英汉化学化工词汇

(第五版)

ENGLISH-CHINESE DICTIONARY OF CHEMISTRY AND CHEMICAL ENGINEERING

(5th Ed.)

科学出版社名词室 编

科学出版社

北京

内 容 简 介

本书是《英汉化学化工词汇（第四版）》的增修订本。修订了其中的错漏之处，删除了与化学化工学科相去甚远的词汇，增补了与化学化工学科密切相关的词汇约 2 万条。全书共收词近 17.5 万条。

本书可供从事化学化工及相关专业的科技人员、生产人员、管理人员和大专院校师生使用。

图书在版编目(CIP)数据

英汉化学化工词汇 = English-Chinese dictionary of chemistry and chemical engineering (5th Ed.)/科学出版社名词室编. –5 版. –北京：科学出版社，2016.1
ISBN 978-7-03-046913-7

Ⅰ. ①英… Ⅱ. ①科… Ⅲ. ①化学–词汇–英、汉②化学工业–词汇–英、汉 Ⅳ. ①O6-61②TQ-61

中国版本图书馆 CIP 数据核字(2015)第 318907 号

责任编辑：张　析／责任校对：蒋　萍
责任印制：肖　兴／封面设计：东方人华

科学出版社 出版
北京东黄城根北街 16 号
邮政编码：100717
http://www.sciencep.com

中国科学院印刷厂 印刷
科学出版社发行　各地新华书店经销

＊

2016 年 1 月第　一　版　　开本：720 × 1000 1/16
2016 年 1 月第一次印刷　　印张：113 1/2
字数：4 358 000

定价：198.00 元
（如有印装质量问题，我社负责调换）

第五版前言

《英汉化学化工词汇》第四版出版已十五年，其中近十年是纸质工具书受网络词典冲击最大的时期，工具书销量逐年下跌，市面上新出版的科技类词典已不多见。传统出版社的优势就在于内容，网络词典虽提供了线索，不能保证知识的准确性。因而，秉承内容为王的理念，我们在艰难的生存环境下，还是决定对我社的品牌项目——《英汉化学化工词汇》进行改版，以期通过精准的内容，在风云变幻的图书市场上，求得一线生机。

本次修订工作中，首先由专家根据学科将《英汉化学化工词汇》第四版中词汇分为13大类，分别由各学科专家逐条审查、修正，特别是删去了大量易于引起法律纠纷的商标及与化学化工专业相去甚远的词汇。并根据《有机化学命名法》(1980)和《无机化学命名法》(1980)对化学物质名称进行了规范。其次，在13大类的基础上，分别就各学科近几年新出现的词汇、术语进行增补。

本次修订不以一味强调增加词条为要，力求定名准确，以全国科学技术名词审定委员会已公布的化学化工标准名词为依据，使词汇定名严谨、准确。

本次修订中参加审稿的有北京大学姚光庆、靳西平、王哲明、刘海超、赵美萍，北京师范大学贾海顺，北京化工大学鲁建民，华南理工大学易筱筠，四川大学刘艳，天津科技大学刘燕红，还有在公司从事与化学化工相关工作的陈红映、姚晖等。

由于化学化工涉及的学科范围较广，不妥和错误之处还会存在，恳请读者在使用中提出宝贵意见，以便将来修订。

科学出版社名词室
2015年1月6日

第四版前言

《英汉化学化工词汇》从第一版的出版到今天已经走过了近四十年的历程。现在已成为科学出版社的一个著名品牌。几十年的修订、再版，使此书的发行总量达到 30 多万册，可以说是科技词汇类图书的成功之作。自 1984 年《英汉化学化工词汇》第三版出版以来，尽管多家出版社也开始出版化学化工专业的词汇，但是《英汉化学化工词汇》第三版仍然是最受广大读者喜爱的科技专业词汇。在第三版出版之后的十余年来，随着化学化工领域各学科的迅猛发展，涌现出了大量新名词、新术语。因此对第三版进行修订、增补，以适应新形势下读者的需要势在必行。为此，我社组织了化学化工行业的部分专家和学者承担了本书的修订工作，历经四年完成了第四版的修订工作。

本书修订工作分三步进行：首先，对第三版词汇按化学化工各分支学科分为 31 大类，分别由各学科专家逐条审查、修正；其次，全面收集化学化工各分支学科近年来出现的新词，特别是第三版中漏收的分支学科，如染料、日用化工等；第三，以全国科学技术名词审定委员会已公布的化学化工标准名词为依据，对所收名词术语进行校订，使所收的名词不仅全、新，而且订名严谨、准确。

本次修订，在第三版基础上增补新词五万余条，使第四版的词汇总数达 17 万条。

我们衷心希望，通过努力，能使第四版成为一部更实用、更具权威性的工具书，同时成为读者们的良师益友！

本书在修订过程中，得到中国化学会、中国化工学会的大力支持，在此表示衷心感谢。

本书虽经一再审订，错误之处仍难免。恳请读者在使用中提出宝贵意见。

科学出版社名词室
1999 年 8 月

第三版前言

《英汉化学化工词汇》(第三版)是在我社 1962 年出版的《英汉化学化工词汇》(再版本)和 1972 年以来陆续出版有关补编的基础上汇编而成的。在汇编过程中，根据中国化学会推荐使用的《无机化学命名原则》(1980)、《有机化学命名原则》(1980)和最近几年来各有关分支学科的名词讨论会的精神，对本书进行了全面校订并增补了新词。本书包括化学化工专业词汇及其有关的科技词汇约 12 万条。

本书由我社王宝瑄汇编校订，并请北京师范大学化学系杨葆昌、无锡轻工业学院化工系苏宜诜审阅，北京医学院有机化学教研室程铁明、李庚新参加了校订工作。

本书在编订过程中得到了北京医学院药学系王序，北京大学化学系张青莲、张滂、张锡瑜，中国科学院化学研究所王葆仁，中国医学科学院药物研究所梁晓天，南开大学王积涛，华东化工学院王承明等的热情指导和帮助，本书再版本出版后还收到了不少读者来信，对收词、订名提出了许多宝贵意见，有些单位和同志还协助提供词汇，谨此一并致谢。

由于化学化工范围很广，编订人员水平有限，本书虽经编、审人员一再校订，但不妥和错误之处还会存在，恳切希望同志们在使用过程中提出增补修订意见，以便再版时收录、改正。

1982 年 10 月

使 用 说 明

1. 词汇正文按英文字母顺序编排。英文复合词看成一个英文词，一律顺排。词汇中斜体字母、阿拉伯数字、希腊文字母、连字符(-)、空格均不参加排序。

2. 同一英文名词，若有数种中文意义时，不同中文意义的译名用①、②、③等分开。

3. 同一中文意义的不同中文译名间用"；"分开，推荐使用的放在最前面。

4. 英文名、中文名中加圆括号的字母及字，在应用时可以替换或省略。但化学物质名词中根据命名原则使用的圆括号不能省略。

5. 中文名的意义或用法，需加以解释说明的，把注释放在六角号"〔〕"内。

6. 中文名后带有"*"者为全国科学技术名词审定委员会公布的定名。

目　　录

第五版前言
第四版前言
第三版前言
使用说明
词汇正文
- A ……………………… 1
- B ……………………… 147
- C ……………………… 239
- D ……………………… 433
- E ……………………… 550
- F ……………………… 628
- G ……………………… 709
- H ……………………… 758
- I ……………………… 829
- J ……………………… 897
- K ……………………… 902
- L ……………………… 915
- M ……………………… 973
- N ……………………… 1083
- O ……………………… 1131
- P ……………………… 1186
- Q ……………………… 1354
- R ……………………… 1365
- S ……………………… 1438
- T ……………………… 1615
- U ……………………… 1721
- V ……………………… 1740
- W ……………………… 1765
- X ……………………… 1793
- Y ……………………… 1798
- Z ……………………… 1801

A

a-　无；缺
aabomycin　阿博霉素
A-addition(=A-polymer)　加聚物
abaca　①蕉麻〔指纤维〕②麻蕉〔指植物〕
Abacavir　阿巴卡韦〔药〕
abacterial　无菌的
abacus　洗金槽；洗矿槽
abadole　阿巴多；2-氨基噻唑；2-噻唑胺
Abar　对溴磷
Abat　双硫磷
Abathion　双硫磷
Abati drying oven　阿巴提干燥(烘)箱
Abbé apertometer　阿贝孔径计
Abbé camera lucida　阿贝明箱
Abbé condenser　阿贝聚光器
Abbé constant　阿贝常数
Abbé number(=Abbé value)　阿贝值
Abbé prism　阿贝棱镜
Abbé refractometer　阿贝折射计
Abbé spectrometer　阿贝光谱仪
Abbé value(=Abbé number)　阿贝值
Abbott bead　单丝钢圈
Abbott (bead winding) machine　单丝钢圈卷成机
abbreviated analysis　简易分析
abbreviated volute　简化螺壳
abbreviation　缩写；略语
Abderhalden's dryer(=drying pistol)　阿布德哈登干燥器；干燥枪
abduct polymerization　包合聚合；包接聚合
Abel apparatus(=Abel flash point tester)　阿贝尔(闪点)试验器
Abel closed tester　阿贝尔闭式试验器
Abel flash point　阿贝尔闪点
Abel flash point tester(=Abel apparatus)　阿贝尔(闪点)试验器
Abel (heat) test　阿贝尔(耐热)试验
Abel-Pensky flash point test　阿贝尔-彭斯基闪点试验
Abel reagent　阿贝尔试剂
Abel tester　阿贝尔(闪点)试验器
abequose　阿比可糖；3-脱氧-D-岩藻糖
aberration　①像差②畸变
aberration constant　像差常数
aberration correction　像差校正
aberrations corrected concave holographic gratings　消像差凹面全息光栅
abherent　阻黏剂；防黏剂；防黏材料
abhesion　脱黏
abhesive　阻黏剂；防黏剂；分离剂
abhesiveness　失黏性
abichite　砷铜矿；光线矿
abienic acid(=abieninic acid)　冷杉酸　$C_{13}H_{20}O_2$
abieninic acid(=abienic acid)　冷杉酸　$C_{13}H_{20}O_2$
abienol　冷杉醇
abies oil　松针油；松香油；冷杉油
abietate　①枞酸酯(盐)②松香酯(盐)
abietene　①枞烯②松香烯　$C_{19}H_{30}$
abietene sulfonic acid　松香烯磺酸
abietic acid　①枞酸②松香酸〔指混合酸〕
abietic anhydride　松香酸酐　$C_{44}H_{26}O_4$
abietic resin　松香酸树脂
abietic-type acid　枞酸型酸
abietin　松香亭
abietinal　枞醛
abietinic　枞树脂的；松香脂的
abietinic acid　松香亭酸
abietinol(=abietyl alcohol)　①枞醇②松香醇
abietinolic acid　松香亭脑酸
abietite　冷杉糖
abietyl　枞酸基
abietyl alcohol　①枞醇②松香醇
abietyl amine　①枞胺②松香胺
abietyl isocyanate　异氰酸松香(醇)酯
ab initio　从头开始
ab initio calculation　从头计算法
ab initio wave function　初始波函数；基于第一性原理的波函数
abiochemistry　①无生化学②无机化学
abiogenesis　自然发生
abiogeny　无生源说；自然发生说
abiologic　非生物学的
abiophysiology　无机生理学
abiuret　不呈缩二脲反应的；无缩三脲反应的
ablation　烧熔；切除
ablation cooling　烧蚀冷却(法)
ablation layer　烧蚀层；消融层
ablation material　烧蚀材料
ablation performance index　烧蚀性能指数；消融性能指数

ablation rate 烧蚀率；消融率
ablation resistance 耐烧蚀性
ablation source 烧蚀源
ablation velocity 烧蚀速率
ablative composite material 烧蚀性复合材料
ablative cooling 烧蚀冷却
ablative photo-decomposition 烧蚀性的光分解
ablative polymer 烧蚀聚合物*
ablative prepreg 烧蚀性预浸料
ablative shield 烧蚀保护层
ablator 烧蚀材料；烧蚀剂
abluent 洗净剂；洗涤剂
ablution ①洗净②洗净液③清除
Abney clinometer 阿布尼测斜器
abnormal 反常的
abnormal addition 反常加成(反应)
abnormal combustion ①反常(异常)燃烧②不规则燃烧
abnormal condition 反常情况
abnormal density 反常密度
abnormal distribution 非正态分布*
abnormal fibre 异状纤维
abnormal glow discharge 异常辉光放电；反常辉光放电
abnormal group 反常基团
abnormality ①反常；不规则②非正态性
abnormal semi-conductor 反常半导体
abnormal settling 反常沉淀；反常凝聚
abnormal structure 异常结构
abnormal test conditions 特殊试验条件
abortive transduction 流产转导；败育转导
above bubble point pressure 高于饱和压力；高于泡点压力
above-critical 临界以上的
above-ground storage tank 地面储藏槽
above proof 超过标准
above-the-melt polymerization 超熔点聚合(作用)
abradability 磨耗性；磨损性；磨蚀性
abradant(=abrasive) 磨料；研磨剂
abraded area 划伤部位(位置)；划伤面积
abraded filament yarn 擦毛长丝〔处于正常长丝和短纤纱之间〕
abrader 磨耗试验机
abrading ①磨耗；磨损；磨蚀②磨光；打磨
abrading agent 磨料；研磨剂
abrading device 磨耗试验机
Abraham consistometer 阿布拉罕稠度计
abrasimeter 耐磨试验仪；磨耗试验机
abrasin oil 次桐油
abrasiometer 磨耗试验机

abrasion 磨耗；磨损；磨蚀
abrasion and wear 磨耗；磨损
abrasion coefficient 磨耗系数
abrasion cycle ①磨蚀周期②磨蚀循环
abrasion disc method 盘式磨蚀(测定)法
abrasion hardness 磨蚀硬度
abrasion index 磨耗指数
abrasion loss 磨耗(减)量
abrasion machine 磨耗试验机
abrasion method of tow conversion 丝束磨断成条法；丝束磨断成纱法
abrasion pattern 磨耗图形
abrasion performance 耐磨(耗)性
abrasion pulsator 间歇式耐磨试验仪
abrasion ratio 磨耗率
abrasion resistance ①磨耗抗力②抗磨性；耐磨性
abrasion resistance improver 耐磨改性剂
abrasion resistance index 抗磨耗指数
abrasion resistance tester 抗磨耗试验机
abrasion resistant compound 耐磨胶料
abrasion-resisting 抗磨损；抗磨耗
abrasion run 磨耗试验
abrasion test 磨耗试验；耐磨试验
abrasion tester 磨耗试验机
abrasion wear 磨耗；磨损；磨蚀
abrasion wheel 磨轮；砂轮
abrasive ①磨料；研磨剂②磨耗的；磨蚀的
abrasive action 磨蚀作用；磨损作用
abrasive action of lubricant 润滑剂磨蚀作用
abrasive belt 磨带
abrasive belt grinding lubricant 磨带润磨油
abrasive binder 磨料黏结剂
abrasive blast equipment 磨料设备
abrasive cleaner 擦洗剂
abrasive cleanser 硬表面去污剂；擦拭清洁剂
abrasive cloth 砂布
abrasive containing lubricant(=abrasive-laden lubricant) 含磨料的润滑油
abrasive disc 磨盘
abrasive erosion 磨蚀性侵蚀；磨蚀性剥蚀
abrasive filler 磨料；研磨剂
abrasive finishing 磨光
abrasive finishing machine 抛光机
abrasive finishing medium 磨光介质
abrasive foam 耐磨泡沫体
abrasive grain 磨料粒；磨光粉
abrasive granule 粒状磨料；粒状摩擦剂
abrasive grinding wheel 磨轮

abrasive hardness　磨料硬度
abrasive-laden lubricant(=abrasive containing lubricant)　含磨料的润滑油
abrasiveness　磨损性
abrasive paper　砂纸；砂磨纸
abrasive powder　磨料；研磨剂
abrasive power　①研磨力②擦洗力
abrasive resistance　耐磨性；耐磨强度
abrasive soap　擦洗皂
abrasive stone　磨石
abrasive substance　研磨材料
abrasive wear　磨耗；磨损；磨粒磨损
abrasive wheel　磨轮
abrasivity　磨蚀度；摩擦损耗度；磨损率
abrastol　萘酚磺酸钙
abrator　喷丸清理；喷砂清理；喷丸清理装置
abraum　①红赭石〔天然氧化铁红〕②层积石
abraum salt　层积盐；废盐
abrazite(=gismondite)　水钙沸石
abriachanite　镁铁青石棉
abric acid　红豆酸；相思豆酸
abridged general view　示意图
abridged goniophotometer　滤色测角光度仪
abridged mashing method　快速糖化法
abridged spectrophotometer　滤光分光光度计
abrin　红豆因
abrine　红豆碱；N-甲基色氨酸
abrodil(=sodium iodomethane sulfonate)　碘甲烷磺酸钠 CH_2ISO_3Na
abromine　昂天莲碱
abrotine　青蒿碱
abrupt change　①陡变②(滴定)突跃
abrupt failure　(塑件)猝裂
abrupt transition　①(挤塑模头)陡变区②突变〔由固态至熔融态的骤然转变〕
abscess　①气孔②脓肿
abscisic acid　脱落酸
abscisin　脱落素
abscissa　横坐标；横轴；横线
abscissa scale　横坐标标度(刻度)；横坐标比例尺
abscissa transform　横坐标变换
absinth　苦艾
absinthe oil　苦艾油
absinthic　①苦艾酒的②苦艾的
absinthic acid　苦艾酸
absinthin　苦艾质；苦艾素
absinthol　苦艾醇
absite　钍钛铀矿

absolute abundance　绝对丰度*
absolute activity　绝对活度
absolute adhesion measurement　绝对附着力测定法
absolute age determination method　绝对年代测定法
absolute alcohol　无水酒精
absolute analysis of GFAAS　石墨炉原子吸收分光光度法绝对分析
absolute asymmetry　绝对不对称
absolute atomic weight　绝对原子量
absolute boiling point　绝对沸点
absolute calibration　绝对校准
absolute calibration curve method　绝对校正曲线法
absolute calibration method　绝对校准法
absolute capacity　绝对容量
absolute colorimetry　绝对比色法
absolute column temperature　绝对柱温
absolute compliance　绝对柔量
absolute configuration　绝对构型
absolute counting　绝对测量*
absolute crystallinity　绝对结晶度
absolute density　绝对密度
absolute detection limit　绝对探测极限
absolute detector　绝对检测器
absolute detector response　检测器绝对响应(值)
absolute detector sensitivity　检测器绝对灵敏度
absolute deviation　绝对偏差*
absolute dielectric constant　绝对介电常数
absolute dry weight　绝对干重
absolute dynamic modulus　绝对动态模量
absolute dynamic viscosity　绝对动态黏度
absolute electric potential difference　电极绝对电位差
absolute elongation　①绝对伸长②绝对伸长率
absolute error　绝对误差*
absolute ethanol　无水乙醇
absolute ether　无水醚
absolute ethyl alcohol　无水乙醇
absolute extract　纯净萃取
absolute frequency　频数
absolute frequency curve　频数曲线
absolute gradient　绝对梯度
absolute hardness　绝对硬度
absolute heating effect　绝对热效率
absolute humidity　绝对湿度
absolute humidity of gas　气体的绝对湿度
absolute isotopic abundance measurement　绝对同位素丰度测量
absolute manometer　绝对压力计
absolute measurement　绝对测量*

absolute methanol 无水甲醇
absolute method 绝对法
absolute method of activation analysis 绝对法活化分析
absolute method of measurement 绝对测量法
absolute mobility 绝对淌度
absolute modulus 绝对模量
absolute oil 净油
absolute permittivity 绝对电容率
absolute potential 绝对电位
absolute precision 绝对精度
absolute pressure 绝对压力
absolute rate constant 绝对速率常数
absolute rate theory(ART) 绝对反应速率理论*
absolute reflectance 绝对反射率；绝对反射比；绝对反射系数
absolute retention time 绝对保留时间
absolute retention volume 绝对保留体积
absolute scattering power 绝对散射本领
absolute sensitivity 绝对灵敏度
absolute solvent power 绝对溶剂力
absolute specific gravity 绝对比重
absolute specificity 绝对专一性
absolute stability constant 绝对稳定常数
absolute structure 绝对结构
absolute temperature 绝对温度
absolute temperature scale 绝对温标
absolute theory of rate process 绝对速率过程理论
absolute units 绝对单位
absolute vacuum 绝对真空
absolute vacuum gauge 绝对真空规
absolute valence 绝对价；最高价
absolute value 绝对值
absolute value of complex modulus 绝对复数模量
absolute velocity 绝对速度
absolute viscosity 绝对黏度
absolute zero 绝对零度
absolute zero potential 绝对零电位
absorbability ①吸收本领②可吸收性
absorbable 可吸收的
absorbable suture （人体）可吸收性缝线〔外科手术用〕
absorbance(A) ①吸光度②吸收度
absorbance accuracy （吸）光度准确度
absorbance index(=absorptivity) 吸收系数；吸光系数
absorbance quotient 吸光商
absorbance ratio 吸光度比值
absorbance ratio method 吸光度比值法
absorbance reproducibility （吸）光度重复性
absorbance spectrum 吸收光谱

absorbance unit 吸收度单位；吸光度单位
absorbance unit full scale(AUFS) 满刻度吸光度单位
absorbance zero stability 零吸收稳定度
absorbancy 吸收度
absorbancy index(=absorbency index) ①吸收系数②吸光系数
absorb band 吸收谱带
absorbed dose 吸收剂量*
absorbed gamma radiation dose γ放射吸收剂量；γ辐射吸收剂量
absorbed layer 吸收层
absorbed water 吸收水
absorbency ①吸收本领②吸水性
absorbency index(=absorbency index) 吸光指数
absorbent ①吸收剂②(有)吸收(本领)的③能吸收的
absorbent bed 吸收层；吸收床
absorbent charcoal 活性炭；吸收性炭
absorbent cotton 脱脂棉
absorbent earth 吸收土
absorbent filter 吸收性滤器
absorbent filtering medium 吸收过滤介质
absorbent oil 吸收油
absorbent packing 吸收性包装
absorbent paper 吸水纸
absorbent power 吸收本领
absorbent solution 吸收剂溶液
absorbent textile fiber 吸收性纺织纤维
absorbent type filter 吸收剂型滤器
absorber ①吸收剂②消振器；缓冲器③吸收器④吸收体
absorber cooler 吸收(器的)冷却器
absorber washer 吸收洗涤器
absorb filler 添加填充剂
absorbility ①吸收能力②吸收率
absorbing 吸收的
absorbing agent 吸收剂
absorbing apparatus 吸收仪器；吸收装置
absorbing band 吸收带
absorbing capacity 吸收本领；吸收能力
absorbing coating 吸收性涂른；防雷达涂层
absorbing column 吸收柱
absorbing liquid 吸收液
absorbing material 吸收剂
absorbing medium 吸收介质
absorbing pipette （气体）吸收球管
absorbing power 吸收本领
absorbing surface 吸收表面
absorbing tower 吸收塔
absorbing trap 吸收阱

absorb moisture　①吸湿度；吸水量②吸湿
absorptance(=absorption factor)　吸收率；吸收比
absorptiometer　①吸光测定计②吸收比色计③(液体)吸气计
absorptiometric method　吸光测定法
absorptiometry　吸光测定法
absorption　吸收*
absorption apparatus　吸收装置
absorption band　吸收谱带
absorption base　吸收基
absorption bottle　吸收瓶
absorption bulb　吸收球管
absorption capacity　吸收本领
absorption cell　吸收池*
absorption chamber　吸收室
absorption coefficient　①吸收系数*②吸光系数
absorption coil　吸收旋管
absorption color　吸收色
absorption column　①吸收柱②吸收塔
absorption compound　吸收化合物
absorption correction　吸收校正
absorption cross section　吸收截面*
absorption curve　吸收曲线
absorption densitometry　吸收光密度法
absorption dose　吸收剂量
absorption edge　吸收边沿；吸收限
absorption edge fine structure(AEFS)　吸收边精细结构
absorption effect　吸收效应
absorption energy　吸收能量
absorption-enhancement effect　吸收增强效应
absorption extraction　吸收提(萃)取
absorption factor(=absorptance)　吸收率；吸收比
absorption fiber probe　吸收纤维探针
absorption filter　吸收滤光片
absorption flask　吸收烧瓶
absorption frequency　吸收频率
absorption funnel　吸收漏斗
absorption gasoline　吸收汽油
absorption hygrometer　吸收湿度计
absorption index　吸收指数
absorption intensity　吸收强度*
absorption intensity expression　吸收强度表示法
absorption jump　吸收陡变；吸收突跃
absorption jump ratio　吸收陡变比
absorption law　吸收律
absorption limit　吸收限
absorption line　吸收线
absorption line profile　吸收线轮廓
absorption line width　吸收(谱)线宽度
absorption liquid　吸收液
absorption loss　吸收(操作)损失
absorption machine　吸收机
absorption maxima wavelength　最大吸收波长
absorption maximum　最大吸收
absorption measurement　吸收测定法
absorption mesh　吸收格(栅)；吸收网格(孔)
absorption meter　①(液体)溶气计②调液厚器
absorption method　①吸收法②吸光(测定)法
absorption of light　光吸收
absorption of moisture　吸湿；吸潮
absorption of radiant energy　辐射能的吸收
absorption oil　吸收油
absorption oil purification process　吸收油精制过程
absorption paper　吸纸
absorption peak　①吸收峰②吸收峰值
absorption peak intensity　吸收峰强度
absorption pipette　吸收球管；吸收移液管
absorption plant　吸收装置
absorption pool of axial direction　轴向吸收池
absorption power　吸收本领；吸收能力
absorption process　吸收法
absorption rate　吸收速率
absorption ratio　吸光比；吸收率
absorption ray　吸收(射)线
absorption reagent　吸收剂
absorption refrigerating machine　吸收制冷机
absorption refrigerator　吸收式制冷机
absorption region　吸收范围
absorption sensitivity　吸收灵敏度
absorption signal　吸收信号
absorption spectrochemical analysis　吸收光谱(化学)分析
absorption spectroelectrochemistry　吸收光谱电化学*
absorption spectrometer　吸收分光计
absorption spectrometry　吸收光谱测定法
absorption spectrophotometry　吸收分光光度法
absorption spectroscopy　吸收光谱法；吸收光谱学
absorption spectrum　吸收光谱*；吸收谱
absorption system　吸收系统
absorption system of refrigeration　制冷吸收系统
absorption test　吸收试验
absorption test of greases　润滑脂吸水性试验
absorption tray　吸收塔盘
absorption tube　吸收管
absorption tube support　吸收管架；吸收管支座
absorption type gage　透过式测厚计

absorption value 吸收值
absorption velocity 吸收速度
absorption vessel 吸收皿
absorptive 吸收的
absorptive capacity 吸收容量
absorptive extraction 吸收提(萃)取法
absorptive silica pigment 吸收性硅类填充剂
absorptive-type filter 吸收型滤器
absorptivity ①吸光系数②吸收性
ABS resin ABS 树脂
abstergent 洗涤剂
abstersion 去垢；洗净
abstersive ①去垢的；有洁净作用的②去污粉
abstract ①提要②抽象的③提取
abstracted factor analysis(AFA) 抽象因子分析
abstract factor 抽象因素
abstraction 夺取(反应)*
abstraction-coupling polymerization 夺取-偶合聚合
abstraction of heat 减热
abstraction reaction 夺取反应；提取反应；抽取反应
abstract rheological concept 抽象流变学概念
absynthol 苦艾醇
abum fixing bath 坚膜定象浴
abundance ①丰度②丰富
abundance of element 元素丰度
abundance of ions 离子丰度
abundance of isotope 同位素丰度
abundance ratio 丰度比
abundance sensitivity 丰度灵敏度
abundance sensitivity correction 丰度灵敏度校正
abundance table 丰度表
abundant nuclei 丰核
abundant ore source 富矿资源
aburamycin 阿布拉霉素
abutilon 白麻
abutment joint 对接接头
abutted surface 相接面；贴合面
abwischen 擦拭
abzyme 抗体酶
acacatechin 类儿茶素
acacetin 金合欢素
acacia ①金合欢②阿拉伯树胶③润滑剂；缓和剂
acacia oil 金合欢油
acaciin ①刺槐苷②金合欢因
acacin 金合欢胶
acacipetalin 树胶含氰苷
academic pigment mixture 非现实混合颜料；学术的理论混合颜料

acadialite 红菱沸石
acadi zeolite 红斜方沸石
acajou balsam 腰果树香脂
acanthaceous 具刺的
acanthite 螺状硫银矿
acanthocytosis 棘红细胞(增多)症
acanthoside 无梗五加苷
acapnia 血液碳酸缺乏；缺碳酸血；血液二氧化碳缺乏
acaprinum 喹啉脲
acarbose 阿卡波糖〔药〕
AC arc 交流电弧
acardite(=diphenyl urea) 二苯脲
acari 螨
acaricide(=miticide) 杀螨剂
acaroid balsam 禾木香脂
acaroid gum(=black boy gum) 禾木胶
acaroid resin 禾木树脂
acaryotic 无核(的)
acatalasia 过氧化物酶缺失症
acaulescent 无梗的
accaroid gum(=yacca gum) 禾木胶
accelerant 促进剂
accelerated ageing ①加速老化②人工时效；人工老化
accelerated ageing test 加速老化试验
accelerated bomb test 加速氧弹试验
accelerated cement 快凝水泥
accelerated clarifier 快速沉降器
accelerated corrosion test 加速腐蚀试验；加速锈蚀试验
accelerated creep 加速蠕变
accelerated cure 快速硫化；加速硫化
accelerated deterioration 加速劣化
accelerated effect 加速效应
accelerated exposure test 加速暴露试验；加速暴晒试验
accelerated flow method 加速流动法*
accelerated gum 速成胶质
accelerated gum test 速成胶质试验
accelerated humidity testing 加速湿度试验
accelerated life 加速老化寿命
accelerated light ageing 加速光老化
accelerated light exposure 加速曝光
accelerated motion 加速运动
accelerated outdoor exposure 加速室外曝置
accelerated outdoor weathering 加速室外天候老化
accelerated oxidation 加速氧化
accelerated oxidation test 加速氧化试验
accelerated ozone ageing 加速臭氧老化
accelerated period 加速期
accelerated porcelain dish test 加速瓷皿试验

accelerated process 加速法；快速法
accelerated resin 速固树脂
accelerated sludge test 加速沉渣试验
accelerated solvent extraction method 加速溶剂萃取法
accelerated soundness test 加速(坚)固度试验
accelerated stock 含促进剂胶料
accelerated stress cracking 加速应力破裂
accelerated sulfur system 促进硫体系
accelerated sulfur vulcanization 促进硫硫化*
accelerated tannage 速鞣法
accelerated test 加速试验
accelerated test for color stability 加速颜色稳定试验
accelerated testing cabinet 加速试验机
accelerated weathering accelerator 加速耐候试验机；加速老化试验机
accelerated weathering exposure 加速风蚀暴晒；加速天候曝置
accelerated weathering resistance 耐加速天候老化性
accelerated weathering test 加速老化试验
accelerating agent ①促进剂②促染剂
accelerating aging 加速老化
accelerating aging by humid-dry cycling 干湿交替加速老化试验
accelerating apparent flow 加速表观流动
accelerating creep 增速蠕变；第三阶段蠕变
accelerating effect ①促进效应②加速效应
accelerating electrode 加速电极；促进电极
accelerating field 加速场
accelerating flow 加速流
accelerating fluid 加速流体
accelerating fluid flow 加速流体流动
accelerating gap 加速隙
accelerating lens 加速透镜
accelerating plastic flow 加速塑性流
accelerating plastic substance 加速塑性物质
accelerating potential 加速电位
accelerating region 加速区
accelerating slit 加速狭缝
accelerating system ①促进系统②加速系统
accelerating tube 加速管
accelerating voltage 加速电压
accelerating voltage scan 加速电压扫描
accelerating well 补偿油井
acceleration ①速化反应②加速作用
acceleration characteristics 加速性能
acceleration inertia load test (动力机械)加速惯性负载试验
acceleration of gravity 重力加速度

acceleration of ions 离子加速
acceleration of ripening 成熟的加速
acceleration period 加速期；促进硫化期
acceleration ramp 加速跳跃
acceleration tensor 加速度张量
acceleration zone 加速区
accelerative thickening 加速稠化
accelerator ①加速剂②促进剂③加速器
accelerator-activator 促进剂-活性剂
accelerator-based activation analysis 使用加速器的活化分析
accelerator dosage 促进剂用量；促进剂含量
accelerator level 促进剂用量
accelerator mass-spectrometry(AMS) 加速器质谱法*
accelerator master batch 加速母炼胶（橡胶）
accelerator pedal 加速踏板
accelerator-produced radionuclides 加速器生产的放射性核素*
accelerator ratio 促进剂(用量)比率；促进剂配比
accelerator retarder 硫化迟延剂
accelerator starvation 促进剂用量不足
accelerator-sulfur cure 硫黄促进剂并用硫化
accelerometer 加速度计
accelerometer array 加速度计阵列
accented term 重点项
acceptability of color matches 允许色差
acceptable daily intake(ADI) 允许日摄取量
acceptable end-product 合格品；正品
acceptable error 容许误差；可接受误差
acceptable explosives 合格炸药
acceptable fiber 合格纤维
acceptable life 有效使用寿命
acceptable limit 允许极限
acceptable quality level(AQL) 合格质量标准
acceptable standard 验收标准
acceptance 验收
acceptance base 受体碱
acceptance level 可接收的水平；验收标准
acceptance of mass filter 滤质器接受容限
acceptance region 接受域*
acceptance sampling 验收取样
acceptance slit 容许狭缝
acceptance tolerance 验收公差
accepted chips 合格木片
accepted stock 合格浆料
acceptor 受体*
acceptor acid 受体酸
acceptor-donor complex 受体-给体(式)复合物

acceptor group　受电子基
acceptor level　受主能级
acceptor of energy　能量吸收器
acceptor residues　受电子体
acceptor-RNA(tRNA)　接受 RNA
acceptor site　接受部位
access　①接近；进入②途径；捷径
access door　检修门；入口门
access hole　入口；入孔
accessibility　①可及度②可及性
accessibility value　可及度值
accessories for lubricating　润滑用辅件
accessory　①附件；辅助设备②附属的；次要的；辅助的
accessory component　附件
accessory factor　补助因素
accessory food factor　补助(食料)因素
accessory food substance　补助食物
accessory ingredient　配合剂；助剂
accessory material　①辅助材料②副矿物
accessory substance　副产物
accident　故障
accidental coagulation　早期凝固
accidental curve　偶然曲线；事故曲线
accidental effect　随机效应；不规则效应
accidental error　偶然误差；随机误差
accident prevention　事故预防；安全措施
acclimation　驯化作用
acclimation period　驯化周期
acclimatization　驯化作用
accommodation　①调节②供应
accommodation coefficient　调节系数
accompaniment　伴生物；伴随物；附属物
accompanying diagram　附图
accompanying element　伴生元素
accompanying mineral　伴生矿物
accompanying substance　外来杂质；夹杂物；异物
accordin mould(=integral mould)　集成模具
accordion interlock fabric　抽条棉毛织物
accountability　可衡算性；可计量性
accountability analysis　可衡算分析；可计量分析
accountability tank　衡算计量槽；(责任)计量槽
accounting procedure　衡算计量程序；计量程序
accounting unit　衡算单位
accretion　增大(作用)
accumulated error　累积误差
accumulating　蓄积的；聚集的；堆集的
accumulation　蓄积；聚集；堆集
accumulation mode　积聚模

accumulation of radioactive waste　放射性废料积存
accumulation wall　富集壁
accumulative　积蓄的；聚集的；堆集的
accumulative column　聚集柱
accumulative crystallization　聚集结晶
accumulative formation　堆集形成
accumulative frequency　累积频数
accumulative pitch error　周节积累误差；齿距积累误差
accumulative probability　累积概率
accumulator　①蓄电池②蓄油器③蓄力器④蓄压器
accumulator box　蓄电(池)槽
accumulator cell　蓄电池
accumulator jar　蓄电瓶
accumulator plate　蓄电(池极)板
accumulator pocket　储气筒袋
accumulator room　蓄电池室
accumulator-separator　蓄电池隔板
accumulator still　缓冲釜
accumulator tank　①储蓄槽②蓄电(池)槽
accumulator tester　蓄电池检验器
accumulator tray　蓄电盘
accumulator tube　储气筒管
accumulator-type blow moulding machine　储料缸型吹塑机
accuracy　准确度*
accuracy control　准确度控制
accuracy grade　准确度级别
accuracy of analysis　分析准确度
accuracy of mass measurement　质量测量准确度
accuracy of measurement　测量精度
accuracy of reading　读数准确度
accurate mass measurement　精确质量测定
accurate mass table　精确质量表
aceanthrene　醋蒽
aceanthrenequinone　醋蒽醌
aceanthrylene　醋蒽烯
acedapsone　双(二)乙酰氨苯砜；醋氨苯砜〔药〕
acedicon　醋底康〔乙酰可待因异构物的商名〕
aceko-black　醋黑
acelation(=acylation)　酰化作用
acenaphth-　苊；次苊基
acenaphthaquinone　苊醌
acenaphthene　二氢苊
acenaphthene quinone　二氢苊醌
acenaphthenequinone monoxime　苊醌单肟
acenaphthenone　二氢苊酮
acenaphthenyl　苊基　$C_{12}H_9-$
acenaphthenylene　1,2-亚苊基　$-C_{12}H_8-$

1-acenaphthenylidene 1-亚苊基 $C_{12}H_8=$
acenaphthylene(=ethylene naphthalene) 苊
acene 并苯*
aceperinaphthane 醋代萘烷
acephenanthrylene 醋菲烯
acerb 涩的
acerbity 涩度
acerdol 高锰酸钙
aceric 槭的
aceric acid 槭汁酸
acerin 槭素
aceritol 槭糖醇
acerous 针状的;无角的;无触须的
acer tannin 槭叶单宁
acescency 酸味
acescent 微酸的
acesulfame 丁磺氨;双氧噁噻嗪
acesulfame potassium 丁磺氨钾;双氧噁噻嗪钾
acet ①次乙基 $CH_3C\equiv$ ②乙酰 CH_3CO-
acetacetate 乙酰乙酸盐(或酯) CH_3COCH_2COOM; CH_3COCH_2COOR
acetacetic acid 乙酰乙酸 CH_3COCH_2COOH
acetacetic ester ①乙酰乙酸酯 CH_3COCH_2COOR ②乙酰乙酸乙酯 $CH_3COCH_2COOC_2H_5$
acetacetic ether 乙酰乙酸乙酯 $CH_3COCH_2COOC_2H_5$
acetal ①乙缩醛;乙醛缩二乙醇 $CH_3CH(OC_2H_5)_2$ ②〔类名〕醛缩醇;缩醛〔醛缩二醇〕 $RCH(OR)_2$
acetalating agent 缩醛剂
acetalation 缩醛化(作用)
acetal copolymer 缩醛共聚物
acetal crosslinker 缩醛交联剂
acetaldazine 乙醛连氮〔—N=N—为偶氮;=N·N= 称为连氮〕 $CH_3CH=NN=CHCH_3$
acetaldehyde 乙醛
acetaldehyde ammonia 乙醛合氨
acetaldehyde-ammonia condensate 乙醛氨缩合物〔促进剂〕
acetaldehyde cyanhydrin 乙醛合氰化氢;2-羟基丙腈
acetaldehyde cyclic phenylethylene acetal 乙醛环苯乙撑缩醛
acetaldehyde disulfonic acid 乙醛二磺酸 $CH(SO_3H)_2CHO$
acetaldehyde oxime 乙醛肟〔防老剂〕
acetaldehyde phenylhydrazone 乙醛苯腙 $C_6H_5NHN=CHCH_3$
acetaldehyde polymer 乙醛聚合物
acetaldehyde resin 乙醛树脂;聚甲醛树脂
acetaldehyde sodium bisulfite 乙醛合亚硫酸氢钠 $CH_3CHO·NaHSO_3$
acetaldol(=aldol) 丁间醇醛;3-羟基丁醛 $CH_3CH(OH)CH_2CHO$
acetaldoxime 乙醛肟
acetal fibre 缩醛纤维
acetalization 缩醛(化)作用
acetalized fiber 缩醛化纤维
acetalizing degree 缩醛化度
acetal phosphatide 缩醛磷脂
acetal plastic 缩甲醛塑料
acetal polymer 缩醛缩聚物;缩醛聚合物
acetal resin 缩醛树脂*
acetal-type nonionics 缩醛型非离子表面活性剂
acetamide(=acetic acid amide) 乙酰胺 CH_3CONH_2
acetamide chloride 二氯代乙酰胺;1,1-二氯乙胺 $CH_3CCl_2NH_2$
acetamidine 乙脒 $CH_3C(=NH)NH_2$
acetamidine hydrochloride 盐酸乙脒 $CH_3C(=NH)NH_2·HCl$
acetamido 乙酰氨基 CH_3CONH-
acetamidoacrylic acid 乙酰氨基丙烯酸
acetamidochloride 二氯代乙酰胺;1,1-二氯乙胺 $CH_3CCl_2NH_2$
2-acetamidofluorene 2-乙酰氨基芴
acetamidoglucal 乙酰氨基葡烯糖
acetamido monoethanol amine 乙酰氨基单乙醇胺
2-acetamido-5-nitrothiazole 2-乙酰氨基-5-硝基噻唑
2-acetamidophenoxazin-3-one 2-乙酰氨基吩噁嗪-3-酮
γ-acetamidopropyl thioacetate 硫代乙酸γ-乙酰氨基丙酯
acetamidopropyl trimonium chloride 乙酰氨基丙基三甲基氯化铵
4-acetamido-trans-azobenzene 4-乙酰氨基反式偶氮苯
acetamine 乙酰胺
acetamino 乙酰氨基
acetaminobenzoic acid 乙酰氨基苯甲酸
acetamino naphthol 乙酰氨基萘酚
acetaminophenol 乙酰氨基苯酚
acetaminosalicylic acid 乙酰氨基水杨酸
acetanil(=acetanilide) N-乙酰苯胺
acetanilide(=antifebrin) N-乙酰苯胺;退热冰〔药〕 $C_6H_5NHCOCH_3$
acetaniside(=methoxyacetanilide) 甲氧基乙酰苯胺 $C_9H_{11}O_2N$
acetanisidine 甲氧基乙酰苯胺
acetanisole 乙酰茴香醚;对甲氧基苯乙酮
acetannin 乙酰单宁
acetarsol(=acetarsone) 乙酰胺胂〔药〕
acetarsone(=acetarsol) 乙酰胺胂〔药〕

acetate(Ac) ①乙酸盐②乙酸酯③乙酸根(基)
acetate butyrate rayon 乙酸丁酸嫘萦
acetate butyrate resin 乙酸丁酸树脂
acetate C-6〔商〕 乙酸己酯
acetate C-8〔商〕 乙酸辛酯
acetate C-9〔商〕 乙酸壬酯
acetate C-10〔商〕 乙酸癸酯
acetate C-12〔商〕 乙酸十二酯
acetate dope 乙酸(纤维)酯涂布漆
acetate dye 醋酯染料
acetate fibre 醋酯纤维；醋酸纤维〔俗〕
acetate film 醋酸纤维素薄膜；纤维素醋酸酯薄膜
acetate from viscose rayon 乙酰化黏胶纤维
acetate hollow filament 醋酯中空长丝
acetate method 乙酸盐法
acetate multifilament 醋酸纤维复丝
acetate rayon 醋酸嫘萦；醋酸纤维；醋纤人造丝
acetate silk 乙酸丝
acetate staple fiber 乙酸短纤
acetate yarn 醋酸纤维
aceto-〔词头〕 乙酸离子〔配盐中的乙酸配位离子〕
acetator 醋化器；酿醋罐
acetazolamide 乙酰唑(磺)胺
acetbromamide(=bromacetamide) 溴乙酰胺
acetbromanilide(=bromacetate) 乙酰溴苯胺
acetene(=ethylene) 亚乙基
acetenyl(=ethynyl) 乙炔基
aceteugenol 乙酰丁子香酚；乙酸丁子香粉酯
acet extract 乙酸浸膏
acethydrazide 乙酰肼 $CH_3CONHNH_2$
acethydroximic acid 乙羟肟酸
acetic acid 乙酸；醋酸〔俗〕
acetic acid amide(=acetamide) 乙酰胺
acetic acid bath 乙酸浴
acetic acid rubber 乙酸凝固橡胶
acetic acid salt spray test 乙酸盐雾试验
acetic aldehyde 乙醛
acetic anhydride(= acetyl oxide) 乙酰化氧；乙(酸)酐 $(CH_3CO)_2O$
acetic ester 乙酸酯 CH_3COOR
acetic ether 乙酸乙酯 $CH_3COOC_2H_5$
acetic hydroperoxide 过乙酸 CH_3COO_2H
acetic oxide(=acetic anhydride) 乙(酸)酐；氧化乙酰 $(CH_3CO)_2O$
acetic peracid(=peracetic acid) 过乙酸
acetic peroxide 过氧化乙酸
acetidin(=ethyl acetate) 乙酸乙酯
acetification 乙酸化作用；醋化作用

acetifier 醋化器
acetimeter 乙酸(比重)计
acetimetry 乙酸测定(法)
acetimide 乙酰亚胺
acetimido ①(=acetylimino)乙酰亚氨基②(=acetimidoyl) 1-亚氨乙基
acetimidochloride 偕氯代乙亚胺；1-氯乙亚胺
acetimidoquinone 乙酰亚氨醌
acetimidoyl 亚氨代乙酰基；1-亚氨乙基
acetin(=glyceryl monoacetate) 甘油单乙酸酯；醋精
acetine(=glyceryl monoacetate) 单乙酸甘油酯
acetin method 乙酸酯法；醋精法〔甘油定量方法〕
acetivenol(=vetiveryl acetate) 乙酸岩兰草酯；乙酸香根酯
acetnaphthalide 乙酰萘胺 $CH_3CONHC_{10}H_7$
aceto ①乙酰②次乙基
aceto-acetamide 乙酰乙酰胺
aceto-acetanilide(=acetyl acetanilide) N-乙酰乙酰苯胺；N-丁间酮酰苯胺 $CH_3COCH_2CONHC_6H_5$
acetoacetate ①乙酰乙酸②乙酰乙酸盐(酯或根)
acetoacetatic ester chelate 乙酰乙酸酯螯合物
acetoacetic acid 乙酰乙酸；丁间酮酸
acetoacetic acid arylide 乙酰乙酸芳基化物；乙酰乙酸芳基酯
acetoacetic anilide N,N-乙酰乙酰苯胺
acetoacetic ester ①乙酰乙酸酯②〔专指〕乙酰乙酸乙酯 $CH_3COCH_2COOC_2H_5$
acetoacetic ester synthesis 乙酰乙酸酯合成(法)
acetoacet-meta-xylidide 乙酰乙酰间二甲苯胺
acetoacet-ortho-anisidide 乙酰乙酰邻茴香胺
acetoacet-ortho-chloranilide N-乙酰乙酰邻氯代苯胺
acetoacet-ortho-toluidide N-乙酰乙酰邻甲苯胺
acetoacet-para-phenetidide 对乙氧基乙酰乙酰苯胺
acetoacetyl 乙酰乙酰基；丁间酮酰基；3-氧丁酰基
5-acetoacetylaminobenzimidazolone 5-乙酰乙酰氨基苯并咪唑酮
acetoacetyl aniline 乙酰乙酰苯胺
2-acetoacetyl pyridine 2-乙酰乙酰吡啶
aceto-aldehyde ammonia 乙醛氨
acetoamidophenol(=acetaminophenol) 乙酰氨基苯酚
acetobenzoic acid anhydride(=acetobenzoic anhydride) 乙酸苯酸酐
acetobenzoic anhydride(=acetobenzoic acid anhydride) 乙酸苯酸酐
acetobrom(=acetbrom) 溴乙酰
acetobromamide 乙酰溴胺 $CH_3CONHBr$
acetobromoglucose 乙酰溴葡萄糖；1-溴-2,3,4,6-四乙酰葡萄糖 $C_{14}H_{19}O_9Br$

acetobutyl alcohol 乙酰丁醇
acetobutyric acid 乙酰丁酸
aceto-carmine 醋酸胭脂红
acetocaustin 三氯乙酸
acetochloral 三氯乙醛；乙氯醛〔俗〕
acetochloroamide 乙酰氯胺 $CH_3CONHCl$
acetocinnamone 乙酰肉桂酮
acetocoumarin 乙酰香豆素
acetocumene 乙酰异丙苯
acetocyclohexane 乙酰环己烷；环己乙酮
acetocyclohexene 乙酰环己烯
acetocyclopentane 乙酰环戊烷
acetocyclopropane 乙酰环丙烷
acetodibromoamide N,N-二溴乙酰胺
acetoethylation 乙酰乙基化(作用)
acetoeugenol(=acetyl eugenol) 乙酰丁子香酚
2-acetofuran 2-乙酰呋喃
acetogenin 多聚乙酰
acetoglyceral 乙环甘油
acetoglyceride 乙酰甘油
acetoguaiacone 乙酰愈创木酮
acetohydroxamic acid 乙酰氧肟酸
acetoin(=3-hydroxy-2-butanone) 乙偶姻〔俗〕；3-羟基-2-丁酮〔-oin 译成偶姻,表示醛的醇酮缩合物〕 $CH_3CHOHCOCH_3$
acetol(=1-hydroxy-2-propanone) 丙酮醇；乙酰甲醇；1-羟基-2-丙酮 CH_3COCH_2OH
acetoluide N-乙酰甲苯胺
acetolysis 乙酸水解；醋解
acetomenaphthone 乙酰甲萘醌
acetometer 乙酸计
aceton acid(=acetonic acid) 乙酮酸
acetonamine 酮胺化物
acetonaphthalene 乙酰萘
acetonaphthalide(=acetyl naphthylamine) 乙酰萘胺
acetonaphthol 乙酰基萘酚
acetonaphthone 乙酰萘；萘乙酮
β-acetonaphthone β-乙酰萘；β-萘乙酮
acetonate 丙酮酸盐(酯)
acetonation(=acetonization) 丙酮化(作用)
acetone(=propanone) 丙酮
acetone acid 醋酮酸〔俗〕；α-羟基异丁酸
acetone alcohol 丙酮醇 CH_3COCH_2OH
acetone amine 丙酮胺 $CH_3COCH_2NH_2$
acetone-benzol dewaxing plant 丙酮苯脱蜡装置
acetone-benzol process 丙酮-苯(脱蜡)法
acetone body(=ketone body) 酮体
acetone chloroform 丙酮氯仿；偕三氯叔丁醇 $CCl_3C(CH_3)_2OH$
acetone cyanohydrin 丙酮合氰化氢；2-甲基-2-羟基丙腈 $(CH_3)_2C(OH)CN$
acetone-diacetic acid 丙酮二乙酸；4-庚酮二酸 $CO(C_2H_4COOH)_2$
acetone-dicarboxylic acid(=β-ketoglutaric acid) 丙酮二羧酸；3-戊酮二酸 $CO(CH_2COOH)_2$
acetone dichloride 二氯代丙酮 $C_3H_4OCl_2$
acetone extract 丙酮提取物
acetone extraction 丙酮提取(法)
acetone-formaldehyde resin 丙酮缩甲醛树脂
acetone-furfural resin 丙酮缩糠醛树脂；糠酮树脂
acetone-methylene chloride solvent 丙酮二氯甲烷溶剂
acetone monochloride 一氯代丙酮
acetone number 丙酮值
acetone oil 丙酮油
acetone oxime 丙酮肟
acetone peroxide 过氧化丙酮
acetone phenylhydrazone 丙酮苯腙
acetone powder 丙酮(制)粉
acetone raffinate 丙酮提余物
acetone resin 丙酮树脂
acetone sodium bisulfite 丙酮合亚硫酸氢钠 $(CH_3)_2CO \cdot NaHSO_3$
acetone tetrachloride 四氯代丙酮 CCl_3COCH_2Cl
acetone titration 丙酮滴定法
acetone trichloride 三氯代丙酮 $C_3H_3OCl_3$
acetone value 丙酮值
acetonic 丙酮的
acetonic acid 醋酮酸〔俗〕；α-羟基异丁酸 $(CH_3)_2C(OH)COOH$
acetonide 丙酮化合物
acetonitrile 乙腈 CH_3CN
acetonization(=acetonation) 丙酮化(作用)
acetonuria(=ketonuria) 酮尿；丙酮尿
acetonyl 丙酮基；乙酰甲基 CH_3COCH_2-
acetonyl-acetone 丙酮基丙酮；2,5-己二酮 $(CH_3COCH_2)_2$
acetonylamine 丙酮基胺
acetonylidene 亚丙酮基 CH_3COCH
acetonylmalonic acid 丙酮基丙二酸
acetonyl urea 丙酮基脲 $NH_2CONHCH_2COCH_3$
acetoolein 乙酰甘油三油酸酯
acetophenanthrene 乙酰菲 $CH_3COC_{14}H_9$
acetophenetide(=acetophenetidine) ①非那西汀〔药〕〔虽未标明位次,但通常指对位〕 $CH_3CONHC_6H_4OC_2H_5$ ②N-乙酰乙氧基苯胺
acetophenetidine(=phenacetin; acetophenetide) ①非那西

汀〔药〕 $C_2H_5OC_6H_4NHCOCH_3$ ②N-乙酰乙氧基苯胺
acetophenine 醋吩宁；2,4,6-三苯基吡啶
acetophenol 乙酰苯酚 $CH_3COC_6H_4OH$
acetophenone(=acetylbenzene; methyl phenyl ketone) 乙酰苯；苯乙酮；甲基苯基(甲)酮 $CH_3COC_6H_5$
acetophenone acetone 乙酰苯基丙酮
acetophenone-formaldehyde resin 苯乙酮-甲醛树脂
acetophenone oxime 苯乙酮肟〔防老剂〕
acetopiperone 乙酰胡椒环
2-acetopyridine 2-乙酰吡啶
acetopyrine 乙酰比林〔药〕
2-acetopyrrole 2-乙酰吡咯
acetosalicylic acid 乙酰水杨酸 $CH_3COOC_6H_4COOH$
acetosulfamine 乙酰磺胺
acetothienone(=thienylmethylketone) 噻嗯基乙酮
acetotoluide(=acetotoluidide) N-乙酰甲苯胺 $CH_3C_6H_4NHCOCH_3$
acetotoluidide(=acetotoluide) N-乙酰甲苯胺
acetous ①乙酸的②醋的
acetovanillon ①乙酰香兰酮②加拿大麻素
acetovanillone(=vanillyl methyl ketone) 乙酰香兰酮；香兰基甲基酮
acetoveratrone 乙酰藜芦酮
acetoxime 丙酮肟 $(CH_3)_2C=NOH$
acetoxy 乙酸基；乙酰氧基 CH_3COO-
acetoxyacetic acid 乙酸基乙酸 CH_3COOCH_2COOH
acetoxyacetone 乙酸基丙酮 $CH_3CO_2CH_2COCH_3$
acetoxy-acetophenone 乙酸基乙酰苯 $C_6H_5COCH_2OCOCH_3$
acetoxy-benzoic acid 乙酸基苯甲酸 $CH_3COOC_6H_4COOH$
acetoxycapric acid 乙酸基癸酸
acetoxycaprylic acid 乙酸基辛酸
acetoxyl 乙酸基；乙酰氧基
acetoxylation 乙酸化(作用)；乙酰氧基化(作用)
acetoxylide 2,4-二甲基乙酰苯胺 $CH_3CONHC_6H_3(CH_3)_2$
acetoxymethylate abietic acid 乙酸基甲基化枞酸
acetoxypregnenolone 乙酸孕烯醇酮
acetoxysilane 乙酰氧基硅烷
acetoxy stearate 乙酸基硬脂酸盐
acetoxytrimethylsilane 乙酰氧基三甲基硅烷
acetoyl butyrate 丁酸-3-丁酯-2-醇酯
acetozone(=acetylbenzoylperoxide) 过氧化乙酰苯甲酰 $C_6H_5COOOCOCH_3$
acetphenetid(=acetophenetide) 乙酰乙氧苯胺
acetphenetidide(=acetophenetide) N-乙酰乙氧苯胺；乙酰氨基苯乙醚 $CH_3CONHC_6H_4OC_2H_5$
acetphenetidin(e) N-乙酰乙氧基苯胺
acettoluide N-乙酰甲苯胺 $CH_3C_6H_4NHCOCH_3$
aceturic acid 醋尿酸；N-乙酰甘氨酸
acetyl 乙酰基
acetylability 可乙酰化性
acetylable 可乙酰化的
acetylacannabinol 乙酰大麻酚
acetyl acetanilide(=aceto-acetanilide) N-乙酰乙酰苯胺 $CH_3COCH_2CONHC_6H_5$
acetylacetate 乙酰乙酸盐(酯)
acetylacetic acid 乙酰乙酸 $CHCOCH_2COOH$
acetylacetic ester ①乙酰乙酸酯 CH_3COCH_2COOR ②〔专指〕乙酰乙酸乙酯 $CH_3COCH_2COOC_2H_5$
acetylacetic ether 乙酰乙酸乙酯 $CH_3COCH_2COOC_2H_5$
acetylacetonate 乙酰丙酮化物
acetylacetonate nickel 乙酰丙酮镍
acetylacetone 乙酰丙酮*
acetylacetone dioxime 乙酰丙酮二肟
acetylacetone extraction 乙酰丙酮萃取
acetylacetone method 乙酰丙酮法
N-acetyl-O-acetylneuraminate N-乙酰-O-乙酰神经氨(糖)酸
acetyl (acid) number 乙酰值
acetylacrifoline(=lycopodium alkaloid 12) 乙酰尖叶(石松)碱
β-acetyl acrolein β-乙酰丙烯醛
acetyladenylate 乙酰腺(嘌呤核)苷酸
acetyl alanine N-乙酰丙氨酸
acetylamidoazotoluene(=azodermine) 乙酰氨基偶氮甲苯
acetylamine(=acetylamino) 乙酰氨基
acetylamino(=acetylamine) 乙酰氨基
acetylaminoacetic acid 乙酰氨基乙酸；N-乙酰甘氨酸 $CH_3CONHCH_2COOH$
acetylamino benzoic acid 乙酰氨基苯甲酸 $CH_3CONHC_6H_4COOH$
acetylaminoglucosidase 乙酰氨基葡萄糖苷酶
1-acetylamino-4-nitronaphthalene 1-乙酰氨基-4-硝基萘
acetylamino salol(=salophen) 乙酰氨基萨罗
acetyl-AMP 乙酰 AMP
acetyl-anisidine N-乙酰茴香胺；N-乙酰甲氧基苯胺 $CH_3OC_6H_4NHCOCH_3$
acetylanthranil 乙酰茴内酐；N-乙酰邻氨基苯酸内酯 $C_9H_7O_2N$
acetylanthranilic acid N-乙酰邻氨基苯甲酸 $C_9H_9O_3N$
acetylate ①乙酰化②乙酰化(产)物
acetylated 乙酰化了的

acetylated castor oil　乙酰(化)蓖麻油
acetylated cellulose　乙酰(化)纤维素
acetylated cobalt borate　乙酰化硼酸钴〔增黏剂〕
acetylated congo　乙酰化刚果(树脂)
acetylated cotton　乙酰化棉〔通过乙酰化作用，制取部分乙酰化的改性棉纤维〕
acetylated monoglyceride　乙酰化单甘油酯
acetylated paper　乙酰纸
acetylated phenolic fiber　乙酰化酚醛纤维
acetylated staple　乙酰化(黏胶)短纤维
acetylating　①乙酰化的②乙酰化作用
acetylating agent　乙酰剂；乙酰化剂
acetylating mixture　乙酰化(混合)剂
acetylating viscose rayon　乙酰化黏胶人造丝
acetylation　乙酰化(作用)
acetylation bath　乙酰化浴
acetylation number(=acetylation value)　乙酰(化)值
acetylation reagent　乙酰化试剂
acetylation value(=acetylation number)　乙酰(化)值
acetylator　乙酰化器；乙酰化装置
acetyl atoxyl　乙酰对氨基苯胂酰
N-acetyl-auramine　N-乙酰基金胺
1-acetylazulene　1-乙酰基薁
acetylbenzene(=acetophenone)　苯乙酮
acetylbenzoate　乙酸苯甲酸酐；苯甲酸乙酰酯 $C_6H_5COOOCH_3$
acetylbenzoic acid　乙酰苯甲酸 $CH_3COC_6H_4CO_2H$
acetyl benzoin　乙酰苯偶姻
acetylbenzoyl(=methylphenyl diketone)　乙酰苯酮；苯丙二酮；丙酮酰苯；甲基苯基二(甲)酮 $CH_3CO \cdot COC_6H_5$
acetylbenzoylperoxide(= acetozone)　过氧化乙酰苯甲酰
acetyl biuret　乙酰基缩二脲
4-acetyl borneol　4-乙酰龙脑
acetyl brilliant blue　乙酰亮蓝
acetyl bromide　乙酰溴 CH_3COBr
N-acetyl butylaniline　N-乙酰丁苯胺
γ-acetyl butyraldehyde　γ-乙酰丁醛
acetyl butyryl　乙酰丁酰
acetyl camphor　乙酰樟脑 $C_{12}H_{18}O_2$
acetyl caproyl(=2,3-octadione)　乙酰己酰；2,3-辛二酮 $CH_3COCO(CH_2)_4CH_3$
acetylcaranine(=bellamarine)　孤挺花碱
acetylcarbamide　乙酰碳酰二胺
N-acetyl carbazole　N-乙酰咔唑 $CH_3CONC_{12}H_8$
acetylcarbinol　乙酰甲醇
acetyl carbromal(=abasin)　亚巴精；对称乙酰基二乙基溴乙酰(基)脲 $(C_2H_5)_2CBrCONHCONHCOCH_3$
o-acetylcarnitine　o-乙酰肉碱

acetyl cedrene　乙酰柏木烯
acetyl cellulose　醋酸纤维素
acetyl chemicals　乙酰化学品
acetyl chloride　乙酰氯 CH_3COCl
acetyl chloride-ferric chloride reagent　乙酰氯-氯化铁试剂〔测聚醚〕
acetylcholine　乙酰胆碱
acetylcholinesterase(AchE)　乙酰胆碱酯酶
acetyl content　乙酰含量
β-acetyl crotonaldehyde　β-乙酰巴豆醛
p-acetyl cumol　对乙酰异丙苯
acetyl cyanide　乙酰氰；丙酮腈 CH_3COCN
2-acetyl-5-cyclohexyl pentane　2-乙酰-5-环己基戊烷
acetyl cyclohexyl-sulfonyl peroxide　乙酰基过氧化磺酰环己烷
acetylcysteine　乙酰半胱氨酸〔药〕
acetyldelcosine　乙酰得可辛
acetyl dichlorohydrin　乙酰二氯丙醇
acetyl-dihydrolycopodine　乙酰二氢石松碱
3-acetyl-2,5-dimethyl furan　3-乙酰-2,5-二甲基呋喃
2-acetyl-3,5-dimethyl pyrazine　2-乙酰-3,5-二甲基吡嗪
acetyl diphenylamine　N-乙酰二苯胺 $(C_6H_5)_2NCOCH_3$
N-acetyldjenkolic acid　N-乙酰黎豆氨酸；N-乙酰亚甲胱氨酸
N-acetyldopamine　N-乙酰多巴胺
acetylene　①乙炔②双亚乙基 $=CHCH=$
acetylene acid　炔属酸
acetylene-air flame　乙炔-空气焰
acetylene alcohol　炔属醇
acetylene black(=Shawinigan black)　乙炔炭黑
acetylene bond　炔键；三键
acetylene bromide　乙炔基溴* $CH≡CBr$
acetylene buoy　乙炔浮标
acetylene burner　乙炔灯
acetylenecarboxylic acid　乙炔基羧酸；丙炔酸 $CH≡CCOOH$
acetylenechloride　乙炔基氯* $CH≡CCl$
acetylene chlorobromide　乙炔化氯溴 $CHCl=CHBr$
acetylene chloroiodide　乙炔化氯碘 $CHCl=CHI$
acetylene converter　乙炔转化器
acetylene coupling(=Glaser coupling)　格拉塞偶合；乙炔偶合
acetylene cylinder　乙炔气瓶
acetylenedibromide　乙炔化二溴；对称二溴代乙烯；二溴乙烯 $CHBr=CHBr$
acetylenedicarboxylic acid　乙炔二羧酸；丁炔二酸 $COOHC≡CCOOH$
acetylenedichloride　乙炔化二氯；对称二氯代乙烯；二

氯乙烯　CHCl=CHCl
acetylenedihalide　乙炔化二卤；对称二卤代乙烯；二卤乙烯　CHX=CHX
acetylenediiodide　乙炔化二碘；对称二碘代乙烯；二碘乙烯　CHI=CHI
acetylenedinitrile　丁炔二腈　CNC≡CCN
acetylenediol　乙炔二醇
acetylene diureine　乙炔二脲〔俗〕
acetylenedivinyl　乙炔基丁间二烯；3,5-己二烯-1-炔　CH_2=CHCH=CHC≡CH
acetylene-enriched solvent　富乙炔溶剂
acetylene flame　乙炔焰
acetylene flasher　乙炔闪光器
acetylene fluorobromide　乙炔化氟溴　CHBr=CHF
acetylene fluorochloride　乙炔化氟氯　CHCl=CHF
acetylene-free ethylene　无乙炔乙烯
acetylene generator　乙炔发生器
acetylenehalide　乙炔基卤　CH≡CX
acetylene hydrocarbon　炔烃
acetylene-hydrogen cyanide process　乙炔-氰化氢工艺
acetyleneiodide　乙炔基碘　CH≡CI
acetylene link(age)　炔键；三键
acetylene monoureine　乙炔一脲
acetylene-nitrous oxide flame　乙炔-氧化亚氮火焰
acetylene-oxygen flame　乙炔-氧气焰
acetylene oxygen hose　乙炔氧胶管；气焊胶管
acetylene series　炔属
acetylene stone　电石
acetylene stripper　乙炔汽提塔；乙炔解吸塔
acetylenetetrabromide　四溴化乙炔　$CHBr_2CHBr_2$
acetylenetetrachloride　四氯化乙炔　$CHCl_2CHCl_2$
acetylenetetrahalide　四卤化乙炔　CHX_2CHX_2
acetylenetetraiodide　四碘化乙炔　CHI_2CHI_2
acetylene unit　乙炔(生产)装置
acetylene urea　乙炔脲　$C_2H_2(CONH_2)_2$
acetylene welding set　乙炔(气)焊装置
acetylenic acid　炔(属)酸
acetylenic alcohol　炔(属)醇
acetylenic bond　炔键；三键
acetylenic carbon　炔键碳
acetylenic compound　炔属化合物
acetylenic glycol　炔(属)烃邻二醇
acetylenic Grignard reagent　炔属格利雅试剂
acetylenic halide　炔属卤化物；卤化炔
acetylenic ketone　炔(属)酮
acetylenic link(age)　炔键；三键
acetylenic polyhalide　炔属多卤化物
acetylenic polymer　乙炔类聚合物*

acetylenic thioether　乙炔型硫醚
acetylenylbenzene　乙炔基苯　CH≡CC_6H_5
acetylenyl carbinol　乙炔基甲醇　CH≡CCH_2OH
acetyl ethanolamine　乙酰基乙醇胺〔增塑剂〕
acetyl ethyl ricinoleate　乙酰蓖麻酸乙酯〔增塑剂〕
acetyl eugenol(=eugenyl acetate)　乙酸丁子香酚酯；乙酰基丁子香酚
acetylferrocene(=ferrocenyl methyl ketone)　二茂铁基甲基(甲)酮；乙酰基二茂铁　$(C_{10}H_9Fe)COCH_3$
acetyl ferrocene polymer　乙酰基二茂铁聚合物
acetyl fibre　乙酰化纤维
acetylfluoride　乙酰氟　CH_3COF
acetyl formazyl　乙酰苯腙基
acetyl formic acid　乙酰甲酸　$CH_3COCOOH$
2-acetyl furan　2-乙酰呋喃
acetyl furan　乙酰呋喃
N-acetylgalactosamine　N-乙酰半乳糖胺
N-acetylgalactosamine diphosphouridine　N-乙酰半乳糖胺二磷酸尿苷
acetyl gasoline　(含)乙炔汽油
acetylglucosamine　乙酰氨基葡萄糖；N-乙酰葡萄糖胺
N-acetylglutamate　①N-乙酰谷氨酸②N-乙酰谷氨酸盐(酯或根)
acetylglycine　N-乙酰甘氨酸　$CH_3CONHCH_2COOH$
acetyl glycocoll(=acetyl glycine)　N-乙酰甘氨酸；乙酰氨基乙酸　$CH_3CONHCH_2COOH$
acetyl group　乙酰基
N-acetylhexosamine　N-乙酰己糖胺
N-acetyl hexosaminidase　N-乙酰氨基己糖苷酶
acetyl hydride　乙酰氢；乙醛
acetylhydroperoxide　乙酰化过氧氢；过醋酸　CH_3COO_2H
N-acetyl-hydroxylamine　N-乙酰羟胺　$CH_3CONHOH$
N-acetyl-hydroxytyramine glucoside　N-乙酰基羟基酪胺葡萄糖苷
acetylide　炔化物*
acetylidene　异乙炔
acetylimino　乙酰亚氨基　CH_3CON=
acetyliminodiacetic acid　乙酰亚氨基二乙酸
N-acetyl indole　N-乙酰吲哚　$CH_3CONC_8H_6$
acetyliodide　乙酰碘　CH_3COI
acetyl ionone　乙酰紫罗兰酮
acetylisable　可乙酰化的
N-acetyl isatin　乙酰靛红　$CH_3CON(CO)_2C_6H_4$
acetyl isobutyryl　乙酰异丁酰
acetyl isoeugenol　乙酰基异丁香酚；乙酸异丁香酚酯
acetyl isopentyryl　乙酰异戊酰

1-acetyl-3-isopropylidene-Δ^1-cyclopentene 1-乙酰-3-异丙叉环戊烯-1

acetyl-isovaleryl 2,3-异庚二酮 $CH_3COCOCH_2CH(CH_3)_2$

acetylith 电石

acetylization 乙酰化作用

acetylization flask 乙酰化用烧瓶

acetylizer 乙酰化器

acetylizing ①乙酰化的②乙酰化作用

acetylizing agent 乙酰化剂

acetyllipoate ①乙酰硫辛酸②乙酰硫辛酸盐(酯或根)

acetyllongifolene 乙酰长叶烯

acetyl-malic acid 乙酰基苹果酸；乙酸基丁二酸 $C_2H_3O_2C_2H_3(CO_2H)_2$

acetylmalonyl enzyme 乙酰丙二酸单酰酶；乙酰丙二酰(基)酶

acetylmannosamine 乙酰甘露糖胺

N-acetylmescaline N-乙酰墨斯卡灵

N-acetyl-3-methoxytyramine N-乙酰-3-甲氧酪胺

acetyl methyl carbinol 乙酰甲基甲醇

N-acetylmuramyl pentapeptide N-乙酰胞壁酰五肽

2-acetylnaphthalene 2-乙酰萘

N-acetyl naphthylamine N-乙酰萘胺

N-acetylneuraminate N-乙酰神经氨(糖)酸；唾液酸

acetyl nitrate 硝酸乙酰酯；硝(酸)乙(酸)酐 $CH_3CO_2NO_2$

acetyl nonanoyl 乙酰壬酰

acetyl number(= acetyl value) 乙酰值

acetylornithine deacetylase 乙酰基鸟氨酸脱乙酰基酶

acetylornithine transaminase 乙酰基鸟氨酸转氨酶

acetyl oxide(=acetic anhydride) 乙酰化氧；乙(酸)酐 $(CH_3CO)_2O$

acetyl pantothenyl ethyl ether 乙酰基泛酰乙醚

acetyl pelargonyl 乙酰壬酰

acetyl pentanoyl 乙酰戊酰

acetylperchlorate 高氯酸乙酰酯；乙酸高氯酸酐 $CH_3CO_2ClO_3$

acetylperoxide 过氧化乙酰 $(CH_3CO)_2O_2$

acetylphenol 乙酰苯酚 $HOC_6H_4COCH_3$

N-acetyl phenylglycine N-乙酰苯氨基乙酸 $C_6H_5N(COCH_3)CH_2CO_2H$

acetylphenylhydrazine 乙酰苯肼

acetylphenyl salicylate 乙酰水杨酸苯酯

acetyl phosphate ①乙酰磷酸②乙酰磷酸盐(酯或根)

acetyl piperidine 乙酸哌啶 $H_3CCONC_5H_{10}$

β-acetylpropionic acid β-乙酰丙酸

acetyl-propionyl 乙酰丙酰；2,3-戊二酮 $CH_3COCOCH_2CH_3$

acetylpropyl alcohol 乙酰丙醇 $CH_3COCH_2CH_2CH_2OH$

acetyl pseudoionone 乙酰假性紫罗兰酮；柠檬叉乙酰丙酮

acetyl pure yellow 乙酰纯黄

acetylpyrazine 乙酰吡嗪

acetylpyridine 乙酰吡啶

2-acetylpyridine 2-乙酰吡啶

3-acetylpyridine 3-乙酰吡啶

acetyl-pyrrole 乙酰吡咯 $CH_3CONC_4H_4$

2-acetyl pyrrole 2-乙酰吡咯

acetyl-resorcin(=euresol; resorcin monoacetate) 乙酰苯间二酚；间苯二酚-乙酸酯；间乙酸基苯酚 $CH_3CO_2C_6H_4OH$

acetyl ricinoleate 乙酰蓖麻酸酯〔增塑剂〕

acetyl-salicylaldehyde 乙酰水杨醛 $C_9H_8O_3$

acetylsalicylate 乙酰水杨酸(盐或酯) $CH_3COOC_6H_4COOR$

acetyl-salicylic acid(=aspirin) 乙酰水杨酸；阿司匹林〔药〕；邻己酸基苯甲酸 $CH_3CO_2C_6H_4COOH$

acetyl salol(=acetylphenyl salicylate) 乙酰萨罗；乙酰水杨酸苯酯 $CH_3CO_2C_6H_4CO_2C_6H_5$

acetyl saponification number 乙酰皂化值

acetyl scarlet 乙酰猩红

acetylserine 乙酰丝氨酸

acetylserotonin N-乙酰-5-羟色胺

acetylsongorine 乙酰准噶尔(乌头)碱

acetylspiramycin 乙酰螺旋霉素〔药〕

acetyl sugar 乙酰化糖

acetylsulfuric acid 乙酰硫酸 CH_3COOSO_3H

acetyltalatisamine 乙酰塔拉(乌头)明

acetyltannic acid 乙酰酸

acetyltannin(=tannigen) 乙酰单宁

2-acetyl thiazole 2-乙酰噻唑

acetylthiohydantoin 乙酰海硫因〔俗〕；1-乙酰基乙内酰硫脲 $CH_3CONCH_2CONHCS$

acetyl thiokinase 乙酰硫激酶

acetyl thiourea 乙酰硫脲 $CH_3CONHCSNH_2S$

acetyl thymol 乙酰百里香酚 $C_{12}H_{16}O_2$

acetyl toluidine N-乙酰甲苯胺 $CH_3C_6H_4NHCOCH_3$

acetyl triallyl citrate 柠檬酸乙酰基三烯丙酯〔增塑剂〕

acetyl tributyl citrate 柠檬酸乙酰基三丁酯〔增塑剂〕

acetyl triethyl citrate 柠檬酸乙酰基三乙酯〔增塑剂〕

acetyl tri-2-ethylhexyl citrate 柠檬酸乙酰基三-2-乙己酯〔增塑剂〕

acetyl trihexyl citrate 柠檬酸乙酰基三己酯〔增塑剂〕

acetyltriphenylsilane 乙酰基三苯基硅烷

acetyltropeine 乙酰托品酯 $CH_3COC_8H_{14}ON$

acetyltryptophan 乙酰色氨酸
N-acetyltyramine N-乙酰酪胺
N-acetyl tyrosine ethyl ester N-乙酰酪氨酸乙酯
acetyl urea 乙酰脲 $CH_3CONHCONH_2$
acetyl valeryl 乙酰戊酰
acetyl value(= acetyl number) 乙酰值
acetyl vanillin 乙酰香兰素
Acheson furnace 阿切孙电炉
Acheson graphite 阿切孙石墨
achilleic acid 蓍草酸；乌头酸
$\quad COOHCH_2C(COOH)=CHCOOH$
achilleine 蓍草碱
achilleline 蓍亭〔蓍草碱的水解产物〕
achilletin 蓍草素
achillin 蓍灵
achiral 非手性(的)*
achiral molecule 非手性分子
achiral reagent 非手性试剂
achirality 非手性型
achmatite achmite 钠辉石；绿帘石
acholest 血清胆碱酯酶测定试纸
achrodextrin(=achroodextrin) 消色糊精
achroic 无色的
achroite 无色电气石
achromatic ①消色差的②消色的
achromatic colours 消光剂
achromatic condenser 消色差聚光器
achromatic indicator 消色指示剂
achromatic lens 消色差透镜
achromatic light 消色光
achromatic magnetic mass spectrometer 消色差磁质谱计(仪)
achromatic objective 消色差物镜
achromatic substage 消色差装置
achromatic system 消色差镜系
achromatism 消色差
achromatization 消色差化
achromic 消色(的)；无色(的)
achromic method 消色测定法
achromic point 消色点
achromycin ①(=tetracycline)四环素②(=puromycin)嘌呤霉素
achrooamyloid 无色淀粉样蛋白
achroodextrin(=achrodextrin) 消色糊精
aci- 〔词头〕酸式
aciclovir 阿昔洛韦〔药〕
aci-compound 酸式化合物
acicular 针状的

acicular goethite 针状水合氧化铁；针状铁黄
acicularity 针状度
acicular particles 针状粒子
acicular pigment 针状颜料
acicular type zinc oxide 针状氧化锌
acid ①酸*②酸的
F-acid F酸；2-萘酚-7-磺酸
γ-acid γ酸；氨基萘酚磺酸
π-acid π酸
acid accelerator 酸性促进剂
acid acceptance 酸容受性；耐酸性；酸稳定性
acid acceptor 酸性接受体；酸性中和剂
acid accumulator 酸性蓄电池*
acid activation 酸活化
acid adipate 酸式己二酸盐(或酯) $COOH(CH_2)_4COOM$
acidaffin 亲酸物
acid alcoholysis 酸式醇解
acid alizarine black 酸性茜素黑
acid alizarine blue 酸性茜素蓝
acid alizarine blue-black 酸性茜素蓝黑
acid-alkali cell 酸碱电池
acid-alkali method of reclaiming 酸碱再生法
acid-alkali-proof hose 耐酸碱胶管
acid alkali reclaiming method 酸碱再生法〔再生胶〕
acid alkylation 酸性烷基化作用
acidamide(=amide) 酰胺 $RCONH_2$
acid amidine 烷基脒 $RC(NH_2)=NH$
acid amidochloride 1,1-二氯代烃胺 $RCCl_2NH_2$
acid analyzer 酸分析器
acid and base resistant rubber gloves 耐酸碱橡胶手套
acid anhydride 酸酐
acid-aniline fuel 酸-苯胺燃料
acid anthraquinone blue 酸性蒽醌蓝
acid arsenate 酸式砷酸盐 MH_2AsO_4；M_2HAsO_4
acid asphalt 酸性地沥青
acidate ①酸化②酰化
acidating ①酸化②酰化
acidating agent 酸化剂
acidation(=acidulation) 酸化；酰化
acid attack 酸侵蚀
acid azelaate 酸式壬二酸盐(或酯) $COOH(CH_2)_7COOM$
acid azide 酰基叠氮 $RCON_3$
acid azo-color 酸性偶氮染料
acid azo-dye 酸性偶氮染料
acid azo pigment 酸性偶氮颜料
acid balling(=acid ball milling) 酸(助球磨)研磨
acid ball milling(=acid balling) 酸(助球磨)研磨

acid barium oxalate 酸式草酸钡 Ba(HC$_2$O$_4$)$_2$
acid-base balance 酸碱平衡
acid-base catalysis 酸碱催化*
acid-base catalyzed reaction 酸碱催化反应*
acid-base determination 酸碱测定(法)
acid-base determination of petroleum products 石油产品的酸碱测定法
acid-base dissociation constants 酸碱解离常数
acid-base equilibrium 酸碱平衡*
acid-base indicator 酸碱指示剂*
acid-base neutralization titration 酸碱中和滴定
acid-base titration 酸碱滴定(法)*
acid-base titration detector 酸碱滴定检测器
acid-base titration method 酸碱滴定法
acid bath 酸浴
acid binding agent 酸结合剂
acid black 酸性黑
acid bleaching 酸性漂白
acid blow case 吹气扬酸箱
acid blue 酸性蓝
acid boiling test 酸煮沸试验
acid bordeaux 酸性枣红
acid brick 耐酸砖
acid brilliant blue 酸性亮蓝
acid bromide 酰基溴 RCOBr
acid brown 萘胺棕
acid buret 酸滴定管
acid calcium carbonate 酸式碳酸钙
acid calcium phosphate 酸式磷酸钙〔1. 磷酸氢钙 CaHPO$_4$; 2. 磷酸二氢钙 Ca(H$_2$PO$_4$)$_2$〕
acid calcium phosphite 酸式亚磷酸钙 CaHPO$_3$
acid capacity 酸容量
acid carbonate 酸式碳酸盐(或酯) MHCO$_3$
acid carboy 酸坛
acid carboy inclinator 酸坛倾斜器
acid catalysed 酸催化的
acid-catalysed cleavage 酸催化裂解
acid-catalysed rearrangement 酸催重排作用
acid catalysis 酸催化*
acid catalyst 酸催化剂
acid catalyst polycondensation 酸催化缩聚(作用)
acid-catalyzed cleavage 酸催裂解
acid-catalyzed polycondensation 酸催化缩聚(作用)
acid catalyzed polymerization 酸(引发)聚合;酸催化聚合
acid catalyzed reaction 酸催化反应
acid centrifugal pump 酸离心泵
acid cerise 酸性樱红

acid chloride 酰基氯;氯化酰基
acid chrome black 酸性铬黑
acid chrome salt 酸性铬盐
acid circulating pump 酸循环泵
acid clay 酸性黏土
acid cleaning 酸洗
acid cleavage 酸裂解
acid coagulant 酸性凝固剂
acid coagulant tank 酸凝槽
acid coagulating bath 酸性凝固浴
acid coagulation 酸性凝固;酸凝固法
acid coke 酸性焦炭
acid color ①酸色②酸颜料
acid complex 酸式配合物〔例如 MHL，其中 M 代表金属，L 代表配位体〕
acid concentration 酸浓度
acid concentrator 酸浓缩器
acid condenser 酸冷凝器
acid condensing agent 酸冷凝剂
acid constituent 酸组分
acid consumption 耗酸量
acid content 酸含量
acid converter 酸性转炉
acid converter process 酸性转炉(炼钢)法
acid cooking 酸法蒸煮
acid cooler 冷酸器;酸冷却器
acid copper arsenite 酸式(性)亚砷酸铜 CuHAsO$_3$
acid corrosion 酸腐蚀
acid corrosion inhibitor 抗酸缓蚀剂
acid cream 酸性霜
acid cresol red 酸性甲酚红
acid crimson〔商〕 酸性绯红
acid cupric arsenite 酸性亚砷酸铜 CuHAsO$_3$
acid curd ①酸凝乳②酸渍
acid cure 酸固化;酸凝
acid-cured epoxy resin 酸固化环氧树脂
acid-cured resin 酸固化树脂
acid-curing 酸固化
acid cut 酸定〔在油漆稀释剂中加入硫酸以测定烃含量〕
acid damaged fibre 酸伤丝;酸烧丝
acid deasphalting 硫酸脱沥青法
acid declining phase 减酸期
acid decomposition 酸性分解
acid-defiber(iz)ing 酸法除纤维〔橡胶〕
acid deficiency 缺酸(条件)
acid-deficient 缺酸(的)
acid-deficient feed 缺酸料液

acid degradation 酸降解
acid degree 酸度
acid deposition 酸沉降
acid desizing 酸退浆
acid diamide 酰二胺
acid digestion ①酸浸法②(钛白)酸解
acid digestion bomb 酸消化罐
acid diol 酸(式)二(元)醇
acid dipping 酸浸渍
acid dip process （表面处理）酸蘸法
acid discharge hose 排酸胶管
acid dissociation 酸式电离
acid dissociation constant 酸解离常数
acid dissociation exponent 酸解离指数
acid drift 变酸倾向
acid dye 酸性染料
acid-dyeable fibre 酸性染料可染型纤维
acid dye colorimetry 酸性染料比色法
acid dyeing 酸性染色
acid dye pigment 酸性染料基颜料；酸性染料型颜料
acid effect 酸效应
acid elevator 升酸器；扬酸器
acid end group 酸端基；羧端基
acid error 酸误差
acid-error of glass pH electrode 玻璃 pH 电极的酸误差
acid ester 酸性酯
acid-ester exchange 酸-酯交换(作用)
acid etch 酸洗；酸浸；酸蚀
acid etching 酸蚀刻
acid ethylenesulfate 酸式硫酸乙烯酯 $C_2H_4(HSO_4)_2$
acid ethyl oxalate 草酸氢乙酯
acid ethylphosphite 酸式亚磷酸乙酯 $C_2H_5OP(OH)_2$
acid ethylsulfate 酸式硫酸乙酯 $C_2H_5OSO_3H$
acid ethylsulfite 酸式亚硫酸乙酯 $C_2H_5OSO_2H$
acid extract 酸性萃取；酸提出物
acid extractor 脱酸机
acid fallout 酸性沉降物
acid-fast stain 抗酸染色法
acid feeder 送酸器；加酸器
acid fluoride 酰基氟 RCOF
acid fog 酸雾
acid-form 酸式
acid former 成酸物质；成酸剂
acid-forming group 成酸基
acid-free ink 无酸墨水
acid-free oil 无酸油
acid-free spinning 无酸纺丝
acid from recovery plant 再生酸

acid fuchsin〔商〕 酸性品红
acid fumarate 酸式富马酸盐(或酯)
  COOHCH=CHCOOM；COOHCH=CHCOOR
acid fume 酸雾
acid gas 酸性气体
acid gas absorption 酸气吸收作用
acid gas extraction 酸性气抽提
acid glutarate 酸式戊二酸盐
acid group ①酸基②酸根
acid halide 酰基卤；卤化酰基
acid hearth 酸性敞炉
acid hose 耐酸胶管
acid hydrazide 酰基肼 $RCONHNH_2$
acid-hydrocarbon emulsion 酸烃乳浊液；酸烃乳状液
acid hydrogen 酸式氢
acid hydrolysis 酸解；加酸水解
acidic ①酸式②酸性③酸(的)
acidic accelerator 酸性促进剂
acidic black 酸性炭黑
acidic carbon black 酸性炭黑
acidic cation exchanger 酸性阳离子交换剂
acidic cleaner 酸性清洗剂
acidic component 酸性组分
acidic deruster 酸性除锈剂
acidic extractant 酸性萃取剂
acidic group 酸性团；酸性类
acidic hydrocarbon 酸性烃
acidic hydrogen 酸性氢
acidic impingement black 酸性接触法炭黑；酸性槽法炭黑
acidic inhibitor(=acidic retardant) 酸性防焦剂
acidic ligand 酸式配位体
acidic material 酸性物质
acidic oxide(=acid oxide) 酸性氧化物；成酸氧化物
acidic phosphate extraction 酸性磷酸酯萃取
acidic potential 酸性电位
acidic process 酸法处理
acidic property 酸性
acidic resin 酸性树脂
acidic retardant(=acidic inhibitor) 酸性防焦剂
acidic-type accelerator 酸性促进剂
acidic waste 酸性废物
acidiferous 含酸的
acidifiable 可酸化的
acidifiable base 可酸化碱
acidification 酸化*
acidified 酸化了的
acidifier ①酸化器②酸化剂

acidifying ①酸化的②酸化(作用)
acidifying agent 酸化剂
acidimeter(=acidometer) 酸度计
acidimetric 酸量滴定的
acidimetric analysis 酸量滴定分析(法)
acidimetric estimation 酸量测定
acidimetric method 酸量滴定法
acidimetric standard 标酸基准；标定酸溶液的基准物
acidimetry 酸量法*
acid imide 酸亚胺；酰亚胺 RC(OH)=NH
acid imidochloride 偕氯代烃亚胺 RCCl=NH
acid inclining phase 增酸期
acid index(=acid number) 酸值
acid indicator 酸性指示剂
acid-in-oil emulsion 油包酸乳液
acid-insoluble lignin 酸不溶木素
acid iodide 酰基碘 RCOI
acid ion 酸基离子
acidite 酸性岩
acidity 酸度*
acidity coefficient 酸度系数
acidity constant 酸度常数*
acidity control 酸度控制
acidity detect 酸度检定
acidity distribution 酸度分布
acidity function 酸度函数
acidity index 酸度指数
acidity in fruits and vegetables 果蔬酸度(测定)
acidity potential 质子亲和势
acidity regulator 酸度调和剂
acidity test 酸度检定
acidization 酸化；酸处理
acidizer 酸性剂；酸处理剂
acid ketone 酮酸
acid-laden fog 酸雾
acid latex 酸性胶乳；阳电荷胶乳
acid leach ①酸沥滤②酸沥滤产物
acidless 无酸的
acidless sulfur 脱酸硫
acid leveling colour 酸性匀染染料
acid light dyestuff 酸性耐光染料
acid lime 酸柠檬；宜母子
acid lining 酸性内衬；酸性炉衬
acid liquor 酸液
acid magenta 酸性品红
acid magnesium citrate 酸式柠檬酸镁 $MgH \cdot C_6H_5O_7$
acid malate 酸式苹果酸盐 $COOHCHOHCH_2COOM$
acid maleate 酸式马来酸盐；酸式顺丁烯二酸盐(或酯) COOHCH=CHCOOM
acid malonate 酸式丙二酸盐 $COOHCH_2COOM$
acid mantle (皮肤的)酸保护层
acid measurer 量酸器
acid medium 酸性介质
acid milling 酸性缩绒
acid milling red 酸性磨红
acid mist 酸雾
acid mist analysis at atmosphere 大气中的酸雾分析
acid modified fibre 酸改性纤维
acid-modified polyester fibre 酸改性聚酯纤维
acid mordant 酸性媒染剂
acid mordant dye 酸性媒染染料
acidness ①酸度②酸性
acid-neutralizing capacity 酸中和容量
acid number(=acid value) 酸值
acid of amber 琥珀酸；丁二酸
acid of ants 甲酸；蚁酸〔俗〕
acid of apples 2-羟丁二酸；苹果酸〔俗〕
acid of milk α-羟基丙酸；乳酸〔俗〕
acid oil 酸性油
acidol 盐酸甜菜碱
acidolysis 酸解*
acidometer(=acidimeter) 酸度计
acidometric conductometric titration 酸量电导滴定(法)
acid open-hearth furnace 酸性敞炉
acidophobe ①疏酸基②疏酸体
acidophobic compound 疏酸化合物
acid orange 酸性橙
acidosis 酸中毒
acidotic 定酸量的
acid oxalate 酸式草酸盐 COOHCOOM
acid oxide(=acidic oxide) 酸性氧化物；成酸氧化物
acid pasting 酸溶法
acid peroxide 酸过氧化物
acid phosphatase 酸性磷酸酶
acid phosphatase staining 酸性磷酸酶染色法
acid phosphate ①酸式磷酸盐②过磷酸钙〔专指〕
acid phosphate den 过磷酸钙(储藏)室
acid phosphate mixer 过磷酸钙混合器
acid phosphate mixing pan 过磷酸钙混合锅
acid phthalate 酸式苯二酸盐 $COOHC_6H_4COOM$
acid pickling 酸渍
acid pigment 酸性颜料
acid pimelate 酸式庚二酸盐 $COOH(CH_2)_5COOM$
acid plant 制酸厂
acid plumping 酸膨胀
acid polymerization 酸(引发)聚合

acid ponceau 酸性丽春红
acid potassium acetate 酸式乙酸钾 $KH(C_2H_3O_2)_2$
acid potassium cobaltous carbonate 酸式碳酸钴钾；碳酸氢钴钾 $KHCO_3 \cdot CoCO_3$
acid potassium oxalate 酸式草酸钾 KHC_2O_4
acid potassium tartrate 酸式酒石酸钾 $COOH(CHOH)_2COOK$
acid precipitation 酸雨
acid pressure leaching 酸法加压浸出
acid producing accelerator 酸性促进剂
acid-proof 耐酸的
acid-proof alloy 耐酸合金
acid-proof brick 耐酸砖
acid proof cast iron 耐酸铸铁
acid-proof cement ①耐酸水泥②耐酸黏合剂
acid-proof concrete 耐酸混凝土
acid-proof enamel 耐酸搪瓷
acid-proof fabric 耐酸织物
acid-proof galosh 耐酸套鞋
acid-proof hose 耐酸软管
acid-proof lining 耐酸内衬
acid-proof material 耐酸材料
acid-proof paint 耐酸涂料；耐酸漆
acid-proof slab 耐酸板
acid proof steel 耐酸钢
acid-proof stoneware 耐酸缸器
acid pump 酸泵
acid purification system 酸净化系统；酸再生系统
acid radical ①酸根②酸基③酰基
acid rain 酸雨
acid ratio 酸比
acid reaction 酸性反应
acid receiver 盛酸器；储酸器
acid reclaim 酸性翻造；酸性再生
acid recovery 酸回收
acid recovery plant 废酸回收厂
acid red 酸性红
acid reduction 酸性还原作用
acid-refined oil 酸洗油
acid refining 酸炼
acid refractory 酸性耐火材料
acid residue 酸性残渣
acid resin 酸性树脂；阳离子交换树脂
acid resistance 耐酸性
acid resistant 耐酸；抗酸
acid resistant coating 耐酸涂层
acid resistant enamelled glass coating 耐酸搪玻璃(涂)层
acid resistant gloves 耐酸手套
acid-resistant paint 耐酸漆
acid resisting 耐酸(的)
acid-resisting brick 耐酸砖
acid-resisting cement 耐酸水泥
acid resisting hose 耐酸胶管
acid-resisting iron 耐酸铁
acid-resisting pump 耐酸泵
acid-resisting steel 耐酸钢
acid-resisting test 耐酸性试验
acid restoring plant 废酸回收工厂
acid rhodamine B 酸性罗丹明B
acid rinse 酸性漂洗
acid rinsing 酸(水)(漂)洗；酸水冲洗
acid rock 酸性岩
acid rubber 酸性橡胶；羧基橡胶
acid salt 酸式盐
acid salt ratio 酸盐比率
acid salt tolerance 酸碱度
acid saponification 加酸皂化
acid scale 酸秤
acid scavenger 酸清除剂；除酸剂；酸中和剂
acid seal 酸封
acid seal paint 酸封漆
acid sebacate 酸式癸二酸盐 $COOH(CH_2)_8COOM$
acid sensitive indicator 酸敏感指示剂
acid sensitive magnesium orthosilicate 酸敏性原硅酸镁
acid-setting resin 酸固化性树脂
acid settler(=acid settling tank) 酸沉降器
acid settling tank(=acid settler) 酸沉降器
acid sight box 窥酸箱
acid site 酸性部位；酸性点
acid slag 酸性炉渣
acid sludge ①酸性污泥②酸渣
acid sludge asphalt 酸渣沥青
acid sludge fuel 酸渣燃料
acid-sludge pitch 酸渣沥青
acid slurry method(=acid swelling) 酸浆法；酸溶胀法
acid smut 酸性煤尘
acid sodium acetate 酸式乙酸钠 $NaH(C_2H_3O_2)_2$
acid sodium tartrate 酸式酒石酸钠 $NaHC_4H_4O_6$
acid soil 酸性土壤
acid-soluble lignin 酸溶木素
acid solution 酸性溶液
acid solvent 酸性溶剂
acid spinning bath 酸性纺丝浴
acid spinning process 酸性(浴)纺丝法；酸性纺丝过程
acid splitting 酸解；加酸分解
acid-sprayed clay 酸喷黏土

acid-stage oil 酸性油
acid stain 酸性染色剂；酸性着色剂
acid steel 酸性钢
acid steeping 酸浸渍；酸退浆
acid storage battery 酸性蓄电池
acid strength 酸强度
acid suberate 酸式辛二酸盐 COOH(CH$_2$)$_6$COOM
acid succinate 酸式丁二酸盐 COOH(CH$_2$)$_2$COOM
acid sulfate 酸式硫酸盐 MHSO$_4$
acid sulfite 酸式亚硫酸盐 MHSO$_3$
acid sulfite process 酸性亚硫酸盐法
acid sulfite semichemical pulp 酸性亚硫酸盐半化学浆
acid swelling(=acid slurry method) 酸浆法；酸溶胀法
acid swollen sole leather 酸胀底革
acid tank 酸槽
acid tar 酸焦油
acid tartrate 酸式酒石酸盐 COOH(CHOH)$_2$COOM
acid tartronate 酸式丙醇二酸盐 COOHCHOHCOOM
acid test 酸性试验
acid thermocoagulation 酸性热凝固法
acid Thorex process 酸式钍雷克斯流程；TBP 萃取过程
acid thymol blue 酸性百里酚蓝
acid tin phosphate 酸式磷酸锡
acid tower 酸塔；制酸塔
acid-treated clay(=acid-treated earth) 酸化黏土
acid-treated distillate 酸化馏出物；酸处理过的馏出物
acid-treated earth(=acid-treated clay) 酸化黏土
acid-treated oil 酸洗油
acid treatment 酸处理
acidulant 酸化剂
acidulated rinsing 酸浴冲洗
acidulating 酸化的；酸化
acidulating agent 酸化剂
acidulation(=acidation) 酸化；酰化
acidulin 盐酸合谷氨酸
acidulous 微酸的；带酸味的
acid value 酸值*
acid-vapor digestion 酸蒸气消化
acid wash 酸洗
acid wash color test 酸洗消色试验
acid washing test 酸洗试验
acid waste 酸性废物
acid waste liquid 酸性废液
acid waste products 酸废料
acid waste water 酸性废水
acid water 酸水
acid weigher 酸秤
acid yellow 酸性黄

acidylable 酰化的
acidylate(=acylate) 酰化
acidylated 酰化了的
acidylating ①酰化的②酰化(作用)
acidylating agent 酰化剂
acidylation(=acylation) 酰化作用
acieral 铝基合金
aci form 酸式*
aci-modification 酸(式变)型
AC impedence method 交流阻抗法*
aci-nitro(=isonitro) 异硝基；硝酸亚基 (HO)ON=
aci-nitro compound (亚)酸式硝基化合物
aci-nitro group (亚)酸式硝基 HOON=
acinose structure 粒状结构
aciolein(=acrylic aldehyde) 丙烯醛 CH$_2$=CHCHO
acipimox 阿昔莫司〔药〕
acitrin 2-苯基辛可宁酸乙酯
acivinil alcohols 不饱和酮醇类
Acker cell(for caustic soda) 阿克(制苛性钠)电池
Ackermann automatic reckoner 埃克曼自动计算盘
Acker process 埃凯法〔一种电解熔盐制氢氧化钠法〕
aclarubicin 阿柔比星〔药〕
Acme burner 顶热灯
Acme screw thread 爱克米螺纹；梯形螺纹
acmite 锥辉石
acne cream 痤疮脂
acne lotion 粉刺化妆水
acocantherin 东非箭毒树苷
acofriose 鼠李糖-3-甲醚
acoin 阿可因〔药〕
acolytine 乌头属碱
aconate 阿康酸盐(或酯)
aconic acid 阿康酸 CH$_2$COOCH=CCOOH
aconine ①乌头原碱②阿康碱
aconitate ①(顺)乌头酸②(顺)乌头酸盐(酯或根)
aconite 乌头
aconite alkaloid 乌头植物碱
aconitic acid 乌头酸；丙烯三甲酸 C$_3$H$_3$(COOH)$_3$
cis-aconitic anhydride 顺-3-羧基戊烯二酸酐；顺-乌头酸酐
aconitine 乌头碱
aconitine arsenate 砷酸乌头碱
aconityl phenetidine 乌头乙氧苯胺
acopyrin 阿科比林〔药〕
acoradiene 菖蒲二烯
acoragermacrone 菖蒲香酮
acorane 菖蒲烷

acorenone 菖蒲烯酮
acoretin 菖蒲脂
acorin 菖蒲苷 $C_{36}H_{60}O_6$
acorn cups 橡碗
acorn flour 浆栎粉
acorn oil 浆栎油
acorn sugar 浆栎糖
acorone 菖蒲酮 $C_{15}H_{24}O_2$
acoronolide 菖蒲内酯
acoroxide 菖蒲醚
acoustic admittance 声导纳
acoustic agitation 声频扰动
acoustical absorption coefficient 吸声系数
acoustical absorption loss 吸声损失
acoustical board 消音板
acoustical fatigue 声致疲劳
acoustical panel 消音板
acoustic capacitance 声容
acoustic coating 吸声涂料；吸音涂料
acoustic coupler 声耦合器〔计〕
acoustic frequency 音频〔物〕
acoustic impedance 声阻抗
acoustic impedance per unit area 单位面积声阻抗
acoustic insulating material 隔音材料*
acoustic irradiation 声波辐射
acoustic maximum 最大音响
acoustic microscope 声学显微镜
acoustic microscopy 声学显微术
acoustic paint 吸声漆；吸音漆
acoustic panel 隔音板；吸音板
acoustic property 声学性能；吸音性
acoustic radiation resistance 声辐射阻尼
acoustic resistant 声抗
acoustics ①声学②音响效应
acoustic wave sensor 声波传感器
acoustoelectric effect 声电效应
acoustooptic effect 声光效应
acousto-optic modulation 声光调制
acousto-optic modulator 声光调制器
acoustooptic tunable filter 声光可调滤光器
acoustooptic tuning 声光调谐
acovenose 毒光药木糖；3-甲基-6-脱氧塔罗糖
AC permeability 交流导电率
AC polarography 交流极谱法*
AC pulse polarography 交流脉冲极谱法
acquired immunodeficiency syndrome(AIDS) 艾滋病；获得性免疫缺陷综合征
acquisition of time-domain signal 时域信号的采集

acquisition time 取数时间*
acraldehyde 丙炔醛
acrichine(=atebrine) 阿的平
acrid ①辛辣的②腐蚀性的
acridan 9,10-二氢化吖啶 $C_{13}H_{11}N$
acridanyl 二氢化吖啶基 $C_{13}H_{10}N-$
acridic acid(=acridinic acid; 2,3-quinoline dicarboxylic acid) 吖啶酸；2,3-喹啉二羧酸 $C_{11}H_7O_4N$
acridine 吖啶 $C_6H_4CHC_6H_4N$
acridine derivatives 吖啶衍生物
acridine dye 吖啶染料
acridine orange 吖啶橙 $NC_{13}H_7[N(CH_3)_2]_2$
acridine red 吖啶红 $CH_3NHC_7H_7O=N(Cl)CH_2$
acridinic acid(=acridic acid; 2,3-quinoline dicarboxylic acid) 吖啶酸；2,3-喹啉二羧酸 $C_9H_5N(COOH)_2$
acridinyl(=acridyl) 吖啶基 $C_{13}H_8N-$
acridol 吖啶酚
acridone(=dihydroketoacridine) 吖啶酮 $C_6H_4COC_6H_4NH$
acridostibine 吖啶锑
acridyl(=acridinyl) 吖啶基 $C_{13}H_8N-$
acriflavine 吖啶黄素
acriflavine hydrochloride 盐酸吖啶黄素
acriflavine neutral 吖啶黄素中性品
acrifoline 尖叶(石松)碱
acrimonious 辛辣的；苦的
acrinyl 羟苄基 $OHC_6H_4CH_2-$
acrinyl isothiocyanate 羟苄基异硫氰酯；异硫氰酸羟苄酯 $OHC_6H_4CH_2NCS$
acritol 吖糖醇
acrivastine 阿伐斯汀〔药〕
acrodynia 肢痛症
acrol 亚烯丙基
acrolactic acid 3-羟基丙烯酸
acroleic acid 丙烯酸 $CH_2=CHCOOH$
acrolein(=acrylaldehyde) 丙烯醛 $CH_2=CHCHO$
acrolein polymer 丙烯醛类聚合物*
acrolein reaction 丙烯醛反应
acrolein resin 丙烯醛树脂
acrol(o)yl 丙烯酰(基)
acromelin 蜈蚣苔灵 $C_{17}H_{16}O_3$
acronidine 山油柑定
acronycidine 山油柑榭定
acronycine 山油柑碱
acropeptides 复氨基酸类
acrose ①吖糖；阿柯糖②合成果糖

acrosite(=pyrargyrite) 硫锑银矿
acryl 丙烯酰基
acrylaldehyde(=acrolein) 丙烯醛
acrylamide(=acrylic amide) 丙烯酰胺
　　$CH_2=CHCONH_2$
acrylamide copolymer 丙烯酰胺共聚物
acrylamide-dodecylamine telomer 丙烯酰胺-十二(烷)胺调聚物
acrylamide gel electrophoresis 丙烯酰胺凝胶电泳
acrylamide monomer 丙烯酰胺单体
acrylamide-vinyl succinimide polymer 丙烯酰胺-乙烯基琥珀酰亚胺共聚物
acrylamidine 丙烯脒
acrylamido methyl propane sulfonic acid 丙烯酰胺甲基丙烷磺酸
acrylamido-terminated polyoxyalkylene 丙烯酰胺基封尾的聚氧链烯
acrylate 丙烯酸盐(或酯)
acrylate-butadiene rubber 丙烯酸丁二烯橡胶
acrylate resin 丙烯酸酯树脂
acrylate rubber 丙烯酸酯橡胶
acrylester-acrylnitril rubber 丙烯酸酯丙烯腈橡胶
acrylester-butadiene rubber 丙烯酸酯-丁二烯橡胶
acrylester-2-chlorovinylether rubber 丙烯酸酯-2-氯乙烯醚橡胶
acrylester rubber 丙烯酸酯橡胶
acryl fibre 丙烯腈纤维〔聚丙烯腈纤维及改性聚丙烯腈系纤维〕
acryl glass 丙烯酸酯类有机玻璃
acrylic 丙烯酸(的)
acrylic acid 丙烯酸 $CH_2=CHCO_2H$
acrylic acid-acrylamide-copolymer 丙烯酸-丙烯酰胺共聚物
acrylic acid-acrylonitril-alkanethiol telomer 丙烯酸-丙烯腈-链烷硫醇调聚物
acrylic acid-butadiene styrene copolymer 丙烯酸-丁二烯苯乙烯共聚物
acrylic acid chloride 氯丙烷
acrylic acid ester 丙烯酸酯 $CH_2=CHCOOR$
acrylic acid fiber 丙烯酸纤维
acrylic acid-grafted poly(vinyl alcohol)fibre 丙烯酸接枝的聚乙烯醇纤维
acrylic acid polymer 丙烯酸聚合物
acrylic acid-polyurethane coating 丙烯酸-聚氨酯涂料
acrylic acid series 丙烯酸系
acrylic aldehyde(=aciolein) 丙烯醛 $CH_2=CHCHO$
acrylic amide(=acrylamide) 丙烯酰胺
acrylic anhydride 丙烯(酸)酐 $(CH_2=CHCO)_2O$
acrylic automotive lacquer 丙烯酸汽车喷漆
acrylic based impact modifier 丙烯酸基冲击性能改性剂
acrylic-based telechelic oligomer 丙烯酸基远螯齐聚物
acrylic bulk top 聚丙烯腈膨体条
acrylic continuous filament 丙烯腈系长丝
acrylic copolymer 丙烯酸系共聚物；丙烯腈系共聚物
acrylic elastomer 丙烯酸酯橡胶；丙烯酸系弹性体
acrylic emulsion 丙烯酸乳液
acrylic emulsion paint 丙烯酸乳液漆
acrylic enamel 丙烯酸瓷漆
acrylic epoxy resin 丙烯酸环氧树脂
acrylic ester 丙烯酸酯
acrylic ester emulsion 丙烯酸酯乳液
acrylic ester polymer 丙烯酸酯聚合物
acrylic ethylene copolymer emulsion 丙烯酸乙烯共聚物乳液
acrylic fiber ①聚丙烯腈纤维；腈纶②丙烯酸(类)纤维
acrylic fibrid 聚丙烯腈纤条体
acrylic filament 丙烯腈系长丝
acrylic glass 丙烯酸酯类有机玻璃
acrylic lacquer 丙烯酸清漆；丙烯酸喷漆
acrylic latex 丙烯酸乳胶；丙烯酸橡胶
acrylic lithin coating 丙烯酸系彩色水泥(砂浆)涂料〔建筑物装饰用〕
acrylic nitrile 丙烯腈 $CH_2=CHCN$
acrylic plastics 丙烯酸类塑料
acrylic polyelectrolyte 丙烯酸聚合电解质
acrylic polymer 丙烯酸(酯)类聚合物*
acrylic primer 丙烯酸(酯)底漆
acrylic resin 丙烯酸(酯)类树脂*
acrylic resin adhesive of second generation(SGA) 第二代丙烯酸酯胶黏剂
acrylic rubber 丙烯酸(类)橡胶
acrylics ①丙烯酸类塑料②丙烯酸类树脂③聚丙烯酸酯类
acrylic size 丙烯酸类树脂胶液；丙烯酸类树脂浆
acrylic thermoplastic resin 热塑性丙烯酸树脂
acrylic tow softening machine 腈纶柔软机
acrylic tow washing machine 腈纶水洗机
acrylic undercoat 丙烯酸中间层
acrylic-vinyl fibre 丙烯腈系-乙烯系共聚物纤维
acrylic-wool blend (聚)丙烯腈系纤维-羊毛混纺纱
acrylic yarn 丙烯腈类纤维纱；丙烯酸类纤维纱
acrylonitrile(=vinyl cyanide) 丙烯腈
acrylonitrile-butadiene-acrylate(ABA) 丙烯腈-丁二烯-丙烯酸酯共聚物
acrylonitrile butadiene copolymer 丙烯腈丁烯共聚物
acrylonitrile-butadiene rubber(NBR) 丁腈橡胶；丙烯腈-

丁二烯橡胶
acrylonitrile butadiene rubber latex　丁腈胶乳
acrylonitrile-butadiene-styrene(ABS)　丙烯腈-丁二烯-苯乙烯共聚物
acrylonitrile-butadiene-styrene plastics　丙烯腈-丁二烯-苯乙烯塑料
acrylonitrile-butadiene-styrene resin　丙烯腈-丁二烯-苯乙烯树脂*；ABS 树脂
acrylonitrile-butadiene-styrene rubber　ABS 橡胶；丙烯腈-丁苯橡胶
acrylonitrile-casein graft copolymer fiber　丙烯腈-酪蛋白接枝共聚物纤维
acrylonitrile-chloroprene rubber　氯腈橡胶
acrylonitrile-ethylene-styrene(AES)　丙烯腈-乙烯-苯乙烯共聚物
acrylonitrile-grafted protein fibre　丙烯腈接枝蛋白质纤维
acrylonitrile-grafted rayon　丙烯腈接枝人造丝
acrylonitrile-isoprene rubber　丙烯腈-异戊二烯橡胶
acrylonitrile-itaconic acid ester copolymer fiber　丙烯腈-衣康酸酯(共聚)纤维
acrylonitrile-methyl methacrylate(AMMA)　丙烯腈-甲基丙烯酸甲酯共聚物
acrylonitrile polymer　丙烯腈聚合物
acrylonitrile-protein fibre　丙烯腈-蛋白质纤维
acrylonitrile resin　丙烯腈树脂
acrylonitrile-starch graft copolymer　丙烯腈-淀粉接枝共聚物
acrylonitrile-styrene-acrylate(ASA)　丙烯腈-苯乙烯-丙烯酸酯(共聚物)
acrylonitrile-styrene-acrylate copolymer　丙烯腈-苯乙烯-丙烯酸酯共聚物
acrylonitrile styrene resin(AS)　丙烯腈-苯乙烯树脂
acrylonitrile-vinyl acetate copolymer　丙烯腈-醋酸乙烯共聚物
acrylonitrile-vinyl chloride copolymer　丙烯腈-氯乙烯(共聚)纤维
acrylonitrile-vinylidene chloride copolymer　丙烯腈-偏二氯乙烯共聚物
acrylonitrile waste　(含)丙烯腈废液
acrylophenone　烯丙酰苯；苯丙烯酮
acryloyl(=acrylyl)　烯丙酰　$CH_2=CHCO-$
acryloyl chloride　烯丙酰氯　$CH_2=CHCOCl$
acrylyl(=acryloyl)　烯丙酰
acrysol　聚丙烯酸酯水乳液
acsinatidine　阿克替定
acsinatine(=(S)-4,7-dimethyl-6,7-dihydro-5H-cyclopenta[c] pyridine)　阿克亭

acsine　阿克辛
acsinidine　阿克定
acterol　一种照射的麦角甾醇
ACTH-like structure　促肾皮素样结构
acticarbon　活性炭〔吸附剂〕
actidione　放线(菌)酮；(戊二酰)亚胺环己酮
actification　再生作用
actified solution　再生溶液
actifier column　再生器；再生塔
actiline(=neomycin B;framycetin)　新霉素 B〔商品名〕
acting force　作用力
actiniasterol　海葵甾醇
actinic　(有)光化(性)的
actinic decomposition　光化分解
actinic degradation　光化降解
actinic glass　闪光玻璃
actinic green　(毒物瓶用的)光化绿玻璃
actinicity　光化度；光化性
actinic light　光合光
actinic rays　光化射线
actinide　锕系元素*
actinide chalcogenide　锕系元素硫族元素化合物
actinide chloride　锕系元素氯化物
actinide contraction　锕系收缩*
actinide elements　锕系元素
actinide halide　锕系元素卤化物
actinide-lanthanide separation　锕系-镧系元素(互)分离
actinide metallide　锕系元素金属化物
actinide oxometallate　锕系元素含氧酸盐
actinide oxyfluoride　锕系元素氟氧化物
actinide sesquioxide　锕系元素倍半氧化物
actinide thiocyanate　锕系元素硫氰酸盐
actinidine　猕猴桃碱
actiniochrome　珊瑚虫色
actinism　射线化学
actinity　光化性；光化度
actinium　锕〔89 号元素，化学符号 Ac〕
actinium A　锕 A　^{215}Po
actinium B　锕 B　^{211}Pb
actinium C　锕 C　^{211}Bi
actinium C′　锕 C′　^{211}Po
actinium C″　锕 C″　^{207}Tl
actinium D　锕 D　^{207}Pb；锕铅
actinium emanation(=actinon)　锕射气　^{219}Em
actinium family　锕系
actinium hydroxide　氢氧化锕　$Ac(OH)_3$
actinium K　锕 K　^{223}Fr
actinium lead　锕铅　^{207}Pb

actinium nitrate 硝酸锕 $Ac(NO_3)_3$
actinium series 锕系〔指89~103号元素〕
actinium tricyclopentadienide 三(环戊二烯)锕 $Ac(C_5H_5)_3$
actinium uranium 锕铀 ^{235}U
actinium X 锕 X ^{223}Ra
actinobacillosis 放线菌病
actinobolin(=actinovorin) 放线菌光素
actinocarcin 放线抗癌素
actinochrysin(=actinomycin C) 放线菌素 C
actinocongestin 海葵毒素
actinodaphnine 放线瑞香宁；黄肉楠碱；六驳碱
actinoerythrin 海葵赤素
actinoflavin 放线菌黄毒
actinoflocin 放线菌团素
actinogan 放线菌制癌素
actinograph (日光)光化力测定器；光化线强度记录器
actinohematin 海葵血红毒
actinoid 锕系元素*
actinoidin 类放线菌素
actinoleukin 放线菌白素
actinolichen 放线菌地衣
actinolite 阳起石
actinolite rock 阳起石岩
actinology ①光化学②射线学
actinolysin 放线菌溶素
actinometer 光能测定仪*
actinometry 光能测定学*
actinomorphy(=radial symmetry) 放线对称；辐射对称性
actinomyce 放线菌
actinomycelin 放线菌丝素
actinomycetales 放线菌目
actinomycetin 白放线菌素
actinomycin 放线菌素
actinon(=actinium emanation) 锕射气
actinone 放线菌酮
actinonin 放线酰胺素 $C_{19}H_{35}N_3O_5$
actinorhodine 放线紫素
actinorubin 放线红素
actinospectacin 放线壮观素；大观霉素
actinotiocin 放线硫肽素
actinouranium 锕铀 ^{235}U
actinouranium decay series 锕铀衰变系*
actinouranium family 锕铀系
actinouranium series 锕铀系
actinovorin(=actinobolin) 放线菌光素
actinoxanthine 放线黄质素
actinozyme 放线菌酶

actinyl 锕系酰*
actinyl fluoride 氟化锕系元素酰基
actinyl group 锕系元素酰基 $AnO_2^{2+}; AnO_2^+$
action ①作用②作用量
action limit 操作极限
actiono(li)te 阳起石
action spectrum 作用光谱
actiphenol 放线菌酚
actithiazic acid 放线噻唑酸
activable tracer 可活化示踪剂
activable tracer technique 可活化示踪技术
activated ①活化的②活性的③激活(了)的
activated adsorption 活性吸附；化学吸附
activated aeration process 活性曝气法
activated aluminium 活性铝
activated aluminium oxide 活性氧化铝
activated atom 活化原子
activated bauxite 活性铁铝氧石；活性铝土矿
activated calcium carbonate 活性碳酸钙
activated carbon ①活性炭〔吸附剂〕②活性炭黑
activated carbon adsorption 活性炭吸附
activated carbon deodorization 活性炭脱臭
activated carbon extraction 活性炭萃取〔增塑剂〕
activated carbon fibre 活化碳纤维
activated carbon filter 活性炭过滤器
activated carbon method 活性炭法
activated-carbon test 活性炭(萃取)试验〔增塑剂〕
activated cellulose 活化纤维素；活性纤维素
activated chalk 活性碳酸钙
activated char 活性炭
activated charcoal 活性炭
activated clay 活性(黏)土；活性瓷土
activated complex 活化络合物*
activated diffusion 活化扩散；激活扩散
activated earth 活性陶土〔补强剂〕
activated fiber 活化纤维
activated fuller earth 活性漂白土
activated glyceride stabilizer 活化甘油酯稳定剂〔四乙铅〕
activated initiation 活性引发
activated low structure black 活性低结构炭黑
activated manganate method 活性锰酸盐法
activated molecule 活化分子
activated monomer 活化单体
activated montmorillonite clay 活性微晶高岭土
activated petroleum coke 活性石油焦
activated polycondensation 活(性)化缩聚
activated polymerization 活化聚合

activated process　活性化法
activated resin　活化树脂
activated silica gel　活化硅胶
activated sludge　活性污泥
activated sludge bulking　活性污泥膨胀
activated sludge digestion　活性污泥消化
activated sludge method　活性污泥处理法
activated sludge plant　活性污泥厂
activated sludge process　活性污泥法
activated sludge tank　活性污泥池
activated-sludge test　活性污泥试验〔测生物降解度用〕
activated state　①活化(状)态②活性状态
activated stock　(含)促进剂胶料
activating accelerator　活化促进剂；助促进剂
activating agent　活化剂
activating bath　活化浴
activating catalyst　活化催化剂
activating group　活化基团*
activating of butane　丁烷活性化；丁烷的脱氢作用
activating oxide　活性氧化物
activating substance　活性物质
activating treatment　活化处理
activation　活化*；活性
activation adsorption　活化吸附(作用)
activation analysis　活化分析*
activation analysis service　活化分析业务
activation autoradiography　活化放射自显影
activation carbon　活性碳
activation center　活化中心
activation control　活化控制
activation cross section　活化截面
activation energy　活化能*
activation-energy of adsorption　吸附活化能*
activation enthalpy　活化焓
activation entropy　活化熵
activation equation　活化方程式
activation grade　活化度
activation grafting　活化接枝
activation heat　活化热
activation ingredient(=activator)　活性剂；活化剂
activation method　活化方法
activation number　活化值
activation overpotential　活化过电位；活化超电势
activation overvoltage　活化超电压
activation parameter　活化参量；活化参数
activation polarization　活化极化*
activation potential　活化电位；活化电势
activation process　激活过程

activation temperature　活化温度
activator(=activation ingredient)　活化剂；活性剂
activator level　活化剂用量
activatory　活化(的)
active　①活性的②有效的③活动的；活泼的④有旋光性的
active absorption volume　有效吸收容积
active acetate　活性乙酸
active acidity　活性酸度
active activation analysis　外加源活化分析；有源活化分析
active addition　充分加入
active adsorption　有效吸附
active agent　活性剂；活化剂
active alkali　活性碱
active amyl　旋性戊基；2-甲基丁基　$CH_3CH(C_2H_5)CH_2—$
active amyl alcohol　旋性戊醇　$C_2H_5CH(CH_3)CH_2OH$
active amyl mercuric bromide　旋性戊基溴化汞　$C_5H_{11}HgBr$
active amyl methyl ketone　旋性戊基甲基(甲)酮　$CH_3(C_2H_5)CHCH_2COCH_3$
active amyl propionate　丙酸旋性戊酯　$C_2H_5CO_2C_5H_{11}$
active amyl stearate　硬脂酸旋性戊酯　$C_{17}H_{35}CO_2CH_2CH=(CH_3)C_2H_5$
active anthracene black　活性蒽炭黑〔补强剂〕
active area　①作用面积②放射性区
active black(=activated carbon)　活性炭〔吸附剂〕；活性炭黑
active cambium　活性形成层
active carbon　活性炭*
active carbon dating method　放射性碳定年代法
active carbon fiber　活性碳纤维
active carbon filter medium　活性炭过滤材料
active cation site　活性阳离子位置
active center　活性中心*
active charcoal　活性炭
active chemicals　活性助剂
active chlorine　活性氯；有效氯
active cocatalyst　活性助催化剂
active condition　活性状态
active constituent　活性组分
active cooling surface　有效冷却面
active coupling　主动耦合
active deposit　活性淀积
active detergent　高效洗涤剂
active earth　活性土
active earth pressure　有效地压
active electrode　活性电极

active emulsion concept　活性乳胶概念；活性乳胶原理
active extreme pressure lubricant　活性特压润滑剂
active filler　活性填料；补强剂
active fire　仍在继续的火灾
active gas　活性气体；腐蚀性气体
active group　活性基团
active hydrogen　活性氢
active hydrogen atom　活性氢原子
active hydrogen compounds　活泼氢化合物*
active hydrogen determination　活泼氢测定
active hydrogen peak　活泼氢谱峰
active immunity　自动免疫性
active ingredient　活性组分
active interrogation　有源探询*
active knife length　切断有效长度
active mass　有效质量
active material　活性物料
active matter　活性物
active matter content　活性物含量
active metal　活性金属
active methylene　活性亚甲基
active methylene group　活性亚甲基
active neutron analysis　外加中子分析
active neutron interrogation　有源中子探询法*
active neutron interrogation method　外加中子探询法；有源中子探询法
active nitrogen　活性氮
active nondestructive assay　外加源无损测试
active oil calculation　可采(有效)石油储量计算
active optical fibre　激活光纤
active output　实际产量
active oxygen　活性氧*
active oxygen method　活性氧法
active oxygen test　活性氧试验
active pigment　活性颜料
active plasticizer　活性增塑剂
active pollution　放射性污染
active polymeric substance　高分子活性物质；活性聚合物
active porosity　有效孔隙度
active potassium dating method　放射性钾定年代法
active primer　活性底漆
active principle　有效要素；有效组分；有效成分
active racemate　活性外消旋盐
active radicals　活性根；活性基团
active region(s)　活性中心；活泼区域；活泼中心(反应)
active rhenium dating method　放射性铼定年代法
active rolling bank　有效的滚动堆积胶

active rubber(=polar rubber)　活性橡胶；极性橡胶
active rubidium dating method　放射性铷定年代法
active screen area　有效筛面积
active site　活性部位；活性点
active site of support　载体的活性部位(点)
active sludge　活性污泥
active solid　活性固体〔可用作吸附剂〕
active solvent　活性溶剂；有效溶剂
active storage　有效存储器
active substance　活性物质
active sulfur　活性硫；腐蚀性硫
active surface　活性表面
active transfer　主动传递；活性运转
active transport　主动转运；活性转运
active valence　有效化合价
active valeric acid　旋(光活)性戊酸
active volume　有效体积
active work in process　实际在制品
active zinc flower　活性锌华；活性氧化锌
active zone　活性区
activin　雄激素诺龙〔商品名〕；活化素〔由垂体合成并由睾丸和卵巢分泌的性激素〕
activity　(放射性)活度*
activity coefficient(=activity quotient)　活度系数
activity limit　活度极限
activity of catalyst　催化剂活性
activity of H ion　氢离子活度
activity of initiator　引发剂活性
activity product　活度积
activity quotient(=activity coefficient)　活度系数
activity rating of catalyst　催化剂的活性评价
activity ratio　活度比率
activity release method　放射性释放法
activity series　活度(顺)序
activity standard　活度标准
activity test　(催化剂的)活性试验
actol　乳酸银
actor　作用物；反应物
actual acidity　实际酸度
actual calorific value　实际热值
actual column　实际塔
actual count　实际支数
actual coverage　①实际覆盖域②实际涂布面积
actual data　①实际数据②实际资料
actual efficiency　实在效率
actual equilibrium constant　真实平衡常数；实际平衡常数
actual factory tread mix　生产用胎面胶

actual filling depth 实际填充深度
actual fineness 实际细度；实际纤度
actual gas 实际气体；真实气体
actual life 实际寿命；有效寿命
actual lift 实在上升
actual load 有效荷载
actual mixing cycle 实际混炼周期〔不包括密炼机进胶排胶时间〕
actual Mooney viscosity 实际门尼黏度
actual octane value 实际辛烷值；纯汽油组分辛烷值
actual plate 实际塔板
actual reelability percentage 实测解舒率
actual reflux 实际回流
actual reserves 实际储量
actual retention volume 实际保留体积
actual road (wear) test 道路试验；实际里程试验〔轮胎〕
actual service life ①实际使用寿命②行驶里程〔轮胎〕
actual service test 实际使用试验
actual solution 实际溶液
actual spot painting 现场涂装
actual stability 实在稳定性
actual stress 实际应力
actual twist 实际捻度
actual useful volume ①实际容积；有效容积②应急容积；安全体积
actual valency(=valence) 实际价
actual viscosity 真实黏度
actuate size 实际尺寸
actuator ①调节器②传动装置
acumycin 针霉素
acute dermal toxicity 急性皮肤毒性
acute dermal toxicity test 急性经皮毒性试验
acute oral toxicity 急性口服毒性
acute oral toxicity test 急性经口毒性试验
acute toxicity 急性毒性
acutumine 尖防己碱
AC voltammetry 交流伏安法*
acyclic 无环的
acyclic compound 无环化合物
acyclic diene metathesis polymerization 非环二烯易位聚合
acyclic elastomer 无环弹性体；无环橡胶
acyclic hydrocarbons 无环烃
acyclic monoterpene 无环单萜(烯)
acyclic polyene chromogens 无环多烯发色体
acyclic stem-nucleus 无环母核〔无环化合物的母链〕
acyclic terpene 无环萜烯
acyl ①酰基②脂酰(基) RCO—

acylability (可)酰化性
acylable 可酰化的
acyl acetic acid 酰基乙酸
acyladenylate 酰基腺苷酸
acylal 缩羰(基)酯
N-acyl-N-alkyltaurate N-酰基-N-烷基牛磺酸盐
acylamide 酰胺
acylamino 酰氨基
acyl amino acid 酰基氨基酸
acyl amino acid ester 酰基氨基酸酯
N-acyl amino alkane acid N-酰基氨基烷酸
acylaminoanthraquinones 酰氨基蒽醌类化合物
acylamino group 酰氨基
acyl amino polyglycol ether sulfate 酰氨基聚乙二醇醚硫酸盐
acylamino yellow 酰氨基黄 $C_{36}H_{20}N_2O_6$
N-acyl anisidine sulfonate N-酰基茴香胺磺酸盐；N-酰化甲氧基苯胺磺酸盐；N-酰化氨基苯甲醚磺酸盐
acyl arginine ester salt 酰基精氨酸酯盐
acylase 酰基转移酶
acylate ①酰化②酰化产物
acylated amino acid 酰基化氨基酸
N-acylated amino sulfonate N-酰氨基磺酸盐
acylated peptide 酰化缩氨酸；酰化肽
acylated phenyl hydrazine sulfonic acid 酰化苯肼磺酸
acylated polymer 酰化聚合物
acylated starch 酰化淀粉
acylating ①酰化的②酰化(作用)
acylating agent 酰化剂
acylation 酰化*
acylation degree 酰化度
acyl azide 酰叠氮*
acyl-azo-compound 酰基偶氮化合物
acyl bromide 酰溴*
acyl carbamyl guanidine 酰替氨基甲酰胍
O-acylcarnitine O-酰基肉碱
acyl cation 酰(基)正离子*
acyl chloride 酰氯*
acyl choline chloride 氯化酰胆碱
acyl cleavage 酰基裂解*
acyl-CoA dehydrogenase 酰基辅酶 A 脱氢酶
acyl-CoA synthetase 酰基辅酶 A 合成酶
acyl cyanide 酰腈*
acyl derivative 酰基衍生物
acyl enzyme 酰(基)酶
acyl fluoride 酰氟*
acyl glutamate 酰基谷氨酸盐
acyl glutamic acid ester 酰基谷氨酸酯

acyl glutamic acid triethanolamine 酰基谷氨酸三乙醇胺
acylglycerol 酰基甘油;甘油酯
acyl guanyl urea 酰基胍基脲
acyl halide 酰卤*
acylhydrazine 酰(基)肼
acylhydrazone 酰腙
acyl imino acetate 酰亚氨基乙酸盐
acylindazolone 酰基二氮杂茚酮
acyl iodide 酰碘*
acyl isethionate 酰基羟乙磺酸盐
acyl isethionic salt 酰基羟乙磺酸盐
acylium cation 酰基阳离子
N-acyllactam N-酰基内酰胺
acyl methyl N-taurate 酰基甲基牛磺酸酯
acyl methyl taurine (乙)酰甲基氨基乙磺酸;(乙)酰甲基牛磺酸
acyl migration 酰基转移作用
acyloin 偶姻*
acyloin condensation 偶姻缩合*
acylolysis 酰基裂解
acylous action 增酸性作用;降碱性作用
acyloxy 酰基;酰氧基 RCOO—
acyloxy-derivatives 酸基衍生物
acyl-oxygen fission 酰氧分裂
acyloxylation 酰氧基化*
acyloxy radical 酰氧自由基
acylpeptide 酰基肽
acylpeptide hydrolase 酰基肽水解酶
acyl peroxide 酰基过氧化物*
acyl phosphonic acid 酰基膦酸
acyl polypeptide 酰化多肽;酰基缩多氨酸
acyl radical 酰基
acyl rearrangement 酰基重排*
acyl sarcosinate 酰基肌氨酸盐;酰基甲替甘氨酸酯
N-acylsarcosine N-酰基肌氨酸
N-acylsphingosine N-酰基(神经)鞘氨醇
acyl stearamine 酰基硬脂胺
N-acyl taurate N-酰基牛磺酸酯
acyl titanate 酰基钛酸酯
acyl tosylate 酰基对甲苯磺酸酐*
acyltransferase 酰基转移酶
acyl urea 酰基脲 $NH_2CONHCOR$
aczoiling (电杆)防腐〔特指使用 aczol 作电杆防腐处理〕
aczol 苯氧化锌及苯氧化铜的含氨溶液〔作木材防腐剂用〕
adalin(=uradal) 阿达林;3-溴-3-乙基丁酰脲
adamant 硬石〔金刚石或刚玉〕
1-adamantanamine 金刚烷胺

adamantane 金刚烷 $C_{10}H_{16}$
adamantane acid 金刚烷二甲酸
adamantine ①金刚石制的②硬如金刚石的③金刚硼〔结晶〕④金刚合金
adamantine acid 金刚烷二甲酸 $C_{10}H_{14}(COOH)_2$
adamantine boron 金刚硼
adamantine spar 刚玉
adamantine type 金刚烷型
adamellite 石英二长石
adamellose 鸽峰岩
adamine(=adamite) 水砷锌矿
adamite(=adamine) 水砷锌矿
Adamkiewicz reaction 亚当凯维奇反应
Adams' catalyst 亚当斯催化剂〔一种氧化铂催化剂〕
adamsite(=phenarsazine chloride) 氯化吩吡嗪;亚当氏毒气 $C_{12}H_9AsCIN$
adanon 阿达农〔药〕;美沙酮
adansonine 猴面包碱
adaptability ①适应性②灵活性〔仪器〕
adapter(=adaptor) ①接管②应接管;接受管③适配器
adapter plate 衬板,垫板〔节段硫化模〕
adapting flange 连接接盘
adaptive analysis method 自适应分析法
adaptive chemical pattern recognition 自适应化学模式识别
adaptive enzyme 适应酶
adaptive immunity 继承免疫性
adaptive immunization 继承免疫作用
adaptive Kalman filter 自适应卡尔曼滤波器
adaptive Kalman filtering method 自适应卡尔曼滤波法
adaptive least square 自适应最小二乘(法)
adaptive linear 自适应线性
adaptive resonance theory 自适应共振理论
adaptor(=adapter) ①接管②应接管;接受管
adaptor block 模头接套
adaptor coextrusion 接套式共挤塑
adaptor elbow 弯接套
adatom 附加原子*
added internal standard method 叠加内标法
added weight 外加重量;重砣
addend 附加物
addendum ①附加物②齿顶高;齿顶
addendum circle 齿顶圆
adder 添加剂
adding machine 计数机
addition 加成;加聚
1,2-addition 1,2-加成
1,4-addition 1,4-加成

3,4-addition　3,4-加成
cis-addition　顺(式)加(成)作用
addition agent　添加剂
additional accelerator　助促进剂
additional amplifier　附加放大器
additional band　附加谱带
additional charge　①补充充电②附加运费
additional crosslinker　助交联剂
additional finishing　补充修饰
additional fuel　附加燃料
additional gas　外加气
additional iterative target transformation factor analysis　加入迭代目标转换因子分析
additional nozzle　辅助喷嘴
additional peak　附加峰
additional polymerization　加成聚合
additional treatment　补充处理
additional vaporization　附加蒸发
additional voltage　补充电压
additional vulcanization　后硫化；二次硫化
addition compound(=additive compound)　加(成化)合物
addition-condensation　加成缩合
addition copolymerization　加成共聚合
addition cyclization polymer　加成环化聚合物
addition-elimination mechanism　加成消除机理*
addition fragmentation chain transfer　加成断裂链转移(反应)
addition method　叠加法*
addition order　加料顺序
addition polycondensation　加成缩聚(作用)〔加成与缩合二者交叉进行的聚合反应〕
addition polymer　加(成)聚(合)物*
addition polymerization　加(成)聚(合)*
addition process　加成法
addition product(=additive product)　加成(产)物
addition reaction　加成反应*
additive　①添加物*②加成的③加和的*
additive blended oil　添加剂调和油
additive capacity　添加剂量
additive color mixture(=additive mixture)　加色法配色；加色法混(合)色
additive color process　加色法
additive complementary colors　加色法补色
additive compound(=addition compound)　加(成化)合物
additive concentration　添加剂浓度
additive copolymerization　加成共聚合
additive depletion　添加剂耗减
additive dimer　加成二聚物

additive dimerization　加成二聚*
additive disk　加色法混色盘
additive effect　①加成效应②加和效应
additive engine oil　含添加剂的机油
additive error　附加误差
additive function　加性函数；可加函数
additive ingredient(=compounding agent)　添加剂；配合剂
additive interference　相加干扰
additive level　添加剂的总含量〔石油产品的〕
additive loss　添加剂损失
additive lubricating oil　含添加剂的润滑油
additive master batch　添加剂母料
additive migration　添加剂迁移
additive mixture(=additive color mixture)　加色法配色；加色法混(合)色
additive motor oil　(含)添加剂机油
additive name　加成名称
additive paint　油漆助剂
additive polymerization　加成聚合
additive primaries　加色法(三)原色〔红、蓝、绿三色〕
additive primary colors　加色法三原色
additive process　加成法；加成工艺
additive product(=addition product)　加成(产)物
additive property　加成性；加合性
additive reaction　加成反应
additive test of lubricating oil　润滑油的添加剂(含量)试验
additive-treated oil　①(含)添加剂润滑油②加添加剂的油品
additive-type oil　含添加剂的油
additivity　加和性；相加性
additivity of degree of freedom　自由度的加和性
additivity of mass spectra　质谱叠加性
additivity of substituted groups　取代基的加和性
additivity of sum of deviation squares　偏差平方和加和性*
additivity of sum of squares　平方和加和性*
additivity of variance　方差加和性
additivity rule　相加规则；相加定律；加成定则
add-on　①填加；加入②加入量；用量③加重率
adduct　加合物；加成化合物
adduct curing agent(=crosslinking agent)　加成固化剂；交联剂
adduct ion　加合离子*
adduct modifier　加合物改性剂
adduct number　加成数
adduct polymerization　加合聚合(作用)

adduct rubber 加合橡胶
adelite 砷钙镁石
adenantherine 海红豆碱
adenase 腺嘌呤(脱氨)酶
adenine(=6-aminopurine) 腺嘌呤；6-氨基嘌呤 $C_5H_5N_5$
adenine deaminase 腺嘌呤脱氨酶
adenosine 腺苷
adenosine deaminase 腺苷脱氨酶
adenosine diphosphate(ADP) 腺苷二磷酸
adenosine hydrolase 腺苷水解酶
adenosine monophosphate(AMP) 腺苷一磷酸；腺苷酸
adenosine triphosphate(ATP) 腺苷三磷酸；三磷酸腺苷
adenylic acid(AMP) 腺(嘌呤核)苷酸；腺苷一磷酸
adeps 动物脂
adeps lanae 羊毛脂
adeps lanae hydrous 含水羊毛脂
adeps suillus(=hog lard) 豚脂
adequate distribution 均匀分布
adequate flow 足够的流动(平)性；合格的流平性
adermin(=vitamin B_6) 抗皮炎素；维生素 B_6
aderosterol 肾上腺甾醇；肾上腺固醇
ader wax 粗地蜡
adfluxion 汇流
adglutinate(=agglutinate) ①烧结②胶结(产)物③凝集
adhatotic acid 鸭嘴花酸
adhere 黏合；黏附
adhere ability 黏合性
adherence(=adherency) 黏合；黏附
adherency(=adherence) 黏合；黏附
adherend 被黏物；黏合体
adherend failure 黏附体破坏
adherent 黏合的；黏附的
adherent water 黏附水
adhering zone 附着区域
adherometer(=adheroscope) 黏合计
adheroscope(=adherometer) 黏合计
adhesion 黏合*；黏着
adhesion agent 黏合剂；胶黏剂
adhesional energy 黏合能
adhesional wetting 附着润湿
adhesion bond 黏合力；密着力
adhesion booster 增黏剂
adhesion contributor 增黏剂；助黏剂
adhesion energy 黏附能
adhesion factor 黏合系数
adhesion failure 黏附破坏；黏合破坏
adhesion force 黏合力；密着力〔橡胶〕
adhesion friction 黏附摩擦
adhesion heat 黏合热；黏附热
adhesion improving agent 胶黏改进剂
adhesion inviting 易黏合的
adhesion layer 黏附层
adhesion mechanism 黏合机理
adhesion of film 薄膜黏附
adhesion power 黏合力；密着力〔橡胶〕
adhesion preventives 防黏剂
adhesion promoter 黏合增进剂；助黏剂
adhesion-promoting ①增加附着力②增黏(作用)
adhesion rate 黏合速率
adhesion ratio 黏力比
adhesion retention 保黏性；黏合保持性
adhesion strength 黏合强度
adhesion strength factor 附着强度系数
adhesion stress 黏附应力
adhesion tension 黏合张力
adhesion test(ing) 密着力试验〔橡胶〕；剥离试验〔橡胶〕
adhesion theory of friction 摩擦黏着理论
adhesion (-type) tyre 抗滑轮胎；雪泥轮胎
adhesion value 黏力值
adhesion work 黏附功；附着功
adhesive 黏合剂*；胶黏剂
adhesive action 黏合作用
adhesive agent ①黏合剂②胶浆
adhesive-assembly 黏(合组)装
adhesive attraction 黏吸作用
adhesive backed 涂满黏合剂的
adhesive backer 胶黏材料
adhesive bar 棒状胶黏剂
adhesive bond 黏附键
adhesive-bonded fabric 无纺(织)布
adhesive bonding ①黏着；黏结；黏合②黏合剂结合法〔非织造织物工艺〕
adhesive bond strength tester 黏合强度试验机
adhesive capacity 黏合度
adhesive cement 胶浆(子)；胶水
adhesive-coated catalyzed laminate 涂胶催化层压板
adhesive-coated uncatalyzed laminate 涂胶无催层压板
adhesive coating 黏合层
adhesive coating technique 黏涂技术
adhesive-dipping 黏合剂浸渍(工艺)
adhesive dispersion 黏合剂分散体
adhesive effect 黏合作用
adhesive-elastomer failure 胶浆橡胶界面剥离
adhesive emulsion 黏合乳剂
adhesive fabric 黏胶布
adhesive failure 密着破坏；脱胶〔橡胶〕；黏附破坏

adhesive film　黏膜
adhesive force　黏合力
adhesive glue　黏合胶
adhesive interface　黏合界面
adhesive interlayer　胶浆夹层
adhesive interlining　热熔衬
adhesive joint　胶接
adhesive label　涂胶标签；不干胶标签
adhesive lamination　①胶黏层压②胶黏复合
adhesive layer　胶层
adhesive linkage　黏结层
adhesive mass　黏合剂
adhesive material　①黏结材料②黏合剂
adhesive meter　黏合计
adhesive method　黏着法
adhesive migration　黏合剂渗移；黏合剂迁移
adhesive mixer　调胶机
adhesive moisture　黏合水分
adhesiveness　①黏合性②黏合度
adhesive phase　黏合相
adhesive powder　粉状黏合剂
adhesive power　黏合力
adhesive primer　底胶
adhesive purpose chloroprene rubber　黏结型氯丁橡胶
adhesive resin　黏合用树脂
adhesive seal　①黏合封口②黏封
adhesive separation　黏附脱层
adhesive spread　黏合剂涂布量
adhesive stick　棒状胶黏剂
adhesive strength　黏合强度
adhesive stress　黏合应力
adhesive tape　胶黏带；胶纸；黏合带
(adhesive) tape test　胶带试验〔附着力检验法之一〕
adhesive tension　黏合张力
adhesive test　黏合试验
adhesive varnish　黏合清漆
adhesive water　吸附水；附着水分
adhesive wear　黏附磨蚀；黏着磨损
adhesivity　黏合性
adhint　黏合接头；胶接
adhumulone　加葎草酮〔啤酒花中的苦味物质之一〕
adiabatic　绝热的
adiabatic and isothermal elastic modulus　绝热和等温弹性模量
adiabatic apparatus　绝热装置
adiabatic calorimeter　绝热式热量计*
adiabatic catalytic cracking　绝热式催化裂化
adiabatic change　绝热变化

adiabatic coating　绝热涂料
adiabatic coefficient of compression　绝热压缩系数
adiabatic column　绝热式(精馏)柱
adiabatic compressibility　①绝热压缩系数②绝热压缩性
adiabatic compression　绝热压缩
adiabatic condition　绝热条件
adiabatic constant　绝热系数
adiabatic contraction　绝热收缩
adiabatic cooling　绝热冷却
adiabatic cooling curve　绝热冷却曲线
adiabatic cooling line　绝热冷却线
adiabatic curve　绝热曲线
adiabatic decomposition temperature　绝热分解温度
adiabatic dehydrogenation　绝热脱氢〔从烷烃产生烯烃〕
adiabatic demagnetization　绝热退磁(作用)
adiabatic diaphragm　绝热隔膜
adiabatic dryer　绝热干燥器
adiabatic drying　绝热干燥
adiabatic efficiency　绝热效率
adiabatic elasticity　绝热弹性
adiabatic electronic transition　绝热电子跃迁
adiabatic equation　绝热方程式
adiabatic evaporation　绝热蒸发
adiabatic expansion　绝热膨胀
adiabatic exponent　绝热指数
adiabatic extruder　绝热挤出机；自热挤出机
adiabatic extrusion　绝热压出；自热压出
adiabatic fast passage　绝热快通过
adiabatic flow　绝热流(动)
adiabatic function　绝热函数
adiabatic head　绝热压头
adiabatic humidification　绝热增湿(作用)
adiabatic index　绝热指数
adiabatic ionization energy　绝热电离能
adiabatic line　绝热线
adiabatic membrane　绝热膜
adiabatic modulus　绝热模量
adiabatic polymerization　绝热聚合
adiabatic process　绝热过程*
adiabatic psychrometer　绝热干湿球湿度计
adiabatic reactor　绝热反应器
adiabatic rectification　绝热精馏
adiabatics　绝热(曲线)
adiabatic saturation temperature　绝热饱和温度
adiabatic scanning calorimetry　绝热扫描量热法
adiabatic single-screw extruder　绝热单螺杆挤塑机
adiabatic system　绝热系统*
adiabatic temperature rise　绝热温升

adiabatic transition 绝热跃迁*
adiabatic wall 绝热壁
adiabatic work 绝热功*
adiactinic 绝射的；不透射线的
adiathermanous body 不透热体
adicity(=valency) 化合价；原子价
adinole 钠长英板岩
adion(=adsorbed ion) 吸附离子
adipaldehyde(=hexanedial) 己二醛
adipamide 己二酰二胺 (CH$_2$CH$_2$CONH$_2$)$_2$
adipate ①己二酸②己二酸盐(酯或根)
adipate plasticizer 己二酸酯增塑剂
adipic ①脂肪的②多脂的
adipic acid(=adipinic acid) 己二酸；肥酸〔俗〕
    COOH(CH$_2$)$_4$COOH
adipic acid ester 己二酸酯
adipic acid-hexamethylene diamine salt 己二酸-己二胺盐；耐纶 66 盐；尼龙 66 盐
adipic aldehyde(=1,6-hexandial) 己二醛
    OHCCH$_2$CH$_2$CH$_2$CH$_2$CHO
adipic anhydride 己二酸酐
adipic chloride(= adipyl chloride) 己二酰二氯
    (CH$_2$CH$_2$COCl)$_2$
adipic dialdehyde 己二醛
adipic diamide 己二酰二胺 (CH$_2$CH$_2$CONH$_2$)$_2$
adipic dihydrazide 己二酸二酰肼
    NH$_2$NH-CO(CH$_2$)$_4$CONHNH$_2$
adipic dinitrile 己二腈 (CH$_2$CH$_2$CN)$_2$
adipimide 己二酰亚胺
adipinic acid(=adipic acid) 己二酸；肥酸〔俗〕
    COOH(CH$_2$)$_4$COOH
adipo-cellulose 含脂纤维素
adipoid 类脂
adipoin 2-羟环己酮 C$_6$H$_{10}$O$_2$
adiponitrile 己二腈
adiposis 肥胖症
adipoyl(=adipyl) 己二酰 —CO(CH$_2$)$_4$CO—
adipyl chloride(=adipic chloride) 己二酰二氯
    (CH$_2$CH$_2$COCl)$_2$
adipyl dihydrazide 己二酰二肼
adjacent ①相邻的；邻位的②交界的
adjacent carbon atom 相邻碳原子
adjacent carbons 相邻碳原子
adjacent chain 邻(接)链
adjacent compound 相邻化合物
adjacent crosslink 邻位交联
adjacent double bonds 相邻双键
adjacent double bonds diene 相邻双键二烯

adjacent element 相邻元素
adjacent hydrogen wag region 邻氢摇摆区
adjacent pitch error 相邻周节误差
adjacent position 邻位；相邻位置
adjacent re-entry 邻近有规进入
adjacent re-entry model 相邻再入模型*
adjacent spot 相邻斑
adjacent surface 邻接面
adjacent-to-end carbon 与末位邻接的碳原子
adjective dye 间接染料
adjoining 邻接
adjoining angle 邻接角
adjoining carbons 邻接碳原子
adjoining position 邻接位置
adjoining region 邻接区
adjunct 添加剂；附属物
adjustable 可调节的
adjustable blade propeller pump 调节叶片螺桨泵
adjustable clamp 调节夹
adjustable curvature 可调曲率
adjustable cutter 可调刀具
adjustable delivery pump 可调输料泵
adjustable discharge gear pump 可调卸料齿轮泵
adjustable frequency 调频传动
adjustable grid 可调栅极
adjustable head extruder 可调口型挤出机
adjustable (-length) V-belt 活络三角带
adjustable limit snap gage 可调式极限卡规
adjustable outrigger collector 可调式外架总管
adjustable resistance 可调电阻
adjustable sampling inspection 调整型抽样检验
adjustable screw(=adjusting screw) 调节螺旋
adjustable sheave 变距槽轮
adjustable siphon 可调虹吸
adjustable slit assembly 可调狭缝组件
adjustable speed drive 变速传动
adjustable tread stock die 可调式面(压出)口型
adjustable vane 可调(节的)叶片
adjustable yarn brake device 可调式丝条填塞装置
adjusted retention time 调整保留时间*
adjusted retention volume 调整保留体积*
adjuster 调节器；校正器
adjusting ①调节的②调节(作用)
adjusting frequency laser 可调谐激光器
adjusting gear 调节齿轮
adjusting screw 调距螺杆
adjusting screw nut 调距螺母
adjustment ①调节；调整②安排

adjustment of mixture 混合物组分的调节
adjustment tank 调配槽
adjust to zero 调整至零点
adjuvant 辅助剂；配料
Adler benzidine reaction 阿德勒联苯胺反应
Adler-Marke powder 阿德勒-马克猎枪药
adlumidine 藤荷包牡丹定
adlumine 紫罂粟碱；山缘草碱
adlupulone 聚蛇麻酮〔啤酒花中的苦味物质之一〕
admicelle 近胶束
admicelle polymerization 近胶束聚合
administration 给药
admiralty brass 海军黄铜
admiralty fuel in kerosene range 煤油沸程的海军燃料油
admiralty fuel oil 海军燃料油
admiralty Redwood viscosimeter 雷乌黏度计〔英国海军〕
admiralty test 干湿陈化试验
admissible error 允许误差
admissible thermodynamic process 容许热力学过程
admission 许可
admission space 装填体积
admission valve ①进气阀②进样阀
admittance 导纳
admixtion ①混合；搅拌②混合物；配合料
admixture ①掺加剂；掺加物②掺加；混合
admixture heat 掺和热
adnephrine(=adrenalin) 肾上腺素
adnic 铜镍系合金；海军镍
adnoral acetate 环己醇乙酸酯
adobe ①风干砖；土砖②灰质黏土；多孔黏土
adogen 甲基三辛基氯化铵
adonic acid 阿东酸 $C_3H_3(COOH)_3$
adonidine 阿东定 $C_{24}H_{40}O_9$
adoniduleite 阿东杜糖
adonilide 侧金盏花内酯
adonin 阿东宁 $C_{24}H_{40}O_9$
adonit(=adonitol) 阿东糖醇；侧金盏糖醇；核糖醇；戊五醇 $HOCH_2(CHOH)_3CH_2OH$
adonite 阿东醇 $C_5H_{12}O_5$
adonitol(=adonit) 阿东糖醇；侧金盏糖醇；核糖醇；戊五醇 $HOCH_2(CHOH)_3CH_2OH$
adonitoxigenin 福寿草毒苷配基；侧金盏花毒苷配基
adonitoxin 福寿草毒苷
adonose 阿东糖
adoption ①采用；选用；沿用②受理；接受；正式通过③实施
adragant 黄芪树胶
adramycin 甲烯土霉素

adrenal 肾上腺
adrenal cortex 肾上腺皮质
adrenal cortex hormone 肾上腺皮质激素；肾皮素
adrenal cortex hormone extract 肾上腺皮质激素萃
adrenal cortical hormone 肾上腺皮质激素
adrenalectomy 肾上腺摘除术
adrenal gland 肾上腺
adrenalin bitartrate 酸式酒石酸肾上腺素；重酒石酸肾上腺素
adrenalin(e)(=adnephrine; epinephrine) 肾上腺素 $C_6H_3(OH)_2(CHOHCH_2NHCH_3)$
adrenalin hydrochloride 盐酸肾上腺素
adrenal medulla 肾上腺髓质
adrenalone 肾上腺酮 $C_6H_5(OH)_2COCH_2NHCH_3$
adrenal virilism 肾上腺性男(性)化症
adrenergic ①肾上腺素(功)能的②肾上腺素(功)能药物
adrenic acid 肾上腺酸；7,10,13,16-二十二碳四烯酸
adrenine(=epinephrine) 肾上腺素
adrenochrome 肾上腺素红 $C_9H_8O_3N$
adrenocorticotrop(h)in 促肾上腺皮质激素
adrenocorticotropic hormone(ACTH) 促肾上腺皮质激素
adrenocortin(=adrenal cortex hormone extract) 肾上腺皮质激素萃；肾上腺皮质液
adrenodoxin (肾上腺)皮质铁氧还蛋白
adrenogenital syndrome 肾上腺生殖系综合症；肾上腺性征异常症
adrenoglomerulotropin 促醛固酮激素
adrenolutine 肾上腺黄素；N-甲基-3,5,6-三羟基吲哚
adrenosterol 肾上腺甾醇
adrenosterone 肾上腺(雄)甾酮
adrenotropic hormone(=corticotropic hormone; adrenotropin; corticotropin) 促肾皮素
adrenotropin(=adrenotropic hormone) 促肾皮素
adriamycin 阿霉素
adronol acetate 乙酸环己酯
adsolubilized monomer 近增溶单体
adsorbability ①吸附性②吸附能力
adsorbability sequence 吸附能力序列
adsorbable 可吸附的
adsorbable organic halogenes(AOX) 可吸附的有机卤化物
adsorbance 吸附量
adsorbed film 吸附膜
adsorbed layer 吸附层
adsorbed material 被吸附物
adsorbed moisture 吸附湿气(水分)
adsorbed oil 吸附油
adsorbed phase 吸附相

adsorbent 吸附剂*
adsorbent activity 吸附剂活性
adsorbent activity function 吸附剂活度函数
adsorbent bed 吸附床
adsorbent-coated glass strip 涂吸附剂玻璃条
adsorbent column 吸附柱
adsorbent deactivator 吸附减活剂
adsorbent filtering medium 吸附剂过滤介质
adsorbent gradient 吸附剂梯度
adsorbent heterogeneity 吸附剂的多相性
adsorbent layer 吸附(剂)层
adsorbent-loaded paper 带吸附剂纸
adsorbent modifier 吸附改性剂
adsorbent reactivation 吸附剂再生(作用)
adsorber ①吸附器②吸附剂
adsorber acid 吸附酸〔淋洒吸附物所用的酸〕
adsorbing ①吸附的②吸附(作用)
adsorbing agent 吸附剂
adsorbing capacity 吸附容量
adsorbing colloid floatation 吸附胶体浮选
adsorbing column 吸附柱
adsorbing gradient 吸附梯度
adsorbing material 吸附剂
adsorbing substance 吸附剂
adsorption 吸附*
adsorption analysis 吸附分析
adsorption applicability 吸附适用性
adsorption balance 吸附平衡；吸附均衡
adsorption band 吸附带
adsorption band indicator 吸附带指示剂
adsorption bed 吸附床
adsorption biodegradation AB 工艺；吸附生物降解工艺
adsorption bond 吸附键
adsorption bubble separation method 吸附气泡分离法*
adsorption by hydrogen bond 氢键吸附
adsorption capability 吸附本领；吸附容量
adsorption catalysis 吸附催化*
adsorption center 吸附中心*
adsorption chromatography 吸附色谱法*
adsorption coefficient 吸附系数
adsorption collection method on carrier column 载体柱吸附收集法
adsorption column 吸附柱
adsorption compound 吸附化合物
adsorption coprecipitation 吸附共沉淀
adsorption current 吸附电流*
adsorption curve 吸附曲线
adsorption cycle 吸附周期

adsorption-desorption of catalyst 催化剂的吸附及脱附性能
adsorption-desorption technique 吸附脱吸技术
adsorption detector 吸附检测器
adsorption displacement 吸附取代*
adsorption dynamics 吸附动力学
adsorption effect 吸附效应
adsorption efficiency 吸附效率；吸附能力
adsorption energy 吸附能
adsorption equation 吸附平衡式；吸附方程
adsorption equilibrium 吸附平衡*
adsorption equilibrium constant 吸附平衡常数
adsorption exponent 吸附指数
adsorption film 吸附膜*
adsorption filtration 吸附过滤
adsorption force 吸附力
adsorption free energy 吸附自由能
adsorption gas chromatography 吸附气相色谱(法)
adsorption gasoline 吸附汽油
adsorption heat 吸附热
adsorption heat detector 吸附热检测器
adsorption hysteresis 吸附滞后(现象)
adsorption index 吸附系数
adsorption indicator 吸附指示剂*
adsorption inhibitor 吸附型钝化剂；吸附型抑制剂
adsorption isobar 吸附等压线*
adsorption isobar line 吸附等压线
adsorption isoster 吸附等容线
adsorption isostere 吸附等量线
adsorption isotherm 吸附等温线*
adsorption isothermal line 吸附等温线
adsorption kinetics 吸附动力学
adsorption layer 吸附层*
adsorption magnetic field 吸附磁场
adsorption measurement 吸附测定法
adsorption mechanism 吸附机理
adsorption medium 吸附介质
adsorption phenomena 吸附现象
adsorption plant 吸附装置
adsorption plate 吸附板
adsorption polymerization 吸附聚合
adsorption potential 吸附电势*
adsorption precipitant 吸附沉淀剂
adsorption precipitation 吸附沉淀*
adsorption premoulding 吸附预成型坯模压法
adsorption process 吸附过程〔自汽油气中生产汽油〕
adsorption process of waste water 废水吸附处理法
adsorption quantity 吸附量*

adsorption rate 吸附速率*；吸附率
adsorption-recovery 吸附法(溶剂)回收
adsorption refining 吸附精制
adsorption separation 吸附分离法
adsorption site 吸附部位；吸附点
adsorption solvent strength parameter 吸附溶剂强度参数
adsorption space 吸附空间
adsorption stripping 吸附分离；吸附气提；解吸
adsorption swelling 吸附膨胀
adsorption test 吸附试验
adsorption theory 吸附理论
adsorption time 吸附时间
adsorption tower 吸附塔
adsorption trap 吸附阱；吸附收集器
adsorption tube 吸附管
adsorption unit 吸附装置
adsorption wave 吸附波*
adsorption zone 吸附带；吸附区
adsorptive(=adsorbate) ①(被)吸附物②吸附的
adsorptive bubble separation method(=adsubble method) 吸附气泡分离法
adsorptive capacity 吸附本领；吸附量
adsorptive catalyst 吸附催化剂
adsorptive clay 吸附白(黏)土
adsorptive complex wave 络合吸附波*
adsorptive effect 吸附效应
adsorptive fiber 吸附性纤维
adsorptive fractionation of lubricating oil 吸附分馏润滑油
adsorptive isotherm 吸附等温线
adsorptiveness 吸附性
adsorptive percolation 吸附渗滤
adsorptive power 吸附力
adsorptive property 吸附性能
adsorptive (stripping) voltammetry 吸附(溶出)伏安法*
adsorptive support 吸附性载体
adsorptivity 吸附性
adstringency 涩度；涩性
adstringent(=astringent) ①收敛剂；涩剂②涩嘴的
adsubble method(=adsorptive bubble separation method) 吸附气泡分离法
A/D transformation 模/数转换
adular 冰长石
adularia 冰长石
adulterant 掺杂剂
adulterated oil 掺杂油
adulteration 掺杂

ADU process 重铀酸铵法〔从 UF_6 制备 UO_2 的方法之一〕
adurol 阿杜酚
adustin 黑刺菌素
adustion of coal 煤的可燃性
advanced composite material(ACM) 高性能复合材料；新型复合材料；先进复合材料
advanced composites 先进复合材料
advanced cracking reactor 先进裂化反应器
advanced 2-fluoride system 高级双氟(化物)体系〔防龋配方〕
advanced package support 前置式筒子架
advanced ribbon breaking winder 新型防叠式卷绕头
advanced stage of cracking 深度裂化阶段
advanced treatment 深度处理
advanced treatment method 高级处理法
advanced wastewater treatment 高级废水处理
advance estimate 预先估计值
advance ignition 提前点火
advance training gasoline 高级教练机汽油〔染成蓝色〕
advancing color 接近色
advancing contact angle 前进接触角*
advancing front 前沿
advection 平流
advertising tape 经纱黏合带
advitant(=vitamin) 维生素
adynerin 无效苷 $C_{30}H_{44}O_7$
AE-cellulose 氨(基)乙基纤维素
aegerite(=wurtzite) 纤维锌矿
aegirine-augite 霓辉石
aegirite 霓石
aenigmatite 三斜闪石
aeolian clay 风成黏土
aeolotropic(=anisotropic; eolotropic) 各向异性的
aeolotropism 各向异性
aeolotropy 各向异性
aeonite(=wurtzite) 纤锌矿
aer- 空气；气
aerated density of catalyst 催化剂的充气比重
aerated filler 疏松填料〔如软木屑、锯木屑〕
aerated flame 充气焰〔气体与空气混合物火焰〕；富空气焰
aerated lagoon 曝气塘
aerated oxidation pond 曝气氧化塘
aerated plastic 泡沫塑料；多孔塑料
aerated pond 曝气池塘
aerated powder bath 沸腾粉末浴；沸腾粉末层
aerated solid 气溶胶；充气固体

aerated water 汽水
aerated yarn 充气纱
aerater 充气器；通气器
aerating agent 充气剂
aerating apparatus 曝气装置
aerating filter 曝气滤池
aerating powder 发泡剂
aeration ①曝气；充气②通气；换气
aeration-agitation 通气搅拌
aeration cell(=oxygen cell) 充气电池；氧气电池
aeration ditch 氧化沟；曝气渠
aeration factor 充气系数；充气因子
aeration lagoon 曝气塘；曝气池
aeration tank 曝气池
aeration test 通气试验
aeration time 通气时间；曝气时间
aerator 曝气器
aerator tank 曝气槽
aereous 青铜色
aerial ①空气的；气体的②空中的
aerial cable 架空电缆；高架线
aerial condenser 空气冷凝器
aerial contamination 空气污染
aerial filter 空气过滤器
aerial oxygen 大气氧气
aerial spraying 空气喷涂
aerial war material 空战材料
aerification 充气；气化；掺气
aero- 〔词头〕①空气②航空
aerobic 需气的；需氧的
aerobic biooxidation 需氧生物氧化
aerobic condition 需氧条件
aerobic degradation 需气降解作用
aerobic digestion 好氧消化
aerobic enzyme 需氧酶
aerobic treatment 好氧处理
aero casing 飞机外胎
aerochemistry 气体化学；空气化学；航空化学
aerocolloidal 气溶胶的
aerocrete 气孔混凝土
aerodynamic ablation 气动烧蚀
aerodynamic(al) 空气动力学的；气动的
aerodynamical interference 气动干扰
aerodynamically-shaped 空气动力状的；流线型的
aerodynamically stretch (空)气动(力)拉伸〔纺黏型非织造布纺丝时的拉伸〕
aerodynamic center 气动中心
aerodynamic deposition 气流凝网〔纤维网成型法之一〕
aerodynamic derivative 气动导数
aerodynamic drag 气动阻力
aerodynamic isotope separation 空气动力学法同位素分离(法)*
aerodynamic lubrication 气体(动力)润滑
aerodynamics 空气动力学
aerodynamics accelerator 气体动力学加速器〔一种化学加速器〕
aeroengine oil 航空润滑剂；航空机油
aerofilter 空气过滤器〔污水净化用〕
aerofloated sulfur 风选硫黄
aerofoil effect 机翼效应
aerofoil fan 机翼型通风机；轴流风扇
aerogel 气凝胶
aerogenesis 产气作用
aerogen gas 空气与汽油蒸气的可燃混合气
aeroglycan 气烯糖
aerograph ①喷雾染色(法)；喷染(法)②气象仪；高空气象计③修版汽笔（印）
aerography 喷染术；喷涂术；高空气象学
aeroklinoscope 浮水气胞
aerolar tissue 蜂窝组织
aerolite ①陨石②哀绕炸药；额那脱〔硝铵、硝酸钾、硫黄炸药〕
aero lubrication (热成型)气垫润滑
aerometer 气流式纤维细度仪
aeronautical rubber 航空用橡胶配件
aero oil 航空汽油
aerophone 扩音器
aerophore 通风面具
aeroplane dope 航空涂料；航空透布油
aeroplane inner tube 飞机轮内胎
aeroplane oil 航空润滑油
aeroplane tyre 飞机轮胎
aeroplankton 空气浮游生物
aeroplastic 航空塑料
aeropulverizer 气流粉碎机；吹气磨粉机
aerosiderite 铁陨石
aerosite(=pyrargyrite) 深红银矿
aerosol 气溶胶*
aerosol abrasive compound 气溶胶磨剂
aerosol analyzer 气溶胶分析器
aerosol container 喷雾剂瓶
aerosol deodorant 气溶胶脱臭剂；气溶胶去臭剂
aerosol filter 气溶胶滤网；气溶胶过滤器
aerosol foam 气溶胶泡沫
aerosol fragrance product 气溶胶芳香产品
aerosol gas ①喷雾(气)②推进剂

aerosol-generation interface 气溶胶发生(器)接口
aerosol grease dispersion 气溶胶分散润滑脂
aerosol lacquer 气溶胶漆
aerosol monitor 气溶胶监测器
aerosol paint 气溶胶漆
aerosol release agent 气喷脱模剂
aerosol sampling 气溶胶取样
aerosol sampling method 气雾剂取样法；气溶胶取样
aerosol silicone release agent 喷用硅油脱模剂
aerosol sprayer 喷雾器
aerosol spraying 气溶胶式喷涂
aerosphere(=atmosphere) 大气
aerostatic lubrication 气体静压润滑
aerostatic press 气(体)压(出)机
aerostatics 空气静力学；气体静力学；航空学
aerotaxis 趋氧作用；趋氧性
aerotherapeutics 空气疗法
aerothermochemistry 空气热化学；气动热化学
aerotolerant 耐氧的
aerotolerant anaerobe 耐氧性厌氧菌
aerotolerant bacteria 耐氧细菌
aeruginosin 铜绿菌素
aerugo ①氧化铜；铜绿②锈
aeschynite 易解石
aescigenin(=escigenin) 七叶配基
aescin(=escin) 七叶素
aescine 七叶皂苷〔药〕
aescinic acid 七叶酸 $C_{24}H_{40}O_2$
aescorcin 七叶酚 $C_9H_8O_4$
aesculetin 七叶亭；6,7-二羟基香豆素
aesculin 七叶灵 $C_{15}H_{16}O_9$
aesculinic acid 七叶灵酸
aether ①以太〔化学史〕②醚
(a)ethyl 乙(基)
aetio(=etio) 本；初
(a)etiocholane 本胆烷
aetiology 病源学；病因学；病理学
(a)etioporphyrin 本卟啉
AFD method AFD(辛烷值测定)法
afenil 氯化钙合四脲 $CaCl_2 \cdot 4(NH_2)_2CO$
afferent 传入的；向心的
afferent phase 传入期
affinage 精炼
affination 精炼法；(离心)洗糖(法)
affined 精炼了的
affine deformation 仿射形变〔元件变形和整体变形一致〕
affine deformation hypothesis 仿射形变假设

affine Gaussian model 仿射高斯模型
affine junction motion 交联点仿射运动
affine transformation 仿射变换
affinin(=spilanthol) 千日菊酰胺
affining ①精炼的②精炼(作用)
affinity ①亲和力②亲和能③亲和性④近似；类似
affinity adsorption 亲和吸附
affinity bond 亲和势键
affinity capillary electrophoresis 毛细管亲和电泳
affinity capture 亲和捕捉法
affinity chromatography 亲和色谱法*
affinity coefficient 亲和系数
affinity constant 亲和常数
affinity curve 亲和曲线
affinity-elution chromatography 亲和洗脱色谱
affinity for metal surface 对金属表面的亲和力〔润滑剂薄膜对金属表面的亲和力〕
affinity interaction 亲和作用
affinity labeling 亲和标记
affinity law 相似定律
affinity membrane 亲和膜
affinity of chemical reaction 化学反应亲和势*
affinity precipitation 亲和沉淀
affinity preference 亲和力次序
affinity residue 亲和残余力(势)；残余亲和力(势)
affinity ultrafiltration 亲和超滤
affixation 添加
affixion 添加；附加
affixture ①添加②添加产物；加成物
afflux(=ad fluxion) 汇流
aflatoxin 黄曲霉毒素
aflatoxin analysis 黄曲霉毒素分析
AFM(atomic force microscope) 原子力显微镜
α-AFP enzyme immuno sensor α-甲胎蛋白酶免疫传感器
African balsam(=illurin balsam) 非洲香脂
African kino 非洲赤胶
African sandalwood oil(=osyris oil) 沙针油
afridol 羟汞甲基苯甲酸钠
after aging 陈化后
after-annealing 后退火
after-bake 后烘
after-baking 后热处理
after-blocking(=reblocking) 后粘连
after-blow 后吹
afterburner 废气再燃器；后燃(烧)室
after-burning 后烧
afterchlorinate 后氯化；补充氯化

after-chlorinated polyvinylchloride fibre 后氯化聚氯乙烯纤维
after-coagulation 后凝固
after-combustion 后燃烧
after condensation 后缩合
aftercondenser 后缩合器
after-contraction 后收缩
after-cooler 后冷却器；二次冷却器
aftercrack 二次裂化
after creaming 后膏化
after-crystallization 后结晶
after cure 后硫化；二次硫化
after-cut 后馏分
after-damp 爆后气
after-depolarization 后去极化
after-drawing 后拉伸
after drip(per) 后流胶乳
after-dripping 喷油后的燃烧〔在喷油嘴内〕
after drying ①后(期)干燥；后(期)固化②再次干燥
after-effect(=elastic after effect) (弹性)后效
after-effect function 后效函数
after-etching 后蚀刻
after-expansion 残余膨胀
after-exposure ①后曝置②后曝光
after-fibrillation 后原纤化(作用)
after-filter 后过滤器
after-filtration (最)后过滤；后滤
afterfire 二次燃烧；后燃烧
after-fixing 后固定
after-flame 余焰
afterflow ①残余塑性变形②续流
after-fractionating tower 二次蒸馏塔
after-glow 后辉；余辉
after-glow depressant 余辉抑制剂；余辉减低剂
after-glow resistance 抗余辉性
after hardening 后固化；后硬化
after heat 余热；衰变热
after-heating 后加热
afterhydrolysis 后水解
after-hyperpolarization 后超极化
after-loading 后负荷
afterloading source 后装源
afterloading technique 后装源技术
afterloading unit 后装源机
after-mature 后熟成
after-mould stress 模塑后应力
after-odour 余臭〔漆膜干后，仍长期放出臭味〕
after polymerization ①后聚合(作用)②(低温辐射聚合中的)二次聚合
afterprocessing 后加工
after-product 后产物
after-purification (最)后净化；补充净化
after-ripen 后熟成
after-saponification 后皂化(作用)
after-scouring 后洗涤；后净化；后煮炼
after-settler 后澄清器；二次澄清器
after-shaping 后成型
after shave lotion 美容水
after shave powder 美容粉
after-shrinkage 后收缩
after-soaping 后皂煮
after-softening (水的)补充软化
after stain(=afterstain) ①互补色；对比色②后着色；后污染
after-strain 后变形
after-stretching 后拉伸
after tack 回黏(性)；返黏性
after-tackiness ①残余黏性；余黏〔油墨〕②回黏现象；回黏性
after-tension 后张力
after the fact testing 事后试验；出厂后试验
after-thickening 返稠
after-treating 后处理
after-treating agent 后处理剂
after-treatment 后处理
after treatment aids 后处理助剂
after treatment bath 后处理浴
after-vulcanization 后硫化作用
after-washing 后水洗
after waxing in sizing 浆纱后上蜡
after-working(=after-effect) 弹性后效
after-yellowing 后黄变〔塑制件经暴晒后发黄现象〕
afzelechin 5,7,4′-三羟基黄烷-3-醇；阿夫儿茶精
against-scale 逆鳞片
agal-agal(=agar) 琼脂；洋菜
agalite 纤滑石
agalma black 寿山黑；阿加马黑
agalmatolite(=pyrophyllite) 寿山石；冻石
agaphite(=turquois(e)) 绿松石
agar(=agal-agal) 琼脂；洋菜
agar-agar 琼脂；洋菜
agarase 琼脂酶
agar block method 琼脂块法
agar chromatography 琼脂色谱(法)〔琼脂作固定相〕
agar cup method 琼脂杯法
agar diffusion 琼脂扩散(法)

agar diffusion technique 琼脂扩散技术
agar embedding method 琼脂埋藏法
agar gel 琼脂凝胶*
agar gel diffusion(AGD) 琼脂凝胶扩散
agar gel electrophoresis 琼脂凝胶电泳(法)
agar glucose 葡萄糖琼脂
agar hanging block 琼脂悬块
agaric 蕈;蘑菇
agaric acid(=agaricic acid) 松蕈(三)酸;2-十六烷基柠檬酸
agaricin 落叶松蕈浸膏
agaricinum(=agaric acid) 松蕈(三)酸;2-十六烷基柠檬酸
agaricol 落叶松蕈醇 $C_{10}H_{16}O$
agar immunoelectrophoresis 琼脂免疫电泳
agaritine 伞菌氨酸;蘑菇氨酸 $C_{12}H_{17}N_3O_4$
agar membrane 琼脂膜
agarofuran 沉香呋喃
agaroid ①类琼脂②琼脂样的
agaroidin 琼脂素
agar oil 沉香油
agarol 沉香醇
agaropectin 琼脂胶
agarose 琼脂糖
agarose gel 琼脂糖凝胶
agarose gel electrophoresis 琼脂糖凝胶电泳
agar plate 琼脂板
agar plate method (微生物菌落数)琼脂平板测定法
agar streak method 琼脂划线法
agar streak test 琼脂涂抹试验
agar tube 琼脂培养试管
agar weed 石花菜
agar wood oil 沉香木油
agarythrine 红蕈碱
agate 玛瑙
agate guide 玛瑙导丝器
agate mortar 玛瑙研钵
agate red 玛瑙红(色)
agateware ①斑纹瓷器②玛瑙器皿③搪瓷器皿
agathen acid(=agathis acid) 贝壳杉酸
agathenedicarboxylic acid 贝壳杉二羧酸
agathic acid(=agathic dicarboxylic acid) 贝壳杉酸;玛瑙酸
agathic dicarboxylic acid(=agathic acid) 贝壳杉酸;玛瑙酸
agathin 阿加赛因〔药〕 $C_{14}H_{14}ON_2$
agathis acid(=agathen acid) 贝壳杉酸
agave decorticating 剑麻纤维制取

agavose 友舌兰糖
agchems(=agrochemicals) ①农用化学品②农产品中提炼的化学品
age ①老化;陈化;熟化②寿命
age concrete 熟化混凝土
age contraction 老化收缩;时效收缩
aged 老化的;陈化的
aged column 老化柱
aged cracked gasoline 陈化的裂化汽油
age determination of the earth 地球年龄测定
aged fission product 老裂变产物;放置过的裂变产物
agedoite 天冬酰胺
aged properties 老化后性能
aged-rubber 老化变质的橡胶
aged viscosity (增塑糊)陈化黏度
aged vulcanizate 老化后硫化胶
age hardening 老化变硬;时效硬化
ageing(=aging) ①老化;陈化;老成②熟化
ageing apparatus 老化试验器
ageing can 老化罐;老成罐
ageing chamber 老化箱;老成箱
ageing effect 老化效应
ageing loss 老化损失
ag(e)ing maturation 熟成
ageing of clay 黏土老化;土壤陈化
ageing of cloth 布匹熟化
ageing of flour 面粉老化;面粉熟化
ageing of polymers 聚合物的老化
ageing of precipitation 沉淀的陈化
ageing of viscose 黏胶丝熟化
ageing oven 老化箱;老化炉
ageing period 老化期
ageing process 老化法
ageing property 老化性质
ageing resistance 抗老化性能
ageing-resistant grease 抗老化润滑脂
ageing shrinkage 老化收缩
ageing stability 经时稳定性;老化稳定性
ageing test 老化试验
ageing time 老化时间;老成时间〔由完成黏合至达到最高黏合强度时间〕
ag(e)ing vessel 老化器
age-inhibiting addition 防老化添加剂
age inhibitor 防老化剂
age instability 老化不稳定性
A gel(=loose gel) 松凝胶
age limit 极限使用寿命
Agene process 埃京法〔一种用三氯化氮漂白面粉的

工艺)
agent ①剂②介质
agent for preventing dirt 阻垢剂
agent-in-oil method (乳化)剂在油中法
agent-in-water method (乳化)剂在水中法
age of catalyst 催化剂寿命
age of minerals 矿石年龄
age overnight 停放过夜
ager 熟化器〔纤维〕
ageratriol 香薷三醇
age resistance 耐老化性
age resister 防老剂
ager tester 老化试验机
age shrink(age) 停放收缩
Agfa vulkameter 爱克发硫化仪
agglomerant ①附聚的②烧结的③烧结工
agglomerate ①附聚物②烧结矿；烧结块③块集岩
agglomerate cell 块集池
agglomerate count 凝集计数法
agglomerated waste 结块(法)废丝〔冷相废丝造粒〕
agglomerating ①附聚的②烧结的③附聚(作用)④烧结(作用)
agglomerating agent ①烧结因素②凝结剂；胶凝剂
agglomeration ①附聚(作用)②烧结(作用)
agglomeration latex 附聚胶乳
agglomerative ①附聚的②烧结的③凝结的；胶凝的
agglutinant ①烧结剂②凝集剂
agglutinating ①凝集的②烧结的；黏结的；胶结的
agglutinating property 烧结性
agglutinating test 烧结试验
agglutinating value (of coal) (煤的)烧结性；黏结性
agglutination ①凝集*②烧结
agglutination assay 凝集反应试验
agglutination latex immunoassay 胶乳凝集免疫分析
agglutination reaction 凝集反应
agglutination test (血清)凝集试验
agglutinative ①凝集的②烧结的
agglutinative absorption 凝集吸收
aggravating 恶化；加重；加剧
aggregate 聚集体*
aggregate capacity 聚集容量
aggregated (particle) structure 聚集(粒子)结构；次结构
aggregate-enhanced Raman scattering 聚集增强拉曼散射
aggregate model 聚集体模型
aggregate model for mechanical anisotropy 力学各向异性的聚集体模型
aggregating ①聚集的②聚集(作用)
aggregation 聚集*

aggregation effect 聚集效应
aggregation-induced emission 聚集诱导发光
aggregation number 聚集数
aggregation number of micelle 胶束聚集数*
aggregation of particles 粒子的聚集
aggregation velocity 聚集速度*
aggregative 聚集的
aggregative flow 聚集流动
aggregative fluidization 聚集式流(态)化
aggressive 侵蚀性的；腐蚀性的
aggressive carbon dioxide 生效二氧化碳
aggressive chemicals 腐蚀性(化学)药品
aggressive marine environment 侵蚀性海洋环境；腐蚀性海洋环境
aggressive tack 干黏性
aging(=ageing) ①老化；陈化*；老成②熟化
aging action 老化作用
aging aids 防护助剂
aging apparatus 老化试验器
aging behavior 老化性能；耐老化性
aging characteristics 老化性能；耐老化性；老化特性
aging coefficient 老化系数
aging condition 老化条件
aging effect 老化效应
aging machine 熟化器〔纤维〕
aging-oven 老化试验箱
aging period ①老化(时)期②放置(时)期③熟化期
aging property 老化性能；耐老化性
aging quality 老化性能；耐老化性
aging resistance 耐老化性；老化性能
aging stability 老化稳定性
aging temperature 老化温度
aging test 老化试验
aging time 老化时间
agitated ball mill 立式搅拌球磨机
agitated batch crystallizer 搅拌式分批结晶器
agitated bed 搅拌床
agitated crystallizer 搅拌结晶器
agitated cylinder dryer 圆筒搅拌干燥器
agitated dryer 搅拌型干燥器
agitated kettle 搅拌锅
agitated tank 搅拌槽
agitating 搅拌
agitating apparatus 搅拌器
agitating cooker 回旋式杀菌釜
agitating reactor 搅动式反应器
agitating tank 搅拌槽
agitating vane 搅拌桨叶

agitation 搅拌(作用)
agitation leach 搅拌沥滤
agitation vat 搅拌瓮
agitator 搅拌器
agitator arm 搅拌杆；搅动叶轮
agitator bath 搅拌(酸)槽〔洗涤石油产品用〕
agitator disc(=agitator disk) 搅拌盘；盘式搅拌器
agitator disk(=agitator disc) 搅拌盘；盘式搅拌器
agitator drier 搅拌干燥器
agitator mill 立式球磨机
agitator-peg mill 销(钉)式搅拌磨
agitator tank 搅拌槽
agitator treating 搅拌洗涤
agitator-type blender 搅拌型掺混机
agitator washer 搅拌式洗涤机
aglucone 配基；配质；(葡萄糖苷的)非糖部
aglycon 苷元*
agmatinase 胍基丁胺酶；鲱精胺酶
agmatine 胍基丁胺；鲱精胺
agnin(=lanolin) 羊毛脂；阿格诺林；阿格宁(含水羊毛脂)
agnosterol 羊毛甾三烯醇；羔甾醇
agon(=prosthetic group) 辅基
agoniadin 槐树皮素；鸡蛋花苷 $C_{21}H_{26}O_{12}$
agonist 兴奋剂；激动剂
agostic 元结；氢混；混氢；抓氢
agostic bond 氢混键；混氢键；抓氢键
agranular 无粒的
agraphitic carbon 非结晶碳；无定形碳；非石墨碳
agreeable odor 愉快香气；愉快气味
agricultural chemistry 农业化学
agricultural emulsion 农用乳(浊)液
agricultural formulation 农业用品配方
agricultural insecticide 农业杀虫剂
agricultural lime 农用石灰
agricultural pesticide 农药
agricultural ply steel 农业层合钢
agricultural rheology 农业流变学；农艺流变学
agricultural salt 农用盐
agricultural surfactant 农用表面活性剂
agricultural varnish 农业用清漆
agriculture ①农业②农学
agrimonia oil 龙牙草油
agrimonine 仙鹤草素
agrimony 欧龙牙草
agrimycin(=agromycin) 农霉素〔链霉素与土霉素的合剂〕
agriplast 农用塑料
agrobacterium 土壤杆菌

agrochemicals(=agchems) ①农用化学品②农产品中提炼的化学品
agrochemistry 农业化学
agroclavine 田麦角碱
agrocybin 田头菇素
agrol fluid 酒精汽油掺混燃料〔78%乙醇+22%汽油〕
agromycin(=agrimycin) 农霉素
agronochemic(al) 农(艺)化学的
agropyrene 冰草炔
agrotextile 农用纺织品
aguamiel 龙舌兰汁
aguilarite 辉硒银矿
aguirin 阿古林〔乙酸可可碱钠的商品名〕
Agulhon's reagent 阿居隆试剂
agustite 磷灰石
ahuaca oil 樟泥油；鳄梨油
Aich's metal 艾奇合金；含铁四六黄铜
aids 助剂
aikinite 白钨形黑钨矿；针硫铋铅矿
ailanthic acid 苦樗酸
aimant(=aimantine; magnetite) 磁铁矿
aimantine(=aimant; magnetite) 磁铁矿
aim colour 目标色
air ①空气②气③风干
air accumulator (压缩)空气储罐；空气储蓄器
air-acetylene flame 空气-乙炔(火)焰
air-actuation 空气起动；空气驱动
air adjusting valve 空气调节阀
air admission ①空气进入②空气入口
air aging 热空气老化；恒温箱老化
air agitation 充气搅拌
air agitator 充气搅拌器
air and fuel mixture 空气燃料混合物
air and gas mixer 空气煤气混合器
air and steam blast 空气蒸汽鼓风
air and water hose 风水胶管
air-arc cutting 空气电弧切割
air aspirating 抽气；吸气
air-assisted electrostatic gun 空气助喷(加速)静电喷枪
air-assisted gun 空气助推喷枪；(空)气助(喷)喷枪
air assist forming 气胀真空成型；气胀成型
air-assist vacuum thermoforming 气助真空热成型
air atomizer 空气雾化器
air atomizing 空气雾化(作用)
air bag 气袋；气胎
airbag base 水胎牙子〔橡胶〕
airbag mold 水胎硫化模〔橡胶〕
air-ballast pump 气镇泵

air balloon cover 气球膜
air barrier 气密层；内衬层
air bath ①空气浴；气锅②干燥室
air bearing 气承；气垫
air-bearing torsion pendulum 空气轴承扭摆
air bed ①(粉末料杯的)气垫；空气层②沸腾床
air bell 气泡；砂眼
air bellow 风箱
air binding 气缚
air bladder 气泡；气囊
air blast 空气鼓风
air-blast dusting machine 鼓风除尘机
air-blast gas burner 鼓风煤气喷灯
air blast process 风选法；空气喷净法
air bleed (=air leakage) ①漏气；空气漏失②空气入口
air bleeder 排气孔；气眼
air bleeding valve 排气阀；排气嘴
air blister 气泡；砂眼
air blow 鼓风
air blower 鼓风机
air blow gun 吹气枪
air blowing 吹气成型
air blowing of asphalt 沥青空气氧化
air blowing oxidation 吹气氧化
air blowing process 吹气氧化过程〔制沥青方法〕
air blowing treatment 吹气处理
air blown ①空气吹制的②空气氧化的
air blown asphalt 氧化沥青
air blown oil 气吹油；氧化油
air blown poke hole 吹气拨火孔
air blown producer (空气)鼓风器
air blow-off 空气喷出法；气吹法
air blow slot 缝隙式吹风口
air bomb aging 空气弹老化
air bomb test 空气弹老化试验
airborne 气飘的；气载的；空中浮游的
airborne activity 气载放射性；大气中放射性
airborne contaminants 气污染物
airborne debris (大气)气载碎片*
air-borne dryer 气流干燥机
airborne dust 气载尘埃
airborne particle 空中浮游粒子
airborne plutonium 气飘钚尘
airborne pollution 大气污染
airborne siliceous material 气漂硅质粉尘
airborne soil 空气媒介污垢
airborne waste 气载废物
air bottle (of pump) (泵的)空气拱室

air brake ①气闸；风闸②气刹车
air brake hose 风闸胶管
air breather ①排气机②通气机
air breathing suit 气衣
air brick 风干砖；砖坯
air bridge fan 气动送风机
air broom 气帚〔吹气扫塔〕
air brush 气刷；喷漆器
air bubble (铸件的)气泡；砂眼
air bubble electrode 空气泡电极；气泡电极
air bubble pitting 气泡点蚀
air-bubble viscometer 气泡黏度计
air-bubble void 气泡空洞
air buffer 空气缓冲器
air buoyancy 空气浮力
air burn 烧焦；过热点
air cap 气帽〔喷漆枪的气帽，用以使漆雾化〕
air-captive tyre 保气式轮胎〔专指双腔式轮胎〕
air car 气垫汽车
air casing (空)气套
air-catalyzed oligomerization 空气催化低聚合(作用)
air cell ①气泡；砂眼②空气电池
air cement gun 气动水泥枪；水泥喷射器
air chamber 空气室；气腔〔无内胎轮胎〕
air change 换气
air checks 气泡；麻孔
air chuck 气动卡盘
air circulation 空气循环；气动循环
air circulation autoclave 空气循环式硫化罐
air circulator 空气循环器
air clamp 气动卡具
air classifier 风力分级器
air classifying mill 气流分级细粉磨
air cleaner 空气净化器
air cleaning 空气净化
air-cleaning facility 空气净化设施
air cock (空)气栓；排气旋塞
Airco-Hoover sweetening 艾尔科-胡佛脱硫法
air collector 空气收集器
air compressing machine 空气压缩机
air compressor 空气压缩机
air compressor oil 空气压缩机油
air compressor pump 空气压缩泵
air condenser ①空气冷凝器②空气电容器
air conditioner 空调器
air conditioning 空调
air conditioning machinery 空调机
air conditioning plant 空调装置

air conditioning unit　空调机组；空调装置
air-conductivity　透气性
air conduit　空气管；风管
air consumption　空气消耗
air contaminant　空气污染物
air contaminate analysis　空气污染分析
air contamination　空气污染
air control valve　空调阀
air convection　空气对流
air conveying　气力输送
air conveyor(=airveyor)　气动运输机
air cooled　空气冷却的
air-cooled condenser　气冷式冷凝器
air cooled cylinder　气冷式汽缸
air cooled gas-cooler　空气冷却的气体冷却器
air-cooled heat exchanger　空气冷却换热器
air-cooled torch tube　空气冷却炬管
air cooler　空气冷却器
air cooling　空气冷却
air-core tire　半实心轮胎；弹性轮胎
Air-Corps method for octane rating　美国空军辛烷值测定法
air craft adhesive　航空黏合剂
air craft coating　飞机涂料
aircraft dope　(飞机)涂布漆
aircraft engine fuel　航空汽油
aircraft engine oil　航空润滑油；飞机(发动机)润滑油
aircraft fuel　航空燃料
aircraft fueler　飞机加油器
aircraft fueling　飞机加油
aircraft gas turbine bearing lubrication　飞机燃气涡轮轴承润滑
aircraft grease　航空润滑脂
aircraft humidity instrument　机载湿度仪
aircraft instrument mounting　航空仪表防震器
aircraft motor gasoline(=aircraft motor spirit)　航空汽油
aircraft motor spirit(=aircraft motor gasoline)　航空汽油
aircraft oil　①飞机油②航空润滑油
aircraft refueler　飞机加油器
aircraft refueling line　飞机加油胶管
aircraft tanker　空中加油飞机
aircraft tire　飞机轮胎
air curing　空气固化；常温固化；自然硫化
air curing type cement　硫化型胶浆；常温硫化胶浆
air current　气流
air curtain(=air screen)　气幕；气帘；气屏
air cushion　①气垫②空气减震器；空气缓冲器
air cushion squeeze　气垫式挤压

air cushion vehicle　气垫汽车
air cylinder　气缸
air-damped balance　(空气)阻尼天平*
air dehydration　风干；晾干
air-depolarized cell　(空气)去极化电池
air-depolarized electrode　(空气)去极化电极
air-deposited clay　风积土
air dew point　空气露点
air diffuser　空气扩散器
air diffuser plate　空气扩散板
air diffusion aerator　空气扩散式曝气装置
air discharge　排气
air discharge cock　排气旋塞；气栓
air disconnect　空气分离
air-displacement method　空气排代法
air distillation　常压蒸馏〔石油产品〕
air distributor　空气分配器
air doctor blade　①空气刮刀②空气刮涂
air dome　充气帐篷
air dome (of pump)　(泵的)气室
air draft　(空气)气流；空气通风
air drain　通风道；通风管
air dried　风干了的；晾干的
air-dried basis　风干体
air-dried brick　风干砖；砖坯
air-dried coating　自干型涂料；气干涂料
air dried coil vanish　自干绕组清漆；自干绕线清漆
air-dried sample　风干试样
air-dried sheet(ADS)　风干(胶)片
air dried size　风干尺寸
air drier　空气干燥器
air-driven attachment　气动附属设备
air drum　空气罐
air dry　风干；气干
air dryer　空气干燥器
air dry fiber　风干纤维
air drying　风干*
air-drying loss　风干失重
air-drying material　气干材料
air-drying time　自干时间；风干时间；常温干时间
air drying type cement　空气干燥型胶浆
air drying varnish　气干清漆；自干清漆
air dry pulp　风干浆粕〔含水约10%〕
air-dry wood　气干材；风干材
air duct　空气管；风管；导风筒；风道
air ejection(=air knockout)　气力顶脱
air ejector　空气喷射器
air electrode　空气电极

air elutriation 空气淘析；空气洗脱
air engine 空气(发动)机
air entangled yarn 空气交络丝；(喷气)交缠丝
air-entrained cement 加气水泥
air-entrained concrete 加气混凝土
air entraining 加气
air-entraining agent 加气剂；携气剂
air-entraining cement 输气水泥
air-entraining concrete 加气混凝土
air entrainment 加气处理
air entrapment 滞留空气；内部气泡
air-entrapping structure 多孔结构
air equilibrium distillation 常压平衡蒸馏
airer 干燥柜(箱)
air escape 泄气
air escape cock 泄气旋塞
air escape valve 放气阀
air evaporation 空气中蒸发
air evaporation gum test 空气蒸发胶质试验〔测定汽油的胶质含量〕
air evaporation sludge test 空气蒸发测沉渣试验
air exhaust 抽气；排气
air exhauster 抽气机
air extraction system 排气系统
air extractor 抽气机
air-feed 送气
air-feeder 送气机
air feed pump 供气泵
air field dump 飞机场废料堆场
air filled plastic 气体填充塑料
air filter 空气过滤器
air filter housing 空气过滤罩
air filter oil 空气过滤器油
air filtration 空气过滤
air-floatation process 充气浮选法；风力浮选法
air floatation table 风选台
air-floated clay 风选陶土
air-floated powder 气浮粉
air-float separator 风选器
air-flow 气流
air flow dryer 气流干燥机；热风烘干机
air-flow meter 气流计
air flow tester 气流式(线密度)测定仪
air flue 风道；烟道
air-flue gas mixture 空气-烟道气混合物
air-foam 气沫；空气泡沫
air-foam cushioning 泡沫胶垫
air-foam rubber 泡沫橡胶

air-foam system 泡沫灭火系统
air foil 翼片
air fractionation system 空气分馏设备
air frame opening seal 飞机舱口密封胶
air free 无气的；没有空气的
air freshener 空气芳香剂
air friction 空气摩擦
air-friendly yarn 包含空气纱；透气纱〔低捻或无捻纱〕
air fuel mixture 空气燃料混合物
air-fuel ratio 空气燃料比率
air-fuel ratio meter 空气-燃料比测定计
air-fuel regulation 空气燃料比例调节
air furnace 鼓风炉
air gage(=air gauge) 空气压力计
air gap 空隙
air gap electrode 气隙电极
air gap temperature 气隙温度
air gas 风煤气〔含空气的煤气〕
air gas attack 毒气空袭
air gauge(=air gage) 空气压力计
air governor 自动空气调节器
air grinder ①风动磨头②风动研磨机；气动砂轮机
air guide (喷枪)空气导管
air gun 空气喷枪
air hardening ①气硬的②空气硬化；气硬
air-hardening lime 气硬石灰
air hardening refractory cement 气硬性耐火水泥
air hardening steel 气冷硬化钢；风钢
air heater 空气加热器
air hoist 空气升液器
air holder 空气储蓄器
air-holding (property) 空气保持性
air hole 气孔
air horn 喇叭状气流(气束)
air hose (压缩)空气胶管；空气压力胶管
air humidifier 空气湿润器；空气增湿器
air humidity 空气湿度；大气湿度
air-hydraulic accumulator 空气水压蓄力器
air-hydrogen flame 空气-氢气火焰
air-impact mill 气动冲击磨
air impermeability 不透气性；气密性
air impermeability test 气密性试验
air-impermeable 不透空气的
air impervious fabric 不透气织物；不透气胶布
air-impervious liner 气密层；内衬层
air inclusion 包含气体；夹气
air infiltration 空气渗透
air inflation 充气

air inflation indicator　充气表
airing　充气
air inhibition　空气阻聚
air inlet　①进气②进气孔
air inlet valve　进气阀；空气进口阀
air input　进气量
air insulation(=air isolation)　空气绝缘
air intake　进气口
air-interlaced yarn　网络变形纱
air interlacing　喷气交缠
air interrupted spark　空气吹断式火花
air ionization　空气电离(化)
air ionization dosimeter　空气电离剂量计
air-ionizing device　空气电离装置；空气离子化装置
air isolation(=air insulation)　空气绝缘
air jacket　空气夹套
air jet　①空气喷嘴②空气喷射
air jet cleaning　空气喷砂除锈
air jet evaporation test　空气喷射蒸发试验〔利用空气喷射蒸发汽油来测定其胶质含量〕
air jet gum　用空气喷射法测定的胶质量〔石油产品〕
air-jet interlaced yarn　气流喷射交缠丝；空气喷射交缠丝
air jet system　空气喷射装置；喷气装置
air jet textured yarn　空气变形纱；喷气变形纱
air knife　风刀；气刀
air knife coating　气刀涂布
air knockout(=air ejection)　气力顶脱
air-laid　气流成网
air-laid fibrous web　气流铺置纤维网
air-laid web　气流成网
air leak　漏气缝
air leakage(=air bleed)　①漏气；空气漏失②空气入口
airless blast　无空气喷砂
airless blast deflashing　冲击除边
airless centrifugal wheel blast　离心轮式无空气喷砂
airless electrostatic spraying　无空气静电喷涂法
airless hot spraying(=hot airless spray)　无空气热喷涂
airless shot blasting installation　无空气抛丸清理机；无空气抛丸清理装置
airless spray　无空气喷涂
airless sprayer　无空气喷雾器
airless spray gun　无空气喷(涂)枪
airless spraying　无空气喷涂
airless spray unit　无空气喷涂装置
air-lift　气提；空气升液器
air-lift agitator　空气升液搅拌器
air-lift bioreactor　气升式生物反应器

air-lift catalytic cracker　气升式移动床催化裂化装置
air-lift cracker　空气提升裂化器
air lift extractor　空气升液萃取器
air-lift (fluid bed) dryer　空气带升式(流化床式)干燥机
air lift mixer settler　空气升液式混合澄清槽
air lift pedestal　气动升降(操作)台
air lift pump　气升泵
air lift tank　空气压送罐
air-lift type agitator　气升式搅拌器；注气搅拌器
air-lift unit　气升(催化剂)装置
air line　空气(谱)线
air line filter　风管过滤器
air line hose　输气胶管
air line lubricator　空气管路润滑器
air liquefaction　空气液化
air-liquid tension　空气液体张力
air liquification　空气液化
air-loaded accumulator　空气荷重式蓄力器
air lock　①锁气室；离毒室②(塑体表面的)气窝；气泡；砂眼③气门；气阻
air loss　风干损耗
air-mail paper　航空邮件纸
air manometer　空气压力计(表)；气压计(表)
air marking　小气泡；麻孔
air marks　麻孔；小气泡
air meter(=anemometer; airometer)　风速计；气流计
air micrometer　空气测微计
air mixture　空气(燃料蒸气)混合物
air moistener　空气湿润器；空气增湿器
air moistening　空气加湿
air monitoring　大气监测
air nozzle　空气喷嘴
air-off　排气
air-oil separator　空气-油分离器
airol　棓酸碘羟铋
airometer(=air meter)　风速计；气流计
air operated　气动的
air-operated controller　气动控制器
air operated lubricator　气动润滑器
air operated orbital sander　气动轨道式喷砂机
air operated power tool　风动工具；气动工具
air operated pump　空气驱动泵；气动泵
air operate valve　气动阀
airosol　气溶胶
air outlet valve　出气阀
air oven　①热空气箱；烘箱；烤箱②老化恒温箱
air oven aging　热空气老化
air oxidation　空气氧化

air oxidation process 空气氧化法
air-painter 喷漆器
air-painting sprayer 压缩空气喷涂机
air-passage ①通气道②通气口
air peak 空气峰
air peener 气动喷丸机；气动喷纱机
air permeability ①透气性；空气透过性②透气度
air permeability test 透气(性)试验
air permeable waterproofing 透气性防水处理
air pervious waterproof fabric 透气雨布
air-petrol mixture 空气-汽油混合物
air pilot valve 空气导向阀
air pipe 空气管
airplane dope (飞机)涂布漆
air plenum 压力通风系统；强制通风；进气增压
air plus steam vulcanization 混气硫化〔空气和蒸气并用〕
air pocket ①残存空气；(帘布层间)气泡②气窝〔蒸汽系统〕③气袋
air pollutant 大气污染物
air pollutant analysis 空气污染分析
air pollution 大气污染
air pollution control system 空气污染控制系统
air pollution emission 大气污染排放(物)
air pollution episode 大气污染事件
air pollution index 大气污染指数
air pollution zone 大气污染区；大气污染带
air preheater 空气预热器
air press 气动压机
air pressure ga(u)ge 气压计(表)
air pressure heat aging 热空气加压老化
air pressure reducer 松压气嘴；空气减压阀
air pressure regulator 气压调节器
air pressure test 气压试验
air-producer gas 空气(发生器)煤气
air-producer gas generator 空气(发生器)煤气发生器
air-proof 不漏气的；气密的
air-proof hose 不漏气胶管
air-propane flame 空气-丙烷火焰
air-pulse gauge 气动脉冲仪
air pulser 空气脉冲器
air pump 气泵
air purge 空气吹扫
air purification 空气净化
air purifier 空气净化器
air quality control region 空气质量控制区(域)
air quality management 空气质量管理
air quality monitoring system 空气质量监测系统

air quenching ①空气急冷②空气骤冷
air quenching chamber 空气骤冷室
air-quenching rate 空气冷却速率
air rate 空气流率
air receiver 空气储蓄器
air recirculation 空气再循环
air refrigerating machine 空气制冷机
air regenerating device ①换气设备②空气换流法
air regulator 空调器
air release agent 脱气剂
air relief valve(=air release valve) 排气阀；放空阀
air removal 排气；除气
air removing roll 除气辊
air reservoir (压缩)空气储罐
air resistance 抗气性；不透气度
air resistance stock 气密性胶料
air-retaining liner 内衬层；气密层
air-retaining wall 内衬层；气密层
air retention (capacity) 空气保持性
air retention time 空气保留时间
air ring 环形垫；环形空气囊
air road 气线
air roll coat 气刀辊涂机
air sampler 空气取样器
air sampling 空气取样
air-sampling filter 大气采样滤膜
air sand blower 吹砂器
air sander 自动打磨机；气动打磨机
air sanitizer 空气消毒剂；空气清净剂
air saturation method 通气饱和法
air saturation value 溶解氧饱和值
air saturator 空气饱和器
air saturator tower 空气饱和塔
air scour 气体冲刷
air screen(=air curtain) 气幕；气帘；气屏
air scrubber 空气洗涤器；空气洗涤塔
air seal 气封
air seasoning 自然干燥；风干
air sectional bag 局部硫化囊
air segmented stream 空气隔断流
air-separated sulfur 风选硫黄
air-separating tank 吹(气分)离罐
air separation 吹(气分)离；风选
air separation plant 吹(气分)离装置
air separator 吹(气分)离器
air separator mill 吹(气分)离磨
air-setting 气干空气固化；空气凝固
air shots 气泡；砂眼〔压延缺点〕

air shrinkage 空气收缩
air shut-off valve 空气关闭阀
air silencer 空气减声器
air silk 空心丝；气泡丝
air sizing 空气定型；空气定径
air slaked 潮解的；空气熟化的
air-slaked lime 潮解石灰
air slaking 潮解；空气熟化
air slide 气阀
air-slip forming (热成型)气胀包模成型
air space 空气室
air spacing 气隙
air sparger 空气分布器
air spinning 气流纺
airspinning-cell 干法纺丝仓
airspinning die 干法纺丝板
airspinning-dope 干法纺丝液
air spots 表面不平〔模制品缺点〕
air spray finishing 空气喷涂法
air spring 空气弹簧
air-steam mixture 空气蒸汽混合物
airstop tube 防刺穿内胎
air strainer 空气滤器
air stream 气流
air stream turbulence 气流湍流
air stripping 气提；空气吹脱
air suit 气衣
air supply 气源
air supply hose 送风胶管
air-supported fabric dome 充气胶布帐篷
air suspension 空气悬挂；空气弹簧
airsweep ①气吸式②气吹
air sweetening 氧化脱硫醇；空气气化脱臭〔使汽油中有臭味的硫醇经空气氧化为没有臭味的二硫化物〕
air switch 空气开关；空气断路器
air tank 气柜；气罐
air tap 空气旋塞
air test 空气试验
air tester(=carbacidometer) 大气碳酸计
air test tube aging 空气试管老化
air-texturing yarn 空气变形丝
air thermometer 空气温度计
air through-put 空气耗用
air-tight 不透气的；气密的
air-tight access door 气密入口门
air-tight container 密闭容器；密闭罐
air-tight joint 气密接头
airtightness 气密性

air-tight seal 气密密封
air tight test 气密性试验
air tire ①空气轮胎②飞机胎
air-to-close 气关式
air-to-fuel ratio 空燃比；空气-燃料比
air-tool oil 气动工具油
air-to-open 气开式
air-town gas flame 空气-煤气(火)焰
air transformer 空气调压器〔净化并加压空气，输出的空气可接喷枪〕
air trap 空气阱
air traps ①气泡；麻孔②气窝
air tube ①空气管②内胎
air tube cooler 管状空气辐射冷却器
air turbine 空气涡轮
air twisting 气流加捻
air tyre 空气轮胎
air valve (空)气阀
air-vapour mixture 空气(石油产品)蒸气混合物
air velocity 空气速度
air velocity gradient 气流速度梯度
air vent 排气口
air vessel 空气罐
airveyor(=air conveyor) 气动运输机
air void 空气空隙；气穴；气泡
air-volume bleeder-gun (空气)容积分压式喷枪；空气分流式喷枪
air vortex 空气涡流(旋涡)；旋风
air vortex blender 空气涡流式混合器
air vortex false twist 空气涡流假捻(法)
air vulcanization 常温硫化；室温硫化
air vulcanizing cement 常温硫化胶浆
air washer 空气洗涤器
air washing 空气洗涤
airwash separator 空气洗涤分离器
air-water cooling 空气-水冷却
air wheel 超低压轮胎
airy ①轻的②通风的；空气的
Aitken counter 艾特肯计数器
Aitken nuclei 艾特肯核
aizumycin 爱图霉素；双环霉素
ajacine 阿芥辛
ajacol 乙氧(基)苯酚
ajawa oil(=ajowan oil) 香旱芹油
Ajax metal 阿贾斯合金〔一种轴承合金〕
Ajax-Northrup furnace 阿贾斯-诺斯拉普高频感应炉
Ajax-Watt furnace 阿贾斯-瓦特电炉
ajmalicine 阿吗碱

ajmalidine 阿吗定
ajmaline 阿吗灵；西萝芙木碱
ajmalinine 阿吗宁
Ajo process (for copper extraction) 阿焦提铜法
ajowan oil(=ajawa oil) 香旱芹油
ajugalactone 筋骨草内酯
Akabori method(=hydrazine method) 肼解法
akazgine 非洲马钱子碱
akcethin 硫乙酸甘油酯
akebigenin 木通配基 $C_{31}H_{50}O_4$
akebine 木通苷 $(C_{35}H_{56}O_{20})_3$
akee oil 阿开木油
akerite 尖晶石
akermanite 镁黄长石
akimycin 秋霉素
akitamycin 秋田霉素
aklavin 阿克拉菌素
Akron abrader 阿克隆磨耗试验机
Akron abrasion tester 阿克隆磨耗试验机
akuammicine 阿枯米辛
akuammine(=vincamajoridine) 阿枯明
akundarol 牛角瓜甾醇
-al〔词尾〕 醛
-al-〔用在词中〕 醛缩
alabandite 硫锰矿
alabaster 雪花石膏
alabaster glass 雪花玻璃；乳色玻璃
alabaster ware 雪花瓷器〔一种骨灰瓷〕
alacreatine 异肌氨酸 $C_4H_9O_2N_3$
alactacid 无乳酸的
alafosfalin 阿拉磷
alamandite(=almandite) 贵榴石
alamethicin 丙甲菌素；丙甲甘肽
alamosite 铅辉石
alanate 铅氢化物
alane 铝烷
alangine 八角枫碱
α-alanine α-氨基丙酸
β-alanine β-氨基丙酸
alanine(Ala) 丙氨酸
alanine aminotransferase 丙氨酸转氨酶
alanine ethyl ether hydrochloride 盐酸丙氨酸乙酯
alanine racemase 丙氨酸消旋酶
alanine transaminase 丙氨酸转氨(基)酶
alanosin 丙氨菌素
alanosine 亚硝基羟基丙氨酸
alant acid anhydride 阿兰酸酐
alant camphor(=helenin) 阿兰脑

alantic acid(=inulic acid; alantolic acid) 旋覆花酸；阿兰酸；土木香酸 $C_{15}H_{22}O_3$
alantin(=inulin) 阿兰粉；土木香素；菊粉
alantol 阿兰醇；土木香油
alantolactone(=helenine) ①堆心菊脑；土木香脑；阿兰内酯②海仑菌素
alantolic acid(=alantic acid) 土木香酸
alant root(=inula) 阿兰根
alant starch 阿兰粉
alanyl 丙氨酰；α-氨基丙酰
alanyl-glycine 丙氨酰甘氨酸
β-alanylhistidine β-丙氨酰组氨酸；肌肽
alarm ①警报②警报器
alarm apparatus 报警(电)器
alarm buzzer 报警蜂鸣器
alarm circuit 警报线路
alarm gauge(=alarm manometer) 警号气压计
alarm indicator 警报指示器
alarm limit 警告限
alarm manometer(=alarm gauge) 警号气压计
alarm signal 警报信号
alarm system 警报系统
alarm thermometer 警号温度计
alaskite 白岗岩
alazopeptin 丙氨肽素
albacol 丙醇
albahaca oil 妥路油
albamycin(=novobiocin) 新生霉素
albarium 大理石灰
albaspidin 三叉蕨素
albedo ①反射率②白色③白果胶
albendazole 阿苯达唑〔药〕
alberene 高级皂石
Alberta oil 阿尔贝塔石油
albertite 阿尔贝塔天然沥青〔加拿大Albert矿所出的一种黑沥青〕
albertol 阿尔贝塔树脂
albescent 浅白色的；带白色的
albimycin 微白霉素
albinism 白化症
albite 钠长石*
albite law 钠长(双晶)律
albitite 钠长岩
albitization 钠长石化(作用)
albizziine 合欢氨酸；脲基丙氨酸
albocarbon(=naphthalene) 萘
albocycline 白环菌素
albomaculine 白斑(网球花)碱

albopannin 绵马(根茎)素 $C_{21}H_{24}O_7$
albumen ①蛋清②蛋清蛋白③清蛋白；白蛋白
albumin 清蛋白；白蛋白
albuminoid ①硬蛋白②类蛋白
albuminometer 清蛋白计
albuminoscope 清蛋白仪；验蛋白器
albuminous ①白蛋白的②蛋白的
albuminous matter 蛋白质
albumin tannate 鞣酸白蛋白
albuminuria 蛋白尿
albumin wave (白)蛋白波
albumosease 清蛋白酶
Alby furnace 阿耳拜电炉
alcahest(=alkahest) 仙丹〔炼丹术〕
alcali(=alkali) 碱；强碱
alcaline 碱的
alcamines 醇胺类
alcapton 尿黑酸；2,5-二羟基苯乙酸
alcaptonuria 尿黑酸尿
alchemical 炼丹的
alchemical reaction ①"炼金术"反应②核反应
alchemist 炼丹家；炼金术士
alchemistic period 炼丹时代
alchemistic symbol 炼丹符号
alchemy 炼丹术*
alchlor 三氯化铝 $AlCl_3$
alchlor process 三氯化铝法〔一种催化裂化法〕
Alcian blue 艾尔西安蓝
Alclad 铝衣合金
alcogas 乙醇汽油混合物
alcogel 醇凝胶
Alco-Gyro cracking process 阿尔柯-杰罗气相裂化过程
alcohol ①醇*②乙醇；酒精
alcohol acid 醇酸；羟基酸
alcohol aldehyde 醇醛；羟基醛
alcohol amide 醇酰胺；羟基酰胺
alcohol amine 醇胺；羟基胺 $R(OH)NH_2$
alcoholase 醇酶
alcoholate ①乙醇化物②醇化物；烃氧基金属
alcoholate ion 烷氧离子 RO—
alcoholature 酊剂
alcohol-benzene extractive 苯醇抽出物
alcohol blast burner 酒精喷灯
alcohol blend 酒精调和液
alcohol burner 酒精灯
alcohol coagulation volume 酒精凝固体积
alcohol dehydrogenase (乙)醇脱氢酶
alcohol error 酒精误差；醇误差

alcohol ester ①羟基酸的酯②醇的酯
alcohol ether ①醚 ROR②羟基醚
alcohol ethoxylate 脂肪醇乙氧基化物
alcohol ethoxysulfate 脂肪醇乙氧基硫酸盐
alcohol fermentation 乙醇发酵；酒精发酵
alcohol fuel 酒精燃料
alcohol-gasoline blends 酒精汽油调和物
alcohol gauge 酒精气压计；酒精测压计
alcohol glycoside 醇(糖)苷；配糖物
alcohol-hydrocarbon mixture 醇-烃混合物
alcoholic ①醇的②乙醇的；酒精的
alcoholic acid 醇酸
alcoholic aldehyde 醇醛
alcoholic ammonia 酒精氨；氨的酒精溶液
alcoholic beverage 酒精饮料
alcoholic determination 定醇法
alcoholic extract ①酒精萃取②酒精提出物
alcoholic extraction 酒精提取
alcoholic fermentation 成(乙)醇发酵；酒精发酵
alcoholic hydroxyl 醇式羟基
alcoholic lye 醇碱液
alcoholic potash 钾碱醇液；氢氧化钾的酒精溶液
alcoholic-potash extraction 钾碱醇液萃取
alcoholimeter 醇定量计
alcohol insoluble matter 乙醇不溶物
alcohol-insoluble substance 乙醇不溶物
alcoholism 酒精中毒
alcoholization 醇化*
alcohol kali 钾碱醇溶液〔氢氧化钾的酒精溶液〕
alcohol ketone 醇酮；羟基酮
alcohol lamp 酒精灯
alcohol lignin 乙醇木素
alcohol motor fuel 酒精马达燃料
alcohol number 醇值
alcohol of crystallization 结晶醇
alcoholometer 酒精比重计；醇比重计
alcoholometry 酒精测定；醇定量法
alcohol oxidation reaction 醇氧化反应
alcohol phenol 醇酚
alcohol phosphate 醇的磷酸酯
alcohol-precipitated starch 醇析淀粉
alcohol purification 酒精净化
alcohol resistance 耐醇性
alcohol soluble absolute 乙醇溶性净油
alcohol-soluble polymer 醇溶性聚合物
alcohol stove 酒精炉
alcohol sulfate 醇硫酸盐
alcohol thermometer 酒精温度计

alcohol titration 酒精滴定法；醇滴定法
alcohol-to-oil ratio 酒精对汽油比率；酒精消耗量
alcohol varnish 醇溶性清漆
alcohol-water blend 酒精水溶液；酒精水调和物
alcoholysate 醇解物
alcoholysis 醇解*
alcoholysis method 醇解法
alcosol 醇溶胶
alcotate 变性剂〔一种用来使有毒气体具有特殊气味或使酒精变性的成分〕
Alco two-stage distillation process 常减压二段蒸馏过程
alcove ①凹室(处)；附室②壁龛③壁橱(柜)
alcoxides 烃氧基金属
alcyl 脂环基
ald- 〔词头〕 醛
aldactone 螺甾内酯
aldalcoketose 醛醇酮糖 $CH_2OH(CHOH)_nCOCHO$
aldamine 乙醛胺(俗)；乙醛合氨；1-氨基乙醇 $CH_3CH(OH)NH_2$
aldaric acid 醛糖二酸
aldarsone 阿耳达宋〔药〕；次硫酸非那肼
aldazine 醛连氮
aldehydase 醛酶
aldehyde ①醛*②乙醛
aldehyde acetal ①缩醛 $RCH(OR)_2$ ②乙缩醛 $CH_3CH(OC_2H_5)_2$
aldehyde alcohol 醛醇
aldehyde-amines 醛胺类
aldehyde-amine (type) accelerator 醛胺类促进剂
aldehyde ammonia 醛氨(加成物)
aldehyde benzoic acid 醛基苯甲酸 $CHOC_6H_4COOH$
aldehyde C-6〔商〕 己醛
aldehyde C-7〔商〕 庚醛
aldehyde C-8〔商〕 辛醛
aldehyde C-9〔商〕 壬醛
aldehyde C-10〔商〕 癸醛
aldehyde C-11〔商〕 十一醛
aldehyde C-12〔商〕 十二醛；月桂醛
aldehyde C-13〔商〕 十三醛
aldehyde C-14〔商〕 桃醛；十四醛〔指γ-十一内酯〕
aldehyde C-16〔商〕 草莓醛；十六醛〔指β-甲基-β苯基缩水甘油酸乙酯〕
aldehyde C-18〔商〕 椰子醛；十八醛〔指γ-壬内酯〕
aldehyde C-19〔商〕 凤梨醛；十九醛〔指己酸烯丙酯〕
aldehyde C-20〔商〕 覆盆子醛；二十醛〔指混合香料〕
aldehyde C-22〔商〕 玫瑰醛；二十二醛〔指混合香料〕
aldehyde C-31〔商〕 银白金合欢醛；三十一醛〔指混合香料〕

aldehyde carbonyl moiety 醛式羰基(部分)〔指醛基酸的羟基〕
aldehydecollidine 5-乙基-2-甲基吡啶
aldehyde condensation 醛醇缩合；醛缩作用
aldehyde dehydrogenase(=aldehyde oxidase) 醛脱氢酶；醛氧化酶
aldehyde group 醛基
aldehyde hydrate 醛水合物*
aldehyde ketone 醛酮 RCOCHO
aldehyde-ketone rearrangement 醛酮重排
aldehyde monoperacetate 醛合单过氧乙酸
aldehyde mutase 醛变位酶；醛歧化酶
aldehydene 乙炔
aldehyde oxidase(=aldehyde dehydrogenase) 醛脱氢酶；醛氧化酶
aldehyde polymer 醛类聚合物
aldehyde resin 醛类树脂
aldehyde tannage 醛鞣(法)
aldehyde tanned leather 醛鞣革
aldehyde tanning 醛鞣
aldehyde tanning agent 醛鞣剂
aldehyde value 醛值
aldehydic acid 醛酸
aldehydic carboxylic acid 醛(基)羧酸
aldehydic hydrogen 醛式氢
aldehydine 乙基甲基吡啶 $C_5H_3N(CH_3)(C_2H_5)$
aldehydo-caprylic acid 壬醛酸；8-醛基辛酸
aldehydo-ester 醛酯 CHOCOOR
aldehydoglyoxylic acid 醛二羟乙酸
2-aldehydoisophthalic acid 2-醛基异邻苯二甲酸
aldehydrase(=xanthine oxidase) 黄嘌呤氧化酶
aldehydrol 水合醛
alder 赤杨；桤木
alder bark 赤杨树皮；桤木皮
alder buckthorn 药炭鼠李
Alder reaction 阿尔德反应〔即双烯合成〕
aldgamycin 阿德加霉素
aldimine 醛亚胺*
aldimine chelate 醛亚胺螯合物
aldimine condensation 醛胺缩合
alditol 糖醇*
aldo- 醛(元)；氧代 O=
aldoalcoketose 醛醇酮糖
aldobionic acid 乙醛糖酸；醛糖二酸
aldobiouronic acid 乙醛糖酸
aldoheptose 庚醛糖
aldohexose 己醛糖
aldoketens 醛烯酮类

aldoketose 醛甾酮糖
aldol 羟醛*
aldolactol 内缩醛
aldol-alpha-naphthylamine(AAN) 2-羟基丁醛-α-萘胺
aldolase(=zymohexase) 醛缩酶；二磷酸果糖酶
aldol condensation 羟醛缩合
aldol group transfer polymerization 醛醇基团转移聚合
aldolisation 缩醛反应；醛醇缩合反应
aldol reaction 醛醇缩合反应
aldol(s) 醇醛缩合物
aldonic acid 醛糖酸
aldopentose 戊醛糖
aldose 醛糖*
aldose reductase 醛糖还原酶
aldosterone 醛甾酮
aldotetrose 丁醛糖
aldotriose 丙醛糖
aldoxime 醛肟*
Aldrin 艾氏剂；氯甲桥萘
alduronic acid 醛糖酸
alebaster 雪花石膏
alembic 蒸馏釜；蒸馏器
alemite grease fitting 压力输送润滑脂的润滑器
alendronate sodium 阿仑膦酸钠〔药〕
aleprestic acid 环戊烯戊酸
alepric acid 环戊烯壬酸
aleprolic acid 环戊烯甲酸
aleprylic acid 环戊烯庚酸
aleuritic acid(=aleutric acid) 紫胶桐酸 $C_{16}H_{32}O_5$
aleurometer （面粉）发力计
aleurone 糊粉
aleurone grain 糊粉粒
aleurone layer 糊粉层
aleutite 闪辉长斑岩
aleutric acid(=aleuritic acid) 紫胶桐酸 $C_{16}H_{32}O_5$
Alexandria paper 亚历山大纸；旃那纸
alexandrite 翠绿宝石；变石
alexandrolite 铬黏土
alexoite 磁黄橄榄岩
alfa 芦苇草
alfacalcidol 阿法骨化醇〔药〕
alfalfa 苜蓿
alfalfa extract 苜蓿浸液
alfalfa-leaf 苜蓿叶
alfalfone 苜蓿酮 $C_{21}H_{42}O$
Alfene synthesis 铝烯合成法〔利用铝催化剂合成α-烯烃方法〕
alferric 含铝铁的

alfin 烯醇
alfin catalyst 醇(碱金属)烯催化剂
alfin initiator 烯醇钠引发剂*
alfin polymer 烯醇(钠)聚合物
alfin rubber 烯醇钠橡胶*
Alfol alcohol 阿尔福醇；铝醇法合成醇
Alfol process 阿尔福法；铝醇合成法
Alfol synthesis 阿尔福合成法；铝醇合成法〔利用铝催化剂合成高级醇的方法〕
Alfrey's rule 阿尔弗雷法则
alfuzosin 阿呋唑嗪〔药〕
alga(e) 藻
algaecide 杀(除)藻剂
algae coal 藻煤
algae preventive effect 防藻效果
algae preventive type antifouling paint 防藻型防污漆
algal bloom 藻花
algarobilla 角豆树；苏方荚
algaroth powder 氯化氧锑 $2SbOCl \cdot Sb_2O_3$
algebraic series 代数级数
algebraic value 代数值
algicidal activity 杀藻活性
algicide 杀藻剂
algin 藻酸铵〔膏化剂〕
alginate 藻酸盐；藻朊酸盐；藻朊酸纤维
alginate calcium 海藻酸钙
alginate fiber 藻(朊)酸纤维
alginate jelly 海藻胶
alginate yarn 海藻纤维纱；藻酸纤维纱
alginic acid 藻酸
alginic (acid) fibre 藻酸纤维；海藻纤维
algistatic activity 抑藻活性
algodonite 微晶砷铜矿
algol 阿果〔染料商名，是还原染料〕
algol blue 阿果蓝
algol bordeaux 阿果枣红
algol brilliant green 阿果亮绿
algology 藻类学
algol olive 阿果橄榄绿
algol printing black 阿果印染黑
algorithm 算法；规则系统
algosol 藻溶胶〔无色瓮染料的溶剂〕
aliased line 混淆线
aliasing phenomenon 假象
alicyclic(=cycloaliphatic) 脂环(族)的
alicyclic acid 脂环酸
alicyclic alkene 脂环烯
alicyclic amine salt 脂环胺盐〔硫化剂〕

alicyclic compound 脂环化合物*
alicyclic hydrocarbon 脂环烃
alicyclic hydrocarbon metabolism 脂环烃代谢
alicyclic nylon fibre 脂环族耐纶纤维
alicyclic polyamine 脂环族多元胺
alicyclic ring 脂环
alicyclic series 脂环系
alicyclic stem-nucleus 脂环母核
alien material 异物
alignment ①对准②校准③定线；定位
alignment chart 列线图
alignment error 安装误差
alignment suppressing agent 取向抑制剂
alinear heating 非线性加热
alinear heating programming 非线性程序加热
aliomycin 棕黄霉素；阿留(棕黄)霉素
aliphatic 脂(肪)族的
aliphatic acid 脂(肪)族酸
aliphatic alcohol 脂族醇
aliphatic aldehyde 脂族醛
aliphatic amine 脂族胺
aliphatic-aromatic polyamide 脂(肪)族-芳(香)族聚酰胺；脂-芳族聚酰胺
aliphatic carboxylic acid 脂族羧酸
aliphatic compound 脂肪族化合物*
aliphatic diamine 脂(肪)族二胺
aliphatic dibasic acid 脂肪族二元酸
aliphatic dicarboxylic acid 脂(肪)族二羧酸
aliphatic diglycidylether 脂肪族二缩水甘油醚
aliphatic diisocyanate 脂肪族二异氰酸酯
aliphatic diolefine 脂族二烯
aliphatic epoxy resin 脂肪族环氧树脂
aliphatic ester 脂族酯
aliphatic ether 脂族醚
aliphatic ether linkage 脂(肪)族醚键
aliphatic group 脂族基
aliphatic halide 脂族卤化物
aliphatic hydrocarbon 脂族烃
aliphatic isocyanate 脂(肪)族异氰酸酯
aliphatic naphtha 脂族石脑油〔溶剂〕
aliphatic nylon fibre 脂(肪)族耐纶纤维
aliphatic olefin 脂族烯
aliphatic oxide 脂族氧化物
aliphatic phosphate 脂族磷酸酯
aliphatic polyamide 脂肪聚酰胺；尼龙
aliphatic polyamide fibre 脂(肪)族聚酰胺纤维
aliphatic polyamine 脂(肪)族多胺
aliphatic polycarbonate 脂肪族聚碳酸酯
aliphatic polyester 脂(肪)族聚酯
aliphatic polysulfide 脂族多硫化物
aliphatic quaternary ammonium inner salt 脂族季铵内盐
aliphatic resin 石油树脂
aliphatics 脂(肪)族化合物
aliphatic saturated hydrocarbon 脂族饱和烃；饱和链烃
aliphatic series 脂肪系；脂族
aliphatic sesquiterpene 脂族倍半萜 $C_{15}H_{24}$
aliphatic solvents 脂族溶剂；脂肪烃溶剂
aliphatic sulfinic acid 脂族亚磺酸 RSO_2H
aliphatic sulfonic acid 脂族磺酸 RSO_3H
aliphatic sulphonyl chloride 脂族磺酰氯
aliphatic sulphoxide 脂族亚砜 RSOR
aliphatic tanning agent 脂族鞣剂
aliphatic terpene 脂肪族萜烯
aliphatic unsaturated hydrocarbon 脂族不饱合烃；不饱合链烃
aliquation 层化；起层
aliquot 等分部分*
aliquot portion 整分部分
aliquot sample 整分试样；等分试样
alit 铝铁岩
alite ①A-水泥石；硅酸三钙石
alizanthrene〔商〕 茜士林
alizaramide 茜酰胺
alizarate 茜酸盐
alizaric acid(=phthalic acid) 茜酸
alizarimide 茜酰亚胺
alizarin(=alizarine) 茜素
alizarin assistant 土耳其红油；太古油〔磺化蓖麻油〕；茜素助剂
alizarin astrol 茜素鲜红
alizarin black 茜素黑
alizarin blue 茜素蓝
alizarin blue-black 茜素蓝黑
alizarin brilliant pure blue 茜素亮纯蓝
alizarin brown 茜素棕
alizarin carmine 茜素胭脂红
alizarin celestol 茜素青
alizarin colors〔复〕 茜素染料
alizarin complexon 茜素氨羧络合剂
alizarin cyanine 茜素菁
alizarin cyanol 茜素蒽醌青
alizarine(=alizarin; 1,2-dihydroxyanthraquinone) 茜素
alizarine dyestuff 茜素染料
alizarine lake 茜素色淀
alizarine yellow C(=2,3,4-trihydroxyacetophenone) 2,3,4-三羟苯乙酮；茜素黄 C $(HO)_3C_6H_2COCH_3$

alizarinic acid(=phthalic acid)　茜酸
alizarin orange yellow　茜素橙黄；3-硝基-1,2-二羟基蒽醌　$C_6H_4(CO)_2C_6H(OH)_2NO_2$
alizarin red S　茜素红 S*
alizarin saphirol　茜素玉醇蓝；茜素蒽醌蓝
alizarin skyblue　茜素青
alizarin sulfonic acid　茜素磺酸
alizarin viridine　茜素青绿
alizarin yellow　茜素黄*
alizarin yellow G(=nitrobenzene azosalicylic acid)　硝基苯偶氮水杨酸；茜素黄 G　$NO_2C_6H_4N = NC_6H_3(OH)CO_2H$
alizarin yellow R(=sodium p-nitrobenzene-azo-salicylate)　茜素黄 R
alizarol brown　茜酚棕
alizarol orange　茜酚橙
alizarol saphirol〔商〕　茜酚玉醇蓝
alkacid process　裂化气净化过程〔先用强碱然后用弱酸洗涤以除去裂化气体中的 H_2S 及 CO_2〕
alkadiene　链二烯
alkadiyne　链二炔
alkahest(=alcahest)　仙丹〔炼丹术〕
alkalamides　碱胺类
alkalescence(=alkalescency)　微碱性
alkalescency(=alkalescence)　微碱性
alkalescent　微碱性的
alkali(=alcali)　碱；强碱
alkali absorption velocity　吸碱速度
alkali-acid desizing　碱酸退浆
alkali-aggregate reaction　碱聚集反应
alkali alcoholate　碱金属醇盐
alkali alkyl　烷基碱金属
alkali amide　氨基碱金属
alkali binding agent　碱性黏合剂；碱结合剂
alkali blue lake　碱性蓝色淀
alkali blue toner　碱性蓝增色剂；碱性蓝调色剂
alkali-catalyzed alcoholysis　碱催化醇解
alkali-catalyzed hydrolysis　碱催化水解
alkali-catalyzed self polymerization　碱催化自聚合(作用)
alkali cellulose　碱纤维素
alkali cellulose cooler　碱纤维素冷却器
alkali chlorine cell　碱氯电池
alkali cleaner　碱脱脂剂；碱性清洗剂
alkali containing glass fibre　(中)碱玻璃纤维
alkali corrosion prevention method　碱蚀防止法；防碱腐蚀法
alkali crack(ing)　碱(致)龟裂；碱(致)开裂
alkali degradation　碱致降解
alkali degreaser　碱(性)脱脂剂
alkali deweighting　碱减量处理
alkali dialyzer　碱液渗析器〔碱液回收装置〕
alkali digester (process)　碱(再生)法
alkali drop test　碱滴试验
alkali earth　碱土〔碱土金属的氧化物〕
alkali-earth metal　碱土金属
alkali extractive　碱抽出物
alkali family　碱族
alkali fastness　抗碱坚牢度；抗碱性
alkaliferous　含碱的
alkaliflable　可碱化的
alkali flame ionization detector　碱火焰电离检测器
alkali formate(=alkali formiate)　碱金属甲酸盐
alkali formiate(=alkali formate)　碱金属甲酸盐
alkali free　无碱的；不含碱的
alkali free glass fibre　无碱玻璃纤维
alkali fusion　碱熔*
alkali fusion reaction gas chromatography　碱熔反应气相色谱(法)
alkaligenous　成碱的
alkali hose　耐碱胶管
alkali hydrolysis diffusion method　碱解扩散法
alkali hydrometer　碱液比重计
alkali hydroxide etchants　碱性氢氧化物腐蚀剂
alkali increase test (for tung oil varnishes)　(桐油漆)加碱后的增稠度试验
alkali ion source　碱离子源
alkali isomerization　碱性异构化
alkali lignin　碱木素
alkali liquor　碱液
alkali loss　碱耗；碱损失
alkali lye　碱液
alkali metal　碱金属*
alkali metal caprolactamate　己内酰胺碱金属盐
alkali metal catalyzed polymerization　碱金属催化聚合
alkali metal dichloroisocyanurate　碱金属二氯异氰尿酸盐
alkali metal flame ionization detector　碱金属火焰电离检测器
alkali metal grease　碱金属润滑脂
alkali metal sensor　碱金属传感器
alkali metal soap　碱金属皂
alkalimeter　碱量计；碳酸定量计；施罗特碳酸定量器
alkali method of spinning　碱法纺丝
alkalimetric　碱量滴定的
alkalimetric analysis　碱量滴定分析(法)
alkalimetric estimation　碱量滴定测定

alkalimetric method 碱量滴定法
alkalimetric standard 标碱基准；标定碱溶液的基准物
alkalimetry 碱量法*
alkaline （强）碱的
alkaline accumulator 碱性蓄电池*
alkaline aqueous suspension 碱性水悬浮体
alkaline bath 碱浴
alkaline black 碱性炭黑
alkaline carbonate 碱金属碳酸盐
alkaline catalyst 碱催化剂
alkaline cell 碱性电池
alkaline char 碱性炭
alkaline chromate treatment 碱性铬酸盐处理法〔铝的表面处理法〕
alkaline cleaner 碱清洗剂
alkaline color 碱色
alkaline constituent 碱组分
alkaline corrosion 碱性腐蚀
alkaline degradation 碱降解
alkaline degreasing(=alkali degreasing) 碱法脱脂；碱法除油
alkaline dropping corrosion resistance test （耐）滴碱腐蚀试验
alkaline dye 碱性染料
alkaline earth 碱土
alkaline earth family 碱土族
alkaline earth metal 碱土金属*
alkaline element 碱性元素；碱性金属元素
alkaline error 碱误差
alkaline-error of glass pH electrode 玻璃pH电极的碱误差
alkaline etching 碱蚀刻；碱腐蚀法
alkaline filler 碱性填料
alkaline fusion 加碱熔化
alkaline granite 碱性花岗岩
alkaline hydrated oxide 碱性氢氧化物
alkaline hydrolysis 碱解；加碱水解
alkaline-induced polymerization 碱引发聚合(作用)
alkaline land 碱性土；碱地
alkaline liquor 碱液
alkaline medium 碱性介质
alkaline metal cleaner 碱性(的)金属清洗剂
alkaline milling 碱性缩绒
alkaline mud 碱性沉渣；碱性泥浆
alkaline neutralization number 碱中和值
alkaline permanganate oxidation 碱性高锰酸盐氧化法
alkaline permanganometry 碱性高锰酸盐滴定法
alkaline polymerization 碱性聚合*

alkaline process ①碱法②碱法制浆
alkaline protease 碱性蛋白酶
alkaline pulp 碱法浆
alkaline purification 碱处理；碱净化
alkaline quench 碱骤冷(法)
alkaline reaction 碱性反应
alkaline reagent 碱性试剂
alkaline reducer(=alkaline reducing agent) 碱性还原剂
alkaline reducing agent(=alkaline reducer) 碱性还原剂
alkaline reduction 碱性还原
alkaline reserve 碱储量；碱藏
alkaline resisting 抗碱的
alkaline rocks〔复〕 碱性岩
alkaline rubber 碱性橡胶〔如丁吡橡胶〕
alkaline saponification 加碱皂化
alkaline semichemical pulp 碱法半化学浆
alkaline-sensitive colored pigment 碱敏性彩色颜料
alkaline silicate 碱性硅酸盐
alkaline sky blue 碱性青
alkaline soaker 碱性浸渍剂
alkaline soil 碱性土
alkaline solution 碱性溶液
alkaline solvolysis 碱性溶离
alkaline splitting 加碱裂解
alkaline steeping agent 碱性浸渍剂
alkaline storage cell 碱性蓄电池
alkaline tower 碱淋塔
alkaline (type) accelerator 碱性促进剂
alkali neutralization number 碱中和值
alkaline wash ①碱洗②碱洗液
alkaline waste water 碱性废水
alkalinity 碱度*
alkalinity of water 水的碱度
alkalinization 碱化(作用)
alkalinous 碱性的
alkalion 负碱性离子；—OH基
alkali phenate(=alkali phenolate) 碱金属酚盐
alkali phenolate(=alkali phenate) 碱金属酚盐
alkali process 碱处理
alkali process of regeneration 碱(再生)法
alkali-proof 耐碱的
alkali purification 碱法纯化
alkali reaction 碱性反应
alkali-reactive aggregate 碱反应性骨料〔与碱性水泥起反应的骨料〕
alkali reclaim 碱回收；碱再生
alkali reclaimed rubber 碱法再生胶
alkali reclaiming process 碱再生法

alkali recovery 碱回收
alkali refine process 碱精制(炼)法
alkali refining 碱精制；碱精炼；碱提纯
alkali regeneration process 碱性再生法
alkali reserve 碱储量；碱藏
alkali resistance 抗碱性
alkali-resistant 抗碱的
alkali-resistant glass fibre AR玻璃纤维；耐碱玻璃纤维
alkali resisting 抗碱的
alkali resisting cellulose 抗碱纤维素
alkali resisting primer 耐碱性底漆
alkali rosinate 松香酸碱金属盐
alkali salt 碱金属盐
alkali-sensitive indicator 碱敏(感)指示剂
alkali-sensitive link 碱敏键
alkali sensitivity 碱敏现象
alkali series 碱性岩系
alkali silicate 碱金属硅酸盐
alkali soil 碱土；碱性土
alkali soluble 碱溶性的；可溶于碱液的
alkali-soluble fibre 碱溶纤维
alkali-soluble resin 碱溶性树脂
alkali source 碱源〔用于碱焰检测器〕
alkali stability 耐碱性
alkali stain 碱斑；碱渍；碱性着色剂(料)
alkali steeping test 碱浸渍试验
alkali-sulfite process 碱性亚硫酸盐法
alkali swelling 碱膨胀
alkali test 碱试法
alkali tolerance ①忍碱性②忍碱度
alkali treatment 碱处理
alkality of gasoline 汽油碱度；汽油碱性
alkali tyre reclaim 碱性轮胎再生胶
alkali wash 碱洗
alkali wash water 碱性洗涤水
alkali waste 废碱
alkali waste liquid 废碱液
alkali waste water 废碱水
alkali weight-loss treatment 碱(法)减量处理
alkali works 制碱厂；苏打工厂
alkali yellowing 碱致泛黄
alkalization 碱化
alkalizing ①碱化的②碱化(作用)；加碱
alkaloid 生物碱
alkaloidal 生物碱的
alkaloidal drug 生物碱药物
alkaloidal reagent 生物碱试剂
alkametric 碱量滴定的

alkamine 氨基醇
alkanal 链烷醛
alkane 烷*；(链)烷烃
alkane amido ethyl sulfosuccinate 烷酰胺乙基磺基琥珀酸盐
alkane dicarboxylic acid 链烷二羧酸
alkanedioic acid 链烷双酸；链烷二羧酸
alkane disulfonate 链烷二磺酸盐
alkane polysulfonate 链烷多磺酸盐
alkane sulfate 链烷硫酸盐
alkanesulfonamide 链烷磺(酰)胺
alkane sulfonate 链烷磺酸盐；烷基磺酸盐
alkanesulfonic acid 链烷磺酸
alkanesulfonyl chloride 链烷磺酰氯
alkanesulphonic acid 烷基磺酸
alkanet 紫草根
alkane tricarboxylate 链烷三羧酸盐(或酯)
alkanisation 链烷化(作用)
alkannan 紫草烷
alkannic acid 紫草酸
alkannin 紫草素
alkannin extract 紫草素提取物
alkannin paper 紫草素试纸
alkanoate 链烷酸酯(或盐)
alkanoic acid 链烷酸
alkanol 链烷醇
alkanolamide 链烷醇酰胺
alkanolamine 链烷醇胺
alkanolester amine (链)烷醇酯胺
alkanoyl 烷酰基
alkanoyloxy 烷酰氧基
alkapton 尿黑酸
alkargen(=cacodylic acid) 卡可基酸；二甲胂酸
alkarsin 卡可基酸氧化物
alkaryl(=alkylaryl) 烷基芳基
alkarylamine 烷芳基胺
alkaryl polyethylene phosphoric acid 烷芳基聚乙烯磷酸
alkasal 水杨酸铝与乙酸钾的混合物
alkatriene 链三烯
alkatriyne 链三炔
Alkazid process 阿尔卡吉特法〔以甲替丙氨酸钾水溶液吸收裂化气体中的H_2S及CO_2的方法〕
alkazid solution 碱性溶液
alkene 烯*
alkene phosphonous acid 烯烃亚膦酸
alkene sulfonate 烯烃磺酸盐
alkene sulfonium 烯烃𬭩
alkenic polymeric monomer 烯类高分子单体

alkeno-derivatives 链烯衍生物
alkenoic acid 链烯酸
alkenone 链烯酮
alkenoxy 链烯氧基
alkenyl 链烯基
alkenyl amine 烯基胺
π-alkenyl anion π-链烯基阴离子
alkenylation 链烯基化(作用)
alkenyl benzene sulfonic acid 烯基苯磺酸
alkenyl halide 卤代烯烃
alkenyl magnesium halide 卤化烯基镁
alkenyl morpholinium 烯基吗啉鎓
alkenyloxy 链烯氧基
alkenyl succinamide 烯基琥珀酰胺
alkenyl succinate 烯基琥珀酸盐；烯基丁二酸盐
alkenyl succinic acid 烯基琥珀酸；烯基丁二酸
alkenyl succinic anhydride 烯基琥珀酸酐
alkermes 胭脂虫；胭脂虫粉；地中海区的一种酒
Alker process 阿尔克法〔由烯烃制乙基苯法〕
alki 掺水酒精
alkine （链）炔烃 C_nH_{2n-2}
alkograph 酒精仪
alkoxide ①醇盐②酚盐
alkoxide-initiated polymerization 醇盐引发聚合(作用)
alkoxy 烷氧基〔有时指烃氧基〕
alkoxy acetic acid 烷氧基乙酸
alkoxyalkyl sulfate 烷氧基烷基硫酸盐
alkoxyamine 烷氧基胺
alkoxyamine salt 烷氧基胺盐
alkoxybenzoyl peroxide 烷氧基苯酰基过氧化物
alkoxycarbonyl 烷氧羰基；烷氧碳酰基
alkoxyethanol 烷氧基乙醇
alkoxy ethoxy sodium sulfate 烷氧基乙氧基硫酸钠
alkoxy group 烷氧基
alkoxyl 烷氧基
alkoxylamine 烷氧基胺 $RONH_2$
alkoxylated ether 烷氧基化醚
alkoxylation 烷氧基化作用
alkoxyl group 烷氧基原子团
alkoxymethyl acrylamide 烷氧基甲基丙烯酰胺
N-alkoxy methyl phosphonic acid 正构烷氧基甲基膦酸
N-alkoxymethyl polyamide N-烷氧甲基聚酰胺
alkoxy methyl pyridinium chloride 烷氧基甲基氯化吡啶鎓
alkoxy methyl sucrose ether 烷氧基甲基蔗糖醚
alkoxy methyl trialkyl ammonium chloride 烷氧基甲基三烷基氯化铵
alkoxynitrile 烷氧基腈

alkoxyphenyl radical 烷氧苯基
β-alkoxy proprionitrile β-烷氧基丙腈
alkoxy propyl amine 烷氧基丙胺
alkoxy propyl amino ethyl sulfonate 烷氧丙胺基乙基磺酸盐
alkoxy propyl sulfonate 烷氧基丙基磺酸盐
alkoxy radical 烷氧自由基
alkoxysilane 烷氧基硅烷
alkoxy-substituted organosilane 烷氧基取代的有机硅烷
alkoxytitanium 烷氧基钛
alkoxytitanium acrylate 丙烯酸烷氧基钛
alkyd 醇酸树脂
alkyd-acrylate copolymer 醇酸-丙烯酸共聚物
alkydal 邻苯二甲酸树脂；醇酸树脂
alkyd emulsion 醇酸乳液；醇酸乳胶
alkyd flat paint 醇酸无光漆
alkyd-modified urea resin 醇酸改性脲醛树脂
alkyd paint 醇酸漆；醇酸涂料
alkyd ratio 醇酸率〔醇酸在树脂中的百分比〕
alkyd resin(=glycerol-phthalic resin) 甘酞树脂；醇酸树脂
alkyd resin enamel 醇酸树脂瓷漆
alkyd resin varnish 醇酸(树脂)清漆
alkyd varnish 醇酸(树脂)清漆
alky gas 酒精和汽油混合燃料
alkyl 烷基〔有时指烃基〕
alkylable (=alkylatable) 可烷基化的
alkyl (aceto) acetal ①烷基(醇)缩醛②烷基(醇)缩乙醛
alkyl acrylate 丙烯酸烷基酯
alkylacyl propyl betaine 烷基酰基丙基甜菜碱
alkyl alkane phosphonate 烷基膦酸烷基酯
4-alkyl-4-alkenyl morpholinium 4-烷基-4-烯基吗啉鎓
N-alkyl-N-alkenyl piperidinium N-烷基-N-烯基哌啶鎓
alkylalkoxy silane (烷基)烷氧基硅烷
alkyl alkoxy solicon ester 烷基烷氧基硅酯
alkyl alkylphosphonate 烷基膦酸单烷基酯
alkyl allyl polyether alcohol 聚烷基烯丙基醚醇
alkylallyl sulfonate 烷基烯丙基磺酸盐
alkylaluminium 烷基铝
alkyl aluminium halide 卤化烷基铝
alkylaluminium sesquichloride 烷基铝倍半氯化物
alkylamide 烷基酰胺
alkyl amido betaine 烷基酰胺甜菜碱
alkyl amido phosphoric acid ester 烷基酰胺基磷酸酯
alkylamine 烷基胺
alkylamine hydrochloride 烷基胺盐酸盐
alkyl aminoacrylate 氨基丙烯酸烷基酯
alkyl amino alkanol 烷基氨基烷醇

alkyl amino carboxylate 烷基氨基羧酸盐
alkyl amino ethanol hydrochloride 烷基氨基乙醇盐酸化物
alkyl amino ethyl maleate 烷基氨乙基马来酸盐
alkyl aminosilane 烷基氨基硅烷
alkyl ammomium chloride 烷基氯化铵
alkyl ammomium dodecylbenzene sulfonate 烷基十二烷基苯磺酸铵
alkylammonium acylate 烷基酰化铵
alkyl ammonium alkanoate 烷基烷酸铵
alkyl ammonium carboxylate 烷基羧酸铵
alkyl ammonium halide 烷基卤化铵
alkylammonium montmorillonite 烷基铵蒙脱石
alkyl ammonium propionate 烷基丙酸铵
alkylaniline rearrangement 烷基苯胺重排作用〔如 $RN \cdot HC_6H_5 \longrightarrow RC_6H_4NH_2$〕
alkylantimony halide 卤化烷基锑
alkylaromatic hydrocarbon disulfide 二硫化烷基芳烃
alkyl aromatics 烷基芳香烃
alkylarsine 烷基胂
alkyl arsine disulfide 二硫化烷基胂
alkyl arsine oxide 氧化烷基胂 $RAs=O$
alkyl arsine sulfide 硫化烷基胂 $RAsS$
alkyl arsine tetrahalide 四卤化烷基胂 $RAsX_4$
alkyl arsonic acid 烷基胂酸 $RAsO(OH)_2$
alkylaryl(=alkaryl) 烷基芳基
alkylarylamine 烷基芳基胺
alkylaryl diglycol ether sulfonate 烷基芳基二甘醇醚磺酸盐
alkyl aryl ether rearrangement 烷芳醚重排作用〔如 $ROAr \longrightarrow RArOH$〕
alkylaryl ethoxy phosphate 烷基芳基乙氧基磷酸盐
alkylaryl ketone 烷基芳基酮
alkyl-aryl p-phenylenediamine 烷基芳基对苯二胺〔防老剂〕
alkyl aryl phosphate 磷酸烷基芳基酯
alkyl aryl phthalate 邻苯二甲酸烷基芳基酯；邻苯二甲酸芳烷基酯
alkylaryl polyethenoxy sulfonate 烷基芳基聚氧乙烯磺酸盐
alkyl aryl polyether alcohol 烷基芳基聚醚醇
alkyl aryl polyether sulfate 烷基芳基聚醚硫酸盐
alkyl aryl polyethyleneglycol ether 聚乙二醇烷基芳基醚
alkylaryl polyglycol ether 烷基芳基聚乙二醇醚
alkylaryl propyl ammonium sulfate 烷基芳基丙基硫酸铵
alkyl aryl recinoleate 蓖麻酸烷基芳基酯
alkylaryl sodium sulfonate 烷芳基磺酸钠
alkylaryl sulfonamide 烷基芳基磺酰胺

alkylaryl sulfonate 烷芳基磺酸盐
alkylatable(=alkylable) 可烷基化的
alkylate ①烷基化②烷基化物
alkylate bottoms 烷基化油蒸馏残液
alkylated 烷基化的
alkylated aromatic hydrocarbons 烷化芳香烃
alkylated bisphenol 烷基化二苯酚〔防老剂〕
alkylated cation 烷基化阳离子
alkylated p-cresol 烷基化对甲酚〔防老剂〕
alkylated cresol 烷基化甲酚
alkylated diphenylamine 烷基化二苯胺〔防老剂〕
alkylate detergent 烷基化物洗涤剂
N-alkylated glucamine N-烷基葡萄糖胺
alkylated hydroquinone 烷基化对苯二酚〔防老剂〕
alkylated mercapto benzimidazole 烷基巯基苯并咪唑
alkylated naphthene 烷基化环烷
alkylated phenol 烷基化苯酚〔防老剂〕
alkylated polymer 烷基化聚合物
alkylated polymethylol aminoresin 烷基(醚)化多羟甲基氨基树脂
alkylated polymethylol melamine 烷基(醚)化多羟甲基三聚氰胺
alkylated styrenated phenol 烷基化苯乙烯化苯酚〔防老剂〕
alkylated succinic acid 烷基化琥珀酸；烷基化丁二酸
alkylated sulfonate 烷基化磺酸盐〔常指烷基苯磺酸盐〕
alkylated tar 烷基化焦油
alkylated thio-bis-phenol 烷基化硫代双苯酚
alkylated urea formaldehyde 烷基化脲甲醛
alkylate fractionator 烷化物分馏塔
alkylate polymer(=alkylate bottoms) 烷化油蒸馏残液
alkylate rerun tower(=alkylate fractionator) 烷化物再馏塔
alkylating ①烷基化的②烷基化作用
alkylating agent 烷基化剂
alkylating phenol 烷基(化)酚
alkylating reagent 烷基化试剂
N-alkylation N-烷基化
O-alkylation O-烷基化*
alkylation ①烷基取代②烷基化
alkylation acid 烷化酸
alkylation gasoline 烷化汽油
alkylation method 烷化法
alkylation plant 烷化工厂
alkylation process 烷化过程
alkylation unit 烷化装置
alkyl azide 烷基叠氮 $R[N_3]$
alkyl azirane 烷基氮杂环丙烷；烷基吖丙啶

alkyl azo-compound 烷基偶氮化合物 RN═NR—
alkyl benzene 烷基苯
alkyl benzene amine 烷基苯胺
alkyl benzene glucoside 烷基苯葡萄糖苷
alkyl benzene sulfonamide 烷基苯磺酰胺
alkyl benzene sulfonate 烷基苯磺酸盐
alkyl benzene sulfonyl halide 烷基苯磺酰卤
alkyl benzene sulphonate(ABS) 烷基苯磺酸
alkyl benzene thiol 烷基苯硫醇
alkyl benzimidazole 烷基苯并咪唑
alkyl benzimidazole sulfonate 烷基苯并咪唑磺酸盐
alkyl benzoate 安息香烷基酯（光聚合引发剂）
N-alkylbenzylamine N-烷基苄胺 $C_6H_5CH_2NHR$
alkyl benzyl dimethyl amine 烷基苄基二甲基胺
alkyl benzyl dimethyl ammonium chloride 烷基苄基二甲基氯化铵
alkylbenzyl dimethylammonium ion 烷基苄基二甲基铵离子
alkylbenzyl dimethylammonium smectites 烷基苄基二甲铵绿土〔蒙脱石〕
alkyl benzyl halide 烷基苄基卤(化物)
alkyl benzyl phthalate 邻苯二甲酸烷基苄酯〔增塑剂〕
alkyl benzyl sucrose ester 烷基苄基蔗糖酯
alkyl benzyl thiosulfate 烷基苄基硫代硫酸盐
alkyl benzyl trimethyl ammonium chloride 烷基苄基三甲基氯化铵
alkyl betaine 烷基甜菜碱
alkyl biphenyl 烷基联苯
alkyl borane 烷基硼烷
alkyl boric acid 烷基硼酸 R_2BOH; $RB(OH)_2$
alkylboron catalyst 烷基硼催化剂
alkylboronic acids 烷基硼酸
alkyl bromide 烷基溴 RBr
alkyl butyral ①烷基(醇)缩丁醛②(=butoxydialkylacetal)(丁氧基型)二烷基(醇)缩丁醛
alkyl carbamate 氨基甲酸烷基酯 NH_2COOR
alkyl carbonate 碳酸烷基酯 $CO(OR)_2$
alkyl catechol 烷基儿茶酚；烷基苯邻二酚
alkyl cellulose ether 烷基纤维素醚
alkyl chain length 烷链长度
alkyl chloride 烷基氯 RCl
alkyl chlorocarbene 烷基氯碳烯
alkyl chlorofluorosilane 烷基氟氯硅烷
alkyl chloroformate 氯甲酸烷基酯 ClCOOR
alkyl chlorosilane 烷基氯硅烷
alkyl chlorosulfite 氯亚硫酸烷基酯 $ClSO_2R$
alkyl compound ①烷基化合物②烃基化合物
alkyl cyanide 烷基腈 RCN

alkyl 2-cyanoacrylate 2-氰基丙烯酸烷基酯 $CH_2C(CN)COOR$
alkyl cyclohexanol 烷基环己醇
alkyl cyclohexene 烷基环己烯
alkyl derivative ①烷基衍生物②烃基衍生物
alkyl diamino glycine hydrochloride 烷基二氨基甘氨酸盐酸化物
alkyl diarylamine 烷基二芳胺〔防老剂〕
alkyl diaryl phosphate 磷酸烷基二芳酯〔增塑剂〕
alkyl diazo-compound 烷基重氮化合物
alkyl diazoimide 烷基叠氮
alkyl diborane 烷基二硼烷
alkyl dihalogenated arsine 烷基胂化二卤；烷基二卤胂 $RAsX_2$
alkyl dimethyl amine oxide 烷基二甲基胺氧化物
alkyl dimethyl benzyl ammonium chloride 烷基二甲基苄基氯化铵
alkyl dimethyl dichlorobenzyl ammonium chloride 烷基二甲基二氯苄基氯化铵
alkyl dimethyl oxazolinium salt 烷基二甲基噁唑啉盐
alkyl diphenyl ether disulfonate 烷基二苯醚二磺酸盐
alkyl disulfide 烷基二硫 RS_2R
alkylene 亚烷基；烯化〔一般指 RCHCHR 型；常指 RCH═ 或 RCHCH— 型基；若其前加希腊字母β，则为间亚烷基如 $RCHCH_2CH$—〕
alkylene carbonate 碳酸亚烃酯
alkylene copolymer 亚烷基共聚物
alkylene diamine 烷撑二胺
alkylene diammonium halide 亚烷基卤化二铵
$α,ω$-alkylene diammonium smectites $α,ω$-亚烷基二铵绿土
alkylene ether sulfide 硫化亚烃醚
alkylene glycol 烷撑二醇
alkylene halohydrin 邻亚烷卤醇 $RCHOHCH_2X$
alkylene imine 烯化亚胺
alkylene oxide 烯化氧
alkylene oxide polymer 环氧烷聚合物；氧化烯类聚合物
alkyleneoxylated bisquaternary ammonium 烯氧基化双季铵
alkylene polyamine 亚烷基多胺
alkylene polysulfide 烯化多硫
alkylene sulfide 烯化硫
alkylene sulfide polymer 硫化烯类聚合物
alkyl epoxide 烷基环氧化物
alkyl epoxy-stearate 环氧硬脂酸烷基酯

alkyl ethanolamine phosphate　烷基乙醇胺磷酸盐
alkyl ethanolamine sulfate　烷基乙醇胺硫酸盐
alkyl ether　烷基醚
alkyl etherified resin　烷基(醇)醚化树脂
alkyl ether sulfate　烷基醚硫酸盐
alkyl ethoxylate sulfate　烷基乙氧基化物硫酸盐
alkyl ethyl dimethyl ammomium halide　烷基乙基二甲基卤化铵
alkyl fluoride　烷基氟　RF
alkyl fluorophosphate　氟磷酸烷基酯
alkyl formal(=methoxy dialkyl acetal)　甲氧基型二烷基(醇)缩(甲)醛；烷基醇缩甲醛
alkyl formamide　N-烷甲酰胺　HCONHR
alkyl glucoside　烷基葡萄糖苷；烷基配糖物
alkyl glyceryl ether sulfonate　烷基甘油醚磺酸盐
alkyl group　①烷基②烃基
alkyl group frequency　烃基振动频率
alkyl halide(=alkylogen)　烷基卤；卤代烷
p-alkylhalobenzene　对烷基卤苯
alkylhalogenobismuthine　烷基卤化铋
alkylhalogenoborine　烷基卤化硼
alkyl halosilane　烷基卤代硅烷〔一种防水整理剂〕
N-alkyl hexahydro azepine　N-烷基六氢氮杂䓬
N-alkylhydrazide　烷基酰肼
alkyl hydrogen sulfate (=alkyl hydrosulfate)　硫酸烷基氢酯
alkyl hydrorubber　烷基氢化橡胶
alkyl hydrosulfate (=alkyl hydrogen sulfate)　硫酸烷基氢酯　$ROSO_3H$
alkyl hydrosulfide　氢硫基烷；烷基硫醇　RSH
alkylhydroxamic acid　烷基异羟肟酸；烷基氧肟酸
alkyl (hydroxyethyl) benzene　烷基(羟乙基)苯
alkyl hydroxyethyl imidazoline　烷基羟乙基咪唑啉
alkyl hydroxyethyl thioether　烷基羟乙基硫醚
alkylide　烷基化物
alkylidene　亚烷基　RCH=
alkylidene bisphenol　次烷基双酚
alkylidene group　亚烷基
alkylidene radical　亚烷基
alkyl imidazole　烷基咪唑
alkyl imidazoline　烷基咪唑啉
alkyl imidazoline hydroxyethyl alkyl amine　烷基咪唑啉羟乙基烷基胺
alkyl imido chloride　烷基氯代亚胺　RCCl=NR
N-alkyl imino diacetic acid　N-烷基亚氨二乙酸
alkyl indole sulfonate　烷基吲哚磺酸盐
alkyl iodide　烷基碘　RI
alkyl isethionate　烷基羟乙基磺酸盐
alkyl isocyanate　异氰酸烷基酯　RNCO

alkyl isocyanide　烷基异氰；烷基胩　RNC
alkyl isorhodanate(=alkylisothiocyanate)　异硫氰酸烷基酯
alkyl isorhodanide(=alkylisothiocyanate)　异硫氰酸烷基酯
alkyl isosulfocyanate(=alkyl isothiocyanate)　异硫氰酸烷基酯
alkyl isosulfocyanide(=alkyl isothiocyanate)　异硫氰酸烷基酯
alkyl isothiocyanate(=alkyl isosulfocyanate;alkyl isosulfocyanide)　异硫氰酸烷基酯　RNCS
alkyl isothiocyanide(=alkyl isothiocyanate)　异硫氰酸烷基酯
S-alkyl isothiourea　S-烷基异硫脲
alkyllammonium　烷基铵
alkyl lead　烷基铅
alkyl lead trihalide　三卤化烷基铅
alkyl lithium catalyst　烷基锂催化剂
alkyl lithium-catalyzed-polybutadiene　烷基锂催化聚丁二烯橡胶；丁锂橡胶
alkyl lithium-initiated polymerization　烷基锂引发聚合
alkyl lithium initiator　烷基锂引发剂
alkyl magnesium bromide　溴化烷基镁　RMgBr
alkyl magnesium chloride　氯化烷基镁　RMgCl
alkyl magnesium halide　卤化烷基镁　RMgX
alkylmercaptan copper　烷基硫醇铜
alkyl mercaptide　烷基硫醇
alkyl mercapto acetic acid　烷基巯基乙酸
alkyl mercuric chloride　氯化烷基汞　RHgCl
alkyl mercuric halide　卤化烷基汞　RHgX
alkyl mercuric hydroxide　氢氧化烷基汞　RHgOH
alkyl mercury　烷基汞
alkyl mercury cyanide　氰化烷基汞　RHgCN
alkyl mercury hydroxide　氢氧化烷基汞
alkyl metal catalyst　烷基金属催化剂
alkyl methacrylate　甲基丙烯酸烷基酯
N-alkyl monoethanol amide　N-烷基单乙醇酰胺
alkyl monoisocyanate　烷基单异氰酸酯
alkyl monosulfide　硫化烷基
alkyl morpholine　烷基吗啉
alkyl naphthalene　烷基萘
alkyl naphthalene sulfonate　烷基磺酸萘〔湿润剂〕
alkyl naphthol　烷基萘酚
N-alkyl nicotinamide　N-烷基烟酰胺
alkyl nitrate　硝酸烷基酯
alkylogen(=alkyl halide)　烷基卤；卤代烷
alkylol amide　烷基醇酰胺
alkylol amine　醇胺；羟基胺

alkylol phosphonamide 烷醇膦酰胺
alkyl orthoformate 原甲酸烷基酯
alkyl oxide 烷基化氧；烷基醚 ROR
N-alkyloxy alkyl compound 正构烷氧基烷基化合物
alkyloxy determination 烷氧基测定
alkyloxy ethylamino propionic acid 烷氧基乙氨基丙酸
alkyl-oxygen fission 烃氧分裂
alkyloyl 烷酰基 RC—$\overset{O}{\underset{\|}{}}$
alkyl paraben 对羟基苯甲酸烷基酯
alkyl peroxide 烷基过氧化物
alkylphenol 烷基(苯)酚
alkyl phenol disulfide 烷基苯酚二硫化物〔硫化剂〕
alkyl phenol ether sulfate 烷基酚醚硫酸盐
alkyl phenol ethoxylate 烷基酚乙氧基化物
alkylphenol-ethylene oxide condensate 烷基酚-环氧乙烷缩合物
alkyl phenol formaldehyde polymer 烷基酚甲醛聚合物
alkylphenol glycidylether 烷基苯酚缩水甘油醚
alkyl phenol monosulfide 一硫化烷基苯酚
alkyl phenol polyether 烷基酚聚醚
alkyl phenol polyethylene glycol ether 烷基苯酚聚乙二醇醚
alkyl phenol polyglycol ether 烷基酚聚乙二醇醚
alkyl phenol polyoxyethylene glycol ether 烷基酚聚乙二醇醚
alkyl phenol sulfonate 烷基酚磺酸盐
alkyl phenoxy butane sulfonate 烷基苯氧基丁烷磺酸盐
alkylphenoxy (poly) ethoxyethyl ammonium salt 烷基苯氧基(聚)乙氧基乙基铵盐
alkyl phenoxy poly (ethyleneoxy) ethanol 烷基苯氧(基)聚氧乙烯乙醇；聚乙二醇-烷基苯醚
alkyl phenyl 烷基苯基；烷代苯基
alkyl phenyl ethylene oxide 烷基苯基环氧乙烷
alkyl phenyl glycerol ether 烷基苯基甘油醚
alkyl phenyl group 烷代苯基
alkyl phosphate 磷酸烷基酯
alkyl phosphine dichloride 二氯化烷基膦 $RPCl_2$
alkyl phosphine oxide 烷基氧膦；烷基膦氧化物
alkyl phosphinic acid 烷基次膦酸 $R_2PO(OH)$
alkyl phosphite 亚磷酸烷基酯
alkyl phosphonic acid 烷基膦酸 $RPO(OH)_2$
alkyl phosphonic diureide 烷基膦酸二酰脲
alkyl phthalic acid 烷基酞酸
alkyl picolinium halide 卤化烷基皮考啉
N-alkyl polyamide N-烷基聚酰胺
alkyl polyamine 烷基多胺

alkyl polyethylene imine 烷基聚乙烯亚胺
alkyl polyglucoside 烷基多葡萄糖苷
alkyl polyglycol ether sulfate 烷基聚乙二醇醚硫酸盐
alkyl polyglycoside 烷基多苷
alkylpolyoxyalkylene phosphate 烷基聚氧烯烃磷酸盐(酯)
alkyl polyoxyethylene 烷基聚环氧乙烷；烷基聚氧乙烯
alkyl polyoxyethylene triphenyl phosphonium chloride 烷基聚氧乙烯三苯基氯化鏻
alkyl polysulfide 烷基多硫化物
alkyl propylene diamine 烷基丙烯二胺
S-alkyl pseudo thiourea S-烷基假硫脲
alkyl pyridinium chloride 氯化烷基吡啶鎓
alkyl pyridinium halide 卤化烷基吡啶鎓
N-alkyl pyridinium halide 卤化 N-烷基吡啶鎓
alkyl pyridinium ion 烷基吡啶鎓离子
alkyl pyridinium salt 烷基吡啶鎓盐
alkyl quaternary ammonium salts 烷基季铵盐
alkyl radical ①烷基②烃基
alkyl residue 烷基(原子团)
alkyl resorcin monoether(=alkyl resorcinol monoether) 烷基间苯二酚单醚
alkyl resorcinol monoether(=alkyl resorcin monoether) 烷基间苯二酚单醚
alkyl rhodanate (=alkyl rhodanide) 硫氰酸烷基酯 RSCN
alkyl rhodanide(=alkyl rhodanate) 硫氰酸烷基酯
alkyl selenide 烷基(化)硒 R_2Se
alkylsilanol 烷基硅醇
alkyl silicate 硅酸烷基酯
alkyl silicone oil 烷基硅酮油
alkylsiloxane 烷基硅氧烷
alkyl sodium sulfate 硫酸烷基(酯)钠 RSO_4Na
alkyl substituted 烷基取代了的
alkyl substituted aza-crown ether 烷基氮杂冠醚
alkyl succinic acid diethyl amide 烷基琥珀酸二乙酰胺
alkyl sucrose 烷基蔗糖
alkyl-sulfamide N-烷基硫酸二酰胺 $R_2NSO_2NR_2$
alkyl sulfaminic acid N-烷(基)氨基磺酸 $RNHSO_3H$
alkyl sulfate 烷基硫酸盐
alkyl sulfhydrate(=alkyl sulfhydryl) (烷基)硫醇 RSH
alkyl sulfhydryl(=alkyl sulfhydrate) 烷基硫醇
alkyl sulfide 硫醚 RSR
alkyl sulfinic acid 烷基亚磺酸 RSO_2H
alkyl sulfoacetate 烷基磺基乙酸盐
alkyl sulfocyanate(=alkyl sulfocyanide) 硫氰酸烷基酯 RSCN
alkyl sulfocyanide(=alkyl sulfocyanate) 硫氰酸烷基酯 RSCN

alkyl sulfonamide 烷基磺酰胺
alkyl sulfonate 烷基磺酸盐 RSO_3M
alkyl sulfonic acid 烷基磺酸 RSO_3H
alkyl sulfonic acid ester 烷基磺酸酯
alkyl sulfonic ester 烷基磺酸酯
alkyl sulfonyl chloride 烷基磺酰氯 RSO_2Cl
2-alkyl sulfopropyl-maleate 2-磺代丙基马来酸烷基酯
alkyl sulfo-propyl succinate 磺基丙丁二酸烷基酯
alkyl sulfosuccinate 烷基磺化琥珀酸盐
alkyl sulfoxide 烷基亚砜 $R_2S=O$
alkyl sulfur compound 烷基硫化物
alkyl sulfuric acid S-烷基硫酸 RSO_4H
alkyl sulfuric ester 烷基硫酸酯
alkyl-sulphate 烷基硫酸盐
alkyl-sulphide radical 烷硫基 $RS—$
alkyl-sulphonate 烷基磺酸盐 RSO_3M
alkylsulphonic acid 烷基磺酸 RSO_3H
alkyl tauride 烷基牛磺酸盐
N-alkyl taurine N-烷基氨基乙磺酸；N-烷基牛磺酸；N-烷基牛胆硷
alkyl tetrahydrophthalate 烷基四氢苯二甲酸酯〔增塑剂〕
alkyl tetrahydro pyranyl sulfate 烷基四氢吡喃硫酸盐
alkyl tetrahydro pyrimidine 烷基四氢嘧啶；烷基四氢间二氮(杂)苯
alkyl tetralin sulfonate 烷基萘满磺酸盐
alkylthioalkyl sulfate 烷基代烷基硫酸盐
alkyl thioborate 硫代硼酸烷基酯
alkyl thiocyanate (=alkyl thiocyanide) 硫氰酸烷基酯 RSCN
alkyl thiocyanide(=alkyl thiocyanate) 硫氰酸烷基酯
alkyl thioglycolic acid 烷基硫代乙醇酸
alkyl thiol acrylate 硫代丙烯酸烷基酯
alkyl thiomorpholine 烷基硫代吗啉
alkyl thiophene sulfonate 烷基噻吩磺酸盐
alkyl thiosulfonic acid 烷基硫代磺酸 RSO_2SH
alkyl thiosulfuric acid O-烷基硫代硫酸 $RSSO_3H$
alkyl thiourea 烷基硫脲 NH_2CSNHR；$CS(NHR)_2$
alkyl thiouronium chloride 烷基氯化硫脲
alkylthiuram disulfide 二硫化烷基秋兰姆〔促进剂〕
alkyl tin 烷基锡
alkyl tin iodide 碘化烷基锡
alkyltin mercaptide 烷基锡硫醇
alkyl titanate 钛酸烷基酯
alkyl toluene sulfonate 烷基甲苯磺酸盐
alkyl tolyl trimethyl ammonium halide 烷基甲苯基三甲基卤化铵
alkyl tosylate 对甲苯磺酸烷基酯
alkyl transfer 烷基转移

alkyl trimethyl ammonium bromide 烷基三甲基溴化铵
N-alkyl trimethylene diamine N-烷基三亚甲基二胺
alkyl trimethylene diamine dioleate 烷基三亚甲基二胺二油酸盐
alkyl tripolyphosphate 烷基三聚磷酸盐
alkyl urea 烷基脲
alkyl vinyl ketone 烷基乙烯基酮
alkyl xanthate 烷基黄原酸盐
alkyl xylene chloride 氯代烷基二甲苯
alkyl zinc halide 卤化烷基锌 $RZnX$
alkymer 烷化油〔烷化汽油的商名〕
alkyne 炔*
alkyne conversion 炔转化
alkyne series 炔系；炔属烃
alkynol 炔醇
alkynyl 炔基
alkynyl compound 炔基化合物
alkynyl magnesium halide 卤化炔镁 $RC\equiv CMgX$
alkyoxycarbonyl 烷酯基 $ROOC—$
alky plant(=alkylation plant) 烷化工厂
allactite 砷水锰矿
allalinite 蚀变辉长岩
allanic acid 脲乙醛酸；尿膜酸
allanite 褐帘石
allantoic acid 尿囊酸；二脲基乙酸 $(NH_2CONH)_2CHCOOH$
allantoin 尿囊素；1-脲基间二氮杂环戊烷-2,4-二酮 $C_4H_6O_3N_4$
allantoxaidine 尿囊毒素 $C_3H_3O_2N_3$
allantoxanic acid 尿囊毒酸 $C_3H_3O_4N$
allanturic acid 亚脲基乙酸
all-around heating 全面加热；均匀加热
all-binder composition 全基料组成
all-binder vehicle 全基料的漆料〔无溶剂漆料〕
all-black stock 全炭黑胶料
all-block nonionics 全嵌段(共聚)非离子表面活性剂
all brass valve 全黄铜阀
all-cascade reactor system 全阶式反应器装置
all core type rayon 全芯型人造丝
all correct ①正确②合格
all cotton fabric 全棉布
allelomorphism(=desmotropism) 稳变异构(现象)
allelotrope 稳变异构体
allelotropism 稳变异构(现象)
allelotype 等位型
allemontite 砷锑矿
allene(=propadiene) 丙二烯 $CH_2=C=CH_2$
allene homologs 丙二烯系同系物 $RCH=C=CH_2$；

$RR'C=C=CH_2$

allene ladder polymer 丙二烯类梯形聚合物
allene polymer 丙二烯聚合物
allenic anion 丙二烯阴离子
allenic compound 丙二烯系化合物
allenic hydrocarbon 丙二烯系烃
Allen-Moore cell 艾伦-穆尔电池
allenolic series 丙二烯系
allergen(=anaphylactin) 过敏素
allergia 过敏性
allergic 过敏
allethrin 丙烯除虫菊酯
alley-stone 矾石
all-failure temperature ①全失效温度②全破坏温度
all-gas furnace 全燃气炉
all glass heated inlet system 全玻璃加热进样系统
all glass solid injector 全玻璃固体进样器
all-heteric nonionics 全混嵌(共聚)非离子表面活性剂
all-hydrocarbon elastomer 全碳氢弹性体
all-hydrogenated 全氢化的
alliaceous 蒜的；葱蒜气息
allicin 蒜素 $(CH_2=CHCH_2S)_2O$
allicinol 蒜醇
allied compound 关联化合物
alligation 和均性
alligator ①鳄鱼②自动卸料池；自动卸料划槽
alligator crack 龟裂；鳄纹
alligator cracking 漆涂层龟裂
alligator effect 鳄皮现象；粗面现象
alligatoring 龟裂；鳄纹
alligatoring lacquer 鳄纹漆；裂纹漆
alligator oil 鳄鱼油
alligator shears 鳄牙剪刀
alligator skin 鳄鱼皮
alligator wrench 管扳手
Allihn condenser 阿林冷凝器
alliin 蒜氨酸；蒜碱 $C_6H_{11}O_3NS$
allin 蒜苷
all iron valve 全铁阀
allistatin 蒜制菌素
allithiamin 蒜硫胺(素)
allithiamine 蒜硫胺素
allitol 蒜糖醇
allizarin yellow A(=2,3,4-trihydroxy-benzophenone) 2,3,4-三羟基苯基苯基(甲)酮；茜素黄 A $C_6H_5COC_6H_2(OH)_3$
all metal catalyst 全金属催化剂
all metal leak valve 全金属泄露阀

all-metal syringe 全金属注射器
allo- 〔词头〕别；异
alloaromadendrene 别香树烯
allobarbital 阿洛巴比妥；二丙烯巴比妥
allobare 异(组)分体
allocation ①分配；配置②存储分配
allochlorophyll 类叶绿素
allocholane 别胆烷
allocholestanone 别胆(甾烷)酮；粪(甾烷)酮
allochroic(=allochromatic) ①变色的；易变色的②非本色的③带假色的
allochroite 粒榴石
allochromasia 变色
allochromatic(=allochroic) ①变色的；易变色的②非本色的③带假色的
allochthonous 移置的
allochthonous theory 移置学说；漂移学说〔地质〕
allocimenol 别罗勒烯醇
allocinnamic acid 别肉桂酸 $C_9H_8O_2$
alloclasite 杂硫铋砷钴矿
allococaine 别可卡因〔可卡因的一种立体异构体〕
allocolloid 同质异相胶
allocortol 别皮(甾)五醇
allocortolone 别皮(甾)酮四醇
allocryptopine 别隐品碱
allocyanine 别花青素；新花青素
allodimer 异二聚物
alloemicymarin 别厄米磁麻苷
alloevodione 别吴茱黄酮
allogibberic acid 别赤霉低酸
allohydroxylysine 别羟基赖氨酸
allohydroxyproline 别羟脯氨酸
all-oil furnace 全燃油炉
alloimperatorin 别前胡精
cis-alloisohumulone 顺-别异葎草酮
alloisoleucine 别异亮氨酸
alloisomerism(=stereoisomerism) 立体异构(现象)
alloisorubijervine 别异玉红杰尔碱
allokainic acid 别红藻氨酸
allolactose 异乳糖
allolupeol 别羽扇醇 $C_{30}H_{50}O$
allomaleic acid 别马来酸；富马酸；别失水苹果酸 $COOHCH=CHCOOH$
allomalenic acid 别马来酸
allomaltol(=3-hydroxy-6-methyl-4-pyrone) 别麦芽酚；3-羟基-6-甲基-4-吡喃酮
allomer 异分同晶质；同构异质
allomerism 异质同晶(现象)

allomerization 叶绿素酐的氧化
allomerized chlorophyll 加氧叶绿素
allomethylose 阿洛甲基糖；6-脱氧阿洛糖
allometric 异速生长的
allomorphism 同质异晶(现象)
allomorphite 贝状重晶石
allomucic acid(=allosaccharic acid) 别黏酸；阿洛糖二酸 $(CHOH)_4(COOH)_2$
allomycin 别霉素
allonic acid 阿洛糖酸 $CH_2OH(CHOH)_4COOH$
allonol(=elemol) 榄香醇
alloocimene 别罗勒烯；2,6-二甲基-2,4,6-辛三烯
alloocimene alcohol 别罗勒烯醇
alloocimenol 别罗勒烯醇
allo-palladium 别钯
alloperiplocymarin 别杠柳苷
alloperiplogenin 别杠柳配基
allopeucenin 别前胡宁
allophanamide(=biuret) 缩二脲
allophanate 脲基甲酸酯
allophane 水铝英石
allophanic acid 脲基甲酸 $H_2NCONHCOOH$
allophanyl 脲羰基；脲基甲酰
allophanylaniline 脲羰基苯胺
alloplasm 异质
allopolar isomerism 变极异构现象
allopregnandiol 别孕烷二醇
allopregnane 别孕(甾)烷
allopseudococaine 别拟可卡因〔可卡因的一种立体异构体〕
allopurinol 别嘌呤醇
allose 阿洛糖
allosteric 变构(象)的
allosteric activation 变构活化
allosteric control 别构控制
allosteric effect 变构效应
allosteric effector 别构效应剂
allosteric enzyme 变构酶
allosteric inhibitor 变构抑制剂
allosteric model 空间变构模型
allosteric site 变构部位
allosteric transition 变构转变；别构转换〔由别构剂引起的蛋白质构象变化〕
allosterism 变构现象
allosteroid 别甾
allostery 变构性；变构现象
allotelluric acid 别碲酸
allotetrahydrocortisol 别四氢皮(甾)醇

allothreo configuration 别苏型(构型)
allothreonine 别苏氨酸 $C_4H_9O_3N$
allotonic 改变张力的
allotonic compound 改变张力的化合物〔能改变水的表面张力，常加入水性涂料中，以改变其流动性〕
allotrope 同素异形体
allotropic 同素异形的
allotropic change 同素异形变化
allotropic substance 同素异形体
allotropic transformation 同素异形变化
allotropism 同素异形(现象)
allotropy 同素异形(现象)
allotypes 同种异型
allouzarigenin 别乌扎配基
all over colour 印花色浆
all-over design 连续花纹
allover method (裱糊时的)满涂胶(浆)法
allowable 允许的；容许的
allowable concentration index 容许浓度指数
allowable crack-per-pass 单程允许裂化率
allowable defect 允许缺陷
allowable deviation 允许偏差
allowable differences in analysis 分析中的容许差
allowable error 允许误差；允差
allowable load 允许负荷
allowable pressure 允许压力
allowable sample size 允许试样量
allowable slip angle 允许偏离角
allowable stress 允许应力
allowable value 允许值
allowance 余量；公差
allowance for finish 加工余量
allowance for machining 机械加工余量
allowed transition 允许跃迁
alloxan(=mesoxalyl urea) 阿脲；四氧嘧啶；2-氧代丙二酰脲 $HN(CO)_3NHCO$
alloxanate 阿脲；四氧嘧啶
alloxan diabetes 阿脲糖尿病；四氧嘧啶糖尿病
alloxanic acid 阿脲酸；脲基丙二酮酸 $C_4H_4O_5N_2$
alloxan monohydrate 水合阿脲；阿脲一水合物 $CONHCONHCOC(OH)_2$
alloxanthin(=alloxantin) 双阿脲；双四氧嘧啶 $C_8H_6O_8N_4 \cdot 2H_2O$
alloxanthoxyletin 别黄木亭
alloxantin(=alloxanthin) 双阿脲；双四氧嘧啶 $C_8H_6O_8N_4 \cdot 2H_2O$

alloxazin 咯嗪 $C_{10}H_6N_4O_2$
alloxazin adenine dinucleotide 咯嗪腺嘌呤二核苷酸
alloxuric bodies 脲环体；嘌呤碱类
alloy 合金；齐
alloyage 合金法
alloy clading 包合金
alloy-coloring alumite 合金着色阳极氧化铝
alloy composition 合金组成
alloyed oil 掺和油；添加植物或动物油的润滑油
alloyed powder 合金粉末
alloy fibre 多组分纤维；合金型熔合纤维
alloy fuel 合金燃料
alloying 制造合金
alloying addition (agent) 掺杂添加(剂)
alloying agent 共混剂
alloying element ①成合金元素②掺杂元素
alloyohimbine 别育亨宾
alloy plastics 塑料合金；合金化塑料
alloy plating 合金电镀
alloy rubber 并用胶；共混胶；橡胶共混物
alloy separation process 合金分离过程
alloy steel 合金钢
alloy superconductor 合金超导体
alloy tool steel 合金工具钢
all-para polymer 全对位聚合物
all-pigment point 全颜料点
all-plastic column 全塑料(制)柱
all point histogram 全位点直方图
all position tyre 全轮位轮胎
all-product line 产品管路〔各种产品管路〕
all program control of quality 全程序质量控制
all-purpose 通用的
all-purpose adhesive 万能胶
all-purpose capillary viscosimeter 通用毛细管黏度计
all purpose cleaner 通用清洁剂
all-purpose detector 通用检测器
all-purpose detergent 通用洗涤剂
all-purpose engine oil 通用机油
all purpose furnace black(APF) 通用炉黑
all-purpose gasoline 通用汽油
all-purpose-grade pine oil 万能(多用)级松油
all-purpose grease 通用润滑脂
all-purpose gum 通用胶
all-purpose instrument 通用仪器
all-purpose rubber 通用(型)橡胶
all purpose washing agent 通用洗涤剂
all-radiant furnace 辐射炉〔无对流加热部分〕
all rag paper 全棉纤维纸

all-reclaimed article 全再生胶制品
all-round adsorbent 万能吸附剂
all-round fastness 全面坚牢度
all-round fibre 全能纤维〔适合于各种用途的纤维〕
all-round performance 全面使用性能
all round (process) 全生产过程
all-round properties 全面性能；综合性能
all rubber article 全胶制品
all-rubber hose 全胶管；软管
all-rubber scraps 全胶废屑〔不含织物等杂质的橡皮屑〕
all-rubber type bearing 全胶式轴承；橡胶轴承
all service gas mask 防毒面具
all-solid state reference electrode 全固态参比电极
all-solvent vehicle 全溶剂漆料
allspice 众香子
allspice oil 药椒油；众香子油
all steel embossing plate 全钢压花板
allulose 阿卢糖；阿洛酮糖
alluvial 冲积的
alluvial layer filter 冲积层式过滤器
alluvium 冲积层
all-volatility reprocessing plant 全挥发法(核燃料)后处理工厂
all-weather 全天候的；全气候性的
all-weather ga(u)ging device 全天候计量设施
all-weather liquefied petroleum gas 全天候液化(石油)气
all-wheel tire 全轮位轮胎
allyl 烯丙基；2-丙烯基 $CH_2=CHCH_2-$
allyl acetate 乙酸烯丙酯 $CH_3CO_2C_3H_5$
allyl acetic acid 烯丙基乙酸；烯戊酸 $CH_2=CHC_2H_4CO_2H$
allyl acetone 烯丙基丙酮；5-己烯-2-酮 $C_3H_5CH_2COCH_3$
allyl acetonitrile 烯丙基乙腈；4-戊烯腈 $C_3H_5CH_2CN$
allyl acrylate 丙烯酸烯丙酯
allyl alcohol 烯丙醇 $CH_2=CHCH_2OH$
allyl alcohol ester 烯丙醇酯类
allyl alcohol polymer 烯丙醇聚合物
allyl aldehyde 丙烯醛
allyl amine 烯丙胺 $CH_2=CHCH_2NH_2$
allyl o-aminobenzoate 邻氨基苯甲酸烯丙酯
allyl-ammonium smectite 烯丙基铵绿土
allyl amyl glycolate 戊基醇酸烯丙酯
p-allyl anisole 对烯丙基茴香醚
allyl anthranilate 邻氨基苯甲酸烯丙酯
allyl arsonic acid 烯丙(基)胂酸 $CH_2=CHCH_2AsO(OH)_2$
allylate ①烯丙基化②烯丙基化物
allylation 烯丙基化(作用)

allyl benzene 烯丙基苯；苯丙烯 $C_6H_5CH_2CH=CH_2$
allyl benzoate 苯酸烯丙酯 $C_6H_5CO_2C_3H_5$
allyl bromide 烯丙基溴 $CH_2=CHCH_2Br$
allyl butyrate 丁酸烯丙酯 $C_3H_7CO_2C_3H_5$
allyl caproate(=allyl hexanoate) 己酸烯丙酯；十九醛〔商〕；凤梨醛〔商〕
allyl caprylate 辛酸烯丙酯
allyl carbinol(=vinyl ethyl alcohol) 乙烯基乙醇；烯丙基甲醇 $C_3H_5CH_2OH$
allyl carbinyl 烯丙基代甲基；3-烯丁基
allylcellulose 烯丙基纤维素
allyl chloride 烯丙基氯；3-氯-1-丙烯 $CH_2=CHCH_2Cl$
allyl cinnamate （肉）桂酸烯丙酯
π-allyl complex mechanism π烯丙型络合机理*
allyl cyanamide(=sinamine) 烯丙基氨基腈；肉桂酸烯丙酯
allyl cyanide 烯丙基腈 $CH_2=CHCH_2CN$
allyl cyclohexaneacetate 环己基乙酸烯丙酯
allyl cyclohexanebutyrate 环己基丁酸烯丙酯
allyl cyclohexanehexanoate 环己基己酸烯丙酯
allyl cyclohexanevalerate 环己基戊酸烯丙酯
allyl cyclohexylcapronate 环己基己酸烯丙酯
allyl cyclohexyl propionate 环己基丙酸烯丙酯；菠萝酯〔商〕
allyl cysteine 烯丙(基)半胱氨酸
δ-allyl cysteine sulfoxide 蒜碱
allyl diglycol carbonate(ADC) 烯丙基二乙二醇碳酸酯
allyldimethylcholrosilane 烯丙基二甲基氯硅烷
allyl disulfide 烯丙基二硫
allylene(=methyl-acetylene) 甲基乙炔；丙炔 $CH_3C\equiv CH$
allylene dichloride 二氯化丙炔 $CH_3CCl=CHCl$
allylene oxide 丙炔化氧 $(CH_3C\equiv CH)_2O$
allyl ester polymer 烯丙酯聚合物
allylestrenol 烯丙雌醇〔药〕
allyl ether 烯丙醚 $(CH_2CH=CH_2)_2O$
allyl 2-ethylbutyrate 2-乙基丁酸烯丙酯
allyl ethyl ether 乙基烯丙基醚 $CH_2CHCH_2OC_2H_5$
2-allyl-2-ethyl propanediol 2-烯丙基-2-乙基丙二醇
allyl formiate(=allyl formate) 甲酸烯丙酯
allyl glycidyl ether 烯丙基缩水甘油醚
allyl group 烯丙基
allylguaiacol 丁香油酚；丁香酸
allyl halide 烯丙基卤 $CH_2CH=CH_2X$
allyl hendecenoate 十一烯酸烯丙酯
allyl heptanoate 庚酸烯丙酯
allyl 2,4-hexadienoate 2,4-己二烯酸烯丙酯
allyl hexanoate 己酸烯丙酯；十九醛〔商〕；凤梨醛〔商〕
allylic 烯丙型(的)*

allylic anion 烯丙位阴离子
allylic bromination 烯丙位溴化作用
π-allylic carbanion π-烯丙位阴(负)碳离子有机
allylic group 烯丙基
allylic halogenation 烯丙型卤化
allylic hydrocarbon 烯丙基烃
allylic hydroperoxylation 烯丙型氢过氧化
allylic migration 烯丙型迁移
allylic polymerization 烯丙基聚合
allylic rearrangement 烯丙位重排(作用)
allylic resin 烯丙基树脂
allylic resonance stabilization 烯丙基共振稳定性
allylic substitution 烯丙位取代〔取代在邻接于双键的C上〕
allylidene 亚烯丙基 $CH_2=CHCH=$
allylidene diacetate 二乙酸亚丙烯酯
allylin 甘油(单)烯丙基醚 $CH_2OHCHOHCH_2OCH_2CH=CH_2$
allyl iodide 烯丙基碘；3-碘-1-丙烯 $CH_2=CHCH_2I$
allyl iodide hexamine 烯丙基碘六胺
allyl α-ionone 烯丙基-α-紫罗兰酮
allyl isoamyl ether 烯丙基异戊基醚；3-异戊氧基丙烯 $C_3H_5OC_5H_{11}$
allyl isocyanate 异氰酸烯丙酯
allyl isonitrile 烯丙异腈 $CH_2=CHCH_2NC$
allyl isopropylacetyl urea 烯丙基异丙基乙酰脲 $(C_3H_7)(C_3H_5)CH=CONHCONH_2$
allyl isorhodanate(=allyl isothiocyanate) 异硫氰酸烯丙酯
allyl isorhodanide(=allyl isothiocyanate) 异硫氰酸烯丙酯
allyl isosulfocyanide(=allyl isothiocyanate) 异硫氰酸烯丙酯
allyl isothiocyanate(=allyl mustard oil) 异硫氰酸烯丙酯；烯丙基芥子油 $C_3H_5N=CS$
allyl isothiocyanide(=allyl isothiocyanate) 异硫氰酸烯丙酯
allyl isovalerate 异戊酸烯丙酯
allyl malonic acid 烯丙基丙二酸 $C_3H_5CH(COOH)_2$
allyl mercaptan 烯丙基硫醇
allyl mercuric iodide 碘化烯丙基汞 $CH_2=CHCH_2HgI$
allyl mercury iodide 碘化烯丙基汞
allyl metal complex 烯丙基金属配合物
allyl methacrylate(AMA) 甲基丙烯酸烯丙酯
4-allyl-2-methoxyphenol 4-烯丙基-2-甲氧基苯酚
allyl methyl disulfide 烯丙基甲基二硫醚
4-allyl-1,2-methylenedioxybenzene 4-烯丙基-1,2-亚甲基二氧苯〔黄樟素的化学名称〕
allyl methyl ether 烯丙基甲基醚 $CH=CHCH_2OCH_3$

allyl methyl trisulfide 烯丙基甲基三硫醚
allyl mustard oil(=allyl isothiocyanate) 烯丙基芥子油；异硫氰酸烯丙酯
allyl nonanoate 壬酸烯丙酯
allyl octanoate 辛酸烯丙酯
allyl oenanthate 庚酸烯丙酯
allyl oxalyl cellulose 烯丙基乙二酰纤维素
allyloxy 烯丙氧基 $CH_2=CHCH_2O-$
allyl oxyglycerol 甘油烯丙基醚；烯丙基甘油 $CH_2OHCHOHCH_2OCH_2CH=CH_2$
allyl pelargonate 壬酸烯丙酯
p-allylphenol 对-烯丙基苯酚；黑椒酚
allyl phenoxyacetate 苯氧乙酸烯丙酯
allyl phenylcinchonimate 苯基辛可宁酸烯丙酯 $C_2H_5N(C_6H_5)CO_2C_3H_5$
allyl phenyl ether 烯丙基苯基醚；3-苯氧基丙烯 $C_3H_5OC_6H_5$
allyl phenyl urea 烯丙基苯基脲 $C_3H_5NHCONHC_6H_5$
allyl plastic 烯丙基塑料
allyl polymer 烯丙基聚合物
allyl polymerization 烯丙基聚合
allyl propionate 丙酸烯丙酯
2-allyl-2-propyl-p-cresol 2-烯丙基-2-丙基对甲酚（防老剂）
allyl propyl disulfide 烯丙基丙基二硫醚
allyl pyridine 烯丙基吡啶 $C_3H_5C_5H_4N$
allyl radical 烯丙基自由基
allyl resin 烯丙基树脂
allyl rhodanate(=allyl rhodanide) 硫氰酸烯丙酯 $CH_2=CHCH_2SCN$
allyl rhodanide(=allyl rhodanate) 硫氰酸烯丙酯 $CH_2=CHCH_2SCN$
allyls ①烯丙基类塑料②烯丙基类树脂
allyl salicylate 水杨酸烯丙酯
allyl sorbate 山梨酸烯丙酯
allyl starch 烯丙基淀粉
N-allyl stearamide N-烯丙基硬脂酰胺
allyl sucrose 烯丙基蔗糖
allyl sulfhydrate 烯丙基硫醇
allyl sulfide 烯丙基化硫；硫化二丙烯 $(CH_2CH=CH_2)_2S$
allyl sulfocarbamide 烯丙基硫脲 $C_3H_5NHCSNH_2$
allyl sulfocyanate(=allyl thiocyanate) 硫氰酸烯丙酯
allyl sulfocyanide(=allyl thiocyanate) 硫氰酸烯丙酯
allylsulfonate 烯丙基磺酸盐
allyl sulphonic acid 烯丙基磺酸
allyl thioaldehyde 烯丙基硫醛
allyl thiocyanate(=allyl thiocyanide) 硫氰酸烯丙酯 C_3H_5SCN
allyl thiocyanide(=allyl thiocyanate) 硫氰酸烯丙酯
allyl thiopropionate 硫代丙酸烯丙酯
allyl thiourea 烯丙基硫脲 $C_3H_5NHCSNH_2$
allyl tiglate 惕各酸烯丙酯；顺芷酸烯丙酯
allyl tolyl ether 烯丙基甲苯基醚
allyl trans-2-methyl-2-butenoate 反-2-甲基-2-丁烯酸烯丙酯
allyltrichlorosilane 烯丙基三氯硅烷（偶联剂）
allyltriethoxysilane 烯丙基三乙氧基(甲)硅烷
allyl trimethyl silane 烯丙基三甲基硅烷
allyl trisulfide 双烯丙基化三硫；三硫二丙烯 $(C_3H_5)_2S_3$
allyl 10-undecenoate 10-十一烯酸烯丙酯
allyl urea 烯丙基脲 $C_3H_5NHCONH_2$
allylveratrol 烯丙基藜芦醚
allysine ε-醛(基)赖氨酸
almadina 大戟树胶
almandine (=almandite) 铁铝榴石；贵榴石
almandite (=almandine) 铁铝榴石；贵榴石
almashite ①绿琥珀②黑琥珀
Almen extreme pressure lubricant testing machine 阿尔门极压润滑剂试验机
Almen (friction) machine 阿尔门(摩擦)试验机
Almen machine(=Almen tester) 阿尔门试验机〔评价润滑油的润滑性能用〕
Almen test 阿尔门试验
Almen tester (=Almen machine) 阿尔门试验机〔评价润滑油的润滑性能用〕
Almen-Wieland tester ALW 试验机
almitrine and raubasine 阿米三嗪萝巴新（药）
almond camphor 杏仁脑
almond flavour 杏仁香精
almond furnace 一种反焰型复熔炉
almond meal 杏仁粉
almond oil 杏仁油
almond paste 杏仁糊
alnico 铝镍钴磁钢
Alniflex solution 阿尔尼夫莱克斯溶液〔一种用在核燃料后处理首端过程的浸取液，组成为 HFHNO$_3$-Al(NO$_3$)$_3$-K$_2$Cr$_2$O$_7$〕
alnuoid 赤杨皮浸膏
alnusenone 赤杨酮；桤木酮
aloatic acid 芦荟酸 $C_{14}H_4O_{10}N_4$
alochrysine 芦荟解素 $C_{15}H_8O_5$
Alocrom pretreatment tank 阿铝克铬姆预处理槽；酸式铬酸盐-磷酸盐预处理槽
aloe-emodin 芦荟大黄素
aloeresic acid 芦荟脂酸 $C_{30}H_{32}O_{14}$

aloes ①芦荟树脂②好望角芦荟
aloetic(al) 芦荟的
aloe-wood 沉香
aloe wood oil 沉香木油
aloin 芦荟素；葡萄糖基蒽酮
alomycin 芦荟霉素
alopecia 脱毛症；秃发症
alos 树脂黄
aloxite 铝砂；人造刚玉磨料
alpaca fabric 羊驼毛织物
alpax 铝硅合金
alpha α〔希腊字母〕
alpha-acetylene α炔
alpha acid α酸；2,8-萘胺磺酸
alpha-activity α放射性
alpha-beta double bond α,β双键
alpha-beta unsaturation α,β不饱和
alpha-branching α分支；α系
alpha-brass α黄铜〔有两种：海军黄铜和铝黄铜〕
alpha-butylene α-丁烯；1-丁烯 C_4H_8
alpha-calibration α射线校准
alpha-carbon α碳
alpha-carbon atom α碳原子
alpha-carotene α胡萝卜素
alpha-cellulose α纤维素
alpha-chlorotoluene α-氯甲苯〔硫化剂〕
alpha cleavage α断裂
alpha coefficient(=side reaction coefficient) α系数；副反应系数
alpha counter α粒子计数器
alpha decay α衰变
alpha (2,4-)dinitrophenol α-(2,4-)二硝基酚
alpha disintegration α蜕变
alpha effect 邻位效应
alpha-emitter α(射线)发射器
alpha-emitting isotope α放射性同位素
alpha-emitting waste α放射性废物
alpha factor α因子
alpha ferrite α铁
alpha fibre α纤维〔纤维素含量在94%以上的纤维〕
alpha-form α形〔蜡晶体〕
alpha-gamma cell α-γ放射性热室
alpha-helix conformation α-螺旋构象
alpha-hydrorubber α-氢化橡胶
alpha-hydroxy acid 果酸；α-羟基酸
alpha-hydroxypropionic acid α-羟基丙酸；乳酸
alpha-methyl-naphthalene α-甲基萘
alpha-methyl styrene polymer α-甲基苯乙烯聚合物

alpha-naphthol α-萘酚 $C_{10}H_7OH$
alpha naphthol test α-萘酚试验
alpha-naphthylamine α-萘胺
alpha-nitroso-beta-naphthol α-亚硝基-β-萘酚
alphanol ①正烷醇②α-取代醇
alpha-oil α油
alpha-olefin α-烯烃
alpha-olefin sulfonate α-烯烃磺酸盐
alpha-oxidation α氧化
alpha-particle α粒(子)；α质点
alpha-particle ionization detector α粒子电离检测器
alpha-particle range α粒(子)射程
alpha-plane α平面
alpha position α位
alpha pulp α纤维素纸浆
alpha-radiography α放射照相法
alpha-ray α射线
alpha-ray-induced α射线诱生的
alpha-ray scattering α射线散射
alpha-ray scintillation α射线闪烁
alpha-ray spectrometer α射线光谱仪；α谱仪
alpha-ray track apparatus α射线检径器
alpha-relaxation α弛豫转变；α松弛转变〔即玻璃化转变〕
alpha-rubber α橡胶；溶胶体橡胶
alpha-slit α狭缝
alpha source α源
alpha spectrometer α谱仪
alpha-substitution α-取代
alpha-tocopherol α-生育酚
alphatopic ①失氢的②差氢的
alphatopic change ①α粒(子)发射变化②失氢变化
alphatron α电离真空规
alpha waste α放射性废物
alphazurine α绿〔指示剂，在 pH=6.0 时，由紫(酸)变绿(碱)〕
alphol(=α-naphthyl salicylate) 水杨酸α-萘酯 $HOC_6H_4CO_2C_{10}H_7$
alphyl 脂苯基
alpinin 良姜素 $C_{17}H_{12}O_6$
alpinone 良姜酮；7-甲氧基-3,5-二羟基黄酮
alprazolam 阿普唑仑〔药〕
alprostadil 前列地尔〔药〕
alquifou 粗粒方铅矿
already-atomised paint cloud (早)已雾化(的)漆雾
Alrok treatment process (重铬酸盐法)氧化铝被膜处理法
Alro quaternaries〔商〕 阿耳绕季盐
Alsace gum 阿耳萨斯胶；糊精

alsol 乙酰酒石酸铝
Alsop drying oven 阿耳索普干燥(烘)箱
alstonidine 鸭脚木定；鸡骨常山次碱
alstoniline 鸭脚木灵；阿斯木碱
alstonine(=chlorogenine) 鸭脚木碱
alstonite 碳酸钙钡矿
altaite 碲铅矿
altering 蚀变
alternant molecular orbital(AMO)method 交替分子轨道法
alternant orbital 交替轨道
alternate ①交替；更迭；相间②区别③间隔的；断续的④替代的
alternate addition method 交替加入法
alternate bond 更迭键
alternate circular ports and annular spaces 交替圆形和环形通道〔换热器的〕
alternate copolymer 交替共聚物
alternate current 交流(电)
alternate double bond 更迭双键
alternate dry and wet test 干湿交替试验
alternate flexural stress 交变弯曲应力
alternate fuel(=alternative fuel) 代用燃料
alternate grafting 交替接枝
alternate immersion test 交替浸渍(腐蚀)试验
alternate layers 轮换层
alternate mechanism 交变机制(机理)；交替机理
alternate motion 交变运动
alternate operating columns 轮换操作塔
alternate operation 轮换运转；轮换操作
alternate phase method 交变相位法
alternate polarity 更迭极性
alternate polarity theory 更迭极性说
alternate stress 交变应力
alternating ①更迭的；交替的；交变的；交错的②交流的
alternating accumulator 交流蓄电池
alternating affinity 更迭亲和力
alternating axes (of symmetry) 更迭(对称)轴
alternating axis 交错轴
alternating axis of symmetry 交错对称轴
alternating bending test 交替弯曲试验；反复弯曲试验
alternating block polymer 交替嵌段聚合物
alternating bond 更迭键
alternating copolymer 交替共聚物
alternating copolymerization 交替共聚合*
alternating current 交流电(流)
alternating current amplifier 交流放大器
alternating current arc 交流电弧

alternating current arc method 交流弧光法
alternating current arc welding 交流电弧焊
alternating current chronopotentiometry 交流计时电位法*
alternating current electrolysis 交流电解
alternating current impedance 交流阻抗
alternating current meter 交流电流计
alternating current motor 交流电动机
alternating current polarization titration 交流极化滴定
alternating current polarography 交流极谱法
alternating current pulse polarography 交流脉冲极谱法
alternating current spark 交流火花
alternating current voltammetry 交流伏安法
alternating damp heat atmosphere 交变湿热环境；湿热交变气氛
alternating deformation 交变形变
alternating displacement 交变位移
alternating double bond 交替双键
alternating humidity atmosphere 交变湿度环境
alternating immersion corrosion test 交替浸渍腐蚀试验
alternating immersion test 交替浸渍试验
alternating intra-intermolecular polymerization 交替分子内-分子间聚合
alternating polarity 交变极性
alternating polymer 交替聚合物
alternating property 交替性质
alternating pulsed field gradient 交变脉冲场梯度
alternating shear 交变剪切
alternating stills 轮换操作分馏塔〔一组蒸馏塔，分别在不同的压力及温度下轮换操作〕
alternating strain 交替应变
alternating strain amplitude 交变应变幅度〔在动态疲劳试验中，最大应变与最小应变代数差的平均值〕
alternating stress 交变应力
alternating stress amplitude 交变应力幅度〔在动态疲劳试验中，最大应力与最小应力代数差的平均值〕
alternating stress load 交变应力载荷
alternating structure (聚合物)交替结构
alternating tendency 交替倾向
alternating twist ①交替加捻〔变形工艺中交替加S捻和Z捻〕②交替捻度〔自捻纺纱条具有的S向和Z向捻度〕
alternating voltage chronopotentiometry 交变电压计时电位法*
alternating voltammetry 交流伏安法
alternation of melting points 熔点的交替
alternative(=alternating) ①更迭的；交替的；交变的②交流的

alternative atom transfer radical polymerization　逆向原子转移自由基聚合
alternative clamp unit　交替锁模装置
alternative copolymerization　交替共聚合(作用)
alternative electro-mechanical system　电-机械变换系统
alternative energy　替代能源
alternative fuel(=alternate fuel)　代用燃料
alternative hypothesis　备择假设*
alternative mathematical method　择一算法
alternative rubber　间聚橡胶
alternative stress　交变应力
alternative synthesis　别种合成法
alternative withdrawal procedure　交替除去法
alternator　交流发电机
althaea root(=althes root)　药用蜀葵根
althea　蜀葵根〔药用〕；蜀癸属植物
altheine(=asparagine)　天冬酰胺
althes root(=althaea root)　药用蜀葵根
althiomycin　异硫霉素
althionic acid(=ethylsulfuric acid)　乙基硫酸
altimeter　高度计
altitude　高度
altitude gage　高度规
altitude grade gasoline　高海拔级汽油
altitude mixture control　海拔调节；高度调节〔调节燃料及空气混合比例以适合海拔高度变更的需要〕
altogether coal　原煤
altretamine　六甲蜜胺〔药〕
altroheptulofuranose　阿卓庚醛呋喃糖
altrose　阿卓糖
alucol　铅溶胶
aludel　梨(状)坛〔冷却水银蒸气用的梨状陶器坛〕
aludel furnace　梨坛炉〔炼汞炉〕
alum　明矾*
alum and salt tanned glove leather　矾盐鞣制手套革
alumatol　额吕马突；硝铵、铝粉炸药
alum cake　矾块
alum earth　明矾土　Al_2O_3
alumel-chromel thermocouple　铝镍-铬镍热电偶
alumen　矾
alum floc　铝矾絮凝剂
alum flour　矾粉
alumian　无水矾石
alumilite process　阳极氧化铝加工处理法
alumilite treatment　阳极氧化铝被膜处理法
alumina　矾土；氧化铝　Al_2O_3
alumina base catalyst　氧化铝载体催化剂
alumina based fibre　氧化铝基纤维
alumina bead　氧化铝珠
alumina-blanc fixe　氧化铝-重晶石粉
alumina boria catalyst　氧化铝-氧化硼催化剂
alumina-boria-silica fibre　铝硼硅(氧化物)纤维
alumina borosilicate glass　硼硅酸铝玻璃
alumina brick　高铝砖
alumina cement　高铝水泥
alumina-ceramic　氧化铝陶瓷；矾土陶瓷
alumina chromatography　氧化铝色谱法
alumina column　氧化铝柱
alumina cream　矾土霜；氢氧化铝
alumina film　氧化铝(薄)膜
alumina for chromatography　色谱(法)用氧化铝
alumina G　氧化铝 G〔氧化铝中加烧石膏〕
alumina gel　氧化铝凝胶
alumina granules　颗粒状氧化铝
alumina hydrate　水合氧化铝
alumina hydrogel　氧化铝水凝胶
alumina-impregnated paper　氧化铝浸渍过的纸
alumina minium　铝红
alumina on silica　硅石上氧化铝
alumina pellets　锭形氧化铝；氧化铝锭片
alumina-silica fibre　铝硅纤维
alumina-silica hydrogel　硅铝水凝胶
alumina silica refractory　硅酸铝耐火材料
alumina silicate　水合硅酸铝；板岩粉〔体质颜料,有灰、黑二色〕
alumina soap　铝皂
γ-alumina spinel　γ型氧化铝尖晶石
alumina supporter　氧化铝载体
aluminate　铝酸盐
aluminate coupling agent　铝酸酯偶联剂
aluminate-nitrile rubber　铝-丁腈橡胶
alumina trihydrate　三水氧化铝；氢氧化铝
alumina whisker　氧化铝晶须
alumina white　矾土白〔不叫铝白,用天然氧化铝作白色颜料〕
alumina white lake　矾土白色淀
alumine　矾土；氧化铝　Al_2O_3
aluminic　①铝的②铝酸的
aluminic acid　①偏铝酸　$HAlO_2$②原铝酸　H_3AlO_3
aluminiferous　①含铝的②含铝土的②含矾的
aluminite　铝氧石；矾石
aluminium(=aluminum)　铝
aluminium acetate　乙酸铝　$(CH_3COO)_3Al$
aluminium acetylacetonate　乙酰丙酮铝　$Al(C_5H_7O_2)_3$
aluminium alginate　海藻酸铝
aluminium alkoxide　烷醇铝；烃氧基铝　$Al(OR)_3$

aluminium alkoxide catalyst 烷醇铝催化剂
aluminium alkoxide reduction 烃氧基铝还原作用
aluminium alkyl 烷基铝
aluminium ammonium sulfate 硫酸铝铵
aluminium amphibole 铝闪石
aluminium anodes 铝阳极
aluminium base 铝基
aluminium base grease(=aluminium soap grease) 铝基润滑脂
aluminium bichromate 重铬酸铝 $Al_2(Cr_2O_7)_3$
aluminium-bitumen paint 铝粉沥青漆；银粉沥青漆〔俗〕
aluminium blind 铝遮阳板
aluminium block heat 铝制组合(块)式加热器；组合式铝块加热器
aluminium boride 硼化铝
aluminium brass 铝黄铜
aluminium bromate 溴酸铝 $Al(BrO_3)_3$
aluminium bromide 溴化铝 $AlBr_3$
aluminium bronze paint 铜粉漆；金粉漆〔俗〕
aluminium bus-bar 铝汇流条
aluminium butoxide 丁醇铝；三丁氧基铝 $Al(C_4H_9O)_3$
aluminium-sec-butylate 仲丁醇铝；仲丁氧基铝
aluminium carbide 碳化铝 Al_4C_3
aluminium carbonate 碳酸铝 $Al_2(CO_3)_3$
aluminium ceramic matrix composite 含铝陶瓷基质复合材料
aluminium cesium sulfate 硫酸铝铯
  $Cs_2SO_4 \cdot Al_2(SO_4)_3 \cdot 24H_2O$; $CsAl(SO_4)_2 \cdot 12H_2O$
aluminium chelate 螯合铝
aluminium chelate compound 铝螯合物
aluminium chlorate 氯酸铝 $Al(ClO_3)_3$
aluminium chloride 氯化铝 $AlCl_3$
aluminium chloride alkylation process 氯化铝烷基化法
aluminium chloride column 氯化铝塔〔分离氯化铝用〕
aluminium chloride hydrocarbon complex 氯化铝-烃配合物
aluminium chloride process 氯化铝法〔利用无水氯化铝精炼润滑油方法〕
aluminium chlorite 亚氯酸铝
aluminium chlorohydrate 水合氯化铝
aluminium chlorohydroxide 羟基氯化铝；碱式氯化铝
aluminium chloroisopropylate 氯异丙醇铝
aluminium citrate 柠檬酸铝
aluminium collapsible tube 铝软管
aluminium electrode 铝电极
aluminium enamel 铝粉瓷漆；银粉瓷漆
aluminium ethide 乙基铝；三乙基铝 $Al(C_2H_5)_3$
aluminium ethoxide 乙醇铝；三乙氧基铝 $Al(C_2H_5O)_3$
aluminium ethyl （三）乙基铝 $Al(C_2H_5)_3$
aluminium ethylate 乙醇铝；三乙氧基铝 $Al(C_2H_5O)_3$
aluminium flake ①薄铝片②高岭土
aluminium fluoride 氟化铝 AlF_3
aluminium fluosilicate 氟硅酸铝 $Al_2(SiF_6)_3$
aluminium foil 铝箔；铝薄片
aluminium foil fibre 铝箔丝
aluminium-foil slitting 铝箔切割(成形)法〔制金属纤维的一种方法〕
aluminium gasket 铝密封垫圈
aluminium gold 铝金〔合金, 含铝黄铜〕
aluminium grease(=aluminium soap grease) 铝皂润滑脂
aluminium hydrate(=aluminium hydroxide) 氢氧化铝
aluminium hydrosilicate 水合硅酸铝
aluminium hydroxide(=aluminium hydrate) 氢氧化铝 $Al(OH)_3$
aluminium hydroxyacetate 乙酸羟铝；碱式乙酸铝 $Al(OH)(C_2H_3O_2)_2$
aluminium hydroxystearate 羟基硬脂酸铝
aluminium iodide 碘化铝 AlI_3
aluminium isobutylate 三异丁醇铝
aluminium isopropoxide 异丙醇铝；三异丙氧基铝 $Al(C_3H_7O)_3$
aluminium isopropylate 异丙醇铝
aluminium-mennige 铝红
aluminium metal coating 金属铝镀层
aluminium methacrylate 甲基丙烯酸铝
aluminium methide 甲基铝；三甲基铝 $Al(CH_3)_3$
aluminium methoxide 甲醇铝；三甲氧基铝 $Al(OCH_3)_3$
aluminium methyl 甲基铝；三甲基铝 $Al(CH_3)_3$
aluminium methylate 甲醇铝；三甲氧基铝 $Al(OCH_3)_3$
aluminium-mono-acetylacetonate bis(ethylacetoacetate) 单乙酰丙酮铝合二(乙酰乙酸乙酯)
aluminium napthenate 环烷酸铝
aluminium nitrate 硝酸铝 $Al(NO_3)_3$
aluminium nitride 氮化铝 AlN
aluminium octoate 辛酸铝
aluminium oleate 油酸铝
aluminium oxide 氧化铝 Al_2O_3
aluminium oxide fibre 氧化铝纤维；Al_2O_3 纤维
aluminium oxide hygrometer 氧化铝湿度计
aluminium oxide isopropylate 三聚异丙氧基氧(化)铝
aluminium oxide octate （三聚）辛酸氧(化)铝
aluminium oxide stearate （三聚）硬脂酸氧(化)铝
aluminium paint 铝涂料
aluminium palmitate 十六碳酸铝；棕榈酸铝
aluminium phenolsulfonate 苯酚磺酸铝
aluminium phenylsulfonate 苯磺酸铝 $(C_6H_5SO_3)_3Al$
aluminium phosphate 磷酸铝 $AlPO_4$

aluminium phosphide 磷化铝 AlP
aluminium polychloride 聚合氯化铝
aluminium polylactyl lactate 聚乳酰乳酸铝
aluminium potassium sulfate 硫酸铝钾；明矾
aluminium powder 铝粉
aluminium propoxide(=aluminium propylate) 丙醇铝；三丙氧基铝
aluminium propylate(=aluminium propoxide) 丙醇铝；三丙氧基铝 $Al(C_3H_7O)_3$
aluminium rhodanate(=aluminium rhodanide) 硫氰酸铝 $Al(SCN)_3$
aluminium rhodanide(=aluminium rhodanate) 硫氰酸铝
aluminium-rimmed screen （熔融纺用）铝(包)边过滤网
aluminium rubidium sulfate 硫酸铝铷 $Rb_2SO_4 \cdot Al_2(SO_4)_3 \cdot 24H_2O; RbAl(SO_4)_2 \cdot 12H_2O$
aluminium secondary butylate 仲丁醇铝
aluminium siding coating 铝壁板用涂料
aluminium silicate 硅酸铝 $Al_2(SiO_3)_3$
aluminium silicate fibre 硅酸铝纤维
aluminium silicate refractory fibrous material 硅酸铝耐高温纤维材料
aluminium soap 铝皂
aluminium soap-carbon black thickened grease 铝皂-碳黑稠化润滑脂
aluminium soap grease(=aluminium base grease; aluminium grease) 铝皂润滑脂
aluminium sodium fluoride 氟化铝钠
aluminium sodium sulfate 硫酸铝钠
aluminium solder 铝焊料
aluminium stearate 硬脂酸铝 $Al(C_{18}H_{35}O_2)_3$
aluminium sulfate 硫酸铝 $Al_2(SO_4)_3$
aluminium sulfide 硫化铝 Al_2S_3
aluminium sulfite 亚硫酸铝 $Al_2(SO_3)_3$
aluminium sulfocyanate(=aluminium sulfocyanide) 硫氰酸铝 $Al(SCN)_3$
aluminium sulfocyanide(=aluminium sulfocyanate) 硫氰酸铝
aluminium sulfophenylate 苯磺酸铝 $(C_6H_5SO_3)_3Al$
aluminium thiocyanate(=aluminium thiocyanide) 硫氰酸铝 $Al(SCN)_3$
aluminium thiocyanide(=aluminium thiocyanate) 硫氰酸铝
aluminium trialkyl 三烷基铝
aluminium tribromide 三溴化铝；溴化铝 $AlBr_3$
aluminium trichloride 三氯化铝；氯化铝 $AlCl_3$
aluminium triethide 三乙基铝；乙基铝 $Al(C_2H_5)_3$
aluminium triethyl 三乙基铝 $Al(C_2H_5)_3$
aluminium trihydrate 三水合铝〔阻燃剂〕
aluminium triiodide 三碘化铝；碘化铝 AlI_3

aluminium trimethide 三甲基铝；甲基铝 $Al(CH_3)_3$
aluminium triphosphate 三聚磷酸铝
aluminium tris(acetylacetonate) 三(乙酰丙酮)铝
aluminium tris(ethylacetoacetate) 三乙基乙酰乙酸铝
aluminium-zinc coating ①铝-锌镀层②铝-锌包膜
aluminized 镀铝的；喷铝的
aluminizing 浸镀铝法；渗铝法
alumino- 〔词头〕铝
alumino-ferric 铝铁剂
aluminon(=aurin tricarboxylic acid) 试铝灵；金精三羧酸
alumino-nickel 铝镍合金
alumino silica gel 硅铝胶
aluminosilicate 硅铝酸盐
aluminosilicate (glass) fibre 硅铝酸盐玻璃纤维
aluminosilicate modification process 硅铝酸盐改良法
aluminosilicophosphate 硅铝磷酸盐
aluminothermics ①铝热法②铝热剂
aluminothermy 铝热法
aluminous ①铝土的②矾的
aluminous cement 高铝水泥
aluminous hornblende 含铝角闪石
aluminous slag 高铝炉渣
aluminum(=aluminium) 铝
aluminum acetate 乙酸铝
aluminum alkoxide 烃醇铝；烷氧基铝
aluminum alkyl 烷基铝
aluminum alloy 铝合金
aluminum alloy ingot 铝合金锭
aluminum arsenite 亚砷酸铝
aluminum bifluoride 复氟化铝 $(Al_2F_6)_3(HF)_4 \cdot 10H_2O$
aluminum borate 硼酸铝
aluminum boride 硼化铝
aluminum borohydride 硼氢化铝；氢化硼铝
aluminum bromide 溴化铝
aluminum bronze 铝青铜
aluminum ceramic 铝陶瓷
aluminum chelate 铝螯合物
aluminum chloride 氯化铝
aluminum-chloride moist 氯化铝雾
aluminum dichromate 重铬酸铝
aluminum diethylmalonate 二乙基丙二酸铝
aluminum diformate （碱式）二甲酸铝
aluminum dipping form 铝模〔浸渍用〕
aluminum dish 铝皿
aluminum di-soap 铝二酸皂
aluminum dodecyl glycinate 十二烷基甘氨酸铝
aluminum dust 铝粉
aluminum ethide 乙基铝；三乙基铝

aluminum ethoxide 乙氧基铝；乙醇铝
aluminum ethyl 乙基铝；三乙铝
aluminum ethylate 乙醇铝；乙氧基铝
aluminum explosive 铝粉炸药
aluminum family 铝族
aluminum flakes ①(=kaolin)陶土〔填充剂〕②薄铝片
aluminum fluosilicate 氟硅酸铝
aluminum foil 铝箔
aluminum foil with paper backing 垫纸铝箔
aluminum-hematoxylin test 铝-苏木精试验
aluminum hydrate 氢氧化铝
aluminum hydroxide 氢氧化铝
aluminum hydroxide gel 氢氧化铝凝胶
aluminum ingot 铝锭
aluminum ink 铝粉墨
aluminum iodate nitrate 碘酸硝酸铝 $Al(IO_3)_2NO_3$
aluminum isopropoxide(=aluminum isoprpylate) 异丙氧基铝；异丙醇铝 $Al(C_3H_7O)_3$
aluminum isopropylate(=aluminum isopropoxide) 异丙醇铝；异丙氧基铝 $Al(C_3H_7O)_3$
aluminum joint ring 铝(接合)密封环
aluminum metal 铝金属
aluminum methoxide 甲氧基铝；甲醇铝 $Al(CH_3O)_3$
aluminum methyl(=trimethylaluminum) 三甲铝
aluminum methylate 甲醇铝；三甲氧基铝
aluminum minerals 铝矿物类
aluminum monobasic stearate 碱式单硬脂酸铝
aluminum muriate 氯化铝
aluminum naphthenate varnish 环烷酸铝清漆
aluminum-β-naphthol-disulfonate β萘酚二磺酸铝
aluminum nitrate 硝酸铝
aluminum oxide 氧化铝
aluminum oxide adsorption 氧化铝吸附法
aluminum oxyhydroxide 碱式氢氧化铝；缩水氢氧化铝
aluminum paint 铝涂料；银灰漆
aluminum palmitate 棕榈酸铝；十六烷酸铝
aluminum paste 铝粉浆〔俗称：银粉浆〕
aluminum phenyl sulfonate 苯磺酸铝
aluminum pot 铝罐；铝锅
aluminum potassium chloride 氯化钾铝
aluminum potassium silicate 铝硅酸钾
aluminum potassium sulfate 硫酸铝钾
aluminum powder 铝粉
aluminum primer 铝粉底漆
aluminum rectifier 铝整流器
aluminum reduction method 铝还原法〔测钛白总钛含量的方法〕
aluminum rhodanate 硫氰酸铝
aluminum rubidium sulfate 硫酸铝铷
aluminum screen 铝幕
aluminum shot 铝粒；铝丸(砂)
aluminum silicate 硅酸铝
aluminum silico-fluoride 硅氟化铝
aluminum sulfate 硫酸铝
aluminum tannage 铝鞣法
aluminum tanning 铝鞣
aluminum tanning agent 铝鞣剂
aluminum thallium sulfate 硫酸铊铝
aluminum triethyl 三乙基铝
aluminum triformate 三甲酸铝
aluminum zinc sulfate 硫酸锌铝；锌矾
alumite 耐酸铝；耐热铝；铝氧化膜
alumite process 阳极氧化铝加工法；氧化铝膜处理法
alumite streak (阳极)氧化铝条痕(条纹)
alumnol β萘酚磺酸铝〔别名〕
alum rock 明矾岩
alum schist(=alum shale) 明矾片岩
alum shale(=alum schist) 明矾片岩
alum speck (明)矾斑(点)
alum steep 明矾液
alum stone 明矾石
alum tannage (明)矾鞣
alum tanned sheep leather for gloves 矾鞣羊皮手套革
alum tanner 矾鞣工
alum tanning 矾鞣
alum tawing 明矾硝制；明矾白鞣
alundum 铝氧粉；刚铝石
alundum boat 刚铝石舟皿
alundum cement 刚铝黏合剂
alundum crucible 刚铝石坩埚
alundum ware 刚铝器皿
alunite 明矾石
alunogen 水硫酸铝石
alurate 阿尿酸(盐或酯)；5-烯丙基-5-异丙基巴比妥酸(盐或酯)
alutaceous(=alutaceus) 淡革色的；淡黄褐色的
alutaceus(=alutaceous) 淡革色的；淡黄褐色的
alvanine 奥万宁
alvite 硅铁锆矿
alypine 阿里品〔药〕
amalgam 汞齐；汞合金
amalgamated 汞齐化了的
amalgamated silver wire cloth 汞齐化银丝布
amalgamated zinc ①汞齐化锌②锌柱
amalgamating ①汞齐化的②汞齐化(作用)
amalgamating plodder 压条式拌和机

amalgamation 汞齐化*
amalgamation half-life 汞齐化半值期
amalgamation process 汞齐法
amalgamation yield 汞齐化率
amalgamator 拌和机；混合机；搅拌机
amalgam cell 汞齐电池
amalgam electrode 汞齐电极
amalgam exchange 汞齐交换
amalgam kryptonate 汞齐氪化物
amalgam reductor 汞齐还原器
amalgum blend 混合胶；并用胶
amalic acid(=amalinic acid) 柔和酸；四甲基双四氧嘧啶 $(CH_2)_4C_8O_8N_4$
amalinic acid(=amalic acid) 柔和酸；四甲基双四氧嘧啶 $(CH_2)_4C_8O_8N_4$
amalthion bordeaux 阿马董枣红〔一种硫化染料的商名〕
amandone colors〔复〕 阿曼董染料〔美国苯胺染料厂靛族还原染料的商名〕
amanil azurine 阿曼尼天青
amanitine 鹅膏蕈碱
amantadine 金刚胺；氨基三环癸烷
amanthrene bordeaux 阿曼士林枣红
amanthrene colors〔复〕 阿曼士林染料〔美国苯胺厂还原染料商名〕
amanthrene dark blue 阿曼士林暗蓝
amanthrene olive 阿曼士林橄榄绿
amaranth ①苋(菜红)②深紫色；紫红色
amarantite 红铁矾
amarbeline 苦杏贝灵
amarbital 苯基乙基巴比妥酸
amargosa(=azadarach;azadaricta) 苦楝树皮
amargosa bark 苦楝根皮
amaric acid 苦杏酸 $C_{23}H_{22}O_3$
amarin(e) 苦杏精；2,4,5-三苯基咪唑啉
amarogentin 苦杏苷 $C_{20}H_{24}O_{10}$
amaroid 苦杏素
amaroidal 微苦的
amaron(=tetraphenylpyrazine) 苦杏碱；四苯基吡嗪 $C_{28}H_{20}N_2$
amaryllidine 朱顶兰定
amatol 额马突〔硝铵、三硝基甲苯炸药〕
amazonite 天河石；绿长石
ambargris 龙涎香
ambelline 安贝灵；颠茄朱顶兰碱
amber ①琥珀②琥珀色
amber acid 琥珀酸；丁二酸 $COOH(CH_2)_2COOH$
amber blanket 琥珀色胶片
amber Cologne soap 琥珀科隆香水皂

amber crepe 琥珀绉片
amber glass 琥珀玻璃
ambergris 龙涎香
ambergris tincture 龙涎香酊
amberite 琥珀炸药；安柏锐特〔一种无烟炸药〕
amberlite 离子交换树脂商品名称
amber oil 琥珀油
amber oxide 琥珀醚
amber oxyaldehyde 琥珀羟基醛
amberplex 离子交换膜
ambers 褐绉片
amber soap 琥珀皂
amber varnish 琥珀清漆；琥珀上光油
amber white 琥珀白
amberwood 酚醛树脂胶合板
amber yellow 琥珀黄
ambident 两可(的)
ambident anions 两可阴离子
ambident ion 两可离子〔如：
$$CH_3\overset{O}{\overset{\|}{C}}CH_2COOEt \longrightarrow H^+ + CH_3\overset{O^-}{\overset{|}{C}}=CHCOOEt$$
$$\longleftarrow CH_3\overset{O}{\overset{\|}{C}} - \overset{-}{C}H - COOEt \, \rbrack$$
ambient 周围的；环境的
ambient air 周围空气；环境空气
ambient cure ①自固化；常温固化②自硫化；常温硫化〔橡〕
ambient fluid 外围流体
ambient moisture 环境湿度；环境水分
ambient noise 环境噪声
ambient operation ①环境条件下操作②室温操作
ambient pollution burden 环境污染负荷
ambient pressure 环境压力
ambient temperature 环境温度；室温；周围温度
ambifunctional adhesion promoter 双功能增黏剂
ambiguity ①双关性②错读
Ambler's apparatus 安布勒气体分析器；气体快速测定仪
amblygonite 磷铝石
amblyslegite 铁苏辉石
ambrain(=ambrein) 龙涎香脂
ambramycin(=tetracycline) 琥珀霉素；四环素
ambreic acid 龙涎裔酸
ambrein 龙涎香醇；龙涎香精
ambreinolide 龙涎香精内酯
ambrette(=musk seed) 黄葵
ambrette seed oil 黄葵油；麝香葵子油
ambrettolic acid(=16-hydroxy-7-hexadecenoic acid) 黄葵

酸；16-羟基十六碳-7-烯酸
ambrettolide(=cyclohexa-decen-7-olide) 黄葵内酯；环十六烯-7-内酯
ambrial 龙涎醛
ambrite 新西兰产的一种沥青产品
ambroid 人造琥珀树脂
ambrosial 动物芳香的
ambrosia oil 土荆芥油〔增香剂〕
ambrosin 豚草素
ambrox 降龙涎香醚〔主要用于香水香精和高档化妆品香精〕
ambroxol 氨溴索〔药〕
ameba(=amoeba) 变形虫；阿米巴
amebacillin 烟曲霉素
amebicide 杀阿米巴药
ameboid 阿米巴样的
amelanotic 无黑(色)素的
ameliaroside 阿美糖苷
amendment ①调理剂②修正
amentoflavone 穗花杉双黄酮；3′,8″-双-4′,5,7-三羟基黄酮；阿曼托黄素
American melting point 美国法(蜡)熔点〔比 ASTM 法测定者高 3°F〕
americate 镅酸盐
americium 镅〔95 号元素，化学符号 Am〕
americium boride 硼化镅 AmB_4; AmB_6
americium carbonate 碳酸镅 $Am_2(CO_3)_3$
americium dioxide 二氧化镅 AmO_2
americium-free plutonium 无镅(的)钚
americium germanate 锗酸镅
americium hydride 氢化镅 AmH_2; AmH_3
americium hydroperoxide 过氧化氢镅
americium hydroxide 氢氧化镅
americium molybdate 钼酸镅 $Am_2(MoO_4)_3$
americium monoarsenide 一砷化镅 AmAs
americium oxalate 草酸镅 $Am_2(C_2O_4)_3$
americium oxide 氧化镅 AmO; AmO_2; Am_2O_3
americium sesquicarbide 三碳化二镅 Am_2C_3
americium sesquitelluride 三碲化二镅 Am_2Te_3
americium sulfate 硫酸镅 $Am_2(SO_4)_3$
americium trichloride 三氯化镅 $AmCl_3$
americyl 镅酰(离子)；双氧镅(离子) AmO_2^+; AmO_2^{2+}
americyl double carbonate 碳酸镅酰复盐〔例如钾盐：$KAmO_2CO_3$; $K_3AmO_2(CO_3)_2$; $K_5AmO_2(CO_3)_3$〕
americyl fluoride 氟化镅酰 AmO_2F_2
amesite 镁绿泥石
amethobottromycin 无甲基波卓霉素
amethocaine 阿美索卡因；丁卡因
amethopterin 氨甲蝶呤
amethyst 紫石英；紫水晶
Amex process ①〔词源 Amine extraction〕阿姆克斯法〔用叔胺萃取法从铀矿浸取液中提取铀〕②阿锔克斯法〔用磷酸三丁酯萃取法从冶金废渣和坩埚料中回收镅〕
amianth(us) 石麻〔与石棉不同，石棉是 $Mg_2CaSi_4O_{12}$, 此物纤维较长〕 $H_4(Mg, Fe)_3Si_2O_{91}$
amiben 3-氨基-2,5-二氯苯甲酸；草灭畏
amic acid 酰氨基酸；酰胺酸
Amice prism 阿米西棱镜；补偿棱镜
amicron 次微(胶)粒；超微粒
amidable 可酰胺化的
amidate ①酰胺化②酰胺化物
amidated 酰胺化了的
amidating ①酰胺化的②酰胺化(作用)
amidating agent 酰胺化剂
amidation 酰胺化(作用)
amidazobenzene 酰胺基偶氮苯
amide ①酰胺②氨化物
amide chloride 二氯代酰胺；偕二氯代烃胺 $RCCl_2NH_2$
amide-epoxy resin 酰胺-环氧树脂
amide-ester polymer 酰胺-酯聚合物
amide exchange reaction ①氨基交换反应②酰胺基交换反应
amide hardener 酰胺固化剂
amide-heterocycle fibre (聚)酰胺-杂环纤维〔指有序芳酰胺-杂环共聚物所制成的纤维〕
amide hydrazone 酰胺腙
amide interchange 酰胺(基)交换
amide linkage 酰胺键
amide nitrogen 酰胺氮
amide oxime 酰胺肟
amide planarity 酰胺平面(法)
amidine 脒
amidine hydrochloride 盐酸脒 $RC(NH_2)=NH \cdot HCl$
amidino(=guanyl) 脒基 $H_2NC(=NH)-$
γ-amidinothiopropyltrihydroxysilane γ-脒基硫代丙基三羟基硅烷〔偶联剂〕
amidinothiourea 脒硫脲
amidinourea 脒脲
amido 酰氨基
amido acid 酰氨基酸；酰氨酸
amido aldehyde 氨基醛 $RNH_2 \cdot CHO$
amido alkyl betaine 酰氨烷基甜菜碱
amido black 酰胺黑
amido bond 酰氨键 —NHCO—
amido colors〔复〕 酰胺染料〔酰胺是德国 MLB 厂的

酸性染料类名〕
amido fast orange 酰胺坚牢橙
amidofluoride mercury(II) 氨基汞化氟
amidogen(=amido) 氨基；酰氨基
amido group 酰氨基
amidohexose 氨基己糖
amidol(=2, 4-diaminophenol dihydrochloride) 阿米酚；阿米多；二氢氯化-2,4-二氨基苯酚
　　　　(NH$_2$)$_2$C$_6$H$_3$OH・2HCl
amido link(=amido lingage) 酰氨键 —NHCO—
amido linkage(=amido link) 酰氨键
amidol reagent 阿米酚试剂
amidolysis 酰氨分解
amidomethylation 酰氨甲基化作用
amidomycin 胺霉素
amidonitrogen 酰氨态氮
amidopurine 氨基嘌呤；腺嘌呤
amidopyrine 氨基比啉；匹拉米童〔药〕
amidosuccinic acid 酰氨丁二酸；琥珀酰氨酸
　　　　CONH$_2$(CH$_2$)$_2$COOH
amido-sulfonic acid 氨基磺酸 H$_2$NSO$_3$H
amidoxalyl(=oxamoyl) 氨基草酰 H$_2$NCOCO—
amidoxim 偕胺肟 —C(=NOH)NH$_2$
amidoxyl 羟氨基 HONH—
amidoxyl acetic acid 羟氨基乙酸 HONHCH$_2$COOH
amido yellow 酰氨黄
amidpulver 酰胺粉；硝铵、硝酸钾、炭末炸药
amidrazone 氨基(某)脒 RC(=NNH$_2$)NH$_2$
amikacin 阿米卡星；丁胺卡那霉素
amiloride 阿米洛利〔药〕
amimycin(=oleandomycin) 竹桃霉素
aminable 可氨化的
aminacrine 氨基吖啶
aminase 氨化酶；氨基酶
aminate ①氨化②氨化产物
aminated 氨化了的
aminating 氨化的②氨化(作用)
aminating agent 氨化剂
amination 氨基化*
amine 胺
amine acetate 乙酸胺
amine adduct 胺加合物
amine-aldehyde complex 胺醛配合物
amine antioxidants 胺类抗氧化剂
amine black green 胺黑绿〔酸性偶氮染料〕
amine blocked acid catalyst 胺封闭的酸催化剂
amine blushing 胺致发白〔漆病〕
amine-carboxyl end-balance 胺-羧端基平衡

amine cellulose 氨基纤维素
amine curing agent 胺固化剂
amine effect 胺效应
amine end concentration 胺端基浓度
amine-ended polymer 端胺基聚合物
amine equivalent 胺当量
amine extraction method 胺萃取法
amine hardener 胺固化剂
amine nitrogen 胺氮
amine nitrogen content 胺式氮含量
amine number 胺值
amine oxidase 胺氧化酶
amine oxide 氧化胺*
amine oxide solvent spun process 氧化胺溶剂纺丝法〔用于再生纤维素纤维制备〕
amine-perchlorate 胺-高氯酸盐〔乙酰化用催化剂〕
amine salt 胺盐
amine salt cationic surfactant 胺盐阳离子表面活性剂
amine soap 胺皂
amine stabilizer 胺稳定剂
amine surfactant 胺型(油溶性)表面活性剂
amine-terminated 胺端(基)的；胺基(链)端位的
amine-terminated polymer 胺端基聚合物
aminetrisulfonic acid(=nitridotrisulfuric acid) 次氨基三硫酸；胺三磺酸 N(SO$_3$H)$_3$
amine type accelerator 胺类促进剂
amine value 胺价；胺(基)值
aminic acid(=formic acid) 甲酸
aminimide 胺化酰亚胺
aminium 铵
aminium salt (某)胺鎓盐 (R$_3$N—)$^+$X$^-$
aminization 胺化(作用)
aminized cotton 胺化棉〔通过胺化作用制取的改性棉纤维〕
amino- 氨基 H$_2$N—
4-amino-4′-chlorobiphenyl 4-氨基-4′-氯代联苯
amino accelerator 氨基促进剂
aminoacetal 氨基乙缩醛 H$_2$NCH$_2$CH(OC$_2$H$_5$)$_2$
aminoacethydrazide 氨基乙酰肼
amino acetic acid 氨基乙酸；甘氨酸 NH$_2$CH$_2$COOH
aminoacetone(=acetonylamine) 氨基丙酮；丙酮基胺 NH$_2$CH$_2$COCH$_3$
o-aminoacetophenone 邻氨基苯乙酮
amino-acid 氨基酸
amino acid acylase 酰化氨基酸水解酶
amino acid amide 氨基酸酰胺
aminoacid analyzer 氨基酸分析器
amino acid composition analyzer 氨基酸组成分析仪

amino acid decarboxylase 氨基酸脱羧酶
amino acid ethyl cellulose 氨基乙基纤维素
amino-acid nitrogen 氨基酸式氮
amino-acid oxidase 氨基酸氧化酶
amino acid phase 氨基酸型固定相
amino acid propyl cellulose 氨基丙基纤维素
amino-acid sequence 氨基酸顺序
amino acid type amphoteric surface active agent 氨基酸型两性表面活性剂
α-aminoacrylic acid α-氨基丙烯酸
aminoacyl 氨酰基
aminoacylase 氨基酰酶
aminoacylhydrazine 氨酰肼
aminoacyl site(=A-site) 氨酰基部位；A 部位
aminoadipaldehyde 氨基己二酸半醛
α-aminoadipic acid α-氨基己二酸
amino-alcohol ①氨基醇②氨基乙醇
aminoaldehexose 氨基己醛糖
amino-aldehyde 氨基醛
aminoalizarin 氨基茜素
amino alkane diphosphate 氨基烷烃二磷酸盐
aminoalkoxy-dioxaborinane β-氨代烷氧基-二氧硼杂环己烷
aminoalkyl acrylate ①氨烷基丙烯酸盐②丙烯酸氨烷基酯
aminoalkylated cellulosic fibre 氨烷基化纤维素纤维
aminoalkyl phosphate 氨基烷磷酸盐；氨烷基磷酸酯
aminoalkylpolyphosphonic acid 氨基烷多膦酸
aminoalkylsilane 氨烷基硅烷
aminoalkyl sulfate 氨烷基硫酸盐；氨烷基硫酸酯
aminoalkyl sulfonate 氨烷基磺酸盐
amino amidate 氨基酰胺化物
aminoanisole(=anisidine) 茴香胺；甲氧基苯胺；氨基苯甲醚 $CH_3OC_6H_4NH_2$
aminoanthraquinone 氨基蒽醌
aminoanthraquinone dye 氨基蒽醌类染料
aminoantipyrene 氨基安替比林
amino-arsenoxide 氨基氧化砷 $NH_2RAs=O$
aminoazobenzene 氨基偶氮苯 $NH_2C_6H_4N_2C_6H_5$
aminoazonaphthalene 氨基偶氮萘 $NH_2C_{10}H_6N_2C_{10}H_7$
2-aminobarbituric acid-N,N'-diacetic acid 2-氨基巴比妥酸-N,N'-二乙酸
22-aminobehenic acid 22-氨基二十二烷酸
aminobenzaldehyde 氨基苯(甲)醛 $NH_2C_6H_4CHO$
aminobenzaldoxime 氨基苯(甲)醛肟
amino-benzene 氨基苯；苯胺
amino-benzene arsonic acid 氨基苯胂酸 $NH_2C_6H_4AsO(OH)_2$

p-aminobenzene sulfonamide(=sulfanilamide) 对氨基苯磺酰胺；磺胺
aminobenzene sulfonic acid 对氨基苯磺酸
4-aminobenzene sulfonyl dodecyl benzene sulfonyl ethylene titanate 4-氨基苯磺酰十二烷基苯磺酰钛酸乙二(醇)酯
aminobenzidine 氨基联苯胺
aminobenzoate 氨基苯甲酸盐(酯)
aminobenzoic acid 氨基苯甲酸 $NH_2C_6H_4COOH$
aminobenzonitrile 氨基苄腈
aminobenzophenone 氨基二苯(甲)酮
aminobenzothiazole 氨基苯并噻唑
amino-benzoyl acetic acid 氨基苯甲酰乙酸 $NH_2C_6H_4COCH_2COOH$
4-aminobenzoyl ethylene isostearoyl titanate 4-氨基苯甲酰异硬脂酰钛酸乙二(醇)酯
amino-benzoyl formic acid 氨基苯甲酰甲酸 $NH_2C_6H_4COCOOH$
aminobenzyl alcohol 氨基苄醇
aminobenzylation 氨苄基化(作用)
aminobenzyl phosphate 磷酸氨苄酯 $O=P(OCH_2C_6H_4NH)_3$
aminobiphenyl(=phenylaniline) 苯基苯胺；氨基联苯 $C_6H_5C_6H_4NH_2$
amino-bonded phase 氨基键合相
amino bonded phase column 氨基键合相柱
amino butyraldehyde 氨基丁醛
amino-butyric acid 氨基丁酸 $CH_3CH_2CHNH_2COOH$
10-aminocapric acid 10-氨基癸酸
6-aminocaproamide 6-氨基己酰胺
amino-caproic acid 氨基己酸 $CH_3(CH_2)_3CHNH_2COOH$
aminocapronitrile 氨基己腈
aminocaprylic acid 氨基辛酸
amino-carbamic acid 氨基氨基甲酸；肼基甲酸 $NH_2NHCOOH$
aminocarbonyl 氨羰基；甲酰胺基 NH_2CO-
aminocarboxylic acid 氨基羧酸
amino carboxylic acid chelate 氨基羧酸螯合物
3-amino-(3-carboxy-propyldimethyl-sulfonium) 二甲锍氨酸；蛋氨酸甲锍盐
aminocellulose 氨基纤维素
amino chemically bonded silica 氨基化学键合硅胶
1-amino-4-chloro-9,10-anthraquinone 1-氨基-4-氯-9,10-蒽醌
4-amino-2-chlorotoluene-5-sulfonic acid 4-氨基-2-氯甲苯-5-磺酸〔亮红胺调色料的化学名〕
5-amino-2-chlorotoluene-4-sulfonic acid 5-氨基-2-氯甲

苯-4-磺酸〔色淀红 C 胺的化学名〕
aminocidin 氨基杀菌素
amino-cinnamic acid 氨基肉桂酸
　　$NH_2C_6H_4CH=CHCOOH$
amino-compound 氨基化合物
amino crosslinker 氨基交联剂
β-amino crotonate β-氨基巴豆酸酯〔增塑剂〕
amino-cumene 氨基异丙苯 $(CH_3)_2CHC_6H_4NH_2$
aminocyclitol 氨基环醇
3-amino-1-cyclohexylamino propane 3-氨基-1-环己氨基丙烷；N-环己基-1,3-丙二胺
1-aminocyclopropane-1-carboxylic acid 氨基环丙烷羧酸
3-amino-3-deoxyadenosine 3-氨基-3-脱氧腺苷
aminodeoxy cellulose 氨基脱氧纤维素
3-amino-3-deoxy-D-glucose 3-氨基-3-脱氧-D-葡萄糖
aminodeoxykanamycin 卡那霉素 B
amino-dicarboxylic acid 氨基二羧酸
amino-diiminoform melamine 氨基二亚氨型三聚氰胺
p-aminodimethylaniline 对氨基二甲基苯胺
aminodiphenylamine 氨基二苯胺
aminodiphenylmethane 氨基二苯甲烷
3-amino-1,2,4-dithiazolidinethion-5 3-氨基-1,2,4-二噻唑烷-5-硫酮〔硫化剂〕
aminodithiophosphate 氨基二硫代磷酸盐〔促进剂〕
amino dodecane 十二胺
12-aminododecanoic acid 12-氨基十二(烷)酸
aminoenanthic acid 氨基庚酸
amino end group 氨端基
aminoethane(=ethylamine) 乙胺；氨基乙烷 $C_2H_5NH_2$
aminoethane sulfonic acid 氨乙磺酸 $H_2NCH_2CH_2SO_3H$
amino ethanethiosulfuric acid(=aminoethylthiosulfuric acid) 氨乙基硫代硫酸
aminoethanol 氨基乙醇
amino ethoxyacetic acid 氨基乙氧基乙酸
aminoethylacetanilide 氨基苯乙酰乙胺
aminoethyl alcohol(=ethanolamine) 乙醇胺；2-羟基乙胺；氨基乙醇；胆胺 $NH_2CH_2CH_2OH$
N-β-(aminoethyl)-γ-aminopropyl methyl dimethoxysilane N-β(氨乙基)-γ-氨丙基甲基二甲氧基硅烷
N-(β-aminoethyl)-γ-aminopropyl trimethoxysilane N-(β-二氨乙基)-γ-氨丙基三甲氧基硅烷〔偶联剂〕
aminoethylbenzene 氨基乙苯
aminoethylcellulose 氨乙基纤维素
S-aminoethylcysteine S-氨乙基半胱氨酸
amino ethyl ethanolamine 氨基乙乙醇胺
N-amino ethyl fatty amide N-氨基乙基脂肪酰胺
(1-aminoethylidene) diphosphonate(AEDP) 1-氨基亚乙基二膦酸 $(NH_2)(CH_3)C(PO_3H_2)_2$

amino ethyl imidazoline 氨基乙基咪唑啉
aminoethyl mercaptan 氨基乙硫醇
S-(2-aminoethyl) phosphorothioic acid S-(2-氨基乙基)硫代磷酸 $(H_2NCH_2CH_2S)PO(OH)_2$
N-aminoethyl piperazine N-氨基乙基哌嗪〔固化剂〕
2-amino-2-ethyl-1,3-propanediol 2-氨基-2-乙基-1,3-丙二醇〔乳化剂〕
aminoethyl pyrazine 氨乙基吡嗪；氨乙基对二氮杂苯
aminoethyl sulfate 硫酸氨乙酯
β-aminoethyl sulfonic acid-N,N-diacetic acid chelate β-氨基乙磺酸-N,N-二乙酸螯合物
3-aminoethyl-3,5,5-trimethylcyclo hexylamine 3-氨乙基-3,5,5-三甲基环己胺〔固化剂〕
amino fluoride 氨氟化物
amino formaldehyde resin 氨基甲醛树脂
amino formic acid 氨基甲酸 NH_2COOH
aminogenesis 氨基的形成；生氨作用
aminoglucose 氨基葡萄糖；葡萄糖胺
amino glutaric acid 氨基戊二酸
　　$COOH(CH_2)_2CHNH_2COOH$
aminoglutethimide 氨鲁米特〔药〕
amino-group 氨基
aminoguanidine 氨基胍 $H_2NC(NH)NHNH_2$
aminoguanidine bicarbonate 氨基胍碳酸氢盐
aminoguanidine carbonate 氨基胍碳酸盐
　　$H_2NC(NH)NHNH_2 \cdot H_2CO_3$
aminoguanidine sulfate 氨基胍硫酸盐
　　$H_2NC(NH)NHNH_2 \cdot H_2SO_4$
amino guanidyl valeric acid 胍基戊氨酸；氨基胍基戊酸
aminoguanyl urea 氨胍基脲
amino halide 氨基卤化物 R_2NX
aminoheptylic acid 氨基庚酸
6-aminohexanol 6-氨基己醇
2-amino-4-hexenoic acid 2-氨基-4-己烯酸
aminohexose 氨基己糖
p-aminohippurate 对氨基马尿酸
amino hydroxy acid 氨基羟基酸
amino hydroxy alkanoic acid amide 氨基羟基烷酰胺
1-amino-4-hydroxyanthraquinone 1-氨基-4-羟基蒽醌
2-amino-4-hydroxy-6-methylpteridine 2-氨基-4-羟基-6-甲基蝶啶
amino hydroxy octyl fatty ester 氨基羟基辛基脂肪酸酯
amino hydroxy propionic acid 羟基丙氨酸；氨基羟基丙酸 $CH_2OHCHNH_2COOH$
2-amino-6-hydroxypurine 2-氨基-6-羟基嘌呤
3-aminoisobutyrate 3-氨基异丁酸
aminoisobutyric acid 氨基异丁酸
α-amino-isopropyl alcohol(=isopropanolamine) α-氨基

异丙醇；异丙醇胺　$NH_2CH_2CHOHCH_3$
α-amino-isovaleric acid　α-氨基异戊酸；异戊氨酸；缬氨酸　$(CH_3)_2CHCHNH_2COOH$
4-amino-3-isoxazolidone　4-氨基-3-异氢噁唑酮；环丝氨酰胺；环丝氨酸
α-amino-β-keto adipic acid　α-氨基-β-酮己二酸
aminoketone and hydroxyketone colouring matters　氨基酮和羟基酮染料
aminoketones　氨基酮类
δ-aminol(a)evulinic acid (ALA)　δ-氨基-γ-酮戊酸
12-aminolauric acid　12-氨基十二(烷)酸；12-氨基月桂酸
aminolipid　氨脂质
aminolysis　氨基分解；氨解
amino-maleic acid　马来酰氨酸
　　$CONH_2CH=CHCOOH$
amino mercapto alkylamide　氨基巯基烷基酰胺
aminomercuration　氨汞化
aminomethyl　氨甲基　H_2NCH_2-
amino methyl aniline　氨基甲基苯胺
aminomethylation　氨甲基化(作用)
aminomethylazo　氨基甲基偶氮
aminomethylbenzoic acid　氨甲苯酸〔药〕
γ-amino-α-methylene butyric acid　γ-氨基-α-亚甲基丁酸
amino methylene phosphonate　氨基亚甲基膦酸盐
aminomethyl propanediol　氨甲基丙二醇
aminomethyl propanol　氨甲基丙醇
2-amino-2-methyl-1-propanol　2-氨基-2-甲基丙醇〔分散剂〕
amino methyl valeric acid　氨基甲基戊酸；L-异白氨酸
aminonaphthalene　氨基萘
2-amino-1-naphthalene sulfonic acid　2-氨基-1-萘磺酸
aminonaphthalimide　氨基萘二甲亚胺
amino-naphthol　氨基萘酚　$NH_2C_{10}H_6OH$
amino-naphthol sulfonic acid　氨基萘酚磺酸
　　$NH_2C_{10}H_5(OH)·SO_3H$
amino-nitrile　氨基腈
2-amino-1-nitro-9,10-anthraquinone　2-氨基-1-硝基-9,10-蒽醌
4-amino-2-nitro-diphenylamine　4-氨基-2-硝基二苯胺
amino nitrogen　氨基氮；胺型氮
4-amino-4'-nitro-stilbene　4-氨基-4'-硝基芪
9-aminononanoic acid　9-氨基(正)壬酸；ω-氨基(正)壬酸〔豆油的成分之一，可制尼龙 12〕
aminononanoic acid　氨基壬酸
amino-octane　氨基辛烷
8-aminooctanoic acid　8-氨基辛酸
amino-oxamide(=oxamic hydrazide; semioxamazide)　草氨酰肼；氨基草酰胺；氨羰基甲酰肼
　　$NH_2COCONHNH_2$

2-amino-6-oxypurine　2-氨基-6-氧嘌呤
2-amino-para-benzoquinone　2-氨基对苯醌
9-aminopelargonic acid　9-氨基壬酸
6-aminopenicillanic acid　6-氨基青霉烷酸
aminopentane　氨基戊烷
amino-pentanoic acid　氨基戊酸
amino peptidase　氨肽酶
amino-phenol　氨基苯酚
aminophenol sulfonic acid　氨基苯酚磺酸
　　$C_6H_3(OH)(NH_2)SO_3H$
aminophenyl acetic acid　氨基苯乙酸
aminophenyl acetonitrile　氨基苯乙腈
aminophenyl arsine sulfide　硫化氨基苯胂
　　$NH_2C_6H_4As=S$
aminophenyl arsinic acid　氨基苯次胂酸
aminophenyl arsonic acid　氨基苯胂酸
aminophenylglycine　氨基苯乙氨酸
aminopherase(=transaminase)　转氨酶
aminophthalic acid　氨基邻苯二甲酸；氨基酞酸
aminophylline　氨茶碱；氨基非林〔药〕
4-aminopipecolic acid　4-氨基六氢吡啶羧酸
aminoplast　氨基塑料
aminoplastic　氨基塑料
aminoplast(ic) resin　氨基塑料树脂
aminoplastics　氨基塑料
aminoplast-modified acrylic fibre　氨基树脂改性(的)丙烯腈系纤维
aminopolycarboxylic acid　氨基多羧酸
amino polymer　氨基聚合物
amino polyoxy alkylene sulfate ammonium salt　氨基聚氧亚烷硫酸铵
aminopolypeptidase　氨基多肽酶
1-aminopropane(=n-propylamine)　正丙胺；1-丙胺
　　$CH_3CH_2CH_2NH_2$
2-aminopropane(=isopropyl amine)　异丙胺；2-丙胺
　　$(CH_3)_2CHNH_2$
aminopropanol　氨丙醇
α-aminopropionic acid　α-氨基丙酸
　　$CH_3CH(NH_2)COOH$
aminopropylbenzene　氨基丙苯
N-(3-aminopropyl)-butane-1,4-diamine(=spermidine)　亚精胺；N-(3-氨丙基)-1,4-丁二胺
γ-aminopropyl triethoxysilane　γ-氨丙基三乙氧基硅烷〔玻璃纤维复合材料用的偶联剂〕
γ-aminopropyltrimethoxysilane　γ-氨丙基三甲氧基硅烷
aminoprotease　氨基蛋白酶
aminopterin　氨基蝶呤
aminopyridine　氨基吡啶

aminopyrine 氨基比林；匹拉米洞〔药〕
aminoquinoline 氨基喹啉
aminoquinoxaline 氨基喹噁啉 $NH_2C_8H_5N_2$
amino resin 氨基树脂（氨基化合物或酰胺化合物与醛类物质作用生成的树脂状物质）
aminoresorcinol 氨基间苯二酚 $NH_2C_6H_3(OH)_2$
amino rubber 氨基橡胶
aminosaccharide 氨基糖
aminosalicylic acid 氨基水杨酸
aminosilane 氨基硅烷
aminosilicone 氨基硅氧烷
aminosiloxane 氨基硅氧烷
amino stearic acid 氨基硬脂酸
amino-succinamic acid 氨基琥珀酰胺酸；天冬酰胺 $CONH_2CH_2CHNH_2COOH$
amino-succinic acid 氨基丁二酸 $COOHCH_2CHNH_2COOH$
aminosugar 氨基糖
aminosulfonic acid 氨基磺酸 NH_2SO_3H
amino terminal 氨基末端；N 末端
amino tetrahydro pyran 氨基四氢吡喃
aminothiazole 氨基噻唑
aminothiophene 氨基噻吩
amino toluene 氨基甲苯；甲苯胺 $CH_3C_6H_4NH_2$
aminotoluene sulfonic acid 氨基甲苯磺酸 $C_6H_3(CH_3)(NH_2)SO_3H$
aminotransferase 转氨酶
aminotriacetic acid(=nitrilotriacetic acid) 次氮基三乙酸；氨基三乙酸 $N(CH_2COOH)_3$
amino tri-arylmethane dye 氨基取代的三芳甲烷染料
amino triazine 氨基三嗪；氨基三氮杂苯
aminotriazine resin 三聚氰胺树脂
aminotriazole 氨基三唑
13-aminotridecanoic acid 13-氨基十三(烷)酸 $H_2NCH_2(CH_2)_{11}COOH$
aminoundecane-carboxylic acid 氨基十二(烷)酸；氨基十一(烷)羧酸 $H_2N(CH_2)_{11}COOH$
11-aminoundecanoic acid 11-氨基十一(烷)酸
amino uracil 氨基尿嘧啶
δ-amino valeramide δ-氨基戊酰胺
aminovaleric acid 氨基戊酸
amino-xylene 氨基二甲苯 $(CH_3)_2C_6H_3NH_2$
amiodarone 胺碘酮〔药〕
amiodoxyl benzoate 碘氧苯甲酸铵 $C_6H_4(IO_2)COONH_4$
amital(=isoamyl ethyl barbiturie acid) 阿米他；异戊基乙基巴比妥酸 $(C_2H_5)(C_5H_{11})=CCONHCONHCO$
amitriptyline 阿米替林〔药〕

amlodipine 氨氯地平〔药〕
ammelide(=cyanuramide) 三聚氰酸一酰胺；氰尿酰胺 $(CN)_3NH_2(OH)_2$
ammeline(=cyanurodiamide) 三聚氰酸二酰胺；氰尿二酰胺 $(CN)_3(NH_2)_2OH$
ammeter(=amperemeter; amperometer) 安培计；电流计
ammine ①氨合物②氨(合)
ammine complex 氨络合物
ammine complex salt 氨络物；氨络盐
ammine type adsorption 氨(络)型吸附
ammino 氨合
ammino-complex 氨配位(化合)物
ammino compound 氨配位(化合)物
ammoina corrosion prevention 氨腐蚀防止法
ammonal 阿芒拿；硝铵、铝、炭炸药
ammonate 氨合物
ammonation 氨合(作用)
ammonchelidonic acid(=chelidamic acid) 白屈菜氨酸
ammongelatine 铵胶
ammongelatine dynamite 铵胶炸药
ammonia 氨；阿摩尼亚；氨水
ammonia absorber 吸氨器
ammonia absorption machine 氨气吸收机
ammonia alkali 氨碱
ammonia amalgam 氨汞齐
ammonia-ammonium ion solution 氨水-铵盐缓冲液
ammonia ash 氨法(制的)碱；氨法碳酸钠
ammoniac ①氨的②氨草胶；氨树胶
ammoniacal 氨的
ammoniacal brine （充）氨盐水
ammoniacal liquor 氨液
ammoniacal nitrogen 氨型氮
ammoniacal solution of cupric oxide 氧化铜氨溶液
ammonia chloride(=ammonium chloride) 氯化铵
ammoniac oil(=ammoniacum oil) 氨草油
ammonia complex 氨络合物
ammonia compression refrigerating machine 氨气压缩制冷机
ammonia compressor 氨气压缩机
ammonia condenser 氨气冷凝器
ammonia converter 氨合成塔
ammonia cooler 氨冷却器
ammonia corrosion 氨腐蚀
ammonia corrosion prevention 氨腐蚀防止法
ammoniac silver nitrate solution 氨性硝酸银液
ammoniacum 氨草胶；氨(树)脂
ammoniacum oil(=ammoniac oil) 氨草油
ammonia cure 氨蒸气熟化

ammonia distiller 蒸氨器
ammonia dynamite 铵爆炸药；硝铵炸药
ammonia elimination process 除氨法
ammonia gas sensing electrode 氨气敏电极
ammonia liquor 氨水
ammonia making machinery 制氨机器
ammoniameter(=ammoniometer) 氨量计；氨浓度计
ammonia nitrogen 氨型氮
ammonia oil 氨油
ammonia oxidation 氨(的)氧化
ammonia oxidation converter 氨气氧化炉
ammonia-preserved latex (加)氨(保存)胶乳
ammonia refrigerant 氨冷冻剂
ammonia refrigerating process 氨冷作用
ammonia refrigerator 氨冷冻机
ammonia saturator 氨饱和器
ammonia scrubber 洗氨器；氨气洗涤器；氨吸收器
ammonia selective electrode 氨选择性电极
ammonia sensor 氨传感器
ammonia soda 氨法(制的)苏打；氨法碳酸钠
ammonia soda ash 氨法苏打灰；粗氨制苏打
ammonia-soda process 氨碱法*
ammonia spirit 氨水〔氨的水溶液〕
ammonia still 氨气塔；蒸氨塔
ammonia sulfate(=ammonium sulfate) 硫酸铵
ammonia synthesis 氨合成法
ammonia synthesis converter 氨合成塔
ammonia synthesis gas 氨合成气
ammonia synthesis unit 合成氨设备
ammonia system 氨溶物系
ammonia system baking powder 氨系膨松剂
ammoniate ①氨合物②有机氨肥(料)
ammoniated ①充氨的②含氨的
ammoniated brine (充)氨盐水
ammoniated iron 氯铁酸铵
ammoniated mercury 氨基汞化氯 NH_2HgCl
ammoniated superphosphate 氨化过磷酸钙
ammoniating ①充氨的②充氨(作用)
ammoniating vat 充氨瓮
ammoniation 氨化(作用)
ammonia treatment 氨处理
ammoniatriacetic acid 氨三乙酸
ammonia vapour test 氨蒸气试验
ammonia washer 洗氨器
ammonia water 氨水
ammonification 氨化(作用)；加氨(作用)；生氨(作用)
ammonifying ①生氨②加氨
ammonio ①氨溶的②铵基

ammoniogen 生氨剂
ammoniogenesis 生氨作用
ammoniometer(=ammoniameter) 氨量计；氨浓度计
ammonite ①菊石②干肉粉③阿芒炸药；硝铵二硝基萘炸药
ammonium 铵
ammonium acetate 乙酸铵 CH_3COONH_4
ammonium acid arsenate 酸式砷酸铵〔通常指$(NH_4)_2HAsO_4$〕
ammonium acid carbonate 酸式碳酸铵 NH_4HCO_3
ammonium acid fluoride 氟化氢铵 NH_4HF_2
ammonium acid sulfate 酸式硫酸铵 NH_4HSO_4
ammonium alginate 藻酸铵〔膏化剂〕
ammonium alkylaryl propyl sulfate 烷基芳基丙基硫酸铵
ammonium alkyl naphthalene sulfonate 烷基萘磺酸铵
ammonium alum(=ammonium aluminium sulfate) 铵铝矾；铵矾
ammonium aluminate 铝酸铵 NH_4AlO_2
ammonium aluminium sulfate(=ammonium alum) 铵铝矾；铵矾
ammonium antimonate 锑酸铵 NH_4SbO_3
ammonium antimonyl tartrate 酒石酸氧锑铵 $NH_4(SbO)(C_4H_4O_6)$
ammonium arsenate 砷酸铵〔计有：1.$(NH_4)AsO_3$；2.$(NH_4)_3AsO_4$；3.$(NH_4)_4As_2O_7$，常指2项而言〕
ammonium arsenite 亚砷酸铵〔计有：1.$(NH_4)AsO_2$；2.$(NH_4)_3AsO_3$；3.$(NH_4)_4As_2O_5$，通常指2项而言〕
ammonium aurichloride 氯金酸铵 $NH_4[AuCl_4]$
ammonium auricyanide 氰金酸铵 $NH_4[Au(CN)_4]$
ammonium aurintricarboxylate 金精三羧酸铵
ammonium aurocyanide 氰亚金酸铵 $NH_4[Au(CN)_2]$
ammonium azelate 壬二酸铵
ammonium benzene sulfonate 苯磺酸铵
ammonium benzoate 苯甲酸铵
ammonium biarsenate 砷酸氢铵 $(NH_4)_2HAsO_4$
ammonium biborate 硼酸氢铵
ammonium bicarbonate 碳酸氢铵 NH_4HCO_3
ammonium bichromate 重铬酸铵 $(NH_4)_2Cr_2O_7$
ammonium bifluoride 氟化氢铵 $(NH_4)HF_2$
ammonium biphosphate 磷酸二氢铵
ammonium bisulfate 硫酸氢铵 NH_4HSO_4
ammonium bisulfite 亚硫酸氢铵 NH_4HSO_3
ammonium bisulfite process 亚硫酸氢铵法
ammonium borate 硼酸铵 NH_4BO_3
ammonium borofluoride 氟硼酸铵 $NH_4[BF_4]$
ammonium bromide 溴化铵 NH_4Br
ammonium bromoplatinate 溴铂酸铵 $(NH_4)_2[PtBr_6]$

ammonium caproate 正己酸铵
ammonium caprylate 正辛酸铵
ammonium carbamate 氨基甲酸铵 NH_2COONH_4
ammonium carbonate 碳酸铵 $(NH_4)_2CO_3$
ammonium ceric sulfate 硫酸高铈铵 $(NH_4)_6[Ce(SO_4)_5]$
ammonium cerous sulfate 硫酸铈铵 $NH_4[Ce(SO_4)_2]$
ammonium chlorate 氯酸铵 NH_4ClO_3
ammonium chloraurate 氯金酸铵 $NH_4[AuCl_4]$
ammonium chloride 氯化铵
ammonium chloroacetate 氯乙酸铵 $NH_4C_2H_2ClO_2$
ammonium chlorocuprate 氯铜酸铵 $(NH_4)_2[CuCl_4]$
ammonium chloroiridate 氯铱酸铵 $NH_4[IrCl_6]$
ammonium chloropalladate 氯钯酸铵 $(NH_4)_2[PdCl_6]$
ammonium chloropalladite 氯亚钯酸铵 $(NH_4)_2[PdCl_4]$
ammonium chloroplatinate 氯铂酸铵 $(NH_4)_2[PtCl_6]$
ammonium chloroplatinite 氯亚铂酸铵 $(NH_4)_2[PtCl_4]$
ammonium chlorostannate 氯锡酸铵 $(NH_4)_2[SnCl_6]$
ammonium chlorotitanate 氯钛酸铵 $(NH_4)_2[TiCl_6]$
ammonium chromate 铬酸铵 $(NH_4)_2CrO_4$
ammonium chromic alum(=ammonium chromic sulfate) 铬铵矾;硫酸铬铵
ammonium chromic sulfate(=ammonium chromic alum) 铬铵矾;硫酸铬铵
ammonium citrate 柠檬酸铵
ammonium cobalt thiocyanate 硫氰酸钴铵
ammonium coco alkyl sulfate 椰子油烷基硫酸铵
ammonium cuprate 铵铜合物〔氢氧化铜的含氨溶液〕
ammonium cyanate 氰酸铵 NH_4OCN
ammonium cyanide 氰化铵 NH_4CN
ammonium cyanoplatinite 氰亚铂酸铵 $(NH_4)_2[Pt(CN)_4]$
ammonium cyanovalerate 氰基戊酸铵
ammonium diacid phosphate 磷酸二氢铵;二酸式磷酸铵 $NH_4H_2PO_4$
ammonium dibutyldithiophosphate 二丁基二硫代磷酸铵〔促进剂〕
ammonium dichromate 重铬酸铵 $(NH_4)_2Cr_2O_7$
ammonium diethyldithiocarbamate 二乙基二硫代氨基甲酸铵〔促进剂〕
ammonium dihydric phosphate 磷酸二氢铵 $NH_4H_2PO_4$
ammonium dihydrogen arsenate 砷酸二氢铵 $NH_4H_2AsO_4$
ammonium dihydrogen phosphate 磷酸二氢铵 $NH_4H_2PO_4$
ammonium dineptunate 重镎酸铵 $(NH_4)_2Np_2O_7$
ammonium dioxodithiomolybdate 二氧二硫代钼酸铵
ammonium dithiocarbamate 二硫代氨基甲酸铵
ammonium dithionate 连二硫酸铵

ammonium diuranate 重铀酸铵 $(NH_4)_2U_2O_7$
ammonium eicosafluoroundecanoate 二十氟代十一烷酸铵 $CHF_2(CF_2)_9COONH_4$
ammonium ferric chloride 氯化铁铵 $NH_4[FeCl_4]$
ammonium ferric citrate 柠檬酸铁铵;枸橼酸铁铵
ammonium ferric sulfate 硫酸铁铵
ammonium ferricyanide 铁氰化铵 $(NH_4)_3[Fe(CN)_6]$
ammonium ferrocyanide 亚铁氰化铵 $(NH_4)_4[Fe(CN)_6]$
ammonium ferrous sulfate 硫酸亚铁铵 $(NH_4)_2SO_4 \cdot FeSO_4 \cdot 6H_2O$
ammonium fluoralkanoate 氟代链烷酸铵
ammonium fluoride 氟化铵 NH_4F
ammonium fluoroborate solution 氟硼酸铵溶液〔裂化催化剂〕
ammonium fluoscandate 氟钪酸铵 $(NH_4)_3[ScF_6]$
ammonium fluosilicate 氟硅酸铵 $(NH_4)_2[SiF_6]$
ammonium fluozirconate 氟锆酸铵 $(NH_4)_2[ZrF_6]$
ammonium formate(=ammonium formiate) 甲酸铵 $HCOONH_4$
ammonium formiate(=ammonium formate) 甲酸铵
ammonium hydrogen fluoride 氟化氢铵 NH_4HF_2
ammonium hydrogen phosphate 磷酸氢铵
ammonium hydrogen sulfate 硫酸氢铵 NH_4HSO_4
ammonium hydrogen tartrate 酒石酸氢铵 $NH_4HC_4H_4O_6$
ammonium hydrosulfide 氢硫化铵 NH_4HS
ammonium hydroxide 氢氧化铵 NH_4OH
ammonium hypophosphate 连二磷酸铵 $NH_4H_2PO_2$
ammonium hypophosphite 次磷酸铵
ammonium hyposulfide 硫代硫酸铵
ammonium hyposulfite 连二亚硫酸铵 $(NH_4)_2S_2O_4$
ammonium iodate 碘酸铵 NH_4IO_3
ammonium iodide 碘化铵 NH_4I
ammonium ion 铵离子
ammonium ion selective electrode 铵离子选择电极
ammonium iridichloride 氯铱酸铵 $(NH_4)_2[IrCl_6]$
ammonium iron (Ⅱ) sulfate 硫酸亚铁铵
ammonium iron alum 铁铵矾 $NH_4Fe(SO_4)_2 \cdot 12H_2O$
ammonium iron chloride 氯化铁铵
ammonium laurate 月桂酸铵;十二酸铵〔胶乳稳定剂〕
ammonium magnesium arsenate 砷酸镁铵 $Mg(NH_4)AsO_4$
ammonium magnesium phosphate 磷酸镁铵 $Mg(NH_4)PO_4$
ammonium mercaptoacetate 巯基乙酸铵;硫代乙醇酸铵
ammonium metaborate 偏硼酸铵 NH_4BO_2
ammonium metaphosphate 偏磷酸铵 NH_4PO_3
ammonium metatungstate(=ammonium metawolframate)

偏钨酸铵　(NH_4)$_2W_4O_{13}$
ammonium metavanadate　偏钒酸铵
ammonium metawolframate(=ammonium metatungstate)　偏钨酸铵
ammonium molybdate　钼酸铵
ammonium molybdophosphate(AMP)　磷钼酸铵　(NH_4)$_3Mo_{12}O_{40}$
ammonium monoacid phosphate　磷酸氢二铵；一酸式磷酸铵　(NH_4)$_2HPO_4$
ammonium monohydric phosphate　磷酸氢二铵
ammonium mucate　黏酸铵；半乳糖二酸铵
ammonium muriate　氯化铵　NH_4Cl
ammonium myristate　肉豆蔻酸铵
ammonium nitrate　硝酸铵
ammonium nitrate limestone　硝酸钙铵
ammonium nitrite　亚硝酸铵　NH_4NO_2
ammonium nitrocaproate　硝基己酸铵
ammonium oleate　油酸铵
ammonium oxalate　草酸铵
ammonium palladic chloride　氯化钯铵；氯钯酸铵
ammonium palladous chloride　氯化亚钯铵；氯亚钯酸铵
ammonium perchlorate　高氯酸铵　NH_4ClO_4
ammonium-perchlorate explosive　高氯酸铵炸药
ammonium perchromate　过铬酸铵
ammonium perfluorocaprylate　全氟辛酸铵
ammonium periodate　高碘酸铵
ammonium permanganate　高锰酸铵　NH_4MnO_4
ammonium peroxodisulphate(=ammonium persulphate)　过(氧化)硫酸铵
ammonium peroxydisulfate　过(二)硫酸铵
ammonium peroxysulfate　过硫酸铵〔引发剂〕　(NH_4)$_2S_2O_8$
ammonium persulfate　过硫酸铵
ammonium persulphate(=ammonium peroxodisulphate)　过(氧化)硫酸铵
ammonium phosphate　磷酸铵　(NH_4)$_3PO_4$
ammonium phosphite　亚磷酸铵　(NH_4)$_3PO_3$
ammonium phosphotungstate(=ammonium phosphorwolframate)　磷钨酸铵
ammonium phosphowolframate(=ammonium phosphortungstate)　磷钨酸铵
ammonium picrate　苦味酸铵
ammonium platinic bromide　溴化铂铵；溴铂酸铵　(NH_4)$_2$[$PtBr_6$]
ammonium platinic chloride　氯化铂铵；氯铂酸铵　(NH_4)$_2$[$PtCl_6$]
ammonium plutonyl phosphate　磷酸钚酰铵　$NH_4PuO_2PO_4$

ammonium polyacrylate　聚丙烯酸铵
ammonium polymethacrylate　聚甲基丙烯酸铵〔增稠剂〕
ammonium polyphosphate　聚磷酸铵
ammonium poly(styrene-maleate)　聚(苯乙烯-顺丁烯二酸)铵
ammonium polysulfide　多硫化铵
ammonium polysulphide　多硫化铵
ammonium primary phosphate　磷酸二氢铵；一代磷酸铵　$NH_4H_2PO_4$
ammonium purpurate(=murexide)　红紫酸铵；骨螺紫
ammonium pyromelliate　苯均四酸铵
ammonium pyrophosphate　焦磷酸铵　(NH_4)$_4P_2O_7$
ammonium pyrosulfate　焦硫酸铵　(NH_4)$_2S_2O_7$
ammonium pyrosulfite　焦亚硫酸铵　(NH_4)$_2S_2O_5$
ammonium pyrrolidine dithiocarbamate　吡咯烷二硫代氨基甲酸铵
ammonium reineckate　雷氏铵
ammonium rhodanate(=ammonium rhodanide)　硫氰酸铵　NH_4SCN
ammonium rhodanide(=ammonium rhodanate)　硫氰酸铵
ammonium rhodanilate　四硫氰基二苯胺合铬酸铵
ammonium salt　铵盐
ammonium salt bath　铵盐浴
ammonium salt extraction　铵盐萃取
ammonium sebacate　癸二酸铵
ammonium secondary phosphate　磷酸氢二铵；二代磷酸铵　(NH_4)$_2HPO_4$
ammonium selenate　硒酸铵　(NH_4)$_2SeO_4$
ammonium selenide　硒化铵　(NH_4)$_2Se$
ammonium selenite　亚硒酸铵　(NH_4)$_2SeO_3$
ammonium silicofluoride　氟硅酸铵　(NH_4)$_2$(SiF_6)
ammonium soap　铵皂
ammonium sodium oxalate　草酸钠铵
ammonium stannic chloride　氯化锡铵；氯锡酸铵　(NH_4)$_2$[$SnCl_6$]
ammonium succinate　丁二酸铵　$C_2H_4(COONH_4)_2$
ammonium sulfamate　氨基磺酸铵
ammonium sulfate(=ammonia sulfate)　硫酸铵
ammonium sulfhydrate　硫氢化铵　NH_4HS
ammonium sulfide　硫化铵　(NH_4)$_2S$
ammonium sulfide method　硫化铵法
ammonium sulfite　亚硫酸铵　(NH_4)$_2SO_3$
ammonium sulfite solution　液体亚硫酸铵
ammonium sulfoantimonate　全硫(代)锑酸铵　(NH_4)$_3SbS_4$
ammonium sulfocyanate(=ammonium sulfocyanide)　硫氰酸铵
ammonium sulfocyanide(=ammonium sulfocyanate)　硫氰

酸铵
ammonium super(phosphate)　过磷酸铵
ammonium tartrate　酒石酸铵　$(NH_4)_2C_4H_4O_6$
ammonium tellurate　碲酸铵　$(NH_4)_2TeO_4$
ammonium tertiary phosphate　(三代)磷酸铵　$(NH_4)_3PO_4$
ammonium tetradecyl naphthalene sulfonate　十四烷基萘磺酸铵
ammonium tetradecyl sulfate　十四烷基硫酸铵
ammonium thiocarbonate　全硫碳酸铵　$(NH_4)_2CS_3$
ammonium thiocyanate(=ammonium thiocyanide)　硫氰酸铵　NH_4SCN
ammonium thiocyanide(=ammonium thiocyanate)　硫氰酸铵
ammonium thiostannate　全硫锡酸铵　$(NH_4)_2SnS_3$
ammonium thiosulfate　硫代硫酸铵　$(NH_4)_2SO_3S$
ammonium titanium oxyoxalate　草酸铵合草酸氧钛
ammonium titanyl oxalate　草酸氧钛(二)铵
ammonium tribasic phosphate　(三代)磷酸铵　$(NH_4)_3PO_4$
ammonium tungstate　钨酸铵　$(NH_4)_2WO_4$
ammonium tungstophosphate(AWP)　磷钨酸铵
ammonium undecanol ethoxy sulfate　十一(烷)醇乙氧基硫酸铵
ammonium uranate　铀酸铵　$(NH_4)_2U_2O_7$
ammonium uranylcarbonate　碳酸氧铀铵
ammonium uranylfluoride　氟化氧铀铵
ammonium vanadate　钒酸铵　$NH_4VO_3; (NH_4)_3VO_3$
ammonium zirconyl carbonate　碳酸氧锆铵；碳酸铵合碳酸氧锆
ammoniun dibasic phosphate　磷酸氢二铵；二碱式酸铵　$(NH_4)_2HPO_4$
ammonla alum　氨明矾
ammonobase　氨基金属　MNH_2
ammonocarbonous acid　氢氰酸　HCN
ammonolysis　氨解作用〔化合物被氨分解,有如水解,如:$RX+NH_3 \longrightarrow RNH_2+HX$〕
ammonoxidation process　氨氧化法
ammonoxidative nitrile synthesis　氨氧化合成腈(法)
ammonpulver　铵炸药
ammoresinol　氨草树脂醇
ammoxidation(=oxidative ammonolysis)　氨氧化反应(作用)；氨解氧化(作用)
ammoxidation of p-xylene method　对二甲苯的氨氧化法〔以对二甲苯为原料在氨存在下通过气相空气氧化制取对苯二酸〕
ammunition　弹药
amobarbital　异戊巴比妥；阿米妥
Amoco process　阿莫科工艺〔芳烃氧化制苯酚法〕
amoeba(=ameba)　变形虫；阿米巴
amoebacide　杀变形虫剂
amoeboid movement　类似阿米巴的运动
amoil　酞酸戊酯
amol　阿摩尔
amorpha frutcosa oil　紫穗槐油
amorphene　紫穗槐烯
amorphin　紫穗槐苷
amorphism　无定形；非晶形
amorphization　非晶态化；无定形化
amorpho-crystalline structure　非晶体-晶体结构
amorphous　非晶态的；无定形的
amorphous alloy　无定形合金；非晶形合金
amorphous aluminosilicate gel　无定形硅铝酸盐凝胶
amorphous band　非晶区谱带
amorphous birefringence　无定形双折射
amorphous carbon　无定形碳
amorphous cellulose　无定形纤维素；非晶形纤维素
amorphous chain　非晶态链；无定形链
amorphous chain structure　非晶态链结构；无定型链结构
amorphous change　无定形的改变
amorphous content　非晶区含量
amorphous defect　非晶区缺陷；无定形区缺陷
amorphous fibre　非晶态纤维
amorphous gel　无定形凝胶
amorphous graphite　无定形石墨〔填料〕
amorphous layer　非晶形层〔即贝尔俾层〕
amorphous material　非晶材料
amorphous matrix　非晶基体
amorphous matter　非晶体
amorphous molecular material　无定形分子材料
amorphousness　无定形(现象)；非晶质
amorphous orientation　非晶态取向；非晶取向
amorphous phase　非晶相；无定形相
amorphous plastic　非晶塑料
amorphous polymer　非晶态聚合物；无定形聚合物
amorphous polyolefins　无结晶聚烯烃
amorphous portion　非晶部分
amorphous precipitation　无定形沉淀
amorphous region　非晶形区
amorphous region mobility　非晶区(链段)活动性
amorphous rubber　无定形橡胶；非晶态橡胶
amorphous silica　无定形二氧化硅
amorphous silicate　无定形硅酸盐
amorphous silicon dioxide aerogel(=silica aerogel)　无定形二氧化硅气凝胶
amorphous silicon dioxide hydrogel(=silica hydrogel)　无

定形二氧化硅水凝胶
amorphous solid 非晶态固体
amorphous solvating polymer 非晶态溶剂化聚合物
amorphous state 非晶态
amorphous structure 无定形结构；非晶态结构
amorphous substance 无定形物
amorphous wax 无定形蜡
amortization ①阻尼；缓冲；减振；消音②折旧
amosite 闪石棉；铁石棉〔填充剂〕
amosite asbestos 铁石棉，长纤维石棉
amount of cure 熟化程度；硫化程度
amount of precipitation 降水量
amoxicillin 阿莫西林〔药〕
amoxicillin and clavulanate potassium 阿莫西林克拉维酸钾〔药〕
amoxicillin and sulbactam 阿莫西林舒巴坦〔药〕
amoxy(=pentyloxy) 戊氧基 $C_5H_{11}O-$
amozonolysis 氨(解)臭氧化反应(作用)
ampeloptin 白蔹素；蛇葡萄素
ampere 安(培)
ampere detector 安培检测器
ampere-hour meter 安时计
amperemeter(=ammeter) 安培计
amperometric determination 电流测定
amperometric method 电流分析法
amperometric titration(=amperometry) 电流滴定(法)
amperometry(=amperometric titration) 电流滴定(法)
amperometry with two indicator electrodes 双指示电极电流滴定法
amperostatic coulometry 恒电流库仑法
amphetamine 苯异丙胺；1-苯基-2-丙胺
amphetamine sulfate 硫酸苯异丙胺
amphi- 〔希腊字头〕①对；双；两②指萘环的2,6位
amphibole 闪石
amphibole lattice 闪石晶格；闪石点阵
amphibole-like strand 闪石状束
amphibolite 闪岩
amphichiralty 兼手性
amphi-form 两侧位式
amphigene(=leucite) 白榴石
amphion 两性离子
amphipathic 两亲的〔指亲水又亲油〕
amphipathic adsorption 两亲吸附作用
amphipathic compound 两亲化合物
amphipathic property 两亲性；亲水亲油性
amphipathic solute 两亲溶质
amphipathy 两亲性；亲水亲油性
amphipatic 两亲的

amphiphathic fluorescent dye 两亲性荧光染料
amphiphatic 两亲(溶)性的〔兼有亲水性和亲油性〕
amphiphile 两亲物
amphiphilic 两亲的
amphiphilic block copolymer 两亲(性)嵌段共聚物
amphiphilic graft copolymer 两亲接枝聚合物
amphiphilic polymer 两亲聚合物
amphiphilic surfactant 两亲型表面活性剂
amphipitic 两亲的
amphi position 远位*
amphiprotic 两性的
amphiprotic compound(=amphoteric substance) 两性化合物
amphiprotic solvent 两性溶剂
ampholine electrophoresis 等电聚焦；两性电解质电泳
ampholyte 两性物*
ampholytic 两性的
ampholytic quaternary ammonium compound 两性季铵化合物
ampholytics(=ampholytic surface active agent) 两性表面活性剂
ampholytic surface active agent(=ampholytics) 两性表面活性剂
ampholytic surfactant 两性表面活性剂
ampholytoid(=amphoteric colloid) 两性胶体
amphomycin(=amfomycin) 安福霉素
amphophic 双亲和性的；双嗜性的
amphophile 双嗜(性)物；双亲和体
ampho-surfactant 两性表面活性剂
amphotere 两性元素
amphoteric 两性的
amphoteric ammonioamidate 两性氨(盐)酰胺化物
amphoteric character 两性性质
amphoteric colloid(=amphotytoid) 两性胶体
amphoteric electrolyte 两性电解质
amphoteric emulsifier 两性乳化剂
amphoteric fat liquor 两性加油乳液
amphoteric fatty compound 两性脂肪族化合物
amphoteric fluoro aliphatic compound 两性氟代脂肪(族)化合物
amphoteric hydroxide 两性氢氧化物
amphoteric imidazoline 两性咪唑啉
amphotericin 两性霉素
amphotericin B(=fungizone) 两性霉素B
amphoteric ion 两性离子
amphoteric ion exchange fibre 两性离子交换纤维
amphoteric ion exchange resin 两性离子交换树脂
amphoteric ionic surfactant 两性离子型表面活性剂

amphoteric membrane 两性膜
amphoteric oxide 两性氧化物
amphoteric polyelectrolyte 两性聚合电解质
amphoteric properties 两性性质
amphoteric quaternary ammonium compound 两性季铵化合物
amphoteric reaction 两性反应
amphoteric resin 两性树脂
amphoterics(=amphoteric surface-active agent) 两性表面活性剂
amphoteric solvent 两性溶剂
amphoteric substance(=amphiprotic compound) 两性化合物
amphoteric surface-active agent(=amphoterics) 两性表面活性剂
amphoteric surfactant 两性表面活性剂
amphoteric tannins 两性鞣质
amphoterisation 两性化(作用)
amphoterism 两性现象
amphotropine 安福托品〔药〕
amphotytoid 两性胶体
amphtericin B 二性霉素
amphyl 酚衍生物
ampicillin ω-氨基苄青霉素; 氨苄西林〔药〕
ampicillin and sulbactam 氨苄西林舒巴坦〔药〕
amplidyne 电机放大器; 交磁放大机
amplification 放大(作用)
amplification factor 放大因子
amplification ratio 放大比
amplifier 放大器
amplifier comparator 放大比较器
amplifier drift tester 放大器漂移测定器
amplifying tube 放大管
amplitude ①幅②振幅
amplitude analysis 振幅分析
amplitude limiter 限幅器
amplitude modulation 振幅调制
amplitude of the square wave voltage 方波电压振幅
amplitude ratio 振幅比
ampoul(e) ①安瓿; 针药管②安瓿剂; 壶药③壶腹〔药〕
ampoule filler 安瓿装药机; 针药封装机
ampulla cap 安瓿瓶胶帽
amrad gum 象苹果胶; 阿姆拉特树胶
amsacrine 安吖啶〔药〕
amurensin 黄柏苷; 脱氢黄柏苷
amurine 黑龙江罂粟碱
amycin 阿霉素
amygdala 扁桃; 杏仁核
amygdalate 扁桃酸盐(或酯) $C_6H_5CHOHCOOM$; $C_6H_5CHOHCOOR$
amygdalic acid 扁桃酸; 苯乙醇酸 $C_6H_5CH(OH)COOH$
amygdalin 扁桃苷; 苦杏仁苷
amygdalinic acid 扁桃酸; 苯乙醇酸 $C_6H_5CH(OH)COOH$
amygdaloid 扁桃状的
amygdalose 苦杏仁糖
amygdophenin 阿米多菲宁〔药〕
n-amyl (正)戊基 $CH_3(CH_2)_4-$
neo-amyl 新戊基 $(CH_3)_3CCH_2-$
sec-amyl 仲戊基 $CH_3CH_2CH_2CH(CH_3)-$
tert-amyl(=1,1-dimethylpropyl) 叔戊基 $CH_3CH_2C(CH_3)_2-$
amylaceous 淀粉的
amyl acetate 乙酸戊酯; 香蕉油〔俗〕
n-amyl acetate 乙酸(正)戊酯 $CH_3CO_2C_5H_{11}$
sec-amyl acetate 乙酸仲戊酯 $CH_3CO_2CH(CH_3)CH_2C_2H_5$
tert-amyl acetate 乙酸叔戊酯 $CH_3CO_2C(CH_3)_2C_2H_5$
amyl acetic ester 乙酸戊酯 $CH_3COOC_5H_{11}$
amylacetylene 戊基乙炔; 庚炔 $CH_3(CH_2)_4C\equiv CH$
amylaceum 葡萄糖
amyl alcohol 戊醇
n-amyl alcohol(=1-pentanol) (正)戊醇
sec-amyl alcohol(=2-pentanol) 仲戊醇
tert-amyl alcohol 叔戊醇
n-amylamine (正)戊胺 $CH_3(CH_2)_4NH_2$
tert-amylamine 叔戊胺
p-tert-amyl aniline 对叔戊基苯胺 $(C_2H_5)(CH_3)_2CC_6H_4NH_2$
amyl anisate 大茴香酸戊酯
n-amyl anthranilate 邻氨基苯甲酸戊酯
amylase 淀粉酶
amylbenzene 戊基苯〔通指 n 即正戊基苯〕
n-amylbenzene(=1-phenylpentane) (正)戊基苯; 1-苯基戊烷 $C_6H_5(CH_2)_4CH_3$
tert-amylbenzene 叔戊基苯 $C_6H_5C(CH_3)_2(C_2H_5)$
amyl benzoate 苯甲酸戊酯
amyl benzyl ether 戊基苄醚
amyl borate 硼酸戊酯 $B(OC_5H_{11})_3$
amyl boric acid 戊基硼酸 $C_5H_{11}B(OH)_2$
amyl boron dihydroxide(=amyl boric acid) 戊基硼酸
n-amyl bromide (正)戊基溴 $CH_3(CH_2)_3CH_2Br$
tert-amyl bromide 叔戊基溴 $(CH_3)_2C(Br)C_2H_5$
amyl butyrate 丁酸戊酯 $C_3H_7CO_2(CH_2)_4CH_3$
tert-amyl butyrate 丁酸叔戊酯 $C_3H_7CO_2C(CH_3)_2C_2H_5$
amyl butyric ester(=amyl butyrate) 丁酸戊酯

amyl caproate 己酸戊酯
amyl-*n*-caproate 己酸(正)戊酯 $C_5H_{11}CO_2C_5H_{11}$
amyl caprylate 辛酸戊酯 $CH_3(CH_2)_6CO_2C_5H_{11}$
amyl carbamate 氨基甲酸戊酯 $H_2NCO_2C_5H_{11}$
tert-amyl carbamate 氨基甲酸叔戊酯
$$H_2NCO_2C(CH_3)_2C_2H_5$$
amyl carbonate 碳酸戊酯 $CO(OC_5H_{11})_2$
amyl cellulose 戊基纤维素
amyl chloride 戊基氯 $C_5H_{11}Cl$
n-amylchloride(=1-chloropentane) (正)戊基氯；1-氯戊烷
$$CH_3(CH_2)_3CH_2Cl$$
tert-amylchloride 叔戊基氯 $(CH_3)_2CClC_2H_5$
n-amyl chloroacetate 氯乙酸戊酯 $ClCH_2COOC_5H_{11}$
amyl chlorocarbonate 氯甲酸戊酯 $C_5H_{11}OCOCl$
amyl chloronaphthalene 戊基氯萘〔溶剂〕
amyl cinnamaldehyde α-戊基肉桂醛
$$C_6H_5CH=C(C_5H_{11})CHO$$
amyl cinnamate (肉)桂酸戊酯
n-amylcinnamic alcohol 正戊基肉桂醇
amyl cinnamic aldehyde 戊基肉桂醛；素馨醛；茉莉醛〔俗〕
α-amyl cinnamic aldehyde dimethyl acetal α-戊基(肉)桂醛二甲缩醛
α-amylcinnamyl acetate 乙酸-α-戊基肉桂酯
α-amylcinnamyl formate 甲酸-α-戊基肉桂酯
α-amylcinnamyl isovalerate 异戊酸-α-戊基肉桂酯
amyl citrate(=triamyl citrate) 柠檬酸戊酯
amylcose acetate 乙酸淀粉
amyl cyanate 氰酸戊酯 $C_5H_{11}OCN$
amyl cyanide(=capronitrile) 己腈；戊基氰
$$CH_3(CH_2)_4CN$$
amyl cyclopentenone 戊基环戊烯酮
amyl decyl phthalate 邻苯二甲酸戊癸酯
amyl dimethyl para-aminobenzoic acid 二甲基对-氨基苯甲酸戊酯
amylene ①戊烯 $CH_3CH_2CH_2CH=CH_2$②〔在复合名中指〕1,5-亚戊基 —C_5H_{10}—
amylene alcohol 叔戊醇 $(CH_3)_2COHCH_2CH_3$
β-amylene-α-carboxylic acid β-戊烯-α-羧酸
amylene oxide 四氢吡喃；氧杂环己烷
amylene oxide ring 四氢吡喃环；氧杂环己烷环
amyl ester 戊酯
amyl ether ①戊基醚 ROC_5H_{11}②戊醚 $C_5H_{11}OC_5H_{11}$
amyl ethyl ketone 乙基戊基酮；3-辛酮 $C_2H_5COC_5H_{11}$
amyl fluoride 戊基氟 $C_5H_{11}F$
amyl formate(=amyl formic ester) 甲酸戊酯
amyl formic ester(=amyl formate) 甲酸戊酯
amyl furoate 糠酸戊酯；呋喃甲酸戊酯
$$C_4H_3OCOOC_5H_{11}$$

α-amyl-β-(α-furyl) acrolein α-戊基-β-(α-呋喃基)丙烯醛
amyl furylacrylate 呋喃基丙烯酸戊酯
amyl halide 戊基卤 $C_5H_{11}X$
amyl heptanoate 庚酸戊酯
n-amyl heptylate 庚酸戊酯
amyl hexanoate 己酸戊酯
amyl hydrogen sulfate 硫酸氢戊酯 $C_5H_{11}OSO_3H$
amyl hydroxide 戊醇 $C_5H_{11}OH$
amylidene(=pentylidene) 亚戊基
amylin 糊精；淀粉不溶素
amyline 淀粉粒的纤维素膜
n-amyl iodide(=1-iodopentane) (正)戊基碘；1-碘代戊烷
$$CH_3(CH_2)_3CH_2I$$
sec-amyl iodide(=2-iodopentane) 仲戊基碘；2-碘代戊烷
$$C_3H_7CHICH_3$$
tert-amyl iodide 叔戊基碘 $(CH_3)_2CIC_2H_5$
amylis 戊基〔amyl 的拉丁名称〕
amyl isorhodanate(=amyl isothiocyanate) 异硫氰酸戊酯
amyl isothiocyanide(=amyl isothiocyanate) 异硫氰酸戊酯
n-amyl isovalerate 异戊酸戊酯
amylit 麦芽糖化酶
n-amyl ketone 二(正)戊基(甲)酮；均十一酮 $(C_5H_{11})_2CO$
amyl lactate 乳酸戊酯
amyl laurate 月桂酸戊酯；十二烷酸戊酯
n-amyl malonic acid (正)戊基丙二酸
$$C_5H_{11}CH(CO_2H)_2$$
n-amyl mercaptan (正)戊硫醇 $CH_3(CH_2)_3CH_2SH$
n-amylmercuric chloride 氯化(正)戊基汞 $C_5H_{11}ClHg$
amylmercuric cyanide 氰化戊基汞
$$CH_3(CH_2)_3CH_2HgCN$$
amylmercuric iodide 碘化戊基汞 $CH_3(CH_2)_3CH_2HgI$
amyl-4-methoxycinnamate 4-甲氧基肉桂酸戊酯
n-amyl methyl carbinol 正戊基甲基醇
2-*n*-amyl-3-methyl-cyclopentanone 2-戊基-3-甲基环戊酮；四氢茉莉酮〔俗〕
2-*n*-amyl-3-methyl-Δ²-cyclopentenone 2-戊基-3-甲基环戊烯-2-酮；二氢茉莉酮〔俗〕
amyl methyl ketone 戊基甲基酮
amyl mustard oil(=amyl isothiocyanate) 戊基芥子油；异硫氰酸戊酯
amyl naphthalene 戊基萘〔溶剂〕
amyl nitrate 硝酸戊酯
amylnitrite 硝基戊烷
n-amyl nitrite 亚硝酸(正)戊酯 $C_5H_{11}ONO$
amylo- 淀粉的
amylo-cellulose 淀粉纤维素

amyl octanoate　辛酸戊酯
amyl octine carbonate　辛炔羧酸戊酯
amyl octylate　辛酸戊酯
amylodextrin　淀粉糊精
amyloform　淀粉仿；淀粉甲醛混合物
amylogen　可溶性淀粉
amylograph　（淀粉）黏焙力测量器
amylohemicellulose　淀粉半纤维素
amyloheptose　淀粉庚糖
amyloid　①淀粉状蛋白胶化纤维素②淀粉的③(硫酸)胶化纤维素
amyl oleate　油酸戊酯
amylolytic activity　淀粉分解力
amylolytic enzyme　淀粉(分解)酶
amylon　淀粉；糖原
amylopectase　支链淀粉酶
amylopectin　支链淀粉
amylose　直链淀粉
amylosynthease　淀粉合成酶
amyl oxalate　草酸戊酯
amyl oxide link(age)　戊氧键
amyloxy-isoeugenol　戊氧基异丁子香酚
n-amyl palmitate　棕榈酸戊酯；十六(烷)酸戊酯
amyl pelargonate　壬酸戊酯　$C_8H_{17}CO_2C_5H_{11}$
tert-amyl phenol　叔戊苯酚　$C_5H_{11}C_6H_4OH$
amyl phenylacetate　苯乙酸戊酯
amyl phenyl ether　戊基苯基醚；戊氧基苯　$CH_3(CH_2)_4OC_6H_5$
amyl phenyl ketone　戊基苯基酮；己酰苯　$C_5H_{11}COC_6H_5$
amyl-propiolaldehyde　戊基丙炔醛；2-辛炔醛　$C_5H_{11}C\equiv CCHO$
amyl propionate　丙酸戊酯
2-amyl pyridine　2-戊基吡啶
amyl pyromucate　焦黏酸戊酯　$C_4H_3OCO_2C_5H_{11}$
2-n-amylquinoline　2-正戊基喹啉
amyl rhodanate(=amyl thiocyanate)　硫氰酸戊酯
amyl salicylate　水杨酸戊酯；柳酸戊酯
amyl stearate　硬脂酸戊酯
amyl succinate　琥珀酸戊酯
amyl sulfocyanide(=amyl thiocyanate)　硫氰酸戊酯
amyl tartrate　酒石酸戊酯
amyl thiocyanate(=amyl thiocyanide; amyl sulfocyanide)　硫氰酸戊酯
amyl thiocyanide(=amyl thiocyanate)　硫氰酸戊酯
amyl thioglycollate　巯基乙酸戊酯〔抗氧剂〕
amyl triethyl silicane　戊代三乙硅　$C_5H_{11}Si(C_2H_5)_3$
amyltrimethoxysilane　戊基三甲氧基硅烷（偶联剂）

amyl trimethyl silicane　戊代三甲硅　$C_5H_{11}Si(CH_3)_3$
amylum(=corn starch)　(玉米)淀粉
n-amyl undecylenate　十一烯酸戊酯
tert-amyl urea　叔戊脲　$C_5H_{11}NHCONH_2$
amyl valerate　戊酸戊酯
n-amyl vinyl carbinol　戊基乙烯基原醇
amyl vinyl carbinyl acetate　乙酸戊基乙烯基甲醇酯
amyradiene　白檀二烯
amyranol　香树醇
α-amyrenol(=α-amyrin)　α-脂檀素
amyrenone(=amyrone)　白檀酮
α-amyrenone(=α-amyrone)　α-脂檀酮
amyrin　香树素
α-amyrin(=α-amyrenol)　α-脂檀素
amyris oil　香树油
amyrol　脂檀醇
amyrol acetate　乙酸脂檀酯
amyrolin　脂檀灵
amyrone(=amyrenone)　白檀酮
α-amyrone(=α-amyrenone)　α-脂檀酮
amytal(=amobarbita)　安密妥〔药〕；异戊巴比妥
ana-　〔希腊字头〕　萘环的1,5位
anabasine　假木贼碱；毒藜碱
anabolism　合成代谢
anabsinthin　苦艾质
anacardic acid　槚如酸；漆树酸　$C_6H_3(C_{15}H_{27})(COOH)(OH)$
anacardol　槚如酚
anaclastic　屈折的；由折射引起的
Anaconda-Trail process (for zinc extraction)　安那康达-特雷耳(提锌)法
anacycline　回环豆碱
anadonis green　氧化铬绿
anaerase　厌氧酶
anaerobe　厌氧菌；厌氧微生物
anaerobiase　厌氧分解酶
anaerobic　①厌氧的②绝氧的
anaerobic adhesive　厌氧胶黏剂；厌氧黏合剂
anaerobic biological treatment　厌氧生物处理
anaerobic degradation　厌气降解(作用)
anaerobic digestion　厌氧消化
anaerobic mastication　无氧塑炼
anaerobic organism　厌氧微生物
anaerobic phase　厌氧期；缺氧期
anaerobic polymerization　厌氧聚合
anaerobic sewage treatment　厌氧性污水处理
anaerobic treatment　厌气处理
anaerobic waste treatment　厌氧性废水处理

anaerobion 厌氧菌；厌氧微生物
anaesthesin(e)(=benzocaine) 阿奈西辛〔药〕；苯佐卡因
anaesthetic(=anesthetic) ①麻醉剂②麻醉的
anagen(e) 安纳晶；乙氧基苯酰胺基喹啉
anagyrine 臭豆碱；安纳吉碱
analcime(=analcite) 方沸石
analcite(=analcime) 方沸石；二水硅铝钠
analgen(e)(=8-ethoxy-5-benzamido quinoline) 安纳晶；8-乙氧基-5-苯酰氨基喹啉
analgesic 止痛药
analgesine(=antipyrine) 安替比林
analgetica 止痛药
analgin(=creolin) 诺瓦经〔药〕；安乃近
anallachrom(=esculetinic acid) 七叶亭酸
analog answer 模拟应答
analog chromatogram 模拟色谱(图)
analog electronic integration 模拟电子积分
analog information 模拟显示
analog pH meter 模拟 pH 计
analogs 类似物；同型物
analog signal 模拟信号
analog slope detector 模拟斜率检测器
analog switch 模拟开关
analog(ue) digital converter 模拟数字转换器
analog(ue) digital transformer 模拟数字转换器
analogue instrument 模拟仪表
analogue output scanner 模拟输出扫描器
analogue signal 模拟信号
analogue to digital conversion 模拟-数字转换
analogy 模拟；类拟
analogy model 模拟模型
analyser(=analyzer) 分析器；检偏振器
analyser cabinet 分析器柜
analyser case 分析器罩
analyser house 分析器室
analyser shelter 分析器棚
analyser tube 分析管
analyses of organochlorine residues 有机氯残留量分析
analyses of organophosphorus residues 有机磷残留量分析
analysin 枯草溶菌素
analysing crystal 分光晶体
analysis by absorption of gases 气体吸收分析
analysis by elutriation 淘析法
analysis by radioactivity 放射分析
analysis error 分析误差
analysis formula 解析式
analysis line 分析线

analysis of covariance 协方差分析
analysis of end group 端基分析
analysis of exhaust gas 排气分析；废气分析
analysis of flue gas 烟气分析
analysis of ionic channel open and closed kinetics 离子通道开闭动力学分析
analysis of nuclear grade 核纯分析
analysis of original organism in body fluid 生物体液原态分析
analysis of variance 方差分析
analysis of variance table 方差分析表
analyst 分析员；化验员
analyte 分析物*
analyte subtraction method 分析物减量法
analytical 分析的
analytical approach 分析方法
analytical balance 分析天平
analytical biodevices 分析生物器件
analytical cell cavity 分析池腔
analytical chemicals 分析试剂；分析化学品
analytical chemistry 分析化学
analytical chromatograph 分析用色谱仪
analytical column 分析柱
analytical cycle 分析周期
analytical data 分析数据
analytical distillation 分析蒸馏
analytical extrapolation 解析外推法
analytical function 分析函数
analytical gap 分析间隙
analytical grade reagent 分析级试剂
analytical grade resin 分析用树脂；分析级树脂
analytical information 分析信息
analytical instrument for solid 固体成分(分析)仪
analytical isotachophoresis 等速电泳分析
analytical line 分析线*
analytical liquid chromatograph 分析(用)液相色谱仪
analytically 分析上
analytically pure 分析纯
analytical method standard 分析方法标准
analytical photometry 分析光度学(法)
analytical plate number 分析塔板数
analytical precision 分析精密度
analytical pure 分析纯(试剂)
analytical quality control 分析质量控制
analytical radiochemistry 分析放射化学
analytical range 分析范围
analytical reaction 分析反应
analytical reagent 分析试剂

analytical reagent grade　分析试剂级
analytical separation　分析分离
analytical signal　分析信号
analytical signal interpretation　分析信号解析
analytical signal processing　分析信号处理
analytical spectroscopy　分析光谱学
analytical standard　①分析标准②分析标准物
analytical system　分析系统
analytical type chromatograph　分析型色谱仪
analytical ultracentrifuge　分析超速离心机
analytical unit　分析单元；分析组件
analytical value　分析值
analytical weights　分析砝码
analytic factor　分析因数
analyticity　解析性；可分析性
analytic unit　分析单元；分析组件
analyzer　①分析器②检偏振器
analyzer magnet　分析器磁铁
analyzer pumping system　分析器抽气系统
analyzer tube　分析管
analyzing　①分析的②分析(工作)
analyzing crystal　分析晶体；分光晶体
anamirtin　①安那米亭②印度防己苷
anamorphism　合成变质；复化变质
anamorphoscope　歪像校正镜
anamorphosis　失真；变形
anapaite　斜磷钙铁矿
anaphoresis　阴离子电泳
anaphrodisiac　制欲剂
anaphylactic shock　过敏性休克
anaphylactin(=allergen)　过敏素
anaphylactogen　过敏原
anaphylatoxin　过敏毒素
anaphylaxis　过敏性反应
anaplerosis　回补；补充
anasthol　安那妥耳
anastigmatic　去像散的；正像的
anastomosing　①吻合的②接通的
anastomosis　联接作用
anastrozole　阿那曲唑〔药〕
anatabine　新烟草碱
anatase　锐钛矿
anatase pigmentary　颜料级锐钛型(二氧化钛)
anatase titanium dioxide　锐钛矿型钛白粉
anation　引入阴离子作用
anatoxin　变性毒素
anauxite　蠕陶土；富硅高岭石
anayodin　安纳尤廷〔药〕；安痢生

anchietine　堇根碱〔从巴西植物囊果堇属植物中提取的〕
anchimeric assistance(=neighbouring group participation)　邻位促进；邻位协助
anchoic acid　壬二酸　$COOH(CH_2)_7COOH$
anchor　①锚；锚固；锚地②(底材表面)粗糙性；锚刺〔增加涂层的附着力〕
anchorage　固着；固定
anchor agitator　锚式搅拌器
anchor bolt　地脚螺栓
anchor catalyst　锚定催化剂
anchor coat　结合层；打底胶浆；初层
anchored compound　黏附的化合物
anchor effect　固着效果
anchoring agent　结合剂；增黏剂
anchoring group　结合团；锚定基团
anchoring material　结合剂；增黏剂
anchoring spurs　着力爪
anchoring strength　结合强度；黏附力
anchor mixer　锚式搅拌器
anchor paddle mixer　锚桨式搅拌器
anchor stirrer(=paddle stirrer)　桨式搅拌器
anchovy oil　鲟鱼油；小鳟油
anchusic acid　安刍酸
anchusin　安刍素
ancillary　①助剂；添加剂②辅助的
ancillary equipment　辅助装置
ancillary test　辅助试验
ancymidol　嘧啶醇
andalusite　红柱石
Anderson's disease　遗传性肝糖储积病〔由于分支酶淀粉-1,4-1,6转葡萄糖苷酶缺乏引起的疾病〕
andesine　中长石
andesite　安山岩
andirin　①莱棕皮黄　$C_{40}H_{43}O_3N$②腊胆尼根苷
andogenin　桉豆配基
andorite　硫锑铅银矿
Andrade's equation　安德雷德方程〔描述聚合物溶液中温度对黏度的影响〕
Andrade's viscosity formula　安德雷德黏度(公)式
andradite　钙铁榴石
Andreasen pipet method　安德列森沉降管粒度分级法
androgen(=male hormone)　雄性激素
androgenic hormone　雄性激素
andrographolic acid　雄茸脑酸
andrographolide　雄茸交酯
androkin　雄甾酮
androl　雄醇

andromedotoxine 梫木毒(素)
andrometoxin 石南毒素
androsin 雄素
androst- 雄(甾)
androstane 雄(甾)烷
androstanediol(=dihydroandrosterone) 雄(甾)烷二醇；二氢雄(甾)酮
androstanedione 雄(甾)二酮
3-androstaneone 3-雄(甾)烷酮
androstanol 雄(甾)烷醇
androstene 雄(甾)烯
androstenediol 雄(甾)烯二醇
1,2-androstenedione 1,2-雄烯二酮
androstenone 雄(甾)烯酮
androsterone(=androtin) 雄(甾)酮
androtermone 雄定酮；(单)衣藻定雄性素
androtin(=androsterone) 雄(甾)酮
Andrussow process 安德卢梭法〔以甲烷、氨、空气为原料催化合成氰化氢法〕
-ane〔词尾〕 烷
anechoic room tyre noise test 消音室轮胎噪声试验
anelastic creep 滞弹性蠕变
anelastic deformation 滞弹性形变
anelasticity ①内摩擦力②滞弹性
anelastic spectrum 滞弹性谱
anelectrolyte 非电解质
anemodispersibility 风力分散率
anemometer(=air meter) 风速计；气流计
anemone camphor 银莲花脑
anemonic acid 银莲花酸
anemonin 银莲花素；白头翁素
anemoninic acid 银莲花裔酸
anemonsite 混合长石
anemostat 气流稳定器
aneroid ①无液的②无液气压表
aneroid barometer 无液气压计
aneroid mixture control 燃烧混合物组织高度校正
anesthesin(=ethyl aminobenzoate) 阿奈西辛〔药〕；苯唑卡因
anesthesiophore radical 麻醉基；苯甲酰基〔致麻醉效应的基〕C_6H_5CO
anesthetic(=anaesthetic) ①麻醉剂②麻醉的
anesthetic ether 麻醉(用)醚
anesthetic steroid compound 麻醉性甾族化合物
anethene 茴香烯；莳萝烯
anethole 茴香脑；对丙烯基茴香醚
anethol trithione 茴三硫〔药〕
anethoqinine 茴香奎宁

aneurinase 硫胺素酶；硫胺分解酶
aneurin(e)(=thiamine;vitamin B_1) 抗神经炎素；盐酸硫胺素；维生素 B_1
anfleurage 脂提法〔提取花中香精的一种方法〕
angel's hair ①金银丝〔织造用〕②玻璃(纤维)丝③假发
angelate 当归酸酯
angelica acid 当归酸
angelica herb oil 当归草油
angelical 当归醛
angelica lactone(=angelic lactone) 当归内酯
angelica oil 当归油
angelica root oil 当归根油
angelica seed oil 当归子油
angelicic acid 当归酸
angelicin 当归根素
angelic lactone(=angelica lactone) 当归内酯
angelicone 当归酮
angelol 当归醇
angeloylzygadenine 当归酰棋盘花碱
angel's hair ①金银丝〔织造用〕②玻璃(纤维)丝③假发
Angelus still 安吉拉斯蒸馏器
angico gum 巴西树胶
angioedema 血管神经性水肿
angioneurosin(=nitro-glycerin) 硝化甘油
angioplast 原生质体；成血管细胞
angle ①角②角铁
angle abrader 定角磨耗试验机；角式磨耗试验机
angle beam probe 斜探头
angle beam searching unit 斜探头
angle block gauge 量角规；角度块规
angle branch 弯管；肘管
angle break 粉碎角
angle centrifuge 斜角离心机
angle (crossflow) nebulizer 直角(型)雾化器
angled flash line 斜向溢料缝
angle dispersive electron spectrometer 角色散电子能谱仪
angled reed 斜齿箱〔预取向丝整经拉伸用〕
angled twill 山形斜纹
angle extrusion head 斜角压出机头；Y形机头
angle gauge 角规
angle gauge block 量出规；角度块规
angle grain 斜纹
angle guide pin 定角位销〔压模〕
angle head 角头；和压出机螺杆成角度的压出机头
angle iron 角铁
angle joint 角接接头
angle mold 角式压模；V形压模

angle molding press　直角型平板机
angle of attack　①冲角②作用角
angle of bend　挠曲角
angle of bite　挟角
angle of clearance　间隙角
angle of contact　接触角；包角
angle of crater　自然倾角
angle of deflection　偏转角
angle of deviation　偏向角
angle of diffraction　衍射角
angle of dip　入射角
angle of discharge　出口角
angle of dispersion　散射角
angle of drop　坠落角
angle of emergence　出射角
angle of extinction　消光角
angle of flexure　屈挠角度
angle of free convergence　自由收敛角
angle of friction　摩擦角
angle of incidence　入射角
angle of inclination　倾角
angle of internal friction　内摩擦角
angle of isocline　①消光角补角②等斜角
angle of lag　落后角〔电〕
angle of lay　布线角度
angle of lead　超前角
angle of molecular orientation　分子取向角
angle of nip　挟角〔两滚间的〕
angle of oscillation　摆动角
angle of pipe　弯管；曲管
angle of pitch　曲距角
angle of polarization　极化角
angle of pressure　压力角
angle of reflection　反射角
angle of refraction　折射角
angle of repose(=angle of rest)　静止角
angle of rest(=angle of repose)　静止角
angle of rotation　旋光角
angle of scattering　散射角
angle of shearing resistance　内摩擦角
angle of slide　滑动角
angle of tear　撕裂角
angle of the V-belt　三角带角度
angle of the V-groove　轮槽角度
angle of tilt　倾斜角
angle of torsion　扭转角
angle of valence　价角
angle of valve seat　阀座角度
angle of view(ing)　视角
angle of weave　编织角度
angle of yarn delivery　导纱角
angle of yarn pull　卷取角
angle plate　角板
angle ply laminate　角铺设层合板
angle press(=side angle press)　角(式)压机
angle reciprocating compressor　角式往复压缩机
angle-resolved secondary ion mass spectrometer　角分辨二次离子质谱仪
angle resolved ultraviolet photoelectron spectroscopy　角分辨紫外光电子能谱法
angle rotor countercurrent chromatograph　倾斜角转子逆流色谱仪
angle section　角型材
anglesite　铅矾；硫酸铅矿
angle steel　角钢
angle strain　斜角应变
angle tear　直角撕裂
angle tear strength　直角撕裂强度
angle tear test　直角撕裂试验
angle test piece　直角形试片
angle thermometer　L型温度计；直角温度计
angle valve　角阀；直角形气门嘴
angle varied XPS depth profile analysis　变角XPS深度剖析法
angle-velocity-flux-contour map　角-速度-等流线图*
anglo-continental worsted system　混合式纺纱系统
ango(la) copalolic acid　安哥拉油酸　$C_{23}H_{36}O_3$
angola weed　纺锤染料衣
angolensin　安哥拉紫檀素
angostura　安果斯都拉树皮；南美芸香树皮〔苦味剂〕
angostura alkaloids　安果斯都拉生物碱类
angostura bark oil　安古树皮油
angostura extract　安古树浸提物
angostura oil　安果斯都拉树油
angostura tincture　安古树酊
angosturine　安果斯都拉树皮碱
angstrom　埃〔$=10^{-8}$厘米〕
angular acceleration　角加速度
angular anomaly　角度异常
angular aperture　角孔径
angular-condensed rings　角稠环
angular cut　①斜切口②斜切
angular deflection　角向挠度
angular degree　角度
angular difference　角差
angular dispersion　角色散

angular displacement 角位移
angular distribution 角分布
angular extrapolation light-scattering method 角外推散光法
angular (extruder) head 斜角压出机头；Y 形机头
angular focusing 角聚焦
angular force 旋转力；角力
angular frequency 角频(率)
angular gloss(=sheen) 斜向光泽度
angular head extruder 斜角压出机
angularity ①尖；棱角②成角度；有角性③弯曲度④斜度；曲率
angular methyl 角(上)甲基
angular methylation 角(上)甲基化(作用)
angular methyl group 角甲基
angular misalignment 管子接偏
angular misalignment loss 角向未对准损耗
angular momentum 角动量；动量矩
angular motion 角运动
angular normal stress 角向正应力
angular orientation 角取向；斜向取向
angular overlap model 角重叠模型
angular quantum number 角量子数
angular resolved photoemission electron spectroscopy 角分辨光电子谱
angular rigidity(=torsional rigidity) 角刚度
angular sheen 糙面光泽；掠角光泽
angular spread 角展度
angular spread of the ion beam 离子束的角发散
angular strain(=shear strain) 角应变
angular substituent 角取代基
angular substituted 角上取代的
angular test 屈挠试验
angular velocity 角速度
angular vibration frequency 角向振(动)频(率)
angustifoliol 安古树酚
angustione 安古树酮
anhalamine 无盐掌胺
anhalidine 无盐掌里定
anhalinine 无盐掌里宁
anhalonidine 无盐掌定
anhalonine 无盐掌宁
anhalonium alkaloids 仙人掌生物碱类
anharmonic coupling constant 非谐性偶合常数
anharmonicity constant 非谐性常数
anharmonicity effect 非谐性效应
anharmonic vibration 非谐振动
anhedritite(=anhydrite) 硬石膏；无水石膏

anhidrotic 止汗剂
anhydrase 脱水酶
anhydrating agent 脱水剂
anhydration ①脱水②干化
anhydride 酐
anhydride curing agent 酸酐固化剂
anhydride of santonic acid(=santonin) 山道酸酐；山道年
anhydridization 酐化
anhydrite(=anhedritite) 硬石膏；无水石膏
anhydro- 脱水
anhydrobase 脱水基质
anhydrocamphoronic acid 脱水樟脑酮酸
anhydro cellobiose 脱水纤维二糖
anhydro cellulose 脱水纤维素
anhydro-dihydrolycopodine 脱水二氢石松碱
anhydroecgonine(=ecgonidine) 芽子定；脱水芽子碱
anhydroecgonine hydrochloride 脱水芽子碱盐酸盐 $C_9H_{13}O_2N \cdot HCl$
anhydroenneaheptitol 脱水壬七醇
anhydroformaldehyde aniline 三聚脱水甲醛合苯胺
anhydroformaldehyde-aniline condensate 脱水甲醛苯胺缩合物〔促进剂〕
anhydroformaldehyde-p-toluidine 脱水甲醛对甲苯胺〔促进剂〕
anhydroglu cochloral(=chloralose) 脱水葡(萄)糖缩氯醛
anhydroglucose 葡萄糖酐
anhydroglucose unit 葡萄糖残基；脱水葡萄糖结构单元
anhydrohexitol 失水己糖醇
anhydrohexose 脱水己糖
anhydroleucovorin 脱水甲酰四氢叶酸；次甲基四氢叶酸
anhydrone 无水高氯酸镁 $Mg(ClO_4)_2$
anhydropanaxadiol 脱水人参二醇
anhydrophenylosazone 脱水苯脎
anhydropolyol fatty acid ester 失水多元醇脂肪酸酯
anhydro ring (内)酐环
anhydro-sorbit 脱水山梨糖醇
anhydrosorbitol 失水山梨糖醇；山梨糖醇酐
anhydrosorbitol ester 无水山梨(糖)醇酯
anhydrosugar 脱水糖
anhydrosulfite 亚硫酸酯酐
anhydrosynthesis 缩水合成
anhydrous 无水的
anhydrous acid 无水酸
anhydrous adeps lanae 无水羊毛脂
anhydrous alcohol 无水乙醇
anhydrous alumina 无水氧化铝
anhydrous aluminium silicate 无水硅酸铝〔填料〕
anhydrous ammonia 无水氨

anhydrous borax 无水硼砂
anhydrous calcium sulfate 无水硫酸钙
anhydrous citric acid 无水柠檬酸
anhydrous dentifrice 无水洁齿用品
anhydrous hematite 无水赤铁矿；天然氧化铁红
anhydrous lanelin 无水羊毛脂
anhydrous soap 无水皂
anhydrous soda(=anhydrous sodium carbonate) 无水碳酸钠
anhydrous sodium carbonate(=anhydrous soda) 无水碳酸钠
anhydrous sodium potassium aluminum silicate 无水硅酸钠钾铝；霞石正长岩〔填料〕
anhydrous solvent 无水溶剂
anhydrous vermiculite 无水蛭石
anhydrovitamin A(=axerophthene) 脱水维生素 A
anibine 蔷薇木碱
anil〔复合词中用〕 缩苯胺〔通指某醛或酮缩苯胺〕
anil formation 缩苯胺形成作用
anilide N-(某)酰苯胺
anilidothiobiazole 1,2,3-噻二唑
anilinate 苯胺金属 $NH_2C_6H_4M$
aniline 苯胺 $C_6H_5NH_2$
aniline acetate 乙酸苯胺 $C_6H_5NH_2CH_3OOH$
2-aniline-5-aminobezene sulfonic acid 2-苯胺基-5-氨基苯磺酸
aniline black 苯胺黑；颜料黑
aniline black dyestuffs 苯胺黑染料；苯胺黑类染料
aniline blue 苯胺蓝；水溶对氮蒽蓝 $C_{37}H_{29}N_3 \cdot HCl$
aniline cloud point 苯胺(浊)点
aniline colorimetric method 苯胺比色法
aniline colors 苯胺染料
aniline dye 苯胺染料
aniline dyeing 苯胺染料染色(法)
aniline dyestuff 苯胺染料
aniline equivalent(=aniline number) 苯胺当量；苯胺数〔燃料爆震稳定性的一种指标〕
aniline finish 苯胺涂饰剂
aniline finishing 苯胺涂饰
aniline formaldehyde resin 苯胺甲醛树脂
aniline gravity constant 苯胺比重常数
aniline green 苯胺绿
aniline hydrobromide 氢溴酸苯胺 $C_6H_5NH_2 \cdot HBr$
aniline hydrochloride 盐酸苯胺 $C_6H_5NH_2 \cdot HCl$
aniline leather 苯胺革；清光革
aniline nitrate 硝酸苯胺
aniline number 苯胺数
aniline oil 苯胺油

aniline orange 苯胺橙
aniline point 苯胺点
aniline printing(=flexography) 苯胺印刷；弹性版印刷
aniline process 苯胺法
aniline purple(=mauvein) 苯胺紫
aniline raffinate 苯胺精炼物
aniline red(=fuchsin) 苯胺红
aniline resin 苯胺树脂
aniline salt 苯胺盐〔通常专指：盐酸苯胺 $C_6H_5H_2 \cdot HCl$〕
aniline sulfate 硫酸苯胺
aniline sulfonic acid 氨基苯磺酸
aniline test apparatus 苯胺点测定仪
aniline violet 苯胺紫
aniline yellow 苯胺黄
anilinium decyl sulfate 癸基苯胺硫酸盐
anilinium ion 苯胺离子 $C_6H_5NH_3^+$
anilino- 苯胺基 $C_6H_5NH—$
β-anilinoethanol(=ethoxylaniline) 羟乙基苯胺；苯氨基乙醇 $C_6H_5NH(C_2H_4OH)$
anilino naphthalene sulfonate 苯胺基萘磺酸盐
anilinonaphthalene sulfonic acid 苯胺基萘磺酸
anilinoplast 苯胺塑料
anilipyrine 安替比林乙酰苯胺；阿尼利匹林〔药〕
anilism 苯胺中毒
anilite 安尼炸药〔液态二氧化氮、汽油炸药〕
anilol 酒精苯胺混合物〔一种高辛烷值汽油的掺和组分〕
anils （醛或酮)缩苯胺；苯胺衍生物〔PhN＝型化合物〕
aniluvitonic acid(=methylcinchoninic acid) 甲基金鸡纳酸 $C_{11}H_9O_2N$
animal alkaloid 动物碱
animal amylase 动物淀粉酶
animal (base) adhesive 动物性黏合剂
animal black(=animal charcoal) 兽炭黑；兽炭；骨炭
animal calorimeter 动物量热器
animal charcoal(=animal black) 兽炭黑；兽炭；骨炭
animal cutting oil 动物切削油〔一种润滑油与动物或植物油的乳状液，用于金属切削〕
animal drawn vehicle tire 马车胎；兽力车胎
animal dyes 动物染料
animal fat 动物脂
animal fiber 动物纤维
animal fibril 动物纤维
animal filler 动物性填充剂
animal glue 动物胶
animal glue adhesive 动物胶黏合剂
animalization （纤维素纤维的)动物(性)化
animalized cellulose fibre 动物性化纤维素；羊毛化纤维素纤维

animalized fibre 动物性化纤维素纤维；羊毛化纤维素纤维
animalized viscose fiber 动物(性)化黏胶纤维；黏胶毛
animal lubricant 动物润滑剂
animal membrane 动物膜
animal nitrogenous fertilizer 动物氮肥料
animal note 动物(香料)香韵
animal nuance 动物香香调
animal oil 动物油
animal parchment 兽皮纸
animal protein factor(APF) 动物蛋白因子
animal resin 动物树脂
animal size 动物胶
animal tallow ①动物脂肪②牛脂；牛油
animal tar 动物焦油
animal wax 动物蜡
animal wool 动物毛
anime(=animi resin) 硬树脂
animikite 锑银矿
animi resin(=anime) 硬树脂
animol(=anymol) 苦槛兰醇
anion 阴离子
anion activator 阴离子活化剂
anion active agent 阴离子活性剂
anion active auxiliary 阴离子活性助剂
anion-active dye 阴离子活性染料
anion base 阴离子碱
anion chromatography 阴离子色谱
anion complex 阴离子络合物
anion defect 阴离子缺陷
anion discharge theory 阴离子放电说
anion exchange 阴离子交换
anion-exchange chromatography 阴离子交换色谱(法)
anion-exchange column 阴离子交换柱
anion exchange electrode 阴离子交换电极
anion-exchange fibre 阴离子交换纤维
anion-exchange membrane 阴离子交换膜
anion-exchange packing 阴离子交换填充物
anion exchanger 阴离子交换剂
anion exchange resin(=anion resin) 阴离子交换树脂
anion gap 阴离子隙
anionic 阴离子的
anionic acid 阴离子酸
anionic agent 阴离子剂
anionically polymerized polybutadiene 阴离子催化聚丁二烯；有规立构聚丁二烯
anionic auxiliary 阴离子助剂
anionic band 阴离子谱带
anionic carboxylic acid 阴离子羧酸
anionic catalyst 阴离子催化剂
anionic catalytic polymerization 阴离子(催化)聚合(作用)
anionic-cationic titration 阴离子-阳离子(表面活性剂)滴定法
anionic chelation complex 阴离子螯合(配合)物
anionic cleavage 阴离子裂解
anionic coordination mechanism 阴离子配位机理
anionic copolymerization 阴离子共聚合(作用)
anionic cycloaddition 阴离子环加成
anionic cyclopolymerization 阴离子环化聚合
anionic detergent 阴离子洗涤剂
anionic donor group 给电子阴离子基；阴离子型给电子基
anionic dye 阴离子染料
anionic dyestuff 阴离子染料
anionic electrochemical polymerization 负离子电化学聚合
anionic emulsifier 阴离子乳化剂
anionic exchange membrane 负离子交换膜
anionic fat liquor 阴离子(型)加油(或脂)乳液
anionic initiation 阴离子引发(作用)
anionic isomerization polymerization 负离子异构化聚合
anionic latex 阴离子胶乳
anionic membrane 阴离子膜
anionic oligoamide 阴离子低聚酰胺
anionic pigment dispersant 阴离子型颜料分散剂
anionic polymerization(=anionoid polymerization) 负离子聚合；阴离子聚合
anionic polyurethane dispersion 阴离子型聚氨酯分散体
anionic polyurethane ionomer 阴离子型聚氨酯离聚物
anionics(=anionic surfaceactive agent) 阴离子表面活性剂
anionic self-emulsifying polyurethane emulsion 阴离子自乳化聚氨酯乳液
anionic softening 阴离子软化
anionic surfaceactive agent(=anionics) 阴离子表面活性剂
anionic surface agent 阴离子表面活性剂
anionic surfactant 阴离子型表面活性剂*
anionic Ziegler-type catalyst 阴离子齐格勒型催化剂
anionite 阴离子交换剂
anion living polymer 阴离子活性聚合物
anion masking agent 阴离子掩蔽剂
anionoid 类阴离子
anionoid polymerization(=anionic polymerization) 负离子聚合；阴离子聚合
anionoid reagent 类阴离子试剂

anionoid substitution 类阴离子取代
anionotropic rearrangement 阴离子转移重排
anionotropy 负离子转移*；阴离子移变(现象)
anion-permeable membrane 阴离子透膜
anion radical initiator 负离子自由基引发剂；阴离子-游离基引发剂
anion-radicals 阴离子基
anion resin(=anion exchange resin) 阴离子交换树脂
anion selective electrode 阴离子选择电极
anion selective membrane electrode 阴离子选择性膜电极
anion semipermeable membrane 阴离子半透膜
anion surface active agent 阴离子表面活性剂
anion transport 阴离子运送；阴离子传送；阴离子迁移
aniontropic solvosystem 阴离子移变溶剂系
anion vacancy 阴离子空穴
anion valency 阴离子价
aniracetam 茴拉西坦〔药〕
anisacetone 对甲氧苯基茴丙酮 $CH_3OC_6H_4CH_2COCH_3$
anisal(=ar-methoxy-benzylidene) 甲氧苯亚甲基 $CH_3OC_6H_4CH=$
anisalcetone 大茴香叉丙酮
anisalcohol 茴香醇；对甲氧基苄醇
anisaldehyde 茴香醛；对甲氧苯甲醛
anisaldehyde diethyl acetal 大茴香醛二乙缩醛
anisaldehyde dimethyl acetal 大茴香醛二甲缩醛
anisaldoxime 茴香肟 $C_8H_9O_2N$
anisate 茴香酸盐 $CH_3OC_6H_4COOM$
anise(=anise fruit) 茴香实
anisealcohol 茴香醇；对甲氧基苄醇
anise camphor 茴香樟脑；茴香脑；对丙烯基苯甲醚
anised oil 茴香油
aniseed ①茴香子②大茴香；八角茴香
aniseed oil 茴香油；大茴油
anise fruit(=anise) 茴香实
anise oil 茴香油
anise seed 茴香子
anise spirit 茴香精
anisette 茴香酒
anise water 茴香水
anisic acid 茴香酸；对甲氧基苯甲酸
anisic aldehyde 茴香醛
anisidine(=aminoanisole) 茴香胺；甲氧基苯胺；氨基苯甲醚 $CH_3OC_6H_4NH_2$
anisidine blue 茴香胺蓝
anisidine sulfonate 茴香胺磺酸盐；甲氧苯胺磺酸盐
anisidino- 茴香氨基；甲氧苯氨基
anisil 茴香偶酰；联茴香酰

anisilic acid 茴香醇酸
anisimal〔商〕 乙酸-2-苯基-2-丙烯酯
anisine 茴香碱
anisketone(=anisyl methyl ketone) 大茴香基甲酮；对甲氧基苯乙酮
aniso-〔词头〕 不同；不等；非等同
anisodamine 山莨菪碱〔药〕
anisodesmic structure 异键结构
anisodiametrical 不等直径的
anisodimensional particle 不对称形粒子
anisokinetic sampling 非等速取样
anisole 茴香醚；苯甲醚；甲氧基苯
anisolesulfonphthalein 茴香磺酞
anisomeric 非异构的
anisometric 不等轴的
anisometric micelle 不等轴胶束
anisometry 不等轴；不等容
anisomorphous 类质异晶型的
anisomycin(=flagecidin) 杀鞭菌素；茴香霉素
anisonitrile 茴香腈 $CH_3OC_6H_4CN$
anisotonic 非等渗的；异渗的
anisotopic element 无同位素的元素
anisotropic(=aeolotropic; anisotropical) 各向异性的
anisotropic absorption 各向异性吸收
anisotropic astigmation 异向性像散
anisotropic body 各向异性体
anisotropic composite material 各向异性复合材料
anisotropic dope 各向异性纺丝液
anisotropic filler 各向异性填料
anisotropic fluid 各向异性流体
anisotropic force system 各向异性力系
anisotropic glass-ceramic 各向异性玻璃陶瓷；非均质玻璃陶瓷
anisotropic hyperfine coupling constant 各向异性超精细耦合常数
anisotropic material 各向异性材料
anisotropic membrane 各向异性膜
anisotropic model filament 各向异性原型丝
anisotropic plate 各向异性板
anisotropic polymer 各向异性聚合物
anisotropic pyrolytic graphite 各向异性热解石墨
anisotropic refraction 各向异性折射
anisotropic relaxation 各向异性松弛
anisotropic solid 各向异性固体
anisotropic solution 各向异性溶液
anisotropic substance 各向异性物
anisotropic swelling 各向异性溶胀
anisotropic temperature factor 各向异性温度因子

anisotropic viscosity 各向异性黏性
anisotropisation 各向异性化作用
anisotropism(=anisotropy) 各向异性
anisotropy(=anisotropism) 各向异性
anisotropy turbulence 非均质性湍流
anisoxide 茴香素
anisoyl 茴香酰；甲氧苯(甲)酰
anisoyl chloride 茴香酰氯；对甲苯甲酰氯
anisyl 茴香基；大茴香基〔俗〕
anisyl acetate 茴香醇乙酸酯
anisyl acetone(=p-methoxy benzyl acetone) 大茴香基丙酮
anisyl alcohol(=p-methoxybenzyl alcohol) 茴香醇；对甲氧基苄醇；对甲氧基苯甲醇
anisyl aldehyde 大茴香醛
anisyl butyrate 丁酸茴香酯
anisyl chloride 茴香酰氯 $CH_3OC_6H_4COCl$
2-anisylethyl 2-甲氧苯乙基
anisyl formate 甲酸大茴香酯
anisylidene(=ar-methoxy-benzylidene) 亚茴香基；甲氧亚苄基
anisylidene acetone 大茴香叉丙酮
anisyl methyl ketone(=anisketone; p-methoxyacetophenone) 大茴香基甲酮；对甲氧苯乙酮
anisyl phenylacetate 苯乙酸大茴香酯
anitol(=anytol) 安尼妥
ankerite ①铁白云石②条状闪长岩
anklet fabric 后帮衬布
anklet rubber 后帮胶片
anlagen(=quinalgen) 奎纳昔；乙氧基苯甲酰氨基喹啉
annabergite 镍华 $3NiO·As_2O_5·8H_2O$
annaline (淀积)硫酸钙 $CaSO_4$
annato ①胭脂树②胭脂树萃〔用作染料〕
annatto(=annotta) 胭脂树红；果红〔胭脂树；用作染料，可染丝、棉和羊毛；用做食品着色剂〕
annatto shrub 胭脂树
anneal 回火；退火
annealed 退火的；退了火的
annealed bulk polymer 热処理过的块状聚合物
annealed density (薄膜)退火密度
annealed track 退火径迹；熟炼径迹
annealed zone 重结晶区
annealer ①缓冷器〔熔纺喷丝头下的加热套管；用以使初成形纤维缓慢冷却〕；退火炉②热処理器
annealing ①退火的②退火(作用)
annealing carbon 退火炭
annealing color 退火色
annealing device 缓冷装置；热処理装置
annealing effect 退火效应；热处理效应
annealing furnace 退火炉
annealing hearth 退火敞炉
annealing oil 退火油
annealing point(=annealing temperature) 退火温度；热处理温度
annealing process 退火过程；熟炼过程
annealing shrinkage ①退火收缩②退火收缩率
annealing temperature(=annealing point) 退火温度；热处理温度
annealing treatment 退火处理；热处理
annelation(=annulation) 成环(反应)
annelleted 稠和的
annerödite 黑铀铌钇矿
annidalin 碘化百里香酚
annihilation 消灭；消失；湮灭
annihilation operator 湮灭算符
annihilation photon 湮灭光子
annihilation radiation 湮没辐射
annihilation rays hypothesis 射线消失假说
annihilator(=extinguisher) ①灭火器②湮灭算符
annofoline 经年石松叶碱
annotine 经年石松碱
annotinine 经年石松宁
annotta(=annatto) 胭脂树红
annual cost 年度费用
annual layer 年轮层
annual overhaul 年度检修
annular ①环的②环形的
annular centrifugal contactor 环形离心萃取器
annular chamber 环形腔；环形室
annular column ①环形柱②环隙塔
annular die 环形模口
annular dissolver 环形溶解器
annular distance(=annular ring) 环孔
annular flow 环形流
annular fuel tank 环状油箱
annular furnace 环形炉
annular grid 环状栅
annular opening 圆孔
annular polymer distribution space 环形熔体分配空间〔在喷丝板上的空间〕
annular producer 环形发生炉
annular ribs 纵向花纹；顺条花纹〔轮胎〕
annular ring(=annular distance) 环孔
annular ring valve 环形阀；环状阀
annular section ①环形断面②环形级③环形部分
annular space 环隙

annular spinneret 环状喷丝板
annular tubes 套管
annular-type stripper 环型汽提器
annulation(=annelation) 成环(反应)
annulene (大环)轮烯
annulenone 环烯酮
annuloid 环状的
annuloline 安纽洛灵
annulus ①环形缝隙②内齿轮③环形套筒
annunciator 信号器；报警器
anodal 阳极的
anodal oxidation 阳极氧化
anode 阳极
anode active material 阳极活性物质
anode bag 阳极袋
anode bar 阳极棒
anode carrier (电镀)阳极支架
anode-cathode ratio 阳极-阴极比率
anode chamber(=anode compartment) 阳极室
anode coagulant dip process 阳极凝固浸渍法
anode compartment(=anode chamber) 阳极室
anode consumption 阳极消耗
anode corrosion efficiency 阳极侵蚀效率
anode current 阳极电流
anode drop 阳极(电势)降
anode effect 阳极效应
anode glow 阳极辉光
anode light 阳极射线
anode liquor 阳极(电解)液
anode mud 阳极泥
anode oxidation 阳极氧化
anode packing material 阳极包裹材料
anode potential 阳极电位
anode process 阳极法
anode ray 阳极射线
anode reaction 阳极反应
anode scrap 阳极屑
anode slime 阳极淀渣；阳极附着物
anode slime (sludge) 阳极泥
anode spot 阳极光点；阳极(斑)点
anodic 阳极的
anodically deposit 阳极沉积
anodic bonding 阳极键合
anodic-cathodic polarographic wave 阳-阴极极谱波
anodic-cathodic wave 换极连续(极谱)波
anodic corrosion 阳极腐蚀
anodic current 阳极电流
anodic current density 阳极电流密度

anodic current efficiency 阳极电流效率
anodic depolarization wave 阳极去极化波
anodic depolarizer 阳极去极剂
anodic deposition 阳极沉积
anodic diffusion current 阳极扩散电流
anodic dissolution current 阳极溶解电流
anodic dissolution wave 阳极溶解波
anodic electrochemical polymerization 阳极电化学聚合
anodic impurity 阳极杂质
anodic inhibitor 阳极(溶解)抑制剂；阳极钝化剂
anodic overvoltage 阳极过电压
anodic oxidation 阳极氧化
anodic passivity 阳极钝态
anodic polarization 阳极极化
anodic polishing 阳极(电)抛光
anodic process 阳极过程
anodic protection 阳极保护
anodic stripping 阳极溶出(分析)
anodic stripping voltammetry(ASV) 阳极溶出伏安法
anodic synthesis 阳极合成
anodic treatment 阳极处理
anodised aluminium 阳极氧化铝；阳极极化铝
anodization 阳极(氧)化
anodized finish 阳极化抛光
anodizing 阳极极化；阳极处理
anodizing process 阳极化过程；阳极处理程序
anodyne 止痛药；镇痛药
anodynin 安替比林〔药〕
anodynon 氯乙烷
anogen 对碘苯磺酸亚汞
p-anol(=p-propenylphenol) 对丙烯基苯酚
$CH_3CH=CHC_6H_4OH$
anol-anone mixture 环己醇-环己酮混合物
anolobine 番荔枝叶碱
anolone (环己)醇(环己)酮(混合物)
anolyte 阳极电解液
anolyte compartment 阳极液室
anomalous 反常的
anomalous dispersion 反常色散
anomalous electric absorption 反常电吸收
anomalous field 异常磁场；剩余磁场
anomalous flow property 反常流动性
anomalous glow discharge 异常辉光放电
anomalous magnetic moment 反常磁矩
anomalous mixed crystal 反常混晶
anomalous osmosis 反常渗透
anomalous refraction 反常折射
anomalous relaxation 反常松弛

anomalous scattering 反常散射*
anomalous scattering method 反常散射法
anomalous structure 反常结构
anomalous term 反常项
anomalous transmission technique 反常透射技术
anomalous transmission topography 异常透射形貌术
anomalous viscosity 反常黏度
anomalous water 反常水；超重水〔现已否定其存在〕
anomalous wave 反常波
anomalous weight 反常重量
anomalous Zeeman effect 反常塞曼效应
anomaly 反常
anomer 端基(差向)异构体*
anomeric carbon 异头碳
anomeric effect 端基异构效应
anomerization 正位异构化(作用)
anomers 正位(差向)异构体
anomite 褐云母
anonaceine 番荔枝碱
anonaine 番荔枝碱
anone 环己酮
anorthic(=triclinic) 三斜的
anorthic lattice 三斜晶格
anorthic system 三斜晶系
anorthite(=calciclase) 钙长石
anorthitite(=calciclasite) 钙长岩
anorthoclase 歪长石
anorthose 歪长岩
anorthosite 斜长岩
anorthospiral 平行螺旋
anosovite 黑钛石
anprolene(=ethylene oxide) 环氧乙烷
ansa compound 柄型化合物
ansa-metallocene 柄型茂金属
Anscuutz thermometer 安舒茨温度计
anserine 鹅肌肽
ansoalcohol 茴香醇；对甲氧基苄醇
ANS resin(=acrylonitrilestyrene resin) 丙烯腈-苯乙烯树脂
anstatic agent 抗静电剂
anstratene(=α-pinene) α-蒎烯
antacid ①解酸药②抗酸剂③抗酸的④解酸的
antacidine 葡萄糖二酸钙
antagonism ①拮抗作用②消效作用③反协同作用
antagonist 对抗物；拮抗物；反协同(试)剂
antagonistic 拮抗药
antagonistic action 对抗作用
antagonistic effect 拮抗效应；对抗效应；反协同效应

antagonist reaction 对抗(性)反应
antagonist surfactant 对抗表面活性剂〔测定表面活性剂用〕
antagonist titration 对抗滴定法
antalkaline ①解碱药②抗碱剂③抗碱的④解碱的
antarafacial 异面的；反面的
antarafacial component 异面组分
antarafacial process 异面过程
antarafacial reaction 异面反应
antazolin 安他唑啉〔药〕
anteiso- 反异
"ante-iso" acids "反-异"酸类
antelope finish 仿羚羊皮整理
antelope skin 羚羊皮
antemetic(=antiemetic) 止吐剂〔药〕
anterior lobe normone 前叶激素
anterior pituitary hormone （脑下）垂体前叶激素
anthanthrene 二苯并[cd, jk]芘
anthanthrone 二苯并[cd, jk]芘-5,10-二酮
anthanthrone orange-red 二苯并芘二酮橙红(橘红)
anthelmintic ①驱虫药；打虫药②驱虫的；打虫的
anthelmintic action 驱虫作用；打虫作用
anthelmycin 抗蠕霉素
anthelvencin 萎蠕菌素
anthemane 春黄菊烷
anthemene 春黄菊烯
anthemidine 春黄菊定
anthemis oil 春黄菊油
anthemol 春黄菊脑
anther 花药
antheraxanthin(=zeaxanthin epoxide) 花药黄质；表氧化玉米黄质
anthion 过二硫酸钾
anthocyanidin ①花色素②花青素
anthocyanin ①花色素苷②花青苷
anthocyanogen 花色素原；花色苷
anthophyllite 直闪石
anthoplearin 海葵素
anthopleurin 海葵素
anthorine A 雪上一枝蒿碱甲
anthosin 花精；安妥新〔碱性偶氮染料〕
anthosine lake 碱性偶氮色淀；安妥新色淀
anthoxanthin ①黄酮②花黄色素
anthra brilliant green 蒽素亮绿
anthracene 蒽
anthracene black 粗蒽炭黑
anthracene blue(=alizarin blue) 蒽蓝
anthracene brown〔商〕蒽棕〔直接蒽醌染料〕
anthracene carboxylic acid 蒽甲酸

anthracene crystal scintillation counter 蒽晶体闪烁计数器
anthracene dihydride 二氢化蒽 $C_{14}H_{12}$
anthracene green〔商〕 蒽绿
anthracene nucleus 蒽环
anthracene oil 蒽油
anthracene perhydride 全氢化蒽；蒽烷 $C_{14}H_{24}$
anthracene ring 蒽环
anthracene scavenger 蒽清除剂
anthracene sulfonic acid 蒽磺酸 $C_{14}H_9SO_3H$
anthracene tetrone(=anthradiquinone) 蒽二醌
anthracene violet 蒽紫
anthracene yellow 蒽黄
anthracenol(=anthrol) 蒽酚
anthracenone(=anthranone) 蒽酮
anthracidin 杀炭疽菌素
anthracite(=anthracite coal) 无烟煤
anthracite coal(=anthracite) 无烟煤
anthracite smalls〔复〕 无烟煤粉
anthracitic bitumite 无烟性烟煤
anthracitic coal 无烟(性)煤
anthracolite 沥青灰岩
anthra colors 蒽素染料〔还原蒽醌染料〕
anthraconite 黑方解石；黑沥青灰岩
anthra copper 蒽素铜
anthracyl 蒽基
anthracylene sulfide polymer 亚蒽基硫醚聚合物
anthradiamine 蒽二胺
anthradiol(=anthrahydroquinone) 蒽氢醌
anthradiquinone(=anthracene tetrone) 蒽二醌
anthraflavic acid 2,6-二羟蒽醌 $C_{14}H_6O_2(OH)_2$
anthragallol(=1,2,3-trihydroxyanthraquinone) 1,2,3-三羟基蒽醌；蒽棓酚 $C_6H_4(CO)_2C_6H(OH)_3$
anthraglucorhein 大黄苷
anthraglucosagradin 药鼠李皮苷
anthraglucosennin 番泻苷
anthraglycoside 蒽苷
anthrahydroquinone(=anthradiol) 蒽氢醌
anthrahydroquinone diacetate 蒽氢醌二乙酸酯
anthraldehyde 蒽(甲)醛
anthramine(=anthranylamine) 蒽胺
anthramycin 氨茴霉素
anthranic-N, N-diacetic acid 邻氨基苯甲酸-N,N-二乙酸
anthranilate 氨茴酸盐(或酯)；邻氨基苯甲酸盐(或酯)
anthranilic acid 邻氨基苯甲酸*
anthranilo 氨茴基〔俗〕；苯邻甲内酰氨基
anthranilo nitrile 氨基苯甲腈 $C_7H_6N_2$
anthraniloyl 氨茴酰〔俗〕；邻氨基苯甲酰 o-$H_2NC_6H_4CO-$
anthraniloylanthranilic acid 氨茴酰氨茴酸
anthranol 蒽酚
anthranol chrome blue 蒽酚铬蓝
anthranol colors 蒽酚染料〔酸性铬染料的一种商名〕
anthranone 蒽酮 $C_6H_4COC_6H_4CH_2$
anthranoyl(=anthraniloyl) 氨茴酰〔俗〕；邻氨基苯甲酰
anthranoyl-anthranilic acid 氨茴酰氨茴酸 $NH_2C_6H_4CONHC_6H_4CO_2H$
anthranylamine(=anthramine) 蒽胺
anthra-olive 蒽素橄榄绿
anthra printing black 蒽素印染黑
anthrapurpurin 蒽红紫
anthrapyrimidine yellow 蒽素嘧啶黄
anthraquinol 蒽二酚
anthraquinoline(=naphthoquinoline) 蒽喹啉
anthraquinone 蒽醌
anthraquinone acridine(=naphthacridinedione) 蒽醌吖啶
anthraquinone aldehyde 蒽醌甲醛
anthraquinone blue 蒽醌蓝
anthraquinone-2,6-disulfonic acid 蒽醌-2,6-二磺酸
anthraquinone dye 蒽醌染料
anthraquinone sulfonic acid 蒽醌磺酸
anthraquinonic acid 蒽醌酸；茜素
anthraquinonyl 蒽醌基 $C_{14}H_7O_2-$
anthrarobin 脱氧茜素
anthrarufine 蒽绛酚；1,5-二羟基蒽醌 $C_{14}H_6O_2(OH)_2$
anthrasol 溶蒽素
anthratetrol 蒽四酚
anthrathiazine 蒽并噻嗪
anthrathiophene 蒽并噻吩
anthratriol 蒽三酚
anthraxolite 碳沥青
anthrax-olive 蒽素橄榄绿
anthraxylon 纯木煤
anthrazine 二蒽并[1,2-1′,2′]哒嗪
anthr brilliant green 蒽素亮绿
anthrene blue 蒽烯蓝
anthrene colors 蒽烯染料〔还原染料的商名〕
anthrene dark blue 蒽烯暗蓝
anthrene golden orange 蒽烯金橙
anthrene jade green 蒽烯玉绿
anthrene red-violet 蒽烯红紫
anthrindan 环戊烷并蒽
anthrindandione 环戊烷并蒽醌
anthroic acid 蒽甲酸 $C_{14}H_9COOH$
anthrol 蒽酚

anthrone 蒽酮
anthrone agent 蒽酮(显色)试剂
anthrone colorimetry 蒽酮比色法
anthropodesoxycholic acid 12-脱氧胆酸；鹅(脱氧)胆酸
anthropogenic source 人为源
anthroxan (邻)苯甲内酰胺
anthroxan aldehyde 苯并异噁唑甲醛
anthryl 蒽基〔有3种异构体〕 $C_{14}H_9$—
anthryl amine 蒽胺
anthryl carbinol 蒽甲醇
anthrylene 亚蒽基
anti- 〔希腊字头〕 反；抗；对；解；阻；减
antiablative polymer 耐烧蚀高分子
anti-abrasion additives 抗磨蚀添加剂
antiabrin 抗红豆素
antiacid additive 抗酸添加剂
anti-activator 阻活剂
anti-adherent 防黏剂；隔离剂
anti-adhesion agent 防黏剂
anti-adhesive 防黏剂；隔离剂
antiager 防老剂；抗老剂
antiagglutinin 抗凝集素
antiaggressin 抗侵袭素
anti-aging agent 防老剂
anti-aging protective 防老剂
antialcoholic drug 解酒药
anti-alopecia factor(=inositol) 抗脱毛因子；环己六醇
antiamoebic agent 抗变形虫药
antiamoebin 抗变形虫素
antian(a)emia factor 抗贫血因子
antianaphylactin 抗过敏素
antiandrogen 抗雄激素物质
antiantidote 抗解毒药
antiar 箭毒木毒汁
antiarin 箭毒木苷
antiaromaticity 反芳香性
anti-arsenical 抗吡药〔解砒霜毒性〕；驱砷剂
antiarsenin 抗砒素〔由于经常服用微量砒霜而在体内产生的抗体〕
antiarthritic 治风湿药
antiasthmatic 治喘息药
antiattrition 减少磨损
anti-bacterial fibre 抗菌纤维
anti-bacterial finish 抑菌整理；抗菌整理
anti-ballistic material 防弹材料
antibaryon 反重子
antiberiberi 维生素 B_1
antiberiberi vitamin(=vitamin B_1) 抗脚气病维生素；维生素 B_1
antibilious 治胆病药
antibiotic ①抗菌素②抗菌的
antibiotics 抗菌素；抗生素
antibiotic with a wide spectrum 广谱抗菌素
antiblastin 抗瘟菌素
antiblaze 抗燃剂
anti-bleed sealer 防渗色封闭剂
antiblennorrhagic 治淋病药
antiblock agent 抗阻塞剂
anti-blocking 抗粘连
anti-blocking agent ①防结块剂②防粘连剂③防贴合剂
anti-blowing agent 消泡剂
anti-blushing agent ①防变色剂②抗混浊剂
antibody 抗体
antibody absorption test 抗体吸收试验
antibody cytolytic 抗体细胞溶解
antibody half-life 抗体半衰期
antibond 反键
antibonding 反键(作用)
anti-bonding agent 防黏剂
antibonding (molecular) orbital(AMO) 反键(分子)轨道
antibonding orbital 反键轨道
anti-bonding state 反键态
anti-bubbling agent 消泡剂
antibump rod 防暴沸棒*
anticachectic ①清血药②治营养不良药
anticaking agent 防结块剂
anticalcium 抗钙(脱灰剂)
anticancer 抗癌剂
anti-carrier 反载体
anticatalysis 反催化(作用)
anti-catalyst 反催化剂；负催化剂
anticatalytic property 负催化性
anticatalyzer 负催化剂
anticatarrhals 消炎药
anticathode 对阴极〔X射线管〕
anti-cavitation 防气蚀；防穴蚀
anti-centrifugal splash 抗甩性
antichaotropic anion 离液序列低的阴离子
antichecking agent 防龟裂剂
anti-chemical properties 抗化学药品性能
antichlor 脱氯剂
antichloration 脱氯；去氯
anticholerin 抗霍乱菌素
anticholinergic 抗胆碱能的；抗胆碱能药
anticholinesterase 抗胆碱酯酶剂
anticipation ①预期②预先处理③提前；先期进行

anti-circular development 反圆心式展开
anticlinal 反错〔扭转角 120°〕
anti-clinal conformation 反错构象
anti-clockwise 反时针(方向)旋转的；左旋的
anticlogging agent 防阻塞剂
anticlogging fuel oil compositions 防结渣添加剂〔锅炉燃料油用〕
anti-clogging safety stop motion 防轧装置
anti-clogging separator 防堵塞分离器
anti-clouding agent 防晕剂；防雾剂
anti-coagulant ①阻凝剂；抗凝(固)剂②抗凝血剂
anti-coagulating action 阻凝
anti-coagulating agent 抗凝(固)剂
anticoagulation 抗凝作用
anticoagulin 抗凝质
anticoalescent 抗凝聚剂；阻凝剂
anticoking additive 防焦添加剂
anticoking treatment 防焦处理
anti-configuration 反型；反式构型
anti conformation 反式构象
anticorrosion ①防腐②防锈
anti-corrosion additive 抗腐蚀添加剂
anti-corrosion adhesive 防腐性黏合剂
anti-corrosion agent 防腐剂；防蚀剂
anti-corrosion cleaning compound 防腐蚀清洗剂
anti-corrosion coating 防腐蚀涂饰
anti-corrosion grease 防腐润滑脂
anticorrosion property 抗腐蚀性
anticorrosive ①防腐的；防蚀的②防锈的
anti-corrosive additive 防腐蚀添加剂
anti-corrosive oil 防锈油
anti-corrosive paint 防锈漆
anticorrosive property 防腐蚀性
anticorrosive rating 防腐蚀等级
anticracking agent 抗龟裂剂；防裂剂〔橡胶〕
anti-cratering agent 防缩孔剂
anti-crawling agent 防蠕变剂
anticreaming agent 阻膏化剂
anticrustator 表面硬化防止剂；防结垢剂
anticryptic 侵隐色的
anti-crystallizing rubber 抗结晶橡胶
anticurl （纤维)防蜷缩
antidandruff agent 去头屑剂
antidandruff shampoo 去头屑洗发剂
anti-darkening agent 防黯色剂
antidegradant 防老化剂；抗降解剂
antidepressive 抗抑郁剂〔药〕
antiderivative 反式衍生物

antidermatitis factor(=pantothenie acid) 抗皮炎因子；泛酸
antideteriorant 防坏剂；防老剂
antidetonant 抗爆剂
antidetonating fluid(=antiknock fluid) 抗爆液；乙基液
anti-detonating joint 抗震缝
antidetonating substance 抗爆物质
anti-detonation 抗爆(作用)
anti-detonation fuel 抗爆燃料
antidetonator(=antiknocking compound) 抗震剂；抗震化合物
antidiabetic ①抗糖尿药②抗糖尿的
antidimmer （镜片)抗朦剂；保明剂
antidimming 抗朦；保明
antidimming agent 保明剂
antidimming disc 抗朦片
antidimming ointment 抗朦软膏
antidimming soap 抗朦皂
antidote 解毒药
antidrip agent 防液滴剂
antidulling 抗变深性
anti-dust 去污斑
antidysenteric 抗痢疾药；止痢药
antielectron 反电子，阳电子
antielectrostatic 抗静电的
antielement 反元素〔反粒子构成的元素〕
antiemetic(=antemetic) 止吐剂〔药〕
anti-emulsifier 抗乳化剂
anti-emulsifying property 抗乳化性
antierrhine 止涕药
anti-explosion 防爆(作用)
anti-explosion fuel 防爆燃料
anti-exposure cracking agent 防候化龟裂剂〔橡胶〕
anti-extrusion ring 抗挤胶圈
anti-fatigue agent 抗疲劳剂
antifebrin 退热冰〔药〕；N-乙酰苯胺
antifelting finishing 防毡缩整理
antiferroelectric ceramics 反铁电陶瓷
antiferroelectric crystal 反铁电晶体
antiferroelectric distortion 反铁电畸变
anti-ferroelectricity 反铁电性
antiferro-magnetic 抗强磁性(的)；反铁磁性的
antiferromagnetic compound 反铁磁(性)化合物
antiferromagnetic resonance 反铁磁共振
antiferromagnetism 反铁磁性
anti-fibrillant masterbatch 抗纤维剂
antifield rotation 反场旋转
anti-flaming 防燃

anti-flammability agent 防燃剂
antiflex cracking antioxidant 防折裂抗氧化剂
anti-float agent 防发花剂
anti-floating agent 防泛剂；抗泛剂
anti-flocculating properties 防絮凝性质
anti-flooding agent 防浮剂；抗浮剂
antifoam 防泡沫的；阻泡沫的；消泡
antifoam additive(=antifoaming additive;antifrother) 防沫添加剂；防起泡添加剂
antifoam agent 防沫剂
antifoamer 防沫剂；消泡剂
antifoaming additive(=antifoam additive) 防沫添加剂
antifoaming agent 消泡剂；防沫剂〔抑制泡沫〕
anti-foaming property 抗泡性
antifoam silicone 消泡硅酮
antifoam still head 防沫蒸馏头
antifoggant 防雾剂
antifogging 防雾剂
antifogging agent 防雾剂〔保持玻璃等透明〕
antifongin 抗真菌素
anti-form 反式
antiformin 安替佛民〔消毒药水〕；次氯酸钠
antifoulant 防污剂
antifoulant additive 防污添加剂
anti-fouling 防污的
antifouling agent 防污剂
anti-fouling composition(=antifouling compound) 防污剂
antifouling compound(=anti-fouling composition) 防污剂
antifouling life 防污有效期；防污寿命
antifouling lubricant 防污润滑剂
antifouling paint 防污漆；防污涂料；防虫漆
antifree-radical agent 抗自由基剂；减自由基剂
antifreeze ①防冻②防冻剂
antifreeze additive 防冻剂；阻冻添加剂
antifreeze agent 防冻剂
antifreeze compound 防冻复合料
antifreeze contamination 防冻沾污
antifreeze fluid 防冻液；防冻剂
antifreeze mixture 防冻混合物
antifreezer 防冻剂
antifreeze solution 防冻溶液
antifreezing agent(=antifreezing dope) 防冻剂
antifreezing compound 防冻化合物
antifreezing dope(=antifreezing agent) 防冻剂
antifreezing fluid 防冻液
antifreezing lubricant 防冻润滑剂
antifreezing oil 防冻润滑油〔石油馏分〕
antifriction ①减摩②减摩剂

antifriction alloy 抗摩擦合金
antifriction bearing 减摩轴承；滚动轴承
antifriction bearing grease 减摩轴承润滑脂
antifriction metal 减摩金属
antifriction property 抗磨性；耐磨性
anti-frosting agent 防霜剂
antifrost salve 防霜膏剂
antifroth agent 消沫剂
antifrother(=antifoam additive) 防沫添加剂；防起泡添加剂
antifrothing agent 消泡剂
antifume agent 抗烟雾剂
anti-fungal fibre 抗菌纤维；耐菌纤维
antifungus 防霉的
antifungus paint 防霉漆
anti-fuzzing finish 防起毛整理
antigalactic 抗泌乳剂；制乳药
antigas additive 抗气(体)添加剂；消泡添加剂
antigassing additive 抗气(体)添加剂；消泡添加剂
anti-gelling agent 防胶凝剂
antigen 抗原
antigen analysis 抗原分析
antigen-antibody complex 抗原抗体复合物
antigen antibody reaction 抗原抗体反应
antigenic determinant 抗原决定簇
antigenicity 抗原性
antigenic substance 抗原物质
antigibberellin 抗赤霉素
antiglare 防目眩；防闪光
anti-glare glass 防眩玻璃；太阳镜用玻璃
antiglobulin 抗球蛋白
antigorite 叶蛇纹石
antigradient 逆梯度；负梯度
antigravity filtration 抗重过滤
antigravity screen 抗重筛〔筛料自下而上筛过〕
antigravity system 抗重系统；空气运输催化剂系统
anti-gray-hair factor(=paraaminobenzoic acid) 抗灰毛因子；对氨基苯甲酸
anti-groove crack （轮胎面)抗花纹龟裂
antigum inhibitor 防胶剂
anti-halation 防光晕
antihalation dye 抗晕染料〔照相〕
anti-hard caking agent 防结(硬)块剂
anti-hazard classification 防爆等级
anti-hemorrhagic compound 止血物
anti-hemorrhagic fibre 止血纤维
antihidrotic 止汗药
anti-homomorphism 反同态；反异质同晶(现象)

anti-Hückel system 反休克尔体系
anti-humidity oil 防潮油
antihypo 过碳酸钾
antiicer 防冰器
antiicing additive 防冰添加剂
antiicing agent 防冻剂
antiicing coating 防结冰涂料
antiicing gasoline 防冻汽油
anti-idiotypic antibody 抗独特型抗体
antiimmune 抗锡疫质
anti-incrustant 防水垢剂
anti-inflammatory action 抗炎作用
anti-irritant 抗刺激剂
anti-isomerism 反式同分异构(现象)
anti-isomorphism 反类质同晶*
antijump baffle (双向溢流塔板上)防跃挡板
antiknock 抗爆
antiknock additive 抗爆添加剂
antiknock agent 抗爆剂
antiknock antagonist 抗爆对抗剂
antiknock blending agent 抗爆掺和剂〔高辛烷值组分〕
antiknock characteristics 抗爆性(质)
antiknock component 抗爆组分；高辛烷值组分
antiknock dope 抗爆添加剂
antiknock dope for diesel fuel 柴油抗爆添加剂
antiknock fluid 抗爆液；乙基液
antiknock fuel 抗爆燃料
antiknock gasoline(=antiknock petrol) 抗爆汽油；高辛烷值燃料
antiknocking compound(=antidetonator) 抗震剂；抗震化合物
antiknock petrol(=antiknock gasoline) 抗爆汽油；高辛烷值燃料
antiknock petrol additive 抗爆汽油添加剂
antiknock properties 抗爆性质
antiknock quality 抗爆性质；防震品级
antiknock rating 抗爆率
antiknock rating method 抗爆性评价法
antiknock responsiveness 抗爆感应性
antiknock substance 抗爆剂
antiknock susceptibility 抗爆感应性
antiknock valuation 抗爆评价
antiknock value 抗爆值
anti-Langmuir isotherm 反朗格缪尔等温线
anti leaching 抗浸出
antilithic 治结石药
anti-livering agent 防肝化剂
antilogarithm 反对数

antiluetin 抗梅毒剂
antilyssic 治狂犬病药
antimalarial ①抗疟的②抗疟药
antimalum 治风湿病药
anti-Markovnikov addition 反马氏加成*
antimellin 蒲桃皮苷
antimer 对映体
antimetabolite 抗代谢物；代谢拮抗物
antimetabolite analog 抗代谢类似物
antimetal 抗金属老化
antimicrobaic(=antimicrobial) ①抗菌的；抗微生物的②杀菌剂；制菌剂
antimicrobial(=antimicrobaic) ①抗菌的；抗微生物的②杀菌剂；制菌剂
antimicrobial agent 抗微生物剂；抗菌剂
antimicrobial spectrum 抗菌谱
antimonation 锑酸化作用
antimonial 锑的
antimonial copper glance 硫锑铜矿
antimonial glass 锑玻璃
antimonial lead 锑铅(合金)
antimonial soap 锑皂
antimoniate(=antimonate) 锑酸盐
antimonic 锑(基；根) =Sb≡
antimonic acid 锑酸
antimonic acid anhydride 锑酸酐；五氧化二锑 Sb_2O_5
antimonic chloride 氯化锑 $SbCl_5$
antimonic compound 锑化合物
antimonic fluoride 氟化锑 SbF_5
antimonic oxide 氧化锑 Sb_2O_5
antimonic oxychloride 三氯氧化锑；三氯化氧锑 $SbOCl_3$
antimonic salt 锑盐
antimonii 锑(的)
antimonine 乳酸锑
antimonious(=antimonous) ①亚锑的；三价锑的②含锑的
antimonite 亚锑酸盐
antimonium(=antimony) 锑
antimono- 偶锑基
antimonoacetylene compound 偶锑乙炔化合物
antimonous(=antimonious) ①亚锑的；三价锑的②含锑的
antimonous acid 亚锑酸
antimonous acid anhydride 亚锑酸酐；三氧化二锑 Sb_2O_3
antimonous arsenate 砷酸亚锑
antimonous arsenite 亚砷酸亚锑

antimonous basic chloride 一氯氧化锑；一氯化氧锑 SbOCl
antimonous bromide 溴化亚锑；三溴化锑 SbBr$_3$
antimonous chloride 氯化亚锑；三氯化锑 SbCl$_3$
antimonous compound 亚锑化合物
antimonous fluoride 氟化亚锑 SbF$_3$
antimonous hydride 三氢化锑 SbH$_3$
antimonous iodide 碘化亚锑 SbI$_3$
antimonous nickel 锑镍齐
antimonous oxalate 草酸氧锑 $(SbO)_2C_2O_4$
antimonous oxide 氧化亚锑；三氧化二锑 Sb$_2$O$_3$
antimonous oxybromide 一溴氧化锑 SbOBr
antimonous oxychloride 一氯氧化锑 SbOCl
antimonous oxyiodide 一碘氧化锑 SbOI
antimonous oxysulfide 氧化锑合硫化锑
antimonous salt 亚锑盐；三价锑盐
antimonous sulfate 硫酸亚锑 Sb$_2$(SO$_4$)$_3$
antimonous sulfide 硫化亚锑 Sb$_2$S$_3$
antimony(=antimonium) 锑
antimony anhydride 锑酐
antimony arsenide 砷化锑
antimony ash 锑灰
antimony black 锑黑；硫化锑 Sb$_2$S$_5$
antimony blende 硫氧锑矿
antimony butter 三氯化锑 SbCl$_3$
antimony cinnabar 锑朱砂；硫氧化锑
antimony compound 锑化合物
antimony crocus 锑藏红〔三氧化二锑和三硫化二锑的共熔体〕
antimony flowers 锑花
antimony glance 辉锑矿
antimony glass 锑镜
antimony golden sulfide 金色硫化锑〔即：五硫化二锑〕Sb$_2$S$_5$
antimony hydride 氢化锑
antimonyl 氧锑(根) SbO—
antimony lactate 乳酸锑 (C$_3$H$_5$O$_3$)$_3$Sb
antimonyl aniline tartrate 酒石酸苯胺氧锑
antimonyl bromide 溴化氧锑；次溴酸锑 SbOBr
antimonyl chloride 氯化氧锑；次氯酸锑 SbOCl
antimonyl compound 氧锑化合物
antimony lead 锑铅
antimonyl iodide 碘化氧锑；次碘酸锑 SbOI
antimony liver 锑肝〔全硫(代)锑酸钙和全硫(代)亚锑酸钙的混和物〕
antimonyl mirror 锑镜
antimonyl oxalate 草酸氧锑 $(SbO)_2C_2O_4$
antimonyl potassium tartrate 酒石酸氧锑钾；吐酒石 K(SbO)C$_4$H$_4$O$_6$·1/2H$_2$O
antimonyl sulfate 硫酸氧锑 (SbO)$_2$SO$_4$
antimony minerals 锑矿类
antimony mirror 锑镜
antimony needle 锑针 Sb$_2$S$_2$
antimony ochre 黄锑矿
antimony orange 锑橙〔主要成分为硫化锑〕
antimony ore 锑矿〔1.泛指；2.专指:辉锑矿〕
antimony oxide 氧化锑；锑白
antimony pentachloride 五氯化锑 SbCl$_5$
antimony pentafluoride 五氟化锑 SbF$_5$
antimony pentamethyl 五甲锑
antimony pentasulfide 五硫化二锑 Sb$_2$S$_5$
antimony pentoxide(=antimony peroxide) 五氧化二锑；过氧化锑
antimony pentoxide sol 五氧化二锑溶胶
antimony peroxide(=antimony pentoxide) 五氧化二锑；过氧化锑
antimony persulfide 五硫化二锑；过硫化锑 Sb$_2$S$_5$
antimony polymer 锑聚合物
antimony potassium oxalate 草酸氧锑钾 K(SbO)C$_2$O$_4$
antimony potassium tartrate 酒石酸氧锑钾 K(SbO)C$_4$H$_4$O$_6$·1/2H$_2$O
antimony red 锑红；五硫化二锑〔颜料〕
antimony regulus 锑块〔含90%锑的粗制金属锑〕
antimony rubber tubing 红橡皮管
antimony sulfate 硫酸锑 Sb$_2$(SO$_4$)$_3$
antimony sulfide 硫化锑
antimony sulfide golden 金色硫化锑
antimony sulfuret 三硫化二锑
antimony tetroxide 四氧化二锑 Sb$_2$O$_4$
antimony tribromide 三溴化锑 SbBr$_3$
antimony trichloride 三氯化锑 SbCl$_3$
antimony triethyl 三乙基锑 Sb(C$_2$H$_5$)$_3$
antimony trifluoride 三氟化锑 SbF$_3$
antimony triiodide 三碘化锑 SbI$_3$
antimony trioxide 三氧化二锑〔用作阻燃剂、催化剂〕
antimony triphenyl 三苯基锑 Sb(C$_6$H$_5$)$_3$
antimony triselenide 三硒化二锑 Sb$_2$Se$_3$
antimony trisulfide 三硫化二锑 Sb$_2$S$_3$
antimony vermilion 锑朱〔颜料〕
antimony vermillon(=crimson antimony) 锑朱
antimony white 锑白〔颜料〕
antimony yellow 锑黄〔颜料〕
antimosan(=potassium antimony) 安替莫散〔药〕；二邻苯二酚-3,5-二磺酸锑钠七水合物
anti muonium 反μ子素
antimushing agent 防(肥皂)糊(烂)剂

antinarcotic　抗麻醉药
antineuralgic　治神经痛药
antineuritic factor(=vitamin B_1)　抗神经炎因子；维生素B_1
antineuritic value　抗神经炎值
antineuritic vitamin　抗神经炎维生素；维生素 B_1
anti-nitrite compound　抗亚硝酸化合物
anti-noise paint　消声漆
antinonnin　安替侬宁〔拜耳1892年研制成功〕〔药〕；二硝甲酸
antinosin　碘酞钠〔药〕；安提诺辛〔一种抗菌性散剂，代替碘仿〕
anti-offset agent　防背面粘脏剂；防蹭背剂；防(粘)脏剂〔油墨〕
antiophthalmic factor(=vitamin A)　抗干眼因子；维生素A
antiophthalmic value　抗干眼值
antioxidant　抗氧化剂*
antioxidant additive　抗氧化添加剂
antioxidant carrier　抗氧化剂载体
antioxidant detergent　抗氧化洗涤剂
antioxidant emulsion　乳化防老剂
antioxidative stabilizer　抗氧化稳定剂
antioxidizer　防氧化剂
anti-oxime　反式肟
antioxygen(=antioxidant)　抗氧化剂
antioxygenic activity　抗氧活性
antiozidant　抗臭氧化剂
antiozonant　抗臭氧剂；防臭氧剂
antiozone property　防臭氧性
antiozonidate　抗臭氧剂
anti-packing chemical　除积垢剂
antiparallel　反(向)平行(的)
antiparallel packing　逆平行堆砌
antiparasite(=antiparasitic)　抗寄生虫药
antiparasitic(=antiparasite)　抗寄生虫药
antiparticle　反粒子
antipathetic　相憎的
anti-patterning　防叠
antipellagra factor　抗糙皮病因子
antipellagra vitamin　抗糙皮病维生素；维生素 PP
antipellagric value　抗糙皮病值
anti-percolator　防渗装置
antiperiodic　抗疟的；抗周期性疾病的
antiperiplanar　反迫〔扭转角180°〕
anti periplanar conformation　反叠构象
antipernicious anemia factor(=vitamin B_{12})　抗恶性贫血因子；维生素 B_{12}
antipernicious anemia principle(=vitamin B_{12})　抗恶性贫血因子；维生素 B_{12}

antiperspirant　防汗剂
anti-perspiration finish　防汗渍整理
antiphase trigonal　反相三角形
antiphlogistic　①抗炎的②消炎剂
antiphlogistic drug　消炎药
antiphlogistic theory　反燃素学说
antiphlogiston　①抗炎药②反燃素
anti-pill fibre　防起球纤维
anti-pilling finishing　防起球整理；抗起球整理
anti-pilling tester with circular abrasion　圆轨迹抗起球试验仪
antipinking fuel　高辛烷值汽油；抗爆燃料
anti-piping compound　防缩孔剂
anti-pitting additives　防针孔添加剂〔电镀〕
antiplasticization　反增塑
antiplasticizer　反增塑剂
antipodagric　治痛风药
antipode　对映体
antipoison　消毒剂；解毒剂
antipoisoning　消毒
anti-pollution　防污染；去污染
anti-pollution barrier　防污染栅
anti-pollution device　防污染装置
anti-pollution system　防污染系统
antipolymerizer　阻聚剂；防聚剂
antiposition　反位
antiprecipitant　抗沉淀剂
antipreignition additive　防预燃添加剂
anti-pressing standing velvet　抗拉力绒
anti-press-off device　防脱套装置
antiproton(=negative proton)　反质子
antiprotozoin　抗原虫菌素
antipyonin　四硼酸钠
antipyrene(=antipyrine)　安替比林〔药〕
antipyretic　①退热剂②退热的
antipyrin(=antipyrine)　安替比林〔药〕；退热药
antipyrine(=antipyrene; antipyrin)　安替比林〔药〕
antipyrine and caffeine citrate　米格来宁〔药〕
antipyrine chloral　安替比林氯醛；安替比林合三氯乙醛
antipyrinyl(=antipyryl)　安替比林基
antipyrotic　治灼伤药
antipyroyl　安替比林(甲)酰
antipyryl(=antipyrinyl)　安替比林基
antique finish　古彩涂饰剂
antique finishing　仿古涂装法
antiquinot　安替奎诺〔药〕
antirachitic　治佝偻病药
antirachitic factor　抗佝偻病因子；维生维 D

antirachitic value 抗佝偻病值
antirachitic vitamin 抗佝偻病维生素
antirad 抗辐射剂
antiradiation agent 抗辐射(试)剂
antiradiation effect 防辐射效应
anti-rads 抗射线(老化)剂
antiredepositing power 抗再沉积(能)力
antiredeposition 抗再沉积(作用)
antiredeposition agent 抗再沉积剂
antiredeposition soil carrier 抗再沉积携污剂
anti-reversion agent 抗硫化返原剂
antirheumatic 治风湿药的
anti-rheumatic fibre 防风湿纤维〔具有低导热、高电位性能的纤维〕
antirheumatin 治风湿药
anti-rivelling paint 防起条纹漆
anti-run-down 防脱散
anti-rust additive 防锈添加剂
antirust composition(=antirust compound) 防锈配方
antirust compound(=antirust composition) 防锈配方
antirust grease 防锈(润滑)脂
antirusting agent 防锈剂
antirusting paint 防锈漆
antirust oil 防锈油
antirust paint 防锈漆；防锈涂料
antirust paper 防锈纸
anti-sag agent 防流淌剂；防流挂剂
anti-sagging agent 防流挂剂
anti-sag index 抗流挂指数
anti-sag property 抗塌性能；坚挺性
antiscale ①防垢②防垢剂〔锅炉〕
antiscaling 防垢(的)
antiscaling compound 防垢剂
antiscorbutic 抗坏血病药
antiscorbutic value 抗坏血病值
antiscorbutic vitamin 抗坏血病维生素；维生素C
anti-scorcher 防焦(烧)剂
antiscorch(ing) ①抗焦(作用)②抗焦剂
antiscorching agent 防焦(烧)剂
antiscratch 抗划道；抗划痕
anti-scratch additive 耐刮擦助剂
anti-scuff additive 抗擦伤添加剂
anti-scuff agent 抗磨剂
anti-scuffing 抗磨蚀；抗擦伤；抗刮(划)伤
anti-scuffing paste 抛光膏；研磨膏
anti-scumming agent 抗浮垢剂；防浮沫剂
anti-sedimentation 抗沉积(降)作用
anti-sedimentation agent 防沉(降)剂

antiseize agent 防卡滞剂；防卡模剂
antiseize compound 抗扯裂化合物
antisense nucleic acid 反义核酸
antisense RNA 反义RNA
antisepsin 防腐素
antiseptic 防腐剂；消毒剂
antiseptic agent 防腐剂
antiseptic effect 防腐效应
antiseptic-germicide 防霉剂；抗菌剂
antiseptic process 防腐处理法
antiseptic rubber 防腐橡胶；含防腐剂的橡胶
antiseptics 防腐剂
antiseptic substance 防腐剂
antiseptic treatment 防腐处理
antiseptic wash(=fungicidal wash) 防腐处理(洗)液
antiseptol 碘硫酸辛可宁
anti-sequestrating agent 反螯合剂
anti-settling agent 防沉剂；抗沉降剂
anti-shatter composition 防碎剂
anti-shock mounting 防冲装置；缓冲橡皮
anti-shrink 防缩(的)
anti-shrink finish 防缩整理
anti-shrinking medium 防(收)缩剂
anti-siccative agent 阻干剂
anti-silking agent 防走丝剂；防丝纹剂
anti-sintering number 抗烧结值
antisiphoning rod(=support rod) 反虹吸棒
anti-skid 防滑性
anti-skid design 防滑花纹
anti-skid groove 防滑花纹沟
anti-skid tread pattern 防滑胎面花纹
anti-skimming agent 防结皮剂
anti-skinning paste 防结皮浆(膏)
anti-slip agent 防滑剂
anti-slip effect 防滑效应
anti-slip paint 防滑漆
antisludge additive(=antisludging agent) 抗淤渣添加剂
antisludging agent(=antisludge additive) 抗淤渣添加剂
antisnag 防抽丝
anti-snag agent 防起毛整剂
anti-snarl device 防扭结装置
anti-softener 防软剂
anti-soil 防污；抗污
anti-soil-redeposition property 防尘污再沾着性
antisomorphism 反同形性；反同构性
antispalling agent 抗散裂剂
anti-spark 防(消)火花的
antispasmin 镇痉粉；解痉剂

antispasmodic 镇痉药；解痉药
anti-spatter 防溅器
antispattering agent 防溅剂
anti-spew agent 防胶边形成剂；除胶边剂
anti-squeak 消声器
antistaining 防污染
antistaling agent 保鲜剂
anti-stat(=anti-static agent) 抗静电剂
anti-static 抗静电的
anti-static additive 抗静电添加剂
anti-static agent(=anti-stat) 抗静电剂
anti-static cartridge 防静电芯筒
antistatic copolymer 抗静电共聚物
anti-static device 抗静电装置
antistatic emulsifier 抗静电乳化剂
antistatic fibre 抗静电纤维
anti-static hose 抗静电胶管
antistatic ingredient 抗静电配料；抗静电组分
anti-static rubber 抗静电橡胶
antistatic treatment 抗静电处理
antisterility factor 抗不育因子；维生素E
anti-stick 抗黏结
antisticking agent 抗黏剂
anti-stickness 抗黏着；抗黏性
anti-Stokes atomic fluorescence 反斯托克斯原子荧光
anti-Stokes fluorescence 反斯托克斯荧光
anti-Stokes scattering 反斯托克斯散射
anti-Stokes shift 反斯托克斯位移
antistripping 抗剥离〔沥青及充填物间的黏结力〕
antistripping agent 抗剥离剂
antisubstance 抗体
anti-suckback nozzle 防回料注嘴
antisudorific 止汗药
anti-sun agent 抗日光剂
antisun material 抗光物质；抗日照物质
anti-surge blow off 防喘振排放
anti-surge control system 防喘振控制系统
anti-swelling 抗溶胀性；抗膨润性
anti-swelling finish 防溶胀处理
anti-symmetric 反对称的
antisymmetrical wave function 反对称波函数
anti-symmetric function 反对称函数
anti-symmetric tensor 反对称张量
antisymmetric wave function 反对称波函数
antisymmetrized molecular orbital(ASMO) 反对称分子轨道
antisymmetrized MO method(ASMO) 反对称分子轨道法

antisymmetry 反对称性
antisymmetry principle 反对称原理
anti-synbiosis effect 反共生效应
antisynergism 反协同现象(效应)
antisyphilitic 治梅毒药
anti-syphon rod(=support rod) 反虹吸棒
anti-tack agent 防黏剂
antitackiness agent 抗黏剂
anti-tarnishing agent 防晦暗剂
anti-tarnish paper 防锈纸
anti-tarnish spray 防污喷雾剂；防晦暗喷剂
anti-termite agent 防白蚁剂
antitetrazin 治神经痛药
antithermin 安替瑟明〔药〕
antithiamine 抗硫胺素
anti-thrombocytic fibre 血小板分离用纤维；血浆分离用纤维
antithyroid ①抗甲状腺的②抗甲状腺物
antithyroid agent 抗甲状腺剂
antitode 解毒物
antitoxic ①解毒剂②抗毒素的
anti-toxicity 防毒
antitoxin 抗毒素
antitoxoplasmic substance 抗毒质体物质
anti-tracking chemicals 防漏电剂
anti-tracking lacquer 抗电弧径迹漆；防迹漆
anti-tracking varnish 防迹漆〔在高电压下不起碳黑迹〕
antitubercular compound 抗痨化合物；抗结核药
antituberculosis 抗痨作用；抗结核作用
antitumor monoglyceride 抗癌甘油乙酸酯
antitussin 止咳素；愈创甘油醚
anti-type 反式
anti-type isomerism 反式同分异构(现象)
anti-variants 防差异剂
antivirubin 抗病毒红素
antivirus 抗病毒素
antivirus action 抗病毒作用
antivitamin 抗维生素
antiwear 抗磨损
antiwear additive(=antiwear agent) 抗磨添加剂
antiwear agent(=antiwear additive) 抗磨添加剂
antiwear film 抗磨膜
antiwelding action 抗焊作用
antiwelding compound 抗焊化合物
anti-wetskid property 抗湿滑性
antiwrinkle 防皱
anti-wrinkling agent 抗皱剂
antixerophthalmic factor 抗干眼因子；维生素A

antixerophthalmic vitamin 抗干眼维生素；维生素 A
anti-yellowing agent 防黄变剂
antodyne 苯氧丙二醇
Antonoff rule 安东诺夫规则
antozone 单原子氧
antraquinone 蒽醌
antu(=α-naphthyl-thiourea) 安妥〔杀鼠剂〕；α-萘硫脲
antwerp blue 亚铁氰酸锌粉
anucleate 无核的
anuria 无尿症
anvil-dross 锻渣
anvil roll 支承辊
anyme oil 苦槛蓝油
anymol 苦槛蓝醇
anysin 鱼石脂
anytin 安尼汀
anytol(=anitol) 安尼妥
apa- 阿朴
apaconitine 阿朴乌头碱〔从乌头碱中提取的一种有毒的碱〕
apagallin 汞四碘酚酞
aparaphysate 无侧丝的
apartment kiln 分室干燥窑
apatite 磷灰石
apatropine(=apoatropine) 阿朴阿托品
apenone 附体肉桂酮
aperient 轻泻药；润肠药
aperiodic elongation 非周期伸长
aperiodic polymer 非周期性聚合物
aperiodic response 非周期性响应
aperiodic strain 非周期应变
apertometer 孔径计
aperture ①孔；口；隙②孔径；口径
aperture lens 孔隙透镜
aperture of screen 筛孔
aperture opening (筛网)孔径开度
aperture ratio 孔径比
aperture size 筛孔度
apex 尖；顶点
apex current 顶电流
apex of the sector field 扇形场的顶点
apex point 顶点；钻尖
Apex process (词源:alcoholplutonium nitrate extraction) 阿派克斯过程〔制备二氧化钚溶胶的过程〕
apex rubber 三角胶芯；三角胶条
apex strip 三角胶条；填充胶条
aphalerite 硫锌矿
aphanesite 砷铜矿

aphanicin 阿番素；束生藻色素乙
aphanin 阿番宁；束生藻色素甲
aphanizophyll 阿番叶素
aphelion ①远日点②远核点
aphermate〔商〕 甲酸-2,3,3-三甲基环乙基酯
aphicide 杀蚜虫药
aphins 蚜色素
apholate 环磷氮丙啶
aphosphorosis 乏磷症；磷缺乏症
aphrite 鳞方解石
aphrizite 泡沸电气石；黑电气石
aphrodine 壮阳碱
aphrodisiac 壮阳药；春药
aphrosiderite 铁华绿泥石
aphthitalite 钾芒硝
aphthonite 银铜矿；银黝铜矿
aphyl 烷基
apiezon〔商〕 阿匹松〔真空润滑脂；真空油膏；一种高真空用的黏合剂〕
apiezon grease〔商〕 阿匹松脂
apiezon oil〔商〕 阿匹松油
apiezon wax〔商〕 阿匹松蜡
apigenidin chloride 氯化芹菜定
apigenin 芹菜素
API gravity API 比重〔美国石油学会 API 比重指数〕
API hydrometer API 比重计
apiin 芹菜苷；芹黄苷
API Lubrication Committee 美国石油学会润滑委员会
apinclum(=apinol) 臭松油
apinol(=apinol) 臭松油
apiol 芹菜脑
apiol aldehyde 芹菜脑醛 $C_{10}H_{10}O_5$
apiole(=apopinol) 芹菜脑
apiolic acid 芹菜脑酸
apiolole 芹菜脑醚
apione 芹菜酮
apionic acid 芹菜酮酸
apionol 芹菜脑酚
apiose 芹菜糖
API service oil classification API 润滑油使用分类
API specifications API 技术规范
aplanatic 等光程的；齐明的
aplastic 非塑性的
aplastic anemia 再生障碍性贫血
aplha coefficient(=side reaction coefficient) α系数；副反应系数
aplite 半花岗岩；细晶岩
aplotaxene 十七碳四烯；单紫杉烯

aplysane 念珠藤烷
apoalpinone 阿朴良姜酮
apoaranotin 脱氧珠囊壳素
apoatropine(=atropamine) 阿托胺；阿朴阿托品
apobornylene 从冰片烯；从龙脑烯
apocaffeine 阿朴咖啡因
apocamphane 阿朴莰烷 C_9H_{16}
apocamphor 阿朴樟脑；脱甲樟脑 $C_9H_{14}O_2$
apocamphoric acid 阿朴樟脑酸 $C_7H_{12}(COOH)_2$
apochromatic 消多色差的
apochromatic objective 消多色差物镜
apocodeine 阿朴可待因
apocrenic acid 阿朴白腐酸
apocyanines 阿朴花青类
apocyclene 从环萜烯
apocynamarin 夹竹桃麻苦素；磁麻苷
apocynein 磁麻素
apocynin 磁麻脂
apocynoid 磁麻素类
apodization treatment 截趾处理；变迹处理
apofacial 反面(的)
apofacial reaction 反面反应
apofenchene 从葑烯；从小茴香烯
apogee ①远地点②远核点
apogossypol 阿朴棉子酚
apogossypolic acid 阿朴棉子酚酸
apoionone 从紫罗兰酮
apolar 非极性的
apolar aprotic solvent 非极性非质子溶剂
apolar solvent 非(无)极性溶剂
apolipoprotein 载脂蛋白
A-polymer(=addition polymer) 加聚物
A-polymerization(=addition polymerization) 加聚作用
apolysin 阿朴利辛〔药〕
apomorphine 阿朴吗啡〔药〕
apomorphine hydrochloride 盐酸阿朴吗啡
apomyelin 阿朴髓磷脂
apophyllite 鱼眼石
apopinol(=apiole) 芹菜脑
apoquinamine 阿朴奎胺碱
apoquinine(=homoquinine) 阿朴奎宁
aporetin 大黄根脂
aporphine 阿朴啡〔吗啡的一种衍生物〕
aporrhegma ①裂衍物②蛋白分解毒质
aposafranine 阿朴藏红
aposafranone 阿朴藏红酮
apothecaries 英国药衡制
apotheosis 极点；顶峰

apothesine 阿朴塞辛〔药〕局部麻醉剂
apotoxicarol 阿朴毒灰叶酚
apotropine 阿朴托品
apparatus ①仪器；器械②装置③镜机
apparatus constant 仪器常数
apparatus error 仪器误差
apparatus for extractive distillation 萃取蒸馏器
apparatus for purity determination by static method 静态法纯度测定装置
apparel fabric 服用织物
apparel fibre 服装用纤维；衣着用纤维
apparel flammability modeling apparatus 服装可燃性模拟试验装置
apparent 表观的
apparent acidity 表观酸度
apparent activation energy 表观活化能
apparent aggregation number 表观缔合数
apparent area 表观面积
apparent attenuation ①表观发酵度②外观耗糖作用
apparent available area 表观可用面积；表观可及面积
apparent bulk density 表观松(态)密度；表观容积密度；表观体积密度
apparent bulk modulus 表观体积模量
apparent coefficient 表观系数
apparent coefficient of compressibility 表观压缩系数
apparent concentration 表观浓度
apparent constant 表观常数
apparent contact area 表观接触面积
apparent conversion 表观转化率
apparent count 表观支数
apparent creep 表观蠕变
apparent crystalline stack height 表观晶体堆垛高度
apparent crystal size 晶体表观尺寸
apparent current density 表观电流密度
apparent density 视密度*；表观密度
apparent detergency 表观去污力〔用污布在洗涤前后反射率改变表示的去污力〕
apparent double glass transition 表观双重玻璃化转变
apparent elastic modulus 表观弹性模量
apparent electrophoretic mobility 表观电泳淌度
apparent exchange capacity 表观交换容量
apparent exchange equilibrium constant 表观交换平衡常数
apparent flow 表观流动
apparent flow curve 表观流动曲线
apparent fluidity 表观流动性；表观流度
apparent fluorescence spectra 表观荧光光谱
apparent formation constant 表观形成常数

apparent gravity 表观比重；视比重
apparent half-life 表观半衰期
apparent hardening curve 表观硬化曲线
apparent ignition temperature 表观着火温度
apparent integrated absorption intensity 表观积分吸收强度
apparent ionization coefficient 表观电离系数
apparent ionization efficiency 表观电离效率
apparent ionization yield 表观电离率
apparent maximum shear rate 表观极大剪切速率
apparent melt viscosity 表观熔体黏度
apparent micro-flow 表观微流
apparent mobility 表观淌度
apparent modulus 表观模量
apparent molar mass 表观摩尔质量
apparent molecular weight 表观分子量
apparent plasticity index 表观塑性指数
apparent polymerization rate 表观聚合速率
apparent porosity 表观孔隙度
apparent powder density 粉末表观密度
apparent purity 表观纯度
apparent relaxation 表观弛豫；表观松弛
apparent retention time 表观保留时间
apparent retention volume 表观保留体积
apparent sensitivity 表观灵敏度
apparent shear rate 表观剪切速率
apparent shear viscosity 表观剪切黏度*
apparent solubility 表观溶(解)度
apparent specific gravity 表观比重；视比重
apparent specific heat 表观比热
apparent specific volume 表观比容
apparent spin-draw ratio 表观纺丝拉伸比
apparent stability constant 表观稳定常数
apparent stress 表观应力
apparent stress overshoot effect 表观应力超越效应
apparent synergism 表观协同现象
apparent temperature drop 表观温度降；表观温差
apparent viscometric flow activation energy 表观黏流活化能
apparent viscosity 表观黏度
apparent volume 表观容积
apparent weight 表观重量
apparition 初现
appearance defect 外表缺陷；外观疵点
appearance failure 严重外表损伤
appearance of fracture 断口外观
appearance of persistent cloud 持久浑浊现象〔汽油及煤油的〕
appearance of the spectrum 谱图概貌
appearance potential 出峰电位；始现电位
appearance rating 外观质量评级
appearance standard 外观标准
appearance temperature 出现温度；出峰温度
appearance time 出峰时间
appendage 附属物；附件；备件
appendix number(=ASTM appendix number) 美国材料试验学会附录号
appetite 食欲
apple acid(=malic acid) 苹果酸；羟基丁二酸
apple base 苹果(香精)基
apple essence 苹果香精
apple flavour 苹果香精
apple jack 油桉
apple oil 苹果油
apple-pectin 苹果胶
apple pomace 苹果酱
apple-seed oil 苹果子油
appliance ①用具②应用
appliance finishes ①家电涂料②仪表漆
appliance paint 家用电器漆
application ①点样；加样②涂布③应用
application box 点样匣
application of adhesive 涂胶
application of sample 加样；点样
application program 操作程序；应用程序
application roll 涂施辊；上墨辊
application roller 涂漆辊
application roll spin finish unit 辊式纺丝上油装置
application specific integrated circuit 专用集成电路
application viscosity 应用黏度；加工黏度
applicator ①涂布器②点样器③涂胶机
applicator blade 涂漆刮刀；刮漆刀
applicator syringe 点样注射器
applied 应用的
applied chemistry 应用化学
applied crystallography 应用晶体学
applied elasticity 应用弹性学
applied force 外加力
applied heat evaporator 外加热式蒸发器
applied petroleum refining 应用石油精制(法)
applied potential 外加电位
applied printing 直接印花
applied radiochemistry 应用放射化学
applied reaction kinetics 应用反应动力学
applied stress 外加应力
applied voltage 外加电压

applied voltage electrolytic current curve　外加电压电解电流曲线
applique　镶嵌；贴花
applique embroidery　贴花刺绣
applique seal　(热合)镶嵌封口
applying of cutting oils　应用切削油
apply oil　上油；加润滑油
approach　①近似；接近②方法；手段③机头导入沟〔压出机〕
approach speed　闭合速度
AP process　AP法〔苯酚烷化工艺〕
appropriate command　专用指令
appropriate weight　①指定重量；指定适量②合理加权
approval test　验收试验；检查试验
approximate analysis　近似分析
approximate formula　近似公式
approximate incremental method　近似增量法
approximate procedure　近似程序
approximate value　近似值
approximation　①似法；②近似；逼近
approximation formula　近似(值)公式
appurtenance　附属物；附属装置；附(配)件
apramycin　阿泊拉霉素；硫酸安普霉素
apricot　杏
apricot flavour　杏仁香精
apricot kernel mixed fatty acid　杏仁混合脂肪酸
apricot-kernel oil　杏仁油
apricot kernel water　杏仁水
apricot paper　杏子纸
aprindine　阿普林定〔药〕
a priori computation　演绎计算
apron　①围裙②裙板
apronbelt　①返料带②裙式传送带
apron board　裙板
apron conveyor　裙式运输器
apron doffing　皮圈剥棉装置
apron draft　皮圈牵伸
apron feeder　裙式加料器
apron leather　皮圈革
apron ring　裙圈；活塞下裙部涨圈
aprotic　(对)质子(有)惰性的
aprotic dipolar solvent　非酸碱偶极溶剂；无质子偶极溶剂
aprotic polar organic solvent　非酸碱极性有机溶剂；无质子极性有机溶剂
aprotic polar organo solvent　非极性有机溶剂；无质子极性有机溶剂
aprotic solvent　非质子溶剂*

aprotinin　抑肽酶〔药〕
aprotogenic solvent　非给质子溶剂*
aptitude test　适应性试验
apurinic acid　无嘌呤(核)酸
apyonin　阿比奥宁〔黄色龙胆紫〕
apyron　乙酰水杨酸锂
aq(=aqueous)　水的；含水的；水成的
AQL sampling　允收水准抽样
aqua〔拉丁文〕　水
aqua acuta　硝酸
aqua ammonia　氨水
aqua anethyl　莳萝水
aqua compound　水合物
aquadag　胶体石墨；导电敷层
aqua destillata　蒸馏水
Aquafluor process　水氟化流程〔美国通用电气公司中西部核燃料后处理工厂曾采用的干法水法结合流程〕
aqua fortis　硝酸
aquagel fibre　(含)水凝胶纤维
aquagel state　水凝胶态
aqua ion　水合离子
aquamarine　海蓝宝石
aquamarine glass　海蓝玻璃
aquameter　水分(测定)计；水量计
aquametry　测水(滴定)法*
aquamycin　水霉素
aquapulper　水力碎浆机
aqua pura　纯水
aqua regia　王水
aqua storage tank　储水槽
aqua-system gel column　水系凝胶柱
aquatech　水中技术；水上技术
aquated　水合(了的)
aquated ion　水合离子
aqua tepida　温水
aquatic chemistry　水化学
aquatic group　水族类
aquaticine　水千里光碱
aquatone　水现网版
aqua vitae　①酒精②烈性酒
aquayamycin　水绫霉素
aqueous　①水的②含水的③水成的
aqueous acrylic pressure sensitive adhesive　水性丙烯酸压敏黏合剂
aqueous adhesive　水基胶黏剂；水溶黏合剂
aqueous alcohol　含水酒精
aqueous alcoholic　含水酒精的
aqueous ammonia　氨水

aqueous caustic　苛性碱液
aqueous coagulation bath　水溶液凝固浴
aqueous continuous phase　水连续相
aqueous cooling solution　冷却水溶液
aqueous corrosion　水腐蚀
aqueous deposit　水成沉积
aqueous dispersion　水分散体
aqueous dispersoid of reclaim　再生橡胶水分散体
aqueous distillate　水馏分
aqueous emulsion　水乳浊液
aqueous extract　水提(出)物
aqueous-favoring　亲水相的
aqueous fusion　水熔(作用)〔结晶体在其本身的结晶水中熔化〕
aqueous hydrazine　含水肼
aqueous hydrogen peroxide solution　过氧化氢水溶液；双氧水水溶液
aqueous layer　水层
aqueous liquid　水成液
aqueous medium　水介质
aqueous mobile phase　含水流动相
aqueous phase　水相
aqueous phase partition　水相分配
aqueous phenol　含水酚
aqueous photoresist　液态光阻
aqueous polymer dispersion　聚合物水分散体
aqueous polymerization　水相聚合(作用)
aqueous reactor fuel　水溶液反应堆燃料
aqueous reprocessing　水法后处理*
aqueous salt spinning　盐溶液纺丝(法)
aqueous sample　含水试样
aqueous-slurry polymerization　水相淤浆聚合(作用)
aqueous soluble oil　可乳化油〔冷却切削刀具用〕
aqueous solution　水溶液
aqueous solution growth　水溶液生长
aqueous solution polymerization　水溶液聚合
aqueous spinning bath　水溶液纺丝浴
aqueous stratum　蓄水层
aqueous subphase　纯水亚相
aqueous suspension polymerization　水相悬浮聚合
aqueous tension　水蒸气张力
aqueous tincture　含水酊剂
aqueous two-phase extraction　双水相萃取
aqueous two-phase system　双水相系统
aqueous type adhesive　水溶型黏合剂
aqueous vapour　水汽；水蒸气
aqueous vapour pressure　水汽压
aqueous vapour tension　水汽张力

aqueous varnish(=water thinned varnish)　水性漆；水稀释性清漆
aquiclude　隔水层
aquifer　含水层
aquinite　三氯硝基甲烷；硝基氯仿
aquo　水合(的)；含水的
aquo-acid　水系酸
aquo-base　水系碱
aquocobalamin　水钴胺素；维生素 B_{12}
aquo complex　水合物；水络(合)物
aquo-compound　含水化合物
aquogel　水凝胶
aquo-hydroxo complex ion　水-羟配离子
aquo ion　水合离子
aquoluminescence　水溶发光*
aquo-pentamine cobaltichloride　氯化一水五氨合高钴　$[Co(NH_3)_5 \cdot H_2O]Cl_3$
aquo-system　水系
aquo-system indicator　水系指示剂
Ar　①氩〔18号元素的化学符号〕②芳基
araban　阿(拉伯)聚糖
arabate　阿(拉伯)糖酸盐　$CH_2OH(CHOH)_3COOM$
arabian oil　阿拉伯石油
arabic acid　阿拉伯酸　$HOCH_2(CHOH)_3CO_2H$
arabic gum　阿拉伯树胶
arabic gum tree　阿拉伯胶树
arabin　阿糖胶　$C_{10}H_{18}O_9$
arabinal　阿(拉伯)醛
arabinofuranose　阿拉伯呋喃糖
arabinofuranosyl adenin　阿糖呋喃腺嘌呤
arabinogalactan　阿拉伯半乳聚糖
arabinoglucuronoxylan　阿(拉伯)糖基葡糖醛酸基木聚糖
arabinose　阿(拉伯)糖；阿戊糖
arabinose phenylhydrazone　阿(拉伯)糖苯腙
arabinoxylan　阿(拉伯)糖基木聚糖
arabinulose　阿(拉伯)酮糖
arabin water　阿糖胶水
arabite(=arabitol)　阿(拉伯)糖醇
arabitic acid　阿(拉伯)糖酸　$CH_2OH(CHOH)_3COOH$
arabitol(=arabite)　阿(拉伯)糖醇
araboascorbic acid　阿(拉伯)糖型抗坏血酸；异抗坏血酸
arabogalactan　阿(拉伯)半乳聚糖
araboketose　阿(拉伯)酮糖
arabonic acid　阿(拉伯)糖酸
arabonic-γ-lactone　阿(拉伯)糖酸-γ-内酯
arabopyranose　阿(拉伯)吡喃糖
araboxylan　阿(拉伯)木聚糖
arabulose　阿(拉伯)酮糖

arachic acid(=arachidic acid)　花生酸
arachidic acid(=arachic acid)　花生酸
arachidonic acid　花生四烯酸；二十碳四烯酸
arachno-　蛛状
arachno coordination compound　网式配合物
arachyl alcohol　花生醇；二十烷醇
araeometer(=hydrometer)　（液体）比重计
aragonite　霰石；文石
aragotite　美国加里福尼亚州天然沥青
araldite　〔商〕环氧类树脂；合成树脂黏结剂
araliene　楤木烯
aralin　楤木苷
aralkyl　芳烷基；芳代脂烷基
aralkylated diphenylamine　芳烷基化二苯胺〔防老剂〕
aramid　芳族聚酰胺
aramid fiber　芳族聚酰胺纤维；芳纶；B 纤维
aramid fiber reinforced plastics(AFRP)　芳纶增强塑料
aramid fibre reinforced concrete　聚芳酰胺纤维增强混凝土
aramid sulfone fibre　聚芳砜纤维
aranilide　N-芳酰苯胺
araroba　柯桠粉
arasan　阿拉散
arasapogenin　五加皂草配基
araucaria oil　南洋杉油
arbitrary constant　任意常数
arbitrary free stream velocity　任意自由流速
arbitrary mechanical stability　任意机械稳定性
arbitrary scale　任意标度；任意刻度
arbitration analysis　仲裁分析
arboar　心轴；刀杆
arbomycin　阿鲍霉素
arbor　皮辊芯；芯子
arboral polymer　树形聚合物
arborescence　树木状；树质
arborescent　乔木状的；树木状的
arborescent crystal　树枝状晶体
arborescin　乔木素；蒿萜
arborine(=glycosine)　山柑子碱；山小橘碱
arborinine　山柑子宁
arborinol　山小橘萜醇；乔木萜醇
arborisation　树枝状
arborized　使分叉；使成树木状的
arbor press　心轴平板机；手动平板机
arbusterol　草莓油
arbute seed oil　熊果油
arbutin　熊果苷；对苯二酚葡萄糖苷
arc　①弧②电弧

arcadian nitrate　农用硝酸盐
arcain(e)　魁蛤素；1,4-二胍丁烷
arc air gauging　电弧气焊
arc and spark stand　电弧电花架
arcanite　单钾芒硝
arcanum　①秘方药②硫酸钾
arc column　弧柱
arc comparison method　弧光比较法
arc concentration technique　电弧浓缩法；电弧浓缩技术
arc coth(=inverse hyperbolic cotangent)　反双曲余切
arc cracking　电弧裂化法
arc cutting　电弧切割
arc discharge　电弧放电
arc discharge ion source　电弧放电离子源
arc draft　弧形牵伸；曲线牵伸
arc erosion　电弧腐蚀
arc flame stabilizing agent　弧焰稳定剂
arc flash welding　电弧闪光焊
arc furnace　电弧炉
arc generator　电弧发生器
arch　拱
archaeological specimen　考古学样品
arch brick　拱砖
arched printing plate　拱形印刷胶板
archeochemistry　考古化学
Archibald's method　阿奇博尔德法
archil　①一种地衣②由地衣所得的红紫色染料
Archimedian screw conveyer　阿基米得螺旋运输机
arching　①拱作用②架拱
arching of cylinder　滚筒中高度
archipelago-like bicomponent fibre　海岛型双组分纤维
architectural paint　建筑用涂料
archless kiln　无拱炉
arch-type dryer　拱形干燥器
archyl(=proyl)　丙炔基
arciform　弓形的
arcilla(=argol)　粗酒石
arcing　发弧光
arc lamp　弧光灯
arc light　弧光
arc line　电弧线；电弧束
arc of contact　接触弧
Arco microknife　阿尔科微型划刀〔检验漆膜划痕硬度和附着力的工具〕
ARCO process　〔词源：Alloy Reguline Chlorination Oxidation〕阿尔科过程；合金块氯化氧化法〔核燃料后处理首端过程之一。在氯化铅熔盐中通氯气溶解锆合金燃料元件外壳〕

arc parameter (电)弧参数
arc plasma 电弧等离子体
arc process 电弧法
arc resistance test 抗电弧试验
arc source 电弧光源
arc spectrum (电)弧光谱
arc spectrum analysis 电弧光谱分析
arc spot welding 电铆焊
arc spraying 电弧喷涂
arctic oil 极圈地区用油;加拿大西北部开采的石油
arctic rubber 耐寒橡胶
arctic sperm oil 北极鲸蜡油
arctic tyre 耐寒轮胎
arctigenin 牛蒡配基
arctiin 牛蒡因
arctiopicrin 牛蒡苦素
ar-curcumene 芳姜黄烯
arc wandering 电弧游移
arc welder (电)弧焊机;电焊机
arc welding 电弧焊接;电弧焊
arc welding machine (电)弧焊机;电焊机
ardennite 硅铝锰矿
ardisic acid 紫金牛酸 $C_{36}H_{64}O_{16}$
ardometer(=optical pyrometer) 光测高温计
area density 面(积)密度〔单位面积重量〕
area detector 面积检测器
area factor 面积系数
area flowmeter 截面流量计
areal deformation 表面形变
arealometer 气流式纤维细度测定仪
area measurement of catalyst 催化剂表面积测定
area method 求面积法
areametric analysis 面积测定分析法;面积法
area normalization method 面积归一化法
area of flame 火焰面积
area scanner 二维扫描机;面扫描机
area source 面源
area stability 面积稳定性
area summation method 面积总值法;面积求和法
area-type meter 面积型流量计〔或流速计〕
areca alkaloids 槟榔生物碱类
arecaidine 槟榔啶;水解槟榔碱
arecaine 槟榔因
arecaline(=arecoline) 槟榔碱
arecane(=arecoline) 槟榔碱
arecin 槟榔素
arecolidine 槟榔里定
arecoline 槟榔碱

arecolineserine 槟榔碱毒扁豆碱混合物
arecolone 槟榔酮
arekane(=arecoline) 槟榔碱
arenaceous(=arenarious; arenous) 砂的;多砂的;砂质的
arenaemycin 阿雷纳霉素
arenarin 沙质菌素
arenarious(=arenaceous) 砂的;多砂的;砂质的
arendalite 暗绿帘石
arene 芳烃
areneboronic acid 芳烃硼酸
arene sulfonate 芳烃磺酸盐
arenesulfonyl(=arylsulfonyl) 芳基磺酰
arenium ion 芳(基)正离子
arenolite 人造矿石
arenomycin 沙霉素
arenous(=arenaceous) 砂质的
areometer(=hydrometer) (液体)比重计
areometry(=hydrometry) (液体)比重测定(法)
areopycnometer 稀液比重计
areosaccharimeter 糖液比重计
arepycnometer 稀液比重计
arfvedsonite 钠铁闪石
argatoxyl 对氨基苯基砷酸银
argemone oil 蓟罂粟油〔碘值约120〕
argemonine 蓟罂粟碱
argent ①银的②银白的
argentamine 银胺液
argentation 银化作用;银染法
argentation chromatography 银化色谱法
argentation thin-layer chromatography 银化薄层色谱(法)
argentic 银的
argentic chloride 氯化银
argentic oxide 氧化银
argentic sulfide 硫化银 Ag_2S
argentiferous 含银的
argentiform 银仿
argentimetry 银量法*
argentine ①银的②似银的③银器④珠光石
argentite(=silvered glance) 辉银矿
argentol 银酚;羟基喹啉磺酸银 $C_9H_5NOHSO_3Ag$
argentometer 银盐定量计
argentometric titration 银量滴定法
argentometry 银盐定量法
argentopyrite 含银黄铁矿
argentous 亚银(的)
argentum virum 水银
argil 陶土;白土
argilla 泥土;铝氧土

argillaceous 泥质的
argillaceous marl 泥质泥灰岩
argillaceous rock 泥质岩
argillaceous slate 泥质板岩
argillite ①厚层泥岩②泥板岩
argillo-arenaceous 泥砂质的
argillous 泥质的
arginine 精氨酸；2-氨基-5-胍基戊酸
arginine carboxylyase 精氨酸脱羧酶
arginine hydrochloride 盐酸精氨酸 $C_6H_{14}O_2N_2HCl$
argininosuccinic acid 精氨(基)琥珀酸
arginyl- 精氨酰(基)
arglecin 精亮氨素
argochrome 银蓝素
argoflavine 银黄素
argol(=arcilla) 粗酒石
argomycin 金船霉素；阿格霉素
argon 氩〔18 号元素，化学符号 Ar〕
argon-β-ray ionization cell 氩β射线电离池
argon-arc welder 氩弧电焊机
argon arc welding 氩弧焊
argon cell 氩(气氛)小室
argon detector 氩检测器
argon inductively coupled plasma(=argon ICP) 氩电感耦合等离子体
argon ion gun 氩离子枪
argon ionization detector 氩电离检测器
argon ion laser 氩离子激光器
argon ion-pumped tunable ring dye laser 氩离子泵浦可调环形染料激光器
argon lamp 氩(气)灯
argon-metastable model 亚稳态氩模型
Argonne bubble chamber magnet conductor 阿尔贡气泡室磁(超)导体
argon plasma 氩等离子体
argon-β-ray detector 氩β射线检测器
argon-β-ray ionization cell 氩β射线电离池
argon-β-ray ionization detector 氩β射线电离检测器
argon welding 氩焊
argrinine phosphate 磷酸精氨酸
argvalin 精缬氨素
argyria 银中毒
argyrine 七叶树碱
argyrite(=argyrose) 辉银矿
argyrodite 硫银锗矿
argyrose(=argyrite) 辉银矿
arheol(=santalol) 檀香脑
arhovin 百里香酚苯甲酸酯与二苯胺的化合物

ari 未熟紫胶〔紫胶虫涌散前采收的紫胶〕
aribine(=harman) 哈尔满；阿锐碱
ariboflavinosis 核黄素缺乏症
aricine(=heterophylline) 夹竹桃碱
Aridye pigment padding Aridye 颜料轧染法；颜料树脂固染法
aridyne 埃定〔表示液体干燥剂干燥能力的单位〕
aristeromycin 芒霉素
aristin 马兜铃素
aristol 碘化百里香酚
aristolane 马兜铃烷
aristolene 马兜铃烯
aristolin 马兜铃灵
aristolochic acid 马兜铃酸
aristolochine 马兜铃碱
aristolochinic acid 马兜铃碱酸
aristololactone 马兜铃内酯
aristolone 马兜铃酮
aristoquin 碳酸奎宁
aritasone 土荆芥酮
arithmetical mean 算术(平)均数(值)
arithmetical progression 算术级数
arithmetic average deviation 算术平均偏差
arithmetic mean 算术平均值
arithmetic mean deviation 算术平均(偏)差
arithmetic mean temperature difference 算术平均温度差
arithmetic mean value 算术平均值
arizonite 红钛铁矿
arkansite 黑钛矿
arkite ①阿克特炸药②白榴霞斑岩
arkon〔商〕 抗热及绝缘的浅色脂环饱和烃树脂
arksutite(=chiolite) 锥冰晶石
arm ①臂②杆
armanene 华山松烯
arm-board 起纹板
armentomycin 畜群霉素
armepavine 杏黄罂粟碱
arm graining board 搓纹板；搓花板
arm mixer 桨式搅拌机
arm of mixer 混合机的桨臂
armor 铠装层〔胶管或电缆〕
armored cable 铠装电缆
armored hose 铠装胶管
armoring 加铠装层
armoring machine 电缆铠装机
armorless cable 无铠装电缆
armor-piercing shell 穿甲弹
armoured concrete 钢筋混凝土

armoured glass 装甲玻璃
armoured hose 铠装软管
armo(u)red thermometer 铠装温度计；带套温度计
arm size （多叶形纺丝孔的）叶片尺寸
arm stake 手工刮软
arm stirrer 搅拌机的桨臂
arm straight paddle mixer 直臂桨式混合机
Armstrong acid 阿姆斯特朗酸
Armstrong metal 阿姆斯特朗金
Armstrong's equation 阿姆斯特朗方程式
Army aircraft engine fuel (grade 92 U.S.) （美国）陆军飞机燃料（辛烷值92）
Army all-purpose engine oil （美国）陆军通用机油
Army and Navy number （美国）陆海军汽油品度值
army duck 军用帆布
army grade （陆军）军用级
army pipe line 美国陆军军用输油管
army rating method of aviation fuels （美国陆军）航空燃料辛烷值评价法
army specifications 军用规范
arnatto(=annatto) 胭脂树红
Arnaudon's green 阿尔诺当绿；磷酸铬
Arndt-Eistert synthesis 阿恩特-艾斯特尔特合成
Arndt tube 阿尔恩德特管
arnica oil 山金车油；山菊油
arnica root oil 山金车根油
arnicin 山金车苷
arnicine 山金车花素
arnimite 块铜矿
Arnold's scrubbing bottle 阿诺德洗瓶
Arnold's U-tube 阿诺德U形管
arnotto ①胭脂树②胭脂树萃〔作染料用〕
arochlors 芳氯物
arogel 粉胶
aroly peroxide 芳酰基过氧化物
aroma 香味；香气
aroma chemical 合成香料
aroma constituent 香味组分
aromadendral 桉树属醛
aromadendrane 香树烷
aromadendrene 香橙烯；香树烯
aromadendrin 香树精
aromadendrol 香树醇
aromadendrone 香树酮；香橙酮
aromatic ①芳香的；有香味的②芳(香)族的③芳香剂；香料
aromatic acid 芳香酸
aromatic adsorption index 芳烃吸附指数〔表示催化剂的活性〕
aromatic adsorption method 芳烃吸附法〔测定催化剂表面积〕
aromatic alcohol 芳族醇；芳香醇
aromatic-aliphatic polyoxamide 芳(香)族脂(肪)族聚乙二酰胺(类)
aromatic amine 芳香胺
aromatic azopolyimide 芳族偶氮聚酰亚胺
aromatic azo polymer 芳族偶氮聚合物
aromatic base ①芳香碱②芳香基〔石油〕
aromatic base crude oil 芳香基原油
aromatic based polymer 芳族聚合物；芳族为基础的聚合物
aromatic cacodyl 芳胂基；芳族卡可基〔或指其化合物〕
aromatic carboxylic acid 芳香羧酸
aromatic character 芳香性
aromatic chemicals〔复〕合成香料
aromatic compound 芳香化合物
aromatic content 芳族含量*
aromatic copolyamide microfibre 芳香族共聚酰胺微细纤维
aromatic copolyurea 芳族共聚脲
aromatic cyclization of paraffins 脂族烃的芳构环化
aromatic cyclodehydration 芳化成环脱水作用
aromatic diamine 芳族二胺
aromatic dianhydride 芳族二(酸)酐
aromatic dicarboxylic acid 芳族二羧酸
aromatic diisocyanate 芳族二异氰酸酯
aromatic diprimary amine antioxidant 芳香二伯胺类防老剂
aromatic elixir 香药酒
aromatic ester 芳香酯
aromatic extraction unit 芳烃抽提装置
aromatic-free 不含芳烃
aromatic-free white oil 不含芳烃石蜡油；不含芳烃（白）油
aromatic fuel （含）芳烃燃料(油)
aromatic group 芳(族)基
aromatic halide 芳族卤化物
aromatic homopolyurea 芳族均聚脲
aromatic hydrocarbon 芳(香)烃
aromatic hydrocarbon oil 芳(香)烃油
aromatic hydroxy acid 芳族羟基酸
aromaticin 香菊精
aromaticity 芳香性*
aromaticity content 芳烃含量
aromatic ketone type photoinitiator 芳酮类光引发剂
aromatic macrocyclic oligomer 芳香环状低聚物

aromatic mercurial 芳族汞制剂
aromatic nitration 芳烃硝化(作用)
aromatic nucleophilic substitution 芳香亲核取代
aromatic nylon fibre 芳族聚酰胺纤维；芳族尼龙
aromatic odor 芳香气味
aromatic oxadiozole polymer 芳基噁二唑聚合物
aromatic oxide 芳香醚；芳族氧化物
aromatic para-structure polymer 芳族对位结构聚合物
aromatic petroleum naphtha 芳族石脑油
aromatic plant 芳香植物
aromatic polyamide 芳族聚酰胺；聚芳酰胺
aromatic polyamide acid 芳族聚酰胺酸
aromatic polyamide complex 芳族聚酰胺络合物
aromatic polyamide fibre 芳族聚酰胺纤维
aromatic polyazomethine fibre 芳香族聚甲亚胺纤维；芳香族聚偶氮甲碱纤维
aromatic polybenzoxazole fibre 芳族聚苯并噁唑纤维
aromatic polyester 芳香族聚酯
aromatic polyhydrazide fibre 芳族聚酰肼纤维
aromatic polyimide fibre 芳族聚酰亚胺纤维
aromatic polyisocyanate 芳(香)族多异氰酸酯
aromatic polymer 芳族聚合物
aromatic polymerization 芳(香)族聚合(作用)
aromatic polysulfonamide 聚芳砜酰胺
aromatic radical 芳(族)基
aromatic rearrangement 芳化重排〔取代基自侧链移到芳环上的重排作用〕
aromatic reserve 芳烃储备〔某些燃料中还未被利用的芳烃溶剂能力〕
aromatic rich cut 富芳烃馏分
aromatic ring 芳(族)环
aromatics ①(=aromatic compounds)芳(香)族化合物 ②芳香剂；香料
aromatic series 芳香系
aromatic sextet 芳香六隅*
aromatic solvent-induced shift 芳香溶剂诱导位移
aromatics reformer 芳烃重整装置
aromatic structure 芳族结构
aromatic substance ①芳香物；香料②芳(香)族化合物
aromatic substitution 芳(香)族取代
aromatic sulfinic acid 芳族亚磺酸
aromatic sulfonic acid 芳族磺酸
aromatic sulfonyl azide 芳香磺酰叠氮化物
aromatic sulfuric acid 芳族硫酸
aromatic syrup of rhubarb 大黄香糖浆
aromatic tar 芳香焦油
aromatic tetraamines 芳族四胺类
aromatic tetracarboxylic acid dianhydride 芳族四羧酸二酐（如均苯四甲酸二酐）
aromatic tincture 香酊
aromatic tincture of rhubarb 大黄酊
aromatic type gasoline 芳族型汽油；芳烃汽油
aromatic vinegar 香醋
aromatic water 香水
aromatic wax 芳香蜡
aromatin 香菊亭
aromatization 芳构化
aromatization catalytic process 催化芳构化过程
aromatized polybutadiene 芳构化聚丁二烯
aromatizer 香料；芳化剂
aromatizing agent 芳化剂
aromatizing cracking 芳构裂化
aromatophore 芳香团
aromatous 芳香的
aromoline 阿莫灵
arone 芳酮
aronotta 胭脂树红
Arons chromoscope 阿朗斯验色器
arosolvan process 芳烃溶剂抽提法；N-甲基吡咯烷酮抽提芳烃法
arosorb process 吸附分离芳烃过程〔一种利用硅胶分离芳烃及烷烃的连续生产过程〕
arotinolol 阿罗洛尔〔药〕
aroxy 芳氧基
aroxylamines 芳氧基胺类
aroyl 芳酰基
aroylation 芳酰基化(作用)
aroylorthanilamide 芳酰基邻氨苯磺酰胺
arquerite 银汞齐
arrangement ①设备②布置；排列
arrangement diagram 布置图
array 数组*；阵列
array detector 阵列检测器
array surface acoustic wave sensor 阵列表面声波传感检测器
arrest ①制动②制动装置
arrester 制动室〔用以使气相热裂的裂化反应停止进行〕
arresting ①制动②制动的
arrestive stabilization 抑制性稳定化
arrestment 制动
arrhenal(=new cacodyls) 甲基胂酸二钠
arrhenate 甲基次胂酸盐 MeHAsOOM
arrhenic acid 甲基次胂酸 MeHAsOOH
Arrhenius activation energy 阿伦尼乌斯活化能
Arrhenius equation 阿伦尼乌斯方程*
Arrhenius frequency factor 阿伦尼乌斯频率因素

Arrhenius ionization theory 阿伦尼乌斯电离理论
Arrhenius law 阿伦尼乌斯定律
Arrhenius plot 阿伦尼乌斯曲线图
Arrhenius theory of electrolytic dissociation 阿伦尼乌斯电离理论
Arrhenius viscosity formula 阿伦尼乌斯黏度公式
arrow-poison 箭毒
arrow symbolism 箭头符号
arsa- 砒；砷杂
arsabenzol(=arsphenamine) 砷酚胺
arsacetin(=acetyl atoxyl; acetylarsanilic acid) 对乙酰氨基苯胂酸 $CH_3CONHC_6H_4AsO(OH)_2$
arsaindane 砷杂二氢茚
arsamin 对氨基苯胂酸钠
arsaminol(=arsphenamine) 砷酚胺
arsanilate 阿散酸盐；对氨基苯胂酸盐
arsanilic acid 阿散酸；对氨基苯胂酸 $NH_2C_6H_4AsO(OH)_2$
arsazen 砷氮烯
Arsem furnace 阿尔森炉；碳精棒管式炉
arsena- 砒；砷杂〔用于杂环命名〕
arsenate 砷酸盐(酯)
arsenazo I 偶氮胂 I
arsenazo III 偶氮胂 III
arsenblende(=orpiment) 雌黄
arseneboronic acid 胂硼酸
arsenfahlerz 砷黝铜矿
arseniate(=arsenate) 砷酸盐(酯)
arsenic 砷〔33号元素，化学符号 As〕
arsenic acid 砷酸
arsenic acid anhydride 砷酸酐；五氧化二砷 As_2O_5
arsenical ①含砷的②砷的
arsenical antimony 砷化锑 AsSb
arsenical copper 砷铜合金
arsenical fahlore 砷锑黝铜矿
arsenical iron 砷化铁
arsenical mundic 砷黄铁矿
arsenical nickel(=niccolite) 红砷镍矿
arsenical pyrite 砷黄铁矿；毒砂
arsenical soap 砷皂
arsenical war gases 砷类毒气
arsenic anhydride 砷酸酐；五氧化二砷 As_2O_5
arsenic apparatus 试砷装备；试砷仪器；砷检测装置
arsenic black(=beta-arsenic) 砷黑；β-砷〔砷的同素异形体；黑色无定形固体〕
arsenic bloom 砷华；三氧化二砷 As_2O_3
arsenic bromide 三溴化砷
arsenic butter 砷油；三氯化砷 $AsCl_3$
arsenic chloride ①五氯化砷②三氯化砷
arsenic compound ①砷化合物；含砷化合物②五价砷化合物
arsenic cured 砷处理了的
arsenic dimethyl 二甲砷
arsenic disulfide(=red arsenic) 二硫化二砷；硫化砷；雄黄〔俗〕
arsenic family 砷族
arsenic flowers〔复〕 砷华；三氧化二砷
arsenic fluoride 五氟化砷 AsF_5
arsenic glass 砷玻璃〔玻璃状的三氧化二砷〕
arsenic group 砷组
arsenic halides 砷的卤化物
arsenic hemiselenide 一硒化二砷
arsenic heterocyclic ring 含砷杂环
arsenic hydride 砷化氢 AsH_3
arsenic iodide 五碘化砷 AsI_5
arsenic lime 砷石灰
arsenic lime liquor 硫化砷石灰浸灰液
arsenic minerals 砷矿
arsenic mirror 砷镜
arsenic orange(=orpiment) 雌黄
arsenic oxide 五氧化二砷 As_2O_5
arsenic oxychloride 氧氯化砷
arsenic pentachloride 五氯化砷 $AsCl_5$
arsenic pentafluoride 五氟化砷 AsF_5
arsenic pentaiodide 五碘化砷 AsI_5
arsenic pentaoxide 五氧化二砷 As_2O_5
arsenic pentasulfide 五硫化二砷 As_2S_5
arsenic phosphide 磷化砷
arsenic poisoning 砷中毒
arsenic polymer 砷聚合物
arsenic powder 砷粉；五氧化二砷 As_2O_5
arsenic rhodanate(=arsenic rhodanide) 硫氰化砷 $As(SCN)_3$
arsenic rhodanide(=arsenic rhodanate) 硫氰化砷
arsenic salt 砷盐
arsenic selenide 三硒化二砷
arsenic sequioxide(=arsenic trioxide) 三氧化二砷；砒霜
arsenic stain 砷斑
arsenic sulfide 五硫化二砷 As_2S_5
arsenic sulfide sol 五硫化二砷溶胶
arsenic sulfocyanate(=arsenic sulfocyanide) 硫氰化砷 $As(SCN)_3$
arsenic sulfocyanide(=arsenic sulfocyanate) 硫氰化砷
arsenic test 试砷法
arsenic thiocyanate 硫氰化砷 $As(SCN)_3$
arsenic tribromide 三溴化砷 $AsBr_3$

arsenic trichloride 三氯化砷 $AsCl_3$
arsenic triethide(=arsenic triethyl) 三乙基砷 $As(C_2H_5)_3$
arsenic triethyl(=arsenic triethide) 三乙基砷
arsenic trifluoride 三氟化砷
arsenic trihydride 砷化三氢 AsH_3
arsenic triiodide 三碘化砷 AsI_3
arsenic trimethide(=arsenic trimethyl) 三甲基砷 $As(CH_3)_3$
arsenic trimethyl(=arsenic trimethide) 三甲基砷
arsenic trioxide(=arsenic sequioxide) 三氧化二砷；砒霜
arsenic triphenyl 三苯基砷 $As(C_6H_5)_3$
arsenic triselenide 三硒化二砷 As_2Se_3
arsenic trisulfide 三硫化二砷；雌黄〔天然产〕
arsenic tube 试砷管
arsenic white 砒霜；三氧化二砷 As_2O_3
arsenic yellow 砷黄 As_2S_3
arsenic ylide 砷叶立德
arsenide 砷化物
arsenidine(=arsepidine) 六氢砷吡啶；砷杂环己烷
arseniopleite 红砷铁矿
arseniosiderite 菱砷铁矿
arsenious ①(=arsenous)亚砷的；三价砷的②砷的
arsenious acid 亚砷酸〔药〕
arsenious trisulfide sol 三硫化二砷溶胶
arsenite 亚砷酸盐
arsenite ion 亚砷酸离子
arsenium 砷
arseniuretted hydrogen 砷化三氢 AsH_3
arsenius 亚砷(的)
arseno- 偶砷基 —As≡As—
arseno acetic acid 偶砷乙酸
arseno-benzene 偶砷苯
arsenobenzene-sodium 偶砷苯钠
arsenobenzol(=arsphenamine) 砷酚胺；胂凡纳明；606〔药〕
arsenocholine 砷胆碱
arsenoclasite 水锰砷石
arseno-compound 偶砷化合物
arsenofuran(=arsenophen) 砷呋喃
arsenogen 砷元
arseno-group 偶砷原子团
arsenolite 砷华
arseno-methane 偶砷甲烷
arsenometric titration 亚砷酸滴定法
arsenometry 亚砷酸滴定法
arsenomolybdate 砷钼酸盐
arsenophen(=arsenofuran) 砷呋喃
arseno-phenol 偶砷(苯)酚
arsenopyrite 毒砂；砷黄铁矿

arsenoso- 亚砷酰；氧砷基 $O=As—$
arsenoso alkane 烷基亚砷酰
arsenotungstic acid(=arsenowolframic acid) 钨砷酸 $As_2O_5·24WO_3·7H_2O$
arsenous 亚砷〔三价〕
arsenous acid 亚砷酸
arsenous acid anhydride 亚砷酸酐；三氧化二砷 As_2O_3
arsenous bromide 三溴化砷 $AsBr_3$
arsenous chloride 三氯化砷 $AsCl_3$
arsenous compound (含)三价砷化合物
arsenous ester 亚砷酸酯
arsenous ethide 三乙基砷 $(C_2H_5)_3$
arsenous fluoride 三氟化砷 AsF_3
arsenous hydride 三氢化砷
arsenous iodide 三碘化砷 AsI_3
arsenous methide 三甲基砷 $As(CH_3)_3$
arsenous oxide 三氧化二砷 As_2O_3
arsenous oxychloride 一氯氧化砷 $AsOCl$
arsenous phosphide 一磷化砷 AsP
arsenous selenide 三硒化二砷 As_2Se_3
arsenous sulfide 三硫化二砷 As_2S_3
arsenous thiocyanate 硫氰化砷 $As(SCN)_3$
arsenowolframic acid(=arsenotungstic acid) 钨砷酸 $As_2O_5·24WO_3·7H_2O$
arsepidine(=arsenidine) 六氢砷吡啶
arsimycin 阿尔西霉素
arsinate 次胂酸盐 $R_2AsO(OM)$
arsindole 砷杂茚
arsine 胂*
arsine cadmium complex 胂镉配合物；胂镉复盐
arsine oxide 胂化氧
arsinic acid 次胂酸〔亚砷酸 H_3AsO_3 的有机衍生物〕
arsinico 次胂酸亚基 $(HO)OAs=$
arsinicoarene 二芳基亚胂酸
arsino 胂基 $H_2As—$
arsinoalkane 烷基胂
arsino-ethane 胂基乙烷；乙基胂 $C_2H_5AsH_2$
arsinoline 胂杂萘 C_9H_7As
arsino-oxybenzophenone 氧胂基二苯(甲)酮
arsinous acid 卑胂酸 R_2AsOH
arso 二氧砷基 $O_2As—$
arsonate 胂酸盐 $RAsO(OM)_2$
arsonation(=arsonylation) 胂酸化(作用)
arsonic acid 胂酸*
arsonic acid type resin 胂酸型树脂
arsonium 砷鎓
arsonium chloride 氯化砷鎓 R_4AsCl
arsonium compound 砷鎓化合物

arsonium halide 卤化砷鎓 R_4AsX
arsonium hydroxide 氢氧化砷鎓 R_4AsOH
arsonium ion 砷鎓离子
arsono- 胂羧基 $(HO)_2OAs—$
arsono alkane 烷基胂酸
arsonylation(=arsonation) 胂酸化(作用)
arsphenamine(=arsenobenzol) 砷酚胺；胂凡纳明；606〔药〕
arsycodile(=sodium cacodylate) 二甲次胂酸钠
arsyl 胂基；二氢胂基 $H_2As—$
arsylene 亚胂基 $HAs=$
artabotrine(=isocorydine) 异紫堇定
artabsin 苦艾素
artarine 椒根碱
art bronze 美术青铜；工艺品青铜
artemazulene 蒿薁
artemether 蒿甲醚〔药〕
artemetin 蒿亭
artemisal 蒿醛；2-甲基丙烯醛
artemisane 蒿烷
artemisia ketone 蒿属酮
artemisia oil 苦艾油
artemisic acid 青蒿酸
artemisin 青蒿素
artemisinin 青蒿素〔药〕
artemisol 青蒿醇
artemone 青蒿酮
arterenol 去甲肾上腺素
arteriosclerosis 动脉硬化
artery 动脉
artesunate 青蒿琥酯〔药〕
art glass 美术玻璃
art-gum eraser 炭画橡皮
artichoke root oil 朝鲜蓟根油
artifact pollution 人为污染
artificial 人工的；人造的
artificial abrasive 人造磨料〔如金刚砂、刚玉等〕；人造磨蚀剂
artificial accelerated exposure test 人工加速曝置试验
artificial ageing 人工老化；人工陈化；人工时效
artificial ageing test 人工老化(时效)试验
artificial aging 人工老化；人工陈化；人工时效
artificial aging fiber 人造纤维
artificial almond oil 人造杏仁油
artificial anthracite 人造无烟煤
artificial asphalt 人造沥青；人造柏油
artificial bate 人造软化剂
artificial catgut 人造肠线
artificial cognac oil 人造康酿克油〔商〕；庚酸乙酯

artificial colloid 人造胶体
artificial contamination 人工污染
artificial cooling 人工冷却
artificial corrosion product ①人工锈蚀物②人工锈
artificial cotton 人造棉
artificial disintegration 人工蜕变
artificial drying 人工干燥
artificial element 人造元素*
artificial endonuclease 人工核酸内切酶
artificial fat 人造油脂
artificial fertilizer 人造肥料
artificial fiber 人造纤维
artificial filament ①人造纤维②人造丝
artificial flavour 人造食用香料；调制食用香精
artificial flavouring 人造食用香料
artificial flower oil 人造花油；人工调制花油
artificial food flavouring 人造食品香料
artificial gas 人造气体
artificial gem 人造宝石
artificial gene 人工基因
artificial geranium oil 调制香叶油
artificial graphite 人造石墨〔填料〕
artificial guano 人造(海)鸟粪
artificial hard water 人造硬水
artificial horn 人造牛角
artificial intelligence 人工智能
artificial intelligence system 人工智能系统
artificial iron oxide 人造氧化铁
artificial latex 合成胶乳
artificial leather 人造革
artificial light 人造光
artificially induced nuclear reaction 人工诱导(原子)核反应
artificial manure 人造肥料
artificial mica 人造云母
artificial mustard oil 人造芥子油 $CH_2=CHCH_2NCS$
artificial neroli oil(=methyl anthranilate) 人造橙花油；邻氨基苯甲酸甲酯
artificial neural network 人工神经网络
artificial nitro musk 人造硝基麝香
artificial nuclear disintegration 人工核蜕变
artificial oil iudustry 人造油类工业
artificial petroleum industry 人造石油工业
artificial process 人工法
artificial protein fibre 人造蛋白纤维
artificial puer 人造软化剂
artificial radioactive nuclide 人造放射性核素；人工放射性核素

artificial radioactive soil 人造放射性污垢
artificial radioactivity 人工放射性
artificial (radio) element 人工放射性元素*
artificial resin 人造树脂
artificial rubber ①合成橡胶②水分散橡胶
artificial rubber latex 人造胶乳
artificial rust ①人工锈蚀②人工锈；人工锈蚀物
artificial sea water 人造海水
artificial silk 人造丝
artificial silk black 人造丝染用黑
artificial silk from collodion 火棉胶人造丝；硝酯人造丝
artificial soil 人造污垢
artificial spice oil 人造调味油料
artificial standard 代用标准物质
artificial structure(=secondary structure) 二次结构〔炭黑〕
artificial sweetener 人造甜味剂
artificial vaseline 人造凡士林；人造矿脂
artificial weathering ①人工老化②人工天候
artificial wool 人造毛
artisan tannery 皮坊
artist board 绘画纸板
artists' oil 美术油〔特选的油不发黄〕
art printing paper 美术印刷纸
ar-turmerone 芳郁金酮
artware glaze 美术釉
artwork 底片
arvomycin(=arbomycin) 阿鲍霉素
aryl- 芳(香)基
arylamide 芳基酰胺
arylamine blend 芳香胺混合物〔防老剂〕
arylamines 芳基胺(类)
arylaminonaphthoquinones 芳氨基萘醌类
aryl arsine oxyhalide 卤氧化芳(基)胂 $ArAsOX_2$
aryl arsine sesquisulfide 三硫化二芳(基)胂 $Ar_2As_2S_3$
aryl arsine tetrahalide 四卤化芳(基)胂 $ArAsS_4$
aryl arsonic acid 芳(基)胂酸 $RAsO(OH)_2$
arylate ①芳基化物②芳化
arylated 芳化了的
arylated alkyl 芳脂基
arylating ①芳基化的②芳基化(作用)
arylating agent 芳基化剂
arylation 芳基化*
1-arylazoazulenes 1-芳基偶氮薁
arylazobenzthiazole dyes 芳基偶氮苯并噻唑染料
4-arylazo-5-pyrazolone 4-芳基偶氮-5-吡唑酮
arylazothiazole dyes 芳基偶氮噻唑染料
arylazothiophene dyes 芳基偶氮噻吩染料
aryl bromide 芳基溴

aryl cation 芳正离子
aryl chloride 芳基氯
aryl chlorofluorosilane 芳基氟氯硅烷
arylchlorosilane 芳基氯硅烷
aryl compound 芳基化合物
aryl cyclic sulfonium zwitter ion 芳环锍两性离子
aryl diarsonic acid 芳基二胂酸 $Ar(AsH_2O_3)_2$
aryl diazonium chloride 氯化芳基重氮 ArN_2Cl
aryl diazonium halide 卤化芳基重氮 ArN_2X
aryl-diazonium initiator 芳基重氮型引发剂
aryle 芳基金属(化合物)
arylene 亚芳基 —Ar—
arylene N-alkylhydrazide 芳撑-N-烷基酰肼
arylene-amide fibre (全)芳撑(聚)酰胺纤维；(聚)芳撑酰胺纤维
arylene-heterocyclic type 芳撑杂环型
arylene oxadiazole 芳撑噁二唑
arylene unit 芳撑单元
aryl fluoride 芳基氟
aryl group 芳基
aryl halide 芳基卤
aryl hydrogen sulfate 硫酸氢芳基酯 $ArOSO_3H$
arylhydroxamic acid 芳基异羟肟酸 $RCONAr(OH)$
aryl hydroxyacetic acid 芳氧基乙酸 $ArOCH_2COOH$
arylide ①芳基化物②芳基金属(化合物) ArM
arylidene 亚芳基 $Ar=$
arylide red 芳基红
aryl iodide 芳基碘
aryl isocyanate 异氰酸芳基酯
aryl lithium 芳基锂 $ArLi$
arylmethanol initiator 芳基甲醇引发剂
aryl methyl ketone 甲基芳基(甲)酮
aryl olefin 芳基烯烃
aryl oxide 芳(族)醚
aryloxy 芳氧基
aryloxy acetic acid 芳氧基乙酸 $ArOCH_2COOH$
aryl phosphate 磷酸芳基酯 $(ArO)_3PO$
aryl phosphine 芳基膦
arylphosphinic acid 芳基次膦酸
aryl phosphite 亚磷酸芳酯
arylphosphonic acid 芳基膦酸 $ArPO(OH)_2$
arylpolyamide 芳基聚酰胺
aryl radical 芳基
aryl residue 芳基
aryl silazane 芳基硅氮烷
aryl stearic acid 芳基硬脂酸
aryl stearic acid soap 芳基硬脂酸皂
aryl stearo nitrile 芳基硬脂腈

aryl substituted 芳基取代了的
aryl sulfamic acid 芳代氨基磺酸 ArNHSO$_3$H
aryl sulfonate 芳基磺酸盐
aryl sulfonic acid 芳基磺酸
aryl sulfonyl chloride 芳基磺酰氯 ArSO$_2$Cl
aryl sulphatase 芳基硫酸酯酶
aryl sulphonate 芳基磺酸盐
arylthioacetylene 芳硫基乙炔
arylthioglycol(l)ic acid 芳巯基乙酸；芳基氢硫基乙酸
aryl thiourea 芳基硫脲
aryne 芳炔*
asafetida ①阿魏②阿魏胶
asafoetida ①阿魏②阿魏胶
asafoetida oil 阿魏油
asalob 苏丹蜜酒
asaprol(=abrastol) β-萘酚-α-磺酸钙
asarabacca 欧细辛
asaresinotannol 阿魏树脂鞣醇
asarin 细辛脑
asarinin 细辛素
asarite 粗细辛脑
asarol 细辛醇
asarone (=propenyl-2,4,5-trimethoxy benzene) 细辛脑；2,4,5-三甲氧苯基丙烯
asaronic acid(=asarylic acid) 细辛酸
asarum oil 细辛油
asaryl(=2,4,5-trimethoxyphenyl) 细辛基；2,4,5-三甲氧苯基
asarylaldehyde 细辛醛；2,4,5-三甲氧基苯甲醛
asarylic acid(=asaronic acid) 细辛酸
asbest 石棉
asbestes fiber 石棉纤维
asbestic 石棉的
asbestine 石棉粉；滑石粉
asbestinite 角闪石棉；石棉
asbesto board 石棉纸板
asbestos(=asbestos) 石棉
asbestos alternative 石棉代用品
asbestos board 石棉板
asbestos brick 石棉砖
asbestos cardboard 石棉纸板
asbestos cement 石棉水泥
asbestos center gauze 石棉心铁丝网
asbestos-ceramic fibre composite 石棉-陶瓷纤维复合材料
asbestos cord 石棉绳
asbestos diatomite 石棉硅藻土
asbestos fabric 石棉织物
asbestos fiber 石棉纤维；石棉绒
asbestos fibreboard 石棉纤维板
asbestos fibre reinforced plastics 石棉纤维增强塑料
asbestos filter 石棉过滤器
asbestos filter cloth 石棉滤布
asbestos fines(=asbestos float) 石棉绒
asbestos float(=asbestos fines) 石棉绒
asbestos gasket 石棉垫片
asbestos gauge 石棉网
asbestos gloves 石棉手套
asbestos magnesia mixture 石棉镁氧混合物
asbestos mat 石棉毡
asbestos mechanical goods 石棉工业制品
asbestos-metallic wool 石棉-金属绒
asbestos mica 石棉云母
asbestos millboard 石棉书皮纸板
asbestos mortar 石棉灰浆
asbestos packing 石棉填料
asbestos pad 石棉垫
asbestos paper 石棉纸
asbestos phenolic tape 石棉酚醛带
asbestos plaster 石棉粉饰；石棉灰浆
asbestos plate 石棉板
asbestos pollution 石棉污染
asbestos-reinforced plastics 石棉增强塑料
asbestos-reinforced polypropylene 石棉增强聚丙烯
asbestos-reinforcement 石棉增强作用
asbestos rope 石棉绳
asbestos rubber sheet 石棉胶板；夹胶石棉板
asbestos slate 石棉板
asbestos sliver 石棉条
asbestos tape 石棉(扁)带
asbestos vein 石棉矿脉
asbestos wire gauze 石棉衬(铉)网
asbestos wool 石棉绒
asbestos yarn 石棉线
asbest paper 石棉纸
asbest sheet 石棉板
asbestus(=asbestos) 石棉
asbetumen 石棉沥青
asbolin 松根油
asbolite 钴土
ascaricide 杀蛔虫药
ascaricidin 杀蛔菌素
ascaridic acid 驱蛔酸
ascaridol(e) 驱蛔萜；1,4-过氧对䓝烯
ascaridolic acid 驱蛔脑酸
ascarite(=soda-asbestos) 苏打-石棉；烧碱石棉

ascaryl alcohol 蛔脂二醇
ascarylose 蛔糖；3,6-二脱氧-L-甘露糖
as-cast ①铸态的②铸出后不加工的③铸出后加工但不进行热处理
ascending chromatography 上行色谱(法)；上行层析
ascending-descending chromatograph 升降色层分离(法)；上行下行色层分离法
ascending development 上行展开(法)*
ascending development method 上行展开法
ascending method 上行法
ascending paper chromatography 上行纸色谱法
ascending paper partition 上行纸分配(法)
ascending paper partition chromatography 上行纸分配色谱法
ascending pipe 直管；(泵的)注入管；(泵的)压入管
ascending retrograde technique 上行退回技术；上升逆行技术
ascending technique 上行技术
ascending tube 上行管
ascending-water-tube boiler 上升水管式锅炉
ascension pipe 上行管
ascharite 纤维硼酸镁石
aschistic process (由功)直接生热法
asclepidin 浓萝藦素
asclepidoid 混萝藦素
asclepin 萝藦类脂；马利筋苷
as-coagulated 初凝态的；凝聚态的
ascorbate 抗坏血酸；维生素 C
ascorbic acid(=vitamin C) 抗坏血酸；维生素 C
ascorbic acid oxidase 抗坏血酸氧化酶；维生素 C 氧化酶
ascorbic oxydase 抗坏血酸氧化酶
ascorbimetry 抗坏血酸滴定法
ascorbinometric titration 抗坏血酸滴定(法)
ascorbyl palmitate 棕榈酸抗坏血酸酯
ascorbyl stearate 硬脂酸抗坏血酸酯
ascosterol 子囊甾醇
as-drawn fibre 初拉伸纤维
asebotin 马醉苷；根皮苷
asebotoxin 马醉毒
aseismic design 抗震设计；耐(地)震设计
asellin 鳖肝油碱
asepsin 对溴乙酰苯胺
asepsis ①无菌法②防腐性
aseptol(=sozolic acid) 苯酚-2-磺酸；搔早酸
asferryl 砷酒石酸铁
as-formed fiber 初生纤维
ash 灰分*
ash alkalinity 灰(分)碱度

ash analysis 灰分分析
ash-atomize curve 灰化-原子化曲线
ash bin ①出灰桶②煤灰桶
ash bowl 灰碗
ash coal 多灰分煤
ash composition 灰分组成
ash content 灰分含量
Ashcroft cell 阿希克罗夫特电池
Ashcroft paper tester 阿希克罗夫特验纸器
Ashcroft sodium process 阿希克罗夫特制钠法
ashe chinkapin 灰栗
ashed 灰化了的；成了灰的
ashery ①灰坑②(浸灰制)钾碱厂〔从灰中取碳酸钾〕
ashes〔复〕 灰
ash fire 灰火；余烬
ash-forming impurities 成灰杂质
ash free 无灰的
ash-free basis 无灰计算
ash-free coal 不含灰分煤
ash fusibility 灰分熔度
ash fusion temperature 灰熔温度
ash gray 淡灰色的；灰白色的
ash hole(=ash pit) 灰坑
ashing ①灰化②抛光；磨光
ashing aid 灰化助剂
ashing curve 灰化曲线
ashing method 灰化法
ashing stage 灰化阶段
ashing temperature 灰化温度
ashing time 灰化时间
ashio base I 芦尾碱 I
ash-lagoon 灰分处理池；炉灰处理池
ashlar(=ashler) 琢石；方石
ashlar brick 琢石砖
ashlaring(=ashlering) 砌琢石
ashler(=ashlar) 琢石；方石
ashlering(=ashlaring) 砌琢石
ashless 无灰的
ashless additive 无灰添加剂
ashless detergent 无灰洗涤剂
ashless detergent dispersant 无灰洗涤分散剂
ashless dispersant 无灰分散剂
ashless filter 无灰滤纸
ashless filter paper 无灰滤纸
ash loss 灰分损失〔灰分烧失量〕
ash (of soda) 苏打灰
ash on dry coal 干煤灰分
ash pan 灰盘

ashphalt 地沥青
ash pit(=ash hole) 灰坑
ash pit door 灰坑门
ash plot 灰化曲线
ash removal(=ash removing) 脱灰
ash-rich fuel 多灰燃料
ash roaster 苏打灰煅烧炉
ash-seed oil 梣子油
ash slate 灰板岩
ash specification 灰分
ash structure 火山灰结构
ash temperature 灰化温度
ash test 灰分试验
ash tray 灰盘
ashy ①灰的②含灰的③灰色的
ash zone 灰层
asiaticosid 积雪(草)苷
asiminine 万寿果碱
asiphyl 氨基苯肿酸汞
as-is fibre 原态纤维
asomate 福美砷〔杀菌剂〕
asp 山杨
asparacemic acid 消旋天冬氨酸
  COOHCHNH₂CH₂COOH
asparagic acid 天冬氨酸 COOHCHNH₂CH₂COOH
asparaginase 天冬酰胺酶；L-天冬酰胺酶〔药〕
asparaginate 天冬酰胺
asparagine(=aspargine) 天冬酰胺
asparaginic acid 天冬氨酸
asparaginyl-(=asparagyl) 天冬酰胺酰(基)
asparagosin 天冬果聚糖
asparagus root oil 石刁柏根油
asparagyl 天冬酰(基)
asparamide(=aspartamic acid) 天冬酰胺
aspargine(=asparagine) 天冬酰胺
aspartame 天冬甜素；天冬甜母
aspartamic acid(=asparamide) 天冬酰胺
aspartate ①天冬氨酸②天冬氨酸盐(酯或根)
aspartic acid 天冬氨酸
aspartic acid amlodipine L-天冬氨酸氨氯地平〔药〕
aspartic-β-semialdehyde 天冬氨酸-β-半醛；丁氨醛酸
aspartocin 天冬菌素
aspartylphosphate 天冬氨酰磷酸
aspect ratio 纵横比；长径比；(轮胎)扁平率
aspergillus flavus 黄曲霉
asperolite 富水硅孔雀石
asperula odorata oil 香车叶草油
asphalt(=bitumen) (地)沥青

asphalt adhesive 沥青黏合剂
asphalt aluminium paint 铝粉沥青漆
asphalt base 沥青基
asphalt base crude oil 沥青基原油
asphalt base oil 沥青基油料〔从沥青基石油制得的产品〕
asphalt-bearing shales 含沥青页岩
asphalt-bearing stocks 含沥青原料
asphalt board 沥青纸板
asphalt cement 沥青膏
asphalt clay emulsion 沥青-白土乳化液
asphalt content 沥青含量〔石油产品的〕
asphalt cutback 沥青稀释物；稀释沥青
asphalt emulsifier 沥青乳化剂
asphalt emulsion 沥青乳胶体
asphalt emulsion coating 沥青乳化漆；沥青乳胶漆
asphaltene 沥青质
asphaltene nucleus 沥青质核
asphalt felt 沥青毡；油(毛)毡
asphalt filler 沥青填充物；沥青填料
asphalt flux 沥青助熔剂
asphalt-free oil 不含沥青油类
asphalt furnace 沥青炉
asphalt grouting 沥青化(作用)
asphaltic acid 沥青酸
asphaltic base 沥青基
asphaltic base oil 沥青基油料
asphaltic bitumen 沥青
asphaltic cement 沥青膏
asphaltic hydrocarbons 沥青的烃类
asphaltic limestone 沥青石灰岩
asphaltic liquid 沥青液
asphaltic mastic 沥青砂胶
asphaltic nature 沥青属性
asphaltic petroleum 沥青质石油
asphaltic products 沥青产品
asphaltic pyrobitumen 石油(焦)沥青；沥青岩
asphaltic residue 沥青基残油
asphaltic road oils 沥青质铺路油；铺路沥青
asphaltic sand 沥青砂
asphaltic sandstone 沥青砂岩
asphaltic substances 沥青物质；沥青状物质
asphaltine 沥青质
asphaltite 沥青岩
asphaltization 沥青化(作用)
asphaltization of residue 残渣油沥青化
asphalt jute serving 沥青浸黄麻包扎(被复)防腐(锈)法
asphalt lac 沥青漆；沥青胶
asphalt mastic 沥青砂胶

asphaltogenic　成沥青的
asphalt oil　沥青质油料
asphaltos　地沥青
asphaltous acid anhydrides　沥青酸酐
asphalt paint　沥青漆
asphalt penetration index　沥青透入度指数
asphalt penetration test　沥青透入度试验
asphalt plant　制造沥青装置
asphalt powder　沥青粉末
asphalt prime coat　路面头道沥青；沥青底漆
asphalt primer　沥青透层；路面头道沥青
asphalt processing　沥青加工
asphalt refinery　沥青工厂
asphalt-retaining power　保持沥青能力
asphalt rock　沥青岩
asphalt (roofing) felt　沥青(屋顶)毡；油毛毡〔俗〕
asphalt-rubber road surfacing　沥青橡胶路面
asphalt saturated felt　沥青毡
asphalt sludge　沥青淤渣
asphalt softening point　沥青软化点
asphalt solvent blend　沥青溶剂混合物
asphalt stock　沥青原料
asphalt stone　沥青石；地沥青
asphalt tar　沥青焦油
asphalt testing apparatus　沥青试验器
asphalt thermometer　沥青温度计
asphalt tile　沥青砖
asphaltum　沥青
asphaltum fume analysis　沥青烟分析
asphaltum oil　残渣油；液态沥青
asphalt varnish　沥青漆
asphalt viscosity　沥青黏度
asphalt volatile constituents　沥青挥发分
aspherical beam condenser　非球面束聚光器；非球面束聚光镜
asphogenin　黄花水仙配基
asphonin　黄花水仙苷
asphyxia　窒息
asphyxiant　①窒息剂②窒息的
asphyxiant gas　窒息气
aspic　穗薰衣草
aspic oil　穗薰衣草油
aspidium oleoresin　绵马油树脂
aspidosamine　绵马胺　$C_{22}H_{28}O_2N_2$
aspidospermine　白坚木碱；楮籽碱
aspirated hygrometer　吸气湿度计
aspiration apparatus for urea determination　吸气定脲装置

aspirator　吸气器
aspirator bottle　吸气瓶
aspirator pump　吸气泵；抽气泵
aspirator threading pistol　吸丝枪
aspirin(=acetyl salicylic acid)　阿司匹林；乙酰水杨酸
aspirochyl(=mercuric atoxylate)　氨基苯胂酸汞
Asplund defibrator　阿斯普伦德单动纤维分离机
as received　按来样计算
assafetida(=asafetida)　①阿魏②阿魏胶
assay　①测定②检定
assay balance　试金天平
assay flask　检定瓶
assaying　①试金；分析矿物②测定；分析③鉴定
assaying table　检定板
assay lead　检定铅
assay method　测定法
assay mill　试验研磨机
assay plate　测定平板
assay reagent　试金试剂
assay strain　检定菌株
assay ton　检定吨〔=29.1667 克〕
assay ton system　化验吨制
assay ton weights　检定吨砝码
assay tool　试金用具
assay vessel　试金器(具；皿)
assemble　装配
assembled light source　组合光源
assembled unit　组合单元；组装单元
assembler　汇编程序
assembling department　①成型车间②装配车间
assembling drawing　装配图
assembling solution　贴合用胶浆
assembling table　成型操作台
assembly　①成型；贴合②装配③系集④组合件
assembly adhesive　装配黏合剂
assembly beaming　并轴
assembly department　装配间
assembly diagram　装配图
assembly drum　成型鼓
assembly (for adhesives)　装配件
assembly glue　装配胶
assembly jig　装配架
assembly language　汇编语言
assembly line method　流水作业法
assembly of independent particles　独立粒子系集
assembly of localized particles　定域粒子系集
assembly of nonlocalized particles　非定域粒子系集
assembly parts　组合件

assembly program 汇编程序
assembly room 成型车间
assembly time 装配时间〔从涂胶至黏合完成〕
assertive colours ①醒目色②鲜明色③警觉色
assessment 查定；估计；估价
assessment of atmospheric environment 大气环境评价
assessment of singeing 烧毛评级
assessor 鉴定器；鉴定管
assignment of configuration 构型的排布
assimilatory coefficient(=assimilatory quotient) 同化系数
assimilatory quotient(=assimilatory coefficient) 同化系数
assistant diagnosis 辅助诊断
assistant flux 助熔剂
assistants 助剂
assisted bacterial coagulation 细菌助凝固法
assisted polymerization 协助聚合
assistor ①加力器；助推器②辅助装置
associated 缔合了的
associated dislocation 缔合位错
associated ion 缔合离子
associated liquid 缔合液体
associated molecule 缔合分子
associated reaction 缔合反应
associated substance of cellulose 纤维素缔合物
associating ①缔合的②缔合(作用)
association 缔合(作用)
association colloid 缔合胶体
association complex 缔合络合物
association constant 缔合常数
association degree 缔合度
association-disassociation equilibrium 缔合解离平衡
association force 缔合力
association heat 缔合热
association in organic phase 有机相中的缔合
association number 缔合数
association of defect 缺陷的缔合
Association of Official Analytical Chemists (美国)分析化学家协会
association polymer 缔合聚合物
association reaction 缔合；缔合反应*
associative ability 缔合能力
associative ligand exchange mechanism 配体交换的缔合机理
associative thickener 缔合性增稠剂
assorting 分类
assortment ①品种；花式②分类；分级拣选；拣选；挑拣
assouplissage 生丝半脱胶(工艺)
as-sprayed coating 喷涂态涂层

as-spun fibre 初纺纤维；初生纤维
assumption 假设
assurance factor 安全系数
assured processibility factor(APF) 安全操作系数
assymmetric carbon atom 不对称碳原子
assymmetry 不对称(现象)
assymmetry effect 不对称效应
astacin 虾红素
A-stage 甲阶；A阶段
A-stage resin 甲阶树脂；可熔阶段树脂
astatic 不稳定的；无定向的
astatide 砹化物
astatine 砹〔85号元素，化学符号At〕
astatism 无定向性
astatium 砹
astatotyrosine 砹代酪氨酸
astaxanthin 虾青素
astemizole 阿司咪唑〔药〕
asteriasterol 海星甾醇
asterin(=chrysanthemin) 紫菀苷；花青-3-葡萄糖苷
asterium 假设存在于最热的星球中的一种元素
asteroidal theory 小行星学说
asteromycin 星霉素
asterrubin(=asterubin) 海星红素；二甲胍基乙磺酸
astersaponin 紫菀皂草苷
asterubin 海星红素；二甲胍基乙磺酸
asthenics 弱肌力药
astigmatic aberration 像散(像)差〔像散焦距差〕
astigmatism 像散性；像散现象
astigmatism compensation 像散补偿
ASTM appendix number ASTM临时规范号
ASTM colour standard ASTM标准色泽
ASTM designation 美国材料试验学会标准编号
ASTM end point ASTM终沸点〔根据ASTM法蒸馏石油产品的终沸点〕
ASTM gum test ASTM胶质测定
ASTM melting point of paraffin ASTM石蜡熔点〔根据ASTM方法测定的石蜡熔点〕
ASTM octane number ASTM辛烷值〔马达法〕
ASTM sampling procedures ASTM取样程序
ASTM specifications ASTM技术规范
ASTM standards 美国材料试验学会标准；ASTM标准
ASTM steam emulsion number ASTM水蒸气乳化值〔油类与水蒸气成乳浊体的特性〕
ASTM timer tube 美国材料试验学会式自动定时管
Aston's mass spectrograph 阿斯顿质谱仪
astrakanite(=astrochanite) 白钠镁矾
astral blue 星蓝

astralite 星字炸药；奥司脱拉特〔硝铵、硝化甘油、三硝甲苯炸药〕
astral oil 星油；变质精制石油
astringency ①涩味②收敛性
astringent(=adstringent) ①收敛剂；涩剂②涩嘴的
astringent cream 收敛霜
astringent lotion 收敛露
astringent matter ①收敛剂②鞣剂
astringent otion 收敛洗液
astrochemistry 天文化学
astrophyllite 星叶石
astroquartz fabric 宇航石英织物〔可耐1093℃高温〕
astrotone〔商〕 十三烷二羧酸环乙撑酯；麝香T〔商〕
asym- 不对称；偏
asym-dicyclohexyl-21-crown-7 不对称二环己基并-21-冠(醚)-7
asymmetric 不对称的
asymmetric absorption band 不对称吸收谱带
asymmetrical peak 不对称峰
asymmetrical top 不对称陀螺
asymmetric amino acid 不对称氨基酸
asymmetric atom 不对称原子
asymmetric carbon 不对称碳*
asymmetric carbon atom 不对称碳原子
asymmetric catalyst 不对称催化剂
asymmetric center 不对称中心
asymmetric compound 不对称化合物
asymmetric fission 非对称裂变；不对称裂变
asymmetric focusing 不对称聚焦
asymmetric hollow fibre 非对称性中空纤维
asymmetric hydrolysis 不对称水解
asymmetric induced copolymerization 不对称诱导共聚(作用)
asymmetric induction polymerization 不对称诱导聚合
asymmetric inductive effect 不对称诱导效应
asymmetric ligand 不对称配体
asymmetric magnetic lens 不对称磁透镜
asymmetric molecule 不对称分子
asymmetric polymer 不对称聚合物
asymmetric potential 不对称电势
asymmetric reflection 不对称反射
asymmetric selection polymerization 不对称选择聚合(作用)
asymmetric stereoselective polymerization 不对称立体选择性聚合
asymmetric stretching vibration 不对称伸展振动
asymmetric surface ridge 不对称表面皱纹
asymmetric synthesis 不对称合成

asymmetric system 不对称系统
asymmetric top molecule 不对称陀螺分子
asymmetric unit 不对称单位
asymmetry 不对称(现象)
asymmetry-based method 不对称法
asymmetry factor 不对称因子
asymmetry of peak 峰的不对称性
asymmetry parameter 非对称参数
asymmetry potential 不对称电势
asymptotes 渐近线
at.%(=atomic percent) 原子百分数；原子浓度
atabrine 疟涤平；阿的平
atacamite(=atakamite) 氯铜矿
atactic 无规(立构)的
atactic block 无规立构嵌段
atactic chain 无规(立构)链
atacticity 无规度；无规立构度
atactic polymer 无规立构聚合物*
atactic polymerization 无规聚合
atactic polypropylene(APP) 无规立构聚丙烯
atactic triads 无规立构三单元组
atactosol 无规溶胶〔光学等向性溶胶〕
atakamite(=atacamite) 氯铜矿
-ate〔词尾〕用于由词尾为-ic的酸所成的盐类或酯类的名称
atebrin(e)〔商〕 疟涤平；阿的平
ate complex 酸根型配合物
atenolol 阿替洛尔〔药〕
athamantin 阿塔曼苦素
athermal 无热的
athermal nucleation 无热成核(过程)
athermal solution 无热溶液*
athermancy 绝热性
athermanous 绝热的；不透热的
athermic effect 绝热效应
atherosclerosis (动脉)粥样硬化
atherosperma oil 澳洲黄樟油
atherospermidine 芒籽定
atherosperminine 芒籽宁
atidine 阿替定〔一种新型双萜碱〕
atisine 阿替素；异叶乌头碱
Atkinson hemin test 阿特金桑血晶试验
atlanolin 阿特兰素
Atlantic process(=Atlantic catalytic reforming process) 大西洋(催化重整)法
Atlantic test 大西洋试验〔大西洋炼油厂催化剂活性评价方法〕
atlanton(e) 大西洋(萜)酮

atlapulgite 活性白土
atlas cedarwood oil 大西洋雪松木油
Atlas fedometer 阿特拉斯褪色仪
Atlas powder 阿特拉斯炸药
atlas stitch 经缎组织
atlas stitch with two-needle underlap 隔针经缎组织
at-line 现场
at-line analysis 现场分析
atm ①(=atmosphere)大气②(=atmospheric pressure)(标准)大气压
atmolysis 微孔分气法
atmometer 汽化计
atmophile element 亲气元素
atmos 大气压
atmosphere (=aerosphere) ①大气②(大)气压③气氛
atmosphere pressure ionization source 大气压电离源
atmospheric ①大气的②空气的
atmospheric acidity 大气酸度
atmospheric aging 大气(作用)老化；自然老化
atmospherical ①大气的②空气的
atmospheric ambient factors 周围自然因素；天候因素
atmospheric and vacuum distillation unit 常减压蒸馏装置(设备)
atmospheric background 大气本底值
atmospheric brown clouds 大气棕色云
atmospheric chemistry 大气化学
atmospheric circulation 大气环流
atmospheric circulation type water cooler 空气循环式水冷器
atmospheric circulation water cooler 空气循环凉水器
atmospheric combination and low pressure absorption nitric acid process 综合法硝酸工艺
atmospheric-compartment drier 常压干燥室
atmospheric composition 大气组成
atmospheric condenser 空气冷凝器；常压冷凝器
atmospheric cooling tower 空气冷却塔
atmospheric corrosive agent 大气腐蚀剂
atmospheric crack 风化裂纹；大气(作用)开裂
atmospheric cracking 大气(作用)龟裂；自然龟裂
atmospheric dispersion 大气色散
atmospheric distillation 常压蒸馏
atmospheric distillation plant 常压蒸馏装置
atmospheric drier 空气干燥器
atmospheric drum drier 鼓式常压干燥器
atmospheric drying 大气干燥
atmospheric dynamics 大气动力学
atmospheric environment capacity 大气环境容量
atmospheric evaporation losses 常压蒸发损失〔石油产品〕
atmospheric exposure test 空气暴露试验；耐候性试验
atmospheric fading protective agent 大气褪色防护剂
atmospheric fallout 大气沉降物
atmospheric flash tower 常压闪蒸塔
atmospheric fluidized bed combustion 常压流化床燃烧
atmospheric homeostasis 大气(成分)稳态
atmospheric humidity 大气湿度
atmospheric isolation 大气隔绝
atmospheric marine environment 海洋大气环境
atmospheric moisture 空气湿度
atmospheric monitoring 大气监测
atmospheric nitric acid process 常压硝酸工艺
atmospheric oxidant 大气氧化剂；大气中氧化性污染物
atmospheric oxidation 空气氧化
atmospheric oxygen 大气氧；空气氧
atmospheric particulate matter 大气颗粒物
atmospheric particulates 大气颗粒物
atmospheric photolysis 大气光解作用
atmospheric pipe 通大气管路
atmospheric pollutant 大气污染物〔指大气中的二氧化硫、二氧化氮等污染物〕
atmospheric pollutant monitoring 大气污染监测
atmospheric polluting material 大气污染物
atmospheric pollution 大气污染
atmospheric pollution chemistry 大气污染化学
atmospheric pollution continue automatic monitoring system 大气污染连续自动监测系统
atmospheric pollution control engineering 大气污染防治工程
atmospheric pollution network 大气污染监视网
atmospheric pollution sources 大气污染源
atmospheric pressure 大气压(力)
atmospheric pressure chemical ionization(APCI) 大气压化学电离
atmospheric pressure ionization 大气压电离
atmospheric pressure ionization mass spectroscopy 常压电离质谱；常压电离质谱法；大气压电离质谱法
atmospheric pressure kiln 常压炉；常压加热炉
atmospheric pressure photoionization 大气压光致电离
atmospheric pressure storage tank 常压油罐
atmospheric pressure vulcanization 常压硫化
atmospheric radical 大气自由基
atmospheric radiochemistry 大气放射化学
atmospheric relief value 大气安全阀；泄气阀
atmospheric re-run 常压重馏
atmospheric sampling method without power 大气样品无动力采集法
atmospheric sampling technology 大气样品采集技术

atmospheric sphere　大气圈，气圈
atmospheric steam　常压蒸汽〔一个大气压的水蒸气〕
atmospheric still　常压管式加热炉
atmospheric storage tank　常压储槽
atmospheric suspended matter　大气悬浮物
atmospheric temperature　常温
atmospheric (temperature) overflow dyeing machine　常温溢流染色机
atmospheric topping　常压拔顶(蒸馏)
atmospheric trace gas　大气痕量气体
atmospheric turbulence　大气湍流
atmospheric vacuum distillation process　常减压蒸馏过程
atmospheric valve　空气阀；放空阀
atmotherapy　吸汽疗法
atoleine(=atolin)　液体石蜡
atolin(=atoleine)　液体石蜡
atom　原子
atom arrangement　原子排列
atom cell　原子化池
atom chemistry　原子化学
atom cluster　原子团簇
atom concentration　原子浓度
atom counter　原子计数器
atometer　蒸发速度测定器
atom force microscopy　原子力显微镜
atomic　原子的
atomic absorption analysis　原子吸收分析
atomic absorption coefficient　原子吸收系数*
atomic absorption detector　原子吸收检测器
atomic absorption line profile　原子吸收线轮廓
atomic absorption spectrometer　原子吸收光谱仪
atomic absorption spectrometer based on resonance detector　共振检测器原子吸收光谱仪
atomic absorption spectrometry(AAS)　原子吸收光谱法
atomic absorption spectrophotometer　原子吸收分光光度计
atomic absorption spectrophotometry　原子吸收分光光度法
atomic absorption spectroscopy　原子吸收光谱学
atomic absorption spectrum　原子吸收光谱
atomic abundance　原子丰度
atomical　原子的
atomic angular distribution function　原子角分布函数
atomic arrangement　原子排列
atomic beam　原子束
atomic beam technique　原子束技术
atomic bomb　原子(炸)弹
atomic bond　原子键
atomic bridge　原子桥
atomic building　原子构成
atomic cataclysm　原子激变
atomic change　原子变化
atomic charge　原子电荷
atomic clock　原子钟
atomic cloud　原子(蒸气)云
atomic collision　原子碰撞
atomic compound　原子化合物
atomic concentration　原子浓度
atomic conductivity　原子传导率
atomic configuration　原子构型
atomic core　原子芯
atomic creation　原子创造
atomic damping constant　原子阻尼常数
atomic dipole moment　原子偶极矩
atomic disintegration　原子蜕变
atomic dispersion　原子分散
atomic displacement reaction　原子置换反应
atomic disruption　原子破裂
atomic distance　原子距离
atomic domain　原子领域
atomic emission　原子发射
atomic emission detector　原子发射检测器
atomic emission line profile　原子发射线轮廓
atomic emission spectra　原子发射光谱
atomic emission spectrometry(AES)　原子发射光谱法
atomic emission spectroscopy　原子发射光谱学
atomic emission spectrum　原子发射光谱
atomic energy　原子(核)能
atomic energy level　原子能级
atomic evolution　原子演变
atomic excitation　原子激发
atomic field　原子场
atomic fission　原子核分裂
atomic fluorescence　原子荧光*
atomic fluorescence mercury-measured spectrometer　原子荧光测汞仪
atomic fluorescence quantum efficiency　原子荧光量子效率
atomic fluorescence spectrometer　原子荧光光谱仪
atomic fluorescence spectrometry(AFS)　原子荧光光谱法
atomic fluorescence spectroscopy　原子荧光光谱学
atomic fluorescence spectrum　原子荧光光谱
atomic force microscope　原子力显微镜
atomic force microscopy　原子力显微镜法
atomic formula　原子式；结构式

atomic fraction 原子分数
atomic (fractional) coordinate 原子(分数)坐标
atomic fragment 原子碎片
atomic framework 原子结构
atomic frequency 原子频率
atomic fusion 核聚变
atomic gas laser 原子气体激光器
atomic g factor(=Lande g factor) 朗德 g 因子；谱线劈裂因子
atomic group(=atomic grouping) 原子团
atomic grouping(=atomic group) 原子团
atomic heat 原子热(容)
atomic heat capacity 原子热容
atomic hydrogen torch 原子氢焰
atomic hydrogen welding 原子氢焊
atomic hypothesis 原子假说
atomic integral 原子积分
atomic ionization potential 原子电离电势
atomic ionoluminescence 原子离子化致发光
atomic irradiation 原子辐射
atomicity 原子价
atomic kernel 原子实
atomic lattice 原子晶格；原子点阵
atomic layer 原子层
atomic line 原子(谱)线
atomic line source 原子(谱)线光源
atomic line width 原子(谱)线宽度
atomic link 原子键
atomic linkage 原子键合
atomic magnetic force 原子磁力
atomic mass 原子质量
atomic mass determination 原子质量测定
atomic mass unit 原子质量单位
atomic meter(=angstrom unit) 原子米；埃单位
atomic migration 原子迁移
atomic model 原子模型
atomic-molecular theory 原子(与)分子(理)论
atomic nature ①原子本性②原子性
atomic non-bonding orbital 原子的非键轨道
atomic nuclear structure 原子核结构
atomic nucleus 原子核
atomic number 原子序数*
atomic orbital 原子轨道*
atomic oscillation 原子振动
atomic parachor 原子等张比容
atomic parameter 原子参量
atomic percentage 原子百分率
atomic plane 原子面
atomic polarity 原子极性
atomic polarizability 原子极化性
atomic polarization 原子极化(作用)
atomic potential 原子电位
atomic power 原子动力
atomic probe field ion microscope 原子探针场离子显微镜
atomic quantum mechanics 原子的量子力学
atomic radial distribution function 原子径向分布函数
atomic radius 原子半径
atomic ratio 原子比
atomic ray 原子射线
atomic reaction 原子反应
atomic reactor 原子反应堆
atomic rearrangement 原子重排
atomic refraction 原子折射度
atomic refractivity 原子折射度
atomic resolution microanalysis 原子分辨微区分析
atomic rotation 原子旋光度
atomics 原子论
atomic scattering factor 原子散射因子*
atomic self-consistent field 原子自洽场
atomic separation 原子间距
atomic shell 原子层
atomic species 原子种类
atomic spectroanalysis 原子光谱分析
atomic spectroscopy 原子光谱(分析)法
atomic spectrum 原子光谱
atomic spin 原子自旋
atomic state 原子态
atomic structure 原子结构
atomic susceptibility 原子磁化率
atomic theory 原子学说
atomic thermal motion 原子热运动
atomic transformation 原子转变
atomic transition width 原子跃迁宽度
atomic transmutation 原子嬗变
atomic unit 原子单位
atomic vapour 原子蒸气
atomic vibration 原子振动
atomic volume 原子体积
atomic wear 原子磨损
atomic weight 原子量
atomic weight measurement 原子量测定
atomic weight of physical scale 物理标度的原子量
atomic weight unit 原子量单位
atom-in-molecule (AIM) method 分子中单原子法
atomiser 喷雾器；喷(水)嘴

atomism　原子论
atomization　①原子化②雾化
atomization cell　原子化池
atomization curve　原子化曲线
atomization efficiency　原子化效率*
atomization in graphite crucible　石墨坩埚原子化法
atomization in graphite tube furnace　石墨管炉原子化法
atomization mechanism　原子化机理
atomization period　原子化期(间)
atomization process　①原子化过程②微粒化过程
atomization stage　原子化阶段
atomization steam　原子化蒸汽
atomization temperature　原子化温度
atomization time　原子化时间
atomized aluminium　细铝粉
atomized fuel　雾化燃料
atomized liquid　雾化液体
atomized lubrication　雾化润滑
atomized oil　雾化油
atomized suspension technique　悬浮液雾化法
atomized water spray　喷水雾；水雾喷射
atomizer　原子化器*
atomizer burner　喷雾燃烧器
atomizer cell　原子化池
atomizer chamber(=spray chamber)　喷雾室；雾化室
atomizer cone(=atomizing cone)　雾化锥
atomizer using graphite furnace　石墨炉原子化器
atomizing air　雾化(用)空气
atomizing burner　雾化喷嘴；原子化燃烧器
atomizing chamber　雾化室；喷雾室
atomizing cone(=atomizer cone)　雾化锥
atomizing disc　喷雾盘
atomizing medium　①雾化介质②微粉化介质
atomizing nozzle　原子化喷嘴；雾化喷嘴
atomizing of liquid fuel　液体燃料雾化
atom jump　原子跃迁
atom line　原子(谱)线
atom localization energy　原子电池；阻挡层光电池
atom mechanics　原子力学
atom-meter　原子米
atom model　原子模型
atomology　原子论
atom pairs　原子对
atom percent calculation　原子百分数计算
atom percent excess　超量原子百分数
atom percent (of isotope)　(同位素)原子(超额)百分数
atom polarization　原子极化
atom-probe　原子探针

atom probe field ion microscope　原子探针场离子显微镜
atom recombination　原子复合
atom reservoir　原子储存器
atom size　原子体积
atom smasher　原子击碎器；粒子加速器
atom source　原子源
atom splitting　原子分裂
atom structure　原子结构
atom transfer radical polymerization(ATRP)　原子转移自由基聚合
atom trapping　原子捕获
atom trapping method　原子捕集法
atom trapping technique　原子捕集技术
atom valence　原子价
atophan(=cinchophen; phenylcinchoninic acid)　阿托方；辛可芬　$C_6H_5C_9H_5NCOOH$
atopic　①特应性的②异位的
atopite　氟锑钙石；黄锑钙石
atorvastatin　阿托伐他汀〔药〕
atoxyl　氨基苯胂酸钠
atoxylate　氨基苯胂酸
atoxylic acid　对氨基苯胂酸　$NH_2C_6H_4AsO(OH)_2$
atractylene　苍术烯　$C_{15}H_{24}$
atractylin　苍术灵
atractylis concrete　苍术浸膏；苍术硬脂〔商〕
atractylis oil　苍术硬脂
atractylogenin　苍术苷配基
atractylol　苍术醇
atractylon　苍术酮
atractyloside　苍术苷
atracurium　阿曲库铵〔药〕
atreol　磺化油；一种由石油馏出物(用硫酸)制成的磺化产品
atricide　杀锥虫剂
atroglyceric acid　阿卓甘油酸；2-苯基甘油酸
atrolactic acid(=atrolactinic acid)　阿卓乳酸；苯基乳酸
atrolactinic acid(=atrolactic acid)　阿卓乳酸；苯基乳酸
atromentin　裂盒蕈色素；阿托曼霉素
atronene　苯基二氢萘
atronic acid　阿窗酸　$C_{17}H_{14}O_2$
atropamine(=apoatropine)　阿托胺；阿朴阿托品
atrophia　萎缩
atrophy　萎缩
-atropic　阿托〔词根〕
atropic acid(=2-phenylacrylic acid)　2-苯基丙烯酸；阿托酸　$C_6H_5C(=CH_2)COOH$
atropine(=coromegine)　阿托品
atropine methonitrate　甲硝阿托品

atropine sulfate 硫酸阿托品
atropisomer 阻转异构体
atropisomerism 旋转对映异构(现象);阻转异构现象
atroponitrile 阿托腈
atropoyl(=2-phenylacryloyl) 2-苯基丙烯酰;阿托酰 $C_6H_5C(=CH_2)CO-$
atroscine 莨菪胺;东莨菪碱;消旋东莨菪碱
atrous 暗灰色
atroxindole 阿托吲哚
attached proton test ①依附质子试验〔APT 试验〕②连氢多寡试验
attaching rubber to metal 橡胶与金属结合
attachment ①附件②装置
attachment energy 附着能
attack 化学浸蚀;起化学反应
attacked bark tissue 受伤树皮〔胶树〕
attack polishing method 腐蚀抛光法
attalic thread 金箔包芯线
attalo〔商〕 (=annatto)胭脂树橙
attapulgite 凹凸棒土〔俗〕;绿坡缕石;硅镁土
attar 挥发油;精油
attar of kewda 露兜油
attar of rose(=rose oil) 玫瑰油
attemperating device 温度控制装置;恒温箱
attemperation 温度控制;调节温度
attemperator 恒温器;控制温度用旋管冷却器
attendance cost 维修费
attendant dissolution phenomena 伴溶现象
attendant tailing 附带拖尾
attenllant 稀释剂;冲淡剂
attenuance ①衰减率②稀释
attenuant 稀释剂;冲淡剂
attenuated total reflectance(ATR) 衰减全反射
attenuated total reflectance correction 衰减全反射光校正
attenuated total reflection 衰减全反射光谱
attenuation coefficient 衰减系数
attenuation factor 衰减因子
attenuation index 衰减指数
attenuation measurement 衰减测定
attenuation modulus 衰减模量
attenuation step pilot lamp 衰减分级指示灯
attenuator 衰减器
Attila 阿蒂拉热室〔法国氟化挥发法核燃料后处理用〕
atto- 阿(托)〔10^{-18}〕〔词头〕
attomole 阿摩尔
attractant 吸引剂;引诱剂
attraction ①引力;吸力②吸引

attraction constant 吸引力常数
attraction potential 吸引位;吸引势
attraction power 吸引能力
attractive force 吸引力
attractive interaction 相互吸力;相互引力
attractive note 快感香韵
attractive potential energy surface 吸引型势能面
attractor electrode 吸(引)力电极
attribute 属性
attrital coal 杂质煤
attrited black 球磨炭黑
attrition 磨耗
attritioning 研磨;精碎
attrition loss of catalyst 催化剂的磨耗损失
attrition loss of support 载体磨碎损耗
attrition medium 研磨介质
attrition mill 磨碎机
attrition of catalyst 催化剂磨耗
attrition resistance 耐磨耗性
attrition resistant 耐磨的
attrition test 磨耗试验
attritive powder 微细粉末
attritor 超微磨碎机;立式球磨机
attritor mill 立式球磨
Atwater calorimeter 阿特沃特量热器
Atwood cracking process 阿特伍德裂化过程〔一种包括有重质油蒸气循环的常压裂化过程〕
atypical 非典型
aubepine(=anisaldehyde) 茴香醛;对甲氧基苯甲醛 $CH_3OC_6H_4CHO$
aubepine nitrile〔商〕 茴香腈
aubepinol 茴香醇
AUC process 碳酸铀酰铵法〔从 UF_6 制备 UO_2 方法之一〕
aucubin(e) 珊瑚木苷 $C_{15}H_{22}O_9$
audio frequency 声频
audio frequency oscillator 声频振荡器
audio method 声频法
audio tube 声频管
audricurin 奥居菌素
auerbachite 锆石
Auer burner 欧爱尔灯
auerlite 磷钍矿
Auer metal 奥厄合金
aufbau principle 构造原理
aufs(=absorbance unit full scale) 满刻度吸光度单位
Augelini furnace 奥格林尼电炉
auger 取样钻

auger blender　螺旋掺混机
Auger chemical effect　俄歇化学效应
Auger chemical shift　俄歇化学位移
Auger coefficient　俄歇系数
Auger depth analysis　俄歇深度分析
Auger depth resolution　俄歇深度分辨
Auger effect　俄歇效应
Auger electron　俄歇电子
Auger electron analyzer　俄歇电子分析器
Auger electron appearance potential spectroscopy　俄歇电子出现电位能谱法
Auger electron detector　俄歇电子检测器
Auger electron spectrometer　俄歇电子能谱仪
Auger electron spectroscopy(AES)　俄歇电子能谱学
Auger electron spectroscopy-low energy electron diffraction　俄歇电子能谱-低能电子衍射
Auger electron spectrum　俄歇电子能谱
Auger electron yield　俄歇电子产率
Auger emission spectroscopy　俄歇电子能谱
auger feeder　螺旋式供料器
Auger image　俄歇电子像
Auger kinetic energy　俄歇动能
Auger line　俄歇线
Auger line scan　俄歇线扫描
Auger line shape　俄歇线形
auger machine　螺旋造砖机
Auger map　俄歇像
Auger mapping　俄歇(电子)图
Auger matrix effect　俄歇基体效应
auger meter　螺旋式配量计
Auger parameter　俄歇参数
Auger recombination　俄歇(电子)复合
Auger shower　俄歇(电子)簇射
Auger signal intensity　俄歇信号强度
Auger spectra　俄歇能谱
Auger spectrometer　俄歇能谱仪
Auger spectrometry　①俄歇能谱②俄歇能谱分析法
Auger spectroscopy　俄歇能谱
Auger transition　俄歇(电子)跃迁
Auger yield　俄歇(电子)产额
augite　辉石
augmented flow　增大流量
auicularin　广寄生苷
aupette　自动移液管
auramine(=pyoktanin yellow)　金胺；脓单宁黄
auranetin　酸橙黄酮；3,4,6,7,8-五甲氧基黄酮
auranofin　金诺芬〔药〕
aurantiacin　齿菌橙；橙齿菌色素
aurantiamarin　苦橙苷
aurantiin　橙花苷
aurantin　橙苷；柚苷
aurantine　橙浸膏
aurantiol　橙花素
auraptene　橙皮油素
auraptenol　酸橙素烯醇
aurate　金酸盐
aureolic acid　金霉酸
aureolin(=primulin)　樱草素
aureomycin(=chlorotetracycline)　金霉素
aureomycine(=chlorotetracycline)　金霉素
aureothin　金链菌素
aureothricin　金丝菌素
auri-　金基；(三价)金基　Au≡
auribromide　溴金酸盐；溴金化物
auribromohydric acid　溴金酸
auric　①金的②正金的；三价金的③含金的
auric acid　金酸
auric bromide　溴化金　$AuBr_3$
auric chloride　氯化金　$AuCl_3$
auric chloride acid　氯金酸
auric compound　正金化合物
auric cyanide　氰化金　$Au(CN)_3$
auric cyanide acid　氰金酸　$H[Au(CN)_4]$
aurichalcite　绿铜锌矿
aurichloride　氯金酸盐　$M[AuCl_4]$
aurichlorohydric acid　氯金酸　$H[AuCl_4]$
auric hydroxide　氢氧化金
auric iodide　碘化金　AuI_3
auric nitrate acid　硝金酸；四硝根合(正)金(氢)酸　$H[Au(NO_3)_4]$
auricome　过氧化氢　H_2O_2
auric oxide　氧化金　Au_2O_3
auric potassium bromide　溴化金钾；溴金酸钾　$K[AuBr_4]$
auric potassium chloride　氯化金钾；氯金酸钾
auric potassium cyanide　氰化金钾；氰金酸钾
auric potassium iodide　碘化金钾；碘金酸钾
auric sodium bromide　溴化金钠；溴金酸钠
auric sodium chloride　氯化金钠；氯金酸钠
auric sulfate　硫酸金　$Au_2(SO_4)_3$
auric sulfide　硫化金　Au_2S_3
auricularine　耳草灵
auricyanhydric acid　氰金酸　$H[Au(CN)_4]$
auricyanide　氰金酸盐　$M[Au(CN)_4]$
auriferous　含金的
auriferous banket　含金砾岩层

auriiodide 碘金酸盐 M[AuI$_4$]
aurimycin(=eurymycin) 泛霉素
aurin(=rosolic acid) 金精；玫红酸
aurinol 金精醇
aurin red 金精红
aurintricarboxylate 金精三羧酸盐
aurin tricarboxylic acid(=aluminon) 试铝灵；金精三羧酸
auripigment 金色料
aurite 亚金酸盐 MAuO
auro- 亚金基；一价金基 Au—
auroauric 正亚金(的)〔指含有二价金的，或一价金与三价金各一原子的化合物〕
auroauric bromide 溴化正亚金 Au$_2$Br$_4$
auro-auric chloride 四氯化二金；氯化正亚金 AuCl$_3$·AuCl;Au[AuCl$_4$]
auro-auric compound 正亚金化合物
auro-auric oxide 四氧化四金；氧化正亚金 Au$_2$O·Au$_2$O$_3$
auro-auric sulfide 四硫化四金；硫化正亚金
aurobromide 溴亚金酸盐 M[AuBr$_2$]
aurochin 对氨基苯甲酸奎宁
aurochloride 氯亚金酸盐 M[AuCl$_2$]
aurochrome 金色素
auro compound 亚金化合物
aurocyanide 氰亚金酸盐 M[Au(CN)$_2$]
aurodiamine 亚金联胺 AuNH·NH$_2$
aurone 噢呋；2-次苯甲基苯(并)呋喃酮
aurorium 假设会产生极光的特征线的元素*
aurosulfide 亚金硫化物 AuSM
aurotensine 金黄紫堇碱
aurous 亚金(的)
aurous amminoiodide 碘化氨合亚金 [Au(NH$_3$)]I
aurous bromaurate 溴金酸亚金 AuBr·AuBr$_3$;Au[AuBr$_4$]
aurous bromide 溴化亚金 AuBr
aurous bromide acid 溴亚金酸 H[AuBr$_2$]
aurous chloride 氯化亚金 AuCl
aurous chloroaurate 氯金酸亚金；四氯化二金 AuCl·AuCl$_3$;Au[AuCl$_4$]
aurous compound 亚金化合物
aurous cyanide 氰化亚金 AuCN
aurous hydroxide 氢氧化亚金 AuOH
aurous iodide 碘化亚金 AuI
aurous oxide 氧化亚金 Au$_2$O
aurous potassium cyanide 氰亚金酸钾；氰化亚金钾 K[Au(CN)$_2$]
aurous sulfide 硫化亚金 Au$_2$S
auroxanthin 金黄质；玉黍黄二呋喃素

aurum vegetable 金黄菊根酸
auryl 氧金根 (AuO)
auryl hydrosulfate 硫酸氢氧金；酸式硫酸氧金 (AuO)HSO$_4$
auryl nitrate 硝酸氧金 (AuO)NO$_3$
Au-Si surface barrier detector 金-硅面垒探测器*
austenic steel 奥氏体钢
austinite 砷锌钙矿
australene α-蒎烯
Australian gum 相思树胶
Australian mahogany oil(=Australian rose wood oil) 澳洲玫瑰木油
Australian rose wood oil(=Australian mahogany oil) 澳洲玫瑰木油
Australian sandalwood oil 澳洲檀香木油
australol 澳桉酚；对异丙基苯酚
austricin 奥菊素
autacoid 自身活性物质
authenticity 可靠性；真实性
authentic sample 真实样品；可信样品
authentic specimen 真实样本；可信标本
authoris(z)ed pressure 法定压力；规定压力；容许压力
auto- 〔希腊字头〕①自②自动
autoabsorption 自动吸收
auto-acceleration 自动加速(作用)
auto-acceleration effect 自动加速效应
auto-acceleration period (聚合反应)自加速期
auto-activation 自动活化
autoadhesion 自黏合；自黏作用
autoanalyzer 自动分析器
autocarousel 自动式圆盘进样器
autocatalysis 自催化*
autocatalytic 自催化的
autocatalytic action 自催化作用
autocatalytical reaction 自动催化反应
autocatalytic decomposition 自催化分解
autocatalytic polycondensation 自催化缩聚
autocatalytic reaction(=autocatalyzed reaction) 自动催化反应
autocatalyzed oxidation 自催化氧化作用
auto-centering adjustment device in sectional warping machine 筒子架自动对中装置
autoclave ①压热器②高压釜③高压灭菌器
autoclave cure 硫化罐硫化
autoclaved sterilization 压热灭菌
autoclave for nitrogen dioxide 二氧化氮高压塔
autoclave molding 压热成型法〔碳纤维复合材料成型方法之一〕

autoclave pan 压热锅；硫化罐
autoclave press 立式硫化罐
autoclave process 压热器法
autoclave saponification 压热器皂化(法)
autoclave test 压热试验
autoclaving 压热器处理；高压灭菌
auto clipping apparatus 自动清丝装置
autocoagulation 自动凝结
autocollimation spectroscope 自准直分光镜
auto collimator 自动准直仪
auto colorimeter 自动比色计
autocondensation 自缩聚
autocorrection 自动校正
auto correlation analysis 自相关分析
autocorrelation coefficient 自相关系数
autocorrelation function 自相关函数
autocrimper 自动卷曲机
auto-cutting attachment 自动划料装置
auto-demolding polymer 自脱模聚合物
auto deposition coating 自动沉积涂料
autodestructive alkylation 自裂解烃化
auto-doffing 自动落筒；自动换筒
autodrum 自动成型机头〔鼓式或半鼓式〕
autodynamograph yarn testing machine 纱线强力自动记录试验机
auto electrothermal vaporization accessory 自动电热蒸发附件
auto-exhaust catalyst 汽车尾气催化剂*
auto-extinguishing composition 自熄(性)组分
auto feed 自动进给；自动加料
autoferritage cylinder(=self-shrinking cylinder) 自增强(高压)圆筒；自紧(高压)圆筒
autofining 自动精制
autofining process 自动精炼过程〔一种催化加氢精炼石油产品过程〕
auto-flame control 自动火焰控制
autoflocculation 自絮凝
auto-flowability 自流动性
autofluorogram 自动荧光图
autofluorograph 自动荧光仪
autofluoroscope 自动荧光镜
autofocus 自(动)聚焦
autofrettage 自紧法
autogamma-scintillation spectrometer γ射线自动闪烁分光计
Autogene sulfur blue 奥托精硫(化)蓝
autogenic succession 自发演替
autogenous decomposition 自动分解；自生分解

autogenous extrusion 自热挤出
autogenous ignition 自燃；自动着火
autogenous ignition temperature 自燃温度；自动着火点
autogenous pressure 自生压力；自压
autogenous soldering 自焊
autogenous welding 自熔接
autografting 自体移植
auto-grease pump 自动注油泵
autohemolysis test 自身溶血试验
autohesion ①自黏性②自黏力
autoignite 自燃；自动着火
autoignition 自燃
autoignition point 自燃点
autoignition temperature 自燃温度
autoimmune disease 自身免疫性疾病
autoinhibition 自阻聚
auto-injection 自动注射
auto-injection control 自动注射控制
auto-injection control panel 自动注射控制板
auto-injection panel 自动注射板
autoionization 自体电离(作用)；自(动)电离
auto-ionization process 自电离过程
auto-knockout 自顶脱
autolean mixture 自动贫化燃烧混合物
autolock 自锁定
autol red 阿脱红〔色淀偶氮染料〕
auto-lubrication 自动润滑
auto-luminescent paint 发光涂料
autolysate 自溶物；自溶产物；自解产物
autolyse 自(己)溶(解)
automat 自动机；自动装置
automata 自动机
automated analysis 自动化分析
automated biochemical analyzer 自动生化分析仪
automated gas chromatographic analysis 自动气相色谱分析
automated injection device 自动注射装置
automated integrating amplifier 自动积分放大器
automated production 自动化生产
automated unloading injection press 自动脱模注压机
automatic 自动的
automatic accumulator regulator 蓄电池自动调节器
automatic acquisition and interpretation of mass spectral 质谱数据的自动获取与解释
automatic activation analysis 自动化活化分析
automatically blast-cleaning 自动喷砂除锈法
automatic amino acid analyzer 氨基酸自动分析仪
automatic analysis 自动分析

automatic analyzer 自动分析仪
automatic analyzer for amino acids 氨基酸自动分析仪
automatic anodizing equipment 自动阳极氧化装置
automatic attenuator 自动衰减器
automatic background subtraction technique 本底自动扣除技术
automatic balance 自动天平
automatic balancing 自动平衡
automatic baseline drift correction 自动基线漂移校正
automatic batch mixing 自动分批混合；自动间歇混合
automatic batchwise gas chromatography 自动分批式气相色谱(法)
automatic belt sander 自动带式砂光剂
automatic blow moulding 自动吹塑
automatic bobbin pneumatic expeller 气动推出式自动落筒装置
automatic brightness control 自动亮度控制
automatic broken weft finder 自动找纬机构
automatic buret(te) 自调(零点)滴定管
automatic capping machine 自动压盖机
automatic clutch 自动离合器〔机工〕
automatic coding photomicrography 自动编码显微照相法
automatic colorimetric analysis 自动比色分析(法)
automatic colorimetric determination 自动比色测定
automatic combustion apparatus 自动燃烧仪
automatic compression mo(u)lding 自动压塑
automatic computing equipment 自动计算设备
automatic continuous analysis 自动连续分析
automatic contrast and brightness adjuster 自动对比度和亮度调节器
automatic control device 自动控制装置
automatic controller 自动控制器
automatic cop change loom 自动换纡织机
automatic corrosion protection system 自动防腐蚀装置(体系)
automatic crimp contraction tester 自动卷曲收缩测试仪
automatic cut-out 自动断路控制
automatic data processing 自动数据处理
automatic defect inspecting system 织疵自动监测系统
automatic desk computer 自动台式计算机
automatic de-sludger 自动除渣器
automatic detection(AD) 自动检测
automatic developing machine 自动展开机
automatic development 自动展开
automatic dielectrometer 自动介电分析仪
automatic dielectrometry 自动(电)介质测试法
automatic die swell detector 口型膨胀自动检测仪

automatic differential titrator 自动差示滴定仪
automatic digital calculator 自动数字计算机
automatic disc printer 自动盘式打印机
automatic dissolver 自动溶解器
automatic door actuator 自动式门调节器
automatic drawing-in 全自动穿经
automatic drier 自动干燥机
automatic drill 自动钻
automatic dryer 自动干燥器
automatic dynamometer ①自动功率计②自动强伸度计③自动测力计
automatic error correction 自动校正误差
automatic expansion valve 自动膨胀阀
automatic feed 自动进料
automatic feeder 自动加料器；自动进料器
automatic fillling machine 自动装填机；自动包装机
automatic filtration 自动过滤
automatic fine tuning 自动微调
automatic flat screen printing range 自动平网印花联合机
automatic fraction 自动馏分；自动分馏物
automatic fraction collector 自动馏分收集器
automatic frequency control 自动频率控制
automatic fuel economizing device 自动节省燃料设备〔自动调节燃烧混合物组成比例的设备〕
automatic fuel saving device 自动节省燃烧设备
automatic fuel shut-off 自动停止进入燃料
automatic gain control 自动增益控制
automatic gas analyzer 自动气体分析计
automatic gas chromatograph 自动气相色谱仪
automatic gas chromatography 自动气相色谱(法)
automatic gas sample injection 气样自动注射
automatic gas sample injection module 气样自动注射组件
automatic gas sampling valve 自动气体进样阀
automatic governor ①自动调节器②自动操纵器；自动控制器
automatic grinding 自动磨刀
automatic hand scale 自动手秤；弹簧秤
automatic headspace analyzer 自动顶空分析器
automatic humidity control 自动调湿装置
automatic ignitor 自动点火器
automatic image analyzer 图像自动分析仪
automatic knotter 自动打结机
automatic layboy (桨板)自动折叠机
automatic leak control 自动漏量控制
automatic lean composition 自动贫化燃烧混合物
automatic leather stacker 自动搭马机

automatic linear temperature programming　自动线性升温程序
automatic lubrication　自动润滑
automatic-manual　自动-手动的
automatic marble-feeder　自动加球机
automatic measure　自动测量；自动测定
automatic measurer　自动量器
automatic milk analyzer　自动牛奶分析仪
automatic monitor　自动监控器
automatic monitoring　自动监测
automatic mould　自动塑模
automatic mo(u)lding-opening press　自动启模平板硫化机
automatic muller　自动平磨机
automatic multi-linear temperature programmer　自动多级线性程序升温器
automatic needle positioning device　自动停针位装置
automatic nitrogen analyzer　自动氮分析仪
automatic null balancing instrument　自动调零装置
automatic oil　自动机润滑油
automatic oiling　自动加油
automatic optical inspection　自动光学检测
automatic package analyzer　(筒子丝)自动卷装分析仪
automatic packaging unit　自动包装机
automatic padding machine　自动拭浆机
automatic particle-size measurement　自动粒度测定
automatic paste filler　自动装膏机
automatic peak catching　自动寻峰；自动峰找正
automatic peak matching　自动峰匹配
automatic pipette　自动吸(移)管
automatic pirn winding machine　自动卷纬机
automatic plaiting and winding-down device　自动落布卷取装置
automatic platform scale　自动台秤
automatic pneumatic control　自动气动控制
automatic pneumatic control of flow rate　流量的自动气动控制
automatic powder measuring machine　自动称粉机
automatic powder spray gun　自动粉末喷枪
automatic pressure controller　自动压力调节器
automatic print-out　自动印出；自动打印
automatic print-out device　自动打印机
automatic process controller　过程自动调节器；过程自动控制器
automatic process gas chromatograph　自动化流程气相色谱仪
automatic process gas chromatography　自动化流程气相色谱(法)
automatic programming machine　自动编程机
automatic quiller　自动卷纬机
automatic range control　自动量程控制
automatic range selector　自动量程选择器
automatic recorder　自动记录仪
automatic recording　自动记录
automatic recording detector　自动记录检测器
automatic recording photoelectric polarimeter　自动记录光电偏振仪
automatic recording photoelectric spectropolarimeter　自动记录光电分光偏振仪
automatic recording spectrophotometer　自动记录分光光度计
automatic recording thermobalance　自动记录式热天平
automatic recording thermometer　自动记录式温度计
automatic recording titrimeter　自动记录滴定仪
automatic recording X-ray spectrometer　自动记录X射线分光计
automatic recovery　①自动再生②自动恢复③自动校(矫)正
automatic reeding hook　自动穿筘钩
automatic register　自动记录器
automatic regulation　自动控制；自动调整
automatic regulator　自动调节器
automatic request　自动检索
automatic rewinding　自动复卷
automatic rich　自动富化
automatic rich position　自动富化位置
automatics　①自动学②自动装置③自动化理论
automatic sample handling system　自动进样系统
automatic sampler　自动进样器*
automatic sampling　①自动取样②自动进样
automatic sander　自动砂光机；自动打磨机
automatic scale　自动秤
automatic selectivity control　自动选择性控制
automatic sensing　①自动读出②自动传感
automatic sensitivity control　自动灵敏度控制
automatic shaking feeder　自摇送料器
automatic shutdown option　自动关车选择(方案)
automatic shutdown seal　自动关闭密封
automatic shuttle change loom　自动换梭织机
automatic siphon　自动虹吸管
automatic soap cutter　自动切皂机
automatic soap press　自动压皂机
automatic spindle-driven turret winder　锭轴传动回转式自动卷绕头
automatic spray gun　自动喷枪
automatic spraying　自动喷涂

automatic spring loaded valve 自动弹簧阀
automatic steelyard 自动弹簧秤;自动磅秤
automatic stuff box 自动调浆箱
automatic submerged arc welding 埋弧自动焊
automatic surface area analyzer 自动表面(面)积分析仪
automatic switch grease 自动转换器润滑脂
automatic switching 自动开关;自动转换器
automatic switching device 自动转接装置;自动开关装置
automatic syphon 自动虹吸管
automatic system 自动系统
automatic tangent baseline correction 自动切线基线校正
automatic temperature controller 自动温度控制器
automatic temperature recorder 自动温度记录仪
automatic temperature recorder controller 自动温度记录调节器
automatic thermobalance 自动热天平
automatic thermostat(=selfacting thermostat) 自动恒温器
automatic thin-layer spreader 自动薄层涂布器
automatic threading device 自动生头装置;自动引丝装置
automatic timer 自动定时器
automatic timing device 自动计时装置
automatic titration 自动滴定
automatic titrator 自动滴定计
automatic titrimeter 自动滴定计
automatic traverse spraying head 自动往返式喷头
automatic tube filling machine 自动装管机
automatic turning sander 自动旋转抛光机(打磨机)
automatic tying-in machine 自动结经机
automatic unloader 自动卸荷器
automatic valve 自动阀
automatic voltage regulator 自动稳压器
automatic weigher 自动秤
automatic weighing 自动称量
automatic welding 自动焊
automatic welding machine 自动焊机
automatic winding 自动卷取;自动收卷
automatic X-ray fluorescence spectrometer 自动X射线荧光光谱仪
automatic zero burette 自动调零(点)滴定管
automatic zeroing 自动调零
automiser spray 喷雾容器
automobile antifreeze solution 汽车防冻液
automobile finish 汽车(罩面)漆
automobile fuel 汽车燃料
automobile gasoline 汽车(用的)汽油
automobile oil 车用润滑油
auto(mobile) panel board 汽车仪表盘纸板
automobile polish 汽车上光蜡
automobile transmission gear oil 汽车传动齿轮润滑油
automolite 铁锌尖晶石
automonitor 自动监控器*
automorphic 自形的;自守的;自同构的
automotive fuel 马达燃料;内燃机燃料
automotive manufacturers (viscosity) oil classification 马达润滑油黏度分类法〔美国汽车制造厂家联合会的技术规范〕
automotive spoilers 汽车阻流板;汽车扰流器
automotive stock 汽车配件胶料
autoordering (纤维分子的)自动排列
auto-orientation 自取向
auto-orientation mechanism 自取向机理
auto-oxidation 自氧化
auto-oxidation chemiluminescence 自氧化化学发光
autooxidation initiation 自氧化引发
autooxidation initiator 自氧化引发剂
autophoresis 自动电泳
autophoretic coating 自泳涂装
autopipette 自动移液管
autopolymer 自聚物
autopolymerizable monomer 自聚型单体
auto-polymerization 自(动)聚合
autoprecipitation 自沉淀*
autoprofile cast film 自动(模口)定形法平挤薄膜
autoprofiling 自动(模口)定形
auto-programming 自动程序(控制)
auto-proportional water sampler 水质自动比例采样器
autoprotolysis 自质子解
autoprotolysis constant 质子自递常数
autoprotolysis ranges of solvents 溶剂的质子自递范围
auto pulp charging system 浆粕自动加料装置
autopurification 自净作用
auto-racemization 自动外消旋
autoradiogram(ARGM) 放射自显影图
autoradiograph 放射性自显影仪;自放射照相机
autoradiographic detection 放射自显影检测
autoradiographic technique 放射自显影技术
autoradiography 放射自显影法
autoradiolysis 自辐解*
auto-radiotitrameter 自动放射滴定计
auto-ranging 自动调程的;自动调范围的
autoreaction 自反应
autoreduction 自还原(作用)
autorefrigeration 自(动制)冷作用
autoretardation 自缓聚

autorich 自动富化〔燃烧混合比例〕
autorotation 自转
auto-sampler ①自动进样器②自动取样器
autosampler digital fibre analyzer 自动取样数字式纤维分析仪〔测量纤维长度及分布〕
auto sampler for precipitation 自动降水采样器
autoscanner 自动扫描仪
autoselector 自动选择器
auto-setter 自动定形机
auto-shim 自动匀场
auto-slit control 自动狭缝控制
auto soap 汽车用皂
autostop ①自动停止②自动停车
autosynchronous network 自动同步网络
auto-synergism 自协同作用
autotermination 自终止(作用)
autothermal extrusion 自热挤塑；自热挤出
autothermally 自热地
auto-thermal plastification 自(供)热塑化(作用)
autothermic cracking 自裂解；自热裂化；氧化裂化
autothermic process 自热法；热自动(补偿)法
autotitrator 自动滴定仪*
autotransformer 自耦变压器
autovac 真空箱
autovapor 自汽压缩(法)；自动汽化
autoverify 自动检验
auto-vulcanization 常温硫化
autovulcanizing stock 常温硫化胶料
autoweak 自(动)贫化〔燃烧混合比率〕
autoxidation 自氧化*
autoxidation of gasoline 汽油的自(动)氧化
autoxidation of hydrocarbon 烃的自(动)氧化
autoxidation-reduction process 自身氧化还原反应
autoxidator 自(动)氧化剂
autoxidisable substance 自(动)氧化物
autoxidizable (可)自(动)氧化的
autozero 自动调零
autozero corrector 自动零点校正器
autozero relay 自动调零继电器
autrometer 自动多元素摄谱仪
autumn wood(=late wood;summer wood) 晚木材
autunite 钙铀云母
Auwers-Skita rule 奥沃斯-斯其达规则〔羰基氢化时在酸中得顺式；在中性溶液中得反式化合物〕
auxanogram(s) 生长图
auxanographic method 生长图谱法
auxanometer 生长计
auxchrome 助色团

auxeticity 拉胀性
auxetic polymer 拉胀高分子；拉胀聚合物
auxiliary ①辅助的②(辅)助剂
auxiliary absorber 辅助吸收器
auxiliary agent 助剂
auxiliary air pump 辅助空气泵
auxiliary anode 辅助阳极
auxiliary baseline 辅助基线
auxiliary battery 辅助电池组
auxiliary blowing agent 发泡助剂
auxiliary bond 副键
auxiliary catalyst 助催化剂
auxiliary cathode 辅助阴极
auxiliary cavity 辅流胶沟；环状流胶沟〔压模〕
auxiliary check point 辅助检测点
auxiliary chemicals 化工助剂
auxiliary circuit 辅助电路
auxiliary column 辅助柱
auxiliary complex 辅助配位化合物
auxiliary complex-former 辅助配位剂
auxiliary complexing agent 辅助配位剂
auxiliary condenser 辅助冷凝器
auxiliary curing agent 助硫化剂
auxiliary department 辅助车间
auxiliary device 辅助装置
auxiliary drier 辅助催干剂
auxiliary electrode 辅(助)电极*
auxiliary emulsifying agent 辅助乳化剂
auxiliary equipment 辅助设备
auxiliary facility 附属设备
auxiliary fluid 辅助流体
auxiliary ingredient 辅助配合剂
auxiliary ion chamber 辅助电离室
auxiliary lens 辅助透镜
auxiliary materials ①辅助材料；物料②配合剂
auxiliary permanent magnet 辅助永久磁铁
auxiliary point 辅助点
auxiliary product 辅助材料；物料
auxiliary properties 次要性能
auxiliary reactor 辅助反应器
auxiliary reservoir 辅助储蓄器；备用储蓄器
auxiliary syntan 辅助性合成鞣剂
auxiliary tanning agent 助鞣剂
auxiliary valence 副价*
auxiliary valency 副(原子)价
auxiliary variable 辅助变量
auxiliary wing fuel tank 机翼辅助油箱〔飞机上的, 必要时可以抛去〕

auxilysin 促溶素
auxin(=plant hormone) 苗长素；植物生长素
auxiometer ①透镜放大计②测力计
auxochrome 助色团
auxochrome group 助色团
auxochromic 助色的
auxochromic group 助色团
auxofluore group 助荧光基团
auxofluorogen 助荧光团
auxogluc 助甜团；致甜基
auxotox 助毒基；助毒团
auxotox radical 甲亚氨基
auxotrophy 营养缺陷型；营养缺陷体
availability 有效性；可用性
availability factor ①使用效率②运转系数
availability ratio 利用系数
available ①有效的②可得的；可用的；适合的
available accuracy 有效准确度
available akali 有效碱
available capacity 有效容量
available chlorine 有效氯
available energy 可用能
available fertilizer 有效肥料
available head 可用压差
available heating value 有效热值
available NPSH 有效的净正吸入压头
available phosphate 有效磷酸盐
available phosphoric acid 有效磷酸
available sites 空位
available stock 可用库存；可利用资源
available surface 有效表面；可及表面
available temperature drop 有效温度下降
avalanche 离子雪崩
avalanche photodiode 雪崩光电二极管
avalanching 磨球崩落
avalite 钾铬云母
avenaceous 燕麦的
avenine 燕麦碱
aventurine ①金星玻璃②砂金石
aventurine glass 金星玻璃
aventurine glaze 金星釉
avenyl 大风子油醛制剂
average 平均
average absolute value 平均绝对值
average access time 平均存取时间；平均取数时间
average aviation gasoline 平均航空汽油
average background radiation 平均背景辐射
average boiling point 平均沸点

average carding time 平均梳理时间
average chain length 平均链长
average column pressure 平均柱压
average composite sample 平均组合样品
average conditioned size 生丝平均公量线密度
average conditioned size test 生丝平均公量条分检验；生丝平均公量线密度(纤度)检验
average crystallite dimension 平均晶粒尺寸；平均微晶大小
average degree of polymerization 平均聚合度
average density 平均密度
average deviation 平均偏差
average error 平均误差
average estimated variance 平均估计方差
average extension of molecular chain 分子链平均伸展
average extent of burning (耐燃试验)平均燃烧程度
average fiber length 纤维平均长度
average fiber length grit 平均纤维长度框架
average fibre diameter 纤维平均直径
average fibre number 纤维平均支数；纤维平均纤度
average filling losses 平均充油损失〔储油槽的〕
average fineness 平均细度；平均纤度
average flow rate 载气平均流速
average fraction defectives 平均次品率
average fracture strength 平均断裂强度
average freight rate assessment 评定平均运费率
average functionality 平均官能度
average gradient 平均陡度；平均梯度
average ignition time 平均着火时间
average increment 平均增量
average initial modulus 平均初始模量
average kinetic chain length 平均动力学链长
average latex 平均胶乳；混合田间胶乳
average length 平均长度
average lethal dose 平均致死量
average life 平均寿命
average life of fluorescence molecule 荧光分子平均寿命
average life (of radioactive atom) (放射性原子的)平均寿命
average life period 平均寿命
average ligand number 平均配体数*
average linear density 平均线密度〔指平均单位长度的重量〕
average load 平均负载
average melting point 平均熔点
average method 平均值法
average modulus 平均模数
average (molecular) orientation 平均(分子)取向度
average molecular velocity 平均分子速度

average molecular volume 平均分子体积
average molecular weight 平均分子量
average number 平均数
average operation time 平均操作时间
average outgoing quality 平均检出质量
average output 平均产量
average paraffins 平均链烷；正构及异构烷烃的混合物
average particle diameter 平均粒径；平均粒子直径
average particle size 平均粒子尺寸
average particle size distribution 平均粒径分布；平均粒度分布
average permeability 平均渗透度
average persistence length (链段的)平均相关长度
average plate height 平均塔板高(度)
average pressure 平均压力
average probability 平均概率
averager ①平均器；均衡器②中和器③中和剂
average rate 平均速率
average rate of flow 平均流速
average residence time 平均停留时间
average retention polarity 平均保留极性
average retention time 平均保留时间
average room condition 平均室内条件
average rusting depth 平均锈蚀深度
average sample 平均样品；平均试样
average sample number 平均取样数
average sequence length 平均序列长度；平均链区长度
average service test 正常使用试验
average size ①平均纤度②平均粒度
average stiffness 平均劲度
average stress 平均应力
average surface profile 平均表面剖面图；平均表面光洁度
average tank circumference 平均油罐圆周
average temperature difference 平均温差
average thermal conductance 平均热导
average thickness 平均厚度
average value 平均值
average velocity 平均速度
average viscosity 平均黏度
average weather condition 平均天候条件
average zone velocity 平均区域速度
averaging of rubber 不同等级生胶的混合
avertin 三溴乙醇 CBr_3CH_2OH
averufin 奥佛尼红素
avgas(=aviation gasoline) 航空汽油
aviation alkylate 航空烃化汽油〔航空汽油中的烃化汽油组分〕
aviation blending fuel 掺混航空燃料

aviation fuel 航空燃料
aviation fuel fraction 航空燃料馏分
aviation gas(=aviation gasoline) 航空汽油
aviation gasoline(=aviation gas; avgas; aviation spirit) 航空汽油
aviation grease 航空润滑脂
aviation kerosene 航空煤油；喷气机用煤油
Aviation Method 航空法
aviation mix 航空用乙基液；航空汽油抗爆液
aviation octane method 航空辛烷值测定法〔评价航空汽油抗爆性的方法〕
aviation octane number 航空辛烷值
aviation petrol 航空汽油
aviation safety fuel 安全航空燃料；高闪点航空燃料
aviation spirit(=aviation gasoline) 航空汽油
aviation turbine fuel 航空涡轮燃料
avicel 微晶纤维素
avicelase 微晶纤维素酶
avicel cellulose 结晶纤维素
avicine 勒楗碱
avicularin 萹蓄苷
avidin 抗生物素蛋白
avi process 浸酸光泽法
avirulent 无毒的
avitaminosis 维生素缺乏症
avitaminous 缺乏维生素的
avivage treatment (人造丝)增柔处理
avocado 油梨
avocado oil 鳄梨油
Avogadro constant 阿伏伽德罗常数
Avogadro's law 阿伏伽德罗(定)律
Avogadro's number 阿伏伽德罗数
avoidable error 可避免的误差
avoirdupois 常衡制〔英、美〕
avometer 万用电表
AV polarography 交变电压极谱法
Avrami equation 阿夫拉米方程
awadcharidine 东乌头定
awadcharine 东乌头灵
awaiting-repair time 等待修复时间
A-waste(s) 放射性废物
awning 保护篷
axenomycin 轴霉素
axerophthene(=anhydrovitamin A) 抗干眼烯；脱水维生素 A
axerophthol(=vitamin A) 抗干眼醇；维生素 A
axes crossing device 滚筒交叉装置
axestone 钺石；斧形软玉石
axial ①轴的②轴向的③直立(的)

axial angle　(结晶)轴角
axial angle apparatus　(结晶)轴角器
axial bond　直立键；竖键
axial clearance　轴向间隙
axial compression　轴向压缩
axial compression column　轴向压缩柱
axial compressor　轴流式压缩机
axial conformation　轴向构象
axial diffusion　轴向扩散
axial diffusion casing　轴流式扩散式机壳
axial direction　轴向
axial dispersion　轴向渗透
axial dissymmetry　轴向不对称性
axial eddy diffusion　轴向涡流扩散
axial electrostatic field　轴向静电场
axial extruder head　轴向挤塑机头
axial fan　轴流风扇
axial flow　轴流
axial-flow blower　轴流式通风机；轴流式鼓风机
axial-flow compressor　轴流式压缩机
axial-flow fan　轴流式风机；轴流式风扇
axial-flow impeller　轴流式叶轮
axial flow opener　轴流开棉机
axial flow pump　轴流泵
axial force　轴向力
axial glide plane　轴向滑移面
axial glide plane a,b,c　轴向滑移面 a,b,c
axial grease　轴(用)滑油
axial growth rate　轴向生长率
axial head　直通机头
axial inductively coupled plasma　端视电感耦合等离子体
axialite　轴晶
axial length　①轴长②轴向长度
axial mixing　轴向混合
axial mixing preventer　轴向防混器
axial modulation　轴向调制
axial molecular diffusion　轴向分子扩散
axial motion　轴向运动
axial normal stress　轴向正应力
axial orientation　①沿轴取向②轴向方位
axial outlet　轴出口
axial peak　轴峰
axial period　轴周期
axial plane　轴面
axial preferred orientation　轴向择优取向；轴向优先取向
axial proton　轴向质子
axial ratio　轴比
axial repeating period　(蛋白质)轴向重复周期
axial seal　轴(向密)封
axial slip　轴向滑脱
axial strain　轴向应变
axial strength　轴向强度
axial stress　轴向应力
axial-substitution　轴位取代
axial swelling　轴向溶胀；轴向膨化
axial symmetric(al) flow　轴对称流
axial symmetry　轴对称
axial-temperature gradient　轴向温度梯度
axial thrust　轴向推力
axial thrust balancing　轴向推力平衡
axial thrust bearing　轴向推力轴承
axial vector　轴向向量
axial velocity　轴向速度
axial winding　长丝缠绕；轴向缠绕
axin　虫漆脂
axinic acid　胭脂虫红酸　$C_{18}H_{28}O_2$
axinite　斧石
axiom　公理
axis　轴
axis-direction fatigue test　轴向载荷疲劳性能测试
axis of rotation　旋转轴
axis of strain　应变轴
axis of stress　应力轴
axis of symmetry　对称轴
axisymmetric　轴对称的
axisymmetric flow　轴对称流(动)
axi-symmetric shell　轴对称壳体
axite　阿西炸药；额克敏特〔硝化甘油、硝棉、石油炸药〕
axle grease　轴用(润滑)油
axle oil　轴用(润滑)油
axo-hexahydrobenzonitrile　偶氮六氢化苯腈
axon　轴突；轴索
axonometry　①测晶学②晶体轴线测定
axoplasm　轴浆
AX system　AX 系统
ayamycin　绫霉素
ayapana oil　三脉泽兰油
ayapanin(=7-methoxy coumarin)　7-甲氧基香豆素
ayapin　三脉泽兰素
aza-　〔词头〕　氮杂；吖
azabicyclic alcohol　氮杂双环醇
azacolutin　防霉菌素
azacyanine〔商〕　吖菁；氮(杂)菁〔作为照相减感剂或钝化剂〕
azacyclo-　氮杂环
1-azacyclopropane(=aziridine)　1-氮杂环丙烷；氮丙啶

azafrin　玄参红酸
azaindole　吖吲哚；氮杂吲哚
azalactones　氮杂内酯
azalea oil　杜鹃花油
azaleine　品红
azalomycin(=azaromycin)　阿沙霉素；阿扎霉素
azaniles〔复〕　制乙酸丝(用)不溶性偶氮染料的伯胺类
azanium　铵；氮鎓　N^+R_4
azanol　羟胺　NH_2OH—
azanyl　氮烷基；氨基　H_2N—
azaporphins　吖卟吩；氮杂卟吩
azarin　亮红(料)
azaromycin(=azalomycin)　阿沙霉素；阿扎霉素
azatetracene　氮杂丁省
azathioprine　硫唑嘌呤〔药〕
azathymine　氮(杂)胸腺嘧啶
azatropylidene　吖草；氮杂草
azatryptophan　氮杂色氨酸
azauracil　氮(杂)尿嘧啶
azaxanthene　氮氧杂蒽
azax powder　额闸克斯炸药
azedarine　苦楝根碱；楝树碱
azelaate　壬二酸盐(或酯)
azelaic acid　壬二酸；杜鹃花酸　$(CH_2)_7(CO_2H)_2$
azelaic acid value　壬二酸值
azelaic dinitrile　壬二腈　$(CH_2)_7(CN)_2$
azelaic semialdehyde　壬二酸半醛；壬酸醛
azelain　壬二酸甘油酯
azel aldehyde　壬二醛
azelaoyl　壬二酰　—$CO(CH_2)_7CO$—
azelaoyl chloride　壬二酰氯
azelate　①壬二酸酯②壬二酸盐
azelatin　壬酸甘油酯
azelon　蛋白质纤维
azene　氮烯
azene intermediate　一价氮中间体
azeotrope　共沸(混合)物*
azeotrope destroyer　共沸破坏剂
azeotrope former　共沸生成添加物
azeotrope forming liquid　共沸生成液体
azeotrope tower　共沸蒸馏塔
azeotropic　共沸的；恒沸的
azeotropic condensation　恒组分缩合；恒比缩合
azeotropic copolymer　恒(组)分共聚物*
azeotropic copolymerization　恒(组)分共聚*
azeotropic dehydration　共沸脱水
azeotropic distillation　共沸蒸馏(作用)
azeotropic distillation column　共沸蒸馏塔

azeotropic drying　共沸干燥
azeotropic extraction　共沸萃取法
azeotropic extractive distillation　共沸萃取蒸馏
azeotropic line　共沸线
azeotropic method　共沸法
azeotropic mixture　共沸混合物
azeotropic point　共沸点
azeotropic process　共沸(蒸馏)过程
azeotropic solution　共沸溶液
azeotropic temperature　共沸温度
azeotropism　共沸作用
azeotropy　共沸(性)
azepine　氮杂*
azete　氮杂环丁二烯
azetidine　氮杂环丁烷
azetidine-2-carboxylic acid　铃兰氨酸；氮杂-2-环丁烷羧酸
azetidinone　β丙内酰胺
azetine　氮杂环丁烯
azetinone　氮杂环丁烯酮
azide　氮化物*
azide group　叠氮基
azidinblue　锥蓝
azido-　叠氮基　N_3—
p-azidoacetophenone　对-叠氮苯乙酮
p-azido benzal acetone　对-叠氮亚苄基丙酮
p-azido benzoaldehyde　对-叠氮苯甲醛　$N_3C_6H_4CHO$
azidobenzoic acid　叠氮苯甲酸
azidocarbonyl　叠氮羰基
azidopolymer　叠氮聚合物
azido pyrene　叠氮芘
aziethane(=aziethylene)　重氮乙烷　$C_2H_4N_2$
aziethylene(=aziethane)　重氮乙烷
azimethane(=azimethylene)　重氮甲烷　CH_2N_2
azimethylene(=azimethane)　重氮甲烷
azimid　联重氮亚胺
azimido-　亚叠氮基；偶氮亚氨基
azimido-benzene(=azimino-benzene)　苯并三唑；1,2,3-三氮(杂)茚
azimido-toluene　亚叠氮基甲苯；甲苯并三唑
azimino-　亚叠氮基；偶氮亚氨基
azimino-benzene　苯并三唑；1,2,3-三氮(杂)茚　$C_6H_4N_3H$
azimuth　方位；方位角；(地)平经(度)
azimuthal angle　方位角
azimuthal half-width at half-height　方位角半高宽〔高度一半时的半宽度〕
azimuthal intensity distribution　方位强度分布
azimuthal quantum number　角量子数

azimuth circle　方向盘
azine　吖嗪*
azine colouring matters　吖嗪染料
azine dyes　吖嗪染料；对氮杂蒽型染料
azine group　吖嗪基
azines　①吖嗪染料；对氮杂蒽型染料②氮杂苯类
azino-　连氮基　＝NN＝
azirane　环氮乙烷
aziridine　氮丙啶；1-氮杂环丙烷
1-aziridine ethanol　1-吖丙啶乙醇
aziridinium salt　吖丙啶鎓盐
aziridinyl　吖丙啶基；氮杂环丙烯基
aziridinylmethacrylate　甲基丙烯酸吖丙啶酯
aziridinyl phosphine oxide　氮丙啶膦的氧化物
azirinomycin　阿际霉素
azithromycin　阿奇霉素〔药〕
azlactone　二氢唑酮*
azlactone synthesis　二氢唑酮合成
azlocillin　阿洛西林〔药〕
azlon　人造蛋白质纤维；再生蛋白质纤维
azo　偶氮　—N＝N—
azoamines　偶氮胺类
azoaminobenzene　偶氮氨基苯〔发泡剂〕
azoaniline　偶氮苯胺；二氨基氮苯
azoanisol　偶氮茴香醚；偶氮苯甲醚
azoaryl ether　偶氮芳基醚
cis-azobenzene　顺式偶氮苯
azobenzene(=azobenzide; azobenzol)　偶氮苯
azobenzide(=azobenzene)　偶氮苯
azobenzil(=benzilam)　三苯基噁唑；2,4,5-三苯基噁唑
azobenzoic acid　偶氮苯甲酸
azobenzol(=azobenzene)　偶氮苯
azobiscyanoisovaleric acid　偶氮二氰基异戊酸
1,1-azobis-1-cyclohexanenitrile　1,1-偶氮二-1-环己腈
2,2-azobis(2,4-dimethylvaleronitrile)　2,2-偶氮二(2,4-二甲基戊腈)〔聚合催化剂〕
azobisformamide　偶氮二甲酰胺
azobisisobutylvaleronitrile　偶氮二(异丁基戊腈)
azobisisobutyronitrile(=azodiisobutyronitrile)　偶氮二异丁腈
azo blue　偶氮蓝
azocarmine　偶氮胭脂红
azochloramide　二氯偶氮脒
azocine　吖辛因
azo-compound　偶氮化合物
azo coupling　偶氮偶合
azo-coupling method　偶氮偶合法
azocumene　偶氮枯烯

azocyanide　偶氮氰化物〔含有一价基N＝NCN的化合物〕
azo-cycle compound　氮环化合物〔环中含NH者〕
azodermine(=acetylamidoazotoluene)　乙酰氨基偶氮甲苯
azodicarboamide　偶氮二碳酰胺
azo-dicarbonamide　偶氮二酰胺
azodicarbonic acid　偶氮二甲酸
azodiformate　偶氮二甲酸酯
azodiisobutyronitrile(=azobisisobutyronitrile)　偶氮二异丁腈
azo-dye　偶氮染料
azo dyeing　偶氮(染料)染色法
azo dyestuff　偶氮染料
azoethane　二乙基偶氮
azofermo(=molybdoferredoxin)　固氮铁钼(氧还)蛋白
azofication　固氮作用
azoflavin(e)　偶氮黄素
azo-foaming agent　偶氮类发泡剂
azoformamide　偶氮甲酰胺
azoformamidine　偶氮甲脒
azoformic acid　偶氮(二)甲酸
azofuchsine　偶氮品红
azo-group　偶氮基
azo-hydrazone tautomerism　偶氮腙的互变异构现象
azoic　①无生(物)的②偶氮的
azoic color(=azoic dye)　偶氮染料
azoic dye(=azoic colour)　偶氮染料
azoic dyestuff　偶氮染料
azoic printing composition　偶氮印染配制品；以冰染料制成的印花用配制品
azoimide　叠氮化氢；叠氮酸
azo indicator　偶氮系指示剂
azo-isobutyric dinitrile　偶氮二异丁腈〔发泡剂〕
2,2′-azo-isobutyronitrile　2,2′-偶氮(二)异丁腈
azolactone　噁唑酮
azo lake　偶氮色淀(化)
azole(=pyrrole)　吡咯
azole series　吡咯系
azolite　硫酸钡、硫化锌混合颜料
azolitmin　石蕊素
azolitmin paper　石蕊素试纸；石蕊精试纸
azomercurial　偶氮汞剂
azomethane　偶氮甲烷
cis-azomethane　顺式偶氮甲烷
azomethane esterification　偶氮甲烷酯化
azomethine　偶氮甲碱；甲亚胺　RN＝CHR
azomethine compounds　偶氮甲碱化合物；甲亚胺化合物〔含特性基团—CH＝N—或＝C＝N—〕

azomethine dyes 偶氮甲碱染料；甲亚胺染料
azomethine formation 甲亚胺形成
azomethine-H 偶氮甲碱H
azonaphthalene 偶氮萘
azonia 氮鎓；氮阳离子 R_4N^+
azonine 偶氮宁〔氨基偶氮苯类的酸性染料〕
azonines〔复〕 偶氮宁类染料〔商名，染醋酸丝用〕
azonium 氮鎓〔R_3N_2X 型的化合物〕
azonol A 偶氮脑A
azophenetol 偶氮苯乙醚
azophenol 偶氮苯酚
azophenyl 苯偶氮基
azophloxine (酸性)偶氮红
azophosphine (碱性)偶氮黄
azophosphon 偶氮膦
azo pigment 偶氮颜料
azopolyamide 偶氮聚酰胺
azo polymer 偶氮类聚合物
azorite 锆石
azorubin 偶氮玉红

azosulfime 1,2,4-硫二唑
azotification 固氮作用
azotometer 定氮仪；氮素计
azotometry 氮量测定法
azo type initiator 偶氮(类)引发剂
azo violet 偶氮紫；镁试剂
azoxime 1,3,4-噁二唑
azoxin-C 偶氮喔星C
azoxine 7-芳基偶氮-8-羟基喹啉-5-磺胺〔金属指示剂〕
azoxin-H 偶氮喔星H $CH_3CO_2C_6H_4COOH$
azoxybenzene(=azoxybenzide) 氧化偶氮苯
azoxy compound 氧化偶氮化物
aztreonam 氨曲南〔药〕
azulene ①薁；甘菊环②(洋)甘菊蓝
azulenoid 薁类化合物
azure blue hue 天青蓝色相
azurite 石青；蓝铜矿
azurite blue 天青蓝
azurlite 兰玉髓

B

babassu kernel oil(=babassu oil) 巴巴苏仁油；巴巴苏油
babassu oil(=babassu kernel oil) 巴巴苏仁油；巴巴苏油
Babbit metal 巴比特合金
babbitting 镶巴氏合金
Babcock flask 巴布科克乳油瓶
Babcock tube 巴布科克管
Babcock method 巴布科克法
Babo flask holder 巴波瓶架
B absorption band B 吸收带
babul(=arabic gum tree) 阿拉伯胶树
baby dryer 小(型)烘缸
baby press 预压辊
baby soap 小儿皂
baby square 小方材
baby tower 小型蒸馏塔
bacampicillin 巴氨西林；氨苄青霉素碳酯
bacca(=sap fruit) 浆果
baccatin 浆果赤霉素
baccharine 巴卡林
bacco 烟草
bac-dry (漂白虫胶)全干
B-acid B 酸；1-氨基-8-萘酚-3,5-二磺酸
bacilipin 枯草杆菌脂；杆菌溶素
Bacillus Licheniformis 地衣芽胞杆菌活菌制剂〔药〕
back ①背面②后面③回；反
back aniline 黑苯胺〔苯胺在毛皮的毛上氧化时所生成的一种化学物质〕
backbiting attack 尾咬进攻
backbiting break 尾咬断裂
backbiting transfer 尾咬转移
back boarding 搓软〔皮革〕
backbone ①主链②主要成分③骨干④骨架
backbone chain 主链
backbone mobility 主链运动性
backbone motion 主链运动
backbone polymer 主链聚合物
backbone rearrangement 主链重排(化)
backbone shape burner 脊形燃烧器
backbone structure 主链结构
back center 尾顶尖
back chelometric titration 螯合返滴定
back chipping 铲根
back coating 背面(里面)涂层；背面(里面)涂装
back coordination 反馈配位

back cupping 背面采脂(法)
back diffusion 反扩散
back donating bonding 反馈键
back donation 反馈作用*；反馈(性)
back draft ①后拉伸②倒转；倒车③逆通风
back electromotive force 反电动势
back end 后段*
backer ①衬里；衬料②底基；底材
Backer sodium process 巴克炼钠法
back-extract 反萃取
back-extractant 反萃取剂
back-extracting 反抽提；反萃取
back-extraction 反萃取
back-extrusion 反向挤出
back fabric 衬里织物
back fall 山坡
backfeed 反馈
backfeed loop 反馈回路
back-fill 反向充灌；回填(充)
back-filled 反填充的；回填充的
backfiller ①反填充(物)②反进料器；反加料器③回填机
back-filling 回填充
backfilling gang 回填组
back fill syringe 回填注射器
back fire 回火
back flash 反闪
back flow ①回流；倒流②开模缩裂
backflow factor 反流系数
back flush 逆流洗涤；反向洗涤；反冲〔色谱〕
back-flushable filter 逆冲洗式过滤器
backflush chromatogram 反吹色谱
backflush filter system 逆冲洗式过滤装置；反洗式过滤装置
back flushing 反吹*；反向冲洗
back flushing chromatography 反吹色谱法
backflushing technique 反吹技术
backflush peak 反吹峰
backflush system 反吹系统
backflush unit 反吹单元；反吹组件
backflush valve 反吹阀
back follower 随动件
back gear 后倒齿轮
background ①本底②背景
background absorption 背景吸收

background absorption correction 背景吸收校正*
background attenuation 背景衰减
background conductance 背景电导
background contamination 本底污染
background contribution 背景贡献
background correction 背景校正*
background correction by measurement of time-resolving spectrum 时间分辨光谱法校正背景
background correction by wavelength modulation 波长调制背景校正法
background corrector 背景校正器
background counting rate 本底计数率
background current 本底电流；基流
background electrolyte 背景电解质
background electrolyte effect 本底电解质效应
background equivalent activity(BEA) （天然）本底放射性当量强度
background factor 背景因子*
background hump 本底突峰
background impurity 本底杂质
background ion 本底离子
background level 本底电平
background luminescence 本底发光
background mass spectrum 本底质谱图
background noise 本底噪声
background of cloth 织物背面；布里
background of mass spectrum 质谱本底*
background of water pollution 水污染背景值
background radiation 背景辐射
background signal 本底信号
background source 背景光源
background spectrum 本底谱线图
background subtraction 本底扣除
background suppression 本底抑制
background value of soil pollution 土壤污染背景值
background vapor pressure 本底蒸汽压
backhousia oil 巴荭油
backing ①裱褙②逆行③背材
backing belt ①倒车皮带②后动皮带
backing board 背纸板
backing burr wheel 罗纹滚姆；绒里弯纱轮
backing coat ①(容器)内壁涂层；里面涂层②(卷材)背面涂层
backing dye 防光晕染料
backing fabric 衬里织物
backing film 背衬膜
backing layer 衬里层
backing material 基材；底子料；背衬料

backing off 反馈；退绕
backing paper 背纸；底子纸
backing plate 随动件；背衬板
backing pressure 前级压强
backing pulley 回行皮带盘
backing pump 前级泵
backing ring ①衬圈②支承环
backing roll 承压滚筒
backing run(=sealing run) ①(焊接)焊根②背焊
backing side trap 前级侧阱
backing stage 前级
backing strip 垫板
backing system 前级(真空)系统
backing up 回投；回冲；回行
backing vacuum 前级真空
backing wire 里网
backing yarn ①基经纱；背面经纱②地纱；背面纱线
back knotter 尾筛
backlash ①后退；后冲②间断；侧向间隙
backlight 背光
back light(ing) 背光法
back lining 背衬；底层(硅)衬
back migration 反迁移
back-mix-flow reactor 返混流反应器
backmixing 逆向混合；回混；反混
back mutation 反突变；回复突变
back paint 里面漆；背面漆；背(里)面底漆
backpanels 支撑板
back pressure 反压(力)；背压力
back pressure control 反压控制
back pressure device 反压装置
back pressure evaporator 背压式蒸发器
back pressure gauge 反压计
back-pressure manometer 背压压力计
back pressure operation 背压操作〔压滤机的〕
back pressure regulator 吸入压力调节阀；回压调节阀
back pressure relief port 背压放出口
back-pressure tubing 反压管
back pressure turbine 背压式涡轮机
back pressure valve 止回阀
back propagation(BP) 反向传播法
back propagation algorithm 反向传播算法
backpulsing 反冲式
back purge 反吹
back purge system 反向冲洗装置；反吹系统
back putty(=bed putty) 打底油灰
back raised cloth 毛背织物
back reflection method 背射法

back rinding 开模缩裂
back roll 后滚筒
back roller line 后辊组
back run 逆行
back running 封底焊
backscatter 反向散射；背反射
back-scatter beta gage 反射型(β射线)测厚计
backscattered 反向散射的
backscattered electron 背散射电子
backscattered electron image 背散射电子像
backscattered electron topography image 背散射电子形貌像
backscatter electron image 背散射电子像
backscattering 背散射
backscattering analysis 背散射分析
back scattering image 反向散射像
backscattering spectrometry(BS) 反散射能谱法
backscattering spectroscopy 反向散射光谱
backset ①逆流；涡流②后退；挫折
backshaft drive 后轴传动
back shooting 逆向爆射
back side ①(木材的)内侧材面②后部；(造纸机)传动侧
backside attack 背面进攻
back-sizing 单面上浆；背面涂布；背面涂胶
back-skip 反跳
backstep sequence welding 分段退焊
backstep welding 退焊
back stopper 反流塞(子)；止回阀；止逆阀
back stress 反应力；背应力
back striking 回击
back substitution 回代
back taper(=back draft) 反退拔
back tender 干燥工〔造纸〕
back titration 返滴定(法)
back up ①垫板②备用设备
backup power source 后备电源
backup system 后备系统；备用系统；支持系统
backup trap 后备阱；备用阱；支持阱
back veneer (胶合板的)里板
back view 后视图
backward-bladed impeller 后弯(叶片)叶轮
backward-curved vane 后弯叶片
backward feed 逆流加料
backward flow 回流；反流
backward inclined type of impeller 后弯(式)叶轮
backward leaning vane 后弯叶片
backward reaction 逆向反应
backward scattering 向后散射

backward voltage 反向电压
backward welding 后焊法
back wash 回洗
back washing 回洗(法)
backwash isotherm 反萃取等温线
backwash water 回洗水
back water (纸浆)残水；回水
back water pump 回水泵
backwinding 复绕
baclofen 巴氯芬〔药〕
bacon rubber 烟片〔橡胶〕
bacteria cellulose 细菌纤维素
bacteria corrosion prevention method 防细菌腐蚀法
bacteria counting apparatus 细菌计数器
bacteria filter 滤菌器
bacterial electrode 细菌电极
bacterial luminescence 细菌发光
bacterial membrane electrode 细菌膜电极
bactericide 杀菌剂
bacteriostatic ①抑菌剂②抑菌的
bacteriostatic agent 抑菌剂
bacteriostatic fibre 抑菌性纤维
bacteriostatic plasticizer 抑菌(性)增塑剂
baddeckite 赤铁黏土
baddeleyite 二氧化锆矿；斜锆石
Badger two-stage distillation process 巴杰尔两级蒸馏法
badian oil 八角茴香油
Badische acid(=2-naphthylamine-8-sulfonic acid) 巴迪氏酸；2-萘胺-8-磺酸
Badouin's reagent 巴杜英试剂
baeckeol 岗松酚 $C_{13}H_{18}O_4$
Baer sorter(=Baer stapler) 拜氏纤维长度分析仪
Baer stapler(=Baer sorter) 拜氏纤维长度分析仪
baeumlerite 盐氯钙石
Baeyer indigo synthesis 拜尔靛合成法
Baeyer strain theory 拜尔张力学说
Baeyer test 拜尔试验
Baeyer-Villiger oxidation 拜尔-维利格氧化
baffle 挡板；折流板
baffle arrangement 挡板排列法；折流板排列法
baffle bar 滞流棒(条)；挡板(条)；折流板
baffle chamber 挡板室
baffle column 挡板塔
baffle column mixer 挡板混合塔
baffled decant line 拦水线
baffled evaporator 折流蒸发器
baffled reaction chamber 隔板式反应室
baffled spray column 挡板喷雾萃取塔

baffle gear　板牙齿轮
baffle pan　挡板塔盘〔蒸馏塔内的〕
baffle plate　挡板
baffle-plate column　折流板式塔
baffle-plate column mixer　挡板混合塔
baffler　①挡板；阻层板；烟道隔板②折流器；导流板③消声器；节流阀
baffle separation　折流分离
baffle spray tower　挡板喷雾塔
baffle washer　折流洗涤器
baffling　折流
bag　袋
bagasse　蔗渣
bagasse digester　渗蔗渣器；浸蔗渣器
bagasse elevator　蔗渣提升机
bagasse fibre　(甘)蔗渣纤维
bagasse polarization　蔗汁旋光度
bagasse pulp　蔗渣浆粕
bag conveyer　袋输送器
bag cure　用水胎硫化〔橡胶〕
bag dust filter　滤尘袋
bag elevator　举袋机
bag extractor　袋提取器
bag filter　袋滤器
bag-for-bag　等量
baggasse fiber　蔗渣纤维
bagged conveyer　袋运送器
bagged tire　装有水胎的生胎〔橡胶〕
bagged tyre　上套轴胎
bagger　装水胎机；装囊机
bagginess　①出兜〔压延缺点〕②脱辊(现象)〔开炼〕
bagging　脱辊
bagging machine　包装机；装填机
bagging property　起拱性能
bag hose　袋滤器
bag hose fume　袋滤器烟雾
bagillumbang oil　三子树油〔取自三子树 Aleurites trisperma，产于菲律宾〕
bag in-take hopper　拆袋投料斗；拆包投料斗
bag-like tire　椭圆形轮胎
bag liner　袋衬里
bag machine　装袋机
bag molding　袋模塑
bag mould　袋模
bag moulding(=flexible membrane moulding)　袋模塑；袋压成型
bag paper　袋纸
bagrationite　铈黑帘石

Ba-grease　钡基润滑脂
bag scraper　袋刮刀
bag-tanning　袋鞣法
bag type dust collector　袋式除尘器
bag unloading hopper　拆袋投料斗；拆包投料斗
baguse　蔗渣
bag wall　袋墙
baicalein(=5,6,7-trihydroxyflavone)　贝加因；5,6,7-三羟黄酮
baicalin　贝加灵；7-葡萄糖醛酸-5,6,7-三羟黄酮 $C_{21}H_{18}O_{11}$
Baier thermometer　拜厄温度计
baikalite　易裂钙铁辉石〔矿〕
baikiain　蓓豆氨酸　$C_6H_9O_2N$
baikiaine　蓓豆碱；四氢吡啶羧酸
bail　①半圆形铁环②半圆形拎环；吊环；钩；耳子③水斗；吊桶④夹紧箍
Bailey crucible holder　贝利坩埚座
Bailey furnace　贝利电炉
Bailey-Walker extraction apparatus　贝利-瓦耳克提取器
Bainbridge's mass spectrograph　班布里奇质谱仪
Bainbridge method　班布里奇法
bainite　贝氏体
bain-marie　水浴器
Baird's random tumble pilling tester　贝尔德无规转筒式起球试验仪
baisacki(=bisacki)　拜萨基〔印度4～5月收的鎏基尼品系紫胶〕
bakc paint　里面漆；背面漆；背(里)面底漆
bakeboard　烘板；烘烤面团板
baked　烘过的；烤干了的
baked carbon　炭粗电极；炭极
baked fabricated carbon　电(阻)炭
bakehouse　①面包厂②面包烘房
bakelite　酚醛树脂；酚醛胶木粉；电木粉
bakelite A　酚醛树脂A；可熔酚醛树脂
bakelite B　酚醛树脂B；半熔酚醛树脂
bakelite C　酚醛树脂C；不熔酚醛树脂
baker　①面包师②烘炉
bakerite　纤硼钙石
Baker-Nathan effect　巴克尔-纳山效应
Baker process (for zinc extraction)　巴克尔(提锌)法
baker sag　烘烤流挂
baker's sugar(=corn sugar)　玉米葡糖
baker's wrap　面包纸〔包面包等产品用纸〕
bakery　面包房
bakery product　焙烤食品
bake sand bake process　烘烤-打磨-烘烤法〔再流平工艺〕

bakeshop ①面包铺②面包烘房
baking 烘；焙
baking additive 烘烤添加剂〔尤指烘制食品的添加剂〕
baking black varnish (黑)沥青(清)烘漆；沥青(黑)烤漆
baking coal 黏结煤；结焦煤
baking compartment 焙烤间
baking cycle 烘烤周期；烘烤时间
baking enamel 烘烤搪瓷
baking finish 烘漆；烤漆
baking hammer finish 烘烤型锤纹漆
baking japans 沥青烘漆
baking lacquer 烤漆
baking loss 烘烤损失
baking of bread 烤面包
baking of brick 烧砖
baking of resin 烘烤树脂
baking of varnish 烘烤油漆
baking oven 烘箱；烤炉
baking paint 烘漆
baking powder 焙粉；发(酵)粉
baking press 热压机
baking process (of sulfonation) 烘焙(磺化)法
baking quality (面包)烘焙质量
baking schedule 烘烤工艺规程；烘烤操作规程
baking soda 碳酸氢钠；小苏打
baking stability 烘烤稳定性
baking temperature 干燥温度；焙烘温度
baking test (面包)烘焙试验
baking varnish 清烘漆；清烤漆
baking vehicle 烘干(型)漆料；烘烤型基料；烘烤媒介物
Baku olive oil 巴库橄榄油
balance ①秤；天平②平衡；均衡
balance arm 天平臂
balance beam 天平梁
balance biaxially oriented film 均衡双轴取向薄膜
balance case 天平箱
balanced 平衡的；配平的
balanced action 平衡作用
balanced bridge 平衡电桥
balanced density method 平衡密度法
balanced-density slurry method 平衡密度淤浆法
balanced-density slurry technique 平衡密度糊状装柱法
balanced-density solvent 平衡密度溶剂
balanced diaphragm 平衡式隔膜；等压隔膜
balanced film 均衡(拉伸)薄膜
balanced filter method 补偿滤光片法
balanced gasoline 平衡汽油〔由启动馏分、加速馏分和动力馏分组成的汽油〕

balanced incomplete block design 平衡不完备区组设计
balanced input 平衡输入
balanced laminate 均衡层合板
balanced lubrication 均衡润滑
balanced mechanical seal 平衡型机械密封
balanced mixer 平衡混频器
balanced modulation 平衡调制
balanced mold 对称压模
balanced opposed compressor 对称平衡式压缩机
balanced orientation 平衡取向
balanced pressure regulator 平衡压力调节器
balanced reaction 平衡反应
balanced state 平衡状态
balanced-suspension procedure 平衡悬浮法
balanced tray thickener 平衡式多层增稠器
balanced twist 均衡加捻〔纺〕
balanced valve 平衡阀
balanced viscosity method 平衡黏度法
balanced yarn 不卷缩(的)纱线
balance equation 衡算方程
balance flow 平衡流
balance force 平衡力
balance method 天平法〔测定相对密度〕
balance of heat 热平衡
balance pan 天平盘
balance pan arrest 天平盘托
balance plastometer(=Hoekestra balance plastometer) 何氏可塑计
balance rest 天平座
balance rheometer 平衡流变仪
balance rider (天平)游码
balance rider hook (天平)游码钩
balance room 天平室
balance scale 天平标尺
balance spring 游丝
balance tank 平衡储槽；平衡油罐
balancing 平衡；配平
balancing axial thrust 平衡轴向力
balancing battery 缓冲电池组
balancing controls 调零设备
balancing device 平衡装置
balancing equation 配平方程式
balancing machine 平衡(试验)机
balancing of chemical equations 化学方程式的平衡
balancing resistance 平衡电阻
balancing tank 平衡储槽；平衡油罐
balanophora wax 蛇菰蜡
balas ruby(=spinel) 玫红尖晶石

balata 巴拉塔树胶；铁线子胶
balata latex 巴拉塔胶乳〔天然反式聚异戊二烯〕
Balback-Thum silver process 巴尔巴克-修姆炼银法
balchanolide 布加内酯
Baldo leaf oil 博都叶油
bald spot 秃斑〔漆病,皱纹漆表面局部小面积不起皱或出现异常的光泽〕
balduilin(=bigelovin) 锦菊素
bale ①包②包装
bale breaker 拆捆机；拆包机
baled goods 包装好的物品
baleen 鲸骨
bale opener 拆捆机；拆包机
bale press 包装机
bale pulper 浆板离解机
baler 打包机
bale room 打包工段
balfourodine 巴福定
balfourolone 巴福酮
baling 包装；打包
baling paper 包装纸
baling press 包装机
baling twine 捆包线
balk(=square log) 枋材；大木
ball ①(砂纸打磨)起球〔漆病〕②球；珠
Ball's sledge sortor 包氏纤维长度分析仪
ball and disc integrator 球盘积分仪
ball and disc type integrator 球盘式积分器
ball and ring apparatus 球环软化点测定器
ball and ring method 球环法〔测定树脂熔点〕
ball and ring test 球环法(软化点)试验
ball and roller bearing 滚珠-滚柱轴承
ball and socket joint 球窝接头
ballas 半刚石〔工业用金刚石〕
ballast 压舱物；压载物
ballast float valve tray 重盘式浮阀塔板
ballast valve tray 重盘式浮阀塔板
ball baling press 毛球打包机
ball bearing 球轴承
ball bearing grease 球轴承润滑脂
ball bearing torque test of greases 润滑脂滚珠轴承扭矩试验〔测定润滑脂的稠度〕
ball check nozzle 球形止回喷嘴
ball check valve 球形单向阀；球形止逆阀
ball chuck 球夹
ball clay 球土
ball cock 球(心)阀
ball cock washer 球阀垫片

ball condenser 球形冷凝器
ball constant 球体常数
ball crusher 球磨机
ball-drop method 落球法〔黏度测定法〕
balled iron 铁坯
ball fall seconds 落球秒数〔黏度测定〕
ball (fall) viscosity 落球黏度
ball float 浮球阀
ball float level controller 浮球液面调节器
ball-float trap 球阀凝气瓣
ball-float valve 球阀
ball flow meter 球形流量计
ball heater 球形加热器
ball impact data 落球法冲力值
ball indentor 球式压入器〔硬度测试用〕
balling ①成球②结球
ballistic 弹道的；冲击的
ballistic galvanometer 冲击电流计〔物〕
ballistic method 冲击法
ballistic pendulum 冲击摆
ballistic programmer 弹道程序器
ballistic-resistant fabric 耐冲击织物；防弹织物
ballistics 弹道学
ballistic temperature programming 弹道程序升温器
ballistic test 冲击试验
ballistic testing machine 冲击式强力试验机
ballistite 弹道炸药；波里斯太特〔硝棉、硝化甘油炸药〕
ball-joint 球承；球形接头
ball joint apparatus 球形连接夹
ball-joint manipulator 球承式(剑式)机械手
ball-joint tongs 球承操作钳
ball-like structure 球状结构
ball loadings (球磨机的)装球量
ball making petal machine 瓣式作球机；分段机
ball mill(=bow mill) 球磨机
ball milling 球磨研磨
ball mill pulverizer 球磨粉碎机
ball mill refiner 球磨精制机
ball non-return valve 球形止回阀
balloelectric 雾电荷的
Ballo-electricity 白劳电性；雾电荷性
ball of black-ash 熔结粗苏打
ballometer 雾粒电荷计
ballonet 气室；气囊〔飞艇〕
ballon inflation 气球充气
balloon ①球形瓶②气球③低压轮胎
balloon breaker ①球破碎机②气圈消除器
balloon control ring 气圈控制环

balloon cover 气球壳
balloon fabric 气球布
balloon filter (素烧)圆滤瓶
balloon flask 球形烧瓶
balloon guide 气圈控制器；气圈导丝器
balloon hull 气球壳
balloon-system loss prevention 防止(石油产品)蒸发损失的球囊装置
balloon top angle 气圈顶角
balloon up 膨胀
ballotini ①微球体②小玻璃球〔流动床热媒〕
ballotini reflecting sphere 反光玻璃珠(微球)〔路线漆等用〕
ballototini 多孔玻璃球载体
ball packing 球状填充物
ballpoint delivery system 圆珠式导出系统
ball powder 丸状药
ball pressure test 球压式(硬度)试验
ball punch impact test 球形冲头冲击试验
ball rebound test 落球回弹试验
ball relief valve 球形安全阀；球心安全阀
ball roller bearing grease 滚珠滚柱轴承润滑脂
ball-shaped preform 球形塑坯
ball-spring (chain) model 球-簧链模型
ball stop ①球阀止逆器件②球形止逆阀
ball structure 球状结构
ball tap 球形旋塞
ball tester 球式硬度试验机
ball test(ing) 球试验〔硬度试验〕
ball test of viscosity 落球式黏度测定
ball tipped pipette 安全吸移管；带球吸移管
ball up 滚成球
ball valve 球(心)阀
ball valve oiler 球阀注油器
ball viscometer 落球黏度计
ball viscosimeter 落球式黏度计
ball viscosity 落球黏度
ball warp 球经
ball warping machine 球经整经机
Bally flexometer 巴利挠曲仪
balm(=balsam) ①香脂；香膏②滇荆芥；蜜蜂花
Balmer's generalized formula 广义巴尔末公式
Balmer formula 巴尔末公式
Balmer lines of H atom 氢原子的巴尔末线
Balmer series 巴尔末系
balm oil(=melissa oil) 滇荆芥油；蜜蜂花油
balmy 香脂香的；香膏香的
balopticon(=stereopticon) 投影放大器

balsam 香脂；香膏
balsam fir 香脂冷杉
balsam fir oil 冷杉油〔加拿大型〕
balsamic 香脂的；香膏的
balsamic odour 香脂气味
balsamic tone 香脂香
balsaminasterol 凤仙甾醇；香脂甾醇
balsam of fir 枞香脂；加拿大香脂
balsamous ①有香脂气味的②香脂香的
balsam plasticizer 香脂增塑剂
balsam polar 香脂白杨
balsam seed 香脂豆
baluster(=banister) ①支持物②支柱
balustrade 支持剂；支持物
Baly's tube 巴利管
Baly absorption tube 巴利吸收管
bamboo 竹纤维
bamboo pulp 竹浆
bamboo steel 竹节钢；扎钢〔俗〕
bambuterol 班布特罗〔药〕
banana bond 香蕉键
banana bond orbital 香蕉键轨道
banana essence 香蕉香精
banana flavour 香蕉香精
banana oil 香蕉油〔俗〕；乙酸戊酯
banbury blending 密闭掺混
banbury compounding 密炼配混
banbury cycle 密炼周期
banburyed 密炼的
banbury fluxing 密闭塑炼
banburying 密炼
banbury mastication 密闭塑炼
Banbury mixer 班伯里密炼机
bancoul nut 油桐籽
band ①绳；带②(光)谱带③频带④波带⑤能带
band absorbance (光)谱带吸收率；谱带吸光度
band absorption 光(谱)带吸收
bandage ①帘布筒②实心轮胎
bandamycin 包扎霉素
band application 带状点样
band area 谱带面积
band asymmetry 谱带不对称(性)
band broadening 谱带增宽；谱带展宽
band broadening in space 空间谱带展宽
band broadening in time 时间谱带展宽
band builder ①布筒机②制带机
band conveyer 带式输送机
band conveyor 带式运输器

band curvature　谱带弯曲
band drier　带式干燥机
band-drive　带式传动；绳子传动
band dryer　带式干燥机；带式干燥器
banded　①有条纹的②连结的；结合的
banded coal　条纹煤
banded corrosion　带状腐蚀
band edge　谱带边沿
banded limestone　带纹石灰石
banded precipitation　条纹沉淀
banded spectrum(=band spectrum)　带(状)光谱
banded spherulite　环带球晶
banded texture　条带织构
banded tire　子午线轮胎
band extrusion process　带条挤压法；铸带方法
band filter　带式过滤机
band gap　带隙
bandgap energy　带隙能
band head　谱带头
band height　谱带高度
band impurity　谱带杂质
banding time　包辊时间
band intensity　谱带强度*
band intercept　谱带底宽
band isotope effect　光带同位素效应
band knife　带式刀
band-knife splitting machine　带刀片皮机〔皮革〕
band lugging　花纹块翻新
band of rotation　转动光谱带
band of rotation-vibration　转(动)振(动)光带
bandoline　润发浆；发油；头油
band ore　条纹矿石
band origin　谱带原点；谱带基线
band oven　带式炉
band overlap at apex　(喷路)锯齿形交错重叠(衔接)
bandpass　(光谱)通带
bandpass filter　通带滤光片
bandpass retarding field analyzer　通带减速场分析器
bandpass width　通带宽度
band pipette　谱带点样管
band polymer　带聚物；带型聚合物
band progression　谱带级(数)
band region　带区
band rejection filtering　带阻滤波
band saw　带锯
band sealing　带式封口
band-selective filter　选带滤光片
band sequence　谱带序

band shape　谱带形状
band shift　带移
band spacing　谱带间跨距
band spectrum(=banded spectrum)　带(状)光谱
band spectrum analysis　带(状)光谱分析
band spinning　条料纺丝
band splitting　谱带分裂
band spreading　谱带扩展
band streak　谱带条纹
band structure　能带结构*
bandtail　①能带尾②谱带尾
band tailing　谱带拖尾
band theory　带理论
band tubing　软韧橡皮管
band viscometer　带式黏度计
band width　带宽*
band wire　带(式导)线
bandy clay　带状黏土
bandylite　氯硼铜矿
bane　毒物
Bang's reagent　巴恩试剂
bang-bang control　继电控制
banister(=baluster)　①支持物②支柱
banisterine　扶栏树碱；骆驼蓬碱；哈尔碱
banjo lubrication　放射管式润滑；离心式润滑
banjo oiler　放射管加油器〔润滑油〕
bank　料垄；辊隙存料
banked work in process　预留在制品
banket　护脚〔水〕；含金砾岩层
bank-note paper　钞票纸
bank of eliminators　节流器组；分离器组
bank of staggered pipes　错列管排
bank of tubes　管排
bankomycin　庄行霉素
bankoul nut oil　石栗油〔取自 Aleurites moluccana 的种子〕
bank pass　辊距
bank post paper　银行用纸
banks(=tube bundle)　管束
banks of staggered pipes　错列管排
banks oil　鱼(肝)油
Ban-Lon process　班纶卷曲变形工艺〔使用填塞箱卷曲法原理〕
Banner machine　钢丝圈卷成机
baobab　木棉；猕猴面包树
baobab oil　木棉油
baphiin　巴非英　$C_{12}H_{10}O_4$
baptifoline　野靛叶素

ψ-baptigenetin ψ-野靛配基亭 $C_{15}H_{12}O_5$
ψ-baptigenin ψ-野靛配基 $C_{16}H_{10}O_5$
baptisoid 野靛蓝素
baptitoxine 野靛毒
bar ①巴〔气压单位，1 巴=10^6 达因/厘米^2〕②(金属、木材等的)杆；棒；条③(光、色的)带；线
bar-and-cake cutter 切条块机
baras camphor 冰片；龙脑；莰醇 $C_{10}H_{17}OH$
barb ①倒钩；倒刺②(植物的)芒刺③触须
Barbados tar 巴巴多斯焦油
barbar dryer 干燥室
barbary gum 坝坝树胶；摩洛哥槐(树)胶
barbatic acid 巴尔巴地衣酸；四甲基地衣缩酚酸
barbatinic acid 巴尔巴酸
barbatol 巴尔巴醇
barbatol-carboxylic acid 巴尔巴醇羧酸
barbatolic acid 巴尔巴醇酸
barbecue sauce 烤肉调味浆〔烤烧时涂抹烤物用的酱汁〕
bar bender 弯条机
barberry 小檗
Barbet's process 巴别特(甘油蒸馏)法
Barbey fluidity 巴比流度
Barbey ixometer 巴比流度测定计〔标准单位为巴比流度〕
Barbey orifice viscometer 巴比锐孔黏度计
barbierite 钠正长石
barbing ①(针上)开刺②竿；棒；柱
barbital(=barbitone) 巴比妥；二乙基丙二酰脲
barbital sodium 巴比妥钠；二乙基丙二酰脲钠；溶性巴比妥
barbitone(=veronal; barbital) 巴比妥；二乙基丙二酰脲
barbiturate(s) 巴比妥酸盐
barbituric acid(=malonyl urea) 巴比妥酸；丙二酰脲
      $CH_3CONHCONHCO$
barbotage 起泡作用
barb spacing 刺距
bar cavity 条式模腔
barcenite 锑酸汞矿
bar chart 方框图
bar chromatogram 棒状色谱图；条状色谱图
bar coater(=wire bar applicator) 绕线棒刮涂器〔棒上绕线粗细可控制涂膜厚度〕
Barcol hardness 巴(科尔)氏硬度
Barcol indenter 巴科尔硬度计
bar copper 棒铜
bar cutter 切条机
bardane oil 牛蒡油〔取自牛蒡 Arctium lappa 的种子〕
bar detergent 块状洗涤剂

bare aluminium 裸铝
bare electrode 光焊条；裸露焊条；裸露电极
bare fabric 露布；露白
baregin 黏胶质
bare glass 原玻璃(纤维)；未处理的玻璃纤维
bare ion 裸离子
bareness 胎面缺胶
bare pipe 暴露的管；裸管；光(滑)管
bare pipe-line 露外管线〔没有保暖的〕
bare plaster (墙壁)裸灰泥层
bare rod 裸焊条；无药焊条
bare silver electrode 裸银电极
bare source 裸源
bare substrate 裸底材；纯净的底材
bare surface 露面
bare tank 露外油罐〔没有保暖的〕
barette 黄化机；黄酸化机
bare wire 光焊丝
barge 驳船
baria 氧化钡
baria-alumina-silica fibre 氧化钡-氧化铝-氧化硅纤维
baric ①气压的②钡的
baricalcite 重解石
baric gradient 气压梯度；压力梯度
barie 巴〔压力单位〕
barilla(=pulverin) 草灰；苏打灰；海草灰苏打
barillor(=barilla) 苏打灰；海草灰苏打
bar iron 棒铁
Bari-Sol process 巴里-索尔(脱蜡)法〔以苯及二氯乙烷与润滑油混合，在离心机中把后者的蜡除去〕
barite 重晶石*
barium 钡
barium acetate 乙酸钡 $Ba(C_2H_3O_2)_2$
barium aluminate 铝酸钡
barium arsenate 砷酸钡〔1. 偏 $Ba(AsO_3)_2$；2.原 $Ba_3(AsO_4)_2$；3. 焦 $Ba_2As_2O_7$〕
barium arsenide 砷化钡 Ba_3As_2
barium arsenite 亚砷酸钡〔1.偏 $Ba(AsO_2)_2$；2.原 $Ba_3(AsO_3)_2$；3. 焦 $Ba_2As_2O_5$〕
barium azide 叠氮化钡 $Ba(N_3)_3$
barium azodicarboxylate 偶氮二羧酸钡〔发泡剂〕
barium-base grease 钡基润滑油
barium-base titanox 钛钡白
barium benzosulfonate 苯磺酸钡 $Ba(C_6H_5SO_3)_2$
barium biarsenate 砷酸氢钡
barium bicarbonate 碳酸氢钡 $Ba(HCO_3)_2$
barium bichromate 重铬酸钡 $BaCr_2O_7$
barium bioxalate 草酸氢钡 $Ba(HC_2O_4)_2$

barium bioxide 过氧化钡 BaO_2
barium bisulfate 硫酸氢钡 $Ba(HSO_4)_2$
barium bisulfite 亚硫酸氢钡 $Ba(HSO_3)_2$
barium borate 硼酸钡〔1. 偏 $Ba(BO_2)_2$; 2. 原 $Ba_3(BO_3)_2$; 3. 焦 BaB_4O_7〕
barium boride 六硼化钡 BaB_6
barium borofluoride 氟硼酸钡 $Ba(BF_4)_2$
barium borotungstate 硼钨酸钡 $2BaO \cdot B_2O_3 \cdot 9WO_3 \cdot 18H_2O$
barium bromate 溴酸钡 $Ba(BrO_3)_2$
barium bromide 溴化钡 $BaBr_2$
barium bromoplatinate 溴铂酸钡 $Ba(PtBr_6)$
barium caprate 癸酸钡
barium carbide 二碳化钡 BaC_2
barium carbonate 碳酸钡 $BaCO_3$
barium chloranilate 氯冉酸钡
barium chlorate 氯酸钡 $Ba(ClO_3)_2$
barium chloride 氯化钡 $BaCl_2$
barium-chloride test 氯化钡测硫试验〔测定油类的硫酸或硫酸盐含量〕
barium chlorite 亚氯酸钡 $Ba(ClO_2)_2$
barium chloroplatinate 氯铂酸钡 $Ba(PtCl_6)$
barium chloroplatinite 氯亚铂酸钡 $Ba(PtCl_4)$
barium chromate 铬酸钡 $BaCrO_4$
barium chrome 钡(铬)黄;铬酸钡
barium cinnamate 肉桂酸钡 $Ba(C_6H_5C_2H_2COO)_2$
barium cresylsulfonate 甲苯磺酸钡 $Ba(CH_3C_6H_4SO_3)_2$
barium crown 铬酸钡;钡黄
barium cuprate 铜酸钡 $Ba[Cu(OH)_4]_2$
barium cyanate 氰酸钡 $Ba(CNO)_2$
barium cyanide 氰化钡 $Ba(CN)_2$
barium cyanoplatinite 氰亚铂酸钡 $Ba[Pt(CN)_4]$
barium detergent additive 含钡洗涤添加剂;含钡去垢添加剂
barium dichromate 重铬酸钡 $BaCr_2O_7$
barium diisopropoxide 二异丙氧基钡
barium di-isopropyl salicylate 二异丙基水杨酸钡
barium dioxalate 草酸氢钡 $Ba(HC_2O_4)_2$
barium dioxide 过氧化钡 BaO_2
barium diphenylamine sulfonate 二苯胺基磺酸钡
barium dithionate 连二硫酸钡 BaS_2O_6
barium ester 钡酯〔兼为钡盐和酯〕
barium ethoxide(=barium ethylate) 乙醇钡;乙氧基钡 $Ba(C_2H_5O)_2$
barium ethylate(=barium ethoxide) 乙醇钡;乙氧基钡
barium ethylsulfate 乙硫酸钡 $Ba(SO_4C_2H_5)_2$
barium extended titanium dioxide 钛钡白
barium feldspar 钡长石

barium ferrate 高铁酸钡 $BaFeO_4$
barium ferrite 亚铁酸钡
barium ferrocyanide 氰亚铁酸钡 $Ba_2[Fe(CN)_6] \cdot 6H_2O$
barium flint 钡燧石
barium fluoborate 氟硼酸钡 $Ba(BF_4)_2$
barium fluobromide 氟溴化钡;氟化钡合溴化钡 $BaBr_2 \cdot BaF_2$
barium fluochloride 氟氯化钡;氟化钡合氯化钡 $BaCl_2 \cdot BaF_2$
barium fluoiodide 氟碘化钡;氟化钡合碘化钡 $BaI_2 \cdot BaF_2$
barium fluoride 氟化钡 BaF_2
barium fluosilicate 氟硅酸钡 $Ba(SiF_6)$
barium fluozirconate 氟锆酸钡 $BaZrF_6$
barium fluxing agent 钡熔剂;二号熔剂
barium glass 钡玻璃
barium hexabromo-linolenate 六溴化亚麻酸钡
barium hydrate 氢氧化钡 $Ba(OH)_2$
barium hydride 氢化钡 BaH_2
barium hydrogen arsenate 砷酸氢钡 $BaHAsO_4$
barium hydrogen phosphate 磷酸氢钡 $BaHPO_4$
barium hydrogen sulfate 硫酸氢钡 $Ba(HSO_4)_2$
barium hydrosulfide 硫氢化钡 $Ba(HS)_2$
barium hydroxide 氢氧化钡 $Ba(OH)_2$
barium hypochlorite 次氯酸钡 $Ba(ClO_2)_2$
barium hypophosphate 连二磷酸钡 $Ba_2P_2O_6$
barium hypophosphite 次磷酸钡 $Ba(H_2PO_2)_2 \cdot H_2O$
barium hyposulfate 连二硫酸钡 BaS_2O_6
barium hyposulfite ①连二亚硫酸钡 BaS_2O_4 ②〔俗指〕硫代硫酸钡
barium iodate 碘酸钡 $Ba(IO_3)_2$
barium iodide 碘化钡 BaI_2
barium isopropoxide(=barium isopropylate) 异丙氧基钡;异丙醇钡
barium isopropylate(=barium isopropoxide) 异丙氧基钡;异丙醇钡
barium lactate 乳酸钡 $Ba(C_3H_5O_3)_2$
barium laurate 月桂酸钡〔稳定剂〕
barium mahogany sulfonate 石油磺酸钡
barium malate 苹果酸钡 $BaC_4H_4O_5$
barium manganate 锰酸钡 $BaMnO_4$
barium metaborate 偏硼酸钡 $Ba(BO_2)_2$
barium metaphosphate 偏磷酸钡
barium metasilicate 硅酸钡 $BaSiO_3$
barium metatitanate 偏钛酸钡
barium metatungstate 偏钨酸钡 $BaO \cdot 4WO_3 \cdot 9H_2O$
barium methoxide(=barium methylate) 甲醇钡;甲氧基钡 $Ba(CH_3O)_2$

barium methylate(=barium methoxide)　甲醇钡；甲氧基钡
barium mineral　钡矿(物)
barium molybdate　钼酸钡　$BaMoO_4$
barium monoxide　氧化钡　BaO
barium myristate　肉豆蔻酸钡
barium naphthenate　环烷酸钡
barium nitrate　硝酸钡　$Ba(NO_3)_2$
barium nitride　二氮化三钡　Ba_3N_2
barium nitrite　亚硝酸钡　$Ba(NO_2)_2$
barium oleate　油酸钡　$C_{18}H_{33}O_2Ba$
barium orthosilicate　正硅酸钡　Ba_2SiO_4
barium oxalate　草酸钡　BaC_2O_4
barium oxide　氧化钡　BaO
barium perchlorate　高氯酸钡　$Ba(ClO_4)_2$
barium periodate　高碘酸钡　$Ba_5(IO_6)_2$
barium permanganate　高锰酸钡　$Ba(MnO_4)_2$
barium peroxide　过氧化钡　BaO_2
barium persulfate　过(二)硫酸钡　$BaS_2O_6(O_2)$; BaS_2O_8
barium phenolate　苯酚钡
barium phenylsulfate　苯代硫酸钡　$Ba(C_6H_5SO_4)_2$
barium phosphate　磷酸钡〔1.(正)$Ba_3(PO_4)_2$；2. 偏 $Ba(PO_3)_2$；3. 焦 $Ba_2P_2O_7$〕
barium phosphide　二磷化钡　BaP_2
barium phosphite　亚磷酸氢钡　$BaHPO_3$
barium plaster　钡盐石膏灰浆〔防辐射用〕
barium platinic chloride　氯铂酸钡　$Ba(PtCl_6)$
barium platinic rhodanate　硫氰合铂酸钡　$Ba[Pt(CNS)_6]$
barium platinocyanide(=barium platinous cyanide)　氰亚铂酸钡　$Ba[Pt(CN)_4]$
barium platinous cyanide(=barium platinocyanide)　氰亚铂酸钡
barium polysulfide　多硫化钡　BaS_n
barium potassium chromate　铬酸钡钾
barium potassium ferrocyanide　氰亚铁酸钡钾　$BaK_2[Fe(CN)_6]$
barium pyroborate　焦硼酸钡　BaB_4O_7
barium pyrophosphate　焦磷酸钡　$Ba_2P_2O_7$
barium rhodanate(=barium rhodanide)　硫氰酸钡　$Ba(CNS)_2$
barium rhodanide(=barium rhodanate)　硫氰酸钡
barium ricinoleate　蓖麻醇酸钡〔有机溶胶或塑性溶胶用热稳定剂〕
barium salt　钡盐
barium selenate　硒酸钡　$BaSeO_4$
barium selenide　硒化钡　$BaSe$
barium selenite　亚硒酸钡　$BaSeO_3$
barium silicate　硅酸钡　$BaSiO_3$
barium silicofluoride　氟硅酸钡　$Ba(SiF_6)$
barium soap　钡皂
barium stannate　锡酸钡
barium stearate　硬脂酸钡　$Ba(C_{18}H_{35}O_2)_2$
barium succinate　丁二酸钡；琥珀酸钡　$BaC_4H_4O_4$
barium sulfate(=barium white)　钡白；硫酸钡　$BaSO_4$
barium sulfate extender　硫酸钡体质颜料
barium sulfate precipitated　沉淀硫酸钡；硫酸钡　$BaSO_4$
barium sulfhydrate　硫氢化钡　$Ba(HS)_2$
barium sulfide　硫化钡　BaS
barium sulfite　亚硫酸钡　$BaSO_3$
barium sulfocyanate(=barium sulfocyanide)　硫氰酸钡　$Ba(CNS)_2$
barium sulfocyanide(=barium sulfocyanate)　硫氰酸钡
barium sulfonate　磺酸钡
barium sulfophenylate　苯代硫酸钡　$Ba(C_6H_5SO_4)_2$
barium sulfovinate　乙代硫酸钡　$Ba(C_2H_5SO_4)_2$
barium superoxide　过氧化钡　BaO_2
barium tartrate　酒石酸钡　$BaC_4H_4O_6$
barium tetraborate　四硼酸钡　BaB_4O_7
barium tetrathionate　连四硫酸钡　BaS_4O_6
barium thiocyanate(=barium thiocyanide)　硫氰酸钡　$Ba(CNS)_2$
barium thiocyanide(=barium thiocyanate)　硫氰酸钡
barium thiosulfate　硫代硫酸钡　$BaSO_3S$; BaS_2O_3
barium titanate　钛酸钡　$BaTiO_3$
barium titanate ceramics　钛酸钡陶瓷
barium triphosphate　磷酸钡　$Ba_3(PO_4)_2$
barium trithionate　连三硫酸钡　BaS_3O_6
barium tungstate　钨酸钡　$BaWO_4$
barium uranite　磷酸钡铀矿
barium white(=baryta white;barium sulfate)　钡白；硫酸钡〔着色剂〕
barium wolframate　钨酸钡　$BaWO_4$
barium yellow　钡黄〔主要成分为铬酸钡〕
bark breaker　树皮碎裂机
bark breaking machine　树皮辗碎机
bark crepe　树皮绉片〔低级生胶〕
bark cutter　树皮切碎机
barker　去皮机；剥皮机
barkevicite　棕闪石
bark extract　树皮浸膏
bark fiber　树皮纤维
bark grinding mill　树皮粉碎机
bark hact　割刀；采脂刀
barking　①去皮；剥皮②树皮化〔光老化现象〕
barking drum　筒式去皮机
barking machine　剥皮机；去皮机
bark liquor　树皮鞣液

bark mill　碾(树)皮机
barkometer(=tannometer)　巴克表；鞣液比重计
barkometer degree(=Either degree)　巴克度
barkometer hydrometer　巴克表；鞣液比重计
bark press　压皮机
bark rubber　树皮绉片〔低级生胶〕
bark shaver　刮(皮)刀
bark shaver face　刮(皮)面
bark shredder　树皮破碎机
bark stripping machine　剥皮机；去皮机
bark tanned　树皮鞣的；红鞣的
bark tanned side leather　树皮鞣的鞋面革
bark tanner　树皮鞣料
barky　含有树皮的
barley seed oil　大麦子油
barmatic　棒料自动送进装置
bar moulding machine　压皂机
barn　靶(恩)〔核反应截面单位, 等于 10^{-24} 厘米^2〕
barn paint　①木板房用漆②仓库用漆③底漆；打底用涂料
barograph　气压记录器
barometer　气压计
barometer cistern　气压计槽
barometer tube　气压计管
barometric (=barometrical)　气压的；测定气压的
barometrical(=barometric)　气压的；测定气压的
barometric condenser　气压冷凝器
barometric discharge pipe　大气排泄管〔干式冷凝器〕
barometric formula　气压公式
barometric fuel regulator　气压燃料调节器
barometric height　气压高度
barometric jet condenser　气压喷水冷凝器
barometric leg(=barometric pipe)　气压排液管；大气腿
barometric leg condenser　气压排液管冷凝器
barometric pipe(=barometric leg)　气压排液管；大气腿
barometric pressure　大气压
barometrograph　气压(照像)记录器
barophoresis　压泳(现象)
baroreceptor　压力感受器
baroscope　验压器
barosmin　①布枯叶脂②云香苷
barosmoid　布枯叶素
barostat　气压补偿器
barotaxis　趋压性；向压性
barothermograph　气压温度记录器
barotropism　向压性
bar oven　条炉
bar pressure　大气压力

barrandite　铝红磷铁矿
barras　①含油树脂；软树脂②毛松香
barrate　①转鼓②干法黄化机；干法黄化鼓
barrel　①桶〔石油容积单位, 等于 158.99 升〕②膛〔枪炮〕③机筒；外套
barrel bore　①圆筒内腔；套筒内腔②圆筒内径；套筒内径
barrel bulk　五立方英尺容积
barrel coating　①转鼓涂装(法)②大桶涂料
barrel colouring　①滚筒着色法②滚筒轧漆法
barrel control thermocouple　筒温控制热电偶
barrel diameter　圆筒(滚筒；套筒)直径
barrel elevator　桶升降机
barrel enamelling　转鼓涂漆
barrel end opening　桶端开口
barrelene azulene　桶烯
barrel filler　自动装桶器
barrel filling　装桶
barrel finishing(=tumbling)　①(小件)滚筒内抛光；转鼓内抛光②转筒中除毛刺
barrel handling　装运油桶
barrel head　桶底
barrel heater　机筒加热器
barrel hoop　桶箍
barreling　装入桶内
barreling time　①滚筒(转鼓)处理时间②滚筒着色加工时间
barrel jacket　机筒外套
barrelled　装入桶内的
barrel lifter　①装桶工人②装桶机
barrel liner　桶衬；机筒内套〔可拆换〕
barrelling　①转鼓涂涂②装桶
barrelling station　装桶站
barrel lining　(挤出机)机筒(筒体)衬里
barrel mixer(=drum tumbler)　滚筒式混合机(器)；转鼓混合器
barrel of pipe　管膛
barrel oil pump　油桶手摇泵〔从油桶把油泵出用〕
barrel packing machine　装桶机
barrel polishing　鼓式抛光；鼓式除边
barrel press　弧形框式平板机
barrel pump　桶泵
barrel ratio　圆筒(机筒；滚筒)长径比〔L/D 比〕
barrel roller　圆筒式辊；筒形辊
barrels daily　每日桶数
barrels per calendar day(BPCD)　每日历日桶数
barrels per day(BPD)　每日桶数
barrels per stream day(BPSD)　每开工日桶数

barrel time 滚光时间
barrel tumbler 转动鼓
barrel tumbling 转鼓抛光；转鼓除边
barrel-type micrometer 套筒式测微计
barrel velocity 压辊速度〔压延机〕
barrel zone temperature 机筒(加热区)温度
barren ①空的；空白的②贫瘠的③不结果的
barren spot 空斑；空层
barrier ①屏蔽层；阻挡层②多孔隔板
barrier bottle 防渗性瓶〔包装〕
barrier coat ①防渗涂层②隔离涂层；封闭涂层
barrier container 防渗性容器
barrier effect 势垒效应
barrier envelope (分散剂对颜料的)阻隔性包封；屏蔽性包封
barrier film ①防渗薄膜②隔离膜
barrier layer ①防渗层②隔离层
barrier layer cell 障层电池；硒(光)电池
barrier leakage 势垒穿透；势垒漏泄
barrier paper ①防渗纸②隔离纸
barrier polymer 阻透聚合物；阻隔聚合物
barrier property 抗渗性；隔离性〔抵抗空气、氧气、水蒸气、油脂、酸、碱以及普通溶剂渗透的性能〕
barrier sheet ①防渗板②隔离片
barrier technology 防渗工艺
barringtogenic acid 玉蕊精酸
barringtogenin 玉蕊配基；玉蕊精宁
barringtogenol 玉蕊精醇
barringtogentin 玉蕊精亭
barringtonin 玉蕊宁
barrow 矸石场
bar soap 条皂；块皂
bar steel 棒钢
bar stock 棒料
bar straightener 直条机；调直机
bar tack 打套结
bar tacking machine 加固缝纫机
bar thermometer 棒式温度计
barthrin 胡椒除虫菊〔杀虫剂〕；(B-)熏虫菊
Bartlett's test for equal variances 等方差性巴特利特检验
Barus effect(=melt swell) 巴勒斯效应〔聚合物流出喷丝孔后突然膨胀的现象〕；熔体膨胀效应
bar viscometer 杆式黏度计
barwood 苏木
barye 微巴〔=1 达因/厘米^2〕
barylite 硅钡铍矿
baryon 重子

barysilite 硅铅矿
baryta 氧化钡
baryta coating 钡地涂层
baryta felspar 钡长石
baryta flint 钡燧石
barytage 钡地涂层
barytage paper 钡地纸
baryta green 钡绿〔锰酸钡〕
baryta hydrate 氢氧化钡 $Ba(OH)_2$
baryta lime 含钡石灰
baryta mixture 氧化钡混合物
baryta paper 钡地纸
baryta phosphate crown 磷铬黄
baryta saltpeter 钡硝石
baryta water 氢氧化钡水溶液
baryta white(=barium white) 钡白；硫酸钡〔着色剂〕
baryta yellow 钡黄；铬酸钡
baryt coating 钡地涂层
barytes(=barite) 重晶石
barytin(e)(=barytes) 重晶石
baryton(=meson) 介子
baryum(=barium) 钡 Ba
basal 基板
basal cross section 底截面
basal dislocation 基本位错
basal face 底面
basal layer 基础层
basal medium 基础培养基
basal metabolic rate(BMR) 基础代谢率
basal metabolism 基础代谢
basal surface 底表面
basalt ①玄武岩②玄武岩制品③黑色瓷器
basal tapping V形割胶
basalt clay 玄武土
base 碱*
π-base π碱
baseball leather 棒球革
base blending stock 基本掺和原料
baseboard(=washboard;scrub board) ①护壁板；墙裙②踢脚板
base bullion lead 粗铅锭
base catalysis 碱催化*
base catalyzed polymerization 碱催化聚合
base-catalyzed rearrangement 碱催重排作用
base-catalyzed solution polymerization 碱催化溶液聚合
base-centered monoclinic 底心单斜
base chemicals 化工品
base circle 基圆

base coat 底层；底漆
base coating machine 底涂机
base color 基本色；基色
base constant 碱常数
base constituent of lubricating oils 润滑油基本组分
base current 基流
base depot 基地堆栈；基地仓库
base dissociation constant 碱解离常数
based on crude 根据原油计算
based on fuel 根据燃料计算
based on oil 根据油类计算
base drift 基线漂移
base drift corrector 基线漂移校正器
base electrode 基极电极
base electrolyte 基底电解质
base element 成碱元素
base exchange 碱交换(作用)
base-exchangeable ions 碱交换离子
base exchange capacity 碱交换量
base exchange resin 碱性(离子)交换树脂
base fabric 底基织物
base face 基准面
base forming element 成碱元素
base gasoline 基本汽油
base intercept 基线截断
base ionization current 基始电离电流
base lacquer 底漆
baseless felt 无基布毡毯
base level control 塔底液位控制
baseline 基线*
baseline adjustment 基线调整
baseline correction 基线校正
baseline corrector 基线校正器
baseline drift 基线漂移*
baseline drift corrector 基线漂移校正器
baseline flatness 基线平直度
baseline fluctuation 基线波动
baseline irregularity 基线不规则性
baseline method 基线法*
baseline noise 基线噪声*
baseline offset correction 基线偏转校正
base line performance 基准绩效
baseline resolution 基线分离
baseline resolved peak 基线分离峰
baseline selection method 基线选择法
baseline shift 基线位移
baseline shock 基线抖动
baseline stability 基线稳定性

baseline stabilizer 基线稳定器
baseline stepping 基线(成台阶状)颤动
baseline wander 基线跳动；基线漂移
baseline width 基线宽度
base material ①原(材)料②基层材料；基质
base material thickness 基材厚度
base measuring pressure 校正压力；测压基准
base measuring temperature 校正温度；测温基准
basement 地下室
base metal ①基底金属〔合金中的主要金属〕②贱金属
base mix 原始混合物
base mole 基本摩尔单位量；基本摩尔
base mortar 底座灌浆
base number 碱值
base of column 柱脚
base of crude oil 原油的基类
base of evaporator 蒸发器底部
base of peg (搅拌)销钉根；搅拌齿根
base of petroleum 石油基
base oil 基础油
base pair(bp) 碱基对
base pairing 碱基配对
base paper 原纸
base peak 基峰*
base plate 底座；机座；底盘
base plate of skirt support 裙座垫片
base point 基点
base polymer 原料聚合物
base potential 基电位
base printing 色基印花
base product 基础产物；蒸馏塔底
base pump station 总泵站
base recipe 基本配方
base resin 原料树脂
baser (metal) 较贱金属
base sequence 碱基顺序
base solution 基液；底液
base stacking 碱基堆集
base stock ①基本原料②基本组分③基本油料
base strength 碱强度
base substance 基质
base unit 基本单位；基本单元
base unit weight 基量；基本单位重量
base-wash 碱洗
base weight 基准重
Bash cell 巴希电池
Bashore resiliometer 巴肖氏弹性试验机
basic ①碱的②碱性的③碱式的④基本的

basic accelerator 碱性促进剂
basic acetate 碱式乙酸盐
basic acetate method 碱式乙酸盐法
basic aliphatic tertiaryamine 碱性脂肪族叔胺
basic alkylation 碱性烷基化
basic aluminium acetate 碱式乙酸铝
basic anhydride 碱性氧化物；碱酐
basic Bessemer cast iron(=basic Bessemer pig iron) 碱性转炉铸铁
basic Bessemer pig iron(=basic Bessemer cast iron) 碱性转炉铸铁
basic bismuth gallate 碱式没食子酸铋
basic bismuth nitrate 碱式硝酸铋 $Bi(OH)_2NO_3$; $Bi(OH)(NO_3)_2$;$(BiO)NO_3$;$(BiO)OH$
basic bismuth salicylate 碱式水杨酸铋 $C_6H_4(OH)COO(BiO)$
basic black 碱性黑〔染〕
basic blue K 碱性蓝 K
basic brick 碱性砖
basic brown 碱性棕〔染〕
basic calcium sulfonate gel 碱式磺酸钙凝胶
basic calcium superphospate 沉淀磷酸钙〔一种钙肥料〕
basic capacity 碱容量；碱度
basic carbonate 碱式碳酸盐
basic carbonate white lead 碱式碳酸铅；铅白〔俗〕
basic cerise 碱性樱红〔染〕
basic chromic chloride 碱式氯化铬
basic chromium sulfate 碱式硫酸铬
basic color 碱性颜料
basic complex 碱式配位化合物
basic converter 碱性转炉
basic converter process 碱性转炉作业法
basic converter steel 碱性转炉钢
basic copper acetate 碱式乙酸铜
basic costs 基本费用
basic cupric carbonate 碱式碳酸铜
basic dimethyl halogenated arsine 碱式卤化二甲胂 $[(CH_3)_2AsX]_6[(CH_3)_2As]_2O$
basic dissociation constant 碱解离常数
basic dye 碱性染料
basic dyeing 碱性染色〔染〕
basic dye pigment 碱性染料性颜料；碱性有机颜料
basic dyestuff 碱性染料
basic electrode 碱性焊条；碱性电极
basic ethylene sulfate 羟乙硫酸 $HOC_2H_4OSO_3H$
basic extractant 碱性萃取剂
basic filler 碱性填料
basic filter ①标准滤光片②标准滤波器

basic flux 碱性熔剂
basic formate method 碱式甲酸盐法
basic formula 基本配方
basic frequency absorption band 基频吸收带
basic fuchsin 碱性品红
basic fuchsin magenta 碱性品红；碱性洋红
basic fusion 碱熔
basic gravity 标准比重
basic group 碱性基；碱性原子团
basic hearth 碱性敞炉
basic hole system 基孔制
basic hydrolysis 碱水解
basic indicator 碱性指示剂
basic ion 阳离子；碱离子
basicity 碱度；碱性
basicity constant 碱度常数
basicity grade 碱度；碱级
basicity value 碱度值
basic lake 碱性色淀
basic lead acetate 碱式乙酸铅
basic lead arsenate 碱式砷酸铅
basic lead carbonate 碱式碳酸铅；白铅粉
basic lead carbonate paper 碱式碳酸铅纸
basic lead chromate 碱式铬酸铅；铬红
basic lead chromate silicate 碱式硅酸铬铅
basic lead nitrate 碱式硝酸铅
basic lead silicate 碱式硅酸铅〔活性剂〕
basic lead silicate white 碱式硅酸铅白
basic lead sulfate 碱式硫酸铅
basic lead sulfate white 碱式硫酸铅白
basic lead white 铅白；碱式碳酸铅 $2PbCO_3 \cdot Pb(OH)_2$
basic lining 碱性炉衬
basic magenta 碱性品红；碱性洋红
basic magnesium carbonate 碱式碳酸镁
basic material ①原料②碱性材料
basic metabolism 基础代谢
basic metal 碱性金属
basic nitril 叔胺 R_3N
basic nitrogen 碱性氮(化合物)
basic note 基香；底香
basic open-hearth furnace 碱性平炉
basic open-hearth steel 碱性平炉钢
basic oxide 碱性氧化物；成碱氧化物
basic pig(=basic pig iron) 碱性生铁
basic pig iron(=basic pig) 碱性生铁
basic polyester 碱性聚酯
basic polymer 原聚合物
basic polymer chain (接枝)聚合物主链

basic property 碱性
basic pseudo-boehmite 碱性假勃姆石
basic pulse generator 基准脉冲发生器
basic reaction ①碱性反应②基础反应
basic recipe 基本配方
basic resin 碱性树脂；阴离子交换树脂
basic rhodamine 碱性玫瑰精
basic rocks 碱性岩
basic salt 碱式盐
basic sediment 碱性沉积
basic shaft 主轴
basic shaft system 基轴制
basic silicate white lead 碱式硅酸铅白
basic simplex method(BSM) 基本单纯形法
basic size 基准尺寸
basic slag 碱性熔渣
basic solvent 碱性溶剂
basic standard 基础标准
basic structure 基础结构
basic substituted 碱性基取代了的
basic sulfate white lead 硫酸铅白
basic superphosphate 沉淀磷酸钙〔一种钙肥料〕
basic topping 碱性染料套染
basic-type accelerator 碱性促进剂
basic volumetric weight(=basic density) 公定容重
basic warp tension 上机张力
basic weight 基重
basic wire 主线材
basic zinc chromate 碱式铬酸锌
basification 碱化；提高碱度；提高盐基度
basified viscose 碱化黏胶纤维；羊毛化黏胶纤维
basifier 碱化剂
basifying 碱化
basil(=basilicum) 罗勒
basilicum(=basil) 罗勒
basil oil 罗勒油
basis ①基准；基础②算法
basis heating 主加热
basis ionization current 基始电离电流
basis material 基体材料；底材
basis set 基组*
basis soap 基皂
basis weight 定重
basketball leather 篮球革
basket centrifuge 篮式离心机
basket drier 篮式干燥机
basket electrode 篮形电极
basket extractor 篮式提取器；篮式浸出器

basket filter 篮式滤器
basket grate 篮式栅
basket purl fabric 席纹织物
basket tipper 篮式倾卸器
basket type evaporator 篮式蒸发器
basket type tower 篮式塔
basket willow 蒿柳
basofor 沉淀硫酸钡
basquet ①篮②桶
bass （秃）椴树
basseol 椴树醇
bassia fat 椴树脂
bassic acid 椴树酸
bassisterol 美果甾醇 $C_{27}H_{46}O$
basswood 美洲椴
basswood oil 椴树油
bastard 非标准的；异常尺码的
bastard coal 硬煤
bast fiber 韧皮纤维
bast fiber spinning 麻纺
bastnasite 氟碳铈镧矿
bastose 黄麻的纤维素；麻纤维
bast zone 韧皮层
basylous(=basic) ①碱的②碱性的③碱式的
basylous action 碱性作用
basylous element 碱性元素
bat 衬垫
batatic acid 巴他酸；甘薯黑疤霉酸
Batavia dammar 巴达维亚玛(树)脂
batavite 透鳞绿泥石〔无铁蛭石〕
batch ①分批②每批容量
batch agitator 分批搅拌器；间歇式搅拌器
batch blending 分批掺和
batch centrifuge 间歇式离心机
batch coke still 间歇式焦化(蒸馏)釜
batch crystallization 分批结晶
batch crystallizer 分批结晶器
batch cure 分批硫化
batch desublimer 分批凝华器
batch dissolution 分批溶解；间歇溶解
batch dissolver 分批溶解器；间歇溶解器
batch distillation 分批蒸馏
batch drier 分批干燥器
batch dryer 分批干燥器
batched bread 成批做的面包
batch end point （输油管线中）油品分批点
batcher 送料计量器；进料量斗
batch extraction 分批萃取；间歇萃取

batch extruder 一次投料压出机
batch filtration 间歇式过滤
batch fractionating 间歇精馏
batch fractionating column(=batch fractionating tower) 间歇式分馏柱(塔)
batch fractionating tower(=batch fractionating column) 间歇式分馏柱(塔)
batching 分批；分组；分段
batching in product lines 产品分批沿管线输送
batching of cloth 布的压光
batching off ①片；出料②导出
batching-out unit 卸料装置；排胶装置
batch ion exchange 间歇(操作)离子交换
batch leaching 分批浸出；间歇浸出
batch mass 每批容量
batch method 分批法
batch method of treatment 分批处理法；间歇式处理方法
batch mill 间歇操作磨机
batch mixer 分批混合器
batch number 批号〔产品或原料的〕
batch of concrete 混凝土的分批(拌和)
batch of dough 面团分批(拌和)
batch-off roll ①卷纸辊②分批辊
batch of rubber 橡胶混合物卷
batch oil 翻砂用油；铸造用油〔制造绳缆用油〕
batch operation 分批操作；间歇操作
batch pan 料盘；药盘
batch plant ①定量供料装置②间歇操作装置
batch polymerization 分批聚合；间歇聚合
batch pot dissolution 间歇罐式溶解
batch process 分批法
batch production 分批式生产；间歇式生产
batch purification 分批式精制；间歇式精制
batch reactor 间歇式反应器
batch rectification 分批精馏
batch run 间歇试验(操作)
batch sampling 分批进样；间歇进样
batch scale 批量秤
batch size 批量大小
batch steam distillation 分批蒸汽蒸馏
batch steeping process 间歇浸渍工艺；分批浸渍工艺
batch still 分批蒸馏器
batch system 间歇式系统
batch temperature 批料温度
batch test 分批试验；间歇试验
batch ticket 批量单；分步规划(计划)
batch-to-batch variations 逐批质量差异
batch treating 分批处理

batch-type 间歇式；分批式
batch type bleaching machine 间歇式漂白机
batch type freezer 间歇式冻结器
batch-type production 分批式生产
batch up 卷取
batch vaporization 分批蒸发
batch weighing 分批称重
batch weight 每批重量
batch-wise 分批的；不连续的
batchwise fractional liquid extraction 分批式分级液体抽提
batchwise polymerization 分批聚合；间歇聚合
batea (硝酸钠)晶体收集器
bate drench 麸液软化
bate pits(=bate pricks; bate stains) ①脱斑②脱灰
bate pricks(=bate pits) ①脱斑②脱灰
Bates hydrometer 贝茨比重计
Bates polariscope 贝茨偏振光镜
bate stains(=bate pits) ①脱斑②脱灰
bath ①浴②浴器
bath brick 砂砖；巴士磨石
bath composition ①电泳(槽)液成分(组成)②电解液成分
bath control 电泳(电解)槽控制
bath enamel 浴槽瓷漆
bath-fed tube 凝固浴管；管(式凝固)浴
bath filter 浸液式过滤器
bath fluid 浴液
bath leaving speed 离浴速度〔丝条离开凝固浴的速度〕
bath lubrication 油浴润滑；浸油润滑
bathochromatic effect 向红效应；深色效应
bathochrome 向红基；向红团
bathochromic (向)红移(的)
bathochromic change 红移
bathochromic effect(=red shift) 向红(增色)效应；减频效应；红移效应
bathochromic series 向红系(列)
bathochromic shift (向)红移
bathocuproine(=2,9-dimethyl-4, 7-diphenyl-1, 10-phenanthroline) 铜试剂；2,9-二甲基-4, 7-二苯基-1,10-菲咯啉
batholite 岩基
bathophenanthroline 红菲咯啉；4,7-二苯基-1,10-菲咯啉 $C_{24}H_{16}N_2$
bath powder 浴(后爽身)粉
bath preparation 浴用配制品
bath ratio(=liquor length) 浴比
bath room units 水浴单元；水浴元件
bath salt 浴盐
bath sealant 浴室封闭剂；槽(池)密封胶〔一种内用有机硅密封胶〕

bath soap 浴皂
bath solution (电镀)槽液
bath stretch 浴内拉伸
bath tub ring 浴缸皂垢环纹〔钙皂、镁皂在浴缸内液面上部边缘形成的环状皂垢〕
bath type cooler 冷却水槽
bath voltage (电解)槽电压
batik dyeing 蜡染
bating 酶解软化
bating drum 软化转鼓
bating liquor 软化液
batiste 丝薄织物
Batroxobin (defibrase) 巴曲酶(降纤酶)〔药〕
Batroxobin (hemocoagulase) 巴曲酶(血凝酶)〔药〕
batt ①絮垫；垫褥②(制毡)纤维层
batten 小圆材〔末端直径在28厘米以下者〕；小方材；狭板
battening 打纬
batter ①粉碎②冲击③捣薄④混合
batter brace (桁架)斜撑
battering ①混合②冲击③捶薄④粉碎
battery 电池组
battery acid 电池用酸
battery bench 电池桌
battery box 电池(组的)箱
battery car 电瓶车
battery carbon 电池用碳
battery case 蓄电池箱
battery cell 电池组电池
battery charger 电池充电器
battery circuit 电池电流
battery contact 电池接头
battery container 蓄电池箱
battery glass 电池用玻璃(容器)
battery hut 电池棚
battery jar 电池槽
battery limit 界区
battery mud 电池淤渣
battery ore 电池用(软锰)矿
battery reversing switch 电池转向(电)闸
battery room 电池室
battery separator 蓄电池隔板
battery switch 电池(组)的开关
battery syringe 电池注射器
battery terminal 电池(组)的端钮
battery tester 电池检验器
battery voltage 电池组电压
batting 棉胎；棉絮

batu 巴土(树)脂
batyl alcohol 鲨肝醇；1-十八烷基甘油醚
Bauer (centri-) cleaner 鲍尔(锥形)除渣器
bauerenol 降香醇；鲍尔烯醇
Bauer-McNett (wet) screening test (石棉)鲍尔-麦克内特湿式筛分试验
Bauer-McNett (wet) test (石棉)鲍尔-麦克内特湿式筛分试验
Baumé degree 波美度〔°Bé〕
Baumé gravity 波美比重度
Baumé hydrometer 波美比重计
Baumé scale 波美度标
Baumé system 波美度制
Baumé tester 波美比重计
baumlerite 氯化钙钾石
Bauschanger effect 鲍辛格效应
bauxite(=beauxite) 铝土矿*
bauxite brick 铝土砖
bauxite cement 铝土水泥
bauxite clay 铝质黏土
bauxite process 铝土过程；铝土催化重整过程〔高温裂化〕
bauxite refractory 铝矾土耐火材料
bauxite-treated gasoline 铝土处理汽油
bauxite treater 铝土处理器
bauxitic clay 铝矾土；铝铁黏土
bauxitic laterite 铝矾性砖红壤
bauxitization 铝矾土化；铝土矿化作用
bavalite 硬铁绿泥石；鲕绿泥石
bavenite 硬沸石；硅铍钙石
bawrerenol 包瑞烯醇
bay 月桂
bayberry ①月桂的果实②南部杨梅的果实③众香子的果实
bayberry oil 月桂子油
bayberry tallow 杨梅油〔取自 Myrica 的果子〕
bay-cherry 桂樱
baycovin 焦碳酸二乙酯〔防腐剂〕
baycurine 矶松根碱
baycuru root 矶松根
Bayer acid 拜耳酸；2-萘胺-7-磺酸
Bayer alumina 拜耳氧化铝
bayerite 三羟铝石
Bayer process (for purifying alumina) 拜耳(精炼矾土)法
bay laurel 加州月桂
bayldonite 乳砷铅矿
bay leaf oil(=bay oil; myrcia oil) 月桂叶油；月桂油〔俗〕
bay oil(=bay leaf oil; myrcia oil) 月桂叶油；月桂油〔俗〕
bayonet catch ①锁梢〔电〕②插座

bayonet catch lid 错齿式罐盖〔硫化罐〕
bayonet gauge stick 插入式计量棒
bayonet joint 插接式接头；插接头
bayonet oil level gauge 插入式油面计
bay rum 月桂香水
bay salt 海盐
bay tree 月桂树
bay-tree oil 月桂油
B-B cut(=butane-butene cut) 丁烷-丁烯馏分
B-B fraction(=butane-butene fraction) 丁烷-丁烯馏分
B-centered lattice B 底心格子
bdellium 芳香树胶
beach head dump 滩头堆栈
beach puncture tester （比齐）刺孔试验〔检验纸板强度用〕
bead ①珠；珠粒②撑轮圈；胎圈③凸缘④成珠；熔珠；焊珠⑤焊缝
bead catalyst 颗粒催化剂
bead chain polymer 珠链聚合物
bead core 叶轮心
bead cover(ing) 叶轮罩
bead covering machine 包装机
bead crack 焊缝裂纹
bead cutter 切边机
bead dispersions 粒状分散体
beaded black(=pelletized black) 粒状炭黑
beaded catalyst 颗粒催化剂
beaded face ①圆面②珠状饰面
beaded glass 玻璃珠；粒状玻璃
beaded paint ①涂料②油漆〔特指色漆〕
beaded pearlite 球状珠光体
beader 弯管机
bead flippering 带式固定
bead heel 凸缘后部
beading 泪珠；涨边；卷边；嵌缝胶
beading up 起珠〔漆病〕
bead insulating extruder 凸缘绝缘压出机
beadlike （珍）珠状的
bead machine 压片机；压锭机
bead polymer 粒状聚合物；珠状聚合物〔用悬浮聚合法生成的小粒状聚合物〕
bead polymerization(=pearl polymerization) 珠状聚合；成珠聚合(法)
bead reaction 熔珠反应
bead reinforcement 凸缘加强
bead rimming machine (for dipped goods) (作)凸缘机
bead ring ①钢丝圈②扣(钢丝)圈盘；碰盘③胎圈压环〔翻胎用〕④钢环〔空气弹簧〕
bead-rod model 珠-棒模型*

bead rolling 轧凸缘
bead seal 胎圈密封胶
bead sealing 缘条封焊
bead set of rim 鞍边；凸缘
bead-spring model 珠-簧模型*
bead test 熔珠试验
bead toe 凸缘嘴
bead tube （玻璃珠）精馏管
bead wire ①凸缘线②钢丝圈
bead wire ring 凸缘线圈
bead wire tape 凸缘线带
beak 圆口灯
beaker 烧杯
beaker clamp 烧杯夹
beaker filter 烧杯过滤器
beaker flask 锥形杯
beaker sampling 烧杯取样；烧杯进样
beam ①光线；光束②射线③束；束流④梁〔天平〕
beam and girder construction 主次梁式结构
beam and slab structure 梁板结构
beam aperture ①射束孔②光束截面
beam blanking option （光）束空白选择
beam chemistry （射线）束化学
beam column construction 梁柱结构
beam combiner （光）束合并器
beam condenser （光）束聚焦装置
beam convergence ①电子束会聚②射束会聚；聚束
beam crossing angle 束交叉角
beam cross section （光）束截面
beam cutter 平压切断机
beam deflection 射束偏转(斜)；电子束偏转
beam density 光束密度
beam diameter 光束直径
beam direction （光）束方向；射束方向
beam divergence （光）束散
beam dump 束流收集器
beam dyer 经轴染色机
beam dynamics 束流动力学
beamed radiation 成束辐射
beamed yarn 经轴纱
beamer 经轴整经机
beam expander （光）束扩展器
beam extractor 束流引出装置
beam flux 束流
beam focusing 束聚焦
beam-folding 光束折叠
beamhouse 浸灰间〔皮革〕
beamhouse operation 准备车间操作

beamhouse weight 浸灰重量〔皮革〕
beaming 并轴；整经
beam intensity 束强*
beam knife 刮皮刀
beam loading 梁式载荷
beam machine 刮皮机
beam monitor 束强监测器
beam of balance 天平梁
beam output 射线束放射量
beam profile 束剖面图；光束轮廓
beam source 束源
beam splitter 分束器*
beam splitting (光)束分离
beam spot 射束点；电子束光点
beam steering ①(光)束控制②(光)束转向
beamster 推挤工
beam warping 分批整经
beam width 光束宽度；束宽
beam worker 准备工
bean ①豆②油嘴；喷油嘴
bean back 减少油嘴(直径)
bean oil 豆油
bean-shaped 豆形的
bean up 放大油嘴(直径)
bearded needle straight bar frame 钩针全成型平机
bearded usnea 须松萝
Bear diagram 拜氏纤维长度分布(图)
bearer ①载体②支座；支承；托架
bear fat 熊脂
bearing 轴承
bearing area 支承面积；轴承面积
bearing block 轴承座；轴承箱
bearing blue 蓝铅油；普鲁士蓝〔着色剂〕
bearing bracket 轴承架
bearing brass 轴承巴氏合金；轴承铜衬
bearing bush 轴瓦
bearing capacity 承载能力；承载量
bearing circle 方向盘
bearing grease 轴承脂
bearing housing 轴承套；轴承箱
bearing materials 轴承材料
bearing metal 轴承合金
bearing oil 轴承油
bearing out (喷枪的)支承间距
bearing property 轴承性能
bearing retainer 轴承承托
bearing strength 承载强度；抗压强度
bearing strength test 承载力试验

bearing stress 承载应力
bearing time 产胶期〔胶树〕
α-bearing waste α(放射性)废物*
beatability 打浆性能
beater 打浆机
beater additive 打浆添加剂
beater chamber 打浆室
beater colloid mill 打浆胶体磨
beater drag 打浆计算器
beater dyeing 打浆机(中)染色
beater loading 打浆机装料
beater oiling roller 打浆机油辊
beater process paper 打浆法纸；纸浆法纸
beater roll 打浆辊
beater roller ①打浆辊②回转器
beater roll pressure ①碎解压力②打浆辊压力
beater room 打浆室
beater sizing 打浆机(中)上胶
beater tup 打浆桶
beat frequency 拍频(率)
beating 打浆；搅打
beating degree 打浆度；叩解度
beating engine 打浆机
beating machine 打浆机
beating method ①打浆法；叩解法②原纤化(成形)法
beating schedule 打浆程序
beating service test of lubricants 润滑剂的搅打使用试验
beating tower 打浆塔；塔式碎解机
beating up 打纬
beat-proof quality 耐热性
beat signal 差拍信号
beat spectrum 差拍谱
Beaudouin's reagent 包道因试剂
beauxite(=bauxite) 铝土矿
beaver ①海獭；海狸②刮料工具③轻型及中型飞机加油〔燃料〕
bebeerilene 贝比烯 $C_{34}H_{28}O_6$
bebeerine(=curine) 贝比碱；卑比令碱
bebeerinemethine 贝比碱甲炔
bebirine(=bebeerine) 贝比碱
beccarite 绿锆石
Bechamp reaction 贝尚反应
bechilite 硼钙石
Bechstein photometer 贝克斯坦光度计
beck 槽；缸
Becke-line method 贝克线法〔测定纤维双折射的一种方法〕
beckelite 方钙铈镧矿

beckerite 酚醛琥珀
Becker process 贝克尔过程；喷嘴法〔铀同位素分离〕
Beck hydrometer 贝克比重计〔测量比重比水轻的液体〕
Beckmann rearrangement 贝克曼重排作用
Beckmann thermometer 贝克曼温度计
Beclometasone dipropionate 倍氯米松〔药〕
Becquerel cell 贝可勒尔电池
Becquerel rays 贝可勒尔射线
bedded 搁置的；成层的
bedded structure 层状结构
bed density 床层密度
bed depth 流化床深度；床(层)深(度)
bedding 层积法；被褥；褥草
bedding wood material 垫木
bed filter 滤床式过滤器
bedframe 底座框架；支承结构
bed knife 固定刀；底刀
bed of fuel 燃料层
bed of furnace 炉底；炉床
bed of mortar 灰浆层
bed of packings 填料床
bed plate ①炉底②台板③底刀板；底刀
bedplate bar ①滚板刀②底刀
bed putty(=back putty) 打底油灰
bedstone ①座石；底石②底梁；底木〔设〕
bed tempertaure (流化)床温度
bed volume(=column volume) 柱床体积
bedye ①着色；施彩色②染〔色〕③漆
beech 山毛榉
beech coal 山毛榉煤
beech gall 山毛榉瘿
beech mast(=beech nut) 山毛榉坚果
beech nut(=beech mast) 山毛榉坚果
beech nut oil(=beech tree oil) 山毛榉树油；山毛榉坚果油
beech tree oil(=beech nut oil) 山毛榉树油；山毛榉坚果油
beechwood 米心树
beechwood creosote 山毛榉杂酚油
beech-wood tar 山毛榉焦油
beef fat 牛脂肪；牛(板)油
beegerite 银辉铋铅矿
beehive ①集气架②蜂巢
beehive beeswax 蜂蜡
beehive coke 蜂巢炉焦炭
beehive coke oven 蜂巢式煤焦炉
beehive cooler 蜂巢式冷却器
beehive kiln (蜂)巢式(炭)窑
beehive oven 蜂巢(炼)焦炉
Bee-Lambert's absorption law 比尔-朗伯吸收定律*
beel curve(=bell-shaped curve) 钟形曲线；圆锥形曲线
Beer-Bouguer law(=Beer-Lambert law) 比尔-波格尔定律〔关于光辐射强度与介质浓度关系的定律〕
Beer-Lambert equation 比尔-朗伯方程
Beer-Lambert law(=Beer-Bouguer law) 比尔-朗伯定律〔关于光辐射强度与介质浓度关系的定律〕
Beer's law 比尔定律*
beeswax 蜂蜡；黄蜡〔俗〕
beeswax candle 蜂蜡烛
beet 甜菜
beet cellar 甜菜窖
beetle ①槌布机②搅打机③脲醛树脂〔塑料〕
behavior ①性能；性质；特性；性状②行为
behenamide 山萮酸酰胺〔抗粘连剂〕
behenic acid 山萮酸；二十二(碳)烷酸 $CH_3(CH_2)_{20}CO_2H$
behen oil(=sorinja oil) 山萮油
behenolic acid 山萮萘炔酸；二十二(碳)-13-炔酸
behenolyl 山萮炔酰；二十二碳-13-炔酰
behenonaphthone 山萮萘酮；山萮酰萘 $C_{10}H_7CO(CH_2)_{20}CH_3$
behenone 山萮酮 $CH_3(CH_2)_{20}CO(CH_2)_{20}CH_3$
behenophenone 山萮苯酮 $C_6H_5CO(CH_2)_{20}CH_3$
behenoxylic acid 山萮氧酸；二十二碳-13,14-二酮酸
behenyl alcohol 山萮醇；二十二醇 $CH_3(CH_2)_{21}OH$
beidellite 贝得石〔蒙脱石的异构体〕
beige ①本色的；未漂白的②米黄色③浅灰色④坯布⑤混色线呢
Beilby layer 拜尔比层
Beilstein's test 贝尔斯坦试验
Beilstein flame detector 贝尔斯坦火焰检测器
beilupeimine 白炉贝素；白炉贝母碱
belamcandin 射干定
belching 汽水沸出
B/E linked scan B/E 联扫
belit B 盐；二钙硅酸盐〔水泥〕
belite 斜硅灰石；二钙硅酸盐 $2CaO \cdot SiO_2$
belith 斜方硅灰石；β-硅钙石
beljankite(=creedite) 铝氟石膏
bell ①钟②漏斗③圆锥体
belladine 孤挺花定
belladonna 颠茄〔药〕
belladonna extract 颠茄浸膏
belladonna leaf 颠茄叶
belladonna root 颠茄根
belladonnine 颠茄宁；异性脱水颠茄碱

bellafoline 颠茄叶素
bellamarine(=acetylcaranine) 孤挺花碱
bell and hopper 进料器
bell and plain end joint 平接
bell and spigot joint 套接
bellaradine(=cuskhygrine) 颠茄定；红古豆碱
bell atomizer 钟式(形)雾化器
bell cap 钟帽；泡帽；泡罩〔蒸馏塔塔盘上半圆球盖〕
bell cell 钟状电解池
belled ①套接的②钟形口的
bell end (of pipe) 承插端
bell float 浮钟
bell glass (玻璃)钟罩
Bellier's test 伯利埃试验〔一种检出花生油的试验〕
belling 插口成型；涨口
bellite ①铬砷铅矿②贝尔炸药；白拉特；硝铵、二硝基苯炸药
bell jar 钟罩
bell jar cell 钟式电池
bell manometer 浮钟压力计
Bellmer bleacher(=Bellmer engine) 贝尔默(漂白)机
Bellmer engine(=Bellmer bleacher) 贝尔默(漂白)机
bell mouth ①喇叭口；锥形孔②锥形底
bellow 风箱
bellow leather 风箱革
bellows ①导风筒；折囊②气囊③波纹管④风箱；皮老虎〔俗〕
bellows pressure gauge 风箱型压力表
bellows seal 波纹管式密封；伸缩鼓式密封
bellows type gun 风箱嘴油枪；风箱型喷射器
bellows type mechanical seal 风箱型机械密封；波纹管式机械密封
bellows type pressure gauge 风箱型压力表
bellows valve 波纹管阀
bell reaction 钟罩反应
bell recorder 浮钟计压器
bell-shaped curve(=beel curve) 钟形曲线；圆锥形曲线
bell-shape distribution curve 钟形分布曲线
bell-shaped rotor 钟形转子；圆锥形转子
bell socket 套接；承插接口
bell sprayer 钟(杯)式喷雾器；杯式喷枪
bell-type atomizer 钟形雾化器；钟形喷雾器
bell type autoclave press 钟式热压锅
bell-type coater 杯形涂装器；盅(杯)式喷枪
bell-type low pressure manometer 钟型低压压力计；钟型低压气压计
bell valve 装料钟
belmacamdin 射干定

belmacamgenin 射干配基
belonesite 针镁钼矿
belonite 针锥晶；棒锥晶
Belro resin 深色热塑性酸性树脂〔从明子制备〕
belt 带；缓冲层
belt bias tyre 带束斜交轮胎
belt booth 带式喷粉(喷漆)橱
belt bucket elevator 带式升桶机
belt building machine 平带成型机
belt carcass 带身；带芯
belt compensator 平带松紧器
belt composition 传动皮带润滑剂
belt-conveyor 带式运输机
belt-conveyor drive 带式运输机起动装置
belt-conveyor support 带式运输机的支撑装置
belt-conveyor take-ups 带式运输机收紧器
belt dressing 皮带油〔鞣革加工用〕
belt drive 皮带传动
belt dryer 带式干燥机
belt drying 带式干燥
belt duck M带布〔纺〕
belt dynamometer 带式功率计
belt elevator 带式升降机
belt feed 带式进料
belt feeder 带式给料机
belt filler(=belt lubricant) 皮带油
belt grinder 带式磨机
belt guide 皮带导向装置
belt haulage (用)胶带输送
belting duck M带布〔纺〕
belting leather 带革；轮带
belt lacing 带接头〔机工〕
belt leather 带革；带皮
belt lubricant(=belt filler) 皮带油
belt magnetic separator 带式磁力分离器
Belt path drawing device 传动过程拉伸装置；纳尔逊辊拉伸装置
belt polishing machine 带式磨光机
belt press 压带机
belt prover 带运机
belt pulley 运输带鼓轮
belt sander 胶合板表面砂磨机
belt-sanding 带式砂磨
belt scale 带秤
belt slope tension 皮带倾斜张力
belt spanner 平带拉紧装置
belt stretcher 皮带拉幅机
belt stretching machine 皮带展幅机

belt stripper 卸运带机
belt tension 皮带张力
belt tightener 平带拉紧装置
belt traction take-off machine 履带式接取机
belt transport detector 传送带检测器
belt weight meter 带秤
Bemberg microporous membrane 铜氨纤维微孔膜
Bemberg silk 本伯格(人造)丝
bemsilk 铜氨人造丝；铜氨丝
Benard cell 贝纳尔德旋流"窝"〔引起漆膜发花的一种旋涡现象〕
Benard cell boundary 贝纳尔德旋流窝(边)界面
Benard circulation motion 贝纳尔德环流(运动)
benazepril 贝那普利〔药〕
bench ①组②凳子③实验桌；工作台
bench drill 台钻
bench life(=working life) 使用寿命
bench mark 标线
benchmark concentration 基准浓度
benchmark value 基准值
bench-mounted hardness tester 台式硬度计
bench of burners 炉组
bench photometer 台式光度计
bench roller 台辊
bench scale 工作台规模；台架规模
bench-scale research 扩大试验
bench side 焦面
bench test 小型试验〔比实验大，比中型试验小〕
bench top 长凳面②长工作台面；长操作台面
bench-type modular unit 台式调制单元
bench type spray booth 台式喷涂橱(室；柜)
bench work 钳工工作
bencyclane 苄环烷
bend ①弯头②弯管；曲管③弯曲④半背皮
bendazol 地巴唑〔药〕
bend bar expander 扩幅弯辊
bend-brittle point 弯曲脆化点
bended tube 弯管
bender ①折弯机②弯管机
Bender sweetening 本德脱硫法
bending ①弯曲度②弯头③弯曲④卷刃
bending creep 弯曲蠕变
bending flexure 弯曲
bending jig 折弯胎模
bending length 弯曲长度
bending load 弯曲载荷
bending machine ①折弯机②卷板机③弯管机④弯曲试验机

bending modulus 弯曲模量；弯曲模数
bending motion 弯曲运动
bending radius(=bend radius) 弯曲半径
bending resistance 抗弯曲性
bending rigidity ①抗弯刚性②抗弯刚度
bending roll fibrillator 弯辊式原纤化装置
bending stiffness 弯曲劲度
bending strain 弯曲应变
bending strength 弯曲强度
bending stress 弯曲应力
bending stress tester 弯曲应力试验仪
bending vibration 弯曲振动
bending Young's modulus test 弯曲杨氏模量试验
Bendix time-of-flight mass spectrometer 本德克斯飞行时间质谱仪
bend loop test 弯曲试验
bend radius(=bending radius) 弯曲半径
bend-shatter point (低温下)弯曲破碎点
bend strength 弯曲强度
bend stress 弯曲应力
Benedict's reagent 本尼迪特试剂
Benedict's test 本尼迪特试验〔鉴别糖分、树脂、淀粉〕
benedict root oil(=water avens root oil) 丁字根油
beneficiation ①选矿②精选③富集
benefit analysis ①(工程)受益分析②优势分析
benemid 对二丙氨基磺酰基苯甲酸
Benesi method 贝纳西法〔用于测定颜料表面酸度〕
bengala 三氧化二铁；氧化铁红；铁丹〔俗〕
Bengough-Stuart anodising process 本戈-斯塔特阳极氧化法
bengucopalic acid 本格珀酸 $C_{10}H_{30}O_2$
bengucopalresene 珀树脂素
benguela copal 本格拉珀
benihene 贝尼烯
benihidiol 贝尼黑二醇
benihinal(=d-myrtenal) 贝尼烯醛；d-桃金娘烯醛
benihinol(=d-myrtenol) 贝尼烯醇；d-桃金娘烯醇
benihiol(=myrtanol) 贝尼醇；桃金娘醇
benihione 贝尼酮
Benin copal 贝宁珀
Beni rubber 本尼橡胶
benitoite 蓝锥矿
benjamin(=benzoin) 安息香
Bennert manometer 本纳特流体压力计
Bennett radio-frequency mass spectrometer 贝内特射频质谱仪
ben oil 山萮油
benorilate 贝诺酯〔药〕

benproperine 苯丙哌林〔药〕
bent crystal 弯晶
benthal deposit 海底沉积物
bent lever 曲杆
bentolite clay 胶质陶土（填料）
bentone grease(=bentonite grease) 膨润土润滑脂；皂土润滑油
bentone lubricant(=bentonite grease) 膨润土润滑脂；皂土润滑油
bentonite 膨润土；皂土；浆土
bentonite clay 膨润土；皂土
bentonite grease(=bentone grease; bentone lubricant) 膨润土润滑脂；皂土润滑油
bentonite suspension 膨润土悬浮液；皂土悬浮液
bentonite-thickened grease 膨润土润滑脂
bent pipe(=bent tube) 弯管
bent plate 曲板
bent strip test 弯条试验
bent tube(=bent pipe) 弯管
bent tube boiler 弯水管锅炉
bent tube section 弯管部分〔冷冻机中淹没在冷却箱的部分〕
Benturi cascade tray 本图利阶梯式塔板
benz- 苯基
benzaceanthrylene 苯并醋蒽
benzaconine 苦乌头碱
benzacridine 苯并吖啶
benzafibrate 苯扎贝特〔药〕
benzal(=benzylidene) 亚苄基；苯亚甲基 C_6H_5CH
benzalacetone 亚苄基丙酮 $C_6H_5CH=CHCOCH_3$
benzalacetophenone 亚苄基乙酰苯 $C_6H_5CH=CHCOC_6H_5$
benzalacetophenone dibromide 亚苄基乙酰苯化二溴 $C_6H_5CHBrCHBrCOC_6H_5$
benzalacetylacetone 亚苄基乙酰丙酮 $C_6H_5CH=CHCOCH_2COCH_3$
benzalagen(=analgen) 安乃近〔药〕
benzalaminophenol 亚苄基氨基苯酚
benzalaniline 亚苄基苯胺；苯甲醛缩苯胺 $C_6H_5CH=NC_6H_5$
benzalation 苯甲醛缩醛化
benzalazine 苄连氮；亚苄基吖嗪 $(C_6H_5CH=N)_2$
benzal bromide 亚苄基二溴；二溴甲基苯 $C_6H_5CHBr_2$
benzal chloride 亚苄基二氯；二氯甲基苯 $C_6H_5CHCl_2$
benzal chloroaniline 亚苄基氯苯胺；苯甲醛缩氯苯胺
benzalcohol 苄醇 $C_6H_5CH_2OH$
benzal cyanhydrin 苯基氰基甲醇 $C_6H_5CH(OH)CN$
benzaldehyde 苯甲醛 C_6H_5CHO

benzaldehyde-acetalized fibre 苯甲醛缩醛化纤维
benzaldehyde-carboxylic acid 苯醛(甲)酸
benzaldehyde cyanhydrin 苯甲醛合氰化氢 $C_6H_5CH(OH)CN$
benzaldehyde dimethyl acetal 苯甲醛二甲缩醛
benzaldehyde glycerine acetal 苯甲醛甘油缩醛
benzaldehyde-phenylhydrazone 苯甲醛苯腙
benzaldehyde propylene glycol acetal 苯甲醛丙二醇缩醛
benzal diacetate 亚苄二乙酸酯；苄二醇二乙酸酯 $C_6H_5CH=(O_2CCH_3)_2$
benzaldoxime 苯甲醛肟
benzal ethylamine 亚苄乙胺；N-乙苄亚胺 $C_6H_5CH=NC_2H_5$
benzal fluoride 亚苄二氟；二氟甲基苯 $C_6H_5CHF_2$
benzal hydrazine 亚苄肼；苄腙 $C_6H_5CH=NNH_2$
benzalization 苯甲醛缩醛化
benzalized fibre 缩苯甲醛化纤维
benzal-keton process(=benzol-keton dewaxing process) 润滑油的脱蜡过程〔利用苯及丙酮混合液〕
benzal lactic acid 亚苄基乳酸 $C_6H_5CH=CHCHOHCO_2H$
benzal malonic acid 亚苄基丙二酸 $C_6H_5CH=C(CO_2H)_2$
benzal methylamine 亚苄基甲胺；N-甲亚苄胺 $C_6H_5CH=NCH_3$
benzal nitroaniline 亚苄基硝基苯胺 $C_6H_5CH=NC_6H_4NO_2$
benzal phthalide 亚苄基酞 $C_6H_5CH=CC_6H_4COO$
benzal pinacolone 亚苄基叔己酮 $(CH_3)_3CCOCH=CHC_6H_5$
benzal-toluidine N-亚苄基甲苯胺
benzamarone 苯苦杏酮
benzamide 苯甲酰胺 $C_6H_5CONH_2$
benzamidine 苄脒；苯甲脒 $C_6H_5C(=NH)NH_2$
benzamido- 苯甲酰氨基 $C_6H_5CONH—$
benzamidocinnamic acid 苯甲酰氨基肉桂酸
benzamidothiophenol 苯酰胺硫酚
benzamidoxime 苄胺肟 $C_6H_5C(=NOH)NH_2$
benzaminic acid 氨基苯甲酸 $NH_2C_6H_4COOH$
5-benzaminoanthraquinone-2-sulfonic acid 5-苯甲酰氨基蒽醌-2-磺酸
benzanil black 亚苯基苯胺黑
benzanil blue 亚苯基苯胺蓝
benzanil color 亚苯基苯胺染料
benzanil fast orange 亚苯基苯胺坚牢橙
benzanilide N-苯甲酰苯胺 $C_6H_5NH·COC_6H_5$

benzanil purpurine 亚苯基苯胺红紫
benzanil yellow 亚苯基苯胺黄
benzanisidide N-苯茴香胺
benzanthoene 苯并蒽酮 $C_{17}H_{10}O$
benzanthracene 苯并蒽 $C_{18}H_{12}$
benzanthramycin 苯氮茴霉素
benzanthrene 苯并蒽
benzarsenious acid 羧苯亚胂酸 $HOOCC_6H_4As(OH)_2$
benzarsonic acid 羧苯胂酸 $HOOCC_6H_4AsO(OH)_2$
benzathine benzylpenicillin 苄星青霉素〔药〕
benzaurin 苯并金精 $C_6H_5C(=C_6H_4O)C_6H_4OH$
benzazide 苯甲酰叠氮 $C_6H_5CON_3$
benzazine 氮萘
2-benzazine 2-氮萘
1-benzazole 1-氮茚；吲哚
benzazole 吲哚；氮茚
benzbromarone 苯溴马隆〔药〕
benzchrmone 苯并色酮
benzchrysene 苯并䓛
benzcinnoline(=o-diphenyleneazine) 苯并噌啉
benzdiazole(=indazole) 吲唑
benzdiketohyrindene(=benz[e]indanedione) 苯并[e]二氢茚-1,3-二酮
benzdioxan 苯并二噁烷
benzedrine 1-苯基-2-氨基丙烷
benzein indicator 苯因系指示剂
benzen-1,3-disulphonyl hydrazide 1,3-苯二磺酰肼〔发泡剂〕
benzene 苯 C_6H_6
benzene-alcohol 苯酒精〔苯与酒精的混合物〕
benzene arsonate 苯胂酸盐 $C_6H_5N=NC_6H_5$
benzene arsonic acid 苯胂酸 $C_6H_5AsO(OH)_2$
benzeneazo-arsonic acid 苯偶氮茴香胺 $C_6H_5N=NC_6H_3(NH_2)CH_3$
benzene-azo-benzene 苯偶氮苯 $C_6H_5N=NC_6H_5$
benzeneazo cresol 苯偶氮甲基苯酚
benzeneazodiphenylamine 苯偶氮二苯胺
benzeneazo-naphthylamine 苯偶氮萘胺
benzeneazophenol (苯)偶氮苯酚
benzeneazo resorcinol(=sudan) 苏丹；苯偶氮间苯二酚 $C_6H_5N_2C_6H_3(OH)_2$
benzene-azo-sulfonic 苯偶氮磺酸 $C_6H_5N_2SO_3H$
benzene bottoms 苯残余物〔苯精馏时蒸馏塔下的部分〕
benzene carbon amide 苯甲酰胺 $C_6H_5CONH_2$
benzene carbon amidine 苯(甲)脒 $C_6H_5C(NH_2)=NH$
benzene carbonic acid(=benzene carboxylic acid) 苯甲酸；安息香酸〔俗〕
benzene carboxylic acid(=benzene carbonic acid) 苯甲酸
benzene color test 苯着色试验

benzene derivatives 苯衍生物
benzene diazonium chloride 氯化重氮苯 $C_6H_5N_2^+Cl^-$
benzene diazonium cyanide 氰化重氮苯 $C_6H_5N_2^+CN^-$
benzene diazonium hydroxide 氢氧化重氮苯 $C_6H_5N_2^+OH^-$
benzene diazonium nitrate 硝酸重氮苯 $C_6H_5N_2^+NO_3^-$
benzene diazonium tribromide 三溴化重氮苯 $C_6H_3Br_2N_2^+Br^-$
benzene dibromide 二溴化苯
o-benzenedicarbamic acid 邻苯二氨基二甲酸
benzene dicarbonic acid(=benzene dicarboxylic acid) 苯二甲酸；苯二羧酸 $C_6H_4(COOH)_2$
benzene dicarboxylic acid(=benzene dicarbonic acid) 苯二甲酸；苯二羧酸 $C_6H_4(COOH)_2$
benzene dichloride 二氯化苯 $C_6H_4Cl_2$
benzene diiodide 二碘化苯 $C_6H_4I_2$
benzene-disulfo-chloride 苯二磺酰氯 $C_6H_4=(SO_2Cl)_2$
benzene disulfohydrazide 苯二磺酰肼〔发泡剂〕
benzene disulfonate 苯二磺酸盐 $C_6H_4(SO_3M)_2$
benzene disulfonic acid 苯二磺酸 $C_6H_4(SO_3H)_2$
benzene-1,3-disulphonyl hydrazide 1,3-苯二磺酰肼〔发泡剂〕
benzene dosimeter 苯剂量计
benzene double bond 苯双键
benzene extraction (carbon black) 苯抽出(炭黑)
benzene halide 卤化苯
benzene hexabromide 六溴化苯；六溴环己烷 $C_6H_6Br_6$
benzene hexacarbonic acid(=benzene hexacarboxylic acid) 苯六甲酸；苯六羧酸 $C_6(COOH)_6$
benzene hexacarboxylic acid(=benzene hexacarbonic acid) 苯六甲酸；苯六羧酸 $C_6(COOH)_6$
benzene hexachloride 六六六；六氯化苯 $C_6H_6Cl_6$
benzene hydrasinonaphthalene 苯肼基萘
benzene hydrocarbon 苯系烃
benzene-insoluble 不溶于苯的
benzenemethylal 苯甲醛
benzene monocarbonic acid(=benzene monocarboxylic acid) 苯甲酸
benzene monocarboxylic acid(=benzene monocarbonic acid) 苯甲酸
benzene monosulfonic acid 苯磺酸 $C_6H_5SO_3H$
benzene nucleus 苯核
benzene pentacarbonic acid(=benzene pentacarboxylic acid) 苯五甲酸
benzene pentacarboxylic acid(=benzene pentacarbonic acid) 苯五甲酸 $C_6H(COOH)_5$
benzene polycarbonic acid(=benzene polycarboxylic acid) 苯多甲酸
benzene polycarboxylic acid(=benzene polycarbonic acid)

苯多甲酸
benzene polyhalide 多卤化苯
benzene resistance 耐苯性
benzene ring 苯环
benzene seleninic acid 苯亚硒酸 C_6H_5SeOOH
benzene series 苯系
benzene siliconic acid 苯(基)硅酸 C_6H_5SiOOH
benzenesulfenyl 苯硫基 PhS—
benzene sulfinate 苯亚磺酸盐 $C_6H_5SO_2M$
benzene sulfinic acid 苯亚磺酸 $C_6H_5SO_2H$
benzene sulfinyl(=phenyl sulfinyl) 苯亚磺酰 C_6H_5SO—
benzene sulfochloride 苯磺酰氯 $C_6H_5SO_2Cl$
benzene sulfohydrazide 苯磺酰肼〔发泡剂〕
benzene sulfonamide 苯磺酰胺 $C_6H_5SO_2NH_2$
benzene sulfonamido- 苯磺酰胺基 $C_6H_5SO_2NH$—
benzene sulfonate ①苯磺酸盐 $C_6H_5SO_3M$ ②苯磺酸酯 $C_6H_5SO_2OR$
benzene sulfone amide 苯磺酰胺 $C_6H_5SO_2NH_2$
benzene sulfone chloride 苯磺酰氯 $C_6H_5SO_2Cl$
benzene sulfonic acid 苯磺酸 $C_6H_5SO_3H$
benzene sulfonic amide 苯磺酰胺 $C_6H_5SO_2NH_2$
benzene sulfonic chloride(=benzene sulfonyl chloride) 苯磺酰氯
benzenesulfonyl 苯磺酰基 $C_6H_5SO_2$—
benzene sulfonyl chloride(=benzene sulfonic chloride) 苯磺酰氯
benzene sulfonyl hydrazide 苯磺酰肼 $C_6H_5SO_2NHNH_2$
benzene sulfonyl hydroxylamine 苯磺酰胲 $C_6H_5SO_2NHOH$
benzene sulphonbutylamide 丁基苯磺酰胺〔增塑剂〕
benzene sulphonyl hydrazide 苯磺酰肼〔发泡剂〕
benzene tetrabromide 四溴化苯；四溴代环己烯 $C_6H_6Br_4$
benzene tetracarbonic acid(=benzene tetracarboxylic acid) 苯四甲酸
benzene tetracarboxylic acid(=benzene tetracarbonic acid) 苯四甲酸 $C_6H_2(COOH)_4$
benzene tetrachloride 四氯化苯；四氯代环己烯 $C_6H_6Cl_4$
benzene tetraiodide 四碘化苯；四碘代环己烯 $C_6H_6I_4$
benzenethiol 苯硫醇
benzene thiosulfonic acid 苯硫代磺酸
benzene-toluene-xylene recovery 苯-甲苯-二甲苯回收；芳烃回收
benzene tricarbonic acid(=benzene tricarboxylic acid) 苯三甲酸 $C_6H_3(COOH)_3$
benzene tricarboxylic acid(=benzene tricarbonic acid) 苯三甲酸 $C_6H_3(COOH)_3$

benzene triozonide 三臭氧化苯
benzene trisulfonic acid 苯三磺酸 $C_6H_3(SO_3H)_3$
benzenoid 苯(环)型的
benzenoid form 苯(环)型
benzenoid hydrocarbons 苯型烃类
benzenoid structure 苯型结构
benzenyl(=benzylidyne) 次苄基
benzenyl amidine 苄脒；苄偕胺亚胺 $PhC(NH_2)=NH$
benzenyl amidothiophenol 2-苯基间氮杂硫茚；次苄基氨硫酚〔俗〕
benzenyl amidoxime 次苄基胺肟；1-苯甲胺-1-肟 $C_6H_5C(NOH)NH_2$
benzenyl chloride 次苄基三氯
benzenyl fluoride 次苄基三氟
benzenyl-naphthyl-amidine N-苄萘脒 $C_6H_5C(=NH)NHC_{10}H_7$
benzenyl trichloride 次苄基三氯；三氯甲基苯 $C_6H_5\cdot CCl_3$
benzepine 苯并环庚三烯
benzerythrene 对联四苯
benzestrol(=octofollin) 辛叶素；异辛雌酚
benzethonium chloride 苯索氯铵〔异辛基苯二聚乙二醇醚二甲基苄基氯化铵〕
benzflavone 苯并黄酮
5,6-benzflavonol 苯并[f]黄酮醇
1,2-benzfluorene 1,2-苯并芴；苯并[a]芴
benzfluorene 苯并芴 $C_{17}H_{12}$
benzfluorenol 苯并芴醇 $C_{17}H_{12}OH$
1,2-benzfluorenone 1,2-苯并芴酮；苯并[a]芴酮
benzfuran 苯并呋喃
benzfuro[2,3-f] quinoline 苯并呋喃并[2,3-f]喹啉
benzheterocycle-imide copolymer 苯并杂环-(酰)亚胺共聚物
benzhydrazide 苯酰肼；亚苯基肼
benzhydrindene 苯并[e]二氢茚
benzhydrindone 苯并[e]二氢茚酮
benzhydrol(=diphenyl-carbinol) 二苯基甲醇 $(C_6H_5)_2CHOH$
benzhydroxamic acid 苯氧肟酸 $C_6H_5C(=NOH)OH$
benzhydryl(=diphenyl-methyl) 二苯甲基 Ph_2CH—
benzhydryl amine 二苯甲基胺 $(C_6H_5)_2CHNH_2$
a-benzhydrylbenzhydrol a-二苯甲基二苯甲醇；1,1,2,2-四苯基甲醇
benzhydryl bromide 二苯甲基溴
benzhydrylidene(=diphenylmethylene) 二苯亚甲基 $Ph_2C=$
benzidine 联苯胺*
benzidine base 联苯胺碱 $NH_2C_6H_4C_6H_4NH_2$

benzidine blue　联苯胺蓝
benzidine dicarboxylic acid　联苯胺二甲酸　$(NH_2)_2C_{12}H_6(COOH)_2$
benzidine dihydrochloride　联苯胺二盐酸盐
benzidine-disulfonic acid　联苯胺二磺酸　$[C_6H_3(NH_2)SO_3H]_2$
benzidine hydrochloride　联苯胺盐酸盐　$(H_2NC_6H_4)_2 \cdot 2HCl$
benzidine-monosulfonic acid　联苯胺一磺酸　$H_2NC_6H_4C_6H_3(NH_2)SO_3H$
benzidine orange　联苯胺橙
benzidine orange toner　联苯胺橙色原〔调色料〕
benzidine-pyridine　联苯胺吡啶
benzidine reaction　联苯胺反应
benzidine rearrangement(=hydrazo rearrangement)　联苯胺重排
benzidine salt　联苯胺盐
benzidine sulfate　联苯胺硫酸盐；硫酸联苯胺　$(H_2NC_6H_4)_2 \cdot H_2SO_4$
benzidine test　联苯胺试验
benzidine yellow　联苯胺黄
benzidino-　联苯氨基　$p\text{-}H_2NC_6H_4C_6H_4NH—$
benzil　偶苯酰*
benzilam(=azobenzil)　2,4,5-三苯基噁唑
benzil dioxime　苯偶酰二肟；二苯乙(二酮)二肟　$(C_6H_5C=NOH)_2$
a-benzil dioxime　a-苯偶酰二肟（a 指顺式,β 指反式）　$(C_6H_5C=NOH)_2$
benzilic acid　二苯基乙醇酸　$(C_6H_5)_2C(OH)CO_2H$
benzilic acid rearrangement　二苯基乙醇酸重排*
benzilic rearrangement　二苯乙醇酸重排*
benzilidene　亚苄基
benzilmono (2-pyridyl) hydrazone　苯偶酰单(2-吡啶)腙
benzil monoxime　苯偶酰一肟　$C_6H_5COC(=NOH)C_6H_5$
benzil osazone　苯偶酰脎　$(C_6H_5C)_2(NNHC_6H_5)_2$
benziloyl　二苯乙醇酰基　$(C_6H_5)_2C(OH)CO—$
benzimidazole　苯并咪唑；间二氮茚
benzimidazoline　苯并咪唑啉
benzimidazolone　苯并咪唑酮
benzimidazolyl-　苯并咪唑基　$C_7H_5N_2—$
benzimide　苯甲亚胺
benzimido-(=benzimidoyl-)　亚氨苄基　$C_6H_5C(=NH)—$
benzimidoyl-(=benzimido-)　亚氨苄基　$C_6H_5C(=NH)—$
benzin　汽油；挥发油
benzindan　苯并二氢茚
benz[c]indane　苯并[c]二氢茚
benz[e]indane　苯并[e]二氢茚
benzindanol　苯并二氢茚醇
benzindene　苯并茚
benz[c]indeno[2,1-α]fluorene　苯并[c]茚[2,1-α]芴
benzindopyran　苯并茚并吡喃
benzine　汽油；挥发油
benzine-air mixture　汽油空气混合物
benzine blow lamp　汽油喷灯
benzine-resistant　抗汽油的
benzine resisting hose　耐汽油软管
benzine soap　挥发油皂
benzinum purificatum　石油醚〔相对密度为 0.634~0.660,沸点为 35~80℃〕
benzisoquinoline　苯并异喹啉
benznaphthalide　苯甲酰萘胺
benzo-　苯并
benzoate　苯甲酸盐(或酯)　$C_6H_5COOM; C_6H_5COOR$
benzo-azurine　苯并天青精
benzo-benzidine conversion　联苯胺重排作用
benzo-black-blue　苯并黑蓝
benzo-blue　苯并蓝〔染〕
benzocaine　苯佐卡因；对氨基苯甲酸乙酯
benzochromone　苯(并)色酮
benzo-colors　苯并染料〔染〕
benzo-15-crown-5(B15C5)　苯并 15-冠(醚)-5
benzocyclooctene　苯并环辛四烯；苯并辛因
2,3-benzodiazine(=phthalazine)　2,3-苯并二嗪；酞嗪　$C_8H_6N_2$
benzodiazole　苯并二唑；间二氮茚
benzodihydropyrone　苯并二氢吡喃酮；二氢香豆素
benzodiphenylene oxide　苯并氧芴；苯并咕吨
benzoe　安息香
benzo fast blue　苯并坚牢蓝
benzo-fast-heliotrope　苯并坚牢淡紫〔染〕
benzo-fast-orange　苯并坚牢橙〔染〕
benzo-fast-red　苯并坚牢红〔染〕
benzoflavine　苯并黄素
7,8-benzoflavone(=naphthoflavene)　萘黄酮；7,8-苯并黄酮　$C_{19}H_{12}O_2$
benzofulvene　亚甲基茚　$C_6H_4CH=CHC=CH_2$
benzofuran　苯并呋喃；氧茚
benzofuran derivative　苯并呋喃衍生物〔防老剂〕
benzofuran resin　苯并呋喃树脂
benzofuranyl(=benzofuryl)　苯并呋喃基；氧茚基　$C_8H_5O—$
benzofuryl(=benzofuranyl)　苯并呋喃基；氧茚基
benzoglycols　苯乙二醇

benzo guanamine 苯胍胺〔即 2,4-二氨基-6-苯基间-三氮杂苯〕
benzo-hydroperoxide 过氧苯甲酸 $C_6H_5COO_2H$
benzohydroxamic acid 苯基异羟肟酸；苯氧肟酸 $C_6H_5CONHOH$
benzoic acid 苯甲酸；安息香酸 C_6H_5COOH
benzoic acid anhydride 苯甲酸酐 $(C_6H_5CO)_2O$
benzoic acid metabolism 苯甲酸代谢
benzoic alcohol 苄醇；苯甲醇 $C_6H_5CH_2OH$
benzoic amide 苯甲酰胺 $C_6H_5CONH_2$
benzoic anhydride 苯甲酸酐 $(C_6H_5CO)_2$
benzoic condensation 安息香缩合
benzoic ester 苯甲酸酯 $C_6H_5COOR; C_6H_5COOC_2H_5$
benzoic sulfimide 邻磺酰苯酰亚胺；糖精 $C_6H_4CHO_2NS$
benzoic trichloride 三氯甲基苯 $C_6H_5CCl_3$
benzoid compound 苯型化合物
benzoin 苯偶姻；二苯乙醇酮；安息香〔俗〕 $C_6H_5CH(OH)COC_6H_5$
benzoin acetate(=acetyl benzoin) 苯偶姻乙酸酯；乙酰苯偶姻 $C_6H_5COCH(C_6H_5)O_2CCH_3$
benzoin alkylether 苯偶姻烷基醚
benzoinated lard 加了苯甲酸的猪油；安息香化猪油
benzoin blue 苯偶姻蓝
benzoin brown 苯偶姻棕〔染〕
benzoin n-butylether 苯偶姻正丁醚 $C_6H_5C(O)CH(OC_4H_9)C_6H_5$
benzoin condensation 苯偶姻缩合
benzoin dark-green 苯偶姻暗绿〔染〕
benzo-indigo-blue 苯并靛蓝
benzoin ethyl ether(=ethoxybenzoin) 苯偶姻乙醚；乙氧基苯偶姻 $C_6H_5CH(OC_2H_5)COC_6H_5$
benzoin fast-red 苯偶姻坚牢红
benzoin gum 安息香胶
benzoin methylether 苯偶姻甲醚
benzoin oil 安息香树油
benzoin oxime 苯偶姻肟 $C_{14}H_{12}O = NOH$
benzoin phenylhydrazone 苯偶姻苯腙 $C_6H_5NHN = C(C_6H_5)CHOHC_6H_5$
benzoin iso-propylether 苯偶姻异丙醚
benzoin soap 苯偶姻皂；二苯乙醇酮皂
benzol ①粗苯②苯
benzol-acetone dewaxing process 酮苯脱蜡过程〔润滑油〕
benzol-acetone solvent 苯-丙酮溶剂
benzol equivalent 苯当量
benzo-light rubine 苯并亮玉红
benzo-light yellow 苯并亮黄
benzoline ①(=benzolene)汽油；挥发油；轻馏分油〔包括汽油与石脑油〕②(=petroleum benzine)石油醚〔别名〕
benzolized oil 混合了苯的油；苯化油
benzol-keton dewaxing process 酮苯脱蜡过程〔润滑油〕
benzol mixture 苯混合物
benzol refining 苯精炼
benzol sulfohydrazide 苯磺酰肼
benzonaphthol 苯甲酸萘酯
benzonitrile(=phenyl cyanide) 苄腈；苯基氰 C_6H_5CN
benzo-olive 苯并橄榄绿
benzoperoxide 过氧化苯甲酰 $(C_6H_5CO)_2O_2$
benzophenid 苯甲酸苯酯 $C_6H_5COOC_6H_5$
benzophenol ①苯酚②苯甲酸苯酯 $C_6H_5COOC_6H_5$
benzophenone(=diphenylketone) 二苯(甲)酮；苯酮〔俗〕；苯酰苯 $(C_6H_5)_2CO$
benzophenone-anil 二苯(甲)酮苯胺 $(C_6H_5)_2C = NC_6H_5$
benzophenone arsenious acid 二苯(甲)酮亚胂酸 $C_6H_5COC_6H_4As(OH)_2$
benzophenone arsine oxide 二苯(甲)酮胂化氧 $C_6H_5COC_6H_4As = O$
benzophenone carboxylic acid 二苯(甲)酮羧酸；苯甲酰苯甲酸 $C_6H_5COC_6H_4COOH$
benzophenone dicarboxylic acid 二苯(甲)酮二甲酸 $COOHC_6H_4COC_6H_4COOH$
benzophenone-2,2-dicarboxylic acid 二苯(甲)酮-2,2-二甲酸 $(HO_2CC_6H_4)_2CO$
benzophenone-oxime(=diphenylketoxime) 二苯甲酮肟 $(C_6H_5)_2C = NOH$
benzophenone phenylhydrazone 二苯甲酮苯腙 $C_6H_5NHN = C(C_6H_5)_2$
benzophenone tetracarboxylic dianhydride 二苯甲酮四酸二酐〔环氧固化剂〕
benzophthalidylidene 苯并酞基叉；苯并酞亚基
benzopinacol 苯频哪醇 $(C_6H_5)_2C(OH)C(OH)(C_6H_5)_2$
benzopurpurin(=ozamin) 苯并红紫
benzopurpurine 4B 苯并红紫 4B
benzopyran-5-one 苯并吡喃-5-酮
benzopyranyl 苯并吡喃基 $C_9H_7O—$
benzopyrene 苯并芘
3,4-benzopyrene 3,4-苯并芘
benzopyridine(=quinoline) 喹啉
benzopyrone 苯并吡喃酮
1,2-benzopyrone 1,2-苯并吡喃酮
benzopyrone series 苯并吡喃酮系
benzopyrrole(=indole) 吲哚
benzoquinaldine 苯并喹哪啶
benzoquinoline 苯并喹啉 $C_{13}H_9N$
benzoquinone 苯醌 $C_6H_4O_2$
p-benzoquinone dioxime 对苯醌二肟〔硫化剂〕

benzoquinone monooxime 苯醌单肟
benzoquinonyl(=quinonyl) 醌基 $C_6H_3O_2—$
benzo-red-blue 苯并红蓝〔染〕
benzoresinol 安息香树脂醇
benzo-rhoduline red 苯并若杜林红；苯并碱红〔染〕
benzoselenadiazole 苯并硒二唑
benzo sky blue 苯并青
benzosulfinide(=saccharin) 糖精 $C_6H_4CHO_3NS$
benzosulfonazole 苯并磺酰唑
benzotetrazine 苯并四嗪
benzothiazole 苯并噻唑；间氮(杂)硫茚；1,3-硫氮杂茚 $C_6H_4N=CHS$
benzothiazole disulfide 二硫化苯并噻唑〔促进剂〕
2-benzothiazole-N-dithiomorpholine 2-苯并噻唑-N-二硫代吗啉〔促进剂〕
benzothiazolesulfenamide accelerators 苯并噻唑次磺酰胺类促进剂
α-benzothiazolethiol α-巯基苯并噻唑
N-(2-benzothiazolethio)phthalimide N-(2-苯并噻唑硫代)邻苯二甲酰亚胺〔促进剂〕
benzothiazole-type accelerator 苯并噻唑类促进剂
benzothiazolyl 苯并噻唑基
benzothiazyl disulfide 二硫化苯并噻唑〔促进剂〕
benzothiazyl-2-sulphene morpholide 苯并噻唑-2-亚磺酰吗啉〔促进剂〕
benzothiophene(=thionaphthene) 苯并噻吩
benzothiophene-based chelate 苯并噻吩基螯合物
benzothioxanthene 苯并噻吨
benzotriazole 苯并三唑*
benzotrichloride 三氯甲苯 $C_6H_5CCl_3$
benzotrifluoride 三氟甲苯 $C_6H_5CF_3$
benzoxanthene dye 苯并呫吨染料
benzoxazine 苯并噁嗪；氧氮杂萘
benzoxazinone 苯并噁嗪酮；氧氮杂萘酮
benzoxazinyl- 苯并噁嗪基 $C_8H_6NO—$
benzoxazole 苯并噁唑
benzoxazolinone 苯并噁唑啉酮
benzoxazolyl 苯并噁唑基 $C_7H_4NO—$
benzoxy(=benzoyloxy) 苯甲酸基；苯甲酰氧基 $C_6H_5COO—$
1-benzoxy-1-(2-methoxyethoxy)-ethane 1-苄氧基-1-(2-甲氧基乙氧基)-乙烷
benzoxyperoxycarbonylation 苯酸过氧羰化作用
benzoyl- 苯甲酰基 $C_6H_5CO—$
benzoylacetanilide N-苯甲酰乙酰苯胺
benzoyl acetate 乙酸苯酯；苯甲酸-乙酸(混合)酐 $C_6H_5COOCOCH_3$

benzoyl acetic acid 苯甲酰乙酸 $C_6H_5COCH_2COOH$
benzoyl acetone 苯甲酰丙酮 $C_6H_5COCH_2COCH_3$
benzoyl acetonitrile 苯甲酰乙腈 $C_6H_5COCH_2CN$
benzoyl acetylacetone 苯甲酰乙酰丙酮 $C_6H_5COCH_2COCH_2COCH_3$
benzoylaconine(=isaconitine; picraconitine) 苯甲酰乌头宁
benzoyl acrylic acid 苯甲酰丙烯酸 $C_6H_5COCH=CHCO_2H$
benzoyl alanine N-苯甲酰丙氨酸 $C_6H_5CONHCH(CH_3)CO_2H$
benzoyl amide 苯甲酰胺 $C_6H_5CONH_2$
benzoylaminobenzoic acid 苯甲酰氨基苯酸
benzoyl-anthranilic acid 苯甲酰氨茴酸〔俗〕；邻苯甲酰氨基苯酸 $C_6H_5CONHC_6H_4CO_2H$
α-N-benzoyl arginine amide α-N-苯甲酰精氨酰胺
α-N-benzoyl arginine ethyl ester α-N-苯甲酰精氨酸乙酯
α-N-benzoyl arginine-β-naphthylamine α-N-苯甲酰精氨酰-β-萘胺
benzoylate ①苯甲酰化②苯甲酰化物
benzoylated 苯甲酰化(了)的
benzoylating 苯甲酰化
benzoylating agent 苯甲酰(化)剂
benzoylation 苯甲酰化(作用)
benzoyl auramine 苯甲酰金胺；苯甲酰槐黄
benzoyl azide 苯甲酰叠氮 $C_6H_5CON_3$
benzoylbenzoic acid 苯甲酰苯甲酸 $C_6H_5COC_6H_4COOH$
benzoyl bromide 苯甲酰溴 C_6H_5COBr
benzoyl chloride 苯甲酰氯 C_6H_5COCl
benzoylcholine 苯甲酰胆碱
benzoyl cyanide 苯甲酰氰 C_6H_5COCN
benzoyldiphenylmethylsilane 苯甲酰二苯基甲基硅烷
benzoylecgonine 苯甲酰芽子碱
benzoylene 亚苯甲酰基 $—C_6H_4CO—$
benzoylene urea 亚苯甲酰基脲
benzoyl-eugenol 苯甲酰丁子香粉 $C_6H_5CO_2C_{10}H_{11}O$
benzoyl fluoride 苯甲酰氟 C_6H_5COF
benzoyl formic acid 苯甲酰甲酸 $C_6H_5COCOOH$
1-benzoylglucuronic acid 1-苯甲酰葡萄糖苷酸
benzoyl-glycine(=hippuric acid) ①苯甲酰甘氨酸；苯甲酰氨基乙酸②苯甲酰甘氨酸，马尿酸 $C_6H_5CONHCH_2CO_2H$
benzoyl glycollic acid 苯甲酰乙醇酸；苯酸基乙酸 $C_6H_5COOCH_2COOH$
benzoyl hydrazine 苯甲酰肼 $C_6H_5CONHNH_2$
benzoyl hydroperoxide 过氧苯甲酸 C_6H_5COOOH
benzoylhydroxamic acid 苯并羟肟酸；苯甲酰氧肟酸

benzoyl hydroxylamine 苯甲酰胲 $C_6H_5CONHOH$
benzoylimino- 亚苯酰氨基；苯甲酰亚氨基 C_6H_5CON
benzoyl iodide 苯甲酰碘 C_6H_5COI
benzoyl isocyanate 异氰酸苯甲酯 C_6H_5CONCO
benzoyl lactic acid 苯甲酰乳酸 $CH_3CH(O_2C_7H_5)COOH$
benzoyl leucomethylene blue 苯甲酰无色亚甲蓝
benzoyl α-naphthylamine(=α-benznaphthalide) N-苯甲酰-α-萘胺 $C_6H_5CONHC_{10}H_7$
benzoyl oxide 苯甲酸酐 $(C_6H_5CO)_2O$
benzoyloxy- 苯甲酸基；苯甲酰氧基 C_6H_5COO-
benzoyloxylation 苯甲酸化作用
4-benzoyloxy-2,2,6,6-tetramethyl piperidine 过氧化苯甲酰氧基-2,2,6,6-四甲基哌啶〔光稳定剂〕
benzoyl peroxide(BPO) 过氧化苯甲酰
benzoylphenylhydroxyamine 苯甲酰苯羟胺；苯甲酰苯胺；N-苯甲酰-N-苯基羟胺 $C_6H_5CON(OH)C_6H_5$
benzoyl phthalic acid 苯甲酰邻苯二甲酸 $C_6H_5COC_6H_3(COOH)_2$
benzoyl pyruvic acid 苯甲酰丙酮酸 $C_6H_5COCH_2COCOOH$
benzoyl sulfonic imide 糖精
benzoyl thiokinase 苯甲酰硫激酶
benzoyl thiourea 苯甲酰硫脲 $C_6H_5CONHCSNH_2$
benzoyl toluidine N-苯甲酰甲苯胺
benzoyltrifluoroacetone 苯甲酰三氟丙酮 $C_6H_5COCH_2COCF_3$
benzoyltrimethylsilane 苯甲酰三甲基硅烷
benzoyltriphenylgermane 苯甲酰三苯基锗烷
benzoyltriphenylmethane 苯甲酰三苯基甲烷
benzoyltriphenylsilane 苯甲酰三苯基硅烷
benzoyltropeine 苯甲酰托品因
N-benzoyltyramine N-苯甲酰酪胺
benzoyl urea 苯甲酰脲 $C_6H_5CONHCONH_2$
N-benzoyl xylide N-苯甲酰二甲苯胺
benzphetamine 甲基苯异丙基苄胺
benzpinacol 苯频哪醇
benzpinacolone 苯频哪酮 $(C_6H_5)_3CCOC_6H_5$
benzpinacone 苯频哪醇；四苯代乙邻二醇；四苯代-1,2-乙二醇 $[(C_6H_5)_2COH]_2$
benzpyrene 苯并[c,d]芘
benzsulfamide 苯磺酰胺 $C_6H_5SO_2NH_2$
benztropine 苯托品〔药〕
benzvalene 盆苯
benzvalene form 盆式〔苯的一种结构式〕
benzyl 苄基；苯甲基 $C_6H_5CH_2-$
benzyl abietate 松香酸苄酯
benzyl acetaldehyde 苄基乙醛

benzyl acetamide N-苄乙酰胺 $CH_3CONHCH_2C_6H_5$
benzyl acetate 乙酸苄酯 $C_6H_5CH_2OCOCH_3$
benzylacetic acid 苄基乙酸
benzyl acetoacetate 乙酰乙酸苄酯
benzyl acetoacetic ester 乙酰苄基乙酸酯 $CH_3COCH(C_6H_5CH_2)COOR; CH_3COCH(C_6H_5CH_2)COOC_2H_5$
benzyl acetone 苄基丙酮 $C_6H_5(CH_2)_2COCH_3$
benzyl acetophenone 苄基乙酰苯；苯丙酰苯 $C_6H_5(CH_2)_2COC_6H_5$
benzyl acrylate 丙烯酸苄酯 $C_2H_3CO_2C_7H_7$
benzyladenine 苄(基)腺嘌呤
benzyl alcohol 苄醇；苯甲醇 $C_6H_5CH_2OH$
benzyl alcohol cellulose 苯甲基纤维素
benzyl amine 苄胺 $C_6H_5CH_2NH_2$
benzyl aminoacetic acid 苄氨基乙酸 $C_6H_5CH_2NHCH_2COOH$
benzyl aminophenol 苄氨基苯酚 $C_6H_5CH_2NHC_6H_4OH$
benzyl aminophenol hydrochloride 盐酸苄氨基苯酚
benzyl aniline 苄基苯胺 $C_6H_5CH_2NHC_6H_5$
benzylaniline-azo-benzene 苄基苯胺偶氮苯
benzylaniline-azo-benzene sulfonic acid 苄基苯胺偶氮苯磺酸
benzyl-anthracene 苄基蒽 $C_7H_6CH_3=C_2H_2=C_6H_4$
benzyl arsine oxide 氧化苄胂 $C_6H_5CH_2As=O$
benzyl arsonic acid 苄胂酸 $C_6H_5CH_2AsO(OH)_2$
benzylated 苄化(了)的
benzylating 苄化
benzylating agent 苄化剂
benzylation 苄(基)化作用
benzyl azide 苄基叠氮 $C_6H_5CH_2N_3$
benzyl benzoate ①苯甲酸苄酯 $C_6H_5CO_2CH_2C_6H_5$ ②苄基苯甲酸盐 $C_6H_5CH_2C_6H_4COOM$
benzylbenzoic acid 苄基苯甲酸
benzyl bibromide 二溴甲基苯 $C_6H_5CHBr_2$
benzyl bichloride 二氯甲基苯 $C_6H_5CHCl_2$
benzylbiphenyl 苄基联苯
benzyl boric acid(=benzyl boron dihydroxide) 苄基硼酸
benzyl boron dihydroxide(=benzyl boric acid) 苄基硼酸 $C_6H_5CH_2B(OH)_2$
benzyl bromide 苄基溴 $C_6H_5CH_2Br$
benzyl butanoate 丁酸苄酯
benzyl butyl adipate 己二酸苄丁酯〔增塑剂〕
benzyl-n-butyl adipate 己二酸苄正丁酯〔增塑剂〕
benzyl butyl alcohol 苄基丁醇
benzyl butyl ether 苄基丁基醚
benzyl butyl phthalate ①邻苯二甲酸苄(基)丁(基)酯 ②苄(基)丁基邻苯二甲酸盐
benzyl butyrate 丁酸苄酯 $C_3H_7CO_2CH_2C_6H_5$

benzyl carbamate 氨基甲酸苄酯 $H_2NCO_2CH_2C_6H_5$	benzyl hydrosulfide 氢硫化苄
benzyl carbanion 苄基负碳离子	benzylhydroxylamine 苄基胲
benzyl carbinol 苯乙醇；苄基甲醇	benzylic 苄型(的)
benzyl carbinyl acetate 乙酸苄基甲酯	benzylic cation 苄(基)正离子*
benzyl cellosolve 苄基溶纤剂	benzylidene(=benzal) 亚苄基；苯亚甲基 $C_6H_5CH=$
benzyl cellulose 苄基纤维素	benzylidene-acetone 亚苄基丙酮；1-苯丁烯-3-酮
benzyl chloracetate 氯乙酸苄酯 $ClCH_2CO_2CH_2C_6H_5$	benzylidene acetophenone 苯叉乙酰苯；亚苄基苯乙酮；苄叉苯乙酮
benzyl chloride 苄基氯 $C_6H_5CH_2Cl$	
benzyl chlorophenol 苄基氯酚	benzylidene aniline 苄叉苯胺；亚苄基苯胺
benzyl cinnamate 肉桂酸苄酯	benzylidene-azine 苄连氮；二苄肼；二亚苄基肼
$C_6H_5CH=CHCOOCH_2C_6H_5$	benzylidene bromide 二溴甲基苯 $C_6H_5CHBr_2$
benzyl cyanamide 苄基氨腈 $C_6H_5CH_2NHCN$	3-benzylidene-2-butanone 3-苄叉-2-丁酮
benzyl cyanide 苄基氰；苯乙腈 $C_6H_5CH_2CN$	benzylidene chloride 二氯甲基苯 $C_6H_5CHCl_2$
benzyl dichloride 二氯甲基苯 $C_6H_5CHCl_2$	benzylidene citronellal 苄叉香茅醛
benzyl diethyl carbinyl acetate 乙酸苄基二乙基原酯	benzylidene diacetate 苄叉二乙酸酯；苯亚甲基二乙酸酯 $C_6H_5CH(OOCCH_3)_2$
benzyldimethylamine 苄基二甲胺〔硫化剂〕	
benzyl 2,3-dimethyl-2-butenoate 2,3-二甲基-2-丁烯酸苄酯	benzylidene-ethylamine N-苄乙胺；N-乙基苄亚胺
	1,2-benzylidene glycerol 1,2-苄叉(缩)甘油醚；亚苄基(缩)甘油醚
benzyl dimethyl carbinol 苄基二甲基甲醇	
benzyl dimethylcarbinyl acetate 乙酸苄基二甲基甲酯	benzylidene glycerol 苄亚甲基甘油；苄亚甲基丙三醇
benzyl 2,3-dimethyl crotonate 2,3-二甲基巴豆酸苄酯	2-benzylidene-1-heptanol(=2-amyl cinnamic alcohol) 2-亚苄基-1-庚醇；2-戊基肉桂醇
benzyldimethylketal 苄基二甲醛缩苯乙酮	2-benzylidene hexanal 2-苄叉己醛
benzyl dimethyl octadecyl ammonium chloride 苄基二甲基十八烷基氯化铵	benzylidene malonate 丙二酸亚苄酯〔增塑剂〕
	benzylidene malonic acid 亚苄基丙二酸；苄叉丙二酸
benzyldioctylamine 苄基二辛基胺 $(C_6H_5CH_2)(C_8H_{17})_2N$	2-benzylidene octanal 2-苄叉辛醛
	benzylidene phthalide 亚苄基-2-苯并呋喃酮；苄叉酞
benzyldioctylphosphine oxide(BDOPO) 苄基二辛基氧膦 $(C_6H_5CH_2)(C_8H_{17})_2PO$	α-benzylidene propionic acid α-苄叉丙酸；α-甲基肉桂酸
benzyl diphenylamine 苄基二苯胺 $(C_6H_5)_2NCH_2C_6H_5$	benzylidene rubber 亚苄基橡胶
benzyl dipropyl ketone 苄基二丙基酮	benzylidene succinic acid 苄叉琥珀酸；亚苄基琥珀酸；苯甲叉丁二酸
benzyl disufide(=dibenzyl disulfide) 二苄(化)二硫 $(C_6H_5CH_2)_2S_2$	benzylidine rubber 亚苄基橡胶
benzyl dodecylamine 苄基十二烷基胺	benzylidyne 次苄基
benzylene bromide 二溴甲基苯 $C_6H_5CHBr_2$	benzyl iodide 苄基碘 $C_6H_5CH_2I$
benzylene chloride 二氯甲基苯 $C_6H_5CHCl_2$	benzyl isoamyl alcohol 苄基异戊醇
benzyl ether 二苄醚 $(C_6H_5CH_2)_2O$	benzyl isoamyl ether 苄基异戊基醚
benzyl ethoxyacetate 乙氧基乙酸苄酯	benzyl isobutyl ketone 苄基异丁基酮
benzyl ethyl alcohol 苄基乙醇；γ-苯丙醇	benzyl isobutyrate 异丁酸苄酯
benzyl ethyl carbinyl acetate 乙酸苄基乙基原酯	benzyl isoeugenol 苄基异丁子香酚
benzyl ethylene 苄基乙烯；苯丙烯 $C_6H_5CH_2CH=CH_2$	benzyl isopropyl carbinyl acetate 乙酸苄基异丙基原酯
benzyl ethyl ether 苄基乙基醚	benzyl isorhodanate 异硫氰酸苄酯
benzyl eugenol 丁子香酚苄醚	benzyl isorhodanide 异硫氰酸苄酯
benzyl formate 甲酸苄酯 $HCO_2CH_2C_6H_5$	benzyl isosulfocyanate 异硫氰酸苄酯
benzyl group 苄基*	benzyl isosulfocyanide 异硫氰酸苄酯
benzyl halide 苄基卤；卤代甲苯 $C_6H_5CH_2X$	benzyl isothiocyanate 异硫氰酸苄酯 $C_6H_5CH_2NCS$
3-benzyl-4-heptanone 3-苄基-4-庚酮	benzyl isothiocyanide 异硫氰酸苄酯
benzyl heptine carbonate 庚炔羧酸苄酯	benzylisothiourea 苄(基)异硫脲 $C_7H_7SC(=NH)NH_2$
benzyl hydrazine 苄基肼 $C_6H_5CH_2NHNH_2$	

benzylisothiourea hydrochloride 苄(基)异硫脲盐酸盐
benzyl isovalerate 异戊酸苄酯 $C_9H_9COOC_7H_7$
benzyl ketone 苄基酮
benzyl lactate 乳酸苄酯 $CH_3CHOHCO_2C_7H_7$
benzyl laurate 十二(烷)酸苄酯 $C_{11}H_{23}CO_2C_7H_7$
benzyl magnesium chloride 氯化苄基镁 $C_6H_5CH_2MgCl$
benzyl magnesium halide 卤化苄基镁 $C_6H_5CH_2Mg$
benzyl malonic acid 苄基丙二酸 $C_6H_5CH_2CH(CO_2H)_2$
benzyl menthol 苄基薄荷醇 $C_{17}H_{26}O$
benzyl mercaptan 苄硫醇 $C_6H_5CH_2SH$
2-benzylmercaptobenzothiazole 2-苄基硫醇基苯并噻唑〔促进剂〕
β-benzylmercaptobutylaldehyde β-苄基硫醇基丁醛〔热稳定剂〕
benzyl mercuric chloride 氯化苄基汞 $C_6H_5CH_2HgCl$
benzyl methacrylate 甲基丙烯酸苄酯 $CH_2\!=\!C(CH_3)COOCH_2C_6H_5$
benzyl methoxyacetate 甲氧基乙酸苄酯
2-benzyl-4-methyl-6-t-butyl phenol 2-苄基-4-甲基-6-叔丁基苯酚
benzyl 3-methylbutyrate 3-甲基丁酸苄酯
benzyl-methyl-carbinol 苄基甲基甲醇；2-苯丙醇 $C_6H_5CH_2CHOHCH_3$
benzyl methyl carbinyl acetate 乙酸苄基甲基原酯
1-benzyl-2-methyl imidazole 1-苄基-2-甲基咪唑〔环氧固化剂〕
benzyl 2-methylpropanoate 2-甲基丙酸苄酯
benzyl methylpropanoate 甲基丙酸苄酯；异丁酸苄酯
benzyl methyltiglate 甲基惕各酸苄酯
benzyl mustard oil 苄基芥子油；异氰酸苄酯 $C_6H_5CH_2NCS$
benzyl naphthalene 苄基萘 $C_6H_5CH_2C_{10}H_7$
benzyl nicotinate 烟酸苄酯
benzyl octyl adipate 己二酸辛苄酯
benzyl oxide (二)苄醚 $(C_6H_5CH_2)_2O$
benzyloxy- 苄氧基 $C_6H_5CH_2O-$
benzyl oxyethanol 苄基氧乙醇
benzyloxy-methylation 苄氧甲基化(作用)
benzylpenicillin 青霉素〔药〕
benzylphenol 苄基苯酚
benzyl phenone 二苯甲酮
benzyl phenylacetate 苯乙酸苄酯
benzylphenyl carbamate 氨基甲酸苄苯酯 $C_7H_7C_6H_4O_2CNH_2$
benzyl phthalate 邻苯二甲酸(二)苄酯
benzyl propanoate 丙酸苄酯
benzyl propionate 丙酸苄酯 $C_2H_5CO_2CH_2C_6H_5$

benzyl propyl carbinol 苄基丙基原醇
benzyl-pyridine 苄基吡啶 $C_6H_5CH_2C_5H_4N$
benzyl-pyrrole 苄基吡咯 $C_4H_4NCH_2C_6H_5$
benzyl radical 苄基自由基
benzyl rhodanate(=benzyl rhodanide) 硫氰酸苄酯 $C_6H_5CH_2SCN$
benzyl rhodanide(=benzyl rhodanate) 硫氰酸苄酯
benzyl salicylate 水杨酸苄酯 $HOC_6H_4CO_2CH_2C_6H_5$
benzyl silicone 苄基硅氧烷
benzylsodium 苄基钠 $C_6H_5CH_2Na$
benzyl succinate 丁二酸二苄酯 $(CH_2COOCH_2C_6H_5)_2$
benzylsulfanilamide(=proseptazine;septazine) 苄基磺胺
benzyl sulfhydrate 苄硫醇 $C_6H_5CH_2SH$
benzyl sulfide (二)苄硫醚 $(C_6H_5CH_2)_2S$
benzyl sulfocyanate(=benzyl sulfocyanide) 硫氰酸苄酯；苄基硫氰 $C_6H_5CH_2SCN$
benzyl sulfocyanide(=benzyl sulfocyanate) 硫氰酸苄酯；苄基硫氰
benzyl sulfone 二苄砜；苄砜 $(C_6H_5CH_2)_2SO_2$
benzyl sulfonic acid 苄磺酸 $C_6H_5CH_2SO_3H$
benzyl sulfoxide 二苄亚砜；苄亚砜 $(C_6H_5CH_2)_2S\!=\!O$
benzyl tartronic acid 苄基丙醇二酸 $C_7H_7C(OH)(CO_2H)_2$
benzylthio- 苄硫基；苄甲硫基 $C_6H_5CH_2S-$
benzyl thiocyanate(=benzyl thiocyanide) 硫氰酸苄酯 $C_6H_5CH_2SCN$
benzyl thiocyanide(=benzyl thiocyanate) 硫氰酸苄酯
benzyl thioether (二)苄硫醚 $(C_6H_5CH_2)_2S$
benzyl thiourea 苄硫脲 $C_6H_5CH_2NHCSNH_2$
benzyl-thiuronium 苄锍脲
benzyl thiuronium salt 苄锍脲盐
benzyl tiglate 惕各酸苄酯
benzyl trans-2-methyl-2-butenoate 反-2-甲基-2-丁烯酸苄酯
benzyl trimethyl ammonium chloride 苄基三甲基氯化铵
benzyl trimethyl ammonium hydroxide 氢氧化三甲基苄基铵
benzyl triphenyl phosphonium chloride 苄基三苯基氯化磷
benzyl urea 苄脲 $NH_2CONHCH_2C_6H_5$
benzyl valerianate 戊酸苄酯
benzyl vinyl ether 苄基乙烯基醚
benzyl violet 4B 苄基紫 4B
benzyl viologen 苄基紫精；联苄吡啶
benzyne 苯炔*
benzyne intermediate 苯炔中间体
ber 枣
beramic color 陶瓷颜料

berbamine 小檗胺
berbamunine 小檗宁
berberal 小檗醛 $C_{20}H_{17}O_7N$
berberic acid 小檗酸 $C_8H_8O_4$
berberilic acid 小檗二酸
berberilic anhydride 小檗二酸酐
berberine 小檗碱〔药〕
berberonic acid 小檗三酸
berberrubine 小檗红碱
berbine 小檗因
bercular corrosion 点状腐蚀
Berek compensator 贝瑞克补偿器
berel gear 伞形齿轮
berengelite 脂光沥青
beresovite 铬铅矿
beresowite 碳铬铅矿
bergamal 香柠檬醛
bergamot 香柠檬
bergamotane 香柠檬烷
bergamotene 香柠檬烯
bergamot mint oil 香柠檬薄荷油
bergamot oil 香柠檬油
bergamot petitgain oil 香柠檬叶油
bergamottin 香柠檬亭
bergapten 香柠檬烯
bergaptene 香柠檬脑
bergaptol 香柠檬酚
Bergbau-Forschung process 伯格鲍-福希格法〔一种烟道气脱硫法〕
bergblau 碳酸铜矿
bergenin(e) 虎耳草素〔岩白菜素是从新疆虎耳草科植物厚叶岩白菜提炼精制得到的〕
Bergius process 伯吉尤斯(煤高压加氢)法
Bergmann and Junk test 伯格曼-靳克耐热试验
bergmeal(=bergmehl) 硅藻土
bergmehl(=bergmeal) 硅藻土
Bergsman micro hardness meter 伯格斯曼显微(镜式)硬度计
Bergsoe process (for tin refining) 伯格索伊(炼锡)法
beriberi 脚气病
berillia 氧化铍 BeO
berillium(=beryllium) 铍 Be
Berkefeld filter 伯克菲尔德(素烧)滤筒
berkelium 锫〔97 号元素，化学符号 Bk〕
berkelium oxide 氧化锫 $BkO_2; Bk_2O_3$
berkelium oxybromide 溴氧化锫 BkOBr
berkelium sulfide 硫化锫 Bk_2S_3
berkelium tetrafluoride 四氟化锫 BkF_4
berkelium thiocyanate 硫氰酸锫
berkelium trichloride 三氯化锫 $BkCl_3$
Berkex process 锫开克斯过程〔从辐照锫靶中回收超锫元素过程的一步。分批溶剂萃取法回收锫〕
berlambine(=hydroxyberberine) 小檗浸碱
Berlin black 柏林黑
Berlin blue 柏林蓝；普鲁士蓝；亚铁氰化铁
Berlin red 柏林红；氧化铁红
Berlin white 柏林白〔$Fe(NH_4)_2Fe(CN)_6$，铁蓝生产中的中间产物〕
Berl saddle 弧鞍形填料；贝尔鞍形填料
Berl saddle packing 弧鞍形填料；贝尔鞍形填料
Bernard law 伯纳德定律
berninamycin 伯尔尼霉素
Bernoulli equation 伯努利方程
Bernoulli's law 伯努利法则
Bernoulli's theorem 伯努利定理
bernzoyl peroxide 过氧化苯甲酰 $(C_6H_5CO)_2O_2$
beromycin 别洛霉素；青霉素 V 钾
berry 浆果
Berthelot-Nernst distribution 贝特洛-能斯特分布
berthierine 磁绿泥石
berthierite 辉锑铁矿
Berthollet's tester 柏氏测氯仪
berthollide 贝陀立体；非定比化合物
berthollide compound 非定比化合物
bertrandite 羟硅铍石
beryl 绿柱石
beryl based glass 绿玉基玻璃
beryllia 氧化铍
beryllia crucible 氧化铍坩埚
beryllia fibre 氧化铍纤维
beryllide 铍化物
beryllium 铍〔4 号元素，化学符号 Be〕
beryllium acetate 乙酸铍
beryllium acetyl-acetonate 乙酰丙酮合铍 $Be[CH(COCH_3)_2]_2$
beryllium alkyl 二烷(基)铍 BeR_2
beryllium ammonium phosphate 磷酸铍铵 $Be(NH_4)PO_4$
beryllium bromide 溴化铍 $BeBr_2$
beryllium bronze 铍青铜
beryllium carbide 碳化铍
beryllium carbonate 碳酸铍 $BeCO_3$
beryllium chloride 氯化铍 $BeCl_2$
beryllium copper 铍铜
beryllium ethide(=beryllium ethyl) 二乙铍 $Be(C_2H_5)_2$
beryllium ethyl(=beryllium ethide) 二乙铍

beryllium fluoride　氟化铍　BeF_2
beryllium-glass composite fibre　铍基玻璃复合纤维
beryllium hydride　氢化铍
beryllium hydrophosphate　磷酸氢铍　$BeHPO_4$
beryllium hydroxide　氢氧化铍　$Be(OH)_2$
beryllium iodide　碘化铍　BeI_2
beryllium methide(=beryllium methyl)　二甲铍　$Be(CH_3)_2$
beryllium methyl(=beryllium methide)　二甲铍
beryllium nitrate　硝酸铍　$Be(NO_3)_2$
beryllium nitride　二氮化三铍　Be_3N_2
beryllium oxalate　草酸铍　BeC_2O_4
beryllium oxide　氧化铍　BeO
beryllium oxide fibre　氧化铍纤维
beryllium oxychloride　二氯一氧化铍　$BeOCl_2$
beryllium phosphate　磷酸铍〔1.偏 $Be(PO_3)_2$; 2.正 $Be_3(PO_4)_2$; 3. 焦 $Be_2P_2O_7$〕
beryllium potassium fluoride　氟化钾合氟化铍；氟化钾铍　$BeF_2·(KF)_2$
beryllium selenate　硒酸铍　$BeSeO_4$
beryllium silicate　硅酸铍〔1. 偏$(BeSiO_3)_n$; 2. 原 Be_2SiO_4〕
beryllium sodium fluoride　氟化钠合氟化铍；氟化钠铍　$BeF_2·(NaF)_2$
beryllium sulfate　硫酸铍　$BeSO_4$
beryllon Ⅱ　铍试剂Ⅱ
beryllonite　磷酸钠铍石
berzelianite　硒铜矿
berzeliite　黄砷榴石
Bessel function　贝塞尔函数
Bessemer converter　酸性转炉
Bessemer converting acid process　酸性转炉作业
Bessemer iron(=Bessemer steel)　酸性转炉钢
Bessemer pig(=Bessemer steel)　酸性转炉钢
bessemer process　酸性转炉工艺
Bessemer steel(=Bessemer iron; Bessemer pig)　酸性转炉钢
best anode temperature　最佳阳极温度
best available treatment　最佳可行处理
best base cure　最适基本硫化
best column efficiency　最佳柱效(率)
best cure　最适硫化；正硫化
best fit　最佳拟合
best-fit line　最佳拟合曲线
best membership method　最佳隶属度方法
best power mixture rate　最佳燃烧混合比
best setting　最佳调节〔燃烧混合比例〕
best tracking(=directional stability)　方向稳定性

best unbiased estimator　最佳无偏估计量(值)
beta　β〔希腊字母〕
beta activity　β放射性
beta-arsenic(=arsenic black)　砷黑；β砷〔砷的同素异形体；黑色无定形固体〕
beta backscattering　β射线照射法
BET absorption method　BET 吸收法
betacaine(=betaeucaine)　β优卡因
beta camphor　β樟脑　$C_{10}H_{16}O$
beta carbon　β碳原子
beta cellulose　β纤维素
beta-decay　β衰变
beta-decay synthesis　β衰变合成〔标记化合物〕
beta decontamination　β放射性去污
BET adsorption isotherm　BET 吸附等温线
beta emitter　β(射线)发射体
beta-emitting　β发射
beta-emitting isotope　β放射同位素
betaeucaine(=betacaine)　β优卡因
beta filter　β滤器
beta-form　β形
beta-gamma cell　β-γ放射性热室
beta-gamma coincidence counting　β-γ符合计数
beta-gamma double bond　β-γ碳原子间的双键
beta-gamma unsaturation　β-γ不饱和
beta-gauge(=beta-radiation thickness gauge)　β射线测厚计
beta glass fibre　β-玻璃纤维；超细玻璃纤维
betahistine　倍他司汀〔药〕
betaine　甜菜碱*
betaine aldehyde　甜菜醛；三甲基甘氨醛
betaine amphoteric surfactant　内铵盐两性表面活性剂
betaine ester sulfonic acid　甜菜碱酯磺酸
betaine hydrazide chloride　氯化甜菜碱酰肼
betaine hydrochloride　盐酸甜菜碱　$C_5H_{11}O_2NHCl$
betaine type amphoteric surface active agent　内铵盐型两性表面活性剂
beta-ionization detector　β(粒子)电离检测器
beta-isorubber　乙种异构化橡胶〔天然胶的低聚合度部分〕
betalamic acid　甜菜醛氨酸
betamethasone　倍他米松〔药〕
beta-naphthol　β萘酚；2-萘酚　$C_{10}H_7OH$
betanidin　甜菜(苷)配基
betanin　甜菜苷
beta-nitrostyrene　β硝基苯乙烯
beta oxidation　β氧化
beta-particle　β粒(子)
beta-pinene　β蒎烯

beta-plane　β平面
beta portion　β部分
beta-position　β位
beta-radiation thickness gauge(=beta-gauge)　β射线测厚计
beta radiography　β放射照相法
beta-ray　β射线
beta-ray absorption method　β射线吸收法
beta-ray back scattering　β射线反向散射法
beta-ray gauge　β射线测厚仪
beta-ray ionization detector　β射线电离检测器
beta-rays　β射线
beta-ray spectrograph　β射线摄谱仪
beta-ray spectrum　β射线谱
BET area　BET 法测定的表面积〔催化剂或吸附剂的〕
beta-relaxation temperature　β松弛温度
beta-rubber　β橡胶；凝橡胶
beta-scission　β断裂
beta-slit　β狭缝
beta source　β(射线)源
beta spectrometer　β谱仪
beta spectrum　β(能)谱
beta stability peninsula　β稳定半岛
beta-stability ridge　β稳定脊
beta stable element　β稳定元素
beta-structure　β结构
beta substitution　β取代
beta-teletherapy source　β远距治疗源；β深部治疗源
betatopic　①失电子的②差电子的
betatopic change　电子放射变化；失电子蜕变
beta-transition　β转变
betatron　电子感应加速器
beta uranium　β铀〔铀的一种同素异形体〕
beta-vibration　β振动
betavoltaic conversion　β辐射电转换
betaxin　维生素 B_1
betel leaves oil　蒌叶油〔取自 *Piper betel*〕
betel-nut(=buyo)　槟榔
betel oil　蒌叶油；槟榔油
betel pepper　蒌叶
betel phenol　蒌叶酚
bethanechol(=urecholine chloride)　乌拉胆碱；氨甲酰甲胆碱
BET isotherm(=Brunauer-Emmett-Teller isotherm)　BET (吸附)等温式
betitol　北蒂糖醇〔1901 年取自甜菜糖醇的一种丁糖醇〕
betol(=salinaphthol)　水杨萘酚
beton　混凝土
betonicine　左旋水苏碱；左旋羟脯氨酸二甲内盐

beton paint　混凝土用涂料
BET surface area　BET 比表面积
Bettendorf test　贝滕多尔夫试验〔一种砷与铋、锑化合物共存试验〕
bettering method　一槽法
better oil　优质油
Betti reaction　贝蒂反应
Bett lead-refining　贝特炼铅法
betula camphor(=betulin)　桦木脑；桦木酮
betula oil　桦木油
betulene　桦木烯
betulenene　脱氢桦木烯
betulenol(=betulol)　桦木醇
betulin(=betulinol)　桦木醇
betulinic acid　桦木酸
betulinol(=betulin)　桦木醇
betulol(=betulenol)　桦木醇
betulonic acid　桦木酮酸　$C_{30}H_{46}O_3$
betuloside　桦木糖苷　$C_{16}H_{24}O_7$
between-laboratory errors　(实验)室间误差
between-procedure errors　操作间误差
betyllium chlorate　氯酸铍　$Be(ClO_3)_2$
Beutell pump　波特耳泵
bevatron　高能质子同步稳加速器
bevel　斜面切割
bevel arm piece　斜三通管
bevel drive　斜(齿)动法
beveled carbon-fiber bundle microelectrode　斜削碳纤维束复合微电极
bevel end　坡管口
bevel gauge　曲尺
bevel gear　伞形齿轮；圆锥齿轮
bevel gear drive　伞形齿轮传动
beveling　切斜边；斜边
beveling machine　刨边机
bevel joint　斜坡接合
bevelled　倾斜的
bevel protractor　斜角规
bevel siding(=clapboard)　互搭板壁；披叠板壁
bevel splice　斜坡接头
beverage analysis and inspection　饮料分析与检验
beyerol　贝叶醇
Beyliss turbidimeter　贝利斯浊度计
bezel　①仪表前盖；(仪表)玻璃框②(荧光)屏③遮光板；聚光圈
Bhatta　熔(紫)胶炉〔印〕
bi-　二；两个；双
biacenaphthylidene(=biacene)　联二氢亚苊

$C_{10}H_6CH_2C \equiv\!\!\equiv CCH_2C_{10}H_6$
biacene(=biacenaphthylidene) 联二氢亚苊
biacetyl 联乙酰；2,3-丁二酮 $CH_3COCOCH_3$
biacetylene 联乙炔 $CHC \equiv\!\!\equiv CCH$
biacetylmonoxime 二乙酰单肟
biacidic base 二(酸)价碱
biackstrap 赤糖糊
biadipate ①酸式己二酸盐 $HOOC(CH_2)_4COOM$ ②酸式己二酸酯 $HOOC(CH_2)_4COOR$
bialin 比阿洛青霉素
biamperometric titration(=biamperometry) 双指示电极电流滴定(法)*
biamperometry(=biamperometric titration) 双指示电极电流滴定(法)*
Bianchi's densimeter 比安希氏密度计〔测定粉末物质密度用〕
bianthronyl 二蒽酮基
bi-apron cross lapper 双帘夹持铺网机
biarsine(=biarsyl) 联胂 $R_2As \cdot AsR_2$
biarsyl(=biarsine) 联胂
biaryl 联芳*
biaryl synthesis 联芳基合成
bias ①偏压②偏流③偏差④斜张网布；斜织法
bias control 偏差控制
bias cut 斜裁；斜向裁断
bias cut cloth 斜切布
bias cutter(=bias cutting machine) 斜切机
bias cutting machine(=bias cutter) 斜切机
biased estimation 有偏估计
bias error 系统误差
bias factor 偏差因素
bias laid 斜贴
bias ply 斜交帘布层〔轮胎〕
bias-ply tyre 斜交轮胎；斜交帘布轮胎
bias tyre 斜交轮胎
biatomic ①二原子的②二酸价的
biatomic acid 二元酸
biax ①双轴向延伸的②双轴向的
biax film 双轴向拉伸薄膜
biaxial 双轴的
biaxial birefringence 双轴双折射
biaxial creep 双轴蠕变
biaxial crystal 双轴晶体*
biaxial deformation 双(轴)向变形
biaxial drawing 双轴拉伸；双轴向牵伸
biaxial extension 双轴伸长
biaxial load 双轴向载荷
biaxial orientation 双轴取向*

biaxial rotation 双轴向旋转
biaxial stress relaxation 双轴向应力松弛
biaxial stress-strain measurement 双轴(向)应力应变测
biaxial stress system 双(轴)向应力系统
biaxial stretch 双轴伸长
biaxial stretching 双轴(向)拉伸
biaxial viscosity 双轴向黏性
biaxial warp-knitted composite 双轴向经编复合材料
biaxial winding 双轴向缠绕
bi-ax orientation 双轴向取向
biazelaate ①酸式壬二酸盐②酸式壬二酸酯
bibasic(=dibasic) ①二元的；二碱价的〔指酸〕②二代的〔指盐〕
bibasic acid 二(碱)价酸
bibasic amino acid 二氨基酸
bibasic keto acid 二元酮酸
bibation ①饮②饮料
bib cock ①活塞②嘴管向下弯的水龙头
bibenzil 联苄 $(C_6H_5CH_2)_2$
5,5'-bibenzimidazole 5,5'-双苯并咪唑
bibenzohydrol 1,1,2,2-四苯基-1,2-乙二醇
bibenzoic acid 联苯甲酸
bibenzoyl 联苯甲酰 $(C_6H_5CO)_2$
bibenzyl 联苄*
biberine 贝比碱
bibliometer 吸水性能测定仪
biborane(=boroethane) 乙硼烷
biborate 酸式硼酸盐
bibromide 二溴化物
bibulous paper 吸墨纸；吸水纸
bicalutamide 比卡鲁胺〔药〕
bicarbonate 碳酸氢盐*
bicarbonate hardness 重碳酸盐硬度
bicarbonate of ammonia 碳酸氢铵
bichloride 二氯化物
bichromate 重铬酸盐
bichromate cell 重铬酸盐电池
bichromated dye (stuff) 重铬酸盐处理染料；铬(处理)染料
bichromate solution 重铬酸溶液〔测定石油产品色度用的标准溶液〕
bichrome 双色的
bicine N-二(羟乙基)甘氨酸
bicolorimeter 双筒比色计；双色比色计
bi-component 双组分
bi-component catalyst 双组分催化剂；双金属系催化剂
bicomponent copolymerization 双组分共聚合
bicomponent fiber 双组分纤维；复合纤维

bicomponent film   双组分薄膜
bicomponent film fibre   双组分薄膜纤维
bicomponent hybrid fibre   双组分混杂纤维
bicomponent structure   (纤维的)双组分结构
biconcave lens   双凹(面)透镜
biconical rotor   双锥形转子
biconstituent fibre   双成分(复合)纤维
biconvex lens   双凸(面)透镜
bicrystals   双晶*；孪晶
bicyclic   二环的
bicyclic compound   二环化合物
bicyclic cryptate   双环穴状化合物
bicyclic epoxide   二环环氧化物
bicyclic hydrocarbon   二环烃
bicyclic monoterpene   双环单萜(烯)
bicyclic sulfide   二环硫化物
bicyclic trisulfone   二环三砜
bicyclo-   二环；双环
bicyclo dihydrohomofarnesyl oxide   双环二氢高金合欢醚
bicyclofarnesal   双环金合欢醛
bicyclohexanone   双环己酮
bicyclohomofarnesal   双环高金合欢醛
bicyclol   双环醇〔药〕
bicyclomycin   双环霉素
bicyclooctane   二环辛烷
bicyclo[4.2.0] octane   二环[4.2.0]辛烷 C_8H_{14}
bicycloparaffin   二环脂族烃
bicyclopentadienylidene(=fulvalene)   富瓦烯
biddiblack   比地黑〔天然颜料，采自比德弗德 Bideford〕
bidentate   ①二配位基；双配位基②(有)二齿的
bidentate chelate   二齿螯合物；双配位基螯合物
bidentate ligand   二齿配位体
bidesyl   联二苯乙酮基 $(C_6H_5COCHC_6H_5)_2$
bidirectional   双向的
bidirectional chromatography   两向色谱(法)
bidirectional fabric   双向织物
bidirectional pig   双向清管器
Bieeker method   布利克法〔电解还原钒化合物的方法〕
bi-electrode amperometric flow-through detector   双电极电流流通型检测器
Biemann's shift rule   比曼匀移规则
Biemann-Watson separator(=fritted glass separator)   比曼-瓦特森分离器；多孔玻璃分离器
Bierbaum scratch hardness   比尔鲍姆划痕硬度
bifendate   联苯双酯〔药〕
biferrocene(=biferrocenyl)   联二茂铁

biferrocenyl(=biferrocene)   联二茂铁
bifid   叉形的；裂成两半的
bifilar   双股的；双线的；双(灯)丝的
bifilar bridge   双线电桥
biflex fuel pump   燃料进料泵
biflorin   双花母草素
bifluorenyl(=difluorenyl)   联芴(基) $C_{26}H_{16}$
bifluoride   二氟化合物
bifocal   双焦点的
bifocus   双焦点
biformin   双形蕈素
biforminic acid   双形蕈酸
biformyl(=ethanedial)   乙二醛
bi-fuel propellant   二元燃料组成的火箭推进剂
bi-fuel system   二元燃料系统〔火箭用〕
bifumarate   富马酸氢盐(或酯)   COOHCH=CHCOOM; COOHCH=CHCOOR
bifunctional antibody   双功能抗体
bifunctional catalyst   双官能催化剂*；双功能催化剂
bifunctional exchanger   双官能交换剂
bifunctional extractant   双官能萃取剂
bifunctional initiator   双官能引发剂*
bifunctional ion exchanger   双官能离子交换剂
bifunctionality   双官能度
bifunctional molecule   双官能分子
bifunctional monomer   双官能(基)单体*
bifunctional oligomer   双官能低聚物
bifunctional reactivity   双官能反应性
bifunctional structural unit   双官能团结构链节
bifurcation   ①分枝；分叉②分叉点
bigarade oil   酸橙花油
bigarade orange(=sour orange)   酸橙；苦橙
big batch   大批量；工业批量
big blow mo(u)lding   大型吹塑成型
bigelovin   锦菊素
big end   连杆大头
big gear wheel(=bull wheel)   主齿轮；大齿轮
big inch line   大直径管线(管路)
big inch pipe   大直径管线(管路)
biglutarate   ①戊二酸氢盐②戊二酸氢酯
big repair   大(检)修
bigroot   大根老鹳草；大根香叶〔俗〕
bigtooth aspen   大齿白杨
biguanide   双胍；缩二胍 $NH_2C(=NH)NHC(NH_2)=NH$
bihexyl   联己基
bihoromycin   比奥罗霉素
biindolyl   联吲哚
biindoxyl   联β-羟吲哚

biindyl 联茚
bi-ion active agent 两性离子(表面)活性剂
bi-ionic potential 异种离子间膜电位
Bijvoet pair 毕伐依特衍射对
bikaverin 比卡菌素
bikhaconitine 白乌头碱
bilateral aliphatic chain 对称脂族链
bilateral effect 双侧效应
bilateral fibre 并列型(双组分)纤维;双侧型纤维
bilateral structure 双面结构
bilateral vulcanization 双面硫化
bilayer 双层
bilayer structure 双层结构*
bilberry-seed oil 覆盆子油
bilbery 欧洲越橘
bile pigment(=gall pigment) 胆汁色素
bilge pipe ①船底水管②污水管
bilge pump 舱底泵
bilge truss hoop 舱底中间箍;舱底中间轴环;舱底中间筒夹
bilinear drift correction 双线性漂移校正
bilinear rotational decoupling pulse sequence 双线性旋转去耦脉冲序列
Biller-Biemann algorithm 比勒-比曼算法
billet ①胶粒〔压出用〕②铅
billet cutter (肥皂)切条(块)机
billet cutting (肥皂)切条(块)
billet forging 锭料煅塑
billiard-ball model 台球模型*
billion ①十亿;10^9〔美制〕②万亿;10^{12}〔英制〕
billion becquerel(BBq;GBq) 十亿贝可勒尔〔=10^9Bq〕
billisecond 毫微秒〔等于10^{-9}秒〕
Billiter diaphragm cell 比利特隔膜电池
Billiter-Leykam cell 比利特-来卡姆隔膜电池
Billiter-Siemens cell 比利特-西门子隔膜电池
bill of materials 材料单
billow forming 气胀成型
billowing 波浪形
Billroth's mixture 比耳罗思混合剂〔一种麻醉剂〕
bilobanone 白果酮
bilobol 银杏酚 $C_{21}H_{34}O_2$
bimalate 苹果酸氢盐(或酯) $HOOCCHOHCH_2COOM$; $HOOCCHOHCH_2COOR$
bimaleate 马来酸氢盐(或酯) $HOOCCH=CHCOOM$; $HOOCCH=CHCOOR$
bimalonate 丙二酸氢盐(或酯) $HOOCCH_2COOM$; $HOOCCH_2COOR$
bimetal 双金属;复合钢材

bimetal lamp 双金属灯
bimetallic 双金属(的)
bimetallic catalyst 双金属催化剂
bimetallic electrode method 双金属电极法
bimetallic electrodes 双金属电极*
bimetallic mechanism 双金属机理
bimetallic μ-oxo alkoxides catalyst μ-氧桥双金属烷氧化物催化剂
bimetallic strip 双金属片
bimetallic-strip thermostat 双金属条恒温器
bimetallic temperature regulator 双金属温度调节器
bimetallic temperature sensor 双金属热敏元件
bimetallic thermocouple 双金属热电偶
bimetallic thermometer 双金属温度计
bimetallic thermostat 双金属温度调节器
bimodal 双峰;双峰值
bimodal deformation 双峰形变
bimodal distribution 双峰分布
bimodal distribution curve 双峰分布曲线
bimodal distribution of defect 双峰值缺陷分布
bimodal molecular weight distribution 双峰型分子量分布
bimolecular 双分子的
bimolecular acid-catalyzed acyl-oxygen cleavage 双分子酸催化酰氧断裂*
bimolecular acid-catalyzed alkyl-oxygen cleavage 双分子酸催化烷氧断裂*
bimolecular base-catalyzed acyl-oxygen cleavage 双分子碱催化酰氧断裂*
bimolecular base-catalyzed alkyl-oxygen cleavage 双分子碱催化烷氧断裂*
bimolecular collisions 双分子碰撞
bimolecular combination 双分子结合
bimolecular disproportionation 双分子歧化
bimolecular electrophilic substitution 双分子亲电取代*
bimolecular elimination 双分子消除*
bimolecular elimination reaction 双分子消除反应
bimolecular elimination through the conjugate base 双分子共轭碱消除*
bimolecular elimination with formation of a carbonyl group 双分子羰基形成消除*
bimolecular law 双分子反应定律
bimolecular leaflets 双分子层片
bimolecular mechanism 双分子反应机理
bimolecular nucleophilic substitution 双分子亲核取代反应
bimolecular nucleophilic substitution (with allylic rearrangement) 双分子亲核取代(含烯丙型重排)*
bimolecular quenching 双分子猝灭

bimolecular reaction 双分子反应*
bimolecular reaction of cracking 双分子裂化反应
bimolecular reduction 双分子还原*
bimolecular termination 双分子终止*
bimolecular termination reaction 双分子(链)终止反应
bin 库；仓
-bin〔词尾〕 宾〔生物碱〕
bin aging ①储存老化②胶料停放
binaphthalene 联二萘 $C_{10}H_7·C_{10}H_7$
binaphthol 联萘酚
binaphthyl 联萘
binarite 白铁矿
binary 二元的
binary acid 二元酸
binary alcohol 二元醇
binary alloy 二元合金
binary catalyst 二元催化剂
binary compound 二元化合物
binary copolymer 二元共聚物
binary copolymerization 二元共聚(作用)；二元共聚合
binary copolymer of chloroprene rubber 二元共聚氯丁橡胶
binary counter 二进制计数器
binary electrolyte 二元电解质；二元电解(溶)液
binary eluent 二元洗脱液
binary emulsion 二元乳状液
binary eutectic 二元低共熔物
binary granite 双云母花岗石
binary initiator 二元引发剂；双成分系引发剂
binary liquid mixtures 二元液体混合物
binary mixture 二元混合物
binary molecule 二元分子
binary oxide catalyst system 二元氧化物催化剂体系
binary phase diagram 二元相图
binary salt 二元盐
binary solution 二元溶液
binary solvent system 二元溶剂系统
binary system 二元体系
binary system ceramics 二元系陶瓷
binary vapour cycle 双汽循环
binary viscosity 二元黏度
bin cure 自硫(化)*
bin curing(=shelf curing) 存放自硫化
bindability 黏合性
bind agent ①黏合剂；接合剂②载色剂
binder ①黏合剂；黏料②夹子
binder course 沥青路面的中间层
binder efficiency 基料效率
binder extraction 基料抽出〔油漆调稀故障〕
binder fibre 黏合(用)纤维；纤维状黏合剂
binder granules 粒状黏合剂
binder index 基料指数
binder precipitation 基料沉淀〔油漆调稀故障〕
binder ratio 颜料/基料比；颜基比
binder resin 黏结剂用树脂
binder-rich (letdown) vehicle 富基料的(调)漆料；高树脂分(调)漆料
binder's board 书皮纸板
binder yarn 固纱
binding 束缚
binding ability 结合(内聚)能力；键合能力
binding agent ①载色剂②接合剂③黏合剂
binding coal 结块煤；黏结性煤
binding energies per nucleon(b.e.p.n) 每核子平均结合能
binding energies per proton(b.e.p.p.) 每质子平均结合能
binding energy 结合能
binding energy of nuclei 原子核结合能
binding force 结合力
binding isotherm 结合等温线
binding layer 黏合层
binding material 黏合剂
binding post 接线柱
binding power 黏合力
binding primer 黏结性底漆；黏合性底漆
binding site 结合部位*；结合点
binding tape method 结扎纱带(试验)法〔测润湿力〕
bin discharger 料斗排料口
bin gate 仓门
Bingham body 宾厄姆体*
Bingham flow 宾厄姆流动
Bingham fluid 宾厄姆流体〔塑性流体的一种〕
Bingham liquid 宾厄姆流体；塑性流体
Bingham plastic 宾厄姆塑性；塑性流体
Bingham plastic fluid 宾厄姆塑性流体
Bingham plasticity 宾厄姆塑性
Bingham plastics 宾厄姆塑性体；宾厄姆塑料
Bingham solid 宾厄姆固体〔可塑性固体物〕
Bingham viscometer 宾厄姆黏度计
Bingham yield value 宾厄姆屈服值
bin hang-up 久留仓中物
binifer 双引发-转移剂*
binitro- 二硝基
bin life 储存期
binnite 炭黝铜矿
binodal (有)双节的；双结点的
binodal solubility curve 双结点溶解度曲线

binode ①双结点②双阳极的
binodial 稳定单相极限线
binomial distribution 二项式分布*
binomial series 二项级数
binomial variance 二项式分布方差
binoxalate ①草酸氢盐②草酸氢酯
binoxide(=dioxide) 二氧化物
bin-pressure test 仓压力试验〔测粉体内摩擦角〕
bin stability 储存稳定性；耐储性
bin stock 储料
bin storage 储存罐储藏
bin-stored material 罐存材料
binuclear aromatics 双核芳烃
binuclear complex 双核配合物
binuclear molecule 双核分子
bioactivation 生物激活
bioactive additive 生物活性添加剂
bioactive polymer 生物活性高分子*
bioactivity monitor 生物活性监测器
bioaffinity sensor 生物亲和传感器
bioanalysis 生物分析(法)
bioassay 生物测定
bioassay method 生物测定法
bioassay of pesticide residue 农药残留量生物测定
bioautography 生物自显影法*；微生物显影
bioavailability 生物可利用性
biobarrier 生物屏蔽层；生物保护膜
biobol 银杏二酚
biocatalytic membrane electrode 生物催化膜电极
biocatalyzing electrode 生物催化电极
biochemical(=physiochemical) 生(物)化(学)的；生(理)化(学)的
biochemical analysis 生物化学分析
biochemical cell 生物化学电池
biochemical degradation 生化降解
biochemical fuel cell 生物燃料电池*
biochemical oxygen demand(BOD) 生化需氧量
biochemical oxygen demand sensor 生化需氧传感器
biochemical purification 生化净化
biochemical separation 生化分离
biochemical treatment process 生化处理法
biochip 生物芯片
biochromatography 生物色谱法
biocide 杀生物剂；抗微生物剂
biocompatibility 生物相容性
biocompatible fibre 生物相容性纤维；生物相适应性纤维
bioconcentration 生物浓缩

bioconversion 生物转化
biocopolymer 生物共聚物
bioctyl(=hexadecane) 联辛基；十六(碳)烷
biocytin 生物胞素；N-生物素酰-L-赖氨酸
biodegradable polyester 生物裂解聚酯
biodegradable polymer 生物降解高分子
biodegradation 生物降解
biodegradation of polymer 聚合物生物降解
biodosimetry 生物剂量学
bioelastic 生物弹性体
bioelastomer 生物弹性体
bioelectricity 生物电流
bioelectrode 生物电极
bioerodable polymer 生物可蚀性高分子
biofergate 葡萄糖酸铁
biofibre 生物纤维（特指由微生物如细菌产生的纤维）
bio-flocculation process 生物絮凝法
biofuel 生物燃料；生物柴油
biofunctional luminescence 生物机能性发光
biofunctional rubber 生物功能橡胶
biogas generating pit 沼气池
biogel 凝胶过滤剂〔聚丙烯酰胺凝胶〕
biogeochemical cycle 生物地球化学循环
bioinformatics 生物信息学
bioinorganic polymer 生物无机高分子
biological accumulation 生物积累
biological activity detection 生物活性检测
biological affinity sensor 生物亲和型传感器
biological agent 生物制剂
biological assay 生物测定
biological chemical degumming 生物-化学联合脱胶
biological chromatography 生物色谱
biological cleaning(=biological purification) 生物净化
biological coagulation 生物凝固〔胶乳〕
biological decomposition 生物分解(作用)
biological degradability 生物降解性
biological degradation 生物降解作用
biological degumming 生物脱胶
biological depollution 生物(学)去污染(法)
biological depolymerization agent 生物解聚剂
biological diversity 生物多样性
biological effect 生物效应
biological efficacy 生物效能
biological electrode 生物电极
biological enrichment 生物富集
biological film 生物膜
biological filter 生物滤池
biological filtration process 生物过滤法

biological functional material immobilization　生物功能物质的固定化
biological half-life　生物半衰期*
biological imitation fibre　仿生物纤维
biological index of water pollution　水污染生物指数
biological indicator method　生物学指示法
biologically active agent　生物活性剂
biological magnification　生物放大
biological mass spectrometry　生物学质谱法
biological membrane　生物膜
biological metabolizing sensor　代谢型生物传感器
biological monitoring for atmospheric pollution　大气污染的生物监测
biological monitoring method of pollution　污染的生物监测方法
biological nitrogen fixation　生物固氮
biological oxidation process　生物氧化法
biological oxygen-consumption sensor　生物耗氧传感器
biological oxygen demand(BOD)　生化需氧量；生化耗氧量
biological plastics　生物塑料
biological polymer　生物聚合物
biological product　生物制品
biological proofing　杀虫处理
biological purification(=biological cleaning)　生物净化
biological redox standard state　生物氧化还原标准状态
biological response modifier　生物学反应修饰剂
biologicals　生物制品；生物制剂
biological standard　生物学标准品
biological-tolerance polymer　生物相容高分子
biological treatment　生物处理
biological visualization procedure　生物观测法
biologic assay　生物检定(法)
bioluminescence　生物发光
bioluminescence immunoassay　生物发光免疫分析
bioluminescence sensor　生物发光传感器
biolysis　生物分解(作用)
biomacromolecule　生物高分子；生物大分子
biomarker　生物标记分子
biomass　生物量
biomass energy　生物质能
biomaterial　生物材料
biomedical mass spectrometry　生物医学质谱法
biomedical polymer　生物医用高分子；生物医药用聚合物
biomembrane　生物膜
biomembrane electrode　生物膜电极
biometry　①生物统计学②生物测量学

biomimetic polymer　仿生高分子
biomimetic recognition　仿生模拟识别
biomimetic sensor　仿生传感器
biomimic electrode　仿生传感器
biomolecule　生命分子；原生质
bionics of polymer　高分子仿生学
biophile element　亲生物元素
bioplastics　生物塑料〔由生物制成可被生物降解的塑料〕
biopolishing　生物抛光
biopolymer　生物高分子*
biopolymerization　生物聚合
biopreparate　生物制剂
biopurification　生物净化
bioreacter　生物反应器
bioreactive polymer　生物活性高分子
bioreactor　生物反应器
biorefractive　抗生物(降解)的〔表面活性剂〕
biorefractory　抗生物(降解)的
bioresistance　抗生物(降解)作用
bioresistant detergent　抗生物(降解)洗涤剂
biorheology　生物流变学
biorientation　双轴取向
biose　①乙糖　$C_2H_4O_2$②二碳糖
biosensor　生物传感器*；生物感应器
bioseparation　生物分离
biosorption process　生物吸附法
biospecific　生物特(异)性的
biospecific column　生物特异性柱
biospecific material　生物特(异)性材料
biosphere　生物圈
biostat　生物稳定剂
biostatistics　生物统计学
biosurfactant　生物表面活性剂
biosynthesis　生物合成
biotexin(=novobiocin)　新生霉素
biotic index of pollution　污染生物指数
biotin(=vitamin H)　生物素；维生素 H
biotin-avidin　生物素-抗生素蛋白复合物
biotite　黑云母
Biot number　毕奥准数
biotransformation　生物转化
bioviscoelastic　生物黏弹性的
bioxalate　①草酸氢盐②草酸氢酯
bioxindol　异靛蓝
bioxyl　氯化氧铋　(BiO)Cl
bipentene　双戊烯
Bipex process　铋派克斯过程〔即磷酸铋沉淀法〕
biphase　两相(的)

biphase current 双相电流
biphasic 双相的
biphenyl 联苯*
biphenyl acetadehyde 联苯基乙醛 $C_6H_5C_6H_4CH_2CHO$
biphenylene 亚联苯基 $—C_6H_4C_6H_4—$
biphenylene bisazo 亚联苯重氮基
　　$—N=NC_6H_4C_6H_4N=N—$
biphenylene isocyanate 异氰酸联苯酯
　　$C_6H_5C_6H_4N=CO$
biphenylene oxide 氧化联苯；二苯并呋喃 $(C_6H_4)_2O$
biphenylhydrazine 联苯基肼 $C_6H_5C_6H_4NHNH_2$
biphenyl isothiocyanate 异硫氰酸联苯酯 $C_{12}H_9NCS$
biphenyl mercury 二苯汞 $Hg(C_6H_5)_2$
biphenylyl 联苯基 $C_6H_5C_6H_4—$
biphenylyl carbonyl 联苯羰基 $C_6H_5C_6H_4CO—$
biphenylyloxy 联苯氧基 $C_6H_5C_6H_4O—$
2-(4′-biphenylyl)-6-phenyl-benzoxazole 2-(4′-联苯基)-6-苯基苯并噁唑 $C_6H_5C_7NOH_3C_6H_4C_6H_5$
biphthalate ①苯二甲酸氢盐②苯二甲酸氢酯
4,4′-biphthalic anhydride 联苯四甲酸二酐
bipimelate ①庚二酸氢盐②庚二酸氢酯
bipolar 双极的；偶极的
bipolar cell 双极细胞
bipolar electrode 双性电极
bipolarity 双极性
bipolar membrane 双极性膜；复合膜
bipolar molecule 双极性基分子
bipolaron 双极化子
bipolar staining 双极染色(体)；复合着色
bipolymer 二聚物；二元共聚物
bipotentiometric titration(=bipotentiometry) 双指示电极电位滴定(法)*
bipotentiometry(=bipotentiometric titration) 双指示电极电位滴定(法)*
bipropargyl 联丙炔 C_6H_6
bipropellant 二元组分火箭推进剂
bipropenyl 联丙烯
bipseudoindoxyl 靛蓝
bipyramid 双锥体
2,2′-bipyridine 联吡啶
2,2′-bipyridyl(=dipy) 2,2′-联吡啶 $C_5H_4NC_5H_4N$
biquinoline(=diquinoline) 联喹啉
2,2′-biquinoline(=biquinolyl) 联喹啉
biquinolyl(=2,2′-biquinoline) 联喹啉
biradical 双游离基；双(自由)基
biradical initiation 双游离基引发
biradical state 双游离基态
birch 桦木；桦树
birch bark tar 桦皮焦油
birch black 桦木(炭)黑
birch bud oil 桦芽油
birch camphor 桦木脑
birch carbon 桦木炭
birchen 桦木的；桦(树)的
birch oil 桦木油
birch-seed oil 桦木子油
birch tar 桦木焦油
birch tar oil 桦木焦油
birch water 桦木汁
bird dung bate 鸟粪软化
Bird (film) applicator 伯德式涂膜器
bird lime 粘鸟胶；山车胶
bird's foot oil 鸟足油；假地豆油
Bird-Young filter 伯德-杨过滤器〔一种圆筒形连续过滤器〕
birectification 双(重)精馏(法)
birectifier 双(重)精馏器
bireflectance ①双(向)反射比②双反射
bireflection ①双反射②反射多色(现象)
birefracting 双(重)折射的
birefraction 双折射
birefractive 双折射的
birefringence 双折射
birefringence of flow 流动双折射
birefringence relaxation 双折射松弛
birefringent 双(重)折射的
birefringent polarizer 双折射起偏振器
birefringent texture 双折射织构
Birkeland-Eyde furnace 伯克兰-艾迪电炉
birotation 双旋光
bi-rotor pump 双转子泵
birotor winder 双转子卷绕头；双旋翼卷绕头
biruea 联二脲 $(NH_2CONH)_2$
biruet 缩二脲 $NH(CONH_2)_2$
bis- 〔词头〕两个；双
bisabolane 红没药烷；防风根烷
bisabolene 红没药烯；防风根烯
bisabol myrrh oil(=opopanax oil) 红没药油；防风根油〔商〕
bisabolol 红没药醇；防风根醇
bisacetaldehyde-oxalyldihydrazone 双乙醛草酰二腙
bis(4-acetoxyphenyl) ether 双(4-乙酸基苯基)醚
bisacetylacetone-propylenediimine 双乙酰丙酮合亚丙基二亚胺
bisacki 颗粒紫胶级；粗紫胶级
bis-acrylic 双丙烯酸的
bis (alkylcyclooctatetraenyl) actinide compound 双(烷基代环辛四烯)合锕系元素化合物

bis-*N*-allyl pyromellite imide　二-*N*-烯丙基均苯四酰二亚胺；均苯四酰二烯丙(基)亚胺
bisamide　双酰胺
bisamination　双氨基化*
bis(4-aminocyclohexyl) methane　双(4-氨基环己基)甲烷
bis(4-aminophenyl) disulfide　二硫化双(4-氨基苯酚)
bis-4-aminophenyl sulfone　双对氨基苯砜
bisanhydrorutilantinone　双脱水鲜红菌素酮
bis-arylation　双芳基化(作用)
bisatin　双醋酚汀
bisazimethylene　联氮乙烯基；双氮亚甲基
bisazo(=tetrazo; bisdiazo)　双偶氮　(—N=N—)$_2$
bisazobenzil　双偶氮甲苯
bisazochromotropic acid　双偶氮变色酸
bis-azo compound　双偶氮化合物
bisbenzimidazole　二苯并咪唑
bisbenzothiazole　二苯并噻唑
N, N'-bis(2-benzothiazolethio) cyclohexylamine　*N,N'*-双(2-苯并噻唑硫代)环己胺〔促进剂〕
1,3-bis(2-benzothiazyl mercaptomethyl)urea　1,3-双(2-苯并噻唑硫醇基甲基)脲〔促进剂〕
bis (benzoxazolyl) derivative　双(苯并噁唑)衍生物〔荧光增白剂〕
bis-benzyl hexamethylenediamine　双苄基六亚甲基二胺〔防老剂〕
bis(beta-hydroxyethyl)-gammaaminopropyl-triethoxy silane　双(β-羟乙基)-γ-氨丙基三乙氧基甲硅烷〔偶联剂〕
bis(4-*t*-butylcyclohexyl) peroxy dicarbonate　过氧化二碳酸双(4-叔丁基环己)酯〔引发剂〕
bis(*n*-butylcyclooctatetraenyl) uranium　双(正丁基环辛四烯)合铀　U(C$_8$H$_7$C$_4$H$_9$)$_2$
1,1-bis(*t*-butylperoxy) cyclohexane　1,1-双(叔丁基过氧基)环己烷〔硫化剂〕
bis(*t*-butylperoxy) diisopropylbenzene　双(叔丁基过氧基)二异丙苯〔硫化剂〕
2,5-bis(*t*-butylperoxy)-2,5-dimethylhexane　2,5-双(叔丁基过氧基)-2,5-二甲基己烷〔硫化剂〕
1,4-bis(*t*-butylperoxy) isopropylbenzene　1,4-双(叔丁基过氧基)异丙苯〔硫化剂〕
1,1-bis(*t*-butylperoxy)-3,3,5-trimethyl cyclohexane　1,1-双(叔丁基过氧基)-3,3,5-三甲基己烷〔硫化剂〕
Bischler-Napieralsky synthesis　毕世勒-纳匹拉尔斯基合成法
bis-*p*-chlorobenzoyl peroxide　过氧化双对氯苯甲酰〔硫化剂〕
bis(2-chloroethyl) ether(BCEE)　二(2-氯乙基)醚 (ClC$_2$H$_4$)$_2$O

bis (chloromercuri) ferrocene　二(氯化汞代)二茂铁
bis(*p*-chlorophenoxy) methane　双对氯苯氧基甲烷
2,2-bis(*p*-chlorophenyl)-1,1-dichloroethane　2,2-双对氯苯基二氯乙烷
1,1-bis(*p*-chlorophenyl)-ethanol　1,1-双对氯苯基乙醇
bischofite　水氯镁石
biscuit fire　初次焙烧
biscuit furnace　坯炉
biscuit kiln　坯窑
biscuit porcelain　素磁〔陶〕
biscuits　紫胶二氧化硅饼〔唱片的原料〕
biscyanate ester　双氰酸酯
biscyclohexanone oxalyldihydrazone　双环己酮草酰二腙；铜试剂
biscyclooctadiene　双环辛二烯
bis-cyclooctatetaenyl-uranium　双环辛四烯合铀　U(C$_8$H$_8$)$_2$
bis (cyclopentadienyl) vanadium　双(环戊二烯基)钒
bis (dialkoxyphosphinyl) alkane　双(二烷氧基氧膦基)烷烃　(RO)$_2$PO(CH$_2$)$_n$PO(OR)$_2$
bis (dialkylphosphinyl) alkane　双(二烷基氧膦基)烷烃　R$_2$PO(CH$_2$)$_n$POR$_2$
bisdiazo(=tetrazo; bisazo)　双偶氮　(—N=N—)$_2$
bis-diazo compound　双重氮化合物
bis-diazotized benzidine　双重氮联苯胺
N,N-bis(3,5-di-*t*-butyl-4-hydroxybenzyl) methylamine　*N,N*-双(3,5-二叔丁基-4-羟基苄基)甲胺〔稳定剂〕
4,4'-bis(2,6-di-*t*-butyl phenol)　4,4'-双(2,6-二叔丁基苯酚)〔防老剂〕
bis(2,4-di-*t*-butylphenyl) pentaerythritol diphosphite　二亚磷酸双(2,4-二叔丁基苯基)季戊四醇酯〔防老剂〕
bis (dibutylthiocarbamyl) disulfide　二硫化双(二丁基硫代氨基甲酰)〔硫化剂〕
bis(2,4-dichlorobenzoyl) peroxide　过氧化双(2,4-二氯苯甲酰)〔硫化剂〕
bis (diethylthiocarbamyl) disulfide　二硫化双(二乙基代氨基甲酰基)〔硫化剂〕
bis (diethylthiophosphoryl) trisulfide　双(二乙基硫代磷酰基)三硫化物〔硫化剂〕
bisdiguanide　双联胍〔药〕
bis-di-*n*-hexylphosphinylmethane　双二正己基膦酰代甲烷；亚甲基双(二正己基氧膦)　(C$_6$H$_{13}$)$_2$(O)PCH$_2$P(O)(C$_6$H$_{13}$)$_2$
2,6-bis(2,4-dihydroxy benzyl)-4-chlorophenol　2,6-二(2,4-二羟苄基)-4-氯苯酚〔胶黏剂〕
bis-(diisopropyl thiophosphonyl) disulfide　双(二异丙基硫代膦酰基)二硫化物〔促进剂〕
3,3-bis (*p*-dimethylaminophenyl) phthalide　3,3-双(对二

甲基氨苯基)苯并呋喃酮

1,3-bis (dimethylaminopropyl) thiourea　1,3-双(二甲氨基丙基)硫脲〔防老剂〕

N,N-bis (dimethylaminotrimethylene) thiourea　N,N-双(二甲基氨基三亚甲基)硫脲〔促进剂〕

4,4-bis (α,α-dimethylbenzyl) diphenylamine　4,4-双(α,α-二甲基苄基)二苯胺〔防老剂〕

N,N-bis (1,4-dimethylbutyl)-p-phenylenediamine　N,N-双(1,4-二甲基丁基)对苯二胺〔防老剂〕

bis (2,6-dimethyl-4-heptyl) phosphoric acid　双(2,6-二甲基-4-庚基)磷酸　$(C_9H_{19}O)_2PO(OH)$

N,N-bis (1,4-dimethylpentyl)-p-phenylenediamine　N,N-双(1,4-二甲基戊基)对苯二胺〔防老剂〕

bis (dioctyl pyrophosphate) ethylene titanate　二(焦磷酸二辛酯)钛酸乙二醇酯

bis (dioctyl pyrophosphate) oxyacetate titanate　二(焦磷酸二辛酯)钛酸羟乙酸酯

bis-dodecylisopropylthiuram disulfide　二硫化双十二烷基异丙基秋兰姆〔促进剂〕

bisebacate　①癸二酸氢盐　$COOH(CH_2)_8COOM$ ②癸二酸氢酯　$COOH(CH_2)_8COOR$

bisecting conformation　等分构象*

bisenamines　双烯胺

bisepoxides (=diepoxides)　双环氧化合物

bis (2,3-epoxypropoxy) ethane　双(2,3-环氧丙氧基)乙烷〔纤维交联剂〕

bisergostadienol　联麦角(甾)醇

bis-ethoxyethoxyethyl phthalate　邻苯二甲酸双乙氧基乙氧乙酯

bis (2-ethyihexyl) phthalate　邻苯二甲酸双(2-乙基己)酯

bis (ethylacetoacetate) oxalyldihydrazone　双(乙酰乙酸乙酯)草酰二腙

bis (2-ethylamino-4-diethylamino-6-triazinyl) disulfide　二硫化双(2-乙氨基-4-二乙氨基-6-三嗪基)〔促进剂〕

bis (ethylcyclooctatetraenyl) uranium　双(乙基环辛四烯基)合铀　$U(EtCOT)_2;U(C_8H_7C_2H_5)_2$

bis (2-ethylhexyl) phosphoric acid　二(2-乙基己基)磷酸　$[C_4H_9CH(C_2H_5)CH_2O]_2PO(OH)$

N,N-bis (1-ethyl-3-methyl-pentyl)-p-phenylenediamine　N,N-双(1-乙基-3-甲基戊基)对苯二胺〔防老剂〕

bisethylxanthate　双黄原酸乙基酯

bis-ethyl xanthogen disulfide (BXD)　二硫化双乙基黄原酸酯

bis-exchange polymerization　双交换聚合

bisfuranyl-18-crown-6　二呋喃并 18-冠(醚)-6　$C_{12}O_6H_{16}(C_2H_2)_2$

bis-(2-furfuryl)-disufide　二硫化双(2-糠基)

N,N-bis (furfurylidene) acetone　N,N-双亚糠基丙酮〔交联剂〕

N,N-bis (furfurylidene) hexamethylenediamine　N,N-双亚糠基己二胺〔硫化剂〕

bisgallia acid　双棓酸

bis-glycol terephthalate　对苯二甲酸双乙二(醇)酯；对苯二甲酸双羟乙酯

bisglyoxaline　联咪唑

bis (hexamethyldisilylamine) zinc (II)　双(六甲基二甲硅烷氨基)锌(II)

bis (hexamethylene) triamine　双-(六亚甲基)三胺；二(己撑)三胺

bishomo-γ-linolenic acid　双高-γ-亚麻酸；8,11,14-二十碳三烯酸

bishrinkage yarn　异收缩(变形)丝；异收缩(变形)纱

bishydrazibenzil　联亚肼基甲苯

bishydrazicarbonyl　环二脲

bishydrazide　双酰肼

1,3-bis-(3-hydroxy-4-benzoyl phenoxy)-2-propanol　1,3-双(3-羟基-4-苯甲酰基苯氧基)-2-丙醇〔光稳定剂〕

2,6-bis(2-hydroxy-3-t-butyl-5-methylbenzyl)-4-methyl-phenol　2,6-双(2-羟基-3-叔丁基-5-甲基苄基)-4-甲基苯酚〔防老剂〕

bishydroxycoumarin　双(羟)香豆素

bis(1-hydroxycyclohexyl)peroxide　过氧化二(1-羟基环己基)〔催化剂〕

bis(4-hydroxy-3,5-di-t-butyl phenoxy)-2-n-octylthio-1,3,5-triazine　双(4-羟基- 3,5-二叔丁基苯氧基)-2-正辛硫基-1,3,5-三嗪〔抗氧剂〕

1,4-bis(2-hydroxyethoxybenzene)　1,4-双(2-羟乙氧基苯)〔交联剂〕

N,N'-bis(2-hydroxyethyl) allylamine　N,N-双(2-羟乙基)烯丙胺〔硫化剂〕

bis (hydroxyethyl) azelate　壬二酸双羟乙酯

bis-β-hydroxyethyl terephthalate　对苯二甲酸双-β-羟乙酯

bis-(hydroxy methyl) urea　二羟甲基脲

bis (hydroxyphenyl) carbonate　碳酸双(羟苯基)酯

1,1'-bis(4-hydroxyphenyl) cyclohexane　1,1'-双(4-羟苯基)环己烷〔防老剂〕

2,2'-bis(4-hydroxyphenyl) propane　2,2'-二(4-羟基苯)丙烷

N,N-bis(2-hydroxypropyl aniline　N,N-双(2-羟丙基苯胺)〔交联剂〕

1,4-bis(2-hydroxypropyl)-2-methyl piperazine　1,4-双(2-羟丙基-2-甲基哌嗪)〔硫化剂〕

bisilicate　焦硅酸盐；(一缩)二硅酸盐

bisindenyl　联茚　$C_9H_7C_9H_7$

bis(1-isobutyl isopentyl) phosphate　磷酸双(1-异丁基异戊基)酯

bismal　比斯马耳〔药〕

bismaleimide 双马来酰亚胺
bismaleimide-triazine resin 双马来酰亚胺三嗪树脂
Bismarok brown(=triaminoazobenzene) 三氨基偶氮苯；碱性棕；俾斯麦棕 $NH_2C_6H_4N=NC_6H_3(NH_2)_2$
bismarsen 铋砷剂〔药〕
bis(methacryloxyethyl phosphate) 磷酸二(甲基丙烯酰氧乙基酯)
N,N'-bis (methoxymethyl) urea N,N'-二甲氧甲基脲
N,N'-bis (methoxymethyl) uron(e) N,N'-二甲氧甲基脲酮
2,2-bis(p-methoxyphenyl)-1,1,1-trichloroethane 2,2-双甲氧苯基-1,1,1-三氯乙烷
bis-(2-methyl-3-furyl) tetrasulfide 双(2-甲基-3-呋喃基)四硫
N,N-bis-(1-methyl-heptyl)-p-phenylenediamine 对称双甲基庚基对苯二胺
2,2′-bis(3-methyl-4-hydroxyphenyl) propane 2,2′-双(3-甲基-4-羟基)丙烷〔防老剂〕
bis(2-methyl-2-pyridyl) glyoxaldihydrazone 双(2-甲基-2-吡啶基)乙二醛二腙
bismite 铋华
bismoidz 铋悬液
bismon 胶态氢氧化铋
bismuth 铋〔83号元素,化学符号Bi〕
bismuth acetate 乙酸铋 $Bi(C_2H_3O_2)_3$
bismuth alloy 铋合金
bismuth benzoate 苯甲酸铋 $Bi(C_6H_5COO)_3$
bismuth black 铋黑
bismuth bromide 三溴化铋 $BiBr_3$
bismuth bronze 铋青铜合金
bismuth camphorate 樟脑酸铋 $Bi(C_{10}H_{14}O_4)_3$
bismuth carbolate(=bismuthyl phenolate) 苯酚二羟铋 $Bi(OH)_2OC_6H_5$
bismuth carbonate(=bismuthyl carbonate) 碳酸氧铋 $(BiO)_2CO_3$
bismuth cesium photo cathode 铋-铯光电阴极
bismuth chloride 三氯化铋 $BiCl_3$
bismuth chromate 铬酸铋
bismuth cinnamate 肉桂酸铋 $Bi(C_6H_5CH=CHCOO)_3$
bismuth citrate 柠檬酸铋 $BiC_6H_5O_7$
bismuth dimethyl dithiocarbamate 二甲基二硫代氨基甲酸铋
bismuth furnace 熔铋炉
bismuth glance 辉铋矿
bismuth glycolyl arsanilate 乙醇酰对氨基苯胂酸铋；乙醇酰阿散酸
bismuth hydroxide 氢氧化铋 $Bi(OH)_3;BiO(OH)$
bismuthic ①铋的②五价铋的
bismuthic compound 五价铋化合物

bismuthic oxide 五氧化二铋 Bi_2O_5
bismuthide 铋化物
bismuthiferous 含铋的
bismuthine ①胺；三氢化铋②辉铋矿③胺基 H_2Bi-
bismuth iodate 碘酸铋
bismuthiol 铋试剂
bismuthite(=bismutite) 泡铋矿
bismuth lactate 乳酸铋 $Bi(C_3H_5O_3)_3$
bismuth meal 铋粉
bismuth molybdate 钼酸铋 $Bi_2(MoO_4)_3$
bismuth nickel 铋镍矿
bismuth nitrate 硝酸铋 $Bi(NO_3)_3$
bismuth ochre 赭铋矿
bismuth oleate 油酸铋 $Bi(C_{18}H_{33}O_2)_3$
bismuth organic compound 有机铋化合物
bismuthous 三价铋的
bismuthous compound 三价铋化合物
bismuthous oxide 三氧化二铋 Bi_2O_3
bismuth oxide 氧化铋 $Bi_2O_3;Bi_2O_4;Bi_2O_5$
bismuth oxychloride 氯氧化铋 $BiOCl$
bismuth oxyfluoride 氟氧化铋；氟化氧铋 $BiOF$
bismuth oxyiodide 碘氧化铋 $BiOI$
bismuth pentoxide 五氧化二铋 Bi_2O_5
bismuth perchlorate 高氯酸铋 $Bi(ClO_4)_3$
bismuth permanganate 高锰酸铋 $Bi(MnO_4)_3$
bismuth peroxide 四氧化二铋 $Bi_2O_2(O_2)$
bismuth phenylate 苯氧基二羟铋 $Bi(OH)_2OC_6H_5$
bismuth phosphate 磷酸铋 $BiPO_4$
bismuth phosphate process 磷酸铋(载体沉淀)过程
bismuth potassium citrate 枸橼酸铋钾〔药〕
bismuth potassium thiosulfate 硫代硫酸铋钾 $K_3[Bi(SO_3S)_3]$
bismuth selenide 三硒化二铋
bismuth silicate 硅酸铋 $Bi_2(SiO_3)_3;Bi_4Si_{13}O_{12}$
bismuth silver 铋银
bismuth soap 铋皂〔可用作催干剂〕
bismuth sodium thiosulfate 硫代硫酸铋钠 $Na[Bi(SO_3S)_3]$
bismuth subcarbonate 碱式碳酸铋 $(BiO)_2CO_3$
bismuth subgallate 碱式没食子酸铋 $Bi(OH)_2·C_7H_5O_5$
bismuth subnitrate 硝酸氧铋
bismuth subsalicylate 碱式水杨酸铋 $BiO·C_7H_5O_3$
bismuth sulfate 硫酸铋 $Bi_2(SO_4)_3$
bismuth sulfide 硫化铋 $BiS;Bi_2S_3$
bismuth sulfocarbonate 酚磺酸铋
bismuth tetroxide 四氧化二铋 Bi_2O_4
bismuth tribromide 三溴化铋 $BiBr_3$
bismuth trichloride 三氯化铋 $BiCl_3$

bismuth triethyl 三乙铋 $(C_2H_5)_3Bi$
bismuth trihydride(=bismuthine) 䥨；三氢化铋
bismuth trimethyl 三甲铋 $(CH_3)_3Bi$
bismuth trioxide 三氧化二铋 Bi_2O_3
bismuth triphenyl 三苯铋 $(C_6H_5)_3Bi$
bismuth trisulfide 三硫化二铋 Bi_2S_3
bismuth tungstate 钨酸铋 $Bi_2(WO_4)_3$
bismuth uranate 铀酸铋
bismuth violet 铋紫
bismuth white 铋白
bismuthyl 氧铋基 BiO—
bismuthyl bromide 溴化氧铋 (BiO)Br
bismuthyl carbonate(=bismuth carbonate) 碳酸氧铋 $(BiO)_2CO_3$
bismuthyl chloride 氯化氧铋 (BiO)Cl
bismuthyl dichromate 重铬酸氧铋 $(BiO)_2Cr_2O_7$
bismuthyl fluoride 氟化氧铋 (BiO)F
bismuthyl hydroxide 氢氧化氧铋 (BiO)OH
bismuthyl iodide 碘化氧铋 (BiO)I
bismuthyl nitrate 硝酸氧铋 $(BiO)NO_3$
bismuthyl phenolate(=bismuth carbolate) 苯酚二羟铋 $Bi(OH)_2OC_6H_5$
bismutite 泡铋矿
bisnaphthol 二萘酚
bis (p-nonylated phenol) phenol phosphite 双(对壬基苯酚)苯酚亚磷酯〔稳定剂〕
bisnorfenchone 双降小茴香酮
bisnorsteroids 联降甾族化合物
bis (octyldecylisopropyl) thiuram disulfide 二硫化双十八异丙基秋兰姆〔促进剂〕
bisoprolol 比索洛尔〔药〕
N,N'-bis (oxydiethylene) thiocarbamoyl sulfenamide N,N'-双(氧二亚乙基)硫代氨基甲酰次磺酰胺〔促进剂〕
bispecific antibody 双特异性抗体
bis (pentabromo phenyl) ether 双(五溴苯)醚〔阻燃剂〕
bis (pentaerythritol ester) 双季戊四醇酯
bis-pentamethylene thiuram tetrasulfide 四硫化双五亚甲基秋兰姆〔促进剂〕
bisphenol A 双酚 A；2,2-双对羟苯基丙烷
bisphenol A bis(2-diazido-naphthoquinone(1,2)-5-sulfonate) 双酚 A 双(2-重氮基萘醌(1,2)-5-磺酸酯)
bisphenol A diglycidyl ether 双酚 A 二缩水甘油醚
bisphenol A disalicylate 水杨酸二双酚 A 酯 $(HOC_6H_4COOC_6H_4)_2C(CH_3)_2$
bisphenol A epoxide 双酚 A 环氧化物
bisphenol A epoxy resin 双酚 A 环氧树脂
bisphenol antioxidant 双酚类抗氧化剂
bisphenol A polycarbonate 双酚 A 聚碳酸酯

1,1′-bis (4-phenol) cyclohexane 1,1′-双(4-苯酚)环己烷〔防老剂〕
bisphenol resin 双酚树脂
bisphenol S 双酚 S
bisphenols 双酚类
1,6-bis(phenyl azomonocarbonate) 1,6-双(偶氮单碳酸苯酯)〔发泡剂〕
bis (phthalocyaninato) thorium 二酞菁合钍 $ThPc_2$；$Th(C_{32}H_{16}N_8)_2$
bis (phthalocyaninato) uranium 二酞菁合铀 UPc_2；$U(C_{32}H_{16}N_8)_2$
bis (piperidinothiocarbonyl) tetrasulfide 四硫化双(哌啶基硫羰基)〔促进剂〕
2,6-bis(2-pyridyl)pyridine 2,6-双-(2-吡啶基)吡啶
bisque 素瓷〔陶〕
bis-N-salicyl hexamethylenediamine 双-N-水杨基己二胺〔硫化剂〕
bissubstitution 双取代作用
bistabilized polyester yarn 两次稳定的聚酯(变形)丝；聚酯低弹丝〔第一次指变形时的热定型，第二次指变形后的热处理〕
bistable 双稳态的
bistable laser diode 双稳激光二极管
bister(=bistre) 深褐色颜料〔取自木煤〕
2,2-bis (tetrabromo-4-hydroxyphenyl) propane 2,2-双(四溴-4-羟苯基)丙烷〔阻燃剂〕
bis(2,2,6,6-tetramethyl-4-piperidyl) sebacate 双(2,2,6,6-四甲基-4-哌啶基)癸二酸酯〔光稳定剂〕
bistetrazole 联四唑
bisthioether linkage 双硫醚键(合)
bisthioglycollic acid 双硫羰乙醇酸
bistolylene diisocyanate 二异氰酸二亚苄酯〔硫化剂〕
bistre(=bister) 深褐色颜料〔取自木煤〕
bis-trialkyltin alkylene dioxide 二氧化亚烷基二(三烷基锡)
bis (tri-arylyltin) maleate 顺丁烯二酸二(三芳基锡) $(=CHCOO)_2(SnAr_3)_2$
bis (tribenzyltin) maleate 顺丁烯二酸二(三苄基锡)
bis (tribromophenoxy) ethane 二(三溴代苯氧基)乙烷
bis (tribromophenyltin) fumarate 反-丁烯二酸二(三溴代苯基锡)
bis-tributyltin alkylene dioxide 亚烷基二氧化二(三丁基锡)
bis-tributyltin phenylene dioxide 亚苯基二氧化二(三丁基锡)
bis(3-triethoxy silylpropyl) tetrasulfide 四硫化双(3-三乙氧基硅烷基丙基)〔偶联剂〕
bis (triphenyltin) mesodibromo succinate 中(均)二溴丁

二酸二(三苯基锡)
bisuberate ①辛二酸氢盐 HOOC(CH$_2$)$_6$COOM②辛二酸氢酯 HOOC(CH$_2$)$_6$COOR
bisuccinate ①丁二酸氢盐 HOOC(CH$_2$)$_2$COOM②丁二酸氢酯 HOOC(CH$_2$)$_2$COOR
bisulfate 硫酸氢盐；酸式硫酸盐
bisulfate fusion 酸式硫酸盐熔融
bisulfide(=disulfide) 二硫化物
bisulfite cooking liquor 亚硫酸氢盐纸浆蒸煮液
bisvinyl 双乙烯基
bisvinyl macromonomer 双乙烯基大分子单体
bisymmetric Zeeman modulator 双对称型塞曼调制器
bit ①钻②刀片；刀头③锥④少量⑤二进制数⑥位
bitartrate ①酒石酸氢盐 HOOC(CHOH)$_2$COOM②酒石酸氢酯 HOOC(CHOH)$_2$COOR
bitartronate ①羟丙二酸氢盐 HOOCCHOHCOOM②酸式羟丙二酸氢酯 HOOCCHOHCOOR
bite(=nip) ①滚距②黏着③溶剂侵蚀
bite thickness 滚距厚度
bithionol 硫氯酚〔药〕
bithiophene 并噻吩
biting-in 腐蚀
bitruder 双螺杆挤出机
bitstone 碎硅石
bitter acid 苦酸
bitter almond camphor 苯偶姻；二苯乙醇酮；苦杏仁脑 C$_6$H$_5$CHOHCOC$_6$H$_5$
bitter almond oil 苦杏仁油
bitter almond oil camphor 苯偶姻；二苯乙醇酮；苦杏仁油脑 C$_6$H$_5$CHOHCOC$_6$H$_5$
bitter almond water 苦杏仁水
bitter ash 苦梣
bitter earth 苦土；氧化镁
bittering ①苦味②加入苦味
bittern 盐卤；卤水
bitter orange(=sour orange) 酸橙；苦橙
bitter orange flower oil 苦橙花油
bitter orange leaf oil(=petitgrain bigarade oil) 苦橙叶油
bitter orange oil 苦橙油
bitter orange peer 苦橙皮
bitter principle 苦素
bitter root 苦根
bitter salt 泻盐；七水合硫酸镁 MgSO$_4$·7H$_2$O
bitter spar 白云石
bittiness 起(粗)粒；起块
bit tool 刀头；刀具
bitumastic 沥青的
bitumastic enamel(=bituminous paint) 沥青漆
bitumastic pipe-line coating 管路沥青涂料
bitumastic solution 水柏油
bitumen(=asphalt) (地)沥青
bitumen-based paint 沥青(基)漆
bitumen blowing process 沥青吹制法
bitumen-coated pipe 涂沥青的管
bitumen cutback 溶于石油馏出油的沥青
bitumen emulsion 沥青乳液
bitumen-impregnated 沥青浸渍的
bitumen limestone 沥青石灰石
bitumen plastic 沥青塑料
bitumen resin 沥青树脂
bitumen soluble 沥青可溶的
bitumen sprayer 沥青喷雾器
bitumination 沥青(固)化
bituminic 沥青(质)的
bituminiferous 含沥青的
bituminization 沥青固化*
bituminized fabric 沥青布
bituminized fibre pipe 沥青纤维管
bituminized hessian 绝缘带；电线包布
bituminized paper 沥青纸
bituminized waste 沥青(固)化废物
bituminizing 沥青浸渍
bituminous 沥青的
bituminous basecoat 沥青底漆
bituminous cement 沥青膏
bituminous coal 烟煤
bituminous coal fines 烟煤细粉
bituminous coating 沥青涂料
bituminous concrete 沥青混凝土
bituminous cutback 稀释的(石油)沥青；轻制沥青
bituminous emulsion 沥青乳化液
bituminous epoxy 环氧沥青
bituminous froth 沥青泡沫体
bituminous fuel 沥青质燃料
bituminous grout 沥青浆
bituminous grouting 沥青化
bituminous industry 沥青工业
bituminous limestone 沥青质石灰岩
bituminous material 沥青材料
bituminous paint(=bitumastic enamel) 沥青漆
bituminous peat 沥青泥煤
bituminous pipe coating 管子沥青涂料
bituminous pitch 沥青；柏油
bituminous primer 沥青底漆
bituminous resins 沥青树脂
bituminous retreat 二次沥青处理

bituminous rock　沥青岩
bituminous sand　沥青砂
bituminous sandstone　沥青砂岩；沥青砂石
bituminous shale　沥青页岩；油母页岩
bituminous substances　沥青(状)物质
bituminous-treated　沥青处理的
bituminous treatment　沥青处理
bituminous varnish　沥青清漆
bituminous wood　沥青木〔具有木质外形的褐煤〕
bitumite　烟煤
bitusol　固体分散胶溶沥青
biurate　酸式脲酸盐
biuret(=allophanamide)　缩二脲
biuret crosslink　缩二脲交联
biuret linkage　缩二脲键(合)
biuret method　双缩脲法
biuret reaction　缩二脲反应；双缩脲反应
biuret test　缩二脲试验
bivalence　二价
bivalent　二价的
bivalent atom　二价原子
bivalent cation　二价阳离子
bivalent hydrocarbon radical　二价烃基
bivalent radical　二价基
bivariant system　双变(度)物系
bivariate distribution　二元分布；二维分布
bivinyl rubber　1,3-丁二烯(基)橡胶
bixin　①胭脂树橙②红木素；胭脂素
bixol　胭脂树醇
bixture　夹具；卡具
bixylidene　联二甲苯胺；2,2,6,6-四甲基联苯胺
bixylyl　联二甲苯基；四甲联苯基
Bjorkman's lignin　磨木木素；贝克曼木素
black acids　黑色酸〔酸渣中所含的磺酸盐〕
Blackadder's method　涂刷(显色)法
black alum　黑矾
black analytical system　黑色分析系统
black antimony　黑锑
black ash　①黑灰②粗碱灰
black-ash ball(=black-ash cake)　①熔结粗碳酸钠②黑灰块
black ash furnace　黑灰炉
black ash liquor　黑灰液
black-ash revolver　黑灰旋转炉
black-ash vat　黑灰浸提桶
black-ash waste　黑灰废液
black backing varnish　(地沥青)黑底漆
black balsam　黑香胶

black batch　炭黑母炼胶
black batt　炭质页岩
black beech　黑假山毛榉
blackberry-seed oil　黑莓子油
blackboard paint　黑板漆
black body　黑体
black body coefficient　黑体系数
black body locus　黑体轨迹
black body radiant emittance　黑体辐射密度
black body radiation　黑体辐射
black box　黑箱
black boy　黑儿胶
black boy gum(=acaroid gum)　禾木胶
black burn　黑色燃烧
black carbon　炭黑
black cashmere　青绒
black center　黑心
black chalk　黑垩〔天然颜料〕
black check　树脂囊
black coal　黑煤
black coating　黑涂料
black collector　炭黑收集器
black compound　炭黑胶料
black content　黑色量；含黑量
black copper　黑铜矿
black copper manganite　亚锰酸铜黑〔耐高温颜料〕
black copper oxide(=cupric oxide;copper monoxide)　黑色氧化铜；一氧化铜　CuO
black core　黑心斑
black cottonwood　美国黑杨
black currant concrete　黑茶藨子花浸膏
black currant flower oil　黑茶藨子花油
black damar　黑玳玛
black dyes　黑色染料
black enamel　黑瓷漆
blackening　黑化
black factice　黑油膏
black fibre　碳纤维
black film　黑膜〔一种厚度极薄的平衡膜，由于不反射光线，呈黑色〕
blackfish jaw oil　海豚腭油
black fish oil　黑鱼油
black flux　粗碳酸钾
black glass　黑色玻璃
black grease　黑润滑膏
black-gum　美国紫树
black harness oil　黑色皮革油
black hematite　黑色赤铁矿

black incorporation time	炭黑混入时间
blacking	涂黑(法)；黑色涂料
blacking brush	涂黑刷
blacking machine	刷黑机
blacking mill	黑色涂料碾磨机
blacking mixer	黑色涂料混合机
blacking up	泛黑
black iron	黑铁板；平铁
black iron oxide	氧化铁黑
black iron oxide pigment	黑色氧化铁颜料
blackish green	墨绿色
blackish olive	深橄榄绿(色)
black jack	铁闪锌矿
black japan	(黑)沥青漆
black-koji mould	黑曲霉菌
black lacquer	黑漆〔大漆〕
black lead	石墨
black lead brush	石墨刷
black-lead crucible	石墨坩埚
black lead lubrication	石墨润滑剂
black-lead paint	黑铅粉漆；黑铅油〔俗〕
black-lead spar	黑铅矿
black light lamp	不可见光灯；黑光灯
black lignite	黑色褐煤
black liquor	黑液
black liquor soap	木浆黑液皂
black loading	填加炭黑；炭黑填充量
black lye	黑液
black magnetic ferrosic oxide	磁性氧化铁黑
black magnetic oxide	磁性氧化铁黑
black manganese	黑锰矿
Blackmar oil thief	布拉克马尔油样采集管
black master	炭黑母料
black-mercury oxide	黑色氧化汞
black mint(=English black mint)	英国黑种薄荷
black-mustard oil	黑芥子油
black mustard-seed oil	黑芥子油
blackness	黑度
black nickel oxide(=nickelic oxide)	氧化镍黑，Ni_2O_3
black non-transparent	黑色不透明
black note	底香
black oak	黑栎
black oil	黑油；润滑油
black oil conversion	黑油转化〔重油转化〕
black oil fuel	黑油燃料
black ore	黑矿
black organic fibre	有机(原料的)碳纤维
black out	①暗场〔双折射测定〕②发暗
black-out coating	遮光涂层
blackout paint	遮黑漆（防空用）
black-out paper	双层纸
black oxide	黑色氧化物〔指 UO_2〕
black oxide of iron	氧化铁黑
black oxide removal	黑氧化
black oxide treatment	黑化
black paint	黑涂料；黑漆
black pepper	黑胡椒
black peppermint(=narrow leaved peppermint)	澳洲桉
black-pepper oil	黑胡椒油
black pigment	炭黑
black pine	黑松
black plant	炭黑厂
black points	①黑霉病②(谷物)黑霉点
black pot	炼硫釜
black powder	黑色火药
black precipitate	黑色沉淀
black process	生产气体炭黑的过程
black products	黑色石油产品
black radiation	黑体辐射
black reactor	炭黑反应炉
black reclaim	黑再生胶
black red	黑红；暗红
black rubber	①变黑生胶②炭黑橡胶
black rust	(小麦秆)黑锈病
black salt	黑盐；粗苏打；黑灰，$Na_2CO_3 \cdot H_2O$
black sand	黑砂
black sassafras	橄榄樟
black sesame oil	黑芝麻油
black sheet iron	马口铁；黑铁皮
black shortness	冷脆性
black silver	脆银矿
black sludge	黑色淤渣
black slurry	炭黑水分散体；炭黑浆状液
blacksmith coal	锻冶煤
black soap(=black acids)	黑色酸〔酸渣中所含的磺酸盐〕
black soil	黑土
black spot	黑斑病；黑点；黑斑
black spotting	黑斑
black spruce	黑云杉
black spruce oil	黑云杉油
black stock	炭黑混合物
black stone	黑石；黑色碳酸盐页岩
black strap(=blackstrap molasses)	赤糖糊
blackstrap molasses(=blackstrap)	赤糖糊
black substitute	黑油胶〔橡胶代用品〕
black-sulfite liquor	黑碱液

black sulfuric acid 黑色硫酸；再生硫酸
black summer oil 黑色夏用机械油；夏用黑机油
black tape 黑胶带；绝缘带
black tellurium 黑碲矿
blackthorn 黑刺李
black tin 黑锡矿
black-top 柏油路面
black tube reclaim 黑内胎再生胶
black varnish 黑色假漆
black walnut oil 黑胡桃油
black wattle 黑荆树；柔毛金合欢
black wattle extract 柔毛金合欢栲胶
black wax 黑蜡
black willow 黑柳
blackwood 黑木金合欢
black wood charcoal 黑炭
bladder green(=sap green) ①暗绿色②树液绿〔一种暗绿色天然染料〕
blade ①叶片②刀片③刀口
blade angle 桨角
blade applicator 刮膜器
blade carrier 刀箱；刀室
blade holder 刀架
blade mixer 叶片式搅拌机
bladen(=hexaethyltetraphosphate) 四磷酸六乙酯
blade-paddle mixer(=blade-paddle stirrer) 桨式搅拌机
blade-paddle stirrer(=blade-paddle mixer) 桨式搅拌机
blade thickness 叶片厚度
blade tip ①刮刀尖；刃尖②叶尖；桨尖
blade twist 叶片的扭曲
blae 灰青炭质页岩；苏格兰产蓝灰色黏土
Blagden's law 布拉格登定律
Blair oven 布累尔烘箱
Blaise reaction 布勒斯反应
Blake jaw crusher 布来克颚式压碎机
Blanc's rule 布兰规则
blanc fix(e) 钡白；硫酸钡粉；重晶石粉 $BaSO_4$
blanching 发白
Blanc method 布兰法
blancometer 白度计
blangel 硅胶
blank 空白*
blank analysis 空白分析
blank assay 空白试验
blank background 空白背景
blank compound 空白胶料
blank controller 空白对照组
blank correction 空白校正

blanked deposit 平伏矿床
blanked-off pipe 关闭管
blanket ①壳层；外壳；套②(反应堆)再生区③毛毡
blanket coater (传送)带式刮涂机
blanket crepe 毡绉胶
blanketing 掩蔽；掩盖；盖覆
blanketing atmosphere 覆盖大气层；保护性气层
blanketing effect 隔离效应
blanketing gas 覆盖气体；保护气体
blanketing smoke 掩蔽烟雾
blanketing with gas 用气体覆盖
blanket of nitrogen 氮气层
blanket processing 再生区(燃料)处理
blanket reprocessing 再生区(燃料)后处理
blanket steam 覆盖水蒸气
blanket vein 横层脉
blanket washer 毛毡洗涤器
blank experiment 空白试验
blank flange 死法兰；法兰盲板
blank gasoline 空白汽油；没有添加剂的汽油
blank ignition 空白灼烧；灼烧恒重
blanking 压力裁断；冲切；(模)冲
blanking calender 压型压延机
blanking machine ①平压裁断机；冲切机②冲压机
blanking mold 冲模
blankite 连二亚硫酸钠
blank luminescence 空白发光
blank measure 空白测量值
blank measure fluctuation 空白测量值波动
blank-off volumetric efficiency 零位容积效率
blank panel 空板
blank pipe 空管；管内过滤器
blank powder 演习火药
blank reading 空白读数
blank reagent 空白试剂
blank run ①空白试验②空白分析；空机操作
blank scattering 空白分散
blank signal 空白信号
blank solution 空白溶液
blank stock 控制备料；调节备料
blank test ①空白试验②空试车
blank titration 空白滴定
blank value 空白(试验)值
Blasius equation 布拉修斯方程
blast ①鼓风②爆炸
blast blower 鼓风机
blast burner 喷灯*
blast cabinet 喷砂橱

blast cleaning 喷砂清洁法；喷丸清洁法
blast dryer 空气干燥机
blast enclosure ①封闭式喷射②喷砂密封罩(室)
blast engine 鼓风机
blaster 爆裂药
blast fan (鼓)风扇
blast furnace 高炉；鼓风炉
blast furnace cement 高炉水泥
blast-furnace cinder 高炉矿渣；鼓风炉渣
blast-furnace coal tar 高炉煤焦油；鼓风炉煤焦油
blast-furnace coke 高炉焦炭；鼓风炉焦炭
blast-furnace dust 高炉灰
blast-furnace elevator 高炉升降机；鼓风炉升降机
blast-furnace filler 高炉填料器
blast furnace gas 高炉煤气
blast-furnace hoist 高炉升举器
blast-furnace pitch 高炉沥青
blast-furnace roasting 高炉煅烧
blast-furnace shaft 高炉身
blast furnace slag 高炉矿渣
blast-furnace smelting 高炉冶炼
blast-furnace tar 高炉煤焦油
blast gas 鼓风气
blast gate 排气阀；排气门
blast head 喷砂(机)头
blast-heating apparatus 同流换热器
blast hood 鼓风罩
blastine 轰炸炸药〔高氯酸铵、硝酸钠、三硝基甲苯炸药〕
blasting ①喷砂②鼓风③爆炸
blasting agent 爆炸剂；炸药
blasting bomb 轰炸弹
blasting cap 起爆筒；雷管
blasting charge 轰裂药
blasting compound 爆炸成分
blasting explosive 轰炸药
blasting gelatine 炸胶
blasting oil 爆炸油；硝化甘油
blasting period 鼓风期
blasting powder 爆炸(火)药
blasting saltpeter 炸硝
blasting shot ①喷射②喷丸
blasting treatment 喷砂处理
blast lamp 喷灯
blast line(=blast main) (空)气管；鼓风管
blast main(=blast line) (空)气管；鼓风管
blast media 喷砂介质
blast nozzle (喷)嘴
blast pipe 排气管；乏气管；送风管；放气管

blast pressure 鼓风压力；风压
blast pump 送风泵
blast roasting 鼓风煅烧
blast smelting 鼓风冶炼
blast smelting furnace 鼓风熔炉
blast-supply 充(空)气管
blast tube ①鼓风管；送风管②喷管③爆炸管
Blaud's pill 碳酸铁丸剂
blaue fixe 硫酸钡粉；重晶石粉
blau gas(=blue gas) 蓝煤气〔纯净水煤气〕
blau-gas process 蓝煤气过程〔800℃高温裂化生产蓝煤气的过程〕
Blaw-Knox decarbonizing process 布劳-诺克斯脱碳(焦化)过程
blaze angle 闪耀角
blazed grating 闪耀光栅
blazed grating with two blazing wavelength 双闪耀波长光栅
blaze of grating 光栅闪耀
blaze speciality 闪耀特性
blaze wavelength 闪耀波长
bleachability 漂白率
bleachable absorber 可变色吸收体
bleach activator 漂白活性剂；助漂剂
bleach chamber 漂白室
bleach consumption 漂白消耗
bleached 漂白(了)的
bleached barytes (增)白重晶石；漂白重晶石
bleached beeswax 白蜡
bleached cotton cloth 漂白棉布
bleached lac(=white lac) 脱色紫胶
bleached linter 漂白棉绒
bleached oil 漂白油
bleached pulp 漂白纸浆
bleached shellac 漂白紫胶；漂白虫胶
bleached tallow 漂白脂
bleacher 漂白器
bleachery 漂白间；晒场；漂白工场
bleaching 褪色；脱色；漂白
bleaching action 漂白作用
bleaching agent 漂白剂
bleaching assistant 漂白助剂
bleaching bath 漂白浴
bleaching clay 漂白土
bleaching earth 漂白土
bleaching engine 漂白机
bleaching extract 漂白栲胶
bleaching fastness 漂白坚牢度

bleaching in stages　分段漂白
bleaching kier　漂白桶；漂白用煮布锅
bleaching liquid(=bleaching liquor)　漂白液
bleaching liquor(=bleaching liquid)　漂白液
bleaching machine　漂白机
bleaching materials　漂白剂
bleaching of cloth　布的漂白
bleaching powder　漂白粉*
bleaching-powder chamber　漂白粉室〔石灰氯化室〕
bleaching power　漂白本领
bleaching process　漂白法
bleaching system　漂白装置
bleaching tank　漂白槽
bleaching tower　漂白塔
bleach liquid(=bleach liquor)　漂白液
bleach liquor(=bleach liquid)　漂白液
bleach powder　漂白粉
bleach stability　(增白剂的)漂白稳定性
bleach-stable brightener　耐漂(白)增白剂
bleach tanning agent　漂白鞣剂
blead steam　额外蒸气
bleb　泡
bleb rate　发泡率〔喷丝板面上发泡的频率〕
bled steam　废蒸气
bleed　①渗料〔挤型机〕②渗移〔增塑剂〕
bleed effect　流失效应
bleeder　①输气管放水阀②输气管水冷凝器的连接管③泄放器
bleeder cock　放水龙头；放水旋塞
bleeder line　放水管；排出管
bleeder type condenser　溢流式大气冷凝器
bleeder-type spray gun　分压式喷枪
bleeder valve　排泄阀；放空器
bleeding　洇，洇色；渗色；(涂料)色料扩散
bleeding loss of grease　润滑脂凝胶收缩作用损失
bleeding of additive　添加剂的渗出(现象)
bleeding of dye　染料印流
bleeding off　①流下；流出②除去③印流
bleeding of lubricant　润滑剂的流失；润滑剂凝胶收缩
bleeding of paint　颜料印流〔油墨〕
bleeding resistance　耐渗色性；抗渗色性
bleeding resistant grease(=bleed-resistant grease)　抗凝胶收缩润滑油；抗流失润滑脂
bleed(ing) test　渗色试验
bleeding test of grease　润滑脂凝胶收缩试验
bleeding time　①流失时间②出血时间
bleed-off　除去；取消
bleedout　泛流

bleed rate　流失速率
bleed resistance　抗渗移性〔增塑剂〕
bleed-resistant grease(=bleeding resistant grease)　抗凝胶收缩润滑油；抗流失润滑脂
bleed-through　①渗胶；胶斑②渗色斑〔漆病〕
bleed valve　排出阀；放泄阀
blemish　①缺陷；疵点；毛病〔轻微的〕②次品
blend　①掺和；调和②掺和物
blend composition　掺和组分；混合组成
blende　闪锌矿
blende burner　闪锌矿煅烧炉
blende cinder　闪锌矿灰渣
blended　①混合的②混合性的；混杂的
blended antioxidant(=mixed antioxidant)　混合防老剂
blended asphalt　掺和沥青
blended coal　掺合煤
blended emulsifier　混合乳化剂
blend(ed) fabric　混纺织物
blended fuel　调和燃料；混合燃料
blended fuel oil　掺和锅炉燃料
blended gasoline　调和汽油
blended liquid phase　混合液相
blended liquid phase column　混合液相柱
blended lubricating oil　调和润滑油
blended motor-fuel　发动机用掺和燃料
blended oil　调和油
blended product　调和产品；掺和产品
blended spinning　共混纺丝
blended stock　混合原料；混合料
blended wool yarn　毛混纺纱
blende furnace　闪锌矿炉
blende gas　闪锌矿炉用气
blender　①掺和机，捏和机②搅切器；捣碎器③调和剂；合香剂；协调剂
blender-agglomerator　混合附聚器
blender loader　①配料器；加料斗②(=cyclone hopper)旋风分离器式加料斗
blende roaster　闪锌矿煅烧炉
blende roasting　煅烧闪锌矿
blende roasting furnace　闪锌矿煅烧炉
blending　①掺和；混合；调和②捣胶〔橡胶〕
blending agent　掺合剂；调和剂
blending agent of fuel　燃料掺和剂
blending bunker　混(合)料仓
blending compound　掺和组分
blending constituent　混合成分
blending fuel　调和燃料
blending hopper　掺和(漏)斗

blending house 掺和部
blending in line 管线掺和
blending machine ①掺和机②搅拌机
blending method 调和方法
blending naphtha 调和石脑油
blending octane curve 调和辛烷曲线
blending octane number(=blending octane value) 调和辛烷值
blending octane value(=blending octane number) 调和辛烷值
blending of aviation gasoline component 航空汽油组成的调和
blending of fuel 燃料调和
blending polymer 共混聚合物
blending rating 调和特性；调混辛烷值〔高辛烷值组分的〕
blending ratio 混合比例
blending roll 辊式掺混机
blending sample 粉碎样品
blending stock 调和用储(备)料
blending tank 掺和槽；混合罐
blending value 掺和值
blend latex 共混胶乳
blend level 混合率
blendor 搅切器；捣碎器
blend ore 掺合矿
blend ratio 掺和比；调和比
blennostasin 异辛可宁
bleomycin 博莱霉素；争光霉素
bleomycin A5 平阳霉素〔药〕
blind cavity 盲腔；盲孔
blind coal 无烟煤
blind end 封闭端；盲端
blind flange ①盖板②堵头盲板
blind hole 盲孔
blinding 窒塞
blinding of screen 筛孔堵塞
blind lapping 无效垫纱；空摆垫纱
blind plate 盲板
blind roaster 闭式烤炉
blind spot 盲点
blisger crepe 泡泡纱
blister ①泡②发泡
blister box 起泡试验箱
blister copper 泡铜；疤铜；粗铜
blistered casting 多孔铸件
blister formation （塑件缺陷）起泡
blister house 起泡试验机

blistering ①疱肿②起泡
blistering collodion 发泡火棉胶
blistering gas 疱肿(性)毒气；糜烂性毒气
blister(ing) resistance 抗起泡性
blister steel 泡钢；疤钢
blister test 起泡试验
blister type cracking 起泡型开裂
blistery 起泡的；有泡的
blistery coefficient 起泡系数
BL method(=borderline method) 边界线方法〔评价汽油抗爆性的方法〕
bloat(ing) 发胀；鼓出
bloating tendency 膨胀性
blob 团迹*
blob model 串滴模型
blob of viscose 黏胶块(疵点)
Bloch equation 布洛赫方程*
Bloch function 布洛赫函数
Bloch-Siegert shift 布洛赫-西格特位移
block 嵌段*
blockage 堵盖
block averaging 信息组平均法
blockboard 细木工板
block briquette 块煤砖
block coal 块煤
block coat(=tie coat) 过渡(涂)层；中间(涂)层
block coefficient 填充系数
block condensation 嵌段缩聚
block copolymer 嵌段共聚物*
block copolymerization 嵌段共聚合*
block diagram ①方框图②示意流程图
blocked adduct 封端加合物
blocked copolyester 嵌段共聚酯
blocked crosslinker 封闭型交联剂
blocked curing agent 封存性固化剂
blocked diamine 封闭的二胺
blocked elastic deformation 受阻弹性形变
blocked group 被保护的基团
blocked-in article 模制品
blocked isocyanate(=masked isocyanate) 封闭(型)异氰酸酯
blocked operation 交替操作；轮换操作
blocked position 被保护的位置；闭塞位置
blocked shellac 结块紫胶
blocked sulfoleate 封闭的磺化油酸盐
block electrophoresis 块状电泳
block factor 区组因素
block flow diagram 方块流程图

block furnace 方形电炉
block glue 胶条
block grease 硬质润滑脂
block-heteric nonionics 混嵌(共聚)非离子表面活性剂
blocking 封闭*；粘连
blocking agent 链端封接剂；封端剂
blocking antibody 封阻抗体；促进抗体
blocking condenser 过渡冷凝器
blocking effect 封闭效应
blocking fibre 嵌段纤维
blocking filter 带通型滤光片
blocking group 保护基团
blocking-layer cell 闭塞膜光电池；阻挡层光电池
blocking leakage under pressure 带压密封(堵漏)技术
blocking of metal indicator 金属指示剂的封闭
blocking out 遮挡；遮光
blocking point ①粘连温度②烫印温度
blocking property 嵌段性
blocking temperature 烫印温度
blocking test 分块试验
block ionomer complex micelles 嵌段离聚复合物胶束
block lava 熔岩块
block material flow diagram 物料流程方块图
block matrix least-squares refinement 块角矩阵最小二乘修正
block mutation 区段突变
block of folded chains 折叠链嵌段
blockout 封网
block penetration 块状试样针入度
block polycondensation 嵌段缩聚
block polyesteramide 嵌段聚酰胺酯
block poly(ester ether) 嵌段聚醚酯
block polymer 嵌段聚合物
block polymerization 嵌段聚合*
block press 模压机
block printing ①木板印刷②人工印花〔纺〕
block printing table 木板印刷台
block resistance 抗黏着性
block rosin 松香块；块状松香
block rubber 橡胶块
block scheme 方框图
block shear test 整体剪切试验
block sides 垫铁；挡板〔硫化用〕
block skiving 切片法
block soap 块皂；皂砖
block soda 苏打块
block structure 嵌段结构
block tin 锡块

block-type graphite heat exchanger 方块式石墨换热器
block up 停用；关闭
block valve 截止阀
blocky particle 块状粒子
blodite (bloedite) blodit 白钠镁矾
blood amylase 血(含)淀粉酶
blood anticoagulant 血液抗凝剂
blood apparatus 血内气体测定器
blood cast 血圆柱
blood cell counter 血细胞计数仪
blood coagulation 血液凝固
blood coagulometer 血液凝结计
blood collector 集血器
blood compatibility 血液相容性
blood composition 血组成
blood count 血球计数
blood examination method 血液检查法
blood fat 血脂
blood gas analysis 血内气体分析
blood gas apparatus 血内气体检验器
blood glucose 血糖
blood group 血型
blood group gene 血型基因
blood method 血液法
blood plasma 血浆
blood platelets 血小板
blood pressure 血压
blood serum 血清
bloodstone 赤铁矿
blood sugar 血糖
blood type agglutination inhibition test 血型凝集抑制试验
blood type autogrouper 血型自动分析
blood viscosity determination 全血黏度测定
bloom ①起霜〔漆病〕②白化
bloomery 精炼炉床
bloom-free plasticizer 不喷霜的增塑剂
Bloom gelometer 布卢姆凝胶计
blooming ①喷霜；喷出②起霜；白化〔透明膜表面发生白云现象〕
blooming agent 喷霜剂
blooming effect 弥散反应
bloom inhibitor 防喷霜剂
bloomless oil 不起霜润滑油
bloom of lubricating oil 润滑油的荧光
bloom oil 起霜油〔松香干馏得的油,沸点约为270℃〕
bloom out 起霜；风化
blot 斑渍；斑点；污渍

blotch 斑污；斑渍
blotchy effect ①疙瘩效应②塌渗效应
blotchy surface 疤面
blotter paper 吸墨纸
blotter press 压滤机
blotter strip 闻香纸条
blotting ①浸透；渗出②印迹③揩吸去(油)④污沾吸作用；吸墨纸效应
blotting paper 吸水纸；吸墨纸
blotting test 吸墨试验
blowability 吹成性
blowback 回吹；反吹
blow bending test 冲击弯曲试验
blow-by 漏气
blow case 吹扬器
blow-cock 排气栓
blow cycle 吹塑周期
blow die 吹塑模
blowdown ①泄料；放空②排污
blowdown condenser 放空冷凝器；泄料冷凝器
blowdown drum 排污罐
blowdown piping 泄料管路
blowdown pit 泄料池
blowdown stack 泄料烟道
blowdown stream 排污液流；泄放液流
blowdown system 泄料系统
blowdown tank 泄料箱
blowdown valve 排汽阀
blower 鼓风机；吹气机
blower drain valve 鼓风机放泄阀
blower performance 通风机性能
blower rotor 鼓风机转子
blower thrust 鼓风机推力
blow extrusion 挤出吹胀〔薄膜〕
blow-extrusion film 吹挤薄膜
blow-forming 吹塑成形
blow gas 吹制水煤气
blow hole 气孔；吹孔
blowing ①放料；放锅；喷发②(塔板上)气流吹开(现象)
blowing activator 发泡活性剂
blowing agent 发泡剂；起泡剂
blowing agent promotor 发泡剂促进剂
blowing agent stability 发泡剂稳定性
blowing cavity ①吹塑阴模②吹塑型腔
blowing characteristics 发泡性
blowing duct 冷却甬道；吹风管道
blowing dust 飞尘；飘尘
blowing engine 鼓风机

blowing fan 送风扇
blowing head 吹塑模头
blowing iron 吹(玻璃)用铁管
blowing machine 鼓风机
blowing method ①吹放法②吹制法
blowing mould 吹模
blowing off ①喷射②吹散
blowing period 吹风期
blowing position 吹炼位置
blowing pressure(=blow pressure) ①吹塑压力②发泡压力
blowing process (制造玻璃纤维的)吹制法
blowing promotor 发泡助剂
blowing ratio 发泡比
blowing speed 吹胀速度
blowing still 吹炼釜
blowing (tower tray) 吹液现象
blow-in the furnace 开炉；开炉送风
blow lamp 喷灯
blow latex foam 泡沫胶
blow molding 吹塑*
blow molding cycle 吹塑周期
blow mo(u)lding parison 吹塑型坯
blown ①吹出的②喷出的
blown animal oil 吹制动物油
blown asphalt 吹制沥青；氧化沥青
blown (asphaltic) bitumen 吹制沥青
blown extrusion 吹胀挤塑
blown film 吹塑薄膜；吹膜
blown-film extrusion 吹胀薄膜挤塑
blown glass 吹制(的)玻璃
blown mineral oil 吹制矿物油
blown oil 氧化油；吹制油；气吹油
blown out 吹(干)净
blown petroleum 氧化残渣油；氧化沥青
blown petroleum asphalt 氧化(石油)沥青〔从石油渣油经吹入空气氧化而制得的沥青〕
blown rubber 海绵胶
blown sponge 海绵胶
blown stand oil 氧化聚合油；吹制定油
blow-off ①吹出②吹散③喷射④放气
blow-off cock(=blow-off pet cock) 排气栓
blow-off pet cock(=blow-off cock) 排气栓
blow-off pipe 排气管；废气管；喷放管
blow-off point (塔板)喷出点〔吹开点〕
blow-off proof hose 耐压胶管
blow-off tank 吹卸槽
blow-off valve 排出阀；放泄阀；放空阀；喷放阀

blow-out ①吹出②爆裂
blow-out patch ①轴圈②管接头③补垫；胎垫
blow-out switch 放气开关
blow-period 吹风期
blow-pipe 吹管
blow-pipe analysis 吹管分析
blow-pipe bead 吹管珠
blow-pipe flame 吹管焰
blow-pipe reaction 吹管反应
blow-pipe set 吹管组；吹管分析的设备
blowpipe test 吹管试验*
blowpipe tip 吹管用鱼尾(灯)
blow pit 泄料池
blow point 发泡温度
blow pressure(=blowing pressure) ①吹塑压力②发泡压力
blow ratio(=blow-up ratio) 吹胀比
blow run 鼓风掺气(过程)
blow-run gas 鼓风气〔加入水煤气〕
blow-run method 鼓风掺气法
blow steam through 吹入水蒸气
blow tank 泄料桶
blow test ①爆破试验②吹气试验③冲击试验
blow-time 吹风时间；吹气时间
blowtorch ①喷气灯；喷焰灯②焊灯(枪)
blow-up ①爆破②吹胀③发泡
blow-up pan 蒸气搅拌锅〔粗糖溶液加石灰后用蒸气搅拌〕
blow-up ratio(=blow ratio) 吹胀比
blow valve 送风阀
blow vent 通气口；排气口
blow wash 吹洗；压水冲洗
blubber 鲸脂
blubbering 提取鲸脂
blubber oil 鲸脂油
blue annealing 蓝回火
blue asbestos(=corcidolite) 蓝石棉
blue aviation gasoline 蓝色航空汽油〔加铅并染成蓝色的高辛烷值汽油〕
blue basic lead sulfate(=blue lead) 碱式硫酸铅蓝；蓝铅〔主要成分 $PbSO_4 \cdot PbO$〕
blueberry 笃斯越橘
blue brittleness 蓝脆性〔热至发蓝色时呈脆性〕
blue camphor oil 蓝樟脑油
blue chamomile 母菊
blue copper 蓝铜矿
blue copperas 胆矾；五水(合)硫酸铜；蓝矾 $CuSO_4 \cdot 5H_2O$

blue copper protein 铜蓝蛋白
blue cross 蓝十字毒气
blue dyes 蓝色染料
blue fish oil 海豚油
blue flag 喀罗林鸢尾
blue flame 蓝焰；一氧化碳焰
blue gas(=blau gas) 蓝煤气〔纯净水煤气〕
blue-green 蓝绿
blue gum 蓝桉胶〔取自蓝桉的叶〕
blue gum tree 蓝桉
blue-hard photographic plate 蓝硬型感光板
blue heat scale 蓝色热轧氧化皮
blue hyacinth 蓝色红锆石〔矿〕
blueing(=bluing) 上蓝；蓝变
blueing agent 发蓝剂
blue iron ore 蓝铁矿
blue jack oak 灰栎
blue john 蓝萤石 CaF_2
blue lead 蓝铅；碱式硫酸铅
blue light 蓝光
blue line 蓝线
blue line process 蓝线法
blue lupin oil 蓝羽扇豆油
blue malachite 蓝铜矿
blue mallee 多苞桉
blue mass 汞软膏
blue-mottled soap 蓝花皂
blue mud 蓝淤泥
bluensin(=bluensomycin) 布鲁霉素
bluensomycin(=bluensin) 布鲁霉素
blue oil 蓝色油〔1. 地蜡矿蒸馏得到的重质油和蜡的混合物；2. 苏格兰页岩重油冷榨脱蜡后的油料〕
blue paste 蓝糊；蓝色涂料
blue pigment 蓝色颜料
blue powder 蓝(锌)粉
blue printing paper 蓝晒纸
blue salt 蓝盐〔染料工业〕
blue scale (褪色试验)蓝色色标
blue sensitive phototube 蓝敏光电管
blue shade 带蓝头；蓝相
blue shift 蓝移*
blue shortness 蓝脆性
blue smoke 蓝烟
blue spruce 蓝云杉
blue stain ①(木材)蓝(青)变；蓝斑〔变质现象〕②(陶瓷)蓝色料〔钒锆化合物〕
bluestone ①胆矾②蓝灰砂岩
blue-tinted(=bluish) 蓝光；带蓝(色)头的

blue trim ①蓝色装潢(饰)物②涂饰成淡蓝色
blue verdigris 乙酸铜
blue verditer 铜绿；碱式碳酸铜
blue vitriol 胆矾*
blue water gas 蓝水煤气
bluing 上蓝；蓝变
bluing agent 上蓝剂
bluish(=blue-tinted) 蓝光；带蓝(色)头的
bluish green 蓝绿
bluish red 蓝红
bluish violet 蓝紫
blumea oil 艾纳香油；艾叶油
Blumer separator 布卢默分离器
blunger 圆筒掺和机
blunging 用水搅拌
blurred image 模糊图像
blushing 发白(现象)〔涂料〕
blushing test 发白试验〔涂料〕
blush mark 泛白迹印
blush preventive agent 防白剂
blush resistance 抗发白性；防潮性
blush resistance determination 防发白试验
blush retarder 湿晕防止剂
BN fibre(=boron nitride fibre) 氮化硼纤维
board 板
board coal 纤维质煤
boarder 定形机
boarding (热压)定型；起纹
boarding machine 搓纹机；搓软机
board machine 纸板机
board mill 板材厂
boardness (织物)硬挺度
board type insulant 板状绝热材料
boardy 手感粗硬；硬挺感
boardy product 硬制品；不柔软制品
boat 舟皿*
boat conformation 船型构象*
boat form 船式
boat pan 舟皿
boat reactor 舟形反应器
boat-slide system 滑舟系统
boat varnish 船用清漆
bob 浮子
bobbin 筒管；大管（俗）
bobbin cutter 盘纸分切机
bobbin dyeing 筒染
bobbin-hose 编织胶管
bobbinite 筒管炸药〔商〕；波兵那特；硫铵炸药〔矿

山用〕
bobbin net 罗纱
bobbin oil 锭子油
bobbin size 线轴尺寸
bobbin winder 抱平机；绕线装置
bobbin yarn 管纱
bobierite 白磷镁石
Bock monoscope 博克单像管
Bode plot 波特图
bodied oil 聚合油
bodiness 增稠；增厚
body 稠度；增稠
body brick 炉体砖
body-centered cubic 体心立方
body-centered cubic lattice 体心立方格子
body-centered cubic structure 体心立方结构
body-centered grating 体心(式结晶)格子
body centered lattice 体心点阵
body-centered orthorhombic 体心正交
body-centered teteragonal 体心四方
body colour 体质色料
body core ①母芯②主体泥芯③主模
body dimension 本体尺寸
body feed 主体加料
body gum 体质胶；增稠剂
bodying ①加厚；加重②增加体积③稠化；增稠
bodying agent ①基础剂〔制造润滑脂用的油类〕②增稠剂
bodying liquid 稠化液；(聚合)稠化液
bodying of oil 油引(发)聚合
bodying speed 黏度增加速度
bodying temperature 热炼温度；(油)聚合温度
bodying up ①加厚〔用紫胶漆打底多次搽涂以增加其厚度〕②稠度增加；增稠
bodying velocity 热炼聚合速度；黏度上升速度
body of catalyst 催化剂的主体
body of coal 煤的主体〔沥青物质的主体〕
body of cracking stock 裂化原料的主体
body of oil 油基；润滑油的底质
body of paint 油漆稠度
body paper 底纸
body pigment 底质颜料
body solder ①(=body putty)车身腻子②聚酯腻子
body stock 坯料
body up ①增稠②加重；加厚
body varnish 外用清漆
boehmite 勃姆石
boehmite process 勃姆膜法〔铝的氧化膜处理方法之一〕

boghead(=boghead coal)　烟煤
boghead coal(=boghead)　烟煤
bog iron(=bog iron ore)　沼铁矿
bog iron ore(=bog iron; bog ore)　沼铁矿
bog ore(=bog iron ore)　沼铁矿
bogus kraft paper　仿包皮纸；仿牛皮纸
bogus wrapping　灰色包装纸
Bohemian crystal　波希米亚玻璃
Bohemian glass　波希米亚玻璃
Bohemian glass tube　波希米亚玻璃管
Bohr atom　玻尔原子
Bohr effect　玻尔效应*
Bohr magneton　玻尔磁子*
Bohr model of atom　玻尔原子模型
Bohr orbit　玻尔轨道*
Bohr radius　玻尔半径*
Bohr theory　玻尔理论
boil down　蒸发
boil dry　煮干
boiled grease　熟炼润滑油
boiled linseed oil　熟炼胡麻子油
boiled oil　清油；熟油；沸炼油
boiled-out water　沸过的蒸馏水
boiled tar　熟焦油；脱水焦油
boiled urushi　熟漆；精制漆（大漆）〔生漆经过搅漆和晒漆两步加工之后称熟漆，亦称精制漆〕
boiler　锅炉
boiler antiscaling composition　锅炉防垢剂
boiler blow-down water　锅炉冷凝水；锅炉回水
boiler cock　锅炉用旋塞
boiler compound　锅炉防垢剂
boiler drum(=boiler shell)　煮(沸)鼓
boiler feed-water　锅炉给水
boiler furnace　(汽)锅炉
boiler horse power　锅炉马力
boiler oil　锅炉燃料(油)
boiler package system　快装锅炉系统
boiler plate　锅炉板
boiler scale　锅垢
boiler shell(=boiler drum)　煮(沸)鼓
boiler shell ring　煮鼓圈
boiler steam room　蒸汽锅炉房
boiler tube　锅炉管
boiling bed　沸腾床；沸腾层
boiling bubble stone(=boiling stone)　沸石
boiling bulb　蒸馏锅
boiling control apparatus　控沸器
boiling-down pan　蒸煮锅

boiling endotherm　沸腾吸热
boiling evaporation　沸腾蒸发
boiling fastness　耐煮性
boiling heat　蒸发热
boiling heat transfer coefficient　沸腾传热系数
boiling kier　蒸煮锅；漂煮锅
boiling-liquid-bath reactor　沸腾液恒温反应器
boiling lye　沸碱水
boiling molten pool　沸腾状熔池
boiling number　沸腾数
boiling of soap　肥皂的熬煮
boiling-on-grain　砂炒法
boiling pan　蒸煮锅
boiling plate　均沸片
boiling point　沸点
boiling point apparatus　沸点测定器
boiling point curve　沸点曲线
boiling point depression(=boiling point lowering)　沸点降低
boiling point diagram　沸点图
boiling point elevation　沸点升高*
boiling point-gravity constant(=boiling point gravity number)　沸点-比重常数
boiling point-gravity number(=boiling point-gravity constant)　沸点-比重常数
boiling point index　沸点指数〔在10毫米汞柱下石油馏分平均沸点和黏度的关系〕
boiling point lowering(=boiling point depression)　沸点降低
boiling point method　沸点(升高)法〔利用沸点上升测定分子量〕
boiling point range　沸点范围
boiling point rising　沸点升高
boiling point rule of retention　保留值沸点规律
boiling point test　沸点试验
boiling point-viscosity constant　沸点-黏度常数
boiling range　沸程
boiling resistance　耐蒸煮性
boiling spread　沸点范围〔石油馏分的〕
boiling starch　部分水解淀粉
boiling station　蒸煮间
boiling stone(=boiling bubble stone)　沸石
boiling tank　蒸煮锅；蒸煮桶
boiling test　①煮沸试验②沸水试验
boiling vessel　蒸煮器
boiling water absorption　①吸沸水量②吸沸水性
boiling water sealing　沸水封闭法
boiling water shrinkage　沸水收缩(率)

boiling water test　(耐)沸水试验
boiling water resistance　耐沸水性
boil off　煮掉
boil-off liquor　废皂水〔纺〕
boil-off shrinkage　退浆收缩(率)；精炼收缩(率)
boil out　熬煮；煎煮
boil over　沸溢
boilproof　耐煮的
boil-stain of cloth　布的煮斑〔因煮炼不足所留斑印〕
boil stringproof　煮熬拔丝〔糖〕
boil-up rate　蒸出速率
boivinose　2-脱氧-6-鼠李糖
bold　大块〔天然树脂和珀珀树脂类树脂的大块〕
boldine　波尔定碱
boldoglucin　玻耳多苷
bole　胶块土；红玄武土
boleg oils　氧化了的矿物油
boleite　银铜氯铅矿
boletic acid　别失水苹果酸
Boling burner　玻林燃烧器
Bollman extractor　博尔曼抽提器
Bollman soybean extractor　博尔曼大豆萃取器
Bolnes engine test　博尔内斯发动机试验
Bologna phosphorus　博龙亚磷〔一种发光硫化钡〕
bolognian phosphorus(=bolognian stone)　重晶石
bolometer　(电阻)测辐射热仪
bolster　活动台；下托板〔平板机〕；软垫
bolt　①圆(门，窗)闩②短圆材③螺栓
bolt circle　螺栓圆周
bolted joint　螺栓接合
bolter　筛
bolt head　①螺栓头②长颈烧瓶
bolt (hole) circle　螺栓周圆；安放螺栓的周圆
bolting　①筛选②螺栓联系
bolting cloth　筛布
bolting house　筛选间
bolting machine(=bolting mill)　筛选机；筛粉机
bolting mill(=bolting machine)　筛选机；筛粉机
bolting reel　转筒筛〔选矿〕
bolting silk　①丝筛；绢筛②丝织筛布
boltless door　错齿式罐盖〔硫化罐〕
bolt oil　螺栓油
boltonite　镁橄榄石；假镁橄石；波尔顿石
bolt steel tank　螺栓连接钢罐
Boltzmann constant　玻耳兹曼常数
Boltzmann distribution　玻耳兹曼分布
Boltzmann distribution law　玻耳兹曼分布定律
Boltzmann equation　玻耳兹曼方程
Boltzmann fluid　玻耳兹曼流体
Boltzmann relation　玻耳兹曼关系(式)
Boltzmann statistics　玻耳兹曼统计
Boltzmann superposition principle　玻耳兹曼叠加原理
Boltzmann transport equation　玻耳兹曼输运方程
bolus　红玄武土
bolus alba　高岭土；瓷土
bomb　①弹②高压气体储罐
bomb aging　弹内(加氧)陈化
bombardment　轰击*；冲击
bomb calorimeter　弹式热量计*
bomb calorimetry　弹式量热法
bomb calorimetry study　弹式量热法研究
bomb furnace　封管炉
bombicesterol　蚕茧醇
bombiosterol　蚕茧醇
bomb method　氧弹法〔测定石油产品的硫含量〕
bomb oxidation　弹内氧化
bomb oxidation test　弹内氧化试验
bomb sulfur　氧弹法含硫量
bomb tube　弹管
bombycin　蚕素
BON acid(=3-hydroxy-2-naphthoic acid)　3-羟基-2-萘甲酸；β-萘酚酸
bond　①黏合；黏合体②键；结合③胶接；胶合
bondability　黏合程度；密着程度
bondability index　黏合指数
bond alternation　键交替现象；分子键交变性质
bond angle　键角*
bond-angle deformation　键角变形
bondant　黏合剂；黏着剂
bond area　黏合面
bond bending　键弯曲
bond breakage　键断裂
bond clay　黏土
bond cleavage　键裂
bond dissociation energy　键裂解能
bond distance　键长
bond distribution analysis　键分配分析法
π-bonded adsorption complex　π键吸附的配合物
bonded assembly　胶接件
bonded energy　键合能
bonded fabric　无纺布；黏合织物
bonded ion exchanger　键合型离子交换剂
bonded joint　胶接接头
bonded liquid phase　键合固定相
bonded phase　(化学)键合(固定)相
bonded-phase chromatography　键合相色谱法；化学键合

相色谱
bonded-phase coverage　键合相覆盖率
bonded-phase packing　(化学)键合(固定)相填充物
bonded reversed-phase plate　反相键合相板
bonded stationary phase　(化学)键合固定相
bond energy　键能*
bond enthalpy　键焓
bonder　黏合剂
bonderization　磷酸盐表面处理
bonderizing　磷化处理〔机〕
bond failure(=separation)　脱开
bond fission　键断裂
bond fixation　键固定
bond force　①键合力②黏合力
bond force constant　键力常数
bond formation　键形成
bond hindrance　键受阻
bonding　黏合；密着；结合
bonding action　黏合(砌合)强度；黏结强度(作用)
bonding adhesive　黏合胶
bonding agent　①键合剂②黏合剂
bonding cement　黏合胶泥
bonding clay　结合黏土
bonding coat　黏合涂层
bonding electron　成键电子
bonding fixture　黏合夹具
bonding force　①结合力②键合力
bonding layer　①黏合层；初层②打底胶浆
π-bonding ligand　π成键配体*
σ-bonding ligand　σ成键配体*
bonding material　黏合材料
bonding medium　黏合介质
bonding (molecular) orbital(BMO)　成键(分子)轨道*
bonding of fibrous material　纤维结合法
bonding orbital　成键轨道
bonding plies　黏结层
bonding point　黏结点；黏合点
bonding position　键合位置
bonding power　黏合本领
bonding region　键(合)区*
bonding resin　黏合树脂
bonding state　①黏合态②键合态
bonding strength　黏合强度；密着力；黏附力
bonding technique　胶黏技术
bonding zone　①黏合区②焊接区
bond length　键长*
bond line　黏合(剂)层；胶层
bondline thickness　黏合(剂)层厚度；胶层厚度

bond migration　键迁移
bond moment　键矩*
bond number　键数
bond orbital　键轨道
bond-orbital method　键轨法
bond order　键级*
bond orientation　键取向
bond paper　证券纸
bond plaster　黏结性抹灰层
bond polarity　键的极性
bond polarization　键的极化
bond radius　键半径
bond refraction　键折射
bond refractivity　键的折射性
bond rubber　结合橡胶
bond rupture　键断裂
bond scission　键裂开
bond shifting　键位移
bonds number cleaver　断键数
bond strength　键强度*
bond stress　黏合应力
bond structure　键结构
bond tautomerism　键的互变(异构)
bond testing machine　黏合力试验机
bond twisting　键扭转
bondu　棒杜铝〔一种耐蚀铝合金〕
bonducin　云实豆素
bond valence　键价*
bond valence-bond length correlation　键价-键长关联*
bond-valence theory　键价理论*
bond valency　键价
bone black　骨炭
bone charcoal　骨炭
bone china　骨灰瓷
bone coal　骨炭
bone densitometry source　骨密度测量源
bone dry　干透；完全干燥
bone dry fiber　绝干纤维
bone dry shellac　漂白干紫胶
bone dry strength　纱线干燥强度
bone dry weight　无水干重
bone fertilizer　骨粉肥料
bone glass　乳色玻璃
bone manure　骨粉肥料
bone oil(=jeppel oil)　骨油
bone pitch　骨沥青
bone tar　骨焦油
bone tar oil　骨焦油

bone tar pitch 骨沥青
bone test 刚性试验〔钢丝〕
boninic acid 包宁地衣酸
bonito liver oil 鲣肝油
bonito oil 鲣油
bonnet 阀盖；阀帽
bonnet valve 帽状阀
Bonotto extractor 博诺托抽提器
bookbinders'varnish 书脊清漆
book-binding leather 书面革
bookcase drier 书柜式干燥器
book clay 书页状瓷土；层状黏土
book inventory 账面库存量
booklike particle 层状粒子；书页状粒子〔白瓷土的特殊晶形结构〕
book paper 印书纸
book storage 百页车中存放
boom 水栅〔纤〕
boomerang 干膜测厚仪〔单磁铁；直接读数〕
boon winding 环向缠绕
boony fiber 木质纤维
Boord olefin synthesis 伯德烯烃合成法
boost carburising 强化(增强)渗碳(作用)
boost-diffuse carburising 增强(强化)扩散渗碳(作用)
booster ①爆管②扩爆药；传爆药③助促进剂
booster action ①推动作用；助推作用；辅助作用②加强作用；加速作用；增压(升高)作用
booster action type accelerator 助促进剂
booster amplifier 升压放大器；辅助放大器
booster compressor 增压压缩机
booster diffusion pump 增压扩散泵
booster dose 促升剂量
booster hose 耐压胶管
booster pump 增压泵
booster solvent 助溶剂
boost gauge 增压压力表；升压计
boosting ①升压(增压)作用；增大；增加(作用)②补充；推动(作用)③辅助加热；火焰-电混合加热
boosting accelerator 助促进剂
boosting site 升压部位
boosting voltage 补充电位
boot 进料斗；接受器
boot cream (皮)鞋油
Booth brass furnace 布思黄铜炉
boothite 铜水绿矾
bootlegging ①逐步脱层②屈挠脱层〔平带〕
boot of elevator 靴架
boot pulley 下鼓轮〔升降机〕

boot-topping paint(=water-line paint) 水线涂料；水线漆
boracic 硼的
boracic acid(=boric acid) 硼酸
boracite 方硼石
boracium(=boron) 硼〔5号元素，化学符号B〕
Borad ring 博拉德双层金属网环
borage 琉璃苣
boral 硼酒石酸铝
borane 硼烷*
borate 硼酸盐
borate buffer solution 硼砂缓冲液
borated water 含硼水
borate flux 硼酸盐熔剂
borate tailing 硼酸盐渣
boration 硼酸化
boratofluoric acid 氟硼酸 H[BF$_4$]
borax 硼砂*；四硼酸钠
boraxal 氧化硼铝〔三氧化二硼和铝的复合体，防中子辐射材料〕
borax bead 硼砂(熔)珠
borax bead reaction 硼砂熔珠反应
borax-bead technique 硼砂珠技术
borax-bead test 硼砂珠试验*
boraxen(e) 硼氧烯
borax fusion 硼砂熔融
borax glass 硼砂玻璃
borax lake 硼砂湖
borax-phosphate solution 硼砂-磷酸盐缓冲液
borax soap 硼砂皂
borax stability 硼砂稳定性
borazine(=borazole) 环硼氮烷；硼嗪〔俗〕 B$_3$N$_3$H$_6$
borazine polymer 硼氮聚合物
borazole(=borazine) 硼嗪〔俗〕
borazon (一)氮化硼；人造亚硝酸硼〔硬度接近金刚石〕
borazone 氮化硼半导体
borazynes 硼炔苯类
Borchers cell 博彻电池
bordeaux ①枣红〔色名〕②酸性枣红〔染料〕
bordeaux lake 酸性枣红色淀
bordeaux turpentine(=French turpentine) 枣红松节油；法国松节油
border 边缘；边沿
border curve 界面曲线
bordered pit 具缘纹孔
bordered pit-pair 具缘纹孔对
borderline 分界线；边界线；轮廓线
borderline acid 边缘酸
borderline base 边缘碱

borderline curve 边界线曲线
borderline dilatancy 边界膨胀性
borderline knock curve 边界(线)爆震曲线
borderline mechanism 边界机理*
borderline method 边界线法〔评价汽油抗爆性〕
borderline motor fuel 卡边发动机燃料〔其性质刚合标准〕
borderline sensitivity 边界灵敏度
border motif 界面形色；边界形态
bore ①(喷丝板的)孔②孔内径③钻孔
bored roll 空心滚筒
bore hole 镗孔
borehole flowmeter 锐孔流量计
borer ①钻孔器②镗床
borer hole 虫孔
borer resistance 抗钻虫性
Borghetty-Bergman value 博尔费第-伯格曼值〔表征表面活性剂钙皂分散能力的数值〕
boric acid 硼酸
boric acid ester 硼酸酯 B(OR)$_3$
boric acid latex 硼酸(保存)胶乳
boric anhydride(=boric oxide) 氧化硼；硼酐 B$_2$O$_3$
boric oxide(=boric anhydride) 氧化硼
boric oxide fibre 氧化硼纤维
boric oxide filament yarn 氧化硼长丝
boric spar 方硼石
boride 硼化物
borinammine 硼氮配位化合物；硼胺
borine 烃基硼 BR$_3$
boring 镗(孔)
boring and inserting sleeve method 扩孔镶套法
boring bar 镗杆
boring cutter 镗刀
boring hose 钻控胶管；凿井胶管
boring lathe 镗床
boring machine 镗床
boring machine operator 镗工
boring method 钻取法〔纸浆取样〕
boring-mill work 镗铣加工
boring sample 岩心取样
boring table 镗床工作台
borinoaminoborine 二硼烷基胺 H$_3$BNHBH$_3$
bormethyl 三甲硼
bornadiene 菠二烯
bornane 莰烷；菠烷
2-bornene(=bornylene) 冰片烯；2-菠烯；2-莰烯
borneo camphor(=borneol) 冰片；龙脑；2-莰醇
borneo camphor oil 龙脑香油

borneo camphor tree 龙脑香树
Borneo copal 婆罗洲玷玛脂
borneol(=borneo camphor) 冰片；龙脑；2-莰醇
borneol acetate(=bornyl acetate) 乙酸冰片酯
borneol flakes 片状；冰片
borneol salicylate(=salit) 水杨酸冰片酯；沙里特
Borneo tallow 婆罗洲脂
Born equation 玻恩方程式
bornesite 甲基肌醇
bornesitol 甲基肌醇
Born-Haber cycle 玻恩-哈伯循环*
bornite 斑铜矿
Born-Oppenheimer approximation 玻恩-奥本海默近似*
bornyl 冰片基；2-莰烷基
bornyl acetate(=borneol acetate) 乙酸冰片酯 C$_{10}$H$_{17}$OOCCH$_3$
bornyl alcohol(=borneol) 冰片；龙脑
bornyl amine 冰片基胺；2-莰胺 C$_{10}$H$_{17}$NH$_2$
bornylane 冰片烷
bornyl α-bromoisovalerate α-溴异戊酸冰片酯 (CH$_3$)$_2$CHCHBrCOOC$_{10}$H$_{17}$
bornyl butyrate 丁酸龙脑酯
bornyl chloride 冰片基氯；2-氯莰 C$_{10}$H$_{17}$Cl
p-bornyl cyclohexanol 对龙脑基环己醇
p-bornyl cyclohexanone 对龙脑基环己酮
bornylene 冰片烯；2-莰烯 C$_{10}$H$_{16}$
bornyl formate 甲酸冰片酯；甲酸-2-莰醇酯 HCOOC$_{10}$H$_{17}$
p-bornyl guaiacol 对龙脑基愈创木酚
bornyl halide 冰片基卤 C$_{10}$H$_{17}$X
bornyl isovalerate(=bornyral) 异戊酸冰片酯 (CH$_3$)$_2$CHCH$_2$COOC$_{10}$H$_{17}$
bornyl lactate 乳酸龙脑酯
bornyl methanoate 甲酸龙脑酯
p-bornyl-2-methyl cyclohexanol 对龙脑基-2-甲基环己醇
p-bornyl-3-methyl-cyclohexanone 对龙脑基-3-甲基环己酮
bornyl nonylate 壬酸龙脑酯
p-bornyl phenol 对龙脑基苯酚
bornyl propionate 丙酸龙脑酯
bornyl salicylate 水杨酸冰片酯 o-HOC$_6$H$_4$COOC$_{10}$H$_{17}$
bornyl thiocyanoacetate 氰硫基乙酸冰片酯
bornyral(=bornyl isovalerate) 异戊酸冰片酯
bornyval 异戊酸莰酯
borobutane 丁硼烷
borocaine 硼酸普鲁卡因
borocalcite 硼钙石

boroethane(=biborane)　乙硼烷
borofluorhydric acid(=borofluoric acid)　氟硼酸　H[BF$_4$]
borofluoric acid(=borofluorhydric acid)　氟硼酸
borofluoride　氟硼酸盐*
borofluorin　硼氟剂
boroglyceride　硼酸甘(油)酯
borohydride　氢硼化物
borohydride sodium　氢硼化钠
borol　硼硫酸钾钠
borolon　合成氧化铝；人工氧化铝　Al$_2$O$_3$
boromycin　硼霉素
boron　硼〔5号元素，化学符号B〕
boron alkyl　烷基硼　BR$_3$
boron amide　硼胺；氮化硼
boron-amine complex　硼-胺配合物
boronation　硼化作用；硼化反应
boronatrocalcite　钠硼钙石
boron bromide　溴化硼　BBr$_3$
boron carbide　一碳化六硼；碳化硼　B$_6$C
boron carbide fiber　碳化硼纤维
boron chamber　(充)硼电离室
boron chloride　氯化硼　BCl$_3$
boron congeners　硼的同族元素
boron-containing polymer　含硼聚合物
boron-doping　掺硼
boron etch-stop　硼腐(浸)蚀停止层
boron family　硼族
boron fiber　硼纤维
boron fiber reinforced plastics(BFRP)　硼纤维增强塑料
boron fibre　硼纤维
boron filament　硼丝
boron fluoride　氟化硼　BF$_3$
boron flux　硼熔剂
boron heterocycles　硼杂环类
boron hydride　①氢化硼②硼烷
boron hydride chloride　氢氯化硼
boron hydroxide　氢氧化硼；硼酸　H$_3$BO$_3$
boronia flower oil　波罗尼花油
boron imide　硼亚胺；亚氮化硼　B(NH)$_3$B
boron iodide　碘化硼　BI$_3$
boronisation(=boronization)　渗硼
boron nitride　一氮化硼　BN
boron nitride fibre　氮化硼纤维
boron nitrite　氮化硼
borono-　二羟硼基　(HO)$_2$B—
boron oxide　氧化硼；硼酐　B$_2$O$_3$
boron phenyl difluoride　苯硼化二氟　C$_6$H$_5$BF$_2$
boron phosphate(=borophosphoric acid)　磷酸硼　BPO$_4$
boron phosphide　磷化硼　BP
boron pollution　硼污染
boron polymer　(含)硼聚合物
boron resin　硼树脂
boron silicate crown　冕牌硅酸硼玻璃
boron skeleton　硼骨架；硼骼
boron sulfide　三硫化二硼　B$_2$S$_3$
boron p-tolyl difluoride　二氟化对甲苯基硼　CH$_3$C$_6$H$_4$BF$_2$
boron tribromide　三溴化硼
boron trichloride　三氯化硼　BCl$_3$
boron triethyl　三乙硼
boron trifluoride　三氟化硼　BF$_3$
boron-trifluoride etherate　醚合三氟化硼〔三氟化硼-醚配合物〕
boron trifluoride ethyl ether complex　三氟化硼乙醚络合物　(C$_2$H$_5$)$_2$O·BF$_3$
boron trifluoride-methanol　三氟化硼合甲醇
boron trifluoride monoethylamine　三氟化硼合单乙胺
boron trimethoxy　三甲氧硼
boron trioxide　三氧化二硼　B$_2$O$_3$
boron trisulfide　三硫化二硼
borophenylic acid　苯基硼酸〔按无机命名原则,苯基在此表明代替了一个OH〕　C$_6$H$_5$B(OH)$_2$
borophosphoric acid(=boron phosphate)　磷酸硼　BPO$_4$
borosilicate　硼硅酸盐
borosilicate crown　硼硅(酸盐)冕
borosilicate flint　硼硅(酸盐)燧石
borosilicate glass　硼硅(酸盐)玻璃
borosilicate glass bead　硼硅酸盐玻璃珠
borosiloxane polymer　硼硅氧烷聚合物
borotungstic acid(=borowolframic acid)　硼钨酸　B$_2$O$_3$·9WO$_3$·24H$_2$O
borous acid　亚硼酸
borowolframic acid(=borotungstic acid)　硼钨酸
boroxane　硼氧烷
boroxene　硼氧烯
boroxin　环硼氧烷　B$_3$O$_3$H$_3$
borphenyl　三苯硼
borsal(–borsyl)　硼硅酸钠
Borsche carbazole synthesis　博歇咔唑合成法
borsyl(=borsal)　硼硅酸钠
bort　钻石粒；圆粒金刚石
borthiin　环硼硫烷　B$_3$S$_3$H$_3$
boryl　硼(烷)基　H$_2$B—
borylarsenate　(偏)砷酸氧硼　(BO)AsO$_3$
borylene　亚硼(烷)基　HB⟨

borylidyne　次硼(烷)基　B≡
borylphosphate　(偏)磷酸氧硼　(BO)PO₃
Bosch's fuel injection pump　勃赤燃料喷注泵〔柴油机〕
Bose-Einstein distribution　玻色-爱因斯坦分布*
Bose-Einstein statistics　玻色-爱因斯坦统计
boseimycin　勃赛霉素
bosh　①(鼓风)炉腹②浴；锅；槽；桶
bosh cooling box(=bosh cooling plate)　炉腹冷却器
bosh cooling plate(=bosh cooling box)　炉腹冷却器
boss　①凸起部；凸台②轮毂
boss ratio　内外径比
bostrycin　卷线孢菌素
bostrycoidin　葡萄孢镰菌素
boswellia resin　乳香树脂
boswellic acid　乳香酸
both sides welding　双面焊
botryolite　葡萄硼石
Böttcher chamber　伯特赫尔计数室
bottle capping machine　玻璃封盖机
bottle cleaning machine　洗瓶机
bottle corking machine　压瓶塞机
bottled gas　瓶装气体
bottle filler(=bottle filling machine)　装瓶机
bottle filling machine(=bottle filler)　装瓶机
bottle glass　瓶玻璃
bottle grade chip　(制)瓶级切片
bottle green　深绿色
bottle hood　瓶盖
bottle method　比重瓶法〔测定比重〕
bottleneck isotope　"瓶颈"同位素
bottleneck zone　瓶颈区；肩颈区
bottle-nose oil　槌鲸油
bottle-nose sperm oil　槌鲸蜡油
bottle-nose whale oil　槌鲸蜡油
bottle oiler　瓶式加油器
bottle polymerization　瓶式聚合
bottle rinser　洗瓶机
bottle sampling　瓶选样
bottle soaker　浸瓶机
bottle thickness gauge　瓶类测厚仪
bottle tilter　倾瓶机
bottle washer(=bottle washing machine)　洗瓶机
bottle washing machine(=bottle washer)　洗瓶机
bottle with cap　带盖瓶
bottle wraps　包瓶纸
bottling　装瓶
bottling machine　装瓶机
bottom　①残渣②底板

bottom allowances　底裕度〔储器底的余裕〕
bottom angle　底角
bottom application　底施；底敷用
bottom blowing　(吹塑)下吹
bottom breaker　下缓冲层
bottom brick　底砖
bottom bunching　打底纱
bottom calendar roll　下压延辊
bottom cap　底盖
bottom cavitation　(聚氨酯发泡缺陷)底部空洞
bottom center(=bottom dead center)　下死点
bottom clearance　径向间隙
bottom coat　底涂层
bottom color　底色
bottom dead center(=bottom center)　下死点
bottom discharge　下出料
bottom drain valve　底部放泄阀
bottom draw　底部吸引
bottom drier　底干催干剂；由里及表催干剂
bottom dyeing　底层染色
bottom ejection　下脱模
bottom feed waste oiler　底层给废料加油器
bottom filling of tank cars　油槽车底部装油
bottom fired pan　底热锅
bottom flue　底烟道；小烟道〔设〕
bottom force　底冲床
bottom heat pan　底热锅
bottom irregularity　底部不规整
bottom lacquer　底漆
bottom leather　底革
bottom loading of lank truck　油槽汽车底部装油
bottom millstone　①底石②底梁；底木〔设〕
bottom note　底香；尾香
bottom oil　残油；油脚
bottom paint　船底漆
bottom pan　储漆槽
bottom plate　底(层柱)板
bottom plate of column　分馏塔底塔盘
bottom position　底部位置
bottom product　(釜；塔)底部产物
bottom pulley　下鼓轮〔提升机〕
bottom ram　阴模；下半模
bottom reeling　底缥
bottom roll　①下滚筒②底辊
bottoms　①底沉积物；脚子②油脚
bottom sample　底部试样
bottom sampler　底部采样器
bottom scale　底标尺

bottoms cooler 油脚冷却器
bottom season 底光
bottom sediment ①底部籽晶(生长)法②底沉积物；脚子
bottom sediment and water 油脚与水；底部残渣与水
bottom settling and water 油脚与水；底部残渣与水
bottom settlings 底沉积物；脚子
bottom sizing ①底胶料②涂底
bottom sludge and water 油脚与水；底部残渣与水
bottom steam （引到蒸馏桶底(的)蒸汽
bottom stream 塔底流出物
bottom suspension 下支
bottom unloading of tank car 油槽车底部卸油
bottom valve 底阀
bottom view 底视图
bottom water 底水〔油罐的〕
bottom wire 下网
bottom zone 底层；底区
bottromycin 波卓霉素
botulin 肉毒(杆菌)毒素
botulinus toxin 肉毒毒素
Bouchardat reagent 布卡达特试剂
bougie ①瓷制多孔滤筒②栓剂③探条；探子
Bouguer-Beer law 布格-比尔定律
Bouguer-Lambert law 布格-朗伯定律
Bouin's fluid 布英防腐液
boulangerite 硫锑铅矿
boulder ①圆石②漂石
boulder clay 冰砾泥
bouldery ore 巨砾矿石
bounce back ①反射；反映②反冲；弹回；回跳③(喷雾粒子)反弹〔部分雾化粒子在飞行途中即固化，及至物体表面有反弹现象〕
bounce cylinder 缓冲气缸
bounce impact elasticity 冲击回弹性
bounce impact force 冲击回弹力
bouncing pin (indicator) 弹跳针〔石油〕
bouncing putty 弹性腻子；抗震腻子
bound acrylonitrile 结合丙烯腈
boundary ①边界②界面
boundary areas 边界面积
boundary coefficient 界面系数
boundary condition 边界条件
boundary conductivity ①边界传导性②边界传导率
boundary crack 界面裂纹
boundary effect 边界效应
boundary electrophoresis 界面电泳
boundary film 边界膜
boundary-film-forming material 形成边界膜的物质
boundary film lubrication 边界润滑；界限润滑
boundary friction 边界摩擦
boundary layer 边界层；界面层
boundary layer dilution 边界层稀释
boundary layer reaction(=interfacial reaction) 界面反应
boundary layer theory 边界层理论
boundary layer transition 边界层过渡
boundary lubrication 边界润滑
boundary material 界面物质
boundary member 限制器；停车器；边界构件
boundary penetration 边界渗透
boundary phase 界面相
boundary potential 界面电位
boundary reflectance 界面反射比
boundary strength 界面强度
boundary surface 界面
boundary tension 界面张力；边界张力
boundary value 边界值
boundary value problem 边界值问题
boundary velocity ①界面速度②极限速度
boundary zone 界面区
bound charge 束缚电荷
bound counterion 束缚反离子
bound energy 结合能
bound gas 结合气体
bounding 约束的；边界的
bound moisture(=bound water) 结合水
bound molecule 束缚分子
bound paraffin chains 结合烷链
bound residue of pesticide 农药结合残留
bound rubber 结合橡胶
bound solubilizate 结合加溶物
bound-state 束缚态
bound styrene 结合苯乙烯
bound sulfur 结合硫黄；化合硫黄
bound sulfur fraction 结合硫黄部分
bound water(=bound moisture) 结合水
bourbonal(=3-ethoxy-4-hydroxy benzaldehyde) 波旁醛；3-乙氧基-4-羟基苯甲醛
bourbonene 波旁烯
Bourdon gauge 波登管压力计；弹簧管压力计
Bourdon pressure gauge 波登式压力计
Bourdon tube gauge 波登管式压力计
Bourdon tube manometer 波登管压力表
bournonite 车轮矿
boussingaultite 铵镁矾
Bouveault aldehyde synthesis 鲍维特醛合成法
Bouveault-Blanc reaction 鲍维特-勃朗克反应

bovey coal 褐煤
bovine serum albumin(BSA) 牛血清清蛋白
bow ①纬纱移位〔纺〕②弯曲；扭曲③凹陷
bow camber 弓形弯曲
bowenite(=serpentine) 鲍文玉；蛇纹岩
bowing 板弯
bowl ①浮筒；辊筒②离心机转筒③滚球
bowl centrifuge 转筒离心机
bowl classifier 分级槽；分类槽
bowl grinding machine 滚球研磨机
bowl metall 粗铸锑〔99%纯的模铸锑〕
bowl mill 球磨机
bowl paper 砑光辊(用)纸
bowl-shell wall 离心机篮壁
bowl temperature 滚筒温度
Bowman & Cichelli's relation 鲍曼-西切利关系〔分批精馏时蒸馏曲线形状和最少回流比、最少理论板数的关系〕
bow mill(=ball mill) 球磨机
bow of weave 弓纬
box ①箱②匣③逻辑框
box board 盒纸板
box bridge 匣式电桥
box calf 纹皮；小牛皮鞋面革
boxcar ①矩形波串②箱车
boxcar detector 矩形波串检测器
BOXCAR integration BOXCAR 积分
box cementing 箱的黏固
box chamber 长方室
box condenser 箱式冷却器
box cooler 箱式冷却器
box corrugated board 波纹盒纸板
box distribution function 箱形
boxed dimension 装箱总尺寸
box filter 箱式滤器
boxing 混料；卷取
boxing-glove leather 拳击手套革
box lead chamber 长方铅室
box lining paper 衬箱纸
box lubricator 匣式润滑器
box negative plate 盒式阴极板
box of the spinning machine 纺机箱
box pilling tester 箱式起球试验仪
box section 箱形截面；箱形断面
box spanner 套管螺丝板
box spinning （离心）罐式纺丝
box spinning machine 离心纺机
boxspun yarn 离心罐纺长丝

box structure 盒(箱)式结构
box-toe 皮鞋包头
box-type extractor 箱式萃取器
box-type furnace 箱式炉
box type galvanometer 箱式电流计
box type heater 箱式加热炉
box-type relaxation spectrum 箱型松弛谱
box wrench 套筒扳手
Boyle's law 波义耳定律
Boyle-Marriote's law 波义耳-马里奥特定律
Boyle's temperature 波义耳温度
Boys calorimeter 博伊斯量热器
B powder(=powder B) B 火药
bra 左矢
Brabender plastograph 布雷本德塑性仪
bracelet 帘布筒
brachyaxis 短轴
brachydome 短轴坡面
brachytherapy 近程(放射)治疗*；短程治疗；近距治疗；浅部治疗
bracing of still 管式炉拉条
bracket ①托架〔机工〕；座架、搁脚〔商〕②方括弧
bracketing method(=interior extrapolation method) 交叉法〔一种定量法〕；插入法
bracketing method of mass measurement 插入法质量测定
bracketing technique 高低射标技术
brackish 稍咸的
brackish cooling water 半咸冷却水
brackishness 半咸性
brackish water 微咸水
bracteine 苞罂粟碱
Bradford crude oil 布拉德福德原油〔美〕
Bradley blade 布来得雷刮(涂)膜器
bradykinin (血管)舒缓激肽
Bragg's equation 布拉格方程
Bragg angle 布拉格角*
Bragg curve 布拉格曲线〔α粒子、裂片的射程-能量关系曲线〕
Bragg diffraction 布拉格衍射
Bragg equation 布拉格方程*
Bragg-Gray cavity principle 布拉格-葛瑞空腔原理〔辐射化学〕
Bragg law 布拉格定律
Bragg method 布拉格法〔X 射线晶体分析〕
Bragg reflection 布拉格反射
Bragg scattering 布拉格散射
Bragg scattering angle 布拉格散射角

Bragg's condition of reflection　布拉格(的)反射条件
Bragg spectrometer　布拉格分光计；布拉格光谱仪
Bragg-William's approximation　布拉格-威廉姆斯近似法
braid　①编织②编织物
braided construction hose　编织胶管
braided fabric　编织物
braided nylon hose　尼龙编织软管
braided ply　编织(加强)层
braided ring　编织环
braiding　编织
braiding machine　编织机
braiding yarn shrinkage factor　编结物纱线收缩系数
braid-lamella pattern　菱(形锯)齿花纹〔轮胎〕
braid ply　编织层
brake　①制动；刹车②制动器；刹车；闸
brake band　制动带
brake-block　制动块；闸瓦
brake boot　制动块；闸瓦
brake disc　制动摩擦盘
brake dressing　制动器润滑脂
brake drum　制动鼓
brake fluid　制动液；刹车油
brake horsepower　实际马力；制动马力
brake lining medium　闸衬片介质；制动衬片介质〔石棉和金属丝等〕
brake metering valve　闸板调节阀
brake pedal　刹车踏板
brake-shoe　制动块；闸瓦
brake specific fuel consumption　(在制动试验台上试验发动机时的)比燃料消耗
brake strip　制动摩擦片
bra-ket vectors of dual space　二度空间的左-右矢量
brake valve　制动阀；闸阀
braking　制动减速
braking radiation(=internal bremsstrahlung)　(内)轫致辐射
bramycin　短霉素；妥布霉素
branch　支；枝
branched　分支的
branched alkyl　支链烷基
branched capillary　分支毛细管
branched (carbon) chain　支(碳)链
branched-chain alkene　支链烯烃
branched-chain alkyl　支链烷基
branched chain amino acid　分支氨基酸
branched chain compound　支链化合物
branched chain explosion　支链爆炸*

branched chain fatty acid　支链脂肪酸
branched chain group　支链基
branched chain hydrocarbon　支链烃
branched chain mechanism　支链机理
branched chain polymer　支链聚合物
branched chain reaction　支链反应*
branched compound　支链化合物
branched copolyester fibre　支化共聚(酯)纤维
branched crack　支链裂解
branched cyclic hydrocarbon　支链环烃
branched decay　分支衰变
branched dimer　支化二聚物
branched-fibril　支化原纤
branched group　支链基
branched hydrocarbon　支链烃
branched isomer　支化异构体
branched molecule　含支链分子
branched monomer　支化单体；支链单体
branched oligomer　支链低聚物；支链齐聚物
branched paraffin　支链烃
branched polyethylene　支化聚乙烯
branched polyethylene terephthalate　支化聚对苯二甲酸乙二(醇)酯
branched polymer　支化聚合物*
branched structure　支化结构
branch high polymer　支化高聚物
branching　①分支；支化②出纹；起梗〔表面缺点〕
branching atom　支折原子
branching coefficient　支化系数
branching content　支链含量
branching decay　分支衰变*
branching density　支化密度
branching index　支化系数
branching mechanism　支化机理
branching probability　支化概率
branching ratio　分支比*
branch pipe　支管
branch point　支化点
branch polymer　支化聚合物；支链聚合物
branch tube　支管
branch unit　支化单元
brand　①火印；烙印②商标
branded oil　优质油
branding　打烙印
branner　磨光机
brashiness　脆性；易碎性
brasileic acid　巴西勒酸
　　$Me(CH_2)_7(CHOH)_2(CH_2)_{12}COOH$

brasilein(=brazilein)　巴西红木精　$C_{16}H_{12}O_5$
brasilic acid(=brazilic acid)　巴西酸　$C_{12}H_{12}O_6$
brasilin　巴西灵　$C_{16}H_{14}O_5$
brasq(=brasque)　①衬料；填料②炉衬
brasque(=brasq)　①衬料；填料②炉衬
brass　黄铜*
brass-fitted cast-iron valve　黄铜配制铸铁阀
brass foundry　①黄铜铸造②黄铜铸件③黄铜铸件厂
brassicasterol　菜子甾醇
brassica steryl propionate　丙酸芸苔甾醇酯
brassicin　蔓菁苷；异鼠李黄酮葡糖苷
brassidic acid(=13-dodecosenoic acid)　巴西烯酸；顺芥酸；(Z)-△¹³-二十二碳烯酸　$C_{21}H_{41}CO_2H$
brassidin　巴西精；巴西烯酸甘油酯；反-13-二十二碳烯酸甘油酯
brassilic acid　巴西基酸；十三烷二酸
　　$COOH(CH_2)_{11}COOH$
brassiness　①黄铜质②黄铜色
brassing　黄铜铸件
brass pipe(=brass tube)　黄铜管
brass plate　黄铜压平板
brass-plating　镀黄铜
brass polish　黄铜光亮剂
brass rivet　黄铜铆钉
brass slicker　黄铜刮刀
brass substrate　黄铜基材；黄铜底料
brass syringe　黄铜喷油器
brass-trimmed valve　黄铜配制阀
brass tube(=brass pipe)　黄铜管
brass valve　黄铜阀
brass worm wheel　黄铜蜗轮
brass-yarn weft　黄铜丝纬线
brassylate　巴西酸盐(或酯)；十三烷二酸盐(或酯)
brassylic acid　巴西基酸；十三烷二酸
　　$HOOC(CH_2)_{11}COOH$
brat　不净煤；原煤
brattice　隔布；围板
braunite　褐锰矿
Braun sample grinder　布老恩样品磨研机
Braun's native lignin(BNL)　布朗氏天然木素
bravaisite　漂云母
Bravais lattice　布拉菲晶格
Bravais-Miller indices　布拉菲-密勒指数
brayer roll　刷色滚(辊)；涂墨滚(辊)
brazan　巴西烷；苯并氧芴
braze　铜焊(接)
brazed copper tube　黄铜管
brazed joint　黄铜接头
braze wilding　铜(钎)焊
brazil copal　巴西珀把树脂
brazilein(=brasilein)　巴西红木精〔可用作天然染料〕$C_{16}H_{12}O_5$
brazilian copal　巴西珀把树脂
brazilian rosewood oil　巴西玫瑰木油
brazilian sassafras oil　巴西黄樟油
brazilic acid(=brasilic acid)　巴西酸
brazilin　巴西木红素；苏枋精　$C_{16}H_{14}O_5$
brazil wax　巴西棕榈蜡
brazil-wood　巴西红木
brazing　①铜焊(接)②钎焊③硬钎焊
brazing filler metal　钎料
brazing solder　焊铜
brazo-　巴西并
brazylium　巴西鎓
bread crumber　面包心磨
bread pan coating　面包烤盘涂料
breadth　宽度
breadth of particle　粒子短轴；粒子宽度
breadth of spectral lines　谱线宽度
bread wrapper　面包纸
bread wrapper adhesive　面包包装纸黏合剂
break　①裂缝；破裂口②(滴定)突跃③革面花纹
breakage　破坏；破裂；断裂
breakage cleavage(=cleavage breakage)　破碎
break-away　开模
break-away friction　滑动摩擦
breakaway speed　开放速度
breakdown　破坏；破裂；断裂；击穿；中断
breakdown aid　塑解剂
breakdown cycle　塑炼周期
break-down maintenance　事故维修；停工检修
breakdown mill　碎研磨
breakdown of boundary film　界限膜破裂
breakdown of carbon-carbon bond　碳-碳键断裂
breakdown of coating　涂层破坏
breakdown of flame　火焰中断；熄火
breakdown of fuel　燃料分散
breakdown of gasoline　汽油破坏〔在氧弹中测定汽油诱导期时压力的急剧下降〕
breakdown of greases　润滑脂破坏；润滑脂分油
breakdown of hydrocarbon　烃断裂
breakdown of oil　油澄清
breakdown of oil film　油膜破坏
breakdown point　①破裂点②强度极限③崩溃点
break-down potential　击穿电位
breakdown product　分解产物

breakdown rate ①断裂速率②分解速率
break-down strength 破坏强度；断裂强度
breakdown tension 击穿电压
breakdown test 破坏试验；稳定性试验
breakdown time 破坏期；汽油诱导期
breakdown voltage 击穿电压
breakdown voltage of membrane 膜击穿电压
breaker ①破乳剂②缓冲层③破碎机④断路器
breaker beater 梳解机
breaker plate ①多孔板②分散板③安全片
breaker rubber 缓冲层胶〔轮胎〕
breaker squeegee 缓冲层隔离胶〔轮胎〕
breaker strip ①缓冲层膈离胶〔轮胎〕②缓冲帆布层〔运输带〕
break-free linseed oil 未裂解的亚麻仁油
break-free oil(=non-break oil) 未裂解(的)油〔用机械法精制油〕
break in fine lines (凹版)细线条断裂〔印刷故障〕
breaking ①破坏②断裂③轧碎④断刀
breaking cup 安全片
breaking down 断裂〔分子〕
breaking-down of hydrocarbons 烃断裂
breaking-down of paraffins 烷烃断裂
breaking-down process 断裂过程
breaking-down test 断裂性试验；破坏性试验
breaking elongation 断裂伸长
breaking emulsion 乳液破碎
breaking energy 断裂能
breaking energy density 断裂能密度
breaking extension 致断伸长
breaking force 断裂力
breaking-in period ①试车时期；试产时期②开动期
breaking length 断裂长
breaking load 致断载荷；断裂负载
breaking machine 致断力试验机
breaking of carbon skeleton 碳骨架的破坏
breaking of emulsion 破乳〔乳浊液破坏〕
breaking of grease by water 润滑脂水解
breaking of oil 油澄清
breaking of vacuum 真空下降
breaking of viscosity 减黏裂化
breaking petroleum emulsion 石油破乳
breaking point ①断点②破乳点
breaking reaction(=destructive reaction) 破坏性反应
breaking resilience ①致断回弹②(=breaking energy) 致断回能
breaking specific strength 拉伸断裂比强度
breaking squeal 猛刹车声

breaking strain 破坏应变；断裂应变；断裂变形
breaking strength ①断裂强度②抗断强度〔物〕
breaking stress 断裂应力
breaking tenacity ①断裂强度②扯断力
breaking test 断裂试验；强度试验
breaking weight 破坏荷重
break-in oil 磨合用油〔新车或机器磨合时用的高润滑性润滑油〕
break length 断裂长度；抗断长度
break of oil 油加热后沉淀
break out 从平环内破出
break out device 单纱打断器
break-out material 易磨损材料
break period 试运转时间
break plate 分配板
break point ①转效点②折点③破乳点
breakseal 破坏(式)密封
break strength 断裂强度
break surface 断裂表面
break test 破坏试验〔试验油的一种方法〕
breakthrough ①穿透；突破②技术革新
breakthrough capacity 漏过能量
breakthrough curve 漏出曲线
breakthrough experiment (离子交换)穿透实验；流穿实验
breakthrough method 漏出法
breakthrough point 突破点
breakthrough point analysis 漏出(点)分析
breakthrough volume 漏出(点)容量
break-up of catalyst 催化剂粉碎
break-up of crude oil 原油馏分组成
break-up pattern 崩裂图案；崩裂型式
break-up time 崩裂时间
breaky 折裂
breast roll 中心辊；胸辊
breast wall temperature (玻璃熔窑内的)胸墙温度
breathability 透气性
breathable film ①透气性薄膜②透气性漆膜
breathable laminate 透气(性)层压制件
breathable plastic(s) 透气性塑料
breather 透气材料
breather roof 呼吸顶；油罐浮顶
breather valve 呼吸阀
breathing ①放气②呼吸
breathing a mould 开模排气；开模放气
breathing area of booth 喷漆橱的通气面积
breathing bag 呼吸袋
breathing coating 透气性涂料

breathing dryer 放气干燥机；放气压榨机
breathing fabric ①透气性人造革②透气性漆布
breathing film 透气薄膜
breathing mask 呼吸面具
breathing roof (油罐的)浮顶；呼吸顶
breathing roof tank 浮顶罐
breathing tank 呼吸式油罐；浮顶(弹性顶)油罐
breathing vibration 呼吸振动
breath of tooth 齿宽
breccia 角砾岩
bredinin 布雷青霉素；布累迪宁
Bredt rule 布雷特规则*
bred uranium 增殖(生成的)铀
breeching 烟道；马主退索
breeder reactor 增殖(反应)堆
breeding-fire 自燃
breeze ①煤粉；煤末②矿粉
bregenin 脑氨脂〔一种氨基脂〕
brei 糊；浆
brein 榄香树脂醇；布惹因
breithauptite 红锑镍矿
Breit-Wigner formula 布赖特-维格纳公式
Bremen blue 布勒门蓝〔颜料〕
Bremen green 布勒门绿〔颜料〕
bremsstrahlung 轫致辐射
bremsstrahlung interrogation 轫致辐射探询
bremsstrahlung isochromat spectroscopy 轫致辐射等色体谱法
bremsstrahlung source 轫致辐射源*
bremsstrahlung X-ray 轫致辐射 X 射线
brenzcain 愈创木酚苄酯
brenzcatechin 焦儿茶酚
Bresk 节路顿胶
bretylium tosilate 托西溴苄铵〔药〕
bretylium tosylate 溴苄铵托西酸盐
breunnerite 铁菱镁石
brevifolin 云实素 $C_{12}H_8O_6$
brevifolin carboxylic acid 云实酸 $C_{13}H_8O_5$
brevilagin 云实鞣花素
brewer's pitch 啤酒厂用沥青；啤酒桶涂料〔用松香、油、蜡及其他材料制成的啤酒桶涂料，并非沥青漆〕
Brewster's law of photo-elasticity 布鲁斯特光弹性定律
Brewster angle 布鲁斯特角
brewsterite 锶沸石
brick clay 砖土
brick condenser 砖制冷凝器
brick die 砖模
brick earth 砖土

bricked-in 砖衬的
brick furnace(=brick kiln) 砖窑
brick grease 脂砖；砖块状润滑脂
bricking 砖衬
bricking-up 用砖填塞
brick kiln(=brick furnace) 砖窑
brick lining 砖衬
brick machine 造砖机
brick oil 砖油；压砖机润滑油
brick press 压砖器
brick red 砖红色
brick setting 砖砌体
brickwork 砌砖工作
brickwork casing ①砖工刷灰；砖工整理②砌砖工作
bridge ①电桥②跨接；桥接
bridge agent 桥键(形成)剂
bridge amplifier 桥式放大器
bridge arm 电桥臂
bridge atom (搭)桥原子
bridge bearing plate 桥梁座垫；桥梁支板
bridge bond 桥键
bridge breaker (料斗)防堵器
bridge circuit 桥式线路
bridge clamp 桥式夹
bridge coating 桥梁涂料
bridge current 电桥电流
bridge current indicator 桥流指示器
bridge current regulator 桥流调节器
bridged complex 桥联式络合物；多核金属络合物
bridged group 桥(连)的)基
bridged hydrocarbon 桥烃
bridged impedance 桥接阻抗
bridged ion 桥联离子
bridged metallocene 桥基茂金属
bridged ring 桥环
bridged-ring system 桥环体系*
bridged-surface hydroxyl 桥接表面羟基
bridge electrometer 桥式静电计
bridge element 桥渡元素
bridge formation 桥结(现象)
bridgehead displacement 桥头取代*
bridgehead mercurial 桥头汞剂
bridgehead nitrogen 桥头氮
bridge hydrogen 桥氢
bridge hydroxyl group 桥式羟基
bridge isomerism 桥式异构；桥键异构
bridge ligand 桥形配体
bridge linkage 桥式联接

bridge network 桥接网络
bridge of valve 阀桥〔设〕
bridge out-of-balance 桥流不平衡
bridge oxygen 桥氧
bridge piece 连接件；横梁
bridge test （电)桥(阻)测验
bridge-type crosslink 桥型交联
bridge wall 坝；管式炉的坝墙
bridge wall tube 坝墙管子
bridge washer 阀桥衬垫
bridging （涂膜的)搭接覆盖；搭桥；桥接
bridging agent 桥键形成剂
bridging amplifier 桥式放大器
bridging atom 桥联原子；桥键原子
bridging bond 桥联键
bridging effect 遮蔽作用；跨隙效应
bridging fine crack 遮蔽细裂缝
bridging group 桥连基*
bridging immobilization 架桥化固定法
bridging ligand 桥连配体*
bridging mechanism 桥联机理
bridging theory 桥连理论
Bridgman-Stockbarger method 晶体生长坩埚下降法*
bridle 辊式张紧装置
brige current indicator 桥流指示器
bright 嫩〔颜色)；明亮的
bright coal 亮煤
bright colour 亮色
bright-coloured preservative stain 浅色防腐着色剂
bright deposit 光亮沉积层(膜)
bright dip 浸渍抛光；浸涂磨光
brightener 增白剂；增艳剂；增亮剂
brightening ①纯化②擦亮；上光
brightening agent ①增白剂②增艳剂③光亮剂〔电镀〕
brightening effect 增白效应
bright-field electron microscopy 亮视野电子显微术
bright-field image 明场像
bright finish ①上光②上光剂③光面
bright fracture 亮断面
bright gold(=liquid gold) （陶瓷描金用)装饰金；亮金；金水
bright line spectrum 明线光谱
brightness ①亮度②光泽③白度
brightness contrast 亮度对比；白度对比
brightness flop （随观测方向而异的)亮度变化
brightness in oil 着油亮度〔颜料着油后其颜色的亮度〕
brightness of electron gun 电子枪亮度
brightness reversion 回色；返黄
brightness temperature 亮白温度
brightness test 明度试验
bright rayon 有光人造丝
bright red ion oxide （合成)氧化铁鲜红
bright sap 净面树材；无皮边材
bright stock 光亮油；重质高黏度润滑油料
bright sulfur 纯硫
bright thread 亮丝；紧经纱
bright transparent color 光亮透明彩色漆
bright yellow 嫩黄
briliant fracture 亮断面
brilliance 亮度；(涂料)明度
brilliancy ①光辉②明亮③亮度
brilliant 亮的
brilliant bordeaux 亮枣红
brilliant cresyl blue 亮甲酚蓝
brilliant crimson 亮绯红
brilliant croceine 亮藏花精
brilliant green 亮绿*
brilliant indigo （还原类)亮靛蓝
brilliantine 润发油
brilliant lanafuchsine 亮蓝纳品红
brilliant metallic effect 明亮的金属效应
brilliant oil black base 亮油黑色基
brilliant orange 亮橙；亮橘红
brilliant phosphine 亮金黄
brilliant ponceau 亮丽春红
brilliant spot 亮斑
brilliant violet 亮紫
brilliant wool blue 亮毛蓝
brilliant yellow 亮黄
Brillouin scatter 布里渊散射
Brillouin scattering 布里渊散射
Brillouin spectroscopy 布里渊光谱法
Brillouin theorem 布里渊定理*
brimstone 硫黄
brimstone acid 硫酸
brimstone burner 燃硫炉
brimstone burner gas 燃硫炉气
brimstone furnace 燃硫炉
brimstone gas 燃硫炉气
brimstone pan burner 燃硫锅炉
brin 单丝
brindle 斑纹；虎斑
brine 盐水
brine change 盐洗
brine cooler 盐水冷却器
brine cooling 盐水冷却

brine corrosion 盐水腐蚀
brine corrosion inhibitor 盐水腐蚀缓蚀剂
brine drum 盐水桶
brine header 盐水集管
brine line 盐水管
Brinell figure 布氏硬度数
Brinell hardness 布氏硬度*
Brinell hardness tester 布氏硬度试验机
Brinell test 布氏(硬度)试验
brine pan 蒸盐锅
brine pipe cooling 盐水管冷却
brine pit ①蒸盐锅②盐井
brine pump 盐水泵
brine refrigeration 盐水冷却
brine spring 盐泉
brine tank 盐水池；盐水箱
brine wash 盐洗
brine well 盐井
bring down ①浓缩②使下降③收缩
bringing-up section 加热或辐射段〔管式炉或裂化炉〕
bring on stream 开始通油
brining 浸盐液
brinish ①盐水的②咸的
brinishness 含盐度
briquet(=briquette) ①煤砖②炭砖
briquette(=briquet) ①煤砖②炭砖
briquette method 团固法；压块法
briquetting ①团固②压块；压制成块
briquetting asphalt 煤砖用沥青
briquetting machine 压块机
briquetting press 压块机
brisance 爆炸威力
brisk rub down 快速磨光；急剧抛光
bristle ①鬃丝②短粗纤维；硬纤维
bristle-like monofilament 鬃状单丝；鬃丝
Bristol board 细料纸板
Bristol glaze 窑釉
Bristol paper 图案纸
British barilla 海草灰
British gum 糊精
British water hardness units 英国的水硬度单位
britonite 脆通炸药〔商〕；波利通那特〔硝化甘油、硝酸钾、草酸铵炸药〕
brittle 脆的；脆性的；易碎的
brittle behavior 脆性性能
brittle coating 脆性涂层
brittle cracking 脆性开裂
brittle creep failure 脆性蠕变破坏

brittle ductile transition 脆韧性转变
brittle-ductile transition temperature 脆性-延性转变温度
brittle failure 脆性破坏
brittle fibre 脆(性)纤维
brittle-flexible transition 脆性-挠性转变
brittle fracture 脆(性)断裂
brittle material 脆性材料
brittleness ①脆性②变脆③脆度
brittleness index 脆性指数
brittleness temperature 脆性温度；脆化点
brittle point 脆点
brittle resistance 耐脆化性
brittle rupture 脆性断裂
brittle silver ore 脆银矿
brittle strength 脆化强度
brittle temperature 脆化温度
brittle-tough transition 脆韧转变
Brix degree 糖度〔比重单位〕
Brix hydrometer 白利比重计
Brix scale 白利刻度
Brix spindle 白利糖度计
broach ①拉刀②拉削
broaching 拉削
broaching machine 拉床
broad 扩孔刀具
broad adhesion characteristics 广谱附着性；广谱黏附(结)性
broad band 宽谱带
broad band absorption 宽带吸收
broad band absorption interference 宽带吸收干扰
broad band decoupling 宽带去耦
broad band filter reflectometry 宽带滤光反射法
broad band NMR 宽带核磁共振
broad band pipette 宽带点样管
broad band radiation 宽带辐射
broad bean oil 阔叶蚕豆油
broad classification 粗分类法
broaden color 复色
broadening 增宽；变宽；展宽
broadening correction 加宽校正
broadening correction factor 加宽校正因子
broadening effect 展宽效应；增宽效应
broadening factor 扩张因子；加宽因子
broadening of spectral lines 谱线展宽*
broadening reagent 展宽试剂；增宽试剂
broad knife 铲刀；宽刮刀；填缝刮刀
broad leaf wattle 阔叶相思树
broad leaved peppermint 阔叶桉

broad-leaved wood　阔叶树材
broad peak　宽峰
broad rib fabric　宽罗纹织物
broad-side paint　舷侧漆（船）
broad spectrum　广谱
broad-spectrum antibiotics　广谱抗菌素
broad spectrum antimicrobicide　广谱(性)抗菌剂
broad term　主要项目
broadwise　横向；纬向
brocade　锦(缎)
brochantite　水胆矾；水硫酸铜矿
brockie oil　陶瓷业用润滑油
Brockmann activity grade　布鲁克曼活性级别
Brockmann activity scale　布鲁克曼活度值
bröggerite　铀钍矿
Brokaw method　勃诺考测定法〔测定甘油单酸酯〕
broke　废纸
broke beater　废物磨粉机
broken　破碎的
broken color　复色；配合色
broken colour effect　①(彩色)斑驳辉映效应②复色效应
broken-down hydrocarbon　已断裂的烃
broken-down rubber　塑炼胶
broken emulsion　已破坏的乳化液
broken fiber　破碎纤维
broken-open perspective view　剖开透视图
broken ore　碎矿石
broken rubber　塑炼胶
broken-staple metal fibre　金属短纤维
broken stone　碎石
broken white　近似白色〔稍带黄头或棕头的白色〕
bromacetanilide(=acetbromanilide)　乙酰溴苯胺
bromacetate　溴乙酸盐　$CH_2BrCOOM$
bromacetic acid　溴乙酸　$CH_2BrCOOH$
bromacetol　2,2-二溴丙烷
bromacyl bromide　溴代酰溴
bromacyl chloride　溴代酰氯
bromal(=tribromoacetaldehyde)　三溴乙醛；溴醛〔俗〕　Br_3CCHO
bromal hydrate　溴醛合水；水合三溴乙醛　$Br_3CCH(OH)_2$
bromalin　布鲁马林〔药〕；咪唑立宾
bromalonic acid　溴丙二酸
bromamide　三溴苯胺
bromamphenicol　溴代氯霉素
bromanil(=tetrabromoquinone)　四溴代(对)苯醌　$O=C(CBr)C=O$
bromanilic acid(=2,5-dibromo-3,6-dihydroxy-p-benzoquinone)　2,5-二溴-3,6-二羟对苯醌
bromargyrite　溴银矿
bromarsenazo I　溴偶氮胂 I
bromate　溴酸盐
bromated(=brominated)　溴代的；溴化了的；含溴的
bromate ion　溴酸根离子　BrO_3^-
bromate titration　溴酸盐滴定
bromatimetric titration　溴酸盐滴定
bromatimetry　溴酸盐滴定
bromating　溴化
bromating agent　溴化剂
bromation(=bromination)　①溴化(作用)②溴处理
bromaurate　溴金酸盐　$M[AuBr_4]$
bromaurite　溴亚金酸盐　$M[AuBr_2]$
brombenzamide　溴苯甲酰胺
brombenzene(=phenyl-bromide)　溴苯
brombutyl　溴丁基
bromcamphor　溴代樟脑
bromchlorphenol blue　溴氯酚蓝　$C_{19}H_{10}Br_2Cl_2O_2S$
bromcholine　溴胆碱
bromcresol green(=tetrabromo-m-cresolsulfonphthalein)　溴甲酚绿〔即溴甲酚蓝〕；四溴间甲苯酚磺酞
bromcresol purple　溴甲酚紫
bromelia　2-乙氧萘
bromeosin　曙红
bromethane　①溴甲烷②溴乙烷
bromethol　三溴乙醇
bromethyl(=ethyl bromide)　乙基溴；溴乙烷
bromethyl tolyl ketone　溴乙基甲苯基(甲)酮；2-溴-1-甲苯基乙酮　$CH_3C_6H_4COCH_2Br$
bromhexine　溴己新〔药〕
bromhydrin　溴代醇〔BrROH 型有机化合物〕
bromic acid　溴酸　$HBrO_3$
bromic ether　溴代乙烷　C_2H_5Br
bromide　①溴化物②溴化物乳剂
bromide chloride　一氯化溴　$BrCl$
bromide extraction system　溴化物萃取系
bromide ion-selective electrode　溴离子选择性电极
bromide paper　①溴素纸②放大纸
bromide rubber　溴化橡胶
brominated(=bromated)　溴代的；溴化了的；含溴的
brominated anthanthrone orange　溴化蒽缔蒽酮橘黄(橙)
brominated butyl rubber　溴化丁基橡胶
brominated flame retardant　溴基阻燃剂
brominated hydrocarbon　溴化烃
brominating　溴化
brominating agent　溴化剂
bromination(=bromation)　①溴化(作用)②溴处理

bromination catalyst 溴化催化剂
bromination method 溴化法
bromination reaction 溴化反应
bromination titration method 溴代滴定法
bromindigo 溴靛蓝
bromine 溴〔35号元素，化学符号 Br〕
bromine absorption value 溴吸附值
bromine cyan(=bromine cyanide) 溴化氰 BrCN
bromine cyanide(=bromine cyan) 溴化氰
bromine flask 溴(滴定)瓶
bromine fluoride 氟化溴 BrF
bromine hydrate 十水合溴 $Br_2 \cdot 10H_2O$
bromine hydride 溴化氢 HBr
bromine index 溴指数〔100克油样所耗溴的毫克数〕
bromine in ring 环上溴
bromine in side chain 支链溴
bromine iodide(=bromine monoiodide) 碘化溴 IBr
bromine monochloride 一氯化溴 BrCl
bromine monoiodide(=bromine iodide) 碘化溴
bromine number(=bromine value) 溴值*
bromine pentachloride 五氯化溴 $BrCl_5$
bromine still 蒸溴器
bromine sulfide 二硫化二溴 Br_2S_2
bromine test 溴化试验
bromine trifluoride 三氟化溴
bromine value(=bromine number) 溴值*
bromine water 溴水
bromite ①亚溴酸盐②溴银矿
bromizate 用溴处理
bromizated 溴化了的；溴处理过的
bromizating 溴处理
bromizating agent 溴化剂
bromization ①溴化作用②溴处理
bromlite 碳酸钙钡矿
bromo 溴代；溴基 Br—
1-bromoacenaphthene 1-溴(代)-1,2-二氢苊 $C_{12}H_9Br$
bromoacetal 溴乙缩醛；二乙氧基乙溴 $BrCH_2CH(OC_2H_5)_2$
bromoacetaldehyde 溴(代)乙醛 CH_2BrCHO
N-bromoacetamide N-溴乙酰胺 $CH_3CONHBr$
bromoacetanilide N-乙酰溴苯胺
bromoacetic acid 溴乙酸 $BrCH_2COOH$
bromo-acetoacetic ester 溴代乙酰乙酸乙酯 $CH_3COCHBrCOOC_2H_5$
bromoacetone 溴丙酮 $BrCH_2COCH_3$
bromoacetonitrile 溴乙腈 $BrCH_2CN$
bromoacetyl bromide 溴乙酰溴 $BrCH_2COBr$
bromoacetyl chloride 溴乙酰氯 $CH_2BrCOCl$

bromoacetylene 溴乙炔 $CH{\equiv}CBr$
bromo-acid 溴代酸
bromo-acid amide 溴代酰胺 $RBrCONH_2$
bromo-acrylic acid 溴代丙烯酸
bromo-aliphatic compound 溴代脂肪族化合物
bromoalkane 溴烷烃
bromoalkylbenzene 溴烷基苯
bromo-amine 溴胺
bromo-amino acid 溴代氨基酸 $NH_2RBrCOOH$
N-bromoantipyrine N-溴安替比林 $BrC_6H_4C_5H_7ON_2$
bromo-benzal α,α-二溴甲基苯 $C_6H_2CHBr_2$
bromobenzene 溴苯 C_6H_5Br
bromo-benzene sulfonic acid 溴苯磺酸 $C_6H_4BrSO_3H$
bromo-benzoic acid 溴苯甲酸 BrC_6H_4COOH
bromo-benzoyl bromide 溴苯甲酰溴 BrC_6H_4COBr
bromo-benzoyl chloride 溴苯甲酰氯 BrC_6H_4COCl
p-bromobenzoylpyrazolone(BBPY) 对溴苯甲酰吡唑啉酮
bromobenzylcyanide 溴氰基苄；溴苯乙腈 $BrC_6H_4CH_2CN$
2-bromobutane 2-溴丁烷
1-bromo-2-butanone 1-溴(代)-2-丁酮
2-bromobutene 2-溴丁烯
bromobutyraldehyde 溴丁醛
bromo-butyric acid 溴丁酸 $C_3H_6BrCOOH$
bromocaffeine 溴代咖啡碱 $C_8H_9O_2N_4Br$
bromo-camphor 溴樟脑
bromocaproic acid 溴己酸
bromochloro-acetic acid 溴氯乙酸 $CHBrClCOOH$
bromochlorobenzene 溴氯苯
bromochlorodifluoromethane 溴氯二氟甲烷
bromochloroethane 溴氯乙烷
bromochlorophene 溴氯芬
bromocinnamaldehyde 溴肉桂醛
bromo-cinnamic acid α-溴代肉桂酸 $C_6H_5CH{=}CBrCOOH$
bromocresol 溴甲酚
bromocresol green 溴甲酚绿*
bromocresol purple 溴甲酚紫
bromocriptine 溴隐亭〔药〕
bromocrotonic acid 溴巴豆酸
bromo-cumene 异丙基溴苯 $(CH_3)_2CHC_6H_4Br$
bromo-cyanogen 溴化氰 BrCN
bromocymene 溴代伞花烃 $(CH_3)_2CHC_6H_3(Br)CH_3$
5-bromodeoxyuridine 5-溴脱氧尿苷
bromodichloromethane 溴二氯甲烷 $CHBrCl_2$
bromodiethyl ether (一)溴二乙醚 $BrCH_2CH_2OC_2H_5$
bromodiiodomethane (一)溴二碘甲烷；二碘甲溴

BrCHI$_2$
bromoemimycin 溴血霉素
bromoergocryptine 溴麦角环肽
bromo-ester 溴酯
bromoethane(=ethyl bromide) 乙基溴；溴乙烷 C$_2$H$_5$Br
2-bromoethanol(=ethylene bromohydrin) 2-溴乙醇 BrCH$_2$CH$_2$OH
bromo-ether 溴醚
5-bromo-2-ethoxyacetanilide 5-溴-2-乙氧基乙酰苯胺
bromoethyl acetate 乙酸溴乙酯
bromo-ethyl ester 溴乙基酯 RCOOC$_2$H$_4$Br
bromo-ethyl-methyl ether 溴乙基甲基醚；溴乙氧基甲烷 C$_2$H$_4$BrOCH$_3$
2-bromoethyl phenyl ether 2-溴乙基苯基醚
bromoform 溴仿；三溴甲烷 CHBr$_3$
bromoformin 布鲁马林〔药〕
bromofumaric acid 溴代富马酸；溴代反丁烯二酸；(Z)-溴代丁烯二酸 CH=CBr(CO$_2$H)$_2$
1-bromoheptane 1-溴庚烷
1-bromohexane(=hexyl bromide) 己基溴；1-溴己烷 CH$_3$(CH$_2$)$_4$CH$_2$Br
bromo-hydrocinnamic acid 溴苯丙酸；溴代氢化肉桂酸 C$_6$H$_4$Br(CH$_2$)$_2$COOH
bromohydroquinone 溴氢醌；溴代苯对二酚 BrC$_6$H$_3$(OH)$_2$
bromoil process 釉溴印画法；相纸上墨法〔使相纸的影像洗去，转为用墨或颜料的方法〕
bromoiodomethane 溴碘甲烷 BrCH$_2$I
bromoketone 溴酮 BrCH$_2$COCH$_2$CH$_3$
bromol 2,4,6-三溴酚
bromolauric acid 溴月桂酸；溴代十二(烷)酸
bromomaleic acid 溴代马来酸；(E)-溴代丁烯二酸 CH=CBr(CO$_2$H)$_2$
bromomalonic acid 溴代丙二酸 BrCH(CO$_2$H)$_2$
bromo-mercuriphenol 溴汞基苯酚 HOC$_6$H$_4$HgBr
bromo-mercury-benzene 溴化苯汞；溴汞基苯 C$_6$H$_5$HgBr
bromomesitylene 溴代莱；2-溴-1,3,5-三甲苯 BrC$_6$H$_2$(CH$_3$)$_3$
bromomethane(=methyl bromide) 溴代甲烷 CH$_3$Br
bromomethyl acetate 乙酸溴甲酯 CH$_3$CO$_2$CH$_2$Br
bromomethylation 溴甲基化
bromo-methyl-ether 溴甲基醚 CH$_2$BrOR
bromomethyl β-naphthyl ketone 溴甲基β萘基甲酮；溴乙酰萘 C$_{10}$H$_7$COCH$_2$Br
5-bromo-2-methylpentane(=isohexyl bromide) 异己基溴；5-溴-2-甲基戊烷 (CH$_3$)$_2$CH(CH$_2$)$_2$CH$_2$Br
bromometric equivalent weight 溴量法当量

bromometric titration 溴量滴定
bromometry 溴量法*
bromomonamycin 溴弧霉素
bromonaphthalene 溴(代)萘
bromonaphthol 溴萘酚 BrC$_{10}$H$_6$OH
bromonitrin 溴硝菌素
bromonitromethane (一)溴硝基甲烷 BrCH$_2$NO$_2$
bromonium 溴鎓；三价溴
bromonium-hydrogen-bonded bromide ion pair 溴鎓-氢-键合溴化物离子对
bromonium ion 溴鎓离子
bromononane 溴壬烷
bromooctane 溴辛烷
bromo-organic compound 含溴有机物
1-bromopentadecane(=pentadecyl bromide) 十五(烷)基溴；1-溴十五烷 CH$_3$(CH$_2$)$_{13}$CH$_2$Br
bromopentanone 溴戊酮
bromophenesic acid 二溴酚 C$_6$H$_3$Br$_2$OH
bromophenic acid 一溴酚 C$_6$H$_4$BrOH
bromophenisic acid 三溴酚 C$_6$H$_2$Br$_3$OH
bromophenol 溴苯酚
bromophenol blue(=tetrabromophenol sulfonphthalein) 溴酚蓝*；四溴苯酚磺酞
bromophenyl-hydrazine 溴苯肼
bromophenyl-hydroxylamine 溴苯基胺
bromophenylmercuric chloride 溴苯基汞化氯 BrC$_6$H$_4$HgCl
bromophenyl phenol 溴苯基苯酚 BrC$_6$H$_4$C$_6$H$_4$OH
1-bromo-2-phenyl-propane 1-溴-2-苯基丙烷
bromophosgene 溴化羰 BrCOBr
bromo-phosphonium 溴化鏻 PH$_4$Br
bromophthalic acid 溴代邻苯二甲酸 BrC$_6$H$_3$(CO$_2$H)$_2$
N-bromophthalimide titration N-溴代邻苯二甲酰亚胺滴定
bromopicrin(=nitrobromoform) 溴化苦；硝基溴仿；三溴硝基甲烷 CBr$_3$NO$_2$
bromoplatinic acid 溴铂酸 H$_2$[PtBr$_6$]
bromoprene 溴代丁烯；2-溴-1,3-丁二烯 CH$_2$=CHCBr=CH$_2$
1-bromopropane(=propyl bromide) 丙基溴；1-溴丙烷 CH$_3$CH$_2$CH$_2$Br
2-bromopropane(=isopropyl bromide) 异丙基溴；2-溴丙烷 CH$_3$CHBrCH$_3$
bromo-propionic acid β-溴丙酸 CH$_2$BrCH$_2$COOH
β-bromopropyl alcohol β-溴丙醇
bromopropylene 溴丙烯
bromopropylene oxide 溴代环氧丙烷
3-bromo-1-propyne(=propargyl bromide) 炔丙基溴；3-

溴-1-丙炔　　HC≡CCH$_2$Br
bromoprotocatechuic aldehyde　溴原儿茶醛
bromopyridine　溴吡啶
bromopyrine　溴安替比林〔药〕
bromopyrogallol red　溴(代)邻苯三酚红*
bromopyruvic nitrile　溴丙酮腈
bromoquinoline　溴喹啉
bromosalicylaldehyde　溴水杨醛　Br(OH)C$_6$H$_3$CHO
bromosoval and procainee　溴米因〔药〕
bromostearic ethyl ester　溴代硬脂酸乙酯
bromostyrene(=bromostyrol)　溴代苯乙烯；溴代苏合香烯
bromostyrol(=bromostyrene)　溴代苏合香烯；溴代苯乙烯
bromosuccinic acid　溴代丁二酸　(CH$_2$CHBr)(CO$_2$H)$_2$
N-bromo-succinimide(NBS)　N-溴代丁二酰亚胺
N-bromosuccinimide titration　N-溴丁二酰亚胺滴定
brom(o)sulf (ophth) alein(BSP)　四溴酚酞磺酸钠
bromosulphonazo III　溴磺偶氮 III
bromoterephthalic acid　溴代对苯二酸　BrC$_6$H$_3$(CO$_2$H)$_2$
p-bromo-thiobenzophenone　对-溴代二苯甲硫酮
bromothricin　溴丝菌素
bromothymol blue(=dibromothymol sulfonphthalein)　溴百里酚蓝*；二溴百里酚磺酞　C$_{27}$H$_{28}$O$_5$Br$_2$S
bromotoluene　溴甲苯
bromotoluidine　溴甲基苯胺
bromo-trichloromethane　(一)溴三氯甲烷；三氯甲溴　CBrCl$_3$
bromotrifluoethylene　溴三氟乙烯
bromouracil　溴尿嘧啶
bromous acid　亚溴酸　HBrO$_2$
bromovirus　雀麦花叶病毒
bromphenol blue　溴酚蓝
brom phenol green　溴酚绿
bromphenol red　溴酚红
bromquatrimycin　溴差向四环素
bromstyrol　溴苯乙烯；α-溴苯乙烯　C$_6$H$_5$CBr=CH$_2$
bromthymol blue(=dibromothymol sulfonphthalein)　溴百里酚蓝；二溴百里酚磺酞
bromural　溴梦拉；α-溴异戊酰脲　(CH$_3$)$_2$CHCHBrCONHCONH$_2$
bromyrite　溴银矿
bronchitis　支气管炎
brongniardite　硫锑铅银矿
Bronner's acid(=2-naphthylamine-6-sulfonic acid)　布珑酸；2-萘胺-6-磺酸　NH$_2$C$_{10}$H$_6$SO$_3$H
Brönsted acid　布朗斯台德酸；质子酸
Brönsted acidity　布朗斯台德酸性(度)
Brönsted base　布朗斯台德碱
Brönsted catalysis law　布朗斯台德催化定律
bronze　青铜*
bronze blue　青铜蓝；普鲁士蓝
bronzed aluminium　青铜色铝粉
bronze gold powders　青铜金粉；铜合金金粉〔由金黄色到黄绿色〕
bronze lacquer　①青铜色漆②金粉漆
bronze paint　①青铜漆②金粉漆
bronze paste　金粉浆〔泛指各种金属粉(铝、铜等)的颜料浆〕
bronze powder　青铜粉
bronze red C　金光红 C
bronze scarlet　①金光大红 C②(=Lake red C)色淀红 C
bronze yellow　古铜黄色
bronzine　青铜色的
bronzing　镀青铜
bronzing fluid(=bronzing medium)　金胶〔大漆〕②金粉(清)漆；金银粉漆基③闪光漆基
bronzing liquid　①金胶〔配制金、银粉漆用的漆料〕〔大漆〕②金粉浆
bronzing medium(=bronzing fluid)　①金胶〔大漆〕②金粉(清)漆；金银粉漆基③闪光漆基
bronzing process　擦金工艺；揩金工艺〔印刷〕
bronzite　古铜辉石
bronzy　①闪铜光的；青铜光泽的②金属光泽
brood lac　(紫胶)种胶
Brooker deviation　布鲁克偏离值
Brookfield gelometer　布氏胶凝点测定计
Brookfield LVF viscometer　布氏 LVF 型黏度计〔一种测表观黏度的转子黏度计〕
Brookfield rotational viscometer　布鲁克菲尔德旋转黏度计
Brookfield type viscometer　布鲁克菲尔德型黏度计；B 型黏度计
Brookfield UL adapter　布鲁克菲尔德超低黏度黏度计
Brookfield yield value　布鲁克菲尔德致流值
brookite　板钛矿
broom absolute　鹰爪豆净油；金雀花净油
broom concrete　鹰爪豆浸膏；金雀花浸膏
broom corn　高粱
broom-seed oil　染料木子油
brosyl(=p-bromobenzenesulfonyl)　对溴苯磺酰基
brosylate(=p-bromobenzenesulfonate)　对溴苯磺酸盐(或酯)
brown　棕(色)；褐(色)；黑(色)
Brown's method　布朗方法
Brown's opacity tube　布朗比浊管
brown acetate of lime　褐醋石

brown acids(=brown soap)　棕色酸；棕色皂〔酸洗时生成的油溶性磺酸盐〕
brown bloom　棕色霜〔煤油灯罩上的〕
Brown-Boveri test　布朗-包维瑞试验〔氧化稳定性〕
brown camphor oil　棕樟油
brown coal　褐煤
brown coal pitch　褐煤沥青
brown coal tar　褐煤焦油
brown crepe　褐绉片；棕色绉(橡)胶
brown dyes　褐染料
Browne test　布劳内热(桐油胶化)试验
brown factice　棕色硫化油膏
brown factis　棕色硫化油膏
brown fish oil　棕鱼油
brown gasoline　褐色汽油〔未经脱硫和脱色的〕
brown ground wood　褐浆粕；褐色细木浆粕
brown hematite　褐赤铁矿
Brownian motion　布朗运动
Brownian movement　布朗运动
Brownian rotation　布朗转动
browning　褐变
browning reaction　棕黄反应
brown iron ore　褐铁矿
brownish black　棕黑
brownish red　棕红
brown kerosene　棕色煤油
brown lime　棕石灰
brown mechanical pulp　棕色机械纸浆
brown mustard　黑芥子
Brown-Neil process　布朗-尼耳(提锌)法
Brown nickel process　布朗炼镍法
brown ocher　褐铁矿
brown oil　棕色油
brown oil of vitriol　棕色硫酸
brown oxide　棕色氧化物〔指 UO_2〕
brown packing paper(=kraft paper)　包皮纸；牛皮纸
brown petroleum　褐色石油
brown red antimony sulfide　硫氧锑矿
brown ring test　棕环试验
brown rot　褐腐
brown seaweed　褐海藻
brown size　褐色胶(料)
brown soap(=brown acids)　棕色酸；棕色皂
Brown-Souders equation　布朗-苏德斯方程
brown spar　铁菱镁矿
brown stock　粗浆；未漂浆
brown stock chest　漂前浆池；黄料浆池
brown stone　褐石
brown substitute　棕(色)代胶
brown tipped　顶端褐变
Brown triangle method　布朗三角图形法
brown vitriol　棕色矾
brown wood board　棕纸板
brown wood pulp　褐色磨木浆
browny　带褐色的；带棕色的
brucealin　鸦胆灵
brucine　番木鳖碱；马钱子碱
brucine test　番木鳖碱试验〔用以试验硝酸〕
brucite　水镁石
Bruehl receiver　布吕耳(承)受器
brugnatellite　次碳酸镁铁矿
bruiser　捣碎机
bruise resistance　抗机械损伤性能
bruise roller　压浆辊筒
bruising　擦伤
bruising mill　破碎机
Brunauer-Emmett-Teller adsorption isotherm(=BET adsorption isotherm)　BET 吸附等温式*
Brunauer-Emmett-Teller equation(=BET equation)　布鲁厄-埃米特-特勒方程式；BET 方程式〔BET 吸附等温线方程〕
Brunauer-Emmett-Teller method　布鲁厄-埃米特-特勒法；BET 法〔用吸附法测定微孔粒子比表面积的方法〕
Brunee separator　布鲁尼分离器
Brunel mass detector　布鲁诺质量检测器
bruneomycin　棕霉素
brunswick black　黑沥青(清)漆〔泛指各种沥青溶于松香水或芳烃溶剂制成的清漆〕
brush　①刷②刷涂
brushability　刷涂性
brush application　刷涂施工；刷涂
brush backed sander(=automatic turning sander)　自动旋转研磨机；刷包砂布打磨机〔以硬毛辊筒为衬垫，上包砂布或砂纸〕
brush binder　刷毛黏合剂〔黏合刷上的毛〕
brush cleaner　漆刷清洗剂
brush coating　①刷涂②刷涂涂料
brush damping machine　刷湿布机
brush discharge　刷形放电
brush drag　滞刷；拉不开漆刷〔俗〕〔油漆黏度过大所致〕
brush dyeing　刷染法
brushed drum screen　带刷鼓式过滤筛
brushed fabric　拉绒织物；起绒织物
brushed pile　拉绒绒毛
brush filler　刷涂填孔剂

brush finish ①刷涂面漆；刷面饰②粉刷
brushgear 电刷装置
brush-heap hypothesis 刷堆假设；积刷假设〔热原子化学〕
brush holder 炭刷架子
brushing 清洁；擦光；刷新；刷拂
brushing characteristic 刷涂性能
brushing compound 刷光涂料
brushing consistency 刷涂稠度
brushing glazing 刷釉
brushing lacquer 刷漆
brushing machine ①刷光机②刷毛机；刷绒机③除尘机
brushing-on paint 刷涂用漆
brushing primer 刷涂底漆
brushing property 刷光性质；涂刷性
brushing-stroke velocity 涂刷速度；运刷速度
brushing wheel 刷光砂轮
brushite 透钙磷石
brush lead 炭刷导线
brushless shaving cream 无刷剃须膏
brush-lubricated 油刷润滑的
brush mark 刷痕
brush method 涂刷(显色)法
brush-on 刷涂
brush-on rust remover 刷(上)除锈剂
brush-on-type 刷涂型
brush-out (试验性)小面积刷涂
brush-out leveling 刷涂流平性
brush-out leveling test method 刷涂流平性试验法
brush pilling tester 毛刷式起球试验仪
brush plating 刷镀
brush plating power pack 刷镀电源
brush plating solution 刷镀溶液
brush polish 刷光剂；上光剂
brush polymer 刷状聚合物
brush roll 毛刷滚
brush spreader 刷式涂布机
brush treatment ①(钢)刷除锈②刷涂防腐蚀漆(剂)
brush type bonded phase 刷子型键合(固定)相
brush type bonded support 刷子型(化学)键合载体
brush wheel 磨轮；砂轮
bryamycin(=thiactin) 硫活素；薛霉素
bryoidin 榄香胶素
bryokinin 苔藓(激)动素
bryonane 薯叶烷
bryonin 欧薯苷
bryophyllin 环落地生根素
bryophyte indicator for pollution 苔藓植物指示污染

bu-(=butyl radical) 丁基
bubble ①气泡②水泡
bubble aeration 吹泡汽通
bubble breaker 消泡剂
bubble cap (蒸馏)泡罩
bubble cap absorption column 泡罩吸收塔
bubble cap column 泡罩(蒸馏)塔
bubble-cap fractionating column 泡罩分馏塔
bubble-cap plate 泡罩(塔)板
bubble-cap plate tower(=bubble-cap tower) 泡罩塔
bubble-cap tower(=bubble-cap plate tower) 泡罩塔
bubble cap tray 泡罩塔盘
bubble-chamber 气泡室；起泡室
bubble-cloud reactor 泡雾反应器
bubble coating 微囊涂料；微泡涂料
bubble column 泡罩塔
bubble counter 计泡计
bubble deck 起泡盘；泡罩塔盘
bubble dust removal 泡沫除尘
bubble film 泡膜
bubble flask 起泡瓶
bubble formation 气泡形成
bubble forming 气胀成型
bubble form pool 泡形池
bubble gauge 气泡指示器
bubble generation 气泡发生(作用)
bubble half-life 气泡半衰期
bubble method 气泡法
bubble nucleation 气泡形成
bubble plate 泡罩(塔)板
bubble plate column 泡罩塔
bubble plate extractor 泡罩抽提塔
bubble plate tower 泡罩塔；泡帽塔
bubble point 始沸点；起泡点
bubble point pressure 始沸点压力；泡点压力
bubble point tester 气泡点数式测试仪〔用于滤芯清洁度测试〕
bubble pressure method 气泡压力法〔测表面张力〕
bubble proof 吹验法；吹泡法
bubbler ①扩散器②打泡器
bubble releasing 脱泡剂；放气；消泡
bubble retention 积泡性；气泡不消
bubble rise time (气泡黏度计的)气泡上升时间
bubble rocks 泡石
bubble rubber 海绵胶
bubble scrubber 泡沫塔
bubble seconds 气泡秒数；泡秒
bubble size distribution 气泡大小分布

bubble test 鼓泡试验
bubble time method 气泡计时法
bubble tower 泡罩塔;泡帽塔
bubble tower condensate 泡罩塔冷凝液
bubble tower overhead 泡罩塔顶馏分
bubble tower section 泡罩塔区段
bubble trap 气(泡)阱
bubble tray 鼓泡塔盘
bubble-tube method 气泡法〔测黏度方法〕
bubble tube visco(si)meter 气泡黏度计
bubble-type-flow counter 气泡型流量计数器
bubble viscometer 气泡黏度计
bubbling 起泡〔漆病〕
bubbling absorber 气泡吸收器
bubbling bed 沸腾床
bubbling cap(=bubbling hood) 泡罩;泡帽
bubbling column reactor 鼓泡塔(式)反应器
bubbling device 鼓泡装置
bubbling hood(=bubbling cap) 泡罩;泡帽
bubbling point 起泡点
bubbling polymerization 鼓泡聚合;沸腾聚合
bubbling type of gas mixing 鼓泡式气体混合(法)
bubbling-type pressure regulator 鼓泡式压力调节器
bubulum oil(=neat's-foot oil) 牛脚油
bucco camphor 布枯樟脑;1-甲基-2-羟基-3-氧-4-异丙基环己烯
Bucherer-Bergs synthesis(=Bucherer synthesis) 布歇尔-贝格斯合成法
Bucherer reaction 布歇尔反应
Bucherer synthesis(=Bucherer-Bergs synthesis) 布歇尔合成法
Buchner-Curtius reaction 布赫内尔-库尔修斯反应
Buchner filter 布氏漏斗;瓷漏斗;平底漏斗
Buchner flask 布氏烧瓶
Buchner funnel 布氏漏斗;瓷漏斗
bucholzite(=fibrolite) 夕线石
buchu camphor(=diosphenol) 布枯樟脑;地奥酚
buchu leaves oil 布枯叶油〔*Barosma* 叶油〕
bucinnazine 布桂嗪〔药〕
buck ①灰水②冲;推
buckbean 睡菜
bucker ①粉碎机;碾碎机;破碎机;压碎机②碎矿机③碎铁机
bucket ①斗②桶;吊桶
bucket carrier(=bucket conveyer) 斗式运输器
bucket conveyer(=bucket carrier) 斗式运输器
bucket cover 斗式提升机
bucket elevator 斗式提升机
bucketful (一)满桶
bucketing 用斗式运输器进料
bucket proofer(=bucket prover) 间格斗式运输器
bucket prover(=bucket proofer) 间格斗式运输器
bucket pump gun 勺斗式加油枪〔加润滑油用〕
bucket trap 勺斗阱
bucket valve 活塞阀
bucket wheel 勺轮
bucking 浸渍;浸润
Buckingham equation 白金汉方程
buck leather (雄)鹿革
buckled ply 皱层;折层
buckle leather 带用革
buckle tear 穿孔撕裂
buckling ①纵向弯曲;扭转②起皱;打折③翘曲④失稳
buckling load 折断载荷
buckling of vessel 容器扭曲
buckling resistance 翘曲阻力;抗纵向弯曲力
buckling strain 弯曲应变
buckling strength 翘曲强度;抗纵向弯曲强度
buckling stress 轴向弯曲应力
buck mortar 推式研钵
buckskin 鹿皮〔革〕
buck-thorn oil 鼠李油;鹿棘油
buckwheat coal 荞麦煤
Bucky rays 巴基射线
Budde effect 巴德效应
buddleoflavonoloside 醉鱼草黄酮醇糖苷
buddleoglucoside 醉鱼草葡糖苷
buddling (矿砂)碎淘
budesonide 布地奈德〔药〕
bud grafting 芽接
budiene 丁二烯
Bueche segmental friction theory 比休链段摩擦理论
bufadienolide 蟾蜍二烯羟酸内酯
bufagin 蟾蜍精
bufalin 蟾蜍灵;二羟蟾毒二烯酸内酯
bufandide 蟾烷内酯
bufatrienolide 蟾三烯内酯
buff ①米色的②抛光轮;磨轮③黄色厚革
buffalo 水牛
buffalo hide 水牛皮
buffalo milk fat 水牛酪脂
buff-burned pattern 抛光灼晕
Buff-Dunlop apparatus 布弗-登洛普(循环)抽提器
buffed 磨(了)面的
buffed grain 摩擦革;磨光革
buffed leather 磨面革

buffer ①缓冲*②缓冲器③缓冲溶液
buffer action 缓冲作用
buffer additive 缓冲溶液添加剂
buffer agent 缓冲剂；缓冲液
buffer battery 缓冲电池组
buffer boundary 缓冲边界
buffer capacity 缓冲容量
buffer change valve 缓冲变模阀
buffer coat 缓冲涂面(层)
buffer coating 缓冲涂层
buffer container 缓冲容器
buffer counterion 缓冲配对离子
buffered 缓冲的；含有缓冲剂的
buffered column 缓冲柱
buffered filter paper 缓冲滤纸
buffered silica gel 缓冲硅胶
buffered tower 挡板器
buffer elution 缓冲洗脱
buffer for electrode calibration 电极校正用缓冲溶液；电极校准用缓冲液
buffer index 缓冲指数；缓冲值
buffering ①缓冲的②缓冲作用
buffering agent 缓冲剂
buffering compound 缓冲剂
buffer intensity 缓冲强度
buffer layer 缓冲层
buffer memory ①缓冲存储②缓冲存储器
buffer mixture 缓冲混合液
buffer pair 缓冲偶
buffer plate 缓冲板
buffer range 缓冲范围
buffer salt 缓冲盐
buffer solution 缓冲溶液
buffer spring 缓冲弹簧
buffer stage 缓冲段
buffer storage ①缓冲存储②缓冲存储器
buffer substance 缓冲物质
buffer system 缓冲体系
buffer tablet 缓冲片
buffer tank 缓冲罐
buffer value 缓冲值
buffer vessel 缓冲罐
buff glazed paper 米色道林纸
buffing 软皮抛光；磨光；打磨
buffing compound 磨光剂
buffing machine 磨光机
buffing mark 磨痕
buffing oil 磨光油

buffing paper 磨皮砂纸
buffing sanding flock raise 砂磨起绒
buffing wheel 砂轮
Bufflex process 布弗莱克斯过程〔离子交换-硫酸介质胺类萃取法从铀矿石浸出液中回收铀,同 Eluex法〕
buff-patched appearance 打磨不匀
buff wheel 磨光轮
buflomedil 丁咯地尔〔药〕
bufotalinic acid 蟾毒他灵酸
bufotalinin 蟾蜍他里宁
bufotalone 蟾蜍他酮；蟾毒配基酮
bufotenidine 蟾毒色胺内盐；N-三甲基-5-羟色胺
bufotenine(=mappine) 蟾毒色胺；N-二甲基-5-羟色胺
bufotoxin ①蟾毒素②蟾毒配基 B 二酯
bufovarin 蟾蜍卵素
bug ①沿管线通信的自动电报②清除管子内部表面的刮器③技术性毛病；缺陷
buggies 炭车
buhr 磨石
buhrstone 磨石；白石
buhrstone mill 石磨(机)
buhrstone mill dressing 石磨刻凿
build ①厚膜；厚涂层；干膜厚度②构型；造型
builder ①组分；成膜物②助洗剂；增效助剂
builder of soap 肥皂助洗剂
building action 增效作用
building bag 胶面轴心〔橡〕
building block ①结构单元②高分子(分子)链节③积木(式的)
building block instrument 组合式仪器系统
building block system 积木式结构系统
building block theory 积木理论
building board 建筑纸板
building brick 建筑砖
building cement 建筑水泥
building coat ①建筑涂层②建筑面层
building component 增效组分
building core 胶面轴心〔橡〕
building drum 胶(性)面(用)转鼓
building machine ①配套机②成型机〔橡〕
building material 建筑材料
building method 合成法〔制汽油〕
building mortar 建筑用灰泥
building of refinery unit 炼油厂建筑
building-out condenser 附加电容器；补偿电容器
building paper 建筑纸
building power of deposited paint ①涂(上)的油漆成膜能力②涂(上)的漆结厚膜能力

building ring 工作圈；速动圈
building room 装配间
building rubber compound 配橡胶料；橡胶组合
building strength 成型贴合强度
building unit （高分子的）单体；结构单元
build-in plasticization 内增塑
build-up 贴胎面；咬模
build-up agent 增深剂
build up process 增层法制程
build-up welding technique 堆焊技术
built detergent 复配洗涤剂
built-for-purpose tools 专用工具
built heavy duty product 加助剂的高效洗涤用品
built-in 内装的；嵌上的；镶上的
built-in check 内在校验；自动校验
built-in compressor ratio 内部压缩比
built-in electrical insulation 内部（电）绝缘
built-in flow circuit 内置流路
built-in function ①内在功能②内部函数
built-in interface controller 固定界面调节器
built-in lubrication 内在润滑
built-in microcomputer 内置式微型计算机
built-in plasticization 内在增塑(作用)
built-in potential 内(建)电位；内(建)电势
built-in problem 内在问题；本身存在的问题
built-in processor 内置式数据处理机
built-in side chain 固有侧链；内在侧链
built-in strain 潜存应变
built-in stress 潜在应力
built-in unit 联机设备；线上设备
built-in wetting agent 内在润湿剂；潜在润湿剂
built-on pump(=built-together pump) 连发动机的泵松密度
built soap 复配皂
built-together pump(=built-on pump) 连发动机的泵松密度
built-up 组合
built-up mandrel 组合圆棒
built-up mica 云母板(材)；云母板坯
built-up mould 组合(塑)模；开合压模
built-up sequence 熔敷顺序
built-up soap 复配皂
bulb ①灯泡②球管；球
bulb apparatus 钾碱球管
bulb barometer(=vessel barometer) 球管气压计
bulbiformin 球茎状素
bulbocapnine 褐鳞碱
bulb pipette 球吸管

bulb stopper 球塞
bulb tube 球管
bulb-type glass electrode 球型玻璃电极
bulge extrusion 鼓泡挤塑法
bulge test 打压试验
bulge theory ①隆丘理论〔地〕②膨胀理论
bulging ①打气②膨胀；鼓胀
bulk ①批量；大量②松密度③大部分；主体④整体⑤胀量
bulk additive 填充剂
bulk analysis 全分析
bulk article 大量制品；标准产品
bulk bin 集装罐；集装箱
bulk boat 石油驳船
bulk carrier 散装大船
bulk chemicals 散装化学品
bulk compliance 体积柔量
bulk composition 体相组成
bulk compression 体积压缩
bulk concentration 体相浓度；主体浓度
bulk conductivity 体积电导率
bulk creep compliance 体积蠕变柔量
bulk crystallization 大量结晶
bulk crystallized polymer 本体结晶聚合物
bulk degradation 整体降解
bulk deliveries of gasoline 汽油散装发货〔油管、油船或油槽车〕
bulk density 堆积密度；堆密度；松密度
bulk development 膨松显现(处理)
bulk diffusion 体扩散
bulk dilation 体积膨胀
bulked textured filament 膨化变形长丝
bulked yarn 膨化变形纱；膨体纱
bulk effect 体效应；体积效应
bulk elasticity 体积弹性
bulker 检查舱货容量；检查人
bulk factor 体积因数；紧缩率
bulk fibre 散纤维
bulk fibre drier 散毛烘燥机
bulk filler 增容剂〔为使产品增加容积和降低成本而添加的粉末状填料〕
bulk fission product 总体裂变产物
bulk flow 总体流动
bulk goods 松散材料；粒状材料
bulk hauling(=bulk oil transportation) 散装输油
bulk heterogeneity 体积非均匀性
bulkiness 膨松性；膨松度
bulking 膨胀

bulking agent　填充剂；增量剂
bulking capacity　膨化能力
bulking density　体积密度
bulking filler　疏松填料；填充剂
bulking intensity　膨松度
bulking material　增量性材料；增量剂；填充剂
bulking power　增量能力；增量本领
bulking property　①膨胀性②总体性③松散性〔纤维素及纸用〕
bulking value　重量体积〔油漆原料每磅加仑数〕
bulking volume　松散容积
Bulkley pressure viscosimeter　巴尔克利压力黏度计〔测润滑脂用〕
bulk liquid　散装液体
bulk liquid storage　散装液体储存
bulk longitudinal relaxation modulus　体积纵向松弛模量
bulk longitudinal viscoelasticity　体积纵向黏弹性
bulk loss compliance　体积损耗柔量
bulk loss modulus　体积损耗模量
bulk material　疏松材料；散料
bulk melting point of lubricant　润滑剂整体熔点
bulk meter　膨松度试验仪
bulk method　批量生产法；成批生产法
bulk modulus(=modulus of volume expansion)　体积模量；本体模量
bulk modulus of elasticity　弹性体积模量
bulk moulding compound(BMC)　块状模塑料
bulk of molecule　分子的大小
bulk oil　①散装油②主体油
bulk oil storage　散装储油
bulk oil tank　散装油罐
bulk oil transportation　散装油料运输
bulkometer　膨松度试验仪
bulk pack(ag)ing　松散包装；散装
bulk packing density　堆积密度；松密度
bulk phase　体相
bulk phase concentration　体相浓度
bulk plant　油库
bulk plant pump　油库泵
bulk polymer　本体聚合物
bulk polymerization　本体聚合；整体聚合
bulk polymerizer　本体聚合釜
bulk polystyrene　本体(聚合)聚苯乙烯
bulk precipitation polymerization　本体沉淀聚合
bulk production　批量生产
bulk property　堆积性能
bulk property detector　总体性能检测器
bulk ratio　体积比
bulk recovery　体积回复
bulk resistance　体积电阻
bulk sample　大样；总试样
bulk sampling　本体采样
bulk settling　整体沉降
bulk shielding　整体屏蔽
bulk shipment　①散装运输②整装运输③散装发货量
bulk solution　本体溶液
bulk specific gravity　视比重；堆积比重；体积比重
bulk station　配油站；散装油站
bulk storage　散装储油
bulk storage area　主储存区
bulk storage compliance　储能体积柔量
bulk storage modulus　储能体积模量
bulk strain　体应变
bulk strength　①整体强度②体积强度
bulk stress　体积应力
bulk stretch yarn　膨体弹力丝
bulk technique　本体聚合工艺
bulk temperature　(按体积计算的)平均温度；整体温度
bulk tensile property　体积拉伸性能
bulk tester　膨松度测试仪
bulk thermal polymerization　本体热聚合
bulk transportation　整批运输
bulk trial　批量生产试验
bulk viscoelasticity(=volume viscoelasticity)　体积黏弹性
bulk viscosity　①体积黏度②本体黏度
bulk volume　①总体积②总容积
bulk weight　整重
bulky　①庞大②笨重
bulk yarn　膨体纱
bulky diammonium derivative　膨松二铵衍生物
bulk yielding　①尺寸易变②容积压缩
bulky oxide　膨松的氧化物(层)
bulky paper　绒毛纸
bulky yarn　膨体纱
bullatine A　雪上一枝蒿甲素
bullet-proof glass　防弹玻璃
bullet-sealing cell　防弹油箱
bullet type tank　卧式圆形储罐
bulleyaconitine A　草乌甲素〔药〕
bull gear　主齿轮；从动齿轮
bull gear drive　主齿轮传动；间接传动
bull hide　分牛皮
bullion　①金银块②纯金③纯银
bullion lead　生铅
bull quartz　烟色石英；烟晶
bull ring　研磨圈

bull screen　碎料筛〔造纸〕
bulls-eye condenser　牛眼形聚光器
bull shaker　摇(动)筛
bulls heart　牛心果
bullvalene　瞬烯
bull wheel(=big gear wheel)　主齿轮；大齿轮
bulnesol　布藜醇
bulrush　芦苇
bumble-bee wax　大蜂蜡
bumetanide　布美他尼〔药〕
bumped head　凸形的底
bumper　①保险杠②防冲装置；缓冲器
bumping　暴沸*
bump leveller　吸振器
Bumstead electroscope　邦斯提德验电器
Buna　布纳(橡胶)；丁(二烯)钠(聚)橡胶
Buna rubber　布纳橡胶；丁钠橡胶
bunch　备纱
bunching　(皂胶的)黏滞现象〔由于产生酸性皂而出现的〕
bundle cleaning method　换热器管束清扫法
bundle dyeing　绞纱染色
bundle fibre strength tester　束纤维强力试验机
bundle finishing　束状纹
bundle (-like) crystal　束状结晶
bundle of electrons　电子束
bundle test　成包试验〔去污力试验〕
bundle-thread　束丝
bundling　分包
bundling press　打包压榨机
bung　塞
bung hole　桶(侧)口
bunker　①仓库②储槽③漏斗〔进料用〕
bunker A fuel　船用A级锅炉燃料油〔黏度最小的燃料油〕
bunker B fuel　船用B级锅炉燃料油〔中等黏度〕
bunker C fuel　船用C级锅炉燃料油〔黏度较高〕
bunker coal　船用煤
bunker fuel oil　船用锅炉燃料油
bunkering　燃料仓储；装燃料
bunkering barge　燃料仓驳船
bunker oil　船用油
bunker point(=bunker station)　燃料储藏站
bunker station(=bunker point)　燃料储藏站
bunkie station　泵送站
Bunsen beaker　烧杯
Bunsen burner　本生灯
Bunsen cell　本生电池
Bunsen clamp　本生夹

Bunsen funnel　本生漏斗
Bunsen gas-bottle　本生洗气瓶
bunsenine(=bunsenite)　绿镍矿
bunsenite(=bunsenine)　绿镍矿
Bunsen photometer　本生光度计
Bunsen reaction　本生反应
Bunsen Roscoe law　本生-罗斯科定律；倒易律
Bunsen valve　本生阀
buoyance constant　浮力常数
buoyancy　浮力；浮扬性
buoyancy correction (of weighing)　浮力校正
buoyancy level indicator　浮球液面指示器
buoyant　漂浮的；有浮力的
buoyant density　浮力密度
buoyant effect　浮力效应
buoyant effect of air　空气浮力效应
buoyant force　浮力
buoyant roof　浮顶
buoyant soap　浮水皂
buoy bogie　浮筒车
buoyed weight　浮重
buoy failure report　浮标损坏报告
buphanamine　布蕃胺
buphanidrine(=distichine)　布蕃君
buphanine　布蕃宁
buphanitine　布蕃尼亭
bupivacaine　布比卡因〔药〕
bupleurol　柴胡醇
bupleurumol　柴胡醇　$C_{37}H_{64}O_2$
buprenorphine　丁丙诺啡〔药〕
burbling　气流分离
burbonal(=bourbonal)　乙香草醛
burdock oil　芒壳油
buret(=burette)　①滴定管②量管
buret cap　滴定管盖
buret clamp　滴定管(节流)夹
buret electrode　滴管电极
buret holder　滴定管架
buret reader　滴定管读数镜
buret support　滴定管架
burette(=buret)　①滴定管②量管
burette float　滴定管浮标
buret(te) meniscus reader(=buret(te) reader)　滴定管弯液面读镜
buret(te) reader(=buret(te) meniscus reader)　滴定管弯液面读镜
burette stand　滴定管台
burette support　滴定管架

burette viscometer　滴定管黏度计
Burgers body　伯格斯体〔理想黏弹体〕
Burgers vector　伯格斯矢量
Burgundy mixture　伯更狄混合液〔含多种离子型和非离子型表面活性剂，用以试测在固体表面上的保留时间〕
burgy　细粉；煤屑
burial ground(=waste graveyard)　废物埋藏场
burial tank　(放射性废物)储埋槽
burial trials　土埋试验
buried　浸入的；浸没的
buried oil pipe line　地下油管
buried pipe line　地下管道
buried piping　地下管道
buried storage　地下油池
buried tank　地下储罐
burkeite　碳酸钠矾
burlap　粗麻布
burlap finish　麻布面饰
burley　白莱烟；细纤维烟草
burmese lacquer　缅甸漆
burnable bone　可燃页岩
burnable poison　可燃毒物
burn cream　烧伤膏
burned area　焦烧面积
burned finish(=fiery finish)　火燎糅饰法〔木器作旧仿古糅饰法之一，先将木器表面用喷灯烤焦，然后用硬刷将烧焦的木质打掉，留下硬质部分犹如浮雕，然后涂漆〕
burner　①灯②灯头③(燃烧)炉④燃烧器
burner angle　燃烧器角度
burner attachment　灯用附件
burner barrel　燃烧器筒身
burner cones complete　灯头全套
burner for flame analysis　火焰分析用燃烧器；喷灯
burner gas　炉气
burner-gas cooler　炉气冷却器
burner gauze　顶纱
burner gauze keep ring　顶纱圈
burner guard　护焰罩
burner hearth　炉底；炉床
burner jet　燃烧器喷嘴
burner lining　炉衬
burner oil　燃烧油；喷射燃料
burner opening　燃烧器口
burner piping　烧嘴管线
burner port　燃烧器口
burner rig(=corrosion chamber)　燃烧试验装置；腐蚀室
burner rig test　燃烧室试验〔模拟燃气轮机环境试验〕
burner rim　燃烧器缘
burner setting　炉子的砌体〔设〕
burner's platform　看火台
burner tail pipe　灯头尾管
burner tile　炉瓦；耐火瓦
burner tip　燃烧器尖端
burnettizing　护材作业法
burning　烧焦
burning behavior　燃烧性状
burning gas　燃烧气体
burning in bulk　主体中燃烧
burning-in knife　刮刡
burning kerosene　灯用煤油
burning of chips　木片黑煮〔林产〕
burning of clay　黏土烧成
burning off　①烧掉；烧尽②火焰除漆法③(搪瓷)烧脱底釉〔缺陷〕
burning-off curve　燃烧曲线*
burning-off oil slicks　燃烧除去浮油〔水面上〕
burning of gas　气体的燃烧
burning of lead　铅焊
burning of ore　煅烧矿石
burning of paper　纸的过干作用
burning of rubber　橡胶的培焦
burning oil　燃烧油
burning pit　燃烧坑
burning point　燃(烧)点
burning quality　燃烧质量
burning quality index　燃烧质量指数
burning quality test　燃烧质量试验
burning rate　燃烧速率
burning resistance　阻燃性；耐燃性
burning retardant　阻燃剂
burning shrinkage　烧缩
burning surface　燃烧表面
burning temperature　燃烧温度
burning test　燃烧试验
burning time　燃烧时间
burning velocity　燃烧速度
burning zone　①燃烧层②燃烧段
burnishing　抛光；磨光；砑光
burnish resistance　①抗磨亮性〔保持无光〕②抗摩擦性；耐磨性
burn-leach process　燃烧-浸取法
burn off effect　燃烧效应
burn ointment　烧伤油膏
burnt alum　烧茂；焦茂
burnt blende　闪锌矿灰渣

burnt coal 天然焦炭
burnt deposit 烧焦镀层
burnt gas 燃烧废气
burn through time 烧穿时间
burnt iron 过烧铁
burnt island red 天然氧化铁红
burnt lime 煅石灰;氧化钙
burn to ash 烧成灰分
burnt ochre 煅赭石
burnt oil 煅油
burnt ore 煅烧矿
burnt paper 过干纸〔纸〕
burnt plaster 烧石膏*
burnt potash 氧化钾 K_2O
burnt pyrite 煅烧黄铁矿
burnt rubber 焦橡胶
burnt sienna 富铁煅黄土
burnt stock 焦母分〔橡胶〕
burnt umber 烧褐土
burn-up 燃耗
burnup analysis 燃耗分析
burnup determination of nuclear fuel 核燃料燃耗测定
bur oak 大果栎
bur oil 芒壳油
burr 毛刺;毛口
burr-drill 圆头锉;钻锥
Burrell-Severs extrusion rheometer 伯勒尔-西弗斯挤压流变仪
burring 去毛刺;内缘翻边
burrstone 磨石
burrstone mill 石磨机
burseran 裂榄素
burserine 香橄榄树脂
bursine 荠(菜)碱
burster 炸裂药
burst fire 点射
bursting 爆破
bursting charge 爆裂药
bursting disc 防爆片
bursting disk 防爆膜
bursting polymerization 爆聚;瞬间聚合
bursting pressure 爆破压力
bursting strength 爆破强度
bursting strength tester 爆破强度试验机
bursting test 爆破试验
burstone 磨石
burstone mill 石磨机
burst spectrum 闪光光谱

burst strength 爆破强度
burst stress 爆破应力
bus-bar 汇流条
bus fuel 公共汽车燃料
bush(=bushing) ①衬套②轴衬;轴瓦
bushel 蒲式耳〔容量单位〕
bushing(=bush) ①衬套②轴衬;轴瓦
bushing center 套子中心
bushing plate 钻模板
bush kauri 鲜贝壳松脂
bush principle 轴衬原理;衬套原理
bushwash 石油和水的乳化液;油罐底残渣
bushy deposits (灯芯上的)丛状沉降物
buspirone 丁螺环酮〔药〕
bustle pipe 促动管
busulfan 白消安〔药〕
butacaine(=butyn) 丁卡因
1,2-butadiene 1,2-丁二烯 $CH_3CH=C=CH_2$
butadiene 丁二烯〔通常指 1,3-丁二烯〕 $CH_2=CHCH=CH_2$
butadiene-acrylonitrile cement 丁腈胶浆
butadiene-acrylonitrile copolymer 丁二烯-丙烯腈共聚物
butadiene-acrylonitrile elastomer 丁二烯-丙烯腈弹性体
butadiene-acrylonitrile rubber 丁腈橡胶
butadiene copolymer 丁二烯共聚物
cis-1,3-butadiene core attractive forces 顺-1,3-丁二烯核(骨架)的吸引力
butadiene cyclization 丁二烯成环作用;丁二烯环化(作用)
butadiene dioxide 二氧化丁二烯〔黏合剂用改性剂〕
butadiene elastomer 丁二烯弹性体
butadiene extraction 丁二烯抽出
butadiene fibre 丁二烯纤维
butadiene isoprene copolymer 丁二烯-异戊二烯共聚物
butadiene-methyl styrene rubber 甲基丁苯橡胶
butadiene methylvinyl pyridine rubber 甲基丁吡橡胶
butadiene-nitrile latex 丁腈胶乳
butadiene nitrile rubber 丁腈橡胶
butadiene plant ①丁二烯厂②丁二烯车间
butadiene process 丁二烯法〔制造己二腈的一种方法〕
butadiene rubber 丁二烯橡胶
cis-1,3-butadiene rubber 顺-1,3-丁二烯橡胶
butadiene sodium polymer 丁(二烯)钠聚合物
butadiene-sodium rubber 丁钠橡胶
butadiene-styrene black masterbatch 炭黑丁苯母胶
butadiene-styrene copolymer 丁二烯-苯乙烯共聚物
butadiene styrene latex 丁苯胶乳;丁二烯-苯乙烯胶乳
butadiene styrene resin 丁二烯-苯乙烯树脂;丁苯树脂

butadiene-styrene rubber 丁苯橡胶
butadiene styrene vinyl pyridine rubber latex 丁苯吡胶乳
butadiene trimer 丁二烯三聚物
butadiene trimerization 丁二烯三聚作用
butadiene vinyl copolymer 丁二烯-乙烯共聚物
butadiene vinyl-pyridine rubber 丁吡橡胶
butadiene vinyl-pyridine rubber latex 丁吡胶乳
butadienyl 丁间二烯基；1,3-丁二烯基
butadiyne 丁二炔
butagas 丁烷气
butalanine α-氨基戊酸
butalastic 丁弹体〔合成丁二烯弹性体〕
butamer process 丁烷异构化法
butanal(=butyraldehyde) 丁醛
butanamide 丁酰胺
butane 丁烷 C_4H_{10}
butane air flame 丁烷空气焰
butane-butylene fraction 丁烷-丁烯馏分
butane-2-carboxylic acid 丁烷-2-羧酸
butane cracking 丁烷裂解
butane diacid 丁二酸 $C_2H_4(COOH)_2$
butanedial 丁二醛 $OHCCH_2CH_2CHO$
butanediamide 丁二酰胺 $H_2NOCCH_2CH_2CONH_2$
1,4-butanediamine 1,4-丁二胺 $H_2NCH_2CH_2CH_2CH_2NH_2$
butanediamine 丁二胺
butane dicarboxylic acid 己二酸；丁烷二羧酸 $COOH(CH_2)_4COOH$
butanedinitrile 丁二腈 $NCCH_2CH_2CN$
butanedioic acid 丁二酸 $HOOCCH_2CH_2COOH$
butanedioic anhydride 丁二酸酐 $C_4H_4O_3$
2,3-butanediol 2,3-丁二醇
butanediol 丁二醇
butanediolamine 丁二醇胺
1,4-butanediol diacrylate 二丙烯酸-1,4-丁二酯
butanediol diacrylate 二丙烯酸丁二酯〔硫化剂〕
butanediol dicaprylate 丁二醇二辛酸酯〔增塑剂 BD〕
butanediol dimethylacrylate 二甲基丙烯酸丁二酯〔硫化剂〕
2,3-butanedione 2,3-丁二酮
butanedione 丁二酮
butane engine fuel 丁烷发动机燃料
butane-enriched water gas 富丁烷的水煤气
butane gasifying 丁烷气化
butane isomerization process 丁烷异构化过程
butane-propane gas 丁烷-丙烷气体
butane splitter 丁烷分离塔
butane sulfinate 丁烷亚磺酸盐
butane sultone 丁磺酸内酯

butane tetracarboxylic acid 丁烷(端)四甲酸 $(COOH)_2CH(CH_2)_2CH(COOH)_2$
1,2,3,4-butane tetracarboxylic dianhydride 1,2,3,4-丁烷四羧酸二酐
butanetriol 丁三醇
butanoate 丁酸盐(或酯)
butanoic acid 丁酸 C_3H_7COOH
1-butanol 1-丁醇
butanol 丁醇 C_4H_9OH
butanolamine 丁醇胺
butanolate(=butylate) 丁醇(金属)盐；丁氧基金属
butanolide 丁内酯
butanol modified resol 丁醇改性可熔酚醛树脂
2,3-butanolone 2,3-丁醇酮
butanolone 丁醇酮
2-butanone 2-丁酮
butanone(=methyl ethyl ketone) 甲基乙基(甲)酮；丁酮 $CH_3COC_2H_5$
butanone diacid 丁酮二酸 $COOHCH_2COCOOH$
butanone oxime 丁酮肟〔防老剂〕
butanoyl(=butyryl) 丁酰(基)
1,2,3,4-butantetraol(=erythritol) 赤藓醇；1,2,3,4-丁四醇
butantriol 丁三醇
butaprenes (聚)丁二烯橡胶类
butatriene 丁三烯
butazone 丁氮酮
butcher cuts 剪屑；切屑；剥伤〔皮革〕
butcher's paper 包肉纸
butea gum 紫矿胶
butein 紫铆因 $C_{15}H_{12}O_5$
2-butenal 2-丁烯醛；巴豆醛
butenal 丁烯醛；巴豆醛
butendiol 丁烯二醇
butene 丁烯
butene diacid 丁烯二酸 $C_2H_2(COOH)_2$
butene dicarboxylic acid 丁烯二酸
butene dioic acid 丁烯二酸
cis-butenediol 顺(式)丁烯二醇
3-butene-4-olide-3-carboxylic acid(=aconic acid) 3-丁烯-4-交酯-3-羧酸，阿康酸
butene rubber 丁烯橡胶
butenic acid 丁烯酸
butenoic acid 丁烯酸 C_3H_5COOH
butenol 丁烯醇
butenolide 丁烯羟基内酯
butenyl 丁烯基
2-butenyl(=crotyl) 巴豆基；丁烯基；2-丁烯基 $CH_3CH=CHCH_2-$

1,4-butenylene 1,4-亚丁烯基 —CH$_2$—CHCH—CH—
butenylidene 亚丁-2-烯基 CH$_3$CH=CHCH=
butenylidyne 次丁-2-烯基 CH$_3$CH=CHC≡
butenyne 丁烯炔
butesin 对氨基苯甲酸丁酯 C$_{11}$H$_{15}$O$_2$N
butex β,β-二丁氧基二乙醚
Butex process 布塔克斯过程〔辐照铀燃料后处理法之一〕
buthotoxin 蝎毒
butin 紫铆黄酮
butine 丁炔
1-butine(=ethyl acetylene) 乙基乙炔;1-丁炔
 C$_2$H$_5$C≡CH
Butler-Volmer equation 巴特勒-沃尔默公式
butolic acid 紫铆醇酸;6-羟基十五烷酸
 C$_{14}$H$_{28}$(OH)COOH
butoxide 丁氧金属;丁醇金属 CH$_3$(CH$_2$)$_2$CH$_2$OM
butoxy 丁氧基 CH$_3$(CH$_2$)$_2$CH$_2$O—
sec-butoxy 仲丁氧基 C$_2$H$_5$CH(CH$_3$)O—
tert-butoxy 叔丁氧基 (CH$_3$)$_3$CO—
butoxy dialkyl acetal 丁氧基型二烷基(醇)缩丁醛
2-butoxy ethanol 2-丁氧基乙醇
1-butoxyethoxy-2-propanol 1-丁氧乙氧基-2-丙醇
butoxyethyl acrylate 丙烯酸丁氧基乙酯
butoxyethyl diglycol carbonate 碳酸丁氧基乙基双甘醇酯
butoxyethyl laurate 月桂酸丁氧基乙酯〔增塑剂〕
butoxyethyl oleate 油酸丁氧基乙酯〔增塑剂〕
2-butoxyethyl pelargonate 壬酸丁氧基乙酯〔增塑剂〕
butoxyethyl stearate 硬脂酸丁氧基乙酯〔增塑剂〕
butoxyglycoluril 丁氧基甘脲
4-butoxyphenyl-2,6-diphenyl pyrylium perchlorate 4-丁氧苯基-2,6-二苯基吡喃鎓过氯酸盐
butoxy resin 丁醇醚化树脂
butrin 紫矿春 C$_{27}$H$_{32}$O$_{15}$
butt coupling 对接偶合
butt end 接合端
butterfly burner 蝶形灯头
butterfly lily 黄姜花
butterfly mixer 蝶形混合器;蝶形搅拌机
butterfly throttle valve 蝶形节流阀
butterfly valve 蝶形阀
buttermilks 犊牛皮(革)
butter milk soap 乳酪肥皂;奶酪润肤皂
butternut 胡桃
butter of antimony 三氯化锑 SbCl$_3$
butter of arsenic 三氯化砷 AsCl$_3$
butter of tin 氯化锡;四氯化锡 SnCl$_4$
butter of zinc 氯化锌 ZnCl$_2$
butter tree 牛油树
butter yellow(=p-dimethylamino-azo-benzene) 甲基黄;对二甲氨基偶氮苯
butt-fusion 对焊;对接熔焊
butt joint 对接接头;对接
butt leather 底革
button 钮(子)
buttonhole stitch 锁眼
button lac 钮状紫胶
button mould 钮扣模
button mushroom 顶磨菇
button polish 钮状紫胶清漆
button speck 钮斑
button switch 按钮开关
buttonwood 一球悬铃木;美国梧桐
butt paddle agitator 平桨式搅拌器
buttressing effect 支撑效应
butt roller 底革辊
butt rot 根腐
butt seam 对接(接合);对头接合
butt weld 对接焊缝;对头焊接
butt welded pipe 对缝焊管
butt welding 对缝焊接
buttwood 环孔材
butyl 丁基 CH$_3$(CH$_2$)$_2$CH$_2$—
sec-butyl 仲丁基 C$_2$H$_5$CH(CH$_3$)—
tert-butyl 叔丁基 (CH$_3$)$_3$C—
N-tert-butylacetamide N-叔丁乙酰胺
 CH$_3$CONHC(CH$_3$)$_3$N
n-butyl acetate 乙酸(正)丁酯 CH$_3$CO$_2$CH$_2$CH$_2$C$_2$H$_5$
sec-butyl acetate 乙酸仲丁酯
tert-butyl acetate 乙酸叔丁酯 CH$_3$CO$_2$C(CH$_3$)$_3$
butylacetic acid 己酸;丁基乙酸 C$_5$H$_{11}$COOH
sec-butylacetic acid 仲丁基酸;3-甲基戊酸
 C$_2$H$_5$CH(CH$_3$)CH$_2$CO$_2$H
tert-butylacetic acid 叔丁基乙酸;3,3-二甲基丁酸
 (CH$_3$)$_3$CCH$_2$COOH
butyl acetoacetate 乙酰乙酸丁酯 CH$_3$COCH$_2$CO$_2$C$_4$H$_9$
butyl acetoacetic ester 乙酰丁基乙酸酯
 CH$_3$COCH(C$_4$H$_9$)COOR
p-tert-butyl acetophenone 对叔丁基苯乙酮
n-butyl-acetylene (正)丁基乙炔;1-己炔 C$_4$H$_9$C≡CH
butyl acetyl ricinoleate 乙酰蓖酸丁酯
β-butylacrolein β-丁基丙烯醛
butyl acrylate 丙烯酸丁酯〔黏合剂〕
N,N'-di-n-butyladipamide N,N'-二正丁基己二酰胺
 C$_4$H$_9$NHCO(CH$_2$)$_4$CONHC$_4$H$_9$

n-butyl alcohol(=propyl carbinol) 丙基甲醇；正丁醇
sec-butyl alcohol 仲丁醇 $C_2H_5CHOHCH_3$
tert-butyl alcohol 叔丁醇 $(CH_3)_3COH$
butyl aldehyde 丁醛 C_3H_7CHO
butyl aldoxime 丁醛肟〔防结皮剂〕
butyl amine 丁胺 $CH_3(CH_2)_2CH_2NH_2$
sec-butylamine 仲丁胺 $C_2H_5(CH_3)CHNH_2$
tert-butylamine 叔丁胺 $(CH_3)_3CNH_2$
butylamine cellulose 丁胺纤维素
butyl 2-aminobenzoate 2-氨基苯甲酸丁酯
t-butylaminoethyl methacrylate 甲基丙烯酸叔丁基氨基乙酯
butyl-*p*-aminophenol 丁基对氨基酚〔成胶抑制剂，防胶剂〕
N-butylaniline *N*-丁(基)苯胺 $C_4H_9NHC_6H_5$
n-butyl anthranilate 邻氨基苯甲酸丁酯
n-butylarsonic acid （正）丁胂酸 $C_4H_9AsO(OH)_2$
butylate(=butanolate) 丁醇(金属)盐；丁氧基金属
butylated 丁基化的
butylated bisphenol A 丁基化双酚 A〔防老剂〕
butylated glycoluril 丁基(醇)醚化甘脲
butylated hydroxyanisol 丁基化羟基苯甲醚；丁基化羟基茴香醚〔防老剂〕
butylated hydroxytoluene 2,6-二叔丁基对甲酚；丁化羟基甲苯〔防老剂〕
butylated melamine-formaldehyde resin 丁醇改性三聚氰胺甲醛树脂
butylated melamine resin 丁醇改性三聚氰胺树脂
butylated methylol melamine 丁醇醚化羟甲基三聚氰胺
butylated urea-formaldehyde resin 丁基化脲甲醛树脂；丁醇醚化脲醛树脂
butylated urea resin 丁醇改性脲醛树脂
butylation 丁化作用
butyl benzene 丁(基)苯 $C_{10}H_{14}$
sec-butylbenzene 仲丁基苯
tert-butylbenzene 叔丁基苯 $C_6H_5C(CH_3)_3$
butyl benzoate 苯甲酸丁酯 $C_6H_5CO_2C_4H_9$
N-tert-butyl-2-benzothiazyl sulfenamide *N*-叔丁基-2-苯并噻唑-次磺酰胺〔促进剂 NBBS〕
butyl *o*-benzoylbenzoate 邻苯甲酰基苯甲酸丁酯 $C_6H_5COC_6H_4CO_2C_4H_9$
butyl benzyl cellulose 丁(基)苄(基)纤维素
butyl benzyl ether 丁苄醚；苄氧基丁烷 $C_4H_9OCH_2C_6H_5$
butyl benzyl phthalate 邻苯二甲酸丁苄酯；酞酸丁苄酯
butyl benzyl sebacate 癸二酸丁苄酯〔增塑剂〕
N-butyl-4,4-bis (*t*-butylperoxy) valerate 4,4-双(叔丁基过氧基)戊酸正丁酯

butyl borate 硼酸丁酯 $B(OC_4H_9)_3$
butyl boric acid(=butyl boron dihydroxide) 丁基硼酸
butyl boron dihydroxide(=butyl boric acid) 丁基硼酸
butyl bromide 丁基溴；1-溴代丁烷 C_4H_9Br
sec-butyl bromide 仲丁基溴 $C_2H_5CHBrCH_3$
tert-butyl bromide 叔丁基溴 $(CH_3)_3CBr$
2-butyl-2-butenal 2-丁基-2-丁烯醛
butyl butenal(=ethylidene hexanal) 亚乙基丁醛；丁基丁烯醛
butyl butyrate 丁酸丁酯 $C_3H_7CO_2C_4H_9$
butyl butyryl lactate 丁酰乳酸丁酯
butyl caproate 己酸丁酯 $C_5H_{11}CO_2C_4H_9$
butyl capronate 己酸丁酯
butyl caprylate 辛酸丁酯 $C_7H_{15}CO_2C_4H_9$
butyl carbamate ①氨基甲酸丁酯 $H_2NCO_2C_4H_9$ ②丁氨基甲酸盐 $C_4H_9NHCOOM$
butyl carbamic acid 丁氨基甲酸 $C_4H_9NHCOOH$
butyl carbinol 丁基甲醇；1-戊醇
butyl carbitol 丁基卡必醇；二甘醇一丁醚；二乙二醇丁醚
butyl carbitol acetate 二甘醇一丁醚乙酸酯；二乙二醇丁醚乙酸酯；丁基卡必醇乙酸酯
butyl carbitol formal 二甘醇二乙醚甲醛〔增塑剂〕
butyl-catechol 丁基邻苯二酚〔成胶抑制剂，防胶剂〕
p-tert-butylcatechol 对叔丁基邻苯二酚〔抑制剂〕
butylcellosolve 丁基溶纤剂；乙二醇一丁醚 $C_4H_9OC_2H_4OC_4H_9$
butyl cellosolve adipate(=dibutoxyethy adipate) 己二酸二丁氧基乙酯〔增塑剂〕
butyl cellosolve pelargonate(=dibutoxyethyl pelargonate; dibutoxyethyl phthalate) 壬酸二丁氧基酯〔增塑剂〕
butyl cellosolve phthalate(=dibutoxyethyl pelargonate) 邻苯二甲酸二丁氧基乙酯〔增塑剂〕
butyl cellulose 丁基纤维素
butyl cellusolve 乙二醇单丁醚〔溶剂〕
butylchloral 丁氯醛〔俗〕 $CH_3CHClCCl_2CHO$
butylchloral hydrate 丁氯醛合水〔俗〕 $CH_3CHClCCl_2CH(OH)_2$
n-butyl chloride （正）丁基氯；1-氯丁烷 $C_2H_5CH_2CH_2Cl$
sec-butyl chloride 仲丁基氯 $C_2H_5CHClCH_3$
tert-butyl chloride 叔丁基氯 $(CH_3)_3CCl$
butyl chloroacetate 氯乙酸丁酯 $ClCH_2CO_2C_4H_9$
butyl chlorocarbonate 氯甲酸丁酯 $ClCO_2C_4H_9$
a-butylcinnamaldehyde *a*-丁基肉桂醛
butyl cinnamic aldehyde 丁基肉桂醛
a-n-butyl cinnamic aldehyde *a*-正丁基(肉)桂醛
butyl citrate 柠檬酸丁酯；柠檬酸三丁酯
p-tert-butyl cresol 对叔丁基甲酚

butyl-o-cresyl ether　丁基邻甲苯基醚　$CH_3C_6H_4OC_4H_9$
butyl crotonate　巴豆酸丁酯；丁烯酸丁酯
　　　　$CH_3CH=CHCO_2C_4H_9$
t-butyl cumyl peroxide　叔丁基过氧化异丙苯〔硫化剂〕
butylcyanide(=valeronitrile)　戊腈；丁基氰
　　　　$CH_3(CH_2)_3CN$
o-$tert$-butyl cyclohexanol　邻叔丁基环己醇
2-sec-butylcyclohexanone　2-仲丁基环己酮
p-$tert$-butyl cyclohexyl acetate　乙酸对叔丁基环己酯
butyl cyclohexyl phthalate　邻苯二甲酸丁基环己酯〔增塑剂〕
butyl 2-decenoate　2-癸烯酸丁酯
butyl decylbenzene sulfonate　癸基苯磺酸丁酯
butyl decyl phalate　邻苯二甲酸丁癸酯〔增塑剂〕
p-t-butyldichloroacetophenone　对-叔丁基苯二氯乙酮
N-butyl diethanol-amine　N-丁二(羟乙基)胺；丁二乙醇胺
　　　　$C_4H_9N(CH_2CH_2OH)_2$
butyl-diglycol-methacrylate　甲基丙烯酸丁氧基二甘
　　　　醇酯
butyl-4-dimethylaminobenzoate　4-二甲氨基安息香酸
　　　　丁酯
6-t-butyl-2,4-dimethyl phenol　6-叔丁基-2,4-二甲基苯酚
　　　　〔防老剂〕
butyl disulfide　二硫化二丁基　$(C_4H_9)_2S_2$
butyl dodecanoate　十二酸丁酯
butylene　①丁烯　$C_4H_8$②〔复合词中〕亚丁基
　　　　—C_4H_8—
1-butylene　1-丁烯　$CH_3CH_2CH=CH_2$
2-butylene　2-丁烯　$CH_3CH=CHCH_3$
butylene-chlorohydrin　亚丁基氯醇；氯丁醇
　　　　$CH_3CHClCHOHCH_3$
butylene diamine　亚丁基二胺；丁撑二胺
1,3-butylene dimethacrylate　二甲基丙烯酸-1,3-丁酯
1,2-butylene-glycol　1,2-丁二醇；丁邻二醇
　　　　$CH_3CH_2CHOHCH_2OH$
1,3-butylene-glycol　1,3-丁二醇；丁间二醇
　　　　$CH_3CHOHCH_2CH_2OH$
1,4-butylene-glycol　1,4-丁二醇；丁隔二醇
　　　　$HO(CH_2)_4OH$
2,3-butylene-glycol　2,3-丁二醇；丁二仲醇
　　　　$CH_3CHOHCHOHCH_3$
butylene glycol acrylate　丙烯酸丁二醇酯〔硫化剂〕
1,3-butyleneglycol diacrylate　二丙烯酸 1,3-丁二醇酯
butylene glycol dimethacrylate　二甲基丙烯酸丁二醇酯
　　　　〔硫化剂〕
2,3-butyleneglycol fermentation　2,3-丁二醇发酵
butylene oxide　亚丁基氧；环氧丁烷；四氢呋喃
　　　　$(CH_2)_4O$
butylene oxide ring　亚丁基氧环；环氧丁烷

butylene sulfide　亚丁基硫环；环硫丁烷；四氢噻吩
　　　　$(CH_2)_4S$
butyl epoxy stearate　环氧硬脂酸丁酯〔增塑剂〕
butyl-9,10-epoxy-stearate　9,10-环氧硬脂酸丁酯
butyl ester　丁酯
butyl ether　①丁基醚　$ROC_4H_9$②二丁醚　$(C_4H_9)_2O$
butylethylene(=1-hexene)　1-丁基乙烯；1-己烯
　　　　$CH_3(CH_2)_3CH=CH_2$
butyl ethyl ether　仲丁基乙基醚
butyl ethyl ketene(BEK)　丁基乙基(乙)烯酮
butyl ethyl ketone　丁基乙基酮
butyl ethyl malonate　丙二酸丁乙酯
　　　　$CH_2(CO_2C_4H_9)(CO_2C_2H_5)$
butyl fluoranthene sulfonate　丁基荧蒽磺酸盐
butyl fluoride　丁基氟；1-氟代丁烷　C_4H_9F
butyl fomiate(=butyl formate)　甲酸丁酯
butyl formate(=butyl fomiate)　甲酸丁酯　$HCOOC_4H_9$
n-butyl formate　甲酸(正)丁酯　$HCO_2CH_2CH_2C_2H_5$
sec-butyl formate　甲酸仲丁酯　$HCO_2CH(CH_3)C_2H_5$
butyl furoate　糠酸丁酯　$C_4H_3OCOOC_4H_9$
butyl glycidyl ether　丁基缩水甘油醚〔稀释剂〕
butyl glycidyl ether acrylate　丁基缩水甘油醚丙烯酸酯
butyl glycol　丁基乙二醇
butyl glycol acetate　乙酸丁基乙二醇酯
butyl glycol ether　乙二醇丁醚〔溶剂〕
butyl Grignard reagent　丁基卤化镁；丁基格利雅试剂
　　　　$MgXC_4H_9$
butyl halide　丁基卤；卤代丁烷　C_4H_9X
butyl 10-hendecenoate　10-十一烯酸丁酯
butyl heptanoate　庚酸丁酯
butyl heptylate　庚酸丁脂
butyl hexanoate　己酸丁酯
butyl hydrogen sulfate　丁(代)硫酸；硫酸氢丁酯
　　　　$C_4H_9OSO_3H$
t-butyl hydroperoxide　过氧化氢叔丁基〔硫化剂〕
butyl hydroxyanisole(2,6-dibutyl-4-methoxyphenol)　丁基
　　　　羟基苯甲醚(2,6-二丁基-4-甲氧基苯酚)〔防老剂〕
butyl 2-hydroxypropanoate　2-羟基丙酸丁酯
butylhydroxy toluene　丁基羟基甲苯
t-butyl hypochlorate　次氯酸叔丁酯〔防焦剂〕
butylic fermentation　丁醇发酵
butylidene　亚丁基　$CH_3CH_2CH_2CH=$
sec-butylidene　仲亚丁基　$CH_3CH_2C(CH_3)=$
butylideneacetone　亚丁基丙酮
4,4′-butylidene-p-aminodimethyl aniline　4,4′-亚丁基对
　　　　氨基二甲苯基胺〔防老剂〕
4,4′-butylidene-bis(6-$tert$-butyl m-cresol)　4,4′-亚丁基
　　　　双-6-叔丁基间甲基苯酚

4,4′-butylidene-bis(3-methyl-6-*t*-butyl phenol) 4,4′-亚丁基双(3-甲基-6-叔丁基苯酚)〔防老剂〕
butylidene chloride 亚丁基二氯
butylidene ethylene 亚丁基乙烯
1,3-butylidene glycol dimethacrylate 二甲基丙烯酸 1,3-亚丁基二醇酯〔交联剂〕
butylidene phthalide 正亚丁基邻苯二甲酰胺
n-butylidene phthalide(=ligusticum lactone) 藁木内酯；正亚丁基酞内酯
butylidene propionic acid 亚丁基丙酸
butylidyne 次丁基 $CH_3(CH_2)_2C\equiv$
butyl inner tube reclaim 丁基内胎再生胶
butyl iodide 丁基碘；碘代丁烷 C_4H_9I
n-butyl iodide (正)丁基碘 $C_2H_5CH_2CH_2I$
sec-butyl iodide 仲丁基碘；2-碘丁烷 $C_2H_5CHICH_3$
tert-butyl iodide 叔丁基碘 $(CH_3)_3CI$
tert-butyl isobutyl ketone 叔丁基异丁酮
butyl isocrotonate 异巴豆酸异丁酯；(Z)-丁烯酸丁酯 $CH_3CH=CHCO_2C_4H_9$
butyl isocyanate 异氰酸丁酯
butyl isocyanide 异氰基丁烷 $CH_3(CH_2)_3NC$
butyl isodecyl phthalate 邻苯二甲酸丁基异癸酯〔增塑剂〕
tert-butyl isopentanol 叔丁基异戊醇
t-butyl isopropyl carbonate 叔丁基碳酸异丙酯〔硫化剂〕
butyl isorhodanate(=butyl isothiocyanate) 异硫氰酸丁酯
butyl isorhodanide(=butyl isothiocyanate) 异硫氰酸丁酯
butyl isosulfocyanate(=butyl isothiocyanate) 异硫氰酸丁酯
butyl isosulfocyanide(=butyl isothiocyanate) 异硫氰酸丁酯
sec-butyl isothiocyanate 异硫氰酸仲丁酯 C_4H_9NCS
tert-butyl isothiocyanate 异硫氰酸叔丁酯 $(CH_3)_3CNCS$
n-butyl isothiocyanate(=*n*-butyl mustard oil) 异硫氰酸(正)丁酯 $C_2H_5CH_2CH_2NCS$
butyl isothiocyanide(=butyl isothiocyanate) 异硫氰酸丁酯
butyl isovalerate 异戊酸丁酯
n-butyl isovalerate 异戊酸(正)丁酯 $(CH_3)_2CHCH_2CO_2C_4H_9$
sec-butyl isovalerate 异戊酸仲丁酯
butyl ketone 5-壬酮
butyl lactate 乳酸丁酯
butyl lactic acid 异丁醇酸；α-甲基乳酸 $(CH_3)_2C(OH)COOH$
butyl laurate 月桂酸丁酯〔增塑剂〕
butyl lauryl phthalate 邻苯二甲酸丁基月桂酯〔增塑剂〕

butyl levulinate γ-戊酮酸丁酯 $CH_3CO(CH_2)_2CO_2C_4H_9$
butyllithium 丁基锂 $n\text{-}C_4H_9Li$
butyllithium-initiated polymerization 丁基锂引发聚合(作用)
butyl magnesium bromide 溴化丁基镁 C_4H_9MgBr
butyl magnesium chloride 氯化丁基镁 C_4H_9MgCl
butylmalonic acid 丁基丙二酸 $C_4H_9CH(CO_2H)_2$
n-butyl mercaptan (正)丁硫醇 $C_2H_5(CH_2)_2SH$
tert-butyl mercaptan 叔丁硫醇 $(CH_3)_3CSH$
butylmercuric bromide 丁基溴化汞
butylmercuric chloride 丁基氯化汞
butylmercuric iodide 丁基碘化汞
butyl methacrylate 甲基丙烯酸丁酯 $C_3H_5CO_2C_4H_9$
butyl-*o*-methoxybenzoate 邻甲氧基苯甲(酸)丁酯 $CH_3OC_6H_4CO_2C_4H_9$
n-butyl 2-methylbutyrate 2-甲基丁酸正丁酯
n-butyl 2-methylpropanoate 2-甲基丙酸正丁酯
8-*tert*-butyl-5-methyl-α-tetralone 8-叔丁基-5-甲基-α-萘满酮；8-叔丁基-5-甲基-α-四氢萘酮
butyl mustard oil 异硫氰酸丁酯 C_4H_9NCS
butyl naphthalene sulfonate(BNS) 萘磺酸丁酯〔止链剂〕
butyl nitrite 亚硝酸丁酯 C_4H_9ONO
butyl nonyl phthalate 邻苯二甲酸丁壬酯〔增塑剂〕
butyl octadecanoate 十八酸丁酯；硬脂酸丁酯
butyl octyl phthalate 邻苯二甲酸丁辛酯
butyl oleate 油酸丁酯 $C_{17}H_{33}CO_2C_4H_9$
butyl oxalate 草酸二丁酯
butyl oxamate 草酰氨酸丁酯；氨羰基甲酸丁酯 $H_2NCOCO_2C_4H_9$
butyl oxide 二丁醚 $(C_4H_9)_2O$
butyl oxide link(age) 环氧丁烷
butyl oxytol 乙二醇一丁醚；丁基溶纤剂（俗）
butyl paraben 尼泊金丁酯；对-羟基苯甲酸丁酯
n-butyl pelargonate 壬酸(正)丁酯 $C_8H_{17}CO_2C_4H_9$
n-butyl pentanoate 戊酸正丁酯
t-butyl perbenzoate 过苯甲酸叔丁酯〔硫化剂〕
t-butyl peroxy-acetate 过乙酸叔丁酯
t-butyl per (oxy) benzoate 过氧苯甲酸丁酯
t-butylperoxy isobutylate 过异丁酸叔丁酯
t-butyl peroxyisopropyl carbonate 叔丁基过氧异丙烯碳酸酯〔交联剂〕
t-butylperoxy-laurate 过月桂酸叔丁酯
t-butylperoxyl benzoate 过苯甲酸叔丁酯
t-butylperoxy pivarate 叔丁基过氧化新戊酸酯；过氧新戊酸叔丁酯
t-butylperoxy-3,5,5-trimethyl hexanoate 叔丁基过氧化-3,5,5-三甲基己酸酯
t-butyl phenol 叔丁基苯酚 $(CH_3)_3CC_6H_4OH$

t-butyl phenoxyl ethanol 叔丁苯氧基乙醇 $(CH_3)_3CC_6H_4OCH_2CH_2OH$
butyl phenylacetate 苯乙酸丁酯 $C_6H_5CH_2CO_2C_4H_9$
α-butyl-*β*-phenyl acrolein *α*-丁基-*β*-苯基丙烯醛
p-*n*-butylphenylarsonic acid 对(正)丁苯胂酸
butyl phenylate 丁基苯基醚 $C_4H_9OC_6H_5$
butyl phenyl carbinol 丁基苯基甲醇；偕苯代戊醇 $C_6H_5CHOHC_4H_9$
p-*t*-butyl phenyl disulfide 二硫化对叔丁基苯基〔促进剂〕
butyl phenyl ether 丁基苯基醚；丁氧基苯 $C_4H_9OC_6H_5$
n-butyl phenyl ketone (正)丁基苯基(甲)酮 $C_4H_9COC_6H_5$
N-butyl-*N*-phenyl-*p*-phenylenediamine *N*-丁基-*N*-苯基对苯二胺〔防老剂〕
N-*sec*-butyl-*N*-phenyl-*p*-phenylenediamine *N*-仲丁基-*N*-苯基对苯二胺〔防老剂〕
p-*t*-butylphenyl salicylate 水杨酸对叔丁基苯酯〔光稳定剂〕
butyl phosphate 磷酸丁酯 $(C_4H_9O)PO(OH)_2$; $(C_4H_9O)_2POOH$; $(C_4H_9O)_3PO$
butyl phthalate 邻苯二甲酸丁酯〔增塑剂〕
3-*n*-butylphthalide 3-正丁基苯酞
butyl phthalyl butyl glycolate 丁基邻苯二甲酰基甘醇酸丁酯；丁基邻苯二甲酰基乙醇酸丁酯〔增塑剂〕
n-butyl propionate 丙酸(正)丁酯 $C_2H_5CO_2C_4H_9$
sec-butyl propionate 丙酸仲丁酯
2-butylpropionic acid 2-丁基丙酸
t-butyl pyrocatechol 叔丁基焦儿茶酚 $(CH)_3CC_6H_3(OH)_2$
butyl pyromucate 焦黏酸丁酯；糠酸丁酯 $C_4H_3OCO_2C_4H_9$
N-butyl pyrrol *N*-丁吡咯 $C_4H_4NC_4H_9$
butyl rhodanate(=butyl rhodanide) 硫氰酸丁酯；丁基硫氰 C_4H_9SCN
butyl rhodanide(=butyl rhodanate) 硫氰酸丁酯；丁基硫氰
butyl ricinoleate 蓖酸丁酯 $HOC_{17}H_{32}CO_2C_4H_9$
butyl rubber(=isoprene-isobutylene rubber;isobutylene isoprene rubber) 丁基橡胶
butyl rubber latex(=isobutylene isoprene rubber latex) 丁基胶乳
butyl salicylate 水杨酸丁酯 $HOC_6H_4CO_2C_4H_9$
butyl sebacate 癸二酸丁酯
butyl stearate 硬脂酸丁酯 $C_{17}H_{35}CO_2C_4H_9$
butyl sulfate 硫酸氢丁酯 $C_4H_9HSO_4$
butyl sulfhydrate(=butyl sulfhydryl) 丁硫醇 C_4H_9SH
butyl sulfhydryl(=butyl sulfhydrate) 丁硫醇
butyl sulfide 二丁(基)硫；丁硫醚 $(C_4H_9)_2S$

butyl sulfocyanate(=butyl thiocyanate) 硫氰酸丁酯
butyl sulfocyanide(=butyl thiocyanate) 硫氰酸丁酯
7-*t*-butyl-*α*-tetralone 7-叔丁基-*α*-萘满酮；7-叔丁基-*α*-四氢萘酮
butyl (tetra) titanate 钛酸丁酯；钛酸(四)丁基酯
2-*sec*-butyl thiazole 2-仲丁基噻唑
4-*t*-butyl-*o*-thiocresol 4-叔丁基邻甲苯硫酚〔塑解剂〕
butyl thiocyanate(=butyl thiocyanide) 硫氰酸丁酯；丁基硫氰 C_4H_9SCN
butyl thiocyanide(=butyl thiocyanate) 硫氰酸丁酯；丁基硫氰 C_4H_9SCN
t-butyl thionitrile 叔丁基硫腈〔防焦剂〕
butyl thiourea 丁基(代)硫脲 $C_4H_9NHCSNH_2$
butyl titanate 钛酸丁酯〔偶联剂〕
butyltitanate dimer 丁基钛酸酯二聚体；钛酸丁酯二聚物
butyl titanate zinc dust paint 钛酸丁酯锌粉漆
butyl *p*-toluenesulfonate 对甲苯磺酸丁酯 $CH_3C_6H_4SO_3C_4H_9$
1-butyl-3-tolylsulfonylurea 1-丁基-3-对甲苯磺酰脲
butyl trichlorosilicane 丁基三氯化硅 $C_4H_9SiCl_3$
butyl trimethylsilicane 三甲基丁基硅 $(C_4H_9)Si(CH_3)_3$
butyl 10-undecenoate 10-十一碳烯酸丁酯
butyl undecylenate 十一烯酸丁酯
butyl urea 丁脲 $C_4H_9NHCONH_2$
butyl urethane 丁基尿烷〔俗〕；丁氨基甲酸乙酯 $C_4H_9NHCO_2C_2H_5$
butyl valerate 戊酸丁酯 $CH_3(CH_2)_3CO_2C_4H_9$
butyl valerianate 戊丁酸酯〔硝基纤维素的溶剂有防潮作用〕
butyl vinyl ether 丁基乙烯醚
butyl xanthate 丁基黄原酸盐(或酯) C_4H_9OCSSM
butyn(=butacaine) 丁卡因
butyne 丁炔
2-butyne 2-丁炔；二甲基乙炔 CH_3CCCH_3
butynediol 丁炔二醇
butynoic acid 丁炔酸 C_3H_3COOH
2-butynylene 1,4-亚丁炔基 $—CH_2C\equiv CCH_2—$
butyraceous ①油的；含油的；油性的；油质的②多油的；油腻的③油滑的
butyral ①缩丁醛；醇缩丁醛②聚乙烯醇缩丁醛
butyraldehyde(=butanal) 丁醛 C_3H_7CHO
butyraldehyde dialkyl acetal(=buthoxy dialkyl acetal) 丁醛型二烷基(醇)缩丁醛
n-butyraldehyde di-*n*-butyl acetal 正丁醛二丁缩醛
butyraldehyde oxime 丁醛肟 $C_2H_5CH_2CH=NOH$
butyraldol(=2-ethyl-1,3-hexanediol) 2-乙基-1,3-己二醇
butyral group (醇)缩丁醛基
butyralization 缩丁醛化(作用)

butyralized　缩丁醛化的
butyral resin　丁醛树脂
butyramide　丁酰胺　$C_2H_5CH_2CONH_2$
butyranilide　丁酰苯胺　$C_3H_7CONHC_6H_5$
butyrase　丁酯酶
butyrate　①丁酸②丁酸盐(酯或根)
butyrelite　泥煤脂
butyric acid　丁酸　$CH_3(CH_2)_2COOH$
butyric acid amide　丁酰胺　$C_3H_7CONH_2$
butyric acid chloride　丁酰氯　C_3H_7COCl
butyric acid value　丁酸值
butyric alcohol　丁醇
butyric aldehyde　丁醛　C_3H_7CHO
butyric anhydride　丁(酸)酐　$(C_2H_5CH_2CO)_2O$
butyric ester　丁酸酯　C_3H_7COOR
butyricin　丁酸梭菌素
butyrin　酪脂；丁精；三丁酸甘油酯
　　　　$(CH_3CH_2CH_2CO)_3C_3H_5O_3$
butyrinase　酪脂酶
butyro-　丁酸(的)
butyrobetaine　丁内铵盐；三甲铵基丁内盐
butyrocopal　丁酸珀珀
butyroin　丁偶姻；5-羟基-4-辛酮　$C_3H_7CHOHCOC_3H_7$
butyrolactam　丁内酰胺
butyrolactone　丁内酯
γ-butyrolactone　γ-丁内酯
butyrolactone β-carboxylic acid　丁内酯-β-羧酸；仲康酸
butyrometer　奶油计
butyrone　二丙基(甲)酮
butyronitrile　丁腈　$C_2H_5CH_2CN$
n-butyrophenone(=propyl phenyl ketone)　丙基苯基(甲)
　　酮　$C_2H_5CH_2COC_6H_5$
butyrospermol　丁酰鲸鱼醇
butyroxymethyl dioxolane　丁酰氧甲基二噁烷
butyryl(=butanoyl)　丁酰(基)　$CH_3CH_2CH_2CO—$
butyryl chloride　丁酰氯　C_2H_5COCl
1-butyryl propylene　1-丁酰基丙烯
4-n-butyryl pyrogallol(=2,3,4-trihydroxybutyrophenone)
　　2,3,4-三羟基苯丁酮；4-丁酰焦棓酚
　　$(HO)_3C_6H_2COC_3H_7$
butyryl urea　丁酰脲　$C_3H_7CONHCONH_2$
buxidine　黄杨叶碱
buxine　黄杨碱

Buxton's fluid　巴克斯顿液
buyo(=betel-nut)　槟榔
buzane　联肼
buzylene　①异四氮烯②异四氮烯基
buzzer　①蜂鸣器②磨轮；砂轮
byak-angelicin　白当归素
byakangelicol　白当归脑
byerlyte　①炼焦烟煤②石油沥青〔贝耳勒法产生的〕
by gravity　自流(地)
by-law　章程；规则；说明书
by-pass　旁路；支路；旁通管
by-pass damper　支路气门
by-passed oil　死油〔由于注入水驱扫不到而造成的〕
by-pass electrode　旁路电极
by-pass extruder　旁路挤塑机
by-pass filter　旁通过滤器
by-pass flue　旁通烟道
by-passing　走旁路
by-pass injection　旁通注射
by-pass injector　旁通注射器；旁通进样器
by-pass line　旁通管
by-pass pipe　旁路管；旁通管
by-pass plug　旁通塞；放油塞
by-pass ratio　旁通比
by-pass sample　旁通进样
by-pass sample line　旁路样品管线
by-pass sampler　旁通进样器
by-pass tee　旁路三通
by-pass valve　旁路阀；旁通阀
by-pass vent extruder　旁通气挤出机
by-product　副产物
by-product ammonia　副产氨
by-product coke　副产焦炭
by-product coking　有副产物的炼焦
by-product gas producer　有副产物的气体发生炉
by-product oven　副产物炉
by-product precipitation　副产品沉淀
by-product recovery gas producer　有副产物回收的气体
　　发生炉
by-product suppressant compound　抑制副产品的化合物
by-product suppressor　副产物抑制剂
bythium(=eka-tellurium)　类碲〔假定的硫属元素名〕
bytownite　倍长石

C

C-face centered lattice C 心点阵
Ca-alumina soap 钙-氧化铝皂
Ca-aluminate-silicate base 铝酸钙-硅酸钙载体
cabagin 维生素 U；甘蓝素
cabardine 麝香
cabasite (=chabazite) 菱沸石
cabinet ①橱；(机)箱②小室
cabinet drier 干燥橱
cabinet dryer 橱式干燥机
cabin paint 船舱漆
cable ①缆②电缆
cable bond 电缆接头
cable compound (=cable oil) 电缆油
cable conductor 电缆导线
cable connector 电缆接头
cable corder 帘子线并捻机
cable core 电缆芯线
cable duct ring 电缆导圈
cable duct tube 电缆导管
cabled yarn 缆线
cable-jacketing 包敷电缆外壳
cable lacquer 电缆漆
cable lay 线绳螺旋缠绕
cable log haul 缆索拉木机
cable net 薄纱；大网眼窗纱
cable oil 电缆油
cable paper 电缆纸
cable sheave bracket 绞缆滑车支架〔油罐上摇管用〕
cable stitch 扎花；绞花织物
cable varnish 电缆清漆；电缆皮清漆
cable wax 电缆蜡
cableway 索道
cabling 并捻(工艺)
cabretta ①直毛绵羊皮②直毛绵羊革
cabreuva oil 卡鲁瓦油
cabtyre cable 焊接电缆
cacahuananche oil 柯莞油〔取自 Licania arborea 的种子，产于南美〕
cacaine 可可碱
cacao (=theobroma cacao) 可可
cacao beans 可可豆
cacao butter (=cocoa butter) 可可脂；可可油
cacaorin 可可苷 $C_{16}H_{20}O_6N_8$
cacao-tallow 可可蜡

C-acid C 酸；2-萘胺-4,8-二磺酸
cacodiliacol (=guaiacol cacodylate) 愈创木酚二甲胂酸酯
cacodyl 卡可基；二甲胂基 $(CH_3)_2As-$
cacodylate 卡可酸盐；二甲胂酸盐 $(CH_3)_2AsOOM$
cacodyl chloride 卡可基氯；氯化二甲胂 $(CH_3)_2AsCl$
cacodyl cyanide 卡可基氰；氰化二甲胂 $(CH_3)_2AsCN$
cacodyl disulfide 卡可基化二硫；二甲胂基化二硫 $[(CH_3)_2As]_2S_2$
cacodyl hydride 卡可基氢；二甲胂 $(CH_3)_2AsH$
cacodylic acid 卡可基酸；二甲次胂酸 $(CH_3)_2AsOOH$
cacodyloxide 卡可基氧；双二甲胂基氧 $[(CH_3)_2As]_2O$
cacodyl sulfide 卡可基硫；双二甲胂基硫 $[(CH_3)_2As]_2S$
cacodyl trichloride 卡可基三氯；二甲胂基三氯 $(CH_3)_2AsCl_3$
cacogenin 卡可配基
cacotheline (=nitrobruciquinone hydrate) 卡可西灵
cacoxenite 黄磷铁矿
cactine 仙影拳碱
cactus 仙人掌
cadalene 卡达烯；4-异丙基-1,6-二甲基萘
cadaverine (=pentamethylene diamine) 尸胺；1,5-戊二胺 $H_2N(CH_2)_5NH_2$
cade oil 杜松油〔医用〕
cadinane 杜松烷
cadinene 杜松烯；荜澄茄烯
cadinenol 杜松烯醇
cadinol 杜松醇
cadion 镉试剂*
cadmia 锌壳；锌渣；菱锌矿
cadmic 镉的；二价镉的
cadmic compound 镉化合物
cadmium 镉〔48 号元素，化学符号 Cd〕
cadmium acetate 乙酸镉 $Cd(C_2H_3O_2)_2$
cadmium-ammonia complex 镉氨络合物
cadmium borotungstate 硼钨酸镉
  $2CdO \cdot B_2O_3 \cdot 9WO_3 \cdot 18H_2O$
cadmium bromate 溴酸镉 $Cd(BrO_3)_2$
cadmium bromide 溴化镉 $CdBr_2$
cadmium carbonate 碳酸镉 $CdCO_3$
cadmium cell 镉电池
cadmium chlorate 氯酸镉 $Cd(ClO_3)_2$
cadmium chloride 氯化镉 $CdCl_2$
cadmium coating 镉(系)涂料〔以镉红或镉黄为颜料的

涂料〕

cadmium colors　镉染料
cadmium crimson　镉大红
cadmium cut-off　镉吸收界限
cadmium cut-off energy　镉截止能
cadmium cyanate　氰酸镉　Cd(OCN)$_2$
cadmium cyanide　氰化镉　Cd(CN)$_2$
cadmium deep orange　镉深橙
cadmium deep red　镉深红
cadmium diethyl dithiocarbamate　二乙基二硫代氨基甲酸镉
cadmium dimethyl dithiocarbamate　二甲基二硫代氨基甲酸镉
cadmium dithionate　连二硫酸镉　CdS$_2$O$_6$
cadmium ethide　乙基镉　Cd(C$_2$H$_5$)$_2$
cadmium 2-ethylhexoate　2-乙基己酸镉〔稳定剂〕
cadmium ferricyanide　氰铁酸镉　Cd$_3$[Fe(CN)$_6$]$_2$
cadmium ferrocyanide　氰亚铁酸镉　Cd$_2$[Fe(CN)$_6$]
cadmium filter　镉滤器
cadmium fluoride　氟化镉　CdF$_2$
cadmium fluosilicate　氟硅酸镉　Cd[SiF$_6$]
cadmium green　镉铬绿〔93%水合氧化铬绿和7%镉黄的混合物〕
cadmium hydride　氢化镉　CdH$_2$
cadmium hydroxide　氢氧化镉　Cd(OH)$_2$
cadmium iodate　碘酸镉　Cd(IO$_3$)$_2$
cadmium iodide　碘化镉　CdI$_2$
cadmium ion selective electrode　镉离子选择电极
cadmium lamp　镉光灯
cadmium laurate　月桂酸镉
cadmium lemon yellow　镉柠檬黄
cadmium light orange　浅镉橙
cadmium line　镉(谱)线
cadmium lithopone　镉钡红
cadmium maroon　镉紫红〔主要成分为硫硒化镉〕
cadmium-mercury lithopones　镉汞钡红〔主要成分为硫化镉、硫化汞和硫酸钡的煅烧共沉淀物〕
cadmium-mercury sulfide　硫化镉汞橘红(橙)
cadmium metasilicate　硅酸镉　CdSiO$_3$
cadmium methide　甲基镉　Cd(CH$_3$)$_2$
cadmium minerals　镉矿物
cadmium nitrate　硝酸镉　Cd(NO$_3$)$_2$
cadmium octoate　辛酸镉
cadmium orange　镉橙
cadmium orthosilicate　正硅酸镉　Cd$_2$SiO$_4$
cadmium oxalate　草酸镉　CdC$_2$O$_4$
cadmium oxide　一氧化镉　CdO
cadmium oxybromide　溴氧化镉　BrCd·O·CdBr

cadmium oxychloride　氯氧化镉；二氯一氧化二镉　ClCd·O·CdCl
cadmium oxyiodide　碘氧化镉　ICd·O·CdI
cadmium perchlorate　高氯酸镉　Cd(ClO$_4$)$_2$
cadmium permanganate　高锰酸镉　Cd(MnO$_4$)$_2$
cadmium peroxide　过氧化镉　Cd(O)$_2$
cadmium phosphate　磷酸镉　Cd$_3$(PO$_4$)$_2$
cadmium pigment　镉颜料
cadmium pollution　镉污染
cadmium primorse　镉草黄
cadmium pyrophosphate　焦磷酸镉
cadmium ratio　镉比
cadmium recinoleate　蓖麻酸镉〔稳定剂〕
cadmium red (=selenium cadmiun pigment)　硒镉颜料；镉红
cadmium ricinoleate　蓖麻醇酸镉
cadmium scarlet　镉猩红
cadmium selenate　硒酸镉　CdSeO$_4$
cadmium selenide　硒化镉　CdSe
cadmium selenide red　硒化镉红
cadmium silicofluoride　氟硅酸镉　Cd[SiF$_6$]
cadmium standard cell　镉标准电池
cadmium stearate　硬脂酸镉
cadmium styphnate　收敛酸镉
cadmium suboxide　一氧化二镉　Cd$_2$O
cadmium sulfate　硫酸镉　CdSO$_4$
cadmium sulfide　硫化镉　CdS
cadmium sulfite　亚硫酸镉　CdSO$_3$
cadmium sulfoselenide　硫硒化镉〔颜色由黄到红〕
cadmium telluride　碲化镉　CdTe
cadmium telluride detector　碲化镉探测器
cadmium tungstate　钨酸镉
cadmium yellow　镉黄　CdS
cadmopone　镉钡黄〔硫化镉和硫酸钡的复合颜料〕
cadmous　亚镉〔一价镉的化合物〕
cadmous compound　亚镉化合物
caesium (=cesium)　铯
caesium intercalation　铯添加；铯插入
caesium vapour lamp　铯光灯
cafesterol (=cafestol)　咖啡醇
cafestrol　咖啡甾醇
caffeic acid　咖啡酸；二羟(基)肉桂酸　(OH)$_2$C$_6$H$_3$CH═CHCOOH
caffeine (=coffeine; theine; guaranine; psoraline)　咖啡碱；咖啡因
caffeine and sodium benzoate　苯甲酸钠咖啡因；安钠咖〔药〕
caffeine citrate　柠檬酸咖啡因

caffeine sodio-benzoate 苯甲酸咖啡碱钠
caffeoyl guinic acid (=caffetannic acid) 咖啡酰奎尼酸；咖啡单宁酸
caffetannic acid (=chlorogenic acid) 咖啡单宁酸 $C_{16}H_{18}O_9$
caffolide 咖啡内酯 $C_5H_3O_5N_3$
caffuric acid 咖啡尿酸
cage 笼
cage bag filter 栅式布袋过滤器
cage bar 笼条
cage compound 笼型化合物*
cage coordination compound 笼状配合物*
cage effect 笼效应*
cage mill 笼式磨机
cage mixer 笼式搅拌机
cage model 笼型
cage oil press 笼式榨油机
cage phenylation 笼状苯基化(作用)
cage reaction 笼闭反应
cage rearrangement 笼状重排*
cage spinning 尘笼纺纱
cage structure 笼形结构
cage type 笼型
cage-type degumming machine 笼式精炼机
cage type formula 笼型式；棱柱式
cahincic acid (=caincic acid) 卡亨酸 $C_{40}H_{64}O_{18}$
Cahn-Ingold-Prelog sequence rule 顺序规则*
cahuecit 卡胡炸药
caincic acid (=cahincic acid) 卡亨酸
caincin (=cahincic acid) 卡亨酸
cairngorm 烟晶
cajeputene 白千层萜
cajeput oil (=cajuput oil) 白千层油
cajuputole (=eucalyptole) 白千层脑
cajuput tree 白千层
cake ①饼；块②滤饼③丝饼
cake compressor 压饼机
cake cutter 切饼机
caked 成块的；块结的
cake ice 冰砖
cake indicator 滤饼指示器
cake of alum 明矾块
cake of filter-press 压滤机的滤饼
cake of fusion 熔块
cake of metal 金属锭
cake of nitre 硝块；硫酸氢钠块
cake of press 压饼
cake of thread ①精纺(线)饼②(纺丝)丝饼
cake shampoo 块状洗发剂
cake sizing 丝饼上浆
cake soap 块皂
cake spinning 丝饼式纺丝
cake-wash machine 丝并洗涤机
caking ①结团；结块②黏结
caking capacity 块结性
caking coal 黏结性煤〔矿〕
caking fertilizer 块状肥料
caking of crystals 晶体(的)块结；晶体胶结
caking power 黏结力
caking property 黏结性
caking test 结块试验；结块检查
calaba oil 卡拉巴油
calabarine 卡拉巴碱
calabash 葫芦
calabash-alkaloid 1 葫芦箭毒碱 1
calabashine (=calebassine) 葫芦箭毒碱
calacane 甜旗烷；卡拉烷
calacone 甜旗酮；菖蒲酮
calaconene 甜旗烯
calamary liver oil 枪乌贼肝油
calamene 卡拉烯；菖蒲萜烯 $C_{15}H_{24}$
calamenene 去氢白菖烯；1,6-二甲基-4-异丙基四氢萘
calameneol (=calamenol) 白菖(蒲)醇
calameone (=calamus camphor) 白菖(蒲)脑
calamine 菱锌矿；水锌矿
calamine cream 锌膏
calamintha oil 风轮菜油
calaminthone 风轮菜酮 $C_{10}H_{16}O$
calamol (=allyl trimethoxy benzene) 白菖(蒲)醚；烯丙基三甲氧基苯
calamone 白菖蒲酮
calamonic acid 白菖(蒲)酸
calamus 菖蒲〔调香用药草〕
calamus camphor (=calameone) 白菖(蒲)脑
calamus oil 白菖(蒲)油
calander ①砑光机〔造纸及纺织〕②压延机橡胶
calander department 砑光车间
calander operator 砑光工
calandria ①多效真空蒸发器②排管式加热器
calandria heat exchanger 排管式热交换器
calandria-type vacuum pan 排管式真空釜
calandria vacuum pan 排管式真空釜
calantas wood oil 香椿木油
calarene 白菖油萜
calaverite 碲金矿
calaya 卡拉亚〔*Annesslea febrifuga* 的果实流浸膏，治

疟药〕

calcar 熔(玻璃)炉；锻烧炉
calcar arch 熔(玻璃)炉
calcareobarite 钙重晶石
calcareous ①石灰的②含钙的
calcareous aeolianite 钙质风成岩
calcareous brick 钙质砖
calcareous cement 石灰黏合剂；钙质水泥
calcareous clay 钙质黏土；石灰质黏土
calcareous dolostone 钙质白云岩
calcareous hydraulic binder 水凝灰浆
calcareous phyllite 钙质千枚岩
calcareous rock 钙质岩；石灰岩
calcareous sandstone 钙质砂岩
calcareous sediment 钙质沉积物
calcareous sinter 钙质凝灰岩
calcareous slag 石灰炉渣
calcareous spar 方解石 $CaCO_3$
calcareous spring 钙泉；石灰泉
calcareous tuff 石灰性凝灰岩；石灰华
calcareous water ①石灰水②硬水
calcarious 含钙的
calcariuria 石灰尿；钙盐尿
calcarius 石灰质的
calcarone 炼硫窑
calcedonite 铅蓝矾；碳铅蓝矾
calcedony 玉髓 SiO_2
calcein (=fluorexon) 钙黄绿素；荧光氨羧络合剂
calcene 胶质碳酸钙；白艳华〔补强剂〕
calc granite 钙质花岗岩
calcia 氧化钙
calcia-alumina-silica fibre 氧化钙-氧化铝-氧化硅纤维
calciborite 钙硼石
calcibreccia 钙质角砾岩
calcic ①石灰的②钙的
calcichrome 钙色素；钙显色剂
calciclase (=anorthite) 钙长石
calciclasite (=anorthitite) 钙长岩
calcicoater ①油性无光建筑涂料；钙(脂)油〔俗〕②高黏度钙脂漆基〔植物油加石灰热炼而成的类凝胶皂〕③钙催干剂〔旧〕
calcicoater flat 钙脂油无光漆
calcicoater paint 钙脂油有光建筑漆
calcicosis 灰石沉着病；灰石肺
calcic plagioclase 钙质斜长石
calcic silicate glass 硅酸钙玻璃
calciferol (=vitamin D_2) 钙化醇；骨化醇；维生素 D_2
calcilization 方解石化(作用)

calcimangite 锰方解石
calcimeter 碳酸计
calcimine (=powdered distemper) 刷墙水粉
calciminge paint 水粉墙壁涂料；粉末水浆涂料〔主要成分为硫酸钙〕
calcimonzonite 钙质二长石
calcimurite 钙氯石 $CaCl_2$
calcinate 煅烧；焙烧
calcinated support 煅烧载体
calcination 焙烧*
calcination plant 煅烧车间
calcination ratio 煅烧出产率
calcinator 煅烧炉
calcinatory 煅烧器
calcined 煅烧(过)的；焙烧(过)的；烘好的
calcined alum 煅烧过的明矾
calcined alumina 煅烧氧化铝
calcined aluminosilicate 煅烧硅酸铝
calcined baryta 氧化钡 BaO
calcined china clay 煅瓷土
calcined clay 煅烧黏土；黏土熟料
calcined coke 煅烧焦炭
calcined co-precipitation 煅烧共沉淀；煅烧共沉淀物
calcined dolomite 烧白云石
calcined gypsum (=casting plaster) 烧(煅)石膏；熟石膏
calcined halloysite 煅烧多水高岭土
calcined kaolin 煅烧高岭土
calcined lime 烧石灰
calcined magnesia (煅烧)氧化镁
calcined ore 煅烧矿石
calcined phosphate 烧制磷肥
calcined pigment 煅烧颜料
calcined plaster 煅石膏
calcined potash 氧化钾 K_2O
calcined pyrite 黄铁矿烧渣
calcined soda 纯碱
calciner 煅烧炉
calcining 煅烧
calcining compartment 煅烧室
calcining furnace 煅烧炉
calcining heat 煅烧温度
calcining kettle 焙烧炉
calcining kiln 煅烧窑
calcining oven 煅烧炉
calcining plant 煅烧厂
calcining temperature 煅烧温度
calcinite 卡斯炸药；克斯那特
calcinitre 钙硝石

calcinol (=calcium iodate) 碘酸钙 Ca(IO$_3$)$_2$
calcinosis 钙质沉着(症);石末沉着(症)
calcioaegirine 钙霓石
calcio-akermanite 钙镁黄长石
calcio-ancylite (=calcio-ankylite) 钙锥锶铈矿;碳钙铈矿
calcio-ankylite (=calcio-ancylite) 钙锥锶铈矿;碳钙铈矿
calciobarite 钙重晶石
calcio-biotite 钙黑云母;杂萤石黑石云母
calcioborite 斜硼钙石
calciocancrinite 艳钙霞石
calciocelestine 钙天青石
calciocelestite 钙天青石
calciocelsian 钙钡长石
calcio-chondrodite 钙质硅镁石
calciodialogite 钙菱锰矿;杂菱锰方解石
calcioferrite 钙磷铁矿
calcio-gadolinite 钙硅铍钇矿
calciojarosite 钙钾铁矾
calciolazulith 钙天蓝石
calciolyndochite 钙钍黑稀金矿
calciomalachite 钙孔雀石
calcio-olivine 钙(钙锰;钙镁)橄榄石
calciopalygorskite 杂坡缕方解石
calciorhodochrosite 杂菱锰方解石;钙菱锰矿
calciosamarskite 钙铌钇铀矿;钙稀土矿
calcioscheelite (=scheelite) 白钨矿
calciospessartine 钙锰铝榴石
calciostrontianite 钙菱锶矿;钙碳锶矿
calciotantalite 钙钽铁矿;杂细晶钽铁矿
calciothermy 钙热(还原)法
calcio-thorite 钙钍石〔含有 CaO 等杂质〕 ThSiO$_4$
calciouraconite 水钙铀矾
calciouranoite 钙铀矿;钙钡铅铀矿
calciovolborthite 钙矾铀矿
calcite 方解石*
calcite prism 方解石棱镜
calcitization 方解石化(作用)
calcitonin 降(血)钙素〔药〕
calcitriol 骨化三醇〔药〕
calcium 钙〔20 号元素,化学符号 Ca〕
calcium acetate 乙酸钙 Ca(C$_2$H$_3$O$_2$)$_2$
calcium acetoacetate 乙酰乙酸钙
calcium acetylacetonate 乙酰丙酮钙
calcium acetylide 碳化钙 CaC$_2$
calcium acid malate 酸式苹果酸钙
calcium alginate fiber 海藻酸钙纤维
calcium alginate yarn (海)藻酸钙纤维纱
calcium aluminate 铝酸钙 Ca(AlO$_2$)$_2$

calcium arsenate 砷酸钙
calcium arsenite 亚砷酸钙
calcium base titanox 钛钙白
calcium benzoate 苯甲酸钙 Ca(C$_6$H$_5$COO)$_2$
calcium biarsenate 砷酸二氢钙 Ca(H$_2$AsO$_4$)$_2$
calcium bicarbonate 碳酸氢钙;重碳酸钙
calcium bichromate 重铬酸钙 CaCr$_2$O$_7$
calcium binding protein 钙结合蛋白
calcium bioxalate 草酸氢钙 Ca(HC$_2$O$_4$)$_2$
calcium biphosphate 磷酸二氢钙 Ca(H$_2$PO$_4$)$_2$
calcium bisulfate 硫酸氢钙 Ca(HSO$_4$)$_2$
calcium bisulfite 亚硫酸氢钙 Ca(HSO$_3$)$_2$
calcium bitartrate 酒石酸氢钙 Ca(HC$_4$H$_4$O$_6$)$_2$
calcium bleach 漂白粉
calcium bomb reduction 钙弹还原(法)
calcium borate 硼酸钙 Ca(BO$_2$)$_2$
calcium borosilicate 硼硅酸钙
calcium brine 钙卤水〔氯化钙卤〕
calcium bromate 溴酸钙 Ca(BrO$_3$)$_2$
calcium bromide 溴化钙 CaBr$_2$
calcium bromiodide 溴碘化钙 CaBr$_2$·CaI$_2$;CaBrI
calcium butoxide 丁醇钙 Ca(C$_4$H$_9$O)$_2$
calcium carbide 碳化钙;电石〔俗〕
calcium carbide-acetylene hygrometer 电石-乙炔湿度计
calcium carbide route 碳化钙法;电石法〔制乙炔的方法之一〕
calcium carbolate 苯酚钙 Ca(OC$_6$H$_5$)$_2$
calcium carbonate 碳酸钙 CaCO$_3$
calcium carbonate-ammonium nitrate 碳酸钙合硝酸铵
calcium carbonate extender 碳酸钙体质颜料
calcium caseinate 酪蛋白钙
calcium chlorate 氯酸钙 Ca(ClO$_3$)$_2$
calcium chloride 氯化钙
calcium chloride brine 氯化钙卤水
calcium chloride jar 氯化钙瓶
calcium chloride tube 氯化钙管
calcium chromate 铬酸钙 CaCrO$_4$
calcium cinnamate (=calcium cinnamylate) 肉桂酸钙 Ca(C$_9$H$_7$O$_2$)$_2$
calcium cinnamylate (=calcium cinnamate) 肉桂酸钙
calcium citrate 柠檬酸钙 Ca$_3$(C$_6$H$_5$O$_7$)$_2$
calcium copper phosphate 磷酸铜钙〔工业品含 0.25%磷酸铜和 99.75%磷酸钙〕
calcium cresyl sulfonate 甲苯磺酸钙 Ca(SO$_3$C$_6$H$_4$CH$_3$)$_2$
calcium cuprate 铜酸钙 Ca[Cu(OH)$_4$]$_2$
calcium cyanamide 氰氨(基)化钙
calcium cyanate 氰酸钙 Ca(OCN)$_2$

calcium cyanide　氰化钙　$Ca(CN)_2$
calcium cyanoplatinite　氰亚铂酸钙　$Ca[Pt(CN)_4]$
calcium cylinder　氯化钙管
calcium di(aluminum tetraethyl)　二(四乙基铝)化钙〔丙烯腈溶液聚合用引发剂〕
calcium dimetaphosphate　偏磷酸钙　$Ca(PO_3)_2$
calcium dimethacrylate　二甲基丙烯酸钙
calcium dioxide　①二氧化钙②过氧化钙
calcium disodium edetate　EDTA 二钠钙
calcium dithionate　连二硫酸钙　CaS_2O_6
calcium dodecyl benzene sulfonate　十二烷基苯磺酸钙
calcium ester　钙酯〔又是钙盐又是酯〕
calcium ethoxide　乙醇钙　$Ca(OC_2H_5)_2$
calcium ethylate　乙醇钙　$Ca(OC_2H_5)_2$
calcium ethylsulfate　乙基硫酸钙　$CaSO_4(C_2H_5)_2$
calcium-exchanged silica pigment　钙离子交换型硅颜料；钙交换硅颜料
calcium extended titanium-dioxide　钛钙白
calcium ferricyanide　氰铁酸钙　$Ca_3[Fe(CN)_6]_2$
calcium ferrite　铁酸钙　$Ca(FeO_2)_2$
calcium ferrocyanide　氰亚铁酸钙　$Ca_2[Fe(CN)_6]$
calcium fluoride　氟化钙　CaF_2
calcium fluosilicate　氟硅酸钙　$Ca[SiF_6]$
calcium formate (=calcium formiate)　甲酸钙　$Ca(OOCH)_2$
calcium formiate (=calcium formate)　甲酸钙
calcium gluconate　葡萄糖酸钙
calcium glycerophosphate　甘油磷酸钙
calcium grease (=calcium soap grease)　钙皂脂；钙基脂
calcium hardness　(水的)钙(质)硬度
calcium hexafluorotitanate　六氟钛酸钙〔缩聚用催化剂〕
calcium hydride　氢化钙　CaH_2
calcium hydrogen phosphate (=calcium hydrophosphate)　磷酸氢钙　$CaHPO_4$
calcium hydrophosphate (=calcium hydrogen phosphate)　磷酸氢钙
calcium hydrosilicate　硅酸氢钙
calcium hydrosulfide　硫氢化钙　$Ca(HS)_2$
calcium hydroxide　氢氧化钙　$Ca(OH)_2$
calcium hydroxybenzoate　羟基苯甲酸钙〔稳定剂〕
calcium hypochlorite　次氯酸钙　$Ca(ClO)_2$
calcium hypochlorite method　次氯酸钙（滴定）法
calcium hypophosphate　连二磷酸钙　$Ca_2P_2O_6$
calcium hyposulfite　连二亚硫酸钙　CaS_2O_4
calcium-insensitive　耐钙(离子)的
calcium iodate (=calcinol)　碘酸钙　$Ca(IO_3)_2$
calcium iodide　碘化钙　CaI_2
calcium ion-selective electrode　钙离子选择电极

calcium (ion) stability　(对)钙(离子)稳定性
calcium isopropoxide　异丙醇钙　$Ca(OC_3H_7)_2$
calcium lactate　乳酸钙　$Ca(C_3H_5O_3)_2$
calcium lactobionate　乳糖醛酸钙
calcium laurate　月桂酸钙〔稳定剂〕
calcium levulinate　乙酰丙酸钙
calcium lignin sulphonate　木(质)素磺酸钙
calcium lime　生石灰；未消石灰　CaO
calcium linoleate　亚油酸钙
calcium lithol toner　钙立索调色料；钙立索色原
calcium L-threonate　苏糖酸钙
calcium mahogany sulfonate　石油磺酸钙
calcium malate　苹果酸钙　$CaC_4H_4O_5$
calcium maleate　马来酸钙　$CaC_4H_2O_4$
calcium manganate　锰酸钙　$CaMnO_4$
calcium metaaluminate　偏铝酸钙　$Ca(AlO_2)_2; Al_2(CaO_4)$
calcium metaborate　偏硼酸钙　$Ca(BO_2)_2$
calcium metaphosphate　偏磷酸钙　$Ca(PO_3)_2$
calcium metaplumbate　偏高铅酸钙　$CaPbO_3$
calcium metasilicate　偏硅酸钙　$CaSiO_3$
calcium metatitanate　偏钛酸钙　$CaTiO_3$
calcium metazirconate　偏锆酸钙　$CaZrO_3$
calcium methoxide (=calcium methylate)　甲醇钙　$Ca(OCH_3)_2$
calcium methylate (=calcium methoxide)　甲醇钙
calcium methyl siliconate　甲基硅酸钙
calcium molybdate　钼酸钙　$CaMoO_4$
calcium monophosphate　磷酸氢钙　$CaHPO_4$
calcium monosaccharate　蔗糖酸钙
calcium naphthenate　环烷酸钙
calcium nitrate　硝酸钙　$Ca(NO_3)_2$
calcium nitride　二氮化三钙　Ca_3N_2
calcium nitrite　亚硝酸钙　$Ca(NO_2)_2$
calcium octoate　辛酸钙
calcium oleate　油酸钙　$Ca(C_{18}H_{33}O_2)_2$
calcium orthoarsenite　原亚砷酸钙　$Ca(AsO_3)_2$
calcium orthodisilicate　原二硅酸钙
calcium orthophosphate (=tricalcium orthophosphate)　正磷酸三钙；正磷酸钙　$Ca_3(PO_4)_2$
calcium orthoplumbate　原高铅酸钙　Ca_2PbO_4
calcium orthosilicate　原硅酸钙　Ca_2SiO_4
calcium oxalate　草酸钙　CaC_2O_4
calcium oxide　氧化钙　CaO
calcium oxychloride　次氯酸钙；漂白粉　$Ca(ClO)_2$
calcium pantothenate　泛酸钙
calcium paracaseinate　副酪蛋白钙
calcium pectinate　果胶酸钙
calcium perchlorate　高氯酸钙　$Ca(ClO_4)_2$

calcium permanganate 高锰酸钙 $Ca(MnO_4)_2$
calcium peroxide 过氧化钙 $Ca[O_2]$
calcium perphosphate 过磷酸钙
calcium phenate (=calcium phenolate) 苯酚钙 $Ca(OC_6H_5)_2$
calcium phenolate (=calcium phenate) 苯酚钙
calcium phenolsulfonate 苯酚磺酸钙 $Ca(SO_3C_6H_4OH)_2$
calcium phenoxide (=calcium phenylate) 苯酚钙 $Ca(OC_6H_5)_2$
calcium phenylate (=calcium phenoxide) 苯酚钙
calcium phosphate 磷酸钙〔$1.Ca(PO_3)_2$; $2.Ca_3(PO_4)_2$; $3.Ca_2P_2O_7$, 通常指 2〕
calcium phosphate (acid) 酸式磷酸钙 $Ca(H_2PO_4)_2$; $CaHPO_4$
calcium phosphate gel 磷酸钙凝胶
calcium phosphate primary 一代磷酸钙；磷酸二氢钙 $Ca(H_2PO_4)_2$
calcium phosphate secondary 二代磷酸钙；磷酸氢钙 $CaHPO_4$
calcium phosphide 二磷化三钙 Ca_3P_2
calcium phosphite 亚磷酸氢钙 $CaHPO_3$
calcium plumbate 高铅酸钙〔1. 偏 $CaPbO_3$; 2. 原 Ca_2PbO_4〕
calcium plumbate-coated pigment 铅酸钙包核颜料；包核铅酸钙
calcium plumbate pigment 铅酸钙颜料
calcium plumbite 铅酸钙 C_2PbO_2
calcium polysilicate 多(聚)硅酸钙
calcium polysulfide 多硫化钙
calcium potassium ferrocyanide 氰亚铁酸钙二钾 $CaK_2[Fe(CN)_6]$
calcium propionate 丙酸钙
calcium propoxide (=calcium propylate) 丙醇钙 $Ca(OC_3H_7)_3$
calcium propylate (=calcium propoxide) 丙醇钙
calcium pump 钙泵*
calcium pyrophosphate 焦磷酸钙 $Ca_2P_2O_7$
calcium racemate 外消旋酒石酸钙
calcium resinate 树脂酸钙
calcium rhodanate 硫氰酸钙 $Ca(CNS)_2$
calcium-5-ribonucleotide 5-核糖核苷酸钙
calcium ricinoleate 蓖麻酸钙〔稳定剂〕
calcium rosin soap 松脂钙皂
calcium salicylate 水杨酸钙
calcium salts 钙盐
calcium-saponite (=Ca-saponite) 钙皂石
calcium selenate 硒酸钙 $CaSeO_4$
calcium selenide 硒化钙 $CaSe$

calcium selenite 亚硒酸钙 $CaSeO_3$
calcium (sequestration) value (螯合)钙值
calcium silicate 硅酸钙 $CaSiO_3$
calcium silicide 硅化钙 $CaSi;CaSi_2$
calcium soap 钙皂
calcium soap grease 钙皂脂；钙基脂
calcium-sodium soap grease 钙-钠皂脂；钙-钠基脂
calcium sorbate 山梨酸钙
calcium stearate 硬脂酸钙
calcium suberate 辛二酸钙 $CaC_8H_{12}O_4$
calcium succinate 丁二酸钙 $CaC_4H_4O_4$
calcium sulfarsenite 硫代亚砷酸钙
calcium sulfate 硫酸钙 $CaSO_4$
calcium sulfate hemihydrate 半水合硫酸钙
calcium sulfide 硫化钙 CaS
calcium sulfite 亚硫酸钙 $CaSO_3$
calcium sulfophenate 苯酚磺酸钙 $Ca(SO_3C_6H_4OH)_2$
calcium sulfovinate 乙硫酸钙 $Ca(SO_4C_2H_5)_2$
calcium sulphoaluminate 硫铝酸钙
calcium sulphydrate 硫氢化钙
calcium superoxide 过氧化钙 $Ca[O_2]$
calcium superphosphate 过磷酸钙〔肥料〕
calcium tallate 松浆(油)酸钙；脂肪酸钙
calcium tartrate 酒石酸钙 $CaC_4H_4O_6$
calcium tetraborate 四硼酸钙 CaB_4O_7
calcium thiocyanate (=calcium thiocyanide) 硫氰酸钙 $Ca(CNS)_2$
calcium thiocyanide (=calcium thiocyanate) 硫氰酸钙
calcium thiosulfate 硫代硫酸钙 $CaSO_3S;CaS_2O_3$
calcium tungstate (=calcium wolframate) 钨酸钙 $CaWO_4$
calcium wolframate (=calcium tungstate) 钨酸钙
calcivorous 嗜钙的
calcon 钙试剂；茜素蓝黑
calconcarboxylic acid 钙指示剂*；钙红
calcothar (=colcothar) 铁丹
calc-spar (=calcite) 方解石
calculated ①算清了的②设计负载
calculated capacity 计算容量
calculated cetane index (CCI) (石油燃料)十六烷指数的计算值
calculated octane number 结算辛烷值
calculated twist 计算捻度
calculating board 计算板
calculating graph for standard addition method 标准加入法计算图
calculation factor 换算因子*
calculation of fixed carbon 固定碳计算法

calculation of reserves 储量计算；储量估计
calculation of steps 操作板数计算
calculation program 计算程序
calculator 计算器
calcyanide 氰化钙
caldariomycin 卡尔里霉素
Calder-Fox scrubber 卡尔德-福克斯涤气器
caldo 硝卤水
caldron ①釜②煮皂釜
Caldwell crucible 卡尔德韦尔坩埚
calebashinine (=calebassinine) 葫芦箭毒宁
calebassine (=calabashine) 葫芦箭毒碱
calebassinine (=calebashinine) 葫芦箭毒宁
caledon colors 加里东染料
caledon gold orange 加里东金橙
caledonian brown 天然铁棕
caledonite 铅蓝矾
calefacient ①发暖剂②发暖的
calefacientia 发暖剂〔药〕
calefaction 发暖作用
calefactor 发暖器
calendar grain 压延效应；压延纹理
calender ①砑光机②压延机
calenderability 压延性*
calender blister 压延起泡
calender bowl 压延机滚筒
calender coating 辊压机涂漆
calender coloring 砑光上色
calender drier 轧光干燥机；轧水干燥机
calendered fabric 轧平式网布
calendered film 压延薄膜
calendered goods 压延制品
calendered sheeting 压延薄片
calender effect (=calender grain) 砑光效应
calenderer 砑光工
calender gauge 压延用测厚计；连续测厚计
calender grain (=calender effect) 砑光效应
calender grease 砑光机脂
calendering 压延；轧光
calender knife 压延切边刀
calender line ①压延过程②压延联动装置
calender machine 砑光机
calender roll 砑光辊
calender run ①砑光②压制；压延
calender sheeting 压延薄片
calender shrinkage 压延收缩
calender sizing 压光机施胶
calender speed 压延速度
calender stack ①砑光机（造纸及纺织）②压延机（橡胶）
calender train ①压延联动装置②多台压延机联动③砑光机
calendic acid 十八碳三烯酸
calendula absolute 金盏花净油
calendulin 金盏花素
calglucon 葡萄糖酸钙
calgon 六偏磷酸钠（清洁剂）
calgon-chrome tannage 偏磷酸盐-铬鞣制
calgon tannage 偏磷酸盐鞣制
calgon-vegetable tanning 偏磷酸盐-植鞣制
caliber ①口径②(量)规；测量器
calibrated dial 校准刻度盘
calibrated filter 校正滤光片
calibrated manometer 校准压力计
calibrated normalized resolution product 校正分离度积
calibrated pipette 校准吸移管
calibrating ①校准②标定；刻度③测定口径
calibrating apparatus 校准用器
calibrating burette 校准滴定管
calibrating gas 标准气体
calibrating pipet (te) 校准吸移管
calibration 校准；校正
calibration accuracy 标定精度
calibration chart (=calibration scales) 校准表；校准图
calibration check compound 校准用对照化合物
calibration correction 刻度校准；校准修正
calibration curve 校正曲线；标准曲线；工作曲线
calibration curve method 校正曲线法
calibration factor 校准因子
calibration filter 校准滤光片*
calibration function 校准函数
calibration gas 校准气体
calibration graph 校正图
calibration leak 校准漏孔
calibration line 定标线
calibration mark 校准标记
calibration model 校正模型
calibration of absorbance scale 吸光度标度校准
calibration of conductivity vessels 电导池校准；电导容器校准
calibration of peak area 峰面积校正
calibration of reference fuel 参比燃料校准；标准燃料校准
calibration of viscometer 黏度计标定
calibration of wavelength scale 波长标度校准
calibration of weights 砝码校准
calibration peak 定标峰

calibration procedure 校准程序；校正方法
calibration sample 校准试样
calibration scales (=calibration chart) 校准表
calibration set 校正集
calibration standard 校准标准；校准标准物
calibration standard method 校准标准法
calibration system 校准系统
calibration table (=calibration chart) 校准表
calibrator 校准器
calicene 杯烯
caliche 生硝
calico-printers soap 印花用皂
calico-printing 印花
calico printing machine 印花机
calico rocks 印花岩
californium 锎
californium amalgam 锎汞齐
californium bomb 锎弹
californium dioxide 二氧化锎
californium oxide 氧化锎
californium oxychloride 氯氧化锎 CfOCl
californium sulfide 硫化锎 Cf_2S_3
californium thiocyanate 硫氰酸锎 $Cf(SCN)_3$
californium tribromide 三溴化锎 $CfBr_3$
californium tricyclopentadienide 三茂合锎；三环戊二烯基锎 $Cf(C_5H_5)_3$
californium trifluoride 三氟化锎
Calingaert-Davis equation 卡林盖特-戴维斯公式〔饱和液体蒸气压公式〕
caliomycin (=caryomycin) 核霉素
caliper ①厚度②纸厚度测定器③卡尺；卡钳
caliper type gauge 测径规；内卡规
caliper vale 卡尺
calisaya 黄色金鸡纳皮
calixarene 杯芳烃
calixcrown 杯冠化合物
caliza 石灰岩；石灰石
calk 生石灰；未消石灰 CaO
calking ①填缝；填孔②嵌缝胶；腻子胶
C-alkylation C-烷基化*
callainite 绿磷铝石
callatome 显微镜切片器
callendars gas thermometer 卡式气体温度计
callipers ①双脚规；测圆器②测径器
callistephin 翠菊苷；花葵素-3-葡萄糖苷
callitrol (=d-citronellic acid) d-香茅酸
callitrolic acid 卡里松醇酸
callophane 荧光分析器

callose 愈创葡聚糖；β-D-(1→3)-葡聚糖
callow 无羽毛的
Callow cell 考洛浮选池
calmagite 钙镁指示剂*；1-(1-羟基-4-甲基-2-苯偶氮)-2-萘酚-4-磺酸
calmine (=barbital sodium) 巴比妥钠
calmodulin 钙调蛋白
calmus oil 菖蒲油〔Acorus calamus 的油〕
calnitro 钙硝肥
calobiosis 同栖共生
calodorant 美嗅剂〔煤气添味剂；酒精变性添味剂〕
calomel 甘汞*
calomel cell 甘汞电池
calomel electrode 甘汞电极*
calomel (normal) electrode 甘汞(标准)电极
calomel reference electrode 甘汞参比电极
calophyllolide 红厚壳烯酮内酯；海棠果素
calorescence 热光
calorgas (=treibgas) 卡气；压缩混烃
caloric ①热②热的
caloric effect 热效应
caloric heat unit 卡热单位
caloricity 热值；卡值
caloricity of gas 煤气的热值
caloric power 热值；发热量
caloric power of gas 煤气热值
caloric receptivity 吸热量
caloric requirement 热需要量
caloric unit 热(量)单位
caloric value (=calorific value) 热值；卡值
calorie 卡〔热量单位〕
calorifacient 生热质；产生热(量)的
calorific 热量的；卡的
calorific capacity (=calorific efficiency) 发热量；热效率；热值；卡值
calorific effect 热效应
calorific efficiency (=calorific capacity) 发热量；热值；卡值；热容(量)
calorific intensity 热强度
calorific power 热值；卡值
calorific radiation 辐射热
calorific value (=caloric value) 热值；卡值
calorifier ①发热器；加热器②热风炉
calorigenic 生热的
calorimeter 热量计*；量热器；卡计；量热计
calorimeter for heat of dilution 稀释热量热计
calorimeter of fusion 熔凝量热器
calorimetric 量热的

calorimetric analysis　量热分析
calorimetric block　量热块
calorimetric bomb　量热弹；卡计弹
calorimetric correction　量热校正
calorimetric entropy　量热熵*
calorimetric gas meter　量热煤气表
calorimetric pyrometer　量热高温计
calorimetric titration　量热滴定；温度滴定
calorimetry　量热学*
calorisator　热法浸提器
calorised coating　铝化(处理)涂料
caloristat　恒温器
caloritropic (=thermotropic)　亲热的；向热的；趋热的
calorization　铝化作用
calorizator (=calorizer)　热法浸提器
calorized coating　铝化(处理)涂料
calorizer (=calorizator)　热法浸提器
calorizing　热镀(铝)法
calorose　转化糖
calorstat　恒温器
calory　卡
calotropin　牛角瓜苷〔取自大牛角瓜的叶和茎〕
calotte　帽罩
calpis　乳浊液
calred (=Cal-Red)　钙红
calsomine　刷墙粉
calumba　咖伦巴(根)
calumbic acid (=columbic acid)　考龙酸；咖伦巴酸
calumbin　咖伦巴根苷；非洲防己苦素
calutron　①加利福尼亚大学回旋加速器②电磁(型)同位素分离器
calvatic acid　马勃菌酸
calx　〔拉丁文〕　石灰
calycanthine　腊梅碱
calycotomine　萼卷豆碱
calyomycin (=caryomycin)　核霉素
cam　凸轮〔机工〕；铿
camacite　梁状铁
camarene　马缨丹烯
camber　弯度
cambered bowl　中高辊筒
cambered inwards　向里凹的〔油罐底〕
cambered outwards　向外凸的〔油罐底〕
cambium　形成层；新生层
cambium ring　形成层环
cambium zone　形成层带
cambogia (=Gamboge)　藤黄
cambric finish　细薄布整理
cambric paper　葛纹纸
Cambridge extensometer　剑桥单纤维拉伸试验仪
cam drive　凸轮传动
cameleon　织物闪色效应
cameleon fibre　变色纤维
camel hide　骆驼皮
cameline oil　亚麻荠油
camellia oil　山茶油〔日本茶油〕
camellin　山茶糖苷　$C_{53}H_{84}O_{19}$
camel tree　金合欢
camenthol　樟脑
γ-camera　γ 照相机*
camera lens　暗箱物镜
camite　堪姆炸药；堪马特
camomile oil　春黄菊油
Ca-montmorillonite　钙-蒙脱石
camouflage paint　伪装漆；覆面漆
camouflaging　掩蔽
campaign　①(一)系列(试验)②运动；活动③阶段④一批；一组；一套⑤季节性生产
Campbell's solubility apparatus　坎贝尔溶解度测定仪
Campbell ejector　坎贝尔射出器
Campbell-Hurley's colorimeter　坎贝尔-赫尔利比色计
Campbell process　坎贝尔(直接成条)工艺
Campbell's solubility apparatus　坎贝尔溶解度测定仪
Camp colorimeter　坎普比色计〔测量钢铁中的锰〕
Campden tablets　坎普登片剂
campeachy wood　苏木
campesterol　菜油甾醇
campestrine　野千里光碱
camphane　莰烷　$C_{10}H_{18}$
camphanic acid　莰烷酸；樟脑酸　$C_9H_{13}O_2COOH$
camphanol　莰烷醇
camphanone　莰烷酮
camphanonic acid　莰农酸；1,2,2-三甲基环戊羧酸
camphanyl　莰基〔有3种异构体〕　$C_{10}H_{17}-$
camphene　莰烯　$C_{10}H_{16}$
camphene hydrate　水合莰烯
camphenic acid　莰烯酸　$C_{10}H_{16}O_4$
camphenilane　莰尼烷　C_9H_{16}
camphenilene　莰尼烯　C_9H_{14}
camphenilidene　亚莰尼基　$C_9H_{14}=$
camphenilidene acetone　樟脑叉丙酮
camphenilol　莰尼醇
camphenilone　莰尼酮　$C_9H_{14}O$
camphenilyl carbinol　莰尼甲醇
camphenolic acid　莰烯脑酸
camphenone　莰烯酮　$C_{10}H_{16}O$

camphidine 莰非啶 $C_{10}H_{19}N$
campho acid 莰佛酸 $C_{10}H_{14}O_6$
camphocarboxylic acid 莰佛羧酸 $C_{11}H_{16}O_3$
camphoceenic acid 莰佛烯酸 $C_9H_{14}O_2$
camphogen (=cymene) 莰佛精，伞花烃
camphoic acid 莰佛酸 $C_{10}H_{14}O_6$
camphoid 樟脑火棉浆
camphol (=isoborneol) 龙脑，2-莰烷醇
campholactone 龙脑内酯 $C_9H_{14}O_2$
campholene 龙脑烯；3-异丙基-1-环戊烯 C_8H_{14}
campholenic acid 龙脑烯酸 $C_{10}H_{16}O_2$
campholic acid 龙脑酸；1,2,2,3-四甲基-1-环戊烷甲酸 $C_{10}H_{18}O_2$
campholide α-龙脑内酯
campholytic acid 龙脑基酸 $C_9H_{14}O_2$
camphomycin 樟霉素
camphonanic acid 龙脑烷酸 $C_9H_{16}O_2$
camphononic acid 樟脑酮酸
camphor 樟脑；2-莰酮 $C_{10}H_{16}O$
camphoraldehyde 樟脑醛
camphoramic acid (=camphoramidic acid) 樟脑氨酸；樟脑酰胺酸 $C_8H_{14}(CONH_2)COOH$
camphoramidic acid (=camphoramic acid) 樟脑氨酸；樟脑酰胺酸 $C_8H_{14}(CONH_2)COOH$
camphoranilic acid 樟脑苯胺酸
camphorated oil 樟脑油
camphorated soap liniment 樟脑皂搽剂
camphor ball 樟脑丸
camphorene 樟脑烯 $C_{20}H_{32}$
camphoric acid 樟脑酸 $C_8H_{14}(COOH)_2$
camphoric acid amide 樟脑酸酰胺；樟脑氨酸 $C_8H_{14}(CONH_2)COOH$
camphoric anhydride 樟脑酸酐 $C_{10}H_{14}O_3$
camphorimide 樟脑酰亚胺 $C_{10}H_{15}O_2N$
camphorism 樟脑中毒
camphor liniment 樟脑擦剂
camphor oil 樟脑油
camphorone (=camphorphorone) 樟脑酮；5-甲基-2-异亚丙基环戊酮 $C_9H_{14}O$
camphoronic acid 分解樟脑酸；樟脑三酸；2,2,3-三甲基-3-羧基戊二酸 $C_9H_{14}O_6$
camphor original oil 樟脑原油
camphoroxime 樟脑肟；2-莰酮肟 $C_9H_{16}C=NOH$
camphoroyl 樟脑酰；莰二酰 $C_8H_{14}=(CO)_2$
camphorphorone (=camphorone) 樟脑酮；5-甲基-2-异亚丙基环戊酮 $C_9H_{14}O$
camphorquinone 樟脑醌；莰醌 $C_{10}H_{14}O_2$
camphor silk 樟蚕丝

camphor soap 樟脑皂
camphor spirit 樟脑酊
camphorsulfonic acid 樟脑磺酸 $C_{10}H_{15}OSO_3H$
camphor type basil oil 樟脑型罗勒油
camphor water 樟脑水
camphor wood oil 樟木油
camphoryl 樟脑基 $C_{10}H_{15}$
camphorylidene 亚樟脑基 $C_{10}H_{14}O=$
camphyl 莰基 $C_{10}H_{16}=$
camphylamine 莰基胺
camphylene 樟烯 $C_{10}H_{16}$
camptonite 康煌岩
camptothecin 喜树碱
cam pump 凸轮泵
camsellite 硼镁石
cam shedding 凸轮开口
camwood 非洲红木
Canada asbestos 加拿大石棉
Canada balsam (=Canada turpentine) 加拿大松节油；加拿大香脂
Canada turpentine (=Canada balsam) 加拿大松节油；加拿大香脂
d-canadene d-加拿烯
canadine 坎那定；氢化小蘖碱 $C_{20}H_{21}O_4N$
canadol 坎那油；重石油醚〔相对密度 0.650～0.700〕
canaigre 酸模〔一种主要原产于北美洲西南部的多年生草本植物(膜萼酸模蓼模属)，长有富含鞣质的块状根，以前用于制革和草药〕
canaigre ink 消迹墨
canal ①水道；管道②(炮)膛
canal drier 管道干燥器
canaline 副刀豆氨酸 $H_2NOCH_2CH_2CH(NH_2)COOH$
canal polymerization 管道聚合
canal ray 极隧射线〔物〕
cananga oil 依兰油
canari oil 金丝油
canary dextrin (e) 黄糊精
canary yellow 金丝雀黄
canavanine 刀豆氨酸
canavaninosuccinate 刀豆氨酸(基)琥珀酸
canavanosuccinic acid 刀豆氨酸(基)琥珀酸
can buoy 罐形浮标
cancellous 网眼状的；多孔的
canceration 癌变
cancerocidal 灭癌的
cancerogen 致癌物
cancidin 灭癌素
can coating 罐头漆；罐头涂层

can creel (条筒丝条)集束架
cancrinite 钙霞石
candela 坎德拉；新(国际)烛光〔光强单位，符号 cd，1 国际烛光=1.018cd〕
candelilla wax 小烛树蜡
candelite 坎底来特烟煤
candesartan 坎地沙坦〔药〕
candicin 坎底辛；对羟苯乙基三甲基铵
candle 烛；蜡烛
candle coal 烛煤
candle filter 滤烛；烛式过滤器
candle fish oil 烛鱼油
candle hardener 蜡烛硬化剂
candle moulding machine 制烛机
candle nut oil 烛果油
candle paper 包烛纸
candle power 烛光
candle wax 烛用蜡
candlewicking 松捻粗棉线；烛芯纱
candy maker's wax 糖果蜡
cane brimstone 硫棒；棒状硫黄
caneine 刀豆碱
canella oil 白桂皮油
cane sugar 蔗糖
cane sugar wax 蔗糖蜡
cane trash 蔗渣
cane unloader 卸蔗机
canfieldite （黑）硫银锡矿
can filler 装罐机；罐头装填机
can filling 装罐；装听
cangerana 卡芨拉树
cangerana oil 卡芨拉油
can half-body 罐头半坯
can ice 罐冰；模制冰
canister ①罐②滤毒罐③金属容器
canister resistance detecting apparatus 滤毒罐阻力测量器
canker ①腐蚀②传(感)染
canlay （丝条)装桶
can lining enamel 罐头内壁涂料
cannabamine 大麻胺类
cannabane 大麻烷 $C_{18}H_{22}$
cannabene 大麻烯 $C_{18}H_{20}$
cannabichrome 大麻色素
cannabidiol 大麻二酚 $C_{21}H_{30}O_2$
cannabidiolic acid 大麻二酚酸
cannabin (e) ①大麻苷②大麻碱
cannabinol 大麻酚 $C_{21}H_{26}O_2$
cannabinone 大麻酮 $C_8H_{12}O$

cannabis 大麻
cannabiscetin (=myricetin) 杨梅树皮素
cannabis oil 大麻油
cannaboid 大麻素
canned ①装罐的②罐头的
canned food 罐头(食品)
canned-motor pump 屏蔽电机泵；密封电机泵
canned oil 听装油
canned pump 密封的轻便泵
cannel (=cannel coal) 烛煤
cannel coal (=cannel) 烛煤
canneloid 烛煤质煤
cannery 罐头工厂
cannibal growth 吞并生长
cannibis 印度大麻脂
can (n)ister ①(金属)罐②滤毒罐
cannizarization 坎氏处理法〔用醇溶钾碱处理醛类〕
Cannizzaro reaction 康尼扎罗反应
Cannon-Fenske capillary viscosimeter 坎农-芬斯克型毛细管黏度计
Cannon-Fenske viscometer 坎农-芬斯克黏度计
cannonite 加隆炸药；加隆那特〔硝化纤维、硝化甘油炸药〕
cannon plug 有孔塞栓
Cannon process 坎农汽油精制过程
Cannon-Ubbelohde glass capillary viscometer 坎农-乌贝洛德玻璃毛细管黏度计
cannula (=canula) 套管
canoe filament 舟形灯丝
can of spinning machine 纺丝罐
canola 低芥酸菜子〔芥酸在 2%以下〕
canola oil 低芥酸菜子油
canonical 正则的
canonical coordinate 正则坐标
canonical correlation 典型相关
canonical distribution 正则分布
canonical ensemble 正则系综*
canonical equation 正则方程式
canonical free energy 正则自由能
canonical molecular orbital 正则分子轨道*
canonical partition function 正则配分函数*
canonical structure 正则结构
canonical transformation 正则变换
canophyllic acid 海棠果酸
canopy hood 伞形罩
can paint 罐头漆
can removal 去壳
can rinser 罐头洗涤机

can spinning machine 罐式纺丝机
can spout 油桶的斜槽嘴
can stability 储藏稳定性
cantaloup(e) 棱瓜；有棱甜瓜
cantaloup(e) seed oil 棱瓜子油
cantharene 二氢邻二甲苯
cantharidin 斑蝥素
canthaxanthin 斑蝥黄
canthinone 铁屎米酮
cantilever-balance 悬臂式天平
cantilever beam impact test (=Izod impact test) 悬臂梁式冲击试验
cantilever beam impact test machine 悬臂梁式碰撞试验机
cantilevered dual side construction 双侧悬臂式结构
cantilever softening point 悬臂(法)软化点
cantilever winder 悬臂式收卷机
Canton phosphorus 坎磷〔二分牡蛎壳与一分硫的混合物〕
can top 油听盖
can top crimp test 罐听卷边试验
canula (=cannula) 套管
canvas 帘子布；帆布
canvas conveyer 帆布运输机
canvas disc wetting test 帆布片润湿试验
canvas duck 轮胎布；轮胎帆布
canvas hose 帆布水龙带
canvvas tank 帆布桶
canyon 设备室
canyon-type cell 狭长型热室；狭谷型设备室
canyon-type centrifuge 屏蔽型离心机
canyon-type pump 屏蔽型泵
caoutchene 生胶干馏渣〔生胶干馏所得的高沸点产物〕
caoutchoid 橡胶类似物
caoutchone 橡胶酮
caoutchouc 生橡胶
caoutchoucene 生胶干馏渣
cap 盖；帽；罩
capability ①能力；本领②性能；效率；生产率
capacitance 电容
capacitance curve of electrode double layer 电极电双层电容曲线
capacitance detector 电容检测器
capacitance moisture meter 电容型水分计
capacitance recorder 电容电阻记录器
capacitance type of titration cell 电容式滴定池
capacitance-type regularity tester 电容式均匀度试验机
capacitive boosted filament pyrolyzer 电容放电热丝裂解器
capacitive coupled microwave plasma 电容耦合微波等离子体
capacitive current 电容电流
capacitive discharge pulse heating 电容放电脉冲加热法
capacitive feedback 电容反馈
capacitive humidity sensor 电容湿度传感器
capacitive hygrometer 电容湿度计
capacitively coupled high frequency plasma torch 电容耦合高频等离子体焰炬*
capacitively coupled microwave plasma 电容耦合微波等离子体
capacitively coupled microwave plasma torch 电容耦合微波等离子体焰炬*
capacitively coupled plasma 电容耦合等离子体
capacitive reactance (电)容(电)抗
capacitive sensing 电容传感
capacitor 电容器
capacitor-discharge spotwelder 电容放电式点焊机
capacitor discharge welding 电容放电焊接
capacitor oil 绝缘油；电容器油
capacity ①额；容量②电容(量)③生产额；生产(能)力
capacity coefficient 容量系数；电容系数
capacity factor ①容量因子*②容量比
capacity for heat 热容(量)
capacity for reaction 反应(能)力
capacity-head curve 容量-压差曲线
capacity in tons per hour 每小时生产吨数
capacity lag 容量滞后
capacity of column 柱容量
capacity of ion exchange resin 离子交换树脂容量
capacity of pump 泵的能量；泵的排量
capacity of the accumulator 蓄电池电容量
capacity operation 全容量操作；满载操作
capacity production 生产能力
capacity range (试验机)测程
capacity rating 额定生产率
capacity ratio 容量比
caparrapidiol (= 3,7,11-trimethyl-1,10-dodecadiene-3,7- diol) 卡帕木二醇
cap board 瓶帽纸板
cap (column) 泡罩
capecitabine 卡培他滨〔药〕
caper bush 刺山柑；老鼠瓜
capilin 茵陈素
capillarin 茵陈素
capillarity ①毛细(现象)②表面活性
capillarity attraction 毛细(管)吸力

capillarity constant 毛细常数
capillarity of fiber 纤维毛细管
capillary (=kapillary) ①毛细管②毛细的
capillary absorption 毛细吸收(作用)
capillary across section effect 毛细管横截面效应
capillary action 毛细管作用
capillary action shaping technique (CAST) 毛细管定形生长
capillary active ①毛细活性的②表面活性的
capillary active compound 毛细活性化合物
capillary active indicator ①毛细活性指示剂②表面活性指示剂
capillary activity ①毛细活性②表面活性
capillary adhesion force 毛细附着力
capillary adsorbed water 毛细吸附水
capillary alum 发矾；发盐
capillary analysis (=kapillaranalyse) 毛细(管)分析
capillary-arc discharge source 毛细管电弧放电离子源
capillary array electrophoresis 阵列毛细管电泳
capillary ascent 毛细上升
capillary attraction 毛细管引力
capillary bore 毛细管孔
capillary break (-up) ①毛细管破裂②(喷丝头)细孔断丝
capillary buffer 毛细管缓冲器
capillary bulb 毛细球
capillary buret 毛细滴定管
capillary capacity 毛细管容量
capillary channel 毛细管道〔漆料渗透到颜料层粒子中的细间隙〕
capillary characteristic 毛细管特性
capillary chemistry 毛细化学；表面化学
capillary column 毛细管柱*
capillary column chromatography 毛细管柱色谱法
capillary condensation 毛细管凝结*
capillary constant 毛细管常数*
capillary correction 毛细校正
capillary depression 毛细下降
capillary dimension 毛细管大小
capillary drop analysis 毛细管点滴分析
capillary electrochromatography 毛细管电色谱法
capillary electrode 毛细管电极*
capillary electrokinetic chromatography 毛细管电动色谱法
capillary electrometer 毛细管静电计
capillary electroosmotic chromatography 毛细管电渗色谱法
capillary electrophoresis 毛细管电泳
capillary electrophoresis electrochemical method 毛细管电泳电化学法
capillary electrophoresis-electrospray ionization mass spectrum 毛细管电泳电喷雾质谱联用
capillary elevation 毛细上升
capillary extrusion 毛细管挤出
capillary extrusion rheometer 毛细管挤出流变仪
capillary feed 毛细管给油
capillary filtering 毛细管过滤
capillary flask 毛细瓶
capillary flow 毛细流动；毛细管流
capillary flow data 毛细流数据
capillary flow measuring method 毛细流动测定法
capillary flowmeter 毛细流量计
capillary flow method 毛细流动法〔阴离子聚合时；求取生成速率常数的方法〕
capillary force 毛细力
capillary gas chromatograph 毛细管气相色谱仪*
capillary gas chromatograph/mass spectrometer/Fourier transform infrared spectrometer 毛细管气相色谱-质谱-傅里叶变换红外光谱联用装置
capillary gas chromatography 毛细管气相色谱法*
capillary gel column 毛细管凝胶柱
capillary gel electrophoresis 毛细管凝胶电泳
capillary glass tube 玻璃毛细管
capillary height method 毛细管高度法〔测表面张力〕
capillary hydrodynamics 毛细管流体动力学
capillary inactive 非毛细活性的；非表面活性的
capillary inlet 毛细管入口
capillary ion analysis 毛细管离子分析
capillary ion electrophoresis 毛细管离子电泳
capillary isoelectric focusing 毛细管等电聚焦
capillary isotachophoresis 毛细管等速电泳
capillary jet 毛细管射流
capillary jet viscometer 毛细管射流黏度计
capillary leak 毛细管漏孔
capillary lubricant 毛细管作用润滑剂
capillary luminescence analysis 毛细发光分析
capillary manometric method 毛细管压力法
capillary melting point determination 毛细管熔点测定
capillary meniscus 毛细管弯月面
capillary micellar electrokinetic chromatography 毛细管胶束电动色谱
capillary moisture capacity 毛细管水分容量
capillary non-gel sieving electrophoresis 毛细管无胶筛分电泳
capillary number 毛细管数
capillary oiler 毛细管给油器
capillary opening 毛细管孔口

capillary orifice viscometer 毛细孔黏度计
capillary outlet 毛细管出口
capillary packed column 毛细管填料(色谱)柱
capillary penetration 毛细管穿透作用
capillary phenomenon 毛细现象
capillary pipet 毛细吸管；毛细移液管
capillary plastometer 毛细管塑性仪
capillary pressure 毛细压力；毛细张力
capillary pyrite 针镍矿
capillary rheometer 毛细管流变仪
capillary rise 毛细上升
capillary rise method 毛细管上升法〔测表面张力〕
capillary sample inlet 毛细管进样口
capillary shield screen effect 毛细管的屏蔽效应
capillary spinneret 毛细孔纺丝板
capillary splitter 毛细管分流器
capillary stopcock 毛细活栓
capillary suction time 毛细管吸升时间
capillary supercritical fluid chromatography 毛细管超临界流体色谱法
capillary support 毛细管载体
capillary surface 毛细(表)面
capillary tension 毛细张力
capillary thickening 初生丝粗节
capillary tube 毛细管
capillary tube method 毛细管法〔测表面张力〕
capillary (tube) viscometer 毛细管黏度计
capillary tubing 毛细管
capillary type electron multiplier 毛细管型电子倍增器
capillary type isotachophoresis 毛细管等速电泳(分析)法
capillary viscometry 毛细管测黏法
capillary visco(si)meter 毛细管黏度计
capillary wave 毛细波；界面波；表面张力波
capillary yawp current 毛细管噪声电流
capillary zone electrophoresis 毛细管区带电泳
capillator 毛细管比色计
capillene 茵陈烯
capillin 毛蒿素
capillometer 毛细试验仪
capillon 茵陈酮
capillone 茵陈酮
capillose 针镍矿
capless tyre valve 无帽胎阀
capnoidine 咖诺定
capnometry 烟密度测定
cap of column 塔泡罩；塔泡帽
cap of multiple-effect evaporator 多效蒸发器泡罩
capped edge 挡边

capped nonionics 封端的非离子表面活性剂
capped pipe 加盖的管子
capped polystyrene 封闭聚苯乙烯
capper 封口机；压盖机
capping group 封端基团
capping reaction 封端反应
capraldehyde 癸醛 $C_9H_{19}CHO$
capraldehyde dimethyl acetal 癸醛二甲醇缩醛
capramide 癸酰胺
caprate 癸酸盐(或酯) $C_9H_{19}COOM$; $C_9H_{19}COOR$
caprenin (=caprin) 三癸精；甘油三癸酸酯
n-capric acid (正)癸酸 $CH_3(CH_2)_8Co_2H$
capric acid (=decylic acid) 癸酸
capric acid chloride 癸酰氯 $C_9H_{19}COCl$
capric alcohol 癸醇
n-capric aldehyde (正)癸醛 $CH_3(CH_2)_8CHO$
capric amide (=capramide) 癸酰胺
capric nitrile 癸腈 $C_9H_{19}CN$
n-capric nitrile (正)癸腈 $CH_3(CH_2)_8CN$
caprilate 辛酸盐(或酯) $C_7H_{15}COOM$; $C_7H_{15}COOR$
caprilic acid 辛酸
caprin (三)癸酸甘油酯
caprine (=nor-leucine) 正亮氨酸
caprinitrile 癸腈 $C_9H_{19}CN$
caprinohydroxamic acid (CHA) 癸基异羟肟酸；十碳异羟肟酸 $C_9H_{19}CONHOH$
caprinolactam 癸内酰胺
caprinoyl (=decanoyl) 癸酰
caproaldehyde 己醛 $C_5H_{11}CHO$
n-caproaldehyde (=hexanal) (正)己醛 $CH_3(CH_2)_4CHO$
n-caproamide (正)己酰胺 $C_5H_{11}CONH_2$
n-caproanilide (正)己酰苯胺 $C_5H_{11}CONHC_6H_5$
caproate 己酸盐(或酯) $C_5H_{11}COOM$; $C_5H_{11}COOR$
n-caproate (=n-hexanoate; n-hexoate) 正己酸酯
cap rock 盖保岩；冠岩
caprohydroxamic acid 六碳异羟肟酸 $C_5H_{11}CONHOH$
n-caproic acid (正)己酸 $CH_3(CH_2)_4Co_2H$
caproic acid (=hexanoic acid; capronic acid) 己酸
caproic aldehyde 己醛
caproic aldehyde diethyl acetal 己醛二乙缩醛
n-caproic anhydride (正)己酸酐 $(C_5H_{11}CO)_2O$
caproin 己酸精；甘油己酸酯
ε-caprolactam ε-己内酰胺
caprolactam (=2-oxohexamethylen(e) imine) 2-氧代六亚甲亚胺；己内酰胺
caprolactam polymerization 己内酰胺聚合
caprolactone 己内酯；羟基己酸内酯
caprolactone acrylate 丙烯酰己内酯

caproleic acid 癸烯酸
capronate 己酸酯
capronic acid (=caproic acid) 己酸
capronitrile (=amyl cyanide) 己腈；戊基腈 $CH_3(CH_2)_4CN$
capronyl chloride 己酰氯 $C_5H_{11}COCl$
caproyl (=hexanoyl) 己酰(基) $CH_3(CH_2)_4CO—$
caproyl ethanol 己酰基乙醇
n-caproylpyrazolone (COPY) 正己酰吡唑啉酮 $C_5H_{11}COC_3H_3N_2O$
capryl ①〔按规定=decanoyl〕癸酰基②〔已混用，一般指〕辛基；辛酰基
capryl acetate 乙酸辛酯
capryl alcohol 辛醇
sec-capryl alcohol 仲辛醇
capryl aldehyde 辛醛 $C_7H_{15}CHO$
caprylamide (=decolyamide) 辛酰胺 $C_7H_{15}CONH_2$
caprylate 辛酸盐(或酯) $C_7H_{15}COOM$；$C_7H_{15}COOR$
capryl chloride 辛酰氯 $C_7H_{15}COCl$
caprylene 辛烯
caprylic acid 辛酸
n-caprylic acid (正)辛酸 $CH_3(CH_2)_6CO_2H$
caprylic acid triglyceride 辛酸甘油三酯
caprylic alcohol 辛醇
caprylic aldehyde 辛醛 $CH_3(CH_2)_6CHO$
caprylic anhydride 辛(酸)酐 $(C_7H_{15}CO)_2O$
caprylic nitrile (=heptyl cyanide) 辛腈；庚基氰 $CH_3(CH_2)_6CN$
caprylidene 辛炔 $CH\equiv C(CH_2)_5CH_3$
caprylin (三)辛酸甘油酯
caprylolactone 辛内酯；羟基辛酸内酯
capryloyl (=octanoyl) 辛酰
capryloyl-collagenic acid 辛酰胶原酸
n-caprylpyrazolone (CYPY) 正辛酰吡唑啉酮 $C_7H_{15}COC_3H_3N_2O$
caprylyl (=octanoyl) 辛酰
caprylyl chloride 辛酰氯 $CH_3(CH_2)_6COCl$
capsaicin(e) 辣椒碱；辣椒素
capsanthin 辣椒红；辣椒质
capsanthol 辣椒醇 $C_{40}H_{57}(OH)_3$
cap screw 有帽螺钉〔机工〕
capsic acid 辣椒酸
capsicin 辣椒胶
capsicine 辣椒碱
capsicol 辣椒香精油
capsicum oleoresin 辣椒油树脂
capsicum resin 辣椒树脂
capsicum tincture 辣椒酊

capsochrome 辣椒红呋喃素
capsorubin 辣椒玉红素
capstan 绞盘；起锚机
capstan haul-off 绞盘式卷取装置
cap stopper 帽塞
capsularin 黄麻苷 $C_{22}H_{36}O_8$
capsular(y) ①雷管的②胶囊的
capsulate(d) ①胶囊包裹的②装入雷管的
capsulating process 胶囊包裹法
capsulation 包囊化作用*
capsule ①荚膜②胶囊③小皿④囊；被膜
capsule body 胶囊体
capsule sampling 封囊进样
capsuloesic acid (=aescinic acid) 马栗酸
captax 卡普踏克斯；促进剂M；2-巯基-1,3-硫氮茚
captive processing plant 附带处理工厂；采矿、水冶联合工厂
captive use 自己使用；自产自用
captopril 卡托普利〔药〕
capture 俘获
capture cross-section 捕获截面；俘获截面
capture effect 捕获效应；俘获效应
capture efficiency 捕获效率；俘获效率
capture gamma-ray 捕获γ射线
capture process 俘获过程；捕获过程
capture reaction 捕获反应
carabrone 天名精内酯酮
caracolite 氯铅芒硝
caracurine I 箭头毒I
caradiene 蒈二烯
carageen(an) 鹿角菜胶
carajura 秋海棠红
carajurin 秋海棠色素
caramel ①焦糖②酱色
caramelan 聚焦糖
carameline 焦黑素
caramelization 焦糖化
caramel malt 黑麦芽汁
caramel odor 焦糖香味
caramel test 焦糖色试验
caranday wax 棕榈蜡
carane 蒈(烷) $C_{10}H_{15}$
caraneol (=carol) 蒈烷醇
carapa oil 烛果油〔取自 *Carapa guayanensis* 种子的油〕
carat 开；克拉〔宝石、金刚石的重量单位，等于0.2克〕
carat balance 克拉天平
caraway 蒿
caraway fruit 蒿子

caraway oil 蒿子油
caraway seed 蒿子
caraway seed oil 蒿籽油；葛缕子油
carbacephem 碳头孢烯
carbachol 卡巴胆碱；碳酰胆碱〔药〕；氨甲酰胆碱
carbacidometer (=air tester) 大气碳酸计
carbalkoxy 烷酯基；烷氧羰基
carbalkoxylation 烷氧羰基化*
carbamamidine (=guanidine) 胍
carbamate 氨基甲酸酯*
carbamazepine 氨基甲酰氮杂
carbamic acid 氨基甲酸*
carbamic chloride 氨基甲酰氯 NH_2COCl
carbamide (=urea) 脲*；尿素 NH_2CONH_2
carbamide peroxide 过氧化脲
carbamide resin 聚脲树脂；碳酰胺树脂
carbamidine 胍 $C(NH_2)_2=NH$
carbamido (=ureido) 脲基
carbamidophosphoric acid 脲基磷酸
carbaminate 氨基甲酸盐(或酯) NH_2COOM; NH_2COOR
carbamino 氨甲酰基
carbamino acid 氨基甲酸 NH_2COOH
carbamino alanine 氨甲酰丙氨酸
carbaminobenzoic acid 脲基苯甲酸 $NH_2CONHC_6H_4COOH$
carbaminoh(a)emoglobin 氨甲酰血红蛋白
carbamino protein 氨甲酰蛋白
carbaminoylcholine 氨甲酰胆碱
carbamite 卡巴买特；二乙二苯基脲〔一种固体火箭燃料组分〕
carbamonitrile (=cyanamide) 氨基氰；氨腈
carbamoyl 氨基甲酰基；甲氨酰〔俗〕 $H_2NCO—$
carbamoylation 甲氨酰化(作用) $NH_2CO—$
carbamoyl ethylation 甲氨酰乙基化(作用) $NH_2COCH_2CH_2—$
carbamoylglyoxaline 氨基甲酰甘啉
carbamyl (=carbamoyl) 氨基甲酰；甲氨酰〔俗〕
p-carbamylaminophenylarsonic acid 对脲苯基胂酸
carbamyl aspartate 氨甲酰天冬氨酸
carbamyl chloride 氨基甲酰氯；甲氨酰氯 NH_2COCl
o-carbamyl-D-serine 邻氨甲酰-D-丝氨酸
carbamylglutamic acid 氨甲酰谷氨酸
carbamyl guanazole 氨基甲酰胍唑
carbamylmethylenephosphonate 氨基甲酰亚甲基膦酸酯
carbamyl ornithine 氨甲酰鸟氨酸
carbamyl phosphate 氨甲酰磷酸
carbamylphosphonate 氨基甲酰膦酸酯

carbamyltaurine 氨甲酰牛磺酸
carbanil 异氰酸苯酯
carbanilic acid 苯氨基甲酸 $C_6H_5NHCOOH$
carbanilide N-碳酰苯胺；对称二苯脲 $CO(NHC_6H_5)_2$
carbanilino (=carbaniloyl;henylcarbamoyl) 苯氨(基)甲酰基
carbaniloyl 苯氨基甲酰
carbanion 碳负离子*
carbanion addition 碳负离子加成作用
carbanionic active species 碳负离子活性体
carbanionic initiation 碳负离子引发(作用)
carbanionic polymerization 碳负离子聚合
carbanion polymerization 碳负离子聚合(作用)
carbarsone (=p-carbamylaminophenylarsonic acid) 卡巴胂；对脲苯基胂酸 $H_2NCONHC_6H_4AsO(OH)_2$
carbazic acid 肼基甲酸 $NH_2NHCOOH$
carbazide (=carbohydrazide) 卡巴肼；卡巴脲；二肼羰；均二氨基脲 $(NH_2NH)_2CO$
carbazochrome 肾上腺色素缩氨脲
carbazolacetic acid 咔唑乙酸
carbazole 咔唑；9-氮杂芴
carbazole dioxazine violet 咔唑二嗪紫
carbazole dye 咔唑染料
carbazole violet 咔唑紫
carbazolyl 咔唑基；氮芴基 $C_{12}H_8N—$
carbazone 卡巴腙〔俗〕；缩(对称)二氨基脲 $—N=NCONHNH—$
carbazone diphenyl 二苯卡巴腙 $PhN=NCONHNHPh$
carbazyl (=carbazolyl) 咔唑基
carbazylic acid 甲脒 $RC(=NH)NH_2$
carbene 卡宾*
carbene chemistry 卡宾化学；碳烯化学
carbenes ①碳质沥青②亚碳化物
carbenicillin 羧苄西林；羧苄青霉
carbenium ion polymerization 碳正离子聚合*
carbenium ions (=carbonium ion) 碳鎓离子；碳正离子
carbenoid 卡宾体*
carbenzol 岩焦油酚
carbethoxy (=ethoxycarbonyl) 乙酯基；乙氧甲酰 $C_2H_5OOC—$
carbethoxy carbene 乙酯基碳烯；乙氧羰基碳烯
2-carbethoxycyclopentanone 2-乙酯基环戊酮
carbethoxyl 乙酯基
carbethoxylation 乙酯基化作用；引入乙酯基
carbethylic acid 碳酸
carbic acid (=nadic acid;3,6-endomethylene-1,2,3,6-tetrahydrophthalic acid) 4-降冰片烯-1,2-二酸；3,6-内亚甲基-1,2,3,6-四氢化邻苯二甲酸

carbic anhydride (=norbornene dicarboxylic anhydride; nadic anhydride)　降冰片烯二酸酐
carbic ring　5-降冰片烯环
carbide　①碳化物②碳化钙
carbide alloy　硬质合金
carbide black　碳化黑
carbide brick　碳硅砖
carbide carbon　化合碳〔碳化物中的碳〕
carbide cermet (s)　金属碳化物；金属陶瓷烧结物
carbide chip　硬质合金刀片
carbide-chlorination　碳化物氯化(作用)
carbide crucible　碳化物坩埚
carbided catalyst　碳化催化剂；结焦了的催化剂
carbide fibre　(金属)碳化物纤维
carbide fuel　碳化物燃料
carbide furnace　碳化炉
carbide slag　电石渣
carbide slagging　碳化物造渣(过程)；碳化成渣(过程)
carbide steel cutting tool　碳素钢刀具
carbide tipped cutting tool　硬质合金刀具
carbide tool　硬质合金刀具
carbide to water acetylene generator　电石入水式乙炔发生器
carbidopa　卡比多巴〔药〕
carbimazole　卡比马唑；甲亢平
carbimide　碳酰亚胺；异氰酸
carbinol　甲醇〔英文此字一般用于复合词中〕　CH_3OH
carbinol base (=rosaline)　甲醇碱；蔷薇苯胺的别名；三-对-氨基苯代甲醇
carbithionic acid　硫代羧酸　RCSOH
carbitol　①二甘醇一乙醚；乙氧乙氧基乙醇②卡必醇〔类名〕　$HO(CH_2)_2O(CH_2)_2OC_2H_5$
carbitol acetate　二甘醇一乙醚乙酸酯
carbitol acrylate (=ethoxy ethoxyethyl acrylate)　丙烯酸卡必醇酯；二甘醇一乙醚丙烯酸酯
carbo-　①羰②碳
carbo activatus　活性炭
carboalkoxy　烷氧羰基；烷氧碳酰
carboamidation　氨羰基化*
carbo animalis　动物炭
carboanion　碳阴离子
carboatomic ring　碳环；碳原子环
carboazotine　克布索丁
carbobenzoxy　苄酯基；苄氧羰基　$C_6H_5CH_2OOC-$
carbobenzoxy chloride　苄氧甲酰氯*
carbobenzoxyglycine　苄氧羰基甘氨酸
carbobenzoyl　羧苯甲酰
carbobenzyloxy derivative　苄酯基衍生物；苯甲氧甲酰基衍生物
carbocation　碳正离子*
carbocationic copolymerization　碳正离子共聚反应
carbocationic polymerization　碳正离子聚合*
carboceric acid　二十七(烷)酸　$CH_3(CH_2)_{25}COOH$
carbochain　碳链
carbochain fibre　碳链纤维
carbochain polymer　碳链聚合物
carbocholine　碳胆碱
carbo cinchomeronic acid　2,3,4-吡啶三羧酸　$C_5H_2N(COOH)_3$
carbocisteine　羧甲司坦〔药〕
carbocoal　半焦
carbocoal tar　半焦油；低温焦油
carbo-corundum　碳刚玉
carbocyanine　羰花青
carbocycle　碳环
carbocyclic　碳环的
carbocyclic aromatic polyamide　碳环芳(香)族聚酰胺
carbocyclic compound　碳环化合物
carbocyclic ladder polymer　碳环梯形聚合物
carbocyclic ring　碳环
carbodianion　碳二负离子
carbodiazone　某双偶氮酮〔指含有二价基—N≡NCON≡N—的化合物〕
carbodication　碳二正离子
carbodihydrazide　卡巴肼；二肼羰
carbodiimide　碳二亚胺*
carbodiimide cyanamide tautomerism　碳二亚胺-氨基氰互变异构体
carbodiphenylimide　碳化双苯亚胺　$C_6H_5N=C=NC_6H_5$
carbodithioic acid　二硫代羧酸；荒酸〔俗〕　RCSSH
carbofrax　金刚硅砖料
carbogenes　碳合气
carbohydrase　糖酶
carbohydrate　碳水化合物*
carbohydrazide　碳酰肼；卡巴肼；均二氨基脲　$H_2NNHCONHNH_2$
carboid　油焦质
carbolate (=phenate)　酚盐；石炭酸盐
carbol crystal violet　石炭酸结晶紫
carbol fuchsin　卡宝品红
carbolic acid (=phenol)　石炭酸〔俗〕；苯酚　C_6H_5OH
carbolic oil　酚油
carbolic soap　酚皂；石炭酸皂
carbo lignius　木质炭
β-carboline　β-咔啉；2,9-二氮芴

carbolite ①卡包立②卡包塑料；磺烃(酚醛)塑料
carbolmethyl violet 酚甲基紫
carbolon 卡包纶〔碳化硅商名〕
carboloy〔商〕 碳化钨钴、钽或钛高硬度合金
carbolxylene 酚二甲苯洗液
carbometer 空气碳酸计
carbomethoxy (=methoxycarbonyl) 甲酯基；甲氧甲酰 CH_3OOC-
carbomycin 碳霉素；卡波霉素
carbon 碳
carbon accumulation 碳储积
carbonaceous 碳的；碳质的；含碳的
carbonaceous chondrite 碳质球粒陨石
carbonaceous coal 半无烟煤
carbonaceous dendrite 碳枝状晶体
carbonaceous deposit 碳质沉积
carbonaceous exchanger 碳质交换剂
carbonaceous fiber 碳纤维
carbonaceous filament 碳纤维长丝
carbonaceous filler 碳质填料
carbonaceous foam 含碳泡沫
carbonaceous impurity 碳质杂质
carbonaceous material 含碳材料
carbonaceous matter 含碳物质
carbonaceous organic material 含碳有机物
carbonaceous pitch fibre 碳质沥青纤维
carbonaceous refuse 含碳废弃物
carbonaceous residue 碳质残渣
carbonaceous shale 碳质页岩
carbon(aceous) soil 碳质污垢
carbon acid 碳氢酸*
carbonado 黑金刚石
carbonamide group (碳)酰氨基 $-CONH_2$
carbon apparatus 定碳器
carbon arc 碳弧
carbon arc cutting 碳弧切割
carbon-arc fadeometer 碳弧式褪色仪
carbon arc gouging 碳弧气刨
carbon-arc lamp 碳弧灯
carbon arc weatherometer 碳弧光耐气候试验机
carbon arc welding 碳弧焊
carbon assimilation 碳同化作用
carbonatation 碳酸盐化(作用)；二氧化碳饱和
carbonate ①黑金刚石②碳酸盐(酯)
carbonate buffer solution 碳酸盐缓冲溶液
carbonated 充了碳酸气的
carbonated juice 充(碳酸)气汁
carbonated liquor (=carbonated lye) 碳酸碱液

carbonated lye (=carbonated liquor) 碳酸碱液
carbonate ester additive 碳酸酯添加剂
carbonate flux 碳酸盐熔剂
carbonate-free lime 纯石灰；纯氧化钙
carbonate fusion 碳酸盐熔融
carbonate hardness 碳酸盐硬度；(水的)暂时硬度
carbonate ion 碳酸离子
carbonate leaching 碳酸盐浸取
carbonate medium 碳酸盐介质
carbonate method 碳酸盐法
carbonate minerals 碳酸盐矿物
carbonate rocks〔复〕 碳酸盐岩
carbonating 碳酸化；充碳酸气
carbonating column 碳酸化柱
carbonating tower 碳化塔
carbonation ①碳酸饱和②碳酸盐法③碳化(作用)④(=carboxylation)羧化作用；引入羧基
carbonation juice 充(碳酸)气汁
carbonation juice pump 充(碳酸)气汁泵
carbonation pan 碳酸饱和锅
carbonatization 碳酸盐岩化
carbon atom 碳原子
carbonator 碳酸化器
carbon backbone chain 碳主链
carbon balance 碳平衡
carbon base paste 碳基糊
carbon base polymer 碳基聚合物
carbon battery 碳电池
carbon binders 炭结合料〔胶质、漆皮〕
carbon bisulfide 二硫化碳 CS_2
carbon black 炭黑
carbon black dibutylophthalate absorption number 炭黑邻苯二甲酸二丁酯吸收值；炭黑DBP吸收值
carbon black dispersion 炭黑分散体〔颜料〕
carbon black external surface area 炭黑外表面积
carbon black fibre 炭黑纤维
carbon black filler 炭黑填充剂
carbon black flame 炭黑焰
carbon black gel complex 炭黑凝胶(复合体)
carbon black iodine absorption number 炭黑吸碘值
carbon black plant 炭黑厂
carbon black volatility (=carbon volatility) (活性)炭法挥发度
carbon bleach 炭漂白
carbon block ①炭精块②炭砖
carbon board 复写纸
carbon bond 碳键
carbon bridge 炭桥

carbon bronze 碳青铜
carbon burning 烧炭
carbon-burning load 碳燃烧负荷〔催化剂再生器〕
carbon-carbon bond 碳-碳键
carbon-carbon crosslink 碳-碳交联；碳-碳交联键
carbon-carbon skeletal chain 碳-碳骨架链；碳-碳主链
carbon chain 碳链
carbon chain compound 碳链化合物
carbon chain fibre 碳链纤维
carbon chain polymer 碳链高分子；碳链聚合物
carbon chloroform extract 碳氯仿抽提物
carbon cloth 碳纤维布〔做电极〕
carbon-coated 被碳沉积盖覆的
carbon coating 碳涂层
carbon color test 石油产品中比色定碳试验
carbon comparison tube 定碳比色管
carbon composite 碳纤维复合材料
carbon compound 碳的化合物
carbon congeners 碳的同族元素
carbon contact 炭精接头
carbon contamination 碳污染
carbon content 碳含量
carbon crucible 石墨坩埚
carbon cycle 碳循环
Carbondale chiller 卡尔邦达冷冻石蜡结晶器
Carbondale refrigeration process 卡尔邦达冷冻过程
carbon dating 碳-14 测定年代
carbon decolourizing efficiency 活性炭脱色效率
carbon deposit (=carbon laydown) 碳沉积；积碳
carbon deposit inhibitor 碳沉积抑制剂
carbon deposition 积碳*
carbon dichloride 二氯化碳；四氯乙烯 $CCl_2 = CCl_2$
carbon dioxide 二氧化碳 CO_2
carbon dioxide absorption 二氧化碳吸收(量)
carbon dioxide absorption tube 二氧化碳吸收管
carbon dioxide capacity 二氧化碳容量
carbon dioxide combining power 二氧化碳结合力
carbon dioxide content 二氧化碳含量
carbon dioxide diffusion coefficient 二氧化碳(CO_2)扩散系数
carbon dioxide diffusion resistance 抗(防)二氧化碳扩散性
carbon dioxide electrode 二氧化碳电极
carbon dioxide fire protection nozzle 二氧化碳灭火喷雾器(喷嘴)
carbon dioxide gas 二氧化碳气；碳酸气
carbon dioxide gas sensing electrode 二氧化碳气敏电极
carbondioxide gas shield welding 二氧化碳气体保护焊

carbon dioxide ice 干冰
carbon dioxide laser 二氧化碳激光
carbon dioxide trap 二氧化碳冷阱
carbon diselenide 二硒化碳 CSe_2
carbon distribution 碳沉积；积碳；碳分布
carbon disulfide 二硫化碳 CS_2
carbon dust 碳尘
carbon electric arc 碳电弧
carbon electrode 碳电极；炭精电极
carbon element hygrometer 碳元件湿度计
carbon equivalent 碳当量
carbon evaporation 表面蒸碳
carbonex 炭黑和矿物油混合物〔混合胶的软化剂〕
carbon family 碳族
carbon fiber 碳纤维*
carbon fiber adsorbent 碳纤维吸着剂
carbon fiber micro-disk electrode 碳纤维微盘电极
carbon fiber microelectrode 碳纤维微电极
carbon fiber reinforced plastics (CFRP) 碳纤维增强塑料
carbon fibre adsorbent 碳纤维吸着剂
carbon fibre-carbon matrix composite 碳纤维-碳基质复合材料
carbon-fibre composite 碳纤维复合材料
carbon fibre-plastic composite 碳纤维-塑料复合材料
carbon fibre precursor 碳纤维母体；碳纤维原丝
carbon fibre reinforced plastics 碳纤维增强塑料
carbon fibre reinforced thermoplastic 碳纤维增强热塑性塑料
carbon fibre reinforcement 碳纤维增强材料
carbon filament 碳灯丝
carbon filament atomizer 碳丝原子化器
carbon filament atom reservoir 碳丝原子储存器
carbon filament cloth 碳纤维织物
carbon-filled hollow fibre 充碳中空纤维
carbon filler 碳质填料
carbon filter funnel 碳质过滤漏斗
carbon finishing method 炭研磨法；炭抛光法
carbon-fluorine spin coupling constant 碳-氟自旋耦合常数
carbon formation on catalyst 催化剂上碳形成
carbon forming properties 碳形成性质；积炭生成性质
carbon-free 不含碳的
carbon-free-backbone chain 非碳主链；无碳主链
carbon freezing 二氧化碳冷冻
carbon gel 炭黑凝胶；结合橡胶
carbon gland 石墨压垫盖
carbon-glass hybrid tape 碳纤维-玻璃纤维混杂带
carbon-graphite reinforcing fibre 碳-石墨增强纤维

carbon group analysis	碳族分析；烃类结构族分析
carbon hexachloride	六氯乙烷 C_2Cl_6
carbon hot-reduction	碳热还原法
carbon humidity element	碳湿敏元件
carbon-hydrogen electrode	碳氢电极
carbon hydrogen ratio	碳氢比
carbonic	碳的
carbonic acid	碳酸 H_2CO_3
carbonic acid derivatives	碳酸衍生物
carbonic acid equilibrium	碳酸平衡
carbonic acid gas	碳酸气；二氧化碳 CO_2
carbonic acid gas pump	碳酸气泵
carbonic acid refrigerating machine	碳酸冷却机
carbonic anhydride	碳酸酐；二氧化碳；碳酐 CO_2
carbon ice	干冰
carbonic ester	碳酸酯
carbonic ether	①碳酸酯 $CO(OR)_2$ ②碳酸乙酯 $CO(OC_2H_5)_2$
carbonic oxide	一氧化碳 CO
carbonic oxide cell	一氧化碳电池
carbonide	碳化物〔指金属碳化物〕
carboniferous (=carbonous)	碳的；含碳的
carbonific	成碳化合物
carbonification	碳化作用
carbonifying	碳化的
carboning flame	碳化焰
carbon ink	碳墨
carboniogenesis	①碳化作用②正碳离子源
carbonite	碳质炸药〔硝化甘油、硝酸钾、锯屑炸药〕
carbonitride	氮化碳；碳氮化物
carbo-nitriding	碳-氮共渗法
carbonitrile	腈
carbonium	碳鎓；碳正
carbonium ion	碳鎓离子；阳碳离子；正碳离子
carbonium ion polymerization	碳离子（催化）聚合（作用）
carbonium radical	碳鎓根
carbonization	碳化（作用）
carbonization effluents	碳化污水
carbonization machine	碳化机
carbonization method	碳化法
carbonization of wood	木材炭化
carbonization process	碳化过程
carbonization properties of coal	煤的碳化性质
carbonization temperature	碳化温度
carbonized	碳化了的
carbonized cellulose	碳(化)纤维素
carbonized fibre	碳化纤维
carbonized path	碳化径迹
carbonized polyacrylonitrile fibre	碳化聚丙烯腈纤维；聚丙烯腈系碳纤维
carbonizer	(氯化铝)去纤维素液
carbonizing	①碳化的②碳化
carbonizing apparatus	碳化装置
carbonizing assistant	碳化助剂
carbonizing flame	还原焰；碳化焰
carbonizing of gas	煤气渗碳
carbonizing plant	碳化装置
carbonizing stove	碳化烘焙室
carbonizing treatment	碳化处理
carbon knock	因积碳而引起的爆震
carbon laydown (=carbon deposit)	碳沉积；积碳
carbonless vulcanized elastomer	无炭黑硫化高弹体
carbon light	碳光
carbon lining	碳质内衬
carbon-13 magnetic resonance (^{13}C-NMR)	碳-13(核)磁共振
carbon molecular sieve	碳分子筛*
carbon monosulfide	一硫化碳 CS
carbon monoxide	一氧化碳 CO
carbon monoxide acetate	双乙氧基碳；一氧化碳乙酸酯 $C(OC_2H_5)_2$
carbon monoxide canister	一氧化碳滤毒罐
carbon monoxide converter	一氧化碳转化器
carbon monoxide detector tube	一氧化碳检测管
carbon monoxide electrode	一氧化碳电极
carbon monoxide filter	一氧化碳滤毒罐
carbon monoxide-hydrogen mixture	一氧化碳-氢气混合物
carbon monoxide indicator	一氧化碳检测管
carbon monoxide recorder	一氧化碳记录器〔记录空气中的含量〕
carbon nanotube	碳纳米管
carbon number	碳数
carbon number rule of retention	保留值碳数规律
carbonoid	类碳型
carbonous (=carboniferous)	碳的；含碳的
carbon oxides	碳的氧化物
carbon oxyiodfluoride (COIF)	氟碘氧化碳
carbon oxysulfide	氧硫化碳 COS
carbon packing	石墨填料；碳素填料
carbon paper	复写纸；炭纸〔俗〕
carbon particles	碳粒子
carbon paste electrode	碳糊电极*
carbon paste immobilization	碳糊固定法
carbon pernitride	叠氮化氰 $CN[N_3]$
carbon plate	碳板；炭精板

carbon porous microsphere 碳多孔微球
carbon powder 炭末；炭粉
carbon print 碳纸晒印
carbon prism 碳棱柱体
carbon process 炭黑印画法
carbon-producing 生碳的
carbon radical 碳自由基
carbon ratio 碳比率；碳系数
carbon remover 除积碳器；除积碳剂
carbon residue 碳渣；焦炭残渣
carbon residue increment 残碳增值
carbon residue test 碳渣试验
carbon resistance furnace 碳(极)阻(力)电炉
Carbon-retting process 卡氏沤麻脱胶法
carbon rheostat 碳变阻器
carbon rich waste 富碳有机废物
carbon rod atomizer 碳棒原子化器
carbon sand 碳素砂
carbon selenosulfide 硒硫化碳 CSeS
carbon settling chamber 碳粒沉积室
carbon shale 碳页岩
carbon shoe ①碳素垫块；碳素滑脚 ②碳刷
carbon silicide 硅化碳；金刚砂
carbon skeleton 碳骨架
carbon smut 润滑剂炭黑
carbon soot 炭黑
carbon source 碳源
carbon steel 碳钢；碳素钢
carbon stick 炭精棒
carbon subnitride 丁炔二腈〔不叫: 低氮化碳〕 CNC≡CCN
carbon suboxide 二氧化三碳*
carbon subsulfide 二硫化三碳
carbon sulfide 二硫化碳 CS_2
carbon test 碳含量试验
carbon tetrabromide 四溴化碳 CBr_4
carbon tetrachloride 四氯化碳 CCl_4
carbon tetrachloride activity 四氯化碳活度
carbon tetrafluoride 四氟化碳 CF_4
carbon tetrahalide 四卤化碳 CX_4
carbon tetraiodide 四碘化碳 CI_4
carbon thrust ring 石墨推承环
carbon tissue 复写纸
carbon-to-carbon linkage 碳-碳链合
carbon-to-carbon rupture 碳-碳键断裂
carbon tool steel 碳素工具钢
carbon tracer 示踪碳
carbon track 导电性炭黑路

carbon tribromide 六溴乙烷 C_2Br_6
carbon trichloride 六氯乙烷 C_2Cl_6
carbon triiodide 六碘乙烷 C_2I_6
carbon tube 碳精管；石墨管；石墨化碳管
carbon value 碳值
carbon volatility (=carbon black volatility) (活性)炭法挥发度
carbon whisker 碳须晶
carbon-white 白炭黑〔白色补强剂〕
carbon yarn 碳纤维
carbonyl 羰基；碳酰 =CO
carbonyl acceptor group 羰基受电子基团
carbonyl addition 羰基加成反应
carbonyl-amine reaction 羰基胺反应
carbonylation 羰基化*
carbonyl bond 羰基键
carbonyl bromide 碳酰溴 $COBr_2$
carbonyl chloride 碳酰氯；光气
carbonyl chlorobromide 碳酰氯溴 COClBr
carbonyl cobalt 羰基钴
carbonyl compound 羰基化(合)物
carbonyl-containing contaminant (含)羰基杂质；羰基污染物
carbonyl derivative 羰基衍生物
carbonyl diamide 碳酰二胺；尿素 $CO(NH_2)_2$
1,1'-carbonyldiimidazole 1,1'-羰基二咪唑
carbonyldioxy 碳酰二氧基；碳酸基
carbonyl disulfide 二硫化羰
carbonyl dithiocarbonic acid 二硫代碳酸 $CO(SH)_2$
carbonyldiurea 碳酰二脲 $OC(NHCONH_2)_2$
carbonyl diurethane 羰基二尿烷；羰基二(氨基甲酸乙酯) $OC(NHCOOCH_2CH_3)_2$
carbonyles 羰合物类
carbonyl ferrocyanic acid 五氰一羰合亚铁酸 $H_2[FeCO(CN)_5]$
carbonyl (ferro) heme 碳氧(亚铁)血红素；羰合血红素
carbonyl fluoride 碳酰氟；羰基氟化物 COF_2
carbonyl group 羰基
carbonyl group frequency 羰基振动频率
carbonyl iron 羰基铁〔分解羰基化铁而得的铁〕
carbonyl link(age) 羰基键
carbonyl-methylene condensation 羰基-亚甲缩合作用
carbonyl-monothio-acid 一硫羟碳酸 CO(SH)OH
carbonyl oxygen 羰基(中的)氧
carbonyl pyrrole 碳酰吡咯 $CO(C_4H_4N_2)_2$
carbonyl reaction 羰基反应
carbonyl reagent 羰基试剂
carbonyl reduction 羰基还原

carbonyl stretching frequency 羰基伸展频率
carbonyl sulfide 硫化羰；氧硫化碳 COS
carbonyl value 羰基值〔指羰基数量〕
carbophile 亲碳性
carbophilic rubber 亲有机物质的橡胶
carboplatin 卡铂〔药〕
carboradiant kiln 金刚砂电炉
carborane 碳硼烷
carborane type 碳硼烷型
carborization ①碳化(作用)②碳化法；渗碳
carbo rubber 炭黑橡胶
carborundum (=silicon carbide) 碳化硅；金刚砂
carborundum brick 金刚砂砖
carborundum furnace 金刚砂(电)炉
carborundum grinding wheel 金刚砂轮
carborundum paper (金刚)砂纸
carborundum paste 活塞粉
carborundum raising 金刚砂起绒
carborundum refractory 碳化硅耐火材料；金刚砂耐火材料
carborundum sharpening stone 磨石〔商〕；油石〔俗〕
carborundum stone grip 磨石夹
carborundum tube 碳化硅管
carborundun grit 金刚砂
carbosant 碳酸檀香酯
carboseal 卡波夕耳〔收集灰尘用润滑剂〕
carbosilane 碳硅烷
carbosphere 碳微球〔作为泡沫塑料填充料〕
carbostyrilic acid (=kynuric acid;cynurenic acid) 犬尿酸
carbostyril (=quinolone) 喹诺酮；2-羟基喹啉
carbothialdine 二硫代氨基甲酸二乙胺酯 $NH_2CSSN(CHCH_3)_2$
carbothioic acid 硫代酸 RCSOH; RCOSH
carbothiolic acid 硫羟酸；巯酸 RCOSH
carbothionic acid 硫羰酸；苊酸 RCSOH
carboxamide 氨甲酰；羧酰胺
carboxamido-group 酰胺基
carboxethyl ester 烃氧基丙酸；羧乙基醚 $ROCH_2CH_2COOH$
carboxide ①羰基化物 $M(CO)_n$②羰基 (=CO)
carbox metal 卡波克斯合金
carbox process (词源: carbothermic-reduction oxidation process) 卡尔保克斯过程；碳热氧化还原过程〔用于碳化铀核燃料再循环〕
carboxy 羧基 HOOC—
carboxyalkyl cellulose 羧甲基纤维素
carboxyalkyl sulfonate 羧基烷基磺酸盐
carboxyarsenazo 偶氮胂羧

carboxyazo 偶氮羧
p-carboxy benzhydrol 对羧基二苯甲醇 $C_6H_5CH(OH)C_6H_4CO_2H$
carboxybetaine 羧基甜菜碱
o-carboxy cinnamic acid 邻羧基肉桂酸 $(C_6H_4C_2H_2)(CO_2H)_2$
carboxy dialkyl benzene sulfonamide 羧基二烷基苯磺酰胺
carboxy ethoxy ethyl amino dodecane 羧基乙氧基乙基氨基十二烷
carboxyethyl 羧乙基 $HOOCC_2H_4$—
carboxyethylated cotton 羧乙基化棉〔棉纤维经丙烯腈和浓烧碱处理或由氰乙基化棉纤维钠盐水解而成〕
carboxyethyl carbamylethyl cellulose 羧乙基氨基甲酰乙基纤维素
carboxyethyl cellulose 羧乙基纤维素
carboxyethyl-cyanoethyl cellulose 羧乙基-氰乙基纤维素
N-(β-carboxyethyl) diethylenetriamine tetraacetic acid (CET-TA) N-(β羧乙基)二亚乙基三胺四乙酸
carboxyfluorescein 羧基荧光素
carboxy group 羧基
carboxy hexyl cyclohexene octanoic acid 羧基己基环己烯辛酸
carboxyl (=oxatyl) 羧基
carboxylamine (=carbamic acid) 氨(基)甲酸
carboxylate 羧化物；羧酸盐(或酯)
carboxylate acrylic fibres 羧化丙烯腈系纤维
carboxylate anion 羧酸根阴离子
carboxylated alkyl imidazolium hydroxide 氢氧化羧基烷基咪唑鎓
carboxylated anionic detergent 羧基化阴离子洗涤剂
carboxylated paper 羧(基)化滤纸
carboxylated phenol formaldehyde polymer 羧基化苯酚甲醛共聚物
carboxylated resin 羧化树脂
carboxylated surfactant 羧化型表面活性剂
carboxylate group 羧基
carboxylate ion 羧酸盐离子
carboxylate latex 羧基胶乳
carboxylation 羧基化*
carboxylation agent 羧基化剂
carboxyl content value 羧基含量值
carboxyl-ended oligomer 羧端基低聚物
carboxyl end group 羧端基
carboxylesterase 羧酸酯酶
carboxy(l)-functional polyester 含羧基官能团的聚酯；羧基聚酯

carboxylic acid　羧酸*
carboxylic acid amide　羧酸酰胺
carboxylic acid analyzer　羧酸分析器
carboxylic acid ester　羧酸酯
carboxylic acid functional surface　羧酸官能表面
carboxylic acid substituent　羧酸取代基
carboxylic acid type cation exchange fibre　羧酸型阳离子交换纤维
carboxylic acid type resin　羧酸型树脂
carboxylic acrylonitrile butadiene rubber　羧基丁腈橡胶
carboxylic acrylonitrile butadiene rubber latex　羧基丁腈胶乳
carboxylic chloroprene rubber latex　羧基氯丁胶乳
carboxylic ether (=carboxylic ester)　羧酸酯
carboxylic group　羧基 —COOH
carboxylic latex　羧基胶乳
carboxylic propyl lauryl amine　羧基丙基月桂基胺
carboxylic resin　羧酸型树脂
carboxylic rubber　羧基橡胶
carboxylic soap　羧酸皂
carboxylic styrene butadiene rubber　羧基丁苯橡胶
carboxyl methyl cellulose　羧甲基纤维素
carboxylnitroso rubber (CNR)　羧基亚硝基橡胶
carboxy(l) polyester　羧基聚酯
carboxyl polymethacrylate　含羧基聚甲基丙烯酸酯
carboxyl reaction　羧基反应
3-carboxyl-1-(p-sulfophenyl)-5-pyrazolone　3-羧基-1-(对-磺基苯)-5-吡唑啉酮
carboxyl terminal　羧基末端；C 末端
carboxyltransferase　羧基转移酶
carboxymethyl　羧甲基
carboxy methyl amylose　羧甲基直链淀粉
carboxymethylated cotton　羧甲基化棉
carboxymethylated wool fibre　羧甲基化羊毛(纤维)
carboxymethylation　羧甲基化(作用)
carboxymethyl cellulose　羧甲基纤维素*
carboxymethyl ester　羧甲基醚；烃氧基乙酸 ROCH$_2$COOH
carboxymethyl mercapto-group　羧甲硫基 HOOCCH$_2$S—
carboxymethyl phenyl arsonic acid　羧基甲基苯胂酸 COOH(CH$_3$)C$_6$H$_3$AsO(OH)$_2$
carboxy methyl starch　羧甲基淀粉
carboxynitrazo　偶氮硝羧
carboxy nitrile rubber　羧基丁腈橡胶〔高耐磨制品用〕
carboxypeptidase　羧(基)肽酶
carboxypeptidase method　羧肽酶法
carboxy phenyl arsine oxide　羧苯胂氧 COOHC$_6$H$_4$As=O
carboxy phenylarsonic acid　羧基苯胂酸 (HO)$_2$OAsC$_6$H$_4$CO$_2$H
carboxy phenyl arsonous acid　羧苯亚胂酸 COOHC$_6$H$_4$As(OH)$_2$
carboxy phenyl dihydroxy arsine　羧苯亚胂酸 COOHC$_6$H$_4$As(OH)$_2$
p-carboxy phenylphosphonic acid　对羧苯基膦酸 HO$_2$CC$_6$H$_4$PO(OH)$_2$
carboxyphenyl salicylate　水杨酸羧苯酯〔稳定剂〕
carboxypolypeptidase　羧基多肽酶
carboxy reactivity　羧基反应性
carboxy terminated　端羧基
carboxy terminated nitrile rubber　羧基丁腈橡胶
4-carboxyuracil　4-羧基尿嘧啶；乳清酸
carboy (=demijohn)　①坛②酸坛
carboy hamper　坛篓
carboy inclinator　酸坛倾架
carboy tilter　酸坛倾架
carbromal (=adalin)　阿大林；二乙代溴乙酰脲 (C$_2$H$_5$)$_2$CBrCONHCONH$_2$
carbuncle　红榴石
carburant　增碳剂；渗碳剂
carburation　①渗碳②增热③(内燃机)汽化(作用)
carburator　①汽化器〔内燃机〕②渗碳器
carburator detergent　汽化器清洗剂
carburelant　增碳剂；渗碳剂
carburet　①增碳；渗碳②汽化〔使汽油与空气混合〕
carbureter　①汽化器〔内燃机〕；化油器②渗碳器
carbureting　①增碳；渗碳②汽化
carburetion　①渗碳作用②(内燃机内的)汽化作用
carburetted air　增碳空气
carburetted alcohol　增碳酒精
carburetted gas　渗碳气
carburetted hydrogen　矿坑气
carburetted water gas　增碳水煤气
carburetter (=carburetor; carbureter)　①增碳器②汽化器〔内燃机〕；化油器
carburetting　①增碳；渗碳②汽化
carburetting oil　增碳用油；生产气体烃用的油料；瓦斯油
carburettor (=carburetter)　①增碳器②汽化器〔内燃机〕；化油器
carburised layer　渗碳层
carburite　碳混铁
carburization　渗碳*
carburizer　渗碳剂
carburizing　碳化；渗碳

carburizing agent 渗碳剂
carbyl 二价碳基 —C—
carbylamine (=isonitrile) 异腈；胩
carbylamine reaction (=Hofmann isonitrile synthesis) 成胩反应；霍夫曼异腈合成法
carbylic acid 碳基酸〔酸基中含碳的酸〕
carbyl sulfate 亚乙基磺酸酐 $C_2H_4O_6S_2$
carbyne 碳炔*
carcass building 帘布层贴合
carcass compound 帘布层胶料
carcass fabric 轮胎织物；工业上胶用底布
carcass material 胎体材料；帘布材料〔轮胎〕
carcass plies 帘布层〔轮胎〕
carcass stock compound 帘布层胶料
Carcel unit 卡索(灯)光度单位〔=6.9 英国烛光单位〕
carcinocidin 消癌菌素
carcino-embryonic antigen (CEA) 癌胚抗原
carcinogen 致癌物
carcinogenesis 致癌作用
carcinogenic 致癌的
carcinogenic compound 致癌化合物
carcinogenicity 致癌作用
carcinogenicity test 致癌试验
carcinogenic nitrogen compound 致癌含氮化合物
carcinogenic property 致癌性
carcinogenic substance 致癌物
carcinolysin 溶癌素
carcinolysis 癌溶解
carcinoma 癌
carcinomic acid 癌脂酸
carcinomycin 癌霉素
carcinostatin 制癌菌素
car cleaner 车辆洗净剂
cardamom 小豆蔻
cardamomin 小豆蔻明
cardamom oil 小豆蔻油；砂仁油
cardamom seed 小豆蔻子
cardan joint 万向轴节
cardanol 腰果酚
cardanolide 卡烷内酯
card board 特等纸板；卡(片)纸板
cardboard wax 纸板石蜡
card cotton yarn 粗梳棉纱
carded knitting yarn 粗梳针织纱
carded union yarn 粗梳混纺纱
carded web 梳理成网；机械成网
carded wool yarn 粗纺毛纱
carded yarn 粗梳纱

cardelmycin (=novobiocin) 新生霉素
cardenolide 卡烯内酯
card for reading meniscus 滴定管弯液面读数卡
card grinding 磨针
cardiac ①心的②强心的③强心剂
cardiac muscle cell 心肌细胞
cardiac pacemaker 心脏起搏器
cardiac sedative 强心剂
cardiac stimulant ①强心剂②强心的
cardiamine 强心胺〔药〕
cardiantia 强心剂；心兴奋剂
cardicin 卡氏菌素
cardinene 小豆蔻烯
carding 起毛；起绒；粗梳
carding and brushing machine 刷毛机；刷绒机
carding leather 梳理机革
carding machine ①粗梳机②起毛机
carding oil 梳理油
cardioid condenser 心形聚光器
cardiolipin 心磷脂；双磷脂酰甘油
card leather 梳革
card middle 衬纸
cardol 强心酚〔取自槚如坚果壳液〕
cardon 刺菜蓟
card sliver 生条
card stripping 抄斩花
car dumper 倾倒车
cardy oil 红花油
careful distillation 精馏
careful oxidation 细心氧化
carene 蒈烯 $C_{10}H_{16}$
Carey-Foster bridge 卡雷-福斯特电桥
car frame 车架
cargo 货船
cargo oil 船运油料；载运油料
cargo pump 货船泵；大排量的泵
cargo tank 货船；油舱；油轮
carhydrine (=dihydrocarveyl acetate) 乙酸二氢香芹酯
carica (=papaya) 木瓜
caricin 番木瓜苷
carinogenicity 致癌性
carionlytic compound 隆凸化合物
carissin 假虎刺苷
carissone 假虎刺酮
caritinoid 胡萝卜素
Carius bomb 卡里乌斯瓶
Carius method 卡里乌斯法〔定卤素和硫〕
carlic acid 卡尔酸 $C_{10}H_{10}O_6$

carlina oxide 鲜蓟醚
carlinene 鲜蓟烯
Carlsbad salt 卡尔斯泉盐
carlson pin (卡氏)定位梢
Carlson process (for cyanamide) 卡尔逊(制氰氨化钙)法
Carman-Kozeny equation 卡曼-康采尼方程
carmeloite 伊丁玄武岩
Carmichael copper extraction process 卡麦克尔提铜法
carminative ①驱风剂②排气剂
carmine 胭脂红；洋红；卡红
carmin(e) lake 胭脂红色淀
carmine red 胭脂红；洋红
carmine vermillion ①(=mercuricsulfide)银朱；硫化汞 ②朱红(色)
carminic acid 胭脂红酸 $C_{20}H_{22}O_{13}$
carminite 砷铅铁矿
carmofur 卡莫氟〔药〕
carmustine 亚硝(基)脲氮芥；卡氯芥；卡莫司汀〔药〕
carnallite 光卤石；砂金卤石 $KCl \cdot MgCl_2 \cdot H_2O$
carnation ①淡红色；肉色②肉色的
carnation absolute 康乃馨净油
carnation oil 康乃馨油
carnauba tree 巴西棕榈
carnauba wax 巴西棕榈蜡；加洛巴蜡
carnaubic acid 巴西棕榈酸；二十四(烷)酸
carnaubyl alcohol 巴西棕榈醇；二十四(烷)醇
carnegieite 三斜霞石
carnegine (=pectenine) 卡乃津 $C_{13}H_{19}O_2N$
carnelian 光玉髓；红玉髓
carnic acid 肉酸
carnine 肌苷；次黄嘌呤核苷
carnitine 肉碱
carnosol 鼠尾草酚
Carnot cycle 卡诺循环*
carnotite 钒钾铀矿
Carnots reagent 卡诺氏试剂〔硫代硫酸铋钠的乙醇溶液，测定钾的试剂〕
Carnot theorem 卡诺定理*
carob 角豆
carob gum 角豆树胶
carobic acid 驱梅树叶酸
carobine 驱梅树碱
carobone 驱梅树叶脂
carob-seed gum 角豆树(豆)胶〔用途：油漆、食品稳定剂、增稠剂、乳化剂、浆料、织物整形、化妆品等〕
carob tree 角豆树；稻子豆树；长角豆
car oil 车油
carol (=caraneol) 蒈烷醇 $C_{10}H_{18}O$

carolic acid 肉霉酸；卡若酸 $C_9H_{10}O_4$
carolinic acid 肉灵酸〔取自 Penicillium charlesii〕 $C_9H_{10}O_6$
carolite (=carrollite) 硫铜钴矿
carone 蒈酮；5-蒈酮 $C_{10}H_{10}O_6$
Caro's acid 过一硫酸 $H_2SO_3[O_2]; H_2SO_5$
carotane 胡萝卜烷
carotene (=carotin) 胡萝卜素*；叶红素
carotene epoxide 环氧胡萝卜素
β-carotene oxide (=mutatochrome) β-胡萝卜素氧化物
carotenoid 类胡萝卜素；类叶红素
carotenol 胡萝卜醇；叶黄素
β-carotenone β-胡萝卜酮
carotin (=carotene) 胡萝卜素；叶红素
carotol 胡萝卜醇
carpaine 番木瓜碱
carpenter's glue 木工胶
carpesia-lactone 天明精内酯
carpet fiber 地毯纤维(黏胶)
carpet lining 衬毯纸
carpet shampoo 地毯香波
carpet wool 地毯毛
carpincho ①水猪皮②水猪皮手套革
carp oil 鲤油
carposide (=caricin) 番木瓜苷
carpyrinic acid 鲤鱼酸
carqueja oil 酒神菊油
carquejol 酒神菊醇
carrageenan (=carragheenan) 角叉菜聚糖；卡拉胶
carrageen gum 鹿角菜胶；卡拉胶
carrageenin 角叉(菜)胶；鹿角(菜)精宁
carragenate 角叉(菜)酸盐；鹿角(菜)酸盐
carragenine 角叉(菜)苷；鹿角菜苷
carrag(h)een 角叉菜；鹿角菜
carrag(h)eenin 角叉菜胶
Carrara glass 卡拉拉玻璃〔一种结构玻璃〕
Carrara marble 卡拉拉白大理石
Carrara porcelain ①卡拉拉瓷〔伯利安瓷的别名〕②素烧〔似白大理石〕
carratene 胡萝卜素
carrene 二氯甲烷 CH_2Cl_2
car retort 车辆干馏釜
carriage grease 马车脂；客车脂
carried-down percentage 载带百分率
carrier (=support) 载体*；载运剂
carrier addition method 载体添加法；逆稀释法
carrier automatic painting machine 载漆自动涂漆机
carrier coprecipitation 载体共沉淀*

carrier displacement　载体置换
carrier distillation　载体蒸馏(法)
carrier distillation method　载体分馏法
carrier dyeing　载体染色
carrier effect　载体效应
carrier fibre　伴纺纤维；载体纤维
carrier free　无载体*
carrier free electrophoresis　无载体电泳(法)
carrier free separation　无载体分离
carrier-free tracer　无载体(同位素)示踪物
carrier gas　载气*
carrier gas diffusivity　载气扩散率
carrier gas flow　载气流
carrier gas inversion　载气逆流；载气反吹
carrier gas regulation　载气调节
carrier gas separator　载气分离器
carrier gas stream　载气流
carrier gas supply line　载气供应管线
carrier gas tank　载气罐
carrier liquid　载液
carrier medium　①载色剂〔颜料〕②载体〔催化剂〕
carrier of oxygen　氧载体
carrier-pellet　催化剂载体片
carrier pin　嵌件定件销
carrier plate　载板
carrier precipitation　载体沉淀*
carrier precipitation agent　载带沉淀剂
carrier psychrometric chart　载体湿度图
carrier roll　托滚〔运输带〕
carrier stream　载流
carroerprecipitation　载体沉淀(作用)
carrollite (=carolite)　硫铜钴矿
carrot　胡萝卜
carrotene　胡萝卜素
carrotenoid　类胡萝卜素
carrotenone　胡萝卜素酮
carrotin (=carotene)　胡萝卜素
carrot seed oil　胡萝卜籽油
carry gas cleanser　载气净化器
carrying agent　载体
carrying capacity　①负荷能力；载重量②耐压力〔泡沫胶垫〕
carry over　携带；带出〔蒸气携带液体；轻馏分携带重馏分〕
carry-over analysis　残留分析
car sampling　槽车取样
carstal　卡斯醇；胡萝卜萜醇
carsting　铸件

carstone　砂铁岩
carte paper　地图纸
Carter lead　卡特法铅白〔碱式碳酸铅〕
Carter process　卡特法〔铅白颜料的制法之一〕
cartesian diver　浮沉子
carthamidin　红花定
carthamin　红花素
carthamus oil　红花油
carthamus yellow (=safflower yellow)　红花黄色素
carton ink　纸板墨水
cartoon　①草图；底图②卡通(片)；动画片
cartridge　①(萃取柱)柱体②筒③药筒④(照相)软片卷⑤(反应堆)燃料管；释热元件
cartridge collector　筒式收集器；筒式捕集器
cartridge element　放热元件
cartridge filter　筒式过滤器；过滤筒
cartridge heater　筒形加热器
cartridge paper　火药纸
carubinose　卡如宾糖；δ甘露糖
carvacrol (=isothymol; 2-methyl-5-isopropylphenol)　异百里香酚；香芹酚　$(CH_3)(C_3H_7)C_6H_3OH$
carvacrol-carboxylic acid　香芹酚甲酸
carvacrol methyl ether　香芹酚甲醚
carvacrotinic acid　香芹酚亭酸
carvacrotinic aldehyde　香芹酚亭醛
carvacryl　香芹基
carvacryl acetate　乙酸香芹酯
carvacrylamine (=2-amino-p-cymene)　香芹胺　$H_2NC_6H_3(CH_3)C_3H_7$
carvacryl ethyl ether　香芹基乙基醚
carvedilol　卡维地洛〔药〕
carved lacquer cabinet　刻漆柜〔大漆〕
carvene (=d-limonene)　香芹烯；d-苧烯
carvene oxide　香芹烯氧化物
carvenol　香芹烯醇
carvenone (=3-p-menthen-2-one)　香芹烯酮　$C_{10}H_{16}O$
carveol　香芹醇；葛缕醇
carvestrene　香芹萜
carveterpin (=sylveterpin)　枞萜二醇；香芹萜二醇
carvol (=d-carvone)　香片酮　$C_{10}H_{14}O$
carvomenthane　香芹烷
carvomenthenal　香芹烯醛
carvomenthene　香芹烯　$C_{10}H_{18}$
carvomenthenol　香芹烯醇
carvomenthol　香芹醇
carvomenthone　香芹酮
carvone　香芹酮　$C_{10}H_{14}O$
carvotanacetone　香芹鞣酮

carvoxime 香芹(酮)肟 $C_{10}H_{14}=NOH$
carvyl acetate 乙酸香芹酯
carvyl propionate 丙酸香芹酯
caryin 山核桃碱
caryl 莒烷基 $C_{10}H_{17}$
carylamine 莒烷胺 $C_{10}H_{17}NH_2$
caryomycin 核霉素
caryophyllane 石竹烷；丁香烷
caryophyllanone 石竹酮
caryophyllene 石竹烯 $C_{15}H_{24}$
caryophyllene alcohol 石竹烯醇
caryophyllene alcohol acetate 石竹烯醇乙酸酯
caryophyllene monoepoxide (单)环氧石竹烯
caryophyllene oxide 石竹烯氧化物
caryophyllenic acid 石竹烯酸
caryophyllenol 石竹烯醇 $C_{15}H_{24}O$
caryophyllin (=oleanolic acid) 石竹素 $C_{30}H_{48}O_3$
caryophyllus aromaticus 丁(子)香
Cary recording spectrophotometer 凯利记录式分光光度计
carzinocidin (=carcinocidin) 消癌菌素
carzinomycin (=carcinomycin) 癌霉素
carzinophillin 嗜癌素
carzinostatin 制癌菌素
casamino acid 酪蛋白氨基酸；酪蛋白水解物
cascade ①(轴流式通风机)叶栅级联②串连
cascade aeration 多级跌水曝气法
cascade aerator 阶式曝气装置
cascade alkylation 阶式烷基化
cascade battery ①阶式蒸浓装置②级联电池组
cascade centripetal 多段冲击采样器
cascade chromatography 级联色谱法
cascade concentration plant 阶式蒸浓装置
cascade concentrator 阶式蒸浓器
cascade cycle 阶式循环
cascade drip cooler 阶式水淋冷却器
cascade dryer 阶式干燥器；多段式干燥器
cascade dyeing 冲流染色
cascade effect 级联效应；阶梯效应
cascade-evaporation 串级-蒸发
cascade experiment 串级实验；级联实验
cascade extraction train 级联型萃取装置
cascade flow ①阶式流动②叶栅流动
cascade heat exchanger 阶式换热器
cascade impactor 阶式碰撞取样器
cascade jet impactor 多级喷嘴碰撞器
cascade mass spectrometer 级联质谱计；串级质谱计
cascade oiling 环给油；油杯润滑
cascade polymer 层叠聚合物
cascade press 阶式压榨
cascade process 级联过程
cascade reaction 阶式反应器
cascade residual 串级余核
cascade sampler 阶式取样器；级联取样器
cascade stretch 分级拉伸；阶式拉伸
cascade system 阶式系统；串连系统
cascade theory 级联理论
cascade-tray fractionating column 阶式盘分馏塔
cascade type cooler 阶式冷却器
cascade washer 阶式洗涤槽
cascara ①药鼠李皮；波希鼠李皮②药鼠李
cascara sagrada 药鼠李〔1.药鼠李皮；2.药鼠李〕
cascarilla 卡藜
cascarilladiene 卡藜二烯
cascarilla oil 卡藜油；香苦木油
cascarillic acid 卡藜酸
cascarilline 卡藜灵
cascarillone 卡藜酮
cascarin 药鼠李苷
case ①箱②盒；套③情形④事件；例
caseanic acid 酪烷酸
case hardening 表面硬化；表层硬化；表面淬火
case hardening carbon black 表面硬结炭黑
case-hardening (drying) 表现硬化
case hardening salt 表面硬化盐；淬火盐
case-hardening steel 表面硬化钢；表层硬化钢
casein-acrylonitrile graft copolymer fibre 酪蛋白-丙烯腈接枝共聚纤维
casein adhesive 干酪素胶
casein agar 干酪素琼脂
casein binder 酪蛋白黏结剂；(干)酪素黏结剂(料)
casein cement 酪素黏合剂
casein dyspeptone 乳酪
casein emulsion 酪素乳液〔俗〕
casein fiber 酪蛋白纤维
casein finish 乳酪涂饰剂
casein-formaldehyde resin 酪蛋白甲醛树脂
casein glue 酪蛋白胶
casein paint 酪蛋白漆
casein paste paint 酪蛋白厚漆
casein plastics (CS) 酪素塑料；酪蛋白塑料
casein primer 酪蛋白漆灰地〔大漆〕
casein protein fibre 酪蛋白质纤维
casein-soybean glue 酪素-大豆蛋白胶
case iron retort 生铁甑
case leather 箱革；箱皮

casement ①玻璃窗扇；竖铰链窗扇②孔模
case packer 装箱机
case sealer 封箱机
case wood 箱材
casher box 打口台
cashew 檟如树
cashew apple oil (=cashew nut oil) 檟如(坚)果油
cashew nut 檟如坚果
cashew-nut oil 檟如坚果油
cashew nut shell liquid 檟如坚果壳液
cashew nut shell oil 腰果壳油；檟如坚果壳油
cashew water coat 腰果油水性防锈底漆
cashmere 开士米；开士米织物
cashmerelike 类开士米
casimiroedine 卡西定
casing ①套②汽缸
casing glass 镶色玻璃
casing-head gas 油井气；油田气
casing head gasoline (从油井气中回收的)天然汽油
casing-head plant 油田气处理装置
casing ply 骨架层
casing ring 壳环〔泵〕
casings for pipe-lines 管线套
cask 桶
cask flask (储运放射性物质的)屏蔽容器；屏蔽罐
cask unloading pool (装燃料)容器卸料池
cassaidine 二氢围涎皮次碱
cassaine 围涎皮次碱
cassamine 围涎皮胺
cassava meal (=farinha) 木薯粉
cassawa 木薯
cassawa starch 木薯粉
Cassel brown (=Vandyke brown) 天然棕土；铁棕
Cassel green 卡塞尔绿〔锰酸钡 $BaMnO_4$〕
Cassella's acid 卡塞拉酸；2-萘酚-7-磺酸
Casselmann green 碱式硫酸铜绿
cassel yellow 氯化铅黄 $PbCl_2 \cdot 7PbO$〔黄色颜料〕
casserole 勺皿
Casser tester 凯塞尔试验机〔铜催化乙酸盐雾试验机的简称〕
cassette mutagenesis 盒式突变
cassia ①山扁豆②肉桂皮
cassia bark 山扁豆皮
Cassia flask 卡氏烧瓶
cassia-juice coloring 挂浆上色
cassia leaf oil 肉桂叶油
cassia leaf powder 肉桂叶粉
cassia lignea (=cassia bark) 山扁豆皮
cassia oil ①山扁豆油②肉桂油
cassia oleoresin 玉桂油树脂
cassia scrapped 桂芯
cassic acid 山扁豆酸
cassie 金合欢
cassie absolute 金合欢净油
cassie ketone 乙酸大茴香酯
cassie oil 金合欢油
cassilvsidin 决明皮素
cassilysin 决明皮溶素
cassiterite 锡石
Casson equation 卡森方程(式)〔黏度分布关系式〕
cast 铸塑*
castability (可)浇铸性
castable 可铸塑的；可流延的
cast acrylic sheet 铸制丙烯片(板材)
castalagin 栗木鞣花素
castalin 栗木素
cast alloy 铸合金；铸齐
cast alloy iron 合金铸铁
castanea 齿栗叶
castanin 栗宁〔存在于栗中〕
castanite 褐铁矾
castavaloninic acid 栗酸
cast brass 生黄铜
cast charge 发射剂〔固体火箭燃料〕
cast-coated paper 高光泽印刷纸；涂布美术纸
cast coating 浇注涂装
cast concrete 模铸混凝土；混凝土铸件
castelamarin 堡树苦素
castellated knife 喋形刮刀
cast film (=casting film) ①流延薄膜②平挤薄膜
cast film extrusion (=chill roll extrusion) 平挤薄膜挤塑；骤冷辊挤塑法
cast film technique 铸膜法
cast gate ①浇口；铸口②流道
cast glass 铸玻璃
castile soap 橄榄油皂；马赛皂
castilloa 美胶树
castilloa gum 美洲橡胶树胶
castiloa rubber 美胶树橡胶
cast-in 镶铸的；浇合的；铸入式的
castine 牡荆碱
casting ①铸品；铸件②浇铸；铸型；铸造
casting belt (流延机)流延(钢)带
casting box 砂箱
casting cycle ①铸塑周期②流延周期
casting film (=cast film) ①流延薄膜②平挤薄膜

casting foundry   铸件
casting lip   (玻璃纸)吐胶口狭缝；流延头狭缝
casting machine   流延机；铸塑机
casting mould   铸型
casting paper   流延用纸
casting pig   铸造用生铁
casting plaster (=calcined gypsum)   烧(煅)石膏；熟石膏
casting polymerization   铸塑聚合
casting resin   铸模树脂
casting roll   流延辊；定型辊
casting syrup   铸塑用料浆
casting temperature   浇铸温度
casting urethane   铸塑聚氨酯
cast iron   生铁；铸铁
cast-iron cold welding   铸铁冷焊
cast iron of second melting   二次熔炼生铁
cast iron pan   生铁锅
cast iron pipe   生铁管
cast iron shot   铸铁球；铸铁砂(丸)
cast iron sinker   生铁沉锤
cast lead pan   铸铅盘
castle manipulator   "城堡"式机械手；高架式机械手
cast molding   铸塑成型
Castner-Kellner cell   卡斯讷-凯耳讷电池
cast of oil   油的色泽〔反射色〕；油的荧光
cast on coating film   漆膜泛荧光
castoramine   海狸胺
castor beans   蓖麻籽
castoreum (=castor)   ①海狸②海狸香
castoreum oil   海狸香油
castoric acid   海狸香酸
castorin   海狸香素
castor machine oil   铝皂稠化机械油
castor oil   蓖麻油
castor oil acid (=ricinoleic acid)   蓖麻油酸〔俗〕；蓖麻醇酸
castor oil isocyanate adduct   蓖麻油异氰酸酯加成物
castor pomace   蓖麻油渣
castor seed   蓖麻籽
cast plastic   铸塑塑料
cast plate   铸制板
cast polymer   铸型用聚合物
cast polymerization   铸塑聚合；浇铸聚合
cast polymerized   浇铸聚合的
cast product   铸型制品
cast resin   铸塑树脂
cast scrap   旧铸物
cast steel   铸钢；浇钢

cast steel shot   铸钢球；铸钢砂(丸)
casumin   卡苏菌素
casurin   木麻黄素
cata-   ①萘状环的 1,7 位②向下；在下
catabolic action   分解作用
catabolic reaction   分解(代谢)反应；降解(代谢)反应
catabolin (=catabolite)   分解(代谢)产物；降解(代谢)产物
catabolism   分解代谢；降解代谢
catabolite (=catabolin)   分解(代谢)产物；降解(代谢)产物
cataclastic structure   碎裂结构
cata-condensed hydrocarbon (CCH)   背缩烃类
catalase   过氧化氢酶；触酶
catalase test   过氧化氢酶试验
catalasometer   催化酶活度计
catalimetric titration (=catalytic titration)   催化滴定
catalin   铸塑酚醛塑料
catalog paper   目录纸
catalpic acid   梓油酸 $C_{18}H_{30}O_2$
catalposide   梓苷；梓实糖苷
catalymetric micro determination   催化微量测定(法)
catalysagen   催化剂原；触媒原
catalysant   被催化物
catalysis   催化*
catalyst   催化剂*
catalyst abrasion   催化剂磨蚀
catalyst accelerator   催化加速剂
catalyst activation   催化剂活化
catalyst active center   催化剂活性中心
catalyst activity   催化剂活性
catalyst adjunct   催化剂助剂
catalyst basket   催化剂筐
catalyst bead system   颗粒催化剂体系
catalyst bed   催化剂床
catalyst bulk density   催化剂堆密度
catalyst carrier   催化剂载体
catalyst case   催化剂室；反应器
catalyst chamber   催化剂室；反应器
catalyst charge   催化剂装料
catalyst circulation   催化剂循环
catalyst circulation rate   催化剂循环速度
catalyst complex   催化剂络合物
catalyst component   催化剂组分
catalyst-containing tubular member   催化剂管件
catalyst cooler   催化剂冷却器
catalyst cracker   催化裂化设备
catalyst damage (=catalyst poisoning)   催化剂中毒*
catalyst deactivation   催化剂失活；催化剂钝化
catalyst density under pressure   加压下催化剂密度

catalyst deterioration (=catalyst poisoning)　催化剂中毒
catalyst dispenser　（聚氨酯浇注发泡机）催化剂分配器
catalyst distillation　催化蒸馏
catalyst downcomer　催化剂下导管
catalyst down-flow principle　催化剂下流原理
catalyst efficiency　催化剂效率
catalyst elevator　催化剂升举机
catalyst fines　催化剂粉末
catalyst flow　催化剂流量
catalyst flow control　催化剂流动控制
catalyst flow plate　催化剂流动板
catalyst for shut-down　停工用催化剂
catalyst fouling　催化剂污损；催化剂结垢；催化剂失活
catalyst furnace　催化(剂)炉；接触炉
catalyst gauze　催化(剂)网
catalyst haze　催化剂致浑
catalyst head　催化剂分配头
catalyst hopper　催化剂漏斗
catalyst induction procedure　催化剂引入程序
catalyst level　催化剂用量
catalyst life　催化剂寿命
catalyst line　催化剂管线
catalyst make-up　催化剂补充
catalyst make-up tank　催化剂配制槽
catalyst-oil disengagement　催化剂-油分离
catalyst-oil ratio　催化剂-油比率
catalyst-oil suspension　催化剂油悬浮体
catalyst-packed column　催化剂填充塔
catalyst particle　催化剂粒子
catalyst particle size determination　催化剂粒径(度)测定
catalyst poison　催化剂毒物
catalyst poisoning　催化剂中毒*
catalyst porosity　催化剂孔隙率；催化剂孔隙度
catalyst precursor composition　催化剂前体成分
catalyst pretreatment　催化剂预处理
catalyst promoter　助催化剂
catalyst ratio (=catalyst-oil ratio)　催化剂-油比率
catalyst reaction　催化反应
catalyst reactivation (=catalyst regeneration)　催化剂再生
catalyst recovery (=catalyst regeneration)　催化剂再生
catalyst reducer　催化剂还原器
catalyst reduction (=catalyst regeneration)　催化剂再生
catalyst regeneration (=catalyst reduction; catalyst reactivation; catalyst recovery)　催化剂再生
catalyst residence time　催化剂驻留时间〔移动或流动催化剂在反应器中逗留的时间〕
catalyst screening　催化剂筛选
catalyst scrubber column　催化剂洗涤塔
catalyst selectivity　催化剂选择性
catalyst separating　催化剂分离〔与产物分离〕
catalyst settler　催化剂沉降器
catalyst slurry　催化剂浆〔催化剂与原料或反应溶剂的浆状混合物〕
catalyst space　催化剂空间
catalyst spraying　催化喷涂
catalyst storage bin　催化剂储仓
catalyst stream　催化剂流
catalyst support　催化剂载体
catalyst surface　催化剂表面
catalyst susceptibility　催化剂感受性
catalyst system　催化剂体系
catalyst-to-additive ratio　催化剂对添加剂比
catalyst-to-charge ratio (=catalyst-to-crude ratio)　催化剂对进料比；催化剂对原料比
catalyst-to-crude ratio (=catalyst-to-charge ratio)　催化剂对原料比；催化剂对进料比
catalyst tolerance to oxidation　催化剂抗氧化性
catalyst-to-oil ratio (=catalyst to-charge ratio)　催化剂对油料比；催化剂对进料比
catalyst tower　催化剂塔
catalyst transfer line　催化剂输送线
catalyst unit　催化单元
catalyst up-flow principle　催化剂上流原理
catalyst volume　催化剂体积
catalyst wax　催化剂表面沉积蜡
catalyst weight　催化剂重量
catalytic　催化的
catalytic action　催化作用
catalytic active site　催化活性位*
catalytic activity　催化活性*
catalytic additive　催化添加剂
catalytic afterburner　催化后燃器；催化后燃室
catalytic agent　催化剂
catalytic air-steam reformer　空气-蒸汽催化转化装置；空气-蒸汽催化重整炉
catalytical　催化的
catalytical discoloring spectrophotometry　催化退色分光光度法
catalytic alkylation　催化烃化
catalytically-blown asphalt　催化吹制沥青
catalytically cracked gasoline　催化裂化汽油
catalytically inert　催化惰性的
catalytic analysis　催化分析
catalytic antibody　催化抗体
catalytic apparatus　催化仪器
catalytic asphalt　催化(氧化)沥青

catalytic capability 催化能力
catalytic carrier 催化载体
catalytic chromatography 催化色谱(法)
catalytic cleaner 催化清洁器
catalytic coal gasification 煤催化气化*
catalytic coal liquefaction 煤催化液化*
catalytic coking 催化焦化
catalytic colorimetry 催化比色法*
catalytic combustion 催化燃烧
catalytic composite 复合催化剂
catalytic converter 催化转化器
catalytic cooxidant 辅助氧化催化剂
catalytic cracker 催化裂解炉；催化裂化装置
catalytic cracking 催化裂化*
catalytic curing 催化固化
catalytic current 催化电流*
catalytic cycle oil 催化(裂化)循环油
catalytic cycle stock 催化(裂化)循环油料
catalytic cyclization 催化环化
catalytic decomposition 催化分解
catalytic decomposition effect 催化分解效应
catalytic degradation 催化解聚；催化降解
catalytic dehydration 催化脱水作用
catalytic dehydrogenation 催化脱氢*
catalytic desulfurhydrogenation 催化加氢脱硫*
catalytic desulfurization 催化脱硫
catalytic dewaxing 催化脱蜡
catalytic disproportionation 催化歧化(作用)
catalytic effect 催化效果
catalytic ester exchanger 催化酯交换装置
catalytic esterification 催化酯化(作用)
catalytic exchange 催化交换
catalytic exhaust purifier 废气催化净化器
catalytic fluorimetry 催化荧光法*
catalytic force 催化力
catalytic functional polymer 催化功能高分子
catalytic fusion 催化熔融
catalytic (gas) chromatography 催化(气相)色谱法
catalytic gas oil 催化(裂化)瓦斯油
catalytic gasoline 催化裂化汽油
catalytic gas reforming 催化气体重整〔低热量变高热量〕
catalytic halogenation 催化卤化(作用)
catalytic hardener 催化硬化剂
catalytic heater 催化加热器
catalytic hydration 催化水合作用
catalytic hydrocracking 催化加氢裂化*
catalytic hydrofinishing 催化加氢精制
catalytic hydrogenation 催化氢化*

catalytic hydrogen wave 催化氢波*
catalytic hydroprocessing 催化加氢处理
catalytic hydrotractor 催化加氢处理
catalytic ignition 催化点火
catalytic ionization detector 催化电离检测器
catalytic isomerisation 催化异构化
catalytic kinetic photometry 催化动力学光度法
catalytic metal ion 催化金属离子
catalytic methylation 催化甲基化(作用)
catalytic nonselective polymerization 催化非选择性聚合(作用)
catalytic oxidation 催化氧化(作用)；接触氧化(作用)
catalytic oxygen wave 催化氧(化)波
catalytic plant 催化裂化工厂
catalytic plasticizer 催化增塑剂
catalytic poison 催化毒(物)
catalytic polyforming 催化聚合重整
catalytic polymer 催化聚合物
catalytic polymerization 催化聚合
catalytic power 催化本领
catalytic process 催化过程
catalytic promoter 助催化剂
catalytic protein wave 催化蛋白波
catalytic purification 催化净化
catalytic reaction 催化反应
catalytic reactor 催化反应器
catalytic reclaiming process 催化再生法
catalytic recombiner 催化复合器
catalytic reduction 催化还原
catalytic reduction wave 催化还原波
catalytic refining 催化精制
catalytic reforming 催化重整*
catalytic reforming furnace 催化重整炉
catalytic reforming unit 催化重整装置
catalytic reverse shift reaction 催化逆移反应
catalytic selectivity 催化选择性*
catalytic site 催化部位
catalytic spectrophotometry 催化分光光度法
catalytic subunit 催化亚单位；催化亚基
catalytic surface 催化表面
catalytic thermal treatment 催化热处理；催化热加工
catalytic thermometric titration 催化热滴定
catalytic thermometric titrimetry 催化热滴定法
catalytic titration (=catalimetric titration) 催化滴定
catalytic tower 催化反应塔
catalytic transfer hydrogenation 催化转移氢化
catalytic unit 催化装置；催化单元
catalytic vaporizer 催化汽化器

catalytic wave 催化波*
catalyzator 催化剂
catalyzed alcoholysis 催化醇解
catalyzed chemiluminescence 催化化学发光
catalyzed coating 催化(固化)型涂料
catalyzed conversion 催化转化
catalyzed degradation 催化降解(反应)
catalyzed lacquer 催化固化漆
catalyzed oil 催化(聚合)油
catalyzed oxidation 催化氧化
catalyzed polymerization 催化聚合(作用)
catalyzed reaction 催化反应
catalyzed shock tube reaction 催化冲击波管反应〔制乙炔工艺〕
catalyzed sorption 催化吸着
catalyzer (=catalyst) 催化剂
catalyzer poison 催化(剂)毒物
catalyzing 催化作用
catalyzing deblocking 催化解封闭
catalyzing enzyme 催化酶
cat and dog production 多规格生产；杂乱生产
cataphoresis (=electrophoresis) 电泳；电渗
cataphoretic 阳离子电泳的
cataphoretic coating 电泳涂层
cataphoretic migration speed 阳离子电泳迁移速度
cataphoretic mobility 阳离子电泳迁移率
catapleiite 钠锆石
catapult strip 弹弓橡皮筋
Catarole process 卡泰洛法〔由 catalytic, aromatic, olefins 三字头组成的复合词，在同一装置中既生产含烯烃的裂解气，又生产芳烃含量高的液体产品〕
catastrophe 突变
catastrophic creep 灾变蠕变；致毁蠕变
catastrophic degradation ①突发性老化；突发性降解②灾变性老化；灾变性降解
catastrophic exposure 毁坏性暴露
catastrophic failure 灾害性破坏
catastrophic oxidation 破(毁)坏性氧化
catastrophic propagation (漆膜开裂等)毁坏性扩散
catathermometer 干湿球温度计
catch 捕集*；接受器
catch-all ①截液器；分沫器②总受器
catch-all steam separator 截液器；分沫器
catch basin 沉水池；收集盘〔石油厂的〕
catch box 截液器
catch crop 间作；间作物
catcher (接)受器
catch foil 捕集箔(片)*

catching 捕集；俘获
catch of used oil 废油收集器
catch plate 导夹盘
catch tank 预滤器；凝汽管；收集槽；捕集槽
catch tray 收集盘
catch-up ①(生产力)恢复；(技术水平)赶超②(平版印刷)局部未印上；脱印
cat-cracker 催化裂解装置
catechin 儿茶酸；儿茶素
catechinic tanning materials 儿茶属鞣料
catechol 儿茶酚；邻苯二酚
catecholamin(e) 儿茶酚胺
catecholborane 儿茶酚硼烷
catechol-dibenzothiazy disulfide 邻苯二酚二硫代二苯并噻唑
catechol oxidase 儿茶酚氧化酶
catechol tan 儿茶类鞣料
catechol violet (=pyrocatechol violet) 邻苯二酚紫
catechotannin 儿茶单宁；儿茶鞣质
catechu 儿茶
catechuic acid 儿茶酸
catechu mimosa 儿茶
catechutannic acid 儿茶单宁酸
catelectrode 阴极
catena- 耦合；联结；链条
catenane 索烃*
catenane polymer 套环聚合物
catenary continuous vulcanizer 弯管式连续硫化机
caterpillar 接取装置；引出装置
caterpillar band 履带
caterpillar take-off 履带式接取装置
caterpillar test engine 开特皮拉试验引擎
catforming 催化重整；铂重整；催化转化法
catforming process 催化重整过程〔大西洋公司含硫里格罗因的催化芳构化过程〕
cat gasoline 催化(裂化)汽油
catharahthine 长春碱；长春花碱
catharometer (=katharometer) 热导计；导热析气计；热导池
cathartic ①泻药②泻的
cathartic acid 泻酸
cathartin 泻素
cathepsin 组织蛋白酶
catheter 导管
catheterization 导管插入法
cathetometer 测高仪*；高差计
cathidine (=cathine) 阿(拉伯)茶碱
cathine (=cathidine) 阿(拉伯)茶碱

cathinine 阿(拉伯)茶宁
cathocin (=novobiocin) 新生霉素
cathode 阴极
cathode box 阴极箱
cathode chamber (=cathode compartment; cathode department) 阴极区；阴极室
cathode compartment (=cathode chamber) 阴极区；阴极室
cathode coupled D.C amplifier 阴极耦合直流放大器
cathode dark current 阴极暗电流
cathode density 阴极密度
cathode department (=cathode chamber) 阴极区
cathode deposit 阴极淀积
cathode drainage 阴极漏电
cathode drop 阴极电位降
cathode excitation 阴极激发
cathode fall 阴极势降
cathode fluorescence 阴极荧光
cathode fluorescence image 阴极荧光像
cathode follower 阴极输出器
cathode glow 阴极辉光；阴极电辉
cathode layer arc method 阴极层弧光法
cathode layer enrichment 阴极层富集
cathode layer enrichment method 阴极区富集法*
cathode layer method 阴极层弧光法〔用于金属分析〕
cathode liquor 阴极电解液
cathode luminescence 阴极发光
cathode material 阴极材料
cathode particle 阴极粒子
cathode potential 阴极电势
cathode ray 阴极射线
cathode ray beam 阴极射线束
cathode ray excited emission spectroscopic analysis 阴极射线激发发射光谱分析
cathode ray oscillographic polarography 阴极射线示波极谱法
cathode ray oscilloscope 阴极射线示波器
cathode ray polarograph 阴极射线极谱仪*
cathode ray polarography 阴极射线极谱法
cathode ray tube 阴极射线管
cathode space 阴极区
cathode spot 阴极斑点
cathode sputtering 阴极溅射
cathode sputtering process 阴极溅射镀膜法；阴极溅射沉积法〔真空镀膜法之一〕
cathode stripping voltammetry 阴极溶出伏安法
cathodic 阴极的
cathodic activity 阴极活性
cathodically polarized platinum electrode 阴极极化铂电极
cathodic coating 阴极(保护)膜
cathodic control 阴极控制
cathodic corrosion 阴极腐蚀
cathodic current 阴极电流*
cathodic delamination 阴极脱层(剥离)；阴极脱胶(解黏合)作用
cathodic depolarizer 阴极去极剂
cathodic deposition 阴极淀积
cathodic disbonding 阴极解黏合作用；阴极分离
cathodic disbondment 阴极脱胶(解黏合)作用；阴极脱层
cathodic disbondment test 阴极脱胶试验；阴极脱层试验
cathodic electrodeposition 阴极电沉积
cathodic etcher 阴极蚀刻器
cathodic inhibitor 阴极抑制剂
cathodic polarization 阴极极化*
cathodic process 阴极(反应)过程
cathodic protection 阴极保护*；阴极防腐
cathodic protection parasites 妨碍阴极保护的物质
cathodic protective primer 阴极保护底漆
cathodic protector 阴极保护器
cathodic reaction 阴极反应
cathodic reduction 阴极还原
cathodic reduction dimerization process 阴极还原二聚法
cathodic reduction wave 阴极还原波；阴极波
cathodic sputtering 阴极溅散
cathodic sputtering chamber 阴极溅射室
cathodic stripping 阴极溶出(法)*
cathodic stripping voltammetry 阴极溶出伏安法
cathodic synthesis 阴极合成
cathodic wave 阴极波
cathodoluminescence 阴极射线发光*
cathodoluminographic method 阴极射线发光光谱分析法
cathodoluminography 阴极射线发光谱图分析法
catholic sputtering 阴极溅射法
catholyte 阴极电解液
catholyte compartment 阴极液室
cathomycin (=novobiocin) 新霉素；新生霉素
cation 阳离子；正离子
cation acid (=cationic acid) 阳离子酸
cation active agent 阳离子型活性剂
cation active auxiliary 阳离子活性助剂
cation active high-molecular electrolyte 阳离子活性高分子电解质
cation-adsorption 阳离子吸附
cation-anion interference 阳离子-阴离子干扰
cationation 阳离子化

cation-cation interference 阳离子-阳离子干扰
cation-dyeable 阳离子染料可染的
cation exchange 阳离子交换
cation exchange capacity 阳离子交换量；碱交换量
cation exchange cellulose 阳离子交换纤维素
cation exchange chromatography 阳离子交换色谱(法)
cation exchange electrode 阳离子交换电极
cation exchange liquid 阳离子交换液
cation exchange membrane 阳离子交换膜
cation exchange packing 阳离子交换填充物
cation exchange paper 阳离子交换纸
cation exchanger 阳离子交换剂*
cation exchange resin 阳离子交换树脂*
cation glass electrode 阳离子玻璃电极
cationic 阳离子的
cationic acid (=cation acid) 阳离子酸
cationic activation 阳离子活化(作用)
cationic additive 阳离子添加剂
cationic auxiliary 阳离子助剂
cationic binder 阳离子黏合剂
cationic brightener 阳离子增白剂
cationic carbocyanine 阳离子碳菁
cationic catalyst 正离子催化剂*
cationic collector 阳离子收集极
cationic conditioning agent 阳离子调节剂
cationic copolymerization 阳离子共聚合(作用)
cationic cyclopolymerization 阳离子环化聚合
cationic detergent 阳离子洗涤剂
cationic dispersing agent 阳离子型分散剂
cationic dispersion 阳离子分散体(液)；阳离子分散作用
cationic dye 阳离子染料
cationic dyeable variant 阳离子染料可染型改性体
cationic dyestuff 阳离子染料
cationic emulsifier 阳离子乳化剂
cationic exchange column 阳离子交换柱(塔)
cationic exchange filter 阳离子交换滤水器
cationic exchange membrane 阳离子交换膜
cationic fat liquor 阳离子(型)加油(脂)乳液
cationic flocculant 阳离子絮凝剂
cationic germicide 阳离子杀菌剂
cationic initiator 正离子引发剂*
cationic latex 阳离子胶乳
cationic layer 阳离子层
cationic monomer 阳离子单体
cationic organic compound 阳离子型有机化合物
cationic pigment dispersant 阳离子型颜料分散剂
cationic polyelectrolyte 阳离子型聚电解质
cationic polymerization 正离子聚合*

cationic polyurethane ionomer 阳离子型聚氨酯离聚物
cationic retarding agent 阳离子阻染剂
cationic ring-opening homopolymerization 阳离子(催化)开环均聚合(反应)
cationics (=cation surface active agent) 阳离子剂；阳离子表面活性剂
cationic softener 阳离子软化剂
cationic softening agent 阳离子软化剂
cationic substance 阳离子物质
cationic surface-active agent 阳离子表面活性剂
cationic surfactant 阳离子型表面活性剂*
cationic tertiary amine 阳离子叔胺
cationic thiazine dyestuff 阳离子噻嗪颜料
cationic titrant 阳离子滴定剂
cationic titration 阳离子滴定法
cation interchange 阳离子交换
cation interfacial active agent 阳离子界(表)面活性剂
cationite 阳离子交换剂
cation masking agent 阳离子隐蔽剂
cationogen 阳离子原；阳离子发生物
cationogenic initiator 阳离子引发剂
cationoid 类阳离子
cationoid activity 类阳离子活度
cationoid polymerization 类阳离子聚合
cationoid reaction 阳离子反应；亲电(子)反应
cationoid reagent 类阳离子试剂
cationoid substitution 类阳离子取代
cationotropic rearrangement 阳离子转移重排
cationotropy 阳离子移变(现象)
cation permselective diaphragm 阳离子选择性渗透膜
cation polymer 阳离子聚合物
cation radical 阳离子自由基
cation resin (=cation exchange resin) 阳离子交换树脂
cation-selective electrode 阳离子选择电极
cation-selective glass membrane 阳离子选择性玻璃膜
cation surface active agent (=cationics) 阳离子剂；阳离子表面活性剂
cativic acid 猫茯酸
cativo gum 猫茯胶
catlinite (=pipestone) 烟斗泥
catmint 荆芥
catnip oil 荆芥油；樟脑草油
catopleite 卡托矿石
cat reformer 催化重整器
cats 阳离子表面活性剂
cat salt 猫盐〔自盐卤中提出的精盐〕
cat's eye 猫眼〔薄膜缺陷〕
cat skin 猫皮

cats product 阳离子表面活性剂
cattail 香蒲
CAT titration 氯胺 T 滴定法
cattle hide 牛皮
catwalk (油罐顶上的窄狭)人行站桥
caucha 晒台〔干燥智利硝石用的〕
Cauchy formula 柯西公式
Cauchy strain 柯西应变
caul (压胶合板时所用的)垫板;衬板
cauldron 煮皂锅
caulerpa 蕨藻
caulerpacin 蕨藻素
cauliflower polymer 花(椰)菜状聚合物
cauliflower polymerization 花(椰)菜状聚合(反应);卷心菜式聚合(反应)
caulk ①堵(捻)缝;填隙②沉淀
caulk compound 嵌缝胶;腻子胶
caulking 填缝〔用麻丝填塞船缝等较大的隙缝〕;填(大)孔;堵洞
caulking chisel 抿灰凿;捻缝凿;填隙凿
caulking compound 嵌缝填料
caulking gum 密封橡胶;填缝橡胶
caulking gun 嵌缝胶注射器;嵌缝胶枪
caulking tool 填缝工具;嵌缝工具;捻缝凿
caulk joint ①嵌缝黏合②嵌缝
caulk welding 填缝焊接
caulocaline 促成茎素
caulophyllin 葳严仙脂
caulophylline 葳严仙根碱
caulosapogenin 葳严仙配基
caulosaponin 葳严仙皂草苷
caul plate 隔板;钢板
cauprene 聚戊烯橡胶
caustic 苛性的;腐蚀(性)的
caustic alkali 苛性碱
caustic ammonia 苛性氨
caustic ash 苛性苏打灰
caustic baryta 氢氧化钡 $Ba(OH)_2$
caustic bottoms 〔复〕苛性红
caustic brittleness 碱蚀致脆
caustic contact-tower 碱接触塔
caustic corrosion 碱腐蚀
caustic cracking 碱性脆裂;碱性开裂
caustic curve 焦散曲线
caustic drum 苛性(钠用的)鼓
caustic embrittlement 碱蚀致脆;苛性脆化
caustic flaking machine 苛性钠刨片机
caustic fusion 碱性熔炼

caustic hydride process 苛化氢化法〔一种除垢法〕
caustic in flakes 片状烧碱
causticity 苛性;碱度
causticization 苛化*
causticized ash 苛性苏打灰
causticized black-ash liquor 苛性黑灰液
causticizer ①苛化器②苛化剂
causticizing 苛化
causticizing agent 苛化剂
causticizing agitator 苛化搅动器
causticizing tank ①苛(性)化槽②苛化桶
caustic lime 苛性石灰
caustic lime mud 苛性石灰泥浆
caustic liquor 苛性(碱)液
caustic lye 苛性碱液
caustic lye of soda 氢氧化钠液;烧碱液
caustic methanol solution 苛性钠甲醇溶液
caustic mud 石灰泥浆
caustic neutralizer column 碱中和塔
caustic oil of arsenic 三氯化砷 $AsCl_3$
caustic pickle 腐蚀性的稀酸液
caustic pot 苛性罐
caustic potash 苛性钾;氢氧化钾
caustic potash lye 钾碱液
caustic pretreating 碱洗;碱预处理
caustic refining 碱炼
caustic salt 苛性盐
caustic scrubber 碱洗气器
caustic scrubbing 碱洗
caustic settler 碱沉降器
caustic sludge 石灰泥浆
caustic soda 苛性钠;烧碱;氢氧化钠
caustic soda ash 苛性苏打灰
caustic soda liquor 苛性苏打水
caustic soda lye 苛性苏打碱水
caustic soda powder 苛性苏打粉
caustic soda softening 苛性钠软水法
caustic solubility 碱溶性;碱溶度
caustic solution 苛性碱溶液
caustic stained panel 碱污染样板
caustic stained substrate 碱污染底材
caustic sulfide lime liquor 硫化钠石灰浸灰液
caustic-treated gasoline 碱处理汽油
caustic treater 碱处理器
caustic wash (=caustic washing) 碱洗;氢氧化钠洗
caustic washing (=caustic wash) 碱洗;氢氧化钠洗
caustic wash tower 碱洗塔
caustification method 苛化法

caustobioliths〔复〕 可燃性生物岩
caustophytoliths〔复〕 可燃性植物岩
caustozooliths〔复〕 可燃性动物岩
cavatition 空化作用
cave 屏蔽室；洞穴
caverne 洞穴
cavernous ①素烧(瓷)的②多孔的
cavernous dolomite 孔状白云岩
cavicol 佳味酚〔俗〕；对烯丙基苯酚
cavitation 空化；成穴
cavitation attack 穴化侵蚀；空腔腐
cavitation corrosion 空化腐蚀
cavitation damage 空穴缺陷；空化损伤；气蚀损失
cavitation effect ①空化效应②气蚀效应
cavitation erosion 麻蚀；气蚀腐蚀；空隙腐蚀；凹陷蚀
cavitation number ①气蚀(系)数；涡空②气穴数
cavitation pump 水泵气蚀
cavities collapse 空泡破坏；空穴破坏
cavity ①模槽②空腔池③共振腔；谐振腔
cavity block 阴模
cavity depth ①阴模深度②模腔深度
cavity dimension ①阴模尺寸②模腔尺寸
cavity effect 空穴效应；空腔(空化)效应
cavity gate 模腔浇口
cavity half 半阴模
cavity impression 阴模型腔
cavity insert 阴模镶块
cavity-molding 模孔铸型；浇铸成型
cavity pressure 模腔压力
cavity resonance 空腔共振
cavity resonator 空腔共振器
cavity ringdown laser absorption spectroscopy 激光腔内共振衰减吸收光谱法
cavity side (注塑模)阴模侧；模腔边
cavity space 阴模模腔
cavity strength 阴模强度
cavity (vacuole) formation 形成模腔
cavity well 阴模模腔
cavone 蓍酮
cawk ①氧化钡②重晶石；硫酸钡矿石
cayaponine 巴西瓜碱；泻瓜碱
Cayenne linaloe oil 伽罗木油；玫瑰木油
Cayley-Hamilton theorem 凯莱-哈密顿定理
C band C 谱带
^{14}C dating 碳-14 年代测定*
ceanothine 美洲茶碱
Ceara rubber 西拉橡胶；木薯橡胶
cebur balsam 萨布香脂

cedar ①雪松②柏木〔泛指〕
cedar camphor (=cedrol) 柏木脑
cedar gum 雷松胶
cedar leaves oil 雪松叶油
cedarn 雪松的
cedar-nut oil 雪松坚果油
cedar oil 雪松油
cedar soap 雪松皂
cedar wood campnor 柏木脑
cedar (wood) oil 雪松油；柏木油
cedar wood oil alcohol 柏木油醇；柏木醇
cedrane 柏木烷
cedranone 柏木烷酮
cedran oxide 氧化柏木烷
cedrat(e) ①橘皮②橙皮
cedrat oil (=citron oil) 枸橼油；香橼油
cedrelanol 洋椿醇
cedrela oil (=Indian mahogany wood oil) 烟香椿木油；印度红木油
cedrellone 洋椿酮
cedrene 柏木烯；雪松烯 $C_{15}H_{24}$
cedrenol 柏木烯醇；雪松烯醇
cedrenone 柏木酮；雪松酮
cedrenyl acetate 乙酸柏木酯
cedrin 雪松素
cedrine 雪松的
cedrol 雪松醇 $C_{15}H_{26}O$
cedrolic acid 柏木脑酸
cedron 策桩子〔植〕
cedrone 柏木酮
cedronella (=lemongrass) 柠檬草；香茅草香草
cedronine 策桩宁
cedroxyde 柏木醚
cedryl acetate 乙酸柏木酯
cedryl methyl ether 甲基柏木醚
cedryl propionate 丙酸柏木酯
cefaclor 头孢克洛〔药〕
cefadroxil 头孢羟氨苄〔药〕
cefazolin 头孢唑林〔药〕
cefdinir 头孢地尼〔药〕
cefepime 头孢吡肟〔药〕
cefixime 头孢克肟〔药〕
cefluosil 铈氟硅石
cefmetazole 头孢美唑〔药〕
cefminox 头孢米诺〔药〕
cefoperazone 头孢哌酮〔药〕
cefoperazone and sulbactam 头孢哌酮舒巴坦〔药〕
cefotaxime 头孢噻肟〔药〕

cefotiam 头孢替安〔药〕
cefoxitin 头孢西丁〔药〕
cefprozil 头孢丙烯〔药〕
cefradine 头孢拉定〔药〕
ceftazidime 头孢他啶〔药〕
ceftizoxime 头孢唑肟〔药〕
ceftriaxone 头孢曲松〔药〕
cefuroxime 头孢呋辛〔药〕
ceiling ①天花板；顶棚②最高限度③升限④船底垫板
ceiling temperature 上限温度；规定最高温度*
ceiling value 最高值
ceilling temperature of polymerization 聚合最高温度*
celadon 青瓷；灰绿色
celadon glaze ①青瓷②青瓷釉
celadonite 绿鳞石
celandine oil 白屈菜油
celanese colours〔复〕 纤烷丝染料
celastin 睡菜苷
celastrine 南蛇藤碱
celastrol (=trypterygine) 雷公藤红；雷公藤碱
celatene colours〔复〕 纤拉厅染料
celcon 聚甲撑氧
celcure 铬酸铜〔木材防腐剂〕
celecoxib 塞来昔布〔药〕
celedonite 绿鳞石
celery 芹菜
celery fruits oil (=celery oil) 芹菜油
celery oil (=celery fruits oil) 芹菜油
celery salt 芹子盐
celery seed oil 芹菜子油
celestial blue 天青蓝
celesticetin (=coelesticetin) 天青菌素
celestine (=celestite) 天青石
celicomycin (=coelicomycin) 天蓝霉素
celite ①C 盐②塞里塑料；次乙酰塑料③硅藻土
cell ①晶胞②细胞③电池④盒；槽⑤测定池
cell angle 晶角
cell body 池体
cell capacitance measurement 细胞电容测定法
cell collapse 瘪泡〔泡沫塑料缺陷〕；微孔萎塌
cell compartment 池间隔；吸收池室
cell constant 池常数；电(导)池常数
cell correction 液池校正
cell cover 热室盖(板)
cell crane 热室吊车
cell crystal 晶胞
cell dimension 晶胞大小
cell efficiency 池效率

cell electrochemistry 细胞电化学
cellexl 二乙基氨乙基纤维素
cell for conductivity measurement 电导(率)测定池
cell for conductometric titration 电导滴定池
cell for flowing sample 流体试样管；流体试样池
cell for high frequency titration 高频滴定池
cell for infrared absorption 红外吸收池；红外吸收槽
cell-free extract 无细胞萃取液；无细胞抽提液
cell-holder 吸收池架
cell-in cell-out method 池入-池出法*
cell insulator 电池绝缘器
cell length (吸收)池长度
cell line 细胞系
cell membrane 细胞膜
cell membrane complex 细胞膜配合物
cell morphology （泡沫塑料)泡孔形态(学)
cell nucleation （泡沫塑料)泡核作用
cellobionic acid 纤维二糖
celloboard process 脲酚醛树脂-木屑连续剥板法
cellocidin 灭胞素
Celloco fractionator Celloco 纤维分离器
cellodextrin 纤维糊精
cellogel 乙酸纤维凝胶
celloidin 火棉
celloidin paper 火棉纸
cellon (=tetrachloroethane) 四氯乙烷
cellopentose 纤维五糖
cellophane (=cellophane paper) 玻璃纸；赛璐玢；透明纸〔商〕
cellophane casting machine 玻璃纸成型机；黏胶薄膜成型机；赛璐玢流延机
cellophane paper (=cellophane) 玻璃纸；赛璐玢；透明纸〔商〕
cellophane yarn 赛璐玢薄膜纱；玻璃纸纱
cell organelle sensor 细胞器传感器
cellosolve 溶纤剂*
cellosolve acetate 乙酸溶纤剂 $C_2H_5OCH_2CH_2COOCH_3$
cellotetraose 纤维四糖
cellotex 纤织剂
cellothene 赛璐嗪；透明聚乙烯薄膜
cellotriose 纤维三糖
cellotropin 赛璐托品〔药〕
celloxan 纤落散〔用碱性染料染醋酸丝时的助染剂〕
celloyarn 玻璃纸条；玻璃纸纱
cell parameters 晶胞参数*
cell path length 吸收池厚度；吸收池光程
cell potential 电池电位

cell signaling processor 细胞信号传递处理器
cell size 孔眼大小；孔度
cell strain 细胞株
cell support 电池架
cell switch 电池换向开关
cell terminal 电池线夹
cell thickness (吸收)池厚度
cell transport 细胞传递
cell type heater 管式加热炉
cell-type spinning system 仓式熔体纺丝装置
celluar silica 多孔硅石
cellubitol 纤维二糖醇
cellula adhesive 泡沫胶黏剂
cellular board 蜂窝板
cellular concrete 蜂窝混凝土；气泡混凝土
cellular ebonite 蜂窝硬质胶；微孔硬胶
cellular fibre 多孔纤维，泡沫纤维
cellular filter 蜂窝滤器
cellular insulant (蜂窝状)多孔绝热材料
cellular material 微孔材料*
cellular plastic 微孔塑料
cellular polymer 多孔聚合物；蜂窝状聚合物
cellular profile 微孔型材
cellular rubber 多孔橡胶；海绵橡胶〔开孔或闭孔〕
cellular sole 微孔大底
cellular striation (泡沫塑料)泡孔条痕
cellular structure 微孔结构*
cellular sulant (蜂窝状)多孔绝热材料
cellular texture 蜂窝状结构
cellulase 纤维素酶
celluloid 赛璐珞；假象牙；纤假(象)牙
celluloid lacquer 硝酸纤维素漆
cellulolytic 能水解纤维素的
cellulolytic enzyme 纤维素(水解)酶
cellulosan 纤维聚糖；复合纤维素
cellulosate 纤维素盐
cellulose 纤维素
β-cellulose β-纤维素
cellulose Ⅰ 纤维素Ⅰ〔天然纤维素，如棉花、麻等〕
cellulose Ⅱ 纤维素Ⅱ〔丝光纤维素或在常温时再生的一种溶解性纤维素〕
cellulose Ⅲ 纤维素Ⅲ〔由纤维素氨化物分解后所得的纤维素〕
cellulose Ⅳ 纤维素Ⅳ〔在200℃以上的高温下分解碱纤维素或纤维素黄酸钠后所得的纤维素〕
cellulose acetate 乙酸纤维素*
cellulose acetate bed 乙酸纤维素床
cellulose acetate-benzoate 纤维素乙酸苯甲酸酯；乙酸苯甲酸纤维素
cellulose acetate butyrate 乙酸丁酸纤维
cellulose acetate cigarette filter tow 二醋酸纤维素丝束
cellulose acetate fiber 醋酸纤维；纤维素酯纤维
cellulose acetate film 乙酸纤维素膜
cellulose acetate flake 二醋酸纤维素切片
cellulose acetate-laurate 乙酸十二酸纤维素
cellulose acetate-nitrate 乙酸-硝酸纤维素
cellulose acetate propionate (CAP) 乙酸-丙酸纤维素
cellulose acetate rayon 纤维素醋酸酯人造丝；醋酸纤维素人造丝
cellulose acetate sulfate 乙酸纤维素酯硫酸盐
cellulose acetate sulphate 纤维素醋酸硫酸酯；醋酸硫酸纤维素
cellulose acetobutyrate 纤维素乙酰丁酸酯；乙酰丁酸纤维素
cellulose acetomethacrylate 纤维素乙酰甲基丙烯酸酯；乙酰甲基丙烯酸纤维素
cellulose acetopropionate 乙酰丙酸纤维素
cellulose adhesive 硝酸纤维素黏合剂
cellulose after-xanthation 纤维素后黄(原酸)化(反应)
cellulose alloy fiber 纤维素合金型纤维；纤维素共混纤维
cellulose anion exchanger 阴离子交换纤维素
cellulose anionic exchange system 纤维素阴离子交换装置
cellulose base fiber 纤维素纤维
cellulose butyl benzoate 丁基-苯甲酸纤维素
cellulose butyl laurate 丁基-十二酸纤维素
cellulose butyrate 丁酸纤维素
cellulose butyrate fiber 丁酸纤维素纤维
cellulose caproate 己酸纤维素
cellulose chain 纤维素链
cellulose chemistry 纤维素化学
cellulose-coated plate 涂纤维素板
cellulose cobalt complex 纤维素钴络合物
cellulose column 纤维素柱
cellulose constituent 纤维素组分
cellulose crystallite 纤维素微晶
cellulose degradation 纤维素降解
cellulose diacetate film 二醋酸纤维素薄膜
cellulose dithiocarbonate 纤维素二硫代碳酸酯；纤维素黄(原)酸酯；黄酸纤维素
cellulose ester 纤维素酯
cellulose ester membrane 纤维素酯(隔)膜
cellulose ether 纤维素醚
cellulose extract 亚硫酸纸浆废液(浸)膏
cellulose fatty acid ester 纤维素脂肪酸酯
cellulose fibre 纤维素纤维

cellulose fibril 纤维素原纤维；纤维素微丝
cellulose film 赛璐玢；纤维素薄膜；黏胶薄膜；玻璃纸〔俗称〕
cellulose filter 纤维素滤器
cellulose filter aid 纤维素助滤剂
cellulose formate 纤维素甲酸酯；甲酸纤维素
cellulose gel 微晶纤维素
cellulose glycollate 纤维素乙醇酸酯
cellulose glycollic ether 羧甲基纤维素；纤维素乙醇酸醚
cellulose graft 接枝纤维素
cellulose hydrate 纤维素水合物
cellulose ion exchange powder 离子交换纤维素粉
cellulose ion-exchanger 离子交换纤维素
cellulose lacquer 纤维素喷漆
cellulose lattice 纤维素晶格
cellulose laurate 十二酸纤维素
cellulose liquor 木素磺酸(鞣)液
cellulose lye 亚硫酸盐纸浆废液
cellulose membrane 纤维素膜
cellulose methylene bis-xanthate 亚甲基(二)纤维素黄酸酯
cellulose micelle 纤维素微胞；纤维素胶束
cellulose (mixed ester) plastic 纤维素(混酯)塑料
cellulose mucus 纤维素黏液
cellulose naphthenate 环烷酸纤维素
cellulose naphthene laurate 环烷酸-十二酸纤维素
cellulose nitrate 硝酸纤维素*
cellulose nitroacetate 硝酸乙酸纤维素；纤维素硝酸乙酸酯
cellulose oleate 纤维素油酸酯
cellulose orthophosphate (=phosphorylated cotton) 磷酸纤维素
cellulose oxalate 纤维素草酸酯；草酸纤维素
cellulose paint 纤维素系涂料
cellulose paper 纤维素纸
cellulose paste 纤维素浆料
cellulose pelargonate 壬酸纤维素
cellulose peroxide 纤维素过氧化物
cellulose phosphate 磷酸纤维素
cellulose pigment finish 有色硝纤涂饰剂
cellulose plastic 纤维素塑料
cellulose plate 纤维素板
cellulose powder 纤维素粉
cellulose propionate 丙酸纤维素
cellulose propionate fibre 纤维素丙酸酯纤维
cellulose pulp 纤维素纸浆
cellulose resinate 树脂酸纤维素
cellulose stearate 硬脂酸纤维素
cellulose sulfate 硫酸纤维素
cellulose sulfonate 纤维素磺酸酯；磺酸纤维素
cellulose thickener 纤维素增稠剂
cellulose thin-layer chromatography 纤维素薄层色谱(法)
cellulose thinner (硝基)纤维素稀料；硝基漆稀料
cellulose thiourethane 硫代氨基甲酸乙酯纤维素；硫脲烷纤维素
cellulose triacetate 三乙酸纤维素
cellulose triacetate film 三醋酸纤维素薄膜
cellulose triester 纤维素三酯
cellulose trithiocarbonate 纤维素三硫代碳酸酯；三硫代碳酸纤维素
cellulose trixanthogen 纤维素三黄(原)酸(酯)基
cellulose turpentine oil 纤维素松节油
cellulose type filler in stiff paste 纤维素型稠浆填孔料
cellulose valerate 戊酸纤维素
cellulose varnish 纤维素清漆
cellulose xanthate 黄原酸纤维素
cellulose xanthic acid 纤维素黄(原)酸酯
cellulose xanthogenate 纤维素黄(原)酸酯；黄(原)酸纤维素
cellulosic 纤维素(性的)
cellulosic composite fibre 纤维素复合纤维
cellulosic crystalline modification 纤维素结晶变体
cellulosic exchanger 纤维素类离子交换剂
cellulosic homogenous polycrystal 纤维素同质多晶体
cellulosic-man-made-fibre 纤维素系人造纤维〔以纤维素为原料的化学纤维〕
cellulosic material 纤维素质
cellulosic matrix 纤维素基体
cellulosics 纤维素制品
cellulosic sorbent 纤维素吸附剂
cellulosic substrate 纤维质底材
cellulosine 木粉
cellulosis 纤维素制品
cellulysis 纤维素分解作用
cell uniformity (泡沫塑料)泡孔均匀性
celluose lacquer 纤维素喷漆
celluronic acid 纤维素糖醛酸
Cellu-sizer 压力式纤维筛分器
cellutyl colour 纤基染料
cell voltage 电池电压
cell volume (泡沫塑料)泡孔体积
cell vortex 贝纳德旋涡
cell with one liquid junction 单液体接界电池
cell without transference 无迁移电池
cell with two liquid junctions 双液体接界电池
celmonit 赛芒炸药；赛芒那特

celotex 隔音板；隔音材料
celsian 钡长石
Celsius (=centigrade) 摄氏度；百分度
Celsius thermometer (=centigrade thermometer) 百分温度计；摄氏温度计
cembrene 西柏烯
cement ①胶泥；黏固剂②水泥；士敏土③接合剂；胶黏剂
cementability ①黏结性②黏结能力
cement asbestos coating 水泥石棉盖覆层
cementation ①硬化②黏固作用；胶接作用③渗碳
cementation coating 渗碳层
cementation effect 胶接效应
cementation furnace 渗碳炉
cementation index (水泥)硬化率
cementation steel 渗碳钢
cement coating 水泥涂料
cement colorant 水泥着色剂
cement concrete 水泥混凝土
cement copper 渗碳铜
cement-dipping method 浸胶法；浸浆法
cement dropping 水泥残滴；水泥滴落物
cemented ①渗了碳的②胶接了的
cemented catalyst 胶接催化器
cemented needle 黏合针头
cemented steel 渗碳钢
cement emulsion paint 水泥乳液漆
cement flour 水泥粉
cement grappier 水泥渣
cement grit 粗水泥
cement grout 水泥浆；水泥砂浆
cement grout hose 水泥浆胶管
cement gun 水泥喷枪
cementing 胶接；溶接；黏合
cementing agent 胶接剂
cementing bath 黏合浴
cementing machine 擦胶机
cementing medium 胶接介质；胶接剂
cementing property 胶接性
cementite 渗碳体；碳化铁体
cementitious 胶接(性)的；黏结(性)的
cementitious coating 水泥涂料；水泥黏结涂料
cementitious rendering 水泥刷面；水泥抹灰(底)层
cementitious substrate 水泥底材
cement laying-on machine 刷胶浆机
cement lined pipe 水泥衬里管
cement lining 水泥炉衬
cement lithin paint 彩色水泥(砂浆)涂料

cement mixer 胶泥混合器〔橡胶〕
cement mortar 水泥灰浆
cement of high index 高标号水泥
cement of low index 低标号水泥
cement paint 水泥漆
cement paste 水泥(灰)浆
cement plaster ①水泥砂浆；水泥灰浆②石膏粉饰〔建筑〕
cement rendering 水泥抹面；水泥抹灰层
cement rock 水泥岩
cement setting 水泥凝结
cement slurry 水泥渣
cement solidification 水泥固化*
cement steel 渗碳钢
cement testing 水泥试验
cement-vinyl based filler 水泥-乙烯填缝料
cement water paint 水泥涂料
cement-water ratio 水灰比
cenosphere 空心微球
censphere 空心微球
centaurin 矢车菊苷
centaurine 百花金碱
center beam method 中心光束法
center bowl 中辊
center cut 中间馏分
center drill 中心(孔)钻
center-driven twin mill 中央传动的双台开炼机
centered-packed 中央封填
center filling 中央充装
center fixture 中心装置器
center-fold film 对折薄膜
center-fold sheet 对折薄膜
center-gauge hatch 中央量油口〔油罐〕
center hole 顶尖孔；中心孔
centering 打中心孔；定中心；定圆心
centering core 泥心
centering machine 定心机
centering ring 中心定位圈
centering screw 调心螺丝
centerless grinder 无心磨床〔机〕
centerless grinding 无心磨削
centerless grinding machine 无心磨床
center line 中心线
center matched 中心相配的
center of filter 滤光片主波长
center-of-gravity rule 重心规则
center of inversion 对称中心*
center-of-mass coordinate 质心坐标*
center of symmetry 对称中心

center pole　中央支柱〔油罐〕
center punch　冲子；定准器
center rest　中心托架〔机工〕；中心架
center rod ejection　中心杆式顶出
center tester　中心试验器
center to side baffle　双流式折流板
center wall up-draft heater　中央火墙烟气上行式加热炉
center winder　中心式卷取机
center winding　中心收卷
center yarn　芯线
centi-〔词头〕　厘；百分之一〔10^{-2}〕
centigrade (=Celsius)　摄氏度；百分度
centigrade degree　百分温度；摄氏温度
centigrade scale　百分温标；摄氏温标
centigrade temperature scale　百分温标；摄氏温度刻度
centigrade thermal unit　磅卡；公制热量单位
centigrade thermometer (=Celsius thermometer)　百分温度计；摄氏温度计
centigrade thermometric scale　摄氏温标
centigram (me)　厘克
centigram method　厘克法〔半微量分析法〕
centimeter　厘米
centipoise　厘泊〔黏度单位〕
centistokes　厘沲〔运动黏度单位〕
central　中央的；中心的
central atom　中心原子*
central axis　中心轴
central axis plane　中心轴面
central composite design　中心复合设计
central core　中心层〔运输带〕
central diffuse scattering　中心漫散射
central discharge　中央卸料
central dogma　中心法则
central downtake　中央降液管
central drain　主排汁沟
central drive　中轴传动法
central field approximation　中心力场近似*
central-force field　中心力场
central grate producer　中央炉箅发生炉
central heating　①集中加热②(建筑物)集中采暖
centraline orange G　中央灵橙 G
central ion　中心离子
centralization of hydrocarbons　烃中季碳原子形成
centralized hydrocarbons　富于季碳原子的烃类
centralized lubricating system　集中润滑系统
centralized lubrication　集中润滑
centralized oiling　集中润滑
central knife edge　中央支棱；中刃

central limit theorem　中心极限定理
central line (CL)　中心线*
central ply　中心层
central projection　集中式投射线
central refrigeration plant　中央制冷装置；中心制冷站
central relaxant　中枢松弛剂〔药〕
central reprocessing plant　(核燃料)后处理中心厂〔一厂处理多堆、多站燃料〕
central scattering component　散射中心成分
central spindle　①心轴②中心纺锤体
central tendency　集中趋势
central transportation (terminal) system　原油储藏基地中心运输系统
centre line average (CLA)　中线平均(法)〔检测钢铁表面粗糙度的方法之一〕
centre of inversion　反演中心
centre of mass　质心
centre of oscillation　振动中心
centre of symmetry　对称中心
centrepiece　①十字轴；十字头；十字架②中心件
centre roll　(压延机)中辊
centres　长中板
centres gap　长中板离缝
centric benzene formula　向心苯式
centric formula　向心式
centricleaner　锥形除渣器
centriclone　锥形除渣器
centricon filter　微量离心浓缩滤器；超滤
centric-preparation TLC　离心制备薄层色谱法
centriffler　两段除渣器
centrifugal　①离心的②离心机
centrifugal absorber　离心吸收器
centrifugal air separation　离心空气分离
centrifugal air separator　离心空气分离器
centrifugal analysis　离心分析
centrifugal apparatus　离心仪器
centrifugal atomising force　离心雾化力；离心喷雾力
centrifugal atomization　离心喷雾
centrifugal atomizing force　离心雾化力；离心喷雾力
centrifugal basket　离心机转筒；离心机吊篮
centrifugal basket drier　篮式离心干燥机
centrifugal blast-cleaning　离心喷砂清洁法；离心喷丸清洁法
centrifugal blender　离心式搅拌机
centrifugal blower　离心鼓风机
centrifugal bolting mill　离心转筒筛
centrifugal box-spinning machine　离心罐式纺丝机；离心式纺丝机

centrifugal casting　离心成型；离心浇铸
centrifugal chromatography (=chromatofuge)　离心色谱法
centrifugal clarifier　离心澄清机
centrifugal classifier　离心分级器；离心分粒器
centrifugal cleaner　锥形除渣器；离心净浆器
centrifugal coating　离心涂漆
centrifugal coating machine　离心涂漆(装)机
centrifugal compressor　离心蒸气机
centrifugal concentrate　离心法浓缩液
centrifugal counter-current chromatography　离心逆流色谱
centrifugal counter-current extractor　离心逆流萃取器
centrifugal cream-separator　乳油分离器
centrifugal crusher　离心式破碎机
centrifugal development　离心展开
centrifugal dewatering　离心脱水
centrifugal discharge　离心出料
centrifugal disk atomizer　离心式圆盘雾化器
centrifugal drier　离心干燥机
centrifugal effect　离心效应
centrifugal emulsor　离心乳化机
centrifugal enamelling equipment　离心涂漆装置
centrifugal extractor　离心萃取器*
centrifugal fan　离心通风机
centrifugal filter　离心滤器
centrifugal filtration　离心过滤
centrifugal finishing　离心涂装；离心涂漆；甩涂〔俗〕
centrifugal floatation method　离心浮选法
centrifugal force　离心力
centrifugal forming　离心成型
centrifugal gas cleaner　离心涤气机
centrifugal gas washing fan (=centrifugal gas cleaner)　离心涤气机
centrifugal grinder　离心研磨机
centrifugal hydroextractor　离心脱水机
centrifugal impact mixer　离心冲击混合器
centrifugal impeller mixer　离心叶轮混合器
centrifugalization　离心分离(作用)
centrifugalizing　离心的
centrifugalled　经过离心分离的
centrifugal lubricator　离心润滑器
centrifugally accelerated paper chromatograph　离心加速式纸色谱仪
centrifugal machine　离心机
centrifugal molecular still　离心型分子蒸馏器
centrifugal oil cleaner　离心净油器
centrifugal oiler　离心加油器
centrifugal oil filter　离心滤油器

centrifugal paper chromatography　离心纸色谱(法)
centrifugal polymerization　离心聚合
centrifugal precipitation tube　离心沉淀管
centrifugal pull　离心力
centrifugal pump　离心泵
centrifugal purification　离心净化
centrifugal purifier　离心净化机
centrifugal reel　离心转鼓；离心圆筒筛
centrifugal roll mill　离心滚磨
centrifugal screen　离心筛
centrifugal scrubber　离心涤气机
centrifugal separation　离心分离(法)*
centrifugal separator　离心分离机
centrifugal settling　离心沉降
centrifugal spinning　离心纺丝(法)
centrifugal spray　离心喷雾
centrifugal spray tower　离心喷淋塔
centrifugal stirrer　离心搅拌器
centrifugal supercharger　离心增压机
centrifugal superfractionator　离心增压分馏器
centrifugal tar extractor　离心焦油提取器
centrifugal type stapling machine　离心式切断机
centrifugal viscose grinder　离心(式)黏胶研磨机
centrifugal viscose rayon spinning machine　离心式黏胶人造丝纺丝机
centrifugal washer　机械洗涤机
centrifugal washing　离心洗涤
centrifugate　离心液
centrifugation　离心法；离心过滤
centrifuge　离心机
centrifuge basket　离心机盘
centrifuge blade (=centrifuge scraper)　离心叶片；离心刀片
centrifuge contactor　离心接触器；离心分馏器
centrifuged latex　离心分离胶乳
centrifuge extraction analysis　离心抽提分析
centrifuge field　离心力场
centrifuge head　离心机转头
centrifuge kerosene equivalent　离心煤油当量
centrifuge microscope　离心显微镜
centrifuge plant　离心法(同位素分离)工厂
centrifuge reclaiming　离心机法废润滑油再生
centrifuge refining　离心精制
centrifuge rotor　离心机转子
centrifuge scraper (=centrifuge blade)　离心叶片；离心刀片
centrifuge separator　离心分离器
centrifuge shield　离心管套

centrifuge speed indicator 离心机示速器
centrifuge stock 离心处理原料
centrifuge test 离心机试验〔用离心机测定石油产品中固体残渣〕
centrifuge treating 离心机处理〔粗制或加工〕
centrifuge trunnion 离心管套
centrifuge trunnion carrier 离心管套座
centrifuge tube 离心(机)管
centrifuging 离心法；离心过滤
centriole 中心粒
centripetal ①向心的②从圆周向里(展开)的
centripetal development 向心展开(法)*
centripetal force 向心力
centripetal pump effect 向心泵效应
centripetal tension 向心张力
centripure process 离心纯化法〔连续煮皂〕
centroid ①面心②体心
centromere 着丝点；着丝粒
centromere index 着丝点指数
centron 原子核
centronucleus 中心核
centrosome 中心体
centrosymmetry 中心对称
cephacetrile 头孢赛曲；氰乙酰头孢菌素
cephaeline 九节因；吐根酚碱
cephalanthin 风箱树苷
cephalein 荷草因素
cephaleine 吐根酚碱
cephalexin 头孢氨苄；先锋霉素Ⅳ
cephalin cholesterol flocculation test 脑磷脂胆固醇絮状试验
cephalochromin 头孢色菌素
cephaloglycine (=cefaloglycin) 头孢来星，先锋霉素Ⅲ
cephalolexin (=cephalexin;cefalexin) 头孢氨苄；头孢力新
cephalomycin 头孢菌素
cephaloridine (=cefaloridine) 头孢噻啶
cephalosporin 头孢菌素
cephalotaxine 粗榧碱
cephalothecin 复端孢菌素
cephalothin (=cefalotin) 头孢噻吩
cephamycin 头霉素
cepharanthine 顶花防己碱
cephem 头孢烯
ceraceous 蜡状的
ceramal 陶瓷合金；金属陶瓷
ceramet 陶瓷金属
cerametalic(s) 金属陶瓷

ceramic ①陶瓷的②无机非金属材料
ceramic abrasive 陶瓷磨料
ceramic adhesive 陶瓷胶黏剂；陶瓷黏合剂
ceramic aerator 陶瓷曝气器；陶瓷通气装置
ceramic block 陶瓷均热块
ceramic body 陶瓷体
ceramic bond 陶瓷黏合剂
ceramic cleaning 陶瓷清洗
ceramic coating 陶瓷涂层
ceramic cobalt blue 瓷蓝；钴蓝
ceramic condenser 瓷电容器
ceramic duct 陶瓷管
ceramic entrapment guide 陶瓷夹持式导丝器
ceramic fibre 硅酸盐纤维；陶瓷纤维
ceramic fibre blanket 陶瓷纤维毛毯
ceramic fibre reinforced composite 陶瓷纤维增强复合材料
ceramic filter 陶瓷过滤器
ceramic foam 多孔陶瓷
ceramic former 瓷模
ceramic glaze (陶)瓷釉
ceramic grinding media 陶瓷研磨介质；瓷球
ceramic heater 陶瓷加热器
ceramic humidity sensor 陶瓷湿度传感器
ceramic hygrometer 陶瓷湿度计
ceramic industry 陶瓷工业
ceramic ink 瓷墨
ceramic insulation 陶瓷绝缘
ceramic jet 陶瓷喷嘴
ceramic media 陶瓷研磨介质；瓷球
ceramic membrane electrode 陶瓷膜电极*
ceramic metal 金属陶瓷
ceramic-metal coating (=metallized ceramic coating) 金属陶瓷涂料
ceramic-metal composite system 陶瓷-金属复合材料系统
ceramic mosaic 陶瓷锦砖；陶瓷马赛克
ceramic nuclear fuel 陶瓷核燃料
ceramic oil 陶瓷用油
ceramic oxide 氧化陶瓷
ceramic packing 陶瓷填料
ceramic plasma 陶瓷等离子体
ceramic plate 陶瓷板
ceramic precursor 陶瓷前体
ceramic ring 陶瓷环
ceramic rod 陶瓷棒
ceramics 陶瓷学
ceramic seal 陶瓷封接
ceramic sensor 陶瓷传感器

ceramic source 陶瓷(放射)源
ceramic sponge 陶瓷海绵
ceramic stator 陶瓷定子
ceramic substrate 陶瓷基质
ceramic-to-metal seal 陶瓷金属封接
ceramic tube 陶瓷管
ceramic tube psychrometer 陶瓷管干湿球湿度计
ceramic twist stop wheel 陶瓷止捻轮
ceramic veneer 陶瓷镶面板；陶瓷面砖
ceramic wax 地蜡
ceramic whisker 陶瓷晶须
ceramography 陶瓷相学
ceramoplastic (=ceraplast) 陶瓷纤维增强塑料
ceramsite 陶粒
ceramsite concrete 陶粒混凝土
ceraplast (=ceramoplastic) 陶瓷纤维增强塑料
cerargrite 角银矿
cerargyrite 角银矿
cerasein 野樱脂
cerasin(e) ①野樱素②角苷脂③角铅矿
cerasine red 角铅矿红
cerasite 樱石
cerate ①铈酸盐②蜡剂；蜡膏
cerate oxidimetry 铈酸盐(氧化还原)滴定法
cerate redox method 铈酸盐氧化还原滴定法
ceratophyr(e) 角斑岩
cereal 谷类
cereal grass 禾谷植物
cerealose ①谷物糖②饴糖
cereal straw 谷草
cerebroside 脑苷脂
Cerenkov counter 契伦科夫计数器
Cerenkov radiation 契伦科夫辐射
cereous (纯)地蜡(制)的
ceresin(e) (纯)地蜡
ceresin(e) wax (纯)地蜡；微晶蜡；不定形蜡
cerevisterol (=7,22-ergostadiene-3,5,6-triol) 啤酒甾醇；酒酵母甾醇
cerex process 铈雷克斯过程
Cerf's theory 西夫理论
ceria 铈土*
ceric 高铈的；四价铈的
ceric basic carbonate 碱性碳酸高铈 $Ce(OH)_2(CO_3)_3$; $Ce(CO_3)_2 \cdot CeCO_3(OH)_2$
ceric compound 高铈化合物
ceric fluoride 四氟化铈 CeF_4
ceric hydrophosphate 磷酸氢高铈 $Ce(HPO_4)_2$
ceric hydroxide 氢氧化高铈 $Ce(OH)_4$

ceric hydroxynitrate 硝酸羟高铈 $Ce(OH)(NO_3)_3$
ceric ion method 高铈离子(纤维接枝)法
ceric nitrate 硝酸高铈 $Ce(NO_3)_4$
ceric oxide 二氧化铈 CeO_2
ceric oxychloride 二氯化氧铈 $CeOCl_2$
ceric oxysulfate 硫酸氧化高铈 $CeO(SO_4)$
ceric selenite 亚硒酸高铈
ceric sulfate 硫酸高铈 $Ce(SO_4)_2$
ceric sulfate dosimeter 硫酸铈剂量计
ceric sulfate method 硫酸高铈(滴定)法
ceride 蜡类脂
cerimetric titration 铈(Ⅳ)滴定法；铈(Ⅳ)量法*
cerimetry 铈(Ⅳ)量法*
cerin 2-羟软木三萜酮；蜡素
cerine 脂褐帘石
cerinic acid 蜡酸；二十六(烷)酸
ceriometry 硫酸铈滴定(法)
cerise 樱桃色(的)；鲜红色(的)
cerite 铈硅石
cerium 铈〔58号元素，化学符号Ce〕
cerium aluminide 铝化铈〔1.$CeAl_4$; 2. $CeAl_3$; 3. $CeAl_2$; 4. CeAl; 5. Ce_2Al〕
cerium compounds 铈化合物类
cerium dioxide 二氧化铈 CeO_2
cerium disulfide 二硫化铈 CeS_2
cerium group 铈组
cerium group lanthanide 铈组镧系元素
cerium-lanthanum naphthenate 环烷酸铈镧
cerium magneside 镁化铈〔1.CeMg; 2. $CeMg_3$; 3.$CeMg_2$〕
cerium mineral 铈矿
cerium naphthenate 环烷酸铈
cerium oxide 二氧化铈
cerium oxychloride 一氯氧化铈 CeOCl
cerium perovskite 含铈钙钛矿
cerium pyrophosphate 焦磷酸高铈 CeP_2O_7
cerium salt-persulphate catalyst 铈盐-过硫酸盐催化剂
cerium sesquioxide 三氧化二铈 Ce_2O_3
cerium tribromide 三溴化铈 $CeBr_3$
cerium trichloride 三氯化铈 $CeCl_3$
cerium triiodide 三碘化铈 CeI_3
cermet 金属陶瓷*
cermet coating 金属陶瓷涂层
cerol colors〔复〕 彩诺染料
ceromelissic acid 蜡蜜酸 $C_{33}H_{66}O_2$
ceropic acid 松针酸 $C_{36}H_{68}O_3$
cerosate 蔗蜡酸盐(或酯) $C_{23}H_{47}COOM$
cerosic acid 蔗蜡酸；二十四(烷)酸 $C_{23}H_{47}COOH$
cerosin 蔗蜡；精制地蜡

cerosinyl 蔗蜡烷基；二十四烷基
cerotate 蜡酸盐(或酯) $C_{26}H_{53}COOM$
cerotene 蜡烯；二十六(碳)烯 $C_{26}H_{52}$
cerotic acid (=hexacosanoic acid) 蜡酸；二十六(烷)酸 $C_{25}H_{51}COOH$
cerotin ①蜡精；甘油三蜡酸酯②蜡醇
cerotinic acid 蜡酸
cerotol 蜡醇
cerotone 蜡酮
cerous (正)铈的；三价铈的
cerous acetate 乙酸铈 $Ce(C_2H_3O_2)_3$
cerous bromate 溴酸铈 $Ce_2(BrO_3)_6$
cerous bromide 三溴化铈 $CeBr_3$
cerous carbonate 碳酸铈 $Ce_2(CO_3)_3$
cerous chloride 三氯化铈 $CeCl_3$
cerous compound 亚铈化合物
cerous dioxysulfate 硫酸二氧二铈 $Ce_2O_2SO_4$
cerous dithionate 连二硫酸铈 $Ce_2(S_2O_6)_3$
cerous fluoride 三氟化铈 CeF_3
cerous hydropyrophosphate 焦磷酸氢铈 $CeHP_2O_7$
cerous hydrosulfate 硫酸氢铈 $Ce(HSO_4)_3$
cerous hydroxide 氢氧化亚铈 $Ce(OH)_3$
cerous hypophosphate 次磷酸铈 $Ce(H_2PO_2)_3$
cerous iodate 碘酸亚铈 $Ce(IO_3)_3$
cerous iodide 碘化铈 CeI_3
cerous metaphosphate 偏磷酸亚铈 $Ce(PO_3)_3$
cerous molybdate 钼酸亚铈 $Ce_2(MoO_4)_3$
cerous nitrate 硝酸亚铈 $Ce(NO_3)_3$
cerous orthophosphate (正)磷酸亚铈 $CePO_4$
cerous oxalate 草酸亚铈 $Ce_2(C_2O_4)_3$
cerous oxide 三氧化二铈 Ce_2O_3
cerous phosphate 磷酸亚铈 $CePO_4$
cerous pyrophosphate 焦磷酸亚铈 $Ce_4(P_2O_7)_3$
cerous selenate 硒酸亚铈 $Ce_2(SeO_4)_3$
cerous sulfate 硫酸亚铈 $Ce_2(SO_4)_3$
cerous tartrate 酒石酸亚铈 $Ce_2(C_4H_4O_6)_3$
cerous tungstate 钨酸亚铈 $Ce_2(WO_4)_3$
ceroxylin 蜡椰树素 $C_{20}H_{32}O$
cerris 西里栎〔即：栎，西里是土耳其地名〕
certification 定值
certification of reference material 标准物质定值
certified 检定的；(书面)证明的
certified body 定值部门
certified burette 检定滴定管
certified colors 合法食用染料
certified reference material 认证标准物质；有证标准物质
certified utensil 检定器具

certified value 标准值；保证值
certified volumetric flask 检定量瓶
certified weights 检定砝码
cerulean blue 青天蓝〔主要成分为锡酸钴〕
ceruleum 天蓝料
cerulignone 蓝木醌
cerusa (=cerussa) 铅白
ceruse 碳酸铅白
cerusite 白铅矿
cervantite 锑赭石；黄锑矿
ceryl 蜡基；二十六烷基 $C_{20}H_{53}$
ceryl alcohol (=hexacosyl alcohol) 蜡醇；1-二十六(烷)醇 $C_{26}H_{53}OH$
cerylate 蜡醇盐 $C_{27}H_{55}OM$
cerylic (=ceryl) 二十六烷基(的)
cesiated 铯衣的
cesium acetate 乙酸铯 $CsC_2H_3O_2$
cesium alum (=cesium aluminium sulfate) 铯矾；硫酸铯铝
cesium aluminium sulfate (=cesium alum) 硫酸铯铝 $Cs_2SO_4 \cdot Al_2(SO_4)_3 \cdot 24H_2O; CsAl(SO_4)_2 \cdot 12H_2O$
cesium ammonium bromide 溴化铯铵 $CsBr \cdot NH_4Br$
cesium arc lamp 铯弧灯
cesium azide 叠氮化铯 $Cs[N_3]$
cesium bicarbonate 碳酸氢铯 $CsHCO_3$
cesium bichromate 重铬酸铯 $Cs_2Cr_2O_7$
cesium bisulfate 硫酸氢铯 $CsHSO_4$
cesium bromate 溴酸铯 $CsBrO_3$
cesium bromide 溴化铯 $CsBr$
cesium carbonate 碳酸铯 Cs_2CO_3
cesium chlorantimonate 氯锑酸铯 $Cs_3[SbCl_6]$
cesium chlorate 氯酸铯 $CsClO_3$
cesium chloraurate 氯金酸铯 $Cs[AuCl_4]$
cesium chloride 氯化铯 $CsCl$
cesium chloroplatinate 氯铂酸铯 $Cs_2[PtCl_6]$
cesium chloroscandate 氯钪酸铯 $Cs_3[ScCl_6]$
cesium chlorostannate 氯锡酸铯 $Cs_2[SnCl_6]$
cesium chromate 铬酸铯 Cs_2CrO_4
cesium chromic sulfate 硫酸铬铯 $Cs_2SO_4 \cdot Cr_2(SO_4)_3 \cdot 24H_2O; CsCr(SO_4)_2 \cdot 12H_2O$
cesium cyanide 氰化铯 $CsCN$
cesium dichromate 重铬酸铯 $Cs_2Cr_2O_7$
cesium disulfatoindate 二硫酸根络铟酸铯；硫酸铟铯 $Cs_2SO_4 \cdot In_2(SO_4)_3 \cdot 24H_2O; CsIn(SO_4)_2 \cdot 12H_2O$
cesium disulfide 二硫化铯 $Cs_2[S_2]$
cesium dithionate 连二硫酸铯 $Cs_2S_2O_6$
cesium electrode 铯电极
cesium ferric sulfate 硫酸铁铯

$Cs_2SO_4 \cdot Fe_2(SO_4)_3 \cdot 24H_2O$; $CsFe(SO_4)_2 \cdot 12H_2O$
cesium ferrous sulfate　硫酸亚铁铯
　　$Cs_2SO_4 \cdot FeSO_4 \cdot 6H_2O$
cesium fluoride　氟化铯　CsF
cesium fluosilicate　氟硅酸铯　$Cs_2[SiF_6]$
cesium hydride　氢化铯　CsH
cesium hydrogen sulfate　硫酸氢铯　$CsHSO_4$
cesium hydroxide　氢氧化铯　CsOH
cesium indium alum　铯铟矾；硫酸铟铯
　　$Cs_2SO_4 \cdot In_2(SO_4)_3 \cdot 24H_2O$; $CsIn(SO_4)_2 \cdot 12H_2O$
cesium iodate　碘酸铯　$CsIO_3$
cesium iodide　碘化铯　CsI
cesium ion　铯离子
cesium manganate　锰酸铯　Cs_2MnO_4
cesium mercury(Ⅱ) tribromide　三溴化汞铯
cesium monoxide　一氧化铯　Cs_2O
cesium nitrate　硝酸铯　$CsNO_3$
cesium nitrite　亚硝酸铯　$CsNO_2$
cesium oxalate　草酸铯　$Cs_2C_2O_2$
cesium oxide　氧化铯　Cs_2O
cesium pentasulfide　五硫化二铯　Cs_2S_5
cesium perchlorate　高氯酸铯　$CsClO_4$
cesium periodate　高碘酸铯　$CsIO_4$; Cs_3IO_5; Cs_5IO_6
cesium permanganate　高锰酸铯　$CsMnO_4$
cesium peroxide　过氧化铯；过四氧化二铯　$Cs_2[O_4]$
cesium platinic chloride　氯化铂铯　$Cs_2[PtCl_6]$
cesium rubidium alum (=cesium rubidium aluminium sulfate)　铯铷矾；硫酸铝铷铯
cesium rubidium aluminium sulfate (=cesium rubidium alum)　硫酸铝铷铯　$CsRb(SO_4) \cdot Al_2(SO_4)_3 \cdot 24H_2O$
cesium rubidium bromide　溴化铷铯　$CsBr \cdot RbBr$
cesium salicylate　水杨酸铯　$CsC_7H_5O_3$
cesium selenate　硒酸铯　Cs_2SeO_4
cesium silicate　硅酸铯　Cs_2SiO_3
cesium silicofluoride　氟硅酸铯　$Cs_2[SiF_6]$
cesium sulfate　硫酸铯　Cs_2SO_4
cesium sulfide　硫化铯〔1.Cs_2S; 2.$Cs_2[S_2]$; 3.$Cs_2[S_3]$; 4.$Cs_2[S_5]$〕
cesium tetroxide　四氧化铯；过四氧化二铯　$Cs_2[O_4]$
cesium thiocyanate　硫氰酸铯　Cs(CNS)
cesium tribromide　三溴化铯　$Cs[Br_3]$
cesium triiodide　三碘化铯　$Cs[I_3]$
cesium trinitride　三氮化铯；叠氮化铯　$Cs[N_3]$
cesium trioxide　三氧化二铯　Cs_2O_3
cesium triphenyl cyanoboron　三苯氰硼铯
　　$Cs(C_6H_5)_3CNB$
cesium trisulfide　三硫化二铯　Cs_2S_3
cesium unit (=moonlight unit)　铯单位

cessation reaction　(链)终止反应
cetalkonium chloride　十六烷基二甲基苄基氯化铵
cetane　十六烷；鲸蜡烷　$C_{16}H_{34}$
cetane additive (=cetane improver)　十六烷值增进剂
cetane improver (=cetane additive)　十六烷值增进剂
cetane index　十六烷指数
cetane number (=cetane ratio)　十六烷值
cetane number booster (=cetane improver)　十六烷值增进剂
cetane number improver (=cetane number booster)　十六烷值增进剂
cetane number in borderline　在交界线的十六烷值；卡边的十六烷值
cetane rating　十六烷品级
cetane ratio (=cetane number)　十六烷值
cetane test method　十六烷值测定法
cetane unit　十六烷值单位
cetane value (=cetane number)　十六烷值
cetanol　鲸蜡醇；十六(烷)醇
cetene　鲸蜡烯；1-十六碳烯　$C_{16}H_{32}$
cetene number (=cetene ratio)　十六烯值
cetene ratio (=cetene number)　十六烯值
cetene unit　十六烯单位
cetenylene　鲸蜡炔　$C_{16}H_{30}$
ceteth　十六烷基聚氧乙烯醚
cetic　鲸蜡的；十六(碳)烷的
cetin　鲸蜡素；棕榈酸鲸蜡(醇)酯
cetirizine　西替利嗪〔药〕
cetol　鲸蜡醇　$C_{15}H_{31}CH_2OH$
cetoleic acid　鲸蜡烯酸；二十二(碳)烯酸
α-cetone (=methyl-α-ionone)　甲基-α-紫罗兰酮
cetostearyl alcohol　十六醇十八醇混合物
cetraric acid　冰岛衣酸
cetrarin　冰岛衣素
cetrimonium chloride　十六烷三甲基氯化铵
cetryl (=hexadecyl)　鲸蜡基；十六(烷)基
cetyl acetate　鲸蜡基乙酸酯；乙酸十六(烷)酯
　　$CH_3CO_2C_{16}H_{33}$
cetyl alcohol　鲸蜡醇；十六(烷)醇
cetyl alcohol sulfate　十六烷醇硫酸酯(盐)
cetylate　①棕榈酸盐(或酯)；软脂酸盐(或酯)②鲸蜡醇盐
cetyl bromide　鲸蜡基溴；1-溴十六(碳)烷
　　$CH_3(CH_2)_{14}CH_2Br$
cetyl cellulose　十六烷基纤维素；鲸蜡基纤维素
cetyl dimethyl amine oxide　鲸蜡基二甲基氧化胺；十六烷基二甲基氧化胺
cetyl dimethyl benzyl ammonium bromide　十六烷基二甲基苄基溴化铵
cetyl dimethyl benzyl ammonium halide　十六烷基二甲

基苄基卤化铵
cetylene (=cetene)　鲸蜡烯
cetylic acid　软脂酸；棕榈酸
cetyl iodide　鲸蜡基碘；1-碘十六(碳)烷　$CH_3(CH_2)_{14}CH_2I$
cetyl lactate　乳酸十六醇酯
cetyl myristate　肉豆蔻酸鲸蜡酯
cetyl palmitate　十六酸鲸蜡酯　$C_{15}H_{31}CO_2C_{16}H_{33}$
cetyl pyridinium bromide　十六烷基溴化吡啶鎓
cetyl pyridinium halide　十六烷基卤化吡啶鎓
cetyl sodium sulfate　十六烷基硫酸钠
cetyl sulfate　十六烷基硫酸酯(盐)
cetyl trimethyl ammonium bromide　十六烷基三甲基溴化铵
cetyl trimethyl ammonium chloride (CTAC)　十六烷基三甲基氯化铵　$[CH_3(CH_2)_{15}N(CH_3)_3]Cl$
C. E. unit operation (=chemical engineering unit operation)　化工单位操作
cevacine　瑟瓦辛
cevadic acid (=tiglic acid)　惕各酸；瑟瓦酸；(Z)-2-甲基丁烯酸　$CH_3CH=C(CH_3)COOH$
cevadilline (=sabadilline)　瑟瓦地灵；沙巴底林
cevadine　瑟瓦定〔一种沙巴碱；参阅 sabadine 等〕
cevine　瑟文〔一种沙巴达碱〕　$C_{27}H_{43}O_8N$
cevitamic acid (=vitamin C)　维生素 C；抗坏血酸
ceyssatite　硅藻土
^{252}Cf plasma desorption ionization source　锎-252 等离子解吸电离源
CFR engine (=cooperative fuel research engine)　CFR 爆震试验机〔石油〕
CFR research test method　CFR 研究试验法〔用于测定汽油爆震性〕
chaavicol methyl ether　黑椒酚甲醚
chabazite (=chabasite)　菱沸石
Chaddock burner　恰多克炉
chafer fabric　胎圈包布；子口包布
chaff　①废物②饲料
chaff cutter　饲料切割机
chain　①链②链条③连锁
chain activation　链活化
chain aggregation　链聚集
chain alignment　链排列
chain assembly　链组装
chain attachment　链连接
chain axis　链轴*
chain backbone　主链；链骨架
chain balance　链式天平
chain block　链滑车

chain branching　链支化
chain branching agent　链支化剂
chain break　链断裂
chain-breaker　断链剂
chain breaking　链断裂
chain bridging　链桥
chain bromination　链上溴化
chain bucket elevator　链斗升降机
chain bundle　链束
chain bundle model　链束模型
chain carrier　链载体*
chain cessation　链终止
chain chlorination　链上氯化
chain cleavage　链断裂；链裂解
chain-cluster　(结构性颜料粒子的)链状聚集；链簇
chain coccus　链球菌
chain-coiling　链盘旋
chain combination　链状结合
chain compound　链状化合物
chain configuration　链构型
chain conformation　链构象
chain contour length　①链的伸直长度②链(分子)等同长度
chain conveyor　链式运输机
chain copolymerization　链式共聚合
chain coupling　链形联轴节
chain-coupling agent　链偶合剂；链偶联剂
chain-cutting agent　①断链剂②调节剂
chain decay　链衰变
chain-decoiling　(分子)链解盘旋；链解蜷；链伸展
chain degradation　链降解
chain dimension　链尺寸
chain disintegration　链蜕变
chain disproportionation　链歧化*
chain dissociation　链离解作用*
chain drive　链(条传)动法
chain-drive lubricant　传动链润滑剂
chained particle　链型粒子
chain element　链单元
chain elevator　链式升降机
chain end　链末端
chain ending　链中止
chain end-to-end vector　链端间矢量
chain entanglement　链缠结
chain-equivalent solute viscosity factor　链当量溶质黏度因子
chain expansion factor　链扩张因子
chain extender　扩链剂；链增长剂

chain extending reaction 链增长反应
chain extension (=chain lengthening) 链伸长
chain extension agent 增链剂
chain extension-crosslinking 扩链交联
chain flexibility 链柔性
chain fluorination 链上氟化
chain-folded crystal 折叠链结晶
chain-folded lamella 折叠链片晶
chain-folded segment 折叠链节
chain-folded structure 折叠链结构
chain folding 链折叠*
chain-fold length 折叠链长度
chain-fold packing 折叠链堆砌
chain-fold region 折叠链区域
chain formation 链(的)形成
chain fragment 链段；链断片
chain geometry 链(几何)形态
chain gill box 链条式针梳机
chain grate 链式筛
chain grizzly 链式筛
chain growing radical 链增长基
chain growth 链增长*
chain halogenation 链上卤化
chain hydrocarbon 链烃
chain-immobilization factor 链受阻因子
chain inhibition 链抑制*
chain inhibitor 链抑制剂*
chain initiation 链引发*
chain-initiation reaction 链引发反应
chain initiator 链引发剂
chain interpenetration 链的相互穿透
chain interruption 链中断
chain iodination 链上碘化
chain irregularity 链不规整性
chain isomerism 链异构
chain isomorphism 链的同(晶)形性；链的类质同晶现象
chain joint 链接
chain lattice 链格
chain length 链长*
chain length distribution 链长分布
chain lengthening (=chain extension) 链伸长
chain length regulator 链长调节剂
chainless mercerizing range 直辊丝光联合机
chain-like structure 链状结构
chain link 花板；链链
chain-linking agent 链键合剂；链连接剂
chain loop 链环
chain lubricant 链润滑油
chain macromolecule 链状大分子
chain mechanism 成链历程
chain mobility ①链流动性；链活动性②链迁移率
chain molecule 链型分子；线型分子
chain notation 垫纱数码
chain of custody sample 监管试样通路
chain of fission products 裂变产物链*
chain of random self-avoiding walks 无规自避行走链
chain of stirred tanks 串联搅拌釜
chainomatic balance 链式天平
chain-ordering 链序列；链有序排列
chain orientation 链取向
chain orientational disorder 链取向无序*
chain orientation distribution 链取向分布
chain-packed model 链束模型
chain-packed structure 链堆砌结构
chain-packet model 链束模型
chain packing 链堆砌
chain polycondensation 成链缩聚(作用)
chain polymer 链型聚合物*
chain polymerization 链(式)聚合*
chain process 链式过程
chain propagation 链增长
chain propagation reaction 链增长反应
chain pulley (=chain wheel) 链轮
chain pump 链式泵
chain reaction ①连锁反应②链式反应
chain reaction polymerization 连锁聚合
chain regulating agent 链长调节剂
chain regulator 链长调节剂；链长控制剂
chain relaxation 链松弛
chain repeating distance 链重复距离*
chain rigidity 链的刚度；链的刚性
chain rule 链式法则；链规则
chain rupture 断链；链锁中断
chain scission 断链(作用)
chain-scission degradation 断链降解*
chain segment 链段
chain slippage 分子链相互滑动
chain space 链(占有的)空间
chain starting 链引发(作用)
chain statistics 链统计学
chain stiffness 链僵硬性；链劲度
chain stitch 链式线迹
chain stitcher 链状缝纫机
chain stop 链终止
chain stopped alkyd 链终止型醇酸树脂
chain stopper 链终止剂；止链剂

chain stopping 链终止
chain stopping activity 链终止活性；链终止活动
chain structure 链型结构*
chain substitution 链上取代*
chain terminating agent 链终止剂
chain-terminating moiety 链终止部分
chain termination 链终止*
chain termination agent 链终止剂；止链剂
chain termination reaction 链终止反应
chain terminator 链终止剂
chain threading 链串
chain transfer 链转移*
chain transfer agent 链转移剂
chain transfer coefficient 链转移系数
chain transfer constant 链转移常数*
chain transfer polymerization 链转移聚合
chain transfer reaction 链转移反应
chain transfer to monomer 向单体链转移
chain transfer to polymer 向聚合物链转移
chain transfer to solvent molecule 向溶剂分子链转移
chain twisting mechanism 链扭转机理
chain uncoiling 链伸展
chain-unfolding 链(的)解折叠；链(的)伸展
chain unit 链单元
chain vector length 链矢量长度
chain weave 链形组织〔纺〕
chain wheel (=chain pulley) 链轮
chain wrench 链条管子钳
chain yield 链产额*
chairamine 切尔明
chair conformation 椅型构象*
chair form 椅式
chaksin 山扁豆碱
chaksine 卡克生
chalcacene 白垩省 $C_{30}H_{16}$
chalcanthite 蓝矾；胆矾；五水(合)硫酸铜 $CuSO_4 \cdot 5H_2O$
chalcedony (=white agate) 白玛瑙；玉髓
chalcedonyx 带纹玉髓
chalcocite 辉铜矿
chalcogen 硫属元素*
chalcogen glass 硫属玻璃〔半导体玻璃〕
chalcogenide 硫属元素化物*
chalcogenide glass 硫属(化合物)玻璃
chalcogenide glass electrode 硫属化物玻璃电极
chalcolite 铜铀云母
chalcomorphite 硅铝钙石
chalcone (=cinnamophenone) 苯乙烯基苯基甲酮；查耳酮

chalcophile element 亲铜元素；亲硫元素
chalcophyllite 云母铜矿
chalcopyrite (=copper pyrite) 黄铜矿
chalcopyrrhotite 铜磁黄铁矿
chalcosiderite 磷铜铁矿
chalcostibite 硫铜锑矿
chalk (=creta) 白垩
chalkboard paint 黑板漆
chalking ①白垩处理②粉化③起垩
chalking compound 隔离剂
chalking of exterior paint 外用漆粉化
chalking paint ①粉化涂料②自(清)洁涂料
chalking resistance 抗起霜性
chalk kauri 白粉状贝壳杉脂
chalk masking 粉化层蒙盖变色；粉化变色
chalkogenide 硫属化物
chalkone 查耳酮 $C_{15}H_{12}O$
chalkostibite 硫铜锑矿
chalk pit 白垩矿场
chalk powder 白垩粉；碳酸钙
chalk rating 粉化评价(级；定)；粉化测(检)定；粉化等级
chalk resistance 抗粉化性
chalk resistance grade titanium dioxide 耐粉化级钛白粉
chalk retardation 阻滞粉化；延缓粉化
chalk slate 白垩板岩
chalkstone 白垩块；痛风石
chalk suspension 白垩悬浮液；白垩悬浊液
chalk whiting 碳酸钙
chalky 白垩的
chalky clay 泥灰岩〔地质〕
chalmersite 方黄铜矿
chalybeate ①铁泉②含铁剂
chalybite 球菱铁矿
chamaecin 扁柏素
chamaecypariol 扁柏醇
chamaelirin (=chamelirin) 地百合苷
chamazulene 母菊蓝
chamazulene carboxylic acid 母菊蓝酸
chamazulenogene 母菊精
chamber ①室②展开室；展开槽
chamber acid 铅室酸
chamber bored roll 空心滚筒
chamber burette 球滴定管
chamber collector 室式收集器
chamber crystal 铅室结晶
chamber drier (=compartment drier) 间格干燥器；分室干燥器
chamber (filter) press 箱式压滤机

chamber gas　铅室气
chamber over-saturation　（展开）槽过饱和
chamber pan　铅室底
chamber plant　铅室硫酸厂
chamber plant acid　铅室酸
chamber pressure　内腔压力；研磨筒(室)内压力
chamber process　铅室法
chamber regulator　调节室
chamber saturation　（展开）槽饱和
chamber saturation chamber (=CS-chamber)　饱和式展开槽
chamber-type nebulizer　室式雾化器
chamber white lead　室铅白
chameleon　三色调效应；变色效应
chameleon paint　示温漆；变色漆；温度指示漆
chameleon thermometer　变色温度计
chamelirin (=chamaelirin)　地百合苷
chamene　扁柏烯；柏木烯
chamenol　扁柏酚
chamfer　倒角；倒棱；(柱子、门槛的)面；凹槽
chamfering　切斜边；倒角
chamic acid　扁柏酸
chaminic acid　扁柏次酸
chamois　麂皮；油鞣革
chamois dressed　油鞣的
chamois dressing　油鞣
chamois fat　油鞣用油脂
chamoising　油鞣法
chamoising process　油鞣法
chamois leather　油鞣革；麂皮〔俗〕
chamois skin　麂皮
chamomile absolute　春黄菊净油
chamomile oil (=camomile oil)　春黄菊油
chamomillin　春黄菊素
chamosite　鲕绿泥石
chamotte　熟料；熟耐火土
champaca (flower) oil　黄兰花油；金香木油
champaca leaf oil　黄兰叶油
champacol (=guaiol)　愈创木醇
changeability　易变性
changeable silk　闪光绸
changeant　闪光效应；闪光织物
change can mixer　换罐捏合机；漆浆捏合机
change gear　变速齿轮；变速装置
change of state　物态的变化
change-over　①转换②转换开关
change-over period　换油周期
change-over switch　换向开关
change-over valve　转换阀
change parts　交换零件
changing load　交变荷重
changing-over　转换
changing solvent system　溶剂变换系统
channel　①频道②通道③联箱④波道⑤槽；沟
channel black　槽法炭黑〔补强剂，颜料〕
channel capacity　信道容量
channel drier　烘道干燥器
channeled spectrum　沟槽光谱
channel effect　沟道效应
channel electron multiplier　通道电子倍增器；渠道式电子倍增器
channel electron multiplier array　通道电子倍增器阵列
channel flow　明渠流动；载流
channel fringe　沟槽杂纹
channeling　沟流
channel point　成沟点〔齿轮转动时润滑剂层中形成未充满沟槽的温度〕
channel process　槽法〔炭黑生产方法之一〕
channel spin　沟道自旋
channel steel　槽钢
channel test　沟试验〔齿轮转动时润滑油层中形成未充满沟槽的试验〕
channeltron　通道倍增器*
channeltron electron multiplier array　通道电子倍增器阵列
channel valve　檐阀
chanoclavine (=secaclavine)　裸麦角碱
chanootin　花柏酚亭
chantalmycin　商塔霉素
chaotic distribution　无序分布
chaotic mixer　混沌混合器；无序混合器
chaotropic anion　离液序列高的阴离子
chappe silk yarn　绢丝
char　①烧焦；焦化②碳
character　特性
character diagram　特性图
characteristic　①特性②特性的
characteristic absorption band　特征吸收谱带
characteristic absorption peak　特征吸收峰
characteristic amino group frequencies in IR　氨基的红外特征频率
characteristic band　特征谱带
characteristic component　特性组分
characteristic concentration　特征浓度
characteristic content　特征含量
characteristic curve　特性曲线

characteristic data　特性数据
characteristic energy loss spectroscopy　特征能量损失谱
characteristic frequency　特性频率
characteristic function　特征函数*
characteristic-group method　特性基团法
characteristic impedance　特征阻抗
characteristic infrared group frequency　特征红外基团频率
characteristic ion　特征离子
characteristic line　特征谱线
characteristic line group　特征线组
characteristic odor　特征气味
characteristic pattern　特征图
characteristic peak　特征峰
characteristic property　特有性能
characteristic radiation　特性辐射
characteristic ratio　特征比*
characteristic reagent　特征试剂；特异试剂
characteristic relaxation time　特征松弛时间
characteristic Reynold's number　特征雷诺数
characteristic rotational temperature　转动特征温度*
characteristic scalar　特征标量
characteristic spectrum　特征光谱
characteristic temperature　特征温度*
characteristic test　特性试验
characteristic value (=eigenvalue)　特征值；本征值
characteristic velocity　特性速度
characteristic vibrational temperature　振动特征温度*
characteristic viscoelastic parameter　黏弹性特征参数
characteristic X-ray　特征X射线
characteristic X-ray spectrum　特征X射线光谱
characterization　①特性②表征；鉴定；表征法
characterization factor　特性因素
characterization of adsorption　吸附特性
characterization of surface structure　表面结构表征
characterization techniques　表征技术
character of service　使用性能
character stain　（木器)特征着色法
character tables　特征标表
char area　炭化面积
charcoal　①炭②木炭③活性炭
charcoal adsorption method　活性炭吸附法
charcoal adsorption process　活性炭吸附过程
charcoal bed　木炭床层
charcoal black　木炭黑
charcoal blast-furnace　木炭鼓风炉
charcoal burner　木炭炉
charcoal iron (=charcoal pig)　木炭生铁
charcoal kiln　炭窑

charcoal pig (=charcoal iron)　木炭生铁
charcoal saturation time　木炭饱和时间〔测吸附活性〕
charcoal test method　活性炭吸附测定法〔测石油气体中汽油含量〕
charcoal trap　炭阱
Chardin filter paper　卡丁滤纸
Chardonnet process　卡当纳特成纤工艺
Chardonnet silk　卡当纳特丝〔一种人造丝〕
char filter　炭滤器
char formation　成焦(作用)
charge　①电荷②充气③充电④进料⑤负荷
charge and discharge key　充放电钥；灌放电钥
charge balance　电荷平衡*
charge balance equation　电荷守恒式
charge capacity　充电额
charge carrier migration　电荷体迁移
charge cask　装料容器
charge cavity　加料孔〔递压机〕
charge cloud　电荷云
charge compensation　电荷补偿*
charge competition　电荷竞争
charge conjugation　电荷共轭
charge controlled reaction　电荷控制反应
charge coupled detector　电荷耦合检测器
charge coupled device　电荷耦合器件
charged acid　荷电酸*
charged atom　荷电原子；带电原子
charged cloud　①带电云雾②带电云
charged electrode　带电电极
charge density　电荷密度*
charge density difference plot　电荷密度差图*
charged flask　充满的烧瓶
charged interface　带电界面
charge discrepancy　电荷偏离
charge displacement　电荷位移
charge distribution　电荷分布*
charge distribution of fission products　裂变产物电荷分布*
charged particle accelerator　荷电粒子加速器
charged particle activation analysis (CPAA)　带电粒子活化分析*
charged particle bombardment　荷电粒子轰击
charged particle spectrograph　荷电粒子谱仪
charged particle X-ray excitation technique (CPXE)　带电粒子X射线激发技术
charged particle X-ray fluorescence　带电粒子(致)X射线荧光
charged particle X-ray fluorescence analysis　带电粒子X

射线荥光分析[*]
charged polymer lattice　带电聚合物点阵
charged powder cloud　带电粉末云团
charged powder particles　带电粉末粒子
charge effect　荷电效应
charge efficiency　给料效率；充电效率
charge escaping　电荷漏失
charge exchange ionization　电荷交换离子化；电荷交换电离
charge exchange ionization　电荷交换离子化
charge exchange reaction　电荷交换反应
charge exchange spectrum　电荷转换谱[*]
charge gas　①裂解气②原料气
charge gas compressor　裂解气压缩机
charge gauge　进料表
charge heater　进料加热器
charge-in　进料
charge-injection detector　电荷注入检测器
charge-mass ratio　荷质比
charge metering device　定量供给装置
charge migration　电荷迁移
charge-neutrality principle　电中性原理
charge neutralization　电荷中和作用
charge neutralizer　电荷中和器
charge number　电荷数
charge of electron　电子电荷
chargeometer　(电池用)硫酸比重计
charge permutation reaction　电荷交换反应
charge potential model　电荷势能模型
charge proportion　进料比例
charge pump　进料泵
charger　①装料机②充电器
charge ratio　投料配比
charge reference　荷电参照
charge-remote fragmentation　电荷远程裂解
charge retention　电荷保留
charge rubber　填料橡皮
charge shift electrophoresis　电荷位移电泳
charge-site initiation　电荷中心引发
charge-site stabilization　电荷中心稳定作用
charge step polarography　电荷阶跃极谱法[*]
charge stock　进料；送入原料
charge stock composition　进料组成
charge-to-count convertor　电荷-计数转换器
charge transfer　电荷传递；电荷转移
charge transfer absorption spectrum　电荷转移吸收光谱
charge transfer band　电荷转移带
charge transfer complex　电荷转移络合物[*]；电荷转移复合物
charge transfer detector　电荷转移检测器
charge transfer device　电荷转移器件
charge transfer energy　电荷转移能[*]
charge transfer initiation　电荷转移引发
charge transfer interaction　电荷转移型相互作用
charge transfer interaction force　电荷转移相互作用
charge transfer overpotential　电量传递过电压
charge transfer polymerization　电荷转移聚合[*]
charge transfer process　电荷转移过程
charge transfer spectra　电荷转移光谱
charge transfer spectrophotometry　电荷转移分光光度计
charge-transfer spectrum　电荷转移光谱[*]
charge transfer structure　传荷结构；电荷转移结构
charge transport by migration　电迁移
charge weight　加料重量
charging　①装料；加料②充电
charging and delivery hose　排吸胶管
charging and discharging curve　充放电曲线[*]
charging apparatus　装料设备
charging belt　充电带
charging capacity　①容胶量；容料量；投料量②蓄电量〔电池〕
charging car　装料斗车
charging connection　装料接头
charging current　充电电流[*]
charging curve　充电曲线；炭化曲线
charging density　电荷密度；装载密度
charging dynamo　充电发电机
charging effect　荷电效应
charging electrode　带电电极
charging equipment of banbury mixer　密炼机加料装置
charging face　装料面
charging feed　送入原料
charging flue　添煤道
charging hole　装料口
charging hopper　装料斗
charging hot pump　进料热油泵
charging load　充电负荷
charging lorry　装料斗车
charging machine　装料机
charging material　加(入)料
charging of accumulators　蓄电池充电
charging pipe　装料管
charging potential　电荷(电)势
charging pump　①送料泵②上水泵
charging set　充电装置
charging side　机面

charging stock 进料
charging stock tank 进料罐
charging tank 供应槽
charging tube 装料管
charging unit 充电单元
charging-up ①充电②装料；加料
charging valve 加料(液)阀
chargometer 充电计；充电表
char index 碳化指数
charing 炭化
Chariton white (=lithopone) 锌钡白
charking ①烧炭②焦化
char layer (燃烧试验)焦化层
char length (燃烧试验)焦化长度
charlock 野芥子
Charlotte colloid mill 查洛特胶体磨
Charlton white 查尔顿白
charming odor 美好香气；优雅香气
char oil 焦油
charpie (=lint) 麻布
Charpy bay 却贝试样
Charpy impact machine (=simple beam impact machine) 单梁(式)冲击(试验)机
Charpy impact test 却贝冲击试验；单梁式冲击试验
Charpy impact tester 却贝冲击试验机
Charpy impact test machine 却贝冲击试验机
Charpy impact test specimen 却贝冲击试样
charred coal 焦煤
charred leather 皮炭
charred spot 炭化斑
charring ①炭化②焦化③烧碳
charring-and-spraying treatment 炭化喷涂处理〔木材防腐〕
charring curve 灰化曲线
charring layer 炭化层
charring reagent 炭化试剂
charring spot 炭化斑
charring stage 灰化阶段
charring temperature 灰化温度
charring time 灰化时间
charry 炭化的
charsa skin 黄颈貂皮
chart ①图；图表；表②记录纸
chart drive speed 记录纸运行速度
chart integrator 图形积分器
chart paper ①记录纸②图纸
chart paper inhomogeneity 记录纸的不均匀性；记录纸的不规整(性)

chart recorder 图表记录仪
chartreusin 教酒菌素
chart speed 记录纸速
chart width 记录纸宽度
char value 炭化值
char volume 炭化体积
chase 模套
chaser 螺纹梳刀；梳刀盘
chasing 雕镂；叠层轧光
chasing calender 雕镂压延机
chasing tool 梳刀
chassis 车底盘〔机工〕
chassis absolute 残花净油
chassis concrete 残花浸膏
chassis paint (汽车)底盘漆
chatoyance ①变彩②猫眼效应
chatoyant ①猫眼石；闪光石；变彩宝石②闪光效应
chatter 刀震
chattering ①振动；颤震②颤震(波)纹〔漆病〕
chatter mark 跳痕；跳刀(伤)
chaulmestrol 晃模酸乙酯
chaulmoogra oil 晃模油；大风子油
chaulmoogrene 晃模烯 $C_{18}H_{34}$
chaulmoogric acid 晃模酸；2-环戊烯十三(烷)酸 $C_5H_7(CH_2)_{12}CO_2H$
chaulmoogroyl 晃模酰 $C_5H_7(CH_2)_{12}CO$
chaulmoogryl 晃模基 $C_5H_7(CH_2)_{12}CH_2$—
chaulmoogryl alcohol 晃模醇 $C_{18}H_{33}OH$
chaulmoogryl group 晃模子油基
chaulmugra (=chaulmoogra) 晃模子
chavibetol 菱叶酚；5-烯丙基-2-甲氧苯酚 $CH_2=CHCH_2C_6H_3OCH_3OH$
chavicic acid 佳味酸
chavicine 佳味碱；黑椒素
chavicol 佳味酚；对烯丙基苯酚 $CH_2=CHCH_2C_6H_4(OH)$
chavicol methyl ether 佳味醇甲醚；草蒿脑
cheapener 降(低)价(格)剂（一般指填料）
cheap oil 廉价石油；已蒸去汽油或轻馏分的油
cheat 雀麦
chebulagic acid 诃黎勒鞣花酸
chebulic acid 诃子酸
chebulinic acid ①诃黎勒酸②云实鞣酸③诃子酸
check ①龟裂；裂纹；断层②校核；核对③抑制；抑止
check analysis 校核分析；对照分析；控制分析
checkar berry oil 白珠(树)油
check ball 制逆球
check bar 制逆棒

check baseline　校核基线
check bridge　炉桥
check column　校核柱
checker　①方格；棋盘格(花样)②抑止者③成品检验工
checker brick　格子砖
checker brickwork　方格式砌砖
checkered　方格的
checkering　方格式
checker work　格式装置
check experiment　对照试验；复验
check for tightness　紧密度核验
check for zero　核验零点
check gauge　校验仪表
checking　①校核；检验②龟裂
checking calculation　验算
checking for zero　零位调整；零点校正
checking resistance　耐(抗)细裂性
checking resistance coating　耐细裂涂料；防细裂涂料
check nozzle　自动关闭喷嘴
check nut　防松螺母；保险螺(丝)帽
checkout system　检查系统；测试装置
checkout test set　检(验)测(试)设备
check plate　挡板
check point　检测点
check result　校核结果
check sample　对照试样；核对试样；质控样品
check test　对照试验；核对试验
check up　检查；验算；核对
check valve (=non-return valve)　止逆阀；止回阀
check weighing　测重
cheddite　谢德炸药；谢代特
cheese box still　干酪盒式蒸馏釜
cheese head screw　圆头螺钉
cheese winding machine　络筒机
cheesiness　酪皮〔漆膜病态之一〕
cheesy　漆膜疲软
cheetah skin　猎豹皮
ch(e)ilitis　唇炎
cheilosis　唇损害
cheirantic acid　桂竹香酸
cheirantin　桂竹香苷
cheirinine　桂竹香宁
cheirolin　桂竹素
cheiroline　桂竹香砜　$CH_3SO_2(CH_2)_3 \cdot NCS$
chekan (=cheken)　契昆番樱桃〔树名〕
chekenetin　契昆亭
chekenine　契昆碱
chekenone　契昆酮

chela　螯；钳
chelant　螯合试剂*
chelate (=chelate complex)　螯合物
chelate complex (=chelate)　螯合物
chelate compound　螯形化合物
chelate effect　螯合效应*
chelate extraction system　螯合物萃取体系
chelate fibre　螯合纤维
chelate formation constant　螯合物形成常数
chelate-forming　形成螯合物的
chelate group　螯合基团
chelate initiator　螯合引发剂
chelate polymer　螯合聚合物*
chelate reagent　螯合剂
chelate resin　螯合树脂
chelate ring　螯合环*
chelate separation process　螯合分离法
chelate solvation　螯合溶剂化
chelate solvent extraction system　螯合剂萃取体系
chelate sorbent　螯合吸附剂
chelate stability　螯合稳定性
chelating　螯合
chelating agent (=sequestering agent; sequestrant)　(多价)螯合剂
chelating anion　螯合阴离子
chelating cellulose filter　螯合纤维素过滤器
chelating chromatography　螯合色谱法
chelating-complexing agent　螯合-配位剂
chelating ion chromatography　螯合离子色谱法
chelating ion-exchanger　螯合(型)离子交换剂*
chelating ion exchange resin　螯合树脂；螯合(型)离子交换树脂
chelating ligand　螯合配体*
chelating polymer　螯合聚合物
chelating reagent　螯合试剂*
chelating resin　螯合(型)树脂*
chelating value　螯合值
chelation　①螯环化②螯合作用*
chelation chromatography　螯合色谱法
chelation extraction　螯合萃取
chelation group　螯合基团*
chelatometric estimation　螯合估测(法)
chelatometric indicator　螯合指示剂
chel(at)ometric titration　螯合滴定(法)
chel(at)ometry　螯合滴定(法)*
chelator　螯合剂
cheleritrin　白屈菜子红碱
chelerythrine (=toddaline)　白屈菜赤碱

cheletropic reaction　螯键反应*
chelic polymer　螯形聚合物
chelidamic acid (=ammonchelidonic acid)　白屈菜氨酸 $C_5H_3ON(COOH)_2$
chelidonamic acid　白屈菜氨酸　$C_7H_7O_6N$
chelidonate　白屈菜酸盐
chelidonic acid　白屈菜酸　$C_7H_4O_6$
chelidonine　白屈菜碱　$C_{20}H_{19}O_5N$
chelidonism　白屈菜中毒
chelidonoid　白屈菜素
chelidoxanthine　白屈菜黄质
chell flow comparator　壳式流动比测器
chelometric titration　螯合滴定
chelometry　螯合滴定法
chelon〔商〕　螯合剂
chelonin　蛇头宁
chelonoid　蛇头素
chelsea boot　橡筋高筒靴
chelsea shoe　(小)橡筋鞋
chemecology　化学生态学
chemic　漂液
chemical　①化学的②化学药品
chemical abrasion　化学磨损
chemical absorbent　化学吸收剂
chemical absorption　化学吸收
chemical acceleration　化学加速(作用)
chemical accelerator　①化学加速剂；化学促进剂②化学加速器
chemical acid　化学处理时带进石油产品中的酸
chemical actinometer　化学光量计
chemical actinometry　化学光量测定法
chemical action　化学作用
chemical activation　化学活化*
chemical activity　化学活性*
chemical addition agent (=chemical additive)　添加剂；化学添加剂
chemical additive (=chemical addition agent)　添加剂；化学添加剂
chemical adsorption　化学吸附
chemical affinity　①化学亲和性②化学亲和势③化学亲和能④化学亲和力
chemical after-effect　化学滞后效应
chemical after-treatment　化学后处理
chemical agent　①化学试剂②化学(药)剂
chemical amplification　化学增幅效应
chemical analysis　化学分析*
chemical antidote　化学解毒药
chemical apparatus　化工设备

chemical arithmetic　化学算术
chemical atomic weight　化学原子量
chemical atomizer　化学喷雾器
chemical attack　化学浸蚀
chemical attraction　化学吸引
chemical balance　化学天平；分析天平
chemical behavoir　化学行为
chemical binder　化学黏合剂
chemical blowing　化学发泡
chemical blowing agent　化学发泡剂
chemical blowing process　化学发泡法；泰勒莱发泡法；冷冻胶凝法
chemical bond　化学键*
chemical bonded phase chromatography　化学键合相色谱法
chemical bonded stabilizer　化学键合的稳定剂
chemical bonding　化学成键
chemical bonding process　化学黏合法
chemical-bound antioxidant　化学结合的抗氧化剂
chemical breakdown　化学损坏
chemical brick　化学砖
chemical brightenning　化学抛光；化学擦亮
chemical burns　化学灼伤
chemical calculation　化学计算
chemical carcinogen　化学致癌物
chemical carcinogenesis　化学致癌(作用)
chemical cell　化学电池
chemical change　化学变化
chemical characteristic　化学特性
chemical chop　化学切割；化学剥溶
chemical circulation cleaning　用化学品循环清洗
chemical cladding removal　化学(法)去壳
chemical cleaning　干洗；化学脱垢
chemical coagulation　化学凝聚
chemical coating　化学涂层；化学涂料
chemical coloring process　①化学染色法②化学着色法
chemical combination　化合*
chemical compatibility　化学相容性
chemical composition　①化学成分②化学组成
chemical composition of paint　涂料的化学组成
chemical compound　化合物
chemical condenser　①电解电容器②化学冷凝器
chemical conditioner　(污泥)脱水助剂
chemical conduct　化学行为
chemical consequence　化学后果〔如核反应的化学后果〕
chemical constant　化学常数
chemical constitution　化学组成；化学结构
chemical constitution formula　化学组成式；化学结构式

chemical consumption　药品消耗
chemical control　化学控制；化学过程的调节
chemical conversion treatment　化学转化处理；(金属的)化学法表面处理
chemical cooling　化学冷却
chemical correct fuel-air ratio　理论恰当油气比
chemical corrosion　化学腐蚀*
chemical coulometer　化学库仑计*
chemical coupling　化学偶合
chemical creep　化学蠕变
chemical crimp　化学卷曲；永久卷曲
chemical crossbond　化学交联键
chemical crosslink　化学交联
chemical crosslinked open tubular column　化学交联开管柱；化学交联空心柱
chemical crosslinking　化学交联
chemical crosslinking agent　化学交联剂
chemical cure　化学硫化
chemical deairing　化学除气法
chemical decanning　化学去壳*
chemical decladding　化学去壳*
chemical decomposition　化学分解
chemical decontamination　化学净化
chemical degassing process　化学除气法
chemical degradation　化学降解*
chemical deliming agent　化学脱灰剂
chemical denudation　化学溶蚀*
chemical deoxidizing　化学除氧法
chemical depolymerization agent　化学解聚剂
chemical derivative　化学衍生物
chemical derivatization method　化学衍生法
chemical derusting　化学除锈
chemical desalting　化学脱盐
chemical descaling　化学脱氧化皮
chemical development　①化学显像法②化学发展
chemical dosimeter　化学剂量计*
chemical dosimetry　化学剂量学
chemical dynamics　化学动力学
chemical ecology　化学生态学
chemical education　化学教育
chemical effect　化学效应
chemical efficiency　化学效率；反应或过程的收率
chemical (electric) gilding　化学(电)镀金
chemical element　化学元素
chemical embossing　化学压花；化学刻花
chemical emulsification　化学乳化
chemical energy　化学能*
chemical engineering　化学工程

chemical engineering design　化工设计
chemical engineering science　化学工程学
chemical engineering unit operation　化工单元操作
chemical engine hose　化工软管
chemical entity　化学个体
chemical environment　化学环境
chemical environmental resistance　耐化学性；耐化学环境性
chemical equation　化学反应式；化学方程式
chemical equilibrium　化学平衡*
chemical equilibrium constant　化学平衡常数
chemical equilibrium in a homogeneous system　均相体系化学平衡
chemical equivalence　化学全同*
chemical equivalent　化学当量
chemical equivalent nuclei and chemical equivalence　化学等同核与化学等价
chemical equivalent weight　化学当量
chemical erosion　化学刻蚀；化学浸蚀
chemical error　化学误差
chemical etching　化学浸蚀
chemical evolution　化学进化
chemical exchange　化学交换*
chemical exchange method　化学交换法
chemical excitation efficiency　化学激活效率
chemical exhaust　化学废气
chemical expert system　化学专家系统
chemical explosive　化学炸药
chemical factor　①化学因数②化学因素
chemical fertilizer　化学肥料
chemical fiber　化学纤维*
chemical fiber reinforced plastics　化学纤维增强塑料
chemical fibre spindle　化学纤维锭子
chemical fibre spinning machinery　化学纤维纺丝机
chemical fibre tow　化学纤维丝束
chemical film　化学处理膜
chemical filter　化学过滤器；化学滤毒器
chemical finishing　化学整理；化学修饰
chemical fixation　化学固定(作用)
chemical flame　化学火焰
chemical flocculation basin　化学絮凝池
chemical flow　化学流
chemical flowsheet　化学流程图
chemical flushing　(颜料的)化学挤水法
chemical foam　化学发泡；化学泡沫胶〔如聚氨酯〕
chemical foaming　化学发泡
chemical foaming agent　化学发泡剂
chemical fog　化学雾

chemical force microscopy 化学力显微镜
chemical formula 化学式
chemical formulation 化学(式)表示
chemical formula weight 化学式量
chemical fuel 化学燃料
chemical fume 化学烟雾
chemical glass 化学玻璃
chemical group composition 化学族组成〔汽油的〕
chemical hydrogen consumption 化学耗氢量
chemical hydrometer 化学比重计
chemical imperfection 化学(上的自然)缺陷
chemical incompatibility 化学不相容性
chemical induced dynamic nuclear polarization 化学诱导动态核极化
chemical industry 化学工业
chemical inertness 化学惰性
chemical inert support 化学惰性载体
chemical inhibitor 抗氧剂；化学抑制剂
chemical initiation grafting 化学引发接枝
chemical integrity 化学完整(性)
chemical interaction 化学相互作用
chemical interference 化学干扰*
chemical intermediate 化学中间体；化学半成品
chemical intumescent flame retardancy 化学膨胀阻燃
chemical intumescent flame retardant 化学膨胀型阻燃
chemical ionization (CI) 化学电离*
chemical ionization detector 化学电离检测器
chemical ionization ion source 化学电离源
chemical ionization mass spectrometry (CIMS) 化学电离质谱法
chemical ionization source 化学电离离子源
chemical ion source 化学电离离子源
chemical isotope separation 化学法同位素分离(法)*
chemical jacket removal 化学法去除(反应堆燃料)外壳
chemical kinetics 化学动力学*
chemical lace 烂花花边
chemical laser 化学激光器；化学激光
chemical leavening 化学发泡
chemical linking 化学交联
chemical literature 化学文献
chemical-loaded molecular sieve 化学品填充分子筛
chemical luminescence 化学发光
chemically active 化学活性的
chemically active pigment 化学活性颜料
chemically active plasticizer 化学活性增塑剂
chemically bonded moiety 化学键合部分
chemically bonded packing 化学键合填充物
chemically bonded phase 化学键合相*

chemically bonded phase chromatography 化学键合(固定)相色谱(法)
chemically bonded protective layer 化学结合型保护层
chemically bonded stationary phase 化学键合固定相
chemically bonded support 化学键合载体
chemically-bonded water 化学结合水
chemically bound water 化学束缚水
chemically clean 化学纯的
chemically coagulated sludge 化学法凝聚污泥
chemically combined 化学结合的
chemically combined water 化合水
chemically crosslinked open tubular column 化学交联开管柱；化学交联空心柱
chemically crosslinked rayon fibre 化学交联的人造纤维
chemically cured finish 化学固化型涂层；催化固化型涂层
chemically foamed plastic 化学发泡塑料
chemically heterogeneous blend 化学非均相混合物
chemically induced dynamic electron polarization (CIDEP) 化学诱导动态电子极化
chemically induced dynamic nuclear polarization (CIDNP) 化学诱导动态核极化*
chemically initiated polymerization 化学引发聚合
chemically mechanical stability 化学上的机械稳定性
chemically modification 化学改性
chemically modified carbon electrode 化学修饰碳电极
chemically modified electrode 化学修饰电极
chemically modified fibre 化学改性纤维
chemically modified microelectrode 化学修饰微电极
chemically modified oil 化学改性油
chemically modified optically transparent electrode 化学修饰光透电极
chemically modified paper 化学改性纸
chemically modified polymer 化学改性聚合物
chemically modified rubber 化学改性橡胶
chemically modified starch 化学改性淀粉；化学变性淀粉
chemically neutral oil 化学中性油〔指石油产品〕
chemically oxidized black 化学氧化炭黑
chemically pure (CP) 化学纯*
chemically resistant 耐化学(物质)的
chemically sensitive field effect transistor 化学敏感场效应管
chemically sensitive layer 化学敏感层
chemically sensitive semiconductor device 化(学)敏(感)半导体器件
chemically softened rubber 化学软化橡胶〔如塑解橡胶〕
chemically woven fabrics 无纺织布
chemical make up 补充药品

chemical manifold 化学歧管；化学集合管
chemical mass 化学质量；有效质量；有效份量
chemical matte finishing 化学法消光处理〔使表面粗糙化〕
chemical-mechanical polishing 化学-机械抛光
chemical mechanism 化学机制
chemical method of separation 化学分离法
chemical microscopy 显微镜分析；化学显微法
chemical microsensor 微型化学传感器
chemical milling 化学蚀刻；化学铣削；化学加工
chemical modification 化学修饰；化学改变
chemical modification electrode 化学修饰电极
chemical monitoring 化学监测
chemical mutagen 化学诱变剂；化学诱变因素
chemical mutagenesis 化学诱变
chemical name 化学名称
chemical nature 化学性质
chemical nickel plating 化学镀镍
chemical noise 化学噪声
chemical nomenclature 化学命名法
chemical octane number 化学辛烷值〔加四乙铅汽油的辛烷值〕
chemical oil dispersant 化学油分散剂
chemical oscillation 化学振荡*
chemical oscillation chronoamperometry 化学振荡计时电流法
chemical overlays 化学被复
chemical oxygen consumption 化学耗氧量
chemical oxygen demand(COD) 化学需氧量
chemical passivation 化学钝化(处理)
chemical passivity 化学钝性
chemical pattern recognition 化学模式识别
chemical permanent set 化学(永久)变定
chemical phenomenon 化学现象
chemical physics 化学物理(学)
chemical pickling 化学清洗；化学浸洗
chemical picture 化学图
chemical pipeline 化工管线
chemical plant 化(学)工厂
chemical plasticizer 化学增塑剂；塑解剂；嚼解剂
chemical plasticizing 化学增塑
chemical plating 化学镀*
chemical polarization 化学极化
chemical polishing 化学抛光
chemical polishing agent 化学擦光剂
chemical polishing effect 化学擦光作用
chemical pollutant 化学污染物
chemical porcelain 化学瓷器
chemical postmodification 化学后处理改性

chemical potential (=partial potential) 化学势
chemical potentiometer 化学电位计
chemical precipitation 化学沉淀作用
chemical preparation 化学制备；化学法预处理
chemical pretreatment agent 化学预处理剂
chemical principle 化学原理
chemical process 化学加工；化学方法；化学过程；化工工艺
chemical process design 化学过程设计
chemical promoter 化学促进剂
chemical proofing 耐(化学)药品性
chemical propellant 化学推进剂
chemical property 化学性质
chemical protection 化学防护
chemical protective clothing 化学劳保服；化学防护服
chemical pseudomorphy 化学假同晶
chemical pulp 化学纸浆
chemical pump 化工用泵
chemical pure 化学纯
chemical purification 化学纯化；化学净化法
chemical quenching 化学猝灭
chemical radiation dosimeter 化学辐射剂量计
chemical rate of processes 过程的化学速度
chemical ray 化学射线
chemical reaction 化学反应*
chemical reaction detector 化学反应检测器
chemical reaction engineering 化学反应工程学
chemical reaction film 化学反应膜
chemical reaction isotherm 化学反应等温式*
chemical reaction of caustic corrosion 碱腐蚀的化学反应
chemical reaction of polymers 聚合物化学反应
chemical reaction power 化学反应力
chemical reaction velocity 化学反应速度
chemical reactivity 化学反应性*
chemical reactor 化学反应器
chemical reagent 化学试剂
chemical reduction 化学还原(作用)
chemical refinery 化学炼油厂
chemical refining process 化学精制过程
chemical refrigeration 化学制冷
chemical regeneration 化学再生
chemical relaxation 化学弛豫
chemical repeat distance 化学(结构)重复距离
chemical reprocessing plant (核燃料)化学后处理工厂
chemical resistance 耐化学性；化学耐性
chemical resistance in binder 黏合剂的耐化学性
chemical resistance of coating 涂层的耐化学性

chemical resistance test 耐化学稳定性试验
chemical-resistant coating 耐化学涂料
chemical-resistant finish 耐化学面漆
chemical-resistant paint 耐化学(腐蚀)漆
chemical-resistant pigment 耐化学(腐蚀)颜料
chemical resistant polymer 耐化学(性)聚合物*
chemical resistant polyurethane coating 耐化学聚氨酯涂料
chemical resistant properties 耐化学性；化学稳定性
chemical-resisting 抗化学作用的；耐化学腐蚀的
chemical-resisting material 耐化学材料
chemical retardation 化学阻滞
chemical retting 化学浸渍法；化学沤麻
chemical rubber ①合成橡胶②化学改性橡胶
chemical rust removing 化学除锈
chemicals 化学药品
chemical sanitizer 化学消毒剂
chemical scission 化学断键；化学断链
chemical sensitization 化学增感作用；化学增敏作用
chemical sensor 化学传感器
chemical setting 化学定形；化学变定
chemicals from petroleum 石油化学产品
chemical shift 化学位移*
chemical shift anisotropy 化学位移各向异性*
chemical shift correlated spectroscopy 化学位移相关谱
chemical shift modulation 化学位移调制
chemical shift of active hydrogen 活泼氢的化学位移
chemical shift of NMR 核磁共振化学位移
chemical shift of X-ray spectrum X射线谱化学位移
chemical shift reagent 化学位移试剂*
chemical shift selective imaging 化学位移选择性成像
chemical shim 化学补偿(剂)；化学控制(剂)
chemical shrunk 化学防缩的
chemical silvering 化学镀银
chemical simulation 化学模拟
chemical slide rule 化学计算尺
chemical smoke 化学烟(雾)
chemical softener 化学软化剂
chemical solvent 化学溶剂
chemical specialty 化学专一性
chemical spectrometry 化学光谱学
chemical spinning 化学(反应)纺丝法；反应纺丝法
chemical spot testing 化学沾污试验
chemical spray liquid 化学喷雾液
chemical squeeze technique 化学挤压技术
chemical stability 化学稳定性*
chemical stabilization 化学稳定化(作用)
chemical stain 化学着色

chemical stencil 化学(仪器)模绘板
chemical stimulation 化学刺激(作用)
chemical stoneware 耐酸陶器
chemical strategy 化学战略
chemical stress-decay law 化学应力衰减定律
chemical stress relaxation 化学应力松弛
chemical stretch (织物)化学弹力整理
chemical stripper 化学脱漆剂
chemical stripping 化学法脱漆
chemical structural formula 化学结构式
chemical subtraction 化学去污
chemical suction hose 吸酸碱胶管
chemical surface control 化学法表面处理
chemical symbol 化学符号
chemical synthesis 化学合成(法)
chemical system 化学物系
chemical technology 化学工艺
chemical tendering 化学脆化
chemical terminology 化学术语
chemical thermodynamics 化学热力学*
chemical tower 化学塔；药剂塔〔塔中装有化学药剂〕
chemical tracer 化学指示(剂)；化学示踪物
chemical transformation 化学转换
chemical transport method 化学传递法
chemical treatment 化学处理
chemical type 化学型
chemical-type plasticizer 化学(型)增塑剂
chemical union 化合
chemical unit 化学(结构)单元
chemical unsheathing 化学除套；化学去壳
chemical valence 化合价；化学原子价
chemical vapor deposition (CVD) 化学气相沉积*
chemical vapor test 化学蒸气(腐蚀)试验
chemical vapor transportation 化学气相输运*
chemical vapour test 化学蒸气(腐蚀)试验
chemical vulcanizate 化学硫化胶
chemical vulcanization 化学硫化
chemical warfare 化学战争
chemical warfare service 化学军务
chemical warfare troops 化学战队
chemical war gas 战争毒气
chemical war material 化学战剂
chemical waste 化学废弃物
chemical weapons 化学武器
chemical weathering 化学风化
chemical welding 化学焊
chemical wood pulp 化学木浆
chemical yield 化学产率*

chemic cistern   漂液槽
chemichromatography   化学(反应)色谱(法)
chemick   漂液
chemicking   漂白；漂液处理
chemico-   化学的〔字头〕
Chemico concentrator   开美科牌浓缩器
chemicoluminescence analysis   化学发光分析
chemicophysics   化学物理学
Chemico process   开美科(厂硫酸再生)法〔从酸渣中〕
chemicrystallization   化学结晶(作用)
chemicure   化学硫化
chemi-excitation   化学激发
chemifold   化学歧管；化学集合管
chemigum   丁腈橡胶
chemigum latex   丁腈乳胶
chemi-ionization   化学电离
chemi-ionization detector   化学电离检测器
chemi-ionization mechanism   化学电离机理
chemi-ionization process   化学电离过程
chemi-ionization reaction   化学电离反应
chemiluminescence   化学发光*
chemiluminescence analysis   化学发光分析
chemiluminescence detector   化学发光检测器
chemiluminescence efficiency   化学发光效率
chemiluminescence enhancer   化学发光增强剂
chemiluminescence enzyme-linked immunoassay (CLEIA)   化学发光酶联免疫分析法
chemiluminescence immunoassay   化学发光免疫分析法
chemiluminescence label   化学发光标记
chemiluminescence probe   化学发光探针
chemiluminescence quantum yield   化学发光量子产率
chemiluminescence reaction   化学发光反应
chemiluminescence reaction kinetic curve   化学发光反应动力学曲线
chemiluminescence reagent   化学发光剂
chemiluminescence-redox detector   化学发光-氧化还原检测器
chemiluminescence sensor   化学发光传感器
chemiluminescence system   化学发光体系
chemiluminescent indicator   化学发光指示剂*
chemiluminescent material   化学发光材料
chemimechanical pulp   化学机械浆
chemiosmotic hypothesis   化学渗透假说
chem(i)otherapy   化学疗法
chemipulp   化学浆粕
chemiresistor   化敏电阻器
chemism   化学历程；化学机理
chemisoptive bond   化学吸附键

chemisorbed anionic adduct   化学吸附阴离子加合物
chemisorbed carbon monoxide   化学吸附的一氧化碳
chemisorbed film   化学吸附膜
chemisorbed material   化学吸附物
chemisorbed molecule   化学吸附分子
chemisorption   化学吸附；化学吸着
chemisorptive bond   化学吸附键
chemist   ①化学家；化学师；化学工作者②药剂师③药房
chemistry   化学*
chemistry of complex   络合物化学
chemistry of gum formation   胶质形成化学
chemistry of phosphate coating   磷酸盐处理膜的化学
chemistry of process   过程化学
chemistry of wash primer   磷化底漆(洗涤底漆)的化学
chemi-thermo-mechanical pulp (APMP)   化学预热机械(磨木)浆
Chemithon reactor   凯米松(磺化)反应器
chemmod cotton   化学改性棉
chemo-   化学
chemoceptor (=chemoreceptor)   化学接受体；化学受纳体
chemocoagulation   化学凝固(法)
chemocreep   化学蠕变
chemofining   石油化学合成；石油加工化学
chemography   化学照相法
chemoheterotrophic bacteria   化学异养(细)菌
chemo-immunity   化学免疫性
chemokinesis   化学增活现象
chemoluminescence   化学发光
chemolysis   化学分解；化学溶蚀
chemometrics   化学计量学*
chemomotive force   化学动力
chemonuclear loop   化学核反应回路
chemonuclear processing   化学核反应处理
chemonuclear reactor   化学核反应堆
chemoreceptor (=chemoceptor)   化学接受体；化学受纳体
chemoresistance   抗化学性；耐化学性
chemorheology   化学流变学
chemosensitivity   化学敏感性
chemosetting   化学固化
chemosmosis   化学渗透*
chemosorbent   化学吸附剂
chemosphere   臭氧层；光化(大气)层
chemostat   化学稳定器
chemosterilant   ①化学消毒剂②化学绝育剂
chemosynthesis   化学合成(作用)
chemosynthetic autotrophy   化能合成自养生物
chemotaxis   趋药性
chemotaxonomy   化学分类学

chemotherapeutics 化学疗法
chemotherapy 化学治疗术
chemotron 电化学转换器
chemurgic 农(业)化(学)加工的
chemurgy 农业化学加工；农业化工；实用化学
chenocholic acid 鹅(脱氧)胆酸
chenodeoxycholic acid 鹅(脱氧)胆酸
chenopodin 藜碱
chenopodium oil 土荆芥(子)油；藜油
chenotaurocholic acid 鹅牛磺胆酸 $C_{29}H_{29}O_6NS$
cherry ①樱桃酒②樱桃③樱木
cherry bay 桂樱
cherry-coal 暗煤；软煤；非焦性煤
cherry essence 樱桃香精
cherry flavour 樱桃香精
cherry gum 樱桃树胶
cherry kernel oil 樱子油
cherry-laurel oil 桂樱油
cherry-pit oil 樱子油
cherry-red 樱桃红的
cherry-red heat 樱桃炽热
cherry-stone oil 樱核油
chert 燧石；黑硅石
Cheshunt mixture 切欣特混合液
chess 雀麦
chessylite 石青；蓝铜矿
chest 柜；箱
chest drying machine 多层干燥机
chestnut ①栗木②栗子
chestnut coal 栗状煤〔3.2~4.4cm 的无烟煤块〕
chestnut extract ①栗木栲胶②栗萃；栗提取物
chestnut (size) 栗子大小〔煤〕
chestnut-wood extract 栗木浸膏
chest steaming cottage 间歇汽蒸机
chetomin 黑毛霉素
chevkinite 硅钛酸铈钇矿
chevreaux ①小山羊皮②铬鞣山羊鞋面革
chevrette ①中小山羊皮②铬鞣山羊鞋面革
chevron packing 人字形轴封
chevron type of tread pattern 人字形胎面花纹
chews 中等大小的煤
Chezy equation 切齐方程式
chian turpentine 松节油
chian varnish ①醇溶性松香清漆②松香钙脂清漆
chia oil 墨西哥油
chiastolite 空晶石
chibou 裂榄树脂
Chichibabin pyridine synthesis 齐齐巴宾吡啶合成

Chichibabin reaction 齐齐巴宾反应
chicken manure 鸡粪
chicken pox virus 水痘病毒
chicoric acid 菊苣酸；二咖啡因(基)酒石酸 $C_{22}H_{18}O_{12}$
chief valence 主(要化合)价
chiffon 绡
chigadmarene 奇格马烯
chilaphylin 智利菌素
chile 番椒
chilenite 软铋银矿
Chile nitrate 智利硝石*
Chile nitre 智利硝石；硝酸钠 $NaNO_3$
Chile saltpeter 智利硝石*
chill ①冷却；冷冻②失光③冷铸
chillagite 钨钼铅矿
chill-back 降温材料；激冷剂
chill car 冷藏车
chill-cast film 骤冷平挤薄膜
chilled ①冷却了的②冷凝了的③淬火过的
chilled cast iron (=chilled iron) 冷铸铁
chilled distillate 冷冻的馏液〔脱蜡时〕
chilled iron (=chilled cast iron) 冷铸铁
chilled iron roll 冷铸铁滚筒
chilled joint 冻胶接头；冷凝接头
chilled joint failure 黏合接头冷凝破坏
chilled roll 冷铸铁滚筒
chilled rubber 冻凝橡胶
chilled solvent 冷冻溶剂
chilled steel 淬火钢；硬化钢
chiller (=chilling machine) ①冷凝器②致冷器③冷却器
chill harden 冷硬化
chill hardening 冷硬化
chilling ①冷却②冷凝③淬火
chilling machine (=chiller) ①冷却辊②冷凝器③冷却器
chilling press 冷压机
chilling roll 冷却滚筒
chilling unit 冷却装置
chill mould 冷铸模
chill roll 冷却辊
chill-roll casting 冷辊流延法
chill roller 冷却辊
chill roll extrusion (=cast film extrusion) 平挤薄膜挤塑；骤冷辊挤塑法
chimaphilin 梅笠灵
chimaphiloid 梅笠素
chime 桶的凹槽
chimeric antibody 嵌合抗体
chimney ①烟囱②灯罩③竖坑；竖井

chimney effect　抽吸作用
chimney filter　滤管
chimney stack　烟囱
chimney type cooling tower　烟囱式冷却塔
chimney (water) cooling tower　烟囱水冷却塔
chimonanthine　山腊梅碱
chimyl alcohol　鲛肝醇　$C_{19}H_{40}O_3$
china　瓷(器)
China blue　中国蓝
china clay　陶土；瓷土
china ink　①墨②墨汁
chinaphthol　鸡纳萘酚
china ware　瓷器
China wood oil (=tung oil)　桐油
chinazoline　喹唑啉
chinchona　金鸡纳皮
Chinese bean oil　豆油
Chinese blue　中国蓝
Chinese Chemical Society　中国化学会*
Chinese cinnamon　中国肉桂；玉桂
Chinese cinnamon oil　中国(肉)桂皮油
Chinese gall nut　(中国)棓子；五倍子
Chinese gallotanninic acid　五倍子单宁酸
Chinese gutta percha　杜仲橡胶
Chinese ink　中国墨汁
Chinese lac　中国虫漆
Chinese lacquer　中国漆
Chinese lilac　华丁香；什锦丁香
Chinese mint oil　华薄荷油
Chinese neroli oil　华橙花油
Chinese oil　桐油
Chinese olive　橄榄
Chinese Pharmacopoeia　中国药典
Chinese pine　油松
Chinese red　中国红
Chinese tallow　乌桕油
Chinese tallow tree　乌桕树
Chinese tannin　五倍子单宁
Chinese vegetable tallow　乌桕油
Chinese vermillion　硫化汞；银朱〔俗〕；朱砂
Chinese wax　白蜡；虫蜡；中国蜡
Chinese white　中国白；锌白
Chinese wild pepper　竹叶椒
Chinese wood oil (=China wood oil)　桐油
Chinese yam　薯蓣
chinic acid　奎宁酸　$(OH)_4C_6H_7COOH$
chinine (=quinine)　奎宁
chiniofon　喹碘方〔药〕

chinioidina (=quinoidine)　奎诺酊
chinley coal　高级烟煤
chinoform (=quinoform)　奎诺仿；碘氯羟基喹啉
chinoic acid　醌酸
chinoidine (=quinoidine)　奎诺酊
chinol (=hydroquinol)　醌醇
chinoline (=quinoline)　喹啉　C_9H_7N
chinone　醌
chinopyrine (=quinopyrine)　奎诺比林
chinoral (=quinoral)　奎诺醛；奎宁合氯醛
chinorin　金鸡纳(树皮)苷
chinosol (=quinosol)　奎诺溶　$C_9H_6ONSO_3K \cdot H_2O$
chinothein　奎诺塞因
chinotoxine　奎诺毒
chinotropine　奎诺托品
chinotto orange　香桃叶橙
chinovin　金鸡纳(树皮)苷
chinovose　奎诺糖
chinpeimine　青贝素；青贝母碱
chintz　印花布
chiococcine　卡亨卡根碱
chiolite (=arksutite)　锥冰晶石
chiomomycin　久莫霉素
chionanthin　北美流苏树素
chip　①小片②木片③切为小片
chip-based capillary electrophoresis (=chip-based CE)　芯片毛细管电泳
chip bin　木片库
chip board　粗纸板
chip breaker　①木片破碎机；木片压碎机②石片压碎机③木屑压碎机
chip conveyor　木片运输器
chip crusher　木片压碎器
chip dispersion technique　压片分散技术；轧片分散技术〔颜料分散法之一〕
chip distributor　木片装锅器
chip drier　木片干燥器
chipless machining　无屑加工
chip machine　削片机；削木机
chip mixing machine　①木屑混合机②碎块混合机
chi-potential　χ电位〔界面电势〕
chipped area　破缺面；碎裂面
chipper　①削片机；切片机②切割；削片
chipping machine　削片机
chippy　研碎了的；粉碎了的
chip removal　排屑
chips　切片
chip screen　木片筛

chip soap 皂片；肥皂粉
chip wringer 碎片榨干器；碎片榨干离心机
chiral 手性(的)*
chiral auxiliary (reagent) 手性助剂*
chiral bonded stationary phase 手性键合固定相
chiral building block 手性子*
chiral carbon atom 手性碳原子
chiral catalyst 手性催化剂*
chiral center 手性中心*
chiral chromatography 手性色谱
chiral coordination compound 手性配合物*
chiral cyclodextrin derivatives stationary phase in GC 手性环糊精衍生物气相色谱固定相
chiral derivation method 手性衍生化法
chiral derivatization reagent 手性衍生化试剂
chiral elution 手性洗脱
chiral gas chromatography 手性气相色谱法
chiral group 手性基团
chiral induction 手性诱导*
chirality 手性*
chirality phase 手性固定相
chirality sensing electrode 对映异构体敏感电极
chiral ligand-exchange complex 配体交换型手性添加剂
chiral metal stationary phase in GC 手性金属络合物气相色谱固定相
chiral mobile phase 手性流动相
chiral mobile phase separation 手性流动相拆分法
chiral molecule 手性分子*
chiral monomer 手性单体
chiral peptide 手性缩氨酸
chiral polymer 手性高分子*
chiral reagent 手性试剂*
chiral recognition 手性识别
chiral selection gas chromatography 手性选择气相色谱法
chiral selector 手性选择剂
chiral separation 手性拆分
chiral shift reagent 手性位移试剂*
chiral solid phase separation 手性固定相拆分法
chiral solvating agent 手性溶剂化试剂
chiral solvent 手性溶剂*
chiral stationary phase 手性固定相；手性静止相
chiratin 当药苷
chiron 手性子*
chirotopic 手性
chirp 间歇噪声
chisel test (胶合板)抗凿试验
chisel tong 錾子钳；凿钳

chi square 卡方 χ^2
chi-square distribution χ^2 分布；卡方分布
chi-square test χ^2 检验法
chitenidine 奎特尼定 $C_{19}H_{22}ON_2$
chitenine 奎特宁 $C_{19}H_{22}O_4N_2$
chitin 甲壳质
chitinase 壳多糖酶；几丁(质)酶
chitobiose 壳二糖
chitodextrin 壳糊精
chitosamine (=glucosamine) 壳糖胺；氨基葡萄糖；葡萄糖胺
chitosan 壳聚糖；脱乙酰壳多糖
chittim bark 药鼠李皮
chloanthite 砷镍矿
chlor- 氯
chloracetamide 氯乙酰胺
chloracetate 氯乙酸盐(酯)
chloracetic acid 氯(代)乙酸 $CH_2ClCOOH$
chloracetone 氯丙酮 $CH_2ClCOCH_3$
chloracetonic acid 3-氯-2-羟基-2-甲基丙酸 $CH_2ClC(OH)(CH_3)COOH$
chloracetophenone 氯乙酰苯
chloracetyl- 氯乙酰(基)
chloracetyl bromide 氯乙酰溴 $CH_2ClCOBr$
chloracetyl chloride 氯乙酰氯 $CH_2ClCOCl$
chloracetyl fluoride 氯乙酰氟 CH_2ClCOF
chloracetyl halide 氯乙酰卤 CH_2ClCOX
chloracetyl iodide 氯乙酰碘 CH_2ClCOI
chloracid 含氯酸
chloracrylate 氯丙烯酸盐 $CHCl=CHCOOM$
chloracrylic acid 氯丙烯酸
chloracyl bromide 氯代酰基溴
chloracyl chloride 氯代酰基氯
chloracyl fluoride 氯代酰基氟
chloracyl halide 氯代酰基卤
chloracyl iodide 氯代酰基碘
chloral (=trichloroacetaldehyde) 氯醛；三氯乙醛 Cl_3CCHO
chloral-acetal 乙缩氯醛；二乙醇缩三氯乙醛 $Cl_3CCH(OC_2H_5)_2$
chloral acetamide 氯醛乙酰胺 $CCl_3CH(OH)NHCOCH_3$
chloral-acetone 氯醛丙酮；5,5,5-三氯-4-羟基-2-戊酮 $Cl_3CCHOHCH_2COCH_3$
chloral acetophenone 氯醛乙酰苯 $CCl_3CH(OH)CH_2COC_6H_5$
chloral alcoholate 2,2,2-三氯-1-乙氧基乙醇；端三氯偕乙氧基乙醇 $Cl_3CCH(OH)OC_2H_5$

chloral-ammonia 氯醛氨；2,2,2-三氯-1-羟(基)乙胺；(一)氨化三氯乙醛；端三氯偕羟基乙胺 $Cl_3CCH(OH)NH_2$

chloral cyanohydrin (=trichlorolactonitrile) 氯醛氰化氢；3,3,3-三氯-2-羟基丙腈；三氯丙醇腈 $Cl_3CCHOHCN$

chloral-formamide 氯醛甲酰胺；甲酰-2,2,2-三氯-1-羟乙胺；甲酰(端)三氯乙醇胺 $Cl_3CCHOHNHCHO$

chloral hydrate 水合氯醛；水合三氯乙醛 $Cl_3CCH(OH)_2$

chloralide 氯醛交酯

chloralimide 氯醛酰亚胺 $C_6H_9N_3Cl_9$

chlorallyl 氯烯丙基 C_3H_4Cl

chlorallylene 3-氯丙烯

chloraloin 氯芦荟素

chloralose (=anhydroglucochloral) 氯醛糖；脱水葡(萄)糖缩氯醛

chloralurethane 氯醛尿烷；氯醛合氨基甲酸乙酯 $CCl_3CH(OH)NHCOOC_2H_5$

chlorambucil 苯丁酸氮芥 $C_{14}H_{19}O_2NCl_2$

chloramine 氯胺

chloramine blue 氯胺蓝〔染〕

chloramine-T 氯胺T；N-氯(代)对甲苯磺酰胺钠 $CH_3C_6H_4SO_2NClNa \cdot H_2O$

chloramphenicol (=chloromycetin) 氯霉素

chloranil 氯醌；四氯代苯对醌；四氯苯醌 $C_6O_2Cl_4$

chloranilam (=chloranilamidic acid) 氯冉氨 $C_6H_3O_3NCl_2$

chloranilamide 氯冉酰胺 $C_6H_2O_3N_2Cl_2$

chloranilamidic acid (=chloranilam) 氯冉酰胺酸

chloranilanilide 氯冉酰苯胺 $C_6Cl_2(NHPh)_2O_2$

chloranilate 氯冉酸盐(或酯)

chloranil electrode 氯醌电极

chloranilic acid 氯冉酸*；2,5-二氯-3,6-二羟醌 $C_6Cl_2O_2(OH)_2$

chloraniline 氯苯胺

chloranion 氯酸根离子

chloranisidine 氯代茴香胺；氯代甲氧苯胺 $NH_2C_6H_3(Cl)OCH_3$

chloranol 氯冉醇；四氯醌醇 $C_6H_2O_2Cl_4$

chlorantimoniate 氯锑复盐

chlorantine fast color 氯冉亭坚牢染料〔直接耐光偶氮染料的一个商名〕

chlorapatite 氯磷灰石

chlorargyrite 氯银矿

chlorarsine 二甲胂基氯

chlorastrolite 绿纤石

chlorate ①氯化②氯酸盐 $MClO_3$

chlorated 氯化了的

chlorate explosive 氯酸盐炸药

chlorate liquor 氯酸钾液

chlorate of potash (=potassium chlorate) 氯酸钾

chlorathiazide 氯(苯并)噻嗪

chlorating ①氯化作用②氯化的

chlorating agent 氯化剂

chloration ①氯化作用②加氯作用

chloratit 克洛炸药；克罗替特

chloraurate 氯金酸盐 $M[AuCl_4]$

chlorauric acid 氯金酸 $H[AuCl_4]$

chlorauride 氯化金 $AuCl_3$

chloraurite 氯亚金酸盐 $M[AuCl_2]$

chlorazide 叠氮化氯 $Cl[N_3]$

chlorazol azurine 氯唑天青

chlorazol color 氯唑染料〔直接染料的一种商名〕

chlorazol dyes 氯唑(类)染料

chlorazol fast bordeaux 氯唑坚牢枣红

chlorazol orange brown 氯唑橙棕

chlorazol yellow GK 氯唑黄GK

chlorazotic acid 王水〔盐酸与硝酸的混合物〕

chlorbenzamide 氯苯甲酰胺

chlorbenzene 氯苯

chlorbutadiene 氯丁(间)二烯；2-氯-1,3-丁二烯 $CH_2=CClCH=CH_2$

chlorbutol 偕三氯叔丁醇；三氯甲基叔丁醇 $CCl_3C(CH_3)_2OH$

chlordane (=octa-klor) 氯丹；八氯化甲桥茚

chlorendate 氯菌酸盐(或酯)

chlorendic acid 氯菌酸；六氯降冰片烯二酸；1,4,5,6,7,7-六氯-二环[2.2.1]庚-5-烯-2,3-二甲酸 $C_9H_4Cl_6O_4$

chlorendic anhydride (=hexachloroendomethylene tetrahydrophthalic anhydride) 氯菌酸酐

chlorethanol 氯乙醇

chlorethene 氯乙烯

chlorethyl (=ethyl chloride) 乙基氯

chlorethylene 氯乙烯

chlorethyl ether 氯乙醚 $(ClC_2H_4)_2O$

chloretone 氯茴酮；偕三氯叔丁醇 $CCl_3C(CH_3)_2OH$

chlorex 二氯乙醚

Chlorex process 克劳雷克斯法；氯化-萃取法（二氯乙醚精炼）

chlorhydric acid 盐酸；氢氯酸

chlorhydrin 氯醇〔含Cl的醇〕

chlorhydrocarbon 氯化烃

chloric acid 氯酸 $HClO_3$

chloric ethane 氯代乙烷；乙基氯

chloridate ①氯化②氯化物

chloride 氯化物

chloride accumulator 氯化铅蓄电池

chloride ion (=chlorion)　氯根离子
chloride ion-sensing membrane　氯离子传感膜
chloride of lime　漂白粉
chloride process titanium dioxide　氯化法钛白；氯化法二氧化钛
chloride shift　氯(离子)转移
chloride volatility process　氯化挥发法
chloridization　氯化(作用)
chloridizing　氯化
chloridizing roasting　氯化焙烧
chloridometer (=chlorometer)　氯量计
chlorimetry　氯量滴定法
chlorin　二氢卟酚
chlorinate　①氯化②氯化(产)物
chlorinated　①氯化了的②绿色的
chlorinated additive　氯化添加剂
chlorinated alkyd　氯化醇酸树脂
chlorinated amide　二氯酰胺；氯化酰胺　$RCCl_2NH_2$
chlorinated aromatic　氯化芳烃
chlorinated asphalt　氯化沥青
chlorinated biphenyl　氯化联苯
chlorinated butyl rubber　氯化丁基橡胶
chlorinated camphene (=toxaphene)　氯化莰烯；毒杀芬
chlorinated compound　氯化化合物
chlorinated copper phthalocyanine　氯化酞菁铜
chlorinated cyclodienes　氯代环二烯类
chlorinated degreaser　氯化脱脂剂
chlorinated dimethyl phenol　氯代二甲酚
chlorinated diphenyl　氯化联苯
chlorinated heavy turpentine　氯化松节重油〔增塑剂〕
chlorinated hydrocarbon　氯代烃
chlorinated kerosene　氯化煤油
chlorinated lignin　氯化木素
chlorinated lime　漂白粉
chlorinated lubricant　氯化润滑剂
chlorinated naphthalene　氯化萘〔蜡状物〕
chlorinated octyl terephthalate　氯化对苯二甲酸辛酯
chlorinated paraffin　氯化石蜡
chlorinated paraffin oil　氯化石蜡油〔增黏剂〕
chlorinated paraffin wax　氯化石蜡
chlorinated phenol　①氯化苯酚②氯化酚
chlorinated polyester fibre　氯化聚酯纤维
chlorinated polyether　氯化聚醚
chlorinated polyethylene (PEC)　氯化聚乙烯
chlorinated polyphenyls　氯化多联苯〔阻燃剂〕
chlorinated polypropylene (PPC)　氯化聚丙烯
chlorinated polyvinyl chloride　氯化聚氯乙烯
chlorinated polyvinyl chloride fiber　过氯乙烯纤维

chlorinated resin　氯化树脂
chlorinated rubber　氯化橡胶
chlorinated rubber coating　氯化橡胶涂料
chlorinated rubber paint　氯化橡胶漆〔防水防锈用〕
chlorinated solvent　氯化了的溶剂
chlorinated triphenyl　氯化三苯〔增塑剂〕
chlorinating　氯化
chlorinating agent　氯化剂
chlorinating roasting　氯化焙烧
chlorination　氯化*
chlorination catalyst　氯化催化剂
chlorination furnace　氯化炉
chlorination in sunlight　日光下氯化
chlorination of drinking water　饮用水加氯〔消毒〕
chlorination process　氯化法
chlorinator　①氯化器②加氯杀菌机〔自来水厂用〕
chlorindin　氯代靛素
chlorine　①氯②氯气
chlorine addition　氯加成
chlorine bleaches　含氯漂白剂
chlorine bleaching agent　含氯漂白剂
chlorine bleach liquor　氯漂(白)液；次氯酸钠溶液
chlorine bomb　氯弹
chlorine carrier　氯载体
chlorine cell　制氯电解池
chlorine chamber　氯气室
chlorine comparator　氯(定量)比色计
chlorine cyan (=chlorine cyanide)　氯化氰　ClCN
chlorine cyanide (=chlorine cyan)　氯化氰
chlorine demand　需氯量
chlorine dioxide　二氧化氯　ClO_2
chlorine family　氯族
chlorine fastness　耐氯漂色牢度；耐氯牢度
chlorine fluoride volatility process　氯氟化物挥发法
chlorine gas　氯气
chlorine gas cell　氯气室
chlorine gas chamber　氯气室
chlorine heptoxide　七氧化二氯　Cl_2O_7
chlorine hydrate　八水合氯　$Cl_2 \cdot 8H_2O$
chlorine hydride　氯化氢　HCl
chlorine in chain　链上的氯
chlorine in ring　环上的氯
chlorine in side chain　侧链上的氯
chlorine iodide　碘化氯
chlorine monofluoride　一氟化氯　ClF
chlorine monoxide　氧化氯；氯的氧化物　Cl_2O
chlorine number　氯值〔用于浆粕的漂白性试验〕
chlorine oxide　氯的氧化物〔1.Cl_2O; 2.ClO_2; 3.Cl_2O_6;

4.Cl_2O_7; 5.$(ClO_4)_4$〕

chlorine pentafluoride 五氟化氯 ClF_5

chlorine purifier (氯化法钛白的)脱氯器

chlorine-releasing compound 释氯化合物

chlorine resistance 耐氯性

chlorine still 氯气发生器

chlorine substitution 氯取代

chlorine trifluoride 三氟化氯 ClF_3

chlorine water 氯水

chlorinex process 含钍燃料元件的高温氯化首端后处理过程

chlorinity 氯含量；氯浓度

chlorinolysis 氯解

chloriodic acid 氯碘酸

chloriodide 氯碘化物 $MCl \cdot MI$

chloriodoform 氯碘仿 $CH \cdot Cl_2I$

chlorion (=chloride ion) 氯根离子

chlorisatic acid 氯靛红酸 $C_8H_6O_3NCl$

chlorisatide 氯靛红

chlorisoindolinone orange 氯化异吲哚酮橙

chlorisoindolinone red 氯化异吲哚酮红

chlorite ①亚氯酸盐 $MClO_2$②绿泥石 $H_4(Mg,Fe)_3Si_2O_9$

chlorite-bisulphite redox system 亚氯酸盐-亚硫酸氢盐氧化还原引发体系

chlorition 亚氯酸根离子 ClO_2^-

chloritoid 硬绿泥石

chlorizate ①氯化②氯化产物

chlorizated 氯化了的

chlorizating 氯化

chlorizating agent 氯化剂

chlorization ①氯化作用②加氯作用

chlorknallgas 氯爆鸣气；爆炸性氯气氢气混合物

chlormadione 氯地孕酮

chlormerodrin 氯汞丙脲；汞新醇；3-氯汞基-2-甲氧基丙基脲 $H_2NCONHCH_2CH(OCH_3)CH_2 \cdot HgCl$

chlormethine 氮芥〔药〕

chlormezanone 氯美扎酮〔药〕

chloro- 氯代；氯(基) Cl—

chloroacetal 氯乙缩醛；氯乙醛缩二乙醇，2,2-二乙氧基-1-氯乙烷 $ClCH_2CH(OC_2H_5)_2$

chloroacetaldehyde 氯乙醛 CH_2ClCHO

chloroacetamide 氯乙酰胺 $ClCH_2CONH_2$

chloroacetanilide 氯乙酰苯胺 $ClC_6H_4NHCOCH_3$

chloroacetic acid 氯乙酸 $ClCH_2CO_2H$

chloroacetic anhydride 氯乙酸酐 $(ClCH_2CO)_2O$

chloroacetone 氯丙酮 $ClCH_2COCH_3$

chloro-acetonic acid 氯乙酮酸；α-甲基-β-氯乳酸 $CH_2Cl(CH_3)C(OH)COOH$

chloroacetonitrile (=chloromethyl cyanide) 氯乙腈 $ClCH_2CN$

chloroacetophenone 氯乙酰苯〔在此名中，未指明氯的位次〕

chloro-acetyl bromide 氯乙酰溴 $ClCH_2COBr$

chloro-acetyl chloride 氯乙酰氯 $ClCH_2COCl$

chloroacetylene 氯乙炔 $CH\equiv CCl$

chloro-acetyl fluoride 氯乙酰氟 $ClCH_2COF$

chloro-acetyl halide 氯乙酰卤 $ClCH_2COX$

chloro-acetyl iodide 氯乙酰碘 $ClCH_2COI$

chloro-acid 氯代酸

α-chloroacrylic acid α-氯丙烯酸；邻氯代丙烯酸 $CH_2=CClCO_2H$

β-chloroacrylic acid β-氯丙烯酸；(端)氯代丙烯酸 $ClCH=CHCO_2H$

4-chloro-albofungin 4-氯白真菌素

chloro-aliphatic compound 含氯脂族化合物

chloroallocinnamic acid 氯别肉桂酸 $C_9H_7ClO_2$

3-chloroallylene 3-氯丙炔 $ClCH_2C\equiv CH$

chloro-amine 氯胺

chloro-amino acid 氯代氨基酸

chloro-2-aminothiophenol 氯-2-氨基苯硫酚 $ClC_6H_3(SH)NH_2$

2-chloro-4-amino toluene-6-sulphonic acid 2-氯-4-氨基甲苯-6-磺酸

chloroaniline 氯苯胺 $ClC_6H_4NH_2$

chloro-anisidine 氯代茴香胺 $ClC_6H_3(OCH_3)NH_2$

chloroanisole 氯代茴香醚；氯苯甲醚 $ClC_6H_4OCH_3$

chloroanthracene 氯蒽

chloroanthranilic acid 氯代氨茴酸；氯代邻氨基苯甲酸 $ClC_6H_3(NH_2)COOH$

1-chloroanthraquinone 1-氯蒽醌 $C_6H_4(CO)_2C_6H_3Cl$

2-chloroanthraquinone 2-氯蒽醌

chloro-antimonate 氯锑酸盐 $M[SbCl_6]$

chloroaurate 氯金酸盐

chloroauric acid 氯金酸

chloroazobenzene 氯代偶氮苯 $ClC_6H_4N_2C_6H_5$

chloroazodin 氯脒佐定；二氯偶氮脒〔水性漆防腐败剂〕

chloroben (=o-dichlorobenzene) 邻二氯苯

chloro-benzal 亚苄基二氯；二氯甲基苯 $C_6H_5CHCl_2$

chlorobenzaldehyde 氯苯甲醛 ClC_6H_4CHO

chlorobenzamide 氯苯甲酰胺 $ClC_6H_4CONH_2$

chlorobenzene 氯苯 C_6H_5Cl

chlorobenzene rubber latex 氯苯胶乳

chlorobenzene sulfonate 氯代苯磺酸盐

chloro-benzoic acid 氯苯甲酸 ClC_6H_4COOH

chlorobenzol 氯苯

chlorobenzophenazine 氯苯并吩嗪

chlorobenzotriazole 氯苯并三唑
chloro-benzoyl bromide 氯苯甲酰溴 ClC_6H_4COBr
chloro-benzoyl chloride 氯苯甲酰氯 ClC_6H_4COCl
p-chlorobenzoyl peroxide 对氯过氧化苯甲酰
chlorobiocin 氯新生霉素
chlorobisphenol 氯代双酚
chloroborane 氯硼烷*
chloroboration 氯硼化(作用)
chloroboric acid 氯硼酸
chlorobromacetate 氯溴乙酸盐 $CHClBrCOOM$
chlorobromacetic acid 氯溴乙酸 $C_2H_2O_2ClBr$
chlorobromglycide 3-氯-2-溴丙烯 $ClCH_2CBr=CH_2$
chlorobromide ①氯溴化物②氯溴化物乳剂
chloro-bromo-acetic acid 氯溴乙酸 $ClBrCHCOOH$
chlorobromopropane 氯溴丙烷
3-chloro-1-bromopropane (=trimethylene chlorobromide) 3-氯-1-溴丙烷 $Cl(CH_2)_3Br$
2-chlorobutadiene 2-氯丁二烯
chlorobutadiene rubber 氯丁橡胶
chlorobutanol 氯代丁醇；〔通指〕偕三氯叔丁醇 $Cl_3CC(CH_3)_2OH$
chlorobutenoic acid 氯代丁烯酸
chlorobutylation 氯丁基化(作用)
chloro-t-butylphenol 氯代叔丁基苯酚 $C_4H_9C_6H_3(OH)Cl$
chlorobutyl rubber 氯丁基橡胶
chlorobutyraldehyde 氯丁醛
chloro-butyric acid 氯丁酸 ClC_3H_6COOH
chlorobutyronitrile 氯丁腈
chlorocalcite 氯钾钙石
chlorocamphor 氯樟脑
chlorocarbene 氯代卡宾；氯碳烯
chlorocarbinol 2-氯乙醇
chloro-carbonate 氯甲酸盐(或酯) $ClCOOM$
chloro-carbonic acid 氯甲酸 $ClCOOH$
chloro-carbonic ester 氯甲酸酯 $ClCOOR$
chloro-carbons〔复〕含氯烃
chloro-carbon solvent 含氯有机溶剂
chlorocarbonylation 氯羰基化*
chlorocarbonyl ferrocene (=ferrocenoyl chloride) 二茂铁碳酰氯；二茂铁羰基氯；氯碳酰二茂铁
chlorocholine chloride (CCC) 氯化氯胆碱；矮壮素；2-氯乙基三甲基氯化铵
chloro-chromic acid 氯铬酸 $Cr_2(OH)Cl$
chlorocinnamaldehyde 氯肉桂醛
α-chlorocinnamic acid α-氯肉桂酸 $C_6H_5CH=CClCOOH$
chlorocobalamin 氯钴胺素
chlorocoumarin 氯香豆素

chlorocresol 氯甲酚
chlorocresol green 氯甲酚绿
chlorocrotonaldehyde 氯巴豆醛
chlorocrotonic acid 氯巴豆酸
8-chloroctanoic acid 8-氯辛酸
chloro-cuprate 氯铜酸盐 $M_2[CuCl_4]$
chloro-cuprite 氯亚铜酸盐 $M[CuCl_2];M_2[CuCl_3]$
chloro-cuprous acid 氯亚铜酸 $H[CuCl_2];H_2[CuCl_3]$
chlorocyanamide 氯氰胺；氯代氨腈 $CNNHCl$
chloro-cyanide 氯氰化物 $MCl \cdot MCN$
chloro-cyanogen 氯化氰 $ClCN$
chlorocyanurate 氯氰尿酸盐
chlorocyclohexane 氯代环己烷
2-chloro-4,6-diamino-s-triazine 2-氯-4,6-二氨基均三嗪〔促进剂〕
chlorodibromomethane (一)氯二溴甲烷 $ClCHBr_2$
m-chloro-diethyl aniline azo-p-benzenesulfonic acid 间氯二乙苯胺偶氮对苯磺酸
chlorodifluoroacetic acid 一氯二氟乙酸 ClF_2CCOOH
chlorodifluoromethane 氯二氟(代)甲烷 $CHClF_2$
chlorodiiodomethane (一)氯二碘甲烷 $ClCHI_2$
chlorodinitro-glycerine 氯二硝基甘油
4-chloro-3,5-dinitro malachite green 4-氯-3,5-二硝基孔雀绿
chlorodioxin 氯化二氧杂环己烷
chlorodiphenyl 氯化联苯〔增塑剂〕
chlorododecane 氯十二烷
chloro-ester 含氯酯
chloroethane (=ethyl chloride) 乙基氯；氯乙烷 C_2H_5Cl
chloroethane sulfonate 氯乙烷磺酸盐
chloroethane-sulfonyl chloride 氯乙磺酰氯 $ClC_2H_4SO_2Cl$
chloroethanol 氯乙醇
2-chloroethanol (=ethylene chlorohydrin) 2-氯乙醇 $ClCH_2CH_2OH$
chloro-ether 含氯醚
chloroethylation 氯乙基化(作用)
chloroethyldiethylamine 氯乙基二乙基胺 $ClC_2H_4N(C_2H_5)_2$
chloroethyl dimethyl dodecyl ammonium chloride 氯乙基二甲基十二烷基氯化铵
chloro ethylene 氯乙烯
chloro-ethyl ester 氯乙基酯 $RCOOC_2H_4Cl$
chloroethyl ethyl sulfide 氯乙基乙基硫醚；氯硫基乙烷 $(ClC_2H_4)S(C_2H_5)$
chloroethyl mercury 氯乙基汞
chloroethyl-methyl ether 氯乙基甲基醚；氯乙氧基甲烷 $ClC_2H_4OCH_3$

chloroethyl phosphate 磷酸氯乙酯〔增塑剂〕
chlorofibre 含氯纤维
chloro-fluocarbons〔复〕 含氯氟烃
chloro-fluoride 氯氟化物
chlorofluorination 氯氟化(作用)
chloro-fluorocarbons (=chloro-fluoro-hydrocarbons)〔复〕含氯氟烃
chloro-fluoro-hydrocarbons (=chloro-fluorocarbons)〔复〕含氯氟烃
chloroform 氯仿；三氯甲烷 $CHCl_3$
chloro-formate (=chloro-formiate) 氯甲酸酯 $ClCOOR$
chloroform extract 氯仿提取物
chloro-formiate (=chloro-formate) 氯甲酸酯
chloro-formic acid 氯甲酸
chloro-formic ester 氯甲酸酯 $ClCOOR$
chloroformic solution 氯仿溶液
chloroform of crystallization 结晶氯仿
chloroform spirit 氯仿酊
chloroformyl 氯甲酰；氯羰基 $ClCO-$
chlorofumaric acid 氯代富马酸 $CH=CCl(CO_2H)_2$
chlorogenic acid 绿原酸；咖啡单宁酸
　　$(HO)_2C_6H_3CH=CHCOOC_6H_7(OH)_3COOH$
chlorogenine 绿原碱；鸡骨常山碱
chloroguanide (=proguanil) 氯胍
chloroheptane 氯庚烷
1-chloroheptane (=heptyl chloride) 1-氯庚烷
　　$CH_3(CH_2)_5CH_2Cl$
ω-chloroheptanoic acid ω-氯代庚酸
chlorohexane 氯己烷
1-chlorohexane (=hexyl chloride) 己基氯；1-氯己烷
　　$CH_3(CH_2)_4CH_2Cl$
chlorohydrination 氯醇化作用
chloro-hydrin(e) ①氯乙醇②氯代醇
α-chlorohydrin glycerin 3-氯-1,2-丙二醇；α-氯甘油
　　$CH_2ClCHOHCH_2OH$
chlorohydrin lignin 氯乙醇木素
chlorohydrin rubber 氯乙醇橡胶
chloro-hydrocarbons〔复〕 含氯烃
chlorohydroquinone 氯代氢醌；氯对苯二酚
　　$ClC_6H_3(OH)_2$
5-chloro-hydroxybenzophenone 5-氯羟基二苯甲酮〔防老剂〕
5-chloro-2-hydroxybenzophenone 5-氯代-2-羟基苯甲酮〔紫外线吸收剂〕
3-chloro-2-hydroxypropyl methacrylate 甲基丙烯酸(3-氯-2-羟丙基)酯
chloro-iodoacetic acid 氯(代)碘乙酸 $CHClICOOH$
chloroiodobenzone 氯碘苯

chloro-iodo-carbons (=chloro-iodo-hydrocarbons) 含氯碘烃
chloro-iodo-hydrocarbons (=chloro-iodo carbons) 含氯碘烃
chloroiodo-hydroxyquinoline 氯碘羟基喹啉
　　$ClOHC_6H_3=C_3H_3N$
chloroiodomethane 氯碘甲烷 $ClCH_2I$
chloro-iridic acid 氯铱酸 $H_2[IrCl_6]$
chloro-lactic acid 氯(代)乳酸 $CH_2ClCHOHCOOH$
chlorolube 含氯合成润滑油
chloro-maleic acid 氯代马来酸；(E)一氯代丁烯二酸
　　$CH=CCl(CO_2H)_2$
chloro-malic acid 氯代苹果酸；氯代羟丁二酸
　　$COOHCHOHCHClCOOH$
chloromalonic acid 氯代丙二酸 $ClCH(CO_2H)_2$
chloromercuri- 氯汞基 $ClHg-$
p-chloromercuri benzoate (PCMB) 对氯汞苯甲酸
chloromercuriferrocene 氯汞基二茂铁
chloromercuri phenol 氯汞酚 HOC_6H_4HgCl
chloro-mercury-benzene 氯汞基苯；苯汞化氯
　　C_6H_5HgCl
chloro-mercury-phenol 氯汞基苯酚 HOC_6H_4HgCl
chlorometer (=chloridometer) 氯量计
chloromethane 氯代甲烷
chloromethoxy-methylation 氯甲氧甲基化(作用)
chloromethyl acetate 乙酸氯甲酯 $CH_3CO_2CH_2Cl$
chloromethyl acylate 酰化氯甲基
chloromethylation 氯甲基化*
chloromethyl chloroformate (=chloromethyl chloroformiate)
　　氯甲酸氯甲(基)酯 $ClCH_2OCOCl$
chloromethyl chloroformiate (=chloromethyl chloroformate)
　　氯甲酸氯甲(基)酯
chloromethyl cyanide 氯甲基氰
chloromethyl cymene 氯甲基异丙苯
chloromethyl ether 氯甲基醚 $ClCH_2OR$
chloromethyl ethyl ketone 氯甲基乙基(甲)酮
　　$ClCH_2COC_2H_5$
chloromethyl methyl sulfate 硫酸氯甲基甲酯
　　$(ClCH_2)(CH_3)SO_4$
chloromethyl naphthalene 氯甲基萘
chloromethyl nonyl ketone 氯甲基壬酮
chloromethyl phenol 氯甲酚
chloromethyl phosphonamide 氯甲基膦酰胺〔阻燃剂〕
chloromethyl phosphonic diamide 氯甲基膦酰(二)胺
　　$ClCH_2PO(NH_2)_2$
chloromethyl phosphonic dichloride 氯甲基膦酰(二)氯
　　$ClCH_2POCl_2$
chloromethyl silane 氯甲基硅烷(类)
chloromethyl stearamide 硬脂酰氯甲基胺
chloromethyl sulfone 氯甲砜

chloromycetin (=chloramphenicol)　氯霉素
chloromycin　氯新霉素
chloronaphthalene　氯萘
chloronaphthalene oil　氯萘油
chloronaphthalene wax　氯萘蜡
chloronaphthol　氯萘酚　$C_{10}H_6Cl(OH)$
chloronitraniline red　氯化硝基苯胺红
chloro-nitric acid　①硝酰氯　NO_2Cl②王水
α-chloro-m-nitroacetophenone　α-氯代间硝基苯乙酮〔防霉剂〕
p-chloro-o-nitroaniline red　对氯邻硝基苯胺红
chloronitroethane　氯硝基乙烷
4-chloro-2-nitro-4-methoxy diphenylamine　4-氯-2-硝基-4-甲氧基二苯基胺
chloro-nitronaphthalene　氯硝基萘　$ClC_{10}H_6NO_2$
1-chloro-1-nitropropane　1-氯-1-硝基丙烷〔防凝剂〕
2-chloro-6-nitrotoluene　2-氯-6-硝基甲苯
chloro-nitrous acid　亚硝酰氯　NOCl
chlorononane　氯壬烷
chlorooctane　氯辛烷
chloro-p-oxy-benzoic acid　氯化对羟基苯甲酸
chloro-palladate　氯钯酸盐　$M_2[PdCl_6]$
chloro-palladite　氯亚钯酸盐　$M_2[PdCl_4]$
chloroparaffin　氯代石蜡烃；氯代烷烃
chloropentabromoethane　(一)五溴乙烷　$ClCBr_2CBr_3$
chloro-pentammine-cobalt chloride　二氯化一氯五氨合钴　$[Co(NH_3)_5Cl]Cl_2$
chloro-pentammine-platinic chloride　三氯化一氯五氨合铂　$[Pt(NH_3)_5Cl]Cl_3$
chloropentane　氯戊烷
1-chloropentane　1-氯戊烷
chlorophenasic acid　五氯苯酚　Cl_5C_6OH
chlorophenate　氯苯酚盐
chlorophenesic acid　二氯苯酚　$Cl_2C_6H_3OH$
chlorophenetole　氯苯乙醚
chlorophenic acid　一氯代苯酚　ClC_6H_4OH
chlorophenisic acid　三氯苯酚　$Cl_3C_6H_2OH$
chlorophenol red　氯酚红
chlorophenosic acid　四氯苯酚　Cl_4C_6HOH
chlorophenoxyacetic acid　氯苯氧基乙酸　$ClC_6H_4OCH_2CO_2H$
chlorophenusic acid　五氯苯酚
chlorophenyl arsine oxide　氯苯氧化砷　$ClC_6H_4As{=}O$
chlorophenyl arsonic acid　氯苯胂酸　$ClC_6H_4AsO(OH)_2$
N-chlorophenyldiazothiourea　N-氯苯重氮硫脲
o-chlorophenyldiazothiourea　o-氯苯重氮硫脲
p-chlorophenyl isocyanate　异氰酸对氯苯酯〔硫化剂〕
p-chlorophenylmaleimide　对氯苯基马来酰亚胺〔硫化剂〕

chlorophoenicite　绿砷锌锰矿
chlorophosphonation　氯膦酰化(作用)
chlorophosphonazo I　偶氮氯膦 I
chlorophosphonazo Ⅲ　偶氮氯膦Ⅲ*
chlorophosphonitrile compound　氯磷腈化合物
chloro-phosphonium　氯化(四烃基)鏻　R_4PCl
chlorophosphonylation　氯膦酰化作用
chlorophthalic acid　氯代邻苯二甲酸　$ClC_6H_3(COOH)_2$
N-chlorophthalimide　N-氯代邻苯二甲酰亚胺〔防焦剂〕
chlorophyr (e)　斑英闪长岩
chloropicrin (=nitrochloroform; trichloronitromethane)　三氯硝基甲烷；氯化苦〔俗〕　NO_2CCl_3
chloro-platinate　氯铂酸盐　$M_2[PtCl_6]$
chloro-platinic acid　氯铂(氢)酸　$H_2[PtCl_6]$
chloro-platinite　氯亚铂酸盐　$M_2[PtCl_4]$
chloro-platinous acid　氯亚铂酸　$H_2[PtCl_4]$
chloroprene　氯丁二烯；2-氯代-1,3-丁二烯　$CH_2{=}CClCH{=}CH_2$
chloroprene polymer　氯丁二烯聚合物
chloroprene rubber　氯丁二烯橡胶；氯丁橡胶
chloroprene rubber latex　氯丁胶乳
chloro propanediol　氯代丙二醇
1-chloropropane (=propyl chloride)　丙基氯；1-氯丙烷　$CH_3CH_2CH_2Cl$
2-chloropropane (=isopropyl chloride)　2-氯丙烷；异丙基氯
2-chloro-1-propanol (=primary propylene chlorhydrin)　2-氯-1-丙醇　$CH_3CHClCH_2OH$
1-chloro-2-propanol (=sec-propylene chlorohydrin)　丙氯仲醇；1-氯-2-丙醇　$CH_3CHOHCH_2Cl$
3-chloro-1-propene　3-氯-1-丙烯
chloropropylation　氯丙基化(作用)
chloropropyl functional silane　氯丙官能基硅烷〔偶联剂〕
γ-chloropropyl triethoxysilane　γ-氯丙基三乙氧基硅烷
γ-chloropropyl trimethoxy silane　γ-氯代丙基三甲氧基硅烷〔偶联剂〕
3-chloro-1-propyne (=propargyl chloride)　炔丙基氯；3-氯-1-丙炔　$HC{\equiv}CCH_2Cl$
chloropyridine　氯吡啶
chloropyrimidine　氯嘧啶〔黏合增进剂〕
chloropyrimidine dye　氯间二氮(杂)苯类染料；氯嘧啶类染料
chloroquine　氯喹
chloroquinol　氯醌醇
chloroquinoline　氯喹啉
6-chloro-2-quinoxaline carboxylic acid-1,4-dioxide　6-氯-2-喹噁啉羧酸-1,4-二氧化物
chloros　次氯酸钠

chlorosalol　氯萨罗〔药〕
chlorosamide　氯水杨酰胺〔药〕
chloroselenic acid　氯硒酸　$HClSeO_3$
chlorosilane　氯硅烷
chlorosis　萎黄病
chlorosity　氯度
chlorospinel　绿尖晶石
chloro-stannate　氯锡酸盐　$M_2[SnCl_6]$
chloro-stannic acid　氯锡酸；六氯合锡(氢)酸　$H_2[SnCl_6]$
chloro-stannite　氯亚锡酸盐　$M_2[SnCl_4]$
chloro-stannous acid　氯亚锡酸；四氯合亚锡(氢)酸　$H_2[SnCl_4]$
chloro stearic anilide　氯代硬脂基酰苯胺
chloro-styrene　氯苯乙烯；氯乙烯基苯　$C_6H_5C_2H_2Cl$
chlorosuccinic acid　氯代丁二酸　$(CHClCH_2)(CO_2H)_2$
N-chlorosuccinimide　N-氯代琥珀酰亚胺〔防焦剂〕
chlorosulfenation　氯亚磺酰化*
chlorosulfonated polyethylene　氯磺化聚乙烯；海普隆
chloro-sulfonation　氯磺化作用〔引入 SO_2Cl〕
chlorosulfonic acid　氯磺酸
chlorosulfonylation　氯磺酰化
chlorosulfophenol S　氯磺酚 S*
chlorosulfuric acid　氯磺酸　HSO_3Cl
chlorosulphonation　氯磺化
chlorosulphonyl isocyanate　氯磺酰异氰酸酯
chlorotetracycline (=aureomycin)　金霉素
chlorothen (=chloropyrilene)　氯吡林；氯森〔药〕；氯噻吡胺
chlorothion　氯硫磷杀虫剂
chlorothiophosphate　氯硫化磷酸酯
2-chloro thioxanthone　2-氯噻吨酮；2-氯-9H-噻吨-9-酮
chlorothymol　氯代百里酚　$CH_3(C_3H_7)C_6H_2(OH)Cl$
chloro-toluene　氯甲苯　$CH_3C_6H_4Cl$
chlorotoluidine　氯甲苯胺
chloro-triammine platinous chloride　氯化一氯三氨合亚铂　$[Pt(NH_3)_3Cl]Cl$
chlorotriazine　氯三嗪〔黏合增进剂〕
chlorotribromomethane　(一)氯三溴甲烷　$ClCBr_3$
chlorotriethyl silane　三乙基氯(甲)硅烷
chlorotrifluoroethylene　三氟氯乙烯
chlorotrifluoroethylene polymer　氯三氟乙烯聚合物
chlorotrimethylsilane (TMCS)　三甲基氯硅烷　$(CH_3)_3SiCl$
chlorouranate　氯铀酸盐
chloro-urea　氯脲　$NH_2CONHCl$
chlorous　亚氯的
chlorous acid　亚氯酸　$HClO_2$
chlorovinyldichloroarsine　氯乙烯基二氯胂

chlorovinylidene　偏二氯乙烯
chlorovinyl ketone　氯乙烯基甲酮
chloroxone (=2,4-D)　2,4-滴；2,4-D；2,4-二氯苯氧基乙酸
chloroxyl　盐酸辛可芬
chloroxylene　氯二甲苯
chloroxylenol　氯二甲酚
chloroxylidine　氯二甲苯胺
chloroxylonine　缎木碱
chlorpheniramine (=chlorphenamine)　氯苯那敏；氯屈米；氯曲米通〔药〕
chlorpromazin　氯丙嗪；冬眠灵
chlorpromazine hydrochloride　盐酸氯丙嗪；冬眠灵
chlorprophenpyridamine (=chlorpheniramine)　扑尔敏；氯芬胺〔药,俗名〕
chlorprothixene　氯普噻吨〔药〕
chlorquatrimycin　差向金霉素
chlorsulfonic acid　氯磺酸　$HOSO_2Cl$
chlorsulphonazo Ⅲ　偶氮氯磺Ⅲ
chlorsulphophenol S　氯磺酚 S
chlortetracycline　氯四环素〔即指金霉素〕
chlortetracycline hydrochloride　盐酸金霉素；盐酸氯四环素
chlorthion　氯硫磷〔杀虫剂〕
chloryl　二氯氧基
chlorylene　三氯乙烯
chloryl fluoride　氟化氯氧
CHN analyzer　碳氢氮元素分析仪
CHN elemental analyzer　碳氢氮元素分析仪
chock　①塞子；楔子②垫木
chocked flow　扼流
choke (=choke coil)　①气阻②节流阀③扼流线圈
choke-capacitance coupled amplifier　扼流圈-电容耦合放大器
choke coil (=choke)　①气阻②节流阀③扼流线圈
choke crushing　阻塞压碎
choke damp　窒息气；二氧化碳气
choked screen　塞孔的筛子
choke feeding　滞塞进料
choking　堵塞；淤塞
choking plug　制动栓(塞)
choky　窒息性的
choladienic acid　胆二烯酸
choladienone　胆二烯酮
cholagogue　利胆剂
cholaic acid (=taurocholic acid)　牛胆酸
cholalic acid (=cholic acid)　胆酸
cholamine (=ethanolamine)　胆胺；乙醇胺
cholane　胆(甾)烷

cholane ring 胆烷环
cholanic acid 胆(甾)烷酸 $C_{24}H_{40}O_2$
cholanthrene 胆蒽 $C_{20}H_{14}$
cholate 胆酸盐(或酯)
cholatrienic acid 胆三烯酸
choleate 络胆酸盐(或酯)
cholecalciferol 胆钙化(甾)醇；维生素 D_3
cholera red test 霍乱红实验
cholestadiene 胆甾二烯
cholestadienol 胆(甾)二烯醇
4,6-cholestadien-3-one 4,6-二烯胆甾-3-酮
cholestane 胆甾烷
cholestanol 胆甾烷醇
3-cholestanone 3-胆甾烷酮
cholestantriol 胆甾烷三醇
cholestene 胆甾烯
cholestenol 胆甾烯醇
cholestenone 胆甾烯酮
cholesterase 胆甾醇酯酶
cholesteric homopolymer 胆甾型均聚物
cholesteric liquid crystal 胆甾醇型液晶
cholesterin(e)(=cholesterol) 胆甾醇；胆固醇〔俗〕
cholesterin hydrate 胆甾醇合水 $C_{27}H_{45}OH \cdot H_2O$
cholesterol (=cholesterin (e)) 胆甾醇；胆固醇〔俗〕
cholesterol complex 胆甾醇配合物
cholesterol digitonide 胆甾醇毛地黄皂苷
cholesterol ester 胆甾醇酯
cholesterol sensor 胆甾醇传感器
cholesterone 胆甾酮
cholesteryl 胆甾烯基
cholesteryl acetate 胆甾醇乙酸酯 $CH_3CO_2C_{27}H_{45}$
cholesteryl benzoate 胆甾醇苯甲酸酯 $C_6H_5CO_2C_{27}H_{45}$
cholesteryl carboxylic ester 胆甾醇(羧)酸酯
cholesteryl compound 胆甾型化合物
cholesteryl liquid crystal 胆甾型液晶
cholesteryl phase 胆甾相
cholesteryl propionate 胆甾醇丙酸酯 $C_2H_5CO_2C_{27}H_{45}$
cholesteryl stearate 硬脂酸胆甾醇酯
cholestrophane 二甲基乙二酰脲
cholic acid (=chololic acid; colalin) 胆酸；3,7,12-三羟甾代异戊酸 $C_{23}H_{36}(OH)_3COOH$
cholic acid treatment 胆酸处理
choline 胆碱 $(CH_3)_3N(OH)CH_2CH_2OH$
choline acetate 乙酸胆碱
choline acetylase 胆碱乙酰化酶
choline bases 胆碱类
choline chloride 胆碱盐酸盐
choline dehydrogenase 胆碱脱氢酶

choline ester 胆碱酯
choline esterase 胆碱酯酶
choline esterase electrode 胆碱酯酶电极
choline oxidase 胆碱氧化酶
choline phosphate 磷酸胆碱
cholinephosphotransferase 转磷酸胆碱酶
choline transacetylase 胆碱转乙酰酶
cholla gum 仙人掌胶
cholodinic acid 胆定酸 $C_{24}H_{38}O_6$
choloidanic acid 胆丹酸 $C_{24}H_{36}O_{10}$
chololic acid (=cholic acid) 胆酸；3,7,12-三羟甾代异戊酸 $C_{23}H_{36}(OH)_3COOH$
cholonic acid 胆酮酸 $C_{26}H_{41}O_5N$
chondrilla rubber 粉苞苣橡胶
chondrodendrine 粒枝碱
chondrodite 粒硅镁石
chondrofoline 谷树叶碱
chondus 角叉藻；角叉(藻)胶
chop ①砍碎②碎块
chop-leach process 切断-浸取法
chop-out die 冲模；冲刀
choppability 短切性
chopped fiber pellets 短纤维粒料
chopped glass strand 短切玻璃丝束
chopped strand 短切原丝；短切纤维〔玻璃纤维〕
chopped strand hybrid composites 混杂短切纤维复合材料
chopped strand mat 短切原丝毡；短切原丝片
chopper ①切碎机②断裂剂〔热原子化学〕③切光器；断续器
chopper amplifier 断续放大器
chopper bar 压纱板
chord 翼弦
chord modulus 切弦模数
chorioma 绒毛膜瘤
chorionepithelioma 绒毛膜上皮癌
chorionic gonadotropin 绒毛膜促性腺激素
choromercuri 氯汞基 ClHg—
Christian effect 克利西安效应
christobalite 方英石
chroatol 氢碘合萜；碘水合萜二醇
chroma ①色度②色饱和度③色品
chroma control 色度调整
chromajar 显色缸
chromal blue G 铬蓝 G; C.I.媒介蓝 55
chroma luminance 色度亮度；色品亮度
chromammine 氨络铬
chroman 苯并二氢吡喃

chromanone 苯并二氢吡喃-4-酮
chroma number 色度值
chromanyl 苯并二氢吡喃基 C_9H_9O—
chromarod (预制)色谱棒
chromascope 色质镜
chromatape 色带
chromate 铬酸盐 M_2CrO_4
chromate brilliant brown 铬酸亮棕
chromate brilliant red 铬酸亮红
chromated aluminum 铬(酸盐)化铝；(化学)镀铬铝
chromate film 铬酸盐处理膜
chromate ion 铬酸盐离子
chromate-molybdate derivative 铬酸盐-钼盐的衍生物
chromate of lead (=lead chromate) 铬酸铅〔铬黄的主要成分〕
chromate-phosphate conversion coatings 铬酸盐-磷酸盐(化学处理)转化层(膜)
chromate-phosphate process 磷(酸)-铬酸盐表面处理法
chromate pigment 铬酸盐颜料；铬系颜料
chromate poisoning 铬酸盐中毒
chromate removal 铬盐脱除
chromate rinse 铬酸盐淋洗；铬酸盐浸洗
chromate treating 铬酸盐处理；铬化处理
chromate waste water 含铬废水
chromatic 色的；彩色的
chromatic aberration 色(像)差
chromatic addition 色加合
chromatic circle 色环
chromatic color 有彩色
chromatic difference 色差
chromatic dispersion 色散现象
chromatic dyeing 袖笼绞纱染色
chromaticity (=chromaticness) 色度；色品
chromaticity coefficient 色度系数
chromaticity coordinate 色度坐标
chromaticity diagram 色度图
chromaticity difference diagram 色差图
chromaticity scale 色度标
chromaticness (=chromaticity) 色度；色品
chromatic value 色度值
chromating 铬酸盐处理法；钝化
chromatism 色(像)差
chromato- 〔词头〕色谱
chromatobar (=chromatopencil) 色谱(固定相)棒
chromatocharger 色谱点样器
chromato-diffusion 色谱扩散
chromato-diffusion term 色谱扩散项
chromatodisk 色谱盘

chromatoelectrophoresis 色谱电泳
chromatofuge (=centrifugal chromatography) 离心色谱法
chromatogenic reaction 显色反应
chromatogenic reagent 显色剂
chromatogram 色谱图*
chromatogram analyzer 色谱分析器
chromatogram calculation 色谱图计算
chromatogram electrochemical method 色谱电化学法
chromatogram evaluation 色谱(图)评定
chromatogram homogeneity 色谱均一性
chromatogram map 色谱图
chromatogram scanning 色谱图扫描
chromatograph 色谱仪*
chromatographable 可用色谱(法)分析的
chromatograph chamber 色谱分离室；色谱分离箱
chromatograph console ①色谱仪控制台②色谱仪控制面板
chromatographer 色谱工作人员
chromatographia 色谱学
chromatographic ①色谱(分析)的；色谱法的②色谱仪的
chromatographic adsorption 色谱吸附
chromatographic adsorption analysis 色谱吸附分析；色层吸附分析
chromatographically pure 色谱纯的
chromatographically separable 可用色谱法分离的
chromatographic analysis 色谱分析
chromatographic assay 色谱验定法
chromatographic band 色谱带
chromatographic bed 色谱床
chromatographic cabinet 色谱罐(盒)；色谱箱
chromatographic characterization 色谱特性
chromatographic column 色谱柱*
chromatographic component 色谱组分
chromatographic computation 色谱计算
chromatographic condition 色谱(分析)条件
chromatographic data 色谱(分析)数据
chromatographic data processor 色谱数据处理机
chromatographic detection 色谱检测(法)
chromatographic detector 色谱仪检测器
chromatographic development 色谱展开
chromatographic efficiency 色谱效能
chromatographic effluent 色谱(分析)流出物
chromatographic evaluation 用色谱法评定
chromatographic flow system 色谱流动系统；色谱流路系统
chromatographic fraction 色谱流分
chromatographic fractionation 色谱分级
chromatographic function ①色谱功能②色谱函数

chromatographic grade　①色谱级②色谱纯
chromatographic grade resin　色谱用树脂
chromatographic instrument　色谱仪
chromatographic integral　色谱积分
chromatographic ion analyzer　离子色谱分析仪
chromatographic jar　色谱缸
chromatographic luminous method　色谱发光法
chromatographic means　①色谱方法；色谱手段②色谱工具
chromatographic mechanism　色谱机理
chromatographic mobility　色谱迁移度
chromatographic optimization coefficient　色谱优化系数
chromatographic optimization function　色谱优化函数
chromatographic packing　色谱填充物
chromatographic paper　色谱纸
chromatographic parameter　色谱参数
chromatographic peak　色谱峰*
chromatographic performance　色谱性能
chromatographic polarity　色谱极性
chromatographic polygon　(R_f值)色谱折线图
chromatographic procedure　色谱分析程序
chromatographic process　①色谱过程；色谱流程②色谱方法
chromatographic profile　色谱图形
chromatographic property　色谱性能
chromatographic reactor　色谱反应器
chromatographic refining　色谱提纯；色谱精制
chromatographic resolution (=chromatographic separation)　色谱分离
chromatographic response function　色谱响应函数
chromatographic retention　色谱保留(值)
chromatographic retention method　色谱保留法
chromatographic run　①色谱(仪)运行②色谱操作③色谱展开
chromatographic scan　色谱扫描
chromatographic sense　①(有关)色谱(的)概念②色谱自动检测
chromatographic separation (=chromatographic resolution)　色谱分离
chromatographic sheet (=chromatosheet)　(薄层)色谱板
chromatographic solution　色谱溶液
chromatographic solvent　色谱溶剂
chromatographic species　①色谱分析类型②色谱分析物质
chromatographic spectrum　(R_f值)色谱折线图
chromatographic spot test　色谱斑点试验
chromatographic surface　色谱表面
chromatographic system　色谱系统

chromatographic system analysis　色谱系统分析
chromatographic technique　色谱技术
chromatographic trace　①色谱记录仪上的图录②(在记录纸上画的)色谱图
chromatographic tube　色谱管
chromatographic work(ing) station　色谱工作站
chromatographing　色谱(分析)
chromatograph operator　色谱仪操作人员
chromatograph oven　色谱仪加热炉
chromatography　色谱法；层析法
chromatography-atomic absorption spectrometry　色谱-原子吸收光谱仪器联用技术
chromatography detector　色谱检测器
chromatography evaluate function　色谱评价函数
chromatography hygrometer　色谱法湿度计
chromatography in aqueous media　水介质色谱(法)
chromatography in centrifugal field (=chromatofuge)　离心色谱法
chromatography jar　色谱缸
chromatography of gases　气相色谱法
chromatography of ions　离子色谱(法)
chromatography on paper　纸(上)色谱(法)
chromatography optimization function　色谱优化函数
chromatography response function　色谱响应函数
chromatography tank　色谱(展开)罐(槽；室)
chromatography trough　色谱槽
chromatography with artificial intelligence　智能色谱
chromatography with mobile adsorbent　移动吸附剂色谱(法)
chromatography workstation　色谱工作站
chromatomap　色谱(图)
chromatomembrane method　色谱膜法
chromatometer　比色计；色度计
chromatopack　色谱纸束
chromatopencil (=chromatobar)　色谱(固定相)棒
chromatopile (=paper-pile)　(色谱)纸堆
chromatoplate　(薄层)色谱板
chromato-polarograph　色谱-极谱仪
chromato-polarography　色谱-极谱法
chromatoroll　色谱(纸)圆筒
chromatosheet (=chromatographic sheet)　(薄层)色谱板
chromatospectrophotometric　色谱分光光度法的
chromatostack　色谱堆
chromato-stick　色谱棒
chromatostrip　色谱条
chromatosulfuric acid　铬硫酸；三氧铬合硫酸　H_2CrSO_7;$H_2[CrO_3 \cdot (SO_4)]$
chromato-thermographic　热色谱的；热色层的
chromato-thermographic gradient　热色谱梯度

chromatothermography 热色谱法
chromatotube 色谱管
chromazol violet 铬唑紫
chrome 铬 Cr
chrome acid bath 铬酸浴
chrome alum (=chromic alum) 铬明矾
chrome avanturine 铬质金星玻璃
chrome azurol S 铬天青S*
chrome bleach (=chrome bleaching) 铬漂白
chrome bleaching (=chrome bleach) 铬漂白
chrome board 彩色石印纸板
chrome brick 氧化铬(耐火)砖
chrome calf 铬鞣小牛革
chrome carbonyl 羰基铬 $Cr(CO)_6$
chrome chamois 铬油鞣
chrome colors 铬(处理)染料〔染〕
chrome content 含铬量
chrome cyanine R(=pontachrome blue) 滂铬蓝〔染〕
chromed 镀(了)铬的
chrome determination 铬含量测定
chromed hide powder 铬处理的皮粉
chrome dye 铬媒染料
chrome dyeing 铬染法
chrome dyestuff 铬(处理)染料
chrome fast cyanine 铬坚牢花青
chrome flashing 铬层薄镀
chrome-formate tanning 铬鞣液-甲酸钠蒙囿鞣
chrome garnet 铬榴石
chrome green (=chromium green) 铅铬绿；铬绿〔铅铬黄与铁蓝的混合颜料〕
chrome iron 铬铁
chrome iron ore 铬铁矿
chrome kid 铬鞣羔革
chrome kip 铬鞣中牛革
chromel 镍铬合金〔含 Ni90%, Cr10%〕
chromel-alumel 镍铬-镍铝〔热电偶材料〕
chromel-alumel couple 铬铝热电偶
chromel-alumel thermoelement 铬镍-铝镍热敏元件
chromel-copel thermoelement 铬镍-铜镍热敏元件
chrome leather 铬革
chrome leather blue 铬革蓝
chrome lemon 铬柠檬
chrome liquor 铬汁；铬液〔革〕
chromel triangle 铬若灭三角
chromel wire gauze 铬若灭铉网
chrome molybdenum steel 铬钼钢
chrome mordant 铬媒染剂
chromene 色烯；苯并吡喃

chrome nickel steel 铬镍钢
chrome ochre (=chromium oxide green) 氧化铬绿
chrome orange 铬橙
chrome ore 铬矿石
chrome oxide 氧化铬〔即三氧化二铬〕
chrome oxide green (=chromium oxide green; chrome ochre) 氧化铬绿
chrome oxide green (=chromium oxide green) 氧化铬绿
chrome paper 铜版纸
chrome pigment 铬颜料
chrome plate 镀铬压平板
chrome plated 镀(了)铬的
chromeplated polished tooling (镀)铬抛光工具
chrome plating ①铬鞣②镀铬
chrome polish ①(镀)铬抛光②铬抛光剂
chrome powder 粉状铬鞣剂
chrome printing bordeaux 铬印染枣红
chrome printing orange 铬印染橙
chrome printing yellow G 铬印染黄 G
chrome-pyrazole 铬吡唑
chrome recovery 铬鞣液回收
chrome red 铬(铅)红〔为碱式铬酸铅〕
chrome refractory 铬质耐火材料
chrome scarlet 铬猩红
chrome shavings 铬革屑
chrome steel 铬钢
chrome syntans 铬合成鞣剂
chrome tannage 铬鞣(法)
chrome tanned 铬鞣的
chrome tanned leather 铬鞣革
chrome tanner 铬鞣工〔革〕
chrome tanning 铬鞣(法)
chrome tanning extract 铬鞣萃〔革〕
chrome trim 镀铬装潢件
chrome ulcers 铬溃疡〔铬中毒引起的病〕
chrome ultramarine 铬群青〔一种绿颜料，组成为铬硅酸钠或铬硅酸铝钠〕
chrome uptake 铬吸收
chrome-vanadium steel 铬钒钢
chrome vermilion 铬朱红
chrome vermillion 钼(铬)红〔别名〕
chrome yellow 铬黄*
chromia-alumina catalyst 氧化铬-氧化铝催化剂
chromiate 络合铬酸盐
chromic ①(正)铬的；三价铬的；六价铬的②铬的
chromic acid 铬酸 H_2CrO_4
chromic acid alumite 铬酸(阳极)氧化铝
chromic acid cell 铬酸电池

chromic acid mixture 洗液；铬酸混合液
chromic acid oxidation 铬酸氧化法
chromic acid rinse 铬酸淋(洗)
chromic acid-strong phosphoric acid decomposition method 铬酸-浓磷酸分解法
chromic alum 铬矾
chromic anhydride 三氧化铬；铬酐 CrO_3
chromic aquopentammine halide 卤化一水五氨合铬 $[Cr(NH_3)_5 \cdot H_2O]X_3$
chromic arsenide 砷化铬
chromic bleaching 铬漂白
chromic boride 硼化铬
chromic bromide 溴化铬 $CrBr_3$
chromic carbide 碳化铬 Cr_3C_2
chromic carbonate 碳酸铬
chromic chloride 氯化铬 $CrCl_3$
chromic chloropentammine dichloride 二氯化一氯五氨合铬
chromic chlorosulfate 硫酸氯铬 $CrClSO_4$
chromic color 铬(处理)染料〔染〕
chromic dihydroxychloride 一氯二氢氧化铬 $Cr(OH)_2Cl$
chromic dioxycarbonate 碳酸二氧铬 $OCr(CO_3)CrO$
chromic dye 铬(处理)染料〔染〕
chromic dyeing 铬染(法)
chromic fluoride 氟化铬 CrF_3
chromic hexamminochloride 三氯化六氨铬 $[Cr(NH_3)_6]Cl_3$
chromic hydroxide 氢氧化铬 $Cr(OH)_3$
chromic iodide 三碘化铬 CrI_3
chromic metaphosphate 偏磷酸铬 $Cr(PO_3)_3$
chromic nitrate 硝酸铬 $Cr(NO_3)_3$
chromic orthophosphate (正)磷酸铬 $CrPO_4$
chromic oxide 氧化铬；三氧化二铬 Cr_2O_3
chromic oxide-on-alumina catalyst (氧化)铬-(氧化)铝催化剂
chromic oxychloride 氯氧化铬 $CrOCl$
chromic oxydisulfate 二硫酸一氧化二铬 $Cr_2O(SO_4)_2$
chromic oxyhydroxide 羟氧化铬 $CrO(OH);HCrO_2$
chromic phosphate 磷酸铬 $CrPO_4$
chromic-phosphoric rinse 铬-磷酸淋洗
chromic potassium alum (=chromic potassium sulfate) 铬钾矾；硫酸铬钾
chromic potassium cyanide 氰化铬钾
chromic potassium oxalate 草酸铬钾 $K_3Cr(C_2O_4)_3 \cdot 3H_2O$
chromic potassium sulfate (=chromic potassium alum) 硫酸铬钾；铬钾矾 $K_2SO_4 \cdot Cr_2(SO_4)_3 \cdot 24H_2O; KCr(SO_4)_2 \cdot 12H_2O$
chromic pyrophosphate 焦磷酸铬 $Cr_4(P_2O_7)_3$

chromic rubidium alum (=chromic rubidium sulfate) 铬铷矾；硫酸铬铷
chromic rubidium sulfate (=chromic rubidium alum) 铬铷矾；硫酸铬铷 $Rb_2SO_4 \cdot Cr_2(SO_4)_3 \cdot 24H_2O; RbCr(SO_4)_2 \cdot 12H_2O$
chromic selenite 亚硒酸铬 $Cr_2(SeO_3)_3$
chromic silicide 硅化铬 Cr_3Si_2
chromic sodium alum (=chromic sodium sulfate) 铬钠矾；硫酸铬钠
chromic sodium sulfate (=chromic sodium alum) 铬钠矾；硫酸铬钠 $Na_2SO_4 \cdot Cr_2(SO_4)_3 \cdot 24H_2O; NaCr(SO_4)_2 \cdot 12H_2O$
chromic sulfate 硫酸铬 $Cr_2(SO_4)_3$
chromic sulfate green 硫酸铬绿〔搪瓷和陶瓷用颜料〕
chromic sulfide 硫化铬 Cr_2S_3
chromic tanning 铬鞣(法)
chromic tartrate 酒石酸铬 $Cr_2(C_4H_4O_6)_3$
chromic thiocyanate 硫氰酸铬 $Cr(CNS)_3$
chromic trioxide 铬酸酐
chromicyanide 铬氰基 $Cr(CN)_6\equiv$
chroming ①镀铬②铬鞣
chroming machine ①铬鞣机②镀铬机
chrominum cross-linkage 铬交联(作用)；铬交键(作用)
chromising 渗铬
chromite 铬铁矿*
chromite brick 铬砖
chromium 铬〔24号元素，化学符号Cr〕
chromium acetate 乙酸铬
chromium alum 铬(明)矾
chromium arsenide 一砷化铬 $CrAs$
chromium boride 硼化铬 $CrB;Cr_3B_2$
chromium carbide 碳化铬 $CrC_3;CrC_4$
chromium carbonate 碳酸铬
chromium chloride 氯化铬
chromium coating 镀铬层；铬涂覆；铬镀层
chromium compound 铬化合物
chromium cross-linkage 铬交联(作用)；铬交键(作用)
chromium dibromide 二溴化铬 $CrBr_2$
chromium dichloride 二氯化铬 $CrCl_2$
chromium dioxide 二氧化铬 $CrO_2;Cr_2O_4$
chromium dioxydichloride 二氯二氧化铬 CrO_2Cl_2
chromium glass 铬玻璃
chromium green (=chrome green) 铬绿；铅铬绿〔铅铬黄与铁蓝的混合颜料〕
chromium hemitrioxide 氧化铬；三氧化二铬 Cr_2O_3
chromium hemitrisulfide 三硫化二铬 Cr_2S_3
chromium hydrate 氢氧化铬
chromium hydrate green 水合氧化铬绿

chromium hydroxide 氢氧化铬
chromium hydroxide green 氢氧化铬绿
chromium intensification 铬盐加厚法
chromium metal 铬金属
chromium minerals 铬矿
chromium monochloride 一氯化铬 CrCl
chromium monosulfide 一硫化铬 CrS
chromium monoxide 一氧化铬 CrO
chromium mordant 铬媒染剂
chromium nitride precipitate 氮化铬沉淀物
chromium oxide ①三氧化二铬②铬的氧化物〔1.CrO; 2.Cr_2O_3; 3.CrO_2; 4.CrO_3; 5.CrO_4〕
chromium oxide green (=chrome ochre; chrome oxide green) 氧化铬绿
chromium peroxide 过(四)氧化铬 $CrO_2[O_2]$
chromium phosphide 一磷化铬 CrP
chromium plated 镀(了)铬的
chromium plated goods 镀铬制品
chromium plated sand finish surface 镀铬喷砂表面
chromium plated steel 镀铬钢
chromium plating 镀铬
chromium potassium oxalate 草酸铬钾
chromium potassium sulfate (=chromium potassium sulphate) 硫酸铬钾; 铬钾矾
chromium salt 铬盐
chromium sesquioxide 三氧化二铬 Cr_2O_3
chromium sesquisulfide 三硫化二铬 Cr_2S_3
chromium silicide 硅化铬
chromium soap 铬皂
chromium sodium sulfate (=chromium sodium sulphate) 硫酸铬钠; 铬钠矾
chromium steel 铬钢
chromium sulfate (=chromic sulfate) 硫酸铬
chromium sulfite 亚硫酸铬
chromium tetroxide 四氧化铬 CrO_4
chromium titanide 钛化铬
chromium trichloride 三氯化铬 $CrCl_3$
chromium trifluoride 三氟化铬 CrF_3
chromium trifluoroacetate 三氟乙酸铬
chromium triiodide 三碘化铬 CrI_3
chromium trioxide 三氧化铬 CrO_3
chromium tungstate 钨酸铬 $Cr_2(WO_4)_3$
chromizing 铬化(处理); 渗铬; 扩散镀铬
chromo board 铜版卡〔俗〕
chromo-citronine 铬柠檬黄
chromocytometry 血红蛋白测定法
chromogen ①发色团; 生色团②色原③铬精(染)
chromogen azurin (e) 铬精天青

chromogen colors〔复〕 铬精染料〔一种铬处理丝染料的商名〕
chromogen cyanine 铬精花青
chromogen green 铬精绿
chromogenic agent 显色剂*
chromogenic reaction 显色反应
chromogenic reagent 显色试剂
chromogen indigo 铬精靛蓝
chromohercynite 铬铁尖晶石
chromo-isomer 异色异构物
chromo-isomerism 异色异构(现象)
chromometer 比色计; 色度计
chromometry 亚铬滴定法
chromone 色酮; 苯并-γ-吡喃酮
chromonol 色酮酚
chromophore 生色团*; 发色团
chromophoric 发色的
chromophoric group 发色团
chromophoric property 发色性质
chromophoric theory 发色团学
chromophotometer 比色计
chromophous emulsion 多色乳状液
chromoscan 彩色扫描
chromoscan densitometer 彩色扫描密度计
chromosol 铬溶染料〔一种媒染偶氮染料的商名〕
chromosol brown 铬溶棕
chromosol color 铬溶染料
chromosol yellow 铬溶黄
chromotrope 铬变素
chromotropic acid 变色酸*; 1,8-二羟基-3,6-二磺酸
chromotropy 异色异构(现象)
chromous ①亚铬的; 二价铬的②铬的
chromous acetate 乙酸亚铬 $Cr(C_2H_3O_2)_2$
chromous acid 亚铬酸 $HCrO_2$; H_3CrO_3
chromous bromide 溴化亚铬 $CrBr_2$
chromous carbonate 碳酸亚铬 $CrCO_3$
chromous chloride 氯化亚铬 $CrCl_2$
chromous compound 亚铬化合物
chromous hydroxide 氢氧化亚铬 $Cr(OH)_2$
chromous iodide 碘化亚铬 CrI_2
chromous oxide 一氧化铬 CrO
chromous phosphate 磷酸亚铬 $Cr(PO_3)_2$; $Cr_3(PO_4)_2$
chromous sulfate 硫酸亚铬 $CrSO_4$
chromous sulfide 一硫化铬; 硫化亚铬 CrS
chromoxane cyanine R 铬花青 R
chromoxide green 氧化铬绿
chromtrope 变色素
chromyl ①铬酰 CrO_2—②氧铬基 CrO—

chromyl bromide　铬酰溴；二溴二氧化铬　CrO_2Br_2
chromyl chloride　铬酰氯；二氯二氧化铬　CrO_2Cl_2
chromyl chloride reaction　铬酰氯反应
chromyl chloride rubber　铬酰氯化橡胶
chromyl sulfate　硫酸氧铬　$(CrO)_2SO_4$
chronic　慢性的
chronic pollution　慢性污染；长期污染
chronic toxicity　慢性毒性
chronoamperometry　计时电流法*；计时安培法
chronocoulometry　计时库仑法*；计时电量法
chronogeochemistry　年代地球化学
chronograph　计时器
chronometer　测时器
chronometric analysis　计时分析
chronometric method　计时法
chronophotography　记时摄影
chronopotentiogram　计时电位图；计时电位滴定曲线
chronopotentiometric stripping analysis　计时电位滴定溶出分析
chronopotentiometric titration　计时电位滴定(法)
chronopotentiometry with linear current sweep　线性扫描计时电位法
chronoscope　计时器
chronoteine　高速摄影机
chronovoltammetry　计时伏安法
chrothiomycin　色硫霉素
chrycentrine　黄花(荷包牡丹)碱
chry(s)-〔词头〕　柯
chrysalicic acid　金蛹酸　$C_7H_5O_6N_3$
chrysalis oil　金蛹油
chrysamine　柯胺　$Na_2C_{18}H_{16}O_6N_4$
chrysammic acid (=chrysamminic acid)　柯氨酸；2,4,5,7-四硝基-1,8-二羟基蒽醌　$(OH)_2C_{14}H_2O_2(NO_2)_4$
chrysammidic acid　柯酰氨酸　$NH_4C_7HO_2(NO_2)_2$
chrysamminic acid (=chrysammic acid)　柯氨酸；2,4,5,7-四硝基-1,8-二羟基蒽醌　$(OH)_2C_{14}H_2O_2(NO_2)_4$
chrysanilic acid　柯苯胺酸
chrysaniline　柯苯胺　$C_{19}H_{15}N_3$
chrysanisic acid　柯茴香酸　$NH_2C_6H_2(NO_2)_2COOH$
chrysanthal　菊醛；6-丙基-4-甲基-3-环己烯甲醛
chrysanthanol　菊醇
chrysanthanone　菊酮
chrysanthemane　菊烷
chrysanthemate　菊酸盐
chrysanthemaxanthin　菊黄质
chrysanthemic acid　菊酸
chrysanthemin　紫菀苷；花青-3-葡萄糖苷；菊色素
chrysanthemine　菊胺　$C_{14}H_{28}O_3N_2$

chrysanthemum concrete　菊花浸膏
chrysanthemum dicarboxylic acid　菊二酸
chrysanthemum dicarboxylic acid monomethylester　菊二酸甲酯
chrysanthemumic acid　菊一酸；菊甲酸
chrysanthemum monocarboxylic acid　菊一酸；除虫菊一羧酸；菊甲酸
chrysanthemum-shaped　菊花状的
chrysanthene　菊烯　$C_{18}H_{35}O$
chrysanthenone　菊烯酮
chrysanthine　菊质　$C_{10}H_{13}O_3$
chrysanthone (=α-thujone)　α-苧酮
chrysarobin　柯桠素
chrysarobin triacetate (=eurobin)　三乙酸柯桠素酯
chrysarobol　柯桠醇
chrysatropic acid (=scopoletin)　柯阿托酸；莨菪亭
chrysazin　柯嗪；1,8-二羟基蒽醌　$C_{14}H_8O_4$
chrysazol　柯札醇；1,8-二羟基蒽　$C_{14}H_{10}O_2$
chryseam　柯阿姆　$C_4H_5N_3S_2$
chrysene　䓛　$C_{18}H_{12}$
chrysenequinone　5,6-䓛醌
chrysenic acid　䓛生酸〔α指2-萘基苯甲酸；β指2-苯基萘甲酸〕
chrysenyl　䓛基　$C_{18}H_{11}—$
chrysidine　柯啶；苯并[c]吖啶　$C_{17}H_{11}N$
chrysil rubber　一枝黄橡胶
chrysin (=5,7-dihydroxy-flavone)　柯因；5,7-二羟黄酮　$C_{15}H_{10}O_4$
chryso-〔词头〕　柯；金黄
chrysoberyl　金绿宝石；金绿玉
chrysocolla　硅孔雀石
chrysoeriol　柯伊利素
chrysofluorene　柯芴；11-氢-苯并[a]芴
chrysohydroquinone　柯氢醌；5,6-䓛二醇
chrysoidine　柯衣定；2,4-二氨基偶氮苯；碱性菊橙〔染〕　$(H_2N)_2C_6H_3N_2C_6H_5$
chrysoidine hydrochloride　盐酸柯衣定；碱性菊橙；2,4-二氨基偶氮苯盐酸盐　$C_{12}H_{12}N_4 \cdot HCl$
chrysoketone　柯甲酮；1,2-苯并芴酮
chrysolepic acid (=picric acid)　苦味酸
chrysolite　贵橄榄石
chrysophanic acid　大黄根酸；大黄酚；3-甲基-1,8-二羟基蒽醌　$C_{14}H_5(OH)_2(CH_3)O_2$
chrysophanin　大黄苷　$C_{20}H_{20}O_9$
chrysophanol　大黄酚（即：大黄根酸）
chrysophenine　直接菊黄〔染〕
chrysophenine G　直接菊黄 G
chrysophyscin　蜈蚣苔素

chrysopicrin (=vulpic acid)　金黄苦；苔酸
chrysoprase　绿玉髓
chrysoquinone (=chrysenequinone)　䓛醌
　　$C_{10}H_6(CO)_2C_6H_4$

chrysorrhetin　番泻叶黄苷；山扁豆叶黄苷
chrysotile　温石棉
chrysotile　纤蛇纹石；温石棉
chrysotoxin　金黄毒；麦角黄毒素
chuchuarine　打印果碱
chuck　夹盘〔商〕
chuck gage　气压表
chuck jaw　卡盘爪
chuck lathe　卡盘车床；钢丝圈卷机
Chugaev elimination　丘加叶夫消除法
chunk　敞口模
chunk out　花纹掉块
chunky particle　厚块状粒子
Chuquicamata copper extraction process　丘卡马塔炼铜法
church method　教堂法（测定遮盖力的一种方法）
church oak varnish　短油内用清漆；教堂栎木清漆
churn　黄化鼓；搅乳器；摇转搅拌筒
churn barrette　碱纤维素黄化器
churning　搅乳器；搅拌；搅动
churning time　搅拌时间
chute　斜槽；滑槽；滑运道
chute feed　斜槽进料；滑槽式喂料器
chute feeder　送料槽
chute hopper　斜沟漏斗
chymotrypsin　胰凝乳蛋白酶；糜蛋白酶
chypre　素心兰香型；西普香水
chypre compound　素心兰香精
chypre green　绿土（硅酸盐绿）
ciba black　汽巴黑〔染〕
cibacet colors　汽巴赛特染料〔染〕
cibacet dyes　汽巴赛特染料
ciba colors　汽巴染料〔染〕
cibanone blue　汽巴弄蓝〔染〕
cibanone bordeaux　汽巴弄枣红〔染〕
cibanone colors　汽巴弄染料〔染〕
cibanone gold orange　汽巴弄金橙
cibanone olive　汽巴弄橄榄绿
cibanone orange　汽巴弄橙
cibanone yellow　汽巴弄黄〔染〕
cibazol　磺胺噻唑
cibetone (=civetone)　灵猫酮
cicada wax　蝉蜡
cichoriin　菊苣苷

cichory root　菊苣根
ciclacidin (=cyclacidin)　环杀菌素
ciclamicin (=cyclamycin)　环霉素
ciclosporin　环孢素〔药〕
cicutene　毒芹萜烯　$C_{10}H_{16}$
cicutine　毒芹碱
cicutol　毒芹醇
cicutoxin　毒芹素
cicutoxine　毒芹素
cider　苹果汁
cider brandy　苹果汁白兰地
cigar　雪茄烟
cigar box wood　烟香椿；印度红木
cigarette　卷烟；香烟
cigarette paper (=cigarette tissue)　卷烟纸
cigarette tissue (=cigarette paper)　卷烟纸
cigar flavour　烟草香精；烟用香精
cilazapril　西拉普利〔药〕
cilia　纤毛
ciliatine　氨乙基膦酸
cillimycin (=lincomycin)　林肯霉素；盐酸林可霉素
cimene (=dipentene)　二聚戊烯；双戊烯
cimetidine　西咪替丁〔药〕
cimicic acid　臭虫酸　$C_{15}H_{28}O_2$
cimicidine　臭单枝夹竹桃碱
cimicifugin　升麻素
cimigenol　升麻醇
cinaebene　山道年萜烯　$C_{10}H_{16}$
cinaebene camphor　山道年脑
cinchaine　辛咖因
cinchene　辛可烯　$C_{19}H_{20}N_2$
cinchol　辛可醇　$C_{20}H_{34}O$
cincholepidine (=lepidine)　辛可勒皮啶（即：勒皮啶）
cincholine　辛可灵
cinchomeronic acid　辛可部酸；3,4-吡啶二甲酸
　　$C_7H_5O_4N$
cinchomeronyl　辛可部酰
cinchona　金鸡纳树皮
cinchona alkaloid　金鸡纳生物碱
cinchonamine　辛可胺　$C_{19}H_{24}ON_2$
cinchonane　脱氧辛可宁　$C_{19}H_{22}N_2$
cinchonate　辛可酸盐；金鸡纳酸盐
cinchonic acid　辛可酸；金鸡纳酸
cinchonicine (=cinchotoxine)　辛可尼辛；辛可毒
cinchonidine　辛可尼定　$C_{19}H_{22}ON_2$
cinchonine　辛可宁*
cinchoninic acid　辛可宁酸
cinchophen (=atophan)　辛可芬；阿托方

$C_6H_5C_9H_5NCOOH$
cinchotannic acid 辛可单宁酸
cinchotine 辛可亭
cinchotoxine 辛可毒
cinchovatin 辛可尼定
cinder 炉渣；熔渣；矿渣
cinder notch 渣口
cinder pit 曲结熔渣
cindery ①灰渣的②似熔渣的
cinefaction (=cineration) 灰化；煅灰法
cinematic 自动络筒的
cinene 芷烯
cinenic acid 桉烯酸 $C_9H_{16}O_3$
1,8-cineole (=encalyptol) 白千层脑；桉叶油素
cineole (=eucalyptol; eucalyptole) 桉叶油素；桉树脑 $C_{10}H_{18}O$
cineolic acid 桉树脑酸 $C_{10}H_{16}O_5$
cineration (=cinefaction) 灰化；煅灰法
cinereous 灰的
cinerin 瓜菊酯；瓜叶除虫菊酯
cinerolone 瓜菊醇酮；瓜叶除虫菊醇酮
cineromycin 烬灰霉素
cinerubin 烬灰红菌素
cine substitution* 移位取代*
cinnabar 辰砂；朱砂；硫化汞
cinnamal (=cinnamylidene) 亚肉桂基；苯亚烯丙基 $C_6H_5CH=CHCH=$
cinnamaldehyde (=cinnamic aldehyde) 肉桂醛 $C_6H_5CH=CHCHO$
cinnamaldehyde ethylene glycol acetal 肉桂醛乙二醇缩醛
cinnamalmalonic ester 亚肉桂基丙二酸酯 $C_6H_5CH=CH=CHC(COOR)_2$; $C_6H_5CH=CH=CHC(COOC_2H_5)_2$
cinna(ma)mide (=cinnamic amide) 肉桂酰胺 $C_6H_5CH=CHCONH_2$
cinnamate 肉桂酸盐
cinnamein 肉桂酸苄酯 $C_6H_5CH=CHCOOCH_2C_6H_5$
cinnamene 肉桂烯〔苯乙烯的别名〕
cinnam(en)ol (=styrene) 苯乙烯
cinnamenyl (=styryl) 苯乙烯基 $C_6H_5CH=CH—$
cinnamenyl angelic acid 苯乙烯基当归酸
cinnamic 肉桂的
cinnamic acid 肉桂酸；苯基-2-丙烯酸 $C_6H_5CH=CHCOOH$
cinnamic alcohol 肉桂醇 $C_6H_5CH=CHCH_2OH$
cinnamic aldehyde (=cinnamaldehyde) 肉桂醛 $C_6H_5CH=CHCHO$

cinnamic aldehyde dimethyl acetal (肉)桂醛二甲缩醛
cinnamic amide 肉桂酰胺
cinnamic anhydride 肉桂(酸)酐 $C_{18}H_{14}O_3$
cinnamic ester 肉桂酯
cinnamic ether 肉桂醚
cinnamide 肉桂酰胺
cinnamilidene 亚肉桂基；苯亚烯丙基
cinnamol 苯乙烯
cinnamomic (=cinnamic) 肉桂的；桂皮的
cinnamon 肉桂
cinnamon bark 肉桂皮
cinnamon bark oil (肉)桂皮油
cinnamone 二苯乙烯甲酮
cinnamonin 肉桂地菌素
cinnamon leaves oil 肉桂叶油
cinnamon oil 肉桂油
cinnamonol 肉桂萜醇
cinnamon spirit 肉桂精；肉桂酊
cinnamon water 肉桂汁
cinnamophenone (=chalcone) 苯乙烯基苯基甲酮；查耳酮
cinnamoyl (=3-phenylacryloyl) 3-苯基丙烯酰；肉桂酰 $C_6H_5CH=CHCO—$
cinnamoyloxyphenylurea 肉桂酸基苯脲；肉桂酰氧基苯脲 $C_6H_5CH=CHCO_2C_6H_4NHCONH_2$
cinnamycin 肉桂霉素
cinnamyl ①肉桂基 $C_6H_5CH=CHCH_2—$②〔有时指〕肉桂酰 $C_6H_5CH=CHCO—$
cinnamyl acetate 乙酸肉桂酯
cinnamyl alcohol 肉桂醇 $C_6H_5CH=CHCH_2OH$
cinnamyl anthranilate 氨茴酸肉桂酯；邻氨基苯甲酸肉桂酯
cinnamyl benzoate 苯甲酸(肉)桂酯
cinnamyl butyrate 丁酸(肉)桂酯
cinnamyl chloride 肉桂酰氯 $C_6H_5CH=CHCOCl$
cinnamyl cinnamate 肉桂酸肉桂酯 $C_6H_5CH=CHCOOCH_2CH=CHC_6H_5$
cinnamylcocaine 肉桂酰古柯碱
cinnamyl ethoxyacetate 乙氧基乙酸(肉)桂酯
cinnamyl formate 甲酸肉桂酯
cinnamylic acid 肉桂酸 $C_6H_5CH=CHCOOH$
cinnamylidene (=cinnamal) 亚肉桂基；苯亚烯丙基 $C_6H_5CH=CHCH=$
cinnamylidene-acetophenone 亚肉桂基乙酰苯 $C_6H_5(CH=CH)_2COC_6H_5$
cinnamyl isobutyrate 异丁酸(肉)桂酯
cinnamyl isovalerate 异戊酸(肉)桂酯
cinnamyl methyl ketone (肉)桂基甲基酮

cinnamyl nitrile 肉桂腈
cinnamyl phenylacetate 苯乙酸肉桂酯
cinnamyl propionate 丙酸(肉)桂酯
cinnamyl valeriate 戊酸(肉)桂酯
cinnarizine 桂利嗪〔药〕
cinnoline 喹啉；邻二氮(杂)萘；1,2-二氮杂萘
$$C_6H_4CH=CHN=N$$
cinnolinyl 喹啉基
cinnyl 肉桂酰
cinoxacin 西罗沙星
cinoxate 西诺沙able；对甲氧基肉桂酸乙基己酯
cipolin(o) 云母大理石
ciprofloxacin 环丙沙星〔药〕
circle ①圆；圆圈；圆周；环②周期；循环③范围；界
circle bend 圆曲管
circle cutting machine 圆切机
circlip 定位环；簧环；弹性挡圈
circuit ①电路②周路③环行
circuit breaker 断路器
circuit-breaker oil 电闸油；绝缘油
circuit closer 闭路器
circuit diagram 线路图
circuiting 圆的；循环的
circular 圆的；循环的
circular adsorption chromatogram 圆形吸附色谱图
circular adsorption chromatography 圆形吸附色谱法
circular arc 圆弧
circular blast-main 环式鼓风管
circular cell 圆形热室
circular chromatography 圆形色谱法
circular column 环形柱
circular comb 圆型精梳机
circular cone 圆锥体
circular cylinder 圆柱体；圆筒
circular development 环形展开(法)*
circular development R_f value 环形展开比移值
circular dichroism 圆二色性*
circular dichroism electrochemistry 圆二色谱电化学法
circular die 圆形冲模
circular electrode 盘状电极
circular filler rod 圆焊条
circular filter paper chromatography 圆形滤纸色谱(法)
circular flow 圆周流动；环流
circular furnace 圆炉；圆窑
circular gas chromatography 循环气相色谱(法)
circular goods 圆型制品
circular groove 圆槽

circular guide 圆导轨
circular hollow tape 套带
circularity coefficient 圆形系数
circular kiln 圆窑
circular knife 圆(盘)刀
circular knitting machine 圆型针织机
circular loom 圆型织机；经纬编织机
circularly polarized light 圆偏振光*
circularly polarized RF-electron nuclear double resonance 圆偏振射频场电子-核双共振
circular measure of angle 弧度
circular mil 圆密尔〔面积单位，直径1密尔的圆面积〕
circular nut 圆顶螺母；环形螺母
circular paper chromatography 圆型纸色谱法*
circular pitch (圆)周节(距)
circular plate 圆板
circular polarization 圆偏振
circular profile 圆形剖面
circular radiant type pipestill 圆形辐射式管状炉
circular saw 圆锯
circular saw blade 圆盘锯锯齿；圆形锯片
circular sawing machine 圆盘锯床
circular shear 圆盘剪床
circular shelf drier 圆架干燥器
circular slitting knife 圆(盘)刀
circular slot 圆槽
circular solid fibre 圆(截面)实体纤维〔非空心纤维〕
circular spider type joint 十字叉连接
circular stamping machine 苎麻圆型打纤机
circular technique 环形术
circular thin-layer chromatography 环形薄层色谱(法)
circular tooth gear 圆弧齿轮
circular velocity (圆)周速率
circular washer 圆垫圈
circular weaving machine 圆型织机；经纬编织机
circular wedge mount 圆楔座
circular winding (长丝缠绕)圆周绕程
circulating 环流；循环；运行；传播；流通
circulating connection (冷却水)循环接头
circulating dip tank 循环浸渍槽
circulating load 循环负荷
circulating lubrication 循环润滑
circulating mixer 循环混合机
circulating oil pump 循环油泵
circulating oil system (=circulation lubrication system) 循环润滑系统
circulating pipe 循环管
circulating pump 循环(式)泵

circulating reflux 循环回流
circulating ring 循环圈
circulating stock 循环油料
circulating tank 环流槽
circulating type hank dyeing machine 循环式绞纱染色机
circulating type oil supply 循环式给油
circulating water 循环水
circulation 循环
circulation column 环形柱
circulation flow 循环流动
circulation flow lubrication 环流润滑
circulation gas 循环气
circulation in drops 点滴环流
circulation layer 循环层
circulation lubrication system (=circulating oil system) 循环润滑系统
circulation oven 循环(热气)炉
circulation pattern 循环模式
circulation pipe (=circulation tube) 循环管
circulation pump 循环泵
circulation reflux 回流
circulation supply system 循环供给系统
circulation tube (=circulation pipe) 循环管
circulator 循环(压缩)机
circulatory cooling system 循环冷却系统
circulatory flow 循环流
circulatory stove 循环炉
circumanthracene 循环蒽
circumference 圆周
circumference coefficient 圆周系数
circumferential angle 圆周角
circumferential cracking 周向龟裂
circumferential grate 篮式炉栅
circumferential grate producer 篮式栅发生炉
circumferential lug space 花纹块圆周间隔
circumferential measurement 周长测量
circumferential pitch 周节
circumferential separation 环形断裂；环状断孔
circumferential speed 圆周速度
circumferential stress 周向应力
circumferential velocity 周围速度
circumferential weld 环焊缝
circumflux 环流
circumpolar 绕极的；周极的
cirolerosus 西罗菌素
cirramycin 卷须霉素
cis- 顺式
cis-addition 顺加成

cis-1,3-butadiene rubber 顺(式)-1,3-丁二烯橡胶
cis-butenediol 顺(式)丁烯二醇
cis-configuration 顺式构型
cis-content 顺式含量
cis-effect 顺位效应
cis form 顺式
cis-isomer 顺式异构体*
cisoid 顺向
cisoid conformation 顺向构象*
cisplatin 顺铂〔药〕
cis-1,4-polybutadiene 顺(式)-1,4-聚丁二烯
cis-polybutadiene rubber 顺式聚丁二烯橡胶
cis-1,4-polybutadiene rubber 顺丁橡胶
cis-polyisoprene 顺(式)聚异戊间二烯；顺(式)聚2-甲基丁二烯
cis-1,4-polyisoprene 顺(式)-1,4-聚异戊二烯
cis-1,4-polyisoprene rubber 顺式-1,4-异戊二烯橡胶
cis-polymerr 顺式聚合物
cis-position 顺位
11-cis-retinol 11-顺视黄醇；新视黄醛B
cis-rich polybutadiene 高顺(式)聚丁二烯橡胶
cis-rubber 顺式橡胶
cissampeline 假软齿花碱
cissing (=crawling) 收缩
cis-stereoisomer 顺式立体异构体
cis-tactic 顺式有规的
cistactic polymer 顺式有规聚合物*
cistern 容器；储水器
cistern barometer 槽式气压计
cistern-car 槽车
cis-thioindigo 顺式硫靛(蓝)
cis-trans effect 顺-反式效应
cis-trans equilibrium 顺-反式平衡
cis-trans-isomer 顺反异构体
cis-trans isomerism 顺反异构*
cistron 顺反子
cis-uranium element 铀前元素*
citalopram 西酞普兰〔药〕
citation 引证；指引
cite 引证；指引
citicoline 胞磷胆碱〔药〕
citraconate 顺甲基丁烯二酸盐(或酯)；柠康酸盐(或酯)
citraconic acid 顺甲基丁烯二酸；(Z)-甲基丁烯二酸；柠康酸 HOOCCH=C(CH₃)COOH
citraconic anhydride 顺甲基丁烯二酸酐；柠康酐 CH₃C=CHCOOCO
citraconoyl 顺甲基丁烯二酰基；柠康酰基

—COC(CH₃)=CHCO—(顺式)

citral (=dimethyl octadienal)　柠檬醛；橙花醛；3,7-二甲基-2,6-辛二烯醛
　　(CH₃)₂C=CHCH₂CH₂C(CH₃)=CHCHO
citral diethyl acetal　柠檬醛二乙缩醛
citral dimethyl acetal　柠檬醛二甲缩醛
citral propylene glycol acetal　柠檬醛丙二醇缩醛
citral type basil oil　柠檬醛型罗勒油
citramalic acid　柠苹酸；2-甲基-2-羟基丁二酸
　　HOOC(CH₃)C(OH)CH₂COOH
citramide　柠檬酰胺　$C_6H_{11}O_4N_3$
citrate　①柠檬酸②柠檬酸盐(酯或根)
citrate buffer solution　柠檬酸盐缓冲液
citrate lyase　柠檬酸裂合酶
citrate plasticizer　柠檬酸酯增塑剂
citrate soluble plant food　柠檬酸溶解性植物养料
citraurin　柠乌素
citrazinic acid　柠嗪酸；2,6-二羟基异烟酸；2,6-二羟基-4-吡啶甲酸
citrazinic amide　柠嗪酰胺；2,6-二羟基异烟酰胺；2,6-二羟基-4-吡啶甲酰胺
citrene　柠檬萜烯　$C_{10}H_{16}$
citreoviridin　黄绿青霉素
citresia　酸式柠檬酸镁
citric acid　柠檬酸　$HO_2CCH_2C(OH)(CO_2H)CH_2CO_2H$
citric acid cycle　柠檬酸循环
citric amide　柠檬酰胺　$C_3H_5O(CONH_2)_3$
citridic acid　乌头酸　$COOHCH_2C(COOH)=CHCOOH$
citriflex process　西垂弗莱克斯过程；柠檬酸氟络合法〔一种试验性的核燃料后处理首端过程,基于用柠檬酸和氟化钠反应产生的氟化氢络合锆的作用〕
citrin　柠檬素；维生素 P
citrine　黄晶；茶晶
citrine ointment　柠檬色软膏；硝酸汞软膏
citrinin　橘霉素　$C_{13}H_{14}O_5$
citriodora oil　柠檬桉油
citromyces　柠(檬菌)霉
citromycetin　柠檬菌素
citromycin　柠檬霉素
citron　柠檬；枸橼；香橼
citronella　香茅；雄刈萱
citronella grass　亚香茅；斯里兰卡香茅
citronellal　香茅醛；3,7-二甲基-6-辛烯醛　$C_9H_{17}CHO$
citronellal cycloglycol acetal　香茅醛环乙二醇缩醛
citronellal dimethyl acetal　香茅醛二甲缩醛
citronella oil　香茅油
citronellene　香茅烯
citronellic acid　香茅酸

citronellol　香茅醇
citronellyl　香茅基
citronellyl acetate　乙酸香茅酯　$C_{12}H_{22}O_2$
citronellyl butyrate　丁酸香茅酯
citronellyl cinnamate　(肉)桂酸香茅酯
citronellyl ethoxyacetate　乙氧基乙酸香茅酯
citronellyl formate　甲酸香茅酯　$HCO_2C_{10}H_{19}$
citronellyl hexylate　己酸香茅酯
citronellyl isobutyrate　异丁酸香茅酯
citronellyl isovalerate　异戊酸香茅酯
citronellyl methoxyecetate　甲氧基乙酸香茅酯
citronellyl methyl ether　香茅基甲醚
citronellyl nitrile　香茅腈
citronellyl oxyacetaldehyde　香茅基含氧乙醛
citronellyl phenylacetate　苯乙酸香茅酯
citronellyl propionate　丙酸香茅酯
citronellyl valerate　戊酸香茅酯
citronine A　柠檬素 A
citronine Y　柠檬素 Y
citron leaf oil　枸橼叶油；香橼叶油
citron oil　柠檬油；枸橼油
citron petitgrain oil　香橼叶油
citron scented gum　柠檬桉
citron soap　柠檬皂
citron yellow　柠檬黄
citronyl　香茅油
citrophen (=p-phenetidine citrate)　柠檬芬；柠檬酸化对氨基苯乙醚　$C_8H_{11}ONC_6H_8O_7$
citropten　白柠檬亭
citrostadienol　α_1-谷甾醇
citrovorum factor (CF)　嗜橙菌因子；亚叶酸；N-甲酰-5,6,7,8-四氢叶酸
citrulline　瓜氨酸　$NH_2CONH(CH_2)_3CH(NH_2)CO_2H$
citrus red　橘红〔食用色素〕
citrus seed oil　柑橘核油
citrusy　柑橘的
citryl　柠檬油
citrylidene　柠檬叉
citrylidene acetaldehyde　梨醛
citrylidene acetone　假性紫罗兰酮
civetane　灵猫烷　$CH_2(CH_2)_{13}CH_2CH=CH$
civetene (=civettene)　灵猫烯；环十七-9-烯
civet extract　灵猫香浸液
civetol　灵猫醇
civetone (=cibetone)　灵猫酮；环十七烯-9-酮
　　$CO(CH_2)_7CH=CH(CH_2)_6CH_2$

civettal (=tetrahydro-*p*-methyl quinoline)　四氢对甲基喹啉
13C labeling　碳-13 标记
Clabour pump　克累伯泵
clack valve　瓣阀
clad　①镀过的；镀过金属的②衬里
cladding　①覆盖层②喷镀
cladding process　覆盖法
cladding vault　(废燃料)包壳库
clad fibre　涂层纤维；包层纤维〔皮芯型玻璃光学纤维〕
cladinose　克拉定糖
cladiomycin　萌霉素
clad metal　包夹金属
cladomycin　芽枝霉素
cladonic acid　石蕊酸
cladosporin　枝孢菌素
clad steel plate　复合钢板
clad vessel　衬里容器
clad wire　包覆丝
claim　①要求赔偿损失权②申请
clairecolle (=clearcole)　①打底白垩胶；油灰②(打底)白铅胶③贴金(箔)胶
Claisen flask　克莱森(烧)瓶
Claisen para-rearrangement　克莱森对位重排作用
Claisen rearrangement　克莱森重排作用
clammy　①胶黏的；黏着(性)的②冷湿的；滑腻的
clamp　①钳②压板③夹④夹板
clamp bolt　夹紧螺栓
clamp check　夹紧压脚
clamp connection　锁状联合
clamp coupling　抱合式联管节
clamp daylight　合膜开档
clamper　夹持器；接线板
clamp fastener　持夹器
clamp frame　夹钳
clamp holder　持夹器
clamping　①夹紧②钳位
clamping apparatus　夹具；卡具
clamping bolt　夹紧螺栓
clamping cycle　模压周期
clamping device　夹持机构
clamping force　闭合力〔平板机〕
clamping level　夹杆
clamping pressure　闭合压力
clamping ring　钢板压环〔轮辋〕
clamping screw　紧固螺丝
clamping shuttle　片梭
clamp plate　夹板；压板

clamp ring　夹圈
clamp roll　夹滚
clamp screw　夹螺旋
clamp time　模压时间
clam-shell bucket　蛤壳式戽斗
Clancy process (for gold extraction)　克兰西(炼金)法
Clanex process (=co-lanthanide-actinide nitrate-extraction process)　克兰纳克斯法；锕系、锕系元素硝酸盐共萃取法〔将锕系、锕系元素混合物用胺类萃取法从硝酸盐溶液转换成氯化物溶液，为 Tramex 法制备料液的过程〕
clapboard (=bevel siding)　互搭板壁；披叠板壁
Clapeyron-Clausius equation　克拉贝龙-克劳修斯方程*
Clapeyron equation　克拉贝龙方程*
clappet (=clappet valve)　止回阀
clappet valve (=clappet)　止回阀
clarain　亮煤
clarase　澄解酶
claret　①红(葡萄)酒②酒红色
claret-red　酒红
clarificant　澄清剂
clarification　澄清法
clarification filtration　澄清过滤
clarification house　澄清房
clarification point　澄清点
clarification tank　澄清池
clarified juice　澄清汁
clarified oil　澄清油
clarifier　①澄清器②澄清剂
clariflocculator　澄清絮凝器
clarifying　澄清的
clarifying agent　澄清剂
clarifying basin　澄清池；沉淀池
clarifying capacity　澄清容量
clarity　①透明度②透明性；澄清度
Clark's pH standard solution　克拉克 pH 标准溶液
Clark cell　克拉克电池；汞电池
Clark degree　克拉克度
Clarke's internal electrolysis apparatus　克拉克内电解装置
Clarke's method　克拉克法
Clarke's soap method　克拉克皂测法
Clarke number　克拉克数
Clarke's internal electrolysis apparatus　克拉克内电解装置
Clark flexibility tester　克拉克挠曲性试验机
Clark-Lubs' buffer solution　克拉克-拉布斯缓冲溶液
Clark-Lubs' indicator　克拉克-拉布斯指示剂

Clark oxygen electrode　克拉克氧电极
Clark's pH standard solution　克拉克 pH 标准溶液
claroline　克拉罗林〔生产气体汽油时用作吸收剂的轻质矿物油〕
clary sage absolute　香紫苏净油
clary sage concrete　香紫苏浸膏
clary sage oil　香紫苏油；鼠尾草油
Clash-Berg flexure temperature　克拉什-伯格挠曲温度
Clash-Berg point　克拉什-伯格点
Clash-Berg test　克-伯法试验（测定高分子挠曲温度）
Classen's electrode　克拉森电极
Classen platinum dish　克拉森铂皿
classical adsorbent　经典吸附剂
classical analysis　经典分析*
classical body　经典物体〔指欧几里得固体、帕斯卡液体、虎克固体、牛顿液体等〕
classical high wet modulus fibre　经典高湿模量纤维
classical light scattering　经典光散射
classical mechanics　经典力学
classical problem　经典问题
classical quantum theory　古典派量子(学说)
classical statistics　经典统计(法)
classical theory of radiation　古典派辐射(学说)
classical thermodynamics　经典热力学*
classic diagram　经典图解(法)
classification　①分类②粒度分级
classification factor　分级因数；等级因数
classification of corrosion preventive　防蚀剂分类(等级)
classification of crude oil　原油分类法
classification of ion selective electrodes　离子选择电极分类法
classification of substance　物质的分类
classification of two dimensional NMR spectra　二维 NMR 谱的分类
classification statistics　分类统计
classified　①分了类的②分了级的
classified grinding　分级研磨
classifier　①粒度分级器②分级器〔如石棉纤维的粗细分级〕
classifier mill　带分级器超微粉碎磨
classifying　①分类②分粒③分级
classifying cone　锥形选粒器；锥形分级器
classifying crystallizer　分级结晶器
classifying screen　分级筛
classing　①分类②分粒③分级
class interval　组距*
class limit　组限
class 2nd weights　二级砝码
class of accuracy　精度等级
class 3rd weights　三级砝码
class settling　分级沉淀
class 1st reagent　一级试剂
class 1st weights　一级砝码
classy cotton　上等棉
clastic rocks〔复〕　碎屑岩
clastogen　断裂剂
clathrate　包合物*
clathrate complex　笼形配合物
clathrate compound　笼形(化)物
clathrate polymerization　笼形聚合
clathrate precipitation　笼形化合物沉淀
clathration　包合作用*
Claude process (for ammonia)　克劳德(制氨)法
claudetite　白砷石
clausenane　黄皮烷
Clausius-Clapeyron equation　克劳修斯-克拉贝龙公式〔平衡状态下的纯物质随着温度的微小变化而产生蒸气压变化的公式〕
Clausius-Duhem inequality　克劳修斯-杜亥姆不等式
Clausius inequality　克劳修斯不等式*
Clausius-Mossotti equation　克劳修斯-莫索提方程
Claus process　克劳斯(二段脱硫)法
Claus's blue　克劳斯蓝〔三氧化二铑的苛性碱溶液〕
clausthalite　硒铅矿
clavacin (=clavatin; claviformin)　棒曲霉素
clavatin (=clavacin)　棒曲霉素
clavatine　克拉瓦碱；棒石松碱
clavatol　克拉瓦醇；棒石松醇；2,4-二羟基-3,5-二甲基苯乙酮
clavatoxine　克拉瓦毒；棒石松毒
clavicepsin　麦角菌苷
claviformin (=clavacin)　棒曲霉素
clavine　棒麦角素
claw clutch　爪形离合器
clay　黏土；陶土；白土
clay burner　(回收)黏土炉；煅白土炉
clay burning　黏土煅制
clay burning unit　黏土煅烧炉
clay catalyst　黏土催化剂
clay-catalyzed degradation　黏土催化的降解(作用)
clay chamber　黏土脱色器
clay coated　白土涂布
clay column　黏土柱
clay contacting　黏土接触
clay contact process　黏土接触过程
clay contact rerun process　白土接触再蒸馏法

clay crucible 陶瓷坩埚
clay cutter 混土器；碎土机
clay deasphalting 黏土脱沥青
clayey 含黏土的；泥质的
clay-filtered oil 黏土过滤精制油
clay filter flask 素烧滤瓶
clay grease 黏土润滑脂
claying 煅烧黏土
clayish 含黏土的；泥质的
clayite 高岭石；杂砷黝铜矿
clay kiln 黏土再生窑
clay kneading machine 捏泥机
clay leg 黏土气动升降器
clay marl 泥灰土
clay membrane electrode 黏土膜电极
clay mill 混土器；碎土机
clay mineral 黏土矿物
clay mixer 混土器
clay mortar 黏土灰浆
clay percolation 白土渗滤法
clay pipe 陶土管
clay plate (=porous sheet) 素烧(瓷)板
clay range 黏土粒度范围
clay reactivation (=clay recovery) 黏土再活化；白土再生
clay recovery (=clay reactivation) 黏土再活化；白土再生
clay refining 白土精制(法)
clay regeneration 白土再生(法)
clay revivifying 黏土复原
clay shale 黏土页岩
clay slate 黏土板岩
clay slip ①黏土断层②黏土糊
clay slurry 黏土泥浆
clay suspension 黏土悬浮体
Clayton refining process 克莱登炼油法
clay tower 黏土塔
clay-treated 黏土处理的；白土精制的
clay treating process (=clay treatment) 黏土处理过程；白土精制过程
clay treatment (=clay treating process) 黏土处理过程；白土精制过程
clay triangle 陶(土)制三角架；泥三角
clay vessel 陶土器皿
clay wash 黏土纯化
clay water slurry 黏土泥浆
clay yield 黏土处理收率
clean 纯化；净化；精炼；提纯
Clean Air Act 清洁空气法〔1963年美国联邦政府制定的防大气污染法令〕

clean-burning fuel 完全燃烧燃料
clean circulation 干净循环；馏出油循环〔在裂化系统〕
clean circulation cracking 干净循环裂化；馏出原料的裂化
clean coal 净煤
clean cracking feed ①洁净裂化原料②新鲜裂化原料〔不加回炼油〕
clean cut ①切割良好的馏分②精确切割③爽利的
clean cut separation 精确分离
clean distillate stock ①洁净馏分②新鲜馏分油料〔不加回炼油〕
cleaner ①除垢器②洗净剂；清洁剂
cleaner's naphtha (=cleaner's solvent) 清洁用石油醚；洗涤用石油醚；干洗溶剂
cleaner's solvent (=cleaner's naphtha) 清洁用石油醚；洗涤用石油醚；干洗溶剂
cleaner blade 清洁器刀片
cleaner blade rod 清洁器刀片杆
cleaner blade spacer 清洁器刀片隔套
cleaner disk 滤清器滤盘
cleaner-polishes aerosol 洗净光亮气溶胶
Cleanex process 克林内克斯过程〔用D_2EHPA间歇萃取法从被腐蚀产物等污染的返回料液纯化超钚元素的过程〕
clean floor sealer 地板用封闭清漆
clean-fresh odor 清鲜香气
clean gas ①纯气体②纯煤气
cleaning 洗涤；洗净
cleaning action 清洗作用
cleaning agent ①清洁器②洗净剂；清洁剂
cleaning and brushing machine 洗刷机
cleaning cartridge 净化器
cleaning cloth ①擦布；抹布②除尘器用布③棉纱头
cleaning compound 洗涤剂
cleaning cycle ①捕集周期②清洗周期
cleaning doctor 刮涂料刀
cleaning down ①清理②彻底冲洗
cleaning-drum 洗涤滚筒
cleaning efficiency 净洗效率
cleaning fluid 清洗液
cleaning formulation 清洁剂
cleaning gas 清洗气
cleaning hose 清洗用水龙带
cleaning index 洗净指数；清洁指数
cleaning lotion 清洁露
cleaning media 清洁剂；除垢剂
cleaning mesh 滤净网
cleaning method ①清洗(清理)方法②除锈方法

cleaning milk　清洁蜜
cleaning mixture　洗涤液；洗涤混合液
cleaning of coal　洗煤
cleaning performance　清洗效果
cleaning plant reject　清洗厂废物
cleaning power　①洗净本领②洗净力
cleaning product　清洁品
cleaning reel　清洁辊
cleaning ring　扫塔环
cleaning solution　洗涤液*
cleaning solvent　洗涤溶剂
cleaning stage　净化阶段
cleaning strainer　粗滤清器
cleaning treatment　净化处理
clean-in-place　就地清洗
cleanliness　洁净(度)
cleanliness factor　洁净系数
cleanness (=cleanliness)　①洁白；清洁②清洁度；净化度
clean oil　洁净油料
cleanout　①清除②清焦〔如裂化炉、分馏塔盘〕
cleanout opening　清扫口〔油罐〕
cleanout stave　清除梯杠；清除横条
cleanout time　清除时间
clean recirculation method　净循环法
clean recycle stock　净循环油料
clean room　无尘室
cleanser　①洗净剂；清洁剂②滤清器；清洁器；刮净器
clean ship　运轻质透明油品用的油船
cleansing　纯化；净化；提纯；精炼
cleansing cream　清洁霜
cleansing oil　清洁(用)油
cleansing solution　洗液
cleansing solvent　清洁用石油醚
clean tanker　精制(石油产品)油槽船
clean technologies　清洁技术
cleanup　净化；提纯；清除；清整；澄清
clean up column　净化柱
clean-up (of gas)　(气体的)吸收
clean up procedure　净化手续
clean up stage　净化步骤
clean water basin　清水池
clean white flesh　洁白(皮底)肉〔制革〕
clean wool content　净毛率
clean yield　净毛率
clear　①纯的②净〔指颜色〕③清洁的
clear acrylic emulsion coating　丙烯酸乳胶透明涂料
clear-air turbulence　晴空湍流
clearance　余隙；间隙；公隙

clearance between roll　辊间间隙
clearance factor　余隙系数
clearance fit　间隙配合；活动配合
clearance flow (=leakage flow)　隙流
clearance gauge　量隙规；塞尺
clearance leakage　间隙漏泄
clearance loss　余隙损失
clearance meter　间隙计；余隙计
clearance pocket　补充余隙
clearance pocket in pump　泵的隙囊
clearance space　余隙空间
clearance take-up mechanism　滚筒调距装置
clearance volume　余隙容积
clear aperture　通光口径
clear area　有效截面(面积)；净面积
clear area of screen　筛的有效面；筛的净面
clear blending value　净调和值；不加四乙铅的调和辛烷值
clear boiled soap　抛光皂
clear coat　清漆(涂)层；透明涂层
clear coat-colo(u)r coat　清漆罩光；彩色(透明)面漆
clear coating　透明涂料
clear coating for chemical resistance　耐化学侵蚀清漆(涂层)
clearcole (=claircolle)　①打底白垩胶；油灰②(打底)白铅胶③贴金(箔)胶
clearcut　净切割
clear distance　净距离；净空
clear dope　透明蒙皮漆
clearer　清洁辊；绒辊
clearer cloth　套筒呢
clearer cotton waste　绒辊花
clearer roller　绒辊
clearer spring　绒辊弹簧
clear filtrate　滤液
clear finish　罩面清漆；透明面漆
clear gasoline　净汽油；无铅汽油
clear-glass bottle　透明玻璃瓶
clearing　①清洗②纯化
clearing agent　①澄清器②澄清剂
clearing factor　清除因子
clearing index　净化指数
clearing machine　①绒毛净化器〔纺〕②净化机
clearing power　洗净本领；洗净力
clear lacquer　清喷漆；透明漆；亮光漆
clear lacquer deposits　透明漆状沉积
clear line of demarcation　清晰的分界线
clear liquid shampoo　洗发清液
clear liquor　澄清液

clearness　①透明性②透明度
clearness index　清晰度
clear octane number (=clear octane rating)　不加铅辛烷值
clear octane rating (=clear octane number)　不加铅辛烷值
clear opening　净宽
clear opening of screen　筛的净面
clear overhead phase　上层清澈相
clear pale yellow　净浅黄色
clear point　澄清点*
clear point method　澄清点法
clear sheet　①透明片材②透明板材
clear side-stripe extruder　透明侧窗式挤塑机
cleat　固着楔
cleavability　可解理性
cleavage　①开裂；解离②解理(性)③裂解
β-cleavage　β裂解
cleavage breakage (=breakage cleavage)　破碎
cleavage column　解聚塔；裂解塔
cleavage face　解理面
cleavage of paraffin　石蜡烃裂解
cleavage plane (=cleavage surface)　解理面
cleavage product　分裂产物
cleavage reaction　断裂反应*
cleavage stress　劈裂应力
cleavage surface (=cleavage plane)　解理面
cleavage test　抗裂试验
cleft welding　裂口焊
clematidin　马兜铃素
clematitol　铁线莲苷
Clemen hardness tester　克里门式硬度计
Clemen scratch(ing) tester　克里门式刮痕硬度试验仪
Clemmensen reduction　克莱门森还原
clenbuterol　克仑特罗〔药〕
Clerget constant　克勒格特常数
Clerici solution　克累西溶液
Cleve's acid　克列氏酸；1-萘胺-6-磺酸和1-萘胺-7-磺酸的混合物
Cleve's α-acid (=1-naphthylamine-5-sulfonic acid)　克列夫氏α-酸；1-萘胺-5-磺酸
Cleve's δ-acid (=1-naphthylamine-7-sulfonic acid)　克列夫氏-δ-酸；1-萘胺-7-磺酸
cleveite　钇铀矿
Cleveland condensing humidity cabinet　克利夫兰型冷凝式潮湿箱
Cleveland flash tester　克利夫兰闪点测定器
Cleveland open-cup flash point test　克利夫兰开杯闪点试验
Cleveland open cup flash point tester　克利夫兰开杯式闪点试验器
Cleveland open cup tester　克利夫兰开杯闪点测定器
Cleveland open tester　克利夫兰开放式试验器
Cleveland tester　克利夫兰试验器
clevis　U形夹；马蹄钩
click chemistry　点击化学
clicker　①平压裁断机；冲切机②冲切工
clicker press　平压裁断机；冲切机
clicking　①冲坯②冲切
clicking die　冲模；冲刀
clicking machine　平压裁断机；冲切机
climatic　气候的
climatic change　气候变化
climatic conditions　气候条件
climatic gasoline　季节性汽油
climatic protection　气候防护
climatic test cabinet　气候试验箱
climbing film evaporator　升膜蒸发器
climb milling　顺铣
climb of dislocation　位错攀移
climograph　温度(曲线)图
clinch　结合
clincher bead core　钉绊顶心
clindamycin　克林霉素〔药〕
cling additives　黏着添加剂
cling compound　抗滑
cling property　抗滑性能
cling rubber　抗滑橡胶
cling-test　紧贴性试验；静电吸附试验
cling tread　抗滑胎面
clinical　临床的
clinical analysis　临床分析
clinical biochemistry　临床生物化学
clinical dextran　临床用右旋糖酐
clinically important　临床重要的
clinical test　临床试验
clinical thermometer　体温计
clinker　①熔块②煤渣
clinker bed　熔结块层
clinker brick　硬砖
clinker clew (=clinker clue)　熔块
clinker concrete　熔块混凝土
clinker formation　熔块形成作用
clinkering (of ash)　(灰的)熔结
clinkering point　(熟料)成熟点；熔结点
clinker ring　熟料圈
clinking press　平压裁断机；冲切机
clinkstone　响岩；响石

clinoaxis 斜轴
clinochlore 斜绿泥石
clinoenstatite 斜顽辉石
clinohumite 斜硅镁石
clinoklase 光线矿
clinometer 倾斜计；斜度计
clinoptilolite 斜发沸石
clinopyramid 斜轴锥
clinostat 加转器
clinozoisite 斜黝帘石
clionasterol (=22,23-dihydroporiferasterol) 穿贝海绵甾醇
clip ①夹；箍②纸夹③扣钩④剪去；截去
clip chain mercerizing range 布铁丝光联合机
clip connector （电极）夹连器
clip fastener 封管机〔设〕
clipper 裁切机
clipper service 快捷服务〔直接由油库向分配点发货〕
clipper transport 快捷运输
clip stretcher (=clip stretching machine) 链式展幅机
clip stretching machine (=clip stretcher) 链式展幅机
clock glass 表(面)玻璃
clock oil 钟表油
clock reaction 计时反应*
clockwise 时钟方向；顺时钟方向
clockwise direction 顺时针方向
clockwise rotation 顺时针方向旋转
clockwise sequence 顺时针次序
clockwise turn 顺时针转动
clod 块；大块；岩块
clodronate disodium 氯膦酸二钠〔药〕
clofazimine 氯法齐明〔药〕
clofibric acid 氯贝酸
clofibride 氯贝胺
clofluperol 氯氟哌醇
clog ①堵塞(物)；障碍②黏着；结渣③止动器
clogestone 氯孕酮
clogged filter 阻塞了的滤器
clogged line 阻塞了的管线
clogged screen 阻塞了的筛网
clogging 堵塞
cloky effect yarn 泡泡纱
clomifene 氯米芬〔药〕
clomipramine 氯米帕明〔药〕
clomocycline 氯莫环素；羟甲金霉素
clonal seedling rubber 无性种橡胶
clonazepam 氯硝西泮〔药〕
clone 克隆

cloned enzyme donor immunoassay 克隆酶给体免疫分析
clone enzyme immunoassay 克隆酶免疫分析
clone selection 克隆选择
clonidine 可乐定〔药〕
clopenthixol 氯哌噻吨〔药〕
clorafin 氯化石蜡
clorprenaline 氯丙那林〔药〕
close-boiling mixture 窄沸(点)混合物
close boiling solvent 接近沸点溶剂
close-burning 黏结性的；成焦性的
close circuit grinding 通路研磨
close coiling 紧盘蛇管
close contact adhesive 近缝黏合剂
close-coupled process 紧凑过程；一体化(方)法
close-coupled processing 一体化(核燃料)处理
close cup flash point 闭杯闪点(测定)
close cut distillate 窄馏分馏液
close-cut fraction 窄馏分；精密分馏馏分
close cut separation 窄馏分分离
closed ①封闭；关闭②连接③密合；紧密
closed angle 尖角
closed assembly 叠装
closed assembly time 叠装时间；叠合时间
closed (blind) joint 无间隙接头
closed brine system 封闭式盐水系统
closed cabinet 密闭箱
closed-cell 闭孔
closed cell foam 闭孔泡沫塑料
closed-cell foamed plastics 闭孔泡沫塑料
closed cell sponge 闭孔海绵胶
closed chain 闭链
closed circuit 闭合回路〔通路〕
closed circuit cell 常流电池
closed circuit cooling system 闭路冷却系统
closed circuit crushing (=closed circuit grinding) 闭路研碎
closed circuit grinding (=closed circuit crushing) 闭路研碎
closcd circuit installation 闭路安装
closed-circuit oil system 闭路循环油系统；闭路循环润滑系统
closed circuit operation 闭路操作
closed circuit potential 闭路电势*
closed circuit voltage 闭路电压
closed-coil reflux 封闭盘管回流〔部分冷凝〕
closed cup flash point 闭杯闪点
closed cup method 闭杯法

closed cup tester 闭杯(闪点)试验器
closed cycle 闭合循环
closed cycle transmutation 封闭循环蜕变
closed delivery 暗流式
closed discharge filter press 闭卸式压滤机
closed-edge gutter stabilizer 封缘淌蜡(成沟)稳定器
closed-end buffer 环形带式磨革机
closed flash point 闭杯闪点
closed flyer 封闭式锭翼
closed grained chilled iron 无砂眼冷铸铁〔滚筒〕
closed heater 封闭加热器
closed hydraulic surge drum 密闭式液压平衡筒
closed impeller 封闭式叶轮
closed impeller pump 隐闭叶轮泵
closed lapping 闭口垫纱
closed loop 闭合回路
closed-loop control 闭环控制〔控〕
closed loop stripping 闭路溶出
closed mixer 封闭式搅拌机
closed mo(u)ld 闭式压模
closed moulding 闭模成型〔增强塑料〕
closed packed structure 紧密堆砌结构
closed packing ①密合装填②密堆积
closed pore 封闭孔
closed return bend 闭式回转管
closed-ring 闭环
closed-ring hydrocarbon 闭环烃
close drying 密闭干燥
closed sand filter 密闭砂滤器
closed side creel 封闭式纱架
closed solubility curve 密闭溶度曲线
closed steam coil 闭口蒸汽盘管
closed storage battery 密闭式蓄电池
closed strip electrophoresis 闭条电泳
closed system 封闭系统*
closed tester 闭杯闪点测定器
closed time 最大密着放置时间〔黏合剂〕
closed-tube mercury manometer 压缩压力计
closed-tube test 闭管试验
closed type impeller 闭式叶轮
close fit 紧(密)配合
close fractionation 精密分馏
close furnace 闭式炉
close grain ①密纹(木材)②细晶粒
close-grained ①细粒的②密纹的(木材)③密实的
close grained wood 细纹木材
closely bound goods 紧密织物
closely set rolls 小滚距；薄滚距；薄螺丝〔俗〕

closely set warp 高密经纱
closely spaced 密集的；间距小的
close-meshed 密网眼的
closeness ①密闭②紧密
close nipple 全外牙(管)
close-packed 密堆积
close-packed structure 紧密堆积结构
close packing 紧密装填；密堆积
closer 闭路器
close roaster (=muffle roaster) 闭式烤炉；马弗烤炉
close running fit 紧动配合
close sizing 紧密上填料
close stitch 锁眼
closest packing 最密装填
closest packing of spheres 球体的最紧密堆积
close test 精密试验
close tolerance extrusion 精密压出
close-up ①闭合；关闭；闭路；接通②靠近；接近③特定镜头；精细观察
close wind 平行卷绕
close working fit 紧滑配合
closing cock 封闭塞
closing column end 密闭柱端
closing machine 封口机；压盖机
closing time 闭模时间
closo 闭合型〔笼型〕
closo coordination compound 闭合式配合物*
closure 密封
closure time (被黏件)密合时间
clot 凝块；块凝物
cloth 布
cloth abrasion tester 织物耐磨试验仪
cloth brown 布帛棕〔纺〕
cloth-coating 织物涂料
cloth count 织物经纬密度
cloth counting glass 织物密度镜
cloth crimp tester 织物皱缩试验仪
cloth doubler 层布贴合机；合布机
cloth fast yellow 布帛坚牢黄
cloth filler 布填料
cloth filter 布滤材
cloth grounding 布帛砑光
cloth guide (人造革涂布等)导布器
clothing leather 服装革
clothing wool 粗纺用毛
cloth inspection machine 验布机
cloth-lined paper 衬(有)布的纸；布衬纸
cloth liner 垫布

cloth looking machine 验布机
cloth marks 布纹
cloth oil 纺织用油
cloth-one-side sheet 单面胶布
cloth orange 布帛橙
cloth penetration ①(人造革)底布渗浆度②底布渗浆
cloth pick count 纬密
cloth proofing 布上涂胶
cloth scarlet 布帛猩红〔染〕
cloth softening 布帛软化
cloth strainer frame 布滤架
cloth tensioner 紧布架
cloth-wear testing machine 织物耐磨试验仪
cloth wick 布芯
clot retraction test 血块收缩试验
clotrimazole 克霉唑〔药〕
cloud 发混
cloud agent 起雾剂
cloud chamber 云室
cloud clear point 混浊消失点
cloud density (电子)云密度
clouded agate 云玛瑙
clouded glass 毛玻璃
cloud formation 云的形成
cloudiness ①浊度②浑浊性
clouding (塑料制件弯曲时)发白
clouding cast 乳光
clouding point 浊化点
cloudness 混浊；不透明性
cloud point 雾点；浊度；浊点
cloud track method 雾迹法；雾径法
cloudy (混)浊的
cloudy dyeing 染色云斑
clovane 丁子香烷
clove 丁子香
clove bud oil 丁子香(花)油
clove bud oleoresin 丁子香(花)油树脂；丁香花蕾油树脂
clove leaf oil 丁子香叶油
clovene 丁子香烯
clovenic acid 丁子香酸
clove oil 丁子香油
cloverleaf pattern 三叶草形
clover oil 三叶草油
clove stem oil 丁子香茎油
clove tree 丁子香树
cloxacillin 5-甲基-3-邻氯苯基-4-异唑青霉素；氯唑西林〔药〕
clozapine 氯氮平〔药〕
club sandwich compound 三层夹心型化合物

clunch 硬白垩
clunging 附着力
clupanodonic acid (=docosa -4,7,11-trien-18-ynoic acid) 鳕鱼酸；4,7,11-二十二碳三烯-18-炔酸
clust clearer 黄麻清纤机
cluster 簇*
cluster analysis 聚类分析*
cluster burner 聚口灯头
cluster cardamom 白豆蔻
cluster compound 簇合物*
cluster formation 成簇
cluster ion 簇离子*
cluster ion theory 离子群(学)说
clusters of needles 针状体簇
cluster theory 簇理论〔指辐射引起的激发或电离粒子的点群〕
clutch 离合器；联动器；接合器
clutch coupling 离合器联结
clutch lining 离合器衬层
C-1 method C-1 法〔航空汽油抗爆性测定法〕
cnicin 蓟苦素
cnidicin 蛇床素
cnidide 川芎内酯
cnidium acid 川芎酸
cnidium lactone 蛇床内酯 $C_{12}H_{18}O_2$
cnidium oil 川芎油
^{13}C nuclear magnetic resonance 碳-13 核磁共振*
coabsorption 共吸收
co-accelerator 共促进剂
coacervate ①凝聚层②凝聚
coacervated phase 凝聚相
coacervation 凝聚；团聚
coacervation method 凝聚法
co-acting roller process 对轮法〔拉制玻璃纤维原丝〕
coaction 相互作用
coactivator 共活化剂
coadhesive 共黏剂
coadsorb 共吸附
co-adsorption 共吸附*
coagel 凝聚胶
coagent 活性助剂
coagglutination 同族凝集
co-agglutination reaction 协同凝集反应；共凝集反应
coagulant ①混凝剂；促凝剂②凝血剂
coagulant agent (=coagulating agent) 促凝剂；混凝剂
coagulant aids 助凝剂；辅助凝聚剂
coagulant casting process 凝固(剂)浇铸法
coagulant cut 蚀痕

coagulant dipping　凝固(剂)浸渍法
coagulant dipping process　凝固剂浸渍法
coagulant sedimentation　混凝沉淀法
coagulase test　凝固酶试验
coagulated aerosol　凝聚型气溶胶
coagulated chain　凝固链
coagulated fibre　凝固纤维；凝结纤维
coagulated latex　凝固胶乳
coagulated phase　凝固相；凝结相
coagulated smectite　凝固绿土
coagulating agent (=coagulant agent)　促凝剂；混凝剂
coagulating basin　①凝固槽②混凝池；凝聚池
coagulating bath　①凝集浴②凝集槽
coagulating bath additive　凝固浴添加剂；纺丝浴添加剂
coagulating dip　凝固浸渍法
coagulating-emulsion precipitation　凝乳状沉淀
coagulating environment　凝固环境；凝固条件
coagulating enzyme　凝固酶
coagulating liquid recovering equipment　凝固液回收装置
coagulating pan　凝固槽
coagulating point　凝固点；凝结点
coagulation　①聚沉*②凝聚；胶凝；凝结
coagulation ability　凝固能力
coagulation accelerator　凝固促进剂；促凝剂
coagulation bath　①凝固浴；凝结浴②纺丝浴
coagulation bath composition　凝固浴组成
coagulation (bath) temperature　凝固(浴)温度
coagulation by blowing with air　充气凝固
coagulation by high frequency heating　高频加热凝固
coagulation by mechanical stirring　机械搅拌凝固
coagulation cascade　阶式凝固器
coagulation condition　凝固条件
coagulation equilibrium　凝固平衡；凝结平衡
coagulation factor　凝血因子
coagulation grade　凝固程度〔以喷出的丝条变成不透明时为界限，用离喷丝头的距离表示〕
coagulation heat　凝结热
coagulation mechanism　凝固机理
coagulation medium　凝固介质
coagulation of oil emulsions　乳化油的凝聚
coagulation point　凝固点；凝结点
coagulation power　凝固能力；凝固本领
coagulation treatment　凝结处理
coagulation value (=flocculation value)　絮凝值；凝结值
coagulative nucleation　凝结成核
coagulator　①凝结器②凝结剂
coagulum content　凝固物含量

coal　煤；石炭
coal analysis　煤(的)分析
coalase　凝固酶
coal ash　煤灰
coal bearing　含煤的
coal bed　煤床
coal bin　煤仓
coal blending　配煤
coal brand　小麦黑粉病
coal brass　富碳黄铁矿
coal breaker　碎煤机
coal brick　煤砖
coal bunker　煤仓
coal-burning　烧煤的
coal by-products　煤副产品
coal cake　煤砖；煤饼
coal carbonization (=coal carbonizing)　煤碳化(作用)；煤干馏
coal carbonizing (=coal carbonization)　煤碳化(作用)；煤干馏
coal charger　①装料车②装煤车
coal clay　耐火黏土；底黏土
coal coke　煤焦炭
coal-coking process　炼焦法
coal combustibles　煤可燃分
coal compression machine　压煤机
coal constitution　煤组成
coal conveyor　运煤车
coal crusher　煤压碎机
coal degradation　煤剥蚀
coal desulfurization　煤炭脱硫
coal distillation　煤蒸馏(作用)
coal distribution　煤分布
coal dressing　选煤
coal dryer　煤干燥器
coal dust　煤屑
coal-dust brick　煤屑砖
coal equivalent　煤当量；燃料当量
coalesced filaments　并丝；凝集丝
coalesced resin globule　凝聚的树脂珠
coalescence　凝聚；聚集；聚结
coalescence in emulsion　乳化液的聚结
coalescence of droplets　小滴的聚结
coalescence point　聚结点；融合点；结合点
coalescencer　聚结剂；聚结器
coalescence rate　聚结(速)率
coalescence resistance　抗聚结性
coalescence temperature　聚结温度；结合温度

coalescence time 聚结时间
coalescent 联合的；结合的；组合的
coalescent agent 聚结剂
coalescer 聚结剂；聚结器
coalescing 聚结的
coalescing agent 聚结剂
coalescing aid 聚结添加剂；聚结助剂
coalescing medium 聚结介质
coalescing particles 聚结的粒子
coaleum 煤烃
coal extract 煤萃取物
coal field 煤田
coal field gas 煤田气
coal-fired 烧煤的
coal firing 用煤加热
coal fish 鳕鱼
coal-fish oil 鳕鱼油
coal flotation 煤浮选
coal formation 煤生成
coal forming process 煤形成过程
coal gas (=coal oven gas) 煤气
coal-gas producer 煤气发生炉
coal geology 煤的地质学
coal grading 煤分级
coal grinding 煤磨碎
coal heating 煤加热的
coal hulk 煤船
coal hydrogenation 煤加氢
coalification 煤化(作用)
coal ignitability 煤着火性；煤可燃性
coal industry 煤炭工业
coaling 加煤；装煤
coal in pile 煤堆
coal in solid 整块煤
coalite gasoline 由低温焦油所得的汽油
coalite tar 低温焦油；半焦油
coalization 聚结
coal measure 煤系
coal mine 煤矿
coal mixing plant 配煤厂
coal naphtha ①煤焦油②苯
coal non-combustibles 煤不可燃物
coal oil 煤馏油
coal origin 煤的来源
coal oven gas (=coal gas) 煤气
coal paste 煤糊〔煤与油的糊状物〕
coal petrography 煤岩学
coal preparation 选煤
coal pulverization 煤的粉碎
coal pulverizer 煤炭粉碎机；碎煤机
coal-pulverizing plant 煤炭粉碎车间
coal pyrite 碳质黄铁矿
coal pyrolysis 煤热分解
coal rank 煤品级
coal resources 煤资源
coal sample 煤样
coal sampling 采煤样
coal screening 煤筛选
coal seam 煤层
coal separating plant 选煤厂
coal separator 煤分离器
coal series 煤系
coal shovel 煤铲；煤锹
coal silt 煤泥滓
coal slack 煤屑
coal slime 煤黏泥
coal sludge 煤淤泥
coal slurry 煤泥
coal stone 煤石
coal store 煤库
coal tar 煤焦油
coal-tar base 煤焦油基
coal tar color 煤焦沥青色
coal tar emulsion 焦油乳液；焦油沥青乳液
coal tar enamel 焦油(沥青)瓷漆
coal tar-epoxy coating 环氧沥青涂料；煤焦沥青环氧涂料
coal-tar fuel 煤焦油燃料
coal-tar heavy oil 重焦油；高温煤焦油
coal-tar hydrocarbon 煤焦油烃
coal-tar industry 煤焦油工业
coal-tar middle oil 中焦油；中温煤焦油
coal-tar naphtha 煤焦油石脑油
coal-tar oil 煤焦油
coal-tar oil stain 煤焦油着色剂
coal-tar pitch 煤焦油沥青
coal-tar products 煤焦油产品
coal tar resin 煤焦油树脂
coal-tar solvent 煤焦油溶剂
coal tar water emulsion 焦油沥青水乳化液
coal throughput 煤处理能力〔煤气甑〕
coal washer 洗煤机
coal washery ①洗煤机②洗煤场
coal washing 洗煤
coal washing plant ①洗煤机②洗煤场
coaly ①属煤的②似煤的③含煤的

coaly rashings　软页岩
co-antiknock agent　抗爆助剂
co-area　协面积
coarse　粗的；粗糙的
coarse abrasive　粗磨料
coarse adjustment　粗调
coarse aggregate　粗集料
coarse bark　粗磨树皮〔制革〕
coarse black　粗粒炭黑
coarse bran　粗糠；粗麸
coarse breaking　粗压碎
coarse break-up　粗粒分散体
coarse calico　粗平布
coarse cement　粗水泥
coarse clay　粗黏土
coarse cloth　粗布
coarse coal　粗煤
coarse content　粗粒含量
coarse cossettes〔复〕　粗甜菜丝
coarse count　低支数
coarse cracking　粗粉碎
coarse crusher　粗压碎机
coarse crushing　粗压碎
coarse denier　粗旦；粗纤度
coarse disperse system　粗分散系统*
coarse dispersion　粗分散体
coarse emulsion　粗乳状液*
coarse fiber　粗纤维
coarse fiber structure　粗纤维结构
coarse filter　粗滤过器
coarse flour　粗木粉
coarse grain　粗粒；粗面；粗颗粒
coarse-grained　粗粒的；粗纹理的
coarse-grained coal　粗纹理煤
coarse grained pig iron　粗粒生铁
coarse grained wood　粗纹理木材
coarse-graining　粗粒化
coarse gravel　粗砂砾
coarse grinding　粗磨
coarse knurl insert　粗滚花嵌件
coarse meal　粗粉末
coarse mesh filter　粗筛网滤器
coarse metal　粗晶硅酸铁
coarse-mined　粗采的
coarseness　粗度；粒度
coarsening　（晶粒)粗化；(表面)粗糙化
coarse oil screen　粗(油)滤器
coarse ore　粗矿
coarse para　粗橡胶；低级白拉胶
coarse particle　粗粒子
coarse powder　粗粉
coarse preshredding arrangement　粗预粉碎机
coarse pyrite　粗黄铁矿
coarse reduction　粗粉碎
coarse roll　粗碾辊
coarse sand　粗砂
coarse screen　粗筛；粗滤器
coarse screening　粗筛
coarse separation　粗分离
coarse-size bave　超粗纤度丝
coarse strainer　粗滤器
coarse stuff　粗涂抹料；打底料
coarse suspension　粗粒悬浮体
coarse texture　宏观结构
coarse thread　粗牙螺纹
coarse thread screw　粗(螺)纹螺钉
coarse tolerance　粗公差
coarse vacuum　低真空
coarse vacuum pump　低真空泵
coarse whiting　重质碳酸钙；重钙〔填充剂〕
coarse yarn　粗支纱；低支纱
coarse zero controller　零点粗调控制器
coastal climate resistance　耐海岸气候性；耐近海气候性
coat　①涂渍②涂层；镀层；涂布③层
coat coverage　①涂层②涂布面积
coated　涂渍的；涂铺的
coated abrasive　砂纸；砂布
coated board　涂布纸板
coated capillary　涂层毛细管
coated carbide fuel　涂敷碳化物燃料
coated coil　涂漆卷材
coated fabric　胶布
coated fibre　涂层纤维〔例如涂金属玻璃纤维〕
coated glass fibre products　玻璃纤维涂覆制品
coated grade whiting　涂胶用碳酸钙
coated ion exchanger　包覆型离子交换剂
coated metallic electrode　涂层金属电极
coated packing material　包覆型填料
coated paper (=enamelled paper)　铜版纸〔商〕；印图纸；蜡图纸；蜡光纸；釉纸
coated particles fuel (CPF)　涂敷颗粒(核)燃料
coated pigment　包核颜料
coated silica pigment　碱式硅硫酸铅白〔别名〕；碱式硫酸铅包(二氧化硅)核颜料
coated substrate　①涂布基材②贴面基材
coated sulfur (=pre-dispersed sulfur)　包得硫黄；预分散

硫黄
coated type 核壳型〔复合半导体纳米微粒的结构型式之一〕
coated web ①涂布基料②贴合基料
coated wire electrode 涂丝电极*
coated-wire ion-selective electrode 涂层离子选择性电极
coater ①涂布机②涂料器
coating ①涂料*②涂层；层；镀层；涂布③涂渍
coating action 盖覆作用
coating additive 涂料添加剂
coating agent 涂层剂；涂布剂；涂饰剂
coating appearance 涂料外观
coating application method 涂料施工方法
coating assessment ①涂料评价②涂层评价
coating blade 刮漆刀
coating booth 喷漆室；喷漆橱
coating calender 贴胶压延机
coating cloud 涂敷云；漆雾
coating composition 涂料组分
coating compound ①涂布用浆胶②覆胶胶料
coating damage 涂层损坏
coating efficiency 涂装效率；涂渍效率
coating facet 涂装范畴
coating failure 涂层断裂；涂层破裂
coating fault 涂料本身缺陷；涂料不合格
coating finish ①涂饰剂②涂饰
coating flat sheet 平板(片)材的涂料
coating fluid 涂布液
coating inspector 涂层探伤仪
coating life 涂层寿命
coating line ①涂布生产线②贴合生产线
coating machine 涂覆机
coating material 涂渍物料；表面盖覆剂；表面修饰剂
coating medium ①涂料②贴合料
coating metal 涂覆用金属；金属涂层用金属
coating mortar 水泥砂浆保护层〔用于铁管外壁〕
coating of rolls 辊轴涂面〔设〕
coating operation 涂渍操作
coating paper 着色纸；图画纸
coating performance ①涂布性能②涂层性能③贴合性能
coating pick-up ①涂料涂敷量②(漆刷)涂料蘸着量
coating pigment 涂料
coating pistol 涂胶用喷枪
coating porosity 涂层孔隙率；涂膜孔隙率
coating powders 粉末涂料
coating process ①涂装(工艺)过程②涂装方法
coating property 被覆性能
coating removal 去除涂层；去漆
coating resin 涂料用树脂
coating resistance 涂层坚牢度
coating roll ①涂布辊②贴合辊
coatings additive 涂料助剂；涂料添加剂
coatings and paint raw materials 涂料和油漆原料
coating sealing 涂层封孔
coating sequence ①涂布顺序②贴合顺序
coating solution 涂渍溶液
coating space 涂装空间；涂装(施工)占用空间；涂装占地面积
coating stress 涂层应力
coating surface ①涂布表面②贴合表面
coating technique ①涂布技术②贴合技术
coating test 盖覆试验〔沥青或沥青乳化液在碎石上的盖覆作用〕
coating tester 胶布(静电火花)测定仪
coating thickness ①涂布厚度②涂层厚度③贴合厚度④贴层厚度
coating thickness measuring device 涂层厚度检测仪；漆膜测厚仪
coating varnish 罩光清漆
coating vehicle 漆料
coating weight ①涂布重量②贴合重量
coating weight measuring test 膜重测定〔计算单位面积漆膜或氧化膜重量的方法〕
coat-thickness 涂层厚度
co-axial 同轴的；共轴的
co-axial cable 同轴电缆
co-axial cylinder viscometer 同轴圆筒式黏度计
coaxial cylinder viscometry 同轴圆筒测黏法
coaxial cylinder viscosimeter 同轴圆筒黏度计
co-axial dual-gas interface 共轴双气式连接装置
co-axial electron entrance 同轴电子入射口
coaxial extruded pellet 同轴挤塑粒料
coaxial extrusion 同轴挤塑
co-axial junction 同轴接口
co-axial make-up flow 同轴加入补充液流
co-axial resonance cavity 同轴共振腔
coazervate 聚析液
coazervation 聚析
cob 圆块
cobackwash 共反萃(取)；共反洗
cobalamin (氰)钴胺素；维生素 B_{12}
cobalamine 钴胺素*
cobalt 钴 〔27号元素，化学符号Co〕
cobalt acetate 乙酸钴
cobalt acetylide 乙炔钴

cobalt aluminate　铝酸钴　$Co(AlO_2)_2$; $Al_2(CoO_4)$
cobaltammine　氨合钴
cobaltammine compound　钴的氨合物
cobalt black　钴黑；一氧化钴　CoO
cobalt bloom　钴华；八水合砷酸钴　$Co_3(AsO_4)_2 \cdot 8H_2O$
cobalt blue　钴蓝；瓷蓝
cobalt bomb　钴弹
cobalt-bromide test　溴化钴试验〔丙烷中水分定性试验法〕
cobalt brown　钴棕　$Fe_2O_3 \cdot CoO \cdot Al_2O_3$
cobalt carbonyl　羰基钴〔$1.Co_2(CO)_8$; $2.Co(CO)_3$〕
cobalt chlorides　钴的氯化物类
cobalt complex　钴络合物
cobalt crust　钴华　$Co_3(AsO_4)_2 \cdot 8H_2O$
cobalt cyanide　氰化高钴
cobalt dibutyldithiocarbamate　二丁基二硫代氨基甲酸钴〔黏合剂〕
cobalt dioxide　二氧化钴　CoO_2
cobalt disulfide　二硫化钴　CoS_2
cobalt drier　钴催干剂
cobalt fatty acid salt　脂肪酸钴盐
cobalt Fischer-Tropsch catalyst　费托合成钴催化剂
cobalt glance　辉钴矿
cobalt glass　钴玻璃
cobalt glass dosimeter　钴玻璃(放射)剂量计
cobalt green　钴绿
cobalt hydrosulfide　硫氢化钴　$Co(HS)_2$
cobalt hydroxychloride　氯羟化钴　$Co(OH)Cl$
cobaltic　①钴的②高钴的；三价钴的
cobaltic aquopentammine salt　一水五氨合高钴盐　$[Co(NH_3)_5H_2O]X_3$
cobaltic bromopentammine salt (=cobaltic bromopurpureo salt)　一溴五氨合高钴盐　$[Co(NH_3)_5Br]X_2$
cobaltic bromopurpureo salt (=cobaltic bromopentammine salt)　一溴五氨合高钴盐
cobaltic chloride　氯化高钴；三氯化钴　$CoCl_3$
cobaltic chloropentammine salt (=cobaltic chloropurpureo salt)　氯化五氨合高钴盐　$[Co(NH_3)_5]Cl_3$
cobaltic chloropurpureo salt (=cobaltic chloropentammine salt)　氯化五氨合高钴盐
cobaltic compound　高钴化合物
cobaltic croceo salt　二硝四氨合高钴盐　$[Co(NH_3)_4(NO_2)_2]X$
cobaltic diaquotetrammine salt　二水四氨合高钴盐　$[Co(NH_3)_4(H_2O)_2]X_3$
cobaltic dichloro-aquotriammine salt (=cobaltic dichro salt)　二氯一水三氨合高钴盐　$[Co(NH_3)_3(H_2O)Cl_2]X$
cobaltic dichlorotetrammine salt　二氯四氨合高钴盐　$[Co(NH_3)_4Cl_2]X$
cobaltic dichro salt (=cobaltic dichloro-aquotriammine salt)　二氯一水三氨合高钴盐
cobaltic dinitrotetrammine salt (=cobaltic flavo salt)　二硝四氨合高钴盐　$[Co(NH_3)_4(NO_2)_2]X$
cobaltic flavo salt (=cobaltic dinitrotetrammine salt)　二硝四氨合高钴盐
cobaltic fluoride　氟化高钴(的)；三氟化钴　CoF_3
cobaltic hexammine salt (=cobaltic hexammonate salt)　六氨合高钴盐　$[Co(NH_3)_6]X_3$
cobaltic hexammonate salt (=cobaltic hexammine salt)　六氨合高钴盐
cobaltic hexanitrite salt　六亚硝酸根合高钴盐　$[Co(ONO)_6]X_3$
cobaltichloride　氯化高钴；三氯化钴　$CoCl_3$
cobaltic hydroxide　氢氧化高钴　$Co(OH)_3$
cobaltic hydroxypentammine salt　一羟五氨合高钴盐　$[Co(NH_3OH]X_2$
cobaltic isoxantho salt　亚硝酸根五氨合高钴盐　$[Co(NH_3(ONO)]X_2$
cobaltic luteo salt　六氨合高钴盐　$[Co(NH_3)_6]X_3$
cobaltic nitratopentammine salt (=cobaltic nitratopurureo salt)　硝酸根五氨合高钴盐　$[Co(NH_3)_5(NO_3)]X_2$
cobaltic nitratopurpureo salt (=cobaltic nitratopentammine salt)　硝酸根五氨合高钴盐
cobaltic nitritopentammine salt　亚硝酸根五氨合高钴盐　$[Co(NH_3)_5(ONO)]X_2$
cobaltic nitropentammine salt　硝基五氨合高钴盐　$[Co(NH_3)_5(NO_2)]X_2$
cobaltic oxide　氧化高钴　Co_2O_3
cobaltic potassium cyanide　氰高钴酸钾；氰化高钴钾
cobaltic potassium nitrite　亚硝高钴酸钾
cobaltic praseo salt　二氯四氨合高钴盐　$[Co(NH_3)_4Cl_2]X$
cobaltic roseo salt　一水五氨合高钴盐；高钴玫红盐　$[Co(NH_3)_5H_2O]X_3$
cobaltic roseo tetrammine salt　二水四氨合高钴盐；高钴玫红四氨盐　$[Co(NH_3)_4(H_2O)_2]X_3$
cobaltic sulfate　硫酸高钴　$Co_2(SO_4)_3$
cobaltic sulfatopentammine salt　硫酸根五氨合高钴盐　$[Co(NH_3)_5(SO_4)]X$
cobaltic sulfatopurpureo salt (=cobaltic sulfatopentammine salt)　硫酸根五氨合高钴盐
cobaltic sulfide　硫化高钴　Co_2S_3
cobaltic tetracyanide salt　四氰合高钴盐　$[Co(CN)_4]M$
cobaltic violeo salt　二氯四氨合高钴盐；高钴紫盐　$[Co(NH_3)_4Cl_2]X$
cobaltic xantho salt　硝基五氨合高钴盐；高钴黄盐　$[Co(CH_3)_5(NO_2)]X_2$

cobalticyanic acid 氰高钴酸 $H_3[Co(CN)_6]$
cobalticyanide 氰高钴酸盐 $M_3[Co(CN)_6]$
cobaltine 辉钴矿
cobaltinitrite 亚硝酸根合高钴酸盐 $M_3[Co(ONO)_6]$
cobaltite 辉钴矿
cobalt linoleate 亚油酸钴
cobalt-magnesia-red (=cobalt pink) 钴-氧化镁红；钴桃红
cobalt molybdate 钼酸钴
cobalt-molybdate catalyst 钴钼催化剂
cobalt-molybdenum catalyst 钴钼催化剂
cobalt monoxide 一氧化钴 CoO
cobalt naphthenate 环烷酸钴（黏合剂）
cobalt nickel pyrite 钴镍黄铁矿
cobalt nitrate 硝酸钴
cobalt nitride 一氮化二钴 Co_2N
cobalt nitrite 亚硝酸钴
cobaltocene 二茂钴
cobalto-cobaltic oxide 一氧化钴合三氧化二钴；四氧化三钴 $CoO \cdot Co_2O_3; Co_2(CoO_4)$
cobalt octoate 辛酸钴
cobaltocyanide 氰钴酸盐；六氰合钴酸盐 $M_4[Co(CN)_6]$
cobaltosic oxide 四氧化三钴；一氧化钴合三氧化二钴 $CoO \cdot Co_2O_3; Co_2(Co_4)$
cobaltosic sulfide 四硫化三钴 $CoS \cdot Co_2S_3$
cobalto sulfate 二硫酸根(合)二价钴酸盐 $[Co(SO_4)_2]^{2-}$
cobaltous (正)钴的；二价钴的
cobaltous acetate 乙酸钴 $Co(C_2H_3O_2)_2$
cobaltous ammonium sulfate 硫酸铵钴 $CoSO_4 \cdot (NH_4)_2SO_4 \cdot 6H_2O$
cobaltous arsenate 砷酸钴 $Co_3(AsO_4)_2$
cobaltous benzoate 苯甲酸钴 $Co(C_6H_5COO)_2 \cdot 4H_2O$
cobaltous bromate 溴酸钴 $Co(BrO_3)_2$
cobaltous bromide 溴化钴 $CoBr_2$
cobaltous butyrate 丁酸钴 $Co(C_4H_7O_2)_2$
cobaltous carbonate 碳酸钴 $CoCO_3$
cobaltous chlorate 氯酸钴 $Co(ClO_3)_2$
cobaltous chloride 氯化钴；二氯化钴 $CoCl_2$
cobaltous chromate 铬酸钴 $CoCrO_4$
cobaltous cyanide 氰化钴 $Co(CN)_2$
cobaltous dihydroxycarbonate 碱式碳酸钴 $CoCO_3 \cdot Co(OH)_2; Co_2(OH)_2CO_3$
cobaltous dithionate 连二硫酸钴 CoS_2O_6
cobaltous ferrocyanide 亚铁氰化钴 $Co_2Fe(CN)_6 \cdot xH_2O$
cobaltous fluoride 氟化钴 CoF_2
cobaltous fluosilicate 氟硅酸钴 $Co(SiF_6)$
cobaltous hexahydrate salt 六水合钴盐 $[Co(H_2O)_6]X_2$
cobaltous hexammonate salt 六氨合钴盐 $[Co(NH_3)_6]X_2$
cobaltous hydrophosphate 磷酸氢钴 $CoHPO_4$

cobaltous hydroxide 氢氧化钴 $Co(OH)_2$
cobaltous iodide 碘化钴 CoI_2
cobaltous linoleate 亚油酸钴 $Co(C_{18}H_{31}O_2)_2$
cobaltous nickelous sulfate 硫酸亚镍合硫酸钴 $NiSO_4 \cdot CoSO_4$
cobaltous nitrate 硝酸钴 $Co(NO_3)_2$
cobaltous oxide 氧化钴 CoO
cobaltous perchlorate 高氯酸钴 $Co(ClO_4)_2$
cobaltous phosphate 磷酸钴〔1. $Co(PO_3)_2$; 2. $Co_3(PO_4)_2$; 3. $Co_2P_2O_7$〕
cobaltous selenate 硒酸钴 $CoSeO_4$
cobaltous selenide 硒化钴 $CoSe$
cobaltous silicate 硅酸钴 Co_2SiO_4
cobaltous stannate 锡酸钴 $CoSnO_3$
cobaltous sulfate 硫酸钴 $CoSO_4$
cobaltous sulfide 硫化钴 CoS
cobaltous telluride 碲化钴 $CoTe$
cobaltous thiocyanate 硫氰酸钴 $Co(CNS)_2$
cobaltous tungstate 钨酸钴 $CoWO_4$
cobalt oxide ①氧化钴②钴的氧化物〔1.CoO; 2.Co_2O_3; 3.Co_3O_4〕
cobalt oxydichloride 二氯一氧化二钴 $CoO \cdot CoCl_2$; $ClCo \cdot O \cdot CoCl$
cobalt phosphide 一磷化二钴 Co_2P
cobalt pigment 钴颜料
cobalt pink (=cobalt-magnesia-red) 钴-氧化镁红；钴桃红
cobalt pyrite 硫钴矿
cobalt red 钴红
cobalt resinate 树脂酸钴
cobalt salt (=cobaltous salt) 钴盐
cobalt sesquisulfide 三硫化二钴 Co_2S_3
cobalt soap 钴皂
cobalt suboxide 一氧化二钴 Co_2O
cobalt sulfate (heptahydrate) 七水硫酸钴 $CoSO_4 \cdot 7H_2O$
cobalt tallate 松浆油酸钴
cobalt tetracarbonyl 四羰合钴 $Co(CO)_4$
cobalt toluate 甲苯(甲)酸钴 $(CH_3C_6H_4COO)_2Co$
cobalt tricarbonyl 三羰合钴 $[Co(CO)_3]$
cobalt ultramarine 钴蓝；钴青
cobalt vermiculite 钴蛭石
cobalt violet 钴紫
cobalt yellow 钴黄 $K_3Co(NO_2)_6$
cobaltyl 氧钴根 (CoO)
cobaltyl sulfate 硫酸氧钴 $(CoO)_2SO_4$
cobalt zincate 锌酸钴 $CoZnO_2$
cobamic acid 钴胺酸
cobamic dichloride 二氯化钴胺酸
cobamide 钴胺酰胺

co-batching 同批作业；同批处理
cobble 粗砾；中(号)鹅卵石；中砾
cobcoal 成团煤炭；大圆块煤
cobinamide 钴啉醇酰胺
cobinic acid 钴啉醇酸
co-boiling rectification 共沸精馏
cobric acid 眼镜蛇酸
cobwebbing 网纹涂覆法；喷(拉)丝涂覆法；(喷涂时形成的)网纹；蛛网
cobweb coating 网纹涂料；拉丝涂料
cobweb whisker 蛛丝状晶须
cobyric acid 钴啉胺酸
cobyrinamide 钴啉胺酸
cobyrinic acid 钴啉酸
coca 柯卡；古柯
cocaethylene 可卡乙碱；高古柯碱
1-cocaine 1-可卡因；1-古柯碱
cocaine hydrochloride 盐酸可卡因
cocainidine 柯卡(因尼)定；甲基可卡因
CO-canister (=carbon monoxide canister) 一氧化碳滤毒罐
cocarboxylase 羧化辅酶；硫胺焦磷酸
cocatalyst 助催化剂；辅催化剂
cocatalytic 助催化的
cocatannic acid 可卡单宁酸
coccinellin 胭脂红
coccognic acid 可可各酸
coccognidic acid 可可各酸
coccolith 颗石
cocculin 木防己苦毒素
cocculine 木防己碱；衡州乌药灵
cocerico oil 西瓜子油
cochineal ①胭脂虫红②(干)胭脂虫
cochineal solution 胭脂红溶液
cochinilin 胭脂红酸
cochin oil 精制(级)椰子油
Cochrane's test for variance 柯奇拉法检验方差
Cochrane's test method 柯奇拉检验法*
cochromatograph 共同用色谱(法)分析
cochromatographic 共同用色谱(法)分析的
cochromatographic separation 共同(进行)色谱分离
cochromatography 共色谱分析法
cochrome 钴铬合金
cocinic acid 椰子油脂肪酸；椰油酸
cocinin 椰子油酸酯；椰子硬脂
cock (活)栓；旋塞；考克〔俗〕
Cockcroft-Walton accelerator 高压倍加器
cock-disc 旋塞垫圈

cockling 起皱；皱纹
cockscomb pyrite 白铁矿
cock type pipette 活栓型移液管
coclamine 衡州乌药胺
coclanoline 衡州乌药醇灵
coclaurine 衡州乌药碱
coco 椰子树
cocoa 可可
cocoa bean 可可豆
cocoa butter (=cocoa oil) 可可油；可可脂
cocoa extract 可可浸液
co-coagulate 共凝固物
coco alcohol sulfate 椰油醇硫酸盐
coco alkyl amino propionic acid 椰油烷基氨基丙酸
coco alkyl diethanol amide 椰油烷基二乙醇酰胺
coco alkyl diethanol amine 椰油烷基二乙醇胺
coco alkyl trimethyl ammonium chloride 椰油烷基三甲基氯化铵
cocoa nib 可可粒
cocoanut (=coconut) 椰子
cocoanut active charcoal 椰壳活性炭
cocoanut charcoal 椰子炭
cocoanut fat 椰子脂
cocoanut fiber 椰子纤维
cocoanut flavour 椰子香精
cocoanut husk 椰子外壳
cocoanut oil 椰子油
cocoanut shell 椰子壳
cocoa oil (=cocoa butter) 可可油；可可脂
cocoa powder 可可粉
cocoa seed 可可籽
cocoa shell 可可壳
cocoa tree 可可树
cocodescol 6-甲基香豆素
coco fatty alcohol sulfate 椰油脂肪醇硫酸盐
coco fatty diethanol amide 椰油脂肪二乙醇酰胺
coco monoethanol amide 椰油单乙醇酰胺
cocondensate 共缩(合)物
cocondensation 共缩合*
co-condensation polymer 共缩聚物
coconut (=cocoanut) 椰子
coconut aldehyde 椰子醛
coconut butter 椰子油；椰子脂
coconut cake 椰子油饼
coconut ethanol amide 椰子油乙醇酰胺
coconut fatty acid 椰子油脂肪酸
coconut fiber 椰子绒；椰子纤维
coconut monoethanol amide 椰油单乙醇酰胺

coconut oil diethanol amide　椰子油二乙醇酰胺
coconut oil diethanol amide phosphate　椰油二乙醇酰胺磷酸酯
coconut oil fatty acid　椰子油脂肪酸
cocoon　茧封〔喷涂一层高黏度涂料于工件的表面，形成一层包覆层以防腐蚀〕
cocoon fiber　蚕丝
cocooning　茧状包覆
cocoonization　被覆包装
cocoon layer　茧层
cocoon outer floss　茧衣
cocoon packing　被覆包装
cocoon shell　茧层
cocoon spraying　①作茧喷涂法；喷茧衣；喷茧层②喷丝包装；被覆包装
COCO process　一氧化碳/一氧化碳(碳同位素分离)法
coco sulfate　椰油(醇)硫酸盐
cocoyl isethionate　椰油基羟乙基磺酸盐
co-crosslinking　共交联
cocrystallization　共结晶*
co-cure　共硫化〔两种胶共同硫化〕
cocuring agent　助硫化剂
cocurrent　①直流；同向②并流
cocurrent catalyst flow　同向流动
cocurrent extraction　并流萃取
co-current fashion　直流型；并流型
cocurrent flow　同向流动
cocurrent laminar flow　共层流
cocurrent leaching　并流浸出
co-current string-up　顺向生头
cod　①鳕②鳕鱼
codamine　可旦民碱
code　①密码②编码
codecontamination　共去污*
codecontamination cycle　共去污循环
coded sample　密码样品
codeic acid　可待酸
codeine (=methylmorphine)　可待因〔药〕
codeine methyl bromide (=eucodin)　甲基溴可待因；优可定〔药〕
codeine phosphate　磷酸可待因
codeinone　可待因酮
codeposition　共沉积
codethyline (=ethyl morphine)　可待乙碱；乙基吗啡 $C_{19}H_{23}O_3N$
code wire　分色电线〔按用途分别着色的电线〕
codimerization　共二聚作用〔两种不同单体共聚成二聚物的作用〕
codimers　共二聚体
coding　①编码②编程序③译成电码④标记
coding amino acid　编码氨基酸
codissolved　共溶的
codistillation　共馏(法)
codistilling　共馏
cod-liver oil　鳕鱼肝油
cod oil　鳕鱼油
co-domain　合域
codon　密码子
Coe-Clevenger method　科-克利文格法
coecomycin　互利霉素
coefficient　系数
coefficient corrected for elevation in boiling point　经沸点上升校正的系数
coefficient multiple derivative spectrophotometry　系数倍率导数分光光度法
coefficient of absorption　吸收系数
coefficient of acid effect　酸效应系数
coefficient of admission　装满系数；充满系数
coefficient of angularity　角度系数
coefficient of charge　装载系数；装料系数
coefficient of complexation effect　络合效应系数
coefficient of compressibility　压缩系数
coefficient of concordance　和谐性系数
coefficient of contraction　收缩系数
coefficient of corrosion　腐蚀系数
coefficient of corrosion inhibition　缓蚀系数*
coefficient of cross viscosity　横向黏性系数
coefficient of cubical expansion　体积膨胀系数
coefficient of cubic expansion　容量膨胀系数
coefficient of cyclic variation　循环变动系数
coefficient of damping　阻尼系数
coefficient of differentiation　分化系数；变异系数
coefficient of diffusion　扩散系数
coefficient of digestability　消化吸收率
coefficient of dilution　稀释系数
coefficient of discharge　卸料系数
coefficient of dispersion　分散系数
coefficient of distribution　分配系数
coefficient of dynamic viscosity　动力黏度；动态黏性系数
coefficient of eddy viscosity　涡流黏性系数
coefficient of efficiency　有效系数；效率
coefficient of elasticity　弹性系数
coefficient of elongation　伸长系数
coefficient of energy dissipation　能耗(散)系数
coefficient of expansion　膨胀系数

coefficient of expansion due to heat　热膨胀系数
coefficient of extension　伸长系数
coefficient of extinction　吸光系数；消光系数
coefficient of fluidity　流动性系数；流度系数
coefficient of friction　摩擦系数
coefficient of fullness　装满系数
coefficient of hardness　硬度系数
coefficient of heat conductivity　导热系数
coefficient of heat transfer　传热系数
coefficient of heat transmission　传热系数
coefficient of hybrid effect　混杂效应系数
coefficient of hydration　水化系数
coefficient of interdiffusion　内扩散系数
coefficient of internal friction　内摩擦系数
coefficient of kinematic viscosity　运动黏度；运动黏性系数
coefficient of leakage　漏损系数
coefficient of linear expansion　线膨胀系数
coefficient of linear thermal expansion　线(性)热膨胀系数
coefficient of losses　耗损系数
coefficient of magnification　放大系数
coefficient of mass absorption　质量吸收系数
coefficient of mass transfer　传质系数
coefficient of mean deviation　平均偏差系数
coefficient of opacity　不透明度；不透明系数
coefficient of orientation-induced crystallization rate　取向诱导的结晶速率系数
coefficient of oxidation　氧化系数
coefficient of partial regression　偏回归系数
coefficient of performance　有效系数；效率
coefficient of plasticity　可塑性系数*
coefficient of plastic viscosity　塑性黏度系数*
coefficient of polymerization　聚合系数
coefficient of pressure　压力系数
coefficient of rank correlation　等级相关系数
coefficient of refrigeration　冷冻系数
coefficient of resistance　阻抗系数
coefficient of restitution　恢复系数
coefficient of rigidity　刚性系数
coefficient of roughness　粗糙系数
coefficient of safety　安全系数
coefficient of scatter (=scattering coefficient)　(光)散射系数
coefficient of scattering　散射系数
coefficient of self-induction　自感系数
coefficient of shear　剪切系数；切变系数
coefficient of shear viscosity　剪切黏度；剪切黏性系数
coefficient of solubility　溶解度系数
coefficient of static friction　静摩擦系数
coefficient of stiffness　劲度系数
coefficient of structure stability　结构稳定性系数
coefficient of tensile viscosity　拉伸黏度系数
coefficient of thermal conductivity　导热系数
coefficient of thermal expansion (CTE)　热膨胀系数
coefficient of thermal transmission　传热系数
coefficient of thixotropic breakdown　触变破坏系数
coefficient of thixotropy　触变系数
coefficient of twist　捻系数；捻度系数
coefficient of variation　变异系数
coefficient of virtual viscosity　有效黏度；有效黏性系数
coefficient of viscosity　黏度系数*
coefficient of viscous traction (=Trauton's coefficient)　拉伸黏性系数；黏性曳引系数
coefficient of volume expansion　体积膨胀系数
coefficient of volume thermal expansion　体积热膨胀系数
coefficient of volume viscosity　体积黏度；体积黏性系数
coefficient of vulcanization　硫化系数
coefficient of waste　浪费系数；废弃系数
coefficient of wetness　湿度系数
coelectrodeposition　共电沉积
coelesticetin (=celesticetin)　天青菌素
coelicolorin　天蓝菌素；蓝链丝菌素
coelicomycin　天蓝霉素
coelute　共洗脱
co-eluted components　共洗脱组分
coenzyme　辅酶
coenzyme Ⅰ (NAD; CoI; DPN)　辅酶Ⅰ；烟酰胺腺嘌呤二核苷酸；二磷酸吡啶核苷酸
coenzyme Ⅱ (NADP; CoⅡ; TPN)　辅酶Ⅱ；烟酰胺腺嘌呤二核苷酸磷酸；三磷酸吡啶核苷酸
coenzyme A(CoA)　辅酶A〔药〕
coenzyme Q　辅酶Q；泛醌
coenzyme Q_{10}　辅酶Q_{10}〔药〕
coenzyme R(=vitamin H; biotin)　辅酶R；维生素H；生物素
coercibility　可压凝性
coercible gas　可压凝气体
coercimeter　矫顽磁力计
coercive force　矫顽(磁)力
coeruleine　媒染棓酸绿〔染〕
coeruleolactite　钙绿松石；微晶磷铝石
coerulomycin　浅蓝霉素
coexistence element　共存元素
coexisting　同时存在；共存
coexisting phase　共存相

coextraction 共萃取
coextraction phenomenon 共萃取现象
coextrudable tie resin 共挤压黏合树脂
coextrude 共挤出
coextruded film 复合薄膜
co-extruded parison 共挤塑型坯
coextrusion 复合挤压；共挤压
coextrusion barrier 共挤阻透层
coextrusion blow 共挤吹塑
coextrusion blow molding 共挤吹塑
coextrusion casting 共挤流延
coextrusion coating 共挤贴合
coextrusion die 共挤塑模头
coextrusion laminating 共挤层合
coextrusion molding 共挤出成型
cofactor 辅助因素
coferment 辅酶
coffee bean (=coffee berry) 咖啡豆
coffee beans tincture 咖啡豆酊
coffee berry (=coffee bean) 咖啡豆
coffee berry oil 咖啡油
coffee berry wax 咖啡蜡
coffee extract 咖啡浸液
coffee ground sludge 咖啡末状淤渣
coffee mill 咖啡磨
Coffey still 科菲蒸馏器
coffic acid 咖啡酸
coffin hoist 匣升降机〔由管沟举起管子用〕
cofield rotation 共场旋转
CO-filter (=carbon monoxide filter) 一氧化碳滤毒罐
co-firing 共烧
co-flocculation 絮凝
coflow 同向流动；协流
coformycin 助间型霉素
co-fumed zinc oxide 炉法含铅氧化锌
cofunctionality 共官能性
cog 嵌齿；大齿
cogasin (水煤气)合成油
cogged V-belt 齿形三角带
cognac 白兰地酒
cognac essence 康乃克油
cognac oil 庚酸乙酯；水芹酸乙酯
cognate amino acid 关连氨基酸
cognate inclusion 均匀包含物
cograft 共接枝
cografting 共接枝
cog-wheel grease 嵌齿轮润滑脂
coherence ①内聚现象②联接③黏结④内聚力

coherence effect 相干效应
coherence length 相干长度
coherence strength 内聚强度
coherence transfer 相干转移
coherency ①黏结②内聚力③连接④内聚现象
coherent 黏附的；黏结的
coherent anti-Stockes Raman spectroscopy 相干反斯托克斯拉曼光谱法
coherent anti-Stokes Raman scattering 相干反斯托克斯拉曼散射
coherent anti-Stokes Raman spectrometry 相干反斯托克斯拉曼光谱法
coherent elastic scattering of radiation 辐射的相干弹性散射*
coherent excitation 相干激发*
coherent forward scattering atomic spectroscopy 前向相干散射原子光谱学
coherent radiation 相干辐射
coherent Raman spectroscopy 相干拉曼光谱法
coherent scattering 相干散射
coherent scattering of molecular 分子相干散射
coherent sliver 并丝条
coherent Stokes Raman scattering 相干斯托克斯拉曼散射
coherent Stokes Raman spectroscopy 相干斯托克斯拉曼光谱法
coherent system of unit of measurement 一贯计量单位制
coherer 金(属)屑检波器
cohesible 可黏合的；可黏结的
cohesion ①内聚(现象)②内聚力③黏结力
cohesional entanglement 凝聚缠结
cohesional failure 内聚破坏
cohesional field 内聚力场
cohesional setting 内聚固定
cohesional strength 黏结强度
cohesion energy 内聚能
cohesion energy density 内聚能密度
cohesion failure 内聚破坏
cohesion pressure 内聚压力
cohesive 内聚的；黏合的
cohesive action 黏结作用；内聚作用
cohesive body 黏结体
cohesive characteristic 内聚特性；集束特性
cohesive end 黏性末端
cohesive energy 内聚能
cohesive energy density (CED) 内聚能密度；凝集能密度
cohesive energy density parameter 内聚能密度参数
cohesive failure 内聚衰坏；内聚破坏

cohesive force　①内聚力②黏结力③黏合力
cohesive material　黏结材料
cohesiveness　黏结性
cohesive power　内聚本领；内聚能力
cohesive pressure　黏结压力
cohesive property　黏结性
cohesive set　抱合定形
cohesive strength　黏结强度；内聚强度
cohesive terminus　黏性末端
cohobation　回流蒸馏；复蒸馏法
cohumulinone　副葎草灵酮
cohumulone　辅葎草酮；合葎草酮
cohydrol　石墨的胶态溶液
cohydrolysis　共水解作用
coil　①旋管；盘管②线圈③线材
coil coating　①卷材涂漆法②卷材涂料
coil condenser (=spiral condenser)　蛇形冷凝器
coil density　线团密度
coiled column　盘旋柱
coiled configuration　蜷曲构型
coiled conformation　卷曲构象
coiled expansion pipe　盘旋膨胀管；膨胀旋管
coiled hose insert　金属螺旋线
coiled loop stitch　绞绕组织
coiled macromolecule　线团大分子
coiled molecule　卷曲分子
coiled oval (tube)　椭圆形旋管；椭圆形蛇管
coiled pipe (=coiler)　①线圈②旋管；盘管
coiled pipe cooler　盘管冷却器
coiled radiator　盘旋散热器
coil enamel　线圈瓷漆；绕阻瓷漆
coiler　①旋管；盘管②线圈
coil evaporator　旋管蒸发器
coil exit　旋管出口
coil finishing varnish　(电机)线圈覆盖清漆
coil former　卷线筒
coil-globule transition　线团-球粒转换
coil heater　蛇(盘)管加热器
coil heat exchanger　螺旋换热器；蛇管换热器；盘管换热器
coil impregnating varnish　绕阻浸渍漆
coil-in-box cooler　箱内旋管式冷却器
coiling　盘旋；卷曲；盘曲
coiling machine　盘管机；盘簧机
coiling of the molecule　分子卷曲
coiling soup　卷花
coiling-type polymer　螺旋型聚合物；盘旋型聚合物
coillayer　卷板式(容器)
coillayer pressure vessel　绕板式压力容器
coil paper　盘纸；筒纸；卷纸
coil planet centrifuge　螺旋管行星式离心分析仪
coil spring model　盘簧模型
coil-type pan　旋管锅
coil varnish　绕阻清漆
coil vulcanized　冷(法)硫化的
coinage gold　货币金
coinage metals　货币金属
coinage silver　货币银
coin bronze　古铜色
coincidence　符合；一致；重合
coincidence circuit　符合电路
coincidence counting　同时计数
coincidence loss　同时计数损失
coincidence mass spectrometer　符合质谱仪
coincidence method　重合法；符合法
coincidence technique　同时计测法
coincidence test　一致性检验；重合性检验
coincident flash method　同期闪火
coinging　压印加工；压花
co-initiator　共引发剂*
coinjection　共注射成型
coinjection molding　共注塑
coin marking　金属划痕〔漆病〕
cointerlaced yarn　混络丝
coin test　硬币(附着力)试验
coin thyme　铜钱叶百里香
co-ion　共同离子；伴离子
coion distribution　同离子分布
coions effect　同离子效应
coir　椰纤维
coixenolide　薏苡酯
coixol　薏苡素
cokability　(可)焦化性
coke　焦炭；焦煤〔俗〕
coke bed　底焦；焦床
coke blast-furnace　鼓风炉；高炉；炼铁炉
coke breeze　焦屑
coke burner　①炼焦工长②炼焦炉
coke-burning　烧焦〔由催化剂烧去焦质〕
coke-burning capacity　烧焦本领；烧焦容量〔催化剂再生器〕
coke button　焦块
coke by-products　炼焦副产品
coke chamber　焦炭室；沉焦室
coke cooling chamber　冷焦室
coked　焦结的；炼成焦的

coke deposits	焦沉积
coked pitch	焦沥青
coke-drum	焦炭鼓
coke-filled scrubber	填焦洗涤器
coke filter	焦滤器
coke formation	焦形成
coke-forming hydrocarbon	成焦烃
coke-forming period	成焦期
coke furnace	炼焦炉
coke iron	焦炭生铁
coke knocker	除焦机
coke laydown	焦沉积
coke-like	焦状的
coke-like sludge	焦状淤渣
coke number	焦值
coke number test	焦值试验
coke oven	炼焦炉
coke oven battery	焦炉群
coke-oven coal tar	焦炉煤焦油
coke-oven coal tar pitch	焦炉煤(焦油)沥青
coke oven gas	焦炉气
coke-oven plant	炼焦炉车间
coke oven tar	焦炉焦油
coke-oven tar pitch	焦炉煤(焦油)沥青
coke-packed scrubber	填焦洗涤器
coke-packed tower	焦炭填充塔
coke pig	焦炭生铁
coke platform	焦(炭)台
coke producer	炼焦煤气发生炉
coke pusher (=coke pushing ram)	推焦车
coke pushing ram (=coke pusher)	推焦车
coke quenching	淬焦
coke quenching car	淬焦车
coker	①焦化装置②炼焦器
coke residue	焦炭残余物
coker-gasifier process	焦化-气化加工法
coker gas oil	焦化瓦斯油；焦化粗柴油
cokery	①炼焦炉②炼焦厂
coke screenings	筛焦层
coke screening station	筛焦台
coke scrubber	洗焦器
coke settler	焦炭沉降器
coke side	焦面
coke slurry	焦泥浆
coke still	焦化蒸馏釜
coke suppressing additive	防焦添加剂
coke test	①成焦试验②残炭试验
coke tower	焦炭塔；焦化塔
coke value	焦值
coke wharf	焦炭装卸台
coke yield	焦炭产率
coking	①结焦②焦化
coking behavior	结焦性能
coking capacity	焦化本领
coking chamber	焦化室；炼焦室
coking coal	炼焦煤
coking heater	焦化加热器
coking of heavy residual oil	重残油的焦化
coking of residues	渣油的焦化
coking of tube	管子结焦
coking period	焦化期；炼焦期
coking plant	炼焦厂
coking plant wastes	炼焦厂废料
coking power	①成焦性②成焦率
coking property	成焦性
coking residue	焦化残油
coking retort	炼焦罐
cokings	蒸馏罐中残渣〔煤焦化后〕
coking space	炼焦室
coking still	焦化釜
coking stoker	炼焦加煤机
coking tar	高温焦油；炼焦焦油
coking test	焦化试验〔确定结焦性能〕
coking tower	焦炭塔；焦化塔；炼焦塔
coking value	焦化值
coking wharf	炼焦台
coking yield	焦化产率
cola acuminata	苏丹可乐果；红可粒
colalin (=cholic acid)	胆酸
colamine (=cholamine)	胆胺；2-羟乙胺
colander	滤器
colas	沥青乳化液；沥青乳胶
colasmix	沥青和铺路材料的混合物
colatannin	可拉单宁
colatein	可拉酚
colatin	可拉精；可拉单宁
colation	过滤
colatorium	滤网；滤筛
cola tree	苏丹可乐果；红可粒
colature	粗滤产物
Colburn's analogy between heat and mass transfer	科卡伯恩热量和质量转换变推法
colchiceine	秋水仙裂碱；去甲秋水仙碱
colchicin	秋水仙素
colchicine	秋水仙素；秋水仙碱〔药〕
colchisal	水杨酸秋水仙碱

colchium seed　秋水仙花子
colcothar (=calcothar)　氧化铁〔颜料〕；铁丹
cold　①(塑物表面)无光(彩)②冷的③寒冷
cold acid treatment　冷酸处理法
cold aeration　冷通气
cold agglutinin　冷凝集素
cold airduct　冷风道；冷空气导管；冷空气通风道
cold alkylation　冷冻烷基化
cold-applied coal-tar coating　冷用煤焦油涂料
cold atomic absorption　冷原子吸收(法)
cold-atomic fluorescence method　冷原子荧光光谱法
cold-atom nonchromatic dispersion AFS　冷原子非色散原子荧光光谱法
cold batch bleach　冷卷漂白
cold beam gun　冷束枪
cold bending　冷弯
cold bend test　冷曲试验
cold blast　冷风
cold-blast pig (iron)　冷铸铁
cold blending　冷掺和
cold blocking　冷压印
cold break　冷淀物；冷却残渣
cold breakdown　冷滚
cold catch pot　低温截液罐；低温分离器
cold cathode discharge　冷阴极放电
cold cathode glow tube　冷阴极辉光放电管
cold cathode ionization gauge　冷阴极电离真空规
cold cathode ion pump　冷阴极离子泵
cold cathode ion source　冷阴极离子源
cold cathode tube　冷阴极管
cold chamber　冷却室
cold charge　低温装料
cold check　低温细裂；冷脆
cold-check resistance　抗冻(细)裂性；抗冷脆性
cold check test　冷开裂试验
cold-check test for clear finish on wood　木器漆低温细裂试验
cold closet　冷藏箱；冷藏室
cold color　冷染染料
cold compression molding　冷压模制法
cold correction　冷态矫正
cold crack　冷裂纹
cold-crack resistance　抗冻龟裂性；抗冷裂性
cold-crack resistance temperature　耐冷龟裂温度
cold-crack stability　冷裂稳定性
cold crack temperature　冷裂温度
cold cream　冷霜；冷膏
cold creep　冷蠕变

cold cross-linking　常温交联
cold crystallization　冷晶化
cold cure　①冷硫化②氯化硫溶液硫化
cold (cured) vulcanizate　冷硫化胶
cold-curing　①冷固化②冷熟化
cold cut　冷切
cold cut chisel　冷錾
cold-cut varnish　冷溶清漆
cold cycle　低温循环(试验)
cold damn　冷拔
cold dipping　冷蘸涂
cold distillation　低温蒸馏
cold draw　冷拉
cold draw effect　冷拉效应
cold drawing　冷拉伸*
cold-drawing lubricant　冷拔用润滑剂〔有色金属及其合金用〕
cold drawn　冷拉
cold drawn oil　冷榨油
cold drawn weldless tube　冷延无缝管
cold drying　低温干燥法
cold-edge texturing process　冷刀口变形法
cold efficiency　冷效率〔生成气热值/燃料热值〕
cold emission　冷发射
cold emission ion source　冷发射离子源
cold-engine sludge　冷冻机沉淀
"cold" ensemble　"冷"系综*
cold factice　白油膏〔软化剂，增量剂〕
cold feed　冷加料；冷进料
cold feed extruder　冷进料压出机
cold field emission　冷场发射
cold filling　冷式灌装
cold finger　指形冷冻器；冷指
cold finger inlet　冷指进样口
cold finger (reflux) condenser　指形(回流)冷凝管
cold finger reservoir　冷指储样器
cold flame　冷(火)焰
cold flexibility　冷挠曲性
cold flex temperature　冷挠曲温度
cold flex tester　冷挠曲试验机
cold flow　冷流
cold flow tester　冷流试验机
cold forging　①冷煅②冷煅塑
cold-forming　冷成型
cold galvanizing　冷镀锌
cold glue size　冷胶料；冷水调和胶浆
cold gluing　冷(室温)胶合
cold grinding　冷研磨

cold hardening 冷加工；冷硬化；加工硬化
cold insulation 保冷
cold insulation mastic 冷绝缘(防潮)胶合辅料
cold junction 冷接点〔热电偶的〕
cold junction compensation 冷接点补偿
cold labeling 冷标记；稳定同位素标记
cold latex 冷聚(丁苯)胶乳
cold leaching 冷浸；常温浸出(取)
cold light 冷光
cold line fittings 冷却管管件
cold made soap 冷制皂
cold malt extract 麦芽冷浸出物
cold mark 冷迹〔制品缺陷〕
cold mastication 冷塑炼；低温塑炼
cold material 冷材料
cold method 冷冻法；低温法
cold milling 冷塑炼
cold-mixed grease 低温混合润滑脂
cold mixing 冷混合
cold moulding 冷塑(法)
cold moulding compound 冷模塑料
cold moulding method 冷模压成型法
cold-neck grease 低温颈口润滑脂〔碾轧机床〕
cold neutron 冷中子
cold oil 冷却油；冷凝油
cold on-column injector 冷柱头进样器
cold operation 冷运行；非放射性运行
cold orientation 冷态取向
cold pad-batch dyeing 冷轧卷堆染色法
cold parison 冷型坯
cold plastication 冷塑炼；低温塑炼
cold plastic matrix 冷塑性基料〔常温溶解型基料〕
cold point 冷点；冻点
cold polymer 冷聚(合)物
cold polymerization 冷聚合
cold-polymerized nitrile rubber 冷聚丁腈胶；低温丁腈胶
cold-polymerized rubber 冷聚橡胶
cold polymerized styrene butadiene rubber 冷聚丁苯橡胶
cold position 冷态位置
cold premixing 冷预混
cold prepressing 冷预压
cold pressed lime oil 冷榨白柠檬油
cold pressing 冷压
cold press process 冷压法
cold process 冷法
cold processing 常温加工
cold process soap 冷制皂
cold production 制冷(作用)

cold propane tank 低温丙烷储器
cold pump 低温泵；低温原料泵
cold punching 冷冲
cold reflux 冷回流
cold repair 冷补
cold resistance ①抗冷性②耐冷度
cold ribbon feed 冷条状进料
cold rinse ①冷淋洗；冷冲(漂)洗②冷浸洗
cold roll 冷轧；冷拔
cold rolled steel 冷轧钢
cold rolling 冷轧*
cold-rolling lubricant 冷轧润滑剂
cold roll neck grease (=cold-neck grease) 低温颈口润滑脂〔碾轧机床〕
cold rubber 冷聚合的橡胶〔一般指丁苯橡胶〕
cold run 冷试验*
cold-runner mould 冷流道模具
cold running 冷加工
cold salt inlet 冷盐入口〔固定床催化裂化过程〕
cold saponification 冷法皂化
cold sensitizer 冷敏(感)剂
cold set 冷固化
cold-sett grease 冷法煮剂润滑脂
cold setting 冷固化；冷(变)定
cold setting adhesive 冷固化黏合剂；冷变定黏合剂〔20℃以下变定〕
cold setting cement 冷固性黏合剂
cold setting glue 冷固性胶黏剂
cold-setting lacquer 常温固化漆；常温挥发性漆
cold-settled cylinder oil 经冷冻沉降脱蜡的汽缸油
cold settler 低温澄清器
cold settling 冷淀积
cold-sett process 冷法煮制过程
cold short 冷脆
cold short iron 冷脆铁
cold shortness 冷脆性
cold shot (透明塑体内的)斑瑕
cold shots (=cold slug) 冷料；冷料头
cold slug (=cold shots) 冷料；冷料头
cold sluggish lubrication 低温下流动迟缓的润滑作用
cold smoke 冷烟〔柴油机用〕
cold snap 快速冷却期；骤冷期
cold soap 冷制皂
cold-soda pulp 冷碱纸浆
cold solder joint 冷焊点
cold-soluble extract 冷溶栲胶
cold source 冷源
cold spots 冷部位

cold spray 冷喷涂
cold sprayed film 冷喷漆膜
cold spraying 常温喷涂；冷喷涂
cold springs 冷泉
cold stamping 冷冲压
cold-starting fuel 低温起动燃料
cold state 冷态
cold stirred soap 冷拌皂
cold storage ①冷藏②冷藏库
cold storage room 冷藏室
cold straightening 冷却直
cold stretching 冷拉伸*
cold styrene butadiene 低温(聚合)丁苯橡胶；冷丁苯胶；软丁苯
cold styrene rubber 冷聚丁苯胶；低温丁苯胶
cold surface 冷表面〔在立式裂化炉内管子没有被覆盖的部分〕
cold tapping 冷攻丝
cold tarring 低温焦油化
cold tear strength 冷撕裂强度
cold test 冷试验；非放射性试验
cold test oil 耐冷油
cold tight mill 冷滚筒小滚距
cold trap 冷阱*
cold-treating process 冷处理过程
cold vapor and non dispersion atomic fluorescence spectrum 冷原子非色散原子荧光光谱
cold vapor atomic absorption 冷蒸气原子吸收
cold vapor atomic absorption method 冷蒸气原子吸收法*
cold vapor atomic fluorescence spectrometry 冷蒸气原子荧光光谱法
cold vapour atomic absorption spectrometry 冷蒸气原子吸收光谱法
cold vat 冷染瓮
cold vulcanization 冷硫化(作用)
cold-vulcanizing machine 冷硫化机
cold-water paint 冷水(涂)漆；冷水涂料
cold water test 冷水(浸)试法
cold-weather lubrication 冬季润滑
cold-weather pipelining ①冬季石油产品的管线输送②冬季管线施工
cold-weld 金属粉末冷焊
cold-welding 冷焊
cold work 冷作
cold-working 冷作；冷加工
Cole-Cole's circular arc law 科尔-科尔圆弧定律
Cole-Cole's complex plane plot 科尔-科尔作图
colemanite 硬硼酸钙石
Coleman method 科尔曼法〔测吸油量〕
cole seed oil 菜(子)油
colibacillus 大肠杆菌
coline 可啉
colinear 共线的
colinearity 同线性
colistin 粘菌素〔药〕
collagen 骨胶原*
collagenase 胶原酶；梭菌肽酶 A
collagen fibrils 胶原纤维
collagen membrane 骨胶膜
collagen microfibril 胶原微纤维
collagenous fiber 胶原纤维
collagenous network 胶原网络
collagenous tissue 胶原组织
collagen precipitating agent 胶原沉淀剂
collagen slurry 胶原浆
collagen sugar 胶原糖
colla piscium (=ichthyol) 鱼石脂
collapsed ball model 塌球模型
collapsed drum 可折桶
collapsed storage tank 可收缩油罐；可折叠油罐
collapse-fissure 塌陷裂缝
collapse (in foamed plastics) 瘪泡〔泡沫塑料中〕
collapse point 崩溃点
collapser (薄膜吹塑机)夹膜板〔即人字板〕
collapse strength (微珠)抗压瘪强度
collapsible 可收缩的；可折叠的
collapsible container 可收缩容器
collapsible drum 可收缩鼓；曲折鼓
collapsible fuel tank (=collapsible storage tank) 可收缩油罐
collapsible mould 分片模；分室模
collapsible orifice 可收缩的锐孔
collapsible pallet 可收缩的棘齿轮的掣子
collapsible storage tank (=collapsible fuel tank) 可收缩油罐
collapsible tube 收缩管；软管
collapsible tube filling and closing 装闭收缩管机
collapsing pressure 破坏压力
collapsing treatment (纤维)致密处理
collar ①杆环〔商〕②轴环
collar burner 环焰灯
collargol 胶态银
collaring ①套轭②作凸缘
collaring machine ①曲边机②皱折机
collar leather 马颈圈革
collar oiler 环形注油器；加油环

collateral chain 旁系链
collateral series 旁系；支系
collected stack 集合烟囱
collecting agent 促集剂
collecting box 收集箱；接受器
collecting chamber 集流室
collecting electrode 接收电极
collecting flue 收集烟道
collecting groove 凝聚槽
collecting main 收集主管
collecting pail 收集桶
collecting pipe 收集管
collecting pit 收集池；收集槽；集水池
collecting precipitate 收集沉淀
collecting spool 收集盘(轴)
collecting system 收集系统
collecting tank 收集池
collecting trap 收集阱；捕集阱
collecting vat 收集槽；收集桶
collecting vessel 收集桶；收集容器
collection ①收集②收集品
collection box ①汇水箱②收集箱
collection efficiency 收(集效)率
collection launder 收集槽
collection of ions 收集离子
collection optics 聚光系统
collection procedure 收集程序
collection rate 集液速率
collection system 收集系统
collection vial 收集管形瓶；集液管形瓶
collective 集体的；聚集而成的
collective effect 集合效应；组合效应
collective gas protection 集体防毒
collective property 集体性质
collective protection 集体防护
collective protection room 集体防护室
collector ①收集器②收集剂〔痕量分析〕③浮选促集剂
collector box 集气盒
collector drum 收集筒
collector electrode 收集电极；捕集电极
collector slit 收集隙缝
collector unit 收集单元
collegen filament 胶原初原纤维；胶原结构基体(单位)；胶原纤维
colleseed oil 菜(子)油
collet ①卧式模塑料罐②挡料圈；节流箱③(玻璃纤维加工用的)挠丝轮
collidine 可力丁

β-collidine (=3-ethyl-4-methylpyridine) β可力丁；3-乙基-4-甲基吡啶
γ-collidine (=2,4,6-trimethylpyridine) γ可力丁；2,4,6-三甲基吡啶
collidine synthesis 可力丁合成
collidinium 可力丁鎓；4-乙基-2-甲基吡啶鎓 $(C_2H_5)(CH_3)C_5H_3NH^+$
collidinium molybdophsophate (CMP) 钼磷酸可力丁鎓
collidinium tungstosilicate (CWSi) 钨硅酸可力丁鎓 $[(C_8H_{11}NH)_4SiW_{12}]O_{40}$
colliery 煤井；煤矿
colligation 共价均成〔每个反应物供应一个电子形成共价 A+B→A∶B，参见 homolysis 共价均解〕
colligative method ①依数(性质的)方法②共价均成方法
colligative properties of solution 溶液的依数性*
colligative property 依数性
colligend 非表面活性的分离物
colligend ion 浮选被促集离子
collimated beam 平行光束；准直光束
collimated monochromatic light 平行单色光
collimated plane grating mounting 自准直式平面光栅装置
collimated roving 平绕粗纱
collimater light beam 准直光束；平行光束
collimating grating 准直光栅
collimating lens 准直透镜
collimating mirror 准直光镜；平行光镜
collimating slit 准直狭缝
collimation 平行校正
collimation converter 准直仪；准直管；平行光管
collimation line 准直轴
collimator 准直管；平行光管
collimator tube 准直管；平行光管
Collin's bleacher 科林漂白器
collinear 直线的；共线的
collinear collision 共线碰撞*
collinomycin 覆盖霉素
collision 碰撞
collision activated dissociation 碰撞活化离解
collisional activated dissociation 碰撞活化
collisional activation 碰撞活化*
collisional activation mass spectrometer 碰撞活化质谱仪
collisional broadening 碰撞展宽；碰撞变宽
collisional deactivation 碰撞失活
collisional energy transfer 碰撞能量转移
collisional half-width 碰撞半宽度
collisionally activated dissociation 碰撞活化解吸
collisional quenching 碰撞猝灭*

collisional radiative model 碰撞辐射模型
collision broadening 碰撞展宽；碰撞变宽
collision bumper 保险杠
collision cascade 碰撞串级
collision cross section 碰撞截面*
collision diameter 碰撞直径
collision effect 碰撞效应
collision efficiency 碰撞效率
collision energy transfer 碰撞能量转移
collision free suspension 无碰撞悬浮体
collision frequency 碰撞频率
collision frequency factor 碰撞频率因子
collision gas cell 碰撞气室
collision induced charge inversion 碰撞诱导电荷反转
collision induced decomposition 碰撞诱导分解
collision induced dissociation 碰撞诱导解离*
collision mean free path 碰撞平均自由路程
collision probability 碰撞概率
collision radiation 碰撞辐射
collision rate 碰撞速率
collision theory 碰撞(理)论
collision theory of reaction rate 反应速率的碰撞理论
collochemistry 胶体化学
collodion (=collodium) 火棉胶；硝棉胶
collodion coating color 硝纤涂层色料
collodion cotton (=collodion nitrocotton; collodion wool) 胶棉；低氮硝化纤维素
collodion emulsion ①火棉乳胶液②胶棉乳剂
collodion film 硝棉胶片；火(棉)胶薄片
collodion lacquer 硝纤清漆
collodion membrane 胶棉膜
collodion membrane electrode 火棉胶膜电极
collodion nitrocotton (=collodion cotton) 胶棉；低氮硝化纤维素
collodion-protamine membrane 硝棉胶-鱼精蛋白薄膜〔测定胶束电荷和表面活性剂活度系数〕
collodion silk 胶棉(人造纤维)丝；胶丝
collodion wool (=collodion cotton) 火棉胶；低氮硝化纤维素
collodium (=collodion) 火棉胶；胶棉
colloid 胶体*
colloidal ①胶体的②胶态的
colloidal agglomerate 胶态(体)附聚物；胶态凝聚物
colloidal alumina 胶体氧化铝
colloidal asbestos 胶态石棉
colloidal bismuth pectin 胶体果胶铋〔药〕
colloidal black 胶态(炭)黑
colloidal calcium carbonate 胶态(体)碳酸钙

colloidal calcium silicate 胶态硅酸钙
colloidal carbon 胶态炭
colloidal chemistry 胶体化学
colloidal clay 胶质黏土
colloidal coal fuel 胶态煤燃料
colloidal complex 胶状复合体
colloidal condition 胶(体状)态
colloidal cotton 胶体棉
colloidal crystal 胶体晶体
colloidal dispersion ①胶态分散②胶态分散体
colloidal electrolyte 胶体电解质*
colloidal evaporation process 胶体蒸发法〔无机多晶纤维制造方法之一〕
colloidal fuel 胶态燃料
colloidal gel 胶态凝胶
colloidal gel filter 胶乳体滤纸
colloidal gold 胶态金
colloidal gold marking immunoassay 胶体金标记免疫分析
colloidal graphite 胶态石墨
colloidal graphite lubrication 胶态石墨的润滑作用
colloidal lubricant 胶态润滑剂
colloidal matter 胶质
colloidal medium 胶态介质
colloidal mercury (=hygrol) 胶态汞；汞胶液
colloidal metal 胶态金属
colloidal microcrystal room temperature phosphorimetry 胶态微晶室温磷光法
colloidal mill 胶体磨
colloidal movement 胶粒运动；胶态运动
colloidal network structure 胶体网状结构
colloidal palladium catalyst 胶态钯催化剂
colloidal particle 胶粒*
colloidal particulate silica 胶粒状二氧化硅；二氧化硅胶态微粒
colloidal precipitate 胶体沉淀*
colloidal precipitation 胶体沉淀
colloidal propellant 胶态发射药；胶态推进药
colloidal property 胶体性质
colloidal shock 胶态休克(震挠；倏变)；胶体冲突〔两种不同的漆料相混时产生的浑浊状态〕
colloidal silica 胶态二氧化硅；胶态硅石
colloidal silicate 胶态硅酸盐
colloidal silicious matter 胶态硅质；胶体含硅物质
colloidal silver theory 胶体银说
colloidal sol 溶胶
colloidal solution 胶体溶液*
colloidal state 胶体状态*

colloidal structure 胶体结构
colloidal sulfur 胶态硫
colloidal suspension ①胶(态)悬(浮)②胶(态)悬(浮)体
colloidal suspension method 胶态悬浮法
colloidal system 胶态物系;胶体体系
colloidal thixotropic structure 胶态触变性结构
colloidal titration 胶态滴定
colloidal tubular particle 管状胶体粒子
colloidal zeolite 胶体沸石
colloidal zinc oxide 胶体(态)氧化锌
colloid amphoion 胶态两性离子
colloid bearer 胶态载体
colloid chemistry (=dispersidology) 胶体化学
colloided silica 胶态硅石;胶态氧化硅
colloid electrochemistry 胶体电化学
colloid equivalent 胶体当量;胶态当量
colloid error 胶体误差;胶质误差
colloid form structure 胶状结构
colloidization 胶态化*
colloidizing 胶(态)化
colloidizing agent 胶化剂
colloid mill 胶体磨*
colloid osmotic pressure 胶体渗透压
colloid particle 胶粒;胶体微粒
colloid scintillation counting 胶体闪烁计数
colloid silica 胶体二氧化硅;白炭黑〔补强剂〕
colloid silicic acid 胶体硅酸
colloid solution 胶体溶液
colloid stability 胶体安定性
colloid state 胶体状态
colloid stationary phase 胶体固定相
colloid titration 胶体滴定
collophanite 胶磷矿
collose 木质胶
collosol 溶胶
collotype 珂罗版
collotype ink 珂罗版墨
colloxylin 火胶棉
colocynth ①药西瓜〔一种用作泻药的葫芦〕②(=colocynth pulp)药西瓜浆
colocynthin 药西瓜苷
colocynth pulp 药西瓜浆
cologne ①古龙香型②古龙香水
cologne spirits 香水级乙醇
cologne stick 古龙香酯
cologne water 古龙香水
colombic acid (=calumbic acid) 考龙酸;咖伦巴酸
colonial spirit 甲醇

colophene 松香
colopholic acid (=colophonic acid) 松香酸
colophonic 松香的
colophonic acid (=colopholic acid) 松香酸
colophonic soap 松香皂
colophonite ①土松香②褐榴石
colophonone 松香酮 $C_{22}H_{18}O_2$
colophony (=rosin) 松香;松脂
colophony resin 松香树脂
color ①(颜)色②色料③颜料④染料
color aberration 色差
colorability 着色性能;着色能力
color acceptance 颜料混合(量);颜料容纳力
color additive 染料助剂
coloradoite 碲汞矿
color analyzer 颜色分析器
colorant 色料;着色剂
coloration ①(颜)色②显色③染色;着色
coloration of cloth 织物涂色〔纺〕
color autoradiography (CAR) 彩色放射自显影法
color base 发色母体
color black 着色用炭黑
color bleeding 渗色;泅色
color bloom 起色霜
color body 发色体
color centre 色心*
color change 变色
color change interval 变色区;变色范围
color change potential 变色电位
color chart 比色图表
color check 染色探伤
color chips (标准)色瓶;漆卡;色片
color circle 色轮;色环
color coat 彩饰层
color code 色标;色码
color comparator 比色计
color comparimetry 比色法
color comparison tube 比色管
color contamination 色脏污;彩色混杂
color content 纯色量
color contrast (彩)色对比度
color coordinate 色坐标
color cord 色带
color deepening agent 颜色加深剂
color degradation (=colour deterioration) 变色〔汽油或灯油〕
color density 色密度
color depth 颜色深度
color deterioration (=color degradation) 变色〔汽油或灯油〕

color developing ①显色性②展色性
color developing reagent 显色剂*
color developing sprayer 显色器
color development 显色
color development agent 显色剂
color development reaction 显色反应
color development reagent 显色试剂
color deviation 色差；颜色失真
color diagram 色图
color difference 色差
color difference measurement 色差测定
color difference meter 色差仪
color disc 色盘；色板
color discharge 拔染剂
color drum 染色转鼓
colored 有色的；着色的
colored article 着色制品；彩色制品
colored background coating 印(白)色衬底漆
colored cement 着色水泥；彩色水泥
colored compound 有色化合物
colored fibre 有色纤维
colored glass 有色玻璃
colored glass filter 有色玻璃滤片
colored indicator 显色指示剂
colored ink ①有色油墨②有色墨水
colored intensity 显色强度
colored ion 有色离子
colored lamp boot 彩色灯泡胶套〔指示灯用〕
colored mix 着色胶料
colored oxidation product 有色(彩色)的氧化生成物
colored paint 色漆
colored paper 有色纸；颜色纸
colored pellet 有色片；有色片状催化剂〔用于观察透明模型内的运动〕
colored reclaim 有色再生胶
colored rubber 有色橡胶
colored soil 着色污垢
colored spot (显)色斑
colored spun 色纺纱
colored stock 有色主分
colored substance 有色物
colored varnish 着色清漆；透明色漆〔俗〕
colored warp 有色经纱
colored water 有色水
colored yarn warping machine 花色纱线整经机
colored zinc plated steel 彩色镀锌钢板；彩板
color effect 色效应
color emissivity 发色性

color error 色误差
color fadeness 褪色
color fadeometer 褪色计
color fastness (颜)色坚牢度；颜色耐久性
color filter 滤色器；滤色片
color fixing 定色；色固定
color floating (=floating) (颜色)发花
color gamut 色域
color grader 色度计
color-grinding machine 颜料(研)磨机
color harmony manual 调色手册；配色手册〔按照奥斯瓦德表色系编制〕
color identification of pipe lines 管线颜色标准化
colorimeter 比色计*
colorimetric 比色的
colorimetrical analysis 色度分析；比色分析
colorimetrical coordinate 色坐标
colorimetric analysis 比色分析(法)
colorimetric determination 比色测定
colorimetric disc 比色盘
colorimetric estimation 比色测定
colorimetric method 比色法
colorimetric method of chromium 铬的比色测定法
colorimetric method with ninhydrin 茚三酮比色法
colorimetric method with phenylhydrazine 苯肼比色法
colorimetric method with salicylate acid 水杨酸比色法
colorimetric purity 色彩纯度
colorimetric reaction 比色反应
colorimetric standard illuminant 彩色标准光源；彩色标准发光物
colorimetric standard solution 比色标准溶液
colorimetric standard stock solution 比色标准储备液
colorimetric system 色度体系
colorimetric titration 比色滴定法
colorimetric tube 比色管
colorimetry 比色法*
color improver 颜色改良剂
color index 比色指数
color indicator 颜色指示剂
coloring 着色；染色
coloring agent ①染料；颜料②着色剂
coloring earth 矿物颜料
coloring material 着色剂；色料
coloring matter ①色素②着色剂
coloring pigment 颜料
coloring power 染色本领
coloring tablet 色素片
coloring test (=color test) ①呈色试验②显色试验

color inhibitor for kerosene　灯油色泽稳定剂
color-in-oil　清油(分散)颜料浆；油性色浆
color intensity　颜色强度
color in varnish　清漆(分散)颜料浆
colorist　着色师
coloristical　彩色的
coloristic property　配色性
colority　①颜色②色度
color lake (=lake)　色淀
colorless　无色
colorless adsorbed compound　无色吸附化合物
colorless chromatogram　无色色谱
colorless chromatography　无色色谱(法)
colorless component　无色组分
colorless metal indicator　无色金属指示剂
colorless plate　无色板
colorless substance　无色物质
color mark　颜色标记
color masterbatch　色母粒；浓色体
colormaster differential colorimeter　主色示差比色计
color matching　配色；调色
color matching function　配色函数
color measurement　颜色测定；测色
color migration　色移
color mill　颜料磨
color mixing room　调色室
color mutant　色泽突变的；色泽突变体(株)
color name　色名
color notation　颜色标号；颜色标志
color number　色数〔浓淡指数〕
color of arc image　弧焰颜色
color of transparent liquid　透明液体颜色
color oil　调色油
color order system　颜色序列体系
color oven　铅丹炉
color paste　染料糊
color perception　色感(觉)
color perception attributes　颜色感觉(知觉)属性
color permanence　保色性
color photography　彩色照相术
color plate　显色板
color prime white　原白色〔石油〕
color process　套色
color-producing bodies　生色物质
color-producing reaction　显色反应
color-producing reagent　显色试剂
color purity　色纯度
color quenching　色致猝灭

color range of indicator　指示剂变色范围
color ratio　颜色比例
color reaction　显色反应
color reaction plate　显色反应板
color reagent　显色试剂
color reference standard　彩色参照标准
color-reflecting interference pigment　色反射干涉颜料
color rendering　①彩色再见(重现)；显色性②分色表③彩色发送器
color reproduction　色再现
color response　色感应
color response curve　色感应曲线
color retention　保色性；色泽稳定性
color reversion　回色
colors　①色料②颜料③染料
color saturability　彩色饱和度
color scale　色标
color screen　滤色屏；色帘
color screen law　色屏定律
color sensation　色感；色感觉
color sensitivity　感色灵敏度；感色性
color separation　分色；色分
color shade　①色灰度；颜色深浅②色相
color shift　①色差②色位移(偏移)
color solid　色立体
color space　色空间
color spot　色斑
color stability　颜色稳定度；颜色稳定性
color stability curve　颜色稳定性曲线
color stability index　颜色稳定度指数
color stability test　颜色稳定度试验
color stabilizer　颜色稳定剂
color-stable oil　颜色稳定的油品〔石油〕
color staining　沾色；搭色
color standard　颜色标准
color standard number　色标数；色标号
color standard white　标准白色〔石油〕
color stimulus　色刺激
color stimulus specification　色刺激值
color strength (=tinctorial strength)　①着色强度；着色力②着色颜料的有效体积浓度
color superfine white　最佳白色〔石油〕
color system　色系
color temperature　(颜)色温(度)
color temperature conversion filter　色温转换滤光片
color test (=coloring test)　①呈色试验②显色试验
color theory　色原(学)说
color tolerance　色差容限

color tolerance computer 颜色容许误差计算机
color transition point 色变点；指示剂颜色转变点
color tree 颜色树
color triangle 颜色三角；原色三相图
color tube camera 比色管暗箱
colotype ①珂罗版②美色法
colour (=color) ①颜色；色彩②着色；染色③显色
colour agent 着色剂
colour aid 着色助剂
colourant dispersion ①着色料(颜料)分散体②颜料分散度③颜料分散
colourant distribution 着色剂分布
colourant slurry 水性颜料浆
colour appraisal 评色；颜色鉴定
colouration ①颜色②染色③显色(作用)
colour batch 色母料
colour brightness 色亮度
colour buffing （镜面)抛光；消色
colour carbon black 色素炭黑
colour carrier 载色剂
colour chalking 色料起霜〔制品缺陷〕
colour change phenomena 色变现象
colour check 比色检验(测)
colour check test 着色检查；着色探伤
colo(u)r coating 彩色涂料
colour code 色码
colour-coded 彩色标记
colo(u)r comparator 比色计；比色器
colour comparison tube 比色管
colour consistency 颜色一致性
colour contrast 颜色对比；颜色反差
colour coordinates 色坐标值
colour defect 色疵点
colour degradation (=colour deterioration) ①色(蜕)变；色退化②褪色
colour density 色密度
colour depth 色浓度
colour deterioration (=colour degradation) ①色(蜕)变；色退化②褪色
colo(u)r development 显色
colour difference 色差
colour discrimination 色识别；色分辨
colour distortion 色变
colour draft 色差
colour dynamics 色彩浓淡法；色彩动力学
coloured 有(染)色的
coloured contaminant 有色污染物；有色杂质
colo(u)red discharge 着色拔染

coloured enamel 有色漆；带色釉
coloured glaze ①彩色釉②彩色罩光(涂)层
coloured moiety （分子的)有色部分；显色部分
coloured molecule 有色分子；着色分子
coloured paste stopper 浆状着色填孔剂
coloured scheme ①着色管线图②颜色标志图例
colo(u)red spun yarn 纺染短纤纱；纤维染色纱；色纺纱
coloured stripe band 彩条带
coloured yarn woven fabric 色织布
colour effect 色相；色彩；色光
colo(u)r fastness 染色牢度
colour fastness test 颜色坚牢度试验
colour fastness tester for light and weather 日晒(气候)色牢度仪
colour fastness tester to ironing 熨烫色牢度仪
colo(u)r fastness to laundering 耐机械洗涤(色)牢度
colour filter ①滤色片②滤色器
colour finish 有色涂饰
colour finishing 彩饰
colour flicker effect 彩色闪烁效应；彩色斑点效应
colour floating (=floating) 颜色发花；发花
colour former 成色物质
colour-forming impurity 成色杂质
colour furnisher 给浆辊
colour gamut 色域；全色图
colo(u)r grinding machine 染料研磨机
colo(u)r harmony ①颜色和谐；颜色协调②调色；配色
colourhold 不褪色；保色
colour identification ①色鉴别②显色鉴定
colourimetric number 比色值；色度值
colour index 色指数；比色指数；彩色指数
colour index number 染料索引号
colouring agent 着色剂
colouring aid 着色助剂
colouring intensity 色彩强度；着色强度
colour(ing) stabilizer 色泽稳定剂
colouring value 着色值
colouristic 色彩的；用色的
colourity 色度
colo(u)r lacking uniformity 颜色不均
colour lake 色淀；沉淀色料
colourless 无色的；不染色的
colo(u)rless coupler 无色成色剂
colour masterbatch （配料用)浓色母料；浓色体
colour match 配色；调色
colo(u)r matching 配色；比色；仿色
colour matching test 配色试验
colour measurement 测色

colour migration 色移
colour painting 彩绘〔大漆〕
colo(u)r pick-up 着色量〔印刷〕
colour quality 色质
colour reduction 消色
colour response curve 色感应曲线；彩色特性曲线
colour sense 色觉；色神
colour shade ①色灰度；颜色深浅②色相
colour shift ①色差②色位移；颜色改变
colour strength 色强度
colour striking 颜料共沉(淀)析出
colo(u)r stripper 脱色剂
colour terminology 颜色术语
colour tolerance 色差容限〔样品和公认标准之间的颜色差别〕
colour tone 色调
colour triangle 色三角；原色三色
colo(u)r vision 色视觉；色视力
colour wash ①着色石灰涂料②着色水胶涂料
coloury ①着(上)色的②色彩丰富的；色泽优良的
Colson's ebulliometer 考尔松沸点计
coltskin 小马皮
colts-tail (=erigeron) 飞蓬
colubrine 可鲁比因
columba (=calumba) 咖伦巴(根)；非洲防己素
columbate 铌酸盐〔1. $MNbO_3$; 2. M_3NbO_4; 3. $M_4Nb_2O_7$〕
columbic ①铌的②含铌的
columbic acid 铌酸〔1. $HNbO_3$; 2. H_3NbO_4; 3. $H_4Nb_2O_7$〕
columbic anhydride 五氧化二铌；铌酐 Nb_2O_5
columbic compound 五价铌化合物
columbin (=calumbin) 咖伦宾；咖伦巴根苷
columbite 铌铁矿 $Fe(NbO_3)_2$
columbium 铌〔41号元素，化学符号Nb〕
columbium carbide 一碳化铌 NbC
columbium dioxide 二氧化铌 NbO_2
columbium monoxide 一氧化铌 NbO
columbium nitride 一氮化铌 NbN
columbium oxide 氧化铌〔1. NbO; 2. Nb_2O_3; 3. NbO_2; 4. Nb_2O_5〕
columbium oxytrichloride 三氯氧化铌 $NbOCl_3$
columbium oxytrifluoride 三氟氧化铌 $NbOF_3$
columbium pentachloride 五氯化铌 $NbCl_5$
columbium pentafluoride 五氟化铌 NbF_5
columbium pentoxide 五氧化二铌 Nb_2O_5
columbium sesquioxide 三氧化二铌 Nb_2O_3
columbium tetroxide 四氧化二铌 Nb_2O_4
columbous 三价铌；亚铌
columbous compound 三价铌化合物

columboxy 铌氧基
columboxy group 铌氧团〔三价〕 ≡NbO
columbyl 铌氧基
columella 中轴
columinescence effect 共发光效应
column ①列②柱③塔
column adaptor 柱连接器；柱接头
column adaptor kit 柱(配套)连接器
column adsorption chromatography 柱吸附色谱(法)
column ageing 柱老化
columnal 中柱
columnar ①柱状的②针状的
column area 塔面积
columnar-shaped particle 柱状粒子
column aspect 柱形
column back-pressure 柱反压力
column-bearing plate 承受储器支架的薄板
column behavior 柱行为
column bleeding 柱流失*
column capacity 柱容量
column capacity factor 柱容量因子
column cassette 盒式(色谱)柱
column characteristics 柱特性
column choice 柱选择
column chromatographic method 柱色谱法
column chromatography 柱色谱法*
column classification 柱分类
column cleaning 柱清洗
column coating 柱涂渍
column combination 柱组合
column compressor 柱加压器
column condition 柱条件
column conditioning ①柱调节②柱老化
column configuration 柱构型
column connecting pipe 柱连接管
column connector 柱连接器
column cross-section 柱截面
column crystallization 柱状结晶(过程)
column dead volume 柱死体积
column development chromatography 柱展开色谱(法)
column diameter 柱直径
column dimension 柱尺寸
column dissolver 圆柱型溶解器
column distiller 蒸馏柱
column efficiency 柱效能*
column electrophoresis 柱电泳(法)
column elution program 柱流出程序
column end plug 柱端塞子

column evaporation　柱中浓缩
column evaporator　蒸发柱；浓缩柱
column exit　柱出口
column extra　柱附件
column extractor　萃取柱
column filling　①柱填充②柱填充物
column fitting　柱配件
column flow rate　柱流速
column foot　塔基；塔脚
column fractionation　柱分级
column head sampling　柱头进样
column holder　柱支架
column hold-up　①柱滞留②塔中存料(量)
column inlet　柱入口
column inlet pressure　柱入口压
column internal diameter　柱内径
column interstitial volume　柱隙体积
column joint　柱接头
column kit　柱配套(器件)
column length　柱长
column life　柱寿命
column liquid chromatography　柱液相色谱
column load　柱负荷
column loadability　柱负载能力
column material　①柱材料②柱填充材料
column mixer　混合柱
column of trays　板式塔
column operation　柱操作；柱式法
column outlet　柱出口
column outlet pressure　柱出口压
column oven　柱加热炉
column overload　柱过载；柱超载
column packer　填柱器
column packing　①柱填充②柱填充物
column packing funnel　柱填充漏斗
column packing method　柱填充法
column parameter　柱参数
column partition chromatography　柱分配色谱法
column performance　柱性能
column permeability　柱渗透性
column plate　塔板
column polarity　柱极性
column preconditioning program　柱预老化程序
column press　柱式压机
column pressure　柱压力
column pressure gradient correction factor　柱压梯度校正因子
column-programming　程序换柱

column radiochromatography (CRC)　柱放射色谱法
column radius　柱半径
column regeneration　柱再生
column rejuvenation　柱复活
column resistance　柱阻力
column resistance factor　柱阻力因子
column response　柱响应
column scrubber　洗柱器
column section　柱截面
column selection　柱选择
column shut-off valve　柱关闭阀
columns in series　串联柱；串联塔
column spinner　旋转柱
column stability　柱稳定性
column still　柱馏器
column support　柱载体
column support cage　色谱柱笼
column switching　柱切换；柱转换
column switching technique　柱切换技术
column switching valve　柱切换阀；柱转换阀
column system　柱系统
column tag　柱标签
column temperature　柱温
column temperature gradient　柱温梯度
column test　柱试验
column throughput　柱处理量
column top　柱顶
column tray　塔板
column tube　柱管
column tube material　柱管材料
column tubule　柱小管
column type　柱型
column valve　柱阀
column void volume　柱空体积
column volume　柱容积；柱体积
column wall　柱(内)壁
column washer　洗涤柱
colupulone　合蛇麻酮
colyseptic　防腐的
colza (=colza oil)　菜子油
colza oil (=colza)　菜子油
comanic acid　靠曼酸　C_6H_4O
comb　①鸡冠②蜂巢
comb dispersant　蜂窝状分散体
comb doffing　斩刀剥棉装置
comber　精梳机
comber detaching active length　精梳分离(拔取)工作长度

comber length detached per nip　分离纤维丛长度；精梳分离(拔取)须丛长度
comber set　精梳机
Combes quinoline synthesis　库姆斯喹啉合成
combex system　干法混合(制粉)系统
combinable　①可以化合的②可以结合的
combinableness　①可化合性②可结合性
combination　①化合(作用)②结合③混合
combination band　综合谱带；组频谱带
combination board　合层纸板
combination cracking　联合裂化
combination dresser　联合上浆机
combination drier　复合催干剂
combination effect　组频效应；组合效应
combination electrode　复合电极
combination frequency　组合频率
combination furnace　复式炉
combination fuse　①复式信线②复式引信
combination gas　富天然气〔含大量石油气〕
combination heat　形成热；化合热
combination lines　组合谱线
combination mat　复合毡
combination mill　联(合)磨
combination modes　组合模式
combination oven　联(立)炉
combination plant　联合设备
combination principle　化合原理；结合原则
combination process　联合作业
combination reaction　化合反应
combination tanned leather　结合鞣制的革
combination termination　结合终止*
combination thermal cracking　联合热裂化；混合热裂化
combination tone　组合频
combination topping and cracking plant　拔顶-裂化联合装置
combination tower　联合塔；复合塔
combination treatment　联合处理，混合处理
combination type process　联合作业；联合操作过程
combination unit　联合装置
combination valve　组合阀
combination vibration　组合振动
combination water bath　联合水浴器
combination weft knitted structure　纬编复合组织
combination wrapped and braided hose　夹布编织胶管
combinatorial array　组合阵列
combined　化合的
combined acid　结合酸
combined action　联合作用；复合作用

combined action of accelerator　加速器的联合作用
combined alkali　结合碱；化合碱
combined bonding technique　组合胶黏技术
combined carbon　化合碳；结合碳
combined charge　(新料与循环料混合)总进料
combined column　复合柱
combined cooling　混合冷却法
combined creep　综合蠕变
combined derivative spectrophotometry　组合导数分光光度法
combined drill　双用钻头
combined fatty acids　化合脂肪酸
combined feed　总进料〔石油〕
combined gas-producer and boiler　联用的煤气发生炉和蒸汽锅炉
combined hydrocracking hydrogenation process　联合加氢裂化加氢法
combined hydrofining reforming process　联合加氢精制重整法
combined hydrogen　化合氢
combined lime　结合石灰
combined liquid crystal polymer　混合型液晶高分子
combined lubricating system　组合式润滑系统
combined mass spectrometry　联用质谱法
combined mixer and sifter　联筛混合器
combined nitrogen　结合氮；固定氮
combined oil processing operation　石油联合加工
combined oxidation-esterification　氧化-酯化联合法
combined package　拼件
combined polymerization　共聚合
combined preparation　短流程前处理
combined pressure and vacuum relief valve　联合压力真空排出阀(放空阀)
combined putting-out and wringing machine　平展挤水两用机
combined rib-lug pattern　混合花纹
combined rolling and sliding friction　滚-滑动摩擦
combined rosin　化合松香
combined rotation and axial flow　旋转和轴向组合流动
combined sewage　混合污水
combined size　一次性浆料；即用浆料；组合浆料
combined spinning twisting frame　纺捻联合机
combined standard uncertainty　合成标准不确定度
combined steady and oscillatory shear　稳态和振动组合剪切
combined storage　混合储藏
combined strength　组合强度
combined suction and force pump　联合真空压力泵

combined sulfur　化合硫
combined sulfur dioxide　化合二氧化硫
combined technique of atomic absorption　原子吸收联用技术
combined technologies　联用技术
combined thermal analyzer　混(组)合型热分析仪
combined warping-sizing machine　整浆联合机
combined water　化合水
combined wire　复合焊丝
combing　梳毛
combing leather　梳刷革
combing waste　精梳落纤
combing wool　精纺用毛
combing yarn　梳刷纱
combining　化合(的)
combining affinity　化合亲和势
combining heat　化合热
combining power　化合力
combining proportion　化合比例
combining standard deviation　并合标准(偏)差
combining volume　化合体积
combining weight　化合量
combining zone　化合区(域)
comb-like molecular structure　蜂窝状分子结构
comb polymer molecule　梳形高分子
comb (shaped) polymer　梳形聚合物*
comburent　燃料；助燃物
combust　燃料
combustibility　可燃性
combustible　①可燃的；易燃的②可燃物
combustible cell type oxygenmeter　燃室型氧分析仪
combustible charge　可燃混合物的装料
combustible component (=combustible constituent)　可燃组分
combustible constituent (=combustible component)　可燃组分
combustible gas　可燃气体
combustible gas detector　可燃气体检查器
combustible gas sensor　可燃气体传感器
combustible liquid　易(可)燃液体
combustible matter　可燃物(质)
combustible matter content　可燃物含量
combustible mixture　可燃混合物
combustibleness　可燃性
combustible point　燃点
combustible substance　可燃物
combustible support　(催化剂的)可燃载体
combustible waste　可燃性废物

combustion　燃烧；氧化
combustion additive　燃烧添加剂
combustion adjuvant　燃烧助剂；助燃剂
combustion air inlet　燃烧层空气的入口
combustion analysis　燃烧分析
combustion boat　燃烧舟皿
combustion bomb　燃烧弹
combustion calorimetry　燃烧量热法*
combustion catalyst　燃烧催化剂
combustion cell　燃烧池
combustion chamber　燃烧室
combustion characteristic　燃烧特性
combustion control　燃烧过程控制
combustion curve　燃烧曲线*
combustion cycle　燃烧周期；燃烧循环
combustion diagram　燃烧曲线图
combustion dust　燃烧尘
combustion efficiency　燃烧效率
combustion emissions　燃烧废气
combustion equipment　燃烧设备
combustion expansion ratio　燃烧膨胀比
combustion flue　炉道；烟道
combustion-fluorination　燃烧-氟化
combustion furnace　燃烧炉
combustion gas　燃烧气体
combustion heat　燃烧热
combustion improver　助燃剂
combustion in moving air　在流动的空气中燃烧
combustion in-situ　就地燃烧
combustion intensity　燃烧强度
combustion knock　燃烧爆击
combustion lag　燃烧滞后
combustion-leach process　燃烧-浸取法
combustion limit　燃烧极限
combustion line (=combustion curve)　燃烧曲线
combustion method　燃烧法
combustion nozzle　烧嘴
combustion of oil in-situ　油就地燃烧
combustion pennisula　燃烧半岛*
combustion period　燃烧周期
combustion pipette　燃烧球管
combustion pressure　燃烧压力
combustion principle　燃烧原理
combustion process　燃烧过程
combustion products　燃烧产物
combustion rate　燃烧速率
combustion reaction　燃烧反应
combustion recorder　燃烧记录器

combustion research 燃烧研究
combustion roughness 燃烧强度
combustion space 燃烧室容积
combustion spiral 燃烧旋管
combustion stability 燃烧稳定性
combustion stabilizer 燃烧稳定剂
combustion-supporting gas 助燃气体
combustion temperature 燃烧温度
combustion test 燃烧鉴别试验
combustion theory 燃烧(学)说
combustion time 燃烧时间
combustion train 燃烧装置
combustion tube 燃烧管*
combustion tube furnace 燃烧管炉
combustion value 热值；卡值
combustion zone 燃烧带；燃烧区域
combwebbing （喷枪）拉丝
comb width 梳宽度
comcentric shell 同心液壳
comelt 共熔；共熔物
comenamic acid 考闷安酸；4,5-二羟吡啶-2-羧酸
comendite 流纹岩
comenic acid 考闷酸
comet formation (谱带)成彗星状〔即拖尾〕
come up 指标达到质量标准〔石油产品〕
comfortability 舒适性
comfort cushioning 良好缓冲
comfort zone 有效温度区；有效反应区
comicellization 共胶束化
commelinin 鸭跖草苷
commensalism 共生现象
commensurability ①公度；同量；同单位②通约性；成比例③相称；合适
commercial 商业的；商品的；工业的；工厂的〔过程、设备、装置〕
commercial allowance 商业公定容差；商业允差
commercial analysis 商品分析
commercial and industrial zone atmospheric corrosion 工商业区大气腐蚀
commercial apparatus 商品仪器
commercial application 工业应用
commercial availability 工业效力
commercial benzol 商品苯
commercial blast 工业(化)喷砂；常规喷砂
commercial burner oil 工业用燃烧油
commercial butane 商品丁烷
commercial catalyst 商品催化剂
commercial chemistry 商品化学

commercial detector 商品检测器
commercial efficiency 经济效率
commercial exploitation 工业开发
commercial fertilizer 商品肥料
commercial formulation 实用配方；生产配方
commercial fuel 商品燃料
commercial furnace 工厂炉子
commercial gasoline 商品汽油
commercial grade 商品级
commercial grade fuel 商品级燃料
commercial inorganic filler 工业无机填料
commercial installation (=commercial unit) 工业设备
commercial lubricating oil 商品润滑油
commercially available apparatus 商品仪器
commercially pure 商业纯的
commercially smooth 商品要求光滑度
commercially sweet 商业香的(气味上满足商品的要求)
commercial naphthenic acid 工业用环烷酸；商品脂环酸
commercial natural gas 商品天然气
commercial oven 商业(工业)用炉
commercial paint ①商品油漆②工业用油漆
commercial pitch 工业沥青
commercial plant 工业设备
commercial polymer 商品聚合物
commercial process 工业化生产过程
commercial product 商品
commercial production 工业生产
commercial pure 商业纯的；工业纯
commercial quality 商品品质
commercial rock gas 天然气；石油气
commercial rubber 商品橡胶
commercial run 工业过程；工业方法
commercial sample 商品试样
commercial scale 工业规模
commercial size ①工业规模②工业规格〔制品〕
commercial soap 商业皂
commercial solvent 商业溶剂
commercial specification 商品规格
commercial standard 商品标准
commercial sulphuric acid 工业用硫酸
commercial type organic pigment 商品有机颜料
commercial unit (=commercial installation) 工业设备
commercial use test 实(际使)用试验
commercial value 工业价值
commercial wax 工业石蜡
commercial wax cake 商品蜡块
commercial weight 通用重量
commercial xylene 商品二甲苯；工业二甲苯

commingle 混合；杂混
commingled yarn 混络丝
commingler 混合器
comminuted polymer 粉末聚合物
comminution 粉碎(作用)
comminution granulation 造粒(法)
comminutor 粉碎机；造粒机
commissioning ①试运转；试车②投(入生)产
committed step 关键步骤；关键反应
Committee D-2 D-2 委员会(ASTM)〔负责石油产品及润滑剂方面分析技术规范的议定〕
Committee D-3 D-3 委员会(ASTM)〔负责气体燃料分析技术规范的议定〕
Committee D-4 D-4 委员会(ASTM)〔负责筑路及路面材料分析技术规范的议定〕
Committee D-16 D-16 委员会(ASTM)〔负责工业芳烃分析技术规范的议定〕
commodity inspection 商品检验
commodity polymer 通用高分子*
commodity resins 日用树脂
common ①普通的②共同的；公共的
common alum （钾)明矾；(铝)钾矾
common ash 欧洲白蜡树
common branded coal 无定结构煤
common brick 普通砖
common brick clay 普通砖土
common crown 普通写字纸〔纸〕
common detector 通用型检测器
common difference 公差
common fennel 小茴香
common heart 石南
common horehound (=marrubium) 普通夏至草〔调香药物名〕
common ion 共(同)离子
common ion effect 同离子效应*
common lime 普通石灰
common name 常用名〔即指 trivial name 俗名〕
common parameter array 共同参数矩阵
common pressure liquid chromatography 常压液相色谱法
common resin 松香
common ring 普通环*
common salt 食盐
common slide valve 普通滑阀
common soap 普通肥皂；家用皂
common turpentine 普通松节油
common valeric acid 异戊酸 $(CH_3)_2CHCH_2COOH$
common yew 欧洲紫杉
communality 公因子方差

communicating pipe 连通管
communication equipment 通信设备
commutation relation 交换关系
commutation rule 对易规则
commutator ①换向器②整流器
commutator compound 换向器(整流器)润滑剂〔由石蜡和蜂蜡制成〕
commutator cover 整流器盖
commutator method 替续法
commutator rectifier 换向整流器
commuting operator 对易算符
comodity resin 日用树脂
comolded 共模塑
comonomer 共聚单体*
comonomer ratio 单体配比
compact ①压紧②紧密的
compact agglomerate 紧密附聚
compact-box mixer-settler 紧凑型箱式混合澄清槽
compact counterbalanced sliding carriage 紧凑型平衡移动式自动落筒车
compact double layer 紧密双(电)层
compacted catalyst 压榨的催化剂；成型催化剂
compactedness (=compactness) ①紧密性②紧密度
compact extractor 紧凑型萃取器
compactibility 紧密性
compacting 压紧
compacting factor 压紧因数
compacting machine 压紧机
compaction ①压实②粉料挤粒
compaction degree 密实度
compact layer 紧密层
compactness (=compactedness) ①紧密性②紧密度
compact particles 压实粒子
compact plant 紧凑型工厂；一体化工厂
compact powder 压型粉末
compact pyrochemical process 紧凑高温化学过程
compact solid 致密固体
compact spinning 紧凑纺丝；短程纺丝
compact steel cord 密封结构钢丝帘线
compact volume 紧密体积
compact winding 紧密缠绕
companion flange 成对法兰；结合法兰；配对法兰
companion ion 配对离子
comparable data 可比数据；参照数据
comparative analysis 比较分析
comparative assay 比较测定；比较试验
comparative data 比较数据
comparative design estimate 设计比较估算

comparative measurement to known standards 对已知标准的比较测量
comparative reactivity 比较反应性
comparative result 比较结果
comparative test (=comparison test) 比较试验
comparator ①比较器②比长仪③比较电路
comparator block 比色座
comparator gauge 比较量规
comparison bridge 比较电桥；惠斯通电桥
comparison buffer 比较缓冲剂；比较缓冲液
comparison buffer method 比较缓冲液法
comparison column 参比柱
comparison method 比较法
comparison method with standard 标准比较法
comparison prism 比谱棱镜
comparison solution 比较溶液
comparison spectroscope 比谱分光镜
comparison spectrum 比较光谱
comparison temperature 比较温度
comparison test (=comparative test) 比较试验
comparison tube 比色管
comparoscope 显微比较镜
compartment ①室；间格②层
compartmentalization 区域化
compartmentalization effect 分隔效应
compartmentation 区域化
compartment drier (=chamber drier) 间格干燥器；分室干燥器
compartment furnace (=compartment kiln) 间格窑；间格炉
compartment mill 间格磨；分室磨碎机
compartment tray 间格盘
compatibility 相容性*；共混
compatibility agent 相容性试剂
compatibility limit 相容极限
compatibility of fuels 燃料的配伍性
compatibility test 相容性试验
compatibility with solvent 与溶剂的配伍性
compatibilization 增容作用*
compatibilizer 增容剂*
compatibilizing agent 相容性试剂
compatibilizing ingredient 兼容成分；并存成分；协调成分
compatible 可配伍的；相容的
compatibleness 配伍性
compatible plasticizer 相容性增塑剂
compatible polymer 相容性聚合物
compatible polymer pair 相容聚合物对

compatilizer 相容剂
compelling high elastic state 强迫高弹态
compensated ionization chamber 补偿电离室
compensated spectra 补偿光谱
compensating cation 补偿阳离子
compensating chemicals 补偿化学品；补偿剂
compensating controller 互补调节器
compensating error 抵消误差
compensating eyepiece 补偿目镜
compensating gauge 补偿片
compensating jacket column 带补偿套管的塔
compensating ocular 补偿目镜
compensating planimeter 补偿求积仪
compensating prism 补偿棱镜
compensating process 补偿方法
compensating roller 松紧调节
compensation apparatus (=potentiometer) 电位计
compensation circuit 补偿电路
compensation colour 补色
compensation effect 补偿效应*
compensation method 补偿(对消)法*
compensation principle 补偿原理
compensation spectrum 补偿光谱*
compensative 补偿的；补充的
compensator (=dance roller) 张力调节器；松紧调节辊
compensator circuit 补偿电路
compensator method 补偿法
compensatory (=compensative) 补偿的
competent ①耐久的②充足的③有能力的
competing analysis 竞争分析
competing phase 对抗相；竞争相
competing reaction 竞争反应
competition coordination 竞争络合(反应)
competitive adsorption 竞争吸附
competitive-binding (based) sensor 竞争性结合传感器
competitive coordination 竞争配位(价)
competitive decay 竞争衰变
competitive enzyme linked immunosorbent assay 竞争酶联免疫吸附法
competitive oxidation 竞争性氧化(作用)
competitive protein-binding assay (CPBA) 竞争性蛋白质结合分析法
competitive reaction 竞争反应
comphene 樟脑烯
comphoric anhydride 樟脑酐
compilation of standard spectra 标准光谱集
compiler 编译程序
complanatine 扁平(石松)碱

complement ①补体②补色
complementary 补足的；互补的
complementary action 互补作用
complementary assay 互补试验
complementary base 互补碱基
complementary base sequence 互补碱基序列
complementary color 互补色
complementary deoxyribonucleic acid 互补 DNA
complementary dominant wavelength 互补色主波长
complementary rocks〔复〕 互补岩
complementary strand 互补链
complementary wavelength 补色波长
complement fabric 配套面料
complement-fixation test 补体结合试验
complete alkalimetric titre 总碱量滴定度
complete analysis 全分析
complete antigen 完全抗原
complete baseline separation 在基线处完全分离
complete break （冲击试验）完全断裂
complete combustion 完全燃烧
complete cure 充分硫化
complete decoupling 全去耦
complete dissociation 完全离解
complete equilibrium 完全平衡；不可逆平衡
complete excitation 完全激发
complete expansion 完全膨胀
complete fertilizer 完全肥料
complete Ferund's adjuvent 完全福氏佐剂（油包水乳剂，由液体石蜡、去污剂和已杀死的微生物组成）
complete gasification 完全气化
complete gasification process 完全气化
complete gelation 完全胶凝
complete hiding 全遮盖
complete inflammation 全部着火
complete ionization 完全电离
complete ionization theory 完全电离学说
complete isomeric change 完全异构变化
complete lubricating refinery 完全型炼厂〔石油全部加工及润滑油生产联合工厂〕
completely automatic colorimetric analyzer 全自动比色分析器
completely denatured 完全变性
completely miscible 完全可以混溶的
completely miscible liquids 完全可(溶)混液体
completely miscible system 全混(溶)体系
completely novel crosslinker 全新型交联剂
completely pyrolytic graphite tube 全热解石墨管
completely randomized design 完全随机化设计

completely reversed stress 对称性交变应力
completely silylanization deactivation 全硅烷化去活
completely soluble 完全可溶的
complete manure 完全肥料
complete medium 完全培养基
complete metal extraction 完全萃取金属
complete mint oil 薄荷原油
complete miscibility 完全混溶性
complete neglect of differential overlap approximation 全略微分重叠近似
complete neglect of differential overlap method (CNDO) 全略微分重叠法*
completeness of combustion 燃烧完全度
complete opacity 全不透明性；全阻光性
complete penetration 焊透
complete polarizable electrode 完全可极化电极
complete processing (of crude oil) 原油完全加工
complete protein 完善蛋白质
complete refinery 完全型炼厂〔石油全部加工及润滑油生产联合工厂〕
complete reflection long-pass spectrophotometry 全反射长光路分光光度法
complete shed 全开梭口
complete synthesis 全合成
complete treatment 完全处理
complete vacuum 绝对真空
completion of cure 固化结束；固化完全
complex ①络合物*②络合的③络合基；配位基
complex acceptor group 复杂受电子基团
complex acid 配酸；络酸
complex adsorption 络合吸附
complex anion 配(位)阴离子*
complexant 络合剂*
complexation 络合作用*
complexation chromatography 络合色谱法*
complexation-induced micellization 络合诱导胶束化
complexation reaction 配位反应
complex atomic spectrum 复杂原子光谱
complex binding reactant 络合反应剂；配位反应剂
complex body 复合(模型)体
complex-bounded 在络合物内界络合(结合)的
complex builder 配位组分
complex bulk compliance 复数体积柔量
complex bulk modulus 复数体积模量
complex catalyst 配位催化剂
complex cation 配(位)阳离子*
complex coacernation 配位积并作用
complex column packing 复合物填充柱

complex compliance 复数柔量*
complex compound 配位化合物
complex compound catalyst 配合物催化剂
complex copolymerization 络合共聚合
complex coupling spectrum 复杂偶合谱
complex deformation 复形变
complex dielectric constant 复数介电系数
complex dielectric permittivity 复数介电常数*
complex dissociation constant 络合物离解常数
complex driving-point force 复驱动点力
complex dyestuff 络合染料
complex elastic modulus 复数弹性模量
complex ether 混合醚 ROR′
complex exchange extraction 络合交换萃取
complex fertilizer 复合肥料
complex fibre 复合纤维
complex film 复合膜
complex flow 复合流动
complex fluids 复杂流体〔包括高分子熔体、高分子溶液、液晶、表面活性剂、胶体、微乳液、生物膜等〕
complex formability 配位度
complex formation 络合物形成
complex formation constant 络合物形成常数
complex formation method 配离子生成法
complex formation titration 配合物形成滴定(法)；配位滴定(法)
complex former 络合物形成体
complex function 复合官能
complex hydrides 配位氢化物
complex hydrocarbon mixture 复杂烃混合物
complex hydroxy ion 配位羟离子
complexible 可配位的
compleximetric determination 络合测定
compleximetry (=complexometry; complexometric titration) 络合滴定(法)*
complex indicator 络合指示剂
complexing ①络合②配位化合
complexing agent 络合剂*
complexing counter ion 络合对离子
complexing efficiency 配位效率
complexing reaction 配位反应
complexing reagent 复式试剂
complexing resin 配位树脂
complexing solvent 络合(性)溶剂
complex initiation system 复合引发体系
complex ion 络离子*
complex ion copolymer 复数离子共聚物
complex ion formation analysis (method) 络离子形成分析(法)
complexity 复杂性
complex liquid 复合液
complex liquid gum 复合液体胶
complex mixture 复杂混合物
complex modulus 复数模量
complex molecule ①复杂分子②配分子
complexometric titration 络合滴定(法)*
complexometry 络合滴定(法)*
complexon (=complexone) 配位酮〔氨羧配位剂的商品名〕
complexon Ⅰ 配位酮Ⅰ；次氮基三乙酸
complexon Ⅱ 配位酮Ⅱ；乙二胺四乙酸
complexon Ⅲ 配位酮Ⅲ；乙二胺四乙酸二钠盐
complexon Ⅳ 配位酮Ⅳ；1,2-环己烯二胺四乙酸
complexonate 乙二胺四乙酸盐
complexon B 乙二胺四乙酸钠
complexone 氨羧络合剂*
complex phosphate 复合磷酸盐
complex plant 联合工厂；复杂工厂〔在一个地区内生产两种以上产品的化工厂〕
complex plastic flow ①复合塑性流②复数塑性流动
complex Poisson's ratio 复数泊松比
complex polycyclic space crosslinked system 复合多环体形交联体系；复合多环立体交联体系
complex process 多相过程
complex reaction 络合反应
complex reciprocal modulus 复(数)倒模量；复数可逆模量
complex reducing agent 复合还原剂
complex relaxation modes 复合松弛模式
complex salt 络盐；配盐
complex sample 复杂试样
complex shear modulus 复(数)切变模量
complex sheath-core 复合皮芯型
complex silica 络合二氧化硅
complex-soap base grease 复合皂基脂
complex soap rease 复合皂基(润滑)脂
complex sound velocity 复数音速
complex spectrum 复杂光谱
complex subtractive colorant mixture 复杂减色剂混合物
complex type antioxidant 复合型抗氧剂
complex utilization 综合利用
complex variable 复变数
complex viscosity 复数黏度*
complex Young's modulus 复数杨氏模量
compliance ①柔量②可挠性；柔软性③柔顺性；适应性④弹性变形
compliance in extension 拉伸柔量

compliance in shear　(剪)切柔量
compliant solvent　符合环保法规的溶剂；合法溶剂
complicated shape　①复杂形状②复杂零件
component　①成分；组分②元件；组件
component analysis　组分分析
component distillation　组分蒸馏；共沸蒸馏
component fibre　组分纤维〔用于组成复合纤维〕
component hole　零件孔
component of fuel blend　掺混燃料的组分
component side　零件面
component solvent　混合溶剂；由各种成分组成的溶剂
composed peak　合成峰
composed twill weave　复合斜纹
composed weave　复杂组织
composite　复合材料*
composite active center　复合活性中心
composite analogy　复合模拟
composite catalyst　复合催化剂
composite cathodic-anodic wave　阴阳极联波
composite cellulosic membrane　复合纤维膜
composite coating　复合涂层
composite column　复合柱
composite correction factor　复合校正因子
composite cupel　特用灰皿
composite curve　组合曲线
composite design　复合设计
composite extrusion　①复合挤塑②复合材料挤塑
composite fabric　复合织物
composite fiber　复合纤维
composite fluid　组合流体〔凝结水和石油的混合物〕
composite force　复合力
composite fuel　组合燃料
composite laminates　复合层压材料
composite lattice　复合点阵
composite material　复合材料
composite materials fine mechanics　复合材料细观力学
composite membrane　复合膜
composite micromechanics　复合材料微观力学
composite mold　组合模
composite molding　复合成型
composite motor fuel　复合发动机燃料
composite network　复合网络
composite oxides　复合氧化物
composite pigment　复合颜料
composite plasma polymer　复合等离子体聚合物
composite powder　复合粉末
composite pre-coated steels　复合预涂钢板
composite product　复合材料制品
composite pulse　组合脉冲
composite reaction　组合反应〔石油〕
composite resistance　复合电阻
composite resistivity　复合电阻率
composite sample　复合试样
composite sheet　①复合片材②复合板材
composite signal　复合信号
composite silk　复合生丝
composite species　复合种类；各种复合形式
composite stabilizer　复合稳定剂；混配稳定剂
composite structure　复合结构；混合结构
composite surface　复合面
composite synthetic fibre　复合合成纤维
composite wave　复合波；波组
composite yarn　包芯纱；复合纱
composition　①胶料②配方③组成④成分
compositional heterogenity　组成非均一性*
compositional refinements　①组成改进②配方改进
composition analysis　成分分析；组成分析
composition distribution function　组成分布函数
composition fibre　复合纤维
composition gradient　组分梯度
composition history curve　组成变化曲线
composition material　组合材料
composition of complex　络合物组成
composition of dust　灰尘组成
composition of petroleum　石油的成分
composition of radiance　辐射频谱
composition of well stream　充气石油成分
composition parameter　组成参数
composition profiles　组成分布；成分分布
composition program　复合程序
composition siding　复合板；硬质纤维板
composition surface　接合面
composition variable　组成变量
compost　堆肥
composting　堆肥化处理
compound　①化合物②复合物；混合物③复合；混合④复(合)的
compound 118(=aldrin)　化合物118；艾氏剂〔农药〕
compound 497(=dieldrin)　化合物497；狄氏剂〔农药〕
compound aminophenazone (antondine)　复方氨基比林(安痛定)〔药〕
compound atom　复合原子
compound body　复质；混合体
compound cellulose　复合纤维素；结合纤维素
compound color　混合色；调和色〔染〕
compound compression　多级压缩

compound compressor　①多级压缩机②双级压缩机
compound contactor　复式萃取器〔有洗涤段的萃取器〕
compound curve　复合曲线
compound cutting oil　复合切削油〔用于金属切削〕
compound cylinder　组合式(高压)圆筒
compound department　混合(车)间
compound dislocation　复合位错
compound dynamo　复绕发电机
compounded　混合的；复合的
compounded latex　复合胶乳
compounded lubricating oil　复合润滑油
compounded mineral oil　复合矿物油
compounded mix　复合填料
compounded oil　复合油
compounded perfume oil　复合香料油
compounded protein　复合蛋白
compounded rubber　复合橡胶；填料混炼胶
compounded rubber stock　混炼胶
compound effervescent powder　复合起泡粉
compound electrode　组合电极
compound emulsifying agent　复合乳化剂
compound ether　混合醚
compound failure　复合破坏；复合破裂
compound fertilizer　复合肥料
compound film type inhibitor　复合膜型缓蚀剂
compound formation　化合物形成
compound formation chromatography　反应色谱(法)
compound gauge　真空压力计
compound glass　多层玻璃
compound indicator　复合指示剂
compounding　①配合②配(药)方③配料
compounding agent (=additive ingredient)　添加剂；配合剂
compounding aid　配合助剂
compounding chemicals　配合药品；助剂
compounding ingredient　配合材料
compounding in parallel　并联混合
compounding in series　串联混合
compounding line　配炼车间；准备车间
compounding material　配合剂；原材料
compounding operation　配合工序
compounding plant　(润滑油)配料车间
compounding practice　配合操作
compounding recipe　配方
compounding room　配料间；配方室
compounding technique　配合技术
compounding variable　配混可变因素
compounding variation　①配方差异②混炼差异

compound ion　复合离子
compound light　复合光
compound manure　复合肥料
compound meter　复合流量计
compound middle lamella　复合中层
compound mill　多仓磨；复式磨
compound molecule　复合分子
compound nucleus　复合(原子)核
compound oil　复合油
compound oven　联立炉
compound perfume oil　复合香料油
compound powder　复合粉剂
compound radical　复根
compound rectifying column　复式精馏塔
compound regenerative oven　复合式再生炉
compound regulator　复合自动控制器
compound resistance　复合阻力
compound resistance in series　串联组合阻力
compound rubber　混合生胶
compound-selective detector　化合物选择性检测器
compound shade　复色
compound spirit (=compound tincture)　复方酊剂
compound substance　复合物质
compound sugar　低聚糖
compound tincture (=compound spirit)　复方酊剂
compound trommel　多层转筒筛
compound-twist silk thread　复捻丝线
compound unit　组合单位
compound wall　多层壁
compound warping　复合经纱排列
compreg (=compregnated wood)　胶压木；木材层积塑料
compregnate　浸压
compregnated wood　(渗)胶压(缩)〔用酚醛树脂处理的压缩木材〕
comprehensive speciation　综合形态分析
comprehensive test　综合试验
comprehensive two dimensional gas chromatography　全二维气相色谱
compressed air　压缩空气
compressed air (ballasted) accumulator　压缩空气储器
compressed-air bearing　压缩空气轴承
compressed air burner　压缩空气灯
compressed air cell　压缩空气盒
compressed air dipping device　压缩空气动浸渍装置
compressed air ejection　(注塑)压气顶出；压气脱模
compressed air hammer drill　气动锤钻
compressed air main　压缩空气干线
compressed air require　(匹配)压缩空气流量

compressed air sealing　压缩空气密封
compressed air spraying　压缩空气喷涂
compressed asbestos sheet　压紧的石棉胶板；夹胶石棉板
compressed gas　压缩气体
compressed gas cylinder　压缩气筒
compressed gas manometer　压缩压力计
compressed limit　压缩限度；压缩极限
compressed pat-kiln test　样饼烧试法
compressed preform　压制料坯
compressed propane　压缩丙烷
compressed wood (=improved wood)　压缩木材
compresser cylinder　压缩机气缸
compressetometer　压缩疲劳试验机
compressibility　①压缩系数②压缩性
compressibility coefficient　压缩系数
compressibility correction factor　压缩系数校正因子
compressibility factor　压缩因数
compressibility factor diagram　压缩因数图
compressibility measuring apparatus　压缩系数测定装置
compressibility of rocket propellants　火箭用燃料的压缩性
compressible　可压缩的；可压紧的
compressible cake　压缩性滤渣
compressible flow　可压缩流动
compressible fluid　可压缩流体
compressing　压缩；压紧
compressing barrier　压缩阻片
compressing machine　压缩机；压紧机
compression　压缩
compression-after-impact testing　冲击试件压缩实验
compressional stress　压缩应力
compressional viscosity　压缩黏性；压缩黏度
compression and recovery test　压缩恢复试验
compression annealing　压缩熟炼；压缩退火〔热原子化学〕
compression blowing (=coblow)　压坯吹塑
compression casting　压铸
compression chamber　压缩室
compression coefficient　压缩系数
compression creep　压缩蠕变
compression cycle　压缩循环
compression-decompression　加压-卸压作用
compression deflection　压缩变形；压缩挠曲
compression deflection characteristics　压缩变形特性；压缩致偏特性
compression deformation　压缩变形
compression elasticity　抗压弹性
compression engine　压缩机
compression factor　压缩系数
compression fatigue test　压缩疲劳试验
compression forming　压缩成型
compression gasoline　气态汽油
compression grease cup　压缩加料润滑杯
compression heat　压缩热
compression heat sealing　加压热合
compression ignition　压缩点火
compression indentation procedure　压缩压痕法；压缩压入法
compression load　压缩载荷
compression load deflection　①压缩载荷挠曲②压缩载荷挠度
compression loss　压缩损失
compression method　压缩法
compression modulus　抗压模量
compression modulus of elasticity　压缩弹性模量
compression mold　(直接式)压模
compression molder　平板硫化机
compression molding　①平板硫化(法)②压缩模塑(法)；压模(法)〔塑料〕
compression mould　压(缩)塑模
compression moulding　压(缩)模(塑)(法)；模塑成型
compression mo(u)lding press　压塑机；平板硫化机
compression pad　压缩式防震垫
compression plant　压缩车间
compression plant gasoline (=compression process gasoline)　压缩冷凝法回收的天然汽油
compression process gasoline (=compression plant gasoline)　压缩冷凝法回收的天然汽油
compression pump　压气泵
compression ratio　压缩比(率)
compression resilience　压缩回弹性
compression set　压缩(永久)变形；压缩变定
compression shear apparatus　压缩剪切装置
compression shear strength　压缩剪切强度
compression shock　压力振动
compression strain　压缩应变
compression strength　抗压强度
compression stress　抗压应力
compression stress relaxation　压缩应力松弛
compression stroke　压缩冲程
compression-system of refrigeration　冷冻设备的压缩系统
compression-tension fatigue　压(缩)张(拉)疲劳
compression test　抗压试验
compression testing machine　①缓冲性能试验机②抗压试验机
compression tool　压塑模具

compression type refrigeration unit 压缩式制冷机组
compression-type regularity tester 压缩式均匀度试验机
compression wave 压力波
compression wood 压缩木材
compression work 压缩功*
compression zone 压缩区
compressive buckling 压缩轴向弯曲
compressive bulk module 体积压缩模量
compressive creep 压缩蠕变
compressive deformation 压缩形变
compressive modulus 压缩模量
compressive parallel plate viscometer 压缩平行板黏度计
compressive resistance 抗压力
compressive set 压缩形变
compressive shrinkage mark 预缩痕
compressive strain 压缩应变
compressive strength 压缩强度*
compressive stress 压缩应力
compressive texturing machine 压缩(卷曲)变形机
compressive transient state 压缩瞬变状态
compressive ultimate strength 压缩极限强度
compressive yield point 压缩屈服点
compressive yield strength 压缩屈服强度
compressive yield stress 压缩屈服应力
compressometer 压缩计
compressor 压缩机；压气机
compressor gun 润滑油增压机；加油枪
compressor housing 压缩机气缸
compressor lubricant (=compressor oil) 压缩机油
compressor lubrication 压缩机润滑
compressor oil (=compressor lubricant) 压缩机油
compressor rated point 压缩机额定点
comprognated wood (渗)胶压(缩)[用酚醛树脂处理的压缩木材]
compromise blend 折衷共混物
Compton cross-section 康普顿截面
Compton effect 康普顿效应
Compton electron 康普顿电子
Compton recoil clcctron 康普顿反冲电了
Compton scattering 康普顿散射
Compton wavelength 康普顿波长
Compton-Wu Youxun scattering 康普顿-吴有训散射
compulsory verification 强制检定；强制检验
computation 计算(技术)
computational error 计算误差
computational pharmaceutical analysis 计算机辅助药物分析
computational spectrophotometry 计算分光光度法
computational stability 计算稳定性
computation method 计算方法
computation program 计算程序
computer ①计算机②计量器③计数器
computer aided design (CAD) 电脑辅助设计
computer-aided interpretation of mass spectrum 计算机辅助质谱解释
computer aided manufacturing (CAM) 电脑辅助系统
computer-aided measuring and control 计算机辅助测量和控制
computer-aided molecular modeling 计算机辅助分子模拟
computer-aided pharmaceutical analysis 计算机辅助药物分析
computer aided quality (CAQ) 电脑辅助质量
computer assistant design 计算机辅助设计
computer code 计算机代码
computer colour matching 计算机拼色
computer control 计算机控制
computer-controlled guide-bar shogging system 电子梳栉横移系统
computer data processing 计算机数据处理
computer instruction 计算机指令
computer integrated manufacturing system 电脑集成制造系统
computerized batch process 计算机化的间歇式生产工艺
computerized chromatograph 连计算机的色谱仪；计算机化色谱仪
computerized color matching 电脑配色
computerized image reconstruction 计算机化影像重现
computerized NAA 计算机化中子活化分析
computerized topography (CT) 计算机化断层显像*
computerized transmission topography (CT-CTT) 计算机化透射式断层照相法
computerized TV emission spectrometer 计算机化电视检测发射光谱仪
computer procedure flowchart 计算机过程流程图
computer program 计算机程序
computer retrieval 计算机检索
computer simulation 计算机模拟
computer simulation of liquid 液体的计算机模拟
computer spectrum-stripping 计算机差谱技术
computer supervision 计算机监督；计算机控制
computer tomography 计算机体层摄影术
computer to plate 计算机直接制版
computing gear 计算传动装置

computing integrator　计算积分仪
conamarin　毒芹根苷
conamine　锥丝胺
Conant's titration cell　科南特滴定池
conarrhimine　康丝瑞明
concave (diffraction) grating　凹面(衍射)光栅
concaved roll　中凹滚筒
concave gradient　凹形梯度
concave grating spectrograph　凹面光栅摄谱仪
concave roller　凹面手压辊
concavity factor　凹度系数；凹率
concentrate　①浓缩物②蒸浓；浓缩③富集④精砂
concentrate colouring　母料着色
concentrated　浓的
concentrated bleaching powder　浓缩漂白粉
concentrated combustion gas burner　(发生)浓焰(的)气体燃烧炉
concentrated crystal soda　倍半碳酸钠；二碳酸一氢三钠 $Na_2CO_3 \cdot NaHCO_3 \cdot 2H_2O$
concentrated fertilizer　浓缩肥料
concentrated flower oil　浓缩花油
concentrated gas liquor (=concentrated liquor)　浓缩氨水
concentrated hydrochloric acid　浓盐酸
concentrated latex　浓胶乳
concentrated liquor (=concentrated gas liquor)　浓缩氨水
concentrated load　集中载荷
concentrated matte　浓缩冰铜；富集冰铜
concentrated nitric acid　浓硝酸
concentrated oil　浓缩精油
concentrated oil of vitriol (COV)　浓硫酸
concentrated phase　浓相*
concentrated polymer solution　聚合物浓溶液
concentrated reflection　集中反射；浓缩反射
concentrated solution　浓溶液
concentrated sulfuric acid　浓硫酸
concentrate spraying　高浓度喷射
concentrating　浓缩
concentrating column　浓缩柱
concentrating effect　浓缩效应
concentrating mill　选矿厂
concentrating pan　浓缩锅
concentrating petroleum product　浓缩石油产品
concentrating table　富集台
concentrating tower (=concentration tower)　浓缩塔
concentrating worm　聚集螺杆；压缩螺杆
concentrating zone　浓缩区
concentrating zone layer plate　浓缩区薄层板
concentrating zone thin layer chromatography　浓缩区薄层色谱法
concentration　①浓度②浓缩③蒸浓④提浓⑤富集
concentration-aggregation method　浓度-聚集度法〔测定颜料分散剂需要量的方法〕
concentration cell　浓差电池*
concentration cell without transference　无迁移浓差电池*
concentration cell with transference　有迁移浓差电池*
concentration coefficient method　浓度系数法
concentration coefficient of diffusion　扩散(的)浓度系数
concentration coefficient of sedimentation　沉积(的)浓度系数
concentration column　浓缩塔
concentration constant　浓度常数*
concentration corrosion cell　浓差腐蚀电池
concentration current　浓差电流
concentration detector　浓度检测器
concentration difference　浓度差
concentration diffusion　浓差扩散
concentration effect　浓度效应
concentration equilibrium constant　浓度平衡常数
concentration factor　浓缩系数
concentration gradient　浓度梯度
concentration gradient curve　浓度梯度曲线
concentration gradient imaging detector　浓差梯度成像检测器
concentration index　浓度指数
concentration index of urine　尿缩指数
concentration level of residues　残留物的浓度量级
concentration limit　浓度极限
concentration logarithmic diagram　浓度对数图*
concentration matching technique　浓度匹配技术
concentration method　浓缩法
concentration of crosslink　交联密度
concentration of poisons　毒物浓度
concentration of sulfuric acid　硫酸的浓缩
concentration overpotential　浓差过电位
concentration overvoltage　浓差超电压；浓差过电压
concentration pan　蒸浓锅
concentration polarization　浓差极化*
concentration potential　浓差电势
concentration profile　浓度剖视图
concentration quenching　浓度猝灭
concentration quotient　浓度商；浓度常数
concentration range of color comparison　比色浓度范围
concentration ratio　浓度比
concentration sensitive detector　浓度敏感型检测器*
concentration sensitivity　浓度灵敏度
concentration solubility product　浓度溶度积

concentration (solvent) jump 浓度(溶剂)跃变*
concentration table (=concentrating table) 富集台
concentration technique 浓缩技术
concentration tower (=concentrating tower) 浓缩塔
concentration trap technique 浓缩收集技术
concentration tube 浓缩管
concentration unit 提浓装置
concentrator 浓缩器
concentrator bowl 离心筒；离心套
concentrator with nozzle discharge 有卸料喷嘴的离心浓缩机
concentric 同心(的)
concentric(al) circle 同心圆
concentric column 同心(色谱)柱
concentric crucible assembly 同轴熔炉组合装置
concentric cylinder rotation viscometry 同心圆筒旋转式测黏法
concentric cylinder viscometer 同心柱黏度计*；同轴圆筒式黏度计
concentric cylinder viscosimeter 同心圆筒式黏度计
concentric float 同心浮标
concentric goniometer stage 同心测角台
concentric hemispherical analyzer 半球形分析器
concentric hose 同心软管；同心圆软管
concentric inner hose 同心内管
concentricity 同心度
concentric joint 同心接合
concentric nebulizer 同心雾化器
concentric packed column 同心填充柱
concentric ring 同心环
concentric sheath-core fibre 皮芯同轴纤维
concentric squirrel cage mill 同心笼式粉碎机
concentric tube 同心管
concentric tube column 同心管柱；精密分馏柱
concentric tubular reactor 同心管式反应器
conceptual design 概念设计；方案设计
conceptual phase 初步设计阶段；草图设计阶段
concerted 协调的；协同的
concerted catalysis 协同催化*
concerted displacement 协调排代
concerted effect 多价效应；协同效应
concerted mechanism 协调机理
concerted reaction 协同反应
concha 甲；壳
conchinamine (=epiquinamine;conquinamine) 康奎明
conchinine (=quinidine) 奎尼丁
conchoidal 贝壳状的
conchoidal fracture 贝壳状破裂

conchoporphyrin 贝卟啉
concoction ①调制②调制品③混合④混合物
concomitant analysis 相伴分析；共存分析
concomitant product 副产物；伴生物
concomitants 附随物质
concomitant variable 协变量
concret block (实体或空心)混凝土块体
concrete 混凝土
concrete aggregate 混凝土聚合料
concrete bond plaster (=bond plaster) 混凝土黏结层；黏结性抹灰层
concrete condenser box 混凝土外壳冷凝器
concrete container 混凝土容器
concrete cracking 混凝土开裂
concrete floor 混凝土地面
concrete floor dressings 混凝土地板装修
concrete floor enamel 混凝土地板瓷漆
concrete form oil (=concrete oil) 混凝土模(板用)油
concrete gravel 混凝土(用)石子
concrete gun 混凝土喷枪
concrete hardener 混凝土硬化剂
concrete holder tank 水泥气柜
concrete lined pipe 衬有混凝土的管
concrete lining 水泥衬里
concrete mixer 混凝土(混合)器；水泥拌合器
concrete mixing machine 混凝土混合机；调泥机
concrete oil (=concrete form oil) 混凝土模(板用)油
concrete oil barge 混凝土油驳船
concrete paint 混凝土漆
concrete pipe 混凝土管
concrete-plastic combination 混凝土-塑料复合材料
concrete plastic composite 混凝土塑料复合物
concrete reinforced pipe 钢筋混凝土管
concrete reinforcement bar (粗)钢筋
concrete sinker 混凝土沉锤
concrete tank 混凝土油罐
concrete water-proofing oil 混凝土防水油
concrete water tank 混凝土水箱
concurrent ①并发的②顺流的；并流的
concussion burst 冲击爆破
condensability 冷凝性
condensable gas 可冷凝的气体
condensable gasoline (天然气中)可冷凝的汽油
condensamine 密叶(马钱)胺
condensate ①缩合物②冷凝物；冷凝液③凝液④缩合
condensate collector 凝液收集器；凝液捕集器
condensate drain 冷凝排水
condensate draining 冷凝水槽

condensated water 冷凝水
condensate flow 冷凝物(水)流
condensate line 冷凝线
condensate pump 冷凝泵
condensate stripper 凝液汽提塔〔石油〕
condensate tank 冷凝槽
condensate trap 冷凝阱；冷凝槽
condensate water 冷凝水
condensating agent 缩合剂
condensation ①缩合②凝聚
condensation agent 缩合剂
condensation air pump 凝汞抽气泵
condensation catalyst 缩合催化剂
condensation coefficient 冷凝传热系数
condensation compound 缩合物
condensation copolymerization 共缩聚(作用)
condensation effect 凝聚效应
condensation exhaust 冷凝排气
condensation heat 凝聚热
condensation hygrometer 冷凝型湿度计
condensation method 冷凝法
condensation monomer 缩聚单体
condensation nucleus 凝聚核
condensation point 冷凝点
condensation polymer 缩(合)聚(合)物
condensation polymerization 缩聚反应；缩聚(作用)
condensation process 冷凝过程
condensation product 缩合产物
condensation pump 冷凝泵
condensation reaction 缩合反应
condensation resin 缩合树脂
condensation rubber 缩聚橡胶
condensation substance 缩合物
condensation temperature 凝聚温度
condensation tube 冷凝管
condensation-type polymerization 缩合型聚合(作用)；缩聚(作用)
condensation water 冷凝水
condensed ①稠合的②冷凝了的
condensed aromatics 稠合芳烃
condensed azo-pigment 缩合偶氮颜料
condensed discharge 浓缩放电
condensed film 凝聚膜
condensed fluid 冷凝液体
condensed gas 冷凝气(体)；凝聚气(体)
condensed gas dispersoid 冷凝气体胶体
condensed heteroaromatic system 稠合杂芳系
condensed nucleus 稠环；稠核

condensed oil 稠合油
condensed phase 凝相；缩相
condensed-phase combustion 凝结相燃烧
condensed phase interference 凝聚相干扰
condensed phase isotope effect 凝聚相同位素效应
condensed polymer 缩(合)聚(合)物
condensed ring 稠环
condensed ring system 稠环系
condensed roving frame 搓捻粗纱机
condensed spark 电容火花
condensed spark excitation (高)电容火花激发法
condensed state 凝聚态
condensed steam 冷凝水蒸气
condensed system 凝聚系统*
condensed tannin 缩合鞣质
condensed tannin extract 缩缩类栲胶
condensed type 缩合型
condenser ①冷凝器②聚光器③电容器
condenser arc 电容电弧
condenser box 冷凝器箱；电容器箱
condenser casing 冷凝器外壳
condenser coil 冷凝旋管
condenser current 电容器电流
condenser current compensation 电容电流补偿
condenser damping 电容阻尼
condenser delivery tube 冷凝器导管
condenser discharge 电容放电
condenser duty 冷凝器负荷；冷凝能率
condenser leg (=condenser leg pipe) 冷凝器气压管
condenser leg pipe (=condenser leg) 冷凝器气压管
condenser microphone 静电传声器
condenser pickup 电容拾振器
condenser pipe (=condenser tube) 冷凝管
condenser spark 电容火花
condenser tube (=condenser pipe) 冷凝管
condenser water 冷凝(管用)水
condensifilter 冷凝滤器
condensing ①冷凝②凝聚
condensing agent ①缩合剂(缩合反应催化剂)②冷凝剂
condensing and controlling zone (双区牵伸的)中牵伸区
condensing apparatus ①冷凝器②电容器③聚光器
condensing chamber 冷凝室
condensing coil 冷凝旋管
condensing engine 冷凝机
condensing equipment 冷凝设备
condensing groove 凝聚槽
condensing humidity cabinet 冷凝湿气箱
condensing lens 聚光透镜

condensing mirror 聚光镜
condensing plant 冷凝车间；冷凝厂
condensing process 凝聚过程
condensing rate 冷凝速率
condensing spinning system 集聚纺纱装置
condensing surface 冷凝面
condensing system 冷凝装置
condensing temperature 液化温度；凝温度
condensing tower 冷凝塔
condensing turbine 凝汽式涡轮机
condensing vapour film 冷凝蒸气膜
condensing vessel 凝洗器
condensing works 冷凝装置
condensing worm 冷凝旋管
condenswater separator 冷凝水分离器；疏水器
condistillation 附馏；共(蒸)馏
condition ①条件；状况②调节③老化
conditional acceptable daily intake 限制性每日允许摄入量
conditional constant 条件常数
conditional equilibrium constant 条件平衡常数*
conditional extraction constant 条件萃取常数
conditional formation constant 条件形成常数*
conditional indicator constant 条件指示剂常数
conditional match 条件配色；条件等色
conditional solubility product 条件溶度积
conditional stability constant 条件稳定常数*
conditioned atmosphere 公定温湿度
conditioned elongation 调湿伸长
conditioned linear density 调湿线密度
conditioned modulus 调湿模量
conditioned titre 调湿纤度
conditioner ①调理池②调节剂
conditioning ①老化*②预处理③调湿④调温⑤调节
conditioning agent 调节剂
conditioning tunnel 空调隧道(式装置)；肥皂空调装置
condition number 条件数
conditions of vulcanization 硫化条件
condition solubility product 条件溶度积
conduct ①传导②导管；套管
conductance ①电导②热导③传导
conductance cell 电导池
conductance detection 电导检测法
conductance electrode 电导电极
conductance method 电导法
conductance ratio 电导率
conductance titration 电导滴定
conductance water 电导水

conductex 导电炭黑
conductibility 传导性；导电性
conductimetric analysis 电导分析
conductimetric titration 电导滴定
conducting air 导电空气；离化空气
conducting coat 导电涂膜
conducting-core heterofilament 导电芯型双组分长丝
conducting glass electrode 传导性玻璃电极
conducting hearth 传导敞炉
conducting paint 导电涂料
conducting polymer 导电高分子；导电聚合物
conducting polymer substance 导电高分子物质
conducting power ①传导本领②导电本领
conduction 传导
conduction band 导带*
conduction electron spin resonance 传导电子自旋共振
conduction heating 传导加热
conduction of heat 热传导
conductive adhesive 导电黏合剂
conductive analysis 电导分析法
conductive body ①导体②导电体
conductive channel black 导电槽黑
conductive coating ①导电涂料②导电涂层
conductive compound 导电化合物
conductive conveyor belt 导电传送带
conductive furnace black 导电炉黑
conductive heat flow 传热热流
conductive lacquer 导电清漆
conductive paste 导电胶
conductive plastic 导电塑料*
conductive polymer 导电聚合物
conductive rubber 导电橡胶
conductivitimeter 电导率仪
conductivity 电导率*
conductivity anisotropy 电导各向异性现象〔某些表面活性剂的胶束溶液,在搅拌时呈现的电导变化〕
conductivity apparatus 电导(测定)仪
conductivity bridge 电导测定电桥
conductivity cell 电导池
conductivity cells of tungsten filament 钨丝热导池
conductivity detector 电导检测器*
conductivity group 电导组
conductivity improver 导电助剂；导电增进剂
conductivity measuring cell 电导率测定槽；电导(率)测定池
conductivity meter 电导计
conductivity method 电导法
conductivity mobility 电导迁移率

conductivity moisture meter　电导水分计
conductivity of metals　金属传导性
conductivity sensor　电导传感器
conductivity type flow meter　电导式流速计；变阻差示压力计
conductivity water　电导水
conductometer　①热导计②电导计
conductometric　热导测量的；电导测定的
conductometric analysis　电导分析
conductometric method　电导(滴定)法
conductometric particle-counting　导电粒子计数测量(法)
conductometric procedure　电导测定法
conductometric sensor　电导型生物传感器
conductometric titration　电导滴定(法)*
conductometric titration apparatus　电导滴定仪
conductometric titration cell　电导滴定池
conductometry　电导分析法
conductor　①管理人②避雷针③导体
conductor thickness　导线厚度
conduct resin　接触成型树脂；触压树脂
conduit　①(导)管②导线管
conduit clip　环状管夹
conduit (tube)　导线管
conduritol　牛弥菜醇；环己烯四醇
condurrite　砷铜矿
cone　锥形筒丝；斜筒丝；圆锥体
cone and plate flow　锥板流动
cone and plate rheogoniometer　锥板式流变性测定仪
cone and plate rheometer　锥板流变仪
cone and plate sensor system　锥板传感系统
cone and plate viscometer　锥板式黏度计*
cone and plate viscosimeter (=cone-plate viscosimeter)　锥板式黏度计；锥形平板黏度计
cone angle　锥角
cone-baffle classifier　锥形挡板分级机
cone bearing tree　针叶树
cone belt (=V-belt)　三角带
cone blender　单锥鼓式搅拌机
cone bottom　锥形底
cone bottomed　有锥形底的；底是锥形的
cone calorimeter　锥形量热仪
cone circular　圆锥
cone classifier　锥形选粒器；锥形分级器
cone closure autoclave　锥形密封式高压釜
cone crusher　锥形压碎机
cone defoamer　锥形漏斗消泡器
cone filter　锥形漏斗
cone joint　圆锥接头
cone mill　锥形磨
cone penetration test　针入度试验
cone-plate sensing system　锥板式传感系统
cone-plate viscometer　锥板式黏度计
cone-plate viscosimeter (=cone and plate viscosimeter)　锥板黏度计；锥形平板黏度计
cone pulley　锥形轮
cone pulley drive　塔轮传动
coner　络筒机；锥形筒子络筒机
cone roof tank　锥顶罐
cone separator　锥形分离器
cone settling tank　锥形沉降槽
cone-shaped bottom　锥形底
cone-shaped pulley　锥轮
cone-shaped tube　锥形管〔用于油的离心分离〕
cone-sheet　锥形片
cone sifter　锥形筛
conessidine　康丝定；锥丝定
conessimethine　锥丝甲碱
conessimine　锥丝明
conessine　康丝碱
conessin(e)　止泻木碱
cone tank　锥形桶
cone thread　宝塔线
cone type liquid-film seal　锥形液膜密封
cone-type pole face　圆锥形磁极面
cone valve　锥形阀
cone viscosimeter　锥式黏度计
cone yarn　宝塔纱
confection　①糖果；蜜饯②混合药剂
confectioner　①糖果制造工②糖果店
confectionery　①糖食店②糖果；蜜饯
confidence　置信水平；置信度
confidence band　置信带
confidence coefficient　置信系数*
confidence interval　置信区间；置信度
confidence level　置信水平
confidence limit　①置信限②置信界限
confidence probability　置信概率
confidence region　置信区域
configuration　构型*
configurational disorder　构型无序
configurational elasticity　构型弹性
configurational entropy　构型熵
configurational free energy　构型自由能
configurational partition function　构型配分函数
configurational polydispersity　构型多分散性
configurational randomness　构型混乱度

configurational unit　构型单元
configuration coordinate　位形坐标*
configuration interaction　组态相互作用*；构型作用
configuration of polymer chain　高分子链的构型
confined chain　受限链
confined-growth crystallinity　限制生长结晶度
confined state　受限态
confining liquid　封闭液
confirmatory reaction　证实反应
confirmatory test　证实试验
conflagrant　速燃的
conflagration　快速燃烧；暴燃
conflicting stream　逆流
confocal laser Raman spectrometer　共焦激光拉曼光谱仪
confocal laser scanning microscopy　共焦激光扫描显微镜
conformability　顺应性；适合性；贴身性
conformal coating　保形涂料；保角涂料
conformal ionic solution theory　共形离子溶液(理)论
conformal solution　共形溶液*
conformance　密贴性
conformation　构象*
conformational analysis　构象分析*
conformational array　构象阵列；构象排列
conformational disorder　构象无序
conformational effect　构象效应*
conformational energy　构象能
conformational entropy　构象熵
conformational freedom　构象自由(度)
conformational inversion　构象反转*
conformational isomer　构象异构体
conformational repeating unit　构象重复单元*
conformational transmission　构象传递*
conformation analysis　构象分析
conformation of polymer chain　高分子链的构象
conformation parameter　构象参数
conformation statistics　构象统计
conformation theory　构象理论
conformer　构象异构体
conforming article　合格品
conformity coefficient　适合系数
conformity index　匹配指数
confounded design　混杂设计
congealer (=refrigerator)　①冷却器②冷藏箱
congealing (=congelation)　冻凝(作用)
congealing point　冻(凝)点
congealing temperature (=congelation temperature)　冻凝温度
congelation　冻凝作用*
congelation point　冻凝点
congelation temperature (=congealing temperature)　冻凝温度
congeners　同族元素
conglobation　球形；球形体；团聚
conglomerate　①(使)凝聚成团②砾岩③砾岩的
conglomerone　团集酮
conglutinating complement absorption test　胶固补体吸收试验
conglutination　共凝集作用；黏合；黏附；固结
conglutinin　共凝集素；团集素
congo　①刚果红②刚果
congo brown　刚果棕
congocidine　刚果素
congo copal　刚果玷珀(脂)
congo-copalic acid　刚果玷珀酸
congo-copalolic acid　刚果玷珀油酸
congo ester　刚果脂；玷珀脂
congo ester resin　刚果酯树脂
congo gum　刚果树胶
congolene　刚果烯
congo red　刚果红*
congo red damage test　刚果红(纤维)损伤测定
congo-red test paper　刚果红试纸
congo rubine　刚果玉红〔染〕
congo yellow　刚果黄
congressane　会议烷；五环金刚烷；五环十四烷
congruent melting point　相合熔点〔固液同成分熔点〕
congruent point　(水合物的)同成分点
congruity of parallel test　平行试验的差别
conhydrine　羟毒芹碱
conhydrinone　毒芹羟碱酮；康海君酮
conic acid (=coniic acid)　毒芹酸
conical　(圆)锥形的
conical beaker　锥形烧杯
conical beam　锥形喷束
conical bell mill　锥形球磨机
conical bottom　锥形底
conical bottom tank　锥底槽
conical buoy　锥形浮；圆锥形浮标
conical column　锥形柱
conical disc mill　圆锥磨
conical drum centrifuge　锥鼓离心机
conical dry blender　锥形干混器
conical duct　锥形管
conical expander　圆锥扩展器

conical feeder 圆锥进料器
conical flask 锥形(烧)瓶
conical flow 锥形流动
conical graduate 锥形量杯
conical head without transition knuckle 无折边锥形封头
conical lamp shade 锥形灯罩
conical mandrel flexibility 锥形心轴弯曲柔韧性
conical mandrel tester 锥形轴挠曲试验机
conical model 圆锥模型
conical nozzle 锥形喷嘴
conical orientation 锥形取向
conical quartering 锥形四分(采样)法
conical refiner 锥形(纸浆)精制机
conical refining engine 锥形磨碎机
conical roof 锥形顶
conical rotating screen (=conical trammel) 锥形旋筛
conical screen 锥形筛；锥形滤网
conical screw 锥形螺杆
conical settling tank 圆锥形沉降槽
conical tank 锥形顶储罐
conical thin layer chromatograph 锥形薄层色谱仪
conical thin layer chromatography 锥形薄层色谱(法)
conical trommel (=conical rotating screen) 锥形旋筛
conical-type Stedmen packing 锥形司梯曼填充物
conicein (e) 毒芹瑟碱
conicic acid (=coniic acid) 毒芹酸
conicine (=coniine) 毒芹碱；2-丙基六氢吡啶
conicylindrical viscometer 锥筒黏度计
conidendrin 铁杉内酯
coniferaldehyde 松柏醛 OH(CH$_3$O)C$_6$H$_3$CH=CHCHO
coniferin 松柏苷
coniferin aglycone 松柏配基
coniferin hydrochloride 盐酸松柏苷
　　(HO)$_2$C$_6$H$_3$CH(OH)CH(NH)$_2$CH$_3$·HCl
coniferol 松柏醇 C$_{10}$H$_{12}$O$_3$
coniferous wood ①针叶树②针叶材；松木
coniferyl 松柏基 HO(CH$_3$O)C$_6$H$_3$CH=CH—
coniferyl alcohol 松柏醇 C$_{10}$H$_{12}$O$_3$
coniferyl alcohol reagent 松柏醇试剂
coniferyl aldehyde 松柏醛
　　HO(CH$_3$O)C$_6$H$_3$CH=CHCHO
coniic acid (=conic acid) 毒芹酸
coniine (=conicine) 毒芹碱；2-丙基六氢吡啶
coning 络筒
coning oil 络筒油
coning-quartering method 锥形四分(采样)法
coning (tower tray) 锥流
conioselinum oil 山川芎油

conjoint effect 联合效应；协同效应
conjugate 共轭
conjugate acid 共轭酸*
conjugate acid-base pair 共轭酸碱对*
conjugate acids and bases 共轭酸和碱
conjugate addition 共轭加成*
conjugate angle 共轭角
conjugate base 共轭碱*
conjugate base mechanism 共轭碱机理*
conjugated ①缀合的②共轭的
conjugated alkadiene 共轭链二烯
conjugated bond 共轭键
conjugated chain 共轭链
conjugated compound 共轭化合物
conjugated diene 共轭二烯
conjugated diolefine 共轭二烯
conjugated double bond 共轭双键
conjugated electrochemical reaction 共轭电化学反应
conjugated fibre 共轭纤维；复合纤维
conjugated hydrocarbons 共轭烃
conjugated link(age) 共轭键
conjugated linoleic acid 共轭(化)亚油酸
conjugated lipid 复合脂质
conjugated liquid pair 共轭液偶
conjugated molecule 共轭分子*
conjugated monomer 共轭单体*
conjugated oxidation reduction 共轭氧化还原反应
conjugated pair 共轭偶
conjugated polyene 共轭多烯
conjugated polymer 共轭聚合物
conjugated radicle 轭合基
conjugated residue of pesticide 农药轭合残留
conjugated stabilization energy 共轭稳定能
conjugated system 共轭体系*
conjugated triple bond 共轭三键
conjugated unsaturated bond 共轭不饱和键
conjugate effect 共轭效应
conjugate effect of chromophores 发色团的共轭效应
conjugate elements 共轭元素
conjugate fiber 组合纤维*
conjugate filament 组合丝缕
conjugate gradient method 共轭梯度法
conjugate ion 共轭离子
conjugate kernel 共轭核
conjugate layer 共轭液层
conjugate line 共轭线
conjugate linear operator 共轭线性算符
conjugate match 共轭匹配

conjugate matrices 共轭矩阵
conjugate metric tensor 共轭度量张量
conjugate phases 共轭相*
conjugate points 共轭点
conjugates 轭合物
conjugate solutions 共轭溶液*
conjugate spinning 复合纺丝
conjugate stress 共轭应力
conjugate system 共轭系统
conjugation ①共轭*②结合作用
conjugation energy 共轭能*
conjugation reaction 共轭(键合)反应
conjugative effect 共轭效应
conjugative mechanism 共轭机理
conjugative monomer 共轭单体
conjunctive name 连接名称
conjunct polymer 混合聚合物
conjunct polymerization 混合聚合法
connected facing 连贴边
connected structure 接合结构
connecting arm ①连接支叉②连接臂
connecting bolt 连接螺栓
connecting bulb 连接球管
connecting conduct 导管
connecting hose 连接管
connecting rod 连(接)杆
connecting rod bearing 联杆轴承
connecting rod busher 联杆轴承套
connecting shackle 链扣
connecting tube 连接管；导管
connecting wire 连接线
connect in series-parallel ①按组连接②串并联接
connection ①连接；接合②连接物
connection angle 连接角
connection foreman 领班
connection gang 接管班
connection in parallel 并联
connection in series 串联
connection size standard 连接尺寸标准
connectivity 连通性
connectivity factor 连通因子
connector ①连接管②连接器③接合物
connector tube 连接管
connect to parallel 并联；平联
conogenic group （可)电离基团
conoidal mill 锥形磨
conopharyngine 榴花碱
Conradson carbon residue 康拉逊残炭值〔在油或石油产品内〕
Conradson carbon test 康拉逊残炭试验
Conradson carbon value 康拉逊残炭值
Conradson test 康拉逊(残炭)试验
conrotatory 顺旋*
consanguinity 同族的
consecutive column 连串柱
consecutive peaks 连串峰
consecutive reaction 连串反应*
consecutive stability constant (=stepwise stability constant) 逐级稳定常数
consecutive titration 逐级滴定
conservation ①资源保护区②防腐③守恒；常住；不灭
conservation equation 守恒方程
conservation of energy 能量守恒
conservation of mass 质量守恒
conservation of matter 物量守恒；物质常住
conservation of momentum 动量守恒
conservation of orbital symmetry 轨道对称性守恒*
conservation of resources 储藏量守恒
conservation plant 废料工厂；利用废料生产的工厂
conservative 防腐剂
conservative substitution （构型)保持置换；保存性置换
conservative system 守恒(体)系
conserve ①保存②蜜饯
conserving 保存
conserving agent 保存剂
consignment 委托
consistance ①适合度②稠度；稠性
consistancy gage 稠度计
consistence test 抽针试验
consistency 稠度
consistency coefficient 稠度系数
consistency controller 稠度控制器
consistency cup 稠度杯
consistency curve 稠度曲线
consistency gage 稠度计
consistency index 稠度指数
consistency indicator 稠度指示器
consistency meter 稠度计
consistency number 稠度值
consistency property 相容性质
consistency regulator 稠度调节器
consistent estimate 一致的估计
consistent unit 一致(的)单位
consistometer 稠度计
console ①控制台；仪表板②托架；落地式支架
consolidated oil 硬脂酸；十八(烷)酸

consolidation　固结；压实
consolidometer　严密(性)检验计；固结试验仪
consoline　硬飞燕草碱
consolute　会溶质的
consolute component　会溶组分
consolute solution　会溶质溶液
consolute temperature　临界共溶温度*
conspecific　同种的
const　①(=constant)常数；恒量②(=constituent)组分；成分
constancy of temperature　温度恒定性
constancy of the cell　电池恒定性
constant　①常数；恒量②恒(常)的；恒定的③不变的
constant action　恒常作用
constant air conditioning　恒温恒湿；恒定空调
constantan　康铜合金〔紫铜与镍等的合金〕
constant boiling　恒沸(点)
constant boiling binary mixture　恒沸二元混合物
constant boiling mixture　恒沸(点)混合物
constant boiling point　恒沸点
constant boiling ternary mixture　恒沸三元混合物
constant cell　恒压电池
constant composition elution　恒溶剂成分洗脱
constant copolymerization　恒比共聚
constant current　恒电流
constant current charge　恒流充电
constant current coulometry　恒电流库仑法*
constant current electroanalysis　恒电流电解分析
constant current electrolysis　恒电流电解法*
constant current polarization titration　恒电流极化滴定
constant current potentiometric titration　恒电流电位滴定
constant current potentiometry　恒电流电位法
constant current process　恒电流过程
constant daughter ion scan　恒定子离子扫描
constant deviation prism　恒偏向棱镜
constant deviation spectrometer　等偏分光仪
constant deviation spectroscope　等偏分光镜
constant difference　①常数；恒量②恒定的
constant discharge　恒定流量；恒定出料量
constant-displacement pump　定量送料泵；计量泵
constant drying condition　恒干燥情况
constant energy synchronous fluorescence spectrometry　等能量同步荧光光谱法
constant entropy compression (=isentropic compression)　等熵压缩
constant environment mixer　恒定(操作)条件混合器
constant equilibrium　恒常平衡
constant error　恒误差
constant flow　定量流动；恒流

constant flow chromatography　恒流色谱法
constant-flow lubrication　恒流润滑
constant flow pump　恒流泵*
constant flow rate　恒定流速
constant freezing point　恒冻点；恒冰点
constant head tank　定位槽
constant-length heat treatment　定长热处理
constant-level control　恒定水平控制
constant-level conversion　恒定深度转化
constant-level lubrication　恒定油位润滑
constant-level lubricator　恒定油位润滑器
constant-level regulator　恒位面调节器
constant load　恒定负荷
constant load balance　恒载天平
constant load compression fatigue tester　定负荷压缩疲劳试验机
constant load elongation　恒载伸长
constant load increment　恒载荷递增；等载荷递增
constant mass division　恒量缩分法
constant moistened sample　恒湿样品
constant molal overflow　恒摩尔回流
constant molal vaporization　恒摩尔汽化
constant net loss　恒定净损耗
constant neutral loss scan　恒定中性丢失扫描
constant neutral loss spectrum　恒定中性丢失谱
constant of action　反应速率常数
constant of conductance cell　电导池常数
constant of integration　积分常数
constant of proportionality　比例常数
constant parent ion scan　恒定母离子扫描
constant pitch　①等距螺纹②等节
constant pitch screw　等螺距螺杆
constant potential amperometric titration　恒电位电流滴定法
constant potential coulometry　恒电位库仑法
constant power process　恒功率法
constant pressure　恒压
constant pressure combustion chart　恒压燃烧作业图
constant pressure filtration　恒压过滤
constant pressure gas thermometer　恒压气体温度计
constant pressure micrometer　恒压测微法
constant pressure pump　恒压泵*
constant pressure pumping system　恒压泵送系统
constant pressure valve　恒压阀
constant proportion　定比；恒比
constant rate　①恒速②常产量
constant rate creep　等速蠕变
constant rate filtration　恒速过滤

constant-rate method of distillation　恒速蒸馏法
constant rate of elongation　恒定伸长速率
constant rate of extension in stretch test　等速延伸拉伸测试
constant rate of loading　恒定加载速率
constant rate of production　常产量；稳定产量
constant rate of strain type tester　恒定应变率型试验机
constant rate of traverse　恒定横移速率
constant-rate period　恒(等速)率阶段
constant rate pump　定量泵
constant region　恒定区
constant resolution　恒定分辨率
constant-shear viscosimeter　恒定剪力黏度计
constant slip machine　恒滑式磨耗试验机
constant stirring tank reactor　连续搅拌釜式反应器
constant strain　恒应变
constant strain test　恒定应变试验
constant stress　恒应力
constant stress test　恒定应力试验
constant stress type tester　恒定应力型试验机
constant stretch history　恒定拉伸历史
constant taper screw　等渐缩螺杆；等斜度螺杆；等距变径螺杆
constant temperature　恒温
constant temperature and humidity　恒温恒湿
constant temperature atomization　等温原子化
constant temperature bath　恒温浴
constant temperature chromatography　恒温色谱法
constant temperature line　等温线
constant temperature method　恒温操作
constant temperature oven　恒温烘箱
constant temperature regulator　恒温调节器
constant tension break　恒张力断裂
constant tension windup　定张力卷取装置
constant underflow　恒底流
constant value control　定值调节；恒值调节
constant voltage　恒电压
constant voltage amperometric titration　恒电压电流滴定法
constant voltage charge　恒压充电
constant voltage circuit　恒压电路
constant voltage generator　恒压发电机
constant voltage transformer　恒压变压器
constant volume　定容；恒定体积
constant-volume burning (=constant volume combustion)　恒定体积燃烧
constant volume combustion　恒定体积燃烧
constant volume feeder　恒体积加料器
constant volume gas thermometer　恒容气体温度计
constant volume pump　恒容量泵
constant volume sampling　恒体积取样
constant volumetric flow rate　恒体积流(动)速(率)
constant weight　恒重*
constant wet-bulb temperature line　恒湿球温度线
constant white (=blanc fixe)　钡白
constant worm pich　等距螺纹
constellation (=conformation)　构象
constellation graph　星座图〔统计化学〕
constituent　成分；构成；组分
constituent material　组分材料
constitution　①组织②成分③结构；构造
constitutional diagram　状态图；相图
constitutional features　结构特征
constitutional formula　结构式；构造式
constitutional heterogeneity　组成非均一性*
constitutional repeating unit　重复结构单元*
constitutional sequence　结构序列
constitutional unit　结构单元*
constitution controller　结构控制剂
constitution heterogeneity　组分非均匀性
constitution of synthetic adhesive　合成胶黏剂的成分
constitution water　①结构水②化合水
constitutive correction constant　本构校正常数
constitutive equation (=rheological equation)　本构方程
constitutive functional　本构泛函
constitutive model　本构模型
constitutive property　结构性(质)
constrained condition　约束条件*
constrained geometry catalyst　限制几何构型催化剂
constrained geometry metallocene catalyst　限定几何构型茂金属催化剂
constrained material　受限材料
constrained optimization　约束最优化*
constrained variation method　约束变分法
constraint condition　约束条件
constraint equation　约束方程
constraint manifold　约束流形
constriction　①压缩；收缩②颈缩
construction　建造
construction adhesive　结构黏合剂
constructive total loss　推定全损
consumption　消耗量；使用量
consumption indicator　消耗指示器
consumption of coal　煤消耗；耗煤量
consumption of fuel　燃料消耗(量)
consumption of oil　油消耗；耗油量

consumption of petroleum　石油消耗
consumption test　(燃料或油的)消耗试验
contact acid　接触(法制的硫)酸
contact action　接触作用
contact adhesion　触点黏结
contact adhesive　接触型胶黏剂；压合式黏合剂
contact agent　接触剂
contact angle　接触角*
contact angle hysteresis　接触角滞后
contact application　触施
contact area　接触面积
contact area tester　支持面仪
contact atomic force microscope　接触式原子力显微镜
contact biological filter　接触生物滤池
contact black　接触炭黑；烟道炭黑
contact carrier　催化剂载体
contact catalysis　接触催化
contact catalyst　接触催化剂
contact chamber　接触室
contact clay　接触白土
contact clay treating　接触白土处理
contact coacervant (=direct coacervant)　接触凝聚剂
contact coking　接触焦化
contact column　接触塔
contact condenser　接触冷凝器
contact cooling　接触冷却
contact corrosion (=crevice corrosion)　接触腐蚀
contact cup　接触盅
contact deasphalting method　接触脱沥青法
contact decolorization　接触脱色(法)
contact desulfuriation　接触脱硫(法)；催化脱硫(法)
contact difference　接触势差；接触位差
contact distillation　接触蒸馏
contact drying　接触式烘燥
contact effect　接触作用
contact electrode process (for sodium)　密接电极(炼钠)法
contact electromotive force　接触电动势
contact exchange　接触离子交换
contact failure　接触断裂；接触衰坏
contact feeler　接触式灵敏元件；接触式探针
contact filtration　接触过滤
contact force　接触力
contact furnace　接触炉
contact gap　接点间隙
contact glue　触压胶
contact goniometer　接触测角器
contact hyperfine interaction　接触超精细相互作用
contacting　接触

contacting electrode　接触电极
contacting element　接触单位
contact inhibition　接触抑制
contact interface　接触界面
contact ion pair　紧密离子对
contact-key　接触键
contact laminate　触压层压板
contact laminating　直接层压(法)〔在小压力下压成层〕
contact lay-up　接触成型法
contact layup mo(u)lding　低压铺叠成型；接触层叠模制法
contact leaching antifouling paint　接触渗毒型防污漆
contactless pickup　无触点传感器
contact maintained plant　直接维修工厂
contact maintenance　接触维修；直接维修
contact mass　接触物质
contact mass revivification　接触物质的复活
contact metamorphic action　接触变质(作用)
contact metamorphism　接触变性
contact metasomatose　接触变质
contact method　接触法〔污水处理〕
contact molding　浇铸成型
contact mouldability　手糊附模性
contactor　接触器
contactor centrifuge acid treating plant　离心机接触混酸处理设备
contactor pump　混合泵
contactor treating process (=contact treating)　接触精制法
contact oven　接触炉；催化剂炉
contact oxidation method　接触氧化法
contact percolation　接触渗滤
contact plant acid　接触(法工厂制的)硫酸
contact plant of sulfuric acid　接触法硫酸厂
contact point　①接触点②电(路)接触点
contact poison　接触毒
contact potential　接触电势*
contact potential difference　接触电位差
contact preparation　①接触制备②印刷准备
contact pressure resin　触压固化树脂
contact print method　接触印相法
contact process　接触法
contact pyrometer　接触式高温计
contact radiation therapy　接触辐射治疗
contact radiotherapy　接触放射治疗
contact reaction　接触反应
contact reactor　接触反应器
contact rectification　接触精馏
contact reforming　接触重整；催化重整
contact resin　低压固化树脂

contact resistance　接触电阻
contact rock　接触岩
contact scar　接触性污染；污斑；污色
contact seal　接触密封；机械密封
contact series　接触次序
contact shift　(费米)联系位移
contact stain(ing)　接触性污染；污斑；污色
contact substance　接触物质
contact surface　接触面
contact surface separator　接触表面分离器
contact test　接触试验
contact thermometer　接触温度计
contact time　接触时间
contact tower　接触塔
contact treating (=contactor treating process)　接触精制法
contact treatment with Fuller's earth　(石油产品)用白土接触精制
contact twin　接触孪晶
contained underground burst　封闭式地下爆炸
container　容器
container board　盒纸板
containerized chemicals　集装箱化学药品
container lining　容器(储罐)衬里
container loading　容器装料
container support　集气瓶架
container unloading　容器卸料
containing mark　容量刻度
containment facilities　防事故装置
contaminant　污染物
contaminant detection　污染物检测
contaminant loading　污染负荷(量)
contaminated area　污染面积；玷污区域
contaminated catalyst　沾污的催化剂；脏催化剂
contaminated clothing laundry　污染衣物洗涤处
contaminated gasoline　污染的汽油
contaminated surface　污染表面
contaminated water　污染水
contaminating material　污染物质
contamination　玷污；污染；沾染；弄脏
contamination control　污染控制；污染管制
contamination index　污染指数
contamination material　污染物质
contamination of water　水(质)污染
contamination precipitation　杂质沉淀
contamination product on paint film　漆膜表面的污染物
contamination test　污垢试验
conteben　缩氨基硫脲；氨硫脲
cis-content　顺式含量

content　①含量②内容物③容度
content by volume　按体积计的含量
content gauge　(油或燃料的)水位仪
content gauge supports　(储器内液体)水位指示器支柱
content line　注入容积刻度线；内标线；容度线
content of powder cups　粉末料杯容量；喷粉杯内容物
content uniformity　装量差异
content volume　内容积；注入容积
Continental Oil process　(美国)大陆石油(公司)法〔由乙烯制直链α-烯烃法〕
contingency table　列联表
continual　连续的
continual method　连续法
continuation　延伸；拓展；延拓
continued (=continual)　连续的
continues automatic monitoring technology　连续自动监测技术
continuity　连续性
continuity equation　连续性方程
continuity law　连续性定律
continuity mechanics　连续介质力学
continuity of state　物态连续性
continuous　连续的
continuous absorption　连续吸收
continuous acid treating　连续酸处理
continuous activated sludge test　连续活性污泥试验
continuous ageing　连续老化；老成
continuous air monitoring program　空气连续监测规划
continuous alcoholysis　连续醇解
continuous alkalisation　连续碱化(作用)；连续碱浸渍
continuous analysis　连续分析
continuous analyzer　连续式分析器
continuous and automatic monitoring system for water pollution　水污染连续自动监测系统
continuous apparatus　①连续(式)器械②连续(式)装置
continuous arc method　连续弧光法
continuous automatic staking machine　连续式自动刮软机
continuous band dryer　链带式干燥机
continuous battery still　连续蒸馏锅组
continuous belt ion exchanger　连续离子交换带；带式离子交换器
continuous belt ion exchange system　连续带式离子交换装置
continuous-belt xanthation　(环)带式连续黄化
continuous blow-down　连续排放
continuous-bucket elevator　连续斗式升降机
continuous buffer system　连续缓冲系统
continuous catalytic hydrogenation　连续催化加氢

continuous centrifuge　连续离心机
continuous chain　连续链
continuous chromatographic refining　连续色谱精制
continuous chromatography　连续色谱(法)
continuous circular settler　连续式圆形澄清池
continuous cleaning and painting process　连续清洗和涂装过程
continuous cleaning phosphatizing and painting　连续清洗、磷化和涂漆
continuous coking　连续焦化
continuous coking process　连续焦化过程
continuous-column dissolver　柱式连续溶解器
continuous contact coking　连续接触焦化
continuous countercurrent decantation　连续向流倾析法
continuous countercurrent operation　连续向流操作
continuous crystallizer　连续结晶器
continuous curing　连续硫化
continuous current　连续流；直流
continuous current convertor　直流换流器
continuous current electromotor　直流电动机
continuous current generator　直流发电机
continuous deaerator　连续脱泡器
continuous dependence theorem　连续依赖性定理
continuous development　连续展开(法)*
continuous dewatering centrifuge　连续脱水离心机
continuous diffusion　连续扩散
continuous digester　连续蒸煮器
continuous distillation　连续蒸馏
continuous distillation plant　连续蒸馏车间
continuous distribution　连续分配
continuous drier　连续(式)干燥器
continuous drying machine　连续式干燥机
continuous duty　连续运转
continuous dyeing　连续染色
continuous electrophoresis　连续电泳
continuous emission spectrum　连续发射(光)谱
continuous ester interchange　连续酯交换
continuous extraction　连续浸提
continuous extrusion　连续挤塑
continuous extrusion blowing　连续挤坯吹塑
continuous extrusion cold tube process　连续挤出冷管坯(吹塑)法
continuous extrusion moulding　连续挤出模塑(法)
continuous feed　连续进料
continuous feed plating press　连续式压光机
continuous feed wringing machine　连续式挤水机
continuous fiber winding　长丝缠绕
continuous fibrillation　连续纤化

continuous (field) method　连续磁粉探伤法
continuous filament　连续纤维
continuous filament rayon　长条螺萦
continuous-filament woven fabric　连续纤维机织物
continuous film　连续膜
continuous film degassing　连续薄膜脱泡
continuous filter　连续滤器
continuous finishing　连续整理
continuous flow　持续流(动)
continuous flow analysis　连续流(动)分析
continuous flow calorimeter　续流量热器
continuous flow chemiluminescence measurement　连续流动化学发光测量
continuous flow chromatography　连续流动色谱(法)
continuous flow enzyme detector　连续流动酶检测器
continuous flow fast atom bombardment　连续流快原子轰击
continuous flow fast atom bombardment interface　连续流动快原子轰击接口
continuous flow method　连续流动法*
continuous flow paper electrophoresis　连续流动纸电泳
continuous flow scintillation counter　连续流动闪烁计数器
continuous flow thermocycling PCR chip　连续流式温度循环芯片 PCR
continuous fluid coker　流化床连续焦化设备
continuous fractionation　连续分馏
continuous fractionation column　连续操作分馏塔
continuous function　连续函数
continuous gradient　连续梯度
continuous gradient extruder　连续梯度挤出机
continuous gradient gel　连续梯度凝胶
continuous grain growth　连续晶粒生长
continuous grease production　润滑脂连续生产过程
continuous grinder　连续研磨机
continuous growth　连续生长
continuous heat resistance　耐热持续性；耐热持久性
continuous-hydrolysis reactor　连续水解反应器
continuous hydrolytic polymerization　连续水解聚合
continuous industrial viscometer　连续式工业黏度计
continuous ion exchange operation　连续离子交换法
continuous kiln　连续窑
continuous laser　连续激光
continuous light absorption　光的连续吸收
continuous load　固定荷重；不变荷重
continuous low-temperature polymerization　连续低温聚合
continuous lubricating film　连续润滑膜

continuous lubrication 连续润滑
continuously flawed region 连续疵点区
continuously variable transmission belt 无级变速传动带
continuous machine 连续机
continuous measuring gage 连续测厚计
continuous medium 连续介质
continuous method for calibration with ISE 离子选择电极连续校准法
continuous mill 连续研磨机
continuous mixer 连续混炼机
continuous mixing conveyer 连续混合运输机
continuous monitoring 连续监测
continuous monitoring technique 连续监控技术
continuous multistage distribution 连续多级分配
continuous network 连续网络
continuous neutron activation analysis (CNAA) 连续中子活化分析
continuous oil circulation 连续的油循环
continuous operation 连续操作；连续运转
continuous oven 连续干燥器；连续烘干机
continuous pad dyeing range 连续轧染联合机
continuous paper electrophoresis 连续纸电泳
continuous phase 连续相
continuous plant 连续操作装置
continuous polyester condensation 聚酯连续缩合
continuous polyester polycondensation process 聚酯连续缩聚工艺
continuous polymer filter 聚合物连续过滤器；纺前过滤器
continuous polymerization 连续聚合*
continuous polymerization unit 连续聚合装置
continuous process 连续法
continuous process vulcanizer 连续操作硫化机
continuous production 连续生产
continuous radio scanning method 连续放射扫描法
continuous reactor 连续(式)反应器
continuous-reading electrotitration apparatus 连续读数式电滴定仪
continuous rectification 连续精馏
continuous reforming regeneration 连续重整再生(法)
continuous ripening 连续熟化
continuous rotary filter 连续回转过滤机
continuous running development 连续流动展开
continuous sample 连续试样
continuous sample inlet system 连续进样系统
continuous sampling ①连续取样②连续进样
continuous saponification 连续皂化
continuous shaft kiln 连续竖窑
continuous shear rheometry 连续剪切流变测定法
continuous shooting of gas 毒气连续射击
continuous slurry equipment 浸压粉碎联合机
continuous soaper 连续制皂器
continuous source method for background correction 连续光源背景校正法
continuous spectrum 连续光谱
continuous spinning 连续纺丝
continuous spinning process 连续纺丝过程
continuous steamer 连续蒸煮器
continuous still 连续蒸馏器
continuous still battery 连续蒸馏
continuous stirred tank reactor (CSTR) 连续搅拌釜式反应器
continuous strand mat 连续原丝毡
continuous stress relaxation 连续应力松弛
continuous tank 连续槽
continuous technique 连续成型
continuous thermoforming 连续热成型
continuous thermosol-pad dyeing machine 连续热溶轧染联合机
continuous thickener 连续稠厚器
continuous thick juice sulfitation 连续浓汁亚硫酸化
continuous throughout de-aerator 连续出料(的)除气器
continuous transfer equivalent plate 连续转移等效塔板
continuous translation group 连续平移群
continuous tray method 连续盘式法
continuous treating 连续精制
continuous tunnel drier 连续热道干燥器
continuous tunnel drying 连续隧道干燥
continuous turbo-mixer 连续式涡轮混合器
continuous variation method 连续变化法
continuous vertical retort 连续的立式蒸馏甑
continuous volumetric method 连续容积变化法
continuous vulcanization 连续硫化
continuous warping 连续整经
continuous washing machine 连续洗涤机
continuous wave (CW) 连续波
continuous wave electron spin resonance 连续波电子自旋共振
continuous wave laser (=CW laser) 连续波激光器*
continuous wave NMR 连续波核磁共振
continuous wave NMR spectrometer 连续波核磁共振(波谱)仪*
continuous welding 连续焊接
continuous whirling-layer dryer 连续旋风式干燥机
continuous working kiln 连续窑
continuous xanthation 连续黄(原)酸化(作用)

continuous X-rays　连续X射线
continuous X-ray spectrum　连续X射线光谱
continus paper　卷筒纸
continuum　连续区*
continuum lamp　连续光谱灯
continuum light source　连续光源
continuum source　连续光源
continuum source background corrector　连续光源背景校正器
continuum theory　连续介质理论
contour　弧面；曲面；外形；恒值线；等值线
contour chromatogram　等高线色谱图
contour curve　等值曲线
contoured　起伏状的
contoured furnace tube　仿形(石墨)炉管
contour(ed) surface　①等值面②围道(路)曲面
contour effect　边缘效应；轮廓效应
contour extrusion　异型挤塑
contour graph　等值线图表
contour length　伸直长度*
contour line　等值线；等高线；轮廓线
contour map of charge density　电荷密度等值线图*
contour plate　靠模样板；仿形样板
contour plot　等值线图表；等高线图
contour spectrum　等高线光谱
contraceptive　避孕药
contract-crimping method　收缩卷曲法
contractiometer　收缩性试验仪
contraction　收缩(作用)；缩小；缩短；缩并
contraction flow　收窄流动；收缩流动
contraction of catalyst　催化剂的收缩
contraction on melting　熔融收缩
contraction percentage　收缩率
contraction pyrometer　收缩高温计
contraction schedule　(催化剂的)收缩状态
contraction stress　收缩应力
contraction work　收缩功
contractive color　缩小色
contractor's hose　建设用胶管
contract research　优选法试验
contra-current　逆流；对流
contra-flow　逆流；对流；反流；回流
contra-flow condenser　逆流冷凝器
contraflow drier　逆流式干燥机
contra-hook　倒钩
contraries　原料中的杂质
contrast　反衬度；对比度
contrast color　对比色；对照色
contrast control　反差控制
contrast enhancement layer technique　反差增强层技术
contrast factor　对比系数；阶调系数
contrast gloss　对比光泽度
contrasting colour　对比色
contrast micrometer　反差测微计
contrast of photographic plate　感光板反衬度
contrast phase microscope　相差显微镜
contrast ratio　对比率；对照比
contrast reaction of color　颜色对比反应
contrast stain　对比染色剂
contrast test　对照试验
contravalency (=covalence)　共价
contravariant vector　逆变向量；反变向量〔数〕
control　①控制；管制②对照物
control action　控制作用
control agent　控制剂
control analysis　控制分析
control and monitor subsystem　控制和监控子系统
control apron　控制皮圈
control board　①控制台(板，盘，屏)；操纵台②配电盘
control button　控制按钮
control center　控制中心
control central line　控制中心线
control chart　控制图*；质控图
control chart for average and range　平均值-极差控制图
control chart for average and standard deviation　平均值-标准偏差控制图
control chart for quality　质量控制图
control chart method　质控图法
control chemical analysis　质控化学分析
control console　①控制台；操纵台②控制板
control cure　对比硫化
control device　控制装置
control equipment　控制仪器；控制设备
control error　控制误差
control experiment　对照实验
control foaming　抑制起泡
control grid　控制栅极
control handle　操纵手柄；控制旋钮
control house　控制室
control instruments　控制仪表
control knob　控制钮
controllability　可控性
controllable extensional experiment　可控延伸试验
controllable factor　可控因素*
controllable reaction　可控反应
controllable viscometric flow　可控测黏流动

control laboratory 检验室；化验室
control latex 空白胶乳；对比胶乳
controlled application of heat 热的控制供给
controlled atmosphere 控制气氛
controlled chain growth polymerization 链增长受控聚合(作用)
controlled chilling 控制冷却
controlled condition 一定条件；恒定条件
controlled current coulometry 控制电流库仑滴定法
controlled current electrogravimetric analysis 控制电流电重量分析法
controlled current electrogravimetry 控制电流电重量分析法
controlled current electrolysis 控制电流电解
controlled current polarography 控制电流极谱法
controlled current potentiometric titration 控制电流电位滴定
controlled-cycle distillation 控制循环蒸馏
controlled degradation 可控降解
controlled deviation 控制偏差
controlled diffusion 受控(制)扩散
controlled factor 可控因素
controlled fancy yarn 控制型花式线
controlled filter 控制过滤
controlled foam detergent 控泡洗涤剂
controlled freezing 受控凝固
controlled frequency 控制频率
controlled fusion 受控(核)聚变
controlled humidity drying 控制湿度(下的)干燥
controlled-humidity thinlayer chromatography 控(制)湿(度)薄层色谱(法)
controlled mechanical fibrillation process 受控机械原纤化法；可调机械原纤化法
controlled polymerization 控制聚合(作用)
controlled polymerization rate 受控聚合速度
controlled pore glass granule 可控孔径玻璃珠
controlled porosity glass 可控多孔性玻璃
controlled porosity product 受控多孔性制品
controlled potential analysis 控制电势分析
controlled potential coulometric analysis 控制电位库仑分析
controlled potential coulometric titration 控制电位库仑滴定(法)*
controlled potential coulometry 控制电势库仑法
controlled potential electrogravimetry 控制电位电重量法
controlled potential electrolysis 控制电势电解
controlled potential electrolysis apparatus 控制电位电解仪
controlled potential electroseparation 控制电位电分离
controlled potential method 控制电位法
controlled propagation (链)受控增长
controlled radical polymerization (CRP) 控制自由基聚合；可控自由基聚合
controlled release pesticide 长效农药
controlled-shrinkage annealing 受控收缩热处理
controlled split 均裂式
controlled structure polymer 有规结构聚合物
controlled-sudsers (=controlled sudsing detergent) 控泡洗涤剂
controlled sudsing detergent (=controlled-sudsers) 控泡洗涤剂
controlled surface 可控表面；限制表面
controlled surface pore glass 可控表面多孔玻璃
controlled surface porosity 可控表面孔度
controlled surface porosity support 可控表面孔率载体
controlled temperature 控制温度；支配的温度
controlled-temperature bath 控温浴
controlled temperature furnace atomizer 控温炉原子化器
controlled thermal decomposition 受控热分解
controlled thermohygrostat 恒温恒湿器
controlled thermonuclear fusion reactor 受控热核聚变反应堆
controlled thermonuclear reaction 可控热核反应
controlled variable 控制变量
controlled volume pump 计量泵
controlled weighted centroid simplex method 控制加权形心单纯形法
controller (=controlling apparatus) 调节器；控制器
control lever 控制杆
control limit 控制限
control line 控制线
controlling apparatus (=controller) 调节器；控制器
controlling board 控制盘；控制板；控制屏(台)
controlling factor 控制因素
controlling potentiometer 自动控制的电位差计
controlling sheen 控制糙面光泽
control loop 控制回路
control of blank value 空白值控制
control of blank value in situ 现场空白(值)控制
control of corrosion 腐蚀控制
control of detection limit 检出限控制
control of microorganism 微生物控制
control of purity 纯度控制
control panel 控制盘；控制板；控制屏(台)

control point 控制点
control rod 控制棒
control roll 导滚
control room 控制室
control sample 对照试样
control sampling 控制进样
control storage 控制储藏
control system 控制系统
control technique ①控制技术②(大气或噪声污染的)净化措施；净化技术
control technique guideline ①控制技术指南②(大气或噪声污染)净化技术指南
control test 对照试验；控制试验
control tool 控制工具
control unit 控制单元；控制组件
control valve 控制阀；调节阀
controvalence (=covalence) 共价
conus (=cone)〔拉丁文〕 圆锥；锥
convaflavin 铃兰黄素
convallamaretin 铃兰苦亭
convallamarin 铃兰苦苷
convallaretin 铃兰亭
convallarin 铃兰苷
α-convallarol α-铃兰醇
convallatoxin 铃兰毒
convected coordinate 迁移坐标；轮换坐标
convected derivative 迁移导数；轮换导数
convected differentiation 迁移微分(法)；轮换微分(法)
convected frame of reference 迁移参考系
convected Maxwell model 交换麦克斯韦模型
convected pressure distribution 变动的压力分布
convection 对流
convection agitation 对流搅拌
convectional drier 对流式干燥器；对流式干燥机
convection bank 对流管束
convection chamber 对流室
convection coefficient 对流系数
convection current 对流〔指气流、热流等〕
convection-diffusion cell 对流扩散池
convection-diffusion thermal conductivity cell 对流扩散式热导池
convection drying ①对流干燥②直接干燥
convection effect 对流效应
convection electrode 对流电极
convection furnace 对流炉
convection heat 对流热
convection heater 对流加热器
convection heating system 对流加热装置

convection heat transfer 对流热传导
convection light 集中光束
convection of air 空气对流
convection process 对流过程
convection section 对流部分
convection system 对流加热系统
convection temperature 对流温度
convection tube 对流管
convection zone 对流段
convective chronoamperometry 对流计时电流法
convective chronocoulometry 对流计时库仑法
convective heat flow 对流热流
convective heat transfer 对流传热
convector 对流散热器；对流加热器；换流器
convenience paste (裱墙纸用的)方便浆糊；方便胶浆
convenient synthesis 简便合成法
convention 惯例
conventional 习用的
conventional activated sludge 常规活性污泥法
conventional activated sludge process 标准活性污泥法
conventional atomization source 常规原子化源
conventional base unit (习)惯用基本单位
conventional chemical constant 习用化学常数
conventional design heater 按一般形式设计的加热炉
conventional dissociation constant 习用离解常数
conventional double heater false-twist texturing machine 常规双加热器假捻变形机
conventional efficient 习用效率；实验效率；总效率
conventional entropy 规定熵
conventional formula 习用式；实验(公)式
conventional liquid chromatography 常规液相色谱法
conventional milling 逆铣
conventional oil 习用油；普通油
conventional packed column 常规填充柱；普通填充柱
conventional paint 传统油漆
conventional rubber 通用橡胶
conventional surfactant 常用表面活性剂
conventional symbol 习用符号
conventional test 普通试验；标准试验
conventional true value 约定真值
conventional-type lubricant 普通润滑油；常用润滑油
conventional viscometer 常规黏度计
convention cutting 逆向切削
convention for estimation of single ion activity coefficient 单离子活度系数计算法
convergence (=convergency) 集束性；收敛性
convergence electrode 会聚电极
convergence limit 会聚极限；收敛极限

convergent 会聚；收敛
convergent beam microdiffraction 会聚束微衍射
convergent beam radiation 会聚束辐射
convergent channel 会聚槽
convergent flow 会聚流动
convergent light image 锥光图
convergent nozzle 会聚喷嘴
convergent synthesis 汇集合成*
converging-diverging nozzle 缩-扩喷嘴
converging flow technique 会聚流动技术
converging guide 集束导丝器
converging ion beam 会聚离子束
converging line 收缩线
converging nozzle 收缩喷嘴
converging stream (在生产上)会聚(的质)流
conversion ①转化②换算
conversion catalyst 转化催化剂；烃类转化过程的催化剂
conversion chart 换算图表
conversion coating 转化层
conversion coating technology 转化型涂料工艺
conversion coefficient 换算系数；转化系数
conversion constant (=conversion factor) 换算因数
conversion corrosion 转化型腐蚀
conversion curve 转化曲线
conversion dynode 转换打拿极
conversion efficiency 转换效率；变换效率
conversion electron 转换电子
conversion energy 转化能
conversion factor (=calculation factor) 换算因子*；换算因数
conversion film 转化膜
conversion level 转化深度
conversion method 转化法〔由一种烃转化为另一种烃的方法〕
conversion of flowmeter 消耗量读数的换算
conversion of hydrocarbons 烃类转化
conversion of monomer 单体转化(率)
conversion of viscosity units 黏度单位换算
conversion per pass 单程转化率
conversion pig 炼钢生铁
conversion process 转化过程
conversion rate 转化率
conversion rate curve 转化率曲线
conversion scale 转换标尺；换算比例尺
conversion system 转化装置
conversion table 换算表
conversion temperature 转化温度
conversion test 转化试验
conversion unit 反应设备
conversion zone 反应层；反应区域
converted count 换算支数
converted products 加工产品
converted water gas 已转化的水煤气
converter (=convertor) ①换流器②转炉③转化器④合成塔
converter gas 转化炉气
converter of acid lining 酸性转炉
converter of basic lining 碱性转炉
converter pig 转炉(用)生铁
converter slag 转炉炉渣
converter spinning process 牵切纺纱法；直接成条纺纱法
converter top 直接成条机纤维条
convertibility 转化性；互换性
convertible coating 转化型涂料
convertible film 转化性漆膜
convertible hydrocarbons 容易转化的烃类
convertible paint 转化型漆
converting 转化；改变
convertor (=converter) ①换流器②转炉③转化器④合成塔
convex 凸(出)的
convex corner 凹角
convex edge 凸边；凸面
convexo-concave 凸凹的
convexo-convexe 两面凸的
convexo-plane 平凸的
convex protrusion 凸起物；隆起物
convex rule coating 钢卷尺涂料
convex surface 凸面
conveyance 输送带；搬运器；转运机；输送；供给
conveyance fluid 输送流体
conveyance loss 运输损失
conveyer 运输机；运输装置
conveyer belt 运输带
conveyer dryer 输送式干燥器；带式烘燥机
conveyer loading hopper 运输载料斗
conveyer scale 传送带秤
conveyer screw 螺旋运输机
conveyer table 转运台
conveyer trough 输送槽
conveyer trough drier 输送槽干燥器
conveyer typecoating 抽涂；带式挤涂
conveyer volume meter 输送容量计
conveyer weigher 运输秤

conveyer weight meter　输送重量计
conveying　输送；运输
conveying band　运输带
conveying belt　①运输带②带式运输机
conveying fan　气运风扇
conveying worm　回旋运输机
conveyor (=conveyer)　输送机；运输机
conveyor band link　运输带扣
conveyor belt　运输带
conveyor belt method　输送带法
conveyor cover　运输带覆盖
conveyor dryer　带式干燥机
conveyor drying　传送式干燥
conveyor fabric　运输带帆布
conveyor head roll　运输带前鼓轮
conveyor idler　运输带托滚；托轮
conveyorization　机械化搬运
conveyor roll　运输带鼓轮
conveyor scale　带式自动秤
conveyor tail roll　运输带后鼓轮
conveyor trough　运输带弯槽
conveyor-type belt　运输带式传动带
convicin　伴蚕豆嘧啶核苷
convolamine　旋花胺
convoluted peak　回旋峰
convolution　褶合；卷积
convolution algorithm　卷积运算
convolution curve method　卷积曲线分析法
convolution difference　卷积差分
convolution difference filter　褶积差分滤波
convolution-integral linear sweep voltammetry　卷积积分线性扫描伏安法
convolution integration　卷积积分
convolution length　捻回长；卷曲长
convolution (of cotton fiber)　天丝转曲(棉纤维)
convolution principle　卷积原理*
convolution spectrum　卷积光谱
convolution spectrum method　卷积光谱法
convolution theorem　褶积定理
convolution voltammetry　卷积伏安法*
convolvamine　旋花胺
convolvicine　旋花素
convolvine　旋花碱
convolvulin　旋花苷；旋花灵
convolvulinic acid　旋花酸
convolvulinolic acid　旋花醇酸；11-羟(基)十四(烷)酸
convulsion　惊厥
Conway's diffusion analysis　微量扩散

conylene (=1,4-octadiene)　1,4-辛二烯〔别名〕
conyrine　康尼碱；毒芹分碱；2-丙基吡啶　$C_8H_{11}N$
cooked grease　热制润滑脂
cookeite　锂绿泥石
cooking　熬炼；热炼
cooking acid　蒸煮酸
cooking boiler　蒸煮锅
cooking fermentation mash　蒸煮发酵醪
cooking kettle　调浆桶；煮浆桶
cooking liquor　蒸煮液
cooking oil　厨用灯油
cooking soda　碳酸氢钠
cooking tank　蒸煮锅
cooking temperature　热炼温度
cooking vat　蒸煮锅
cool　①冷却②冷的
coolant　切削剂；冷却剂
coolant clarification　冷却剂澄清
coolant flow　冷却剂流
coolant oil　冷却油
coolant thermostat　冷却恒温槽
cool color　冷色
cool down　冷却
cooled　冷却了的
cooler　①冷却剂②冷却器
cooler casing　冷却器外壳
cooler condenser　冷却冷凝器
cooler crystallizer　冷却结晶器
cooler drum　冷却转鼓
cooler pan　冷却盘
cool flame　冷焰*
cool flame emission　冷焰发射
cool flue gas　冷烟道气
Coolidge's quartz fiber manometer　考利基石英丝压力计
Coolidge tube　考利基管；热阴极电子管
co-oligomer　共齐聚物
co-oligomerization　共齐聚
cooling　冷却
cooling agent　冷却剂
cooling air　冷却空气
cooling air baffle　冷却空气挡板
cooling apparatus (=cooler)　冷却器
cooling area　冷却面积
cooling bath　冷却浴
cooling blast　冷却通风
cooling block　冷却块
cooling calender　冷却滚压机
cooling capacity　冷却本领

cooling chamber　冷却室
cooling coil　冷却旋管
cooling curve　冷却曲线*
cooling cylinder　冷缸
cooling draught　冷却气流；冷却通风
cooling drum　冷却滚筒
cooling effect　冷却效应
cooling fins　散热片
cooling frame　冷板框
cooling hearth　冷却炉床
cooling installation　冷却装置
cooling jacket　冷却套管
cooling jig　冷却装置
cooling mandrel　冷却芯型
cooling medium　冷却介质
cooling medium pump　冷却介质泵；输送冷却介质的泵
cooling method　冷却法
cooling mixture　冷却混合物
cooling of evaporation　汽化致冷(冷却)
cooling off　冷却
cooling oven　冷却室
cooling pan　冷却锅
cooling peak　冷却峰
cooling pipe　冷却管
cooling pond　冷却池
cooling power　冷却本领；冷却能力
cooling press　①冷却压榨机②冷却压滤机
cooling pump　冷却泵
cooling-rate curve　降温速率曲线
cooling roll　冷却辊
cooling roller　冷却辊
cooling space　冷却(套管的)空间
cooling stack　冷却烟道
cooling surface　冷却面
cooling system　冷却系统
cooling tank　冷却槽；冷却桶
cooling telethermometer　冷却(系统的)遥测温度计
cooling theory　冷却理论
cooling time　冷却时间
cooling tower　凉水塔；冷却塔
cooling tower suction sump　冷却塔空吸池
cooling tray　冷却盘
cooling trough　冷却水槽
cooling tube　冷却管
cooling tunnel　冷却隧道；隧道式冷却装置
cooling water　冷却水
cooling water circulation　冷却水循环
cooling water pipe　冷却水管

cooling water treatment　冷却水处理
cooling water tube　冷却水供应管
cooling worm　冷却旋管
cooling zone　(再生炉内)冷却层
cool note　凉香韵
cool wash　冷洗法
coom　煤烟；炭黑
coomassie brilliant blue G_{250}　考马斯亮蓝 G_{250}
coopal powder　苦拔炸药
cooper's wood　桶材
cooperage　①桶匠工作场②桶匠工艺
cooperation complex　共作用配合物
cooperation test　协作(协同)试验
cooperative action　协同作用
cooperative adsorption　协同吸附(作用)
cooperative effect　合作效应*
cooperative feedback inhibition　合作反馈抑制；协同反馈抑制
Cooperative Fuel Research Committee　燃料研究协会〔是石油和马达制造的联合委员会，起源于 1922 年〕
cooperative motion　协同运动
cooperative site　协同部位
cooperativity　协同效应
cooperite　硫砷铂矿
coopery　桶匠业
coordinate　①配位的；配价的②坐标
coordinate axis　坐标轴
coordinate bond　配位键
coordinate complex　配位化合物
coordinate complex salt　配位复盐
coordinate-covalent bond　配位共价键*
coordinated-anionic mechanism　配位阴离子机理
coordinated anionic polymerization　配位阴离子聚合(作用)
coordinated cationic polymerization　配位阳离子聚合(作用)
coordinated ionic polymerization　配位离子聚合(作用)
coordinated metal complex　配位金属配合物
coordinated polymer　配位聚合物
coordinated water　配位水
coordinate electrovalent bond　配位电价键
coordinate factor　配位因子
coordinate formula　配位式
coordinate link (=coordinate bond)　配位键
coordinate linkage (=coordinate link)　配位键
coordinately linked　配位地；键合的
coordinate motion　协调动作
coordinate paper　坐标纸
coordinate plane　坐标平面；标轴面
coordinate polymerization　配位聚合

coordinate repression	①电价阻遏②配位阻遏；协同抑制
coordinate scale	坐标标度(刻度)；坐标(比例)尺
coordinate system	坐标系
coordinate valence	配位价
coordinate valence force	配位价力
coordinating atom	配位原子
Coordinating Fuel and Equipment Research Committee	燃料及设备研究协调委员会
Coordinating Lubricant and Equipment Research Committee	润滑剂及设备研究协调委员会
coordinating polyhedron	配位多面体*
coordination	配位作用*
coordination agent	配位剂*
coordination anion	配(位)阴离子*
coordination bond	配位键*
coordination catalysis	配位催化*
coordination catalyst	配位催化剂
coordination cation	配(位)阳离子*
coordination center	配位中心*
coordination chemistry	配位化学*
coordination complex	配位化合物
coordination compound	配位化合物*
coordination effect	配位效应
coordination formula	配位式；配价式
coordination group	配位基
coordination ion	配离子*
coordination isomerism	配位异构*
coordination lattice	配位格子；配位晶格
coordination link (=coordination linkage)	配位键
coordination linkage (=coordination link)	配位键
coordination number	配位数*
coordination polyhedron	配位多面体
coordination polymer	配位聚合物*
coordination polymerization	配位聚合*
coordination position isomerism	配位位置异构(现象)
coordination reaction	配位反应*
coordination shell	配位壳(层)
coordination sphere	配位层
coordination theory	配位理论
coordination type initiator	配位型引发剂
coordination valence (=coordinative valency)	配(位)价
coordinative activity	配位活性
coordinative polymerization isomerism	配位聚合异构*
coordinative valency (=coordination valence)	配(位)价
coorongite	弹性藻沥青
coot fat	黑鸭脂
cooxen	钴辛〔氢氧化钴乙二胺络合物，$Co(en)_3(OH)_2$，式中 en 代表乙二胺，可做纤维素溶剂〕
cooxidant	辅助氧化剂
co-oxidation	共氧化*
cop	纬管纱；纡子
copaene	珂珀烯
copaiba (=copaiba balsam)	珂珀香脂
copaiba balsam (=copaiba)	珂珀香脂
copaiba oil (=copaiva oil)	珂珀油
copaiva oil (=copaiba oil)	珂珀油
copaivic acid	珂珀酸 $C_{20}H_{30}O_2$
copal	珂珀(树脂)
copal ester	珂珀酯
copal gum	珂珀树胶；珂珀树脂
copalic acid	珂珀酸；黄脂酸
copalinic acid	珂珀脂酸
copal oil	珂珀油
copalolic acid	珂珀油酸
copal varnish	珂珀清漆
copane	胡椒烷
cop dyeing	管纱染色
copel	铜镍合金
copellidine	2-甲基-6-乙基二氢吡啶 $C_8H_{17}N$
copernik	镍铁合金
copiamycin	丰富霉素
copiapite	叶绿矾；铁矾石
coping	顶层；挡板
copious oil supply	溢流润滑作用
coplanar	共(平)面的
coplanar displacement	共(平)面位移
coplanar hydrogen bond	共(平)面氢键
coplanarity	共(平)面性
coplasticizer	辅增塑剂*
copoiva (=copaiba)	珂珀香脂
copolyalkenamer	共聚烯烃
copolyalkylene oxide	共聚氧化烯
copolyamide	共聚多酰胺
copolyamide fibre	共聚酰胺纤维
copolycondensation	共缩聚*
copolyester	共聚酯
copolyesteramide	共聚酯酰胺
copolyester elastomer	聚酯橡胶
copolyesterification	共聚酯化(作用)
copolyether	共聚醚
copolyether ester fibre	共聚醚酯纤维
copolyether ester polymer	醚酯共聚物；共聚醚酯聚合物
copolyhydrazide	共聚酰肼
copolyhydrazide amide	共聚酰肼酰胺
copolyimide	共聚多酰亚胺
copolymer	(二元)共聚物*

copolymer composition 共聚物组成
copolymer composition curve 共聚物组分曲线
copolymer composition diagram 共聚物组分图
copolymer effect 共聚物效应
copolymer fibre 共聚物纤维
copolymeric antistatic agent 共聚(型)抗静电剂
copolymerisate 共聚(合)物
copolymer isomorphism 共聚物同晶型(现象)*
copolymerization 共聚(反应)*
copolymerization activity 共聚合活性
copolymerization behaviour 共聚合行为
copolymerization equation 共聚合方程*
copolymerization kinetics 共聚反应动力学
copolymerization parameter 共聚参数
copolymerization rate 共聚(反应)速率
copolymerization ratio 共聚比
copolymerization reactivity ratio 共聚合(单体)竞聚率；共聚合单体反应性比率
copolymerization temperature 共聚(反应)温度
copolymerized oil 共聚油
copolymer micelle 共聚物胶束*
copolymer of acrylate and butadiene 丙烯酸酯与丁二烯共聚物
copolymer oil 共聚物油
copolymer paint 共聚物型漆；树脂型漆
copolymer rubber 共聚型橡胶；合成橡胶
copolyolefine 共聚烯烃
copolyoxamide 共聚草酰胺
copolyoxymethylene 共聚甲醛
coppe-epoxy type EMI shielding coating 铜-环氧型电磁干扰屏蔽涂料
copperferrite 铁酸铜
copper 铜；紫铜；红铜
copper accelerated acetic acid salt spray test 铜催化乙酸盐雾试验
copper acetate 乙酸铜 $Cu(CH_3COO)_2$
copper aceto-arsenite 翠绿；乙酸铜合亚砷乙酸铜
copper activity 铜活性〔润滑剂侵蚀铜零件的倾向〕
copper aftertreatment 铜后处理
copper alloys 铜合金类
copper alum 铜明矾
copper amalgam 铜汞齐
copper ammonia fibre 铜氨纤维
copper ammonia silk 铜铵丝
copper ammonium acetate process 乙酸铜铵法
copper ammonium chloride 氯化铜铵
copper and manganese contamination 铜锰污染〔生胶〕
copper arsenide 砷化铜 $Cu_3As;Cu_5As_2$

copper arsensite 亚砷酸铜
copperas (=green vitriol) 绿矾；七水(合)硫酸亚铁 $FeSO_4 \cdot 7H_2O$
copper-asbestos gasket 铜皮石棉垫
copperas red (绿)矾制铁红
copper avanturine (=copper aventurine) 铜砂金石
copper azide complex 叠氮络铜盐
copper-base alloy 铜基合金
copper beaker 铜烧杯
copper bearing 含铜的
copper blast-furnace 铜鼓风炉
copper blue 铜蓝 CuS
copper bond 黄铜焊接
copper borate 硼酸铜
copper borate-coated silica 包核硼酸铜
copper bromide 溴化铜
copper bronze ①青铜；古铜②青铜粉；铜青铜〔指有金属光泽的铜合金粉颜料〕
copper cake 铜饼
copper calorimeter 铜量热器；铜热量计
copper carbonate 碳酸铜
copper chloride 氯化铜
copper chloride-oxygen sweetening 氯化铜-氧脱硫醇〔石油〕
copper chloride process 氯化铜精制法
copper chloride solution 氯化铜溶液
copper-chrome black 铜铬黑(颜料)
copper chromite 亚铬酸铜
copper chromite black 亚铬酸铜黑
copper chromite catalyst 亚铬酸铜催化剂
copper-clad conductor 镀铜(超)导体
copper clad laminate 覆铜板
copper coating 铜镀层
copper-constantan couple 铜-康铜热电偶
copper-constantan thermocouple 铜-康铜热电偶
copper-converter gas 铜转气炉
copper corrosion index 铜腐蚀指数
copper corrosion-inhibited grease 防铜腐蚀润滑脂
copper corrosion test 铜板腐蚀试验
copper coulometer 铜(极)电量计
copper crucible 铜坩埚
copper degradation 铜降解
copper dialkyldithiocarbamate 二烷基二硫代氨基甲酸铜〔促进剂〕
copper di-n-butyldithiocarbamate 二正丁基二硫代氨基甲酸铜〔促进剂〕
copper dimercaptobenzothiazole 二硫醇基苯并噻唑铜〔促进剂〕

copper dioxide 过氧化铜 Cu[O₂]
copper dish evaporation test (=copper dish gum test) 铜皿胶质试验
copper dish gum 铜皿胶质
copper dish gum test (=copper dish evaporation test; copper dish residue test) 铜皿胶质试验
copper dish residue test (=copper dish gum test) 铜皿胶质试验
copper dish stability （聚丙烯）铜碟法稳定度
coppered 铜化的；镀铜的；包铜的
copper emerald 透视石；绿铜矿
copper-epoxy type EMI shielding coating 铜-环氧型电磁干扰屏蔽涂料
copper ethylacetoacetate 乙酰乙酸乙酯络铜〔杀菌剂〕
copper ethylphenyldithiocarbamate 乙基苯基二硫代氨基甲酸铜〔促进剂〕
copper family elements 铜族元素
copper ferrocyanide 亚铁氰化铜 $Cu_2Fe(CN)_6 \cdot 7H_2O$
copper flask 铜烧瓶
copper foil 铜箔
copper froth 铜泡石
copper fulminate 雷酸铜
copper fungicide 铜杀霉剂；铜杀(霉)菌剂
copper funnel 铜漏斗
copper furnace 冶铜炉
copper gauze 铜丝网；铜铉网
copper glance 辉铜矿
copper-glass composite fibre 铜-玻璃复合纤维
copper green 铜绿
copper (greening) inhibitor 铜抑制剂；防铜老化剂
copper group 铜组
copper head ①铜头②铜斑
copper hydride 一氢化铜 CuH
copper hydroxy carbonate 碱式碳酸铜
copper index 铜值；铜指数
coppering 镀铜
copper inhibitor 防铜老化剂；铜抑制剂
copper in lubricating oil 润滑油铜含量
copper intensification 铜盐加厚法
copper-iodine anti-oxidant system 铜-碘抗氧剂体系
copper ion selective electrode 铜离子选择电极
copper liquor 铜(水)溶液
copper matte 冰铜
copper mercaptide 硫醇铜 (RS)₂Cu
copper mercaptobenzothiazole 硫醇基苯并噻唑铜〔促进剂〕
copper mordant 铜媒染剂
copper naphthenate 环烷酸铜

copper-nickel-chromium plated steel 镀铜镍铬钢
copper nitride 一氮化三铜 Cu_3N
copper number 铜值
copperon 试铜铁剂
copper ore 铜矿
copper orthosilicate 硅酸铜矿；透视石
copper oxalate 草酸铜 CuC_2O_4
copper oxide 一氧化铜；氧化铜 CuO
copper oxide and zinc accumulator 氧化铜和锌蓄电池
copper oxide cell 氧化铜电池
copper oxide rectifier 氧化铜整流器
copper packing 铜填料
copper paint 含铜船底漆；木船船底漆
copper-PAN indicator 铜-PAN 指示剂；铜潘酚指示剂
copper peroxide 过氧化铜 $Cu[O_2] \cdot Cu(OH)_2$
copper phosphide 磷化铜 Cu_3P
copper phthalo blue 铜酞菁蓝
copper phthalocyanine 铜酞菁
copper phthalocyanine blue 铜酞菁蓝
copper phthalocyanine green 铜酞菁绿
copper plate ①铜板②铜版
copper plated 镀铜的
copper-plated balls 镀铜小球
copper plate singeing machine 铜板烧毛机
copper plating 镀铜
copper powder 铜粉
copper printing roller 紫铜印染辊筒
copper protoxide (=copper suboxide) 氧化亚铜
copper pyrite 黄铜矿
copper 8-quinolinolate 8-羟基喹啉铜
copper ratio 铜值
copper rayon 铜氨螺萦
copper refining 铜精炼
copper resinate 树脂酸铜
copper rhodanide 硫氰酸铜；硫氰化铜
copper rope 铜丝绳
copper ruby glass 铜玉红玻璃
copper salicylimine 水杨基亚胺铜〔硫化剂〕
copper salt antioxidant 铜盐抗氧剂
copper salt method 铜盐法
copper sensitized thermionic detector 铜敏化热离子检测器
copper sequestrating agent 铜抑制剂
copper sesquioxide 三氧化二铜 Cu_2O_3
copper shavings 铜屑
copper sheet 铜片；薄铜板
copper silicide 一硅化二铜 Cu_2Si
copper smelter (=copper smelting furnace) 炼铜炉
copper smelting 炼铜

copper smelting furnace (=copper smelter) 炼铜炉
coppersmith 铜匠
coppersmith's forming iron 锅型铁
copper soap 铜皂
copper solution 铜溶液
copper spark method 铜火花法
copper sponge 海绵(状)铜
copper staining ①铜锈②铜污染；铜斑
copper storage battery 铜蓄电池
copper strip test 铜条试验
copper sub-oxide (=cuprous oxide) 氧化亚铜
copper sulfate 硫酸铜
copper sulfate pentahydrate 五水硫酸铜 $CuSO_4 \cdot 5H_2O$
copper sulfate test 硫酸铜水溶液腐蚀试验
copper sulphate solution mordant 硫酸铜溶液酸洗剂
copper sweetening process 氯化铜脱硫法
copper tack 紫铜钉
copper-tolerant latex 耐铜老化胶乳
copper toning 铜调色法
copper tool 铜刀
copper tubing 铜管；紫铜管
copper turnings 铜刨花
copper uranite 铜铀云母
copper value 铜值
copper-vanadium catalyst 铜-钒催化剂
copper voltameter 铜伏安计
copper welding rod 铜焊条
copper whisker 铜须晶
copper wire 铜丝；铜铉
copper (wire) gauze 铜铉网；铜丝网
coppery ①似铜的②含铜的③铜的
copper zeolite Y 铜 Y 型沸石
copper-zinc accumulator 铜锌蓄电池
Coppet's law 科佩特定律
copping 卷纬
copra 干椰子仁
copra oil 椰子油
copraol 椰子脂
coprecipitated catalyst 共沉淀催化剂
coprecipitation 共沉淀*；共沉淀现象
coprecipitation method 共沉淀法
coprecipitation separation method 共沉淀分离法
coprecipitator 共沉淀剂
coprinin 4-甲氧甲苯醌；鬼伞菌素
co-processing 共处理；同时处理
coproduct 联产品
coprolite 粪化石
coprolithus 粪石

coproporphyrin 粪卟啉
coproporphyrinogen 粪卟啉原
coprostane 粪(甾)烷
coprostanol (=coprosterol) 粪(甾)醇
coprostanone 粪(甾)酮
coprostene 粪(甾)烯
coprostenol 粪(甾)烯醇
coprostenone 粪(甾)烯酮
coprosterol (=coprostanol) 粪(甾)醇
coprosterone 粪(甾)酮
cops ①堆；顶②绕丝轴③圆锥形盘绕纱线
coptine 黄连次碱
coptisine 黄连碱
cop-to-bobbin winding machine 管纱络筒机
copulant 同形结合体
copulated compound 连系化合物
copulation 连系
cop winder 卷纬机
cop winding tension 卷纬张力
copy error 复制错误
copying 仿形切削
copying cutting 仿形切削；靠模切削
copying ink 复写墨
copying machine 仿形机床；靠模机床
copying milling machine 仿形铣床
copying paper (=copying tissue) 复写纸；拷贝纸
copying tissue (=copying paper) 复写纸；拷贝纸
copy machine 仿形机床
copyrine 2,7-萘啶；2,7-二氮杂萘 $C_8H_6N_2$
copyrolysis 共裂解
coquina (介)壳灰岩；贝壳岩
coralinomycin 珊瑚霉素
corallin ①玫红酸②珊瑚精
corallinate (=rosolate) 玫红酸盐
corallite 珊瑚石；珊瑚色大理石
coral ore 珊瑚矿
coral red 珊瑚红（碱性铬酸铅色料）
coral-red glaze 珊瑚红釉
coral reefs 珊瑚矿脉
coramine 尼可刹米；可拉明〔药〕
Corbin chlorate process 科宾氯酸盐制造法
corbisterol 蚬甾醇
corchorin 黄麻因
corchoritin 黄麻亭
corchortoxin 黄麻毒
corcidolite (=blue asbestos) 蓝石棉
cord ①帘布；帘线②条痕③绳④科德〔一种木材体积单位〕

cord body　帘布层；胎体；胎身
cord count　帘布密度
cord dipping　帘子线浸胶；帘线浸胶
cord drive　绳传动
corded fabric　灯芯布
cord fabric　帘子布；帘布
cord factor　索因子
cordial　①强心药②浸果酒
cordierite　堇青石
cordierite-based glass fibre　堇青石基玻璃纤维
cord-in-rubber test　橡胶内帘子线试验
cordite　柯达炸药；柯戴特〔硝棉、硝化甘油、石油脂炸药〕
cord lock-ups　帘布卷边
cordol　帘布酚；水杨酸三溴苯酯　$C_6H_4OHCOOC_6H_2Br_3$
cord ply　帘线层；帘布层
cord roller　导纱辊
cord tyre fabric　轮胎布，帘布
cord-wood　层积材〔以科德（即128立方英尺）为单位出售的小木料和枝材等〕
cordycepic acid　虫草酸　$C_7H_{12}O_6$
cordycepin　虫草品〔取自冬虫夏草属的胃虫草〕
cordycepose　虫草糖　$OHCCHOHCH(CH_2OH)_2$
core　①心层②铁心③原子实④(中)心⑤心板〔层压木板的〕
coreactant　共反应剂
coreaction　共反应〔指三种以上反应物同时反应〕
coreactivity　共反应性
coreagent　共反应剂
core alignment machine　铁心矫正器
core binder (=core oil)　型心黏结剂
core binder pitch　型心黏结沥青
core box　芯盒
core building machine　筒制机；筒模接合(橡胶物品)机
core building method　筒制法；筒模接制法
core-corona structure　核冠结构
core diameter　心直径
cored mold　芯式塑模
coreduction　同时还原；共还原
core ejector　脱芯机
core electron　芯电子·
core flow　核心流
core gap　芯板离缝
core hardness tester　核心硬度试验机
core insert　模芯嵌件
corelborin　噻根苷
core material　芯材
core mercerization　透芯丝光；纤(维)芯(子)丝光
core method　筒制法；筒模(制轮胎)法

core mould process　(泡沫塑料)夹芯成型法
core of paste　泥心〔陶瓷〕
core oil (=core binder)　型心黏结剂
core pigment　包核颜料
core-pin　成孔销
corepressor　辅阻遏物
core print　砂芯头；泥心头
core resistance　中心流体的阻力；(湍流)核心阻力〔物〕
core-sheath method　皮芯法〔复合纤维成形法之一〕
core sheet　塑料贴面板芯
core shell copolymer　核-壳共聚物
core-shell-corona structure　核-壳-冠结构
core-shell emulsion　芯-壳乳液(胶)
core-shell emulsion polymerization　芯-壳乳液聚合
core shell latex polymer　壳胶乳聚合物
core-shell model　核-壳模型
core-shell polymerization　核-壳聚合
core-shell type　核壳型〔复合半导体纳米微粒的结构型式之一〕
core-skin nucleus model　核心表皮核模型
core type latex (=core-shell type latex)　芯-壳型胶乳
core varnish　硅钢片清漆；磁心清漆
core wire　铁丝芯骨
core wood　薪柴
coriaceous　①如革的②革的
coriamyrtin　马桑内酯
coriander　芫荽〔调香药物名〕；香菜
coriander oil　芫荽油
coriander oleoresin　芫荽油树脂
coriander seed oil　芫荽子油
coriandrol (=linalool)　芫荽(萜)醇
coriarine　马桑素
Cori ester (=glucose-1-phosphate)　柯里酯葡萄糖-1-磷酸酯〔酒精发酵过程中的中间产物〕
corilagin　鞣料云实素；1-棓酰-3,6-二羟基联苯二甲酰基葡萄糖　$C_{27}H_{22}O_{18}$
coring　抽芯
corinnic acid　柯扔酸
Coriolis splitting　科里奥利谱线分裂
cork　①软木塞②软木③塞
cork acid　软木酸；辛二酸〔俗名〕
cork arm graining board　软木搓花板
cork black　软木(炭)黑
cork board　软木板
cork borer　软木钻孔机
cork borer sharpener　穿孔器削锋刀
cork boring machine　穿孔机
cork brick　多孔砖；软木块

cork-composition gasketing　软木胶垫片
cork drill　(木塞)钻孔器
cork dust　软木粉〔防冷凝液填料〕
cork elm　栓皮榆
corker　封塞机；木塞压紧机
cork extractor　拔(软木)塞器
cork fender　软木护板
cork flour　软木粉
cork gauge　塞径规
cork-insulated metal pedestal　填软木的隔热金属架(底座)
cork joint　软木结合
cork lacquer　软木漆
cork mat(t)　软木垫
cork-packed gland　用软木填充的衬垫
cork press　压(软木)塞器
cork presser　压(软木)塞器
cork puller　拔(软木)塞器
cork ring　软木环
cork screw　螺旋拔塞器
corkscrew curl fibre　螺旋形卷曲
corkscrew weft weave　纬面螺旋纹织物
cork sheet　软木
cork stopper　(软)木塞
cork tile　软木砖
cork tree bark　黄柏
corkwood　黄蘗属；黄柀椤属；软木
corn cob　玉米穗轴
corn cob grits　玉米棒屑
corncob particles　玉米穗轴碎块；碎玉米穗轴
corn couckle　麦仙翁
corn distillery　玉米酒厂
cornelian　光玉髓
corner　①角；隅；棱②弯头；弯管
corner angle　顶角；棱角
corner defect　棱角缺陷；缩边露角〔漆病〕
cornering　(轮胎)耐急转弯性能
corner joint　角接接头
corner pockets　死角
corner sample　切角样品
corner sampling　切角取样
cornerslick　角光子
corner weld　角焊
cornerwise　对角的；斜交的
Cornet forceps　科尔内特镊
corn fatty acid　玉米油脂酸
corn fibre　玉米蛋白质纤维
corn flour mill　玉米粉磨
cornic acid (=cornin)　梾木酸

corniferous rock　角页岩；中部泥盆层岩
cornin (=cornic acid)　梾木酸
cornine　梾木苷
corning　制成粒
Corning filter　康宁滤光片
Cornish distribution　科尼希分配
Cornish powder　科尼希(炸)药
Cornish stone　科尼希石
cornit　硫化煤黑〔防锈颜料〕
cornite　柯恩炸药；柯珞那特〔氯酸钠、三硝基甲苯炸药〕
corn mill　玉米磨
corn oil　玉米油
corn oil fatty acid　玉米油脂肪酸；玉蜀黍油脂肪酸
corn protein fiber　玉米蛋白质纤维
corn protein plastic　玉米蛋白质塑料
corn remover　脱鸡眼剂；脱老茧皮剂
corn spar　结晶方解石
corn stalk　玉米杆
corn starch (=amylum)　(玉米)淀粉
Cornu mounting　考纽装置
cornuoid　梾木合素
Cornu prism　考纽棱镜
cornutine　低麦角碱
cornutine citrate　柠檬酸低麦角碱
cornutol　麦角流浸膏
coromat　包在管外防止腐蚀的玻璃丝
coromegine (=atropine)　阿托品
corona　①电晕②日冕③冠
coronadite　铅硬锰矿
coronal　①花环；冠②冠(状)的
corona plasma　电晕等离子体
corona resistance　①电晕电阻②耐电晕性
coronaridine　冠狗牙花定
coronarine　狗牙花碱
corona voltage　电晕电压
coronene　蒄；晕苯　$C_{24}H_{12}$
coronillin　小冠花苷
coronising　科伦奈津整理
corossol　刺果番荔枝
corotoxigenin　冠花毒配基
corpaverine　柯杷(魏)碱
corporin (=progesteron)　激孕甾酮
corps　精制脂肪
corpse light　危兆光；鬼火〔俗称〕；磷光
corpuscle　①微粒；粒子②血球
correct boiling range　校正的馏程；校正的沸程
correct count　修正支数
correct cure　最适硫化；正硫化

correct depth　正确深度
correct dipstick level　(计量板上)标准水准的记号
corrected area　校正面积
corrected fibre number　修正纤维支数〔折合成标准回潮率的支数〕
corrected fluorescence excitation spectrum　校正荧光激发光谱
corrected fluorescence spectrum　校正荧光光谱
corrected gas hold-up volume　校正气体保留体积
corrected grain　修正粒面
corrected oil　合格油
corrected retention distance　校正保留距离
corrected retention time　校正保留时间*
corrected retention volume　校正保留体积*
corrected reticulocyte count　网织红血球数校正值
corrected specific fuel consumption　校正比燃料消耗率
corrected temperature　校正温度
corrected yarn number　修正纱线支数
correcting factor (=correction factor)　校正因子
correction　校正*
correction by flame　火焰矫正
correction by sensitiveness　灵敏度校正
correction chart　校正图表
correction factor (=correcting factor)　校正因子
correction factor for pressure　压力校正因子
correction for buoyancy　浮力改正
correction formula　改正公式
correction method　校正(方)法
correction of fixed bias　固定偏倚的校正
correction of relative bias　相对偏倚的校正
correction rate　校正率
correction temperature　校正温度
correction term　修正项
correction value　校正值；修正值
correct level　校正水准；标准水准
correct mixture　标准气体混合物
correctness　正确度
correct oil　合格油
corrector roll　校正辊
correct size　公量纤度
correlated　相关的；对比的
correlated (color) temperature　相关色温
correlated electron pair approximation (CEPA)　相关电子对近似
correlated index　相关指数
correlated molecular motion　相关分子运动
correlated sweep excitation　相关扫描激发
correlation　相关性；相符性
correlation analysis　相关分析*
correlation coefficient　相关系数*
correlation coefficient in population　总体相关系数
correlation coefficient in sample　样本相关系数
correlation data　对比数据
correlation diagram　相关图
correlation factor　相关因子
correlation function　相关函数*
correlation index　对照索引；相关索引
correlation matrix　相关矩阵*
correlation method　对比法；比较法
correlation spectroscopy　关联能谱法
correlation test　相关性检验*
correlation time　相关时间*
correlator　相关器
correlogram　相关(曲线)图
correspondence　①对应②对比③相当
correspondence factor analysis (CFA)　对应因子分析
correspondence principle　对应原理
corresponding point　对应点；相应点
corresponding pressure　对比压力
corresponding rating　对比等级；对应等级
corresponding solution　对应溶液*
corresponding state　对应状态*
corresponding temperature　对比温度
corresponding viscoelastic state　对比黏弹态
corresponding volume　对比体积
corrigent　①矫正的②矫正药
corrin　咕啉*
corrinoid　类可啉
corroded crystal　溶蚀(斑)晶；蚀缘晶体
corroded iron　锈蚀铁
corroded metal coating　腐蚀的金属涂层
corroded surface　①腐(锈)蚀的表面②受伤的表面
corrodent　腐蚀剂
corrodibility　可腐蚀性
corrodible　可被腐蚀的
corroding　腐蚀
corroding agent　腐蚀剂
corroding chamber (=corroding house)　腐蚀室
corroding electrode　腐蚀电极
corroding house (=corroding chamber)　腐蚀室
corrod(o)kote paste　腐蚀膏〔金属腐蚀试验用〕
corrod(o)kote test　腐蚀膏试验
corronel　耐蚀镍钼铁合金
corrosion　腐蚀；侵蚀
corrosion accelerator　腐蚀加速剂
corrosion agent　腐蚀剂

corrosion allowance　腐蚀余度
corrosion at initial boiling point　初沸点腐蚀性〔样品温度等于初沸点温度时对铜片的腐蚀性〕
corrosion behavior　腐蚀行为
corrosion by common salt　食盐腐蚀
corrosion by gas　气体腐蚀
corrosion by non-freezing solution　不冻液腐蚀
corrosion by oxygen blowing　吹氧腐蚀
corrosion by restoration water　回收水腐蚀
corrosion by saturated dissolved hydrogen　吹氢腐蚀；饱和氢溶液腐蚀
corrosion cell　腐蚀电池
corrosion chamber (=burner rig)　燃烧试验装置；腐蚀室
corrosion contaminant　腐蚀污染物
corrosion cracking　腐蚀龟裂；腐蚀断裂
corrosion current　腐蚀电流*
corrosion depth　腐蚀深度
corrosion due to welding　焊接腐蚀
corrosion during distillation　蒸馏时(铜片上)的腐蚀
corrosion embrittlement　腐蚀脆变
corrosion environment　腐蚀环境
corrosion fatigue　腐蚀疲劳
corrosion fatigue cracking　腐蚀疲劳龟裂
corrosion fatigue limit　腐蚀疲劳极限
corrosion fatigue test method　腐蚀疲劳试验法
corrosion figure　蚀图；蚀像
corrosion film　腐蚀膜
corrosion form　腐蚀形(状)
corrosion gas　腐蚀性气体
corrosion gauge　腐蚀计
corrosion ga(u)ge point　腐蚀检测点；腐蚀测定点
corrosion index　腐蚀指数
corrosion-inhibited fuel　防腐蚀燃料
corrosion-inhibiting　耐腐蚀的
corrosion-inhibiting coating　防腐蚀涂层
corrosion-inhibiting film　腐蚀抑制膜；缓蚀膜；防腐蚀膜
corrosion-inhibiting layer　缓蚀层；腐蚀抑制层；防锈层
corrosion inhibition　腐蚀抑制(作用)
corrosion inhibitive function　腐蚀抑制功能；缓蚀功能
corrosion inhibitive pigment　腐蚀抑制性颜料
corrosion inhibitor　缓蚀剂*
corrosion loss　腐蚀损耗
corrosion monitoring　腐蚀监测
corrosion-oxidation attack　腐蚀氧化侵蚀
corrosion pacifier　缓蚀剂
corrosion penetration　腐蚀深度
corrosion pit　腐蚀斑点；侵蚀点
corrosion plate target　腐蚀板极(阳极)靶

corrosion potential　腐蚀电势*
corrosion prevention　腐蚀防护
corrosion preventive　①防腐蚀②防腐剂
corrosion preventive compound　防腐化合物
corrosion-prone　易于腐蚀的
corrosion-proof　耐腐蚀的
corrosion protection　防腐*
corrosion protection value　防腐蚀值
corrosion rate　①腐蚀速率②腐蚀率
corrosion rate numbering　腐蚀速率计数系统
corrosion rate-ocean current diagram　海水流速-腐蚀速度图
corrosion rate of zinc　锌腐蚀率
corrosion remover　防腐剂
corrosion resistance　抗蚀性；耐蚀型
corrosion resistance measurement　耐腐蚀性测定
corrosion resistant (=corrosion resisting)　耐腐蚀的
corrosion resistant coating　耐腐蚀涂料；防腐蚀涂料
corrosion-resistant finishes　①耐腐蚀漆②耐腐蚀涂层
corrosion-resistant water-reducible alkyd coatings　水稀释性醇酸耐(防)腐蚀涂料
corrosion resisting (=corrosion resistant)　耐腐蚀的
corrosion resisting alloy　抗腐蚀合金
corrosion resisting steel　抗腐蚀钢；耐蚀钢
corrosion resistivity　耐蚀性
corrosion spool　腐蚀试片
corrosion stability　耐腐蚀性
corrosion strength　腐蚀强度；腐蚀力
corrosion system　腐蚀体系
corrosion-tank finishing process　(酸)浸蚀工艺〔碱式碳酸铅的生产工艺〕
corrosion target　腐蚀电极；腐蚀阳极靶
corrosion test　腐蚀试验
corrosion test for oils　油类腐蚀试验
corrosion testing of lubricating oils　润滑油腐蚀试验
corrosion type　腐蚀类型
corrosion weight loss　腐蚀减重
corrosiron　耐(腐)蚀铸铁；科伦西朗耐腐蚀硅钢
corrosive　①腐蚀剂②腐蚀(性)的
corrosive atmosphere　腐蚀性空气
corrosive chemical fume　腐蚀性化学烟雾
corrosive enamel　防腐磁漆
corrosive film　腐蚀(性)膜
corrosive mercuric chloride　氯化汞；升汞　$HgCl_2$
corrosive nature　腐蚀性能；腐蚀作用
corrosiveness　腐蚀性
corrosive poison　腐蚀毒物；腐蚀抑制
corrosive sample　腐蚀性试样

corrosive sublimate 升汞；氯化汞
corrosive sulfur 腐蚀性硫黄
corrosive wear 腐蚀损耗；蚀耗
corrosivity 腐蚀性
corrosometer 腐蚀性测定计；测蚀计
corrugate ①起皱；起波纹②波(纹)状的
corrugated 波纹的；起皱的
corrugated board 波面纸板；瓦楞纸
corrugated box board 波纹盒纸板
corrugated buff 瓦楞形抛光轮
corrugated container board 瓦楞箱纸板
corrugated edge 荷叶边
corrugated expansion joint 波形伸缩接头
corrugated gauze packing 波纹金属网填充物
corrugated hose 螺纹胶管
corrugated iron 瓦楞铁皮
corrugated metal joint ring 波形金属接头环
corrugated mixing rolls 皱纹混合磨
corrugated panel 皱纹板；瓦楞板
corrugated pipe 波纹管
corrugated plate interceptor (CPI) 波纹板拦截器
corrugated ring 波纹环
corrugated roll 槽纹滚筒
corrugated roll crusher 皱纹滚碎机
corrugated sheet 波纹板
corrugated sheet iron 波纹铁板
corrugated stitcher 波式缝纫机
corrugated tube 波形管
corrugated-type expansion joint 波形膨胀节
corrugating medium 瓦楞夹心原纸
corrugation 起皱
corrugation former 瓦楞成型机；瓦楞成型模
corrutgated box board 波纹盒纸板
cortepinitannic acid 皮松单宁酸 $C_{32}H_{34}O_{17}$
cortex 皮质；皮层
cortexolone 11-脱氧皮(甾)醇
corticinic acid 栓皮酸 $C_{12}H_{10}O_6$
corticotrop(h)in 促肾上腺皮质(激)素
corticotropic hormone (=adrenotropic hormone) 促肾皮素
corticotropin releasing factor (CRF) 促(肾上腺)皮质(激素)释放因子；促皮质释放素
cortisone (=11-dehydro-17-hydroxycorticosterone) 11-脱氢-17-羟皮质(甾)酮；可的松；皮质酮
cortisone acetate 乙酸可的松
cortisone-oral glucosetolerance test 可的松口服葡萄糖耐量实验
corundum 刚玉*；金刚砂
corundum abrasive 刚玉磨料

corundumite 刚玉
corundum ware 刚玉(制)实验仪器
coruscation 闪烁；闪光
corvoline 碱性乌鸦黑
corybulbine 紫堇鳞茎碱；紫堇球碱
corycavamine 紫堇胺
corycavidine 紫堇维定
corycavine 紫堇文碱
corydaline 延胡索碱；紫堇碱
corydine 紫堇丁
coryfin 可力芬；乙基乙醇酸薄荷酯
corynantheidine 柯楠碱
corynantheine 柯楠因
corynanthic acid 柯楠酸
corynanthidic acid 柯楠低酸
corynanthidine 柯楠定
corynanthine (=rauhimbine) 柯楠质
coryneine 棍掌碱
corynine (=yohimbine) 可立宁；育亨宾
corypalline 紫堇杷灵
corytuberine 紫堇块茎碱
cosanates 二十级(烷)酸盐(或酯)
cosanes 念烷系〔C_{20} 到 C_{29} 的链烷〕
cosanic acid 念酸〔由 C_{20} 到 C_{29} 的二十级(烷)酸〕
cosanols 二十级(烷)醇
cosaprin 对乙酰氨基苯磺酸钠
cosegment 共聚链段
coseparation 共分离；同时分离
coset 陪集；旁系
cosine 余弦
coslettising process 科斯赖特法；磷酸铁处理法
coslettizing 磷酸铁护铁法；磷化
cosmane 波斯菊烷
cosmene 波斯菊萜；2,6-二甲基-1,3,5,7-辛四烯
cosmetic ①化妆品；润肤剂②美容的
cosmic ①宇宙的②广大无边的
cosmic abundance 宇宙丰度
cosmic inventory 宇宙万物
cosmic iron 陨铁
cosmic particle 宇宙(微)粒
cosmic radiation 宇宙辐射
cosmic ray (=ultra-gamma ray) 超 γ 射线；宇宙线
cosmochemistry 宇宙化学
cosmogenic 宇宙发生的；由宇宙射线产生的；来自宇宙线的
cosmogenic radionuclide 宇生放射性核素
cosmos oil 波斯菊油
CO_2 snow (=dry ice) 干冰

cosolubility 共增溶性
cosolubilization 共加溶(作用)
co-solubilizer 辅助加溶剂
cosolute 共溶质
cosolvency ①共溶性②共溶本领；共溶度
cosolvent 共溶剂；助溶剂
cosolvent effect 共溶效果
cosolvent-rich solution 富(含)助溶剂溶液
cospinning 共纺*
cossaite 致密钠云母
cossette 甜菜丝
cost 成本；投资；费用
co-stabilizer 共稳定剂
cost accounting 成本核算
costaclavine 肋麦角碱
cost aluminum pressure head 铸铝加压头；铸铝压力分配头
cost and freight 生产费用和总运输费
cost control 成本控制
costen 木香烯 $C_{15}H_{24}$
costene 广木香烯
cost estimating 成本估算
cost-freight-insurance 生产费用、总运输费及保险费
costing formula 成本配方
cost of development 开发费用；开发成本
costol 木香醇 $C_{15}H_{24}O$
costrip 共反萃
costunolide 木香烃内酯
costus absolute 广木香净油
costusic acid 木香酸 $C_{15}H_{22}O_2$
costuslactone 木香内酯
costus root oil 广木香根油
cosulfonate 共磺化(物)
cosurfactant 辅助表面活性剂
cosy 保温套
cosyl 二十级(烷基
cotarnic acid 可他酸
cotarnin(e) 可他宁
cotarnone 可他酮
cotectic 共结的；低共熔线
cotelomer 共调聚(合)物
coticula 砥石
coto 柯桃〔或指其树皮而言〕
coto bark 寇托木皮
cotogenin 柯桃配基
cotoin 柯桃因
cotruder 双螺杆挤出机
cottage rose 白玫瑰

cotton ①棉②棉的
cotton and wool mixtures 棉毛混纺织物；棉毛交织物
cotton asbestos 石棉
cotton balls〔复〕 硼钠钙石
cotton blue 棉染蓝〔染〕
cotton blushing 棉致发白现象；硝基漆发白〔漆病〕
cotton cake 棉子饼
cotton classification room 原棉分级室
cotton detergent 棉织品洗涤剂
cotton dull thread 棉无光缝纫线
cotton dye 棉染料
Cotton effect 卡滕效应*
cotton fabric 棉织物；棉布
cotton fiber 棉纤维
cotton fibre fineness and maturity tester 二次压缩法棉纤维细度成熟度测定仪
cotton filter tube 棉絮过滤管
cotton flock 棉绒
cotton-free 不黏棉花〔一种表面干燥情况〕
cotton gin 轧棉机
cotton glazed thread 棉蜡光缝纫线
cotton gum 棉胶
cotton gum tree 胶皮糖香树
cotton hydrocellulose 水解棉纤维素
cottonin 棉化麻纤维；棉化(韧皮)纤维
cottonizing ①用棉花包裹②棉化
cotton lamp wicks 棉灯芯
cotton linter 棉绒
cotton mercerized thread 棉丝光缝纫线
cotton mixing 和花；混棉
cotton oil 棉油
cotton orange 棉染橙
cottonous (=cottony) 似棉的；柔软的；棉的
cotton packing 棉绳辫填料；棉质衬垫；棉质密封件
cotton plant oil 棉叶油
cotton plug 棉絮塞
cotton plugged 棉絮塞住的
cotton press 榨棉机
cotton printing 棉布印花
cotton printing mill 棉布印花(工)厂
cotton rag (擦漆用)棉擦布；抹布
cotton-rayon blend 棉-人造棉混纺
cotton red 棉染红
cotton scarlet 棉染猩红
cotton scrim strip 棉质纱布胶带
cotton seed 棉子
cotton seed cake 棉子饼
cotton seed fatty acid 棉籽脂酸〔活性剂、分散剂、乳

化剂〕
cotton seed hull　棉子壳
cotton seed hull ash　棉子壳灰
cotton seed meal　棉子粉
cotton seed oil　棉子油
cotton seed oil soap　棉油皂
cotton seed oil stearin　棉籽油硬脂精；棉硬脂
cotton silk union　棉丝联织物
cotton-spun acrylic　棉型丙烯腈系纤维
cotton stopper　棉絮塞
cotton waste　废棉(纱)；废花
cotton waste filter　废棉滤器
cotton wax　棉蜡
cotton weed oil　棉田除草油
cotton wool　棉绒
cotton-wool filter　棉绒滤器
cottony (=cottonous)　似棉的；柔软的；棉的
cotton yellow　棉染黄〔染〕
Cottrell's ebulliometer　科特雷尔沸点计；沸点酒精计
Cottrel(l) dehydrator　科特雷尔脱水器
Cottrel(l) gas filter (=Cottrel(l) gas cleaner)　科特雷尔净气器；气体电滤器
Cottrell molecular weight determination apparatus　科特雷尔分子量测定仪
Cottrel(l) precipitator　科特雷尔沉淀器；科特雷尔聚尘器
Cottrel(l) process　科特雷尔法；静电气体净制法
Cottrell pump　科特雷尔泵
Coubrough process　考伯劳过程
couch　①层②床③压出
couch board　多层纸板
coucher　伏工
couch press　横式挤压机
couch roll　伏辊
couch together　层叠
Couette's law　库爱特定律
Couette correction　库爱特改正
Couette flow　库爱特流动
Couette-Hatschek cup/bob rheometer　库爱特-哈谢克杯/摆锤式流变仪
Couette loss　库爱特损失
Couette-Margules' law　库爱特-玛古累斯定律
Coulogravimetric analysis　库仑重量分析
Coulomb's friction　库仑摩擦
Coulomb's law　库仑定律
Coulomb's yield criterion　库仑屈服判据
Coulomb barrier　库仑势垒；库仑位垒
Coulomb explosion　库仑爆裂
Coulomb hole　库仑穴*

Coulombian force　库仑力
Coulombic attraction　库仑引力
Coulombic barrier energy　库仑能障
Coulombic energy　库仑能*
Coulombic force　库仑引力
Coulombic interaction　电荷相互作用
Coulombic repulsion　库仑斥力
Coulomb integral　库仑积分*
Coulomb interaction energy　库仑相互作用能
coulombmeter　库仑计；电量计
coulometer (=voltameter)　库仑计；电量计
coulometric analysis　库仑分析；电量分析
coulometric analysis with constant current　恒电流库仑分析
coulometric detector　库仑检测器
coulometric determination　库仑定量法
coulometric moisture meter　库仑水分计；电量水分计
coulometric monitor　库仑监测器
coulometric titration　库仑滴定；电量滴定
coulometric titration with externally generated reagent　外生试剂库仑滴定；外生试剂电量滴定
coulometric titrator　库仑滴定装置
coulometry　库仑法；电量法
coulopotentiography　库仑电势谱法
coulostatic method　恒电量法*
coumachlor　氯杀鼠灵
coumalic acid　阔马酸　C_6H_4O
coumalin　香豆灵；邻吡喃酮
coumaraldehyde　香豆醛　$OHC_6H_4CH=CHCHO$
coumaran　香豆冉；苯并二氢呋喃　C_8H_8O
coumarandione　香豆冉二酮；苯并二氢呋喃二酮
coumaranone　香豆冉酮；苯并二氢呋喃-3-酮
coumaric acid (=coumarinic acid)　香豆酸；羟苯基丙烯酸　$HOC_6H_4CH=CHCOOH$
coumaric aldehyde　香豆醛　$HOC_6H_4CH=CHCHO$
coumarilic acid　香豆基酸；苯并呋喃-2-甲酸　C_8H_5OCOOH
coumarin　香豆素
coumarin-3-carboxylic acid　香豆素-3-甲酸
coumarinic acid　香豆酸；羟苯基丙烯酸　$HOC_6H_4CH=CHCOOH$
coumarinic anhydride　香豆酸酐
coumarinic lactone　香豆素
coumarone (=cumarone)　苯并呋喃；香豆酮
coumarone-indene resin　苯并呋喃-茚树脂*
coumarone resin　苯并呋喃树脂
coumaroyl guinic acid　香豆酰奎尼酸　$C_{16}H_{18}O_8$
coumermycin (=cumermycin)　香豆霉素
coumothiazone　香豆噻嗪

count ①编织密度②支数；号数〔纺〕③计算计数
count conversion factor 支数换算系数
counter ①鞋后帮②反③计数器
counteractant 冲消剂〔除恶臭〕
counteracting ①反作用；抵消②平衡；消解
counter anion 抗衡负离子
counter balance 托盘天平
counterbalance cell 平衡池
counterbalance valve 平衡阀
counter blade 对刃
counterbore ①沉孔②埋头钻；钳口钻；平底扩孔钻
counter-clockwise 反时针方向的
counter-clockwise rotation 反时针回转
counter-clockwise turning 反时针方向旋转
counter-clockwise twist 反手捻；Z 捻
counter-current 逆流；反流
counter-current absorber 逆流吸收塔
counter-current action 逆流效应
counter-current azeotropic distillation 逆流共沸蒸馏
counter-current cascade 逆流级联
counter-current cell 逆流电池
counter-current centrifuge 逆流离心机
counter-current chromatograph 反流色谱仪
counter-current chromatography 反流色谱法*
counter-current condenser 逆流冷凝器
counter-current decantation 逆流倾析
counter-current decantation system 逆流倾析系统
counter-current distribution 反流分配*
counter-current distribution apparatus 反流分配仪；反流分布装置
counter-current distribution method 逆流分配法
counter-current drying 逆流干燥
counter-current electrophoresis 对流电泳*
counter-current extraction 逆流萃取*
counter-current flow 逆流流动
counter-current fractionation 逆流分级
counter-current heat exchanger 逆流热交换
counter-current immunoelectrophoresis 反流免疫电泳
counter-current ion exchange 反流离子交换
counter-current ionphoresis 反流离子电泳(法)
counter-current jet condenser 逆流淋凝器
counter-current multiple contact operation 反流多级接触操作
counter-current multiple extractor 反流多级抽提装置
counter-current multi-stage extraction 反流多级萃取；逆流多级萃取
counter-current nitration 反流硝化
counter-current operation 反流运行；反流操作

counter-current pipe exchanger 反流管状换热器
counter-current principle 逆流原理
counter-current process 反流过程；逆流过程
counter-current scrubbing 逆流洗涤
counter-current separation process 反流分离过程
counter current soap boiling 逆流(洗涤)煮皂(法)
counter-current stripper 逆流气提塔；逆流解吸塔
counter-current tower 逆流塔
counter-current tray 逆流(塔)盘
counter-current treatment 逆流处理
counter-current washing 逆流洗涤
counter-current-wise 逆流地
counter-diffusion 逆扩散
counter electrode 对电极；反电极
counter-electromotive force 反电动势
counterelectrophoresis (CE; CEP) 对流电泳
counter extraction 逆萃取；反抽提
counter flange 对接法兰；过渡法兰
counter flow 对流；反流；倒流
counter-flow condenser 逆流冷凝器
counter-flow cooling tower 逆流式凉水塔
counter-flow drier 逆流干燥器
counter-flow heat exchanger 逆流式换热器
counter flow jet 对流喷嘴
counterflush 反向冲洗；反向洗涤
counter force 反作用力
counter (gegen) ion 反荷离子*
counterimmunoelectrophoresis (CIEP) 对流免疫电泳
counter ion 抗衡离子；相反离子；反离子；对离子
counterion effect 反离子效应
counter irritant 抗刺激剂
counter motion 反向运动
counterpart 配对物；对应物
counterpoise 配衡(体)
counterpoised balance 配重天平
counterpoise weighing 置换称量法
counter pressure 反压
counter propagation 反向传播
counter reaction 逆反应
counter-rotating peddle 逆转式桨叶
counter-rotating propeller 逆向转动螺旋桨；逆向转动推进器
counter-rotation 反转
countershaft 副轴；天轴
countersink 喇叭孔
counter solvent 反萃溶剂
counter spectrometer ①计数器能谱仪②计数器光谱仪
counter spectrum 等位光谱

counterstain 复染色；对比染色
countersunk rivet 隐头铆钉
counter tube 计数管
counterweight ①平衡(重)锤；配重②砝码
counting assembly 计数装置
counting cell 计数池
counting dish 计数盘
counting efficiency 计数效率
counting error 计数误差
counting glass 织物分析镜；织物密度镜
counting instrument ①计数器②计量器
counting loss 计数损失
counting rate 计数率*
counting rate meter 计数率计
counting room 计数室
counting time 计数时间
countra-flow 逆流；回流；反流
count rate 计数率；计数比
country rock 原岩；主岩
country tar (=kiln burned pine tar) 松焦油；土窑松焦油
counts corrected for chemical yield (CCY) 化学产额校正计数
counts corrected for coincidence (CCC) 符合校正计数
counts per minute 每分钟计数
count wool 支数毛
Coupier's blue 考皮尔蓝
couplage 偶联
couplant 偶联剂
couple ①偶；对②偶联③力偶；电偶
couple action 电偶作用
couple agent 偶联剂
coupled ①偶合的；耦合的②成对的③联结的④偶联的
coupled action 偶联反应(作用)
coupled column 联用柱；多级柱
coupled dimer 偶联二聚物
coupled electron 偶合电子
coupled instrument system 联用仪器系统
coupled length 带管接头的长度〔胶管〕
coupled oxidation 偶联氧化*
coupled oxidation-reduction 偶合氧化还原
coupled-pair many-electron theory (CPMET) 耦合电子对多电子理论*
coupled phosphorylation 偶联磷酸化(作用)
coupled reaction 偶合反应
coupled simultaneous technique 偶合联用技术
coupled stress 偶合应力
coupled type 偶合型〔复合半导体纳米微粒的结构型式之一〕
couple-electron pair approximation (CEPA) 耦合电子对近似*
coupler ①偶合剂②成色剂③连接器
couple stress fluid 偶合应力流体
coupling 偶合*
coupling agent 偶联剂*
coupling amine 偶合胺
coupling amplifier 耦合放大器
coupling compound 偶合化合物
coupling constant 偶合常数*
coupling dyestuff 偶合染料
coupling effect 偶合效应
coupling factor 偶联因子
coupling finish 偶联处理剂
coupling frequency 耦合频率
coupling joint ①联轴节②联管节
coupling of line 管路偶联管
coupling of momentum 角动量的耦合
coupling pair 偶合线对
coupling polymerization 偶联聚合*
coupling process 偶合法
coupling program 联合程序
coupling reaction 偶联反应
coupling reaction chemiluminescence 偶合反应化学发光
coupling reagent 偶合试剂
couplings 连接器
coupling scheme 偶合图像
coupling shaft 联接轴联轴
coupling size 增强型浸润剂；偶联型浸润剂
coupling solvent 偶合溶剂
coupling spanned space 跨越空间的耦合
coupling technique 联用技术
coupling termination 偶合终止*
coupling value 偶合值；偶联值
coupling viscosity 偶合黏度
coupon 取样管
coupon location 试样点
course 过程
course of reaction 反应过程
course plate 导流板；流向板
course spacing 圈高
courtzilite 一种沥青变态物
covalant carbides 共价型碳化物
covalant character 共价特性
covalant hydrides 共价型氢化物
covalant oxides 共价型氧化物
covalence 共价
covalency 共价(性)

covalent 共价的
covalent bond (=covalent link; covalent linkage) 共价键
covalent carbide 共价碳化物
covalent combination 共价(键)化合
covalent complex 共价配合物
covalent compound 共价化合物
covalent coordination bond 共价配位键*
covalent crystal 共价晶体*
covalent force 共价力
covalent formula 共价式
covalent interaction 共价相互作用
covalent link (=covalent bond) 共价键
covalent linkage (=covalent bond) 共价键
covalently bound surface modifying agent 共价键表面改性剂
covalently linked 以共价键联系的
covalent molecule 共价分子
covalent radius 共价半径*
covalent ring structure 共价环状结构
covalent union 共价结合；共价(键)化合
covar 〔商〕 科伐(铁镍钴)共膨胀合金
covariance 协方差*
covariance matrix 协方差矩阵
covariant derivative 协变导数
covariant differentiation 协变微分
covariation 相关变异；并发变异；协变
covelline (=covellite) 铜蓝；靛铜矿
covellite (=covelline) 铜蓝；靛铜矿
cover 盖；套
coverage 覆盖度
coverage density ①覆盖密度②表面电荷密度
coverage factor 覆盖因子
coverage rate 覆盖率；涂布面积；遮盖力
cover coat 表护层；保护层
cover coat enamel ①面层珐琅；面釉②罩面瓷漆
covered ①涂抹了的②覆盖了的；掩盖了的
covered arc welding 手工电弧焊
covered elastic yarn 包纱橡筋线
covered yarn 固纱
cover glass ①保护玻璃②盖玻片
covering ①掩盖②套
covering agent 遮盖剂
covering flux ①覆盖熔剂②覆盖焊药
covering machine 包线机
covering power 遮盖力；覆盖能力
covering strip 盖板；盖条
cover layer (=cover lay) 覆盖层；包封膜
cover light ①涂层过薄②过轻
cover mould 盖模；前模
cover paper 书面纸
cover plate 盖板
coverplate oven 盖片炉
cover sheet (塑料贴面板透明)保护层；层压板面板；覆盖薄片
cover slip 条状盖板
covert 细斜
covolume 协体积
covulcanizability 共硫化性
cow "母牛"〔可从中提取短寿命放射性同位素的母体同位素〕
Cowan screen 寇文(离心式)圆筛
cowhide 牛皮
cow leather 牛皮
Cowper blast air heater 热风炉
Cowper stove 考珀(鼓风)炉
cowrie 胶珇钯；玛瑙贝
cowslip 西洋樱草；莲香花
Cox chart 柯克斯蒸气压图
coxcomb 鸡冠状纹；梳状纹
cozy 保温套
cozymase 辅酶
C-polymer 缩聚物
C-polymerization (=condensation polymerization) 缩聚作用
CPVC of multicomponent pigment system 多成分(组分)的颜料体系的临界颜料体积浓度
CPVC of resin emulsion paint 树脂乳胶漆的临界颜料体积浓度
crab ①蟹②蟹肉③(野生)酸苹果；酸苹果树
crabbing machine 煮呢机
crab oil 山楂油；烛果油〔取自 Carapa guayncnsis 的种子〕
crab wood oil 山楂树油
crack ①龟裂；裂纹②裂化；裂解
crackability 可裂化性；热裂容易度
crackability of stock 燃料的裂化度
crack appearance 龟裂显现
crack arrest 滞止裂纹
crackate 裂化产物；裂解产物
crack crude 原油裂解
crack detection 裂纹检查
crack detector 裂纹控测仪；探伤仪
crack-drawing 裂膜拉伸
cracked ①有裂缝的②裂化的；裂解的
cracked asphalt 裂化沥青
cracked clear gasoline 裂化净汽油〔不含铅水的〕
cracked constituent 裂化成分

cracked distillate 裂化馏分
cracked-distillate rerun plant 裂化馏分再蒸馏装置
cracked fuel 裂化燃料
cracked fuel dilution 裂化燃料稀释作用
cracked fuel oil 裂化燃油
cracked gas 裂化气
cracked gasoil 裂化瓦斯油
cracked gasoline 裂化汽油
cracked leaded gasoline 裂化含铅汽油
cracked motor fuel 裂化发动机油
cracked naphtha 裂化石脑油
cracked oil 裂化油
cracked pitches 裂化沥青
cracked product 裂化产物；裂解产物
cracked residuum (=cracked residue) 裂化渣油
cracked spirit 裂化汽油
cracked still 裂化炉
cracked still gas 裂化炉气
cracked stock 裂化原料
cracked wax olefin 石蜡裂化烯烃
crackene 裂化烃
cracker ①裂化器；裂化装置②碾碎辊③脆点心
crack extension 缝隙扩张
crack extension force 裂纹扩展力
crack failure 龟裂破损
crack filler 填缝料；裂缝填充物
crack formation 龟裂形成
crack grain 裂面
crack growth 龟裂增长
crack growth rate 龟裂增长率
cracking ①裂化；裂解②断裂③破裂④裂纹
cracking activity (催化剂的)裂化活性
cracking capacity 裂化设备的生产量
cracking case 裂化反应器
cracking case vapour line 裂化反应器的蒸气管道
cracking catalyst 裂解催化剂
cracking chamber 裂化塔
cracking characteristic of catalyst 催化剂的裂化特性
cracking coil 裂化旋管；裂化炉管
cracking condition 裂化条件
cracking cycle 裂解周期；裂化周期
cracking cycle efficiency 裂化周期效率
cracking distillation 裂化蒸馏
cracking efficiency 裂化效率
cracking equipment 裂化设备
cracking fractionator 裂化设备的精馏塔
cracking furnace 裂化炉
cracking gas 裂化气

cracking heater (=cracking furnace) 裂化炉
cracking in liquid phase 液相裂化
cracking intensity 裂化强度
cracking in vapour phase 气相裂化
cracking in various media 在各种(惰性)介质中裂化
cracking lacquer 裂纹漆
cracking level 裂化深度
cracking load 破坏载荷
cracking-off 剥脱；脱落
cracking-off stand 碎折(用)支柱
cracking of oil 石油裂化
cracking pattern 裂化谱图；裂片图；裂解谱图
cracking per pass 单程裂化
cracking plant 裂化厂
cracking process 裂化法
cracking reaction 裂化反应
cracking residue (=cracking residuum) 裂化渣油
cracking residuum (=cracking residue) 裂化渣油
cracking resistance 抗碎裂性
crackings〔复〕 脆脂
cracking severity 裂解深度；裂化强度
cracking stability 裂化稳定性
cracking still ①裂化罐②裂化(筒式)炉
cracking still gas 裂化气
cracking stock 裂化原料
cracking tar 裂化焦油
cracking test 抗裂试验；裂缝试验
cracking tester (屈挠)龟裂试验机
cracking tube 裂化炉的管子
cracking unit 裂化装置
cracking unit evaporator 裂化装置的蒸发塔
cracking unit fractionator 裂化装置的精馏塔
cracking zone 裂化层
crack initiation 龟裂诱发
crack initiation energy 裂纹发生的能量
crackle 网状细裂纹；龟裂；碎裂花纹
crackle coating 裂纹漆
crackle finish 裂纹漆
crackle forming 龟裂；裂纹
crackle glaze 裂纹瓷釉；爆瓷釉
crackle lacquer 裂纹漆
crackle varnish 裂纹漆
crackle ware 碎(纹陶)瓷
cracklin (=crackle) ①碎裂花纹②小裂缝
crack meter 龟裂探测仪；裂纹(缝)探测仪
crack nucleation 碎裂成核(过程)
crack performance 龟裂现象
crack-per-pass 单程裂化量

crack propagation　裂纹扩展
crack resistance　抗龟裂性
crack sealer　①填缝剂；填缝料②裂纹封闭剂；龟裂均闭剂
crack sensitivity　裂缝敏感性
crack test　裂缝试验
cradle　台架；支架；皮圈架
cradle feeder　托架送料机
craft　①技巧；技艺；手艺；工艺技术；技能②飞机；飞船；飞行器③手工业；行业；工种
craft bag　工艺包
crafter　气泡孔；针眼
craft of gilding　镀金工艺；涂金工艺
crag　岩石碎块
cragged　岩多的；崎岖的；峭壁多的
Craig distribution apparatus　克雷格分配仪
Craig extraction　克雷格萃取
Craig extraction apparatus　克雷格萃取仪
craignurite　英闪玢岩
Cram's rule　克拉姆规则*
Cramer's rule　克莱姆法则
crammer feeder　填塞喂料机
cranberry-seed oil　红莓子油
cranberry wax　红莓子蜡；蔓越橘蜡
crane　起重机
crank　曲柄
crank arm　曲柄臂
crank axle　曲(柄)轴
crankcase　曲轴箱
crankcase breather　曲轴箱通气管
crankcase catalyst　曲轴箱(内油的氧化)催化剂
crankcase conditioning oil　曲轴箱冲洗用油
crankcase explosion　曲轴箱爆炸〔由于可燃物蒸气的积聚着火〕
crankcase lubrication　曲轴箱润滑
crankcase mayonnaise　曲轴箱低温沉淀物
crankcase oil　曲轴箱润滑油
crankcase oil additive　曲轴箱润滑油添加剂
crankcase oil classification　曲轴箱用油分类
crankcase oil dilution test　曲轴箱润滑油稀释试验
crankcase oil foaming　曲轴箱润滑油起泡沫
crankcase oil foaming test　曲轴箱润滑油泡沫试验〔使空气在75°F下通过油样〕
crankcase oil oxidation　曲轴箱润滑油氧化
crankcase oil types　曲轴箱润滑油等级
crank center　曲柄中心
crank detection　裂纹检查；探伤
crank detection test　探伤试验
crank handle (=crank lever)　手摇曲柄
cranking　摇动；起动
crank journal　曲(柄)轴颈
crank lever (=crank handle)　手摇曲柄
crank mechanism　曲柄机构
crank pin　曲柄销
crank pin bearing　曲柄销轴承
crankpin grease　连杆润滑脂
crankpin oiler　蒸汽机内用于连杆销子的离心加油器
crank press　曲轴式压机
crank radius　曲柄半径
crank shaft　曲轴
crank shaft bearing　曲轴轴承
crank shaft grinder　曲轴磨床
crankshaft grinding machine　曲轴磨床
crank shaft lathe　曲轴车床
crankshaft mechanism　曲轴机理
crankshaft rotation　曲轴旋转
crank stock　曲轴制动器
crank throw　曲柄行程
crank web　曲柄臂
cranny　裂缝
crapemyrtle　紫薇；百日红〔俗〕〔其炭用于漆器的推光〕
crape structure　绉纱结构
craping defect　绉疵；绉纹不匀
craping machine　起绉机
crash door　防冲门
crash pad　防冲垫
crataegin　山楂素
crataegolic acid　山楂酸
crate　①机箱；插件②篓；筐
crater　弧坑；坎；陷口
cratereee　缩孔受体
craterifermycin　火山口霉素
crateriform　火山口状
cratering　①缩孔(露底)；陷穴〔漆病〕②印品起泡〔印刷故障〕
crawler track　履带
crawling　①表面涂布不均②收缩〔涂漆后，形成小滴或小球致使漆膜龟裂〕
crawl model　爬行模型
craze　银纹*
craze matter　银纹体
craze resistance　耐微裂性
crazing　银纹；起裂纹
crazing mill　碎矿机；碎锡矿机
crazing-of-conformal coating　敷形涂层微裂纹
cream　①乳油；稀奶油②水浆③膏

cream base 膏基；膏(用)底物
cream caustic soda 苛性碱膏〔含水固体苛性碱〕NaOH·H₂O
creamed latex 膏化法(浓缩)胶乳
creamer 杂化材料
creamery 乳油制造厂
creamery hose 酪业胶管
cream formation ①乳状液形成②乳油形成
creaming 乳状液分层*
creaming agent 乳状液分层剂
creaming machine 乳油搅打机；甩奶油机
creaming of emulsion 乳液分层；乳液的破乳
creaming of latex 胶乳的澄清
creamlike consistency 奶油状稠度；膏状稠度
cream of lime 石炭浆
cream of tartar 酒石
creamy ①似乳油的②含奶油的
creamy consistency 奶油状稠度
creamy ground （木浆糊墙纸的)乳白色底(子)
creamy lather 似乳油泡沫
creamy paste ①乳白色浆②奶油状浆料
creamy white 乳白
creasability 耐折绉性；防绉性
crease 折痕
crease and shrink resistant finish 防皱防缩处理
creased bag 打折硫化胶囊；起绉胶囊
creased casing 胎体打折
crease folding 皱褶折叠
creaseproof 不皱的；防皱的
crease recovery 折皱弹性；皱褶回复
crease recovery tester 抗皱性测定机；折皱弹性测定仪；折皱复原测定机
crease resist 防皱；抗皱
crease resistance 耐皱性；抗皱性
crease-shedding 抗皱
crease shy 抗皱；除皱
crease whitening 折叠泛白
creasing ①折叠②卷边
creasing angle 卷边角铁模板
creasing strength 耐折叠强度
creasote (=creosote) 杂酚油
creatase 肌酶
creatine 肌酸；甲胍基乙酸
creatine phosphate 磷酸肌酸
creatine phosphoric acid 肌磷酸；磷酸肌酸
H₂PO₃·NHC(=NH)N(CH₃CH₂COOH)
creatininase 肌酸酐酶

creatinine 肌酸酐 NH=CH(CH₃)CH₂CONH
creatinine standard solution 肌酸酐标准液
creedite (=beljankite) 铝氟石膏
creel 粗纱架；筒子架；经轴架〔纺〕
creep 蠕变*
creepage ①蠕变②蠕动
creepage track 蠕变痕迹；塑性变形纹
creep back 蠕变回缩
creep compliance 蠕变柔量
creep corrosion 裂隙腐蚀
creep curve 蠕变曲线
creep effect 蠕变效应
creep elongation ①蠕变伸长②蠕变伸长率
creep extension ①蠕变延伸②蠕变延伸率
creep flow 蠕变流动
creep fluidity 蠕变流度；蠕变流动性
creep fracture 蠕变破坏
creep function 蠕变函数
creeping 蠕升
creeping effect 蠕动作用
creeping film 蠕升膜
creeping flow 蠕动流
creeping lateral contraction ratio 侧向蠕变收缩比
creeping traverse mechanism 滞缓横动机构
creeping willow 匍匐柳
creep limit 蠕升极限
creep modulus 蠕变模量
creepocity 易蠕变性
creep path （电花)径迹
creep rate 蠕变速率
creep recovery 蠕变回复
creep relaxation 蠕变松弛
creep resistance ①蠕变阻力；抗蠕变力②蠕变强度
creep rupture 蠕变破裂
creep rupture strength 蠕变破裂强度
creep rupture test 蠕变破坏试验
creep strain 蠕变变形；蠕变应变
creep strength 抗蠕变强度
creep stress 蠕变应力
creep test 蠕升试验
creep tester 蠕变试验仪
creep time curve 蠕变时间曲线
creep under constant pressure 定压蠕变
creep yield time 蠕变屈服时间
cremeomycin 乳脂霉素
cremnitz white 碱式碳酸铅；铅白
cremometer (=creamometer) 乳油计

crenellated leveling blade 锯齿状流平刮刀
crenic acid 白腐酸；克连酸
crenulation 细圆齿；细圆齿状物
creolin (=analgin) 诺瓦经〔药〕；安乃近
creosol (=methoxycresol) 甲氧甲酚；4-甲基-2-甲氧基苯酚 $CH_3OC_6H_3(CH_3)OH$
creosote (=creasote) 杂酚油
creosoted roll rubber 含杂酚油的生胶片
creosote oil 杂酚油
creosote oleate 油酸杂酚(油)
creosote valerate 杂酚油戊酸酯；戊酸杂酚油
crepe ①绉②绉片③绉网
crepenynic acid 还阳参油酸；顺(式)十八碳-9-烯-12-炔酸；(Z)-十八碳-9-烯-12-炔酸
crepe paper 绉纹纸；卷绉纸〔俗〕
crepe rayon fabric 嫘萦绉绸
crepe rayon yarn 嫘萦绉丝；紧捻人造丝
crepe sole 绉胶底
crepe twisting 绉线加捻；紧捻
crepe varnish 皱纹清漆
crepe yarn 皱皮纱
crepin 黄鹌菜素
creping 起绉(工艺)
creping machine 制绉绸机
crepitation 作碎裂声
crepon finish ①皱纹漆②皱缩整理
cresalol (=cresyl salicylate) 水杨酸甲苯酯
cresatin 乙酸间甲苯酯
crescent pump 新月形齿轮泵
crescent tear test 月牙型撕裂试验
cresidine 甲酚定；3-氨基对甲苯甲醚 $CH_3C_6H_3(OCH_3)NH_2$
cresoform 甲酚甲醛
cresol 甲(苯)酚
p-cresol 对甲酚
cresolase 甲(苯)酚酶
cresolbenzein 甲酚苯因
cresol-carboxylic acid 甲酚羧酸 $CH_3C_6H_3(OH)COOH$
p-cresol 2-diazidonaphthoquinone (1,2)-5-sulfonate 2-重氮基萘醌(1,2)-5-磺酸对甲酚酯
cresol-formaldehyde (CF) 甲酚-甲醛(树脂)
cresol-formaldehyde resin 甲酚-甲醛树脂
cresol liquor 甲酚溶液
cresol-novolac epoxy resin 甲酚-线型(热塑性)酚醛环氧树脂
cresolphthalein 甲酚酞 $C_6H_4COOC= (C_6H_3CH_3OH)_2$
cresol purple 甲酚紫*

cresol red (=o-cresolsulfonphthalein) 甲酚红；邻甲酚磺酞
cresol resin 甲酚(-甲醛)树脂
cresolsulfonic acids 甲酚磺酸类 $C_7H_8O_4S$
o-cresolsulfonphthalein (=cresol red) 邻甲酚磺酞；甲酚红
cresorcin 2,7-二甲基异荧光黄
cresorcinol 2,4-甲苯二酚
cresorcyl 二羟甲苯基 —$C_6H_2(OH)_2CH_3$
cresotic acid (=cresotinic acid) 甲酚酸；甲基水杨酸 $CH_3C_6H_3(OH)COOH$
cresotinic acid (=cresotic acid) 甲酚酸；甲基水杨酸 $CH_3C_6H_3(OH)COOH$
cresotyl 甲酚酰；邻羟间苯甲酰；2-羟-3-甲基苯甲酰 $HO(CH_3)C_6H_3CO—$
cresoxyacetic acid 甲苯氧基乙酸
m-cresoxy acetic acid 间甲苯氧基乙酸
cresoxy (=tolyloxy) 甲苯氧基〔邻位、间位或对位〕
crest ①峰值②峰顶③顶点
crest of bands 谱带顶
crest of flame 焰峰；焰波
crest of peak 峰顶
crest of weir 堰口
cresyl ①羟甲苯基②(=tolyl)甲苯基〔邻位、间位或对位〕
cresyl acetate 乙酸甲苯酯
p-cresyl acetate 乙酸对甲酚酯
cresylate 甲酚盐 $CH_3C_6H_4OM$
p-cresyl benzyl ether 对甲酚苄醚；茉莉醚
cresyl diglycol carbonate 碳酸甲苯双甘醇酯
cresyl diphenyl phosphate 磷酸甲苯联苯酯
p-cresyl dodecanoate 十二(烷)酸对甲酚酯
cresylene (=tolylene) 亚甲苯基
cresyl ethyl ether 甲酚乙醚
cresyl-2-glyceryl ether 邻甲苯基-2-甘油醚
cresyl glycidyl ether 甲酚缩水甘油醚
o-cresylic acetate 乙酸邻甲苯酯
cresylic acid 甲苯基酸〔俗名〕
cresylic blue 甲酚蓝〔氧化还原指示剂〕
cresylic phenolics 甲酚类酚醛塑料
cresylic resin 混甲酚树脂
p-cresyl isobutyrate 异丁酸对甲酚酯
cresylite 甲苯炸药
p-cresyl methyl ether 对甲酚甲醚
cresylol 甲(苯)酚
p-cresyl phenylacetate 苯乙酸对甲酚酯
cresyl phosphate 磷酸甲苯酯 $CH_3C_6H_4O \cdot PO(OH)_2$; $(CH_3C_6H_4O)_2 \cdot PO(OH)$; $(CH_3C_6H_4O)_3 \cdot PO$; $(CH_3C_6H_4O)_3 \cdot PO$
cresyl salicylate (=cresalol) 水杨酸甲苯酯
cresyl p-toluene sulfonate 对甲苯磺酸甲苯酯

p-cresyl valerate　戊酸对甲酚酯
creta (=chalk)　白垩
cretaceous　白垩的
crevasse crack (=crevasse)　（大）裂缝；龟裂
crevet　熔壶
crevice　裂隙
crevice corrosion　裂隙腐蚀
crevice corrosion speed　缝隙腐蚀速度
crevice oil　裂隙油（位于页岩矿内的石油）
crew-cut micelle　平头胶束
crew-cut polymer　平头状高分子
crewel work　绒线刺绣
CRGI-Glidden miniviscometer　CRGI-格里登微型黏度计
cribble　①粗筛②粗粉③粗的
cribriform　筛状的
crichtonite　尖钛铁矿；锶钛铁矿
cricondenbar　临界冷凝压力
cricondentherm　临界冷凝温度
Criegee's oxidation　克里吉氧化(反应)
crimpability　(可)卷曲性；卷曲能力
crimp contraction　卷缩率；卷曲缩率；皱缩率
crimped fiber　卷曲纤维
crimped rayon staple　卷曲嫘萦短纤
crimp effect　蜷缩效应
crimp energy　卷曲能
crimper　卷曲机；折皱器
crimper doffer　假捻机自动落筒装置
crimp extension　卷曲延伸
crimp fabric　泡泡纱；绉纹织物
crimp force　卷曲力
crimping　皱缩；起皱；收缩；卷曲
crimpness　卷曲度；绉缩性
crimp-proof　不皱的
crimp-proof fabric　不皱布
crimp-proof finish　防皱整理〔纺〕
crimp stripe　泡泡纱条纹；绉条纹
crimp tester　(纤维)卷曲度测定仪
crimson　绯红
crimson antimony (=antimony vermillon)　锑朱
crimson glory rose concrete　墨红玫瑰浸膏
crimson lake　绯红色淀〔染〕
crin　克林；马毛；粗丝
crinine　文殊兰碱
crinkle crepe　泡泡纱
crinkle finish (=ripple finish)　皱纹(罩面)漆；波纹面饰
crinkling (=wrinkling)　起皱；皱缩〔漆病〕
crioscopic method　冰点降低法〔测分子量〕
cripping load　破坏荷重；折断荷重

cripping test　弯折试验
crispature　卷缩
crisp handle　挺爽手感
crispness　挺爽性
cristobalite　方英石
criteria stability　准则稳定性
criterion　判据；判别准则
crith　克瑞〔气体质量单位，即标准状况下一升氢的质量，约合 0.0896 克〕
crithmene (=γ-terpinene)　海茴香烯；γ-松油烯
critical　临界的；中肯的
critical absolute temperature　绝对临界温度
critical aggregation concentration　临界聚集浓度
critical angle　临界角
critical angle refractometer　临界角折射计
critical area (s)　危险部位；危险面；锈蚀严重的表面
critical chain length　(大分子)临界链长
critical coefficient　临界系数
critical compression ratio　临界压缩率
critical concentration　临界浓度
critical concentration of micelle formation　临界胶束浓度
critical condensation temperature　临界冷凝温度
critical condition　①临界情况②临界状态
critical consolute temperature　临界共溶温度
critical constants　临界常数*
critical cooling velocity　临界冷却速度
critical correlation coefficient　临界相关系数
critical coupling　临界偶合；临界交联
critical crazing strain　临界微裂应变
critical current capacity　临界载流量；临界电流容量
critical current density　临界电流密度
critical curve　临界曲线
critical damage　临界破损
critical damping　临界阻尼
critical degree of crosslinkage　临界交联度
critical density　临界密度
critical deposition potential　临界沉积电位
critical diameter of particle separation　粒子分离临界直径；粒子分离中肯直径
critical difference　临界差
critical dimension　临界尺寸
critical dissolution temperature　临界溶解温度
critical dissolving concentration　临界溶解浓度
critical dissolving time　临界溶解时间
critical draft　临界牵伸
critical energy　临界能
critical energy of reaction　反应临界能*
critical entanglement molecular weight　临界缠结分子量

critical equation 临界方程
critical excitation potential 临界激发电势
critical extension ratio 临界拉伸比
critical extent of reaction 临界反应程度
critical extinction point 临界熄火点
critical factor 临界因素
critical fibre length 临界纤维长度
critical fragment length 滑脱长度
critical fusion frequency 临界(视觉)停闪频率;临界融合频率
critical heat 临界热
critical humidity 临界湿度
critical illumination 临界照明度
critical inlet pressure 临界入口压
criticality 临界性;临界状态
criticality alarm system 临界报警系统
critical length 极限长度;断裂长度;致断长度
critical length of filament 纤维临界长度
critical load 临界负荷;危险负荷
critical longitudinal stress 临界纵向应力
critical mass 临界质量
critical material 关键材料
critical micelle concentration (CMC) 临界胶束浓度*
critical micelle point 临界胶束点〔即临界胶束浓度〕
critical miscibility temperature 临界溶混(性)温度
critical moisture content 临界水分(量)
critical molar volume 临界分子体积
critical molecular weight 临界分子量
critical nucleation 临界成核(现象)
critical nuclide 关键核素
critical oxygen determinator 临界氧浓度测试仪〔耐燃性试验用〕
critical parameter 临界参数
critical path method (CPM) 临界途径法;关键途径法〔工厂施工及检修进度合理安排方法〕
critical path scheduling 临界途径法;关键途径法〔工厂施工及检修进度合理安排方法〕
critical phenomenon 临界现象
critical pigment volume concentration (CPVC) 临界颜料体积浓度
critical point 临界点*
critical point dryer 临界点干燥器
critical position 临界位(置)
critical potential 临界电势
critical pressure 临界压力*
critical processing temperature 临界操作温度
critical radius 临界半径
critical radius of a cluster 簇的临界半径

critical range 临界范围
critical region 临界区域;拒绝域;否定域
critical relative humidity 临界相对湿度
critical relaxation time 临界松弛时间
critical resistance 临界电阻
critical Reynolds number 临界雷诺数
critical sampling mass 临界取样量
critical settling point 临界沉降点
critical shearing stress 临界剪应力
critical shear rate 临界剪切速率
critical shear stress 临界剪切应力
critical solubility 临界溶度
critical solution point 临界溶点
critical solution temperature 临界共溶温度*;临界溶解温度
critical solution temperature method 临界溶解温度法
critical speed 临界速率
critical stability temperature 临界稳定温度
critical state 临界(状)态
critical strain 临界应变;临界变形
critical stress 临界应力
critical stress intensity factor 临界应力强度因子
critical supersaturation ratio 临界过饱和比
critical surface tension 临界表面张力
critical temperature 临界温度*
critical temperature of wetting 临界润湿温度
critical transition point 临界转折点
critical value 临界值*
critical variable 临界变量
critical velocity 临界速度
critical volume 临界体积*
critical volume fraction 临界体积分数
critical wind velocity 临界风速
crivaporbar 临界蒸气压力
crizzling 表面微裂纹
croceic acid 藏红花酸;2-萘酚-8-磺酸
crocein 藏花精
crocein orange 藏花橙
croceocurine 藏花箭毒素
croceomycin 藏花霉素
croceous 藏花色的
crocetin 藏花酸 $C_{20}H_{24}O_4$
crochet 钩编织物
crocic acid (=croconic acid) 克酮酸;邻二羟环戊烯三酮 $C_5H_2O_5$
crocidolite 青石棉
crocidolite asbestos 青石棉
crocin 藏花素

crock　①烟炱②瓮
crockemeter　渗色仪；摩擦色牢度仪
crockery　陶器
crocking　①染污②喷霜③摩擦掉色
crocking fastness　颜色抗摩擦牢度
crocking meter　沾色试验仪
crocking tester　耐磨(色)牢度试验机
crock-meter (=crockmeter)　摩擦掉色测定器；耐摩擦牢度试验机
crock resistance　抗摩擦掉色性
crocodile skin　鳄鱼皮
crocodile tanned skin　鳄鱼革
crocodiling　龟裂；鳄纹
crocoisite　铬铅矿
crocoite　铬铅矿
croconic acid　克酮酸；邻二羟环戊烯三酮　$C_5O_3(OH)_2$
crocose　藏花糖　$C_6H_{12}O_6$
crocus　①擦粉②藏红花
crocus metallorum　锑藏花红〔三氧化二锑和三硫化二锑的共熔体〕
croisure　丝鞘；捻鞘
croloy　〔商〕铬基合金〔特种不锈钢，C-Cr-Fe-Mo-V 合金〕
cronak process　常温溶液浸渍法
crook　①弯曲；挠曲②钩子
crooked log　弯曲材
crookesite　硒铊铜银矿
Crookes tube　克鲁克司管
crospovidone (=polyvinylpolypyrrolidone)　聚乙烯聚吡咯烷酮
cross aldol condensation　交叉羟醛缩合*
Cross and Bevan cellulose　克贝尔维素；C.B.纤维素
cross-banded plies　直交组合
cross bar gasket　加横条管状封圈
cross beam　横梁
cross beam technique　交叉束技术*
cross beater mill　十字型锤磨机
cross belt　交叉传送带；交叉带
Cross-Bevan color reaction　克罗斯-贝文呈色反应
cross-blending　交叉混合
cross bombardment　交叉轰击
cross bond　交联键
cross bonding　交联
cross-breaking strength (=flexural strength)　挠曲强度
cross bred wool　杂交种羊毛
cross-bridge　横桥；交联桥
cross-bridging　交联
cross-bridging agent　交联剂
cross-channel flow　跨螺槽流(动)

cross classification　交叉分组*
cross coating　交叉刷涂法；纵横涂漆法
cross conjugation　交叉共轭*
cross contamination　交叉污染；交互沾染
cross conveyor　横式运输机
cross-coupling reaction　交叉偶联反应*
cross crack　横裂纹
cross current　错流
cross-current solvent extraction　串联的溶剂抽提
cross cut　①划方格；交叉切割②横切；横割
cross cut circular saw　横裁圆锯
cross-cut file　横割纹锉
cross cutter　横裁禁断机；卧式裁断机
cross cut test　①划格法附着力试验②划十字法盐雾试验
cross cutting　横向切削；模切
cross direction　横向
cross direction elongation　横向伸长率
cross direction tear strength　横向撕裂强度
cross direction tensile strength　横向抗张强度
crossed beam technique　交叉束技术〔化学加速器〕
crossed beam thermal lens　交叉束热透镜
crossed belt　交叉皮带；合带
crossed coil meter　环形计器
crossed contamination　交叉污染*
crossed detection technique　交叉检测法
crossed dispersion　交叉色散
crossed double bond　横交双键
crossed drive　交叉传动
crossed electric-magnetic field electron-capture ion source　正交电磁场电子捕获离子源
crossed electric-magnetic fields mass spectrometer　正交电磁场质谱计
crossed ends　绞头
crossed immunoelectrophoresis　交叉免疫电泳
crossed laser-molecular beam technique　激光-分子束交叉技术*
crossed molecular beam　交叉分子束*
crossed molecular-ion beam tandem TOFMS　交叉分子-离子束串级飞行时间质谱
crossed nicol　交叉棱镜
crossed polarizer　正交偏振棱镜
crossed polar system　参差极化系统
cross elastic effect　交叉弹性效应；横向弹性效应
cross elasticity　交叉弹性；横向弹性
cross feed　横给；横向进给
cross-feed line　相互供应输送管
cross-feed system　相互供应系统

cross field 井字区；交叉场
cross-field flow fraction 交叉场流分级
cross figure 交叉图案
cross flow 交叉流动；横向流动
cross-flow adsorber 错流吸附器
cross-flow filtration 交叉流过滤
cross-flow quench 横吹冷却；侧向冷却
cross flow tray 单溢流塔板
cross force 交叉力；横向力
Cross furnace 克劳斯裂化装置的裂化炉
cross-grafted copolymer 交叉接枝共聚物
cross grain 斜纹理
cross-grained 交叉转位的
cross grained plywood 横纹胶合板
cross-grain modulus 横纹(弹性)模量
cross-hatch 交叉线；截面线
cross-head ①十字头②横梁③(=T-head)直角机头〔电线压出机〕
cross head bolt bearing 十字头销轴承
cross head guide 十字头导槽
cross head pin 十字头销
cross head pin bearing 十字头销轴承
cross head pin oiler 用于连杆十字头的加油器〔火车头〕
crossing 十字交叉刷涂法
crossing effect 交叉效应
crossing streams 叉流
crossing thread 纬线
crossite 青铝闪石；铝铁闪石
cross-laid web 交叉铺置纤维网
cross laminate 正交层压制品
cross laminated 交向(纹理)层压
cross lapping 交叉铺网
cross-lining 横衬；交叉衬里
crosslink ①交联②交联键
cross-linkable 交联性的；可交联的
cross-linkable monomer 交联性单体
crosslinkable plasticizer 交联型增塑剂
cross linkage ①交联②交键
cross-link bond 交联键
crosslink cure 交联硫化
crosslink density 交联点密度
cross-link dextran 交联葡聚糖
crosslinked acrylic acid polymer 交联丙烯酸聚合物
cross-linked agent 交联剂
cross-linked capillary 交联毛细管柱
cross-linked cellulose fiber 交联纤维素纤维〔黏胶〕
cross linked chain 交联链
cross-linked elastomer 交联弹性体

cross-linked gel 交联凝胶
cross-linked insertion 交叉嵌入
cross-linked network 交联网络
crosslinked point 交联点
cross-linked polyacrylate fibre 交联聚丙烯酸酯纤维
cross linked polymer 交联聚合物
cross-linked polystyrene bead 交联聚苯乙烯珠
cross-linked region 交联区
cross-linked rubber 交联胶；硫化胶
cross-linked system 交联体系
cross-linked thermosetting coating 交联性热固化涂料；热固化交联涂料
cross linker 交联剂
crosslink finishing 交联整理
cross-link index 交联指数
crosslinking 交联*
crosslinking action 交联反应
crosslinking agent (=adduct curing agent) 加成固化剂；交联剂
crosslinking catalyst 交联催化剂
cross-linking coating 交联型涂料
crosslinking coefficient 交联系数
cross-linking comonomer 交联共聚单体
cross-linking copolymer 交联共聚物
crosslinking degree 交联度
crosslinking density 交联密度
crosslinking efficiency 交联效率
crosslinking index 交联指数*
crosslinking indicator 交联指示剂
cross-linking level 交联度
crosslinking ligand 交联配体
cross-linking mechanism 交联机理
crosslinking of polymer 聚合物交联反应
crosslinking plasticizer 交联性增塑剂
cross-linking point 交联点
cross-linking properties 交联性能
cross-linking radical acceptor 交联基团接收剂
crosslinking rate 交联速度
cross linking reaction 交联反应
cross-linking set 交联点
crosslinking side reaction 交联副反应
crosslinking site 交联点
cross-linking stage 交联阶段
crosslinking structure 交联结构
cross-linking system 交联体系
cross linking yield 交联收率
cross-link intensity 交联强度
cross-link low-density polyethylene 交联低密度聚乙烯

crosslink mobility　交联键移位
cross mark　(珩磨)网纹
cross-member　横向构件
cross metathesis　交叉易位(反应)
cross-migration　交互渗移；相互迁移
cross-miss interlock　双罗纹交错浮线织物
cross motion　横向运动
cross nicol　正交尼科耳
cross-over　交叉(污染)
cross-over bend　横跨弯头
cross-over experiment　交叉实验
cross-over flue　横跨焰道
crossover gasoline valve　重叠的汽油阀
crossover line　重叠的输送管
cross-over oxidation　交叉氧化
crossover point　横跨点
crossover reaction　交叉反应
cross peak　交叉峰
cross piece　四通
cross pipe　十字管
crossplate　十字板
cross-plated rib fabric　交错添纱罗纹织物
cross-ply　斜交帘布层
cross polarization　交叉极化*
cross profile　截面；横截面
cross propagation　交叉增长*
Cross reaction chamber　克劳斯裂化装置的反应室
cross-recovery method　交叉回收方法
cross reference　交叉参照
cross relaxation　交叉弛豫*
cross-section　截面
cross-sectional area　截面面积
cross-sectional picture　截面图
cross-section detector　截面积检测器
cross-section ionization detector　截面积电离检测器
cross-shake (=radial shake)　径裂
cross shear　横向剪切力
cross slide unit　横滑组
cross solvent extraction　逆流溶剂萃取
cross staining　交互污染
cross strain　交叉应变；横向应变
cross stress　交叉应力；横向应力
cross talk　呼应作用
cross termination　交叉终止*
cross termination reaction　交叉终止反应〔两种不同单体的游离基在共聚反应中产生不活性分子,此种反应称为交叉终止反应〕
cross traverse spooling　交叉卷绕

cross tube　①十字管②四通管
cross tuck　珠地网眼织物
Cross unit　克劳斯(裂化)装置
cross validation　交叉校验
cross validation method　交叉校验法
cross valve　十字阀
cross-vein　交错矿脉；交错脉
cross viscosity　交叉黏性；横向黏性
cross viscosity coefficient　第二黏性系数；横向黏性系数
cross viscous effect　交叉黏性效应；横向黏性效应
cross wall　横墙；隔板
cross warping skein　十字绞
cross weld　横向焊缝
cross winding　络交
crosswise laminate　正交层压制品
crosswise lamination　①交向层压②交向层合
croton　①巴豆②巴豆油
crotonal (=crotonylidene)　亚巴豆基　$CH_3CH=CHCH=$
crotonaldehyde　巴豆醛；丁烯醛　$CH_3CH=CHCHO$
crotonaldehyde-aniline　丁烯醛苯胺〔促进剂〕
croton aldehyde process　丁烯醛选择精制(润滑油)过程
crotonamide　巴豆酰胺；丁烯酰胺　$CH_3CH=CHCONH_2$
crotonate　巴豆酸酯；2-丁烯酸酯
crotonbetaine　巴豆甜菜碱
croton chloral　巴豆氯醛；丁烯氯醛
crotonic acid (=β-crotonic acid)　巴豆酸；丁烯酸〔取自巴豆 Croton tiglium〕　$CH_3CH=CHCO_2H$
crotonic aldehyde　巴豆醛；2-丁烯醛　$CH_3CH=CHCHO$
crotonic anhydride　巴豆酸酐；丁烯酸酐　$(CH_3CH=CHCO)_2O$
crotonic nitrile (=propenyl cyanide)　巴豆腈；丙烯基腈　$CH_3CH=CHCN$
crotonization　丁烯醛化(作用)
crotonoid system　巴豆烯共轭系统
croton oil　巴豆油
crotonol　巴豆油醇
crotonolactone　巴豆酰内酯
crotonolic acid　顺式-2-甲基-2-丁烯酸；甲基巴豆酸
crotononitrile (=crotonic nitrile)　丁烯腈；巴豆腈
crotonoside　巴豆苷　$C_{10}H_{13}O_5N_5$
crotonoyl　巴豆酰；丁烯酰　$CH_3CH=CHCO-$
croton resin　巴豆脂
crotonyl　①(=crotonoyl)巴豆酰基　$CH_3CH=CHCO-$　②巴豆基；丁烯基　$CH_3CH=CHCH_2-$
crotonyl alcohol　巴豆醇；丁烯醇　$CH_3CH=CHCH_2OH$
crotonyl chloride　巴豆酰氯；丁烯酰氯　$CH_3CH=CHCOCl$
crotonyl-CoA　巴豆酰辅酶 A；丁烯酰辅酶 A

crotonylene 巴豆炔；2-丁烯炔 $CH_3C\equiv CCH_3$
crotonylidene (=crotonal) 亚巴豆基 $CH_3CH=CHCH=$
crotonyl mustard oil 巴豆基芥子油；异硫氰酸γ-丁烯酯
crotoxin 响尾蛇毒素
crotyl (=2-butenyl) 巴豆基；丁烯基；2-丁烯基 $CH_3CH=CHCH_2-$
crotyl alcohol 巴豆醇；2-丁烯醇 $CH_3CH=CHCH_2OH$
crotylamine 巴豆胺 $CH_3CH=CHCH_2NH_2$
crotyl bromide 巴豆基溴；丁烯基溴 C_4H_7Br
crotyl chloride 巴豆基氯；丁烯基氯 $CH_3CH=CHCH_2Cl$
crotyl halide 巴豆基卤；丁烯基卤 C_4H_7X
crotyl iodide 巴豆基碘；丁烯基碘 $CH_3CH=CHCH_2I$
crotyl mustard oil 巴豆基芥子油；硫氰酸巴豆基酯 $C_4H_7\cdot NCS$
crow's foot cracking 鸡爪裂
crowbar 铁挺；橇棍〔机工〕
crowding 加密；浓缩；聚集
crowding effect 集群效应；拥挤效应
crowding factor 群集因子
croweacin 克罗葳素
Crowell pump 克罗韦唧筒(泵)
crowfoot checking 爪状细裂纹
crowfoot crack 爪状裂纹；皱裂
crown ①帽②窑顶
12-crown-4 12-冠(醚)-4
18-crown-6 18-冠(醚)-6
27-crown-9 27-冠(醚)-9 $C_{18}H_{36}O_9$
30-crown-10 30-冠(醚)-10
crown bowl 中高滚筒
crown compound 冠状化合物
crowned roll ①中凸辊②有顶盖的轴
crown ether 冠醚*
crown ether stationary phase 冠醚固定相
crown filler 上等填料
crown form 冠式
crown gall 冠瘿
crown gear 冕状齿轮；伞形齿轮
crown glass 冕玻璃；无铅玻璃
crown root 冠根
crown top 灯帽
crozzling 鳄鱼皮状裂纹
Cr-rich zone 富铬区
crucible 坩埚*
crucible assay 坩埚试金法；坩埚分析
crucible cast steel 坩埚(铸)钢
crucible cover 坩埚盖
crucible disc 坩埚片

crucible furnace 坩埚炉
crucible holder 坩埚座
crucibleless zone melting 无坩埚区熔法*
crucible oven 坩埚炉
crucible steel 坩埚钢
crucible swelling number 坩埚膨胀序数
crucible tongs 坩埚钳
crucible triangle 坩埚(用)三角架
crucible type solid source 坩埚型固体离子源
cruciform baffle 十字形挡板
cruciform jet 十字形喷丝头
cruciform joint 十字接头
crud ①(腐蚀)沉淀物；沉积物②(界面)污物
crude ①原油②粗的③天然的
crude analysis 原油分析
crude antimony ①粗锑②辉锑矿
crude arsenic 粗砷；粗氧化砷
crude asbestos 粗石棉
crude asphalt 粗沥青
crude asphaltic petroleum 沥青基原油
crude battery 原油(蒸馏)装置
crude beet juice 粗甜菜(渗)汁
crude benzole 粗苯
crude bottom 粗脚子
crude carbolic acid 粗酚
crude charging capacity 原油处理量
crude cracker 原油裂化设备
crude cresylic acid 粗甲酚
crude desulfurization 原油(加氢)脱硫
crude distillation 原油蒸馏
crude distillation tower 原油蒸馏塔
crude emulsion 乳化原油
crude evaluation 原油评价
crude fat 粗脂(肪)；原脂(肪)
crude fiber 粗纤维
crude fiber digestion 原纤维消化
crude fish serap 粗制鱼渣
crude flashing 原油闪蒸
crude flash tower 原油闪蒸塔
crude fractionating column 原油分馏塔
crude fuel 原燃料；原煤
crude gas 原煤气；不纯煤气
crude gas liquor 原煤气水
crude gasoline 粗汽油
crude glycerine 粗甘油
crude heavy solvent 粗挥发油
crude heavy solvent naphtha 粗的重溶剂石脑油
crude hydrocarbon 粗制烃

crude lac (=seed lac)　粗紫胶
crude light solvent naphtha　粗的轻溶剂石脑油
crude lube stock　润滑油原料
crude material (s)　原料
crude matte　半冶金属
crude metal　粗金属
crude mineral oil (=crude oil)　石油；原油
crude mint oil　薄荷原油
crude naphtha　粗石脑油；粗挥发油
crude odor　粗糙香气；粗糙气味
crude oil (=crude petroleum; crude mineral oil)　石油；原油
crude oil cyclization　原油环化
crude oil emulsion　原油乳状液
crude oil remove　原油除去剂
crude oil still　原油蒸馏锅
crude oil storage tank　原油储罐
crude oil unit　原油蒸馏装置；常减压装置
crude oil upgrading　原油预处理
crude oil working tank　原油日用储罐
crude peppermint oil　薄荷原油
crude petroleum (=crude oil)　石油；原油
crude petroleum fuel oil　石油燃料油
crude pine tar　粗松焦油
crude product　粗制品
crude production　原油的开采(量)；采油(量)
crude protein　粗蛋白质
crude pyroligneous acid　粗木醋液
crude resources　原油(天然)资源
crude rubber　①(原料)生胶②天然橡胶〔专指三叶橡胶〕
crudes　①天然胶②原矿〔未选过的矿〕③生胶
crude sampler　原油取样器
crude scale　粗石蜡
crude scale wax　①(鳞状)粗汗蜡〔石油〕②一次发汗蜡〔石油〕
crude separation　①粗分离②原油分离
crude settling tank　原油沉降罐
crude sewage　原污水
crude shale oil　粗页岩油
crude solvent　粗挥发油
crude solvent naphtha　粗溶剂汽油
crude spirit　粗汽油〔英国名〕
crude splitter column　原油分裂塔；粗品分割塔
crude stabilization　原油稳定
crude still　原油蒸馏锅
crude storage　原油库
crude tar　原焦油
crude test　粗糙试验；最简单试验

crude topper　原油拔顶装置
crude turpentine　粗松节油
crude unit　原油蒸馏装置
crude wax　原石蜡
crude wood naphtha　粗木精
crude wood spirit　粗木精
crude wood turpentine　粗木制松节油
crud removal　①腐蚀沉积物去除②界面污物去除
cruentine A　瓜叶菊碱甲
cruise rating octane number　贫油混合物辛烷值；巡航条件下的辛烷值
cruising range　航程；巡航半径
cruium trifluoride　三氟化锔　CmF_3
crumbled-sheet model　碎屑状薄片模型
crumbliness　(可)破碎性；脆性
crumbling surface　风化的表面
crumbly mass　脆性物质
Crum Brown and Gibson rule　克拉姆·布朗和吉布森规则
crumb rubber　①粒状生胶②废胶末
crumbs　①粒状生胶②废胶末
crumpled　①皱的；起皱纹的②盘曲的
crumple pattern　皱纹
crumpling resistance　耐揉性
crup leather　马臀革
crushed coke　碎焦炭
crushed levant grain　熨光绉纹粒面
crushed solid　碎固体
crusher　压碎机；药碾
crusher gauge　(气体)压缩压力计；铜柱压力计
crusher oil　压碎机用油
crushibility　可碎性
crushing　压碎；粉碎；破碎
crushing apparatus　压碎器
crushing boulders　球磨(用)圆石
crushing capacity　压碎本领；碾碎本领
crushing force　压碎力
crushing machine　破碎机
crushing mills　轮碾机
crushing resistance　抗碾性
crushing roll　压碎辊；辊筒压碎机
crushing sand　粉碎砂(粒)
crushing strength　抗碎强度
crushing stress　①压碎应力②(容器)压瘪应力
crushing test　碾碎试验
crushing tester　碾碎试验器
crushing yield load　(容器)压瘪屈服载荷
crush test　压裂试验

crust　①外皮②壳
crust forming agent　表皮形成剂；结皮剂
crusting　结硬皮
crustization　结壳
crust leather　坯革；半硝革
crust of cobalt　钴壳；钴华
crust of iron　①铁渣②氧化皮
crust sorter　坯革挑选工
cryobox　低温箱
cryochemistry　低温化学
cryoconcentration　低温浓缩
cryodesiccation　(冷)冻干(燥)
cryodrying　低温干燥；深冷干燥
cryofocusing　冷集；低温富集
cryoformaldehyde tanning　低温甲醛鞣制
cryogen　冷冻剂；制冷剂
cryogenic　低温的
cryogenic distillation　低温蒸馏
cryogenic dry etching　低温干法腐蚀
cryogenic grinding　冷冻粉碎；深冷磨碎
cryogenic microgrinding　深冷微磨碎
cryogenic operation　低温操作
cryogenic pump　低温抽气泵
cryogenic pumping　低温抽气
cryogenic purification　深冷净化
cryogenic relaxation　深冷松弛
cryogenics　低温实验法（通常低于$-100°C$）
cryogenic separation　深冷分离
cryogenic stripping　冷冻脱漆；冷冻脱膜
cryogenic superconductor　低温(深冷)超导体
cryogenic surface　低温(体)表面
cryogenic tank　低温储罐
cryogenic target　低温靶
cryogenic temperature　深冷温度〔$-150°C$以下〕
cryogenic upgrading　深冷提纯
cryogenic work　低温操作
cryogenine (=m-benzoamino-semicarbazide)　间氨甲酰基苯氨基脲；冷却精　$H_2NCONHNHC_6H_4CONH_2$
cryohydrate (=cryosel)　冰盐；低共熔冰盐合晶
cryohydrate point　低共熔冰盐结晶点；冰盐点
cryohydric　冰盐的；低共熔冰盐结晶的
cryohydric point　冰盐点；低共熔冰盐结晶点
cryolite　冰晶石*
cryolite glass　冰晶玻璃；冰晶石
cryolithionite　锂水晶石
cryometer (=kryometer)　低温计
cryomicroscope　低温显微镜
cryomycin　冷霉素

cryophilic　嗜寒的；喜低温的；喜冰雪的
cryophilic salt minerals　喜冰无机盐
cryophorus　冰凝器〔物〕
cryoprotective agent　防冻剂
cryopulverizer　深冷粉碎机
cryopump　低温泵
cryoscope　冰点测定器
cryoscopic method　冰点(降低)法
cryoscopic solvent　冰点降低溶剂
cryoscopic titration　冰点降低滴定(法)
cryoscopy　冰点降低测定(法)
cryosel (=cryohydrate)　冰盐；低共熔冰盐合晶
cryosol　冰胶体溶液；冰溶胶〔冰为分散相的胶体溶液〕；低温溶胶〔仅在低温下才稳定的胶体〕
cryosorption pump　低温吸附泵
cryosorption pumping　低温吸附抽气
cryostat　深冷恒温器；低温恒温器
cryostatic stabilization　低温稳定(作用)
cryosystem　低温系统
cryoultramicrotome　冰冻超薄切片机
cryptal　桉油萜醛；对异丙基环己烯甲醛　$C_{10}H_{16}O$
cryptand　穴状配体*
cryptate　窝穴体
cryptate　穴状化合物
cryptate compound　穴(状化)合物*
cryptaustoline　厚壳桂碱
cryptene　隐烯
cryptocaria bark　厚壳桂皮
cryptocarpine　隐卡品；厚壳桂品
cryptocavine (=cryptopine)　隐品碱
cryptochlorophaic acid　隐氯地衣酸
cryptocrystalline　隐晶的〔矿〕
cryptocrystalline quartz　隐晶石英；潜晶质石英
cryptocrystalline silica　隐晶二氧化硅
cryptocyanine　隐花青〔染料〕
cryptoflavin　隐黄素
cryptohalite　方氟硅铵石
cryptojaponol　柳杉酚
cryptol　隐醇　$C_9H_{16}O$
cryptolepine　白叶藤碱
α-cryptomerene　α-柳杉烯
cryptomeridiol　柳杉二醇
cryptomeriol　柳杉醇
cryptomerione　柳杉烯酮
cryptomerone　柳杉酮
cryptometer　遮盖(力)计；遮盖力测定仪
cryptomorphic　隐形的
cryptomycin　隐球霉素

cryptone 隐酮；异丙基环己烯酮
cryptopimaric acid 柳杉酸
cryptopine (=cryptocavine) 隐品碱 $C_{21}H_{23}NO_5$
cryptopinone 隐蒎酮
cryptopleurine 小穗苎麻素；侧厚壳桂碱
cryptoporphyrin 隐卟啉
cryptopyrrole 隐吡咯；2,4-二甲基-3-乙基吡咯
cryptopyrrole carboxylic acid 隐吡咯羧酸
cryptoscope (=fluorescope) 荧光镜
cryptosterol 隐甾醇
cryptotaenene 隐腾烯 $C_{10}H_{16}$
cryptotanshinone 隐丹参酮
cryptovalence (=cryptovalency) 隐价；异常价
cryptovalency (=cryptovalence) 隐价；异常价
cryptowolline 厚壳桂灵
cryptoxanthin 隐黄质；玉米黄质
cryptoxanthine 隐黄质
crystal 晶体*；结晶
crystal aggregate 结晶聚集(体)
crystal alcohol 结晶醇
crystal analysis 晶体分析
crystal angle 晶角
crystal anisotropy 晶体各向异性
crystal axis 晶轴*
crystal birth 晶体产生
crystal boiling 结晶煮沸〔糖〕
crystal bonding 晶体键合；晶体结合
crystal carbonate 晶碱；苏打结晶 $Na_2CO_3 \cdot H_2O$
crystal cell ①晶胞②晶体光电池③晶格
crystal chemistry 晶体化学*
crystal chloroform 结晶氯仿
crystal class 晶类
crystal clear 无色透明的；清澈透明的
crystal clearity 晶状透明度
crystal clear syrup 清澈透明的浆状物
crystal clear transparency 无色清澈透明度
crystal coating 晶纹涂料
crystal configuration 晶体结构
crystal conglomerate 众晶
crystal control producer 晶体控制发生器
crystal coordinates 晶体坐标
crystal defect 晶体缺陷*
crystal defect diffraction 晶体缺陷衍射
crystal detector 晶体检测器
crystal diffraction 晶体衍射
crystal druse 晶簇
crystal edge 晶棱*
crystal elasticity 晶体弹性

crystal enamel 晶纹瓷漆
crystal engineering 晶体工程
crystal ether 结晶醚
crystal face 晶面*
crystal face corrosion speed ratio 晶面腐蚀速度比
crystal field 晶体场
crystal-field effect 晶场效应
crystal field model 晶体场模型
crystal field splitting 晶体场分裂*
crystal field splitting parameter 晶体场分裂参数
crystal field stabilization energy (CFSE) 晶体场稳定能
crystal field theory 晶体场理论*
crystal fineness 晶体细度
crystal finish 晶纹面漆
crystal form 晶形
crystal formation 晶体形成；晶体生长
crystal-forming polymer 成晶聚合物
crystal glass 结晶玻璃；富铅玻璃；水晶玻璃
crystal glaze 结晶釉
crystal grain 晶粒
crystal grating 晶格
crystal growth 晶体生长*
crystal habit 晶癖*；晶体习性
crystal habit modifier 晶体习性改变剂
crystal imperfection 晶体非理想性
crystal-ionic radius 结晶离子半径
crystal lacquer 结晶纤维喷漆
crystal lacquer finish 挥发性晶纹面漆
crystal lattice (结)晶格(子)
crystal lattice defect 晶格缺陷
crystal lattice energy 晶格能
crystalline 结晶的；晶状的
crystalline aggregate 晶状聚集体
crystalline-amorphous transition 晶态-非晶态转变*
crystalline band 结晶谱带；晶带
crystalline block 晶体区域
crystalline bloom 起霜；晶霜〔漆病〕
crystalline cellulose 晶态纤维素
crystalline cellulose powder 结晶纤维素粉
crystalline complex 晶体复合物
crystalline content 晶体含量
crystalline density 晶体密度
crystalline diffraction 晶体衍射
crystalline dimension 晶体尺寸
crystalline dispersion 晶体分散
crystalline domain 结晶畴；晶畴
crystalline effect 结晶效应
crystalline electric field (=crystalline field) 晶体场

crystalline electrode　晶体电极
crystalline element　晶胞
crystalline fibre　晶态纤维
crystalline field (=crystalline electric field)　晶体场
crystalline flour (=silica flour)　晶粉；硅粉〔填料〕
crystalline fold period　晶体折叠周期
crystalline form　结晶形状；晶形
crystalline forms of wax　蜡的晶形
crystalline fracture　结晶断面
crystalline glaze　结晶釉
crystalline globe　球晶
crystalline grade　结晶(程)度
crystalline grain　晶粒
crystalline-granular texture　结晶粒状结构；晶粒结构
crystalline granule　(结)晶粒
crystalline hydrate　结晶水合物；水合结晶
crystalline imperfection　晶体非理想性
crystalline indice　结晶指数
crystalline lamellae　晶态薄层；片晶
crystalline lattice　晶格
crystalline liquid　晶性液体
crystalline material　结晶性材料
crystalline matrix　结晶母体；结晶基体
crystalline melting point　晶体熔点
crystalline metal　晶体金属
crystalline micelle　晶体微胞；晶状微胞
crystalline monochrometer　晶体单色仪
crystalline morphology　结晶形态学
crystalline order　晶态有序
crystalline orientation　结晶取向；晶体取向
crystalline perfection　晶体完整性
crystalline phase　晶相
crystalline phosphate coating　结晶(的)磷酸盐被膜
crystalline phosphate layer　结晶磷酸盐层(膜)
crystalline polymer　结晶聚合物*
crystalline polyolefins　结晶聚烯烃类
crystalline polysilicic acid　结晶聚硅酸
crystalline poly (vinyl chloride)　结晶聚氯乙烯
crystalline portion　结晶部分
crystalline precipitate　晶形沉淀*
crystalline precipitation　晶形沉淀
crystalline region　晶区
crystalline rubber　结晶橡胶
crystalline rupture　结晶状断口
crystalline sequence　晶体序列
crystalline silica　①结晶二氧化硅②结晶硅石
crystalline silica-aluminate　晶体硅铝酸盐；分子筛；人造泡沸石

crystalline silicate　结晶硅酸盐
crystalline size　晶粒大小*
crystalline slag　晶性熔渣
crystalline slip　晶面滑移
crystalline solid　结晶固体
crystalline state　晶态*
crystalline structure　晶体结构
crystalline substance　结晶物质
crystalline sulfuric acid　结晶硫酸
crystalline temperature　结晶温度
crystalline texture　晶体织构
crystalline transducer　晶体传感器
crystalline transition　结晶转变
crystalline versus amorphous ratio　晶区-非晶区比率
crystalline wax　晶态蜡
crystalline zinc oxide　结晶氧化锌
crystalling dish　结晶皿
crystallinic acid　结晶酸
crystallinity　结晶性；结晶度
crystallinity index　结晶度指数
crystallisate　结晶物
crystallite　微晶；晶粒
crystallite alignment　微晶排列；微晶取向
crystallite dimension　微晶尺寸
crystallite length　微晶长度
crystallite orientation　微晶取向
crystallite reticulum　微晶网状结构
crystallite size　微晶大小
crystallizability　可结晶性
crystallizable　(可)结晶的
crystallization　结晶*；晶化
crystallization effect　结晶效应
crystallization half-life　结晶半衰期
crystallization heat　结晶热
crystallization-inhibited rubber　抑制结晶橡胶
crystallization inhibitor　结晶抑制剂〔阻止石蜡晶体在油内增长的混合物〕
crystallization initiation time　结晶诱导期
crystallization interval　结晶间歇
crystallization kinetic parameter　结晶动力(学)参数
crystallization kinetics　结晶动力学
crystallization modifier　结晶调节剂
crystallization morphology　结晶形态学
crystallization of amorphous solid　非晶晶化法
crystallization parameter　结晶参数
crystallization phase　结晶相
crystallization point　结晶点；结晶温度
crystallization rate　结晶速率

crystallization rate characteristic 结晶速率特性
crystallization rate constant 结晶速率常数
crystallization sensitive band 结晶敏感谱带
crystallization temperature 结晶温度
crystallization vat 结晶桶
crystallization velocity 结晶速度
crystallized glass fibre 微晶玻璃纤维；晶化玻璃纤维
crystallized region 结晶区；结晶段
crystallized soda 晶碱
crystallizer 结晶器
crystallizer pan 结晶(蒸发)器；结晶盘
crystallizer tank 结晶桶；结晶槽
crystallizing 结晶
crystallizing agent 结晶剂；形成晶体的媒介物
crystallizing boiler 结晶蒸发器
crystallizing dish 结晶皿
crystallizing evaporator 结晶蒸发器
crystallizing field 结晶区；沉积区
crystallizing finish 晶纹(罩面)漆；晶纹面饰〔非病态〕
crystallizing pan 结晶盘
crystallizing point 结晶点；结晶温度
crystallizing tank 结晶桶
crystallizing varnish 晶纹清漆
crystalloblastesis 晶质改变作用
crystalloblastic texture 变晶结构
crystallochemical analysis 结晶化学分析
crystallochemical reactivity 结晶化学反应性
crystallogram 晶体衍射图
crystallographic 结晶(学)的
crystallographic axis (结)晶轴
crystallographic component 结晶组分
crystallographic data 晶体学数据*
crystallographic dimension 晶体大小
crystallographic direction 晶体学方向
crystallographic interplanar spacing (纤维结构中)晶(学)的平面间距
crystallographic parameter 晶体参数
crystallographic plane 晶面
crystallographic plane groups 晶体学平面群*
crystallographic point group 晶体学点群
crystallographic register 结晶学定位；结晶配准
crystallographic shear 结晶学切变*
crystallographic site 晶格位置
crystallographic slip mechanism 晶面滑移机制
crystallographic symmetry 晶体学对称性*
crystallographic texture 晶体织构*
crystallographic unit cell 晶体学单元晶格；结晶晶胞
crystallography 晶体学*

crystalloid ①晶样的②类晶体
crystalloid solution 晶体溶液
crystallo-luminescence 结晶发光；晶体发光
crystallomycin 晶霉素
crystallon 晶核；籽晶
crystal mass 结晶状物
crystal membrane electrode (单)晶膜电极
crystal model 结晶模型
crystal modifier 结晶改良剂
crystal modulus 晶体模量
crystal nucleation 晶核化(作用)
crystal nucleus 晶核*
crystal offsetting 晶体偏置
crystal oil 结晶油；未脱蜡的油
crystalon 刚晶
crystal order 晶序
crystal orientation 晶体取向
crystal orientation function 晶体取向作用；晶体取向功能
crystal paper varnish (糊墙)纸用晶纹清漆
crystal pattern 晶体图案
crystal perfection index 晶体完整性指数
crystal phase 晶相
crystal phosphor(s) 结晶发光体；结晶荧光体
crystal plane 晶面
crystal plane slip 晶面滑移
crystal plasticity 晶体塑性
crystal point 结晶点
crystal ponceau 结晶丽春红〔染〕
crystal powder 晶体粉末
crystal pulling method 晶体生长提拉法*
crystal refractometer 结晶(晶体)折射仪；结晶(晶体)折光仪
crystal region 结晶区
crystal repeating unit 晶体重复单位
crystal rotating method 旋转结晶法
crystal seeds 晶种
crystal shape 晶形
crystal size 晶粒大小
crystal-size distribution (CSD) 晶体大小分布
crystal skeleton 晶体间架
crystal spectrometer 晶体分光仪
crystal structure 晶体结构*
crystal structure analysis 晶体结构分析
crystal structure determination 晶体结构测定
crystal structure of wax 蜡的晶体结构
crystal symmetry 晶体对称性
crystal system 晶系*

crystal texture　晶体组织
crystal tilting　晶体倾斜
crystal transformation　晶体变换
crystal twin　双晶*
crystal varnish　结晶清漆；烘干晶纹清漆
crystal violet　结晶紫*；龙胆紫
crystal violet lactone　结晶紫内酯
crystal water　结晶水
crystal yarn　水晶纱
cryst (crystalline)　①结晶的；晶状②晶体的
crystobalite　白硅石〔石英的一种晶态，产生于硅酸铝催化剂加热过程中〕
crytate　穴状化合物
crythrite　钴华；八水合砷酸钴　$Co_3(AsO_4)_2 \cdot 8H_2O$
Csrex process　铯锶雷克斯过程〔词源:cesium Strontium Rare Earth Extraction，铯锶稀土提取法，用酸性磷酸酯和取代酚类萃取剂〕
C-stage　丙-阶段
C-stage resin　酚醛树脂C；不熔阶段树脂
cstk (=centistoke)　厘泊〔运动黏度单位〕
C-terminal　C末端；羧基末端
C-T plane grating mounting　C-T平面光栅装置
cubane　立方烷
cubanite　方黄铜矿
cube　①立方体②立方；三次幂
cubeb　荜澄茄
cubeb camphor　荜澄茄脑
cubebene　荜澄茄油烯　$C_{15}H_{26}$
cubebic acid　荜澄茄酸　$C_{29}H_{14}O_7$
cubebin　荜澄茄素
cubeb oil　荜澄茄油
cubebol　荜澄茄醇
cubeb pepper　荜澄茄
cubebs oil　荜澄茄油
cube dicer　方粒切粒机
Cu-Be electron multiplier　铜铍电子倍增器
cube gambier　立方块槟榔膏
cubelike molecule　立方体状分子
cube method　立方法〔用以测定沥青物的熔点〕
cube spar　硬石膏；无水石膏　$CaSO_4$
cubic(al)　①立方体的②立方晶系③立方的
cubical dilatation (=volume dilatation)　体积膨胀
cubic(al) elasticity　体积弹性
cubic(al) grating　立方晶格
cubical packing pattern　(粒子的)正六面体堆积方式
cubical particle　正六面体粒子
cubical-shaped　立方形的
cubic arrangement　立体排列

cubic average boiling point　立方平均沸点
cubic axis　立方轴
cubic capacity　容积；容量；空间；体积
cubic close packing　立方密堆积*
cubic crystal　立方晶体；等轴系晶体
cubic curve　三次曲线
cubic dilation　体积膨胀
cubic elasticity　体积弹性；抗压缩性
cubic fermi　立方费米〔费米为长度单位，等于10^{-13}厘米〕
cubic lattice　立方格子
cubic lattice cell　立方点阵晶胞
cubic niter　钠硝石；智利硝石
cubic powder　方体炸药
cubic spline interpolation　三次样条插值法
cubic stacking　立方堆积
cubic strain　方体张力
cubic system (=isometric system)　等轴晶系
cubic trapped ion cell　立体离子阱
cubitainer (=cubical container)　方容器；方罐
cubozols　溶蒽素；溶性蒽系还原染料
cucaivite　硒银铜矿
cucoline (=sinomenine)　汉防己碱
cucumber alcohol　黄瓜醇
cucumber oil　黄瓜油
cucumber-seed oil　黄瓜子油
cucumber tree　锐叶木兰
cucumis melo　甜瓜
cucurbit　①葫芦②葫芦形蒸馏瓶
cucurbitacin　葫芦素
cucurbitacine　葫芦素(类)
cucurbitin　南瓜子氨酸
cucurbitine　南瓜子碱
cucurbitol　南瓜子醇　$C_{24}H_{40}O_4$
cuddleoside　醉鱼草糖苷
cudesmol (=selinenol)　蛇床烯醇；瑟灵烯醇
cue　①嵌入②滴定度③信号④晶质因子⑤尾接指令(计)
cuff　胶管管头
cularimine　枯拉明
cularine　枯拉灵
cull　残胶〔塑化筒内〕；剩料
culled wood　废材；边材
cullet　碎玻璃
cull pick-up　残料沟
culm　①麦杆②小块无烟煤③煤屑
culture　培养
culture bottle　培养瓶
cultured broth　培养液
cultured cell electrode array　培养细胞电极阵列

culture dish (=double dish) 培养皿
culture flask 培养瓶
culture medium 培养基；培养介质
culture tube 培养管
culture yeast 培养酵母
cumal (=cuminal) 亚枯茗基；对异丙亚苄基
cumaldehyde (=cumic aldehyde) 枯醛；对异丙基苯甲醛
　　$(CH_3)_2CHC_6H_4CHO$
cumar gum 香豆树脂
cumaric acid (=coumaric acid) 香豆酸
　　$OHC_6H_4CH=CHCOOH$
o-cumaric aldehyde methyl ether 邻香豆醛甲醚
cumarone (=coumarone) 苯并呋喃
cumar resin 香豆树脂
cumberlandite 钛铁长橄岩
cumene (=isopropyl benzene) 枯烯；异丙基苯
　　$C_6H_5CH(CH_3)_2$
cumene hydroperoxide 氢过氧化枯烯
cumene process 枯烯法
cumene sulfonate 异丙基苯磺酸盐；枯烯磺酸盐
cumengeite 铜氯铅矿
cumengite 铜氯铅矿
cumenol 枯烯醇；对异丙基苄醇
　　$(CH_3)_2CHC_6H_4CH_2OH$
cumenuric acid 枯烯尿酸
cumenyl 枯烯基；异丙基苯基（邻,间或对）$(CH_3)_2CHC_6H_4-$
p-cumenyl acetic acid 对异丙苯乙酸
p-cumenyl acrylic acid 对枯烯基丙烯酸；对异丙苯基丙烯酸
cumermycin (=coumermycin) 香豆霉素
cumic acid 枯酸；对异丙苯甲酸
　　$(CH_3)_2CHC_6H_4COOH$
cumic alcohol 枯醇；对异丙基苄醇
　　$(CH_3)_2CHC_6H_4CH_2OH$
cumic aldehyde (=cumaldehyde) 枯醛；对异丙基苯甲醛
　　$(CH_3)_2CHC_6H_4CHO$
cumic amide 枯酰胺；对异丙基苯甲酰胺
　　$(CH_3)_2CHC_6H_4CONH_2$
cumidic acid 枯二酸 $C_{10}H_{10}O_4$
cumidine 枯胺 $C_9H_{13}N$
cumidino 枯氨基；对异丙苯氨基
　　$p-(CH_3)_2CHC_6H_4NH-$
cumin ①枯茗；欧蒔萝②异丙苯
cuminal (=cumal) ①枯茗基；对异丙亚苄基②枯茗醛
cuminalacetic acid 亚枯茗基乙酸
cuminalcohol 枯醇；对异丙基苄醇
　　$(CH_3)_2CHC_6H_4CH_2OH$
cuminaldehyde 枯茗醛；对异丙基苯醛
　　$(CH_3)_2CHC_6H_4CHO$
cuminalmalonic acid 亚枯茗基丙二酸
　　$(CH_3)_2CHC_6H_4CH=C(COOH)_2$
cuminamic acid 枯茗氨酸
cuminamide 枯茗酰胺 $(CH_3)_2CHC_6H_4CONH_2$
cumine hydroperoxide 异丙苯化过氧氢
cuminic acetaldehyde 枯茗乙醛
cuminic acid 枯茗酸；对异丙基苯甲酸
　　$(CH_3)_2CHC_6H_4COOH$
cuminic acid amide 枯茗酰胺
cuminic acid nitrile (=cumonitrile) 枯茗腈；对异丙基苯甲腈
cuminic alcohol 枯茗醇 $(CH_3)_2CHC_6H_4CH_2OH$
cuminic aldehyde 枯茗醛 $(CH_3)_2CHC_6H_4CHO$
cuminil 枯茗偶酰 $C_9H_{11}COCOC_9H_{11}$
cuminilic acid 枯茗基酸 $C_{20}H_{24}O_3$
cumin nitrile 枯茗腈
cumin oil 枯茗油
cuminoin 枯茗偶姻 $C_{20}H_{24}O_2$
cuminol 枯茗醇 $C_{10}H_{14}O$
cumin-seed oil 枯茗子油
cuminuric acid 枯茗尿酸 $C_{12}H_{15}O_3N$
cuminyl (=p-isopropylbenzyl) 枯茗基；对异丙苄基
　　$(CH_3)_2CHC_6H_4CH_2-$
cuminyl acetate 乙酸枯茗酯
cuminylacetic acid 枯茗基乙酸
cuminyl alcohol 枯茗醇 $C_{10}H_{14}O$
cuminylamine 枯茗基胺 $C_{10}H_{15}N$
cuminylidene (=p-isopropyol benzylidene) 亚枯茗基；对异丙亚苄基
cummingtonite 镁铁闪石
cumobenzyl alcohol 三甲苯甲醇 $C_6H_2(CH_3)_3CH_2OH$
cumohydroquinone 枯氢醌
cumol 异丙苯〔别名〕
cumonitrile 异丙苯腈 $C_{10}H_{11}N$
cumoquinone 枯醌
cumoyl (=p-isopropylbenzoyl) 枯酰；对异丙苯甲酰
cumulative 累积的
cumulative absolute frequency 累积频数
cumulative constant (=gross constant) 累积常数；总常数
cumulative copolymer composition 累积共聚组成
cumulative damage 累积损坏；总损坏
cumulative distribution 累积分布
cumulative distribution diagram 累积分布图
cumulative distribution function 累积分布函数
cumulative distribution under-size particle 累积筛下粒度分布
cumulative double bond 累积双键

cumulative drop impact method　累计坠落冲击法
cumulative error　累积误差
cumulative feedback inhibition　累积反馈抑制
cumulative frequency　累积频率
cumulative graph　累积曲线图
cumulative index　累积索引
cumulative ionization　累积电离
cumulative mean　累积平均(值)
cumulative oversize　累积筛上物；累积筛余物
cumulative particle size distribution　累积粒度分布
cumulative production　累计产量；总产量
cumulative property curve　累积特性曲线
cumulative residence time distribution function　累积逗留时间分布函数
cumulative screen analysis　累积筛析
cumulative stability constant　累积稳定常数
cumulative throughflow　累积流量
cumulative time　总时间；累积时间
cumulative toxic action　累积中毒作用
cumulative toxicant　累积性毒物
cumulative volume　总体积
cumulative weight percentage　累计重量百分比
cumulative yield　累积产额
cumulene　累积多烯
cumyl (=cumenyl)　枯基；枯烯基；异丙苯基
cumylacetic acid　枯基乙酸
cumyl alcohol (=cumic alcohol; isopropylbenzyl alcohol)　枯基醇；枯醇；对异丙基苄醇；对异丙(基)苯甲醇 $(CH_3)_2CHC_6H_4CH_2OH$
p-cumyl benzoate　苯甲酸对异丙苄酯
cumyl t-butyl peroxide　叔丁基过氧化异丙苯〔硫化剂〕
cumylene　亚枯茗基；对异丙亚苄基；枯茗醛
cumyl hydroperoxide　枯基过氧化氢；过氢氧化异丙苯
cumylic acid (=durylic acid)　枯基酸；2,4,5-三甲基苯甲酸 $C_6H_2(CH_3)_3COOH$
cumyloxy radical　枯氧游离基
cumyl peroxide　枯基过氧化物；过氧化二异丙苯
cumylphenate　枯基酚盐
cumylphenol　对异丙基酚；枯基酚
p-cumylphenol　对枯基酚
cuoxam　古克瑟姆〔碱性铜离子氨配合物，可作纤维素溶剂〕
cup and ball viscosimeter　杯球黏度计〔带杯球的黏度计〕
cup-and-cone　杯锥〔装填器〕
cup anemometer　杯形气流计；杯形风力计
cuparane　花侧柏烷
cuparene　花侧柏烯
cuparone　花侧柏酮
cuparophenol　花侧柏酚
cupel　烤钵；灰皿
cupel furnace　烤钵炉
cupellation　①烤钵冶金法②烤钵试金法
cupellatrion process　烤钵冶金法
cupelling　烤钵冶金
cupelling furnace　烤钵炉
cupferrate　铜铁试剂盐
cupferron　铜铁试剂；N-亚硝基-β-苯胲铵
cupferronate　铜铁试剂盐
cup flow figure (=cup flow index)　杯模流动度；杯模法流动指数
cup flow index (=cup flow figure)　杯模流动度；杯模法流动指数
cup flow test　杯溢法流动试验
cup grease　稠结润滑(脂)膏
cup leather　制动缸皮碗
cup lump　杯凝胶；胶团
cup of the mold　模巢；模腔
cupola　化铁炉；圆顶(鼓风)炉
cupola (blast) furnace　化铁(鼓风)炉
cupped gasket　杯形衬垫
cupping　采脂；杯吸法；杯突(过程)
cupping tester (=Erichsen tester)　杯突试验仪；埃里克森试验仪
cup plate method　杯皿法
cupral (=Na DDC)　二乙基二硫代氨基甲酸钠；铜试剂
cupraloy　铜铬银合金
cuprammonia　铜氨液
cuprammonium　铜铵
cuprammonium cellulose　铜铵纤维素
cuprammonium cellulose complex　铜铵纤维素络合物
cuprammonium hydroxide　氢氧化铜铵
cuprammonium-ion　四氨合铜离子
cuprammonium process　铜铵法〔人造丝制造方法〕
cuprammonium rayon　铜铵嫘萦；铜氨纤维
cuprammonium silk　铜铵丝
cuprammonium solution　铜铵溶液
cuprammonium sulfate　硫酸铜铵；硫酸四氨合铜 $[Cu(NH_3)_4]SO_4$
cuprammonium viscosity　铜铵(溶液)黏度
cupraol　椰子脂
cuprase　铜酶〔一种无毒氢氧化铜胶体，用以治皮下癌〕
cupra silk　铜铵丝
cuprate　铜酸盐　$M_2[CuO_2]$
cuprate silk　铜铵丝
cuprea bark　铜色树皮
cupreine　铜色树碱；脱甲奎宁

cuprene 聚炔 $(C_7H_6)_x$
cupressene 柏烯〔取自 *Cupressus macrocarpa* 大果柏〕$C_{20}H_{32}$
cupressin ①柏(树)油②(=delapril hydrochloride)地拉普利
cupric (正)铜的；二价铜的
cupric abietinate 松香亭酸铜 $Cu(C_{19}H_{27}O_2)_2$
cupric acetate 乙酸铜 $Cu(C_2H_3O_2)_2$
cupric acetylacetonate 乙酰丙酮酸铜〔催化剂〕
cupric acetylide 乙炔化铜
cupric aluminate 铝酸铜
cupric ammine 氨络铜；氨基铜
cupric ammonium chloride 氯化铜铵；氯化铜合氯化铵 $CuCl_2 \cdot 2NH_4Cl \cdot 2H_2O$
cupric anhydride 三氧化二铜 Cu_2O_3
cupric arsenate (正)砷酸铜 $Cu_3(AsO_4)_2$
cupric arsenite 亚砷酸铜 $Cu_3(AsO_3)_2$
cupric bichromate 重铬酸铜 $CuCr_2O_7$
cupric bioxalate 草酸氢铜 $Cu(HC_2O_4)_2$
cupric bitartrate 酒石酸氢铜 $Cu(HC_4H_4O_6)_2$
cupric bromide 溴化铜 $CuBr_2$
cupric cement (含)铜(补齿)泥
cupric chlorate 氯酸铜 $Cu(ClO_3)_2$
cupric chloride 氯化铜 $CuCl_2$
cupric chromate 铬酸铜 $CuCrO_4$
cupric compound 正铜化合物
cupric cyanide 氰化铜 $Cu(CN)_2$
cupric diamminochloride 氯化二氨铜 $[Cu(NH_3)_2]Cl_2$
cupric diamminohydroxide 氢氧化二氨铜 $[Cu(NH_3)_2](OH)_2$
cupric dichromate 重铬酸铜
cupric dimethyldithiocarbamate 二甲基二硫代氨基甲酸铜〔促进剂〕
cupric dithionate 连二硫酸铜 CuS_2O_6
cupric ethylenediamine tetraacetate 乙二胺四乙酸铜〔铜抑止剂〕
cupric ferrocyanide 氰亚铁酸铜 $Cu_2Fe(CN)_6 \cdot 7H_2O$
cupric fluoride 氟化铜 CuF_2
cupric fluosilicate 氟硅酸铜 $Cu[SiF_6]$
cupric hydroxide 氢氧化铜 $Cu(OH)_2$
cupric hyposulfite 连二亚硫酸铜 CuS_2O_4
cupric iodate 碘酸铜 $Cu(IO_3)_2$
cupric iodide 碘化铜 CuI_2
cupric ion 铜离子
cupric mercaptobenzothiazole 硫醇基苯并噻唑铜〔促进剂〕
cupric metaborate 偏硼酸铜 $Cu(BO_2)_2$
cupric nitrate 硝酸铜 $Cu(NO_3)_2$
cupric nitrite 亚硝酸铜 $Cu(NO_2)_2$
cupric nitroprussiate 亚硝铁氰化铜 $Cu[Fe(CN)_5NO]$
cupric nitroprusside 硝普酸铜
cupric nucleinate 核酸铜
cupric octoate 辛酸铜〔催化剂〕
cupric oleate 油酸铜 $Cu(C_{18}H_{33}O_2)_2$
cupric oxalate 草酸铜 CuC_2O_4
cupric oxide 氧化铜；氧化正铜 CuO
cupric oxide plate 氧化铜片
cupric oxychloride 氯化铜合氧化铜 $CuO \cdot CuCl_2$
cupric perchlorate 高氯酸铜 $Cu(ClO_4)_2$
cupric phosphate 磷酸铜 $Cu_3(PO_4)_2$
cupric phosphate (acid) 磷酸氢铜 $CuHPO_4$
cupric phosphide 二磷化三铜 Cu_3P_2
cupric potassium chlorate 氯酸铜合氯酸钾；氯酸铜钾 $Cu(ClO_3)_2 \cdot 2KClO_3$
cupric potassium chloride 氯化铜合氯化钾；氯化铜钾 $CuCl_2 \cdot 2KCl \cdot 2H_2O$
cupric potassium cyanide 氰化铜合氰化钾；氰化铜钾 $Cu(CN)_2 \cdot 2KCN \cdot 2H_2O$
cupric potassium ferrocyanide 氰亚铁酸铜钾；亚铁氰化铜钾 $K_2CuFe(CN)_6 \cdot H_2O$
cupric potassium tartrate 酒石酸铜钾 $K_2Cu(C_4H_4O_6)_2$
cupric pyrophosphate 焦磷酸铜 $Cu_2P_2O_7$
cupric rhodanate (=cupric rhodanide) 硫氰酸铜 $Cu(SCN)_2$
cupric rhodanide (=cupric rhodanate) 硫氰酸铜
cupric salicylimide 水杨基亚胺铜〔硫化剂〕
cupric salt of 2-mercaptobenzothiazole 2-硫醇基苯并噻唑铜盐〔促进剂〕
cupric selenate 硒酸铜 $CuSeO_4$
cupric selenide 一硒化铜 $CuSe$
cupric silicate 硅酸铜 $CuSiO_3$
cupric silicofluoride 氟硅酸铜 $Cu(SiF_6)$
cupric sodium chloride 氯化铜合氯化钠；氯化铜钠 $CuCl_2 \cdot 2NaCl \cdot 2H_2O$
cupric subacetate 碱式乙酸铜 $CuO \cdot 2Cu(C_2H_3O_2)_2$
cupric subcarbonate 碱式碳酸铜 $Cu_2(OH)_2CO_3$; $CuCO_3 \cdot Cu(OH)_2$
cupric sulfate 硫酸铜 $CuSO_4$
cupric sulfide 硫化铜 CuS
cupric sulfocyanate 硫氰酸铜 $Cu(SCN)_2$
cupric sulfophenate 苯磺酸铜 $Cu(SO_3C_6H_5)_2$
cupric tartrate 酒石酸铜 $CuC_4H_4O_6$
cupric tetramminochloride 氯化四氨(合)铜 $[Cu(NH_3)_4]Cl_2$
cupric tetramminohydroxide 氢氧化四氨(合)铜 $[Cu(NH_3)_4](OH)_2$

cupric tetramminosulfate　硫酸四氨(合)铜　[Cu(NH$_3$)$_4$]SO$_4$
cupric tetrathionate　连四硫酸铜　CuS$_4$O$_6$
cupric thiocyanide　硫氰酸铜　Cu(SCN)$_2$
cupric thiosulfate　硫代硫酸铜　CuSO$_3$S;CuS$_2$O$_3$
cupric wolframate　钨酸铜
cupricyanide　氰铜酸盐
cupriethylenediamine solution　铜乙二胺溶液
cupriethylenediamine viscosity　铜乙二胺黏度
cupriferous pyrite　含铜黄铁矿
cuprifluoride　氟铜酸盐　M$_2$[CuF$_4$];M[CuF$_3$]
cuprisone　铜试剂
cuprite　赤铜矿
cupri-thioformin　铜硫甲酰素
cuprizone　双环乙酮草酰双腙
cupro-　亚铜
cuproadamite　含铜水砷锌矿
cuproammonium fiber　铜氨丝
cuprobromide　溴亚铜酸盐　M[CuBr$_2$];M$_2$[CuBr$_3$]
cuprocompound　亚铜化合物
cuprocopiapite　铜叶绿矾
cuprocupric　正亚铜的；含一价铜和二价铜的化合物
cuprocupric compound　正亚铜化合物；一价铜与二价铜化合物
cuprocupric cyanide　氰化正亚铜；铜氰酸亚铜　Cu(CN)$_2$·Cu(CN)$_2$;Cu$_2$Cu(CN)$_4$
cuprocyanide　氰亚铜酸盐　M$_3$[Cu(CN)$_4$]
cuprodescloizite　矾铜铅矿
cuprohemol　铜血粉
cuproine (=2,2'-biquinoline)　亚铜试剂；2,2'-联喹啉
cuprol　核酸铜
cupro lead　铅铜合金
cuprometric titration　亚铜盐滴定
cuprometry　亚铜盐滴定法
cupron (=α-benzoinoxime)　铜试剂；安息香肟
　　C$_6$H$_5$CHCC$_6$H$_5$
　　　|　||
　　　OH NOH
cupron cell　氧化铜电池
cupronickel　铜镍合金；白铜
cuprophane　铜氨薄膜；铜纺〔一种人造丝〕
cuproscheelite　(杂)铜白钨矿
cupro silicon　铜硅合金
cuprosulfate　硫酸铜复盐
cuprous　亚铜的；一价铜的
cuprous acetate　乙酸亚铜　CuC$_2$H$_3$O$_2$
cuprous acetate ammoniacal (=copper ammonium acetate)　乙酸亚铜铵
cuprous acetylide　乙炔化亚铜　Cu$_2$C$_2$
cuprous amminobromide　溴化氨亚铜　[Cu(NH$_3$)]Br
cuprous bromide　溴化亚铜　Cu$_2$Br$_2$
cuprous carbonate　碳酸亚铜　Cu$_2$CO$_3$
cuprous chloride　氯化亚铜　Cu$_2$Cl$_2$
cuprous compound　亚铜化合物
cuprous cyanide　氰化亚铜　Cu$_2$(CN)$_2$
cuprous ethanolamine chloride solution　氧化亚铜乙醇胺溶液
cuprous ethylxanthate　乙黄原酸亚铜
cuprous fluoride　氟化亚铜　Cu$_2$F$_2$
cuprous fluosilicate　氟硅酸亚铜　Cu$_2$SiF$_6$
cuprous formate　甲酸亚铜
cuprous hydrosulfate　硫酸氢亚铜　CuHSO$_4$
cuprous hydroxide　氢氧化亚铜　CuOH
cuprous iodide　碘化亚铜　Cu$_2$I$_2$
cuprous ion　亚铜离子　Cu$^+$
cuprous nitride　氮化亚铜　Cu$_3$N
cuprous oxide　氧化亚铜；一氧化二铜　Cu$_2$O
cuprous oxychloride　氧氯化亚铜
cuprous phosphide　磷化亚铜　Cu$_3$P
cuprous potassium cyanide　氰化钾亚铜；氰化钾合氰化亚铜　CuCN·3KCN; KCu(CN)$_2$
cuprous pyridine acetate solution　吡啶乙酸亚铜溶液
cuprous rhodanide　硫氰酸亚铜
cuprous selenide　一硒化二铜　Cu$_2$Se
cuprous sulfate　硫酸亚铜　Cu$_2$SO$_4$
cuprous sulfide　硫化亚铜　Cu$_2$S
cuprous sulfite　亚硫酸亚铜　Cu$_2$SO$_3$
cuprous sulfocyanide　硫氰酸亚铜
cuprous telluride　碲化亚铜　Cu$_2$Te
cuprous thiocyanate　硫氰酸亚铜　Cu(SCN)
cuproxam lignin　氧化铜氨液(制)木素
cuproxide　铜氧矿
cup screen　圆柱筛
cup-shaped vane　筒形叶片
cup type electrostatic paint sprayer　杯式静电喷漆机
cup-type pycnometer　杯式比重瓶
cupu oil　壳斗油；大叶可可油〔取自 *Theobroma* 的种子〕
cup viscometer　杯式黏度计
cup weight per gallon　(每)加仑重量杯；加仑重量换算杯〔容量为83.3毫升的黄铜杯，用以换算每加仑的液体重量〕
cup wetting method　杯状电极湿润法
curability　固化性能；硫化性能
curable polymer　可固化聚合物
curacao peel　库拉索皮
curameter　硫化计；硫化程度测定计

curamycin 居拉霉素
curangin 枯苒苷；苦兰加苦苷
curarine 箭毒碱
curarisant alkaloids 箭毒植物碱
curative 固化剂；硬化剂；熟化剂
curative agent 固化剂
curative carrier 硫化剂载体
curative concentration 固化剂浓度
curative system 硫化体系
curative treatment 修补处理
curcas oil 麻风(树)油〔取自 Jatropha curcas 的种子〕
curcuma oil 姜黄油
curcuma (test) paper 姜黄试纸
curcumene 姜黄烯
curcumenol 姜黄烯醇
curcumin 姜黄(素) $C_{21}H_{20}O_6$
curcumine 姜黄素
curcumol 姜黄醇
curcumone 姜黄酮
curd ①液体凝结物②凝乳
curd fiber phase 皂粒纤维相
curd fiber(s) 皂粒纤维
curding out 凝结析出
curd tension 凝乳张力
curd tension meter 乳凝张力计
curdy ①多凝乳的②似凝乳的③凝结的
curdy precipitate 凝乳状沉淀
cure ①二次加热〔硅橡胶〕②熟成〔胶乳〕③熏烟〔烟片〕④硫化；硬化⑤凝固〔胶乳水泥〕
cure accelerator (硫化)促进剂
cure activating ingredient 硫化活性剂
cure activator (硫化)活性剂
cure agent (=curing agent) ①固化剂；硬化剂②硫化剂③熟化剂
cure bag 蒸煮室；硫化室
cure coefficient 硫化系数
cure conditions 硫化条件
cure curve 硫化曲线
cure cycle (=curing cycle) 固化周期
cured ①治愈②硫化③熏制好的④腌干的⑤熟化的
cured crumb 自硫胶粒
cured natural rubber 天然(胶)的硫化胶
cured phenolic resin fibre 固化(的交联)酚醛树脂纤维
cured product 硫化成品
cured properties 硫化胶性能
cured resin 凝固树脂；硬树脂
cured scrap 硫化碎片胶
cure index 固化指数

cure-in-place 现场固化
cure kinetics characterization 固化动力学特性
cure kinetics parameter 固化动力学参数
curelastometer 硫化弹性仪
cure mechanism 固化机理
curemeter 硫化仪
curemeter curve 硫化(仪)曲线
cure model 固化模型
cure oven 硫化罐
cure period 固化时间；固化(周)期
cure point 硫化温度
cure process 固化工艺
cure process monitoring 固化工艺监控
cure profile 固化曲线
cure range 硫化(平坦)范围
cure rate 硫化速度
cure reaction (=curing reaction) 固化反应
cure retarder 硫化迟延剂
cure shrinkage 固化收缩
cure stock 硫化胶
cure temperature 固化温度
cure time 固化时间
cure under pressure 加压固化
curia 二氧化镉 CmO_2
curide 镉系元素〔96~103 号元素〕
curie (Ci) 居里〔放射性强度单位，$=3.7\times10^{10}$ 次衰变/秒〕
Curie's constant 居里常数
Curie's law 居里定律
Curie electroscope 居里验电器
curie-equivalent 居里当量
curiegraph 镭疗照片
Curie magnetic susceptibility 居里磁化率
Curie point (=Curie temperature) 居里温度；居里点
Curie point pyrolysis 居里点裂解
Curie point pyrolysis-mass spectrometer 居里点裂解质谱仪
Curie point pyrolyzer 居里点裂解器
Curie point temperature 居里点温度
curiescopy (体素)镭注射试验
Curie temperature (=Curie point) 居里温度；居里点
curietron 居里治疗机
Curie-Weiss law 居里-韦斯定律
curine 箭毒碱
curing 固化*
curing accelerator ①固化促进剂②硫化促进剂〔橡胶〕
curing agent 熟化剂；硫化剂〔在橡胶工业中 curing 意指硫化〕；固化剂；腌制剂
curing agent blush 固化剂(致)泛白〔环氧涂料用胺类固

化剂,如湿度大,则易泛白〕
curing aid dispersions 预分散硫化剂
curing arm 熟化心轴
curing bag 熟化室
curing behaviour 固化特性
curing bladder 硫化胶囊
curing catalyst 硫化促进剂
curing chamber 熟化室
curing characteristics 固化特性
curing core 熟化心轴
curing cycle 硫化周期
curing degree ①固化度②硫化度③熟化度
curing diaphragm 硫化隔膜
curing exotherm 固化放热
curing fixture 固化夹具
curing fluid 硫化流体〔指硫化介质〕
curing level 硫化(程)度
curing mechanism 固化机理;固化机制
curing medium 硫化介质
curing oven 固化烘箱;固化炉;固化烘干炉
curing parameter 硫化参数
curing period 养护期
curing polyol 固化用多元醇
curing press 平板(硫化)机
curing rate 硫化速度;固化速度
curing speed 硫化速度;固化速度
curing stress 固化应力
curing system 硫化体系
curing temperature ①熟化温度②固化温度③硫化温度
curing test 熟化试验
curing time ①熟化期②硫化期
curing time lag 硫化时间滞后
curing tube ①薄壁水胎②硫化管道
curing tunnel ①隧道式固化炉②隧道式塑化箱
curite 铀铅矿
curium 锔
curium aluminate 铝酸锔 $CmAlO_3$
curium carbonate 碳酸锔 $Cm_2(CO_3)_3$
curium hydride 氢化锔
curium hydroxide 氢氧化锔 $Cm(OH)_3$
curium nitrate 硝酸锔 $Cm(NO_3)_3$
curium oxalate 草酸锔 $Cm_2(C_2O_4)_3$
curium oxide 氧化锔 $CmO; Cm_2O_3; CmO_2$
curium oxychloride 氯氧化锔 $CmOCl$
curium oxysulfate 硫酸氧锔 $Cm_2O_2SO_4$
curium phosphate 磷酸锔 $CmPO_4$
curium recovery facility (CRF) 锔回收设施
curium sesquioxide 三氧化二锔 Cm_2O_3
curium tetrafluoride 四氟化锔 CmF_4
curl ①蜷曲;卷毛;翘曲;旋转;涡流②成卷;成螺旋状③纸张起皱
curled buff 蜷曲式布抛光轮;布蜷抛光轮
curled edge 卷边
curled selvage 卷曲边
curling ①(=coilingsoup)卷涂垄纹〔施工缺陷〕②翘曲;扭曲〔木材〕③(=crawl)收缩变形④卷曲;卷缩⑤缩釉〔陶瓷〕
curling factor (=griseofulvin) 灰黄霉素
curling joint 卷边接缝
curly 卷曲的
curly cotton 萝卜丝
curly grain 卷纹;旋涡纹
curly schist 纤维片岩;卷曲片岩
curoid 锔系元素
curometer 硫化计〔硫化度测定计〕
current analysis 电流分析法
current-applied voltage curve 电流-电压曲线
current-cessation chronopotentiometry 电流中止计时电位法
current collector 集流体;集电体
current color 流行色
current compensation 电流补偿
current cost 市价;时价
current density 电流密度*
current detector 检流器
current efficiency 电流效率
current feedback 电流反馈
current integrator 积分电流计
current intensity tunnelling spectroscopy 正向隧道电流强度能谱法
current jump 电流突跃(现象)
current measurement 电流测定
current meter (冲)流速计;测流计
current noise 电流噪声
current ratio 流动比率
current rectifier 整流器
current regulator 节流器;电流调整器
current reversal 电流回扫*
current reversal chronopotentiometry 电流回扫计时电位法*
current-sampled polarography 电流取样极谱法
current-sampled voltammetry 电流取样伏安法
current scanning polarography 电流扫描极谱法*
current-scan voltammetry 电流扫描伏安法
current source 电源
current step 电流阶跃*

current step chronopotentiometry 电流阶越计时电位法
current supplyline 供电导线
current sweep 电流扫描*
current-time curve 电流-时间曲线
current-voltage curve 电流-电压曲线
current yield 日产量；现时产量
currier's knife ①刨②刮刀
currier's oil 骨(质)油；鞣革用油
currier's ink 铁黑；灯黑
currying ①梳洗加色②加脂法
currying work ①修整间②加脂操作〔皮革〕③修整
curry leaf tree oil 克尼九里香油
curtain 幕；帘；帷；挡
curtain coated finish 幕涂剂
curtain coater 幕涂机
curtain coating 幕涂
curtain gas 气帘
curtaining 垂落〔涂后漆膜形成较大面积的下垂〕
curtain wall penel ①间壁墙板②壁板③围墙板
Curtius conversion 克尔提斯转变〔酰肼成酰叠氮的转变〕
Curtius reaction 克尔提斯反应〔$RCON_3 \rightarrow RNCO+N_2$〕
curua palm oil 棕榈油
curvature 曲度
curvature effect 曲率效应
curve break 曲线转折点
curve correction 曲线校直
curve crystal monochromater 弯晶单色器
curved adapter 弯曲应接管
curved arrow symbol 弯箭头符号
curved blade open turbine agitator 开启弯叶涡轮式搅拌器
curved coiler tube 曲线斜管
curved crystal 弯(曲)晶(体)
curved crystal spectrometer 弯曲晶体分光计
curved expander 弧形拉幅机〔纺〕
curved flow （液体)在弯曲的管内流动
curved pipe 弧形管
curved plywood 曲形胶合板
curved rim 变形轮辋
curved roll ①弧形开幅辊②曲辊；弧形辊
curved shell roof 曲面薄壳屋顶
curved-sided weir 曲线堰
curved surface 弯曲表面
curve fitting 曲线拟合*
curve fitting method 曲线拟合法〔求单体反应性比值的方法之一〕
curvemeter 曲度计

curve smoothing 曲线平滑
curvilinear coordinate 曲线坐标
curvilinear figure 曲线图形
curvilinear regression 曲线回归
curvilinear shear flow 曲线剪切流动
curvilinear system 曲线(体)系
curvity 曲度
curvularin 苦乌素 $C_{16}H_{20}O_5$
curzerenone 蓬莪术酮
cuscamidine 库斯柯米定
cuscamine 库斯柯明
cusco bark 库斯柯皮
cuscohydrine 库斯柯液碱
cusconidine 库斯柯尼定
cusconine 库斯柯宁
cus-cus 岩兰草；香根
cuscus oil 岩兰草油；香根油
Cushings syndrome 柯兴氏综合症
cushion ①缓冲垫层②垫③缓冲器④胶垫
cushion block 垫块
cushion breaker 缓冲层
cushion coat 垫层
cushion cover 垫子套
cushioned channel valve 气垫槽阀
cushion factor 缓冲因数
cushion gas 气垫
cushioning ①缓冲；减震②软垫；缓冲器
cushioning action 缓冲作用
cushioning capacity 缓冲能力
cushioning effect ①减震效应②缓冲效应③缓冲作用
cushioning material 缓冲材料；减震材料；衬垫材料
cushion ply (=cushion rubber) 垫层(橡胶)
cushion rubber (=cushion ply) 垫层橡胶
cusohygrine 红古豆碱；古柯液碱
cusp 会切点
cusparine 库柏碱；西花椒碱
cusped inclusion compound 尖形包合物
cuspidine 枪晶石
cuspidite 枪晶石
CUSP process (词源: Concentrated Urania Sol Preparation) 浓二氧化铀溶胶制备过程
custard apple 南美番荔枝
custody transfer 密闭输送
custom-built extruder 非标准挤压机；定制(做)的挤出机
custom chemicals 专门定购化学品
custom coater 委托喷漆厂；定约油漆施工单位
custom color ①(客户)指定色；专配的色②定制漆；(零售店应顾客要求)专配的漆

customer-designed   非定型设计的；专门设计的
custom fabricated   按订货条款制造的
custom irradiation   委托照射
custom labeling   委托标记
custom-made   定做的
custom (-made) molding   特制模制品；订制配件
custom parts   特制模制品；订制配件
custom preparation   委托制备
custom processing   定货加工
custom source   委托(放射)源；定做源；定制源
custom synthesis   委托合成
cut   ①克特〔紫胶浓度单位，磅(紫胶)/加仑(酒精)〕②切；切断，切割③分流；切取馏分；分割比④油分〔100磅树脂中加入油的加仑数〕
cutal   硼单宁酸铝
cut-and-trial method   渐近法；试凑法；逐步接近法
cut and try method   试探法；逐次近似法；误差逐次缩小法
cut-and-try work   试凑试验；实验工作
cut and weigh method   剪下称重法
cut area   切断面
cut-away picture (=cut-away view)   (局部)剖视图；内部结构图
cut-away view (=cut-away picture)   (局部)剖视图；内部结构图
cut back   ①稀释②稀释产物③稀释的
cut-back asphalt   稀释沥青〔溶于石油馏出物的沥青〕
cut-back bitumen   稀释沥青〔软化剂〕
cutback oil   稀释油
cut-back product   ①稀释产物②轻馏分〔石油〕
cut-back tank   稀释的容器
cutch   儿茶
cut chenille   绳绒线
cut clear   乳化稀释浆
cutdrafil direct spinner   牵切直接纺丝机
cut film   薄膜切片
cut filter   截止滤波器
cut fraction   馏分
cut glass   雕玻璃
cut house   (石油工厂的)分类专业小组
cutic acid   角质酸〔植〕
cuticle   表皮层；外皮；角质层
cuticle cream   表皮润滑膏；指甲油
cut in   ①接通②插入③加入
cutinic anthracite   角质无烟煤
cut lamination   开料
cut-layer (s)   示层面；(层压材料)露层
cut length   切断长度；切段长度

cut linter   棉籽绒
cut log   ①块状原木②木段③木块
cut-middles method   中段切断称重法
cut noodle   切面
cutocellulose   角质纤维素
cut off   ①关闭②断流③切去④切断⑤截止⑥(塑体表面的)模缝脊
cut-off bias   截止偏压
cut-off blade   遮光片
cut-off clamp   断流夹
cut-off cock   切断旋塞；断流开关
cut-off device   断流器
cut-off energy   截止能量
cut-off filter   截止滤波器
cut-off level   截止(限制)电平；门限(坎)电平
cut-off mould (=flash mould)   溢出式塑模；溢出式压模
cut off the gas   断气
cut-off valve   截止阀；逆止阀
cut-off wavelength   截止波长
cut oil   乳化(石)油
cut-open view   剖视图
cut out   ①安全器〔电〕②切断
cut-out dump bell   冲切式哑铃试片
cut-out plug   断流栓
cut-over   接入；开动；开通；转换
cut-paper weighing method   剪纸称重法
cut plush fabric   割圈毛圈织物；割圈长毛绒织物
cut point   ①馏出温度②分馏点③切割点
cut resistance test (=cut-through resistance test)   抗切割试验；抗磨(穿)性试验；耐切削性试验
cuts   馏
cut selvage   破边；切割边
cut-set   割集
cut size   分级粒径；极限粒径
cut sole   切成型的底革
cut squirt   切断(用)喷水
cut strand   割断原丝
cut-switch sampling   切换进样
cutter   ①切刀②切断器③分馏器；切割机
cutter bar   刀杆
cutter depth   切削深度
cut-through resistance test (=cutresistance test)   抗切割试验；抗磨(穿)性试验；耐刮性试验
cutting   ①切；切割②切削
cutting abrasion   切削性磨耗
cutting action   切割作用
cutting agent   切削剂
cutting angle   切削角

cutting block 冲模；冲头
cutting compound (=cutting emulsion) 切削油〔切削工具的润滑和冷却用〕
cutting die 刀模
cutting disc 圆盘刀
cutting edge ①刃口；刀刃②切(削)刃③切割边
cutting emulsion (=cutting compound) 切削液〔切削工具的润滑和冷却用〕
cutting fluid 切削液；乳化切削油
cutting fluid (oil) 切削液(油)
cutting hardness 切削硬度
cutting loss 裁切损耗
cutting lubricant 切削(润滑)液
cutting mill 切割磨
cutting off ①断路②割断
cutting oil 切削油
cutting oil additive 切削油添加剂
cutting out ①切断②断路
cutting paste 切削油膏
cutting plane algorithm 割平面法
cutting process 切削方法
cutting remnant 筒脚纱；管脚纱
cutting resistance 抗切割性
cutting ring method 环刀法
cuttings 切屑；刨屑
cutting table for soap 切皂机；切皂台
cutting tip ①切削刀片②割嘴；割尖〔氧割器的〕
cutting tool 切削工具〔机工〕；刨刀
cutting tool coolant oil 切削工具冷却油
cutting tool lubricant 切削工具润滑剂；刨刀润滑剂
cutting torch 割矩
cutting uniformity 裁断均匀性
cutting unit 切割装置
cutting velvet 割绒
cuttle fish 乌贼
cuttling 折布
cut-top lozenges 截顶菱形面
cut up 粉碎；切碎(断)
cut-up mill 切造车间
cuvee 酒桶内容物
cuvette (=cell) ①电池②小池③比色杯
CW laser 连续波激光器
cyamelide 氰白；三聚异氰酸 $(CNOH)_3$
cyameluric acid 氰白尿酸
cyameluric chloride 氰白尿酰氯
cyamurodiamide 氰尿二酰胺
cyan〔词头〕 氰基 CN—
cyanacetic acid 氰基乙酸

cyan-acetic ester ①氰基乙酸酯 $CNCH_2COOR$ ②〔专指〕氰基乙酸乙酯 $CNCH_2COOC_2H_5$
cyan alcohol 氰醇
cyanaldehyde 氰基乙醛
cyanamid 氰氨；氨基氰
cyanamide 氨腈*；氨基腈
cyanamide chloroformoxime 氨基腈氯甲醛肟
cyan-amide nitrogen 氰氨式氮
cyanamide process 氰氨法
cyananilide (=phenylcyanamide) 苯氨腈 C_6H_5NHCN
cyanate 氰酸盐 MOCN
cyanated 氰化了的
cyanating 氰化
cyanation ①氰化作用②氰化法
cyanato- 氰酰；氰氧基 $N \equiv CO—$
cyanaurite 氰亚金化物 $MAu(CN)_2$
cyan bromide 溴化氰 BrCN
cyancarbamic acid 氰代氨基甲酸 $C_2H_2O_2N_2$
cyan carbonic acid 氰基甲酸 CNCOOH
cyan chloride 氯化氰 ClCN
cyancoumarin 氰基香豆素 $C_9H_5O_2CN$
cyanethine 氰乙碱〔俗〕；2,6-二乙基-5-甲基-4-氨基嘧啶 $C_9H_{15}N_3$
cyan etholin 氰酸乙酯 C_2H_5OCN
cyanex process 萨亚乃克斯过程〔偕醇腈化学交换法分离碳同位素〕
cyan fluoride 氟化氰 FCN
cyanformate 氰(代)甲酸盐(或酯) CNCOOM(R)
cyanformic acid 氰(代)甲酸 CNCOOH
cyanguanidine 氰(基)胍
cyan halide 卤化氰 XCN
cyanhydrin 氰醇
cyanhydrin synthesis 氰醇合成法
cyanic acid (=hydrocyanic acid) 氢氰酸 HOCN
cyanidation 氰化*
cyanide 氰化物
cyanide bath 氰化物浴
cyanide fusant 氰熔体 $Ca(CN)_2 \cdot NaCl$
cyanide gold-refining 氰化物炼金法
cyanide green (氰化)铁绿；铁氰化铁
cyanide nitrogen 氰基氮
cyanide process 氰化物法
cyanide selective electrode 氰离子选择电极；氰根选择电极
cyanide titration 氰化物滴定
cyanidin chloride 氯化氰定 $C_{15}H_{11}ClO_6$
cyaniding 氰化
cyanidum 氰化物

cyanilide 苯氨腈 C_6H_5NHCN
cyanine ①花青〔染料类名〕；菁②(=quinolineblue)喹啉蓝 $C_{29}H_{35}B_2I$
cyanine dyes 花青染料
cyanite 蓝晶石
cyanmethine 2,6-二甲基-4-氨基嘧啶；氰甲碱〔俗〕 $C_6H_9N_3$
cyano- 氰基 $N{\equiv}C{-}$
cyano acceptor group 氰基受电子基团
cyanoacetaldehyde 氰基乙醛 $NCCH_2CHO$
cyanoacetamide 氰基乙酰胺 $NCCH_2CONH_2$
cyanoacetanilide 氰基乙酰苯胺 $C_6H_5NHCOCH_2CN$
cyano-acetate ①氰基乙酸盐 $CNCH_2COOM$ ②氰基乙酸酯 $CNCH_2COOR$
cyanoacetic acid 氰基乙酸 HO_2CCH_2CN
cyanoacetophenone 氰基苯乙酮 $(CN)C_6H_4COCH_3$
cyanoacetylene 丙炔腈 $HC{\equiv}CCN$
cyanoacrylate 氰基丙烯酸酯
cyanoacrylate adhesive 氰基丙烯酸酯黏合剂
β-cyanoalanine β-氰丙氨酸
ω-cyanoalkyl 端氰烷基；氰烷基
cyano-aniline (=cyananilide) 苯氨腈
cyanobacteria 蓝藻细菌
o-cyanobenzamide 邻氰基苯(甲)酰胺
cyanobenzene 苯甲腈
cyanobenzoic acid 氰基苯甲酸
cyanobenzyl 氰苄基 $CNC_6H_4CH_2-$
cyanobenzylchloride 氰苄基氯
cyano benzyl cyanide 氰基苯乙腈 $CNC_6H_4CH_2CN$
cyano-bonded phase 氰基键合相
cyano butyric acid 氰基丁酸 CNC_3H_6COOH
cyanocarbon 氰碳化合物
cyanocarbonic acid 氰(代)甲酸
cyanocobalamin 氰钴胺素；维生素 B_{12}
cyano derivative 氰基衍生物
cyanoethanol 氰乙醇
2-cyanoethyl acrylate 丙烯酸 2-氰基乙酯
cyanoethylated fibre 氰乙基化纤维
cyanoethylated polyvinyl alcohol 氰化基化聚乙烯醇
cyanoethyl cellulose 氰基纤维素
cyano-ethylene 丙烯腈
cyanoethyl ether of cellulose 纤维素氰乙基醚
1-cyanoethyl-2-ethyl-4-methylimidazole 1-氰乙基-2-乙基-4-甲基咪唑
1-cyanoethyl-2-ethyl-4-methylimidazole trimellitate 1-氰乙基-2-乙基-4-甲基咪唑合偏苯三酸酯
1-cyanoethyl-2-isopropylimidazole 1-氰乙基-2-异丙基咪唑
1-cyanoethyl-2-methylimidazole 1-氰乙基-2-甲基咪唑
1-cyanoethyl-2-methylimidazole trimellitate 1-氰乙基-2-甲基咪唑合偏苯三酸酯
1-cyanoethyl-2-phenyl-4,5-di (cyanoethoxymethyl) imidazole 1-氰乙基-2-苯基-4,5-二(氰乙氧甲基)咪唑
1-cyanoethyl-2-phenylimidazole 1-氰乙基-2-苯基咪唑
1-cyanoethyl-2-undecylimidazole 1-氰乙基-2-十一基咪唑
1-cyanoethyl-2-undecylimidazole trimellitate 1-氰乙基-2-十一基咪唑合偏苯三酸酯
cyanoform 氰仿；三氰基代甲烷 $CH(CN)_3$
cyanoformate (=cyanoformic ester) 氰基甲酸酯
cyanoformic acid 氰基甲酸 $CNCOOH$
cyanoformic ester (=cyanoformate) 氰基甲酸酯 $CNCOOR$
cyanogas 氰钙粉
cyanogen (=dicyanogen) 氰 $(CN)_2$
cyanogenation 氰化作用
cyanogen azide 叠氮化氰 CNN_3
cyanogen bromide 溴化氰 $BrCN$
cyanogen bromide method (=von Braun reaction) 溴化氰法
cyanogen chloride 氯化氰 $ClCN$
cyanogenetic 生氰的
cyanogen fluoride 氟化氰 FCN
cyanogen halide 卤化氰 XCN
cyanogenic 生氰的
cyanogen iodide 碘化氰 ICN
cyanogen-oxygen flame 氰氧焰
cyanogen sulfide 硫化氰 $(CN)_2S$
cyanoguanidine 氰基胍
cyanohydrin 羟腈*
cyanomaclurin 木波罗单宁 $C_{15}H_{12}O_6$
cyanomalonic ester 氰基丙二酸酯
cyano-manganese-complex 氰(络)锰酸盐
cyanometer 天空蓝度测定仪；青度计
cyanomethylation 氰甲基化*
cyano (methylmercuri) guanidine 甲基汞替氰基胍
cyanometric titration 氰滴定法
cyanomycin 蓝霉素
cyanophoric glycoside 氰基苷
cyanopropionic acid 氰基丙酸
cyanopropylene oxide 氰基丙烯化氧
cyanosensor 氰基传感器
cyano-silicone rubber 氰基硅橡胶
cyanosorbic acid 氰基山梨酸
cyano-styrene 氰基苯乙烯
cyanotype 氰印画法；蓝晒法

β-cyanovaleramide β-氰基戊酰胺
cyanovaleric acid 氰戊酸
cyanoximide 某氰肟 NCC(=NOH)—
cyanoximido-acetic acid 氰肟乙酸 NCC(=H)COOH
cyanozonolysis 氰臭氧分解
cyanphenine 2,4,6-三苯基-1,3,5-三嗪；2,4,6-三苯基均三嗪
cyanpyrrole 氰(代)吡咯
ω-cyan-undecanoic acid ω-氰基十一酸 CNCH$_2$(CH$_2$)$_9$COOH
cyanuramide 氰尿酰胺
cyanurate 氰尿酸酯
cyanurdiamide (=ammeline) 氰尿(酸)二酰胺；三聚氰酸二酰胺
cyanuric acid 氰尿酸；三聚氰酸 NHCCNHCONH—CO—
cyanuric acid ethyl ester 氰尿酸乙酯；三聚氰酸乙酯
cyanuric chloride 氰尿酰氯 C$_3$N$_3$Cl$_3$
cyanuric dye 氰尿染料
cyanuric ester 2,4,6-三羟基-1,3,5-三嗪酯；2,4,6-三羟基-均三嗪酯
cyanuric triazide 氰尿酰三叠氮；三叠氮代氰尿酰；三聚氰酰三叠氮
cyanurin 氰尿蓝
cyanuro 1,3,5-三嗪基；均三嗪基
cyanurodiamide 氰尿二酰胺
cyanurtriamide (=melamine) 氰尿酸三酰胺
cyanuryl chloride 氰尿酰氯；三聚氰酰氯 C$_3$N$_3$Cl$_3$
cyathin 蛋巢菌素
cybernetics 控制论
cyboma 集散微晶
cybotactates 群聚体
cybotactic 群聚的
cybotactic groups 群聚体
cybotactic region 群聚区
cybotactic state 群聚状态
cybotaxis 群聚；非晶体分子立方排列
cycart 循环筒(式)
cycart collector 循环过滤筒式集尘器
cycart filter 循环筒式过滤器
cycasin 苏铁素
cyclacidin 环杀菌素
cyclamal (=cyclamen aldehyde) 仙客来醛
cyclamen 仙客来；兔耳花〔俗〕
cyclamen aldehyde (=cyclamal) 仙客来醛
cyclamen aldehyde diethyl acetal 仙客来醛二乙缩醛
cyclamen aldehyde-2-methyl-2,4-pentandiol acetal 仙客来醛-2-甲基-2,4-戊二醇缩醛
cyclamen concrete 仙客来浸膏
cyclamen flower oil 仙客来花油
cyclamic acid (=cyclohexane sulfamic acid) 环己烷氨基磺酸 C$_6$H$_{11}$NHSO$_3$H
cyclamidomycin 赛拉霉素；环氨霉素
cyclamin 仙客来苷
cyclamiretin 仙客来亭
cyclamycin (=troleandomycin) 竹桃霉素三乙酸酯；环霉素
cyclanoline 轮环藤酚碱
cyclanone 环烷酮
cyclator 浓缩絮凝器；絮凝增稠器
cyclazine 环吖嗪
cycle ①循环②周期③环
cycle amplitude 循环振幅
cycleanine (=methyl-isochondodendrine) 轮环藤宁
cycle compound 环化合物
cycle controller 循环进程调节器
cycle counter 周波计数器
cycle efficiency 循环效率
cycle efficiency of cracking process 裂化过程的循环效率〔裂化连续运转时间对整个操作周期的百分比〕
cycle length 循环时间
cycle log (自动化)程序控制器；程序调整器；程序装置
cycle matrix 循环矩阵
cyclene 环烯
cycle of operation (裂化过程中)连续运转周期
cycle oil 循环油；催化裂化油
cycle parameter 周期参数
cycle penalty 循环障碍
cycle period 循环周期
cycle rate 循环速度(率)
cycle selector 循环选择机
cycle speed 循环速度
cycle stock 循环油；回炼油
cycle time 循环时间
cycle-time efficiency 连续运转效率
cycle timer 循环定时器
cycle tube valve 循环管阀
cyclic ①环状的②循环的
cyclic action 环化作用
cyclic activation 循环活化
cyclic AC voltammetry 循环交流伏安法
cyclic addition 环化加成；环状加成；成环加成
cyclic adenosine monophosphate 环磷腺苷〔药〕
cyclic adsorption refining 循环吸附精制
cyclical 环的；环状的；循环的

cyclical chromatography　①环形色谱(法)②循环色谱(法)
cyclic alcohol (=cycloalcohol)　环(状)醇
cyclic alkanol　环烷醇
cyclically ordered pulse sequence　循环指令脉冲序列
cyclical operation　循环操作
cyclic alternating current voltammetry　循环交流伏安法
cyclic(al) twin　轮式双晶
cyclic amide　环酰胺
cyclic amine　环胺
cyclic anhydride　环酐
cyclic base　环(状)碱
cyclic bond　环(内)键
cyclic borate　环硼酸盐
cyclic carbonate　环状碳酸酯
cyclic carbonate compound　环状碳酸酯化合物
cyclic chronopotentiometry　循环计时电位法
cyclic compound　环(状)化合物
cyclic compound with side chain　含有侧链的环状化合物
cyclic compression　反复压缩
cyclic condensation　环(状)缩合
cyclic condensation polymer　环状缩聚物
cyclic coordinate　环坐标
cyclic crown ether　环状冠醚；环式冕醚
cyclic diester (=lactide)　环(状)二酯；交酯
cyclic dimer　环(状)二聚物
cyclic dimerization　环化二聚(作用)
cyclic dimethyl polysiloxane　环(状)二甲基聚硅氧烷 $[(CH_3)_2SiO]_n$
cyclic diolefine　环二烯
cyclic diureide　环二酰脲
cyclic electronic current effect　环电流效应
cyclic ester　环(状)酯
cyclic ether　环醚
cyclic ethylene urea　环乙撑脲
cyclic fluidized-bed combustion boiler　循环流化床锅炉
cyclic formula　环状式
cyclic guanosine monophosphate　环鸟苷酸
cyclic hexaglycine　环六甘氨酸；环六氨基乙酸
cyclic hydrocarbon　环烃
cyclic imide　环状亚胺
cyclic inner ether　环状内醚
cyclic ketone　环酮
cyclic ketone resin　环酮树脂
cyclic ketoxime　环酮肟
cyclic lactone　环内酯
cyclic link (=cyclic linkage)　环(内)键
cyclic linkage (=cyclic link)　环(内)键

cyclic loading　周期载荷
cyclic memory　循环存储器
cyclic methyl phenylpolysiloxane　环状聚甲基苯基硅氧烷；环状甲基苯基聚硅氧烷　$[(CH_3)Si(C_6H_5)O]_n$
cyclic molecule　环状分子
cyclic monomer　环状单体
cyclic motion　循环运动
cyclic octamethyltetrasiloxane　环状八甲基四硅氧烷 $[(CH_3)_2SiO]_4$
cyclic olefine　环烯
cyclic olefinic bond　环内双键
cyclic oligoamide　环状低聚酰胺
cyclic oligoester　环状低聚酯
cyclic oligo (ethylene isophthalate)　环低聚间苯二甲酸乙二(醇)酯
cyclic oligomer　环(状)低聚物
cyclic phosphonate ester　环膦酸酯
cyclic phosphonitrilic dihalide trimer　环状二卤磷腈三聚物
cyclic plasticizer　环状增塑剂
cyclic polycondensation　环化缩聚(作用)
cyclic polyene chromogen　环多烯发色体
cyclic polymer　环状聚合物
cyclic polymerization　环化聚合(作用)
cyclic polyolefine　环状(结构的)聚烯
cyclic polysilazane　环状聚硅氮烷
cyclic potential-sweep voltammetry　循环电位扫描伏安法
cyclic regeneration　连续再生
cyclic regenerative mechanism　循环再生机理
cyclic-ring structure　环状结构
cyclic rubber　环化橡胶
cyclics　环状化合物
cyclic scan　循环扫描
cyclic staircase voltammetry　循环阶梯伏安法
cyclic stationary electrode voltammetry　循环静止电极伏安法
cyclic stress　周期应力；交变应力
cyclic stress energy dissipation　周期应力能量耗散
cyclic stress relaxation　周期应力松弛
cyclic stress-strain　周期应力-应变
cyclic structure　环状结构
cyclic substituent　环式取代基
cyclic sulfide　环状硫化物
cyclic surface　圆纹曲面
cyclic system　环路系统
cyclic terpene　环式萜烯；环萜(烃)
cyclic tertiary base　环式叔碱

cyclic tetracarboxylic dianhydride 环四羧酸二(酸)酐
cyclic time 循环时间
cyclic transition state 环化过渡态
cyclic triangular wave polarography 循环三角波极谱法
cyclic triangular wave voltammetry 循环三角波伏安法
cyclic trimer 环状三聚物
cyclic trimerization 环化三聚(作用)
cyclic voltammetry 循环伏安法
cyclic voltammogram 循环伏安图
cyclic wetting and drying test 干湿循环试验
cyclinder gas 钢瓶气体
cycling plant 回收地层内气体的天然石油厂
cyclite 赛克炸药；赛克拉特；二溴苄
cyclitol 环多醇
cyclization 环化
cyclization of olefines 烯烃环化
cyclization of paraffins 烷烃环化
cyclization polymerization 环化聚合
cyclization process 环化过程
cyclized oligomer 环化低聚物
cyclized polymer 环化聚合物
cyclized rubber 环化橡胶
cyclized rubber resin 环化橡胶树脂
cyclized structure 环化结构；环形结构
cyclizing 环化〔橡胶表面处理〕
cyclo…ene ①环…烯②芳…环〔按 IUPAC 有机命名法规定在此词之前若用 benzo-等词头时，此词代表芳环如: benzocyclooctene 苯并芳辛环,指苯并环辛间四烯〕
cycloaddition 环加成
cycloaddition polymerization 成环加成聚合(作用)
cycloaddition reaction 环加成反应
cycloalcohol 环式醇
cycloaliphates 脂环烃
cycloaliphatic (=alicyclic) 脂环(族)的
cycloaliphatic epoxy 脂环系环氧
cycloaliphatic ketoximes 环脂(肪)酮肟类
cycloalkane 环烷
cycloalkanoates 环烷金属化合物
cycloalkanol 环烷醇
cycloalkanone 环烷酮
cycloalkapolyene 环(烷)多烯
cycloalkene 环烯
cycloalkene polymerization 环烯聚合
cycloalkenyl group 环烯基
cycloalkyl 环烷基
cycloalkylation 环烷化(作用)
cycloalkyl benzyl ether 环烷基苄醚

cycloalkyne(s) 环炔(烃类)
cycloalliin 环蒜氨酸
cycloamination 环胺化(作用)
cycloaminium 环铵
cycloartenol 环阿屯醇；9,19-环-24-羊毛甾-3β-醇
cycloartenone 环阿屯酮；9,19-环-24-羊毛甾-3-酮
cyclobutadiene 环丁二烯 C_4H_4
cyclobutadipyrimidine 环丁二嘧啶
cyclobutane-carboxylic acid 环丁烷羧酸
cyclobutanedinone 环丁(烷)二酮
cyclobutane (=tetramethylene) 环丁烷
cyclobutanol 环丁醇 $CH_2CH_2CH_2CHOH$
cyclobutanone 环丁(烷)酮 $(CH_2)_3CO$
cyclobutene 环丁烯
cyclobutyl 环丁基 C_4H_7-
cyclobutyl methyl ketone 环丁基甲基甲酮
cyclobuxine 环布辛；环黄杨星
cyclochlorotin 环氯霉素
cyclocitral 环柠檬醛
cyclocitronellol 环香茅醇
cyclocitrylidene acetic acid 环柠檬叉乙酸
cyclocolorenone 环色烯酮
cyclo compounds 环状化合物
cyclocondensation 环化缩合(作用)
cyclocooligomerization 成环低共聚合
cyclocopolymer 环状共聚物；成环共聚物
cyclocopolymerization 成环共聚合(作用)
cyclocta sulfur 八环硫黄
cyclodeca- 芳癸并〔指环癸间五烯并〕
cyclodecanone 环十酮
cyclodehydration 环化脱水作用；脱水成环作用
cyclodehydrogenation 环化脱氢作用
cyclodepolymerization 环化解聚反应
cyclodepsipeptide 环缩酚肽
cyclodextrin 环糊精
cyclodextrin derivative gas chromatography 环糊精衍生物气相色谱
cyclodextrin electrokinetic chromatography 环糊精电动色谱(法)
cyclodextrin stationary phase 环糊精固定相
cyclodihydromyrcene 环二氢月桂烯
cyclodimerisation 环二聚
cyclododecalactam 环十二烷内酰胺〔尼龙 12 的原料〕
cyclododecane 环十二烷〔尼龙 12 的中间体〕
cyclododecanol 环十二烷(基)醇
cyclododecanone 环十二酮

cyclododecanone dimethyl ketal 环十二酮二甲缩酮
cyclododecatriene 环十二碳三烯
cyclododecene 环十二烯
cyclododecenol 环十二烯醇
cyclododecyl ethyl ether 环十二烷(基)乙醚
cycloethylene brassylate 十三烷二羧酸环乙撑酯；麝香T
cyclofenchene 环葑烯 $C_{10}H_{16}$
cyclogeranic acid 环牻牛儿酸 $C_{10}H_{16}O_2$
cyclogeraniol 环香叶醇
cyclogeraniolene 环香叶烯
cyclogeraniolene alcohol 环香叶烯醇
cyclogeranyl acetate 乙酸环香叶酯
cyclohepta- 芳庚并（指环庚三烯并）
cycloheptaamylose 环庚直链淀粉；β-环状糊精
cycloheptadecanone (=dihydrocivetone) 二氢灵猫酮；环十七(烷)酮
cycloheptadecenone 环十七烯酮
9-cycloheptadecen-1-one 9-环十七烯-1-酮
cycloheptadiene 环庚二烯 C_7H_{10}
1,2-cycloheptaenedionedioxime (=heptoxime) 1,2-环庚二酮二肟 $C_7H_{10}(NOH)_2$
cycloheptane (=suberane) 环庚烷；软木烷
$CH_2(CH_2)_5CH_2$
cycloheptane aldehyde 环庚烷(基)甲醛
cycloheptane-spirocyclo-propane 环庚烷-螺环丙烷；螺[2.6]壬烷
cycloheptanol (=suberol) 环庚醇；软木醇
cycloheptanone (=suberone) 环庚酮；软木酮
$CH_2(CH_2)_5CO$
cycloheptasulfur 环七硫
cycloheptatriene 环庚三烯
cycloheptatrienium ion 环庚三烯锑离子
cycloheptatrienylium cation 环庚三烯正离子
cycloheptene (=suberene) 环庚烯；软木烯
$CH_2(CH_2)_4CH=CH$
cycloheptyl 环庚基 $C_7H_{13}-$
cycloheptylamine 环庚胺
cyclohexaamylose 环己烷直链淀粉；α-环状糊精
cyclohexadecanolide (=dihydro ambrettolide) 二氢黄葵内酯；环十六内酯
cyclohexadecanone 环十六酮
1,3-cyclohexadiene 1,3-环己二烯 C_6H_8
1,4-cyclohexadiene 1,4-环己二烯
cyclohexadiene dioxide 二氧化环己二烯；二环氧基环己烷
cyclohexadienyl 环己二烯基

cyclohexadienylene 亚环己二烯基 $—C_6H_6—$
cyclohexadienylidene (=phenylidene) 环己二烯亚基
2,5-cyclohexadienylidene (=phenylidene) 2,5-环己二烯亚基
cyclohexadienyne (=benzyne) 环己二烯一炔；苯炔
1,4-cyclohexalene diisocyanate 1,4-亚环己基二异氰酸酯；1,4-环己撑二异氰酸酯
cyclohexaline 环己醇
cyclohexan-diol 环己二醇
cyclohexane 环己烷 $CH_2(CH_2)_4CH_2$
cyclohexaneacetic acid 环己烷基乙酸
cyclohexane-carboxylic acid (=hexahydrobenzoic acid) 环己烷羧酸；六氢化苯甲酸 $CH_2(CH_2)_4CHCOOH$
cyclohexanediamine chelate 环己二胺螯合物
cyclohexanediamine tetraacetic acid 环己二胺四乙酸
cyclohexanedimethanol 环六烷二甲醇
1,4-cyclohexane dimethanol 1,4-环己烷二甲醇
cyclohexane dimethanol succinate 丁二酸环己烷二甲醇酯
cyclohexanediol 环己二醇
cyclohexanehexol (=inositol;cyclohexanhexanol) 环己六醇；肌醇 $(CHOH)_6$
cyclohexane number 环己烷值
cyclohexane oxidation process 环己烷氧化法
cyclohexane oxide 环己烷氧化物
cyclohexanepentol 环己五醇
cyclohexane-rich 富环己烷的
cyclohexane sulfamic acid (=cyclamic acid) 环己烷氨基磺酸 $C_6H_{11}NHSO_3H$
cyclohexanethanol 环己基乙醇
1,3,5-cyclohexane trione trioxime 环己烷间三肟；环己三酮肟
cyclohexanhexanol (=inositol;cyclohexanehexol) 环己六醇；肌醇 $(CHOH)_6$
cyclohexanhexanone 环己六酮 $(CO)_6$
cyclohexanol (=hexalin) 环己醇 $CH_2(CH_2)_4CHOH$
cyclohexanol acetate 乙酸环己酯
cyclohexanol phenylacetate 苯乙酸环己酯
cyclohexanone 环己酮 $CH_2(CH_2)_4CO$
cyclohexanone carboxylic acid 环己酮羧酸
cyclohexanone-formaldehyde resin 环己酮-甲醛树脂
cyclohexanone-oxime 环己酮肟 $CH_2(CH_2)_4C=NOH$
cyclohexanone oxime route 环己酮肟路线
cyclohexanone peroxide 过氧化环己酮〔引发剂〕

cyclohexanpentol (=quercitol)　环己五醇；栎醇　$CH_2(CHOH)_4CHOH$

cyclohexanyl (=cyclohexyl)　环己基　C_6H_{11}—

cyclohexapyrazine　环己基并吡嗪

cyclohexasilanyl　环己硅烷基　$SiH_2(SiH_2)_4SiH$—

cyclohexatriene　环己三烯

cyclohexene　环己烯　$CH_2(CH_2)_3CH=CH$—

cyclohexene hydroperoxide　氢过氧化环己烯　$C_6H_{10}O_2$

cyclohexene oxide　氧化环己烯

cyclohexenyl　环己烯基　C_6H_9—

cyclohexenylene　亚环己烯基　—C_6H_8—

cycloheximide　放线菌酮；环己酰亚胺

cyclohexitol　环己六醇类化合物

cyclohexyl　环己基　C_6H_{11}—

cyclohexyl acetaldehyde　环己基乙醛

cyclohexyl acetate　乙酸环己酯　$CH_3CO_2C_6H_{11}$

cyclohexyl acetic acid　环己基乙酸

cyclohexyl acrylate　丙烯酸环己酯

cyclohexylamine　环己胺　$C_6H_{11}NH_2$

cyclohexylamine acetate (=fixing salt)　凝固盐；乙酸环己胺〔凝固剂〕

cyclohexylamine lactate　乳酸环己胺

cyclohexylamine salt of 2-mercaptobenzothiazole　2-硫醇基苯并噻唑环己胺盐〔促进剂〕

cyclohexyl aminoalkane carboxylic acid　环己氨基链烷羧酸

cyclohexyl o-aminobenzoate　邻氨基苯甲酸环己酯

cyclohexylammonium chloride　氯化环己铵〔促进剂〕

p-cyclohexyl-anisole　对环己基茴香醚　$CH_3OC_6H_4C_6H_{11}$

cyclohexyl-azocarbonitrile　环己基偶氮碳腈〔发泡剂〕

4-cyclohexyl benzaldehyde　4-环己基苯甲醛

N-cyclohexyl-2-benzothiazole sulfenamide　N-环己烷基-2-苯并噻唑次磺酰胺

cyclohexyl bromide　环己基溴　$C_6H_{11}Br$

cyclohexyl butyrate　丁酸环己酯

cyclohexylcarbinol　环己基甲醇

cyclohexyl chloride　环己基氯　$C_6H_{11}Cl$

cyclohexyl cinnamate　肉桂酸环己酯

2-cyclohexyl cyclohexanol　2-环己基环己醇

2-cyclohexyl cyclohxanone　2-环己基环己酮

2-cyclohexyl dithiobenzimidazole　2-环己基二硫代苯并咪唑

2-cyclohexyl dithiobenzothiazole　2-环己基二硫代苯并噻唑〔促进剂〕

cyclohexylene　亚环己基　—C_6H_{10}—

cyclohexylene-1,2-diisocyanate　亚环己基-1,2-二异氰酸酯

β-cyclohexyl ethanol　β-环己基乙醇

N-cyclohexyl-p-ethoxyaniline　N-环己基对乙氧基苯胺〔防老剂〕

cyclohexyl ethyl acetate　乙酸环己基乙酯

cyclohexyl ethyl alcohol　环己基乙醇

cyclohexyl ethylamine　环己基乙基胺〔促进剂〕

N-cyclohexylethylammonium cyclohexyl ethyl dithiocarbamate　环己基乙基二硫代氨基甲酸 N-环己基乙基铵〔促进剂〕

cyclohexyl formate　甲酸环己酯　$C_6H_{11}COOH$

cyclohexyl hydroperoxide　环己基过氧化氢

cyclohexylidene　亚环己基　$CH_2CH_2CH_2CH_2CH_2C$=

cyclohexylidene methylamine　亚环己基甲胺

cyclohexyl iodide　环己基碘　$C_6H_{11}I$

cyclohexyl isovalerate　异戊酸环己酯

cyclohexyl mercaptan　环己硫醇　$C_6H_{11}SH$

cyclohexyl methacrylate　甲基丙烯酸环己酯　$CH_2=C(CH_3)C(O)OC_6H_{11}$

N-cyclohexyl-p-methoxyaniline　N-环己基对甲氧基苯胺〔防老剂〕

cyclohexyl oxalate　草酸环己酯

cyclohexyl oxypropionitrile　环己氧基丙腈

cyclohexyl peroxide　环己基过氧化物

cyclohexyl β-phenylacrylate　β-苯丙烯酸环己酯

N-cyclohexyl-p-phenylenediamine　N-环己基对苯二胺〔防老剂〕

N-cyclohexyl-4-phenyl-2-thiazolesulfenamide　N-环己基-4-苯基-2-噻唑次磺酰胺〔促进剂〕

cyclohexyl phthalate　苯二甲酸环己酯

cyclohexyl polysulfide　环己基多硫化物

cyclohexyl propionate　丙酸环己酯

N-cyclohexyl pyrrolidone　N-环己基吡咯烷酮〔溶剂〕

cyclohexyl ring　环己基环

cyclohexyl stearate　硬脂酸环己酯

cyclohexyl sulfamic acid　环己氨基磺酸

N-cyclohexyl taurine　N-环己基牛磺酸　N-$C_6H_{11}NHCH_2CH_2SO_3H$

cyclohexyl tetralin sulfonate　萘满磺酸环己酯

N-(cyclohexylthio)-o-benzoic sulfimide　N-(环己硫代)邻苯硫酰亚胺〔促进剂〕

cyclohexyl thiocarbamyl disulfide　二硫化环己基硫代氨基甲酰〔促进剂〕

N-cyclohexylthiophthalimide　N-环己基硫代邻苯二甲酰亚胺〔防焦剂〕

cycloid　摆线

cycloidal blower　摆旋鼓风机
cycloidal gas meter　摆旋气表；摆线式气表
cycloidal ion path　回旋离子轨迹；摆线离子轨道
cycloidal mass spectrometer　摆线质谱仪*
cycloidal pump　摆旋泵
cycloid gear　摆线齿轮
cycloisomerisation　环异构(化)
cyclol　环醇
cyclolavandulal　环薰衣草醛
cyclolavandulol　环薰衣草醇
cyclomethicone　环(二)甲基硅酮
cyclomethyl hexanol　甲基环己醇
cyclomethyl hexanone　甲基环己酮
cyclomone (=2-n-heptyl-3-methyl cyclopentanone)　2-庚基-3-甲基环戊酮
cyclomonoolefin　环单烯(属)烃
cyclomycin (=tetracycline)　四环素
cyclone　①旋风分离器②旋风除尘器③环酮〔通指：四芳基环戊二烯酮，专指：四苯基环戊二烯酮〕
cyclone bed　旋流床
cyclone burner　旋风燃烧器
cyclone catalyst collector　旋风催化剂收集器
cyclone collector　旋风收集器
cyclone combustion chamber　旋风式燃烧室
cyclone dust collector　旋风除尘器
cyclone dust separator　旋风分尘器
cyclone evaporator　旋风蒸发器
cyclone filter　旋风过滤器
cyclone furnace　旋风炉
cyclone hopper　旋风式加料斗
cyclonene (=piperonyl cyclonene)　环酮烯；胡椒基环己烯酮
cyclone overflow　旋流分离器溢流
cyclone reclaim powder intake　旋风分离器回收粉入口
cyclone screener　旋风筛分机
cyclone scrubber　①旋风涤气器②旋风洗涤器③旋风集尘器
cyclone separated diatomaceous　旋风分离硅藻土
cyclone separator　旋风分离器
cyclone thickener　旋风增稠器
cyclone type electrostatic spraying apparatus　旋风式静电喷漆装置
cyclone underflow　旋流分离器底流
cyclonic current　旋流
cyclonic gas scrubber　旋风涤气器
cyclonic spray tower　旋风喷淋器
cyclonite　旋风炸药；三次甲基三硝基胺
cyclonol　三甲基环己醇

cyclonona-　芳壬并〔指环壬间四烯并〕
cyclononadecanone　环十九酮
cyclononane　环壬烷　$CH_2(CH_2)_7CH_2$
cyclononanone　环壬酮
cyclon thickener　旋风增稠器
cycloocta-　芳辛并〔指环辛间四烯并〕
cyclooctaamylose　环辛烷直链淀粉；γ环状糊精
cyclooctadiene　环辛二烯
cyclooctane　环辛烷　$CH_2(CH_2)_6CH_2$
cyclooctatetraene　环辛四烯
cyclooctatetraenyl compound　环辛四烯化合物
cyclooctene　环辛烯
cyclooctyne　环辛炔
cycloolefines　环烯
cycloolefin polymer　环烯烃聚合物
cyclooligomerization　环齐聚合(作用)
cyclooxahexadecanolide　氧杂环十六内酯
cycloparaffin　环烷
cycloparaffinic hydrocarbons　环烷烃
cyclopean concrete　块石混凝土
cyclopenta-　环戊二烯并
cyclopentadecane　环十五烷　$CH_2(CH_2)_{13}CH_2$
cyclopentadecanol　环十五(烷)醇
cyclopentadecanone　环十五(烷)酮　$CH_2(CH_2)_{13}CO$
cyclopentadiene　环戊二烯
1,3-cyclopentadiene　1,3-环戊二烯
cyclopentadienide anion　环戊二烯负离子
cyclopentadienyl　环戊二烯基　C_5H_5-
cyclopentadienyl complex compound　环戊二烯(基)配合物；环戊二烯基配位化合物；茂基配位化合物
cyclopentadienylidene　亚环戊二烯基　$CH=CHCH=CHC=$
cyclopentadienyl potassium　环戊二烯合钾　C_5H_5K
cyclopentadienyl sodium　环戊二烯合钠　C_5H_5Na
cyclopentamine　环戊胺
cyclopentane (=pentamethylene)　环戊烷
cyclopentane-carboxylic acid　环戊烷甲酸　$(CH_2)_4CHCOOH$
cyclopentane-1,1-dicarboxylic acid　环戊-1,1-二羧酸
cyclopentane-1,2-diol　环戊-1,2-二醇
cyclopentane-1,2-dione　环戊-1,2-二酮
cyclopentane tetracarboxylic acid dianhydride　环戊烷四甲酸二酐

cyclopentane tetracarboxylic mono anhydride　环戊烷四甲酸一酐
cyclopentanethiol　环戊硫醇
cyclopentano-　环戊(并)
cyclopentanol　环戊醇
cyclopentanone (=dumasin)　环戊酮
cyclopentanone oxime　环戊酮肟
cyclopentanoperhydro-phenanthrene　甾环；环戊环多氢菲
cyclopenta[c]pyridine　阿克亭
cyclopentene　环戊烯
cyclopentenone　环戊烯酮
cyclopentenyl　环戊烯基　C_5H_7—
cyclopentenylidene　亚环戊烯基
1-cyclopentenyl vinyl methyl ketone　1-环戊烯基乙烯基甲酮
cyclopentyl　环戊基　C_5H_9—
cyclopentyl carbinol　环戊基甲醇
cyclopentylene　1,5-亚环戊基　—C_5H_8—
1,2-cyclopentylene dinitrilo tetraacetic acid (CPDTA)　1,2-亚环戊基二次氮基四乙酸　$C_5H_8[N(CH_2COOH)_2]_2$
cyclopentylidene　亚环戊基　$CH_2CH_2CH_2CH_2C{=}$
cyclopentyl mercaptan　环戊硫醇
cycloperoxene　环过氧烯〔类名,过氧桥加在链式共轭烯1,4-位上或环二烯上〕
cyclophane　环芳
cyclophosphamide　环磷酰胺〔药〕
cyclopin　圆弧素
cyclo-polycondensation reaction　环化缩聚反应
cyclopolymerization　环化聚合
cyclopolyolefin　环状聚烯烃
cyclo (poly) siloxane　聚环硅氧烷
cyclopropanaphthalene　环丙烷萘
cyclopropane-carboxylic acid　环丙烷甲酸
cyclopropane dicarboxylic acid　环丙烷二甲酸　$C_3H_4(COOH)_2$
cyclopropane-1,1,2-tricarboxylic acid　环丙烷-1,1,2-三羧酸
cyclopropane (=trimethylene)　环丙烷
cyclopropanone　环丙酮
cyclopropene　环丙烯　C_3H_4
cyclopropenyl cation　环丙烯阳离子
cyclopropenyl radical　环丙烯基
cyclopropyl　环丙基　C_3H_5—
cyclopropylamine　环丙胺
cyclopropyl-carbinol　环丙基甲醇
cyclopropylene　环丙烯

cyclopropylidene　亚环丙基
cycloreversion　裂环(作用)
cycloreversion reaction　裂环反应
cyclorubber　环化橡胶
cycloserine　环丝氨酸
cyclosidic linkage　环(式)侧键
cyclosilane　环硅烷
cyclosilazane　环硅氮烷；环硅氨烷
cyclosilicate　环状硅酸盐
cyclosiloxane　环硅氧烷　$(SiH_2O)_n$
cyclosiloxane polymerization　环硅氧烷聚合
cyclosilthiane　环硅硫烷
cyclosubstituted　环取代的
cyclotetradecanone　环十四酮
cyclotridecanone　环十三酮
cyclotrimerization　环化三聚合(作用)
cyclotrimethylene trinitramine (=onit)　渥尼脱；翁尼特；环三亚甲基三硝胺
cyclotrisiloxane　环丙硅氧烷；三聚硅氧烷　$(SiH_2O)_3$
cyclotristannothiane　环三锡(杂)硫烷；环三硫杂锡己烷
cyclotron　回旋加速器
cyclotron resonance mass spectrometer　回旋共振质谱仪
cycloundecanone　环十一酮
cycloversion catalytic cracking process (=cycloversion process)　固定床矾土催化裂化过程
cycloversion desulphurization　固定床矾土催化脱硫过程
cycloversion process (=cycloversion catalytic cracking process)　固定床矾土催化裂化过程
cycloversion reforming process　固定床矾土催化重整过程
cyclural　环己烯基巴比妥
cydonic acid　木瓜酸；十七(烷)酸
cydonin　榲桲胶糖　$C_{18}H_{28}O_{14}$
cylinder　①圆筒②量筒③钢瓶；钢筒④气缸；汽缸⑤机筒
cylinder and dial machine　双面圆型针织机
cylinder block　气缸柱；气缸座
cylinder body　气缸体
cylinder bore　气缸直径
cylinder borer　镗缸机
cylinder boring machine　镗缸机
cylinder bridge　圆筒电桥
cylinder cap　气缸盖
cylinder cast iron　气缸铸铁
cylinder clearance　气缸余隙
cylinder coordinate　柱面坐标
cylinder cover　气缸盖
cylinder drier　筒式干燥器；干燥筒

cylinder drying　滚筒干燥
cylinder drying machine　筒式干燥机
cylinder extension　滚筒伸长部分
cylinder fillet　大辊筒针布
cylinder head　气缸盖
cylinder (head) gasket　气缸床；气缸垫片
cylinder heating　圆筒加热
cylinder lapping machine　气缸研磨机
cylinder liner　气缸衬里；气缺套
cylinder lubricator　(蒸馏制动器的)气缸润滑器
cylinder machine　圆网(造纸)机
cylinder mould　圆网
cylinder mould machine　圆网抄浆机
cylinder oil　气缸油
cylinder-piston viscometer　圆筒活塞式黏度计
cylinder plate method　筒碟法；杯碟法〔测抗菌素等用〕
cylinder roaster　筒式煅烧炉
cylinder roasting　筒内煅烧
cylinder singeing machine　柱式燎烤机
cylinder sleeve　气缸套
cylinder still　直立筒式蒸馏锅
cylinder stock　气缸油
cylinder stock oil　气缸油
cylinder stock solution　气缸油溶液
cylinder temperature　①汽缸温度②筒身温度
cylinder undercasing　大漏底
cylinder valve　气缸阀
cylinder vat　圆网机浆槽
cylinder wall　气缸壁
cylinder water jacket　气缸水套
cylinder wear　气缸磨损
cylinder yankee machine　圆网单缸造纸机
cylind lapping machine　气缸研磨机
cylindrial polar coordinate system　柱面极坐标系
cylindrical　圆筒(形)(的)
cylindrical bulb　长圆球
cylindrical can　圆桶；圆形罐
cylindrical capillary tube　圆筒形毛细管
cylindrical cell　圆柱池；圆筒池
cylindrical condenser　①管形(圆柱形)电容器②筒形冷凝器③柱形聚光器
cylindrical container　圆筒形储罐
cylindrical coordinate　柱面坐标
cylindrical deflector analyzer　圆(柱)形偏转分析器
cylindrical diaphragm　隔膜筒
cylindrical drier　圆筒干燥机
cylindrical fibre　圆(截面)纤维；圆柱形纤维
cylindrical gauze oxidizer　网筒氧化器
cylindrical graphite furnace　管状石墨炉
cylindrical mandrel apparatus　圆柱轴试验装置
cylindrical micelle　圆柱形胶束
cylindrical mill　柱式磨
cylindrical mirror analyzer　筒镜分析器
cylindrical model　圆筒模型
cylindrical piston water meter　筒式活塞水表
cylindrical retort　圆(筒形)甑
cylindrical roaster　柱式煅烧炉
cylindrical rodlike spindle　(黏度计的)圆柱棒状测量轴
cylindrical screw　圆柱螺旋
cylindrical shell section　筒节
cylindrical tank　圆(筒形)槽
cylindrical test piece　圆柱形试片
cylindrical tube　柱形管
cylindrical void　①条形孔隙②条孔〔纤维缺陷〕
cylindrical washer　洗涤筒；筒式洗涤器
cylindrical worm　圆柱形蠕虫状〔胶束形状〕
cylindric water jetting　滚筒式射流喷网
cylindrulite　柱晶
cylindrulite (row structure)　(圆)柱晶
cymaric acid　磁麻酸
cymarigenin　磁麻配基〔取自磁麻 Apocynum cannabium〕
cymarin　磁麻苷
cymarose　磁麻糖　$C_7H_{14}O_4$
cymbopol　茅属醇
cymene　伞花烃；甲基异丙基苯　$CH_3C_6H_4CH(CH_3)_2$
o-cymene　邻伞花烃；邻异丙基苯甲烷
p-cymene　对伞花烃；对异丙基苯甲烷
cymene alcohol　伞花烃醇
p-cymene hydroperoxide　对甲基异丙基苯过氧化物
cymenol　伞花醇
2-p-cymenol　2-对伞花醇
p-cymen-8-ol　对伞花-8-醇
cymenyl (=cymyl)　伞花基
cymic acid　甲基异丙基苯甲酸
cymidine　伞花碱；氨基伞花烃
cymogene　粗丁烷；近乎纯的丁烷
cymyl (来自 cymene)　伞花基；甲基异丙苯基
p-cymyl propanal　对伞花基丙醛
cynanchogenin　牛皮消苷元
cynarin　洋蓟酸；二咖啡酰奎尼酸
cynnematin (=cephalosperin N)　头孢菌素
cynoctonine　北乌头碱
cynodontin　长蠕孢犬牙素；四羟基甲基蒽醌
cynoglossine　倒提壶碱
cynoglossophine　倒提壶碱
cynotoxin　磁麻毒

cynurenic acid (=carbostyrilic acid)　犬尿酸
cyperenal　莎草烯醛；香附烯醛
cyperene　莎草烯；香附烯
cyperol　莎草醇；香附醇
cyperone　莎草酮　$C_{15}H_{22}O$
cyperus oil　莎草油
cypral　丝柏油醛
cypress camphor　柏木脑
cypressene　丝柏油烯
cypress lavender-cotton oil　檀香艾油
cypress oil　柏油
cyprian vitriol　塞浦路斯矾；硫酸铜与硫酸锌复盐　$CuSO_4 \cdot 3ZnSO_4 \cdot 28H_2O$
cyprite　辉铜矿
cyproheptadine　赛庚啶〔药〕
cyrtolerinetin　曲特素
cyrtomine　贯众明
cyrtominetin　贯众亭
cyrtopterinetin　冷蕨亭

cysteine-carbazole reaction　半胱氨酸-咔唑反应
cysteine sulfuric acid reaction　半胱氨酸-硫酸反应
cystine (=dithiobisalanine)　胱氨酸
cytarabine　阿糖胞苷〔药〕
cytisine　野靛碱；金雀花碱
cytochrome C　细胞色素 C〔药〕
cytochrome C electrochemistry　细胞色素 C 电化学
cytochrome oxidase (=indophenol oxidase)　靛酚氧化酶
cytometer　血细胞计数仪
cytomimetic biomaterial　细胞膜仿生材料
cytomixis　细胞融合
cytosensor　细胞传感器
cytosine riboside　胞(嘧啶核)苷
cytoskeleton　细胞骨架
cytospaz　莨菪碱
cytoxan　环磷酰胺
Czerny-Turner mounting　切尔尼-特纳装置
Czochralski method　晶体生长提拉法*

D

d〔符号〕比重
D- 右型
d- 右旋
dabber ①(半球形)软毛漆刷②(飞金，烫金用的半球形)飞金刷③(=inking ball)着墨器
dabbling 灌注的
dab oil 比目鱼油
dacarbazine 达卡巴嗪〔药〕
dacite 石英安山岩
dacrene 泪柏二萜
Dacron color-sealed black staple 大可纶色封型黑色短纤维〔指熔体染色型纤维〕
dacron fibre 聚(对苯二甲酸乙二)酯纤维
dacron terephthalate fiber 达可纶
dacryagogue 催泪剂
dacrydene 泪柏烯
Dacthal(=dimethyl tetrachloroterephthalate) 四氯代对苯二甲酸二甲酯
dactinomycin(=actinomycin D) 放线菌素 D
dactylarin 四氢三羟基甲氧基甲基甲基蒽二酮；指孢霉素
dactylin 鸭茅黄素；鸭茅灵
dado joint 槽接接头
dag 石墨粉
dagincoleic acid 达金醇酸〔熔点 170℃，取自婆罗洲哒玛〕$C_{22}H_{44}O_4$
dagingenoleic acid 达金醇裂酸〔熔点 125～126℃，取自婆罗洲玛〕$C_{13}H_{26}O_3$
dahmenite 达门炸药；达门那特
daidzein 黄豆苷原；7,4'-二羟基异黄酮
daidzin 黄豆苷〔即：异黄酮苷 isoflavone glucoside〕
daily capacity 每日产量
daily crude capacity 每日原料加工能力
daily flow 日产量；日流量
daily inspection 每日检验
daily loss 每日损失
daily output 日产量
daily production 日产量
daily requirement 每日需要
dairy industry 乳品工业；制酪工业
dairy machinery 制酪机
dairy product 乳制品
dairy salt 食盐
Dakin reaction 达金反应
Dakin's solution 达金溶液〔次氯酸盐与过硼酸钠的混合物〕
dalacin(=streptovaricin) 曲张链菌素
dalbergin 黄檀素
dalton 道尔顿〔质量单位，等于一氧原子量的十六分之一，约为 1.65×10^{-24} 克；1 克约为 6×10^{23} 道尔顿〕
Daltonian compound 道尔顿式化合物；定比化合物
daltonide 道尔顿体*
Dalton's law 道尔顿定律
Daly detector 戴里检测器
Daly electrode 戴里电极
Daly multiplier detector 戴里倍增检测器
dam ①坝②挡板
damage accumulation 损坏累积
damaged 损伤了的；损坏了的
damaged tire 废(轮)胎
damage resistant 耐划伤(的)
damage tolerance 损伤容限
damaging deformation (落锤冲击试验)损坏形变
damaging force (落锤冲击试验)损坏力
damarin 大马烯
damascenine 大马(士草)宁
damascenone 大马烯酮〔旧称突厥酮〕
4-damascol 4-大马醇
damascone 二氢大马酮；突厥酮
dam-board 挡板
dambonite 橡胶素；二甲基肌醇
dambose 不旋肌醇；橡胶糖
dame's violet oil 夜堇油
damiana 达迷草
dammaradienol 琥玛烷二烯醇
dammarenediol 琥玛烯二醇
dammarenolic acid 琥玛烯醇酸
dammar gum 琥玛树脂
dammarolic acid 琥玛醇酸
dammar resin(=damar resin) 琥玛树脂
dammar varnish 琥玛清漆
dammar wax 琥玛蜡
dammarylic acid 琥玛基酸 $C_{36}H_{60}O_3$
damourite 变白云母；水白云石
damp and hot effect 湿热效应
damped 阻尼的；减震的
damped balance 阻尼天平
damped elastic response(=delayed elasticity) 迟延弹性响应
damped harmonic oscillator 阻尼简谐振子

damped oscillation　阻尼振荡
damped torsional oscillator　减振扭转振荡器；阻尼扭转振荡器
dampening　回潮
dampening solution(=fountain solution)　①湿润液②润版液；润版药水
damper　①烟道挡板；调节风门②阻抑器；减震器；缓冲器③气门；气闸
damper plate　气闸
damper regulator　气闸调节器
damper vane　调气车翼
damping　①缓冲②减震③阻尼④回潮
damping capacity　阻尼容量
damping coating　阻尼涂料
damping coefficient　阻尼系数；减振系数
damping constant　阻尼常数
damping device　阻尼装置
damping effect　阻尼效应；减振作用
damping factor　阻尼因子
damping factor analysis　阻尼因子分析
damping fluid　缓冲液
damping force　减振力
damping machine　回潮机
damping material　阻尼材料
damping oil　制动油
damping peak　阻尼峰*
damping period　阻尼期；激后复原期
damping property　阻尼性能
damping ratio　阻尼比率
damping resistance　阻尼电阻
damping vat　汽蒸桶
damping viscosity　阻尼黏性
dampness　①湿度②潮湿
damp-proof　防潮的
dampproofer　防潮剂
damp-proof fungicidal paint　防潮防霉漆
damp-proofing fluid　防潮液
damp-proofing liquid　防潮液；防潮涂剂
damp-proof installation　防潮设备
damp-proof membranes　防潮膜；防潮层
damp-proof paint　防潮漆
damp surface　潮湿表面
damsine　达牡素
dan　小车；空中吊动车；杓；瓢；桶
danaite　钴毒砂
danalite　铍石榴子石
danazol　达那唑〔药〕
danburite　赛黄晶

dancer arm　松紧调节臂
dancer roll　调布辊；松紧调节辊；张力调节辊
dancer tension control device　浮动辊式张力控制装置
dancing　跳动的
Danckwerts theory　丹克沃茨理论
daN(deca-newton)　10 牛顿〔等于 1.019716 千克力〕
dandered coal　天然焦
dandruff　皮屑
dandy roll　压绞辊
dangerously explosive　极易爆炸的
dangerous oils　易燃石油产品〔闪点低于 23℃〕
dangling chain end　悬链端
Daniel flow technique(=Daniel's flow point method)　丹尼尔流点技术；丹尼尔流点法
Daniell battery　丹尼耳电池组
Daniell cell　丹尼耳电池*
Daniell standard cell　丹尼耳标准电池
Daniel's flow point (method)　丹尼尔流点法
Daniel wet point　丹尼尔湿点
danish agar(=furcellaran)　红藻胶
Danish concentrator　丹氏浓缩器
danks　黑色炭质页岩
dannemorite　锰铁闪石
dansyl(=1-dimethylaminonaphthalene-5-sulfonyl)　1-二甲氨基萘-5-磺酰；丹酰
dansyl chloride　丹磺酰氯；5-二甲氨基萘磺酰氯
Dapex method　达佩克斯法；二烷基磷酸萃取法
Dapex process　达佩克斯过程〔词源: dialkylphosphate extraction,①从铀矿石浸出液用 D2EHPA 和 TBP 萃取法提取铀的过程②提取超钚元素的 D2EHPA 萃取过程〕
daphnarcine　瑞香水仙碱
daphne　瑞香
daphne concrete　瑞香浸膏
daphne oil　瑞香油
daphnetic acid　瑞香酸
daphnetin(=7,8-dihydroxy coumarin)　瑞香素；7,8-二羟香豆素
daphnin　瑞香苷
daphniphylline　虎皮楠碱
daphnite　铁绿泥石
daphnoline(=trilobamine)　瑞香醇灵
dapsone　氨苯砜〔药〕
daptomycin　达托霉素
darapskite　钠硝矾
darcy　达西〔渗透力单位, 厘米/秒〕
Darcy friction factor　达西阻力因数；达西摩擦因数
Darcy number　达西数

Darcy's law 达西定律
Darex process 达雷克斯过程〔核燃料后处理首端过程的一种。用王水溶解不锈钢燃料元件外壳〕
dark ①黑暗的②暗（指颜色）
dark adaptation 暗适应
dark atom 暗原子；无放射性原子
dark band 暗带
dark color 暗色；深色〔色光发暗〕
dark-colored paint 深色漆
dark-colored stocks 暗色原料
dark current 暗电流；暗流
dark discharge 无光放电
dark distiller's grain 暗酒糟
darkening 发黑；暗黑；变黑
darkening of oil 油品(的颜色)变暗
dark-field condenser 暗场聚光器
dark-field image of weak beam 弱束暗场像
dark-field microscope 暗场显微镜*
dark-field microscopy 暗场显微术
dark flame 暗焰；无光焰
dark grain 暗酒糟
dark green 暗绿
dark-ground illuminator 暗视野照明器；暗场照明器
dark heat 暗热
dark light 暗光；不可见的光线
dark line 暗线
dark line spectrum 暗线光谱
dark petroleum oils 深色石油油料
dark reaction 黑暗反应
dark red 暗红
dark-red silver ore 硫锑银矿
dark repair 暗修复
dark room 暗室
darkroom clock 暗室钟
dark selvedge 糊边
dark substitute 黑油膏
Darling sodium process 达林炼钠法
darnel 毒麦
dart ①镖②突进；突发
dart drop test 落镖试验〔冲击试验〕
dart flight 落镖行程
dart impact strength 落镖冲击强度
dart union 活络管子节
Darzen condensation(=glycidic ester condensation) 缩水甘油酯缩合；α,β-环氧酸酯缩合；达参缩合
Darzen dichloroacetate synthesis 达参二氯乙酸酯合成
dash adjustment 缓冲调节
dash board ①仪表控制板；仪表盘②挡水(泥)板
dashboard panel 仪表板
dash coat 泼涂(层)；溅涂
dash-control 缓冲控制
dash current 冲击电流；超值电流
dash plate 缓冲板
dashpot ①控制品②缓冲槽③黏壶；阻尼延迟器〔模拟流体黏性〕
dash thermometer 缓冲温度计
dasycarpine 厚果(红豆树)碱
dasymeter 炉热消耗计；气体密度测量器
data ①数据②资料③基线
data accumulation 数据累积
data acquisition 数据获取
data acquisition system 数据获取系统
data analysis 数据分析
data balancing 数据平衡
database 基本数据；数据库
data base management system 数据库管理系统
data collection 数据收集
data collection terminal 数据收集终端
data enhancement 数据增强
data fitting 数据拟合
data handling 数据处理
data-in 数据输入
data logger 数据记录器；巡回检测器
data logging ①数据记录②巡回检测
data matrix 数据矩阵
data message 数据信息
data-out 数据输出
data output 数据输出
data processing 数据处理
data processing system 数据处理系统
data quality objective 数据质量目标
data recording 数据记录
data reduction 数据还原
data report (试验)数据报告
data restoration 数据复原
data retrieval recorder 数据检索记录器
data sampling 数据采集
data scatter 数据散布
data sets 数据组
data sheet 一览表；记录表
data smoothing 数据平滑
data system 数据系统
date kernel oil 枣椰油
date palm 海枣
dating 年代测定；年龄测定
dating of fossils 化石年代测定

datiscetin 橡精
dative bond 配价键
datolite 硅钙硼石
datugen 曼陀罗精
datugenin 曼陀罗配基
datum level 基准面
datum line 基(准)线
datum point 基准点
datum temperature 基准温度
daturic acid(=margaric acid) 十七(烷)酸
daturine 曼陀罗碱
daub ①胶泥②底色〔革〕③抹胶④涂；抹⑤打底色〔革〕
daub coat 漆革第一层涂料
daubreeite 铋土
daubreelite 陨硫铬铁
dauby 胶黏的
daucane 胡萝卜烷
daucarine 胡萝卜子素
daucene 胡萝卜烯
daucine 胡萝卜碱
daucol 胡萝卜醇
daucosterol 胡萝卜苷
daughter 子体
daughter cell 子细胞
daughter element 子元素
daughter ion 子离子*
daughter ion spectrum 子离子谱
daughter isotope 子体同位素
daughter nuclide 子核；子体核素*
daughter polymer 聚合物子体
daunomycin 柔红霉素；道诺霉素；柔毛霉素
daunorubicin 柔红霉素〔药〕
daunorubicinol 柔红霉素醇
dauricine 北豆根碱；蝙蝠葛碱
davanafuran 印蒿呋喃
davana oil 印蒿油
davanone 印蒿酮
daviesite 柱氯铅矿
Davis-Bruning Colorimeter 戴维斯-布鲁宁比色计
Davis-Gibson filter 戴维斯-吉普森滤波器
davit 吊柱
Davydov spliting 达维多夫分裂
Davys' lamp 戴维(安全)灯
Dawson gas 道森煤气；半水煤气
dawsonite 片钠铝石
daylight ①板距②台距③日光
daylight filter 日光滤波器
daylight filter glass 日光过滤玻璃

daylight fluorescent pigment 白昼荧光颜料
daylight lamp 日光灯
daylight opening (压机)压板间距
day output 日产量
day tank furnace 日槽熔炉
day-to-day loss control 日常(燃料)损耗控制
day-to-day test 逐日试验
D band D 谱带
DBBP lanthanide-actinide process 丁基膦酸二丁酯萃取分离镧系-锕系元素过程
d-block element d 区元素
DC amplifier 直流放大器
DC arc 直流电弧
DC arc welding(=direct current arc welding) 直流电弧焊
d-character factor d 特性因子
DC naphtha(=Stoddard solvent) 干洗溶剂汽油
DC polarography 直流极谱法*
DC resin(=silicone resin) 硅氧烷树脂
DC welder 直流电焊机
DDPA process 十二烷基磷酸萃取过程〔从铀矿石酸浸液中提取铀，类似 Dapex 法〕
de- 脱；去；除；解；减；消；反；止
11-deacetoxywortmamin 11-去乙酸基渥曼青霉素
deacetylated cellulose acetate 脱乙酰基醋酸纤维素
deacetylated chitin 脱乙酰基甲壳质
deacetylating ①脱乙酰的②脱乙酰作用
deacetylation 脱乙酰作用
deacetylprotoveratrine 脱乙酰原藜芦碱
deacetylstrychnospermine 脱乙酰裸子(马钱子)明
deacidification 脱酸作用
deacidizing 还原；脱氧脱酸
Deacon catalyst 迪肯催化剂
Deacon process 迪肯制氯法
Deacon's furnace 迪肯炉
deactivated catalyst 失活催化剂
deactivated polymer 无活性聚合物；链终止聚合物
deactivated silica gel 去活化硅胶
deactivated zeolite 去活化沸石
deactivating agent 去活化剂
deactivating group 钝化基团*
deactivation 失活作用*；钝化作用
deactivation capacity 去活化能力
deactivation column 减活柱；去活化柱
deactivation effect 减活化效应；去活化效果
deactivation of support 载体的钝化
deactivation period 惰化期
deactivation rate 去活率；钝化率
deactivator 减活化剂；去活化剂

deactive gel　钝化凝胶
dead　①暗的②死的③不活泼的
dead air pocket　滞留空气；存气
dead air space　(多孔物质的)闭塞空间(空隙)；封闭空隙；滞留空气区
dead angle　死角
dead band　死谱带
dead beating　重力打浆〔在打浆机中打得过度，以致纸张脆弱〕
dead Borneo　低级天然橡胶
dead burned　僵烧的
dead burned calcium sulfate　烧僵硫酸钙〔完全无水的硫酸钙〕
dead-burned dolomite　僵烧白云石
dead-burned gypsum　僵烧石膏
dead-burned magnesia　僵烧氧化镁
dead burned plaster　僵烧石膏
dead burning　僵烧
dead burnt　僵烧的
dead catalyst　废催化剂
dead center(=dead centre)　①(冲程的)死点②(车床的)死顶点
dead color　暗色
deadcoloring　呆色；死色
dead-end batch reactor　死端式间歇反应器
dead-end convertor　死端式转化器〔一种加氢设备〕
dead-ended filter　终端过滤器
dead-end inhibitor　终端抑制剂
dead-end polymerization　死端聚合*；无活性端聚合
dead-end pressure　死点压力
deadener　①隔音涂料②消声器
deadening　①隔音的②隔音(作用)
deadening felt　隔音毡
deadening felt paper　隔音纸；垫物纸
dead entry　断电进槽〔电泳〕
dead fiber　死纤维
dead flat　①绝对无光；完全无光(泽)②完全无光涂层
dead halt　完全停机
dead head　冒口
dead hearth generator　死膛发生炉
dead knife　固定刀；死刀
dead knots　死节
dead-leg　死区
dead light　探光玻璃；舷窗玻璃
dead lime　死灰池；用过的灰池(革)
dead line　①停用管线②环行管线③截止时间
dead load　静重；静载荷
dead lustre　无光

deadly compound　致命化合物
deadly nightshade root　颠茄根
dead macromolecule　无活性大分子
dead milled　过炼
dead milled rubber　过炼橡胶
dead milling　塑炼过度；过炼
dead oil　重油
dead plate　障热板
dead point　死点
dead polymer　无活性聚合物
dead retention time　死保留时间
dead roasted　僵烧了的
dead roasting　①僵烧的②僵烧
dead rock　空岩〔不含矿的岩石〕
dead rolled rubber　过炼橡胶
dead rolling　塑炼过度
dead rubber　①过炼胶②缺填料橡胶〔混炼不匀现象〕
dead small　(煤、矿石等)小块料；粉末；细末
dead soft steel　极软钢〔碳含量少于0.1%〕
dead space　死空间；死体积
dead spot　死角
dead stall　小块煤
dead steam　废汽
dead-stop circuit　永停电路；死停电路
dead-stop end point　永停终点；死停终点
dead-stop end point method　永停终点法；死停终点法
dead-stop process　永停终点(滴定)法
dead-stop titration(=biamperometric titration)　永停滴定；双指示电极电流滴定
dead-storage oxidation　久存老化
dead time　死时间*
dead time loss　死时间漏记
dead volume　死体积
dead weight　静重
dead-weight capacity (tonnage)　负载容量(吨位)
dead weight gauge　静重仪
dead zone　死区域
DEAE-cellulose(=diethylaminoethyl cellulose)　二乙氨基乙基纤维素
deaerated biological treatment　脱气生化处理
deaerating agent　①脱气剂②(糊料制备)脱泡剂
deaerating tank　脱气槽；除气槽
deaeration　脱气
deaeration of viscose　黏胶脱泡
deaerator　脱气器
DEAE-Sephadex　二乙氨基乙基交联葡聚糖(凝胶)
deaf ore　哑矿〔含矿很少的矿〕
deagglomeration　解凝集(作用)

deaggregating effect 解聚集效应
deaggregation 解聚作用
de-air 去气；除气；排气
dealcoholization 脱醇(作用)
dealkalization 脱碱作用
dealkalization softening 脱碱软化
dealkylation 脱烷基化；去烃(作用)
dealkylation catalyst 脱烷基化催化剂
dealuminization 脱铝作用
deamidated 脱去酰氨基的
deamidating ①脱酰氨基的②脱酰氨基(作用)
deamidation 脱酰氨基(作用)
deamidization 脱酰氨基(作用)
deaminase 脱氨酶
deaminated 脱氨基的
deaminating ①脱氨基的②脱氨基(作用)
deamination ①脱氨基②脱氨(基)作用
deaminizating 脱氨基
deaminizing ①脱氨基的②脱氨基(作用)
deammoniation 去氨
deapogossyool 脱阿朴棉子醇
deaquation 脱水的；脱水作用
dearated 脱气的；脱氧的
dearomatization 脱芳构化(作用)
de-aromatized solvent 脱芳构溶剂
de-arsenicated oil of vitriol(DOV) 脱砷硫酸
dearsenicator 脱砷塔
de-ashing 脱灰〔催化剂的洗涤工序〕
deashing device 除灰设备
deasphalt 脱沥青
deasphalted oil 脱沥青油
deasphalting 脱沥青；脱沥青的
deasphalting agent(=deasphalting medium) 脱沥青剂
deasphalting medium(=deasphalting agent) 脱沥青剂
deasphalting method 脱沥青法
deasphalting process 脱沥青过程
deasphalting solvent 脱沥青溶剂
deasphalting tower 脱沥青塔
deasphaltizing ①脱沥青②脱沥青的
death ray (致)死光(线)
debeading 去钢丝圈
debenzolized oil 脱苯油
debenzylation 脱苄基作用
debenzyloxy carbonylation 脱苄氧羰基作用
debiteuse 土制浮标；槽子砖
debituminization 脱沥青(作用)
deblocking 解封(闭)
deblooming (石油产品)去荧光

deblooming agent 消荧光剂
de Boer process 德·保尔法〔碘化物气相分解法〕
debond 脱胶
debonding 剥离；脱胶〔橡胶〕
Deborah number 德博拉数〔松弛时间与实验观测时间的比值〕
deboration 脱硼作用
de-branding 去烙印
debris 碎片；碎屑；爆片
de Broglie relation 德布罗意关系式
de Broglie wave 德布罗意波
debubbling 脱泡；消泡
debug 调试
debugging on-line 联机调试
deburring 修边；除去毛边
deburring device 毛纺除草装置
de-burring machine 除草籽机
debutanization 脱丁烷〔作用〕
debutanized gasoline 脱丁烷汽油；无丁烷汽油
debutanizer 脱丁烷塔
debutanizing 脱丁烷
debutanizing column(=debutanizing tower) 脱丁烷塔
debutanizing tower(=debutanizing column) 脱丁烷塔
debutylizing ①脱丁基的②脱丁基(作用)
debye 德拜〔偶极矩的单位，deb〕
Debye-Bueche's viscosity equation 德拜-比歇黏度方程
Debye characteristic temperature 德拜特性温度
Debye equation 德拜方程
Debye force(=induction force) 德拜力；感应力
Debye-Hückel limiting law 德拜-休克尔极限定律*
Debye-Hückel theory of strong electrolyte 德拜-休克尔强电解质理论*
Debye-Onsager theory of electrolytic conductance 德拜-昂萨格电导理论
Debye photograph ①德拜照片；粉末X射线照片②德拜摄影；粉末X射线摄影
Debye-Scherrer method 德拜-谢乐(X射线检验)法；粉末照相法
Debye-Scherrer ring 德拜-谢乐环
Debye's equation 德拜公式*
debye temperature 德拜温度
debye unit(D) 德拜单位〔电偶极矩单位，等于10^{-18}静电单位·厘米，或3.336×10^{-30}库仑米〕
Debye-Waller factor 德拜-沃勒因子
deby red(=chrome red) 镉铅铬绿；铬绿
deca- 〔词头〕十；癸
decaborane 癸硼烷
decabromodiphenyl 十溴代联苯 $(C_6Br_5)_2$

decabromodiphenyl oxide(DBDPO)　十溴二苯醚〔阻燃剂〕
decabromophenyl oxide　十溴二苯醚〔阻燃剂〕
decachlorobiphenyl　十氯代联苯　$Cl_5C_6C_6Cl_5$
decacyclene　十环烯　$C_{36}H_{18}$
decade　十倍程
decade resistance box　十进电阻箱
decadienal　癸二烯醛
1,3-decadiene　1,3-癸二烯；癸间二烯　$CH_3(CH_2)_5CH=CHCH=CH_2$
decadienoic acid　癸二烯酸　$C_9H_{15}COOH$
2,4-decadienoyl isobutylamide　2,4-癸二烯酰异丁酰胺
decaffeination　除去咖啡因
decagram　十克
decahedron　十面体
decahydrocarbazole　十氢咔唑；十氢化氮芴　$C_{12}H_{19}N$
decahydrodihydroxy-methoxyfluorenone　十氢二羟基甲氧基芴酮
decahydro-β-naphthaldehyde　十氢-β-萘醛
decahydro naphthalene(=decalin)　十氢化萘；萘烷
decahydro-α-naphthol　十氢-α-萘酚　$C_{10}H_{18}O$
decahydro-β-naphthol　十氢-β-萘酚
decahydro-β-naphthyl acetate(=β-decalol acetate)　乙酸-β-萘烷酯
decahydro-β-naphthyl ethyl ether　十氢-β-萘乙醚
decahydro-β-naphtyl formate　甲酸十氢-β-萘酯
decahydropyrazino pyrazine　十氢吡嗪并吡嗪〔硫化剂〕
decahydroquinoline　十氢喹啉　$C_9H_{17}N$
decal　①待复制图纸②贴花法〔陶瓷玻璃装饰法〕
decalactone　癸内酯
decalcification　脱钙(作用)
decalcified　脱钙的
decalcifying agent　脱钙剂
decalcomania　印花釉法
decalcomania paper　印刷转写纸
decalescence　(钢条)吸热
decalescent　钢条吸热的
decalin(=decahydro naphthalene)　萘烷；十氢化萘　$C_{10}H_{18}$
decaliter　十升
decalol　萘烷醇
α-decalol acetate　乙酸-α-萘烷酯
β-decalol acetate(=decahydro-β-naphthyl acetate)　乙酸-β-萘烷酯
decalone　萘烷酮
decamethonium　十烷双铵；十甲季铵
decamethylene diamine　癸二胺
decamethylene diammonium adipate　己二酸癸二铵

decamethylene-dimethyl azodicarboxylate　偶氮二羧酸十亚甲基二甲酯〔硫化剂〕
decamethylene-glycol　1,10-癸二醇　$HOCH_2(CH_2)_8CH_2OH$
4,5-decamethylene pyrimidine　4,5-十亚甲基嘧啶
decanal　癸醛
decanal dimethyl acetal　癸醛二甲缩醛
decane　癸烷　$C_{10}H_{22}$
1,10-decanediamine　1,10-癸二胺　$H_2N(CH_2)_{10}NH_2$
decane dicarboxylic acid　癸烷二羧酸　$C_{10}H_{20}(COOH)_2$
decanedioic acid　癸二酸　$C_8H_{16}(COOH)_2$
decanedioyl　癸二酰基　$-CO(CH_2)_8CO-$
decanoate　①癸酸②癸酸盐(酯或根)
decanohydroxamic acid　十碳异羟肟酸　$C_9H_{19}C(O)NHOH$
decanoic acid　癸酸　$C_9H_{19}COOH$
decanoin　癸酸精；(三)癸酸甘油酯
1-decanol　1-癸醇　$CH_3(CH_2)_8CH_2OH$
decanol　癸醇
decanone　癸酮
decanoyl(=caprinoyl)　癸酰基　$CH_3(CH_2)_8CO-$
decanoyl chloride　癸酰氯
decanoyl peroxide　过氧化二癸酰〔催化剂〕
Decan process(=decanting process)　邓凯法；滗析法
decantation　滗；倾析
decantation tank　滗析槽
decanter　滗析器；倾析器；倾析槽
decanter centrifuge　沉降式离心机
decanting　①滗析的②滗析
decanting bottle　滗析瓶
decanting cylinder　滗析筒
decanting flask　滗析瓶
decanting glass(=decanting jar)　滗析瓶
decanting jar(=decanting glass)　滗析瓶
decanting tank　滗析槽
decantor　①滗析器②油水分离器
decanyl acetate　乙酸癸酯
deca-n-octyl heptabutylene octaphosphine oxide　十正辛基七亚丁基八氧膦
decapitation　断头
decaploid　十倍体
decarbidizing　脱碳沉积；脱焦炭
decarbonater　脱碳酸气塔
decarbonization　脱碳(作用)
decarbonizer　脱碳剂
decarbonizing　①脱碳的②脱碳(作用)
decarbonylation　脱羰*
decarboxyamidation　脱羧酰胺化(作用)

decarboxylating ①脱羧基的②脱羧基(作用)
decarboxylating agent 脱羧剂
decarboxylation 脱羧*
decarboxylation temperature 脱羧温度
decarboxylative acylation 脱羧基酰化作用
decarboxylative elimination 脱羧消除
decarboxylative halogenation 脱羧卤化*
decarboxylative nitration 脱羧硝化*
decarboxylizing ①脱羧基的②脱羧基(作用)
decarburizating ①脱碳的②脱碳(作用)
decarburization (钢铁)脱碳(作用)
decarburizing 脱碳
2,4,6,8-decatetraene 2,4,6,8-癸四烯
decationizing ①除去阳离子的②除去阳离子(作用)
decatizing ①汽蒸的②汽蒸(作用)
decatizing fastness 汽蒸坚牢度
decatone 癸酮
decatyl(=decyl) 癸基 $CH_3(CH_2)_8CH_2-$
decay chain 衰变链
decay coefficient 衰变系数
decay constant 衰变常数；衰变常量
decay curve 衰变曲线
decay curve analysis 衰变曲线分析
decay daughter 衰变子体
decay energy 衰变能量；蜕变能量
decay family 衰变系；放射系
decay heat 衰变热
decay law 衰变定律；衰减定律；衰变律
decay modulus 衰变模量；衰变系数
decay of afterglow 余辉衰减
decay of luminescence 发光衰变
decay period 衰变期
decay probability 衰变概率
decay product 衰变产物
decay rate 衰减率
decay scheme 衰变图式；衰变纲图
decay sequence 衰变序列
decay series 衰变系；放射系
decay stress 衰变应力
β-decay synthesis β衰变合成*
decelerated particle 被减速粒子
decelerating electrostatic field 减速静电场
decelerating flow 减速流
decelerating fluid flow 减速流体流
decelerating plastic flow 减速塑性流
decelerating plastic substance 减速塑性物质
deceleration 减速
deceleration period 减速期

decenal 癸烯醛
2-decenal(=decylenaldehyde) 癸烯-2-醛
decene 癸烯 $C_{10}H_{20}$
decene dicarboxylic acid 癸烯二羧酸；十二碳烯二酸 $C_{10}H_{18}(COOH)_2$
decenedioic acid 癸烯二酸 $C_8H_{14}(COOH)_2$
decenoate(=decylenate) 癸烯酸盐(或酯)
decenoic acid 壬烯二酸；十一碳烯二酸 $C_9H_{16}(COOH)_2$
decenone 癸烯酮
decenoyl 癸烯酰(基)
decentralized and coordinated control 分散集中型控制
deceptively simple spectrum 假象简单图谱*
deceptive spectrum 假象图谱
dechalking 擦去粉剂；除去隔离剂
dechenite 红钒(酸)铝矿
dechine 双绉类织物
dechlorinating ①脱氯的②脱氯(作用)
dechroming 褪铬；铬鞣革脱鞣
dechromisation(=dechromization) 去铬；除铬
dechromization(=dechromisation) 去铬；除铬
deci- 〔拉丁字头〕 分；十分之一
decibel 分贝
decidigit 十进位数
decigram 分克；十分之一克
decigram method 分克法〔常量法〕
deciliter 分升；十分之一升
decimal ①十进的②小数；十进分数
decimal point 小数点
decimal rheostat 十进变阻器
decimal weigher balance 十进天平
decimation (数据)抽取
decimeter 分米
decimilligram analysis 超微量分析
decimolar 十分之一摩尔的
decine 癸炔
decinormal solution 十分之一当量溶液
deci-normal table (标准液)十分之一当量表
decision level 判断水平
decision rule 决策规则
decision-tree approach 判别树法
decision variable 决策变量；决策变数
decisive effect 决定作用
decitex 分特〔纤维的线密度单位〕
decked 装了甲板的
decker 稠料器；脱水机；浓缩机；圆网浓缩机
deckering 纸料增稠；脱水
deckle 框

deckle band　定边带
deckle board　框板
deckle frame　(制模)框架
deckle strap　框带
deck valve　台阀
decladding　去壳*
decline phase　衰亡期；死亡期
declining curve　下降曲线
declomycin(=demethytetracycline)　去甲基四环素
decloxizine　去氯羟嗪〔药〕
decocta　煎剂
decoction　煎(煮)
decoctum　煎剂
decoding　解码
decoherence　剥离；裂开
decohesion　剥离；裂开
decoic acid　癸酸
decoil　开卷；展开卷料
decoking　除焦；清焦
decolor　脱色；掉色
decolorant　脱色剂
decoloration　脱色(作用)
decolorimeter　脱色计
decoloring clay　漂白土*
decoloriser(=decolorizer)　脱色剂
decolorization　脱色(作用)
decolorizer(=decoloriser)　脱色剂
decolorizing　脱色的
decolorizing agent　脱色剂；退色剂
decolorizing carbon (=decolorizing char(coal))　脱色炭
decolorizing char(coal)(= decolorizing carbon)　脱色炭
decolorizing earth　脱色土
decolorizing efficiency　脱色效率
decolorizing resin　脱色树脂
decolorizor　脱色剂
decolour　脱色
decolouring　脱色
decolouriser(=decolourizer)　脱色剂
decolourized shellac　脱色紫胶片
decolourizing agent　脱色剂
decommissioning　(设施)退役；停运；核设施退役
decomplexation　解络(作用)
decomposability　可分解性
decomposed coal　分解的煤
decomposer　分解器
decomposing　①分解的②分解(作用)
decomposing furnace　分解炉〔Dickon 法中氯化氢的氧化炉〕

decomposing pot　分解釜
decomposition　分解*
decomposition catalyst　①分解催化剂②由分解得到的催化剂
decomposition constant　分解常数
decomposition course　分解过程
decomposition efficiency　分解效力；分解效率
decomposition hazard index　分解危险指数
decomposition heat　分解热
decomposition inhibitor　分解阻化剂
decomposition per pass　单程分解
decomposition point　分解点
decomposition potential　分解电位
decomposition principle　分解原理
decomposition product　分解产物
decomposition rate　分解速率
decomposition reaction　分解反应
decomposition (resolving) voltage　分解电压
decomposition stage　分解阶段
decomposition temperature　分解温度
decomposition tension　分解电压
decomposition value　分解值
decomposition voltage　分解电压
decompression　降压
decompression section(=decompression zone)　释压段
decompressor　减压器
deconjugation　早期解离
decontaminability　可去污性
decontaminable　可去污的；易去污的
decontaminant　去污剂*
decontaminating　除污染
decontaminating agent　净化剂；去污剂*
decontamination　净化；去杂质(作用)；纯化；去污(作用)；消毒〔放毒气后〕
decontamination agent　净化剂；除污染剂
decontamination bleach mixture　消毒漂粉浆
decontamination factor　去污系数；净化系数
decontamination index　净化指数；去污指数
decontamination plant　净化装置
decontamination rate of urban refuse　城市垃圾无害化处理率
decontamination squad　去污班；去污小组
deconvolution　去褶合；解卷积
deconvolution method　解卷积法
deconvolution technique　解卷积技术
decorating　美化
decorating fire　彩烧；彩烤
decorating foil　①彩饰箔②装潢膜

decoration ①装饰②施彩(作用)
decorative chromium coating 装饰性镀铬层
decorative emulsion paint 装饰性乳胶漆
decorative epoxy powder coating 装饰性环氧粉末涂料
decorative painting 装饰性涂装(漆)
decorative plywood 装饰胶合板
decorative porcelain 彩瓷
decorative powder coating 装饰性粉末涂料
decorative rib 装饰线
decorporation 排出；促排
decorticator 去皮机
decoupling mode 去耦方式
decoyl(=capryl) 辛酰
decoylamide(=caprylamide) 辛酰胺
decreased logarithmic phase 对数减少期
decrease passages 毛纺缩道
decreasing drying rate period 减速干燥阶段
decreasing-lead screw(=decreasing pitch screw) 减螺距螺杆
decreasing pitch screw 收敛式螺杆
decreasing worm-pitch 不等螺距
decrement 递减；衰减量
decrement gauge 减量气压计
decrepitation 烧爆作用*
decrystallization 解晶作用
decrystallized 非晶化的；消晶化的
decyanation 脱氰基*
decyanoethylation 脱氰乙基*
decyclization 解环作用
decyl(=decatyl) 癸基 $CH_3(CH_2)_8CH_2-$
decyl acetate 乙酸癸酯 $CH_3CO_2C_{10}H_{21}$
decyl alcohol 癸醇
n-decyl alcohol (=1-decanol) (正)癸醇；1-癸醇 $CH_3(CH_2)_8CH_2OH$
decyl aldehyde 癸醛
decyl aldehyde dimethyl acetal 癸醛二甲缩醛
decyl amide 癸酰胺
decyl amine 癸胺 $CH_3(CH_2)_8CH_2NH_2$
decylate 癸酸酯
decyl benzene sulfonate 癸基苯磺酸盐
decyl bromide 癸基溴 $CH_3(CH_2)_8CH_2Br$
decyl butyl phthalate 邻苯二甲酸癸丁酯〔增塑剂 DBP〕
n-decyl butyrate 丁酸癸酯
decyl chloride 癸基氯 $CH_3(CH_2)_8CH_2Cl$
n-decyl cinnamate (肉)桂酸癸酯
n-decyl decylate 癸酸癸酯
n-decyl dicresyl phosphate 磷酸正癸基二甲苯酯〔增塑剂〕
decyl dimethyl benzyl ammonium halide 癸基二甲基苄基卤化铵
decyl dimethyl imidazole 癸基二甲基咪唑
decylenate(=decenoate) 癸烯酸盐(或酯)
α-decylene(=1-decene) 1-癸烯 $CH_2=CH(CH_2)_7CH_3$
decylenic acid 癸烯酸 $C_{10}H_{18}O_2$
decylhexyloctyl phthalate 邻苯二甲酸癸己基辛酯〔增塑剂〕
decylic acid(=capric acid) 癸酸
decylic aldehyde 癸醛
decyl iodide(=1-iododecane) 癸基碘 $CH_3(CH_2)_8CH_2I$
decyl methyl ether 癸基甲基醚
decyl nitrate 硝酸癸酯 $C_9H_{19}CH_2ONO_2$
decyl nitrite 亚硝酸癸酯 $C_9H_{19}CH_2ONO$
n-decyl-n-octyl phthalate 邻苯二甲酸正癸正辛酯〔增塑剂〕
decyl phenyl ether 癸基苯基醚
n-decyl propionate 丙酸癸酯
decyl tridecyl phthalate 邻苯二甲酸癸基十三酯〔增塑剂〕
decyl-triethyl-silicane 癸基三乙基硅 $(C_{10}H_{21})Si(C_2H_5)_3$
decyl-trimethyl-silicane 癸基三甲基硅 $(C_{10}H_{21})Si(CH_3)_3$
decyne 癸炔
decyne carboxylic acid 癸炔羧酸
decynoic acid 癸炔酸 $C_9H_{15}COOH$
decynyl 癸炔基
dedicated 专用的
dedicated computer 专用计算机
dedicated instrument 专用仪器
dedust 除尘
dedusted coal 去掉粉末的煤
dedusting ①除灰的②除灰
dedusting agent 除尘剂
de-electrifying 去电
de-electronating agent 去电子剂〔即氧化剂〕
de-electronation 去电子(作用)〔即氧化作用〕
Deeley friction machine 迪来摩擦试验机〔用于评定油性或油膜强度〕
deemulsification 破乳化(作用)
deemulsifier 去乳化剂；破乳化剂
deemulsifying agent(=demulsifying agent) 破乳剂
deenergized ①不带电的；切断电源的②去激励的；去能的
deentrainment 去夹带；除雾末
deentrainment column 除雾末柱
deentrainment filter 除雾末过滤器
deentrainment tower 除雾塔
deep ①深〔指颜色〕②深(度)的
deep agar method 厚层琼脂法

deep bluish-tinge (带)深蓝色相
deepcolor 深色；饱和色
deep-colored alkyd flat paint 深色醇酸无光漆
deep colour ①深色②浓色
deep cooling 深冷
deep cracks 深裂
deep drawing lubricant 金属深拔润滑剂
deep drawing quality 深拉性；深冲性；高延伸性
deep earth burial 地下深部埋藏〔放射性废物〕
deeper cracking 深度裂化
deep-etch litho 平凹板
deep filter 深度过滤器
deep flight worm 深(螺)纹螺杆
deep forming 深度成型〔热成型〕
deep freeze resistance 耐冷冻性
deep freezing 深度冷凝
deep fuel bed 深燃料层
deep hole drill 深孔钻头
deephole fractionating process 深孔分馏法
deep penetration electrode 深熔焊条
deep penetration welding 深熔焊
deep pile fabric 长毛织物
deep reactive ion etching 深度反应离子腐蚀；深度反应离子刻蚀
deep refrigeration 深度冷冻
deep-seated 深成的；深嵌的
deep seated rocks〔复〕 深成岩
deep seated stain 深沟腐蚀
deep self-burial 地下深部自埋藏(放射性废物)
deep shade 深色；浓色
deep stone 深石色；深灰色
deep tint 深色
deep tone 深色
deep tread 厚胎面；高花纹
deep UV region 远紫外区
deep vacuum 高真空
deep well pump 深井泵
deep well pumping unit 深井泵
deer skin 鹿皮
deer-skin filter 鹿皮滤器
deer's tongue 香蛇鞭菊
deethanation(=deethanization) 脱乙烷(作用)
deethanization(=deethanation) 脱乙烷(作用)
deethanizer 脱乙烷塔
deethanizing 脱乙烷
deethanizing absorber 脱乙烷吸收塔
deethanizing column 脱乙烷塔
de-ethylation 脱乙基作用

de-ethyleneglycol reaction 脱乙二醇反应
de-excitation 退激；去激法
defatting ①脱脂的②脱脂(作用)
defecated juice 澄清汁
defecating pan 澄清盘
defecation 澄清(作用)
defecation pan 澄清釜〔糖〕
defecator 澄清槽
defect ①缺乏②缺陷
defect cluster 缺陷簇*
defective fuel 有缺损的燃料
defective goods 不合格品；次品
defective insulation 不良绝缘
defective patch ①有缺陷的斑点(污点)〔漆病〕②找补缺陷
defective percentage 次品率
defective rate 次品率
defective spray 不良喷雾；缺陷喷雾
defect lattice 缺陷晶格
defectoscope 探伤仪
defects of gas accumulation (阳极氧化膜)蓄气性发花(起泡)
defect solid chemistry 缺陷固体化学
defects on spun silk yarn 糙节
defect structure 缺陷结构
defect theory 缺陷理论；损伤理论
defensive gas 防御用毒气
defensoat(=dephenzoat) 蒜素制剂
defervescence 止沸；退热
defferential dilatometry 差示热膨胀法
defiber 分离纤维
de-fibering 分离纤维；脱去纤维
defibrator 纤维分离机
defibrator-chemipulper 纤维分离管式连续蒸煮器
defibrillation 解原纤；原纤分离(作用)
defibrinated 脱纤维蛋白的
defibrinated blood 脱纤血液
defibrination ①脱纤维作用②磨(制木)浆；磨木制浆
deficiency 缺乏
deficient oil supply 油供给不足
defilement 污损；污染物
defined solid angle method 固定立体角法
defining 去飞边
definite 固定的
definite chemical composition 固定(化学)组成
definite melting point 固定熔点
definite proportion 定比
definite spot 明显斑点

deflagrability 爆燃性
deflagrating mixture 爆燃混合物
deflagrating spoon 爆燃匙
deflagration 爆燃(作用)
deflagrator 爆燃器
deflasher 修边机；除边机
deflashing 切除胶边；修边
deflated 排气的
deflation ①放(出空)气；排气②漏气
deflation opening 排气孔
deflation valve 放气阀
deflecting nozzle 折向喷嘴
deflection ①偏转；偏斜②挠曲；挠度③挺度④偏差⑤曲折
deflection angle 偏转角
deflection electrode 偏转电极
deflection function 偏离函数*
deflection refractive index detector 偏转折光指数检测器
deflection refractometer detector 偏转(示差)折光检测器
deflection separator 折流分离器
deflection surface 挠曲面
deflection system 偏转系统
deflection temperature 挠曲温度〔高分子物质的软化点温度〕
deflection test 挠曲试验
deflection type 偏转型
deflection type differential refractometer 偏转型差示折射计
deflectometer 挠度计
deflector 折转板；偏转器；挡板
deflectorless air pattern gun 无换向器(空)气流喷枪
deflector nozzle 无换向器喷嘴(喷管)
deflector-type separator (evaporator) 折板除沫器
deflegmation(=dephlegmation) 分凝
deflexion ①挠曲②挠度③偏斜④偏斜度
deflexion coil 偏转线圈
deflocculant(=deflocculating agent) 抗絮凝剂；(黏土)悬浮剂
deflocculated graphite 胶态石墨〔在润滑脂上〕
deflocculated sludge 抗絮凝淤渣
deflocculating 消絮凝
deflocculating agent(=defloculant) 抗絮凝剂；(黏土)悬浮剂
defloculation 抗絮凝(作用)
defloculation agent 抗絮凝剂
defloculator 抗絮凝离心机
De Florez cracking process 德-弗劳瑞兹(汽相)裂化过程
De Florez cylindrical heater(=De Florez upshot heater) 德-弗劳瑞兹筒式直立加热炉
De Florez predrilling method 德-弗劳瑞兹钻眼检查法〔用以检查设备的腐蚀速度〕
De Florez unit 德-弗劳瑞兹(汽相)裂化装置
De Florez upshot heater 德-弗劳瑞兹筒式直立加热炉
defluorinating ①脱氟的②脱氟(作用)
defluorination 脱氟作用
defoamant 消泡剂
defoamer 消泡剂
defoamer agent 去沫剂；消泡剂
defoaming ①去沫的②去沫(作用)；消泡
defoaming agent 消泡剂；抗泡剂；消沫剂
defoaming effectiveness 消泡效率
defoaming surfactant 消泡表面活性剂
defocusing 散焦；去聚焦
defocusing technique 去聚焦技术
Defo durometer 迪福硬度计
Defo elasticity 迪福弹性
Defo hardness 迪福硬度
defoliant 脱叶剂
Defo plastometer 迪福塑度仪
deform 变形；畸形
deformability 可变形性
deformable body 形变体；柔体
deformable dielectric material with memory 可变形电介质记忆材料
deformable magnetic material with memory 可变形磁性记忆材料
deformation 形变；变形
deformational nonlinearity 形变非线性
deformation analysis 形变分析
deformation at break 断裂点形变
deformation at constant volume 恒体畸变
deformation at yield 屈服点形变
deformation band (滑移)形变带
deformation behavior 形变(特)性
deformation coordinate 形变坐标
deformation free energy 形变自由能
deformation frequency 形变频率
deformation gradient 形变梯度
deformation interferometer 变角干涉仪
deformation of bond angles 键角形变
deformation orientation 形变取向
deformation range 形变范围；变形区
deformation rate 形变速率
deformation rate tensor 形变速率张量
deformation ratio 形变比率
deformation resilience 形变回弹
deformation resistance 抗变形性

deformation set 永久形变；变定
deformation structure 形变结构
deformation tensor 形变张量
deformation under torsion 扭力形变
deformation vibration 形变振动
deformeter 形变测定器；应变仪
deforming stress 形变应力
Defo value 迪福值
De Frise ozonizer 德-弗莱斯臭氧器
defrost 除雾
defroster 除雾器
defrosting 除霜；除冰
defrother ①消泡剂②消泡器
deg ①浸(润)湿②喷雾；雾化
degas 脱气
degasification 脱气(作用)
degasifier 脱气器
degassed crude 脱气原油
degassed oil 脱气石油
degassed silicone rubber 脱气硅橡胶〔已除去低分子挥发物的硅橡胶〕
degassed solution 除气溶液；脱气溶液
degasser 脱气装置
degassing ①排气；脱气；放气②解毒气
degassing column 脱气塔
degassing mould 排气塑模
degassing opening 排气孔
degassing orifice 排气孔
degassing tank 脱气罐
degassing temperature 排气温度
degassing tower 脱气塔
degassing vessel 脱气罐；脱气槽
de-gate 浇口料切除
degelatinized bone dust 脱胶骨粉
degeneracy ①退化性②简并性③简并
degenerated branched chain reaction 退化支链反应
degenerate state 简并态
degenerate vibration 简并振动
degenerative oxidizing condition 降级氧化条件
degenerative transfer 衰减转移
degerming 除菌；消毒
degerming agent 除菌剂
deglossing liquid 消光液
deglycerinizing(=deglycerizing) 除去甘油
degradability 降解性
degradable carbon 降解碳
degradable polymer 降解性高分子
degradation ①降解②老化

degradation catalyst 降解催化剂
degradation chain transfer 退化链转移
degradation curve 降解曲线
degradation (degradative) chain transfer 退化链转移
degradation delayer 降解延缓剂
degradation mechanism 降解机理
degradation peak 降解峰
degradation product 降解产物
degradation product analysis 降解产物分析
degradation rate 降解速率
degradation reaction 降解反应
degradation-resistant 抗降解的
degradation-resistant fibre 抗降解纤维
degradation theory 降解(学)说；降解理论
degradative benzoylation 降解性苯甲酰化
degradative chain transfer 降解性链转移
degradative chain transfer reaction 降解链转移反应；裂解链转移反应
degradative metathesis 易位降解反应
degradative reaction 降解反应
degraded cellulose 脆化纤维素；降解纤维素
degraded colour 暗色
degraded gelatin 降解明胶
degrading adduct 降解加合物
degrading shear rate (引起聚合物)裂解(的)剪切速率
degras 氧化鱼油；油鞣余脂
degreasant 脱脂的；去脂的
degreased 脱(了)脂的
degreaser 脱脂(垢)剂；除油剂
degreasing ①脱脂②除去油腻
degreasing agent 脱脂剂
degreasing effect 脱脂效果
degreasing machine 脱脂机；去除油脂机
degreasing power 脱脂力
degree ①程度②度
degree Baumé 波美度
degree Celsius 摄氏度
degree centigrade 摄氏(温)度数〔符号℃〕
degree-day 度-日
degree Engler 恩氏黏度
degree Fahrenheit 华氏(温)度数〔符号℉〕
degree Kelvin 开氏温度
degree of absorption 吸附程度
degree of acetalization 缩醛(化)度
degree of acetylation 乙酰化度
degree of acid 酸度
degree of action 作用(程)度
degree of activation 活化度

degree of activity 活度
degree of addition-condensation 加成缩合度
degree of adhesion 黏合度
degree of admission 容许程度
degree of adsorption 吸附度
degree of anchor (表面)粗糙度
degree of anisotropy 各向异性度
degree of association 缔合度
degree of attention 保养程度
degree of beating 打浆度
degree of birefringence 双折射率
degree of bleaching 漂白度
degree of blushing 发白度
degree of branching 支化(程)度
degree of brightness 亮度；明度
degree of brilliance 光亮度
degree of bulging 膨胀度
degree of chemisorption 化学吸附度
degree of clarity 透明度
degree of cleaning 清洁度；清洗程度
degree of coiling 盘曲度；盘旋度
degree of compactness 密实度；紧密度
degree of compression 压缩度
degree of confidence 置信度
degree of conversion 转化度
degree of cooking 蒸解度
degree of creaming 膏化度
degree of crimp 卷曲度
degree of crosslinking 交联度
degree of crystallinity 结晶度
degree of cure 硫化程度*
degree of degeneration 退化(程)度
degree of degradation 降解度
degree of depolarization ①去极化度②消偏振度
degree of depolymerization 解聚度
degree of disorder 无序度
degree of dispersion 分散度；色散度
degree of dissociation 离解度
degree of distortion 畸变度
degree of drawing 拉伸度
degree of drying 干燥程度
degree of dullness 不活泼度；钝度
degree of efflorescence 起霜程度
degree of electrolytic dissociation 电离度
degree of elongation 伸长度
degree of esterification 酯化度
degree of etherification 醚化度
degree of excitation 激发度

degree of exhaustion 抽空度
degree of expansion 膨胀度
degree of fermentation 发酵度
degree of fiber openness 纤维开松度
degree of fineness 细度
degree of finish 光洁度
degree of fixation 固定度；固着度
degree of flatting 消光度
degree of flexibility 挠曲程度；柔韧程度
degree of flocculation 絮凝度
degree of fluidity 流度
degree of formalization 缩甲醛(化)度
degree of freedom 自由度*
degree of freedom in system 系统自由度
degree of functionality 官能度
degree of grafting 接枝度
degree of grind 研磨程度
degree of grinding 粉碎度
degree of hardness 硬度
degree of hydration 水化程度
degree of hydrolysis 水解度
degree of immaturity 未成熟(程)度
degree of imperfection (结晶)缺陷(程)度
degree of inertness (to chemicals) (对化学药品)惰性度
degree of internal order (分子)内部有序度
degree of ionization 电离度*
degree of irregularity 紊乱度；不规则度
degree of isomerization 异构化程度
degree of isotacticity 全同(立构)规整度
degree of liquefaction 液化度
degree of loading 载荷度
degree of maturity 成熟度
degree of metamerism 条件配色度；条件等色度；位变异构度
degree of mixedness 混匀程度
degree of mixing 混合度
degree of molecular orientation 分子取向度
degree of nitration 硝化(程)度
degree of nitriding 氮化度；渗氮度
degree of nitrification 硝化度
degree of oil modification 油度；油长
degree of opalescence 乳光度
degree of order 有序度
degree of orientation 取向(程)度
degree of overlapping 重叠度
degree of photodegradation 光致解聚度
degree of plasticification 塑化度
degree of plasticity 可塑度

degree of polarity 极性度
degree of polarization ①偏光度②极化(程)度
degree of polycondensation 缩聚度
degree of polymerization 聚合度*
degree of porosity 孔隙度；气孔度
degree of preorientation 预取向度
degree of purification 净化度
degree of quaternization 季碱(化)度
degree of radial uniformity 径向均匀度
degree of rancidity 度；酸败度
degree of randomization 无规化度；随机化(程)度
degree of randomness 无规度
degree of reaction 反应度
degree of refining 精炼度
degree of relaxation 松弛度
degree of reproducibility 再生产力
degree of ripeness 成熟度〔棉纤维〕；熟成度〔黏胶溶液〕
degree of saturation 饱和度
degree of setting 定形度
degree of shrinkage 收缩度
degree of sizing 施胶度
degree of soiling 染污度
degree of specialization 专业化程度
degree of specific gravity 比重度
degree of sphericity 球度；圆球度
degree of staining 着色度
degree of structuration 结构度
degree of substitution 取代度
degree of swelling 溶胀度；泡胀度
degree of syndiotacticity 间同(立构)规整度
degree of tacticity 构型规整度
degree of tanning 鞣度系数；鞣透度
degree of thoroughness 清洗(除锈)彻底程度
degree of unsaturation 不饱和度
degree of vacuum 真空度
degree of viscosity 黏度
degree of water retention 水滞留度；水分保持度
degree of wetness 润湿度
degree of wetting 润湿度
degree of whiteness 白度
degree of xanthation 黄(原)酸酯化度；黄化度
degree of yellowness 发黄度
degree Plato 柏拉图度
degree Rankine 兰金度数
degree Réaumer 雷默温度
degree (s) Schopper-Riegler (SR) 打浆度
degree Twaddell 特瓦德尔度〔液体比重表示法之一〕
degression 渐减；递降

degritting 溢流分选
deguelin 鱼藤素
degumed silk 脱胶丝
degummed hemp 大麻精麻
degummed ramie 脱胶苎麻
degummed ramie grouping 苎麻分磅
degummed silk 脱胶丝
degumming 脱胶(的)
degumming agent 脱胶剂
degumming bath 脱胶浴
degumming of oil 油脱胶
degumming of silk 丝脱胶
degumming soap 脱胶皂
de-hairing 去毛
dehalogenating ①脱卤的②脱卤(作用)
dehalogenation 脱卤*
dehalogenation reaction 去卤反应
deHan salt 德海因盐 $SbF_3 \cdot (NH_4)_2SO_4$
deheptanizer 脱庚烷塔
dehesion 减黏
dehexanizer 脱己烷塔
dehexanizing 馏除己烷
dehexanizing column 脱己烷塔
dehumidification 减湿(作用)；干燥(作用)
dehumidifier ①减湿剂②减湿器；吸湿装置
dehumidifying ①减湿的②减湿(作用)
dehumidizer 减湿剂
dehusking (谷类)脱皮
dehydrant 脱水剂
dehydrase 脱水酶
dehydratase 脱水酶
dehydrate ①脱水②脱水物
dehydrated 脱了水的
dehydrated alcohol 脱水酒精；无水酒精；绝对酒精
dehydrated aluminosilicate 脱水硅铝酸盐
dehydrated castor oil 脱水蓖麻油
dehydrated castor oil fatty acid 脱水蓖麻油脂肪酸
dehydrated food 脱水食物
dehydrated gypsum 脱水石膏；无水石膏
dehydrated silica 脱水二氧化硅
dehydrated tar 脱水焦油
dehydrated weight 干重；脱水重量
dehydrater(=dehydrator) ①脱水剂②脱水器
dehydrating ①脱水的②脱水(作用)
dehydrating agent 脱水剂
dehydration 脱水*
dehydration accelerator 脱水(作用)促进剂
dehydration agent 脱水剂

dehydration box　脱水箱
dehydration catalyst　脱水催化剂
dehydration column　脱水塔
dehydration column reboiler　脱水塔再沸器
dehydration desalting process　脱水脱盐法
dehydration-oxidation reaction　脱水-氧化反应
dehydration peak　脱水峰
dehydration polymerization　脱水聚合
dehydration property　脱水性
dehydration tank　脱水槽
dehydration transition point　脱水转变点
dehydrator(=dehydrater)　①脱水剂②脱水器
dehydro　脱氢
dehydroabietic acid　脱氢枞酸
dehydroabietinol　脱氢枞醇
dehydroabietylamine　脱氢枞胺
dehydroacetic acid　脱氢乙酸〔俗〕；a,r-二乙酰基乙酰乙酸　$CH_3COCH_2COCH(COOH)COCH_3$
dehydroadiponitrile　脱氢己二腈
dehydroandrosterone　脱氢雄甾酮
dehydroangustione　脱氢安钩酮
dehydroannulenes　脱氢轮烯
dehydroascorbate　脱氢抗坏血酸
dehydroascorbic acid　脱氢抗坏血酸
dehydrobenzene(=benzyne)　脱氢苯；苯炔
dehydro benzoyl-acetic acid　脱氢苯甲酰乙酸
$C_6H_5COCHCOCH=C(C_6H_5)OCO$
dehydrobilirubin(=uteroverdin)　胆绿素
α-dehydrobiotin　α-去氢生活素
dehydrobromination　脱去溴化氢
dehydrobufotenine　脱氢蟾蜍特宁
dehydrocamphenilic acid(=tricyclenic acid)　三环萜酸
dehydrocarbylation　脱烃基化(作用)
dehydrochlorination　脱去氯化氢
7-dehydrocholesterol　7-脱氢胆甾醇
dehydrocholic acid　脱氢胆酸
dehydrocondensation　脱氢缩合
11-dehydrocorticosterone　11-脱氢皮质(甾)酮
dehydrocostuslactone　脱氢广木香内酯
dehydrocyclization　脱氢环化(作用)
dehydrocyclization catalyst　脱氢环化催化剂
dehydrocyclocitral　脱氢环柠檬醛
dehydro-cyclo-dehydrogenation　脱氢-环化-脱氢
dehydrodibenzodianthronyl　脱氢二苯并二蒽酮基
dehydrodigallic acid　脱氢双没食子酸
dehydroepiandrosterone(=dehydroisoandrosterone)　脱氢表雄(甾)酮；脱氢异雄(甾)酮

dehydrofluorination　脱氟化氢作用
dehydrogallic acid　脱氢棓酸
dehydrogenated　脱了氢的；去了氢的
dehydrogenated agent　脱氢剂
dehydrogenated oil　脱氢油
dehydrogenated rosin　脱氢松香
dehydrogenating　①脱氢的②脱氢(作用)
dehydrogenating agent　脱氢剂
dehydrogenation　脱氢*
dehydrogenation catalyst　脱氢催化剂
dehydrogenation polymerization　脱氢聚合(作用)
dehydrogenation reactor　脱氢反应器
dehydrogenative condensation　脱氢缩合
dehydrogenization　脱氢(作用)
dehydrogeranic acid　脱氢香叶酸
dehydrohalogenation　脱卤化氢*
dehydroherbarin　脱水叶蜡菌素
dehydrohomocamphoric acid　脱氢高樟脑酸
dehydrohydantoic acid　脱氢海因酸
11-dehydro-17-hydroxycorticosterone(=cortisone)　11-脱氢-17-羟皮质(甾)酮；可的松；皮质酮
dehydroiodination　脱碘化氢作用
dehydroisoandrosterone(=dehydroepiandrosterone)　脱氢表雄(甾)酮；脱氢异雄(甾)酮
dehydrolinalool　脱氢芳樟醇
dehydrolysis　脱水(作用)
dehydrolyzing agent　脱水剂
dehydromucic acid　脱氢黏酸；呋喃-2,5-二羟酸
β-dehydroperilla ketone　β-脱氢紫苏酮
dehydroperillic acid　①脱氢苏子酸②脱氢紫苏酸
3,4-dehydroproline　3,4-脱氢脯氨酸
5-dehydroquinate　①5-脱氢奎尼酸②5-脱氢奎尼酸盐
dehydroquinic acid　脱氢奎尼酸
7-dehydrositosterol　7-脱氢谷甾醇
dehydrotachysterol　脱氢速甾醇
dehydrothiamine　脱氢硫胺素
dehydrothiophen　脱氢噻吩
dehydrothio-p-toluidine　脱氢硫代对甲苯胺
dehydrotransandrosterone(=dehydroepi an-drosterone)　脱氢反雄甾酮
dehydroxylation　脱羟基作用
deicer　防冰剂；碎冰器
deicing agent　防冻剂；防冰剂
deicing fluid　防冻液
Deijaguin-Landau and Verwey-Overbeek theory of flocculation　DLVO絮凝理论*
de-inking　脱墨〔自纸上脱去印刷的颜料〕
de-inking solution　脱色溶液

deiodination 脱碘作用
deionisation(=deionization) 去离子作用
deionization(=deionisation) 去离子作用
deionized water 去离子水*
deionizer 去离子器；脱离子器
deionizing 除去离子
deisobutanizer 脱异丁烷塔
deisopropyl dehydroabietate 脱异丙基脱氢枞酸盐(或酯)
dejacket 脱壳；去壳；除壳
dejacketer 脱皮(除壳)装置
dejacketing solution 脱壳(用)液
dekatron 十进制计数管；十进管
deknitting 解编
delaminated clay (千)层状瓷土
delaminated kaoline 层状高岭土
delaminating 层离；脱层
delamination 层离；脱层
delamination resistance (层压制品)抗分层性；抗脱层性
delamination tank 层离罐；层解罐；粒度分级罐
delanium graphite 人造石墨
delanium graphite safety disc 高纯度压缩石墨安全盘
delavaconitine 紫草乌碱甲；紫草乌碱 $C_{30}H_{41}O_6N$
De Laval centrifugal extractor 德-拉伐尔离心萃取器
De Laval centrifuge 德-拉伐尔离心机
De Laval oil purifier 德-拉伐尔石油产品净化器
De Laval zinc process 德-拉伐尔炼锌法
delay 滞后
delay action 延迟作用
delayed action accelerator 延迟促进剂*
delayed-action activator 延迟剂
delayed action chemicals 延迟作用化学剂
delayed coagulant 缓凝剂；延迟凝固剂
delayed coagulation 缓凝(作用)
delayed coincidence counting method 延时同时计数法
delayed coke 延迟焦化
delayed coking 延迟焦化
delayed coking process 延迟焦化过程
delayed compliance 缓发柔量
delayed cracking 延迟裂纹
delayed crazing 延迟龟裂
delayed cure finish 延迟焙烘整理
delayed deflection 延迟挠曲
delayed deformation 延迟形变
delayed delamination 后期脱层
delayed drying varnish 干性滞后清漆；干性迟缓的漆
delayed elastic deformation 迟延弹性形变
delayed elasticity(=damped elastic response) 迟延弹性响应
delayed elastic recovery 迟延弹性回复
delayed elastic solid(=Kelvin solid) 开尔文固体
delayed expansion 延迟膨胀
delayed extraction 延迟引出；延迟排出
delayed fluorescence(DF) 迟滞荧光；缓发荧光
delayed fracture 延迟断裂
delayed hypersensitivity 迟发过敏性
delayed ignition 延迟引燃
delayed knock 延迟爆击
delayed modulus 缓发模量；迁延系数
delayed neutron 缓发中子
delayed neutron counting 缓发中子计数
delayed neutron logging 缓发中子记录
delayed neutron monitor 缓发中子监测器
delayed quench 徐冷；延迟冷却
delayed recovery 延迟回复
delayed sagging 延迟流挂
delayed time after pulse 脉冲后延迟时间
delayed timing 滞后成圈
delayed type of recovery (涂料胶体结构的)推迟型恢复
delaying basin 滞留池
delay period ①滞燃期②延迟期〔自燃、爆发或闪光〕
delay quenching collar 徐冷环套
delay time ①延迟时间②滞燃时间
delay zone 延迟区
delcosine 硬飞燕草次碱；翠雀胺
deleading reagent 脱铅剂
deleafing ①脱漂浮性；脱叶展性②漂浮性消失；叶展性消失(丧失)
deleafing action (铝粉)脱漂浮作用；解漂浮作用
Delepine's amine synthesis 德勒平氏胺合成法
delessite 铁叶绿泥石
deleterious 有毒的；有害的
deleterious change 有害的变化；恶化
deleterious effect 有害效应；有害影响
deletion 缺失；删去
delicate colour 嫩色；柔和光泽
delicate fabric detergent 纤细织物洗涤剂
delignification 去木质作用
de-liminate ①脱层；分层②脱胶③起鳞
deliming 脱灰
deliming agent 脱灰剂
deliquation 冲淡；稀释；潮解
deliquescence 潮解*
deliquescent (容易)潮解的；容易吸收湿气的
deliquescent crystal 易潮解的晶体
deliquescent effect 吸湿效应；吸湿性
deliquium 潮解物

deliriant 谵妄药
delivering 输送；供应
delivery 分送；运送
delivery capacity 生产额；交货额；排量
delivery conduit 导(出)管
delivery end 卸料端
delivery flask 分液瓶
delivery head(=delivery lift) 水头；压力差
delivery hose 导(出)管；输油软管
delivery lift(=delivery head) 水头；压力差
delivery line ①导(出)管②排量标线；排出量刻度线
delivery nozzle 输送管嘴；排出喷嘴
delivery phase 远送相；分送相
delivery pipe 送水管
delivery point 交货地点
delivery pump 输油泵；给料泵
delivery rate 给料速度；输料速度
delivery ratio 排出比；输送比
delivery stroke 输送冲程
delivery tube 导(出)管
delivery value 输送本领
delivery valve 导出阀；排气阀
delivery volume 给料体积；倾出体积
delocalization 非定域作用；离域作用
delocalization energy(DE) 非定域能；离域能
delocalization of electrons 电子的非定域化
delocalized 非定域的；离域的
delocalized bond 离域键*
delocalized-bond model 非定域键模型；离域键模型
delocalized energy 非定能；离域能
delocalized molecular orbital 离域分子轨道*
delphamine 翠雀胺
delphatine 翠雀亭
delphelatine 翠雀拉亭
delpheline 翠雀灵
delphine blue 噁嗪蓝
delphinic acid(=isovaleric acid) 异戊酸
delphinidin 翠雀素；飞燕草色素；花翠素
delphinidin chloride 氯化翠雀啶
delphinin 翠雀苷；花翠苷；飞燕草色素苷
delphinine 翠雀宁
delphinoidine 翠雀碱
delphisine 翠雀素 $C_{31}H_{50}O_7N$
delphocurarine 翠雀混碱
delphonine 翠雀芳宁
delsoline 翠雀灵
delta acid δ酸；2-萘酸-7-磺酸〔染料中间体〕
delta-blade mixer 三角形叶片混合器

delta-carbon δ位碳原子
delta cross section 三角形截面
delta-cyanovaleramide δ-氰基戊酰胺
delta-function δ函数
deltaline(=eldeline;delphelatine) 翠雀它灵；德它灵
deltamine(=eldelidine) 翠雀它明〔白升麻中含有的一种化学成分〕；匹莫林〔异名〕；盐酸二甲双胍〔商品名〕；德它明
delta-oxidation δ位氧化
delta-position δ位〔对酸及杂环指的是第5位〕
delta ray δ射线
delta-response δ响应
delta-ring Δ环；三角垫
delta-substitution δ位取代
delug 切除花纹
deluge system 集水系统
deluster 去光剂
delusterant 去光剂
delustered fibre 消光纤维；无光纤维
delustering 除去光泽
delustering agent 去光剂
delustering pigment 消光颜料
delustrant 消光剂
delustred yarn 无光丝
delustring 除去光泽
delvauxite 胶磷铁矿
delves cup 凹杯
Delves sampling cup 德尔夫取样杯
demagnetization 去磁作用；消磁
demand capacity lag(=demand lag) 需容滞后〔石油加工中温度控制过程〕
demand lag(=demand capacity lag) 需容滞后
demand side 需要面
demangenization 除锰作用；脱锰作用
demargarination 脱奶油化作用
demasking 解蔽
demasking agent 解蔽剂
demecolcine 脱羰秋水仙碱
dementholized peppermint oil 薄荷素油
demercuration 脱汞(作用)
demetalization 脱金属(作用)
demetallated 脱金属的
demethanation(=demethanization) 脱甲烷(作用)
demethanization(=demethanation) 脱甲烷(作用)
demethanizer 脱甲烷塔
demethanizing 脱甲烷
demethanizing column(=demethanizing tower) 脱甲烷塔
demethanizing tower(=demethanizing column) 脱甲烷塔

5-demethoxy-9-ubiquinone 5-脱甲氧-9-泛醌
demethylating ①脱甲基的②脱甲基
demethylation 脱甲基化*
3,4-demethyl benzoic acid 3,4-二甲基苯甲酸
demethylcelesticetin 去甲基天青菌素
2-demethyldeacetylcolchicine 2-脱甲基脱乙酰基秋水仙碱
demethylhomolycorine 脱甲基高石蒜碱
Demet process 德梅特过程〔一种裂化催化剂脱金属过程〕
demetric acid 去实酸
demicellization 解胶束化
demidovite 天青硅孔雀石
demijohn(=carboy) ①酸坛②坛
demineralization 去矿化*
demineralized water 软化水;脱矿质水
demineralizer 脱盐装置;软水器
demineralizing plant (水的)软化装置
demi-sec spinning 半干纺精纺工艺
demissidine 垂茄定
demissin 狭茄素
demissine 垂茄碱
demister 除沫器;除雾器
de-misting agent 去雾剂
demixing 分层〔指混合液分成两层〕
demixing area 混合物分层的区域
demixing point 分层点
demixion 分层
Demjanov rearrangement 德姆雅诺夫重排〔可用于扩环〕
demodulation polarography 解调极谱
demoisturizer 干燥塔;干燥器
demoisturizing column 干燥塔
demolding 脱模;出模;下模
demolition 炸毁
demolization 过热分散(作用)
demonomerization 脱单体(作用);除单体(作用)
demonstrated film performance 已(得到)验证的漆膜性能
demonstrating apparatus ①示教仪器②表演仪器
demonstration ①表演②示教③验证
demonstration plant ①实验厂;样板厂②示范装置
demo plant 示范工厂;验证工厂
demoulding 脱模
demountable ①活的;活络的;可以拆卸的②可以换装的
demountable hollow cathode lamp 可拆卸空心阴极灯
demountable X-ray tube 可拆卸 X 射线管

Dempster's mass spectrometer 登普斯特质谱计
Dempster type mass spectrometer 登普斯特质谱计
demulcent 润药;缓和的
demulsibility ①反乳化性②反乳化率
demulsibility test 反乳化试验
demulsibility tester 乳化度检验器
demulsifiable lubricant composition 反乳化润滑剂成分
demulsification 反乳化(作用);破乳(作用)
demulsification number 反乳化值
demulsification test 反乳化试验
demulsifier ①反乳化剂;破乳剂②乳液澄清器
demulsifying ①反乳化的②反乳化(作用)
demulsifying agent(=deemulsifying agent) 破乳剂
demulsifying compound 反乳化剂
den ①(=diethylentriamine) 二亚乙基三胺;二(β-氨乙基)胺 $H_2N(CH_2)NH(CH_2)_2NH_2$②窖;小储藏室
Denaby powder 登纳比(炸)药;铵硝化钾炸药
denamycin 德纳霉素
denaturant 变性剂
denaturated alcohol 变性酒精
denaturating ①变性的②变性(作用)
denaturation 变性(作用)
denatured alcohol 变性酒精;工业酒精
denatured fuel 变性燃料;改性燃料
denatured molasses 变性糖蜜
denatured salt 变性食盐
denaturing agent 变性剂
dendrimer 树枝状高分子
dendrimer chemistry 树状高分子化学
dendrimerization 树枝状聚合
dendrite 树枝(状)晶体*
dendritic(al) 树枝状的
dendritic algorithm 树图算法
dendritic copper powder 枝状铜粉
dendri(ti)c crystal 树枝状晶体
dendritic metal 枝状金属
dendritic polymer 树枝(状)高分子
dendritic salt support 枝状盐载体;枝晶盐载体
dendrobine 石斛碱
dendrogram 谱系图
dendrolasin 榧素
denesting 不套叠
de-nib 除粗粒;磨光
denier 旦*〔纤度单位,9000 米纤维重1克为1旦〕
denier control system 生丝纤度控制系统
denierer 定纤感知器;纤度感知器
denier indicator 纤度感知器
denier irregularity 纤度不匀率

Denier process 德尼耶法〔聚氯乙烯糊二氧化碳发泡法〕
Denige's reagent 德尼格试剂
denitrated acid 脱硝酸〔去了硝酸盐的酸〕
denitrated cellulose nitrate 脱硝基硝酸纤维素
denitrated collodion 脱硝基火棉胶
denitrated nitrocellulose 脱硝基硝酸纤维素
denitrated zeolite 脱硝沸石
denitrating column(=denitrating tower) 脱硝(酸盐)塔
denitrating tower(=denitrating column) 脱硝(酸盐)塔
denitration 脱硝(酸盐)(作用)
denitrator 脱硝(酸盐)器
denitridation 脱氮化层(作用)〔炼钢〕
denitrification ①脱氮(作用)②反硝化(作用)
denitrification percent 脱氮率
denitrifying ①脱硝的②脱硝(作用)③反硝化的
denitrogenation 脱氮(作用)
denitrogeneration 脱氮(作用)
Dennestedt furnace 丹尼斯特(电热燃烧)炉
denomination 面额；(度量衡的)单位
denotation impulse 指示脉冲
de novo 从新；从头
de novo synthesis 从头合成；全程合成
dense (稠)密的；浓厚的
dense bed 紧密的(催化剂)床层
dense cake 密致结块
dense gas chromatography 高密度气相色谱法；超临界流体色谱法
dense media process 重液选矿
dense media separation 重液分离；重介质分离
dense membrane 密度膜
dense packing 紧密包装；密实堆积
dense phase 密相
dense phase fluidized bed 密相流化床
dense-phase lifting of catalyst 催化剂密相提升
dense phase suspension 密相悬浮体
dense-phase transporting system 密相输送系统
dense smoke 浓烟
dense soda ash 重苏打灰〔吕布兰法制得的苏打〕
densification ①增浓作用；稠化(作用)②密封；封；封严
densified laminated wood 加重层积材；强化层积材
densifier 稠化剂
densiflorene 赤松烯
densify fluid 黏滞流体
densimeter ①光密度计②比重计
densimetry ①密度法②比重分离法
densite 登斯炸药；登煞特；硝胺、硝酸钾、三硝基甲苯炸药
densi-tensimeter (蒸气)密度-压力计
densitometer 光密度计*；测微光度计
densitometer comparator (显像、光)密度计比较仪
densitometric assay 光密度分析法
densitometric scanning of thin-layer 光密度薄层扫描(法)
densitometric titration 光密度滴定
densitometry 光密度分析法
density ①密度②浓度③强度④黑度
density in situ (裂化产品)在过程中的密度
density balance 密度天平
density bottle 密度瓶
density control valve 调浓阀；浓度控制阀
density crystallinity 密度结晶度
density distribution 密度分布
density fluctuation 密度波动；密度涨落
density fluids 密度液〔测定矿物粉末密度所用的液体〕
density fractionation 密度分离法
density function 密度函数*
density functional theory 密度泛函理论
density gauge ①密度计②比重计
density gradient 密度梯度
density gradient centrifugation 密度梯度离心(分离)法
density gradient column 密度梯度柱；密度梯度管
density gradient electrophoresis 密度梯度电泳(法)
density gradient method 密度梯度测试法
density gradient sedimentation 密度梯度沉降
density gradient sedimentation method 密度梯度沉降法
density gradient separation 密度梯度分离法
density gradient tube 密度梯度管
density index 密度指数
density matrix 密度矩阵*
density matrix theory 密度矩阵理论
density meter 密度计；比重计
density method 密度法
density of catalyst 催化剂密度
density of latex 胶乳密度
density of packing 填充密度
density of spectral line 谱线黑度*
density of states(DOS) 态密度*
density operator 密度算符*
density operator (matrix) method of NMR theory NMR理论的密度算符(矩阵)方法
density recorder 自记密度计；比重记录器
density separation ①密度分离(法)②比重分离法
density temperature coefficient 密度温度系数
density transformation 黑度换值
density variation 密度偏差
densograph 黑度曲线；密度曲线自动描绘仪
densography X射线照片密度检定法

densometer ①密度计②纸透气度测定仪
dent ①凹穴；凹陷②切口③记号
dental alloy 补齿合金
dental gas 笑气；一氧化二氮 N_2O
dentalite 补齿合金
dental paste 牙膏
dental plaster 牙科石膏
dental pulp 牙髓
dental resin 牙科用树脂
dental soap 牙用皂
dentate 配位基
denticotic acid 十二碳-5-烯酸
dentifrice(=dentifrice powder) 牙粉
dentifrice powder(=dentifrice) 牙粉
dentifrice soap 牙膏(粉)用皂
dentifrice water 牙水
dentin(e) 牙质
denting ①压凹②刻齿
dentists' amalgam 牙医用汞剂〔Hg 70%, Cu 30%〕
dentition ①出牙②牙列
dentral 树图算法
dent resistance 耐冲击性
dentrite 树枝晶
dentron 单枝；树枝化单元
dentronized polymer 树枝化聚合物
denture plastics 补齿塑料
denudation ①去肥；瘠化②溶蚀
denuded oil 去垢油
denuding ①溶蚀②去垢③剥裸
deodar cedarwood oil 雪松木油
deodarone 雪松酮
deodorancy ①除臭②除臭作用③祛臭剂
deodorant 除臭剂；芳香剂
deodorant cream 除臭膏
deodorant liquid 除臭液
deodorant paste 除臭糊
deodorant soap 除臭皂；抑臭皂
deodorization 除臭；脱臭；香化
deodorization column 脱臭(除臭)柱(塔)
deodorizer ①除臭机②除臭剂
deodorizing ①除臭的②除臭
deodorizing agent 脱臭剂；去臭剂；矫臭剂
deodorizing composition 脱臭剂
deodorizing compound 祛臭剂
deodorizing solvent 脱臭溶剂；无臭溶剂
de-oiling ①去油的②去油
deolation 解配聚
deorientation 解取向(作用)

deos(=deodorant) 祛臭剂
deoxidant 脱氧剂
deoxidation 脱氧(作用)
deoxidiser(=deoxidizer) 脱氧剂
deoxidized copper 脱氧铜
deoxidizer 脱氧剂
deoxidizing ①脱氧的②脱氧(作用)
deoxidizing agent 脱氧剂
deoxy(=desoxy) 脱氧
deoxycellulose 脱氧纤维素
deoxygenated water 脱氧水
deoxygenation 脱氧*
deoxygenator 脱氧器；除氧器
deoxyhexose 脱氧己糖
deoxynivaenol 脱氧(雪)腐镰刀菌烯酮
deoxynucleoside 脱氧核苷
deoxynucleotide 脱氧核苷酸
deoxynybomycin 脱氧尼博霉素
deoxypentose 脱氧戊糖
deoxyriboaldolase 脱氧核糖醛缩酶
deoxyribonuclease(DNase) 脱氧核糖核酸酶
deoxyribonucleic acid(DNA) 脱氧核糖核酸
deoxyribonucleoside 脱氧核(糖核)苷
deoxyribonucleotide 脱氧核(糖核)苷酸
deoxyribose 脱氧核糖
deoxyriboside 脱氧核(糖核)苷
deoxyribotide 脱氧核(糖核)苷酸
deoxytetracycline(=doxycycline) 多西环素；强力霉素
deoxythymidine(T; dT) 脱氧胸(腺嘧啶核)苷
deoxythymidine monophosphate 胸腺嘧啶脱氧核苷酸
deoxythymidine triphosphate 脱氧胸苷三磷酸
deoxythymidylic acid(TMP; dTMP) 脱氧胸苷酸
deoxyuridine 脱氧尿苷
deoxyuridylic acid 脱氧尿苷酸
deoxyvasicinone 脱氧鸭嘴花碱酮
deozonization 脱臭氧(作用)
depainting 脱漆
deparaffinating 脱蜡
department ①车间②部门
departure 偏差
depassivation 去钝化
dependent variable 因变量；因变数
depentanizer 脱戊烷塔
deperoxidation 脱过氧化(作用)
deperoxidation catalyst 脱过氧化催化剂
dephenolization 脱酚(作用)
dephenolizer 脱酚剂
dephenolizing ①脱酚的②脱酚(作用)

dephenzoat   蒜素制剂
dephlegmating   ①分凝的②分凝(作用)
dephlegmation   分凝；分馏
dephlegmation tower(=dephlegmator)   ①分馏塔②分馏柱③分凝器
dephlegmator(=dephlegmation tower)   ①分馏塔②分馏柱③分凝器
dephlogistication   脱燃素(作用)；减轻发炎
depickling   脱酸〔革〕
depigmentation   脱色素(作用)
depilating   ①脱毛的②脱毛(作用)
depilating agent   脱毛剂
depilation   脱毛(作用)
depilator   脱毛机
depilatory   脱毛剂
depilatory paste   脱毛浆
depilatory powder   脱毛粉
depinker   抗爆剂
deplating   退镀
depleted   废弃的；变质的〔石油储藏〕；消耗的
depleted brine   废盐水
depleted fraction   贫化馏分
depleted reservoir   枯竭矿层；空矿层
depleted uranium   贫化铀*
depletion   ①取金〔冶〕②贫化；放空；耗尽③减少体液〔医〕
depletion effect   贫乏效应
depletion layer   耗尽层
depletion of additive   添加剂的消耗
depletion of film thickness   漆膜厚度枯瘦(瘠薄)化
depolariser   ①去极化剂②消偏振镜
depolarization   ①解偏振作用*②去极化*
depolarization degree   消偏(振光)度
depolarization factor   消偏振因素；退极化因子
depolarization ratio   退偏比值；消偏振度
depolarized electrode   去极化电极*
depolarizer   去极化剂*
depollution   去污染；除污染
depollution of tannery waste water   制革废水处理法
depolyalkylation   解聚烷基化(作用)
depolymerase   解聚酶
depolymerization   解聚*
depolymerizator   解聚(合)器；解聚(合)机
depolymerized rubber   解聚橡胶
depolymerizer   解聚(合)器；解聚装置
depolymerizing agent   解聚剂
depolymerizing reactor   解聚(合)反应器
deposit   ①沉积；淀积②沉积物

deposit accumulation   沉积物的聚积
deposit attack   沉积侵蚀
deposit build-up(=deposit accumulation)   沉积物的聚积
deposit corrosion   沉积腐蚀
deposited activity   沉积放射性
deposited coating   积附涂层
deposited film   沉淀膜
deposited matter   沉积物；沉淀物
deposited metal   熔敷金属
deposited metal film   蒸镀金属膜；敷金薄膜
deposit film   沉积膜
deposit formation   沉积物的形成
deposit-free   无沉积物
deposit ga(u)ge   沉积计
deposit-induced ignition   沉积物所引起的着火〔内燃机中〕
depositing velocity   沉积速度
deposition   ①沉积(作用)②沉积物
deposition chamber   淀积室
deposition efficiency   喷涂沉积效率
deposition of solids   固体沉积
deposition polymerization   沉积聚合
deposition potential   沉积电势；析出电势
deposition pressure(=deposition tension)   沉积张力
deposition product   析出产物；沉淀产物
deposition rate   沉积速度
deposition sedimentation   沉积
deposition tension(=deposition pressure)   沉积张力
deposition thickness   (真空镀膜)积附厚度
deposit modifier   改变沉积物结构的添加剂
depositor   沉积器
deposit seeds   沉淀颗粒
deposit tank   沉积槽；沉积桶
deposit thickness   沉积层厚度
depot   储存；基地
depreciation   减值；折旧
depreciation charge   折旧费
depreciation rate   折旧率
depressant   抑制剂；镇静剂
depressant of surface tension   表面张力降低剂
depressed crown moulding   压缩胎冠造型
depressed design   压花；压印花纹
depressimeter   凝固点降低计
depressing agent   抑制剂
depression   降低；减低
depression constant   凝固点降低常数
depression effect   抑制效应*
depression gate   (注塑模)沉陷式浇口

depression of freezing point 凝固点降低
depression of pour point 倾点降低
depression tank 真空小油箱
depressor ①抑制剂；阻化剂②缓冲剂③阻浮计
depressor bar recorder 抑制记录棒
depressor effect 抑制效应
depropagation ①负增长反应②链断裂作用*
depropagation reaction 减增长反应；反增长反应
depropanization 脱丙烷
depropanizator(=depropanizing column; depropanizer tower; depropanizing tower) 脱丙烷塔
depropanizer 脱丙烷器
depropanizer tower(=depropanizing column; depropanizator; depropanizing column) 脱丙烷塔
depropanizing 脱丙烷
depropanizing column(=depropanizator;depropanizer tower; depropanizing tower) 脱丙烷塔
depropanizing tower(=depropanizing column; depropanizator; depropanizer tower) 脱丙烷塔
depropylation 脱丙基(作用)
deproteinization 脱蛋白作用
deproteinizing agent 脱蛋剂
deprotonated ion 失质子离子
deprotonation 脱质子化作用
deprotonation equilibria 脱质子平衡
depside 缩酚(羧)酸
depsidone 缩酚酸环醚
depsipeptide 缩酚肽
depsiphore 鞣性团
depth ①深度②色泽浓度
depth analysis 深度分析
depth filter 深层滤器
depth gauge 深度计
depth measuring instrument 测深仪
depth of ball charge (球磨机)装球量深度
depth of burial(DOB) 埋深
depth of colo(u)r 色浓度
depth of conversion 转化深度
depth of corrosion 腐蚀深度
depth of cut 切削深度
depth of field 场深；景深
depth of film(=seeing into the film) 漆膜深度(厚度)；漆膜表观深度(厚度)
depth of foam 泡沫厚度；发泡层厚度
depth of focus 焦深
depth of immersion 浸(渍)深(度)
depth of masstone 浓色深度；自身颜色深度
depth of penetration 插入深度

depth of pit 蚀孔深度
depth of rib 棱片长度
depth of shade 色浓度
depth of threat 牙深
depth profiling 深度剖(面分)析
depth resolution 深度分辨率
depuration 纯化；净化；提纯；精炼
derbylite 锑钛铁矿
derbyshire spar 萤石 CaF_2
deresination 脱胶脂(作用)；脱树脂(作用)
deresination agent 脱树脂剂
deresined lube stock 脱胶脂润滑油料
deresining 脱胶脂；脱树脂
deresining plant 脱胶脂装置
derfaz process 德尔法兹过程〔用活性铁和沸石净化放射性废液法〕
derivant 衍生物
derivating agent 衍生剂
derivation 衍生；导出
derivative ①衍生物②导(函)数
derivative absorption spectroscopy 导数吸收光谱法
derivative alternating current polarograph 导数交流极谱仪
derivative anodic stripping voltammetry 导数阳极溶出伏安法
derivative autotitrator 导数自动滴定仪
derivative behavior 衍生物的性能
derivative chronopotentiometry 导数计时电位法*
derivative curve 导数曲线
derivative differential thermal analysis 微分差热分析
derivative dilatometry 微商热膨胀法；导数热膨胀法
derivative double-loop stitch 变化重经组织
derivative fluorescence 导数荧光
derivative formation 形成衍生物
derivative method 衍生(物)法
derivative normal pulse polarography 常规导数脉冲极谱法
derivative plain weave 平纹变化组织
derivative polarogram 导数极谱图
derivative polarograph 导数极谱仪
derivative polarography 导数极谱法*
derivative potentiometric titration 导数电位滴定
derivative pulse anodic stripping voltammetry 导数脉冲阳极溶出伏安法
derivative pulse polarography 导数脉冲极谱(法)*
derivative ratio method 导数比率法
derivative ratio spectrophotometry 比值导数分光光度法
derivative ratio spectrum method 比值导数谱法

derivative spectrophotometry　导数分光光度法
derivative spectrum　导数光谱*
derivative synchronous fluorescence　导数同步荧光
derivative synchronous fluorescence method　导数同步荧光法
derivative synchronous fluorimetry　导数同步荧光法
derivative thermodilatometry　导数热膨胀法
derivative thermogravimetric analysis(DTG)　导数热重量分析；导数热重分析；微分热重分析
derivative thermogravimetric curve　导数热重量分析曲线
derivative thermogravimetry(DTG)　导数热重量分析法；微商热重法
derivative titration method　微分滴定法
derivative variable-angle synchronous fluorescence spectroscopy　导数可变角同步荧光光谱法
derivative warp-knitted stitch　变化经编组织
derivatization　衍生(作用)
derivatization gas chromatography　衍生气相色谱法
derivatization method　衍生法
derivatization reaction　衍生化反应
derivatization reaction detector　衍生反应检测器
derivatization reagent　衍生化试剂
derivatization room temperature phosphorimetry　衍生室温磷光法
derivatizing reaction vial　衍生反应小管
derivatograph　①示差热分析仪②测偏仪
derivatography　示差热分析法；导数热谱法；微分热谱法
derived fossils〔复〕　转生化石
derived high polymer　衍生高聚物
derived lipid　衍生脂类
derived protein　衍生蛋白
derived quantity　导出量
derived rubber　衍生橡胶
derived unit　导出单位
derma　真皮
dermadin　木菌素
dermatan sulfate　硫酸皮肤素
dermateen　漆布；布质假皮
dermatitant　刺激皮肤物
dermatol　桔酸铋
dermatoscope　双筒显微镜
dermatosome　微纤维素
dermics　皮肤病药
dermiformer　生皮剂
dermol　大黄根酸铋
dermostatin　制皮菌素

derric acid　鱼藤酸
derrick　①提升重量的绞盘(绞车)②塔式起重机③井架
derrid　鱼藤脂
derrin　鱼藤酮
derris root　鱼藤根
deruster　除锈剂
derusting　除锈
derusting method　除锈法
derusting of section　分段除锈
derusting with strippable film　(用)可剥性薄膜除锈法
derustit　电化学除锈法
7-desacetoxyhelvolic acid　7-去乙酸基蜡黄酸
desactivation　钝化；不活性化
desaeration　脱气
desalgin　氯仿的胶态溶液
desalicetin　去水杨天青菌素
desalination　脱盐(作用)
desalination of sea water　海水淡化
desalination plant　淡化工厂
desalinator　海水淡化厂
desalinization　①脱盐(作用)②(海水)淡化
desalinized water　脱盐水
desalt　脱盐
desalted water　脱盐水
desalter　脱盐设备
desaltification　脱盐(作用)
desalting　脱盐(作用)
desalting agent　脱盐剂
desalting reactor　(海水)脱盐反应器
desamidation　脱酰胺(作用)
desamidization(=deamidization)　脱酰胺作用
desaspidin　异鳞毛蕨素
desaturation　去饱和(作用)
desaulesite　硅锌镍矿
descaling　脱垢；去垢
descaling chisel　除锈凿；除鳞(片)凿
descaling hammer　除氧化皮锤；除锈锤
descaling pistol　除氧化皮枪
descendant　子系体；后代
descending chromatogram　下行色谱图
descending chromatography　下行色谱法
descending development　下行展开(法)*
descending development method　下行展开法
descending method　下行法
descending paper chromatography　下行纸色谱(法)
descending paper partition chromatography　下行纸分配色谱法
descending paper strip chromatography　下行纸条色谱(法)

descending slurry 下行淤浆
descending technique 下行术
descending vertical flue 下降式竖(式)烟道
deschlorothricin 脱氯丝菌素
descloizite 钒铅锌矿
describing function 描述函数
description 性状
description of product 产品说明书
descriptive literature 说明书
descriptor 描述符
desdamethine 甲硫吡丙菌素
desdanine(＝cyclamidomycin) 赛拉霉素
desealing tongs 起封钳
deselenization 脱硒*
desensibilization 去敏化作用
desensitivity 减感性
desensitization 减感(作用)
desensitizer 减感剂
desensitizing agent 脱敏剂
desensitizing development 减感显影
desensitizing dye 减感染料
desensitizing effect 减感效应
deserpidine(=canescine;recanescine;raunormine) 地舍平
desertification 荒漠化；沙漠化
desertomycin 沙漠霉素
desetting 反变定；去定形；解凝固
desheating 去除外壳(或外套)；脱壳
deshielding 去屏蔽
deshielding effect 去屏蔽效应
deshydro 脱氢
deshydroxy 脱羟
desiccant(=drying agent) 干燥剂*
desiccant gel 干燥剂(胶)
desiccated balance 干燥平衡
desiccating agent 干燥剂
desiccation 干燥；脱水
desiccative ①干燥的②干燥剂
desiccator(=exsiccator) 干燥器；保干燥
desicchlora 燥钡盐
design 设计
designability of material properties 材料性能可设计性
designated size ①指定粒度②设计尺寸；指定尺寸
designation ①指明；指定；选定②名称；符号；牌号；标志③目标；目的地
designation number 标准指数
design capacity 设计(生产)能力
design center size 设计中心线密度
design compression ratio 设计压缩机比

design criteria 设计规范；设计标准；设计准则
design data 设计数据
design drawing 设计图
design instruction 设计说明书
design loading 设计负荷
design of experiment 试验设计法
design of polymer 高分子设计
design paper 方格纸；绘图纸
design point 设计点
design roller coating 辊涂压花法
design specifications 设计规范
design stress 设计应力
design variable 设计参数
desilication 脱硅(酸)作用
desilicification 脱硅(作用)
desilver 脱银
desilverisation 脱银(作用)
desilylation 脱甲硅基作用
desintegration ①裂变；蜕变②分裂③去整合(作用)
desintegrator 磨碎机
desirable animal nuance 愉快的动物香调〔指天然动物香料〕
desired alkylbenzene 目的烷基苯
desired coating 理想涂层
desivac process 冰冻干燥过程
desizability 退浆能力；退浆性
desized product 脱浆制品
desizing 脱浆；退浆；除浆
desizing agent 退浆剂；脱浆剂
desk 实验台；工作台
desk calculator 台式计算器
desk centrifuge 台式离心机
desk-electrophoresis 台面电泳法
desk fan 台风扇
desk for glass blowing 吹制玻璃(操作)台
desk lamp 台灯
desktop ①台式的；桌式的②案头的；桌上的
desliming 底流分离；除去矿泥
deslubbing roll 除粗节丝辊；去纱节辊
desludging 除去淤渣
de-smear 除胶渣
Desmet extractor 德斯梅抽提器束沸石
desmoenzyme 不溶酶；结合酶
desmolase 碳链(裂解)酶
desmopressin 去氨加压素〔药〕
desmotrope 稳变异构体
desmotropic 稳变异构的
desmotropic compound 稳变异构物

desmotropic form 稳变异构体
desmotropism 稳变异构(现象)
desmotropy 稳变异构(现象)
desmycosin 脱碳霉糖泰乐菌素〔泰乐菌素 tylosin 的水解产物〕
desodoration 除臭；香化
desodorization 除臭；脱臭
desolvating nebulizer 去溶剂雾化器
desolvation 去溶剂化*
desolvation interference 去溶干扰
desolventizer 脱溶剂器
desolventizing 脱(去)溶剂
desorber 解吸塔
desorption 脱附(作用)*
desorption chemical ionization 解吸化学电离
desorption chemical ionization source 解吸化学电离源
desorption coefficient 解吸系数；脱附系数
desorption column 解吸塔
desorption curve 解吸曲线；脱附曲线
desorption efficiency 解吸效率；脱附效率
desorption factor 解吸因子；脱附因子
desorption ionization 解吸电离
desorption isotherm 解吸等温线；脱附等温线
desorption of gases 气体的解吸(脱附)
desorption of moisture 脱湿(作用)；减湿(作用)
desorption potential 脱附电势
desorption spectrum 解吸谱
desorption tower 解吸塔
desosamine 德糖胺〔红霉素的降解产物〕 $C_8H_{17}NO_3$
desoxidation 脱氧
desoxyalizazin 3,4,9-蒽三酚
desoxybenzoin(=phenylbenzyl ketone) 脱氧苯偶姻；二苯乙酮；苯基苄基(甲)酮 $C_6H_5COCH_2C_6H_5$
desoxyribonucleic acid 脱氧核糖核酸
desoxyribose(=ribodesose) 脱氧核糖
desoxystreptomycin 脱氧链霉素
dessicant 干燥剂
destabilization 去稳定作用
destabilizer 去稳定剂
destabilizing agent 去稳定剂
destabilizing effect 失稳效应；去稳效应
destain 脱色
destarch 脱浆；去淀粉
destarched 退(了)浆的
destaticization 消静电作用
destaticizer 去静电剂
destemming 去梗；除梗
desthiobiotin 脱硫生物素

destinker 去味器
destomycin 越霉素
destruction gas chromatography 破坏性气相色谱(法)
destruction of ozone layer 臭氧层的破坏
destruction test 破坏性试验；断裂试验
destructive ①破坏的；毁坏的 ②破坏性
destructive alkylation 破坏烷基化；破坏烃化
destructive analysis 损毁分析；破坏性分析
destructive chlorination 破坏氯化作用
destructive detector 破坏性检测器
destructive distillation ①破坏蒸馏；分解蒸馏 ②干馏
destructive distillation of wood 木材干馏
destructive distillation turpentine 干馏(木)松节油
destructive gas chromatography 破坏性气相色谱(法)
destructive hydrogenation 破坏加氢
destructive inspection 破坏(性)检查
destructive interference 相消干扰
destructive interference coating 反雷达涂层
destructive method 有损探伤法
destructive oxidation 破坏性氧化
destructive process 破坏过程〔在此过程中原料的分子结构发生变化〕
destructive reaction(=breaking reaction) 破坏性反应
destructive test ①破坏性试验 ②爆破试验
destructive vibration 破坏性震动
destructor (垃圾)焚烧炉
destructurization 变构(作用)
destruxin 黑僵菌素；腐败菌素
desublimation 凝华(作用)；去升华(作用)
desugar 脱糖；提出糖分
desugaring process 脱糖方法
desugarization 脱糖(作用)
desugarizing 脱糖
desulfate 脱硫；脱硫酸盐
desulfating 脱硫酸盐
desulfation 脱硫(作用)；脱硫酸盐(作用)
desulfidation 脱硫(作用)
desulfonation 脱磺酸基*
desulfurase 脱硫酶
desulfurated 脱硫的；无硫的
desulfurating ①脱硫的 ②脱硫(作用)
desulfuration 脱硫(作用)
desulfurization 脱硫*
desulfurization chamber 脱硫室
desulfurization-hydrogenation 脱硫加氢(作用)
desulfurized fuel 脱硫燃料
desulfurized naphtha 脱硫石脑油
desulfurizer ①脱硫剂 ②脱硫装置

desulfurizing 脱硫
desulfurizing agent 脱硫剂
desulfurizing furnace 脱硫炉
desulphurase 脱硫酶
desulphur(iz)ation 脱硫(作用)
desuperheating 脱过热；过热下降
deswelling 退(溶)胀(作用)
desyl(=α-phenylphenacyl) 二苯乙酮基；α-苯甲酰苯甲基
detachability (可)脱渣性
detachable 可拆卸的；可移专的
detachable core type top roller 活皮辊；活心式皮辊
detachable device 可拆装置
detachable needle 可拆卸针头；活动针头
detached caustic soda 块状苛性苏打
detaching 拆卸；分离；移去
detaching roller 分离罗拉
detachment 除去；劈去；移去
detackifier 防黏剂
detail 详图
detail drawing ①详图②零件图
detailed balancing principle 详细平衡原理
detailed-estimated design 细估设计
detailed list 明细表
detail paper 誊写纸
detail scan 窄扫描
detanning 脱鞣
detar 脱焦油
detar column 脱焦油塔
detarrer 脱焦油器
detarring ①脱焦油的②脱焦油(作用)
detarring precipitator (脱)焦油沉降器
detearing ①沥水；沥干②除滴；除液(漆)滴
detectability 检测性；可检测性
detectable 可检测的
detectable limit 可测下限
detectable peak 可检测峰
detectable sample 可检测试样
detecting element 检测元件
detection 检测
detection agent 检定剂；检测剂
detection device 检测装置
detection level 检测水平
detection limit 检出限*
detection limit of spectral analysis 光谱分析检出限
detection line 检测线*
detection of analytical signal 分析信号检测
detection of interfere and restrain conductivity 干扰抑制电导率检测
detection of laser-induced light beam intervene 激光诱导光束干涉检测
detection of laser-induced light heat and deflexion 激光诱导光热光偏转测量
detection of shellfish toxins 贝类毒素检测
detection period 检测期*
detection sensitivity 检测灵敏度
detection system 检测系统
detection tube for acetylene 乙炔检测管
detection tube method 检定管法
detectivity 检测能力
detector 检测器*
detector aspect 检测器型式
detector block 检测器(金属)块
detector cell 检测器池
detector cleaner 检测器清洗器
detector compartment 检测器室
detector cycle 缫丝探索周期
detector detectability 检测器检测性
detector electrode 检测器电极
detector geometry 检测器几何形状
detector heater 检测器加热器
detector linear range 检测器线性范围
detector noise 检测器噪声
detector of dual-beam difference 双束差分检测器
detector output 检测器输出
detector oven 检测器加热炉
detector response 检测器响应(值)
detector sensitivity 检测器灵敏度
detector signal 检测器信号
detector temperature 检测器温度
detector time constant 检测器时间常数
detector tube 检查管；视镜管
detector tube method 检测管法
detention period 停留时间；延迟时间
detention time 滞留时间
detergency ①去污力②洗净；脱垢
detergency promoter 助洗剂
detergent 洗涤剂*
detergent action 去垢作用
detergent additive 去垢添加剂
detergent alcohols 洗涤剂用醇类〔指$C_{12} \sim C_{18}$脂肪醇〕
detergent alkylate 洗涤剂用烷基化物〔常指$C_{12} \sim C_{14}$烷基苯〕
detergent auxiliary 洗涤助剂；去垢助剂
detergent base 洗涤剂基料
detergent builder 洗涤剂助剂；助洗剂

detergent characteristic 洗涤剂特性
detergent cleaning agent 去污剂
detergent dispersant 洗涤分散剂；清洁分散剂
detergent gas 洗净气
detergent liquid 液体洗涤剂；洗涤液
detergent oil 去垢油；洗涤油
detergent paste 膏状洗涤剂；洗衣膏
detergent perfume 洗衣粉香精
detergent phosphate 洗涤剂用磷酸盐〔指缩聚磷酸盐〕
detergent power 去垢本领
detergent range alcohol 洗涤剂(用)醇
detergent release 除垢性脱模剂
detergent resistance 耐洗涤剂性
detergent-resistant 耐洗涤剂的
detergent resistant coating 耐洗涤剂涂料
detergent solution 去污液
detergent surfactant 除垢表面活性剂
detergent tablet 片状洗涤剂
detergent wax 净石蜡
deterioration ①恶化(作用)；变坏(作用)②退化(作用)③变质④磨损
deterioration accelerator 加速老化试验机
deterioration failure 劣化失效
deterioration of cracked gasoline 裂化汽油变质
determeter (往复式)去污试验仪
determinable error 可测误差
determinant ①(抗原)决定簇②行列式③决定子④欲测物
determinantion wave function 行列式波函数
determination ①判定；测定②注定；确定
determination limit 测定限*
determination of acid value 酸值的测定〔油脂分析〕
determination of alkalinity of ash 灰分的碱度测定〔皮粉分析〕
determination of biochemical oxygen demand 生化需氧量测定
determination of blood cholesterol 血液胆甾醇定量法
determination of chlorophyll in plant 植物叶绿素测定
determination of coagulation factor activity 凝血因子活性测定
determination of dust by light scattering measurement 散射光测尘法
determination of fat 脂肪含量的测定
determination of filler (润滑脂内)填料含量的测定
determination of flavacin 黄曲霉素测定
determination of minimum inhibitory concentration 最低抑制浓度测定法
determination of molecular weight 分子量测定
determination of nitrogen in soil 土壤中氮素的测定
determination of organics in soil 土壤有机质测定
determination of plasminogen activated inhibitor 纤溶酶原激活抑制物测定
determination of platelet life span 血小板寿命测定
determination of protein by potentiometer 蛋白质电位滴定测定法
determination of protein by UN-spectrometry 紫外光度法测定蛋白质
determination of red cell electrophoresis time 红细胞电泳时间测定
determination of soap (润滑脂内)皂量的测定
determination of soil air 土壤空气测定
determination of soil moisture 土壤自然含水量测定
determination of soil nutrients 土壤营养成分测定
determination of tinting strength 着色力测定
determination of tissue plasminogen activator 组织型纤溶酶原激活物测定
determination of vitamin 维生素测定
determination of volatile matter 挥发物测定
determination of water of crystallization 结晶水定量分析
determination with equal numbers 等重复次数测定
determination with unequal numbers 不等重复次数测定
determinism 决定论
deterministic system 确定性系统
deterrent ①妨碍的；阻止的②阻滞剂；灭菌剂
detersive 洗涤(的)
detersive efficiency 洗涤效率；去污效率
detersive power 去污力
detinning 脱锡
Detol process 德托尔法〔催化脱烷基化法〕
detonating ①爆炸的②爆炸(作用)
detonating agent 发爆剂；爆炸剂
detonating cap 发火帽；雷管
detonating explosive 爆轰炸药
detonating fuel 爆震燃料
detonating fuse 爆炸信线；爆炸导火索
detonating gas 爆鸣气
detonating gas coulometer 爆鸣气库仑计；爆鸣气电量计
detonating primer 引爆雷管
detonating relay 继爆管
detonating velocity 爆燃速度；起爆速度
detonation 爆震；爆炸(作用)；轰发
detonation characteristic 爆震特性
detonation flame spraying 爆炸喷涂
detonation front 爆震正面
detonation meter 爆震计

detonation point 爆震点
detonation pressure 爆(轰)压(力)
detonation propagation 爆炸传播
detonation temperature 爆炸温度
detonation time 爆炸时间
detonation transmission 爆炸传递
detonation velocity 爆炸速度
detonation wave 爆(炸)波
detonative power 爆炸力
detonator ①发爆剂；爆炸剂；引爆剂②雷管；发火管；发爆器
detonics 爆炸学
detour roll 导丝辊
detoxicant 去毒剂
detoxicated gas 解毒的气体
detoxication 解毒(作用)
detoxification 解毒(作用)；去毒(作用)
detoxifier 解毒剂
detrital 碎屑的
detritiation 除氚*
detrition 磨耗
detritus ①腐质②碎屑
detritylation 脱三苯甲基作用
detrusion 冲移
detuning 失调*
deuterated 氘化的
deuterated antioxidant 氘化抗氧剂
deuterated cellulose 氘代纤维素
deuterated polyethylene 氘化聚乙烯 $(CD_2)_n$
deuterated reagent 氘代试剂
deuterated solvent 氘代溶剂*
deuterated water 重水 D_2O
deuterating petroleum product 氘化的石油产品
deuteration ①氘化②氘化作用〔氘代氢〕
deuteric acid 含氘酸
deuteride 氘化物*
deuteriocarbon 碳化氘；氘烃
deuteriochloroform 氘(代)氯仿
deuterioorganic compound 含氘有机化合物
deuterium(=deuteronium) 氘〔氢的同位素符号 ^{2}H 或 D〕；重氢
deuterium arc background correction 氘弧灯背景校正
deuterium arc lamp 氘弧灯
deuterium bond 氘键；重氢键
deuterium bromide 溴化氘 DBr
deuterium discharge lamp 氘放电灯
deuterium exchange 氘交换*；重氢交换
deuterium isotope effect 氘同位素效应
deuterium-labelled 氘标记的
deuterium labelling technique 氘标记技术
deuterium lamp 氘灯
deuterium lamp background correction ①氘灯校正背景②氘灯背景校正法
deuterium lock 氘锁
deuterium-oxide 氧化氘；氘水；重水 D_2O；HDO
deuterium-peroxide 过氧化氘 D_2O_2
deuterium stabilization 氘稳场；氘核稳场
deuterium-tritide 氚化氘 DT
deutero ①次；第二②衍生的③含氘的
deuteroacetic acid 含氘乙酸
deuteroacetone 含氘丙酮
deuteroacetoprotic acid 氘乙酸 CD_3COOH
deutero-aetioporphyrin 次初卟啉
deutero-albumose 次胨
deuteroammonia 含氘氨
deuteron(=deuton) 氘核*
deuteronium(=deuterium) 氘
deuteroparaffin 氘化烃
deutero phosphoric acid 氘磷酸
deuteroporphyrin 次卟啉
deuteroproteose 次胨
deutero rubber 氘橡胶；重氢橡胶
deuteroxide 重水；氧化氘
deuteroxyl 氘氧基 DO—
deutomycin 丢托霉素
deuton(=deuteron) 氘核
deutration 氘化反应
Deutschland hardness 德国硬度
devaporation 止汽化(作用)
Devarda alloy 德瓦达合金
developed chromatographic strip 已展开色谱条
developed cylinder 显影滚筒
developed dyes 显色染料
developed ore 开采矿
developer ①显影剂②展开剂③显色剂
developer solution 显影液
developing ①显影②展开
developing agent 显影剂
developing bath ①显影浴②显色浴
developing boundary layer 发展中边界层
developing dye 显色染料
developing factor 显影因数
developing power ①显影本领②展开能力
developing radical 显影基
developing solution ①展开液②显影液
developing solvent 展开剂*

developing tank　①展开罐(槽)②显影罐
developing time　①显影时间②展开时间
development　①展开②显影③显色④进展
development chamber　展开室
development distance　展开距离
development duration　展开时间
development of gas　①放出气体②放出毒气
development of heat　放热；生热
development period　展开时间
development rate　展开速度
development system　展开系统
development tank　展开罐
development time　显影时间
devernalization　去春化作用
deviation　偏差*
deviation adjustment　偏差调整
deviation compensation　偏差补偿
deviation monitor assembly　偏差监测系统
deviation of weight　重量偏差
deviations from Beer's law　偏离比尔定律
deviatoric state of stress　偏应力状态
deviatoric stress tensor　偏应力张量
device　①装置；设备②手段③设计
device for radioactivity measurement　放射性测定器
device for repeated diffraction　二次衍射装置
device for scanning paper chromatogram　纸色谱扫描器
devil liquor(=devil water)　废液
devil water(=devil liquor)　废液
devitrification　反玻璃化
devitrified glass fibre　微晶玻璃纤维；失透玻璃纤维
devitrifying　消失透明；失透(作用)；反玻璃化
devolatilization　脱挥发分(作用)
devolatilizer　①(螺杆反应器)排气室；排气段②脱挥发组分装置
devolatilizing chamber　(螺杆反应器)脱挥腔；排气室
devulcanization　脱硫
devulcanized product　再生胶
devulcanized (waste) rubber　再生胶
devulcanizer　脱硫器；反硫化器
devulcanizing　脱硫
devulcanizing agent　脱硫剂
devulcanizing pan　脱硫釜
dew　露
Dewar benzene　杜瓦苯
Dewar form　杜瓦式
Dewar formula　杜瓦式
Dewar's bottle　真空瓶；杜瓦瓶
Dewar's rules　杜瓦规律

Dewar structure　杜瓦结构
Dewar-type polymerization reactor　杜瓦瓶型聚合反应器
Dewar (vacuum) flask(=vacuum flask)　真空瓶；杜瓦(真空)瓶
Dewar vessel　杜瓦瓶；真空保温瓶
dewatered slurry　脱水料浆
dewaterer　脱水器
dewatering　①脱水的②脱水(作用)
dewatering box　脱水箱
dewatering centrifuge　脱水离心机
dewatering conveyor　脱水传送带
dewatering tank　脱水槽；澄清槽
dewater unit　脱水装置
dewaxed damar　脱蜡哒玛
dewaxed oil　脱蜡油
dewaxed oil stripper　脱蜡油汽提塔
dewaxed shellac　脱蜡虫胶
dew corrosion　露点腐蚀
dew cycle　露循环；(蒸馏)液滴循环
dew cycle corrosion tester　露点循环腐蚀试验机
dewetting　半润湿
deweylite　水蛇纹石
dew of death　致死露
dew point　露点
dew point corrosion test　露点腐蚀试验
dew point curve　露点曲线
dew point hygrometer　露点湿度计
dew point hygrometry　露点测湿法
dew point meter　露点计
dew point method　露点法；露点测定法
dew point temperature　露点温度
dew retting　露浸渍
dewy　(潮)湿的；回湿(潮)的
dewy freshes　清鲜香气
dexamethasone　地塞米松；9-氟-16-甲基脱氢皮质(甾)醇
dexanthation　脱黄(原酸)化作用
dexiotropic　右旋的
dexpanthenol(=D-pantothenyl alcohol)　D-泛醇；右泛醇
dextrorotatory form　右旋型
dextral　右方的；向右的；右旋的
dextral motion　右旋运动
dextran　葡聚糖*
dextranase　葡聚糖酶
dextran gel electrophoresis　葡聚糖凝胶电泳
dextran sulfate　硫酸葡聚糖
dextrin　糊精*
dextrin adhesive　糊精黏合剂
dextrinase　糊精酶

dextrin glue 糊精胶
dextrinization 液化；糊精化
dextro 右旋的
dextrochrysin 右金菌素
dextro-compound 右旋(化合)物
dextroform 右旋型
dextroglucose(=dextrose) 右旋糖；葡萄糖
dextrogyrate(=dextrogyric) 右旋的
dextrogyric(=dextrogyrate) 右旋的
dextro isomer 右旋异构体*
dextromethorphan 右美沙芬〔药〕
dextromycin 右霉素
dextronic acid 右旋糖酸；葡萄糖酸
dextropimaric acid(=pimaric acid) (右旋)海松酸
dextropimarinal 右香醛
dextrorotary 右旋的
dextrorotary crudes 右旋原油
dextrorotation 右旋
dextrorotatory 右旋的
dextrorotatory compound 右旋化合物
dextrorotatory isomer 右旋异构体
dextro-rotatory substance 右旋物质
dextrosan 聚右旋糖
dextrosazone 右旋糖脎
dextrose(=dextroglucose) 右旋糖；葡萄糖
deyamittin 假软齿花苷
dezincification 脱锌〔腐蚀〕
dezincification preventive substance 脱锌防止剂
dezincing 脱锌
dg 分克
D glass fibre D 玻璃纤维；低介电玻璃纤维
DHD process DHD 法〔高压脱氢过程〕
Dhobi marking nut oil(=marking nut oil) 印果油
dhurrin 蜀黍苷
di- ①二；双〔指基的数目，如: dinitro 二硝基，指复杂基时应加圆括号〕②联(二)〔指两个基以一价相联，如 diphenyl 联(二)苯〕③双〔指两个单体相结合，如 dipyrrole 双吡咯, diurea 双脲, disalicylic acid 双水杨酸〕
diabantite 辉绿泥石
diabase 辉绿岩
diabase-aplite 辉绿细晶岩
diabasic texture 辉绿岩结构
diabasite 辉绿玢岩类
diaboline 达波灵碱
diacetamide 二乙酰基胺 $(CH_3CO)_2NH$
diacetanilide N,N-二乙酰苯胺 $C_6H_5N(COCH_3)_2$
diacetate 双乙酸盐(或酯)
diacetate fibre 二醋酯纤维

diacetic acid ①二乙酰乙酸 $(CH_3CO)_2CHCOOH$ ②乙酰乙酸 $CH_3CH_2COCH_2COOH$
diacetic ester ①二乙酰乙酸酯 $(CH_3CO)_2CHCOOR$ ②〔专指〕二乙酰乙酸乙酯 $(CH_3CO)_2CHCOOC_2H_5$
diacetin(=glyceryl diacetate) 甘油二乙酸酯；二醋精 $(CH_3COOCH_2)_2CHOH$
diacetine(=glyceryl diacetate) 二乙酸甘油酯
diacetone(=acetylacetone) 双丙酮；乙酰丙酮 $CH_3COCH_2COCH_3$
diacetone-acryloamide(DAA) 双丙酮丙烯酰胺
diacetone alcohol 双丙酮醇 $(CH_3)_2COHCH_2COCH_3$—
diacetone-amine 双丙酮胺 $(CH_3)_2C(NH_2)CH_2COCH_3$—
diaceto-succinic ester 二乙酰丁二酸酯 $ROCO(CHCOCH_3)_2COOR$
diacetoxyl 双乙酸基
diacetoxylphenylisatin 二(乙酸基苯基)靛红 $C_6H_4N=COHC=(C_6H_2O_2CCH_3)_2$
diacetoxysciroenol(=anguidin) 蛇形菌素
diacetyl ①联乙酰化合物；丁二酮②二乙酰基
diacetyl acetic acid 二乙酰乙酸 $(CH_3CO)_2CHCOOH$
diacetyl acetic ester 二乙酰乙酸酯 $(CH_3CO)_2CHCOOR$
diacetyl acetone 二乙酰基丙酮；庚间三酮 $(CH_3COCH_2)_2CO$
diacetyl amide 二乙酰胺 $(CH_3CO)_2NH$
diacetylamino 二乙酰氨基 $(CH_3CO)_2N$—
diacetyl anilide N,N-二乙酰苯胺 $C_6H_5N(COCH_3)_2$
diacetylated 二乙酰的；二乙酰基取代了的
diacetyl benzidine N,N-二乙酰联苯胺 $(CH_3CONHC_6H_4)_2$
diacetyl-carbinol 二乙酰甲醇 $(CH_3CO)_2CHOH$
diacetyldioxime 丁二酮肟*
diacetyl diphenyl octane 二乙酰基二苯基辛烷〔防老剂〕
diacetyl diphenyl osazone 二乙酰基二苯脎〔抗臭氧剂〕
diacetyl disulfide 二乙酰二硫 $(CH_3CO)_2S_2$
diacetylene 联乙炔；丁二炔 $HC≡CC≡CH$
diacetylene-benzene 二乙炔基苯 $C_6H_4(C≡CH)_2$
diacetylene-dicarboxylic acid 联乙炔二羧酸；己二炔二酸 $HOOCC≡CC≡CCOOH$
diacetylene glycol(=hexadiindiol) 联乙炔二醇；己二炔二醇 $HOCH_2C≡CC≡CCH_2OH$
diacetylene polymer 二乙炔聚合物*
diacetylenic acid 二炔酸
2,4-diacetyl-fluoroglucine 2,4-二乙酰荧光甜素
diacetylhydrazine 二乙酰基肼 $(CH_3CONH)_2$
diacetyl monomethoxime 双乙酰一甲氧肟；丁二酮一甲氧肟 $CH_3COC(=NOCH_3)CH_3$
diacetylmonoxime(=isonitrosomethyl ethyl ketone) 肟甲基乙基(甲)酮 $CH_3COC(=NOH)CH_3$

diacetyl morphine(=heroine)　二乙酰吗啡；海洛英
diacetyl oxide　二乙酰化氧；乙酸酐　$(CH_3CO)_2O$
diacetyl peroxide　过氧化二乙酰　$(CH_3CO)_2O_2$
diacetyl succinic acid　二乙酰丁二酸　
　　$COOHCH(COCH_3)CH(COCH_3)COOH$
diacetyl tannic acid　二乙酰单宁酸
diacetyl tannin(=tannigen)　二乙酰单宁；单宁精
diacetyl tartaric acid　二乙酰酒石酸
　　$COOH[CH(OCOCH_3)]_2COOH$
diacetyl urea　二乙酰脲　$(CH_3CONH)_2CO$
diachylon plaster　油酸铅硬膏
diacid　二酸〔类名〕
diacid amide　二酰胺　$(RCO)_2NH$
diacid arsenate　二酸式砷酸盐；砷酸二氢盐　MH_2AsO_4
diacidic base　二元碱
diacid salt　二酸式盐；二氢盐〔分子中仍含有两个氢离子〕
diaclases　压力裂缝
diacolation　渗萃；渗漉
diacrylamine　二丙烯酰胺〔防老剂〕
diacryl ethylene titanate　钛酸二丙烯酰乙二(醇)酯
diactetoxyseripenol　二乙烯草镰刀菌烯醇
diactinic　透光化线的
di-active amyl succinate　丁二酸二旋性戊酯
　　$(CH_2CO_2C_5H_{11})_2$
diacyl　二酰基
diacyl compound　二酰基化合物
diacyl iodoglycerol　二乙酰碘代甘油
diacyl peroxide　二酰基过氧化物
diad　二单元组*
diad axis　二重轴
diadduct　二(基团)加成物；二元加成物
diadem colours　二登染料〔商名,染毛铬处理类染料〕
diadipate　①己二酸氢盐　$COOH(CH_2)_4COOM$②己二酸氢酯　$COOH(CH_2)_4COOR$
diadochite　磷铁华
diad prototropy　二素质子移变作用(现象)
diagenesis　成岩作用
diagenism　沉积变质作用
diageotropism　横向地性
diagglomerate latices　双附聚胶乳
diagnosis　诊断
diagnostic biochemistry　诊断生物化学
diagnostic observation　诊断观察；判断性观察
diagnostic peak　鉴定峰
diagnostic performance of a laboratory result　实验结果的诊断效能
diagnostic program　诊断程序

diagnostic reaction　诊断反应
diagnostic reagent　诊断剂
diagnostic routine analysis　诊断常规分析
diagnostic structural analysis　诊断结构解析
diagnostic test　诊断试验
diagonal　①对角线②对角的
diagonal brace　斜撑
diagonal chromatographic technique　对角线色谱技术
diagonal chromatography　对角线色谱(法)
diagonal grain　斜纹；斜纹理
diagonal hybrid orbital　直线型杂化轨函数
diagonal least-squares refinement　对角最小二乘修正
diagonal matrix　对角矩阵
diagonal montage　对角线剪辑(表)
diagonal peak　对角线峰
diagram　图；简图；线图；图表
diagram loom timing　织机工作圆图
diagrammatic　图解的；图表的
diagrammatic drawing　示意图
diakinesis　(生殖细胞分裂)的丝球期〔指母染色体和父染色体在核中配对的时期〕；终变期
dial　①二醛②刻度盘；标度③拨号盘
dial balance　刻度盘天平
di-alcohols　二元醇类
dialdehyde　二醛
dialdehyde cellulose　二醛基纤维素
dial flow meter　刻度盘流速计
dial gauge　直读式厚度计
dialin　二氢化萘
dial indicator　①刻度指针②千分表
dialkanol piperazine ester　二烷醇哌嗪酯
dialkene(=diolefine)　二烯烃
dialkoxy　二烷氧基
dialkoxyaluminum acylate　酰化二烷氧基铝
dialkoxyaluminum methacrylate　甲基丙烯酸二烷氧基铝
dialkoxy diamino silane　二烷氧基二氨基硅烷
dialkoxy phosphazine　二烷氧膦嗪
dialkoxyphosphinyl　二烷氧膦基
dialkyl　二烷基(的)
dialkyl adipate　己二酸二烃酯
dialkyl alkylene diphosphonic acid　亚烷基二烷基双膦酸
dialkyl alkylphosphonate　烷基膦酸二烷基酯　$RPO(OR)_2$
dialkylaminobenzene azoanilino functional group　二烷基氨基苯偶氮苯胺官能团
dialkylamino-biphenylamino group　二烷基氨-联苯基氨基团
dialkyl arsine　二烷基胂　R_2AsH

dialkyl arsine oxide 二烷基胂化氧 $(R_2As)_2O$
dialkyl arsine sulfide 二烷基胂化硫 $(R_2As)_2S$
dialkyl arsine trihalide 二烷基胂化三卤 R_2AsX_3
dialkyl arsinic acid 二烷基次胂酸 $R_2AsO \cdot OH$
dialkyl-aryl peroxide 二烷基-芳基过氧化物
dialkyl aryl thiourea 二烷基芳基硫脲
dialkylate 二烷基化合物
dialkylated 二烷基化了的
dialkylated diphenylamine 二烷基化二苯胺〔防老剂〕
dialkyl benzene 二烷基苯
dialkyl benzene sulfonate 二烷基苯磺酸盐
dialkyl cyano arsine 二烷基胂化氰 R_2AsCN
dialkyl-N,N-dialkylcarbamyl phosphonate N,N-二烷基氨基甲酰膦酸二烷基酯 $R_2NCOP(O)(OR)_2$
dialkyl dihydrogen methylenebisphosphonate 亚甲基双膦酸二烷基酯;亚甲基双(膦酸单烷基酯) $CH_2[P(O)(OR)(OH)]_2$
dialkyl dimethylammonium chloride 二烷基二甲基氯化铵
dialkyl dimethylammonium halide 二烷基二甲基卤化铵
dialkyl dimethylammonium ion 二烷基二甲基铵离子
dialkyl dimetyl ammonium bromide 二烷基二甲基溴化铵
dialkyl dithiophosphate 二硫代磷酸二烷基酯
dialkylene 二烯基
dialkyl ether 二烷基醚
dialkyl halogenated arsine 二烷基卤化胂 R_2AsX
dialkyl imino imidazolidine 二烷基亚氨基咪唑啶
dialkyl indan 二烷基茚满
dialkyl isocyanate 异氰酸二烷基酯〔硫化剂〕
dialkyl maleate 马来酸二烷基酯
dialkyl mercury 二烃基汞 HgR_2
dialkyl methyl phosphonate 甲基膦酸二烷基酯 $CH_3PO(OR)_2$
dialkyl morpholinium alkyl sulfate 烷基硫酸二烷基吗啉鎓
dialkyl phenol 二烷基酚
dialkyl phenol sulfide 二烷基酚硫化物〔防老剂〕
dialkyl phosphinic acid 二烷基次膦酸 $R_2PO(OH)$
dialkyl phosphonate 二烷基膦酸酯
dialkyl phosphoric acid extraction 二烷基磷酸萃取法
dialkyl phthalate 邻苯二甲酸二烷基酯
dialkyl potassium imino diphosphate 二烷基亚氨二磷酸钾
dialkyl pyrophosphoric acid 二烷基焦磷酸 $(RO)(HO)P(O)(O)P(OH)(OR)$
dialkyl sebacate 癸酸二烷基酯
dialkyl selenide dihalide 二烷基二卤化硒
dialkyl sodium sulfosuccinate 二烷基磺基琥珀酸钠

dialkyl sulfate 二烷(基)硫酸盐
dialkyl sulfide 二烷基硫醚 R_2S
dialkyl sulfone 二烷基砜 RSO_2R
dialkyl sulfosuccinate 磺基琥珀酸二烷基酯
dialkyl sulfosuccinic salt 琥珀酸二烷基酯磺酸盐
dialkyl sulfoxide 二烷基亚砜
dialkyl tetralin 二烷基萘满
dialkyl thiamorpholinium dioxide 二烷基硫代吗啉鎓二氧化物
dialkyl thiourea 二烷基硫脲〔促进剂〕
dialkyl tin 二烷基锡 SnR_2
dialkyl tin sulfide trimer 二烷基锡化硫(环)三聚物
dialkyl zinc 二烃基锌 ZnR_2
diallage 异剥石
diallyis ①聚二烯丙酯②聚二烯丙酯(系)树脂
diallyl ①1,5-己二烯 $CH_2=CH(CH_2)_2CH=CH_2$ ②二烯丙基 $(CH_2=CHCH_2—)_2$
diallyl amine 二烯丙基胺 $(CH_2=CHCH_2)_2=NH$
diallyl aniline N,N-二烯丙基苯胺 $C_6H_5N(CH_2CH=CH_2)_2$
diallyl azelate 壬二酸二烯丙酯〔增塑剂〕
diallyl barbituric acid 二烯丙基巴比妥酸
diallyl chlorendate 氯菌酸二烯丙酯
diallyl-p-cresol 二烯丙基对甲酚〔防老剂〕
diallyl cyanamide 二烯丙基氰胺 $(C_3H_5)_2NCN$
diallyl disulfide 二硫化二烯丙基 $[CH_2=CHCH_2S]_2$
diallyl ether 二烯丙基醚〔硫化剂〕
diallyl fumarate 富马酸二烯丙酯〔硫化剂〕
diallylidene pentaerythrite 二亚烯丙基季戊四醇酯〔硫化剂〕
diallyl isophthalate 异酞酸二烯丙酯;间苯二酸二烯丙酯
diallyl maleate 顺丁烯二酸二烯丙酯;马来酸二烯丙酯
diallyl oxalate 草酸二烯丙酯 $(COOC_3H_5)_2$
diallyl oxide (二)丙基醚 $(CH_2=CHCH_2)_2O$
diallyl phthalate 邻苯二甲酸二烯丙酯〔增塑剂〕
diallyl phthalate resin 苯二甲酸二烯丙酯树脂
diallyl polymer 二烯丙基聚合物
diallylsilane 二烯丙基硅烷
diallyl sulfide(=diallyl thioether) 二烯丙基硫醚
diallyl thioether(=diallyl sulfide) 二烯丙基硫醚
diallyl thiourea 二烯丙基硫脲 $CS(NHC_3H_5)_2$
diallyl trisulfide 二烯丙基三硫化物 $(C_3H_5)_2S_3$
dially phthalate 苯二甲酸二烯丙酯
dial micrometer 直读式测微仪;厚度千分表
dialozite 菱锰矿 $MnCO_3$
dialphanol adipate 己二酸二脂族醇酯〔增塑剂〕
dialphanyl phthalate(=dialphanol phthalate) 邻苯二甲酸

二脂族醇酯〔其中 alphanol 是指 $C_7 \sim C_9$ 脂族醇〕
dial plate　①刻度板②刻度盘
dial section　针盘
dial setter　盘式设定器；盘式调节器
dial thermometer　刻度温度计
dial type　度盘式
dial type feed mechanism　转盘式加料器
dial type micrometer　指针读数测厚仪
dialuramide(=uramil)　尿咪；5-氨基巴比妥酸
dialuric acid(=5-hydroxybarbituric acid)　径尿酸〔俗〕；5-羟基巴比妥酸　$C_4H_4O_4N_2$
dialysate　渗析液
dialysed　渗析的；渗出的
dialyser　渗析器
dialysis　渗析*；渗透
dialysis bag　渗析袋
dialysis culture　透析培养
dialysis membrane　渗析膜
dialysis potential　渗析势(位)
dialysis tube　渗析管
dialytic　渗析的
dialytic coefficient　渗析系数；透析系数
dialyzate　渗析液
dialyzator　渗析器
dialyzer　①渗析器②渗析膜
dialyzing paper　渗析纸
diamagnetic　抗磁(性)的
diamagnetic anisotropy　抗磁各向异性
diamagnetic complex　抗磁性配合物
diamagnetic compound　抗磁化合物
diamagnetic contribution　抗磁性作用
diamagnetic dication　抗磁性双阳离子
diamagnetic exciton　反磁激子
diamagnetic material　反磁性物质
diamagnetic nickel chelate　抗磁性镍螯合物〔聚烯烃纤维的光稳定剂〕
diamagnetic ring current　抗磁环电流*
diamagnetic screening　抗磁屏蔽
diamagnetic shielding　抗磁屏蔽；反磁性屏蔽
diamagnetic shielding effect　抗磁性屏蔽效应
diamagnetic shift　抗磁位移*
diamagnetic substance　反磁性物质；抗磁性物质；抗磁体
diamagnetic susceptibility　抗磁化率
diamagnetism　抗磁性
diamalt　麦芽浸出液
diamantin　金刚铝
diameter　直径
diameter expansion　扩径
diameter limit cupping　树径采(脂)法
diameter of pigment agglomerate　颜料附聚物直径
diameter of primary pigment particle　原始颜料粒子直径
diameter ratio scale　直径比标尺
diameter run-out　径向跳动
diametral grate　直径栅
diametral grate producer　直径栅式(煤气)发生器
diamide　肼；联氨　$NH_2 \cdot NH_2$
diamidine　①联脒②(某)二脒
4,4-diaminodiphen-oxy-propane(=propamidine)　二脒二苯氧基丙烷
diamido　二(酰)氨基
diamidogen　①肼；联氨　$NH_2NH_2$②二胺　NH_2RNH_2
diamidosuccinic acid　丁二酰胺　$CONH_2(CH_2)_2CONH_2$
diamine　①肼；联氨②—(元)胺③双胺染料〔德国 Cassella 染料厂偶氮直接染料类名〕
diamine blue　双胺蓝
diamine blue-black　双胺蓝青
diamine brilliant blue　双胺亮蓝
diamine bronze　双胺铜青
diamine brown　双胺棕
diamine cyanine　双胺花青
diamine dark-green　双胺暗绿
diamine deep black　双胺深黑
diamine dyes〔复〕　双胺染料〔德国 Cassella 厂偶氮染料商名〕
diamine fast scarlet　双胺坚牢猩红
diamine gold yellow　双胺金黄
diamine pure blue　双胺纯蓝
diamine rose　双胺玫红
diamino　二氨基
diamino acid　二氨基(甲)酸
3,6-diaminoacridine sulfate(=proflavine)　硫酸原黄素；硫酸-3,6-二氨基吖啶　$C_{13}H_{11}N_3H_2SO_4$
2,4-diamino-*trans*-azobenzene　2,4-二氨基反式偶氮苯
2,5-diamino-1,4-benzoquinone　2,5-二氨基-1,4-苯醌
diaminobutane　二氨基丁烷
diaminochloromethyl thiopropyl silanetriol　二氨基氯甲烷合硫丙基硅三醇
1,2-diaminocyclohexane tetraacetic acid(DCTA)　环己二胺四乙酸
1,10-diaminodecane　1,10-癸二胺　$H_2N(CH_2)_{10}NH_2$
diamino dicarboxylic acid　二氨基二羧酸
4,4'-diaminodicyclohexylmethane　4,4'-二氨基二环己基甲烷
2,2-diaminodiethylether-*N,N,N',N'*-tetraacetic acid (DEETA)　2,2-二氨基二乙醚-*N, N, N', N'*-四乙酸
2,2-diaminodiethylsulfide-*N,N,N',N'*-tetraacetic acid

(DESTA) 2,2-二氨基二乙硫醚-N,N,N',N'-四乙酸
diamino dihydroxy arsenobenzene 二氨基二羟偶砷苯 $NH_2(OH)C_6H_3As=AsC_6H_3(OH)NH_2$
1,10-diamino-4,7-dioxadecane 1,10-二氨基-4,7-二氧杂癸烷
2,4-diaminodiphenylamine 2,4-二氨基二苯胺
4,4'-diaminodiphenylether 4,4'-二氨基二苯醚
p,p'-diaminodiphenyl methane p,p'-二氨基二苯甲烷
diaminodiphenyl oxide 二氨基二苯醚
diaminodiphenyl sulfide(=thioaniline) 硫苯胺〔俗〕；二氨基二苯硫醚 $(NH_2C_6H_4)_2S$
4,4'-diaminodiphenyl sulfone 4,4'-二氨基二苯砜〔交联剂〕
diaminodiphenyl sulfone(DDS) 二氨基二苯砜
diamino diphosphatides 二氨基二磷脂类
1,2-diaminoethane 1,2-乙二胺；1,2-二氨基乙烷
diaminogen dye 双胺精染料
1,6-diaminohexane 1,6-二氨基己烷；1,6-己二胺
diamino maleonitrile 二氨基顺丁烯二腈
3,6-diamino-10-methylacridium chloride(=acriflavine) 3,6-二氨基-10-甲基吖啶氯化物
p-di-(aminomethyl) benzene 对二氨甲基苯 $C_6H_4(CH_2NH_2)_2$
diamino monocarboxylic acid 二氨基一(元)羧酸
2,8-diaminophenoxathiin 2,8-二氨基吩噁噻蒽；2,8-二氨基氧硫杂蒽
diamino-phenylacetic acid 二氨基苯乙酸 $(NH_2)_2C_6H_3CH_2COOH$
diaminophosphazine 二氨基膦嗪 $N=P(NHR)_2N=P(NHR)_2N=P(NHR)_2$
diaminopimelic acid 二氨基庚二酸
diamino polyacetic acid 二氨基多乙酸
1,2-diaminopropane(=propylene diamine) 丙邻二胺；1,2-二氨基丙烷 $CH_3CH(NH_2)CH_2NH_2$
1,2-diaminopropane-tetraacetic acid 1,2-丙二胺四乙酸
diaminostilbene 二氨基芪；二氨基-1,2-二苯乙烯
diaminostilbene disulfonic acid 二氨基芪二磺酸 $[H_2NC_6H_3(SO_3H)CH]_2$
diamino sulfonic acid 二氨基磺酸 $RNH·SO_2NH_2$
2,4-diaminothiodiazole 2,4-二氨基-1-噻-3,4-二唑
5,5'-diamino-$trans$-thioindigo 5,5'-二氨基反式硫靛
2,4-diaminotoluene 2,4-二氨基甲苯〔防老剂〕
diaminotriazine 二氨基三嗪
2,5-diaminovaleric acid 2,5-二氨基戊酸；乌氨酸
diammine palladous chloride 二氯化二氨钯 $[Pd(NH_3)_2]Cl_2$
diammonium ①联铵②二铵 $NH_4·NH_4$
diammonium glycyrrhizinate 甘草酸二铵〔药〕

diammonium hydrogen phosphate 磷酸氢二铵〔防老剂〕
diammonium N-octyldecyl sulfosuccinate N-十八烷基磺化琥珀酸二铵〔发泡剂〕
diammonium orthophosphate (正)磷酸氢二铵 $(NH_4)_2HPO_4$
diammonium phosphate 磷酸氢二铵
diamond 金刚石；金刚钻〔俗〕；钻石〔俗〕
diamond black 金刚黑
diamond cutter 玻璃割刀
diamond cutting 金刚石切割
diamond flavine 金刚黄〔偶氮铬处理染料〕
diamond glass cutter 金刚石切玻刀
diamond grid (tower) 钻石格子填料
diamond grinding wheel 金刚石砂轮
diamond knurled insert 菱纹滚花嵌件
diamond-like pattern ①菱形图案②金刚石花型
diamond mortar(=Plattner mortar) 冲击钵；钢研钵
diamond petal 菱形花瓣(状)
diamond point burr 金刚石尖点纹
diamond pyramid hardness 金刚石棱锥压入硬度
diamond shaped metal pieces 多棱金属碎块
diamond shaped needle 棱形针
diamond spar 刚石；钢玉
diamond type lattice 金刚石型晶格
diamorph 聚合体
diamorphine hydrochloride 盐酸海洛因；盐酸二乙酰吗啡
diamorphism 二形现象
diamyl ①二戊基②联戊基
diamylamine 二戊胺 $(C_5H_{11})_2NH$
diamyl amyl phosphonate(DAAP) 戊基膦酸二戊酯 $(C_5H_{11}O)_2PO(C_5H_{11})$
diamyl benzene 二戊苯 $(C_5H_{11})_2C_6H_4$
diamyl carbonate 碳酸二戊酯 $(C_5H_{11}O)_2CO$
diamyl disulfide 二戊基(化)二硫 $(C_5H_{11}S)_2$
diamylene ①癸烯②萜烯
diamyl ester 二戊酯 $C_5H_{11}OCORCOOC_5H_{11}$
diamyl ether 二戊醚 $C_5H_{11}OC_5H_{11}$
di-t-amylhydroquinone 二叔戊基对苯二酚 $[C_2H_5C(CH_3)_2]_2C_6H_2(OH)_2$
diamylose 双多糖类
diamyl oxalate 草酸二戊酯〔增塑剂〕
di-t-amylpheoxyethanol 二叔戊基苯氧基乙醇
diamyl phthalate 邻苯二甲酸二戊酯 $C_6H_4(CO_2C_5H_{11})_2$
diamyl succinate 丁二酸二戊酯 $(CH_2CO_2C_5H_{11})_2$
diamyl sulfite 亚硫酸二戊酯 $(C_5H_{11}O)_2SO$
diamyl sulfosuccinate 磺基琥珀酸二戊酯
diamyl tartrate 酒石酸二戊酯〔增塑剂〕

dian(=bisphenol A)　双酚 A；二酚基丙烷
dianemycin　猎神霉素
dianhydrohexitol stearate ester　硬脂酸双无水己糖酯
dianil azurine　双苯胺青
dianil black　双苯胺黑
dianil blue　双苯胺蓝
dianil bordeaux　双苯胺枣红
dianil chrome brown　双苯胺铬棕
dianil fast red　双苯胺坚牢红
dianil garnet　双苯胺紫酱
dianiline　双苯胺
dianiline dye　双苯胺染料
dianilinomethane　双苯胺基甲烷
dianil ponceau　双苯胺丽春红
dianil pure yellow　双苯胺纯黄
dianil yellow　双苯胺黄
dianin　狄安宁
danion　二价阴离子
dianionic surfactant　双阴离子表面活性剂
o-dianisidine　邻联(二)茴香胺　$(CH_3OC_6H_3NH_2)_2$
dianisidine blue　联茴香胺蓝
dianisidine diisocyanate　联甲氧基苯胺二异氰酸酯〔硫化剂〕
dianisidine red　直接偶氮红
dianodic method　双阳极抑制剂法
danthoside　石竹皂苷
dianthracene　双蒽
dianthranide　联(二)蒽　$C_{28}H_{18}$
dianthraquinone　联(二)蒽醌　$(C_6H_4COC_6H_4C=)_2$
dianthrene blue　双蒽蓝〔还原靛类染料〕
dianthus saponin　瞿麦皂苷
diantimony trisulfide　三硫化二锑〔促进剂〕
4,4'-diantipyrinylmethane(=diantipyrylmethane)　二安替比林甲烷*
diantipyrylmethane(=4,4'-diantipyrinylmethane)　二安替比林甲烷*
diaparene chloride　氯化迪阿帕伦
diapason　①范围；水平②射域③音域
diapers　①尿布②兜布〔一种测去污力的布样〕
diaphaneity　透明度
diaphoretic　①发汗药②发汗的
diaphorite　辉锑铅银矿
diaphragm　①横隔膜②光阑③隔膜；膜片
diaphragm-actuated regulator　薄膜调节器
diaphragm actuator　薄膜(式)执行机构
diaphragm box level controller　膜合式料位控制器
diaphragm cell　隔膜电池
diaphragm current　隔膜电流
diaphragm direct liquid introduction interface　薄膜式液体直接进样连接装置
diaphragm expansion joint　膜片胀缩接合
diaphragm fuel pump　膜片燃料泵
diaphragm gas meter　隔膜煤气计
diaphragm gate　隔膜形浇口
diaphragm gauge　薄膜压力计
diaphragm leather　隔膜革；隔膜皮
diaphragmless cell　无隔膜电解池
diaphragmless injector　无隔片进样器
diaphragm manometer　薄膜压力计
diaphragm meter　膜片计量器
diaphragm-operated　膜片的；鼓膜的
diaphragm packing　隔膜封填
diaphragm pressure gauge　膜片压力计
diaphragm pump　隔膜泵；膜式泵
diaphragm screen　膜筛
diaphragm type　隔膜式
diaphragm type compressor　膜式压缩机
diaphragm valve　隔膜阀
diaphthol(=quinaseptol)　迪阿索耳；间磺酸邻氧喹啉〔尿道消毒剂〕；奎色醇
diaporthin　腐皮壳菌素
diapositive　透明正片
diaquooxonium ion　二水合氧鎓离子；三水合氢离子〔$H_7O_3^+$; $H_3O^+ \cdot 2H_2O$〕
diarabinose　二阿拉伯糖；二阿戊糖
diarachin　甘油二花生酸酯
diaroyl peroxide　过氧化二芳酰基
diarrhea　泻肚；腹泻
diarsenate　二砷酸盐
diarsenide　二砷化物
diarsenite　二亚砷酸盐
diarsenous acid　焦砷酸；三缩二原砷酸
diarsine　联胂　$R_2As \cdot AsR_2$
diarsonium(=cacodyl)　二甲胂基
diarsyl(=biarsine)　联胂
diarylamines　二芳基胺(类)
1,4-diarylaminoanthraquinone　1,4-二芳氨基蒽醌
diaryl arsenious acid　二芳基亚胂酸　R_2AsOOH
diaryl arsine oxyhalide　二芳基胂化氧卤　R_2AsOX
diaryl arsine trihalide　三卤化二芳胂　R_2AsX_3
diarylguanidine　二芳胍〔促进剂〕
diarylhydrazine rearrangement(=benzidine rearrangement)　二芳基肼重排作用
diaryl hydroxy arsine　二芳基亚胂酸　R_2AsOOH
diarylide yellow　二芳基黄；联苯胺黄

diaryl mercury 二芳基汞 HgR$_2$
diarylmethane 二芳甲烷
diaryl-*p*-phenylenediamine 二芳基对苯二胺〔防老剂〕
diaryl secondary amine antioxidant 二芳基仲胺类防老剂
diaryltin dihalide 二卤化二芳基锡
diaryl zine 二芳基锌
diascope 透明玻片；反射幻灯机；投影机
diasone(=disodium formaldehyde-sulfoxylate diamino-diphenylsulfone) 亚磺氨苯砜钠；大艾松；地阿宋
diasonograph 超声诊断仪
diasphaltene 脱沥青
diaspirin 双阿司匹林
diaspore 水铝石〔矿〕；水矾石；水矾土
diasporite 硬水铝石
diastase 淀粉酶制剂
diastasic action 淀粉糖化作用
diastasimetry 糖化力测定
diastereoisomer 非对映异构体
diastereoisomeric 非对映异构的
diastereoisomeride 非对映异构体
diastereoisomerism 非对映异构(现象)
diastereoisomerization 异向异构
diastereomer 非对映(异构)体*
diastereomeric 非对映的
diastereomeric excess 非对映体过量*
diastereomeric form 非对映形
diastereosequence distribution 非对映链段分布
diastereotopic 非对映(异构)的
diastrophism 地壳变动
diatactic polymer 双无规立构聚合物
diathermal 透热的
diathermancy 透热性
diathermanous 透热的
diathermanous body 透热体
diathermic 透热的
diathermic membrane(=diathermic wall) 绝热膜；绝热壁
diathermic wall(=diathermic membrane) 绝热壁；绝热膜
diathermometer 导热计
diathermous 透热的
diathermy 透热法
diatol 双元油；碳酸二乙酯 CO(OC$_2$H$_5$)$_2$
diatom 硅藻
diatomaceous calcite 硅藻方解石
diatomaceous earth(=diatomaceous silica) 硅藻土
diatomaceous shale 硅藻页岩
diatomaceous silica(=diatomaceous earth) 硅藻土
diatomaceous support 硅藻土载体
diatom earth 硅藻土

diatomic 双原子的
diatomic acid 二价酸〔不能叫二元酸〕
diatomic alcohol 二元醇；二羟醇
diatomic base 二价碱
diatomic molecule 双原子分子*
diatomics-in-molecule(DIM) method 分子中双原子法
diatomine 硅藻素
diatomite 硅藻土
diatomite calcium carbonate 硅藻碳酸钙
diatomite support 硅藻土载体
diatoms 硅藻土
diatom structure 硅藻结构
diatoxanthin 硅藻黄质
diatretyne 穿孔蕈炔素
diatropic interference 纬向干扰
diatropic plane 纬向面
diaxial addition 双直键加成*
diaxial orientation 双轴取向
2,3-diazabicyclo[2.2.1]hept-2-ene 2,3-二氮杂双环[2.2.1]-庚-2-烯
1,8-diazabicyclo-[5.4.0] undec-7-ene 1,8-二氮杂双环-[5.4.0]-十一-7-烯
1,4-diazabicylo[2.2.2]-octane 1,4-二氮杂二环[2.2.2]-辛烷
diazacyclo 〔词头〕二氮杂环
5,6-diaza-2,4,6,8-decatetraene 5,6-二氮杂-2,4,6,8-癸四烯
diazamerocyanines 二氮杂部花青
diazamine blue 重氮胺蓝
diazamine bordeaux 重氮胺枣红
diazamine brilliant black 重氮胺亮黑
diazamine colors 重氮胺染料〔重氮胺是法国国家染料厂可行重氮化的直接染料商名〕
diazamine fast red 重氮胺坚牢红
diazamine orange 重氮胺橙
diazamine violet 重氮胺紫
diazane 二氮烷；肼 H$_2$NNH$_2$
1,1-diazanediyl 1,1-二氮亚烷基；1,1-亚肼基 H$_2$NN=
1,2-diazanediyl 1,2-二氮亚烷基；1,2-亚肼基 —NHNH—
diazanetetrayl 二氮亚烷二基；1,2-肼双亚基
diazanorcarane 二氮杂原蒈烷
diazanyl 二氮烷基；肼基 H$_2$NNH—
diazapolyoxamacrobicyclic ligand 二氮多氧大双环配体
diazelate ①壬二酸氢盐 COOH(CH$_2$)$_7$COOM ②壬二酸氢酯 COOH(CH$_2$)$_7$COOR
diazene 二氮烯 HN=NH
diazenediyl 二氮亚烯基 —N=N—

diazenyl 二氮烯基 HN—N—
diazepam 地西泮〔药〕
diazepine 二氮杂䓬〔俗〕
diazete 二氮杂环丁二烯 $C_2N_2H_2$
diazetidine 二氮(杂)环丁烷 $C_2N_2H_6$
diazetine 二氮杂环丁烯 $C_2N_2H_4$
diazido- 二叠氮基
4,4′-diazido chalcone 4,4′-二叠氮基查耳酮
diazidoethane 二叠氮基乙烷
2-diazido-naphthoquinone(1,2)-5-sulfonic acid sodium salt 2-重氮基萘醌(1,2)-5-磺酸钠盐
diazine 二嗪*
diazine blue 二嗪蓝
diazine brown 二嗪棕
diazine colors 二嗪染料〔美国南星染料厂可行重氮化的直接染料的商名〕
diazine fast orange 二嗪坚牢橙
diazine fast red 二嗪坚牢红
diazine garnet 二嗪紫酱
1,2-diazine(=pyridazine) 哒嗪；1,2-二嗪
1,3-diazine(=pyrimidine) 嘧啶；1,3-二嗪
1,4-diazine(=pyrazine) 吡嗪；1,4-二嗪
diazine ring 二嗪环
diazinon 二嗪磷；二嗪农〔杀虫剂〕
diazo 重氮基 —N＝N—; N≡N＝
diazo acetate 重氮基乙酸盐(或酯) N_2＝CHCOOM
diazo acetic acid 重氮基乙酸 N_2＝CHCOOH
diazo acetic ester 〔专指〕重氮基乙酸乙酯 $N_2CHCOOC_2H_5$
diazoacetonitrile 重氮乙腈 NN＝CHCN
diazoalkane 重氮烷*
diazoalkane polymerization 重氮链烷聚合
diazo amino 重氮亚氨基 —N＝NNH—
diazo aminobenzene 重氮氨基苯 C_6H_5N＝$NNHC_6H_5$
diazoaminobenzoic acid 重氮氨基(二)苯甲酸
diazoamino compound 重氮氨基化合物
diazoamino rearrangement 重氮胺重排作用
diazoaminotoluene 重氮氨基(二)甲苯
diazoanhydride 重氮酐
diazoate 重氮酸盐 ArNNOOM
diazobenzene ①苯重氮酸②重氮苯
diazo benzene acid 苯重氮酸 C_6H_5N＝NOH
diazo benzene chloride 氯化重氮苯 $C_6H_5N_2Cl$
diazo benzene hydroxide 氢氧化重氮苯 $C_6H_5N_2OH$
diazobenzene sulfonic acid 重氮苯磺酸
diazo black 重氮黑
diazo brilliant green 重氮亮绿
diazocine 二氮芳辛；二氮杂环辛间四烯

diazo color 重氮染料
diazo compound 重氮化合物*
diazo coupling 重氮偶合(作用)
diazo coupling reaction 重氮偶合反应
4-diazo diphenylamine sulfonate 4-重氮二苯胺磺酸盐
diazo ester 重氮甲酸酯
diazoethane 重氮乙烷
diazo fast blue 重氮坚牢蓝
diazo fast yellow 重氮坚牢黄
diazofuran 二氮杂呋喃
diazogen black 重氮精黑
diazogen blue 重氮精蓝
diazogen bordeaux 重氮精枣红
diazogen colors 重氮精染料
diazogen red 重氮精红
diazogram 重氮化色谱图
diazohydrates 重氮醚类〔指含有二价基—N＝NO—的化合物类〕
diazohydroxide 重氮氢氧化物*
diazoic acid 重氮酸 RN_2OH
diazoimide(=hydrazoic acid) 叠氮酸
diazoimido compound 重氮化合物
diazo indigo blue 重氮靛蓝
diazoketone rearrangement(=Wolff rearrangement) 重氮甲酮重排作用
diazoketones 重氮酮
diazol black ERN 重氮盐黑 ERN
diazol bordeaux 重氮盐枣红；显色盐枣红
diazol colors 重氮盐染料；显色盐染料
diazole 二唑〔吡唑或咪唑〕
diazol fast orange 重氮盐坚牢橙
diazol fast scarlet 重氮盐坚牢猩红
diazolidinyl urea 二偶氮利定脲；二偶氮烷基脲
diazo light bordeaux 重氮浅枣红
diazol light orange 重氮盐浅橙
diazol light yellow 重氮盐浅黄
diazols 重氮盐类
diazomalonic acid 重氮丙二酸
diazomethane 重氮甲烷 CH_2N_2
diazomethane synthesis 重氮甲烷合成
diazomethane via methoxyl-end-group analysis 重氮甲烷分析法〔测定聚酰胺数均分子量的方法〕
4-diazo-3-methoxy-diphenylamine sulfonate 4-重氮-3-甲氧基-二苯胺磺酸盐
diazomycin(=duazomycin) 偶氮霉素
diazonitrophenol 重氮硝基酚
diazonium 重氮(化) N(≡N)—
diazonium chloride 氯化重氮(物) $RN_2^+Cl^-$

diazonium compound　重氮化合物
diazonium coupling　重氮偶联*
diazonium halide　卤化重氮(物)　$RN_2^+X^-$
diazonium hexafluorophosphote　六氟磷酸重氮盐　$[RN_2]^+PF_6^-$
diazonium hydroxide　氢氧化重氮(物)　$RN_2^+OH^-$
diazonium ion　重氮鎓离子　$N≡N^+—$
diazonium salt　重氮盐*
diazonium tetrafluoroborate　四氟硼酸重氮盐　$[RN_2]^+BF_4^-$
diazo olive　重氮橄榄绿
diazo paper　重氮复印纸
diazoparaffins　重氮烷(属烃)
diazo-phenol　氢氧化重氮苯酚　$HOC_6H_4N=NOH$
diazo plate　重氮基板
diazopropyl azodicarboxylate　偶氮二羧酸重氮丙酯〔发泡剂〕
diazo-reaction　重氮化反应
diazosalicylic acid　重氮水杨酸
diazo salt　重氮盐
diazosoplit　重氮分解物
p-diazosulfanilic acid　对重氮苯磺酸
diazosulfide　苯并噻二唑
diazotability　重氮化本领
diazotable dye　重氮染料
diazotate　重氮酸盐　ArNNOM
diazo-test　重氮试验
diazotetrazole　重氮四唑
diazotic acid　重氮酸
diazotisation　重氮化(作用)
diazotizable　可重氮化的
diazotizating　重氮化的
diazotization　重氮化*
diazotization titration method　重氮化滴定法
diazotized　重氮化了的
diazotized tobias acid　重氮化-2-氨基-1-萘磺酸；重氮化吐氏酸
diazotizing　①重氮化的②重氮化(作用)
diazotizing colours　重氮染料〔可以重氮化的染料〕
diazotizing dyes(=diazotizing colours)　重氮染料
diazotol rapid fast colours　重氮酚快速坚牢染料；偶氮酚类
diazo transfer　重氮基转移*
diazotroph　固氮生物
diazotype　重氮印象法
diazo-type paper　重氮型复印纸
diazoum chloride　氯化重氮盐　$N≡N(R)^+Cl^-$
diazouracil　重氮尿嘧啶　$\overline{CONHCONHCH}=CN=NOH$
diazouracil test　重氮尿嘧啶试验
diazo violet　重氮紫
diazoxide　二氮嗪〔药〕
diazoxy　重氮氧基　$—N(=O)=N—$
diazthines　噻二嗪类
Diban(=dibasic aluminium nitrate)　二碱式硝酸铝　$Al(OH)_2NO_3$
Diban process　德班法；二碱式硝酸铝法
dibasic　①二元的；二碱价的〔指酸〕②二代的〔指盐〕
dibasic acid(=diprotic acid)　二元酸
dibasic acid esters(=di-esters)　双酯类〔合成润滑油〕
dibasic alcohol　二元醇；二羟醇
dibasic aminoacid　氨基二酸
dibasic calcium phosphate(=secondary calcium phosphate)　二代磷酸钙；磷酸氢钙
dibasic carboxylic acid　二(价)羧酸
dibasic ester　二价酸酯
dibasic keto acid(=dibasic ketonic acid)　二价酮酸；酮二酸
dibasic ketonic acid(=dibasic keto acid)　二价酮酸；酮二酸
dibasic lead phosphate　二元磷酸铅〔稳定剂〕
dibasic lead phosphite　二盐基亚磷酸铅；亚磷酸氢铅
dibasic magnesium phosphate　磷酸氢镁；二代磷酸镁
dibasic phenol acid　二价酚酸；酚二酸
dibasic polyploid　二基数多倍体
dibasic potassium phosphate　磷酸氢二钾
dibasic salt(=secondary salt)　二代盐
dibasic sodium phosphate　磷酸氢二钠
dibbling　穴植；点播
dibehenolin　二山萮精；甘油二山萮酸酯
dibenzalacetone　二亚苄基丙酮　$(C_6H_5CH=CH)_2CO$
dibenzamide　二苯甲酰胺　$(C_6H_5CO)_2NH$
2,2'-dibenzamido-diphenyl disulfide　2,2'-二苯甲酰氨基二苯基二硫
dibenzanthracene　二苯(并)蒽
dibenzanthrone green　二苯并蒽酮绿
dibenzanthrone violet　二苯并蒽酮紫
dibenzenazoresorcin　二苯偶氮间苯二酚　$(C_6H_5N_2)_2=C_6H_2(OH)_2$
dibenzenyl　①二苯基乙炔②二亚苄基
dibenzhydroxamic acid　二苯甲氧肟酸　$C_6H_5CONHOCOC_6H_5; C_6H_5C(OH)=NOCOC_6H_5$
dibenzimidazole disulfide　二硫化二苯并咪唑〔促进剂〕
dibenzimide　二苯甲酰亚胺
dibenzo-18-crown-6(DB18C6)　二苯并-18 冠(醚)-6

$C_4H_4C_{12}O_6H_{16}C_4H_4$

dibenzodioxine 二苯并二噁烯
dibenzofuran 氧芴；二苯并呋喃
dibenzopyridine(=acridine) ①二苯并吡啶②吖啶
dibenzosuberone 二苯并环庚酮
dibenzothiazole disulfide 二硫化二苯并噻唑〔促进剂〕
dibenzothiazole monosulfide 一硫化二苯并噻唑〔硫化剂〕
dibenzothiazyl-dimethyl-thiourea 二苯并噻唑基二甲基硫脲
dibenzothiazyl-disulfide 二硫化二苯并噻唑基
dibenzothiocarbocyanine 二苯并硃花青
dibenzothiophen 硫芴
dibenzoyl ①联苯甲酰②二苯(甲)酰(基) $(C_6H_5CO)_2$
dibenzoyl-acetone 二苯甲酰丙酮 $(C_6H_5CO)_2CHCOCH_3$
dibenzoyl disulfide 二苯甲酰二硫 $(C_6H_5CO)_2S_2$
dibenzoyl-ethylene 二苯甲酰乙烯 $(C_6H_5COCH=)_2$
dibenzoyl-ethylene diamine N,N'-二苯甲酰乙二胺 $(C_6H_5CONHCH_2)_2$
dibenzoyl hydrazine 二苯甲酰肼 $C_6H_5CONHNHCOC_6H_5$
dibenzoyl ketone 二苯甲酰基(甲)酮 $(C_6H_5CO)_2CO$
dibenzoyl methane 二苯甲酰甲烷 $(C_6H_5CO)_2CH_2$
dibenzoyl ornithine(=ornithuric acid) 鸟尿酸；N,N'-二苯甲酰鸟氨酸
dibenzoyl peroxide 过氧化二苯(甲)酰 $(C_6H_5CO)_2O_2$
p,p-dibenzoylquinone dioxime 对对二苯甲酰醌二肟
2,4-dibenzoyl resorcinol 2,4-二苯甲基苯二酚〔紫外线吸收剂〕
dibenzoyl resorcinol 二苯甲酰间苯二酚〔紫外线吸收剂〕
dibenzoyl sulfide 二苯甲酰硫化物〔塑解剂〕
dibenzoyl thiourea 二苯甲酰硫脲 $(C_6H_5CONH)_2CS$
dibenzphenanthrene 二苯并菲
dibenzpyrenequinone 二苯并芘醌
dibenzthioxine(=phenoxthine) 夹氧硫杂蒽
dibenzyl ①联苄基 $C_6H_5CH_2CH_2C_6H_5$②二苄基 $(C_6H_5CH_2)_2$
dibenzyl acetic acid 二苄基乙酸 $(C_6H_5CH_2)_2=CHCO_2H$
dibenzyl adipate 己二酸二苄酯〔增塑剂〕
dibenzyl amine 二苄胺 $(C_6H_5CH_2)_2NH$
dibenzyl-p-aminophenol 二苄基对氨基苯酚
dibenzyl aniline 二苄苯胺 $(C_6H_5CH_2)_2NC_6H_5$
dibenzyl azelate 壬二酸二苄酯〔增塑剂〕
dibenzyl disulfide 二苄(化)二硫 $(C_6H_5CH_2S)_2$
dibenzyl dithiocarbamate 二苄酯二硫代氨基甲酸酯
dibenzyldithio-oxamide 二苄基二硫代草酰胺
dibenzylene diisocyanate 二异氰酸二亚苄酯〔硫化剂〕

dibenzyl ether 二苄醚 $(C_6H_5CH_2)_2O$
dibenzyl fumarate 富马酸二苄酯 $(=CHCO_2CH_2C_6H_5)_2$
dibenzylidene 二亚苄基
dibenzylidene mannitol 二亚苄基甘露醇
dibenzylidene sorbitol 二亚苄基山梨糖醇
dibenzylidene xylitol 二亚苄基木糖醇
dibenzyline 苯氧苄胺
dibenzyl ketone 二苄基(甲)酮 $(C_6H_5CH_2)_2=CO$
dibenzyl maleate 马来酸二苄酯 $(=CHCO_2CH_2C_6H_5)_2$
dibenzyl-mercury 二苄基汞 $C_{14}H_{14}Hg$
1,3-dibenzyl-2-methyl imidazolium chloride 1,3-二苄基-2-甲基咪唑鎓氯化物
dibenzyl oxalate 草酸二苄酯 $(CO_2CH_2C_6H_5)_2$
dibenzyl phthalate 邻苯二甲酸二苄酯 $C_6H_4(CO_2C_7H_7)_2$
dibenzyl sebacate 癸二酸二苄酯〔增塑剂〕
dibenzyl succinate 丁二酸二苄酯 $(CH_2CO_2C_7H_7)_2$
dibenzyl sulfanilate 二苄(基)对氨基苯磺酸盐
dibenzyl sulfide 二苄硫 $(C_6H_5CH_2)_2S$
dibenzyl sulfone 二苄砜 $(C_6H_5CH_2)_2SO_2$
dibenzyl sulfoxide 二苄基亚砜 $(C_6H_5CH_2)_2SO$
dibenzyl tartrate 酒石酸二苄酯 $(CHOHCO_2C_7H_7)_2$
dibenzyl thiourea 二苄基硫脲 $(C_6H_5CH_2NH)_2CS$
2,5-dibiphenylyl-oxazole(BBO) 2,5-二联苯基噁唑 $(C_6H_5C_6H_4)_2 \cdot C_3NOH$
diblock copolymer 二嵌段共聚物*
diborane 乙硼烷；二硼烷 B_2H_6
diborane diammine 二氮合乙硼烷
dibrom- 二溴
dibromated 二溴化的
dibrom-dimethyl ether 二溴二甲醚
dibromide 二溴化物
dibrominated(=dibromizated) 二溴化的
dibromizated(=dibrominated) 二溴化的
dibromoacetaldehyde 二溴乙醛 $CHBr_2CHO$
dibromoacetamide 二溴乙酰胺 $Br_2CHCONH_2$
dibromoacetic acid 二溴乙酸 Br_2CHCO_2H
dibromoacetone 二溴丙酮
dibromo-acetyl bromide 二溴乙酰溴 $CHBr_2 \cdot COBr$
dibromoacetylene 二溴代乙炔 $BrC≡CBr$
dibromoanthranilic acid 二溴邻氨基苯甲酸 $Br_2C_6H_2(NH_2)CO_2H$
dibromoarsenazo II 二溴偶氮胂II
dibromobarbituric acid 二溴巴比妥酸 $Br_2C(CONH)_2CO$
dibromo-benzene 二溴(代)苯 $C_6H_4Br_2$
dibromobenzoic acid 二溴苯甲酸
dibromobenzyl peroxide 二溴苄基过氧化物〔塑解剂〕

dibromo-butyric acid 二溴丁酸 $C_3H_5Br_2COOH$
dibromo-cinnamic acid 二溴肉桂酸
　　$C_6H_5CBrCBrCOOH$
dibromo compound 二溴化合物
dibromo-*o*-cresol sulfonphthalein(=bromocresol purple)
　　溴甲酚红紫
1,2-dibromo-1,2-diphenyl ethane 1,2-二溴-1,2-二苯乙烷
　　〔螯合剂〕
dibromodiphenyl ether 二溴二苯醚 $(BrC_6H_4)_2O$
dibromodiphenylsilane 二苯基二溴(甲)硅烷
dibromo ester 二溴酯
dibromo ether 二溴醚
dibromoethyl alcohol 二溴乙醇 $CHBr_2CH_2OH$
di-(1-bromoethyl) sulfide 双(1-溴乙基)硫 $C_4H_8Br_2S$
dibromofumaric acid 二溴富马酸 $(=CBrCO_2H)_2$
dibromogallic acid(=gallobromol) 二溴棓酸
　　$C_7H_4O_5Br_2$
5,7-dibromo-8-hydroxyquinoline 5,7-二溴-8-羟基喹啉
dibromomaleic acid 二溴马来酸 $(=CBrCO_2H)_2$
dibromomalonic acid 二溴丙二酸 $Br_2C(CO_2H)_2$
dibromomalonyl bromide 二溴丙二酰溴
　　$COBrCBr_2COBr$
dibromomethane(=methylene bromide) 二溴甲烷
dibromomethylethylketone 二溴丁酮
dibromonitromethane 二溴硝基甲烷 Br_2CHNO_2
dibromophenol 二溴苯酚〔阻燃剂〕
dibromophenolphthalein 溴酚红
dibromophenyl propionic acid 二溴苯基丙酸
dibromopolybutadiene 二溴聚丁二烯
2,3-dibromopropyl acrylate 2,3-二溴丙基丙烯酸酯
dibromo pyranthrone 二溴皮蒽酮
dibromopyrogallolsulfonaphthalene(DBPSN) 二溴代焦
　　棓酚磺萘
dibromopyrogallolsulphophthalein 二溴连苯三酚磺基酞
dibromopyruvic acid 二溴丙酮酸 $CHBr_2COCO_2H$
2,6-dibromoquinone chlorimide 2,6-二溴醌氯亚胺
dibromo-succinic acid 二溴丁二酸
　　$COOH(CHBr)_2COOH$
dibromothymol sulfonphthalein(=bromothymol blue) 二
　　溴百里酚磺酞；溴百里蓝 $C_{27}H_{28}O_5Br_2S$
dibromotichromin 二溴酞色敏
3,5-dibromotyrosine 3,5-二溴酪氨酸
α,α'-dibromo-*p*-xylene α,α'-二溴对二甲苯
dibucaine hydrochloride 盐酸辛可卡因；盐酸路佩卡因；
　　苏夫卡因
di(butan-3-one-1-yl) sulfide 二(3-丁酮-1-基)硫醚
dibutene 二丁烯；聚二丁烯
β,β-dibutoxy diethyl ether(=Butex; dibutylcarbitol) β,β-
　　二丁氧基二乙醚 $(C_4H_9OC_2H_4)_2O$
di(butoxyethoxyethyl) adipate 己二酸二(丁氧基乙氧基
　　乙)酯〔增塑剂〕
di(butoxyethoxyethyl) glutarate 戊二酸二(丁氧基乙氧基
　　乙)酯〔增塑剂〕
di(butoxyethoxyethyl) sebacate 癸二酸二(丁氧基乙氧基
　　乙)酯〔增塑剂〕
dibutoxyethyl adipate 己二酸二丁氧基乙酯〔增塑剂〕
di(β-butoxyethyl) ether(=Butex) 二(β-丁氧乙基)醚
　　$(C_4H_9OC_2H_4)_2$
dibutoxyethyl glutarate 戊二酸二丁氧基乙酯〔增塑剂〕
dibutoxyethyl phthalate 邻苯二甲酸二丁氧基乙酯(增塑剂)
dibutoxyethyl sebacate 癸二酸二丁氧基乙酯
1,1-dibutoxy-2-phenylethane 1,1-二丁氧基-2-苯基乙烷
dibutoxytitanium-bis(octylene glycolate) 二丁氧基钛二
　　(辛二醇酯)
dibutyl- ①二丁基②联丁基
di-*n*-butylacetic acid 二丁基乙酸 $(C_4H_9)_2CHCOOH$
dibutyl acid phosphate 酸式磷酸二丁酯
dibutyl adipate 己二酸二丁酯〔增塑剂〕
2,6-di-*t*-butyl-4-allyl phenol 2,6-二叔丁基-4-烯丙基苯酚
　　〔防老剂〕
dibutylamine 二丁基胺〔促进剂〕
dibutylaminopropylamine 二丁氨基丙胺
　　$(C_4H_9)_2NC_3H_6NH_2$
dibutyl ammonium dibutyl dithiocarbamate 二丁基二硫
　　代氨基甲酸二丁铵
dibutyl ammonium oleate 油酸二丁铵
N-dibutylaniline *N*-二丁苯胺 $C_6H_5N(C_4H_9)_2$
di-*n*-butyl azelate 壬二酸二正丁酯〔增塑剂〕
dibutyl butylphosphonate(DBBP) 丁基膦酸二丁酯
　　$(C_4H_9O)_2PO(C_4H_9)$
4,4-dibutyl-γ-butyrolactone 4,4-二丁基-γ-丁内酯
dibutyl carbitol(DBC) 二丁基卡必醇 $(C_4H_9OC_2H_4)_2O$
dibutyl carbonate 碳酸二丁酯 $(OC_4H_9)_2CO$
2,6-di-*t*-butyl-*p*-cresol(=2,6-di-*t*-butyl-4-methyl phenol)
　　2,6-二叔丁基对甲酚
di-butyl cyanamide 二丁氨腈 $(C_4H_9)_2NCN$
dibutyl decanedioate 癸二酸二丁酯
dibutyl diethyl carbamoylphosphonate(DBDECP) 二乙基
　　氨基甲酰膦酸二丁酯 $(C_2H_5)_2NC(O)P(O)(OC_4H_9)_2$
dibutyl-*N,N*-diethyl-carbamylmethylene phosphonate
　　(DBDECMP) *N,N*-二乙基氨基甲酰亚甲基膦酸二丁
　　酯 $(C_2H_5)_2NC(O)CH_2P(O)(OC_4H_9)_2$
dibutyl dihydrogen methylenebisphosphonate(DBMDP)
　　亚甲基双膦酸二丁酯；亚甲基双(膦酸单丁酯)
　　$CH_2[PO(OC_4H_9)(OH)]_2$
dibutyl disulfide 二丁二硫 $C_4H_9SSC_4H_9$

dibutyl dithiocarbamate 二硫代氨基甲酸二丁酯
dibutyl ester 二丁酯 R(COOC$_4$H$_9$)$_2$
dibutyl ether ①二丁醚 C$_4$H$_9$OC$_4$H$_9$ ②二丁基醚 R(OC$_4$H$_9$)$_2$
dibutylethyl azelate 壬二酸二丁基乙酯〔增塑剂〕
dibutyl ethylene diphosphonic acid(DBEDP) 亚乙基二膦酸二丁酯 C$_2$H$_4$[P(O)(OH)(OC$_4$H$_9$)]$_2$
dibutyl fumarate 富马酸二丁酯
3,5-di-t-butyl-4-hydroxybenzyl acrylate 3,5-二叔丁基-4-羟基苄基丙烯酸酯〔防老剂〕
3,5-di-t-butyl-4-hydroxybenzyl stearyl ester 3,5-二叔丁基-4-羟基苄基硬脂酰酯〔防老剂〕
di-(3-t-butyl-4-hydroxy-6-methyl phenyl) sulfide 二(三叔丁基-4-羟基-6-甲基苯基)硫化物〔防老剂〕
3,5-di-t-butyl-4-hydroxy-toluene 3,5-二叔丁基-4-羟基甲苯〔防老剂〕
dibutyl hydroxy toluene 二丁基(对)甲酚
dibutyl itaconate 衣康酸二丁酯〔增塑剂〕
dibutyl ketone 二丁基(甲)酮;5-壬酮 (C$_2$H$_5$CH$_2$CH$_2$)$_2$CO
dibutyl malate 苹果酸二丁酯 C$_4$H$_4$O$_5$(C$_4$H$_9$)$_2$
dibutyl maleate 马来酸二丁酯 (=CHCO$_2$C$_4$H$_9$)$_2$
dibutyl malonate 丙二酸二丁酯 CH$_2$(CO$_2$C$_4$H$_9$)$_2$
2,6-di-t-butyl-4-methoxyphenol 2,6-二叔丁基-4-甲氧基苯酚〔防老剂〕
dibutyl methylene bis-thioglycolate 二丁基亚甲基双硫代乙二醇酯〔增塑剂〕
dibutyl methylene diphosphonic acid(DBMDP) 二丁基亚甲基双膦酸 CH$_2$[P(O)(OH)(OC$_4$H$_9$)]$_2$
2,6-di-t-butyl-4-methyl phenol 2,6-二叔丁基-4-甲基苯酚
dibutyl methyl phosphate(DBMP) 磷酸二丁基甲酯
dibutyl 1,8-octanedicarboxylate 1,8-辛烷二羧酸二丁酯
dibutyl oleamide 二丁基油酰胺〔增塑剂〕
dibutyl oxalate 草酸二丁酯 (CO$_2$C$_4$H$_9$)$_2$
di-t-butyl peroxide 过氧化二叔丁基〔硫化剂〕
1,1-di-t-butyl peroxy cyclohexane 1,1-二(叔丁基过氧基)环己烷
di-sec-butyl peroxy dicarbonate 过(氧)二碳酸二仲丁基酯
$α,α$-di(t-butylperoxy) diisopropylbenzene $α,α$-二(叔丁基过氧基)二异丙苯〔硫化剂〕
di-(t-butylperoxy) isophthalate 二(叔丁基过氧基)间苯二酸酯
1,3-di(t-butyl-peroxyisopropyl) benzene 1,3-双(叔丁基过氧异丙基)苯
4,4-di-t-butyl peroxy-n-pentyl butytate 4,4-二叔丁基过氧丁酸正戊酯〔硫化剂〕
2,6-di-t-butyl phenol 2,6-二叔丁基苯酚〔防老剂〕

2,4-di-t-butylphenyl 3,5-di-t-butyl-4-hydroxy benzoate 3,5-二叔丁基-4-羟基苯甲酸-2,4-二叔丁基苯酯〔防老剂〕
N,N-di-sec-butyl-p-phenylenediamine N,N-二仲丁基对苯二胺〔防老剂〕
2,6-di-t-butyl-4-phenyl phenol 2,6-二叔丁基-4-苯基苯酚〔防老剂〕
dibutyl phenyl phenol sodium disulfonate 二丁基苯基酚二磺酸钠
2,6-di-t-butyl phenyl-p-phenylenediamine 2,6-二叔丁基苯基对苯二胺〔防老剂〕
dibutyl phenyl phosphate 磷酸二丁基苯酯 (C$_4$H$_9$C$_6$H$_4$O)$_2$POOC$_6$H$_5$
dibutyl phenylphosphonate(DBPP) 苯基膦酸二丁酯 (C$_4$H$_9$)$_2$(C$_6$H$_5$)PO
di-sec-butyl phenylphosphonate(DSBPP) 苯基膦酸二仲丁酯 [CH$_3$CH$_2$CH(CH$_3$)O]$_2$PO(C$_6$H$_5$)
dibutyl phenyl polyglycol ether 二丁基苯基聚乙二醇醚
1,1-dibutyl-4-phenyl thiosemicarbazide 1,1-二丁基-4-苯基硫代氨基脲〔防老剂〕
dibutylphosphate(DBP) 磷酸二丁酯 (C$_4$H$_9$O)$_2$PO(OH)
dibutyl phthalate 邻苯二甲酸二丁酯;增润剂〔俗〕 C$_6$H$_4$(CO$_2$C$_4$H$_9$)$_2$
$α,α$-di(t-butylproxy) diisopropylbenzene $α,α$-二(叔丁基过氧基)二异丙苯〔硫化剂〕
dibutylpyrophosphoric acid(DBPP) 焦磷酸二丁酯 O[P(O)(OC$_4$H$_9$)(OH)]$_2$
di-t-butyl quinone 二叔丁醌〔防老剂〕
dibutyl sebacate 癸二酸二丁酯 [(CH$_2$)$_4$CO$_2$C$_4$H$_9$]$_2$
dibutyl succinate 丁二酸二丁酯 (CH$_2$CO$_2$C$_4$H$_9$)$_2$
dibutyl sulfate 硫酸二丁酯 (C$_4$H$_9$O)$_2$SO$_2$
dibutyl sulfide 二丁基硫 (C$_2$H$_5$CH$_2$CH$_2$)$_2$S
dibutyl sulfite 亚硫酸二丁酯 (C$_4$H$_9$O)$_2$SO
dibutyl sulfone 二丁砜 (C$_4$H$_9$)$_2$SO$_2$
dibutyl tartrate 酒石酸二丁酯 (CHOHCO$_2$C$_4$H$_9$)$_2$
dibutyl terephthalate 对苯二甲酸二丁酯〔增塑剂〕
dibutyl thiophosphite(DBTP) 硫代亚磷酸二丁酯 (C$_4$H$_9$O)$_2$PSH
N,N'-dibutyl thiourea N,N'-二丁基硫脲〔促进剂〕
dibutyl thiourea 二丁基硫脲
dibutyl tin bis(2-mercaptobenzothiazole) 双(2-硫醇基苯并噻唑)二丁锡〔稳定剂〕
dibutyl tin bromide 二丁基锡化二溴 Sn(C$_4$H$_9$)$_2$Br$_2$
dibutyl tin dilaurate 二月桂酸二丁基锡〔稳定剂〕
dibutyl tin dimethacrylate 二甲基丙烯酸二丁基锡
dibutyl tin laurate 月桂酸二丁基锡;二丁基月桂酸锡
dibutyl tin maleate 顺丁烯二酸二丁基锡
dibutyl tin oxide 氧化二丁锡〔稳定剂〕

dibutyl urea 二丁脲
dibutyl xanthate disulfide 二硫化黄原酸二丁酯〔促进剂〕
dibutyl xanthogen disulfide 二硫化二丁基黄原酸
dibutyrin 二丁精；甘油二丁酯
dibutyryl 联丁酰基；二丁酰基
dic- 狄克〔音译，生物碱〕
dicacodyl 双卡可基；双二甲肼
dicaesium hexachlorplutonate 六氯钚酸二铯 Cs_2PuCl_6
dicalcium phosphate 磷酸二钙
dicaprolactam disulfide 二硫代二己内酰胺〔硫化剂〕
dicapryl adipate 己二酸二辛酯〔增塑剂〕
dicapryl phthalate 邻苯二甲酸二辛酯
dicapryl sebacate 癸二酸二辛酯〔增塑剂〕
dicarbazyl 联咔唑基
dicarbocyanines 二羰菁
dicarbonate ①碳酸氢钠；小苏打②碳酸氢盐；重碳酸盐 $MHCO_3$
dicarbonyl compound 二羰基化合物
dicarboximide 二羧酰亚胺 $R(COO)_2NH$
dicarboxyarsenazo III 二羧偶氮胂III
dicarboxy arseno benzene 偶砷苯酸 $COOHC_6H_4As \cdot AsC_6H_4COOH$
4,4'-dicarboxydiphenyl ether 4,4'-二羧基二苯醚
dicarboxyl 联羧基；乙二酸；草酸
dicarboxylcellulose 二羧基纤维素
dicarboxyl chloride 二元酰氯
dicarboxylic acid 二元羧酸
dicarboxylic acid chloride 二羧酰氯
dicarboxylic acid ester 二元羧酸酯
dicarboxylic ester 二羧酸酯
di-(p-carboxymethoxy phenyl) oxide 氧化二(对羧甲氧基苯)
dicaryotic nuclear division 双核分裂
dication 二价阳离子
dicationic monomer 双阳离子单体
dice 小方块；小片
diced material 方粒状材料
dicentrine 荷苞牡丹碱
dicer 切成粒机；切粒机
dicerotin 二蜡精；甘油二蜡酸酯
dicetyl(=n-dotriacontane) （正）三十二（碳）烷；联十六烷基 $CH_3(CH_2)_{30}CH_3$
dicetyl ether 双十六烷醚〔润滑剂〕
dichain polyamide 双链聚酰胺
dichain polymer 双链聚合物
dicharging agent 脱色剂；漂白剂；拔染剂
dichlofluanid 苯氟磺胺；抑菌灵
dichlor- 二氯

dichloracetyl chloride 二氯乙酰氯 $CHCl_2 \cdot COCl$
dichloralide 二氯交酯
dichloramine-T 二氯胺-T；对甲苯磺酰二氯胺
dichlorated 二氯化的
dichlorethyl ether 二氯乙醚 $ClCH_2CH_2OCH_2Cl$
dichloride 二氯化物
dichlorinated(=dichlorizated) 二氯化的
dichlorizated(=dichlorinated) 二氯化的
dichloro- 二氯
dichloroacetal 二氯缩醛；二乙醇缩二氯乙醛 $Cl_2CHCH(OC_2H_5)_2$
dichloro acetaldehyde 二氯乙醛 Cl_2CHCHO
dichloro acetamide 二氯乙酰胺 $Cl_2CHCONH_2$
dichloroacetic acid 二氯乙酸；二氯醋酸
sym-dichloroacetone 对称二氯丙酮 $(ClCH_2)_2CO$
unsym-dichloroacetone 不对称二氯丙酮 $Cl_2CHCOCH_3$
ω,ω-dichloroacetophenone ω,ω-二氯乙酰苯
dichloroacetyl chloride 二氯乙酰氯 $Cl_2CHCOCl$
dichloroacetylene 二氯乙炔 C_2Cl_2
dichloroazodicarbonamide(=azochloramide) 二氯偶氮胩 $[ClN=C(NH_2)N=]_2$
dichlorobarbituric acid 二氯巴比妥酸 $Cl_2C(CONH)_2CO$
dichlorobenzene 二氯(代)苯 $C_6H_4Cl_2$
o-dichlorobenzene 邻二氯苯
p-dichlorobenzene 对二氯苯
3,3'-dichlorobenzidine 3,3'-二氯联苯胺
4,4'-dichlorobenzophenone 4,4'-二氯代二苯甲酮
dichlorobenzothiazole-2-morpholine 二氯苯并噻唑-2-吗啉〔硫化剂〕
5,6-dichlorobenzoxazolone 5,6-二氯苯并噁唑酮〔防霉剂〕
dichlorobenzoyl peroxide 过氧化二氯苯甲酰〔硫化剂〕
3,3'-dichloro-4,4'-biphenylene diisocyanate(CDI) 3,3'-二氯 4,4'-亚联苯基二异氰酸酯
dichlorobutane 二氯丁烷
dichlorobutyric acid 二(代)丁酸 $C_3H_5Cl_2COOH$
dichlorocarbene 二氯卡宾 $Cl_2C=$
dichloro compound 二氯化合物
3,3'-dichloro-4,4'-diamino diphenylmethane 3,3'-二氯-4,4'二氨基二苯甲烷〔硫化剂〕
dichlorodibromomethane 二氯二溴甲烷 Cl_2CBr_2
dichlorodiethyl ether 二氯二乙醚 $(C_2H_3Cl_2)_2O$
dichlorodiethyl formal 二氯二乙缩甲醛〔溶剂〕
dichlorodiethyl sulfide 二氯二乙硫醚；芥子气
dichlorodifluoromethane 二氯二氟甲烷
dichlorodiiodomethane 二氯二碘甲烷 Cl_2CI_2

dichlorodimethyl ether 二氯二甲醚 ClCH$_2$OCH$_2$Cl
1,3-dichloro-5,5-dimethyl hydantoin 1,3-二氯-5,5-二甲基乙内酰脲〔防焦剂〕
dichlorodinitromethane 二氯二硝基甲烷 Cl$_2$C(NO$_2$)$_2$
4,4'-dichloro-diphenylsulfone 4,4'-二氯二苯砜
dichloro-diphenyl-trichloroethane(=2,2-bis(p-chlorophenyl)-1,1,1-trichloroethane) 二氯二苯基三氯乙烷；双(对氯苯基)三氯乙烷
dichlorodipropyl ether 二氯二丙醚
dichlorodivinyl chloroarsine 二氯二乙烯氯胂
dichloro ester 二氯酯
dichloroethane 二氯乙烷
dichloro ether ①〔类名〕二氯代醚②二氯乙醚〔俗名，指：α,β-二氯代乙醚 CH$_2$ClCHCOC$_2$H$_5$〕
di(2-chloroethoxy) methane 二(2-氯乙氧基)甲醚〔溶剂〕
dichloroethyl alcohol(=2,2-dichloroethanol) 二氯乙醇；2,2-二氯乙醇 Cl$_2$CHCH$_2$OH
dichloroethylene 二氯乙烯
dichloroethylene carbonate(DCEC) 碳酸二氯乙烯酯
dichloroethyl ether 二氯乙醚
dichloroethyl sulfide 二氯二乙硫醚；芥子气 (CH$_2$Cl·CH$_2$)$_2$S
dichlorofluorescein 二氯荧光黄*
α-dichlorohydrin(=glycerol dichlorohydrin) α-二氯甘油；1,3-二氯-2-丙醇 CH$_2$ClCHOHCH$_2$Cl
dichlorohydrine 二氯丙醇
dichloroindophenol 二氯靛酚
dichloroiodomethane 二氯碘甲烷 Cl$_2$CHI
dichloroisocrotonic acid 二氯异巴豆酸 CH$_3$(CHCl)$_2$CO$_2$H
dichloroisocyanuric acid 二氯异氰脲酸 OCNClCONClCONH
dichloroisopropyl ether 二氯代异丙醚
dichloromaleic acid 二氯马来酸 (=CClCO$_2$H)$_2$
dichloromaleic anhydride 二氯马来酸酐〔硫化剂〕
dichloromalic acid 二氯苹果酸 COOHCCl$_2$CHOHCOOH
dichloromethane(=methylene chloride) 二氯甲烷
dichloromethylation 双氯甲基取代作用〔引入两个氯甲基〕
dichloromethyl p-chlorophenyl ketone 二氯甲基对氯苯基(甲)酮 ClC$_6$H$_4$COCHCl$_2$
dichloromethylene diphosphonate 二氯亚甲基二膦酸
1,3-dichloro-5-methyl-5-isobutyl hydantoin 1,3-二氯-5-甲基-5-异丁基乙内酰脲〔防焦剂〕
dichloromethyl phenylsilane 甲基苯基二氯(甲)硅烷
dichloro methyl silane 二氯甲基硅烷
dichloronitroethane 二氯硝乙烷

dichloronitrohydrin(=dichloropropyl nitrate) 硝酸二氯丙酯 ClCH$_2$CHClCH$_2$ONO$_2$
1,5-dichloropentane 1,5-二氯戊烷
dichloropentane 二氯戊烷〔溶剂〕
dichloropentenyl methacrylate 甲基丙烯酸二氯戊烯酯
dichlorophen(=2,2-dihydroxy-5, 5-dichlorodiphenyl-methane) 二氯芬；二羟二氯二苯甲烷
dichlorophenarsine hydrochloride 盐酸合二氯酚胂
dichlorophenol dihydrochloride 二氢氯化二氯苯酚
2,6-dichlorophenol indophenol 2,6-二氯酚靛酚
dichlorophenol sulfonphthalein 氯酚红；二氯苯酚磺酞
2,4-dichlorophenoxyacetic acid(=2,4-D) 2,4-滴；2,4-二氯苯氧乙酸
α,α-dichloro-4-phenoxy acetophenone α,α-二氯-4-苯氧基苯乙酮
di(p-chlorophenyl)-dichloroethane(=2,2-bis(p-chloro-phenyl)-1,1-dichloroethane) 双氯苯二氯乙烷；双(对氯苯基)二氯乙烷
3,3'-dichlorophenyl diisocyanate 3,3'-二氯苯基二异氰酸酯〔硫化剂〕
3-(3,4-dichlorophenyl)-1,1-dimethylurea(DCMU) 二氯苯(基)二甲脲
1,3-dichloro-p-phenylene diisocyanate 1,3-二氯对亚苯基二异氰酸酯〔硫化剂〕
di(p-chlorophenyl)-methylcarbinol(=1,1-bis(p-chlorophenyl)-ethanol) 双(对氯苯基)甲基甲醇；双(对氯苯基)乙醇
dichlorophenyl phenyl phosphate 磷酸双氯苯基苯酯 C$_6$H$_5$OPO(ClC$_6$H$_4$O)$_2$
dichlorophenyl sulfonic acid 二氯苯肼磺酸 Cl$_2$C$_6$H$_2$(SO$_3$H)NHNH$_2$
dichlorophenyl p-toluene sulfonate 对甲苯磺酸二氯苯酯 CH$_3$C$_6$H$_4$SO$_3$C$_6$H$_3$Cl$_2$
dichlorophosphazine(=phosphorous nitride dichloride trimer) 二氯化氮磷三聚物；二氯膦嗪
dichlorophosphination 二氯膦化作用
dichlorophosphinyl 二氯膦基
dichloropropanol α-二氯丙醇
dichloropropylene 二氯丙烯
dichloropropyl nitrate 硝酸二氯丙酯
dichlorostyrene 二氯苯乙烯
dichlorosuccinic acid 二氯代丁二酸 COOH(CHCl)$_2$COOH
dichlorosulfonphthalein 二氯磺酞
dichloro tetrafluoro ethane 二氯四氟乙烷
dichlorotetraglycol 二氯四甘醇〔glycol 在复合名 di~, tri~, tetra~ 中译甘醇〕 (ClCH$_2$CH$_2$OCH$_2$CH$_2$)$_2$O
N,N'-dichloro-p-toluene sulfenamide N,N'-二氯对甲苯次磺酰胺〔促进剂〕

dichlorotriglycol 二氯三甘醇 Cl(CH₂CH₂O)₂CH₂CH₂Cl
dichlorourea 二氯脲 CO(NHCl)₂
dichloro-*p*-xylene 二氯对甲苯
dichloro xylenol 二氯二甲苯酚
dichrin 常山英
dichrograph 二色性测定仪
dichroic 二色性的
dichroic dyestuff 二向色染料
dichroic filter 二色性滤色器
dichroic IR 双色红外线；二色性红外线
dichroic mirror 二向色反射镜；分色镜
dichroic polarizer 二向色偏振器
dichroic ratio 二色性比
dichroine 常山碱
dichroism 二色性；二向色性
dichromate 重铬酸盐
dichromate cell 重铬酸电池
dichromate ion 重铬酸根离子 $Cr_2O_7^{2-}$
dichromate titration 重铬酸钾(滴定)法*
dichromatic 二色性的
dichromatic effect (结晶)二色(性)效应
dichromatism 二色性
dichrometer 二色仪
dichromic acid 重铬酸 $H_2Cr_2O_7$
dichromism (感官)二色现象；(结晶)二向色性
dichromotomous sampler 双分道采样器
dichroscope 二色镜
dichrostachinic acid *S*-琥珀基半胱氨酸；二辛可宁酸
dicinchonine 双金鸡宁；双辛可宁
dicing cutter 切粒机
dicing machine 切粒机
dicinnamalacetone 二苯基壬四烯酮；双亚肉桂基丙酮
N,N'-dicinnamylidene-1,6-hexamethylenediamine *N,N*'-二亚肉桂基-1,6-己二胺〔硫化剂〕
dickite 地开石；迪凯石
diclofenac 双氯芬酸〔药〕
dicloran(=2,6-dichloro-4-nitroaniline) 2,6-二氯-4-硝基苯胺
dicocoalkyl dimethyl ammonium chloride 二椰油烷基二甲基氯化铵
dicodeine (四)聚可待因
dicodid(e) 二氢化可待因酮
diconchinine(=diquinidine) 双奎尼定
diconical 双锥形的
dicopac 维生素B_{12}吸收双同位素试验(Schilling 试验) 药箱〔Nycomed Amersham 生产〕
dicoumarin ①3,3'-亚甲基-4-双羟香豆素②败坏翘摇素
dicoumarol(=dicoumarin) 败坏翘摇素

dicresol disulfide 二甲酚二硫化物〔防老剂〕
di-*o*-cresol monosulfide 二邻甲酚一硫化物〔防老剂〕
dicrotaline 二猪屎豆碱
dicroton(=dicroton aldehyde) 二聚丁烯醛
dictagymnin 白藓醚
dictam 白藓
dictamine 白藓胺；狄克胺
dictamnal 白藓醛
dictamnin 白藓碱
dictamnolactone 白藓脑内酯
dictamnolic acid 白藓脑内酸
dictamnolide 白藓交酯
dictam oil 白藓油
dictyosome (分散)高尔基体
dicumyl 二枯基
dicumyl peroxide 过氧化二异丙苯
dicumyl peroxide vulcanized elastomer 过氧化二枯基硫化的弹性体
dicyan ①氰②二氰(基) (CN)₂
dicyanamide 二氰胺 (CN)₂NH
dicyandiamide(=cyanoguanidine) 双氰胺；二聚氨基氰；氰基胍 $H_2NC(=NH)NHCN$
dicyandiamidine(=guanylurea) 脒基脲；胍基甲酰胺 $H_2NC(=NH)NHCONH_2$
dicyanin(e) 双花青
1,4-dicyanobutane 1,4-二氰基丁烷
dicyanogen(=cyanogen) 氰 (CN)₂
dicyanopropane 氰基丙烷
1,2-dicyanopropane(=propylene cyanide) 丙邻二腈；1,2-二氰基丙烷 $CH_3CH(CN)CH_2CN$
dicyanovinyl acceptor group 二氰乙烯基受电子基团
dicyclic 双环的
dicyclic compound 双环化合物
dicyclic hydrocarbons 双环烃类
dicyclic ring 双环核
dicyclo- 双环〔不是二环 bicyclo〕
dicyclohexyl ①(=decahydrobiphenyl)双环己基②二环己基 $[CH_2(CH_2)_4CH]_2$
dicyclohexyl adipate 己二酸二环己酯 $(CH_2CH_2CO_2C_6H_{11})_2$
dicyclohexyl amine 二环己基胺 $(C_6H_{11})_2NH$
dicyclohexyl amine nitrite 亚硝酸二环己胺〔气相防锈剂〕
dicyclohexylammonium benzothiazole mercaptide 硫醇苯并噻唑二环己铵〔促进剂〕
dicyclohexyl ammonium nitrite 亚硝酸二环己基铵〔气相防锈剂〕 $(C_6H_{11})_2H_2NNO_3$
dicyclohexyl azelate 壬二酸二环己酯〔增塑剂〕

N,N'-dicyclohexyl-2-benzothiazolesulfenamide　N,N'-二环己基-2-苯并噻唑次磺酰胺〔促进剂〕

dicyclohexylcarbodiimide(DCC; DCCI)　二环己基碳二亚胺

dicyclohexyl-18-crown-6(DC18C6)　二环己基并-18-冠(醚)-6　$C_4H_8C_{12}O_6H_{20}C_4H_8$

N,N'-dicyclohexyl dithio-oxamide　N,N'-二环己基二硫代草酰胺〔催化剂〕

dicyclohexyl ketone　二环己基甲酮　$(C_6H_{11})_2CO$

dicyclohexyl maleate　马来酸二环己酯　$(=CHCO_2C_6H_{11})_2$

dicyclohexyl methane diisocyanate　二环己基甲烷二异氰酸酯

dicyclohexyl oxalate　草酸二环己酯

dicyclohexyl peroxy dicarbonate　过氧化二碳酸二环己酯〔引发剂〕

N,N'-dicyclohexylphenyl-p-phenylenediamine　N,N'-二环己基苯基对苯二胺〔防老剂〕

dicyclohexyl phthalate　邻苯二甲酸二环己酯　$C_6H_4(CO_2C_6H_{11})_2$

N,N'-dicyclohexyl-1,2,4,5-tetrathia-3,6-diazine　N,N'-二环己基-1,2,4,5-四硫代-3,6-嗪〔硫化剂〕

dicyclohexylurea(DCU)　二环己(基)脲

dicyclooctatetracene-neptunium　二环辛四烯合镎　$Np(C_8H_8)_2$

dicyclooctatetracene-plutonium　二环辛四烯合钚　$Pu(C_8H_8)_2$

dicyclopentadiene　二聚环戊二烯　$(CH=CHCH_2CH=CH)_2$

dicyclopentadienyl　联环戊二烯

dicyclopentadienyl berkelium chloride　氯化二茂合锫　$Bk[(C_5H_5)_2Cl]_2$

dicyclopentadienyl beryllium　二茂合铍　$Be(C_5H_5)_2$

dicyclopentadienyl cobalt(=cobaltocene)　二环戊二烯(基)钴；二茂钴　$(C_5H_5)_2Co$

dicyclopentadienyl iron(=ferrocene)　二茂铁

dicyclopentadienyl-metal　二环戊二烯基金属；(二价)金属茂　$(C_5H_5)_2M$

dicyclopentadienyl metal halide　二环戊二烯基卤化金属；卤化(二茂金属)

dicyclopentadienyl nickel(=nickelocene)　二环戊二烯(基)镍；二茂镍

dicyclopentadienyl tin　二茂锡

dicyclopentadienyl titanium dichloride(=titanocene dichloride)　二氯化环戊二烯(基)钛；二氯二茂钛

dicyclopentadienyl zirconium dichloride(=zirconocene dichloride)　二氯二环戊二烯基锆；二氯二茂锆

dicyclopentamethylene thiuram tetrasulfide　四硫化双环五亚甲基秋兰姆〔促进剂〕

$1,1'$-dicyclopropyl uranocene　$1,1'$-二环丙基代双环辛烯合铀　$U(C_8H_7 \cdot C_3H_5)_2$

dicycuring agent　双氰胺固化剂

didanosine　去羟肌苷〔药〕

didecanoyl peroxide　过氧化二癸酰

didecyl adipate　己二酸二癸酯

didecylamine　二癸基胺

didecyl disulfide　二癸基二硫化物

didecyl ether　二癸醚〔润滑剂〕

didecyl glutarate　戊二酸二癸酯〔增塑剂〕

didecyl ketone　二癸基(甲)酮；11-廿一(碳)烷酮　$CH_3(CH_2)_9CO(CH_2)_3CH_3$

didecyl phosphite　亚磷酸二癸酯〔增塑剂〕

didecyl phthalate　邻苯二甲酸二癸酯

didepside　二缩酚酸

dideuteride　二氘化物

dideuterio acetylene　二重氢乙炔

dideutero-p-aminobenzoic acid　二氘对氨基苯甲酸　$NH_2C_6H_2D_2COOH$

dideuteroethylene　二氘(代)乙烯

1,1-dideuteropropene　1,1-二氘化丙烯

N,N'-di(1,3-dimethylbutyl)-p-phenylenediamine　N,N'-二(1,3-二甲基丁基)对苯二胺〔防老剂〕

N,N'-di(1,5-dimethylpentyl)-p-phenylenediamine　N,N'-二(1,5-二甲基戊基)对苯二胺〔防老剂〕

didimolite　钙蓝石

di(dioctylphosphato) ethylene titanate　二(二辛基磷酸氧基)钛酸乙烯酯〔偶联剂〕

di(dioctyl pyrophosphato) ethylene titanate　二(焦磷酸二辛酯)钛酸乙烯酯

di-diol bond　双二醇键

didiploid　双二倍体

didodecenyl dimethyl ammonium nitrate　双十二碳烯基二甲基硝酸铵　$(C_{12}H_{23})_2(CH_3)_2N \cdot NO_3$

didodecyl disulfide　双十二烷基二硫

didodecyl phthalate　邻苯二甲酸双十二酯〔增塑剂〕

didymia　氧化钕镨　Di_2O_3

didymic acid　苔酸

didymium　钕、镨混合物〔曾误认为一种元素〕

didymium chloride　氯化钕镨

didymium filter　钕镨滤器；钕镨滤波器

didymium nitrate　硝酸钕镨

didymium oxide　氧化钕镨

didymium sulfate　硫酸钕镨

die　①口型；型板；压出板②冲模；硬膜；塑模衰减曲线

die annulus　环形模口

die blade　模唇
die block hardener　冲模硬化剂
die-burn　烧伤
die casting　型铸；模铸
die casting zinc alloy　膜铸锌合金
Dieckmann reaction　迪克曼反应
die connection part　模头连接件
die constant　模子常数
die cut　①冲切；修边②裂口；缺口；裂缝
die cutter　压力裁断刀；冲刀
die cutting　压力裁断；冲切；冲剪
die design　模具设计
die-face cutter　模口切料刀
die glazing　出模上光
die jaw　模唇
die land　口模成型面
Dieldrin(=compound 497)　狄氏剂；氧桥氯甲桥萘；化合物497
Dieldrin-attapulgite mixture　狄氏剂-凹凸棒土混合物
dielectimetry　介电常数分析法
dielectric　①电介质；电介体②电介的
dielectric absorption　介电吸收
dielectric absorption spectra　介电吸收谱
dielectric after effect　介电后效；电解质后效应
dielectric analysis　介电分析(法)
dielectric anisotropy　介电各向异性
dielectric breakdown　介电击穿
dielectric breakdown strength　介电击穿强度
dielectric breakdown test　绝缘击穿试验
dielectric capacity　介电容量
dielectric characteristics　介电特性
dielectric constant　介电常数*
dielectric constant detector　介电常数检测器
dielectric constant of free space　自由空间的介电常数
dielectric curing　(热固性塑料)介电固化
dielectric depolarization　介电退极化
dielectric dispersion　介电色散
dielectric dispersion curve　介电色散曲线
dielectric displacement　介电位移
dielectric dissipation factor　介电损耗因子*
dielectric dryer　高频干燥机
dielectric drying　高频干燥
dielectric expansion　介电发泡；高频发泡
dielectric fatigue　介电疲乏
dielectric flux　介电通量
dielectric glass fibre　D玻璃纤维；低介电玻璃纤维
dielectric heater(=electronic heater)　电子加热器；介电加热器
dielectric heating　介电加热
dielectric heat sealer　介电热合机；高频热合机
dielectric insulation　介电绝缘
dielectric loss　介电损耗
dielectric loss angle　介电损耗角
dielectric loss angle tangent　介电损耗角正切
dielectric loss constant　介电损耗常数*
dielectric loss factor　介电损耗因子
dielectric loss index　介电损耗指数
dielectric loss tangent　介电损耗角正切
dielectric material　电介体；绝缘材料
dielectric modulus　介电模量
dielectric moisture meter　介电水分计
dielectric monitoring　介电监控
dielectric oil　介电油；绝缘油
dielectric phase angle　介电位相角
dielectric phase difference　介电相差
dielectric polarization　电介质极化
dielectric power factor　介电功率因数
dielectric power test　介电本领试验
dielectric preheating　介电预热；高频预热
dielectric property　介电性质
dielectric relaxation　介质弛豫*
dielectric relaxation spectrum　聚合物的介电松弛谱
dielectric relaxation time　介电弛豫时间*
dielectric saturation　介电饱和
dielectric spectra　介电谱
dielectric stability　(高频热合)介电稳定性
dielectric strength　绝缘强度；介电强度
dielectric stress　介电应力
dielectric substance　介电质；介电质材料
dielectric test　介电强度试验
dielectric thermal analysis　介电热分析
dielectric thermal analyzer　介电热分析仪
dielectrometer　介电计；介电测试器
dielectrometric titration(=dielectrometry)　介电(常数)滴定(法)*
dielectrometry(=dielectrometric titration)　介电(常数)滴定(法)*
dielectrophoresis　介电泳
Diels-Alder adduct(s)　狄尔斯-阿尔德(反应)加成物
Diels-Alder reaction　狄尔斯-阿尔德反应*
die mandrel　模芯；芯模
dien(=diethylene triamine)　二亚乙基三胺；二(2-氨基乙基)胺　$(NH_2CH_2CH_2)_2N$
diene　双烯*
diene component(=dienophile)　亲二烯物；亲双烯体
diene copolymer　二烯聚合物

diene hydrocarbon elastomer 二烯(基)弹性体
diene interpolymer 二烯系共聚物〔二烯系单体与乙烯、丙烯或异戊间二烯等的共聚物〕
diene isomerism 二烯(同分)异构(现象)
diene monomer 双烯单体；二烯单体
diene number 二烯值
dienephilic group 亲二烯基(团)
diene polymer 二烯系聚合物
diene polymerization 双烯类聚合*
diene rubber 二烯(基)橡胶
diene series 二烯系
dienes with adjacent double bonds 邻二烯属〔具有邻双键的二烯属〕 RC＝C＝CR
dienes with conjugated double bonds 共轭二烯属〔具有共轭双键的二烯属〕 RHC＝CHCH＝CHR
dienes with independent double bonds 独立二烯属〔具有远离独立双键的二烯属〕 RCH＝CH(CH$_2$)$_n$CH＝CHR
diene synthesis 双烯合成*
diene synthetics 二烯(基)合成橡胶
diene value 二烯值
dienoic acid 二烯酸
dienol 二烯醇
dienomycin 二烯霉素
dienone 二烯酮
dienone-phenol rearrangement 二烯酮-酚重排作用
dienophile 亲双烯体*
dienophile curative 亲双烯硫化剂
dienophilic compound(=dienophile) 亲二烯物；亲双烯体*
die of stamp 捣矿砧
die opening 口型槽口
die orifice 口型槽口
die passage 模口通道
diepoxides(=bisepoxides) 双环氧化合物
die pressing 模压
die pressure 口型内压力
diergol 双组分火箭燃料
die ring 口模圈；口模套圈
dierucin 二芥精；甘油二芥酸酯
diesel cetane method 柴油(机燃料)十六烷值测定法
diesel-dope 柴油(机燃料)的添加剂
diesel engine 柴油机；狄塞尔机
diesel engine fuel oil 柴油机(燃料)油〔低速柴油机用〕
diesel-fuel 柴油机燃料
diesel-fuel additive 柴油(机燃料)添加剂
diesel-fuel carbon 柴油(机燃料)残炭
diesel-fuel cetane number 柴油(机燃料)的十六烷值
diesel-fuel distillation test 柴油(机燃料)蒸馏试验
diesel-fuel distribution 柴油(机燃料在燃料系统中)的分布
diesel-fuel dribbling 柴油(机燃料)流入(燃烧室内)
diesel-fuel end point 柴油(机燃料)的终馏点
diesel-fuel flash point 柴油(机燃料)的闪点
diesel-fuel gravity test 柴油(机燃料)比重的测定
diesel-fuel ignition 柴油(机燃料)的着火
diesel-fuel ignition quality 柴油(机燃料)的自燃特性
diesel-fuel improver 提高柴油(机燃料)十六烷值的添加剂
diesel-fuel index number 柴油指数
diesel fuel oil 柴油；柴油机用柴油
diesel-fuel pour point 柴油(机燃料)的倾点
diesel generator 柴油发电机
diesel index 柴油指数；狄塞尔指数
diesel knock 柴油爆击
diesel number 柴油值
diesel oil 柴油；狄塞尔燃料
diesel starting fuel 柴油机起动燃料
die shoe 模瓦
die shrinkage 模压收缩率；计算收缩
die sinking press 冷模压型机
diesoline 轻柴油
die space 型腔；模腔；模槽
die spider 辐架
die-spinning nozzle 喷丝嘴
diester 二酯
diester-diurea polymer 二酯二脲聚合物
diester oil 合成双酯润滑油
diester plasticizer 双酯增塑剂
di-esters(=dibasic acid esters) 双酯类〔合成润滑油〕
die swell(=extrudate swelling) 挤出物胀大
die swell correction 喷丝孔膨胀校正
die swell ratio 挤出物胀大比率
diet 饮食；膳食
dietary ①饮食的 ②食谱
dietary standard 饮食标准；食谱标准
dietary supplement 营养强化剂
dietetics 膳食学
diethacetic acid 二乙基乙酸 $CH(C_2H_5)_2COOH$
diethanolamine 二乙醇胺(俗)；二羟乙基胺 $HN(CH_2CH_2OH)_2$
diethanolamine methoxy cinnamate 甲氧基肉桂酸二乙醇胺
diethanolaniline 二乙醇苯胺〔增塑剂〕
diethazine 二乙吖嗪
diethenoid 二烯系的
diethenoid tetracyclic triterpene alcohol 二烯系四环三萜烯醇

diethoxalic acid 4-羟基己酸；二乙草酸〔俗〕 C₂H₅CHOH(CH₂)₂COOH
diethoxy acetic acid 二乙氧基乙酸 CH(OC₂H₅)₂COOH
2,2-diethoxyacetophenone 2,2-二乙氧基苯乙酮
3,4-diethoxy benzyl alcohol 3,4-二乙氧基苄醇
1,1-diethoxyethane 1,1-二乙氧基乙烷
diethoxyethyl adipate 己二酸二(乙氧基)酯
di(2-ethoxyethyl) peroxy dicarbonate 过(氧)二碳酸双(2-乙氧基乙基)酯
diethoxy methane 二乙氧基甲烷
di(ethoxy-thiocarbonyl) disulfide 二(乙氧基硫代羰基)二硫化物〔促进剂〕
diethyl 二乙基的
diethyl acetal 二乙基乙缩醛
diethylacetaldehyde 二乙基乙醛 (C₂H₅)₂CHCHO
N,N-diethyl acetamide N,N-二乙基乙酰胺；乙酰二乙胺 CH₃CON(C₂H₅)₂
diethylacetic acid 二乙基乙酸 (C₂H₅)₂CHCOOH
diethyl acetoacetic ester ①〔通指〕二乙基乙酰乙酸酯 CH₃COC(C₂H₅)₂COOR ②〔专指〕二乙基乙酰乙酸乙酯 CH₃COC(C₂H₅)₂COOC₂H₅
diethylacetonitrile 二乙基乙腈 (C₂H₅)₂CHCN
diethylacetylene 二乙基炔 C₂H₅C≡CC₂H₅
diethyl acetylmalonate 乙酰丙二酸二乙酯 CH₃COCH(CH₂C₂H₅)₂
diethyl adipate 己二酸二乙酯〔增塑剂〕
diethyl alkylaspartate 烷基天冬氨酸二乙酯
diethyl aluminium hydride 氢化二乙基铝；二乙基铝烷
diethylaluminum chloride 氯化二乙基铝
diethylamine 二乙胺 (C₂H₅)₂NH
7-diethylamino-3-chlorofluorane 7-二乙氨基-3-氯荧烷
diethylaminoethanol 二乙氨基乙醇
diethylaminoethyl 二乙氨乙基
diethylaminoethylcellulose 二乙氨乙基纤维素〔含有(C₂H₅)₂NCH₂CH₂—置换基的纤维素〕
diethylaminoethyl-cotton 二乙氨乙基棉；二乙氨基乙基纤维素
diethylaminoethyl dextran 二乙氨基葡聚糖
diethylaminoethyl methacrylate 甲基丙烯酸二乙氨基乙酯
diethyl aminoformyl phosphonate 氨基甲酰膦酸二乙酯
1-(N,N-diethylaminomethyl) benzothiazolyl-thione-2 1-(N,N-二乙氨甲基)-2-苯并噻唑基硫酮〔促进剂〕
diethylaminopropylamine 二乙氨基丙胺 (C₂H₅)₂NC₃H₆NH₂
diethylammonium diethyldithiocarbamate(DDDC) 二乙基二硫代氨基甲酸二乙基铵 (C₂H₅)₂NCSSN(C₂H₅)₂
diethylammonium dimethylthiocarbamate 二甲基二硫代氨基甲酸二乙基铵〔促进剂〕

diethylammonium dithiocarbamate 二硫代氨基甲酸二乙基铵
diethylaniline 二乙苯胺
diethylaniline orange(=ethyl orange) 乙基橙〔不必另定名为二乙苯胺橙〕
diethyl arsine 二乙胂
diethyl arsinic acid 二乙次胂酸 (C₂H₅)₂AsOOH
diethylated 二乙基的
diethyl azodicarbozylate 偶氮二羧酸二酯〔发泡剂〕
diethyl azodiformate 偶氮二甲酸二乙酯〔发泡剂〕
diethyl barbituric acid(=veronal) 佛罗那；二乙基巴比妥酸 C₈H₁₂O₃N₂
diethyl-2-benzothiazole-sulfenamide 二乙基-2-苯并噻唑次磺酰胺
diethyl-bis(3-mercaptopropyl) silane 二乙基双(3-硫醇基丙基)硅烷
diethyl butanedioate 丁二酸二乙酯
di-2-ethyl butyl azelate 壬二酸二-2-乙基丁酯壬二酸二己酯〔增塑剂〕
1,1-diethyl-4-butylthiosemicarbazide 1,1-二乙基-4-丁基硫代氨基脲〔防老剂〕
diethyl camphorate 樟脑酸二乙酯
diethylcarbamazine 乙胺嗪〔药〕
diethyl carbitol 二甘醇二乙醚；二乙基卡必醇〔俗〕
diethyl carbonate 碳酸二乙酯
diethyl chloroarsine 二乙氯胂
diethylchlorophosphate 二乙基磷酰氯；氯代磷酸二乙酯 (C₂H₅O)₂P(O)Cl
1,1-diethyl-4-cyclohexyl thiosemicarbazide 1,1-二乙基-4-环己基硫代氨基脲〔防老剂〕
diethyl decanedioate 癸二酸二乙酯
diethyl 3,5-di-t-butyl-4-hydroxybenzyl phosphonate 3,5-二叔丁基-4-羟基苄基膦酸二乙酯〔防老剂〕
diethyl 2,3-dihydroxy butanedioate 2,3-二羟基丁二酸二乙酯
diethyl dihydroxylamine 二乙基二羟胺
diethyl 2,3-dihydroxy succinate 2,3-二羟基琥珀酸二乙酯
diethyl-α,β-diketone 二乙基-α,β-二酮
diethyl dioxide 过氧化二乙基 (C₂H₅)₂O₂
diethyldiphenylthiuram disulfide 二硫化二乙基二苯基秋兰姆〔促进剂〕
diethyldiphenylthiuram monosulfide 一硫化二乙基二苯基秋兰姆〔促进剂〕
diethyl diphenyl urea 二乙基二苯基脲
diethyl dithiocarbamate 二乙基氨基二硫代羧酸盐 (C₂H₅)₂NCSSM
diethyl dithiophosphoric acid 二乙基二硫代磷酸

$(C_2H_5O)_2PSSH$
diethylene ①二亚乙基②环丁烷
diethylenediamine(=piperazine) 哌嗪
diethylene glycol 二甘醇 $O(CH_2CH_2OH)_2$
diethylene glycol abietate 松香酸二甘醇酯
diethylene glycol adipate 己二酸二甘醇酯
diethylene glycol diacrylate 二丙烯酸二甘醇酯
diethylene glycol dimethacrylate(DEGDMA) 二甲基丙烯酸二甘醇酯
diethylene glycol dimethyl ether(=diglyme) 二甘醇二甲醚 $O(CH_2CH_2OCH_3)_2$
diethylene glycol dinitrate 二硝酸二甘醇酯
diethylene glycol ethyl ether 二甘醇乙醚
diethylene glycol laurate 月桂酸二甘醇酯
diethylene glycol monobutyl ether 二甘醇一丁基醚；(一缩)二乙二醇单丁醚
diethylene glycol monobutyl ether acetate 二甘醇单丁(基)醚乙酸酯；(一缩)二(聚)乙二醇-丁乙酸酯
diethylene glycol monoethyl ether acetate 二甘醇一乙醚乙酸酯
diethylene glycol monomethyl ether 二甘醇一甲醚；(一缩)二乙二醇单甲醚；甲基卡必醇
diethylene glycol monomyristate 一肉豆蔻酸二甘醇酯
diethylene glycol monooleate 一油酸二甘醇酯
diethylene glycol monopalmitate 一棕榈酸二甘醇酯
diethylene glycol monostearate 一硬脂酸二甘醇酯
diethylene glycol myristate 肉豆蔻酸二甘醇酯
diethylene glycol oleate 油酸二甘醇酯
diethylene glycol palmitate 棕榈酸二甘醇酯
diethylene glycol stearate 硬脂酸二甘醇酯
diethylene glycol succinate(DEGS) 丁二酸二乙二醇酯
diethylene oxide 二乙烯化(二)氧 $C_4H_8O_2$
diethylene tetramine 二亚乙基四胺〔硫化剂〕
diethylene triamine 二亚乙基三胺
diethylenetriamine hydroxypropylate N,N-二羟丙基二亚乙基三胺
diethylenetriamine pentaacetic acid(DTPA) 二亚乙基三胺五乙酸
diethylenetriamine pentamethylenophosphonic acid (DTPP) 二亚乙基三胺五亚甲基膦酸
diethylenetriamine pentamethyl phosphinic acid(DTPPA) 二亚乙基三胺五甲基次膦酸
γ-(diethylenetriamino) propyl trimethoxysilane γ-(二亚乙基三氨基)丙基三甲氧基硅烷
diethyl ester〔泛指〕二乙酯 $R(COOC_2H_5)_2$
diethyl ethanedicarboxylate 乙烷二羧酸二乙酯
diethylethanolamine 二乙基乙醇胺〔催化剂〕
diethyl ether(=ethyl ether) (二)乙醚

1,1′-diethyl ferrocenoate 1,1′-二茂铁羧酸二乙酯
N,N-diethylformamide(DEF) N,N-二乙基甲酰胺
N,N-diethylhexamethylene diamine N,N'-二乙基己二胺 $C_2H_5NH(CH_2)_6NHC_2H_5$
di(2-ethylhexyl) azelate 壬二酸二(2-乙基己基)酯
di(2-ethylhexyl) chloromethylphosphonate(DEHCLMP) 一氯甲基膦酸二(2-乙基己基)酯
di-(2-ethylhexyl)-4,5-epoxy tetrahydrophth-alate 4,5-环氧四氢邻苯二甲酸二(2-乙基己基)酯〔增塑剂〕
di-2-ethylhexyl fumarate 富马酸二-2-乙基己酯；富马酸二辛酯〔增塑剂〕
di-2-ethylhexyl hexahydrophthalate 六氢邻苯二甲酸二-2-乙基己酯；六氢邻苯二甲酸二辛酯〔增塑剂〕
di-2-ethylhexyl maleate 马来酸二-2-乙基己酯；马来酸二辛酯〔增塑剂〕
di(2-ethylhexyl) peroxy dicarbonate 过(氧)二碳酸双(2-乙基己基)酯
di(2-ethylhexyl) phosphoric acid 二(2-乙基己基)磷酸
di-2-ethylhexyl phthalate 邻苯二甲酸二-2-乙基己酯；邻苯二甲酸二辛酯〔增塑剂〕
di-2-ethylhexyl sebacate 二(2-乙基己基)癸二酸酯〔一种合成润滑物质，增塑剂〕
di-2-ethylhexyl succinate 丁二酸二-2-乙基己酯；丁二酸二辛酯〔增塑剂〕
di-2-ethylhexyl terephthalate 对苯二甲酸二-2-乙基己酯；对苯二甲酸二辛酯〔增塑剂〕
di(2-ethylhexyl) tetrahydrophthalate 四氢邻苯二甲酸二(2-乙基己基)酯
diethyl hydrine 甘油-1,2-二乙醚 $CH_2OEtCHOEtCH_2OH$
diethyl hydroxysuccinate 羟基琥珀酸二乙酯
diethylidene ①二亚乙基②2-丁烯
N,N-di(N'-ethylideneanilino) aminobenzene N,N-二(N'-亚乙基苯胺基)氨基苯〔防老剂〕
diethylin 二乙氨基 —$N(C_2H_5)_2$
diethyline 二乙精；甘油-1,2-二乙醚 $CH_2OEtCHOEtCH_2OH$
diethylisopropoxy aluminum 二乙基异丙氧基铝 $(CH_3)_2CHOAl(C_2H_5)_2$
diethyl itaconate 衣康酸二乙酯
diethyl ketone 二乙基甲酮
diethyl malate 苹果酸二乙酯 $C_2H_5OCOCH_2CHOHCOOC_2H_5$
diethyl maleate 马来酸二乙酯 $C_2H_5OCOCH=CHCOOC_2H_5$
diethyl maleic acid 二乙基马来酸 $COOHC(C_2H_5)=C(C_2H_5)COOH$
diethyl malonate 丙二酸二乙酯 $C_2H_5OCOCH_2COOC_2H_5$

diethyl malonic acid 二乙基丙二酸 COOHC(C_2H_5)$_2$COOH
diethyl malonic ester ①〔通常泛指〕二乙基丙二酸酯 ROCOC(C_2H_5)$_2$COOR ②〔专指〕二乙基丙二酸二乙酯 C_2H_5OCOC(C_2H_5)$_2$COOC_2H_5
diethyl-malonyl urea(=barbitone) 二乙基丙二酰脲；巴比妥〔俗〕
diethyl meso-2,3-dibromosuccinate 中-2,3-二溴琥珀酸二乙酯
diethylmethylammonium diethyldithiocarbamate 二乙基二硫代氨基甲二乙基甲酸基铵〔促进剂〕
N,N'-di(1-ethyl-3-methyl)-p-phenylened-iamine N,N'-二(1-乙基-3-甲基)对苯二胺〔防老剂〕
2,3-diethyl-5-methylpyrazine 2,3-二乙基-5-甲基吡嗪
N,N'-diethyl-p-nitrosoaniline N,N'-二乙基对亚硝基苯胺〔防老剂〕
N,N'-diethyl-p-nitrosophenylamine N,N'-二乙基对亚硝基苯胺〔防老剂〕
diethyl 1,8-octanedicarboxylate 1,8-辛烷二羧酸二乙酯
diethylol dimethyl 二羟乙基二甲基
diethyl oxalate 草酸二乙酯；乙二酸二乙酯
diethyl oxaloacetate 草乙酸二乙酯〔俗〕；丁酮二酸二乙酯 C_2H_5OCOCOCH_2COOC_2H_5
diethyl oxaloacetic ester ①〔通常泛指〕二乙基草乙酸酯 ROCOCOC(C_2H_5)$_2$COOR ②〔专指〕二乙基草乙酸二乙酯 C_2H_5OCOCOC(C_2H_5)$_2$COOC_2H_5
diethyl peroxide 过氧化二乙基 (C_2H_5)$_2$O$_2$
diethyl phosphinic acid 二乙基次膦酸 (C_2H_5)$_2$POOH
diethyl phosphoric acid 磷酸二乙酯 (C_2H_5O)$_2$POOH
diethyl phthalate 邻苯二甲酸二乙酯
2,2-diethyl-1,3-propanediol(=prenderol) 甫任德醇〔俗〕；2,2-二乙基-1,3-丙二醇
2,3-diethylpyrazine 2,3-二乙基吡嗪
diethylpyrocarbonate 焦碳酸二乙酯
diethyl sebacate 癸二酸二乙酯〔增塑剂〕
diethyl selenide 二乙基硒 C_2H_5SeC_2H_5
diethylsilane 二乙基甲硅烷；二乙基硅(甲)烷
diethylsilanediol 二乙基硅(甲)烷二醇
diethyl stilbestrol 己烯雌酚；乙芪酚
diethyl succinate 丁二酸二乙酯 C_2H_5OCOCH_2CH_2COOC_2H_5
diethyl succinic acid 二乙基丁二酸 COOH(CHC$_2H_5$)$_2$COOH
diethyl sulfate(=ethyl sulfate) 硫酸二乙酯
diethyl sulfourea 二乙基硫脲 CS(NHC$_2H_5$)$_2$
diethyl sulphate 硫酸二乙酯
diethyl tartrate 酒石酸二乙酯 C_2H_5OCO(CHOH)$_2$COOC_2H_5

diethyl terephthalate 对苯二甲酸二乙酯〔增塑剂〕
2,5-diethyltetrahydrofuran 2,5-二乙基四氢呋喃
diethyl thionamic acid 二乙氨磺酸 (C_2H_5)$_2$NSO$_3$H
N,N-diethyl thiourea N,N-二乙基硫脲 CS(NHC$_2H_5$)$_2$
diethyl tin 二乙基锡 Sn(C_2H_5)$_2$
diethyl tin bromide 溴化二乙基锡 Sn(C_2H_5)$_2$Br$_2$
diethyltin dihalide 二卤化二乙基锡
diethyl tin sulfide trimer 二乙基锡化硫(环)三聚物
1,1'-diethyluranocene 1,1'-二乙基代双环辛四烯合铀 U($C_8H_7C_2H_5$)$_2$
diethyl urea 二乙基脲 CO(NHC$_2H_5$)$_2$
diethyl xanthate disulfide 二硫化黄原酸二乙酯〔促进剂〕
diethyl zinc 二乙基锌 Zn(C_2H_5)$_2$
dietician 饮食学专家
dietics 饮食学
die-to-roll gap 模辊距
die torpedo 鱼雷形分流芯
dietotherapy 膳食疗法；营养疗法
dietzeite 碘铬钙石
Dietzel silver refining 迪茨耳炼银法
difatty amido alkoxylated ammonium halide 二脂肪酰氨基烷氧基卤化铵
difference ①差(数)②差别；差异；区别
difference absorption spectrum 吸收差谱
difference band 差频谱带
difference equation 差分方程
difference Fourier method 差值傅里叶法*
difference Fourier synthesis 差值傅里叶合成；差值电子密度合成
difference gauge 极限量规
difference mode 差分模式
difference of fundamental frequencies 差频
difference of temperature 温度差
difference spectroscopy 差谱
difference spectrum 示差谱*
difference value ①酯化值②差值
differential ①微分②微分的③差示的
differential absorption curve 示差吸收曲线
differential aeration cell 差动充电电池；微分氮气电池
differential alternating current polarograph 示差法交流极谱仪
differential amperometry 示差电流滴定(法)*
differential amplifier 差示放大器
differential analysis 差示分析(法)
differential anodic stripping voltammetry 微分阳极溶出伏安法
differential calorimeter 示差量热计

differential calorimetry 微分量热法；示差量热法
differential capacity 微分电容*；微分容量
differential cell 示差液池
differential centrifugation 差速离心；差示沉降离心法
differential chromatography 示差色谱法
differential coefficient 微分系数；微商
differential colorimeter 差示比色计
differential condensation 微分冷凝
differential conductometric titration 差示电导滴定
differential contact extractor 连续接触式萃取设备
differential detection 差示检测
differential detector 差示检测器；微分型检测器
differential diaphragm 差压隔膜
differential dilatometer 示差膨胀计
differential dilatometry 示差热膨胀测量法
differential distribution curve 微分分布曲线
differential distribution of crosslinking 交联微分分布
differential double column 示差双柱
differential dyeing 差异染色
differential electrolytic potentiometry 微分电解电位法
differential enthalpic analysis 差示热焓分析
differential equation 微分方程式
differential expansion 不均匀膨胀；局部膨胀
differential extraction 微分萃取
differential Faradaic rectification polarography 微分法拉第整流极谱法
differential fiber 改性纤维*
differential flow 微分流量
differential flow controller 差示流量控制器
differential fractionation 微分分馏
differential gage 示差测厚计
differential galvanometer 差绕电流计
differential gear 差动齿轮；差速传动
differential head 差压头；示差水头
differential head meter 差高计
differential heat(=partial heat) 微分热；定浓热
differential heating 差分加热；局部加热
differential heat of adsorption 微分吸附热*
differential heat of dilution 微分稀释热；定浓稀释热
differential heat of solution 微分溶解热*；定浓溶解热
differential heat of sorption 微分吸收热
differential in-gel electrophoresis 胶上差示电泳
differential kinetic chemiluminescence analysis 示差动力学化学发光分析
differential laser absorption radar technology 示差吸收激光雷达技术
differential liquid level gauge 水位差流速计
differential manometer 差示压力计

differential mass distribution curve 微分质量分布曲线
differential-melting behaviour 差示熔融性状
differential method 示差法；微分法；差示法
differential migration 微差移动
differential molecular weight distribution 微分分子量分布
differential number distribution curve 微分数分布曲线
differential operation 差示操作
differential partial condensation 部分冷凝的蒸馏分离
differential piston compressor 级差式压缩机
differential polarogram 差示极谱图
differential polarographic titration 示差极谱滴定
differential polarography 示差极谱(法)*
differential potentiometric titration 差示电位滴定
differential potentiometric titrator 示差电位滴定仪；微分电位滴定仪
differential potentiometry 示差电位法
differential potentiotitration 差示电位滴定法
differential precipitation 示差沉淀
differential pressure 差压
differential pressure flow meter 差压流速计
differential pressure indicator 差压指示计
differential pressure indicator controller 差压指示调节器
differential pressure level controller 水平式差压计
differential pressure meter 差压计
differential pressure recorder 差压记录器
differential protection 差动保护(装置)
differential pulse anodic stripping voltammetry 差示脉冲阳极提溶伏安法
differential pulse polarography 差示脉冲极谱法
differential pulse voltammetry 示差脉冲伏安法
differential pump 差动泵
differential pumping 差压抽气
differential reaction cross section 微分反应截面*
differential reaction-rate kinetic analysis 速差动力学分析(法)
differential reactor 微分反应器
differential refraction detector 差示折光检测器
differential refractive index detector 示差折光率检测器*
differential refractometer 差示折光计
differential refractometer detector 差示折光检测器
differential relay 差示继电器
differential respirometer 示差呼吸计
differential rewinding 差速复卷
differential saturation technique 差分饱和法
differential scanning calorimeter 差示扫描量热计
differential scanning calorimetric curve 差示扫描量热曲线
differential scanning calorimetry(DSC) 示差扫描量热法*

differential scanning calorimetry method　差热分析法
differential scanning photocalorimetry　光差量热扫描分析法
differential scattering cross section　微分散射截面*
differential screen analysis　示差筛析；特异筛析
differential sedimentation　差示沉降
differential sensitivity　差示灵敏度
differential signal　差示信号
differential spectra　示差光谱
differential spectrometry　示差光谱法
differential spectrophotometry　示差分光光度法*
differential spectroscopy　微分谱
differential spectrum　示差光谱*
differential sputtering　差动溅射
differential staining　对比着色
differential surface speed　线速比〔滚筒〕
differential taking-up mechanism　差微式卷取装置
differential temperature　示差温度
differential tensimeter　示差张力计
differential thermal analysis(DTA)　差热分析*
differential thermal analysis curve　差热分析曲线
differential thermal analyzer　差热分析仪
differential thermal curve　差热曲线
differential thermal gravimetric analysis　差热重量分析
differential thermocouple　差示热电偶
differential thermodilatometry　差示热膨胀法；示差热膨胀法
differential thermodynamic parameter　差热力学参数
differential thermogravimetric analysis　微分热重分析；导数热重分析
differential thermogravimetric curve　热重曲线
differential thermogravimetry　差热重量测定法；导数热重量测定法
differential thermometer　差示温度计
differential time　微分时间
differential titration　差示滴定
differential titration method　示差滴定法
differential transformer　差示变压器
differential type detector　微分型检测器*
differential valve　差动阀
differential vapor pressure thermometer　差示蒸气压温度计
differential velocity　差示速度；差动速度
differential velocity in horizontal plane　水平差示速度；水平差速
differential viscosity　差示黏度；特异黏度
differential voltammetry　微分伏安法
differential weight distribution　微分重量分布(函数)
differential winder　差速卷纸器
differential X-ray absorptiometry　示差X射线吸收法；X射线吸收线光谱法
differentiated rocks〔复〕　分化脉岩；分异岩
differentiating effect　分辨效应；区分效应
differentiating nonaqueous titration　差示非水滴定
differentiating solvent　区分(性)溶剂；辨别溶剂
differentiating titration　示差滴定
differentiation　①微分(法)②辨别；区别
differentiation of analytical signal　分析信号求导
differentiator　①微分器②微分电路③示差装置
different-size particles　不同粒径的粒子
diffracted intensity　衍射强度
diffracting mask　衍射幕；衍射掩模
diffraction　衍射
diffraction angle　衍射角
diffraction arc　衍射弧；衍射条纹
diffraction broadening　衍射增宽
diffraction contrast　衍射对比(度)；衍射反差
diffraction crystal　衍射晶体
diffraction effect　衍射效应
diffraction formula　衍射公式
diffraction grating　衍射光栅*
diffraction grating monochrometer　衍射光栅单色器
diffraction intensity　衍射强度
diffraction of X-rays　X射线衍射
diffraction order　衍射级
diffraction pattern　衍射花样
diffraction peak　衍射峰
diffraction phenomenon　衍射现象
diffraction spectrum　衍射光谱
diffraction spot　衍射斑
diffraction streak　衍射条纹
diffraction symbol　衍射群
diffraction symmetry　衍射对称性*
diffractive lens　衍射镜头
diffractometer　衍射仪
diffractometry　衍射学
diffusate　渗出液
diffuse　①渗出②扩散③漫射的
diffuse band absorption　漫射光带吸收
diffuse coefficient　扩散系数
diffuse correction　漫射修正
diffused　①漫射(的)②扩散的；弥漫的
diffused cloud　扩散云团
diffused light　漫射光
diffusedness　扩散性
diffuse double layer　扩散双层*

diffused reflection 漫散反射〔光学〕
diffused transmission 漫(散)透射〔光学〕
diffused zinc coating 扩散渗锌涂料(层)
diffuse flow 扩散流
diffuse flux 扩散通量
diffuse haze 漫射光雾
diffuse interface 扩散界面
diffuse layer 扩散层
diffuse peak 离散峰；扩散峰
diffuse-porous wood 散孔材
diffuser ①洗料器②漫射体③扩散器④浸提器
diffuse reflectance ①扩散反射率②漫反射
diffuse reflectance IR 漫反射红外光谱
diffuse reflectance spectroscopy 漫反射光谱法
diffuse reflection(=Fourier transform infrared spectroscopy) 漫反射红外光谱
diffuse reflection principle 漫反射原理
diffuse relaxation time 扩散弛豫时间
diffuser micropump 扩散器微泵
diffuser plate 扩散板；弥散板
diffuser priming of pump 扩散器启动泵
diffuser series 漫射系
diffuser tube 扩散管
diffuse scattering 漫散射
diffuse tint 渗流色〔糖〕
diffuse transmission density 漫透射密度
diffuse transmission factor 漫透射因数
diffuse transmittance 漫射透光率；漫射透射比
diffuse X-ray peak 弥散X射线峰
diffusibility 扩散本领
diffusible 可扩散的
diffusible calcium 可扩散钙
diffusible ion 可扩散离子
diffusibleness 扩散本领
diffusing (colour) coupler 扩散型成色剂
diffusing surface 漫射表面；扩散表面
diffusion ①扩散；弥漫②渗滤③漫射
diffusion agent 扩散剂
diffusional creep 扩散蠕动
diffusional flow 扩散流动
diffusional interchange 扩散交换
diffusional resistance 扩散阻力
diffusional solute flux 扩散溶质助熔剂
diffusion analysis 扩散分析
diffusion angle 扩散角
diffusion barrier 扩散膜[*]
diffusion battery ①浸提器组②扩散池组
diffusion blotting 扩散印迹

diffusion boundary layer 扩散边界层
diffusion capacity 扩散容量
diffusion cell ①(糖)渗滤池②扩散池
diffusion cell electrolytic hygrometer 扩散池电解湿度计
diffusion coating 扩散层
diffusion coefficient 扩散系数[*]
diffusion combustion 扩散燃烧
diffusion constant 扩散常数
diffusion control 扩散控制
diffusion controlled current 扩散控制电流
diffusion controlled reaction 扩散控制的反应[*]
diffusion controlled termination 扩散控制终止[*]
diffusion couple 扩散偶
diffusion current 扩散电流[*]
diffusion current compensation apparatus 扩散电流补偿装置
diffusion current constant 扩散电流常数[*]
diffusion double layer 扩散双电层
diffusion effect 扩散效应
diffusion electrode 扩散电极
diffusion equation 扩散方程
diffusion film coefficient 扩散膜系数
diffusion flame 扩散焰
diffusion flame breakdown 扩散火焰的破坏
diffusion flame system 扩散火焰系统
diffusion flow 扩散流
diffusion gradient 扩散梯度
diffusion heat 扩散热
diffusion hygrometer 扩散湿度计
diffusion index 扩散指数
diffusion juice 浸出汁〔糖〕
diffusion knife 切刀；切丝刀片〔糖〕
diffusion law 扩散(定)律
diffusion layer 扩散层
diffusion length 扩散长度
diffusion light 散射光
diffusion-limited current 极限扩散电流
diffusion limiting current 极限扩散电流
diffusion loss 扩散损失；扩压损失
diffusion model 扩散模型
diffusion of gases 气体扩散
diffusion of silver salt 银盐扩散(法)
diffusion overpotential 扩散超电势[*]
diffusion overvoltage 扩散过电压；浓差过电压
diffusion path 扩散程
diffusion photometer 漫射光度计
diffusion potential 扩散电位[*]
diffusion pressure 扩散压(力)

diffusion process	①扩散法②扩散过程
diffusion process for coating	扩散镀敷法；扩散渗金法
diffusion pump	扩散泵*
diffusion rate	扩散速率
diffusion ratio	扩散比(率)
diffusion-recombination	扩散与复合
diffusion resistance	耐(抗)扩散性；防扩散性
diffusion resistance coefficient	防扩散性系数
diffusion ring	扩散环
diffusion rubber	纯化橡胶
diffusion separation technique	扩散分离技术
diffusion vacuum pump	扩散(真空)泵
diffusion value	(光学)漫射值
diffusion vane	扩散(器)叶片
diffusion velocity	扩散速度
diffusion washing (filter cake)	扩散洗涤
diffusion water	浸提用水；浸用水〔糖〕
diffusiophoresis	扩散电泳
diffusive	①扩散的②浸出的〔糖〕
diffusive mixing	扩散混合
diffusive odor	逸发香气
diffusive wear	扩散磨损
diffusivity	扩散系数；扩散能力
difluorated	二氟化的
difluorene	联芴 $C_{26}H_{18}$
difluorenyl(=bifluorenyl)	联芴(基) $C_{26}H_{16}$
difluorenylene(=difluorylene)	联亚芴(基) $C_{26}H_{18}$
difluoride	二氟化物
difluorinated(=difluorizated)	二氟化的
difluorizated(=difluorinated)	二氟化的
difluoro	二氟
difluoro-benzene	二氟代苯 $C_6H_4F_2$
difluorocarbene	二氟卡宾 CF_2
difluoro compound	二氟化合物
difluorodimethylsilane	二甲基二氟甲硅烷
difluoro ester〔泛指〕	二氟酯
difluoro ether〔泛指〕	二氟醚
difluoromethane(=methylene fluoride)	二氟甲烷
difluorophosphoric acid	二氟磷酸
difluorophthalic anhydride	二氟邻苯二甲酸酐
difluorylene(=difluorenylene)	联亚芴(基) $C_{26}H_{18}$
diformin(=glycerin diformate)	甘油二甲酸酯；二甲精 $(HCO_2)_2C_3H_6O$
di(2-formyl phenyl) ethylene titanate	二(2-甲酰苯基)钛酸乙二(醇)酯
difumarate	①富马酸氢盐 $COOHCH=CHCOOM$ ②富马酸氢酯 $COOHCH=CHCOOR$
difunctional	双官能的
difunctional additive	双官能添加剂
difunctional caster oil	双官能团蓖麻油
difunctional comonomer	双官能共聚单体
difunctional compound	双官能化合物
difunctional epoxy compound	双官能环氧化合物
difunctional epoxy resin	双官能(度)环氧树脂
difunctional exchange resin	双官能团交换树脂
difunctional initiator	双官能团引发剂
difunctionality	双官能度；双官能性
difunctional monomer	双官能单体
difurfuroyl	①二糠酰②联糠酰
difurfuroyl disulfide	二硫化二糠酰
difurfuryl disulfide	二糠基二硫；二糠基二硫醚
difurfuryl ether	二糠基醚
difurfuryl sulfide	二糠基硫醚
N,N'-difurfuryl thiourea	N,N'-二糠基硫脲
difuroyl peroxide	过氧化二糠酰
digallic acid	双没食子酸；鞣酸 $C_6H_2(OH)_3COOC_6H_2(OH)_2COOH$
digalloil	双棓酸酯；鞣酸酯
digenic acid(=kainic acid)	海人草酸
digentisic acid	缩二龙胆酸；一缩双 2,5-二羟苯甲酸 $C_6H_3(OH)_2COOC_6H_3(OH)_2COOH$
digermane	乙锗烷
digested sludge	消化污泥
digester	①浸煮器；蒸煮器；蒸煮罐②消化器；熟化槽
digester devulcanizing method	水油脱硫法
digester liquor	蒸解液
digester process	蒸煮再生法
digester room	浸煮室；蒸煮室；蒸煮锅
digester silo	锅顶料仓
digestibility	可消化性
digestibility coefficient	消化系数
digestible	易消化的
digesting	①消化的②煮解的
digesting shelf	蒸煮架
digestion	①消化；消解②老化③浸提④蒸煮
digestion process	蒸解(再生)法〔碱法〕
digestion tank	陈化器
digestion time	陈化时间
digestive enzyme	消化酶
digestive ferment	消化酶
digestive gas	消化气体
digestor	①浸煮器②老化器③消解容器
digging machine	掘凿机
digicitrine	毛地黄黄酮
digifolein	毛地黄叶英
digilanid	毛地黄苷

diginin 毛地黄宁
diginose 2-脱氧毛地黄糖
digipoten 毛地黄同效苷
digitable 数据表格
digital-analog conversion 数-模转换
digital-analog converter 数-模转换器
digital answer 数字答案
digital clock 数字钟
digital coding 数字编码
digital control system 数字控制系统
digital converter 数字转换器
digital counter 数字计数器
digital display 数字显示
digital feature 数值特征
digital feature of random variable 随机变量数值特征
digital fibrograph 计数式纤维长度测定仪
digital filter 数字滤波
digital hologram ①数字全息图②数字全息照片
digitalin 毛地黄苷
digital indicator 数字指示器
digitaline nativelle 原汁毛地黄他灵〔药;地芰毒粗制剂〕
digital ion activity meter 数字式离子活度计
digital ion meter 数字式离子计
digitalis 毛地黄
digitalis glycoside 毛地黄苷
digital microwave hygrometer 数字式微波湿度计
digital monitoring system 数字监测系统
digital multimeter 数字式万用表;数字式多量程测量仪表
digitalose 毛地黄糖
digital phase shifter 数字式相移器
digital pH meter 数字式pH计
digital printer 数字打印器
digital programmer 数字程序器
digital quadrature detection ①数字正交检波②数字正交检测
digital readout 数字读数
digital response 数字响应(值)
digital set programmer ①数字程序装置;数字程序设计器②数字程序设计员
digital shaft 数字轴
digital signal 数字信号
digital signal processor 数字信号处理器
digital simulation 数字模拟
digital slope detector 数字斜率检测器
digital subtraction 数字减法
digital temperature control 数字温度控制器
digital temperature indicator 数字温度指示器

digital temperature programmer 数字程序升温器
digital thermohydrometer 数字温差比重计
digital thermohygrometer 数字式温湿度计
digital to analog converter 数模转换器
digital-tone 数字色调
digital voltmeter 数字电压表
digital weigh-feeder 数字式称量加料器
digitanol 毛地黄烷醇
digitiser 数字转换器
digitization of mass spectra 质谱的数字转换
digitogenin 毛地黄皂苷配基
digitol 毛地黄酊
digitonide 毛地黄皂苷化物
digitonin 毛地黄皂苷
digitoxigenin 毛地黄毒苷配基;β-(丁烯酸内酯)-14-羟(基)甾醇
digitoxin 毛地黄毒苷
digitoxose 毛地黄毒素糖
digitoxoside 毛地黄毒糖化物
diglutarate ①戊二酸氢盐 $COOH(CH_2)_3COOM$ ②戊二酸氢酯 $COOH(CH_2)_3COOR$
diglyceride 二脂酰甘油酯;甘油二酯
diglycerin 双甘油
diglycerol 双甘油;一缩二甘油 $(CH_2OHCHOHCH_2)_2O$
diglycerol polyglycidylether 一缩二甘油多缩水甘油醚
diglycidyl ether 二环氧甘油醚
diglycidyl ether of bisphenol A 双酚A二环氧甘油醚
diglycidyl terephthalate 对苯二甲酸二环氧丙酯
diglycol 二甘醇;一缩二乙二醇 $HOCH_2CH_2OCH_2CH_2OH$
diglycol aldehyde 二甘醇醛 $OHCCH_2OCH_2CHO$
diglycolamidic acid 亚氨基二乙酸 $NH(CH_2COOH)_2$
diglycolamine 二甘醇胺
diglycolic acid 二甘醇酸;一缩二乙二醇酸
diglycolide 乙交酯
diglycol laurate 月桂酸二甘醇酯;月桂酸一缩二乙二醇酯
diglycollic acid 二甘醇酸
diglycol oleate 油酸二甘醇酯
diglycol phthalate 邻苯二甲酸二乙二醇酯;酞酸二乙二醇酯〔俗〕
diglycol ricinoleate 蓖麻酸二甘醇酯〔增塑剂〕
diglycol stearate 硬脂酸二甘醇酯
diglycol terephthalate 对苯二甲酸(二)乙二(醇)酯
diglycosyl diglyceride 二糖基二甘油酯
N,N'-diglycylethylene diamine(DGEN) N,N'-二氨基乙酰乙二胺
diglyme(=diethylene glycol dimethyl ether) 二甘醇二甲

醚 O(CH$_2$CH$_2$OCH$_3$)$_2$
digol(=diethylene glycol) 二甘醇
digoxigenin 地谷新配基;洋地黄毒苷
digoxin 地高辛〔药〕;异羧基洋地黄毒苷原
diguanide 缩二胍
dihalide 二卤化物
dihalo 二卤的
dihaloalkane 二卤(代)烷
dihalogen acid 二卤代酸
dihalogenated 二卤代的
dihalogenated acetone 二卤代丙酮 CHX$_2$COCH$_3$; CH$_2$XCOCH$_2$X
dihalogenated benzene 二卤代苯 C$_6$H$_4$X$_2$
dihalogenated ester 二卤代酯
dihalogenated ether 二卤代醚
dihalogeno-benzene 二卤代苯 C$_6$H$_4$X$_2$
dihalogenophenyl borine 二卤化苯基硼;苯基二卤化硼
dihalomethylenation 二卤亚甲基化(作用)
dihedral angle 双面夹角
dihedral reflector 二面反射镜
N,N'-diheptyl-p-phenylenediamine N,N'-二庚基对苯二胺〔防老剂〕
diheptyl phthalate 邻苯二甲酸二庚酯〔增塑剂〕
dihexyl ①二己基②十二烷
dihexyl adipate 己二酸二己酯〔增塑剂〕
dihexylamine 二己胺〔硫化剂〕
dihexyl azelate 壬二酸二己酯〔增塑剂〕
dihexyl-N,N-dibutyl carbamylmethylene phosphonate N,N-二丁基氨基甲酰亚甲基膦酸二己酯
di-n-hexyl phosphoric acid(HDHXP) 二正己基磷酸
dihexyl phthalate 邻苯二甲酸二己酯〔增塑剂〕
dihexyl sebacate 癸二酸二己酯〔增塑剂〕
dihexylsulfoxide(DHXSO) 二己基亚砜
dihydracrylamic acid 双乳胺酸;双乳酰胺 CH$_3$CHOHCOOCH(CH$_3$)CONH$_2$
dihydracrylic acid 双乳酸〔俗〕;乳酰乳酸 CH$_3$CHOHCOOCH(CH$_3$)COOH
dihydranol 庚基间苯二酚
dihydrate 二水(合)物
dihydric 二羟基的
dihydric acid 二价酸
dihydric alcohol 二羟醇;二元醇
dihydric arsenate 砷酸二氢盐 MH$_2$AsO$_4$
dihydric phenol 二羟酚;二元酚
dihydric phosphate 磷酸二氢盐 MH$_2$PO$_4$
dihydric salt 二氢盐〔分子中仍含有两个氢离子〕
dihydride 二氢化物
dihydrite 假孔雀石;斜翠绿磷铜矿

dihydro- 二氢(化)
dihydroabietic acid 二氢枞酸
dihydroabietyl alcohol 二氢枞醇 C$_{19}$H$_{31}$CH$_2$OH
dihydroactinidiolide 二氢猕猴桃(醇酸)内酯 C$_{11}$H$_{16}$O$_2$
dihydroalantolactone 二氢土木香内酯
dihydroambrate 二氢龙涎香酯
dihydro ambrettolide(=cyclohexadecanolide) 二氢黄葵内酯;环十六内酯
dihydroandrosterone(=androstanediol) 雄(甾)烷二醇;二氢雄(甾)酮
dihydroanethole 二氢茴香脑
dihydroanhydrovitamin A 二氢脱水维生素 A
dihydroartemisinin 双氢青蒿素〔药〕
dihydroascorbate 二氢抗坏血酸
dihydrobenzaldehyde 二氢苯甲醛
dihydrobenzene 二氢苯 C$_6$H$_8$
dihydroberberine 二氢小檗碱
dihydrobilirubin 二氢胆红素
dihydrobromide 二氢溴化物 R(HBr)$_2$
dihydrocalciferol 二氢钙化甾醇
dihydrocarveol 二氢香芹醇;二氢葛缕醇
dihydrocarvone 二氢香芹酮;二氢葛缕酮
dihydrocarvyl acetate 乙酸二氢香芹酯
dihydrochalcone 二氢查耳酮;芳基丙酰芳烃
dihydrochloride 二氢氯化物;二盐酸化物 R(HCl)$_2$
dihydrocholesterin 二氢胆甾醇
dihydrocholesterol 二氢胆甾醇
dihydrocinnamic aldehyde 二氢(肉)桂醛;苯丙醛
dihydrocitronellol(=tetrahydrogeraniol) 二氢香茅醇;四氢香叶醇
dihydrocitronellyl acetate 乙酸二氢香茅酯
dihydrocivetol 二氢灵猫醇;环十七(烷)醇
dihydrocivetone(=cycloheptadecanone) 二氢灵猫酮;环十七(烷)酮
dihydrocodeine 双氢可待因〔药〕
dihydro-compound 二氢化合物
dihydrocostunolide 二氢广木香内酯
dihydrocostuslactone 二氢广木香内酯
dihydrocoumarin 二氢香豆素
dihydrocyclocitral 二氢环柠檬醛
dihydrocyclolavandulol 二氢环薰衣草醇
dihydrodaunomycin 二氢道诺霉素
dihydrodeguelin 二氢鱼藤素
dihydrodiethylstilbestrol 己(烷)雌酚
dihydroequilenin 二氢马萘雌(甾)酮
dihydroeremophilol 二氢雅槛兰醇
2,2-dihydroergosterol 2,2-二氢麦角甾醇
dihydroergotamine 双氢麦角胺〔药〕

dihydroergotoxine 二氢麦角碱〔药〕
dihydroerysodine 二氢刺桐定
dihydroeudesmol 二氢桉叶醇
dihydroeugenol 二氢丁子香酚
dihydrofarnesol 二氢金合欢醇
dihydroflavonol(=flavanonol) 黄烷酮醇
dihydrofluoride 二氢氟化物 R(HF)$_2$
dihydrofolic acid 二氢叶酸
dihydrogen ammonium phosphate 磷酸二氢铵
dihydrogen phosphate 磷酸二氢盐
dihydrogladiolic acid 二氢唐菖蒲青霉酸
dihydroiodide 二氢碘化物 R(HI)$_2$
dihydroionone 二氢紫罗兰酮
dihydroisopimaric acid 二氢异海松酸
dihydrojasmone 二氢茉莉酮
dihydrojasmone lactone 二氢茉莉酮内酯
dihydroketoacridine 吖啶酮 C$_6$H$_4$COC$_6$H$_4$NH

dihydrol 二聚水〔即(H$_2$O)$_2$〕
dihydrolichesterinic acid 二氢苔甾酸
dihydrolinalool 二氢芳樟醇
dihydromyrcene 二氢月桂烯〔链转移剂〕
dihydromyrcenol(=2,6-dimethyl-7-octen-2-ol) 二氢月桂烯醇；2,6-二甲基-7-辛烯-2-醇
dihydronancimycin 二氢南锡霉素
1,4-dihydronaphthalene 1,4-二氢萘〔调节剂〕
dihydronitidine 二氢光花椒碱
dihydronootkatone 二氢诺卡酮；二氢圆柚酮
dihydronopaldehyde 二氢诺卜醛
dihydronorsclareol oxide 二氢降香紫苏醚
dihydropalustric acid 二氢长叶松脂；二氢右旋松香酸
dihydro perfluoro alkanoic acid ester 二氢全氟烷酸酯
dihydrophytol(=phytanol) 植烷醇 C$_{20}$H$_{41}$OH
dihydropimaric acid 二氢海松酸
dihydropyrane 二氢吡喃
dihydroquercetin 二氢栎精
dihydroriboflavin 二氢核黄素
dihydrorobinetin 双氢刺槐亭
dihydrorotenone 二氢化鱼藤酮
dihydrorotundifolone 二氢圆叶薄荷酮
dihydrosamidin 二氢萨米定
dihydrosclareol 二氢香紫苏醇
2,2-dihydrostigmasterol 2,2-二氢豆甾醇
dihydrostreptomycin 双氢链霉素
dihydrotachysterol 二氢速甾醇
dihydrotagetone(=2,6-dimethyl-7-octen-4-one) 二氢万寿菊酮；2,6-二甲基-7-辛烯-4-酮
dihydro-α-terpineol 二氢-α-松油醇

dihydrotheelin(=estradiol) 二氢雌酮；雌二醇
1,2-dihydro-2,2,4-trimethyl-6-ethoxyquinoline 1,2-二氢-2,2,4-三甲基-6-乙氧基喹啉〔防老剂〕
1,2-dihydro-2,2,4-trimethyl quinoline 1,2-二氢-2,2,4-三甲基喹啉〔防老剂〕
dihydrotrypacidin 双氢杀锥曲菌素
dihydro-β-vetivol 二氢-β-岩兰醇
dihydro-β-vetivone 二氢-β-岩兰酮
dihydrowogonin 二氢汉黄芩素；双氢次黄芩素
dihydroximic acid 联(甲)羟肟酸
dihydroxy 二羟(基)
dihydroxy acetic acid 二羟基乙酸 CH(OH)$_2$COOH
dihydroxy acetone 二羟基丙酮 CH$_2$OHCOCH$_2$OH
dihydroxyacetone phosphoric acid 磷酸二羟丙酮
2,4-dihydroxy acetophenone(=resacetophenone) 雷琐苯乙酮〔俗〕；2,4-二羟基苯乙酮 (HO)$_2$C$_6$H$_3$COCH$_3$
dihydroxy acid 二羟酸
dihydroxy alcohol 二羟醇；二元醇
dihydroxy anthracene 二羟蒽 C$_{14}$H$_8$(OH)$_2$
1,2-dihydroxyanthraquinone(=alizarine) 1,2-二羟基蒽醌；茜素
dihydroxyarachidic acid 二羟花生酸；二羟(基)二十(烷)酸
dihydroxy-arseno-benzene 二羟偶砷苯 OHC$_6$H$_4$As·AsC$_6$H$_4$OH
dihydroxy-arsino-benzophenone 二羟胂苯基苯基(甲)酮；苯甲酰苯亚胂酸 C$_6$H$_5$CO·C$_6$H$_4$As(OH)$_2$
dihydroxy-azo-benzene 二羟代偶氮苯 OHC$_6$H$_4$N$_2$C$_6$H$_4$OH
o,p-dihydroxy-azo-p-nitrobenzene(=magneson) 试镁灵
dihydroxy-benzene 二羟基苯；苯二酚 C$_6$H$_4$(OH)$_2$
dihydroxy benzhydrol 二羟基二苯甲醇
dihydroxy benzophenone 二羟基二苯(甲)酮
2,5-dihydroxybenzoquinone 2,5-二羟基苯醌〔防老剂〕
dihydroxy-benzyl alcohol 二羟苄醇 (OH)$_2$C$_6$H$_3$CH$_2$OH
dihydroxy-bis (lactato) titanate 二羟基二乳酸钛；二羟基钛二乳酸酯
3,7-dihydroxy cholanic acid 3,7-二羟胆酸、鹅(脱氧)胆酸
1,25-dihydroxy cholecalciferol 1,25-二羟胆钙化(甾)醇；1,25-二羟维生素 D$_3$
dihydroxy compound 二羟基化合物
6,7-dihydroxycoumarin(=esculetin) 七叶亭；6,7-二羟香豆素 C$_9$H$_6$O$_4$
dihydroxy cyclobutenedione(=squaric acid) 方形酸；二羟基环丁烯二酮
dihydroxy dibasic acid 二羟二酸 HOOCR(OH)·R(OH)COOH
dihydroxy dibenzanthracene 二羟二苯并蒽

1,2-dihydroxy-1,2-dihydroanthracene 1,2-二羟基-1,2-二氢蒽
2,2′-dihydroxy-4,4′-dimethoxybenzophenone 2,2′-二羟基-4,4′-二甲氧基二苯甲酮〔紫外线吸收剂〕
2,2′-dihydroxy-4,4′-dimethoxy-5-sodium sulfonate benzophenone 2,2′-二羟基-4,4′-二甲氧基-5-磺酸钠二苯甲酮〔紫外线吸收剂〕
4,4′-dihydroxydiphenyl 4,4′-二羟基联苯〔防老剂〕
4,4′-dihydroxydiphenylcyclohexane 4,4′-二羟基二苯基环己烷〔防老剂〕
dihydroxydiphenyl disulfide 二羟基二苯基二硫化物〔引发剂〕
dihydroxydiphenyl ether 二羟基二苯基醚
dihydroxy diphenylsulphone 二羟基二苯砜
di(hydroxyethyl) resorcinol 二(羟乙基)间苯二酚〔硫化剂〕
di-2-hydroxyethyl terephthalate 对苯二甲酸(二)乙二(醇)酯
5,7-dihydroxyflavone 5,7-二羟黄酮
dihydroxyfluoboric acid 二羟(基)氟硼酸
dihydroxyisobutylamine 二羟基异丁胺
dihydroxyl 二羟基
dihydroxylated acid 二羟(基)酸
dihydroxyl derivative 二羟衍生物
dihydroxy-maleic acid 二羟马来酸
　　COOHCOH＝COHCOOH
3,4-dihydroxy mandelic acid 3,4-二羟(基)苦杏仁酸
dihydroxy menthofuran 二羟基薄荷呋喃
2,6-dihydroxymethyl-4-chlorophenol resin 2,6-二羟甲基-4-氯代苯酚树脂〔硫化剂〕
dihydroxy monobasic acid 二羟一(碱)价酸
　　R(OH)R(OH)COOH
dihydroxy naphthalene 二羟萘 $C_{10}H_6(OH)_2$
1,8-dihydroxynaphthalene-2,4-disulfonic acid(=2S-acid) 2S 酸;1,8-二羟基萘-2,4-二磺酸
5,8-dihydroxynaphthoquinone 5,8-二羟基萘醌
2,2′-dihydroxy-4-octyloxy benzophenone 2,2′-二羟基-4-辛氧基二苯甲酮〔光稳定剂〕
1,6-dihydroxyphenazine 1,6-二羟基吩嗪
dihydroxyphenyl acetic acid 二羟苯乙酸
3,4-dihydroxy phenylalanine(DOPA) 3,4-二羟苯丙氨酸;多巴
3,4-dihydroxy-β-phenylethylamine(=hydroxytyramine; dopamine) 3,4-二羟基-β-苯乙胺
3,4-dihydroxyphenyl glycol 3,4-二羟(基)苯乙二醇
dihydroxyphenyl propane 二羟苯基丙烷;双酚A〔防老剂〕
2,5-dihydroxyphenylpyruvic acid 2,5-二羟(基)苯丙酮酸
α,ω-dihydroxy-poly(methyl methacrylate) α,ω-二羟基聚甲基丙烯酸甲酯

dihydroxy propane sulfonic acid 二羟基丙烷磺酸
dihydroxypropionic acid(=glyceric acid) 甘油酸;2,3-二羟基丙酸 $HOCH_2CHOHCOOH$
dihydroxypropyl methacrylate 甲基丙烯酸二羟丙酯
dihydroxy stearic acid 二羟硬脂酸 $C_{17}H_{33}(OH)_2COOH$
2,3-dihydroxy succinic acid(=tartaric acid) 2,3-二羟丁二酸;酒石酸 $COOH(CHOH)_2COOH$
dihydroxytartaric acid 二羟(基)酒石酸;四羟丁酸
dihydroxy tetrabasic acid 二羟基四(碱)价酸
　　$(COOH)_2R(OH)·R(OH)(COOH)_2$
dihydroxy tetrachloro diphenyl thioether 二羟四氯二苯基硫醚
dihydroxy tribasic acid 二羟基三(碱)价酸
　　$COOHR(OH)·R(OH)(COOH)_2$
dihydroxy trihydroxybenzoyloxy benzoic acid 二羟基三羟基苯甲酰氧基苯甲酸
1,25-dihydroxyvitamin D_3 1,25-二羟维生素 D_3
diimidazole-anthraquinone polymer fibre 二咪唑-蒽醌聚合物纤维
diimide ①二酰亚胺 ②偶氮 HN＝NH
di-imido dicarboxylic acid 二亚氨基二羧酸
diimine 二亚胺
diimino 二亚氨基
diimino succinonitrile 二亚氨基丁二腈
3,5-diiminotriazole(=guanazole) 3,5-二亚氨基(二氢化)三唑;胍唑
diindogen(=indigotin) 靛蓝
diine 二炔
diiod(=diiodo) 二碘(代)
diiodated 二碘化了的
diiodide 二碘化物
diiodinated(=diiodizated) 二碘化了的
diiodizated(=diiodinated) 二碘化了的
diiodo(=diiod) 二碘(代)
diiodo-acetic acid 二碘乙酸 CHI_2COOH
diiodo-benzene 二碘苯 $C_6H_4I_2$
diiodo-compound 二碘化合物
diiodo-ester 二碘代酯
diiodo-ether 二碘代醚
diiodo-ethyl ether 二碘乙醚;不对称二碘代二乙醚 $C_2H_3I_2OC_2H_5$
di-iodofluorescein 二碘荧光素
diiodoform 四碘乙烯
diiodohydroxyquinoline 双碘喹啉〔药〕
diiodomethane(=methylene iodide) 二碘甲烷
diiodo-methyl-arsonic acid 二碘甲基胂酸 $CHI_2AsO(OH)_2$
3,5-diiodothyronine 3,5-二碘甲腺氨酸

diisoamyl ①二异戊基②联异戊基；2,7-二甲基辛烷
diisoamyl ester 二异戊酯 R(COOC$_5$H$_{11}$)$_2$
diisoamyl ether 二异戊醚 R(OC$_5$H$_{11}$)$_2$;C$_5$H$_{11}$OC$_5$H$_{11}$
diisoamyl phosphate(DIAP) 磷酸二异戊酯
diisobutyl ①二异丁基②联异丁基；2,5-二甲基己烷
diisobutyl adipate 己二酸二异丁酯〔增塑剂〕
di-isobutyl aluminum chloride 氯化二异丁基铝
diisobutyl azelate 壬二酸二异丁酯〔增塑剂〕
diisobutyl carbinol(DIBC) 二异丁基甲醇
diisobutyl carbinyl acetate 乙酸二异丁基原酯
diisobutyl carbinyl pelargonate 壬酸二异丁基原酯
diisobutylene 二异丁烯〔表面活性剂〕
diisobutyl ketone 二异丁基甲酮
diisobutyl maleate 马来酸二异丁酯〔增塑剂〕
N,N'-diisobutyl-p-phenylenediamine N,N'-二异丁基对苯二胺〔防老剂〕
diisobutyl phthalate 邻苯二甲酸二异丁酯
diisobutyl sodium sulfosuccinate 二异丁基磺基丁二酸钠；琥珀酸二异丁酯磺酸钠
diisobutyrin 二异丁精；甘油二异丁酸酯
diisobutyryl peroxide 过氧化二异丁酰
diisocyanate ①二异氰酸酯②二异氰酸盐
diisocyanate rubber 二异氰酸橡胶；聚氨酯橡胶
2,4-diisocyanatoluene(2,4-TDI) 2,4-甲苯二异氰酸酯
di(p-isocyanatophenyl) methane 二(对异氰酸苯基)甲烷〔硫化剂〕
diisodecyl adipate 己二酸二异癸酯
diisodecyl glutarate 戊二酸二异癸酯〔增塑剂〕
diisodecyl phthalate 邻苯二甲酸二异癸酯；酞酸二异癸酯
diisoeugenol 二异丁子香酚
diisoheptyl phthalate 邻苯二甲酸二异庚酯〔增塑剂〕
diisononyl adipate 己二酸二异壬酯〔增塑剂〕
diisononyl phthalate 邻苯二甲酸二异壬酯〔增塑剂〕
diisooctyl adipate 己二酸二异辛酯〔增塑剂〕
diisooctyl azelate 壬二酸二异辛酯〔增塑剂〕
diisooctyl decyl phthalate 邻苯二甲酸二异辛基癸酯〔增塑剂〕
4,4'-diisooctyl diphenylamine 4,4'-二异辛基二苯胺〔防老剂〕
diisooctyl dodecanedioate 十二双酸二异辛酯〔增塑剂〕
diisooctyl fumarate 富马酸二异辛酯〔增塑剂〕
diisooctyl isophthalate 间苯二甲酸二异辛酯〔增塑剂〕
diisooctyl maleate 马来酸二异辛酯〔增塑剂〕
diisooctyl monoisodecyl trimellitate 偏苯三酸二异辛基一异癸酯〔增塑剂〕
di(isooctyl phenol) disulfide 二硫化二(异辛基苯酚)〔硫化剂〕
N,N'-diisooctyl-p-phenylenediamine N,N'-二异辛基对苯二胺〔防老剂〕
diisooctyl phthalate 邻苯二甲酸二异辛酯〔增塑剂〕
diisooctyl sebacate 癸二酸二异辛酯
diisopentyl phthalate 邻苯二甲酸二异戊酯
diisopropanolamine 二异丙醇胺
diisopropenyl ①二异丙烯基②2,3-二甲基丁二烯
diisopropyl ①二异丙基②2,3-二甲基丁烷
diisopropyl adipate 己二酸二异丙酯
diisopropylamine 二异丙胺〔促进剂〕
diisopropylamine diisopropyldithiocarbamate 二异丙基二硫代氨基二异丙胺〔促进剂〕
diisopropylaminoethanol 二异丙基氨基乙醇
diisopropyl azodiformate 偶氮二甲酸二异丙酯〔发泡剂〕
diisopropylbenzene(DIPB) 二异丙苯
di-isopropylbenzene hydroperoxide 过氧化氢二异丙苯〔ABS、合成橡胶等聚合引发剂〕
diisopropylbenzensulfonate 二异丙基苯磺酸盐
N,N'-diisopropyl-2-benzothiazolesulfenamide N,N'-二异丙基-2-苯并噻唑次磺酰胺〔促进剂〕
diisopropylcarbinol 二异丙基甲醇
di-isopropylene glycol salicylate 二异丙二醇水杨酸酯〔紫外线吸收剂〕
diisopropyl ester 二异丙酯 R[COOCH(CH$_3$)$_2$]$_2$
diisopropyl ether 二异丙醚 R[OCH(CH$_3$)$_2$]; (CH$_3$)$_2$CHOCH(CH$_3$)$_2$
diisopropyl ethylamine 二异丙基乙胺
diisopropyl fluorophosphate(DFP) 二异丙基氟磷酸(或酯)
diisopropylidene acetone(=phorone) 二异亚丙基丙酮；佛尔酮〔俗〕
diisopropyl ketone(=isobutyrone) 二异丙基(甲)酮 [(CH$_3$)$_2$CH]$_2$CO
diisopropyl methylphosphonate(DIMP) 甲基膦酸二异丙酯
diisopropyl naphthalene 二异丙基萘
diisopropyl naphthalene sulfonate 二异丙基萘磺酸盐
diisopropyl peroxydicarbamate 过氧基二氨基甲酸二异丙酯〔硫化剂〕
di-isopropyl peroxydicarbonate 过氧化二碳酸二异丙酯〔催化剂〕
3,5-diisopropyl phenol 3,5-二异丙基苯酚〔硫化剂〕
3,5-diisopropyl phenolformaldehyde 二异丙基苯酚甲醛〔硫化剂〕
2,4-diisopropyl phenylacetaldehyde 2,4-二异丙基苯乙醛
diisopropyl pyrophosphoric acid(DIPPP) 二异丙基焦磷酸 O[P(O)(OC$_3$H$_7$)(OH)]$_2$
diisopropyl thiophosphonyl disulfide 二异丙基硫代膦酰二硫〔促进剂〕
N,N'-diisopropyl thiourea N,N'-二异丙基硫脲〔促进剂〕

diisopropylthiuram disulfide 二硫化二异丙基秋兰姆〔促进剂〕
diisopropyl xanthate disulfide 二硫化黄原酸二异丙酯〔促进剂〕
diisopropyl xanthogenate disulfide 二硫化黄原酸二异丙酯
diisosafrole 二异黄樟(油)素
diisostearoyl ethylene titanate 二异硬脂酰基钛酸亚乙酯〔偶联剂〕
diisotactic 双全同立构的
diisotactic polymer 双全同立构聚合物
diisotridecyl 邻苯二酯二异十三酯
diisotridecyl phthalate 邻苯二甲酸二异十三酯〔增塑剂〕
dika fat 地咖脂；加蓬依苦木脂
dika oil 地咖油；加蓬依苦木油〔取自 *Irvingia gabonensis* 的果实〕
diketen 双烯酮
diketo 二酮(基)
diketoalcohol 二酮醇
diketocoriolin 二酮革盖菌素
2,3-diketogulonate 2,3-二酮古洛糖酸(或酯)
diketogulonic acid 三酮古洛糖酸
diketone 二酮
diketopiperazine 哌嗪二酮*
diketopiperazine ring 哌嗪二酮环
diketo pyrrolo-pyrrole pigment(DPP pigment) 二酮基吡咯并吡咯颜料〔红色耐高温有机颜料〕
diketotriazolidine(=urazole) 尿唑
dilactamic acid 双乳胺酸；双乳酰胺 $CH_3CHOHCOOCH(CH_3)CONH_2$
dilactic acid 双乳酸 $CH_3CHOHCOOCH(CH_3)COOH$
dilatability 膨胀性
dilatable (可以)膨胀的
dilatancy ①膨胀性②胀流性；触稠性
dilatancy index 膨胀指数
dilatant 胀流型体；触稠体
dilatant consistency 膨胀性稠度
dilatant dispersion 胀流型分散体
dilatant flow 胀流型流动
dilatant fluid 胀流型流体
dilatant liquid 胀流性流体
dilatant thickener 胀流增稠器
dilatant vehicle 膨胀性漆料
dilatation 膨胀
dilatational strain 膨胀应变
dilatational strain energy gradient 膨胀应变能(量)梯度
dilatational viscosity 膨胀黏度
dilatent 胀流型流体

dilation coefficient 膨胀系数
dilatometer 膨胀计*
dilatometric 膨胀测定的
dilatometric method 膨胀测定法
dilatometric thermometer 膨胀温度计
dilatometry 膨胀计测定法；热膨胀法
dilaurin 二月桂精；甘油二月桂酸酯
dilauroyl peroxide 过氧化二月桂酰
dilaurylamine(DLA) 二月桂胺 $(C_{12}H_{25})_2NH$
dilaurylformamide(DLE) 二月桂基甲酰胺 $HCON(C_{12}H_{25})_2$
dilauryl phosphate 磷酸二月桂酯
dilauryl phthalate 邻苯二甲酸二月桂酯；酞酸二月桂酯
dilauryl thiodipropionate 硫代二丙酸二月桂酯
N,N'-dilauryl thiourea N,N'-二月桂基硫脲〔促进剂〕
dilead trioxide 三氧化二铅
dilevulinic acid 联4-氧戊酸；癸二酮-4,7-二酸
dilinolein(=dinolin) 二亚油精；甘油二亚油酸酯
dilithium initiator 双锂引发剂
dillapiole 莳萝脑
dill oil 莳萝油
Dillon extrusion plastometer 狄朗挤出塑度仪
dill seed oil 莳萝子油
diltiazem 地尔硫〔药〕
diluent 稀释剂；冲淡剂
diluent naphtha 石脑油溶剂；石脑油稀释剂
diluent solvent 稀释溶剂
dilutability (可)稀释度
dilute acetic acid 稀乙酸
dilute acid 稀酸
dilute acid bath 稀酸槽
dilute hydrochloric acid 稀盐酸
dilute juice 稀汁
dilutent modifier 稀释剂改性剂
dilute nuclei 稀核
dilute phase 稀相*
dilute phase lifting 稀相提升〔催化剂的气提升输送法〕
dilute phase suspension 稀相悬浮体
dilute phase system 稀相系统
dilute phase zone 稀相区
dilute-Purex process 稀钚雷克斯流程
diluter 稀释器
dilute solution 稀溶液
dilute solution theory 稀溶液理论
dilute solution viscosity(DSV) 稀薄溶液黏度
dilute sulfuric acid 稀硫酸
diluting agent 稀释剂
diluting ratio 稀释比

dilution ①稀释；冲淡②稀释度；冲淡度
dilution additive 稀释剂
dilution colorimeter 稀释比色计
dilution correction 稀释(误差)校正
dilution discharge 稀释排放法
dilution effect 稀释效应
dilution error 稀释误差
dilution factor 稀释因子
dilution function ①稀释函数②冲淡函数
dilution heat 稀释热；冲淡热
dilution law 稀释定律
dilution limit 稀释极限
dilution metering 稀释测定(法)
dilution method 稀释法
dilution plate method 稀释平板分离法
dilution ratio 稀释率；稀释比(例)；稀释比值
dilution stability 稀释稳定性
dilution table 稀释表
dilution test 稀释试验
dilution test of fuel 燃料稀释试验
dilution trap 稀释槽
dilution value 稀释值
dilution viscometer 稀释黏度计
dimalate ①苹果酸氢盐 $COOHCHOHCH_2COOM$ ②苹果酸氢酯 $COOHCHOHCH_2COOR$
dimaleate ①马来酸氢盐 $COOHCH=CHCOOM$ ②马来酸氢酯 $COOHCH=CHCOOR$
dimalonate ①丙二酸氢盐 $COOHCH_2COOM$ ②丙二酸氢酯 $COOHCH_2COOR$
dimalonic acid 亚乙基四酸；双丙二酸 $(COOH)_2CHCH(COOH)_2$
dimazon 二乙酰氨基偶氮甲苯 $C_{18}H_{19}N_3O_2$
dimedone(=dimethyl cyclohexanedione;dimethone) 双甲酮（俗）
dimedone dioxime 双甲酮双肟
dimedone test 双甲酮试验
dimefline 二甲弗林〔药〕
dimelissin 二蜂酸精；甘油二蜂酸酯
dimenhydrinate 茶苯海明〔药〕
dimension ①维；度②大小；尺寸③量纲；因次
dimensional 因次的；量纲的
dimensional accuracy 尺寸精度
dimensional analysis 量纲分析；因次分析
dimensional change 尺寸变化
dimensional deformation 外部变形
dimensional formula 因次式
dimensional homogeneity 量纲齐次性
dimensional instability 易变形性
dimensionality ①维数②因次性
dimensionally stable anode(DSA) 尺寸稳定阳极；形稳阳极
dimensional sensitivity to moisture 湿敏尺寸；尺寸对湿敏感性
dimensional stability 尺寸稳定性
dimensional strength 耐变形力
dimensional tolerance 尺寸公差
dimensional uniformity 尺寸一致性
dimensional variance 尺寸变化
dimension and tolerance 尺寸与公差
dimensioned 有因次的
dimensioning 尺寸示法；定尺度；量尺寸
dimensionless 无因次的；无量纲(的)
dimensionless coordinate 无因次坐标
dimensionless glass transition 无因次玻璃化转变
dimensionless group 无因次群
dimensionless local memory time 无因次局部记忆时间
dimensionless number 无因次数
dimensionless parameter 无因次参数
dimensionless ratio 无因次比(率)
dimensionless scale 无因次系数；无因次比尺
dimensionless specific speed 无因次比转速
dimensionless term 无维量项；无因次项
dimensionless variable 无因次变数；无因次变量
dimension limit 极限尺寸
dimension-lumber(=timber stock) 规格材
dimension of polymer chain 高分子链的大小
dimension of quantity 量纲
dimension of radical 基的大小
dimension of surface tension force 表面张力因次
dimension stability 尺寸安定性
dimentionless quantity 无因次量；无量纲量
dimer 二聚体*
dimercaprol(BAL) 二巯基丙醇
dimercaptoacetate 二巯基乙酸
dimercaptoethyl maleate 顺丁烯二酸二巯基乙酯
dimercaptoethyl phthalate 邻苯二甲酸二巯基乙酯
dimercaptopropanol(BAL) 二巯基丙醇
dimercaptosuccinate 二巯基丁二酸
dimercaptosuccinic acid(DMSA) 二巯基丁二酸；二巯基琥珀酸
2,5-dimercapto-1,3,4-thiadiazole disulfide 二硫化-2,5-二硫醇基-1,3,4-噻二唑〔促进剂〕
dimercaptothiodiazole 二巯基噻二唑
dimercurammonium 汞氮基 $HgN-$
dimercurousammonium 氨汞基 NH_2Hg-
dimeric 二聚的

dimer(ic) acid　二聚酸
dimeric compounds　二聚化合物
dimeric dibasic acid　二聚二元酸
dimeric form　二聚型
dimeric polymer　二聚物；二分子聚合物
dimeric 2,4-toluene diisocyanate　2,4-二异氰酸甲苯酯的二聚体〔防老剂〕
dimerisation(=dimerization)　二聚
dimerising dye　二聚染料
dimerization(=dimerisation)　二聚
dimerization reaction　二聚反应
dimerized linoleic acid　二聚亚油酸
dimerized tolylene-2,4-disocyanate　二聚-2,4-甲苯二异氰酸酯
dimer theory　二聚体学说
Di-Me solvent dewaxing　二氯乙烷-二氯甲烷溶剂脱蜡(法)
dimeso-periodic acid　二中高碘酸　$H_4I_2O_9$
dimetalation　二金属取代(作用)
dimethacetic acid　二甲基乙酸　$CH(CH_3)_2COOH$
dimethacryl ethylene titanate　二甲基丙烯酰钛酸乙二(醇)酯
dimethano　二甲桥
dimethicone　二甲硅油〔药〕
dimethicone copolyol　聚二甲基硅氧烷共聚醇
dimethicone copolyol stearate　二甲基硅氧烷共聚醇硬脂酸酯
dimethiconol　聚二甲基硅氧烷醇
dimethone(=dimedone)　双甲酮〔俗〕
dimethoxy　二甲氧(基)
dimethoxyanilino-diisocyanate　二甲氧基苯胺二异氰酸酯〔硫化剂〕
p-dimethoxybenzene　对二甲氧基苯〔防老剂〕
dimethoxy benzene　二甲氧基苯
2,5-dimethoxy benzoquinone　2,5-二甲氧基苯醌
3,4-dimethoxy benzyl alcohol　3,4-二甲氧基苄醇
dimethoxybiphenyl　二甲氧基联(二)苯
3,3′-dimethoxy-4,4′-biphenyl diisocyanate　3,3′-二甲氧基-4,4′-联苯(撑)二异氰酸酯
di(3-methoxybutyl) peroxy dicarbonate　过(氧)二碳酸双(3-甲氧基丁基)酯
4,6-dimethoxy coumarin　4,6-二甲氧基香豆素
di-p-methoxy-diphenylamine　二对甲氧基二苯胺
dimethoxy-ethane　二甲氧基乙烷
dimethoxyethoxy ethyl azelate　壬二酸二甲氧乙氧基乙酯
α,α-di(methoxyethoxy)-α-phenylacetophenone　α,α-二(甲氧乙氧基)-α-苯基苯乙酮
dimethoxyethyl adipate　己二酸二甲氧基乙酯〔增塑剂〕

3,3′-dimethoxy-4,4′-isocyanic diphenyl　3,3′-二甲氧基-4,4′-二异氰酸酯联苯〔胶黏剂〕
di(3-methoxyiso propyl) peroxy dicarbonate　过(氧)二碳酸双(3-甲氧基异丙基)酯
1,3-dimethoxy-10-methylacridone　1,3-二甲氧基-10-甲基吖啶酮
3,4-dimethoxyphenethyl　3,4-二甲氧苯乙基
3,4-dimethoxyphenylacetyl　3,4-二甲氧苯乙酰
1,1-dimethoxy-2-phenylethane　1,1-二甲氧基-2-苯基乙烷
1,1-dimethoxy-2-phenylpropane　1,1-二甲氧基-2-苯基丙烷
3,4-dimethoxyphthalic acid　3,4-二甲氧基邻苯二甲酸
3,4-dimethoxy styrene　3,4-二甲氧基苯乙烯
dimethoxy succinic acid　二甲氧基丁二酸　$COOH(CHOCH_3)_2COOH$
α,α-dimethoxy toluene　α,α-二甲氧基甲苯
dimethoxy trityl chloride　二对甲氧三苯甲基氯
3,4-dimethoxy-1-vinylbenzene　3,4-二甲氧基-1-乙烯基苯
dimethyl　①二甲基②乙烷
N,N-dimethylacetamide(DMA)　N,N-二甲基乙酰胺
dimethyl acetic acid　二甲基乙酸　$CH(CH_3)_2COOH$
dimethyl acetoacetic ester　①二甲基乙酰乙酸酯　$CH_3COC(CH_3)_2COOR$ ②〔专指〕二甲基乙酰乙酸乙酯　$CH_3COC(CH_3)_2COOC_2H_5$
dimethyl acetophenone　二甲基苯乙酮
2,5-dimethyl-3-acetyl furan　2,5-二甲基-3-乙酰基呋喃
2,4-dimethyl-5-acetylthiazole　2,4-二甲基-5-乙酰基噻唑
2,3-dimethylacrolein　2,3-二甲基丙烯醛
N,N-dimethylacrylamide　N,N-二甲基丙烯酰胺　$CH_2=CHCON(CH_3)_2$
β,β-dimethyl acrylate　丙烯酸-β,β-二甲酯
dimethylacrylic aldehyde(=tiglic aldehyde)　二甲基丙烯醛；惕各醛　$CH_3CH=C(CH_3)CHO$
dimethylacylamide　二甲酰胺化合物
2-N-dimethyladenine　2-N-二甲腺嘌呤
dimethyl adipate　己二酸二甲酯〔增塑剂〕
dimethyl allene(=2,3-pentadiene)　二甲基丙二烯；2,3-戊二烯
dimethyl allyl carbinyl acetate　乙酸二甲基烯丙基原酯
γ,γ-dimethylallyl pyrophosphate　焦磷酸γ,γ-二甲(基)烯丙酯
dimethylallyl transferase　二甲烯丙基转移酶
dimethyl amination　二甲基胺化(作用)
dimethylamine　二甲胺
dimethylamine sulfate　硫酸二甲胺
dimethylamine sulfonate　磺酸二甲胺〔乳化剂〕

dimethylamine unhairing 二甲胺脱毛
cis-4-dimethylaminoazobenzene 顺-4-二甲氨基偶氮苯
p-dimethylamino-azo-benzene(=butter yellow; benzene-azodimethylaniline) 甲基黄；对二甲氨基偶氮苯
dimethylamino benzaldehyde 二甲氨基苯甲醛
p-dimethylaminobenzalrhodanine 银试剂；对二甲胺基亚苄基罗丹宁
4-dimethylamino benzoic acid 4-二甲氨基安息香酸
2-dimethylamino-1,4-benzoquinone 2-二甲氨基-1,4-苯醌
N-(4-dimethylamino benzylidene)-p-nitroaniline N-(4-二甲氨基苄亚甲基)对硝基苯胺
4-dimethylamino cinnamaldehyde 4-二甲氨基肉桂醛
2,3-dimethylamino dicarbonic acid(=diphenylamine-2,3-dicarboxylic acid) 2,3-二甲氨基二碳酸
dimethylamino ethanol 二甲氨基乙醇
dimethylaminoethyl acrylate 丙烯酸二甲氨基乙酯〔黏合剂〕
dimethylamino ethyl benzoate 安息香酸二甲氨基乙酯
dimethylamino ethyl methacrylate 甲基丙烯酸二甲氨基乙酯〔硫化剂〕
2-(N,N-dimethylamino) ethyl methacrylate 甲基丙烯酸2-(N,N-二甲氨基)乙酯
dimethylamino methylphenol 二甲氨基甲基苯酚
2-dimethylamino-2-methyl-1-propanol 2-二甲氨基-2-甲基-1-丙醇〔乳化剂〕
4-dimethylamino-2-nitro-$trans$-azobenzene 4-二甲氨基-2-硝基反式偶氮苯
4-dimethylamino-4-nitro-stilbene 4-二甲氨基-4-硝基芪
3-(p-dimethylaminophenyl)-3-(1,2-dimethyl indole-3-yl) phthalide 3-(对二甲氨基苯基)-3-(1,2-二甲基吲哚-3-基)苯酞
dimethylamino phenyl methane 二甲氨苯基甲烷
dimethylamino propionitrile 二甲氨基丙腈
3-(N,N-dimethylamino) propyl acrylate 丙烯酸3-二甲氨基丙(基)酯
dimethyl aminopropylamine 二甲氨基丙胺〔硬化剂〕
4-dimethylamino-α,β,β-tricyano styrene 4-二甲氨基-α,β,β-三氰基苯乙烯
dimethylammonium dimethylmonothiocarbamate 二甲基一硫代氨基甲酸二甲铵〔促进剂〕
dimethylammonium hydrogen isophthalate 间苯二甲酸氢二甲铵〔促进剂〕
dimethylammonium salt of 2-mercaptobenzothiazole 2-硫醇基苯并噻唑二甲铵盐〔促进剂〕
dimethylaniline 二甲基苯胺 $C_6H_5N(CH_3)_2$
N,N-dimethyl-p-anisidine N,N-二甲基对甲氧基苯胺
dimethyl anthranilate(=methyl N-methyl anthranilate) N-甲基邻氨基苯甲酸甲酯

dimethyl arsenic chloride 氯化二甲胂 $(CH_3)_2AsCl$
dimethyl arsenic cyanide 氰化二甲胂 $(CH_3)_2AsCN$
dimethyl arsenious oxide 氧化二甲胂 $[(CH_3)_2As]_2O$
dimethyl arsine 二甲胂 $(CH_3)_2AsH$
dimethyl arsine disulfide 二硫化二甲胂 $[(CH_3)_2As]_2S_2$
dimethyl arsine ethylsulfide 乙硫化二甲胂
dimethyl arsine oxide 氧化二甲胂 $[(CH_3)_2As]_2O$
dimethyl arsine sulfide 硫化二甲胂 $[(CH_3)_2As]_2S$
dimethyl arsine trichloride 三氯化二甲胂 $(CH_3)_2AsCl_3$
dimethyl arsinic acid 二甲次胂酸 $(CH_3)_2AsOH$
dimethylated 二甲基化的
dimethylation 二甲基化作用；二甲基取代作用
dimethyl azelate 壬二酸二甲酯 $CH_3OOC(CH_2)_7COOCH_3$
dimethyl azulene 二甲基薁
dimethyl benzene 二甲苯 $C_6H_4(CH_3)_2$
dimethylbenzidine(=tolidine) 二甲基联苯胺；联甲苯胺
dimethylbenzimidazole ribosidephosphate 二甲(基)苯并咪唑核苷磷酸
2,3-dimethylbenzoquinone 2,3-二甲基苯醌
N-dimethyl-2-benzothiazolesulfenamide N-二甲基-2-苯并噻唑次磺酰胺〔促进剂〕
dimethyl benzoyl 二甲苯酰
dimethyl benzyl carbinol 二甲基苄基原醇
dimethyl benzyl carbinyl acetate 乙酸二甲基苄基原酯
dimethyl benzyl carbinyl butyrate 丁酸二甲基苄基甲酯
4,5-dimethyl-2-benzyl-1,3-dioxolane 4,5-二甲基-2-苄基-1,3-二氧杂环戊烷
2,5-dimethyl-2,5-bishexane 2,5-二甲基-2,5-双己烷
dimethyl butadiene 二甲基丁二烯 $CH_2=C(CH_3)C(CH_3)=CH_2$
dimethylbutadiene rubber 二甲丁二烯橡胶；甲基橡胶
dimethylbutane 二甲基丁烷
N-(1,3-dimethylbutyl)-N-phenyl-p-phenylenediamine N-(1,3-二甲基丁基)-N-苯基对苯二胺〔防老剂〕
dimethyl cadmium 二甲基镉 $Cd(CH_3)_2$
dimethyl carbate 卡百酸二甲酯〔蚊虫忌避剂〕；驱蚊灵
dimethylcarbinol 异丙醇
dimethyl cellosolve 二甲基溶纤剂
dimethyl cellulose 二甲基纤维素
dimethyl chlorendate 氯菌酸二甲酯〔防霉剂〕
dimethyl chloro-arsine 二甲氯胂 $(CH_3)_2AsCl$
dimethylchlorosilane 二甲基氯硅烷
dimethyl-crackene 二甲基裂化烯〔改进油品荧光的添加剂〕
dimethylcyanamide 二甲基氰胺
dimethyl cyano-arsine 二甲氰胂 $(CH_3)_2AsCN$
dimethyl cyclohexanedione(=dimedone;dimethone) 双甲酮〔俗〕

1,3-dimethyl cyclohexane-2-one 1,3-二甲基环己烷-2-酮
2,2-dimethyl cyclohexanol 2,2-二甲基环己醇
dimethyl cyclohexanyl adipate 己二酸二甲基环己酯
1,2-dimethyl cyclohexene 1,2-二甲基环己烯
dimethyl cyclohexyl phthalate 邻苯二甲酸二甲基环己酯
3,4-dimethyl-1,2-cyclopentanedione 3,4-二甲基-1,2-环戊二酮
dimethyl diallylammonium 二甲基二烯丙基铵
2,5-dimethyl-2,5-dibenzoylperoxy hexane 2,5-二甲基-2,5-二(苯甲酰过氧基)己烷
2,5-dimethyl-2,5-di(t-butyl peroxy) hexane 2,5-二甲基-2,5-二(叔丁基过氧基)己烷〔硫化剂〕
(S)-4,7-dimethyl-6,7-dihydro-5H-cyclopenta[c]pyridine (S)-4,7-二甲基-6,7-二氢-5H-环戊烷并吡啶;阿克亭
dimethyl dichlorosilane(DMCS) 二甲基二氯硅烷
dimethyl dihydroresorcinol(=dimedon) 5,5-二甲基二氢化间苯二酚
dimethyl diisocyanate 二异氰酸二甲酯〔硫化剂〕
dimethyl diketone 丁二酮 $CH_3COCOCH_3$
N,N'-dimethyl-N,N'-di(1-methylpropyl)-p-phenylenediamine N,N'-二甲基-N,N'-二(1-甲基丙基)对苯二胺〔防老剂〕
N,N'-dimethyl-N,N'-dinitrosoterephthalamide N,N'-二甲基-N,N'-二亚硝基对苯二甲酰胺〔发泡剂〕
dimethyl dioctadecylammonium 二甲基二(十八烷基)铵
dimethyl dioctadecylammonium chloride 二甲基二(十八烷基)氯化铵
2,5-dimethyl-2,5-di(peroxybenzoyl)-3-hexyne 2,5-二甲基-2,5-二(苯甲酰过氧基)-3-己炔
2,5-dimethyl-2,5-di(peroxy-2-ethylhexanoyl) hexane 2,5-二甲基-2,5-二(过氧化-2-乙基己酰)-己烷
4,4'-dimethyl diphenylmethane-2,2',5,5'-tetraisocyanate 4,4'-二甲基二苯基甲烷-2,2',5,5'-四异氰酸酯
dimethyl diphenyloxysilane 二甲基二苯氧基(甲)硅烷 $(C_6H_5O)_2Si(CH_3)_2$
dimethyl disulfide 二甲基二硫醚
dimethyldithiocarbamate 二甲基二硫代氨基甲酸盐〔促进剂〕
dimethyl dithiocarbamic acid 二甲基二硫代氨基甲酸;二甲基氨磺酸
dimethyl dithiol ethylene copper compound 二甲基乙二硫铜配位化合物
dimethyl dithiourethane 二甲基二硫代氨基甲酸乙酯;二甲(基)氨磺酸乙酯
2,2-dimethyl dodecylaldehyde 2,2-二甲基十二醛
dimethylene ①二亚甲基②乙烯
dimethyl ester 二甲酯 $R(COOCH_3)_2$
dimethylethanolamine 二甲基乙醇胺
dimethyl ether 二甲醚 CH_3OCH_3; $R(OCH_3)_2$

dimethyl ethyl acetic acid 二甲基乙基乙酸;邻二甲代丁酸 $C(CH_3)_2(C_2H_5)COOH$
2,6-dimethyl-3-ethyl pyrazine 2,6-二甲基-3-乙基吡嗪
dimethyl ethynyl carbinol 二甲基乙炔基原醇
1,3-dimethyl ferrocene 1,3-二甲基二茂铁 $(C_5H_5)Fe[C_5H_3(CH_3)_2]$
1,1'-dimethyl ferroceneoate(=1,1'-ferrocene dicarboxylic acid dimethylester) 1,1'-二茂铁二羧酸二甲酯
dimethyl formamide 二甲基甲酰胺
N,N-dimethyl formamide(DMF) N,N-二甲基甲酰胺 $HCON(CH_3)_2$
dimethyl glutarate 戊二酸二甲酯〔增塑剂〕
dimethyl glycol phthalate 酞酸二甲基乙二醇酯;邻苯二甲酸二甲基乙二醇酯
dimethyl glyoxal 二甲基乙二醛
dimethyl glyoximate 丁二酮肟盐
dimethyl glyoximato cobalt 丁二酮肟合钴
dimethyl glyoxime(=diacetyl dioxime) 丁二酮肟
dimethyl guanidine 二甲基胍
2,6-dimethyl heptanal 2,6-二甲基庚醛
2,6-dimethyl-4-heptanol 2,6-二甲基-4-庚醇
di-1-methylheptyl-1-methylheptyl phosphonate 1-甲庚基膦酸二(1-甲庚基)酯 $(C_8H_{17})PO(OC_8H_{17})_2$
N,N-dimethyl histamine N,N-二甲基组胺
dimethyl hydantoin formaldehyde 二甲基乙内酰脲甲醛
dimethyl hydantoin formaldehyde resin 二甲基乙内酰脲甲醛树脂
dimethyl hydrine 甘油二甲醚
dimethyl hydroquinone 二甲基对氢醌
dimethylin 二甲氨基 $-N(CH_3)_2$
1,1-dimethy lindane 1,1-二甲基-1,2-二氢化茚
N,N'-dimethyl-indigo N,N'-二甲基靛蓝
dimethyline 二甲灵;甘油二甲醚
dimethyl inositol 二甲基肌醇
dimethyl irigenin 二甲基鸢尾配基
6,7-dimethyl isoalloxazine 6,7-二甲基异咯嗪;感光黄素
2,2-dimethyl-3-isobutenylcyclopropane-1-carboxylic acid 菊一酸;菊甲酸
dimethyl isobuty(roy)l chloride 二甲基异丁酰氯
dimethyl isophthalate 间苯二甲酸二甲酯
dimethyl itaconate 衣康酸二甲酯〔增塑剂〕
dimethyl ketol 二甲基乙酮醇
dimethyl ketone(=acetone) 丙酮
dimethyl maleate 马来酸二甲酯〔增塑剂〕
dimethyl maleic acid 二甲基马来酸 $COOHC(CH_3)=C(CH_3)COOH$
dimethyl malonate 丙二酸二甲酯 $CH_3OCOCH_2COOCH_3$

dimethyl malonic acid　二甲基丙二酸　COOHC(CH$_3$)$_2$COOH

dimethyl malonic ester　①〔泛指〕二甲基丙二酸酯　ROCOC(CH$_3$)$_2$COOR　②〔专指〕二甲基丙二酸二乙酯　C$_2$H$_5$OCOC(CH$_3$)$_2$COOC$_2$H$_5$

dimethyl mercury　二甲汞　Hg(CH$_3$)$_2$

2,6-dimethyl-2-methoxy-7-octanol　2,6-二甲基-2-甲氧基-7-辛醇

2,2-dimethyl-3-methylene norbornane　2,2-二甲基-3-亚甲基降莰烷

dimethyl methylene norbornane(=camphene)　二甲基亚甲基降莰烷；莰烯

6,6-dimethyl-2-methylene norpinane　6,6-二甲基-2-亚甲基蒎烷

dimethyl methyl ether　二甲基甲醚

dimethyl monothiocarbamate　一硫代氨基甲酸二甲酯〔促进剂〕

2-(2,6-dimethyl-4-morpholinothio) benzothiazole　2-(2,6-二甲基-4-吗啉基硫代)苯并噻唑〔促进剂〕

3,3-dimethyl naphthoquinone　3,3-二甲基萘醌〔硫化剂〕

N,N-dimethyl-2-naphthylamine　N,N-二甲基-2-萘胺〔防老剂〕

dimethyl nonanal　二甲基壬醛

4,8-dimethyl nonyl acetate　乙酸4,8-二甲基壬酯

6,6-dimethyl-2-norpinene-2-aldehyde　6,6-二甲基-2-降蒎烯-2-甲醛

dimethyl octadienal(=citral)　二甲基辛二烯醛；柠檬醛

dimethyl octadiene　二甲基辛二烯

3,7-dimethyl-2,6-octadiene nitrile　3,7-二甲基-2,6-辛二烯腈；柠檬腈〔商〕

dimethyl octadienol(=linalool)　二甲基辛二烯醇；芳樟醇

dimethyl octanol　二甲基辛醇

dimethyl octanyl acetate　乙酸二甲基辛酯

3,7-dimethyl-6-octenoic acid　3,7-二甲基-6-辛烯酸

3,7-dimethyl-6-octen-1-ol　3,7-二甲基-6-辛烯-1-醇

3,7-dimethyl-6-octenyl acetate　乙酸3,7-二甲基-6-辛烯酯

N,N-di(1-methyloctyl)-p-phenylenediamine　N,N-二(1-甲基辛基)对苯二胺〔防老剂〕

dimethyloctyne diol(=3,6-dimethyl-4-octyne-3,6-diol)　二甲基辛炔二醇；3,6-二甲基-4-辛炔-3,6-二醇〔表面活性剂〕

dimethylol adipic acid amide　二羟甲基己二酰二胺

N,N-dimethylol carbamate　氨基甲酸二羟甲酯　(HOCH$_2$)$_2$NCOOR

dimethylol dimethyl　二甲羟基二甲基

dimethylol ethyl dicarbamate　二羟甲基乙二氨基甲酸酯〔纤维交联剂〕

dimethylol ethylene urea　二羟甲基亚乙基脲〔催化剂〕

dimethylol propionic acid　二羟甲基丙酸

dimethylol urea　二羟甲基脲　HOCH$_2$NHCONHCH$_2$OH

dimethyl oxalate　草酸二甲酯

2,4-dimethyl pentane　2,4-二甲基戊烷

2,4-dimethyl-2-pentenoic acid　2,4-二甲基-2-戊烯酸

dimethylphenol butyraldehyde　二甲基苯酚丁醛〔防老剂〕

2,4-dimethyl phenylacetaldehyde　2,4-二甲基苯乙醛

dimethyl phenyl arsenite　苯亚胂酸二甲酯　C$_6$H$_5$As(OCH$_3$)$_2$

dimethyl phenyl arsine oxide　二甲基苯基胂化氧　(CH$_3$)$_2$C$_6$H$_3$As=O

2,4-dimethyl phenylbutyraldehyde　2,4-二甲基苯丁醛

dimethyl phenylcarbinol　二甲基苯甲醇

dimethyl phenyl carbinyl acetate　乙酸二甲基苯基原酯

dimethyl phenylene diamine　二甲(基)苯二胺

3,5-dimethyl phenylethyl acetate　乙酸3,5-二甲基苯乙酯

3,5-dimethyl phenylethyl alcohol　3,5-二甲基苯乙醇

dimethyl phenylethyl carbinol　二甲基苯乙基原醇

dimethyl phenylethyl carbinyl acetate　乙酸二甲基苯乙基原酯

3,3'-dimethyl phenylmethane　3,3'-二甲基苯甲烷

dimethyl phenyl-para-cresol　二甲基苯基对甲酚

2,5-dimethyl phenyl propionaldehyde　2,5-二甲基苯丙醛

dimethyl phenyl silanol　二甲基苯(甲)硅(烷)醇

dimethyl phosphinic acid　二甲基次膦酸　(CH$_3$)$_2$POOH

dimethyl phosphite　亚磷酸二甲酯　HPO(OCH$_3$)$_2$

dimethyl phosphonate　膦二甲酯　RPO(OCH$_3$)$_2$

dimethyl phthalate　邻苯二酸二甲酯

dimethyl polysiloxane　聚二甲基硅氧烷〔隔离剂〕

N,N'-dimethyl-1,3-propanediamine　N,N'-二甲基-1,3-丙二胺

2,2-dimethyl-1,3-propanediol　2,2-二甲基-1,3-丙二醇

1,1-dimethylpropyl　叔戊基

dimethyl propylene urea　二甲基亚丙基脲

dimethylpropyl-p-phenylenediamine　二甲基丙基对苯二胺〔防老剂〕

dimethyl purine　二甲基嘌呤

2,6-dimethyl pyrazine　2,6-二甲基吡嗪

3,5-dimethyl pyrazole　3,5-二甲基吡唑

dimethyl pyridine　二甲基吡啶

5,5-dimethyl-1-pyrroline　5,5-二甲基-1-吡咯啉

4,6-dimethyl resorcinol(=xylorcinol)　4,6-二甲基间苯二酚；木间二酚

dimethyl resorcinol ether　间苯二酚二甲醚

dimethyl sebacate　癸二酸二甲酯　(CH$_2$)$_8$(CO$_2$CH$_3$)$_2$

dimethyl selenide(=selenium dimethyl)　二甲硒　(CH$_3$)$_2$Se

N-dimethyl serotonin　N-二甲(基)-5-羟色胺；蟾毒色胺

dimethyl silane(=dimethyl silicane)　二甲(基)甲硅烷　$(CH_3)_2SiH_2$
dimethyl silicane(=dimethyl silane)　二甲(基)甲硅烷　$(CH_3)_2SiH_2$
dimethyl silicone oil　二甲基硅油〔脱模剂〕
dimethyl silicone polymer fluid　二甲基硅氧烷聚合液〔油压传动系统用油〕
dimethylsiloxane polymer　二甲基硅氧烷聚合物
dimethyl siloxane rubber　二甲基硅(氧烷)橡胶
N,N-dimethyl stearylamine　N,N-二甲基硬脂胺　$CH_3(CH_2)_{16}CH_2N(CH_3)_2$
dimethyl succinate　丁二酸二甲酯　$(CH_2CO_2CH_3)_2$
dimethyl succinic acid　二甲基丁二酸　$(CH_3CH)_2(CO_2H)_2$
dimethyl sulfate　硫酸二甲酯　$(CH_3O)_2SO_2$
dimethyl sulfide　二甲硫；甲硫醚　$(CH_3)_2S$
dimethyl sulfite　亚硫酸二甲酯　$(CH_3O)_2SO$
dimethyl sulfone　二甲砜　$(CH_3)_2SO_2$
dimethyl sulfonium methylide　二甲基亚甲基锍　$(CH_3)_2S=CH_2$
dimethyl sulfourea(=dimethyl thiourea)　二甲基硫脲
dimethyl sulfoxide　二甲亚砜　$(CH_3)_2SO$
dimethyl sulfoxonium methylide　二甲基亚甲基氧锍
dimethyl sulphate test　硫酸二甲酯试验
dimethyl sulphonazo Ⅲ　二甲基磺偶氮Ⅲ
dimethyl sulphoxide(DMSO)　二甲基亚砜
dimethyl tartrate　酒石酸二甲酯　$(CHOHCO_2CH_3)_2$
dimethyl telluride　二甲碲　$(CH_3)_2Te$
dimethyl terephthalate　对邻苯二甲酸二甲酯　$C_6H_4(CO_2CH_3)_2$
dimethyl tetrahydroindene　二甲基四氢茚
4,5-dimethylthiazole　4,5-二甲基噻唑
N,N-dimethyl thioacetamide　N,N-二甲基-硫代乙酰胺
dimethyl thioether　二甲基硫醚
dimethyl thiourea　二甲基硫脲　$CS(NHCH_3)_2$
dimethylthiuram disulfide　二硫化二甲基秋兰姆〔促进剂〕
dimethyl tin oxide　氧化二甲基锡；二甲基氧化锡
5,8-dimethyl tocol　5,8-二甲基母酚酐；β生育酚
dimethyl trisulfide　二甲基三硫
dimethyl trithiocarbonate　三硫代碳酸二甲酯
N,N-dimethyl tryptamine(=nigerine)　N,N-二甲基色胺
dimethyl urea　二甲脲　$CO(NHCH_3)_2$
sym-dimethyl urea　对称二甲脲　$(CH_3NH)_2CO$
unsym-dimethyl urea　不对称二甲脲　$(CH_3)_2NCONH_2$
dimethyl uric acid　二甲基尿酸　$C_5N_4(CH_3)_4$
2,4-dimethyl-5-vinylthiazole　2,4-二甲基-5-乙烯基噻唑
3,7-dimethyl xanthine　可可碱；3,7-二甲(基)黄嘌呤
dimethyl yellow(=methyl yellow)　甲基黄〔不另叫二甲基黄〕
dimethyl zinc　二甲锌　$Zn(CH_3)_2$
p-dimethyoxy benzene　对二甲氧基苯
dimethyoxyethyl phthalate　邻苯二甲酸二甲氧基乙酯〔增塑剂〕
dimetric　正方的；四边形的
diminishing angle　递减角度
diminution　减少
diminution factor　衰减率
dimister　除雾器
dimixo-octyl phthalate(=diisooctyl phthalate)　邻苯二甲酸二异辛酯〔增塑剂〕
dimmer　调光器；光度调整器；减光器
dimolecular　双分子的
dimolecular reaction　双分子反应
dimolecular reduction　双分子还原
dimorphic　双晶的
dimorphism　双型；双晶现象
dimorpholino ethane　二吗啉乙烷
dimorpholinyl disulfide　二硫化二吗啉〔硫化剂〕
dimorphous　双晶的
dimorphous substance　双晶物质
dimple effect　涟漪效应
dimples　①表面砂眼②窝穴；凹部③压成十字形
dimyrcene　二聚月桂烯
dimyristyl ether　双肉豆蔻醚〔防静电剂〕
dinaphthalene　联萘；联二萘　$C_{10}H_7 \cdot C_{10}H_7$
dinaphtho　二萘并
dinaphthol　联萘酚
dinaphthyl　①二萘基②联萘
dinaphthylamine　二萘胺
di-β-naphthyl disulfide　二-β-萘二硫
dinaphthylene　二亚萘基
dinaphthyl ether　二萘醚
dinaphthyl ketone　二萘(甲)酮
dinaphthyl methane　二萘甲烷
dinaphthyl sulfide　二萘硫
DIN color difference equation　DIN色差方程式
DIN color system　DIN色(体)系
dincopentyl acctic acid　双新戊基乙酸　$[(CH_3)_3CCH_2]_2CHCOOH$
dineric　二液界面的
Dingler's green　①磷酸铬绿②丁勒绿
dingy　(色彩)黯淡
dinicotinic acid　烟碱二酸；吡啶二甲酸
dinitraniline　二硝基苯胺
dinitraniline orange(=dinitro aniline orange)　二硝基苯胺橙

dinitration 二硝化作用〔引入了两个硝基〕
dinitric acid 二硝酸
dinitrile fibre 聚偏氰乙烯纤维
dinitro-〔词头〕 二硝基
4,6-dinitro-2-aminophenol 4,6-二硝基-2-氨基苯酚
2,4-dinitro anisole 2,4-二硝基茴香醚
dinitro benzene 二硝基苯 $C_6H_4(NO_2)_2$
p-dinitrobenzene-azo-naphthol 对二硝基苯偶氮萘酚；镁试剂
2,4-dinitro benzene dimethyl dithiocarbamate 二甲基二硫代氨基甲酸-2,4-二硝基苯酯
dinitro benzoylene urea 二硝基邻亚苯甲酰脲 $C_8H_4O_6N_4$
dinitrocellulose 二硝酸纤维素
dinitro chloro benzene(=dinitro chloro benzol) 二硝基氯苯
dinitro chloro benzol(=dinitro chloro benzene) 二硝基氯苯
dinitro compound 二硝基化合物
4,6-dinitro-o-cresol 4,6-二硝基邻甲酚
4,6-dinitro-o-cyclohexyl phenol 4,6-二硝基邻环己基苯酚
dinitro diazophenol 二硝基重氮酚
dinitro diphenylamine 二硝基二苯基胺
4,4'-dinitro diphenylmethyl carboanion 4,4'-二硝基二甲基负碳离子
dinitro fluoro benzene 二硝基氟苯
dinitrogen 双氮*
dinitrogen tetroxide 四氧化二氮 N_2O_4
dinitrogen trioxide 三氧化二氮 N_2O_3
dinitro glycerine 二硝基甘油；甘油二硝酸酯
dinitro glycerine explosive 二硝基甘油炸药；甘油二硝酸酯炸药
dinitroglycol(=glycol dinitrate) 二硝酸乙二醇酯；硝化乙二醇 $(O_2NOCH_2)_2$
2,5-dinitro hydroquinone 2,5-二硝基氢醌
dinitro hydroquinone acetate 二硝基氢醌乙酸酯
dinitro hydroxyazo 二硝基羟偶氮
2,4-dinitro-Meisenheimer complexes 2,4-二硝基-迈森海默配合物
dinitro methane 二硝基甲烷 $(NO_2)_2CH_2$
dinitro naphthalene 二硝基萘
dinitro naphthalene disulfonic acid 二硝基萘二磺酸 $(NO_2)_2C_{10}H_4(SO_3H)_2$
dinitro naphthalene sulfonic acid 二硝基萘磺酸 $(NO_2)_2C_{10}H_5SO_3H$
2,4-dinitro-1-naphthol-7-sulfonic acid(=flavianic acid) 黄萘酸；2,4-二硝基-1-萘酚-7-磺酸 $(NO_2)_2C_{10}H_4(OH)SO_3H$

dinitro pentamethylene tetramine 二硝基五亚甲基四胺〔发泡剂〕
dinitro phenamic acid 二硝基氨基苯酚；苦氨酸 $NH_2C_6H_2(OH)(NO_2)_2$
dinitro phenol(DNP) 二硝基苯酚
dinitro phenylation 二硝基苯基化；DNP 化
2,4-dinitrophenyl dimethyl dithiocarbamate 二甲基二硫代氨基甲酸-2,4-二硝基苯酯
2,4-dinitro phenylhydrazine 2,4-二硝基苯肼
dinitro phenyl hydrazine method 二硝基苯肼法〔聚氧乙烯分析〕
dinitro phenylhydrazone 二硝基苯腙
2-(2,4-dinitro phenylthio) benzothiazole 2-(2,4-二硝基苯硫基)苯并噻唑
dinitroso 二亚硝基
p-dinitrosobenzene 对二亚硝基苯〔硫化剂〕
N,N'-dinitroso-N,N'-dimethylterephthalamide N,N'-二亚硝基-N,N'-甲基对苯二甲酰胺〔发泡剂〕
dinitrosodiphenylguanidine 二亚硝基二苯胍〔塑解剂〕
dinitroso pentamethylene tetramine 二亚硝基五亚甲基四胺
dinitrosophenyl benzothiazyl disulfide 二硫化二硝基苯基苯并噻唑〔促进剂〕
2,4'-dinitrosophenyl-2-mercaptobenzothiazole 2,4'-二硝基苯基-2-硫醇基苯并噻唑〔促进剂〕
dinitrosophenyl thiobenzothiazole 二亚硝基苯基硫代苯并噻唑〔促进剂〕
2,4-dinitroso-resorcinol 2,4-二亚硝基间苯二酚〔塑解剂〕
dinitroso terephthalamine 二亚硝基对苯二甲酰胺
dinitro thiobenzoic acid 二硝基硫代苯甲酸
dinitro toluene 三硝基(甲)苯 $CH_3C_6H_3(NO_2)_2$
dinking machine 平压切断机〔橡〕
dinolin(=dilinolein) 二亚油精；甘油二亚油酸酯
dinonyl adipate 己二酸二壬酯〔低温增塑剂〕
4,6-dinonyl-o-cresol 4,6-二壬基邻甲酚〔防老剂〕
dinonyl furmarate 富马酸二壬酯〔增塑剂〕
n-dinonyl ketone 二壬基(甲)酮；10-十九(碳)烷酮 $(C_9H_{19})_2CO$
dinonyl maleate 马来酸二壬酯〔增塑剂〕
dinonylnaphthalene disulfonic acid(DNN-DSA) 二壬基萘二磺酸
dinonylnaphthalene sulfonic acid(DNS;DN-NS) 二壬基萘磺酸
dinonyl phthalate 邻苯二甲酸二壬酯；酞酸二壬酯
dinonyl sebacate(DNS) 癸二酸二壬酯〔增塑剂〕
Dinoseb(=2-(1-methyl-n-propyl)-4,6-dinitro phenol) 2-(1-甲基-正丁基)-4,6-二硝基苯酚〔农药〕
dinuclear 两核的；两环的

dinuclear aromatics 两核芳烃
dinuclear complex 双核配合物
dinucleotide 二核苷酸
dinucleotide-fold 二核苷酸折叠
dinyl 联(二)苯；二苯基
dinyl kettle 联苯炉
dioctahedral analogue 双八面体同型物
dioctahedral type 双八面体型
dioctanoyl peroxide 过氧化二辛酰
dioctyl ①二辛基②十六烷
N-di-sec-octyl acetamide(DOAA) N,N-二仲辛基乙酰胺 $CH_3CON(C_8H_{17})_2$
dioctyl acetic acid 二辛基乙酸 $[CH_3(CH_2)_7]_2CHCO_2H$
dioctyl adipate 己二酸二辛酯
dioctyl azelate 壬二酸二辛酯
dioctyl decyl phthalate 邻苯二酸辛癸酯
dioctyl ether (二)辛醚 $[CH_3(CH_2)_7]_2O$
dioctyl fumarate 富马酸二辛酯〔增塑剂〕
dioctyl isophthalate 间苯二甲酸二辛酯〔增塑剂〕
dioctyl itaconate 衣康酸二辛酯〔增塑剂〕
dioctyl ketone 二辛基(甲)酮；9-十七(碳)烷酮 $[CH_3(CH_2)_7]_2CO$
dioctyl maleate 马来酸二辛酯〔增塑剂〕
di-n-octyl-2-oxo-propanphosphonate 2-氧代-丙基膦酸二正辛酯 $CH_3COCH_2PO(OC_8H_{17})_2$
dioctyl-p-phenylenediamine 二辛基对苯二胺〔防老剂〕
dioctyl phenyl phosphate 磷酸二辛基苯酯〔增塑剂〕
di-p-octylphenyl phosphoric acid 二(对辛基苯基)磷酸
dioctyl phosphate 二辛基磷酸盐
dioctyl phosphinic acid(HDOP) 二辛基次膦酸
dioctyl phosphite 亚磷酸二辛酯〔稳定剂〕
dioctyl phthalate 邻苯二甲酸二辛酯
dioctyl sebacate 癸二酸二辛酯
dioctyl sodium sulfosuccinate 琥珀酸二辛酯磺酸钠
dioctyl sodium sulphosuccinate 琥珀酸二辛酯磺基钠
dioctyl succinate 丁二酸二辛酯；琥珀酸二辛酯〔增塑剂〕
dioctyl sulfosuccinate 琥珀酸二辛酯磺酸盐
dioctyl sulfoxide(DOSO) 二辛基亚砜 $(C_8H_{17})_2SO$
dioctyl terephthalate 对苯二甲酸二辛酯〔增塑剂〕
di-n-octyl tetrahydrophthalate 四氢邻苯二甲酸二正辛酯〔增塑剂〕
di-n-octyltin-S,S'-bis-(iso-octyl mercaptoacetate) 二-正辛基锡-S,S'-双(巯基乙酸异辛酯)
diode 二极管
diode array detector 二极管阵列检测器
diode array spectrophotometer 二极管阵列分光光度计
diode laser 二极管激光器
diode temperature sensor 二极管温度传感器

dioform 1,2-二氯乙烯
diogenite 古铜无球陨石
-dioic acid 双酸；二酸
diol 二元醇
-diol 二醇；二酚〔芳香族〕
diol bond 二醇键
diol (bonded) phase 二醇基(键合)相
diol diacid polyester resin 二醇-二酸聚酯树脂
dioleate 二油酸酯
diolefine(=dialkene) 二烯烃
diolefinic 二烯属的
diolefinic acid 二烯酸
diolefins 二烯属烃 C_nH_{2n-2}
diolefin series 二烯系
diolein 二油精；甘油二酸酯
dioleostearin 二油一硬脂精；甘油一硬脂酸二油酸酯
dioleoyl ethylene titanate 二油酰基钛酸亚乙酯〔偶联剂〕
dioleyl adipate 己二酸二油基酯
dioleyl ethylene titanate 二油酰基钛酸亚乙酯〔偶联剂〕
dioleyl phosphite 亚磷酸二油基酯〔稳定剂〕
-dione〔词尾〕 二酮
Dionic water tester 戴氏测水器
dionin(=ethylmorphine hydrochloride) 盐酸乙基吗啡；地昂宁
diophane 玻璃纸
diopside 透辉石
dioptase 绿铜矿；透视石
diopter 屈光度；折光度
dioptrics 折光学
dioptry 折光度
diorgano siloxane 二有机基硅氧烷
diorite 闪长岩
dioritite 闪长细晶岩
diortho-periodic acid 二原高碘酸 $H_{12}I_2O_{13}$
diorthotolyl guanidine 二邻甲苯胍
dioscin 薯蓣素
dioscorea-sapogenin 薯蓣皂草配基
dioscorea-sapotoxin 薯蓣皂草毒
dioscorine 薯蓣碱；地奥碱
diosellinic acid 薯蓣酸；地奥酸
diosgenin 薯蓣皂苷配基
diosmetin 地奥亭；洋芫荽黄素；香叶木素
diosmin 地奥司明〔药〕
diosphenol 地奥酚
diosphenol acetate 乙酸地奥酚酯
diospyrol 柿酚
dioxadiene 二噁二烯
dioxalate 草酸氢盐(或酯)

dioxanate 二㗁烷盐
1,4-dioxan(e) 1,4-二㗁烷；二氧杂环己烷 $(CH_2)_4O_2$
dioxane 二㗁烷；二氧杂环己烷 $(CH_2)_4O_2$
dioxane lignin 二㗁烷木素；二氧杂环己烷木素
dioxane polycarboxylate 二㗁烷多羧酸酯
dioxazine 二㗁嗪 $C_3H_3O_2N$
dioxazine violet pigment 二㗁嗪紫颜料〔无金属杂环颜料之一，耐候性极优〕
dioxetane 二氧杂丁烷
dioxetene 二氧四环烃
dioxide 二氧化物
dioxime 二肟
dioxime of furoxane dialdehyde 呋㗁烷二醛二肟 $C_2H_2O_3(CH=NOH)_2$
dioxin 二㗁英；二氧杂苊
dioxine 二㗁烯；二氧杂环己烯
3,4-dioxohexane 3,4-己二酮
dioxolane 二㗁茂烷；二氧戊环
dioxole 间二氧杂环戊烯
dioxolone 二氧杂戊二烯酮
dioxopromethazine 二氧丙嗪〔药〕
dioxosiloxane 二氧代二乙硅醚
dioxotetrafluoro molybdate 二氧四氟合钼酸盐
5-(2,5-dioxotetrahydrofuryl)-3-methyl-3-cyclohexene-1,2-dicarboxylic anhydride 5-(2,5-二氧代四氢呋喃基)-3-甲基-3-环己烯-1,2-二羧酸酐
dioxy- 〔词头〕 ①二氧②二羟
dioxyarylene unit 二氧代亚芳香基单元
4,4′-dioxydiphenyl 4,4′-二氧二苯
dioxydisulfotungstate 二氧二硫钨酸盐 $M_2WO_2S_2$
dioxygen 双氧*；分子氧
dioxygenyl hexafluoroantimonate 六氟锑酸双氧基 $O_2^+SbF_6^-$
dioxystearic acid 二羟基硬脂酸
dioxytartaric acid 二羟基酒石酸 $COOHC(OH)_2C(OH)_2COOH$
dipalmitin 二棕榈精；甘油二棕榈酸酯
dipalmito-olein 二棕榈一油精；甘油一油酸二棕榈酸酯
dipalmito-stearin 二棕榈硬脂精；甘油一硬脂酸二棕榈酸酯
dip Alocrom plant 阿铝克铬姆浸渍处理装置〔铝表面化学处理装置之一〕
dip-and-dry product 浸渍制品
dip-and-scrap (浸渍)蘸刮
dip-and-squeeze (浸渍)蘸挤
dip application 浸涂施工
dipara-periodic acid 二仲高碘酸 $H_8I_2O_{11}$
dip bonding 浸渍黏合(法)

dip can 选样器
dip cement 浸渍胶浆子〔橡〕
dip-coated strip 浸涂样板条
dip coating 浸泡涂膜
dip composition 浸渍合剂；浸渍组分
dip compound 浸渍混合物
dip-drain technique 浸渍-流干法
dipentadecyl-carbinol 双十五基甲醇；16-三十(碳)烷醇 $[CH_3(CH_2)_{14}]_2CHOH$
dipentadecyl ketone 双十五基(甲)酮
dipentaerythrite(=dipentaerythritol) (一缩水)二季戊四醇；聚二季戊四醇
dipentaerythritol 二(聚)季戊四醇；(一缩)二季戊四醇
dipentaerythritol hexanitrate 六硝基二季戊四醇；六硝酸二季戊四醇酯
dipentamethylenethiuram hexasulfide 六硫化双五亚甲基秋兰姆〔促进剂〕
dipentene 二聚戊烯；双戊烯；松油精；苎烯
di-n-pentyl phosphoric acid(HDAP;DAP) 二正戊基磷酸 $(C_5H_{11}O)_2PO(OH)$
1,3-dipenylbut-1-ene 1,3-二苯基丁-1-烯
dipeptide 二肽
dipeptide phase 二肽型固定相
diperacid 双过酸 $H_2(R[O_2])_2$
diperchromic acid 二过铬酸 H_3CrO_7
diperoxy brazilic acid 二过氧巴西酸
dip forming process 浸渍成形法〔复合多晶纤维成形法之一〕
dip gum 浸油树胶
diphase 两相的；二相的
diphase cleaner 两相洗净剂
diphase rubber 两相橡胶
diphase system 两相系；二相系
diphasic titration 两相滴定法
dip hatch(=dip hole) 计量口
diphemanil(=diphenmethanil) 二苯马尼
diphenamic acid 联苯甲酰胺酸；苯基苯甲酰胺酸 $C_6H_5C_6H_3(CONH_2)COOH$
diphenate ①邻苯二甲酸氢盐 $(C_6H_4)_2(COOM)_2$ ②邻苯二甲酸氢酯 $(C_6H_4)_2(COOR)_2$
diphenazyl 二苯甲酰乙烷
diphenhydramine 苯海拉明〔药〕
diphenic acid 联苯甲酸
diphenimide 联苯四酰环亚胺
diphenmethanil(=diphemanil) 二苯马尼
diphenol ①联苯酚 $(C_6H_4OH)_2$ ②〔类名〕二酚
diphenolic acid(=4,4-bis-4-hydroxyphenyl pentanoic acid) 双酚酸

diphenoxy phosphazine 二苯氧基膦嗪
diphenyl ①联(二)苯 ②二苯基 $(C_6H_5)_2$
N,N'-diphenyl acetamidine N,N'-二苯乙脒 $CH_3C(NHC_6H_5)=NC_6H_5$
diphenyl acetic acid 二苯基乙酸 $(C_6H_5)_2CHCO_2H$
diphenyl acetylene(=tolane) 二苯乙炔 $C_6H_5C\equiv CC_6H_5$
diphenyl alkane disulfonate 二苯基烷二磺酸盐(或酯)
diphenylamine 二苯胺 $(C_6H_5)_2NH$
diphenylamine blue 二苯胺蓝*
diphenylamine-2,3-dicarboxylic acid 二苯胺-2,3-二羧酸
diphenylamine reaction 二苯胺反应
diphenylamine sulfate 二苯胺硫酸 $(C_6H_5)_2NHH_2SO_4$
diphenylamine sulfonate 二苯胺磺酸盐 $C_6H_5NHC_6H_4SO_3M$
diphenylamine sulfonic acid 二苯胺磺酸 $C_6H_5NHC_6H_4SO_2H$
diphenylamino-azo-m-benzene sulfonic acid 二苯胺偶氮间苯磺酸
diphenylamino-azo-p-benzene sulfonic acid(=tropeolin OO) 金莲橙；二苯胺偶氮对苯磺酸
diphenylamino chlorarsine 二苯胺氯胂
diphenylamino cyanarsine 二苯胺氰胂
diphenyl antimony chloride 氯化二苯基锑
diphenyl antimony cyanide 氰化二苯基锑
diphenyl arsenious acid 二苯亚砷酸 $(C_6H_5)_2AsOOH$
diphenylarsine 二苯胂 $(C_6H_5)_2AsH$
diphenyl arsine oxide (双)二苯胂化氧 $[(C_6H_5)_2As]_2O$
diphenyl arsine oxychloride 二苯(基)胂化氧氯 $(C_6H_5)_2AsOCl$
diphenyl arsine sulfide 二苯胂黄硫 $(C_6H_5)_2As=S$
diphenyl arsine trichloride 三氯化二苯胂 $(C_6H_5)_2AsCl_3$
diphenyl arsinic acid 二苯亚胂酸 $(C_6H_5)_2AsOOH$
diphenyl arsonic acid 二苯胂酸
diphenylate 二苯氧化物 $R(OC_6H_5)_2$
N,N'-diphenylbenzidine N,N'-二苯基联苯胺 $(C_6H_5NHC_6H_4)_2$
diphenylbenzidine 二苯基联苯胺
diphenyl bibenzoate 联苯二甲酸二苯酯
diphenyl black 联苯黑
diphenyl-bromomethane 二苯溴甲烷；二苯甲基溴 $(C_6H_5)_2CHBr$
1,4-diphenyl butane 1,4-二苯基丁烷 $(C_6H_5CH_2CH_2)_2$
diphenyl t-butyl phosphate 磷酸二苯基叔丁酯 $C_4H_9C_6H_4=OPO=(OC_6H_5)_3$
diphenyl carbamyl chloride 二苯氨甲酰氯 $(C_6H_5)_2NCOCl$
diphenylcarbazide 二苯卡巴肼*
sym-diphenylcarbazide 对称二苯卡巴肼

diphenylcarbazone 二苯卡巴腙*
sym-diphenyl carbazone 对称二苯基卡巴腙〔俗〕；苯肼羰基偶氮苯 $C_6H_5N=NCONHNHC_6H_5$
diphenyl carbene 二苯基卡宾
diphenyl carbinol 二苯基甲醇 Ph_2CHOH
diphenyl carbonate 碳酸二苯酯 $(C_6H_5O)_2CO$
diphenyl chloroarsine 二苯氯胂；二苯基氯胂 $(C_6H_5)_2AsCl$
diphenyl chloromethane 二苯氯甲烷；二苯甲基氯 $(C_6H_5)_2CHCl$
diphenyl cresyl phosphate 磷酸二苯基甲苯酯〔增塑剂〕
diphenyl cresyl phthalate 邻苯二甲酸二苯甲苯酯〔增塑剂〕
diphenyl cyanoacrylate 氰基丙烯酸二苯酯
diphenyl cyanoarsine 二苯基氰胂 $(C_6H_5)_2As(CN)$
diphenyl decyl phosphate 亚磷酸二苯基癸基酯〔抗氧剂〕
diphenyl decyl phosphite 亚磷酸二苯基癸酯〔稳定剂〕 $(C_6H_5O)_2(C_{10}H_{21}O)P$
diphenyl diacetylene 二苯基联乙炔；二苯基丁二炔 $PhC\equiv CC\equiv CPh$
diphenyl dicarboxylic acid 联苯二羧酸 $(C_6H_4)_2(COOH)_2$
diphenyl dichloromethane 二苯基二氯甲烷 $(C_6H_5)_2CCl_2$
diphenyl dichloro silane 二苯基二氯(甲)硅烷
diphenyl di-n-dodecylsilane 二苯基二-正十二烷基甲硅烷
diphenyl diethylene 二苯基联乙烯；二苯基丁二烯 $PhCH=CHCH=CHPh$
diphenyl diisocyanate (连)二苯基二异氰酸酯 $(C_6H_4NCO)_2$
diphenyl diketohexane 二苯基己二酮 $PhCO(CH_2)_4COPh$
diphenyl diketone 二苯基二(甲)酮 $C_6H_5COCOC_6H_5$
diphenyl diketooctane 二苯基辛二酮 $PhCO(CH_2)_6COPh$
diphenyl dimethoxy silane 二苯基二甲氧基甲硅烷 $(C_6H_5)_2Si(OCH_3)_2$
diphenyl dimethyl ethane 二苯基二甲基乙烷
diphenyl diphenoxysilicane 二苯基二苯氧基硅 $(C_6H_5)_2Si(OC_6H_5)_2$
diphenyl disulfide 二苯二硫 $(C_6H_5)_2S_2$
diphenyl disulfoxide 二苯二砜；联(二)苯亚砜 $(C_6H_5SO)_2$
diphenylene ①二亚苯基 (—C_6H_4—)₂ ②联亚苯基 —$C_6H_4C_6H_4$— ③联二亚苯 $C_6H_4=C_6H_4$
diphenylene disulfide (夹)二硫杂蒽
diphenylene imide(=carbazole) 咔唑

diphenylene-oxide 二苯并呋喃 $(C_6H_4)_2O$
diphenylenimide(=carbazole) 咔唑
diphenylenimine(=carbazole) 咔唑
diphenyl ester 二苯酯 $R(COOC_6H_5)_2$
unsym-diphenylethane 不对称二苯基乙烷 $(C_6H_5)_2CHCH_3$
diphenyl ether （二）苯醚 $(C_6H_5)_2O$
diphenyl ether calorimeter （二）苯醚量热计
diphenylethylene 二苯基乙烯
1,1-diphenylethylene 1,1-二苯基乙烯
N,N-diphenyl-ethylene diamine N,N-二苯亚乙基二胺
di-β-phenylethyl ether 二-β-苯基乙醚 $(C_6H_5CH_2CH_2)_2O$
N-diphenylformamide N-二苯基甲酰胺
N,N-diphenyl-formamidine N,N-二苯基甲脒；对称二苯甲脒 $HC(NC_6H_5)NHC_6H_5$
diphenyl glyoxalone(=diphenylimidazolone) 二苯基甘噁酮；二苯基咪唑酮 $C_6H_5C=C(C_6H_5)NHCONH$
diphenyl guanidine 二苯胍；促进剂D $(C_6H_5NH)_2C=NH$
diphenylguanidine acetate 乙酸二苯胍〔促进剂〕
diphenylguanidine dimethyldithiocarbamate 二甲基二硫代氨基甲酸二苯胍〔促进剂〕
diphenylguanidine oxalate 草酸二苯胍〔促进剂〕
diphenyl guanidine phthalate 邻苯二甲酸二苯胍
diphenylguanidine tartrate 酒石酸二苯胍〔促进剂〕
diphenyl heating kettle 联苯加热釜
diphenylheptyl phosphine oxide(DPHPO) 二苯基一庚基氧膦 $(C_6H_5)_2(C_7H_15)PO$
diphenyl hydantoin 苯妥英〔俗〕；二苯基乙内酰脲 $CONHCONHC=(C_6H_5)_2$
diphenyl hydramine 二苯羟基胺；二苯醇胺
diphenyl hydrazine 二苯肼
diphenyl hydrazobenzene 四苯肼 $C_{24}H_{20}N_2$
diphenyl hydroxy arsine 二苯亚胂酸 $(C_6H_5)_2AsOOH$
diphenyl-imidazolone 二苯咪唑酮
diphenyline rearrangement(=benzidine rearrangement) 联苯胺重排作用
diphenyl iodonium hydroxide 氢氧化二苯碘 $(C_6H_5)_2IOH$
diphenyl iodonium iodide 碘化二苯碘 $(C_6H_5)_2II$
di(4-phenyl isocyanate) methane 二 4-苯基异氰酸甲烷
diphenyl ketene 二苯乙烯酮 $(C_6H_5)_2C=CO$
diphenyl ketone(=benzophenone) 二苯(甲)酮；苯酰苯
diphenyl-ketoxime 二苯(甲)酮肟
diphenyl malonate ①丙二酸二苯酯 $CH_2(CO_2C_6H_5)_2$ ②二苯基丙二酸盐 $(C_6H_5)_2C(COOM)_2$

diphenyl mercury 二苯汞 $(C_6H_5)_2Hg$
diphenyl methane 二苯甲烷 $(C_6H_5)_2CH_2$
diphenyl methane colouring matters(=ketone imines) 二苯基甲烷染料〔即：酮亚胺类〕
diphenyl methane-4,4'-diisocyanate 二苯甲烷-4,4'-二异氰酸酯
diphenyl methane dye 二苯甲烷染料
diphenyl methyl 二苯甲基 $(C_6H_5)_2CH-$
diphenyl methyl arsine hydroxybromide 溴羟化二苯基甲基胂 $CH_3(C_6H_5)_2AsBrOH$
diphenylmethylation 二苯甲基化(作用)
diphenyl methyl bromide 二苯甲基溴
diphenyl methyl chlorosilane 二苯甲基氯硅烷 $(C_6H_5)_2CH_3SiCl$
diphenyl methylene 二苯亚甲基 $(C_6H_5)_2C=$
diphenyl mono-o-xenyl phosphate 磷酸二苯基单邻-联苯酯 $(C_6H_5O)_2(C_6H_5C_6H_4O)PO$
diphenyl-nitrosamine 二苯亚硝胺 $(C_6H_5)_2NNO$
diphenyl octyl phosphate 磷酸二苯辛酯〔增塑剂〕
diphenylol propane 二酚基丙烷；双酚A〔俗〕
2,5-diphenyloxazole(PPO) 2,5-二苯基噁唑
diphenyl oxide （二）苯醚；二苯基氧 $C_6H_5OC_6H_5$
diphenyl oxide-4,4'-disulfonyl hydrazide 4,4'-二磺酰肼代二苯醚
diphenyl pentaerythritol phosphite 亚磷酸二苯基季戊四醇酯〔稳定剂〕
diphenyl peroxide 过氧化二苯
diphenyl-p-phenylene diamine 二苯基对苯二胺
diphenyl phosphate 磷酸二苯酯 $(C_6H_5O)_2POOH$
diphenyl phosphine 二苯膦 $(C_6H_5)_2PH$
diphenyl phosphinic acid 二苯基次膦酸 $(C_6H_5)_2POOH$
diphenyl phthalate 邻苯二甲酸二苯酯 $C_6H_4(CO_2C_6H_5)_2$
diphenylpicrylhydrazine 二苯基苦基肼
diphenylpicrylhydrazyl 二苯基苦味酰肼基
4,4'-diphenylpropane diisocyanate 4,4'-二苯基丙烷二异氰酸酯
diphenylpropane-1,3-dione(DPPO) 二苯基-1,3-丙二酮；二苯酰甲烷 $C_6H_5COCH_2COC_6H_5$
2,6-diphenyl pyrylium perchlorate 2,6-二苯基过氯酸吡喃鎓
diphenyl selenide 二苯硒；苯硒醚 $(C_6H_5)_2Se$
diphenyl selenium dichloride 二氯化二苯硒 $(C_6H_5)_2SeCl_2$
diphenyl silanodiol 二苯基(甲)硅烷二醇
diphenyl silicon diisocyanate 二苯基甲硅烷二异氰酸酯
diphenyl steam boiler 联苯蒸气(锅)炉
diphenyl succinate 丁二酸二苯酯 $(CH_2CO_2C_6H_5)_2$

diphenyl succinic acid 二苯基丁二酸
　　COOH(CHC$_6$H$_5$)$_2$COOH
diphenyl sulfide 二苯硫；苯硫醚　(C$_6$H$_5$)$_2$S
diphenylsulfon-3,3'-disulfonyl hydrazide 3,3'-二磺酰肼
　　二苯砜〔发泡剂〕
diphenyl sulfone 二苯砜　(C$_6$H$_5$)$_2$SO$_2$
diphenyl sulfourea 二苯基硫脲　CS(NHC$_6$H$_5$)$_2$
diphenyl sulfoxide 二苯亚砜　(C$_6$H$_5$)$_2$SO
diphenylsulphone-3,3'-disulfonyl hydrazide 二苯砜-3,3'-二磺酰肼
diphenyl thiocarbazone(=dithizone) 双硫腙；二苯基卡巴腙　C$_6$H$_5$N=NCSNHNHC$_6$H$_5$
diphenyl thiocyanoarsine 二苯硫氰胂　(C$_6$H$_5$)$_2$AsCNS
diphenyl thiourea 二苯基硫脲　CS(NHC$_6$H$_5$)$_2$
diphenyl tin 二苯基锡　(C$_6$H$_5$)$_2$Sn
diphenyl tin oxide 氧化二苯锡；二苯基氧化锡
diphenyl-*m*-tolylmethane 二苯基间甲基甲烷
　　CH$_3$C$_6$H$_4$CH(C$_6$H$_5$)$_2$
diphenyl tolyl phosphate 磷酸二苯基甲苯酯〔增塑剂〕
diphenyl triazole 二苯基三唑
diphenyl tridecyl phosphite 亚磷酸二苯基十三烷基酯
　　(C$_6$H$_5$O)$_2$(C$_{13}$H$_{27}$O)P
diphenyl-triketone(=dibenzoyl ketone) 二苯基三(甲)酮；二苯甲酰基(甲)酮　(C$_6$H$_5$CO)$_2$CO
unsym- diphenylurea 不对称二苯脲　(C$_6$H$_5$)$_2$NCONH$_2$
diphenyl urea 二苯脲　CO(NHC$_6$H$_5$)$_2$
sym-diphenylurea(=carbanilide) 对称二苯脲
　　(C$_6$H$_5$NH)$_2$CO
diphenylurethane 二苯基尿烷；二苯氨基甲酸乙酯〔俗〕
　　(C$_6$H$_5$)$_2$NCO$_2$C$_2$H$_5$
diphenylvinyl phosphine 二苯基乙烯基膦
　　(C$_6$H$_5$)$_2$P(CH=CH$_2$)
diphenylxylenyl phosphate 磷酸二苯基二甲苯酚〔增塑剂〕
dip hole(=dip hatch) 计量口
diphosgene 双光气；氯甲酸三氯甲酯
diphosphane 二膦；二磷烷　PH$_2$PH$_2$
diphosphanetetroic acid 连二磷酸　(HO)$_2$(O)PP(O)(OH)$_2$
diphosphate 二磷酸
diphosphatidyl glycerol 双磷脂酰甘油；心磷脂
diphosphatomanganate 二磷酸根络锰酸盐
diphosphine 二膦；二磷烷　PH$_2$PH$_2$
diphosphine dioxide (亚烷基)双氧膦
　　R$_2$PO(CH$_2$)$_n$POR$_2$
1,3-diphosphoglyceraldehyde 1,3-二磷酸甘油醛
1,3-diphospho glycerate 1,3-二磷酸甘油(酯)
diphospho-2-methyl-1,4-naphthohydroquinone 二磷酸-2-甲基-1,4-萘氢醌
diphosphoric acid 焦磷酸　H$_4$P$_2$O$_7$

diphosphorous acid 二亚磷酸；焦亚磷酸
　　(HO)$_2$POP(OH)$_2$
diphthalate 苯二酸氢盐(或酯)
diphtheria 白喉
diphtheria antitoxin 白喉抗毒素
diphtheria toxin 白喉毒素
diphtherinic acid 白喉菌酸
dip-hydro technique 水溶液浸渍整理工艺
diphyl 联(二)苯
diphyl cycles 联苯循环系统
diphyl heated piping 联苯加热管道
diphyl steam 联苯蒸气
diphyl vapour generator 联苯蒸气发生器；联苯锅炉
dipicolinate 二皮考啉酸盐
dipicolinic acid 吡啶二羧酸
dipicrylamine(=hexanitrodiphenylamine) 二苦胺；六硝基二苯胺　[(NO$_2$)$_3$C$_6$H$_2$]$_2$NH
dipicryl sulfide(=hexanitrodiphenyl sulfide) 二苦硫；六硝基二苯硫　[(NO$_2$)$_3$C$_6$H$_2$]$_2$S
dipimelate 庚二酸氢盐(或酯)
dipinene 双蒎烯
2,5-dipiperidino-*p*-benzoquinone 2,5-二哌啶基对苯醌
dipiperonalacetone 二亚胡椒基丙酮
　　(CH$_2$O$_2$C$_6$H$_3$CH=CH)$_2$CO
dip iron 刮脂刀
dipivaloylmethane(DPVM) 二叔戊酰甲烷；二(三甲基乙酰基)甲烷　(CH$_3$)$_3$CCOCH$_2$COC(CH$_3$)$_3$
dipleg 浸入管
diplococcin 双球菌素
diplococcus 双球菌
diplococcus pneumoniae 胸膜肺炎双球菌
diplodnabactivirus 双脱噬菌体；双DNA噬菌体
diplogen(=deuterium) 氘；重氢
diploicin 双球标氏衣素
diploid 二倍体
diploidy 二倍性
diplomycelium 二倍体菌丝体
diplomycin 双球霉素
diplonema 双线
diplotene stage 双线期
dip lubricating system 浸入润滑系统
dip method 浸染法
dip-mix 浸渍剂
dip moulding 浸渍成型；蘸塑
dipolar 偶极的；两极的
dipolar shift(=pseudo-contact shift) 偶极位移
dipolar addition 偶极加成
dipolar adsorbent 偶极性吸附剂

dipolar aprotic solvent 偶极非质子溶剂
dipolar bond 偶极键
dipolar complexes 二极络合物
dipolar decoupling 偶极去偶
dipolar decoupling-magic angle spinning-nuclear magnetic resonance 去偶-魔角旋转-核磁共振
dipolar electret 偶极驻极体
dipolar energy 偶极能
dipolar fluid 偶极流体
dipolar force 偶极力
dipolar interaction 偶极相互作用
dipolar ion 偶极离子
dipolar loss 偶极损耗
dipolarophile 亲偶极的
dipolar protophilic solvent 偶极亲质子溶剂
dipolar protophobic solvent 偶极疏质子溶剂
dipolar relaxation 偶极弛豫
dipolar shift reagent （偶极）位移试剂
dipolar solvent 极性溶剂
dipolar transition 偶极跃迁
dipole 偶极
dipole moment(=moment of dipole) 偶极矩
dipole attraction 偶极引力
dipole coupling 偶极偶合
dipole-dipole broadening 偶极-偶极展宽
dipole-dipole interaction 偶极-偶极相互作用
dipole-dipole relaxation 偶极-偶极弛豫
dipole disorientation 偶极解取向
dipole effect 偶极效应
dipole-elastic loss 高弹偶极损耗；链段偶极损耗
dipole-induced-dipole interaction 偶极-诱导-偶极相互作用
dipole interaction force 偶极力
dipole ion 偶极离子
dipole layer 偶极层
dipole molecule 偶极分子
dipole moment 偶极矩
dipole orientation 偶极取向
dipole polarization 偶极极化
dipole-quadrupole interaction 偶极-四极相互作用*
dipole-radical loss 侧基偶极损耗
dipole strength 偶极强度
dipole transition moment 偶极跃迁矩
dipolymer 二聚物
dipolymerization 二聚(作用)
dipotassium compound 二钾化合物
dipotassium hydrogen phosphate 磷酸氢二钾 K_2HPO_4
dip painting 浸涂漆

dipped article ①浸渍制品②无缝制品〔橡〕
dipped beam(=low beam) 弱光
dipped cord 浸胶帘线；浸渍帘子线
dipped electrode method 浸渍电极法
Dippel's oil 骨焦油；地柏油
dip-pen 浸笔技术
dipper ①收脂工；刮脂工②浸渍机
dipper sampling 勺取样
dipperstick 水位指示器；测量尺；量油尺
dipping ①浸浴②浸涂③浸渍④收脂；刮脂
dipping agent 浸渍剂
dipping bath 浸渍浴
dipping black 浸涂沥青漆
dipping compound 浸渍混炼胶
dipping counter 浸液计数器
dipping electrode 浸液电极
dipping enamel 浸涂瓷漆
dipping form(=dipping former) 浸渍模
dipping former(=dipping form) 浸渍模
dipping glazing 浸渍釉
dipping lacquer 浸漆
dipping latex 浸渍用胶乳
dipping machine 浸渍机
dipping mandrel 插入芯杆
dipping mangle 浸渍器
dipping mix 浸渍混炼胶
dipping mold 浸渍用模(型)
dipping platinum microelectrode 浸入式铂微电极
dipping polish 浸渍抛光
dipping process 浸渍过程
dip (ping) process reclaim 压出法再生胶
dipping rack 浸渍挂架
dipping refractometer 浸液折射计
dipping solution 浸渍溶液
dipping tank 浸渍槽
dipping technique 浸渍技术
dipping test 浸涂试验
dipping time 浸渍(涂)时间
dipping varnish 浸渍清漆
dip pipe 浸渍管
dippnictide 二磷族元素化物
dip polishing 浸渍抛光
di-primary 二伯的
dip rod 浸量尺；水位指示器；量油尺
dipropargyl ①联炔丙基；二炔丙基②1,5-己二炔 $(CH\equiv CCH_2)_2$
dipropenyl(=2,4-hexadiene) 联丙烯基；2,4-己二烯
diprophylline 二羟丙茶碱〔药〕

dipropoxy-bis (acetylacetonato) titanium 二丙氧基二(乙酰丙酮)钛
dipropoxytitanium-bis(ethylacetoacetate) 二(乙基乙酰乙酸)二丙氧基钛；二丙氧基钛二(乙基乙酰乙酸酯)
dipropoxytitanium-bis(lactate) 二丙氧基钛二乳酸酯
dipropoxytitanium-bis(triethanolaminate) 二(三乙醇氨)钛酸二丙(氧基)酯
dipropyl 二丙基 $(C_3H_7)_2$
dipropyl-acetic acid 二丙基乙酸 $(C_3H_7)_2CHCO_2H$
dipropyl adipate 己二酸二丙酯 $[(CH_2)_2CO_2C_3H_7]_2$
dipropyl amine 二丙胺 $(C_2H_5CH_2)_2NH$
dipropyl amino-benzaldehyde 对二丙氨基苯(甲)醛 $(C_3H_7)_2NC_6H_4CHO$
dipropyl barbituric acid 二丙基巴比妥酸 CONHCONHCOC=$(C_3H_7)_2$
dipropyl carbanilide-4,4'-dicarboxylate 均二苯脲-4,4'-二甲酸二丙酯 $CO(NHC_6H_4CO_2CH_2C_2H_5)_2$
dipropyl carbonate 碳酸二丙酯 $(C_2H_5CH_2O)_2CO$
dipropyl disulfide 二丙基二硫醚 $(C_2H_5CH_2S)_2$
dipropylene glycol 双丙甘醇；一缩二丙二醇
dipropylene glycol dimethacrylate 二甲基丙烯酸二丙二醇酯
dipropylene glycol salicylate 水杨酸二丙二醇酯
dipropylenetriamine 二亚丙基三胺 $H_2NC_3H_6NHC_3H_6NH_2$
dipropyl ether (二)丙醚 $(C_2H_5CH_2)_2O$
dipropyl-ethyl-phenylsilicane 二丙基乙基苄基甲硅烷 $(C_3H_7)_2(C_2H_5)(C_6H_5CH_2)Si$
dipropyl-hexylmethane 二丙基己基甲烷；4-丙基癸烷 $(C_3H_7)_2CHC_6H_{13}$
dipropyl ketone 二丙基(甲)酮；4-庚酮 $(C_2H_5CH_2)_2CO$
dipropyl maleate 马来酸二丙酯 $(=CHCO_2C_3H_7)_2$
dipropyl malonate 丙二酸二丙酯 $CH_2(CO_2C_3H_7)_2$
dipropyl-mercury 二丙基汞 $(C_2H_5CH_2)_2Hg$
dipropyl naphthalene sulfonate 二丙基萘磺酸盐
dipropyl nitrosamine 二丙亚硝胺 $(C_3H_7)_2NNO$
dipropyl oxalate 草酸二丙酯 $(CO_2C_3H_7)_2$
di-n-propyl peroxydicarbonate 过(氧)二碳酸二丙酯
dipropyl phthalate 邻苯二甲酸二丙酯 $C_6H_4(CO_2C_3H_7)_2$
dipropyl polyoxyethylene glycol 二丙基聚氧乙烯(乙)二醇
dipropyl succinate 丁二酸二丙酯 $CH_2CO_2C_3H_7)_2$
dipropyl sulfate 硫酸二丙酯 $(C_2H_5CH_2O)_2SO_2$
dipropyl sulfide 二丙硫；丙硫醚 $(C_2H_5CH_2)_2S$
dipropyl sulfite 亚硫酸二丙酯 $(C_2H_5CH_2O)_2SO$
dipropyl sulfone 二丙砜 $(C_2H_5CH_2)_2SO_2$
dipropyl tartrate 酒石酸二丙酯 $(CHOHCO_2C_3H_7)_2$

dipropyl trisulfide 二丙基三硫
sym-dipropylurea 对称二丙脲 $(C_2H_5CH_2NH)_2CO$
unsym-dipropylurea 不对称二丙脲 $(C_3H_7)_2NCONH_2$
dipropyonyl peroxide 过氧化二丙酰
diprotic 双质子的
diprotic acid(=dibasic acid) 二元酸
diproton 双质子
dip sampler 浸入取样器；插入取样器
dip solution 浸渍溶液
dip-spin coating technique 旋转浸涂技术
dip-spin technique 浸渍-离心涂布技术；浸渍离心涂装技术
dip stick 浸量尺；水位指示器；测量尺；量油尺
dip stool 浸量管；计量管
dip tank 浸渍槽；浸浆槽
di-p-tolyl lead chromate 二(对甲苯基)铬酸铅
dip tray 浸涂托板
dip treating 浸渍处理
dip-tumbler 浸渍转筒
dip-type cell 浸入型电池
dip varnish 浸渍漆
dip weight 浸量锤；测量海洋深度的锤
dipy(=2,2'-bipyridyl) 邻联吡啶
dipyrazolone 二吡唑酮
dipyridine 联吡啶
1,3-di(pyridine-4)propane 1,3-二(吡啶-4)丙烷〔硫化剂〕
di-quaternary ammonium cellulose halide exchanger 双季铵纤维素卤化盐(离子)交换剂
di-quaternary ammonium thiocyanate 硫氰酸双季铵
di-quaternary exchanger 双季铵盐(离子)交换剂
di-quaternization 双季碱化；成双季碱反应
diquinidine(=diconchinine) 双奎尼定
diquinoline(=biquinoline) 联喹啉
diquinolyl 联喹啉
diquinone(s) 二醌(类)
Dirac delta function 狄拉克δ函数〔物〕
Dirac notation 狄拉克符号
diradical 双(自由)基
diradical initiation 双基引发(作用)
direactive glyceride 二反应基甘油酯；单甘油酯
direct-acting pump 直接作用泵
direct acting steam pump 直动泵；蒸汽直接作用泵
direct-acting valve 直动阀
direct ammonia process 直接制氨法
direct ammonia recovery 氨气直接回收法
direct analysis(=direct method of analysis) 直接分析法
direct analysis of daughter ion 子离子直接分析
direct anodic stripping voltammetry 直接阳极溶出伏

安法
direct antiglobulin test 直接抗球蛋白试验；直接湿片检查
direct bright fast blue 直接鲜坚牢蓝
direct bromination 直接溴化作用
direct brown 直接棕
direct chelatometric titration 直接螯合滴定
direct chemical ionization 直接化学电离
direct chemical ionization source 直接化学电离源
direct chlorination 直接氯化作用
direct coacervant(=contact coacervant) 接触凝聚剂
direct colo(u)r 直接染料
direct comparison method of measurement 直接比较测量法
direct condenser 回流冷凝器
direct(-connected) drive 直接传动
direct cooking process 直接蒸煮法
direct cooling 直接冷却法
direct cotton dye(=substantive azo dye) 直接偶氮染料；直接棉染料
direct count 定长制细度
direct-counting method 直接计数法
direct-coupled amplifier 直接耦合放大器
direct current(=direct electric current) 直流电(流)
direct current amplifier 直流放大器
direct current arc 直流电弧
direct current arc light source 直流电弧光源
direct current arc method 直流电弧法
direct current arc welding(=DC arc welding) 直流电弧焊
direct current converter 直流换流器
direct current coulometric detector 直流库仑检测器
direct current electro-motor 直流电动机
direct current generator 直流发电机
direct current integrating ammeter 直流积分电流计
direct current plasma 直流等离子体
direct current plasma jet 直流等离子体喷焰
direct current plasma self oscillator 直接等离子体自激振荡发生器
direct current plasma source 直流等离子光源
direct current polarogram 直流极谱图
direct current polarograph 直流极谱仪
direct current polarography 直流极谱法
direct current power source 直流电源
direct dark green 直接暗绿
direct desulfurization 直接脱硫
direct digital control 直接数字控制
direct dilution method 直接稀释法
direct dipping process 直接浸渍法

direct displacement 直接取代
direct distillation method 直接蒸馏法
direct-distilled 直馏的
direct dye(=salt dye;substantive dye) 直接染料
direct dyeing 直接染色
direct dyestuffs 直接染料
directed aldol condensation 定向醛醇缩合
directed inter-esterification 定向酯交换
direct effect 直接效应
direct electric current(=direct current) 直流电(流)
direct emulsion 直接乳胶
direct esterification 直接酯化
direct esterification-polycondensation process 直接酯化缩聚法
direct evaluation 直接计算法；直接估算法
direct evaporation 直接蒸发
direct expansion 直接发泡
direct expansion cooling 直接膨胀冷却
direct fast yellow 直接坚牢黄
direct fertilizer 直接肥料
direct filtration 直接过滤
direct fire 活火头；活火
direct fired 活火烧了的
direct fired heater 直接火力加热器；直焰炉
direct fired rotary drier 直烧旋转干燥器
direct fire heating 活火加热
direct firing 直接烧；直接用火加热
direct firing furnace 直接火焰炉；直焰炉
direct flow indicator 直接流量指示器
direct fluid interface 直接流体接口
direct fluorescence method 直接荧光法
direct fusion 直接熔化
direct gravimetric method 直接重量(分析)法
direct halogenation 直接卤化作用
direct heat drier 直接加热干燥器
direct-heated mold 直接加热压模
direct heat exchange 直接热交换
direct heating 直接加热
direct heat oven 直接加热炉
direct humidifier 直接增湿器
direct hydration 直接水合
direct IDA 直接同位素稀释分析法
direct imaging mass analyzer 直接成像质量分析器
direct immuno-fluorescence technology 直接免疫荧光技术
direct impact strength 直接冲击强度；正面冲击强度
directing effect of group 取代基的定向作用
directing group 定向取代基
direct injection 直接注射

direct injection burner 直接喷入燃烧器；直接喷射燃烧器
direct injection enthalpimetry 直接注射焓测定(法)
direct inlet probe 直接进样探头；直接进样杆
direct inlet system 直接进样系统
direct iodination 直接碘化作用
direction ①方向②指导
directional freeze 定向凝固
directional intermolecular force 定向分子间力
directional(ity) effect 定向效应；方向性效应
directional lay up moulding 定向铺设模压法
directionally ordered structure 指向有序结构
directionally oriented structure composites 定向结构经编复合材料
directional motion 定向运动
directional non-random heterogeneity 非随机方向的不均匀性
directional property 方向性
directional reflectance 定向反射率；定向反射系数
directional reflectance factor 定向反射系数
directional solidification 取向性凝固
directional stability(=best tracking) 方向稳定性
directional valency 定向(方向性)原子价
direction distribution 取向分布
direction focusing 方向聚焦
direction of affinity 亲和方向
direction of illumination 光照方向；照射方向
direction of twist 捻向
direction property 各向异性；方向性
direct isotope dilution method 直接同位素稀释法
directive action 定向作用
directive effect 定向效应
direct labor 普通工；非熟练工人
direct latex casting 胶乳直接铸型〔橡〕
direct line 直线
direct line atomic fluorescence 直跃线原子荧光
direct line fluorescence 直跃线荧光
direct liquid introduction 直接液体进样
direct liquid introduction interface 直接液体进样接口
directly chemical combination 直接化学结合法
directly heated tube 直热型管
directly interjected sample introduction 直接插入固体进样
directly jet combustor 直接喷入式燃烧器
direct maintenance 直接维修
direct measurement 直接测量法
direct metal solution theory 金属直接溶解学说
direct method 直接法

direct method of analysis(=direct analysis) 直接分析法
direct method of determination 直接测定法
direct method of measurement 直接测量法
direct nitration 直接硝化作用
direct nitric method 直硝法
direct operating expense 直接操作费用
direct oxidation 直接氧化
direct oxidation test 直接氧化试验
direct particle(=primary particle) 原始粒子
direct photometric titration 直接光度滴定
direct piping 直接(到输送管的)干线
direct plate method 直接平板法
direct polymerization 直接聚合
direct polymer to websystem 聚合物直接成网法
direct polyphosphate process 直接多磷酸法
direct positive process 直接阳图加工法
direct potentiometry 直接电位(分析)法
direct printing 直接印染
direct print plywood 直接印刷胶合板
direct probe 直接进样探头；直接进样探针；直接进样探测
direct production cost 直接生产成本
direct pyrolysis method 直接热解法
direct reacting bilirubin 直(接反)应胆红素
direct reaction 直接反应
direct reading 直接读数
direct reading analytical balance 直读分析天平
direct reading balance 直读天平
direct reading instrument 直接读数仪
direct reading optical-emission spectrometer 直读光发射光谱仪
direct reading pH meter 直读(式)pH计
direct reading spectrometer 直读光谱仪
direct reading system ①直读式②直(接)读(数)系统
direct reading thermometer 直读式温度计
direct recording polarograph 直接记录式极谱仪
direct reeling machine 直缫缫丝机
direct reflectance ①定向反射比②直接反射比
direct replacement 直接置换
direct roll coating 同向辊涂装法
direct roller coat 同向辊涂
direct roving 直接无捻粗纱
direct sample introduction device 直接进样装置
direct sample probe 直接进样探头
direct sampling 直接取样
direct sampling leak valve 直接进样(漏孔)阀
direct sampling system 直接取样系统
direct sky blue 直接青

direct smelting 直接熔化
direct spin draw 直接纺丝拉伸(法)
direct spinning 直接纺纱
direct sprue gate 直接注口式浇口
direct steam 直接蒸汽；新汽；活汽
direct substitution 直接取代
direct sulfonation 直接磺化作用
direct synthesis strong nitric acid process 直接合成浓硝酸工艺
direct titration 直接滴定
direct transmitted beam 直接透射的光束
direct ultraviolet absorbance measurement 直接紫外吸光(度)测量(法)
direct union 直接联合
direct valency 直接化合价
direct viewing 直观
direct viscose 直接(法)黏胶
direct vision prism 直视棱镜
direct vision spectroscope 直视分光镜〔物〕
direct visual comparison 直观比较(法)
direct warping 直接整经
direct weighing method 直接称量法
direct welding 双面点焊
direct welt 毛起头
direct yarn number 直接制纱线线密度
diresorcin 联间苯二酚
diresorcinol 联间苯二酚 $C_{12}H_6(OH)_4$
diresorcylic acid 一缩双 3,5-二羟苯甲酸
Direx process 直接萃取法
diricinoleic acid 双蓖酸
diricinoleidine 二蓖精；甘油二蓖酸酯
dirt ①杂质；夹杂物②污垢；废屑
dirt-carrying 携污；载污
dirt catcher 滤尘器；除尘器
dirtiness resistance 污垢热阻〔在热交换器管壁上〕
dirt particles(=impurity) 杂质
dirt pick-up resistance (石料建筑漆)抗积尘性；防吸尘性
dirt pick-up test ①吸垢试验②积垢试验
dirt-removing power 去污能力
dirt repellent 防污剂
dirt-repellent treatment 防污处理
dirt resistance 防污性；防尘性
dirt resistance coatings 防尘涂料
dirt retention 积垢；积尘性
dirt settling (漆膜)积垢；附着物
dirt tester 杂质测定器
dirty 污〔指颜色〕
dirty charge stock 重残油进料；污油进料
dirty factor 污垢系数
dirty hole 孔内异物
dirty oil 污油
dirty ship 运输黑色石油产品的油(槽)船
dirty spot 污斑
dirty tanker 运输黑色石油产品的油(槽)船
dis- 〔词头〕①无；非；不②除(脱)去③相反；反对④分离；分开
disaccharide(=disaccharose) 二糖；双糖
disaccharose(=disaccharide) 二糖；双糖
disagglomeration 瓦解(作用)
disaggregation 解聚作用
disagreeable odor 不愉快气息
disalicylatethilendiamine 双水杨酸二乙烯二胺〔减活化剂〕
disalicylic acid(=salicyl salicylic acid diplosal) 双水杨酸；水杨酰水杨酸 $HOC_6H_4CO_2C_6H_4CO_2H$
disalicylide(=salosalicylide) 双水杨酸内酯
disappearance 消失
disappearing fibre 溶解性纤维
disassembled equipment 解体设备；拆卸的装置
disassembly 拆卸
disassembly saw 拆卸锯
disassembly table 拆卸台
disassimilation 异化(作用)
disassociation 离解(作用)
disaturated (glyceride) 二饱和的(甘油酯)
disazo compound 二重氮化合物
disazo dyes 双偶氮染料
disbonded region 脱胶(脱层)区；剥离区
disbondment radius 脱胶(脱层)半径；剥离半径
disc ①(圆)盘②片
discard molasses 〔复〕废糖蜜
disc bellow 盘式风箱
disc bowl centrifuge 碗盖式离心机
disc bowl clarifier 碗盖式澄清器
disc bowl type centrifuge 碗盖式离心机
disc chromatogram 圆盘形色谱图
disc chromatography 圆盘色谱(法)
disc column 圆盘塔
disc comparator 圆盘比色计
disc conveye(o)r flame ionization detector 传动盘式火焰电离检测器
disc crusher 盘式压碎机
disc electrophoresis 圆盘电泳(法)；盘状电泳(法)
discending paper partition chromatography 下行纸分配色谱(法)
disc enzyme electrophoresis 盘状酶电泳(法)
disc evaporator 盘式蒸发器

disc filter 盘滤机
disc finishing machine 圆盘磨光机
disc fungi 盘菌
disc grizzly 盘式筛
discharge 流量
dischargeability 拔染性
dischargeable capacity ①排送能力②(储罐的)有效容积
dischargeable color(=dischargeable dye) 拔染染料
dischargeable dye(=dischargeable color) 拔染染料
discharge angle 出口角
discharge area 出口切面面积
discharge atomizer 排出雾化器
discharge capacity 排出量
discharge cask 卸料容器
discharge cell 放电室
discharge chute 卸料斜槽；卸料溜槽
discharge cock 排水旋塞
discharge coefficient ①放电系数②流量系数
discharge curve 流量过程线
discharged ①放电的②泻出的
discharge detector 放电检测器
discharge end 出料端
discharge equipment 卸出装置
discharge gas 废气
discharge gate 给料输送管上的阀门；卸料阀门
discharge grating 卸料炉栅
discharge gutter 排出沟
discharge hatch 排放口；出料口
discharge head 压头；(压缩机)压力的高度
discharge header 排出集合管
discharge head (extruder) 压出机头
discharge hole 排出孔
discharge hopper 卸料斗
discharge hose 排水胶管
discharge ionization 放电电离
discharge knockout drum 出口分离器
discharge lamp 放电灯
discharge leader 排出管
discharge line 卸出线
discharge liquid 排出液
discharge loss 卸出损失
discharge manifold (泵的)卸出支管；卸出连接管
discharge material 放电材料
discharge nozzle(=discharge opening) 卸料口
discharge opening(=discharge nozzle) 卸料口
discharge orifice 排放孔
discharge pipe 泄水管
discharge pipe line 排放管线
discharge pipe loss 输送管内(水力)的损失
discharge piping 卸出输送管
discharge plasma 放电等离子体
discharge polarography 放电极谱法
discharge polymerization 放电聚合
discharge port 排出口
discharge porting 排气口
discharge potential 放电电位
discharge pressure 排出压力
discharge printing agent 拔染剂
discharge printing paste 拔染印色浆
discharge pump 排出泵
discharger 放电器
discharge ram 卸料推杆
discharge rate 放电速率
discharge rate number 排出速率数
discharge roller 出料辊
discharge side 卸料面
discharge spectrum 放电光谱
discharge spectrum detector 放电光谱检测器
discharge spout 漏嘴
discharge stack ①卸料管②排风管；(空气)净化管
discharge tank 出料桶
discharge temperature 排出温度
discharge tension 放电电压
discharge through gas 气体放电
discharge tube 卸料管；排出管
discharge valve 排出阀；卸料阀
discharge velocity 输送速度；排料速度
discharge velocity triangle (叶轮)出口速度三角形
discharge voltage 放电电压
discharge water 排放水；废水
discharging ①拔染②放电③卸料
discharging agent 拔染剂
discharging color(=discharging dye) 拔染染料
discharging current 放电电流
discharging dye(=discharging color) 拔染染料
discharging performance 排出性能
discharging potential 放电电势
discharging print 拔染印色
discharging pump 出料泵；排气泵
discharging rate 放电率
discharging side 卸料面
discharging time 出料时间
disc heat interchanger 盘式换热器
Dische's reaction 联苯胺反应
disc holder 圆盘支座
disc immunoelectrophoresis 盘状免疫电泳

disc integrator　盘式积分仪
disc meter　盘式流量计
disc mill　圆盘式粉碎机
disc mixer　圆盘式混合机
disc-needle tester　盘针式(耐磨)试验仪
discoloration　脱色；褪色；变色；漂白
discoloration method　色变法〔热稳定试验〕
discolored　脱(了)色的；褪(了)色的
discolored polymer　泛色聚合物
discolor hole　孔黑；孔灰；氧化
discoloring　脱色的；褪色的
discoloring agent　脱色剂
discoloring antioxidant　脱色性防老剂
discoloring clay(=discoloring earth)　漂白土；脱色土
discoloring earth(=discoloring clay)　漂白土；脱色土
discoloring type anti-oxidant　变(脱)色型防老化剂
discolor on aging　老化退色(变)
discoloured polymer　泛色聚合物；变色聚合体
discolourization　褪色
discolour spectrophotometry　褪色分光光度法
disconnect centrifuge　分离式离心机
discontinuity　不连续性；间断性
discontinuity layer　①不连续层②跃变层
discontinuity stress　间断性应力
discontinuous　不连续的；间断的；间歇的
discontinuous band absorption　不连续带吸收
discontinuous buffer system　不连续缓冲系统
discontinuous centrifuge　间歇式离心机
discontinuous countercurrent distributor　非连续式逆流分配器
discontinuous crystallizer　间歇式结晶器
discontinuous cyanamide process　间歇式氰胺盐制造法
discontinuous degassing　间歇脱泡；静止脱泡
discontinuous distribution　不连续分配
discontinuous extrusion　不连续挤塑
discontinuous fiber　定长纤维
discontinuous filter　间歇式过滤器
discontinuous gradient　不连续梯度
discontinuous-metallic filament　金属短纤维；不连续金属丝
discontinuous microelement grinder　间歇式超微细离心粉碎机
discontinuous operation　间歇操作
discontinuous orientation　不连续取向
discontinuous pasteurizer　间歇式巴氏杀菌器
discontinuous phase　不连续相
discontinuous polycondensation　间歇缩聚；分批缩聚
discontinuous presteaming　不连续酒料蒸煮法
discontinuous process　间歇过程
discontinuous rectifying plant　间歇精馏厂
discontinuous simultaneous differential thermal analysis and gas chromatography　差热分析-气相色谱间歇联用法
discontinuous simultaneous technique　间歇联用技术
discontinuous sterilization　间歇消毒；分步消毒
discontinuous welding　断续焊(接)
discordant polar system　参差极性系统
discotic phase　盘状相
discount present value　贴现值
discous　盘状的
disc pulverizer　盘式粉磨机
discrasite　锑银矿
discrete　个别的；分立的；分在的
discrete adsorption　选择吸收；离散吸收
discrete aggregate　离散的聚集体(凝聚体)
discrete amount　个别量
discrete distribution　分散分布；离散分布
discrete dynode　分立倍增极
discrete dynode multiplier　分立式电极倍增器
discrete energy level　离散能级
discrete fast Fourier transform　离散快速傅里叶变换
discrete flaw　分散性疵点；分散裂缝
discrete Fourier transform　离散傅里叶变换
discrete globule　分散小球；离散小球
discrete material　分散(型)物料
discrete maximum principle　离散系统最大值原理
discrete method　分段法；非连续法；离散法
discrete molecular energy level　分立分子能级
discrete particle　个别微粒
discrete peak　离散峰
discrete phase　分散相；不连续相
discrete point　分立点；离散点
discrete random variable　离散型随机变量
discrete relaxation　不连续弛豫(谱)；离散弛豫(谱)
discrete relaxation time　离散弛豫时间
discrete spectral interval　离散光谱间隔
discrete spectrum　离散光谱
discrete spot　离散斑
discrete stage contact　分级接触
discrete structure　个别结构
discrete system　离散系统；分立系统
discrete variational (molecular orbital) method(DVM)　离散变分(分子轨道)法
discrete variation method(DVM)　离散变分方法
discrete viscoelastic spectra　离散黏弹谱
discriminant analysis　判别分析*
discriminant function　判别函数

discriminating threshold　甄别阈
discrimination test　鉴别试验；甄别试验
discriminator　甄别器；鉴别器
discriminator circuit　甄别电路
disc-ring reactor　盘环型(搅拌缩聚)反应器
disc rotor　盘形叶轮
disc ruling　盘划线
disc sander　砂轮机
disc-sanding(=disk-sanding)　砂纸圆盘打磨；圆盘砂光
disc saw　圆盘锯
disc-shaped micelle　盘状胶束
disc tower　蝶式塔
disc-type atomizer　盘式雾化器；盘型喷雾器
disc-type impeller　盘式高速搅拌机
discutient　①消肿药②消肿的
disc valve　盘形阀
disease yeasts〔复〕　致病酵母
disebacate　癸二酸氢盐(或酯)
di-secondary　二仲的
diselenide　联硒化物　RSe·SeR
disengaged vapor　释放汽
disengagement　分离；分相
disengaging gear　齿轮离合器；脱开装置
disengaging zone　分离区
disentanglement　解开；解缠结
disentrainment　①雾沫分离②卸下
disentrainment section　除雾末(夹带)段
diseptal(=uleron)　二甲氨磺酰磺胺
diseptal C(=disulon)　磺酰磺胺
disfigurement　外貌损伤；损形；瑕疵
disgorging　除去沉淀物
disgregation　分散(作用)
dish　碟；皿；盘；盆
dish baffle　盘形挡板
dished　凹陷；表面下凹；凹状扭曲
dished bottom(=dished head)　碟形底；碟形头
dished-bottom tank　带碟形底的储罐
dished head(=dished bottom)　碟形底；碟形头
dished lid　碟形盖
dished surface　凹面
dish-ended　碟形底的
dish gas holder　湿式气柜
dish roaster　焙烧碟；盘式焙烧炉
dishwasher　洗碟机
dishwasher detergent　洗碟机用洗涤剂
dish washing agent　洗碟剂
dishwashing (foam) test　碟洗(泡沫)试验
dishwashing test　碟洗试验

disilane　乙硅烷　Si_2H_6
disilanoxy　乙硅烷氧基　H_3SiSiH_2O-
disilanyl　乙硅烷基　H_3SiSiH_2-
disilanylamino　乙硅烷氨基　H_3SiSiH_2NH-
disilanylene　亚乙硅烷基　$-SiH_2SiH_2-$
disilanylthio　乙硅烷硫基　H_3SiSiH_2S-
disilazane　二硅烷基胺；二硅氮烷
disilazanoxy　二硅氮烷氧基；甲硅烷氨基甲硅烷氧基　$H_3SiNHSiH_2O-$
disilazanyl　二硅氮烷基；甲硅烷氨基甲硅烷基　$H_3SiNHSiH_2-$
disilazanylamino　二硅氮烷氨基；二甲硅烷氨基　$H_3SiNHSiH_2NH-$
disilicic acid　焦硅酸　$(H_2Si_2O_6)_n$
disilico(n) ethane　乙硅烷
disilmethylene　亚甲基二硅(形)
disiloxane　二硅氧烷；二甲硅醚　$(SiH_3)_2O$
disiloxanoxy　二硅噁烷氧基；甲硅醚氧基；甲硅烷氧代甲硅醚氧基　$H_3SiOSiH_2O-$
disiloxanyl-　甲硅醚基；甲硅烷氧代甲硅烷基　$H_3SiOSiH_2-$
disiloxanylamino　二硅噁烷氨基；甲硅醚氨基；甲硅烷氧代甲硅醚氨基　$H_3SiOSiH_2NH-$
disiloxanylene　甲硅醚亚基　$-SiH_2OSiH_2-$
disiloxanylthio　二硅噁烷硫基；甲硅醚硫基　$H_3SiOSiH_2S-$
disilthiane　二硅硫醚　$H_3SiSSiH_3$
disilthianoxy　二硅噻烷氧基；甲硅硫醚氧基　$H_3SiSSiH_2O-$
disilthianyl　二硅噻烷基；甲硅硫醚基　$H_3SiSSiH_2-$
disilthianylthio-　二硅噻烷硫基；甲硅硫醚基　$H_3SiSSiH_2S-$
disilver salt　二银盐
disilyldisilanyl　三甲硅烷代硅基　$(H_3Si)_3Si-$
disinfectant　①消毒剂；灭菌剂②消毒的
disinfectant soap　消毒皂
disinfecting action　消毒作用
disinfecting agent　杀菌剂
disinfecting apparatus　消毒器
disinfection　消毒(作用)
disinfector　消毒器
disinfestation　灭(昆)虫法
disinomenine(=dehydrosinomenine)　双青藤碱；双汉防己碱
disinsection　灭(昆)虫法
disintegrability　崩解能力；可崩解性
disintegrated chalk　干磨碳酸钙；重质碳酸钙
disintegrating ability　分裂本领

disintegrating agent　崩解剂
disintegrating machine　粉碎机；研磨机；切片机
disintegration　①分裂；分解②蜕变
disintegration chain　蜕变链
disintegration coefficient　蜕变系数
disintegration constant　蜕变常数
disintegration energy　蜕变能
disintegration product　蜕变产物
disintegration rate　蜕变率
disintegration scheme　蜕变方式
disintegration series　蜕变系(列)
disintegration theory　蜕变(学)说
disintegrator　①粉碎机；切片机②清棉机；开毛机〔纺〕
disintegrator gas washer　喷散式涤气机
disintoxication　解毒
disjoining pressure　分离压
disk(=disc)　①(圆)盘②片
disk bottom　碟型封头
disk bursting test　轮盘破裂试验
disk centrifuge　碟式分离机
disk colourimeter　圆盘比色计
disk comparator　旋盘比色计
disk-controlled flame　(用流线型)圆控制(的火)焰
disk conveyer　圆盘式输送机
disk crusher　盘式破碎机
disk-disk rheometer　双盘式流变仪
disk dryer　圆盘(式)干燥器
disked bottom　碟形底盖
disked closure　碟形顶盖
disked end　碟形封头
disk electrode　圆盘电极
disk electrophoresis　圆盘电泳
disk feeder　圆盘加料机
disk filter　圆板过滤器
disk friction　轮盘摩擦；圆盘摩擦
disk friction loss　轮盘摩擦损失
disk grinder　圆盘研磨机
disk laser　圆盘(形)激光器
disklike molecule　盘状分子
disk meter　圆盘计量器
disk pre-filter　圆盘预滤器
disk-ring electrode　盘环电极
disk-ring reactor　盘环型(搅拌缩聚)反应器
disk selecting mechanism　圆齿片式选针机构
disk-shaped　圆盘形的
disk sprayer　盘式喷涂机；盘形喷涂机
disk stirrer　圆盘搅拌器
disk storage　磁盘储存器

disk type electrostatic sprayer　转盘式静电喷涂机
disk-type fibre opener　盘式纤维开松机
disk type friction false twister　圆片式摩擦假捻器
disk valve　盘形阀
disk viscometer　盘式黏度计
dislocation　位错*
dislocation array　位错列
dislocation damping　位错阻尼
dislocation glide　位错滑移
dislocation line　位错线
dislocation loop　位错环
dislocation motion　位错运动
dislocation nucleation　位错成核(过程)
dislocation rearrangement　位错重排
dislocation theory　位错理论
dislodged sludge　沉积泥渣
dislodger　沉积槽
disluster　失光；失去光泽
dismantling　拆除；拆卸
dismounting　卸下；拆卸
dismulgan Ⅲ　狄司摩根Ⅲ〔石油破乳化剂〕
dismutase　歧化酶
dismutation　歧化(作用)
di-soap　二酸皂
disodium acetylene　乙炔二钠　C_2Na_2
disodium compound　二钠化合物
disodium EDTA　乙二胺四乙酸二钠盐
disodium ethylene bisdithiocarbamate　亚乙基二(二硫代氨基甲酸钠)　$[NaSSCNH_2CH_2]_2$
disodium ethylene diamine tetraacetate　乙二胺四乙酸二钠〔稳定剂〕
disodium hydrogen phosphate　磷酸氢二钠　Na_2HPO_4
disodium methyl arsonate　甲基胂酸二钠　$CH_3AsO(ONa)_2$
disodium monoalkyl phosphate　单烷基磷酸二钠
disodium N-octadecyl sulfosuccinate　N-十八烷基磺基琥珀酸二钠〔稳定剂〕
disodium phosphate　磷酸(氢)二钠
disodium salt　二钠盐
disodium succinate　琥珀酸二钠
disodium sulfonate　磺酸二钠
disodium sulfosuccinate　磺基琥珀酸二钠
disodium tetraborate　四硼酸二钠　$Na_2B_4O_7$
disomatic　二晶质的
disopyramide　丙吡胺〔药〕
disorder　无序
disordered　无序(态)的；不规则的
disordered carbon　无序碳；不正常碳

disordered chain 不规则链；不整齐链
disordered chain propagation 不规则链增长
disordered fold surface 不规则折叠面
disordered orientation 无序取向
disordered region 无序区
disordered state 无序状态
disordered structure 无规结构；无序结构
disordering agent 破序剂
disordering effect 无序化效应
disorganized fiber (分子)排列紊乱的纤维
disorganized form 无规状态
disorientation 解取向；乱取向〔物〕
disorientation time 解取向时间
disoriented polymer 解取向聚合物
disoxidant(=disoxidation agent) 脱氧剂
disoxidation(=dioxidation; deoxidation) ①脱氧(作用)②还原(作用)
disoxidation agent(=disoxidant) 脱氧剂
disparking machine 剥皮机
dispatching 装运；分配；装货
dispenser ①分配器②给料器③自动售货机④药剂师
dispensing 配药(方)
dispensing balance 药剂天平
dispensing equipment 配料设备；分配油的装置
dispensing pump 配剂泵；分配泵
dispensing section 分配班；分配段
dispensing station 分配站
dispensing test (油和燃料)分配器的检验
disper 分散机
dispergation 解胶；胶液化(作用)
dispergator 解胶剂
dispermatle 高速变速搅拌机
dispersal mill 分散式研磨机
dispersancy 分散力；分散性
dispersancy index 分散指数
dispersant 分散剂
dispersant agent 分散剂
dispersant-coated particle 分散剂包覆粒子
dispersant efficiency 分散剂效率
dispersant layer 分散剂层
dispersate 分散质
dispersed 分散(了)的
dispersed aerosol 分散型气溶胶
dispersed color 分散颜料
dispersed component 分散组分
dispersed dyestuff 分散(性)染料
dispersed light 弥散光；色散光
dispersed medium 分散介质

dispersed particle 分散粒子
dispersed phase 分散相
dispersed substance 分散介质
dispersed sulfur dye 分散硫化染料
dispersed system 分散体系
dispersed weave (玻璃纤维)分散织法
disperse dyes 分散染料
disperse dyestuffs 分散染料
disperse medium 分散介质
disperse mill ①乳化机；乳化磨②分散磨
disperse phase 分散相*
disperser ①(蒸馏塔中的)泡罩②分散混合器③分散剂
disperser blade 分散叶片
disperse reinforcement 分散性强化复合材料
disperser hood (蒸馏塔中)泡罩
disperse system 分散系统*
dispersibility 分散性；分散能力
dispersible 可分散的
dispersible particle 分散性粒子
dispersidology(=colloid chemistry) 胶体化学
dispersimeter (光学)微粒计
dispersing ①分散的②分散(作用)
dispersing additive 分散添加剂
dispersing agent 分散剂
dispersing aid 分散助剂
dispersing auxiliary 分散助剂
dispersing coefficient ①色散系数②分散系数
dispersing element 扩散元件；散射元件
dispersing lens 发散透镜
dispersing medium ①分散剂②分散介质
dispersing power 分散能力
dispersing-type spectroradiometer 色散型分光辐射计
dispersion ①分散(作用)②分散体③色散；色散率④离差
dispersion agent 分散剂
dispersion aids 助分散剂
dispersion analysis 分散分析
dispersion coating 分散性涂料
dispersion coefficient 分散系数
dispersion colloid 分散胶体
dispersion color 分散色料
dispersion component 色散分量
dispersion contactor 分散混合器〔使被混合的液体喷射成雾状〕
dispersion correction ①色散校正②分散度校正
dispersion curve 色散曲线；散射曲线
dispersion degree 分散度
dispersion effect 分散效应

dispersion efficiency　分散效率
dispersion factor　分散因子
dispersion force　色散力*
dispersion impeller　分散盘(叶轮)
dispersion index　①分散指数②色散指数
dispersion interaction force　色散力
dispersion law　①色散定律②分散定律
dispersion length　色散长度
dispersion machine　分散机
dispersion medium　分散介质
dispersion method　分散法
dispersion micelle　分散胶束
dispersion mill　分散研磨机；分散磨
dispersion mode　色散模式
dispersion of distribution　分布宽度
dispersion of refractive index　折射率分散；折射率色散；扩散折射率
dispersion of specific rotatory power　旋光色散
dispersion paint　分散性涂料；乳化漆
dispersion phase　分散剂；分散外相
dispersion phenomena of viscoelastic body　黏弹体色散现象
dispersion polymerization　分散聚合(作用)
dispersion potential　分散电位
dispersion prism　色散棱镜
dispersion process　分散过程
dispersion rate　分散速率
dispersion ratio　分散率
dispersion reclaiming process　分散(再生)法
dispersion resin　分散树脂
dispersion rubber　(水)分散橡胶
dispersion sensitive property　分散敏感性
dispersion signal　色散信号
dispersion spectrum　色散谱
dispersion stabilization　分散稳定性
dispersion strengthened coating material　弥散强化材料
dispersion system　分散体系
dispersion technology　分散技术
dispersion test　分散实验
dispersion torque　分散转矩
dispersity(=dustability)　分散度
dispersity ratio　①色散度比②弥散度比
dispersive　分散的
dispersive ability　分散能力
dispersive action　分散作用
dispersive capacity(=dispersive power)　分散本领；分散力
dispersive element　色散元件

dispersive infrared spectrometer　色散型红外光谱仪
dispersiveness　分散度
dispersive power(=dispersive capacity)　分散本领
dispersive Raman scattering　色散型拉曼散射仪
dispersive-slewed scan　色散旋转扫描
dispersive spectrometry　色散分光法
dispersive system　色散系统
dispersivity　分散性
dispersivity test　分散性试验；分散度测定
dispersoid　分散(胶)体；分散质
dispersol yellow　散胶黄〔乙纤丝的染料〕
disphenoid　双半面晶体
displaceable　可排代的；可置换的
displaced　排代的；代替的
displaced term　置换项；排代项
displaced volume　排出体积；置换体积
displacement　①排代；置换；取代②位移
displacement analysis　顶替法；置换分析
displacement angle　①(介电损耗)位移角②(长丝缠绕)进缠角
displacement chromatography　顶替色谱(法)
displacement component　位移分量
displacement crosslinking　置换交联(作用)
displacement current　置换电流
displacement development　顶替展开(法)
displacement effect　顶替效应
displacement flux　位移通量
displacement gradient　位移梯度
displacement heat exchanger　容积式换热器
displacement law　①排代(定)律；置换定律②位移定律
displacement liquid　①顶替溶液②置换液③封闭液
displacement nitration process　排代硝化法
displacement oil pump　排代油泵；活塞式油泵
displacement polymerization　置换聚合
displacement process　排代法；置换法
displacement ratio　置换率
displacement reaction　置换反应*
displacement rule　排代规则
displacement series　排代(次)序；置换顺序
displacement substoichiometry　排代亚化学计量法
displacement titration　排代滴定法；置换滴定法
displacement type level transmitter　位移式液位传感器
displacement-type liquid meter　根据排代原理运转的流量计
displacement-type pump　活塞泵
displacement volume　排代体积
displacement washing　置换洗涤
displace meter　变位流量计

displacer 浮筒；平衡浮子；置换剂；置换器
displacing agent 顶替剂
displacing development 顶替展开法
displacing elution 顶替洗脱
displacing liquid 顶替液
displacing solution 顶替溶液
display ①显示；显像②显示器；指示器
display board 显示屏；指示盘
display instrument 显示仪表
display stands 陈列柜
display system 显示系统
display unit 显示部件；显示单元；显示装置
disposable ①一次性的；用完扔掉的②可处理的③可自由使用的
disposable item ①废品②一次应用制品
disposable nonwovens 用即弃型非织造布
disposable packaging 不回收包装；可弃包装
disposal 排列；配置；处理；整理
disposal by land 陆地处置
disposal by sea 海洋处置
disposal facilities〔复〕 排放设备
disposal of sewage 污水处理
disposal site 垃圾处理场
disposition 安排；布置
disposition of cavity 模槽排列
disproportionated rosin 歧化松香
disproportionation 歧化反应*
disproportionation catalyst 歧化催化剂
disproportionation condensation 歧化缩合作用
disproportionation of hydrogen 氢的重新分配
disproportionation reaction 歧化反应
disproportionation termination 歧化终止
disproportionative condensation 歧化缩合作用
disrotation 对旋
disrotatory 对旋
disruption 破裂(作用)
disruptive 破裂的
disruptive coloration ①混杂色②虎斑色③斑驳辉映彩绘法(晕色法)
disruptive discharge 击穿放电〔电〕
disruptive disk method 裂盘法〔用于测定爆震力〕
disruptiveness 破裂性
disruptive oxidation 破坏氧化(作用)
disruptive selection 歧化选择；分裂选择
disruptive strength 破坏强度
disruptive voltage 击穿电压
disrupture ①破裂②毁坏
disrupture force 破裂力

dissemination ①散布(作用)；传播②分散(作用)
dissimilar thread 夹花丝
dissimilation 异化作用
dissimilation plasmid 降解质粒
dissipation 散逸(作用)；分散
dissipation coefficient 损耗系数
dissipation factor 损耗因子；分散因子
dissipation loss 介电损失
dissipation of energy 能量散逸
dissipation of heat 热分散；热散逸(作用)
dissipative effect 损耗效应
dissipative energy 散逸能
dissipative structure 耗散结构
dissociant (微生物)变异株
dissociated 解离的
dissociating ①离解的②离解(作用)
dissociating force 离解力
dissociating group 离解基团
dissociating medium 离解介质
dissociating method 离解法
dissociating power 离解本领；离解能力
dissociating solvent 离解性溶剂
dissociation 离解*
dissociation channel 解离通道
dissociation chemisorption 离解化学吸附作用
dissociation constant 离解常数
dissociation constants of acids 酸的解离常数
dissociation degree 离解度
dissociation energy 解离能*
dissociation energy of bond 键解离能
dissociation enhanced lanthanide 离解增强镧系元素
dissociation equilibrium 解离平衡
dissociation extraction 解离萃取
dissociation factor 解离因子
dissociation heat 离解热
dissociation kinetics 离解过程动力学；溶解过程动力学
dissociation limit 解离极限
dissociation of molecules 分子离解
dissociation pressure 离解压力
dissociation process 离解过程
dissociation product 离解产物
dissociation tension 离解张力
dissociation threshold 解离阈值
dissociative 离解的
dissociative chemisorption 离解化学吸附
dissociative equation 离解方程(式)
dissociative ligand exchange mechanism 配体交换的解离机理

dissolubility ①溶(解)度②溶(解)性；(可)溶性
dissoluble 可溶的；能溶(解)的
dissoluble resin content 可溶性树脂含量
dissolution 溶解(作用)
dissolution boiler 溶化锅
dissolution heat 溶解热
dissolution method 溶样方法
dissolution wave 溶解波
dissolvability ①溶(解)度②溶(解)性；(可)溶性
dissolvable 可溶的；能溶(解)的
dissolvane 溶解烷
dissolvant 溶剂；溶媒
dissolved 溶解的
dissolved acetylene 液化乙炔〔在钢瓶中溶于丙酮〕
dissolved-air floatation (加压)气浮法
dissolved bone 溶解的骨粉
dissolved bone compound 溶骨化合物
dissolved gases 溶解的气体
dissolved impurity 溶解的杂质
dissolved matter 溶解物
dissolved oxygen 溶解氧
dissolved oxygen analyzer 溶解氧分析器
dissolved oxygen detector 溶解氧检测器
dissolved oxygen determination 溶解氧测定
dissolved oxygen meter 溶解氧计
dissolved-oxygen sag curve 溶解氧下垂曲线
dissolved solid 溶解固体
dissolved sulfide 溶存硫化物
dissolved tar 溶解焦油
dissolvent(=solvent) 溶剂
dissolver 溶解器
dissolver basket 溶解器的篮筐
dissolving ①熔化〔冶〕②溶解③溶解的
dissolving capacity 溶解力
dissolving drum 溶解鼓
dissolving metal reduction 溶解金属还原
dissolving pan 溶解锅
dissolving power 溶解力
dissolving pulp 溶解纸浆；化学纸浆
dissolving salt B 苄氨基苯磺酸钠；溶解盐 B
dissolving tank 溶解槽；溶解池；溶解罐
dissolving vat 溶解瓮
dissolving vessel 溶解罐(槽)
dissonance 不谐和；非谐振
dissymmetric 不对称的
dissymmetrical structure 不对称结构
dissymmetric molecule 不对称分子
dissymmetric peak 不对称峰

dissymmetry 不对称现象
dissymmetry coefficient 不对称系数
dissymmetry of scattering 散射的非对称性
distacin(=distacyne) 偏端菌素
distacyne(=distacin) 偏端菌素
distamycin 偏端霉素
distance matrix 距离矩阵
distance piece 位置调节片；垫片
distance recorder 遥测记录器
distance ring 定距环
distance velocity lag 距速滞后〔石油加工过程自动控制的术语〕
distance washer 垫片
distannic compound 二(正)锡化合物
distannic ethide ①三乙锡(游基)②六乙二锡 $Sn_2(C_2H_5)_6$
distannoxane 二锡氧烷
distant (reading) thermometer 遥测温度计
distarch glycerol 甘油双淀粉
distaval(=thalidomide) 反应停；沙立度胺〔镇静药,由于引起婴儿畸形；1962 年停售〕
distearin 二硬脂精；甘油二硬脂酸酯 $(C_{17}H_{35}COO)_2C_3H_5OH$
distearodaturin 二硬脂一苷精；甘油一苷酸二硬脂酸酯
distearolin 二硬脂精；甘油二硬脂酸酯
distearoylethylenediamine 二硬脂酰乙二胺
distearyl amine 二硬脂胺；双十八(烷)基胺
distearyl dimethyl ammonium chloride 二硬脂基二甲基氯化铵；双十八烷基二甲基氯化铵
distearyl ether 二硬脂醚〔脱模剂〕
distearyl pentaerythritol diphosphite 二硬脂酰二亚磷酸季戊四醇酯〔防老剂〕
distearyl thiodipropionate 硫代二丙酸二硬脂酰酯〔防老剂〕
distemper 刷墙粉；水浆涂料
distemper colors 墙粉色(颜)料；水浆涂料色(颜)料
distensibility 伸长性；扩张性
disthene 蓝晶石
distillable 可蒸馏的
distillable fraction 可蒸馏的部分
distilland 被蒸馏物；蒸馏液
distillate 馏出液
distillate cooler 蒸馏冷却器
distillate curve 馏出物(曲)线
distillate fuel 馏出的燃料
distillate fuel oil 馏出燃料油
distillate oil 馏出油
distillate wax 石蜡馏分

distillating ①蒸馏的②蒸馏(作用)
distillating still 蒸馏釜
distillation 蒸馏(作用)
distillation apparatus 蒸馏器
distillation bench 蒸馏装置组
distillation cascade 蒸馏级联
distillation chamber 蒸馏室
distillation characteristic 蒸馏特性
distillation column 蒸馏柱
distillation column network 蒸馏塔网络
distillation curve 蒸馏曲线
distillation cut 馏分
distillation effect 蒸馏效应
distillation end point 蒸馏终点
distillation equipment 蒸馏装置
distillation film 蒸馏薄膜
distillation flask 蒸馏瓶
distillation gas 蒸馏气
distillation loss 蒸馏损失
distillation method 蒸馏法〔确定馏分组成的方法〕
distillation plant 蒸馏设备
distillation plate calculation 蒸馏塔板的计算
distillation plus loss activity 馏出量加损失活性〔催化剂〕
distillation range 馏程；沸程；蒸馏范围
distillation rate 蒸馏速率
distillation reactor 蒸馏反应器
distillation refluence ratio 蒸馏回流比
distillation residue 蒸馏残渣
distillation still 蒸馏釜
distillation tailings 尾馏分
distillation test 蒸馏试验
distillation test residue 蒸馏后的残渣
distillation tower 蒸馏塔
distillation tray 蒸馏塔板
distillation tube 蒸馏管
distillation under pressure 加压蒸馏
distillation value 蒸馏值
distillation yield 馏出产量
distillation zone 蒸馏区
distillator 蒸馏器
distillatory ①蒸馏的；蒸馏用的②蒸馏器
distillatory kettle 蒸馏锅
distillatory vessel 蒸馏器
distilled 蒸馏过的；馏出的；馏得的
distilled fatty acid 蒸馏脂肪酸
distilled gasoline 直馏汽油
distilled lime oil 蒸馏白柠檬油
distilled oil 馏出油
distilled oil of vitriol 精馏的硫酸
distilled oleine 馏过的甘油三油酸酯；馏油精〔俗〕
distilled pyroligneous acid 蒸馏木醋液
distilled-to-dryness 蒸干
distilled water 蒸馏水
distilled water corrosion 蒸馏水腐蚀
distiller 蒸馏器
distiller condenser 蒸馏冷凝器
distiller's dried grain 干酒糟
distiller's grain 酒糟
distiller's malt 酿酒麦芽
distiller's solubles 酒糟
distiller's wort 酿酒麦芽汁
distiller's yeast 酒曲；酿酒酵母
distillery ①蒸馏室②(造)酒厂
distilling ①蒸馏的②蒸馏(作用)
distilling apparatus 蒸馏装置
distilling column ①蒸馏柱②蒸馏塔
distilling flask 蒸馏瓶；分馏烧瓶
distilling furnace 蒸馏炉
distilling head 蒸馏头
distilling machinery 蒸馏机(器)
distilling plant 蒸馏装置
distilling tower 蒸馏塔
distilling tray 蒸馏塔板
distilling tube 蒸馏管
distilling vacuum flask 真空蒸馏瓶
distil overhead 馏过头
distinctive temperature 特色温度；相异温度
distinctness ①差别②清晰度
distinguishable electron method(DEM) 可区分电子法
distinguishing test 区别试验
distort ①扭(畸)变，变形，翘曲②失真
distorted lattice 畸变晶格
distorted loops 三角眼
distorted peak 畸峰*
distorted-wall column 扭变壁管蒸馏塔
distorted-wave Born approximation(DWBA) 畸变波玻恩近似(法)
distortion 扭变；畸变；变形；失真
distortion energy theory 歪形能理论
distortion failure condition 畸变衰坏条件
distortion point 扭变点；畸变点
distortion silk 变形生丝
distortion temperature 热变形温度
distortion under heat 热扭变
distortion wave(=standing wave) 驻波

distributed charge 分布电荷
distributed control system 集散控制系统
distributed source 分布源
distributer(=distributor) ①分配器②配电盘
distributing 分配的
distributing agent 分散剂
distributing board 配电盘
distributing box 分配槽
distributing channel(=distributing duct; distributing flue) 配水渠
distributing duct(=distributing channel) 配水渠
distributing flue(=distributing channel) 配水渠
distributing gutter(=distributing trough) 配水槽
distributing main ①配电干线〔电工〕②(流体)输送干线
distributing manifold 分配的干线
distributing oil box 润滑油中央加油器
distributing plate 分布板
distributing reservoir 分配池
distributing trough 配水槽
distributing well 配水井
distribution ①分配；分布②配电
χ^2-distribution χ^2分布
distribution board 配电盘
distribution chromatography 分配色谱
distribution coefficient(=distribution number;distribution ratio) 分配系数
distribution constant 分配常数
distribution curve 分布曲线
distribution density 分布密度
distribution depot 分配仓库
distribution diagram 分布图
distribution fraction 分布分数
distribution function 分布函数
distribution function of chain conformations 链构象的分布函数
distribution function of lag time 滞留时间分布函数
distribution header 分配总管
distribution hopper 分配料斗
distribution isotherm 分配等温线
distribution law 分配定律*
distribution license 配给执照
distribution line ①配电线路②配料线
distribution moment 分布矩
distribution number(=distribution coefficient) 分配系数
distribution octane number 分布辛烷值
distribution of degree of polymerization 聚合度分布
distribution of droplet mass 滴量分布
distribution of molecular weight 分子量分布
distribution of particle size 粒度分布
distribution of polymerization degree 聚合度分布
distribution of pore size 孔径分布
distribution of radiant heat 辐射热分布
distribution of reflux (塔内)回流分布
distribution of relaxation times 弛豫时间分布
distribution of residence-time(DRT) 保留时间分布
distribution of retardation time 推迟时间分布
distribution of sizes 粒度分布
distribution of velocity 速率分布
distribution on number basis 按数量分布
distribution on weight basis 按重量分布
distribution parameter 分布参数
distribution pattern (颜料粒径)分布方式
distribution principle 分配定则；分配原理
distribution ratio(=extraction coefficient;distribution coefficient) 萃取系数；(萃取)分配比
distributive channel 配水渠(槽)
distributor(=distributer) ①分配器②配电盘
distributor block 纺丝箱；分配箱
distributor roller 分печатногоприпуска 分散辊；传墨辊
disturbance 扰动；干扰
disturbance flow 扰动流；干扰流动
disturbing ①流云拉毛涂装法②扰动；(无规律地)搅乱
disturbing factor 干扰因素
disturbing frequency 干扰频率
distyrene 联苯乙烯 $(C_6H_5CH{=}CH)_2$
distyryl 联苯乙烯
disuberate 辛二酸氢盐(或酯)
disubstituted 双取代的；二基取代的
disubstituted carbinol 仲醇；二代甲醇 R_2CHOH
N,N'-disubstituted dithiocarbamate N,N'-取代二硫代氨基甲酸酯
N,N'-disubstituted oxalamide N,N'-双取代草酰胺
N,N-disubstituted sodium sulfanilate N,N-双取代对氨基苯磺酸钠
disubstitution 双取代作用；二基取代作用
disubstitution product 双取代物
disuccinate 丁二酸氢盐(或酯)
disulfanilamide 双磺胺；磺酰磺胺 $H_2NC_6H_4SO_2NHC_6H_4SO_2NH_2$
disulfide 二硫化物
disulfide bond 二硫键
disulfide linkage 二硫键
disulfide oil 含二硫化物的油〔从轻质石油馏分中抽出的硫醇氧化而成的二硫化物，以油状自抽出物中分出〕
disulfine blue 二硫蓝；专利蓝
disulfiram(=tetraethylthiuram disulfide) 二硫化四乙基秋

兰姆〔促进剂〕
disulfo-〔词头〕 二硫
disulfo-benzoic acid ①二磺基苯甲酸 $COOHC_6H_3(SO_3H)_2$ ②二硫代苯酸；苯荒酸 C_6H_5CSSH
disulfo-cyanic acid 二聚硫代氰酸 $(HCNS)_2$
disulfoisophthalic acid 二磺基间苯二甲酸
disulfole(=dithiole) 二噻环戊二烯
disulfo-metholic acid 甲二磺酸 $CH_2(SO_3H)_2$
disulfo-naphtholic acid 萘酚二磺酸 $OHC_{10}H_5(SO_3H)_2$
disulfonate 二磺酸盐（或酯）
disulfone 二砜
disulfonic acid 二磺酸 $R(SO_3H)_2$
disulfonic acid amide 二磺酰胺 $R(SO_2NH_2)_2$
disulfonic acid chloride 二磺酰氯 $R(SO_2Cl)_2$
disulfosuccinic acid alkyl ester sodium salt 二磺基琥珀酸烷基酯钠盐
disulfoxide 二亚砜 RSOSOR
disulfuric acid(=pyrosulfuric acid) 焦硫酸；一缩二(正)硫酸
disulon(=disulfanilamide;diseptal C) 磺酰磺胺；双磺胺
disulphonazo Ⅲ 二磺偶氮Ⅲ
disultone 二磺内酯
disvolution 退化；变形
disymmetrical 二对称的
disymmetry constant 双对称常数
disyndiotactic 双重间同立构的
disyndiotactic polymer 双间同立构聚合物；双间规聚合物
dit 小(孔)沙眼
ditactic 构型的双中心规整性
ditacticity 双有规立构度
ditactic polymer 双有规立构聚合物
ditaine(=echitamine) 面条树皮碱；鸡骨常山毒碱；艾奇明
ditallow amide ethylene ammonium acetate 二牛油酰氨亚乙基乙酸铵
ditallow dimethyl ammonium chloride 二牛油基二甲基氯化铵
ditan 二苯甲烷
ditane 二苯甲烷
ditartrate 酒石酸氢盐（或酯）
ditartronate 丙醇二酸氢盐（或酯）
ditch ①沟；渠②开沟；修沟
diterminal oxidation 双末端氧化
diterpene 二萜*
diterpenoid resin acid 二萜树脂酸
diterpenoid(s) 双萜（类）
di-tertiary 二叔的〔胺；醇〕

di-tertiary butyl peroxide 二叔丁基过氧化物
ditetrafurfuryl propyl phthalate 邻苯二甲酸双四糠丙酯〔增塑剂〕
ditetrahydrofurfuryl adipate 己二酸双四氢糠(醇)酯〔增塑剂〕
di-tetrahydrofurfuryl maleate 马来酸双四氢糠酯 $(CH_2CO_2CH_2C_4H_7O)_2$
ditetrahydrofurfuryl phthalate 邻苯二甲酸双四氢糠(醇)酯〔增塑剂〕
di-tetrahydrofurfuryl succinate 丁二酸双四氢糠酯
dithiadiazole 二噻二唑
1,4-dithiadiene 1,4-二噻二烯；对二硫杂环己二烯
dithiane 二噻烷*
dithiazine 二噻嗪
dithiazole 二噻唑
dithiene 二噻烯
dithienyl methane 二噻吩基甲烷
dithio ①联硫基 —SS—②二硫代
dithioacetal 二硫缩醛
dithioacetic acid 乙二硫代酸；二硫代乙酸 CH_3CSSH
dithioacid 二硫代羧酸；荒酸（俗） RCSSH
dithiobenil 双硫代苯偶酰素；二苯基二硫乙酮基
dithiobenzoic acid 二硫代苯甲酸 C_6H_5CSSH
dithiobisalanine(=cystine) 胱氨酸
dithio-bis-benzothiazole 二硫代二苯并噻唑〔促进剂〕
dithiobis-ethylenenitrilo-tetraacetic acid(BADDS) 二硫双亚乙基次氮基四乙酸
dithiobisxylene 二硫代双二甲苯〔再生活化剂〕
dithiobiuret 二硫代缩二脲 $NH_2CSNHCSNH_2$
dithiocarbamate(s) 二硫代氨基甲酸盐类
dithiocarbamic acid 二硫代氨基甲酸 $CS(NH_2)SH;C(NH)(SH)_2$
dithiocarbaminocarboxylic acid 脲二硫代羧酸 $NH_2CONHCSSH$
dithiocarbonic acid 二硫代碳酸 $CO(SH)_2;CS(OH)SH$
dithiocarboxylic acid 二硫代羧酸 RCSSH
dithiocyanogen 二硫代氰
dithiodiglycollic acid 亚二硫基二乙酸
dithioerythritol 二硫赤藓糖醇
1,2-dithio-glycerine 1,2-二硫代甘油 $C_3H_8OS_2$
dithioglycol 乙二硫醇 $C_2H_6S_2$
dithio-hydroquinone 二硫代氢醌；1,4-苯二硫酚 HSC_6H_4SH
dithiol(=toluene-3,4-dithiol) 二硫酚；甲苯-3,4-二硫酚
dithiolane 二硫戊环
dithiole(=disulfole) 二噻环戊二烯
dithionate ①连二硫酸盐 $M_2S_2O_6$ ②〔罕用，泛指〕二硫代羧酸盐 RCSSM

dithione　①二硫酮②硫特普
dithionic acid　①连二硫酸　$H_2S_2O_6$　②〔罕用，泛指〕二硫代羧酸　RCSSH
dithionite　连二亚硫酸盐
6,8-dithio-*n*-octanoic acid　6,8-二硫正辛酸；硫辛酸
dithionous acid　连二亚硫酸　$H_2S_2O_4$
dithio-oxamide(=rubeanic acid)　二硫代草酰胺；红氨酸　$(C=SNH_2)_2$
dithiophosphate　二硫代磷酸酯(盐)
dithiophosphinic acids　二硫代次膦酸
dithiophosphonate　二硫代膦酸酯
dithio-resorcin　二硫代间苯二酚；间苯二硫酚　HSC_6H_4SH
dithio-salicylic acid　二硫代水杨酸　HOC_6H_4CSSH
dithiosemicarbazone　二硫代缩氨(基)脲
dithiosuccinimide　二硫代琥珀酰亚胺
dithiothreitol(DTT)　二硫苏糖醇
dithiourazole　二硫代脲唑
dithiourethane　二硫代氨基甲酸乙酯
dithizonate　双硫腙盐
dithizone(=diphenyl thiocarbazone)　双硫腙；二苯基硫卡巴腙　$C_6H_5N=NCSNHNHC_6H_5$
dithizone extraction method　双硫腙抽提法
dithranol　1,8,9-蒽三酚
dititanate　一缩二亚钛酸盐
ditolyl　①联甲苯②二甲苯基　$(CH_3C_6H_4)_2$
di-*o*-tolyl carbodiimide　二(邻甲苯基)碳化二亚胺
ditolyl diketone　联甲苯酰；二(甲苯基)二甲酮　$CH_3C_6H_4C(O)C(O)C_6H_4CH_3$
ditolylene　联甲代亚苯基　$-C_6H_3(CH_3)C_6H_3(CH_3)-$
ditolylene diisocyanate　联亚甲苯基二异氰酸酯〔防老剂〕
di-*o*-tolylene diisocyanate　4,4-联二甲代亚苯基二异氰酸酯
di-*o*-tolyl-ethylene-diamine　二邻甲苯基乙二胺
ditolyl formamide　联甲苯甲酰胺
di-*o*-tolyl guanidine　二邻甲苯基胍
ditolylmethane　二甲苯基甲烷　$(CH_3C_6H_4)_2CH_2$
ditolyl suifide　二甲苯基硫；甲苯硫醚　$(CH_3C_6H_4)_2S$
ditolyl sulfone　二甲苯砜　$(CH_3C_6H_4)_2SO_2$
di-*p*-tolyl tin　二对甲苯锡
di (tridecyl) amine　双十三烷基胺　$(C_{13}H_{27})_2NH$
ditridecyl phthalate　邻苯二甲酸二(十三)酯〔增塑剂〕
ditridecyl thiodipropionate　硫代二丙酸二(十三)酯〔防老剂〕
di (3,5,5-trimethyl hexyl) adipate　己二酸二(3,5,5-三甲基己基)酯
di (trimethyl) silazane　二(三甲基)硅氮烷
ditropolonyl　联草酚酮
dittany　岩薄荷；白藓〔调香药草〕
dittany root　白鲜根
ditto　同前；同上
diumycin　久霉素
diundecyl ketone　双十一基(甲)酮；12-三烷酮　$(C_{11}H_{23})_2CO$
diundecyl phthalate　苯二酸十一烷酯
diuranate　重铀酸盐　$M_2U_2O_7$
diurea　双脲；环二脲；四氮杂己环对二酮　$CO=(NHNH)_2=CO$
diuresis　多尿(症)
diurethane　二氨基甲酸酯；二异氰酸酯
diuretic(s)　①利尿的②利尿剂
diurnal cycle　昼夜循环
diurnal heating　昼夜加热
diurnal load　每日负载；昼夜负载
divalent　二价的
divalent alcohol　二元醇；二羟醇；二价醇
divalent cation　二价阳离子
divalent cation exchange　二价阳离子交换
divalent element　二价元素
divalent radical　二价基
divalerin　甘油二戊酸酯
divalomic acid　双瓦落酸
divanadyl　二氧二钒根〔四价根〕　V_2O_2
divariant system　二变系；双变量系统
divaricatic acid　柔扁枝衣酸
divarication　分离；分散
divergence(=divergency)　①分流性；分散性②发散；散度
divergence angle　张角；扩散角；开度角
divergency(=divergence)　①分流性；分散性②发散；散度
divergent　发散的
divergent die　分流口模；扩张式模头
divergent slit　发散狭缝
divergent streams　分支流动
divergent structure　发散结构
diverse ion effect　异离子效应
diverse salt effect　异盐效应
diversine　青藤碱
diversion box　转移箱
diversity　①多样性；相异(性)②参差；分集
diversity factor　不等率；差异度
diverter　分流器
divided circle　分度盘
divided-flow purification　分流精制
divided manifold　分支管
divided rim　对开式轮辋

divided threadline extrusion 双丝束纺丝
dividend 被除数
divider 分规；双脚规；分离器
divider and rounder unit 搓圆机
divider wall 隔离墙；分水墙
dividing box 分块机
dividing head 分度头
dividing surface 分割表面
dividual draw 分区穿综法
divinol 芸实醇；低分子聚丁二烯
divinyl ①联乙烯；丁二烯；1,3-丁二烯；丁间二烯 $CH_2=CHCH=CH_2$ ②二乙烯基 $(CH_2=CH-)_2$
divinylacetylene 二乙烯基乙炔 $CH_2=CHC\equiv CCH=CH_2$
divinylbenzene 二乙烯基苯
divinylene imine(=pyrrole) 环二亚乙烯基亚胺；吡咯
divinylic monomer 丁二烯单体
divinyl rubber 丁二烯橡胶
divinyl sulfide 二乙烯基硫；乙烯硫醚 $(CH_2=CH)_2S$
divinyl thioether 二乙烯基硫醚
divinyltrichlorobenzene 二乙烯基三氯苯
1,1'-divinyluranocene 1,1'-二乙烯基代双环辛四烯合铀 $U(C_8H_7CH=CH_2)_2$
divisibility 解理性；可劈性〔晶〕
divisible inactive 可分性不旋光的〔指外消旋的〕
dixanthyl urea 二黄质基脲 $(C_{13}H_9ONH)_2CO$
Dixon packing 狄克松环
Dixon ring 狄克松环；金属网θ环〔填料〕
Dixon's test 狄克松检验
Dixon's test method 狄克松检验法
dixylene disulfide mixture 二甲苯基二硫化物混合物〔塑解剂〕
dixylyl disulfide 二甲苄基二硫化物〔塑解剂〕
diyne 二炔
dizirconyl 三氧二锆根〔二价根〕 Zr_2O_3
dizirconyl bromide 二溴三氧化二锆 $(Zr_2O_3)Br_2$
dizirconyl chloride 二氯三氧化二锆 $(Zr_2O_3)Cl_2$
dizirconyl nitrate 硝酸三氧化二锆 $(Zr_2O_3)(NO_3)_2$
djalmaite 钽钛铀矿
djave butter 毒雾冰草油
djenkol bean 今可豆；羊公豆
dl 分升
dm 分米
DNA amplifier DNA 扩增仪
DNA array DNA 阵列
DNA engineering DNA 工程
DNA enzyme DNA 酶
DNA microarray DNA 微集阵列
DNA microchip DNA 微集芯片
DNA polymerase DNA 聚合酶
DNA repair DNA 修复
DNA replication DNA 复制
DNA sensor 脱氧核糖核酸电化学传感器
DNA transcription processor DNA 转录处理器
DNP derivative 2,4-二硝基苯衍生物
DNP-hydrazine 2,4-二硝基苯肼
DNP-hydrazone 2,4-二硝基苯腙
dobby loom 多臂织机
dobby machine 多臂机；多臂织机
Dobereiner lamp(=Dobereiner match box) 多伯临纳发火器
Dobereiner match box(=Dobereiner lamp) 多伯临纳发火器
doble flame furnace 返焰炉
Dobson unit 多布森单位〔臭氧柱浓度单位〕
dobutamine 多巴酚丁胺〔药〕
docetaxel 多西他赛〔药〕
dockage 船渠费
docking (煤的)灰分评价
dockside fender 码头防冲装置
docosahexenoic acid 二十二碳六烯酸
docosandioic acid 二十二烷二酸；二十二烷双酸 $C_{20}H_{40}(COOH)_2$
n-docosane (正)二十二(碳)烷 $CH_3(CH_2)_{20}CH_3$
docosanoic acid 二十二烷酸；山萮酸 $C_{21}H_{43}COOH$
docosanol 二十二(烷)醇
docosatetraenoic acid 二十二碳四烯酸
docosendioic acid 二十烯双酸；二十碳烯二酸 $C_{20}H_{38}(COOH)_2$
docosene dicarboxylic acid 二十二烯二羧酸 $C_{22}H_{42}(COOH)_2$
docosenoic acid 二十二烯酸 $C_{21}H_{41}COOH$
cis-13-docosenoic acid(=erucidic acid;erucid acid) 芥酸；顺(式)13-二十二(碳)烯酸
docosoic acid 二十二烷酸；山萮酸 $C_{21}H_{43}COOH$
docosyl 二十二(烷)基 $CH_3(CH_2)_{20}CH_2-$
docosyl alcohol 二十二(烷)醇 $CH_3(CH_2)_{20}CH_2OH$
docosyl sulfate 二十二烷基硫酸盐
doctor-bar 刮片
doctor-blade application 刮涂施工
doctor blade coating(=knife coating) 刮刀涂漆法；刮涂法
doctor blade method 刮涂法
doctor kiss coater 辊刮涂机；轻触刮涂器
doctor knife 刮刀
doctor line(=doctor mark) 刮(刀)痕；刮(刀)条纹
doctor mark(=doctor streak; knife mark) 刮(刀)痕；刮(刀)

条纹〔刮刀口有伤，刮涂时留下的条状痕迹〕
doctor negative 低硫的
doctor positive 高硫的
doctor process(=doctor treatment) 博士法精制过程；亚铅酸钠精制过程
doctor roll 涂胶调节辊；涂胶量控制辊
doctor solution 博士溶液；亚铅酸钠溶液
doctor streaker 刮刀条痕〔刮涂缺陷〕
doctor sweetener 博士法(脱硫醇)精制装置；亚铅酸钠精制装置
doctor sweetening 博士法脱硫醇(或脱臭)；亚铅酸钠脱硫醇(或脱臭)
doctor sweet gasoline 博士试验阴性的汽油
doctor sweet product 博士试验阴性的油品
doctor test 亚铅酸钠试验〔检测硫〕
doctor treatment 博士法精制；亚铅酸钠精制
document paper 公文纸
dodder oil 菟丝子油
dodeca〔词头〕十二
dodecaboride 十二硼化物
dodecadienoic acid 十二碳二烯酸 $C_{11}H_{19}COOH$
dodecafluoroheptylacrylate 丙烯酸十二氟庚酯
dodecafluorocyclohexane 十二氟代环己烷 C_6F_{12}
dodecagon 十二边形
dodecahedrane 十二面烷
dodecahedron 十二面体
dodecahexaene 十二碳六烯
dodecalactone 十二内酯
δ-dodecalactone δ-十二内酯
dodecamethylene diamine 十二亚甲基二胺 $H_2N(CH_2)_{12}NH_2$
dodecamethylene diammonium suberate 辛二酸十二亚甲基二铵盐
dodecanal 十二(烷)醛
dodecandioic acid 十二双酸 $C_{10}H_{20}(COOH)_2$
n-dodecane (正)十二烷 $CH_3(CH_2)_{10}CH_3$
dodecane dicarboxylic acid 十四(烷)双酸 $C_{12}H_{24}(COOH)_2$
dodecanedioic acid 十二烷二酸
1,12-dodecanediol 1,12-十二烷二醇
dodecane sizing 十二烷油剂
dodecane sulfonate 十二烷(基)磺酸盐
dodecane sulfonic acid 十二烷磺酸
dodecanic acid 十二酸 $C_{11}H_{23}COOH$
dodecanoate 十二(烷)酸盐
dodecanoic acid 十二(烷)酸
dodecanol 十二烷醇
dodecanoyl 十二(烷)酰 $CH_3(CH_2)_{10}CO-$

dodecapentaene 十二碳五烯
dodecatriyne 十二碳三炔
dodecene 十二碳烯
dodecene dicarboxylic acid 十二碳烯二甲酸 $C_{12}H_{22}(COOH)_2$
dodecenedioic acid 十二碳烯双酸 $C_{10}H_{18}(COOH)_2$
dodecenoic acid 十二碳烯酸 $C_{11}H_{21}COOH$
dodecenylsuccinic anhydride 十二碳烯丁二酸酐；十二碳烯琥珀酸酐
dodecenyltrialkylmethylamine 十二碳烯基三烷甲基胺 $(C_{12}H_{23})(R_3C)NH$
dodecoic acid 十二酸 $C_{11}H_{23}COOH$
12-dodecosenoic acid 顺(式)-12-二十三碳烯酸；(Z)-$\triangle^{12}$-二十三碳烯酸
dodecyl 十二烷基 $CH_3(CH_2)_{10}CH_2-$
dodecyl acetate 乙酸十二(烷)酯 $CH_3CO_2C_{12}H_{25}$
dodecyl acrylate 丙烯酸十二(基)酯
dodecyl alanine 十二烷基氨基丙酸
dodecyl alcohol 十二(烷)醇
dodecyl aldehyde 十二醛；月桂醛〔俗〕
dodecyl amine 十二烷胺 $C_{12}H_{25}NH_2$
dodecylamine hydrochloride 十二胺盐酸盐
dodecyl amine thiosulfate 硫代硫酸十二烷基胺
dodecyl amino propionic acid 十二烷基氨基丙酸
dodecylammonium bentonite 十二烷基铵膨润土
dodecyl ammonium caprylate 十二烷基辛酸铵
dodecyl ammonium carboxylate 十二烷基羧酸铵
dodecyl anisidine 十二烷基茴香胺；十二烷基甲氧基苯胺
dodecylate 十二(烷)酸酯
dodecylated diphenyl ether disulfonate 十二烷基二苯醚二磺酸盐
dodecylbenzene 十二烷基苯
dodecylbenzene sulfonate 十二烷基苯磺酸盐
dodecylbenzene sulfonyl chloride 十二烷基苯磺酰氯
dodecylbenzylamine(DBA) 十二烷基苄基胺 $(C_{12}H_{25})(C_6H_5CH_2)NH$
dodecylbenzyl dimethyl ammonium bromide 十二烷基苄基二甲基溴化铵
dodecyl bromide(=lauryl bromide) 十二烷基溴 $CH_3(CH_2)_{10}CH_2Br$
dodecyl butyl polyvinyl pyridinium salt 十二烷基丁基聚乙烯吡啶鎓盐
dodecyl caproate 己酸十二(烷)酯 $C_5H_{11}COO(CH_2)_{11}CH_3$
dodecyl chloride(=lauryl chloride) 十二烷基氯 $CH_3(CH_2)_{10}CH_2Cl$
dodecyl cyanide 十二烷基氰；十三(碳)烷腈

$CH_3(CH_2)_{10}CH_2CN$

dodecyl dichlorobenzyl dimethyl ammonium chloride 十二烷基二氯(代)苄基二甲基氯化铵

dodecyl dimethyl phosphine oxide 氧化十二烷基二甲基膦

dodecylene 十二碳烯 $C_{12}H_{24}$

α,β-dodecylenic acid α,β-十二烯酸

dodecyl gallate(=lauryl gallate) 没食子酸十二烷基酯；没食子酸月桂酯

dodecyl isoquinolinium bromide 溴化十二烷基异喹啉鎓

dodecyl mercaptan 十二烷硫醇

dodecyl mercapto acetic acid 十二烷基巯基乙酸

1-dodecyl-2-methyl-3-benzylimidazolium chloride 氯化1-十二基-2-甲基-3-苄基咪唑鎓

dodecyl morpholine oxide 氧化十二烷基吗啉

4-dodecyloxy-2-hydroxybenzophenone 4-十二烷氧基-2-羟基二苯甲酮〔防老剂〕

dodecylphosphoric acid(DDPA) 十二烷基磷酸 $(C_{12}H_{25}O)PO(OH)_2$

dodecyl polyoxyalkanol 十二烷基聚氧烷醇

dodecyl polyoxy ethylene glycol 十二烷基聚氧乙烯(乙)二醇

dodecyl polyoxy ethylene sulfate 十二烷基聚氧乙烯硫酸盐

dodecyl polyvinyl pyridinium salt 十二烷基聚乙烯吡啶鎓盐

α-n-dodecyl-α-propyl-β-aminopropionic acid α-正十二烷基-α-丙基-β-氨基丙酸

dodecyl pyridinium halide 十二烷基卤化吡啶鎓

dodecyl pyridinium pentachlorophenate 十二烷基五氯苯酚吡啶鎓

dodecyl salicylate 水杨酸十二烷酯〔紫外线吸收剂〕

dodecyl sulfate 十二烷基硫酸盐

dodecyl taurine 十二烷基牛磺酸；十二烷基氨基乙磺酸

dodecyltriethylphosphonium 十二烷基三乙基膦鎓

dodecyltrimethyl ammonium chloride 氯化十二烷基三甲铵〔表面活性剂〕

6-dodecyl-2,2,4-trimethyl-1,2-dihydroquinoline 6-十二烷基-2,2,4-三甲基-1,2-二氢喹啉

dodecyne 十二(碳)炔 $C_{12}H_{22}$

dodecynoic acid 十二碳炔酸 $C_{11}H_{19}COOH$

Dodge crusher 道奇压碎机

Dodge jaw crusher 道奇(颚式)压碎机

Dodge-Metzner friction factor correlation 道奇-梅茨纳摩擦因子关系式

Doebner-Miller quinoline synthesis 多布纳-米勒喹啉合成

Doebner reaction 多布纳反应

doeglic acid 十九碳烯酸

doegling oil 真甲鲸油

Doerner-Hoskins distribution 德尔纳-霍斯金斯分配

doeskin ①羊皮绒面手套革②鹿皮绒面革

doff-assist 落筒辅助装置

doffer 道夫

doffer transfer ratio 道夫转移率

doffing (薄膜)落卷；(收卷)松脱

doffing digital time signalization 数字式落丝时间信号

dogbone shaped cross section 骨形截面；哑铃形截面

dog button 马钱子；番木鳖子

dog clutch 爪形离合器

dog eared 卷边

dog fat 狗脂(肪)

dog fennel oil 犬茴香油

dog-fish liver oil 角鲛鱼肝油

dog grass 匐匍冰草

dog holes 排钉孔

dog house 原料预热室

dog-leg (bend) 反转弯曲

dog leg (ged) piping 弯曲管路

dog rose 犬蔷薇

dog's grass 鸭茅

dog skin 橘皮；狗皮纹

dogskin leather 狗皮革；狗皮制的革

dogwood oil 山茱萸油

doily 补强片

do-it yourself paint 家用涂料

dolabradiene 罗汉柏二烯；斧松二烯

dolabrine 罗汉柏素

dolabrinol 罗汉柏酚

dolerite 粗玄岩；辉绿岩

dolerophanite 褐铜矾

dolichol 多萜醇

doling 干削

doling machine 干削机

dolomite 白云石*

dolomite cement 白云石水泥

dolomite limestone 白云石灰岩

dolomitization 白云石化作用〔地〕

dolomol 硬脂酸镁粉

dolomold 白云石晶模

dolphin oil 海豚油

domain 微区；畴

domain morphology 微区形态学

domain of order 有序畴

domain structure 微区结构

Do-malt continuous malting system 都马尔特连续制麦芽系统

dome ①拱顶〔蒸馏釜的〕②坡面〔结晶〕

dome center manhole 储罐中部入孔
domed 拱凸；表面凸起
dome innage 圆顶盖的容积
dome manhole 圆顶入孔
dome-shaped(=dome-type) 圆顶状的
domestic burner 民用炉；家用炉
domestic coke 家用焦炭
domestic detergent 民用洗涤剂
domestic fuel 国产燃料；民用燃料；家用燃料
domestic fuel oil 家用柴油（即：煤油）
domestic fungus 食用菌
domestic gas 民用气体；家用气体
domestic grade 民用规格
domesticine 南天竹碱〔取自 Naudani domestica〕
domestic method 常用方法
domestic oil 国产石油
domestic sewage 生活污水
domestic soap 家用皂；洗衣皂
domestic washing-up liquid 家用洗涤剂
domestic waste water 生活污水
domestomycetes 木腐菌
dome-type(=dome-shaped) 圆顶状的
dome-type tank 拱顶油罐
domeykite 砷酮矿
dominant bath 主浴(式)〔三氧化硫磺化方式〕
dominant bath type 主浴式
dominant color 主色
dominant hue 主色相
dominant lamella 主片晶
dominant peak 主峰
dominant reaction 主反应
dominant surface 主表面
dominant wavelength 主波长
domingite 硫锑铅矿
domperidone 多潘立酮〔药〕
donarite 多纳炸药〔硝铵、TNT、硝化甘油等混合炸药〕
donating bond 给予键；施予键；供键
donation 给予(作用)
donator(=donor) 给体；供体
dona tub 种子罐
donaxine(=gramine) 克胺
dongola leather 山羊革
dongola tannage 铝-槟榔膏结合鞣法
donicity 给体性；给体数
donkey pump 蒸气往复泵〔锅炉进水泵〕
donkey skin 驴皮
Donnan equation 唐南方程式
Donnan equilibrium 唐南平衡*

Donnan exclusion 唐南排斥
Donnan potential 唐南电位；膜电位
Donnan's membrane theory 唐南膜理论
Donnan softening 唐南软化〔渗析法除水中阳离子〕
donor 给体*
donor-acceptor 给电子-受电子体
donor-acceptor chromogen 给电子-受电子发色体
donor-acceptor complex polymerization 给体-受体复合物聚合
donor-acceptor copolymer 给体-受体共聚物
donor-acceptor system 给体-受体体系
donor atom 供电子原子；配位原子
donor group 供电子(原子)团
donor impurity 施主杂质
donor level 施主能级
donor molecules 供电子分子
donor residues 给电子体；给电子基
donor site 供体部位
donor strength 给电子能力
donor-substituted quinones 给电子基取代的醌类化合物
Doolittle equation 杜利特尔方程
Doolittle viscosimeter 杜利特尔扭力黏度计
door extractor(=door lifting machine) 起门机
door lifting machine(=door extractor) 起门机
dopachrome 多巴色素；红悲素
dopamine 多巴胺〔药〕
dopant 掺杂剂
dopant-selective etching 掺杂选择腐蚀剂
dope ①蒙布漆；飞机翼涂料；涂布漆②添加剂；掺杂
dope brushability 蒙布涂刷性
dope bucket 装封口油灰的斗〔管子接头配件〕
dope can (发动机开动时)加油用油枪
dope cotton 高黏度硝化纤维
doped crystal 掺杂晶体*
doped envelope 涂漆层；涂漆外壳
doped fuel 含添加剂的燃料
doped oil 含添加剂的油
doped rutile 金红石型晶格颜料
doped titanias 掺杂钛颜料；钛系晶格颜料
doped type 掺杂型〔复合半导体纳米微粒的结构型式之一〕
dope dye ①纺丝原液染料②调色染料〔用量极少，用以调整色相〕
dope-dyed fiber 着色纤维
dope header 纺丝液集合管
dopentacontane 五十二烷 $C_{52}H_{106}$
dope pot 稀释封口润滑油用壶〔接合输送管用〕
doper 油枪；润滑脂枪

doping 掺杂*
doping agent 掺杂剂
doping gradient 掺杂梯度
Doppler breadth 多普勒宽度
Doppler broadening 多普勒宽化*；多普勒变宽
Doppler effect 多普勒效应
Doppler half width 多普勒半宽度
dopplerite ①一种天然沥青②腐殖质块；灰色泥炭
Doppler linewidth 多普勒线宽
Doppler shift 多普勒位移
Doppler shift frequency 多普勒频移
Dore bead 多尔熔珠
Dore metal 多尔合金
doremol 氨草醇
doremone 氨草酮
doremyl acetate 乙酸氨草酯
doricin 多丽菌素
dormant coagulation 潜伏凝固
dormant species 休眠种
Dorr agitator 多尔搅拌器
Dorr balanced-tray thickener 多尔型平衡盘式增稠器；料盘式增稠器
Dorr clarifier 多尔澄清器
Dorr classifier 多尔分粒器；多尔分级器
Dorrco filter 多尔科滤机
Dorr system of leaching 多尔沥滤法
dosage ①用量②配料③剂量
dosage bunker 配料槽；剂量槽
dosage effect 剂量效应
dosage mortality curve 剂量死亡曲线
dosage pump 剂量泵
dosage rate 剂量率
dose 剂量；药量
dose equivalent(DE) 剂量当量〔单位为雷姆〕
doser 定量给料器
dose rate (放射)剂量率
dose-response curve 剂量-反应曲线
dose-response relationship 剂量-反应关系
dosimeter 剂量计；剂量仪器
dosimetry (放射)剂量测定法
dosing ①定量给料②剂量给药
dosing feeder 计量进料装置
dosing hopper 计量漏斗
dosing machine 计量机
dosing plant ①定时供料装置②间歇操作装置
dosing pump 计量泵
dosing station 计量台；进料计量台
dosing unit 计量装置；定量装置；配料装置

dosing vessel 计量槽；配料槽
dot blot 斑点印迹
dot pattern 点花样
dot product 点积
dot resolution 点分辨能力
n-dotriacontane(=dicetyl) (正)三十二(碳)烷；联十六基 $CH_3(CH_2)_{30}CH_3$
dotriacontyl 三十二(烷)基 $CH_3(CH_2)_{30}CH_2-$
dotting recorder 打点式记录器
double ①(成)双的②两重的；二倍的③加倍
double acid phosphate 双料过磷酸钙
double acting ①双作用；往复②双动(式)
double acting compressor 双作用压缩机
double acting cylinder 双作用气缸；往复式气缸
double acting press 双动压力机
double acting pump 双动泵；双向(作用)泵
double addition method 两次添加法
double alkalization 二次碱化；二次浸碱
double and draft 并合与牵伸
double arc furnace method 双重电弧炉法
double arc light method 双弧光法
double arc method 双电弧法*
double Auger process 双俄歇过程
double axial flow blower 双吸轴流风机
double bacillus 双杆菌
double balsam fir 福莱瑟胶枞
double-base propellant 双元燃料火箭推进剂
double-bath dyeing 双浴染色
double battery switch 双电池开关
double beam 双光束
double beam atomic absorption spectrometer 双光束原子吸收光谱仪
double beam atomic absorption spectrophotometer 双光束原子吸收分光光度计；双光束原子吸收光谱仪
double beam densitometer 双光束光密度计
double beam double-focusing mass spectrometer 双束双聚焦质谱计
double beam electric null infrared spectrometer 双光束电(调)零点红外分光计
double-beam fast scan instrument 双光束快速扫描仪
double beam grating spectrophotometer 双光束光栅分光光度计
double beam mass spectrometer 双光束质谱仪
double-beam microinterferometer 双光束干涉显微镜
double beam optical null infrared spectrometer 双光束光(调)零点红外分光计
double beam optical system 双光束光学系统
double beam scanning spectrophotometer 双光束扫描式

　　　　　　分光光度计
double beam spectrometer　双光束分光计；双光束光谱仪
double beam spectrophotometer　双光束分光光度计*
double beam type　双光束(方)式；双光束型
double beater　双效打浆机
double bed burner(=double bed furnace)　双膛炉
double bed deionization　复床式脱盐；复床式去离子
double bed furnace(=double bed burner)　双膛炉
double bed system　复床式
double bend　双支管
double bevel-groove　K型坡口
double blade mixer　双桨搅拌机
double blind performance evaluation sample　双盲性能评价试样
double bob method　双浮子法
double bond　双键
double bonded hydrocarbons　双键烃类
double bond migration　双键转移
double-bond polymerization　双键聚合
double bottom　双底；夹底
double buff　双折布抛光轮
double calender　双重碾光机
double caliper　双测径器；双卡(尺)
double cantilever beam cleavage test　双悬臂梁开裂试验
double carbide　复碳化物
double-casing godet　夹套式导丝辊
double chain conveyor　双链运输机
double-chain polymer　双链型聚合物
double chamber electrode　双腔电极
double channel atomic absorption spectrometer　双通道原子吸收光谱仪
double channel plate　双通道板
double channel screw　双螺槽螺杆
double-charge ion　双电荷离子
double check　双重校对；双重保险
double cloud point　双浊点
double coated film　双面涂覆胶片；双面涂漆薄膜
double-coated paper　双层涂料纸
double collector　双接收器
double-column qualitative method　双柱定性法
double composition grease　复合润滑脂
double compound　复合物
double-cone classifier　双锥选粒机
double-cone gasket　双锥面垫圈
double-cone-impeller mixer　双锥形桨式混合机
double-cone mixer　双锥混合机
double-cone rotary vacuum dryer　双锥形回转真空干燥器

double-cone seal ring　双锥密封环
double cone viscometer　双锥(式)黏度计
double conjugation　双共轭(化)
double contact controller　自动控制器〔物〕
double containment　复式包容；双包容；双密封；双重封闭
double contingency principle　两种可能性原则；双重事故原则
double covalence　双共价
double-covered cylinder　双壁滚筒
double covering　两层包纱防蚀法
double crimp wire screen　双皱筛布
double crystal topography　双晶形貌术
double deck classifier　双层分级机
double-decked　双层的；铺两层的
double decker oven　双层炉
double-deck floating roof　(油罐的)双层浮顶
double deck screen　双台筛
double-deck twister　双层加捻机；双层捻线机
double-deck winder　双层络筒机
double decomposition　复分解
double-decomposition grease　复分解润滑脂
double decomposition reaction　复分解反应
doubled edge　卷边
double derivative IDA　双衍生物同位素稀释分析法
doubled goods〔复〕　双料货物
double diagonal brushing machine　锯齿形刷毛机；锯齿形起绒机
double diffusion test　双向扩散试验
double dilution method　二重稀释法
double dipsizing　两次浸渍上浆
double disc electrophoresis　两向盘状电泳；双盘电泳
double-discharge gear pump　双出口齿轮泵
double dish(=culture dish)　培养皿；北德利氏皿
double-double effect　双偶效应；四素组效应
double-double hypothesis　双偶效应假说
double-doughnut　双环形；双圆环
doubled pipe exchanger(=double pipe heat exchanger)　套管换热器
double-drum dryer　双转鼓干燥器
double-duty lube truck　两用润滑油运输货车
double-duty plant　两用设备
double duty regulator　双作用调节器
double effect　双效
double effect evaporator　双效蒸发器
double electrode layer　双电层
double electrovalence　双电价
double-end fired furnace　两头烧火加热炉

double extraction   双重抽提
double extra-heavy(=double extra-strong)   特厚壁的〔管〕
double extra-strong(=double extra-heavy)   特厚壁的〔管〕
double faced knitted fabric   双面针织物
double-faced paper   双面纸
double fastener   双夹
double filler lap joint   对面塔焊接
double fillet welded T joint   双面 T 字形焊接
double-film theory   双膜理论
double-fired heater   带两个燃烧室的加热炉
double flash evaporation   两级闪蒸
double flight feed screw   双叶螺旋加料器
double floor kiln   双层焙炉
double flow   ①双流②双程错流〔塔板〕
double flow centrifugal compressor   双吸离心压缩机
double fluid cell   双液电池
double focusing   双聚焦
double focusing analyzer   双聚焦分析器
double focusing at all mass   全质量双聚焦
double focusing for one mass only   单质量双聚焦
double focusing mass spectrograph of Mattauch-Herzog type   马-赫型双聚焦质谱仪
double focusing mass spectrometer   双聚焦质谱计
double focusing sector spectrometer   双聚焦扇形质谱计
double folding   双叠式；再摺式；双摺式
double formation   双晶；孪晶
double fronting   双前沿
double funnel   ①双膛炉②二重漏斗
double-gap   双间隙(的)
double glass transition   双玻璃化转变
double godet machine   双导丝盘(离心)纺丝机
double-graded oil   双特性油
double grade engine oil   双级(发动)机油
double groove   双面坡口
double group   双值群
double-headed nail   双头钉
double hearth burner(=double hearth furnace)   双膛炉
double hearth furnace(=double hearth burner)   双膛炉
double-heart wood   复心材
double heel suture   复式跟缝
double helical gearing   人字齿轮传动装置
double helical mixer   双螺桨混合器
double helix   双螺旋
double helix hypothesis   双螺旋假说
double helix model   双螺旋模型
double heterojunction laser   双异质结晶激光器
double high input impedance amplifier   双高输入阻抗放大器

double high input impedance ion meter   双高输入阻抗离子计
double-humped   双峰的
double IDA   双同位素稀释分析法
double indicator titration   双指示剂滴定法
double injection   双注射
double injector   复式喷射器
double inlet centrifugal blower   双吸离心式风机
double inlet system   双进样系统
double ionization   双电离
double ionization chamber   双电离室
double irradiation   双照射
double isotope derivative method   双同位素衍生物法
double isotope dilution method   双同位素稀释法
double-jacket   双层夹套
double junction reference electrode   双液接参比电极
double junction salt bridge   双液接盐桥
double labeling(=double-tagging)   双重标记
double labeled compound   双标记化合物
double lap joint   双搭接接头
double layer   ①双层②电偶层
double layer capacitance   双电层电容
double layer capacity   双层电容
double layer compression   双(电)层压缩
double layer current   双电层电流
double layer potential   双层电位
double layer twisting machine   双层加捻机
double layer warp-knitted geotextile   双层经编土工布
double lifter multiple shuttle box   双侧升降式多梭箱
double line method for background correction   双线法校正背景
double linkage   双键
double linked carbon   双键碳原子
double-loop tricot stitch   重经平组织
double-magic nuclei   双幻数核
double magnetic prism   双重偏转磁棱镜
double mantle column oven   双罩柱加热炉
double-mash decoction mashing method   双醪煮出糖化法
double mechanical seal   双端面机械密封
double melting   双重熔化
double melting point   复熔点；双重熔化点
double-mercerizing method   双浸碱法
double monochromator   双单色器
double motion agitator   双动搅拌器
double motion paddle mixer   双动桨式混合器
double-needle overlapping loop   重经线圈
double of doublets   双二重峰

double offset expansion U-bend　双效 U 形补偿器
double orientation　双(向)取向〔指薄膜拉伸取向〕
double orientation rotation　双旋转
double paper backing sheet　双层背纸板;双层背纸型
double parison technique　双型坯技术
double-pass drier　双路干燥器
double pass internal reflection element　双程内反射元件
double pass monochromator　双光路分光器;双光路单色器
double pass tray　双流塔板
double path system　双路系统
double pipe　套管
double pipe chiller　套管冷冻器;套管结晶器
double pipe condenser　套管冷凝器
double pipe cooler　套管冷却器
double pipe cooler crystallizer　套管冷却式结晶器
double pipe crystallizer　套管结晶器
double pipe heat exchanger(=doubled pipe exchanger)　套管换热器
double pipe heat interchanger　套管换热器
double pipette　双吸移管
double pique stitch　点纹组织
double-piston pump　双活塞泵
double-pole doublethrow switch　双刀双掷开关
double-pole singlethrow switch　双刀单掷开关
double potential step chronoamperometry　双电位阶跃计时电流法
double potential step chronocoulometry　双电位阶跃计时库仑法
double potential step chronopotentiometry　双电位阶跃计时电位法
double pressed stearic acid　二压硬脂酸
double-pressing　双重压榨
double proofed fabric　胶灰布
double pulsed chronoamperometry　双脉冲计时电流法
double pulsed field gradient spin echo method and NOE spectroscopy　双脉冲场梯度自旋回波法-核欧沃豪斯效应谱
double quadrupole mass spectrometer　双四极质谱仪
double quadrupole resonance　双四极共振
double quantum　双量子
double quantum coherence　双量子相干
double quantum filtered correlation spectroscopy　双量子滤波相关谱
double quantum transfer experiment　双量子转移实验
double quantum transition　双量子跃迁;双级跃迁
doubler　①层布贴合机②合布机③拈线机〔纺〕
doubler adhesion　双料黏结;双料黏合

double-recoil method　双反冲法
double reduction unit　双级减速机
double refracting　双折射的
double refraction　双折射
double replacement　双重取代;互(置)换
double resonance　双共振
double resonance difference spectroscopy　双共振差式波谱法
double resonance via level crossing　能级交叉双共振
double resonance with coupled multiplets　多重耦合双共振
double reverse bend　双反折
double (roll) coater　双辊施涂机;双辊涂漆机
double roll crusher　双辊破碎机
double-roller mill　双辊研磨机
double roller mixer　双辊混合器
double salt　复盐*
double samples method　双样(品)法
double samples plot　双样(品)图
double seal　双重水封
double sectioned column　两节柱
double sector cell　扇形双池
double-service rack　双效栈桥;两用栈桥
double sheathed needle　双套针管
double shell digester　双层煮解器
double shell rotary drier　双壳旋转干燥器
double side　双面板
double side coating　①双面上胶②双面涂漆
double-sided loop pile fabric　双面毛圈起绒织物
double size　①双层胶黏②双倍尺寸
double sizing　双面涂胶
double-skeleton catalyst electrode　双骨架催化电极
double slide valve　双动滑阀
double slit　双狭缝
double solvent　复溶剂
double solvent extraction　双溶剂萃取;双溶剂抽提
double solvent process　复溶剂选择精制过程
double spanner　双头螺丝板
double spinneret block　双喷丝板组合体;双喷丝板组合件
double spot　(色谱)双斑点
double spread　双面涂布(胶);相对面涂布
double spreading　两面贴胶
double staining　双重着色
double standard interpolative method　双标准内插法
double steaming　二度蒸煮
double-steeping　二次浸渍
double strand　①(绳、线等)双股②双丝;双纤维;双排

double-stranded 双链的
double-stranded chain 双重链;双股链
double stranded DNA 双链 DNA
double stranded helix 双(股)螺旋
double-stranded polymer fiber 双股型聚合物纤维〔梯型聚合物纤维〕
double strand polymer 双股链聚合物;梯形链聚合物
double stretching 双拉伸
double sulfate method 硫酸复盐法
double superphosphate 双料过磷酸钙
double sweep tee 双弯三通
doublet ①电子对;电子偶②偶极子③双重(谱)线④双峰⑤双电子键
double-tagging(=double labeling) 双(重)标记
double tandem 双串联式(的);双串列(的)
double thread 双头螺纹
double thread-up open-width washing machine 回形穿布平幅水洗机
double toggle locking system 双肘节锁模装置;双肘节锁模方式
double tone polarography 双叠波极谱法
double toning 双重色调
double-tracer clearance technique 双指示剂清除技术
double-tracer technique 双指示剂法;双示踪技术
double transfer paper 复转印纸
double-tube converter(=double-tube reactor) 套管反应器
double-tube fixed bed reactor 双套管固定床反应器
double-tube reactor(=double-tube converter) 套管反应器
double tuned coil 双调谐线圈
double tuned signal coil probe 双调谐信号线圈探头
double-twist twisting machine 倍捻加捻机
double U-groove 双面 U 型坡口
double upshot furnace 直立双膛(烟气上行)加热炉
double uranyl acetate reagent 乙酸铀酰试剂
double V butt joint 双面 V 型对焊接;X 型焊接
double V-groove V 型坡口
double-walled 夹壁的
double-walled annulus 双壁圆环
double-walled can 复壁罐;两层器壁的桶
double-walled kettle 二重釜
double walled tank 双壁(储)罐
double wall funnel 夹壁漏斗;双层漏斗
double water glass 双料水玻璃〔即钾钠均有〕
double-wavelength spectroscope 双波长光谱法
double weighing 交换称量法
double withdrawal procedure 两相除去法
double zone gas producer 两段煤气发生炉
double zoning 成双区

doubling ①重合;并合〔纺〕②夹胶
doubling calender 重合砑光机
doubling generation time 倍增时间
doubling machine 复制机
doubling-over test 折叠试验;180°折弯试验
doubling roller 贴合辊
doubling time 倍增时间
doubly bonded(=doubly linked) 双键连接的
doubly charged ion 双价离子
doubly-closed-shell isotope 双闭合壳层同位素
doubly label(l)ed soil 双标记污垢
doubly linked(=doubly bonded) 双键连接的
doubly linked carbon 双键连接的碳原子
doubly refractive 双折射的
dough 捏塑体
dough batch 打面机
dough brake 碾面机;面团辊轧机
dough chute 面团滑放槽
dough fermentation 面团发酵
dough inflation 面团发起
doughing ①揉面的②揉面
doughing machine 揉面机
doughing of malt 麦精面团
dough maker(=dough-making machine) 揉面机
dough-making machine(=dough maker) 揉面机
dough maturity 面团成熟
dough mill ①揉面机②调浆机〔橡〕
dough mixer 揉面机
dough moulding compound(DMC) 团状模塑料
doughnut 环状物;环状结疤
doughnut effect 环流效应(分散)
doughnut fiber 中空纤维;环形截面纤维
doughnut-section kiln 环形断面的再生器
doughnuts-shaped pattern 环形喷束
doughnut tyre (超)低压轮胎
dough raising 面团发起
dough raising powder 发粉
dough recovery (面团)醒发
dough roller 面团碾
dough room 捏制面团间
dough-slackening 面团松弛现象
dough softening 面团软化
dough trough 面团槽
dough truck 面团车
doughy 面团状的
douglas fir oil 花旗松油
doumasin 环戊烯酮
dovetail 燕尾

dovetail groove 燕尾槽
Dow-Badische process 道-巴底斯法〔由乙炔、一氧化碳及水反应制丙烯酸的方法〕
Dow cell (for bromine) 道氏(制溴)电池
dowel 定位销
dowel bush 定位销套
dowel hole 定位销孔
dowel joint 套接接头
dowel pin 合缝销；导钉；两尖钉
downcast 下风井
down-channel flow 顺螺槽料流；顺螺槽流(动)
down-channel pressure 顺螺槽压力
down-channel pressure profile (挤塑)顺螺槽压力分布
down-channel velocity (挤塑)顺螺槽流速
downcomer 下水管；落水管；降液管
down-convection pipe still 烟道气向下流动的管式炉
down corner (精馏柱的)溢流管
downcurve 下降曲线
down detergent 绒毛清洗剂
down draft(=down draught) 倒风；倒焰
down-draft furnace 倒风炉；倒焰炉；烟道气下行的加热炉
down-draft kiln 倒风窑；倒焰窑
downdraft metier 顺流型干式纺丝机
down-draught(=down draft) 倒风；倒焰
down-draught producer 倒风发生器；下抽式发生炉
downfield 低磁场
down-flop 随角异色效应
down-flow 下流；向下流动
down flow apron 降液挡板；降液裙板
down-flow fixed bed 下流式固定床
down-flow fluid-catalyst unit 下流式催化装置
down-flow fraction 贫化部分
downflow-pipe 下水管
down-flow principle 下流原则
down-flow reactor 下流式反应器；降液式反应器
down-flow spout 降液管；溢流管
down-flow system 顺流式；下流式
down-flow weir 溢流堰
downgauge(=downgage) (薄膜)减厚
downhill pipe line 下倾的输送管线；位于斜坡上的输送管线
downhill pumping installation 斜坡上的泵装置
down-off ①下流量②塔底流出量
down-pipe 下流管；水落管；降液管
down-pipe collect box 下流管集合箱
down-run 向下吹气
downshot-type furnace 烟道气下行式加热炉

downspout 降液管；溢流管
downspout conductor 下流管
downstream 下流；下游；顺流
downstream line 下游管线
downstream process 下游过程
downstream processing 下游处理
downstream side 下游侧
downstream solute migration rate 流出溶质迁移速率
down stroke 下冲(程)
down stroke hydraulic press 下冲式水压机
downstroke press 下压式压机
downstroke twisting 下行捻线
downtake 下导管；降液管
downtake feed header 下流供应的总管
downtake pipe 下导管；降液管
down tank 下流槽
down-time ①下落时间②停车时间；故障期
Downton pump 当通(手摇)泵
downtwisting arrangement 下加捻装置
downward 向下
downward displacement 向下排代
downward extrusion 向下挤塑(法)
downwash 下沉洗流；冲洗下沉污染物
Dow process 道法〔十二烷基磷酸萃取精制铀再转换成四氟化铀的过程〕
dowsonite 片钠铝石
dowtherm 道氏热载体〔一种换热剂；联(二)苯及二苯氧化物的混合物〕
dowtherm heating fluid 导热姆加热液
dowtherm process 道氏热载体法〔利用联(二)苯及二苯氧化物混合物换热剂的过程〕
dowtherm separator 导热姆分离器
dowtherm service facility 道生热载体供应装置
dowtherm vent trap 导热姆放空捕集器
doxapram 多沙普仑〔药〕
doxazosin 多沙唑嗪〔药〕
doxepin 多塞平〔药〕
doxifluridine 脱氧氟尿苷〔药〕
doxofylline 多索茶碱〔药〕
doxorubicin 阿霉素〔药〕
doxycycline(=deoxytetracycline) 多西环素；强力霉素
dozen (一)打〔=12 个〕
drachenfels trachyte 云母粗面岩
drac(h)orhodin 龙血树深红素
draconis sanguis 血竭
dracoresene 血竭黄脂
dracorubin 龙玉红
dracylic acid(=benzoic acid) 安息香酸；苯甲酸

draff 渣滓；糟粕
draft ①气流②通风③斜度④牵伸〔纺织〕
draft chamber 通风室；通风橱
draft flue 通风管
draft furnace 通风炉
draft gauge 通风计；差式风压计
draft indicator(=draft gauge) 通风计；差式风压计
draft irregularity 污气缺陷
draft ratio 拉伸比
draft roll 拉伸辊
draft temperature 拉伸温度
draft test 抗污气流试验
draft tube 通流管
draft tube baffle crystallizer DTB 结晶器；引流管缓冲结晶器
draft tube mixer 引流管混合器；导管混合器
drag ①下垂②阻力③牵引④刮板
drag-chain conveyer 链式运输器
drag-chain conveyor 拖链式运输机
drag classifier 刮板(式)分粒机；耙式分粒机
drag coefficient 曳力系数；阻力系数
drag conveyer 刮板(式)运输机
drag flow 黏性流；有阻力流
drag flow capacity (挤塑)推进流流量
drag flow constant (挤塑)推进流常数
drag force 曳力
drag-force flow sensor 拉力流动传感器
dragging 咬模；模型擦伤
drag line 拉伸纹〔热成型品缺陷〕
drag-line scraper 带式刮土机
drag-link conveyor 联杆式运输机
drag-out 拖带〔电镀〕
drag reducer 减阻剂
drag reducing fluid 减阻流体
drag reduction 减曳(作用)
drag screen 刮板筛
drag spot 浆疙瘩
drag viscometer 拖拉黏度计；曳力黏度计
drain ①泄水；排水；排泄〔机工〕②排放口
drainage ①浚(电)流②排水③排水法
drainage bunker 泄水斗
drainage effect (挤塑)穿流效应
drainage error 滴沥误差
drainage error in capillary viscometer 毛细管黏度计的流出体积误差
drainage hopper(=drain cup) 排液漏斗；排水漏斗
drainage pipe(=drain pipe) 排泄管
drainage pump ①排水泵②排油泵
drainage rate 叩解度；打浆度
drainage tray 排水槽
drainage tubing 排水管
drain board 排水板
drain box 排水箱
drain catch pot 排放捕集槽
drain cock 放水旋塞；排水栓
drain conduit 排水管；出水管
drain connection 排水管
drain contact 漏极接头
drain cup(=drainage hopper) 排液漏斗；排水漏斗
drain elbow 排水肘管
drainer ①排水器②储浆池③滴干板
drain filter 排(油、水)污过滤器
drain hole 排泄孔；放出口
draining ①排水；泄水；排泄②漏滴；流尽；滴干；沥
draining area 排泄面
draining box 沥箱
draining dish 漏(滴)皿
draining effect 穿流效应
draining funnel 放液漏斗
draining period(=drain period) ①换油周期②排放期
draining profile 流淌的侧影
draining pump 排水泵
draining table 排水台
draining test(=working viscosity test) 涂刷黏度试验
draining tunnel 排泄隧道；排泄洞
draining valve 排(放)泄阀
drain intervals 换油期限
drain line 排水管线；泄水管线
drain mark 水渍
drain outlet 排水出口
drain pan 排水盘
drain period(=draining period) ①换油周期②排放期
drain pipe(=drainage pipe) 排泄管
drain plate 排油口的盖
drain plug 泄出塞
drain port 泄口；排污口
drain pressure 排出管线内的压力
drain rack 干燥架
drain sample 排出的试样
drain sleeve 冷凝管；排出套管
drain tank 排水槽；储油槽〔在润滑油循环系统内〕
drain time 排液时间
drain trap 排水阱；沉淀池
drain tunnel 排风洞〔排溶剂蒸气装置〕
drain valve 排泄阀
drape and vacuum forming 包模真空成型

drape assist frame　助压型框
drape-flex test　(织物)悬垂挠曲试验
drape forming　(热成型)包模成型
drape molding　包模成型
drastic cracking　深度裂化
drastic deasphalting　深度脱沥青
drastic extraction　深度抽提
drastic reduction　①深度还原②大幅度削减
draught　①气流②通风③吸饮
draught cupboard　烟橱；通风橱
draught gauge　风力计；风压表
draught hood　烟橱通风罩
draught principle　通风原理
Draves test method　德拉夫斯试验法〔测润湿力〕
Draves wetting time　德拉夫斯润湿时间
dravite　镁电气石
draw　拉伸
drawback　①缺点；缺陷；瑕疵；障碍(物)②回火
draw beam　拉软床〔革〕
draw-crimping　拉伸卷曲
draw depth　(热成型)撑压深度
draw direction　拉伸方向
draw down　缩小横切面；轧扁，锤平；收缩
drawdown bar　刮板；刮规
draw-down blade　刮刀
draw-down guide　刮涂导杆
drawdown rating　刮板流平等级
draw-down ratio　拉缩比；垂伸比；牵伸比
draw down rod　①刮漆棒②压延棒〔铺展凹印或苯胺印的低黏度油墨可用的工具〕
drawdown sheet　遮盖力试验纸；刮样纸
draw frame　拉伸机
draw frame and lap machine combined　条并卷联合机
draw hole　拉孔
draw in　吸入；引入；流入
draw-induced crystal　拉伸诱发晶体
drawing　①图画②绘画③抽；拉④拉伸
drawing and coiling apparatus　拉盘(毛细管)装置
drawing camera　绘图室
drawing compound　拉丝化合物
drawing depth　拉延深度
drawing godet　拉伸(导丝)盘
drawing grain　皱面
drawing-heat treating sequence　拉伸-热处理顺序
drawing machine　(玻璃或毛细管)拉制机
drawing machines series　牵伸机系列
drawing number　图号
drawing paper　绘图纸；画图纸

drawing radius　拉延半径
drawing ratio　①(挤塑)牵引比②(纤维)牵伸比
drawing-spinning machine　纺丝拉伸(联合)机
drawing stress　拉伸应力
drawing temperature　回火温度
drawing tolerance　(热成型)撑压公差
drawing zone　拉伸区
draw mould　铸(塑)模；薄壁模子
drawn fiber　并纺纤维
drawn grain　皱面
drawn needle　抽针
drawn pin type insert　空心嵌销
drawn shell type insert　空心壳型嵌件
drawn sliver　熟条
drawn tow　(无捻)拉伸丝束
drawn tube　拉制管；冷拔管
draw-off　①泄水②曳除
draw-off cock　放水旋塞；排水栓
draw-off godet　导丝盘；拉伸盘
draw-off juice　浸出汁〔糖〕
draw-off level　排泄孔的水平
draw-off pan(=draw-off tray)　泄流板
draw-off pump　①抽出泵②排液泵
draw-off tray(=draw-off pan)　泄流板
draw-off valve　排水阀；泄水阀
drawout　(涂料、油墨)刮平；刮匀；铺展开
draw pan　拉伸槽
drawplate oven　拉板炉；活底炉
draw ratio　拉伸比
draw-relax-anneal process　拉伸-弛豫-热处理工艺
draw resonance　①纺丝程共振②拉引共振
draw rod(=tie rod)　拉杆
draw roll　拉伸辊
draw shoes　拉伸导向板
draw speed　牵伸速度；抽出速度
draw texturing machine　牵伸变形机
draw texturizing machine　拉伸卷曲机
draw-tube　伸缩管
draw twister　牵伸加捻机
draw twisting　拉伸加捻
draw-twist machine　拉伸加捻机
draw winder　牵伸卷绕机
Drechsel gas washing bottle　玻璃煤气洗涤瓶
Drechsel washer　玻璃煤气洗涤器
dredge　①挖泥机②拖网
dredge ore　贫矿石
dredge pump　吸泥泵；排污泵
dredging box　戽斗

dregginess 含渣量
dreggy 含渣的；多渣的
dregs 渣
drench ①浸润②脱灰〔革〕③浸麸液〔革〕④敷料缸
drenching 浸麸液
drenching apparatus 灌水机
drench pit 麸液槽〔革〕
dressed masonry 细石工；镶面坊工
dressing ①修饰②调味品〔香料〕
dressing agent 修饰剂
dressing disinfection 服装除毒
dressing jar 药膏缸
dressing leather 饰革
dressing vat 鞣制瓮
dressing-works 选矿厂
Drexel bottle(=Drexel washer) 煤气洗涤瓶
Drexel washer(=Drexel bottle) 煤气洗涤瓶
dribble blending 一滴一滴地混合
dribbled fuel 没有蒸发的燃料
dribbling 滴掉；滴下
dried alum 烧明矾
dried aluminium hydroxide gel 干氢氧化铝凝胶
dried blood (干)血粉
dried catalyst 干催化剂
dried emulsion 干乳状液
dried fish 干鱼粉
dried joint 干燥接头
dried malt 干麦芽
dried meat 干肉
dried milk 奶粉
dried milk albumin 干燥乳品蛋白
dried pulp 干蔗渣〔糖〕
dried rumen bacteria 干燥瘤胃菌(制剂)
dried sample 干样品
dried skimmed milk 脱脂奶粉
dried yeast 干酵母
drier 催干剂
drier felt 干燥毛布；干燥毡
drierite 燥石膏
drier white 吐渣
drift ①漂移②弹性后效
drift action 漂移作用
drift current 漂移电流
drift distance 漂移距离
drift length 漂移长度
drift mobility 漂移迁移率
drift rate 漂移速率
drift region 漂移区
drift space 漂移空间
drift time 漂移时间
drift tube mass spectrometry 漂移管质谱分析
drift velocity 漂移速度
driftwood 流送木材
drikold(=dry ice) 固体二氧化碳；干冰
drill ①钻(子)②钻头③钻床
drill bit 钻头；钻嘴
drilled passage 钻道
drill fit 牢配合；紧配合
drilling 钻孔
drilling fluid 钻孔液体
drilling machine 钻床；钻孔机
drilling machine operator 钻工
drilling mud 钻探泥浆
drilling operating 钻孔；操作；钻孔工序
drill point 钻针
drill press 穿孔机；打眼机
drills 斜纹布
drimenene 辛辣木素
drimenol 八氢三甲基萘甲醇；辛辣木醇
drimenone 辛辣木酮
drinking paper 吸水纸
drinking water 饮用水
drinking water supply 饮用水供应
driography 无水胶印
drip ①引管②滴口③采酸管④滴⑤滴注(法)
drip bead 珠状流挂
drip catcher 液滴捕集器
drip chamber 排水室；沉淀池
drip cock 滴(降)栓
drip collector (of lead chamber) (铅室的)采酸管
drip condenser 水淋冷凝器
drip cooler 水淋冷却器
drip drying 滴干
drip feed 滴给；逐滴供给；一滴一滴的供给
drip feed lubricator 滴给加油器
drip feed oiling system 滴油式润滑系统
drip furnace 液体燃料燃烧用炉
drip gasoline 液滴汽油
dripless paint ①不流挂的涂料②触变性涂料
dripless seal 无液滴封闭
drip lubrication 滴油式润滑(作用)
drip lubricator 滴油润滑器
drip mat 滴流垫
drip oil 裂解冷凝油
dripping ①液滴②滴下
dripping time 滴流时间

drip pipe　凝液排出管
drip pockets　凝液收集袋〔在输气管内〕
drip point(=initial point)　初馏点
drip pot　气体分液罐
drip-proof　防滴式的；不透水的
drip rail　防滴横挡；挡水横条
drip return tube　液滴回流管
drip ring　润滑环
drip-rinsing process　淋洗法
drip-saver pan　液滴捕集器
drip tank　收集桶〔收集流动的油类或其他液体〕
drip tin　铅滴管
drip tray　(接)滴漆盘
drive　①传动〔机工〕②推进
drive assembly　传动装配
drive clutch　传动离合器
drive fit　紧配合；打入配合
drive gear　大齿轮；驱动齿轮
drive gear guard　大齿轮罩
drive line　传动体系
driven equilibrium　驱动平衡
driven machine　从动机
drive nozzle　压力喷嘴；高压远射喷嘴
driven roll　从动辊；被动辊
driven roller conveyer　辊式传动运输机
driven shaft　从动轴
driven sheave　从动槽轮
driven wheel　从动轮；被动轮
drive pinion　驱动小齿轮
drive power　驱动功率
driver(=driving mechanism)　驱动器
drive roll　后滚筒
drive shaft　①传动轴②主动轴
drive-type oil cup　压力加油器
drive unit　传动装置
drive wheel　传动轮；主动轮
driving belt　传动皮带
driving force　推动力
driving-point velocity　策动点速度
driving power　驱动力
driving pulley　主动滑轮
driving roll　后滚筒
driving shaft　传动轴
driving sheave　主动槽轮
driving test　驾驶试验
drop　①滴②降落
drop analysis　点滴分析
drop bar grizzly　落棒筛

drop black　煅骨碳；落黑
drop-burette　滴量-滴定管
drop center rim　深式轮辋
drop counter　计滴器
drop counter-current chromatography　液滴逆流色谱
dropdown　(吹塑型坯)下垂
drop electrode　滴(液)电极
drop eliminator　液滴去除器；水分去除器
droperidol　氟哌利多〔药〕
drop error　滴误差
drop feed oiler(=drop lubricator)　液滴加油器
drop height　①(冲击试验)落锤高度②(坠落试验)落高
drop impact test　落锤冲击试验
drop indicator　点滴指示剂
drop knocker　滴落敲击器
droplet　滴
droplet countercurrent chromatograph　液滴反流色谱仪
droplet countercurrent chromatography　液滴反流色谱法*
droplet generator　液滴发生器
droplet satellite　余滴；终了液滴
droplet size　液滴体积
droplet-size distribution　液滴大小分布
droplet test　点滴试验
drop line integration　垂线积分
drop lubrication　液滴润滑(作用)
drop lubricator(=drop feed oiler)　液滴加油器
drop meter　滴量计
drop method　点滴法
drop number method　滴数法
drop oiler　液滴加油器
drop out　吊装孔
drop-out line　放空线；排泄线
drop out point　滴点；析出点
drop-out sump　排泄器
dropper　滴管
dropping　①滴落②点滴③滴液
dropping amalgam electrode　滴汞齐电极
dropping ball viscometer　落球黏度计
dropping bottle　滴瓶
dropping cock　滴(降)栓
dropping electrode　滴液电极
dropping funnel　滴液漏斗
dropping liquid　滴液
dropping mercury cathode　滴汞阴极
dropping mercury electrode(DME)　滴汞电极
dropping pipette　滴液吸移管
dropping pipet type electrode　滴管型电极
dropping sphere method　落球法

drop pipette 滴液吸移管
drop ply 空层；漏层
drop point 滴点
drop qualitative test 点滴定性试验
drop reaction 点滴反应
drop scale method 滴量方法
drop softening-point method 软化点滴测法
drop test 点滴试验
drop-testing machine 点滴试验机
drop time 滴下时间*
drop value 滴值
drop-volume method 滴体积法〔测表面张力〕
drop weight apparatus 滴重器
drop-weight impact test 落锤冲击试验
drop-weight method 滴重法
drop weight tensiometer 滴重量法表面张力测定仪
drop-weight test 落重试验
dropwise 逐滴地；一滴滴地；滴状
drop-wise condensation 滴状冷凝
dropwort 欧合叶子
droserone 茅膏(菜)酮；甲基二羧萘醌
drosophilin 脆柄菇素
dross ①浮渣②渣滓
dross coal 渣煤；不黏(结性)煤
drossing 撇渣
drossing oven 撇渣炉
drossing process 造渣过程
drossy ①浮渣的②渣滓的
drossy coal 渣煤；不纯的低级煤
drowned pipe 淹没(在水内)的管子
drowned pump 淹没的泵；深井泵
drritol 鱼藤醇
drug 药
drug-fast 抗药的；耐药的
drug-fast strain 抗药性菌株
druggist ①药商②调剂员
druggist balance 调剂天平；毒物天平
druggist rubber sundries 〔复〕医用橡胶制品
drug-habit 药瘾
drug metabolism 药物代谢
drug mill 药磨
drug room ①药房②配料间
drug screening assay 药物筛选试验
drug standard 药品标准
drug sundries 卫生用品
drum ①鼓；转鼓；转筒②桶
drum agitator 转鼓搅拌器；圆筒式搅拌机
drum barker 鼓式剥皮机
drum beet cutter 鼓式甜菜切丝机〔糖〕
drum beet slicer(=drum beet cutter) 鼓式甜菜切丝机
drum coater 转鼓式涂漆机
drum cooler 转鼓式冷却器；冷却转鼓
drum crystallizer 鼓式结晶器
drum cylinder crystallizer 圆筒式结晶器
drum director system 转鼓自控系统
drum door 鼓门
drum drafter 滚筒式牵伸机构
drum-drawing mat-forming aggregate 滚筒薄毡机组
drum dry 圆筒烘干
drum dryer 转鼓式干燥机
drum drying 转鼓式干燥
drum dyeing 转鼓染色
drum feeder 转鼓加料机
drum filler 装桶机〔将产品装入桶内的设备〕
drum film dryer 转鼓真空过滤干燥器
drum filter 转鼓真空过滤机；滤鼓
drum gudgeon 转鼓轴孔
drum handling truck 运桶用的小车
drum head 桶底
drum heater 转鼓式加热器
drum iron 转鼓支架
drum kiln 鼓式焙炉
drum leather 鼓皮
drum liner 圆筒(桶)衬里机；滚筒衬里
drum malting 鼓式发芽
drummed oil 桶装油
drum method 鼓式法
drum mill 桶式研磨机
drumming 转鼓加工(法)
drumming process 转鼓加工法
drum mixer 鼓式混合机
drum moulder 鼓式塑形机〔面包〕
drum pallet 桶衬垫(支架)
drum racking 滚筒(的)移位
drum reactor 鼓式反应器
drum rinser 洗桶器
drum sander ①鼓形抛光轮②滚筒砂磨机③砂鼓磨床
drum scourer ①转筒洗刷机②转筒剥壳机③转筒精炼机
drum screen 转筒筛
drum screening 转筒筛选
drum separator 转鼓式分离器
drum setting-out machine 鼓型平展机
drum shaft 成型鼓主轴
drum sieve 滚筒筛
drum sifter 转筒筛

drum sifting　转筒筛选
drum skimmer　转鼓撇油器
drum stability　装桶稳定性；桶藏稳定性
drum stuff　鼓中加脂〔革〕
drum stuffing　鼓中加脂(法)
drum tannage(=drum tanning)　转筒鞣制法
drum tanning(=drum tannage)　转筒鞣制法
drum tumbler(=barrel mixer)　滚筒式混合机(器)；转鼓混合器
drum-type drying machine　鼓式烘燥机
drum type vacuum filter　转鼓式真空过滤机
drum washer　转筒洗涤机
drum washing　转筒洗涤
drum washing machine　鼓式洗涤器
druse　晶簇
drusen　硫黄颗粒
drusy structure　晶簇结构
dry acid treatment　干酸处理
dry adhesion　干燥黏合
dry aerosol　干气溶胶
dry ageing　干老化
dry-air curing　干空气固化
dry-air pump　干风泵；干(燥空)气泵
dry analysis　干法试验；干法分析
dry and wet bulb hygrometer　干湿球湿度计
dry and wet bulb psychrometer　干湿球温度计
dry application　干施
dry-area(=dry spot)　干斑
dry ashing　干法灰化
dry ashing method　干灰化法
dry assay　干法试金
dry assaying　干法试金
dry atmospheric corrosion　干大气腐蚀；干腐蚀
dry attapulgite　干(燥)凹凸棒石；硅镁土
dryback　(油墨)干后暗色；干后变色
dry base exchange　干法碱交换
dry basis　干基*
dry-basis moisture content　干基湿含量
dry battery　干电池组
dry blast　鼓风；送风
dry blast-cleaning　干喷砂除锈(法)
dry-blend　干混剂
dry blending　干掺和；干混合
dry bottom producer　干底发生器
dry box　①干燥吸收剂②干燥吸收器③干箱；手套箱
dry brightness　干亮度
dry brush　干刷法
dry bulb temperature　干球温度

dry bulb thermometer　干球温度计
dry bulk　①干燥体积；松散体积②干的散装货物
dry burning coal　低级的不结块煤
dry capacity　干额；干容量
dry catalyst　干催化剂
dry cell　干电池
dry cell battery　干电池
dry chemical extinguishing method　干式化学灭火法
dry-chlorinated　干式氯化(防缩处理)
dry churn　干法黄化机；干式黄化鼓
dry classification　干式分级
dry clay　干燥白土
dry cleanable leather　耐干洗革
dry cleaner's naphtha(=dry cleaner's solvent)　干洗溶剂；干洗挥发油
dry cleaner's solvent(=dry cleaner's naphtha)　干洗溶剂；干洗挥发油
dry cleaning　干洗(净)
dry cleaning agent　干洗剂
dry cleaning composition　干洗剂
dry cleaning detergent　干洗(洗涤)剂
dry cleaning gasoline　干洗汽油
dry cleaning process　干洗法
dry cleaning resistance　耐干洗性
dry cleaning soap　干洗(用)肥皂
dry cleaning solvent　干洗溶剂
dry coal　干煤；长焰煤
dry-collapse　干燥致密化
dry coloring　干法染色
dry colo(u)r　干颜料
dry column　干柱
dry column chromatography　干柱色谱法
dry column-packing　干法装柱
dry combustion method　干式燃烧法；干热法
dry compression　干压缩
dry concentration　干选；风选
dry condenser　①干冷凝器②干电容器
dry cook　干煮；不完全蒸煮
dry cooling　干冷却
dry corrosion　干腐蚀
dry crosslinking　干交联
dry crushing　干(法压)碎
dry crushing mill　干碎机
dry curd　干皂粒
dry cylinder liner　干式气缸衬垫
dry decatizing　干法汽蒸
dry deposition　干沉降
dry dip　干底革再鞣

dry distillation 干馏
dry doubling ①干并〔纺〕②干贴〔胶〕
dry dust collector 干灰收集器
dry elutriation 干淘(洗); 空气选粒
dry elutriator 干淘机
dryer(=drier) ①干燥剂②干燥器③烘缸
dryer charge gas 干裂解气
dryer feed 干进料
dryer felt 造纸干燥毡毯
dryer part 干燥部; 煤干部
dry etching 干法腐蚀
dry ether 干醚
dry expansion evaporator 干膨胀式蒸发器
dry extract 干提出物
dry fall 过喷干燥
dry fallout 干散落物
dry felt 干毛毯
dry filler 干填充剂
dry film 干膜
dry-film brightness 干膜亮度
dry film lamination 压膜
dry film lubricant 干膜润滑剂
dry film lubrication 干膜润滑(作用)
dry filter 干(燥过)滤器
dry filtration 干法过滤
dry finish 干浆料
dry finishing 干浆; 干法上浆
dry fire retardant finish 干式防火整理
dry flex ①干挠曲②干挠曲强度
dry flushing (颜料)干挤(水)法
dry friction 干摩擦
dry galvanizing 干法镀锌
dry gas ①干气; 贫气②干煤气
dry gas fuel 干煤气燃料
dry gas-holder 干式储气柜
dry gas meter 干式气表
dry gas purification ①干气净化②干法气体净化
dry gas purifier ①干气净化器②干法气体净化器
dry gluing 干式胶合
dry gluten ①干麸②干燥面筋
dry grinder 干磨机
dry grinding 干磨
dry ground 干法粉碎; 干磨
dry ground muscovite mica 干磨白云母粉; 干法云母粉; 干磨硅铝酸钾
dry-ground whiting 干磨碳酸钙
dry hard 硬干
dry-hard time 硬干时间

dry-heat cure 热空气硫化
dry heat exposure 干热态曝置
dry heating curing 干热熟化
dry heat-setting 干热定形
dry heat stability ①干热稳定性②干热稳定度
dry heat sterilization ①干热灭菌②热空气消毒
dry heat treatment method 干热处理法
dry heat vulcanization 干热硫化
dry hide 干皮
dry hide powder 干皮粉
dry honing equipment 干燥抛光设备
dry hydrate 氢氧化钙; 熟石灰粉
dry ice 干冰; 固体二氧化碳
dry incineration 干式灰化
drying 干燥
drying agent 干燥剂*
drying and stretching frame 干燥展开架
drying apparatus 干燥器; 干燥设备
drying basin 干燥皿
drying battery 干电池; 化学干燥器
drying bed 干化场; 干燥场
drying block (元素分析)微量干燥装置
drying cabinet 干燥箱
drying chamber 干燥室; 烘房
drying characteristic curve 干燥特性曲线
drying chest 干燥箱
drying column 干燥塔
drying condition 干燥条件
drying cupboard 干燥柜; 橱式干燥机
drying cycle 干燥周期
drying cylinder 干燥筒; 干燥鼓
drying cylinder in sizing 浆纱烘筒
drying drum 干燥鼓
drying filter 干滤器
drying floor 干燥地板
drying frame 干燥架
drying hopper 干燥箱
drying index 干燥指数
drying in frozen 冷冻干燥
drying ingredient 干燥配合剂
drying inhibitor 阻干剂
drying in the open 自然干燥
drying line 干燥联动装置
drying loft 干燥室
drying machine 干燥机
drying mark 干燥(黑)斑
drying medium 干燥介质
drying meter (漆膜)干燥计; 测干仪

drying oil 干性油
drying oil fatty acid 干性油脂肪酸
drying oven 干燥箱；烘箱
drying pistol 干燥枪
drying plant 干燥设备
drying power 干燥能力
drying press 榨干机
drying rack 干燥架；烘架
drying rate 干燥速率
drying reel 干燥丝框
drying retarder 干燥抑制剂
drying room 干燥室
drying saturation degree 干燥饱和度
drying shed 烘干室
drying stand 干燥台；干燥架
drying steam 干燥用(蒸)汽
drying stove 干燥箱
drying table 干燥台
drying temperature 干燥温度
drying test 干燥试验
drying thermal efficiency 干燥器的热效率
drying time 干燥时间
drying time automatic recorder(=drying time recorder) 干燥时间自动记录仪
drying time meter 干燥时间测定器
drying time recorder(=drying time automatic recorder) 干燥时间自动记录仪
drying tower 干燥塔
drying tray 干燥盘
drying tube 干燥管
drying tumbler 干燥鼓；烘干转筒；转鼓式烘干机
drying tunnel 烘道
drying-up point 干燥点
dry jet wet spinning 喷湿法纺丝；干喷湿纺法
dry kiln 干燥窑
dry laminate 干式叠层〔塑料〕
dry liming 干灰澄清作用；干法澄清作用〔糖〕
dry loft 坯革堆放间
dry lubricant 干膜润滑剂
dry lubricity 干燥润滑性
dry lute 干封泥
dry main 干总管；干馏气管
dry matter 干燥物质
dry matter content 干料含量；干物质含量
dry measure 干测量
dry method 干法
dry-milled clay 干法黏土；干磨细瓷土
dry milling 干滚

dry mixer 干式混合机
dry mixing 干(法)混合
dry naphtha 无水石脑油
dry natural gas 干燥天然气
dryness 干燥
dryness fraction 干燥度；烘干度
dryness tester 干度检验器
dry neutralization 干中和；(接触精制时)白土对酸性油的中和
dryobalanops camphor 龙脑；冰片
dryocrassin 粗蕨素；东北贯众素
dry-off moisture 除去湿气；除掉水分
dry-off oven 烘干炉
dry oil 干性油；无水石油
dry operation 干场作业
dry ore 干矿石
dry-out sample 无水试样
dry-packing 干(法)填充
dry paint 干油漆；干漆膜
dry painting 干粉涂漆(法)；涂粉末涂料
dry pan 干盘
dry pan mill(=dry pans) 干盘磨机
dry pans(=dry pan mill) 干盘磨机
dry pelletizing method 干式造粒法
dry phoshating 干法磷化处理
dry pickup 干燥附着量
dry piece 干(麦芽)料
dry pigmentation 干着色；干混法〔颜料不经研磨分散直接加入大量漆料，分散成色漆的过程〕
dry pipe 蒸汽收集器；过热蒸汽输送管
dry plate 干片
dry point 干点
dry powder continuous foam generator 连续式干燥泡沫发生器
dry powder generator 干燥粉末发生器
dry powder paint 干粉涂料；粉末涂料
dry press brick 干压砖
dry press brick machine 干压压砖机
dry pressing 干压
dry pressure drop （气液接触塔)无回流时的压力损失
dry process 干法成型
dry processed chalk(=pulverized chalk) 干磨碳酸钙〔填料〕
dry process enameling 干法搪瓷
dry pulp 干纸浆
dry purification 干法提纯
dry purification process 干式(气体)净化过程
dry quenching (of coke) 干法熄焦

dry reaction　干反应
dry refluxing　干回流
dry rendering　干法炼油
dry residue　干残渣
dry ring spinning　环锭干纺
dry-rot　干腐
dry rubber content　干胶含量
dry rubber substance　干橡胶
dry rubbing　干打磨
dry-run protection　防无液运转
dry-run tar　干馏焦油
dry salted hide　干腌皮
dry screening　干(法过)筛
dry-seal gasholder　干式气柜
dry-seal lifting roof　干封呼吸顶
dry separation　干选；风选
dry setting　干热定形
dry shrinkage　干热收缩；干燥收缩
dry skin lotion　干皮洗涤剂
dry sludge　干淤渣
dry soap　干皂
dry-solvent receiver　无水溶剂接受器
dry species analysis　干法形态分析
dry spinning　干纺
dry spinning cell　干纺甬道；干纺丝室
dry-spinning column　干纺甬道；干纺丝室
dry spots　干摩擦点；未被润滑的部分
dry spray　干喷〔过喷或喷逸〕
dry stage　干燥阶段
dry steam　干(燥)蒸汽
dry steam distillation　干(蒸)汽蒸馏(法)
dry stock　干料
dry strength　干燥强度
dry strippable paper　(可)干剥纸
dry substance　干燥物质
dry-sump lubrication　干沉淀式润滑作用
dry support　干燥载体
dry-suspension process　干态悬浮法
dry tack　干黏着性
dry tack adhesion　成型黏着性
dry tack free　表面干燥；指触干燥
dry tannage　干鞣法
dry temperature　干燥温度
dry tenacity　干燥强度
dry tensile strength　干拉力；干抗拉强度
dry test　①排空试验②干试法
dry test gas meter　干试煤气用量表
dry through　全干

dry time　干燥时间
dry tolerance　耐干性
dry toluene　无水甲苯
dry to tack free　不黏干燥
dry-to-touch　指触干燥
dry trimming(=dry refluxing)　干回流〔用密封的盘管部分冷凝器调节回流〕
dry tube mill　干管磨机
dry tumbling　干滚滚(转)抛(光)
dry tumbling drum　干滚转鼓
dry vacuum　干真空
dry vacuum distillation process　真空干馏法
dry vacuum pump　干式真空泵
dry vanning　干式摇选法
dry way　干法
dry web impregnation　干卷料浸渍
dry-web process　干纸饱和法〔浸渍纸〕
dry wet spinning　干湿法纺丝
dry winding　干法缠绕
dry xanthation　干法黄化；干黄(原酸)化作用
dual　二元的；二重的
dual adsorption site　双吸附中心
Dualayer distillate process　杜莱伊尔馏分油碱洗电沉降法
Dualayer gasoline process　杜莱伊尔汽油碱洗法
dual ball check valve　双球止逆阀
dual beam spectrophotometer　双光束分光光度计
dual beam UV monitor　双光束紫外监测器
dual burner　可用两种燃料的燃烧器
dual-catalyst　双功能催化剂
dual cavity mould　双槽(单塞)塑模
dual cell　双池结构
dual chamber furnace　双室炉；双膛炉
dual-channel atomic absorption spectrophotometer　双道原子吸收分光光度计
dual-channel flame photometric detector　双道火焰光度检测器
dual-channel screen changer　双流道换网机
dual-circulation steam generator　双循环系统的蒸汽发生器
dual-coating system　双涂层体系
dual collector　双接收器
dual column　双柱
dual column chromatography　双柱色谱(法)
dual column ion chromatography　双柱离子色谱法
dual column system　双柱系统
dual crossing　双交叉〔管路及管线上的〕
dual curing technology　双重固化工艺〔紫外线热固〕
dual decay　双衰变

dual-die extrusion system 双模头挤塑装置
dual disc set point 双转盘组合点
dual drive 双传动
dual effect compressor 双效压缩机
dual emulsion 二元乳化液(乳浊液)〔由二相组成〕
dual evaporation 二级蒸发
dual filter system 双联过滤装置
dual firing （煤与石油）混合加热
dual flame detector 双焰检测器
dual flame ionization detector 双火焰电离检测器
dual-flow oil filter 双流油过滤器
dual flow path 双流路
dual fluorescence 双重荧光
dual focusing mass spectrometer 双聚焦质谱仪
dual functional 双功能的
dual functional catalyst 双官能催化剂*
dual-furnace cracking unit 双炉裂化装置
dual graft 双接枝
dual initiator system 双引发剂体系
dual installation 复式装置；双重装置
dual ion beam detector 双离子束探测器
dualistic 二元的
dualistic formula 二元式
dualistic nature 二象性
dualistic theory 二元(学)说；二元论
duality 二象性
dual-jacketed lamp 双夹套灯
dual-line pipe extrusion 双列管材挤塑
dual matched column 双匹配柱
dual mechanism 双重机理
dual mounts 双(轮)胎
dual-pack(=two-pack) 两罐装；双组分
dual pen recorder 双笔记录仪
dual plunger mould 双塞塑模
dual pressure controller 高低压控制器
dual pressure nitric acid 双加压硝酸工艺
dual property 二象性
dual-purpose additive 双效添加剂
dual-purpose column 两用(萃取)柱
dual-purpose oil 双效油〔具有润滑和冷却的作用〕
dual purpose tire 公路越野两用轮胎
dual rated tread rubber 双速硫化胎面胶
dual reservation mechanism 双保留机理
dual-solvent 25 process 双溶剂25过程〔高浓缩铀核燃料后处理流程，一循环用TPB，二、二循环用MIBK〕
dual solvent system 二元溶剂体系
dual-temperature exchange separation 双温交换分离(法)
dual-temperature process 双温过程

dual temperature separation 双温分离(法)
dual tire vulcanizer 双模化胎硫化机
dualumin 坚铝〔铝基合金〕
dual wavelength-factor spectrophotometry 双波长系数分光光度法
dual wavelength spectrophotometer 双波长分光光度计*
dual wavelength spectrophotometry 双波长分光光度法
dual wavelength TLC scanner 双波长薄层色谱扫描仪
dual worm extruder 双螺杆挤出机
duamycin 二重霉素
duazomycin(=diazomycin) 偶氮霉素
dubbin(g) 软化与防水油脂〔革〕
Dubbs cracking process 杜布斯式热裂化过程
Dubbs residuum 杜布斯式热裂化残油
dubhium〔符号：Db, 1916年埃德(Eder) 经游离而得的元素的假定名，后经证明即元素镱(ytterbium)〕
Du Bois cell 杜·波伊斯电池
Du Bois thermobalance 杜·波伊斯热天平
Duboscq colorimeter 杜波斯克比色计
Dubrovai process 杜布罗瓦过程；气相氧化裂化过程
duck 轻帆布
duclauxin 杜克拉青霉素
duct 导管
ductile 延性的
ductile base oil 优等延性油
ductile-brittle transition 韧脆转变
ductile coating 延展性涂层
ductile-crack-growth fracture 延性裂纹增长断裂
ductile fracture 延性破裂*
ductile iron 延性铁
ductileness(=ductility) 延性
ductilimeter 延性计
ductility(=ductileness) 延性
ductility machine 延性试验机
ductility retention ①延性保留②延性残留率
ductility test 延展性试验
ductility transition temperature 塑性转变温度
ductilometer 延展性试验计
duct(ing) lining 风道衬里
ducting powder 撒粉；扑粉
ductor(=ductor roller) 墨斗辊
duct thermostat 温度调节器
duff 煤屑
Duff reaction(=formylation reaction) 达夫反应；甲酰化作用
dufrenite 绿磷铁矿
dufrenoysite 硫砷铅矿
dugong oil 儒艮油

Duhem-Margules equation　杜亥姆-马居尔方程
dulcamarin　苦茄苷；蜀羊泉苷
dulcification　加甜
dulcin　①卫矛(己六)醇②对(位)乙氧基苯脲；甘素〔俗〕
dulcite(=dulcitol)　卫矛(己六)醇
dulcitol(=dulcite;dulcose)　卫矛(己六)醇；半乳糖醇　$HOCH_2(CHOH)_4CH_2OH$
dulcose(=dulcitol)　卫矛(己六)醇
dull　无光的
dull-appearing　外观发暗的
dull Berlin black　无光黑(沥青)漆
dullbox　无光鞋面革；多脂无光泽修正面大牛鞋面革
dull coal　暗煤
dull corrugated roll　暗瓦垅轧辊
dull fiber　暗纤维
dull film　无光膜
dull finish　①消光剂②消光修饰③消光
dull-finish lacquer　无光硝基面漆
dull hard coal　暗硬煤
dulling　①发浑(指清漆)②发暗〔指颜料〕③消光
dulling agent(=flating agent)　消光剂
dull-lustered rayon yarn　无光人造纤纱；无光人造丝
dull lustre　无光泽；暗光
dullness　①无光度；消光度②晦暗
dull polish(ing)　消光；打毛；磨成无光
dull rayon　无光嫘萦
dull red heat　暗红炽热
dull surface　无光面；黯面
Dulong and Petit's rule　杜隆-珀替规则
Dulong formula for heating value　杜隆热值公式
dumasin(=cyclopentanone)　环戊酮
Dumas method　杜马法
dumb-bell model　哑铃(状)模型
dumb bell sample　哑铃试样；哑铃状样件
dumbbell-shaped pattern　哑铃形喷束
dumbbell specimen　哑铃形试验片；哑铃形样板
dumbbell spray pattern　哑铃形喷束
dumb-bell test piece　哑铃试验片
dumbwaiter　小型升降机
dummy level　拟水平
dummy plate　隔板
dummy ring　填密环
dummy source　假放射源
dummy variate　虚拟变量；假变量
Dumont's blue　钴蓝
dumortierigenin　杜母配基
dumortierite　蓝线石
dump　①渣坑②垃圾堆

dump bin　废料箱
dump car　①倾卸车②废料车
dump chest　卸料池
dumped　废弃的；弃扔的；堆积的
dumped packing　堆积填充物
dumping　卸料的；倾卸的；倾倒的
dumping device　卸料装置
dump oil　桶装油
dump point　塔板〔蒸馏塔〕；倾流点
dump tank　接受器
dump valve　倾泄阀〔安全阀〕
dundasite　碳酸铅铝矿；白铅铝矿
dung bate(=dung bath)　粪(脱灰)液〔革〕
dung bath(=dung bate)　粪(脱灰)液〔革〕
dunging process　除媒染剂法
dung salt　粪盐；砷酸氢二钠　Na_2HAsO_4
dunite　纯橄榄岩
dunking　浸泡
Dunlop tripsometer　邓录普摆锤式弹性仪
dunnione　董尼酮
dunnite　董炸药；董那特；苦味酸铵炸药
Dunouy (ring) tensiometer　杜诺依(环法)表面张力仪
duodenin　肠降血糖素
duohydrate(=dihydrate)　二水合物
duomycin(=aureomycin)　金霉素
duoplasmatron ion source　双等离子体离子源
duoplasmatron source　双等离子体源
duo-sol extraction　双溶剂提取
duo-sol process　双溶剂润滑油精制过程
duotal　碳酸愈创木酚酯
duo-trio test　(味觉)对比试验
dupatorin　半齿泽兰素
DU penetrator(=depleted uranium penetrator)　贫化铀(制)穿甲弹
duplet　①电子对；电子偶②对
duplet bond　双键；偶键
duplet electron　偶电子
duplex　双的；加倍的
duplex alloy　二相合金
duplex baking powder　双焙粉；复合膨松剂
duplex ball mill　双行型球磨机
duplex classifier　双重分粒器
duplex color ink　复色墨
duplex double acting pump　双缸复动泵
duplex film　双重膜
duplex filter　复式过滤器
duplex grate producer　双箅发生器
duplexite　硬沸石

duplex nickel plating 双层(电)镀镍
duplex nozzel 双联喷嘴
duplex paper 双层纸
duplex polymer filtration 双联聚合物过滤
duplex printing machine 双印机〔纺〕
duplex pump 双缸泵；双筒泵
duplex reciprocating pump 双缸往复泵
duplex refiner （肥皂)双联压炼机
duplex regulator 双效调节器
duplex type 复式
duplex vacuum plodder 双联真空压条机
duplex winding 双联缠绕(法)
duplicate 双份*
duplicate cavity plate 双联模板；滑动模槽板
duplicate injection 重复注射
duplicate stained 二重染色
duplicate stripper-plate 双联脱模板
duplicate test 重复试验
duplicating paper 双层纸
duplication method 复制法
duplicator ①复印机；复制机②复制者③二倍器；倍增器；倍加器
Du Pont flexural tester 杜邦屈挠试验机；杜邦弯曲试验机
Du Pont impact tester 杜邦冲击试验机
Du Pont process 杜邦法〔全循环制尿素〕
duprene 氯丁橡胶
duprene rubber 氯丁橡胶
Dupre's equation 杜普雷公式*
durability 耐久性；耐用年限
durability test 寿命试验
durable ①耐久的；经久；坚牢的；有持久力的②耐用物品
durable antistatic finish 持久性抗静电整理
durable aroma 持久香气
durable-press fabrics 耐久性压烫织物
durable product 耐用品
durable soil-release 持久去污性
durable years 耐久年限；使用年限
durain 暗煤(硬块)
Duraloy 杜拉洛伊铁铬合金
duralumin 硬铝
duraluminum 铝合金；轻合金
duramen(=heart wood) 心材
duramycin 都拉霉素；耐久霉素
durangite 橙砷钠石
duranol 杜南醇
duranol black 杜南醇黑
duranol blue 杜南醇蓝
duranol colors 杜南醇染料〔英国 BDC 染料厂醋酸纤维染料商名〕
duranol orange 杜南醇橙
duranthrene black 杜南士林黑
duranthrene blue 杜南士林蓝
duranthrene brilliant violet 杜南士林亮紫
duranthrene colors 杜南士林染料〔英国 BDC 染料厂还原染料商名〕
duranthrene golden orange 杜南士林金橙
duranthrene red violet 杜南士林红紫
duration 持续时间*
durdenite 碲铁矿
durene(=1,2,4,5-tetramethylbenzene;durol) 杜烯；1,2,4,5-四甲苯；均四甲苯 $(CH_3)_4C_6H_2$
durene carboxylic acid 杜烯羧酸；2,3,5,6-四甲基苯甲酸
durene diacetic acid 均四甲苯(基)二乙酸
durenol 杜烯酚 $C_{10}H_{14}O$
duridine 杜啶；2,3,5,6-四甲基苯胺
duriron 硅铁
duro 硬度计
durocalibrator 硬度计校正器
duroelasts 硬弹体；刚弹体
durohydroquinone 杜氢醌；四甲对苯二酚
durol(=durene;1,2,4,5-tetramethylbenzene) 杜烯；1,2,4,5-四甲苯；均四甲苯 $(CH_3)_4C_6H_2$
duromer 硬质体；刚性体
durometer 硬度计；硬度测验器
durometer-hardness(=Shore hardness) 肖氏硬度
durometer level ①硬度范围②硬度级
durometer number 硬度计读数
duroplasts 硬塑料
duroquinol 杜氢醌；四甲基氢醌；四甲基对苯二酚
duroquinone 杜醌；四甲基对苯醌
Durran's melting point 达兰(水银法)熔点
Durran's mercury method 达兰水银法〔测树脂熔点〕
duryl(=2,3,5,6-tetramethylphenyl) 杜基；2,3,5,6-四甲(基)苯基
duryl aldehyde 杜基醛；2,4,5-三甲基苯甲醛
durylene(=tetramethyl-p-phenylene) 亚杜基；四甲亚苯基
durylic acid 杜基酸；2,4,5-三甲基苯酸 $(CH_3)_3C_6H_2COOH$
dussertite 绿砷钡铁石
dust ①(灰)尘②屑；末；粉③粉剂
dust abating 除尘(的)
dustability(=dispersity) 分散度
dust agglomeration 粉末附聚(作用)
dust-arrester installation 集尘装置

dust blow-away preventive　尘粉飞扬防止剂
dust borne gas　含尘气体
dustborne radioactivity　放射性尘埃
dust cage　尘笼
dust car　废料车
dust catcher　集尘器
dust chamber　除尘室
dust cloud　尘雾
dust collecting　除尘；集尘
dust collecting efficiency　集尘效率
dust collecting electrode　集尘电极
dust collection hose　吸尘胶管
dust collector　集尘器
dust concentration　粉尘浓度
dust condensing flue　沉灰烟道
dust content　含尘量
dust corrosion experiment　尘埃腐蚀实验
dust diluent　粉剂稀料
dust disposal　粉尘处理
dust-dry　不沾尘干燥；脱尘干燥〔一种表面干燥情况〕
dust emission　扬尘
duster　①除尘器；除尘机②撒粉器③集尘器
dust exhauster　除尘器
dust-exhaust system　除尘系统
dust exploding　粉尘爆炸
dust explosion　粉尘爆炸
dust extractor　集尘器
dust fall　降尘；落尘
dust fastness　耐尘性
dust filter　滤尘器
dust-fired　粉末燃烧的
dust free　无尘
dust-free blasting　无尘喷砂法
dust free chamber　无尘室
dust-free operating space　无尘操作区
dust gauze　滤灰网
dust guard　防尘盖；防尘罩
dust hazard　(粉)尘害
dust hood　吸尘罩
dustiness　尘污
dusting　①撒粉；涂粉②除尘③粉尘飞扬
dusting agent　隔离剂
dusting bronzes　①敷青铜粉；涂青铜粉漆②揩金〔印刷〕
dusting clinker　粉脆熔块
dusting flour　(扑)撒用粉
dusting machine　除尘器
dusting material　腌料〔用于皮革腌鞣〕；隔离剂〔橡胶〕
dusting process　撒粉印画法

dust-laden air　有尘空气
dust laying　除尘(的)；降尘(的)
dust-laying agent　消尘剂
dust laying oil(=dust palliative)　防尘油
dustlike　尘状的；粉状的
dust load　粉尘负荷
dust loss　粉尘损失
dust mask　防尘面具
dust palliative(=dust laying oil)　吸尘油
dust precipitator　集尘器
dust-proof　防尘的
dust-proof lighting fitting　防尘照明装置
dust proof suede　防尘绒面革
dust removal installation　除尘设备
dust-removal system　除尘系统
dust remover　除尘器
dust respirator　防尘面具
dust retention　积尘；积垢
dust sampler　尘埃取样器
dust sampling tube　尘粒取样器
dust scrubber　洗尘器
dust separating　灰尘分离的；除尘的
dust separation　灰尘分离；除尘
dust separator(=dust settler)　集尘器
dust settler(=dust separator)　集尘器
dust settling　除尘的；集尘的
dust settling chamber　除尘室
dust settling pocket　除尘袋
dust shield　防尘罩
dust stop　密封装置
dust sucking plant　吸尘装置
dust suctor　吸尘器
dust suppression　粉尘消除；除尘措施
dust-tight　绝尘
dust trap　集尘器；除尘器
dusty　①灰尘的②多尘的
dusty fuel　粉状燃料
Dutch brick　荷兰砖；高温烧结砖
Dutch cheese　荷兰干酪
Dutch cleanser　荷兰去垢粉
Dutch cyclone　荷兰旋风分离器；水力旋风分离器
Dutch gold(=Dutch metal)　荷兰金；假金；高锌黄铜合金〔锌12%～20%，其余为铜，金箔代用品〕
Dutch liquid(=Dutch oil)　二氯乙烷
Dutch oil(=Dutch liquid)　二氯乙烷
Dutch oven　荷兰炉子
Dutch stack process(=Dutch process)　荷兰堆积法；堆垛法

Dutch white lead 荷兰法铅白；碱式碳酸铅
Dutch white metal(=Dutch pewter) 荷兰白〔铜：锑：锡=10：9：81〕
duterium rubber 氘橡胶；重氢橡胶
duty 本领；生产量；负荷；功；功用
duty cycle capacity 断续负载容量
duty factor ①占空因数；占空系数②工作系数；利用因数
duty horse power 报关马力；额定马力
duty ratio ①功能比；功率比；负载比②(脉冲)占空率；占空系数
duxite 杜克炸药；杜克煞特
D-value D值；中和延缓值〔加改性剂后黏胶中和时间与不加改性剂的黏胶中和时间的比值〕
dvi- ①类；似〔元素〕②〔梵文〕第二的
dvicesium(=francium) 类铯；钫 Fr
dvielement 类元素〔指周期表中尚未发现的元素〕
dvimanganese(=rhenium) 类锰；铼 Re
dvitellurium(=polonium) 类碲；钋 Po
dwelling 滞留保压
dwelling period 停留期
dwell tack (黏合)持续黏性
dwell time 停留时间；采样时间*
dyad ①二价的②二数的③二价元素④双；对
dyadic ①并矢式；并向量②二(价元)素的；二数(进)的；双值的；二元的
dyadic notation 并向量符(号)
dyadic numbers 二进数
dydimic acid 石蕊酸
dydrogesterone 地屈孕酮〔药〕
dye ①染料②染(色)
dyeability ①染色性②染色度
dyeability modifier 可染性能改良剂
dyeable 可染色的
dye absorption spectra method 染料吸收光谱法〔测临界胶束浓度〕
dye-acceptance 染料接受性；染料容纳性
dye-acceptancy 染料接受性；染料容纳性
dye accessibility 染料可及性
dye adsorption 染料吸附(作用)
dye affinity 染料亲和力
dye bath 染色浴
dyebath stability 染浴稳定性；染液稳定性
dye beck 染槽；染色桶
dye binding method 染料结合法
dye bleeding 染料析水
dye cation 着色阳离子
dye check 着色探伤；染色探伤

dyed gasoline 着色汽油
dye dichroism 染料二向色性
dye diffusion velocity 染料扩散速度
dyed-in 渗入；染进
dyed-in screener technique 掩蔽剂染进技术；屏蔽剂渗入技术
dyed in the mass 原液染色；纺液着色
dyed yarn 染色纱
dye exhausting rate 染料吸净率；染料吸尽率
dye exhaustion 染料吸净(作用)
dye fastness 染料坚牢度
dye fixatives 固色剂
dye fixing 染料的固定
dye-fixing agent 固色剂
dye-forming oxidation 成色氧化(作用)
dye-forming reaction 成色反应
dye gigger 精染机
dye house 染坊
dye indicator 染料指示剂
dyeing ①染色的②染色(作用或工程)
dyeing ability 染色本领
dyeing accelerant 促染剂；导染剂
dyeing affinity 染色亲和力
dyeing assistant 染色助剂
dyeing auxiliary 染色助剂
dyeing beck 染色桶
dyeing beck stick 染色桶棒
dyeing behavior 染色性能
dyeing capacity 染色力
dyeing defect 染疵；色花
dyeing drum 染色转鼓
dyeing equilibrium 染色平衡
dyeing fastness 染色坚牢度
dyeing house 染坊
dyeing inhibitor 染色防胶剂〔树脂的形成〕
dyeing intensification 染料加厚法
dyeing intensifier 染色增强剂
dyeing in width 宽幅染色
dyeing kinetics ①染色动力学②染色速度
dyeing liquor 染色液；染料溶液
dyeing machine 染色机
dyeing machinery 染色机
dyeing paste 染色浆
dyeing-sizing combined technique 染浆联合工艺
dyeing solution 染色溶液；染料溶液
dyeing speck 染色斑点；染色瑕疵
dyeing strip 染色条
dyeing uniformity 染色均匀性

dye intermediate　染料中间体
dye laser　染料激光器
dye level　染色水平；染色深度
dye leveller　均染剂
dye levelling　染料均染性
dye liquor　染料溶液；染色液
dye migration preventing agent　染料迁移防止剂
dye number　染色值（加溶）
dyeometer　上色率测验器
dye paste　染料浆
dye pasting　染料制浆
dye penetrant　染色渗透液
dye penetration constant　染料渗透常数
dye-polymer association process　染料-聚合物缔合过程
dye printing　印染；印花
dye printing paste　印染浆
dyer　染色工人；染匠
dye receptiveness　染料吸取性
dye-receptive polar group　吸取染料极性基团
dye receptive site　染料吸取位置
dye-receptivity　染料吸取性
dye receptor　染料吸取体
dye resisting agent　防染剂；拒染剂
dye retarding agent　染色阻滞剂；缓染剂
dyer's bath　染浴器
dyers greenweed　金雀花
dye sensitified polymerization　染料增感聚合(法)
dye sensitization　染料增感作用
dye-sensitized photoinitiation　染料敏化光引发(作用)
dye sensitized polymerization　染料增感聚合(作用)
dye solubilization　染料增溶
dye solution　染料溶液
dye sorption　染料吸着作用
dye sprayer　染料喷枪
dyestain　染渍
dye stain test　(染料)着色试验〔染色法鉴定氧化膜气孔的封闭程度〕
dye strength　染料浓度
dyestuff　①染料；颜料②着色剂（橡胶）
dyestuff fixing　染料的固定(固着)
dyestuff former　染料形成体；染料前体
dyestuff intermediate　染料中间体
dyestuff penetration　染料渗透(作用)
dye toning　染料调色法
dye transfer agent　染料传递剂
dye uptake　上色率；上染率；吸色率
dye vat　染色瓮
dye vatting　瓮染法

dye wood　染料木
dyewood-extract　染料木萃
dying affinity　染色亲和力
dying out　(反应)停止；衰灭消失
dylene　聚苯乙烯
dynad　原子内场
dynagraph　内应力测定仪
dynamagnite(=nitromagnite)　硝化甘露醇
dynamic(=dynamical)　①动态的②动力的③有力的
dynamic adsorption　动态吸附
dynamical(=dynamic)　①动态的②动力的③有力的
dynamical angle of repose　动态休止角
dynamical behavior　动态行为
dynamical boundary condition　动力边界条件
dynamical compliance　动态柔量
dynamical correlation　动态相关
dynamical creep　动态蠕变
dynamical deformation　动态形变
dynamical elasticity　动态弹性
dynamical equation　动力学方程
dynamical loss shear compliance　动态剪切耗能柔量
dynamical loss shear modulus　动态剪切耗能模量
dynamic allotropy　动态同素异形
dynamically formed membrane　动态形成膜
dynamically modified silica chromatography　动态改性硅胶色谱法
dynamical theory of strength　强度动力学理论
dynamical variable　动力学变数
dynamic analysis　动态分析；动力学计算；动力特性分析
dynamic angle of friction　动摩擦角
dynamic behavior　动态性状；动态行为
dynamic bending modulus　动态弯曲模量
dynamic bioanalysis　动态生化分析
dynamic birefringence　动态双折射
dynamic braking　动力制动
dynamic buckling　动态变形
dynamic calibration　动态校准
dynamic characteristic　动态特性
dynamic check　动态检查；动态校验
dynamic chemorheology　动态化学流变学
dynamic coated method　动态涂渍法
dynamic coating　动态包覆；动态涂渍
dynamic compensation　动态补偿；动态校正
dynamic complex ion exchange model　动态复合离子交换模型
dynamic compliance　动态柔量
dynamic condenser　动电容器
dynamic condenser amplifier　动电容放大器

dynamic condenser electrometer 动电容静电计
dynamic condition 动态条件
dynamic creep 动态蠕变
dynamic deactivation 动态去活
dynamic deactivity 动态脱活
dynamic density functional theory 动态密度泛函理论
dynamic dielectric analysis 动态介电法
dynamic differential thermal analysis(DDTA) 动态差热分析
dynamic diffusion 动态扩散
dynamic drape 动态悬垂性
dynamic drift (硬度计指针)动态漂移
dynamic elasticity 动态弹性
dynamic elastic modulus 动态弹性模量
dynamic electrical behavior 动电性能
dynamic elevated temperature-pressure test 动态升温-升压试验
dynamic elevated temperature test 动态升温试验
dynamic equilibrium ①动态平衡②动力平衡
dynamic error 动态误差
dynamic FAB 动态快原子轰击(接口)
dynamic fatigue 动态疲劳
dynamic fatigue test 动态疲劳试验
dynamic flexing 动态曲挠
dynamic foam height 动泡沫高度
dynamic force 动力
dynamic force spectroscopy 动态力光谱
dynamic formula 动(态)式
dynamic frequency 动态频率
dynamic friction 动摩擦
dynamic friction(al) coefficient 动(力)摩擦系数
dynamic hardness 冲击硬度
dynamic headspace analysis 动态顶空分析法
dynamic heat distortion 动态热变形
dynamic hygroscopicity 动态吸湿性
dynamic hysteresis loop 动态滞后回线
dynamic infrared dichroism 动态红外二色性
dynamic infrared spectroscopy 动态红外光谱法
dynamic instrument 动态仪器
dynamic interfacial tension 动界面张力
dynamic ion exchange 动态离子交换
dynamic ion exchange chromatography 动态离子交换色谱法
dynamic ion exchange model 动态离子交换模型
dynamic isomeride 动态异构休
dynamic isomerism 动态异构现象
dynamic light scattering 动态光散射
dynamic liquid chromatographic column 动态液相色谱柱

dynamic loop-target 流动环路靶
dynamic loss 动态损耗
dynamic loss factor 动态损耗因子
dynamic loss shear modulus 动态损耗剪切模量
dynamic loss tensile modulus 动态拉伸耗能模量
dynamic mass spectrometer 动态质谱仪*
dynamic mechanical analysis 动态机械法
dynamic mechanical analyzer of polymer 高聚物动态力学分析仪
dynamic mechanical loss tangent 动态力学损耗正切
dynamic mechanical spectroscopy(DMS) 动态力(学)谱学
dynamic mechanical thermal analysis 动态热机械分析
dynamic mechanical thermal analyzer 动态热机械分析仪
dynamic method 动态法；动态测定法
dynamic mixer 动态混合器
dynamic modification 动态改性
dynamic modulus 动态模量
dynamic nuclear polarization 动态核极化
dynamic oscillatory measurement 动态振动测量
dynamic oxidation test 动力学氧化试验
dynamic packing 动态密封
dynamic packing technique 动态(柱)填充术
dynamic potential 动电势
dynamic potentiometry 动态电位滴定法
dynamic pressure 动压力
dynamic programming 动态程序设计；动态规划
dynamic quenching 动态猝灭
dynamic range 动态范围
dynamic reflectance spectroscopy 动态反射光谱法
dynamic relaxation 动态松弛
dynamic reserve 动态储备
dynamic resilience 动态弹性回弹
dynamic resistance 动态阻力；动态电阻
dynamic resolving power 动态分辨本领
dynamic response 动态响应
dynamic response of ISE 离子选择电极动态响应
dynamic reversibility 动态可逆性
dynamic rheometer 动态流变仪
dynamic rigidity 动态刚性度
dynamic scan 动态扫描
dynamic screening (砂磨机)动态过滤
dynamic seal 动态密封
dynamic secondary ion mass spectrometry 动态二次离子质谱法
dynamic separation 动态分离
dynamic separation of the pH gradient pH 梯度动态分离
dynamic shear compliance 动态剪切柔量

dynamic shear viscosity 动态切变黏度
dynamic shrinkage 动态收缩
dynamic similarity 动力相似
dynamic similarity principle 动力学相似原理
dynamic spin packet 动态自旋波包
dynamic split system 动态分离系统
dynamic spring analysis 动(态弹)簧分析
dynamic sputtering 动态溅射
dynamic stability 动态稳定性
dynamic state 动态
dynamic state simulation 动态模拟
dynamic steam devulcanization 废胶动态再生法
dynamic stiffness ①动态刚性度；动力劲度②动模数③动态稳定度
dynamic storage shear compliance 动态剪切储能柔量
dynamic storage shear modulus 动态剪切储能模量
dynamic storage tensile modulus 动态拉伸储能模量
dynamic strain optical coefficient 动态光弹(性)系数
dynamic stress 动态应力
dynamic suction lift 动式吸入高度
dynamic surface tension 动态表面张力*
dynamic test 动态试验〔如疲劳试验、蠕变试验、冲击试验〕
dynamic tester 动平衡测定仪
dynamic testing 动态试验
dynamic theory of continuous medium 连续介质动力学理论
dynamic thermomechanical analysis 动态热机械分析法
dynamic thermomechanical analyzer 动态热机械分析仪
dynamic thermomechanometry 动态热机械分析(法)
dynamic titration technique 动态滴定法；三角波程控滴定法
dynamic transition 动态转变
dynamic valve 动力阀
dynamic viscoelasticity 动态黏弹性
dynamic viscoelastic measurement 动力黏弹性测量
dynamic viscoelastometer 动态黏弹仪
dynamic viscoplasticity 动态黏塑性
dynamic viscoplasticoelasticity 动态黏塑弹性
dynamic viscosity 动态黏度
dynamic vulcanization 动态硫化
dynamic X-ray technique 动态 X 射线技术
dynamic Young's modulus 动态杨氏模量
dynamite 达纳炸药；达纳马特
dynamite glycerol 达纳炸药用甘油
dynammon 达纳猛炸药；硝铵-炭炸药
dynamo ①电机②发电机
dynamograph 自动记录测力计
dynamometer ①测力计②功率计
dynamo oil 电机油
dyne 达因
dynobel 达诺贝尔
dynode 倍增电极；打拿极
dypnone 缩二苯乙酮
dysanalite(＝dysanalyte) 钙铌钛矿
dysbiosis 生态失调
dyschromasia 变色
dyscrasite 锑银矿
Dysonian line shape 戴森线型
Dyson's interference microscope 戴森干涉显微镜
dysoprosium phosphate 磷酸镝 $DyPO_4$
dysoxylonene 樫木烯
dysprosia 氧化镝 Dy_2O_3
dysprosium 镝〔66 号元素，化学符号 Dy〕
dysprosium bromide 溴化镝 $DyBr_3$
dysprosium carbonate 碳酸镝 $Dy_2(CO_3)_3$
dysprosium chloride 氯化镝 $DyCl_3$
dysprosium hydroxide 氢氧化镝 $Dy(OH)_3$
dysprosium nitrate 硝酸镝 $Dy(NO_3)_3$
dysprosium oxide 氧化镝 Dy_2O_3
dysprosium oxychloride 氯氧化镝 DyOCl
dysprosium sulfate 硫酸镝 $Dy_2(SO_4)_3$
dyssophotic 弱光性
dystectic 高熔的
dystectic mixture 高熔混合物
dystectic polymer 高熔(点)聚合物
dystomespar 硅硼钙石

E

Eagle mounting 伊格尔装置
Eagle-Picher blister box 伊格尔-皮切尔型起泡试验箱
eagle wood oil 沉香木油
EAP filter(=extended area pack filter) 面积扩展的组合过滤器
early barrier 早势垒*；前势垒
early growth phase 早期生长相；生长前阶段
early strength cement 早硬水泥；快硬水泥
early turbulence 初期湍流
early wood(=spring wood) 早材；春材
ears 灯花〔灯芯上的积炭〕
earth ①泥；土②地(球)③接地
earth alkali metal 碱土金属
earth bearing strength 接地耐力
earth burner 烧土炉
earth burning 烧土
earth cable 接地电缆
earth color 矿物颜料
earth conductor cable 接地电线
earth connection 接地
earth-covered pile 土堆窑〔干馏木桩制松焦油用〕
earth crust 地壳
earthed electrode 接地电极
earthed metal shield 接地金属护罩
earthed surface 接地表面
earthed system 接地系统；接地装置
earthen ①陶制的；土的；土制的②土地的
earthen container 陶制容器
earthen storage tank 陶制油罐
earthenware 陶器
earthenware clay 陶器土
earthenware coil 陶器旋管
earthenware pipe 陶器管
earthenware pot 陶器桶
earthenware ring 陶器圈；瓦圈
earthenware tower 陶器塔；瓦塔
earthenware vessel 陶器器皿
earth family element 土族元素
earth-fault finder 接地故障探测器
earth fault relay 接地故障继电器
earth filter 土滤器
earth flax 石棉
earth-free 不接地的
earthing 接地
earthing installation 接地装置
earth kiln 土(炭)窑
earth leakage 漏电
earth metal 土金属
earthnut oil 花生油
earth oil 石油
earth pigment 无机颜料；土性颜料
earth potential 地电位；地电势
earthquake load 地震载荷
earth refining 陶土精制
earth regeneration 陶土再生
earth regeneration oven 陶土再生炉
earth rubber 泥胶
earth-type filter 陶制过滤器
earth wax 地蜡
earthy brown coal 土状褐煤
earthy cobalt 钴土矿
earthy element 土族元素
earthy odor 壤香
ease 简易；轻便
ease of addition 加成容易程度；加成本领
ease of coloring 易着色性
ease of dispersion 易分散程度；分散容易程度
ease of ignition 易燃性
Eastern perfume 东方型香精
Eastman-Kodak process 伊斯门-科达法〔由甲烷-丁烷制乙炔法〕
easy access 容易接近；便于(检修)
easy-bleaching pulp 易漂白浆粕
easy-care finish 易保养涂饰剂；免烫整理
easy-care leather 易保养革
easy-care property 易保养性能
easy cation-dyeable polyester 阳离子染料易染型聚酯；易染型改性聚酯
easy fit 轻转配合
easy flow 易流动；高流动性
easy open convenience can 易拉罐；易开罐
easy processing agent 操作助剂
easy push fit 轻推配合；滑动配合
easy servicing 小修
easy slide fit 轻滑配合；滑动配合
easy starter 简易启动装置
easy-to-blend 易掺混的
easy-to-degrade organic pollutant 易降解有机污染物

easy-to-disperse pigment　易分散型颜料
easy-to-dye　易染的
easy-to-handle　①易于处理的②易于操纵的
eatable　①可食用的②食品
eau-de-Javel　惹芙耳溶液〔次氯酸钠或次氯酸钾溶液〕
eau-de-Labarraque　拉巴腊溶液〔次氯酸钠溶液〕
eave-frosting　起凸形霜纹〔漆病〕
E band　E 谱带
ebb-flow cycle　涨落周期；盛衰周期；脉动周期
EBC process　电子束固化法
Eberhard effect　埃伯哈德效应
Eber's reagent　埃伯尔试剂
Eberstadt method　埃伯施塔特测定法〔测定醋酸纤维素中乙酰基 CH_3CO ——含量的非均相皂化法〕
Ebert-Fastie mounting　艾伯特-法斯第装置
ebonite　硬质胶*
ebonite accumulator box(=ebonite accumulator jar)　硬质胶蓄电池箱
ebonite accumulator jar(=ebonite accumulator box)　硬质胶蓄电池箱
ebonite compound　硬(质)胶胶料
ebonite dust　硬质胶粉；胶木粉
ebonite fittings　胶木配件
ebonite powder　胶木粉；硬质胶粉
ebonite wax　乌木蜡
ebony　乌木
ebucin　葡萄糖酸钙
ebullated bed　沸腾床；流化床
ebullator　沸腾器
ebullience　沸腾
ebullient　沸腾的
ebullient fluidized solid bed　沸腾流化固体床
ebulliometer　①沸点升高计②沸点酒精计
ebulliometry　沸点升高测定法
ebullioscope　①沸点计②沸点酒精计
ebullioscope method　沸点酒精度测定法
ebullioscopic constant　沸点升高常数
ebullioscopic equation　沸点升高公式
ebullioscopic method　沸点升高法
ebullioscopic solvent　沸点升高溶剂
ebullioscopy　沸点升高(测定)法
ebullition　沸腾
eburicoic acid　齿孔酸
eburnamenine　象牙(洪达木)烯宁
eburnamine　象牙(洪达木)胺
eburnamonine　象牙(洪达木)酮宁
ECB(electrically controlled birefringence)　电控双折射
ecbalin(=elateric acid)　喷瓜酸

ecboline　麦角碱
EC carbon acrylic type EMI shielding coating　导电炭黑丙烯酸型电磁干扰屏蔽涂料
eccentric　偏心的
eccentric cylinder rheometer　偏心圆筒式流变仪
eccentric disk　偏心盘
eccentric factor　偏心因子
eccentricity　①偏心率②偏心性③偏心距
eccentric oiler　偏心注油器
eccentric pivot　偏心轴
eccentric press　偏心式压机
eccentric reducer　偏心大小头；偏心变径管
eccentric shaft　偏心轴
eccentric strap　偏心环
eccentric toggle lever press　偏心肘杆式压机
ecdysial fluid　蜕皮液
ecgonidine(=anhydroecgonine)　芽子定；脱水芽子碱
ecgonidine methyl ester　芽子定甲酯
ecgonine　芽子碱
ecgoninic acid　芽子碱酸　$C_7H_{11}O_3N$
echelette　红外光栅
echelette grating　红外光栅
echelle　阶梯光栅
echelle grating spectrometer　中阶梯光栅光谱仪
echelon　梯阵；阶梯
echelon grating　阶梯光阑
Echesle　爱克斯尔〔糖度比重计名〕
echiceric acid　艾奇蜡酸
echicerin　艾奇蜡精
echinacosid　海胆苷
echinenone　海胆酮
echinenone　海胆酮；β 胡萝卜素-4-酮
echinochrome　海胆色素
echinocystic acid　刺囊酸
echinopanacene　日参烯
echinopanacol　日参醇
echinopsine　蓝刺头碱
echinuline　①海胆灵②灰绿曲霉素
echitamidine　鸡骨常山定碱
echitamine(=ditaine)　鸡骨常山碱；艾奇明
echitenine　艾奇宁
echitin　鸡骨常山酸　$C_{32}H_{52}O_2$
echiumine　蓝蓟碱
echo　①回声②回波
echo detection　回波检测
echo-planar imaging　回波二维成像
echugin(=echujin)　箭毒苷
echujin(=echugin)　箭毒苷

Eckert number 埃克尔特数
eclipsed 重叠的
eclipsed conformation 重叠构象*
eclipsed form 重叠式
eclipse period (增殖)隐蔽期；晦暗期
eclipse phase 隐蔽期
eclipsing effect 重叠效应*
eclipsing strain 重叠张力*
eclogite 榴辉岩
ecological agriculture 生态农业
ecological balance 生态平衡
ecological effect 生态效应
ecological efficiency 生态效率
ecological fiber 生态纤维；无污染纤维
ecological sustainability 生态持续性
ecology 生态学
econometer 炉气碳酸计
economical apparatus 廉价仪器
economically recoverable oil 工业性可采石油储量
economical ore 经济矿藏
economic balance 经济核算
economic gain 利润；赢利
economic value 经济价值
economize energy emulsion 节能乳化液
economizer ①省热器②节油器③省煤器
economizer bank (空气)预热管；节热器排管
economy ①经济②节俭
ecotextile 生态纺织品
ecracite 厄拉炸药；伊煞特；三硝基甲酚铵
ectype ①异常型②复制品
EDA complex 电子给体-受体复合物
eddy 涡流
eddy-conductivity 涡流传导性
eddy current ①涡流②涡电流
eddy current heater 涡电流加热器
eddy current heating 涡电流加热；感应加热
eddy current inspection 涡流探伤
eddy current thickness meter 涡(电)流(法)测厚仪
eddy diffusion 涡流扩散*
eddy diffusion factor 涡流扩散因子
eddy diffusion term 涡流扩散项
eddy diffusivity 涡流扩散率
eddy effusion 涡流喷射
eddy flame 湍流焰；扰动焰；紊流火焰
eddy generation 发生涡流
eddying 涡流；涡流的形成
eddying effect 涡流效应
eddying flow 涡流流动

eddying motion 涡流运动
eddying resistance 涡流阻力
eddy kinematic viscosity 涡流运动黏度
eddy losses 涡流损失
eddy motion 涡流运动
eddy resistance 涡流阻力
eddy shedding 涡流分离
eddy thermal diffusivity 涡流热扩散系数
eddy transfer 涡流传递
eddy velocity 涡流速度
eddy viscosity ①涡流黏度②涡黏性
eddy viscosity coefficient 涡流黏性系数
edeine 伊短菌素
Edeleanu extract 埃德林抽提物
Edeleanu extraction 埃德林抽提
Edeleanu raffinate 埃德林精炼物
Edeleanu refining process 埃德林精炼法〔石油〕
Edeleanu urea dewaxing process 埃德林尿素脱蜡法
edestin fiber 麻仁纤维
edetate 乙二胺四乙酸盐；EDTA 盐
edetic acid(EDTA) 乙二胺四乙酸；乙底酸（俗）
edge 棱；边；边缘
edge angle 边缘角；偏角
edge breaks 边缘破裂
edge compressive strength 棱边抗压强度
edge coverage 棱(边)角覆盖力；边角涂覆力(性)
edge crack 边缘裂缝
edge-crack test 边缘分层拉伸测试
edge crimped yarn 刀口卷曲变形丝
edge crimping process 刀口卷曲(工艺)
edge crumbling 边缘缺损
edge cutting machine 切边机
edge dislocation 刃型位错
edge distance 边距
edge doctor 刮边刀；边缘涂料刮刀
edge drainage (液膜)棱边排液
edge effect 边缘效应
edge excitation red shift 边缘激发红移
edge feeler 探边器
edge filter 流线式滤器
edge filtration(=streamline filtration) 流线式过滤
edge fold ①折边②卷边
edge forming 卷边
edge frequency 截止频率
edge gluing 棱边胶合
edge joint 边缘焊接；纵拼
edge jointing adhesive ①(成角)边接黏合剂②边缘接缝黏合剂

edge lighting 侧缘折光
edge mill(=edge runner) 轮碾机
edge nucleation 晶棱成核
edge phenomena 边缘现象
edge planer 刨边机
edge planer machine(=edge planing machine) 刨边机；板边刨床
edge point 端接接头
edge rim gate(=side edge gate) 齐边浇口
edge runner(=edge mill) 轮碾机
edge runner mill(=edge-running mill) 轮碾机
edge sealing 端封
edge spacing 板边空地
edge stone 磨石〔机〕
edge straightener 边缘碾平器
edge stress 棱边应力
edge substitution 边缘取代
edge-textured yarn 刀口法变形丝
edge texturing 刀口变形(工艺)
edge-to-edge 边对边
edge to edge resolution 边边分辨率
edge tool ①削边刀②修饰边缘的工具③有刃口的刀具
edge-trimming cutter 切边刀
edge trimming device 切料装置
edge-trim regrind 边料回用；回用边料
edge weld 端面焊缝；对边焊
edgewise （层压)沿边(的)；侧向(的)
edgewise compression 侧面压缩
edgewise compressive strength 侧压强度
edgewise load 沿边载荷
edgewise pressure 沿边压力
edging profile 镶边用型材
edgy field 尖端场
edibility 食用价值
edible ①食用的；食品的②营养的
edible fat 食用脂肪
edible function fiber 食用功能纤维
edible gelatin 食用明胶
edibleness 食用价值
edible oil 食用油
edible pigment 食用色素
edible tubular collagen casing 食用胶原肠衣
edingtonite 钡沸石
edinol 盐酸对氨基羟苯甲醇
Edison accumulator 爱迪生电池；铁镍蓄电池
Edison cell 爱迪生电池
edisonite 金红石
Edison storage battery(=iron nickel accumulator) 爱迪生蓄电池；铁镍蓄电池
editic acid(EDTA) 乙二胺四乙酸；乙底酸
Edmister's equation 埃德明斯特公式〔多组分气体吸收装置的吸收率计算式〕
Edmond's balance 埃德蒙天平〔测定气体相对密度用〕
edrophonium chloride 依酚氯铵〔药〕
EDTA titration EDTA 滴定；乙二胺四乙酸滴定法
educational chromatograph 教学(用)色谱仪
educt 离析物
eduction 离析
eduction pipe 放气管；排气管
eductor 喷射器
edudesmin 桉树胶素
edulcorant 加甜剂；甜味剂
edulcoration 洗净；除杂；纯化
edulcorative 加甜的
edulcorator 加甜剂
Edward's bottle 爱德华气瓶
Edward's gas density balance 爱德华气体密度天平
eel-liver oil 鳗鱼肝油
eel oil 鳗鱼油
efavirdine 依非韦伦〔药〕
effect ①效应；效；效果②结果③影响
cis-effect 顺位效应
cis-trans effect 顺-反式效应
α-effect α效应
effect confounding 效应混杂
effective 有效的
effective adsorption capacity 有效吸附容量
effective alkali 有效碱
effective amount 有效量
effective aperture 有效孔径
effective aspect ratio 有效长宽比
effective atmosphere 有效(大)气压；表压
effective atomic number 有效原子数
effective atomic number rule 有效原子序数法则
effective bandwidth 有效带宽；谱带半宽度
effective bond length 有效键长
effective capacity(=effective power) 有效动力;有效功率
effective carbon number 有效碳(原子)数
effective catalyst composition 有效催化剂组成
effective charge 有效电荷
effective charge of defect 缺陷的有效电荷
effective chart width 有效(记录)纸宽度
effective concentration 有效浓度
effective concentration of network chains 网络链的有效浓度
effective constant 有效常数

effective crosslink　有效交联
effective cross section　有效截面
effective cross section of ionization　有效电离截面
effective current　有效电流
effective density　有效密度
effective diameter　有效直径
effective diffusion coefficient　有效扩散系数
effective diffusivity　有效扩散系数；有效扩散率
effective dipole moment　有效偶极矩
effective dose　有效剂量
effective efficiency　效率
effective entanglement density　有效缠结密度
effective enthalpy　有效焓
effective estimator　有效估计值
effective factor　有效因数
effective fiber length　实际纤维长度；有效纤维长度
effective field　有效场
effective film　有效膜
effective film thickness　有效膜厚度
effective fixative　有效定香剂
effective fractional binder volume　有效基料体积分数
effective functionality　有效官能度
effective half-life　有效半衰期
effective interfacial area　有效相界面积
effective length　有效长度
effective lethal phase　致死效应期；有效致死期
effective level　有效用量
effective life　有效使用期；有效寿命
effective mass　有效质量
effective mobility　有效淌度
effective molecular weight　有效分子量
effectiveness　效力
effectiveness of pesticide　农药药效
effective network chain　有效网链
effective nuclear charge　有效核电荷
effective number of plate　有效塔板数
effective number of replication　有效重复数
effective output　有效生产量
effective peak number　有效峰数
effective permeability　有效渗透性
effective plate height　有效塔板高度
effective plate number　有效塔板数
effective plate volume　有效塔板容积
effective porosity　有效孔隙率
effective power(=effective capacity)　有效动力；有效功率
effective pressure　有效压力
effective quantum number　有效量子数
effective radius　有效半径

effective reaction time　有效反应时间
effective size of grain　有效粒度
effective size of pattern　喷束的有效尺寸
effective slip velocity　有效滑动速度
effective slit width　有效狭缝宽度；有效缝隙宽度
effective stack height　有效烟囱高度
effective stress　有效应力
effective sulfur content　有效硫含量
effective surface　有效表面
effective temperature　有效温度
effective temperature difference　有效温度差
effective thickness　有效厚度
effective value　有效值
effective viscosity　有效黏度
effective viscosity of oil wedge　楔内油的有效黏度
effective viscosity value　有效黏度值
effective volatility　有效挥发度
effective volume　①有效体积②有效容积
effective volume factor　有效体积因子
effective width　有效宽度
effect lacquer　美饰漆；真空涂漆
effect of increased concentration　增浓效应
effect of interaction　相互作用效应；交互效应
effect of light　光的作用；光效应
effect of order　有序效应
effect of orientation　取向效应
effect of rapid deformation　快速变形作用
effect of rapid impact　快速冲击作用
effect of signal averaging　信号平均效应
effect of soil drying　干土效应
effector　试验器；操纵装置；效应器
effector phase　效应期
effect pigment　随角异色效应颜料
effects of annular strain　环的张力效应
effect yarn　饰纱
effervescence　泡腾；起泡(沫)
effervescent　泡腾的；起泡的
effervescent bath table　泡腾浴台
effervescent beverages〔复〕　泡腾饮料；充气饮料
effervescent mixture　发泡剂
effervescent phosphate　泡腾磷酸盐〔酸性磷酸盐和碳酸盐的混合物〕
effervescent potassium citrate　泡腾柠檬酸钾
effervescent salt　泡腾盐
efficacy　功效；效验
efficiency　效率
efficiency curve　效率曲线；功率曲线
efficiency factor　有效因子；效率

efficiency of atomization 原子化效率
efficiency of column 柱效率
efficiency of cracking unit 裂化设置效率；裂化设置的处理量
efficiency of dispersion (样品的)气散效率
efficiency of electro-acoustic transducer 声电换能器效率
efficiency of fractional separation 分级分离效率
efficiency of grafting 接枝效率
efficiency of heat engine 热机效率
efficiency of initiation 引发效率
efficiency of initiator 引发剂效率
efficiency of nebulization 雾化效应
efficiency of sample utilization 样品利用率
efficiency test 效率试验
efficient ①效率高的②有效的
efficient estimator 有效估计量
efficient recovery 有效回收
efficient statistics 有效统计量
efficient vulcanization 有效硫化
efflorescence(=effloresceny) ①风化*②开花③起霜
efflorescency(=efflorescence) ①风化②开花③起霜
efflorescent ①风化的②花状③起霜的
efflorescent action 风化作用
efflorescent test 风化试验
effluence 流出
effluent ①流出液*②流出物
effluent brine 废盐水
effluent disposal 废水处理
effluent fraction 流出的馏分；流出物馏分
effluent fractioning 流出液分级
effluent gas 废气；烟道气
effluent gas analysis 废气分析；烟道分析
effluent gas treatment system 废气处理系统
effluent limitation ①排污限度；排放物限度②三废排放限度
effluent oil 流出油
effluent pit 废水坑
effluent purification 污水净化
effluent quality standard 排放质量标准
effluent sample valve 流出物采样阀；试样流出阀
effluent segregation system 废水分离系统
effluent settling chamber 废水沉淀池；废水沉降池
effluent splitter 流出物分流器
effluent stream 流出气流；流出液流
effluent temperature 流出物温度
effluent treatment 废水处理
effluent treatment plant 污水净化装置；污水处理站
effluent treatment unit 排出物处理装置；污水处理装置
effluent water 废水
effluogram 液流图
efflux ①流出；射流②流量
efflux coefficient 流出系数
efflux cup 流出杯；流出黏度计
efflux cup method 流杯法〔测黏度〕
efflux cup viscometer 流出杯黏度计
efflux drain time (杯中液体的)流干时间
efflux flow 排出液流
effluxion 流出；射流
efflux method 流出法〔测黏度〕
efflux time 流出时间
efflux-type viscometer 流出式黏度计
efflux velocity 流出速度
efflux viscometer 流出式黏度计
efflux volume 流出体积
effusiometer 隙透计；(气体)扩散计
effusion ①泻流②渗透；隙透③渗出液
effusion cell 泻流室
effusion method 隙透法
effusion separator 泻流分离器；渗透型分离器
effusive 流出的；射流的
effusive beam source 隙流束源*
effusive electrode 喷汞电极
effusiveness 喷发性
effusive rocks〔复〕 喷发岩；喷出岩
E fiber glass E 玻璃纤维；无碱玻璃纤维
efsiomycin(=fluvomycin) 河流霉素
egesta 排泄物
egg 蛋；卵
egg albumin fiber 卵清蛋白纤维
egg-counting method 虫卵计数法
egg end 蛋形底
egging 加蛋黄
egg oil ①涂蛋油②蛋黄油
egg packer's oil 涂蛋油
egg-plant type flask 茄形烧瓶
eggshell enamel 半光瓷漆；蛋壳光瓷漆
eggshell falt 蛋壳光；半光
egg-shell finish 蛋壳面饰
eggshell gloss 蛋壳光〔60°光泽计 20%～35%〕
egg-size 蛋块〔指煤级〕
eglestonite 氯汞矿
Egloff's equation 埃格劳夫方程式
egomaketone(=β-dehydroperillaketone) β-脱氢紫苏酮
egress ①出去；出外②出口；出路
egress opening 流出口；排出口
Egypt green 硅酸铜绿

Egyptian blue 埃及蓝〔主要成分为硅酸铜〕
Egyptian jasper 埃及碧玉
ehrichin 艾霉素
Ehrlich diazo reaction 埃尔利希重氮反应
ehrlichin 艾氏菌素
Ehrlich's reagent 埃尔利希试剂
eicolin 石南素
eicosaborane 二十硼烷 $B_{20}H_{16}$
eicosanoid 类二十烷酸;类花生酸
eicosene diacid(=icosendioic acid) 二十碳烯二酸
eigen 本征
eigenfunction(=characteristic function) 本征函数;特征函数
eigenfunction orbital 本征轨道
eigenstate 本征态
eigen value(=characteristic value) 本征值;特征值
eigen value eigen vector 特征值与特征向量
eigenvalue's equation 特征方程
eigenvector 特征向量*
eighteen-membered ring(=eighteen ring) 十八元环
eighteen ring(=eighteen-membered ring) 十八元环
eight-membered ring(=eitht ring) 八元环
eight ring(=eight-membered ring) 八元环
eikonometer 光像测定器
einstein 爱因斯坦〔能量的单位,$E=6.06×10^{23}$ 量子〕
Einstein coefficient 爱因斯坦系数
einsteinium 锿〔99 号元素,化学符号 Es〕
einsteinium amalgam 锿汞齐
einsteinium oxychloride 氯氧化锿 EsOCl
einsteinium sesquioxide 三氧化二锿 Es_2O_3
einsteinium trichloride 三氯化锿 $EsCl_3$
Einstein mass-energy equation 爱因斯坦质能方程
Einstein model of flow 爱因斯坦流动模型
Einstein's law 爱因斯坦定律
Einstein's theory for flow of suspension 爱因斯坦悬浮体流动理论
Einstein's viscosity equation 爱因斯坦黏度方程
einzel lens 单透镜
Eirich crusher 艾里希(浆粕连续)粉碎机
E isomer E 异构体*
Either degree(=barkometer degree) 巴克度
ejecta 喷出物;抛出物;排出物
ejecting outline blanking die 顶出式外形冲切模
ejecting plug 推顶杆;推顶钉
ejecting press 挤压机
ejection 挤出;喷出;推顶;推出;排出
ejection bar 推顶销
ejection cycle ①顶出周期②顶出时间

ejection pin 推顶销;脱模销
ejection pin assembly 脱模销组合件
ejection temperature 脱模温度
ejector ①喷射器②脱模销〔塑料〕
ejector air pump 喷射泵
ejector condenser 喷射冷凝器
ejector drier 喷射干燥器
ejector housing 注塑模顶杆空间
ejector jet pump 喷射器;注射泵
ejector mixing 喷射式混合器;喷射式搅拌器
ejector-operating mechanism 顶杆操作机构
ejector pin 推顶销;脱模销〔塑料〕
ejector plate return pin 脱模板回程销
ejector priming 喷射器;启动泵
ejector pump 喷射泵
ejector ram 推顶活塞
ejector rod 推顶柱
ejector rod clamp screw 顶出固定螺丝
ejector spring 脱模弹簧
ejector vacuum pump 喷射真空泵
ejector water air pump 喷水空气泵
eka- ①类〔元素〕②〔梵文〕第一
eka-actinium 类锕〔第 121 号元素〕
eka-aluminum 类铝〔即:镓〕
eka-astatine 类砹〔第 117 号元素〕
eka-bismuth 类铋〔第 115 号元素〕
ekaconjugated polymer 准共轭聚合物
ekaconjugation 准共轭(作用)
eka-element 待寻元素〔历史上对周期表中尚缺的元素的叫法〕
eka-francium 类钫〔第 119 号元素〕
eka-gold 类金〔第 111 号元素〕
eka-hafnium 类铪〔第 104 号元素〕
eka-iodine 类碘〔即: 砹〕
eka-iridium 类铱〔第 109 号元素〕
eka-lead 类铅〔第 114 号元素〕
eka-lutetium 类镥〔第 103 号元素, 铹〕
eka-mercury 类汞〔第 112 号元素〕
eka-osmium 类锇〔第 108 号元素〕
eka-platinum 类铂〔第 110 号元素〕
eka-polonium 类钋〔第 116 号元素〕
eka-radium 类镭〔第 120 号元素〕
eka-rakon 类氡〔第 118 号元素〕
eka-rhenium 类铼〔第 107 号元素〕
eka-silicon 类硅〔即: 锗〕
eka-tantalum 类钽〔第 105 号元素〕
eka-thallium 类铊〔第 113 号元素〕
eka-tungsten 类钨〔第 106 号元素〕

eka-ytterbium 类镱〔第102号元素；锘〕
ekmolin 鱼素
elaboration 加工作用
elaboration products 加工产物
elaeomargaric acid(=elaeostearic acid;eleostearic acid) 桐酸；9, 11, 13-十八碳三烯酸 $C_{18}H_{30}O_2$
elaeometer(=elaiometer) 验油比重计
elaeostearic acid(=eleostearic acid;elaeomargaric acid) 桐酸；9, 11, 13-十八碳三烯酸 $C_{18}H_{30}O_2$
elaeostearin 桐酸精（俗）；甘油三桐酸酯
elaidiate 反式油酸盐
elaidic acid 反油酸；反(式)9-十八碳烯酸 $C_{17}H_{33}CO_2H$
elaidic acid test 反油酸试验
elaidin 反油酸精；甘油三反油酸酯
elaidinization 反油酸重排作用；反油酸转位
elaidinization oil 反油酸化油
elaidin reaction 反油酸反应
elaidin test 反油酸检验
elaidodistearin 反油酸二硬脂精；甘油反油酸二硬脂酸酯
elaidyl alcohol 反(式)十八烯醇
elaiometer(=elaeometer) 验油比重计
elaiomycin 洋橄榄霉素；油霉素
elaiophilin(=elaiophylin) 洋橄榄叶素
elaiophylin(=elaiophilin) 洋橄榄叶素
elaoptene 油萜
elastance 倒电容〔等于1/C〕
elastic 弹性的
elastica 橡胶
elastic after effect(=after-effect) （弹性）后效
elastic after-working(=elastic after-effect) 弹性后效
elastically isotropic 弹性各向同性的
elastic anisotropy 弹性各向异性
elasticator 增弹剂
elastic behavior 弹性
elastic body 弹性体
elastic break down pressure 弹性失效压力
elastic cell 弹性细胞
elastic coefficient 弹性系数
elastic collision 弹性碰撞
elastic compliance 弹性柔量
elastic constant 弹性常数
elastic contribution 弹性基值；弹性作用
elastic core 弹性核心
elastic coupling 弹性联轴节
elastic covered yarn 弹性包(芯)线
elastic deformation 弹性形变
elastic deformation limit 弹性变形极限

elastic displacement 弹性移位
elastic distortion 弹性畸变
elastic dumb-bell model 弹性哑铃模型
elastic effect 弹性效应
elastic elongation 弹性伸长
elastic energy correction in capillary 毛细管弹性能改正
elastic entropy 弹性熵
elastic extension 弹性延伸
elastic failure 弹性破坏；永久变形
elastic fatigue 弹性疲劳
elastic fiber 弹性纤维
elastic filler 弹性填料
elastic fluid 弹性流体
elastic force 弹(性)力
elastic fore-effect 弹性前效
elastic gel 弹性凝胶
elastic gum 生橡胶；弹性胶
elastic hysteresis 弹性滞后
elastic isotropy 弹性各向同性
elasticity 弹性
elastic(ity) modulus 弹性模数
elasticity number 弹性值
elasticity of bulk 体积弹性
elasticity of compression 压缩弹性
elasticity of elongation 伸长弹性
elasticity of flexure 弯曲弹性
elasticity of shape 形状弹性
elasticity of torsion 扭转弹性
elasticity of volume 体积弹性
elasticity tensor 弹性张量
elasticity test 弹性试验
elasticizer 弹性剂；增弹剂
elastic lag 弹性滞后
elastic light scattering 弹性光散射
elastic limit 弹性极限
elastic liquid 弹性液体
elastic loss factor 弹性损耗因子
elastic manometer 弹性压力计
elastic-mechanical inversion 弹性-力学转化
elastic melt swelling 弹性熔体膨胀
elastic memory 弹性记忆(效应)
elastic modulus 弹性模量
elastic net 弹性网眼
elastic neutron scattering 中子弹性散射
elastic normal stress coefficient 弹性正应力系数
elastic nylon 弹性锦纶；弹性尼龙
elastic oscillation 弹性振动；弹性振荡
elastico-viscosity ①弹黏性②弹性黏度

elastico-viscous fluid 弹黏性流体
elastico-viscous liquid 弹黏性液体
elastico-viscous solid(=Kelvin body) 弹黏性固体
elastic paper 伸性(牛皮)纸
elastic peak 弹性峰
elastic-plastic ①弹性塑料②弹塑性的
elastic-plastic behavior 弹塑性性质
elastic polymer 弹性聚合物
elastic polymer liquid 弹性聚合物液体
elastic potential 弹性势
elastic properties 弹性特性〔指弹性、抗张力、伸长性、永久变形〕
elastic quartz 弹性石英
elastic reactance 弹性阻抗
elastic reaction 弹性后效；弹性反应
elastic recoil 弹性反冲；弹性回弹
elastic recovery 弹性回复；弹性回复率
elastic relaxation 弹性松弛
elastic relaxation time 弹性松弛时间
elastic resilience 回弹性
elastic resisting force 弹性阻力
elastic response 弹性响应
elastic restoring force 弹性恢复力
elastic ribbon 松紧带
elastic scaler 弹性尺
elastic scattering 弹性散射*
elastic scattering electrons 弹性散射电子
elastic shear 弹性剪切
elastic shear modulus 弹性剪切模量
elastic slip 弹性滑移
elastic sol 弹性溶胶
elastic solid(=Hookean body) 弹性固体
elastic sols 弹性溶胶
elastic stage 弹性状态
elastic stiffness 弹性刚度
elastic strain 弹性应变
elastic strain energy 弹性应变能
elastic straining 弹性应变
elastic strain recovery 弹性应变回复
elastic stress 弹性应力
elastic stretching 弹性伸长
elastic sulfur(=plastic sulfur) 弹性硫
elastic synthetics 弹性合成物
elastic tape 弹性带；松紧带
elastic tensile modulus 弹性拉伸模量
elastic thread 弹性线
elastic thread covering machine 弹性线包覆机
elastic torsion 弹性扭力；弹性扭转
elastic turbulence 弹性湍流
elastic vibration 弹性振动
elastic-viscoplastic body 弹-黏塑(性)体
elastic-viscosity ①弹黏性②弹性黏度
elastic-viscous system 弹黏(性)体系
elastic washer 弹性垫圈
elastic wave 弹性波
elastic wrinkle resistant fabric 弹性抗皱织物
elastic yarn 胶芯纱线
elastification 弹性化
elastirotor extruder 无螺杆挤出机
elastodurometer 弹性硬度计
elastodynamic screwless extruder 弹性动力式无螺杆挤出机
elastogel 弹性凝胶
elasto-hydrodynamic lubrication 弹性流体力学润滑
elastomer 高弹体*
elastomer adduct 高弹体加成物
elastomer film 高弹体膜；弹性体膜
elastomer foam 高弹体泡沫料
elastomer gel 弹性体凝胶；凝胶弹性体
elastomeric 弹性体的；高弹体的
elastomeric adhesive 弹性体黏合剂
elastomeric compound 胶料；混炼胶；弹性体型混合料
elastomeric fibre 弹性(体)纤维〔聚氨基甲酸酯弹性纤维的统称〕
elastomeric foam 泡沫胶；橡胶海绵
elastomeric paint 弹性漆
elastomeric plastic 弹性体塑料
elastomeric polyester 弹性体聚酯
elastomeric polymer 弹性聚合物
elastomeric stamp 弹性印章
elastomeric state 高弹态
elastomeric yarn warping machine 弹性纱线整经机
elastomer pigmentation 弹性体着色
elastomer powder 合成橡胶粉
elastomer sequence 高弹性分子链序列
elastometer 弹性计
elasto-optical coefficient 弹光系数
elasto-osmometry 渗透压(高)弹性测定法
elastoplast 弹性塑料
elastoplastic polymer 弹塑性聚合物
elasto-plastic range 弹塑性范围
elastoplastics 弹性塑料
elastoplastic state 弹塑态
elastoplastic system 弹塑体系
elastopolymer 弹性高聚物
elastoprene 二烯橡胶

elasto viscometer 弹黏计
elasto-viscosimeter 弹性黏度计；黏弹计
elasto-viscous 弹黏的
elastoviscous polymer 弹黏性聚合物
elasto-viscous system 弹黏(性)体系
elateric acid(=ecbalin) 喷瓜酸 $C_{24}H_{28}O_5$
elaterin 喷瓜素
elaterium 喷瓜汁
elaterometer(=elatrometer) 气体密度计
elaterone 喷瓜酮 $C_{24}H_{30}O_5$
elatigenin 高飞燕草配基；厄拉剔配基
elatrometer(=elaterometer) 气体密度计
elayl ①乙烯②亚乙基
elbaite 锂电气石
elbow ①肘管；弯管②肘
elbow cover opener 弯管开盖器
elbow feeder 肘管送料器
Elbs persulfate oxidation 埃尔布斯过硫酸盐氧化
Elbs reaction 埃尔布斯反应
elcometer (gage) 干膜厚度仪；测厚仪
eldeline 接骨灵
elderberry oil 接骨木果油
elder flower absolute 接骨木花净油
elder flower concrete 接骨木花浸膏
elder flower oil 接骨木花油
eldrin(=rutin) 芸香苷；芦丁
eleagnine 胡颓子碱
elecampane 土木香
elecampane oil 土木香油；菊油
elecampin 土木香粉
elective affinity 有择亲和势
electret 驻极体；永久极化的电介质
electret thermal analysis 驻电体热分析
electric 电的
electric absorption 电的吸收
electric accumulator 蓄电池
electrical 电的
electrical analog 电模拟
electrical arc discharge 电弧放电
electrical arc photosource 电弧光源
electrical arc spectrum 电弧光谱
electrical bonding power 电子键合力
electrical breakdown 电击穿
electrical breakdown tension 击穿电压
electrical capacity-type moisture meter 电容式测湿仪
electrical centering 静电法调整中心
electrical characteristic 电特性
electrical charge 电荷
electrical computer 电动计算机
electrical conductivity 导电性；电导系数；电导率
electrical conductivity analyzer 电导率分析器
electrical conductivity detector 电导检测器
electrical conductivity fibre 导电性纤维
electrical conductor 导(电)体
electrical discharge degradation 电火花降解
electrical discharge detector 放电检测器
electrical discharge pyrolyzer 放电热解器
electrical discharge spectrum detector 放电光谱检测器
electrical dissipation 电力耗散
electrical double layer 电双层；双电层
electrical effect 电场效应*
electrical efficiency 电机效率；电效率
electrical energy 电能
electrical equivalent circuit 等效电路
electrical erosion 电场腐蚀
electrical excitation 电激发
electrical faults 漏电；走电
electrical field flow fraction 电场流分级
electrical field magnified injection 场放大进样
electrical fittings 电子配件；电气配件
electrical forces 电力；电场力
electrical glass fibre (低碱)电绝缘玻璃纤维；E 型玻璃纤维
electrical grade 电气级；电工级
electrical heated thermostat 电热恒温器
electrical immersion heater 浸没式电(加)热器
electrical impedance 阻抗
electrical inductance 电感
electrical induction heating 电感应加热
electrical instrument 电气仪表
electrical insulating coating 电绝缘涂料
electrical insulating varnish 电绝缘清漆
electrical insulation compound ①电绝缘胶(料)②电绝缘混合料
electrical insulator 电绝缘体；非导电体
electrical interference 电干扰
electrical isolation fibre 电绝缘纤维
electrical knife 电(热)刀
electrical loss factor 电损耗因子
electrical loss modulus 电损耗模量
electrical loss spectra 电损耗谱
electrically conducting coating 导电涂料
electrically conductive fibre 导电纤维
electrically fired bomb 电燃式弹
electrically-heated degreasing plant 电热式脱脂装置
electrically heated grid 电热炉栅；电热熔栅

electrically heated thermocouple 电热温差电偶
electrically heated thermostat 电热恒温器
electrically-operated platform truck 电动装卸车；电池车
electrically operated valve 电动(调节)阀
electrical-mechanical analogy 电-力学模拟
electrical neutrality of solution 溶液的电中性
electrical neutral zones 不带电区；电中性区
electrical null 电零点；零电点
electrical operating parameter 电操作参数
electrical parameter 电参数
electrical potential 电位；电势
electrical-potential gradient 电位梯度；电势梯度
electrical precipitator 电沉器
electrical property 电性能
electrical pulse 电脉冲
electrical resistance heater 电阻加热器
electrical resistance method 电阻法
electrical resistance test 耐电压试验
electrical resistivity 电阻系数；电阻率
electrical shock ①电击；触电②电休克
electrical signal 电信号
electrical stability 电稳定性
electrical static charge 静电荷
electrical static eliminator ①静电消除器②静电消除剂
electrical stitcher 电动压合滚
electrical stop motion 电气自停装置
electrical strip heater 电热丝式加热器
electrical surface resistivity 表面电阻率
electrical tape 绝缘带；电线包布
electrical treating 电处理〔尤指电晕处理〕
electrical wiring 电气布线；配线；电线线路
electrical work 电功
electric analysis 电分析
electric arc 电弧
electric arc furnace 电弧炉
electric arc metallizing process 电弧法金属涂覆工艺
electric arc reaction 电弧反应
electric arc welding 电弧焊
electric audibility method of knock rating 汽油爆震性电声学测定法
electric battery 电池组
electric bell 电铃
electric birefringence 电场致双折射
electric bleaching 电漂白
electric block 电动葫芦
electric blue ①钢青色②电火花蓝(灰蓝色)
electric breakdown strength 击穿电压
electric bulb 电灯泡

electric calamine 电异极矿
electric calorimeter 电量热器；电热量计；电卡计
electric capacity(=electrostatic capacity) 电容；静电容量
electric cell 电池
electric chain-block 电动葫芦
electric charge 电荷
electric charge density 电荷密度
electric charge dissipating finish 电荷消散整理
electric circuit 电路
electric circuit analogy 电路模拟
electric clock 电钟
electric computer 电动计算机
electric concentration 电热蒸浓(法)
electric condenser 电容器
electric condenser oil 电容器油
electric conductance 电导
electric conductance determination 电导测定
electric conduction 导电；电导
electric conduction paint 导电漆
electric conductive adhesive 导电胶黏剂
electric conductivity ①电导率②电导性
electric conductor 电导体
electric contact thermometer 电接点温度计
electric convection 电(流)对流
electric cracking 电裂解
electric current 电流
electric current density 电流密度
electric decantation 电滗析；电倾析；电渗析
electric dehydration 电脱水(作用)*
electric depolarization 通电去极化
electric desalter 电(透析)脱盐装置
electric desalting 电气脱盐
electric detarring precipitator 电力脱焦油沉淀器
electric detonator 电(发)爆管
electric dipole 电偶极子
electric dipole moment 电偶(极)矩
electric dipole transition 电偶极跃迁
electric discharge 放电
electric discharge ionization 放电电离
electric discharge machining 放电加工
electric discharge pyrolyzer 放电热解器
electric displacement 电(位)移
electric dissociation 电离
electric dissociation exponent 电离指数
clcctric doublc laycr 双电层*
electric drying apparatus 电干燥器；电烘箱
electric dust precipitator 电集尘器
electric element 电池

electric emulsion treater　电乳浊液处理器
electric endosmosis　电内渗；电渗
electric energy　电能
electric equivalent　电当量
electric exhaust fan　电(动)排气扇
electric field　电场
electric field directed layer-by-layer assembly technique　电场诱导的静电组装技术
electric field gradient(EFG)　电场梯度
electric field induced infrared absorption　(电)场致红外吸收
electric field intensity　电场强度
electric-field jump　电场跃迁
electric field scanning　电场扫描*
electric fire(s)　电炉
electric flow meter　电动流量计；电量计
electric fork lift　叉式电池车
electric fuel gauge　电测燃料仪〔燃料消耗量的电测量器〕
electric furnace　电炉
electric fuse　①电引信②电信线
electric glow discharge method　辉光放电法
electric gun　电子枪
electric heated roll　电加热轧辊
electric heater　电热器；电炉
electric heating element　电热元件
electric hot plate　电热板
electric hygrometer　电湿度计
electric hygrometry　电测湿法
electric ignition device　电点火装置
electric impulse　电脉冲
electric induction furnace　电感炉
electric insulating enamel　(电)绝缘瓷漆
electric insulating oil　(电的)绝缘油；变压器油
electric-insulating paper　电绝缘纸
electric insulating varnish　(电)绝缘清漆
electric interlock　电接合；电连接
electric iron　①电熨斗②电烙铁
electricity meter　量电计
electricity potential detector　电位检测器
electric jacket　电热套
electric kinetic potential　电动势
electric light bulb　电灯泡
electric mantle　电热套
electric moisture meter　电水分计；电气湿度测定计
electric motor　电动机；马达
electric noise　电噪声
electric oiler　电动加油器
electric oscillation　电振荡

electric osmosis　电渗(现象)
electric oven　电烘箱
electric panel　配电板；电气仪表盘
electric panel general　总配电板
electric pipe precipitator　电管除尘器
electric platen　电热板
electric plate precipitator　电板除尘器
electric-pneumatic converter　电动-气动变换器
electric polarization　电极化
electric porcelain　电瓷
electric potential　电势；电位
electric potential difference　电位差
electric power plants　发电厂
electric precipitation　电力沉淀；电除尘
electric precipitator　电力沉淀器
electric pressure　电压
electric pressure gauge　电学压力计
electric protection　电保护；电防腐
electric pulse　电脉冲
electric quadrupole moment　电四极矩
electric quadrupole radiation　电四极辐射
electric quantity　电量
electric radiation　电辐射
electric relay　继电器
electric resistance　电阻
electric resistance furnace　电阻炉
electric resistance hygrometer　电阻式湿度计
electric resistance moisture meter　电阻式水分计
electric resistance pyrometer　电阻高温计
electric resistance thermometer　电阻温度计
electric resistant manometer　电阻测压计
electric resistivity　电阻率
electric rod-curtain precipitator　棒帘式电除尘器
electric rot　电腐蚀
electric sector　电扇形
electric separator　电力分离器；电滤器
electric solenoid operation　电磁操作
electric spark　电火花
electric spark machining　电火花加工
electric stage　电热台
electric static spraying　静电喷涂
electric steel　电炉钢
electric steel furnace　电炉钢炉
electric stitcher　电动压合器
electric stop motion　电气自停装置
electric strength　耐电强度
electric substation　变电站
electric supply　电源

electric surface density 表面电子密度
electric susceptibility 电极化率
electric switch oil 电键油
electric tannage 电鞣(法)
electric tape cloth 电线包布
electric telemeter 电力遥测计〔电远距测定器〕
electric tension 电压
electric tube furnace 管式电炉
electric twist meter 电动捻度计
electric valve 整流管
electric vector 电矢(量)
electric vibrator 电振(动)器
electric volume resistivity 体积电阻率
electric washer 电动洗涤机；电动洗衣机
electric washing machine 电动洗涤机；电动洗衣机
electric wave 电波
electric welder 电焊机
electric welding 电焊
electric welding rod 电焊条
electric wide and narrow device 电气宽窄调幅装置
electric wiring 电气布线
electric work 电功
electric zone method 电域(敏感)法
electrifiable 可起电的；可带电的
electrification ①带电②起电
electrification detector 起电侦查器
electrified body 带电体
electrified culture 通电培养
electrifying 起电
electrion oil (=electrochemically treated oil) 高压放电合成油
electrion process 高压放电法
electro- 〔词头〕 ①电气②电化③电动④电力⑤电解
electroactive polymer 电活性聚合物
electroactive species 电活化物种
electroactive substance 电化学活性物质
electro adsorption 电吸附
electroaffinity ①电亲和性②电亲和力
electro-analysis 电解分析
electroanalytical chemistry 电分析化学
electroanalytical detector 电分析检测器
electroanalytical technique 电分析技术
electroantennogram 触角电图
electro-attracting 吸电子的
electrobalance 电平衡
electro-bath 电镀浴；电镀槽
electro-beam 电子束
electrocalorio effect 电热效应

electro capillarity 电毛细(管)现象
electro-capillary adsorption 电毛细(管)吸附
electrocapillary curve 电毛细管曲线
electrocapillary effect 电毛细效应
electrocapillary maximum 电毛细管极大*
electrocapillary phenomenon 电毛细现象*
electrocapillary phoresis 电毛细泳
electrocapillary zero 电毛细零点
electrocapillary zero potential 电毛细零点电位
electrocarbonization 电法炼焦
electrocatalysis 电催化*
electrocharge effect 电荷效应
electrochemical 电化(学)的
electrochemical acetylene analyzer 电化学乙炔分析仪
electrochemical action 电化作用；电化学作用
electrochemical activation 电化学活化(作用)
electro-chemical additive 电化添加剂
electrochemical analysis 电化(学)分析
electrochemical biosensor 电化学生物传感器
electrochemical buffer solution 电化学缓冲溶液
electrochemical constant 电化(学)常数
electrochemical corrosion 电化学腐蚀*
electrochemical crystal structure 电化学晶体结构
electrochemical deposition 电化学沉淀；电化学淀积
electrochemical desalination process 电化学脱盐(淡化)法
electrochemical desorption 电化学脱附
electrochemical desulfuration 电化脱硫作用
electrochemical detector 电化学检测器
electrochemical diffused collector phototransistor 电化学扩散集电极光电晶体管
electrochemical disintegration 电化分解
electrochemical DNA sensor 电化学 DNA 传感器；电化学脱氧核糖核酸传感器
electrochemical double layer 电化学双电层
electrochemical duplication 电化学偶联化(反应)
electrochemical effect 电化学效应
electrochemical equilibrium 电化(学)平衡
electrochemical equivalent 电化当量
electrochemical etching 电化学刻蚀；电化学抛光
electrochemical etch modulation 电化学腐蚀调节
electrochemical etch-stop 电化学腐蚀停止
electrochemical flow-through detector 电化学流通检测器
electrochemical force of corrosion 腐蚀的电化学力
electrochemical free energy 电化学自由能
electrochemical immunoassay 电化学免疫分析
electrochemical industry 电化工业
electrochemical integrating device 电化(学)积分装置

electrochemical intercalation 电化嵌入(作用)；电化掺杂(作用)
electrochemical kinetics 电化学动力学
electrochemical luminescence 电化学发光
electrochemically actuated mercury pump 电化学致动汞微泵
electrochemically initiated polymerization 电化学引发聚合
electrochemically treated oil(=electrion oil) 高压放电合成油
electrochemical machining 电化学加工
electrochemical masking 电化学掩蔽
electrochemical mass 电化质量
electrochemical method 电化学分析法
electrochemical oscillation 电化学振荡*
electrochemical overpotential 电化学超电势*
electrochemical oxidation 电化学氧化*
electrochemical polarization 电化学极化
electrochemical polishing bath 电化学抛光浴
electrochemical polymerization 电化学(引发)聚合
electrochemical potential 电化学势(位)
electrochemical probe 电化学探针
electrochemical process 电化法
electro-chemical protection 电化学保护
electrochemical quartz crystal microbalance 电化学石英晶体微量天平
electrochemical quartz crystal microweighing 电化学石英晶体微天平
electrochemical reaction 电化学反应
electrochemical reduction 电化学还原
electrochemical reflection spectrum 电化学反射光谱
electrochemical relaxation technique 电化学弛豫法
electrochemical sensor 电化学传感器
electrochemical separation method 电化学分离方法
electrochemical series (元素)电化序；电位序
electrochemical spectrum 极谱波
electrochemical steady-state 电化学稳态
electrochemical synthesis 电化学合成*
electrochemical transient 电化学暂态
electrochemical transistor 电化学晶体管
electrochemical treatment 电化学处理
electrochemical valence 电(化学)化(合)价
electrochemiluminescence 电致化学发光
electrochemiluminescence immunoassay 电化学发光免疫分析
electrochemiluminescent 电化学致光的；电致化学荧光的
electrochemistry 电化学*

electrochemograph 记录式极谱仪
electrochemotherapy 电化学疗法
electrochromatographic technique 电色谱技术
electrochromatography 电色谱法
electrochromic display device(ECD) 电致变色显示器；电致显色装置
electrochromic polymer 电致变色聚合物
electrochromism 电着色
electrochromism film 电致着色异常薄膜；电镀膜着色
electrocleaning 电清洗
electro-coagulation 电凝固(作用)；电凝结(作用)
electrocoat 电沉积涂料；电泳漆
electrocoating 电涂；电泳涂漆
electrocoating bath 电泳槽
electrocoating paint 电沉积涂料；电泳漆
electrocoat tank 电泳槽；电泳涂漆槽
electroconcentration 电浓缩法
electroconductibility 电导率；导电性
electroconductiometric gas analyzer (溶液)电导式气体分析计
electroconductive 电导性的
electroconductive agent 导电剂
electroconductive coating 导电镀层
electro-conductive fiber 导电纤维
electro-conductive material 导电材料
electroconductive paste 导电胶
electroconductive polymer 导电高分子
electroconductive printing ink 导电印刷油墨
electroconductive resin 导电树脂
electroconductive synthetic fibre 导电合成纤维
electroconductivity 电导率
electrocorrosion 电腐蚀
electrocratic 电稳的
electrocremage 电滗析；电倾析
electrocrystallization 电结晶
electrocure process 电固化法；电处理法
electrocuring 电子束固化
electrocyclic reaction 电环化反应；π键环化反应
electrocyclic rearrangement 电环(化)重排*
electrocyclization 电环化
electrode 电极*
electrode array 电极阵列
electrodecantation 电倾析
electrode capacity 电极电容*
electrode carbon 电极炭
electrode chamber 电极室；电极腔体
electrode concentration cell 电极浓差电池
electrodecontamination 电去污；电净化

electrode deposit 电极沉积
electrode function potential 功能电位
electrode gap 电极隙；电极间距
electrode holder 电极(固定)架
electrode impedance 电极阻抗
electrode interfacial phenomenon 电极界面现象
electrode interference 电极干扰
electrode kinetics 电极(过程)动力学
electrodeless 无电极的
electrodeless discharge 无极放电
electrodeless discharge detector 无(电)极放电检测器
electrodeless discharge lamp 无极放电灯*
electrodeless discharge tube 无极放电管
electrodeless heating 无极加热
electrode of first kind 第一类电极
electrode of second kind 第二类电极
electrode of third kind 第三类电极
electrode of zeroth kind 第零类电极
electrode pair 电极对
electrode polarization 电极极化作用
electrodeposit 电解淀积；电镀
electrodeposited cadmium coating 电沉积镉涂层；电镀镉膜
electrodeposited chromium 电沉积铬；电镀铬
electrodeposited coating(=electrophoretic coating) 电沉积涂覆法；电泳涂覆法
electro-deposited film 电沉积膜
electrodeposited paint film 电沉积漆膜；电泳漆膜
electrodeposition 电沉积*
electrodeposition coating 电泳涂料
electrode potential 电极电位
electrode potential under electrolysis 电解(电极)电位
electrode probe 电极探测器
electrode process 电极过程*
electrode reaction 电极反应
electrodesalting 电气脱盐
electrode stand 电极架
electrode stem 电极杆
electrode storage 电极保存
electrode surface 电极表面
electrodetarrer 焦油静电沉降器
electrode time constant 电极时间常数
electrode vessel 电极室
electrode voltage 电极电压
electrodialysed water 电渗析水
electrodialyser 电渗析器
electrodialysis 电渗析
electrodialytic 电透析的；电渗析的

electrodialytic cell 电渗析池
electrodialytic membrane 电渗析膜
electrodialytic technique 电渗析技术
electrodialyzed culture 电渗析培养
electrodialyzer 电渗析器；电渗析装置
electro-diffusion 电扩散
electrodip process 电沉积法
electrodischarge machining(EDM) 放电加工
electrodisintegration 电破坏；电分裂；电致衰变
electrodispersion 电分散(作用)
electro-driven chromatography 电驱动色谱法
electro-driven spindle spinning frame 电锭细纱机
electrodynamics 电动力学
electrodynamometer ①电功率计②力测电流计
electroendosmosis 电内渗(现象)
electro-endosmosis membrane control 电内渗膜控制
electroendosmotic flow 电渗流
electroendosmotic flow marker 电渗流标记物
electroendosmotic mobility 电渗流淌度
electroengraving machine 电动刻花(纹)机
electroerosive wear 电浸蚀磨损
electro-exchange polymer 电子转移聚合物
electroextraction 电萃取
electrofax process 电子摄影法
electrofilter 电过滤器
electrofixer 电热固色器
electro-flotation method 电解浮选法；电解浮渣法
electrofluid mechanics 电流体力学
electrofluorination 电解氟化
electrofocusing 电聚焦
electrofocusing electrophoresis 聚焦电泳(法)
electroformed cavity 电铸模腔
electroformed sieve 电铸筛
electroforming 电铸
electrofuge 离电体*
electrofuse 电熔融
electrofusion 电融合
electrofying 电光整理
electro-galvanized steel 热浸镀锌钢；电镀锌钢
electrogenerated chemiluminescence(ECL) 电致化学发光
electrogenerated reagent 电解制备的试剂
electrograph ①电图②电谱③电版机
electrographic 电图分析的
electrographic analysis 电图分析；电谱分析
electrographic method 电谱分析法；电图分析法
electrographite 电炉石墨；人工(造)石墨
electrography 电谱法*

electrogravimetric analysis 电重量分析
electrogravimetric trace analysis 电重量痕量分析法
electrogravimetry 电重量法*
electrograving 电蚀刻
electrohydraulics 电动水力学
electrohydrodimerization 电解氢化偶联作用；电解还原二聚作用
electrohydrodynamic 流体电动力型(的)
electrohydrodynamic ionization 流体电动力学电离
electroimmunoassay(EIA) 电免疫分析
electroimmunodiffusion 电免疫扩散(法)
electroinactive substance 无电活性物质；电化学非活性物质
electro-induction 电感应
electroinitiated polymerization 电引发聚合
electroinitiation 电引发(作用)
electro-inorganic chemistry 电无机化学
electro-insulating coating 绝缘涂料
electroinsulating finishing varnish 电绝缘覆盖漆
electroinsulating property 绝缘性能
electroinsulation varnish 电绝缘清漆
electroionic process(=electroion process) 高压放电法
electroionization(=ionization) 电离(作用)
electroion process(=electroionic process) 高压放电法
electrokinetic chromatography 电动色谱法
electrokinetic effect 电动效应
electrokinetic force 电动力
electrokinetic method 界面电动势测定法；界面电位差测定法
electrokinetic phenomena 电动现象
electrokinetic potential 动电势；动电位
electrokinetic property 动电性能
electrokinetics 动电学
electrokinetic theory 电动力学说；动电学理论
electrokinetic ultrafiltration analysis 动电超滤分析
electroleaching 电沥滤；电浸取
electroless coating 无电涂敷；化学电泳涂装法
electroless deposition 无电沉积；化学沉积
electroless deposition of metallic coating 无电沉积金属涂层(涂料)
electroless plating 无电镀膜法；化学镀膜法
electroluminescence 电致发光*
electroluminescent devices (电)场致发光器件
electroluminescent material (电)场致发光材料
electroluminescent panel (电)场致发光板
electroluminescent polymer 电致发光聚合物
electrolyser 电解池；电解槽
electrolysis 电解*

electrolysis at constant current 恒电流电解
electrolysis at controlled current 控制电流电解
electrolysis at controlled potential 控制电位电解
electrolysis bath 电解槽
electrolysis humidity sensor 电解法湿度传感器
electrolysis in fused salt 熔盐电解
electrolysis in molten salt 熔盐电解
electrolysis polymerization 电解(引发)聚合
electrolysis vessel 电解瓶
electrolyte ①电解质②电解(溶)液
electrolyte conductivity detector 电解质电导检测器
electrolyte hydrometer 电液比重计
electrolyte impedance humidity sensor 电解质阻抗湿度传感器
electrolyte number 电解质值
electrolyte solution 电解质溶液
electrolyte strength 电解质强度
electrolyte transition range 电解质转变范围
electrolytic 电解的
electrolytic agent 电解剂
electrolytical(=electrolytic) 电解的
electrolytic analysis 电解分析(法)
electrolytic anodising process 电解阳极氧化法
electrolytic apparatus 电解仪器
electrolytic assay 电解检验
electrolytic bath 电解槽
electrolytic bleaching 电解漂白
electrolytic caustic regeneration 苛性钠电解再生法
electrolytic caustic soda 电解苛性钠；电解氢氧化钠
electrolytic cell 电解池*
electrolytic chlorine 电解氯
electrolytic chromatography 电解色谱法
electrolytic cleaning 电解净化
electrolytic coating ①电泳涂料②电镀层
electrolytic common salt process 电解食盐法
electrolytic conductance cell ①电解(质)电导电池②电解(质)电导管
electrolytic conduction 电解导电
electrolytic conductivity ①电解(质)电导率②电解质导电性
electrolytic conductivity detector 电解(质)电导检测器
electrolytic conductor 电解导电体
electrolytic copper 电(解)铜
electrolytic corrosion 电解腐蚀
electrolytic corrosion test 电腐蚀试验
electrolytic coupling 电解偶合(工艺)
electrolytic current 电解电流
electrolytic degreasing 电解脱脂

electrolytic deposition 电解沉积
electrolytic dimerization 电解二聚化(反应)
electrolytic dissociation 电离(作用)
electrolytic dissociation constant 电离常数
electrolytic dissociation equilibrium 电离平衡
electrolytic dissolution 电解溶解
electrolytic double layer 双电层
electrolytic equilibrium 电解平衡
electrolytic etching 电解浸蚀
electrolytic etching technique 电解浸蚀技术〔可用于制金属长丝〕
electrolytic extraction 电解萃取
electrolytic fluorescence spectrophotometry 电解荧光分光光度法
electrolytic furnace 电解炉
electrolytic gas 电解气；爆鸣气
electrolytic gravimetry 电解重量分析(法)
electrolytic grinding 电解磨削
electrolytic hydrodimerization 电解加氢二聚化(作用)
electrolytic hygrometer 电解湿度计
electrolytic industry 电解工业
electrolytic initiation 电解引发
electrolytic ion (电解)离子
electrolytic ionization 电解电离
electrolytic labeling 电解标记
electrolytic method 电解法
electrolytic moisture analyzer 电解(法)水分测定仪
electrolytic nebulizer 电解雾化器
electrolytic oxidation 电解氧化
electrolytic pickling 电解酸洗
electrolytic polarization 电解极化(作用)
electrolytic polishing 电解抛光
electrolytic polymerization 电解聚合*
electrolytic potential 电解势
electrolytic preparation 电解制备
electrolytic process 电解法
electrolytic reaction 电解反应
electrolytic rectifier 电解整流器
electrolytic reduction 电解还原
electrolytic refining 电解精炼
electrolytic regeneration 电解再生
electrolytic scrubbing 电解洗涤
electrolytic separation 电解分离
electrolytic separation factor 电解分离因子
electrolytic soda process 电解制碱法
electrolytic solution 电解溶液
electrolytic solution pressure(=electrolytic solution tension) 电解(溶液)压(力)
electrolytic solution tension(=electrolytic solution pressure) 电解(溶液)压(力)
electrolytic solution tension theory 电解压理论
electrolytic synthesis 电解合成
electrolytic tension 电压；电解溶液张力
electrolytic tin plate 电镀锡薄板；电镀马口铁
electrolytic titration 电势滴定
electrolytic transport 电解迁移法
electrolytic voltage 电(槽)电压
electrolytic water 电解水
electrolytic water analyzer 水分电解分析器
electrolytic white lead 电解铝白
electrolytic winning 电解提取；电解冶金法
electrolyzable 可以电解的
electrolyzer 电解池
electrolyzing 电解
electromagnet ①电磁铁②电磁体
electromagnetic 电磁的
electromagnetic adhesive 电磁黏合剂
electromagnetic apparatus 电磁仪器
electromagnetic circulation 电磁循环
electromagnetic clutch 电磁离合器
electromagnetic enhancement model 电磁增强模型
electromagnetic field 电磁场
electromagnetic flowmeter 电磁流量计
electromagnetic force 电磁力
electromagnetic force balance 电磁力天平
electromagnetic image shift 电磁像位移
electromagnetic induction 电磁感应
electromagnetic interference(EMI) 电磁干扰
electromagnetic lens system 电磁透镜系统
electromagnetic pump 电磁泵
electromagnetic radiation 电磁辐射
electromagnetic radiation polymerization 电磁辐射(引发)聚合
electromagnetic radiation X-ray excited fluorescence analysis 电磁辐射激发X射线荧光分析
electromagnetic separation 电磁分离(法)*
electromagnetic separator 电磁分离器；磁选器
electromagnetic spectrum 电磁谱
electro-magnetic strain gauge 电磁应变仪
electromagnetic switch 电磁开关；电磁阀
electromagnetic theory 电磁(学)说
electromagnetic thermal analysis 电磁热分析
electromagnetic thickness gage 电磁式测厚仪
electromagnetic transducer 电磁换能器
electromagnetic unit 电磁单位
electromagnetic valve 电磁阀

electromagnetic vibratory feeder 电磁振动加料器
electromagnetic wave 电磁波
electromagnetism 电磁；电磁学
electromagnetism lens 电磁透镜
electromechanical coupling constant 机电偶合常数
electromechanical crystal structure 电机结晶构造
electromechanical integrator 电-机械积分器
electromechanical potentiometer 电机电势计
electromechanical printer 电机打印器
electromechanical torque measuring system 机电转矩测量系统
electromechanical transducer 机电换能器
electromembrane process 电隔膜法
electromer 电子异构体
electromeric change 电子异构变化
electromeric effect 电(子)移(动)效应
electromeric form 电子异构体
electromeric migration 电子异构移动
electromerism 电子异构(现象)
electromerization 电子异构(作用)
electrometallurgy 电冶(金)学
electrometer 静电计
electrometer amplifier 静电计放大器
electrometer tube 静电计电子管
electrometer valve 静电计电子管
electrometric analysis 量电分析
electrometric determination (of valence) (化合价的)电测定
electrometric method 电测法；量电法
electrometric migation 电子异构迁移
electrometric pH indicator pH 电测指示仪
electrometric titration 电化学滴定
electrometric titration outfit 电化学滴定装置
electrometry 量电法；电测法
electromigrated paper 电迁移(用)纸
electromigration 电迁移
electromigration injection 电迁移进样
electromolecular propulsion 带电分子推进分离法
electromotance 电动势
electromotion 电动
electromotive equilibrium 电动平衡
electromotive force 电动势*
electromotive force series 电动序
electromotor 电动机；马达
electromoulding 电铸
electron 电子
π-electron π电子
σ-electron σ电子*
electron absorption detector 电子吸收检测器

electron accelerating voltage 电子加速电压
electron acceptability 电子接受能力
electron acceptable monomer 电子接受性单体
electron-accepting group 受电子基团
electron-accepting molecule 电子接受分子
electron acceptor(EA) 电子受体
electron affinity 电子亲和势*
electron affinity detector 电子亲和检测器
electron affinity substance 电子亲和性物质
electron affinity yield 电子亲和率
electronating agent 增电子剂；还原剂
electronation 增电子(作用)〔即还原作用〕
electronation reaction 增电子反应；还原反应
electron atmosphere 电子云
electron attachment coefficient 电子附着系数
electron-attacting group(=electrondrawing group) 吸电子基团
electron attractive group 电子吸引基团
electron attractivity 电子吸引性
electron attractor 电子吸引体
electron avalanche 电子雪崩
electron band energy 电子带能
electron beam 电子束
electron beam accelerator 电子束加速器
electron beam analysis 电子束分析
electron beam atomizer 电子束原子化器
electron beam collimation 电子束准直
electron beam crosslinking 电子束交联
electron beam curing 辐射交联；电子束固化
electron beam current 电子束电流
electron beam damage 电子束损伤
electron beam divergence 电子束发散
electron beam evaporation 电子束蒸发法
electron-beam extractor 电子束引出装置
electron beam gun 电子束枪
electron beam melting 电子束熔化
electron beam microprobe analysis 电子束微探针分析
electron beam polymerization 电子束聚合
electron beam probe 电子束探针
clectron beam surface probe 电子束表面探针
electron beam vulcanization 电子束硫化
electron-beam welding 电子束焊接
electron binding state 电子结合态
electron bombard 电子轰击
electron-bombardment 电子轰击
electron bombardment ionization 电子轰击电离
electron bombardment ionization source 电子轰击源
electron bombardment ion source 电子轰击离子源

electron capture 电子俘获
electron capture chemical ionization 电子捕获化学电离
electron capture detector(ECD) 电子捕获检测器
electron capture dissociation 电子捕获裂解
electron capture gas chromatography 电子俘获气相色谱(法)
electron capture gas liquid chromatography 电子捕获气液色谱法
electron capture ion source 电子捕获离子源
electron capture labelling 电子俘获标记
electron capture negative ion chemical ionization 电子捕获负离子化学电离
electron capture process 电子俘获过程
electron capture supply 电子俘获电源
electron capturing compound 电子俘获化合物
electron carrier 电子载体
electron channel effect 电子通道效应
electron channel image 电子通道图
electron channeling pattern 电子通道图
electron channel pattern 电子沟道图样
electron chronometer 电子计时器
electron cloud 电子云*
electron collimating magnet 电子准直磁铁
electron collision 电子碰撞
electron concentration 电子浓度
electron configuration 电子组态
electron contributing group 电子供给基团
electron correlation 电子相关
electron crystallography 电子晶体学
electron curing 电子射线固化
electron current image method induced by electron beam 电子束感生电流像法
electron-defect compound 缺电子化合物
electron deficiency 缺电子
electron-deficient bonding 缺电子键合
electron-deficient compound 缺电子化合物
electron density 电子密度
electron density distribution 电子密度分布
electron density map 电子密度图〔电子云的等密度线图〕
electron density projection 电子密度投射
electron density synthesis 电子密度合成
electron detachment 电子脱离
electron diffraction 电子衍射*
electron diffraction diagram 电子衍射图
electron diffraction investigation 电子衍射研究
electron diffraction pattern 电子衍射图像
electron diffractometer 电子衍射仪
electron dipole-electron dipole interaction 电子与电子的偶极-偶极相互作用
electron displacement ①电子位移②电子排代
electron donability 电子给予性；电子供给性
electron-donating center 给电子中心
electron-donating group 给电子基团*
electron donating solvent 供电子溶剂
electron donation 给电子能力
electron donor 电子给体
electron donor-acceptor(EDA) complex 电子给(体)受体络合物
electron donor solvent 给电子溶剂
electron-dot formula 点电子式；路易斯式
electron double resonance 电子双共振
electron drift 电子漂移
electron drift detector 电子漂移检测器
electron drift velocity detector 电子漂移速度检测器
electronegative 电负的；带负电的
electronegative deoiling 阳极除油
electronegative element 电负性元素
electronegative gas 负电性气体
electronegative ion 负(电性)离子
electronegative potential 阴电势
electronegative radical 阴(电)性根
electronegative substituent 电负性取代基
electronegativity 电负性*
electronegativity difference 电负差
electronegativity scale 电负(度)标；电负性标度
electronegativity value 电负值
electron ejection 电子射出
electron-electron double resonance 电子-电子双共振
electron emission 电子发射
electron emission current 电子发射电流
electron emission spectra 电子发射能谱
electron emissive coating 电子发射涂覆
electron energy loss spectrometer 电子能量损失谱仪
electron energy loss spectroscopy(ELS) 电子能量损失谱(学)*
electron energy selector 电子能量选择器
electron entrance hole 电子入口孔
electron escape depth 电子逃逸深度
electroneutral 电中性的
electroneutrality condition 电中性条件
electroneutrality relation 电中性关系
electron exchange 电子交换
electron exchange chromatography 电子交换色谱(法)
electron exchange interaction constant 电子交换相互作用常数
electron-exchanger 电子交换剂

electron exchange resin 电子交换树脂
electron excited X-ray fluorescence analysis 电子激发 X 射线荧光分析
electron feedback 电子反馈
electron focusing 电子聚焦
electron gas 电子气
electron gun 电子枪
electron hole 电子空穴
electron-hole pair 电子-空穴对
electron-hole recombination 电子空穴复合
electronic 电子的
electronic absorption band 电子吸收带
electronic absorption spectroscopy 电子吸收光谱
electronic absorption spectrum 电子吸收光谱
electronic action 电子作用
electronically controlled AC arc 电子控制交流电弧
electronically controlled spark 电子控制火花
electronic balance 电子天平
electronic band gap 电子能带距离(范围)
electronic charge 电子电荷
electronic chronometer 电子计时器
electronic circuitry 电子电路系统
electronic cloud 电子云
electronic colour scanner 电子彩色扫描器
electronic computer 电子计算机
electronic conduction 电子导电
electronic conductor 电子导体
electronic configuration 电子排布；电子构型
electronic control 电子控制
electronic coulometer 电子库仑计
electronic coupling 电子偶合
electronic cure 高频硫化
electronic current 电子流
electronic data processing 电子数据处理
electronic degree of freedom 电子自由度
electronic dehydration 高频加热干燥
electronic denier monitor 电子式旦数监测仪；电子纤度监测仪
electronic densitometer 电子(光)密度计
electronic detonation meter 电子爆震计
electronic digital integrator 电子数字积分器
electronic digital stopwatch 电子数字秒表
electronic displacement ①电子位移②电子排代
electronic double layer repulsion 双电(子)层排斥
electronic drift 电子漂移；电子仪器漂移
electronic effect 电子效应
electronic-electronic energy transfer 电子-电子能量传递

electronic emission 电子发射
electronic energy 电子能
electronic energy level 电子能级
electronic equation 电子方程式
electronic equilibrium 电子平衡
electronic excitation 电子激发(作用)
electronic feeding wheel 电子式喂纱轮
electronic fineness tester 电子细度测定仪
electronic flow-meter 电子流量计
electronic flow-meter system 连续式电子流量计
electronic formula 电子式
electronic glueing 电子胶合
electronic heater(=dielectric heater) 电子加热器；介电加热器
electronic heating 电子加热
iso-π-electronic hydrocarbon 等 π 电子烃
electronic hypothesis 电子假说
electronic integrator 电子积分器
electronic interconnect 电子互换
electronic Lande factor 电子朗德因子
electronic level 电子能级
electronic magnetic interference 电磁干扰
electronic mass 电子质量
electronic measuring machine 电子量革机
electronic micrography 电子显微摄影
electronic microscope 电子显微镜
electronic microscopy 电子显微测定法
electronic migration 电子迁移
electronic moisture meter 电子水分计
electronic noise 电子噪声
electronic nose 电子鼻
electronic number 电子数
electronic open-width circular knitting machine 开幅电子圆机
electronic orbit 电子轨道
electronic orbital angular momentum 电子轨道角动量
electronic paramagnet resonance(EPR) 电子顺磁共振
electronic parameter 电子参数
electronic particle counting 电子微粒计数法
electronic partition function 电子配分函数
electronic polarizability 电子极化率
electronic polarization 电子极化
electronic potentiometer 电子电位计
electronic power supply 电子电源
electronic preheating 高频(率)预热
electronic probe 电子探针
electronic processing 电子处理
electronic programmer 电子程序器

electronic quadrupole moment　电四极矩
electronic quenching　电子猝灭*
electronic ratio　电子比率
electronic readout section　电子(仪)读出部分
electronic recorder　电子记录仪
electronic regulator　电子调节器；电子控制器
electronic scale　电子秤
electronic scanner　电子扫描器
electronic self-propelled stitch frequency welding　电子自动推进缝编高频熔接
electronic shell　电子层；电子壳
electronic smoothness evaluator　电子(织物)皱纹测验器
electronic spectroscopy　电子光谱(法)
electronic spectrum　电子光谱*
electronic speed probe　电子测速器
electronic spin angular momentum　电子自旋角动量
electronic spin resonance(ESR)　电子自旋共振
electronic structure　电子结构
electronic susceptibility　电子磁化率
electronic temperature controller　电子温度控制器
electronic tension meter　电子张力(测定)仪
electronic theory　电子(学)说
electronic timer　电子计时器
electronic titer control　电子纤度控制
electronic tongue　电子舌
electronic transfer process　电子转移过程
electronic transition　电子跃迁
electronic trouble-shooting　电子故障检验
electronic-vibrational-rotational band　电子-振动-转动谱带
electronic-vibrational-rotational transition　电子-振动-转动跃迁
electronic view of valency　化合价的电子观
electronic voltage regulator　电子管式电压调节器；电子管式调压器
electronic voltmeter　电子伏特计
electronic vulcanizer　高频硫化装置
electronic warp length counter　电子经纱计长器〔用于整经拉伸上浆系统〕
electronic weighing system　电子称重系统
electronic work function　电子功函数
electron impact(EI)　电子碰撞；电子轰击
electron impact-chemical ionization source　电子轰击-化学电离复合源
electron-impact degradation　电子冲击降解(作用)；电子碰撞降解(作用)
electron impact desorption(EID)　电子碰撞解吸
electron-impact-induced fragmentation　电子碰撞诱导碎裂(作用)
electron impact ion source　电子碰撞离子源
electron injector　电子注入器
electron interaction　电子相互作用
electron ionization　电子电离
electron ionization ion source　电子电离源
electron ionization mass spectrum　电子电离质谱
electron ionization source　电子离子化源
electron-ion recombination　电子-离子复合
electron isomerism　电子异构现象
electron jump　电子跃迁
electron-lattice interaction　电子晶格相互作用
electron lens　电子透镜
electron level　电子能级
electron magnetic moment　电子磁矩
electron magnetic resonance　电子磁共振
electron magnetic resonance spectroscopy　电子磁共振波谱法
electron metallography　电子金相学
electron-microautoradiography(EMAR)　电子显微放射自显影法
electron microdiffraction　电子显微衍射(法)
electron micrograph　电子显微图
electron micrography　电子显微摄影
electron microprobe(EMP)　电子微探针
electron microprobe analysis　电子微探针分析
electron microscope　电子显微镜*
electron microscope autoradiography(EMAR)　电子显微(镜)放射自显影法
electron microscope microanalyzer(EMMA)　电子显微镜微量分析器
electron microscope photomicrograph　电子显微镜(显微)照片
electron microscopic analysis　电(子显微)镜分析
electron microscopic examination　电子显微镜检验法
electron microscopy　电子显微术
electron migration　电子移动(作用)
electron mobility detector　电子迁移率检测器
electron multiplication　电子倍增作用
electron multiplier　电子倍增器*
electron multiplier detector　电子倍增检测器
electronnegativity rule　电负性规则
electron-nuclear dipole interaction　电子与核的偶极-偶极相互作用
electron-nuclear double resonance(ENDOR)　电子-核双共振*
electron nuclear double resonance spectroscopy　电子-核双共振谱学

electron-nuclear-nuclear triple resonance 电子-核-核三重共振
electron number of electrode reaction 电极反应的电子数
electronography 静电印刷术
electron optics 电子光学
electron orbit 电子轨道
electron-osmosis 电渗
electron pair 电子对；电子偶
electron pair acceptor(EPA) 电子对受体
electron pair bond 电子对键
electron pair donor(EPD) 电子对给体
electron pairing method 电子配对法
electron pair production 电子对产生
electron pair repulsion 电子对互斥
electron paramagnetic resonance(=electron spin resonance) 电子顺磁共振；电子自旋共振
electron paramagnetic resonance spectrometer 电子顺磁共振谱仪；电子自旋共振谱仪
electron paramagnetic resonance spectroscopy (电子)顺磁共振(波)谱法；电子自旋共振(波)谱法
electron pencil 电子锥；电子束
electron-phonon collision 电子声子相互作用*
electron photographic marking-off method 电子照相画线法
electron photomicrograph 电子显微摄影；电子显微照片
electron photomicrography 电子显微摄影法
electron plasma 电子等离子体
electron polarization 电子极化
electron population analysis 电子群体分析
electron potential difference ga(u)ge 电子电势差计
electron probe 电子探针
electron probe micro-analysis 电子探针微量分析*
electron probe microanalyzer 电子探针微量分析器；电子探针显微分析仪
electron probe micro-region analysis 电子探针微区分析
electron probe surface mass spectrometry 电子探针表面质谱分析
electron probe X-ray microanalyzer 电子探针X射线微量分析仪
electron radiation 电子辐射
electron ray 电子射线；电子束
electron-ray drying 电子射线干燥
electron-ray luminescence 电子射线发光
electron ray titration assemble spectrometer 电子管式电位差滴定装置
electron reaction 电子反应
electron relay 电子中继体
electron-releasing group 释电子基团

electron repeller 电子推斥极
electron repelling 电子排斥
electron replacement 电子增减；电子置换
electron repulsion energy 电子排斥能
electron rich 富电子的
electron scanning micrograph 扫描电子显微照片
electron scattering 电子散射*
electron scattering power 电子散射能力
electrons cluster 电子簇；电子雾
electron sensitive film 电敏胶片
electron shake-off 电子震脱
electron sharing 电子共用
electron shell 电子层；电子壳
electron spectrometer 电子能谱仪
electron spectrometer for chemical analysis 化学分析用电子能谱仪
electron spectroscopy(ES) 电子能谱(学)*
electron spectroscopy for chemical analysis 化学分析用电子能谱法
electron spectrum 电子能谱
electron spin 电子自旋
electron spin echo 电子自旋回波
electron spin echo-electron nuclear double resonance 电子自旋回波电子-核双共振
electron spin echo envelope modulation 电子自旋回波包络线调制
electron spinning 电子自旋
electron spin resonance(=electron paramagnetic resonance) 电子自旋共振；(电子)顺磁共振
electron spin resonance absorption 电子自旋共振吸收；(电子)顺磁共振吸收
electron spin resonance imaging 电子自旋共振成像；顺磁共振成像技术；电子自旋共振成像技术
electron spin resonance oxidimetry 电子自旋共振氧定量法
electron spin resonance spectroscopy 电子自旋共振波谱仪；电子自旋共振波谱学
electronspray 电喷雾
electronspray mass spectrometry 电喷雾质谱法
electron stimulated desorption 电子受激脱附
electron structure 电子结构
electron suppressor slit 电子抑止狭缝
electron temperature 电子温度
electron theory 电子(学)说
electron titration 电子滴定
electron transfer 电子转移*
electron transfer chain 电子转移链
electron transfer protein 电子传递蛋白
electron-transfer reaction 电子转移反应

electron transition 电子跃迁
electron transit time 电子跃迁时间
electron transmission micrograph 电子透射显微照相
electron-transport chain 电子迁移链
electron transport system 电子传递体系
electron trap 电子阱
electron tube 电子管；真空管
electron tube voltmeter(=vacuum tube voltmeter) 电子管伏特计
electronuclear breed 电核法增殖〔核燃料〕
electron velocity selector 电子速度选择器
electron volt 电子伏特
electron wave 电子波
electron-withdrawing group 吸电子基团*
electron-withdrawing power 吸电子能力
electron-withdrawing substituent 吸电子取代基
electrooptical anisotropy 电学光学各向异性现象
electrooptical detector 电光学检测器
electrooptical effect 光电效应
electrooptical ion detector 电光式离子检测器
electrooptic method 电光法
electrooptic multichannel detection 电光学多道检测
electroorganic synthesis 电有机合成
electroosmose process 电渗析法
electroosmosis 电渗
electroosmotically driven chromatography 电渗流驱动色谱法
electroosmotic mobility 电渗迁移率
electrooxidation 电(解)氧化
electropaint 电沉积涂料；电泳漆
electropainting 电涂
electropainting installation 电泳涂漆设备
electropainting plant 电泳涂漆装置
electropermissibility 电通透性
electrophile(=electrophilic reagent) 亲电子试剂；亲电体
electrophilic 亲电子的
electrophilic addition 亲电加成*
electrophilic aromatic substitution 亲电芳香取代
electrophilic atom localization energy 亲电子原子定域能
electrophilicity 亲电性
electrophilic reaction 亲电子反应
electrophilic reactivity 亲电子反应性
electrophilic reagent(=electrophile) 亲电子试剂；亲电体
electrophilic rearrangement 亲电重排*
electrophilic substitution 亲电取代*
electrophobic 疏电子的
electrophobic reagent 疏电子试剂*
electrophoresis 电泳(法)

electrophoresis analysis 电泳分析
electrophoresis apparatus 电泳仪
electrophoresis chromatogram 电泳色谱
electrophoresis pattern 电泳图(谱)
electrophoresis strip 电泳条
electrophoresis tank 电泳槽
electrophoretic 电泳的
electrophoretic analysis 电泳分析
electrophoretic behavior 电泳性能
electrophoretic blotting 电泳印迹
electrophoretic buffer 电泳缓冲液
electrophoretic coating(=electrodeposited coating) 电沉积涂覆法；电泳涂覆法
electrophoretic column 电泳柱
electrophoretic current 电泳电流
electrophoretic effect 电泳效应
electrophoretic fingerprint 电泳指纹(图)
electrophoretic focusing 电泳聚焦
electrophoretic force 电泳力*
electrophoretic injection 电迁移进样
electrophoretic light scattering 电泳光散射
electrophoretic medium 电泳介质
electrophoretic method 电泳法
electrophoretic migration 电泳迁移
electrophoretic mobility ①电泳淌度②电泳迁移
electrophoretic painting 电泳涂漆；电泳涂装
electrophoretic potential 电泳电位
electrophoretic primer 电泳底漆
electrophoretic property 电泳性能
electrophoretic resolution 电泳分辨率
electrophoretic separation 电泳分离
electrophoretic velocity 电泳速度
electrophoretogram 电泳图
electrophorus 起电盘
electrophotographic method 电子照相法；光电摄影法
electrophotography 光电摄影法；电子照相法
electro-physical machining 电物理加工
electroplated coating 电镀(涂)层
electroplating 电镀
electroplating bath 电镀浴
electroplating technology 电镀工艺
electroplating template 电镀模板
electropneumatic converter 电动-气动转换器
electropneumatic valve 电控气动阀
electropolishing 电抛光*
electropolymerization 电引发聚合
electropolymerization macromolecule embedment 电聚合高分子包埋法

electroporation 电穿孔；电打孔
electropositive 电正的；阳电的
electropositive deoiling 阴极除油
electropositive element 电正性元素
electropositive ion 阳(电性)离子
electropositive metal 电正性金属；电阳性金属
electropositive potential 阳电势
electropositive radical 阳(电性)根
electropositivity ①电正性；阳电性②电正度
electropotential 电极电位
electroprecipitator ①电力沉淀器②静电集尘器
electro-psychrometer 电子湿度计；电测湿度计
electro-pulse column 电脉冲柱
electropure process 低频交流电导杀菌(法)
electropyrometer 电阻高温计
electroquartz 电造石英
electroradioimmunoassay(ERIA) 电放射免疫分析
electro ray 电子射线
electroreduction 电(解)还原
electrorefining 电解精炼
electroregulator 电热调节器
electroresponse 电(流)感应
electroretardation filtration 电阻滞过滤(作用)
electrorheological fluid 电流变液
electrorheophoresis 电流变泳
electroscope 验电器
electroscopy 验电法；气体电离检定法
electrosensitive paper 电敏纸
electroseparation 电分离
electro-slag welding 电渣焊
electrosparking deposition and build up 电火花沉积和堆焊
electrosparking hardening method 电火花放电硬化法
electro-spin 静电纺纱
electro spot test 电图分析法；电谱法；电点滴试验
electrospray 电雾化
electrospray deposition 电喷镀沉积
electrospray interface 电喷雾接口
electrospray ionization 电喷雾离子化；电喷雾电离；电喷雾电离技术
electrospray ionization mass spectrometry 电喷雾电离质谱(法)
electrospray ion source quadrupole mass spectrography 电喷雾离子源四极质谱法
electrospray method 电喷雾法
electrostatic 静电的
electrostatic adsorption 静电吸附(作用)
electrostatic aerosol coating 静电气溶胶涂装法

electrostatic agglomeration 静电附聚(作用)
electrostatically spraying 静电喷射〔制超细纤维的成形方法之一〕
electrostatic analyzer 静电分析器
electrostatic atomizer 静电雾化器；静电喷雾器
electrostatic attraction 静电引力
electrostatic barrier 静电能垒
electrostatic binding (污垢的)静电结合
electrostatic blade coater 叶片式静电喷(刮)涂机
electrostatic bond 静电键
electrostatic bunching (粉末干筛时的)静电絮凝
electrostatic capacitance 静电容
electrostatic capacitance thickness meter 静电电容(法)测厚仪
electrostatic capacity(=electric capacity) 电容；静电容量
electrostatic charge 静电荷
electrostatic charge density 静电荷密度
electrostatic clinging property 静电黏附性
electrostatic cloud 静电云团
electrostatic coating 静电涂装；静电喷涂
electrostatic deflection 静电偏转
electrostatic deposition 静电沉积法
electrostatic detearing 静电除漆；静电沥水〔浸涂或电泳涂装时将多余的漆除掉〕
electrostatic disc atomizer 静电盘雾化器；盘式静电喷雾器
electrostatic dissipative compound 静电消除配混料
electrostatic dry spraying 静电干粉喷涂法
electrostatic duster 静电式喷粉机
electrostatic effect 静电效应
electrostatic energy 静电能
electrostatic energy barrier 静电能垒
electrostatic field 静电场
electrostatic filter 静电过滤器
electrostatic flocking 静电植绒
electrostatic flock printing 静电植绒印花
electrostatic fluidized bed 静电沸腾床；静电流动床
electrostatic-fluidized bed coating 静电流化床涂装；静电沸腾床涂装；静电流态化涂装
electrostatic focusing 静电聚焦
electrostatic force 静电力
electrostatic force microscopy 静电力显微镜
electrostatic free energy 静电自由能
electrostatic generator 静电发电机
electrostatic heat 静电加热
electrostatic induction 静电感应
electrostatic interaction 静电相互作用
electrostatic ion reflector 静电离子反射器

electrostatic law 静电律
electrostatic lens 静电透镜
electrostatic mass filter 静电滤质器
electrostatic mirror 静电反射镜
electrostatic nebulizer 静电雾化器
electrostatic oil filter 静电油过滤器
electrostatic painting promote agent 静电涂装助剂
electrostatic particle guide 静电粒子导向装置
electrostatic potential 静电势*
electrostatic potential gradient 静电势梯度
electrostatic powder coating 静电粉末涂装；静电粉末喷涂
electrostatic powder spraying 静电粉末喷涂法
electrostatic power sources 静电发生器；静电源
electrostatic power unit 静电发生器
electrostatic precipitation 静电沉淀
electrostatic precipitator 静电除尘器；静电沉积器
electrostatic repulsion 静电推斥
electrostatics 静电学
electrostatic sample collection trap 静电试样收集阱
electrostatic separator 静电离析器
electrostatic setting 静电电压定位值；静电电压校准
electrostatic shield 静电屏蔽
electrostatic spinning 静电纺丝
electrostatic spray 静电喷雾
electrostatic spray coating 静电喷涂
electrostatic spray equipment 静电喷漆机
electrostatic sprayer 静电喷涂机
electrostatic spraying 静电喷涂
electrostatic spraying voltage 静电喷涂电压
electrostatic tension 静电压
electrostatic unit 静电单位
electrostatic valence rule 电价规则*
electrostatic viscous filter 静电黏性过滤器
electrostatic wrap-around 静电迂回(环抱)性
electrostenolysis 膜孔电(淀)积(作用)
electrostriction 电致伸缩*
electrostrictive 电致伸缩的
electrostrictive compliance 电致伸缩柔量
electrosynthesis 电合成*
electrotaxis 趋电性；向电性
electrothermal analysis(ETA) 电热分析(法)
electrothermal atomic absorption spectrometry 电热原子吸收光谱法
electrothermal atomization 电热原子化
electrothermal atomization atomic absorption spectrometry 电热原子化原子吸收光谱法
electrothermal atomizer 电热原子化器
electrothermal device 电热装置
electrothermal effect 电(致)热效应
electrothermal equivalent 电热当量
electrothermal furnace 电热炉
electrothermal graphite 电加热石墨炉
electrothermal graphite atomizer 电加热石墨炉原子化器
electrothermal mantle 电热罩
electrothermal metallurgy 电热冶金
electrothermal vaporization 电热蒸发
electrothermic industry 电热工业
electrothermics 电热学
electrotinplate 电镀锡
electrotitration 电滴定
electrotransfer reaction 电子传递反应
electrotropic action 向电作用
electrotropic change 向电变化
electrotropism(=electrotropy) 向电性
electrotropy(=electrotropism) 向电性
electrotype 电铸版
electrotype mould 电铸版胎模
electrotyping 电铸*
electrotyping process 电铸(印板)法
electro-ultrafiltration 电超过滤
electrovalence 电价
electrovalency 电价；电化价
electrovalent bond 离子键
electrovalent compound 离子化合物
electrovalent coordination bond 电价配位键*
electrovalent crosslink 电价(键)交联
electroviscosity 电黏性
electro-viscous effect 电黏效应；电滞效应
electrovisualisation 电观测法
electrowinning 电解提取*；电解沉积
electro-zinc 电镀锌(层)
electrum ①琥珀金；镍银②银金矿
elektrion oil(=electrion oil) 高压放电润滑油
Elektron metal 镁铝合金；埃雷克特龙镁合金
elema 硅碳棒
elemane 榄香烷
elemazulene 榄香薁
elemecin 异榄香素
elemene 榄香烯
elemenic acid 榄香烯酸
elemenol 榄香烯醇
elemenolic acid 榄香烯脑酸
elemenone 榄香烯酮
elemenonic acid 榄香烯酮酸
element ①元素②单体③要素④元件；零件
elemental abundance 元素丰度

elemental analysis 元素分析*
elemental analyzer 元素分析仪
elemental atomic ratio 元素原子比
elemental chromatography 元素色谱法
elemental depth profiling 元素深度剖(面分)析
elemental enrichment factor method 元素富集因素法
elemental line scan 元素线扫描
elemental semiconductor 元素半导体
elemental sensitivity factor 元素灵敏度因子
elemental sulfur(=primary sulfur) 元素硫
elementary ①元素的②基础的③初等的
elementary analysis 元素分析
elementary body 原粒体
elementary cell 单位晶格
elementary charge 基本电荷
elementary color 原色；基色
elementary composition 元素组成
elementary count 单纤维支数
elementary fibril 初级原纤
elementary fibril bundle 初级原纤束
elementary microfibril 初级微原纤；基元微原纤
elementary organic analysis 元素有机分析
elementary particle 基本粒子
elementary polymer 元素高分子
elementary process 单元过程；基础过程；基元反应
elementary reaction 基元反应
elementary rheology 初等流变学
elementary substance 单质*
elementary sulfur(=elemental sulfur) 元素硫
elementary volume 体(积)元
elementary weave 原组织
element chlorine free bleaching 无元素氯漂白
element effect 元素效应
element of filter 滤芯
element of volume 微元体积
elementoorganic fiber-forming polymer 元素有机成纤聚合物
elemento-organic polymer 元素有机聚合物
elementosiloxane polymer 硅氧烷元素有机聚合物
element polymer 元素高分子
element purge 部件清洗
element selection device 元素选择器件
element selective detector 元素选择性检测器
elements of symmetry 对称要素
elemi ①榄香②榄香脂
elemic acid 榄香酸
elemicin 榄香素
elemi oil 榄香脂油

elemi resin 榄香树脂
elemol 榄香醇
elemolactone 榄香内酯
elemolic acid 榄香醇酸
elemonic acid 榄香酮酸
eleolite 脂光石
eleonorite 簇磷铁矿
eleoptene 油萜〔精油中的液体萜〕
eleostearate 桐酸酯
eleostearic acid(=elaeostearic acid;elaeomargaric acid) 桐酸；9,11,13-十八碳三烯酸 $C_{18}H_{30}O_2$
eletrect 驻极体
eletrokinetic potential 电动电势*
eletroosmosis 电渗
eleutherin 艾榴英
eleutherinol 艾榴醇
eleutherol 艾榴脑
elevated ①升高的；提高的；高的②高位的；架空的
elevated dosing vessel 高位计量槽
elevated jet condenser 气压冷凝器
elevated pressure 高压
elevated tank 高架储罐；高位槽
elevated temperature 升温；增高温度；高温
elevating apron 提升皮带
elevating screw 升降螺旋桨
elevation ①海拔；高度②正视图
elevation back 后视图
elevation drawing 正面图
elevation of boiling point 沸点升高
elevation side 侧视图
elevator 提升机
elevator boot 提升机底槽；升降机的滑车箱
elevator-cup 提升机箕斗
elevator-platform 升降台
elevtrovalent 电价的
α-elgenone α-埃金酮
cis-elimination 顺位消除
α,β-elimination α,β消除*
γ-elimination γ消除*
β-elimination β消除*
elimination-addition 消除-加成
elimination of compound 消去化合物
elimination of group 消去取代基
elimination of hydrogen embrittlement 消除氢脆
elimination of variables 消元法
elimination of water 脱水；消去水
elimination polymerization 消除聚合
elimination reaction 消除反应*

eliminator ①空气净化器②液滴分离器
eliminator baffle plate 除湿折流板；除水珠折流板
eliminator stack ①空气净化管②漆雾分离管；液滴分离管(系统)
eliminostat 静电消除器
eliminostatic 静电消除
elinvar 铁镍铬恒弹性钢
elipse head 椭圆封头
elixir 酏剂；甘香酒剂
elk side leather 多脂铬铝鞣(牛皮半张)鞋面革
ell 弯头(管)
ellagene 鞣花烯
ellagic 鞣花的
ellagic acid(=gallogen) 鞣花酸；棓原 $C_{14}H_2O_4(OH)_4$
ellagitannic acid 鞣花单宁酸 $C_{14}H_{10}O_{10}$
ellagitannin 鞣花单宁；鞣花鞣质
Eller stress relaxometer 爱勒应力弛豫仪
elleryone 吉莉酮
elliotinoic acid 湿地松酸
elliotinol 湿地松醇
Elliot test 埃利奥特(闪点测定)试验
Elliot tester 埃利奥特(闪点测定)试验器
ellipsoid 椭圆体
ellipsoidal flowmeter 椭圆流量计
ellipsoidal head 椭圆形封头；椭圆形顶盖
ellipsoidal mirror 椭圆棱镜
ellipsometer 椭偏仪
elliptic acid 鱼藤酸
elliptical and circular polarized light 椭圆和圆偏振光
elliptical cell rotation method 椭圆池旋转法
elliptical cross-section 椭圆截面
elliptical fiber 椭圆形(截面)纤维
elliptically polarized light(=elliptic polarization) 椭圆(偏)振光
elliptically vibrating screen 椭圆振动筛
elliptical polarization 椭圆偏振
elliptical-shaped spray pattern 椭圆形喷束
elliptical spot 椭圆形斑
ellipticity 椭圆度
elliptic orbit 椭圆轨道
elliptic polarization(=elliptically polarized light) 椭圆(偏)振光
elliptol 鱼藤醇
elliptolone 鱼藤醇酮
elliptone 鱼藤酮
elliptonone 鱼藤酮酮
Ellis model for flow 埃利斯流动模型
Ellis' mortar 埃利斯研钵；(大型)冲击钵

ellispsometry 椭圆光度仪
elm 榆树
elm bark 榆树皮
Elmendorf tearing tester 埃尔曼多夫扯裂试验仪
Elmendorf tear strength 埃尔曼多夫撕裂强度
Elmendorf test 埃尔曼多夫试验
Elmore continuous centrifuge 埃尔摩连续离心机
El Nino phenomenon 厄尔尼诺现象
elomag process 镁合金阳极氧化法
elongation 伸长；延伸
elongational flow 伸长流动；拉伸流动
elongational viscosity 拉伸黏度
elongation at break 断裂伸长；致断伸长
elongation at constant load 定载伸长
elongation at failure 衰坏伸长〔致衰坏的伸长〕
elongation at rupture 破坏伸长〔致破坏的伸长〕
elongation at specified load 定负荷伸长(率)；特定载荷伸长(率)
elongation at yield 屈服伸长〔达屈服点的伸长〕
elongation at yield point ①屈服点伸长②屈服点伸长率
elongation factor 延伸因子
elongation in knot test 结节伸长(率)试验
elongation in loop test 勾接伸长(率)试验；环扣伸长(率)试验
elongation measurement 伸长率测量
elongation percentage 延伸率；伸长率
elongation rate 伸长率
elongation ratio 伸长比
elongation retention ①伸长保留②伸长保留率
elongation scale 伸长率标度
elongation set 伸长变定；永久伸长
elongation strain 伸长应变
elongation stress 张应力
elongation tensor 伸长张量
elongation ultimate 极限延伸率
elongator 伸出件；伸出臂
Eloxal process 铝表面阳极氧化法〔以草酸为电解液〕
elpidite 钠锆石
Elrepho brightness 厄利福白度
Elrod-Maron-Krieger equation for shear rate 埃尔罗德-马伦-克里格剪切速率方程
elsholtzaldehyde 香薷醛 $C_6H_6O_2$
elsholtziaketone 香薷酮 $C_{10}H_{14}O_2$
elsholtzia oil 香薷油
elsholtzic acid 香薷酸 $C_6H_6O_3$
elsholtzione 香薷酮 $C_{10}H_{14}O_2$
Elsner green 锌酸钴绿；钴绿
eluant 洗脱剂；洗脱液

eluant applied in sequence　按顺序加的洗脱剂
eluant component　洗脱剂组分
eluant composition　洗脱剂组成
eluant gas　洗脱(用)气体；载气
eluant series　洗脱液系列
eluant strength　洗脱剂浓度
eluant strength gradient　洗脱剂强度梯度
eluant stripper column　洗脱剂清洗柱
eluate　洗出液*
elucidation of structure　阐明结构式
eluent(=elutriant)　洗脱液；洗脱剂
eluent gas　洗脱(用)气体；载气
eluent gradient　流动相梯度
eluent ion　淋洗离子
Eluex-Amex process　埃留克斯-阿姆克斯过程〔铀的联合精制过程〕
Eluex process　埃留克斯过程〔离子交换-叔胺溶剂萃取联合回收铀的过程〕；淋萃流程
eluotropic　洗脱的
eluotropic series　洗脱序*
eluotropic strength　洗脱强度
elutant　洗出液
eluting　①洗脱②流出
eluting agent　洗脱剂
eluting effect　洗脱效应
eluting gas　流出气体
eluting order　洗脱顺序
eluting peak　①洗脱峰②流出峰
eluting power　洗脱能力
eluting sequence　洗脱顺序
eluting temperature　洗脱温度
elution　①洗脱(法)②流出③淋洗
elution analysis　洗脱分析
elution characteristic　洗脱特性
elution chromatography　洗提色谱法；洗脱色层(分离)法；洗脱色谱法
elution curve　①洗脱曲线②流出曲线
elution development　洗脱展开(法)
elution-extracting resin　萃淋树脂
elution fractionation　洗脱分级；淋洗分级
elution gas chromatography　洗脱气相色谱(法)
elution method　洗脱法；洗提法；冲洗法
elution order　①洗脱顺序②流出顺序
elution process　①洗脱过程②洗脱方法
elution program　洗脱程序
elution range　流出范围
elution requirement　洗脱条件
elution system　洗脱系统

elution time　①洗脱时间②流出时间
elution volume　①洗脱容积②流出体积
elutive power　洗脱能力
elutriant(=eluent)　洗脱液；洗脱剂
elutriating　淘析；淘选
elutriating apparatus　淘析器
elutriating cylinder　淘析柱
elutriating flask　淘析瓶
elutriating funnel　淘析漏斗
elutriation　淘析；淘洗；淘净
elutriation analysis　淘析法
elutriation test　淘析试验；沉淀试验
elutriator　淘析器
elutriator-centrifuge　淘析离心机
elutropic　洗脱的
eluviation　淋滤作用
elwotite　硬钨钛合金
elymoclavine　野麦角碱
eman　埃曼*〔大气放射性单位，等于 10^{-10} 居里/升〕
emanation　①(略 Em)射气；氡的同位素②发射；放射
emanation chamber　射气箱
emanation electrometer　放射性静电计
emanation electroscope　放射性验电器
emanation test　氡试验；射气试验
emanation thermal analysis　放射热分析；射气热分析
emanium(=emanon)　射气
emanon(=emanium)　射气
Embden ester　恩布登酯
Embden-Meyerhof-Parnas pathway(EMP)　EMP 途径；糖酵解途径
embedded coordinate　嵌入坐标
embedded ion　嵌入离子；包埋离子
embedded temperature detector　嵌入式检温计；埋置温度计
embedding　埋置*；包埋
embedding compound　镶嵌化合物
embedding medium　包埋介质；埋封介质
embedment　包埋；浇铸封闭
embeliate　恩贝酸盐
embelic acid(=embelin)　恩贝酸；恩贝灵；3-十一烷基-2,5-二羟基-1,4-苯醌
embelin(=embelic acid)　恩贝酸；恩贝灵；3-十一烷基-2,5-二羟基-1,4-苯醌
emblem oil　巴林油〔石油〕
emblic extract　余甘栲胶
embolite　氯溴银矿
emboss　①浮雕②压花；压纹
embossed　①浮雕的②压纹的

embossed bonding　轧纹法
embossed cigarette tissue　绸纹香烟纸
embossed coating surface　凹凸花纹涂覆表面
embossed colo(u)red lacquer　堆彩漆器
embossed finish　压花涂饰剂
embossed glass　浮雕玻璃
embossed leather　压花革
embossed paper　绸纹纸；轧花纸
embosser　压纹机；压花机
embossing　压花；压纹；压型
embossing calendar　压纹机；压花机
embossing cylinder　压花辊筒
embossing die　压花模；印模
embossing machine　压花机；压纹机
embossing pad　压花垫
embossing roller　压花辊；压纹辊
embossment　浮雕；凸雕
embritite　硫锑铅矿
embrittlement　变脆；发脆；脆化
embrittlement point(=embrittlement temperature)　脆变点；脆变温度
embrittlement temperature(=embrittlement point)　脆变点；脆变温度
embrittlement time　脆变时间
embryo oil　胚芽油
embryos　晶胚
emerald　①绿宝石；翡翠；纯绿柱石；祖母绿〔矿〕②翡翠绿色
emerald green　翡翠绿；巴黎绿
emeraldine　翠绿亚胺〔苯胺黑氧化染色的中间产物〕
emeraldite　绿辉石
emerald oxide of chromium　氧化铬翠绿〔水合氧化铬绿〕
emerald shiner　绿宝石；艳绿色钻石
emergency break-off　紧急停车装置
emergency construction　防险建筑物
emergency control　紧急控制
emergency cooling　急冷；骤冷
emergency generator　备用发电机
emergency material　应急(用)备料
emergency pipe line　紧备备用输送管
emergency pump　紧急备用泵
emergency repair　紧急修理
emergency system　应急系统；事故系统
emergency valve　事故阀
emergency venting　紧急排放
emergent stem correction　水银柱露出部分的校正
emerin　伊默菌素

emerizing　起绒(工艺)
emersion　浮出
Emersol process　爱默索法〔一种精制脂肪酸方法〕
Emerson beater　爱默生打浆机
Emerson calorimeter　爱默生量热器
emery　金刚砂；刚砂；刚玉粉
emery board　刚砂板
emery cake　刚砂块
emery cloth　(刚)砂布
emery fillet　金刚砂布带
emery paper　(刚)砂纸
emery roller　金刚砂磨辊
emery stick(=emery stone)　刚砂石；油石
emery stone(=emery stick)　刚砂石；油石
emery wheel　①金刚砂轮②砂轮机
emetamine　吐根胺
emetic　催吐药
emetic war gas　呕吐性军用毒气
emetine　吐根碱
emetine　依米丁〔药〕
emetine hydrochloride　盐酸吐根碱；盐酸依米丁
emicin　厄迷僧；榄香素
Emila rotary viscometer　埃米拉旋转黏度计
emimycin　放霉素
emiratin　安镰孢菌素
EMI shielding coating　电磁干扰屏蔽涂料；防电磁干扰涂料
emission　发射
emission cathodes　发射阴极
emission detector　发射检测器
emission efficiency　发射效率
emission electron microscope　发射式电子显微镜
emission frequency　发射频率
emission interference　发射干扰*
emission line　发射谱线
emission line spectrum　发射线光谱
emission monitoring　排放监视
emission monochromator　发射光单色器；荧光单色器
emission of light　光线的发射
emission regulation　发射调节
emission spectra　发射光谱
emission spectral analysis　发射光谱分析
emission spectral analysis of isotopes　同位素发射光谱分析
emission spectrochemical analysis　发射光谱(化学)分析
emission spectrometer　发射光谱分析仪；发射分光仪
emission spectrometric analysis　发射光谱分析*
emission spectroscope　发射分光镜

emission spectrum 发射光谱
emission wavelength 发射波长
emission X-ray analysis 发射 X 射线分析
emissive power(=emittance) 发射本领
emissivity 发射率；辐射系数
emittance(=emissive power) 发射本领
emitted energy 发射能量
emitter 发射体
γ-emitter γ(射线)辐射体；γ(射线)辐射源
emitter follower 发射极输出器；发射极跟随器
emitting 发射
β-emitting 辐射β粒子的
emitting electrode 发射电极
emitting surface 发射面
emmenagogue 通经药
emodic acid 大黄酸；1,8-二羟基-3-羧基蒽醌
emodin 大黄素
emoline oil 艾摩林油
emollescence 软化作用
emolliate 软化
emollient ①软化剂②软化的③润肤剂④润滑药
emphysema 气肿
empire ①电绝缘漆②绝缘③支配
empire cloth 油基绝缘漆布
Empire powder 恩派尔炸药
empirical adhesion measurement 实验(经验)附着力测定法
empirical coefficient 经验系数
empirical constant 经验常数
empirical curve 经验曲线
empirical distribution 经验分布
empirical equation 经验方程(式)
empirical formula ①实验式；经验式；成分式②经验公式
empirical law 经验(定)律
empirically determined constant 经验定义常数；经验测定常数
empirical method 实验方法
empirical parameter 经验参数
empirical rule 经验规则
emplaster 灰膏
emplastrum 硬膏剂
emplectite 硫铜铋矿
empressite 碲银矿
empty column 空柱
empty dyeing 上色浅淡
emptying fittings 排空的零件
empty space 真正空间
empty tower 空塔
empty tube column 空管柱
empyreumatic ①烧焦了的②焦臭的
emulgator ①乳化剂②乳化器
emulgent 乳化剂
emulsator 乳化器
emulsibility 乳化性
emulsible oil 乳化油
emulsic oil 乳浊状油〔切削液〕
emulsicool 乳浊状油〔切削液〕
emulsifiability 乳化性
emulsifiable 可以乳化的
emulsifiable concentrate(s) 可乳化的浓缩物；浓缩乳剂(俗)
emulsifiable oil 可乳化的油
emulsifiable paint 乳化漆
emulsifiable paste 乳膏
emulsifiable solution 乳油
emulsification 乳化作用
emulsification apparatus 乳化(试验)器
emulsification column 乳化塔
emulsification of lubricating oils 润滑油的乳化(作用)
emulsification stability 乳化稳定性
emulsification test 乳化试验
emulsified ammonium nitrate 乳化硝酸铵〔燃料助燃剂〕
emulsified aqueous wax polish 水-蜡乳液光亮剂
emulsified asphalt 乳化沥青
emulsified asphalt varnish 乳化沥青漆
emulsified bitumen 乳化沥青
emulsified crude oil 乳化原油
emulsified hydrocarbon fuel 乳化烃燃料
emulsified oil 乳化油
emulsified petroleum solvent 乳化石油溶剂
emulsified rubber 乳化橡胶
emulsifier(=emulsifying agent) 乳化剂
emulsifier free emulsion polymerization 无乳化剂乳液聚合
emulsifier-free latex 无乳化剂的胶乳
emulsifier of the oil-in-water type 水包油型乳化剂
emulsifying 乳化的
emulsifying ability 乳化力
emulsifying agent(=emulsifier) 乳化剂
emulsifying dispersivity 乳化分散性
emulsifying efficiency 乳化效率
emulsifying hydrocarbon 乳化烃
emulsifying mixer 乳化混合器
emulsifying power 乳化效率；乳化本领
emulsifying property 乳化性
emulsifying stability 乳化稳定性

emulsifying tower 乳化塔
emulsifying wax 乳化蜡
emulsin 苦杏仁酶
emulsion 乳状液*
emulsion adhesive 乳化黏合剂；乳胶黏合剂
emulsion adjuvant 乳化助剂
emulsion asphalt ①乳化沥青②乳剂用沥青
emulsion band 乳化带；乳化区
emulsion-based decorative enamels 乳液型装饰漆
emulsion-based filler 乳液型填孔剂
emulsion binder ①乳液基料②乳液黏结剂
emulsion breaker ①破乳器②破乳剂
emulsion breaking 破乳；乳胶分解
emulsion calibration curve 乳剂校准(特性)曲线
emulsion chamber 乳胶室；乳化室
emulsion characteristic 乳剂特性
emulsion characteristic curve 乳剂特性曲线
emulsion cleaner(s) 乳液型清洗剂；乳化清洗剂
emulsion cleaning 乳液净洗
emulsion coatings 乳胶涂料；乳液涂料
emulsion colloid 乳胶体
emulsion copolymerization 乳液共聚
emulsion cosmetic 乳液化妆品
emulsion cutting oil 乳化切削油
emulsion degreasion 乳化脱脂
emulsion film 乳胶漆膜
emulsion flash spinning method 乳液闪蒸纺丝法
emulsion flash spinning process 乳液闪蒸纺丝法
emulsion flavour 乳液香精
emulsion fuel 乳化燃料
emulsion inhibitor 乳胶阻化剂；脱乳胶剂
emulsion inverse 乳液转相；乳液逆转
emulsion-like dispersion 乳胶状分散体；乳液状分散体
emulsion machine 乳化机
emulsion oil-in-water 水包油型乳液
emulsion opal glasses〔复〕乳光玻璃
emulsion paint 乳化漆
emulsion particle 乳胶(微)粒
emulsion polymer 乳液聚合物
emulsion polymerization 乳液聚合*
emulsion polymerization technique 乳液聚合工艺
emulsion polymerized butadiene styrene rubber(ESBR) 乳聚丁苯橡胶
emulsion (polymerized) rubber 乳液聚合橡胶；乳聚橡胶
emulsion process 乳化法；乳化过程
emulsion pump 乳胶输送泵
emulsion putty 乳液腻子
emulsion reactor 乳化反应釜；乳化反应器(罐)

emulsion resin 乳液法树脂
emulsion resistance 抗乳化
emulsion resolving 乳胶分解；脱乳化
emulsion sensitivity 乳胶灵敏度
emulsion sludge 乳胶淤渣沉淀〔齿轮箱内低温淀〕
emulsion spinning 乳液纺丝
emulsion splitter 破乳剂
emulsion stability 乳胶稳定度
emulsion stabilization 乳(状)液稳定作用
emulsion stabilizer 乳化稳定剂
emulsion strength 乳化强度
emulsion stripper ①乳胶漆脱漆剂②糊墙纸剥离剂
emulsion sulfiding process 乳液黄化法
emulsion texture 乳液状组织〔一种矿石组织〕
emulsion thickener 乳液增稠剂
emulsion treater 乳化处理器；破乳器
emulsion type 乳状液类型
emulsion-type alkyd 乳液型醇酸；醇酸树脂乳化液
emulsion (type) cleaner 乳液(型)洗净剂
emulsion-type cleaner-polish 乳液型去污抛光剂(膏)
emulsion-type degreasing agent 乳液型脱脂剂
emulsion-type membrane 乳化膜
emulsion viscosity 乳液黏度
emulsion xanthation 乳液黄化
emulsive 乳化的
emulsoid 乳胶(体)
emulsoid partical 乳胶(微)粒
emulsor ①乳化剂②乳化器
enalapril 依那普利〔药〕
enamel ①搪瓷；珐琅②瓷漆③釉质
enamel bead 搪瓷珠
enamel blue 搪瓷青
enamel clay 搪瓷土
enamel coating ①瓷漆②涂搪瓷；上釉
enamel color 搪瓷颜料
enamel dye 搪瓷颜料
enameled leather 漆皮
enamel furnace 搪瓷窑
enamel glass 釉彩玻璃
enamel holdout(=hold-out) 瓷漆不渗性
enamel kiln(=enamel furnace) 搪瓷窑
enamel knotting 木节封闭瓷漆
enamel lacquer (纤维素)瓷漆
enamel leather 漆皮
enamelled 搪瓷的
enamelled hard board 漆面硬质纤维板
enamelled paper(=coated paper; enamel paper) 铜版纸；印图纸；蜡图纸

enamelled tile 釉瓷瓦；琉璃瓦
enamel lined 搪瓷的
enamel(l)ing 搪瓷
enamel(l)ing furnace 搪瓷窑
enamel lining 搪瓷衬里
enamel paint 瓷漆
enamel painting 搪瓷画；搪瓷彩饰
enamel paper(=enamelled paper) 铜版纸；印瓷纸；蜡图纸
enamel sticky 亮油发黏
enamel thickness gage 瓷漆厚度仪
enamelum 釉质
enamel varnish 瓷漆用漆料；瓷漆料
enamel wash basin 搪瓷脸盆
enamel white 锌钡白
enamic form 烯胺式
enamine 烯胺
enanthal(=heptaldehyde; enanthic aldehyde) 庚醛
enanthal cyclic glycerol-1,3-acetal 庚醛环甘油-1,3-缩醛
enanthaldehyde 正庚醛
enanthal diethyl acetal 庚醛二乙缩醛
enanthal dimethyl acetal 庚醛二甲缩醛
enanthic acid(=heptanoic acid; heptylic acid) 庚酸 $CH_3(CH_2)_5CO_2H$
enanthic alcohol 庚醇
enanthic aldehyde(=enanthal; heptaldehyde) 庚醛
enanthine 庚炔
enanthol 庚醇
enanthotoxin 水芹毒素
enanthyl(=heptanoyl) 庚酰
enanthylic acid(=enanthic acid) 庚酸
enantioasymmetric polymerization 对映(体)不对称聚合
enantiomer 对映(异构)体
enantiomer selective chemical sensor 对映体选择性化学传感器
enantiomorph 对映体
enantiomorphism 对映形态
enantiomorphous 对映(结构)的
enantiomorphous crystal 对映结晶
enantioner 光学异构对映体
enantioselectivity 对映选择性
enantiosymmetric polymerization 对映(体)对称聚合
enantiotopic 对映异位(的)*
enantiotropic 互变性的；双变性的
enantiotropic change 对映性变化
enantiotropic liquid 互变液体
enantiotropy 对映现象
enargite 硫砷铜矿

encaline 恩卡菌素
encalyptol(=1,8-cineole) 白千层脑；桉叶油素
encapsulated 包胶的；用胶囊包起来的
encapsulated adhesive 胶囊型胶黏剂
encapsulated coating 包囊涂料；微胶囊涂料
encapsulated source 密封源；包封源
encapsulating 包胶；封入
encapsulating compound 电器外封胶
encapsulating polymer 包囊聚合物
encapsulating sample 胶囊试样
encapsulation ①密封；封装②包胶；封入胶内
encapsulation emulsion 胶囊乳剂
encapsulization 胶囊化；包膜
encased knot 镶嵌节；内含节
encasing ①包壳；装箱(盒)；包装②镶框；饰面③(刀、剑)入鞘
encaustic ①蜡彩画法②烧彩釉(的)
encaustic brick 琉璃砖；彩(釉)砖；釉面砖
encaustic painting ①蜡画②(上)釉烧(的)画
encaustic tile 烧彩瓦；琉璃丸
enclosed 被围的；封闭了的
enclosed dynamic screening (砂磨机)密闭式动态过滤
enclosed impeller 闭式叶轮
enclosed jigger 封闭式卷染机
enclosed knot 隐节
enclosed member 闭合构件〔箱、槽形的壳体〕
enclosed phase 被围相；封闭相
enclosed scale thermometer 内标温度计
enclosed single-shell drier 封口单壳干燥器
enclosed small media mill 密闭式细介质磨；密闭砂磨机
enclosed source 封闭式(离子)源
enclosed strip electrophoresis 闭条电泳
encloser 外壳；罩壳
enclosing cover (叶轮)轮盖；紧密盖
encode 编码
encounter complex 遭遇络合物
encrimson 红色涂料
encrust 包的皮壳；结壳
end ①终点；尾端②终了③结局
end absorption 末端吸收
end assay method 终点测定法
end-block 端基封闭〔高分子〕
end-block copolymer 封端共聚物；端基封闭共聚物
end-blocked 端基封闭的；端接的
end boiling point 终沸点
end boiling point analyzer 终馏点分析仪
end bond 端键
end cap 管端盖板；管子堵头

end-cap molecule 封端分子
end capped 封端的
end capping 封端(反应)*
end-capping reagent 封端剂；端接剂
end-caps 封端化合物；端接化合物
end cell 末端电池；补充电池
end-centered orthorhombic 底心正交
end check 端轮裂；端环裂
end closure(=end cover) 端盖
end-column 端柱
end correction 末端改正；终端校正
end count 单丝支数；经密
end cover(=end closure) 端盖
end cuts 尾馏分；最后的馏分
end-delivery head 直通(压出)机头
end-down of heart yarn 芯线断头
endecenoic acid 十一碳烯酸
end effect 末端效应
endellite 多水埃洛石；多水高岭土
endemic goiter 地方性甲状腺肿
endergonic reaction 吸能反应
end face seal 端面密封
end-fed 尾端进料；端面进刀
end force 端面压力
end gas 终结气体；尾气
end gas reaction 终结气体反应
end grain block 横纹木块
end group 端基
end group analysis 端基分析
end group concentration 端基浓度
end group determination 端基测定
end group distribution 端基分布
end group measurement 端基测定(法)
end group protected oligoamide 端基保护(的)低聚酰胺
end group titration 端基滴定
endiaron 恩台阿咙〔药〕；二氢氯喹〔通用名〕
ending mark 结头不良
end-initiation 末端引发
endiometer 气体容量分析管
end isomer 末端异构体
end item 成品
end joint 端拼
endless abrasive belt 环形砂带
endless belt 环状带；循环皮带
endless belt buffing paper 环形磨革砂纸
endless belt-puller 循环抽丝带；循环带式拉丝器
endless-belt splice 环形带式接头
endless blanket 环形衬布

endless chain 环形链
endless filter belt 循环过滤带
endless oil groove 环状油槽
endless screw 蜗杆
endless vulcanization 连续硫化
end liner 底衬
end link 端键
end mill head 立铣头
end missing 断经
end monomerization 链端单体化
end motion 端极运动；轴向运动
end movement 端极运动
endo- ①内②桥〔环内桥接〕
endo addition 内型加成
endocamphene 内莰烯
endocarp 内果皮
endo-compound 桥(环)化合物
endo-configuration 内向构型
endocrine function test 内分泌功能试验
endocyclic 桥环的
endocyclic bond 桥环键
endocyclic compound 桥环化合物
endocyclic double bond 桥环双键
endocytosis 胞饮；内吞；内摄
endoelectrical reaction 收电反应
endoergic 吸热的；吸能的
endoergicity 获能度
endoergic reaction (核反应中的)收能反应；吸能反应
endo-exo configuration 内(向)-外(向)构型
endo-exo isomerization 内(向)-外(向)异构化
endogenous inhibitor 内源性抑制剂
endo isomer 内型异构体
endokrocin 黄肾盘衣素
endolysin 内溶菌素；内溶素
endo-methylene group 内亚甲基
3,6-endomethylene-1,2,3,6-tetrahydro phthalic anhydride
 (=carbic anhydride) 3,6-内亚甲基-1,2,3,6-四氢邻苯二甲酸酐；降冰片烯二酸酐
endomycin 涂霉素
endomycopsis 拟内孢霉
end-on complex 端连配合物
end-on coordination 端向配位
end opening 下口
endo receptor 内向型受体
end organ 灵敏元件；传感器
endorphin 内啡肽
endosmometer 电渗计
endosmosis 内渗

endotherm 吸热；吸热量
endothermal 吸热的
endothermal change 吸热变化
endothermal decomposition 吸热分解
endothermal reaction 吸热反应
endotherm area 吸热面积
endothermic 吸热
endothermic(al) reaction 吸热反应
endothermic catalytic reaction 吸热催化反应
endothermic change 吸热变化
endothermic compound 吸热化合物
endothermic conversion 吸热转化
endothermic decomposition process 吸热分解过程
endothermic degradation reaction 吸热降解反应
endothermic dehydration 吸热脱水
endothermicity 吸热性
endotherm(ic) peak 吸热峰
endothermic reaction 吸热反应
endothermic transition 吸热转变；吸热转换
endotoxoid 类内毒素
end out (纺织)断经；缺经
end-over-end drum tumbler 竖转式鼓式拌和机
end-over-end winding 极向卷绕
endoxan 环磷酰胺
endoxy- 桥氧
end peak 尾峰
end place 盖板；端板
end plate 封头；端盖
end plate current(EPC) 终板电流
end play 轴端余隙；轴向间隙
end point ①终点②终馏点
end point correction 终点校正
end point error 终点误差*
end point gasoline 终点汽油
end point indicator 终点指示器；终点指示剂
end point of distillation 终馏点
end point of fraction 终点馏分
end point yield curve 终馏点收率曲线
end-product ①最终产物②最终结果
end properties 成品性能
end reaction 终反应；尾反应
end-reactive polymer 末端活性聚合物
end reinforcement 端部加强层
end requirement (成品)使用要求
end-result 最后结果
end ring 端环；机头护环
endrop 果树喷洒油〔一种石油产品〕
ends ①(纱)线②经纱；经线〔帘布〕

end shake 端轮裂；端环裂
ends in warp 总经根数
end split 端劈裂
end stopper (of chain) (链的)终止剂
end stopping (of chain) (链的)终止
end structure (链)端结构
end temperature 最后温度
end terminal 端线夹；极靴
end thrust 轴端推力
end-to-end ①端基对端基；头-尾结合②从一端到另一端③头对头；头顶头(连接)
end-to-end binding 首尾结合；端基对端基
end-to-end distance 末端距*
end-to-end distance of polymer chain 高分子链的末端距
end-to-end vector 末端间矢量
end unit 链端单元；端基
enduracidin 持久杀菌素
endurance 耐久；耐劳
endurance crack 疲劳裂缝
endurance failure 疲劳破坏
endurance life 耐久寿命
endurance life test 持久性试验；耐久试验
endurance limit(=fatigue limit) 疲劳极限
endurance strength(=endurance limit) 耐久强度
endurance test(=endurance trial) 耐久试验
endurance trial(=endurance test) 耐久试验
end-use 产品用途；最终用途
end-use allocation 最终利用分配
end-use performance 使用效能
end-use property 使用性能
end user 产品使用者
end-use temperature (成品)使用温度
end-use test 使用期试验
end view 端视图
end-window X-ray tube 端窗式 X 射线管
endwise skew 端部向上弯曲
endy 裂断丝绞
-ene (词尾) 烯
-enediol (词尾) 烯二醇 $RC(OH)=C(OH)R$
enema 开塞露〔药〕
ene reaction 烯反应
energetic atom 高能原子*
energetic gamma emitter 高能γ发射体
energetic particle 高能粒子
energetic plasma 高能等离子体
energetic ray 高能射线
energetics ①唯能论②能力学
energizing ①激发②供能③通电④施力

energon 能子
energy 能*
energy absorber coating 电磁能吸收涂层；反雷达涂层
energy-absorbing resin mat 吸能树脂毡片
energy absorption 能量吸收
energy analyzing microscope 能量分析显微镜
energy analyzer 能量分析器
energy balance 能量衡算；能量平衡
energy balance in flow 流动能量平衡
energy band 能带
energy barrier 能障；能垒
energy chain 能链
energy conservation 能量守恒
energy consumption 能量损耗*
energy content(=internal energy) 能含量；内能
energy conversion device 换能器
energy conversion polymer 能量转换高分子
energy crisis 能源危机
energy criterion for stability 稳定性的能量判据
energy datum 能量数据
energy deficiency 能量缺乏
energy diagram 能级图
energy dilemma 能源危机
energy dispersion 能量色散
energy-dispersion X-ray fluorescence spectrometer 能量色散 X 射线荧光光谱
energy dispersion X-ray fluorescence spectrometry 能量色散 X 射线荧光光谱法
energy dispersive detector 能量色散探测器
energy dispersive X-ray analysis(EDX) 能量分散 X 射线分析
energy dispersive X-ray fluorescence(EDXRF) 能量分散 X 射线荧光(分析)
energy dispersive X-ray fluorescence spectrometry 能量分散式 X 射线荧光光谱法
energy dispersive X-ray spectrometer 能量色散 X 射线光谱仪
energy dissipation 能量耗散
energy distribution 能量分布
energy distribution curve 能量分布曲线
energy diverged X-ray fluorescence spectrometry 能量色散 X 射线荧光光谱法
energy divergence X-ray diffractometer 能量色散 X 射线衍射仪
energy effect 能量效应
energy efficiency 能量效率
energy elastic eigenstress 能量弹性特征应力
energy elasticity 能量弹性

energy equation 能量方程
energy equivalent 能量当量
energy filter 能量过滤器
energy fluence 能注量；能量流量
energy flux density 能通量密度
energy focusing 能量聚焦
energy gap ①能量曲线中断处②禁(能)带宽度；能量范围③能域；能级距离
energy gradient 能量梯度
energy input ①能量输入②能耗量
energy irradiation 能量辐照
energy level 能级
energy level diagram 能级图
energy loss 能量损耗
energy loss spectrometer 能量损耗光谱仪
energy of absolute zero 绝对零点能
energy of activation 活化能
energy of adhesion 黏合能
energy of desorption 解吸能
energy of flow 流动能
energy of retraction 弹性复原能
energy of rupture 破坏能
energy of solvation 溶剂化能
energy of wetting 润湿能
energy output 能量输出；输出能量
energy-poor bond 低能键
energy poor phosphate bond 低能磷酸键
energy profile 能线图
energy quantum 能量子
energy randomization 能量随机化
energy recovery 能量回收
energy relationship 能量关系
energy release 能的放出；能量释放
energy resilience 回弹能
energy resolution 能量分辨*
energy resolved mass spectrum 能量分辨质谱
energy-rich bond 高能键
energy-rich phosphate 高能磷酸化物
energy rich phosphate bond 高能磷酸键
energy saving 节能
energy sink ①吸能源；散能源；散能装置②潜能
energy source 能源
energy spectrum 能谱
energy spread 能量发散
energy spread of the ion beam 离子束的能量发散
energy state 能态
energy storage capacity 能(的)储量
energy-to-break 断裂能

energy to fracture 破裂能
energy transfer 能量传递*
energy transfer chemiluminescence 能量转移化学发光
energy-transfer model 能转移模型
energy unit 能量单位
energy utilization patterns 能源利用方式
energy without pollution(=pollution-free energy) 无污染能源
ene synthesis 单烯合成
eneyne 烯炔
enfleurage 冷吸
enfleurage absolute 冷吸净油
enfleurage process 冷脂吸法
enflurane 恩氟烷〔药〕
engagement 接合
engine 发动机；引擎
engine cleansing agent 发动机洗涤剂
engine conditioning oil 发动机清除油
engine distillate 发动机用(石脑)油馏出物
engine dyeing 机染
engineered safety system 工程安全系统
engineered storage 工程加固储存
engineering 工程(学)
engineering chemistry 工程化学
engineering compounded plastics 工程复合塑料
engineering cost 工程成本
engineering design 工程设计
engineering economics 工程经济学
engineering feasibility 技术(上的)可能性
engineering flow sheet 工艺流程图
engineering material 工程材料
engineering plastic 工程塑料
engineering properties 工程性质
engineering stress 工程应力
engineering thermodynamics 工程热力学
engineering thermoplastics 工程热塑性塑料
engineering unit of viscosity 黏度的工程单位
engineering unit system 工程单位制
engine fuel 发动机燃料
engine kerosene 发动机煤油
engine loading 发动机负载
engine lubricating oil 发动机(润滑)油
engine oil 机油；发动机(润滑)油
engine room 机房
engine sizing 机器上胶
engine solar oil 发动机瓦斯油；索拉油(太阳油)
engine testing of lubricating oils 润滑油的发动机试验
Engler curve 恩氏(蒸馏)曲线

Engler degree(=Engler number; Engler unit) 恩氏(黏)度数
Engler distillate of gasoline (汽油的)恩氏(蒸馏)馏分
Engler distillation 恩氏蒸馏
Engler-distilled 恩氏蒸馏的
Engler distilling flask 恩氏蒸馏瓶
Engler flask 恩氏长颈瓶
Engler number(=Engler degree) 恩氏(黏)度数
Engler orifice viscometer 恩氏锐孔黏度计
Engler second 恩式秒
Engler unit(=Engler degree) 恩氏(黏)度数
Engler viscosimeter 恩氏黏度计
Engler viscosity 恩氏黏度
English crystal 铅玻璃
English crystal glass 全铅晶质玻璃
English elm 大叶榆
English gallon 英制加仑〔＝4.546 升〕
English heating test 润滑脂加热试验
English melting point 英国熔点
English powder 氯氧化锑
English red 含硫酸钙铁红
English system 英制
English white ①白垩②胡粉；蛤灰
engobe 釉底料
engobe coating 上釉
engobing 上釉底料
engrain 深染
engram 记忆印迹
engraved copper roller 刻花铜辊
engraved cylinder 刻花辊
engraved lacquer 雕漆〔大漆〕
engraved roll 花滚筒
engraved roll coating 槽辊涂装；槽辊涂布；凹印辊涂布
engraved scale thermometer 刻标温度计
engraved stem thermometer 刻标棒状温度计
engraving 雕刻；镂蚀
engraving machine 镌板机〔设〕
engraving wax 刻度用石蜡
enhanced 增强的
enhanced analytical agent 增效分析试剂
enhanced corrosion resistance 增强的耐腐蚀性；强化耐腐蚀性
enhanced sensitivity manostat 增高灵敏度的冲压器
enhancement 增强
enhancement effect 增感效应
enhancement effect of fluorescence 荧光增强效应
enhancement factor 增强因子
enhancer 增强器

enhydros 玉髓晶洞
enhydrous 含水的
enigmatite 三斜闪石
enimic form 烯亚胺式
enimization 烯亚胺化作用
enlargement ①扩大；放大；拓张②增补
enlargement factor 放大因数；扩增因数
enlargement in section 截面扩大（槽或管的）
enlargement loss 拓张损失
enniatin(=enniatine) 恩镰孢菌素
enodis 无节的
enolate 烯醇化物
enol ester 烯醇酯
enol ether 烯醇醚
enol form 烯醇式
enolic ester type 烯醇型酯 $R_2C=C(OH)R$
enolizability 烯醇化程度
enolization 烯醇化
enol lactone 烯醇内酯
enology 酿造学
enol phosphate 烯醇磷酸酯
enol phosphopyruvic acid 烯醇磷酸丙酮酸
enomycin 伊诺霉素
enone 烯酮
enoyl 烯酰(基)
enradin 恩拉鼎〔是一种多肽类抗生素预混剂，有效成分为恩拉霉素〕
enramycin 恩拉霉素
enriched flour 加料面粉
enriched gas 富煤气
enriched gas oil 浓缩粗柴油
enriched isotope 浓缩同位素
enriched metal stable isotope 浓缩金属稳定同位素
enriched mixture 浓缩混合物
enriched oil 浓缩油
enriched target 富集靶
enriched uranium(EU) 浓缩铀；加浓铀
enricher 浓缩器
enriching ①浓缩；富集②选集；选矿
enriching column 浓缩塔
enriching section （塔中）浓缩段；提浓段；精馏段
enriching stage 浓缩阶段
enrichment 浓缩；富集
enrichment by flotation 浮(沫)选(集)
enrichment factor 富集因子；浓缩因子
enrichment meter （同位素）浓缩度计
enrichment of mixture 混合物的浓缩
ensemble ①总体〔数〕②系综

ensemble current 总效应电流
ensemble effect 集团效应
ensemble noise 集合噪声分析法
ensile 保藏秣草
enstatite 顽辉石；顽火石
entagenic acid 内根酸；藤子酸
entangled thread 乱纱；乱丝
entanglement 缠结
entanglement crosslink 缠结交联点
entanglement network 缠结分子网
entanglement of polymer chain 高分子链的缠结
entanglemeter 纤维缠结测试仪
entangler 交络器；交络喷嘴
entangling degree of air-textured yarn 空气变形纱交缠度
entatic state 张力态
enteramine(=5-hydroxytryptamine; serotonine) 5-羟基色胺
enteric coating(=encapsulated coating) 包囊涂料；微囊涂料
entering end 入口
entering pin 穿心销
enterocin 肠道菌素；促肠活动素
enterococcin 肠道球菌素
enterococcus 粪链球菌；肠道球菌
enterocrinin 促肠液(激)素
enterogastrone 肠抑胃素
enteromycin 肠霉素
enteromycin carboxamide 肠霉素酰胺
enteropathogenic bacteria 肠道致病菌
enterotoxin 肠毒素
enterovirus 肠道病毒
enthallite(=euthalite) 密方沸石
enthalpic effect 热函效应
enthalpic elasticity 焓弹性
enthalpic method 热焓法
enthalpimetric 量焓的
enthalpimetric analysis 量热分析
enthalpimetric titration 量热滴定
enthalpimetry 测焓；热焓测量
enthalpogram 焓测定(曲线)图
enthalpy(=heat content;heat function) 热函；焓
enthalpy change 焓变；热函变化；热函改变
enthalpy-concentration diagram 焓浓度图；热函浓度图
enthalpy-controlled 焓控制的
enthalpy correction 修正焓
enthalpy-entropy compensation 焓-熵补偿
enthalpy function 焓函数
enthalpy of activation 活化焓
enthalpy of adsorption 吸附热函

enthalpy of humid air 湿空气的焓
enthalpy of melting 熔融热函；熔融焓
enthalpy of polymerization 聚合焓
enthalpy of relaxation 弛豫焓
enthalpy recovery factor 焓恢复系数
enthalpy titration(=thermometric titration) 热函滴定；温度滴定
entire split system 全分流系统
entity 本体
entrained catalyst （被气体)带走的催化剂
entrained oil 夹带的油
entrainer ①夹带剂②(共沸蒸馏中加入的)共沸剂
entraining agent 夹带剂
entrainment 夹带；输送
entrainment separator(=entrainment trap) 夹带物分离器
entrainment trap(=entrainment separator) 夹带物分离器
entrance 入口；进口
entrance correction （挤塑模头)流入校正
entrance effect （挤塑模头)流入效应
entrance effect in capillary flow 毛细管流动入口效应
entrance optics 入射光学系统
entrance pressure 进压
entrance pressure drop 入口压(力)降(低)
entrance slit 入射狭缝；入口狭缝
entrapment 截留〔一种表面改性方法〕
entrapped air 残存空气；存气
entrapped bubble 残存空气；存气；内部气泡
entrapped moisture 夹带水分；残留水分
entrapped slag 夹渣
entrapping 截留
entrapping method 包埋法
entromycin 内透霉素〔链霉素和杆菌肽及次甲基水杨酸盐的混合制剂〕
entropically elastic component 熵弹性分量
entropic configuration 熵构型
entropic deformation mechanism 熵形变机理
entropic effect 熵效应
entropic elasticity 熵弹性
entropic force 熵力
entropic interaction 熵相互作用
entropic stabilization 熵稳定性
entropic-stabilized dispersion 熵稳定分散
entropy 熵*
entropy balance 熵平衡
entropy change 熵变
entropy contribution 熵基值〔数〕
entropy effect 熵效应
entropy-elastic deformation 熵弹性形变

entropy-elasticity 熵弹性
entropy-elastic stress 熵弹性应力
entropy flow 熵流
entropy flux 熵流
entropy increase 熵的增加
entropy of absolute zero 绝对零点熵
entropy of activation 活化熵
entropy of dilution 稀释熵
entropy of disorientation 消向熵
entropy of dissolution 溶解熵
entropy of elasticity 弹性熵
entropy of evaporation 汽化熵
entropy of melting 熔融熵；熔化熵
entropy of mixing 混合熵
entropy of network formation 网络组成熵
entropy of superheating 过热熵
entropy parameter 熵参数
entropy production 熵增量
entropy source strength 熵源强度
entropy spring 熵跃
entry ①入口；进入②通路③表值
entry accumulator 进入储器(塔)
entry effect （挤塑)进口效应
entry roll 输入滚筒；喂入辊
entry-stand 喂入托架
entry zone 入口区
envelope ①外壳；封皮②包络线③封套(式)④被膜；胞质鞘
envelope conformation 信封(型)构象
envelope curve 包络线；包迹
envelopment 包封；包裹
envenomation （表面)毒化
environment 环境
environmental analytical chemistry 环境分析化学
environmental attack （遭)环境侵蚀(破坏)
environmental auditing 环境审计
environmental background 环境背景值
environmental background value 环境背景值
environmental barrier 环境壁垒
environmental cabinet 人造环境柜〔创造必要的温度、湿度、压力等环境条件的工作柜〕
environmental capacity 环境容量
environmental carcinogenesis 环境致癌(作用)
environmental cell 环境(模拟)室
environmental chamber 环境模拟箱
environmental chemistry 环境化学
environmental complex 环境复合体
environmental conservation 环境保护

environmental contamination 环境污染
environmental control system 环境控制系统
environmental corrosion (周围介质)接触腐蚀;环境腐蚀
environmental costs 环境成本
environmental cracking 环境诱发龟裂
environmental crisis 环境危机
environmental criteria 环境标准;环境准则
environmental destruction 环境破坏
environmental deterioration 环境恶化
environmental disruption 环境破坏
environmental economics 环境经济学
environmental effect 环境效应
environmental element 环境要素
environmental endocrine disrupting chemicals 环境内分泌干扰物
environmental endocrine disruptor 环境内分泌干扰(破坏)物
environmental error 环境误差
environmental evaluation 环境评价
environmental factor 环境因素;外界(条件)因素
environmental function zoning 环境功能区划
environmental health standard 环境卫生标准
environmental hormone 环境激素
environmental hygiene 环境卫生(学)
environmental impact assessment 环境影响评价
environmental impact statement 环境影响报告
environmental index 环境指数法
environmentalism 环境保护论;环境论
environmental labeling 环境标志
environmental management 环境管理
environmental medical monitoring 环境医学监测
environmental monitoring 环境监测
environmental monitoring of pesticide 农药环境监测
environmental mutagenesis 环境诱变
environmental planning 环境规划
environmental pollution 环境污染
environmental pollution chemistry 环境污染化学
environmental pollution monitoring 环境污染监测
environmental pollution standard analysis method 环境污染标准分析方法
environmental-polymer interface 环境-聚合物界面
environmental pressure 环境压力
environmental protection 环境保护
environmental purification 环境净化
environmental quality 环境质量
environmental quality assessment 环境质量评价;环境质量评估
environmental quality assessment index 环境质量评估(价)指数
environmental quality criteria 环境质量标准;环境质量准则
environmental quality indicator 环境质量指标
environmental quality pattern 环境质量模型
environmental quality standard 环境质量标准
environmental reference materials 环境标准物
environmental remote sensing 环境遥测;环境遥感
environmental resistance 耐环境性能〔如臭氧、氧、热、氧化性药品〕
environmental risk assessment 环境风险评估
environmental sample 环境样品
environmental scanning electron microscope 环境扫描电子显微镜
environmental self-purification 环境自净能力
environmental stability 环境稳定性
environmental standard 环境(保护)标准
environmental standard material 环境标准物质
environmental standard reference material 环境标准参考物质
environmental stress cracking 环境致裂
environmental subsidy 环境补贴
environmental surveillance 环境监视
environmental temperature 环境温度
environmental test facility 环境模拟试验设备
environment and environmental element 环境和环境要素
environment cracking 环境致裂
environment friendly textile 环保纺织品
environment hormone 环境激素
environmetrics 环境计量学
enyne 烯炔
enynic acid 烯炔酸
enzymatic 酶催的
enzymatic conversion ①酶糖化②酶转化
enzymatic diagnosis 酶法诊断
enzymatic hydrolysis 酶促水解
enzymatic inhibition method 酶抑制法
enzymatic laundry powder 加酶洗衣粉
enzymatic pathway 酶催化途径
enzymatic polymerization 酶聚合作用
enzymatic reaction 酶促反应
enzymatic thermistor 酶热敏电阻
enzyme activation 酶激活
enzyme activator 酶活化剂
enzyme activity ①酶活性②酶活度
enzyme activity unit 酶活性单位
enzyme allosteric effect 酶的别构效应

enzyme-catalytic kinetic spectrophotometry 酶催化动力学分光光度法
enzyme-containing detergent 加酶洗涤剂
enzyme detergent 加酶洗涤剂
enzyme electrode 酶电极
enzyme electrophoresis 酶电泳
enzyme imitative polymer 类酶高分子
enzyme immunoassay 酶免疫分析法；酶免疫测定
enzyme-immuno sensor 酶-免疫传感器
enzyme inhibition 酶抑制
enzyme inhibitor 酶抑制剂
enzyme labeling 酶标记
enzyme-labeled electrochemical immunoassay 酶标电化学免疫分析法
enzyme like polymer 模拟酶高分子
enzyme-linked fluorescence immunoassay 酶联荧光免疫分析
enzyme-linked immunosensor 酶联免疫传感器
enzyme-linked immunosensor of α-fetoprotein α-甲胎蛋白酶免疫传感器
enzyme-linked immunosorbent assay 酶联免疫吸附测定；酶联免疫分析
enzyme multiplied immunoassay technique 酶放大免疫分析技术
enzyme precursor 酶原
enzyme preparation 酶制剂
enzyme radiochemical assay 酶的放射化学测定法
enzyme receptor assay 酶受体分析
enzyme selective detector 酶选择性检测器
enzyme sensor 酶传感器
enzyme steeping 酶浸渍退浆
enzyme-substrate complex 酶-底物复合体
enzyme substrate electrode 酶底物电极；酶电极
enzyme system 酶系统
enzymic quantitative analysis 酶定量分析法
enzymic resolution ①酶法离析②酶法拆分
enzymic structure analysis 酶结构分析
enzymology 酶学
eolotropic(=aeolotropic) 各向异性的
eosin(=tetrabromofluorescein) 曙红；四溴荧光素 $C_{20}H_8O_5Br_4$
eosine 曙红；酸性曙红
eosine lake 曙红色淀
eosinophyll 叶曙红素
eosin sodium salt 曙红钠(盐) $C_{20}H_6O_5Br_4Na_2$
EP additive(=extreme pressure additive) (耐)特压添加剂；极压添加剂
epandrosterone 表雄(甾)酮

eperisone 乙哌立松〔药〕
ephedine 麻黄定
ephedrine 麻黄碱〔药〕；麻黄素
epi-〔词头〕 ①表；外②表示萘的16-位③指糖类α-位置上的立体异构④表示桥式结构
epiberberine 表小檗碱
epibolic stress 松弛消失应力
epiborneol 表冰片；3-莰烷醇
epiboulangerite 块硫锑铅矿
epicamphor 表樟脑；3-氧代莰烷 $C_{10}H_{16}O$
epicatechin 表儿茶素
epichitosamine 甘露糖胺；类葡萄糖胺
epichlorite 次绿泥石
epichlorohydrin 表氯醇〔俗〕；3-氯-1,2-环氧丙烷；氯甲代氧丙环 OCH_2CHCH_2Cl
epichlorohydrin ethyleneoxide rubber 共聚氯醇乙烯化氧橡胶
epichloro hydrin rubber 表氯醇橡胶；均聚氯醇橡胶
epicholestanol 表胆甾烷醇；表二氢胆甾醇
epicholesterol 表胆甾醇
epicoprosterol 表粪甾醇
epicurerenone 表蓬莪术酮
epicyanohydrin(=cyanopropylene oxide) 表氰醇〔俗〕；3,4-环氧丁腈；氰甲代氧丙环 OCH_2CHCH_2CN
epicycle reduction gear train 套接式周圆减速齿轮
epicyclic screw mixer 周转螺杆式混合机
epicyclic train 行星齿轮系；周转轮系
epidermin 表皮纤维
epididymite 板晶石
epidihydrocholesterol 表二氢胆甾醇；表胆甾烷醇
epidioxy-〔词头〕 环二氧；桥二氧 —OO—
epidithio-〔词头〕 环二硫；桥二硫 —SS—
epidote 绿帘石
epidotization 绿帘石化作用
epiethylin 表乙灵；二乙氧甲烷 $C_2H_5OCH_2OC_2H_5$
epigallocatechin 表儿茶素
epigenesis 后生说
epigenetic deposit 后成矿床
epigenite 砷硫铜铁矿
epihalohydrin 表卤代醇
epihydric acid 2,3-环氧丙酸
epihydric alcohol(=epihydrin alcohol) 缩水甘油
epihydrin 1,2-环氧丙烷
epihydrin alcohol(=epihydric alcohol) 缩水甘油
epihydrin carboxylic acid 缩水甘油甲酸

epihydrinic acid 缩水甘油酸 $CH_2CHCOOH\ \backslash O /$

epiindolinedione 表吲哚啉二酮
epiiodohydrin 表碘醇；3-碘-1,2-环氧丙烷 OCH_2CHCH_2I

epilaccishelbolic acid 表紫胶壳脑酸
epilaccishellolic acid 表紫草茸虫胶酸〔为紫草茸的提取物之一〕
epilaksholic acid 表壳脑醇酸
epilamens 油膜的表面活性
epilating agent 脱毛剂
epillumination 落射光
epimidin 淫羊藿定
epimino 环亚胺；桥亚胺 —NH—
epinephrine(=adrenine) 肾上腺素
epinine 麻黄宁〔合成的麻黄素代用药〕；N-甲基-3,4-二羟基苯乙胺 $(HO)_2C_6H_3CH_2CH_2NHCH_3$
epioxybiotin 表氧生物素
epipastic 粉状
epi-phase 表相
epiphasic 表相性的〔在用分液漏斗提取时，趋向存在于表层液体中的性质〕
epiphenylin 2,3-环氧-1-苯氧基丙烷
epipolic 荧光(性)的
epi-position 表位
epiquinidine 表奎尼定
epiquinine 表奎宁
epirubicin 表柔比星〔药〕
episcotister 光栅；截光盘；斩光盘
epishellolic acid 表紫胶酸；表虫胶酸
episome 附加体；游离体
epispastic ①发疱的②疱烂剂
epistasis(=epistasy) ①液膜；上位②强性
epistephamiersine 表千金藤默星碱
epistephanine 表千金碱
episteroid 表(式)甾族化合物
epistilbite 柱沸石
episulfide 环硫化物
epitaxial 外延的；(晶体)取向生长(附生)的
epitaxial crystallization 外延结晶；附生结晶
epitaxial growth 外延晶体生长；附生晶体生长
epitaxial growth reaction 外延生长反应
epitaxy 外延；取向生长〔晶体〕
epitaxy mechanism 取向机制
epitestosterone 表睾(甾)酮
epitetracyclin 差向四环素

epithermal 超热的
epithermal capture 超热捕获；超热俘获
epithermal neutron activation analysis 超热中子活化分析
epithio- 〔词头〕 环硫；表硫；桥硫 —S—
epithioamide 环硫酰胺
epithioethyl benzene 环硫乙基苯
EP lubricant (耐)特压润滑剂；极压添加剂
epoformin 环氧型菌素
epon ester 环氧酯
epoxidase 环氧酶
epoxidation(=oxirane formation) 环氧化作用
epoxide 环氧化物
epoxide content 环氧化物含量
epoxide equivalent 环氧当量
epoxide group 环氧基
epoxide number 环氧值
epoxide resin 环氧树脂
epoxide ring 环氧环
epoxide stabilizer 环氧化物稳定剂
epoxidized 环氧化的
epoxidized butyl ester 环氧化丁酯
epoxidized cyclic acetal 环氧化环状缩醛
epoxidized linseed oil 环氧化亚麻仁油
epoxidized plasticizer 环氧化增塑剂〔指环氧化不饱和脂肪酸〕
epoxidized soybean oil 环氧化豆油
epoxidizing agent 环氧化剂
epoxy 环氧；表氧；桥氧基 —O—
epoxy resin 环氧树脂
epoxy acrylate 环氧丙烯酸酯
epoxy adduct 环氧加合物
epoxy adhesive 环氧树脂胶黏剂
1,2-epoxyalkane 1,2-环氧烷烃
epoxy alkane 环氧烷烃
epoxy alkyd resin 环氧(改性)醇酸树脂
epoxy alkyd varnishes 环氧醇酸清漆
epoxyalkylsilane 环氧烷基硅烷
epoxy-amine resin 胺固化环氧树脂
8,9-epoxycedrane 8,9-环氧柏木烷
epoxy chloropropane 环氧氯丙烷
epoxy coating 环氧(树脂)涂料
epoxy compound 环氧化合物
epoxy cresol novalac resin(ECN) 环氧甲酚醛树脂；环氧酚醛树脂
epoxy cross-linking chemistry 环氧交联化学
epoxy curing agent 环氧交联剂
β-(3,4-epoxycyclohexyl) ethyl trimethoxysilane β-(3,4-环

氧环己基)乙基三甲氧基硅烷
epoxy derivative　环氧衍生物
epoxydon　顶环氧菌素
epoxy equivalent (weight)　环氧当量
epoxy ester　环氧酯
epoxy ester undercoat　①环氧酯底漆②环氧酯中间涂层
epoxy ethane　环氧乙烷
epoxy foam　环氧(树脂)泡沫塑料
epoxy glass substrate　环氧玻璃基板
epoxy glyceride　环氧甘油酯
epoxy glycerite　环氧甘油〔润滑剂〕
epoxy group(ing)　环氧基；表氧基〔环氧基易和杂环命名相混；故用表氧基较好〕
epoxy isoimperatorin　环氧异欧前胡素
epoxylinalool(=linalool monoxide)　环氧芳樟醇
epoxylinalyl acetate　乙酸环氧芳樟酯
2,6-epoxy-6-methyl heptane-2,5-dicarboxylic acid　2,6-环氧-6-甲基庚烷-2,5-二羧酸，桉树脑酸
epoxy-modified lacquer　环氧改性清漆
epoxy-modified novolak filament　环氧改性的酚醛树脂长丝
epoxy molding material　环氧树脂模塑(材)料
epoxy monomer　环氧单体
epoxy moulding powder　环氧树脂模型粉
epoxyn　环氧树脂类黏合剂
epoxy novolac　环氧酚醛
epoxy novolac adhesive　线型酚醛环氧黏合剂
epoxy-novolac resin　环氧-线型(热塑性)酚醛树脂
epoxy number　环氧值〔每100克环氧化合物中所含环氧基的克当量值〕
epoxyoleic acid　环氧(化)油酸
epoxy oxygen　环氧基氧；环氧态氧
epoxy plasticizer　环氧增塑剂
epoxy plastics　环氧塑料
epoxy-polyamide blend　环氧聚酰胺混合物
epoxy-polyamide coatings　环氧-聚酰胺涂料
epoxy-polyamide resin　环氧-聚酰胺树脂
epoxy polymer　环氧聚合物
epoxy prepreg　环氧树脂预浸料
1,2-epoxy propane　1,2-环氧丙烷
1,3-epoxypropane　1,3-环氧丙烷
2,3-epoxy propanol　2,3-环氧丙醇；2,3-缩水甘油
epoxy propionic acid　环氧丙酸；缩水甘油酸
epoxypulegone　环氧胡薄荷酮
epoxy putty　环氧(树脂)腻子
epoxy resin (based) powder coatings　环氧树脂(系)粉末涂料
epoxy resin cure　环氧树脂固化
epoxy resin derivatization　环氧树脂的衍生作用
epoxy resin ester　环氧(树脂)酯
epoxy resin varnish　环氧(树脂)清漆
epoxy ring　环氧环
epoxysiloxane　环氧硅氧烷
epoxy stabilizer　环氧稳定剂
epoxy stearate　环氧硬脂酸酯
epoxystearic acid　9,10-环氧硬脂酸
epoxy substrate　环氧树脂底材(基底)
epoxythio　环氧硫；桥氧硫　—OS—
epoxy tooling　环氧模具
epoxy value　环氧值〔每100克环氧化合物中所含环氧基的克当量值〕
epoxy zinc rich primer　环氧富锌底漆
eppendorf micropipette　微量取液器；微量进样器
Epprecht viscometer　埃普雷切特黏度计〔一种圆筒式旋转黏度计〕
epristeride　爱普列特〔药〕
EPR spectrometer　顺磁共振(波谱)仪
epsilon-amino caproic acid　ε-氨基己酸
epsilon potential　ε势能
epsomite　泻盐矿
eptatretin　粘盲鳗素；盲鳗心血管素〔从粘盲鳗的鳃、心、肝和肌肉内分离所得的一种芳胺类物质，具有显著刺激起搏点的强烈心脏兴奋作用〕；八目鳗鱼丁〔心脏跳动剂〕
EP test　(耐)特压试验；极压试验
Epton (two-phase) titration　埃普顿(两相)滴定法
epuration　提纯；精炼
equal-arm balance　等臂天平
equal cracking　等分裂化；对称裂化
equal energy spectrum　等能量光谱
equal energy synchronous fluorescence method　等能量同步荧光光谱法
equalization　①均衡法②均涂作用③均涂性④均染性⑤均匀比
equalized pressure　均衡压力
equalizer　均匀管；平衡管
equalizer rod　均绕线材涂漆机
equalizing　均衡；调整
equalizing abrasion　均匀磨耗
equalizing agent　匀染剂
equalizing effect　均衡效应；调整效应
equalizing line　均衡管；平衡管
equalizing tank　调节池；均衡槽
equally accurate measurement　等精度测量
equal pitch　等螺距
equal probability　等概率

equal settling velocities　等沉降速度
equal tension　(长丝缠绕)等张力
equal tension shed　等张力梭口
equal turbidity method　等浊滴定法
equal turbidity titration　等浊滴定
equal twill　双面斜纹
equal variance　等方差性；方差齐性；方差一致性
equal variance test　等方差检验；方差齐性检验；方差一致性检验
equant alpha form of copper phthalocyanine pigment　等径α形铜酞菁颜料
equation for approximating curve　近似曲线方程
equation of continuity　连续性方程
equation of filtration　过滤方程
equation of motion　运动方程
equation of polarographic wave　极谱波方程式
equation of state　物态方程式；状态方程
equation of state for tension　张力状态方程(式)
equator　赤道线
equatorial　平展(的)
equatorial arc like reflection　赤道弧状反射
equatorial bond　平(伏)键*
equatorial conformation　平伏构象
equatorial position　平伏位置
equatorial reflection　赤道反射
equi〔词头〕等；同等；均；平均
equiasymptotical stability　等度渐近稳定性
equiaxed　等轴的；各方等大的
equiaxed crystal　等轴晶体
equiaxial　等轴的
equiaxial grains　等轴粒子
equiaxial polygonal grain　等轴多面晶粒
equibinary polymer　等二元聚合物
equicohesive temperature　等内聚温度；等凝聚温度
equi-composition method　等成分法；等组成法
equi-composition standard　等成分标准；等组成标准
equidirectional　同方向的
equidistant　等距的
equienergy map　等能(量)图
equiflux heater　双面辐射式加热炉
equigranular texture　均匀粒状结构；等粒状组织
equi-interference method　等干涉法
equi-interference standard　等干涉标准
equil　平衡
equilenin　马萘雌(甾)酮　$C_{18}H_{18}O_2$
equilenin benzoate　苯甲酸马萘雌酮酯
　　　$C_{18}H_{17}OO_2CC_6H_5$
equilibrant　平衡力

equilibrated　平衡的；补偿的
equilibrated adsorption　平衡吸附
equilibration　①(展开剂蒸气)饱和②平衡化
equilibration of stationary phase　固定相平衡
equilibration system　平衡装置
equilibration tube　平衡管；补偿管
cis-trans equilibrium　顺-反式平衡
equilibrium　平衡
equilibrium adsorption model　平衡吸附模型
equilibrium air distillation　平衡空气蒸馏〔在排除空气的情况下；把燃料试样与其蒸气相接触进行蒸馏〕
equilibrium boiling point　平衡沸点
equilibrium box　平衡箱
equilibrium coefficient　平衡系数
equilibrium compliance　平衡柔量
equilibrium composition　平衡成分
equilibrium concentration　平衡浓度
equilibrium condensation curve　平衡冷凝曲线
equilibrium condition　平衡状态；平衡条件
equilibrium constant　平衡常数
equilibrium constants for isotope-exchange reactions　同位素交换反应的平衡常数
equilibrium contact angle　平衡接触角
equilibrium-controlled reaction　平衡控制反应
equilibrium conversion　平衡转化
equilibrium copolymerization　平衡共聚(作用)
equilibrium curve　平衡曲线
equilibrium data　平衡(状态时的)数据
equilibrium dew point　平衡露点
equilibrium diagram　平衡图(表)
equilibrium dialysis　平衡透析
equilibrium dialysis method　平衡渗析法
equilibrium distillation　平衡蒸馏
equilibrium distribution　平衡分布
equilibrium distribution of sizes　(分子)大小的平衡分布
equilibrium drainage　稳定泄油；均衡排泄
equilibrium electrode potential　平衡电极电位
equilibrium elution　平衡洗脱
equilibrium energy of wetting　润湿平衡能
equilibrium equation　平衡方程式
equilibrium establishment　建立平衡
equilibrium exchange current density　平衡交换电流密度
equilibrium exclusion theory　平衡排斥理论
equilibrium flash still　平衡闪蒸锅
equilibrium flash vaporization　平衡闪蒸；平衡急骤蒸发
equilibrium flash vaporizer　平衡闪蒸设备
equilibrium gradient separation method　平衡梯度分离法
equilibrium hygroscopicity　平衡吸湿性

equilibrium hypothesis　平衡假设
equilibrium law　平衡定律
equilibrium melting point　平衡熔点
equilibrium mixture　平衡混合物
equilibrium mode distribution length　稳态模式分布长度
equilibrium modulus　平衡模量
equilibrium moisture　平衡湿度
equilibrium moisture content　平衡湿量；平衡含水率
equilibrium moisture regain　平衡回潮率
equilibrium moisture regain value　平衡回潮(率)值
equilibrium oil　平衡油
equilibrium osmotic pressure　平衡渗透压
equilibrium phase　平衡相
equilibrium point　平衡点
equilibrium polycondensation　平衡缩聚(作用)
equilibrium polymer　均衡聚合物
equilibrium polymerization　平衡聚合
equilibrium potential　平衡电势
equilibrium pressure　平衡压力
equilibrium ratio　平衡比
equilibrium regain　平衡回潮率
equilibrium relationship　平衡关系
equilibrium relative humidity　平衡相对湿度
equilibrium response function　平衡响应函数
equilibrium saturation　平衡饱和度
equilibrium segregation coefficient　平衡分凝系数
equilibrium shear compliance　平衡剪切柔量
equilibrium shear modulus　平衡剪切模量
equilibrium shifting method　平衡移动法
equilibrium solubility　平衡溶解度
equilibrium stage　平衡级
equilibrium state　平衡态
equilibrium still　平衡蒸馏锅
equilibrium surface structure　平衡表面结构
equilibrium surface tension　平衡表面张力
equilibrium swelling　平衡溶胀
equilibrium system　平衡系统
equilibrium temperature　平衡温度
equilibrium theory　平衡理论
equilibrium time　平衡时间
equilibrium unit　平衡单元
equilibrium value　平衡常数
equilibrium vaporization　平衡蒸发
equilibrium water　平衡水
equilibrium water content　平衡水分
equilin　马烯雌(甾)酮　$C_{18}H_{20}O_2$
equimolal counter diffusion　等摩尔反方向扩散
equimolar　等克分子的；克分子数相等的

equimolar compound　等克分子化合物
equimolar response　等摩尔响应(值)
equimolar series method　等摩尔系列法
equimolecular　等分子的
equimolecular counter diffusion　等分子对向扩散
equimolecular quantity　等分子量
equimolecular surface　等分子表面
equine gonadotropin　雌马促性腺(激)素
equipartition　均分
equipartition of energy　能量均分
equipartition principle　均分原则；等分原则
equipment　设备；装备
equipment appurtenance　设备附件
equipment configuration　①设备装置②设备构造型式
equipment design　设备设计
equipment for rain water collection　采雨装置
equipment investment　设备投资
equipment layout　设备布置
equipotential　等电位的
equisetic acid　乌头酸
equisetine(=palustrine)　沼泽(木贼)碱
equisetonine　木贼宁
equisetum rubbing　木贼磨光；锉草打磨〔大漆〕
equisolvation point　等溶剂化点
equispaced　等距的
equispaced baffles　等间隔的挡板
equitactic polymer　全同体同等量聚合物*
equitransference　等传输
equitransferent filling solution　等传输充填液
equivalence(=equivalency)　①等价；化合价相等②相等③相当
equivalence factor　当量因子
equivalence point　等当点；等效点
equivalence potential　等当点电势
equivalence principle　等效原则；等效性原理
equivalency(=equivalence)　①等价；化合价相等②相等③相当
equivalency point　等当点；等效点
equivalent　①当量②当量的③相当的
equivalent chain　等效链
equivalent chain length　等效链长
equivalent concentration　当量浓度；规度；规定浓度
equivalent conductance　当量电导
equivalent conductivity　当量电导(率)
equivalent cure　等效硫化
equivalent DFT　同等通风(条件)
equivalent diameter　等量直径
equivalent efflux time　(黏度杯的)等效流出时间

equivalent electric conductivity   当量电导率
equivalent electron circuit   等效电子电路
equivalent fractional overall porosity   当量总孔隙率分数
π-equivalent heteroatoms   π等价杂原子
equivalent length   等量长度
equivalent loading   等量填充；等量配合；等量负荷；等量荷载
equivalent Munsell value notation differences   等价孟塞尔值差
equivalent number   当量值
equivalent octane blends   当量辛烷掺和物〔按抗爆性与标准燃料相当的〕
equivalent orbital(EO)   等价轨道
equivalent orbital method   等价轨道法
equivalent particle diameter   当量粒径
equivalent percent   当量百分率
equivalent point   当量点
equivalent point potential   等当点电位
equivalent position   等效点
equivalent reflection   等效反射
equivalent sound level   等效声级
equivalent sphere   等效球体
equivalent spherical diameter   (粒子)当量球体直径
equivalent steam setting temperature   等效蒸气定形温度
equivalent stiffness   等效劲度；等效刚度
equivalent strain   当量应变
equivalent strength   等效强度
equivalent stress   当量应力
equivalent thickness   等值厚度
equivalent volume   当量体积
equivalent vulcanization   等效硫化
equivalent weight   当量
equiviscous temperature   等黏温度
equivocation   可疑度
equi-wet-bulb-temperature line   湿球温度线
equus   马脂
eradicant   铲除剂
eradicator   去污剂
eraser   擦字橡皮
erasibility   耐擦性(能)
erasing knife   刮刀
erasing rubber   擦字橡皮
erbia   氧化铒 Er_2O_3
erbium   铒
erbium chloride   氯化铒 $ErCl_3$
erbium hydroxide   氢氧化铒 $Er(OH)_3$
erbium nitrate   硝酸铒 $Er(NO_3)_3$
erbium oxalate   草酸铒 $Er_2(C_2O_4)_3$

erbium oxide   氧化铒 Er_2O_3
erbium oxychloride   氯氧化铒 $ErOCl$
erbium sulfate   硫酸铒 $Er_2(SO_4)_3$
erdin   欧菌素
erection   装配
erection drawing   装配图
erector   毛囊筋
eremacausis   慢性氧化；缓慢氧化
eremeyevite   硼铝石
eremophiladienone   雅槛兰二烯酮
eremophilane   雅槛兰烷
eremophilatrienone   雅槛兰三烯酮
eremophilone   雅槛兰酮
eremy   金刚砂
erg   尔格〔能量单位，$=10^{-7}J$〕
ergastic substance   后含物
ergine   麦碱
ergobasine   麦角巴生
ergobasinine(=ergometrinine)   麦角异新碱
ergocalciferol   麦角钙化(甾)醇；维生素 D_2
ergochrysin   麦角黄素
ergoclavine   麦角棒碱
ergocornine   麦角考宁
ergocristine   麦角日亭
ergocristinine   麦角日亭宁
ergocryptine(=ergomolline)   麦角隐亭
ergocryptinine(=ergomollinine)   麦角隐宁
ergoflavin   麦角黄素
ergoline   麦角灵
ergometer   尔格计；尔格测量器
ergometric method of test   尔格(漂白土)试验法
ergometrine(=ergobasine; ergotocine; ergostretine; ergonovine)   麦角新碱
ergometrinine(=ergobasinine)   麦角异新碱
ergomolline(=ergocryptine)   麦角隐亭
ergomollinine(=ergocryptinine)   麦角隐宁
ergonovine   麦角诺文
ergopinacol   联麦角(甾)醇
ergosine   麦角僧
ergosinine   麦角僧宁
ergostane   麦角甾烷
ergostanol   麦角甾烷醇 $C_{28}H_{50}O$
ergostenol   麦角甾烯醇 $C_{28}H_{48}O$
ergosterin   麦角甾醇；麦角固醇
ergosterol   麦角甾醇 $C_{28}H_{44}O$
ergostetrine   麦角新碱
ergot   麦角
ergota   麦角

ergotamine 麦角胺
ergotamine caffeine 麦角胺咖啡因〔药〕
ergotaminine 麦角胺宁；麦角异胺
ergothioneine(=thiohistidine-betaine) 硫组氨酸甲基内盐；麦硫因
ergotic acid 麦角酸 $C_{15}H_{30}O_{15}(NH_2)SO_3H$
ergotine 麦角碱
ergotinine 麦角亭宁；麦角异毒碱
ergotocine 麦角新碱
ergotoxin 麦角毒素
ergotoxine 麦角毒
ericamycin 欧石南霉素
Erichsen testing machine 埃里克森试验机
ericinol 喇叭茶醇
ericolin 石南素
erie black 伊利黑；酸性黑
erigrisin(=erygrisin) 红灰菌素
erinite 铁蒙脱石；翠绿砷铜矿
erio alizarine blue 羊毛茜素蓝
erio alizarine violet 羊毛茜素紫
erio anthracene blue 羊毛蒽蓝
eriochrome black A 铬黑 A*
eriochrome blue-black 铬蓝黑
eriochrome blue black B 铬蓝黑 B
eriochrome blue black R 铬蓝黑 R
eriochrome cyanine R 铬花青 R*
eriochrome violet B 铬紫 B
eriodictin 圣草苷
eriodictyol 圣草酚；毛纲草酚
eriodictyon oil 圣草油
eriodin 毛纲草素
erio fast floxine 羊毛坚牢根皮红
erioglaucine A 罂红 A*
eriometer 纤维细度测定仪；衍射测微器
erionite 毛沸石
erizomycin 埃里兹霉素
erkensator 离心式浆粕净化机；立式离心除纱机
erlangen blue(=prussian blue) 铁蓝
Erlenmeyer flask 锥形瓶*
erodent 腐蚀剂
erometer 气体比重计
erosion 磨耗；侵蚀；烧蚀；消蚀；剥蚀
erosion control 侵蚀控制
erosion-control agent （土壤）防侵蚀剂；侵蚀控制剂
erosion-corrosion ①侵蚀腐蚀；磨(损)腐蚀；溃蚀②冲击(刷)腐蚀③紊流；腐蚀
erosion rate 侵蚀速率；腐蚀速率
erosion resistance 耐侵蚀性

erosive 磨耗的；侵蚀的
erosive wear 冲刷磨损；侵蚀损耗
erosone 豆薯酮 $C_{20}H_{16}O_6$
erratic 无定的
erratic baseline 不稳定基线；漂移基线
erratic cell structure 不均匀孔眼结构
erratic composition 不规则结构；不固定组成
erratic flow 涡流；扰动流
erratic relaxation shrinkage 弛豫(状态)异常收缩率；异常弛豫回缩率
erratic stitch pattern 花纹失真
errhine 引涕药；引嚏药
error allowance 容许误差
error analysis 误差分析
error compensator 误差补偿器
error correct feed back method 误差修正反馈法
error curve 误差曲线
error factor 误差因素
error function 误差函数
error propagation 误差传递
error sum of square 误差平方和
error variance 误差方差
eruciate 芥酸盐(或酯)
erucic acid(=cis-13-docosenoic acid;erucidic acid) 芥酸；顺(式)13-二十二(碳)烯酸 $CH_3(CH_2)_7CH=CH(CH_2)_{11}COOH$
erucic acid oil 芥酸油〔主要酸为芥酸的油〕
erucicamide 芥酸酰胺
erucidic acid(=erucic acid; cis-13-docosenoic acid) 芥酸；顺(式)13-二十二(碳)烯酸 $CH_3(CH_2)_7CH=CH(CH_2)_{11}COOH$
erucin 芥酸精；甘油三芥酸脂
erucylacetic acid 瓢儿菜基乙酸；二十四碳烯酸；神经酸 $CH_3(CH_2)_7CH=CH(CH_2)_{13}COOH$
erucyl alcohol 瓢儿菜醇
erycin 刺桐素
erygrisin(=erigrisin) 红灰菌素
erysimin 糖芥苷
erysipelas 丹毒
erysipeloid 类丹毒
erysocine 刺桐素
erysodine 刺桐定
erysonine 刺桐宁
erysopine 刺桐平
erysothiopine 刺桐硫平
erysothiovine 刺桐硫碱
erysovine 刺桐碱
erytauin 百金花苷

erythema 红斑
erythraline 刺桐灵
erythramine 刺桐胺 $C_{18}H_{21}O_3N$
erythratine 刺桐亭
erythrene 刺桐烯；丁间二烯 $CH_2=CHCH=CH_2$
erythric acid(=erythritic acid) 赤藓(糖)酸
$CH_2OH(CHOH)_2COOH$
erythricine(=gentianine) 龙胆碱；秦艽甲素
erythrin 赤藓素；地衣红(素)
erythrinan 刺桐烷
erythrine 刺桐碱
erythrinine 刺桐宁
erythriphileine(=erythrophloeine) 围涎树碱
erythrite(=erythritol) 赤藓醇
erythritic acid(=erythric acid) 赤藓(糖)酸
erythritol(=erythrite;1,2,3,4-butantetraol) 赤藓醇
erythritol laurate 月桂酸赤藓酯
erythritol oleate 油酸赤藓酯
erythritol palmitate 棕榈酸赤藓酯
erythritol tetranitrate 四硝酸赤藓酯 $C_4H_6(ONO_2)_4$
erythro- 赤
erythroaphin 蚜红素
erythrocin(=erythromycin) 红霉素
erythro complex compound 赤配合物
erythro configuration 赤型构型
erythrodextrin 红糊精
erythro-diisotactic 叠(同)双全同立构〔是指双全同立构中两个不对称中心是可以叠同的，即构型相同(通过 C—C 键的内旋转)〕
erythro-diisotactic polymer 赤型双全同立构聚合物
erythro-di-iso-*trans*-tactic 叠(同)双反式全同立构
erythrodiol 高根二醇
erythro-disyndiotactic 叠(同)双间同立构
erythro-disyndiotactic polymer 赤型双间同立构聚合物
erythro form 赤型；赤式
erythrogenic acid 生红酸；赤原酸；十八碳烯炔酸 $C_{18}H_{26}O_2$
erythroglaucin 毛罂红
erythroglucin 赤丁四醇
erythroidin 刺桐若定
erythroidine 刺桐定
erythro isomer 赤型异构体
erythrol 赤醇；3-丁烯-1,2-二醇
$CH_2=CHCH(OH)CH_2OH$
erythrolaccin 红紫胶素
erythrol tetranitrate 四硝酸赤藓酯 $C_4H_6(NO_3)_4$
erythromycin(=erythrocin) 红霉素〔药〕
erythromycin ethylsuccinate 琥乙红霉素〔药〕

erythronic acid 赤糖酸
erythronolactone 赤酮酸内酯
erythrophloeine(=erythriphileine) 围涎树碱
erythropoietic porphyria 生血性卟啉症
erythropoietic protoporphyria 生血性原卟啉症
erythropoietin 促红细胞生成素
erythro-polymer 赤型聚合物
erythropsin 视紫红(质)
erythroresinotannol 赤脂单宁醇
erythro salt 赤配位盐
erythrose 赤藓糖；丁糖；四碳糖
erythrosiderite 红钾铁盐
erythrosine(=tetraiodofluorescein) 赤藓红；四碘荧光素；新品酸性红〔染〕
erythrosine sodium salt 赤藓红钠盐 $C_{28}H_6O_5I_4Na_2$
erythroskyrine 红醌茜菌素
erythroxyanthraquinone 赤氧基蒽醌
erythroxyline 古柯碱
erythroxylon 古柯
erythrulose 赤藓酮糖
escape depth 逃逸深度
escape orifice 排泄口；放出口
escape peak 小峰〔X 射线分析〕；逃逸峰
escape pipe 排气管
escape route 逃逸通道
escaping coefficient 逃逸系数；逸出系数
escaping tendency 逃逸趋势
escaping vapor 逃逸蒸汽；逸出蒸汽
escharotic 苛性剂
Eschka mixture for sulfur determination 埃斯卡测硫混合剂
Eschka's method 埃斯卡混熔法
Eschka's mixture 埃斯卡混合剂
escholerine 厄邵(藜芦)碱
Eschweger soap(=mottled soap) 爱氏肥皂；斑纹皂；兰花皂〔俗〕
Eschweiler-Clarke reaction 埃谢韦勒-克莱克反应
escigenin(=aescigenin) 七叶配基
escin(=aescin) 七叶素
esciorcin(=escorcin) 七叶酚
escorcin(=esciorcin) 七叶酚
esculetin(=6,7-dihydroxycoumarin) 七叶亭；6,7-二羟香豆素 $C_9H_6O_4$
esculetinic acid 七叶亭酸 $C_9H_8O_5$
esculin(=polychrom) 多色素；七叶灵
eseridine 金丝碱；氧化毒扁豆碱
eserine 毒扁豆碱
eserine salicylate 水杨酸毒扁豆碱 $C_{15}H_{21}N_3O_2C_7H_6O_2$

Eshka method 埃史卡测硫法〔石油产品〕
esmeraldaite 杂褐铁矿
esmolol 艾司洛尔〔药〕
esomeprazole 埃索美拉唑〔药〕
esotoxin 内毒素
espartille oil 羽状草油
esparto 西班牙草
esparto wax 西班牙草蜡
esperin 埃斯波素
espinomycin 针棘霉素
ESR age ESR 年龄
ESR dating ESR 年代测定
ESR dosimetry ESR 剂量测定法
ESR measurement 电子自旋共振测量
ESR microscopy ESR 显微镜
ESR spectrometer 电子自旋共振(波)谱仪
ESR spectrum 电子自旋共振；顺磁共振光谱
ESR tube 电子自旋共振管
essange oil 埃散油
essence ①香精②精油③香料④香气
essence of Mirbane 硝基苯；密斑油
essence of resin 松香精
essential ①必需的②本质的；主要的③酯的；酯化的
essential amino acid 必需氨基酸
essential colour 主色
essential component 主要成分
essential element 必需元素*
essential fatty acid 必需脂肪酸
essential group 主要原子团
essential ingredient 主要成分
essential number 酯化值
essential oil 精油*
essential oil microemulsion (香)精油微乳液
essential tone 基本色调
essential trace element 必需微量元素
essential value 酯化值
essential water 组成水*
essexite 碱性辉长岩；厄塞岩
essonite 钙铝榴石；肉桂石
establishment ①建立；确定②组织③基础
estate rubber 庄园橡胶；种植橡胶
estazolam 艾司唑仑〔药〕
ester 酯*
ester acid 酸式酯
ester alkyl 酯烷基
ester aminolysis 酯氨解
esteratic site 酯部位
ester-bentonite grease 酯-膨润土润滑脂

ester bond 酯键
ester n-butyl acetate 乙酸(正)丁酯
ester sec-butyl acetate 乙酸仲丁酯
ester carbonyl group 酯羰基
ester condensation 酯缩合(作用)
esterdiol 酯二醇
ester ether 醚酯〔是醚又是酯〕
ester exchange(=interesterification) 酯交换
ester exchange catalyst 酯交换催化剂
ester exchange polycondensation 酯交换型缩聚
esterfied resin 酯化树脂
ester gum 酯胶；松香甘油酯
ester gum oxidation 酯胶氧化
esterification 酯化*
esterification catalyst 酯化催化剂
esterification column 酯化塔
esterification condensation 酯化缩合(反应)作用
esterification degree 酯化度
esterification equivalent weight 酯化当量
esterification law 酯化律
esterification number 酯化值
esterification on polymers 高聚物酯化反应
esterification-polymerization process 酯化聚合反应过程
esterification rate 酯化速度
esterification reagent 酯化剂
esterification waterproofing 酯化防水整理法
esterified 酯化了的
esterified cellulose 酯化纤维素
esterified silica 酯化二氧化硅
esterified silica support 酯化硅质载体；酯化硅胶载体
esterifier 酯化反应器
esterifying 酯化
esterifying agent 酯化剂
esterifying alcohol 酯化醇
esterifying catalyst 酯化催化剂
esterifying reagent 酯化(试)剂
ester interchange(=transesterification) 酯交换(作用)
ester-interchange catalyst 酯交换催化剂
ester interchange-polymerization 酯交换聚合(作用)
ester interchange vessel 酯交换釜
esterization 酯化(作用)
ester linkage 酯键(合)
ester lubricant 酯润滑剂
esterlysis 酯解(作用)
ester montanyl montanate 褐煤酸二十八酯；二十八酸二十八酯；蒙旦酸蒙旦酯 $C_{27}H_{55}COOC_{28}H_{57}$
ester number 酯(化)值；酯价
ester of halogenated acid 卤代酸酯

ester oil 酯油〔由酯组成的合成油〕
ester oil grease 酯润滑油〔以酯为主的润滑油〕
ester oxycellulose 氧化纤维素酯
ester plasticizer 酯增塑剂
ester rubber 酯类橡胶
ester value 酯化值
esthetics 美感性；美观性
estimate 估计量
estimated value 估计值；测定值
estimated viscosity 测定的黏度
estimating of crude oil 原油评价
estimation ①估定；估计②测定
estimation by eye 目测；目测法
estimator 估计量；估计值
estimator of variance 方差估计值
estin 棘霉素
estoral 硼酸蓝基酯
estradiol(=dihydrotheelin) 二氢雌酮；雌二醇 $CH_3C_{17}H_{19}(OH)_2$
estradiol benzoate 苯甲酸雌二醇〔药〕
estradiol valerate 戊酸雌二醇〔药〕
estragole 蒿脑；对甲氧基苯丙烯；对烯丙基茴香醚 $CH_2=CHCH_2C_6H_4OCH_3$
estragon oil 龙蒿油
estrane 雌(甾)烷
estrane diol 雌(甾)烷二醇 $C_{18}H_{30}O_2$
estranol 雌烷酚 $C_{18}H_{30}O$
estratriene 雌三烯 $C_{18}H_{24}$
estrin 雌激素
estriol(=oestriol;theelol) 雌(甾)三醇；16,17-二羟甾酚 $C_{18}H_{24}O_3$
estriol glucuronate 雌(甾)三醇葡萄苷酸
estrogen 雌激素
estrogenic hormones(=female hormones) 雌(情)激素(类)
estron 乙酸纤维素纤维
estrone(=oestrone;theelin) 雌(甾)酮 $CH_3C_{17}H_{18}O(OH)$
Estrup-Wolfgang theory 埃斯特鲁普-沃尔夫冈理论
esulin 七叶苷
Et(=ethyl)〔符号〕乙基
etabetacin 庚二菌素
etalon 标准具；校准器
etamsylate 酚磺乙胺〔药〕
etamycin 宜他霉素
-etane〔词尾〕 表示四节环
Etard reaction 埃塔反应
etchant 蚀刻剂；浸蚀剂
etch cone 刻蚀锥
etch depth 蚀刻深度

etched figure 刻蚀像(图)
etched glass beads 经刻蚀的玻璃珠
etched mark ①蚀刻痕②刻度线
etcher's wax 蚀刻用的石蜡
etch figure 蚀象
etch hole 蚀刻孔
etching 蚀刻；侵蚀
etching agent 蚀刻剂
etching effect 侵蚀效应
etching figure 蚀刻(图)像
etching-interfering method 腐蚀干涉法〔测玻璃球的均匀度〕
etching machine 浸蚀机
etching mechanism 腐蚀机理
etching needle 蚀刻针
etching of concrete surface 混凝土表面的侵蚀
etching off 蚀去
etching powder 磨毛粉
etching reagent 蚀刻剂
etching resist 抗蚀阻剂
etching solution ①磷化处理液；洗涤液②浸蚀液；蚀刻液
etching tank 浸蚀槽；腐蚀槽
etching test 蚀刻试验
etching waste liquor ①蚀刻废液②磷化(处理)废液
etching wear 浸蚀磨损
etch pattern 腐蚀图形
etch pit 刻蚀坑
etch-pit counting method 蚀刻坑计数法
etch pitting 浸蚀蚀痕(法)〔晶体结构完善度鉴定法之一〕
etch-polish 蚀刻抛光
etch primer 磷化底漆；反应性底漆
etch rate 浸蚀速度
etch solution test 颜料耐蚀试验〔油墨〕
etch-stop 腐蚀停止
eteline 四氯乙烯
ethacetic acid (正)丁酸
ethal 十六烷醇
ethalic acid 十六烷酸
ethambutol 乙胺丁醇〔药〕
ethamine 乙胺
ethanal 乙醛
ethanamide 乙酰胺
ethanamidine 乙脒
1,2-ethandiol 1,2-乙二醇
1,2-ethandithiol 1,2-乙二硫醇
ethane 乙烷 CH_3CH_3

ethane cracking heater 乙烷裂解加热炉
ethane diacid(=ethanedioic acid) 乙二酸；草酸 HOOCCOOH
ethane dicarboxylic acid 乙烷二甲酸 $C_2H_4(COOH)_2$
ethanedioic acid(=ethane diacid) 乙二酸；草酸 HOOCCOOH
ethanediol 乙二醇
1,2-ethanedithiol(=ethylene mercaptan) 1,2-亚乙基二硫醇；1,2-乙二硫醇 $HSCH_2CH_2SH$
ethane-ethylene fraction 乙烷-乙烯馏分
ethane selenonic acid 乙基硒酸 $C_2H_5SeO_3H$
ethanesulfenyl chloride 乙硫基氯
ethane tellurinic acid 乙基亚碲酸 $C_2H_5TeO_2OH$
ethane telluronic acid 乙基碲酸 $C_2H_5TeO_2OH$
ethane tetracarboxylic acid 乙烷四甲酸 $(COOH)_2CHCH(COOH)_2$
ethanetetrayl 联(二)亚甲基
ethanethiol 乙硫醇 CH_3CH_2SH
ethanethioyl 硫代乙酰基 $CH_3CS—$
ethano 桥亚乙基〔指—CH_2CH_2—基跨在环中的词头〕
ethanoic acid 乙酸；醋酸〔俗〕 CH_3COOH
ethanol 乙醇 C_2H_6O
ethanolamine(=aminoethyl alcohol;cholamine) 乙醇胺；2-羟基乙胺；氨基乙醇；胆胺 $NH_2CH_2CH_2OH$
ethanolate(=ethylate) 乙醇盐；乙氧化(基)金属
ethanolato(=ethoxy) ①乙醇化物②乙氧基
ethanolato-titanium ①乙氧基钛②钛酸乙酯
ethanolic extract 乙醇萃取物
ethanol morpholine ester 乙醇吗啉酯
1-ethanol-2-thiol 2-巯基-1-乙醇
ethanol-toluene azeotrope 乙醇甲苯恒沸法
ethanolysis 乙醇解*
ethanolysis method 乙醇分解法
ethanoyl(=acetye) 乙酰(基)
ethene(=ethylene) ①乙烯 $C_2H_4$②1,2-亚乙基 —C_2H_4—
etheno 亚乙基 —CH＝CH—
ethenoid 乙烯型的
ethenoid resin 乙烯树脂
ethenol(=vinyl alcohol) 乙烯醇
ethenone(=ketene) 乙烯酮
ethenoxy group 氧乙烯基
ethenoxylated nonionic surfactant 乙氧(基)化非离子表面活性剂
ethenoxylation reaction 乙氧(基)化反应
ethenoxy unit(s) 氧乙烯单位；环氧乙烷单位
ethenyl(=ethylidyne;vinyl) ①乙烯基②次乙基
ethenylene(=vinylene) ①1,2-亚乙烯基 —CH＝CH—
②1,1,2-次乙基 —CH_2CH＝
ethenylidene(=vinylidene) 亚乙烯基 CH_2＝C＝
ethenylidene monomer 亚乙烯基系单体；偏乙烯系单体 CH_2＝CX_2
ethenyl testosterone 乙烯基睾丸甾酮
ether 醚*
ether acid 醚酸
ether alcohol 醚醇 ROROH
ether alcohol sulfate 醇醚硫酸盐
etherate ①醚合物②乙醚配合物
etherate of trifluoroboron 三氟化硼合乙醚
ether bond content 醚键含量
ether cleavage 醚裂开
ethereal 醚的
ethereal extract 乙醚萃取物
ethereal fruit oil 果香油
ethereal liquid 醚状液
ethereal odor 轻飘气味
ethereal oil 精油
ethereal salt(=ester) 酯
ethereal sludge (含)醚浆；醚渣
ethereal solution 醚溶液
ethereal sulfate 硫酸乙酯
ethereous(=etheric) 醚的
ether ester 醚酯〔是醚又是酯〕
ether extract 乙醚提取物
ether extraction 醚类萃取
ether from petroleum 石油醚
etheric(=ethereous) 醚的
etheric acid 乙酰乙酸
etheride 酰卤化物〔含有—COX〕
etherification 醚化(作用)
etherification reaction 醚化反应
etherification uniformity 醚化均匀度
etherification waterproofing 醚化防水整理法
etherified 醚化了的
etherified cellulose 醚化纤维素
etherified urea resin 醚化脲树脂
etherifying 醚化
etherifying agent 醚化剂
ether index 乙醚指数
ether inhibitor 醚抑制剂
ether ketone 酮醚 RCOROR
ether link(=ether linkage) 醚键 ROR
ether linkage(=ether link) 醚键
ether link content 醚键含量
ether linking 醚键合；醚(键)结合
ether of cellulose 纤维素醚

ether of crystallization 结晶醚
ether oil 醚油；醇精油
etherol solution 醚化溶液
etheron 能媒子
etherophosphoric acid 磷酸一乙酯 $C_2H_5H_2PO_4$
etherosulfuric acid 乙(基)硫酸；硫酸氢乙酯 $C_2H_5SO_4H$
ether oxygen atom 醚氧原子
ether peroxide 过氧化二乙醚
ether plasticizer 醚增塑剂
ether resin ①醚树脂②环氧树脂
ether ring 醚环；含氧环
ether-soluble 溶于醚的
ether-splitting 醚裂
ether sulfate 醚硫酸盐
ether thioether 醚硫醚〔增塑剂〕
ethesdanine 乙硫吡丙菌素
ethide 乙基化物 MC_2H_5
ethidene 亚乙基
ethidium bromide 溴化乙锭；溴化 3,8-二氨基-5-乙基-6-苯基菲啶鎓
ethine(=acetylene) 乙炔
ethinyl(=ethynyl) 乙炔基 $CH≡C—$
ethinyl estradiol 乙炔基雌(甾)二醇
ethionic acid 乙二磺酸
ethionine 乙(基)硫氨酸
ethiops mineral 黑硫汞矿
ethisterone 17-乙炔睾(甾)酮
ethocel 乙基纤维素
ethodin 依沙吖啶；利凡诺尔〔杀菌药〕
etholide 羟酸链酯 $RCH(OCOR)CH_2COOH$
ethosuximide 乙琥胺〔药〕
ethotoin(=3-ethyl-5-phenylhydantoin) 3-乙基-5-苯基乙内酰脲
ethovan(=ethyl vanillin) 乙基香兰素
ethoxal 乙(酯)草酸
ethoxalyl 乙草酰(基) $C_2H_5OOCCO—$
ethoxide 乙醇盐；乙氧基金属 C_2H_5OM
ethoxy(=ethanolato) ①乙醇化物②乙氧基 $C_2H_5O—$
ethoxyacetic acid 乙氧基乙酸 $C_2H_5OCH_2COOH$
ethoxyaniline 乙氧基苯胺 $C_2H_5OC_6H_4NH_2$
p-ethoxybenzaldehyde 对乙氧基苯甲醛
ethoxybenzaldehyde 乙氧基苯(甲)醛 $C_2H_5OC_6H_4CHO$
ethoxybenzidine 乙氧基联苯胺
ethoxybenzoic acid 乙氧基苯甲酸 $C_2H_5OC_6H_4CO_2H$
ethoxybenzoin 乙氧基苯偶姻
ethoxybiphenyl 乙氧基联(二)苯 $C_2H_5OC_6H_4C_6H_5$
ethoxycarbonyl 乙酯基；乙羰基 $C_2H_5OOC—$
ethoxy cellulose 乙氧基纤维素

ethoxy cinnamic acid 乙氧基肉桂酸 $C_2H_5OC_6H_4CH=CHCOOH$
ethoxy dialkyl acetal 乙氧基型二烷基(醇)缩乙醛
6-ethoxy-1,2-dihydro-2,2,4-trimethyl quinoline 6-乙氧基-1,2-二氢-2,2,4-三甲基喹啉
2-ethoxy ethanol 2-乙氧基乙醇
ethoxy ethoxyethyl acrylate(=carbitol acrylate) 丙烯酸乙氧乙氧基乙酯；丙烯酸卡必醇酯
ethoxyethyl acetate(EA) 乙酸乙氧基乙酯；乙二醇一乙醚乙酸酯
ethoxy ethylene 乙氧基乙烯
ethoxyethyl laurate 月桂酸乙氧基乙酯
2-ethoxyethyl methacrylate 甲基丙烯酸 2-乙氧基乙酯
ethoxy-ethyl ricinoleate 蓖麻酸乙氧基乙酯；乙二醇一乙醚蓖麻醇酸酯 $C_{17}H_{32}(OH)COOCH_2CH_2OCH_2CH_3$
ethoxy group 乙氧基
ethoxyl- 羟乙基 $HOC_2H_4—$
ethoxylaniline(=β-anilinoethanol) 羟乙基苯胺；苯氨基乙醇 $C_6H_5NH(C_2H_4OH)$
ethoxylate 乙氧基化物
ethoxylated alkyl betaine 乙氧基化烷基甜菜碱
ethoxylated alkylcresol 乙氧基化烷基甲酚
ethoxylated amine 乙氧化胺
ethoxylated amylphenol 乙氧基化戊基酚
ethoxylated benzylcresol 乙氧基化苄基甲酚
ethoxylated carbinamine 乙氧基化甲醇胺
ethoxylated coco amine 乙氧基化椰子胺
ethoxylated ester 乙氧基酯
ethoxylated fatty acid 乙氧基脂肪酸
ethoxylated fatty amide 乙氧基(化)脂肪族酰胺
ethoxylated fatty monoamine 乙氧基化脂肪族一元胺
ethoxylated fatty polyamine 乙氧基化脂肪族多胺
ethoxylated hydroxy benzene sulfonamide 乙氧基化羟基苯磺酰胺
ethoxylated naphthenic amine 乙氧基化环烷胺
ethoxylated nonylphenol 乙氧基化壬基酚
ethoxylated octadecanol 乙氧基化十八(烷)醇
ethoxylated oleic acid 乙氧基化油酸
ethoxylated pentaerythritol ether 乙氧基化季戊四醇醚
ethoxylated quaternary 乙氧基化季铵盐化合物
ethoxylated quinoline sulfonate 乙氧基化喹啉磺酸盐
ethoxylated sorbitan ester 乙氧基化山梨糖醇酐酯
ethoxylated sorbitol fatty acid ester 乙氧基化山梨糖醇脂肪酸酯
cthoxylation 乙氧基化
ethoxyl dialkyl acetal 乙氧基型二烷基(醇)缩(乙)醛
ethoxylin(=epoxy resin) 环氧树脂
ethoxyline resin 环氧树脂

ethoxylized fatty alcohol 乙氧基化脂肪(族)醇
ethoxylized nonylphenol 乙氧基化壬基酚
ethoxyphenyl(=phenelyl) 乙氧苯基
p-ethoxyphenyl-succinimide(=pyrantin) 吡喃亭
　　$C_2H_5OC_6H_4N(COCH_2)_2$
ethoxy phenyl urea 乙氧基苯脲
　　$C_2H_5OC_6H_4NHCONH_2$
ethoxy propenylphenol 乙氧基丙烯基苯酚
ethoxy propyl acetate(EPA) 乙酸乙氧基丙酯
ethoxyquin 乙氧喹啉
2-ethoxythiazole 2-乙氧基噻唑
ethoxy-triethyl silicane 三乙基硅氧基乙烷
　　$(C_2H_5)_3SiOC_2H_5$
ethoxytrimethyl silane 乙氧基三甲基(甲)硅烷
ethyl 乙基　CH_3CH_2-
ethyl abietate 松香酸乙酯　$C_{19}H_{29}CO_2C_2H_5$
ethyl acetanilide 乙替乙酰替苯胺〔增塑剂〕
ethyl acetate 乙酸乙酯　$CH_3CO_2C_2H_5$
ethyl acetic acid 丁酸；乙基乙酸　$C_2H_5CH_2COOH$
ethyl acetoacetate 乙酰乙酸乙酯
ethyl acetoacetate chelate 乙酰乙酸乙酯螯合物
ethyl acetoacetic acid 乙基乙酰乙酸；2-乙基-3-丁酮-1-酸　$CH_3COCH(C_2H_5)COOH$
ethyl acetoacetic ester ①乙基乙酰乙酸酯
　　$CH_3COCH(C_2H_5)COOR$②〔专指〕乙基乙酰乙酸乙酯　$CH_3COCH(C_2H_5)COOC_2H_5$
ethylacetoacetone(EAA; HEAA) 乙基乙酰丙酮
　　$CH_3COCH(C_2H_5)COCH_3$
ethyl acetylene(=1-butine) 乙基乙炔；1-丁炔
　　$C_2H_5C\equiv CH$
ethyl acetyl glycoate 乙酰乙醇酸乙酯
1-ethyl-2-acetyl pyrrole 1-乙基-2-乙酰基吡咯
ethyl acetyl ricinoleate 乙酰蓖麻酸乙酯
ethyl acetyl salicylate 乙酰水杨酸乙酯
ethyl acid sulfate 酸式硫酸乙酯
ethyl aconitate 乌头酸乙酯
ethyl acrylate 丙烯酸乙酯
2-ethyl acrylic aldehyde 2-乙基丙烯醛
ethyl active amyl ether 乙基旋性戊基醚
ethylal 乙醛
ethyl alcohol(=ethyl hydroxide) 乙醇；酒精〔俗〕
ethyl aldehyde 乙醛　CH_3CHO
ethyl allophanate 脲基甲酸乙酯　$NH_2CONHCO_2C_2H_5$
ethyl allylacetylacetate 烯丙基乙酰乙酸乙酯
　　$CH_3COCH(C_3H_5)CO_2C_2H_5$
ethyl allyl ether 乙基烯丙基醚　$CH_2=CHCH_2OC_2H_5$
ethyl alpha-hydroxyisobutyrate α-羟基异丁酸乙酯
　　$(CH_3)_2COHCOOC_2H_5$

ethylaluminium dichloride 二氯化乙基铝〔催化剂〕
ethylaluminium sesquichloride 三氯化乙基铝〔催化剂〕
ethyl aluminum dichloride 二氯乙基铝〔烯烃类聚合用催化剂〕
ethylamine(=aminoethane) 乙胺；氨基乙烷　$C_2H_5NH_2$
ethylamine hydriodide 乙胺化氢碘　$C_2H_5NH_2HI$
ethylamine hydrobromide 乙胺化氢溴　$C_2H_5NH_2HBr$
ethylamine hydrochloride 乙胺化氢氯　$C_2H_5NH_2HCl$
ethylamino 乙氨基　C_2H_5NH-
ethyl aminoacetate 氨基乙酸乙酯　NH_2CH_2COOEt
ethyl 2-aminobenzoate 2-氨基苯甲酸乙酯
ethyl β-aminocrotonate β-氨基巴豆酸乙酯
　　$CH_3C(NH_2)=CHCO_2C_2H_5$
ethylaminoethanol(=ethylaminoethyl alcohol) 乙氨基乙醇　$C_2H_5NHCH_2CH_2OH$
ethylaminoethyl alcohol(=ethylaminoethanol) 乙氨基乙醇
ethyl 7-aminoheptanoate 7-氨基庚酸乙酯
　　$H_2N(CH_2)_6COOC_2H_5$
2-ethyl-1-amino-hexane 2-乙基-1-氨基己烷
　　$C_4H_9(C_2H_5)CHCH_2NH_2$
ethylammonium ions 乙基铵离子
ethylammonium smectite 乙基铵绿土
ethylammonium vermiculite 乙基铵蛭石
ethyl amyl carbinol 乙基戊基原醇
ethyl amyl carbinyl acetate 乙酸乙基戊基原酯
ethyl amyl ketone 乙戊基(甲)酮
ethyl-n-amyl ketone 乙戊酮；3-辛酮
ethylane glycol monobutyl ether 乙二醇单丁醚
ethylaniline phosphate 乙基苯胺磷酸盐
ethyl anisate 茴香酸乙酯；对甲氧基苯甲酸乙酯
ethyl anthranilate 邻氨基苯甲酸乙酯
2-ethyl anthraquinone 2-乙基蒽醌
ethyl-arsenic acid 乙基砷酸　$C_2H_5AsO(OH)_2$
ethyl arsenite 亚砷酸乙酯　$(C_2H_5)_3AsO_2$
ethyl arsine 乙胂　$C_2H_5AsH_2$
ethyl arsine disulfide 二硫化乙胂　$C_2H_5AsS_2$
ethyl arsine oxide 乙胂化氧　$C_2H_5As=O$
ethylate(=ethanolate) 乙醇盐；乙氧化(基)金属
ethylated ①乙基化了的②含乙基的
ethylated gasoline 乙基化汽油
ethylating 乙基化；乙基化的
ethylating agent 乙基化剂
ethylation 乙基化*
ethylation in the cold 冷时乙基化
ethyl azelaate 壬二酸二乙酯
　　$C_2H_5OCO(CH_2)_7COOC_2H_5$
2-ethylaziridine(=ethyl ethyleneimine) 乙基吖丙啶

1-ethylaziridinyl methacrylate 甲基丙烯酸-1-乙基吖丙啶酯；甲基丙烯酸-吖丙啶乙酯
ethyl azodicarboxylate 偶氮二羧酸乙酯〔调节剂〕
ethyl benzalacetoacetate 亚苄基乙酰乙酸乙酯
ethyl benzenesulfonate 苯磺酸乙酯 $C_6H_5SO_3C_2H_5$
2-ethyl benzene thiol 2-乙基苯硫醇
ethyl benzoate 苯甲酸乙酯 $C_6H_5CO_2C_2H_5$
ethyl benzoylacetate 苯甲酰基乙酸乙酯
ethyl benzoylacetoacetate 苯甲酰乙酰乙酸乙酯 $C_2H_3OCH(C_7H_5O)CO_2C_2H_5$
ethyl o-benzoyl benzoate 邻苯甲酰苯甲酸乙酯
1-ethylbenzoylene urea 1-乙基亚苯甲酰基脲 $C_{10}H_{10}O_2N_2$
ethyl benzoylformate 苯甲酰甲酸乙酯
ethyl benzylacetoacetate 苄基乙酰乙酸乙酯
ethyl benzylbenzene 乙基苄基苯 $C_2H_5C_6H_4CH_2C_6H_5$
ethyl benzyl cellulose 乙基苄基纤维素
ethyl benzyl ether 乙基苄醚
ethyl benzyl ketone 乙基苄基(甲)酮；丙酰甲苯 $C_2H_5COCH_2C_6H_5$
ethyl-bismuthine 三乙铋 $Bi(C_2H_5)_3$
ethyl-borate 硼酸乙酯 $B(OC_2H_5)_3$
ethyl-boric acid(=ethyl-boron dihydroxide) 乙基硼酸〔基表示取代 OH〕 $C_2H_5B(OH)_2$
ethyl-boron dihydroxide(=ethyl-boric acid) 乙基硼酸
ethyl brassylate 十三烷二酸乙酯
ethyl bromide(=bromoethane) 乙基溴；溴乙烷 C_2H_5Br
ethyl bromoacetate 溴乙酸乙酯 $BrCH_2CO_2C_2H_5$
ethyl bromoacetoacetate 溴乙酰乙酸乙酯
2-ethyl butanol 2-乙基丁醇
2-ethyl butyl acetate 乙酸异己酯
2-ethyl-3-butylacrolein 2-乙基-3-丁基丙烯醛
ethyl butyl carbinol 乙基丁基甲醇
ethyl butyl carbonate 碳酸乙丁酯
ethyl butyl cellulose 乙基丁基纤维素
ethyl butyl ether 乙基丁基醚 $C_2H_5OC_4H_9$
ethyl butyl ketone 乙基丁酮
ethyl butyl malonate 丙二酸乙丁酯
ethyl butyl sulfide 乙基丁基硫；乙丁硫醚；乙硫基丁烷 $C_2H_5SC_4H_9$
ethyl butyrate 丁酸乙酯 $C_3H_7CO_2C_2H_5$
α-ethyl butyric acid α-乙基丁酸
2-ethyl butyric acid 2-乙基丁酸
ethyl butyric acid sulfate 乙基丁酸硫酸盐
ethyl butyrolactone 乙基丁内酯
ethyl cacodylic acid 乙基卡可酸；二乙次胂酸 $(Et)_2AsOOH$
ethyl-cadmium 乙基镉 $Cd(C_2H_5)_2$

ethyl caprate 癸酸乙酯 $CH_3(CH_2)_8CO_2C_2H_5$
ethyl caproate 己酸乙酯
ethyl caprylate 辛酸乙酯
ethyl carbamate(=urethane) 尿烷〔俗〕；氨基甲酸乙酯 $NH_2CO_2C_2H_5$
ethyl carbazole 乙基咔唑
ethyl carbitol 二甘醇一乙醚
ethyl carbonate 碳酸乙酯 $CO(OC_2H_5)_2$
ethyl carbostyrile 乙基喹诺酮；乙基-2-羟基喹啉 $C_6H_4CH=C(C_2H_5)NHCO$
ethyl-2-carboxyglutaconate 2-羧基戊烯二酸乙酯
ethyl carbylamine(=ethyl isonutrile) 乙胩；乙基异腈 C_2H_5NC
ethyl carvacrol 乙基香芹酚
ethyl carvacryl ether 乙基香芹醚
ethyl cellosolve 乙基溶纤剂；乙二醇一乙醚
ethyl cellulose 乙基纤维素
ethyl chaulmoograte 晁模酸乙酯 $C_5H_7(CH_2)_{12}CO_2C_2H_5$
ethyl chloride(=chloroethane) 乙基氯；氯乙烷 C_2H_5Cl
ethyl chloroacetate 氯乙酸乙酯 $ClCH_2CO_2C_2H_5$
ethyl chlorocarbonate(=ethyl chloroformate) 氯甲酸乙酯 $ClCO_2C_2H_5$
ethyl chloroformate(=ethyl chlorocarbonate) 氯甲酸乙酯 $ClCO_2C_2H_5$
3-ethyl-3-chloromethoxy-cyclobutane 3-乙基-3-氯甲氧基环丁烷
ethyl chlorophyllide 叶绿素乙酯
ethyl cinnamate (肉)桂酸乙酯
α-ethyl cinnamic aldehyde α-乙基(肉)桂醛
ethyl cinnamyl ether 乙基(肉)桂醚
ethyl citral 乙基柠檬醛
ethyl citronellate 香茅酸乙酯
ethyl coumarin 乙基香豆素
ethyl crotonate 巴豆酸乙酯 $C_3H_5CO_2C_2H_5$
ethyl p-cumenylacetate 对异丙基苯乙酸乙酯
ethyl cyanacetate 氰基乙酸乙酯 $NCCH_2CO_2C_2H_5$
ethyl cyanacetic acid 乙基丙腈酸 $CNCH(C_2H_5)COOH$
ethyl-cyanacetic ester 乙基氰基乙酸酯；2-氰基丁酸酯 $C_2H_5CH(CN)COOR$
ethyl cyanamide 乙基氨腈 C_2H_5NHCN
ethyl cyanate 氰酸乙酯 $(C_2H_5)OCN$
ethyl cyanide 乙基氰；丙腈 EtCN
ethyl-2-cyano-3,3-diphenyl acrylate 2-氰基-3,3-二苯基丙烯酸乙酯〔紫外线吸收剂〕
ethyl cyanoformate 氰基甲酸乙酯 C_2H_5OOCCN
ethyl cyanuramide 乙基氨腈 EtNHCN

ethyl cycloheptane 乙基环庚烷 $C_2H_5C_7H_{13}$
ethyl cyclohexane 乙基环己烷 $C_2H_5C_6H_{11}$
ethyl cyclohexanepropionate 环己烷基丙酸乙酯
ethyl 3-cyclohexyl propanoate 3-环己烷基丙酸乙酯
ethyl cyclohexylpropionate 环己基丙酸乙酯
3-ethyl-2-cyclopenten-2-ol-1-one 3-乙基-2-环戊烯-2-醇-1-酮
ethyl decanoate 癸酸乙酯
ethyl decine carbonate 癸炔碳酸乙酯
ethyl decylate 癸酸乙酯
ethyl 2-decynoate 2-癸炔酸乙酯
ethyl-diacetate 双乙酸乙酯
ethyl diacetoacetate 二酰乙酸乙酯 $(C_2H_3O)_2CHCO_2C_2H_5$
ethyl diacetylacetate 二酰乙酸乙酯 $(CH_3CO)_2CHCOOC_2H_5$
2-ethyl-1,4-diazine 2-乙基-1,4-二嗪
ethyl diazoacetate 重氮基乙酸乙酯 $N_2CHCO_2C_2H_5$
ethyl dibenzyl-glycolate 二苄基乙醇酸乙酯 $(C_6H_5CH_2)_2COHCO_2C_2H_5$
ethyl dibromoacetate 二溴乙酸乙酯 $Br_2CHCO_2C_2H_5$
ethyl dibromoarsine 乙基二溴胂；二溴化乙基砷 $C_2H_5AsBr_2$
ethyl dibromocinnamate 二溴肉桂酸乙酯
ethyl dibutylcarbamate 二丁氨基甲酸乙酯 $(C_4H_9)_2NCO_2C_2H_5$
ethyl dichloroacetate 二氯乙酸乙酯 $Cl_2CHCO_2C_2H_5$
ethyl dichloroarsine 乙基二氯胂；乙胂化二氯 $C_2H_5AsCl_2$
ethyldichlorophosphate 乙基磷酰二氯；二氯磷酸乙酯 $(C_2H_5O)Cl_2PO$
ethyl diethylacetoacetate 二乙基乙酰乙酸乙酯 $CH_3COC(C_2H_5)_2CO_2C_2H_5$
ethyl diethylcyanoacetate 二乙基氰基乙酸乙酯 $CNC(C_2H_5)_2COOC_2H_5$
ethyl diethylmalonate 二乙基丙二酸乙酯 $C_2H_5OCOC(C_2H_5)_2COOC_2H_5$
ethyl dihydroxypropyl *p*-aminobenzoic acid 二羟丙基对氨基苯甲酸乙酯
ethyl dimethylacetoacetate 二甲基乙酰乙酸乙酯 $CH_3COC(CH_3)_2CO_2C_2H_5$
ethyl 4-dimethylaminobenzoate 4-二甲氨基安息香酸乙酯
ethyl dimethyl dioxolane acetate 2,4-二甲基-1,3-二噁烷-2-乙酸乙酯
2-ethyl-3,5-dimethylpyrazine 2-乙基-3,5-二甲基吡嗪
ethyl dinitrobenzoate 二硝基苯甲酸乙酯 $(NO_2)_2C_6H_3CO_2C_2H_5$

ethyl disulfide 乙二硫醚 $C_2H_5SSC_2H_5$
ethyl dodecanoate(=ethyl laurate) 十二烷酸乙酯；月桂酸乙酯
ethyl dodecylate 十二烷酸乙酯
ethylenation 乙烯化作用
ethylene ①乙烯 $CH_2=CH_2$ ②1,2-亚乙基 —CH_2CH_2— ③1,3-二氧戊环-2-酮
ethylene acetate(=glycol diacetate) 二乙酸乙二醇酯 $CH_3COOCH_2CH_2OOCCH_3$
ethylene-acetic acid 环丙基甲酸 $(CH_2)_2CHCOOH$
ethylene-acrylate copolymer 乙烯-丙烯酸酯共聚物
ethylene-acrylate rubber 乙烯丙烯酸酯橡胶
ethylene-air mixture 乙烯-空气混合物
ethylene bis-dimethacrylate 双二甲基丙烯酸乙二酯
ethylene bis-ethylazodicarboxylate 双乙基偶氮二羧酸亚乙酯〔硫化剂〕
ethylene bisoleoamide 亚乙基双油酰胺〔防黏剂〕
ethylene bis-oxyethylenenitrilo tetraacetic acid(EGTA) 亚乙基双(氧亚乙基次氮基)四乙酸
ethylenebisstearamide 亚乙基双硬脂酰胺〔防黏剂〕
ethylene brassylate 十三碳二酸乙二醇酯；巴西酸乙二醇酯；麝香T〔商〕
ethylene bromide 溴化乙烯；1,2-二溴乙烷 CH_2BrCH_2Br
ethylene bromohydrin(=2-bromoethanol) 2-溴乙醇 $BrCH_2CH_2OH$
ethylene-butadiene copolymer 乙烯-丁二烯共聚物
ethylene-butylene polymer 乙烯-丁烯聚合物
ethylene carbonate 碳酸亚乙酯；乙二醇碳酸酯；碳酸乙二酯
ethylene carboxylic acid 丙烯酸 $CH_2=CHCOOH$
ethylene chlorhydrin 2-氯乙醇
ethylene chloride 氯化乙烯 CH_2ClCH_2Cl
ethylene chlorobromide 氯乙基溴；1-氯-2-溴代乙烷 CH_2ClCH_2Br
ethylene chlorohydrin(=2-chloroethanol) 2-氯乙醇 $ClCH_2CH_2OH$
ethylene concentrator 乙烯提浓器
ethylene cyanohydrin 3-羟基丙腈 $HOCH_2CH_2CN$
ethylene diacetate 二乙酸乙二酯 $CH_3COO(CH_2)_2OOCH_3$
ethylene diamide 亚乙基二酰胺
ethylenediamine 1,2-乙二胺 $NH_2CH_2CH_2NH_2$
ethylenediamine adduct 乙二胺加合物
ethylenediamine-bisisopropylphosphonic acid (EDIP; H_4EDIP) 乙二胺双异丙基膦酸
ethylenediamine-bis-methylenephosphonic acid (EDMP; H_4EDMP) 乙二胺双亚甲基膦酸

ethylene diamine chlorate 氯酸化乙二胺
ethylene diamine dinitrate 二硝酸化乙二胺
ethylene diamine dipropionic acid 乙二胺二丙酸
ethylenediamine hydrate 乙二胺合一水 $(CH_2NH_2)_2H_2O$
ethylenediamine hydroxypropylate NN-二羟丙基乙二胺
ethylenediamine tetraacetic acid(EDTA) 乙二胺四乙酸；乙底酸〔俗〕
ethylenediamine tetrahydroxy propylene 乙二胺四羟丙烯
ethylenediamine tetramethylphosphinic acid(EDTPA) 乙二胺四甲基次膦酸
ethylenediamine tetrapropionic acid(EDTP) 乙二胺四丙酸
ethylenediamine triacetic acid 乙二胺三乙酸
ethylenediammonium uranyl chloride 乙二铵氯化铀酰
ethylene dibromide 二溴化乙烯
ethylene dicarboxylic acid 丁二酸；乙二甲酸；琥珀酸〔俗〕 $COOHCH_2CH_2COOH$
ethylene dichloride 二氯化乙烯；亚乙基二氯；1,2-二氯乙烷
ethylene dichloride process 二氯乙烯法〔用以制取氯乙烯单体〕
ethylene difluoride 二氟化乙烯；亚乙基二氟 CH_2FCH_2F
ethylene dihalide 二卤化乙烯 CH_2XCH_2X
ethylene diiodide 二碘化乙烯 CH_2ICH_2I
ethylene-dimalonic acid 亚乙基双丙二酸；丁烷四甲酸 $(COOH)_2CH(CH_2)_2CH(COOH)_2$
ethylene dimethacrylate 二甲基丙烯酸乙二醇酯
ethylene dinitrate 二硝酸乙二醇酯
ethylene dinitroamine 亚乙基二硝胺
ethylenedioxy 亚乙二氧基 —OCH_2CH_2O—
ethylenedioxy bis(propylamine) 亚乙二氧基双丙胺 $H_2N(CH_2)_3O(CH_2)_2O(CH_2)_3NH_2$
ethylenedioxy titanium diisostearate 二异硬脂酸亚乙二氧基钛
ethylene diphenate(=ethylene diphenyl ether) 乙二醇二苯醚 $(C_6H_5OCH_2)_2$
ethylene diphenyl ether(=ethylene diphenate) 乙二醇二苯醚 $(C_6H_5OCH_2)_2$
ethylene diphenyl sulfone 亚乙基二苯二砜 $(C_6H_5SO_2CH_2)_2$
ethylene distearamide 二硬脂酰乙二胺
ethylene disulfhydrate 1,2-乙二硫醇 $HSCH_2CH_2SH$
ethylene disulfonic acid 乙基二磺酸
ethylene epoxide 氧丙环；氧化乙烯；环氧乙烷 $(CH_2)_2O$
ethylene-ethylacetate elastomer 乙烯乙酸乙酯橡胶

ethylene-ethyl acrylate 乙烯-丙烯酸乙酯共聚物
ethylene fluoride 氟化乙烯；亚乙基二氟 CH_2FCH_2F
ethylene glycol 1,2-亚乙基二醇；乙二醇 $HOCH_2CH_2OH$
ethylene glycol adipate(EGA) 己二酸乙二醇酯
ethylene glycolate 乙醇酸亚乙酯；乙二醇二乙醇酸酯
ethylene glycol bath 乙二醇浴
ethyleneglycol bis(2-aminoethylether) tetraacetic acid (EGTA) 乙二醇双(2-氨基乙醚)四乙酸
ethylene glycol bisthioglycolate(=glycol dimercaptoacetate) 乙二醇二(巯基乙酸酯)〔交联剂；环氧固化促进剂〕
ethylene glycol butyl ether 乙二醇一丁醚；丁基溶纤剂〔俗〕
ethylene glycol diacetate 二乙酸乙二醇酯
ethylene glycol dibenzoate 二苯甲酸乙二醇酯〔增塑剂〕
ethylene glycol dibutyl ether 乙二醇二丁醚〔高沸点特种溶剂〕
ethylene glycol dibutyrate(=glycol dibutyrate) 二丁酸乙二醇酯
ethylene glycol diethyl ether 乙二醇二乙醚
ethylene glycol diformate 二甲酸乙二醇酯
ethylene glycol diglycidyl ether 乙二醇二缩水甘油醚
ethyleneglycol diglycidyl ether acrylate 乙二醇二缩水甘油醚丙烯酸酯
ethylene glycol dilaurate 二月桂酸乙二醇酯
ethylene glycol dimethacrylate 二甲基丙烯酸乙二醇酯
ethylene glycol dimethyl ether(=glyme) 甘醇二甲醚；乙二醇二甲醚
ethylene glycol dinitrate 二硝酸乙二醇酯
ethylene glycol dipropionate 乙二醇二丙酸酯〔增塑剂〕
ethyleneglycol ethyl ether 乙二醇乙醚
ethylene glycol-isophthalic acid-terephthalic acid copolymer 乙二醇-间苯二甲酸-对苯二甲酸共聚物
ethylene glycol monoacetate 单乙酸乙二醇酯 $HOCH_2CH_2OOCCH_3$
ethylene glycol monoallyl ether 乙二醇单烯丙基醚
ethylene glycol monobenzyl ether 乙二醇单苄醚〔醋酸纤维素用溶剂〕
ethylene glycol monobutyl ether 乙二醇一丁醚〔溶剂〕
ethylene glycol monodecanoate 单癸酸乙二醇酯
ethylene glycol monododecyl ether 乙二醇单十二烷基醚
ethylene glycol mono-ether 乙二醇单醚
ethylene glycol monolaurate 单月桂酸乙二醇酯
ethylene glycol monoleate 单油酸乙二醇酯
ethylene glycol monomyristate 单豆蔻酸乙二醇酯
ethylene glycol monostearate 单硬脂酸乙二醇酯
ethylene halide 卤化乙烯 CH_2XCH_2X
ethylene halohydrin 卤代乙醇

ethylene hydride(=ethane)　乙烷
ethylene hydroxy sulfuric acid　羟乙基硫酸　$OHC_2H_4 \cdot HSO_4$
ethylene imide　吖丙啶；氮丙环
ethylene imine　①吖丙啶②哌嗪
ethyleneimino　亚乙基氨基　$CH_3CH=N-$
ethylene iodide　碘化乙烯　ICH_2CH_2I
ethylene iodohydrin　碘代乙醇
ethylene-lactic acid　羟丙酸；3-羟丙酸　CH_2OHCH_2COOH
ethylene-malonic acid(=vinaconic acid; cyclopropane-dicarboxylic acid)　酒康酸；环丙烷二甲酸　$CH_2CH_2C(COOH)_2$
ethylene mercaptan(=1,2-ethanedithiol)　1,2-亚乙基二硫醇；1,2-乙二硫醇　$HSCH_2CH_2SH$
ethylene-methacrylic acid　乙烯-甲基丙烯酸共聚物
ethylene monoacetate(=glycol monoacetate)　一乙酸乙二醇酯
ethylene naphthalene(=acenaphthylene)　苊
ethylene nitrate　硝酸亚酯
ethylene nitrite　亚硝酸亚乙酯；乙二醇二亚硝酸酯　$(ONOCH_2)_2$
ethylene oligomerization　乙烯低聚(作用)
ethylene oxide　环氧乙烷；氧丙环　CH_2CH_2O
ethylene oxide adduct　环氧乙烷加成物；氧化乙烯加成物
ethylene oxide chain　环氧乙烷链
ethylene oxide condensate emulsifier　环氧乙烷缩合乳化剂
ethylene oxide condensation　环氧乙烷缩合(作用)
ethylene oxide hydroxybenzamide　环氧乙烷羟基苯甲酰胺
ethylene oxide nonylphenol　环氧乙烷壬基酚
ethylene oxide oligomer　环氧乙烷低聚物
ethylene oxide rubber　乙烯氧化橡胶
ethylene oxide telomer　环氧乙烷调聚物
ethylene oxide unit(s)　环氧乙烷单位
ethylene oxidic ester　缩水甘油酯　RC_2H_2OCOOR
ethyleneoxy　乙烯氧基
ethylene perbromide　全溴乙烯；四溴代乙烯　$CBr_2=CBr_2$
ethylene perchloride　全氯乙烯；四氯代乙烯　$CCl_2=CCl_2$
ethylene periodide　全碘乙烯；四碘代乙烯　$CI_2=CI_2$
ethylene-petroleum oil copolymers　乙烯石油共聚物〔一种合成润滑油〕

ethylene polyamine　乙烯多胺
ethylene polymer　乙烯高聚物
ethylene polymer oil　乙烯聚合油
ethylene product　乙烯产品
ethylene-propylene copolymer　乙烯-丙烯共聚物
ethylene-propylene-diene(EPD)　乙烯-丙烯-二烯三元共聚物
ethylene-propylene-diene elastomer　三元乙丙橡胶
ethylene propylene diene monomer(EPDM)　三元乙丙橡胶；乙丙三元橡胶
ethylene-propylene diene terpolymer　乙烯-丙烯-二烯三元共聚物
ethylene propylene monomer(EPM)　二元乙丙橡胶；乙丙二元橡胶
ethylene-propylene polymer(EPM)　乙烯-丙烯共聚物
ethylene propylene rubber(EPR)　二元乙丙橡胶；乙丙二元橡胶
ethylene propylene rubber latex　乙丙胶乳
ethylene propylene terpolymer(EPT)　三元乙丙橡胶；乙丙三元橡胶
ethylene-propylene terpolymerisate rubber　三元乙丙橡胶
ethylene recycle　循环乙烯
ethylene rhodanate(=ethylene rhodanide; ethylene sulfocyanate)　硫氰酸亚乙酯；乙二醇二硫氰酸酯　$C_2H_4(SCN)_2$
ethylene rhodanide(=ethylene rhodanate)　硫氰酸亚乙酯；乙二醇二硫氰酸酯　$C_2H_4(SCN)_2$
ethylene-rich gas　富乙烯气体
ethylenes　乙烯属烃
ethylene sulfite　亚硫酸乙二醇酯
ethylene sulfocyanate(=ethylene rhodanate)　硫氰酸亚乙酯；乙二醇二硫氰酸酯　$C_2H_4(SCN)_2$
ethylene-sulfonic acid　亚乙基二磺酸　$C_2H_4(SO_3H)_2$
ethylene telomer　乙烯调聚物
ethylene terephthalate　对苯二甲酸亚乙酯〔增塑剂〕
ethylene terephthalate polymer　对苯二酸-乙二醇缩聚物
ethylene tetrabromide　四溴(化)乙烯　$CBr_2=CBr_2$
ethylene-tetracarboxylic acid　乙烯四甲酸　$(COOH)_2C=C(COOH)_2$
ethylene tetrachloride　四氯(化)乙烯　$CCl_2=CCl_2$
ethylene-tetrafluoroethylene(ETFE)　乙烯-四氟乙烯共聚物
ethylene-tetrafluoroethylene copolymer　乙烯-四氟乙烯共聚物
ethylene-tetrafluoroethylene copolymer fibre　乙烯-四氟乙烯共聚纤维
ethylene thiocyanate(=ethylene thiocyanide)　硫氰酸亚乙酯　$(CH_2SCN)_2$

ethylene thiocyanide(=ethylene thiocyanate) 硫氰酸亚乙酯
ethylene thioketal 乙二硫醇缩酮
ethylene thiourea(=2-imidazolidine thione) 亚乙基硫脲
ethylene tribromide 三溴代乙烯 $CHBr=CBr_2$
ethylene trichloride 三氯代乙烯 $CHCl=CCl_2$
ethylene triols 烯属三醇
ethylene urea(=2-imidazolidinone) 亚乙基脲；2-咪唑啉酮 $(CH_2NH)_2CO$
ethylene urea formaldehyde resin 亚乙基脲醛树脂
ethylene-vinyl acetate(EVA) 乙烯-乙酸乙酯共聚物
ethylene vinyl acetate copolymer(EVA) 乙烯-乙酸乙烯酯共聚物
ethylene-vinyl acetate rubber 乙烯乙酸乙烯酯橡胶
ethylene-vinyl alcohol(EVAL) 乙烯-乙烯醇共聚物
ethylene-viny lalcohol copolymer 乙烯-乙烯醇共聚物
α-ethylenic acid α-烯酸
α,β-ethylenically-unsaturated α,β-烯键式不饱和的
ethylenically-unsaturated carboxylic acid ester 烯键式不饱和羧酸酯
ethylenic bond(=ethylenic link) 烯键
ethylenic homopolymer 乙烯型均聚物
ethylenic link(=ethylenic bond;ethylenic linkage) 烯键
ethylenic linkage(=ethylenic link) 烯键
ethylenimine(=aziridine) 氮丙啶；氮杂环丙烷 $NHCH_2CH_2$
ethylenimine polymer 1,2-亚乙基亚胺聚合物
ethyl erucate 芥酸乙酯 $C_{21}H_{41}CO_2C_2H_5$
ethyl ester 乙酯
ethyl ether ①(二)乙醚 $C_2H_5OC_2H_5$ ②乙基醚 ROC_2H_5
ethyl ether diaminetetraacetic acid 二乙醚二胺四乙酸
ethyl ethylacetoacetate 乙基乙酰乙酸乙酯 $CH_2COCH(C_2H_5)CO_2C_2H_5$
ethyl ethyleneimine(=2-ethylaziridine) 乙基吖丙啶
ethyl ethylidene acetone 乙基亚乙基丙酮
ethyl eugenol 乙基丁子香酚
ethyl farnesenate 金合欢酸乙酯
ethyl ferrocenoate(=ferrocene carboxylic acid ethyl ester) 二茂铁羧酸乙酯
ethyl fluid 乙基液〔汽油(四乙铅镇震)精〕
ethyl fluid blending 乙基液掺和
ethyl fluoride(=fluoroethane) 乙基氟；氟代乙烷 C_2H_5F
ethyl fluorosulfonate 氟磺酸乙酯
ethyl formate 甲酸乙酯 $HCOOC_2H_5$
ethyl fumarate 富马酸乙酯；反-丁烯二酸乙酯
ethyl α-furancarboxylate(=ethyl pyromucate) α-呋喃甲酸乙酯；焦黏酸乙酯 $C_4H_3OCOOC_2H_5$
ethyl 2-furanpropionate 2-呋喃基丙酸乙酯
ethyl furfurylacetate 糠基乙酸乙酯
ethyl furfuryl hydracrylate 羟基丙酸乙基糠酯
ethyl furoate 糠酸乙酯；α-呋喃甲酸乙酯 $C_4H_3OCOOC_2H_5$
α-ethyl-β-furyl acrolein α-乙基-β-呋喃基丙烯醛
ethyl-β-furyl glycidate β-呋喃基缩水甘油酸乙酯
ethyl 3-(2-furyl)-propanoate 3-(2-呋喃基)丙酸乙酯
ethyl furylpropionate 呋喃基丙酸乙酯
ethyl gallate 棓酸乙酯 $(HO)_3C_6H_2CO_2C_2H_5$
ethyl gasoline 乙基汽油
ethyl geranate 香叶酸乙酯
ethyl geranyl acetone 乙基香叶基丙酮
ethyl glutarate 戊二酸二乙酯 $C_2H_5OCO(CH_2)_3COOC_2H_5$
α-ethylglutaric acid α-乙基戊二酸 $C_2H_5C_3H_5(COOH)_2$
ethyl glycerate 甘油酸乙酯 $(HO)_2C_2H_3CO_2C_2H_5$
ethylglycinate 甘氨酸乙酯
ethyl glycine 乙基甘氨酸 $C_2H_5NHCH_2CO_2H$
ethyl glycol(=cellosolve) 乙基乙二醇
ethyl glycol acetate 乙基乙二醇乙酸酯
ethyl glycolate 乙醇酸乙酯 $HOCH_2CO_2C_2H_5$
ethyl-Grignard-reagent 乙基卤化镁 $MgXC_2H_5$
ethyl guaiacol 乙基愈创木酚
ethyl halide 乙基卤 C_2H_5X
ethyl heptanoate 庚酸乙酯
ethyl heptine carbonate 庚炔羧酸乙酯
ethyl heptylate 庚酸乙酯 $C_6H_{13}CO_2C_2H_5$
ethyl heptyl ether 乙基庚基醚；乙氧基庚烷 $C_2H_5OC_7H_{15}$
ethyl-2-heptynoate 2-庚炔酸乙酯
ethyl hexane 乙基己烷
2-ethyl-1,3-hexanediol 2-乙基-1,3-己二醇
ethyl hexanoate 己酸乙酯
ethylhexanol 乙基己醇
ethylhexene 乙基己烯
2-ethylhexoate ①2-乙基己酸酯；异辛酸酯②2-乙基己酸盐；异辛酸盐
ethylhexoate drier 乙基己酸盐催干剂；异辛酸盐催干剂
2-ethylhexoic acid 2-乙基己酸；异辛酸
2-ethyl hexyl acetate 乙酸-2-乙基己酯
ethyl hexylate 己酸乙酯
ethyl hexyl carbinol 乙基己基原醇
ethyl n-hexyl carbinyl acetate 乙酸乙基己基原酯
ethylhexyl cyano diphenyl acrylate 氰基二苯基丙烯酸乙基己酯
2-ethylhexyl diphenyl phosphate 2-乙基己基二苯基磷酸

酯〔增塑剂〕
2-ethyl-1,3-hexylene 2-乙基-1,3-己二烯
2-ethylhexyl epoxytallate 环氧妥尔酸-2-乙己酯〔增塑、稳定联合剂〕
2-ethyl hexyl ethanoate 乙酸异辛酯；乙酸-2-乙基己酯
ethyl hexyl ether 乙基己基醚；乙氧基己烷
 $C_2H_5OC_6H_{13}$
ethyl hexyl ketone 乙基己基酮
ethylhexyl methoxy cinnamate 甲氧基肉桂酸乙基己酯
2-ethylhexyl phosphonic acid(H_2EHP) 2-乙基己基膦酸
 $(2-C_2H_5C_6H_{12})PO(OH)_2$
ethylhexyl salicylate 水杨酸乙基酯
ethylhexyl sulfosuccinate 琥珀酸乙基己基磺酸盐
ethyl hinokiate 扁柏酸乙酯
ethyl hydrate 乙醇
ethyl hydrazine 乙肼 $C_2H_5NHNH_2$
ethyl hydrocinnamate 氢化肉桂酸乙酯
 $C_6H_5(CH_2)_2CO_2C_2H_5$
ethyl hydrocyanic ether 乙基氰；丙腈；氢氰酸乙酯
 C_2H_5CN
ethyl hydrogen peroxide 乙基过氧化氢
ethyl hydrogen sulfate 乙基硫酸；硫酸乙氢酯
 $C_2H_5SO_4H$
ethyl hydroperoxide 乙过醇；乙基过氧化氢 C_2H_5OOH
ethyl hydroselenide(=ethyl selenomercaptan) 乙硒醇；乙基硒氢 C_2H_5SeH
ethylhydrosiloxane 乙基氢硅氧烷
ethyl hydroxide(=ethyl alcohol) 乙醇
ethyl p-hydroxybenzoate 对-羟基苯甲酸乙酯；尼泊金乙酯
ethyl 3-hydroxybutanoate 3-羟基丁酸乙酯
ethylhydroxylamine 乙胺
5-ethyl-3-hydroxy-4-methyl-2(5H)-furanone 5-乙基-3-羟基-4-甲基-2(5H)-呋喃酮
ethyl hydroxy methyl ketone 乙基羟甲基酮
ethyl α-hydroxypropionate α-羟基丙酸乙酯
2-ethyl-2-hydroxy-4H-pyran-4-one 2-乙基-2-羟基-4H-吡喃-4-酮
ethyl hydroxy pyrone(=ethyl maltol) 乙基麦芽酚；乙基羟基吡喃酮
ethyl hypochlorite 次氯酸乙酯 C_2H_5OCl
ethylidene 亚乙基 $CH_3CH=$
ethylidene acetobenzoate 乙酰苯甲酸亚乙酯〔溶剂〕
ethylidene acetone(=methyl propenyl ketone) 亚乙基丙酮；甲基丙烯基(甲)酮 $CH_3CH=CHCOCH_3$
ethylidene bichloride 亚乙基二氯；1,1-二氯乙烷
 CH_3CHCl_2
ethylidene bromide 亚乙基二溴；1,1-二溴代乙烷
 CH_3CHBr_2
ethylidene chloride 亚乙基二氯；1,1-二氯乙烷
 CH_3CHCl_2
ethylidene cyanhydrin 亚乙基氰醇 $C_2H_4(OH)CN$
ethylidene diacetate 二乙酸亚乙基酯；1,1-二乙酸基乙烷
ethylidene dibromide 亚乙基二溴；1,1-二溴代乙烷
 CH_3CHBr_2
ethylidene dichloride 亚乙基二氯；1,1-二氯乙烷
 CH_3CHCl_2
ethylidene difluoride 亚乙基二氟；1,1-二氟乙烷
 CH_3CHF_2
ethylidene dihalide 亚乙基二卤 CH_3CHX_2
ethylidene diiodide 亚乙基二碘；1,1-二碘乙烷
 CH_3CHI_2
ethylidene diisocyanate 二异氰酸亚乙酯
ethylidene dimethylether 亚乙基二甲基醚
ethylidene diurethane(=ethylidene urethane) 亚乙基脲烷〔俗〕；亚乙基二氨基甲酸乙酯
ethylidene ether ①乙缩醛；乙醛缩乙二醇②醛缩醇；缩醛
ethylidene fluoride 亚乙基二氟；1,1-二氟乙烷
 CH_3CHF_2
γ-ethylidene glutamic acid γ-亚乙基谷氨酸
ethylidene glycol 亚乙基二醇 $CH_3CH(OH)_2$
ethylidene halide 亚乙基二卤 CH_3CHX_2
ethylidene hexanal(=butyl butenal) 亚乙基己醛；丁基丁烯醛
ethylidene iodide 亚乙基二碘；1,1-二碘乙烷 CH_3CHI_2
ethylidene lactic acid(=lactic acid) 乳酸
 $CH_3CHOHCOOH$
5-ethylidene-2-norbornene 5-亚乙基-2-降冰片烯
ethylidene perbromide 亚乙基二溴；1,1-二溴代乙烷
 CH_3CHBr_2
ethylidene perchloride 亚乙基二氯；1,1-二氯乙烷
 CH_3CHCl_2
ethylidene perfluoride 亚乙基二氟；1,1-二氟乙烷
 CH_3CHF_2
ethylidene periodide 亚乙基二碘；1,1-二碘乙烷
 CH_3CHI_2
ethylidene peroxide 亚乙基过氧(化物)
ethylidene-succinic anhydride 亚乙基丁二酸酐
 $CH_3CH=C_4H_2O_3$
ethylidene-urea 亚乙基脲；亚乙脲 $CH_3CH=NCONH_2$
ethylidene urethane(=ethylidene diurethane) 亚乙基脲烷〔俗〕；亚乙基二氨基甲酸乙酯
ethylidine(=ethylidyne) 次乙基 $CH_3C\equiv$
ethylidyne(=ethylidine) 次乙基 $CH_3C\equiv$
ethylin 乙灵〔多元醇的乙醚或甘油的乙醚〕

ethyl iodide(=iodoethane) 乙基碘；碘乙烷
ethyl iodoacetate 碘乙酸乙酯 $ICH_2CO_2C_2H_5$
ethyl ionone 乙基紫罗兰酮
ethyl ionylidene acetate 乙酸乙基紫罗兰叉酯
ethyl isoamyl carbinyl acetate 乙酸乙基异戊基原酯
ethyl isoamyl ether 乙基异戊基醚 $C_2H_5CO_5H_{11}$
ethyl isoamyl ketone 乙基异戊酮
ethyl isobutyl ether 乙基异丁基醚 $C_2H_5COH_2CH(CH_3)_2$
ethyl isobutyl ketone 乙基异丁基(甲)酮 $C_2H_5COC_4H_9$
ethyl isobutyrate 异丁酸乙酯
ethyl isocyanate 异氰酸乙酯 C_2H_5NCO
ethyl isocyanide 乙基肼 CH_3CH_2NC
ethyl isonitrile(=ethyl carbylamine) 乙肼；乙基异腈 C_2H_5NC
ethyl isopropyl ether 乙基异丙基醚；异丙氧基乙烷 $C_2H_5OCH(CH_3)_2$
ethyl isopropyl ketone 乙基异丙基(甲)酮；3-异丙酮 $C_2H_5COCH(CH_3)_2$
ethyl isorhodanide(=ethyl isothiocyanate) 异硫氰酸乙酯
ethyl isothiocyanate(=ethyl isorhodanide) 异硫氰酸乙酯
ethyl isovalerate 异戊酸乙酯
ethylization(=ethylizing) 乙基化
ethylized fuel 乙基汽油
ethylizer 乙基化器〔将气缸加乙基油的装置〕
ethylizing(=ethylization) 乙基化
ethyl α-ketopropionate α-酮基丙酸乙酯
ethyl lactate 乳酸乙酯 $CH_3CHOHCO_2C_2H_5$
ethyl laurate(=ethyl dodecanoate) 十二烷酸乙酯；月桂酸乙酯
ethyl lead susceptibility 乙基铅敏感性〔汽油的铅感性〕
ethyl levulinate γ-戊酮酸乙酯 $CH_3CO(CH_2)_2CO_2C_2H_5$
ethyl linalool 乙基芳樟醇
ethyl linalyl acetate 乙酸乙基芳樟酯
ethyl magarate 十七烷酸乙酯 $CH_3(CH_2)_{15}CO_2C_2H_5$
ethyl-magnesium-bromide 溴化乙基镁 $MgBrC_2H_5$
ethyl-magnesium-chloride 氯化乙基镁 $MgClC_2H_5$
ethyl-magnesium-halide 卤化乙基镁 $MgXC_2H_5$
ethyl malate 苹果酸乙酯
ethyl maleate 马来酸乙酯；顺丁烯二酸(二)乙酯
ethyl malonate ①乙基丙二酸盐(或酯) $C_2H_5CH(COOM)$ ②丙二酸二乙酯 $CH_2(COOEt)_2$
ethyl malonic acid 乙基丙二酸 $C_2H_5CH(CO_2H)_2$
ethyl malonic ester 乙基丙二酸酯
ethyl maltol(=ethyl hydroxy pyrone) 乙基麦芽酚；乙基羟基吡喃酮
ethyl mandelate 扁桃酸乙酯 $C_6H_5CHOHCO_2C_2H_5$
ethyl menthol 乙基薄荷醇 $C_{12}H_{24}O$

ethyl mercaptan 乙硫醇 C_2H_5SH
β-ethylmercaptopropionaldehyde β-乙基硫醇丙醛〔热稳定剂〕
ethyl 2-mercaptopropionate 2-巯基丙酸乙酯
ethyl mercuric chloride 氯化乙基汞 $HgClC_2H_5$
ethylmercuric hydroxide 氢氧化乙基汞 C_2H_5HgOH
ethyl mercuric-iodide 碘化乙基汞 $HgIC_2H_5$
ethyl mercuric phosphate(=ethyl mercury phosphate) 磷酸乙基汞 $(C_2H_5HgO)_2PO(OH)$
ethyl mercuride(=ethyl mercury) 二乙汞
ethyl mercury(=ethyl mercuride) 二乙汞 $Hg(C_2H_5)_2$
ethylmercury hydrogen phosphate 磷酸乙基汞
ethyl mercury phosphate(=ethyl mercuric phosphate) 磷酸乙基汞 $(C_2H_5HgO)_2PO(OH)$
ethyl methacrylate 甲基丙烯酸乙酯；异丁烯酸乙酯 $C_3H_5CO_2C_2H_5$
ethyl methane sulfonate(EMS) 甲磺酸乙酯
ethyl method for gum content 胶质含量的乙基测定法〔汽油中胶质测定法〕
α-ethyl-p-methoxyphenyl propionaldehyde α-乙基对甲氧基苯丙醛
ethyl methyl acetic acid 乙基甲基乙酸 EtCHMeCOOH
ethyl N-methyl antranilate N-甲基邻氨基苯甲酸乙酯
ethyl 2-methylbutyrate 2-甲基丁酸乙酯
ethyl methyl dioxolane acetate 乙酸乙基甲基二噁烷酯
2-ethyl-4-methylimidazole 2-乙基-4-甲基咪唑〔催化剂〕
ethyl methyl ketone 甲基乙基甲酮；甲乙酮
ethyl methyl phenylglycidate 草莓醛；十六醛
3-ethyl-2-methylpyrazine 3-乙基-2-甲基吡嗪
1-ethyl-4-methylquinolinium iodide 1-乙基-4-甲基喹啉碘
ethyl morphine(=codethyline) 可待乙碱；乙基吗啡 $C_{19}H_{23}O_3N$
ethylmorphine hydrochloride(=dionin) 盐酸乙基吗啡；地昂宁
N-ethyl morpholine N-乙基吗啉〔硫化剂〕
ethyl mustard oil 乙基芥子油；异硫氰酸乙酯 C_2H_5NCS
ethyl myristate 肉豆蔻酸乙酯；十四(烷)酸乙酯 $C_{13}H_{27}CO_2C_2H_5$
ethyl myristinate 肉豆蔻酸乙酯；十四(烷)酸乙酯
ethylnaphthalene 乙萘 $C_2H_5C_{10}H_7$
ethyl α-naphthyl ether 乙基α-萘基醚 $C_{10}H_7OC_2H_5$
ethyl nicotinate 烟酸乙酯 $C_5H_4NCO_2C_2H_5$
ethyl nitrate 硝酸乙酯 $C_2H_5ONO_2$
ethyl nitrite 亚硝酸乙酯 C_2H_5ONO
ethyl nitroacetate 硝基乙酸乙酯 $NO_2CH_2CO_2C_2H_5$
ethyl nitrobenzoate 硝基苯甲酸乙酯

$NO_2C_6H_4CO_2C_2H_5$
ethyl nitrocinnamate 硝基肉桂酸乙酯
　　$NO_2C_6H_4CH=CHCO_2C_2H_5$
ethyl nitrolic acid 乙硝肟酸 $CH_3C(NOH)NO_2$
ethyl nonanoate 壬酸乙酯
ethyl nonylate 壬酸乙酯
ethyl 2-nonynoate 2-壬炔酸乙酯
ethyl number 乙基值
ethyl 9-octadecenoate 9-十八烯酸乙酯
ethyl octine carbonate 辛炔羧酸乙酯
ethyl octylate 辛酸乙酯
ethyl octyl ether 乙基辛基醚 $C_2H_5OC_8H_{17}$
ethyl octyl ketone 乙基辛基(甲)酮; 3-十一(烷)酮 $C_2H_5COC_8H_{17}$
ethyl octynecarbonate 辛炔酸乙酯
ethyl 2-octynoate 2-辛炔酸乙酯
ethyl oenanthate 庚酸乙酯; 酒醚〔俗〕
ethylogen 乙烯原
ethyloic 羧甲基 —CH_2COOH
ethylol 羟乙基 —CH_2CH_2OH
ethylolacrylamide 羟乙基丙烯酰胺
ethyl oleate 油酸乙酯 $C_{17}H_{33}CO_2C_2H_5$
ethylolmethacrylamide 羟乙基甲基丙烯酰胺
ethyl orange(=sodium diethylaniline-azobenzene sulfonate) 乙基橙; 二乙基苯胺偶氮苯磺酸钠
ethyl orthoacetate 原乙酸乙酯 $CH_3C(OC_2H_5)_3$
ethyl orthocarbonate 原碳酸乙酯 $C(OC_2H_5)_4$
ethyl orthoformate 原甲酸乙酯 $HC(OC_2H_5)_3$
ethyl orthosilicate 正硅酸乙酯 $Si(OC_2H_5)_4$
ethyl oxalacetate 草乙酸乙酯〔俗〕; 2-丁酮二酸二乙酯
ethyl oxalacetic acid 乙基 2-氧(代)丁二酸
　　$COOHCOCH(C_2H_5)COOH$
ethyl oxalate 草酸乙酯〔溶剂〕
ethyl oxamate(=ethyl oxamide) 草氨酸乙酯〔俗〕; 氨羰基甲酸乙酯 $H_2NCOCO_2C_2H_5$
ethyl oxamide(=ethyl oxamate) 草氨酸乙酯〔俗〕; 氨羰基甲酸乙酯 $H_2NCOCO_2C_2H_5$
ethyl oxide (二)乙醚 $C_2H_5OC_2H_5$
ethyl oxindole 乙基羟吲哚 $C_8H_6ONC_2H_5$
ethyl 3-oxobutanoate 3-氧代丁酸乙酯
ethyl 4-oxopentanoate 4-氧代戊酸乙酯
ethyl 2-oxopropanoate 2-氧代丙酸乙酯
ethyl oxydithiocarbonic acid 乙氧基二硫代羧酸; 乙黄原酸 C_2H_5OCSSH
ethyl oxyhydrate 老姆醚
ethyl palmitate 十六(烷)酸乙酯; 棕榈酸乙酯 $C_{15}H_{31}CO_2C_2H_5$
ethyl pelargonate 壬酸乙酯 $C_8H_{17}CO_2C_2H_5$

ethylpentane 乙基戊烷 $(C_2H_5)_3CH$
ethylpentanol 乙基戊醇
ethylpentene 乙基戊烯
ethyl perbromide 全溴乙烷 C_2Br_6
ethyl perchlorate 高氯酸乙酯 $C_2H_5ClO_4$
ethyl perchloride 全氯乙烷; 六氯乙烷 C_2Cl_6
ethyl periodide 全碘乙烷 C_2I_6
ethyl peroxide 过氧乙醚 $(C_2H_5)_2O_2$
ethyl perselenide 二乙基化二硒 $C_2H_5SeSeC_2H_5$
ethyl persulfide 二乙基化二硫 $C_2H_5SSC_2H_5$
ethyl petrol 乙基石油醚; 含四乙铅汽油
ethylphenol 乙基苯酚 $C_2H_5C_6H_4OH$
o-ethyl phenol 邻乙基苯酚
p-ethyl phenol 对乙基苯酚
ethyl phenolate(=phenetole) 苯乙醚; 乙氧基苯 $C_2H_5OC_6H_5$
ethyl phenoxyacetate 苯氧基乙酸乙酯
ethyl phenoxybutyrate 苯氧基丁酸乙酯 $C_2H_5OCH_2(CH_2)_2CO_2C_2H_5$
p-ethyl phenylacetaldehyde 对乙基苯乙醛
ethyl phenylacetate 苯乙酸乙酯
ethyl β-phenylacrylate β-苯基丙烯酸乙酯
ethyl phenylbarbiturate 乙基苯基巴比妥酸盐
ethyl phenylbutyrate 苯丁酸乙酯
ethyl phenylcarbinol 乙基苯基甲醇 $C_2H_5CHOHC_6H_5$
ethyl phenyl carbinyl butyrate 丁酸乙基苯基甲酯
ethyl phenyl ether 乙基苯基醚 PhOEt
ethyl phenylglycidate 苯环氧丙酸乙酯
ethyl 3-phenylglycidate 3-苯基缩水甘油酸乙酯
α-ethylphenylhydrazine α-乙基苯基肼 $C_2H_5NHNHC_6H_5$
2-ethyl phenyl mercaptan 2-乙基苯硫醇
ethylphenyl-nitrosamine(=N-nitroso-ethylaniline) N-亚硝基-N-乙基苯胺 $C_6H_5N(NO)C_2H_5$
ethyl phenylpropiolate 苯丙炔酸乙酯
ethyl phenyl sulfide 乙基苯基硫; 乙硫基苯; 苯乙硫醚 $C_6H_5SC_2H_5$
ethyl phenyl sulfone 乙基苯基砜 $C_2H_5SO_2C_6H_5$
ethyl phenyl triethylsilicane 乙苯基三乙基硅 $C_2H_5C_6H_4Si(C_2H_5)_3$
ethyl phenyl urea 乙基苯基脲
ethyl phosphate 磷酸三乙酯 $(C_2H_5O)_3PO$
ethyl phosphine 乙膦 $C_2H_5PH_2$
ethyl phosphinic acid 乙次膦酸 $C_2H_5PO(H)(OH)$
ethyl phosphoacid(=ethyl phosphonic acid) 乙膦酸; 乙基磷酸 $C_2H_5PO(OH)_2$
ethyl phosphonic acid(=ethyl phosphoacid) 乙膦酸; 乙基磷酸 $C_2H_5PO(OH)_2$

ethyl phosphoric acid 乙代磷酸；磷酸一乙酯
ethyl phosphorous acid 乙基亚磷酸 $C_2H_5PO(OH)_2$
ethyl phthalate(DEP) 邻苯二甲酸乙酯〔溶剂〕
ethyl phthalyl ethyl glycollate 乙基邻苯二甲酰基甘酸乙酯〔增塑剂〕
ethyl pinonate 漀酮酸乙酯
N-ethyl piperidine N-乙基哌啶〔硫化剂〕
ethyl polysilicate 聚硅酸乙酯
ethyl potassium 乙基钾 KC_2H_5
1-ethyl-1-propanol 1-乙基-1-丙醇
ethyl propargyl ether 乙基炔丙基醚；2-丙炔乙醚 $CH≡CCH_2OC_2H_5$
ethyl propenoate 丙烯酸乙酯
ethyl propiolate 丙炔酸乙酯
ethyl propionate 丙酸乙酯
ethyl propionylpyrurate 丙酰基丙酮酸乙酯
α-ethyl-propyl acetate 乙酸α-乙基丙酯 $CH_3CO_2CH(C_2H_5)_2$
ethyl propyl-acetoacetate 丙基乙酰乙酸乙酯 $CH_3COCH(C_3H_7)CO_2C_2H_5$
ethyl propyl acrolein 乙基丙基丙烯醛 $C_3H_7CH=C(C_2H_5)CHO$
ethyl propyl carbinol 乙基丙基甲醇
ethyl propyl-dibenzylsilicane 乙基丙基二苄基硅 $(C_2H_5)(C_3H_7)(C_6H_5CH_2)_2Si$
ethyl propyl ether 乙基丙基醚；乙氧基丙烷 $C_2H_5OCH_2C_2H_5$
ethyl propyl ketone 乙基丙基酮
ethyl pseudoionone 乙基假性紫罗兰酮
ethyl pyrazine 乙基吡嗪
3-ethyl-2-pyrazinyl methyl ketone 3-乙基-2-吡嗪基甲基酮
ethylpyridine 乙基吡啶 $C_2H_5C_5H_4N$
ethyl pyromucate(=ethyl α-furancarboxylate) α-呋喃甲酸乙酯；焦黏酸乙酯 $C_4H_3OCOOC_2H_5$
ethyl pyruvate 丙酮酸乙酯 $CH_3COCO_2C_2H_5$
ethyl quinolinium iodide 碘化 N-乙基喹啉鎓
ethyl rhodanate(=ethyl rhodanide) 硫氰酸乙酯；乙基硫氰 C_2H_5SCN
ethyl rhodanide(=ethyl rhodanate) 硫氰酸乙酯；乙基硫氰 C_2H_5SCN
ethyl rhodinol 乙基玫瑰醇
ethyl ricinoleate 蓖麻酸乙酯 $HOC_{17}H_{32}CO_2C_2H_5$
ethyl rubber 乙基橡胶
ethyl safranate 藏红花酸乙酯
ethyl salicylate(=sal ethyl) 水杨酸乙酯
ethyl sebate 癸二酸乙酯
ethyl selenide 二乙硒 $(C_2H_5)_2Se$
ethyl seleninic acid 乙基亚硒酸 C_2H_5SeOOH
ethyl selenomercaptan(=ethyl hydroselenide) 乙硒醇 C_2H_5SeH
ethyl silicane 乙基甲硅烷 $C_2H_5SiH_3$
ethyl silicate(=tetraethyl orthosilicate) 硅酸(四)乙酯
ethyl-siliconic acid 乙基硅酸 C_2H_5SiOOH
ethyl silicon oil 乙基硅油
ethyl sodio benzyl malonate 钠代苄基丙二酸乙酯 $C_2H_5OCOCNa(C_6H_5CH_2)COOC_2H_5$
ethyl sorbate 山梨酸乙酯
ethyl stannic acid(=ethyl stannonic acid) 乙基锡酸
ethyl stannonic acid(=ethyl stannic acid) 乙基锡酸
ethyl stearate 硬脂酸乙酯 $C_{17}H_{35}CO_2C_2H_5$
ethyl stibamine 乙胺
ethyl styryl ketone 乙基苯乙烯酮
ethyl suberate 辛二酸乙酯 $C_2H_5OCO(CH_2)_6COOC_2H_5$
ethyl succinate 丁二酸乙酯；琥珀酸乙酯 $C_2H_5OCO(CH_2)_2COOC_2H_5$
ethyl succinic acid 乙基丁二酸 $COOHCH_2CH(C_2H_5)COOH$
ethyl sulfate(=diethyl sulfate) 硫酸乙酯 $(C_2H_5)_2SO_4$
ethyl sulfide 乙硫醚；二乙硫 $(C_2H_5)_2S$
ethyl sulfinic acid 乙亚磺酸 C_2H_5SOOH
ethyl sulfocarbonate 硫代碳酸乙酯
ethyl sulfocyanate(=ethyl sulfocyanide) 硫氰酸乙酯；乙基硫氰 C_2H_5SCN
ethyl sulfocyanide(=ethyl sulfocyanate) 硫氰酸乙酯；乙基硫氰 C_2H_5SCN
ethyl sulfonamide 乙磺酰胺 $C_2H_5SO_2NH_2$
ethyl sulfone 乙基砜；二乙砜 Et_2SO_2
ethyl sulfonic acid 乙磺酸 $C_2H_5SO_2OH$
ethyl sulfonyl chloride 乙磺酰氯 $C_2H_5SO_2Cl$
ethyl-sulfoxide 乙基亚砜；二乙亚砜 $(C_2H_5)_2S=O$
ethyl sulfuric acid 乙基(代)硫酸；硫酸一乙酯；硫酸氢乙酯 $C_2H_5OSO_2OH$
ethyl-sulfurous acid 乙基(代)亚硫酸 $C_2H_5OSO_2H$
ethyl sulfuryl chloride(=ethylchlorosulfonate) 乙基硫酰氯；氯磺酸乙酯 $C_2H_5OSO_2Cl$
ethyl-tartrate 酒石酸二乙酯 $C_2H_5OCO(CHOH)_2COOC_2H_5$
ethyl-telluride 二乙碲 $Te(C_2H_5)_2$
ethyl-2-thiazolyl ether 乙基-2-噻唑基醚
ethylthio 乙硫基 CH_3CH_2S-
ethyl thioacetate 硫代乙酸乙酯 $CH_3COSC_2H_5$
S-ethyl thiocarbamate(=thiourethane) 硫代氨基甲酸 S-乙酯；硫尿烷
O-ethyl thiocarbamate(=thioxanthamide) 硫代氨基甲酸 O-乙酯

ethyl thiocarbonate 硫代碳酸乙酯
ethyl thiocyanate(=ethyl thiocyanide) 硫氰酸乙酯 C_2H_5SCN
ethyl thiocyanide(=ethyl thiocyanate) 硫氰酸乙酯
ethyl thioether 乙硫醚；二乙基硫 $(C_2H_5)_2S$
ethyl thioglycolate 巯基乙酸乙酯 $HSCH_2CO_2C_2H_5$
ethyl thioglycollic acid 乙硫基乙酸 $C_2H_5SCH_2CO_2H$
ethyl thionamic acid 乙氨基磺酸 $C_2H_5NHSO_3H$
ethylthiophene 乙基噻吩
2-ethylthiophenol 2-乙基硫代酚
ethyl thiosulfonic acid 乙硫基磺酸 $C_2H_5SSO_3H$
ethyl-thiourea 乙硫脲 $NH_2CSNHC_2H_5$
ethyl tiglate 惕各酸乙酯
ethyl tin 四乙基锡 $Sn(C_2H_5)_4$
ethyl tin chloride 三乙基氯化锡 $SnCl(C_2H_5)_3$
ethyl tin dichloride 二乙基二氯化锡 $SnCl_2(C_2H_5)_2$
ethyl tin monochloride 三乙基氯化锡 $SnCl(C_2H_5)_3$
ethyl tin tribromide 三溴化乙基锡
ethyl tin trichloride 乙基三氯化锡 $SnCl_3(C_2H_5)$
ethyl toluate 甲苯甲酸乙酯 $CH_3C_6H_4CO_2C_2H_5$
ethyltoluene 乙基甲苯 $C_2H_5C_6H_4CH_3$
ethyl tolyl ether 乙基甲苯基醚 $C_2H_5OC_6H_4CH_3$
ethyl trans-2-butenoate 反-2-丁烯酸乙酯
ethyl trans-2-cis-4-decadienoate 反-2-顺-4-癸二烯酸乙酯
ethyl trans-2,3-dimethylacrylate 反-2,3-二甲基丙烯酸乙酯
ethyl triazine resin 乙基三嗪树脂
ethyl triazoacetate 叠氮基乙酸乙酯 $N_3CH_2CO_2C_2H_5$
ethyl tribromacetate 三溴乙酸乙酯 $Br_3CCO_2C_2H_5$
ethyl trichloroacetate 三氯乙酸乙酯 $Cl_3CCO_2C_2H_5$
ethyl trichloro silicane 乙基三氯硅 $C_2H_5SiCl_3$
ethyl triethoxy silicane 乙基三乙氧基硅 $C_2H_5Si(OC_2H_5)_3$
ethyl trifluoroacetate 三氟乙酸乙酯 $CF_3COOC_2H_5$
ethyl trimethyl silicane 乙基三甲基硅 $C_2H_5Si(CH_3)_3$
ethyl undecylate 十一(碳)烷酸乙酯 $C_{10}H_{21}CO_2C_2H_5$
ethyl undecylenate 十一烯酸乙酯
ethyl undecyl ketone 乙基十一基(甲)酮；3-十四(烷)酮 $C_2H_5COC_{11}H_{23}$
ethyl urea 乙脲 $C_2H_5NHCONH_2$
ethyl urethane 氨基甲酸乙酯 $NH_2COOC_2H_5$
ethyl valerate 戊酸乙酯
ethyl value 乙基值
ethyl vanillate 香兰酸乙酯
ethyl vanillin(=ethovan) 乙基香兰素
ethyl vanillylalcohol 乙基香兰醇
ethyl vinyl carbinol 乙基乙烯基甲醇 $CH_2=CHCHOHC_2H_5$
ethyl vinyl ether 乙基乙烯基醚〔溶剂〕
ethyl vinyl ketone 乙基乙烯基甲酮〔溶剂〕
ethyl vinyl sulfide 乙基乙烯基硫；乙硫基乙烯
ethyl xanthate(=ethyl xanthogenate) 黄原酸乙酯；乙黄原酸乙酯 $C_2H_5OCSSC_2H_5$
ethyl xanthic acid 乙基黄原酸；乙氧基荒酸 C_2H_5OCSSH
ethyl xanthogenate 黄原酸乙酯
ethynation(=ethynylation) 乙炔化(作用)
ethyne 乙炔
ethynyl 乙炔基 $CH\equiv C—$
ethynylation(=ethynation) 乙炔化(作用)
ethynyl carbinol 乙炔基甲醇 $HC\equiv CCH_2OH$
ethynylene 亚乙炔基 $—C\equiv C—$
17β-ethynylestradiol 17β-乙炔雌(甾)二醇
ethythrin 乙基除虫菊〔杀虫剂〕
etianate 因烷酸酯
etianic acid 因烷酸
etidine 因二烯
etidronate disodium 羟乙膦酸钠〔药〕
etidronic acid 羟乙磷酸
etimicin 依替米星〔药〕
etio(=aetio) 本；初
etiocholane 本胆烷
etiocholanedione 本胆烷二酮
etiocholanolone 本胆烷醇酮
etiocholanolone fever 本胆烷醇酮热
etiocobalamin 本钴胺
etiophyllin 初卟啉合镁盐
etnylenetrithiocarbamate 三硫代氨基甲酸亚乙酯〔促进剂〕
etomidate 依托咪酯〔药〕
etoposide 依托泊苷〔药〕
etruscomycin(=lucensomycin) 意北霉素；鲁斯霉素
ettringite 钙矾石；钙铝矾
eualpinol 优良姜醇
eubacteria 真细菌
eubiosis 生态平衡
eucaine 优卡因〔药〕
eucairite 硒铜银矿
eucalin 优卡林
eucalyptene 桉树烯 $C_{10}H_{16}$
eucalyptin 桉树素
eucalyptol(=cineole) 桉树油素
eucalyptole(=cajuputole) 桉树脑；顺(式)-1,8-萜二醇内醚；白千层脑 $C_{10}H_{18}O$
eucalyptolene 桉树脑烯 $C_{10}H_{16}$

eucalyptus oil ①桉树油②桉叶油
eucarvane 优香芹烷
eucarvone 优香芹酮；优葛缕酮
eucatropine 优卡托品〔药〕
eucazulene 桉蓝烃 $C_{15}H_{18}$
euchinine 优奎宁〔药〕
euchlorine 优氯〔氯与二氧化氯的混合物〕
euchroite 翠砷铜矿
euclase(=euclasite) 蓝柱石
euclasite(=euclase) 蓝柱石
Euclidean body(=rigid solid) 欧几里得体；刚性体
eucodal 二羟可待因酮；优可达〔药〕
eucodin(=codeine methyl bromide) 甲基溴可待因；优可定〔药〕
eucol 愈创木酚乙酸酯
eucolloid 真胶体
Eucommea rubber 杜仲胶
eucrytite 锂霞石
euctolite 橄金云斑岩
eudalene 优达烯；1-甲基-7-异丙基萘
eudesmane 桉叶烷
eudesmene(=selinene) 桉叶烯 $C_{15}H_{24}$
eudesmin 桉素
eudesmol 桉叶油醇
eudesmyl acetate 乙酸桉叶酯
eudialite 异性石
eudiaron 优地酮
eudiometer 量气管；容积变化测定管
eudiometry 空气纯度测定法；气体测定(法)
euflavine 优黄素
eugallol(=pyrogallol monoacetate) (一)乙酸焦棓酚 $CH_3COOC_6H_3(OH)_2$
eugenel type basil oil 丁香罗勒油
eugenic acid ①(=eugetinic acid)丁子香酸②(=eugenol)丁子香酚
eugenin 丁子香色酮
eugenitin 丁子香亭
eugenitol 丁子香脑
eugenol 丁子香酚
eugenol acetate 乙酸丁子香酚酯 $CH_3CO_2C_{10}H_{11}O$
eugenol benzoate 苯甲酸丁子香酚酯
eugenol benzyl ether 丁子香酚苄醚
eugenol cinnamate 肉桂酸丁子香酚酯 $C_8H_7CO_2C_{10}H_{11}O$
eugenol formate 甲酸丁子香酚酯 $C_{11}H_{12}O_3$
eugenol methyl ether 丁子香酚甲醚
eugenol phenylacetate 苯乙酸丁子香酚酯
eugenol type of basil oil 丁子香酚型罗勒油

eugenone 丁子香酮
eugentiogenin 优龙胆配基
eugenyl 丁子香基
eugenyl acetate(=acetyl eugenol) 乙酸丁子香酚酯；乙酰基丁子香酚
eugenyl benzoate 苯甲酸丁子香酚酯
eugenyl cinnamate (肉)桂酸丁子香酚酯
eugenyl formate 甲酸丁子香酚酯
eugenyl methyl ether 丁子香酚甲基醚
eugetic acid(=eugetinic acid) 丁子香酸 $C_{11}H_{12}O_4$
eugetinic acid(=eugetic acid) 丁子香酸 $C_{11}H_{12}O_4$
euhedral 正体面的；同面体型；全形的；自形的
euionone 优紫罗兰酮
eukairite 硒铜银矿
eukodal(=oxycodone) 盐酸羟可待酮；优可达〔药〕
eulachon oil 烛鱼油
Euler equation of motion 欧拉运动方程
Euler head 欧拉压头；理论压头
Euler number 欧拉准数
Euler's expansion formula 欧拉展开式
eulicin 美菌素
eulytine(=eulytite) 闪铋矿；硅铋矿；硅铋石
eumelanin 正常黑素
eumenol 当归浸膏；白芷
euminycin 弓霉素
eumycin 优霉素
euonic acid 卫矛酸
euonymin(=euonymus) 卫矛苷
euonymit 卫矛醇
euonymus(=euonymin) 卫矛苷
euosmite 木脂石
eupafolin 楔叶泽兰素
euparin 泽兰素
eupatene 泽兰烯
eupatilin 异泽兰黄素
eupatol 泽兰醇
eupatolide 泽兰内酯
eupatolin 泽兰苷
eupatolitin 泽兰苷配基
eupatoretin 泽兰亭
eupatoria oil 泽兰油
eupatorine 泽兰碱
eupatorinin 泽兰苷
eupatoriopicrin 泽兰苦质
euphadienol(=-euphol) α-大戟脑；大戟二烯醇
α-euphol(=euphadienol) α-大戟脑；大戟二烯醇
euphorbadienol 大戟根二烯醇
euphorbia A 大戟 A

euphorbia B 大戟 B
euphorbia C 大戟 C
euphorbic acid 大戟酸 $C_{24}H_{30}O_6$
euphorbin 大戟素
euphorbium 大戟(乳)脂
euphorbol 大戟醇
γ-euphorbol(=euphadienol) γ-大戟醇；大戟二烯醇
euphorbone 大戟酮 $C_{30}H_{48}O$
euphorin 苯基氨基甲酸乙酯
euphthalmine 优卡(加)托品
euphyllite 钠钾云母
eupittone(=eupittonic acid) 优皮酮；优皮酮酸
eupittonic acid(=hexamethylaurine;eupittone) 优皮酮；优皮酮酸；六甲金精 $C_{19}H_8(OCH_3)_6O_3$
euporphin(e) 溴甲烷阿朴吗啡
eupyrin(e) 优剖林〔药〕
euquinine(=quinine ethyl carbonate) 无味奎宁；碳酸乙酯奎宁
Eurex process 尤雷克斯过程〔词源: enriched uranium extraction，意大利发展的高浓铀燃料胺类萃取后处理流程〕
eurhodine 二氨吖嗪；优若定
eurhodole 二羟吖嗪；优若哚
eurimycin(=eurymycin) 泛霉素
eurobin(=chrysarobin triacetate) 三乙酸柯桠素酯
eurocidin 优洛杀菌素
europia 氧化铕 Eu_2O_3
europium 铕
europium chloride 三氯化铕 $EuCl_3$
europium dichloride 二氯化铕 $EuCl_2$
europium hydroxide 氢氧化铕 $Eu(OH)_3$
europium nitrate 硝酸铕 $Eu(NO_3)_3$
europium oxide 氧化铕 Eu_2O_3
europium oxychloride 氯氧化铕 $EuOCl$
europium sesquioxide 氧化铕 Eu_2O_3
europium sulfate 硫酸铕 $Eu_2(SO_4)_3$
europous chloride 二氯化铕 $EuCl_2$
europous sulfate 硫酸亚铕 $EuSO_4$
eurotin 败菌素
Eurowatt process 磷酸三丁酯-煤油处理过程
eurymycin(=aurimycin) 泛霉素
eurythrol 牛胆浸膏
eusantonenic acid 优山道年烯酸
eusantonine 优山道年
euscopol 优斯可朴
eusintomycin 硬脂酸合霉素
eustenin 优斯登宁
eutacticity 理想的构型规整性
eutannin 优单宁
eutaxitic structure 共融斑状结构
eutectic ①共晶；共晶体②易熔的；低共熔的③易熔质；低共熔体(点)；低共熔(混合物)
eutectic alloy 低共熔合金
eutectic behavior 低共熔行为
eutectic colonies 共晶群体
eutectic crystal 共晶；低共熔晶体
eutectic graphite 共晶石墨
eutectic melting point 低共熔点；共晶熔点
eutectic mixture 低共熔(混合)物*
eutectic point 低共熔点*
eutectic polymer 低共熔聚合物
eutectics ①低共熔性②低共熔混合物
eutectic salt(=low melt salt) 低共熔盐；易熔盐
eutectic side 低共熔体侧
eutectic state 低共熔态
eutectic temperature 低共熔温度
eutectoid 低共熔体；共析体
eutectoid point 低共熔点
eutectoid steel 共析钢
eutectophyric structure 流状结构
euthalite(=enthallite) 密方沸石
eutrophication 富营养化
eutrophic lake 富营养湖
eutropic 异序同晶的
eutropic series 异序同晶系
eutropy 异序同晶(现象)
euvitrain 纯镜煤
euxanthic acid(=euxanthinic acid) 优黄酸 $C_{19}H_{16}O_{10}$
euxanthin 优黄质
euxanthinic acid(=euxanthic acid) 优黄酸
euxanthonic acid 优呫吨酮酸 $C_{13}H_{10}O_5$
euxantogen 优黄原
euxenite 黑稀金矿
evacuant 排除药
evacuated chamber 真空室
evacuated space 稀疏空间
evacuated vessel 抽空容器
evacuating equipment 抽空设备
evacuation ①抽(成真)空②撤离
evacuation hull 浸取后的包壳
evacuation of viscose 黏胶脱泡；黏胶去气
evaluating of crude oil 原油评价
evaluation 估定；评定；评价
evaluation method 评价方法
evaluation of catalyst microstructure 催化剂显微结构的测定

evaluation test 评价试验
evaluation test of column 柱评价试验
e value e 值
evanescent elasticity 渐消(失)弹性
evanescent wave 消失波
evanescent wave induced fluorescence spectroscopy 逝波诱导荧光光谱法
Evans diagram 埃文斯图
evansite 核磷铝石
evaporable 可蒸发的；挥发的
evaporated 蒸发的
evaporated crystallizer 蒸发结晶器
evaporated salt 食盐
evaporated skimmed milk 脱脂奶粉
evaporated volume 蒸发体积
evaporating 蒸发；汽化
evaporating apparatus 蒸发器
evaporating column 浓缩柱；蒸浓柱
evaporating dish 蒸发皿
evaporating field (海水)蒸发场
evaporating pan 蒸发锅
evaporating pipe 浓缩管；蒸发管
evaporating pot 蒸发罐
evaporating tower 浓缩塔；蒸浓塔
evaporation 蒸发(作用)
evaporation area 蒸发(表)面；汽化(表)面
evaporation coefficient 蒸发系数
evaporation coil 蒸发旋管；蒸发器
evaporation cooling 蒸发冷却
evaporation cooling effect 蒸发冷却效应
evaporation curve 蒸发曲线
evaporation drying 蒸发干燥
evaporation efficiency 蒸发效率
evaporation gum test 蒸发胶质试验〔油品〕
evaporation heat 蒸发热
evaporation latent heat 蒸发潜热
evaporation losses 蒸发损失
evaporation method 蒸发法
evaporation number 蒸发数；蒸发值
evaporation rate 蒸发率；蒸发速率；蒸发比率；挥发率
evaporation rate of solvent 溶剂蒸发速率
evaporation residue 蒸发残留物
evaporation retardant 蒸发抑制剂
evaporation suppressor 蒸发抑制剂
evaporation surface 蒸发面
evaporation test 蒸发试验
evaporation test of gasoline 汽油的蒸发试验
evaporation to dryness 蒸干
evaporation tower 蒸发塔
evaporative 蒸发的
evaporative capacity 蒸发能力
evaporative condenser 蒸发冷凝器
evaporative cooler 蒸发冷却器
evaporative cooling 蒸发冷却
evaporative crystallizer 蒸发(式)结晶机
evaporative light-scattering detector 蒸发光散射检测器
evaporative power 蒸发本领
evaporative surface condenser 蒸发表面冷凝器
evaporativity 蒸发度
evaporator 蒸发器
evaporator body 蒸发器身
evaporator coil 蒸发器旋管
evaporator room 蒸发室
evaporator scale 蒸发器垢层
evaporator station 蒸发站
evaporator tails stream 蒸发器残余流；蒸发器尾料流
evaporator tower 蒸发塔
evaporimeter(=evaporometer) 蒸发计
evaporites 蒸发(后剩余)残渣
evaporometer(=evaporimeter) 蒸发计
evase exhaust stack 缩颈排气(风)管(器)
evasion lapping 闪避垫纱；躲避垫纱
even ①偶(数)的；双(数)的②均匀的
even AH(=even alternate hydrocarbon) 偶交替烃
even blend 掺混均匀
even calender 等速压延机
even color 均匀的颜色
even density reeding 等密度穿筘法
even dye 均染染料；均匀染料
even-dyeing colour 匀染(性)染料
even-electron ion 偶电子离子
even-electron rule 偶电子规则
even-even 偶-偶的
even-even nuclei 偶-偶核〔核内质子数和中子数均为偶数〕
even-even polyamide 偶-偶聚酰胺〔组分二羧酸及二胺的碳原子数均为偶数〕
even function 偶函数
even heating 均匀加热
even load 均匀载荷
even mass 偶质量
evenness 均匀性
evenness defect test 丝条斑检验
evenness of dye 染料的均染性
evenness standard photograph 匀度标准样照
evenness transducer 均匀度传感器

even-numbered 偶数的；偶碳原子数的
even-odd 偶-奇的
even-odd nuclei 偶-奇核〔核内质子数为偶数，中子数为奇数〕
even soap 偶碳原子数皂
even surface 平滑表面
even tension (长丝缠绕)张力均匀
eventual cracking 最后的开裂；可能发生的开裂
even wear 均匀磨耗
ever-cut finish 烂化整理
evericin 埃维菌素
Everitt's salt 埃弗立特盐 $K_2Fe[Fe(CN)_6]$
evermonte structure 双罗纹集圈空气层织物
evernesic acid(=everninic acid) 扁枝衣酸；甲基羟基甲氧基苯甲酸
evernic acid 扁枝衣二酸 $C_{18}H_{16}O_7$
everniine 扁枝衣因 $C_{16}H_{14}O_3$
everninic acid(=evernesic acid) 扁枝衣酸；甲基羟基甲氧基苯甲酸
everninocin 扁枝衣菌素
everninomicin 扁枝衣霉素
evocating agent 引发剂
evocator(=evocating agent) 引发剂
evodene 吴茱萸烯 $C_{10}H_{16}$
evodiamine 吴茱萸碱
evodin 吴茱萸定
evodinic acid 吴茱萸啶酸
evodinone 吴茱萸啶酮 $C_{26}H_{32}O_9$
evodione 吴茱萸二酮
evodionic acid 吴茱萸二酸
evodionol 吴茱萸脑
evodol 吴茱萸醇 $C_{26}H_{28}O_9$
evodone 吴茱萸酮 $C_{10}H_{12}O_2$
evogin 吴茱萸精 $C_{24}H_{28}O_8$
evolution ①发出②进化；演变；释放
evolutionary operation(EVOP) ①调优运算②开发计划；发展计划
evolution period 发展期
evolved gas analysis(EGA) 逸析气体分析法
evolved gas detection(EGD) 逸出气检测法
evolving factor analysis(EFA) 渐进因子分析
evonymin(=euonymin) 卫矛苷
evosin 茶萸地衣素
evotion 侵蚀
evoxanthidine 椒吴萸萸定
evoxine 吴茱萸素
Ewald's construction 埃瓦尔德图解法
Ewald's diffraction sphere 埃瓦尔德衍射球

Ewald's sphere of reflection 埃瓦尔德反射球〔纤维X射线衍射图用语〕
exact mass 准确质量
exact mass measurement 精确质量测定
exactness 正确性；精确性；严密性
exaggerated test 超常试验〔在特别不利的条件下试验〕
exalgin N-甲基乙酰苯胺
exaltation ①炼浓；精炼②超加折射③升高
exaltolide(=muscolide) 环十五内酯
exaltone 环十五烷酮
examination 检验(法)
examination at random ①随机抽查②随机试验
examination of genetically modified organisms foods 转基因食品检验
examination of waste water 废水检验
exanol 轻聚油〔汽油轻馏分的聚合物〕
excavation ①采掘；挖掘；掘进②土方工程
excavator 挖掘器；挖掘机
excellent scratch resistance 高抗刮性
Excer process 埃克萨法〔从铀矿浸出液等用离子交换和电解还原法制UF_4的过程〕
excess ①过量②余物
excess air 过剩空气
excess air number 空气过剩系数
excess chemical potential 超额化学势
excess electron 多余电子；过剩电子
excess enthalpy 超额焓*
excess entropy 超额熵*
excess free energy 超额自由能*
excess function 超额函数
excess heat ①过热②余热
excessive bleed 过量流失
excessive gum formation 过量树脂的形成
excessive handling 处理过度
excessive heating ①过热②过热的
excessive load (测试)过载
excessive material 剩余材料；余料
excessive pressure 超压；过压
excessive soil 过度污染
excessive sulfur content 过高的硫含量
excessive wear 磨损过度
excess quantity 过剩量
excess Rayleigh ratio 超瑞利比
excess stabilizer reflux (来自)稳定塔的过量回流
excess surface tension 逾限表面张力
excess (thermodynamic) function 超额(热力学)函数；过量(热力学)函数
excess (thermodynamic) property 过剩(热力学)性质

excess volume 超额体积*
excess water 过量水分
exchange 交换；互换
exchangeable 可互换的
exchangeable cation 可交换阳离子
exchangeable ion 可交换离子
exchange acidity 置换酸度；交换酸度
exchange adsorption 交换吸附
exchange capacity 交换容量
exchange chromatography (离子)交换色谱(法)
exchange column 交换柱
exchange current 交换电流
exchange current density 交换电流密度
exchange decomposition 交换分解(作用)
exchange density 交换密度
exchanged heat 交换热
exchange energy 交换能
exchange equilibrium 交换平衡
exchange equilibrium constant 交换平衡常数
exchange extraction 交换萃取
exchange filter 切换式过滤器；交换式过滤机
exchange force 交换力
exchange group 交换基(团)
exchange half-time 半交换期*
exchange integral 交换积分
exchange interaction 交换相互作用
exchange ion 交换离子
exchange isotherm 交换等温线
exchange kinetic 交换反应速度；交换动力学
exchange matrix 交换基质
exchange membrane 交换膜
exchange narrowing 交换变窄
exchange of know-how 技术交流
exchange polarization 交换极化
exchanger 离子交换剂
exchange rate 交换速率
exchange reaction 交换反应
exchange repulsion 交换排斥
exchange resin (离子)交换树脂
exchange site 交换点
exchange syntan 代替性合成鞣剂
exchange zone 交换带；交换区
excimer ①激基缔合物*②受激原子③受激分子
excimer fluorescence 激基缔合物荧光
excimer laser 准分子激光器
excipient 赋形剂
excipient vehicle 赋形剂
exciplex 激基复合物*

exciplex fluorescence 激基复合物荧光
excitability 激发性；兴奋性
excitants 刺激药；兴奋剂
excitation ①激发②兴奋
excitation-complex mechanism 激发复合机制
excitation condition 激发条件
excitation curve 激发曲线
excitation-decomposition 激发分解
excitation-emission matrix 激发-发射矩阵
excitation energy 激发能
excitation formula 激发式
excitation function 激发函数
excitation generator 励磁发电机
excitation labeling 激发标记
excitation level 激发(能)级
excitation light source 激发光源*
excitation mechanism 激发机理
excitation monochromator 激发单色器
excitation of X-rays X射线激发
excitation peak 激发峰
excitation potential 激发电位
excitation potential of spectral line 谱线激发电位
excitation probability 激发概率
excitation process 激发过程
excitation purity 激发纯度
excitation signal 激发信号
excitation source 激发光源
excitation spectrum 激发光谱
excitation wavelength 激发波长
excited atom 受激原子
excited configuration 激发组态
excited ethylene 受激乙烯(分子)
excited ion 受激离子
excited level 受激能级
excited molecule 受激分子
excited polymer 受激聚合物
excited singlet state 激发单重态
excited state 激发态*
excited state chemistry 激发态化学
excited state quencher 激发态抑制剂；激发态；猝灭剂
excited triplet state infrared spectrometry 受激三重态红外线光谱法
exciter 激发机；主控振荡器；辐射器
excite state 激态
exciting agent 激发剂；引发剂
exciting current 励磁电流
exciting line 激发光线
exciting moment 激发矩

excitomotors 激动药
exciton 激子*
exciton-phonon coupling 激子-声子耦合
excluded volume 排除体积*；已占体积
excluded volume effect 已占体积效应；已占空间效应
exclusion 摈斥；排斥
exclusion area 禁区
exclusion chromatography 排阻色谱法；筛析色谱法
exclusion limit ①排阻限②排除极限
exclusion limit molecular weight 排斥极限分子量
exclusion principle 不相容原理
exclusion seal 挤压型密封
exclusion TLC 排阻薄层色谱法
exclusive correlation spectroscopy 专一化学位移相关谱
excrementum pteropi 五灵脂
excreta 排泄物
excretion 排泄
excursion 漂移；偏差；振幅
exdenning ①卸货②卸料
execution drawing 施工图
executive routine 检验程序
executive system ①操作系统②行政系统③执行系统
exemestane 依西美坦〔药〕
exempt concentration 最大允许浓度
exempt quantities 最大允许量
exempt solvent 豁免溶剂
exergenic reaction 放能代谢反应
exfocal 焦距外的
exfoliatin 脱叶菌素
exfoliation 剥落；片落；剥离；剥落物
exfoliation corrosion 片状剥落腐蚀
exhalation valve 吸气阀
exhauser 喷射器
exhaust 排气〔机工〕；抽空
exhaust air cleaning 排放空气净化处理
exhaust and intake manifold 排气及进气歧管
exhaust blower 排风机；抽风机
exhaust boiler 废热锅炉
exhaust cam 排气门突轮
exhaust chamber 抽气箱
exhaust chimney 排气烟囱
exhaust duct 排风管
exhausted 用过的；无用的；废的
exhausted cossettes〔复〕 沥滤渣；沥滤过的甜菜切片
exhausted liquid 废液
exhausted lye 废碱液
exhausted molasses 橘蜜〔糖〕；废糖蜜
exhausted pulp 沥滤渣；沥滤过的甜菜切片

exhausted solution 废溶液
exhausted water 废水
exhaust end 卸料端
exhauster 排气机
exhaust fan 排气扇；排风机
exhaust flow 废气流
exhaust gas 废气
exhaust gas analyzer 废气(组成的)分析器
exhaust gas (catalytic) converter 废气(催化)转化器
exhaust gas desulfurization 排烟脱硫；废气脱硫
exhaust gas recirculation 排气循环
exhaust gas temperature 排气温度
exhaust hood 排气罩(橱)
exhaust hose 排气胶管
exhausting 抽空；排气
exhausting agent 吸尽剂〔印染〕
exhausting chemiluminescence sensor 消耗型化学发光传感器
exhausting column ①回收塔②再生塔③汽提塔
exhausting section 极度贫化区
exhaustion 抽出；排(气)；抽空；耗尽
exhaustion of earth 漂白土耗费程度〔石油产品精制后〕
exhaustion plate 排气板
exhaustion rate 上染率；上色率
exhaustive 彻底的；完全的
exhaustive analysis 全分析；彻底分析
exhaustive bromination 完全溴化作用
exhaustive extraction 极限抽提
exhaustively chlorinated 完全氯化了的
exhaustive methylation 彻底甲基化*
exhaustive nitration 完全硝化作用
exhaustive oxidation 彻底氧化
exhaustive test 彻底试验
exhaust line 排废线
exhaust manifold 排气集管；排气歧管
exhaust muffler 排气消声器
exhaust opening 排气孔
exhaust outlet 排气口
exhaust pipe 废气管
exhaust plenum ①废气充满②排气集气室
exhaust port 排气口；排出孔
exhaust purification 废气净化
exhaust resistance 排放阻力；流出阻力
exhaust silencer 排气消声器
exhaust steam 废蒸汽
exhaust-steam separator 废气分离器
exhaust trunking 排(抽)气管线
exhaust tube 排气管

exhaust valve 排气阀；废气门
exhaust vent 排气口
eximine(=dicentrine) 荷包牡丹碱
existent corrosion 原在腐蚀〔油品中固有的物质造成的腐蚀〕
existing gum(=potential gum) 原在胶
exit 出口
exit accumulator 出口储料器(塔)
exit dose (射线)放射量
exit gas 出(去)气
exit loss 出口(压头)损失
exit pressure 出口压力
exit seal 出口密封
exit slit 出射狭缝；出口狭缝
exit tube 出口管
exit velocity triangle 出口速度三角形
exit window 出射窗
exo 外；挂〔取代在侧链上〕
exo addition 外型加成*
exocondensation 外缩成环；支链缩合
exoconfiguration 外向构型
exocyclic 外向环
exocyclic diene 环代二烯
exocyclic double bond 环外双键
exocytosis 胞吐作用；胞泌作用
exoelectrical reaction 放电反应
exoelectron 外(逸)电子
exoelectron emission(EEE) 外(逸)电子发射
exoenergic 放能的；放热的
exoenergic reaction 放能反应；释能反应
exoergicity 释能度
exogenous toxin 外毒素
exograph X射线底片〔检验焊接情况〕
exohorone 外激素
exo isomer 外型异构体
exon 编码顺序；外显子
exophthalmic goiter 突眼性甲状腺肿
exophthalmotropic hormone 促突眼激素
exosmosis 外渗
exosphere 外大气圈；外大气层
exotherm ①放热量②放热反应曲线图
exothermal 放热的
exothermal change 放热变化
exothermal peak 放热峰；发热峰
exothermal reaction 放热反应
exothermic 放热的
exothermic catalyst 放热催化剂
exothermic channel 放热通道

exothermic composite material 放热性复合材料
exothermic compound 放热化合物；发热剂
exothermic contribution 放热贡献
exothermicity 放热性
exothermic peak 放热峰
exothermic phase 放热相
exothermic reaction 放热反应
exothermic transition point 放热转变点
exotic atom chemistry(=new atom chemistry) 奇异原子化学〔包括正子素化学、介子素化学、介子原子化学等〕
exotoxin 外毒素
expandable bead 可发珠粒料
expandable graphite 可膨胀石墨〔阻燃体系〕
expanded-bag method 扩张袋法
expanded bed 膨胀床
expanded ebonite 多孔硬质胶
expanded film 扩展膜*
expanded flow 扩张流
expanded graphite 膨胀石墨
expanded metal 网形铁〔机工〕
expanded metal packing (tower) 延展金属填料
expanded monolayer 扩张单层
expanded perlite 膨胀珍珠岩；膨胀珠光体
expanded phase 扩张相
expanded phenol-formaldehyde resin 多孔酚醛树脂
expanded plastics 泡沫塑料
expanded polystyrene(EPS) 发泡聚苯乙烯
expanded polystyrene veneer 膨胀型聚苯乙烯胶合板
expanded rubber 海绵橡胶
expanded-scale indirect photometry 扩展标尺间接光度法
expanded-scale spectrophotometry 扩展标尺分光光度法
expanded-tube method 管材扩张制膜法
expanded uncertainty 扩展不确定度；总不确定度
expanded vinyl leather (cloth) 聚氯乙烯泡沫人造革
expander ①骤冷器②撑模器③扩张器④增充剂
expandibility ①膨胀性②发泡性
expanding 扩张；展开
expanding agent 发泡剂
expanding bed chromatography 膨胀床色谱技术
expanding drill 扩孔钻
expanding foam 发泡
expanding machine ①空气定型机②压力定型机
expanding mandrel 胀开心轴；可调式心轴；可胀式心轴
expanding material 增量剂；增容剂
expanding property 发泡性
expanding roller 展幅机；拉幅机
expanding volume 膨胀体积

expand joint　①伸缩接合②伸缩缝
expand test　胀管试验
expansibility(=expansiveness)　膨胀性
expansible bend　膨胀弯管
expansible joint　膨胀接合
expansibleness(=expansibility)　膨胀性
expansin　苹果菌素
expansion　扩展*
expansion bend　膨胀弯管；补偿器
expansion chamber　膨胀室
expansion coefficient　膨胀系数
expansion coil　膨胀旋管〔冷却机〕
expansion compensator　膨胀补偿器
expansion cracks　①发泡开裂②膨胀龟裂
expansion cushion　膨胀垫；补偿器
expansion drying　膨胀干燥；闪蒸干燥
expansion equation　膨胀方程式
expansion factor　溶胀因子
expansion flow　膨胀流
expansion insert　胀入式嵌件
expansion joint　膨胀节
expansion loop　膨胀环〔输送管内〕
expansion molding technique　膨胀成型技术〔碳纤维复合材料成型法之一〕
expansion pipe　膨胀管；补偿管
expansion rate　①发泡速率②膨胀速率
expansion ratio　膨胀率
expansion ring　伸缩环；膨胀圈
expansion roof of tank　储(油)罐的膨胀顶；储(油)罐的呼吸顶
expansion stroke　膨胀冲程
expansion test　膨胀试验
expansion theorem　展开定理
expansion thermometer　膨胀温度计
expansion trap　膨胀(式)阱
expansion U-bend　膨胀 U 形管；补偿器
expansion valve　膨胀阀；安全阀
expansion volume　膨胀容积
expansion wave　膨胀波
expansion work　膨胀功
expansive　(可以)膨胀的
expansive color　放大色〔视觉放大〕
expansiveness(=expansibility)　膨胀性
expansivity　膨胀性
expectation of variance　预期方差
expectation value　期望值
expected life　预期寿命
expected service life　预期使用寿命
expected value　期望值
expected variance　预期方差
expedient　有利的
expedition pump　远程泵
expellant gas　排出的气体
expelled air-vapour mixture　排出的空气-蒸气混合物
expeller　螺杆脱水机
expendable　消耗品
expenditure　费用；消耗
expensive floral　珍贵花香
experiment　实验；试验
experimental　实验的
experimental apparatus　实验装置
experimental chemistry　实验化学
experimental data　试验数据
experimental design　实验设计*
experimental drum　试验转鼓
experimental error　实验误差
experimental evidence　实验证据
experimental index　试验指标
experimental plate height　实验塔板高度
experimental run　试(生)产；实验性生产
experimental temperature　实验温度；试验温度
experimentation　实验术
experiment control　实验控制
experiment table　实验台；实验桌
expert system　专家系统
expert system of gas chromatography　气相色谱专家系统
expiration date　先效期
expiration date of reference material　标准物质的有效期
explicit　显明的；明白的
explicit programming　显式编程
exploded pulp　爆炸法纸浆
exploded view　展开图；部件分解图
exploder　爆炸剂；爆轰剂
exploding　爆炸；爆发
exploding atom　爆裂原子
exploisve mixture　爆炸混合物
exploitation　利用
exploratory research　探索性研究
exploratory run　试(生)产；实验性生产
explosibility　爆炸性
explosimeter　爆炸浓度计；气体爆炸危险浓度指示剂；爆炸计
explosion　爆炸
explosion concentration　爆炸浓度
explosion door　爆炸安全门
explosion gas pipette　爆炸气管

explosion hatches　防爆门〔储油罐的〕
explosion heat　爆炸热
explosion limit　爆炸极限；爆炸界限
explosion method　爆炸法
explosion peninsula　爆炸半岛
explosion pipette　爆炸球管
explosion product　爆炸产物
explosion-proof　防爆；抗爆
explosion-proof chromatograph　防爆色谱仪
explosion-proof enclosure　防爆外壳
explosion-proof equipment　防爆装置
explosion-proof motor　防爆电动机；防爆马达
explosion-proof structure　防爆结构
explosion-proof tank　防爆桶；防爆储罐〔石油〕
explosion range　爆炸范围；爆炸极限
explosion ratio　爆炸浓度
explosion reaction　爆炸反应
explosion seismology　爆炸地震学
explosion temperature　爆炸温度
explosion tube　爆炸管；爆燃管
explosion wave　爆炸波
explosive　①炸药②爆炸(性)的
explosive agent　爆炸剂
explosive atmosphere　爆燃性空气
explosive bonding (welding)　爆炸焊接
explosive brisance　猛度
explosive bulge test　爆破试验
explosive bullet　爆炸子弹
explosive charge　爆炸装填物
explosive compound　爆炸化合物
explosive force　爆炸力
explosive forming　爆炸成形
explosive fuel　爆炸烧料；易燃燃料
explosive gas　爆炸气
explosive limit　爆炸极限
explosive limits of hydrocarbons　烃类的爆炸极限〔与空气混合物〕
explosive limits of mixture　混合物爆炸极限
explosive mixture of gases　气体的爆炸混合物
explosiveness　爆炸性
explosive oil　爆炸油；硝化甘油
explosive pyrite　爆性黄铁矿
explosive range　爆炸范围；爆炸极限
explosive reaction　爆炸反应
explosive train　爆炸串列
explosive view　爆炸图
explosive welding　爆炸焊
explosivity　爆炸性

explosivity limit　爆炸限度
exponent　①指数②说明者
exponential　指数的
exponential calculation method　指数计算法
exponential curve　指数曲线
exponential decay constant　指数衰减常数；指数下降常数
exponential decay law　指数衰变律
exponential dilution flask　指数稀释瓶
exponential dilution method　指数稀释法
exponential extrapolation　指数外推
exponential filter　指数滤波
exponential flow　指数流动
exponential function　指数函数
exponential gradient　指数梯度
exponential gradient device　指数梯度装置
exponential growth　指数增长
exponential growth constant　指数增加常数；指数生长常数；指数上升常数
exponential interpolation　指数内插
exponential phase　指数期
exponential regression　指数回归
export kerosene　出口煤油
exposed stem correction　露出柱段的校正
exposed tank　危险油罐〔在有着火危险距离内的油罐〕
exposing lamp　曝光灯
exposing process　爆煮法
exposition　①曝光②暴露③陈列
exposure　①暴露②曝光③照射量；曝射量〔以伦琴为单位的物理量〕
exposure angle　①曝置角②暴晒角
exposure area　①曝置面积②暴晒面积
exposure boiling deposition　暴沸淀积
exposure condition　暴晒条件；暴露条件
exposure cracking　自然龟裂；暴露龟裂
exposure data　暴晒数据
exposure distance　危险距离〔有着火危险的距离〕
exposure dose　①暴晒剂量②照射剂量；辐照剂量③暴露剂量
exposure labeling　曝射标记
exposure meter　曝光表；曝光计
exposure of pipe line　输送管的暴露
exposure parameter　①暴露参数②曝光参数
exposure period　①曝置期②曝光期
exposure rack　暴晒架
exposure rate　曝射率；照射率
exposure scale　①暴晒标度②曝光(露)标度
exposure test　暴露试验

exposure time 曝光时间
exposure to Atlas Weatherometer 阿特拉斯人工老化机暴晒
exposure to gasoline 汽油浸渍；浸汽油
exposure to tropical weather(ing) 热带天然暴晒
exposure to ultraviolet 紫外线照射
exposure to weather(ing) 天然暴晒
exposure zone 暴露区
expressed 榨出的；压榨的
expressed lime oil 压榨白柠檬油
expressed oil 榨出油；压榨油
expressed orange oil 榨出的橘油
expressible oil 可榨出的油
expression ①压榨②式③表现
expression proteomics 表达蛋白质组学
ex-proof 防爆
exquisite lift 优雅散发香；细腻散发香
exsiccant ①干燥剂②干燥的
exsiccated alum 烧明矾
exsiccated sodium phosphate 干燥的磷酸钠
exsiccation 干燥(作用)；除湿(作用)
exsiccator(=desiccator) 保干器；干燥器
Exsolex process 溶剂萃油法〔用溶剂萃取棉子油或豆油的工艺过程〕
exsolution 脱溶；出溶；出溶作用
extend cure 延续硫化
extended 拉长的；伸长的
extended absorption edge fine structure 扩展吸收边精细结构
extended aeration 延时曝气
extended bond 伸直键；伸展键
extended chain 伸直链
extended chain configuration 伸直链构型
extended-chain crystal 伸展链晶体*
extended-chain crystallite 伸直链微晶
extended chain filament 伸直链(晶)丝
extended-chain lamellae 伸直链片晶；伸直链晶层
extended chain length 伸直链长度；伸展链长度
extended chain model 伸直链模型；伸展链模型
extended cone and plate flow 扩展(型)锥板流动
extended conjugation 扩大的共轭体系
extended diagram 展开图
extended dislocation 扩展位错
extended Hartree-Fock method 推广的哈特莱-弗克法
extended heating surface 伸长的换热面
extended Hückel method(EHM) 推广的休克尔法
extended Hückel molecular orbital method(EHMO) 推广的休克尔分子轨道法

extended Hückel theory 推广的休克尔理论
extended length 伸展长度
extended linear macromolecule 伸直线型大分子
extended mandrel method 长心轴法
extended rubber ①增容橡胶②油充橡胶③拉伸的橡胶
extended surface tube 扩展表面管〔如翅片管、钉头管等〕
extender 补充剂；增充剂；增量剂
extender paste 填料浆；填料膏
extender pigments(=pigment extender) 体质颜料
extender plasticizer 补充增塑剂；助增塑剂
extender polymer 增量聚合物
extensibility ①伸长性②伸长度
extensile(=extensible) 可伸长的
extensimeter 伸长计
extensiometer(=extensometer) ①伸长仪②应变仪③张量计④拉伸试验器
extension ①伸长；延伸；拉伸；伸展；扩大②增加③广度④伸出部
extensional flow(=elongational flow) 拉伸流动；伸长流动
extensional motion 拉伸运动
extensional shearing flow 拉伸剪切流动
extensional viscosity 拉伸黏度
extension at break 断裂伸长
extension elongation ①拉伸延长；拉长②拉伸长度；伸出长度
extension modulus 伸长模量；张拉模量
extension ratio 拉伸比
extension rod 伸长杆
extension speed 伸长速率
extension strain 伸长变形
extension thermometer 长管温度计；长梗温度计
extension-type 伸缩式；伸缩自如的
extensively used 广(泛使)用的
extensive property 广度性质*
extensive variable 广延变数；外延变量
extensometer(=extensiometer) ①伸长仪②应变仪③张量计④拉伸试验器
extent ①广延②延伸(程度)
extent of corrosion 腐蚀程度
extent of crosslinking 交联度
extent of grafting 接枝度
extent of hydration 水化程度
extent of polymerization 聚合程度
extent of reaction 反应进度*
extent of re-equilibration 再平衡程度
exterior ①外面②外部的；外用的

exterior charging　①外电荷②(喷)枪外充电；枪外带电
exterior coating of pipe line　输送管的外套
exterior durability　室外耐久性
exterior exposure test　户外暴晒试验
exterior finish　外用罩面漆
exterior-grade preservative　外用级防腐剂
exterior latex paint　外用乳胶漆
exterior paint　外用漆
exterior plywood　室外用胶合板
exterior solvent-borne paint　外用溶剂型涂料
exterior stucco　外墙拉毛粉刷
exterior trim paint　外用装饰漆
exterior type plywood　(室)外用型胶合板；防湿胶合板
exterior varnish　外用清漆
exterior weatherability　户外耐候性
exterior wood finish　户外木器漆
external　①外(面)的②外部
external antistatic agent　外用抗静电剂
external antistatics　外用抗静电剂
external auxiliary column　外部辅助柱
external calandria　外加热排管
external catalyst return　催化剂外回路
external circuit　外电路
external coating　外涂层
external column　外柱
external comparison method　外标准法
external compensation　①外消旋(作用)②外补偿
external compensation method　外部补偿法；外对消法
external condensation process　外冷凝法
external conversion of energy　能量外转换
external crack　外部龟裂
external crosslinking　外交联(作用)
external cutting　外圆切削；外圆车削
external diameter　外径
external diffusion　外扩散
external dimension　外部尺寸
external effectiveness factor　粒外有效因子
external equalizer　外平衡管
external exposure　外照射
external factor　外界因素
external fibrillation　表面纤丝化；表面带化
external force　外力
external friction　外摩擦
external gear oil pump　外啮合齿轮油泵
external grinding　外圆磨削
external haze　表面混浊
external heat exchanger　外置式换热器
external heating　外热法

external heavy atom effect　外重原子效应
external ignitor　外接点火器
external impregnation　外部浸渍
external indicator　外(用)指示剂*
external interheater　外部中间加热器
external labeling　外标记
external lock　外锁
externally compensated compound　①外消旋化合物②外补偿化合物
externally cooled Fischer-Tropsch process　外部冷却的费-托合成过程
externally fed spray gun　外送料喷枪
externally fired retort　外热(干馏)釜
externally heated　外面加热的
externally heating　外面加热
external mix　(喷)枪外混合
external mix air cup　外混(合)式喷雾嘴
external mix spray gun　外混式喷枪
external mix type gun　外混(合)式喷枪
external modulation　外调制
external (mold) lubricant　外施脱模剂
external multitubular heater　外多管加热器
external phase　外相
external phase of emulsion　乳液外相〔连续相〕
external photoelectric effect　外光电效应
external plasticization　外增塑(作用)
external plasticizer　外增塑剂
external pressure　外压
external pressure vessel　外压容器
external reference electrode　外参考电极；外参比电极
external reflectance　外反射
external reflux　外回流〔塔的冷却回流；从外面送入塔内的回流〕
external reflux of column　分馏柱的外回流
external releasing agent　外脱模剂
external resistance　外电阻
external ring　外环
external screw thread　外螺纹
external sheath　外层包覆
external shell-and-multitube cooler　外部夹套多管冷却器
external standard　外标*
external standard method　外标法
external strain　外应变
external stress　外应力
external surface　外表面
external symmetry　外对称
external temperature　室外温度；外部温度
external thermometer　外标尺式温度计

external thread 外螺纹；阳螺纹
external variable 外变量
external work 外功
external wort boiling heater 体外麦汁煮沸加热器
external wort boiling system 体外麦汁煮沸系统
external wort boiling vessel 体外麦汁煮沸器
extinction ①消光②熄灭；消灭
extinction angle 消光角
extinction coefficient 消光系数
extinction contrast 消光衬比
extinction distance 消光距
extinction exponent 吸光指数
extinction modulus 吸光率
extinction pattern 消光花样；消光图案
extinction position 消光位
extinction quotient 吸光商
extinction rule of refraction(=rule of absent reflection) (折射)消光规则
extinction time 熄灭时间〔耐燃试验〕
extinction turbidimeter 消光浊度计
extinction value 消光值
extinct radionuclide 熄灭的放射性核素〔曾在地球上存在过，但现已衰变完而消灭〕
extinguisher(=annihilator) ①灭火器②湮灭算符
extinguishing method 灭火方法
extinguish material 灭火料
extinguishment 熄灭
extirpation 摘除；切除
Exton reagent 埃克斯顿试剂
extra ①额外的②非常的；特优的③额外附加物
extra bed factor 床外因素
extra carbon atom 额外碳原子
extra carrier gas 外加载气
extra check 特别对照；额外对照
extra-cluster 外簇
extra-column 附加柱
extra-column effect 柱外效应
extra column variance 柱外变化量
extra-column volume 柱外体积
extra conductive furnace black 超导电炉黑
extract(=extractum) ①提取；萃取；抽提②提取液；萃取液；浸膏
extractability 萃取率
extractable(=extractible) 可提取的
extractable acid 可萃取酸
extractable softener 抽出性软化剂
extractable species 可萃取物
extract and oxidation sweetening 抽提物氧化脱硫(法)

extractant(=extracting agent) 提取剂；萃取剂
extract atmospheric flash tower 抽出液常压闪蒸塔
extracted complex 萃取络合物
extracter 抽提器；萃取机；脱水机
extract exchanger 提取热交换器
extract furnace 提取加热炉
extractibility (可)萃取性；(可)提取性
extractible(=extractable) 可提取的
extracting (植鞣革)滚浓鞣液
extracting agent(=extractant) 萃取剂；提取剂
extracting power 萃取能力
extract(ing) reagent 萃取剂；抽提剂
extracting solvent 萃取溶剂
extracting stage 萃取级
extracting tower 萃取塔
extraction 浸提(法)；萃取(法)；提取(法)
extraction agent 萃取剂；提取剂
extraction apparatus 抽提器；提取器；萃取器
extraction bank 萃取槽组
extraction bath 萃取浴；抽提浴
extraction battery 抽提器组；萃取组
extraction cartridge 萃取柱柱体(柱身)
extraction-catalytical kinetic spectrophotometry 萃取催化动力学光度法
extraction chamber 萃取室
extraction chromatography 萃取色谱(法)
extraction coefficient(=distribution ratio) 萃取系数；(萃取)分配比
extraction column 抽提塔；萃取塔
extraction constant 萃取常数
extraction-contamination problem 萃取沾污问题；提取污染问题
extraction curve 萃取曲线
extraction cycle 萃取周期
extraction data 提取数据
extraction disk 抽取碟片
extraction distillation column 萃取蒸馏塔
extraction distribution coefficient 萃取分配系数
extraction efficiency 萃取效率
extraction equilibrium 萃取平衡；抽提平衡
extraction equilibrium constant 萃取平衡常数
extraction equipment 浸提(取)设备；萃取装置
extraction factor 萃取因素
extraction feed 萃取原料
extraction flask 提瓶；萃取瓶；浸提瓶
extraction floatation 萃取浮选法
extraction flow 排气流量
extraction fractionation 萃取分级*

extraction gravimetry　萃取(抽提)重量分析法
extraction indicator　萃取指示剂
extraction-inhibition kinetic spectrophotometry　萃取阻抑动力学光度法
extraction isotherm　萃取等温线
extraction line　①出料管；排出管②抽提管路
extraction mechanism　萃取机理
extraction metallurgy　萃取冶金学；提取冶金学
extraction method　萃取法
extraction of sugar　糖提取法〔糖〕
extraction packed column　填充抽提塔
extraction-photometric method　萃取光度法
extraction pump　排气泵
extraction raffinate　萃余液
extraction rate　萃取率；浸提率；出粉率
extraction section　萃取段
extraction separation　萃取分离(法)；提取分离(法)
extraction separation method　萃取分离法
extraction shell　提取筒壳
extraction solvent　萃取溶剂；抽提溶剂
extraction spectrophotometric method　萃取分光光度法
extraction stage efficiency　萃取级效率
extraction suppression　萃取抑制
extraction system　萃取系统；抽提系统
extraction thimble　抽提套管
extraction titration method　萃取滴定法
extraction tower　抽提塔
extraction tube　提取管；抽提管
extraction yield　萃取率；提取率
extractive　①抽提的②抽出物；萃；浸膏
extractive distillation　萃取蒸馏
extractive electrode　萃取电极
extractive metallurgy　萃取冶金学；提取冶金学
extractive reaction　萃取反应；抽提反应
extractive substance　浸提(萃取)物
extractive titration　萃取滴定
extract layer　提取层
extract liquor　浸提(萃取)液
extract of glycyrrhiza(=extract of licorice root; extract of liquorice)　甘草浸膏；甘草萃；甘草提出物
extract of licorice root(=extract of glycyrrhiza)　甘草浸膏；甘草萃；甘草提出物
extract of liquorice(=extract of glycyrrhiza)　甘草浸膏；甘草萃；甘草提出物
extract oil　提取油
extractor　萃取器；提器；抽提器；萃取机；脱水机
extractor basket　提取器筐
extractor regime　间歇式萃取器周围模式

extract phase　提取相；提取层
extract prediction　浸出物预测
extract pressure flash tower　提取加压蒸发塔
extract printing　拔染印花
extract-separation　抽提物分离
extract storage　提出物储器
extract stripper　抽出液汽提塔
extract sweetening　抽提物脱硫(法)
extract tanning　浸膏鞣(制)
extractum(=extract)　抽出物；浸膏
extract wool yarn　再生毛纱
extract yeast　酵母膏
extra deep tread type　超深胎面花纹轮胎
extra entropy production　额外熵产生
extra-heat-endurance　耐超温性；耐过热性
extra heavy　特强
extra heavy pipe　特强管；厚壁管
extra heavy size　特重浆
extra high modulus　特高模量
extra high molecular polymer　超高分子量聚合物
extra high pressure　超高压
extrait　①花净油②花香精
extra-lemon pale　超柠檬白〔石油产品标准颜色的标志〕
extra light　①超轻(质)的②特浅(色)的
extra light calcined magnesia　超轻质氧化镁；高活性氧化镁
extra-light drive fit　轻压配合
extra-limy slag　特强石灰炉渣
extra long staple　超长短纤
extra loss　额外损失
extramolecular　分子外的
extraneous air separator　外部空气离析器
extraneous component　①外加组分②额外分量〔数〕
extraneous cracking　外部裂化〔轻质碳氢化合物的裂化〕
extraneous factor　外界因素
extraneous material　外来杂质；异物
extraneous matter　外来杂质；异物
extraneous peak　外来峰
extraneous residue limit　再残留限量；外来残留限量
extranuclear　核外的
extranuclear electron　核外电子
extra-nuclear structure　核外结构
extra-orange pale　超橙黄白〔石油产品标准颜色的标志〕
extraordinary chemical activity　超高化学活性
extraordinary ray　非常射线；非常光线
extra-pale　超苍白〔石油产品标准颜色的标志〕
extrapolated base line　外推(起始)基线
extrapolated curve　外推曲线

extrapolated method 外推法；外插法
extrapolated octane number 外推辛烷值
extrapolated onset temperature 外延起始温度
extrapolated range 外推范围
extrapolated viscosity 外推黏度
extrapolation 外推(法)
extrapolation error 外推误差
extrapolation technique 外推法
extra pure 特纯的
extra-pure grade 超纯级
extrasin 胸腺素
extra slack running fit 松动配合
extra stabilization 超常稳定(作用)
extra steam 额外蒸气
extra-stress 附加应力；额外应力
extra-stress tensor 附加应力张量
extra-strong pipe 特强管；非常坚固的管子；粗管
extra strong stand oil 过稠原油
extra super fine wool 特超细羊毛
extraterrestrial 地(球)外的；行星际的
extraterrestrial chemistry 地外化学；行星(际)化学
extraterrestrial disposal 地(球)外(放射性废物)处理(法)；外层空间(放射性废物)处置(法)
extra thick oval brush 超厚椭圆形漆刷
extra warp 特加经纱
extra weft (特)加纬
extreme dimension 极限尺寸
extreme light weight hide 最轻磅皮
extremely strong odor 极强烈香气；极强烈气味
extremely viscous 特黏的；非常黏的
extreme narrowing limit 极窄化极限
extreme pressure 特(高)压(力)；极压
extreme pressure addition agent(=extreme pressure additive) (耐)特压添加剂
extreme pressure additive(=extreme pressure addition agent) (耐)特压添加剂
extreme pressure characterisitic (耐)特压性质
extreme pressure compound(=extreme pressure additive) (耐)特压添加剂
extreme pressure dope(=extreme pressure additive) (耐)特压添加剂
extreme pressure film 极压膜
extreme pressure gear oil (耐)特压齿轮油
extreme pressure grease (耐)特压润滑脂
extreme pressure lubricant (耐)特压润滑剂
extreme pressure lubricant tester 润滑剂(耐)特压试验器
extreme-pressure lubrication 极压润滑
extreme pressure machine (耐)特压试验机〔油类耐高压性能试验机〕
extreme pressure material (耐)特压添加剂
extreme pressure oil (耐)特压油〔含耐特高压力添加剂的油〕
extreme pressure properties 极压性能
extreme pressure soap 特压皂〔润滑脂生产用皂〕
extreme pressure turbine oil (耐)特压涡轮油
extreme trace analysis 超痕量分析
extreme ultraviolet 远紫外(区)；超紫外
extreme ultraviolet ray 远紫外线
extremum principle 极值原理
extremum value 极值*
extrinsic 非本征的；外赋的
extrinsic contaminant 外来杂质；外来沾污
extrinsic defect 杂质缺陷
extrinsic insoluble 外来不溶物
extrinsic photoconductor 本征光导器件
extrinsic semiconductor 含杂质半导体；外赋半导体
extropolated rise time(ERT) 外延起发时间〔发泡〕
extrudability 可挤出性
extrudant 压出物；压出半成品
extrudate 压出物；压出型材；挤出物
extrudate coiling 挤出物盘绕
extrudate distortion 挤出物畸变
extrudate post swell 挤后膨胀
extrudate swelling(=die swelling) 挤出物膨胀
extrudation 挤出；压出(作用)
extruded aluminum product 冲压铝制品
extruded article 挤压制品
extruded-bead sealing 挤出熔体熔接
extruded catalyst 挤压催化剂
extruded coating 挤塑贴面；挤塑涂层
extruded column 顶出柱
extruded melt 挤塑熔体
extruded net 挤塑网
extruded pipe 挤塑管材
extruded plastic mesh 挤塑塑料网
extruded profile 挤塑型材
extruded profile shape 挤塑异型材
extruded sheet 压出胶片
extruded stock 挤出料
extruded tape 挤塑带材
extruded velocity 挤出速度
extruder 压出机；挤压机
extruder and cooling train 压出冷却联动装置
extruder attitude 挤塑机型式
extruder base 压出机机座
extruder blanks 压出坯料

extruder bore 挤塑机机膛；挤塑机内径；挤塑机口径
extruder-calender 压出压延机
extruder constant 挤塑机常数
extruder core 压出机芯型
extruder cylinder liner 压出机机筒内衬
extruder diameter 挤塑机直径
extruder die 压出机口型
extruder grain ①挤压造粒②挤出取向(性)效应
extruder head 压出机机头
extruder head core 压出机机头芯型
extruder length to diameter ratio 挤压(出)机长(度)(直)径比(L/D)
extruder output 压出量
extruder-pelletizer 挤出切粒机
extruder pressure 压出(机)压力
extruder pump 挤压出料泵；挤出泵
extruder rate 压出速度
extruder rheometer 挤出式流变仪
extruder rifled barrel 挤塑机来复线机筒
extruder screen 挤出机过滤网
extruder screen pack 挤出机筛网组；挤塑机滤网叠
extruder screw 压出机螺杆
extruder size 压出机型号
extruder stand 压出机机座
extruding 挤压；挤出；压出(程序)
extruding aids 压出助剂
extruding-head 压出机机头
extruding machine 压出机；挤压机；挤出机
extruding press ①压出机②挤出机
extruding profile 挤塑型材
extruding test 压出试验
extrudometer 压出计
extrusion 挤出*
extrusion aid 助挤剂
extrusion blow molding 挤出吹塑
extrusion casting 挤出流铸
extrusion coater 挤涂机
extrusion coating(=substrate extrusion coating) 挤压涂装
extrusion coloring 挤出(压)着色挤塑着色
extrusion device 挤出装置
extrusion die 挤出模板；挤出模头
extrusion die swell 挤出胀大
extrusion direction 挤压(出)方向
extrusion draw blow molding 拉吹塑成型
extrusion-dried rubber 压出干燥法橡胶
extrusion dryer 螺杆脱水机
extrusion grade 挤出级
extrusion head 压出机机头

extrusion index 压出指数
extrusion lamination 挤塑层合；挤压黏合(法)；挤压贴膜(法)
extrusion land 压出口型唇口
extrusion line 压出(生产)流水线
extrusion machine 压出机；挤出机
extrusion mandrel 挤塑模芯
extrusion manifold 挤出料道
extrusion mark 挤塑印迹
extrusion method 顶出法
extrusion molding 挤出成型
extrusion output 挤出量
extrusion packing 挤压填充
extrusion plastometer 挤压式塑性计
extrusion pressure 挤出压力；压出力
extrusion quality 压出质量
extrusion rheometer 挤出式流变仪
extrusion rheometry 挤出流变测定法
extrusion screw 挤出机螺杆
extrusion stress 挤压应力〔塑性变形〕
extrusion stretched blow moulding 挤拉吹塑
extrusion swelling 挤出胀大
extrusion technique 压出工艺；顶出技术
extrusion technology 挤塑工艺学；挤出(压)工艺学
extrusion temperature 挤出温度
extrusion theory 挤塑理论
extrusion type injection machine 压出式注压机
extrusion type plastometer 挤压式塑性计
extrusion variable 挤塑变量
extrusion viscometer 挤出式黏度计
exudate 渗出物
exudation 渗出
exudation mark 渗出迹印〔制品缺陷〕
exudation of pigment 颜料渗出；颜料渗色
exudation pressure 渗流压力
exuding ①(颜料)渗色②渗出
ex vivo 动态半体外
eye ①小孔②眼圈③滴眼器④眼状垫环⑤芽眼〔胶树〕
eye bolt 有眼螺栓
eyebright 小米草〔用于制洗眼液〕
eye dropper 滴管
eyed structure 网眼结构
eye-hole 观察孔
eye irritation 眼睛刺激性
eyelash pomade 睫毛膏
eyelet(=eyelet-hole) ①小孔；孔眼②鸟眼③观察孔④金属环
eyelet-hole(=eyelet) ①小孔；孔眼②鸟眼③观察孔④金

属环
eye liner 眼线笔；眼睑墨
eyepiece 目镜
eyepiece micrometer 目镜测微计；目镜测微尺
eye-rest color 悦目色〔指绿光波附近的色〕
eye shield 护眼罩；护目罩；防护眼镜

eyes protector 护目镜
Eykman equation 埃克曼方程
Eyring model for flow 埃林流动模型
Eyring theory of viscous flow 埃林黏性流动理论
Eyring viscosity 埃林黏度
ezomycin 鲡霉素

F

Faber viscosimeter 法伯尔黏度计
fabric ①织物②结构；组织
fabricability 加工性
fabric air permeability tester of mid-low pressure 织物低压透气量仪
fabricated caravan 装配式大篷车
fabricating 二次加工
fabricating technique 制造技术
fabrication ①制造②捏造③成形
fabrication on site 现场制作
fabric backing 织物反面；织物背面
fabric back stay 后跟加强布；里后跟布
fabric base coat 布基涂层
fabric belt 纤维传送带；布传送带
fabric breaker 帘布缓冲层
fabric bursting test 织物顶破强力试验
fabric bursting tester 织物胀破强度
fabric cell 用胶布作成的气球；用胶布作成的袋子
fabric coating 织物涂胶；织物覆胶
fabric coating unit 织物覆胶装置
fabric construction 织物构造
fabric count 织物密度
fabric cover 织物总紧度
fabric covering 用胶布作成的外壳
fabric dipping unit 织物浸胶装置
fabric dryer 烘布机
fabric element 毡的滤波器元件
fabric enter-storage standard-piece rate 织物入库正品率
fabric expanding 织物扩幅
fabric fall 织物悬垂性
fabric feeder 供布器
fabric-filled moulding compound 布填模塑料
fabric filler 织物填料
fabric filter 织物过滤器
fabric-free reclaim 无纤维再生胶
fabric fuel tank 软燃料箱
fabric geometry serial 结构相
fabric hybrid composites 混杂织物复合材料
fabric impact test 织物冲击强力试验
fabric luster streak 织物亮纹（纤维截面受压变形所致）
fabric off-loom standard-piece rate 织物下机正品率
fabricometer （电子式）织物形稳定性测试仪
fabric penetration 织物渗透
fabric porosity 织物紧密性；织物空隙度；织物透孔性
fabric proofing 布上涂胶；织物覆胶
fabric puller 帘布伸展器
fabric pushing tester 钢球式顶破强力仪
fabric seal 织物封
fabric sheen 织物光泽
fabric shock test 织物冲击强力试验
fabric slippage resistance 织物纰裂强度
fabric slitting 织物裁断
fabric softener 织物软化剂
fabric take-off 织物卷取
fabric tank 软桶
fabric tearing tester 织物撕裂仪
fabric thickness 织物厚度
fabric treating unit 织物处理装置
fabric tyre 帘布轮胎
fabric water-bag test 织物兜水测试
fabric with different edges 差边布〔纺〕
fabrillation 原纤化
Fabry-Perot etalon 法布里-珀罗标准具
Fabry-Perot interferometer 法布里-珀罗干涉仪
Fabry's disease 费勃莱氏病；酰基鞘氨醇己三糖苷酶缺乏症
FAC(Fat Analysis Committee) color FAC 色度
face ①面板②正面；割面；采脂面〔样胶〕
face bend test 表面弯曲试验
face brick 面砖
face brick clay 面砖土
face-centered crystal 面心晶体
face-centered cube 面心立方体
face-centered cubic lattice 面心立方格子
face-centered cubic structure 面心立方结构
face-centered grating 面心格子
face centered lattice 面心点格*
face-centered orthorhombic 面心正交
face-centred cell 面心晶格
face-centred cubic structure 面心立方结构
face-centred lattice 面心点阵；面心晶格
face-coating 表面涂层；正面涂层
face color(=top color) ①面色②直视色〔观测者的视线几乎垂直于漆膜表面所见的颜色〕
face cream 润肤香脂；面霜
face cutting 车平面；端面车削
faced plywood 贴面胶合板
face finish ①表面修饰②表面涂饰剂

face glazing 面釉
face grinding 平面磨削
face guard(=face mask) 面具
face-lift 上油漆
face mask(=face guard) 面具
face of crystal 晶面
facepiece (of mask) 面罩〔防毒面具〕
face plate 花盘
face powder 香粉
face protector(=face shield) 防护面罩
face seal 表面密封
face shield(=face protector) 防护面罩
face side 正面〔设〕
face soap 洗脸皂
facet 小平面
face-to-face adhesion 面对面附着
facette 水面；水平面
face value 面值
face velocity 面速度
face veneer （胶合板的)面板
face warp 表经
face width 表面宽度
facial isomer 面式异构体
facial mask 美容面具
F-acid F 酸；2-萘酚-7-磺酸
facile hydrolysis 易水解
facility ①方便；灵活②可能(性)③设备；器材
facing ①面料②面饰
facing brick 贴面砖；饰面砖
facing concrete 护面混凝土
facing rubber 衬面胶；衬里胶
facing stone 贴面石；饰面砖
facing tile 饰面砖
facing up ①表面加工；表面找平；刮平表面②贴面；镶面；表面装修
factice(=factis) (硫化)油膏；油胶
factis(=factice) (硫化)油膏；油胶
factitious 人为的；人造的
factograph 断口组织(显微镜)照片
factor ①系数；因数；因子②因素
factor analysis 因素分析；因子分析
factor analysis spectrophotometry 因子分析分光光度法
factor group analysis 因子群分析
factorial 阶乘；阶乘积
factorial analysis 因数分析
factorial development 系数显影法
factorial effect 因素效应*
factorial experiment 析因实验

factorial experiment design 析因试验设计
factor in precipitation 沉淀系数
factor interaction 因素间交互作用
factorization 因子分解
factor loading 因子载荷
factor of adhesion 黏附系数
factor of expansion 热膨胀系数
factor of merit ①优良因素；质量因素②灵敏值
factor of porosity 孔隙率
factor of safety ①安全因素②安全系数
factor of shrinkage 收缩系数
factors of cracking 裂解系数
factor system 因数方式；因数法
factor weight ①因数重量②因数砝码③因数权重
factor weight system 因数重量法
factory 工厂
factory coatings 工业用涂料
factory effluent 生产废水
factory formula 生产配方
factory noise 工厂噪声
factory processing 工厂条件下加工
factory research 工厂研究
factory runs 大量生产；成批生产
factory seconds 成品的副号
factory test 工厂试验
factory testing 工厂条件下试验；成批试验
factory trials 工厂条件下试验；成批试验
faded 褪了色的
fade meter 褪色试验机；褪色计
fadeness test 褪色试验
fadeometer 褪色计
fadeometer life 褪色计中寿命
fade resistance 抗褪色性
fade-resistant color 耐褪色颜料
fading 褪色
fading memory 衰退记忆
fading of gasoline 汽油褪色
fading test 褪色试验
fadometer 褪色仪
fagaramide 崖椒酰胺 $C_{14}H_{17}O_3N$
fagaric acid 崖椒酸 $C_7H_{11}O_5N$
fagaric aldehyde 崖椒醛 $C_{12}H_{14}O_4N$
fagarine 崖椒碱 $C_{21}H_{23}O_5N$
fagarol 崖椒醇 $C_{20}H_{18}O_2$
Fagergren flotation cell 费格临浮选池
fag(g)ot ①一束②连接；联系
fag(g)oted iron 束铁
fag(g)oting 打捆

fag(g)ot iron   束铁
faggot wood   柴；薪材
fagine   水青冈碱
fagopyrin   荞麦碱
fahl ore   黝铜矿
Fahrenheit   华氏(温度)
Fahrenheit degree   华氏温度
Fahrenheit scale   华氏温标
Fahrenheit scale of temperature   华氏温标
Fahrenheit thermometer   华氏温度计；华氏寒暑表
faience   彩陶器；釉陶
faience tile   釉陶面砖
failed test sample   不合规格的样品
failure(=brittle; fracture; rupture)   破坏；衰坏
failing load   破坏荷重
failure analysis   失效分析；破坏分析；故障分析
failure condition   破坏条件
failure criterion   衰坏判据；破坏判据
failure envelope   破坏包络线；断裂点轨迹
failure load   破坏载荷
failure mechanism   破坏机理
failure mode   衰坏方式；破坏方式
failure stress   破损应力
failure surface   衰坏面；破坏表面
failure time   破损时间
faint   ①暗淡的〔指色〕②弱的
faint blue   暗蓝色
faintly acid   弱酸(性)的
faintly alcaline reaction   弱碱性反应
faintly colored   暗淡色
faint odor   微弱香气
fair-faced concrete   磨光盖面混凝土
fairing   边；缘；侧线
Fajans method   法扬斯法
Falcon engine rust test   福尔肯发动机生锈试验
fales grain   闭砂
Falex machine   法列克斯摩擦机〔用于油类润滑性能的评价〕
Falex test   法列克斯试验〔在法列克斯摩擦机上评价润滑油的润滑性能〕
fall-back pipe   回降管道
fall-ball method   落球法〔测黏度方法〕
fallen calfskins   死小牛皮〔革〕
fall head   压头高度
falling abrasive tester   落砂法试验器
falling ball impact test   落球冲击试验
falling ball method   落球法
falling ball resilience tester   落球式弹性计

falling ball viscometer   落球黏度计
falling ball viscosity   落球黏度
falling body viscosimeter   落体黏度计
falling coaxial cylinder viscometer   同轴圆筒下落式黏度计
falling cylinder viscometer   圆筒下落式黏度计
falling-dart impact test   落锤冲击试验
falling drop   落滴
falling drop method   落滴法〔测定液体密度〕
falling drying rate period   减速干燥阶段
falling film diluter   降膜稀释器
falling film evaporator   降膜(式)蒸发器
falling film molecular still   降膜分子蒸馏器
falling film reactor   降膜式反应器
falling film sulfonation unit   降膜式磺化装置
falling liquid film   下落液膜；降(液)膜
falling needle viscosimeter   落针黏度计
falling-off phenomenon   降变现象
falling-pendulum apparatus   落锤式织物撕破强力测试仪
falling-rate period (drying)   降速阶段
falling sand abrasion test   落砂磨损试验
falling sand apparatus   落砂试验器
falling solution   再浸液〔皮革〕
falling-sphere damage test   落球(法)损伤测定
falling sphere viscometer   落球式黏度计*
falling temperature technique   降温技术
falling weight test   冲击试验
fall into   分解(为几部分)
fall off factor   渐疏因子
fall-out   ①落下②(核爆炸)散落物；(放射性)沉降物；落尘
fall out deposition   (核爆炸)散落物沉降
fall out transport   (核爆炸)散落物运输
fallow   ①休耕之地②休耕的；荒芜的③淡棕黄色的
fallowing   休闲制
fall sphere test   落球试验
fall tube   排水管
false   假；不可靠的
false beech(=mountain beech)   山地假山毛榉
false body   假稠性；假黏度
false bottom(=false floor)   假底
false bottom retort   假底甑
false equilibrium   假平衡
false feedback inhibition   假反馈抑制
false flax oil   亚麻荠油
false floor(=false bottom)   假底
false fusion   假熔
false grain   细糖晶粒

false indigo　假靛蓝；野靛蓝
false line　假线
false Mooneys　假门尼黏度〔黏度不随分子量增加而增加〕
false Mooneys viscosity　假门尼黏度〔黏度不随分子量增加而增加〕
false morels　类羊肚菌
false ring　假年轮；伪年轮
false truffle　①腹菌②大团囊菌
false twist　假捻
false twist spindle　转子假捻器
false-twist texturing　假捻变形(工艺)
false wave　假波；第二极大波
famciclovir　泛昔洛韦〔药〕
family　①族〔周期表〕②科〔生物分类〕
famotidine　法莫替丁〔药〕
fan　①风扇②扇(子)
fanal blue　法哪蓝
fanal colors　法哪染料〔色淀染料的一种商名〕
fanal pink　法哪桃红
fanal yellow-green　法哪黄绿
fan angle　(喷束)扇形角
fan belt　风扇带
fan-blast　通风机的鼓风
fanchinin　粉己宁
fancy air textured yarn　花色喷气变形丝；空气变形花式纱
fancy coal　上等煤；精选煤
fancy finish　美术涂饰剂
fancy leather　美术革
fancy lump coal　上等大块煤
fancy paint　装饰漆；装潢漆
fancy perfumes　想象型香精
fancy yarn　花色纱
fan discharge　通风机排量
fan-driven generator　风(机传)动发电机
fan-drying　排风干燥
fangchinoline　防己醇灵
fanner　扇风机
fanning　①通风②扇形
Fanning equation　范宁公式
fanning machine(=fanning mill)　风车；簸扬机
fanning mill(=fanning machine)　风车；簸扬机
Fanning's equation for turbulent motion　湍流运动的范宁方程式
fanno line　法诺线
fan-tail die　扇尾形口模
fan-type thin layer chromatograph　扇形薄层色谱仪
farad　法拉〔电容单位〕

faradaic current　法拉第电流*
Faradaic demodulation current　法拉第型解调电流
Faradaic rectification current　法拉第型整流电流
faraday　法拉第〔电量单位〕
Faraday balance　法拉第天平
Faraday cage　法拉第笼
Faraday cage effect　法拉第笼蔽效应
Faraday caging　法拉第笼蔽
Faraday constant　法拉第常数
Faraday cup　法拉第杯
Faraday cup detector　法拉第杯检测器
Faraday cylinder　法拉第筒
Faraday dark space　法拉第暗区
Faraday effect　法拉第效应
Faraday impedance　法拉第阻抗
Faraday rectification　法拉第整流
Faraday rotation　法拉第旋转
Faraday's law　法拉第定律
faradiol　款冬二醇　$C_{30}H_{50}O_2$
faradism　感应电流
faradization　电疗
farcinicin　皮疽菌素
fard　脂粉〔香料〕；胭脂
far field diffraction pattern　远场衍射图
fargite　钠沸石
farinaceous　面粉的
farinaceous size　淀粉浆液
far infrared　远红外*
far-infrared absorption　远红外(线)吸收
far infrared absorption spectrum　远红外吸收光谱
far infrared beam-splitter　远红外分束器
far infrared detector　远红外检测器
far infrared dryer　远红外干燥器
farinfrared fibre　微元生化纤维
far-infrared heater　远红外线加热器
far-infrared heating　远红外加热
far-infrared heating plate　远红外线平面加热器；远红外线加热板
far-infrared heating stove　远红外加热炉
far-infrared lamp　远红外灯
far-infrared oven　远红外炉
far infrared photoconductor　①远红外光电导体②远红外探测器
far-infrared radiation　远红外辐射
far infrared ray　远红外线
far infrared region　远红外区
far infrared scanner　远红外扫描仪(装置)
far infrared source　远红外光源

far infrared spectrometer　远红外光谱仪；远红外线分光计
far infrared spectroscopy　远红外光谱法；远红外光谱学
far infrared spectrum　远红外光谱
farinha(=cassava meal)　木薯粉
farinograph　面粉试验仪
farinometer　谷粉测量计
farinose　淀粉纤维素
farm　①田庄②农场③耕种
Farmer's reducer　法麦尔减薄剂
farmiglucin(=paromomycin)　巴龙霉素
farm manure　场地肥料
farnesal　法呢醛；金合欢醛
farnesane　法呢烷；金合欢烷
farnesene　法呢烯；金合欢烯
farnesenic acid　法呢烯酸；金合欢烯酸
farnesiferol　阿魏脂醇
farnesol　法呢醇；金合欢醇　$C_{15}H_{26}O$
farnesyl-　法呢基
farnesyl acetate　乙酸金合欢酯
farnesyl acetone　金合欢基丙酮
farnesylpyrophosphate　焦磷酸法呢酯
farnoquinone　法呢醌；金合欢醌；维生素 K_2
farrerol　发热醇；杜鹃素
far-super heavy nuclei　远超重核
far ultraviolet absorption spectroscopic detector　远紫外吸收分光检测器
far ultraviolet ray　远紫外线
far ultraviolet region　远紫外区
fascinated-compound silk　包芯复合生丝
fasculation　缩聚
fashioning　精加工
Fason powder　法逊炸药
fast　①快②牢固③紧的④坚牢的；不褪色的⑤禁食；绝食
fast accelerator　快速促进剂
fast analysis　快速分析
fast atom bombardment　快原子轰击
fast atom bombardment ion source(FAB)　快速原子轰击离子源
fast atom bombardment mass spectroscopy　快原子轰击质谱法
fast atom bombardment source　快原子轰击(离子)源
fast atom gun　快原子枪
fast axis　快轴
fast-breeder reactor(FBR)　快中子增殖(反应)堆
fast chemistry　快化学
fast colo(u)r　坚牢染料；坚牢色泽
fast cure　(快)速硫化；短时硫化
fast curing cement　快硫化胶浆；常温硫化胶浆

fast curing property　快速硫化特性
fast curing rubber　快速硫化橡胶
fast detector　快速检测器
fast dispersibility　(快)速分散性
fast dissolving　速溶的
fast-draining film　快排水膜
fast-dry floor sealer　快干地板封闭底漆
fast-dyed　不褪色的
fast dyed yarn　不褪色纱线
fastener　夹持器；接线柱；紧固件
fastening　①紧固(的)②紧固物③支撑设备
fastening screw　紧固螺丝
fast extrusion furnace　快速压出炉
fast flow rate paper　高流速纸
fast Fourier transform　快速傅里叶变换
fast freezing　速冻
fast green　坚牢绿
fast-high-temperature cure　快速高温硫化
fasting　禁食；绝食；断食
fast insert probe　快速插入探针
fast inversion recovery　快速反转恢复
fast inversion recovery Fourier transform　快速反转恢复傅里叶变换
fast linear sweep voltammetry　快速线性扫描伏安法
fast loop　快速回路
fastness　坚牢度；不褪色性
fastness of dye　染料的坚牢度〔染〕
fastness of dyeing　染色坚牢度〔染〕
fastness rate　耐光度；着色坚牢度
fastness to alkali　耐碱度
fastness to bleaching　耐漂(白)度
fastness to boiling　耐煮度
fastness to decatizing　耐蒸度
fastness to dust　耐(灰)尘度
fastness to efflorescence　抗风化性
fastness to ironing　耐熨度
fastness to light　耐光度
fastness to perspiration　耐汗度
fastness to rubbing　耐磨度
fastness to washing　耐洗度
fastness to water　耐水度
fastness to wear　耐磨耗性；抗磨能力
fast neutron　快中子
fast neutron activation analysis　快中子活化分析
fast neutron detector　快中子探测器
fast oil orange　坚牢油橙
fast partial least square method　快速偏最小二乘法
fast particle bombardment　快粒子轰击(电离)

fast passage 快通过
fast protein liquid chromatography 快速蛋白质分析液相色谱法
fast radiochemical separation 快放射化学分离
fast radiochemistry 快速放射化学〔指短寿命同位素的快速化学分离〕
fast reaction 快速反应
fast reaction kinetics 快速反应动力学
fast reactor 快(中子反应)堆
fast red 坚牢红〔染〕
fast relaxation 快松弛；快弛豫
fast response 快响应
fast roll 快速滚筒
fast scan 快速扫描
fast setting 快速固化
fast shadow (色谱斑的)快速阴影
fast sky blue 坚牢天蓝；磺化铜酞菁
fast solvent 常温快速挥发溶剂
fast test reactor(FTR) 快中子试验堆
fast thawing 速熔
fast to light 耐光的；不变色的；不褪色的
fast violet ①锰紫②坚牢紫
fast wool yellow 坚牢羊毛黄
fast yellow 坚牢黄〔染〕
fasudil 法舒地尔〔药〕
fat acid 脂肪酸
fatal 致死的
fatal dose 致死剂量
fatality 致死率
fat asphalt 肥沥青〔含大量铺路用沥青的混合物〕
fat clay 富黏土；可塑性黏土；油性黏土
fat coal 肥煤
fat colors 油溶染料
fat dye 油溶染料
fat dyestuffs 油溶染料
fate 历程；去向；走向
fat edge 厚边；溜边；淤边〔漆膜病态〕
fat-free extraction paper 脱脂提取纸
fat-free filter paper 脱脂滤纸
fat gas 油气；肥气
fat grease 动物润滑脂〔用动物脂肪做成的润滑脂〕
fat hardening 油脂硬化
fat hydrolysis 油脂水解
fat hydrolyzing process 油脂水解法
fatigue 疲劳
fatigue break-down 疲劳破坏
fatigue breaking 疲劳断裂
fatigue corrosion crack 疲劳腐蚀开裂

fatigue crack 疲劳裂缝；疲劳断裂
fatigue curve 疲劳曲线
fatigue data ①疲劳数据②疲劳值
fatigue endurance 耐疲劳性
fatigue endurance limit 疲劳持久极限
fatigue failure 疲劳衰坏；疲劳破坏
fatigue life 疲劳寿命
fatigue limit(=endurance limit) 疲劳极限
fatigue loss 疲劳损耗
fatigue machine 疲劳试验机
fatigue mechanism 疲劳机理
fatigue notch factor 疲劳缺口因数
fatigue notch sensitivity 疲劳缺口敏度
fatigue-proof rubber 防疲劳橡胶
fatigue protective agent 抗疲劳剂
fatigue ratio 疲劳比
fatigue resistance ①耐疲劳性；抗疲劳性②动强度
fatigue strength 疲劳强度
fatigue test 疲劳试验
fatigue tester 疲劳试验机
fatigue tester of flexural property 挠曲性能疲劳试验仪
fatigue under flexing 挠曲疲劳；弯曲疲劳
fatigue under scrubbing 挠曲疲劳；搓擦疲劳
fatigue wear 疲劳磨损
fat iodine value 油脂碘值
fat lime 肥石灰
fat liquor 加油(或脂)乳液
fat liquored 乳液加油(或脂)的
fat liquoring 上脂；上油
fat-lub test 液体中含油量测定
fatlute 油泥
fat melter 熔脂器
fatness ①多脂②油腻③油性④润滑性
fat oil 饱和油
fat ore 富矿石
fat rendering 炼脂；脂肪熔炼
fat resistance 防油性；不透油性
fat sand 富沙
fat saponification 油脂皂化(作用)
fat-soluble 脂溶性的
fat-soluble vitamin 脂溶性维生素
fat solution 油脂溶液
fat solvent 油脂溶剂
fat splitter 裂脂器
fat-splitting agent 油脂水解(催化)剂
fat splitting plant 脂肪分裂工厂
fat substitute 脂肪代用品；代用油脂
fat tannage 油脂鞣

fat tanned 油脂鞣(了)的
fattening (油漆储存期间的)增稠；返稠〔漆病〕
fatting agent 加脂剂
fat turpentine 脂化松节油；稠化松节油
fatty ①脂(肪)的②多脂的；油的③脂(肪)族的；链烃类的
fatty acid 脂肪酸
fatty acid alcohol amide 脂肪酸烷醇酰胺
fatty acid alkanol amine 脂肪酸烷醇胺
fatty acid alkyl amide 脂肪酸烷基酰胺
fatty acid amide 脂肪酸酰胺
fatty acid amide sulfonate 脂肪酸酰胺磺酸盐
fatty acid ammonium salt 脂肪酸铵盐
fatty acid anhydride 脂肪酸酐
fatty acid cationic ester 脂肪酸阳离子酯
fatty acid chloride 脂肪酸氯
fatty acid ethanol amide 脂肪酸乙醇酰胺
fatty acid ethylene oxide condensate 脂肪酸环氧乙烷缩合物
fatty acid fractionation 脂肪酸分馏
fatty acid glyceride 甘油脂肪酸酯
fatty acid hydrazide 脂肪酸酰肼
fatty acid inositol ester 脂肪酸环己六醇酯
fatty acid lakes 脂肪酸色淀
fatty acid mannitol ester 脂肪酸甘露醇酯
fatty acid methyl ester 脂肪酸甲酯
fatty acid methylol amide 脂肪酸羟甲酰胺
fatty acid monoethyl polyoxyethylene ester 脂肪酸单乙基聚氧乙烯酯
fatty acid monoglyceride 脂肪酸单甘油酯
fatty acid monoglyceride monosulfate 脂肪酸单甘油酯单硫酸盐
fatty acid nitrile 脂肪酸腈
fatty acid pentaerythritol ester 脂肪酸季戊四醇酯
fatty acid peroxidase 脂(肪)酸过氧物酶
fatty acid phenylethyl amide 苯乙基脂肪酰胺
fatty acid pitch 脂肪酸沥青
fatty acid polyethylene glycol ester 脂肪酸聚乙二醇酯
fatty acid polyglycol ester 脂肪酸聚乙二醇酯
fatty acid polyvinyl alcohol ester 脂肪酸聚乙烯醇酯
fatty acid process 脂肪酸法
fatty acid quaternary ammonium salt 脂肪酸季铵盐
fatty acid separation 脂肪酸分离
fatty acid sorbitan ester 脂肪酸失水山梨糖醇酯
fatty acid sulfate 脂肪酸硫酸酯〔阴离子表面活性剂如红油或酯化油〕
fatty acid triethanol amine 脂肪酸三乙醇胺
fatty acid triglyceride 三脂肪酸甘油酯

fatty acyl alkylol amide 脂肪酰烷醇酰胺
fatty acyl amino alcohol 脂肪酰氨基醇
fatty acyl amino carboxylic acid 脂肪酰氨基羧酸
fatty acyl amino diphenyl ether sulfonic acid 脂肪酰氨基二苯醚磺酸
fatty acylated glucamine 脂肪酰葡萄糖胺
fatty acyl benzene sulfonate 脂肪酰苯磺酸盐(或酯)
fatty acyl carbazole sulfonic acid 脂肪酰咔唑磺酸
fatty acyl guanidine 脂肪酰胍
fatty acyl guanyl urea 脂肪酰脒基脲
fatty acyl hydrazide 脂肪酰肼
fatty acyl methyl metanilic acid 脂肪酰甲基间氨基苯磺酸
fatty acyl naphthalene sulfonate 脂肪酰萘磺酸盐(或酯)
fatty acyl phenol 脂肪酰基酚
fatty acyl phenoxyethyl amide 脂肪酰苯氧乙酰胺
N-fatty acyl-sphingosine N-脂(肪)酰(神经)鞘氨醇
fatty acyl sulfanilic acid 脂肪酰氨基苯磺酸
fatty acyl taurine 脂肪酰牛磺酸
fatty acyl urea 脂肪酰脲
fatty alcohol 脂肪族醇
fatty alcohol adduct 脂肪醇加合物
fatty alcohol chloroacetate ester 脂肪醇氯乙酸酯
fatty alcohol ether sulfosuccinate 脂肪醇醚磺基琥珀酸盐
fatty alcohol ethoxylate 脂肪醇乙氧基化物
fatty alcohol glucoside 脂肪醇葡糖苷
fatty alcohol phosphate 脂肪醇磷酸酯
fatty alcohol polyglycol ether 脂肪醇聚乙二醇醚
fatty alcohol sulfate 脂肪醇硫酸酯
fatty alcohol sulfonate 脂族醇磺酸酯
fatty alcohol sulfosuccinate ester 脂肪醇磺基琥珀酸酯
fatty aldehyde 脂肪(族)醛
fatty alkanol amide 脂肪酰胺；(脂族)链烷醇酰胺
fatty alkyl sulfoacetate 磺基乙酸(脂族)烷基酯
fatty amide 脂肪酰胺
fatty amide oxide 脂肪酰胺氧化物
fatty amide sulfonate 磺酸化脂肪酰胺
fatty amine 脂肪族胺
fatty amine basicity 脂肪(族)胺的碱度
fatty amine hydrochloride 脂肪胺盐酸盐
fatty amine oxide 脂肪胺氧化物；氧化脂肪胺
fatty amine polyalkylene glycol 聚亚烷基二醇脂肪胺；聚撑二醇脂肪胺
fatty amine salt 脂肪胺盐
fatty arsine oxide 脂肪胂氧化物
fatty compound 脂肪族化合物
fatty cutting oil 脂肪切削油〔用于金属切削〕

fatty degeneration 脂肪变质
fatty diethanol amide 脂肪二乙醇酰胺
fatty ether 脂肪族醚
fatty glyceride 脂肪酸甘油酯
fatty group ①脂肪族②脂肪(族)基
fatty liver 脂肪肝
fatty mercaptan 脂肪硫醇
fatty monoamine polyglycol 聚乙二醇脂肪一元胺
fatty monoethanol amide 脂肪酸单乙醇酰胺
fatty monoglyceride 脂肪酸单甘油酯
fatty nitrile 脂肪腈
fatty odor 油脂气息
fatty oil 脂肪油
fatty paint 油性漆
fat type soil 脂肪类污垢
fatty phosphine 脂肪膦
fatty polyamide 脂肪族聚酰胺
fatty polyamine polyglycol 聚乙二醇脂肪多胺
fatty series 〔复〕 脂肪系
fatty spew 脂肪斑〔皮革〕
fatty sulfonamide 脂肪磺基酰胺
fatty thioether 脂肪硫醚
fatwood 明子；多脂材
fat wrinkles〔复〕 油皱
faucet 龙头管
faucet end 龙头管末
faucet joint 龙头接嘴
faujusite 八面沸石
fault ①故障；毁损②缺陷
fault checker 故障检验器
fault detector 毁损指示器
fault diagnosis 故障诊断
fault finder(=fault indicator) ①故障指示器②探伤仪
fault indicating lamp 毁损位置指示灯
fault indication 故障指示
fault indicator(=fault finder) ①故障指示器②探伤仪
fault in seam 接缝缺陷
faultless 无缺陷的；完美无缺的
fault localization 故障位置测定法；探伤定位法
fault location 毁损位置的测定
fault locator 毁损定位器
fault tress analysis 故障树形图分析
faulty 有毛病；错误的
faulty acid treatment 不合规定的硫酸精制
faulty coal 劣质煤
faulty finish 不合格涂层(装)
faulty lubrication 不合规定的润滑(作用)
faulty operation 操作错误

faulty shearing 误行剪毛
faulty wood 疵伤木材
Fauser process 佛瑟法〔烃类部分氧化制合成气法〕
Faversham powder 法佛斯哈姆炸药
Favier explosive 法维尔(炸)药
Faville-La Vally tester 法维-拉瓦利测定器〔测定油膜破裂时的负荷〕
Favorskii reaction 法沃斯基反应
Favorskii rearrangement 法沃斯基重排作用
favoured transition 有利跃迁
fawcettiine 洛芙林；佛石松碱
fayalite 铁橄榄石
faying surface 接触面；吻合面
f-block element f区元素
F-centered lattice 面心格子
F-1 clear octane rating F-1 净辛烷率〔无乙基液研究法辛烷值〕
F-criterion F 检验
F-distribution F 分布*
feasibility study 可行性研究
feather 羽毛
feather alum 铁明矾；毛矾石
Feather analysis 费瑟分析*
Feather analyzer 费瑟分析器
feather checking 微裂；细裂
feather crystal 羽毛状晶体
feather-edge 薄边；羽状薄边
feathered lead 羽毛状铅
feathered tin 羽(毛)状锡
feather grass 羽状草
feathering ①羽毛；羽状物②釉上硫花
feathering pump 螺旋桨式泵〔当泵静止时叶面与流体垂直〕
feather key 滑键
feather length 毛圈长
feather-like precipitate 羽状沉淀
feather ore 羽毛矿
Feather plot 费瑟曲线(法)
feather shag 长毛绒
feather spray 羽状喷雾
feather-weight paper 轻磅纸
feathery needles 羽毛针状体
feat position welding 平焊
feature ①零件；部件②特征
feature selection 特征选择
febrifacient 发热药
febrifuge 退热剂
febrifugin(e) 退热碱；常山碱

feces 粪(便)
fecula ①细淀粉②淀粉渣；酒糟
feculence(=feculency) 混浊
feculency(=feculence) 混浊
feculent 混浊的
feculose (冰醋酸处理淀粉制成的)混合酯
feeble hydraulic 弱水力的
feeble magnetism 弱磁性
feed ①进料；送料②喂；饲③进给④饲品；饲料
feed adjustment 调速；料液调节
feed analysis (牲畜)饲料分析
feed apron 裙板送料器
feedback ①回料②反馈
feedback circuit 反馈电路
feedback control 反馈控制
feedback control system 反馈控制装置
feedback inhibition 反馈抑制
feedback loop 反馈回路
feedback mechanism 反馈机制
feedback principle 反馈原理
feedback ratio 反馈比
feedback-type proportioning controller 反馈型比例控制器
feed bank ①(三辊磨辊子旋转时的)料堆；漆堆；墨堆②喂料垄〔塑料〕
feed belt 进料皮带
feed bin ①供应仓库②进料斗
feedblock coextrusion 供料头式共挤塑
feed bush(=sprue bushing) 注道衬套
feed cable 电源电缆
feed channel 进料道
feed charge 装料
feed charging meter 进料计
feed chute 进料槽
feed clarification 料液澄清
feed cock 给水龙头
feed compartment 加料室
feed composition 进料组成
feed control 进料控制
feed conveyer 进料运输机
feed cylinder(=heating cylinder) 料筒
feed delivery pipe 进料泄出管
feed disk 盘式进料器
feed distributing plate 进料分布板
feed drum 筒式进料机
feed end 加料端
feed entrance point 进料位置
feeder ①进料器；加料器②馈(电)线③塞尺

feeder auger screw 供料机螺旋进料器
feeder cable 电力干线
feed exchanger 进料换热器
feedforward control 前馈控制
feedforward inhibition 前馈抑制
feed funnel 进料斗
feed gas 原料气
feed glass 进料玻璃；控制玻璃〔注油器的〕
feed head 进料口；进料头
feed heater 进料加热器
feed heat exchanger 进料热交换器
feed hopper 进料斗
feed hose 供粉软管；供漆软管
feeding ①进料；送料；给料②喂养
feeding belt 进料带
feeding culture 流加培养
feeding engine ①进料泵②上水泵
feeding equipment 进料装置；加料设备
feeding experiment ①送料试验②饲料试验；喂养试验
feeding hopper 进料斗
feeding neck 进料口
feeding platform 进料平台
feeding pump 进料泵
feeding rate 喂丝速率；进料速率
feeding roller 进料辊
feeding screw 螺旋进料器
feeding throat 进料口
feeding trough 喂料槽
feeding up 稠化；发满〔俗〕
feed inlet 进料口
feed line 原料输送管；原料供给线
feed mechanism ①进料机构〔机工〕②给水机构〔机工〕
feed mill 进料磨
feed naphtha 石脑油原料
feed nozzle 喷料嘴
feed opening 下料口
feed-out shaver 单面进出削匀机
feed pipe ①进料管②给水管
feed plate 进料板
feed point 供给点
feed port 进料孔
feed preheater 进料预热器
feed preparation technique 进料预处理方法
feed preparation unit 进料预处理装置
feed-product exchanger 进料-产品热交换器
feed pump ①进料泵②上水泵
feed rack (给)齿条
feed rate 加料速度

feed regulator 进料调节器
feed roll 进料辊
feed roller 加料辊
feed shaft 进给杆；进刀杆
feed solution 萃取进料
feed spool 馈送盘(轴)
feed spout 给料斜槽
feed sprue 注料口
feed steam 进汽
feed stock 原料
feed stock conversion 原料转化率
feed strainer 进料吸滤器
feed stream 原料流
feedstuff analysis 饲料分析
feed suction pipe 进料泵的吸入管
feed supplement 饲料添加剂
feed surge bin 进料斗；进料缓冲斗
feed surge drum 进料平衡罐
feed system 进料系统
feed table 进料盘
feed tank 供应槽
feedthrough line 供应管线
feed timer 进料定时器
feed tray 进料盘
feed tube 进料管
feed-type 供(加，送)料类型；给料方式
feed valve 进料阀
feed vaporizer 原料蒸发器
feed water ①(锅炉)给水②饮用水
feed weighing 供料称量
feed well 给水井
feed yeast 饲料酵母
feeler 测隙规；厚薄规；隙片
feeler blade 测隙片；隙片
feeler gauge 测隙规；厚薄规
feeler microscope 接触式测微显微镜
feel flow draw of block 方框流程图
feeling 手感
feel test of lubricant 润滑剂的触觉试验
feet of head 压头英尺数
Fehling reagent 费林试剂
Fehling solution 费林溶液
Fehling's reaction 费林反应
Fehling test 费林试验
feldspar 长石
feldspar box 长石箱
feldspathic porcelain 长石瓷
feldspathic rock 长石岩

feldspathoides〔复〕 似长石
felite ①F 盐；己盐②F 岩〔一种 belite B 岩〕
fell 生皮
Fellgett advantage 费尔盖特效益
felling stitch 折缝
fellmonger ①去毛皮商②去毛
fellmongered pelt 去毛绵羊皮
fellmongering 去毛
fellmongery 去毛作坊
felloe band 钢带；载重带
Fell relationship 费尔关系(曲线)〔研究漆膜厚度与遮盖力关系的曲线〕
felodipine 非洛地平〔药〕
felon herb 艾蒿
felsite 霏细岩
felsitic rock 霏细岩
felsophyre 霏细斑岩
felspar 长石
felspathic 长石的
felt (毛)毡
feltability 缩绒性；毡合性
felt backing 毛毡里；(毛)毡地
felt buff 毡轮；毡抛光轮
felt-cloth 毡布；毛布
felt collar 呢绒绷带
felt conditioner ①呢绒洗涤器②毛布洗涤器
felt covering 毛毡衬面
felt cylinder 毛毡圆筒；毛毡滤心
felt disc 毡轮；盘式毡磨轮
felt drier ①呢绒干燥剂②毛布干燥剂
felt drying machine 包毡滚筒烘干机
felt dust packing 毛毡填充物
felt element 毛毡滤心；毛毡过滤装置
felt filter 毡滤器
felt for paper making machine 造纸机用毡
felt gasket 毡垫；毛毡垫圈
felt guide 呢绒导带
felting 缩绒；毡合；毡化
felting down ①毡垫打毛②(漆膜上的)黏毛
felting polishing disc 毡质抛光轮
felting property 缩绒性
felting soap 洗毛皂；缩绒皂
feltless wet machine 无毛毯湿抄机
felt packing 毛毡衬垫
felt pad 毛毡坐垫
felt polishing disk 毡抛光轮
felt roll 毡辊；毛毯辊
felt roll oiler 毡辊注油器

felt seal 毛毡封
felt side 正面；毛布面
felt suction box 毡吸箱
felt whipper 净毡器
felt widening roll 起动辊〔设备〕
feltwork 致密纤维
felt wrapped roll 压光辊；绒辊
female ①雌的；女性的②女性③雌性〔动植物〕
female die 阴模；下半模
female flange 凹面法兰
female hormone 雌性激素
female mold 阴模；下半模
female rotor 阴转子
female screw 阴螺丝；内螺丝
female section 阴模；下半模
female thread 阴螺纹；内螺纹
femic 铁镁质
femic minerals〔复〕 铁镁质矿物
feminization 女性化；雌性化
femto- 尘；飞〔词头〕；毫微微（10^{-15}）
femtometer 毫微微米（10^{-15}米）
femtomole 飞摩尔
femtosecond fluorescence up-conversion 飞秒荧光直接转化技术
-fen〔词尾〕 含硫的烃
fence fitting 护栅器材
fenchane 莰烷；小茴香烷
fenchanol 莰醇；小茴香醇
fenchanone 莰酮；小茴香酮
fenchene 莰烯；小茴香烯
fenchenic acid 莰烯酸
fenchlorphos 皮蝇硫磷〔农药〕
fenchocamphorone 茴香樟脑；脱甲樟脑；莰樟酮 $C_9H_{14}O_2$
fenchol 莰醇；小茴香醇
fencholenic acid 莰烯酸
fencholic acid 莰酸；小茴香酸
fenchone 莰酮；小茴香酮
fenchone oxime(=fenchoxime) 莰肟；莰酮肟
fenchoxime(=fenchone oxime) 莰肟；莰酮肟
fenchyl 莰基
fenchyl alcohol 莰醇；小茴香醇
fenchylamine 莰胺 $C_{10}H_{19}N$
fenchylchloride 莰基氯 $C_{10}H_{17}Cl$
fenchyl thiocyanoacetate 氰硫基乙酸莰酯
fender ①护板②护冲器
fenestrane 四环壬烷
feniculin 小茴香灵；对丙烯基苯酚异戊烯醚

fenite 长霓岩
fenks 鱼油渣
fennel 小茴香
fennel oil 小茴香油
fennel-seed oil 小茴香油
fennel water 小茴香水
fenofibrate 非诺贝特〔药〕
Fenske equation 芬斯克方程
Fenske helix packing 芬斯克螺旋型填料
Fenske packing 芬斯克填料
fentanyl 芬太尼〔药〕
fenthion 肟硫磷〔农药〕
fento-〔词头〕毫沙；微纤（10^{-15}）
Fenton reagent 芬顿试剂
fenugreek 葫芦巴
fenugreek oil 葫芦巴油；灵香草油
feochromin(=phaeochromin) 暗色菌素
ferbam(=ferric dimethyldithiocarbamate) 二甲基二硫代氨基甲酸铁；二甲氨基荒酸铁；福美铁
ferberite 钨铁矿
ferganite 磷钒铀矿
fergusonite 褐钇铌矿
fermator 生化反应器；发酵器
Fermat principle 费马原理
ferment ①酵素；酶②发酵
fermentating tube 发酵管
fermentation 发酵
fermentation alcohol 发酵酒精
fermentation amyl alcohol 发酵戊醇
fermentation butyl alcohol ①发酵(法)丁醇②异丁醇
fermentation chamber 发酵室
fermentation chemistry 发酵化学
fermentation coefficient 发酵系数
fermentation cylinder 发酵罐
fermentation diagram 发酵曲线；发酵图谱
fermentation glycerol 发酵(法)甘油
fermentation inhibitor 发酵抑制剂
fermentation in sealed vessel 密闭(式)发酵法
fermentation lactic acid 发酵乳酸
fermentation liquid 发酵(溶)液
fermentation liquor 发酵液
fermentation lock ①水封②发酵闸
fermentation media 发酵介质
fermentation milk 发酵乳
fermentation plant 发酵工厂
fermentation recovery 发酵回收率
fermentation residue 发酵残渣
fermentation test 发酵试验

fermentation tube 发酵管
fermentation valve 发酵阀
fermentation vat ①发酵池②瘤胃②浸泡池
fermentative 发酵的
fermenting 发酵
fermenting cellar 发酵窖
fermenting plant 发酵工厂
fermenting room 发酵室
fermentogen 酶原
fermentograph 发酵测定仪〔测定发酵面团产气能力〕
fermentometer 发酵计
fermentum 酵母
fermi 费米〔长度单位，$=10^{-13}$ 厘米〕
Fermic golden rule 费米黄金定则
Fermi contact interaction 费米接触相互作用
Fermi contact shift 费米接触位移
Fermi-Dirac distribution 费米-狄拉克分布
Fermi-Dirac statistics 费米-狄拉克统计
Fermi hole 费米穴*
Fermi level 费米能级
Fermi resonance 费米共振
Fermi surface 费米表面
fermium 镄〔100 号元素，化学符号 Fm〕
fermium amalgam 镄汞齐
fermium bromide 溴化镄
fern ①蕨(类植物)；羊齿②香薇香型
Fernbach flask 费氏烧瓶
fern extract 香蕨浸剂
fern oil 羊齿油
fern soap 香薇型香皂
ferramido-chloromycin 铁铵氯新霉素
Ferranti coaxial cylinder viscometer 费兰蒂同心轴圆筒黏度计
Ferranti furnace 费兰蒂电炉
Ferranti portable viscometer 费兰蒂轻便式黏度计
Ferranti-Shirley cone-plate viscometer 费兰蒂-雪莉锥板黏度计
ferrate 高铁酸盐
ferrazite 磷钡铅矿
ferri- 铁
ferriammonium chloride 氯铁酸铵 $NH_4[FeCl_4]$
ferriammonium chromate 铬酸铵铁 $NH_4[Fe(CrO_4)_2]$
ferriammonium sulfate 铁铵矾〔硫酸铁与硫酸铵形成的水合复盐〕
ferribacterium 铁杆菌
ferric (正)铁的；三价铁的
ferric acetate 乙酸铁 $Fe(C_2H_3O_2)_3$
ferric acid 高铁酸 H_2FeO_4

ferric alum(=iron alum) 铁明矾
ferric ammonium chromate 铬酸铵铁
ferric ammonium citrate 柠檬酸铁铵
　　$(NH_4)_3C_6H_5O_7 \cdot FeC_6H_5O_7$; $(NH_4)_3Fe(C_6H_5O_7)_2$
ferric ammonium oxalate 草酸铁铵 $(NH_4)_3Fe(C_2O_4)_3$
ferric ammonium tartrate 酒石酸铁铵
ferric bedellite 铁贝得石
ferric bromide 溴化铁 $FeBr_3$
ferric carbonate 碳酸铁 $Fe_2(CO_3)_3$
ferric chlorate 氯酸铁 $Fe(ClO_3)_3$
ferric chloride 氯化铁 $FeCl_3$
ferric chromate 铬酸铁
ferric citrate 柠檬酸铁 $FeC_6H_5O_7$
ferric compound 正铁化合物
ferric cyanide 氰化铁
ferric diethyldithiocarbamate 二乙基二硫代氨基甲酸铁〔促进剂〕
ferric dimethyldithiocarbamate(=ferbam) 二甲基二硫代氨基甲酸铁；二甲氨基荒酸铁；福美铁
ferric ferrocyanide 亚铁氰化铁
ferric fluoride 氟化铁 FeF_3
ferric formate(=ferric formiate) 甲酸铁 $Fe(HCO_2)_3$
ferric formiate(=ferric formate) 甲酸铁
ferric hydroxide 氢氧化铁 $Fe(OH)_3$
ferric hypophosphite 次磷酸铁 $Fe(H_2PO_2)_3$
ferricinium ion 二茂铁离子 $[Fe(C_5H_5)_2]^+$
ferric iodide 碘化铁 FeI_3
ferric ion 铁离子
ferric ion-selective electrode 高铁离子选择电极
ferric lactate 乳酸铁 $Fe(C_3H_5O_3)_3$
ferric malate 苹果酸铁 $Fe(C_4H_4O_5)_3$
ferric metasilicate 硅酸铁
ferric nitrate 硝酸铁 $Fe(NO_3)_3$
ferri-compound 正铁化合物
ferric oxalate 草酸铁 $Fe_2(C_2O_4)_3$
ferric oxide 氧化铁；三氧化二铁
ferric peptonate 胨合铁
ferric phosphate 磷酸铁 $FePO_4$
ferric phosphate sludge 磷酸铁泥渣
ferric potassium alum 铁钾矾
ferric potassium ferrocyanide 亚铁氰化铁钾；钾盐铁蓝
ferric potassium sulfate 铁钾矾〔硫酸钾和硫酸铁的水合复盐；硫酸铁钾〕
ferric pyrophosphate 焦磷酸铁 $Fe_4(P_2O_7)_3$
ferric pyrosulfate 焦硫酸铁 $Fe_2(S_2O_7)_3$
ferric pyrothioarsenate 焦硫代砷酸铁 $Fe_4(As_2S_7)_3$
ferric rhodanate(=ferric rhodanide) 硫氰酸铁 $Fe(SCN)_3$
ferric rhodanide(=ferric rhodanate) 硫氰酸铁

ferric rubidium alum(=ferric rubidium sulfate) 铁铷矾
ferric rubidium sulfate 铁铷矾〔硫酸铷和硫酸铁的水合复盐;硫酸铷铁〕
ferric succinate 丁二酸铁
ferric sulfate 硫酸铁 $Fe_2(SO_4)_3$
ferric sulfide 硫化铁 Fe_2S_3
ferric sulfocyanate(=ferric thiocyanate) 硫氰酸铁
ferric tannate 单宁酸铁
ferric thiocyanate(=ferric sulfocyanate; ferric thiocyanide) 硫氰酸铁
ferric thiocyanide(=ferric thiocyanate) 硫氰酸铁
ferricyanic acid 氰铁酸;六氰合铁氢酸 $H_3[Fe(CN)_6]$
ferricyanide 氰铁酸盐;铁氰化物 $M_3[Fe(CN)_6]$
ferricyanide titration 铁氰化物滴定
ferricyanometry 铁氰化物滴定法
ferriferro cyanide 氰亚铁酸铁;亚铁氰化铁〔铁蓝的化学成分〕
ferriferrous 正(铁)亚铁的
ferriferrous chloride 氯化正亚铁;五氯化二铁 $FeCl_2 \cdot FeCl_3$
ferriferrous compound 正(铁)亚铁化合物
ferriferrous gneiss 铁质片麻岩
ferriferrous hydroxide 氢氧化正亚铁 $2Fe(OH)_3 \cdot Fe(OH)_2$
ferriferrous oxide 氧化正亚铁;四氧化三铁 Fe_3O_4
ferriferrous sulfate 硫酸正亚铁 $FeSO_4 \cdot Fe_2(SO_4)_3$
ferriferrous sulfide 硫化正亚铁;四硫化三铁 $FeS \cdot Fe_2S_3$
ferrigluconate 葡萄糖酸高铁盐
ferriin 菲咯啉(正)铁配离子
ferrimagnetism 亚铁磁性
ferrimanganese 铁锰齐
ferrimanganic 铁锰(根)〔指兼含三价铁 Fe^{3+} 及三价锰 Mn^{3+}〕
ferrimanganic pyrophosphate 焦磷酸铁锰
ferrimetric titration 铁(Ⅲ)盐滴定
ferrimetry 铁(Ⅲ)盐滴定法
ferrimycin 高铁霉素
ferriporphyrin 铁卟啉
ferripotassium 铁钾(根)〔根含有钾及三价铁〕
ferripotassium citrate 柠檬酸铁钾
ferriprotoporphyrin(=heme) 正铁血红素
ferripyrin(=ferropyrin) 铁比林
ferrisodium 铁钠(根)〔指含有钠及三价铁〕
ferrite ①铁酸盐②纯粒铁③铁素体;铁氧体
ferrite matrix 铁氧体(铁素体)晶体;铁氧体基质
ferrite red (合成)铁红
ferrite steel 铁素体钢

ferrite yellow 铁黄
ferrithioformin 铁硫甲酰素
ferritic-austenitic alloy 铁素奥氏体合金
ferritic based compound plating layer 铁基复合镀层
ferritic steel 铁素体钢
ferritin assay 血清铁蛋白测定
ferro- 〔词头〕亚铁
ferro-alloy 铁合金
ferroaluminium 铁铝合金;铝铁合金 Fe_2Al_5
ferrobilin 三氯铁胆青盐
ferrocarbon titanium 铁碳钛齐
ferrocene 二茂铁*
ferrocene carboxylic acid ethyl ester(=ethyl ferrocenoate) 二茂铁羧酸乙酯
1,1-ferrocenedicarboxylic acid 1,1-二茂铁二羧酸
ferrocene polymer 二茂铁聚合物
ferrocenoate 二茂铁(二)羧酸酯(盐)
ferrocenoyl chloride(=chlorocarbonyl ferrocene) 二茂铁碳酰氯;二茂铁羰基氯;氯碳酰二茂铁
ferrocenyl 二茂铁基
ferrocenyl methyl ketone(=acetylferrocene) 二茂铁基甲基(甲)酮;乙酰基二茂铁 $(C_{10}H_9Fe)COCH_3$
ferrocenylmethyl methacrylate 甲基丙烯酸二茂铁代甲酯
ferrocenyltriphenylsilane 茂铁三苯硅烷
ferrocerium 铁铈齐
ferrochrome 铬铁合金
ferro-chromium 铬铁(合金)
ferro-compound 二价铁化合物
ferroconcrete 钢筋混凝土
ferrocyanic acid 氰亚铁酸;亚铁氰酸 $H_4[Fe(CN)_6]$
ferrocyanide 氰亚铁酸盐;亚铁氰化物
ferrocyanide process 氰亚铁酸盐法;亚铁氰化物法
ferrocyanide titration 亚铁氰化物滴定
ferrocyanometry 亚铁氰化物滴定法
ferrodolomite 铁白云石
ferroelectric 铁电的
ferroelectric ceramics 铁电陶瓷
ferroelectric ceramic thin films 铁电陶瓷薄膜
ferroelectric copolymer 铁电共聚物
ferroelectric crystal 铁电性晶体*
ferroelectric effect 铁电效应
ferroelectric hysteresis loop 电滞回线
ferroelectricity 铁电性*
ferroelectric polymer 铁电聚合物
ferroeletric domain 铁电畴
ferroferric 亚铁正铁(根)
ferroferric compound 亚(铁)正铁化合物

ferroferric oxide 四氧化三铁 $FeO \cdot Fe_2O_3$
ferroferricyanide 氰铁酸亚铁；(正)铁氰化亚铁 $Fe_3[Fe(CN)_6]_2$
ferroferrocyanide 氰亚铁酸亚铁；亚铁氰化亚铁 $Fe_2[Fe(CN)_6]$
ferrofining 铁剂精制
ferrograph 铁粉记录图；铁磁示波器
ferrogum 橡胶磁铁
ferroheme (亚铁)血红素
ferroin 邻菲咯啉亚铁离子*；亚铁试剂
ferromagnesium ①铁镁齐②亚铁镁(根)
ferromagnesium sulfate 硫酸亚铁镁 $FeSO_4 \cdot MgSO_4 \cdot 6H_2O$
ferromagnetic 铁磁的
ferromagnetic polymer 铁磁聚合物
ferromagnetic resonance 铁磁共振
ferromagnetic resonance absorption 铁磁共振吸收
ferromagnetism 铁磁性
ferro-manganese 铁锰(合金)
ferromanganic 亚铁锰(根)〔指含有二价铁与三价锰〕
ferromanganous 亚铁亚锰(根)〔指兼含二价铁与二价锰〕
ferromanganous chloride 氯化亚铁亚锰 $MnFeCl_4$
ferromanganous sulfate 硫酸亚铁亚锰 $FeMn(SO_4)_2 \cdot 12H_2O$
ferrometric method 亚铁量法
ferrometric titration 亚铁量滴定
ferrometry 亚铁量法〔滴定法〕
ferro-molybdenum 钼铁(合金)
ferromycin 菲洛霉素
ferron(=7-iodo-8-hydroxyquinoline-5-sulfonic acid) 7-碘-8-羟基喹啉-5-磺酸；试铁灵
ferro-nickel 镍铁(合金)
ferro-phosphorus 磷铁
ferropotassium 亚铁钾(根)
ferro-silico-aluminum 硅铝铁(合金)
ferro-silico-manganese 硅锰铁(合金)
ferrosilicon 高硅铸铁
ferro-silicon alloy 硅铁(合金)
ferro-silico-nickel 硅镍铁(合金)
ferro-silicon process 硅铁法
ferro-silico-titanium 硅钛铁(合金)
ferrosi sulfate 硫酸亚铁
ferrosoferric compound 亚(铁)正铁的化合物
ferro-titanium 钛铁(合金)
ferro-tungsten 钨铁(合金)
ferro-uranium 铀铁(合金)
ferrous 亚铁的；二价铁的

ferrous abrasives 铁磨料
ferrous acetate 乙酸亚铁 $Fe(C_2H_3O_2)_2$
ferrous acid 亚铁酸
ferrous ammonium phosphate 磷酸亚铁铵
ferrous ammonium sulfate 硫酸亚铁铵
ferrous arsenide 二砷化三铁 Fe_3As_2
ferrous bicarbonate 碳酸氢亚铁 $Fe(HCO_3)_2$
ferrous bromide 溴化亚铁 $FeBr_2$
ferrous carbonate 碳酸亚铁 $FeCO_3$
ferrous chloride 氯化亚铁 $FeCl_2$
ferrous compound 亚铁化合物
ferrous disulfide 二硫化铁 FeS_2
ferrous dosimeter 亚铁剂量计
ferrous ferricyanide 氰铁化亚铁
ferrous fluoride 氟化亚铁 FeF_2
ferrous fluosilicate 氟硅酸亚铁 $FeSiF_6$
ferrous fumarate 富马酸亚铁
ferrous gluconate 葡萄糖酸亚铁
ferrous hydrogen carbonate 碳酸氢亚铁 $Fe(HCO_3)_2$
ferrous hydroxide 氢氧化亚铁 $Fe(OH)_2$
ferrous hypophosphite 次磷酸亚铁 $Fe(H_2PO_2)_2$
ferrous hyposulfite 连二亚硫酸亚铁 FeS_2O_4
ferrousimetry 亚铁滴定法
ferrous iodide 碘化亚铁 FeI_2
ferrous lactate 乳酸亚铁 $Fe(C_3H_5O_2)_2$
ferrous line group 铁谱线组
ferrous metal 铁类金属；黑色金属
ferrous metallurgy 钢铁冶金；黑色冶金
ferrous metasilicate 硅酸亚铁 $FeSiO_3$
ferrous nitrate 硝酸亚铁 $Fe(NO_3)_2$
ferrous orthosilicate 正硅酸亚铁 $FeSiO_4$
ferrous oxalate 草酸亚铁 FeC_2O_4
ferrous oxide 氧化亚铁 FeO
ferrous phosphate 磷酸亚铁 $Fe_3(PO_4)_2$
ferrous phosphide 磷化铁〔指 Fe_3P_4、Fe_2P_3 等〕
ferrous platinichloride 氯铂酸亚铁 $Fe[PtCl_6]$
ferrous pyrothioarsenate 硫代焦砷酸亚铁 $Fe_2As_2S_7$
ferrous rhodanate 硫氰酸亚铁 $Fe(SCN)_2$
ferrous selenide 一硒化铁 $FeSe$
ferrous sulfamate 氨基磺酸亚铁 $Fe(NH_2SO_3)_2$
ferrous sulfate 硫酸亚铁 $FeSO_4$
ferrous sulfate dosimeter 硫酸亚铁剂量计
ferrous sulfide 硫化亚铁
ferrous sulfocyanate(=ferrous sulfocyanide) 硫氰酸亚铁 $Fe(SCN)_2$
ferrous sulfocyanide(=ferrous sulfocyanate) 硫氰酸亚铁
ferrous tartrate 酒石酸亚铁 $FeC_4H_4O_6$
ferrous telluride 碲化亚铁 $FeTe$

ferrous thiocyanate(=ferrous thiocyanide)　硫氰酸亚铁 Fe(SCN)$_2$
ferrous thiocyanide(=ferrous thiocyanate)　硫氰酸亚铁
ferrous thiosulfate　硫代硫酸亚铁　FeSO$_3$S; FeS$_2$O$_3$
ferrous titanate　钛酸亚铁　FeTiO$_3$
ferrous tungstate　钨酸亚铁　FeWO$_4$
ferrovanadium　钒铁(合金)
ferroverdin　绿铁(合金)
ferroxyl indicator　铁锈指示剂
ferroxyl test　(镀层及漆膜的)孔隙率试验；铁锈(指示剂)试验
ferrozine　二茂铬铁
ferruginol　铁锈醇；弥罗松酚〔取自植物铁罗汉〕
ferruginous　含铁的
ferrule　①衬套；套管②嵌环③管箍；管接头
ferrum　铁
ferry　①驳船②(F-)菲瑞铜镍合金
Ferry-Williams' approximation　费里-威廉斯近似式
fertile　可转换的；可再生的；可增殖的
fertile absorber　有效吸收剂
fertile mycelial filament　结果菌丝
fertile mycelium　结实菌丝体
fertile nuclide　可转换核素*
fertile-to-fissile ratio　可转换的与易裂变的核素的比
fertiliser(=fertilizer)　(化学)肥料
fertility　①肥(沃)度；肥力②肥沃性③能生产性
fertility vitamin　生育维生素；维生素E
fertilizer(=fertiliser)　(化学)肥料
fertilizer analysis　肥料分析
fertilizer element　肥料要素
fertilizer law　肥料法
fertilizer sawer　撒肥机
ferulaic acid(=ferulic acid)　阿魏酸
ferulaldehyde　阿魏醛
ferulene　阿魏烯
ferulic acid(=ferulaic acid)　阿魏酸
feruloyl quinic acid　阿魏酰奎尼酸
fervanote　水钒铁矿
fervenulin　热诚菌素
Féry glass prism　费里玻璃棱镜
Féry prism　费里棱镜
Féry pyrometer　费里高温计
Féry refractometer　费里折射仪；费里折光仪
fesogenin　沤索配基
festoon　热成型片材
festoon cooler　悬挂式冷却装置
festoon drier　浮花干燥器
festoon dryer　长环悬挂干燥机；悬挂式干燥机
festooning　在浮花干燥器中进行干燥
festooning oven　垂悬式烘箱
festoon rack　蓄布架；挂布架
fetich bean oil　神豆油
fettle　修边
Feulgen reaction　福尔根氏反应
Feulgen stain　福尔根染色法
Feussner spark　费氏火花
Feussner spark source　费氏火花源
feverbark　治热树皮
F-factor　F因子
fiber　纤维
fiber accumulation　(玻璃纤维)纤维集束
fiber alignment　纤维定向
fiber attrition　纤维磨损
fiber bonding　纤维黏合
fiber bundle　纤维束
fiber can　①纤维板桶②棉条桶
fiber container　纤维容器
fiber content　纤维含量
fiber count　纤维支数
fiber-covered plywood　纤维面层(压木)板
fiber crimpness　纤维蜷曲度
fiber density　纤维密度
fiber diagram　纤维图
fiber diameter　纤维直径
fiber (diffraction) pattern　纤维(衍射)图
fiber dispersion　纤维分散
fiber fineness　纤维细度
fiber finishing　纤维表面处理
fiber forming　成纤
fiber-forming polyamide　成纤聚酰胺；拉丝聚酰胺
fiber furnish　纤维配比
fiberglas(s)　玻璃纤维；玻璃丝
fiber glass(=spun glass)　玻璃纤维
fiber glass coating　玻璃纤维套
fiberglass drum　玻璃钢转鼓
fiber glass packing　玻璃纤维垫
fiber glass reinforced plastic　玻璃纤维增强塑料
fiber-grade monomer　化纤单体
fiber grease(=fibrous grease)　纤维状润滑脂
fiber hybrid composites　混杂纤维复合材料
fiberized　纤维化了的
fiberizer　石棉毛纺机
fibcrizing　纤维解离
fiber junction　纤维液接
fiber length　①纤维长度②平均纤长
fiber length control　(润滑脂)纤维长度的控制

fiberlike   纤维状〔胶束形状〕
fiberlike texture   纤维状结构
fiber optical sensor   ①光纤传感器②光纤探测器
fiber optic-based chemical sensor   光纤化学传感器
fiber optic biosensors   光纤生物传感器
fiber optic chemical sensor   光纤化学传感器
fiber optic humidity sensor   光纤湿度传感器
fiber optic light tube   光纤光管
fiber optic probe   光纤探头
fiber optics   纤维光学
fiber optic sensor   光纤传感器
fiber orientation   纤维取向；纤维定向
fiber packing   (硬化)纸板衬垫
fiber pattern   ①纤维纹理②纤维组织图
fiber period   纤维(轴向)等同周期
fiber profile   纤维纵切面图
fiber protrusion   纤维突出
fiber recovery machine   纤维回收机
fiber reinforced plastics   纤维增强塑料
fiber reinforcement   纤维强化热固性；纤维强化复合材料
fiber saturation point   纤维饱和点
fiber separator   纤维分选器
fiber show   露丝〔增强塑料疵点〕
fiber spin-dying   化纤原液染色
fiber spinning   纤维纺丝
fiber staining   纤维着色
fiber strength   纤维强度
fiber stress   纤维强度
fiber structure   纤维结构
fiber wax   纤维蜡
fiber wetness   纤维浸润性
fibr-   〔词头〕纤维
fibrage   纤维编织；纤维层
fibrated asphalt coating   含纤维沥青涂料；纤维增强沥青涂料
fibration   纤维形成；纤维化
fibre(=fiber)   ①纤维②纤维纸板
fibre acetylated staple   乙酰化短纤维
fibre aftertreatment   纤维后处理
fibre agglomerate   纤维团；纤维结块
fibre array   纤维束；纤维排列
fibre assemblies   纤维束；纤维团
fibre axis   纤维轴
fibre bending fatigue tester   纤维弯曲疲劳测试仪
fibre blend   纤维混合(物)
fibre-building polyester   成纤用聚酯；制造纤维的聚酯
fibre bunching   玻纤离析

fibre coating technique   纤维涂层技术
fibre cohesion   纤维抱合力
fibre composite   纤维复合材料
fibre compressive elasticity tester   纤维压缩弹性试验机
fibre configuration   纤维构造；纤维构型；纤维形态
fibre crystallinity   ①纤维结晶度②纤维结晶性
fibred   ①纤维状的②纤维质的
fibre density analysis   纤维密度分析
fibre diagram   纤维长度分布图；纤维结构图
fibre diameter video analyzer   纤维直径视频分析仪；纤维直径图像分析仪
fibre disc sander   布轮磨光机
fibre drag   纤维间分离阻力
fibre exposure   露纤
fibre finish   ①纤维整理②纤维整理油剂
fibre flock   绒屑；纤维屑
fibre fluorescence tester   纤维紫外线荧光测试法
fibre forming compound   成纤(维)化合物
fibre-forming condition   成纤条件
fibre-forming crystallite   成纤微晶
fibre-forming material   成纤材料
fibre-forming polyamide   成纤聚酰胺
fibre-forming polymer   成纤聚合物
fibreglass   玻璃纤维；玻璃丝
fibre glass casement fabric   玻璃丝窗帘布
fibre glass reinforced laminate   玻璃纤维增强层合塑料
fibreglass reinforced thermoplastic   玻璃纤维增强热塑性塑料
fibre-grade monomer   纤维级单体
fibre-grade purity   纤维级纯度
fibre laying   纤维平伏性
fibre length array diagram   纤维长度-根数排列图
fibrelike polymer crystal   纤维状聚合物晶体
fibre lubrication   纤维润滑(作用)
fibre mass content   纤维质量含量
fibre matting   纤维缠结；纤维消光
fibre melting point microscope   纤维熔点显微镜
fibre melting point tester   纤维熔点仪
fibremeter   (光电式)纤维定量喂给控制装置
fibre modification   纤维改性(作用)；纤维变性(作用)
fibre modifier   纤维改性剂；纤维变性剂
fibre moisture regain tester   纤维回潮测试仪；纤维回潮率测试仪
fibre morphometer   纤维形态测视仪
fibre nep   纤维结〔纤维疵点〕
fibre network   纤维网络
fibre number   纤维纤度
fibre orientation   纤维取向；纤维定向

fibre parallelization 纤维平行化(作用)；纤维平行排列
fibre path(=filament path) 丝路
fibre period 纤维(链段)重复长度；纤维等同周期
fibre porosity 纤维的多孔性
fibre prosthesis 纤维(制)假体〔人造器官〕
fibre-reactive dye 纤维用活性染料
fibre reinforced concrete and cement 纤维增强混凝土
fibre reinforced metal composite 纤维增强金属复合材料
fibre reinforced plastics 纤维增强塑料
fibre reinforced thermoplasticity plastics 纤维增强热塑性塑料
fibre-related biotechnology 纤维相关生物技术
fibre resin 纤维(用)树脂；纤维级树脂
fibre-resin composite 纤维-树脂复合材料
fibre sealing medium 纤维封闭介质
fibre section 纤维切片；纤维截面
fibre setter 纤维定形机
fibres-from-film technology 裂膜(法)纤维工艺学；薄膜(法)纤维工艺学
fibre show(=fibre prominence) 露丝
fibre spicules （石棉)针状纤维
fibre strand ①纤维辫子②并丝〔化纤疵点〕
fibre strength 纤维强度
fibre-strengthened 纤维增强的
fibre strength tester 纤维强力测试仪
fibre structure(=fibre texture) 纤维结构
fibre taper 纤维轴向不均匀性；纤维拔斜率
fibre-tendering 纤维损伤
fibre texture 纤维结构；纤维组织
fibre thickening 纤维变粗
fibre thickness 纤维粗度
fibre torsion fatigue tester 纤维扭转疲劳测试仪
fibre tow 纤维束
fibre transfer 纤维转移(作用)
fibre variant 纤维变体；纤维变种
fibre volume content 纤维体积含量
fibre waste reclaiming equipment 废丝再生装置
fibre weight 纤维量〔以每厘米重若干毫克计算纤维细度〕
fibre whiteness 纤维白度
fibrid 类纤维；沉析纤维
fibril 原纤*
fibrilation 原纤化
fibrilia 麻纤维；麻织物
fibrilization during drawing 拉伸纤维化
fibrilla 小纤维
fibrillar alumina 丝状氧化铝

fibrillar crystal 原纤晶体
fibrillar crystallization 原纤晶(作用)
fibrillar structure 原纤(维)结构
fibrillar unit 原纤单元
fibrillar(y) ①原纤的②纤丝的
fibrillate 纤化
fibrillated fibre 原纤化纤维
fibrillated yarn 纤化丝
fibrillating roll 纤化辊
fibrillation 帚化；纤丝化；细纤维化
fibrillator 纤化机
fibrillen 小纤维
fibrillous ①原纤的②纤丝的
fibrinogen 血纤维蛋白原
fibrocement 纤维水泥
fibroelasts 韧弹性体
fibroferrite 纤铁矾
fibroid ①纤维状的；纤维性的②由纤维组成的
fibroillar(y) 微丝的
fibrolite 硅线石
fibrolysin 溶纤维剂；柔瘢药
fibroplast 韧塑性体
fibrosity 微丝性；原纤维性
fibrous 纤维状的
fibrous alumina 纤维状氧化铝
fibrous architecture 纤维状构造
fibrous asbestos 纤维状石棉
fibrous carbon 纤维状碳；碳纤维
fibrous cellulose powder 纤维状纤维素粉
fibrous composite 纤维复合材料
fibrous crystal 纤维晶
fibrous crystallization 纤维状结晶
fibrous filler 纤维填料
fibrous fleece 纤维网
fibrous fracture 纤维裂痕
fibrous glass 玻璃纤维；玻璃丝
fibrous glass mat 玻璃纤维席
fibrous grease(=fiber grease) 纤维状润滑脂
fibrous insulant 纤维绝热材料
fibrous iron 纤维断口铁
fibrous long spacing(FLS) 纤维长间距
fibrous magnesium silicate 硅酸镁纤维；石棉纤维
fibrous material 纤维质材料
fibrous metal 纤维状金属；金属纤维
fibrous micelle 纤维状胶束；纤维微胞
fibrous molecule 纤维状分子
fibrous morphology 纤维形态学
fibrous packing 纤维填料

fibrous peat 泥碳纤维
fibrous polymer ①成纤性聚合物②纤维状聚合物
fibrous product 纤维状制品
fibrous protein 纤维状蛋白
fibrous red iron ore 纤维赤铁矿
fibrous reinforcement 纤维增强料；纤维补强
fibrous reticulum 纤维网
fibrous rock 纤维状岩石
fibrous seed crystal 纤维状子晶
fibrous silicate 纤维状硅酸盐
fibrous structure 纤维状结构
fibrous tactoid 纤维状类晶团聚体〔纤维结构中不同长度的线型分子所形成的有规棒状胶体〕
fibrous talc 纤维型滑石粉
fibrovascular fibre 微管束纤维
ficcoxanthin 岩藻黄素
fichtelite 朽松木烷
Fick's first law 菲克第一定律
Fick's law of diffusion 菲克扩散定律
Fick's second law 菲克第二定律
ficoceryl alcohol 榕蜡醇
ficocerylic acid 榕蜡酸
ficosapentenoic acid 二十碳五烯酸
fictitious field 虚构场
fictitious peak 假峰
fictitious strain energy 虚拟应变能
fidelity 重现精度；精确(度)；保真(度)
field ①场②范围③田野④现场；工地
field-aided zone melting (电)场辅助的区熔法
field ammoniation 田间加氨
field amplified injection 场放大进样
field analysis and monitoring 现场分析与监测
field analytical chemistry 现场分析化学
field angle ①张角②视场角
field antipoisoning 战地消毒
field assay 野外测试；现场分析
field assembly 现场组装
field broadening 场(致)变宽
field calculation 现场计算
field coating 现场涂装涂料
field concentration 战地浓度
field condition 开采条件
field desorption(FD) 场解吸
field desorption ionization source 场解吸电离源
field desorption ion source 场解析离子源
field desorption mass spectrometry(FD-MS) 场解吸质谱法
field desorption source 场解吸源
field distribution 场分布
field effect 场效应
field effect biosensor 场效应生物传感器
field effect transistor(FET) 场效晶体管
field emission 场致发射
field emission detector 场(致)发射检测器
field emission electron gun 场(致)发射电子枪
field emission electron microscopy 场发射电子显微镜
field emission electron source 场发射电子源
field emission gun 场(致)发射枪
field emission ion source 场致发射离子源
field emission microscope(FEM) 场致发射显微镜
field-emission scanning electron microscopy 场发射扫描电子显微镜
field emission spectroscopy(FES) 场致发射光谱法
field emitter 场致发射
field emitter tip 场发射器尖顶
Fielden drymeter 菲尔登湿度测定器
field equation 场方程
field erected 现场安装的
field experience ①野外经验②现场经验
field experiment ①野外试验②现场试验
field fabrication 现场制作
field flow fraction 场流分级
field flow fractionation(FFF) (重力)场流分级
field flow fractionator 场流分离仪
field focused nuclear magnetic resonance 磁场聚焦核磁共振
field focusing 场聚焦
field-free region 无场区
field frequency control 场频控制
field frequency lock 场频联锁
field frequency stabilization 场频稳场
field gradient 场梯度
field increment 场增量
field induced emission gun 场发射枪
field inspection 现场检验
field intensity 场强
field ion emission 场(致)离子发射
field ionization 场致电离
field ionization ion source 场致电离离子源
field ionization mass spectroscopy 场电离质谱法
field ionization source 场电离源
field ion mass spectroscopy 场离子质谱法
field ion microscope(EIM) 场致离子显微镜；场离子显微镜检查法
field ion microscopy(FIM) 场离子显微镜检查法
field jump 电场跃变
field method 现场方法

field mint　薄荷
field modulation　场调制
field mushroom　草菇；野菇
field of activity　活性范围〔分子链〕
field of application　应用范围
field of force　力场
field of view　视场；视野
field of vision　视场
field performance　野外使用特性
field personnel training　现场人员训练
field poisoning　战地布毒
field pump　①野外用泵②油矿用泵
field result　野外试验结果
field service　①野外使用②现场使用
field service compressor　现场用压缩机；移动式压缩机
field storage　①野外储存②现场储存油库
field survey　现场调查
field sweep　扫场
field tank　①野外储罐②现场储罐
field test(=field trial)　①野外试验②现场试验
field trash　夹杂物
field trial(=field test)　①野外试验②现场试验
field usage　①野外使用②现场使用
field weld　现场焊接
field work　现场分析
fierce braking　急剧制动；猛刹车
Fiesers' rules　菲泽规则
fiffle　格槽分样器
fifteen-membered ring　十五元环
figged soap　纹饰皂
figging　(肥皂的)结晶化
fight back　顶回；弹回
fighting grade gasoline　军用航空汽油
fig smut　黑曲霉
fig soap　纹饰皂
figure　①图(形)②数字③数值④形状⑤位数⑥花纹
figured cloth　花布
figured effect　花纹效应
figured silk　提花丝织物
figured stripe chiffon　条花绸
figure imitative deviation rate　花形仿制差异率
figure of merit curve　质量曲线
Fikentscher's viscosity formula　菲肯切尔黏度式
fil　①长丝②单丝〔指复丝中的单根细丝〕③灯丝〔阴极〕
filamatic process　连续纺丝法
filament　①(纤)丝②灯丝③丝极
filamentar(y)　丝的；纤细的
filamentary cermet　金属陶瓷丝〔陶瓷与金属的复合材料〕

filamentary fibril　丝状原纤维
filamentary material　纤维材料；丝状物料
filamentary structure　纤维状结构；细丝状结构
filamentary superconductor　线状超导体；丝状超导体
filament assembly　热丝组件
filamentation　成丝
filament axis　纤维轴
filament balance　灯丝平衡
filament battery　灯丝电池；灯丝蓄电池
filament carbonization　灯丝碳化
filament crystal　丝(状结)晶
filament current　灯丝电流
filament deflector guide　导丝棒；长丝转向钩
filament denier　单纤维旦数
filament detector　热丝检测器
filament emission　灯丝发射
filamentization　成丝(作用)
filament number　单丝纤度
filamentose(=filamentar(y))　丝的；纤细的
filamentous(=filamentar(y))　丝的；纤细的
filamentous capsid　丝状衣壳
filamentous fungi　丝状(真)菌
filamentous yeast　丝状酵母
filament pullout test　纤维拔出试验
filament pyrolysis　热丝热解
filament pyrolyzer　热丝热解器
filament supply　灯丝电源
filament temperature　灯丝温度
filament-type pyrolyzer　热丝裂解器
filament warping sizing and beaming technology　长丝整浆并工艺
filament weight ratio　(复合材料)长丝重量比
filament winding　长丝维卷绕法
filament winding process　长丝卷绕工艺
filament yarn　长丝纱线
filament yarn sizing machine　长丝浆丝机
filar　丝状的
filar eyepiece　细丝测微目镜
filariomycosis　丝霉病
filar micrometer　纤条测微计
filature　巢丝；巢丝机
file　锉；锉刀
file finishing　锉削
filer content　填料含量
filicic acid　绵马酸
filicin　绵马精
filicinic acid　绵马精酸　$C_8H_{10}O_3$
filiform　丝状的

filiform corrosion 丝状锈蚀；起红丝〔俗〕
filiform corrosion resistance 耐丝状锈蚀；耐红丝腐蚀
filiform molecular structure 线状分子结构
filimarisin(=filipin) 菲律宾菌素
filing (锉)屑
filing fixture 锉工夹具
filipin 菲律宾菌素
filitannic acid 绵马单宁酸；绵马鞣酸
filix 绵马
filix extract 绵马浸剂
filixic acid 绵马根酸
fill ①填充②纬纱
fill-and-wipe 填抹上色
filled 填充了的；填了料的
filled can dump 油桶堆栈
filled capacity 满装量
filled column 填料柱
filled plastics 填充塑料
filled polyisoprene rubber 填充聚异戊二烯橡胶
filled rubber 填充橡胶
filled soap 填充皂
filled styrene butadiene rubber 填充丁苯橡胶
filled system thermometer 封入式温度计
filled tower 填料塔
filled vulcanizate 填料硫化胶
filled zero technique 添零技术
filler ①填料；填充物②体质颜料③装罐机
filler/fiber conductive 导电填充剂/导电纤维
filler absorption 填料混入
filler acceptance 填料混入
filler bars 花纹沟槽护胶
filler block 衬块
filler bowl 滤杯
filler coat (底漆)涂层
filler content 填料含量
filler-free vulcanizate 无填料硫化胶；纯硫化胶
filler gum 填充胶料
filler-inset 过滤(器)芯子
fillerless process 无纬组织
filler loaded compound 填料(填充)胶料
filler loading 填充(剂用)量
filler metal 填焊金属；填充金属
filler-neck hose 进燃料胶管〔汽车〕
filler opening(=filling hole) 填充孔；注入孔
filler orientation 填料取向
filler plate 填料板〔塑料〕
filler reinforcement 填充补强
filler reinforcement effect 填料加强效应
filler reservoir 填料储器
filler retention 填料保持
filler ring 衬环
filler rod ①焊条②嵌条
filler seam 填密缝
filler sheet 垫片
filler speck 填料斑瑕
filler test of grease 润滑脂内填料量的测定
filler yarn 纬纱；纬线〔纺〕
fillet 胶瘤〔填充在两被黏物交角处的那部分胶黏剂(如蜂窝夹芯与面材胶接时，夹芯端部所形成的胶黏剂圆角)〕
filleted corner 圆角
fillet radius 圆角半径
fillet weld 角焊缝
fillet welding 角焊
fill in ①填充物②纬纱
filling ①装填；填充②装满③装灌④填腻子
filling agent 填充剂；填料
filling capacity 填充能力；填充性能
filling compound 填料
filling connection 装油软管与石油产品储槽之间的接合
filling density 纬密；纬纱密度
filling device 装填机；装填装置
filling direction (织物)纬向
filling equipment 装填设备；装油设备
filling evaporation loss 装油时的蒸发损失
filling face twill 纬面织物
filling factor 填充系数；占空因数
filling fittings 装填零件；注油零件
filling flange (供粉筒的)加料法兰(接口)
filling funnel 注液漏斗
filling gun 注油枪
filling hole(=filler opening; filling opening; filling orifice) 填充孔；注入孔
filling house 装油房
filling layer 衬层
filling line 装油管线
filling machine 灌注机器
filling main 装油干线
filling mark 灌注记号
filling material 填充材料
filling neck 填充颈；注油软管
filling nozzle 装填管嘴；注入短管
filling of coated abrasives 砂布(砂纸)沾满(粉)屑
filling of mold 装模；填模
filling (of pie) ①馅子②底纱；筒脚〔筒管上的残纱〕
filling of tank cars 油槽车装油

filling opening(=filling hole)　①填充孔②注入孔
filling orifice(=filling hole)　①填充孔②注入孔
filling piece　填隙片
filling pipe　装灌管；注油管
filling plant　装油车间；注油装置
filling point　装油站
filling pressure　(注塑)充模压力
filling primer　填充底漆
filling pump　注油泵
filling riser　灌装鹤管
filling sleeve　灌装软管
filling soap　填料皂
filling solution　充填液
filling space　①灌装空间②充填空间
filling station　加油站
filling tank　供应槽
filling thread　纬线；纬纱
filling time　灌装时间
filling trunk　装油干线；灌注干线
filling tube　灌装管；加料管
filling up　填充
filling valve　进油阀
fillingwise　纬向；横向
fillister　凹刨
fillister head screw　有槽凸圆头螺钉
fill line　充填输送管
fillmass(=massecuite)　糖膏
fillmass chute　糖膏斜槽
fillmass emptying gate　糖膏卸膏门
fillmass pump　糖膏泵
fill orifice　①填充孔②注入孔；灌注孔
fill thread　纬线
fill-type insulation　疏松隔热层；填充型隔热层
fill up　填充
film　①膜；薄膜②(照相)软片
film action　润滑膜性状
film adhesive　薄膜胶黏剂；黏合(薄)膜；薄膜胶胶带
film aging　漆膜老化
film applicator coating　贴膜法〔将油漆用成膜器预制成薄膜然后转贴在施工物表面上的一种方法〕
film badge　胶片剂量计
film balance　膜天平
film bearing　油膜轴承
film blowing　吹塑成膜
film boiling　薄膜沸腾
film boiling range　薄膜沸腾区
film breakup　膜破裂
film bubble　(挤出吹胀)薄膜膜泡

film-builder　成膜物质
film building　①构膜的②膜的构成
film caster　漆膜制备器
film casting　薄膜铸塑
film clarity　薄膜透明度
film coagulating speed　成膜速度；薄膜凝固速度
film coalescence　①聚结成膜②膜粘连；膜熔合
film coalescence aid　成膜助剂
film coating　涂膜
film coefficient　膜系数
film coefficient of heat transfer　传热膜系数
film compensator　薄膜补偿器
film condensation　膜状冷凝(作用)
film conversion　薄膜再加工
film curing　漆膜固化
film curvature　薄膜曲率
film defect　漆膜病态；漆膜缺陷
film degradation　漆膜降解；涂膜(层)降解
film density　(氧化)膜密度
film destruction　(泡沫)薄膜破坏
film-detector　漆膜检测仪
film diffusion　薄膜扩散
film discontinuity　漆膜不连续性
film distillation apparatus　液膜蒸馏器〔蒸馏时被蒸馏的液体形成薄膜〕
film downgaugeability　薄膜减厚性
film-drainage transition temperature　膜排水性转变温度
film drier　薄膜干燥机
film durability　漆膜耐久性
film effect　薄膜效应
film elasticity　膜弹性
film erosion　漆膜浸蚀
film evaporator　薄膜蒸发器
film extraction　薄膜萃取(法)
film extrusion　挤出成膜；薄膜挤出
film-fiber　膜裂纤维；薄膜纤维
film flow　薄膜流动
film formation　薄膜形成；成膜
film-former　成膜剂
film forming　成膜的
film-forming agent　成膜剂
film forming aid　膜辅助剂
film forming characteristics　成膜特性
film forming compound　成膜化合物
film forming emulsifier　成膜乳化剂
film-forming ingredient　成膜配料；成膜剂
film forming material　成膜物
film forming matter　成膜物质

film forming molecule　成膜分子
film forming process　成膜工艺
film forming properties　成膜性能
film-forming resin　成膜性树脂
film forming speed　成膜速度
film forming surfactant　成膜表面活性剂
film gate　膜状浇口
film gauge　膜厚仪；漆膜测厚仪
film grade　薄膜级
film growth　薄膜生长
film-hardening agent　漆膜固化剂
film hardness　①薄膜硬度②漆膜硬度
film healing　膜修复
film-inspector　漆膜检测仪
film integrity　漆膜完整性；漆膜连续性
film lamination　薄膜层合
film lubrication　液体润滑(作用)
film membrane　薄膜
filmogen　成膜物质〔总称〕
filmometer　漆膜抗张强度测定器
film orientation　薄膜分子定向
film penetration theory　薄膜浸透理论
film plasticizer　薄膜增塑剂
film potential　膜电位
film pressure　膜压力
film puncher　胶片打孔机
film reactor　膜式反应器
film resilience　膜回弹性
film rupture　膜破裂
film spreading device　薄膜扩展装置
film stage　成膜期
film stiffness　漆膜劲度；漆膜刚度；漆膜刚性
film strength　膜强度
film strength additive　增加(润滑油)膜强度的添加剂
film strength grease additive　增加润滑脂膜强度的添加剂
film stripping test　漆膜剥离强度试验
film sulfonator　膜式磺化器
film temperature　薄膜温度
film test　膜试验
film tester　漆膜试验计
film theory　薄膜理论
film thickness　膜厚度
film thickness gauge　膜厚仪
film thickness measurement　漆膜厚度测定
film thickness test　(漆)膜厚度检验
film-to-fiber process　薄膜成纤工艺
film toughness　膜强度
film type boiler　膜式汽化炉

film type condensation　膜式冷凝
film type evaporator　液膜式蒸发器
film type inhibitor　(薄)膜式缓蚀剂；膜式防锈剂；缓蚀(防锈)膜
film type reaction　膜式反应
film-type rectifying section　膜式精馏段
film uniformity　膜均匀性；膜厚均匀性
film vacuum evaporator　薄膜真空蒸发器
film viscosity　(油)膜黏度
film web　①膜幅②膜状物
film wise condensation　膜状冷凝
film wise operation　膜式操作；膜式运行
filmy　薄膜的
film yarn　薄膜纱；薄膜丝
film-yarn rope　裂膜(纤维)纱绳；裂膜丝绳；薄膜(纤维)纱绳；薄膜纤维绳
filobactivirus　丝状噬菌体
filter　滤色片*
filterability　①过滤本领；滤过额；滤过率②过滤性
filterability constant　过滤性常数
filterable　滤过的
filter aid　助滤剂
filter alum　明矾滤层
filter area　过滤面积
filteration　过滤(作用)
filteration resistance　过滤阻力
filter bag　滤袋
filter basin　滤池
filter bed　滤垫；滤层
filter bell　滤钟
filter belt　过滤带
filter board　滤板
filter bottom　滤底
filter bottom tank　滤底桶
filter bowl(=filter casing)　滤罩
filter box　滤箱
filter cake　滤饼
filter candle　滤烛；过滤烛管
filter cartridge　①芯形过滤器②过滤器芯子
filter casing(=filter bowl)　滤罩
filter cell　滤槽
filter chamber　过滤室
filter circuit　滤波器线路
filter clogging　过滤器堵塞
filter cloth　滤布；滤网
filter coke　过滤用焦炭〔精制〕
filter compressor　滤饼压榨器
filter cone　锥形滤器；滤锥

filter conveyer blade 过滤机输送带刮刀〔用于石蜡分离〕
filter cotton washing machine 过滤棉洗涤机
filter crucible 滤埚
filter cycle 过滤周期
filter dehydration 过滤脱水(法)
filter diaphragm 滤膜
filter diaphragm cell 滤膜电池
filter disc 滤片
filter dish 滤皿；滤盘；滤碟
filter disk assembly 滤片组
filter distributor 过滤分布器
filter drum 滤鼓
filter durability 滤毒罐持久性
filter dust collector 过滤式集尘器
filter dust separator 过滤式集尘机
filtered 滤过的
filtered cylinder oil 过滤汽缸油
filtered cylinder stock 过滤汽缸油料
filtered flue gas 滤过的烟道气
filtered juice 滤汁；滤液
filtered linseed oil 过滤亚麻籽油
filtered noise field 过滤噪声场
filtered stock 过滤油料
filtered tar 过滤焦油
filter effect ①过滤效果②滤波效应
filter efficiency 过滤效率
filter element 滤心；过滤元件
filter feed tank 过滤原料槽
filter flask(=filtering flask;suction bottle) 吸滤瓶
filter fly 滤程
filter frame 滤框
filter freezing 过滤冷凝
filter funnel 过滤漏斗
filter gauze 滤布
filter glass 滤色(光)玻璃；滤光镜；玻璃滤器
filter house 滤过部
filtering ①过滤②过滤的③滤波
filtering accessory 过滤装置
filtering agent 过滤剂
filtering apparatus 过滤仪器；过滤设备
filtering area 过滤面积
filtering bag 滤袋
filtering bed 滤床
filtering candle 滤烛；滤用空烛
filtering cartridge 滤筒
filtering centrifuge 过滤式离心机
filtering clay(=filtering earth) 过滤用白土
filtering crucible 滤埚
filtering device 过滤设备
filtering disk 滤片
filtering earth(=filtering clay) 过滤用白土
filtering effectiveness 过滤效率
filtering element 滤心
filtering flask(=suction bottle; filter flask) 吸滤瓶
filtering frame 滤框
filtering funnel 滤液漏斗
filtering jar 滤缸
filtering layer 滤层
filtering medium 滤质；过滤介质
filter(ing) medium resistance 滤材阻力；过滤介质阻力
filtering of analytical signal 分析信号滤波
filtering rate 过滤速率
filtering sintered metal 过滤用烧结金属板
filtering stock 过滤母液
filtering surface 滤面；过滤面
filtering velocity 滤速
filter-insert 过滤(器)芯子
filter juice 滤出汁；滤液
filter leaf 滤叶
filter liquor 滤(出)液
filter loading 过滤器负荷；滤池负荷
filter loading rate 过滤塞满速度
filter mantle ①过滤罩；滤罩②微孔陶瓷过滤管
filter material 过滤材料
filter media 过滤器介质
filter medium (过)滤介体；过滤介质
filter medium resistance 过滤介质阻力
filter membrane 滤膜
filter off(=filter out) 滤出
filter of graded density 分级比重过滤
filter operation 过滤操作
filter out(=filter off) 滤出
filter pack 过滤组合件
filter pad 过滤垫；滤板
filter panel (过)滤板
filter paper 滤纸
filter paper analysis 滤纸分析
filter paper chromatography 滤纸色谱法
filter paper for chromatography 色谱法用滤纸
filter paper pulp 滤纸浆
filter paper test 滤纸反应；滤纸试验
filter photoelectric photometer 滤光片式光电光度计
filter photometer 滤光光度计
filter plant 过滤设备
filter plate 滤板

filter plugging test 过滤堵塞试验
filter pocket 滤袋
filter pond 滤池
filter press 压滤机；压滤器
filter press cake 压滤饼
filter press cell 压滤机型电解槽
filter press cloth 压滤布
filter press dressing 压滤机整装
filter pressing 压滤
filter press plate 压滤机板
filter press station 压滤站
filter pulp ①过滤用纸浆②滤棉
filter pump 滤泵
filter range 滤光范围；滤波范围
filter residue 滤渣
filter ruffle 滤套
filter sack 滤袋
filter screen 滤网
filter sheet 滤片；过滤层
filter sieve 过滤筛；过滤网
filter spectrophotometer 滤光片式分光光度计
filter stand 滤架；漏斗架
filter stencil 漏印模
filter stick 滤棒
filter stoppage 过滤器堵塞
filter support 滤网支撑器
filter thickener ①过滤增稠剂②过滤增稠器
filter tow 过滤用长丝
filter tube 滤管
filter twill 过滤斜纹布
filter wash 洗滤
filter washer 洗滤器
filter washing 洗滤
filth 污物；垢
filtrability 过滤性；可滤性
filtrable virus 过滤性病毒
filtrate 滤液
filtrated air 无菌空气；过滤空气
filtrated stock 过滤母液
filtrate factor 滤液因素
filtrate pipe 滤液管
filtrate receiver 滤液接收器
filtrate tank 滤液桶
filtrate vacuum receiver 滤液真空接收器
filtrating 过滤
filtrating equipment 过滤装置；过滤设备
filtration 过滤
filtration adjuvant 过滤助剂

filtration aid 助滤剂
filtration area 过滤面积
filtration capacity 过滤能力
filtration-chlorination 过滤-氯化法
filtration cycle 过滤周期
filtration drying 过滤干燥
filtration enrichment 过滤浓缩法
filtration factor 过滤因数
filtration naphtha 过滤石脑油；里格罗因溶液
filtration packing 过滤填充
filtration pump 过滤泵
filtration rate 过滤速率
filtration resistance 过滤阻力
filtration stand 漏斗架
filtration under diminished pressure 减压过滤
filtration under reduced pressure 减压过滤
filtration velocity 过滤速度
filtration yield 过滤(得)量
Filtrol clay catalyst 费尔特洛尔粉状白土催化剂〔美〕
Filtrol (fractionation) process 费尔特洛尔白土接触精制蒸馏过程
filtros 滤石
filum 丝
fimbria ①菌毛②纤毛③伞毛
fimbriate colony 缫边菌落
fimbriation 菌毛形成；伞毛形成
fin 毛刺；飞边；(模)缝脊；模脊
final acid separator 最终酸分离器
final ascendant 末端上升
final attenuation 最终发酵度；消糖度
final boiling point 终馏点；终沸点；干点
final coat 末道漆；末道涂层
final condenser 最终冷凝器
final cooler 最后冷却器
final cure 后硫化；二次硫化
final curing ①终固化②终塑化
final dimension 成品尺寸
final end-product 成品
final evaporation rate 最终蒸发速率
final filter 末级过滤器
final finish coat 末道面漆
final hardness 最终硬度
final lacquer 末道漆
final mashing temperature 最终糖化温度
final mixing 终炼；二段密炼
final mother-liquor 终结母液
final moulder 最后滚边机
final polishing 终抛光

final polycondensation reactor 最终缩聚反应釜；后缩聚釜
final polymer 成品聚合物
final polymerization 终聚合(作用)
final product 最终产物
final purification plant(=final purifier) (合成气体)最终净化装置
final purifier(=final purification plant) (合成气体)最终净化装置
final quantity ①最终量②答数
final refining 最后精制
final residuum 最后残渣；渣油
final retention time 后期保留时间
final retention volume 后期保留体积
final sample 最终样品
final saturation 最后饱和
final set 最后凝固；最后贯入度
final settling tank 后沉淀池；最终沉淀池
final sizing 最后筛分；精筛
final spreading coefficient 终了铺展系数
final state 终态；终止状态
final state structure 终态结构
final temperature 终点温度
final treatment 后处理；最终处理
final velocity 最后速度
final water saturation 最终含水饱和度
final yield 最后产率
finasteride 非那雄胺〔药〕
find ①新油田②发现〔物〕
finder ①探测器②(显微镜)监视器
Findlay cell (for caustic soda) 芬德来制苛性钠电池
fin drum dryer 翅片转鼓干燥器
fine ①细的；薄的；精密的②微粉碎的；细磨的③优质的；高级的
fine adjustment 精密校正；细调；微调
fine aggregate 细集料
fine bore column 细孔柱
fine-bore tube 细芯管
fine breaking 精细压碎
fine break-up 细粒分散
fine breeze 煤粉；煤尘
fine ceramics 精细陶瓷；精制陶瓷
fine chemicals 精细化学品
fine chemicals industry 精细化学品工业
fine clay 细泥
fine clay slurry 细致的黏土悬浮物
fine cloth 精制的金属丝网
fine coal 粉煤
fine coke 细炭粉末；焦粉
fine composition 优质香精
fine control 微调
fine course 紧密横列
fine crack 微龟裂
fine crusher 细碎机
fine crushing 精细压碎
fine cut ①精加工②细隔距
fine dispersion 精细分散
fine drafts mixing 精棉配棉
fine earth 细漂白土粉
fine emulsion 细(滴)乳状液
fine fibered 细绒的；纤细的
fine fibrous 细纤维状的
fine file 细(纹)锉(刀)
fine filter 精制过滤器
fine fine velvet 极细砂
fine finishing cut 精加工
fine focus X-ray tube 细焦 X 射线管
fine furnace black 细炉黑
fine gold 纯金
fine grain 细粒
fine grain development 微粒显像法
fine grained salt 细粒盐
fine grained solid support 细粒载体
fine grained wood 细纹木材
fine graphite 细石墨
fine grinder 细研磨机
fine grinding 精细研磨；精磨
fine grists 细粉；精白粉
fine grit wheel 细砂轮
fine hair removal 除净绒毛
fine hard para 高级硬巴拉胶
fine heating element 微加热元件
fine instrument lubricant 精密仪表润滑剂
fine iron ore 细铁矿
fine laundering 轻洗〔指洗涤丝绸、毛织品和合成纤维〕
finely cleaned 精制的
finely disintegrated fuel(=finely pulverized fuel; finely subdivided fuel) 磨成细粉的燃料
finely divided 细碎的
finely divided catalyst 磨碎的催化剂
finely divided ice 粉碎的冰〔油脱蜡时过滤的促进剂〕
finely divided resin 细粒树脂；微粒树脂
finely divided suspension 微细悬浮液(体)
finely granular 细粒状的
finely ground clay 磨得很细的黏土
finely ground particle 微粒；小颗粒

finely pulverized fuel(=finely disintegrated fuel)　磨成细粉的燃料
finely pulverized powder　微细粉末
finely subdivided fuel(=finely disintegrated fuel)　磨成细粉的燃料
fine melt　全熔(炼)；强(热)裂解
fine melt copal　全熔珀玾(树脂)
fine melt resin　全熔树脂
fine-meshed　带细孔的；细筛孔的
fine-mesh filter(=fine-mesh serene)　带细孔的滤网
fine-mesh sereen(=fine-mesh filter)　带细孔的滤网
fine mesh solid support　细筛目载体
fine-mesh wire　细孔丝网用金属丝
fine-mesh wire cloth　带细孔的金属丝网
fine mesh wire gauze　细网目丝网；细孔金属丝网
fine metal　纯金属
fine micron mill　超微粉碎机(研磨机)
fine needles　细针状体
fineness　①粒子细度；粒度②细度③纯度
fineness gauge　细度计；刮板细度计
fineness-maturity tester　细度成熟度试验仪
fineness modulus　细度系数
fineness of dispersion　分散细度；分散粒度
fineness of finish　精加工光洁度
fineness of grind　研细度
fineness of grind gauge　(颜料)细度计；研磨细度计
fineness of grind (of paint)　研细度〔染料、涂料〕
fineness of grind test　研磨细度试验
fineness of powder　粉末细度
fineness ratio　粒度比(例)
fineness tester　纤度测定器
fine number　(纤维)细支
fine orifice　细孔
fine papers〔复〕　高级纸张
fine particle　细粒
fine particle calcium carbonate　细粒子碳酸钙；轻质碳酸钙；轻钙
fine particles fluidization　细粒流态化
fine perfumery　上等香料
fine polymer　精细高分子
fine-pored　细孔的
fine porosity　微孔性
fine powder　细粉
fine pulverizer　精细磨粉机
fine purification　精制
fine purification plant　(合成气体的)精制装置
fine-quality　优质
fine reinforcing fiber　微细增强纤维
finery　精炼炉
fines　微细粉末〔指金属〕；细屑；碎屑
fine salt　细盐
fine-scale composite　微型复合材料
fine screen　精筛
fine screening　精筛
fine setting　精密调整
fine sheet　薄片材
fine sieve　细筛
fine sizing　精筛
fine splitting　精细分裂
fines roaster　矿粉煅烧矿
finestill　精馏〔高度纯水重蒸馏〕
finestiller　精馏塔；精馏器
fine structure　精细结构
fine structure broadening　精细结构谱线变宽
fine structure distribution　精细结构分布
fine structure of NMR　核磁共振的精细结构
fine structure splitting　精细结构分裂
fine suspension　①微悬浮体②微悬浮液
fine texture　小颗粒结构；细密组织
fine thermal black　细热裂黑
fine thread　细牙螺纹
fine-titred　细纤度的
fine tuning　精细调谐
fine vacuum(=high vacuum)　高真空
fine velvet　细砂
fine wool　细毛
fine zero control　零点微调控制
Fingal process　芬哥尔法〔英国研究的强放射性废物玻璃固化法〕
finger　测厚规；指针
finger buff　指形抛光轮
finger cot　指套
finger cracking test　指形抗裂试验
finger-dabbing　姆指按印
finger-form extending　手指状伸展；手指状延伸
finger grip　指状夹具
finger joint　指接
finger jointed plywood　指接胶合板
finger nail test　刻划硬度试验
finger print　指纹图谱；酶解图谱；指纹；指印
fingerprint chromatogram　指纹色谱(图)
fingerprint figures of thermoanalysis　热分析的指纹特征谱图
finger printing　指纹法
fingerprint mark　指纹标记
fingerprint method　指纹法

fingerprint pyrogram 指纹热解图
finger print region 指纹区域〔红外光谱中〕
finger print removal test 指印防锈试验
fingerprint smudging 核材料(同位素)组成不明
fingerprint spectrum 指纹谱
fingers 指粒〔手指状颗粒〕
finger tab-out test 指黏试验
finger test 手指试验
finger-tip control 按钮控制
finger valve （压缩机）单槽阀
finimeter ①储量计②储氧计
fining 澄清
fining agent 澄清剂
fining pot 精炼罐
finish ①完结②整理；整饰；修饰③涂漆④涂饰剂⑤罩面漆；末道漆
finish coat 罩面层
finish-coat paint 罩面漆
finish cut 精加工
finished blend 混成油
finished board 成品
finished cement 水泥成品
finished cure 后硫化；二次硫化
finished distillate 成品油中间馏分〔包括灯油和柴油〕
finished enamel ①末道瓷漆②成品瓷漆
finished form 成品形状
finished fuel 商品燃料
finished goods 成品；产品
finished goods warehouse 成品仓库
finished hose 胶管成品
finished item 成品；完成项目
finished leather 成革；整饰过的皮革
finished liner 精制垫布；加工垫布
finished market product 成品商品
finished ore 精选矿石
finished polymer 成品聚合物
finished primer 末道底漆
finished product 成品
finished proprietary item 专用制品
finished state 成品形状
finished stock 成品
finished stock storage 成品储藏
finished wax 制成的蜡
finisher 后缩聚器；整理厂
finisher exhaust system 后缩聚器排放系统
finisher-reactor 后缩聚反应器
finish grinding 最后磨碎
finishig gilding 飞金；涂金；镀金

finishing ①修饰；整理；完工②涂装
finishing agent 修饰剂；整理剂
finishing allowance 加工余量；加工留量
finishing apparatus ①后缩聚设备②后处理设备
finishing bath tray 上油槽
finishing beater 成浆机
finishing brush 画图的润饰
finishing calender 完工砑光机
finishing change 完工变化
finishing coat 罩光漆；罩面漆；末道涂层
finishing column ①成品塔②后处理塔
finishing composition 整理剂；油剂
finishing cure 后硫化；二次硫化
finishing cut 精加工；完工切削（机工）
finishing department ①整理车间②完成车间
finishing lacquer ①挥发性面漆②硝基面漆
finishing metal 精炼金属
finishing mill ①精制机②精轧机
finishing operation 最后工序；完成工序
finishing paint 面漆；罩面漆
finishing plant ①后缩聚设备②整理车间③后处理车间
finishing powder （烫金、飞金用）金银粉
finishing reactor 后缩聚反应器
finishing rolls 整理机
finishing room 成品车间
finishing sequence 整理程序；后处理程序
finishing smooth 光面整理
finishing strip 补强胶条
finishing system 涂装系统
finishing test 底面漆配套试验
finishing varnish 罩面清漆
finishing wheel 精磨砂轮；抛光砂轮
finishing work 修整厂
finish insert 瓶颈成型嵌具
finish ironing 成品烫
finish machining 精加工
finish mark 加工符号
finish mix 制成(的)料
finish needling 末道针刺
finish pass 定型辊隙
finish plate 光泽整理；光泽表面
finish presser 整理熨烫机
finish roll 给油辊
finite concentration 一定浓度
finite-difference method 有限差分法
finite elastic-plastic theory 有限弹塑性理论
finite elastoviscoplasticity 有限弹黏塑性
finite linear viscoelasticity 有限线性黏弹性

finite perturbation theory(FPT)　有限微扰理论
finite population　有限总体
finite reflux　有限回流
finite reflux operation　（塔)有限回流运转
finite spin　有限自旋
finite strain　有限应变
Finkelstein reaction　芬克尔斯坦反应
finned coil　有翅盘管
finned heat exchanger　翅片换热器
finned pipe　翅片管
finned surface　翅面
finned tube　翅片管
finned tube condenser　翅管冷凝器
finned tube exchanger　翅管换热器
finocchio　甜茴香
fins　飞边；毛边
Finsen lamp　芬生灯
fir balsam　加拿大香脂
fire alarm system　消防警报设备
fire annihilator　灭火器
fire assaying　火试金法；火试金分析
fire-assay procedure　火试金法
fire back　回火
fire bank　隔火墙
fire box(=firebox)　燃烧室
fire break　防火间隔
fire brick　（耐)火砖
firebrick support　耐火砖载体
fire bulkhead　防火壁；防火隔板
fire clay　（耐)火泥；耐火土
fire clay brick　火泥砖；耐火砖
fire coal　取暖用煤
fire control　消防；防火
fire control unit　消防装置；防火装置
fired　①发火②发射③喷射
fire damp　沼气；甲烷
fired ceramic coating　烧成陶瓷涂料；烧结陶瓷涂料
fired coil　用火加热的旋管
fire detector　火灾探测器；爆炸性气体测定器
fired heater　火焰加热器；裂解炉
fire-dike　防火堤
fired in burst　点射
fire-distillation　用火加热的蒸馏
fire door　炉门；火门
fire engine hose　消防胶管；水龙带
fire evaluation test　燃烧评定试验
fire extinction　灭火
fire extinguishant　灭火剂

fire extinguisher　灭火器
fire extinguishing agent　灭火剂
fire extinguishing equipment　灭火设备
fire extinguishing pump　消防用泵
fire-fighting　火喷
fire fighting equipment　灭火设备
fire fighting foam　灭火沫
fire (fighting) hose　消防胶管；水龙带
firefly fluorescein　萤火虫荧光素
fire foam　灭火沫
fire foam producing machine　泡沫发生器
fire gases　可燃气体；易燃气体
fire goods　易燃物
firegrate　炉箅
fire-hazardous　易着火的
fire intensity　燃烧强度
fire kiln　直接火式焙炉
fire lighter　①火焰增光剂②点火剂③点火器
fire mask　消防面具
fire opal　火蛋白石
fire performance　耐燃性
fire pit test　燃痕试验
fireplace　①炉②壁炉③火室
fire point　燃点；着火点
fire polishing　火琢
fire precaution　防火措施
fire prevention　防火；防火措施
fireproof　耐火的；防火的
fireproof agent　防燃剂
fireproof alloy　耐热合金
fireproofed paper　防火纸
fireproof fibre　防火纤维；不燃纤维
fireproof finishing　防火加工
fireproof gasoline tank　不怕火的汽油箱
fire-proof grease　耐火润滑脂
fireproofing　防火的
fireproofing agent　防燃剂；防火剂
fire proof(ing) finish　防火整理
fire-proof mat　耐火席子〔用特殊组分浸过的〕
fireproof material　耐火材料
fire proofness　耐火性
fire propagation　火焰蔓延
fire property　可燃性
fire red　颜料火红；氯化对位红
fire resistance　耐火性
fire resistance rating　耐火性等级
fire-resistant　耐高温的
fire resistant conveyor belt　防火运输带

fire-resistant finish 防火整理；抗燃整理
fire resistant plywood 耐火胶合板
fire resisting construction 防火结构；耐火结构
fire-resisting finish 防火罩面漆
fire resistive 耐火的；抗燃的
fire retardance 抗延燃性；阻燃性
fire retardancy test 阻燃试验
fire retardant 阻燃剂；防火剂
fire-retardant additive 阻燃添加剂
fire-retardant coating 防火涂料
fire-retardant fibre 阻燃纤维；难燃纤维
fire-retardant finish 阻燃整理；阻燃处理
fire retardant plasticizer 防火增塑剂；阻燃性增塑料
fire-retardant shingle stain 板墙防火色浆；板墙防火涂料
fire retarder 阻燃剂
fire retarding agent 防火剂；阻燃剂
fire risk 火险；着火危险
fire room 锅炉间；火室
fire sighting 验火
fire still 用火加热的蒸馏锅
Firestone extrusion plastometer 费尔斯通挤出式可塑计
Firestone flexometer 费尔斯通曲挠试验机
Firestone steel ball rebound tester 费尔斯通落球式弹性计
fire suppressor 防燃剂
fire test ①着火点测定②灼烧度试验
fire test on fire proofed wood 防火木材试验法
fire tube 火管
fire tube boiler(=multitubular boiler; tubular boiler) 火管锅炉
fire wall 隔火墙；绝热隔板；防火隔板
fireworks 焰火
firing 射击；放枪
firing chamber 火室
firing equipment 加热设备
firing pin 撞针〔军〕
firing point 燃点；着火温度
firing pot 火加热罐
firing range 着火范围〔指着火温度的范围〕
firing resistance 耐燃性能
firing ring 测热圈
firing shrinkage 烧缩
firing temperature 燃点；着火温度
firing time test 燃烧速度试验
firing voltage 点火电压；开始放电电压
firm ①坚固的②坚定的
firm cure 充分硫化；彻底硫化
firm feel 厚实手感
firming 硬化；固化

firming agent 固化剂
firmness ①坚固性②稳定性③硬度；坚实度
firmo viscosity(=viscoelasticity) 黏弹性
firm time 变硬时间；硬化时间
firmware 固件
fir needle oil 杉针油；杉油
fir seed oil 杉子油
first acidaffins 第一亲酸性
first approximation （第)一级近似
first bath 初染浴
first betti number 网络的秩
first break draft 预牵伸
first class grade chemical 一级试剂
first clear flour 头等净粉
first CMC 第一临界胶束浓度
first coat ①底漆；底涂层；头道漆②封闭涂层
first comonomer 第一共聚(用)单体
first cost 投资
first cure ①(=semicure)半硫化；定型硫化②第一段硫化〔运输带〕
first cut 初馏分；酒头〔俗〕
first derivative spectrum 一级导数光谱
first dimension 第一(展开)方向
first discolouration 初期变色(褪色)
first draft 第一类牵伸
first electroviscous effect 第一电黏效应
first evaporator 第一级蒸发器
first fillmass 初糖膏
first filtration 头道过滤
first grade 头等的；高级的
first kind oxidation-reduction electrode 第一类氧化还原电极
first latex crepe 头等皱胶
first latex rubber 头等橡胶
first law of absorption 光吸收第一定律
first law of thermodynamics 热力学第一定律
first lees ①初生酒泥②粗沉淀物
first level ionic line 一级离子线
first level ionization 一级电离
first messenger 第一信使
first normal stress coefficient 第一正应力系数
first normal stress difference 第一正应力差
first order ①一级；一次②第一级〔数〕
first order configuration 一级组态
first order coupling spectrum 一级耦合谱
first order derivative titration 一级导数滴定
first order differential titration 一次微分滴定；一级导数滴定

first order directional focusing mass spectrometer 一级方向聚焦质谱计
first order direction focusing 一级方向聚焦
first order focusing 一级聚焦
first order nucleation 一级成核作用
first order phase transformation 一级相变
first order phase transition 一级相变*
first order reaction 一级反应
first order reduced density matrix 一阶约化密度矩阵
first order spectrum 一级光谱；一级图谱
first order spin pattern 一级自旋图形
first-order transition ①一级跃迁②一级转变〔指高分子物从高弹态到黏流态转变〕
first order transition point ①一级转变点②一级跃迁点
first-order transition temperature 一级转变温度
first order velocity focusing 一级速度聚焦
first overtone 第一泛音
first pass effect 首过效应；第一关卡效应
first patent flour 一级专利粉
first-ply-failure 最先一层失效
first polycondensation reactor 前缩聚(反应)器
first quality 优质
first resonance line 第一共振线
first runnings（复） 初馏物
first screen 头道筛选
first stage cure 第一阶段硫化
first stage of decreasing drying rate 减速干燥第一阶段
first-stage polymerization 初级聚合；前段聚合
first strike 初打
first sugar 一级糖
first-surface painting ①正面彩绘②正面涂漆；外表面绘画
first sweating 一次发汗〔蜡〕
first Virial coefficient 第一维里系数
first wort 初麦芽汁
firware 固件
Fischer's acid 费歇尔酸；2-萘酚-6-磺酸
Fischer's method 费歇尔法〔人造石油制法〕
Fischer indole synthesis 费歇尔吲哚合成法
Fischer iron-refining process 费歇尔炼铁法
Fischer osazone reaction (费歇尔)成脎反应
Fischer phenylhydrazone reaction (费歇尔)苯腙反应
Fischer projection 费歇尔投影式*
Fischer reagent 费歇尔试剂〔盐酸苯肼与乙酸钠的混合液〕
Fischer-Speier esterification 费歇尔-斯皮尔酯化作用
Fischer's potassium nitrite method 费歇尔亚硝酸钾法
Fischer's yellow 费歇尔黄；硝酸钴钾

Fischer synthesis 费歇尔合成
Fischer-Tropsch catalytic process 费-托催化过程
Fischer-Tropsch hydrocarbons 费-托法合成的烃类
Fischer-Tropsch process 费-托法〔用 CO 和 H_2 合成烃类的过程〕
Fischer-Tropsch reaction 费-托反应
Fischer-Tropsch synthesis 费-托合成法
Fischer-Tropsch wax 费-托合成蜡
fiscus rubber 印度榕橡胶
fisetic acid 非瑟酸〔即: 非瑟酮〕
fisetin 非瑟酮；漆树黄酮；3,7,3′,4′-四羟(基)黄酮
fisetinidin 非瑟酮定
fisetinidol 非瑟酮醇；7,3′,4-三羟基黄烷-3-醇
fish backs spreader 鱼背展幅机（橡）
fish bean 鱼(毒)豆
fish-bellied beam 鱼腹式梁
fish berry 印度防己
fish bone method 鱼骨割法
fisherman's soap 渔民皂〔能去鱼腥〕
Fisher Shell sorptometer 费歇尔-谢尔吸着测定仪
fish gelatine(=fish glue) 鱼胶
fish glue(=fish gelatine) 鱼胶
fish guano 鱼渣粉
fishhook symbolism 鱼钩符号
fishing ①捕鱼②捞盐
fishing lines 渔网线
fishing pan 蒸盐锅
fishing salt 腌鱼盐
fishing twine 渔网绳；钓鱼线
fish liver oil 鱼肝油
fish oil 鱼油
fish-oil stuffing 鱼油加脂
fish-oil tannage 鱼油鞣(法)
fish paper 鱼纸；青壳纸〔商〕
fish poison 鱼毒
fish protein fiber 鱼蛋白纤维
fish scaling 片落；剥落；剥离；起鳞〔搪瓷〕
fish-scaling effect (纸张涂层的)鳞爆现象〔漆病〕
fish stearine 鱼油硬脂
fish tail 鱼尾灯头；扩焰器
fish tail burner 鱼尾状灯；鱼尾状燃烧器
fish tail head ①鱼尾灯头②鱼尾机头
fishtail sheet die 鱼尾形挤板口模
fishtail-type kneader 鱼尾型捏合机
fish tallow 鱼脂
fish top 鱼尾式灯头
fisser 可裂变物质
fisser material 易裂变物质

fissia 氧化裂片合金
fissible 可裂变的(物质)
fissile ①片状的；页状的②(易)裂变的(物质)
fissile isotope 裂生同位素
fissile material 易裂变物质
fissile nucleus 裂变核
fissile nuclide 易裂变核素
fissility 分裂性；可裂变性
fissiogenic isotope 裂生同位素；裂变成因同位素
fission 裂变
fissionability 可裂变度
fissionability parameter 可裂变度参数；可裂变性参数 (Z^2/A)
fissionable 可裂变的
fissionable material 可裂变物质〔指 ^{233}U、^{235}U、^{239}Pu等〕
fissionable nuclide 可裂变核素
fission barrier 裂变位垒
fission chamber 裂变室
fission chemistry 裂变化学
fission counter 裂变计数器
fission cross-section 裂变截面
fission energy 裂变能
fission fragment 裂变碎片
fission fragment ionization desorption 裂变碎片电离解吸
fission gas 裂变气体〔产物〕
fission monitor 裂变检测体；裂变检测核
fission multiplicity detector 裂变倍增探测器
fission product 裂变产物；裂片(元素)
fission product burner 裂变产物燃烧器
fission product chemistry 裂变产物化学
fission products 裂变产物
fission radiochemistry 裂变放射化学
fission track 裂变径迹
fission track age 裂变径迹年龄
fission track dating 裂变径迹年代测定
fission trigger 裂变触发器
fission yeast 裂殖酵母
fission yield 裂变产额
fission yield monitor 裂变产额监测体
fissiparity 分裂繁殖
fissium 裂变产物合金；裂片合金〔指辐照燃料中除去裂变材料、放射性气体、铯、锶、稀土等后，剩余贵金属裂变产物的总称；或指组成与其相同的模拟合金〕
fissium oxide 裂变产物合金氧化物
fissochemistry 原子核分裂化学；核裂化学
fissure 裂缝；裂口；裂隙

fissured 裂缝的
fissuring ①裂隙②节理
fistelin 菲瑟配基
fistula 瘘；瘘管
fit ①配合②装配
fitness 拟合度
fitoncide 植物杀菌素
fitoncidin 葱素
fit quality 配合等级
fitter 装配工；钳工
Fittig reaction(=Wurtz-Fittig reaction) 武慈-维悌希反应
fitting ①装配②配件；零件③接头
fitting ink 磨合油墨；刮研涂料
fitting lubricant 安装(用)润滑剂
fitting room 装配车间；成型车间
fitting surface 配合面
fit tolerance 配合公差
fit value 适合值
Fitz chilsonator 造粒机；造粒器；费兹造粒机
Fitzgerald-Thomson furnace 裴兹杰惹-汤姆逊电炉
Fitz mill 中碎机；费兹粉碎机(磨)
Fitz-Simons viscosimeter 费兹-西蒙斯黏度计
five bowl universal calender 五滚筒式砑光机
five carbon ring naphthene 五碳环烷
five-grade xylene 五级二甲苯
five-membered ring(=five-ring) 五元环
five-pole cell 五极池
five-ring(=five-membered ring) 五元环
five-roller(=five-roll mill) 五辊滚压机
five-roll mill(=five-roller) 五辊滚压机
fixanal 分析用配定试剂
fixation ①定像；定影；凝固②定香〔香料化学〕③固定
fixation bath 固着浴；固化浴
fixation fluid 固定液
fixation method 固定法
fixation of gas 气体的(热)稳定
fixation of nitrogen 固氮作用
fixation pair 固定线对
fixation point 定点
fixation reaction 固定反应
fixation test 固定试验
fixative ①固定剂②定香剂〔香料〕
fixative solution 固定液；定影液
fixed 固定的
fixed acid 固定酸
fixed acidity 不挥发酸度
fixed adsorbent 固定吸附剂
fixed adsorbent bed 固定式吸附剂床

fixed agent　固定剂
fixed air　固定空气；二氧化碳
fixed alkalies　固定碱〔在正常条件下不蒸发的碱〕
fixed ammonia　固定氨；结合氨
fixed analyzer transmission　固定通能分析器
fixed analyzer transmission mode　固定通能模式分析器
fixed-angle　(转子)固定角度式
fixed ash　固定灰分
fixed barrier　固定阻片
fixed base　固定碱
fixed bed　固定床
fixed bed adiabatic reactor　固定床绝热反应器
fixed bed catalyst　固定床催化剂
fixed bed catalyst chamber(=fixed bed catalytic reactor)　固定床催化反应器
fixed bed catalytic cracking　固定床催化裂化
fixed bed catalytic process　固定床催化过程
fixed bed catalytic reactor(=fixed bed catalyst chamber)　固定床催化反应器
fixed bed contactor　固定床接触器
fixed bed hydroforming　固定床(临)氢重整
fixed bed ion exchange　固定床离子交换
fixed bed process　固定床过程
fixed bed reactor　固定床反应器
fixed bed unit　固定床设备
fixed bias　固定偏倚
fixed capital　固定资本；固定资产
fixed carbon　固定碳
fixed catalyst　固定(床层)催化剂
fixed charge method　①固定消耗量法②定电荷计数法
fixed charges　固定消耗量
fixed coke burning capacity　(再生时)固定烧焦本领
fixed configuration　固定构型
fixed controller　(指标)固定的控制器
fixed coordinate　固定坐标
fixed count method　定数计时法
fixed cutting oil　动植物油切削油〔金属切削时用作润滑冷却用的动植物油〕
fixed die blade　定模唇
fixed die plate　(固)定模板
fixed dosage　固定的剂量(用量)；规定的用量
fixed dye　固定染料
fixed-effect model　固定效应模型
fixed eluant　固定洗脱液
fixed equipment　固定设备
fixed error　固定误差
fixed factor　固定因素
fixed form　定形

fixed fuse　①固定引信②固定导火线③保险丝
fixed gas　固定气体〔在常压和常温下不冷凝的气体〕
fixed gauge　固定规
fixed grate producer　固定炉篦煤体发生炉
fixed grease　固定油制的润滑脂
fixed group　固定基团
fixed hearth furnace　固定床膛火炉
fixed hydrocarbon　非挥发性
fixed interference method　固定干扰法
fixed ion　固定离子
fixed ion concentration　固定离子浓度
fixed jaw (of breaker)　(压碎机的)固定颚
fixed light　定光灯
fixed liquid phase　固定液相；液体固定相
fixed mold　固定式压模
fixed mold press　固定模平板机
fixed mould　固定压模
fixed nitrogen　固定氮
fixed nitrogen fertilizer　固定氮肥
fixed oil　固定油〔在通常情况下指不挥发性的油，在石油中指动植物性油〕
fixed oil identifying test　固定油的鉴定
fixed oil radicals　固定油基
fixed oils　固定油类
fixed orifice　固定孔
fixed oxygen　固定氧；化合氧
fixed partition cell　固定隔板沉淀池
fixed path length cell　固定光程长吸收池
fixed pattern noise　固定模式噪声
fixed pellets　固定丸剂
fixed phase　固定相
fixed point　固定点
fixed point of temperature　固定温度点
fixed primary ion method　固定主要离子法
fixed pulley　固定皮带轮
fixed pump　固定泵
fixed-rate division　定比缩分法
fixed residue　固定残渣
fixed routine　固定程序
fixed schedule　固定的目录；固定的一览表
fixed set point control　定值调节；恒值调节
fixed shunt　固定分流器
fixed-sieve jig　定筛簸海机
fixed size reeling　定纤巢丝
fixed solution with hardener　坚膜定影液
fixed spreader　固定式涂板器
fixed substance　固定物质；具有一定性质的物质
fixed tan　结合鞣质

fixed temperature 固定温度
fixed temperature and humidity 恒温恒湿
fixed tray 固定盘；固定板
fixed tube sheet exchanger 固定管板换热器
fixed tube sheet heat exchanger 固定管板式(列管)换热器
fixed tube sheets 固定管格(换热器)
fixed variable 固定变量*
fixed weigher 固定天平
fixed white 硫酸钡
fixer 保香剂
fixer solution 定影液
fixing 定位；固着
fixing agent ①固定剂②定影剂③固色剂
fixing bath ①定影液②固着浴
fixings〔复〕 ①调味品②调味
fixing salt(=cyclohexylamine acetate) 凝固盐；乙酸环己胺〔凝固剂〕
fixing solution 固定液；定影液
fixity 不挥发性；稳定性
fix sized moving window evolving factor analysis 固定尺寸移动窗口因子分析
fixture ①装置品；装置器②卡具；夹具
fizz 充气饮料
fizzum(Fz) 富锆裂片合金
Flade potential 弗拉德电位；活化电位
flag ①扁石②菖蒲③旗
flagecidin(=anisomycin) 杀鞭菌素；茴香霉素
flagging 石板
flag leaf 菖蒲叶；香蒲叶
flagstaffite 柱晶松脂石
flagstone 铺街石；板石
flairred(=redflair) (蓝色)红光位移；向红位移
flajolotite 锑铁土
flake 薄片
flake aluminum 薄铝片
flake brass 片状黄铜粉
flake caustic 片状烧碱
flake chipping 表皮落屑（轮胎）
flake graphite （鳞）片状石墨
flake hematite 片状赤铁矿；天然云母氧化铁
flake lac 片状紫胶；虫胶片〔俗〕
flake lead 铅华；碱式碳酸铅
flake litharge 鳞状氧化铅；鳞状密陀僧
flake metal powder 片状金属粉
flake-off 剥落；片落
flake orientation (金属粉)鳞片(排列)取向
flake powder 片状炸药
flaker 刨片机

flake shellac 片(紫)胶
flake soap 皂片
flakes of mica 云母片
flake white 铅白；碳酸铅白
flakiness 成片性
flaking ①片落；剥落②刨片
flaking machine 刨片机
flaking mill 薄片机；压扁机
flaking resistance (涂膜)抗片落性；耐片落性
flaky 片状的；鳞状的
flaky grain 片状颗粒
flaky graphite 薄片状石墨；鳞状石墨
flaky particle 薄片状粒子
flaky texture 薄片状结构
flamability 可燃性
flamboyant coating 闪光涂层施工法
flamboyant finish 闪光漆
flame 火焰
flame ablation 火焰烧蚀；熔化烧蚀
flame absorption 火焰吸收
flame adapter 火焰接管；火焰应接管
flame analysis 火焰分析
flame analytical equivalent 火焰分析当量
flame anchor 火苗
flame arrester 灭火器
flame atomic absorption spectrometry 火焰原子吸收光谱法
flame atomic emission spectrometry 火焰原子发射光谱法
flame atomic fluorescence spectrometry 火焰原子荧光光谱法
flame atomization 火焰原子化*
flame atomizer 火焰原子化器
flame axis 焰轴
flame background 火焰背景
flame blow-off 火焰吹灭
flame burning velocity 火焰燃烧速度
flame calorimeter 火焰量热计
flame chamber 燃烧室
flame chipping 烧剥；火焰剥皮
flame cleaning 火焰清洁法
flame coal 焰煤
flame coloration 焰色
flame color test 焰色试验
flame cone(=inner cone) 焰心
flame couple 热电偶
flame cracking ①火焰裂化②火焰开裂
flame curing 火焰固化
flame cutting 火焰切割

flame damper(=flame arrester) 灭火器
flame deposition process 火焰沉积法〔复合多晶纤维成形法之一〕
flame-descaling ①火焰除锈②火焰除鳞
flame detector 火焰检测器；自动防火器
flame emission 火焰发射
flame emission detector 火焰发射检测器
flame emission spectrometry 火焰发射光谱法
flame emission spectrum 火焰发射光谱
flame envelope 火焰包围物
flame exposure device （马丁-平斯基闪点及燃点测定器的)点火装置
flame flash-back 火焰通路；火焰闪灭
flame formation 火焰形成
flame front 火焰锋；火焰前缘
flame front stability 火焰正面的稳定度
flame fuel ①火焰燃料②纵火油料③喷火油料
flame fusion method 晶体生长焰熔法
flame gas 火焰气
flame geometry interference 火焰形状干扰
flame gouging 火焰气刨
flame gun spray （粉末)火焰喷涂法
flame hardening ①火焰硬化②火焰淬火
flame holder 火焰稳定器
flame hydrolysis 火焰水解(法)；高温水解法
flame ignition 点火
flame ignitor 点火器
flame impingement 火焰(对火墙的)冲击
flame ionization 火焰电离；火焰离子化
flame ionization detector(FID) 火焰离子化检测器
flame ion mass spectrometry 火焰离子质谱法
flame jet ①火焰喷射②火焰喷口
flame laminating 火焰层合
flame lamp 火焰灯
flameless 无焰的
flameless atomization 无焰原子化
flameless atomizer 无焰原子化器
flameless ionization detector 无火焰电离检测器；无火焰离子化检测器
flameless powder 无焰火药
flameless procedure 非火焰法；无焰法
flame movement 火焰移动
flame normalizing 火焰正火
flame oxidation 火焰氧化
flame path 火道
flame photometer 火焰光度计
flame photometric analysis 火焰光度分析
flame photometric detector 火焰光度检测器

flame photometric determination 火焰光度测定
flame photometric titration 火焰光度滴定
flame photometry 火焰光度法
flame photometry detector 火焰光度检测器
flame photosource 火焰光源
flame polish （火)焰抛光
flameproof 耐火的
flameproof bulkhead 防火壁
flameproof enclosure 防火罩
flame-proofer 耐焰剂
flameproof fiber 防燃纤维；防火纤维
flame-proof finishing ①防火涂装②防火加工；防火处理
flame proofing 防火的；耐火的
flame-proofing agent 耐焰剂；耐火剂
flameproofing organic fiber 防燃有机纤维；防火有机纤维
flame-proofness 耐焰性；耐火性
flame propagation 火焰(在燃料混合物内)的传播
flame propagation mode 火焰传播法
flame-quenching polymer 自熄性聚合物；火焰淬熄性聚合物
flame radiation 火焰辐射
flame reaction 焰色反应
flame reactor 火焰(喷射)反应器
flame resistance 耐火性
flame resistance rubber 耐火橡胶
flame resistance test 耐火性试验
flame resistant 耐燃性(的)
flame resistant fiber 抗燃纤维；防火纤维
flame-resistant resin 抗燃树脂；防火树脂
flame resisting 耐火的；防火的
flame resisting agent 阻燃剂
flame resistivity 阻焰性；耐火性
flame retardance 防延燃性；阻燃性
flame retardancy 防延燃性；阻燃性
flame retardant 阻燃剂
flame-retardant additive 阻燃添加剂
flame retardant agent 阻燃剂
flame retardant coating 阻燃涂层(料)
flame-retardant fiber 阻燃纤维
flame retardant grade 阻燃剂等级；防火阻燃级
flame retardant lubricant 阻燃滑润(油)剂
flame retardant paint ①防火漆②阻燃漆
flame-retardant plasticizer 阻燃增塑剂
flame-retardant plywood 阻燃胶合板
flame retardant property 阻燃性能
flame retardant rating 阻燃等级

flame retardant resistance　阻燃性
flame retarder　阻燃剂
flame retarding polymer　阻燃聚合物
flame rod　火柱
flame sensitizer　火焰光度增感剂；焰光敏化剂
flame sensor　火焰传感器
flame source　火焰光源
flame spectrometer　火焰光谱仪
flame spectrometric analysis　火焰光谱分析*
flame spectrometry　火焰光谱法
flame spectrophotometer　火焰分光光度计
flame spectrophotometry　火焰分光光度法
flame spectroscopy　火焰光谱学
flame spectrum　火焰光谱
flame speed　火焰速度
flame spray coating　火焰喷涂〔金属或塑料粉末涂层〕
flame-sprayed ceramic　火焰喷涂陶瓷
flame-sprayed plastic coating　火焰喷涂塑料覆盖层
flame spray gun　火焰喷枪；(金属)熔融喷枪
flame spraying　火焰喷涂法；火焰喷射法
flame spray powder coating　粉末火焰喷涂
flame spread　展焰性
flame spreader　火焰扩张器〔试验灯油标准灯的〕
flame spread rating　火焰扩散等级
flame spread test　火焰蔓延试验
flame stability　火焰稳定性
flame stabilization　火焰的稳定
flame stabilizator　火焰稳定剂
flame sterilization　火焰灭菌(法)
flame straightening method　火焰校正法
flame structure　火焰结构
flame temperature　火焰温度
flame temperature detector　火焰温度检测器
flame test　焰色试验*
flame thermionic detector　火焰热离子检测器
flame thermocouple detector　火焰热电偶检测器
flame thrower　火焰投射器
flame-tight　不漏火焰的
flame tip　焰舌
flame trace　火焰痕迹
flame transmission　火焰传动
flame trap　火光防止筒(罩)；防焰器
flame travel　火焰的移动
flame treating　火焰处理
flame tube　燃烧管
flame tube cooling　燃烧管冷却
flame tunnel test　火焰穿孔试验
flame velocity　火焰速度

flame vibration　火焰摆动；火焰晃动
flame wedge　火焰楔
flame zone　火焰带
flaming combustion　有焰燃烧
flammability　耐燃性；可燃性
flammability classification　可燃性分级
flammability index　可燃性指数
flammability limit　自燃的极限；可燃性极限
flammability point　燃(烧)点
flammability rating　可燃性等级；易燃性评价
flammability standard　可燃性标准
flammability test　易燃性试验；燃烧性试验
flammability tester　可燃性试验仪
flammable　可燃的
flammable gas　可燃气体
flammable inhibitor　发火抑止剂
flammable limit　可燃极限
flammable liquid　可燃液
flammable range　易燃范围
flammable solvent　易燃溶剂
flammable vapor　易燃性蒸气
flammulin　火菇菌素
flanch(=flange)　法兰；凸缘；接盘
Flanders stone　石墨
flange(=flanch)　法兰；凸缘；接盘
flange coupling　法兰式联轴节
flange crimping　(橡胶)凸缘卷曲法；法兰卷曲法
flanged　装有法兰的
flanged cast-iron pipe　法兰铸铁管
flanged connection　法兰连接
flanged coupling　法兰连接；法兰接头
flanged edge weld　卷边焊
flanged end　法兰形端部
flanged hose coupling　胶管法兰接头
flanged joint　法兰连接
flanged packing　法兰衬垫
flange gasket　法兰垫
flange gasket pre-stress　法兰垫初期应力
flange gasket residual stress　法兰垫残余应力
flange gasket stress　法兰垫(承受的)应力
flange joint　法兰连接
flange packing　法兰垫
flanging　折边
flanging machine　折边机
flanging test　管子在法兰上的试验
flank　腹边皮
flannel disk　法兰绒磨光盘
flap　衬带；垫带；压条

flap-covered grease fitting 活盖润滑脂注入器
flap-covered lubricator 活盖注油器
flapper 活瓣；挡板
flap sander 叠层砂光机
flap tray 活门塔板
flap valve 片状阀；瓣阀
flap wheel(=lamellar buff) 叠层抛光轮
flare ①火舌；火苗②火炬；火把③钟口④火舌管⑤漏斗
flare back 火舌回闪
flared 扩口的；漏斗式的
flared joint 扩口连接
flare line 火舌管；(石油)废气燃烧管路
flare opening 漏斗口；漏斗扩大的部分
flare point 燃烧点；着火点
flare stack ①火炬烟囱；火炬管②放空烟囱
flare stack burner 火炬烟囱燃烧器
flare stack ignition device 火炬烟囱点火装置
flare system 火炬系统
flare tip 火舌尖
flare up 骤燃；闪光
flaring 骤燃
flaring cup 展口杯；喇叭形杯
flaring exit 圆锥形出口；喇叭形出口
flaring pail 圆锥形提桶
flaser gabbro 鳞状辉长岩
flaser texture 鳞状组织
flash ①闪(光)②一闪③(模塑)溢料④(模)缝脊⑤闪蒸法
flash arrestor 灭火器；消焰器
flash back ①反闪②(=fire back)回火
flash-back criterion 反闪准则
flash-back fire 反闪火焰
flash bottoms 闪蒸残渣
flash chamber 闪蒸室
flash chromatography 闪式色谱法；急骤色谱法
flash clearance 溢料间隙
flash column 闪蒸塔
flash condenser 闪蒸冷凝器
flash cup 闪点杯；测定闪点的杯子
flash-cure 快速固化
flash curve 闪蒸曲线
flash deaerator 瞬间脱泡器；快速脱泡器
flash desorption 闪脱；急骤解吸
flash distillation 闪蒸馏；急骤蒸馏
flash distillation column 闪蒸塔
flash distillation plant 闪蒸装置
flash down 闪降；压力迅速下降
flash drum 闪蒸槽
flash dryer 气流干燥器；急骤干燥器

flash drying 急骤干燥
flashed glass 镶色玻璃
flashed vapour 急骤蒸气
flasher ①闪蒸器②闪光器
flash evaporating injection 闪蒸进样法
flash evaporation 闪蒸
flash evaporator 闪蒸器
flash exposure 瞬时暴露
flash factor 凝液膨胀系数
flash film concentrator 急骤薄膜浓缩器
flash film evaporator 急骤薄膜蒸发器
flash fin 溢料
flash fire 急骤燃烧；暴燃
flash fire test 突燃试验；闪火试验
flash gas 闪蒸气体
flash gasoline 闪蒸汽油
flash gas refrigeration 气体闪蒸冷冻
flash grinder 砑光机
flash heater 急速加热器
flash ignition temperature(FIT) 闪(骤)燃温度
flashing ①急骤蒸发②玻璃镶色③闪光
flashing chamber(=flash chamber) 闪蒸室
flashing column(=flash column) 闪蒸塔
flashing indicator 闪光指示器
flashing lamp 闪光灯
flashing light 闪光
flashing off 急骤馏掉
flashing point 闪点
flashing reactor 闪蒸反应器；快速反应器
flashing test 闪点测定；高压绝缘试验
flashless 无闪光的；无焰的
flashless mold 不溢式压模
flashless molding 不溢式模压法
flash liberation 瞬时释放；一次分离
flash light 闪光
flash light bomb 闪光弹
flash method 闪蒸法
flash mixer 快速混合器
flash molding 平板硫化法〔橡胶〕
flash mould 急骤模；溢出式塑模
flash off 闪蒸出
flash-off time ①闪干时间；晾干时间②闪蒸时间
flash out 闪蒸排出
flash-over(=flash-through) ①击穿；诱爆②闪弧；飞弧
flash overflow mold 溢(出)式压模；开放式压模
flash period 闪光周期
flash photolysis 闪光光解*
flash photolysis lamp 闪光光解放电管

flash photolysis method 闪光光(分)解法
flash pipe 闪燃管
flash point 闪点；闪燃点
flash point apparatus 闪点试验器
flash point in closed cup 闭杯闪点
flash point in open cup 开杯闪点
flash point test 闪点试验
flash (point) tester 闪点测定器
flash point yield curve 闪点产率曲线
flash polymerization 闪发聚合*
flash pot 闪蒸罐
flash pyrolysis 闪热解*；快速热解
flash pyrolysis reaction 急骤热解反应
flash resistance 耐着火性
flash ridge 绉纹边缘
flash roasting 飘悬焙烧
flash rust inhibitor 闪锈抑制剂；锈抑制剂
flash section 闪蒸部分
flash separation 石油接触脱气
flash-signal lamp 闪光信号灯
flash spectroscopy 闪光光谱法
flash spectrum 闪光光谱
flash spinning 瞬时纺丝法；急骤纺丝法
flash spinning method 闪(蒸)纺(丝)法〔超细纤维成形法之一〕
flash subcavity mold 溢式多巢模
flash tank 闪蒸罐
flash temperature ①闪蒸温度②闪点温度
flash test 闪点试验
flash tester 闪点测定仪
flash thickness 飞边厚度
flash-through(=flash-over) ①击穿；诱爆②闪弧③飞弧
flash tower 闪蒸塔
flash trapping stage (油气分离器中)一级分离
flash trimming 修边；除边
flash type mold 溢(出)式压模
flash vaporization 闪蒸
flash vaporization curve 闪蒸曲线
flash vaporization inlet 闪蒸入口
flash vaporization point 闪蒸温度
flash vaporizer 闪蒸器
flash vessel 闪蒸器；急骤蒸发器
flash vulcanization 超速硫化
flash welding 闪光焊
flash yield curve 闪蒸曲线
flash zone 闪蒸段
flask 烧瓶；瓶
flask brush 烧瓶刷
flask chamber 逃胶腔；跑胶孔
flask charge 烧瓶内已称过的剂量
flask combustion method 烧瓶燃烧法
flask constant 容器常数
flask heater 烧瓶加热器
flask holder 烧瓶夹
flask neck 烧瓶颈
flask opener 开瓶器
flask with side arm 具支管烧瓶
flask with three necks 三口烧瓶
flat ①无光的②平坦的③扁平的
flat abrasion 平磨
flat and round steel-chain 平圆式钢链
flat annular disk 扁环式圆盘
flat arch (裂化炉内的)扁平拱门
flat band method 平板法〔橡胶〕
flat-bar knitting machine 平机；横机
flat baseline 平坦基线；稳定基线
flat base rim 平底式轮辋
flat bearing (天平)支棱(承接)面
flat bed chromatogram 平板色谱(图)
flat belt 平带
flat billet 平错齿饰
flat blade 平叶片
flat bottom 平底
flat bottomed 平底的
flat bottom flask 平底烧瓶
flat bundle test 片状纤维束强力试验；平列纤维束强力试验
flat burner flame 燃烧器的扁平火焰
flat butted seam 对头接缝
flat cell 扁平池
flat compression strength 平压强度
flat conveyor belt 平型运输带
flat crepe 平绉
flat crystal spectrometer 平面晶体光谱仪
flat cure ①平坦性硫化；平坦效应②平铺硫化
flat curve 扁平曲线
flat deck roof 平顶盖
flat disk turbine agitator 圆盘平直涡轮式搅拌器
flat drill 平钻
flat duck 工业用帆布
flat edge (刮刀的)扁平刃
flat enamel 无光瓷漆
flat-ended horizontal cylindrical 平底卧式圆筒形罐
flat fiber 扁丝
flat file 扁锉；平锉
flat finish 消光

flat finishing ①涂无光涂料②平面涂装法
flat flame 无光焰
flat flow 扁平流(型)
flat gauze oxidizer 平网氧化器
flat grain 平纹〔切向纹理〕
flat grate 平面光栅
flat grating 平面光栅
flat grizzly 扁平条栅
flat hearth generator 气化平炉
flat iron 扁铁
flat lacquer ①无光挥发性漆；无光喷漆②无光硝基漆
flat latex grinds 无光乳胶研磨料
flat modulus 平坦定伸(强力)
flat multi-shed loom 平型多梭口织机
flatness 平直度
flat open turbine agitator 开启平直涡轮式搅拌器
flat-operated relief valve 浮筒操纵安全阀
flat paint 无泽油漆；晦漆
flat panel 平板；仪器面板
flat passage 平滑的通道；平滑的导管
flat peak 平顶峰
flat plate 平板
flat plate diffuser 平板扩散器
flat-plate heat exchanger 平板式换热器
flat plate X-ray pattern 平板 X 射线图像
flat polarization 平面偏振
flat position welding 平焊
flat product 平型制品
flat purl machine 平板机
flat rib top knitting machine 自动罗纹辅料横机
flat rotating screen 旋转平筛
flat rubber belting 平带
flat rubber product 平型制品
flat screen 平板筛
flat screen printing range 平网印花联合机
flat seaming 平缝
flat sheet 平板
flat sheet membrane 平片膜
flat sieve 平筛
flat site of peak 峰平坦区域
flat slate 扁平的石板
flat spirals 光滑旋管
flat spotting 胎面压扁；平点
flat spray type 平喷型
flat steel 扁钢
flat stock 平形工作物
flat stone mill 平面磨
flat strainer 平筛；平板筛浆机

flat strip steel 平钢带
flat suction box 平板吸水箱
flat surface(=plane) 平面
flattened square head bolt 平顶方头螺栓
flattening ①平化；变平②压平
flattening oven 平板(玻璃)炉
flattening test 压扁试验
flat tension strength 平拉强度
flatting ①平化；变平②消光
flatting agent 平光剂〔涂料〕
flatting effect 消光效应
flatting efficiency 消光效率
flatting extender 消光性体质颜料
flatting mill 压平机
flatting oil 消光油
flatting paint 平光油漆
flatting putty 油灰；涂料
flatting silica 消光粉；消光二氧化硅
flatting varnish 平光清漆涂料
flat-tipped spatula 平头调刀；平头刮刀
flat-toothed belt 齿形(平型)传动带
flat-top peak 平顶峰
flat-top sealed can 带平顶盖的密闭罐〔润滑油用〕
flat transmission belt 平型传动带
flat-tread profile 扁平胎面
flat tread rubber tire 扁平胎面轮胎
flat under taker-in 刺辊分梳板
flat valve 平阀
flat varnish 平光清漆
flat vibrating screen 平板振动筛
flat wafer 平晶片
flat wall brush 扁平涂墙刷
flat wall paint 无光墙(壁)漆
flat ware 浅皿；盘碟
flat washer 平垫圈
flat welding 平焊
flat width (吹塑薄膜)平折宽度
flatwise compression 贯层压缩
flat yarn(=flat fiber) 扁丝
flav〔词头〕 黄
flavacid 黄杀菌素
flavacin 黄曲霉素
flavandiols 黄烷双醇
flavane 黄烷 $C_{15}H_{14}O$
flavaniline 黄苯胺；2-对氨苯基-4-甲基喹啉 $NH_2C_6H_4C_9H_5NCH_3$
flavanol 黄烷醇 $C_{15}H_{10}O_3$
flavanone 黄烷酮

flavanonol(=dihydroflavonol) 黄烷酮醇
flavanthrene 黄烷士林
flavanthrone 黄烷士酮
flavanthrone paper 黄蒽酮试纸
flavaspidic acid 黄绵马酸
flavaspidinol 黄绵马定醇
flavatin 黄曲霉菌素
flavensomycin 菌虫霉素；黄烯霉素酸；黄质霉素
flaveolin 浅黄链丝菌素；浅黄菌素
flavescence 变黄；黄白病
flavianate 黄安酸盐 $C_{10}H_4O_8N_2SM$
flavianic acid(=2,4-dinitro-1-naphthol-7-sulfonic acid) 黄萘酸；2,4-二硝基-1-萘酚-7-磺酸 $(NO_2)_2C_{10}H_4(OH)SO_3H$
flavicidin(=penicillin F) 青霉素 F
flavicin(=penicillin F) 青霉素 F
flavin 黄素
flavin mononucleotide(FMN) 黄素单核苷酸
flavipin 黄柄曲菌素
flavo〔词头〕黄
flavochrome 黄色素；叶红呋喃素
flavofungin 黄抗霉素
flavogallol 黄棓醇
flavol 黄醇；1,3-蒽二酚
flavomycin 黄霉素
flavomycoin 黄霉菌素
flavone 黄酮*
flavone glycoside 黄酮糖苷
flavonoid 类黄酮；黄酮类化合物
flavonol 黄酮醇
flavophenine(=chrysamine) 黄芬宁；柯胺
flavophosphine 黄䣀
flavopurpurin 黄红紫素
flavor ①香味②食用香料③调料
flavorant 食用香料；食品香料
flavor chemistry 香味化学；香料化学
flavor component 香味组分
flavor composition ①食用香精②香味组分
flavored 调味；加香
flavor enhancer 香味增强剂；风味增强剂
flavorhodin 紫菌红素甲
flavoring ①香味②调香③调味剂
flavoring agent ①增香剂②调味剂
flavoring composition ①食用香精②风味组分
flavoring oil 食用油质香精
flavoring tablet 食用片剂香料
flavoring tincture 食用香酊
flavorist 食品调香者
flavor material 食用香料
flavor potentiator 香味增强剂；风味增强剂
flavor profile method 风味剖析法
flavor substance ①食用香料②调料
flavor terminology 风味术语
flavor threshold 香味阈值；口味阈值
flavor threshold concentration 香味阈值浓度；口味阈值浓度
flavoskyrin 黄天精
flavothebaone 黄蒂巴酮
flavo(u)r ①味；香味；风味；滋味②食用香料③食用香精④调料
flavour enhancers 增味剂
flavouring ①香味②调味剂③食用香料
flavouring agent ①增香剂②调味剂③食用香料
flavouring emulsion 乳液食用香精
flavouring extract 食用香料浸剂
flavouring powder 粉状香料
flavourless 没味道的；无味的
flavourous 有香味的；有滋味的；味浓的
flavoviridomycin 黄绿霉素
flavoxanthin 黄黄质；叶黄呋喃素
flavoxate 黄酮哌酯〔药〕
flavucidin 杀黄球菌素
flavylium 黄锌盐
flavylium ion 花(色)锌；2-苯基苯并吡(喃)锌
flaw ①缺陷②裂纹
flaw detection 探伤
flaw detector 探伤仪
flawed article 次品；副号
flawless 完好的；无缺陷的
flax breaker card 亚麻梳麻机
flax degumming 亚麻脱胶
flax dressing 打麻；清理亚麻
flax fiber 亚麻纤维
flax oil 亚麻油
flax retting 亚麻浸洗；亚麻脱胶
flaxseed 亚麻子
flax seed coal 碎粒煤
flaxseed mucilage 亚麻子黏液
flax-seed oil 亚麻油；麻子油
flax tow 亚麻短纤；亚麻落麻
flax wax 亚麻蜡
flaying 剥皮〔皮革〕
FLC(ferroelectric liquid crystal) 铁电液晶
fleabane oil 飞蓬油
fleat 平泥板
flecks streaks 斑条

fleece 羊毛
fleeting angle 传动角度
Fleischman brew buffer 弗雷斯克曼发酵液缓冲剂〔面团改良剂〕
fleroxacin 氟罗沙星〔药〕
flesher 仿麂皮皮革
flesh finish 肉面涂饰
fleshing beam 刮皮板
fleshing cylinder 去肉机刀轴
fleshing knife 刮肉刀
fleshing machine 刮肉机
fleshings〔复〕 肉面〔皮革〕
flesh side 肉面；皮里
flesh-side dubbing 肉面揩油液
flesh to flesh 肉面相对
Fletcher bleacher 弗利歇漂白机
fleur 鸢尾
flex 挠曲；屈挠
flexamine 蛇叶(尼润)胺
flex brittle test 挠曲脆性试验
flex crack 挠曲龟裂
flex cracking 挠裂；挠曲开裂
flex-cracking inhibitor 抗挠曲龟裂剂
flex-crack(ing) resistance 耐挠曲龟裂性
flex cycle 挠曲周期；曲挠循环
flexer 疲劳生热试验机；挠曲试验机
flex fatigue 挠曲疲劳
flex foam 软质泡沫塑料
flexibility 挠性；鞣性；揉曲性；柔韧性
flexibility chain 柔性链
flexibility of operation 操作灵活性；加工适应性
flexibility of polymer chain 高分子链的柔性
flexibility point ①挠曲点②软化点
flexibility retention ①柔软度保留②柔软度保留率
flexibility test 弯曲试验；柔韧性试验
flexibilized grade 增韧级
flexibilizer 增韧剂
flexible 挠性的；易揉曲的；柔韧的；韧性的；曲挠性的
flexible active site mechanism 易变活性中心机理
flexible alkyd 柔性醇酸树脂
flexible back-bone 柔性骨架；挠性骨架
flexible-bag laminating 膜袋层合(法)
flexible bond 柔性键
flexible chain 柔性链
flexible chain polymer 柔性链聚合物
flexible coating 挠性涂层；韧性涂层
flexible collodion 挠性火棉胶
flexible connection 挠性连接；活动连接
flexible connector 挠性联接器；挠性接头
flexible container 挠性容器；软质(包装)容器
flexible copper tubing 挠性铜管；软铜管
flexible core 软线
flexible coupling 挠性联轴节
flexible electron multiplier 挠曲型电子倍增器
flexible fiber 挠性纤维
flexible film 软(质)薄膜
flexible fitting (石)挠性零件
flexible fuel tank 挠性燃料桶
flexible glue 揉曲性黏合剂
flexible hose 柔软管；挠性软管
flexible hose connection 柔软管连接
flexible hose pump ①挠性软管泵②无脉动泵
flexible joint 挠性接头
flexible linear macromolecule 挠性线型大分子
flexible lubricator 挠性润滑器
flexible manufacturing system(FMS) 可调加工系统
flexible material 柔性材料
flexible membrane moulding(=bag moulding) 袋模塑；袋压成型
flexible metal hose 软金属管
flexible metallic hose 挠性金属管
flexible metal shoe (油罐浮顶水封的)挠性金属密封套
flexible metal tube 挠性金属管；金属软管
flexible pipe 挠性管
flexible pipeline pig 挠性管道清洗器
flexible plastics(=flexiplastics) 软(质)塑料
flexible plastic sleeve 软塑料套筒
flexible rule 卷尺
flexible shaft level controller 挠性轴的水准控制器
flexible shaft level indicator 挠性轴的水准指示器
flexible sheet ①挠性板；软(质)板(材)②软片
flexible side group 柔性侧基
flexible spacers 柔性间隔基
flexible substrate 挠曲性底材；柔性底材
flexible tank 折叠式油罐
flexible transport 无轨运输
flexible tube 挠性管
flexible tubing 挠性管路
flexible urethane foam 软质聚氨酯泡沫塑料
flexibly jointed chain 柔性联结链
flexicoking 灵活焦化
flexicracking 灵活裂化
flexine 蛇叶(尼润)辛
flexing 挠曲
flexing cycle 挠曲周期
flexing elasticity 挠曲弹性

flexing fatigue 挠曲疲劳
flexing fatigue life 抗挠疲劳寿命；耐挠曲疲劳性
flexing life 挠曲寿命
flexing machine 挠曲机；致柔机
flexing resistance 抗挠性
flexing test 挠曲试验
flexing tubing 软管；纯胶管
flexinine 蛇叶(尼润)碱
flexion 挠曲；弯曲
flexion zone 屈挠区
flexiplast 挠性塑料
Flexitray 弗莱克斯圆盘形浮阀塔板
flexixanthin 屈曲黄素
flexizer 增韧剂
flex life 抗挠寿命
flex life test 抗挠寿命试验
flex lip extruder 活模唇挤塑机
flexographic ink 苯胺油墨；胶印油墨；橡皮凸版油墨
flexographic press 苯胺印刷机；胶版印刷机
flexographic printing 苯胺印刷；胶(版)印(刷)
flexography 胶版印刷；胶印
flexomer 挠性聚合物
flexometer 挠度计
flexometer for hysteresis test 滞后试验挠曲计
flex point ①拐点②挠曲点；软化点
flex stiffness 挠曲劲度；抗挠曲性
flex strength 挠曲强度
flex temperature 软化点；挠曲点
flex test 挠曲试验；弯曲试验
flexuose(=flexuous) 锯齿状的；之字形的
flexural cracking 挠曲龟裂
flexural creep 挠曲蠕变
flexural deflection 挠曲畸变
flexural elongation ①挠曲伸长②挠曲伸长率
flexural failure 挠曲损坏
flexural glide 弯曲滑移
flexural modulus 弯曲模量
flexural modulus of elasticity 弹性挠曲模量；挠曲弹性模量
flexural moment 挠曲力矩；弯矩
flexural offset yield strength 弯曲偏置屈服强度
flexural rigidity 抗弯刚度
flexural shock 挠曲冲击
flexural stiffness 挠曲劲度；弯曲硬挺度
flexural strain 挠曲应变；弯曲应变
flexural strength 弯曲强度
flexural stress 挠曲应力；弯曲应力
flexural vibration 弯曲振动

flexural yield strength 挠曲屈服强度
flexure 挠曲；揉曲；弯曲
flexure temperature 弯曲温度；柔软温度
flexure test 挠曲试验；弯曲试验
flex-whiten 挠曲发白
flicker ①闪光；闪烁；闪变②摇晃
flickering flame 闪烁火焰
flickering of spray 雾化气流脉动
flick separator 弹动分离器
Flieger's equation 弗利格方程式
flight 刮板；抄板
flight conveyor 刮板运送器
flight depth 螺纹深度
flight hole 通气口
flight land 螺纹棱面
flight lead 螺纹导程
flight path 飞行路径
flight pitch 螺(纹间)距
flight scrimp rail 螺纹扩幅板
flight tube 飞行管
flight velocity 飞行速度
flimsy cloth 稀布；松织布〔纺〕
flindersin(e) 弗林辛；二甲吡喃并喹啉酮
flinkite 褐水砷锰矿
flint 燧石〔矿〕；打火石〔商〕
flint ball 燧石
flint clay 燧土；燧石状土
flint dry hide 僵板皮
flint glass 燧石玻璃；火石玻璃
flint glazed paper 蜡光纸
flint glazing (用)燧石砑光
flint glazing calender 燧石砑光机
flintiness 坚硬度
flint lining 燧石衬
flint pebble 燧石子
flint plaster 燧石石膏
flint stone 燧石；打火石
flint wheat 硬质小麦
flinty ①燧石的②坚硬的
flinty crust 硬壳
flinty slate 燧石板岩
flinty-type sludge 很硬的淤渣
flip angle 倾倒角
flip-flop 金属闪光效果；触发器
flip-flop column system 回转柱系统
flipper 挡泥板
flipping machine 挡泥板制造机
flip tone 闪光面色

fliter agent　（过）滤剂
flitter　①片状金粉②闪光颜料
float　①浮②浮标；浮筒③磷灰石粉④浮线
floatability　浮游性；浮游本领；可浮选性
float-actuated level controller　浮标水准控制器
float-actuated tape　浮筒动作测深度表尺
floatater　浮选机
floatation　浮选*
floatation agent　浮选剂
floatation analysis　浮选分析
floatation by precipitation adsorption　沉淀吸附浮选
float charge　浮动充电
float circuit　浮筒线路〔汽化器水准控制用〕
float controlled valve　浮子阀
float displacement　浮标排水
floater　①浮标②浮渣
float gauge　浮标；浮规
float glass　浮法玻璃
float head heat-exchanger　浮头式(列管)换热器
floating　①浮行②浮雕③发花
floating agent　浮选剂
floating arm　浮臂
floating barge　(海洋)钻井浮船；自浮钻船
floating barrier　浮动挡板；浮油栅
floating baseline　浮动基线
floating bath soap　浮水浴皂
floating bell manometer　浮钟压力计
floating bottom　浮底
floating chase　活络模套
floating chase mould　浮套塑模；双压塑模
floating coat　①二道抹灰层②二道漆；中涂层
floating controller　浮动控制器
floating cover　浮顶
floating crosshead pin　浮动十字头销
floating diffuser　浮式扩散器
floating end cover　浮头盖(换热器)
floating filling station　浮动的加油站
floating foam　浮动的泡沫
floating gate electrode　浮置控制极
floating head　浮头；浮盖
floating head exchanger　浮头式换热器
floating hydrocarbon　浮烃
floating knife coater　浮动刮刀涂布机；浮刀刮涂机；活动刮涂机
floating oil　浮油
floating oil boom　浮油栅
floating oil filter　浮(在表面上的)油过滤器
floating packing　活动填充

floating pipe　浮动管道；活动管道
floating plate　中间热板〔平板机〕
floating platen　浮动压板；中间压板
floating pollutants　浮动污物；漂浮污物
floating reclaim　熔融的再生橡胶
floating refinery　浮动炼油厂；船上炼油厂
floating reservoir　浮式油罐
floating ring　浮动圈
floating-ring seal　浮环密封
floating roll　浮动辊
floating roof　(油罐的)浮顶
floating roof drain　浮顶油罐的排水系统
floating roof hydrocarbon bank　浮顶油罐
floating roof reservoir　浮顶储罐
floating roof seal　浮顶封
floating roof tank　浮顶油罐
floating skimmer　浮式撇取器；浮式撇油器
floating sleeve　(输油软管的)浮套
floating soap　浮水皂
floating spherical Gaussian function(FSGO)　浮动球高斯函数
floating-tongue type jet tray　浮舌塔
floating trough　浮槽
floating tube sheet cover　浮管盖片；管式换热器的活动盖
floating type thermometer　浮式温度计；刻度在里面的温度计
float(ing) valve　浮球阀
floating-vane type jet tray　浮动喷射塔
floating weight　上顶栓
floating zone melting　悬浮区熔法
float length　液比
float level controller　浮动式液面调节器
float meter　浮尺
float needle　浮针
float on line　线路上的浮标
float-operated relief valve　浮球控制安全阀
floator　浮游选矿机
float product　浮选产物
floats　①铰接②磷灰石粉
float spindle　浮针
float steam trap　浮行蒸气冷凝器
float-stone　浮石
float stone grinding　(用)浮石粉抛光
float suction control　浮筒吸气控制
float switch　浮筒开关〔俗〕；浮控开关
float tank　浮箱；浮选箱
float test　浮杯试验

float tester (沥青的)浮子测定器
float type area flowmeter 浮子式面积流量计
float type flowmeter 浮子式流量计
float type tank gauge 浮子式油罐液面计
float-type viscometer 浮标式黏度计
float valve 浮阀
float valve needle 浮动阀的针
float valve tray 浮阀塔板
float viscosimeter 浮标黏度计
float viscosity 浮测黏度；在浮标黏度计内测定的黏度
floc 絮凝物
flocbed 絮凝层
floccose 丛卷毛的；羊毛状的
flocculability 絮凝性
flocculant 絮凝剂
floccular 絮凝的
flocculate 絮凝物
flocculated colloid 凝聚胶体
flocculated pigment 絮凝颜料
flocculated sludge 絮凝渣
flocculating 絮凝化
flocculating agent 絮凝剂
flocculating aids 助凝剂
flocculating colloid 絮凝胶体
flocculating constituent 絮凝体
flocculating effect 絮凝效应
flocculating settling 絮凝沉降
flocculating yeast 絮凝酵母
flocculation 絮凝；凝结
flocculation accelerator 絮凝促进剂
flocculation basin 絮凝池
flocculation concentration 絮凝浓度
flocculation gradient 絮凝梯度
flocculation kinetics 絮凝动力学
flocculation number 絮凝值
flocculation oil removing method 石油絮凝脱水法
flocculation rate 絮凝速度
flocculation structure 絮凝结构(构造)
flocculation value(=coagulation value) 絮凝值；凝结值
flocculator 絮凝器
floccule 絮凝粒；絮状物
flocculence ①棉絮状②絮凝性
flocculent(=flocky) 絮凝的
flocculent curd 絮状(乳)块
flocculent deposit 絮凝状沉淀
flocculent gel 絮凝胶
floccules 絮凝粒
flock 短纤维(填料)；绒屑

flock coating 栽绒涂料；植绒涂料
flocked fabric 植绒织物
flocked suede 植绒仿鹿皮
flock filler 绒屑填充剂
flock finish 植绒涂装法；涂栽绒面漆
flock finishing 栽绒涂装
flock gun 植绒(涂层的专用)喷枪
flocking ①植绒(工艺)；静电植绒②短绒；棉束；毛束
flocking machine 植绒机
flocking settling 凝聚沉降
flock printing 植绒印刷
flock spraying 绒屑喷布〔喷布在已涂有黏合剂的表面上可形成毛毡状〕
flocky(=flocculent) 絮凝的
floc point 絮凝点
flocs〔复〕 棉絮
flocs unit 絮凝体
floc test 絮凝试验〔石油〕
floded molecule 折叠分子
flokite 针丝光沸石
flood 溢流；涨溢；淹没；沉入
flood coating 幕式淋(流)涂；流涂；浇涂
flood cooling 泛流冷却
flooded burner 溢流燃烧器
flooded continuous dissolver 液流式连续式溶解器
flooded evaporator 溢流式蒸发器
flooded lubrication(=flood lubrication) 淹没润滑
flooded zone 溢流区
flooding ①溢流②液泛〔分馏时的液阻现象〕③泛色〔涂料〕；泛浮；浮色
flooding agent 溢流剂
flooding fine particle 溢流微粒
flooding head (水幕)喷头
flooding nozzle 冲洗喷嘴
flooding pipe 溢流管
flooding point 泛液点
flooding point in column 塔内溢流点
flooding rate ①溢流速率②液泛速率
flooding solution plating 流动镀
flooding valve 溢流阀
flooding velocity ①溢流速度②液泛速度
flood light 泛光照明；强力照明；浮光灯
flood lubrication 溢流式润滑
Flood reaction 弗鲁德反应〔氯代三烃基硅烷合成法〕
floor ①底(面)②地板
floor cleaner 地板洗净剂
floor conveyor 地面传送带
floor heater 地龙式加热器

flooring ①地面材料②地面
flooring plaster 地板(料)石膏
floor paint 地板油漆
floor price 最低价格；最低出口价格
floor section 炉底管段
floor space (设备的)占地面积
floor temperature of polymerization 下限聚合温度
floor tile 铺地瓷砖
floor-to-floor conveyor 楼上下运输机；斜坡运输机
floor tube (管状炉)靠近底部的管子
floor type grinding wheel 落地砂轮机
floor-type hand punch 落地式手扳冲刀
floor varnish 地板清漆
floor wax 地板蜡
flop 随角异色
flop color 侧视色；(水)平视色〔观测角几乎与漆膜表面平行时所见的颜色〕
floppy disk 软盘
floral 花的；花香的
floral background 花香尾韵
floral base 花香基
floral bouquet 百花香；百花型
floral character 花香特征
floral composition ①花香香精②花香组分
floral compound 花香香精
floral essence 花萃；花中提出物
floral perfume ①花制香料②花香香精
floral scent 花香
floral suggestion 花香韵调
floral type 花香型
floral water 香水；花露水
floral wax 花蜡
florence flask 平底烧瓶
Florentine lake 佛伦汀色淀；绯红色淀〔别名〕
flores ①花〔植物〕②华〔升华所得的化学品〕
flores martiales 氯化铁铵
floribundine 多花(罂粟)碱
florida clay 漂白土
florida earth 凹凸棒土〔别名〕
floridean starch 花淀粉
florimycin 弗罗里霉素；紫霉素
floripavidine 多花罂粟定
floripavine 多花罂粟因
florist and twisting tissue 扎花纸
Flory dynamometer 弗洛里常数
Flory-Fox constant 弗洛里-福克斯常数
Flory-Huggins equation 弗洛里-哈金斯方程〔有关高分子溶液的统计热力学理论所推导的方程式〕

Flory-Huggins theory 弗洛里-哈金斯理论
Flory-Stockmayer's theory 弗洛里-施托克迈尔理论〔关于高分子凝胶化的理论〕
flory temperature(=θ-temperature) 弗洛里温度；θ温度
floss ①絮状物②丝棉③浮渣
flotation ①浮〔物〕②浮选
flotation agent 浮选剂
flotation cell 浮选池
flotation concentrate 浮选精矿
flotation line 浮线；水线
flotation medium 浮选介质；悬浮介质
flotation method 浮选法
flotation oil 浮选油
flotation process 浮选法
flotation promoter 促浮剂
flotation spectrophotometry 浮选分光光度法
flotation test 浮选测试法
flour 粉末纤维；(纤维)细屑；细料
flour blender 面粉掺和机
flour blending machine 面粉掺和机
flour bolt 面粉筛
flour dough 面团
flour dredge(r) 面粉掺和机
flouring of mercury 汞的乳化作用
flour mill 面粉磨
flour mite 面蛆
flour of powder 细(粉)末
flourohalocarbon 氟卤烃
flourohydrocarbon 氟代烃
flourometer 澄清器；澄粉器
flour (storage) room 面粉仓库
flour tester 面粉细度试验器
flour weevil 面象虫
flouve 黄花草
flouvo oil 黄花草油
flow ①流；流动②流量
flowability 流动性
flow-activation energy 流动活化能
flowage 流
flow agent 流动剂；流平剂
flowage structure 流程结构
flow aid 流动性助剂；助流剂
flow analysis 流动分析法
flow area 流动面积
flow-back 逆流；回流
flow behavior 流动特性
flow behavior index 流动特性指数
flow birefringence 流动双折射

flow boundary　流动边界
flow box　流料箱
flow calorimeter　(气体)流(动)量(热)计
flow capacity　流量；流动能力；泄水能力
flow casting　中间铸型法
flow cell　流动池；流通池
flow cell compartment　流动池室
flow cell technique　流动池技术
flow characteristic　流动特征
flow chart(=flow sheet)　①流程图②程序框图
flow circuit diagram　流路图
flow coat　流涂；浇涂〔涂料〕
flow coating　流涂；浇涂〔涂料〕
flow coating machine　流涂机
flow coating machine for radiator　暖气片用流涂机
flow coat method　流涂法；浇涂法〔涂料〕
flow coefficient　流量系数
flow colorimeter　流动式比色计
flow component　分流
flow condition　流动条件
flow contribution　流速分额
flow control　流量控制；进料控制
flow controller　流量控制器
flow control system　流量控制系统
flow control valve　流量控制阀
flow cracks　流动痕迹
flow cup　流杯〔黏度计的一种〕
flow cup viscometer　流出杯黏度计
flow curtain electrophoresis　流幕电泳
flow curve　流动曲线
flow cuvette　流动吸收池
flow cytometry　流动式细胞光度法；流式细胞术
flow data　流动值
flow detectors in radiochemistry　放射化学流通检测器
flow diagram(=flow sheet)　流程图；流动图解
flow dichroism　流动二向色性
flow diffuser　流量扩散器
flow directing device　直流装置
flow direction　流向
flow distributing system　流量分配系统
flow disturbance　流体扰动
flow diverter　流路换向阀
flow-down burning　下行燃烧
flow elasticity　流动弹性
flow electrolysis and thin layer method　流动电解和薄层技术
flow-equalization screen　均流筛；气流均衡筛
flow equalizer　均流筛；气流均衡筛

flow equation　流动方程
flower　①华〔升华所得的化学品〕②花〔植物〕
flower absolute　花(制)净油
flower bud　花蕾
flower concrete　花(制)浸膏
flower extract　花浸提物；花浸液
flowering hormone　成花激素
flower mark　花斑〔制品缺陷〕
flower of sulfur　硫华
flower oil　花(制)油
flower ointment　花香软膏；花香油膏
flowers of antimony　锑华；氧化锑
flow factor　流量因数
flow feedback control　流量反馈控制
flow feeder　流动送料机
flow field　流场
flow fluctuation　流量波动
flow friction characteristics　流动摩擦特性
flow gauge　流量计；测流规
flow glass　流动观察玻璃
flow graph　流向图
flow homogenizer　流量均化器
flow in　流入
flow in corner　绕角流动
flow in critical region　在临界范围内流动
flow index　流动指数
flow indicator　流量指示器；进料量指示器
flow indicator alarm　流量指示警报器
flow-induced crystallization　流动诱导结晶
flow inducer　导料器
flowing　流动的
flowing afterglow technology　流动余辉技术
flowing chromatogram　流动色谱(图)
flowing colorimeter　流动比色计
flowing fluid ratio　液体比〔流液中各种流体的流量比〕
flowing furnace　熔化炉
flowing junction　流动接界
flowing liquid colorimetric detector　流液比色检测器
flowing mercury electrode　流汞电极
flowing out　流出
flowing pressure　流动压力
flowing property　流动性
flowing temperature　流动温度
flowing-through chamber　穿流室
flowing varnish　流平性清漆
flow injection analysis(FIA)　流动注射分析(法)
flow injection analysis gradient technique　流动注射分析梯度技术

flow injection analyzer 流动注射分析器
flow injection-atomic absorption spectrometry 流动注射-原子吸收光谱仪联用技术
flow injection chemiluminescence measurement 流动注射化学发光测量
flow injection immunoassay 流动注射免疫分析
flow injection-renewable surface technique 流动注射-可更新表面技术
flow injection-spectrophotometry 流动注射分光光度法
flow injection system 流动注射系统
flow in three dimension 三维流动
flow in vortex 涡流
flow irregularities 流动不规则性
flow length (模具)流程
flow limiting passage 限制泵供给的通路
flow line ①流送管②由油井至油气分离器的管路③流水线
flow line pressure 流送管内压力
flow line production 流水作业的生产法
flow line technique 流水作业法
flow mark 波纹
flow meter 流量计；流速计
flowmeter in low pressure 低压流量计
flow method 流动法
flow microcalorimetry 流动微量热法
flow minder 流量指示器
flow mixer 流体混合器
flow model 流动模型
flow molding 传递模压法；流动成型
flow-monitoring device 流量监控装置
flow network 流路网络
flow nipple 油嘴；节流嘴
flow noise 流动噪声
flow nozzle 测流嘴
flow number 流速；流量数值
flow off 流(下)
flow of lubricant 润滑油的流动；润滑油流
flow of momentum 动量流
flow of powder 粉末流
flow of stream 流动
flow operation 流水作业
flow orientation 流动取向
flow out 流出
flow outside boundary 边界外流动
flow over ①溢出；满出②(挠曲产生的)内裂纹
flow path 流路
flow pattern 流型；流动方式
flow pipe 压力水管；送水管

flow point(=pour point) 倾点；流(动)点；倾倒点；浇铸点
flow-pressure diagram 流量-压力图
flow production 流水作业
flow program 流量程序
flow programmed chromatography 程序变流色谱法
flow programmer 程序变流器
flow programmer panel 程序变流盘
flow programming 程序变流
flow programming high performance liquid chromatography 程序变流高效液相色谱法
flow promoter 流动性促进剂
flow promoting agent 流动性促进剂
flow promotor 流动助剂
flow rate ①流量②流率；流速③比移值〔指色谱中 R_f 值〕
flow rate meter 流量计
flow rate of carrier gas 载气流速*
flow ratio 流(量)比
flow ratio controller 流量比率控制器
flow reactor 流动反应器*
flow recorder 流量记录器
flow recording controller 流量记录控制器
flow region 流动区域
flow regulating valve 节流阀
flow resistance 流阻；水力摩阻
flow-restrictive screw 分流螺杆
flow reversal FIA 逆流动注射分析
flow rule 流动法则
flow schematic diagram 流程示意图
flow scheme(=flow sheet) 流程图；流程表
flow sensitive detector 流量敏感检测器
flow sensor 流动传感器
flow sheet(=flow chart; flow diagram) 流程图；流动图解
flow sheet simulation system chemical technology 化工流程模拟系统
flow snubber 限流器
flow splitter 分流器
flow spoiler 扰流器
flow spreader 多孔辊；匀浆辊；整流辊
flow stability 流动稳定性
flow stress 流动应力
flow summarizer 流量累积器
flow switch 流量开关；流向转换器
flow system 流动控制系统
flow table 流盘；流动稠度试验台
flow tank 沉淀池
flow temperature(=flux temperature) 流动温度
flow temperature test 流动温度测定试验
flow test 流动试验

flow tester　流动试验仪
flow-through cell　流通池
flow-through coil planet centrifuge　流通型螺旋管行星式离心(分离)仪
flow-through colorimeter　流通式比色计
flow-through detector　直通型检测器
flow-through direct current discharge emission detector　直通型直流放电发射检测器
flow-through electrophoresis　流通电泳(法)
flow-through ionophoresis　流通离子电泳(法)
flow-through method　直流法
flow-through orifice　流量孔板；孔板流量计
flow thru electrode　流通电极
flow time　流平时间
flow tube　测流管
flow tube reactor　流动管反应器
flow-up burning　上行燃烧
flow velocity　流速
flow visualization　流动形象化
flow void　流空
flow work　流动功
floxin acetate　乙酸α-戊基肉桂酯
floxin isovalerate　异戊酸α-戊基肉桂酯
Floyd tester　弗洛依德试验器〔评价齿轮润滑剂极压性质的一种携带式的仪器〕
fluavil　古塔树脂〔古塔波胶所含树脂〕
fluconazol　氟康唑〔药〕
fluctuating　波动；起伏
fluctuating flow　波动流动；脉动流动
fluctuating nature　波动性能
fluctuating strain　波动应变
fluctuating stress　波动应力
fluctuation　波动；起伏
fluctuation noise level　涨落噪声电平
fluctuation of cencentration　浓度波动
fluctuation of temperature　温度波动
fluctuation scattering　(溶液)浓度起伏散射；浓差散射
flucytosine　氟胞嘧啶〔药〕
fludarabine　氟达拉滨〔药〕
fludrocortisone　9-氟皮质(甾)醇；氟氢可的松
flue　①烟道②焰道
flue ash　烟道灰
flue collector　主烟道
flue-cured tobacco　烟熏烟
flue curing　烟熏〔烟草〕
flue dust　烟道尘
flue gas　烟道气
flue gas analysis　烟道气体分析

flue gas analyzing apparatus　烟道气分析装置
flue gas blower　烟道气鼓风机
flue gas desulfurization　烟气脱硫
flue gas holder　烟道气气柜
flue gas purifier　烟道气净化器
flue gas recirculation　烟道气再循环
flue gas recirculation furnace　烟道气循环炉
flue gas return　烟道气的回炉
flue gas treatment　烟道气处理
flue gas vent　烟道气的放空口
fluellite　氟铝石
fluence　能流*
fluence rate　积分通量率；流量率
fluent granular solid　流化床粒状催化剂
fluff　①磨里②起毛
fluffer　纤维分离机；疏解机
fluffiness　①起毛现象②蓬松性
fluffing　磨里
fluffing machine　磨里机；磨面机
fluff packing　疏松堆积
fluffy cake　毛茸丝饼；松散丝饼
fluffy carbon　容易破碎的炭粒；松散的炭粒
fluffy soda　发面碱；碱
fluid　流体
fluidal　液体的
fluidal structure　流纹构造
fluid analyzer unit　流体分析装置
fluid bed　流化床
fluid bed burner　流化床燃烧炉
fluid bed catalytic cracking unit　流化床催化裂化设备
fluid bed combustion　流化床燃烧
fluid bed dryer　流化床干燥器
fluid bed flowmeter　流化床流量计
fluid bed process　流化(沸腾)床法
fluid bed vulcanization　流动床硫化
fluid behaviour　流体特性
fluid body　流体
fluid carbon　挥发性碳
fluid catalyst　流化状态的催化剂
fluid catalyst process　流化床催化过程
fluid catalyst reactor　流化催化(剂)反应器
fluid catalytic cracker　流化床催化裂化设备
fluid catalytic cracking process　流化床催化过程
fluid catalytic hydroforming　流化床催化(临)氢重整
fluid channcl　流道
fluid char adsorption process　活性炭流化吸附法
fluid choke　流体流阻器
fluid coke　流化焦炭

fluid coker 流化焦化器
fluid coking 流化焦化
fluid coking-gasification 流化焦化气化
fluid coking process 流化床焦化过程
fluid coking-steam gasification 流化焦化-蒸气气化
fluid combustion 液体燃料的燃烧
fluid contacting apparatus 流体接触装置
fluid contacting unit 流体接触装置
fluid container 液体容器
fluid continuous cracking unit 流化床连续裂化设备
fluid core 流体核心
fluid core electrolytic cable 液体芯线电缆
fluid cracking catalyst 流化床裂化催化剂
fluid deformation 流体形变
fluid degradation 流体降解
fluid density 流体密度
fluid density meter 流体密度计
fluid distributing apparatus 流体分配装置
fluid draw-texturing unit 流体拉伸变形部件；流体拉伸变形装置
fluid dynamics 流体(动)力学
fluide bed burner 硫化床燃烧炉
fluid elastic deformation 流体弹性形变
fluid end of pump 泵的流体端；泵的压力室
fluid energy mill 气流粉碎机；流能磨
fluid energy milled mica 气流粉碎云母粉
fluid exposure 流体暴露
fluid extract 流体提出物
fluid film 流体薄膜
fluid film lubrication 流体薄膜润滑作用；边界膜润滑作用
fluid filter 流体过滤器
fluid flow 流体流动
fluid flow indicator 流体流动指示器；液体流量指示器
fluid flow pattern 液流图
fluid fracture 流体破裂
fluid friction 流体摩擦
fluid handling 流体的处理；液体的运送
fluid head 流体高差；液压头
fluid heating furnace 流体加热炉
fluid heating medium 流体加热介质
fluid hose 流体管；油漆(软)管
fluid hydroformer 流化床(临)氢重整设备
fluid hydroforming 流化床(临)氢重整
fluidic 流体的
fluidication 流化
fluidic channel 流体管道
fluidic computer 流体计算机
fluidic inductance 流体感应系数

fluidic logic sampling 射流逻辑进样
fluidic microsystem 流体微系统
fluidic oscillator 流体振荡器
fluidic resistance 流(体)阻(力)
fluidics 流控技术；射流技术；流体学
fluidifier 塑化剂
fluidifying 流(体)化
fluidimeter(=fluidometer; fluidity meter) 流度计
fluid inelastic deformation 流体非弹性形变
fluid injecting atomizer 射流喷雾器
fluid issuing 流体出口
fluidity 流动性；流度；流值
fluidity at low temperature 低温下的流动性
fluidity at rest 静止时流动性
fluidity coefficient 流度系数
fluidity function 流动性函数
fluidity improver 流动性改进剂
fluidity index 流度指数；流动性指数
fluidity meter(=fluidmeter) 流度计〔测定不同温度下油品黏度的黏度计〕
fluidity ratio 流度比〔螺杆进料点与出料点熔体的比流度之比〕
fluidizable particle size 可流体化的颗粒大小
fluidization 流化(作用)
fluidization column 流(态)化塔
fluidization dip coating (粉末)流化床浸涂法；沸腾床浸涂法
fluidization flow 流化(物料)流；流(态)化流动
fluidization number 流(态)化数
fluidization of solid 固体流态化
fluidization quality 流化品质；流态化质量
fluidization technique 流化技术
fluidized 流态化的
fluidized bed(=liquid bed) 流体床；沸腾床
fluidized bed calciner 流化床煅烧器
fluidized bed catalytic cracker 流化床催化裂化装置
fluidized bed denitration 流化床脱硝
fluidized bed denitrator 流化床脱硝器
fluidized bed dipping 流化床浸涂法；沸腾床浸涂法
fluidized bed dryer 流化床干燥器
fluidized bed fluoride volatility process(FBFVP) 流化床氟化挥发法
fluidized bed furnace 流化床加热炉
fluidized bed granulation 流化床造粒(法)；沸腾床造粒(法)
fluidized bed incinerator 流化床焚烧炉
fluidized bed method 沸腾床法；流动床法
fluidized bed of adsorbent 吸附剂流化床
fluidized bed plant 流化床设备

fluidized bed process 流化床法；沸腾床法
fluidized-bed reactor 流化床反应器
fluidized bed roasting 流化床焙烧；沸腾床焙烧
fluidized bed seal leg 流化床料腿封
fluidized bed technique 流动床硫化法
fluidized carbonization 流化床焦化
fluidized catalyst 流化催化剂
fluidized catalytic cracker 流化催化裂化装置
fluidized circulation 流化循环
fluidized coal beds 流化煤层
fluidized column 流化柱
fluidized distillation 流化床蒸馏
fluidized fixed bed 流化的固定床层
fluidized gasification 流化床气化
fluidized granulation 流动造粒
fluidized iron catalyst process 流体铁催化法〔以 CO 与 H_2 在流化铁催化剂上合成烃类的过程〕
fluidized layer 流化床；沸腾床
fluidized particle bed 流化粒床
fluidized point 流(态)化点
fluidized powder 流态化粉末
fluidized powder bed 粉末沸腾床；粉末流化床
fluidized reactor 流化床反应器
fluidized roaster 流化床焙烧炉；沸腾床焙烧炉
fluidized solid 流化固体
fluidized solid bath 流化固体浴；流动固体加热炉
fluidized solid oxidation 流化粒氧化；流化固体氧化(作用)
fluidized solid process 固体流化法
fluidized solid roaster 流化床焙烧炉；沸腾床焙烧炉
fluidized state 流态化状态
fluidized state polymerization 流化态聚合
fluidizer ①流化剂②流化床装置
fluidizing 流化的
fluidizing agent 流(态)化剂；流化气体
fluidizing air 流动空气
fluidizing cooler 流化式冷却器
fluidizing dryer 流化式干燥器
fluidizing grid 流动栅；流化分配栅
fluidizing medium 流化介质
fluidizing mixer 液体化搅拌机
fluidizing velocity 流化速度
fluid jet 流体射流
fluid jet mill 气流粉碎机
fluid layer 流体层
fluid level gauge 液位指示器
fluid level gauge rod 测定液位的指示杆
fluid-like state (粉末的)类液态

fluid loss 失水量
fluid lubrication 流体润滑
fluid mechanics 水力学；液体力学
fluid medium 流体；流质；流体介质
fluidmeter(=fluidity meter) 流度计〔测定不同温度下油品黏度的黏度计〕
fluid milling technique 流(液)体研磨分散技术
fluid mobility 流体淌度；流体流动性
fluid needle 针形喷阀
fluidness 流动性；流度
fluid oil 润滑油；液态油
fluidometer 流度计
fluid-operated controller 流体传动控制器
fluid orifice 射流孔；射流锐孔
fluid ounce 流体盎司
fluid paraffin 液体石蜡
fluid point 流点〔增塑糊转变为黏性膜层时的温度〕
fluid pressure 流体压力；流体静力学压力
fluid pressure booster 流体压力加强鼓风机
fluid-pressure moulding 液压模塑法；(膜)袋(模)塑(法)
fluid process 流化床过程
fluid property number 流体性质数；流体特性值
fluid-pump 液压泵；冲洗液泵
fluid relief 内液压缓冲
fluid resin 流体树脂
fluid resistance 流体阻力
fluid rheology 流体流变学
fluid rubber 液体橡胶
fluid seal 液封
fluid seed method 流体接种法；气流接种法
fluid sensor 流体传感器
fluid separation device 流体分离装置
fluid shearing 流体剪切
fluid slag 流动熔渣
fluid-solid chromatography 液固色谱(法)
fluid speed meter 流速计
fluid state 流态
fluid statics 流体静力学
fluidstatio pressure 静水压力
fluid strainer 滤液器
fluid technology 射流技术
fluid-tight 不漏流体的；液体密封
fluid ton 流体吨〔体积为 0.906 立方米〕
fluid-type shale retorting 流化床页岩干馏
fluid unit 流化床设备
fluid-valve 液流阀；水力阀
fluid wax 液体石蜡
fluid wedge 流体楔；油楔

fluid with deformable microelements 具可变形微元素的流体
fluid with rigid non-spherical microelements 具刚性非球形微元素的流体
flume 斜槽；流斜槽；侧流槽
flumerin 二钠-2-羟汞荧红〔药〕
fluming 水的输送
fluming water 输送水
flunarizine 氟桂利嗪〔药〕
fluo- 〔词头〕 ①氟 ②荧
fluobenzene 氟(代)苯 C_6H_5F
fluobenzoic acid 氟苯甲酸 FC_6H_4COOH
fluoborate 氟硼酸盐 $M[BF_4]$
fluoboric acid 氟硼酸 $H[BF_4]$
fluoborine 氟碳铈矿
fluoborite 氟硼镁石
fluocarbon 氟代烃
fluocerite 氟铈矿
fluochromate 氟基铬酸盐 $M[CrO_3F]$
fluochromic acid 氟基铬酸〔基字不宜略去；因为"氟基"指代"—OH"而氟某酸则为氢配位酸的一种〕 $H[CrO_3F]$
fluocin 荧光极毛菌素；荧光菌素
fluocolumbate 氟铌酸盐 $M_2[NbF_7]$
fluodensitometry 荧光密度测定法
fluoflavin 荧黄素
fluoform 氟仿
fluogermanate 氟锗酸盐 $M_2[GeF_6]$
fluogermanic acid (氢)氟锗酸 H_2GeF_6
fluohafnate 氟铪酸盐 $M_3[HfF_7]$
fluohydric 氢氟化的
fluohydric acid(=hydrofluoric acid) 氢氟酸
fluohydrocarbon 氟代烃
fluomethane 氟代甲烷 CH_3F
fluooxycolumbate 氟氧铌酸盐
fluoplumbic acid (氢)氟铅酸 H_2PbF_6
fluoprot(o)actinate 氟镤酸盐 $M_2[PaF_7]$
fluopsin 荧假胞菌素
fluor 氟〔9号元素，化学符号F〕
fluoracyl chloride 氟乙酰氯
fluoracyl fluoride 氟乙酰氟
fluoracyl halide 氟乙酰卤
fluoradiography 荧光照相法
fluorandiol(=fluorescein) 荧烷二醇；荧光素
fluorane 荧烷 $C_{20}H_{12}O_3$
fluoranthene 荧蒽 $C_{16}H_{10}$
fluoranthenequinone 荧蒽醌 $C_{16}H_8O_2$
fluor apatite 氟磷灰石

fluorated 氟化的
fluorating 氟化
fluorating agent 氟化剂
fluoration 氟化作用
fluorbenzene(=fluobenzene) 氟苯
fluorboric acid 氟硼酸 HBF_4
fluorchrome 荧光染料
fluoremetry 荧光测定
fluorenamine 芴胺
fluorene 芴
fluorene alcohol(=fluorenol) 芴醇
fluorene analogue 芴同系物
2,7-fluorenediamine-pyromellitic dianhydride copolymer 2,7-芴二胺-苯均四酸二酐共聚物
fluorenic acid 芴酸 $C_{14}H_{10}O_2$
fluorenol(=fluorine alcohol) 芴醇
fluorenone 芴酮
fluorenone-oxime 芴酮肟 $C_{12}H_8=C=NOH$
fluorenyl 芴基 $C_{13}H_9-$
fluorenylidene 亚芴基 $C_{13}H_8=$
fluorescamine 荧光胺
fluorescein 荧光黄*
fluorescein-bromine method 荧光溴法
fluorescein diacetate 二乙酸荧光素酯 $C_{24}H_{16}O_7$
fluorescein dye 荧光素染料
fluorescein isothiocyanate 异硫氰酸荧光素
fluorescein paper 荧光素试纸
fluorescein sodium 荧光素钠
fluorescein test 荧光素试验〔检定醇酸树脂的存在〕
fluorescence 荧光
fluorescence amine derivation 荧光胺衍生化法
fluorescence analysis 荧光分析
fluorescence chromatogram 荧光色谱(图)
fluorescence chromatography 荧光色谱(法)
fluorescence color 荧光色
fluorescence correlation spectroscopy 荧光相关光谱法
fluorescence cycle 荧光循环
fluorescence depolarization 荧光去极作用；荧光消偏振作用
fluorescence derivatization 荧光衍生
fluorescence detection 荧光检测(法)
fluorescence detector 荧光检测器*
fluorescence efficiency 荧光效率*
fluorescence excitation 荧光激发
fluorescence excitation spectrum 荧光激发光谱
fluorescence extinction method 荧光熄灭法
fluorescence immunoassay 荧光免疫分析法
fluorescence indicator 荧光指示剂

fluorescence intensity 荧光强度*
fluorescence in ultraviolet light 紫外线内的荧光
fluorescence label 荧光标记
fluorescence lifetime 荧光寿命
fluorescence line narrowing 荧光线变窄现象
fluorescence method 荧光法
fluorescence microscope 荧光显微镜
fluorescence microscopy 荧光显微法*
fluorescence microwave double resonance(FMDR) 荧光微波双共振法
fluorescence monitor 荧光监测器
fluorescence peak 荧光峰
fluorescence photon-burst spectroscopy 荧光光子脉冲光谱法
fluorescence polarimetric method 荧光偏振法
fluorescence polarization 荧光偏振
fluorescence polarization immunoassay 荧光偏振免疫分析
fluorescence probe 荧光探针
fluorescence quantum counter 荧光量子计数器
fluorescence quantum efficiency 荧光量子效率
fluorescence quantum yield 荧光量子产额
fluorescence quencher 荧光猝灭剂
fluorescence quenching 荧光猝灭
fluorescence quenching analysis 荧光猝灭分析
fluorescence quenching method 荧光猝灭法*
fluorescence radiant flux 荧光辐射通量
fluorescence radiation 荧光辐射
fluorescence reagent 荧光试剂
fluorescence self-absorption effect 荧光自吸收效应
fluorescence signal 荧光信号
fluorescence solar collector 荧光太阳能聚光器
fluorescence spectrophotometer 荧光分光光度计
fluorescence spectrophotometry 荧光分光光度法
fluorescence spectroscopy 荧光光谱法
fluorescence spectrum 荧光光谱
fluorescence spot-out method 荧光点滴定法
fluorescence standard substance 荧光标准物*
fluorescence state 荧光态
fluorescence thin layer 荧光薄层
fluorescence thin layer plate 荧光薄层板
fluorescence titration 荧光滴定
fluorescence titrimetric method 荧光滴定法
fluorescence yield 荧光产额*
fluorescent 荧光的
fluorescent additive 荧光添加剂
fluorescent agent 荧光(增白)剂
fluorescent antibody inhibition 荧光抗体抑制反应
fluorescent-antibody technique 荧光抗体技术

fluorescent bleacher 荧光增白剂
fluorescent brightener 荧光增白剂
fluorescent brightening agent 荧光增白剂
fluorescent coating 荧光涂料
fluorescent derivative 荧光衍生物
fluorescent dye 荧光染料
fluorescent fault detector 荧光探伤仪
fluorescent film technique 荧光薄膜技术
fluorescent indicator 荧光指示剂*
fluorescent indicator adsorption 荧光指示剂吸附
fluorescent indicator adsorption method 荧光指示剂吸附法〔测定石油产品中的族组成〕
fluorescent indicator adsorption technique 荧光指示剂吸附术
fluorescent ink 荧光油墨
fluorescent lamp 荧光灯；日光灯〔俗〕
fluorescent layer 荧光层
fluorescent light fittings 荧光灯装置
fluorescent lighting 荧光照明
fluorescent light reflector 荧光反光板(罩)；日光灯反光板
fluorescent marker method 荧光标记法
fluorescent microscope 荧光显微镜
fluorescent microscopy 荧光显微镜检验法；荧光显微术
fluorescent oil 荧光油
fluorescent penetrant 荧光渗透剂
fluorescent photometer 荧光光度计
fluorescent pigment 荧光颜料
fluorescent plate technique 荧光板技术
fluorescent polyester fiber 荧光(增白)聚酯纤维
fluorescent probe technique 荧光探针技术
fluorescent pumping energy 荧光抽运能
fluorescent pumping source 荧光抽运源
fluorescent reaction 荧光反应
fluorescent reagent 荧光试剂
fluorescent scanning technique 荧光扫描技术
fluorescent screen 荧光屏
fluorescent screen alignment device 荧光屏微调装置
fluorescent solid 荧光体
fluorescent substance 荧光物质
fluorescent thin layer plate 荧光薄层板*
fluorescent tracer 荧光示踪剂
fluorescent tracer technique 荧光示踪法
fluorescent treponemal antibody absorption test 荧光密螺旋体抗体吸收试验
fluorescent tube 荧光管
fluorescent whitener 荧光增白剂
fluorescent whitening agent 荧光增白剂
fluorescent X-ray X射线荧光

fluorescent X-ray spectroscopy　荧光 X 射线光谱学
fluorescent yellow　荧光黄
fluorescent yield　荧光产额
fluorescer　荧光增白剂
fluorescope(=cryptoscope)　荧光镜
fluorescopy　①荧光学②荧光检查
fluorexon(=calcein)　钙黄绿素
fluorexone　荧光素配位剂；荧光素络合剂
fluorhydric acid　氢氟酸　HF
fluorhydrocarbon　氟代烃
fluoric　①氟的②氟代的
fluoric acid　氟酸〔常误用以指氢氟酸〕
fluoric ether　①氟代酸的酯②氟代烷烃　RF③〔专指〕氟代乙烷　C_2H_5F
fluoridation　氟化反应
fluoride　氟化物
fluoride amine　氟化胺
fluoride extraction system　氟化物萃取体系
fluoride fusion　氟化物熔融
fluoride ion selective electrode　氟离子选择电极
fluoride slagging　氟化造渣
fluoride waste　含氟废水
fluorigenic labeling technique　荧光生成标记技术；致荧光标记技术
fluorigenic reaction　荧光发生反应
fluorimeter(=fluorometer)　①氟量计②荧光计
fluorimetric detector　荧光检测器
fluorimetric determination　荧光测定
fluorimetric labelling technique　荧光计量法标记技术
fluorimetry　荧光测定法
fluorin　荧菌素
fluorinate　氟化
fluorinated　氟化的
fluorinated acrylic ester　氟化丙烯酸酯
fluorinated alkanol　氟化链烷醇
fluorinated alkyl compound　氟化烷基化合物
fluorinated amidoamine oxide　氟化酰氨基氧化胺
fluorinated carboxylic acid　氟代羧酸
fluorinated compound　氟化物〔合成润滑脂〕
fluorinated elastomer　含氟弹性体
fluorinated emulsifying agent　氟化乳化剂
fluorinated ether　氟化醚
fluorinated ethylene propylene　(聚)氟化乙烯丙烯；聚全氟乙烯丙烯（即四氟乙烯与六氟丙烯的共聚物树脂）
fluorinated hydrocarbon propellant　氟化烃推进剂
fluorinated hydrophobe　氟化疏水基
fluorinated paraffins　氟化链烷烃
fluorinated polyester　氟化聚酯

fluorinated polymer(=fluoropolymer)　氟化高聚物
fluorinated silicone rubber　氟硅橡胶
fluorinated siloxane copolymer　氟化硅氧烷共聚物
fluorinated smectite　氟化绿土
fluorinated tenside　含氟表面活性剂
fluorinated triazine rubber　三嗪氟橡胶
fluorinating agent　氟化剂
fluorination　氟化*
fluorination ash　氟化灰
fluorination process　氟化过程；氟化作业
fluorinator　氟化器
fluorine　氟〔9 号元素，化学符号 F〕
fluorine-contained polymerisate fibre　含氟聚合物纤维
fluorine containing lubricant　含氟润滑剂
fluorine-containing polymer support　含氟聚合物载体
fluorine (-containing) rubber　氟橡胶
fluorine crown　氟铬黄
fluorine cyanide　氟化氰　FCN
fluorine emission　含氟发射物
fluorine fluxing agent　氟熔剂〔$NH_4BF_4 \cdot NH_4F$ 和 HF 的混合物〕
fluorine hydride　氟化氢　HF
fluorine iodine　氟化碘〔总名，具体物名应为几氟化碘〕
fluorine ion-selective electrode　氟离子选择电极
fluorine pollution　氟污染
fluorinion(=fluorion)　氟离子
fluoriodate　氟碘酸盐　$R \cdot IO_2F_2$
fluor iodine　氟化碘〔总名，具体物名应为几氟化碘〕
fluorion　氟离子
fluorite(=fluorspar)　萤石；氟石
fluorizated　氟化的
fluorizating　氟化
fluorizating agent　氟化剂
fluorization　氟化作用
fluoro-　氟代；氟(基)
fluoroacetate ions　氟乙酸根离子
fluoroacetic acid　氟乙酸　FCH_2CO_2H
fluoro-acetic chloride　氟代乙酰氯　CH_2FCOCl
fluoro-acetic fluoride　氟代乙酰氟　CH_2FCOF
fluoro-acetic halide　氟代乙酰卤　CH_2FCOX
fluoro-acid　氟代酸
fluoro-acid amide　氟代酸酰胺　$RFCONH_2$
fluoroacrylate　氟化丙烯酸酯(类)
fluoro-aliphatic compound　氟代脂肪族化合物
fluoroalkoxysilane　氟烷氧基硅烷
fluoroalkylpolysiloxane　氟烷基聚硅氧烷
fluoroalkylsiloxane polymer　氟代烷基硅氧烷聚合物
fluoroamino acid　氟代氨基酸　$NH_2RFCOOH$

fluoroaniline 氟苯胺 $C_6H_4(F)NH_2$
fluoroanisole 氟茴香醚；氟苯甲醚 $FC_6H_4OCH_3$
fluorobenzamide 氟苯甲酰胺 $C_6H_4FCONH_2$
fluorobenzene 氟(代)苯 C_6H_5F
fluorobenzoic acid 氟苯甲酸 $FC_6H_4CO_2H$
fluoroborate 氟硼酸(盐)
fluoroborate ion selective electrode 氟硼酸根离子选择电极
fluoroboric acid 氟硼酸 HBF_4
fluorobromobenzene 对氟溴苯 FC_6H_4Br
2-fluoro-1,3-butadiene(=fluoroprene) 氟丁二烯
fluorobutane 丁基氟；1-氟代丁烷 C_4H_9F
fluorocarbohydrate 氟代碳水化合物
fluorocarbon 碳氟化合物
fluorocarbon additive 碳氟(化合物)添加剂
fluorocarbon agent 碳氟剂
fluorocarbon coating 氟烃树脂涂料
fluorocarbon dye dispersion 碳氟(化合物)染料分散体
fluorocarbonnitrogen compound 含氮的碳氟化合物
fluorocarbon oil 氟代烃油；氟油
fluorocarbon polymer 氟碳聚合物
fluorocarbon radical 氟碳基
fluorocarbon resin 氟碳树脂*
fluorocarbon rubber seal 氟碳橡胶密封
fluorocarbon silane 碳氟(基)硅烷
fluorocarbon silicone rubber 氟硅橡胶
fluorocarbon type surface active agent 氟烃系表面活性剂
fluorocarbon wax 碳氟蜡
fluorocarboxylic acid 氟代羧酸
fluorochemical extender 氟化物增量剂
fluorochemical finish 氟化物整理
fluorochlorobenzene 氟氯苯 FC_6H_4Cl
fluorochlorobromomethane 氟氯溴甲烷 $CHClBrF$
fluorochlorohydrocarbon 氟氯烃
fluorochloromethane 氟氯甲烷 CH_2ClF
fluorochloroparaffins 氟氯链烷烃
fluorochrome 荧光染料；荧色物
fluorocitric acid 氟代柠檬酸
fluorocomplex 含氟络合物
fluorocurine 荧光箭毒素
fluorocurinine 荧光箭毒宁
fluoro-cyanogen 氟化氰 FCN
fluorocyclopropylsilane 氟代环丙基硅烷
fluorodensitometer 荧光光密度测定的
fluorodensitometry 荧光光密度分析法
1-fluoro-1,2-dibromoethane 1-氟-1,2-二溴乙烷 $FCHBrCH_2Br$
fluorodinitrobenzene 氟二硝基苯
fluoroelastomer 氟橡胶；含氟弹性体
fluoroester 氟代酯
fluoroethane(=ethyl fluoride) 乙基氟；氟代乙烷 C_2H_5F
fluoroether 氟代醚
fluoroether rubber 氟醚橡胶
fluoroethylene polymer 氟(化)乙烯聚合物
fluoroethylene resin 氟树脂
fluoroethyl ester 氟代乙酯 $RCOOC_2H_4F$
fluorofiber 含氟纤维
fluoroform 三氟甲烷；氟仿 CHF_3
fluoroformol 氟仿液〔含28%氟仿的溶液〕
fluorogen 荧光团
fluorogram 荧光谱图
fluorography 荧光自显影
fluoro gum 氟橡胶
fluoro gum I 氟橡胶 I
fluoro gum II 氟橡胶 II
fluoro gum III 氟橡胶 III
fluoro hectorite 氟水辉石；含氟锂蒙脱石
fluoroimmunoassay 荧光免疫分析
fluoroleum 荧光油〔使用呈现颜色的添加剂的商品名称〕
fluorolube 氟碳润滑油
fluorolubricant 氟化碳润滑油
fluorometer(=fluorimeter) ①氟量计②荧光计
fluoromethane 甲基氟；氟代甲烷 CH_3F
fluoromethylation 氟甲基化
fluorometric analysis 荧光分析
fluorometric bituminological analysis 荧光沥青分析
fluorometric detector 荧光检测器
fluorometric determination 荧光测定法；荧光分析法
fluorometric titration 荧光滴定
fluorometry(=fluorimetry) 荧光分析法
fluoronaphthalene 氟萘 $C_{10}H_7F$
fluorone 荧光酮；3-异呫吨酮 $C_{13}H_8O_2$
fluoroneptunate 氟镎酸盐
fluoronitrobenzene 氟硝基苯 $NO_2C_6H_4F$
fluoronitrobenzoic acid 氟硝基苯甲酸 $C_6H_3F(NO_2)COOH$
1-fluorooctane 1-氟辛烷
fluoroolefins 氟代烯烃(类)
fluoroorganic substituent 有机氟取代基
fluoropentabromoethane (一)氟五溴乙烷 Br_3CCBr_2F
fluoropentachloroethane (一)氟五氯乙烷 Cl_3CCCl_2F
fluorophenol 氟苯酚 FC_6H_4OH
fluorophlogopite 氟金云母
fluorophore 荧光团

fluorophosphoric acid 氟磷酸
fluorophotometer 荧光光度计
fluorophotometric analysis 荧光光度分析
fluorophotometric titration 荧光光度滴定
fluoroplastic film 含氟树脂薄膜
fluoroplastics 氟塑料
fluoroplutonate 氟钚酸盐
fluoropolyene 氟多烯
fluoropolymer 氟聚合物
fluoropolymer block 含氟聚合物嵌段
fluoropolymer grease 含氟聚合物的润滑脂
fluoropolysiloxane based grease 氟代聚硅氧烷基酯
fluoroprene(=2-fluoro-1,3-butadiene) 氟丁二烯
fluoro-propane 氟(代)丙烷 C$_3$H$_7$F
fluoroprotactinate 氟镤酸盐
fluororesin 氟树脂
fluoro rubber 氟橡胶
fluorosilicate 氟硅酸盐
fluorosilicic acid 氟硅酸 H$_2$[SiF$_6$]
fluoro silicone rubber 氟硅橡胶
fluorosilicon oil 氟硅油
fluorosis 氟中毒
fluorospectrophotometer 荧光分光光度计
fluorosulfuric acid 氟磺酸 HSO$_3$F
2-fluoro-1,1,1,2-tetrabromoethane 2-氟-1,1,1,2-四溴乙烷 Br$_3$CCHBrF
fluorotoluene 氟代甲苯 CH$_3$C$_6$H$_4$F
1-fluoro-1,1,2-tribromoethane 1-氟-1,1,2-三溴乙烷 BrCH$_2$CFBr$_2$
fluoro trichloroethylene (一)氟三氯乙烯 FClC=CCl$_2$
fluoro trichloromethane (一)氟三氯甲烷 Cl$_3$CF
fluorouracil 氟尿嘧啶〔药〕
fluorous 氟的
Fluorox process 弗鲁罗克斯过程〔从铀浓缩溶液和固体制备 UF$_4$ 的过程〕
fluorozirconate (六)氟锆酸盐
fluorspar(=fluorite) 萤石；氟石
fluorubin 荧红环
fluosilicate 氟硅酸盐 M$_2$[SiF$_6$]
fluosilicate lead refining process 氟硅酸盐炼铅法
fluosilicic acid 氟硅酸 H$_2$[SiF$_6$]
fluosolid(=fluidized solid) 流动化物质〔沸腾床用〕
fluostannate 氟锡酸盐 M$_2$[SnF$_6$]
fluostannic acid 氟锡酸 H$_2$[SnF$_6$]
fluostannous acid 氟亚锡酸 H$_2$[SnF$_4$]
fluosulfonic acid 氟磺酸 HSO$_3$F
fluotantalate 氟钽酸盐 M$_2$[TaF$_7$]
fluotitanic acid (氢)氟钛酸 H$_2$TiF$_6$

fluoxetine 氟西汀〔药〕
flupentixol 氟哌噻吨〔药〕
fluphenazine 氟奋乃静〔药〕
flurazepam 氟西泮〔药〕
Flurex process 弗卢雷克斯过程〔Excer 法的另一方案，从硝酸铀酰制备四氟化铀复盐的方法。用离子交换膜电解槽〕
fluroescence 荧光
fluroplast-4 氟塑料-4；聚四氟乙烯
fluroresin 氟树脂
flush ①冲洗；冲水②平(齐)的③变红
flush air 吹洗用空气
flush away 洗涤；洗去
flush-back filter 回冲式过滤器
flush bead ①平的串珠线脚②平的焊缝
flush connection 溢流接口
flush distillation 一次蒸发
flushed blue pigment 挤水蓝颜料
flushed color paste 挤水色浆；挤水颜料浆
flushed colors 〔复〕底色
flushed paste 挤水色浆
flusher 净化器；冲洗器
flush filter plate 平槽压滤板
flushing ①冲洗②(沥青)泛油
flushing agent 挤水剂
flushing line 冲洗管线
flushing oil 洗涤油；洗液
flushmounted pressure transducer 同平面装配压力传感器
flush-off type remover 冲洗型脱漆剂
flush out 清洗；吹洗
flushout valve 冲出阀
flush parting line 齐面分模线
flush plate 平槽滤板
flush production 初期产量；最盛期产量
flush trimmer 剔除器
flush welding 闪光焊
flutamide 氟他胺〔药〕
fluted (带有)槽纹的
fluted cylinder 槽纹辊
fluted funnel 棱沟漏斗
fluted roll 槽纹辊
fluted roller(=fluted roll mill) 槽纹辊
fluted roll mill(=fluted roller) 槽纹辊
fluticasone 氟替卡松〔药〕
fluticasone and salmeterol 氟替卡松沙美特罗〔药〕
fluting 开槽；切槽
flutter 颤动；振动

flutter valve　翼形阀
fluvastatin　氟伐他汀〔药〕
fluvomycin(=efsiomycin)　河流霉素
flux　熔剂*
flux-calcined　热碱处理的
flux density　通量密度
fluxed asphalt　加过稀释剂的沥青
flux gradient　通量梯度
fluxibility　①熔性②熔度
fluxing　稀释；冲淡
fluxing agent　融合剂
fluxing oil　沥青稀释油
fluxing point　①熔融点②塑化点
fluxing temperature　熔化温度
fluxion　①流②熔③流动液体
fluxional　微分的；不断变化的
fluxional molecule　立体易变分子
fluxional structure　循变结构*
fluxion structure　流纹构造
flux lime　石灰石
flux method　助熔剂法
flux monitor　通量监测体
flux oil　①沥青稀释油②天然沥青
flux-reaction technique　助熔剂反应法
flux stabilizer　磁通稳定器
flux stone　助熔石
flux temperature(=flow temperature)　流动温度
fly ash　飘尘
flyback　①回(扫)描；(扫描)回程②倒转③反馈
fly bar　辗齿；刀片；飞刀
flyer twisting ply machine　翼锭捻线机
flying knife　飞刀；回转刀
flying shear cutter　定长切断机
flying sponge　飞绵
flying spot scanner　飞点扫描器
fly powder　喷蝇粉〔含三氧化二砷〕
fly spray　喷蝇油；灭蝇喷射油
fly spray test　喷蝇试验；杀虫液体和油的杀虫特性试验
fly wheel　飞轮
FNH powder　FNH 炸药；无光无湿炸药
foalskins〔复〕　小牛皮
foam　泡沫*
foamability　发泡性；发泡能力
foamable composition　泡沫剂
foamable polystyrene　发泡性聚苯乙烯
foam adhesive　泡沫黏合剂
foam agent　起泡剂
foam analysis　起泡分析

foam application　泡沫的应用〔石油储器内〕
foam article　泡沫制品
foam asphalt　泡沫沥青
foam-back　泡沫背衬
foam beater　发泡剂；起泡剂
foam blanket　泡沫层
foam bonding　泡沫黏合法
foam booster　泡沫促进剂
foam boosting agent　发泡助剂
foam breaker　破泡剂*
foam casting　泡沫塑料铸塑
foam catcher　泡沫捕集器
foam chamber　泡沫发生室
foam-coated fabric　泡沫胶布
foam collapse　泡沫崩溃；泡沫结构破坏
foam column　泡沫塔
foam connection　软管与(桶内)撒沫器的联结器
foam control　气泡控制；泡沫控制
foam control agent　控泡剂
foam crown　泡沫塑料拱面
foam curing　泡沫交联
foam degumming　泡沫脱胶
foam density　泡沫密度
foam depressant　抑泡剂
foam drainage　泡沫排放
foamed adhesive　泡沫胶黏剂
foamed carbonaceous char　泡沫状碳质炭
foamed concrete　泡沫混凝土
foamed condition　(油在齿轮箱内)起泡状态
foamed impregnation　泡沫浸轧法〔黏合法制非织造布工艺〕
foam(ed)-in-place process　现场发泡法
foamed latex　泡沫胶乳
foamed plastics　泡沫塑料；多孔塑料
foamed polyethylene　泡沫聚乙烯
foamed polymer　泡沫聚合物
foamed polystyrene　泡沫聚苯乙烯
foamed polyurethane coating　聚氨酯泡沫涂料
foamed rubber　泡沫橡皮；海绵橡胶
foamed sandwich structure　泡沫夹层结构
foamed thermoplastic resin　泡沫热塑树脂
foam end point　泡沫终点
foamer　发泡剂；起泡剂
foamer inhibitor　泡沫抑制剂
foam fabric　泡沫胶布
foam fabrication　泡沫塑料再加工；泡沫塑料裁切
foam fermentation　起泡发酵
foam floating roof　泡沫浮层

foam flotation 泡沫浮选；泡沫分离
foam flow 泡沫流
foam formation 起泡；发泡
foam fractionation 泡沫分级分离；泡沫精馏
foam generating device 起泡装置〔泡沫测定仪的部件〕
foam generator 泡沫发生器
foamglass 泡沫玻璃
foam glue 发泡黏合剂
foam gluing 发泡黏合
foam height 泡沫高度*
foam hydrant 灭火器泡沫的给水栓
foamicide 破泡剂；削泡剂
foam improver 增泡剂
foaminess 发泡性
foaming 发泡；发生泡沫；起泡
foaming ability 发泡能力；发泡性
foaming adhesive 发泡胶黏剂
foaming agent 起泡剂*
foaming analysis 起泡分析
foaming effect 发泡作用；发泡效果
foaming oil 容易发泡的油
foaming power 发泡能力
foaming process ①泡沫法②发泡过程
foaming regulator 泡沫调节剂
foaming space 泡沫空间
foaming stability test 泡沫稳定性试验
foaming structure 泡沫结构
foaming substance 发泡性物质；起泡物
foaming system 泡沫系统
foaming test apparatus 起泡试验设备
foam inhibiting agent 消泡剂；阻泡剂
foam inhibition 泡沫抑制
foam inhibitor 抑泡剂
foam-in-mould 模内发泡
foam-in-place compound 现场发泡混合料
foam-in-situ 就地发泡
foamite 灭火药沫；泡沫灭火剂
foamite extinguishing method 泡沫灭火法
foam killer 消泡剂；消沫剂；破泡剂
foam-killing agent 消泡剂；消沫剂；破泡剂
foam killing oil 消泡油剂
foam-latex 泡沫胶
foam life 泡沫寿命
foam liquid 起泡液
foam maker 发泡器
foam mat drying 泡沫干燥
foam meter 泡沫仪
foam mixing chamber 起泡组分混合室；泡沫发生室

foam modifier 泡沫调节剂
foam molding 泡沫塑料成型
foam number 泡沫值
foam oil 泡沫油；制纸用灯油
foam overblow 发泡过度；过膨胀
foam plastic 泡沫塑料
foam powder 发泡粉
foam preventer 防泡剂
foam profile 泡沫分布型〔指泡高-温度曲线或泡高-洗涤时间曲线〕
foam promoter 泡沫促进剂
foam-promoting builder 泡沫促进剂
foam reducing composition 抗泡沫剂
foam rise 发泡
foam rubber 泡沫胶
foam-rubber backing 泡沫(塑料)-橡胶底衬
foam-rubber hardness tester 海绵橡胶硬度计
foam rubber sealing 泡沫橡胶密封
foams 泡沫体
foam separation 泡沫分离
foam solution 泡沫塑料溶液
foam sponge 多孔海绵
foam sponge indentation tester 泡沫胶耐压试验机
foam sponge mold 泡沫胶硫化模
foam stability 泡沫稳定性
foam stabilization 泡沫稳定(作用)
foam stabilizer 稳泡剂*
foam-stabilizing action 泡沫稳定作用
foam suppressing agent 抑泡剂
foam suppressor 消泡剂；抑泡剂
foam time 消泡时间
foam trap 泡沫收集器
foam value 泡沫值
foam volume 泡沫体积；泡沫量
foamy (多)泡沫的；泡沫似的
foamy structure 泡沫结构
foamy virus 泡沫病毒
focal 焦点的
focal distance 焦距
focalization 焦距调整；聚焦；对光
focal length 焦距
focal plane 焦平面
focal plane array 焦平面阵列
focal plane detector 焦(平)面检测器
focal point 焦点
focus 焦点
focused electron diffraction 会聚束电子衍射
focused open-vessel microwave digestion 聚焦敞口微波

消解
focusing 聚焦
focusing chromatography 聚焦色谱(法)
focusing electrode 聚焦电极
focusing equation of concave grating 凹面光栅的聚焦方程
focusing lens 聚焦(透)镜
focusing of ion beam 离子束聚焦
focus lens 聚光镜
fodder sweetener 饲料用甜味剂
Foehr and Fenske method for structural group analysis 费约尔和芬斯基结构族分析法
foeniculin 茴香苷
fog ①雾②生雾③被翳④雾面
fog cabinet 雾室
fog chamber (盐)雾箱
fog density 雾翳黑度
fog-foam unit 雾状泡沫设备
fogged metal 晦面金属
foggers 润湿器〔使灰尘从气体中沉降下来〕
fogging 雾化
fogging oil 雾化油
fogging spraying 喷雾法
fog marking 雾痕
fog nozzle 喷雾嘴
fog resistance (漆膜)抗结雾性
foil 箔；薄片
foil coating 金(属)箔涂料
foil coating lacquer 金箔罩光漆；金箔漆
foil colorimeter 金属箔比色计
foil decorating ①彩箔装潢②彩纸装潢
foil finishing 彩箔装璜
foil free 不粘箔干
foil gauge 应变片
fold 折叠*
-fold〔词尾〕 倍；重
foldability 可折叠性
fold back 返折〔注塑制品缺陷〕
fold back effect 折叠(效)应
foldback peak 折叠峰
fold domain 折叠微区*
folded buff 折叠抛光轮
folded chain 折叠链
folded-chain configuration 折叠链构型
folded-chain crystal 折叠链晶体
folded-chain crystallite 折叠链微晶
folded-chain fringed micellar grain model 折叠链缨状胶束粒子模型

folded chain lamellae 折叠链晶片
folded chain model 折叠链模型
folded-chain structure 折叠链结构
folded conformation 折叠(链)构象
folded container 可叠合容器
folded crystal 折叠晶体
folded filter 折纸漏斗〔漏斗上放着折过的滤纸〕
folded length of molecular chain 分子链的折叠长度
folded plate structure 折板结构
folded structure 折叠结构
folder 织物折断强度试验器
fold filling 软管装油
fold filter 折叠滤纸
folding 折叠
folding back 捣胶；翻胶
folding boat 橡皮船
folding bobbin 并纱筒子；并丝筒子；并线筒子
folding box board 摺合纸板
folding edge 弯折边缘
folding endurance ①耐折性②耐折度
folding length 折叠链长度
folding machine 折叠机
folding period 折叠周期
folding plane 折叠面
folding press 折叠机
folding resistance 耐折叠性
folding strength 耐折强度
folding tester 耐折叠试验
foldover of peak 谱峰折叠
foldovers 起折〔膜卷缺陷〕
fold plane 折叠面
fold surface 折叠表面
foliaceous(=foliate) ①层状的②叶状的
foliate(=foliaceous) ①层状的②叶状的
foliated coal 层状褐煤
foliation ①成层②成片
folicanthine 叶坎生
folimycin 多叶霉素
Folin's method 福林法
Folin's reaction 福林反应
Folin's reagent 福林试剂
Folin urea apparatus 福林定脲器
Folin urea bulb 福林定脲球管
folione(=methyl 2-octynoate) 辛炔酸甲酯
follow ①追随②生效
follower 随动件；从动部〔机工〕
follower rest 随动中心架
following reaction 后行反应

follow-up pressure 随动加压
follow-up pulley 从动轮；随动轮
fomecin 层孔菌素
fondant 软糖料〔糖〕
fondu ①(颜色)全混合的②高铝水泥
Foner magnetometer 福纳磁强计
font (煤油灯的)油座；油壶
fontactoscope （矿）泉放射性检测计
food 食品；食物
food additive analysis 食品添加剂分析
food additives 食品添加剂
food allergy 食物过敏
food analysis 食品分析
Food and Drug Administration （美国)食品与药物管理局
food antiseptics 食品防腐剂
food chain 食物链
food color 食品着色剂；食品色素
food dye 食用染料
food filming agent 食品被膜剂
food flavor(ing) ①食品香味②食品香精③食品香料 ④食品调味剂
food flavour 食用香精
food manufacturing agent 食品制造用添加剂
food poison 食物毒
food pollution 食物污染
food preservative analysis 食品防腐剂分析
food processing aid 食品加工助剂
food soil 食物类污垢
food starch 食用淀粉
foodstuff rubber 食品工业用橡胶
food wrapper 食品包装纸
foolproof 安全装置
fool's gold(=pyrite) 黄铁矿
foot ①脚②英尺
football-helmet leather 橄榄球头盔革
football leather 足球革
foot bellow(=foot blower) 脚踏风箱
foot blower(=foot bellow) 脚踏风箱
foot board 脚踏板
foot-candle 英尺-烛光；尺烛光〔照度单位〕
footing 脚子〔糖浆下脚的结晶〕
foot lambert 英尺-朗伯〔亮度单位〕
foot measure 用英国长度单位测量
foot-operated air pump 脚踏打气筒
foot-powder 足粉
footprinting 足印法
foot pump 脚踏泵
foots ①脚子；下脚②渣滓③油脚

foots oil 脚子油；油脚子
footstep bearing 立轴承；托杯轴承
foot switch 脚踏开关
foot-ton 英尺吨〔=2240 英尺-磅〕
foot valve 脚阀
footwear reclaim 胶鞋再生橡胶
foraminous 有孔的；多孔的
foraminous die plate 多孔模板
forbidden 禁阻
forbidden band 禁带
forbidden explosives 禁运爆炸物
forbidden line 禁(忌)线
forbidden radiative transition 禁阻辐射跃迁
forbidden region 禁区
forbidden transition 禁阻跃迁
f orbitals f 轨道
force ①力②强制；强迫
force and splash lubricating 强制及击溅润滑
force-area curves 力-表面积曲线
force-at-yield 屈服力
force balance 力平衡
force balance-type manometer 力平衡式压力计
force close 气闭式
force constant 力常数
force constant f 力常数 f
forced air circulation 强制空气循环
forced circulation 强制循环
forced circulation evaporation 强制循环蒸发
forced circulation evaporator 强制循环蒸发器
forced circulation type evaporator 强制循环蒸发器
forced convection 强制对流
forced convection oven 强制对流炉(烘箱)
forced crystallization 强制结晶
forced dehydration 强制脱水
forced draft 强制通风
forced draft cooling 强制通风冷却
forced draft cooling tower 强制通风凉水塔
forced draft fan 强制通风风扇
forced draught 强制通风
forced draught (water) cooling tower 强制通风(水)凉水塔
forced drying 强制干燥
forced emulsification 强制乳化
forced feed 强制进料；加压进料
forced feed lubrication 强制供油润滑(作用)；加压供油润滑(作用)
forced feed lubricator 强制供油润滑器
forced feed system 强制进料系统
forced filtration 强制过滤

forced-flow 强制流动
forced flow leaching 强流沥滤法
forced ignition 强制点火
forced lubrication 强制润滑
forced oscillation 强制振荡；强迫振荡
forced rubbery deformation 受迫橡胶态形变
forced ventilation 强制通风
forced vibration 强制振动；强迫振动
forced vibration method 强制振动法
forced vibration torsion pendulum 强制振动扭摆
forced vibrator 强制(式)振动器
forced vortex 强制旋涡
forced vortex blade 强制涡流叶片
forced vortex zone 强制旋涡区
force fit 压配合
forceful arc 强电弧
force lift pump 增压泵
force modulation microscopy 力调制显微镜
force of adhesion 黏附力
force of cohesion 内聚力
force of flocculation 絮凝力
force of friction 摩擦力
force of inertia 惯性力
force of periphery 切线方向的力
force open 气开式
force oscillation 强制振荡；强迫振动
force piston 模塞；阳模
force plug 模塞；阳模
force plunger 模塞；阳模
forceps 镊子
force pump 压力泵
forcer ①压出机②挤出机
force-to-draw 拉伸力
force-to-stretch 拉伸力
forcing 强制；加压
forcing hose 加压软管
forcing machine ①压出机②挤出机
forcing pipe 加压管
forcing pump 强制泵；压力泵
forcing spindle 压力轴
forcite 福斯炸药；福煞特
Ford cup 福特(黏度)杯
Ford viscosity cup 福特黏度杯
fore and aft traction 前后方向牵引力
forechamber 前室
forecooling 预冷却
fore-distillation 预蒸馏
fore flow 前驱流(动)

foreflow section 前驱流区间；辅助流区间
forehearth (玻璃窑炉的)喂液器
foreign atom 局外原子；杂质原子；外来原子
foreign body ①混合物；掺合物②杂质
foreigness 外来物质
foreign impurity 外来杂质；夹杂物；异物
foreign labeling 外标记
foreign material 外来杂质；夹杂物；异物
foreign matter 外来杂质；夹杂物；异物
foreign matter content 杂质含量
foreign object 外物
foreign odor 不适的气味；异臭
foreign pigment 有色杂质
foreign solids 外来的机械杂质
foreign substance 外来杂质；夹杂物；异物
foreign taste 异样口味；异味
forel 绵羊皮纸
foreline trap 前级真空冷阱
forensic activation analysis(FAA) 法医活化分析
forensic analysis 法医检定法
fore-prism 前置棱镜
forepump 前级真空泵
fore-run 初馏物
forerunning 初馏
foreshot 初馏物
fore-spreading 头道延展
forestine 佛氏乌头亭
forevacuum 前级真空
forevacuum gauge 前级真空规
forevacuum pressure 前级真空压强
forevacuum pump 前级真空泵
forevacuum trap 前级真空冷阱
foreward welding 左焊法
forewarmer 预热器
forgeability 可锻性
forgeable 可锻的；延性的
forged 锻造的
forged flange 锻接盘；锻法蓝
forged sheffield plate 锻造包银铜板
forged steel 锻钢
forged steel roller 锻钢滚筒
forge hearth 锻铁炉
forgenin 甲酸四甲铵 $HCOON^+(CH_3)_4$
forge pig 锻铁
forge scale 锻鳞
forgetting factor analysis 遗忘因子分析法
forgetting factor method 遗忘因子法
forge welding 锻焊

forging 锻造
forging iron 锻铁
forging steel 锻钢
fork 叉(子)
fork chain(=forked chain) 支链
forked group(=fork group) 支基
fork group(=forked group) 支基
fork lift 叉式起重车
forklift truck 铲车
fork-shaped agitator 叉形搅拌器
fork tongs 叉式钳
fork truck 叉车
form ①形(式)②成形③形成；生成；造成
cis-form 顺式
formability 易成型性
formacyl 甲酰基
formal ①克式量的②式符③缩甲醛④二甲醇缩甲醛
formal bond 形式键〔在式中理论上可写出而实际上太远的键〕
formal charge 形式电荷
formal concentration 克式量浓度
formaldehyde 甲醛 HCHO
formaldehyde-acetamide 甲醛合乙酰胺
formaldehyde ammonia ethyl chloride condensate 甲醛氨乙基氯缩合物
formaldehyde determination 甲醛测定；甲醛鉴定
formaldehyde dialkyl acetal(=methoxy dialkyl acetal) 甲醛型二烷基醇缩甲醛
formaldehyde dimethyl acetal 甲醛缩二甲醇 $CH_2(OCH_3)_2$
formaldehyde oxime 甲(醛)肟 $H_2C=NOH$
formaldehyde sodium bisulfite 甲醛化亚硫酸氢钠 CH_2OHSO_3Na
formaldehyde sodium hydrosulfite 甲醛化连二亚硫酸钠 $CH_2ONa_2S_2O_4$
formaldehyde sodium sulfoxylate 甲醛化次硫酸钠 CH_2OHSO_2Na
formaldehyde tanning 甲醛鞣料
formaldehyde p-toluidine condensate 甲醛对甲苯胺缩合物〔促进剂〕
formaldehyde treated waste(FTW) 甲醛处理过的(放射性)废物
formaldehyde treating of gasoline 汽油的甲醛处理〔去掉形成树脂的组分〕
formaldehyde urea copolymer 脲醛共聚物
formaldoxime 甲醛肟 $H_2C=NOH$
formal glycerine 甲醛缩甘油 $H_2C=C_3H_6O_3$
formal group 缩甲醛基

formalin 甲醛水；甲醛水溶液；福尔马林
formality 克式浓度
formalization 缩甲醛化
formalization treatment 缩甲醛化处理
formalized 缩甲醛化的
formalized PVA fiber 聚乙烯醇缩甲醛纤维；维尼纶
formalizing 缩甲醛化
formaloin 甲醛缩芦荟素
formal potential 表观电位；克式电位；式量电位
formal synthesis 中继合成*
formal titration 羟甲基滴定法
formal toxoid 液体类毒素
formal treatment 形式处理
formal variable 形式变量
formamide 甲酰胺 $HCONH_2$
formamide-formaldehyde resin 甲酰胺-甲醛树脂
formamidine 甲脒 $HN=CHNH_2$
formamidine disulfide 甲脒化二硫 $[(NH=)(NH_2)CS]_2$
formamidinesulphinic acid 甲脒亚磺酸
formamidino(=amidino) 脒基 $H_2NC(=NH)-$
formamido 甲酰氨基 $OHCNH-$
5-formamido imidazole-4-carboxamide ribotide 5-甲酰氨基-4-氨甲酰咪唑核苷酸
formamidoxime 氨基甲肟 $H_2NCH=NOH$
formamine(=hexamethylene-tetramine) 甲醛胺；六亚甲基四胺
formamyl 氨基甲酰基 $-CONH_2$
form-and-spray 预成型喷塑
formanilide N-甲酰苯胺 C_6H_5NHCHO
formate ①甲酸②甲酸盐(酯或根)
formation ①形成②生成
formation constant 形成常数
formation curve 形成曲线
formation function 形成函数
formation heat 形成热；生成热
formation of carbon 积碳的形成
formation of cracking 龟裂
formation of deposits 沉积的形成
formation of gas 气体的形成；气化
formation of two zones 成双区
formation reaction 生成反应
formative substance 成型物质
formazane 甲䐶
formazan reaction 甲䐶反应
formazyl 甲䐶基；偕苯偶氮基；偕苯肼基代甲基
form birefringence 形状双折射
form constancy 形状恒定性；形状持久性
form copying 仿形加工法；靠模加工法

form cutter 成型刀具
form drag 形阻力
formed asphalt 加工(的)沥青
formed plywood 成型胶合板
form effect 形状效应
former ①模型②成型机头
former sleeve 定型套；定径管
formestane 福美坦〔药〕
form factor 形态因子
formhydrazide(=formyl hydrazine) 甲酰肼 $H_2NNHCHO$
formhydroxamic acid 甲异羟肟酸
formic 甲酸的
formic acid 甲酸；蚁酸〔俗〕
formic aldehyde 甲醛
formic anhydride 甲酸酐 $(HCO)_2O$
formic ether ①甲酸甲酯 $HCOOCH_3$ ②甲酸酯
formimido chloride 氯甲亚胺 $CHCl=NH$
formimido ether 亚氨代甲基醚；烃氧基甲亚胺 $CH(OR)=NH$
formimidoyl 亚氨代甲基 $HC(=NH)-$
formimino 亚胺甲基 $-CH=NH$
formiminoether 亚胺甲基醚 $NH=CHOR$
formin 甲酸精；甘油甲酸酯
forming 形成；成形；成型
forming pad 金属片压型用胶板
forming process 成型加工
forming roll 型辊
forming rubber 压型用胶
forming shop 制模厂
forming table 网案；网架
form-ink-roller 涂墨滚
form lacquer (混凝土模板)防黏着涂料
form of chart expression about two-dimensional spectra 二维谱图形表示法
formohydrazide 甲酰肼
form oil 定形油〔用于木材〕
formol(=formaldehyde) ①甲醛②甲醛防腐剂
formolation 甲醛化
formolite 硫酸甲醛
formolite number 硫酸甲醛值〔石油〕
formolite reaction 硫酸甲醛反应〔用于油品测试〕
formolite test 甲醛试验〔测定硫酸甲醛值的方法〕
formol test 甲醛滴定法；甲醛试验
formol titration 甲醛滴定〔测定蛋白质水解产物中的氨基〕
formol titration method 甲醛滴定法
formolysis 甲醛分解作用

formonitrile 甲腈；氢氰酸 HCN
formonitrolic acid 甲硝肟酸 $CH_2O_3N_2$
formo(o)nonetin 7-羟(基)-4-甲氧异黄酮
formoterol 福莫特罗〔药〕
formoxime 甲醛肟
formoxy〔词头〕 醛氧(基)
formoyl 甲酰基 HCO—
form release agent 脱模剂
form tool 样板刀
formula ①式；公式②(化学)式③药方④配方⑤分子式
formula adjustment 配方调整
formula calculation 配方计算
formula change 配方变更
formula ingredients 配合用(原材)料
formula ratio 配方规定(用)量
formular conductivity 式量传导系数
formulary 处方集；配方手册
formula-symbol 结构符号
formulated oil 调配(润滑)油
formulating ①配方研究(设计)②配制
formulating of recipe 配方设计
formulation ①列出公式②公式化
formulation of cleaner 清洗剂(去污剂)配方
formulation parameter 配方参数
formulator 配方设计师
formula unit 分子式单元
formula variation 配方差异
formula weight (化学)式量；分子量
formycin 间型霉素
formyl 甲酰基 HCO—
formyl acetic acid 甲酰乙酸 $HCOCH_2COOH$
formyl acetone 甲酰丙酮 $HCOCH_2COCH_3$
formylate 甲酰化
formylated 甲酰化的
formylating 甲酰化
formylating agent 甲酰(化)剂
formylation 甲酰化*
formylation number 甲酰数
formylation ratio 甲酰比值
formylation reaction(=Duff reaction) 达夫反应；甲酰化作用
formylation value 甲酰值
formyl benzofuran 甲酰基苯并呋喃
formyl chloride 甲酰氯；氯化甲酰 HCOCl
formylcyclododecane 甲酰基环十二烷
formyl fluoride 甲酰氟；氟化甲酰 HCOF
formyl halide 甲酰卤 HCOX
formyl hydrazine(=formhydrazide) 甲酰肼

H₂NNHCHO
formyl-hydroxybenzoic acid 甲酰基羟基苯甲酸
HOC₆H₃(CHO)COOH
formylic acid 甲酸；蚁酸〔俗〕HCOOH
α-formylisoglutamine α-甲酰异谷氨酰胺
formylmerphalan 氮甲〔药〕
formylmethionine 甲酰甲硫氨酸
formyloxy 甲酸基；甲酰氧基 HCOO—
formylphenate 甲醛酚盐
formyl violet 甲酰紫
foromacidin 福罗杀菌素
forsterite 镁橄榄石
Forsterling prism system 费斯特林棱镜系(统)
forsythin 连翘质
forsythiogenol 连翘精醇
fortified phenol(=hindered phenol) 受阻(苯)酚；取代(苯)酚
fortified rosin 强化松香
fortified wine 强化酒
fortifier 增强剂；补强剂
fortifying fiber 增强纤维
fortoin 甲醛缩可土树皮素
forward ①向前；前进②前部的
forward action 向前作用；前向作用
forward bias 正向偏压
forward curved vane 前弯叶片
forward extraction （正向)萃取
forward feed 顺流进料
forward feeding 顺流送料；向前进料
forward flow ①顺流；向前流动②前滑
forward library search 正向谱库检索
forward motion 前向移动
forward pilot 前端指示器
forward process 热芳构化过程
forward reaction 正向反应
forward scattering 向前散射
forward speed 前进速度
foscarnet sodium 膦甲酸钠〔药〕
fosfomycin(=phosphonomycin) 磷霉素
fosforus(=phosphorus) 磷
fosinopril 福辛普利〔药〕
fossil ①化石②化石的
fossil copal 化石砝玻(树)脂；矿物砝玻脂
fossil flax 石棉〔填充剂〕
fossil flour 硅藻土
fossil fuel 化石燃料；矿物燃料
fossillin(=vaselin) 凡士林
fossil meal 化石粉

Foster chart 福斯特色差图
Foster-Wheeler furnace 福斯特-威洛炉
Foster-Wheeler process 福斯特-威洛(蒸气转化)法
fotemustine 福莫司汀〔药〕
fougere perfumes 馥奇香精
fougere scents 馥奇香调
foul 污秽的
fouled catalyst 中毒的催化剂
fouled sulfuric acid 不纯的硫酸；沾污了的硫酸
foul gas 秽臭气；不凝性气体
fouling 挂胶；结皮；结垢
fouling factor 污垢系数
fouling inhibitor 污垢抑制剂
fouling of catalyst 催化剂的污损；催化剂结垢〔包括结焦〕
fouling products 污垢物
fouling rate ①(催化剂)中毒率②污垢率
fouling resistance 污垢热阻
fouling resistance rating 耐污(染)等级
fouling tendency 结垢性；结垢倾向
foul main ①焦油总管②总管
foul odor analysis method 恶臭物质分析法
foul smelling oil 秽臭的(粗)石油产品
foul solution 秽臭溶液；污液；废液
found 铸造
foundation ①基础②地基
foundation bolt 地脚螺钉；地脚螺栓
foundation coke oven 地基炼焦炉
foundation emulsion 粉底(乳)液
foundation fabric 底布；地布
foundery(=foundry) ①铸造车间②铸造
founding 铸造
foundry(=foundery) ①铸造车间②铸造
foundry coke 铸(造)用焦炭
foundry cupola 铸造圆顶炉
foundry facing 铸件涂料；石墨涂料
foundry flask 砂箱
foundry iron 铸(造)用生铁
foundry scrap 铸造废料
fountain solution(=dampening solution) ①湿润液②润版液；润版药水
fouramine(=oxidation bases) 四胺〔染〕；氧化色基
four-ball bearing apparatus(=four-ball machine) 四球试验机〔用于评价润滑油的润滑性能〕
four-ball extreme pressure tester 四球极压试验机
four-ball machine(=four-ball bearing apparatus) 四球试验机〔用于评价润滑油的润滑性能〕
four-ball tester 四球式试验机

four-ball test of lubricants　润滑油的四球机试验
four-ball top　四球角锥〔四球摩擦机〕
four-bond motion　四键运动
four-center photopolymerization　四中心光聚合反应
four center polymerization　四中心聚合
four-center reaction　四中心反应
four-center transition state　四中心过渡状态
four-center (type) reaction　四中心反应
four channel magnetic mass spectrometer　四通道磁式质谱仪
four circle diffractometer　四圆衍射仪
four circle goniometer　四圆测角仪
four circle single crystal diffractometer　四圆单晶衍射仪
four-coordinate complex　四配位(数)复盐；四配价复盐
fourdrinier　长网成形器；长网造纸机
fourdrinier machine　长网(造纸)机
fourdrinier paper machine　长网造纸机
fourdrinier part　长网部
fourdrinier type drying machine　长网干燥机
fourdrinier wire　长网线
four-effect evaporator　四效蒸发器
four-electron ligand　四电子配体
four-element model　四要素模型〔黏弹性力学模型的一种〕
four element theory　四元素(学)说
four-end draw texturizing　四头拉伸变形工艺
four-fold axis　四重轴
four-fold monopole mass spectrometer　四重单极质谱仪
four hearth burner(=four hearth furnace)　四膛炉；四室炉
four hour varnish　四小时快干漆
Fourier analysis-synthesis　傅里叶分析合成
Fourier law　傅里叶定律
Fourier number　傅里叶准数
Fourier series　傅里叶级数
Fourier space　傅里叶空间
Fourier spectrometer　傅里叶谱仪
Fourier synthesis　傅里叶合成
Fourier transform AC voltammetry　傅里叶变换交流伏安法
Fourier transformation electron spin resonance　傅里叶变换电子自旋共振
Fourier transform electron paramagnetic resonance　傅里叶变换顺磁共振；傅里叶变换电子自旋共振
Fourier transform infrared　傅里叶变换红外
Fourier transform infrared detector　傅里叶变换红外检测器
Fourier transform infrared photoacoustic spectrometer　傅里叶变换红外光声光谱仪
Fourier transform infrared photoacoustic spectroscopy　傅里叶变换红外光声光谱
Fourier transform infrared spectrometer(FTIR)　傅里叶变换红外光谱仪
Fourier transform infrared spectroscopy　傅里叶变换红外光谱法
Fourier transform ion cyclotron resonance　傅里叶变换离子回旋共振
Fourier transform ion cyclotron resonance mass spectrometer　傅里叶变换离子回旋共振质谱仪
Fourier transform ion cyclotron resonance mass spectroscopy　傅里叶变换离子回旋共振质谱法
Fourier transform ion cyclotron resonance spectrometry　傅里叶变换离子回旋共振谱法
Fourier transform mass spectrometer(FTMS)　傅里叶变换质谱仪
Fourier transform near infrared spectroscopy　傅里叶变换近红外光谱法
Fourier transform nuclear magnetic resonance(FTNMR)　傅里叶变换核磁共振(法)
Fourier transform Raman spectrometer　傅里叶变换拉曼光谱仪
Fourier transform Raman spectroscopy　傅里叶变换拉曼光谱法
Fourier transform spectroscopy(FTS)　傅里叶变换光谱(法)
Fourier zeugmatography　傅里叶变换投影重建法
four-jaw chuck　四爪卡盘
four-membered ring　四元环
four parameter model　四参数模型；线性黏弹性流变模型
four-pen recorder　四笔记录仪
four-pi beta counter　$4\pi\beta$计数器
four-pin method　四针法
four-point adjustment　四点调节
four-point control method　四点控制法
four probe method　四探针法
fourrine D　四灵〔商〕；毛皮对氨黑
four-ring　四元环
four-roll crusher　四辊碾碎机
four-roll mill　四辊磨
four-stage mass spectrometer　四级质谱计
four stage zinc phosphating systems　四步法磷酸锌处理系统
fourteen-membered ring(=fourteen-ring)　十四元环
fourteen-ring(=fourteen-membered ring)　十四元环
fourth evaporator　第四效蒸发器
four way plug valve　四通旋塞阀
four-way valve　四通阀
four wire connector　四线联接
foward-backward scattering　向前-向后散射

fowlerite 锌锰辉石
fox fat 狐狸脂
foxglove 毛地黄叶
foxy ①赤褐色的；有褐斑的②(颜料晶型变化引起的)色相变化
foxy batch 凡尔登红染槽
foyaite 流霞正长岩
Fraas breaking point 弗拉斯沥青破裂点(或温度)
Fraas tester (测定沥青强度的)弗拉斯试验器
fractal 分形
fraction ①部分②馏分③级分④分数
fractional ①分数的；小数的②分成几份的
fractional analysis 分馏分析
fractional centrifugation 分部离心分离法
fractional charge 分数电荷
fractional coagulation 分级凝固；分凝(作用)
fractional collector 馏分收集器
fractional column ①分馏塔②分馏柱
fractional combustion 分级燃烧
fractional composition 馏分组成
fractional condensation 分凝(作用)；分级冷凝
fractional condenser 分级冷凝器
fractional condensing tube 分凝管
fractional condensing unit 分级冷凝装置
fractional conversion 部分转换
fractional correction 分馏校正
fractional crystallization 分级结晶
fractional cumulative yield 分级累积产额
fractional decomposition 分级分解
fractional dissolution 分级溶解
fractional distilling tube ①分馏柱②分馏塔；部分冷凝器
fractional distillation 分馏(作用)
fractional distillation column ①分馏柱②分馏塔
fractional distillation effect 分馏效应
fractional distillation process 分馏过程
fractional distilling flask 分馏(烧)瓶
fractional distribution 馏分分布；分级分布
fractional error ①分步误差②部分误差
fractional extraction 分馏萃取
fractional factorial design 部分析因设计
fractional filtration 分步过滤
fractional free volume 自由体积分数
fractional hydrolysis 分级水解；分步水解
fractional independent yield 分级独立产额
fractional melting 分(步)熔化〔用于油的脱蜡〕
fractional melting apparatus (分)熔化设备
fractional neutralization 分级中和
fractional oil content 部分石油含量
fractional order reaction 分数级反应；非整数级反应
fractional porosity 部分孔隙率；相对孔隙率
fractional precipitation 分步沉淀*
fractional pressure 分压
fractional release (放射性)释放比率
fractional replication 分数配置
fractional saturation 部分饱和
fractional separation 分级分离
fractional solid volume 固体体积分数
fractional solution 分级溶解
fractional sublimation 分级升华
fractional transfer 转移分数值
fractional void 空隙度〔空隙体积与粒子体积的和〕
fractional volatilization 分级挥发
fractional wax crystallization 石蜡分步结晶
fractional weight 小数砝码
fractional weight loader 小数砝码输载器
fractionate ①分馏②分级
fractionated ①分馏的②分级的
fractionated by coacervate extraction 凝聚析出分级法；凝聚抽提分级法
fractionated oil 分级油
fractionated ratio (同位素)分馏比
fractionating ①分馏②分级
fractionating column 分馏柱
fractionating condenser 分级冷凝器
fractionating device 分馏装置
fractionating diffusion pump 分馏型扩散泵
fractionating distillation 分馏
fractionating efficiency 分馏效率
fractionating flask 分馏(烧)瓶
fractionating plate 分馏(塔)的塔盘
fractionating process 分馏法
fractionating screen analysis 分选；筛分
fractionating system 分馏系统
fractionating tower 精馏塔
fractionating tray 分馏(塔)的塔盘
fractionating tube 分馏管
fractionation 分级*
fractionation by adsorption 吸附分离
fractionation by Brownian diffusion 布朗扩散分级(法)
fractionation by distillation 分馏；蒸馏分离
fractionation by sedimentation velocity 沉降速度分级(法)
fractionation by selective adsorption 选择吸附分级(法)
fractionation column ①分馏塔②分级柱
fractionation correction 分馏校正
fractionation crystallization 分级结晶法
fractionation due to capillary leak 毛细管漏孔的分馏

fractionation efficiency of tower 塔的分馏效率
fractionation precipitation 分级沉淀
fractionation range 分馏范围；分级分离范围
fractionation tower 分馏塔
fraction atomized 原子化度
fractionator 分馏器
fraction collection trap 馏分收集阱
fraction collector 馏分收集器*
fraction defective 次品率
fraction desolvated 去溶度
fraction exchange 同位素交换率
fraction lift 馏程
fraction number 馏分号码
fraction of distillate 馏分
fraction of sample in gas (or liquid) phase 试样在气(液)相中所占分数
fraction of saturation value 饱和值分数
fractograph 断裂面显微镜图像；断口组织的(显微镜)照片
fractorite 分级炸药；福来克赖特
fracture ①破碎②断面③骨折④断裂⑤破坏
fracture initiation test 开裂试验
fracture length 裂断长
fracture mechanics 断裂力学
fracture plane 破裂面；断面
fracture propagation 裂纹蔓延
fracture strength 断裂强度
fracture stress 断裂应力
fracture surface 断面；破裂面
fracture surface energy 断裂表面能
fracture toughness 断裂韧性*
fracture transition temperature 断裂转变温度
fracturing 压裂
fracturing fluid (油井)压裂液
fradicin 弗氏菌素*
fradiomycin 弗氏霉素；新霉素
fragile 脆的；脆性的；易碎的
fragile support 脆性载体
fragile target 脆靶；易碎靶
fragility ①脆性②脆度
fragin 脆假胞菌素
fragment 碎片；片段
fragmental medulla 碎片型毛髓
fragment antigen binding 抗原结合片段
fragmentary system 碎裂体系
fragmentation 碎裂*
fragmentation curve 裂解曲线
fragmentation mapping 碎裂图解

fragmentation pathway 碎裂途径
fragmentation pattern 碎裂方式；碎裂过程图形
fragmentation process 碎裂过程
fragmentation reaction 碎裂反应
fragmentation rule 碎裂规则
fragment ion 碎片离子
fragmentography 碎片谱(法)
fragment peak 碎片峰
fragrance ①芳香②香味；香气③香料
fragrance material 香料
fragrance product 芳香制品
fragrance raw material 香原料
fragrance retention 留香性
fragrant 香的；芳香的
fragrant body 含香物
fragrant oil 香精油
fragrant substance 芳香物；香料
framberry 树莓；悬钩子
frame ①框；架②格架③构架④机身；机架
frame accumulator 框式蓄电池
framed 模框的
frame(d) building 构架建筑
frame dried hide 撑干皮
framed soap 框制皂
frame filter 板框式过滤器
frame filter press 框式压滤机
frame plate 框板；框式压滤板
frame side member 边架
frame spinning 环锭纺
framycetin 弗氏菌丝素
framycin 弗拉霉素；护乐霉素
Francis equation 弗兰西斯公式
Francis formula 弗兰西斯公式
francium 钫[87号元素，化学符号Fr]
Franck-Condon principle 弗兰克-康登原理
franckeite 辉锑锡铅矿
Franck-Hertz experiment 弗兰克-赫兹实验
francolite 细晶磷灰石
frangibility ①脆性；可碎性②脆度
frangipanni 红鸡蛋花
frangula-emodin 大黄素
frangule 泻鼠李皮
frangulin 泻鼠李皮苷
franguloside(=frangulin) 泻鼠李皮苷
Frank-Care cyanamide process 弗兰克-卡尔氰氨(基)化钙制造法
frankincense 乳香
frankincense oil 乳香油

franklinic(=static)　静电的
franklinite　锌铁尖晶体
Frank-Rabinowitch effect　弗兰克-雷平诺维奇效应〔即笼效应〕
Frary metal　弗雷里合金
Frasch process　弗拉施法〔石油产品在氧化铜上用蒸馏法脱硫〕
fraueneis　透明石膏　$CaSO_4 \cdot 2H_2O$
Fraunhofer diffraction　夫琅和费衍射
Fraunhofer line　夫琅和费谱线
fraxetin　白蜡树亭
fraxin　白蜡树苷
fraxinellone　白蜡树酮　$C_{14}H_{16}O_3$
fraxinine　白蜡树宁
fraxinol　白蜡树酚
fraxitannic acid　白蜡树皮单宁酸　$C_{26}H_{22}O_{14}$
fraxitin　白蜡树亭
fray counter　毛羽计数仪
fraying　纱线滑溜；纱线位移；纤维磨散
Frazier permeameter　弗雷泽织物透气性试验仪
freak stocks　非商品性的石油产品；中间产品
free　①单体的②自由的；游离的
free acid　游离酸
free acidity　游离酸度
free affinity　自由亲和力
free air　大气；大气层的空气
free air temperature　大气温度；大气层空气的温度
free air thermometer　大气温度计
free alkali　游离碱
free alkalinity　游离碱度
free alkali test　(石油产品中)游离碱度的测定
free alongside ship　船侧就岸交货
free amine group　游离氨基
free amino end group　链端自由氨基
free amino group　自由氨基
free ammonia　游离氨
free and total chlorine　游离氯和总氯
free anhydride　游离酐〔指硫酸酐〕
free area　有效截面
free asphalt　游离沥青
free atom　单体原子；自由原子
free beating　游离打浆
free bed　自由床
freeboard of fluidized bed　稀相区
free boundary electrophoresis　自由界面电泳
free burning coal　单烧煤〔不结焦烟煤〕
free carbon　单体碳；游离碳
free carboxyl group　自由羧基

free cellulose xanthic acid　游离纤维素黄(原)酸
free cementite　游离渗碳体；游离碳化铁体
free-chalking pigment　易粉化型颜料
free charge　自由电荷
free chlorine　游离氯
free column　空柱
free column volume　空柱体积
free convection　自然对流
free corrosion potential　自然腐蚀电位
free counterion　自由反离子
free cross-section　空隙截面；自由横切面
free crushing　自由压碎
free cutting　高速切削
free damping oscillation vibrometer　自由衰减振荡示振仪
free diffusion　自由扩散
free discharge　自由流出
freedom　①自由②自由度
freedom from cracking　抗龟裂；无色裂现象
free-draining　自由穿流*
free-draining model　自由穿流模型
free-draining molecule　自由穿流分子
free electron　自由电子；游离电子
free electron gas　自由电子气
free electron laser　自由电子激光器
free electron method(FEM)　自由电子法
free-electron model　自由电子模型
free electrophoresis　自由电泳(法)
free element　单体元素
free end　自由端
free energy　自由能
free energy change　自由能变化
free energy function　自由能函数
free energy functional　自由能泛函
free energy of activation　活化自由能
free energy of formation of hydrocarbon　烃生成的自由能
free enthalpy　自由焓
free expansion　自由膨胀
free fatty acids　游离脂肪酸
free feed　自由进料
free fiber in viscose　黏胶中不溶解纤维
free film　游离漆膜
free flame　活火头；活火；自由火焰
free-flight　滑脱；松脱
free floating factor analysis　自浮因子分析法
free floating target transform factor analysis　自由浮动目标转换因子分析法

free flow	自由流动
free-flow agent	助流动剂
free-flowing	流动性良好的；自由流动的
free flowing granule	自由流动(的)颗粒〔指流动性良好〕
free-flowing material	①流动性材料②松散性
free flowing powder	自由流动(的)粉
free-flowing property	自由流动性
free-flowing softener	易流动软化剂；低黏度软化剂
free flow valve	易流阀
free from acid	无酸的；不含酸的
free from corrosive substances	无腐蚀物的
free gas	游离气体；自由气体
free gas space in column	柱中游离气体空间〔气体滞留〕
free gas volume	自由气体容积；死容积
free gauge	自由区隔距
free hand sketch	徒手草图
free hydroxyl group	游离羟基
free-hydroxyl-group concentration	自由羟基浓度
free induction decay	自由感应衰减
free internal rotation	自由内旋转
free ion	自由离子
free isocyanate	游离异氰酸酯
free jet	自由气流
free length	净长度
free lime	游离石灰
free linkage	自由键合；自由连接
freely extruded velocity	自由挤出速度
freely-jointed chain	自由连接链*
freely-rotating chain	自由旋转链
freely settled density of catalyst	催化剂自由堆积密度
freely soluble	易溶(解)的
free machining steel	快削钢
free mobility	自由迁移率；自由流动性
free moisture	自由湿气；游离水分
free moisture content	自由湿含量
free molecule	自由分子
freeness	①游离度②打浆度
freeness number	排水系数
freeness test	打浆度试验〔造纸〕
freeness tester	打浆度测验器
free nitrogen	单体氮
free oil	游离油〔未反应油〕
free on board(FOB)	①就船交货价格②离岸价格
free organic acid	游离有机酸
free oscillation	自由振荡
free oxide	游离氧化物
free paraffin chains	自由烷烃链
free paraffins	游离石蜡〔被分离出来的石蜡〕
free path	自由程
free path of particle	质点自由程
free phenol	单体酚；游离酚
free phosphate detergent	无磷洗涤剂
free position	任意状态
free precession signal	自由进动信号
free radical	自由基
free radical addition	自由基加成
free radical catalyst	自由基催化剂
free radical chain degradation	自由基链降解
free radical chain polymerization	自由基连锁聚合
free-radical chemistry	自由基化学
free-radical crosslinking agent	自由基交联剂
free radical curing	自由基硫化
free radical generator	自由基发生器；游离基发生器
free-radical initiation	自由基引发
free-radical initiator	自由基引发剂；游离基引发剂
free radical intermediate	自由基中间体
free radical isomerization polymerization	自由基异构化聚合
free radical lifetime	自由基寿命
free radical mass spectrometry	自由基质谱分析
free-radical mechanism	游离基(自由基)机理
free-radical oxidation	自由基氧化
free radical polymerization	自由基聚合；游离基聚合
free radical reaction	自由基反应*
free radical scavenger	自由基清除剂*
free radical spectrum	自由基光谱
free radical termination	游离基终止
free radical terminator	游离基终止剂
free radical trap	自由基捕集器
free radical trapping agent	自由基捕集剂
free radical type	自由基型
free resin	游离(单体)树脂
free resonance vibration method	自由共振法
free rinsing detergent	易漂净洗涤剂
free roller	张力调节辊
free rosin	游离松香
free rosin size	白色松香胶
free rotating chain	自由旋转链
free rotation	自由转动；不受限自旋
free running frequency	固有频率
free running generator	自激振荡器
free running liquid	易流动液体
free salt	游离盐
free sand blasting	无遮喷砂(处理)法
free settling	自由沉降
free settling ratio	自由降落系数

free settling tank classifier　自降分级器
free silica　①游离二氧化硅②(黏土、硅藻土等中的)杂质二氧化硅
free sintering　自由熔结(即无模烧结)
free solubilizate　自由加溶物
free solution capillary electrophoresis　自由溶液毛细管电泳
free solution electrophoresis　自由溶液电泳(法)
free space　自由空间
free spin　自由转动
free spiro union　自由螺接
free standing cubicle　自立式柜
free state　游离状态；单体状态
free stock　游离状浆
freestone　易切石
freestream turbulence　自由流湍流
free sulfur　单体硫
free sulfur cure　游离硫硫化；元素硫硫化
free sulfur dioxide　自由二氧化硫
free surface energy　自由表面能
free surface instability　自由面不稳性
free surface spanning　自由面纺丝
free surface velocity　自由表面速度
free time of flight　自由飞行时间
free valence　自由价
free valency　自由价
free variability　自由变异性
free velocity　自由速度
free vibration　自由振动
free vibration method　自由振动法
free volume　自由体积
free volume theory　自由体积理论
free volume (tower)　空隙容积
free vortex blade　自由涡流叶片
free vulcanization　无模硫化
free water　自由水分；游离水分
free water content　游离水含量；自由水分含量
freeze coagulation　冻结
freeze concentration　冷冻浓缩
freeze desalination　冰冻脱盐
freeze desalting process　冷冻脱盐法
freeze drying　冷冻干燥
freeze etching　冷冻蚀刻(法)
freeze fracturing　冷冻破碎
freeze-gel process　冷冻胶凝法
freeze grinding　冷冻研磨
freeze mill　冷冻研磨
freeze-off time　冻结时间
freeze out　冻结
freeze point　①冰点②凝固点
freeze-proof　①防冻的②防冻性
freezer(=congealer; refrigerator)　①制冷器②冷藏箱
freeze resistance　耐寒性；抗霜性
freeze-thaw　冰冻-融化
freeze-thaw cycle　①冻-融周期②冻-融循环
freeze-thaw resistance　抗冻融性
freeze-thaw stability　冻融稳定性
freeze-thaw stability controller　冻融稳定控制剂
freeze-thaw stabilizer　冻融稳定剂
freeze-thaw stable latex　抗冻胶乳
freeze-thaw test(=freezing and thawing test)　冻融试验
freeze-thaw valve　冻融阀
freezing　凝固；冻结
freezing and thawing test(=freeze-thaw test)　冻融试验
freezing constant　凝固点降低常数
freezing curve　冷凝曲线
freezing-in　冻结；凝入
freezing microtome　冻结切片机
freezing mixture　冷却剂；冷冻混合物；致冷混合物
freezing out　结冰；冻结出来
freezing point　①凝固点②冰点
freezing point apparatus　凝固点测定器
freezing point depressant　防冻剂
freezing point depression　凝固点降低*
freezing point depression analysis　冰点降低分析
freezing point method　冰点法
freezing point test　冰点试验
freezing polymerization　冷冻聚合(作用)
freezing polymer stop device　冷冻聚合体截流装置；冷冻阀
freezing resistance　耐寒性；抗霜性
freezing salt　冷冻用盐
freezing test　①冰点的测定②乳浊液均匀性和冰恒性的测定
freezing thawing cycle　冻融循环
freezing trap　冷阱
freezing way　冷冻法
freibergite　银黝铜矿
Fremont impact test piece　弗莱蒙冲击试验片
Fremy's salt　费里米盐
French blue(=ultramarine)　群青；佛青
French chalk　滑石
French folio　稿纸；存根纸
French milling　肥皂研磨(法)
French polishing　①紫胶清漆②醇溶性清漆
French process (for zinc extraction)　法兰西提锌法

French rosemary oil　法国迷迭香油
French sweet basil oil　法国罗勒油
French turpentine(=bordeaux turpentine)　枣红松节油；法国松节油
French varnish　揩涂高光泽清漆
French veronese green　水合氧化铬绿
French white　法国白；滑石粉
Frenkel defect　弗仑克尔缺陷*
frequency　频数*
frequency array detection　频率阵列检测
frequency band　频带
frequency bandwidth　频带宽度
frequency control unit　频率控制单元
frequency counter　频率计数器
frequency curve　频数分布曲线
frequency difference　频率差；差频
frequency difference radiofrequency detector　差频射频检测器
frequency discriminator　鉴频器
frequency dispersion　频(率分)散
frequency distribution　频数分布*
frequency distribution curve　频率分布曲线
frequency domain　频域
frequency domain signal and time domain signal　频域信号与时域信号
frequency domain spectra　频域谱
frequency drift　频率漂移
frequency effect　频率效应
frequency factor　频率因子
frequency meter　频(率)计
frequency modulation　频率调制
frequency of beat　拍频；振频
frequency offset unit　频率补偿装置
frequency particle distribution　频率粒度分布
frequency ratio　频比（通常指强制频率与自然频率的比）
frequency relaxation spectrum　频率松弛谱
frequency response　频率响应
frequency shift of scattering　散射光频率的位移
frequency spectrograph　频谱仪
frequency spectrum　频谱
frequency sweep　扫频
frequency synthesizer　频率合成器
frequency temperature reducibility　频率-温度换算法则
frequent cleaning　频繁清洗；经常清洗
frequentic acid　常见青霉酸
frequentin　常见青霉菌素
fresco　①壁画；湿壁画②湿灰泥墙面涂彩色水浆涂料
fresh　①新鲜的②清凉的③淡的④鲜（指颜色）
fresh air　新鲜空气
fresh charge　新鲜进料
fresh cider　新鲜苹果汁
fresh cracked gas　新鲜裂化气
freshener　①溶剂表面增黏剂②增鲜剂
freshening　①回炼；复炼②热炼③塑炼④半成品刷汽油
fresh feed　新鲜原料
fresh feed pump　进料泵
fresh feed surge drum　新鲜原料收集器
fresh hide　鲜皮
fresh keeping agent　保鲜剂
fresh-keeping method　（产品）保鲜法
fresh latex　新鲜胶乳
fresh lime　新灰
fresh lime liquor　新灰液
freshly set mortar　新凝泥浆
freshly smoked magnesium oxide　新制的气相法氧化镁
freshness　(新)鲜度
freshness date　质量保证期
freshness determination　鲜度测定
freshness storage life　鲜度保藏期
freshness test　鲜度检验
fresh note　清鲜香韵
fresh oil　新油
fresh rubber　新(鲜)橡胶
fresh soda　新鲜碱
fresh spirit　石脑油
fresh surface　新生表面
fresh water　淡水
fresh water condenser　淡水冷凝器
freshwater ecosystems　淡水生态系统
fresh water pump　清水泵
freshwater resources　淡水资源
fresnel　兆兆周；兆兆赫〔兆兆为 10^{12}〕
Fresnel coefficient of reflection　菲涅耳反射系数
Fresnel diffraction　菲涅耳衍射
Fresnel fringe　菲涅耳条纹
Fresnel reflectance　菲涅耳反射系数
Fresnel reflection　菲涅耳反射(现象)
Fresnel reflection law　菲涅耳反射律
Fresnel refractometer　菲涅耳折光计
Fresnel refractometer detector　菲涅耳折光检测器
frettage　摩擦腐蚀
fretting　流水磨蚀；流水侵蚀
fretting corrosion　摩擦磨蚀
fretting corrosion mechanism　摩擦磨蚀机理；磨蚀机理
fretting corrosion preventive　摩擦腐蚀防止法；磨蚀防

止法
fretting fatigue　磨蚀疲劳
fretting oxidation　磨蚀氧化
Freudenberg's lignin　铜铵木素
Freudenreich culture flask　弗罗伊登里奇培养烧瓶
Freudenreich flask　弗罗伊登里奇培养(烧)瓶
Freundlich adsorption equation　弗罗因德利奇吸附公式
Freundlich adsorption isotherm　弗罗因德利奇吸附等温式
Freundlich isotherm　弗罗因德利奇等温线
Freund reaction　弗罗因德(闭环)反应
freyalite　硬铈钍矿
friability　脆性；易碎性
friable　脆的；易碎的
friable surface　易碎的表面；松散剥落的表面
friar's cap　乌头根
Fricke dosimeter　弗里克剂量计；硫酸亚铁化学剂量计
friction　摩擦(力)
frictional characteristic of lubricant　润滑剂的(抗)摩擦特性
frictional coefficient　摩擦系数
frictional damping　摩擦阻尼
frictional drag　摩擦阻力
frictional drop　摩擦差
frictional effect　摩擦效应
frictional energy　摩擦能
frictional flow　摩擦流
frictional force　摩擦力
frictional heat　摩擦热
frictional loss　摩擦损失
frictional pressure loss　摩擦压力损失
frictional reducer　减摩剂
frictional resistance　摩擦阻力
frictional stiffness　摩擦劲度
frictional stress　摩擦应力
frictional testing machines　摩擦试验机
frictional theory　摩擦理论
frictional torque　摩擦力矩
frictional wear　摩(擦损)耗
frictional work　摩擦功
frictionating　①摩擦②擦胶〔橡胶〕
friction board　耐擦纸板
friction brake-drum　摩擦制动鼓
friction bush device　套筒型摩擦装置
friction calender　异速压延机
friction calendering　擦胶压延〔橡胶加工〕
friction clutch　摩擦离合器
friction coat　擦胶胶片
friction compound　擦胶剂
friction disc　摩擦片
friction drive　①摩擦传动②异速传动
friction factor　摩擦系数
friction field distribution　摩擦力界分布
friction gear　摩擦传动装置
friction glazed paper　磨光纸
friction glazing　擦釉
friction goods　摩擦制品
friction head　摩擦压头
frictioning　擦胶；刮胶
frictioning calender　异速砑光机
frictioning ratio　异速比例
frictionless flow　无摩擦流
frictionless fluid　无摩擦流体；无黏性流体；理想流体
friction loss　摩擦损失
friction machine　①摩擦上光机〔织染〕②摩擦(生)电机
friction-motion speed　不同速度；异速
friction of motion　运动摩擦
friction of rest　静止摩擦
frictionometer　摩擦系数测定器
friction plate　摩擦片
friction pressure　摩擦压力
friction pull　密着力
friction pulley　摩擦滑轮
friction pump　摩擦泵
friction ratio　滚筒速比
friction reducing agent　减摩剂
friction resistance　摩擦阻力
friction ring　摩擦环；摩擦圈
friction rocks　擦胶碎块〔橡胶〕
friction separator　摩擦分离器
friction speed　不同速度；异速
friction speed calender　异速压延机
friction squeezer　摩擦榨取机
friction starching mangle　摩擦上浆机
friction stock　擦胶剂
friction surface　摩擦面
friction surface belt　毛面运输带
friction tape　胶布
friction test　摩擦试验
friction top　压紧盖
friction value　摩擦值；摩擦系数
friction welding　摩擦焊接
Friedel-Crafts catalyst　傅瑞德尔-克拉夫茨催化剂〔烃化催化剂〕
Friedel-Crafts-Karrer nitrile synthesis　傅瑞德尔-克拉夫茨-卡勒成腈合成法
Friedel-Crafts polymerization　傅瑞德尔-克拉夫茨聚合

Friedel-Crafts reaction　傅瑞德尔-克拉夫茨反应
Friedel-Crafts synthesis　傅瑞德尔-克拉夫茨合成法
friedelin　无羁萜；软木三萜酮
Friedländer quinoline synthesis　傅瑞德兰德喹啉合成法
Friele-MacAdam-Chickering color difference equation　弗利尔-麦克亚当-奇克林色差方程
frieseite　杂硫银铁矿
Fries migration　傅瑞斯移动
Fries reaction　傅瑞斯反应
Fries rearrangement　傅瑞斯重排反应
Fries rule　傅瑞斯规则
frigorific　冰冻的
frigorific mixture　冰冻混合物；制冷混合物
Frigorifico hide　弗利哥利非克生皮〔拉丁美洲卤腌皮〕
frigorific unit　致冷单元
frigorimeter　低温计
frigory　1千卡〔冷冻热量单位〕
fringe　条纹
fringed fibril　缨状原纤；毛边原纤
fringed-lamellae structural model　缨状片晶结构模型
fringed micelle　缨状微束
fringed-micelle model　缨状微束模型
fringed micelle structure　缨状微胞结构；毛边胶束结构
fringe micelle　缨状微束
fringe order　缨状有序
fringe stitch　流苏组织
fringe tree bark　流苏树皮
fringing field　弥散场
fringing field effect　弥散场效应
frit　玻璃料；釉料
frited glass separator　多孔玻璃分离器
fritillarine　贝母属碱；弗列惕拉令碱
fritilline　贝母灵
fritimine　贝母素丙
fritiminine　贝母素丁
frit inlet　多孔入口
frit outlet　多孔出口
fritted　①烧结的；熔结的②多孔的
fritted disc　烧结盘；烧结板
fritted glass(=sintered glass)　烧结玻璃；多孔玻璃
fritted glass crucible　烧结玻璃(过滤)坩埚
fritted glass filter　多孔玻璃过滤器
fritted glass filter plate　多孔玻璃滤板
fritted glass separator(=Biemann-Watson separator)　比曼-瓦特森分离器；多孔玻璃分离器
fritted glass ware　烧结玻璃器皿
fritted glaze　熟釉
fritted porcelain　烘炙陶瓷

fritted rocks〔复〕　烤变岩；烧岩
fritted ware　烧结器皿
fritting　烘炙〔陶瓷〕；烧结；熔结
fritting furnace　烧结炉
frizing　①刮油②去粒面
frog-suit　气衣；防毒衣
front　①前沿②前部③前面的；前部的
frontal analysis　迎头分析；前沿分析
frontal chromatography　迎头色谱法；前沿色谱法
frontal method　前沿法
frontal orbital　前沿轨道
front barrel flange　机筒前端法兰
front brick　面砖；正面砖
front elevation　正面(图)；正视(图)
front end　前段；前端
front-end processor　前端处理机
front end volatility　(燃料)轻馏分挥发性
front factor　前置因子
front ferrule　前箍；前套圈
front guide bar　前梳栉
frontier electron density　前沿电子密度
frontier electron theory　前沿电子理论
frontier molecular orbital　前沿分子轨道
frontier orbital　前沿轨道*
front line　前沿线
front-loading washing machine　前装式洗衣机
front mould　前模
front of developer　展开剂前沿
front of flame　火焰正面
front outline　①正视图②垂直投影
front roll　前滚筒
fronts〔复〕　颈皮〔皮革〕
front scavenge pipe　前清除管
front shoe　前固定板；喷嘴板〔塑料〕
front side　操作面前侧；操作侧面
front stringer　底座
front surface　正面；锋面
front surface attachment　前表面附件
front view　正视图
frost　霜
frosted　①盖有霜的②除去光泽的③冻伤的
frosted cotton　霜后花
frosted finish　毛化整理；磨砂
frosted glass　毛玻璃；磨砂玻璃
frosted glass plate　毛玻璃片
frosted plastic　无光塑料
frosted roll　无光辊；消光辊
frosted rusk　起霜干面包

frosting ①起霜②霜纹③去光泽；消光
frosting phenomenon 起霜现象；消光现象
frost line ①冰冻线②深冷线③霜白线〔塑料〕
frost patterns 霜花图案
frostproofer 防冻剂
frost resistance 防起霜性〔指漆膜变态性质〕
frostwork 霜花
froth 泡沫
froth chromatography 泡沫色谱(法)
frothed latex 泡沫胶乳
frother ①起沫剂②发泡机
froth flotation 泡沫浮选
froth flotation method 泡沫浮选法
froth flotation process 泡沫浮选法
frothiness 起泡沫性
frothing 起沫
frothing agent 起沫剂
frothing aid 发泡(助)剂
frothing machine 起泡机；泡沫发生器
froth-over ①沸腾②冒泡；逸出
froth promotor 泡沫促进剂
froth rubber 泡沫胶
froth stability 泡沫稳定性
froth suppressor 泡沫抑制剂；消泡剂
frothy 泡沫状的；多泡沫；起泡的
Froude number 夫劳德数
frozen motion 冻结运动
frozen polymerization 冻结聚合(作用)〔液状单体在熔点以下冷却至发生结晶时进行的聚合〕
frozen rubber 冷冻(橡)胶
frozen strain 冻结应变
frozen stress 冻结应力
frozen target 冰冻靶
frozen-wall continuous fluorinator 冰冻壁连续氟化器
fructigenin 产果镰孢菌素
fructose diphosphate sodium 果糖二磷酸钠〔药〕
frue vanner 淘矿机
fruit ①水果②果实③果香
fruit acid 水果酸
fruital 果香
fruitcake 水果蛋糕
fruit coat oil 水果肉油
fruit essence 果香香精
fruit fibre 果实纤维
fruit flavour 果香香精
fruit glaze agent 被膜剂
fruit juice 果汁
fruit odor 果香

fruit packer's oil 水果装罐用油
fruit pulp 果酱
fruit-scent 果香
fruit section yarn 橘瓣形截面长丝
fruits tissue 水果包装纸
fruit syrup 果子露
fruit tissue paper 水果包裹纸
fruity 果香
fruity odour 水果香；水果味
Frumkin correction 弗鲁姆金修正
frustoconical 截头锥形的
frustrated multiple internal reflectance 多次内反射装置
Fry's theory(=alternate polarity theory) 弗莱氏理论；交替极性说
F-test F检验*
FTV(flavor threshold values) 香味阈值
fuchsin(=magenta red) 品红；洋红
fuchsin-aldehyde reagent 品红醛试剂
fuchsin red 品红
fuchsisenecionine 富斯千里光宁
fuchsite 铬云母
fuchsone 品红酮；对二苯亚甲基环己二烯酮
fucolipid 岩藻糖脂
fucosan 岩藻聚糖
fuel 燃料
fuel acids 燃料酸〔燃料燃烧后产生的酸类〕
fuel additive 燃料添加剂
fuel-air mixture analyzer 燃料-空气混合物分析器
fuel-air ratio 燃料-空气比
fuel-air ratio indicator 燃料-空气比指示器
fuel alcohol 燃料酒精；动力酒精
fuel analysis 燃料分析
fuel and air mixture 燃料和空气混合物
fuel antiknock quality 燃料抗爆性
fuel ash corrosion 燃料灰腐蚀
fuel ash deposition 燃料灰沉积
fuel assembly cask 燃料组件罐〔核〕
fuel atomization(=fuel atomizing) 燃料雾化
fuel atomizing(=fuel atomization) 燃料雾化
fuel battery 燃料电池
fuel bay 燃料存放池
fuel bed 燃料层
fuel bed combustion 燃料层着火燃烧
fuel bell 燃料钟
fuel blend ①混合燃料②掺杂汽油
fuel breeding 燃料增殖(核)
fuel brick 燃料块
fuel bundle 燃料棒束

fuel bunker　燃料仓
fuel calorimeter　燃料量热计
fuel capacity　燃料容量
fuel cartridge　燃料元件
fuel cassette　燃料盒
fuel cell　燃料电池
fuel cell bottom　燃料电池底
fuel cell detector　燃料电池检测器
fuel centralizer　燃料集中分配器
fuel charge　燃料供送
fuel charger　燃料供送泵
fuel charging means　燃料供送器
fuel clad　燃料包壳
fuel coefficient　燃料(消耗)系数
fuel collecting pipe　燃料收集管
fuel composition　燃料成分
fuel consumption　燃料消耗
fuel consumption curve　燃料消耗曲线
fuel consumption meter　燃料消耗计
fuel consumption rate　燃料消耗(速)率
fuel consumption test(=fuel consumption trial)　燃料消耗试验
fuel consumption trial(=fuel consumption test)　燃料消耗试验
fuel content gauge　燃料含量指示计
fuel contribution value　燃料贡献值；燃料影响值
fuel control unit　燃料供给控制器
fuel corrosion　燃料腐蚀
fuel cycle cost　燃料循环成本
fuel delivery system　燃料输送系统
fuel depot　燃料仓库
fuel depth　燃料深度；燃料厚度
fuel dilution　燃料稀释
fuel dilution test　燃料稀释试验
fuel discharge　燃料放出；燃料排出
fuel distillation bell　燃料蒸馏罩
fuel dope　燃料添加剂
fuel duty　燃料供应量
fuel economizer　节油器
fuel economy　燃料经济(机工)
fuel efficiency　燃料效率
fuel element　燃料元件
fuel element case　燃料元件盒
fuel element flask　燃料元件运输容器
fuel encapsulating machine　燃料封装机
fuel engineering　燃料工艺
fuel feed　燃料供给
fuel feed control　燃料供给控制；燃料供给调节系统

fuel feeder　燃料加料器；喷雾器
fuel feeding bell　燃料进口钟
fuel feed system　燃料供给系统
fuel filling column　燃料柱；燃料塔；加油柱
fuel film　燃料膜
fuel filter　燃料过滤器；滤油器(机工)
fuel flow meter　燃料流速计
fuel flow rate　燃料流速
fuel fog　燃料雾
fuel gas　可燃气体
fuel gas system　燃料气系统
fuel gas treatment　燃料气精制
fuel gauge　量油计；油规；燃料液面计
fuel handling　燃料储运
fuel heater　燃料加热器
fuel hopper　燃料斗
fuel hose　燃料管
fuel ignition quality　燃料的自燃性
fuel indicator　燃料液面指示器
fueling　加燃料；燃料转注
fueling connection　(液体)燃料转注连接管(船舰)
fueling gear　燃料接受装置(船舰)
fueling main　燃料转注干线(船舰)
fueling oil hose　燃料油管
fuel injection　燃料喷射
fuel injection nozzle　燃料喷嘴
fuel injection pump　燃料油喷射泵
fuel kernel　燃料芯核
fuel-lean flame　贫燃火焰
fuel level indicator　燃料油位指示器
fuel line　燃料管；油管
fuel manifold　燃料歧管
fuel mixer　燃料混合器
fuel mixture　燃料混合物；可燃混合物
fuel mixture indicator　燃料性质指标
fuel of high (anti) knock rating　高抗爆性燃料；高辛烷值燃料
fuel of nuclear reactor　核(反应堆)燃料
fuel oil　燃料油；液体燃料
fuel oil additive　燃料油添加剂
fuel oil analysis　燃料油分析
fuel oil barge　石油驳船；油驳
fuel oil burner　燃料油喷嘴
fuel oil compensating system　燃料油补偿系统
fuel-oil consumption ratio　燃料-润滑油消耗比
fuel oil distillate　燃料油馏出物
fuel oil drum　燃料油罐
fuel oil filling system　燃料油装料系统

fuel oil filter 燃料油过滤器；滤油器
fuel oil flash tower 燃料油闪蒸塔
fuel-oil-gas tar 燃料油气焦油
fuel oil meter 燃料油计
fuel oil product 燃料油成品
fuel oil residue(=fuel oil residuum) 燃料油残渣；重油；石油残油
fuel oil residuum(=fuel oil residue) 燃料油残渣；重油；石油残油
fuel oil sampling 燃料油取样
fuel oil service pump 燃料油泵
fuel oil service tank 燃料油储罐
fuel oil settling tank 燃料油沉淀箱
fuel oil slag 燃料油炉渣
fuel oil stabilizer 燃料油稳定剂
fuel oil supply 燃料油供给
fuel oil tank 燃料油储罐
fuel oil yield 燃料油产率
fuel outlet 燃料出口管
fuel particle 燃料颗粒
fuel performance 燃料性质；燃料鉴定
fuel pin 燃料细棒状元件
fuel-proof 防油的；耐汽油的
fuel pulverization 燃料雾化
fuel pump 燃料泵
fuel rating 燃料性质评价
fuel rating engine 燃料试验发动机；燃料性质评价发动机
fuel ratio 燃烧比率
fuel reprocessing （核）燃料后处理
fuel requirement 燃料需要
fuel resin 燃料树脂
fuel resistance 耐油性
fuel-resistant rubber 耐油橡胶
fuel-rich flame 富燃火焰*
fuel rod 油柱
fuel segregation 燃料分离
fuel sensitivity 燃料敏感性
fuel servicing truck 燃料供应车；油槽车
fuel sheath 燃料包壳
fuel ship 油船；油轮
fuel slippage 燃料流动性
fuel soot 燃料烟黑
fuel spray 燃料喷雾
fuel spreader 装料钟
fuel station 燃料站
fuel storage 燃料储藏；燃料库
fuel storage hopper 燃料仓斗

fuel storage pool 燃料储存池
fuel store 燃料库
fuel strainer 滤油器
fuel tank 油箱
fuel tanker ①油槽车②飞机油箱
fuel tester 燃料试验器
fuel testing 燃料试验
fuel thermal efficiency 燃料热效率
fuel thickener 燃料稠化剂；凝油剂
fuel value (of food) (食物的)燃烧值
fuel vaporization 燃料蒸发
fuel vapour 燃料蒸气
fuel volatility 燃料挥发性
fuel volatility adjustment 燃料挥发度调节
fuelwood 薪材；薪炭材
fugacity 逸度；逸性；有效压力
fugacity coefficient 逸度系数
fugillin(=fumagillin) 夫马洁林；烟曲菌素
fugimeter 耐晒牢度试验仪
fugin(=fugutoxin) 河豚毒
fugitive 短效的
fugitive accelerator 短效加速剂
fugitive binder ①短效黏结剂②短效成膜物
fugitive dye 短效染料
fugitive lubricant 挥发性润滑剂
fugitiveness ①不稳定性②挥发性
fugitiveness to light 光褪色性；不耐光性
fugitive plasticizer 短效增塑剂
fugitive tint 短效色泽；易褪色泽
fugitometer(=fadeometer) 褪色度试验计；褪色计
fugutoxin(=fugin) 河豚毒
Fujiex 螺旋状倾斜管萃取器
fulcin(=griseofulvin) 灰黄霉素
fulcrum 支点
fulgenic acid 俘精酸；烟菌酸
$$R_2C=C(COOH)C(COOH)=CR_2$$
fulgide 俘精(酸)酐 $R_2C_6O_3R_2$
fulguration 闪熔；电干燥法
fulgurator 盐溶液雾化器
fulgurite 闪电熔岩
fuligonic acid 炱黏菌酸
fuligonin 烟垢黏液菌素
full and clothe feel 丰满手感
full annealing 全退火
full aperture drum 活底桶
full aroma 饱满香气
full automatic 全自动的
full blast 全风

full bleaching 完全漂白
full-blown sponge rubber 充分发泡海绵胶
full-bodied 高黏度的；高增容的
full-bodied paint 高黏稠漆
full-boiled process 全沸煮法〔制皂〕；热法〔煮皂〕
full-boiled soap 全沸(煮)皂
full boiling point 终沸点〔全部馏分蒸出时的温度〕
full charge 完全装料；全负荷
full chrome side 全铬鞣牛皮半张鞋面革
full circle goniometer 全圆测角仪
full coat ①厚涂层；厚膜②全涂；满涂
full color ①纯色；饱和色〔未冲淡的色〕②彩色；全色
full color image 全色像图
full crystal 全晶玻璃〔含铅晶玻璃〕
full cure 充分硫化
full curve 完全曲线
full detergency oil 含最大量去垢剂的油
full dip infiltration 全部浸入浸渍法
full dip process 全浸法(涂装)
full-dull 全无光；全消光
fuller ①填料工〔纸〕②毡合工〔纺〕
fullerene 球壳状碳分子
Fuller-Lehigh mill 富勒-利弗研磨机
fuller's earth 漂白土
full-face gasket 宽敞面法兰垫；带螺栓孔的法兰垫
full flashing 一次闪蒸；一次骤蒸馏
full-flash operation 一次闪蒸操作
full flight length (螺杆的)螺纹总长度
full-floating mechanical packing 全浮动机械填料
full fluoride protection 全氟保护作用
full-fresh oiling 新油润滑
fullgas 全燃气
full gloss 全光；高光泽
full gloss potential 最大光泽潜力
full goniophotometer 全色测角光度仪
full graphic panel 全测量系统图示板
full hardness 全干硬度
full hydrodynamic lubrication 完全流体润滑
full identification of pesticide 农药全分析
fulling 缩绒(过程)
fulling agent 缩绒剂
fulling assistant 缩绒助剂
fulling clay 缩绒黏土
fulling mill 漂洗机
fulling soap 缩呢皂；缩绒皂
fulling stock 揉皮器
full killed steel (完全)镇静钢
full lean mixture 全贫混合物

full load ①全负荷②全填充
fullmasse(=massecuite) 糖膏
full matrix least-squares refinement 全角矩阵最小二乘修正
full mercerized finish 全丝光(整理)
full milk bread 全乳面包
full-molded type rubber bearing 整体模制式橡胶轴承
fullness ①完全②丰满度
fullness of shade 色的饱和度；色纯度
full overlap method 全重叠法
full-plush jacquard fabric 高密度提花毛圈织物
full potential 全电势(位)
full pressure circulating lubrication system 全压循环润滑系统
full pressure lubricating system 全压润滑系统
full pressure lubrication 全压润滑；强制润滑
full pressure ratio 全压比率
full production 全能力生产
full range 满量程
full regeneration 完全再生
full-rich position 全富位置
full scale 实际尺寸；实物大小；满刻度
full scale deflection 满刻度偏转
full scale model 实尺模型
full scale plant ①成套设备②工业装置
full scale production 全部生产
full scale reading 满刻度读数
full scale response 满刻度响应
full scale test 实物试验
full scan chromatogram 全扫描色谱图
full sensitivity contrast 全灵敏度对比
full shade 饱和色
full-sized 实比；与原物大小一样
full sized brick 整砖
full-sized plate 标准尺寸的塔盘
full skin fiber 全皮(层)纤维
full-step mode 整步模式
full thread 全螺纹
full tinting strength 全着色力
full view 全视图
full-wave phase control 全波相位控制
full weight 总重(量)；全重
full wet coat 充分润湿涂层
full width at half maximum(FWHM) 半峰宽度〔极大值一半处的全宽度〕
fully acetylated cotton 全乙酰化棉纤维
fully-additive process 全加成法
fully automatic 全自动的

fully automatic doffing device　细纱全自动集体落纱装置
fully automatic dual-spindle transfer　全自动双锭子换筒装置
fully continuous line　全连续生产线
fully developed flow　完全展开流
fully eclipsed form　全重叠式
fully extended chain　完全伸直链
fully-fashioned flat knitting machine　全成型横机
fully fluorinated paraffins　完全氟化的烷烃
fully hydrolyzed　全水解的
fully oriented yarn　全取向丝
fully positive mold　溢(出)式压模；开放式模
fully refined wax　充分精制的蜡〔颜色、含油量、气味等合乎特殊要求的结晶石蜡〕
fulmargin　银胶溶液
fulmenite　俘门炸药；福明那特
fulminate　雷酸盐*
fulminating　雷爆(的)
fulminating cap　雷帽；雷汞爆管
fulminating gold　雷爆金；亚金联胺
fulminating powder(=percussion powder)　雷爆火药
fulminic acid　雷酸　C=NOH
fulminurate　雷尿酸盐　(C=NOM)$_3$
fulminuric acid(=isocyanuric acid)　雷尿酸；异氰尿酸；异三聚氰酸　NCCH(NO$_2$)CONH$_2$
Fulton furnace　弗尔顿电炉
Fulton zinc process　弗尔顿炼锌法
fulvalene(=bicyclopentadienylidene)　富瓦烯
fulvene　富烯*
fulvic acid(=fulvinic acid)　灰黄霉酸
fulvicin(=griseofulvin)　灰黄霉素
fulvinic acid(=fulvic acid)　灰黄霉酸
fulvoplumerin　蛋花杏素
fumagillin(=fumidil)　夫马洁林；烟曲霉素
fumanomycin　烟霉素
fumaramic acid　富马酰胺酸　NH$_2$COCH=CHCOOH
fumaramide　富马酰胺　(=CHCONH$_2$)$_2$
fumaraniloyl　反式苯氨甲酰丙烯酰基　C$_6$H$_5$NHCOCH=CHCO—
fumarate　①富马酸；反(式)丁烯二酸②延胡索酸盐(酯或根)；反丁烯二酸盐(酯或根)
fumarhydrazide　富马酰肼　C$_4$H$_6$O$_2$N$_4$
fumaria　紫堇
fumaric acid　富马酸；延胡索酸
fumaric resin　反丁烯二酸(酐)树脂；富马酸(酐)树脂
fumarimide　富马酸亚胺　C$_4$H$_2$O$_2$N$_2$
fumarine　富马碱；原鸦片碱
fumaroid　富马型；反丁二烯型；反式异构化合物

fumaroid form　富马型；反丁烯二酸型
fumarol　(喷)气孔
fumarol acid(=boric acid)　硼酸　H$_3$BO$_3$
fumarol gases　火山气
fumaroyl acetoacetate　富马酰乙酰乙酸
fumarprotocetraric acid　富马原冰岛衣酸
fumaruric acid　富马尿酸
fumaryl　反(式)富马酰；反(式)丁烯二酰　—COCH=CHCO—
fumarylacetoacetic acid　富马酰乙酰乙酸
fumaryl chloride　富马酰氯；反(式)丁烯二酰氯　(=CHCOCl)$_2$
fumatorium　熏蒸消毒室
fume　烟(雾)
fume cupboard　烟橱；通风橱
fume cure　氯化硫气体硫化
fume disposal　①排烟②废气处理
fumed silica(=pyrogenic silica)　热解法二氧化硅；火成二氧化硅
fume extraction　抽吸烟雾(作用)
fume extraction system　雾化萃取器
fume extractor　抽烟器；风抽子〔俗〕
fume hood(=fuming cupboard)　烟橱；通风橱
fume incinerator　微粒污染物焚烧炉
fumeless dissolution　无烟溶解
fume loss　烟雾损失
fume-proof enamel　耐烟雾瓷漆
fume-proof paint　耐烟雾漆
fume red lead　烟法红丹
fume-resistant　防烟的
fume-resistant paint　耐烟雾漆
fume stack　烟雾排放管；烟囱
fumid　烟色的
fumidil(=fumagillin)　夫马洁林；烟曲霉素
fumigachlorin　刺烟氯菌素
fumigacin(=helvolic acid)　蜡黄酸；烟曲霉酸　C$_{32}$H$_{44}$O$_8$
fumigant　熏蒸剂
fumigatin　烟曲霉醌
fumigating　熏蒸
fumigating pastille　薰香锭
fumigation　熏蒸法；烟熏法
fumigator　烟熏器
fuming　发烟的
fuming acid　发烟酸
fuming cupboard(=fume hood)　烟橱；通风橱
fuming hood(=fuming cupboard)　烟橱；通风橱
fuming hydrochloric acid　发烟盐酸
fuming liquid　发烟液体

fuming nitric acid　发烟硝酸
fuming-off point　烟化点
fuming-off temperature　爆燃温度
fuming oil of vitriol(=fuming sulfuric acid)　发烟硫酸
fuming sulfuric acid(=fuming oil of vitriol)　发烟硫酸
function　①官能；功能②函数
functional　官能的；功能
functional-analytical group　分析官能团；分析功能基
functional coating　功能性涂料；功能性涂层
functional compound　官能化合物
functional design　功能设计；性能设计
functional efficiency　功能效率
functional electrode　功能电极
functional extender　功能性增量剂；功能性体质颜料
functional fiber　功能纤维
functional filler　功能性填料
functional finish　功能性整理剂
functional genomics　功能基因组学
functional group　功能基；官能团
functional group analysis　官能团分析
functional group chromatogram　官能团色谱图
functional group labelling　官能团标记
functional group retention index　官能团保留指数
functional inspection　作业检查
functionality　官能度
functionality evaluate analysis of cosmetics　化妆品功能性评价分析
functional joint　伸缩接合
functional liquid　官能流体；官能液
functionallization　官能作用
functionally gradient material　功能倾斜材料；功能递变材料
functional membrane　功能膜
functional molecule　官能分子
functional monomer　功能性单体
functional nano material　功能纳米材料
functional pigment　功能型颜料
functional plug-in module　功能插入组件
functional polymer　功能高分子*
functional proteomics　功能蛋白组学
functional representation of behaviour　特性的泛函表示
functional retention index　官能团保留指数
function flow　函数流
function group　官能团
function group frequency　官能团频率区
function isomerism　官能异构
function of complex variable　复变函数
function of mutual information　相关信息函数

function of particle density　粒子密度函数
function of state　(状)态函数
function value　函数值
Funda filter　回转水平滤板过滤机
fundamental　①基本的②原理③基频
fundamental band　基频峰
fundamental chain　母链；主链
fundamental characteristics　基本性能
fundamental color　基(本)色
fundamental cycle　基本环路
fundamental factor　①基本因素②基波系数
fundamental flame speed(=fundamental flame velocity)　火焰基本速度
fundamental flame velocity(=fundamental flame speed)　火焰基本速度
fundamental frequency　基频
fundamental frequency band　基频谱带
fundamental metric tensor　基本度量张量〔数〕
fundamental parameter　基本参数
fundamental principle　基本原理
fundamental procedure　基本操作；基本步骤
fundamental secular frequency　特征基频
fundamental series　基系
fundamental test　基本试验；深入的试验
fungicidal paint　防霉漆
fungicidal properties　防霉性能
fungicidal wash　①杀(真)菌处理液；杀(真)菌药水②防霉底漆
fungicide　杀(真)菌剂
fungicide sprayer　杀菌剂喷洒器
fungicidin(=nystatin)　制霉菌素
fungimycin　真菌霉素；培里霉素
funginert　①不生霉的；不发霉的②防霉的
funginertness　抗霉力；防霉性
funginon　真菌油素
fungi pellet　真菌菌丝球
fungi-proof　防霉性
fungistatic agent　抑(霉)菌剂
fungistatic plasticizer　防霉增塑剂
fungistatin　制真菌素
fungisterol　菌甾醇
fungizone(=amphotericin B)　两性霉素 B
fungocin　杆菌制霉素
fungoid　类真菌的；真菌样的
fungous　真菌的；霉菌性的
fungous enzyme　霉菌酶
fungusproofing　①防霉处理②防霉的
fungus resistance　防霉性

fungus-resistant paint 防霉涂料；防霉漆
fungus resistant plasticizer 抗霉增塑剂
funiclarin(=funicularin) 索芽孢菌素
funiculosin 索青霉素
funiculus 菌丝索
funnel 漏斗
funneled pipe 带漏斗的管子
funnel flow 漏斗状流
funnel spinning process 漏斗式纺丝法
funnel stand(=funnel support) 漏斗架
funnel support(=funnel stand) 漏斗架
funnel tube 长梗漏斗
funnel type viscosity test 漏斗式黏度试验〔胶浆〕
funnel with filter plate 多孔板漏斗
funtumia 绢丝橡胶
funtumine 丝胶树碱
fur ①毛皮；裘皮②锅垢
furac 夫拉克〔呋喃甲基二硫代羧酸铅及锌的俗名〕
furacrolein(=furfuracrolein) 呋喃丙烯醛
fural 呋喃亚甲基
2-furaldehyde 2-呋喃甲醛
furaldehyde 糠醛 C_4H_3OCHO
furamide 糠酰胺
furan 呋喃
furanacrylic acid(=furfuracrylic acid) 呋喃丙烯酸
   $C_4H_3OCH=CHCO_2H$
furancarbinol(=furfuryl alcohol) 呋喃甲醇；糠醇
α-furancarboxylic acid(=pyromucic acid) α-呋喃甲酸；焦黏酸
β-furancarboxylic acid(=furoic acid) 糠酸；β-呋喃甲酸
furandione 呋喃二酮
furan formaldehyde 呋喃甲醛
furanidine 呋喃烷；四氢呋喃
2-furanmethanethiol 2-呋喃甲烷硫醇
furan nucleus 呋喃环
furanodienone 呋喃并二烯酮
furanoeremophilone 呋喃雅槛蓝酮
furanoid 呋喃型化合物
furanomycin 呋喃霉素
furanopetasin 呋喃蜂斗菁
furan plastics 呋喃塑料
furan resin 呋喃树脂
furan resin adhesive 呋喃树脂黏合剂
furazano[b]pyridine 呋咱并[b]吡啶
fur dressing 毛皮修整
fur dyes 毛皮染料
furfuracrolein(=furacrolein) 呋喃丙烯醛
furfuracrylic acid(=furanacrylic acid) 呋喃丙烯酸
   $C_4H_3OCH=CHCO_2H$
furfural ①(=furfurol)糠醛 C_4H_3OCHO ②(=furfurylidene)亚糠基；呋喃亚甲基
   $OCH=CHCH=CCH=$
furfural acetone 糠醛丙酮
furfuralcohol(=furyl alcohol) 呋喃基甲醇；糠醇
α-furfuraldehyde α-糠醛
furfural diacetate 糠醛二乙酸酯
furfural dimethyl mercaptal 糠醛二甲缩硫醛
furfural glycerine 糠醛缩甘油 $C_5H_4O \cdot C_3H_6O_3$
furfural grease 糠醛润滑脂
furfural number 糠醛值
α-furfural oxime α-糠醛肟 $C_4H_3OCH=NOH$
furfural phenol resin 糠醛苯酚树脂
furfural phenylhydrazine 糠醛苯肼
furfural phenylhydrazone 糠醛苯腙
   $C_4H_3OCH=NNHC_6H_5$
furfural point 糠醛点
furfural process 糠醛萃取法
furfural refining process 糠醛精制法〔润滑油或柴油〕
furfural resin 糠醛树脂
furfural resin adhesive 糠醛树脂黏合剂
furfural solvent extraction process 糠醛溶剂抽提法
furfural storage 糠醛储器
furfural treating tower 糠醛抽提塔
furfuramide(=hydrofuramide) 糠醛胺；二氨缩三糠醛
   $(C_5H_4O)_3N_2$
furfuran(=furan) 呋喃 $CH=CHCH=CHO$
furfurine 糠醛碱；2,4,5-三呋喃基咪唑 $C_{15}H_{12}O_3N_2$
furfuroin(=furfuryl fural) 糠基糠醛
   $C_4H_3OCOCH_2OC_4H_3$
furfurol(=furfural) 糠醛 C_4H_3OCHO
furfuryl 糠基；呋喃甲基 $OCH=CHCH=CCH_2-$
furfuryl acetate ①乙酸糠酯 $C_4H_3OCH_2O_2CCH_3$ ②糠基乙酸盐(或酯)
furfuryl-acetic acid 糠基乙酸 $C_4H_3OCH_2COOH$
furfuryl acetone 糠基丙酮
furfuryl alcohol(=furancarbinol) 呋喃甲醇；糠醇
furfuryl alcohol resin 糠醇树脂
furfuryl amine 糠胺 $C_4H_3OCH_2NH_2$
furfuryl butyrate 丁酸糠酯 $C_3H_7CO_2CH_2C_4H_3O$
α-furfuryl caprylate 辛酸-α-糠酯
furfuryl chloride 糠基氯 $C_4H_3OCH_2Cl$
furfuryl disulfide 糠基二硫
furfuryl ether 二糠基醚；糠醚

furfuryl fural(=furfuroin)　糠基糠醛
　　$C_4H_3OCOCH_2OC_4H_3$
furfurylidene(=furfural)　亚糠基；呋喃亚甲基
　　$OCH=CHCH=CCH=$
furfurylidene-acetone　亚糠基丙酮
　　$C_4H_3OCH=CHCOCH_3$
furfurylidene-acetophenone　亚糠基乙酰苯
　　$C_4H_3OCH=CHCOC_6H_5$
N-furfurylidenebenzothiazole sulfenamide　N-亚糠基-2-苯并噻唑次磺酰胺〔促进剂〕
furfuryl isopropyl sulfide　糠基异丙基硫醚
furfuryl isovalerate　异戊酸糠酯
furfuryl mercaptan　糠基硫醇
furfuryl 2-methyl butanoate　2-甲基丁酸糠酯
furfuryl methyl ether　糠基甲基醚
α-furfuryl octanoate　辛酸-α-糠酯
α-furfuryl pentanoate　戊酸-α-糠酯
furfuryl propionate　丙酸糠酯
furfuryl pyromucate　焦黏酸糠酯
　　$C_4H_3OCO_2CH_2C_4H_3O$
furfuryl salicylate　水杨酸糠酯
furfuryl sulfide　糠基硫醚
furfuryl thioacetate　硫代乙酸糠酯
2-furfurylthio-3-methylpyrazine　2-糠基硫代-3-甲基吡嗪
furfuryl thiopropionate　硫代丙酸糠酯
α-furfuryl valerate　戊酸α-糠酯
furil(=α,α-difurfuroyl)　糠偶酰；联糠酰；α,α-联呋喃甲酰　$(C_4H_3OCO)_2$
furil-dioxime　糠偶酰二肟　$(C_4H_3OC=NOH)_2$
furnace　①加热炉②熔炉
furnace air　炉内空气
furnace annealing　炉内退火
furnace arch　炉顶
furnace atomic absorption spectrometry　高温炉原子吸收光谱法
furnace bottom　炉底；炉床
furnace bridge wall　管式炉隔墙
furnace capacity　炉容量；有效热负荷〔裂化炉〕
furnace charge　炉内装料
furnace charge pump　往炉内装料泵
furnace clinker　炉渣结块
furnace coal　冶金煤；炉煤
furnace coke　冶金焦炭
furnace cooling　随炉冷却
furnace discharge　①炉出料②从(管式或裂化)炉中出来的热油料
furnace distillate　炉馏出物

furnace door　炉门；窑洞
furnace draft　炉子抽力
furnace dust　炉灰
furnace floor　炉底
furnace fuel　炉用燃料
furnace gas　炉气
furnace gas condenser　炉气冷凝器
furnace gas scrubber　炉气洗涤器
furnace hearth　炉底；炉床
furnace inlet stream　炉入口流体
furnace oil　炉油
furnace operating curve　炉子操作曲线
furnace outlet stream　炉出口流体
furnace pot　①杯；盘②蒸发皿；溜槽
furnace pyrite ore　炉用黄铁矿
furnace rear　炉后
furnace treatment　炉热处理
furnace tube　炉管
furnace tube spring hanger　炉管弹簧吊架
furnacing　焙烧；煅烧
furnish　①供给②配料
furnishing　①调成②供给③置备
furniture　家具；器具
furniture finish lacquer　家具涂装用硝基漆
furniture lacquer　家具漆
furniture leather　家具革
furniture polish　家具擦光漆
furniture varnish　家具清漆
furoate　糠酸盐(或酯)
furocoumarin(e)　呋喃香豆素；呋喃并香豆精
furodiazole(=oxdiazole)　呋二唑；噁二唑
furogen colors　呋精染料
furogen French gray　呋精法国灰
furogen olive　呋精橄榄绿
furogen sand　呋精砂
furogen yellow　呋精黄
furoic acid(=β-furancarboxylic acid)　糠酸；β呋喃甲酸
furoin　糠偶姻；联糠醛　$C_4H_3OCHOHCOC_4H_3O$
furol　糠醛　C_4H_3OCHO
furol viscosity test　赛氏重油黏度测定
furonic acid　糠基乙酸；呋喃基丙酸　$C_4H_3OCH_2COOH$
furosemide　呋塞米〔药〕
furostan　呋甾烷
furoxane　N-氧化噁二唑
furoyl　糠酰；呋喃甲酰　$CH=CHOCH=CCO=$
furoyl chloride　呋喃甲酰氯
furoyltrifluoroacetone(FTA; HFTA)　呋喃甲酰三氟丙酮

$C_4H_3OCOCH_2COCF_3$
fur polishing machine 毛皮上光机
furrier 毛皮商
furrow ①畦②沟槽；皱纹③作沟槽
furrow application 条施
fur scouring agent 毛皮洗涤剂
furskins〔复〕 带毛皮
further tearing 连续撕裂
furyl 呋喃基 $OC_4H_3—$
furyl acetone 呋喃基丙酮
furylacrolein 呋喃基丙烯醛 $C_4H_3OCH=CHCHO$
α-furyl acrolein α-呋喃丙烯醛
furylacrolein oxime 呋喃基丙烯醛肟
$C_4H_3O(CH)_3NOH$
furylacrylamide 呋喃基丙烯酰胺
$C_4H_3OCH=CHCONH_2$
furyl alcohol(=furfuralcohol) 呋喃基甲醇；糠醇
1-(2-furyl)-heptane 1-(2-呋喃基)庚烷
1-(2-furyl)-1-hexanone 1-(2-呋喃基)己酮
furylidene 亚呋喃基〔通常指β-亚呋喃基〕
$CH=CHOCH_2C=$
2-furyl methanethiol 2-糠基甲硫醇
3-furylmethyl 3-呋喃甲基 $CH=CHOCH=CCH_2—$
2-furyl methyl ketone 2-呋喃甲基酮
furyl propionaldehyde dimethyl acetal 呋喃丙醛二甲缩醛
fusafungine 镰孢真菌素
fusain 丝炭
fusamarin 镰菌生素
fusanin 镰孢菌素
fusanol 穗檀醇
fusaric acid(=fusarinic acid) 镰孢菌酸；萎蔫酸；5-丁基-2-吡啶甲酸
fusarin 镰菌素
fusarine 镰孢素；镰刀菌素；萎蔫素
fusarinic acid(=fusaric acid) 镰孢菌酸；萎蔫酸；5-丁基-2-吡啶甲酸
fusariocin 珠镰孢菌素
fusariogenin 镰刀霉毒素
fusarium species 镰酸盐(酯)
fusarubin 镰红菌素
fuscin 暗褐菌素
fuscomycin 褐霉素
fuse ①引信②熔化③保险丝④导火线
fused ①熔融的②稠(合)的
fused acetate ①醋酯纤维硬粒②醋酯纤维黏结丝条

fused (calcium magnesium) phosphate 钙镁磷肥
fused (calcium magnesium phosphorus) potash fertilizer 钙镁磷钾肥
fused catalyst 熔融催化剂
fused caustic 熔融的苛性碱
fused cement 熔融水泥
fused copal 热熔炼珀把
fused corundum 电熔刚玉
fused drier 熔融(法)催干剂
fused electrolytic cell 熔质电池
fused heterocycle 稠杂环
fused peak 重叠峰
fused polycyclic system 稠合多环系
fused quartz 熔融石英
fused resinate 熔融树脂酸盐
fused ring 稠环
fused ring compound 稠环化合物
fused rocket electrophoresis 融化火箭电泳
fused salt 熔盐
fused salt chemistry 熔盐化学
fused salt electrolysis process 熔盐电解法〔复合多晶纤维成形法之一〕
fused salt electrorefining 熔融盐电精制
fused salt extraction 熔融盐萃取
fused salt-liquid metal extraction 熔盐-液相金属萃取
fused salt liquid phase 熔盐液相
fused salt polarography 熔盐极谱法
fused silica 熔凝硅石
fused silica open tubular column 熔融石英开管柱；熔融石英空心柱
fused silica tube 熔融石英管
fused six-membered rings 稠合六元环
fused temperature 熔解温度
fusel oil 杂醇油
fuse primer 导火管
fuse wire 保险丝
fusibility ①熔性②熔度
fusible ①易熔的②可熔化的
fusible alloy 易熔合金
fusible clay 易熔黏土
fusible cone (示温)熔锥
fusible metal 易熔金属
fusibleness ①熔性②熔度
fusible plug (易)熔塞
fusible powder 可熔性粉末
fusible salt 易熔盐 $NaNH_4 \cdot HPO_4$
fusicoccin 壳梭孢(菌)素
fusidic acid(=fusidin) 梭链孢酸

fusidin(=fusidic acid)　梭链孢酸
fusidinic acid　梭链孢酸
fusiform　纺锤形
fusiform ray　纺锤形射线
fusile　①可聚变的；可聚变物质②可熔的
fusing　熔化
fusing energy　熔化能
fusing oven　熔融(化)炉
fusing point　熔(化)点
fusing soldering　熔焊
fusing sprayed deposit　熔融喷镀沉积(层)
fusion　①(核)聚变②熔融*
fusion bond coating　熔融黏附涂料(涂层)
fusion-bonded epoxy coating　熔融黏附环氧涂料
fusion bonding　熔融黏结
fusion-bonding process　熔融黏结法
fusion cake　熔块
fusion casting　熔铸
fusion chemistry　聚变化学
fusion coating　熔融涂覆；热熔涂装
fusion electrolysis　熔(融)盐电解
fusion enthalpy　熔化焓
fusion-fission reaction　聚变-裂变反应
fusion-free fiber　不熔(化)纤维
fusion fuel　聚变燃料
fusion heat　熔化热
fusion method　熔化法
fusion mixture　熔化混合物
fusion name　稠合名称
fusion point　熔点；核聚变温度；塑化温度
fusion point of ash　灰的熔点；灰分熔点
fusion pore　融合孔
fusion pot　熔罐
fusion product　聚变产物
fusion rate　塑化速率
fusion reactor　聚变反应堆
fusion temperature　熔化温度

fusion time　①熔融时间②(增塑糊)塑化时间③熔凝时间
fusion tube　熔管
fusion viscosimeter　熔融黏度计
fusion welding　熔焊
fusion with alkali　加碱熔化
fusion zone　(母材)熔化区
Fuso black body　平板式远红外线加热炉
fustic　黄颜木
fustic extract　黄颜木浸膏
fustin　黄颜木素；3,7,3′,4′-四羟基双氢黄酮
fuze　引信
fuzz　毛丝；微毛；绒毛
fuzzball　灰蘑菇；马勃菌
fuzzy clustering analysis　模糊聚类分析
fuzzy cluster-partial least squares regression spectrophotometry　模糊聚类-偏最小二乘光度法
fuzzy comprehensive evaluation　模糊综合评判
fuzzy comprehensive index　模糊综合指数法
fuzzy control　模糊控制
fuzzy decision　模糊决策
fuzzy equivalence relation　模糊等价关系
fuzzy hierarchial clustering　模糊系统聚类法
fuzzy information　模糊信息
fuzzy logic　模糊逻辑
fuzzy matrix　模糊矩阵
fuzzy nonhierarchial clustering　逐步模糊聚类法；动态模糊聚类法
fuzzy operation　模糊运算
fuzzy orthogonal design　模糊正交设计
fuzzy pattern recognition　模糊模式识别
fuzzy probability　模糊概率
fuzzy reasoning　模糊推理
fuzzy set　模糊集合
fuzzy simulation　模糊模拟
fuzzy theory　模糊理论

G

gabardine 华达呢
gabbro 辉长岩
gabbromycin(=paromomycin) 巴龙霉素
gabbro norite 辉长苏长岩
gabbro pegmatite 辉长伟晶岩
gabbro sienite(=gabbro syenite) 辉长正长岩
gabbro syenite(=gabbro sienite) 辉长正长岩
gabexate 加贝酯〔药〕
gabianol 页岩油
Gabon copal 加蓬茈杷脂
Gabriel isoquinoline synthesis 盖布瑞尔异喹啉合成法
Gabriel synthesis 盖布瑞尔合成法
gabrite 脲甲醛聚合物
G-acid(=2-naphthol-6,8-disulfonic acid) G 酸；2-萘酚-6,8-二磺酸
gad 沉胶渣
gadelaidic acid 反(式)二十碳烯酸
gadget 机件
gadiometer 磁强梯度计
gadoleic acid 顺(式)9-二十碳烯酸
gadolinia 氧化钆 Gd_2O_3
gadolinite 硅铍钇矿
gadolinium 钆〔64 号元素，化学符号 Gd〕
gadolinium bromate 溴酸钆 $Gd(BrO_3)_3$
gadolinium bromide 溴化钆 $GdBr_3$
gadolinium carbonate 碳酸钆 $Gd_2(CO_3)_3$
gadolinium chloride 氯化钆 $GdCl_3$
gadolinium fluoride 氟化钆 GdF_3
gadolinium hydroxide 氢氧化钆 $Gd(OH)_3$
gadolinium nitrate 硝酸钆 $Gd(NO_3)_3$
gadolinium oxalate 草酸钆 $Gd_2(C_2O_4)_3$
gadolinium oxide 氧化钆 Gd_2O_3
gadolinium oxychloride 氯氧化钆 GdOCl
gadolinium sesquioxide 氧化钆；三氧化二钆 Gd_2O_3
gadolinium sulfide 硫化钆 Gd_2S_3
gadolinium zeolite 钆沸石
gadose 鳕肝油脂
gaff 带钩阀〔可以使石油管在固定位置上移动〕
gafrinin 格菲宁
gagat 软煤
gage(=gauge) ①表；计②规；量规③压力计④样板⑤计量；测量
gage cock 玻璃液位表旋塞；油表旋塞
gage controller 厚度控制器〔压延用〕
gage distortion 厚薄不匀
gage glass 玻璃液面计
gage mark ①标线〔示片〕②计量标记
gage-pole 量油杆
gage pressure 表压；计示压力
gage thickness ①滚距②线规粗度〔钢丝〕
gage variation 尺寸变化
gaging(=gauging) ①测压(的)②计量(的)
gaging hatch 量油孔
gahnite 锌光晶石
gaidic acid 2-十六(碳)烯酸
gain ①增益②放大
gain control 增益控制；增益调节
gain curve 增益曲线
gaine ①盒子②套③罩④箱
gain factor 增益(放大)因数
gain in weight(=weight gain) 重量增加；增重
gain in yield (产品)产率增加
gain of leather 得革率
gaize 生物蛋白岩
gal 伽〔重力加速单位，等于 1 厘米/秒²〕
galactal 半乳醛
galactan 半乳聚糖 $(C_6H_{10}O_5)$
galactaric acid 半乳糖二酸；黏酸
galactinol 肌醇半乳糖苷
galactite 乙基半乳糖
galactitol 半乳糖醇；卫矛醇
galactoaraban 半乳糖阿拉伯糖胶
galactochloral 半乳糖氯醛
galactococcus 乳球菌
galactoflavin 半乳糖黄素
galactofuranose 呋喃半乳糖
galactoglucomannan 半乳糖葡萄糖甘露聚糖
galactometer 乳(比)重计
galactomethylose 岩藻糖
galactose 半乳糖 $C_5H_{11}O_5CHO$
galactotoxin 乳毒素
galacturia 乳糜尿(症)
galam butter 牛油树油
galanga 高良姜
galangal 良姜
galangal oil(=galange oil) 高良姜油
galangal root 良姜；红豆蔻
galange oil 高良姜油

galangin 高良姜精
galanginidin 高良姜精定
galanthamine(=lycoremine) 雪花胺；加兰他敏
galanthidine(=lycorine) 石蒜碱
galanthine 雪花碱
galantin 甘丙氨菌素
galaxolide 加乐麝香
galbanol 格蓬醇
galbanum 古蓬香脂；波斯树脂
galbanum fruit oil 格蓬果油
galbanum gum(resin) 古蓬树胶(树脂)
galbanum oil 格蓬油
galegine 山羊豆碱
galena 方铅矿
galenical 草药
galenite 方铅矿
galenobismuthite 辉铅铋矿
Galigher tilting filter 加林格尔斜式过滤器
galiium sulfate 硫酸镓 $Ga_2(SO_4)_3$
Galileo number 伽利略数
galiosin 茜素酸苷
galipeine 尬梨因
galipene 尬梨烯
galipidine 尬梨定
galipine 尬梨频
galipoidine 尬梨波定
galipol 尬梨醇 $C_{15}H_{26}O$
galipoline 尬梨波灵
galipot 海松树脂
galipot gum 海松树胶
galipot resin 海松树脂
galirubin 加利红菌素
galitannic acid 猪殃单宁酸 $C_{14}H_{16}O_{10}H_2O$
gall ①棓子；没食子②胆汁
gallacetophenone(=alizarin yellow) 2,3,4-三羟基苯乙酮；茜素黄 $C_6H_2(OH)_3COCH_3$
gallal(=aluminum subgallate) 碱式棓酸铝
gallamide 棓酰胺；3,4,5-三羟基苯甲酰胺 $(HO)_3C_6H_2CONH_2$
gallane 镓烷
gallanilide(=gallanol) 棓酰苯胺；棓酚；3,4,5-三羟基苯甲酰苯胺 $C_{13}H_{11}NO_4$
gallanol(=gallanilide) 棓酰苯胺；棓酚；3,4,5-三羟基苯甲酰苯胺 $C_{13}H_{11}NO_4$
gall apple(=gall nut) 棓子
gallate ①棓酸盐 $C_6H_2(OH)_3COOM$②棓酸；没食子酸③镓酸盐 $MGaO_2$
gallate propyl 没食子酸丙酯

gallerte 凝胶体
galley furnace(=gallery furnace) 长廊炉
gallic ①正镓的；三价镓的②棓子的
gallic acid ①(=3,4,5-trihydroxybenzoic acid)棓酸；五倍子酸；没食子酸②镓酸
gallic acid equivalent(=GAE) 棓酸当量；没食子酸当量
gallic acid ester 棓酸酯；没食子酸酯；五倍子酸酯
gallic compound 正镓化合物
gallic hydroxide 氢氧化镓 $Ga(OH)_3$
gallicin(=methyl gallate) 棓酸甲酯 $(HO)_3C_6H_2CO_2CH_3$
gallic oxide 三氧化二镓 Ga_2O_3
gallin 棓灵
galling ①磨损；擦伤②卡住；黏住
gallinol 棓酚
gallion 试镓灵
gallipot 软膏壶〔药〕
gallium 镓〔31号元素，化学符号Ga〕
gallium dichloride 二氯化镓
gallium family 镓族
gallium family element 镓族元素
gallium hydroxide 氢氧化镓 $Ga(OH)_3$
gallium monoxide 一氧化镓 GaO
gallium nitrate 硝酸镓 $Ga(NO_3)_3$
gallium potassium alum(=gallium potassium sulfate) 镓钾矾；硫酸镓钾
gallium potassium sulfate(=gallium potassium alum) 硫酸镓钾；镓钾矾 $K_2SO_4 \cdot Ga_2(SO_4)_3 \cdot 24H_2O$；$KGa(SO_4)_2 \cdot 12H_2O$
gallium selenate 硒酸镓 $Ga_2(SeO_4)_3$
gallium sesquioxide 三氧化二镓 Ga_2O_3
gallium trichloride 三氯化镓 $GaCl_3$
Gall-Montlaux cell 戈尔-芒劳克斯电池
Gall-Montlaux process 戈尔-芒劳克斯氯酸制造法
gall nut 棓子；五倍子；没食子
gallnuts extract 棓子浸膏
gallobenzophenone 2,3,4-三羟基苯基(甲)酮 $C_6H_2(OH)_3COC_6H_5$
gallobromol(=dibromogallic acid) 二溴棓酸 $C_7H_4O_5Br_2$
gallocatechin 棓儿茶酸
gallocyanin 棓花青；媒染棓酸青 $C_{15}H_{12}O_5N_2$
gallocyanine 棓花青；媒染棓酸青〔染〕
galloflavin 棓黄素
gallogen(=ellagic acid) 鞣花酸；棓原 $C_{14}H_2O_4(OH)_4$
gallol 棓酚；N-3,4,5-三羟苯基苯胺 $C_6H_5NHC_6H_2(OH)_3$
gallon 加仑
gallonage 加仑量
gallon-degree 加仑-度〔美国采用的冷却单位〕

gallon-octane 加仑-辛烷值〔加仑量和辛烷值的乘积〕
gallophenine GD 棓吩宁 GD
gallotannic acid 棓单宁酸；鞣酸
gallotannin 棓单宁
gallous 亚镓的；二价镓的
gallous chloride 氯化亚镓 $GaCl_2$
gallous compound 亚镓化合物
gallous oxide 氧化亚镓 GaO
galloyl 棓酰；3,4,5-三羟苯甲酰 $(3,4,5)-(OH)_3C_6H_2CO-$
gall pigment(=bile pigment) 胆汁色素
galmey 异极(锌)矿
galvanic 电流的
galvanic action 动电(化学)作用；电池作用
galvanic anode 自发阳极；流电阳极；动电阳板；牺牲阳极
galvanic battery 蓄电池组；原电池组
galvanic cell 伽伐尼电池；原电池
galvanic corrosion 原电池腐蚀
galvanic current 动电电流；直流电流
galvanic deposit 电沉积
galvanic effect 电化(学)效应
galvanic electricity 动电；流电
Galvanic electric potential difference 伽伐尼电位差；绝对电位差
galvanic element (原)电池
galvanic pile(=voltaic pile) 电堆
galvanic polarization 电极极化作用
galvanic potential 伽伐尼电位；电转电位
galvanic protection ①防原电池腐蚀；防电流腐蚀②阴极保护
galvanic series 电势序
galvanic series in sea water 海水中腐蚀电位序表
galvanised steel ①镀锌钢②镀锌钢材
galvanization 镀锌(作用)
galvanized 镀锌的；电镀的
galvanized iron 白铁〔商〕；镀锌铁
galvanized (iron) plain sheet 白铁片；平铁片〔商〕
galvanized sheet iron 镀锌铁皮；白铁(皮)〔俗〕
galvanizer ①电镀工②电镀器
galvanizing 镀锌
galvanizing flux 镀锌液
galvanizing kettle 镀锌锅
Galvano-chemistry 电化学
galvanograph 电流记录图；电镀版
galvanogustometer 电味觉计
galvanolysis 电解
galvanometer 电流计；检流计
galvanometer recorder 检流计记录器

galvanometric torque-twist tester 扭转式电流(测定)仪
galvanometry 电流测定法
galvanoplasty 电铸
galvanoscope 验电流器
galvanostat 恒电流计
galvanostatic method 恒电流法
galvanostat method 恒电流法
galvanotaxis 趋电性
galvanotropism 向电性；趋电性
galvano-voltameter(=voltameter) 伏安计
galvano-voltammeter(=voltammeter) 伏安计
galyl 伽利耳〔一种含砷剂〕
gamabufagin 和蟾蜍精〔取自日本产蟾蜍〕
gamabufogenin 和蟾蜍配基
gamabufotalin 和蟾蜍他灵
gamabufotoxin 和蟾蜍毒
gama-vulcanizate (丙种射线)辐射硫化胶
gambier(=gambir) 黑儿茶
gambine ①邻亚硝基萘酚②媒染绿③毛皮绿
gambir(=gambier) 黑儿茶
gambirine 黑儿茶碱
gamboge 藤黄
gambogia 藤黄
gambogic acid 藤黄酸
gamet 砾岩
game theory 博弈论
gamma absorptiometer γ(射线)吸收计〔测铀、钚等浓度用〕
gamma absorption analysis γ(射线)吸收分析(法)
gamma absorptometer γ(射线)吸收计
gamma acid 伽马酸；2-氨基-5-萘磺酸 $NH_2C_{10}H_5SO_3H$
gamma-activation 用γ射线活化
gamma-aminopropyl triethoxy silane γ-氨丙基三乙氧基甲硅烷〔偶联剂〕 $NH_2CH_2CH_2CH_2Si(OC_2H_5)_3$
gamma camera γ照相机
gamma carbon γ碳原子
gamma cellulose γ纤维素
gamma distribution model γ分布模型
gamma excited X-ray fluorimeter γ射线激发X射线荧光(分析)仪
gamma ferrite γ铁
gamma function γ函数〔数〕
gammagraph γ射线照相
gamma-iron oxide γ氧化铁
gamma lay γ层
gamma nonyl lactone γ-壬内酯；椰子醛〔商〕；十八醛〔商〕
gamma number γ值；酯化度
gamma oxidation γ氧化

gamma portion　γ部分
gamma position　γ位
gamma-radiation　γ辐射
gamma-radiation-induced polymerization　γ辐射诱导聚合(作用)
gamma-radiator　γ辐射体；γ辐射源
gamma-ray　γ射线
gamma-ray absorption　γ射线吸收
gamma-ray counter　γ射线计数器
gamma-ray densimeter　γ射线密度计
gamma-ray detector　γ射线探测器
gamma-ray equipment　γ射线设备
gamma-ray-initiated polymerization　γ射线引发聚合(作用)
gamma-ray laser　①γ射线激光器②γ激光
gamma-ray nuclear resonance spectrofluorimetry　γ射线核共振分光荧光法
gamma-ray orbitography　γ射线眼科照相法
gamma-ray penetration type liquid density meter　γ射线穿透式液体密度计
gamma-ray photon　γ光子
gamma-ray polymerization　γ射线(引发)聚合
gamma-ray spectrometer　γ射线分光计
gamma ray spectroscopy　γ射线光谱学
gamma source　γ(射线)源
gamma space　γ空间
gamma spectrometer　γ谱仪
gamma spectrometry　γ谱法
gamma spectroscopy　γ谱学
gamma substitution　γ位取代
gamma undecalactone　γ-十一内酯；桃醛〔商〕；十四醛〔商〕
gamma value　γ值；酯化度
gamophen　六氯酚
gamut　①整个范围；全部②色域
ganciclovir　更昔洛韦〔药〕
gancidin(=cancidin)　灭癌素
gangaleoidin　节枝定
gang drive　集体传动；成组传动
gang-driven　同轴传动
gang mold　多巢模
gangtokumycin(=gangtomycin)　甘托克霉素
gangtomycin(=gangtokumycin)　甘托克霉素
gangue　脉石；矿渣；尾矿
g-anisotropy　g值各向异性
ganister　致密硅岩
ganmycin(=carcinomycin)　癌霉素
ganomalite　硅钙铅矿

ganophyllite　辉叶石
gaoline filter　汽油过滤器
gap　①(间)隙；缝②(翼)隔〔机工〕
gap filler　裂缝填充物；缝隙填料
gap-filling adhesive　空隙充填性黏合剂
gap-filling cement　填缝胶泥
gap-filling properties　间隙填充性能；填缝性能
gap joint　带缝接头
gapless superconductivity　无隙超导性
gapped cuts　跨切馏分
gap setting　①辊隙调定②(挤塑模)模口间隙调定
gap size　间隙尺寸
gap test　空隙检验
garage fuel-trap　停车场的燃料过滤器
garage poison　汽车库毒
garancin　茜色淀
garantose(=saccharin)　糖精
garbage　(食用)下脚
garbage grease　下脚油脂
garcia nutans oil　(产于墨西哥等地的)大戟科植物干性油
garcinin　藤黄皮素
garcinolic acid　藤黄酸
garden celandine　白屈菜干
garden cress　积鸡菜
garden cress oil　独行菜油；园芹油
garden heliotrope　缬草
gardenia absolute　栀子花净油
gardenia compound　栀子香精
gardenia concrete　栀子花浸膏
gardenia flower oil　栀子花油
gardenin　栀子宁
garden rocket oil　园岩油
garden rue oil　夏芸香油
Gardner bubble viscometer　加德纳气泡黏度计
Gardner-Coleman oil absorption point　加德纳-柯尔曼吸油点
Gardner-Coleman rub-out oil absorption method　加德纳-柯尔曼刮刀混合吸油量测定法
Gardner color　加德纳颜色
Gardner color difference meter　加德纳色差计
Gardner color scale　加德纳色标
Gardner color standard　格氏色标准；加德纳色标准
Gardner color standard number　加德纳色标号
Gardner consistency　加德纳稠度
Gardner drying meter　加氏(漆膜)干燥计
Gardner drying time recorder　加德纳干燥时间记录器
Gardner engine　加德纳引擎；润滑油试验用 L-2 型单缸

发动机
Gardner gloss meter　加德纳光泽计
Gardner-Holdt bubble tube　加氏(气泡)管
Gardner-Holdt viscosity tube　加德纳-霍尔特黏度管
Gardner impact strength　冲击强度
Gardner letter viscosity　加德纳字母黏度
Gardner mark　加氏符号；格氏记号〔格氏管黏度的表示符号〕
Gardner mobilometer　格氏淌度计；加德纳淌度计
Gardner-Park permeability cup　加德纳-帕克渗透杯
Gardner-Parks adhesion test　加德纳-帕克斯黏接试验
Gardner-Park's tensile strength meter　加德纳-帕克抗张强度试验机
Gardner straight line wash ability machine　加德纳直线式耐洗刷性试验机
Gardner variable impact tester　加德纳可调冲击试验器
Gardner vertical viscometer　加德纳竖式黏度计
gard technique　沙门氏菌单相分离技术
gargle　含漱剂
gargoylism　软骨代谢障碍病
garlandosus　加得菌素
garlicin　大蒜素
garlic oil　蒜油
Garlock packing　加洛克衬垫
garment grading sketch　服装放码图；服装推档图；推档总图
garment leather　衣服革；皮革服装
garment wax printing　服装蜡染
garmin　肉叶芸香碱
garnet　石榴子石*
garnet abrasive　石榴石磨料
garnetiferous skarn　石榴硅卡岩
garnet lac　深红色的精制紫胶
garnet laser　石榴石激光器
garnet paper　①(金刚)砂纸②红晶色纸
garnet red　紫酱红
garnet shellac　石榴紫胶
garnetted stock　回收纤维
garnierite　硅镁镍矿
Garrique evaporator　加立克蒸发器
garryine　加山萸碱
garter-type spring　环形弹簧
gas　①气体②气态③煤气④毒气
gas absorbent　气体吸收剂
gas absorbent bed　气体吸收床
gas absorber oil　气体吸收油；洗油
gas absorption　①气体吸收②毒气的吸收
gas adsorbent bed　气体吸附床

gas-adsorbent carbon　吸附气体用碳
gas-adsorbent coal　吸附气体的焦炭
gas adsorption　①气体吸附②毒气吸附
gas adsorption chromatography　气体吸附色谱法
gas adsorption technique　气体吸附技术
gas aided injection molding　气辅注塑
gas-air flame　气体-空气焰
gas-air interface　气体-空气界面
gas alarm　毒气警报
gas amplification　气体膨胀
gas analysis　气体分析
gas analysis apparatus　气体分析器
gas analyzer　气体分析器
gas and oil separator　油矿的油气分离器
gas-assist injection molding　气体辅助注射成型
gas-atomizing nozzle　气体雾化喷嘴
gas attack　①毒气攻击②气体腐蚀
gas bacillus　产气杆菌
gas bag　气袋
gas balance　气体天平
gas balloon　称气瓶；气体比重瓶
gas-barrier property　气体屏蔽性；阻气性
gas bath　气浴
gas battery　气体电池组
gas behavior　气体行为
gas bell　气柜钟罩
gas black　气黑；气烟末
gas blanket　气(体)层
gas blanket centrifuge　气体保护式离心机
gas bleaching　气体漂白
gas blow mixing　气泡搅拌
gas blown bubble　气发泡孔〔泡沫塑料〕
gas blown foam　气发泡沫塑料
gas blow off　放气；排气
gas blow-out　气体吹出
gas blue　铁蓝〔指由煤气副产氰化物废液生产的亚铁氰酸盐〕
gas bomb　①气体钢瓶②毒气(炸)弹
gas-booster　辅助压缩设备〔气体输送〕
gas bottle　洗气瓶
gas bubble　气泡
gas burette　气体量管
gas burner　①燃气喷嘴②煤气喷灯
gas by-products　气体工业的副产品
gas calorimeter　煤气量热器
gas capacity　气体储器容积
gas carbon　气碳
gas carburization　气体渗碳

gas cartridge 蓄气筒
gas cavity 气孔
gas cell ①气极电池②毒气帐篷
gas cementation 气体渗碳
gas centrifuge process 气体离心法
gas certificate 储器中气体的检定
gas chamber 毒气室
gas characteristic 气体特性
gas checking 晶纹；气裂
gas check valve 闭气阀
gas chromatogram 气相色谱(图)
gas chromatograph 气相色谱仪*
gas chromatograph-gas chromatography 二维气相色谱仪
gas chromatographic analysis 气相色谱分析
gas chromatographic method 气相色谱法
gas chromatographic technique 气相色谱技术
gas chromatograph-ion trap detector 气相色谱-离子阱检测器联用
gas chromatograph-mass spectrometer system 气相色谱仪-质谱计系统
gas chromatograph-quadrupole mass spectrometer 气相色谱四极质谱仪
gas chromatography(GC) 气相色谱法
gas chromatography hygrometer 气相色谱湿度计
gas chromatography-infrared technique 气相色谱-红外联用技术
gas cleaner 气体净化器
gas cleaning 气体净化
gas cleaning unit 气体净化装置
gas cloud 毒(气)云
gas coal 气煤〔矿〕
gas coke 煤气焦炭
gas collecting main 煤气聚焦总管；煤气总管
gas collecting tube 煤气聚集管
gas collector 煤气聚集器；集气器
gas column 气柱
gas combustion 气体燃烧
gas-combustion pipette(=slow combustion pipette) 缓燃(烧)吸移管；气体燃烧吸移管
gas compartment 气体空间；气体室
gas compressibility correction factor 气体压缩系数校正因子
gas compressor 压气机
gas condensate 凝析油
gas condenser 气体冷凝器
gas conditioning 气体处理
gas conduit 气体导管

gas constant 气体常数
gas container 容气器
gas content 气体含量
gas conversion process 气体转化过程
gas cooled fast breeder reactor(GCFBR) 气冷快中子增殖(反应)堆
gas cooler ①气体冷却器②煤气冷却器
gas cooling 气冷的
gas core reactor 气芯(反应)堆
gas corrosion 气体腐蚀
gas coulometer 气体电量计
gas crazing 细裂(见底)〔漆病〕
gas cure (用)气体硫化
gas cutting 氧炔割；气割
gas cylinder ①毒气筒②集气筒
gas defense 毒气防御
gas defense for animals 畜类的毒气防御
gas degeneration 气体简并
gas delivery ①气体分离②气体输送
gas densitometer 气体密度计
gas density balance 气体密度天平
gas density balance detector 气体密度天平检测器
gas density meter 气体密度计
gas density recorder 气体密度记录器
gas depoisoning 气体消毒
gas desorption 气体解吸
gas detection work 毒气侦查勤务
gas detector 气体检测器〔检查漏气用〕
gas-development agent 发泡剂
gas dewatering 气体脱水
gas diffusion column 气体扩散柱
gas diffusion electrode 气体扩散电极
gas diffusion process 气体扩散过程
gas diffusion unit 气体扩散装置
gas discharge 气体放电
gas discharge detector 气体放电检测器
gas discharge ion source 气体放电离子源
gas discharge lamp 气体放电灯
gas discharge plasma 气体放电等离子体
gas discharge tube 气体放电管
gas disease 毒气病
gas dispersoid 气溶胶；气态分散体
gas distillate 气体馏出物；富含汽油烃类的天然气
gas distributing system 煤气管道网
gas distributor 气体分布器；分气盒
gas-driven displacement 气动置换
gas-driven displacement pump 气动置换泵
gas dry filter trap 气体干燥滤阱

gas drying bottle	气体干燥瓶
gas drying unit	气体干燥塔
gas duct	烟道
gasdynamics	气体动力学
gaseity	气态
gas electrode	气体电极
gas engine	煤气发动机
gas engine oil	气机润滑油
gas entrainment method	加气法〔聚氨酯发泡〕
gas entrainment type fermentor	气体吸入式发酵罐
gaseous	气体的；气态的
gaseous cement	气体渗碳剂
gaseous combustion	气体燃烧
gaseous conductance	①气体传导②气体电导
gaseous contaminant	气态污染物
gaseous corrosion	气相腐蚀；气体腐蚀
gaseous detonation	气体爆炸
gaseous diffusion	气体扩散
gaseous diffusional separation	气体扩散分离(法)
gaseous diffusion separator	气体扩散分离器
gaseous discharge	气体放电
gaseous dissociation	气体离解
gaseous effluent	排气；废气
gaseous effluent cooler	废气冷却器
gaseous electronic detector	气态电子检测器
gaseous emission	气体发射；气体排放
gaseous enclosure	气体留藏
gaseous equilibrium	气相平衡
gaseous escape	气体逸出；漏气
gaseous exchange	换气
gaseous film	气态膜
gaseous fluid	气态流体
gaseous fluidization	气态流化作用
gaseous fuel	气体燃料
gaseous fuel calorific value	气体燃料的热值
gaseous fuel dew point	气体燃料的露点
gaseous fuel specific gravity	气体燃料的比重
gaseous fuel water content	气体燃料的水分含量
gaseous gap	气隙
gaseous hydrocarbons	气态烃类
gaseous impurities	气态杂质；气体杂质
gaseous inclusion	气体留藏
gaseous mixture	气体混合物
gaseousness	气态
gaseous oxygen	气态氧
gaseous phase	气相
gaseous pollutant	气态污染物
gaseous polymerization	气相聚合(法)
gaseous product	气体产物
gaseous quenching device	气体骤冷装置
gaseous radwaste	气态放射性废物
gaseous reagent	气体试剂
gaseous sample	气体样品
gaseous solubility	气体溶解性
gaseous solution	气溶体
gaseous state	气态
gaseous steam	气态蒸汽；过热蒸汽
gaseous sterilization	气体灭菌
gaseous tension	气体压力；气体张力
gaseous titanium tetrachloride	气态四氯化钛
gaseous waste	废气
gas equation	气体方程式；气体公式
gaser	γ射线激光(器)
gas evolution	①气体发生②气体逸出；气体释放
gas evolution analysis	流出气体分析
gas evolution method	气体发生法
gas expanded rubber	①微孔橡胶②闭孔泡沫胶
gas expansion	气体膨胀
gas explosion tube	气体爆燃管
gas exposure labeling	气体曝射标记(法)
gas factor	①油气比②气体产率
gas-fading inhibitor	气(烟)熏褪色抑制剂
gas family	石油气体；烃族
gas fastness	耐气体坚牢度
gas field	天然气田
gas-filled detector	充气检测器
gas-filled lamp	灌气(电)灯泡
gas-filled thermometer	充气温度计
gas-filled tube	充气管
gas-filler incandescent lamp	充气白炽灯泡
gas-filling station	充气站
gas film	气膜
gas film coefficient	气膜系数
gas film controlling	气膜控制
gas film mass transfer coefficient	气膜传质系数
gas filter	滤气器
gas-fired	燃烧煤气的
gas-fired furnace	烧煤气的炉子
gas-fired heater	燃气加热炉
gas fired infrared drying oven	煤气加热红外线干燥炉
gas firing	燃气供暖法
gas fittings	气管接头
gas flame	煤气火焰
gas flame thermopile	气焰热电偶
gas flow	气流
gas flow controller	气流控制器

gas flow counter 气(体)流(量)计数器
gas flow impedance detector 气流阻抗检测器
gas flow meter 气流计
gas flow path 气体流路
gas flow rate 气体流速
gas flue 烟道
gas flush syringe 气冲注射器
gas-foam analysis 起泡分析
gas for flame analysis 火焰分析用气体
gas-forming material 成气物料；形成气体的物质
gas free 不含气的
gas-free certificate 石油储槽中气体检定器
gas-freed tank 无油品蒸气的储罐
gas fuel 气体燃料
gas fume 气熏
gas furnace 煤气炉
gas gathering system 气体收集系统
gas gauge (流体)压力计；压力表
gas-gel chromatography 气体凝胶色谱(法)
gas generating bottle 气体发生瓶
gas generation agent 气体发生剂；发泡剂；发孔剂
gas generator 气体发生器
gas gouging 气割
gas graphite reactor 气冷石墨(反应)堆
gas grid 气体供应网
gash 裂口
gas handling system 进气系统
gas heater 煤气暖炉
gas heating 气体加热
gas heating phosphating tank 煤气加热磷化处理槽
gas holder ①储气器；气柜②煤气储柜；气体储罐
gas-holder bell 储气器的钟罩
gas-holder foundation 气柜基础
gas-holder grease 储气器润滑脂
gas-holder operation 气柜操作
gas holder tank 储气器
gas hold-up 气体滞留
gas hold-up time 气体滞留时间
gas hold-up volume 气体保留体积
gas horsepower 气体马力
gas hourly space velocity 气时空速
gas house 煤气房
gas house coal tar(=gas house tar) 煤气房煤焦油
gas house tar(=gas house coal tar) 煤气房煤焦油
gas hydrate (石油)气体水化物
gasifiable 可气化的
gasificating desulfurization 气化脱硫
gasification 气化(作用)

gasification latent heat 气化潜热
gasification (of coal) in place (煤的)地下气化
gasification zone 气化带
gasiform 气态的；(形成)气体的
gasifying 气化
gasifying reactor 气化反应器
gas ignition 气体着火
gas impermeability test 不透气性试验
gas-impermeable cloth 不透气布
gas-impervious 不透气的
gas impurity 气体杂质
gas industry 煤气工业
gas infrared spectrograph 气体红外光谱仪
gasing 充气
gas insertion probe 插入气体探头
gas in solution 溶解气
gas interferometer 气体干扰仪
gas ion 气体离子；气态离子
gas ionization chamber 气体电离室
gas ionization detector 气体电离检测器
gasitication 厌氧气体发酵法；(污水)甲烷发酵法
gas jet 煤气嘴
gas-jet method 气体喷射法
gasket ①垫圈；垫片②接(合)垫料
gasket coating 密封涂料；密封涂层
gasket cutter 切垫片机；垫片旋切车床
gasket groove 按放垫片环槽
gasketing material 衬垫料
gasket joint 垫圈接头
gasket packing 板式填料；密封垫圈
gasket ring 密封圈；密封环
gasket seat 垫圈座
gasket splicer 密封圈接头机
gas lamp 煤气灯
gas law 气体定律
gas law correction factor 气体定律校正因素(因子)
gas law equation 气体定律方程式
gas leak 漏气
gas leakage 气体泄漏
gas leakage test 气密性试验
gas-lift 气升
gas-lift unit 气升循环装置
gas light ①煤气灯光②煤气灯
gas-like phase 类气相
gas lime 煤气石灰〔其中含有氢硫化钙等杂质〕
gas line 煤气管
gas line dehydrator 气体管道脱水器
gas liquefaction 气体液化

gas-liquid chromatography(=gas-partition chromatography)　气液色谱法
gas-liquid partition chromatography　气液分配色谱法
gas-liquid partition coefficient　气液分配系数
gas liquid reaction　气(体)液(体)反应
gas-liquid-solid chromatography　气液固色谱法
gas liquor　煤气水
gas-load accumulator　气压蓄力器
gas lock　气封；气塞；气栓
gas loss　气体损失
gas main　煤气总管
gas-making　煤气制造
gas-making log book　煤气制造记录簿
gas-making period　气化周期
gas making retort　生气曲颈甑
gas manometer　气体压力计
gas marks　气泡；麻孔
gas mask　防毒面具；毒气面具
gas mask canister　防毒面具罐
gas mask charcoal　防毒面具用炭
gas mask for civilian　民众防毒面具
gas mask with valve expiration　带阀呼吸面具
gas meter　①气量计；气表②煤气表
gas meter leather　气表革
gas microanalysis　气体微量分析
gas mixer　气体混合器
gas mixture　①气体混合物②毒气混合物
gas mortar　毒气迫击炮
gas motor fuel　气体动力燃料
gas multiplication　气体放大
gas naphtha　气体石脑油
gas nipple　气嘴
gas nozzle　排气口；喷气嘴
gaso(=gasoline)　汽油
gas odorant　气体加味剂
gas odorizer　气体臭味鉴定器
gasogene　①木炭燃气②汽水制造机
gas oil　瓦斯油；粗柴油；汽油
gas oil cracking　粗柴油裂解
gas oil cracking process　瓦斯油裂化法
gas-oil exchanger　油气热交换器
gas-oil fluid viscosity　含溶解气石油的黏度
gas-oil interface　油气分界面
gas-oil mixture　气态和液态石油产品混合物
gas-oil ratio(GOR)　油气比
gas-oil recycle stock　循环瓦斯油
gasol　气体油
gasolene(=gasoline)　汽油

gasoline(=gaso;gasolene)　汽油
gasoline acidity test　汽油酸度测定
gasoline additive(=gasoline dope)　汽油添加剂
gasoline-air mixture　汽油-空气混合物
gasoline alkylate　烷基化汽油
gasoline antifreeze mixture　汽油防冻混合物〔作为冷冻液体的汽油混合物〕
gasoline anti-icing additive　汽油防冻添加剂
gasoline barge　汽油驳船
gasoline blast burner　汽油喷灯
gasoline breakdown　汽油深度氧化
gasoline burner　汽油灯
gasoline can　汽油桶；汽油罐
gasoline carrier　①汽油储箱〔运送用〕②汽油槽车
gasoline chamber　汽油室
gasoline coalescer　汽油凝结剂
gasoline condenser　汽油冷凝器
gasoline consumption　汽油消耗；油耗
gasoline content of natural gas　天然气的汽油含量
gasoline corrosion　汽油腐蚀
gasoline corrosion test cup　汽油试蚀杯
gasoline detergent additive　汽油清洁添加剂
gasoline distillation　汽油蒸馏
gasoline dope(=gasoline additive)　汽油添加剂
gasoline engine　汽油机
gasoline filling station　加汽油站
gasoline fraction　汽油馏分
gasoline freezing point　汽油凝固点
gasoline gas　汽油气
gasoline gauge　汽油表
gasoline gauge dial　汽油表标度板(盘)
gasoline grade　汽油级别
gasoline gravity　汽油比重
gasoline gum　汽油胶质
gasoline gumming test cup　汽油试胶杯
gasoline hydrocarbon　汽油烃
gasoline knocking　汽油爆震性
gasoline leak　汽油漏失
gasoline level　汽油液面
gasoline level gauge　汽油液面计
gasoline line　汽油管路
gasoline loss control　汽油损失控制
gasoline mercaptan absorber　汽油中硫醇的吸收塔
gasoline meter　汽油计量表
gasoline mileage　汽油英里程〔每加仑汽油行驶的距离〕
gasoline mixture　汽油混合物
gasoline octane number(=gasoline octane rating)　汽油辛烷值

gasoline octane rating(=gasoline octane number)　汽油辛烷值
gasoline oil(=gasoline)　汽油
gasoline-oil consumption ratio　汽油-油消耗比
gasoline oxidation　汽油氧化
gasoline performance　汽油(在发动机中)的行为
gasoline pick-up fraction　拾获汽油馏分
gasoline plant　(气体)汽油厂
gasoline pool　汽油池〔将不同质量汽油掺合成一定规格的产品汽油〕
gasoline power fraction　汽油主馏分
gasoline precipitation test　汽油沉淀试验
gasoline pressure gauge　汽油压力计
gasoline production　汽油生产
gasoline-proof grease　防汽油的润滑脂
gasoline pump　汽油泵
gasoline pump room　(转注)汽油的泵室
gasoline receiver　汽油接受器
gasoline refining　汽油精制
gasoline resistance　耐汽油性
gasoline scale　汽油秤
gasoline separator　汽油离析器
gasoline shut-off　汽油管阀
gasoline soap　汽油皂
gasoline splitter　汽油分馏塔
gasoline stabilizer　①汽油稳定塔②汽油稳定剂
gasoline startability　汽油起动性质
gasoline station　(加)汽油站
gasoline storage　汽油库
gasoline strainer　汽油过滤器
gasoline substitutes　汽油代用品
gasoline suction lift　吸提汽油
gasoline sulfur test　汽油中硫含量测定
gasoline sweetener　汽油脱硫装置
gasoline system　加汽油的系统
gasoline tank　汽油罐；汽油箱
gasoline tanker　汽油运输船
gasoline tank gauge　汽油罐液面指示器；汽油箱液面指示器
gasoline testing outfit　汽油试验装置
gasoline tetraethyl lead test　汽油中四乙铅含量测定试验
gasoline trap　汽油捕集器；汽油阱
gasoline treating process　汽油精制过程
gasoline upgrading　汽油提(高)质(量)；汽油改质
gasoline value　汽油值
gasoline vapour　汽油蒸气
gasoline vapour pressure　汽油蒸气压
gasoline volatility test　汽油挥发性试验

gasoline-water solution　汽油-水溶液
gasoline yield　汽油产率
gasoloid　气溶胶；气胶溶体
gasometer　①气量计；气表②气柜；储气罐
gasometer flask　气量计；气量瓶
gasometer leather　气表革
gasometric　气体定量的
gasometric analysis　气体定量分析
gasometric titration　气压滴定法
gasometry　气体定量分析法
gas-operated　用煤气操作的
gasoscope　气体检验器
gas outlet　排气管；导气管
gas output　煤气输出量
gas oven　煤气炉
gas-partition chromatography(=gas-liquid chromatography)　气液色谱法
gas path　气路
gas pellets　粒状发泡剂
gas permeability　透气性
gas permeation　气体透过；透气
gas-per-mile gauge　汽油每英里耗量计；每英里路程燃料消耗的计量器
gas-pervious　透气的
gas phase　气相
gas phase chemiluminescence　气相化学发光
gas phase coulometry　气相库仑法；气相库仑滴定法；气相电量滴定法
gas phase electron adsorption　气相电子吸附
gas phase permeation　气相渗透
gas phase polymerization　气相聚合
gas phase protein sequenator　气相蛋白质序列分析仪
gas-phase reaction　气相反应
gas-phase suspension process　气相悬浮过程
gas phase titration　气相滴定
gas phase xanthation　气相黄化
gas pin　(气)针孔
gas pipe　①气管②煤气管
gas pipe line　气体管线
gas pipette　验气球管
gas plant　煤气厂
gas planum　充气室
gas pocket　气囊；气孔；气窝〔塑料〕
gas pressure regulator　气体调节器；气压控制器
gas producer　煤气(发生)炉
gas-producing factor　气体生成因数
gas projector　毒气掷射器
gas-proof　①防(毒)气的②不透气的

gas proofness　不透气性；气密性
gas-proof shelter　避毒所
gasproof test　①防污气性试验②不透气性试验
gas propellant　①(火箭发动机)气体燃料②(气溶胶)喷射剂
gas proportioner　气体比例调节器
gas pulsation　气体脉动作用
gas pump　气体泵
gas purger　放气器；不凝性气体排除器；气体排放器
gas purging　气体排除；气体排放
gas purification　气体精制；气体净化
gas purifier　①气体净化器②煤气净化器
gas purifying　①气体净化(作用)②煤气净化(作用)
gas radiation　气体辐射
gas range　煤气灶
gas reaction　气体反应
gas receiver　储气桶；气柜
gas recovery　气体回收
gas recovery system　气体回收系统
gas recycle process　气体循环过程
gas recycle pump　气体循环泵
gas reformer　煤气体转化器
gas reforming　气体转化；煤气转化〔从高卡煤气转化为低卡煤气的过程〕
gas refractometer　气体折光计
gas regulator　气体调节器
gas release　①发泡②发泡量〔发泡剂〕
gas-release blowing agent　释气类发泡剂
gas relief line　①排气线②放气管线③减压管道
gas resistance detector　气阻检测器
gas retaining property　气体保持本领
gas retort　①干馏甑②干馏炉
gas retort carbon　煤气甑碳
gas reversion　裂化改质〔气态烃的转化〕
gas reversion process　气态烃裂化改质过程
gas reversion reaction　裂化改质反应
gas ring burner　环形气体燃烧器
gas rock　含天然(煤)气岩石
gas sample　气体试样
gas sample bulb　气体试样球管
gas sample scrubber　气样洗涤器
gas sample tube　气体取样管
gas sampling　气体取样
gas sampling tube　气体取样管
gas sampling valve　气体进样阀
gas saturation　气体饱和率
gas scattering effect　气体散射效应；气体扩散效应
gas scrubber　气体洗涤器

gas scrubber column　气体洗涤塔
gas scrubbing　涤气过程
gas scrubbing oil　涤气油
gas seal　气封
gas seepage　气体渗出
gas sensing electrode　气敏电极*
gas sensing probe　气敏探头
gas sensing sulphur dioxide probe　气敏二氧化硫探头
gas sensitive chromatograph　气敏色谱仪
gas sensitive chromatography　气敏色谱法
gas sensitive reflection spectrum　气敏反射光谱
gas sensitive semiconductor detector　气敏半导体检测器
gas sensor　①气体传感器②气敏元件
gas separate pot　气体分离罐
gas separation unit　气体分离装置
gas separator　气体分离器
gasser　(煤)气井
gas service pipe　气体分配管
gas shield　气体保护；气体防护；气体防护设备
gas shielded arc welding　气体保护电弧焊
gas singeing machine　煤气烧毛机
gassing factor　充气系数
gassing rubber　泡沫胶
gassing tendency　气体生成趋势
gas-solid adsorption　气固吸附(法)
gas-solid chromatography　气固色谱法
gas-solid contactor　气-固接触设备
gas-solid film　气固膜
gas sorption analyzer　气体吸着分析仪
gas space　气体空间
gas spirit　气态汽油
gas standard　气体参数
gas storage holder　储气器；储气罐
gas stove　煤气炉
gas stream　气流
gas stream atomizer　气流喷雾器
gas stripping　气提法
gas supply line　供气管线
gas-supply pipe　输气管
gas sweetening　气体净化；气体脱硫
gassy　气体的；气态的
gas syringe　气体注射器
gas tank　①气槽②煤气槽
gas-tanker(=gas-tank truck)　气槽车
gas-tank truck(=gas-tanker)　气槽车
gas tar　煤气焦油
gas tar emulsion　煤气焦油乳液
gas tension　气体张力

gasteromycetes 腹菌
gas thermometer 气体温度计
gas-tight 不透气的；气密的
gas-tight joint 不透气接合
gas tightness 气密性；不透气性
gas-tight shielded enclosure 气密屏蔽室
gas-tight syringe 气密注射器
gas-tight test 气密性试验
gas titration 气体滴定
gas trap 气体分离器；气阱；气体收集器
gas treating process 气体净化法；气体处理法
gas treating system 气体净化系统；气体净化装置
gastric acidity 胃液酸度
gastricism 消化障碍
gastric juice 胃液
gastric juice test 胃液检查
gastrodine 天麻素〔药〕
gas tube 气体管线；气柜
gas tungsten arc welding(GTAW) 钨电极惰性气体保护焊
gas turbine 燃气轮机；燃气透平
gas turbine engine lubricant 燃气轮机轮机润滑剂
gas turbine fuel 燃气轮机燃料
gas turbine liquid-fuel burner 燃气轮机液体燃料燃烧器
gas type mass spectrometer 气体型质谱计
gas vacuole 气泡
gas-vapour mixture 蒸气-空气混合物
gas velocity 气体速度
gas vent 气体出口；通气口
gas voltameter 气体电量计
gas volume detector 气体体积检测器
gas volumeter 气体体积计
gas volumetric analysis 气体容量分析
gas volumetric chromatography 气体体积色谱(法)
gas volumetric method 气体体积分析法
gas vulcanization (用)气体硫化
gas washer 湿煤气净化器
gas washing 洗气；气体洗涤；煤气洗涤
gas washing bottle 洗气瓶
gas wash tower ①煤气洗涤塔②气体洗涤塔
gas water 涤气用水
gas water-heater 煤气烧水炉
gas water ratio 气-水比
gas weld 气焊
gas welder 气焊机
gas welding 气焊
gas well 火井；天然气井
gas works 煤气厂

gasworks coal tar(=gasworks tar) 煤气(厂)焦油
gasworks coke 煤气(厂)焦炭
gasworks liquor 煤气(厂)水
gasworks pitch 煤气厂沥青
gasworks tar(=gasworks coal tar) 煤气(厂)焦油
gasworks tar pitch 煤气厂焦油沥青
gas yield 煤气产量
gas-yielding polymer 析出气体聚合物
gatavalin 谷缬菌素
gatch 蜡饼；含油蜡
gate ①大门②闸门③堰板
gate current 门控电流
gated beam 门控光束
gated decoupling 门控去偶
gated injection 门式进样法
gated photodetector 门控光电检测器；门控光电探测器
gated trapping 门捕获
gate feed hopper 框式加料斗
gate mixer 框式混合器
gate paddle agitator 框桨式搅拌器
gate stirrer 框式搅拌器
gate tensiometer 门式张力器；梳形张力器
gate type agitator 框式搅拌器
gate valve 闸阀；闸门阀
gatherer 收集器
gathering 收集；富集
gathering agent (螯合金属)搜集剂
gathering comb 分股集束器
gathering line 收集管线；输送管
gathering pipeline ①集油管线②集气管线
gathering station 收集站
gathering system 收集系统
gathering tank ①集油罐②收集罐
Gathurst powder 伽瑟斯特(炸)药
gatifloxacin 加替沙星〔药〕
Gattermann aldehyde synthesis 加特曼醛合成法〔一种甲酰基化法〕
Gattermann-Koch reaction 加特曼-科霍反应〔一种芳环甲酰化法〕
Gattermann reaction 加特曼反应
gauche 非对称的；左方的
gauche conformation(=skew conformation) 邻位交叉构象
gauche form 左右式；旁式
gauche-*trans* conformational transition 旁-反式构象转变
gauche-*trans* energy difference 旁-反式(构象)能量差
gauffer calender 浮花压制机
gauffered board 波面纸板

gauffered cloth 绉纹布
gauffer machine 压纹机
gauffrage 压纹
gauge(=gage) ①表②规；量规③计④样板
gaugeable 能量的；可计量的
gauge band 暴筋〔薄膜收卷缺陷〕
gauge block 块规
gauge bob 测深锤
gauge cock 水位表旋塞
gauged 校正的；计量的；标记的
gauged burette 标准滴定管
gauged cement 标定水泥
gauged lime 标准石灰；石灰浆
gauged oil （经过沉淀除去水和杂质)再计量的原油
gauged orifice 计量孔；检验孔
gauge feeler 探视
gauge glass 计液玻管；锅炉水位指示玻璃管
gauge hatch 计量口
gauge hole 计量孔
gauge invariant 规模不变量
gauge length 计算的样品长度
gauge line 计量管
gauge mark 计号
gauge nipple 计量口
gauge point 计量基准点
gauge pressure 计示压力；表压
gauger 计量器
gauge reading 计量仪器读数
gauge rod(=gauge stick) 探测杆；计量杆；表尺
gauge seal 表封；计量孔盖
gauge stick(=gauge rod) 探测杆；计量杆；表尺
gauge table 计量(图)表；校正表
gauge tank 计量槽
gauging ①量压②计量
gauging device 计量装置
gauging glass 计水玻管；锅炉水位指示玻璃管；玻璃水位计
gauging hatch(=gauge hatch) 计量口
gauging loss （石油产品)计算时的损失
gauging nipple(=gauge nipple) 计量口
gauging pass 定厚辊隙
gauging tank 计量桶；计量储槽
gaultheria oil 白珠油；冬青油〔俗〕
gaultheric acid 水杨酸甲酯
gaultherilene 白珠木烯
gaultherin 白珠木苷
Gauss cell 高斯电池
Gauss chain 高斯链

Gauss error curve 高斯误差曲线
Gauss(ian) approximation 高斯近似
Gaussian chain 高斯链
Gauss(ian) concentration profile 高斯浓度图
Gaussian curve 高斯线型*
Gaussian distribution 高斯分布；正态分布
Gaussian elimination method 高斯消元法
Gaussian elution band 高斯洗脱谱带
Gaussian equation 高斯方程
Gaussian error function 高斯误差函数
Gaussian function 高斯函数
Gaussian line shape 高斯线型
Gaussian network 高斯网络
Gaussian peak 高斯峰
Gaussian-shaped concentration distribution 高斯型浓度分布
Gaussian spring 高斯弹簧
Gaussian-type orbital(GTO) 高斯型轨道
Gaussmeter 高斯计
Gauss's double weighing method 高斯交换称量法
Gauss-Seidel iteration method 高斯-赛德尔迭代法
gauze 网；纱(布)
gauze and leno weave 纱罗组织
gauze cone 网状锥体
gauze element 网状滤心
gauze filter 网状滤器
gauze filter tray 网状过滤盘
gauze packing 网状填充物
gauze platinum electrode 铂网电极
gauze sieve 网筛
gauze strainer 网状滤器
gauze top burner 网头灯
gauze wire cloth 钢丝布
gauze with asbestos 石棉网
Gay-Lussac's method 盖吕萨克方法；等油(滴定)法
Gay-Lussac tower 盖氏塔；盖吕萨克塔
Gay-Lussac type 盖吕萨克型
gaylussite 单斜钠钙石
gazogene ①煤气发生器②配气机③木炭燃气
G-component G 组分
GC-TLC coupling 气相色谱-薄层色谱联用法
gear ①齿轮；牙轮②传动装置；齿链
gear backlash 齿轮隙
gear box 齿轮箱；传动箱
gear box oil(=gear case oil) 齿轮箱油；传动箱用润滑油；传动油
gear case 齿轮箱；传动箱
gear case oil(=gear box oil) 齿轮箱油；传动箱用润滑油

传动油
gear compound 齿轮油；传动油；复齿轮
gear coupling 齿轮联轴器
gear cutting machine 齿轮(加工)机床
gear drive 齿轮传动
gear flowmeter 齿轮流量计
gear grease 齿轮润滑油
gear guard 齿轮罩
gear head motor 齿轮减速电动机
gear hobbing 滚削；滚齿
gear hobbing machine 滚齿机
gear hole 齿轮孔
gear increaser 齿轮增速机
gearing ①(齿轮)传动②传动装置
gearlex coupling 齿轮式弹性轴接
gear lubricant 齿轮润滑剂
gear lubricating mineral grease 齿轮用矿物润滑脂
gear lubrication 齿轮润滑
gear oil 齿轮油
gear oil additive 齿轮油添加剂
gear pump(=gear rotary pump; gear wheel pump) 齿轮泵
gear pump metering system 齿轮泵计量系统
gear ratio 齿轮比；传动比
gear reducer 齿轮减速器
gear reduction unit 齿轮减速机
gear roller 齿辊
gear rotary pump(=gear pump) 齿轮泵
gear shaker cutting 刨齿机切削
gear shaking 刨齿
gear shaving 刨削
gear spinning pump 纺丝齿轮泵
gear stopper 齿轮制动器
gear-type coupling 齿轮联轴节
gear unit 齿轮箱；传动箱
gear wheel pump(=gear pump) 齿轮泵
geat ①浇口；铸口②流道
GE-cellulose 胍(基)乙基纤维素
gedanite 脂状琥珀
gedda wax(=ghedda wax) 甘达蜂蜡
gedrite 铅直闪石
geepound 机磅〔计重的一种单位〕
Geeraerd cell 吉雷德电池
Geer oven 吉尔老化试验箱；吉尔恒温老化箱
gefarnate 吉法酯〔药〕
gegenion 反离子
gehlenite 钙黄长石
Gehman creep apparatus 吉曼蠕变试验仪
Gehman torsional apparatus 吉曼转扭试验机；吉曼低温

屈挠试验机
geic acid(=ulmic acid) 赤榆酸
Geiger counter 盖(革)氏计数器计算机(系统)
Geiger counter detector 盖革计数器检测器
Geiger counter tube 盖革计数管
geigerin 格忌素
Geiger-Müller counter 盖革-缪勒计数管
Geiger-Müller counting tube 盖革-缪勒计数管
Geiger-Müller tube 盖革-缪勒计数管
Geiger-Nutall equation 盖革-努塔尔方程
geijerene 吉枝烯
geijerin 吉枝素
geikielite 镁钛矿
gein 水杨梅苷
Geissler pycnometer 盖斯勒比重瓶
Geissler tube 盖斯勒管
geissoschizoline(=pereirine) 缝籽木早灵；缝籽碱
geissospermine 缝籽碱
gel 凝胶*
gel-anatase 胶锐钛矿
gelasin 琼脂素
gelata 凝胶剂
gelate 胶凝
gelatification 胶凝作用；凝胶化作用
gelatin 明胶；动物胶
gelatination(=gelation) 胶凝作用；凝胶化(作用)
gelatin dynamite 明胶炸药
gelatin(e) capsule 胶囊
gelatin filter 胶膜滤光片
gelating 凝胶化
gelating agent ①胶凝剂②胶化剂
gelatinization ①胶凝(作用)②明胶化(作用)
gelatinization point ①胶凝点②糊化温度
gelatinization temperature 胶凝温度；胶化温度
gelatinization test 胶化试验
gelatinized cellulose acetate 胶化乙酸纤维素
gelatinizer 胶凝剂；胶化剂
gelatinizing ①胶凝化②胶体化
gelatinizing agent 胶凝剂
gelatin liquification test 明胶液化试验
gelatin medium 明胶培养基
gelatinoid ①(凝)胶状的②胶状物；胶体；胶化物
gelatinous 凝胶状的；胶质的；胶状的
gelatinous fiber ①凝胶纤维②胶质木纤维
gelatinous floater 凝胶状漂浮物
gelatinous precipitate 胶状沉淀*
gelatin sponge 海绵胶
gelatin-veronal buffer 明胶佛罗那缓冲液

gelation(=gelatination)　胶凝作用；凝胶化(作用)
gelation dose　(辐射)胶凝化剂量
gelationization　胶凝化
gelation particle　胶化颗粒
gelation point　胶凝点
gelation reaction　凝胶反应
gelation-stimulant　促(胶)凝剂
gelation temperature　胶凝温度
gelation time　胶凝时间
gelatum　①胶冻②凝胶体
gelbecidin　盖伯杀菌素
gel buildup　(聚合反应器的)凝胶质结垢
gel cellophane membrane　凝胶赛璐玢膜
gel-cement　胶质水泥
gel chromatograph　凝胶色谱仪
gel chromatography　凝胶色谱法
gel coat　凝胶涂层；凝胶涂料
gelcoat finish　凝胶面层；亮光面层加工
gel coating　凝胶涂料
gel coat surface　凝胶涂层表面
gel column　凝胶柱
gel consistency　凝胶稠度
gel content　凝胶含量
geldanamycin　格尔德霉素
gel deposition　凝胶沉积(作用)
gel diffusion　凝胶扩散
gel diffusion chromatography　凝胶扩散色谱法
gel-dipping　凝胶(体)浸涂(法)；凝胶漆浸涂(法)
gel-dipping process　凝胶(体)浸涂法；凝胶漆浸涂法
gel effect　凝胶效应
gel electrophoresis　凝胶电泳
gelemeter　凝胶时间测定计
gel exclusion chromatography　凝胶排阻色谱(法)
gel fibre　凝胶纤维
gel filtration　凝胶过滤(法)
gel filtration chromatography　凝胶过滤色谱(法)
gel filtration media　凝胶过滤介质
gel filtration method　凝胶过滤法
gelfischerite　胶水磷铝石
gel formation　凝胶形成
gel-formation power　成胶能力
gel-forming agent　胶凝剂
gel-forming material　胶凝剂
gel fraction　胶凝部分
gel fractionator　凝胶分段分离器
Ge-Li detector　锗-锂探测器
gelification　胶凝作用；凝胶化
gelignite　葛里炸药；吉里那特〔硝化甘油、硝化纤维、硝酸钾、木粉炸药〕
gel inhibitor　凝胶抑制剂
gel inner volume　凝胶内体积
gel interstitial volume　凝胶外体积
geliomycin(=heliomycin)　日光霉素
gel isoelectrofocusing　凝胶等电点聚焦
gelkielite　镁钛矿
gellable aviation fuel　可胶凝航空燃料
gel lacquer　凝胶型涂料；凝胶漆
Gellan gum　结冷胶
gellant　胶凝剂；胶化剂
gelled acid　加胶的酸；浓缩酸
gelled alcohol　胶态醇
gelled crude　胶凝石油
gelled fiber　胶化纤维；凝胶纤维
gel(led) filament　凝胶丝
gelled flat　胶化失光
gelled fuel　胶凝燃料；胶状油
gelled hydrocarbon　胶态烃
gelled materials　胶化物质
gel-like consistency　凝胶状稠度
gel-like constituent　凝胶状成分
gel-like soap　类凝胶皂
gel-like structure　凝胶结构；胶状结构
gelling　胶凝作用；凝胶化作用
gelling action　胶凝(化)作用
gelling agent(=gelling solvent)　胶凝剂；胶凝溶剂
gelling machine　胶化机；聚氯乙烯增塑糊烘炉
gelling point　胶凝点
gelling process　胶凝法
gelling solvent(=gelling agent)　胶凝剂；胶凝溶剂
gelling-vulcanizing　胶凝硫化
gel melting point　凝胶熔点
gel network　凝胶网络
gelometer　胶凝计
gelose(=galactan)　半乳聚糖
gel paint　凝胶漆
gel particle　胶粒〔黏胶溶液〕
gel permeation chromatography　凝胶渗透色谱(法)
gel permeation chromatography-low angle laser light scattering　凝胶渗透色谱-小角度激光散射法
gel phenomena　凝胶现象
gel point　胶凝点
gel polymer　胶凝聚合物
gel pore　凝胶孔(隙)；凝胶微孔
gel precipitation process　凝胶沉淀过程
gel rigidity　胶冻坚度
gel rubber　凝(胶体)橡胶

gel-rutile 胶金红石〔矿〕
gelsamine 钩吻胺
gel scintillator 凝胶闪烁体
gelsedine 钩吻定
gelsemic acid(=scopoletin) 钩吻酸
gelsemicine 钩吻素
gelsemin 钩吻明
gelsemine 钩吻碱
gelseminic acid(=scopoletin) 钩吻酸
gelseminine 钩吻宁
gelsemium 钩吻
gelsemoid 钩吻剂
gelsemoidine 钩吻未定
gelsevirine 钩吻绿碱
gel sieving 凝胶筛分
gel slicing machine 凝胶切片机
gel spinning 凝胶纺(丝)
gel sponge 凝胶海绵
gel staining method 凝胶染色法
gel strength 胶凝强度
gel structure 凝胶结构
gel swelling 凝胶溶胀度
gel swelling factor 凝胶溶胀系数
gel swelling value 凝胶溶胀值
gel test 凝胶试验；胶化试验
gel thin layer chromatography 凝胶薄层色谱法
gel time 胶凝时间；胶化时间
gel type resin 凝胶型树脂
gelutong 节路顿(树)脂
gel vehicles ①胶衣基料②胶凝性漆基〔触变性厚膜聚酯漆用〕
gelzircon 胶锆石〔矿〕
gem 宝石；玉
gem-(=geminate) 偕〔取代在同一碳原子上〕
gemcitabine 吉西他滨〔药〕
gem-difluoroalkane 偕二氟烷烃
gem-dimethyl effect 偕二甲基效应
gem-dinitro compound 偕二硝基化合物
gemdinitroparaffin 液体二硝基石蜡
gemfibrozil 吉非罗齐〔药〕
geminal 成对的；孪位的
geminal coupling 偕偶
geminal function 孪函数
geminal substituted hydrocarbon 双取代的烃
geminate(-gem) 偕〔取代在同一碳原子上〕
geminative 成双的
geminimycin 双生霉素
geminus ①双结构②双子座

gemmatin 埃蕈色素 $C_{17}H_{12}O_7$
genalkaloids 氧化生物碱类〔氨基变为氨氧基的生物碱〕
genatropine 颓托品
gene 基因
genealogy 谱系；家系
gene amplification 基因增殖
gene analyzer 基因分析仪
genechip 基因芯片
gene diagnosis 基因诊断
gene dosage 基因数量；基因剂量
gene expression 基因表达；基因表现
gene manipulation technique 基因操作技术
gene map 基因图
gene mutation 基因突变
gene pool 基因库
general ①一般的②普通的
general assembly 总装配
general characteristic 通性；一般特性
general chemistry 普通化学
general chromaticity coordinate 普通色度坐标
general chromaticity diagram 普通色度图
general corrosion 均匀腐蚀；全面腐蚀
general drawing 总图
general electric recording spectrophotometer 普通电子记录分光光度计
general facilities 通用设施；公用工程
general formula 通式
general identification test 一般鉴定试验
generalization 通则；概说
generalized ①推广的；广义的②概括的；综合的
generalized Bingham body 广义宾汉体
generalized body 广义(物)体
generalized coordinate 广义坐标
generalized Hooke's law 广义虎克定律
generalized least square method 广义最小二乘法
generalized liquid model 广义液体模型
generalized Maxwell model 广义麦克斯韦模型
generalized Newtonian fluid 广义牛顿流体
generalized Newtonian liquid 广义牛顿液体
generalized Reynolds number 广义雷诺数
generalized solid model 广义固体模型
generalized solubility parameter 广义溶度参数；广义可溶性参数
generalized space 广义空间
generalized standard addition method 广义标准加入法*
generalized valence bond(GVB) method 广义价键法
generalized variance 广义方差
generalized Voigt model 广义沃伊特模型

general layout drawing 总布置图；总平面图
generally labeling 一般标记；普通标记
general molded goods 一般模制品
general overbaul 大修
general polarity 普通极性
general polystyrene 普通聚苯乙烯
general principle 普通原理
general purpose 通用型
general-purpose adhesive 通用型黏结剂；万能胶〔俗〕
general purpose calender 通用压延机
general purpose chloroprene rubber 通用型氯丁橡胶
general purpose computer 通用计算机
general purpose detergent 通用洗涤剂
general-purpose-grade pine oil 通用级松油
general-purpose grease 通用润滑脂
general purpose rubber 通用(型)橡胶
general purpose varnish 通用清漆〔内用、外用均可〕
general radiation 一般辐射
general rubber 通用橡胶
general transport equation 一般输运方程
general-use article 一般制品
general utility 正常利用率
general view 总图；全视图
general yield fracture mechanics 全面屈服断裂力学
generated reagent 发生试剂
generating electrode 发生电极
generating medium 发生介质
generation ①发生②代
generation of internal heat 内部发热
generation rate 发生(速)率
generative fuel 再生燃料
generator ①发电机②发生器③发烟器
generator electrode 发生电极
gene recombination technique 基因复合技术；基因再结合技术
gene redundancy 基因重复
gene replication 基因复制
generic term 通称；总称
generic test 定属试验
geneserethol 金丝脑
geneserine 金丝碱；氧化毒扁豆碱
geneserolene 金丝烯
geneseroline 金丝灵
genesis ①起源；来源②生殖
genet absolute 鹰爪豆净油；金雀花净油
genet concrete 鹰爪豆浸膏；金雀花浸膏
genetic algorithm 遗传算法
genetic code 遗传密码
genetic engineering 基因工程
genetic information 遗传信息
genetic operation 遗传操作
genetic relation 母子体关系
genet oil 金雀花油
Geneva nomenclature 日内瓦命名法
gengshengmeisu 更生霉素
genimycin 世霉素
geniposide 金尼泊苷
genistein 染料木黄酮；5,7,4′-三羟(基)异黄酮
genisteine(=l-α-isoparteine) l-α-异鹰爪豆碱
genistin 染料木苷
genkwanin 芫花素；5,4′-二羟基-7-甲氧基黄酮
genome 基因组
genomic 染色体组的
gentamicin 庆大霉素；艮他霉素
gentamycin 庆大霉素〔药〕
genthite 水硅镁镍石
gentiamarin 龙胆苦苷
gentian 黄龙胆
gentiana lutea 黄龙胆
gentianic acid(=gentisinic acid) 龙胆酸
gentianidine 龙胆定；秦艽乙素
gentianine(=erythricine) 龙胆宁；秦艽甲素
gentianose 龙胆三糖
gentian root 龙胆根
gentian violet 龙胆紫；结晶紫
gentienin 龙胆恩宁
gentiin 龙胆苷
gentiobiose 龙胆二糖；β-(1→6)葡二糖
gentiobiuronic acid 龙胆二糖醛酸
gentiodextrin 龙胆糊精
gentiogenin 龙胆苦配基
gentiopicrin 龙胆苦苷
gentisaldehyde 2,5-二羟苯甲醛 (OH)$_2$C$_6$H$_3$CHO
gentisein 龙胆赛因
gentisic acid 2,5-二羟(基)苯甲酸
gentisic aldehyde 2,5-二羟苯甲醛
gentisin 龙胆黄素；1,7 二羟(基)-3-甲氧呫吨酮
gentisin alcohol 龙胆素醇
gentisinic acid(=gentianic acid) 龙胆酸
gentisyl alcohol(=saleripol) 龙胆醇
gentle oxidation 细心氧化
gentle turbulence 轻度湍流；缓慢的湍流
gently curved shape 轻微弯曲的形状；平缓弯曲的形状
genuine 真正的；真的
genuine soap 皂基
geocerinic acid 地蜡酸

geocerinone 地蜡酮
geocerite 地蜡；硬蜡
geochemical prospecting sample 化探样品
geocomposite 土工复合材料
geocronite 砷硫锑铅矿
geodesic line 短程线
geodesic line winding 短程缠绕
geodin 地曲菌素
geodistomycetes 耐干木腐菌
geoffrayin(=rhatanin) 娜擅宁
geological dating method 地质定年代法
geological thermometer 地质温度计
geom 吉纶〔聚氯乙烯树脂的一种商名〕
geomat 土工垫
geometric 几何(学)的
geometrical 几何(学)的
geometrical conversion 几何转变
geometrical effect 几何效应
geometrical equivalence 几何等效
geometrical factor 几何因子；几何因数
geometrical inversion 几何转位；几何转化
geometrical isomer 几何异构体
geometric(al) isomeride 几何异构体
geometrical isomerism 几何异构(现象)
geometrical optics 几何光学
geometrical regularity 几何规整度
geometrical shape 几何形状
geometrical similarity 几何相似性
geometric(al) stereoisomer 立体几何异构体
geometrical symmetry 几何对称
geometric axis 几何轴线
geometric configuration 几何构型
geometric isomer 几何异构体
geometric isomerism(=rotamerism) 几何异构(现象)
geometric mean 几何平均数；等比平均数
geometric metamerism 几何(光学)条件配色；条件等色；几何位变异构
geometric pattern 几何图形
geometric programming 几何规划
geometric series 几何级数
geometric standard deviation 几何标准(偏)差
geometric stiffness 几何刚度
geometric volume 几何体积
geometry optimization 几何优化
geomycin 土霉素；地霉素
geonet 土工网
Geon process 吉洪法〔烃混合物抽提法〕
geopotential 重力势

georgette 乔其纱
geoside 水杨梅苷
geosynthetic clay liners 土工合成材料黏土衬层
geothermal energy 地热能
geothermometer 地质温度计
geotropism 向地性
gerade 对称
geranial 牻牛儿醛；香叶醛
geranic acid 牻牛儿酸；香叶酸；3,7-二甲基-2,6-辛二烯-1-酸
geraniene 牻牛儿萜烯
geranine 直接猩红〔染料〕
geraniol 牻牛儿醇；香叶醇
geraniolene 香叶烯
geranium lake 天竺葵红色淀〔酸性染料和氯化钡、硝酸铅沉淀而得〕
geranium oil 老鹳草油；香叶油
geranium over rose oil 香叶和玫瑰花共蒸油
geranium rose oil 玫瑰香叶油
geranonitrile 香叶腈
5-geranoxy-7-methoxycoumarin 5-香叶氧基-7-甲氧基香豆素
8-geranoxy psolarene 8-香叶氧基补骨素
geranyl 牻牛儿基；香叶基 $C_{10}H_{17}$—
geranyl acetate 乙酸牻牛儿酯 $CH_3CO_2C_{10}H_{17}$
geranyl acetoacetate 乙酰乙酸香叶酯；乙酰乙酸牻牛儿酯
geranyl acetone 香叶基丙酮
geranyl anthranilate ①邻氨基苯甲酸香叶酯②邻氨基甲酸牻牛儿酯
geranyl benzoate 苯甲酸香叶酯
geranyl butyrate 丁酸牻牛儿酯 $C_3H_7CO_2C_{10}H_{17}$
geranyl caproate 己酸香叶酯
geranyl caprylate 辛酸香叶酯；辛酸牻牛儿酯
geranyl cinnamate (肉)桂酸香叶酯
geranyl ethoxyacetate 乙氧基乙酸香叶酯
geranyl 2-ethyl butyrate 2-乙基丁酸香叶酯
geranyl ethyl ether 香叶基乙基醚
geranyl formate 甲酸牻牛儿酯 $HCO_2C_{10}H_{17}$
geranyl geranylidene acetic acid 香叶基香叶叉乙酸
geranyl heptylate 庚酸香叶酯
geranyl hexanoate 己酸香叶酯
geranyl hexylate 己酸香叶酯
geranyl isobutyrate 异丁酸香叶酯
geranyl β-kctobutyrate β-酮基丁酸香叶酯
geranyl methoxyacetate 甲氧基乙酸香叶酯
geranyl methyl ether 香叶甲醚
geranyl 3-oxobutanoate 3-氧代丁酸香叶酯

geranyl oxyacetaldehyde 香叶基含氧乙醛
geranyl pelargonate 壬酸香叶酯
geranyl phenylacetate 苯乙酸香叶酯
geranyl propionate 丙酸香叶酯
geranylpyrophosphate 牻牛儿(基)焦磷酸；3,7-二甲基-2,6-辛二烯焦磷酸
geranyl tiglate 惕各酸香叶酯
geranyl α-toluate α-甲基苯甲酸香叶酯
geranyl valerate 戊酸香叶酯
Gerard's pine oil 格雷松油
Gerber method 盖勃氏法
gerhardtite 铜硝石
germacrane 大根香叶烷
germacranolide 大根香叶内酯
germacrene 大根香叶烯
germacrol 大根香叶脑
germacrone 大根香叶酮
germanate 锗酸盐〔1. 偏 M_2GeO_3；2. 正 M_2GeO_4〕
German black 德国黑
germander 石蚕
germanic ①(正)锗的；四价锗的②含锗的
germanic chloride 四氯化锗 $GeCl_4$
germanic compound (正)锗化合物
germanicol 日耳曼醇
germanic oxide 氧化锗 GeO_2
germanite 亚锗酸盐；二价锗酸盐 M_2GeO_2
germanium 锗
germanium chloride 氯化锗 $GeCl_2$; $GeCl_4$
germanium chloroforme 三氯甲锗烷；氯锗仿 $GeHCl_3$
germanium dioxide 二氧化锗〔聚酯聚合催化剂〕
germanium ethide 乙基锗 $Ge(C_2H_5)_2$; $Ge(C_2H_5)_4$
germanium hydride 氢化锗
germanium hydride chloride 氢氯化锗
germanium methide 甲基锗〔二甲基锗 $Ge(CH_3)_2$ 或四甲基锗 $Ge(CH_3)_4$〕
germanium monoxide 一氧化锗 GeO
germanium oxide 氧化锗 GeO; GeO_2
germanium rectifier 锗整流器
germanium sulfide 硫化锗 GeS; GeS_2
germanium tetrachloride 四氯化锗 $GeCl_4$
germanium tetraethyl 四乙锗 $(C_2H_5)_4Ge$
German kieselguhr 德国硅藻土
germanoformic acid 甲锗酸 HGeOOH
germanomolybdate 锗钼酸盐
German orange 德国橙；酸性橙
germanous 亚锗的；二价锗的
germanous chloride 二氯化锗 $GeCl_2$
germanous compound 亚锗化合物

germanous oxide 一氧化锗 GeO
German saltpetre 德国硝石
German silver 白铜〔一种铜镍锌合金〕；德银
germ-carrying 带菌的
germ-free box 无菌箱
germicidal 杀(细)菌的
germicidal soap 杀菌皂
germicide 杀菌剂
germidine 胚芽定
germifuge 杀菌剂
germ-killing efficiency 杀菌效率
germ processed oil 加有动植物油脂的复合润滑油
germ-resistant ①防腐的；杀菌的②防腐剂；杀菌剂
germyl 甲锗烷基；三氢锗基 H_3Ge-
α-germyl ketones α-锗烷基酮类化合物
germylsilans 锗烷基硅烷类
gerobriecin 截短菌素
geronaldehyde 葛让醛
geronic acid 葛让酸；2,2-二甲基-6-氧化庚酸
gerontine(=spermine) 精素；精胺
gersdorffite 辉砷镍矿
gerty 燕麦粉
gestagen 助孕剂
gestrinone 孕三烯酮〔药〕
getter ①吸气剂；消气器②吸气剂
getter ion pump 吸气离子泵
getter pump 吸气泵
geyserite 硅华
g-factor g 因子
ghatpot 绫
ghatti gum 茄替胶
ghedda wax(=gedda wax) 甘达蜂蜡
gheddic acid 三十四(烷)酸
ghee 印度酪脂
ghee oil 牛酪油
G-H mixtures(=G-H solvents) G-H 混合物；G-H(混合)溶剂
ghost 空胞
ghost-causing 引起假峰
ghost-cut 消除假峰
ghost image 阴影
ghosting 双茎；粗细茎
ghosting effect 假峰效应
ghost line 鬼线
ghost peak 假峰
ghost peak of X-ray X 射线假峰
ghost spot 假斑
G-H solvents(=G-H mixtures) G-H 混合物

giant magnetoresistive effect 巨磁阻效应
giant micelle 巨胶束
giant molecule 巨分子
giant particle(GP) (核爆炸沉降物中)强放射性粒子
giant pulse 巨脉冲
giant salamander oil 鲵鱼油
gibbane 赤霉素烷
Gibb chlorate process 吉布氯酸盐制造法
gibberella 赤霉菌
gibberellane 赤霉素烷
gibberellenic acid 赤霉烯酸
gibberellic acid 赤霉酸;赤霉素
gibberellin 赤霉素;赤霉酸
gibberene 赤霉芴;1,7-二甲基芴
gibberic acid 赤霉酸
gibberone 赤霉素酮
Gibbs adsorption equation 吉布斯吸附公式
Gibbs convention 吉布斯规定
Gibbs-Duhem equation 吉布斯-杜亥姆方程
Gibbs-Duhem relation 吉布斯-杜亥姆关系
Gibbs elasticity 吉布斯弹性
Gibbs ensemble 吉布斯系综
Gibbs free energy 吉布斯自由能*
Gibbs free energy of activation 活化吉布斯自由能
Gibbs function 吉布斯函数
Gibbs indophenol test 吉布斯靛酚试验
gibbsite 三水铝石*
Gibbs-Konovalow's rule 吉布斯-孔诺伐洛夫定律
Gibbs' method 吉布斯氏法
Gibbs process (醇酸树脂)吉布斯法
Gibbs' reagent 吉布斯氏试剂
giberellin 赤霉素
gieseckite 绿霞石
Gieseler fluidity 吉氏流动度
giga- 〔词头〕京;千兆;十亿〔符号 G, 10^9〕
giga-electron-volt(GeV) 千兆电子伏特;十亿电子伏特〔10^9eV〕
giganteol 牛角瓜醇
gigantic acid 大曲酸
gigantine 大曲菌素
gigantolite 青块云母
gigging (织物)起毛
Gilamite 吉氏石脂
gild ①镀②镀金
gilding 饰金;涂金;飞金
gilding with gold 鎏金
Gilead balsam(=Mecca balsam) 麦加香脂
gill drawing 绢纺并条;绢纺练条

gillenia 美吐根
Gillespie's bicolorimeter 吉莱斯皮双色比色计
Gillespie's still 吉莱斯皮蒸气压测定装置
Gillett-Rhoads furnace 吉勒特-罗兹电炉
gill fungi 伞菌
Gilliland correlation 吉利兰特公式〔理论塔板与回流比关系式〕
gilsonite(=uintahite; uintaite) 硬沥青
gilt 镀金;烫金
Giltner tube 吉耳特讷无氧培育管
gin ①杜松烧酒(商)②轧棉(子)机③打桩机
gingelly oil(=sesame oil) 芝麻油
ginger ale 姜麦酒;姜味较淡的姜汁汽水
ginger beer 姜啤酒
ginger bread 姜料面包
ginger grass oil 姜草油
gingerin 姜油脂
ginger oil 姜油
gingerol 姜醇
ginger oleoresin 姜油树脂
ginger snap 姜汁饼干
ginger tincture 姜酊
gingivitis 龈炎
gingkol 银杏酚
gingkolic acid 银杏酚酸
gin house 清花厂
ginkgetin 银杏黄素;银杏黄酮苷;银杏黄酮
ginkgol 银杏酚
ginkgolic acid 银杏酚酸
ginkgotoxin 银杏毒
ginning 轧棉;轧花
ginning outturn 衣分
ginnol 银杏醇
gin pole 中央立柱〔油罐〕
ginseng 人参
ginseng oil 人参油
ginsenin 人参宁
giobertite 菱镁矿
gipsite γ-三羟铝石;三水铝石
Girard-Street graphite process 吉拉德-斯特里特石墨制造法
girbotol absorber 乙醇胺法吸收器〔以乙醇和二乙醇溶液精制气体用〕
girbotol process 乙醇胺法
girder 横梁
girdle ring 腰环〔葫芦形空气弹簧〕
Girod furnace 吉罗德电炉
giroflee oil 桂竹香油;罗兰花油

girth sheets 圈板
girth welding 环缝焊接
gismondite(=abrazite) 水钙沸石
git ①浇口；铸口②流道
gitaligenin 芰他配基
gitalin 芰他灵
gitatoxigenin 芰他毒配基
gitatoxin 芰他毒
githagenin 剪秋罗配基
githagin 剪秋罗苷
gitin 芰苷
gitogenin 芰皂配基 $C_{27}H_{44}O_4$
gitonin 芰皂苷
gitoxigenin 芰皂配基
gitoxin 芰皂素
glabra lactone 异白芷内酯
glabric acid 平滑酸；光甘草酸
glacial ①冰的②冰河的
glacial acetic acid 冰醋酸〔俗〕
glacial acrylic acid 冰丙烯酸
glacial clay 冰黏土
glacial dye 冰染染料
glacial phosphoric acid 冰磷酸
glacier 冰河
gladiolic acid 剑霉酸；唐菖蒲青霉酸
glagerite 多水高岭土
glallizing 冲淡法〔果汁加工〕
glance coal 亮煤
glance pitch 辉沥青
glancing angle 掠射角
glancing angle X-ray diffraction 掠射角X射线衍射法
gland （密封)压盖；密封套；填料盖
gland box 填料函
glanded cottonseed 有腺棉籽
glandless cottonseed 无腺棉籽
gland packing 压盖填料
glarimeter （纸面)光泽计
Glaser coupling(=acetylene coupling) 格拉塞偶合；乙炔偶合
glaserite 钾芒硝
Glasgow-type generator 格拉斯哥式水煤气发生炉
glass ①玻璃②镜③玻盅；玻(璃)杯④玻片
glass adhesive 玻璃胶黏剂
glass antidimmer 玻璃防雾剂
glass atomizer 玻璃雾化器
glass backing-plate 玻璃衬板
glass-base laminated material 玻璃基层压制品
glass bead 玻璃珠

glass bead column 玻璃珠柱
glass beads support 玻璃球载体
glass beaker 玻璃烧杯
glass black 灯黑
glass blower 玻璃吹制工
glass blower's burner 吹玻璃用喷灯
glass blower tool 吹玻璃用具
glass blowing 玻璃吹制
glass bonded mica 玻璃云母
glass bonded refractory coating 玻璃质耐火涂层
glass bubbles 玻璃小孔；玻璃泡
glass capillary 玻璃毛细管
glass capillary gas chromatography 玻璃毛细管气相色谱法
glass capillary viscometer 玻璃毛细管黏度计
glass cell 玻璃吸收池；玻璃比色池；玻璃比色杯
glass cement 玻璃胶泥
glass ceramic 玻璃陶瓷
glass-ceramic fiber 玻璃-陶瓷纤维
glass-clad platinum iridium thermal conductivity detector 玻璃封裹铂铱热导检测器
glass cloth 玻璃布
glass coating 搪玻璃
glass colors 玻璃(着色用的)颜料
glass color standard 玻璃颜色标准
glass column 玻璃柱
glass cover-plate 玻璃盖板
glass crucible 玻璃坩埚
glass cullet 碎玻璃
glass cutter 玻璃割刀
glass cutting 玻璃切割
glass cylinder 玻璃杯；玻璃量筒
glass diaphragm manometer 玻璃薄膜压力计
glass dish evaporation test(=glass dish gum test) 玻璃皿蒸发试验；玻璃皿胶质试验
glass dish gum test(=glass dish evaporation test) 玻璃皿蒸发试验；玻璃皿胶质试验
glass dosimeter 玻璃(制放射性)剂量计
glass drawing 拖玻璃
glass drops 玻璃珠
glassed vessels 覆盖有玻璃的钢质容器
glass effect 玻璃化效应
glass electrode 玻璃电极*
glass enamel 搪玻璃
glass fabric 玻璃布
glass felt 玻璃毡
glass fiber 玻璃纤维；玻璃丝；玻璃毛
glass fiber coating 玻璃纤维涂料

glass fiber lubricant　玻璃纤维润滑剂
glass fiber paper　玻璃纤维纸
glass fiber reinforced plastic(GFRP)　玻璃纤维增强塑料；玻璃钢〔俗〕
glassfibre　玻璃纤维
glass fibre cloth　玻璃纤维布
glass fibre coupled　玻璃纤维黏结的
glass fibre covering　玻璃丝被覆；玻璃丝包缠
glass fibre-epoxy laminate　玻璃纤维-环氧(树脂)层压物
glass fibre filter　玻璃纤维过滤器
glass fibre membrane　玻璃纤维膜
glass-fibre reinforced polyamide　玻璃纤维增强聚酰胺
glass fibre reinforcement　玻璃纤维增强材料
glass filter plate　玻璃滤板
glass flake　玻璃(鳞)片
glass flake coatings　玻璃叶片(鳞片)涂料
glass flour　玻璃粉〔磨料〕
glass-former　成玻璃材料；玻璃形成体
glass forming substance　玻璃形成物
glass frame　玻璃架；玻璃框
glass funnel　玻璃漏斗
glass furnace　熔玻璃炉
glass gauge　玻璃管规；玻璃液位表
glass glazed　浓釉的〔釉得很浓，以至于很亮〕
glass glazing　浓釉
glass grinding media　玻璃研磨介质；玻璃球
glass helices　玻璃单环填充物
glass house　玻璃工厂；温室
glass hydrometer　玻璃比重计
glass-indicating electrode　玻璃指示电极
glassine(=glassine paper)　半透明玻璃纸
glassine paper(=glassine)　半透明玻璃纸
glassiness　玻璃状态；玻璃质
glassing　磨光；打光
glassing jack(=glassing machine)　磨光机
glassing machine(=glassing jack)　磨光机
glass ink　玻璃墨水〔在玻璃上写字用〕
glass inlet port　玻璃入口
glass insert　玻璃插入物
glass introduction port　玻璃进样口
glassivation　①玻璃钝化；玻璃保护层②玻璃熔封(密封)
glass jar　玻璃缸(瓶)〔实验室用〕
glasslike polymer　类玻璃聚合物
glass lined　搪玻璃的；搪瓷的
glass-lined kettle　搪玻璃釜；玻璃衬里反应釜
glass-lined knotting jar　搪玻璃的木节封闭剂专用罐
glass lined reaction vessel　搪玻璃反应器
glass-lined steel　搪玻璃钢；搪瓷钢

glass lined tubing　玻璃内衬管
glass lining　搪玻璃；搪瓷内衬
glass making　玻璃制造
glass marker　玻璃的刻度器
glass mats　玻璃纤维垫
glass measure　(玻璃)量杯
glass melt　熔融玻璃
glass melting　玻璃熔融
glass membrane electrode　玻璃膜电极
glass membrane ion-selective electrode　玻璃膜离子选择电极
glass metal　玻璃金属
glass microballoon　玻璃微珠
glass microelectrode　微型玻璃电极
glass microfiber filter media　玻璃微纤维过滤介质
glass monofilament　玻璃单丝
glass paddle　玻璃搅拌桨
glass paper　玻璃砂纸
glass pearls　玻璃珠
glass pear-shaped bulb　梨形玻璃灯泡
glass pencil　玻璃铅笔〔在玻璃上写字的铅笔〕
glass pipe　玻璃管
glass pipe line　玻璃制输送管
glass plate　玻璃片；玻璃板
glass plate level gauge　玻璃板液面计
glass point　玻璃点
glass pot　玻璃罐；玻璃坩锅
glass powder　玻璃粉
glass pressure reactor　玻璃耐压反应器
glass-reinforced　玻璃(纤维)增强的
glass-reinforced polypropylene　玻璃纤维增强聚丙烯
glass retort　玻璃甑
glass rod　玻璃棒
glass roving　玻璃粗纱
glass roving cloth　玻璃纤维无捻纱布
glass-rubber transition　玻璃-橡胶态转变
glass-rubber transition point　玻璃态-橡胶态转变点；二级转变点
glass-rubber transition temperature　玻璃态橡胶态转换温度；二级转变温度
glass sampling bulb　玻璃进样球(形管)
glass sand　玻璃砂
glass sheet　玻璃片
glass-silicide coating　玻璃-硅化物涂层
glass silk　玻璃丝
glass slicker　玻璃刮刀
glass spatula　玻璃刮铲
glass spinning　玻璃丝纺

glass spreading rod 玻璃涂布棒
glass state 玻璃态*
glass state temperature 玻璃化温度；二次转变温度
glass-stem thermometer 玻璃温度计
glass stopper 玻璃塞
glass-stoppered bottle 玻璃塞瓶
glass tank 玻璃桶
glass tiff 方解石
glass tile 玻璃瓦
glass-to-metal seal 玻璃-金属封接
glass transition 玻璃化转变
glass transition point 玻璃化点
glass-transition temperature 玻璃化转变温度*
glass trap 玻璃分水器
glass tube 玻璃管
glass tube level gauge 玻璃管液面计
glass tubes of chromometer 比色计玻管
glass tubing gauge 玻管量规
glass-type ceramic coating 玻璃质陶瓷涂层
glass valve 玻璃阀
glass veil 覆盖毡
glassware ①玻璃器皿②玻璃仪器
glass wool 玻璃棉；玻璃绒；玻璃毛
glass wool extractor 玻璃棉抽出器
glass wool filter 玻璃棉滤器
glass wool inserter 玻璃棉塞
glass work 玻璃制造(业)
glasswort 玻璃草〔自其灰中可取得碱〕
glass writing pencil ①刻玻笔②蜡笔
glassy 玻璃状的；透明的
glass yarn 玻璃丝
glassy carbon 玻璃化炭黑
glassy carbon composite 玻璃态碳素复合材料
glassy carbon electrode 玻(璃)碳电极
glassy carbon tube 玻(璃状)碳管
glassy compliance 玻璃态柔量
glassy corn 玻璃质粒
glassy deformation 玻璃态形变
glassy elastic deformation 玻璃态弹性形变
glassy phosphate 玻璃状磷酸盐(偏磷酸盐)
glassy polymer 玻璃状聚合物
glassy shear modulus 玻璃态剪切模量
glassy solvent 玻璃状溶剂
glassy state 玻璃态
glassy tensile modulus 玻璃态拉伸模量
glassy yellow 透明黄色
glassy zone 玻璃(态)区域
glauberite 钙芒硝（矿）

glauber salt 芒硝*
glaucine 海罂粟碱
glauco-(希文 glauko-) 戈烙〔音译〕；霜灰色
glaucobilin 胆蓝素
glaucochroite 钙锰橄榄石
glaucodot 铁硫砷钴矿
glauconite 海绿石
glaucophane 蓝闪石
glaucophanite 蓝闪岩
glaucophyllin 绿叶二酸 $C_{31}H_{32}N_4Mg(COOH)_2$
glaukonine 海绿石
glaze 釉；釉料
glaze brick 釉面砖；玻璃砖
glazed 上(了)釉的；打光的
glazed board 高光泽纸板
glazed brick 玻璃砖；面砖
glazed finish 打光涂饰剂
glazed horse 亮马皮
glazed imitation parchment 釉光仿羊皮纸
glazed joint failure (黏合接头)玻璃状破坏
glazed kid 亮小山羊革
glazed paper 釉纸；蜡光纸
glazed porcelain casserole 上釉(带把)瓷皿
glazed powder (石墨)打光了的火药
glazed printing paper 道林纸
glazed tile 琉璃瓦
glaze firing 上釉焙烧；烧釉
glazer ①上光机②上光烘箱
glaze stains 釉着色剂
glazier's putty 镶玻璃用腻子；玻璃工(用)腻子；镶嵌玻璃用的油灰
glazier's salt 硫酸钾 K_2SO_4
glazing ①施釉②砑光；打光③釉化
glazing agent 上光剂
glazing calender 压光机
glazing cask 砑光桶
glazing colour 施釉色料；透明色料
glazing glass 打光玻璃辊
glazing jack 打光机
glazing kiln 上釉窑
glazing machine 砑光机；磨光机；打光机；抛光机
glazing pottery 上釉陶器
glazing roll 上光辊
glcerin tricaprate(=tricaprin) 甘油三癸酸酯；三癸精 $(C_9H_{19}CO_2)_3C_3H_5$
glcerol-α-monoallyl ether 甘油 α-单烯丙醚
Gleason four-square test 格来逊试验〔长时间试验条件下测定润滑油的方法〕

glebomycin 土块霉素
gledigenin 皂荚配基
gledinin 皂荚宁
gleditschine 皂荚碱
gleditsin 皂荚素
glessite 棕琥珀；圆粒树脂石
glibenclamide 格列本脲〔药〕
gliclazide 格列齐特〔药〕
glide lamella 滑移层
glide plane 滑移面
glide reflection 滑移
glimepiride 格列美脲〔药〕
gliorosein 蔷胶霉菌素
gliotoxin 胶霉毒素
glipizide 格列吡嗪〔药〕
gliquidone 格列喹酮〔药〕
glistening 闪耀的
glistening colony 闪光菌落
Glitsch distillation tray 格利希蒸馏塔板
glitter ①闪光②闪光颜料；闪光物
glitter effect 闪光效应
glittering 光泽；亮度；闪光；放光
global chain orientation （分子）链大尺度取向
global distillation 全球蒸馏效应
global environment monitoring 全球环境监测
global environment monitoring system 全球环境监测系统
global optimal area 全局最优化区域
global optimization 全局最优化
global warming 全球变暖
globar 碳化硅(炽)热棒；硅碳棒
globar infrared source 硅碳棒红外光源
globar resistance heater 碳(化)硅棒电阻式发热元件
globe ①球②地球
globe digester 球形蒸煮釜；蒸球
globe fish liver oil 河豚鱼肝油
globe fish poison 河豚毒
globe lining 球形衬
globe-roof 球形顶；圆顶
globe-roof tank 球形顶油罐
globe rotary bleacher 球形旋转漂白器
globe valve 球阀
globicin 格鲁比型杆菌素
globomycin 球霉素
globucid 球磺胺；磺胺乙基硫二氮杂环戊二烯
globulane 蓝桉烷
globular 球状的
globular-chain crystal 球状链晶体

globularetin 卷柏分苷
globularin 卷柏苷
globular interface 球形界面
globular micelle 球状胶束
globular molecule 球状分子
globular polymer 球型聚合物
globular powder 球粒剂；球状粉末
globular protein 球状蛋白
globular swelling 球状溶胀
globule ①小球；小珠②液滴；油珠；汞珠③小丸
globule arc technique 球形(熔珠)电弧法
globule method of arcing 球形电弧法
globulimeter 血球计算器
γ-globulin γ球蛋白；丙种球蛋白
globulin zinc insulin 球蛋白锌胰岛素
globulite 球晶
globulol 蓝桉醇
globulus 球状体
glocken cell 钟式电解池
gloeocapsa 黏球藻
glonoin(=nitroglycerol) 硝化甘油
glory-hole 火焰窥孔
glory scattering 辉光散射
gloss ①光彩；光泽②使有光泽③上釉④浅饰
gloss agent 光泽剂
gloss beam(=specular reflectance beam) 镜面反射光束
gloss coat 光泽涂膜
gloss control 光泽控制；光泽调整
gloss control agent 光泽控制助剂
gloss decatizing machine 压光蒸呢机
gloss difference 光泽差
gloss emulsion paint 有光乳化漆；有光乳胶漆〔俗〕
gloss enhancer 光亮剂
gloss finish 光泽整理；丝光整理
glossimeter 光泽机
gloss increaser 光泽剂
glossiness ①光泽性②砑光度③光泽度
glossing agent 抛光剂
glossing up ①抛光；出亮；擦亮②泛光；出光斑〔无光漆膜出现不该有的光泽〕
gloss ink 亮光油墨
gloss ink vehicles 亮光油墨调墨油
gloss level 光泽度；光泽标准(等级)
gloss loss 失光
glossmeter 光泽计
gloss number 光泽值
gloss oil 光泽油；松香清漆
glossometer 光泽仪

gloss paint 光泽涂料
gloss promoter 增光剂
gloss rating ①光泽度(等级)②光泽评定
gloss reading 光泽读数
gloss reduction 消光；光泽减少；失光〔漆病〕
gloss retention 保光性
gloss shifting 光泽不匀；发花
gloss stability 光泽稳定性
gloss topcoat 有光面漆
gloss varnish 有光清漆；亮油
gloss white 光泽白〔硫酸钡和矾土白共沉淀物〕
glossy ①砑光的光泽〔革〕②整光的光泽〔纤、纺〕
glossy coating 有光涂层
glossy finish(ing) 最后抛光
glossy ink 亮油墨；有光油墨
glossy paper 釉纸；蜡光纸
glossy polyester 有光聚酯
glossy surface 光泽面
glossy white finish 有光白面漆
glost ①釉②上釉的瓷器
glost fire 烧釉
glost firing 烧釉
glost kiln 釉烧窑
glost ware 釉皿；上釉制品
glove 手套
glove box 手套箱*
glove kid 手套革；手套皮
glove leather 手套皮
Glover acid 格洛弗塔酸
Glover tower 格洛弗塔
glow ①辉光；发光②灼热
glow corona 辉光电晕
glow current 发光电流
glow curve 辉光曲线
glow discharge 辉光放电
glow discharge detector 辉光放电检测器
glow discharge lubricant 辉光放电所得润滑剂
glow discharge mass spectrometry 辉光放电质谱法
glow-discharge modification 辉光放电改性
glow discharge optical emission spectrometry 辉光放电光学发射谱
glow-discharge polymerization 辉光放电聚合(作用)
glow discharge source 辉光放电源
glow discharge sputtering 辉光放电溅射
glow discharge voltage regulator tube 恒电压放电管；稳压管
glowing 辉光；灼热的
glowing ceramic rod 辉光磁棒

glowing combustion 灼热燃烧
glow-in-the-dark paint 暗处发光漆；夜光漆
glow lamp 辉光灯
glow plasma 辉光等离子体
glow resistance 抗灼热性；抗发光性
glow-resistant 抗灼热的
glow-to-arc transition 辉光-电弧转换
glow transfer tube 计数放电管
glucic acid 丙烯醇酸 $CH_2=COHCOOH$
glucin 甜素〔一种甜味剂〕
glucinum 元素铍〔别名〕
glucinum alkyl 二烷基铍 BeR_2
glucinum ethide 二乙基铍 $Be(C_2H_5)_2$
glucinum ethyl 二乙基铍
glucinum methide 二甲基铍 $Be(CH_3)_2$
glucitol(=sorbitol) 葡萄糖醇；山梨醇
gluckauf 格溜考夫〔炸药〕
glucoalyssin 葡配庭荠精
glucobiogen 葡萄糖酸钙
glucochloral 葡萄糖缩氯醛 $Cl_3CCH=O_2=C_6H_7O(OH)_3$
glucocholic acid(=glycocholic acid) 甘胆酸
glucocinin 激糖素
glucocorticoid 糖皮质激素
glucodialohexose 葡二醛己糖
glucoferrum 葡萄糖铁
glucofuranose 呋喃(型)葡萄糖
glucofurone 葡萄糖酸γ-内酯
glucogallic acid 棓酸葡萄糖苷
glucogallin 葡萄糖棓苷
glucoprotein(=glycoprotein) 糖蛋白
α-D-glucopyranosido-β-D-fructofuranoside α-D-吡喃葡萄糖(苷基)-β-D-呋喃果糖苷；蔗糖
glucosan 葡聚糖
D-glucose D-葡萄糖；D-型葡萄糖
α-glucose(=phlorose) 根皮糖；α-葡萄糖
glucose biosensor 葡萄糖生物传感器
glucose oxidase-peroxidase method 葡萄糖氧化酶-过氧化物酶比色法
glucose sensor 葡萄糖传感器
glucuronic acid 葡糖醛酸
glue 胶黏剂
glue digester 化胶器
glued wood 胶合板
glue enamel plate (热)胶版
glue etching (玻璃表面)胶膜蚀刻法
glue film(=glue line) 胶层
glue gun 黏合剂用喷枪；涂胶枪

glueing 黏合；胶接；胶黏；胶合
glue in pearls 颗粒胶
glue laminate 胶合层压制件
glue line(=glue film) 胶层
glue peneration ①透胶②透胶率
glue pot 熬胶锅
glue putty 胶质油灰
gluer 制胶水器
glue sizing ①(木器)打胶地②上胶；施胶
glue spread 涂布量
glue spreader 涂胶辊；涂胶器
glue spreading 涂胶
glue stock 胶料
gluey 黏合的；胶黏的
glueyness 黏合性
gluing 黏合；胶黏
glumamycin 颖霉素
glumitocin 谷催产素；4-丝-8-谷酰胺催产素
gluside(=saccharin) 糖精 $C_7H_5O_3NS$
glutaconaldehyde 戊烯二醛
glutaconanilic acid 戊烯苯胺酸
glutaconate ①戊烯二酸②戊烯二酸盐(酯或根)
glutaconic acid 戊烯二酸 $HO_2CCH_2CH=CHCOOH$
glutaconic anhydride 戊烯二(酸)酐
$$OCCH_2CH=CHCO$$
glutaconic nitrile 戊烯二腈 $NCCH_2CH=CHCN$
glutaconimide 戊烯二酰亚胺
glutaconyl ①戊烯二酰(基)②戊烯二酸单酰(基)
glutamic-oxal(o)acetic transaminase(GOT) 谷(氨酸)草(酰乙酸)转氨酶
glutamic-pyruvic transaminase(GPT) 谷(氨酸)丙(酮酸)转氨酶
glutamic pyruvic transaminase sensor 谷(氨酸)丙(酮酸)转氨酶传感器；GPT 传感器
glutaminyl- 谷酰氨基 $H_2NCOCH_2CH_2CH(NH_2)CO-$
glutamoyl- 谷酰基 $-COCH_2CH_2CH(NH_2)CO-$
glutamycin 戊二霉素
glutaraldehyde 戊二醛
glutaraldehyde tanning 戊二醛鞣(制)
glutaramic acid 戊酰胺酸 $CONH_2(CH_2)_3COOH$
glutaramide 戊二(酸)酰胺 $CONH_2(CH_2)_3CONH_2$
glutarate ①戊二酸②戊二酸盐(酯或根)
glutaric acid 戊二酸
glutaric anhydride 戊二(酸)酐
glutaric dialdehyde 戊二醛 $CHO(CH_2)_3CHO$
glutarimide 戊二酰亚胺
glutaronitrile 戊二腈 $CH_2(CH_2CN)_2$

glutaryl- 戊二酰(基)；戊二酸单酰(基) $-CO(CH_2)_3CO-$
gluten test 面筋含量测定法
glutin ①明胶蛋白②谷胶酪蛋白
glutinene 黏霉烯
glutinic acid 戊炔二酸 $HOOCC\equiv CCH_2COOH$
glutinol 黏霉醇
glutinone 黏霉酮
glutinosin 黏霉菌素
glutinousness ①黏(滞)性②黏(滞)度
glutol 明胶(合)甲醛
glutton fat 獾脂
glutynic acid 戊炔二酸
glycemia 糖血症
glyceraldehyde 甘油醛；2,3-二羟基丙醛 $HOCH_2CHOHCHO$
glyceramine 甘油胺；1,3-二羟基-2-丙胺 $CH_2OHCHNH_2CH_2OH$
glycerate 甘油酸盐 $C_3H_5O_4M$
glyceric acid(=dihydroxypropionic acid) 甘油酸；2,3-二羟基丙酸 $HOCH_2CHOHCOOH$
glyceric aldehyde 甘油醛 $CH_2OHCHOHCHO$
glycerin(=glycerine) 甘油
glycerin abietate(=glyceryl abietate) 甘油松香酸酯
glycerinate 甘油酸盐(或酯)
glycerinated 含甘油的
glycerinating 甘油处理
glycerin α-chlorohydrin α-氯甘油；3-氯-1,2-丙二醇 $CH_2ClCHOHCH_2OH$
glycerin β-chlorohydrin β-氯甘油；2-氯-1,3-丙二醇
glycerin chlorohydrin 氯甘油
glycerin dibromohydrin 二溴甘油
glycerin α-dibromohydrin α-二溴甘油；1,3-二溴-2-丙醇 $(CH_2Br)_2=CHOH$
glycerin β-dibromohydrin β-二溴甘油；2,3-二溴-1-丙醇 $BrCH_2CHBrCH_2OH$
glycerin 1,3-dibutyl ether 甘油-1,3 二丁醚 $(C_4H_9OCH_2)_2=CHOH$
glycerin dichlorohydrin 二氯甘油
glycerin α-dichlorohydrin α-二氯甘油；1,3-二氯-2-丙醇 $(CH_2Cl)_2=CHOH$
glycerin β-dichlorohydrin β-二氯甘油；2,3-二氯-1-丙醇 $ClCH_2CHClCH_2OH$
glycerin diformate(=diformin) 甘油二甲酸酯；二甲精 $(HCO_2)_2C_3H_6O$
glycerin diglycidyl ether 甘油二缩水甘油醚
glycerin α-diiodohydrin α-二碘甘油；1,3-二碘-2-丙醇 $(CH_2I)_2CHOH$

glycerin 1,3-diisoamyl ether 甘油-1,3-二异戊基醚 $(C_5H_{11}OCH_2)_2CHOH$

glycerin 1,3-dimethyl ether 甘油-1,3-二甲基醚 $(CH_3OCH_2)_2CHOH$

glycerin dinitrate 甘油二硝酸酯

glycerin 1,3-diphenyl ether 甘油-1,3-二苯醚 $(C_6H_5OCH_2)_2CHOH$

glycerine(= glycerin) 甘油

glycerine alpha-monoallyl ether 甘油α-单烯丙醚〔交联剂〕

glycerin(e) cement 甘油(型)胶泥〔油性漆的脂肪酸酯与铁盐反应生成 $R(OH)_3$ 型化合物〕

glycerine litharge cement 甘油氧化铅黏合剂

glycerine monofatty ester 甘油单脂肪酸酯；脂肪酸单甘油酯〔俗〕

glycerin epoxy 甘油环氧基(树脂)

glycerin ester 甘油酯

glycerin ether 甘油醚

glycerin foots 甘油下脚

glycerin heptylate(=triheptin) 甘油三庚酸酯；三庚精 $(C_6H_{13}CO_2)_3C_3H_5$

glycerin iodohydrin(=3-iodopandiol) 碘甘油；3-碘-1,2-丙二醇 $CH_2ICHOHCH_2OH$

glycerin monoacetate(=monacetin) 甘油一乙酸酯；一醋精 $CH_3CO_2CH(CH_2OH)_2$

glycerin α-monobutyl ether 甘油α-一丁醚 $CH_2OHCHOHCH_2OC_4H_9$

glycerin α-monobutyrate 甘油α-一丁酸酯 $C_3H_7CO_2CH_2CHOHCH_2OH$

glycerin monoformate(=monoformin) 甘油一甲酸酯；一甲精 $HCO_2CH_2CHOHCH_2OH$

glycerin monoguaiacol ether 甘油一愈创木酚醚 $CH_3OC_6H_4OC_3H_5(OH)_2$

glycerin α-monoisoamyl ether 甘油α-一异戊醚 $CH_2OHCHOHCH_2OC_5H_{11}$

glycerin α-monomethyl ether(=monomethylin) 甘油α-一甲醚；单甲甘油醚；一甲灵 $CH_3OC_3H_5(OH)_2$

glycerin mononitrate 甘油一硝酸酯

glycerin monooleate(=glyceryl monooleate) 甘油单油酸酯 $C_{17}H_{35}CO_2CH_2CHOHCH_2OH$

glycerin β-monostearate 甘油β-一硬脂酸酯 $C_{17}H_{35}CO_2CH=(CH_2OH)_2$

glycerin of tragacanth 黄蓍胶甘油

glycerin soap 甘油皂

glycerin still 蒸甘油锅

glycerin treatment 用甘油处理

glycerin triacetate(=triacetin) 甘油三乙酸酯；三醋精 $(CH_3CO_2)_3C_3H_5$

glycerin triarachidate(=triarachidin) 甘油三花生酸酯；三花生精 $(C_{19}H_{39}CO_2)_3C_3H_5$

glycerin tribenzoate(=tribenzoin) 甘油三苯甲酸酯；三苯精 $(C_6H_5CO_2)_3C_3H_5$

glycerin tributyrate(=tributyrin) 甘油三丁酸酯；三丁精 $(C_3H_7CO_2)_3C_3H_5$

glycerin tricaproate(=tricaproin) 甘油三己酸酯；三己精 $(C_5H_{11}CO_2)_3C_3H_5$

glycerin tricaprylate(=tricaprylin) 甘油三辛酸酯；三辛精 $(C_7H_{15}CO_2)_3C_3H_5$

glycerin triformate(=triformin) 甘油三甲酸酯；三甲精 $(HCO_2)_3C_3H_5$

glycerin triglycidyl ether 甘油三环氧丙醚
$$C_3H_5(OCH_2-\underset{O}{CH}-CH_2)_3$$

glycerin triisovalerate(=triisovalerin) 甘油三异戊酸酯；三异戊精 $(C_4H_9CO_2)_3C_3H_5$

glycerin trilaurate(=trilaurin) 甘油三(十二酸)酯；三月桂精 $(C_{11}H_{23}CO_2)_3C_3H_5$

glycerin trimyristate(=trimyristin) 甘油三(十四酸)酯；三肉豆蔻精 $(C_{13}H_{27}CO_2)_3C_3H_5$

glycerin trinitrate(=nitroglycerin) 甘油三硝酸酯；硝化甘油 $(O_2NO)_3C_3H_5$

glycerin trinitrite 甘油三亚硝酸酯 $(ONO)_3C_3H_5$

glycerin trioleate(=triolein) 甘油三油酸酯；三油精 $(C_{17}H_{33}CO_2)_3C_3H_5$

glycerin tripalmitate(=tripalmitin) 甘油三(十六酸)酯；三棕榈精 $(C_{15}H_{31}CO_2)_3C_3H_5$

glycerin tripropionate 甘油三丙酸酯 $(C_2H_5CO_2)_3C_3H_5$

glycerin tristearate(=glycerol tristearate;tristearin) 甘油三硬脂酸酯；三硬脂酸甘油酯

glycerin trivalerate(=trivalerin) 甘油三戊酸酯；三戊精 $(C_4H_9CO_2)_3C_3H_5$

glycerinum 甘油

glycerin value 甘油值

glycerite of starch 淀粉甘油〔混合物〕

glycerol- ①(=glyceryl)甘油基；丙三基 $-CH_2CHCH_2-$ ②丙三氧基 $-OCH_2CHCH_2O-$

glycerogel 甘油凝胶
glycerogelatin 甘油胶冻
glycerol 甘油；丙三醇 CH₂OHCHOHCH₂OH
glycerol acetate 甘油乙酸酯
glycerol α-allyl ether α-烯丙基甘油醚
glycerol bath 甘油浴
glycerol 1,3-bisisooctyl ether 甘油-1,3-二异辛醚
glycerol boriborate 甘油硼砂硼酸酯
glycerol dichlorohydrin(=α-dichlorohydrin) α-二氯甘油；1,3-二氯-2-丙二醇
glycerol dioleate 甘油二油酸酯
glycerol 1,3-distearate(=glyceryl 1,3-distearate) 甘油-1,3-二硬脂酸酯
glycerol ether acetate 甘油醚醋酸酯〔增塑剂〕
glycerol laurate 甘油月桂酸酯
glycerol lead subacetate 碱式乙酸铅甘油(溶液)
glycerol liquor 甘油水
glycerol monoester 甘油单(脂酸)酯
glycerol monoester of salicylic acid 水杨酸单甘油酯
glycerol monolaurate(=glyceryl monolaurate) 甘油单月桂酸酯
glycerol monooleate 甘油单油酸酯〔增塑剂〕
glycerol monostearate(=glyceryl monostearate; monostearin) 甘油单硬脂酰酯
glycerol oxyethylated-oxypropylated ester 甘油乙氧基化-丙氧基化酯
glycerol phthalate(=glyceryl phthalate) 邻苯二甲酸甘油酯；甘油邻苯二甲酸酯
glycerol-phthalic resin(=alkyd resin) 甘酞树脂；醇酸树脂
glycerol polyglycidylether 丙三醇多缩水甘油醚
glycerol ricinoleate 甘油蓖麻醇酸酯
glycerol sorbitan ester 甘油失水山梨糖醇酯
glycerol sorbitol ester 甘油山梨糖醇酯
glycerol-starch composition 甘油-淀粉润滑剂
glycerol tartrate 甘油酒石酸酯
glycerol tetrahydropyranyl ether 甘油四氢(化)吡喃基醚
glycerol triacetate 甘油三乙酸酯〔增塑剂〕
glycerol tributyrate(=glyceryl tributyrate) 甘油三丁酸酯；三丁精
glycerol trinitrate 甘油三硝酸酯 C₃H₅(NO₃)₃
glycerol tristearate(=glycerin tristearate;tristearin) 甘油三硬脂酸酯；三硬脂酸甘油酯
glycerol trithioglycollate 甘油三巯基乙酸酯
glycerolysis 甘油醇解
glycero-monoricinoleate(=glyceryl monoricinoleate) 甘油单蓖麻(醇)酸酯
glycerophosphatase 甘油磷酸酶

glycerophosphate 甘油磷酸盐
α-glycerophosphate cycle α-甘油磷酸循环
glycerophosphorylethanolamine 甘油磷酰乙醇胺
glycerosazone 甘油醛脎
glycerose 甘油糖 C₃H₆O₃
glycerosone 甘油酮
glycero-tristearate 甘油三硬脂酸酯
glyceroyl 甘油酰；2,3-二羟丙酰 HOCH₂CH(OH)CO—
glyceryl 甘油基；丙三基 —CH₂CHCH₂—
glyceryl abietate(=glycerin abietate) 甘油松香酸酯
glyceryl acetate 甘油乙酸酯
glyceryl alcohol 丙三醇；甘油
glyceryl amine 甘油基胺
glyceryl aminobenzoate 甘油氨基苯甲酸酯
glyceryl p-aminobenzoate 甘油对氨基苯甲酸酯；对氨基苯甲酸甘油酯
glyceryl p-aminobenzoic acid 甘油对氨基苯甲酸酯
glyceryl borate 甘油硼酸酯
glyceryl α-chlorohydrin(=chlorohydrin, α-chlorohydrin) α-氯甘油；3-氯-1,2-丙二醇
glyceryl diacetate(=diacetin) 甘油二乙酸酯；二醋精 (CH₃COOCH₂)₂CHOH
glyceryl dichlorohydrin(=alphadichlorohydrin) α-二氯丙醇；1,3-二氯丙醇
glyceryl 1,3-distearate(=glycerol 1,3-distearate) 甘油-1,3-二硬脂酸酯
glyceryl ditolyl ether 甘油二甲苯醚
glyceryl ester 丙三酯；甘油酯；三甘油酯〔俗〕
glyceryl glycyrrhetinate 甘油甘草亭酸酯
glyceryl maleate polyester 顺丁烯二酸甘油酯聚酯
glyceryl monoacetate(=acetin) 甘油单乙酸酯；醋精
glycerylmonoamine 甘油基一胺；丙三醇一胺
glyceryl monolaurate(=glycerol monolaurate) 甘油单月桂酸酯
glyceryl monooleate(=glycerin monooleate) 甘油单油酸酯
glyceryl monooxalate 甘油一草酸酯
glyceryl monophthalimide(=glycerin monophthalimide) 甘油单邻苯二甲酰亚胺；甘油-酞酰亚胺
glyceryl monoricinoleate(=glycero-monoricinoleate) 甘油单蓖麻(醇)酸酯；单蓖麻酸甘油酯〔消泡剂〕
glyceryl monostearate(=glycerol monostearate; monostearin) 硬脂酸甘油酯
glyceryl oleate 甘油油酸酯
glyceryl palmitate 棕榈酸甘油酯；十六烷酸甘油酯
glyceryl phosphatide 甘油磷脂
glyceryl phthalate(=glycerol phthalate) 邻苯二甲酸甘油

酯;甘油邻苯二甲酸酯
glyceryl phthalimide 甘油邻苯二甲酰亚胺
glyceryl salicylate 甘油水杨酸酯
glyceryl stearate 甘油硬脂酸酯
glyceryl thioglycolate 甘油巯基乙酸酯
glyceryl triacetate(=triacetin) 甘油三乙酸酯;三乙酸甘油酯;三醋精
glyceryl triacetylricinoleate 三乙酰(化)蓖麻醇酰甘油酯
glyceryl tribenzoate 三苯甲酸甘油酯
glyceryl tributyrate(=glycerol tributyrate) 三丁酸甘油酯;三丁精
glyceryl trichloride 甘油基三氯;1,2,3-三氯丙烷 $CH_2ClCHClCH_2Cl$
glyceryl triethyl ether 甘油三乙醚;1,2,3-三乙氧基丙烷 $C_3H_5(OC_2H_5)_3$
glyceryl trimellitate 偏苯三酸甘油酯
glyceryl trimyristate(=trimyristicin) 三肉豆蔻酸甘油酯
glyceryl trinitrate 三硝酸甘油酯;三硝基甘油
glyceryl trioleate 三油酸甘油酯
glyceryl tripalmitate(=tripalmitin) 甘油三棕榈酸酯;三棕榈酸甘油酯;三棕榈精
glyceryl tripropanoate 三丙酸甘油酯
glyceryl triricinoleate 甘油三蓖麻(醇)酸酯;(三)蓖麻醇酸甘油酯
glyceryl tristearate(=stearin) 甘油三硬酯酸酯;硬脂精
glycidamide 环氧酰胺
glycidate ①缩水甘油酸酯②缩水甘油酸盐
glycide(=2,3-epoxy-1-propanol) 缩水甘油;甘油内醚;2,3-环氧-1-丙醇
glycidic acid 环氧丙酸*
glycidic alcohol 缩水甘油醇 $C_2H_3OCH_2OH$
glycidic ester 缩水甘油酸酯 C_2H_3OCOOR
glycidic ester condensation(=Darzen condensation) 缩水甘油酯缩合;α,β-环氧酸酯缩合;达参缩合
glycidol 缩水甘油 $C_2H_3OCH_2OH$
glycidol ether 缩水甘油醚 $C_2H_3OCH_2OR$
glycidoxy〔词头〕 环氧丙氧基
γ-glycidoxypropyltrimethoxy silane γ-环氧丙氧丙基三甲氧基硅烷
glycidyl 缩水甘油基
glycidyl acrylate 丙烯酸缩水甘油酯
glycidyl alkyl ether polymer 缩水甘油烷基醚聚合物
glycidyl allyl ether 烯丙基缩水甘油醚;缩水甘油烯丙醚
glycidyl ester 缩水甘油酯
glycidyl ether 缩水甘油醚
glycidyl ethyl ether 缩水甘油乙醚 $C_2H_3OCH_2OC_2H_5$
glycidyl methacrylate 甲基丙烯酸缩水甘油酯
glycidyl phenyl ether 苯基缩水甘油醚;缩水甘油苯醚

glycidyl polymer 缩水甘油聚合物
glycinamide 甘氨酰胺
glycinamide ribonucleotide 甘氨酰胺核苷酸
glycinamidine 甘氨脒
glycinate 氨基乙酸盐(或酯)
glycine 甘氨酸
glycoal 烯糖〔脱去了两个 HO 的糖〕
glycoalkaloid 配糖(生物)碱
glycocholic acid(=glucocholic acid) 甘胆酸
glycocoll 甘氨酸;氨基乙酸 CH_2NH_2COOH
glycoconjugate 糖缀合物;缀合糖;复合糖
glycocyamidine 胍基乙酸内酰胺 $C_3H_5ON_3$
glycocyamine(=guanidinoacetic acid) 胍基乙酸 $HN=C(NH_2)NHCH_2CO_2H$
glycogen staining method 糖原染色法
glycol ①1,2-亚乙基二醇;甘醇〔俗〕;1,2-乙二醇 ②二元醇的类名
glycol adipate 己二酸乙二醇酯
glycolal 乙醇醛 CH_2OHCHO
glycolaldehyde 乙醇醛;羟基乙醛 $HOCH_2CHO$
glycolaldehyde diethyl acetal 乙醇醛缩二乙醇 $HOCH_2CH=(OC_2H_5)_2$
glycolamide 2-羟乙酰胺
glycolate 甘醇酸酯
glycol borate 硼酸乙二醇酯
glycol-catalyst solution 催化剂乙二醇溶液
glycol cellulose 纤维素乙二醇醚
glycol concentration 乙二醇浓度;乙二醇浓缩
glycol condenser 乙二醇冷凝器
1,2-glycol content 1,2-二醇含量
glycol decanoate 癸酸乙二醇酯
glycol diacetate(=ethylene acetate) 二乙酸乙二醇酯 $CH_3COOCH_2CH_2OOCCH_3$
glycol dibenzyl ether 乙二醇二苄基醚 $(C_6H_5CH_2OCH_2)_2$
glycol dibutyrate(=ethylene glycol dibutyrate) 二丁酸乙二醇酯 $(C_3H_7CO_2CH_2)_2$
glycol dicaprate 二癸酸乙二醇酯 $(C_9H_{19}CO_2CH_2)_2$
glycol dicaproate 二己酸乙二醇酯 $(C_5H_{11}CO_2CH_2)_2$
glycol dicaprylate 二辛酸乙二醇酯 $(C_7H_{15}CO_2CH_2)_2$
glycol dicarboxylate 二羧酸乙二醇酯
glycol diformate 二甲酸乙二醇酯 $(HCO_2CH_2)_2$
glycol diguaiacic ether 乙二醇双愈创木酚醚 $(CH_3OC_6H_4OCH_2)_2$
glycol dilaurate 双十二(烷)酸乙二醇酯 $(C_{11}H_{23}CO_2CH_2)_2$
glycol dimercaptoacetate 乙二醇二巯醇基乙酸酯〔促进剂〕
glycol dimercaptopropionate 乙二醇二巯基丙酸酯〔交联

剂、环氧固化促进剂〕

glycol dimethyl ether(=dimethoxy-ethane)　乙二醇二甲醚；二甲氧基乙烷　$(CH_3OCH_2)_2$

glycol dimyristate　双十四(烷)酸乙二醇酯　$(C_{13}H_{27}CO_2CH_2)_2$

glycol dinitrate(=dinitroglycol)　二硝酸乙二醇酯；硝化乙二醇　$(O_2NOCH_2)_2$

glycol dipalmitate　双十六(烷)酸乙二醇酯　$(C_{15}H_{31}COOCH_2)_2$

glycol dipropionate　二丙酸乙二醇酯　$(C_2H_5COOCH_2)_2$

glycol distearate　二硬脂酸乙二醇酯　$(C_{17}H_{35}COOCH_2)_2$

glycol dodecyl ether　乙二醇十二烷基醚

glycol ester　乙二醇酯

glycol ether　乙二醇醚；二甘醇；一缩二乙二醇

glycol ether acetate　乙二醇醚乙酸酯

glycol ether content　乙二醇醚含量

glycoletherdiaminotetraacetic acid(GEDTA)　乙二醇醚二胺四乙酸

glycol ether ester　乙二醇醚酯(类)

glycol ether of cellulose　纤维素乙二醇醚

glycol ethyl ether　乙二醇乙基醚

glycol ethylidene-acetal　乙二醇缩乙醛
　　$CH_3CH=(OCH_2)_2$

glycol fatty acid stearate　脂肪酸硬脂酸二醇酯

glycol head tank　乙二醇高位槽

glycolic acid phenyl ether　苯氧基乙酸

glycolic antimony oxide　三氧化二锑乙二醇悬浮液

glycolic manganese acetate　乙酸锰乙二醇悬浮液

glycolic suspension　乙二醇悬浮液

glycolide(=glycollide)　乙交酯；乙醇酸交酯
　　$CH_2CO_2CH_2COO$

glycolipid　糖脂

glycolipide(=glycolipin)　糖脂

glycolipin(=glycolipide)　糖脂

glycol isopropyl ether　乙二醇异丙醚　$C_3H_7O(CH_2)_2OH$

glycollate　乙醇酸盐

glycollic acid　乙醇酸；羟基乙酸　$HOCH_2CO_2H$

glycollic aldehyde　乙醇醛；羟基乙醛　CH_2OHCHO

glycollic amide　乙醇酰胺　$HOCH_2CONH_2$

glycollic anhydride　乙醇酸酐　$(OHCH_2CO)_2O$

glycollic nitrile　乙醇腈　$HOCH_2CN$

glycollide(=glycolide)　乙交酯；乙醇酸交酯
　　$CH_2CO_2CH_2COO$

glycol lubricant　乙二醇润滑剂

glycol-modified polyethylene terephthalate(PETG)　醇化聚酯

glycol monoacetate(=ethylene monoacetate)　一乙酸乙二醇酯

glycol monobenzyl ether　乙二醇一苄醚　$C_7H_7OC_2H_4OH$

glycol monoethyl ether(=cellosolve)　乙二醇一乙醚〔溶纤剂〕　$HOCH_2CH_2OC_2H_5$

glycol monoformate　一甲酸乙二醇酯　$HCO_2CH_2CH_2OH$

glycol monomethyl ether　乙二醇一甲醚　$CH_3O(CH_2)_2OH$

glycol monopalmitate　单棕榈酸乙二醇酯

glycol monosalicylate　一水杨酸乙二醇酯

glycolonitrile　乙醇腈　$HOCH_2CN$

glycoloyl(=glycolyl)　乙醇酰　$HOCH_2CO-$

glycol recovery　乙二醇回收

γ-glycols　γ-二醇类；遥二醇类

glycol terephthalate　乙二醇对苯二甲酸酯〔增塑剂〕

glycol urethane　乙二醇脲烷

glycoluric acid　脲基乙酸

glycoluril(=glyoxaldiurene)　乙醛缩二脲；甘脲

glycolyl(=glycoloyl)　乙醇酰　$HOCH_2CO-$

glycolylguanidine　乙醇酰胍

glycolylurea(=hydantoin)　乙内酰脲；海因〔俗〕

glycopolymer　糖聚合物

glycoprotein(=glucoprotein)　糖蛋白

glycosa(e)-6-phosphate　葡萄糖-6-磷酸酯〔增塑剂〕

glycosamine　葡萄糖胺；氨基葡萄糖

glycosamino glucan　葡萄糖胺基葡聚糖；葡萄糖胺多糖

glycosuria　糖尿

glycuresis　糖尿；糖尿症

glycyl　甘氨酰；氨基乙酰　H_2NCH_2CO-

glycyl alcohol(=glycerol)　甘油；丙三醇

glycylglycine dipeptidase　甘(氨酰)甘(氨酸)二肽酶

glycyramarin　甘草苦苷

glycyrol　甘草醇

glycyrrhetate　甘草酸盐

glycyrrhetin　甘草亭

glycyrrhetinic acid(=glycyrrhetic acid)　甘草亭酸

glycyrrhiza　甘草

glycyrrhizic acid(=glycyrrhizinic acid)　甘草酸

glycyrrhizin　甘草甜

glycyrrhizin ammoniated　甘草甜素铵盐

glycyrrhizinic acid(=glycyrrhizic acid)　甘草酸　$C_{42}H_{62}O_{18}$

glyicine type amphoterics　甘氨酸型两性表面活性剂

glyme(=ethylene glycol dimethyl ether)　甘醇二甲醚

glyoxal　乙二醛　$O=CHCH=O$

glyoxalase　乙二醛酶

glyoxal bis(hydroxyanil)　乙二醛缩双羟苯胺

glyoxal dioxime 乙二醛二肟；乙二肟
glyoxaldiurene(=glycoluril) 乙二醛缩二脲；甘脲
glyoxal ethyline(=2-methyl imidazole) 2-甲基咪唑
glyoxalic acid 乙醛酸 CHOCOOH
glyoxalic acid hydrate(=glyoxylic acid) 二羟乙酸；水合乙醛酸
glyoxalin(=imidazole) 咪唑；1,3-二氮杂茂
glyoxalinyl(=glyoxalyl;imidazolyl) 咪唑基；甘噁啉基
glyoxal sulfate 硫酸缩乙二醛
glyoxalyl(=glyoxalinyl; imidazolyl) 咪唑基；甘噁啉基
glyoxime 乙二肟；乙二醛二肟 HON=CHCH=NOH
glyoximic acid 乙醛酸肟；肟基乙酸
glyoxylaldehyde 乙二醛
glyoxylase(=glyoxalase) 乙二醛酶
glyoxylate ①乙醛酸②二羟乙酸③二羟乙酸盐
glyoxylate bypass 乙醛酸循环；乙醛酸支路
glyoxylate cycle 乙醛酸循环；乙醛酸支路
glyoxyl carboxylic acid 乙醛酰羧酸 HOOCCOCHO
glyoxylic acid(=glyoxalic acid hydrate) 二羟乙酸；水合乙醛酸 (OH)₂CHCOOH
glyoxylic acid oxime(=glyoximic acid) 乙醛酸肟；肟基乙酸
glyoxyloyl(=glyoxylyl) 乙醛酰 OHCCO—
glyoxylurea(=allanturic acid) 乙醛酸脲；脲叉乙酸
glyoxylyl(=glyoxyloyl) 乙醛酰
glyoxysome 乙醛酸循环体
glyptal 醇酸树脂；丙三醇-邻苯二甲酸树脂
glyptal resin 甘酞树脂；丙三醇邻苯二甲酸树脂
glysantine 各里散亭〔一种使水结冰以降温的物质〕
G.M.counter(=Geiger-Müller counter) 盖革-缪勒计数器
gmelinite 钠菱沸石
gmelinol 石梓醇
gnarl ①木节②木瘤
gneiss 片麻岩
gneiss-granite 片麻花岗岩
gneissic structure 片麻结构〔岩〕
gneissose granite 片麻花岗岩
gnoscopine 格诺莨菪品
goa 柯桠粉
goa butter ①羚羊奶油②藤黄油
goaf(=gob) 空岩；不含矿的岩石
go-and-return ①两端间②来回的
goat hair 山羊毛
goat skin 山羊皮
goat tallow 山羊脂
gob(=goaf) 空岩；不含矿的岩石
gobelin blue 暗青绿色
godang wax(=gondang wax) 榕树蜡

godetless spinning machine 无导丝盘纺丝机
godet roller 导丝轮
godet wheel 导丝轮
go-devil(=go-on-devil) 清管刮刀
go-devil tracing (输送管中)刮刀痕迹
godown 堆栈；仓库
goethite 针铁矿
goetzenite 氟硅钙钛矿
goffered paper 皱纹纸
goffering 形成浮花；形成皱纹
GO-fining and RESID fining 瓦斯油加氢精制及渣油加氢精制(法)
go-gauge 过端量规
goggles 眼罩；护目眼镜
goingin(=sinkage) 下陷；渗入
goiter 甲状腺肿
Golay cell 红外线指示器
Golay coil 戈雷探测器；戈雷线圈
Golay column 戈雷柱〔毛细管柱〕
Golay detector 戈雷检测器；红外线检测器
Golay equation 戈雷方程*
Golay pneumatic detector 戈雷气动池检测器；戈雷(气动)池
gold ①(黄)金②(黄)金(色)的
gold balance 金秤〔贵重金属天平〕
gold bath 金盐液
gold bromide 溴化金
gold bronze 金色铜粉
gold chloride 氯化金
gold cyanide 氰化金
gold-cyanide complex 金氰络合物
gold dichloride 二氯化金
gold dioxide 二氧化金 Au₂O・Au₂O₃
gold electrode 金电极
golden ①(黄)金的②金制的；含金的③金色的
golden cut method 黄金分割法
golden germander 金石蚕
golden glow 金辉黄
golden lacquer inlay 金漆镶嵌
golden red 金红色
golden rod 一枝黄〔一种橡胶草〕
goldenrod oil 一枝黄花油
golden sulfide (of antimony) 五硫化二锑 Sb₂S₅
golden wattle 金荆树；密花相思树；密花金合欢
golden yellow 金黄色的
golder lacquer inlay 金漆镶嵌
gold evaporation 表面蒸金
gold flat yarn 片金线

gold foil 金箔
gold-foil electroscope type Kundsen gauge 克努森金箔验电器型真空计
gold gasket 金密封垫圈
gold hydrocyanic acid 氰化金酸 $H[Au(CN)_4]$
gold hydrogen nitrate 硝金酸 $H[Au(NO_3)_4]$
gold ink 金色油墨
goldinodex 高迪菌素
gold iodide 碘化金〔AuI、AuI_3等〕
gold lacquer 淡金水〔一种涂于铝箔时有金色效果的紫胶漆〕
gold leaf 金叶
gold leaf electroscope 金箔验电器
gold monobromide 一溴化金；溴化亚金 AuBr
gold monochloride 一氯化金；氯化亚金 AuCl
gold monoiodide 一碘化金；碘化亚金 AuI
gold monoxide 一氧化金；氧化亚金 Au_2O
gold needle engraving 戗金〔漆艺〕
gold number 金值
gold orange 金橙〔商〕；酸性橙
gold paint 金粉漆；金黄色漆
gold perchloride 三氯化金 $AuCl_3$
gold pigment ①金粉颜料②金黄色铜粉颜料
gold plating 镀金
gold point 金点
gold potassium bromide 溴化金钾；溴金酸钾 $K[AuBr_4]$
gold potassium chloride 氯化金钾；氯金酸钾 $K[AuCl_4]$
gold potassium iodide 碘化金钾；碘金酸钾 $K[AuI_4]$
gold powder 金粉
gold preheating loop 预热金环管
gold ribbon 金带；金带饰
gold ruby(=lake red D) 色淀红 D
gold ruby glass 金红玻璃
gold salt 氯金酸钠 $Na[AuCl_4]$
gold (silver) filigree and inlay 金(银)丝镶嵌〔漆艺〕
gold silver jeweleries ①贴金漆；描金胶；涂金胶②镶嵌珠宝的金银工艺品
gold size(=oil gold size) ①油性金胶；金胶油②金漆；金胶；贴金用漆
gold-size paste 金胶糊
gold-size surface 金胶漆表面〔涂金胶漆的表面〕
Goldsmith process 戈德史密斯法
gold sodium bromide 溴化金钠；溴金酸钠 $Na[AuBr_4]$
gold sodium chloride 氯化金钠；氯金酸钠 $Na[AuCl_4]$
gold sol 金溶胶
gold stamp(ing) 烫金
gold stoving varnish 金色清烘漆

gold-tin purple 金锡紫
gold-tinted 黄光；带黄(色)头的
gold toning 金调色法
gold tooling 烫金工具
gold tribromide 溴化金；三溴化金 $AuBr_3$
gold trichloride 氯化金 $AuCl_3$
gold tricyanide 氰化金 $Au(CN)_3$
gold triiodide 碘化金 AuI_3
gold trioxide 氧化金 Au_2O_3
gold trisulfide 硫化金；三硫化二金 Au_2S_3
gold tubular reactor 金管反应器
gold wire seal 金丝密封圈
Golgi apparatus 高尔基(氏)体
Goly column 戈雷柱
Gomberg-Bachmann-Hey synthesis 冈伯格-巴赫曼-黑氏合成法
Gomberg reaction 冈伯格反应
gome 润滑油的积炭
gonadorelin 戈那瑞林〔药〕
gonane 甾烷
gondang wax(=godang wax) 榕树蜡
gondoic acid 巨头鲸鱼酸；二十碳-11-烯酸
gonecystolith 精囊石
goniochromatism 角异色性
goniometer ①(晶体)测角计②测向器
goniometer head 测角头
goniophotometer 测角光度计
goniophotometry 测角测光法
goniospectrophotometer 测角分光光度计
goniospectrophotometry 测角分光光度法
go-no-go 一定口径
gonystilol 线桂醇
Gooch crucible 古氏坩埚*
Gooch filter 古氏滤器；古奇滤器
Gooch funnel 古氏漏斗
good aging 老化性能良好
good agricultural practice 良好农业措施
good cure 适度硫化；适中硫化
Goodeve's rotational viscometer 古迪夫旋转黏度计
good fixative 良好定香剂
good knitting 融合良好
Goodloe packing 古德洛填料
goodness of fit 拟合优度*
goodness of fit test 拟合优度检验
good oil 精制油；提纯油
Goodrich cord testing vibrator 古德里奇型帘子线测试用振动仪
Goodrich rotor flexer 古德里奇型转子挠曲试验机

Goodrich viscurometer 古德里奇黏度硫化计
goods in bulk 集装物
good sized beam rate 浆纱好轴率
good solvent 良溶剂
goods piece length 成品匹长
Goodyear angle type abrader 固特异变角度型磨耗试验机
Goodyear ozone fatigue tester 固特异臭氧疲劳试验机
go-on-devil(=go-devil) 清管刮刀
gooseberry 圆醋栗；需俱果
gooseberry flavor 圆醋栗香精
goose fat 鹅脂
goosefoot oil 美洲土荆芥油
goose neck 鹅颈管〔蒸馏釜上的汽相馏出管线〕
goose skin 鹅皮〔漆病〕
gorlic acid 环戊烯十三碳烯酸；大风子烯酸
goserelin 戈舍瑞林〔药〕
goslarite 皓矾
gossypetin 棉子皮亭
gossypetonic acid 棉子皮酮酸
gossypin 棉纤维素
gossypitol 棉子皮醇
gossypitone 棉子皮酮
gossypoid 棉根皮素
gossypol 棉子酚
gossypolic acid 棉子酚酸
gossypose 棉子糖
gossypyl alcohol 棉子基醇
gothite 针铁矿
gouache ①水胶颜料打底法②水胶颜料打底作画法
gouache paint 水胶颜料〔树胶与水彩的调和物〕
goudron 焦油；沥青
gougerotin 谷氏菌素
gougeroxymycin 谷氏氧霉素
gouging abrasion 碰撞磨损；冲击磨损
gouging edge ①半圆形凿刃②弧口状(周)边；新月形边
Gouin accumulator 高因蓄电池
Goutal formula for heating value 高塔耳热值公式
goutine 痛风粉
Gouy cell 古依电池
Gouy-Chapman double layer 古依-查普曼双层
Gouy method 古依法
governor ①调节器②调速器
governor control 调速器控制
governor valve 调节阀
goyol 告衣醇
grab 抓斗
Graco Hydra-spray process 格拉克液压喷涂法

gradation of aggregate (颜料粒子)凝聚体的级配(级别)
gradation of color 颜色的深浅程度(等级)；颜色的浓淡程度(等级)
grade ①级；度②级〔生物分类〕
graded 分了级的；分了类的；选过了的
graded aggregate 分级粒料〔经过一定筛目的石子与沥青的混合物，铺路用〕
graded coal 筛过煤
graded crushing 分级压碎
graded hydrolysis 分段水解
graded ply belt 阶梯式运输带
graded potential 分段电势
grade efficiency of separation 分级分离效率
grade estimation 质量评定
grade hydrolysis 分段水解
grade line (管路的)坡度线
grade number 熔体(流动)指数；熔融指数
grade of fit 配合等级
grade of reagent 试剂品级
grader ①分类器②平路机
grades ①分级；级别②坡度
grade sampler of air particulate matter 大气颗粒物分级采样器
grades of rosin 松香(的颜)色(等)级
grade standard 等级标准；品级标准
gradient 梯度；陡度
gradient block copolymer 梯度嵌段共聚物
gradient centrifugation 梯度离心
gradient column (密度测定)梯度管〔测定尼龙、塑料的密度〕
gradient cooling 分级冷却；逐步冷却
gradient copolymer 梯度共聚物
gradient curve 梯度曲线
gradient development 梯度展开
gradient device 梯度装置
gradient elution 梯度洗脱
gradient elution adsorbent system 梯度洗脱吸附剂系统
gradient elution chromatography 梯度洗脱色谱法
gradient elution device 梯度洗脱器
gradient elution partition chromatography 梯度洗脱分配色谱(法)
gradient elution separation 梯度洗脱分离
gradient enhanced NOE spectroscopy 梯度增强核欧沃豪斯效应谱
gradienter 测梯度仪；倾斜测定器
gradient freezing 梯度冷冻
gradient gel electrophoresis 梯度凝胶电泳
gradientless contactor 无梯度接触器

gradientless reactor　无梯度反应器
gradient liquid chromatograph　梯度液相色谱
gradient liquid chromatography　梯度液相色谱法
gradient method　梯度法
gradient mixed applicator　梯度混合涂板器
gradient mixed spreader　梯度混合涂板器
gradient mixing device　梯度混合装置
gradient morphology　梯度形态
gradient penetration　梯度渗透
gradient polymer　梯度聚合物
gradient ratio calibration　梯度比校正
gradient reversal imaging　梯度回波成像
gradient scanning　梯度扫描
gradient search　梯度寻优*
gradient shape　梯度形状
gradient slope　梯度坡度
gradient solvent　梯度溶剂
gradient start　梯度起点
gradient steepness　梯度倾斜度
gradient technique　梯度技术
gradient thin layer chromatography　梯度薄层色谱(法)
gradient thin layer plate　梯度薄层板
gradient titration　梯度滴定
gradient type　梯度型式
grading　①分级②粒度的确定
grading analysis　粒度分析
grading machine　平土机；分级机
grading of support size　载体粒度分级
grading room　检选室；分级室
grading sieve　分级筛
grading sifter　分级筛；分选筛
gradiomanometer　压差密度计
gradocal filter　分级滤器
gradocal membrane　分级滤膜
gradual change　渐变
gradual condensation　逐级缩合(作用)
gradual contraction　截面逐渐缩小
gradual development　逐级展开
gradual enlargement　截面逐渐扩大
gradual migration　逐步迁移
graduate　①量筒；量杯②刻度；划分度数
graduated　刻(了)度的
graduated bottle　刻度瓶
graduated centrifuge tube　刻度离心管
graduated colour　(彩色)层次渐变的颜色
graduated cylinder　量筒
graduated dial　指(度)盘；分度盘；刻度盘
graduated flask　量瓶

graduating　刻度；分度
graduation　刻度；分度
graduation error　刻度误差
graduation house(=graduation tower)　梯塔
graduation mark　刻度线
graduation of thermometer　温度计刻度
graduations in degrees　按度分刻度
graduation tower(=graduation house)　梯塔
graduator　刻度器；分度器
Graebe-Ulmann carbazole synthesis　格雷伯-乌尔曼咔唑合成法
Graesser rain-bucket extractor　格雷泽雨斗萃取器
Graetz number　格雷数
graft　①嫁接②移植物
graftable　①可接枝的②可移植的
graft copolymer　接枝共聚物
graft copolymer hollow fibre　接枝共聚物中空纤维
graft copolymerization　接枝共聚合*
graft copolymerization by ultraviolet radiation　紫外线辐射(法)接枝共聚
graft copolymerization on the surface of polymers　聚合物表面接枝反应
graft degree　接枝度
graft dispersant　接枝分散剂
grafted　接枝的
grafted cellulose　接枝纤维素
grafted chain　接枝链
grafted fibre　接枝纤维
grafted hollow fibre　接枝中空纤维
grafted monomer　接枝单体
graft effect　接枝效应
graft elastomer　接枝弹体
graft hybridization　嫁接杂交
graft implantation　接枝植入
grafting degree　接枝度
graft(ing) efficiency　接枝效率
grafting knife　芽接刀
grafting monomer　接枝(用)单体
grafting reactant　接枝反应物
grafting site　接枝点
graft(ing) yield　接枝率
graft modification　接枝改性
graftomer　接枝聚合物
graft polymer　接枝聚合物
graft polymerization　接枝聚合
graft polymerization yield　接枝聚合率
graft-promoting additive　接枝促进剂
graft ratio　接枝比

graft rayon staple　接枝人造短纤维
graft reaction　接枝反应
graft rejection　移植排斥
graft rubber　接枝橡胶
graft side chain　接枝侧链
graham bread　黑面包
graham flour　黑面粉〔未去麸的面粉〕
grahamite　脆沥青
Graham salt　格雷姆盐
Graham's law　格雷姆定律
Graham's law of diffusion　格雷姆扩散定律
grail　砾石；鹅卵石
grain　①粒②粒面③谷粒④胭脂虫〔染料〕⑤木材纹理⑥(沙)颗粒　⑦格令〔重量单位，=0.065 克〕⑧晶体
grain alcohol　乙醇；酒精〔专指以粮食为原料〕
grain alignment　(片材)纹理对齐
grain boundary　晶粒间界
grain boundary carbide　晶界碳化物
grain boundary density　晶粒间密度
grain boundary diffusion　晶粒间界扩散
grain boundary precipitate　晶(粒边)界沉淀物
grain cast-iron roll　未淬火的铸铁滚筒
grain density　晶核密度
grain distillery　酒精厂
grain drier　谷物干燥机
grained　粒状的
grained catalyst　粒状催化剂
grained soap　粒状皂；盐析皂
grained tin　粒状锡
grain effect　压纹效应；晶粒效应
grain embossing　粒面压花
grainer　①脱毛器②起纹器③蒸发器
grain formation　晶粒的形成
grain-free　无晶核的
grain growing　晶粒生长
grain growth　晶粒长大
grain growth inhibitor　晶粒生长抑制剂
graininess　①粒度②粒性③搓纹；搓花〔皮革〕
graining　①粒化②皂析③起纹
graining board　压纹板；搓纹板；搓花板
graining change　(皂胶的)盐析(阶段)；成粒(阶段)
graining colour　①木纹着色料②彩色木纹透明漆〔描绘木纹用的有色半透明漆〕
graining liquid　木纹清漆；木纹油
graining machine　木纹机；搓纹机
graining of tin　锡的粒化
graining out　①皂析②加盐分离③盐析
graining paint　木纹漆(描绘木纹用)

graining paste　(搪瓷绘花)颜料浆
graining point　皂析点；成颗粒点；粒化点
graining table　搓花台；搓纹台〔皮革〕
grain lac　粒状紫漆
grain layer　粒面层
grain leather for gloves　手套羔皮
grain-like composite fibre　木纹状复合纤维
grain mill　磨坊
grain of crystallization　结晶中心；晶核；晶籽
grain of soap　皂粒；皂核
grain of wood　木材纹理
grain oil　杂醇油
grain printing　木纹印刷
grain propagation　晶粒扩展
grain refining　晶粒细化
grain-refining agent　晶粒细化剂
grain roller　粒面滚；压花辊
grain side of leather　皮革粒面
grain size　①粒积②晶粒大小
grain soap　皂粒
Grain stain　革兰染色
grain staking　干革拉软
grain stillage　酒糟；沉渣
grain surface　粒面
grain tester　谷粒检验器
grain tin　粒状锡
grain type cracking　粒形开裂
grainule size　粒度
grainy　粒状的；颗粒的
grainy sludge　粒状残渣
gram(=gramme)　克
gramamycin(=glumamycin)　颖霉素
gram atom　克原子
gram atomic weight　克原子量
gram calorie　克卡；卡；小卡〔相对于大卡即千卡〕
gram centimeter　克厘米
gram equivalent　克当量
gram force　克力；克重
gram formula weight　克(化学)式量
gramicidin　短杆菌肽
gramine　芦竹碱；2-二甲氨甲基吲哚
graminic acid　短杆菌酪酸
graminifoline　禾叶(千里光)碱
graminomycin　禾霉素
gram ion　克离子
gram mass　克质量
gramme(=gram)　克
gram method　常量分析

grammol 克分子
gram molecular solution 克分子溶液
gram molecular volume 克分子体积
gram molecular weight 克分子量
gram molecule 克分子
gram particle 克粒
gram particle weight 克粒量
gram-rem 克雷姆〔积分吸收剂量单位〕
Gram's iodine solution 革兰氏碘液
Gram stain 革兰氏染色
Gram-stained 革兰氏染色的
gram thorium unit 克钍单位
granatal 石榴皮醛
granatane 石榴皮烷
granatanine 石榴皮单宁
granataninol 石榴皮单宁醇
granatenine 石榴皮碱
granatic acid 石榴皮酸
granaticin 榴菌素
granatin 石榴皮亭
granatoline 石榴皮灵
granatonine 石榴皮宁
granatotannic acid 石榴皮单宁酸
granatum 石榴皮
granatylamine 石榴皮基胺
grandcanonical ensemble 巨正则系综*
grandcanonical partition function 巨正则配分函数
granddaughter nuclide 第三代子体核素
grand ensemble 巨(正则)系综
grane ambra 龙涎香
granegillin(=hydroxyaspergillic acid) 谷霉菌素〔即：羟基曲霉酸〕
Gran function 格兰函数
granisetron 格拉司琼〔药〕
granite 花岗石
granite linoleum 花岗石油毡
granite paper 花岗石纹纸
granite-ware 花岗岩石器
granitic plaster 仿花岗石面饰
granitide 黑云花岗岩
granitoid 花岗类岩
granodiorite 花岗闪长岩
granodolerite 花岗粗粒岩
granolithic toppings 人造石饰面(层)
granophyre ①花斑岩②文象斑岩
granophyric texture 花斑(岩)结构
granoplasm 颗粒性原浆
Gran's plot 格兰作图法；格氏作图法

Gran's plot paper 格兰作图纸；格氏作图纸
grantianine 石榴皮宁
granula (颗)粒
granular 粒状
granular activated carbon 颗粒活性炭
granular aluminum powder 粒状铝粉
granular ash 苏打粒；粒状苏打
granular bed 粒状床
granular boundary 颗粒界面
granular calcium chloride 粒状氯化钙
granular carbon 粒状活性炭
granular catalyst 粒状催化剂
granular charcoal 粒状木炭
granular-crystalline structure 粒状结晶的结构
granular finish ①粒状涂饰；粗饰面②粒状面漆
granular form 粒状；粒形
granular fracture 粒状破裂
granular gypsum 粒状石膏 $CaSO_4 \cdot 2H_2O$
granular ion exchange resin 粒状离子交换树脂
granular iron 粒状铁
granularity ①粒度②粒性
granular layer 颗粒层
granular magnesium silicate 粒状硅酸镁
granular mass 颗粒质量
granular membrane 筛网过滤器
granular pearlite 粒状珠光体
granular polymer 粒状聚合物
granular polymerization 成粒聚合(作用)
granular rubber 粒状(橡)胶
granular size 粒度
granular soda ash 苏打粒；粒状苏打
granular structure 粒状结构
granular tapioca 木薯粉片
granular yeast 粒状酵母
granular zone 粒料区
granulate 粒化
granulate cooling 粒料冷却
granulated 成粒的；粒状的
granulated catalyst 粒状催化剂
granulated cork 软木屑
granulated fertilizer 颗粒肥料
granulated gel 粒状凝胶
granulated glass sphere 玻璃珠
granulated rough (黏有砂粒等)粗糙不平；粗涩
granulated rubber 粒状(橡)胶
granulated tin 粒状锡
granulating 造粒；粒化过程
granulating agent 成粒剂

granulating machine　成粒机
granulating mill　磨谷机
granulation　成粒(作用)
granulation polymerization　成粒聚合(作用)
granulator　成粒机
granule　粒；颗粒〔无定形的〕
granule dry hopper　粒料烘干储槽
granule manufacture　颗粒制造；造粒
granule size　粒度(大小)
granules of catalyst　催化剂颗粒
granulesten　大豆磷脂
granulimetry　粒度测定法
granulite　白粒岩
granulometer　粒度计；颗粒测量仪
granulometric composition　颗粒的组成
granulometry　粒度测定法；粒度分析
granulophyre　微花斑岩
granulous　粒状的
grapefruit neroli oil　圆柚花油
grapefruit oil　圆柚油
grapefruit petitgrain oil　圆柚叶油
grape-fruit seed oil　柚子油
grapefruit yellow　葡萄柚黄
grape hyacinth　总状麝香兰
grape juice　葡萄汁
grape-kernel oil(=grape seed oil)　葡萄子油
grape seed oil(=grape-kernel oil)　葡萄子油
grape sugar　葡萄糖　$C_6H_{12}O_6$
graph　①(曲线)图；标绘图②〔词尾〕仪；机
graphic(=graphical)　图解的
graphical(=graphic)　图解的
graphical analysis　图解分析
graphical calculation board　计算板
graphical classification　(坐标)图解分类
graphic(al) copying　复印
graphical differentiation　图解微分
graphical integration　图解积分
graphical method　图解法
graphical plotting　图解法
graphical representation　图解表示
graphical-statistical analysis　图解统计分析
graphical weeping point(GWP)　图示漏液点
graphic arts coatings　标牌涂料；指示牌涂料
graphic formula　图(解)式
graphic granite　文象花岗岩
graphic instrument　图解仪；自动记录仪
graphic integration　图解积分
graphic interpretation　图示法；图解法

graphic investigation　图解研究
graphic language　图形语言
graphic method　图解法
graphic panel　图像化仪表板；测量系统图示板
graphic recorder　图解记录器
graphic symbol　图解符号
graphic texture　图解结构
graphite　石墨
graphite anode　石墨阳极
graphite arc　石墨电弧
graphite atomic absorption spectrometry　石墨炉原子吸收光谱法
graphite braid atomizer　石墨带原子化器
graphite carbon　石墨碳
graphite corrosion　石墨腐蚀
graphite crucible　石墨坩埚
graphite cup electrode　石墨杯电极
graphite cuvette atomizer　石墨坩埚原子化器
graphited　①石墨化的②涂石墨剂的
graphited fiber　石墨纤维
graphited throughout　全部用石墨剂处理的
graphite electrode　石墨电极
graphite ferrule　石墨卡套
graphite fibre　石墨纤维
graphite filament atomizer　石墨丝原子化器
graphite furnace　石墨炉
graphite furnace atomic absorption spectrometry(GFAAS)　石墨炉原子吸收光谱法
graphite grease　石墨(润滑)脂
graphite heat exchanger　石墨换热器
graphite intercalation compound　石墨插层化合物；石墨层间化合物
graphite-lined platinum crucible　衬以石墨的白金坩埚
graphite lubricant　石墨润滑剂
graphite lubricating oil　石墨润滑油
graphite lubrication　石墨润滑
graphite moderated reactor　石墨减速反应堆
graphite monochromator　石墨单色器
graphite network　石墨网
graphiteoxide　石墨氧化物
graphite packing　石墨填料
graphite paint　石墨漆
graphite platform　石墨平台
graphite powder　石墨粉
graphite reinforced synthetic material　石墨增强合成材料
graphite resistance heater　石墨电阻加热器
graphite rod atomizer　石墨棒原子化器
graphite spark method　石墨火花法

graphite tube burner   管式石墨炉
graphite tube coated with refractory metal carbide   难熔金属碳化物涂层石墨管
graphite tube lined with metal foil   衬金属箔石墨管
graphite whisker   石墨搅拌器
graphitic   石墨的
graphitic acid   石墨酸
graphitic carbon   石墨碳
graphitic mica   石墨云母；硅酸钾铝
graphitic rock   石墨岩
graphitiferous   (含)石墨的
graphitisable   可石墨化的
graphitizability   石墨化能力；可石墨化性
graphitizable substrate   可石墨化(的)基质
graphitization   石墨化作用
graphitized carbon   石墨化的碳
graphitized carbon black   石墨化炭黑；导电炭黑
graphitized carbon column packing   石墨化炭黑柱填料
graphitized carbon fibre   石墨化碳纤维
graphitized electrode   石墨敷面电极
graphitized fibre   石墨化纤维
graphitizing   ①石墨化作用②披覆石墨
graph of molecular orbital   分子轨道图形
grapholite   石墨片岩
graph paper   坐标纸；方格纸；作图纸
graph theory   图论
graph theory of molecular orbital   分子轨道图形理论
graser   γ射线激光(器)
Grashof number   格拉斯霍夫数
grass bleaching   (草场)暴晒漂白法
Grasselli abrasion machine   格拉西里磨耗试验机
grasseriomycin   蚕病霉素
grass fiber(=straw fiber)   草纤维
grass hopper   ①准备焊接管子用的修正和联接工具②输送设备
grass hopper conveyor   跳动运输器
grass hopper pipe coupling method   导管装配的分组法
grassing bleach   草场暴晒漂白法
grass pollen   青草花粉
grass-root refinery   地面炼油厂
grass rubber   草胶；草本橡胶
grate   炉箅；炉栅；炉条
grate bar   炉栅
grate coal   筛选的煤
grateless producer   无炉箅气体发生器
grate mill   栅条磨
grater   磨光机
grate size   筛块（煤级50～100毫米）

graticule   ①标线(片)；十字线；量板②方格图③网格
grating infrared spectrometer   光栅红外分光计
grating spectrograph   光栅光谱仪
grating spectrometer   光栅光谱仪
grating spectrophotometer   光栅分光光度计
grating spectroscope   光栅光谱仪
grating spectrum   光栅光谱
gratuitous inducer   安慰诱导物
gravel   砂砾；石子；小石
gravelliferon(e)   芸香内酯；芸香酚内酯；芸香酮
gravel mill   卵石磨(机)
gravelometer   砂石磨损试验器(仪)；卵石磨损试验器
gravel-projecting machine   砾石喷射装置；喷砾石机
gravel stones〔复〕   卵石
graveoline   芸香灵；芸香碱
graveolinine   芸香宁；山小橘碱
graves〔复〕   金属渣
gravimeter   重差计
gravimeter bottle   比重瓶
gravimetric   (测定)重量的；重量分析的
gravimetric analysis   重量分析
gravimetric burette   重量(分析)滴定管；称量滴定管
gravimetric determination   重量测定
gravimetric factor   重量分析因数
gravimetric hygrometer   重量法湿度计
gravimetric method   重量(分析)法*
gravimetric separation   重量法分离
gravitation   ①引力②重力；地心引力
gravitational   重力的；引力的
gravitational constant   万有引力常数
gravitational fall   重力沉降
gravitational field   重力场
gravitational force   重力
gravitational packing method   重力填充法
gravitational potential   引力势(位)；重力势(位)
gravitational sedimentation   重力沉降
gravitational separation   重力分离
gravitational settling   重力沉降
gravitation constant   重力加速度；万有引力常数
gravitation filter   重力滤器
gravitation tank   重力供料罐
gravitol(e)   格拉维妥(药)
gravitometer   重差计〔测定比重用〕；比重计
gravity   ①重力②美国石油学会制定的比重
gravity accumulator   重力储蓄器
gravity balance   比重秤
gravity Baumé   波(美)氏比重
gravity bottle   比重瓶

gravity bucket conveyor　重力斗式运输机
gravity cell　重力电池
gravity chute　重力滑槽
gravity circulating oil system(=gravity lubricating system)　重力自流润滑系统
gravity circulation　重力循环
gravity circulation lubrication　重力循环润滑
gravity closing　自重闭合
gravity concentration　重力浓度
gravity convection　重力对流
gravity conveyer　重力运输机
gravity correction　重力校正
gravity determination　重力测定〔测定比重〕
gravity dewaxing　重力脱蜡
gravity drainage　重力下水；自然下水
gravity feed　重力自流进料
gravity feed cup spray gun　重力送料杯式喷枪
gravity-feed line(=gravity-feed pipe)　重力自流进料管
gravity-feed lubricator　重力自流润滑器
gravity-feed pipe(=gravity-feed line)　重力自流进料管
gravity-feed tank　重力自流进料罐
gravity feed type gun　重力送料式喷枪
gravity filter　重力滤器
gravity flow　重力流动
gravity flowing bed　重力流动床
gravity force　重力
gravity fuel feed　燃料重力自流进料
gravity fuel system　燃料重力自流进料系统
gravity gasoline tank　汽油重力自流进料箱
gravity head　重力压头
gravity-head feeder　压力进料机
gravity line　重力自流管路
gravity loading　重力装载
gravity lubricating system(=gravity oil system)　重力自流润滑系统
gravity meter　重差计〔测比重用〕
gravity-mid per cent curve　比重-蒸出50%曲线
gravity mill　捣碎机
gravity nulifying effect　重力抵消作用
gravity oil feed　重力自流加油
gravity oil system(=gravity lubricating system)　重力自流润滑系统
gravity packing　重力填充法
gravity petrol tank　自动加汽油用的油箱
gravity purity　重力纯度〔糖〕
gravity roller conveyor　重力辊式运输机
gravity scale　重度标度〔按重度计量柴油的经验式〕
gravity segregation　按比重分层

gravity separation　按比重分离
gravity separator　重力沉降分离器
gravity settler　重力沉降槽
gravity settling　比重沉降法
gravity settling basin　重力沉降池
gravity settling tank　重力沉降槽
gravity stamp battery　捣磨机
gravity stamp mill　捣碎机
gravity system　重力自流进料系统
gravity-system water-cooling　冷却的热虹吸系统
gravity take-up　重力式收紧器
gravity tank　自动进料罐
gravity tank truck　自动加油的油槽车
gravity-temperature correction graph　比重-温度校正线图
gravity test　比重测定
gravity thickening　重力浓缩；重力(沉降)增浓
gravity water　重力水
gravity wave　重力波
gravity weight　铅锤
gravity-weight measurement　根据体积和比重的重量测定
gravi-volumeter　重量容积计
gravure　照相凹板〔印刷〕
gravure coater　凹凸涂布机
gray(=grey)　①灰(色)的②戈瑞〔吸收剂量单位〕
gray acetate of lime　灰醋石
grayanotoxin　木藜芦毒素；灰安毒
gray antimony　辉锑矿
gray balance(=gray scale balance)　灰度平衡；灰色平衡
gray body　灰体
gray body radiation　灰体辐射
gray cast iron　灰铸铁；灰口铁
Gray catalytic desulfurization　格雷催化脱硫法
Gray chiller　格雷石蜡结晶器
Gray clay treating process　格雷白土处理法
gray cleaning　原色布的清洗工程；粗布清洗工程
gray cobalt　辉钴矿　CoAsS
gray copper(=gray copper ore)　黝铜矿
gray copper ore(=gray copper)　黝铜矿
gray corrosion　灰色斑点
Gray desulfurization process　格雷脱硫法
graying　泛灰(色)
gray iron(=gray pig iron)　灰口铁；灰铸铁
grayish-green　灰绿色
gray maintenance coating　灰色修补涂料
gray masks　灰色遮挡框；灰度掩模
gray mold　灰霉

grayness 灰(色)度
grayness after rubbing 擦后变质〔家具涂饰时用油打光过早而发灰〕
gray-off 染灰；染色发灰
gray oil 灰色油
gray paint 灰色漆
gray pig iron(=gray iron) 灰口铁；灰铸铁
gray pine oil 灰松油
gray platinum 灰色铂
gray powder 灰色粉
gray room 粗布帐篷
gray scale(=grey scale) ①灰度标；灰度(色)级谱；灰色标度②亮度色标
gray scale balance(=gray balance) 灰度平衡；灰色平衡
gray scale chart ①灰色级谱图(表)②灰度表
gray scale histogram 灰度直方图
gray slag 铅熔渣
Grayson-Wolf separator 格雷森-沃尔夫分离器
Gray tester 格雷试验器
gray thread 原色纱
Gray tower 格雷塔
Gray treated distillate 格雷法精制的馏分
graywacke 杂砂岩
graywake 杂砂岩
gray wrap 灰色包装纸
graze collision 擦边碰撞
grazed relief tile 釉面浮雕瓷砖；浮雕琉璃砖
grazing angle 入射余角；掠射角；擦地角
grazing incidence mounting 掠角入射式装置
grease ①脂膏②润滑脂〔石油〕③动物脂④油腻
grease additive 润滑脂添加剂
grease alkali 润滑脂碱〔制造润滑脂皂基用的碱〕
grease analysis 润滑脂分析
grease antioxidant 润滑脂的抗氧化剂
grease apparent viscosity 润滑脂的视黏度
grease-barrier property 不透脂性
grease base 润滑脂的基
grease bleeding 润滑脂分油
grease box(=grease chamber) 润滑脂罐
grease burnishing 脂膏抛光
grease cartridge lubricator 弹筒式润滑器
grease chamber(=grease box) 润滑脂罐
grease channeling 润滑脂(中生成的)沟流
grease characteristic 润滑脂性质
grease classification 润滑脂分类
grease cock 润滑脂栓
grease composition 润滑脂组成
grease compounding 润滑脂组分混合

grease compounding plant 制造润滑脂工厂
grease consistency 润滑脂稠度
grease consistometer 润滑脂稠度计
grease contamination 润滑脂沾污；油污
grease coupler 润滑过的管接头；润滑过的管件
grease creeping 润滑脂蠕升
grease cup 杯状润滑器；油杯
greased 加过润滑脂的
grease dispensing test 润滑脂配制试验
grease dropping point 润滑脂滴点
grease dye 润滑脂染料
grease elasticity 润滑脂弹性
grease evaporation loss 润滑脂蒸发损失
grease filler 润滑脂填料
grease filter 润滑脂过滤器
grease flow test 润滑脂流动性测定
grease gelling 润滑脂胶凝
grease grades 润滑脂品级
grease gun(=grease pump; grease squirt; grease syringe) 润滑脂枪
grease gun fitting 润滑脂枪喷嘴
grease gun hose 黄油枪胶管
grease gun lubrication 用润滑脂枪润滑
grease gun nipple 润滑脂枪螺纹接套
grease hardening resistance 润滑脂抗硬化安定性
grease kettle 润滑脂锅
grease lubricant 润滑脂
grease-lubricated 润滑脂润滑的
grease lubrication 润滑脂润滑
grease lubrication fittings 润滑脂润滑用零件
grease lubricator 加润滑脂用润滑器
grease-making plant 润滑脂制造厂
grease manufacture 润滑脂制造
grease mark(=lubricant bloom) 油迹；油斑
grease melting point 润滑脂熔点
grease oil 润滑油
grease organic filler 润滑脂有机填料
grease pail 润滑脂桶
grease paint(=nondrying coating) 不干性油膏
grease pan 润滑脂盘
grease pencil 油彩画笔；石印笔
grease penetrameter 润滑脂针入度测定器
grease penetration test 润滑脂针入度测定
grease penetrometer 润滑脂针入度测定器
grease perfume 润滑脂香料
grease plant 润滑脂工厂
grease pocket 润滑脂杯
grease-proof 防油的；耐油的

grease-proof coating　耐油脂涂料
grease proofness　①防油性②防油度
grease-proof paper　防油纸
grease proof wrapping　防油包装纸
grease pump(=grease gun)　润滑脂枪
greaser　①加润滑脂工人②加润滑脂设备
grease remover　除润滑脂物
grease-removing agent　脱脂剂
grease resistance　耐油(脂)性；抗油性；耐润滑油性
grease-resistant coating　防油脂涂料
grease seal　润滑脂封
grease separation　润滑脂分油
grease separator　分油器
grease solidification　润滑脂固化
grease-spoiled oil　润滑脂污染的油料
grease spot　油渍
grease spray lubrication　润滑脂喷射润滑
grease squirt(=grease gun)　润滑脂枪
grease stability　润滑脂安定性
grease surface　油面
grease swelling　润滑脂膨胀
grease syringe(=grease gun)　润滑脂枪
grease testing　润滑脂试验
grease thickener　①润滑脂增稠剂②增稠器
grease thickening　润滑脂增稠
grease trap　润滑脂分离器
grease tube　润滑脂管
grease way　润滑沟纹
grease-well lubrication　润滑脂槽润滑
grease working machine　润滑脂搅和设备；润滑脂搅和机器
greasiness　油脂性
greasing　润滑(过程)
greasing equipment　润滑用设备
greasing oil　润滑油
greasing station　加润滑剂站
greasing substance　润滑剂；有润滑性的性质
greasy　油脂的
greasy fabric　毛织坯料
greasy flame　油脂焰
greasy lustre　脂状光辉〔煤〕
greasy property　润滑性质
greasy wool　汗羊毛
great calorie　大卡；千卡
great coal　大煤；精选的煤；块煤
greater galangal　大高良姜
greaves〔复〕　金属渣
Greek fire　希腊火药

green　①绿(色)的②新鲜的③未成熟的
green acid　绿酸
green acid soap　绿酸皂〔油不溶性石油磺酸皂〕
green adhesion　半成品密着力；胶料黏着性
green bacteria　绿色细菌
green barrier　绿色壁垒
green belt　带坯
green bloom oil(=green cast oil)　绿油
green blue　竹绿色
green brick　砖坯
green budding　绿色芽接
green bulk density　熟化前堆积密度；熟化前压缩密度
green camphor oil　樟脑蓝油
green cast oil(=green bloom oil)　绿油
green cinnabar　绿辰砂；氧化铬绿〔别名〕
green clear　鲜绿
green compound　未硫化胶
green concrete　新浇混凝土；未硬化混凝土
green copperas(=iron copperas)　绿矾；七水(合)硫酸亚铁
green copper ore　绿铜矿；硅孔雀石
green cross　绿十字
green density　未硬化密度；未熟化密度
green earth(=terre verte)　绿土
green emerald　①巴黎绿；乙酸亚砷酸铜；砂绿〔商〕②翡翠
green enamel　绿瓷漆
green floral　鲜花香
green food　绿色食品
green-for-safety mining product　安全矿用品〔如防火运输带〕
green gases〔复〕　绿色气体
green glue stock　生胶料
green heart　绿心
green hellebore　绿藜芦
green hide　鲜皮
green house aerosol　温室气溶胶
greenhouse effect　温室效应
greenhouse gases　温室气体
greening　发绿
greening lacquer　古铜色漆〔文物修复用仿古涂料的一种〕
greenish black　墨绿
greenish blue　绿蓝
greenish cast　绿痕〔木器涂饰中的缺陷〕
green lacquer hammer finish　绿色硝基锤纹漆
green lake　绿色淀〔铁蓝与黄色淀的复合物〕
green-leaf scent　青叶子香；鲜叶子香
green liquor　绿液；碱化液

green lumber　主材
green mallee　绿桉
green manure　绿肥
green manure crop　绿肥作物
green masonry　未硬化的(砖、石)砌体
green molasses　原蜜
green mold　绿霉
green nickel oxide　氧化镍绿　NiO
green note　青香韵
green nuance　青香
greenockite　硫镉矿；天然硫化镉
green oil　原油；绿(石)油
green ore　原矿；未选过的矿
greenoxane　青叶噁烷；2-(3-庚基)-1,3-二氧戊环〔一种人工合成的液体香料〕
green petroleum coke　绿色石油焦；新出炉的石油焦
green phospho-molybdic acid toner　磷钼酸绿调色剂(料)；磷钼酸绿色原
green pigment　绿(色)颜料
green pinch　硫化前互黏
green potato　鲜土豆香
green product　绿色产品
green ramie　青苎麻
green resin　生树脂；未经熟化的树脂
green-rot　青腐
Green rotational viscometer　格林旋转黏度计
green rubber　粗制生胶；未再炼生胶
green rum　①初步试运转②初步试产；临时生产
green salt　绿盐；四氟化铀　UF_4
green salted glue stock　新腌肉面〔皮革〕
green salted hide　盐鲜皮
green sand　绿砂
green seal　绿印(级)
green sirup(=green syrup)　绿蜜糖；未精制糖浆
green soap　绿肥皂；软皂
green stock　生料；原始混合物；生胶料
Green's viscometer　格林氏黏度计
green syrup(=green sirup)　绿蜜糖；未精制糖浆
green tack　成型黏着性；半成品黏着性
green tar　绿焦油；巴巴多斯石油
green technology　绿色工艺学
Green test　格林试验〔汽油中胶质测定〕
green tire　生胎
green topnote　青头香
green T-substance　绿 T 料；氯甲酸-氯甲酯
green ultramarine　群青绿
green vegetable　青菜香
green verditer　铜蓝〔颜料〕

green vitriol(=copperas)　绿矾；七水(合)硫酸亚铁　$FeSO_4 \cdot 7H_2O$
green ware　生陶器〔未烧过的陶器〕
green-yellow　绿黄
gregorite(=ilmenite)　钛铁矿　$FeO \cdot TiO_2$
greige cloth　原坯布；本色布
greige yarn　原色纱
greisen　云英岩
grenz ray　跨界射线
Grethen's weight bottle　格雷森(挥发液)称量瓶
Grevet chromic cell　格雷维特铬酸电池
grevillol　银桦酚；十三烷(基)苯二酚
grey　灰(色)的
grey aluminium sealer　灰色铝粉封闭剂
grey analytical system　灰色分析系统
grey (cloth)　原坯布；本色布
grey clustering analysis　灰色聚类分析
grey correlation analysis　灰色关联分析
grey filter　灰色减光板
greyish-green　灰绿色
grey model　灰色模型
grey rot fungus　灰色腐败真菌
grey scale(=gray scale)　①灰度色标；灰度级谱；灰色标度②亮度色标
grey scaly　灰卡
grey wedge　灰色光劈
grey yarn　原色纱
grid　①栅极②栅条；栅板；格栅
grid accumulator　栅条蓄电池
grid battery　栅极电池
grid current　栅极电流
grid deck　弦栅填料
grid distributor　分布板
grid electrode　栅(形电)极
grid electrostatic paint spray apparatus　栅极式静电喷涂装置
grid leak　栅漏(阻)
grid packed tower　栅条填充塔
grid packing　栅格填料
grid pressure　格栅压
grid reception station　栅式接受站；气体分配站的接受站
grid resistor　栅极电阻器
grid tower　栅条填充塔
grid tray　栅条盘
grid tray column　栅条盘分馏柱
grid valve tray　链条形浮阀塔盘
grid voltage　栅极电压
Griess reaction　格里斯(脱氨基)反应

Griess'reagent 格里斯试剂
Griess test 格里斯试验
Griffin beaker 格里菲烧杯
Griffin mill 格里菲磨机
Griffith's fracture criterion 格里菲思断裂判据
grifolin 奇果菌素
grifulvin(=griseofuvin) 灰黄霉素
Grignard compound 格利雅化合物
Grignard nitrile synthesis 格利雅腈合成法
Grignard reaction 格利雅反应
Grignard's reagent 格氏试剂*；格利雅试剂
grille 格栅
grille spectrometer 栅格谱仪
Grillo process 格里洛法
Grimm discharge lamp 格里姆放电灯
Grimm discharge spectroscopic analysis 格里姆放电光谱分析
Grimm glow discharge source 格里姆辉光放电光源
Grimm lamp 格里姆(放电)灯
grindability 可磨度
grind-burn-leach process 磨碎-燃烧-浸取过程
grind edge 磨棱边
grindekol 胶草醇
grindelia 胶草
grindelic acid 胶草酸
grinder ①研磨机②碎木机③粉碎机④砂轮机
grinder buffing (粗)砂轮打磨
grinder cylinder 碎木机压力缸
grinder-mixer 粉碎搅拌机；粉碎混合机
grinder pocket 碎木机盛木装置
grind fineness of paint 颜料研细程度
grind gage 细度计；刮板细度计
grinding (研)磨
grinding aids 磨料；研磨剂
grinding allowance 磨削余量
grinding charge (of mill) 被磨物
grinding consistency 研磨稠度
grinding disc 砂轮
grinding drum 压碎机；磨碎机
grinding dust 研磨粉尘
grinding fluid 研磨液；用于金属研磨的冷却液
grinding japan 压浆用快干漆料〔常含紫胶〕
grinding knife machine 磨刀机
grinding lubricant 研磨液
grinding machine 研磨机
grinding machine operator 磨工
grinding material 磨料；研磨剂
grinding media charge ①研磨介质装填②磨料装填量
grinding medium 磨料；研磨剂
grinding mill 研磨机
grinding miller 研磨轮
grinding oil 研磨油；润磨油
grinding pan 磨盘
grinding pigmentation 研磨着色法；轧(色)浆法〔先将颜料用少量漆料研磨分散成一次粒子的色浆过程〕
grinding ratio 研磨比
grinding rotor 粉碎转子；研磨转子
grinding slab (颜料捏合用)平板(石板或玻璃板)
grinding stock 磨料；研磨剂
grinding-type resin 研磨型树脂
grinding vehicle 研磨漆料；压浆用漆料
grinding vessel 研磨筒
grinding wheel 磨轮；砂轮〔商〕
grind-leach process 磨碎-浸取过程
grindmeter 研磨试验器
grind of paint 颜料研碎
grindometer 细度计；刮板细度仪
grindstone 磨石
grindstone dresser 磨石刻槽器
grindstone pit 磨石井
grinning through (遮盖力不足引起的)露底〔漆病〕
grip 夹头
grip coat (搪瓷)底层
griper 河上运煤船
griphtie 暧昧石
grip nut 夹紧螺母
grip-pad 履带板〔履带式接取机〕
grippage 抓握性；抓握能力
gripper bar 叼纸牙排
gripping holder 夹紧装置
gripping ring 夹紧环；卡环
grisamine 灰霉胺
Griscom-Russell evaporator 格里斯科姆-鲁塞尔蒸发器
grisein 灰霉素
griselimycin 浅灰霉素
griseococcin 灰球菌素
griseofagin 灰榉菌素
griseoflavin 灰黄菌素
griseofulvin(=fulcin;fulvicin;grisovin;grifulvin) 灰黄霉素
griseolutein 灰藤黄菌素
griseolutic acid 灰藤黄酸
griseolutin 灰藤黄素
griseomycin 灰色霉素
griseorhodin 灰紫红菌素
griseoviridin 灰绿菌素
grisic acid 灰霉酸

grisine(=grizin)　灰菌素
Grison dynamito　格里森炸药
grisonomycin　灰诺霉素
grisorixin　灰争菌素
grisounite　格桑炸药；格锐烧那特〔硝酸甘油、硫酸镁、棉花炸药〕
grisoutite　格搔炸药；格锐烧太特〔硝铵、三硝基萘、硝酸钾混合炸药〕
grisovin(=griseofulvin)　灰黄霉素
grist　谷粉
grit　①硬渣②粗砂
grit blasted roll　喷砂处理轧机
grit-blast(ing)　喷铁砂
grit chamber　粗砂(沉淀)室
grit content　粗粒含量；粗粒度
grit number　筛目数；粒度
grit soap　砂皂
gritstone　砂岩
grittiness　砂性
gritty particle　杂质粒子
grizin(=grisine)　灰菌素
grizzle　低级煤
grizzly　栅筛
grizzly screen　栅筛
grizzly screening　栅筛析
groats　去壳谷粒
grog　陶渣
grog brick　耐火砖
grommet　垫圈；金属孔眼；绝缘孔圈；环管
grommet-type V-belt　活络三角带
groove　①槽②坡口〔焊接〕
groove angle　槽角；坡口角度
groove corrosion　槽腐蚀；焊接部位腐蚀
grooved bowl　压型辊
grooved cylinder　槽纹滚筒
grooved drum winder　槽筒式络筒机
groove depth　槽深；坡口高度
grooved liner　机筒槽纹内衬
grooved pulley　槽轮
grooved roll　槽纹辊
grooved roller conveyor　槽辊式输送器
grooved sheave　槽轮
groove face　沟面
groove flange　槽面法兰
groove gasket　嵌入式垫圈
groove welding　坡口焊
grooving　切槽；开槽
gross　①过失误差*；粗差②全部的③总数

gross assets　投资总额
gross atomic population　总原子布居
gross calorific power(=gross calorific value)　总热值
gross calorific value(=gross calorific power)　总热值
gross capacity　总容量
gross constant(=cumulative constant)　总常数；全常数；累积常数
gross distortion　严重变形；大变形
gross effect　有效功率
gross efficiency　总效率
gross error　过失误差；粗差
gross fission product　总裂变产物
gross heating value(=gross thermal value)　总热值
gross heat of combustion　总热值；总燃烧热
gross polymer　粗制聚合物
gross rubber　①充油充炭黑胶料②胶料；混炼胶
gross sample　总样品；大样
gross structure　整体结构；宏观结构
gross thermal value(=gross calorific value)　总热值
gross tolerance　总公差
gross tonnage　总吨位
grossul rite　钙铝榴石
gross volume　毛体积
gross weight　毛重
Grotthuss-Draper's law　格罗图斯-德雷珀定律
ground　①地②地面③底(子)④研磨⑤磨碎的；磨细的⑥磨口的⑦接地
ground-and-washed chalk　重质碳酸钙
ground calcium carbonate　重质碳酸钙
ground caustic　①粉状腐蚀剂②(苛)碱粉
ground chalk　重质碳酸钙
ground clearance　离地高度
ground coat　(搪瓷)底层；底漆
ground coat paint　底漆
ground colour　①地色；底色②暗色
ground dolomite　白云石粉〔含镁碳酸钙〕
ground dried hide　缩干皮
ground dyeing　底施颜料
ground effect machine　气垫汽车
ground finish　精磨加工
ground glass　磨砂玻璃
ground glass apparatus　磨口玻璃仪器
ground glass joint　磨口玻璃接头；玻璃磨片
ground glass stopper　磨口玻璃塞；毛玻璃塞
ground glass stoppered flask　磨口玻璃瓶
ground-in　磨口的
grounding　接地
ground ivy oil　活血丹叶油

ground joint 磨口接头
ground lead 接地(引)线
ground limestone 重质碳酸钙
ground line 基线
ground litharge 黄(铅)丹
ground mass 基本重量
ground material 磨料
ground noise 本底噪声
ground nut fiber 花生蛋白纤维
groundnut oil 花生油
ground oyster shells 蛤灰〔大漆〕；蛤粉〔天然碳酸钙填料〕
ground paper 原纸；纸坯
ground paste (研磨过的)色浆；(色漆)料浆
ground pattern 底样
ground peg 桩子
ground perlite 珍珠岩粉；膨胀珍珠岩
ground phosphate (rock) 磷酸矿石
ground pigment 粉末填充剂
ground pipe 地下管道
ground plasm 基质
ground pulp(GP) 磨木浆
ground pumice 浮石粉〔可作磨用粉〕
ground rubber ①废胶末②再生胶粉
grounds 沉淀；渣滓
ground scrap 废胶末
ground shape 研磨断面形〔压延滚筒〕
ground silica (结晶)二氧化硅粉；石英粉；硅石粉
ground slurry 研磨浆料
ground springs 无纤维废胶末
ground state 基态
ground stopper 毛玻璃塞；玻璃磨塞
ground stopper bottle 磨口玻璃塞瓶；毛玻璃塞瓶
ground storage 地上仓库
ground temperature 地面温度
ground tire 废胎胶粉
ground vulcanized scrap 废硫化胶末
ground warp 地经〔纺〕
ground waste 磨碎的废料
ground waste rubber 磨碎废胶
ground water 地下水
ground water level 地下水位
ground weft 地纬〔纺〕
ground whiting 重质碳酸钙
ground wood pulp 细(磨)木浆
groundwork 基础
group ①类〔周期表〕②族〔周期表〕③基④团⑤组⑥群〔生物分类〕

group agglutination reaction 组凝集反应
group analysis 基团分析法〔一种测定分子量的方法〕
group connection 综合接续〔电〕
group drive 成组传动
group element 同组元素
group frequency 基团频率*
grouping ①类〔周期表〕②基③(原子)团④组〔分析化学〕
group-migration polymerization 基团转移聚合
group moment 基矩
group number 基数〔计算HLB值〕
group orbital 群轨道
group precipitant(=reagent for group precipitation) 组沉淀试剂
group precipitation 组沉淀
group property 同组通性
group reaction 组反应
group reagent 组试剂
group refraction 基团折射
group representation 群的表示
group retention index 基团保留指数
group sampling 分组抽样
group separation (按)组分离
group switch 综合开关
group theory 群论
group transfer 基团转移
group transfer polymerization 基团转移聚合*
group transfer reaction 基团转移反应
group type analysis 基团类型分析
group weight method 分组称重法
grout 薄浆〔水泥〕
grout additive 砂浆添加剂；(水泥)薄浆添加剂
grout filler ①灌浆填料②灌缝砂浆
grout hose 泥浆胶管
grouting 灌浆；灌浆法
grouting compound 稀浆
growing chain 生长链
growing chain end 生长链末端
growth ①生长②增加
growth coefficient 增殖系数
growth of crystals 晶体生长
growth of defect 缺陷增长
growth of stress 应力生长
growth orientation 生长取向
growth peak 生长峰
growth rate 增长率；增长速度
growth rate coefficient 生长速度系数
growth regulating substance(=growth regulator) 生长调节素

growth regulator(=growth regulating substance) 生长调节素
growth spirals 生长蜷线
growth stimulation 生长刺激(作用)
growth striation 生长层；生长条纹
growth substance 助长剂
growth suppressor 生长阻遏剂
Grubbs'test method 格鲁布斯检验法
grubilin 格鲁菌素
grummet 眼圈；垫环
Grundmann aldehyde synthesis 格伦德曼醛合成法
grunerite 铁闪石
Grunwald-Winstein equation(=Winstein-Grunwald equation) 温斯坦-格伦瓦尔德公式
GS-moisture chamber (控制)湿度展开槽；调湿展开槽
G solvency G 溶解本领
G solvent G 溶剂
guaiac 愈创木(树)脂
guaiacetin 瓜亚西丁；羟苯氧基乙酸钠 $C_8H_7O_4Na$
guaiacin 愈创木粉
guaiacol 愈创木酚；邻甲氧基苯酚 $CH_3OC_6H_4OH$
guaiacol acetate 愈创木酚乙酸酯 $CH_3OC_6H_4O_2CCH_3$
guaiacol benzoate 愈创木酚苯酸酯 $CH_3OC_6H_4O_2C_7H_5$
guaiacol cinnamate 愈创木酚肉桂酸酯 $CH_3OC_6H_4O_2C_9H_7$
guaiacol enzylicether(=pyrobain) 派鲁卡因；愈创木酚苄醚
guaiacol salicylate 愈创木酚水杨酸酯 $C_{14}H_{12}O_4$
guaiacol valerate 愈创木酚戊酸酯 $CH_3(CH_2)_8CO_2C_6H_4OCH_3$
guaiaconic acid 愈创木酯酸
guaiac resin 愈创木(树)脂
guaiac(um) 愈创木脂
guaiacwood acetate 乙酸愈创木酯
guaiac wood oil 愈创木油
guaiacyl(=o-methoxyphenyl) 愈创木基；邻甲氧苯基
guaiacyl acetate 乙酸愈创木酯
guaiacyl phenylacetate 苯乙酸愈创木酯
guaiadiene 愈创木二烯
guaiamar 愈创木甘油醚
guaiane 愈创木烷
guaianolide 愈创内酯
guaiaretic acid 愈创木酸
guaiazulenic acid 愈创木䓬酸
guaiol(=champacol) 愈创醇
guaiol acetate 乙酸愈创木酯
guaioxide 愈创木醚
guaiyl butyrate 丁酸愈创木酯

guamycin 瓜霉素
guanajuatite 硒铋矿
guanamine 胍胺；三聚氰二胺
guanamine resin 三聚氰二胺树脂；胍胺树脂
guanamycin 瓜那霉素
guanazole(=3,5-diiminotriazole) 3,5-二亚氨基(二氢化)三唑；胍唑
guanghuimeisu 光辉霉素
guanidine(=carbamamidine) 胍 $(H_2N)_2C=NH$
guanidine accelerator 胍类促进剂
guanidine acetate 乙酸胍 $CH_5N_3HC_2H_3O_2$
guanidine acetic acid 胍基乙酸 $NH_2C(=NH)NHCH_2COOH$
guanidine aldehyde resin 胍醛树脂
guanidine carbonate 碳酸胍 $(CH_5N_3)_2H_2CO_3$
guanidine chlorate 氯酸胍
guanidine derivative 胍基衍生物
guanidine hydrochloride 盐酸胍 $CH_5N_3 \cdot HCl$
guanidine nitrate 硝酸胍 $CH_5N_3 \cdot HNO_3$
guanidine octadecyl sulfate 十八烷基硫酸胍
guanidine perchlorate 高氯酸胍
guanidine petroleum sulfonate 石油磺酸胍
guanidine picrate 苦味酸胍
guanidine salt 胍盐
guanidine sulfate 硫酸胍 $(CH_5N_3)_2H_2SO_4$
guanidine thiocyanate 硫氰酸胍 CH_5N_3HCNS
guanidino(=guanidyl) 胍基 $H_2NC(=NH)NH-$
guanidinoacetic acid(=glycocyamine) 胍基乙酸 $HN=C(NH_2)NHCH_2CO_2H$
guanidino-acid 胍基酸
2-guanidino propionic acid 2-胍基丙酸
guanidoethyl cellulose 胍乙基纤维素
guanidotaurine 胍基牛磺酸
guanidyl(=guanidino) 胍基 $H_2NC(=NH)NH-$
guanite 鸟粪石
guanyl(=amidino) 脒基 $H_2NC(=NH)-$
guanyl guanidine 脒基胍；缩二胍 $NH_2C(=NH)NHC(NH_2)=NH$
guar 瓜耳(树)胶〔膏化剂〕
guarana 瓜拉那〔药〕
guarana tannin 泡林藤单宁
guaranine(=caffeine) 咖啡碱；咖啡因
guaranteed reagent 保证试剂
guaranteed reagent-grade 保证试剂级
guarantee reagent 保证试剂
guarantee test 保证试验
guard board 防护纸板
guard catalyst 保护催化剂

guard chamber ①防护室；保护塔②(催化剂)防淀积器
guard column 保护柱*
guard hair 针毛
guard mask 防护罩
guard reactor 保护反应器
guard ring 防护圈
guard tube 防护管
guar gum 瓜耳(树)胶
guayule rubber 银菊橡胶
gudgeon 耳轴
gudgeon pin 活塞销
guejarite 柱辉铜锑矿
Guerbet reaction 格尔伯特反应
guest 客体*
guest compound 客体化合物
guest molecular recognition 客体分子识别
guest molecule 客体分子
guhr 硅藻土
guhr dynamite 硅藻土炸药
Guichard's thermobalance 盖查德热天平
guidance ①引导；控制②导槽，导板；导轨
guide (向)导；指导
guide arrangement 引导装置
guide bar gaiting 对梳规律
guide bar gaiting diagram 穿纱对梳图
guide bar lapping 垫纱规律
guide bar nesting 梳栉集聚
guide bend test 靠模弯曲试验
guide bush(=dowel bush) 导销套
guide coat 标志涂层；制导涂层；磨合涂膜
guide end (三辊磨)档板端
guide eye 导丝头；绕丝头
guide head 导丝头；绕丝头
guide line 分度线
guideline level 指导性限量
guideline test (稳定性)预测试验
guide pin 导丝棒；导销
guide pin bushing 导销套
guide plate 导向隔板
guide ring 导流环
guide rod 导杆
guide roll 导辊
guide roller 导辊；导轮
guide shell 导流筒
guide threading 穿纱器
guide tip (挡板的)挡尖
guide tube 导管
guide vane 导叶；导流叶片

guiding channel 管道片
Guignet's green 吉勒特绿
Guild colorimeter 吉尔德比色计
Guillery impact machine 回旋打桩机
guillotine 横切机；截切机
guillotine knife ①切断机刀②铡胶刀
guillotining ①铡断；切割②切生胶
guillotining equipment 铡刀式剪切设备
guinea green B 基尼绿 B
Guldberg-Waage's law 质量作用定律
gulf 轮槽〔三角带轮〕
gulf red 铁红
gulomethylose 6-脱氧古洛糖
gulonate 古洛糖酸
gulonic acid 古洛糖酸 $CH_2OH(CHOH)_4COOH$
gulonolactone 古洛糖酸内酯
gulose 古洛糖
gum ①树胶②龈③纯橡胶(料)
gum acacia 阿拉伯树胶
gum accroides(=accroides) 禾木胶
gum-bichromate process 树胶重铬酸盐印画法
gum bloom 起霜〔漆病〕
gum blushing 树脂致白〔漆病〕
gumbrine 胶盐土〔漂白用〕
gum camphor 樟脑
gum cap 胶帽
gum cement 橡胶黏合剂
gum check 脂囊
gum content 树胶含量
gum cover 胶套
gum dipping 浸胶；浸浆
gum duct 树脂道
gum elastic 弹性树胶
gum filler 填充胶条
gum ghatti 印度胶
gum guaiacum 愈创木胶
gum inhibiting index 胶质阻抑指标；生胶抑制剂质量指标
gum inhibitor 胶质抑制剂
gum latex 橡浆
gum level 胶质含量
gumlike material 类胶物质
gummase 漆酵素
gummed label 涂胶标签
gummed paper 胶纸
gummed paper method 胶纸(集尘)法
gummed tank 衬胶槽
gummed yarn 已浆纱

gummeline 糊精
gummi 树胶
gumminess 黏合性；胶黏性
gumming 涂胶
gumming dirt 胶泥
gumming tendency 生成胶质的倾向
gumming test 黏合试验
gummosity 黏合性
gummous 黏合的
gummy 黏合的
gummy bottoms (石油中)胶质脚
gummy fibre 树胶状纤维
gummy oil 胶质油〔含大量胶质的油〕
gummy residue 胶质残余
gum phase 胶状(固定)相
gum pot 熔胶锅
gum process 树胶印画法
gum-producing substance 胶质形成物〔促使汽油生胶的物质〕
gum residue 胶质残余
gum resin 脂松香；树胶脂
gum rosin 松香
gum rubber tubing 纯胶管；软管
gum running 树胶熔炼；树胶热裂解〔某些天然树胶经高温熔融裂解后易溶于油〕
gum sandarac 山达脂；桧树胶；杜松胶
gum seam 全胶接缝
gum senegal 阿拉伯树胶
gum spirit cement 松节油灰泥
gum stock ①纯胶料②擦胶胶料③热炼胶
gum sugar 胶糖；阿拉伯糖
gum test 胶质(含量)测定
gum thus 松脂；松香
gum tolerance 胶质容忍量
gum tragacanth 黄(树)胶；龙(须)胶
gum tragasol 豆胶
gum turpentine 脂松节油
gum water 胶水
gun ①枪②油枪③润滑油泵
gun adapter 油枪嘴
gun-applied caulk 枪喷填缝料；枪注嵌缝料
gun-applied furnace insulation (油)枪喷(涂)炉绝缘〔一种可以利用混凝土枪来喷射的加热炉保温材料〕
gun barrel(=gun barrel tank) ①沉淀罐②气体分离器
gun barrel tank(=gun barrel) ①沉淀罐②气体分离器
guncotton 火棉；纤维素六硝酸酯
gun graphite 枪用石墨
gun grease 枪用润滑脂

gunite 喷浆〔一种涂在反应室内的防水防油物质〕
gunite lining of tank 喷浆罐衬〔一种衬在油罐内的防水防油物质〕
gunite material 喷浆材料〔一种水、水泥及砂的混合物，可以用喷枪喷至混凝土上防水防油用〕
guniting 喷涂；喷浆；喷镀
gun metal 炮铜
gunny cloth (bag) 黄麻布(袋)
gun oil 枪油
gunpowder 火药
gun powder-residue detection 火药残渣鉴定〔活化分析用于破案〕
gun recoil oil 枪炮制退油
gun slot 喷枪口
Gunzberg test 冈斯伯尔格试验
guoethol 乙氧苯酚
Guoy balance 古依天平
gurjun 古芸香胶
gurjunazulene 古芸薁
gurjun balsam oil 古芸香胶油
gurjunene 古芸烯
gurjunene ketone 古芸烯酮
gurjuresene 古芸素
Gurney-Lurie chart 葛尼-鲁利传热(计算)图
gusher 喷穴
gusset 筋板；角牵板
gustation ①尝味②味觉
gustatory 味觉的
gustiness 阵风性；湍流度
gut 油管内加热用的水蒸气小管
gutta-percha 古塔波胶；杜仲胶
gutter ①沟；槽②导脂器
gutter stabilizer 窄槽稳定器
guttiferin 藤黄素
Gutzeit test 古蔡试验*
Guye furnace 盖伊电炉
guy wires 牵索；长绳；链
G-value G 值〔辐射化学中，每吸收 100 电子伏特的辐射所发生的分子变化数〕
gym finish 溶剂型高光泽地板面漆
gymnemic acid 森林匙羹藤酸
gymnemin 武靴叶素
gynergon(=estradiol) 雌甾二酚
gynesine(=trigonelline) 葫芦巴碱
gynocardia 大风子〔植〕
gynocardia oil 大风子油
gynocardic acid 大风子酸；环戊基十三酸 $C_{17}H_{33}CO_2H$
gynocardin 大风子定

gynocardinic acid　大风子定酸　$C_{12}H_{19}O_9COOH$
gynoval　异冰片基戊酸酯
gypse　石膏
gypseous　石膏的
gypsine　砷酸铅　Pb_3AsO_4;$Pb_2H_2(AsO_4)_2$
gypsite　土(状)石膏
gypsous　石膏的
gypsum　石膏*
gypsum cement　石膏水泥
gypsum compact　雪花石膏
gypsum concrete　石膏混凝土
gypsum earth　土(状)石膏
gypsum filler　石膏填孔料
gypsum filling　石膏填孔
gypsum-free layer　不含石膏层
gypsum kettle　烧石膏锅
gypsum lath　石膏板条
gypsum lime mortar　石膏灰浆
gypsum plaster　石膏粉饰〔建筑〕
gypsum plaster slab(=plaster slab)　石膏板
gypsum pottery plaster　(制陶器用)石膏灰泥；石膏陶土泥；陶器成形用烧石膏
gypsum product　石膏制品
gypsum sand　石膏粒
gypsum slag cement　石膏矿渣水泥
gypsum stains〔复〕　石膏斑
gypsum tile　石膏瓦(板)
gyrating breaker　回转碎裂机
gyrating crusher　回转压碎机
gyration　回转
gyration (shaft)　涡形回转(轴)
gyratory　旋转的
gyratory breaker　回转碎裂机
gyratory cone crusher　回转锥形压碎机
gyratory crusher　回转压碎机
gyratory milling　回转磨碎
gyratory motion sieve　(三元)回转振动筛
gyratory paddle　回转桨〔搅拌机〕
gyratory rock-breaker　回转碎岩机
gyratory screen　旋转筛
gyratory shaker　转动摇床
gyratory sieve　回转筛
Gyro cracking　杰罗裂化法
Gyro gasoline　杰罗(法制得的)汽油
gyro-horizon　回转水
gyrolite　白钙沸石
gyro-magnetic ratio　磁旋比
gyromytrin　鹿花菌素
gyrophoric acid　三苔色酸；石茸酸
Gyro process　杰罗过程；杰罗气相裂化过程
gyroscope　陀螺仪；旋转仪
gyroscope lubricant　回转仪用润滑剂
gyroscopic effect　回转效应
gyro speed　旋转速度
gyrotron　电子回旋脉泽；回旋管
gyrotropy　回旋磁性
Gyro unit　杰罗设备；杰罗法气相裂化设备

H

Haake rheometer　哈克流变仪
Haake RV3 viscometer　哈克 RV3 型黏度计
Haanel-Heront furnace　哈内耳-赫朗特电炉
Haas-Oettel cell　哈斯-奥提耳电池
Haber ammonia process　哈伯制氨法
Haber's process　哈伯氏固氮法
habotai　纺
hachimycin (=trichomycin)　曲古霉素；抗滴虫霉素
hachure　刻线；影线；晕线
H-acid　H 酸；1,8-氨基萘酚-3,6-二磺酸
hackle clamp　夹麻器
hacksaw　弓锯
Hacoba bulking system　哈科巴膨化变形法
haco oil　鲸鱼油
hadacidin　杀腺癌菌素
Hadamard transform infrared spectrometer　阿达马变换红外光谱仪
Hadamard transform Raman spectroscopy　阿达马变换拉曼光谱法
Hadamard transform spectroscopy　阿达马变换光谱
haddock liver oil　鳕鱼肝油
hadromal (=ferulaldehyde)　阿魏醛
hadron　强子〔强相互作用粒子〕
hadronic atoms chemistry　强(子)原子化学
haemacytometer (=hemacytometer)　血球计；血细胞计数器
h(a)emagglutination　血凝集(作用)
haemaglobinometer (=hemoglobinometer)　血红蛋白计
haemanthamine　网球花胺
haemanthidine (=pancratine)　网球花定
haemanthine　血安生
haematimeter　血球计
haematine (=hematine;hematoxylin)　苏木精；苏木紫
haematine crystal (=hematoxylin)　苏木精
haematinometer　血红素计
haematocrit value　血细胞比率值；血球比率值
h(a)ematoglobin　血红蛋白
h(a)ematological pipet(te)　血液吸量管
h(a)ematoporphyrin　血卟啉
haematoxylin (=hematoxylin)　苏木精
h(a)emocyte　血细胞；血球
haemocytometer　血球计；血细胞计数器
haemostat　①止血剂②止血钳
haemproteins　血蛋白

hafnifluoride　氟铪酸盐；氟铪化合物 $M_2[HfF_6];M_3[HfF_7]$
hafnium　铪〔72 号元素，化学符号 Hf〕
hafnium hydroxide　氢氧化铪〔1. 原 $Hf(OH)_4$；2. 偏 $HfO(OH)_2$〕
hafnium oxide　二氧化铪
hafnium oxychloride　二氯一氧化铪　$HfOCl_2$
hafnium sulfate　硫酸铪　$Hf(SO_4)_2$
hafnocene　二茂铪
hafnyl chloride　氯化氧铪；二氯一氧化铪　$HfOCl_2$
Hagg carbide　海格碳化铁
Hagstrum type mass spectrometer　哈氏质谱仪
Hahn echo　哈恩回波
haidingerite　砷钙石
hair　毛发
hair bulb　毛球
hair cell　毛细胞
hair color　毛发染料
hair coloring preparation　染发剂
hair crack　发裂；毛状细裂纹；微细裂纹
hair cracking　毛细裂纹
hair-destroying process　毁毛法〔用于去毛〕
hair-dissolving process　溶毛法〔用于去毛〕
hair follicle　毛囊
hair grease　毛填料润滑脂
hair hygrometer　毛发式湿度计
hair line finishing　精细直线打磨法
hair-on tanning　带毛鞣制
hairpin flue　发夹烟道；U 形烟道
hairpin structure　发夹结构；U 形结构
hairpin tube　发夹管；U 形管
hair restorer　头发修整机
hair salt　发盐；铁明矾；羽明矾
hair sieve　马尾筛
hair washing machine　洗毛机
hairy acrylic fibre　毛状丙烯腈系纤维
hairy micelle　毛发形胶束
halation　晕光作用
halation of line　线晕
halazone　卤胺宗；哈拉宗；4-(N,N-二氯胺磺酰基)苯甲酸
hal-crystallization time　半结晶期
haleite (=EDNA)　二硝基乙胺
Halex process　哈莱克斯过程〔磷酸三丁酯-四氯化碳萃取辐照核燃料后处理法〕

half adjust 四舍五入
half aldehyde 醛酸；半醛 CHORCOOH
half aldehyde of succinic acid 丁醛酸 CHO(CH$_2$)$_2$COOH
half amide 半酰胺；单酰胺
half amide of malonic acid 丙二酸半酰胺
    CONH$_2$CH$_2$COOH
half-angle 半角
half angle of divergence 半张角
half auto 半自动式
half band width 半峰宽度
half bleaching 半漂白
half-blocked diisocyanate 半封闭的二异氰酸酯
half boiled process 半沸煮法；热法〔煮皂〕
half boiled soap 半沸煮皂；热法皂
half bond 半键
half cell 半电池
half cell potential 半电池电位
half-chair conformation 半椅型构象*
halfcystine 半胱氨酸
half-decay time 半衰期
half-dressed warp 经纱黏并
half-electron model 半电子模型
half element 半电池
half equation 半方程式
half ester 半酯
half-field transition 半场跃迁
half-filled-shell effect 半充满(电子)壳层效应
half-finished goods 半成品；半制品
half foam life period 泡沫半衰期
half-frequency spin wave 半频自旋波
half-full 半满
half-hard rubber 半硬质胶
half-hydrate 半水合物
half inert 半抑制试验〔测定纤维的不匀率〕
half-intensity width 半强度宽度
half intersecting gill box 半开式针梳机
half jersey 双经平组织；双梳经平组织
half leather 半皮面装订；半革装
half-lethal dose 半致死剂量
half life 半衰期
half life period 半衰期
half-lives of actinides 锕系的半衰期
half mask 半罩面具
half matt gloss 半光(泽)
half-maximum intensity 半峰强度〔强度最大值的一半〕
half-maximum line breadth 半峰线宽
half mean width 半平均宽度
half mercerizing 半碱化；半丝光处理

half metallocene 单茂金属化合物
half meter bridge 半米电桥
half milano rib 三平；罗纹半空气层
half mold 半模；敞口模
half-neutralization potential 半中和(点)电势
half normal 半正常；半标准的
half-peak breadth 半峰宽度
half-peak potential 半峰电位
half-peak width 半峰宽度
half period ①半衰期②半周期
half period method ①半衰期法②半周期法
half-pint can 半品脱(漆)罐
half polymer 低聚物
half quantum number 半量子数
half-quenching concentration 半猝灭浓度
half-reaction 半反应
half-reaction potential 半反应电势
half-reduction in amplitude (条痕)振幅半衰
half-round file 半圆锉
half-round steamer system 半圆汽蒸箱式练漂联合机
half saturated 半饱和的
half-scouring of raw silk 生丝半练法
half-second butyrate 半秒(级)丁酸纤维素〔半秒乙酸丁酸纤维素的简称〕
half second cellulose butyrate 半秒丁酸纤维素(酯)
half shade device 半阴装置；半影装置
half shadow polarimeter 半阴旋光计
half-shell pressure vessel 瓦片式压力容器
half-shell split repair clamp 半壳分离修理夹
half silk 半丝
half soling 半圆形(保护槽的)衬垫
half spent lye 半废液；碱析水
half-step height 半阶高
half stock (=half stuff) 半成品(纸浆)
half stuff (=half stock) 半成品(纸浆)
half-sum method 升降中点法〔测定渗透压〕
half thickness 半值层厚度
half-time 半衰期
half-time of crystallization 半结晶期
half tint 中间色调；中间色
half tone 半色调；中间色调
halftone black ink 网目版黑墨
half-tone dot ①网目版网点②半色调点
half-tone ink 铜板墨
half-tone method 浓淡色调法；半色调法
half-tone news 照相印报纸
half-tone tint 半色调
half transformation point 半转变点

half-truck vehicle 半履带式车
half-unit cell composition 半晶胞组分
half-value layer (HVL) 半值层
half-value period 半值期
half-wash 半洗
half-wavelength plate 半波长板
half-wave plate 半波片
half-wave potential 半波电位*
half-way unit 半工业装置
half-white oil 半亮油
half-width 半宽度
half-width of spectral line 谱线半宽度
half wool 半毛
halide (=halogenide) 卤化物*
halide disc 卤化物片
halide end group 卤素端基
halide ion electrode 卤离子电极
halide ions 卤化物离子
halide slagging 卤化(物)造渣
halide volatilization process 卤化物挥发法
halite 石盐；岩盐 NaCl
Hall aluminum process 霍尔制铝法
Hall conductivity detector 霍尔电导检测器
Hall effect 霍尔效应
halleflinta 长英角岩
Hall electrolytic conductivity detector 霍尔电解电导检测器
Haller-Bauer reaction 哈勒-鲍尔反应
Hallikainen capillary viscometer 哈利凯南毛细管黏度计
Hallikainen rotating-disk viscometer 哈利凯南转盘式黏度计
Hallikainen sliding-plate viscometer 哈利凯南滑板式黏度计
hallow cylinder method 拉筒法〔测玻璃表面张力〕
halloysite 多水高岭土 $Al_2Si_2O_7 \cdot nH_2O$
Hall probe 霍尔探头
Hall probehead 霍尔探头
Hall process 霍尔法〔电解还原制备金属铀块法〕
halluoinogen 致幻剂〔药〕
Hall voltage 霍尔电压
halo ①〔词头〕卤②晕圈
haloalcohol 卤代醇
haloalkylation 卤烷基化*
haloalkylphosphate 磷酸卤代烷基酯
haloalkylphosphine 卤代烷基膦
haloalkylsilane 卤烷基硅烷
halocarbon 卤代碳
halochromism 加酸显色；卤色化（作用）
halochromy 加酸显色现象

halo effect （堆版的)光晕效应
halo fibre 光晕纤维
haloform 卤仿；三卤甲烷 CHX_3
haloform reaction 卤仿反应
halogen 卤(素)
halogen acetone 卤代丙酮
halogen acid (=haloid acid) 氢卤酸；卤化氢
halogen acid amide 卤代酰胺
halogen acyl bromide 卤代酰溴
halogen acyl chloride 卤代酰氯
halogen acyl fluoride 卤代酰氟
halogen acyl halide 卤代酰卤
halogen acyl iodide 卤代酰碘
halogenate 卤化
halogenated ①卤代的②卤化的
halogenated acid 卤代酸
halogenated acrylic ester 卤代丙烯酸酯
halogenated alcohol 卤代醇
halogenated aliphatic acid 卤代脂肪酸
halogenated amine 卤代胺
halogenated anthanthrone 卤化二苯并芘二酮；卤化蒽蒽醌（橙色有机颜料）
halogenated aralkyl-aryl ether 卤化芳烷基-芳基醚（白色有机颜料）
halogenated aromatic compound 卤化芳族化合物
halogenated aryl-arsonic acid 卤代芳胂酸 $XRAsO(OH)_2$
halogenated benzoic acid 卤代苯甲酸
halogenated butyl rubber 卤化丁基橡胶
halogenated carboxylic acid 卤代羟酸
halogenated compound 卤代化合物
halogenated diphenyl ether 卤化二苯基醚
halogenated ethylenic compound 卤化乙烯化合物
halogenated hydrocarbon monitoring method 卤代烃的监测方法
halogenated isoviolanthrone 卤化异紫蒽酮
halogenated ketone 卤代酮
halogenated paraffin 卤代链烷烃
halogenated rubber 卤化橡胶
halogenated solvent 卤化溶剂
halogenating ①卤代②卤化
halogenating agent 卤化剂
halogenating reaction 卤化反应
halogenation 卤化*
halogenative decarboxylation 卤化脱羧作用
halogen azide 叠氮化卤 $X[N_3]$
halogen benzoic acid 卤代苯甲酸 XC_6H_4COOH
halogen carrier 卤载体

halogen compound	卤素化合物
halogen-containing	含卤素(的)
halogen-containing acrylonitrile polymer	含卤素丙烯腈聚合物
halogen containing monomer	含卤素单体
halogen cyan 卤化氰 CNX	
halogen derivative	卤素衍生物
halogen ester	卤代酯
halogen ether	卤代醚
halogen ethyl ester 卤乙酯 $RCOOC_2H_4X$	
halogen family	卤族
halogen-free flame retardant	无卤阻燃剂
halogen hydride	卤化氢
halogenide (=halide)	卤化物*
halogen in ring	环上的卤
halogen in side chain	支链中的卤
halogen lamp	卤素灯
halogen-metal interconversion	卤素-金属互换作用
halogeno-acid	卤代(羟)酸
halogeno-aliphatic compound	卤代脂肪化合物
halogenoamine rearrangement	卤胺重排作用
halogeno-benzene	卤代苯
halogenocations	卤代阳离子
halogeno-cyanogen 卤代氰 CNX	
halogenodecarboxylation	卤代脱羧作用
halogenohydrin	卤醇
halogenoid	类卤基
halogen organic acid	卤代有机酸
halogeno-sugar	卤代糖
halogenosulfanes	卤代硫烷
halogenous	含卤的
halogen quenching counter	卤素猝灭计数管
halogen refrigerant	卤素冷冻剂；用卤素衍生物操作的冷冻剂
halogen sediment	卤化沉积
halogen-silver salt reaction	卤素-银盐反应
halogen substituted	卤素取代了(的)
halohydrin	卤代醇
halohydrin thioether	卤代醇硫醚
halohydrocarbon	卤代烃
haloid	卤(族)的
haloid acid (=halogen acid)	氢卤酸；卤化氢
haloid element	卤族元素
haloid ether	卤酸酯
haloid fluoride	卤素氟化物
haloing	白边；白圈；晕圈
haloklasite	哈洛克拉炸药
halometallic lactam	卤金属内酰胺
halometer	盐量计
halomethyl	卤甲基
halomethylation	卤甲基化作用
halomicin	卤霉素
halomonomer content	卤化单体含量
haloolefin	卤代烯烃
haloperidol	氟哌啶醇〔药〕
halophenol	卤酚
N-halophenyl-leucoauramine	N-卤化苯基-无色金胺
halophile	亲卤素的
halophosphoric acid	卤代磷酸
halophosphorus compound	卤代磷化合物
halophyte	盐土植物；耐盐植物
haloplankton	海水浮游生物
halopolymer	卤代聚合物；卤聚物
halo rubber	卤化橡胶
halosolvent	卤溶媒；含卤素化合物溶媒
halotrichite	铁明矾
halowax	卤蜡〔β-氯代萘〕
halt instruction	停机指令
hamamelin	金缕梅灵
ham ameli tannin	金缕梅单宁
hamamelose	金缕梅糖
hamartite	氟碳铈矿
hambergite	硼铍石
Hametag process	哈默塔克法〔铝粉干(球)磨法〕
Hamilton-Cayley theorem	哈密顿-凯莱定理
Hamiltonian	哈密顿(算符)*
Hamiltonian coordinate	哈密顿坐标
Hamiltonian operator	哈密顿算符
Hamilton method	哈密顿法〔漆漠的机械性能检验方法之一〕
hammer	锤
hammer crusher	锤(式压)碎机
hammered iron	锻铁
hammer finish	锤纹(罩面)漆
hammering	锻铁；打铁
hammer mill	锤磨机
hammer oil	气锤用润滑油
hammer paint	锤纹漆
hammer slag (=hammer scale)	铁屑
hammer swing mill	锤磨机
hammer swing sledge mill	锤击磨
hammer test	锤击试验〔评定润滑脂的纤维结构〕
hammer tone	锤纹
hammer tone silver	锤纹银粉漆；锤纹铝粉漆
hammer welded pipe	锻造管
Hammett acidity	哈米特酸度

Hammett acidity function 哈米特酸度函数*
Hammett adsorption indicator 哈米特吸附指示剂
Hammett equation 哈米特方程式
Hammett indicator 哈米特酸标指示剂
Hammett relations 哈米特关系*
Hammett's acidity function 哈米特酸度函数
Hammett scale 哈米特酸(度)标(度)；哈米特标度
Hammett's coefficient 哈米特系数
Hammett sigma correlation 哈米特 σ 相关
Hammett's rule 哈米特法则〔苯环上邻位对位取代基衍生物的反应性与结构关系的经验法则〕
Hammett-Zucker postulate 哈米特-朱克假定
Hammick-Illingworth rule 哈米克-伊林沃思规则
Hammond postulate 哈蒙德假说*
hampered-flow model 受阻流动模型
hamycin 哈霉素
hanadamine 哈哪达明；花田碱
hand ①手②手工；人工
hand adjustable reamer 手调铰刀
hand-ashed producer 手工除尘气体发生炉
hand ashing 手工除尘
hand atomizer (=hand atomiser) 手工喷雾器
hand bellow 手风箱
handboard 硬质纤维板
hand builder 柔软剂；改进手感的整理剂
hand-building 手工成形
hand-built product 手工(制)品
hand burner 手控喷灯；手控燃烧器
hand calciner 手工锻炉
hand centrifuge 手摇离心机
hand cleaner 洗手剂
hand cleaning ①手工除锈②(重涂时)表面处理等级
hand control 手工控制
hand control valve 手动控制阀
hand cream 润手香脂
hand crusher 手摇压碎机
hand-dipping 用手工测量罐内油量
hand drill 手(摇)钻
hand driven screw press 手动平板机
handedness 手型性；螺旋性
hand expansion valve 手动膨胀阀
hand-fed 人工加料的
hand-fed producer 手工加料气体发生炉
hand feed 人工加料
hand feeding 人工给料
hand feel and drape 手感舒适
hand-filling 人工包装
hand-finishing 手工修饰

hand fit 压入配合
hand furnace 人工炉
hand glassing 手工推光
hand-grainer 搓纹工；搓花工
hand graining 手工搓纹；手工搓花
hand grit soap 擦手砂皂
hand gun 手动注射器
handhole 手孔
handianol 汉地醇
handianolic acid 汉地醇酸
handianone 汉地酮
handing-in 手工穿经
handkerchief polymer 片型聚合物；二维聚合物
hand lay-up 手糊成形
hand lay-up method 手糊成型
handle ①柄；把手②手感
handler 铺鞣池
handler vat 平鞣池
handling 处理；装卸；输送
handling ease 加工容易；便于加工
handling equipment 装卸设备；输送设备
handling facilities 装卸设备
handling machinery 装卸机；输送器
handling operation 装卸工作
handling property ①处理性能②手感性
handling speed 周转速度
hand lotion 润手乳液
hand lubrication 手工润滑
hand lubricator (=hand oiler) 手提加油器
hand lugging 拉带法
hand machine ①钻床②手摇机器
hand-made 手工制造
handmade article 手工(制)品
hand-made overshoe 手工套鞋
hand made paper 手工(制)纸
handmade specialities 特种手工品
hand-manipulated 用手操作的
hand mould 手工压模
hand moulding 手工模塑
hand moulding press 手工模压机
hand oiler (=hand lubricator) 手提加油器
hand opener 手工开瓶器
hand-operated 用手操作的
hand-operated valve 用手操作的阀
hand operation 手动操作
hand pasting 手糊成型法
hand picking 人工挑选
hand plunger 搅灰器；木笛

hand plunger grease pump 手动杠杆式上油泵
hand-poked producer 手工出灰气体发生器
hand poking 手工出灰
handpress 手动压机
hand-propelled tripper 手摇推进式倾料器
hand pump 手泵
hand punch 手扳冲刀
hand-rabbled furnace 人工炉
hand-rabbled roaster 人工烤炉
hand rabbling 人工扒动
handrailing 扶栏
hand-raked furnace 人工炉
hand-raked roaster 人工烤炉
hand raking 人工扒动
hand regulation 手动调整〔机工〕
hand roller 手压滚
hand sampling 手工取样；手工进样
hand saver 保护手套；手护
hand-screw vulcanizing press 手旋硫化机
hand scrubber 手刷〔商〕
hand scudding 手工推挤
hand semi-rotatory pump 手摇半转泵
hand setting 手工推平
hand shear cutting machine 手扳铡断机
handsheet 手抄纸
hand sheet machine 手工抄纸器
hand slaking 人工熄灭
hand sleeve 手套
hand slicker 推皮刮刀
hand slicking 手工推平
hand spectrophotometer 手提式分光光度计
handspike method 推杆法
hand spray ①手动喷雾②手工喷涂；人工喷涂
hand stop valve 手动关闭阀
hand-stuff 人工填充
hand stuffing 手工加脂法
hand (superphosphate) den 人工过磷酸盐穴
hand swabbing 手工刷浆；手工刷色
hand (tear) test 手撕试验
hand test 手上试验；在手掌上蒸发测定汽油的挥发性
hand washing 手洗；人工洗涤
hand wheel 手轮
hand worked furnace 人工炉
hand worked roaster 人工烤炉
handy-size 小型；轻型
hanfangchine 汉防己碱
Hanford thermite process 汉福特铝热过程〔将高放射性废物转变成不溶性的硅酸盐或硅铝酸盐的高温法〕

hanger 吊架
hange stripping 挂具脱漆；吊具剥漆
hanging bell manometer 悬钟压力计
hanging bracket 吊架；悬托架
hanging clip 晾皮夹子
hanging drier 悬挂式干燥器
hanging drop 悬滴
hanging drop cell 悬滴槽
hanging drop culture 悬滴培养
hanging drop test 悬滴试验
hanging filter 滤棒
hanging mercury drop electrode 悬汞电极*；滴汞电极
hanging mercury electrode 悬汞电极；滴汞电极
hanging panel 挂屏〔大漆〕
hanging paper 裱糊纸
hanging up 工作架
hang-up 挂模；挂料；挂模斑
hank 纱线；绞丝
hank drier 绞纱烘燥机
hank shellac 绞状漂白紫胶
hanksite 碳酸芒硝
hank spraying washing machine 丝绞淋洗机
hank-to-bobbin winding machine 绞纱络筒机
Hansa monoazo orange 单偶氮汉萨橙；单偶氮橙颜料
Hansa monoazo yellow 单偶氮汉萨黄；单偶氮黄颜料
Hansa-Mühle soybean extractor 汉萨-米勒大豆萃取器
hansa yellow G〔商〕 汉萨黄 G；耐晒黄 G
Hantzsch pyridine synthesis 汉栖吡啶合成法
Hantzsch-Widman name 汉栖-魏德曼名称
Hanus iodine number 哈纳斯碘值
Hanu's method 哈纳斯法
haphazard arrange ①随机排列②杂乱排列
haploperine 尖叶(芸香)碱
haplophytine 单枝(夹竹桃)碱
hapten 半抗原
hapten radioimmunoassay 半抗原放射免疫测定(法)
hapto 配位点；络合点
harbor barge 码头驳船
harbor equipment 码头装卸设备
hard 硬的
hard acid 硬酸
hard aggregate 坚固结集
hard alloy 硬质合金
hard and soft acid and base (HSAB) 软硬酸碱*
hard anodic oxidation coating 硬质阳极氧化膜
hard ash coal 硬灰煤
hard asphalt 硬沥青；硬质沥青
hard base 硬碱

hard black	硬黑；硬烟末〔商〕
hard-block	刚性嵌段；硬嵌段
hard-block segment polymer	刚性嵌段共聚体；硬嵌段聚合体
hard boiling	强煮；硬熬
hard borosilicate glass	硬质硼硅酸玻璃；耐火玻璃
hard burned	硬烧的；高温焙烧过的
hard burr	硬草刺
hard-caked sediment	①硬(结)沉淀(物)②硬底〔油漆储存缺陷〕
hard carbon	硬碳；固体碳粒
hard (carbon) black	硬质炭黑；补强炭黑
hard cast-iron cylinder	冷铸铁滚筒
hard (china) clay	硬质陶土
hard cloth	粗布
hard coal	硬煤；无烟煤
hard cook	硬煮；过度蒸煮
hard copal	硬(树脂)
hard dry	硬干
hard elastic fibre	弹性硬纤维
hard elasticity	硬弹性
hard elastic material	硬弹性材料
hard elastic region	硬弹性区域
harden	硬化
harden ability	可淬(硬)性；淬透性
Harden and Young ester	哈杨二氏酯；果糖-1,6-二磷酸
hardened bush	硬化防护套管
hardened fat	硬化脂
hardened iron	硬化铁；冷铸铁
hardened oil	硬化油
hardened resin	硬树脂；凝固树脂
hardened rosin	硬化松香
hardened rubber	硬化橡胶
hardened steel	硬化钢
hardener	①硬化剂②坚膜剂
Harden furnace	哈顿电炉
hardening	硬化；淬火
hardening action	硬化作用
hardening agent	①硬化剂②坚膜剂
hardening bath	硬化浴；淬火浴
hardening carbon	硬化碳
hardening curve	硬化曲线
hardening furnace	硬化炉；淬火炉
hardening kiln	硬化窑
hardening liquid	硬化液；淬火液
hardening of fat	油脂的加氢作用
hardening of grease	润滑脂的硬化；润滑脂的固化
hardening of resin	树脂硬化；树脂凝固
hardening oil	硬化油；淬火油
hardening resistance of grease	润滑脂抗硬化
hardening strain	硬化应变
hardening stress	硬化；应力
hardening treatment	硬化处理
hardening yield value	硬化屈服值
hard facing	表面硬化
hard fat	硬脂
hard fiber	硬纤维
hard fibre board	硬质纤维板
hard film rust preventives	硬膜防锈剂
hard finish	硬化整理；硬挺整理
hard finish plaster	缓硬石膏
hard-fired ware	硬烧陶瓷；高温烧成陶瓷
hard floor paint	地板漆〔商〕
hard flow	低流动性
hard gel	硬(质)凝胶
hard glass	硬质玻璃
hard gloss paint	硬光漆
hard-grained roll	硬铸铁滚筒
hard grease	硬脂膏
hard gum	硬树胶
hard head	①硬渣②硬头
hardimeter	水质硬度测试器
Hardinge mill	哈丁磨机
harding temperature	硬化温度
hard lacquer	硬漆
hard lac resin	硬紫胶树脂
hard lead	硬铅
hard lump	硬块
hard metal	硬质合金
hard mortar	硬性灰浆
hardness	①硬度②硬性
hardness creep	硬度蠕变
hardness degree	硬度标度
hardness index of catalyst	催化剂硬度指数
hardness ions	硬性离子〔指水中的钙、镁等离子〕
hardness meter	硬度计
hardness number	硬度值；硬度指标
hardness of water	水的硬度
hardness of X-rays	X射线硬性；X射线硬度
hardness scale	硬度标(度)
hardness test	硬度试验
hardness tester	硬度试验仪
hardness testing machine	硬度实验机
hard oil putty	硬质油性腻子
hardometer	硬度计
hard paraffin	硬石蜡；固体石蜡

hard pitch 硬质沥青
hard polymer 硬弹性聚合物
hard porcelain 硬瓷
hard processing channel black 难加工槽黑
hard putty ①(用黄丹或铅白固化的)油性腻子；油灰②快干水性石膏腻子
hard radiation 硬(性)辐射
hard ray 硬(性)射线
hard residue 硬性残留物〔未降解的表面活性剂〕
hard resin 硬树脂
hard rock phosphate 硬岩磷酸盐
hard rubber 硬质胶；硬橡胶；胶木
hard rubber article 硬质胶制品；胶木制品
hard rubber scraps 硬质胶屑
hard rubber sheet 硬胶板；胶木板
hards 硬煤
hard settling 硬结；结块
hard shell sphere model 硬壳球体模型
hard size 重施胶
hard sludge 硬淤渣；焦状渣
hard soap 硬肥皂；钠皂
hard-soft acid-base principle 硬软酸碱原理
hard solder 硬焊料；硬钎料
hard soldering 硬焊接；用焊料焊接
hard-sphere fluid 硬球形流体
hard spun yarn 紧捻纱；强捻纱
hard steel 硬钢
hard stock 硬浆
hard stopping ①快干水性石膏腻子②硬填孔料
hard surfacing 表面淬火
hard-textured pigment 硬质(地)颜料〔难分散的颜料〕
hard time 固化时间
hard-to-degrade organic pollutant 难降解有机污染物
hard tough film 硬而坚韧的(漆)膜
hardware ①实物；成品；装备；附件；元件；金属构件②硬件；计算机主体
hard water 硬水
hard water resisting soap 抗硬水皂
hard water soap 硬水肥皂
hard wax 硬蜡
hard-wearing 耐磨损的
hard wood 阔叶材；硬材
hard-wood charcoal 阔叶材木炭；硬材木炭
hard wood distillation 阔叶材干馏；硬材干馏
hardwood flour 硬木粉〔填料〕
hard wood tar 阔叶材焦油；硬材焦油
hard wood tar pitch 阔叶材(焦)沥青；硬材(焦)沥青
hard X-ray 硬 X 射线

Hardy-Rand-Rittler PIC plate test 哈迪-兰德-里特尔伪等色图试验
hardystonite 锌黄长石
hard zinc 硬锌
Hargreaves-Bird cell 哈格里夫斯-伯尔德电池
harimycin 下里霉素
Harker section 哈克尔截面
harl 麻类长纤维；麻丝
harmalan 哈梅蓝
harmaline 骆驼蓬灵
harmalol 骆驼蓬酚
harman 哈尔满
harmful carbon deposition 有害的碳沉积
harmine 哈尔明；骆驼蓬碱
harminic acid 骆驼蓬酸 $C_{10}H_8O_4N_2$
harmless carbon deposition 无害的碳沉积
harmol 哈尔酚
harmonic ions 谐振离子
harmonic mean 调和平均
harmonic motion 谐和运动
harmonic oscillator ①谐振子②谐振荡器
harmonic peak 谐振峰
harmonic test 谐波试验〔指动态模量实验〕
harmonic vibration 谐波振动
harmonic vibration correction 谐振校正
harmonic wave 谐波
harmonizing color 调和色；和谐色
harmony 谐和；调和；协调；融洽
harmotome 交沸石
harmyrin 哈母仁
harness draft plan 穿综图
harness leather 马具革
harness oil 皮革油
harness twine 通丝
harp 竖琴式管子结构加热炉
harpoon model 鱼叉模型
Harries reaction 哈里斯反应
harringtonine 三尖杉酯碱
Harris vibrating screen 哈里斯摆(动)筛
harsh 粗糙
harsh concrete 干硬性混凝土
Harshel's demulsibility test 哈希尔抗乳化性试验
harshest condition 最苛刻的条件
harsh feeling 粗糙感
harsh grain 粗粒革
harsh mix ①粗糙搅拌；不均匀搅拌②干硬性混合料
Hart condenser 哈特冷凝器
hartell (=orpiment) 雌黄

Hart impact tester 哈特式冲击强度试验机
hartite 晶蜡石
Hartley band 哈特莱吸收带
Hartley's test 哈特莱检验
Hartmann diaphragm 哈特曼光阑
Hartmann formula 哈特曼公式
Hartmann-Hahn matching condition 哈特曼-哈恩匹配条件
hart moulding scraper 圆形造型刮板
hartree 哈特里〔能量单位,1哈特里=27.211652电子伏特〕
Hartree-Fock approximation 哈特里-福克近似
Hartree-Fock limit 哈特里-福克极限*
Hartree-Fock method 哈特里-福克法
Hartree-Fock-Roothaan equation 哈特里-福克-罗特汉方程*
Hartree-Fock SCF method 哈特里-福克自洽场方法*
hartsalz 硬盐〔钾石盐与钾盐镁矾的混合物,用作肥料〕
hartshorn 鹿角精;氨水
hartshorn carbon 鹿角粉;鹿角灰〔鹿角烧制的炭粉,漆器推光用〕
hartshorn oil 骨油
hartshorn salt 碳酸铵
Harvest 英国工厂规模高放射性废液玻璃固化过程
harvester oil 收割机油
harzburgite 斜方辉橄岩
Hasenclever lead pan 哈森克累弗铅锅
hashab ①塞内加尔金合欢②精制级阿拉伯树胶
hashish 大麻粉
hash mark 条痕〔电泳病态〕
Hass process (for zinc refining) 哈斯(炼锌)法
hasubanonine 莲花宁碱
haswelite 豪威炸药
hatch ①人孔;升降孔②闸门;炉门
hatchettite 伟晶蜡石
hatching-out 以火星点燃混合物
hat leather 制帽革
hat orifice 圆柱形锐孔
hat packing 帽形轴封
hat sweatband leather 帽圈革
Hatta number 八田数
Hatton green 哈顿绿〔磺化酞菁蓝与钛镍黄的共沉淀物〕
haulage 调动;输送;供给;曳引
haul-off gear 导出装置;接取装置
haul-off tension 引出张力
hausmannite 黑锰矿
hauyne 蓝方石
hauynite 蓝方石
Haüy's law of crystallization 霍伊结晶定律

hawking 液下浸轧染色
Hawkins cell 铁镍电池
hawkite 豪卡炸药;豪卡特
Haworth methylation 霍沃思甲基化作用
hawser 粗绳;缆
hawthorn seed oil 枳子油
Hayashi rearrangement 林氏重排作用
hayatin 海牙亭
hayatinin 海牙亭宁
Hayden process (for copper refining) 海敦(炼铜)法
Hayem solution 海姆试液
hay filter 枯草过滤器〔用于过滤含油废水〕
Haynes-Engle process 哈奈斯-英格尔过程〔用氢氧化钠从碳酸盐浸出液中沉淀铀的过程〕
hay tank 高架油罐
hazard index 危害指数
hazardous area 危险区
hazardous chemicals 危险药品
hazardous material 有害物质
hazardous solvent 有害溶剂
hazardous substance 危险品
hazardous wastes 危险废物
haze 灰霾;薄雾
haze formation in gasoline 汽油的混浊
hazelnut oil 榛子油
hazemeter (薄膜)混浊度测量仪;雾度测定仪
Hazen number 黑(曾)氏色值
Hazen platinum-cobalt standard 哈森铂钴标准〔用氯铂酸钾和氯化高钴水溶液比色法测定近无色液体颜色的标准〕
Hazen unit 哈森单位〔测定色度用〕
haze tallow (=sumach tallow;Japan wax; haze tallow) ①漆脂;皮油(俗)②(误称)日本蜡;木蜡〔得自野漆树坚果的果皮,主要成分为棕榈酸甘油酯和少量其他脂肪酸甘油酯,并非为蜡〕
haze value 雾度值;浊度值
haze wax (=haze tallow;Japan wax; sumach wax) ①漆脂;皮油〔俗〕②(误称)日本蜡;木蜡
haziness 混浊性;混浊度
haziness of gasoline 汽油浊度
hazing ①(漆膜)起雾②(清漆)轻度挥发
H-bonding 氢键合
H cell H型电解池
H-clinoptilolite 正-斜发沸石
H-component H组分
H-D analyzer 氢-氘分析器
head ①高差②头;前部③压头;盖④水头
head box 网前箱;流浆箱

head-capacity curve　压头-排量曲线；压头-输送量曲线
head coal　精煤
head coefficient　压头系数
head-driven conveyor　前(鼓轮)驱动式运输带
head end　首端
head-end plant　道端过程工厂(车间)
head-end process　首端过程*
header　①集管②联管箱
header box　高位箱；高位槽
header pipe　总管
headers　①头部；顶盖；端板②横梁；帽木
header suport　总管管架
header tank　高位槽
head fraction　头馏分；初馏分
headgroup　首基
head helmet　头兜
heading　①作头②露头
headlight oil　头光油〔煤油在65℃以上的部分〕
head loss　压头损失
head meter　压头(测流)计
head of band　谱带头
head of delivery　压头；扬程
head of mill　下磨；磨底
head of pump　泵压头
head of (spectral) band　(光)谱带头
head of tender　泵送石油产品的主泵组
head-on collision　迎头碰撞
head pressure　压头
head pressure ga(u)ge　入口压力计；泵口压力计
head product　初产物；初馏物
head pulley　①前鼓轮〔运输带〕②(=top pulley)上鼓轮〔提升机〕
heads　头馏分
head space　(容器中)液面上(部)空间；顶部空间
head space analysis　顶空分析
head space concentrating injector　顶空浓缩进样器
head space gas chromatography　顶空气相色谱法*；液(面)上(部)气体分析
head space sampler　液面上取样器
headspace sampling　①液面上取样②顶空进样
head stage of microelectrode amplifier　微电极放大器前级
head-tail sequence　头尾顺序
head tank　压力槽；压头槽
head-to-head　头-头接；头接头
head to head addition　头-头加成
head-to-head bond　(聚合物)头-头键合
head-to-head linkage　头-头键合

head-to-head polymer　头-头聚合物*
head-to-head sequence　头-头序列
head-to-head unit　头-头接单元
head-to-tail linkage　头-尾键合
head-to-tail polymer　头-尾接聚合物
head-to-tail polymerization　头-尾聚合
head truss hoop　吊桶钩
head variation in glass capillary viscometer　玻璃毛细管黏度计压头变化
head vat　头池〔鞣池组中最浓的鞣池〕
healant　修补剂
healed varnish　综清漆
healing rate constant　恢复速率常数
healing stone　屋顶岩板
healing time　恢复时间
Healograph　希洛多强力试验机
heap　堆
heap roasting　堆摊焙烧
heart carrier　鸡心卡头〔车床〕
heart check (=rift crack)　心裂〔木材干裂〕
heart cut　①中心馏分②中心切割
heart cutting　中心切割
hearth　炉膛
hearth furnace　床炉
heart sugar　环己六糖；肌糖
heart wood (=duramen)　心材
heat　热*
heatable reservoir inlet　加热储槽进样器
heat absorbent　热吸收剂；吸热剂
heat-absorbing　吸热的
heat-absorbing reaction　吸热反应
heat absorption　热吸收
heat absorption capacity　吸热容量；吸热量
heat abstraction　热去除；热抽吸
heat accumulation　生热性；发热性
heat-activable tanning agent　热激活性鞣剂
heat activated adhesive　热活化胶黏剂
heat-adherable　热黏附性的
heat-affected zone　热影响区
heat-affected zone crack　热影响区裂纹
heat-age discoloration　热老化变色
heat ageing　加热老化
heat ageing additive　防热老化剂
heat ageing inhibitor　防热老化剂
heat alarm　热警报；温度报警信号
heat application　热供应
heat balance　①热(量)平衡②热(量)差额表③热量衡算
heat band　加热带；电热丝

heat barrier　热障
heat block　加热块
heat-blocking　热黏着性
heat-bodied oil　厚油；(热)聚合油
heat-bodying　热炼；热聚合；加热稠化
heat bodying of oil　油的热黏化
heat bonding　热黏结
heat booster　加热器；预热器
heat break-free oil　未热炼油；未裂解油
heat build-up　生热(性)；发热(性)
heat calculation　热平衡计算
heat capacity　热容*
heat capacity at constant pressure　等压热容
heat capacity at constant volume　等容热容
heat-carrying agent　热载体；载热剂
heat change　热交换
heat coagulant　热凝固剂
heat coil　加热盘管
heat collapse　热破坏；热变形
heat color　①耐热染料②火色〔冶金〕
heat compensating jacket　热补偿夹套
heat conduction calorimeter　热导式热量计*
heat conduction relaxation function　导热松弛函数
heat conductivity　①导热性②热导率
heat conductor　导热体
heat consumption　耗热量
heat content (=enthalpy; heat function)　热函；焓
heat content of gas　气体热含量
heat control　热控制；温度调节
heat convection　热(的)对流
heat conversion factor　热换算因子；热当量
heat-convertibility　热转化性；热固性
heat convertible　热固(性)的；热变定的
heat convertible resin　热固(性)树脂
heat conveyance system　传热系统
heat crack　热龟裂
heat cross-linking　(加)热交联
heat-curable　热固性的
heat cure　热硫化
heat cured system　热固化体系
heat-cured urethane coating　热固型聚氨酯涂料
heat curing　热硫化
heat curing accelerator　热硫化促进剂
heat curing system　热硫化法
heat current　热流
heat cycle　加热(模塑)周期
heat deflection temperature (HDT)　热挠曲温度
heat deflection temperature under load　热变形温度

heat deformation　热致形变
heat-degradation　热降解(作用)
heat degree　热度
heat desizing　热脱浆
heat deterioration　加热劣化变质；热老化
heat diffusivity　①导热性②热导率
heat disorientation　热致去取向
heat dispersion　热分散；热散逸；散热性能
heat dissipation　热消散；热散逸
heat distortion　热畸度；热扭变
heat distortion point　热变形点；软化点
heat distortion temperature　热畸变温度
heat distortion test　热变形试验
heat drop　热降；温度降落
heat durability　耐热性
heat duty of pipe still　管式炉热负荷
heated　热的；加热的
heated absorption tube　(加)热吸收管
heated bar　测热棒
heated capillary spinneret　喷孔加热式纺丝板
heated chamber　外加热加热室
heated drum　鼓式干燥机
heated effluent　热废液
heated gas sampler　加热气体进样器
heated graphite atomizer　加热石墨原子化器
heated graphite furnace atomizer　电热石墨炉原子化器
heated inlet system　加热进样系统
heated nebulization chamber　(加)热雾化室
heated roll　热辊
heated rubber　过热胶片〔生胶缺点〕
heated scrub　加热洗涤液
heated snubbing pin　加热制动销
heated spot　热迹
heated storage　加热的储藏器
heated-tool welding　热烙铁焊接
heat effect　热效应
heat efficiency　热效率
heat elimination　散热
heat embossing　热压花
heat embrittlement　热硬化
heat emission　热(的)发射
heat endurance　耐热性
heat-endurance test　耐热试验
heat energy　热能
heat equivalent　热当量
heat equivalent of work　热功当量
heater　①加热器②硫化罐③硫化机
heater band　加热带；电热丝〔压出机〕

heater block 加热(组合)块
heater bridge wall couple 加热炉隔墙测温热电偶
heatercase 电热箱
heater effluent 裂解产物
heater flue couple 加热炉废气(或出气口)的热电偶
heat erosion 热侵蚀
heater outlet couple 炉出口处热电偶
heater plate 加热板
heater press 焙燥机
heater reclaim 蒸汽法再生
heater (reclaiming) process 蒸汽(再生)法；油再生法
heater stack couple 出口气体的热电偶
heat evolution 热析出；放热
heat exchange 热交换
heat exchange coefficient 热交换系数
heat-exchange equipment 换热设备
heat-exchange fluid 液体载热剂
heat exchange medium 热交换介质
heat exchange pebbles 石子载热体
heat exchanger 换热器；热交换器
heat exchanger coil 热交换盘管(蛇形管)
heat exchanger network 换热器网络
heat exchanger tube sheet 换热器管片
heat exchanger wall 换热器壁
heat expansion 热膨胀
heat extraction 热除去
heat fast 耐热的
heat fastness 热牢度
heat flexibility 热柔韧性
heat flexibility test 热挠曲试验
heat flow 热流
heat flow meter 热流计
heat flow modulus 热流系数
heat flow rate 热流速率
heat fluctuation 热波动
heat flux 热流；热通量
heat flux detector 热流检测器
heat flux differential scanning calorimetry 热流型差示扫描量热法
heat flux vector 热通量向量；热流向量
heat forming 热成型
heat function (=enthalpy; heat content) 热函；焓
heat fusing 热熔融；热熔化
heat gain 热增益；热增加
heat gel test apparatus 热胶凝试验器
heat generated by fuel 燃料发生的热
heat-generating body ①发热体②高频加热件
heat generating reaction (=heat generation reaction) 放热反应
heat generation 生热(性)；发热(性)
heat generation in flow 流动发热
heat generation reaction (=heat generating reaction) 放热反应
heat gradient 热梯度
heat gun 空气加热枪
heat-hardenable resin 热固性树脂
heather effect 混色效应
heather mixture 混色毛纱
heat history 热史；热历程；累积热
heat history equivalent 热史等价系数
heat index (=heat number) 热指数
heat indicating coating 示温涂料
heat indicating paint 示温涂料
heat-indicating pigment (=temperature-indicating pigment) 示温颜料
heat-induced laser beam deflection analysis 热诱导激光束偏转分析法
heat inertia 热惯性
heating ①加热的②加热
heating aeration 加气气化法
heating agent 热媒
heating aging 热老化；快速老化
heating apparatus 加热装置
heating area 加热面积
heating arrangement 加热装置
heating bath 热浴
heating block 加热块
heating body 加热体；散热体
heating boiler 供暖锅炉
heating capability 升温能力
heating capacity 热容量；热值
heating chamber 加热室；燃烧室
heating circuit ①加热电路②加热回路③加热系统
heating coil 加热旋管
heating coil set 加热盘管组
heating-cooling balance 加热冷却平衡
heating-cooling cycle 加热-冷却循环
heating cord 电炉线
heating curve 热曲线；升温曲线
heating curve determination 加热曲线测定
heating cycle 加热周期
heating cylinder (=feed cylinder) 料筒
heating drum 加热(转)鼓；加热滚筒
heating duct 热风管
heating effect 热效应
heating efficiency 加热效率

heating electrode welding　加热电极焊接
heating element　热体；热媒；加热元件
heating equipment　加热设备
heating facilities　加热用设备
heating flue　供暖烟道
heating function　加热功效；加热功能
heating funnel　保温漏斗
heating furnace　加热炉
heating gas　可燃气体
heat(ing) generator　①热发生器②高频加热器
heating grade natural gas　加热用天然煤气
heating installation　①暖气设备②加热设备
heating intensity　加热强度
heating in the open　敞口加热
heating in water bath　水浴加热
heating jacket　加热夹套；罩式电热器；加热帽
heating load　加热负载
heating lock　热扣合法
heating loss　加热减量
heating mantle　加热夹套；罩式电热器；加热帽
heating medium　载热体
heating medium for high temperature　高温载热体
heating member　加热体
heating mortar　加热乳钵；加热研钵
heating oil　燃料油
heating oven　加热炉；烘箱
heating pad　加热板
heating peak　加热峰
heating pickling tank　热酸浸洗池(槽)
heating pin　加热销
heating pipe　加热管
heating plate　加热板
heating power　燃烧热；热值
heating radiator　散热器
heating rate　加热速率
heating rate curve　加热速率曲线
heating red　赤红加热
heating section　加热部分
heating space　加热空间
heating spray chamber　加热雾化室
heating steam　加热蒸汽
heating surface　加热面
heating system　加热系统
heating tube　加热管
heating under diminished pressure　减压加热
heating under reduced pressure　减压加热
heating unit　加热装置；加热元件
heating up　加热

heating value　热值；卡值
heating wall　火墙
heating zone　加热段；加热区
heat input　热量耗费；热量输入
heat-insulated　绝热的；不传热的
heat insulating ability　①绝热性；保温性②绝热效率
heat insulating coat　隔热涂层
heat insulating material　隔热材料；保温材料
heat insulation　热(的)绝缘
heat insulator　①隔热材料；保温材料②热绝缘器
heat intensity　热强度
heat interchange　热交换
heat interchanger　热交换器；换热器
heat jacketed drum　夹套加热鼓
heat-labile　不耐热的；热不稳定的
heat lag　热滞后；热延迟
heat laminating　加热复合〔泡沫塑料与织物〕
heat leak　热渗透
heat leakage　热渗漏；热漏损
heat lens absorbance detection　热透镜光谱检测法
heat lens detection of intersect laser-induced　相交束激光诱导的热透镜测量
heat liberation　放热
heat life　热寿命
heat load　热负荷
heat loss　热消耗
heat mark　热伤
heat medium　热媒；加热介质
heat medium oil　载热油
heat modification　(加)热改性；热变性
heat-modified　热改性的；热变性的
heat number (=heat index)　热指数
heat of ablation　烧蚀热；消融热
heat of activation　活化热
heat of adhesion　黏附热
heat of admixture　混合热
heat of adsorption　吸附热
heat of association　缔合热
heat of bond formation　成键热
heat of coagulation　凝结热
heat of combination　化合热
heat of combustion　燃烧热
heat of compression　压缩热
heat of condensation　冷凝热
heat of conversion　转化热
heat of cracking　裂化热
heat of crystallization　结晶热
heat of decomposition　分解热

heat of detonation　爆轰热；爆炸热
heat of diffusion　扩散热
heat of dilution　稀释热*
heat of dissociation　离解热
heat of dye absorption　染料吸收热
heat of emersion　浸润热*
heat of evaporation　蒸发热
heat of explosion　爆炸热
heat of formation　生成热
heat of friction　摩擦热
heat of fusion　熔化热
heat of hydration　水合热
heat of immersion　浸湿热
heat of ionization　电离热
heat of isomerization　异构热
heat of linkage　键合热
heat of liquefaction　液化热*
heat of liquid　液体热含量
heat of micellization　胶束形成热*
heat of mixing (=heat of mixture)　混合热
heat of mixture (=heat of mixing)　混合热
heat of neutralization　中和热
heat of oxidation　氧化热
heat of polymerization　聚合热
heat of precipitation　沉淀热
heat of racemization　(外)消旋热
heat of radiation　辐射热
heat of reaction　反应热
heat of reaction detector　反应热检测器
heat of reduction　还原热
heat of solidification　固化热
heat of solution　溶解热
heat of solvation　溶剂化热
heat of sorption detector　吸着热检测
heat of spreading　铺展热
heat of sublimation　升华热
heat of swelling　溶胀热
heat of transition　转化热；转变热；转换热
heat of vaporization　汽化热
heat of vulcanization　硫化热
heat of wetting　润湿热
heat output　燃烧热；热值
heat passage　热(的)传递
heat penetration　加热深度
heat pipe　热管
heat pipe exchanger　热管换热器
heat plasticization　热塑炼；高温塑炼
heat polymer　热聚物

heat polymerization　热聚合
heat polymerization rubber　热聚橡胶
heat polymerized oil　热聚合油
heat preservation cloth　保温布
heat processing　热加工
heat producing reaction　放热反应
heat production　放热；热的产生
heat-proof　抗热的；隔热的
heat-proof glass　耐热玻璃
heat-proof material　耐热材料
heat-proof porcelain　耐热瓷器
heat propagation　热传导；热扩散
heat-protection layer　防热层
heat prover　废气和排出气体的分析器
heat pump　蒸汽泵
heat quantity　热量
heat quantity unit　热量单位
heat radiating equipment　热辐射设备
heat radiation　热辐射
heat rate　加热速率
heat rating　热功率；耗热率
heat ray　热(射)线
heat-reactive phenolic resin　热反应性酚醛树脂；热固性酚醛树脂
heat reactive resin　热固化树脂
heat reactivity　热反应性
heat receiver　收热机
heat reclamation system　热量回收系统
heat-recoverable material　热可逆材料
heat recovery　热量回收
heat-recovery condenser　热回收冷凝器
heat recovery system　热量回收系统
heat refining　(油)热炼
heat-reflecting body　热反射体
heat reflecting paint　反射热的涂料
heat reflow coating　热熔再流平涂料
heat-regenerating fiber　热(量)再生纤维〔含远红外辐射陶瓷纤维〕
heat regeneration　交流换热；热量再生
heat regenerator　交流换热器
heat regulator　热调节器
heat rejection　热消耗
heat relaxing apparatus　热松弛(处理)装置
heat-relaxing treatment　热松弛处理
heat release　放热
heat release kinetics　放热动力学
heat removing agent　去热剂
heat requirement　需热量

heat reserving material　保温材料
heat reservoir　储热器；热库
heat resistance　①耐热性②耐热度〔可耐的最高温度〕
heat resistance paint　耐高温漆
heat resistance test　耐热试验
heat resistant　耐热的
heat resistant coating　耐热涂料
heat resistant finish　①耐热面漆②耐热涂装
heat resistant furnace　耐热炉
heat resistant paint　耐热漆
heat resistant pigment　耐热颜料
heat resistant polymer　耐热性聚合物
heat-resistant quality　耐热性；热安定性
heat resister　防热老化剂
heat resisting　耐热
heat resisting alloy　耐热合金
heat resisting alloy steel　耐热合金钢
heat-resisting durability　耐热持久性
heat resisting enamel　①耐热瓷漆②耐热搪瓷
heat resisting glass　耐热玻璃
heat resisting steel　耐热钢
heat-retaining　储热的；保持热的
heat retention　储热能力；保温能力
heat rise　生热；发热
heatronic　高频(率)电热
heatronic moulding　高频(率电热)模塑(法)
heat sag value　热下垂度值
heat scale　热轧(钢)氧化皮
heat-seal　熔焊；熔接；热封
heat sealability　热封性；熔焊性
heat-seal coating　热封涂层
heat sealer　热合性；封口机
heat sealing　熔焊；熔接；热封
heat-sealing temperature　热封温度
heat sensing element　热敏元件
heat sensitive adhesive　热敏黏合剂
heat sensitive adhesive agent　热敏黏合剂
heat sensitive latex　热敏性胶乳
heat-sensitive material　热敏性物料
heat-sensitive paint (=temperature-indicating paint)　示温漆
heat sensitive process　热敏法
heat sensitive sensor　热敏传感器
heat sensitivity　热敏性
heat sensitization　热敏化(作用)
heat sensitized mixing　热敏混炼
heat sensitizer　热敏化剂
heat sensitizing agent　热敏化剂

heat sensor　热传感器；热传感元件
heat service　高温下使用
heat set　热变定
heat set ink　热固着油墨
heat-set stretch yarn　热定形弹力丝
heat settability　热定形性；热变定性
heat setting　热定形；热固定
heat-set vehicles　热固着连接料〔油墨〕
heat shield　挡热板；挡热罩
heat shield material　隔热材料；保温材料
heat shock　热震荡
heat-shrink　热收缩
heat-shrinkable　热缩的
heat-shrinkable fibre　热收缩纤维
heat-shrinkage　①热收缩②热收缩率
heat sink　吸热层；吸热设备
heat sink method　吸热法；热吸收法
heat sink plane　散热层
heat-softened resin　热软化树脂
heat softening　热塑炼；热软化
heat source　热源
heat stability　热稳定性；耐热性
heat stability test　(涂料黏度的)热稳定性试验
heat stability test of greases　润滑脂热安定性试验
heat-stabilized　热稳定的；耐热的
heat stabilizer　热稳定剂*
heat stabilizing additive　热稳定添加剂
heat-stable　耐热的；热稳定的
heat stable antioxidant　热稳定防老剂；防热老化剂
heat stable conditioning agent　热稳定整理剂；耐热整理剂
heat-stable inhibitor　热稳定剂
heat stable material　耐热材料
heat stable resin　耐热树脂
heat storage capacity　热容量
heat storage well　储热井
heat straightening　热校直
heat stress　热应力
heat stretch　热伸张
heat summation　热量总和
heat supply　热供应
heat test　①耐热试验②热试法
heat tint　回火颜色
heat transfer　传热；热传递
heat transfer agent　传热介质；载热剂
heat transfer bracket　传热托架
heat transfer by conduction　传导传热
heat transfer by convection　对流传热

heat transfer by forced convection　强制对流传热
heat transfer by radiation　辐射传热
heat transfer coefficient　传热系数
heat transfer coefficient for deposits　污垢系数
heat transfer density　传热密度
heat transference　传热
heat transferer　传热剂；热载体
heat transfer factor　传热因数
heat-transfer film　传热膜
heat-transfer fluid　传热流体
heat transfer index　传热指数
heat transfer lag　传热滞后
heat transfer medium　传热介质
heat transfer number　传热值
heat-transfer oil　传热油
heat transfer perimeter　传热周边
heat transfer printing　热转移印刷；热转移印花
heat transfer rate　传热速率
heat transferring　传热
heat transferring agent　传热剂
heat transfer salt　传热盐
heat transfer surface　传热面
heat transfer with constant temperature　恒温传热
heat transfer with variable temperature　变温传热
heat transmission　传热
heat transmission coefficient　传热系数
heat transmission oil　传热油
heat-treated oil　热处理油
heat-treated wood　热处理(木)材
heat treating　热处理
heat treating temperature　热处理温度
heat treatment　热处理
heat treatment after welding　焊后热处理
heat treatment controller　热处理控制器
heat treatment oil　热处理用油
heat under reflux　回流加热
heat unit　热(量)单位
heat up　加热
heat-up time　升热时间；加热时间
heat utilization　热的利用
heat value　热值；卡值
heat value of mixture　混合物的热值
heat vulcanizate　热硫化胶
heat vulcanization　热硫化
heat waste　热损失
heat wave　热波
heat-yellowing test　加热泛黄试验
heavier body　比较黏的原料；高黏稠性原料

heavier component　重组分
heavier in viscosity　较黏的
heavier petroleum fraction　重质石油馏分
heaviest overhead oil　最黏馏出油
heavily cracked　深度裂化的
heavily-laden conveyor　重负传送带
heaving shales　重页岩；膨胀页岩
heavy　重的
heavy addition　大量添加
heavy alkylate　①重质烷基化物②重烷基化油；烷基化油蒸馏残液
heavy and crude oil engine　重油(与)原油发动机
heavy aromatics　重芳烃
heavy aromatic solvent　重质芳香族溶剂；高沸点芳香族溶剂
heavy asphalt　重沥青；重质沥青
heavy asphaltic crude　沥青基重质石油
heavy asphaltic oil　重质沥青油
heavy asphaltic residues　重质沥青残渣
heavy atom effect　重原子效应
heavy atom method　重原子法*
heavy atom quenching　重原子猝灭
heavy benzol　重苯
heavy bitumen　重质沥青
heavy bodied　黏稠性质的；高黏稠的
heavy bodied oil　黏稠油品
heavy body　高黏度；高容量
heavy building　结构坚固；构造结实
heavy-burned magnesia　重质氧化镁；低活性氧化镁
heavy-calcined magnesia　重质氧化镁
heavy carrier gas　重载气
heavy charge　重质原料
heavy chemical　重化学品
heavy coated electrode　厚涂层电极
heavy component　(精馏的)较重组分；高沸点组分
heavy condensation products　①重质冷凝产物②重质缩合产物
heavy corrosion　重腐蚀
heavy crude　重质石油
heavy crystalline phosphate layer　重结晶磷酸盐膜
heavy current　强电流
heavy current surge　强电流冲击
heavy cut　重馏分
heavy cycle oil　重循环油
heavy denier　粗纤度
heavy dip coating　厚浸渍涂层
heavy distillate　重馏分
heavy duck　厚帆布

heavy duplex wallpaper　重质双层糊墙纸
heavy duty　重型；载重；重级
heavy duty "B" battery　大容量乙电池组
heavy-duty black　重防腐蚀(强防腐)沥青漆
heavy-duty brush　重役漆刷〔粗糙表面涂漆用〕
heavy-duty coating　强防腐蚀涂料；重防腐涂料
heavy-duty compressor　重型压缩机
heavy-duty corrosion inhibiting　重防蚀(涂装)的；高效防蚀(涂装)的；重役阻蚀的
heavy duty detergent　重役型洗涤剂；重垢型洗涤剂
heavy-duty diesel oil　重型柴油
heavy-duty dispersion impeller　重载分散盘
heavy-duty drawframe　重型拉伸机
heavy-duty engine oil　重型发动机油
heavy-duty lubricating grease　重型润滑脂
heavy-duty oil　重型油
heavy-duty paste mixer　厚浆混合机
heavy-duty portable　重型移动式
heavy-duty primer　重防腐蚀底漆
heavy duty samming and setting machine　重型平展匀湿两用机
heavy duty stirring blades　重载搅拌桨叶
heavy-duty supplement 2 oil　高添加级系列 2(重负荷)机油〔含约 11%的各种添加剂〕
heavy duty system　重防腐蚀体系；重防腐蚀涂料
heavy duty tube shield　重型管罩
heavy electron (=meson)　重电子；介子
heavy element　重元素
heavy element chemistry　重元素化学
heavy end　重尾馏分
heavy ends tower　重质尾部馏分(分馏)塔
heavy feed stock　重质原料
heavy film　厚层；厚膜〔油〕
heavy flash　厚胶边；厚毛边
heavy flint glass　重火石玻璃
heavy flow　强裂油流
heavy fluid separation　重流体分离
heavy foam　重质泡沫胶
heavy force fit　重压配合
heavy fraction　重馏分
heavy fuel　高黏度燃料；重质燃料
heavy-gage rubber sheet　厚胶片；胶板
heavy-gage rubber thread　粗胶丝
heavy gas oil　重瓦斯油；重柴油
heavy gasoline　重质汽油
heavy gel-like soap　高黏度类凝胶皂
heavy-grade black　高抗性黑沥青漆；重质黑沥青漆
heavy-gravity　①浓的；黏稠的②重质的

heavy grease　重质润滑油
heavy grease film　润滑脂厚膜
heavy hydrocarbon　重(质)烃
heavy hydrocarbon gases　重质烃类气体
heavy hydrocarbon oil　重质烃油
heavy hydrogen　重氢
heavy ice　重冰〔固态重水〕
heavy impurity　重质杂质；重组分杂质
heavy industry　重工业
heavy ion accelerator　重离子加速器
heavy ion beam implosion　重离子束内爆
heavy ion bombardment　重离子轰击
heavy ion induced MS　重离子诱导质谱法
heavy ion linear accelerator　重离子线性加速器
heavy ion nuclear chemistry　重离子核化学*
heavy ion probe　重离子探针
heavy ion sputtering　重离子溅射
heavy-isotope buildup　重同位素积累
heavy keying fit　重型固定配合
heavy laundering　强效性洗涤〔指洗涤污垢较重的棉织品〕
heavy leather　重革
heavy leather rolling machine　底革滚压机
heavy liquid　重质液体
heavy liquid residuum　重质液体残油
heavy lubricating oils〔复〕　厚黏度润滑油
heavy media separation　重介质分离
heavy metal　重金属
heavy mineral sand　重矿砂
heavy mineral spirit (=heavy petroleum spirit)　重质溶剂油；重石油醚
heavy naphtha　重石脑油
heavy natural gasoline　重质天然汽油
heavy neutral oil　重质中性油
heavy oil　重油；重柴油；粗石油
heavy oil cracking　重油裂化
heavy oil film　厚油膜
heavy oil fraction　重油馏分
heavy oil heater　重油加热炉
heavy organic soil　黏稠的有机污物
heavy paraffin hydrocarbon　重烷烃
heavy paraffinic impurity　重(链)烷烃杂质；重石蜡基杂质
heavy paste　稠糊
heavy petrol　重汽油
heavy petroleum oil　重石油
heavy petroleum spirit　重质溶剂油；重石油醚
heavy petroleum wax　重石油蜡
heavy phase　重(比重)相；重(液)相

heavy phase weir 重相堰
heavy platinum metal 铂类重金属
heavy polymer 重聚合物；高分子聚合物
heavy pressure 高压
heavy product 重(组合)产品；重质产品
heavy production 大规模(量)生产
heavy rare earths 重稀土(元素组)
heavy recoil oil 重反冲油；重后座油
heavy recycle stock 重质循环油
heavy repair 大修
heavy residual stocks 重残油
heavy residue 重残渣
heavy resinous petroleum residue 含胶质重石油残渣
heavy ring 承力环
heavy sand 重砂
heavy scale 厚氧化皮
heavy section 厚壁〔制品〕
heavy shade 饱和色
heavy side 重侧
heavy soda 重苏打
heavy soda ash 重苏打灰
heavy solids 机械杂质粗粒〔废油中〕
heavy solution 重液
heavy solvent naphtha 重质溶剂石脑油
heavy spar 重晶石
heavy still bottoms 重质残油
heavy straight-run gasoline 重直馏汽油
heavy straw board 厚草纸板
heavy syrup 浓缩糖浆
heavy tar distillate 重质焦油馏出物
heavy vessel 重型容器
heavy virgin naphtha 直馏重汽油馏分
heavy-walled 厚壁的
heavy water 重水
heavy weight hide 重磅皮；重量皮
heavy weight rubber product 重型(橡胶)制品
heavy wine 烈酒
heavy wire 粗钢丝
heavy wool grease 重质羊毛润滑脂；重质毛填料润滑脂
heberling 中小山羊皮
hebiscetin 木槿黄酮
heckling yarn 精梳纱
hecogenic acid 核柯精酸
hecogenin 核柯配基
hecogenoic acid 核柯精酸
hecogenone 核柯精酮
hecololactone 核柯内酯
hecto- 〔希腊字头〕百

hectogamma 百微克
hectogram 百克
hectograph 胶板誊写版〔印刷〕
hectolambda 百微升
hectoliter 百升
hectometer 百米
hectorite 锂蒙脱石
hectorite grease 水辉石润滑脂
hedamycin 赫达霉素
heddle (=heald) ①综丝；综线②综片
hedenbergite 钙铁辉石
hederagenin 常春(藤苷)配基 $C_{30}H_{48}O_4$
hederin 常春藤素
hedge mustard oil 篱芥子油
hedge radish oil 篱莱菔油
hedonal (=methyl-n-propyl carbinol urethane) 氨基甲酸-2-戊酯 $H_2NCO_2CH(CH_3)CH_2CH_2CH_3$
hedrites 多角晶
hedyotine 耳草碱
heel and toe wear 倒角磨损；跟趾磨耗
heel clearance 轴封踵隙
heeling (=heel leather) (鞋)后跟皮
heel leather (=heeling) (鞋)后跟皮
heel lift 鞋跟皮层
heels 残余料；下脚料；留底
Hegman fineness 赫格曼细度
Hegman gage 赫格曼细度计
Hegman grind gage 赫格曼细度计；刮板细度计
Hegman scale 赫格曼细度标度
Hegman scale reading 赫格曼细度标度读数
Hehner cylinder 亥讷(比色)筒
Hehner number (=Hehner value) 亥讷值〔不溶脂酸及不皂化物(总)值〕
Hehner test 亥讷检验(法)〔墨水中检验甲烷法〕
Hehner tube 亥讷管
Hehner value (=Hehner number) 亥讷值〔不溶脂酸及不皂化物(总)值〕
heifer 中牛皮
height equivalent to an effective plate 有效等(塔)板高度
height equivalent to an effective theoretical plate 有效理论塔板等效高度
height equivalent to an overall transfer unit 总传递单元高度
height equivalent to a theoretical plate (HETP) 理论塔板高*；理论塔板等效高度
height equivalent to a theoretical stage 理论塔板的等量高度
height equivalent to mass transfer unit 传质单元等效

高度
height equivalent to one theoretical plate　理论塔板等效高度
height equivalent to transfer unit　传质单元高度
height of an effective plate　有效塔板高度
height of a transfer unit (=height equivalent to transfer unit)　传质单元高度
height of catalytic unit　催化装置高度
height of center　中心高
height of packing equivalent to one theoretical plate　理论板等板填充高度
height of reaction unit　反应装置高度
height per transfer unit　传质单元高度
height-times-width method　高乘宽法；H. W 法
Heilter-London approximation　海特勒-伦敦近似
Heisenberg representation　海森堡表象
Heisenberg-Slater function　海森堡-斯莱特函数
Heisenberg uncertainty principle　海森堡测不准原理
Heisenberg uncertainty relation　海森堡测不准关系
Heitler-London method　海特勒-伦敦法
helamycin (=hilamycin)　喜霉素
helcosol　焦棓酸铋〔别名〕
helenalin　堆心菊灵；土木香灵〔取自 Helenium 堆心菊或 Inula helenium 土木香〕
helenien　堆心菊素；土木香素
helenine (=alantolactone)　①堆心菊脑；土木香脑；阿兰内酯②海仑菌素
helenite　弹性地蜡
helianthate　半日花酸盐
helianthin(e)(= methyl orange)　半日花素；甲基橙；酸性黄
helianthrene　半日花烯
helianthrone　半日花酮
heliarc unit　氩弧焊具
heliarc welding　氩弧焊
heliarthemin　半日花苷
helical agitator　直螺旋(式)搅拌器
helical blade　螺旋叶片
helical blade stirrer　螺旋形片状搅拌器
helical burr　螺纹
helical chain conformation　螺旋形链构象
helical coil　螺旋状管
helical condenser　螺旋形冷凝器
helical configuration　螺旋构型
helical conformation　螺旋型构象
helical-conveyer centrifuge　螺旋卸料离心机
helical dislocation　螺旋位错
helical drive　斜齿轮传动
helical fin　螺旋翅片

helical flow　螺旋流
helical flow column　(螺)旋流(动)柱
helical-lobe compressor　螺杆压缩机
helical molecule　螺旋型分子*
helical-path mass spectrometer　螺线质谱计
helical pipe　螺旋管
helical polymer　螺旋形聚合物
helical ribbon mixer　螺旋带搅拌机
helical screw pulp conveyer　螺旋纸浆传送器
helical stirrer　螺旋搅拌器
helical structure　螺旋结构
helical whisker　螺旋须晶；蜷线须晶
helicene　螺旋烯*
helices　单环
helicidin　螺杀菌素
helicine　绣线菊苷
helicity　螺旋性
helicobasidin　桑卷担子素
helicocerin　卷角孢菌素
helicoid (=helicoidal)　螺旋状的；螺旋面
helicoidal (=helicoid)　螺旋状的；螺旋面
helicoidal configuration　螺旋构型
helicoidal flow　螺旋流动
helicoidal groove　螺旋形槽
helicoidal sifting machine　螺旋转筒筛
helicoid screw conveyor　(无缝)螺旋运输器
helicoid-type mass spectrometer　螺旋型质谱仪
helicon　大喇叭固定壁离心分离法〔铀同位素分离法之一，又称 UCOR 法〕
helide　氢化物〔假想物〕
Heli-Grid packing　希里-格来得填充物
helindone blue BB〔商〕　赫林顿蓝 BB
helindone fast scarlet C〔商〕　赫林顿坚牢猩红 C
helindone orange R〔商〕　赫林顿橙 R
helindone red 3B〔商〕　赫林顿红 3B
helindone violet J2R〔商〕　赫林顿紫 J2R
helindone yellow 3GN〔商〕　赫林顿黄 3GN
helio fast yellow 6GL〔商〕　日光坚牢黄 6GL
heliography　日光胶版法
heliolite　日长石
heliomycin　日光霉素
heliophyllite　日叶石；斜方氯砷铅矿
heliopsine　赛菊芋碱
heliosupine　天芥菜平
heliotric acid (=heliotrinic acid)　天芥菜酸
heliotridane　天芥菜烷　$C_8H_{15}N$
heliotridene　天芥菜烯　$C_8H_{13}N$
heliotridine　天芥菜定　$C_8H_{15}O_2N$

heliotridylamine 天芥菜基胺
heliotrine 天芥菜碱
heliotrinic acid (=heliotric acid) 天芥菜酸
heliotrope 鸡血石
heliotrope gray 鸡血石灰(色)
heliotrope oil 天芥菜油
heliotropic acid 天芥菜酸
heliotropin ①天芥菜精②(=piperonal)3,4-亚甲二氧基苯甲醛；胡椒醛
heliotropine 天芥菜精
Heli-Pack packing 亥里-派克填料
helisterol 半日花甾醇
helium 氦
helium atoms diffraction 氦原子衍射
helium density of catalyst 催化剂的氦密度
helium detector 氦检测器
helium family element 氦族元素
helium gas 氦气
helium ionization detector 氦电离检测器
helium leak detection mass spectrometer 氦质谱探漏仪
helium leak detector 氦探漏仪
helium-neon gas laser 氦-氖气体激光器
helium photoionization detector 氦光电离检测器
helix 螺旋形
helix agitator 螺带式搅拌器
helix angle 螺旋角
helix chain 螺旋链
helix column 螺旋形柱
helix countercurrent chromatograph 螺旋形逆流色谱仪
helix scale 螺旋弹簧秤
helix structure 螺旋形结构
helleborein 嚏根草苷
helleboresin 嚏根草素
helleboretin 嚏根草亭
helleborin 嚏根草因
hellebrigenin 嚏根草苷配基
hellebrin 嚏根草苷
helleoboretin 嚏根草亭
Hellmann-Feynman theorem 赫尔曼-费恩曼定理
Hell-Volhard-Zelinsky halogenation 赫尔-乌尔哈-泽林斯基卤化作用
helmet 盔兜
Helmholtz cell 亥姆霍兹电池
Helmholtz double layer 亥姆霍兹双层
Helmholtz equation 亥姆霍兹方程
Helmholtz free energy 亥姆霍兹自由能
Helmholtz principle of minimum energy dissipation 亥姆霍兹最小能量损耗原理
Helmholtz relation for solid deformation 亥姆霍兹固体形变关系式
helminthic 驱虫药
helminthosporal 长蠕孢醛
helminthosporin 长蠕孢素；三羟基甲基蒽醌
helminthosporol 长蠕孢醇
helper drive 辅助传动
helsinkite 钠长绿帘岩
Helvetia leather 油脂鞣革
helvite 日光榴石
helvolic acid (=fumigacin) 蜡黄酸；烟曲霉酸 $C_{32}H_{44}O_8$
helvolinic acid 羟蜡黄酸
helvomycin 蜡黄霉素
hemachate 血点玛瑙
hemacytometer (=haemacytometer) 血球计；血细胞计数器
hemafibrite 红纤维石
hemagglutination inhibition 血凝抑制
hemanthine 网球花碱
hemartine extract 苏木浸膏
hematein 苏木因；氧化苏木精
hematic acid 血酸
hematine (=haematine; hematoxylin) 苏木精；苏木紫
hematine crystal 苏木精晶体
hematinic 补血药
hematinic acid 血色酸
hematite 赤铁矿
hematite pig 由赤铁矿炼成的生铁
hematolithe 红砷锰矿
hematomancy 验血诊断法
hematoporphyrin 血卟啉 $C_{34}H_{38}O_6N_4$
hematoxylin (=haematine; hematine) 苏木精；苏木紫
hematoxylineosin staining 苏木素-伊红染色法
hematoxylon 洋苏木
hematoxylone 苏木酮
heme 血红素；亚铁原卟啉
hemellitene 半莱；连三甲苯；1,2,3-三甲苯
hemellitic acid (=2,3-xylic acid) 半莱酸；2,3-二甲基苯甲酸 $(CH_3)_2C_6H_3COOH$
hem flange 折边(卷边)法兰
hemiacetal 半缩醛
hemiacetal group 半缩醛基
hemiacetal linkage 半缩醛键(合)
hemicellulase 半纤维素酶
hemicellulose 半纤维素
hemicellulose level 半纤维素含量
hemicolloid 半胶体
hemicolloid structure 半胶体结构

hemicrystalline 半晶状的；半结晶的
hemicyclic 半环的
hemicyclic double bond 半环状双键
hemiformal 半缩甲醛 CH$_2$(OH)OR
hemihedral 半面的
hemihedral form 半面晶形
hemihedral symmetry 半面型对称
hemihedry 半面体
hemihydrate 半水合物
hemi-hydrated plaster 半水合石膏
hemiketal 半酮缩醇
hemilignin 半木素
hemimellitene (=1,2,3-trimethylbenzene) 半莱〔俗〕；连三甲苯；1,2,3-三甲苯 (CH$_3$)$_3$C$_6$H$_3$
hemimellitic acid (=benzene-1,2,3-tricarboxylic acid) 半莱酸〔俗〕；苯连三酸；1,2,3-苯三酸 C$_6$H$_3$(COOH)$_3$
hemimercaptol 半硫代半缩醛 R$_2$C(SR)OH
hemimetamorphosis 半变态
hemi-micelle 半胶束
hemimorphism 半对称形；异极性〔地〕
hemipic acid (=hemipinic acid) 半蒎酸；3,4-二氧基邻苯二甲酸
hemipinic acid (=3,4-dimethoxyphthalic acid; hemipic acid) 半蒎酸；3,4-二氧基邻苯二甲酸 (CH$_3$O)$_2$C$_6$H$_2$=(COOH)$_2$
hemipolymer 半聚体
hemiprismatic 半棱晶的
hemiquinoid 半醌型
hemispherical analyzer 半球形分析器
hemispherical meniscus （水银柱)半球形弯月面；(水银柱)半球形弯液面
hemispherical reflectance 半球面反射率
hemispheroid 半球形
hemitactic polymer 半立构规整聚合物
hemiterpene 半萜 C$_5$H$_8$
hemitrisulfide 三硫化二物
hemitrope 半体双晶
hemitropic solid 半迷向固体
hemlock ①毒芹②铁杉
hemlock bark extract 铁杉树皮浸膏
hemlock extract 铁杉(树皮)栲胶
hemlock oil 铁杉油〔取自 Tsuga canadensis〕
hemmed 折边的
hemming 缝边
hemocyanin 血蓝蛋白
hemodialysis fibre 血液透析纤维
hemoglobin 血红蛋白

hemolytic test 溶血试验
hemolytic thermal scission reaction 均裂性热断(链)反应；热均裂反应
hemoprotein 血红素蛋白
hemopyrrole 血吡咯；2,3-二甲基-4-乙基吡咯

HC=C(C$_2$H$_5$)C(CH$_3$)=C(CH$_3$)
　　　　　　NH

hemorheology 血液流变学
hemostasis 止血
hemostatic ①止血剂②止血的
hemostatic agent 止血剂
Hempel analysis (=Hempel distillation) 亨佩耳蒸馏
Hempel distillation (=Hempel analysis) 亨佩耳蒸馏
Hempel flask 亨佩耳蒸馏瓶
Hempel gas apparatus 亨佩耳气体分析器
Hempel gas pipette 亨佩耳验气球管
Hempel's column 亨佩耳(蒸馏)柱
hempen 大麻
hempen core 麻芯线
hemp seed oil 大麻子油
hem type 摆型
henbane 天仙子
henbane extract 天仙子萃取物
hencicosanedioic acid 二十一烷二酸 C$_{19}$H$_{38}$(COOH)$_2$
Hencky strain (=logarithmic strain; natural strain) 固有应变
Hencules high-shear viscometer 亨库利斯高剪切黏度计
hendecadienoic acid 十一碳二烯酸 C$_{10}$H$_{17}$COOH
hendecanal 十一(烷)醛
hendecane (=undecane) ①十一(碳)烷②十一(碳)(级)烷 C$_{11}$H$_{24}$
hendecane diacid 十一烷二酸 C$_9$H$_{18}$(COOH)$_2$
hendecane dicarboxylic acid 十一烷二羧酸；十三碳二酸 C$_{11}$H$_{22}$(COOH)$_2$
hendecanedienoic acid 十一碳二烯酸 C$_{10}$H$_{17}$COOH
hendecanedioic acid 十一烷二酸 C$_9$H$_{18}$(COOH)$_2$
hendecanoic acid 十一(烷)酸 C$_{10}$H$_{21}$COOH
hendecanol 十一烷醇
hendecanone 十一烷酮
hendecene 十一碳烯 C$_{11}$H$_{22}$
hendecene diacid 十一碳烯二酸 C$_9$H$_{16}$(COOH)$_2$
hendecene dicarboxylic acid 十一碳烯二羧酸 C$_{11}$H$_{20}$(COOH)$_2$
hendecene dioic acid 十一碳烯二酸 C$_9$H$_{16}$(COOH)$_2$
hendecenoic acid 十一碳烯酸 C$_{10}$H$_{19}$COOH
10-hendecenoic acid 10-十一烯酸
10-hendecenyl acetate 乙酸 10-十一碳烯-1-醇酯
hendecoic acid 十一碳(烷)酸 C$_{10}$H$_{21}$COOH

hendecyl (=undecyl) 十一(烷)基
hendecyne 十一碳炔 $C_{11}H_{20}$
hendecyne diacid 十一碳炔二酸 $C_9H_{14}(COOH)_2$
hendecyne dicarboxylic acid 十一碳炔二羧酸 $C_{11}H_{18}(COOH)_2$
hendecyne dioic acid 十一碳炔二酸 $C_9H_{14}(COOH)_2$
hendecynoic acid 十一碳炔酸 $C_{10}H_{17}COOH$
heneguen 剑麻〔一种龙舌兰纤维〕
3,6-heneicosandiol 3,6-二十一烷二醇；3,6-二十一双醇 $CH_3(CH_2)_{14}CHOH(CH_2)_2CHOHCH_2CH_3$
heneicosane (正)二十一(碳)烷
heneicosane diacid 二十一烷二酸 $C_{19}H_{38}(COOH)_2$
heneicosane dicarboxylic acid 二十一烷二羧酸 $C_{21}H_{42}(COOH)_2$
heneicosanic acid (=heneicosoic acid) 二十一烷酸 $C_{20}H_{41}COOH$
heneicosanoic acid 二十一(碳)烷酸 $CH_3(CH_2)_{19}COOH$
heneicosene 二十一碳烯 $C_{21}H_{42}$
heneicosoic acid (=heneicosanic acid) 二十一(烷)酸
heneicosyl 二十一(烷)基 $CH_3(CH_2)_{19}CH_2-$
Henkel process 亨克尔法〔由苯甲酸或邻苯二甲酸制对苯二甲酸〕
henna 指甲花；散沫花染剂〔棕红色，尤用于染发和涂饰皮肤〕
Henneberg's solution 亨内伯格氏溶液
Henning process (for nickel extraction) 亨宁(炼镍)法
henpentacontane 五十一烷 $C_{51}H_{104}$
henry 亨利〔电感单位〕
Henry's constant 亨利常数
Henry's law 亨利定律*
Henschel mixer 亨舍尔混合机〔高速混合机的一种〕
hentetracontane 四十一烷 $C_{41}H_{84}$
hentriacontane 三十一(碳)烷 $CH_3(CH_2)_{29}CH_3$
hentriacontanol 三十一烷醇
hentriacontyl 三十一(烷)基 $CH_3(CH_2)_{29}CH_2-$
cis-3-henxenol (=leaf alcohol) 叶醇；顺式己烯-3-醇
hepa filter 空气滤清器
hepar antimony 锑肝〔氧化锑、亚锑酸钾、硫代锑酸钾与硫酸钾的混合物〕
hepar calcis 钙肝〔硫化钾和多硫化钾的混合物〕
heparin 肝素
heparinase 肝素酶
heparin sodium 肝素钠
hepar sulfuric reaction 硫肝反应
hepatic sulfur reaction 硫肝反应
hepatocuprein 肝铜蛋白；超氧物歧化酶
HEPES-buffered salt solution HEPES 缓冲盐溶液
Hepex process 〔词源:heavy element partitioning by extraction〕海派克斯过程〔用酸性膦酸酯萃取分离重元素(超铀元素)的过程〕
heplulose 庚酮糖
heptabasic alcohol 七元醇
heptacene 并七苯
heptachlor 七氯
heptachlor-1-naphthol 七氯-1-萘酚
heptachlorodiethyl ether 七氯二乙醚
heptachloropropane 七氯丙烷 $Cl_3CCCl_2CHCl_2$
heptacontane (正)七十(碳)烷
heptacosane 二十七(碳)烷
heptacyclic compound 七环化合物
heptacyclic ring 七核环
heptad ①七价物②七价的
heptadecadienoic acid 十七碳二烯酸 $C_{16}H_{29}COOH$
heptadecandioic acid (=heptadecane diacid) 十七烷二酸 $C_{15}H_{30}(COOH)_2$
heptadecane (正)十七(碳)烷
heptadecane diacid (=heptadecandioic acid) 十七烷二酸 $C_{15}H_{30}(COOH)_2$
heptadecane dicarboxylic acid 十七烷二羧酸 $C_{17}H_{34}(COOH)_2$
heptadecanoic acid 十七(烷)酸 $C_{16}H_{33}COOH$
9-heptadecanol 9-十七(烷)醇 $[CH_3(CH_2)_7]_2CHOH$
heptadecanol 十七(烷)醇 $C_{17}H_{35}OH$
heptadecanone 十七烷酮 $C_{17}H_{34}O$
heptadecanoyl 十七(烷)酰 $CH_3(CH_2)_{15}CO-$
heptadecendioic acid (=heptadecene diacid) 十七碳烯二酸 $C_{15}H_{28}(COOH)_2$
heptadecene 十七碳烯 $C_{17}H_{34}$
heptadecene diacid (=heptadecendioic acid) 十七碳烯二酸 $C_{15}H_{28}(COOH)_2$
heptadecene dicarboxylic acid 十七碳烯二羧酸 $C_{17}H_{32}(COOH)_2$
heptadecenoic acid 十七碳烯酸 $C_{16}H_{31}COOH$
heptadecoic acid (=heptadecylic acid) 十七(烷)酸 $C_{16}H_{33}COOH$
heptadecy iodide 碘代十七烷 $CH_3(CH_2)_{15}CH_2I$
heptadecyl 十七(烷)基 $CH_3(CH_2)_{15}CH_2-$
heptadecylaldehyde (=margaric aldehyde) 十七(烷)醛 $CH_3(CH_2)_{15}CHO$
heptadecyl-amine 十七(烷)胺 $C_{17}H_{35}NH_2$
heptadecyl bromide 十七基溴 $CH_3(CH_2)_{15}CH_2Br$
heptadecyl chloride 十七基氯 $CH_3(CH_2)_{15}CH_2Cl$
heptadecylic acid (=heptadecoic acid) 十七(烷)酸 $C_{16}H_{33}COOH$
2-heptadecylimidazole 2-十七基咪唑
heptadecyl tetrapropenyl succinic acid 四丙烯基丁二酸

十七烷基酯
heptadecynoic acid 十七碳炔酸 $C_{16}H_{29}COOH$
2,4-heptadienal 2,4-庚二烯醛
heptadiene 庚二烯
heptadienoic acid 庚二烯酸 C_6H_9COOH
heptadiyne 庚二炔
heptafulvalene 庚富瓦烯
heptafulvene 庚富烯
heptafungin 七烯真菌素
heptahydric alcohol (=heptahydroxy alcohol) 七元醇
heptahydroxy alcohol (=heptahydric alcohol) 七元醇
γ-heptalactone γ-庚内酯
heptaldehyde (=enanthal;enanthic aldehyde) 庚醛 $CH_3(CH_2)_5CHO$
heptaldehyde oxime (=heptaldoxime) 庚肟；庚醛肟 $CH_3(CH_2)_5CH=NOH$
heptaldoxime (=heptaldehyde oxime) 庚肟；庚醛肟 $CH_3(CH_2)_5CH=NOH$
heptalene 庚间三烯并庚间三烯；庚搭烯
heptalenium 并庚间三烯离子
heptaline 甲基环己醇
heptamer 七聚物
heptamethylene ①环庚烷②七亚甲基
heptamethylene diamine 庚二胺；七亚甲基二胺
heptamethylene glycol 庚二醇；七亚甲基二醇
heptamide 庚酰胺 $CH_3(CH_2)_5CONH_2$
heptamycin 庚霉素
heptanal 庚醛
heptanamido (=hexanecarboxamido) 庚酰氨基 $CH_3(CH_2)_5CONH-$
heptanaphthenic acid 庚环烷酸；环己甲酸
heptandioic acid (=heptane diacid) 庚二酸 $C_5H_{10}(COOH)_2$
heptandiol 庚二醇
heptane 庚烷
heptane diacid (=heptandioic acid) 庚二酸 $C_5H_{10}(COOH)_2$
heptane dicarboxylic acid 壬二酸 $C_7H_{14}(COOH)_2$
heptanediol 庚二醇
2,3-heptanedione 2,3-庚二酮
heptanedioyl 庚二酰 $-CO(CH_2)_5CO-$
heptane number 庚烷值
heptane-octane mixture 庚烷-辛烷混合物
heptaner 庚烷塔
heptane-xylene equivalent 庚烷-二甲苯当量
heptanoate ①庚酸②庚酸盐(或酯)
heptanoic acid (=enanthic acid;heptylic acid) 庚酸 $CH_3(CH_2)_5CO_2H$
heptanoic anhydride (=heptylic anhydride) 庚酸酐 $(C_6H_{13}CO)_2O$
heptanol 庚醇
heptanone 庚酮
2-heptanone (=methyl amyl ketone) 甲基戊基(甲)酮；2-庚酮 $CH_3(CH_2)_4COCH_3$
heptanoyl (=enanthyl) 庚酰 $CH_3(CH_2)_5CO-$
heptanthiol 庚硫醇
heptantriene 庚三烯 C_7H_{10}
heptaoxacycloheneicosane 七氧代环二十一烷；21-冠(醚)-7 $C_{14}O_7H_{28}$
heptaphene 庚芬；蒽并四并苯 $C_{30}H_{18}$
heptaploid 七倍体
heptatomic ①七原子的②七元的
heptatomic ring 七元环
heptavalent 七价的
heptavalent element 七价元素
2-heptenal 2-庚烯醛
4-heptenal diethyl acetal 4-庚烯醛二乙醇缩醛
heptene 庚烯 C_7H_{14}
heptene diacid 庚烯二酸 $C_5H_8(COOH)_2$
heptene dicarboxylic acid 庚烯二羧酸；壬烯二酸 $C_7H_{12}(COOH)_2$
2-heptenic aldehyde 2-庚烯醛
heptenoic acid 庚烯酸 $C_6H_{11}COOH$
1-hepten-4-one 1-庚烯-4-酮 $C_3H_7COCH_2CH=CH_2$
heptenone 庚烯酮
heptenyl 庚烯基
heptenyl methyl carbonate 碳酸庚烯甲酯
heptet 七重峰
heptine 庚炔 C_7H_{12}
α-heptine carboxylic acid α-庚炔羧酸
heptitol 庚糖醇
heptoic acid (=heptylic acid) 庚酸 $CH_3(CH_2)_5COOH$
heptonic acid 庚糖酸 $CH_2OH(CHOH)_5COOH$
heptonitrile (=hexyl cyanide) 己基氰；庚腈 $CH_3(CH_2)_5CN$
heptose 庚糖
heptoxide 七氧化物
heptoximate 庚肟盐
heptoxime (=1,2-cycloheptaenedionedioxime) 1,2-环庚二酮二肟 $C_7H_{10}(NOH)_2$
heptulose 庚酮糖
hepturonic acid 庚糖醛酸 $OHC(CHOH)_5COOH$
n-heptyl (正)庚基
heptyl 庚基 $CH_3(CH_2)_5CH_2-$
heptyl acetate 乙酸庚酯 $CH_3CO_2C_7H_{15}$
heptyl alcohol 庚醇

heptyl aldehyde　庚醛　$C_6H_{13}CHO$
heptyl amine　庚胺　$C_7H_{15}NH_2$
heptylate　①庚酸②庚酸盐(酯或根)
heptyl benzene　庚基苯　$C_7H_{15}C_6H_5$
heptyl bromide (=1-bromoheptane)　1-溴庚烷　$CH_3(CH_2)_5CH_2Br$
γ-heptyl butyrolactone　γ-庚基丁内酯
heptyl chloride (=1-chloroheptane)　1-氯庚烷　$CH_3(CH_2)_5CH_2Cl$
n-heptyl cinnamate　肉桂酸庚酯
n-heptyl dimethylacetate　二甲基乙酸正庚酯
heptylendioic acid　庚烯二酸　$C_5H_8(COOH)_2$
heptylene　庚烯　C_7H_{14}
heptylene diacid　庚烯二酸　$C_5H_8(COOH)_2$
heptylene dicarboxylic acid　壬二酸　$C_7H_{12}(COOH)_2$
heptyl esters　庚酯
heptyl ether　庚醚　$(C_7H_{15})_2O$
heptyl formate　甲酸庚酯　$HCO_2C_7H_{15}$
2-heptylfuran　2-庚基呋喃
heptyl heptylate　庚酸庚酯　$C_6H_{13}CO_2C_7H_{15}$
heptylic acid (=heptanoic acid; enanthic acid)　庚酸　$CH_3(CH_2)_5CO_2H$
heptylic anhydride (=heptanoic anhydride)　庚酸酐　$(C_6H_{13}CO)_2O$
n-heptyl isobutanoate　异丁酸正庚酯
n-heptyl isovalerate　异戊酸正庚酯
heptyl malonic acid　庚基丙二酸　$C_7H_{15}CH(CO_2H)_2$
2-n-heptyl-3-methyl-cyclopentanone　2-庚基-3-甲基环戊酮
heptyl methyl ether　庚基甲醚；甲氧基庚烷　$CH_3O(CH_2)_6CH_3$
5-heptyl-3-methyl-2(3H)-furanone　5-庚基-3-甲基-2(3H)-呋喃酮
n-heptyl 2-methyl propanoate　2-甲基丙酸正庚酯
heptyl nitrite　亚硝酸庚酯　$CH_3(CH_2)_6ONO$
n-heptyl octanoate　辛酸正庚酯
heptyl 3-phenyl propenoate　3-苯基丙烯酸庚酯
heptyl propionate　丙酸庚酯　$C_2H_5CO_2C_7H_{15}$
heptyl-trimethyl-silicane　庚代三甲硅　$(C_7H_{15})Si(CH_3)_3$
heptyndioic acid (=heptyne diacid)　庚炔二酸　$C_5H_6(COOH)_2$
heptyne　庚炔
heptyne diacid (=heptyndioic acid)　庚炔二酸　$C_5H_6(COOH)_2$
heptyne dicarboxylic acid　庚炔二羧酸　$C_7H_{10}(COOH)_2$
heptynoic acid　庚炔酸　C_6H_9COOH
herapathite (=quinine iodosulfate)　碘硫酸奎宁
herbicidal oil　除草油

herbicides　除草剂
herb rubber　草胶；草本橡胶
Hercolt aluminum process　赫科特制铝法
Hercules drop method　海格立斯滴点法〔测熔点〕
Hercules powder process　海格立斯火药公司法〔液态烃催化裂化法〕
Hercules stone (=magnetite)　磁铁矿
herculin　鹤枯灵；棒状花椒酰胺
hercynin(e)　组氨酸三甲基内盐
herderite　磷铍钙石
hereditary functional　追忆性泛函
hereynite　铁尖晶石
Herman's orientation function　赫曼取向因子；赫曼取向函数
hermetic　气密的
hermetic art　炼金术
hermetic centrifuge　密封离心机
hermetic closure　密封；密闭
hermetic seal　①密封；气密封接②密封封接
hermetic separator　密封分离器
Hermex process　赫尔美克斯过程；汞提取法〔用汞萃取净化辐照金属铀或金属废屑的方法〕
Hermite process (for hypochlorite)　赫密特(次氯酸盐制造)法
Hermitian operator　厄米算符
herniarin　7-甲氧(基)香豆素
heroin (=heroine;diacetyl morphine)　海洛因；二乙酰吗啡
heroine (=heroin;diacetyl morphine)　海洛因；二乙酰吗啡
herpestine　核佩斯亭
herpes virus　疱疹病毒
herqueichrysin　郝金青霉素
herquein　郝青霉素
herrerite　铜菱锌矿
herring bone gear (=double helical spur gear)　人字齿轮
herring-bone method　鱼骨(式)法
herring-bone profile　人字花纹〔轮胎〕
herring-bone wrap　人字包缠式
herring oil　鲱鱼油
Herry furnace (for calcium carbide)　赫里(制电石)电炉
Hersch cell detector　赫希池检测器
Herschel demulsibility number　赫谢尔反乳化数值
herschelite　碱菱沸石
hertz (Hz)　赫(兹)〔周/秒〕
Hertzian wave　赫兹波
Herz reaction　赫茨反应
hesion　吸收力
hesperetin　橙皮素；5,7,3'-三羟基-4'-甲氧基黄烷酮　$C_{16}H_{14}O_6$

hesperetin chalcone 查耳酮橙皮素
hesperetol 橙皮酚 $CH_2=CHC_6H_3(OH)OCH_3$
hesperidene (=d-limonene) 橙皮烯；苧烯
hesperidin 橙皮苷
hesperidine 橙皮碱〔前胡中含有〕；前胡碱
hesperitinic acid 橙皮酸 $C_{10}H_{11}O_4$
hessian ①砂坩埚②粗麻屑
hessian-backed wallcovering 粗麻布夹衬的裱墙材料
hessian crucible 砂坩埚
hessite 碲银矿
hessonite 钙铝榴石
Hess's law 赫斯定律*
HET acid (=hexachloro endoethylene tetrahydrophthalic acid) 氯桥酸
hetaryne 杂芳炔；脱氢杂环
hetarylation 杂芳化作用
heteric-block nonionics 混嵌(共聚)非离子表面活性剂
hetero〔希腊字头〕 杂；不同
heteroannular diene 异环二烯
heteroantagonism 异型拮抗作用
heteroaromatic compound 芳香杂环化合物
heteroaromatics 杂芳族化合物
heteroatom 杂原子
heteroatomic 杂原子的
heteroatomic ring 杂(原子)环
heteroauxin (=3-indoleacetic acid) 异植物生长素；3-吲哚基乙酸
heteroazeotrope 非均相共沸混合物
heterobaric 异(原子)量的
heterobaric heterotope 异量异序(元)素；异量异位元素
heterobetulin 杂桦木脑
heterobonding ①杂键(合)②杂黏合
heterocatenary polymer 杂链聚合物
heterocedasticity 非方差齐性
heterochain (=heterogeneous chain) 杂链
heterochain fibre 杂链纤维
heterochain polymer 杂链聚合物*；杂链高分子
heterochelate 混合配位体螯合物
heterochinine (=heteroquinine) 杂奎宁
heterochirality 异手性
heterochromatic light 多色光
heterochromosome 异染色体
heterochrosis 变色
heterocoagulation 杂凝聚
heterocomplex 杂合物
heteroconjugate 复共轭对配合物
hetero conjugation 复共轭
heterocumulenes 杂累接双键化合物系〔如 $S=C=N—$,
$=C=C=N—,O=C=N—,—N=C=N—$等结构〕
heterocycle 杂环
heterocycleamide 杂环酰胺
heterocycle-amide fibre 杂环(聚)酰胺纤维
heterocycle-containing polyamide 含杂环(的)聚酰胺
heterocycle polymer 杂环聚合物
heterocyclic 杂环的
heterocyclic atom 杂环原子
heterocyclic bridged analogue 杂环桥联同系物
heterocyclic chemistry 杂环化学
heterocyclic compound 杂环化合物*
heterocyclic diamine 杂环二胺
heterocyclic ladder polymer 杂环梯形聚合物
heterocyclic nitrogen compound 杂环氮化(合)物
heterocyclic nucleus 杂环核
heterocyclic oxonol 带杂环基团的氧杂菁
heterocyclic polyamine 杂环多(聚)胺
heterocyclic polymer 杂环高分子*
heterocyclic radical 杂环基
heterocyclic ring 杂环
heterocyclics 杂环族化合物
heterocyclic spiran 杂环螺烷
heterocyclic stem nucleus 杂环母核
heterocyclic tetraacid 杂环四酸
heterocyclization 杂环化(反应)
heterocyclo addition 杂环加成
heterocyclyl ethylation 杂环乙基化(作用)
heterodesmic 杂键(的)
heterodesmic structure 杂键结构
heterodisperse 杂分散(的)；外分散
heterodispersity 杂分散性；外分散性
heterodynamics 非均相动力学
heterodyne 外差法
heterodyne method 外差法
heteroenoid system 杂烯系
hetero epitaxy 异质外延
heteroexcimer 异激态原子
hetero fiber 异质复合纤维
hetero-filament 异质丝
heterofilm 异质薄膜
heterofil yarn 异芯丝包圈纱
heteroflocculation 杂絮凝；不均匀絮凝
heterofunctional condensation 杂官能缩合
heterogel 杂凝胶
heterogeneity 不均匀性；多相性
heterogeneity index 不均匀性指数；多分散性指数
heterogeneity of liquid film 液膜不均匀性

heterogeneity test 多相性测定〔沥青物质的〕	heteromorphism 多晶型现象
heterogeneous 不均匀的；多相的	heteronuclear ①杂环的②杂核；异核
heterogeneous azeotrope 非均匀共沸混合物	heteronuclear chemical shift correlated spectroscopy 异核化学位移相关谱
heterogeneous catalysis 多相催化*	
heterogeneous catalysis action 多相催化作用	heteronuclear correlation spectroscopy via long range coupling 远程偶合异核相关谱
heterogeneous catalyst 多相催化剂	
heterogeneous chain compound 杂链化合物	heteronuclear decoupling 异核去偶*
heterogeneous chain polymer 杂链聚合物	heteronuclear diatomics 异核双原子分子
heterogeneous electrode 非均相电极	heteronuclear double resonance 异核双共振
heterogeneous enzyme immunoassay 非均相酶免疫分析	heteronuclear lock 异核联锁
heterogeneous equilibrium 多相平衡*	heteronuclear Overhauser effect spectroscopy 异核欧沃豪斯效应相关谱
heterogeneous grafting 多相接枝	
heterogeneous hydrogenation 多相氢化*	heteronuclear relayed coherence transfer spectroscopy 异核接力相关谱
heterogeneous index 不均匀指数	
heterogeneous ion-exchange membrane 非均相离子交换膜	heteronuclear substitution 杂环取代反应
	heteronuclear total correlation spectroscopy 异核总相关谱
heterogeneous laminate ①非均质层压材料②非均质层合材料	
	heteronucleus 杂环核
heterogeneous material 不均匀材料；多相材料	heterooligosaccharide 杂寡糖
heterogeneous membrane 非均态交换膜	heterophase chemiluminescence 异相化学发光
heterogeneous membrane electrode 非均相膜电极；多相晶膜电极	heterophase polymerization 多相聚合
	heterophylline (=aricine) 夹竹桃碱
heterogeneous nitration 多相硝化作用	heterophyllolite 杂晶片岩；杂粗晶纤维状片岩
heterogeneous phase polymerization 非均相聚合	heteropolar 异极的；有极的
heterogeneous polyblend 非均相共混物	heteropolar binding 异极键联
heterogeneous polymerization 非均相聚合*	heteropolar bond 异极键
heterogeneous reactor 非均相(反应)堆	heteropolar compound 异极化合物
heterogeneous ring compound 杂环化合物	heteropolarity 异极性
heterogeneous sample 多相试样；不均匀试样	heteropolar link (=heteropolar linkage) 异极键
heterogeneous slurry polymerization 非均相悬浮聚合(作用)	heteropolar linkage (=heteropolar link) 异极键
	heteropolar liquid 异极性液体
heterogeneous state 非均匀态	heteropolar valence 异极化合价
heterogeneous system 非均相系统*	heteropolyacid 杂多酸*
heterogeneous vulcanization 不均匀硫化*	heteropolyanion 杂多阳离子
heterogenetic 多相的；不均匀的	heteropolybase 杂多碱*
heterogenicity 异质性；不均一性	heteropoly blue 杂多蓝
heterogenity factor 非均相因子	heteropolycondensation 杂缩聚作用；杂缩聚反应
heterogenize 多相化；非均一化	heteropolymer 杂聚物
heteroglycan 杂聚糖；杂多糖	heteropolymerization 杂聚合(作用)
heterograph ink 复印墨	heteropolymer latex 杂聚物胶乳
heterohexosan 异聚己糖	heteropolymolybdate 杂多钼酸盐
heteroion (混)杂离子	heteropoly molybdic acid 杂多钼酸
heterolactic acid bacteria 杂异乳酸菌	heteropolynuclear coordination compound 杂多核配合物*
heterolipid 杂脂	
heterolupeol 杂羽扇醇 $C_{30}H_{50}O$	heteropolyoxometallate 杂多金属氧酸盐*
heterometric titration 比浊滴定	heteropolyoxyalkylene chain (混)杂聚氧化烯链
heterometry 比浊滴定法	heteropolysaccharidase 杂多糖酶
heteromorphic (=polymorphic) 多晶(型)的	heteropolysaccharide 杂多糖

heteroproteose 杂朊
heteroquinine (=heterochinine) 杂奎宁
heterosaccharide 异糖化物
heterosegment 杂(原子)链段
heteroside ①葡萄(糖)苷②苷配糖物
heterosteric 异(型空间)配(位)的
heterosubstituted compound (混)杂取代化合物
heterosugar 异糖
heterotactic 非均态有规立构
heterotactic placement 非均态有规立构排列
heterotactic polymer 杂同立构聚合物*
heterotactic unit 杂同(立构)单元
heterotope ①异(原子)序元素②(同量)异序(元)素
heterotopic 异位(的)*
heterotopic atom 异序原子
heterotrophic bacteria 异养细菌
heteroxanthine 7-甲(基)黄嘌呤
heteroyarn 异质复合丝；异质复丝
heterozygosis 杂合；异型接合
heterpolar attachment energy 异极附着能
hetoform 肉桂酸铋〔别名〕
hetol 肉桂酸钠
heulandite 片沸石
Heumann indigo synthesis 霍伊曼靛合成法
heuristic method 启发式方法
hevea 三叶橡胶
hevea latex 天然胶乳
hevea rubber 天然橡胶
heveene 黑非因〔天然胶的分解蒸馏产物〕
hevein 橡胶蛋白
hevelian halo 淡晕
hewettite 针钒钙石
hex 六氟化铀〔俗〕
hexa〔希腊字头〕 六
hexaalkylated 六烃基化的
hexa-atomic ①六元的②六价的
hexabasic ①六(碱)价的；六元的②六价的
hexabasic acid 六元酸；六(碱)价酸；六价酸
hexabasic alcohol 六元醇；六羟基醇
hexabasic carboxylic acid 六元羧酸
hexabasic ester 六元酸酯
hexabasic salt 六代盐；六碱盐
hexaborane 六硼烷 B_6H_{10}
hexabromated 六溴化的
hexabromide 六溴化(合)物
hexabromide method 六溴化合物法〔油和脂肪酸检验法之一〕
hexabromide number (=hexabromide value) 六溴值

hexabromide test 六溴化(合)物试验
hexabromide value (=hexabromide number) 六溴值
hexabrominated (=hexabromizated) 六溴化的
hexabromo- 六溴(代)
hexabromo-benzene 六溴(代)苯 C_6Br_6
hexabromo-biphenyl 六溴联苯
hexabromo-cyclododecane 六溴环十二烷〔阻燃剂〕
hexabromo-cyclohexane 六溴环己烷；六溴化苯 $C_6H_6Br_6$
hexabromoethane 六溴乙烷 Br_3CCBr_3
hexabromophenol 六溴代苯酚 C_6Br_6O
hexabutoxymethyl melamine 六丁氧甲基三聚氰胺
hexabutylphosphoric triamide (HBPT) 六丁基磷酸三酰胺 $[(C_4H_9)_2N]_3PO$
hexacarbonylmolybdenum 六羰基钼 $Mo(CO)_6$
hexacarboxylic acid 六羧酸
hexacene 并六苯
hexachlorethane 六氯乙烷 Cl_3CCCl_3
hexachloride 六氯化合物
hexachlorinated (=hexachlorizated) 六氯化的
hexachloro 六氯(代)
hexachloroacetone 六氯丙酮
hexachloroamericate 六氯镅酸盐 M_3AmCl_6
hexachloroanthracene 六氯蒽 $C_{14}H_4Cl_6$
hexachlorobenzene 六氯(代)苯 C_6Cl_6
hexachlorobutadiene 六氯丁二烯
hexachloro-cyclohexane 六六六；六氯化苯
hexachlorocyclopentadiene 六氯环戊二烯〔阻燃剂〕
hexachlorodiethyl sulfide 硫化六氯双乙烷
hexachloroendomethylene tetrahydrophthalic acid 氯菌酸；六氯代降冰片烯二酸
hexachloroendomethylene-tetrahydrophthalic anhydride (=chlorendic anhydride) 氯菌酸酐
hexachloroethane 六氯乙烷 Cl_3CCCl_3
hexachloroethane mixture 六氯乙烷混合剂
hexachloronaphthalene 六氯(代)萘 $C_{10}H_2Cl_6$
hexachlorophenol 六氯苯酚 C_6OCl_6
hexacontane 六十(碳)烷 $CH_3(CH_2)_{58}CH_3$
hexacontyl 六十(烷)基 $CH_3(CH_2)_{58}CH_2—$
hexacoordinated aluminum-hydroxy 六配位的铝-羟基
hexacosandiacid 二十六烷二酸 $C_{24}H_{48}(COOH)_2$
hexacosandiendioic acid 二十六碳二烯二酸 $C_{24}H_{44}(COOH)_2$
hexacosandienoic acid 二十六碳二烯酸 $C_{25}H_{47}COOH$
hexacosandioic acid 二十六烷二酸 $C_{24}H_{48}(COOH)_2$
hexacosane (正)二十六(碳)烷
hexacosane dicarboxylic acid 二十六烷二羧酸 $C_{26}H_{52}(COOH)_2$

hexacosanic acid 二十六(烷)酸 $C_{25}H_{51}COOH$
hexacosanol 二十六(烷)醇
hexacosendiacid (=hexacosendioic acid) 二十六碳烯二酸；二十六烯双酸 $C_{24}H_{46}(COOH)_2$
hexacosendioic acid (=hexacosendiacid) 二十六碳烯二酸
hexacosene dicarboxylic acid 二十六碳烯二羧酸 $C_{26}H_{50}(COOH)_2$
hexacosenoic acid 二十六碳烯酸 $C_{25}H_{49}COOH$
hexacosoic acid 二十六烷酸 $C_{25}H_{51}COOH$
hexacosyl 二十六(烷)基 $CH_3(CH_2)_{24}CH_2—$
hexacosyl alcohol 二十六(烷)醇
hexacyanobutadiene 六氰基丁二烯
hexacyanogen 六聚氰 C_6N_6
hexacyclic compound 六环化合物
hexacyclic ring 六核环
hexad ①六价物②六价的③六面的④一列六个
hexadecadienedioic acid 十六碳二烯二酸 $C_{14}H_{24}(COOH)_2$
hexadecadienoic acid 十六碳二烯酸 $C_{15}H_{27}COOH$
hexadecandioic acid (=hexadecane diacid) 十六烷二酸 $C_{14}H_{28}(COOH)_2$
hexadecane (=bioctyl) 联辛基；十六(碳)烷
hexadecane diacid (=hexadecandioic acid) 十六烷二酸 $C_{14}H_{28}(COOH)_2$
hexadecane dicarboxylic acid 十六烷二羧酸 $C_{16}H_{32}(COOH)_2$
hexadecan octaene 十六碳八烯
hexadecanoic acid 十六(烷)酸 $C_{15}H_{31}COOH$
hexadecanol 十六(烷)醇
hexadecanoyl 十六(烷)酰 $CH_3(CH_2)_{14}CO—$
hexadecatrienoic acid 十六碳三烯酸
hexadecendioic acid (=hexadecene diacid) 十六碳烯二酸 $C_{14}H_{26}(COOH)_2$
hexadecene 十六碳烯
hexadecene diacid (=hexadecendioic acid) 十六碳烯二酸
hexadecene dicarboxylic acid 十六碳烯二羧酸 $C_{16}H_{30}(COOH)_2$
ω-7-hexadecenlactone ω-7-十六烯酸内酯
hexadecenoic acid 十六碳烯酸 $C_{15}H_{29}COOH$
9-hexadecenoic acid (=palmitoleic acid) 棕榈油酸；鳖酸；9-十六(碳)烯酸
hexadecine 十六碳炔 $C_{16}H_{30}$
hexadecoic acid 十六(烷)酸 $C_{15}H_{31}COOH$
hexadecyl 十六(烷)基 $CH_3(CH_2)_{14}CH_2—$
N-hexadecylacetoacetamide 十六烷基乙酰基乙酰胺
hexadecylamine-linoleic acid complex 十六胺亚油酸络合物〔腐蚀抑制剂〕
hexadecyldimethyl benzyl ammonium 十六烷基二甲基苄基铵(离子) $[C_{16}H_{33}(CH_3)_2NCH_2C_6H_5]^+$
hexadecylendioic acid (=hexadecylene diacid) 十六碳烯二酸
hexadecylene 十六(碳)烯
hexadecylene diacid (=hexadecylendioic acid) 十六碳烯二酸 $C_{14}H_{26}(COOH)_2$
hexadecylene dicarboxylic acid 十六碳烯二羧酸 $C_{16}H_{30}(COOH)_2$
hexadecylenic acid 十六碳烯酸 $C_{15}H_{29}COOH$
hexadecylic acid 十六(烷)酸 $C_{15}H_{31}COOH$
α-n-hexadecyl-α-propyl-β-aminopropionic acid α-正十六烷基-α-丙基-β-氨基丙酸
hexadecyl trichlorosilane 十六烷基三氯硅烷
hexadecyl trimethyl ammonium bromide 溴化十六烷基三甲铵
hexadecyl trimethyl ammonium derivative 十六碳烷基三甲铵衍生物
hexadecyndioic acid (=hexadecyne diacid) 十六碳炔二酸
hexadecyne 十六(碳)炔 $C_{16}H_{30}$
hexadecyne diacid (=hexadecyndioic acid) 十六碳炔二酸 $C_{14}H_{24}(COOH)_2$
hexadecyne dicarboxylic acid 十六碳炔二羧酸 $C_{16}H_{28}(COOH)_2$
hexadecynoic acid 十六碳炔酸 $C_{15}H_{27}COOH$
7-hexadecynoic acid (=palmitolic acid) 棕榈炔酸；7-十六(碳)炔酸 $CH_3(CH_2)_7C≡C(CH_2)_5COOH$
hexadentate ligand 六配位体；六配位基
hexadiendioic acid (=hexadiene diacid) 己二烯二酸
1,5-hexadiene (=diallyl) 1,5-己二烯 $(CH_2=CHCH_2)_2$
2,4-hexadiene (=dipropenyl) 2,4-己二烯；联丙烯基
hexadiene diacid (=hexadiendioic acid) 己二烯二酸 $C_4H_4(COOH)_2$
hexadienedioic acid 己二烯二酸
hexadienic acid (=hexadienoic acid) 己二烯酸 C_5H_7COOH
2,4-hexadienoate 山梨酸酯
hexadienoic acid (=hexadienic acid) 己二烯酸
hexadiindiol 己二炔二醇
hexadiine 己二炔
hexadiyne 己二炔
hexaethylated 六乙基(化)的
hexaethylbenzene 六乙基苯 $(C_2H_5)_6C_6$
hexaethyldilead 六乙基二铅 $Pb_2(C_2H_5)_6$
hexaethyl disilazine 六乙基二(甲)硅氮烷
hexaethyl disiloxane (=triethyl-silicon oxide) 双氧化(三乙基)硅 $[(C_2H_5)_3Si]_2O$
hexaethylditin (=triethyltin) 三乙基锡；六乙基二锡

[(C₂H₅)₃Sn]₂ → $[(C_2H_5)_3Sn]_2$

hexaethylene glycol　六甘醇　$HOCH_2(CH_2OCH_2)_5CH_2OH$
hexaethyltetraphosphate (HETP)　四磷酸六乙酯　$(C_2H_5O)_6P_4O_7$
hexafluorated (=hexafluorinated; hexafluorizated)　六氟化的
hexafluorinated (=hexafluorated)　六氟化的
hexafluorizated (=hexafluorated)　六氟化的
hexafluoro　六氟(代)
hexafluoroacetone　六氟丙酮
hexafluorobenzene　六氟(代)苯　C_6F_6
hexafluoroethane　六氟乙烷　C_2F_6
hexafluoroisopropanol　六氟异丙醇〔聚酯纤维溶剂〕
hexafluoroisopropyl alcohol　六氟代异丙醇
hexafluorophosphoric acid　氟磷酸；六氟合磷氢酸
hexafluoropropane-2,2-diol　六氟丙烷-2,2-二醇
hexafluoropropylene　六氟丙烯
hexafluoropropylene oxide　六氟环氧丙烷
hexafluoropropylene trimer　六氟丙烯三聚物
hexafluorozirconate　六氟锆酸盐
hexagon　六方形；六角形
hexagonal　六方形的
hexagonal array pulse field gel electrophoresis　六方阵列脉冲电场凝胶电泳
hexagonal close-packed lattice　六方密集格子
hexagonal close-packed structure　六方密装结构
hexagonal close packing　六方密堆积*
hexagonal crystal　六方晶
hexagonal crystal system　六方晶系
hexagonal lattice　六方点格*
hexagonal mesomorphic form　六方介晶型
hexagonal nut　六角螺母
hexagonal plate　六方形片状体
hexagonal platelet　六方形片晶
hexagonal recess　(蒙脱石晶体结构中的)六角凹穴
hexagonal system　六方晶系*
hexahalogenated　六卤代的
hexahalogenated benzene　六卤代苯　C_6X_6
hexahedron　六面体
hexahelicene　六螺烯
hexahydrate　六水合物
hexahydric　六羟的
hexahydric acid　六元酸
hexahydric alcohol　六元醇；六羟基醇
hexahydro　六氢化
hexahydroanthracene　六氢化蒽　$C_{14}H_{16}$
hexahydrobenzaldehyde　六氢化苯(甲)醛；环己基甲醛　$CH_2(CH_2)_4CHCHO$
hexahydrobenzene　六氢化苯；环己烷　C_6H_{12}
hexahydrobenzoic acid　六氢化苯甲酸；环己基甲酸　$CH_2(CH_2)_4CHCOOH$
hexahydrobenzyl alcohol (=cyclohexylcarbinol)　六氢化苯甲醇；环己基甲醇　$C_6H_{11}CH_2OH$
hexahydrobilin　六氢后胆色素
hexa-hydrocannabinol (=pyrahexy)　六氢大麻醇
hexahydrocresol　甲基环己醇；六氢化甲酚
hexahydrocumene (=isopropylcyclohexane)　六氢化枯烯；异丙基环己烷　$C_3H_7C_6H_{11}$
hexahydro-mellitic acid　六氢化苯六甲酸　$C_6H_6(COOH)_6$
hexahydro-mesitylene　六氢化莱；六氢化均三甲苯　$C_6H_9(CH_3)_3$
hexahydronaphthalene　六氢化萘　$C_{10}H_{14}$
hexahydrophenol　环己醇；六氢化苯酚
hexahydrophenol acetate　乙酸环己酯
hexahydro-phthalic acid　六氢化邻苯二甲酸；环己烷邻二甲酸　$C_6H_{10}(COOH)_2$
hexahydrophthalic anhydride　六氢邻苯二甲酸酐
hexahydroretene　六氢惹烯
hexahydrosalicylic acid　六氢化水杨酸　$HOC_6H_{10}COOH$
hexahydroterephthalate　六氢化对苯二甲酸酯
hexahydroterephthalic acid　六氢化对苯二甲酸　$C_6H_{10}(COOH)_2$
hexahydroxy　六羟基
hexahydroxy acid　六羟基酸
hexahydroxy alcohol　六羟基醇
hexahydroxybenzene　六羟基苯；苯六酚　$C_6(OH)_6$
hexahydroxy dibasic acid　六羟基二元酸
hexahydroxy diphenic acid　六羟基联苯甲酸
hexahydroxy monobasic acid　六羟基一元酸
hexahydroxy stearic acid　六羟基硬脂酸
hexaiodated (=hexaiodizated)　六碘化的
hexaiodizated (=hexaiodated)　六碘化的
hexaiodo　六碘(代)
hexaiodobenzene　六碘代苯　C_6I_6
hexakis〔词头〕　六个
hexakis-methoxy methyl melamine　六甲氧甲基三聚氰胺；六(个)甲氧甲基三聚氰胺
γ-hexalactone (=tonkalide)　γ-己内酯；香豆内酯
hexalin (=cyclohexanol)　环己醇
hexalin acetate　乙酸环己酯
hexamer　六聚物
hexametapol　六甲基磷酰三胺　$[(CH_3)_2N]_3PO$
hexamethonium bromide　六甲溴铵
hexamethoxy diphenic acid　六甲氧基联苯甲酸

hexamethoxymethyl melamine 六甲氧甲基三聚氰胺
hexamethylacetone (=pivalic ketone) 六甲基丙酮 $(CH_3)_3CCOC(CH_3)_3$
hexamethylated 六甲基化的
hexamethyl-aurine 六甲金精
hexamethyl-benzene 六甲基苯 $C_6(CH_3)_6$
hexamethyldisilane (HMDS) 六甲基二硅烷*
hexamethyldisiloxane (HMDSO) 六甲基二硅醚*
hexamethylenamine 六亚甲基四胺；乌洛托品 $(CH_2)_6N_4$
hexamethylene ①1,6-亚己基；六亚甲基 —$CH_2(CH_2)_4CH_2$—②环己烷
hexamethyleneadipamide unit 己二酰己二胺单元〔耐纶66的结构单元〕
hexamethylene diadipamide 双己二单酰己二胺；六次甲基二酰胺
hexamethylene diamine 六亚甲基二胺；1,6-己二胺 $NH_2(CH_2)_6NH_2$
hexamethylene diamine bis(phenylphosphonate) 己二胺双苯基膦酸盐
hexamethylene diamine tetraacetic acid (HMDTA) 亚己基二胺四乙酸
hexamethylene diammonium azelate 壬二酸己二胺盐；锦纶69盐；尼龙69盐
hexamethylene dicyanide 辛二腈 $NC(CH_2)_6CN$
hexamethylene-diisocyanate 1,6-己二异氰酸酯 $OCN(CH_2)_6NCO$
hexamethylene dinitrilo-tetraacetic acid 己二胺四乙酸；亚己基二次氮基四乙酸 $(HOOCCH_2)_2N(CH_2)_6N(CH_2COOH)_2$
hexamethylene formal 甲醛缩己二醇
hexamethylene-glycol (=1,6-hexandiol) 1,6-己二醇 $HO(CH_2)_6OH$
hexamethylene imine 六亚甲基亚胺
hexamethylene isophthalamide unit 间苯二甲酰己二胺(结构)单元
hexamethylene-tetramine (=urotropine) 六亚甲基四胺；乌洛托品 $(CH_2)_6N_4$
hexamethylene triamine 六亚甲基三胺
hexamethylolmelamine (HMM) 六羟甲基三聚氰胺
hexamethylpararosaniline 结晶紫；龙胆紫
hexamethylpararosaniline hydrochloride (=crystal violet) 六甲碱性副品红盐酸盐〔即结晶紫〕
hexamethylphosphoramide (HMPA) 六甲基磷酰胺 $[(CH_3)_2N]_3PO$
hexamethyl phosphoric triamide 六甲基磷酰三胺
hexamethyl tetracosahexene 六甲基二十四碳六烯；角鲨烯 $C_{30}H_{50}$

hexamethyl tetracosane (=squalane) 六甲基二十四烷；异三十烷；角鲨烷
hexamine (=hexamethylenetetramine) 六胺；乌洛托品
hexammine 六氨合物
hexamminecobalt (III) chloride 氯化六氨合高钴
hexammine cobaltrichloride 三氯化六氨合钴 $[Co(NH_3)_6]Cl_3$
hexammine platinic chloride 四氯化六氨铂 $Pt[(NH_3)_6]Cl_4$
hexamycin 己霉素
hexanal 己醛
hexanamide 己酰胺
hexandial 己二醛
1,6-hexandial (=adipic aldehyde) 己二醛 $OHCCH_2CH_2CH_2CH_2CHO$
hexandioic acid (=hexane diacid) 己二酸
hexandiol 己二醇
1,6-hexandiol (=hexamethylene-glycol) 1,6-己二醇
hexandione 己二酮
hexane (正)己烷
hexaneate ①己酸②己酸盐(酯或根)
hexanecarboxamido (=heptanamido) 庚酰氨基
hexane diacid (=hexandioic acid) 己二酸 $C_4H_8(COOH)_2$
hexanedial 己二醛
hexanediamide 己二酰二胺
hexanediamine 己二胺
hexane dicarboxylic acid 辛二酸；己烷二羧酸 $C_6H_{12}(COOH)_2$
hexanediol 己二醇
1,6-hexanediol diacrylate 1,6-己二醇二丙烯酸酯〔硫化剂〕
1,6-hexanediol diglycidylether 1,6-己二醇二缩水甘油醚
1,6-hexanediol dimethacrylate 二甲基丙烯酸1,6-己二酯
2,3-hexanedione 2,3-己二酮
hexanedione 己二酮
hexanedioyl 己二酰 —$CO(CH_2)_4CO$—
hexanenitrile 己腈
hexanepentol 己五醇
hexanethiol 己硫醇
1,2,6-hexanetriol 1,2,6-己三醇
hexane upgrading 己烷提级；己烷改性
hexan-hexol 己六醇 $CH_2OH(CHOH)_4CH_2OH$
hexanitrato complex 六硝酸根配合物
hexanitrin 己六醇六硝酸酯
hexanitro 六硝基
hexanitrodiphenyl 六硝基联苯
hexanitrodiphenylamine (=dipicrylamine) 二苦胺；六硝基二苯胺 $[(NO_2)_3C_6H_2]_2NH$

hexanitrodiphenyl ester 六硝基二苯酯
hexanitrodiphenyl sulfide (=dipicryl sulfide) 二苦硫；六硝基二苯硫 $[(NO_2)_3C_6H_2]_2S$
hexanitrodiphenyl sulfone 六硝基二苯砜
hexanitroethane 六硝基乙烷 $(NO_2)_3CC(NO_2)_3$
hexanitro-mannite 六硝酸甘露醇酯
hexanitro-mannitol (=mannitol hexanitrate) 六硝基甘露醇；甘露醇六硝酸酯 $(O_2NOCH_2)_2[CH(ONO_2)]_4$
hexanitro-oxanilide 六硝基乙二酰苯胺
n-hexanoate 正己酸酯
hexanoic acid (=caproic acid) 己酸
hexanol 己醇〔通常指: 1-己醇〕
hexanolactam 己内酰胺
1,4-hexanolide 1,4-己内酯
hexanone (=oxohexane) 氧代己烷；己酮
2-hexanone (=methyl butyl ketone) 甲基丁基(甲)酮；2-己酮 $CH_3COC_4H_9$
3-hexanone 3-己酮
hexanoyl 己酰 $CH_3(CH_2)_4CO—$
2-hexanoyl furan 2-己酰基呋喃
hexaoctyltributylene tetraphosphine oxide (HOTB-TPO) 六辛基三亚丁基四氧膦
hexaoxacyclooctadecane 六氧杂环十八烷；18-冠(醚)-6 $C_{12}O_6H_{24}$
hexaoxadiamine macrobicyclic complex 六氧二胺大双环配合物
hexaoxadiaza bicyclohexacosane 六氧二氮杂双环十八烷 $C_{18}H_{36}N_2O_6$
hexaphene 己芬；萘并四并苯 $C_{26}H_{16}$
hexaphenol resin 六(苯)酚树脂
hexaphenoxy tungsten 六苯氧基钨
hexaphenyl 六苯基
hexaphenylethane 六苯乙烷
hexaphyranose 吡喃己糖；六环己糖
hexaploid 六倍体
hexapropyldisiloxane 六丙基二硅氧烷；三丙基乙硅醚
hexaric acid 己糖二酸
hexasaccharide 多聚己糖
hexasilane 己硅烷 Si_6H_{14}
hexasubstitution product 六取代产物
hexatetrahedron 六四面体
hexathionic acid 连六硫酸 $H_2S_6O_6$
hexatomic ①六元的②六价的
hexatomic acid 六价酸
hexatomic alcohol 六元醇；六羟基醇
hexatomic base 六价碱
hexatomic ring 六元环
hexatriacontane 三十六(碳)烷 $C_{36}H_{74}$

hexatriene 己三烯
hexavalence 六价
hexavalent 六价的；六元的
hexavalent alcohol 六元醇；六羟基醇
hexavalent chromium 六价铬
hexavanadate 六聚钒酸盐 $M_4[V_6O_{17}]$
hexenal 己烯醛
hexendioic acid (=hexene diacid) 己烯二酸
hexene 己烯
hexene diacid (=hexendioic acid) 己烯二酸 $C_4H_6(COOH)_2$
hexene dicarboxylic acid 己烯二羧酸；辛烯二酸 $C_6H_{10}(COOH)_2$
4-hexene-3-one 4-己烯-3-酮
hexenic (=hexenoic) 己烯的
α,β-hexenic acid α,β-己烯酸
hexenic acid 己烯酸 C_5H_9COOH
hexenoic (=hexenic) 己烯的
hexenoic acid 己烯酸
hexenoic aldehyde 己烯醛
hexenol 己烯醇
5-hexen-2-ol 5-己烯-2-醇 $CH_2=CH(CH_2)_2CHOHCH_3$
hexenone 己烯酮
hexenoyl 己烯酰
hexenyl 己烯基 $C_6H_{11}—$
4-hexenyl alcohol 4-己烯醇
hexestrol (=hexoestrol) 己雌酚
hexethal sodium 己妥钠；己基巴比妥钠
hexides 己糖二酐〔脱二水己六醇类〕
hexin 己菌素
hexine (=hexyne) 己炔
hexinic acid (=hexinoic acid) 己炔酸 C_5H_7COOH
hexinoic acid (=hexinic acid) 己炔酸 C_5H_7COOH
hexiology 生物环境学
hexitan 己糖醇酐〔脱一水己六醇〕
hexitol 己糖醇 $CH_2OH(CHOH)_4CH_2OH$
n-hexoate 正己酸酯
hexobarbital soluble 可溶性环己烯巴比妥〔药〕
hexobarbitone sodium 环己烯巴比妥钠〔药〕
hexobiose 己二糖
hexoestrol (=hexestrol) 己雌酚
hexogen (=trimethylene trinitramine) 六素精；三亚甲基三硝基胺
hexoic acid 己酸 $C_5H_{11}COOH$
hexokinase 己糖激酶
hexon(e) 异己酮；2-异己酮；甲基异丁基(甲)酮 $(CH_3)_2CHCH_2COCH_3$
hexone base 六碳碱

Hexone-23 process 己酮-23 过程〔从辐照钍中提取铀-233，用甲基异丁基酮作萃取剂〕
Hexone-25 process 己酮-25 过程〔从辐照高浓铀中回收纯化铀-235，用甲基异丁基酮作萃取剂〕
hexonic acid 己糖酸 $CH_2OH(CHOH)_4COOH$
hexonit 海宋炸药；RDX-NG-PETN 混合药
hexopentosan 己戊聚糖
hexosamine 己糖胺；氨基己糖
hexosan 己聚糖
hexosazone 己糖脎
hexose 己糖 $C_6H_{12}O_6$
hexosediphosphatase 己糖二磷酸(酯)酶
hexosediphosphate 己糖二磷酸；二磷酸己糖
hexosediphosphoric acid 己糖二磷酸 $C_6H_{10}O_4(OPO_3H_2)_2$
hexose gallic acid ester 己糖棓酸酯
hexoseisomerase 己糖异构酶
hexosemonophosphate 己糖磷酸；磷酸己糖
hexosemonophosphate pathway 磷酸己糖途径
hexose monophosphate shunt 磷酸己糖支路；磷酸己糖途径
hexose monophosphoric acid 己糖一磷酸 $C_6H_{11}O_5OPO_3H_2$
hexose phosphate 磷酸己糖
hexosephosphate dehydrogenase 磷酸己糖脱氢酶
hexosidase 己糖酶
hexotriose 己三糖
hexoxide 六氧化物
hexoylene 2-己炔
hexulose 己酮糖
hexuronic acid 己糖醛酸
hexyl ①己基 $CH_3(CH_2)_4CH_2-$ ②(=hexanitrodiphenylamine) 海西尔；六硝基二苯胺；六硝炸药
hexyl acetate 乙酸己酯 $CH_3CO_2C_6H_{13}$
2-hexyl-4-acetoxy tetrahydrofuran 2-己基-4-乙酰氧基四氢呋喃
n-hexyl acetylene (=1-octyne) 正己基乙炔；1-辛炔 $CH_3(CH_2)_5C\equiv CH$
3-hexyl acrolein 3-己基丙烯醛
n-hexyl alcohol 正己醇
n-hexyl aldehyde 正己醛
hexyl amine 己胺 $CH_3(CH_2)_5NH_2$
hexyl benzene 己基苯 $C_6H_{11}C_6H_5$
hexyl bromide (=1-bromohexane) 己基溴；1-溴己烷 $CH_3(CH_2)_4CH_2Br$
n-hexyl butyrate 丁酸己酯
γ-n-hexyl-γ-butyrolactone γ正己基-γ丁内酯
hexyl chloride (=1-chlorohexane) 己基氯；1-氯己烷 $CH_3(CH_2)_4CH_2Cl$
α-n-hexyl cinnamic aldehyde α-己基(肉)桂醛
hexyl cyanide (=heptonitrile) 己基氰；庚腈 $CH_3(CH_2)_5CN$
2-n-hexyl cyclopentanone 2-己基环戊酮
2-n-hexyl-Δ^2-cyclopentenone 2-己基环戊烯-2-酮；二氢异茉莉酮
hexyldecyldiisocyanate 十六烷二异氰酸酯
n-hexyl-n-decyl phthalate 邻苯二甲酸正己正癸酯〔增塑剂〕
hexyl-(2-dioctylphosphinylethane)-phosphinic acid (HDOPEP) 己基-(2-二辛基膦酰乙基)次膦酸 $(C_6H_{13})[(C_8H_{17})_2POC_2H_4]POOH$
hexylene 己烯 C_6H_{12}
α,β-hexylene aldehyde α,β-己烯醛；叶醛
hexylene glycol 己二醇
α,β-hexylenic acid α,β-己烯酸
hexyl ether 己醚；二己基醚 $C_6H_{13}OC_6H_{13}$
hexyl formate 甲酸己酯 $HCO_2C_6H_{13}$
2-hexyl-4-hydroxy-1,3-dioxan 2-己基-4-羟基-1,3-二噁烷
hexylic acid 己酸 $C_5H_{11}COOH$
hexylidene 亚己基 $CH_3(CH_2)_4CH=$
hexylidyne 次己基 $CH_3(CH_2)_4C\equiv$
hexyl iodide (=1-iodohexane) 己基碘；1-碘己烷 $CH_3(CH_2)_4CH_2I$
n-hexyl isovalerate 异戊酸己酯
hexyl mercaptan 己硫醇 $CH_3(CH_2)_4CH_2SH$
hexyl methacrylate 甲基丙烯酸己酯
hexyl 2-methylbutyrate 2-甲基丁酸己酯
hexyl nitrite 亚硝酸己酯 $C_6H_{13}ONO$
n-hexyl-n-octyl-n-decyl phthalate 邻苯二甲酸正己正辛正癸酯〔增塑剂〕
hexyl oleate 油酸己酯
hexyl phenyl carbinol (=α-phenyl-n-heptanol) 己基苯基甲醇；α-苯基庚醇 $C_6H_{13}CHOHC_6H_5$
N-hexyl-N'-phenyl-p-phenylenediamine N-己基-N'-苯基对苯二胺〔防老剂〕
n-hexyl propionate 丙酸己酯
hexyl resorcin 己基间苯二酚 $C_6H_{13}C_6H_3(OH)_2$
hexyl-triethyl-silicane 己基三乙基硅 $C_6H_{13}Si(C_2H_5)_3$
n-hexyl trimelitate 偏苯三酸正己酯〔增塑剂〕
hexyl-trimethyl-silicane 己基三甲基硅 $C_6H_{13}Si(CH_3)_3$
n-hexyl valerate 戊酸己酯
hexyndioic acid (=hexyne dacid) 己炔二酸 $C_4H_4(COOH)_2$
hexyne (=hexine) 己炔
hexyne diacid (=hexyndioic acid) 己炔二酸 $C_4H_4(COOH)_2$

hexyne dicarboxylic acid 己炔二羧酸；辛炔二酸 $C_6H_8(COOH)_2$
hexynic acid (=hexynoic acid) 己炔酸 C_5H_7COOH
hexynoic acid (=hexynic acid) 己炔酸 C_5H_7COOH
hexynol 己炔醇
Hey reaction (=Gomberg-Bachmann-Hey synthesis) 黑氏合成法
H-film H膜*
HF process 氢氟酸法〔在水法热铀处理过程中加入氟氢酸合锆以改进对锆等裂变产物净化的过程〕
HF-vapor etching 氟化氢蒸气腐蚀
hialomycin 透明霉素
hibar ①高压②高压的
hibar pre-packed column 高压预填柱
hibbsite 三水矾土；三水合氧化铝 $Al_2O_3 \cdot 3H_2O$
hibschite 水榴石
hibulking 高膨体(变形)；高膨体变形(工艺)
hickory nut oil 山核桃油
hickory oil 山核桃油〔取自 Carya〕
hidden flaws 暗伤；暗疵
hiddenite 翠绿锂辉石
hidden layer 隐含层
hide ①生皮②塑料板坯
hide cellar 生皮窖
hide fiber 皮纤维
hide finishes 皮革涂饰剂
hide glue 皮胶〔皮革〕
hide house 生皮仓库
hide inspection bureau 生皮检验局
hide-marking hammer 生皮打号锤
hide powder 皮粉
hide powder method 皮粉方法
hides〔复〕 皮革
hide sider 生皮对剖器
hide sorter 生皮挑选工
hide stamping machine 生皮打印机
hide substance 皮革物质
hiding chart (=hiding power chart) 遮盖力试验纸；黑白格纸
hiding economy 遮盖经济性〔遮盖面积/元〕；遮盖力经济学
hiding extender 遮盖性填料；盖底力填料
hiding lack 遮盖力不足
hiding paint ①盖底漆②隐蔽色涂料
hiding pigment 盖底颜料
hiding power 遮盖力；覆盖力；盖底力；被覆力；遮盖本领
hiding power chart (=hiding chart) 遮盖力试验纸；黑白格纸
hiding power efficiency 遮盖力效率
hiding power value 遮盖(力)值
hidrotic 发汗药
hielmite 钙铌钽矿
hierachical chromatography response function 串行色谱响应函数
hierachical clustering method 系统聚类法
hierarchial cluster analysis 系统聚类分析；谱系聚类分析
hierarchial design 分层设计
hierarchical self-assembly 等级自集
hiflash 高闪点油
hi-flash naphtha (=high-flash naphtha) 高闪点石脑油
hi-flash solvent 高闪点溶剂
Higbie theory 希格比理论
Higgins contactor 希金斯接触器〔一种动态离子交换柱〕
Higgins ion contactor 希金斯离子交换柱
high ①高的②高级的
high abrasion furnace black (HAF) 高耐磨炉黑
high abrasion goods 耐磨制品
high abundance sensitivity mass spectrometer 高丰度灵敏度质谱计
high activity electrode 高活性电极
high added acid concentration 高浓度的酸添加剂
high additive oil 高添加剂含量的油品
high affinity 高亲和能(力)
high air pollution potential 高空气污染潜势
high alkalinity additive 高碱性添加剂
high altitude LP gas 高丁烷含量的液化石油气体
high alumina 富矾土的
high alumina cement 高铝水泥
high alumina clay 高铝黏土
high alumina refractory 高铝耐火材料
high alumina shale 高铝页岩
high amide frequency 高酰胺键出现率
high analysis fertilizer 高成分肥料
high antiknock fuel 高抗爆燃料
high antiknock rating base fuel 高抗爆性基础油
high assay 高指标样品
high back pressure 高背压
high bake 高温烘烤
high beam ①高光束；强光②上方光束
high bloomery furnace 高温炼铁炉
high boiler 高沸点化合物
high boiling 高沸点的
high boiling component (精馏的)高沸点组分

high-boiling fraction 高沸馏分
high-boiling hydrocarbons 高沸点烃类
high-boiling petroleum products 高沸点石油产品
high boiling point 高沸点
high boiling solvent 高沸点溶剂
high branched chain 高支化链
high build 高成膜性；成厚涂层性；高固体性
high bulk fiber 高膨体纤维
high bulking 高膨体化变形(工艺)
high calcium lime 高钙石灰
high capacity tire 高负荷轮胎
high-carbon 高碳的
high-carbon alloy steel 高碳合金钢
high-carbon graphite 高碳石墨
high-carbon residue oil 高碳残渣油
high carbon steel 高碳钢
high carbon tar 高碳焦油
high-cetane fuel 高十六烷值燃料
high-clarity ①高透明性②高透明度
high colour channel black 高色素槽法炭黑
high compressed steam 高压蒸汽
high concentration chloroprene rubber latex 高浓度氯丁胶乳
high conductance valve 高导通阀
high conductivity 高电导率
high-consistency 高稠度
high consistency rotational viscosimeter 高黏度用黏度计
high-consistency viscometer 高稠度黏度计
high-count fabric 精细织物；高支纱织物
high-coverage (铝粉)高覆盖能力
high crosslink density 高交联密度
high-cycle efficiency 工作周期的高生产率
high cycle fatigue 高周疲劳
high degree of accuracy 高精密度
high degree of dispersion 高分散度
high density 高密度；高浓度
high density bleacher 高密度漂白器
high density gas chromatography 高密度气相色谱法
high density glass 高密度玻璃珠（研磨介质）
high-density linear polyethylene 高密度线型聚乙烯
high density lipoprotein (HDL) 高密度脂蛋白
high-density matter 高密度物质
high-density overlay 高密度复层胶合板
high density polyethylene (HDPE) 高密度聚乙烯*
high-density polyethylene plastics (HDPE) 高密度聚乙烯塑料
high-destruction thermal cracking 高度破坏的热裂化
high dilution principle 高度稀释原理

high dispersivity 高分散性
high-distillation thermometer 高温蒸馏用温度计
high dropping point grease 高滴点润滑脂
high-durability 高耐久性
high durometer rubber 高硬度橡胶
high duty (=heavy duty) 重型；载重
high-duty boiler 高压锅炉
high-duty fuel 高辛烷值燃料；高功率燃料
high duty gas producer 重压煤气发生器
high early strength cement 高快固水泥
high-early-strength portland cement 早强硅酸盐水泥
high efficiency 高效率；高生产率
high efficiency liquid chromatography 高效液相色谱(法)
high efficiency open-width water mangle 高效平幅轧水机
high efficiency packing tower 高效填料塔
high efficiency particulate air filter 高效空气微粒过滤器
high efficiency silver nitrate liquid chromatography 高效硝酸银液相色谱法
high efficient column 高效柱
high elastic deformation 高弹形变
high elasticity 高弹性
high elastic rubber 高弹性橡胶
high electrolyte 强(性)电解质
high elongation 高相对伸长
high elongation furnace black 高伸长炉黑
high end point 高终沸点
high energy adsorption site 高能吸附点
high energy aviation fuel 高能航空燃料
high energy ball mill 高能球磨法
high energy battery 高能电池(组)
high energy chemistry 高能化学
high energy collision 高能碰撞
high energy electron 高能电子
high energy electron diffraction (HEED) 高能电子衍射*
high energy fragmentation 高能碎裂
high energy ion scattering spectroscopy 高能离子散射谱；高能离子散射能谱法
high energy ion scattering spectrum 高能离子散射谱
high energy irradiation 高能辐照〔例如用紫外线辐照〕
high-energy mill 高能磨
high energy particle 高能粒子
high-energy phosphate bond 高能磷酸键
high energy photo-electron spectroscopy 高能光电子能谱学
high-energy radiation 高能辐射
high energy radiotherapy 高能放射治疗
high energy spallation 高能散裂
high energy throughput 高能通量

high-enriched uranium (HEU)　高浓(缩度)铀
higher　①较高的②高级的；较高级的
higher acetylenes stripper　高炔属汽提塔
higher actinide　重锕系(元素)
higher alcohols　高碳醇；高级醇
higher alkyl sulfate　高烷基硫酸酯
higher aromatics　高级芳烃
higher-boiling compound　高沸点化合物
higher chloride　高级氯化物
higher derivatives　高级(化学)衍生物
higher fatty acid　高级脂肪酸
higher fatty alcohol sulfate　高级脂肪醇的硫酸酯
higher harmonic AC polarography　多阶谐波交流极谱法*
higher homolog　高级同系物
higher homologue　高级同系物
higher hydrocarbon　高级烃
higher member　高级物
higher member of the series　同系物的高级物
higher order compound　高级化合物
higher order reaction　高级次反应
higher order structure　高次结构；高序结构
higher polymer　高聚物
higher preorientation　高预取向
higher-way classification　多种方式分组
highest attainable vacuum　极限真空
highest occupied molecular orbital (HOMO)　最高占据(分子)轨道*
highest practical molecular weight　最高实用分子量
highest useful compression ratio (HUCR)　最高有效压缩率
highest vacuum　高度真空
high expansion material　高增容材料
high explosive　高级炸药
high fidelity　高真实性
high-field superconducting magnet material　高磁场超导磁体材料
high filler loading capacity　高填充量
high film build　①厚膜②高成膜性
high finish　高度光泽；非常光泽
high fired　高温焙烧过的
high-fixed carbon hydrocarbons　分子中碳原子数多的烃
high-flash fuel　高闪点燃料
high-flash naphtha (=hi-flash naphtha)　高闪点石脑油
high-flash oil　高闪点油
high flash point　高闪点
high-flash solvent　高闪点溶剂
high flux beam reactor (HFBR)　高通量(中子)束堆

high flux isotope reactor (HFIR)　高通量同位素(生产)堆
high-foamers (=high foaming detergent)　高泡洗涤剂
high foaming　高泡的
high foaming detergent (=high-foamers)　高泡洗涤剂
high free rosin size　高游离松香胶
high frequency alternator　高频振荡(发生)器
high frequency analysis　高频分析
high frequency bonding　高频胶接
high-frequency concentration meter　高频浓度计
high frequency conditioning　高频调湿
high frequency conductance　高频电导
high frequency conductometric titration (=high frequency conductometry)　高频电导滴定(法)*
high frequency conductometry (=high frequency conductometric titration)　高频电导滴定(法)*
high frequency current　高频(率)电流
high frequency dielectric constant　高频电容率；高频介电常数
high frequency discharge　高频放电*
high frequency electron spin resonance　高频电子自旋共振
high frequency gluing　高频胶合
high frequency hardening　高频熔接
high frequency heating　高频加热
high-frequency induction furnace　高频电感(加热)炉
high frequency induction plasma spraying　高频等离子喷涂
high frequency induction spraying　高频喷涂
high frequency method　高频法
high frequency moisture meter　高频水分计
high frequency noise　高频噪声
high frequency paramagnetism　高频顺磁(性)
high frequency plasma　高频等离子体
high frequency plasma torch　高频等离子体炬
high frequency polarography　高频极谱法
high frequency pyrolysis　高频热解
high frequency seal machine　高频熔接机
high frequency spark ion source　高频火花离子源*
high frequency spark method　高频火花法
high frequency spark source　高频火花光源
high frequency spectrum　高频(率)光谱
high frequency titration (=oscillometric titration)　高频滴定
high frequency titrator　高频滴定计
high frequency titrimeter　高频滴定计
high frequency torch　高频(火)炬
high-frequency vulcanization　高频(率)硫化
high frequency welding　高频焊接

high-gloss black 特(级)亮炭黑；高光泽黑
high gloss coating 高光泽涂层
high gloss finish 高度光泽；非常光泽
high-grade 高级的；优等的；高纯(试剂)
high-grade cast iron 高级铸铁
high-grade fuel 高级燃料
high-grade lacquer 高级硝基漆
high-grade rubber 高等级橡胶
high-grade sugar 高级糖
high-gravity filler 重质填料
high-gravity fuel 高比重燃料
high-gravity oil 高比重油品
high gum compound 高含胶率胶料
high heating power (=high heating value) 高热值
high heating rate DTA 高升温速率差热分析(器)
high heating value (=high heating power) 高热值
high heating value gas 高热值煤气
high-heat-resistant aluminum coating 耐高温铝粉涂料
high-hiding color 高被覆力着色剂
high-impact (HI) 高抗冲(的)
high impact copolymer 高弹态共聚物
high impact material 耐冲击材料
high impact plastic 耐冲塑料
high impact polyethylene 耐冲击性聚乙烯
high impact polystyrene (HIPS) 高抗冲聚苯乙烯*
high impact polystyrene rigidity 高冲击性聚苯乙烯
high impact rayon 高冲击强度人造丝
high in aromatics 富芳烃的
high input impedance 高输入阻抗
high intensity hollow cathode lamp 高亮度空心阴极灯
high intensity lamp 高亮度灯
high iron oxide type electrode 氧化铁型焊条
high iron portland 高铁波特兰水泥
high iron portland cement 高铁波特兰水泥
high-knock rating gasoline 高辛烷值汽油
high-lateral order 高侧序
high leg boot 长靴
high level cave 高放(射性)热室
high level production 高水平生产
high level switch 高液位开关
high-level waste 高放废物*
high lift pump 高压泵
high-lightness 高亮度；辉亮
high limit of tolerance 容许上限；上限公差
high-load ①高载荷(的)②高含量(的)
high-loaded column 高负荷柱
high loading 高填充(剂用)量
high-load melt index 高负荷熔体流动指数

high-loft 高膨松度；高膨体
high-low temperature cycles test 高低温交变试验
highly acid 高酸的；强酸性的
highly acid slag 高酸渣
highly aggressive 高侵蚀性
highly alkyl-substituted aromatics 多烃基取代了的芳族化合物
highly angled twill 急斜纹
highly basic 高碱的；强碱性的
highly basic slag 高碱渣
highly branched 多支的；多枝的
highly branched chain 多支链
highly branched compound 多支链化合物
highly combustible 高度易燃的；高度易燃品
highly compressed 高压的
highly compressed steam 高压蒸汽
highly concentrated hydrogen peroxide 高度浓缩的过氧化氢
highly cracked gasoline 高度裂化汽油
highly detergent oil 高(度)清净(分散)性(能)的润滑油
highly dilution method 高度稀释法
highly efficient regeneration 高效再生；完全再生；高度再生
highly elastic material 高弹性材料
highly hydrogenated 高度氢化的
highly inflammable 高易燃的
highly olefinic gasoline 富(含)烯烃汽油
highly oxidized 高度氧化的
highly pigmented film 高颜料分漆膜
highly polymerized compound 高聚合度化合物
highly refined oil 高度精制油
highly repetitive sequence 高度重复序列
highly stripped heavy ion 高度剥离的重离子；深剥重离子
highly twisted rayon 多捻螺萦
highly volatile 易挥发的
high magnesium lime 高镁石灰
high melting 高熔点的；难熔的
high melting glass 难熔玻璃
high melting glass fiber 高熔点玻璃纤维
high melting glaze 高温釉
high melting grease 高熔点润滑脂
high melting metal 难熔金属
high-melting plastics 高熔点塑料
high methane gas 富甲烷气
high-modification-ratio 高改性比；高变性比
high-modulus 高模量(的)
high-modulus and low-shrinkage fiber 高模低缩纤维

high modulus furnace black　高模数炉黑
high modulus glass fiber　M 玻璃纤维；高模量玻璃纤维
high modulus polymer　高模量聚合物*
high modulus yarn warping machine　高模量纱线整经机
high molecular　高分子的
high molecular compound　高分子化合物
high molecular polymer　高聚物；高分子聚合物
high molecular type surfactant　高分子(型)表面活性剂
high molecular waste　高分子废物
high molecular weight (HMW)　①高分子量②高分子量的
high molecular weight amine　高分子胺
high molecular weight compound　高分子(量)化合物
high-molecular-weight hydrocarbon　高分子量烃
high-molecular weight polymer　高分子量聚合物
high molecule　高分子；大分子
high-nitrogen stock　高氮原料
high nucleation density　高成核密度
high octane　高辛烷值
high octane compound　高辛烷值化合物
high-octane fuel　高辛烷值燃料
high-octane gasoline　高辛烷值汽油
high-octane motor fuel　高辛烷值发动机燃料
high octane number component　高辛烷值组分
high octane rating　高辛烷值
high-ohmic input resistance　高欧姆输入电阻
high-ohmic resistance　高欧姆电阻
high-oil-absorption pigment　高吸油量颜料
high oiliness　高油性；好润滑性
high overpotential electrode　高过电位电极
high-performance　优越性能；高性能；高效的
high performance capillary electrophoresis　高效毛细管电泳
high performance column　高效柱
high performance composite　高性能复合材料
high performance detector　高性能检测器
high performance electrophoresis chromatograph　高效电泳色谱仪
high-performance fuel　高热值燃料；优质燃料
high performance ion exchange chromatography　高效离子交换色谱法
high performance liquid chromatograph　高效液相色谱仪
high performance liquid chromatography (HPLC)　高效液相色谱法*
high performance polymer　高性能高分子*
high-performance random card　高性能无规梳棉机；高性能无规梳理机
high performance resin　高性能树脂
high performance thin layer chromatography　高效薄层色谱法
high-performance vehicle　高性能漆料
high pick fabric　高纬密织物
high pigment loading　(体质)颜料高填充量
high-pitch cone roof　高倾度圆锥形顶盖〔油罐〕
high-plasticity mixer　高塑性混合器
high point petroleum　高沸点石油
high polar region　高极性区
high polish　高度光泽
high polymer　高聚物*
high polymer chemistry　高分子化学
high polymeric compound　高聚物
high polymeric polyelectrolyte　高聚物电解质
high-polymer-technology　高分子工艺学；高聚物工艺学
high-porosity polymer material　高孔隙率聚合物材料
high potential condenser　高压容电器
high potential gradient　高电位梯度；高电位差
high-potential quinone　高效醌；高能醌
high-pour (test) oil　高倾点油
high-precision calender　高精密度压延机
high precision isotope ratio mass spectrometer　高精度同位素比质谱计
high pressure　高压
high-pressure absorber　高压吸收塔
high-pressure alkylation　高压烷基化；高压烃化
high pressure area　高压区域
high pressure capillary　高压毛细管
high pressure capillary rheometer　高压毛细管流变仪
high-pressure catalytic unit　高压催化设备
high pressure cell　高压池
high-pressure charging　高压加料
high pressure chemical ionization source　高压化学电离源
high pressure chemistry　高压化学
high-pressure coal hydrogenation　煤的高压加氢
high pressure column　高压柱
high pressure compressor　高压压缩机
high pressure crystallography　高压晶体学*
high pressure diamond cell　高压金刚石池
high pressure differential thermal analysis　高压差热分析
high pressure evaporation tower　高压蒸发塔
high pressure flash　高压闪蒸
high pressure float valve　高压浮球阀
high pressure flow cell technique　高压流通池技术
high pressure Fourier transform infrared spectroscopy　傅里叶变换高压红外光谱法
high pressure gas chromatography　高压气相色谱法
high pressure gradient　高压梯度

high-pressure grease 高压润滑脂
high pressure hydraulic airless spray 高(压)液压无空气喷涂
high-pressure hydrogenation 高压氢化；高压加氢
high pressure ion exchange (HPIX) 高压离子交换
high pressure ion source 高压离子源
high pressure jet cleaning 高压喷射清洗
high-pressure jet cutting technique 高压喷射切割技术
high pressure laminate 高压层积材
high pressure liquid chromatography 高压液相色谱(法)
high pressure lubricant 高压润滑剂
high pressure mass spectrometry 高压质谱分析
high pressure mercury arc lamp 高压汞弧(光)灯
high pressure nitric acid process 高压法硝酸工艺
high-pressure oil 高压油
high pressure phase 高压相*
high-pressure point of oil wedge 油楔入的最高压力点
high pressure polyethylene (HPPE) 高压聚乙烯
high pressure pump 高压泵
high pressure pumping system 高压泵送系统
high pressure reactor 高压反应器
high pressure safety cut-out 高压安全切断器
high pressure seal 高压密封；高压封口
high pressure separator 高压分离器
high-pressure side ①高压侧②高压旋管〔冷冻机中〕
high pressure solvent pump 高压溶剂泵
high pressure spectroscopy 高压光谱学
high pressure spectrum 高压光谱
high-pressure spinning 高压纺丝
high pressure storage 高压储存
high pressure syringe 高压注射器
high pressure X-ray 高压 X 射线
high productive chromatography 高产色谱(法)
high protrusion barb 高突刺
high pulp 高糖渣
high pumping value 高泵送值；高泵汲值
high purity carbonaceous material 高纯度碳质材料
high-purity cellulose 高纯度纤维素
high purity chemicals 高纯度药品
high purity-grade 高纯度级
high purity substance 高纯度物质
high purity terephthalic acid 高纯度对苯二甲酸
high-quality burning oil 优质煤油
high quality cast iron 优质铸铁
high-rate activated-sludge process 加速活性污泥法；加速曝气法
high rate of inflammation 高着火速度
high rate of oxidation 高氧化速度

high-rate-of shear viscosimeter 高剪速黏度计
high-rate trickling filter 高速淋水滤床
high ratio reduction gear 高速比减速机
high-reading thermometer 高温(用)温度计
high reflux ratio 高回流比
high reliability 高可靠度
high-relief wallcovering 高浮雕型糊墙材料
high repetition rate source 高重复率光源
high resilience 高回弹力
high resilient rubber 高弹性橡胶
high resistance alloy 高电阻合金
high resistance pyrometer 高阻高温计
high resolution ①高分辨②高分辨率
high resolution chromatographic separation 高分辨色谱分离
high resolution column 高分辨柱
high resolution data processing 高分辨数据处理
high resolution electron microscope 高分辨电子显微镜
high resolution electron microscopy (HREM) 高分辨电镜*
high resolution gas chromatography 高分辨气相色谱法
high resolution grating 高分辨光栅
high resolution image 高分辨像
high resolution interferometer 高分辨干涉仪
high resolution liquid chromatography 高分辨液相色谱(法)
high resolution mass spectrometer 高分辨质谱仪(计)
high resolution mass spectrometry 高分辨质谱法
high resolution mass spectrum 高分辨质谱
high resolution matrix 高分辨矩阵(点阵)
high resolution NMR spectra 高分辨率核磁共振谱
high resolution spectroscopy 高分辨光谱法
high rubber mixing 高橡胶混合(法)
high selective sensitivity 高选择性灵敏度
high sensitive gas analysis 高灵敏度气体分析
high sensitivity detector 高灵敏度检测器
high sensitivity galvanometer 敏感检流计
high separating efficiency 高分离效能
high shear 高剪切(力)
high shear force 高剪切力；高切力
high shear grinding 高剪切研磨；高切力研磨
high-shear mixer 高剪切混合器
high shear rate 高剪切速率；高切速
high shear viscometry 高剪切测黏法
high side roller mill 高边轮碾机
high silica glass fiber 高硅氧玻璃纤维
high silica zeolites 高硅沸石
high silicon cast iron 高硅铸铁
high silicon iron 高硅铁
high-slip polystyrene 高滑爽性聚苯乙烯

high solid　高固体分
high solid alkyd-melamine system　高固体分醇酸-三聚氰胺(涂料)体系
high solid lacquer　高固体分喷漆
high solid latex　高固物胶乳
high-solid level　高固体分
high solid prepolymer　高固体(分)预聚物
high solvating plasticizer　高溶剂化增塑剂；高熔合性增塑剂
high-solvency naphtha　高溶解力石脑油
high-solvent-content vehicle　高溶剂含量漆料
high specific activity　高比放射性
high speed　高速的
high speed analysis　高速分析
high speed centrifugation　高速离心
high speed centrifuge　高速离心机
high speed chain-type mercerizing range　高速布铗丝光联合机
high speed chromatography　快速色谱法
high speed cooking　快速蒸煮
high speed counter current chromatograph　高速反流色谱仪
high speed counter current chromatography　高速逆流色谱法；高速反流色谱法
high speed cure　快速硫化
high speed cutting　高速切削
high speed detector　高速检测器
high-speed Diesel fuel　高速柴油
high speed digital computer　高速数字计算机
high speed disk impeller　高速轮盘(成)搅拌器
high speed dispersor　高速分散机
high speed dissolver　高速溶解器
high speed evaporator　高速蒸发器
high speed gas chromatography　高速气相色谱(法)
high-speed impingement mill　高速冲击磨
high-speed kneader (=high-speed kneading machine)　高速捏和机
high-speed kneading machine(=high-speed kneader)　高速捏和机
high-speed line mill　高线速磨〔即胶体磨〕
high speed liquid chromatography　高速液相色谱(法)
high speed liquid partitioning column　高速液相色谱分配柱
high-speed lubrication　高速(下)润滑
high speed (membrane) osmometer　高速(薄膜)渗透计
high-speed mixer　高速混合机
high-speed photography　高速照像术
high speed plasma chromatography　高速等离子色谱(法)
high speed reversed-phase liquid chromatography　高速反相液相色谱(法)
high speed scrubber　高速涤气器
high speed spin-drawing　高速纺丝拉伸
high-speed spinning stretching crimping cutting　高速纺丝拉伸卷曲切断一步法
high speed spring　高速弹簧
high speed steel　高速钢
high speed temperature programmed　快速程序升温
high speed turbine stirrer　高速涡轮式搅动器
high-speed vulcanization　快速硫化
high speed wind-up　高速卷绕；高速卷取
high-spin　高自旋
high-spin complex　高自旋配合物
high streaming velocity arc method　高流速弧光法
high-strength　①高强度的②高着色力的
high-strength cast iron　强力铸铁
high strength glass fibre　S玻璃纤维；高强玻璃纤维
high-strength hydrogen peroxide　高浓度过氧化氢
high strength portland　高强度波特兰水泥
high styrene resin　高苯乙烯树脂
high styrene rubber　高苯乙烯橡胶
high-sudsers (=high sudsing detergent)　高泡洗涤剂
high sudsing detergent (=high-sudsers)　高泡洗涤剂
high-sulfur crude　高硫原油
high-sulfur crude oil　高硫原油
high-sulfur fuel　高硫燃料
high-sulfur oil　高硫油
high surface area heat dissipation　高表面热散逸
high-tech　高技术
high-tech fiber　高技术纤维
high temperature　高温(度)
high temperature ageing　高温老化
high temperature and high pressure yarn dyeing machine　高温高压染纱机
high temperature annealing　高温退火
high temperature ashing　高温灰化法
high temperature carbonization　高温干馏；高温炭化
high temperature chemistry　高温化学
high-temperature coatings　高温涂料
high temperature coke　高温焦炭
high temperature composite　高温复合材料
high-temperature conversion coating　高温转化涂层
high temperature cooling　高温冷却；用高沸点液体冷却
high temperature corrosion　高温腐蚀
high temperature decomposition　高温分解
high temperature destructive hydrogenation　高温破坏加氢
high temperature distillation　高温蒸馏
high temperature dyeing　高温染色

high temperature electrode　高温电极
high temperature electrodialysis　高温电渗析
high temperature epoxy (HTE)　高温树脂
high temperature gas chromatography　高温气相色谱(法)
high temperature gel chromatography　高温凝胶色谱法
high temperature grease　高温润滑脂
high temperature heating medium　高温载热体
high temperature-high pressure dyeing machine　高温高压染色机
high temperature hydrocarbon aromatization　(烷)烃的高温芳构化
high temperature hydrogenation　高温加氢
high temperature hygrometer　高温湿度计
high temperature initiator　高温引发剂
high temperature inlet system　高温进样系统
high temperature lacquer　高温漆
high temperature life　高温下使用寿命
high temperature lubricant　高温润滑剂
high temperature mass spectrometry　高温质谱法
high temperature modification　①高温型②高温改进
high temperature overflow dyeing machine　高温溢流染色机
high temperature oxidation　高温氧化
high temperature performance of greases　润滑脂的高温性质
high temperature plasticizer　耐高温增塑剂
high temperature polymerization　高温聚合；热聚合
high temperature-pressure cheese dyeing machine　高温高压筒子染色机
high temperature-pressure continuous open-width steaming range　高温高压平幅连续汽蒸设备
high-temperature pressure dyeing　高温高压染色
high temperature probe　高温探头
high-temperature processing　高温加工
high temperature propane tank　高温丙烷储罐
high temperature pyrolysis deactivation with polysiloxane　聚硅氧烷高温裂解去活
high temperature reflectance spectroscopy　高温反射光谱法
high-temperature resistant varnish　耐高温清漆
high temperature-resistant wire lacquer　耐高温漆包线漆
high temperature sample dyeing machine　高温样品染色机
high temperature scale　高温氧化皮
high temperature service　高温(下)作业
high temperature shale oil　高温页岩油
high temperature silanizing deactivation　高温硅烷化去活
high temperature sludge　高温渣
high temperature solution polycondensation　高温溶液缩聚(作用)
high-temperature spot　高温辉点；高温斑点
high temperature stability　高温稳定性
high temperature-steel　耐高温钢
high temperature superconductivity　高温超导性
high temperature tar　高温焦油
high temperature tempering　高温回火
high temperature thermometer　高温(用)温度计
high temperature varnish　高温漆
high temperature viscosimeter　高温黏度计
high temperature X-ray diffraction analysis　高温X射线衍射分析
high tempering　高温回火
high temp superconductivity　高温超导性
high-tenacity　高强度；高韧度；高强力
high tenacity rayon　高强嫘萦
high tenacity viscose rayon spinning machine　黏胶强力丝纺丝机
high tensile brass　高拉力(强度)黄铜
high tensile steel　高级钢
high tensile wire　高强力钢丝
high tension　高压
high tension cable　高压电缆；高压线
high tension condenser spark method　高压电容火花法
high tension current　高压电流
high tension electrostatic field　高压静电场
high tension spark method　高压火花法
high tension voltage　高(电)压
high tension voltmeter　高压伏特计
high test bleaching powder　高效漂白粉；漂粉精〔商〕
high-test cast iron　高级铸铁
high-test cement　高级水泥
high-test gasoline　高级汽油
high-throughput combinatorial method　HT组合法
high titania type electrode　氧化钛型焊条
high-titer soap　高凝固点皂
high-tonnage plastics　大宗塑料
high-torque extruder　高扭矩挤塑机
high-touch fiber　高触感(性)纤维；触感优良的纤维
high-vacuum (=fine vacuum)　高(度)真空
high vacuum apparatus　高真空设备
high-vacuum desorption　高真空解吸
high vacuum distillation　高真空蒸馏
high-vacuum pump　高度真空泵
high-vacuum pyrolysis　高真空热解
high-vacuum residue　高真空蒸馏残渣
high vacuum system　高真空系统
high-vacuum valve　高真空阀

high velocity combustion　高速燃烧
high velocity electron　快速电子
high velocity steam arc method　高速蒸汽电弧法
high velocity thermal field flow fraction　高速热场流分级
high vinyl polybutadiene rubber　高乙烯基聚丁二烯橡胶
high viscosity　高黏度
high viscosity index lubricating oil　高黏度指数润滑油
high viscosity nitrocellulose　高黏度硝化棉
high viscosity oil　高黏性油
high-visibility paint　高能见度漆；发光漆
high-visibility pigment　高能见度颜料；荧光颜料
high volatile coal　高挥发分煤
high voltage　高(电)压
high voltage alternating current arc　高压交流电弧
high voltage atmospheric ionizer　高压大气电离器
high voltage cable　高压电缆；高压线
high voltage capacitor spark　高压电容火花
high voltage capillary electrophoresis　高压毛细管电泳
high voltage capillary zone electrophoresis　高压毛细管区带电泳
high voltage electrode　高压电极
high voltage electrodialysis　高压电渗析
high voltage electron microscopy　高压电子显微镜检术
high voltage electrophoresis　高压电泳*
high voltage glow-discharge ion source　高压辉光放电离子源*
high voltage paper electrophoresis　高(电)压纸电泳(法)
high voltage paper ionophoresis　高(电)压纸离子电泳法
high voltage rectified spark　高压整流火花
high voltage spark source　高压火花光源
high voltage supply　高压电源
high voltage zone electrophoresis　高(电)压区带电泳(法)
high volume air sampler　高容量采气器
high volume product　大量生产制品
high volume sampler　高容积取样器
high-warp loom　立经式织机
high wet-modulus fiber　高湿模量纤维
high wet modulus rayon　变化型高湿模量黏胶纤维；高湿模量黏胶纤维
high work-factor oil　高工作系数的油
high yield　高产率
hikizimycin　引地霉素
hilamycin (=helamycin;hiramycin)　喜霉素
Hildebrand function　希尔德布兰德函数
Hildebrand value　希尔德布兰德(溶解)值
Hill filter　希尔过滤器
Hill plot　希尔图
Hill reaction　希尔反应

Hill's measuring apparatus of vapor pressure　希尔蒸气压测定仪
hilow bulked yarn　高低膨体混纺纱〔用收缩率不同的纤维混纺〕
α-himachalene　α-雪松烯
d-himenol　d-红香茅醇
himgravine　喜瑞文〔生物碱〕
Hi-mixer　高效混合器；静态混合器〔使用时无声，无振动〕
hindered internal rotation　受阻内转动
hindered isocyanate　受阻异氰酸酯〔加热时能分解成异氰酸酯的化合物〕
hindered motion　受阻运动
hindered phenol (=fortified phenol)　受阻(苯)酚；取代(苯)酚
hindered phenolic antioxidant　受阻碍酚性抗氧剂
hindered rotation　位阻旋转
hindered rotation potential　受阻旋转位能
hindered segmental motion　受阻链段运动
hindered settling　阻滞沉降
hindered settling classification　阻滞沉降分级；干涉沉降分级
hindered settling ratio　受阻沉降比例
hindered shrinkage　受阻收缩
hindered transition　受阻跃迁
hindered translational motion　受阻平移运动
hind leg　后腿
hindrance　阻碍
hinelight　高强度荧光灯
hinesol　茅苍术醇
hinge　铰链〔机工〕
hinge area　铰链区；关节区
hinge characteristics　强制机械特性
hinged copolymer　铰链型共聚物
hinged split mo(u)le　合页式对开模
hinokitiol (=β-thujaplicin)　β-苧侧素；4-异丙基苧酚酮；日扁柏素
Hinsberg method　兴斯堡法
Hinzke's sulfur burner　欣兹克烧硫炉
hiochic acid (=mevalonic acid)　甲瓦龙酸〔俗〕；3,5-二羟基-3-甲基戊酸；袂瓦龙酸
hip oil　枳子油
hippeastrine　小星蒜碱
hippo　吐根
hippocastanum (=horse-chestnut)　马栗树；七叶树
hippopotamus　河马
hippulin　马尿灵；异马烯雌(甾)酮
hippuran　碘马尿酸钠

hippurate 马尿酸盐；苯甲酰氨基乙酸盐 $C_6H_5CONHCH_2COOM$
hippurazide 马尿酰叠氮 $C_6H_5CONHCH_2CON_3$
hippuric acid (=benzoyl-glycine) 马尿酸；苯甲酰氨基乙酸 $C_6H_5CONHCH_2CO_2H$
hippuric acid test 马尿酸实验
hippuroyl (=hippuryl) 马尿酰；苯甲酰氨基乙酰 $C_6H_5CONHCH_2CO—$
hippuryl (=hippuroyl) 马尿酰；苯甲酰氨基乙酰 $C_6H_5CONHCH_2CO—$
hiptagin 狗角藤苷
hirame liver oil 比目鱼肝油
hiramycin (=hilamycin) 喜霉素
Hirschburger continuous bulking system 希施布格尔连续膨化装置
Hirsch funnel 赫尔什漏斗
hirsutic acid 多毛酸
hirsutidin 三甲花翠素；报春色素
hirsutin 三甲花翠苷；报春色素苷
hirudin 水蛭素
hirudo 药用水蛭
hisingerite 硅铁土
histamine 组胺
histamine hydrochloride 盐酸组胺 $C_5H_9N_3·2HCl$
histidinal 组氨醛
histidine 组氨酸 $C_3H_3N_2CH_2CH(NH_2)CO_2H$
histidine dihydrochloride 二盐酸组氨酸 $C_6H_9O_2N_3·2HCl$
histidinol 组氨醇
histidomycin 组氨霉素
histidyl 组氨酰(基) $N_2C_3H_3CHCH_2CO—$
histochemical dyes 组织化学染料
histocompatibility test 组织相容性试验
histophotometry 显微组织分光光度法；显微分光测定法
historadioautography 组织放射自显影法
historadiography 组织放射显影法；组织放射照相法
history of deformation 形变历程；形变史
history of deformation gradient 形变梯度历程
histosol 有机土
hit and miss endways 偶然性的通道
hit and miss (experiment) 尝试试验
Hittorff's method 希托夫法*
hjeavu ends〔复〕 重尾馏分
HLB value HLB 值；亲水-亲油平衡值
HM carbon fiber 高模量碳纤维
HM limed Poly-pale〔商〕 石灰聚合松香
H-number H 值；H 数；Hf 值〔溶剂质量指标之一,定义为每 10^9 升溶剂萃取的铪 Hf 分子数〕
hoangnan 马钱子皮

hob ①切压母模②滚铣③齿轮滚铣
hobbing ①切压制模(法)②滚铣③齿轮滚铣
hobbing machine 滚齿机
hobbing press 切压机
hob-sinking 切压(制模)
Hochst-Wacker process 赫希斯特-瓦克法〔乙烯直接氧化制乙醛〕
hodorine 百部碱 II
hodydamycin 郝迪达霉素
Hoekestra balance plastometer (=balance plastometer) 何氏可塑计
Hoepfner copper process 霍普弗讷炼铜法
Hoepfner nickel process 霍普弗讷炼镍法
Hoepfner zinc process 霍普弗讷炼锌法
Hoesch synthesis 霍西合成法
Hofmann coulometer 霍夫曼电量计
Hofmann degradation 霍夫曼降解
Hofmann isonitrile synthesis (=carbylamine reaction) 霍夫曼异腈合成法
Hofmann-Martius rearrangement 霍夫曼-马修斯重排作用
Hofmann pinch cock 霍夫曼弹簧夹；霍夫曼节流夹
Hofmann reaction 霍夫曼反应
Hofmann's method 霍夫曼方法
Hofmann's rule 霍夫曼规则*
Hofmeister series (=lyotropic series) 感胶离子序
hof-stage ①(显微)熔点测定器②热板
hog 重型多刀粉碎机
hog fat 猪(脂)油
hogged 碎的
hog gum 苹婆胶
hog lard (=adeps suillus) 豚脂
hog machine 回转鱼鳞刀切碎机
hogskin 野猪皮〔指 carpincho 及 peccary〕
hog skins〔复〕 猪皮〔皮革〕
hog still 蒸馏塔
hohlraum 空穴
hohmannite 碱性水铁矾
hoist ①升举器；起动机②卷扬机③起重机④升降机
holarrhenine 止泻木雷宁
holarrhessimine 止泻木西明
holarrhidine 止泻木定
holarrhimine 止泻木明
holarrhine 止泻木碱
hold-back 重馏分所含的轻馏分
holdback carrier 反载体*
hold-down bars 固定棒
hold-down grid 填料压板〔填料塔〕

hold-down support　固定支座
holdenite　红砷锌锰矿
holder　①储蓄器②座③架④夹
holder (of gas)　①储气罐②夹③座；架
holding bin　储料筒
holding capacity　容量；容积
holding device　夹具
holding oxidant　支持氧化剂
holding reductant　支持还原剂
holding tank　收集槽；存储槽
holding time　保留时间；占用时间
hold out　①保持性②不渗透性
hold paint　船舱漆
hold time　停留时间
hold-up　滞留量；塔藏量；停止
hold-up capacity　塔藏量；滞留容量
hold-up tank　接受槽；储存槽
hold-up time　①停留时间②维持时间③停工时间
hold-up volume　滞留体积
hole　空穴*
hole base system　基孔制
hole breakout　孔破
hole bubble　孔穴状气泡
hole capture　空穴俘获
hole current (=positive hole current)　空穴电流
hole diameter　孔径
hole equivalence　空穴等价性
hole forming pin　穿孔针
hole location　孔位
hole plate method　杯碟法
hole-pressure error　孔压误差
hole theory　空穴理论
hole void　孔壁空洞
holey bread　整麦面包
holiday　①空胶；露布；露白〔胶布缺点〕②露胶〔隔离剂涂布不匀〕
holiday detection　漏涂点检查
holiday detector　漏涂点检测仪
holistic approach　整体分析
holland (cloth)　充亚麻棉布
hollander　打浆机；荷兰式打浆机
hollander tub　打浆槽
Hollemann's rules　霍莱曼氏规则
Holley-Mott contactor　霍利-莫特萃取器
hollocellulose　全纤维素
hollow　凹陷
hollow acetate fibre　中空醋酯纤维
holloware (=hollow ware)　空心器皿；凹形器皿

hollow article　中空制品；空心制品
hollow bead　空心颗粒
hollow beam　(喷涂的)空心射束；环状射束
hollow-body dried lacquer　无胎漆器；脱胎漆器
hollow-bowl clarifier (=hollow-bowl centrifuge)　空杯澄清机；空杯离心机
hollow braid　空心带
hollow casting　中空铸型法
hollow cathode　空心阴极
hollow cathode atomizer　空心阴极原子化器
hollow cathode discharge　空心阴极放电
hollow cathode discharge emission spectroscopic analysis　空心阴极放电发射光谱分析
hollow cathode discharge lamp　空心阴极灯；空心阴极放电管
hollow cathode discharge spectroscopic analysis　空心阴极放电光谱分析*
hollow cathode discharge tube (HCDT)　空心阴极放电管；空心阴极灯
hollow cathode duoplasmatron　空心阴极双等离子体源
hollow cathode lamp　空心阴极灯*
hollow cathode plume　空心阴极羽
hollow concrete　空心混凝土
hollow cone　①空心锥体②白度测定器
hollow cored fiber　中空纤维；空心纤维
hollow draw-in spindle　空心内拉轴
hollow drill　空心钻
hollow fiber　中空纤维*
hollow fiber-filter dialysis　空心纤维过滤器透析
hollow fiber membrane cartridge　空心纤维膜滤架
hollow fiber reverse-osmosis modules　中空纤维反渗透率；中空纤维逆渗透系数
hollow fiber suppressor　中空纤维抑制器
hollow fibre gas permeator　中空纤维气体渗透器
hollow-fibre membrane　中空纤维膜
hollow-fibre ultrafiltration device　中空纤维超滤装置
hollow filling (=hollow packing)　空心填充
hollow flatten fibre　中空扁平纤维
hollow float　空心浮球；空心浮标
hollow flocking filament　空心植绒长丝
hollow glass fibre　中空玻璃纤维
hollow needle　空心针(阀)
hollowness　①空心；多孔性②空心度
hollowness of fibre　纤维中空度
hollow optical fibre　空心光纤
hollow packing (=hollow filling)　空心填充
hollow piece　中空制品；空心制品
hollow plywood　空心胶合板

hollow polyester fibre　中空聚酯纤维
hollow prism　空心棱镜
hollow profiled fibre　中空异形纤维
hollow profile tire　高花纹轮胎
hollow pyramids　空心棱锥(晶体)
hollow rayon fibre　中空人造纤维
hollow shelf　空心隔板
hollow space　空洞
hollow sphere　空心球
hollow synthetic fibre　中空合成纤维
hollow tile　空心瓦
hollow ware (=holloware)　空心器皿；凹形器皿
hollow ware glass　空心玻皿
hollyhock-seed oil　蜀葵子油
Holmes Manley cracking process　荷姆斯-曼莱裂化过程
holmia　氧化钬　Ho_2O_3
holmic　钬(的)〔指三价 Ho 原子的化合物〕
holmic oxide　氧化钬
holmium　钬〔67号元素，化学符号 Ho〕
holmium chloride　氯化钬　$HoCl_3$
holmium hydroxide　氢氧化钬　$Ho(OH)_3$
holmium nitrate　硝酸钬　$Ho(NO_3)_3$
holmium oxalate　草酸钬　$Ho_2(C_2O_4)_3$
holmium oxide　氧化钬　Ho_2O_3
holoaxial　全轴
holoblast　全变晶
holocaine (=phenacaine)　哈洛卡因；芬那卡因
holocellulose　全纤维素
holocrystalline　全晶的
hologram image　全息图像
hologram memory fibre　全息记忆纤维
holographic　全息的；全息照相的
holographically ruled grating　全息光栅
holographic diffraction grating　全息衍射光栅
holographic grating　全息光栅*
holographic microscope　全息照相显微镜
holographic microscopy　全息照相显微术
holographic optical element　全息光学元件
holographic recording　全息记录
holography　全息照相术
holohedral　全面的；全对称(晶)形的
holohedral form　全面(晶)形；全对称(晶)形
holohedron　全面体
holohedry　全面像；全晶形
hololens　全息透镜
holometer　测高计
holomorphism　全面形；全对称形态
holomycin　全霉素

holonics　全息学
holopolar form　全极式
holosiderite　全陨铁
holotactic　全规整
holotactic polymer　全规整聚合物
holothin (=desacetyl holomycin)　去乙酰全霉素
holothurin　海参素
holotoxin　全毒素
holotype　主模式株
Holtsmark broadening　霍尔兹马克变宽
holystone　浮石
homatropine　后马托品
home-made　①自制的②国产的
homeokinesis　均等分裂
homeomorphism (=homomorphism)　异质同晶(现象)；同态
homeopathia　顺势医疗论
homeostasis　内稳态*
homeotherms　同热剂〔一种有机磷杀虫剂〕
homeotypic division　同型分裂
home petroleum industry　本国石油工业
home rubber　国产橡胶
homilite　硅硼钙铁矿
homo　高；同(型)
homoallylic alcohol　高烯丙醇*
homoallylic system　高烯丙系
homoannular diene　同环二烯
homoantipyrine　高安替比林
homoarbutin　高熊果苷
homoarecolin　高槟榔碱
homoarginine　高精氨酸
homoaromaticity　同芳香性*
homoatomic chain　同素键；纯键〔同种元素原子间的键〕
homoatomic ring　同素环；同种原子环
homoborneol　高冰片；高莰醇
homocamphor　高樟脑
homocamphoric acid　高樟脑酸〔高指多一个—CH_2—〕$COOHC_8H_{14}CH_2COOH$
homocaryophyllenic acids　高石竹烯酸
homocentric　共心的
homocentric rays　共心射线
homocerebrin　类脑素
homo-chain polymer　均链聚合物；碳链聚合物
homochelidonine　高白屈菜碱
homochemical compound　二(型)原子化合物
homochiral　纯手性(的)*
homochirality　同手型
homochroman　高苯并二氢吡喃

homochromatography 同系色谱法
homochromo-isomer 同色异构体
homochromo-isomerism 同色异构(现象)
homocinchonidine 高辛可尼丁
homocitric acid 高柠檬酸
homocitrullyl-amino-adenosine 高瓜氨酰氨基腺苷
homocoagulation 同粒凝结(作用)
homoconjugate 共轭对配合物；共共聚物
homoconjugation 匀共轭；隔碳共轭
homoconjugation effect 匀共轭效应
homocycle ①碳环②同素环
homocyclic ①碳环的②同素环的
homocyclic compound ①碳环化合物②同素环化合物
homocyclic nucleus ①碳环核②同素环核
homocyclic ring ①碳环②同素环
β-homocyclocitral β-环高柠檬醛
homocysteine 高半胱氨酸；同型半胱氨酸
homocystine 高胱氨酸；同型胱氨酸
homocystinuria 高胱氨酸尿
homodesmic 均键(的)
homodesmic structure 均键结构
homodiene 高二烯
homodimerization 均二聚作用
homodisperse 均匀分散；均相分散
homodyne 零差；自差法
homodyne spectroscopy 均差波谱(学)；均差波谱法
homoenergetic 同能的；匀能的
homoenolate 高烯醇化物
homoentropic ①同熵的；匀熵的②高熵的
homoeoblastic texture 等粒(变晶状)结构
homoepicamphor 高表樟脑
homoepitaxy 均相外延
homoequilenin 高雌马甾酮
homo-fibre 单组分纤维；均质纤维
homofil 单组分丝；单组分纤维
homogeneity 均一性；均匀性〔黏胶液〕
homogeneity limit 均匀性极限
homogeneity of reference material 标准物质均匀性
homogeneity of variance 方差齐性
homogeneity spoiling pulse (HSP) 均匀性破坏脉冲
homogeneity spoil pulse 均匀性破坏脉冲
homogeneity test 结构均一性测定
homogeneity test of reference material 标准物均匀性检验
homogeneity to the eye 目测均匀性
homogeneization 均(一)化作用
homogeneous 均匀的；均相的
homogeneous azeotrope 均匀共沸混合物

homogeneous body 均匀体
homogeneous bottle neck model 同质瓶颈模型
homogeneous broadening 均匀增宽
homogeneous catalysis 均相催化*
homogeneous catalyst 均相催化剂
homogeneous catalytic hydrogenation 均相催化氢化
homogeneous chain compound 同素链化合物
homogeneous chemical reaction 均相化学反应
homogeneous combustible mixture 均匀可燃混合物
homogeneous combustion 均匀燃烧
homogeneous design 均匀设计*
homogeneous design of experiment 均匀试验设计
homogeneous dispersion 均匀分散(体)
homogeneous displacement gradient 均匀位移梯度
homogeneous distribution law 均匀分配定律
homogeneous dye 均一染料；同性染料(染色)
homogeneous electrode 均相电极
homogeneous enzyme immunoassay 均相酶免疫分析
homogeneous equation 齐次方程
homogeneous equilibrium 均相平衡*
homogeneous explosive mixture 均匀爆炸混合物
homogeneous extraction 均相萃取
homogeneous fluid 均匀流体
homogeneous graft 均相接枝
homogeneous-heterogeneous 均相-多相催化(作用)
homogeneous hydrogenation 均相氢化*
homogeneous hydrolysis 均相水解
homogeneous ion exchange membrane 均相离子交换膜
homogeneous laminate 均质层压板
homogeneous light 单色光
homogeneous magnetic field 均匀磁场
homogeneous material 均匀物料；均质材料
homogeneous medium 均匀介质
homogeneous membrane electrode 均相膜电极*
homogeneous metallocene catalyst 均相茂金属催化剂
homogeneous mixture 均匀混合物
homogeneous network 均匀网络
homogeneous nucleation 均相成核*
homogeneous phase 均相
homogeneous phase flame chemiluminescence 均相火焰化学发光
homogeneous polyblend 均相共混物
homogeneous polymerization 均相聚合*
homogeneous precipitation 均匀沉淀*；均相沉淀
homogeneous precipitation method 均相沉淀法
homogeneous ray 单色光
homogeneous reaction 均相反应*
homogeneous reactor 均相(反应)堆

homogeneous ring compound 碳环化合物
homogeneous sludge 油中均一淤渣
homogeneous solution polymerization 均相溶液聚合（作用）
homogeneous state 均态
homogeneous strain 均匀应变
homogeneous stress 均匀应力
homogeneous system 均相系统*
homogeneous thermal expansion 均匀热膨胀
homogeneous tube 均匀管
homogeneous turbulence 均匀湍流
homogeneous X-ray 单色 X 射线
homogenization 均化(作用)法
homogenized grease 均化的润滑脂
homogenizer 均化器
homogenizing 均化(作用)
homogenizing agent 均化剂
homogentisate 尿黑酸；2,5-二羟苯乙酸
homogentisic acid 尿黑酸；2,5-二羟苯乙酸；高龙胆酸 $(OH)_2C_6H_3CH_2COOH$
homogeranic acid 高香叶酸
homogneous complex 均匀配合物
homoharringtonine 高三尖杉酯碱〔药〕
homohydratropic acid 高氢化阿托酸
homo-hydroquinone 2-甲基-1,4-苯二酚
homo-ion 同离子
homo-ionic mineral 同离子矿物
homo-ionic solution 同离子溶液
homo (io)thermic 恒温的；调温的
homo (io)thermism 温度恒定；温度调节
homoisocitric acid 高异柠檬酸
homoisohydric 同酸(根)等氢离子的
homoisoleucine 高异亮氨酸
homolactic acid bacteria 同型乳酸菌
homolanthionine 高羊毛氨酸
homolevulinic acid 4-氧代己酸
homologen 同系化合物
homologisation 均裂作用
homologous line 匀称谱线
homologous pair 对应线
homologous series 系〔有机化合物〕；同系列
homolog(ue) 同系物
homology 同系(现象)
homolycorine (=narcipoetine) 高石蒜碱
homolysis 均裂*
homolytic 均裂的
homolytic cleavage 均裂
homolytic fission 均裂

homolytic mechanism 均裂机理
homolytic process 均解过程
homolytic reaction 均裂反应*
homolytic substitution 均裂取代〔即自由基或原子取代一个 H〕
homolytic thermal scission reaction 均裂性热断(链)反应；热均裂反应
homomartonite 高马炸药；溴丁酮
homomenthone 高薄荷酮
homometric 同 X 光谱的
homomixer 均相混合机
homomixis 同源融合；同宗接合
homomorph effect 异质同晶效应
homomorphic 同态的
homomorphism (=homeomorphism) 异质同晶(现象)；同态
homomycin 匀霉素
homonuclear decoupling 同核去偶*
homonuclear decoupling chemical shift correlated spectroscopy 同核去耦化学位移相关谱
homonuclear diatomic molecule 同核双原子分子
homonuclear double resonance 同核双共振
homonuclear gated decoupling 同核门控去耦
homonuclear lock 同核联锁
homonuclear phase sensitive chemical shift correlated spectroscopy 同核相敏化学位移相关谱
homonuclear spin echo J modulation 同核自旋回波 J 调制
homonuclear stabilization 同核稳场
homonuclear two dimensional chemical shift correlation spectrum 同核二维化学位移相关谱
homonuclear two dimensional J resolved spectroscopy 二维同核 J 分辨(解)谱
homopantoyltaurine 高泛酰牛磺酸
homophase 同相
homophoric acid 高樟脑酸
homophorone 高佛尔酮
homophthalic acid 高邻苯二甲酸；邻羧基苯乙酸 $HO_2CC_6H_4CH_2COOH$
homopimanthrene 高海松烯
homopinol (=pinene hydrate) 水合蒎烯；高蒎醇 $C_{10}H_{18}O$
homopiperonyl (=3,4-methylenedioxyphenethyl) 高胡椒基；3,4-亚甲二氧苯乙基
homopolar 无极的
homopolar adsorption 无极吸附
homopolar bond 无极键
homopolar compound 无极化合物

homopolar crystal　无极晶体
homopolar dipole　无离子键偶极子
homopolar link (=homopolar linkage)　无极键
homopolar linkage (=homopolar link)　无极键
homopolar polymer　无极聚合物；同极性聚合物
homopolar valency　无极价
homopolyacrylonitrile fibre　均聚丙烯腈纤维
homopolyamide　均聚酰胺
homopolycondensation　均相缩聚*
homopolyester fibre　均聚酯纤维
homopolymer　均聚物*
homopolymerization　均聚反应*
homopropagation　同种增长
homopter(o)carpin　高紫檀素
homopyrocatechol　高儿茶酚；甲苯间二酚
　　$CH_3C_6H_3(OH)_2$
homoquinine　高奎宁
homoretene　高惹烯
homoricinstearolic acid　高蓖硬酸
homorph　同态(象)；异质同晶(现象)
homosalicylic acid　高水杨酸
homosapogenin　高皂角配基
homoscedastic　同方差的
homoscedasticity　方差齐性*；同方差性
homoserine　高丝氨酸
homoseryl　高丝氨酰基
homosigmatropic rearrangement　同 σ 迁移重排*
homospin decoupler　同核自旋去耦器
homospoil technique　均匀扰动法
homostephanoline　高千金藤诺灵
homosteroid　高甾化合物
homosynergism　均态增效(作用)
homoterpenylic acid　高萜酸　$C_7H_{11}O_2CH_2COOH$
homothermal process　匀温过程
homothetic work-hardening　(同)位(相)似加工硬化
homothujyl alcohol　高苧酚；高崖柏醇
homotonia　等渗性；等压性
homotope　同族素
homotopic　①等位(的)②同伦的*
homotrilobine (=isotrilobine)　异三裂碱
homotropine　升托品
homovanillic acid　高香草酸
homovanillin　高香草醛；3-甲氧基-4-羟基苯乙醛
　　$HO(CH_3O)C_6H_2CH_2CHO$
homoveratric acid　高藜芦酸
homoveratroyl (=3,4-dimethoxyphenylacetyl)　高香草酰；
　　3,4-二甲氧苯乙酰
homovitamin A　高维生素 A

hondamycin　宏大霉素
Honda's thermobalance　本多氏热天平
hondrometer　粒度计；微粒特性测定计
hone　磨石
honey-comb　蜂窝；蜂巢
honeycomb clinker　蜂窝状烧结块
honeycomb composite　蜂窝状复合材料
honeycomb construction　蜂窝式结构
honeycomb core　蜂窝芯
honeycomb cracks　网状龟裂；蜂窝状龟裂
honey-combed　蜂窝结构
honeycomb filter　蜂窝式过滤器
honeycombing　蜂窝状层合材料
honeycomb-like　蜂窝状的
honeycomb radiator　蜂窝式微热器
honeycomb sandwich　蜂巢型芯材
honeycomb structure　蜂窝状结构
honeycomb structure laminate　蜂窝状层压品
honeycomb substrate　蜂窝状载体
honeycomb support　蜂窝状载体*
honeydew　甘汁；蜜露
honeystone　蜜蜡石
honeysuckle　忍冬
honing plating　电镀珩磨
hood　①通风橱②盖
hooded impeller　闭式叶轮；闭式轮
ho oil　芳(香)樟(树)油
hook　外弧
hook comb reed　钩形梳筘
Hookean body (=elastic solid;Hookean solid)　弹性固体；
　　胡克体
Hookean elastic behavior　胡克弹性行为
Hookean elasticity　胡克弹性
Hookean region　胡克区；弹性区域
Hookean solid (=elastic solid; Hookean body)　胡克体
Hookean spring　胡克弹簧
Hookean strain　胡克应变；弹性应变
Hooke-Rauchy elasticity equation　胡克-柯西弹性方程
hook joint strength　钩接强力
hook-shape　钩形
Hook's law　胡克定律
hoolamite　浮石
hoolamite indicator　胡拉麦脱一氧化碳显示器
hoolamite test　胡拉麦脱一氧化碳检测
hoop　箍
hoop stress　周向应力；环向应力
hoop tension　周张力
hootenanny　拆除水管清洗污水装置

Hoover automatic miller　胡佛自动研磨计
Hoover muller　胡佛式研磨机；平磨机
hop　(啤)酒花；蛇麻；草花；忽布
hop-back　酒花浸取槽
hopcalite　①浩咖炸药；霍加拉特②一种催化剂〔锰、钴、银的氧化物混合物,用于气体防毒面具〕
hop drier　酒花干燥器
hopeite　磷锌矿
hop oil　酒花油
hopper　①进料斗；加料(漏)斗②储料斗
hopper-cooled　连续水冷却的
hopper door　进料斗盖
hopper dryer　斗式干燥器
hoppered bottom　锥形底
hopper klep　计量漏斗
hopper lining　装料斗衬里
hopper scale　库秤
hopping　库秤
Höppler rolling ball viscosimeter (=Höppler viscosimeter)　霍普勒落球黏度计
Höppler viscosimeter (=Höppler rolling ball viscosimeter)　霍普勒落球黏度计
hop seed oil　酒花子油
hops oil　酒花油
hoptanedioic acid　庚二酸　HOOC(CH$_2$)$_5$COOH
horbachite　硫镍铁矿
hordenine　大麦芽碱；对二甲氨乙基苯酚
horizon-symmetric plane grating mounting　水平对称平面光栅装置
horizontal　横(平)的；横卧的；水平的
horizontal balanced reciprocating compressor　对称平衡往复压缩机
horizontal-bed glazing jack　水平式打光机
horizontal belt conveyor　卧式胶带运输机；卧式运输带
horizontal boring machine　卧式镗床〔水平钻孔用〕
horizontal burette　横式滴定管
horizontal centrifugal screen　卧式离心筛
horizontal-chamber furnace　卧式炉
horizontal-chamber oven　卧式炉
horizontal chromatography　水平色谱(法)
horizontal circular chromatography　水平圆形色谱法
horizontal collecting area　水平收集面
horizontal compressor　卧式压缩机
horizontal coordinate　横坐标；水平坐标
horizontal cylinder drying machine　卧筒干燥机
horizontal cylindrical tank　卧式圆形油罐
horizontal development　水平展开
horizontal diaphragm cell　平膜电解池
horizontal digester　卧式蒸煮器
horizontal drying oven　卧式干燥炉
horizontal exchanger　卧式热交换器
horizontal extruder　卧式挤出机
horizontal filter paper chromatography　水平滤纸色谱法
horizontal flow　水平流
horizontal flow pipe reactor　平流管式反应器
horizontal flue　横(式)气道
horizontal-flued oven　卧卤炉
horizontal gas coke　卧炉焦
horizontal grizzly　水平格筛
horizontal hydroextractor　卧式脱水机
horizontal injection　①水平注塑②水平注射
horizontal leno　横罗
horizontal level　水平线
horizontally split　水平剖分
horizontally split casing　水平剖分式机壳
horizontal mill　卧式研磨机
horizontal nip embossing　水平辊隙压花
horizontal opposed reciprocating compressor　对称平衡压缩机；卧式往复压缩机
horizontal paper electrophoresis　水平滤纸电泳
horizontal plane　水平面
horizontal position welding　横焊
horizontal pulse extractor　卧式脉冲萃取器
horizontal pump　卧式泵
horizontal retort　卧式(干馏)釜
horizontal roller　卧式辊涂机
horizontal rotary furnace　卧式转炉
horizontal rotating cylinder　水平旋转圆筒筛
horizontal rotating disc　水平(旋)转盘
horizontal rotatory dryer　卧式回转干燥器
horizontal sand grinder　卧式砂磨机
horizontal screw finisher　卧式螺旋后缩聚器
horizontal shear　卧式剪床
horizontal slab gel electrophoresis　水平板凝胶电泳法
horizontal sole leather roller　卧式底革滚压机
horizontal spindle pump　卧式心轴泵
horizontal spinning extruder　卧式挤出纺丝机
horizontal spray chamber　横式喷雾室
horizontal storage　平放储存
horizontal table filter　平盘(真空)过滤机；平面过滤机
horizontal tube evaporator　平管蒸发器
horizontal tubular absorber　水平管吸收器
horizontal zone melting　水平区域熔化
hornblende　角闪石
hornblende rock　角闪岩
Horne beater　霍恩打浆机

hornfels structure 角页岩结构
horn frefoil oil 角状车轴草油
hornification 角化；角质化
hornito 熔岩滴丘
horn lead 角铅
horn scoop 角勺
Horn's dry lead subacetate method 霍恩干碱式乙酸铅法
horn silver 角银 AgX; AgCl
hornskin 漆皮
hornstone 角岩
Horouf continuous bulking machine 霍劳夫连续膨化变形机
horse 支架；木马〔皮革〕
horse box leather 马箱革
horse chestnut oil 马栗油
horse chestnut tannin 马栗单宁
horse fat 马脂
horse grease 马脂
horse hair 马鬃
horse hide 马皮
horse marrow fat 马髓脂
horse mask 马面具
horse mint oil 马薄荷油
horse oil 马油
horse power 马力
horse radish 辣根
horse radish oil 辣根油
horse's foot oil 马足油
horse-shoe magnet 马蹄型磁铁
horse-shoe main 马蹄形总管
horse-shoe manifold 马蹄形管道
horse-shoe mixer 马蹄式混合机
horse-shoe stirrer 马蹄式搅拌机
horse-shoe type mixer 马蹄式混合器
horse stacker 搭马机
horsfordite 锑铜矿
horsing 搭马
Horton multispheroid 霍尔顿型多弧(滴)形油罐
hortonolite 镁铁橄榄石
Horton sphere 霍尔顿型球形压力储罐
Horton spheroid 霍尔顿型滴形压力储罐〔挥发性液体用〕
Hortvat's tube 离心分离管
hose cap 管帽；管堵
hose carrier 携管器；软管运送车
hose cart 软管运送车
hose clamp (=hose clip) 软管夹
hose clip (=hose clamp) 软管夹

hose cock 软管钩
hose connection 软管连接
hose connector 软连接管
hose cord 水龙带合股纱线
hose coupling 软管偶联器
hose davit 软管吊车
hose expansion 管体膨胀
hose hawser 软管吊绳；软管缆
hose joint 软管连接头
hose line (=hose pipe) 软管管线
hose liner 软管套头
hose lining 内胶；内层胶；内径胶〔胶管〕
hose machine 软管机
hose nipple 软管短节
hose pipe (=hose line) 软管管线
hose pump 软管泵
hose reducer 异径接管
hose reel 软管卷盘
hose supporting clip 软管挂环
hose test 软管试验
hose thread 软管螺纹
hose tube 软管
hose union 软管接合器
hose wrench 软管扳手
hosho oil 芳(香)樟(树)油
hosiery 针织品；针织物
host 主体*
Hostapon process 霍斯塔庞法〔制烷基磺酸盐〕
host compound 主体化合物
host crystal 基质晶体
host-guest complex 主宾络合物*
host-guest complexation 主体-客体配位作用
host-guest coordination 主体-客体配位作用
host-guest coordination compound 主客体配合物*
host-guest molecular recognition 主-客体分子识别
host lattice 主晶格
hot 热的
hot acid polymerization process 热(硫)酸聚合过程
hot acid treatment 热酸处理
hot air aging 热空气老化
hot-air (and) steam cure 混汽硫化〔热空气与蒸汽并用〕
hot-air band dryer 热风带式干燥器
hot air cabinet hot-air circulating system 热空气循环系统
hot air cabinet 热气烘箱
hot air circulating system 热空气循环系统
hot air convection 热风(空气)对流
hot-air cure 热空气硫化

hot air drier 热风干燥器
hot air drying 热风干燥
hot air funnel 热气漏斗；双层漏斗
hot air hand dryer 热风带式干燥器
hot airless spray (=airless hot spraying) 无空气热喷涂
hot air oven 热风炉
hot-air paint stripper 热空气喷射除漆机
hot air sizing machine 热风式浆纱机
hot-air steam cure 热气蒸汽硫化
hot air stenter 热风拉幅机
hot air sterilization 干热灭菌；热空气消毒
hot air sterilizer 干热灭菌器
hot-air stripper 热空气喷射除漆机
hot-air stuffing mill 热风加脂转鼓
hot-air vulcanization 热空气硫化
hot-alkylation 热烷基化
hot and cold tube 冷热管
hot applied bitumen 热施工沥青
hot applied coal tar coating (=hot coal-tar coating) 热煤焦油覆盖
hot applied enamel 热施工瓷漆
hot atom 热原子*
hot atom annealing 热原子退火*
hot atom chemistry 热原子化学*
hot atom reaction 热原子反应*
hot band 热谱带
hot bath 热浴
hot-blast 热(鼓)风
hot-blast furnace 热风炉
hot-blast heater 热风炉
hot-blast main 热风总管
hot-blast pig iron 热风生铁
hot-blast stove 热风炉
hot body 供热体
hot break 加热消蚀
hot breakdown 热打浆
hot break juicing 热法榨汁〔制糖〕
hot-briquetting 热团矿
hot capping 热挂背
hot-cast 热铸〔塑〕
hot cathode 热阴极
hot cathode hydrogen discharge tube 热阴极氢放电管
hot cathode ionization gauge 热阴极电离规
hot cathode low-pressure mercury lamp 热阴极低压汞灯
hot cathode X-ray tube 热阴极 X 射线管；热阴极电子管
hot cell ①热室*②放射性小室
hot charge 热原料
hot chemical vapour deposition 热化学蒸气沉积；化学蒸气热沉积
hot chromatography 热色谱(法)
hot clear end-point 热透明点；醇解终点
hot coal-tar coating (=hot-applied coal tar coating) 热煤焦油覆盖
hot coining 热后压铸
hot column stabilizer 稳定塔
hot compression molding 热压成型(法)；热压模塑法〔碳纤维复合材料成型法之一〕
hot crack 热裂
hot cure 热硫化
hot desert test 热砂试验
hot die-face pelletizer 热模面切粒机
hot dip 热浸法
hot dip coating 热浸涂
hot dip galvanized (process) 热浸镀锌(法)
hot dipped tinning 热浸镀锡
hot dipping method 热浸法
hot-dip plastisol coating 热浸塑料溶胶
hot-doctor treatment (汽油)含硫处理
hot drawing 热拉
hot-drawn tube 热拉管
hot drip gum test 热滴胶质试验〔在热空气流中蒸发汽油滴以测定汽油中胶质含量〕
hot embossing 热压法；热压型
hot enamel coating 热搪瓷涂盖
hot-engine sludge 高温残渣
hot extrusion 热挤法
hot extrusion molding 热挤模塑
hot feed 热进料；热加料
hot filament ionization gauge 热丝电离真空计
hot filter 保温漏斗
hot flame 高温(火)焰
hot floor 平底干燥器
hotflue 热烟道
hot-fluid 热流体
hot fluid jet texturing process 热气流喷射变形工艺
hot foiling 热烫彩箔(工艺)
hot formed 热成形的
hot funnel 保温漏斗
hot gas line for defrosting 融霜热气管线
hot gas purification 热气净化
hot-gas recycle process 热气循环过程
hot-gas welding 热气焊接
hot glue size 热胶料；热水调胶浆
hot house 燃烧室〔炭黑〕
hot industry 高温工业
hot jacket 热外套

hot jet welding　热风焊接；热风熔接
hot joining　热接
hot lab　热实验室；放射性实验室
hot laboratory　热实验室*；放射性实验室
hot lacquer　热喷漆
hot-laid mixture　热拌混合料
hot machine　高温机器
hot mastication　热塑炼；高温塑炼；高温软化
hot matched die press mo(u)lding　对模热压成型
hot material conveyor belt　耐热运输带
hot melt　热熔体
hot-melt adhesive　热熔胶黏剂；热熔性黏合剂
hot-melt application　热熔施工
hot melt bonded nonwovens　热风片
hot melt coating　热融涂层〔遇热就融化〕
hot melt process　热熔法
hot melts　热熔体
hot metallurgy laboratory (HML)　强放射性冶金实验室；热冶金实验室
hot metal working lubricant　用于(轻)金属热加工的润滑剂
hot methylation　热甲基化
hot milling　热混炼
hot-mix asphalt　热沥青混合物
hot mold　热模
hot-moulded　热模塑
hot mo(u)lding　热模塑
hot-neck grease　(轧钢机)滚棒轴头润滑脂
hot oil　热馏出物
hot oil centrifugal pump　热油离心泵
hot oil distillation　热油蒸馏
hot oil duct　热油导管
hot oil expression　热压油法
hot oil machine　热油机
hot oil pump　热油泵
hot particle (HP)　(核爆炸沉降物中)强放射性粒子
hot-pen type recorder　热笔式记录仪
hot piping　热管道输送
hot pit　热鞣池
hot plate　电热板*
hot plate chromatography　热板色谱法
hot plate test　热板试验
hot point　热点
hot polymer　热聚物
hot-polymerizable mastic adhesive　热聚合型膏状黏合剂
hot polymerization　热聚作用
hot polymerized styrene butadiene rubber　热聚丁苯橡胶
hot press　热压机

hot pressed oil　热榨油
hot pressing　热压
hot-process　热处理
hot radical　热自由基
hot reflux　热回流
hot reflux condenser　热回流管
hot repair　热补；汽补
hot resistance　耐热性
hot retreading　高温胎面翻新；热翻新
hot rolled　热轧的
hot rolled tube　热轧管
hot rolling　热轧
hot roll neck grease　轧钢机热滚棒轴头润滑脂
hot room　加温室；烘胶房
hot run　热试验*
hot runner　热流道
hot-runner mould　热流道模具
hot semiwork　"半热"工厂；热的半负载试运转工厂
hot-set　热固化；热变定
hot setting　高温固化
hot-setting adhesive　热固性黏合剂〔100℃以上硬化〕；热硬化胶黏剂
hot setting machine　热定形机
hot setting resin　热固性树脂
hot short　不耐热的；热脆性的
hot-short iron　热脆铁
hot slot arrangement　热槽板；热槽装置
hot spark ion source　热火花离子源
hot spot　①热点；热部位②过热部位；腐蚀点
hot spot application　热点利用
hot spot in pipestill　管式炉热点
hot spot model　热斑模型；热点模型
hot spotting　预热
hot spray　热喷涂
hot spray coating　热喷涂涂料
hot spraying　热喷涂
hot spraying technique　热喷涂技术；热喷涂工艺
hot-spray lacquer　热喷漆
hot spray process　热喷涂法
hot stage　①热板；热台②(显微)熔点测定器
hot stage melting point determination　热台熔点测定
hot stage microscope　湿阶显微镜
hot stamping　烫印；热压印
hot start　高电流起弧
hot steam funnel　热蒸汽漏斗
hot strength　高温强度
hot strength retention　热态强度保留率
hot stretch　热拉伸

hot-stretched cord 热拉伸的帘子线
hot stuffing 热加脂(法)
hot styrene rubber 热聚丁苯橡胶
hot surface 热表面〔反常的吸收剂表面〕
hot tack 热黏性
hot tarring 热沥青铺路
hot tear resistance 抗高温撕裂性
hot tensile (strength) 高温抗张(强度)
hot test 热试验
hot-tip gate mo(u)lding 热点式浇口注塑
hot top 保温帽；热顶
hot tube furnace 热管炉
hot-vapour lines 热蒸气线路
hot velocity 热原料速度
hot vinyl ①热塑性乙烯树脂②热塑性乙烯系船底漆
hot vulcanization 热硫化(作用)
hot-vulcanized silicone rubber 热硫化硅橡胶
hot waste 强放射性废料
hot wastewater 高温废水
hot water boiler 热水锅炉
hot water cure 热水处理
hot water funnel 热水漏斗
hot-water glue size 热水稀释胶料
hot water gravity circulation 热水重力循环
hot water heating ①热水加温②过热水(加压)硫化
hot water outlet piping 热水流出管道系统
hot-water paste (裱糊墙纸用的)热水调和胶浆
hot water resistance 耐热水性
hot water tank 热水槽
hot water test 热水试验
hot well 热水井
hot wet strength 热态湿强度
hot wire ammeter 热丝安培计
hot-wire anemometer 热线风速计
hot wire cell 热丝池
hot wire detector 热丝检测器
hot wire flowmeter 热丝流量计
hot wire glass tubing cutter 热线玻管切断机
hot wire thermal conductivity detector 热丝热导检测器
hot wire type gas analyzer 热线型气体分析计
hot working 热加工
hot work in the battery 电池组中的热功
hot zone 热区
Houben-Fischer nitrile synthesis 霍本-费希尔腈合成法
Houben-Hoesch synthesis (=Hoesch synthesis) 霍本-赫西合成法
houdriflow catalytic cracking 胡得利催化裂化
houdriflow catcracker 胡得利移动床催化裂化装置

houdriformer 胡得利催化重整装置
houdriforming process 胡得利催化重整过程
Houdry adiabatic dehydrogenation process 胡得利固定床绝热催化脱氢过程
Houdry butane dehydrogenation 胡得利丁烷脱氢(法)
Houdry catalytic cracking process 胡得利催化裂化过程
Houdry catalytic naphtha 胡得利催化裂化汽油馏分
Houdry catalytic reforming process 胡得利催化重整过程
Houdry cracking case 胡得利固定床催化裂化反应器
Houdry-Daikyo hydrodesulfurization 胡得利-大协和加氢脱硫(法)
Houdry fixed-bed unit 胡得利固定床催化剂的裂化设备
Houdry flow catalytic cracking 胡得利移动床催化裂化(法)
Houdryforming 胡得利重整
Houdry pellets 胡得利片状催化剂
Houdry process (=Houdry catalytic cracking process) 胡得利固定床催化裂化过程
Houdry process beads 胡得利球状催化剂
Hounsfield tensometer 洪斯费尔德张力计
hourly space velocity 时空间速度
hours on stream (=hours to stream) 操作时间
hours to stream (=hours on stream) 操作时间
house brand 工厂标号
household chemical resistance 耐日用化学品性
household chemicals 日用化学品
household detergent 家用洗涤剂
household fuel 家用燃料
household fuel gas 家用煤气
household lino paint 家用亚麻仁油醇酸树脂漆
household oil 家庭用润滑油
household sewage 住户污水；生活污水
household soap 家用皂
housekeeping 车间整理；车间管理
houseman 石油加工工厂控制室操作者
house paint 房屋漆；民用漆〔建筑漆〕
housing ①套②罩③壳④套管
housing type press 罩式压机
Howard-Bridge ozonizer 霍华德-布里奇臭氧器
Howard crystallizer 霍华德式结晶器
Howard process 霍华德法〔用氯精制汽油法〕
Howe-Baker electrical desalting 豪-贝克电气脱盐(法)
howlite 硅硼钙石
hoxamycin 贺克沙霉素
Hoyer's solution 霍耶氏(氮源需求测定)溶液
H-point standard addition method H 点标准加入法
hR_f value hR_f 值〔$100 \times R_f$ 值〕
H-section steel (=I-section steel) 工字钢

H-shift　氢位移
H solvency　H 溶解本领
H solvent　H 溶剂
hub　①轮毂②中心轴（压出机螺杆）③冲压阴模
Hubach test　胡伯克氏(测定氰化物及低铁氰化物)法
hubbed flange　高颈法兰
hubbing (=hobbing)　切压制模(法)
hubbing press　冷模压型机；切压机
hub disk　（叶轮)轮盘〔轮毂〕
Huber process　休伯法〔沉淀二氧化硅的一种制造方法〕
Hübl number　胡伯值〔脂肪或油的碘值〕
Hübl solution　胡伯溶液〔测定碘值的试剂〕
Hübl-Waller method　胡伯-瓦勒碘值测定法
hubnerite　钨锰矿
hub ratio　(叶轮)轮毂比
Hückel molecular orbital method (HMO)　休克尔分子轨道法*
Huckel's rule　休克尔规则*
Hudson's lactone rule　赫德逊内酯规则
hue　①色彩②色调；深浅浓淡
hue circle　色相环；色(彩)环；色(调)环
Huey test　晶间腐蚀试验；不锈钢腐蚀试验
Huff electrostatic separator　胡夫静电分离器
Huggins coefficient　哈金斯系数
Huggins equation　哈金斯公式
hull　①薄膜②壳；皮
hull cell　哈氏槽
huller　脱壳机
hull paint　船壳漆；船体漆
hull rubber　薄膜橡胶
human and biosphere　人与生物圈
human chorionic gonadotrophin sensor　人绒毛膜促性腺素传感器
human error　人为误差
human factor　人为因素
human genome project　人类基因组计划
human hair　人发；头发
humatin (=paromomycin)　巴龙霉素
humectant　保湿剂
humectant-plasticizer　湿润性增塑剂
humectants　水分保持剂；增湿剂
hum filter　交流声滤除器
humic acid　①腐殖酸②黑腐酸
humic coal　腐殖煤
humic compounds (=humic substances;humics)　腐殖质
humics (=humic compounds;humic substances)　腐殖质
humic substances (=humic compounds;humics)　腐殖质
humic sulfur　腐殖硫

humid　湿的
humid ageing　湿态老化
humid air　湿空气
humidenthalpy　湿焓；湿热函
humid ether　湿醚；含水醚
humid gas　湿气(体)
humid heat　湿比热
humidification　增湿(作用)
humidification by contacting with heated liquid　热液增湿
humidification by mixing of gas　混合增湿
humidification process　增湿过程
humidifier　增湿器；湿润器
humidifying　加湿
humidifying chart　湿度表
humidifying equipment　增湿设备
humidiometer　湿度计
humidistat　恒湿器
humidity　湿度
humidity aging　潮湿老化
humidity blush (=blushing)　湿致白斑
humidity cabinet　湿热试验箱；潮湿箱；(调)温(调)湿箱
humidity cabinet test　湿润箱试验
humidity chart　湿度图
humidity controller　湿度控制器
humidity distribution　湿度分布
humidity drier　湿度调节干燥器
humidity indicator　湿度指示器
humidity meter　湿度计
humidity pickup　①吸湿②吸湿量
humidity ratio　湿度率；湿度比
humidity recorder　湿度记录器
humidity resistance　防潮性能
humidity sensor　湿度传感器
humidity-temperature index (=humiture)　温湿指数
humidity-thermal cycle cracking　湿热循环开裂
humidity tolerance　湿度允许偏差
humidizer　增湿剂
humidor　①恒湿器；保湿器②蒸汽饱和室
humidostat　①恒湿(调节)器②保湿箱
humid tropical condition　高温高湿状态；湿热状态
humid volume　湿容积
humin　腐黑物；腐殖质
humin substances　腐黑物
humite　硅镁石
humitector　测湿仪
humiture (=humidity-temperature index)　温湿指数
hummer　哼声器〔用于电导测定〕
hummer screen　电磁簸动

humoral immunity 体液免疫性
hump ①峰②巅峰值
Humphrey's generator 汉姆弗利式水煤气发生炉
Humphries' sulfur burner 汉弗莱斯烧硫炉
hump strip 护圈胶
humulane 葎草烷
humulene 葎草烯
humulinone 葎草灵酮
humulone 葎草酮
humulo tannin 葎草单宁
humus 腐殖质
humus coal 腐殖质煤
humus sludge 腐殖质污泥
humycin (=paromomycin) 巴龙霉素
Hund-Mulliken-Hückel method 洪德-马利肯-休克尔法
hundred 百
hundred million 万万；亿
hundred volume total formula 体积百分率配方
hundred weight 二十分之一吨〔=112英磅或100美磅〕
Hund's rule 洪德规则*
hungry 欠鞣皮〔革〕
hungry area 空胶；露布；露白〔胶布缺点〕
hungry surface (=starved surface) ①(漆膜)丰满度差的表面②干瘪的表面③塌渗的表面
Hunsdiecker reaction 汉斯迪克反应
Hunter multipurpose reflectometer 亨特万能反射(白度)计
Hunter whiteness 亨特白度
hunting 探求
hunting leather 粗绒面革
huntite 碳酸钙镁石
huperzine A 石杉碱甲〔药〕
hurdle 栅格
hurdle scrubber 栅格式气体洗涤器
hurling pump 旋转泵
hurricane drier 风干室
hush cloth 吸音绒布
husk 壳
husked 脱壳的
hutch 选矿箱
Huttig temperature 许蒂希温度*
HV-cable ①高压电缆②高压电缆接口
HV generator (粉末喷涂)高压静电发生器
hyacinth 红锆石
hyacinthin (=phenylacetaldehyde) 苯乙醛；风信子质
hyacinth oil 风信子油
hyaline ①透明的②(=hyalin)透明素
Hyalite (=Müller's glass) 玻璃蛋白石

hyalo- 〔词头〕 ①玻璃②透明
hyalobasalt 玻璃玄武岩
hyalography 玻璃蚀刻术
hyaloidin 玻璃糖质
hyalophane 钡冰长石
hyalophyr 玻基斑岩
hyalosiderite 透铁橄榄石
hyalotekite 硼硅钡铅矿
hyaluronic acid (=mucinase) 透明质酸
Hybinette process (for nickel extraction) 海宾尼特(炼镍)法
hybrid ①杂种②杂化物
hybrid-base oil 混合基石油
hybrid base petroleum 混合基石油
hybrid biosensor 混合生物传感器
hybrid complex 杂化配合物
hybrid composite 混杂复合材料
hybrid curing technology 混合式固化工艺
hybrid effect 混杂效应
hybrid effect of fiber fracture strain 纤维断裂应变混杂效应
hybrid electronic balance 混合力电子天平
hybrid fabric 混杂织物
hybrid flowsheet 混杂流程；杂化流程
hybrid grinding and dispersion device 研磨分散并行(联合)磨；混合式研磨分散设备
hybrid interface 混杂界面
hybridization 杂化*
hybridization in situ 原位杂交
hybridization of atomic orbitals 原子轨道的杂化
hybridized orbital 杂化轨道
hybrid mass spectrometer 混合型质谱仪
hybrid material 杂化材料
hybrid mill 混合型研磨机
hybrid milling and dispersion machine 研磨分散并行(联合)磨
hybrid orbital 杂化轨道*
hybrid oriented nematic 混合取向向列
hybrid polymer 混合聚合物；杂化聚合物
hybrid polymerization 杂化聚合
hybrid powder coating 混合粉末涂料〔专指环氧-聚酯树脂混合粉末涂料〕
hybrid process 混杂流程；杂化流程
hybrid resistor 混合电阻
hybrid rocks 混杂岩
hybrid structure 混杂结构
hybrid volume ratio 混杂体积比
hybrimycin 杂交霉素

hycar 丁腈橡胶
hydantoic acid 脲基乙酸；海因酸 $H_2NCONHCH_2CO_2H$; $COOHCH_2NHCONH_2$
hydantoin (=glycolylurea) 乙内酰脲；脲基醋酸内酰胺
hydantoin acetic acid 海因乙酸；乙内酰脲基代乙酸 $C_3H_3N_2O_2CH_2COOH$
hydantoin-derivative-modified polyester 乙内酰脲衍生物改性聚酯
hydantoin distearate 二硬脂酸乙内酰脲〔润滑剂〕
hydantoin propionate 丙酸乙内酰脲
hydatogenous rocks 液成岩
hydnocarpic acid 副大风子酸；2-环戊烯十一烷酸 $C_5H_7(CH_2)_{10}COOH$
hydnocarpoyl 副大风子酰〔来自 hydnocarpic acid〕 $C_5H_7(CH_2)_{10}CO-$
hydnocarpus oil 大风子油
hydnocarpyl 副大风子油基〔来自 hydnocarpyl alcohol〕 $C_5H_7(CH_2)_{10}CH_2-$
hydnocarpyl alcohol 副大风子醇
hydous salt 水合盐
hydr-〔词头〕 ①氢；氢化②水
hydrabarker 水力剥皮机
hydrability 水化性
hydracetin (=pgrodin) 乙酰苯肼〔药〕
hydracid 氢酸
hydraclone 连续除渣器
hydracrylic acid 羟基丙酸 CH_2OHCH_2COOH
hydracrylonitrile 3-羟丙腈
hydrafiner 水化精磨机；高速精浆机
hydragel 水凝胶
hydragog(ue) 利尿药
hydralazine 肼屈嗪〔药〕
hydram 氨(胺)化水溶紫胶
hydramine 羟基胺；醇胺
hydrangenol 绣球酚
hydrangin 八仙药精
hydranginic acid 绣球酸；二羟(基)苊甲酸
hydrant (水)龙头
hydrapulpter 水力碎浆机
hydrargillite ①水铝氧；水铝矿 $Al(OH)_3$②银星石
hydrargotin 鞣酸亚汞
hydrargyria 汞中毒
hydrargyrum〔拉丁名〕 汞
Hydrar process 海德拉尔法〔苯催化氢化制环己烷法〕
hydrastic acid 北美黄连酸 $C_9H_6O_6$
hydrastine 白毛茛碱；北美黄连碱
hydrastinine 白毛茛分碱
hydratability ①水合本领②水合性

hydratable 能水合的
hydrate 水合物*
hydrated 水合的
hydrated alumina 水合氧化铝
hydrated aluminium silicate 水合硅酸铝
hydrated aluminum oxide 水合氧化铝；氢氧化铝
hydrated antimony pentoxide (HAP) 水合五氧化二锑
hydrated barta 氢氧化钡 $Ba(OH)_2$
hydrated calcium sulfate 水合硫酸钙
hydrated cation 水合阳离子
hydrated cellulose 水合纤维素
hydrated chromium green 水合铬绿
hydrated chromium oxide 水合氧化铬(绿) $Cr_2O_3·2H_2O$
hydrated electron 水合电子*
hydrated form 水合式
hydrated grease 水化润滑脂
hydrated halloysite 水埃洛石；多水高岭土
hydrated ion 水合离子
hydrated lime (=slaked lime) 熟石灰
hydrated magnesium aluminum silicate 水合硅酸铝镁〔凹凸棒土的主要成分〕
hydrated magnesium silicate 水合硅酸镁；滑石粉
hydrated metal cation 水合金属阳离子
hydrated micelle 水化胶束
hydrated nitrocellulose 水合硝酸纤维素
hydrated oxide 水化氧化物
hydrated peroxide 水合过氧化物
hydrated proton 水合质子
hydrated radius (离子的)水化半径
hydrated salt 水合盐
hydrated sheath 水化层
hydrated shell 水化层
hydrated soap 含水皂
hydrate inhibitor 水化抑制剂
hydrate of aluminium 氢氧化铝 $Al(OH)_3$
hydrate of barium 氢氧化钡 $Ba(OH)_2$
hydrate storage 水化法储存
hydrate water 水合水；结合水
hydrating 吸水
hydration 水合*
hydration energy 水合能*
hydration heat 水合热
hydration isomerism 水合同分异构(现象)
hydration layer 水化层
hydration number 水合数*
hydration of ion 离子(的)水合作用
hydration of olefines 烯烃的水化作用
hydration plant 水化设备

hydration shell　水合外层*
hydration value　水合值
hydration water　结合水
hydratisomery　水合同分异构(现象)
hydrator　水化器
hydratropic acid　氢化阿托酸
hydratroponitrile　氢化阿托腈　$C_6H_5CH(CH_3)CN$
hydratropoyl　氢化阿托酰；邻苯丙酰　$C_6H_5CH(CH_3)CO-$
hydraulic　水力的；水压的；液压的
hydraulic accumulator　水压储蓄器
hydraulic aided gun (=hydraulic assisted gun)　液压助喷喷枪
hydraulic airless spray　液压无空气喷涂
hydraulic assisted gun (=hydraulic aided gun)　液压助喷喷枪
hydraulic atomizing　液压雾化；液压喷雾
hydraulic atomizing principle　液压雾化原理
hydraulic barrel lifter　水力桶状升降机
hydraulic binder　水力结合器
hydraulic bolting device　液压上紧(螺钉)装置
hydraulic brake fluid　水力制动用液体
hydraulic bronze　耐蚀铅锡黄铜
hydraulic bursting test　水压爆破试验
hydraulic cement　水硬水泥
hydraulic characteristic of column　精馏柱的水力特性
hydraulic classification　水力分粒法
hydraulic classifier　水力分粒机
hydraulic conveying　水力输送
hydraulic coupling　水力联轴器；液力联轴器
hydraulic cyclone (=hydrocyclone)　旋液分离器
hydraulic cylinder　水筒；水压力缸
hydraulic decoking　水力法出焦
hydraulic disperser　液压分散机
hydraulic drive　液(压传)动
hydraulic dynamometer　水力测功计
hydraulic elevator　液压升降机
hydraulic elutration　水选；水力淘析；水簸
hydraulic extruder　液压式压出机；柱塞式压出机
hydraulic flow　湍流
hydraulic fluid　水力(系统用)流体
hydraulic fluid-lift platform　激动升降台
hydraulic fluid pressure　①液压油压力②液压系统压力
hydraulic foam tower　水力泡沫器
hydraulic force　水力
hydraulic fracturing　水力破碎
hydraulic gradient　水力梯度
hydraulic head　液压压头
hydraulic hoist　水力提升机

hydraulic hose　水力制动用软管
hydraulic index　水硬率
hydraulic jack　液压千斤顶
hydraulic jig　水簸机
hydraulic joystick　液压操纵杆；水力操纵杆
hydraulic leather　水压机革〔密封用革〕
hydraulic lift　水力起重机
hydraulic lime　水硬石灰
hydraulic loss　水力损失
hydraulic machine　水压机
hydraulic main　水压主管
hydraulic mangle　液压浸轧机
hydraulic mining　水力采矿法
hydraulic modulus　水硬系数
hydraulic mortar　水凝灰浆
hydraulic motor　液压马达
hydraulic oil　液压油
hydraulic operation　液压动作
hydraulic packing　液压封垫；水封
hydraulic pipe　水力管；输水管
hydraulic pipe-line　输水管路
hydraulic planer　液压刨床
hydraulic press　液压机
hydraulic press cylinder　平板机液压筒
hydraulic pression extruder　液压柱塞式压出机
hydraulic press oil　液压机用油
hydraulic pressure　液压
hydraulic pressure gradient　液压梯度
hydraulic pressure head　液压压头
hydraulic pressure system　液压系(统)
hydraulic pressure test　液压试验
hydraulic pump　水力泵；液压泵
hydraulic rabbit capsule　液动跑兔盒；液动传递(靶)盒
hydraulic radius　水力半径；液压半径
hydraulic ram　压力扬汲机
hydraulic ram press　液压平板机
hydraulics　水力学；液压
hydraulic separation　水力离析
hydraulic separator　水力离析器
hydraulic shear　水力剪切机
hydraulic slip　液体漏失
hydraulic spraying　液压喷涂；无空气喷涂
hydraulic spring grease cup　水力弹簧油杯
hydraulic squeezing arrangement　液力压轧装置
hydraulic test　水压试验
hydraulic transmission fluid　水力传动系统用流体；液压传动液
hydraulic turbine　水轮机

hydraulic universal staking machine 液压立式铆软机
hydraulic variable speed drive 液力变速传动
hydraulic vulcanizing press 水压硫化机
hydraulic water 水压用水
hydrautic cyclone 水力旋流分粒机
hydrazi 1,2-亚肼代；环亚联氨基
hydrazi-acetic acid 1,2-亚肼代乙酸
hydrazide 酰肼*
hydrazi-methylene 1,2-亚肼代甲烷
hydrazine 肼；联氨 NH_2NH_2
hydrazinecarboxylate 肼羧酸酯
hydrazine carboxylic acid 肼羧酸 $NH_2NHCOOH$
hydrazine derivative 肼的衍生物
hydrazine hydrate 水合肼
hydrazine hydrochloride 盐酸肼 $NH_2NH_2 \cdot HCl$
hydrazine hydrogen chloride 氯化氢肼；盐酸肼
hydrazine-isophthalic acid-terephthalic acid copolymer 肼-间苯二酸-对苯二酸共聚物
hydrazine method 联氨法
hydrazine nitrate 硝酸肼
hydrazine reaction 肼反应
hydrazine sulfate 硫酸肼
hydrazine sulphonate 磺酸肼
hydrazine-terephthalic acid copolymer 肼-对苯二甲酸共聚物
hydrazinium 铵 $[NH_2NH_3]$；$[NH_3NH_3]$
hydrazino 肼基；联氨基 $H_2NNH—$
hydrazino-benzoic acid 肼基苯甲酸 $H_2NNHC_6H_4CO_2H$
hydrazino-borane 肼基硼烷
hydrazinolysis 肼解(作用)
hydrazo 1,2-亚肼基；联氨基 $—NHNH—$
hydrazoate 叠氮化物 $M[N_3]$
hydrazo-benzene 1,2-亚肼基苯；二苯肼 $C_6H_5NHNHC_6H_5$
hydrazobenzoic acid 1,2-亚肼基苯甲酸 $(NHC_6H_4COOH)_2$
hydrazo compound 氢化偶氮化物*
hydrazo derivative 1,2-亚肼基衍生物
hydrazo-dicarbonamide 联二脲；亚肼基二甲酰胺 $(H_2NCONH)_2$
hydrazoic acid 叠氮酸
hydrazoindol 亚肼基吲哚；1,2-二吲哚肼
hydrazonaphthalene 1,2-二萘肼
hydrazone 腙*
hydrazone reaction 腙反应
hydrazone tautomer 腙式互变异构体
hydrazonic acid 腙酸 $RC(OH)=NNN_2$

hydrazonium 铵
hydrazonium salt 铵盐
hydrazono 亚肼基；亚联氨基 $H_2NN=$
hydrazo rearrangement (=benzidine rearrangement) 联苯胺重排
hydrazotoluene 1,2-二甲苯肼
hydrazulmine 氢氮明 $C_4H_6N_4$〔一种氰和氨的反应物〕
hydrazy 偕腙肼 $RC(=NNH_2)NHNH_2$
hydremia 稀血症
hydric (含)羟的
hydride 氢化物
hydride-abstraction 夺氢化物(作用)
hydride atomic absorption spectrometry 氢化物原子吸收光谱法
hydride generation 氢化物发生(法)
hydride generation-atomic absorption spectrometry 氢化物发生-原子吸收光谱法
hydride generation-atomic fluorescence spectrometry 氢化物发生-原子荧光光谱法
hydride generator 氢化物发生器
hydride ion 氢负离子
hydride of arsenic 砷的氢化物；砷化(三)氢；胂 AsH_3
hydride separation-atomic absorption method 氢化物分离原子吸收法
hydride shift 氢负离子转移
hydride-shift polymerization 氢化物位移聚合
hydride shift reaction 氢负离子转移反应
hydrides of nitrogen 氮的氢化物〔如 NH_3; N_2H_4; N_2H_2 和 $H[N_3]$等〕
hydride transfer 氢负离子转移
hydride-transfer reaction 氢负离子转移反应
hydridometallocarborane 氢化金属碳硼烷
hydrin〔词尾〕醇〔取代的醇〕
hydrindanol 茚烷醇
hydrindene (=indane) 二氢化茚
hydrindenic acid 二氢化茚酸
hydrindone 二氢化茚酮
hydriodic acid 氢碘酸
hydriodide 氢碘化物
hydrion ①氢离子②质子
hydrion paper pH 试纸
hydro ①氢化的②氢的③水
hydroabietic acid 氢化松香酸
hydroabietyl alcohol 氢化枞醇
hydroabietylamine ①氢化枞胺②氢化松香胺
hydroacylation 加氢酰化*
hydroalkylation 加氢烷基化
hydroammonolysis (=reductive ammonolysis) 氢化氨解

(作用)
hydroaromatic 氢化芳族的
hydroaromatic compound 氢化芳族化合物
hydroaromatic series 氢化芳香系
hydrobalata 氢化巴拉塔树胶
hydrobenzamide (=tribenzal diamine) 三苯甲醛缩二氨；氢化苯酰胺〔俗〕 $C_6H_5CH=(N=CHC_6H_5)_2$
hydrobenzoin 氢化苯偶姻；对称二苯基乙二醇 $(C_6H_5CHOH)_2$
hydrobiology 水生生物学
hydrobiotite 水黑云母
hydroblasting 水力清砂；水力清理
Hydrobon 催化加氢精制
Hydrobon catalyst 加氢精制催化剂；加氢脱硫催化剂〔以氧化铝为载体的钴钼系统催化剂〕
hydroborated 硼氢化的
hydroborates (=boron hydrides) 硼氢化合物
hydroboration 硼氢化*
hydroborons 氢化硼
hydrobromic acid 氢溴酸；溴化氢 HBr
hydrobromic ether 溴代烃 C_2H_5Br
hydrobromination 溴氢化作用
hydrobromo-auric acid 溴金酸 [$HAuBr_4$]
hydrocal plaster 高强度石膏水泥
hydrocaoutchouc 氢化橡胶
hydrocarbon 碳氢化合物；烃
hydrocarbon alcohol 烃基醇
hydrocarbon analysis 烃类分析
hydrocarbon analyzer 烃分析器
hydrocarbonate 碳酸氢盐 $MHCO_3$
hydrocarbonate of sodium 小苏打 $NaHCO_3$
hydrocarbon-based fuel 烃基燃料
hydrocarbon-based solution 烃基溶液
hydrocarbon black 烃黑；炭黑
hydrocarbon chain 烃链
hydrocarbon coking 烃焦化
hydrocarbon compound 碳氢化合物
hydrocarbon conversion 烃类转化
hydrocarbon conversion process 烃类转化过程
hydrocarbon core 烃内核
hydrocarbon cracker 烃裂化器
hydrocarbon cracking 烃裂化
hydrocarbon decoking 烃脱焦
hydrocarbon desulfurization 烃脱硫
hydrocarbon detector with alarm 烃类报警探测仪
hydrocarbon distribution 烃类分布
hydrocarbon elastomer 烃类弹性体
hydrocarbon extraction 烃类抽提

hydrocarbon flame 烃火焰
hydrocarbon-formaldehyde resin 烃-甲醛树脂
hydrocarbon fuel 烃类燃料
hydrocarbon functional groups 烃官能团
hydrocarbon gas 烃气；气态烃
hydrocarbon gas-turbine fuel 燃气轮机烃类燃料
hydrocarbon gels 烃类胶质
hydrocarbon grease 烃类润滑脂
hydrocarbon group analysis (=hydrocarbon type test) 烃类分析
hydrocarbon hydrate 烃水合物
hydrocarbon hydroisomerization 烃加氢异构化(作用)
hydrocarbon in air-vapour mixture 蒸气-空气混合物中的烃(类蒸气)
hydrocarbon isomerization 烃异构化(作用)
hydrocarbon mixture 烃混合物
hydrocarbon oil 烃油
hydrocarbon oil plasticizer 烃油增塑剂
hydrocarbon oligomer 烃齐聚物
hydrocarbon performer furnace 烃类转化炉；烃类加工用炉
hydrocarbon phase 烃相
hydrocarbon polymer 碳氢聚合物；烃类聚合物
hydrocarbon (polymer) elastomer 烃高聚物弹性体
hydrocarbon pressure reforming 烃加压重整
hydrocarbon processing 烃加工
hydrocarbon products 烃类产品
hydrocarbon pyrolysis 烃裂解
hydrocarbon reforming 烃重整
hydrocarbon resin 烃类树脂*
hydrocarbon ring assembly 集合烃环
hydrocarbon solubility 烃类溶解度
hydrocarbon-soluble cobalt compound 烃溶钴化物
hydrocarbon-soluble metal salt catalyst 烃溶金属盐催化剂
hydrocarbon solvent 烃类溶剂
hydrocarbon spectrum 烃谱
hydrocarbon steam conversion 烃蒸气转化
hydrocarbon steam reforming 烃蒸气重整
hydrocarbon stripper 烃汽提塔
hydrocarbon succinic amide 烃基琥珀酰胺
hydrocarbon sulfones 烃砜
hydrocarbon sweetening 烃脱硫
hydrocarbon synthesis 烃类合成
hydrocarbon type analysis (=hydrocarbon group analysis) 烃类族分析
hydrocarbon type content 烃型含量
hydrocarbon type oil 烃类油

hydrocarbon type test (=hydrocarbon group analysis)　烃类分析

hydrocarbon vapor monitoring method　烃类气体监测方法

hydrocarbon vapour　烃蒸气

hydrocarbon water gas　烃类制得的水煤气

hydrocarbon water gas process　烃类制水煤气法

hydrocarbon with condensed nuclei (=hydrocarbon with condensed rings)　稠环烃

hydrocarbon with condensed rings (=hydrocarbon with condensed nuclei)　稠环烃

hydrocarbon with separated nuclei (=hydrocarbon with separated rings)　集合环烃

hydrocarbon with separated rings (=hydrocarbon with separated nuclei)　集合环烃

hydrocarbonylation　加氢甲酰化(作用)

hydrocarbostyril　氢化喹诺酮　C_9H_9ON

hydrocarboxylation　氢羧基化

hydrocarbyl　烃基

hydrocarbylation　烃基化(作用)

hydrocarbyl hexahydropyrimidine　烃基六氢嘧啶

hydrocarbyl tetracarboxylic acid　烃基四羧酸

hydrocatalytic reforming　临氢催化重整

hydrocellulose　水解纤维素

hydrocellulose nitrate　水解纤维素硝酸酯

hydrocephalus　脑积水；水脑

hydrochelidonic acid　氢化白屈菜酸；(对称)庚酮二酸　$CO(CH_2CH_2COOH)_2$

hydrochemistry　水质化学

hydrochinidine (=hydroconchinine;hydroquinidine)　氢化奎尼定；氢化奎宁

hydrochloric acid　盐酸；氢氯酸　HCl

hydrochloric acid spray test　盐酸喷雾试验

hydrochloric ether　氯代烃

hydrochloride　氢氯化物；盐酸化物

hydrochlorinated rubber　盐酸橡胶

hydrochlorination　氢氯化反应

hydrochlorithiazide　氢氯噻嗪〔药〕

hydrochloro-auric acid　氯金酸　$H[AuCl_4]$

hydrocinnamaldehyde　氢化肉桂醛；苯基丙醛　$C_6H_5CH_2CH_2COH$

hydrocinnamamide　氢化肉桂酰胺；苯丙酰胺　$C_6H_5(CH_2)_2CONH_2$

hydrocinnamic acid (=phenylpropionic acid)　氢化肉桂酸；苯基丙酸　$C_6H_5CH_2CH_2CO_2H$

hydrocinnamoyl　氢化肉桂酰基；苯丙酰基　$C_6H_5CH_2CH_2CO-$

hydrocinnamyl　氢化肉桂基；苯丙基　$C_6H_5CH_2CH_2-$

hydrocinnamyl alcohol　氢化肉桂醇；苯丙醇　$C_6H_5(CH_2)_2CH_2OH$

hydroclassifier　水力分粒器

hydroclassifying　水力分粒法

hydroclave　液压釜；蒸缸

hydroclone　水力旋流器

hydrocodimer　氢化共二聚体

hydrocol naphtha fraction　铁催化剂流化床合成的石脑油馏分

hydrocol process　铁催化剂流化床合成过程

hydroconchinine (=hydroconquinine)　氢化奎宁

hydrocondensation　加氢缩合反应

hydrocone crusher　液压式锥形破碎机

hydroconquinine (=hydroconchinine)　氢化奎宁

hydro-conversion　加氢转化(作用)；加氢转换(作用)

hydro-cooling　用水冷却

hydrocortisone　氢化可的松；皮质(甾)醇

hydrocotarnine　氢化可塔宁

hydrocracker　加氢裂化器

hydrocracking　氢化裂解

hydrocracking gas oil　加氢裂化粗柴油

hydrocracking process　加氢裂化法；加氢裂解法

hydrocracking unit　加氢裂化装置

hydrocyanation　氢氰化(作用)

hydrocyanic acid (=cyanic acid;hydrogen cyanide)　氢氰酸；氰化氢　HCN

hydrocyanic acid fumigant　氢氰酸蒸熏剂

hydrocyanic ester (=hydrocyanic ether)　氰酯；腈　RCN

hydrocyanic ether (=hydrocyanic ester)　氰酯；腈

hydrocyanide　氢氰化物

hydrocyclone (=hydraulic cyclone)　旋液分离器

hydrocyclone separator　旋液分离器

hydrocyclo-rubber　氢(化)环(化)橡胶

hydrodealkylation　加氢脱烷基化(作用)

hydrodecyclization　氢化开环作用

hydro-denitrification　加氢脱氮(作用)

hydro-denitrogenation-hydrocracking process　加氢脱氮-加氢裂化联合法

hydrodepolymerization　加氢解聚(作用)

hydroderivating　加氢衍生(作用)

hydrodesulfurization (=hydrodesulfurization process)　加氢脱硫过程

hydrodesulfurization process (=hydrodesulfurization)　加氢脱硫过程

hydrodesulfurizing　加氢脱硫

hydrodesulphurization　氢化脱硫作用

hydrodimerization　加氢二聚化；加氢偶聚作用

hydrodimerized　加氢二聚的

hydrodisproportionation 加氢歧化
hydrodynamic ①流体动力学②流体动力(的)
hydrodynamically equivalent sphere 流体力学等效球
hydrodynamic chromatography 流体动力学色谱法
hydrodynamic diffusion 流体动力学扩散
hydrodynamic drag 流体动力学阻力
hydrodynamic effect 流体效应
hydrodynamic equilibrium system 流体动力学平衡体系
hydrodynamic film 流体动力薄膜
hydrodynamic film formation 流体动力薄膜的形成
hydrodynamic friction 流体动摩擦
hydrodynamic interaction 流体动力学相互作用
hydrodynamic length (大分子)流体动力学长度；流线型长度
hydrodynamic lubrication 流体润滑
hydrodynamic lubrication region 流体润滑范围
hydrodynamic orientation 流体动力学取向；流线型取向
hydrodynamic pressure 流体压力
hydrodynamic radius 流体动力学半径
hydrodynamics 流体动力学
hydrodynamic screening 流体力学屏
hydrodynamic shear 流体剪切力
hydrodynamic stability 流体动力学稳定性
hydrodynamic theory of fluid lubrication 流体润滑的流体动力学理论
hydrodynamic voltammetry 流体动力学伏安法
hydrodynamic volume 流体力学体积*
hydrodynamometer 流速计
hydroelectric 水电的
hydroelectrical cell 水电池〔电化〕
hydroelectric power 水电
hydroentanglement 水刺；水力缠结
hydroergotinine ①麦角毒素②氢化麦角素
hydro-exhaustion 抽水；吸水
hydroexpansivity 水膨胀性
hydro-extraction ①脱水②水力提取
hydroextractor 脱水机；脱水器
hydroferricyanic acid 氰铁酸；六氰合铁氰酸 $H_3[Fe(CN)_6]$
hydrofined oil 加氢精制油
hydrofining 加氢精制；氢化提纯
hydrofining gasoline 氢化汽油
hydrofining-hydrocracking process 加氢精制-加氢裂化联合法
hydrofining of cracked distillates 裂化馏出油加氢精制
hydrofining solvent 加氢精制溶剂
hydrofinishing 加氢精制
hydrofluoaluminic acid 氟铝酸；$H_3[AlF_6]$
hydrofluoric acid (=fluohydric acid) 氢氟酸
hydrofluoric acid alkylation process 氢氟酸烷基化法
hydrofluoric ether 氟代烃
hydrofluoride 氢氟化物
hydrofluorinating 氢氟化
hydrofluorination 氢氟化作用
hydrofluorinator 氟氢化器
hydrofluosilicic acid 氟硅酸；六氟合硅氢酸 $H_2[SiF_6]$
hydroformate 临氢重整生成物；加氢重整汽油
hydroformer bottoms (=hydroformer heavy aromatics; hydroformer polymer) 临氢重整的残渣
hydroformer gasoline 临氢重整汽油
hydroformer heavy aromatics (=hydroformer bottoms) 临氢重整的残渣
hydroformer polymer (=hydroformer bottoms) 临氢重整的残渣
hydroformer vessel 临氢重整反应器
hydroforming ①临氢重整②临氢重整的
hydroforming process 临氢重整过程
hydroform method 液压成形法
hydroformylation 加氢甲酰基化*
hydroformylation of olefins 烯烃的醛化
hydrofracture 水力破碎
hydrofuge 防湿的；不透水的
hydrofuramide (=furfuramide) 糠醛胺；二氨缩三糠醛
hydrogasification 加氢气化
hydrogasoline 加氢汽油
hydrogel 水凝胶*
hydrogen ①氢②氢气
hydrogenable 可以氢化的
hydrogen-absorbing alloy 储氢合金
hydrogen abstraction reaction 夺氢反应
〔R·+HR ⟶ RH+R·〕
hydrogen acceptor 氢受体
hydrogen acid 氢酸
hydrogen acid ester (=hydrogen acid ether) 氢酸酯
hydrogen addition 加氢
hydrogen adsorption 氢吸附
hydrogen analyzer 氢分析计
hydrogenant ①氢化的②还原的
hydrogenant agent 氢化剂
hydrogen arc lamp 氢弧灯
hydrogen arsenide 砷化(三)氢；胂 AsH_3
hydrogenatable 可氢化的
hydrogenate ①氢化②氢化物
hydrogenated 氢化(了)的
hydrogenated alkylated tar 氢化烷化焦油
hydrogenated aviation fuel 加氢的航空燃料

hydrogenated butadiene-acrylonitrile rubber　氢化丁腈橡胶
hydrogenated coal tar oil　加氢的煤焦油
hydrogenated fat　氢化脂；硬化脂
hydrogenated fatty acid　氢化脂肪酸
hydrogenated fuel　加氢燃料
hydrogenated gasoline　加氢汽油
hydrogenated methyl abietate　氢化松香酸甲酯〔增塑剂〕
hydrogenated motor spirit　氢化汽油
hydrogenated naphtha　氢化石脑油
hydrogenated naphthalene　氢化萘
hydrogenated oil　氢化油；硬化油
hydrogenated oil and fat　氢化油脂
hydrogenated polybutene　氢化聚丁烯
hydrogenated polymer　氢化聚合物
hydrogenated products　加氢产品
hydrogenated propylene tetramer (HPT)　氢化丙烯四聚物
hydrogenated residuum　加氢后残渣
hydrogenated resin　氢化树脂
hydrogenated ring　加氢的环
hydrogenated rosin　氢化松香
hydrogenated rubber　氢化橡胶
hydrogenated terphenyl　氢化三联苯
hydrogenating desulfurization　加氢脱硫
hydrogenating of coal　煤的氢化
hydrogenation (=hydrogenization)　氢化*
hydrogenation apparatus　氢化装置
hydrogenation catalyst　氢化催化剂
hydrogenation cracking　氢化裂解
hydrogenation gas chromatography　氢化气相色谱(法)
hydrogenation gases　加氢过程中生成的气体
hydrogenation gasoline　氢化汽油
hydrogenation of carbon dioxide　二氧化碳的加氢
hydrogenation of carbon monoxide　一氧化碳的加氢
hydrogenation of coal　煤的氢化
hydrogenation of oil　油的氢化(加氢)作用
hydrogenation of shale oil　页岩油加氢
hydrogenation of tar　焦油氢化
hydrogenation plant　加氢装置；加氢设备
hydrogenation pressure　加氢压力
hydrogenation process　加氢过程
hydrogenation reaction chamber　加氢反应器
hydrogenation reactor　氢化反应器
hydrogenation residues　加氢残渣
hydrogenation solvent　氢化溶剂
hydrogenation unit　加氢装置；加氢设备
hydrogenation vessel　氢化器皿
hydrogenative cleavage　氢化裂开

hydrogenator　氢化器
hydrogen attack (=hydrogen damage)　氢蚀
hydrogen balance　氢平衡
hydrogen blanket　氢气层
hydrogen blistering　氢致起泡
hydrogen bomb　氢弹
hydrogen bond　氢键
hydrogen bond acceptor　氢键受体
hydrogen-bonded　氢键键合的
hydrogen-bonded adduct　氢键加成物
hydrogen-bonded hydrolyzate　氢键合(的)水解物
hydrogen bonder　氢键键合剂
hydrogen bond index　氢键指数
hydrogen bonding　氢键合；氢键键合
hydrogen-bonding affinity　氢键(键合)亲和性
hydrogen bonding agent　氢键键合剂
hydrogen bonding energy　氢键合能
hydrogen bound interaction force　氢键相互作用力
hydrogen bridge　氢桥
hydrogen-bridged ion　氢桥离子
hydrogen brittleness　氢脆
hydrogen bromide　溴化氢
hydrogen-bromine reaction　氢溴反应
hydrogen-carbon link　氢碳键
hydrogen-carbon ratio　氢碳比率
hydrogen carrier　氢载体
hydrogen-cascade process　阶式加氢法
hydrogen chloride　氯化氢
hydrogen chloride column　氯化氢塔
hydrogen-chlorine cell　氢氯电池
hydrogen coil　氢气旋管
hydrogen complex　氢配合物(配合物 ML 中 M 为 H 时)
hydrogen-containing　含氢的
hydrogen content of hydrocarbons　烃类中的氢含量
hydrogen cracking　加氢裂化；氢压下裂化
hydrogen cyanide (=hydrocyanic acid)　氢氰酸；氰化氢 HCN
hydrogen cyanide sensor　氰化氢传感器
hydrogen damage (=hydrogen attack)　氢蚀
hydrogen deficient fuel　贫氢燃料〔含氢少的燃料〕
hydrogen delayed crack　氢致潜伏开裂(龟裂)；氢气滞后破裂
hydrogen desulfurization　加氢脱硫
hydrogen difluoride ion　二氟化氢离子
hydrogen discharge lamp　氢(放电)灯
hydrogen discharge tube　氢放电管
hydrogen donator　授氢体
hydrogen donor　氢供体

hydrogen donor diluent visbreaking process　氢气稀释剂；
　　减黏裂化法
hydrogen doping　掺氢的；氢掺杂
hydrogen electrode　氢电极*
hydrogen embrittlement　氢脆
hydrogen end group　氢端基
hydrogen equivalent　氢当量
hydrogen evolution　氢形成；氢逸出；氢析出
hydrogen evolution thermal conductometry　逸氢热导法
hydrogen exchange　氢交换
hydrogen exponent (=hydrogen ion exponent)　氢离子
　　指数
hydrogen extraction apparatus　氢提取装置
hydrogen flame detector　氢火焰检测器
hydrogen flame ionization detector　氢火焰电离检测器
hydrogen flame temperature detector　氢火焰温度检测器
hydrogen fluoride　氟化氢
hydrogen fluoride alkylation process　氟化氢烷(基)化法
hydrogen fluoride extraction　氟化氢抽提
hydrogen form　氢型
hydrogen gas　氢气
hydrogen gas cylinder　氢气钢瓶
hydrogen ghydrogen generator　氢气发生器
hydrogen halide　卤化氢
hydrogen halide acceptor　卤化氢接受体
hydrogen-halides reaction　氢卤反应
hydrogen induced cracking (HIC)　氢致开裂；氢致龟裂
hydrogen-in-petroleum test　石油中氢含量测定
hydrogen iodide　碘化氢
hydrogen iodide test　氢碘酸试验
hydrogen ion　氢离子
hydrogen ion activity　氢离子活度
hydrogen ion concentration　氢离子浓度
hydrogen ion determination apparatus　氢离子测定器；pH
　　测定器
hydrogen ion exponent (=hydrogen exponent)　氢离子
　　指数
hydrogen ion index　氢离子指数
hydrogen ion indicator　氢离子指示剂；酸碱指示剂
hydrogen ion selective electrode　氢离子选择电极
hydrogen isomerization process　(加)氢异构化过程(法)
hydrogenium　金属氢
hydrogenization (=hydrogenation)　加氢(作用)
hydrogenized　氢化了的
hydrogenizing　氢化
hydrogen lamp　氢灯
hydrogen-like atom　类氢原子*
hydrogen microflare detector　氢微闪检测器；氢(火)焰温
　　度检测器
hydrogen migration　氢转移
hydrogen migration polymerization　氢转移聚合(作用)
hydrogen molecular ion　氢分子离子
hydrogen molecule ion　氢分子离子
hydrogen nitrate　硝酸　HNO_3
hydrogen nitride　氨　NH_3
hydrogen nucleus　氢核〔即质子〕
hydrogen number　氢值
hydrogenolysis　氢解*
hydrogenolytic cleavage　氢(化裂)解(作用)
hydrogenous　①含氢的②氢的
hydrogenous coal　褐煤；含水量高的煤
hydrogen overpotential　氢过电位；氢超电压
hydrogen overvoltage　氢超电压
hydrogen-oxygen coulombmeter　氢氧库仑计；水解电
　　量计
hydrogen-oxygen fuel cell　氢氧燃料电池*
hydrogen pentasulfide　五硫化二氢　$H_2[S_5]$
hydrogen permeability test　渗氢(性)试验
hydrogen peroxide　过氧化氢　$H_2[O_2]$
hydrogen peroxide decomposition　过氧化氢分解
hydrogen peroxide freezing process (=Tallalay process)
　　过氧化氢冷冻法
hydrogen peroxide-permanganat system　过氧化氢-高锰
　　酸盐系统
hydrogen peroxide solution　过氧化氢溶液
hydrogen persulfide　过硫化氢
hydrogen phosphate　磷酸氢盐
hydrogen phosphide　磷化氢　PH_3
hydrogen plasma stream　氢等离子流
hydrogen polarization　氢极化
hydrogen preferential oxidation　氢优先氧化
hydrogen process　应用氢的(石油加工)过程
hydrogen refining　加氢精制
hydrogen reforming　临氢重整
hydrogen revivification　(催化剂)用氢再生
hydrogen-rich　富氢的
hydrogen-rich flame　富氢焰
hydrogen-rich gas　富氢气
hydrogen-rich recycle　富氢的循环气
hydrogen salt　酸式盐
hydrogen scale　氢标度
hydrogen selenide　硒化(二)氢　H_2Se
hydrogen sensor　氢传感器
hydrogen shift　氢移位
hydrogen shift polymerization　氢位移聚合
hydrogen-silver palladium separator　氢-银钯分离器

hydrogen soldering 氢焊
hydrogen stream 氢气流
hydrogen sulfate ①硫酸氢〔即: 硫酸〕②硫酸氢盐 MHSO$_4$
hydrogen sulfate of sodium 硫酸氢钠 NaHSO$_4$
hydrogen sulfide 硫化氢
hydrogen sulfite ion 硫化氢离子
hydrogen telluride 碲化氢 H$_2$Te
hydrogen tetrasulfide 四硫化二氢 H$_2$[S$_4$]
hydrogen thermometer 氢温度计
hydrogen transfer (=hydrogen transference) 氢转移
hydrogen transference (=hydrogen transfer) 氢转移
hydrogen transfer polymerization 氢转移聚合*
hydrogen transfer reactions 氢转移反应; 氢传递反应
hydrogen transportase 氢转移酶
hydrogen trisulfide 三硫化二氢 H$_2$[S$_3$]
hydrogen tritide 氚(化)氢
hydrogen upgrading 氢气提浓
hydrogen value 氢值
hydrogen wave 氢波*
hydrogermanation 锗氢化作用
hydroglucan 水解葡聚糖
hydrograph 过程线
hydrogutta-percha 氢化杜仲胶; 氢化古塔坡树胶
hydrohalic 氢卤的
hydrohalide (=hydrohalogen) 氢卤化物
hydrohalogen (=hydrohalide) 氢卤化物
hydrohalogenating agent 氢卤化剂
hydrohalogenation 氢卤化作用; 加上卤化氢
hydrohematite 水赤铁矿
hydroherderite 水磷铍钙石
hydroheterolite 水锌锰矿
hydroilmenite 水钛铁矿
hydroiodic acid 氢碘酸 HI
hydroiodination 碘氢化反应
hydroiodo-auric acid 碘金酸; 四碘合金氢酸 H[AuI$_4$]
hydroisomerisation (=hydroisomerization) 氢化异构化(作用)
hydroisomerizing 加氢异构
hydrokinetics 流体动力学
hydrol 二聚水分子〔即 H$_4$O$_2$〕
hydrolabil 对水不稳定的; 非水稳的
hydrolactometer 乳比重计
hydrolapachol 氢化拉帕醇
hydrolinc 吹制油
hydrolith 氢化钙
hydrolose 甲基纤维素
hydrolube 氢化润滑油

hydrolysable tan 水解类鞣质
hydrolysable tannin extracts 水解类栲胶
hydrolysable tannins 水解(类)单宁; 水解(类)鞣质
hydrolysate (=hydrolyzate) 水解产物; 水解液
hydrolysed polyvinyl acetate 水解(的)聚醋酸乙烯酯
hydrolysis 水解*
hydrolysis constant 水解常数*
hydrolysis current 水解电流
hydrolysis effect 水解效应
hydrolysis hydrogenation 水解氢化
hydrolysis of polymers 高聚物的水解
hydrolysis polycondensation 水解缩聚
hydrolysis process 水解工艺
hydrolysis product 水解产物
hydrolysis-resistant carboxylate surafce 耐水解的羧酸盐表面
hydrolyst 水解催化剂
hydrolyte 水解质
hydrolytic 水解的
hydrolytic action 水解作用
hydrolytical 水解的
hydrolytic complex-formation titration 水解络离子形成滴定
hydrolytic constant 水解常数
hydrolytic deaminization 水解脱氨
hydrolytic decomposition 水解
hydrolytic degradation 水解降解
hydrolytic dissociation 水离解
hydrolytic elimination 水解排除(作用)
hydrolytic enzyme 水解酶
hydrolytic instability 水解不稳定性
hydrolytic polycondensation 水解缩聚(作用)
hydrolytic polymerization 水解聚合
hydrolytic precipitation titration 水解沉淀滴定
hydrolytic reagent 水解剂
hydrolytic resistance 抗水解性
hydrolytic scission 水解分裂
hydrolytic stability 水解稳定性
hydrolyzable substituent 可水解取代基
hydrolyzable tannin 水解单宁; 水解鞣酸
hydrolyzate (=hydrolysate) 水解产物; 水解液
hydrolyzed 水解了的
hydrolyzed polyacrylamide 水解的聚丙烯酰胺
hydrolyzed shellac 水解紫胶
hydrolyzed titanium dioxide 水解二氧化钛
hydrolyzer 水解器
hydrolyzing 水解的
hydromagnesite 水菱镁矿

hydromechanics	流体力学
hydrometallation	氢金属化*
hydrometallurgy	湿法冶金学
hydrometamorphism	水热变质
hydrometer (=araeometer;areometer)	（液体）比重计
hydrometer calibration	（液体）比重计校准
hydrometer degree	比重计标度
hydrometer method	比重法〔测定石油产品的比重〕
hydrometric (=hydrometrical)	测定比重的
hydrometrical (=hydrometric)	测定比重的
hydrometry (=aerometry)	液体比重测定（法）
hydromodulus	流量模数
hydronamic orientation	流体力学取向
hydronaphthoquinone	萘二酚
hydron blue R〔商〕	海昌蓝 R
hydrone	①钠铅合金②（单体）水分子
hydronephelite	水霞石
hydronephrosis	肾积水
hydronitric acid	（氢）叠氮酸 $H[N_3]$
hydronitrous acid (=nitroxylic acid)	亚硝酸 HNO_2
hydronium ion (=hydroxonium ion)	水合氢离子；水合质子 H_3O^+
hydrooligomer	氢化低聚物
hydroperoxidation	氢过氧化（作用）
hydroperoxide	氢过氧化物*
hydroperoxide decomposition	氢过氧化物分解
hydroperoxide formation	氢过氧化物形成
hydroperoxide initiated polymerization	氢过氧化物引发的聚合作用
hydroperoxy bond	氢过氧键
hydroperoxyl radical	过氧羟基
hydrophane	水蛋白石
hydrophil balance	亲水平衡
hydrophile	①亲水物②亲水胶体
hydrophile-lyophile balance (HLB)	亲水-亲油平衡*
hydrophile-lyophile balance number	亲水-亲油平衡值
hydrophilic (=hydrophylous)	亲水的
hydrophilic adsorbent	亲水吸附剂
hydrophilic bicomponent yarn	亲水(性)双组分丝
hydrophilic bis-biuret	亲水性二缩二脲
hydrophilic centre	亲水中心
hydrophilic chain	亲水链
hydrophilic colloid	亲水胶体*
hydrophilic ether bond	亲水醚键
hydrophilic filler	亲水(性)填料
hydrophilic finish	亲水性整理
hydrophilic gas electrode	亲水气体电极
hydrophilic gel	亲水胶
hydrophilic group	亲水基
hydrophilic hollow fibre	亲水性中空纤维
hydrophilic isocyanate	亲水性异氰酸酯
hydrophilic isocyanate terminated prepolymer	亲水性异氰酸酯封端预聚物
hydrophilicity	亲水性；亲水程度
hydrophilic-lipophilic balance	亲水性亲油性比率；亲水性亲油性平衡
hydrophilic monomer	亲水(性)单体
hydrophilic nature	亲水性
hydrophilic oleomicelle	亲水非极性芯胶束
hydrophilic polyether chain segment	亲水性聚醚链段
hydrophilic polymer	亲水聚合物
hydrophilic side chain	亲水侧链
hydrophilic sol	亲水溶胶
hydrophilic substance	亲水物质
hydrophilite	氯钙石
hydrophility	亲水性
hydrophilization process	亲水化法〔一种用表面活性剂溶液的分离工艺〕
hydrophilizing agent	亲水剂
hydrophilous (=hydrophilic)	亲水的
hydrophite	水铁蛇纹石
hydrophobe	①疏水物②疏水胶体
hydrophobe base	疏水基
hydrophobic	疏水的；憎水的
hydrophobic activation	疏水活化
hydrophobic adsorbent	疏水吸附剂
hydrophobic agent	疏水剂
hydrophobic association	疏水缔合*
hydrophobic bond	疏水键
hydrophobic chain	疏水链
hydrophobic chromatography	疏水色谱法
hydrophobic coating	疏水涂层
hydrophobic colloid	疏水胶体*
hydrophobic core	疏水内芯
hydrophobic dye	疏水染料
hydrophobic fibre	疏水性纤维
hydrophobic force	疏水力
hydrophobic gas electrode	疏水气体电极
hydrophobic group	疏水基
hydrophobic hydration	疏水水合作用
hydrophobic interaction chromatography	疏水作用色谱法
hydrophobicity	疏水性；疏水程度
hydrophobic membrane	疏水膜
hydrophobic microcapillary vent	憎水性微毛细管通气口

hydrophobic moisture transportation 疏水导湿性
hydrophobic oil 疏水油
hydrophobic organic polymer 疏水性有机聚合物
hydrophobic polyelectrolyte 疏水性聚合电解质
hydrophobic polymer 疏水聚合物
hydrophobic porous membrane 疏水多孔膜
hydrophobic pumice 疏水浮石；憎水浮石
hydrophobic residue 疏水残基
hydrophobic rubber 疏水(性)橡胶
hydrophobic segment 疏水链段
hydrophobic sol 疏水溶胶*
hydrophobic substance 疏水物质
hydrophobic surface modifier 疏水表面改性剂
hydrophobic synthetic fibre 疏水性合成纤维
hydrophobing agent 疏水剂
hydrophobization 疏水化
hydrophosphate 磷酸氢盐
hydrophylic property 亲水性
hydrophylous (=hydrophilic) 亲水的
hydroplaning 打滑行为；轮胎水膜滑行
hydroplastic 水增塑塑料
hydroplasticity 湿塑性
hydroplumbation 铅氢化作用
hydropneumatic accumulator 水气储压器
hydro-pneumatic clamp 液压气动合模装置；液压气动夹具
hydropolymer 氢化聚合物
hydropolymerization 氢化聚合作用
hydropress 水压平板机
hydropretreating 加氢预处理
hydroprocessing 加氢操作；加氢处理
hydropteridine 氢化蝶啶
hydropyrolysis process 水热解法
hydroquinidine (=hydroconchinine; hydrochinidine) 氢化奎尼定
hydroquinine (=hydrochinine) 氢化奎宁
hydroquinol (=chinol) 醌醇
hydroquinone 氢醌*；对苯二酚
hydroquinone diacetate 氢醌二乙酸酯；对苯二酚二乙酸酯
hydroquinone dibenzyl ether 氢醌二苄基醚；对苯二酚二苄醚 $(C_6H_5CH_2O)_2C_6H_4$
hydroquinone dihydroxyethyl ether 氢醌二羟基乙醚；1,4-二(β-羟基乙氧基)苯
hydroquinone dimethyl ether 氢醌二甲基醚；对苯二酚二甲醚 $C_6H_4(OCH_3)_2$
hydroquinone-krypton elathrate (HQ-Kr) 氢醌-氪笼形包合物

hydroquinone monobenzyl ether 氢醌一苄基醚；对苯二酚一苄醚 $C_6H_5CH_2OC_6H_4OH$
hydroquinonesulfonic acid 氢醌磺酸
hydroquinophthalein 氢醌酚酞 $C_{20}H_{12}O_5$
hydrorefining 加氢精制
hydroreforming (=hydroforming) 临氢重整
hydroresorcinol 1,3-环己二酮 $C_6H_8O_2$
hydrorubber 氢化橡胶
hydrorubeanic acid 氢化红氨酸；二巯基乙二胺 $NH_2CH(SH)(SH)CHNH_2$
hydrosafroeugenol 氢化黄樟丁香酚
hydroscope 验湿器
hydroscopic 收湿的
hydroscopicity 吸湿度；吸湿性；吸水性
hydroscopic moisture 湿存水
hydroscopic property 吸湿性；吸水性
hydroscopic substance 吸湿物
hydroseparator 水力分离器；沉降槽
hydroset 湿定形
hydrosetting process 湿定形工艺
hydrosilicate catalyst 氢化硅酸盐催化剂
hydrosilication 硅氢化作用*
hydrosilicocarbons 氢硅碳化合物
hydrosilicofluoric acid 氟硅酸；六氟合硅氢酸 $H_2[SiF_6]$
hydrosilicons〔复〕 硅氢化合物
hydrosilylation 氢化硅烷化
hydroskimming refinery 轻度加氢炼油厂
hydrosol 水溶胶*
hydrosoluble 水溶的
hydrosolvent 水溶剂
hydrosorbic acid 氢化山梨酸；氢化己二烯酸
hydros resin〔商〕 脱氢松香树脂
hydrostable 对水稳定的；抗水的
hydrostannation 锡氢化作用
hydrostatic(al) 流体静力的
hydrostatic balance (=specific gravity balance) 比重秤
hydrostatic bell 水压钟
hydrostatic capillary viscometer 流体静压毛细管黏度计
hydrostatic compression 围压压缩；静水压缩
hydrostatic creep 流体静力蠕变；静水力学蠕变
hydrostatic gauge 流体静压压力计
hydrostatic injection 静力进样
hydrostatic lubrication 流体静压(力)润滑
hydrostatic machine 水压机
hydrostatic method (=Westphal balance method) 韦氏比重秤法
hydrostatic oiler 水压加油器
hydrostatic pressure ①(流体)静压②(静)水压(力)

hydrostatic pressure injection 流体静压进样
hydrostatic pressure test 流体静压试验；水压试验
hydrostatics 流体静力学
hydrostatic sink 静流容器
hydrostatic state of stress 静水力学应力状态
hydrostatic stress (=isotropic stress) 各向同性应力；静水应力；流体静应力
hydrostatic tensile component 静水拉力分量
hydrostatic test 水压实验
hydrostatic weighing 静流称重
hydrosulfate ①硫酸氢盐 $MHSO_4$②硫酸化物 $R·H_2SO_4$
hydrosulfate of sodium 硫酸氢钠
hydrosulfide 氢硫化物 MSH
hydrosulfide group 巯基；氢硫基 HS—
hydrosulfide of ammonia 氢硫化铵 $(NH_4)HS$
hydrosulfite ①亚硫酸氢盐 $MHSO_3$②连二亚硫酸盐 $M_2S_2O_4$
hydrosulfite bleaching 亚硫酸氢盐漂白
hydrosulfite of sodium 亚硫酸氢钠 $NaHSO_3$
hydrosulfite vat 亚硫酸(氢盐)釜
hydrosulfo 巯基；氢硫基
hydrosulfuric acid 氢硫酸；硫化氢
hydrosulfuryl 巯基；氢硫(基) HS—
hydrosulphite ①亚硫酸氢盐②连二亚硫酸盐
hydrosulphite ager 还原蒸箱
hydrotalcite 水滑石
hydrotalcite-like compound 类水滑石
hydrotaxis 向水性
hydroterpin 氢化松节油
hydro-tester 水分测定器
hydrotetramer 氢化四聚物
hydrotherapy 水疗(法)
hydrothermal ①水热的；湿热的②热液的
hydrothermal contraction 水热收缩；热液收缩
hydrothermal crystallization 水热结晶
hydrothermal method 晶体生长水热法*
hydrothermal process 热液处理
hydrothermal stability 湿热稳定性
hydrothermal synthesis 水热合成法
hydrothermal transformation 水热转变
hydrothermal yellowing 湿热泛黄
hydrotimeter 水硬度计
hydrotimetric burette 水硬度滴定管
hydrotimetric flask 水硬度量瓶
hydrotorting 加氢干馏
hydrotreated 氢化处理的；加氢处理的
hydrotreater 加氢器

hydrotreating 加氢处理
hydrotreating facility 氢化处理设备
hydrotreating-hydrocracking process 加氢处理-加氢裂化联合法
hydrotreatment 加氢处理
hydrotrimer 氢化三聚物
hydrotrope 水溶助长剂
hydrotropic 水溶助长剂的
hydrotropic action 水溶助长作用
hydrotropic agent 水溶助长剂
hydrotropic extraction 水溶抽提
hydrotropic pulping 水溶助长性制浆法
hydrotropic solution 水溶溶液
hydrotropic solvent 水溶溶剂
hydrotropism 向水性；趋湿性
hydrotropy 水溶助长性
hydrotylosin (=relomycin) 氢泰乐菌素〔即：雷洛霉素〕
hydrourushiol 氢化漆酚
hydrous ①含水的②水合的；水化的③水状的
hydrous aluminum potassium silicate 水合硅酸铝钾；云母
hydrous calcium sulfate 含水硫酸钙；石膏
hydrous iron oxide 水合氧化铁；羟基氧化铁
hydrous lanolin 含水羊毛脂
hydrous magnesium silicate 含水硅酸镁；滑石粉
hydrous material 含水材料
hydrous oxide 水合氧化物
hydrous titanium oxide (HTO) 水合氧化钛
hydrous water (配)合水
hydrous wool fat 含水羊毛脂
hydrous zirconium oxide (HZO) 水合氧化锆
hydroviscose 氢化黏胶
hydroxamic acid 异羟肟酸
hydroxamino (=hydroxyamino) 羟氨基 HONH—
hydroxide 氢氧化物
hydroxide anion 氢氧离子；氢氧阴离子
hydroxide ion 氢氧离子；羟离子
hydroxide ion concentration 氢氧离子浓度；羟离子浓度
hydroxide radical ①羟基②氢氧根 HO—
hydroxidion 羟离子 HO—
hydroximic acid 羟肟酸 $RC(OH)=NOH$
hydroximino (=hydroxyimino) 肟基 $HON=$
hydroxocobalamin(e) 羟钴胺素；维生素 B_{12a}
hydroxo complex 羟配合物
hydroxo group 配位羟离子
hydroxo ligand 羟配位体
hydroxonium ion (=hydronium ion) 水合氢离子 H_3O^+
hydroxy 羟基 HO—

hydroxy-acetaldehyde 乙醇醛 CH_2OHCHO
hydroxyacetic acid (=oxyacetic acid) 羟基乙酸；乙醇酸 $HC_2OHCOOH$
hydroxy-acetone 羟基丙酮 CH_3COCH_2OH
hydroxyacetonitrile 羟基乙腈
hydroxy-acetophenone 羟苯乙酮；乙酰苯酚 $OHC_6H_4COCH_3$
hydroxy-acetylenic acid 羟基炔酸
hydroxy-acid ①含氧酸②羟(基)酸；醇酸
hydroxy-acid amide 羟基酰胺 $ROHCONH_2$
hydroxy-acid bromide 羟代酰溴
hydroxy-acid chloride 羟基酰氯
hydroxy-acid fluoride 羟基酰氟
hydroxy-acid halide 羟基酰卤
hydroxy-acid iodide 羟基酰碘
hydroxy-acid lithium soap 羟基酸锂皂润滑脂
hydroxy acrylic acid 羟基丙烯酸 $CHOH=CHCOOH$
hydroxy aldehyde 醇醛；羟基醛
hydroxy-aldehydic acid 醇醛酸 $CHOR(OH)COOH$
hydroxyalkanoic acid 羟基烷酸
hydroxyalkyl 羟烷基
N-(2-hydroxyalkyl) amide N-(2-羟烷基)酰胺
hydroxy-alkyl amine 羟基胺 $OHRNR_2$
hydroxyalkylation 羟烷基化*
hydroxyalkyl ether 羟烷基醚
hydroxyalkyl mercaptan 羟烷基硫醇
2-hydroxy-4-alkyloxy-benzophenone 2-羟基-4-烷氧基二苯甲酮〔光稳定剂〕
hydroxyalkyl phosphate 磷酸羟烷基酯
hydroxyallysine aldol 羟赖氨醛醇；联赖氨酸
hydroxy amide 羟基酰胺 $R(OH)CONH_2$
hydroxy amine 羟基胺 $R(OH)NH_2$
hydroxyamino (=hydroxamino) 羟氨基 $HONH-$
hydroxy-amino-acid 羟基氨基酸 $ROHRNH_2COOH$
hydroxy-amino-butyric acid 羟基氨基丁酸 $NH_2C_3H_5OHCOOH$
3β-hydroxy-20α-aminopregnene-5-glucoside 3β-羟基-20α-氨基孕烯-5-葡萄糖苷
hydroxyandrostenedione 羟雄(甾)烯二酮
hydroxy-anhydride 羟基酸酐
3-hydroxyanthranilic acid 3-羟基-2-氨基苯甲酸
hydroxyanthraquinone 羟基蒽醌
hydroxyapatite 羟基磷灰石*
hydroxyapatite implant fibre 掺入羟基磷灰石纤维
hydroxyaphylline 羟基无叶(假木贼)碱
hydroxyaralkyl radical 羟代芳烷基
hydroxyarene sulfonate 羟基芳烃磺酸盐
hydroxyaryl 羟芳基
hydroxyarylmethane 羟基芳甲烷
hydroxyaspergillic acid 羟基曲霉酸
hydroxyazobenzene 羟基偶氮苯
hydroxyazobenzene sulfonate 羟基偶氮苯磺酸盐 $HOC_6H_4N=NC_6H_4SO_3M$
5-hydroxy-barbituric acid 5-羟基巴比妥酸
hydroxy-benzalacetophenone (=hydroxy chalcone) 羟基亚苄基乙酰苯；羟基查耳酮 $C_6H_5COCH=CHC_6H_4OH$
o-hydroxy-benzaldehyde (=salicylaldehyde) 水杨醛；邻羟基苯醛 HOC_6H_4CHO
o-hydroxybenzoic acid (=salicylic acid) 水杨酸；邻羟基苯甲酸 $HOC_6H_4CO_2H$
1-hydroxybenztriazole 1-羟基苯并三唑
p-hydroxy benzyl acetone 对羟基苄基丙酮
hydroxy-benzyl alcohol 羟苄醇 $OHC_6H_4CH_2OH$
hydroxy-benzyl chloride 羟苄基氯 $OHC_6H_4CH_2Cl$
hydroxy-benzyl halide 羟苄基卤 $OHC_6H_4CH_2X$
hydroxy-benzyl isorhodanate (=hydroxy-benzyl isothiocyanate) 异硫氰酸羟基苄酯
hydroxy-benzyl isorhodanide (=hydroxy-benzyl isothiocyanate) 异硫氰酸羟基苄酯
hydroxy-benzyl isosulfocyanate (=hydroxy-benzyl isothiocyanate) 异硫氰酸羟基苄酯
hydroxy-benzyl isosulfocyanide (=hydroxy-benzyl isothiocyanate) 异硫氰酸羟基苄酯
hydroxy-benzyl isothiocyanate (=hydroxy-benzyl isothiocyanide) 异硫氰酸羟基苄酯 $OHC_6H_5CH_2NCS$
hydroxy-benzyl isothiocyanide (=hydroxy-benzyl isothiocyanate) 异硫氰酸羟基苄酯
hydroxy-benzyl mustard oil (=hydroxy-benzyl isothiocyanate) 异硫氰酸羟基苄酯
p-hydroxybenzylpenicillin 对苄羟青霉素；青霉素X
hydroxybetaine 羟基内铵盐〔抗静电剂〕
2-hydroxy-1,3-bis (4-phenylaminophenyloxy) propane 2-羟基-1,3-双(4-苯氨基苯氧基)丙烷〔防老剂〕
3-hydroxy-2-butanone 3-羟基-2-丁酮
hydroxy butanone (=oxobutanol) 氧代丁醇；羟基丁酮
hydroxy butyl methyl cellulose 羟丁基甲基纤维素
2-(2-hydroxy-3-t-butyl-5-methylphenyl)-5-chlorobenzotriazole 2-(2-羟基-3-叔丁基-5-甲基苯基)-5-氯苯并三唑〔紫外线吸收剂〕
2-hydroxy butyraldehyde 2-羟基丁醛
β-hydroxybutyrate cycle β-羟丁酸循环
β-hydroxybutyric acid β-羟(基)丁酸
hydroxycamptothecin 羟喜树碱〔药〕
α-hydroxycaproic acid α-羟(基)己酸
hydroxycarbamide 羟基脲〔药〕

2-(α-hydroxy-α-carboxyethyl)-thiamine pyrophosphate 2-(α-羟基-α-羧乙基)硫胺焦磷酸；活性丙酮酸
hydroxycellulose　氧化纤维素
hydroxy chalcone (=hydroxy-benzalacetophenone)　羟基亚苄基乙酰苯；羟基查耳酮 $C_6H_5COCH=CHC_6H_4OH$
2-hydroxy-3-chloropropyl acrylate　丙烯酸-2-羟基-3-氯丙酯
hydroxychloroquine　羟氯喹〔药〕
25-hydroxycholecalciferol　25-羟胆钙化(甾)醇；25-羟维生素 D_3
7-hydroxycholesterol　7-羟胆甾醇
6-hydroxychroman　6-羟基苯并二氢吡喃；6-羟基苯并氢化吡喃
p-hydroxycinnamic aldehyde　对羟基(肉)桂醛
10-hydroxycodeine　10-羟基可待因
3-hydroxy coronaridine　3-羟基冠狗牙花定
17-hydroxycorticosteroid　17-羟皮(质甾)醇
7-hydroxycoumarin (=umbelliferone)　7-羟基香豆素；伞形酮
1-hydroxycyclohexylhydroperoxide (HCH)　1-环己醇过氧化氢
1-hydroxy cyclohexyl phenylketone　1-羟基环己基苯甲酮
2-hydroxy-p-cymene　2-羟基对伞花烃
ω-hydroxydecanoic acid　ω-羟基癸酸；端羟基癸酸
17-hydroxy-11-dehydrocorticosterone　17-羟基-11-脱氢皮(质甾)酮；皮质素
17-hydroxydeoxycorticosterone　17-羟脱氧皮(质甾)酮
2-(2-hydroxy-3,5-di-t-butylphenyl) benzotriazole　2-(2-羟基-3,5-二叔丁基苯基)苯并三唑〔光稳定剂〕
2-(2-hydroxy-3,5-di-t-butylphenyl)-5-chlorobenzotriazole　2-(2-羟基-3,5-二叔丁基苯基)-5-氯苯并三唑〔紫外线吸收剂〕
2-(2-hydroxy-3,5-di-isopentyl phenyl) benzotriazole　2-(2-羟基-3,5-二异戊基苯基)苯并三唑〔光稳定剂〕
4-hydroxy-3,5-dimethoxybenzoic acid (=syringic acid)　丁香酸；4-羟基-3,5-二甲氧基苯甲酸 $HO(CH_3O)_2C_6H_2CO_2H$
4-hydroxy-3,5-dimethoxy benzyl alcohol (=syringyl alcohol)　丁香醇；4-羟基-3,5-二甲氧基苄醇
p-hydroxydiphenylamine　对羟基二苯胺〔防老剂〕
2-hydroxy-4-dodecyloxybenzophenone　2-羟基-4-十二烷氧基二苯甲酮〔紫外线吸收剂〕
hydroxy end unit　羟端基
25-hydroxyergocalciferol　25-羟麦角钙化(甾)醇；25-羟维生素 D_2
hydroxyethoxylation　羟乙氧基化

2-hydroxyethyl acrylate　丙烯酸-2-羟乙基酯
2-hydroxyethyl acryloyl phosphate　磷酸丙烯酰 2-羟乙酯
2-hydroxyethyl amide　2-羟基乙酰胺
hydroxyethylated　羟乙基化的
hydroxyethylated cotton　羟乙基化棉
hydroxyethyl cellulose　羟乙基纤维素*
hydroxyethyl ether cellulose　羟乙基纤维素醚
N-hydroxyethyl-ethylenediamine　N-羟乙基乙二胺 $HOCH_2CH_2-NHCH_2CH_2NH_2$
hydroxyethyl ethylenediamine triacetic acid　羟乙基乙二胺三乙酸
2-hydroxy-4-(2-ethyl hexyoxy) benzophenone　2-羟基-4-(2-乙基己氧基)二苯甲酮〔光稳定剂〕
β-hydroxyethyl hydrazine　β-羟乙基肼
(1-hydroxyethylidene) diphosphonate (HEDP)　1-羟基亚乙基二膦酸
1-hydroxy-ethylidene-1,1-disodium phosphonate (HEDSP)　1-羟基亚乙基-1,1-双膦酸钠 $CH_3C(OH)[P(O)(OH)(ONa)]_2$
N-hydroxyethylimino diacetic acid (HIMDA)　N-羟乙基亚氨基二乙酸
2-hydroxyethyl methacrylate　甲基丙烯酸-2-羟酯
hydroxyethyl methacrylate-methyl methacrylate copolymer　甲基丙烯酸羟乙酯-甲基丙烯酸甲酯共聚物
5-hydroxyethyl-4-methylthiazole　5-羟乙基-4-甲基噻唑
2-α-hydroxyethyl thiamine pyrophosphate　2-α-羟乙基硫胺焦磷酸；活性乙醛
2-hydroxy-4-heptoxy benzophenone　2-羟基-4-庚氧基二苯甲酮〔光稳定剂〕
hydroxyimino (=hydroximino)　肟基　HON=
hydroxy-indole (=oxyindole)　羟(基)吲哚
hydroxyisophthalic acid　羟基间苯二酸
δ-hydroxy-γ-keto-norvaline　δ-羟基-γ-酮正缬氨酸
3-hydroxykynurenine　3-羟基犬尿氨酸
hydroxyl acid　羧基羧酸
hydroxyl acrylic　羟基丙烯酸
hydroxylamine　羟胺
hydroxylamine disulfonate　胲基二磺酸盐；羟氨基二磺酸盐
hydroxylamine-phosphate oxime process　羟胺-磷酸盐肟法
hydroxylammonium acetate　乙酸羟铵
hydroxylammonium phosphate　磷酸羟铵
hydroxylated contaminant　羟基(化的)污染物
hydroxylation　羟基化*
hydroxyl-bearing solvent　含羟基的溶剂
hydroxyl-ended oligomer　羟端基低聚物
hydroxyl end group　羟端基

hydroxyl equivalent 羟基当量
hydroxyl group frequencies 羟基振动频率
hydroxylic gibbsite-like surface 羟基三水铝型表面
hydroxylic oxide 羟基氧化物
hydroxylic solvent 羟基溶剂
hydroxyl-terminated 羟基封端(的)
hydroxyl-terminated poly(ethylene oxide) 羟基端接的聚氧乙烯
hydroxyl value 羟基值；羟基数
hydroxylysine 羟(基)赖氨酸
hydroxymalonic acid (=tartronic acid) 丙醇二酸；羟基丙二酸　$HOCH(CO_2H)_2$
p-hydroxymercuri-benzoic acid 对羟基汞基苯甲酸　$HOHgC_6H_4COOH$
hydroxymethane sulfinate 羟甲基亚磺酸盐
2-hydroxy-5-methoxy acetophenone 2-羟基-5-甲氧基苯乙酮
3-hydroxy-4-methoxybenzaldehyde (=isovanillin) 异香兰素；异香草醛；3-羟基-4-甲氧基苯甲醛　$(CH_3O)C_6H_3(OH)CHO$
2-hydroxy-4-methoxy benzophenone 2-羟基-4-甲氧基二苯甲酮〔紫外线吸收剂〕
7-hydroxy-6-methoxy coumarin (=scopoletin) 7-羟基-6-甲氧基香豆素；莨菪亭　$C_{10}H_8O_4$
2-hydroxy-4-methoxy-5-sulfon benzophenone 2-羟基-4-甲氧基-5-磺基二苯甲酮〔光稳定剂〕
hydroxymethylated 羟甲基化的
hydroxymethylation 羟甲基化*
hydroxymethyl cellulose xanthate 纤维素黄酸羟甲酯
hydroxymethyl cyclodecane 羟甲基环癸烷
5-hydroxymethyl cytosine (HMC) 5-羟甲基胞嘧啶
4-hydroxymethyl-2,6-di-t-butyl phenol 4-羟甲基-2,6-二叔丁基苯酚〔防老剂〕
5-hydroxymethyl furfural 5-羟甲基糠醛
β-hydroxy-β-methyl glutaryl-CoA(HMG-CoA) β-羟基-β-甲基戊二酸单酰 CoA
2-hydroxy methylmercapto benzene 2-羟基甲硫基苯
2-hydroxy methyl-5-norbornene (=5-norbornene-2-methanol) 5-降冰片烯-2-甲醇；2-羟基-5-降冰片烯〔涂料用聚合物改性剂〕
3-(hydroxymethyl)-2-octanone 3-(羟甲基)-2-辛酮
2(2-hydroxy-5-methylphenyl) benzothiazole 2(2-羟基-5-甲基苯基)苯并噻唑〔紫外线吸收剂〕
4-hydroxy-N-methyl-proline 4-羟基-N-甲基脯氨酸
2-hydroxy-2-methyl-propiophenone 2-羟基-2-甲基苯丙酮　$(CH_3)_2C(OH)C(O)C_6H_5$
N-hydroxymethyl pyridine N-羟甲基吡啶
3-hydroxy-2-methyl-4-pyrone 3-羟基-2-甲基-4-吡喃酮；麦芽酚
6-hydroxy-1-methylquinolinium 6-羟基-1-甲基喹啉鎓
N-hydroxymethyl tetrahydrofolate N-羟甲基四氢叶酸
α-hydroxy-β-naphthaldehyde α-羟基-β-萘醛
2,3-hydroxynapharylamide 2,3-羟萘芳酰胺
5-(2-hydroxy-3-naphthoylamino) benzimidazolone 5-(2-羟基-3-萘甲酰胺)苯并咪唑酮
N-(4-hydroxynaphthyl) pyridinium N-(4-羟基萘基)吡啶鎓
5-hydroxy-4-octanone 5-羟基-4-辛酮
2-hydroxy-4-n-octoxybenzophenone 2-羟基-4-正辛氧基二苯甲酮〔紫外线吸收剂〕
hydroxyorganosilane 羟基有机硅烷
α-hydroxyoxime α-羟基肟
γ-hydroxy-β-oxobutane γ-羟基-β-氧代丁烷
ω-hydroxypentadecanoic acid ω-羟基十五酸；端羟基十五烷酸
2-hydroxy-2-phenylacetophenone 2-羟基-2-苯基乙酮
p-hydroxyphenylalanine 对羟苯丙氨酸；酪氨酸
2-hydroxyphenylamine 2-羟基苯胺
2-hydroxy-phenyl benzotriazole 2-羟苯基苯并三唑〔紫外线吸收剂〕
p-hydroxyphenyl butanone 对羟基苯丁酮
N-p-hydroxyphenyl-β-naphthylamine N-对羟苯基-β-萘胺〔防老剂〕
4-hydroxypipecolic acid 4-羟基六氢吡啶羧酸
hydroxypivalic acid 羟基新戊酸；羟基三甲基乙酸
hydroxypolyamide 羟基聚酰胺
17-hydroxyprogesterone 17-羟孕(甾)酮
hydroxyprogesterone caproate 己酸羟孕酮〔药〕
2-hydroxypropane-1,3-diaminotetraacetic acid (HPDTA) 2-羟基丙烷-1,3-二氨基四乙酸
2-hydroxypropanic acid 2-羟基丙酸
1-hydroxy-2-propanone 1-羟基-2-丙酮
α-hydroxy-propionaldehyde (=lactic aldehyde) α-羟基丙醛；丙醇醛；乳醛
hydroxy-propionic acid 羟基丙酸　C_2H_5OCOOH
hydroxypropionitrile 羟基丙腈
2-hydroxypropyl acrylate 丙烯酸 2-羟丙基酯
hydroxy propyl cellulose (HPC) 羟(基)丙基纤维素
hydroxypropyl methacrylate (HPM) 甲基丙烯酸羟丙酯
hydroxypropyl methyl cellulose (HPMC) 羟丙基甲基纤维素
γ-hydroxypropyl methyl sulfide γ-羟丙基甲基硫醚
9-hydroxyprostenoate 9-羟前列腺烯酸
6-hydroxypurine (=hypoxanthine) 6-羟基嘌呤；次黄质　$C_5H_4N_4O$
hydroxy-pyruvic acid 羟基丙酮酸　$CH_2OHCOCOOH$
8-hydroxyquinoline (=oxine) 8-羟基喹啉

hydroxy radical 羟自由基*
hydroxysilane (=silicol) 羟基硅烷 R₃SiOH
hydroxy silicone oil 羟基硅油
d-hydroxysparteine (=17-oxosparteine) d-羟基鹰爪豆碱
10-hydroxy-stearic acid 10-羟基硬脂酸
hydroxystearic ethylene diamide 羟基硬脂酸亚乙基二酰胺
hydroxystearic tetraethylene pentamide 羟基硬脂酸四亚乙基五酰胺
hydroxystearic triethylene tetramide 羟基硬脂酸三亚乙基四酰胺
17-hydroxysteroid 17-羟甾类
hydroxysulfobetaine 羟磺基内铵盐〔抗静电剂〕
hydroxy-terminated 端羟基的
hydroxy-terminated polybutadiene 端羟基聚丁二烯
hydroxy-terminated poly(ethylene oxide) 羟基端接的聚氧乙烯
o-hydroxythiobenzoic acid 邻羟基硫代苯甲酸〔防焦剂〕
2-hydroxytrimethylene dinitrilo tetraacetic acid (HPDTA) 2-羟基亚丙基二次氮基四乙酸
5-hydroxytryptamine (=serotonine;enteramine) 5-羟基色胺
5-hydroxytryptophan (e) 5-羟色氨酸
3β-Hydroxy-12-ursen-28-oic acid 熊果酸
hydroxyvaline 羟基缬氨酸
hydroxy value 羟基值
9-hydroxy-xanthene (=xanthydrol) 9-羟基呫吨
HOCH=(C₆H₄)₂O
hydroxyzine 羟嗪〔药〕
hyenic acid (=tricosylacetic acid) 二十五(碳)烷酸
CH₃(CH₂)₂₃CO₂H
hyetometer 雨量表
hygral change 湿度变化
hygral expansion 湿膨胀
hygral expansion index 湿膨胀指数
hygroautometer 自记湿度计
hygrochasy 逐湿性
hygrodeik 图表湿度计；图示湿度计
hygrograph 湿度计
hygrokinesis 感湿性
hygrol (=colloidal mercury) 胶态汞；汞胶液
hygrology 湿度学
hygrometer 湿度计
hygrometry 测湿法；湿度测定法
hygronom 测湿仪
hygroscopic degree 吸湿度
hygroscopic expansion 吸湿膨胀
hygroscopic ion-exchange membrane 吸湿性离子交换膜

hygroscopicity 吸湿性
hygroscopic moisture 湿存水
hygroscopic property 吸湿性；吸水性
hygroscopic synthetic fibre 吸湿性合成纤维
hygroscopic water 湿存水*
hygroscopic water content 吸水量；平衡湿量
hygroscopy 测湿法；湿度测定法
hygrosensor 湿度探测器
hygrostat 恒湿器*
hygrotaxis 趋湿性
hygrotemper 通风式增湿装置
hygrothermal effect 湿热效应
hygrothermograph 温湿仪；温湿自记器；湿度温度记录仪
hygrothermoscope 温湿仪
hylotropy ①恒熔性②恒沸性
hyperbolic scan 双曲线扫描
hyperboloid 双曲面；双曲线体
hyperboloidal gear 双曲线齿轮
hyperbranched polymer 超支化聚合物
hyperchromatism 着色过度
hyperchromicity 增色性
hyperchromism 增色现象
hypercolor 温变色
hyper conjugation (=quasi-conjugation) 超共轭效应；似共轭效应
hyper-dispersant 超分散剂
hyperelastic 超弹性的
hyperelectronic polarization 超电子极化
hyperergy 高反应性
hyperfiltration membrane 超细滤膜；纳滤膜
hyperfine 超精细的
hyperfine coupling constant 超精细偶合常数*
hyperfine coupling mechanism 超精细偶合机理
hyperfine interaction 超精细相互作用
hyperfine splitting 超精细(结构)分裂
hyperfine splitting constant 超精细分裂常数
hyperfine structure 超精细结构*
hyperfine sublevel correlation spectroscopy 超精细子能级相关谱
hyperfine tensor 超精细张量
hyperfine transition 超精细(结构)跃迁
hyperformer reactor 移动床重整反应器；超重整反应器
hyperforming 移动床重整；超重整
hypermolecular 超分子的
hypermultiplet 超多重线
hyperometer 沸点测定仪
hyperpanchromatic emulsion 高全色乳剂

hyperplane 超平面
hyperpressure gas chromatography 超高压气相色谱法
hyperpure 超纯的
hyper Raman effect 超拉曼效应
hyper Raman scattering 超拉曼散射
hyper-Rayleigh scattering 超瑞利散射
hypersensitive 超灵敏的；过敏的
hypersensitized 超高灵敏度的
hypersonic flame spraying 超音速火焰喷涂
hyperstoichiometric 超化学计量的
hyperthermometer 超高温度计
hyperviscosity 超黏度
hyphenated technique 联用技术
hypo 海波*
hypoboric (acid) ester 连二硼酸酯
hypoborous acid 次硼酸 $HOBH_2$
hypo (chlorite) bleaching 次氯酸盐漂白
hypochlorite of lime 次氯酸钙；漂粉
hypochlorite titration 次氯酸盐滴定
hypochorite-iodide reaction 次氯酸盐-碘化物反应
hypochromicity 减色性；缺色性
hypochromism 缺色性；少色性
hypoelastic body 亚弹性体；次弹性体
hypoelasticity 亚弹性
hypoepistephanine 次表千金藤碱
hypogoeic acid 8-棕榈烯酸；8-十六碳烯-1-酸 $C_{15}H_{29}COOH$

hypohalous 次卤(酸)的
hypohalous acid 次卤酸
α-hypophamine α-垂体胺；后叶催产素
β-hypophamine β-垂体胺；后叶加压素
hypothesis 假说
hypothesis about means 总体均值的假设检验
hypothesis test 假设检验*
hypothetical evaporation pattern 假想蒸发模式
hypothetical model 假定模型
hypothetical structure 假拟结构*
hypoxanthine (=6-hydroxypurine) 6-羟基嘌呤；次黄质 $C_5H_4N_4O$
hypsochrome 浅色基*
hypsochromic 向蓝移(的)
hypsochromic effect 蓝移效应；向蓝效应
hypsochromic shift 向蓝移；向短波长(位)移
hypsometer ①沸点测定器②沸点测高器
hysteresis 滞后现象*；滞后
hysteresis curve 磁滞曲线
hysteresis effect 滞后效应
hysteresis loss 磁滞损耗；滞后损耗
hysteresis of adsorption 吸附滞后*
hysteresis set 滞后变定；滞后定形
hysteretic modulus 磁滞模量
hysterometer 滞后试验仪
Hyswing ball mill 希斯温球磨机；四滚筒球磨机
hythergraph 温湿度关系图

I

iatrochemistry 医疗化学
ibogaine 伊菠因〔抗抑郁药〕
ibotenic acid 鹅膏蕈氨酸
iboxygaine 伊菠氧碱
ibuprofen 布洛芬〔药〕
ibuprofen and codeine 洛芬待因；布洛芬可待因〔药〕
icariin 淫羊藿苷 $C_{33}H_{42}O_{16}$
ice 冰
ice bank evaporator 结冰式蒸发器
ice-bath ①冰浴器②冰浴
ice-bear fat 北极熊油
ice-blocking 结冰
ice blue 冰染蓝
ice box 冰箱
ice calorimeter 冰量热器
ice chain 冰链
ice chest 冰柜
ice chopper 碎冰机
ice color(=ice colour) 冰染料
ice color process 冰染法
ice colour(=ice color) 冰染料
iced water 冷冻水；冰水
ice dye 冰染料
ice dyeing 冰染
ice dyestuff 冰染料
ice fat 冰脂；卡丽托脂
Iceland crystal 冰洲石
Iceland moss 冰岛藓
Iceland spar 冰洲晶石
ice-machine oil 制冰机油
ice making machine 制冰机
ice making machinery 制冰机(器)
ice mill 磨冰机
ice mould 冰铸
ice of land origin 陆冰；陆(原)冰
icephobic coating 防结冰涂料
ice point 冰点
ice point depression 冰点下降
ice spar 冰晶石
ice trap 冰阱；用冰冷却的接受器
ice-water bath 冰水浴
ichnography ①(用圆规和直尺作的)装饰图案艺术②平面图(法)
ichthalbin 白蛋白合鱼石脂〔药〕

ichthammol 鱼石脂〔药〕
ichthargan 银鱼石脂〔药〕
ichthoform 鱼石脂〔药〕
ichthyol(=colla piscium) 鱼石脂
ichthyol oil 鱼石油
ichthyol sulfonate 鱼石脂磺酸盐
ICI cone and plate viscometer ICI 锥板黏度计；ICI 平底锥形黏度计
icing beater 冰糖搅棍机〔制糖〕
icing machine 结糖机；冰糖机〔制糖〕
icos(a)- 〔词头〕二十
icosadecaene 二十碳十烯
icosadienoic acid 二十碳二烯酸
icosandioic acid(=icosane diacid) 二十烷二酸
icosane 二十(碳)烷 $C_{20}H_{42}$
icosane diacid(=icosandioic acid) 二十烷二酸
icosane dicarboxylic acid 二十烷二甲酸
icosanic acid 二十(烷)酸
icosanoic acid 二十(烷)酸；花生酸 $CH_3(CH_2)_{18}COOH$
icosanol 二十(烷)醇 $CH_3(CH_2)_{18}CH_2OH$
icosapentaenoic acid 二十碳五烯酸
icosendioic acid(=eicosene diacid) 二十碳烯二酸
icosene diacid 二十碳烯二酸
icosene dicarboxylic acid 二十碳烯二甲酸；二十二烯二酸
icosenoic acid 二十碳烯酸 $C_{20}H_{38}O_2$
icosinene 地蜡烯 $C_{26}H_{38}$
icosoic acid 二十(烷)酸
icosyl 二十烷基 $CH_3(CH_2)_{18}CH_2—$
icosylene 二十碳烯
ICP emission spectrometer 电感耦合等离子体发射光谱仪
ICP-MS(inductivity coupled plasma mass spectrometry) 电感耦合等离子体质谱
ICPP electrolytic dissolver 爱达荷化学处理工厂电解溶解器
ICP quantometer 电感耦合等离子体光量计
ICP spectrograph 电感耦合等离子体摄谱仪
icterogenin 黄疸配基 $C_{35}H_{52}O_6$〔取自 Lippia renmanni 过江藤〕
icterus 黄疸
icy 凉香
icyl blue 衣色蓝
icyl colors 衣色染料〔染黏胶嫘萦用的一种染料商名〕

icyl orange　衣色橙
idaein　越橘色苷
-ide〔词尾〕　某化物
ideal　①理想②理想的；完美的
ideal body　理想体
ideal cascade　理想级联
ideal chain　理想链
ideal column　理想塔
ideal conformation　理想构象
ideal copolymerization　理想共聚合*
ideal crystal　理想晶体*
ideal cycle　理想循环
ideal dispersion　理想分散
ideal elastic body　理想弹性体
ideal elasticity　理想弹性
ideal fluid　理想流体
ideal focusing field　理想聚焦场
ideal focusing magnetic field boundary　理想聚焦磁场边界
ideal gas　理想气体
ideal gaseous film　理想气态膜
ideal gas equation　理想气体方程；气体定律
ideal gas law　理想气体定律
idealized radiator　理想辐射体
ideal lubricant　理想润滑剂
ideally elastic behaviour　理想弹性特性
ideal mixing　理想混合
ideal network　理想交联网
ideal nonpolarized electrode　理想的非极化电极
ideal operating temperature　理想操作温度
ideal orientation　理想取向
ideal permselective membrane　理想选择透过膜
ideal plasticity　理想塑性
ideal plate　理想塔板
ideal polarized electrode　理想极化电极
ideal refrigeration cycle　理想冷冻循环
ideal scale color notation　理想颜色标度；颜色标号理想标度
ideal solution　理想溶液*
ideal solvent blend　理想混合溶剂〔指不相互作用的混合溶剂〕
ideal stage　理想级
ideal state　理想态
ideal temperature　理想温度〔弗洛里温度或θ温度〕
ideal viscous fluid　理想黏性流体
-idene　亚基〔加于任何一个基名上；表示在联结处为双键〕
identical　相同的；恒等的

identical period　等同周期
identification　鉴定*；鉴别
identification limit　鉴定极限
identification mark(=identifying mark)　标志
identification of hydrocarbons　烃的鉴定
identification paint　标志用漆
identification test　鉴定试验；识别试验；鉴别试验
identification threshold　鉴定阈；识别阈
identifier　①鉴定器②鉴定剂③鉴定人；检验人
identifying mark(=identification mark)　标志
identifying of oil　油的鉴定；油牌号的确定
identity element　①同一元素②恒等参数
identity matrix　单位矩阵
identity period　等同周期*
idiochromatic　自色的；本色的
idiotypes　独特型
iditol　艾杜糖醇　$CH_2OH(CHOH)_4CH_2OH$
idle　①空闲的②空转的
idle boiler　空闲锅炉
idle capacity　储备容量
idle current　无功电流
idle hours　停工小时数；窝工小时
idle mixing time　①停止混合(作用)时间②闲置时间
idler　惰轮
idler carrier　托滚(支)座
idle roller　空滚子
idler pulley　托滚
idler roll　托滚
idler rubber　①剪余碎片②碎硫化胶片；硫化胶边
idle space　有害空间；无效空间
idle time　闲空时间；停歇时间
idle time cost　停工费用
idle unit　闲置设备
idling　空运转
idocrase　符山石
idonic acid　艾杜糖酸　$CH_2OH(CHOH)_4COOH$
idosaccharic acid　艾杜糖二酸
idose　艾杜糖
idoxuridine　碘苷；5-碘脱氧尿苷
iduronase　艾杜糖醛酸酶
iduronate　①艾杜糖醛酸②艾杜糖醛酸盐(酯或根)
iduronic acid　艾杜糖醛酸
-idyne〔词尾〕　次(某)基
ifosfamide　异环磷酰胺〔药〕
Igclite　聚氯乙烯塑料
ignavine　惰碱
igneous breccia　火成角砾岩
igneous concentration　煅烧富集

ignitability 可燃性；着火性
ignitable 可燃的
ignited basic silica 灼烧的基料二氧化硅；灼烧的原料二氧化硅
ignited residue 烧余残渣
igniter(=ignitor) ①点火器②传火药
igniter fuse 点火引线
igniter pawl 点火掣
ignitibility 可燃性
ignitible 可燃的；可着火的
ignition 灼烧
ignition accelerator 着火加速剂
ignition cable 发火电缆
ignition center 着火中心；发火中心
ignition coil 发火线圈
ignition controller ①着火控制剂②着火控制器
ignition delay 着火延迟
ignition delay angle 着火迟延角
ignition dope 自燃促进剂
ignition emission detector 热发射检测器
ignition energy ①点火能(量)；引爆能量②燃烧能
ignition fuel 引燃燃料
ignition heat 着火热
ignition inhibitor 着火抑制剂
ignition lag 延迟发火
ignition limit 燃烧极限
ignition line 着火线；燃烧点火线
ignition loss 灼烧损失；灼减
ignition point 着火点；燃点
ignition quality 着火性
ignition quality improver 发火性增进剂〔改进柴油着火性质用添加剂〕
ignition range 着火范围
ignition residue 灼烧残渣
ignition source 火源；着火源；起火源
ignition temperature 着火温度
ignition test 灼烧试验
ignition time 着火时间；点火时间
ignition transformer 引火变压器
ignition tube 灼管
ignitor(=igniter) ①点火器②传火药
ignitor fuse 点火引信
ignitron 点火器；引燃管
igntion wire 点火线
ikarugamycin 斑鸠霉素
ikpan oil 西瓜子油
ikutamycin 生田霉素
ilamycin 岛霉素

ilangen 衣兰质
ilexanthin 冬青黄质
ilexin A 冬青苷 A $C_{19}H_{28}O_{10}$
ilexin B 冬青苷 B $C_{33}H_{54}O_{10}$
ilicic acid 冬青酸
ilicic alcohol 冬青醇 $C_{30}H_{50}O$
ilicin 冬青(苦)素
ilicyl alcohol 冬青酰醇 $C_{22}H_{38}O$
Ilkovic equation 伊尔科维奇方程(式)
ill-conditioned 病态的；坏条件的
ill-conditioned system 病态体系
ill-conditioning 病态方程组
illicium oil 莽草油
illite 伊利水云母；伊利石
ill-prepared surface 处理不当的表面；处理得不干净的表面
illudin 隐杯伞素
illuminance (光)照度；照明度
illuminant ①发光物②照明剂
illuminant B B光源
illuminating apparatus 照明装置
illuminating attachment 照明设备
illuminating gas 照明气
illuminating glass 照光玻璃
illuminating oil 灯油
illuminating spot 光斑
illumination ①照明②照明学③照度
illumination level ①照度水平②照度级
illuminator ①发光器②照灯；照明装置
illuminometer 照度计
illuminophore 发光团
illumi yarn 光亮丝
illuric acid 衣卢酸
illurin balsam(=African balsam) 非洲香脂
illustrator's azo paper 画家偶氮纸
illustrator's special paper 画家特种纸
ilmenite 钛铁矿 $FeO \cdot TiO_2$
ilmenite concentrate 钛铁矿富集物；钛精矿
ilmenite sand 钛铁矿砂
ilmenitite 钛铁岩
ilmenorutite 钛铁金红石
ilotycin(=erythromycin) 红霉素
ilvaite 黑柱石
imabenzil 衣马苯甲酰
image 像
image analysis 镜像分析
image analysis system 影像分析系统
image analyzing computer 影像分析计算机〔测纤维细

度用〕
image blackness 影像黑度
image broadening 像加宽
image confusion 弥散像；像弥散；影像紊乱
image current detector 像电流检测器
image displacement 像位移
image dissecting ionprobe 图像解剖离子探针
image dissector 析像管
image distance 像距
image energy 镜像能
image field curvature 像场弯曲
image formation 成像
image formation system 成像系统
image-forming dyes 成像染料
image gloss 映像光泽
image intensifier gamma-ray camera 影像增强γ射线照相机
image of adsorption current 吸收电流像
image of transmission electron 透射电子像
image plane ①平面像②像平面
image reconstruction computer algorithm 图像重整计算机算法
image restoration 像复原
image retention 像保留
image-reversal process 影像反转过程
image rotation prism 成像旋转棱镜
image sensor 图像传感器
image sharpness 像清晰度
image-superimposable molecule 镜像重合分子
image synchronization 像同步
image transfer 影像转移
image treatment 图像处理
image width 像宽
imaginary modulus 虚(数)模量
imaginary viscosity 假黏度
imaging agent 显像剂*
imaging atom probe 成像原子探针
imasatin 衣氨酸内酰胺；3-羟基-2-氧代-3-(2-氧代-3-吲哚亚氨基)吲哚
imbalance 不平衡；不相等；不稳定(性)
imbedded agent 嵌接剂〔纤维分子链内嵌段单元〕
imbedded particle 嵌入粒子
imbedding 镶铸*
imbedibility 嵌入性；埋入性；吸入性
imbedment 包埋；埋封
Imbert zinc process 因伯特炼锌法
imbibing 吸液
imbibition ①吸移②加水

imbibition pressure 吸胀压；溶胀压
imbricated scale structure 鳞片状叠覆结构
imbrication 鳞片状
imbue ①浸渍②熔化(固体油脂)；加脂③浸染
Imconel (nickel alloys) 因康镍合金；铬镍铁合金
imecic alcohol 衣袂醇 $C_{22}H_{38}O$
imesatin 3-亚氨基-2-吲哚酮；3-亚氨基-2-氧代-1,3-二氢吲哚〔辛可芬的中间体〕
imflammable gas 易燃气体；可燃性气体
imflammable solvent 可燃性溶剂
imidapril 咪达普利〔药〕
imidazo- 咪唑并
imidazole(=glyoxalin) 咪唑；1,3-二氮杂茂
4-imidazoleacetic acid 4-咪唑乙酸
α,β-imidazoledicarboxylic acid α,β-咪唑二羧酸
imidazole lactic acid 咪唑乳酸
imidazolidine 咪唑烷；四氢化咪唑
2,4-imidazolidinedione 2,4-咪唑烷二酮
2-imidazolidine thione(=ethylene thiourea) 亚乙基硫脲
2-imidazolidinone (=ethylene urea) 2-咪唑啉酮
imidazolidinyl(=imidazolidyl) 咪唑烷酮 $C_3H_7N_2—$
imidazolidone 咪唑烷酮 $C_3H_6N_2O$
imidazolidone caproic acid 咪唑酮己酸
imidazolidyl(=imidazolidinyl) 咪唑烷基
imidazoline 咪唑啉
imidazoline phosphonic acid 咪唑啉膦酸
imidazoline phosphoramide 咪唑啉膦酰胺
imidazoline propyl triethoxysilane 咪唑啉丙基三乙氧基硅烷
imidazolinium 咪唑啉鎓盐
imidazolinium betaine 咪唑啉鎓甜菜碱
imidazolinium compounds 咪唑啉鎓化合物
imidazolinium surfactant 咪唑啉型表面活性剂
imidazolinol 咪唑啉醇
imidazolinone 咪唑啉酮
imidazolinyl 咪唑啉基 $N_2C_3H_5—$
imidazolium 咪唑盐
imidazolium chloride 氯化咪唑
imidazolium compounds 咪唑鎓化合物
imidazolone 咪唑啉酮
imidazolyl 咪唑基 $C_3H_3N_2—$
imidazolyl acrylic acid 咪唑基丙烯酸
β-imidazolylpyruvic acid β-咪唑(基)丙酮酸
imidazopyrazine 咪唑并哒嗪
1H-imidazo [b] pyrazine 1H-咪唑并[b]哌嗪
imidazopyridine 咪唑并吡啶
1H-imidazo [1,2-a] pyridinium 1H-咪唑并[1,2-a]吡啶鎓
imide 二酰亚胺*

imide chloride　偕氯代亚胺　RCCl＝NH
imide salt　亚胺(基)盐
imidic acid　亚氨(基)酸　RC(＝NH)OH
imidic ester　亚氨(基)酸酯
imidization　(酰)亚胺化
imido　亚氨(基)　HN＝
imido-acetic acid　亚氨(二)乙酸　(CH$_2$COOH)$_2$NH
imidoamine salt　亚氨基胺盐
imidoate　亚氨酸酯；偕亚氨醇盐
imido-carbonic acid　亚氨碳酸　(OH)$_2$C＝NH
imido-carbonic ester(=imido-carbonic ether)　亚氨碳酸酯　(RO)$_2$C＝NH
imido-carbonic ether(=imido-carbonic ester)　亚氨碳酸酯
imido-chloride　偕氯代亚胺　RCCl＝NH
imidodicarboxylic acid　亚氨二羧酸　NH(COOH)$_2$
imidodiphosphate　亚氨二磷酸盐
imidodiphosphoric acid　亚氨二磷酸　H$_4$P$_2$O$_6$NH
imidodisulfamide(=imidodisulfuramide)　亚氨二磺酰胺　H$_2$NSO$_2$NHSO$_2$NH$_2$
imidodisulfate　亚氨二硫酸盐
imidodisulfuramide(=imidodisulfamide)　亚氨二磺酰胺
imidodisulfuric acid　亚氨二硫酸
imido-ester(=imidoate; imido-ether)　亚氨酸酯　RC(OR)＝NH
imido-ether(=imido-ester)　偕亚氨醚；亚氨酸酯
imidogen　亚氨基　HN＝
imido-group　亚氨基
imido-halide　偕卤代亚胺　RCX＝NH
imido-hydrogen　亚氨型氢
imidole(=pyrrole)　吡咯
imidophosphonic acid　亚氨膦酸
imidopolyphenyl oxide　亚氨(基)多苯(基)氧化物
imidotrimetaphosphoric acid　亚氨三偏磷酸
imidotriphosphoric acid　亚氨三磷酸
N-imidoylamidine　N-亚氨脒
imine　亚胺*
imine-enamine tautomerism　亚胺-烯胺互变异构*
imino　亚氨基　HN＝
imino-acetic acid　亚氨(二)乙酸　(CH$_2$COOH)$_2$NH
imino-aceto-acetic acid　亚氨乙酰乙酸　CH$_3$C＝(NH)CH$_2$COOH
imino-acid　亚氨基酸　NH(RCOOH)$_2$
imino-base　亚氨碱　R$_2$NH
imino-bis-propylamine　亚氨基二丙胺
imino-choride　偕氯代亚胺　RCCl＝NH
imino-compound　亚氨基化合物
iminodiacetic acid(IDA)　亚氨二乙酸　NH(CH$_2$COOH)$_2$

iminodiacetonitrile　亚氨基二乙腈　NH＝(CH$_2$CN)$_2$
iminodicarboxylic acid　亚氨基二羧酸
iminodiformic acid　亚氨基二甲酸
β,β'-iminodipropionitrile　β,β'-亚氨二丙腈
imino-ester(=imidoate)　亚氨酸酯　RC(OR)＝NH
imino-ether(=imino-ester)　亚氨醚；亚氨酸酯
imino-ether hydrochloride　盐酸化亚氨基醚　RC(OR)＝NH·HCl
imino-formmelamine(=isomelamine)　亚氨型三聚氰胺；异三聚；氰胺
imino-formyl chloride　亚氨基甲酰氯；氯甲亚胺　CHCl＝NH
iminoglutaric acid　亚氨基戊二酸
imino-halide　偕卤代亚胺　RCX＝NH
imino-nitrogen　亚胺型氮
iminopropinate　亚氨基丙酸盐
imino-urea　胍　C(NH$_2$)$_2$＝NH
imipenem and cilastatin sodium　亚胺培南西司他丁钠〔药〕
imipramine　丙咪嗪〔药〕
imitated grain　假粒面
imitation　①模仿②仿造的；假(造)的③仿造品
imitation flavor　人造食用香料；调制食用香精
imitation leather　人造皮革
imitation method　模仿法
imitation parchment　假羊皮纸
imitation pearl paint　仿珠光漆
imitation red lead　假红丹
immedial green　直达绿〔染〕
immediate deformation　瞬时形变
immediate elastic deformation　瞬间弹性变形
immediate oxygen demand　瞬间需氧量；直接需氧量
immediate reading　瞬间读数；瞬(间指)示
immediate set　瞬间变定
immersant　浸渍剂
immersed displacer　浸入式排出器
immersed strip electrophoresis　浸条电泳
immersion　浸(渍)；浸入
immersion alkali cleaning　浸渍式碱洗；碱浸洗涤(去污)法
immersional wetting　浸润作用；浸入润湿
immersion bath　浸渍浴；浸渍槽
immersion battery　浸液电池
immersion cleaning　浸洗净化法
immersion coating　浸涂
immersion cooler　沉浸冷却器
immersion cooling　沉浸冷却；浸没冷却
immersion counter　浸液型计数器
immersion electrode　浸液电极

immersion extractor 浸泡式浸出器
immersion freezer 浸液致冷器；浸没式冷冻机
immersion heater 浸液加热器
immersion jig 液下卷染机
immersion length 浸长；浸没长度；浸浴长度
immersion lens 浸没透镜
immersion line 浸入线〔温度计上〕
immersion liquid 浸渍液
immersion method 液浸法；浸入法
immersion objective 浸没物镜
immersion of thermometer 温度计浸入深度
immersion oil 浸渍油
immersion phosphating plant 浸渍式磷化装置
immersion refractometer 浸入式折光仪
immersion resistance 耐浸渍性
immersion roller 浸液辊
immersion set 瞬间永久变形；初始永久变形
immersion test 浸没试验
immersion thermocouple 浸入式热电偶
immersion thermometer 浸没式温度计
immersion type thermocouple 浸渍型热电偶
immersion ultramicroscope 浸渍(式)超显微镜
immiscibility 不溶混性
immiscible ①不溶混的②不混合的
immiscible elastico-viscous liquid 不溶混弹黏液体
immiscible liquid 不溶混的液体
immiscible phase 不混溶相
immiscible solution 不溶混溶液
immiscible solvent 不溶混溶剂
immobile 固定的
immobile interface 固定界面
immobile liquid 固定液
immobile phase 固定相
immobilization ①固定；定位②固化
immobilization factor 固定因子
immobilized 固定的；静止的；不动的
immobilized enzyme 固定化酶；固相酶
immobilizing membrane 固定化膜
immortal polymerization 不死的聚合
immortelle concrete 蜡菊浸膏
immortelle oil 蜡菊油 A1
immune 免疫的
immune body 免疫体
immune cytolysis 免疫胞溶
immune elimination 免疫消除
immune protein 免疫蛋白质
immune response 免疫响应；免疫反应
immune serum 免疫血清
immunity 免疫性
immunity analysis of capillary electrophoresis 毛细管电泳免疫分析
immunity electrode 免疫电极
immunization 免疫
immunizing agent 免疫剂
immunnoelectropherogram 免疫电泳图
immunoadsorbent 免疫吸附剂
immunoadsorption 免疫吸附
immunoaffinity chromatography 免疫亲和色谱法；免疫亲和层析法
immunoassay 免疫分析法
immunoassay kits 免疫分析试剂盒
immunobiology 免疫生物学
immunoblotting 免疫印迹
immunochemistry 免疫化学
immunocomplex 免疫复合物
immunodeficiency 免疫缺乏
immunodiffusion(ID) 免疫扩散
immunoelectrofocusing 免疫电聚焦
immunoelectroosmophoresis(IEOP) 免疫电渗电泳
immuno electrophoresis 免疫电泳*
immunoenzymatic technique 免疫酶技术
immunoenzymology 免疫酶学
immunofiltration 免疫过滤
immunofixation 免疫固定
immunofluorescence 免疫荧光
immunofluorescent test 免疫荧光试验
immunogen 免疫原
immunogenicity 免疫原性
immunoglobulin 免疫球蛋白
immunoglobulin G(IgG) 免疫球蛋白 G
immunolabelling technique 免疫标记技术
immunological enhancement 免疫封阻
immunologically competent cell 免疫活性细胞
immunological memory 免疫记忆
immunological reaction 免疫反应
immunological rejection 免疫排斥
immunological suppression 免疫抑制
immunological surveillance 免疫监督
immunological tolerance 免疫耐受性
immunological unresponsiveness 免疫无响应性；免疫无反应性
immunology 免疫学
immunoluminometric assay 发光免疫分析
immuno-nephelometry 免疫浊度测定法
immunoprecipitation 免疫沉淀
immunoprophylaxis 免疫预防

immunoradioautography 免疫放射自显影
immunoradiometric assay(IRMA) 免疫放射分析*
immunoreaction 免疫反应
immuno sensor 免疫传感器
immunosorbent 免疫吸附剂
immunosuppression 免疫抑制
immunotherapy 免疫疗法
impact 冲击
impact adhesive 感(触)压黏合剂；压敏黏合剂
impact bending strength 冲击弯曲强度
impact bending test 冲击弯曲试验
impact break 冲击破裂
impact cross section 冲击断面
impact crushing 冲击粉碎
impact damage 碰撞损伤；撞坏
impacted compression strength 冲击后压缩强度
impact elasticity 冲击弹性
impact elasticity tester 冲击试验机
impact erosion 冲击腐蚀
impact fatigue 冲击疲劳
impact fatigue test 冲击疲劳试验
impact force 冲击力
impact fracture 冲击破坏
impact fusion 冲击熔融
impact grade 抗冲击等级
impact grinder 冲击粉碎机
impact grinding 冲击粉碎
impact hardness 冲击硬度
impact head 冲击高差
impact idler 缓冲托滚
impact ionization 碰撞电离
impact ionization coefficient 碰撞电离系数
impact load 冲击载荷
impact machine (for testing) 冲击机
impact mill 冲击研磨机
impact-model 撞击模型
impact modifier 抗冲改性剂*
impact molding 冲压成型
impactometer 碰撞(空气)取样器
impact parameter 撞击参数*
impact plies 缓冲层
impact polystyrene 耐冲击聚苯乙烯
impact pressure 冲击压力
impact puncture penetration 冲击穿透力
impact resilience 抗冲(击)性
impact resilience rubber 抗冲(击)橡胶
impact resilience tester 抗冲(击)试验仪
impact resiliometer 冲击回弹性试验机；抗冲强度试验机
impact resistance 耐冲击强度；耐冲击性
impact resistance test 抗冲击性试验
impact resistant plastics 耐冲击塑料
impact-resistant polystyrene(IPS) 耐冲击聚苯乙烯
impact retention ①残留冲击力②残留耐冲击值
impact screen 振动筛
impact sound reduction 冲击消音
impact spectrum 冲击谱
impact strength 冲击强度*
impact stress 冲击应力
impact tear test 冲击撕裂试验
impact tension machine 打桩机
impact test 冲击试验
impact tester 冲击试验机
impact testing machine 冲击试验机
impact value 冲击值
impact viscosity ①冲击黏性②冲击黏度
impact wear 冲击磨损
impact weight 冲击荷重
impalpable powder 细微粉末
impasto 厚涂颜料法
impedance 阻抗
impedance bridge 阻抗电桥
impedance-coupled amplifier 阻抗耦合放大器
impedance method 阻抗法
impedence spectroscopy 电化学阻抗波谱
impede settling 滞沉
impedimeter 阻抗计
impedimetric cell 阻抗滴定池
impedimetric titration 阻抗滴定
impedimetry 阻抗滴定法
impedimetry cell 阻抗滴定池
impedometer 阻抗计
impeller ①叶轮②高速搅拌机
impeller approach 叶轮入口
impeller blasting 抛丸除锈
impeller breaker 叶轮式粉碎机
impeller channel 叶轮流道
impeller cleaning 抛丸除锈法
impeller cut 叶轮切割
impeller dish 叶轮盘
impeller hub 叶轮轮毂
impeller mounting ring 叶轮固定环
impeller output 叶轮排出量
impeller pump 叶轮泵
impeller selection 叶轮选择
impeller tip speed 叶轮外缘速度

impeller vane 叶轮叶片
impeller wear ring 叶轮衬环
impelling action 推进作用
impelling agent 促进剂
impelling ratio 推进比
impelling strength 冲击韧性
impending plastic flow 急湍塑性流动
impenetrability 不可渗透性
imperatorin 前胡醚；白茅苷
imperfact crystal formation 不完全晶型；无定形结晶
imperfect ①不完全的②不够好的〔表面加工〕③非理想的〔气体〕
imperfect combustion 不完全燃烧
imperfect crystal 缺陷晶体*
imperfect crystal formation 不全晶型；无定形结晶〔石蜡〕
imperfect gas 非理想气体；真实气体
imperfectly elastic behaviour 不完全弹性特性
imperfect vacuum 部分真空
imperforated 无孔隙的；无气孔的；不穿孔的
imperial gallon 英国加仑〔为4.546升〕
imperial green 翡翠绿；巴黎绿
imperialine 西贝母碱；西贝素
imperial smelting furnace 铅锌鼓风炉
impermeability 不透性
impermeable(=impervious) 不透性的；不可渗透的
impermeable clay 不渗透黏土
impermeable coil 不透性线圈
impermeable film 不渗透性漆膜
impermeable graphite 不透性石墨
impermeable molecule 不渗透分子
impervious(=impermeable) 不透性的；不可渗透的
impervious carbon 不渗透性碳
impervious graphite 不渗透性石墨
impervious lining 不渗透性内衬
impervious material 不透水材料
impervious nature 不透水特性；不渗透性
imperviousness 不渗透性
impervious surface 抗渗表面
impervious to petrol 不透汽油的
impingement ①冲击；撞击；打击②震③水锤
impingement area(s) 碰撞面；冲击面
impingement attack 冲击腐蚀；滴蚀；水锤浸蚀
impingement baffles 撞击挡板
impingement black 接触法炭黑；槽法炭黑
impingement collection 碰撞采集
impingement separator 撞击式气液分离器
impinger 碰撞取样器

impinging corrosion 冲击腐蚀；水锤腐蚀
implement 工具〔机工〕；用具
implication ①含意②牵连
imporosity 非多孔性；紧密结构
imported ①导入的②输入的
imported effect 导入效应
imported polar effect 导入极化效应
imposite 整板
impounding reservoir 蓄水池
impoverished fuel 贫化燃料
impreg ①浸渍木料②浸渍材料
impregnability 浸渗本领
impregnable 可渗透的
impregnant 浸渍剂；渗透剂
impregnated 浸(渍)过的；浸制(了)的
impregnated and compressed wood(=impreg) 浸脂(压缩)木材
impregnated catalyst 浸渍催化剂
impregnated cloth 浸渍布
impregnated densified wood 浸渍压缩木材
impregnated glass-fiber cloth 浸渍玻璃布；玻璃漆布
impregnated granules 浸渍颗粒〔催化剂〕
impregnated paper 浸渍滤纸
impregnated paper chromatography 浸渍纸色谱(法)
impregnated thin layer chromatography 浸渍薄层色谱法
impregnated web 浸渍基料〔层压塑料〕
impregnating 浸渍
impregnating agent 浸渍剂
impregnating bath 浸渍浴
impregnating compound(=impregnation compound) 浸渍化合物
impregnating equipment 浸胶机
impregnating insulating varnish 浸渍绝缘清漆
impregnating machine 浸渍机
impregnating process ①浸渍程序；浸渍过程②浸渍处理法
impregnating resin 填充性树脂
impregnating time 浸渍时间
impregnating varnish 浸渍清漆
impregnation 浸渍*；浸胶
impregnation compound(=impregnating compound) 浸渍化合物
impregnation method 浸渍法
impregnator 浸渍器；浸浆机；浸胶机；浸浆槽
impression ①印痕；压痕②模槽③印刷
impression bolster 型腔固定板
impression cylinder 压花滚筒
impression moulding(=contact pressure moulding) 触压

成型
impression resin 触压(固化)树脂
impression throw-off 滚筒脱开
impression throw-on 滚筒合拢
impression top 防滑花纹面；(粗)糙面；毛面〔运输带〕
impringement attack 冲击腐蚀
imprint 压痕
imprinter 打印机
improper 不合适的
improper axis 非真轴
improper fuel 不合适的燃料
improper fuel combustion products 不合适的燃料产品
improperly wound （卷取）成形不良
improper rotation 非真转动
improved drying oil 改良干性油〔合成干性油〕
improved four-site model 改进型四位置模型
improved wood(=compressed wood) 压缩木材
improver 改进剂
impulse 脉冲；冲动；冲击
impulse action 推进作用；冲力作用
impulse counter 脉冲计数器
impulse glow discharge 脉冲辉光放电
impulse impeller 推进叶轮；冲力叶轮
impulse response 脉冲响应
impulse sealer 高频熔接机〔塑料〕
impulse source 脉冲源
impulse welding 脉冲焊
impulse-wheel meter 脉冲轮计
impulsive model 冲击模型
impure 不纯的
impurity(=dirt particles) 杂质
impurity absorption 杂质吸收
impurity analysis by spark source 火花源杂质分析
impurity atoms 杂质原子
impurity-containing waste polyester 含杂质废聚酯
impurity content 杂质含量
impurity defect 杂质缺陷
impurity element 杂质元素
impurity level 杂质量
imputrescence 防腐性
imputrescibility 防腐性
imuran 咪唑硫嘌呤
-in 〔词尾〕碱；因
inaccuracy 不准确性；不精确度
inaction period 钝化周期；不作用时间
inactivating ①钝化②使失效
inactivation 钝化；失活
inactivator 钝化剂

inactive ①不活泼的②钝性的③不旋光的
inactive base 钝性载体〔催化剂的〕
inactive black 非活性炭黑
inactive filler 钝性填料；非活性填充剂
inactive form ①钝性体②不旋光体
inactive gas 不活泼气体；惰性气体
inactive oil 不旋光石油〔光学上的〕
inactive pigment 惰性颜料
inactive positions of alternant system 交替体系的不活泼位置
inactive solvent 惰性溶剂
inactone 钝酮素
inadequacy 不适宜
inadequate lubrication 不适宜的润滑
inadherent 不黏结的
inadhesion ①不密着；不黏合②缺乏黏性
inadhesive 不能黏结的
inamycin(=novobiocin) 新生霉素
in-and-out construction of tank 阶梯式储罐
in-beam electron impact mass spectrum 电子束轰击(固体)质谱
in-beam electron ionization 在束电子电离
in-beam spectroscopy 射谱能光谱法
in-between-type reinforcing material 中间型补强材料
inborn error 先天性障碍
inborn factor 先天因素
in bulk 成块；成堆
incandescence 白热；白炽
incandescent 白炽的；白热的
incandescent burner 白炽灯
incandescent lamp 白炽灯
incandescent light ①白炽灯②白热光
incandescent wire 炽热灯丝
incanine 灰(毛束)草碱
in-can preservative 罐内防腐剂
incarnadine 淡红色；肉色
incarnatin 翘摇苷
incarnatyl alcohol 翘摇醇
incendiary 燃烧的
incendiary agent 燃烧剂
incendiary bomb 燃烧弹
incendiary material 燃烧材料
incendiary oil 纵火油料
incendiary pencil 燃烧管
incendiary shell 燃烧炮弹
incense ①香②香气③熏香
incher 小管
inches of head 压头英寸〔用英寸表示的压头〕

incident 入射的
incident angle 入射角
incident beam 入射(光)线；入射束
incident beam collimator 入射束准直管
incident energy 入射能
incident flux 入射光通量
incident intensity 入射光强度
incident light 入射光
incident light source 入射光源
incident radiation 入射辐射
incident ray 入射线
incinerating 煅烧的
incineration ①焚烧；焚化②煅烧
incineration dish 煅烧盘
incineration of radwaste 放射性废物煅烧*
incinerator ①焚烧炉；焚化炉②煅烧炉
incipient 开始的；起首的
incipient cure ①初期硫化②早期硫化；焦烧
incipient decay 初腐
incipient fusion 垂熔
incipient melting 初熔
incipient plastic flow 初塑性流动；早期塑性流动
incipient scorch 早期硫化；焦烧
incipient settling 早期沉降
incipient sintering 初期熔结作用
incitement 激动；激发
inclination 倾斜
inclined 倾斜的
inclined clarifier 斜式沉降器
inclined conveyor 倾斜运输机
inclined flow meter 倾斜流速计
inclined gauge 斜管液面计
inclined grizzly 斜置栅筛
inclined hide processor 倾斜式生皮加工器
inclined manometer 倾斜压力计
inclined plane tester 斜面式强力试验机
inclined plane viscometer 斜面黏度计
inclined retort 斜式甑
inclined step grate producer 斜级炉
inclined T joint 斜接接头
inclined tube evaporator 斜管蒸发器
inclined tube gauge 斜管气压计
inclined tube manometer 倾斜管式压力计
inclinometer 倾斜式织物静摩擦测试仪
included gas ①吸留气体②溶解在石油中的气体
included sapwood 内函边材
inclusion ①包含；包含②夹杂物
inclusion analysis in metal 金属中夹杂物分析

inclusion assay 夹杂分析
inclusion body 包合体
inclusion complex 包合配合物
inclusion complexation 包结络合
inclusion compound 包合物；包含化合物；包结物
inclusion constant 包结常数
inclusion-forming substance 包合形成物
inclusion migration 包合物迁移
inclusion polymerization 包合聚合
inclusion reaction 包结反应
inclusion X-ray tube 封入X射线管
incoagulable 不可凝结的
Incoby nickel alloys 因科比镍合金；耐热镍合金
incoercible 不能压缩的
incoherence 不相干性
incoherent 不相干的
incoherentness 不相干性
incoherent scattering 不相干散射
incohesion 不相干性
incoliy 因科铬伊；耐高温镍铬铁合金
incombustibility 不燃性
incombustible 不燃(烧)的
incombustible conveyor belt 不燃性运输带；防火运输带
incombustible fabric 不燃纺织品
incombustible matter 不燃物
incombustible mixture 不燃混合物
incombustible substance 不燃物
incoming beam 入射光束
incoming charge 进料；进料费用
incoming gas 进气
incoming mixture 进料混合物
incoming signal 输入信号
incoming stock 进料；进入原料
incommensurability 不相称；不均衡性；不成比例
incommensurate crystal 无公度晶体*
in commercial quantity 工业规模地
incompatibility ①配伍禁忌②不相容性
incompatible ①禁忌的②不相容的
incompatible elasticity theory 不相容弹性理论
incompatible salts 不相容盐类
incomplete 不完全的
incomplete adhesion 欠黏
incomplete antibody 不完全抗体
incomplete channel 不完整孔道
incomplete chemical combination 不完全化合
incomplete combustion 不完全燃烧
incomplete equilibrium 不完全平衡
incomplete fertilizer 不完全肥料

incomplete Ferund's adjuvent 不完全福氏佐剂〔不含微生物〕
incomplete fusion 未焊透
incomplete hiding 不完全遮盖(的)
incomplete oxidation 不完全氧化
incomplete penetration 未焊透
incomplete protein 不完全蛋白质
incomplete reaction 不完全反应
incomplete reduction 不完全还原
incomplete root penetration 根部未焊透〔焊接〕
incomplete singeing 不完全烧焦
incomplete vulcanization 不完全硫化；部分硫化
incompressibility 不可压缩性
incompressible 不可压缩的
incompressible simple fluid 不可压缩简单流体
incompressible viscous fluid 不可压缩黏性流体
incondensable 不能缩合的
incondensable gas 不冷凝气体
inconel 铬镍铁耐热蚀合金；因科镍合金
inconel drum 铬镍铁合金转鼓
incongealable 不冻(结)的
incongruence 异元性；不相容性
incongruent 异元的；不相容的
incongruent melting point 异元熔点；(液固)异成分熔点
incongruity 异元性；不相容性
incongruous 不一致的；不合式的
inconstancy 无常；不定
in control 在控制内的
incontrollable 不能控制的
inconvertibility ①不可逆性②不能转变性
inconvertible ①不可逆的②不能转变的
incorporating mill 并和机
incorporation 结合；掺和；掺入
incorporation into 并入
incorporation time 混合时间
incorrect consistency (油墨的)稠度失调
incorrodible 不腐蚀的；抗腐蚀的
increased yield value 增加的屈服值；屈服值增量
increasing creep 渐增蠕变
incredible natural abundance double quantum transition experiment(INADEQUATE) 自然丰度双量子特殊跃迁试验
increment ①增量②份样
incremental charge polarography 增电荷极谱法
incremental draw process 递增拉伸工艺；增量拉伸工艺
incremental plastic theory 增量塑性理论
increment percent 增长率；增量(百分)率
increment sample 增量样品

increment sampling 增量取样法
incretin 肠降血糖素；肠促胰岛素
incretion 内分泌
incretory 内分泌的
incrustation ①表面硬化②结壳③沉垢
incrustation-proof pipe 防表面硬化纺丝管
incubation ①保温，孵化，培育②潜伏
incubation period(=induction period) 诱导期；感应期
incubation time 潜伏期；诱导期
incubator 恒温箱；恒温器；孵化箱
indaconitine 印乌头碱
indalone 避虫酮
indamine 吲达胺 $HN=C_6H_4=NC_6H_4NH_2$
indan 茚满
indandione 2,3-二氢-1,3-茚二酮 $C_6H_4COCH_2CO$
indane(=hydrindene) 二氢化茚 $C_6H_4CH_2CH_2CH_2$
indan(e) musk 茚满型麝香
indanol 茚满醇
indanol acetate 乙酸茚满酯
β-indanone 2,3-二氢-2-茚酮 $C_6H_4CH_2COCH_2$
indanone(=indone) 2,3-二氢-1-茚酮
indanthrone 阴丹酮
indanthrone blue 阴丹酮蓝
indanyl 2,3-二氢化茚基 C_9H_9-
indapamide 吲达帕胺〔药〕
indazole 吲唑 $C_6H_4CH=NNH$
indazolinium compound 吲唑䓬化合物；吲唑季铵化合物
indazolone 吲唑酮
indazolyl 吲唑基 $C_7H_5N_2-$
indefinite 不定的
indene- 〔词头〕茚并
indene 茚 $C_6H_4CH_2CH=CH$
indeneacetic acid 茚(基)乙酸
indene-coumarone resin 茚-香豆酮树脂
indene resin (聚)茚树脂
indenofluorene 茚并芴
indenoindene 茚并茚
indeno [1,2-b] indole 茚并[1,2-b]吲哚
indenoisoquinoline 茚并异喹啉
indenol 茚酚
indenone(=indone) 二氢茚酮
indentation ①压痕；凹陷；凹穴②针入度〔硬度计〕

indentation force 压入力
indentation hardness 压痕硬度*
indentation index ①压痕指数②针入度指数
indentation load 压陷载荷
indentation recovery 压痕复原
indentation test 压痕试验
indented V-belt 齿形三角带；风扇带
indenting ball 硬度计球
indenting load 压陷载荷
indention 成穴
indentor 压头〔硬度计的〕
indentor point 凹刻针头；压针头
indenture(=indention) 成穴
indenyl 茚基〔有7种异构体〕 C_9H_7—
independent 独立的
independent arc furnace 独立电弧炉
independent chain(=uncrosslinking chain) 独立链；未交联链
independent component 独立组分
independent control 单独控制；单独调整
independent double bonds(=independent double links) 孤立双键
independent double links(=independent double bonds) 孤立双键
independent electron approximation 独立电子近似
independent electron pair approximation(IEPA) 独立电子对近似*
independent oiler 单独加油器
independent random variable 独立随机变量
independent variable(s) 自变数；自变量
independent yield 独立产额*
in-depth distribution 深度分布
in-depth resolution 深度分辨
inderite 多水硼镁石
indestructibility 不灭性
indeterminable 不可测的
indeterminate 未定的
indeterminate error 未定误差；不可测误差
indeterminate principle 测不准原理
indeterminism 测不准论
index ①指数②索引③指标④分度头
index compound 母体化合物；索引化合物
index error 误差指数
index for oiliness 油性指数
index for routine analysis for atmospheric pollution 大气污染常规分析指标
index gradient 指数梯度
indexing transfer 定位传送器
index number 指数；平均沸点〔汽油的〕
index of activity 活度指数
index of cementation 渗碳指数
index of notch sensitivity(=notch factor) 缺口敏感度指数
index of polarity 极性指数
index of refraction 折光指数
index of routine analysis for atmospheric pollution 大气污染常规分析指标
index of routine analysis for water pollution 水污染常规分析指标
index of whiteness 白度指数
index of yellowness 黄度指数
index plate ①分度盘；标盘②说明牌
index thermometer 有刻度的温度计
Indiana isopentane process 印第安纳(公司)异戊烷法
Indiana oxidation test 印第安纳氧化法
Indian beeswax 印度蜂蜡
Indian grass oil 印度草油
Indian gum 印度树胶
Indian ink 印度墨水；黑墨水
Indian laurel oil 印度月桂油
Indian mahogany wood oil 烟香椿木油；印度红木油
Indian marigold oil 印度万寿菊油；孔雀草油
Indian melissa oil 印度蜂花油
Indian mustard seed oil 印度芥子油
Indian rape oil 印度菜油
Indian red 印度红；三氧化铁；氧化正铁
Indian red ocher 印度赭红；天然氧化铁红
Indian valerian oil 印度缬草油
Indian wintergreen oil 印度冬青油
Indian yellow 印度黄；酸性黄
India paper(=bible paper) 圣经纸
India rubber 天然橡胶〔最早的天然橡胶名称〕
India rubber tubing(=rubber tubing) 橡皮管
indican ①尿蓝母；β-吲哚硫酸钾〔动物〕②β-吲哚葡萄糖苷〔植〕
indicanometer 靛蓝母计
indicanuria 尿蓝母尿
indicarmine 靛胭脂；酸性靛蓝
indicated 标明的；指明的
indicated efficiency 指示效率
indicated horse power 指示马力；指示功率
indicated hydrogen 指示氢
indicated mean effective pressure 平均指示压力
indicated pressure 指示压力
indicated value 指示值
indicated weight 标(明的)重(量)

indicated work　指示功
indicating　指示
indicating controller　指示(器的)控制仪
indicating device　指示装置
indicating electrode　指示电极
indicating error　指示误差
indicating flow meter　指示流量计
indicating gauge　指示器
indicating head　指示头
indicating instrument　指示仪表
indicating liquid level gauge　液面指示仪表
indicating meter　指示计
indicating micrometer　指示测微计
indicating pressure gauge　指示压力表
indicating thermocouple　指示热电偶
indication　①指出②表示③征候
indication of quality　质量指数
indicative factor　标示因素；标示因子
indicator　指示剂*
indicator acid　酸性指示剂
indicator analysis　指示剂分析(法)
indicator base　碱性指示剂
indicator blank　指示剂空白*
indicator board　指示器板
indicator card　①指示器卡片②指示图③示功图
indicator constant　指示剂常数*
indicator correction　指示剂校正
indicator current　指示电流
indicator diagram　①指示器图表②示功图
indicator diagram of work　示功图
indicator dial　指示盘
indicator electrode　指示电极
indicator element　示踪元素
indicator error　指示剂误差
indicator exponent　指示剂指数
indicator for diazotization　重氮化指示剂
indicator for non-aqueous acid-base titration　非水酸碱滴定指示剂
indicator for nonaqueous system　非水体系指示剂
indicator for nonaqueous titration　非水滴定指示剂
indicator lamp　指示灯
indicator method　指示剂法
indicator motion　指示器动作
indicator organism for atmospheric pollution　大气污染指示生物
indicator paper　试纸〔浸有指示剂的纸〕
indicator photometric titration　指示剂光度滴定
indicator piston　指示活塞

indicator pressure　指示压力
indicator range　指示幅度；变色范围；变色区
indicator test　指示剂试验
indicator titration method　指示剂滴定法
indicator transition point　指示剂变色点
indicator tube　检测管
indicator tube method　检测管法
indicator value　指示值
indicator work　指示功
indicatrix　指示面；指示量；指示线；特征曲线
indicaxanthin　梨果仙人掌黄质
indicia　标记
indicolite　蓝电气石；蓝碧玺
indifference　中性；惰性
indifferent　①惰性的②不重要的
indifferent electrode　惰性电极；参比电极；参考电极
indifferent electrolyte(=supporting electrolyte)　支持电解质；惰性电解质
indifferent gas　惰性气(体)〔指零族气体元素或不起作用的气体〕
indifferent oxide　惰性氧化物
indifferent salt　惰性盐；中性盐
indifferent solvent　惰性溶剂
indiffusible　不扩散的
indiffusible ion　不扩散离子
indigo(=indigotin)　靛蓝
indigo-attapulgite complex　靛蓝-凹凸棒土配合物
indigo blue　靛蓝
indigo blue printing fabric　蓝印花布；靛蓝花布
indigo bottom　靛蓝打底
indigo carmine　靛蓝胭脂红
indigo copper　铜蓝
indigo dicarboxylic acid　靛蓝二羧酸　$C_{18}H_{10}O_6N_2$
indigo disulfonic acid　靛蓝二磺酸　$C_{16}H_8O_2N_2(SO_3H)_2$
indigofera oil　槐蓝油
indigoid　靛青类
indigoid colors　靛类染料
indigoid colouring matters　靛染料〔含有—COC═CCO—结构〕
indigoid dyes　靛类染料
indigoide colors〔复〕　靛类染料
indigometer　靛蓝计
indigo monosulfonate　靛蓝一磺酸盐*
indigo monosulfonic acid　靛蓝一磺酸　$C_{16}H_9O_2N_2(SO_3H)$
indigo paste　靛蓝糊
indigo process　靛蓝法
indigosol　溶靛素〔染料〕

indigosol brilliant rose 溶靛素亮玫瑰红
indigosole dye 溶靛素染料
indigosol golden yellow 溶靛素金黄
indigosol orange 溶靛素橙
indigosol printing black 溶靛素印染黑
indigosol printing blue 溶靛素印染蓝
indigosol red-violet 溶靛素红紫
indigo tetrasulfonate 靛蓝四磺酸盐*
indigo tetrasulfonic acid 靛蓝四磺酸
indigotin(=diindogen; indigo) 靛蓝
indigo trisulfonate 靛蓝三磺酸盐
indigo vat 靛蓝瓮
indigo vatting 靛蓝的瓮化
indigo violet 靛蓝紫
indigo white 靛白
indinavir 茚地那韦〔药〕
indirect 间接的
indirect ammonia recovery 间接氨回收
indirect analysis 间接分析
indirect arc furnace 间接电弧炉
indirect atomic absorption spectrometry 间接原子吸收光谱法
indirect calorimetry 间接量热法
indirect condenser 间接冷凝器
indirect cooking process 间接蒸煮法
indirect cooling surface 间接冷却面
indirect coulometer 间接电量计
indirect coulometric analysis 间接库仑分析
indirect cure 间接硫化
indirect desulfurization 间接脱硫
indirect detection 间接检测
indirect drive 间接传动；大齿轮传动
indirect EDTA titration 间接 EDTA 滴定法
indirect evaporation 间接蒸发
indirect fertilizer 间接肥料
indirect fired drier 间接火热干燥器
indirect fluorescence detection 间接荧光检测
indirect fluorescence method 间接荧光法
indirect fluorescent antibody technique 间接荧光抗体技术
indirect gas heating 煤气间接加热
indirect gravimetric analysis 间接重量分析
indirect heated mold 间接加热模型
indirect heat exchange 间接热交换
indirect heating 间接加热；外部加热
indirect hemagglutination 间接凝集反应
indirect humidifier 间接润湿器
indirectly attributable expense 间接成本
indirect method 间接法*
indirect method in atomic absorption spectrometry 间接原子吸收光谱法
indirect method of determination 间接测定法
indirect method of measurement 间接测量法
indirect optical sensor 间接光学传感器
indirect oxidation 间接氧化
indirect photometric chromatography 间接光度(检测)色谱法
indirect photometric ion chromatography 间接光度(检测)离子色谱法
indirect photometric titration 间接光度滴定
indirect pirn 间接纡
indirect process zinc oxide 间接法氧化锌
indirect produced strong nitric acid process 间接法生产浓硝酸工艺
indirect recoil effect 间接反冲效应
indirect reduction 间接还原
indirect regeneration 间接再生系统；循环再生系统
indirect sample introduction method 间接进样法
indirect spectrophotometric method 间接分光光度法
indirect spectrophotometry 间接分光光度法
indirect steam 间接蒸汽
indirect substitution 间接取代
indirect sulfonation 间接磺化
indirect titration 间接滴定
indirect ultraviolet detection 间接紫外检测
indirect weighing method 间接称量法
indirect welding 单面点焊
indirubin 靛玉红〔药〕
indiscriminate 无差别的
indispensable amino acid 必需氨基酸
indissolubility 不溶(解)性；不均(匀)性
indissoluble 不溶(解)的
indissolvable 不溶(解)的
indissolvableness 不溶(解)性
indistinguishability 不可区分性
indium 铟〔49 号元素，化学符号 In〕
indium anomaly 铟反常
indium bromide 三溴化铟 $InBr_3$
indium chloride 氯化铟 $InCl; InCl_2; InCl_3$
indium dibromide 二溴化铟 $InBr_2$
indium dichloride 二氯化铟 $InCl_2$
indium diiodide 二碘化铟 InI_2
indium flame detector 铟火焰鉴定器
indium generator 铟(同位素)发生器
indium hydrosulfide 氢硫化铟 $In(HS)_3$
indium hydroxide 氢氧化铟 $In(OH)_3$

indium monobromide 一溴化铟 InBr
indium monochloride 一氯化铟 InCl
indium monoiodide 一碘化铟 InI
indium monosulfide 一硫化二铟 In_2S
indium nitrate 硝酸铟 $In(NO_3)_3$
indium oxide 氧化铟 In_2O_3
indium oxybromide 溴氧化铟 InOBr
indium oxychloride 氯氧化铟 InOCl
indium-sensitized flame photometric detector 铟敏(化)火焰光度检测器
indium sesquioxide 三氧化二铟 In_2O_3
indium sulfate 硫酸铟 $In_2(SO_4)_3$
indium sulfide 硫化铟 In_2S_3
indium tribromide 三溴化铟 $InBr_3$
indium trichloride 三氯化铟 $InCl_3$
indium trifluoride 三氟化铟 InF_3
indium triiodide 三碘化铟 InI_3
indium trioxide 三氧化二铟 In_2O_3
indium trisulfide 三硫化二铟 In_2S_3
individual 独特的*；单一的；个体；个别
individual analysis 分别分析
individual coefficient of heat 个别热系数
individual curing ①单独硫化②单独处理
individual deviation 单独偏差
individual drive 单独传动
individual fiber 单根纤维
individual film coefficient 单项膜系数〔传热时〕
individual film coefficient of mass transfer 分传质系数
individual film coefficient on a volume basis 分体积膜系数
individual liquid sample 个别液体样品
individual maintenance cost 单项维修成本
individual metal lamp 单金属灯
individual nuance 独特香调
individual point lubrication 分散润滑
individual protection ①个别防毒②个别防护
individual vulcanizer 单人硫化机
indivisible inactive compound 不可分的不旋光化合物
indizating agent 碘化剂
indo 碘代 I—
indoaniline 靛苯胺
indocarbon 靛炭〔一种硫化染料商名〕
indocentoic acid 茚百酸
indodiazole 茚并二唑
indoferron 吲哚铁试剂
indoform 靛仿
indogen 靛元基
indogenide 靛元化物

indoine 吲哚因
indoine blue 吲哚因蓝
indolal 吲哚醛
indole 吲哚*
3-indoleacetic acid(=heteroauxin) 异植物生长素；3-吲哚基乙酸
indole alkaloid 吲哚生物碱
indolenine(=pseudo-indole) 假吲哚 C_8H_7N
indoline 二氢吲哚
α-indolinone α-吲哚满酮
indolinyl 二氢吲哚基 C_8H_8N—
indolinylidene 亚二氢吲哚基 $CH_2NHC_6H_4C=$
indolizine(=pyrindole) 中氮茚；吲嗪 C_8H_7N
indolmycin 吲哚霉素
indolocarbazole 吲哚并咔唑；氮茚并氮芴
indolol 吲哚酚
indolone 吲哚酮
indolyl(=indyl) 吲哚基；氮杂茚基〔有7种异构物〕 NC_8H_6—
indolylacetic acid 吲哚乙酸
indolylethylamine 吲哚乙胺
indolylidene 二氢亚吲哚基
3-indolylpyruvic acid 3-吲哚基丙酮酸
indometacin 吲哚美辛〔药〕
α-indone 二氢-α-茚酮 $C_6H_4CH_2CH_2CO$

indone(=indanone) 2,3-二氢-1-茚酮
indoor air pollution 室内空气污染
indoor application 户内用；室内用
indoor paint 内用漆
indophenine 靛吩咛 $(C_{12}H_7NOS)_2$
indophenol 靛酚 $C_{12}H_9O_2N$
indophenol blue 靛酚蓝
indophenol blue colorimetry 靛酚蓝比色法
indophenol dye 靛酚染料
indophenol method 靛酚法
indophenol oxidase(=cytochrome oxidase) 靛酚氧化酶
indophenol test 靛酚试验
indophenol titration method 靛酚滴定法
indospicine α-氨基-ε-脒基己酸
indoxyl(=3-indolol) 3-吲哚氧基；吲羟 $HNCH=C(OH)C_6H_4$

indoxyl-β-glucoside(=indican) 羟基吲哚-β-葡萄糖苷；尿蓝母
indoxylglucuronic acid 葡萄糖醛基吲哚苷酸
indoxylic acid(=3-hydroxy-2-indole carboxylic acid) 3-羟

基-2-吲哚羧酸　$C_6H_4C(OH)C(CO_2H)NH$

indoxyl-sulfuric acid　羟基吲哚硫酸
indraft　吸入；流入
in-draw texturing machine　内拉伸法变形机；同时拉伸变形机
in-drum mixing　桶内混合
induced　①诱导的②感生的；感应的
induced absorption　诱导吸收
induced activity　①诱导活动性②诱导放射性；感生放射性
induced bioluminescence　诱发的生物发光
induced charge　感生电荷
induced circular dichroism　诱导圆二色性
induced color　感应色；衍生色
induced current　感生电流；感应电流
induced decomposition　诱导分解
induced deformation　诱发形变
induced degradation　诱导降解
induced deposition　诱导沉积
induced dipole　诱导偶极
induced draft　压力通风；人工通风
induced-draft fan　烟泵；排烟机
induced draught　人工通风
induced draught cooling tower　排风式凉水塔
induced electrical polarity　诱导极性
induced electron emission　感应电子发射
induced emission　诱导发射
induced enzyme　诱导酶
induced exhaust　引风管；抽风管
induced fission　诱发裂变*
induced-fit　诱导匹配
induced-flow heater　感生流加热器
induced force　感应力
induced grafting　诱导接枝
induce dipole moment　诱导偶极矩
induced magnetic moment　感生磁矩
induced (molar) polarization　(摩尔)诱导极化
induced neutron counting　感生中子计数
induced oxidation　诱导氧化(作用)
induced precipitation　诱导淀积；诱导沉淀
induced radiation　感应辐射；诱导辐射
induced radioactivity　感生放射性
induced radioisotope　感生放射性同位素
induced reaction　诱导反应*
inducer　①诱导物；诱发物②导流片
inducing catalyst　诱导催化剂
inducing charge　施感电荷
inducing combustion　诱导燃烧
inducing interaction force　诱导力
inductance　①电感②感应系数
inductance and eddy current technique　感应和涡流技术
inductance bridge　电感电桥
inductance coil(=induction coil)　电感线圈；感应线圈
inductance type of titration cell　电感式滴定池
inducted mixture　引进的混合物
induction　①诱导②感应③归纳(法)
induction bridge　感应电桥
induction coil(=inductance coil)　电感线圈；感应线圈
induction drip melting　感应滴熔
induction force(=Debye force)　德拜力；感应力
induction-free　无感应
induction furnace　感应电炉
induction fusing　感应熔化；感应熔融
induction heater　感应加热器
induction heating　感应加热
induction heating reactor　感应加热反应器(釜)；电感加热反应釜
induction melting　感应熔化；电感熔融
induction method　感应方法
induction motor　感应电动机
induction period(=incubation period)　诱导期；感应期
induction period in gum formation　(汽油中)胶质生成诱导期
induction period of crystallization　结晶诱导期；晶化诱导期
induction period of drying　干燥诱导期
induction period of gasoline　汽油的诱导期
induction period test　诱导期试验
induction pipe　①送水管②送气管
induction regulator　感应调压器
induction stability　诱导稳定性
induction surface hardening　感应表面硬化
induction time　诱导时间
induction valve　①送汽阀②送水阀
induction voltage　感应电压
induction welding　感应焊接
inductive　①感应的②诱导的
inductive cleavage　诱导断裂
inductive effect　诱导效应*
inductive effect index　诱导效应指数
inductively coupled high frequency plasma torch　电感耦合高频等离子体焰炬*
inductively coupled microwave plasma torch　电感耦合微波等离子体焰炬
inductively coupled plasma　感应耦合等离子体

inductively coupled plasma atomic emission spectrometry (ICP-AES)　电感耦合等离子体原子发射光谱法*
inductively coupled plasma atomic emission spectrum　电感耦合等离子体原子发射光谱
inductively coupled plasma atomic fluorescence spectrometry　电感耦合等离子体原子荧光光谱法
inductively coupled plasma ionization source　电感耦合等离子体电离(离子)源
inductively coupled plasma mass spectrometer(ICP-MS)　电感耦合等离子体质谱仪
inductively coupled plasma mass spectrometry　电感耦合等离子体质谱
inductively coupled plasma source　电感耦合等离子体光源
inductively couple plasma　电感耦合等离子体
inductive phase　诱发期
inductive reactance　电感电抗
inductive separation　感应分选
inductive transducer　感应换能器
inductivity　诱导率
inductivity heated plasma　感应加热等离子体
inductometer　电感计
inductor　①感器②诱导物③感应物④感应体
inductothermy　感应电热器
indulin　纯松木素
induline　引杜林(染料)；吩嗪蓝；对氮蒽型蓝
induline base B　引杜林色基B〔可作汽油防胶剂〕；对氮蒽蓝B
induline fat-soluble　油溶引杜林
induline spirit-soluble　醇溶引杜林
indurated fiber　硬纤维
induration　硬化；变硬；固化；凝固
industrial　工业的
industrial absorbent　工业用吸收剂
industrial alcohol　工业酒精；变性酒精
industrial aromatic hydrocarbons　工业芳香烃
industrial atmospheric corrosion　工业大气腐蚀
industrial chemistry　工业化学
industrial chromatograph　工业色谱仪
industrial chromatography　工业色谱(法)
industrial chromium plating　工业镀铬
industrial coating　工业涂装
industrial crystallization　工业结晶；大量结晶
industrial effluent　工业废液；工业废水
industrial equipment lubricant　工业装置用润滑油
industrial explosive　工业炸药
industrial filtration aids　工业助滤剂
industrial fume　工业烟雾

industrial furnace　工业炉
industrial gas　工业煤气
industrial gas chromatographic method　工业气相色谱(法)
industrial gasoline　工业汽油
industrial gasses　工业气体
industrial grade　工业级
industrial 90-grade benzene　90级工业苯〔沸点范围782～120℃〕
industrial grade toluene　工业级甲苯
industrial grade xylene　工业级二甲苯
industrial intermediates　工业中间体
industrial ion meter　工业离子计
industrial irradiator　工业辐照器
industrial leather　工业用革
industrial lubricant　工业润滑油
industrial lubrication　工业装置的润滑
industrial maintenance paint　工业维修漆；工业保养漆
industrial mask　工厂用面具
industrial methylated spirit　工业用甲醇变性酒精
industrial naphtha　工业石脑油
industrial odorant　工业用香料
industrial oil　工业用油
industrial organic finishing　工业有机涂装；工业有机涂料
industrial petroleum chromatography　石油工业色谱
industrial pH meter　工业pH计
industrial primer　工业底漆
industrial processing cloth　工业筛网
industrial product　工业产品
industrial property　工业特性
industrial public nuisance　工业公害
industrial pumping　工业泵送；厂内泵送
industrial pumping equipment　工业泵送设备；厂内泵送设备
industrial radiography　工业放射照相法
industrial rapid balance　工业快速秤
industrial refrigerating plant　工业制冷装置
industrial rheology　工业流变学*
industrial rubber　商品橡胶
industrial safety　工业安全
industrial salt　工业盐；原盐
industrial scale　工业规模
industrial sewage　工业污水
industrial soap　工业(用)皂
industrial (solid) tyre　工业实心胎
industrial solvent　工业溶剂
industrial specialties　工业特制品；工业专用品
industrial spirit　工业酒精

industrial stoichiometry 工业化学计算(法)
industrial surfactant 工业用表面活性剂
industrial topcoat 工业面漆
industrial-use functional fibre 产业用功能纤维
industrial waste 工业废物；工业废液
industrial waste treatment 工业废物处理；工业废液处理
industrial waste water 工业废水
industrial waste water treatment 工业废水处理
industrial water 工业用水
industrial white oil 工业白油；深度精制油
industry 工业
indyl(=indolyl) 吲哚基；氮(杂)茚基
indylidene 二氢亚吲哚
-ine〔词尾〕 ①炔②因；碱
inearnatyl alcohol 三十四(烷)醇
inedible fat 非食用脂
inedible oil 非食用油
inelastic collision of ions 离子非弹性碰撞
inelastic deformation 非弹性形变
inelasticity 非弹性
inelastic neutron scattering spectroscopy(INS) 中子非弹性散射能谱学
inelastic scattering 非弹性散射*
inene(=α-gurjunene) 荜烯；α-古芸烯
inequality 不等式
inert 惰性的
inert additive 惰性添加剂
inertance 惰性
inert anode 不溶性阳极〔惰性阳极〕
inert atmosphere 惰性气氛
inert-atmosphere glove box 惰性气氛手套箱
inert-atmosphere processing cell 惰性气氛处理热室
inert base ①惰性基质②(涂料的)非活性固体分
inert black 非活性炭黑
inert carrier 惰性载体
inert carrier gas 惰性载气
inert coating 惰性膜
inert complex 惰性配合物
inert conductor 惰性导体
inert constituent 惰性组分
inert coordination compound 惰性配合物*
inert diluent 惰性稀释剂
inert diluent effect 惰性稀释剂效应
inert diluent gas 稀释用惰性气体
inert electrode 惰性电极
inert electrolyte 支持电解质；惰性电解质
inert element 惰性元素
inert filler 惰性填料*

inert-free gas 无惰性气体；不含惰性成分的气体
inert gas 惰性气体
inert-gas arc welding 惰性气体保护焊
inert gas atom 惰性气体原子
inert gas configuration 惰性气体结构
inert gas deposition 惰性气体沉积法
inert gas generator 惰性气体发生器
inert gas ion gun 惰性气体离子枪
inert gas plasma 惰性气体等离子体
inert gas welding 惰性气体焊接(法)
inert group 惰性基团
inert heat carrier 惰性热载体
inertia ①惯性②惯量
inertia force 惯性力
inertial dust collection 惯性除尘
inertial effect 惯性效应
inertial effect in fluid 液体中惯性效应
inertial element 惯性单元
inertial separator 惯性集尘器；惯性分离器
inertial settling 惯性沉降
inertia mass 惯性质量
inertia shaking force 摇控惯性力
inert material 惰性物料
inert media 惰性介质
inertness ①惰性②惯性
inert packing 惰性填料
inert peak 稳定峰
inert pigment 惰性颜料；体质颜料
inert plasticizer 惰性增塑剂
inert polymer 惰性聚合物
inert powder 惰性粉料
inerts 惰性组分；惰性气体
inert sample 惰性试样
inert solvent 惰性溶剂*
inert substance 惰性物质
inert support 惰性载体
inert to deformation 抗变形性
inexhausted 未用尽的
inexplosive 不爆炸的
inextensible 不可延伸的
inextractable 不被萃取的
inextricable mixture 不可分混合物
infallible powder 确发炸药
infection 传染(性)；感染
infectious 传染的
in-feeding 进料；加料
infer-coat 二道底漆
inferior coal 低质煤

inferior fuel 低质燃料
in-field use 就地使用；现场使用
infiltration ①渗(入过)滤②渗透；渗入
infiltration rate 渗入速率；渗滤速率
infinite 无限的；无穷的
infinite amount of vaporization 无限蒸发〔分馏时无限汽化〕
infinite diameter column 无限直径柱
infinite diameter effect 无限直径效应
infinite dilution ①无限冲淡②无限淡度；无限稀度
infinite dilution activity coefficient 无限稀释活度系数
infinite fluid viscometer 无限流体黏度计
infinitely thick layer 无限厚度(层)
infinitely variable speed transmission 无限度变速传动
infinite networks 无限网状结构
infinite population 无限总体
infinite reflux 无限回流；全回流
infinite reflux operation (蒸馏塔)无限回流操作
infinite reflux rate 无限回流(速)率
infinite reflux ratio 无限回流比率
infinitesimal ①无限小的；无穷小的②无限小(量)
infinitesimal deformation 无穷小形变
infinitesimal strain 无穷小应变
infinites network 无限交联网
infinite swelling 无限溶胀
infinite thickness 无限厚度
inflame 燃烧
inflamed gases 燃烧(着的)气体
inflaming retarding 阻燃*
inflammability 易燃性
inflammability point 燃点；着火温度
inflammability test 易燃性试验
inflammable 易燃的
inflammable air 可燃空气〔历史上名称〕；氢气
inflammable gas 易燃气体
inflammable liquid 易燃液(体)
inflammable minerals〔复〕 可燃性矿物
inflammableness 易燃性
inflammable store 易燃物储藏库
inflammable substance 易燃物
inflammable vapour 易燃蒸气
inflammation 着火；燃烧
inflammation point 着火点
inflatable 可膨胀的；可充气的
inflatable bag 气囊
inflatable bag method 充气袋模法
inflatable cushion 气垫
inflatable seat 充气座椅

inflatable skeleton 充气骨架；充气房屋
inflated 膨胀了的；打了气的
inflated article 充气制品
inflating 打气；装气
inflating agent(=inflating medium) 发泡剂；生气剂
inflating medium(=inflating agent) 发泡剂；生气剂
inflating valve 气门嘴
inflation 充气吹胀*
inflation film 吹胀薄膜
inflation film process 吹塑薄膜法
inflation method 吹胀法
inflation pressure 充气压力
inflation process 充气法
inflation sleeve 打气筒
inflatoplane 充气橡皮飞机
inflection ①屈折②内弯
inflection point 拐点*；转折点
inflection temperature 拐点温度
inflexible 不可弯曲的；僵硬的；刚性的
inflexible polymer chain 刚性聚合物链
inflexion ①弯曲；屈曲②拐折；拐点
inflexion point ①屈折点②拐点
in-flight refuelling hose 空中加油胶管
inflow 流入
influence exploder 非触发引信；感应雷管
influence function 影响函数；感应函数
influence of a third element 第三元素影响
influence quantity 影响量
influent 流体；液体；流入液
influent gradient 流入液梯度
information 信息*
informational macromolecules 信息大分子
information content 信息容量*
information content of a spectrum 谱图的信息量
information data 信息数据
information document 情报资料
information efficiency 信息效率*
information encoding 信息编码
information gain 信息增益
information gathering network 情报收集网
information handing 信息处理
information pattern 信息图
information polymer 信息聚合物〔生物体内传递信息高分子化合物的总称〕
information profitability 信息效益*
information retrieval 信息检索；情报检索
information storage 信息存储器
information value of reference material 标准物质参考值

informative pattern 可提供数据的图形
informing power 供信能力；信息参量
informosome 信息体
infra-red 红外(线)的
infra-red absorption 红外线吸收
infrared absorption band 红外吸收(谱)带
infrared absorption cell 红外吸收池
infrared absorption curve 红外吸收曲线
infrared absorption intensity 红外吸收强度
infra-red absorption method 红外线吸收分析法
infrared absorption moisture meter 红外吸收水分计
infrared absorption spectroscopy 红外吸收光谱(法)
infrared absorption spectrum 红外吸收光谱
infrared-active 红外活性的
infra-red analysis 红外线分析法
infrared analytical instrument 红外分析仪器
infra-red analyzer 红外线分析器
infrared baking 红外线烘干；红外线烤干
infra-red band 红外(线)谱带
infrared beam condenser 红外光束聚光器*
infrared cell 红外吸收池
infrared detector 红外检测器*
infrared dichroism 红外二向色性
infrared dryer 红外线干燥器
infrared drying 红外线干燥
infrared electronic feeler 红外电子探测器
infrared emission spectroscopy 红外发射光谱
infrared emulsion 红外型乳剂
infrared filter 红外滤波器
infrared Fourier transform spectrometry 傅里叶变换红外光谱法
infrared frequency 红外线频率
infrared gas analyzer 红外气体分析器
infrared generator 红外线发生器
infrared heater 红外线加热器
infrared heating 红外线加热
infrared heating oven 红外线加热炉
infrared heat lamp 红外加热灯
infrared hygrometer 红外湿度计
infrared inactive 不显红外的；非红外活性的
infrared interference pattern 红外干涉图
infrared lamp 红外线灯
infrared laser radiation 红外线激光辐射
infrared light 红外光
infrared light source 红外光源
infrared measurement 红外(光谱)测定
infrared microscope 红外显微镜
infrared microscopy 红外显微技术
infrared microspectrometer 红外显微分光仪
infrared moisture analysis 红外水分分析
infrared moisture meter 红外水分计
infrared noise level 红外噪声水平
infrared penetration cure 红外(线)硫化
infrared photoelectric self-spreading and centering apparatus 红外光电自动扩展对中装置
infrared photometer 红外光度计
infrared polarization spectrum 红外偏振光谱
infrared polarizer 红外偏振器*
infrared projector 红外线发射体
infrared pyrometer 红外线高温计
infrared radiation 红外辐射
infrared radiator 红外线辐射器
infrared range 红外线范围
infrared rapid-scan monochromator 红外快扫描单色仪；快扫描红外单色仪
infrared ray 红外线
infrared ray drying 红外线干燥
infrared ray lamp 红外线灯
infrared reference spectrum 红外吸收参比图谱
infrared reflectance curve 红外反射曲线
infrared reflecting camouflage paint 红外线反射伪装涂料
infrared reflection-absorption spectroscopy 红外反射-吸收光谱法
infrared region 红外区
infrared scanner 红外扫描器；红外检查器
infrared sensor 红外传感器
infrared signature 红外特征
infrared solvent 红外溶剂
infrared source 红外光源*
infrared spectra 红外光谱图
infrared spectroelectrochemistry 红外光谱电化学法
infrared spectrogram 红外光谱图
infrared spectrograph 红外摄谱仪
infrared spectrometer 红外分光计
infrared spectrometer detector 红外光谱检测器
infrared spectrometry 红外光谱法
infrared spectrophotography 红外分光摄谱学
infrared spectrophotometer 红外分光光度计*
infrared spectrophotometry 红外分光光度法
infrared spectroscopic analysis 红外光谱分析
infrared spectroscopy(IR) 红外光谱法*
infrared spectrum 红外光谱
infrared spectrum analysis 红外光谱分析
infrared spectrum remote-sensing analysis 红外光谱遥感分析
infrared standard spectrum 红外标准图谱

infra red stoving 红外线烘干
infrared stoving finish 红外线烘漆；红外线烤漆
infrared thermography 红外热成像法
infrared thermometer 红外测温计
infrared-transmitting material 红外透明材料
infrared transmitting solvent 透过红外线的溶剂
infrared two-colour detector 双色红外检测器
infrared two-colour moisture meter 双色红外水分计
infrared unit 红外线装置
infrared vulcanization 红外(线)硫化
infrared wave band 红外波段范围
infrared window material 红外窗片材料
infrasizer 超微粒空气分级器
infrastructure 下部结构；底层结构；基础结构
infusibility ①不熔性②难熔的
infusible ①不熔的②难(能)熔的
infusible precipitate 不熔性沉淀；难熔性沉淀
infusion ①灌输②浸剂③浸渍
infusion kettle 浸渍锅
infusorial earth 硅藻土
ingate 入口孔
ingenol 巨大戟二萜醇
ingestion 摄取
ingot 锭；块；坯料；锭料
ingot furnace 热锭炉；铸锭炉
ingot iron 铁锭；生铁；铸铁
ingot metal 金属锭；铸金属
ingrain 纱染〔纱先染,然后织〕
ingrain color 显色染料〔一类染料的类名〕
ingrain dye 显色染料
ingrain dyestuff 显色染料
ingramycin 英格拉霉素；白环霉素
ingredient ①组分②成分③配合剂；拼料
ingress ①进口②进口处
ingress pipe 导入管
inhalable particles 可吸入颗粒物
inhalation 吸入；吸入物；吸入法
inhaler ①吸气器②滤气器
inherent 固有的；原有的
inherent astigmatism 固有像散
inherent colour 固有颜色；本身颜色
inherent dispersibility 固有分散性
inherent flaw 内部缺陷；固有缺陷
inherent instability 固有的不稳定性
inherent limitation 固有限度；原有限度
inherently flameproof 本质防燃的；内在防燃的
inherent moisture ①固有水分②结合水
inherent quality 内在质量
inherent regulation 内在调节；自调节
inherent toughness 固有柔韧性
inherent variability 固有偏差
inherent viscosity 比浓对数黏度*
inhibited alkaline cleaner 阻蚀性碱清洗剂
inhibited oil 抑制油；抗氧化剂油
inhibited oxidation 抑制氧化〔抗氧化剂存在下的氧化〕
inhibited polymerization 受抑聚合
inhibiting degradation 受抑降解
inhibiting solution 缓蚀液
inhibiting value 抑制值；安定性增高值
inhibition 阻聚作用*
inhibition analysis 抑制分析
inhibition constant 阻聚常数
inhibition discoloring spectrophotometry 抑制褪色光度法
inhibition index 抑制指数
inhibition kinetic-spectrophotometry 阻抑动力学光度法
inhibition mechanism 抑制机理
inhibition period 抑制期
inhibition ratio 抑制比率
inhibitive priming paint 防锈底漆
inhibitive substance ①阻聚剂②防老剂③迟延剂
inhibitive wash 缓蚀洗液；阻蚀洗液
inhibitor 阻聚剂*；抑制剂
inhibitor constant 抑制剂常数
inhibitor content 抑制剂含量
inhibitor dye 抑制剂染料
inhibitor efficiency 腐蚀抑制效率
inhibitor of oxidation 氧化抑制剂
inhibitor of polymerization 聚合阻止剂；阻聚剂
inhibitor response 抑制剂感应
inhibitor solution 抑制剂溶液
inhibitor susceptibility(=inhibitor response) 抑制剂感应
inhibitor sweetening 抑制剂脱硫(法)
inhibitory concentration 抑制浓度
inhibitory effect 抑制作用
inhibitory phase 抑制相
inhibitory quotient 抑菌商数
inhomogeneity 不匀一性；多相性
inhomogeneous 不匀的；多相的
inhomogeneous broadening 非均匀展宽
inhomogeneous field magnet 非均匀场磁铁
inhomogeneous field magnet technique 非均匀磁场法
inhomogeneous force system 不均匀力系
inhomogeneous magnetic field mass spectrometer 非均匀磁场质谱计
inhomogeneous reaction 非均相反应*
inhomogenous 不匀的；多相的；杂拼的

inifer 引发-转移剂*
iniferter 引发-转移-终止剂*
inimer 引发剂单体
initial (最)初的；开始的
initial action 初反应；起始反应
initial adhesion 初黏合性
initial adjustment 初调；零度调正
initial allowance 机械加工余量
initial balance 初始平衡
initial band 初始谱带
initial boiling point 初沸点；初馏点
initial bond strength 初期结合力
initial cell wall 初生胞壁
initial charge 初充电
initial collapse 初始坏裂；初始断裂
initial colour 初色；最初的颜色
initial come-up time of unit 设备从开工到正常操作的时间
initial concentration 起始浓度
initial cost 基建投资；原始成本
initial creep 初蠕变
initial crusher 初级压碎机
initial cure 起始硫化
initial daily production 最初日产量〔设备开工后第一天的日生产量〕
initial deflection ①初始挠曲②初始挠度
initial detonating agent 起爆剂
initial elasticity 初弹性；瞬间弹性
initial evaporation rate 初始蒸发速度
initial fill material 电泳槽初投涂料；电泳原漆
initial fixing image 初始定影
initial flexural strength 初始弯曲强度
initial foam height 起始泡沫高度；当时泡沫高度
initial geometrical defect 初始几何缺陷
initial gloss 初始光泽；(暴晒前的)原始光泽
initial grab 初期黏附力
initial hydrogen blistering 初期氢致起泡
initial ignition 起始着火
initial isobaric period 初始等压时间
initial level 初始能级
initial lubrication 起动润滑
initial modulus 起始模量；初始模量
initial Newtonian viscosity 起始牛顿黏度
initial orientation 初始取向
initial oxidant 初始氧化剂
initial point(=drip point) 初馏点
initial polymerization 初始聚合；预聚合
initial pressure 初压

initial pure whiteness 原始纯白(色)度
initial rate constant 初速常数
initial raw plasticity 塑炼前可塑性
initial refining 捏炼；一次精炼〔再生胶〕
initial retention time 初期保留时间
initial retention volume 初期保留体积
initial rubber 原料(生)胶
initial set 初凝
initial set-up 初期定型
initial slope 初始斜率
initial speed 初速度
initial spot 初始斑
initial spreading coefficient 初始铺展系数*
initial strain 初始应变
initial stress 起始应力；初应力
initial stress defect 初始应力缺陷
initial surface 初始表面；原始表面
initial surface protection 原始表面保护
initial tangent modulus 起始切线模量
initial tear 初撕裂
initial tear strength 起始撕裂强度
initial temperature 起始温度
initial true boiling point 最初实沸点；最初真沸点
initial vapour pressure 初蒸气压
initial velocity 初速度
initial viscosity 初始黏度
initial vulcanization step 硫化开始阶段；硫化初期
initial wet modulus 初始湿模量
initial yellowing 初期黄变
initial yield value 起始屈服值
initiating accident sequence 引发源
initiating agent 引发剂
initiating capability 引发能力
initiating explosive 起爆药
initiating power 引发能力
initiating radical 引发基
initiating species 引发形式；引发种类
initiating stage ①初期②引发阶段
initiation 引发*
initiation code 起始密码
initiation complex 起始复合物
initiation of chain 链的引发
initiation of detonation 爆炸的引发
initiation of flame 火焰的发生
initiation rate 引发速率
initiation rate constant 引发速率常数
initiation reaction 引发反应
initiation temperature 引发温度

initiative ①初步②主动③创造力
initiator 引发剂*
initiator activity 引发剂活性
initiator codon 起始密码子
initiator efficiency 引发剂效率
initiator of polymerization 聚合引发剂
initiator radical 引发(剂)基
initiator system 引发剂体系
initiator transfer agent 引发-转移剂
initiator transfer agent terminator 引发-转移-终止剂
initiator transfer constant 引发剂转移常数
injectability 注射能力
injected fuel spray 注射燃料硫
injecting septum 进样隔膜胶垫
injection 注射；注入；喷射
injection band spreading 注射谱带扩展
injection block 注射块
injection blow moulding 注射吹塑
injection channel 注样通道
injection compression molding 注塑压缩成型
injection condenser 喷射冷凝器
injection cooling 注入冷却
injection depth 注射深度
injection efficiency 注射效率
injection error 注射误差
injection heater 注射加热器
injection hotspot 注射加热部位
injection laser 注入型激光器
injection life 注射寿命
injection mold 射模
injection-molded item 注压制品
injection molder 注射模型成形机
injection molding 注射成型*
injection molding machine ①注压(硫化)机②注(射)模(塑)机〔塑料〕
injection mould 注(射)塑)模
injection-moulded parison 注塑型坯
injection moulding 注(射)模(塑)法
injection nozzle 注射管嘴
injection orifice 注射孔
injection piston 注射柱塞
injection plunger 注射柱塞
injection port 注射口；进样口
injection pressure 注射压力
injection pump 注射泵；喷射泵
injection splicer 注压(制品)接头机
injection syringe 注射管；注射器
injection technique 注射技术
injection temperature 注射温度
injection torpedo 注射分流梭
injection valve 进样阀
injection volume 注入体积
injection water 注射水
injection welding 注塑焊接
injector 注射器；喷射器
injector fuel pipe 注射燃料管
injector pump 注射泵
injector tube apparatus 软管通风防毒器
injector wall 注射器(管)壁
inject time 注射时间
ink ①墨水②(油)墨③墨汁
ink blue 碱性蓝〔别名〕
ink-eraser ①擦墨水的橡皮②消字水涂改剂；消字灵
ink(ing) feed roller 注墨辊；加墨辊〔印刷机〕
inking roller 注墨辊；加墨辊〔印刷机〕
ink-jet printing 喷墨印刷技术
inkometer 油墨汁
ink process 紫黑印画法
ink receptivity 吸墨性(能)
ink receptor 受墨物；吸墨剂〔如纸张、塑料中的填料等多孔物〕
ink-repellant ①疏墨的；不沾墨的②抗油墨性的
ink resistance 抗墨性(能)
ink stick (中国)墨
ink stone ①水绿矾〔矿〕②砚台
ink tablet 墨
ink transparency 油墨透明度
inlaid lacquer ware 镶嵌漆器
inlaid linoleum 镶嵌油毡
inlay ①镶嵌物；衬纬；垫纱②镶嵌
inlay method 镶补法
inlay printing 镶印
inleakage ①哺乳②漏入
inlet ①入口；进口②入口管
inlet capacity 吸入容量；吸入口容量
inlet coefficient 入口系数；引入系数
inlet flow rate 入口流量；入口流速率
inlet orifice 进料口
inlet pressure 入口压力
inlet pressure control 入口压力控制
inlet sample system 进样系统
inlet splitter 入口分流器；进样分流器
inlet stroke 吸入冲程
inlet system 入口系统
inlet temperature 入口温度
inlet-to-outlet system 从入口到出口系统

inlet valve ①入口阀②供给阀；进气阀
inlet velocity 入口速度
in-line (在)管线内；流线
in-line analysis 流线分析；线上分析
in-line arrangement 顺排
in-line blending 在管道中进行(流体)掺和；管道混合
in line coater 流水线涂装机
in-line continuous production 流水连续生产
in-line embosser 串联压花机
in-line extrusion-stretch-blow 串联挤坯拉伸吹塑
in-line filter 在管线中的滤器
in-line meter 在线仪表；在管道中的仪器
inline procedure 联机程序
in-line product blending 在线掺混产品
in-line pump 管线泵
in-line pyrolysis gas chromatography 管线内热解气相色谱(法)
in-line sampling by flow injection 流动注射在线进样
in-line sheet-fed thermoformer 串联单片喂料式热成型机
in-line stretch-blow 串联式拉伸吹塑
in-mould coating 模塑涂料；模内涂料
in mould polymerization 模具内聚合
inner 内部的；里面的
inner anhydride 内酐
inner bark 内(树)皮
inner complex 内配合物；螯合物
inner complex salt 内络盐
inner condensing tube 冷凝器内管
inner cone(=flame cone) 焰心
inner coordination sphere 内配位层*
inner crystal crack(=transcrystallic cracking) 穿晶裂纹
inner cylinder 内圆筒
inner dead point 内止点
inner diameter 内径
inner electric potential 内电势
inner electron 内层电子
inner equilibrium 内平衡
inner ester 内酯
inner ether 内醚
inner face 内表面
inner filter effect 内过滤作用
inner flame 内焰
inner granular crack 穿晶裂纹
inner granular stress corrosion crack 穿晶应力腐蚀开裂；晶内应力腐蚀开裂
inner heat 蓄热；内部发热
inner hemiacetal 内半缩醛
inner liner 内衬

inner loop ①内圈②近核圈
inner-mesh 内网
inner molecular reaction 分子内反应*
innermost electron shell 最内电子壳层
inner orbital 内层轨道*
inner-orbital complex 内轨型配合物
inner-orbital configuration 内轨型(构型)
inner orbital coordination compound 内轨配合物*
inner orbital mechanism 内轨机理*
inner phase 内相
inner potential 内电位
inner quality 内在质量
inner quantum number 内量子数
inner reamer 内扩孔锥；绞刀
inner reference electrode 内参比电极
inner root sheath 内根鞘
inner salt 内盐*
inner section 内部断面；内空断面；内剖视图
inner shell 内壳层*
inner-shell ionization 内壳(层)电离(作用)
innersole 内底
inner sphere complex 内层配合物
inner sphere complexation 内层配位*
inner sphere mechanism 内层机理*
inner stress 内应力
inner structure 内部结构
inner transformation 内部转变
inner transition element 内过渡元素*
inner tube reclaim 内胎再生橡胶
inner tube rubber 内胎胶(料)
inner Ventury 内文丘里管
inner volume 内体积
inner wall 内壁
inner width 内隙；内部宽度
innocuous 无害的
innocuous substance 无害物质
innovation 革新
innovator 革新者；改革者
innoxious 无害的
innoxious substance 无害物质
inoculating crystal 晶种；籽晶
in oil 荫油
inorfil(=inorganic filament) 无机(长)丝
inorganic 无机的
inorganic accelerator 无机促进剂
inorganic acid 无机酸
inorganic acidity 无机酸度
inorganic active filler 活性无机填料

inorganic addition agent　无机添加剂
inorganic additive　无机添加剂
inorganic adhesive　无机胶黏剂
inorganic alloys　无机合金〔金属和非金属的合金〕
inorganically gelled grease　无机物稠化的润滑脂
inorganic analysis　无机分析
inorganic base　无机碱
inorganic binder　①无机漆基②无机黏结剂；无机黏合剂
inorganic building material　无机建材
inorganic cation　无机阳离子
inorganic chemistry　无机化学*
inorganic chloride　无机氯化(合)物
inorganic chromatography　无机色谱(法)
inorganic coating　无机涂料
inorganic colloid　无机胶体
inorganic colloid-gelled grease　无机胶体凝胶润滑脂
inorganic color　无机颜料
inorganic color(ed) pigment　无机彩色颜料
inorganic compound　无机化合物
inorganic coprecipitant　无机共沉淀剂
inorganic degreaser　无机脱脂剂
inorganic deruster　无机除锈剂
inorganic digestion　无机物消解
inorganic exchanger　无机离子交换剂
inorganic fiber　无机纤维
inorganic fibre reinforce　无机增强纤维
inorganic filler　无机填料；矿质填充剂
inorganic gel-thickened grease　无机稠化润滑脂
inorganic indicator　无机指示剂
inorganic inhibitor　无机缓蚀剂；无机防腐蚀剂；无机防锈剂；无机抑制剂
inorganic insoluble test　无机不溶试验
inorganic ion effect　无机离子效应
inorganic ion exchange　无机离子交换
inorganic ion exchange paper　无机离子交换纸
inorganic ion exchanger　无机离子交换剂
inorganic ionic polymer　无机离子型聚合物
inorganic ion qualitative detection　无机离子定性测定
inorganic ion tolerance　(洗涤剂的)耐无机离子量
inorganic mass spectrometer　无机质谱仪
inorganic mass spectrometry　无机质谱法
inorganic materials adsorbing combination　无机材料吸附结合法
inorganic nitrogenous fertilizer　无机氮肥
inorganic non-metallic nanomaterial　无机非金属纳米材料
inorganic organic polymer　无机-有机高分子
inorganic-organic property balance　无机性-有机性平衡值
inorganic paint　无机涂料
inorganic particulate　无机粒子
inorganic phosphorescent paint　无机夜光漆；无机磷光漆
inorganic pigment　无机颜料
inorganic pollutant　无机污染物
inorganic polymer　无机高分子*
inorganic precipitant　无机沉淀剂
inorganic qualitative analysis　无机定性分析
inorganic reagent　无机试剂
inorganic rubber　无机橡胶
inorganic rust preventive film　无机防锈膜
inorganic sediments　无机沉积物
inorganic substance　无机物(质)
inorganic theory　无机学说
inorganic thickening agent　无机增稠剂
inorganic zinc coating　无机(富)锌涂料
inorganic zinc rich paint　无机富锌漆
inosamine　肌醇胺；环己六醇胺
inose　肌醇；环己六醇
inosine(I)　次黄(嘌呤核)苷；肌苷
inosine acid　肌苷酸；次黄嘌呤核苷酸　$OHC_5H_2N_4C_5H_8O_3OPO_3H_2$
inosite(=inositol; cyclohexanhexaol)　肌醇；环己六醇　$(CHOH)_6$
inositol hexaphosphoric acid(=phytic acid)　肌醇六磷酸
inositol nicotinate　烟酸肌醇酯
inositol triphosphoric acid　肌醇三磷酸
inosose　肌醇单酮
inoxidability　不可氧化性
inoxidable　不可氧化的
inoxidizability　不可被氧化性
inoxidizable　不可被氧化的
inpaint　补涂
in parallel　并联
inpasto　厚涂颜料法
in-phase　同相的
in-phase component　同相分量
in-place foaming property　现场发泡性能
in-place mo(u)lding　就地成型；就地模塑
in-place pipe cleaning　管线原地清洗；管线不拆除的清洗
in-plane bending vibration　面内弯曲振动
in-plane deformation vibration　面内形变振动
in-plane retardation　面内延迟
in-plane switching mode　面内切换模式
in-plant assay　厂内测定；厂内分析；厂内检验
in-plant equipment　厂内设备

in-plant handling　厂内搬运
in-plant stock handling　半成品搬运
in-plant transportation　内部运输
in-pot melting　罐内熔融
in-processing test　生产(条件下)试验
in-process inventory　过程中投入量
in-process material　加工用原材料
in-process product　半成品
input　①输入②进料③输入量
input accountability　物料投入统计计量；投料衡算计量
input attenuation　输入衰减
input capacitance　输入电容
input command decoder　输入指令译码器
input current　输入电流
input data　输入数据
input impedance　输入阻抗
input layer　输入层
input material　原料；进料
input of column　蒸馏塔进料
input pressure　输入压力
input rate　进料率
input resistor　输入电阻(器)
input speed　进料速度
input stage　输入级
input terminus　输入端
input transformer　输入变压器
input voltage　输入电压
inquartation(=inquarting)　四分法〔炼金术〕
inquarting(=inquartation)　四分法〔炼金术〕
in-reactor loop　堆内回路
in-reactor process　反应釜内法
in-register　互相对准
in-rubber degradation　橡胶内(帘线)降解
in running condition　①在运转情况下②运转正常
in-running rolls　对向回转的滚筒
insect farm　杀虫剂车间〔石油炼厂内〕
insect flower　除虫菊
insect flower oil　除虫菊花油
insecticidal coating　杀虫涂料
insecticidal materials　杀虫材料；杀虫剂；杀虫药物
insecticidal oil(=insecticide oil; insect oil)　杀虫油
insecticidal paint　杀虫漆
insecticidal plasticizer　杀虫增塑剂
insecticidal solution　杀虫液
insecticide　杀虫剂；杀虫药
insecticide base　杀虫剂主剂
insecticide dust　杀虫剂(粉)
insecticide oil(=insecticidal oil; insect oil)　杀虫油

insecticide paint　杀虫漆
insect oil(=insecticide oil; insecticidal oil)　杀虫油
insect-spotted thread　虫点丝
insensitive nuclear enhanced by polarization transfer　不灵敏核的极化转移增强法
insensitive nucleus enhancement by polarization transfer (INEPT)　不灵敏核极化转移增益法
inseparability　不可分离性
in series　串联
in-series continuous production　顺序连续生产
insert　①插入物②插入
insert Dewar bottle　插入型杜瓦瓶
inserted tooth method　镶齿法
insert holder pin　插夹针
insertion　插入*
insertion-decomposition　嵌入-分解
insertion-excitation-decomposition reaction　嵌入-激发-分解反应〔热原子化学〕
insertion group　嵌入基团
insertion polymerization　插入聚合*
insertion reaction　插入反应；嵌入反应
insertion translocation　着丝区；插入易位
insert molding　夹物模压
insert pyrometer　插入式高温计
insert tube　插入管
in-service inspection　工作中检查；运行中检查
in-service job　临时修理；紧急修理
in-service test　实用试验
in-service use　实际使用
insiccation　干燥
inside　①内部②在内③内部的
inside box section　箱形内截面
inside casing line　胎体内轮廓线
inside coating　内壁涂层
inside corrosion　内表面腐蚀
inside diameter　内径
inside drop stitch　里漏针
inside girth　内周长
inside gradient　内梯度
inside helper　正式工人
inside indicator　内指示剂
inside-out filter　外流式过滤器
inside radius　内半径
insipid　无味的
in-site polymerization　就地聚合；模具内聚合
in situ　原位；现场；原地
in situ activation analysis　现场活化分析
in situ analysis　原位分析

in situ blending 原位共混
in situ coal gasification 煤的就地气化
in situ composite 原位复合材料
in situ concentration sampler 原位富集进样器
in situ concentration sampling 原位富集进样
in-situ curing 现场硫化法；注入硫化法
in-situ density 原地密度〔不取样测定的密度〕
in situ electrolysis 原位电解
in situ epoxidation process 原位环氧化方法
in-situ growth （结晶体)原地生长
in situ hybrid composite 现场硫化法；注入硫化法
in situ hybridization 原位杂交
in situ initiator 原地引发剂
in situ leaching 现场浸取；就地浸取
in situ method 现场法；就地法
in situ neutron activation 现场中子活化(分析法)
in situ neutron activation analysis 现场中子活化分析*
in situ polycondensation 原位缩聚
in situ polymerization 原位聚合
in situ polymerized film 原地聚合薄膜
in situ precipitation 原处沉淀；原位沉淀
in situ preparation 现场制备；就地制造
in situ pretreatment 原位预处理*
in situ quantitation 现场定量；原位定量；(薄层)板上定量
in situ recovery 现场回收；就地回收
in situ retorting 现场干馏；就地干馏
in situ tracer 原位示踪剂
in-skin foamed article 带皮泡沫制品
in-skin foaming 带皮发泡法〔泡沫胶外包一层聚氯乙烯薄皮〕
insol ①(=insoluble)不溶的②(=insolubility)不溶(解)性
insolation 暴(晒)；晒
insole 内底
insolubility 不溶(解)性
insolubilization （使)不溶解；降低可溶性
insolubilized enzyme 固相酶
insolubilizer 不溶黏料
insolubilizing 不能溶解的
insolubilizing organic pigment 不溶型有机颜料
insoluble 不溶(解)的
insoluble ash 不溶灰分
insoluble azo color 不溶偶氮染料
insoluble bromide number(=insoluble bromide ratio; insoluble bromide value) 不溶溴值
insoluble bromide ratio(=insoluble bromide number; insoluble bromide value) 不溶溴值
insoluble bromide value(=insoluble bromide number;

insoluble bromide ratio) 不溶溴值
insoluble dry pigment 不溶性干颜料
insoluble enzyme 固相酶
insoluble layer 不溶解层
insoluble matter 不溶物质
insoluble monolayer 不溶性单层
insoluble phosphate 不溶性磷酸盐
insoluble residue 不溶残渣
insoluble resin 不溶树脂；不溶胶质
insoluble sludge 不溶残渣
insoluble substance 不溶物(质)
insoluble tar 沉淀(木)焦油
insolvent-based packaging ink 非溶剂性包装油墨
in-source decay 源内裂解
in-source fragmentation 源内裂解
inspecting engineer 验收工程师
inspecting standard 检验标准
inspection 检查；监督；检验
inspection gauge 检验规
inspection of refinery equipment 炼油厂设备检查
inspection plug 检查用堵头〔检查液面时可将它打开〕
inspection rejects 挑除废品；剔废
inspection with magnetic powder 磁粉探伤
inspiration ①进气；吸气②启发；灵机
inspirator ①呼吸器②喷气注水器
inspissated ①凝结作用②浓缩作用
inspissated juice 浓缩汁
inspissated oil 浓缩石油
inspissation 蒸浓(法)
inspissator 蒸浓器
instability ①不稳定性②不稳定度
instability constant 不稳定常数*
instability criterion 不稳定度准数
installation ①装置②安装③计算法
installation diagram 安装图
installation plan 安装图
installation specification 安装检修规程
instantaneous 瞬时的；立刻的
instantaneous adhesive 瞬时黏合剂
instantaneous compliance 瞬时柔量
instantaneous coupling modulus 瞬时偶合模量
instantaneous current 瞬时电流
instantaneous deflection 瞬时挠度
instantaneous deformation 瞬时形变
instantaneous elastic deformation 瞬时弹性形变
instantaneous elasticity 瞬时弹性
instantaneous elastic recovery 瞬时弹性回复
instantaneous elongation ①瞬时伸长②瞬时伸长率

instantaneous extension 瞬时延伸
instantaneous modulus 瞬时模量
instantaneous molal concentration 瞬时摩尔浓度
instantaneous nucleation 瞬时成核
instantaneous rate of deformation 瞬时形变速率
instantaneous rate of flow 瞬时流速
instantaneous set 瞬时变定；瞬时固定
instantaneous specific heat 瞬时比热
instantaneous strain 瞬时应变
instantaneous stress 瞬时应力
instantaneous surface tension 瞬时表面张力
instantaneous tangent modulus 瞬时正切模量
instantaneous value 瞬时值
instantaneous velocity of reaction 瞬时反应速度
instanteneous dipole moment 瞬时偶极矩
instant grinding ①即时打磨；瞬时抛光②即时研磨（粉碎）
instant productivity 瞬时产量；瞬时产率
instant set polymer 瞬时(快速)固化聚合物
instant steam 随手蒸汽
instant vaporization 瞬间汽化
in statu nascendi 生成瞬间〔化学反应进行中产品的生成〕
instream analysis 物流内分析
Instron tensile tester 英斯特朗张力试验仪
instruction case ①毒气试样盒②指导(书)盒
instruction set 指令系统
instrumental analysis 仪器分析*
instrumental charged-particle activation analysis(ICPAA) 带电粒子仪器活化分析
instrumental constant 仪器常数
instrumental correction 仪器校正
instrumental deviation 仪器偏差
instrumental deviations from Beer's law 偏离比尔定律的仪器偏差〔指仪器响应值偏离比尔定律产生的偏差〕
instrumental error 仪器误差
instrumental fast neutron activation analysis(IFNAA) 快中子仪器活化分析
instrumental method 仪器分析法
instrumental neutron activation analysis(INAA) 中子活化仪器分析*
instrumental photon activation analysis(IPAA) 光子活化仪器分析(法)
instrumentation ①仪器应用；仪表化②成套仪器装置
instrumentation amplifier 仪表放大器；数据放大器
instrument board 仪表盘；仪表屏板
instrument constant 仪器常数
instrument effect 仪器效应
instrument error 仪器误差
instrument for hollow cathode discharge spectroscopic analysis 空心阴极放电光谱分析装置
instrument for laser micro-spectrochemical analysis 激光显微光谱化学分析装置
instrument of pressing sliver 麻条压条器
instrument panel 仪表盘；仪表屏板
instruments for scattering analysis 散射分析仪器
instrument symbol 仪器符号
insuccation 浸渍(法)
insufficient lubrication 润滑不良
insufficient turn length 圈长不足〔纱线疵点〕
insufficient vehicle 劣质漆料
insulamine 英色胺
insulant ①绝缘材料②绝热材料
insulated ①绝缘的②保温的
insulated column 保温蒸馏塔
insulated material 绝缘材料
insulated paint 绝缘漆
insulated panel 绝缘板
insulated paper 绝缘纸
insulated roof 保温顶盖
insulated side wall 保温罐壁
insulated water bottle 绝缘采水瓶
insulated wire 电磁线；漆包线〔俗〕
insulating ①绝缘的②保温的
insulating board 隔音板；隔热板
insulating brick 保温砖
insulating cement 绝缘水泥
insulating ceramic 绝缘陶瓷
insulating ceramic foam 绝缘泡沫陶瓷
insulating coat 绝缘涂层
insulating core transformer 绝缘芯变压器
insulating efficiency 保温效率
insulating foam 绝缘泡沫胶
insulating head 绝缘喷头
insulating jacket 绝热夹套；隔热夹套
insulating layer 绝缘层；保温层
insulating lining 保温衬里
insulating material ①绝缘材料②保温材料
insulating medium 隔离剂
insulating oil 绝缘油；变压器油
insulating oil acid and alkalies test 绝缘油的酸碱试验
insulating oil dielectric strength 绝缘油的介电强度
insulating oil emulsion test 绝缘油的乳化试验
insulating oil flash point 绝缘油的闪点
insulating oil gas content 绝缘的气体含量

insulating oil organic chloride and sulfate test　绝缘油有机氯化物及硫酸盐试验
insulating oil pour point　绝缘油的倾点
insulating oil resistibility　绝缘油的介电阻力
insulating oil saponification number　绝缘油的皂化值
insulating oil sludge　绝缘油的沉淀
insulating packing　绝缘填料
insulating paint　绝缘漆
insulating paper　绝缘纸
insulating pipe　保温管
insulating plastic　绝缘塑料
insulating power　绝缘能力
insulating property　绝缘性质
insulating quality　绝缘性；绝缘质量
insulating refractory　隔热耐火材料
insulating shaft　(喷枪)绝缘枪身
insulating sheath　绝缘护套
insulating strip　绝缘带
insulating tape　绝缘胶带；电线胶布
insulating tube　绝缘管
insulating varnish　绝缘清漆
insulating wax　绝缘蜡
insulation　①绝缘②保温
insulation against heat　热绝缘
insulation can　保温箱
insulation class　绝缘等级(种类)
insulation coating　绝缘涂料
insulation compound　①绝缘混合剂；绝缘涂料；绝缘材料②保温(隔热)混合物
insulation course(=insulation layer)　保温层
insulation donating agent　电绝缘性配合剂
insulation lacquer　(挥发性)绝缘漆
insulation layer(=insulation course)　保温层
insulation paper　绝缘纸；隔电纸〔商〕
insulation resistance　①绝缘强度②绝缘电阻
insulation silicone type coating　绝缘性硅烷型涂料
insulation tank　绝缘槽
insulation tape　绝缘胶带
insulation tube　绝缘管；隔电管
insulative water paint　水性绝缘漆
insulatng paste　绝缘膏
insulator　①绝缘体②绝缘器；绝缘子③隔离剂
insullac　绝缘漆
insurance rating　保险等级
insurance risk　保险事故
INS value　INS 值〔皂化价减去碘价〕
intact　完整的；整体的
intact mill scale　完整(无损)的轧制氧化皮

intaglio　凹板印刷
intaglio ink　凹板墨
intake air　进气(管口)
intake charge　进气；充气
intake chute　加料槽；导液槽
intake pressure　进气压力
intake zone　进料区
Intalox saddle　英特洛克斯鞍形填料；槽鞍形填料；矩鞍形填料
Intalox saddle packing　英特洛克斯鞍形填料；槽鞍形填料；矩鞍形填料
in-tank mixing　罐内混合
in-tank-solidification(ITS)　罐内固化；槽内固化
intarvin　甘油十七酸酯
integerrimine(=squalidine)　全缘(千里光)碱
integral　①完全的②积分③积分的
integral absorption　①积分吸收②总吸收(量)
integral absorption coefficient　积分吸光系数
integral accuracy　积分精度*
integral blend　整体掺混
integral capacity　积分电容；积分容量；界面容量
integral color　整体颜色；总色泽
integral colouring　整体着色
integral cooling system　完整冷却系统；整体化冷却系统
integral curve　积分曲线
integral denitrogenation-dehydrogenation　联合脱氮-脱氢(法)
integral density　合成密度
integral detector　积分型检测器
integral distribution　积分分布
integral distribution curve　累积分布曲线
integral dose　总剂量
integral drive　直接传动
integral extrusion　整体挤塑
integral fixed-bed reactor　固定床积分反应器
integral flexibility　(高分子)内部柔韧性
integral function　①积分函数②整函数
integral heat(=total heat)　积分热；变浓热；总热
integral heat of adsorption　积分吸附热*
integral heat of dilution　积分稀释热
integral heat of solution　积分溶解热*
integral insulation　整体绝缘
integrally stiffened skin　整体加强蒙皮
integral method　积分法
integral mobility　积分离子淌度
integral mould(=accordin mould)　集成模具
integral oil filter　积分滤油器
integral part　①组成部分②重要部分
integral peak　完整峰

integral plasticization　全增塑(作用)；内增塑(作用)
integral property detector　整体性质检测器
integral pump　积分泵
integral reactor　积分反应器
integral sensitivity　积分灵敏度
integral skin rigid urethane foam　结皮聚氨酯硬泡沫塑料
integral spectroscopy　直接谱；积分谱
integral thickness　总厚度
integral type detector　积分型检测器*
integral type flange　整体法兰
integral voltammetry　积分伏安法
integral water proof(ing)　总体防水〔指整体不透水性〕
integral weight　整数砝码
integral weight distribution　积分重量分布(函数)
integral weight distribution curve　积分重量分布曲线
integrated absorption method　积分吸收法*
integrated biological-chemical process　生物-化学联合法
integrated chip-based capillary electrophoresis　集成芯片毛细管电泳
integrated circuit　集成电路
integrated control　综合防治
integrated control of water pollution　水污染综合防治
integrated dissipation inequality　积分耗散不等式
integrated hydrocarbon stripping process　联合烃气提法〔两种烃气流同时加工〕
integrated hydrofining-hydrocracking　联合加氢精制-加氢裂化(法)
integrated hydrotreating-hydrocracking　联合加氢处理-加氢裂化(法)
integrated intensity method　累积强度法
integrated line　①综合生产流程②成套设备
integrated microconduit　集成微管道
integrated mill　大型工厂；联合工厂
integrated oil company　大型石油公司
integrated prevention and control of environment　环境污染综合防治
integrated processing　统一集中处理
integrated radiation dose　累积辐射剂量
integrated refinery　联合炼油厂
integrated rubber　集成橡胶
integrated system　一体化体系；一体化系统
integrated tower　有反应器的蒸馏塔〔与反应器并在一起的蒸馏塔〕
integrating amplifier　积分放大器
integrating capacitor　积分电容器
integrating detector　积分型检测器
integrating function　积分功能

integrating instrument(=integrating meter)　积分仪
integrating meter(=integrating instrument)　积分仪
integrating motor　积分电动机
integrating scan　积分扫描
integrating type photoelectrometer　积分型光电光度计
integration　①整合(作用)②集成③积分(法)
integration accuracy　积分精度
integration constant　积分常数
integration filter　集成滤波器
integrative approach　积分方法
integrative suppression　整合校正
integrator　①积分仪②整合器
integrator pulser　积分仪脉冲器
intelligent polymer　智能高分子材料；智能聚合物
intelligent transducer　智能转换器
intended life　预定使用寿命
intense　正〔指颜色〕；强
intense floral odor　强烈花香
intense ion beam　强离子束
intense odor　强烈香气；强烈气息
intensified silicon intensified target　增强硅增强靶
intensifier　增强器
intensifying agent　强化剂
intensity　强度
intensity anomaly　强度异常(现象)
intensity coefficient　强度系数
intensity coefficient method　强度系数法
intensity distribution　强度分布
intensity factor　①强度因素②强度因数
intensity fluctuation spectroscopy(IFS)　(谱线)强度涨落波谱(法)
intensity mark method　强度标度法
intensity modulation　强度调制
intensity of absorption line　吸收线强度
intensity of current　电流强度
intensity of field　场强度
intensity of illumination　照度；光照强度
intensity of magnetization　磁感应强度
intensity of packing　填充物强度
intensity of precipitation　降水强度
intensity of radiation　辐射强度
intensity of spectral line　谱线强度
intensity of suction　吸入强度
intensity ratio　强度比
intensity ratio of line pair　线对强度比*
intensity rule for multiplet　多重线强度定则
intensive drying　充分干燥
intensive factor　强度因数

intensive filtration 强化过滤
intensive internal mixer 强力密炼机
intensive mixing 充分混合
intensiveness 密集性
intensive property 强度性质*
intentional pollution 有意污染
inter-〔拉丁字头〕 ①(在)中(间); 在中②互相; 合; 一起
interaction 相互(作用); 交互作用
interaction between light and substance 光与物质相互作用
interaction coefficient 相互作用系数
interaction force 相互作用力; 干扰力
interaction force between molecules 分子间相互作用力
interaction of atom pairs 原子对的相互作用
interaction parameter 相互作用参数
interaction potential 相互作用势
interactive graphics terminal 人机对话式图形终端
interactive image editing 交互式图像编辑
interactively fitted polynomial baseline 交互拟合多项式基线
inter-and intra-molecular oxidation 分子内和分子外的氧化作用
interatomic 原子间的
interatomic attractive force 原子间吸引力
interatomic bond ①原子间结合②原子间键
interatomic bond energy 原子间键能
interatomic diamagnetic shielding 原子间抗磁屏蔽
interatomic distance 原子间距离
interatomic force 原子间力
inter-attraction 相互吸引
interband 间纹
interband region 带间区
interbedded 间歇的
interblend 相互掺混
intercalated material 嵌入材料
intercalating agent 嵌加剂; 插入剂
intercalation 插层反应*
intercalation chemistry 插层化学*
intercalation compound 嵌入化合物
intercalation compounding 插层复合
intercalation coordination compound 夹层配合物*
intercalation hybrid 插层复合法
intercalation polymerization 插层聚合
intercalibration 相互校准
intercell liquid transfer 热室间液体输送
intercept ①截距②截取; 切断③相交; 交叉
interception 遮光(作用)
interchain (高分子)链间
interchain bond 链间结合; 链间交联

inter-chain bonding 链间键合
interchain interaction 链间相互作用
interchain spacing 链间距
inter-chamber tower 中间塔〔硫酸〕
interchange 互换; 交换
interchange ability (可)互换性
interchangeable (可)互换的; 可拆的
interchangeable column system 可互换的柱系统
interchangeable element 可互换的元件
interchangeable gear 可互换的齿轮
interchangeable grades 互换级别〔石油产品〕
interchangeable ground 标准玻璃磨口
interchangeable ground glass joint 可互换的玻璃磨口接头
interchangeable injector 可互换进样器
interchangeableness (可)互换性
interchangeable oil filter 互换滤油器
interchangeable parts 互换(性)配件
interchangeable seal 可互换密封
interchange esterification 酯交换
interchange mechanism 互换机理*
interchange of heat 热交换
interchanger 换热器
interchange reaction 互换作用
inter coagulation 互凝聚
inter-coat adhesion 涂层间附着(力)
intercoat contamination 层间污损
inter-coiled loop stitch 互绞组织
intercommunicating tube 连接管
intercomparison 相互比较
interconal gases to flame 火焰的内锥气体
interconal zone 锥间区
intercondensation polymer 共缩聚物
intercondenser 中间冷凝器
interconnected capillary 互连毛细管
interconnecting cable 互连电缆
interconnecting group 连接官能团
interconnection 互连; 互相连接
interconversion 互变
interconvertible 可互换的; 可调换的
intercooler 中间冷却器
intercooling 中间冷却
intercooling device 中间冷却设备
intercross 相互交叉
intercrosslinking 链间交联
intercrystal boundary 晶间界面
intercrystalline 晶(粒片)间的
intercrystalline amorphous region 晶间无定形区

intercrystalline bond   晶间键
inter-crystalline corrosion   晶间腐蚀
intercrystalline crack   晶间裂纹
intercrystalline fibril   晶间原纤
intercrystalline fracture   晶间破裂
intercrystalline swelling   晶间溶胀
interdiffusion   互扩散
interdiffusion zone   扩散带
interdimer   内二聚体；共二聚体
interelectrode distance   电极间距
interelement correction   共存元素校正；第三元素校正
interelement effect   共存元素效应
interelevenaldehyde   2-十一烯醛
interenin(=adrenal cortex hormone extract)   (肾上腺)皮质激素制剂
interesterification(=ester exchange)   酯交换
interesting effect   满意效果；重要作用
interesting flowery   满意花香
interesting note   满意香韵
interetherification   相互醚化
interface   ①界面*；交界②接口③连接器④(两种仪器)联用
interface activity   表面活性
interface analysis   界面分析
interface bus   接口总线
interface condensation   界面缩聚
interface control   界面控制
interface extension   界面宽广度
interface film   界面膜
interface level indicator   界面指示计
interface modifier   界面改性剂
interface polymerization   界面聚合
interface reaction   界面反应
interface reaction constant   界面反应常数
interface reaction order   界面反应级数
interface structure   界面结构
interface surface energy   界(表)面能
interface temperature   界面温度
interface transfer coefficient   界面传递系数
interface wave   界面波
interfacial   界面的；面际的
interfacial activity   界面活性
interfacial adhesion   ①界面黏附②界面黏力
interfacial adsorption   界面吸附
interfacial agent   界面活性剂
interfacial aging   界面老化
interfacial angle   界面角
interfacial area   界面面积

interfacial bond   界面黏合；界面键
interfacial complex(IFC)   (两相)界面络合物
interfacial concentration   界面浓度
interfacial electric double layer   界面双电层
interfacial electrochemistry   界面电化学
interfacial eletrokinetic potential   界面动电势
interfacial energy   界面能
interfacial esterification   界面酯化
interfacial excess   界面超额*；界面过剩
interfacial failure   界面破坏
interfacial film   界面膜
interfacial force   界面力
interfacial free energy   界面自由能
interfacial graft polymerization   界面接枝聚合(作用)
interfacial layer   界面层
interfacial migration   界面迁移；界面转移
interfacial misfit strain   界面错配应变
interfacial phase   界面相
interfacial phenomenon   界面现象
interfacial plasticization   界面增塑(作用)
interfacial polarization   界面极化
interfacial polycondensation   界面缩聚*
interfacial polymerization   界面聚合*
interfacial potential   界面势
interfacial pressure   界面压力
interfacial reaction(=boundary layer reaction)   界面反应
interfacial reaction method   界面反应法
interfacial resistance   界面阻力
interfacial rheology   界面流变学
interfacial shear viscosity   界面剪切黏度
interfacial solute adsorption   界面溶质吸附
interfacial stress   界面应力
interfacial surface area   界面表面积
interfacial synthesis   界面合成
interfacial tensimeter   界面张力计
interfacial tension   界面张力*
interfacial tension lowering   界面张力降低
interfacial tension ring method test   界面张力环法试验
interfacial thermodynamics   界面热力学
interfacial work   界面功
interfacing   ①交界②联用③接口
interfacing unit   连接器件
interference   ①干涉②干扰③噪声
interference band   干扰谱带
interference color   干扰色
interference-colour chart   干涉色谱图〔测定双折射〕
interference effect   干扰效应
interference element   干扰元素

interference figure　干扰图
interference filter　干涉滤光片*
interference fringe　干涉条纹
interference microscope　干涉显微镜
interference modulated spectrometer　干涉调制光谱仪
interference polarizer　干涉偏光仪
interference potential　干扰电位
interference preventant　防干扰剂
interference refractometer　干涉折光计
interference spectroscopy　干扰光谱仪；干涉分光仪
interference spectrum　干扰光谱
interferential　干扰的
interfering　干扰的
interfering component　干扰组分
interfering diffusion current compensator　干扰扩散电流补偿器
interfering element correction　干扰元素校正
interfering ion　干扰离子
interfering line　干扰线
interfering material　干扰物质
interfering substance　干扰物质
interferogram　干涉图
interferometer　干涉仪
interferometric method　干涉测量法；干涉量度法
interferometric microscopy　干涉显微术
interferometry　干涉分析法
interferon inducer　诱起剂
inter-fibrillar H-bonding　原纤间氢键(结)合
interfibrillar substance　纤维间质
interfibrillar surface　原纤间表面
interfibrillar swelling　原纤间溶胀
interfibrillar void　原纤间空隙
interfused　①混合的；混熔的②熔合的
interfusion　①渗透②混合；熔合
inter-generational compensation　代际补偿
inter-generational equity　代际公平
intergranular　晶间的
intergranular attack　晶间浸蚀
intergranular corrosion　晶间腐蚀
intergranular corrosion test solution　晶间腐蚀试验液
intergranular cracking　晶间开裂
intergranular diffusion　晶间扩散
intergranular texture　晶(粒)间结构
intergrown knots　活节
interhalogen compound　卤间化合物*
interhalogen fluorination　卤间氟化*
interim-233 process　暂定铀-233 提取过程〔TBP 萃取过程；类似 TBP-233 过程〕

interim storage　暂时储藏；临时储藏
interindustry emission control　工业间排放控制
interionic　离子间的
interionic attraction theory　离子互吸(理)论
interior　①内部；里面②内部的
interior dry wall primer　内用干墙底漆
interior emulsion paint　内用乳胶漆
interior enamel　内用瓷漆
interior extrapolation method(=bracketing method)　交叉法〔一种定量法〕；插入法
interior finish　①室内装饰②内用涂料；内墙面漆
interior finishing varnish　①内用末道清漆②内用罩光清漆
interior flat wall paint　平面内墙漆
interior grade wood preservative　内用级木材防腐剂
interior paint　内用漆
interior phase　内相
interior polyurethane varnish　内用聚氨酯清漆
interior pressure　内压力
interior primer-sealer　内用封闭底漆
interior type plywood　室内用胶合板
interior varnish　内用(清)漆
interlab comparison　实验室间比较
interlaboratory　研究室间
inter-laboratory error　(实验)室间误差
inter-laboratory repeatability error　(实验)室间重复性误差
interlace　交织
interlaced fabric　交织织物
interlacement　交织；组织〔纺〕
interlacing　交织的
interlacing fabric　交织织物
interlacing line　①交织线②交叉线
interlamellar　层间
interlamellar adsorption　界面薄层吸附
interlamellar cation　层间阳离子
interlamellar deformation　层间形变
interlaminar liquid　(泡沫)层间液体
interlaminar shear　层间剪力
interlaminar shear strength　层间剪切强度
interlaminar strength　层间剥离强度
interlaminar stress　层间应力
interlaminar tensile strength　层间拉伸强度
interlayer　夹层；中间层
interlayer adhesion　层间附着
interleaving paper　隔离纸；剥离纸
interlining　夹层；中间层
interlinking　互连

interlock bourrelet 双罗纹凸条织物
interlock gating 棉毛配置
interlocking 联锁；咬合；啮合
interlocking labyrinth seat 联锁迷宫式密封
interlocking lead brick 联锁型铅砖；嵌入式铅砖
interlocking phase 中间相
interlocking rotor 联锁式转子
interlocks 联动装置
interlock switch 互锁开关
intermass 衬料
intermedeol 臭根醇
intermediary ①中间的；媒介的②中间体；半成品；媒介(物)
intermediary metabolism 中间代谢
intermediate 中间体*
intermediate agent 中间剂
intermediate alloy 中间合金
intermediate base crude 中间基原油
intermediate coat 二道底漆
intermediate (coat) peeling 层间脱落
intermediate color 间色〔介于原色和复色之间的颜色〕
intermediate compound 中间化合物；中间体
intermediate condenser 中间冷凝器
intermediate cooler 中间冷却器
intermediate crusher 中级压碎机
intermediate crushing 中级压碎
intermediate cut 中间馏分
intermediate density lipoprotein 中密度脂蛋白
intermediate depot 中间堆栈
intermediate detector 中间检测器
intermediate distillate 中间馏分
intermediate element 过渡元素
intermediate equipment 中间型设备
intermediate fluid 中介液体
intermediate formation 中间形成
intermediate fraction 中间馏分
intermediate image spectrometer 中间成像光谱仪
intermediate-level waste 中放废物*
intermediate loading 中等填充量；中量填充
intermediate material 中间产品
intermediate metal 中间金属
intermediate modulus 中间模量〔拉伸进程中任一定伸长的模量〕
intermediate octane rating 中间辛烷值
intermediate oil 中间馏分油
intermediate oxide 两性氧化物
intermediate pan 中间(体)盘
intermediate peeling 层间脱落

intermediate phase 中间相
intermediate-pressure reactor 中压反应器
intermediate pressure system 中间加压系统
intermediate product 中间体；中间产物
intermediate proof 中间证明
intermediate reaction product 中间反应产物
intermediate reaction tower 中间反应塔
intermediate receiver 中间接受器
intermediate receptacle 中间容器；中间接受器
intermediate reed 中间筘；二道筘
intermediate refinery stock 炼厂中间产品
intermediate roll 中滚筒
intermediate scale wax(=intermediate sweat wax) 中间蜡
intermediate screening 中间筛
intermediate stage 中间阶段
intermediate steam 间断蒸汽
intermediate stiffener 中间加强杆
intermediate storage 中间储存
intermediate storage tank 中间储罐
intermediate strut 中间支撑
intermediate sweat wax(=intermediate scale wax) 中间蜡
intermediate tank 中间储罐
intermediate temperature adhesive(=warm-setting adhesive) 中温硬化黏合剂
intermediate temperature deposits 中温沉积物
intermediate temperature glue 中温(固化)胶
intermediate temperature setting 中温固化
intermediate temperature setting adhesive 中温硬化黏合剂〔31~99℃硬化〕
intermediate temperature sludge 中温沉淤
intermediate temperature varnish 中温清漆
intermediate tone 中间色调
intermediate warehouse 中间产品仓库
intermedium 中间体；媒介物
intermeshing flight 啮合螺纹
intermetallic 金属间的
intermetallic compound 金属间化合物*
intermetallics 金属互化物
intermicellar equilibrium 胶束间平衡
intermiscibility 相(互)溶(混)性
intermittency effect 断续照明效应；间歇曝光效应
intermittent 间歇的；间断的
intermittent-arc method 间歇电弧法
intermittent atomic contact force microscope 断续接触式原子力显微镜
intermittent carbonation 间歇充(碳酸)气
intermittent column 间歇(操作)蒸馏塔
intermittent-contact recorder 间歇接触记录机

intermittent discharge 间歇放电
intermittent drier 间歇干燥器
intermittent dryer 间歇式干燥机
intermittent duty 间歇操作
intermittent filter 间歇过滤器
intermittent fixed-bed process 间歇式固定床过程
intermittent fractionation 间歇式精馏
intermittent illumination 间歇照光(法)
intermittent irradiation 间歇照射；断续照射
intermittent kiln 间歇窑
intermittent lubrication(=intermittent oiling) 间歇式润滑
intermittent motion 间歇移动
intermittent oiling(=intermittent lubrication) 间歇式润滑
intermittent operation 间歇操作
intermittent polymerization 间歇聚合
intermittent process 间歇操作法；分批操作法
intermittent pumping 间歇泵送
intermittent reaction 间歇反应
intermittent salt spray test 间歇式(喷)盐雾试验
intermittent sampling 间歇取样
intermittent sealing 间歇热合
intermittent stress 断续应力
intermittent stress relaxation 间断应力松弛
intermittent take-off 间歇出料
intermittent tensile stress 间歇抗张应力
intermittent transfer 间歇传送
intermittent vertical retort 间歇竖立蒸馏器
intermittent welding 断续焊
intermixer 密闭式混炼机
intermixing 混合
intermixing in pipe lines 在管线中混合
intermixture 混合物
intermode resonance 内模共振
intermodulation polarography 互调极谱法
intermodulation voltammetry 互调伏安法
intermolecular 分子间
intermolecular acetalization 分子间缩醛化
intermolecular attraction 分子间引力；分子内引力
intermolecular bond 分子间键
intermolecular bonding 分子间键合
intermolecular bonding energy 分子间键合能
intermolecular cohesion 分子间内聚力
intermolecular comonomer distribution 分子间共聚单体分布
intermolecular condensation 分子间缩合
intermolecular cross-linking 分子间交联
intermolecular energy transfer 分子间能量传递
intermolecular excluded volume 分子间排除体积
intermolecular force 分子间作用力
intermolecular hydrogen bonding 分子间氢键
intermolecular interaction 分子间相互作用
inter-molecular linkage 分子间键
intermolecular migration 分子间转移作用
inter-molecular nucleation 分子间成核过程
intermolecular orbital 分子间轨道
intermolecular order 分子间有序(状态)
intermolecular oxidation 分子间氧化作用
intermolecular penetration 分子间渗透
intermolecular polymerization 分子间聚合
intermolecular potential 分子间势能
intermolecular reaction 分子相互反应
intermolecular rearrangement(=intermolecular transposition) 分子间重排作用
intermolecular relaxation 分子间弛豫
intermolecular repulsion 分子间斥力；分子内斥力
intermolecular respiration 分子间呼吸
intermolecular second order moment 分子间二级矩
intermolecular selectivity 分子间选择性
intermolecular strain 分子间应变
intermolecular termination 分子间终止
intermolecular transposition(=intermolecular rearrangement) 分子间重排作用
internal 内(部)的；内件
internal absorbance 内吸光度
internal absorbent method 内吸收法
internal addition 内位加成
internal alcoholysis 内醇解
internal anhydride 内酐
internal antistatic agent 内用抗静电剂
internal barrier 内势垒
internal bremsstrahlung(=braking radiation) (内)轫致辐射
internal carbon referencing 碳内标
internal catalyst bed 内催化剂床
internal charging gun 机内充电喷枪；内部带电喷枪
internal check valve (油罐)内单向阀
internal circuit 内电路
internal coagulant 内施凝固剂
internal coating 内涂剂
internal cohesive energy 内(凝)聚能
internal combustion 内燃(烧)
internal combustion engine 内燃机
internal compensation 内消旋(作用)
internal compensation method 内消旋法
internal conductance system 内导体系
internal content 内含(量)

internal conversion　内转换*
internal conversion coefficient　内转换系数
internal conversion electron　内转换电子
internal cooling　内冷却
internal cracking　内部龟裂
internal crosslinker　内交联剂
internal crosslinking　内交联(作用)
internal crystal framework　结晶内部格架；内部晶格
internal crystal structure　晶体内部结构
internal damping　内阻尼
internal degree of freedom　内自由度
internal diameter　内径
internal diameter method　内径法
internal diffusion　内扩散*
internal double bond　内双键
internal drum filter　内滤鼓式过滤机；内滚筒压滤机
internal drying test　内部干燥试验
internal dying　纺前染色；内染色
internal electrically energizable heater element　内供电式加热器元件
internal electrogravimetry　内电解法*
internal electrolysis　内电解法
internal emulsifying agent　内乳化剂
internal energy　①内能*②内部能量
internal ester(=internal ether)　内酯
internal ether(=internal ester)　内酯
internal expanded coupling　内胀式管接头
internal factor　内在因素
internal fibrillation　内部帚化；内部纤维化
internal filling solution　内充溶液
internal filling solution of sensing electrode　敏感电极的内充溶液
internal firebox boiler　内火室锅炉
internal fissure　①内开裂②内裂纹
internal fluid friction(=internal friction of liquids)　液体内摩擦
internal force　内力
internal friction　①内摩擦②内耗
internal friction of liquids(=internal fluid friction)　液体内摩擦
internal furnace ion source　内炉离子源
internal furnace source　内炉源
internal gas-heated retort　内热(干馏)釜
internal haze　内混浊
internal heated runner　内加热流道
internal heating　内加热
internal heating effect　内加热效应
internal heat of evaporation　内蒸发热

internal heavy atom effect　内重原子效应
internal heteronuclear D lock　异核氘内锁
internal indicator　内指示剂
internal interference　内干扰；内干涉
internal ionization　内部电离作用
internal latent heat　潜热
internal lock　内锁*
internally blocked crosslinker　内封闭交联剂
internally compensated inactive compound　内消旋无旋光性化合物
internal mixer　密炼机
internal morphology　内部形态学
internal nonuniformity　内部不均匀性
internal normalization　内标归一化
internal nucleophilic substitution　分子内亲核取代*
internal olefin　内烯烃
internal oleyl sulfate　内油醇硫酸酯
internal oxidation　分子内氧化作用
internal partition function　内分配函数
internal penetration　内部渗透；内渗透性
internal phase　内相
internal-phase ratio　内相比
internal plasticization　内增塑作用*
internal plasticizer　内增塑剂
internal (position) isomer　内(位)异构体
internal reaction　内反应
internal rearrangement　内部重排(作用)
internal redox indicator　液内氧(化)还(原)指示剂
internal reduction　分子内还原作用
internal reference　内标；内参比；内参考
internal reference electrode　内参比电极
internal reference element　内参比元素
internal reference element technique　内参比元素技术
internal reference line　内参比线
internal reflectance　内反射
internal reflectance spectroscopy　内反射光谱法
internal reflection　内反射光谱
internal reflection element　内反射元素
internal reflux　内回流；热回流
internal reflux ratio　内回流比
internal relative deviation　内相对偏差
internal relaxation　内部松弛；内部弛豫
internal releasing agent　内脱模剂
internal ring　内环
internal rotation　内旋转*
internal rotation angle　内旋转角
internal rubber mixer　橡胶混合器
internals　①内部构件；内部零件②本质(性)

internal salt 内盐
internal screw pump 内螺旋泵
internal separation 分离
internal sizing 内(部)施胶
internal stabilizer 内稳定剂
internal standard ①内标*②内标物③内标法
internal standard area 内标面积
internal standard element 内标元素*
internal standard line 内标线*
internal standard method(=marker method) 内标法*；标示法
internal standard normalization 内标归一化(法)
internal standard sample 内标试样
internal standard substance 内标物
internal strain 内应变
internal stress 内应力
internal stress due to electroplating 电镀内应力
internal stress relaxation 内应力松弛
internal stripping section 内汽提段
internal surface 内表面
internal surface area 内表面积
internal target 内靶*
internal thread 内螺纹；阴螺纹
internal-type ball float 内在型浮球
internal variable 内变量
internal vibration 内振动
internal view 内部剖视图
internal volume 内体积
internal waviness 内波纹(度)
international ampere 国际安培
international atomic weight 国际原子量
international biological standard 国际生物标准
International Bureau of Weights and Measures 国际度量衡局
international candle 国际烛光〔照度单位；符号 int.c〕
International Conference on Thermal Analysis(ICTA) 国际热分析协会
international nomenclature 国际命名法
international oil industry 国际石油工业
International Organization of Flavor Industry(IOFI) 国际香料工业组织
international prototype of kilogram 国际千克原器
international prototype of meter 国际米原器
international standard 国际标准
International Standardization Organization(ISO) 国际标准化组织
international standard substance 国际标准品
international steam table 国际蒸汽表

international sugar scale 国际砂糖标度
international system of units 国际单位制
international temperature scale 国际温(度)标
International Union of Pure and Applied Chemistry (IUPAC) 国际纯粹与应用化学联合会*
international unit 国际单位
internode 波腹；节间
internuclear coupling 核间耦合作用
internuclear cyclization 核间成环作用
internuclear dipole-dipole interaction 核间偶极相互作用
internuclear double resonance(INDOR) 核间双共振*
internucleotide linkage 核苷酸间键合
interparticle 粒子间的
interparticle distance 粒子间距
interparticle double layer 粒间双电层
interparticle force 粒子间(作用)力
interparticle pore 粒间孔隙
interparticle porosity ①颗粒间多孔性②颗粒间孔隙率
interparticle potential 粒间相互作用势，粒间相互作用电位
interpenetrating 互(相贯)穿的
interpenetrating networks 互穿网络
interpenetrating network system 互穿网络体系
interpenetrating polymer 互穿聚合物
interpenetrating polymer networks(IPN) 互穿聚合物网络*
interpenetration 互穿；互相渗透
interphase 界面；际面；间期〔细胞分裂〕
interphase interaction 相间(的)相互作用
interphase precipitate 界面沉淀；相际沉淀
interplanar distance (晶)面间距
interplanar spacing 晶面间距*
interplaner space 星际空间
inter-plant handling 厂间运输
interplay hybridization (异种纤维)隔层杂化
interply fiber hybrid composites 混杂纤维复合材料
interpolar 极间的；两极之间的
interpolation 内推(法)；内插(法)
interpolation technique 内插法；内推法
interpolative relation analysis 内插法相对分析
interpolyamide 均共聚酰胺
interpolymer 共聚体
interpolymer complex 高分子络合物；复合物
interpolymerizable 可均共聚的
interpolymerization 共聚作用
interreaction 相互反应；相互作用
interrelation 相互关系
interrogation source 探询源

interrupted 断续的；间歇的
interrupted arc 断续光弧
interrupted arc method 断续电弧法
interrupted elution 间断洗脱
interrupted elution chromatography 间断洗脱色谱(法)
interrupted extraction 间歇萃取；断续萃取
interrupted oil supply 间歇供油
interrupted polymerization 间断聚合(作用)
interrupted-ramp strain test 断续斜线上升应变试验
interrupted stepwise polymerization 间断逐步聚合
interrupter 断续器
interrupter method 间歇法
interruption of chain 链的断裂
intersecting flow curve 交叉流动曲线
intersection chart 网络图；交织图；计算图表
intersection point 交点
inter snubber 中间缓冲器
inter space ①间隙；间隔②间隔带
interstage cooler 中间冷却器；级间冷却器
interstage cooling 中间冷却；级间冷却
interstage flow 级间流(量)
interstice ①细隙；间隙；裂缝②间隔
interstitial 空隙间的
interstitial air space 孔隙空间；填隙空间
interstitial atom 填隙原子
interstitial carbides 填充型碳化物
interstitial compound 填隙化合物
interstitial defect 间隙缺陷*
interstitial deposit ①填隙沉积物②填隙矿床
interstitial fraction 隙间部分
interstitial liquid 空隙液体
interstitial opening 填隙孔；晶隙孔
interstitial polymerization 间隙聚合
interstitial ratio 间隙比
interstitial solution 填隙液
interstitial stagnant fluid model 间隙性滞流模型
interstitial structure 填隙结构
interstitial vacancy defect 间隙空位缺陷*
interstitial velocity 间隙速度
interstitial void 晶格间隙*
interstitial volume 间隙体积*
interstitial water 间隙水
interstratified layer 间层
interstructural 结构间的
interstructural plasticizer 结构间增塑剂
intersystem crossing 系信窜越*
intersystem crosslinking 体系间交联
intertangled lamellae 互相缠结薄层

inter-telomerization(=telocopolymerization) 共调聚反应*；调节共聚(作用)
intert-gas atoms 惰性气体原子
intert-gas configuration 惰性气体结构
intertraction 吸浓作用
interval 区间；间隔
interval basicity 碱度差值
interval estimation 区间估计*
intervals of oil changing 换油时间
inter-vulcanizability 共硫化性
interweaving 交织；组织〔纺〕
interwoven crystal 混杂晶体；组合晶体
interzonal zone 中间区
interzone ①地区间；地带之间②晶带间
intimate admixture 紧密混合物；紧密掺和物；均匀混合物
intimate ion-pair 紧密离子对
intimately mixed continuous fiber hybrid composites 分散型连续混杂纤维复合材料
intimate mixture 紧密混合物
intisy 印蒂〔一种橡胶植物〕
intolerable concentration 不可耐浓度
intolerable limit 不可耐限度
intolerance 不可耐
Intolox saddle 英驼洛克斯鞍形填料
intoxican ①毒剂②麻醉剂
intoxicant ①毒物②致醉物
intoxicating ①致毒②致醉
intoxication ①中毒②喝醉
intoxication of catalyst 催化剂中毒
intra- 内
intra-annular 环内的
intra-annular tautomerism 环内互变异构(现象)
intra-atomic 原子内的
intra-atomic energy 原子内能
intrachain (高分子)链内
intrachain force 链内作用力
intra-chain potential energy 链内位能
intrachain reaction (分子)链内反应
intraclass mixture 同属混合物
intra-cluster 内簇
intra-column injection system 柱内注射系统
intracrystalline 晶体内的
intracrystalline imperfection 晶内缺陷
intracrystalline swelling 晶内溶胀
intraester 内酯
intra-intermolecular polymerization 分子内-分子间聚合*
intra-inter ply fiber hybrid composites 层内-层间混杂纤

维复合材料
intra-laboratory comparison 室内对照
intra-laboratory duplicate 实验室内的双份测定
intra-laboratory repeatability error (实验)室内重复性误差
intramolecular 分子内的
intramolecular acetalization 分子内缩醛化
intramolecular acetal ring 分子内缩醛环
intramolecular acyl 分子内酰基
intramolecular anhydride 内酐
intramolecular arylation 分子内的芳基化
intramolecular charge transfer 分子内电荷转移
intramolecular condensation (分子)内缩合(作用)；成环(作用)
intramolecular crosslinking 分子内交联
intramolecular cyclization 分子内环化(作用)
intramolecular elimination reaction 分子内消除反应
intramolecular energy transfer 分子内能量传递*
intramolecular force 分子内力
intramolecular hydrogen bond 分子内氢键
intramolecular labelling 分子内标记
intramolecular migration 分子内迁移作用；分子内重排作用
intramolecular oxidation and reduction 分子内部氧化还原(作用)
intramolecular polymerization 分子内聚合(作用)
intramolecular potential 分子内势能；分子内位能
intramolecular reaction 分子内反应
intramolecular rearrangement(=intramolecular transformation; intramolecular transposition) 分子内重排作用
intramolecular relaxation 分子内弛豫*
intramolecular repulsion 分子内斥力
intramolecular second order moment 分子内二级矩
intramolecular selectivity 分子内选择性*
intramolecular stiffness 分子内硬挺性
intramolecular strain 分子内应变
intramolecular transfer 分子内转移
intramolecular transformation(=intramolecular rearrangement; intramolecular transposition) 分子内重排作用
intramolecular transposition(=intramolecular rearrangement; intramolecular rearrangement) 分子内重排作用
intranuclear (原子)核内的
intranuclear cascade 核内串级(过程)
intranuclear tautomerism 核内互变异构；环内互变异构
intrapair energy 分子对内的能量
intraparticle diffusion 粒子内部扩散
intraperiod band 内周期带
intraplant transport 厂内输送

intraply 布层内的
intraply fiber hybrid composites 层内混杂纤维复合材料
intraply hybrid (增强)层内(异种纤维)杂化
intraresin reaction 树脂内反应
intratomic 原子内的
intratomic force(=valence force) 原子间力；价力
intratransguanilation 内转移胍化作用
intricacy 错综；复杂；缠结
intricate casting 复杂的铸件
intricate detail 复杂花纹〔模制品〕
intrinsic 内在的；本质的；内禀的；固有的
intrinsic acidity 固有酸度*
intrinsic activity 固有活性
intrinsically flame retarded 本质抗燃的；本质阻燃的
intrinsically safe system 本质安全系统
intrinsic anisotropy 内在各向异性
intrinsic barrier 内禀能垒*
intrinsic basicity 固有碱度*
intrinsic birefringence 特性双折射
intrinsic breakdown 特性击穿
intrinsic breaking energy 内在断裂能；特性断裂能
intrinsic constant 固有常数
intrinsic contaminants 内在杂质；固有杂质
intrinsic corrosion resistance 固有的耐腐蚀性
intrinsic defect 本征缺陷*
intrinsic dissymmetry 特性不对称性
intrinsic dynamic viscosity 固有动力黏度
intrinsic energy(=internal energy; energy content) 内能；蕴能；能含量；能藏
intrinsic error 基本误差；固有误差
intrinsic factor 内(源)因素
intrinsic fluidity 内在流动性；固有流动性
intrinsic heat 内热；蕴热
intrinsic mechanical property 内在机械性能；内在力学性能
intrinsic mobility 内在流动性
intrinsic nature 本质，禀性；特征
intrinsic orientation constant 固有取向常数
intrinsic oxygen index 特性氧指数
intrinsic photoconductor 本征光电导器件
intrinsic pressure 内压；蕴压
intrinsic reactivity 本征反应性；本质反应性
intrinsic rigidity 固有刚性
intrinsic safety 本征安全
intrinsic semiconductor 本征半导体；纯半导体
intrinsic solubility 固有溶解度*
intrinsic tracer 内部示踪物
intrinsic viscosity 特性黏数*

intrinsic viscosity number 特性黏数
intrinsic viscosity of polymer 聚合物特性黏度〔聚合物由分子量决定的黏度〕
introduction ①进样②注射③进入
introduction pneumatic valve 进气阀
introduction valve 注射阀；进样阀
introfaction 加速浸泡(作用)；浸透化
introfier 浸泡剂
intron 内含子
intronium surface active agent 内季铵盐型表面活性剂
intrusion 侵入作用
intumescence 膨胀
intumescent 膨胀现象的
intumescent coating 发泡型防火涂料；膨胀型防火涂层
intumescing 膨胀
inturgescence 膨胀
intussusception 吸受；摄取
inulenin 旋覆花糖 $(C_6H_{10}O_5)_x$
inulic acid(=alantic acid) 旋覆花酸；阿兰酸
inulin 旋覆花粉；菊粉 $(C_6H_{10}O_5)_n$
inuline(=anthranoyllycoctonine) 安茴酰牛扁碱
inuloid 类木香素；类旋覆花粉
inulol 木香油
inundator 浸泡器
inuroyleanol 罗复酚
in-use condition 使用条件
in-use testing 实用试验
inusterol A(=taraxasterol) 旋覆花甾醇 A
inusterol B 旋覆花甾醇 B
inusterol C 旋覆花甾醇 C
invar 不胀钢；因钢；因伐合金
invariable interference interval 不变干涉区域
invariance requirement 不变量要求
invariant ①不变(更)的；无变度的②不变式
invariant equilibrium 无变度平衡；不变(更)平衡
invariant of strain tensor 应变张量不变量
invariant system 无变度(体)系；不变(更)系(统)
invasin ①扩散因子②透明质酸酶
inventory 渣油
inventory change report(ICR) 库存变更报告
inventory time 库存时间
inverse aqua regia 逆王水
inverse chronopotentiography 反向计时电位法
inverse cooling-rate curve 冷却速率倒数曲线
inverse derivative potentiometric titration 反向导数电位滴定
inverse dilution method 逆稀释法
inverse dispersion polymerization 反相分散聚合

inversed phase chromatography 反相色谱(法)
inversed solubility curve 逆溶度曲线
inverse emulsion polymerization 反相乳液聚合
inverse experiment 反转(反向)实验
inverse gas chromatography 反气相色谱法
inverse gas liquid chromatography 反气液色谱法
inverse gated decoupling 反(转)门控去耦
inverse heating rate curve 加热速率倒数曲线
inverse IDA 逆同位素稀释(分析)法
inverse impact strength 反(面)冲击强度
inverse isotope dilution method 反同位素稀释法
inverse isotope effect 逆同位素效应*
inverse lattice(=reciprocal lattice) 倒易晶格
inverse matrice 逆矩阵
inverse metamorphism 逆变形
inverse method 反逆法
inverse miniemulsion polymerization 反相微乳液聚合
inverse phase ion pair chromatography 反相离子对色谱法
inverse polarography 逆极谱法
inverse-power fluid 逆幂型流体
inverse Raman scattering(IRS) 反拉曼散射
inverse ratio 反比例
inverse-spinel type structure 反尖晶石晶型结构
inverse substitution 相反取代(作用)
inverse suspension polymerization 反相悬浮聚合
inverse voltage 反电压
inverse voltammetry 逆向伏安法*
inversion ①反转*②倒反③转化
inversion axis 反演轴
inversion barrier 反演势垒
inversion center 对称中心；倒反中心
inversion of phases ①相变型②相转化
inversion phase emulsion 转相乳化(作用)
inversion point 转化点
inversion recovery 反转恢复法
inversion recovery method 反转恢复法
inversion recovery spin echo 反转恢复自旋回波
inversion symmetry 反演对称
inversion temperature 转化温度
inversion voltammetry 逆向伏安法
inversive 转化的
inverted 转化了的；逆转了的
inverted evaporation 向下蒸发；反向蒸发
inverted filter 反滤层
inverted gated decoupling 反转门控去偶
inverted orthoflow 反正流〔把再生器放在反应器之上的催化裂化装置〕
inverted peak 反峰；倒峰

inverted plasticity 膨胀性；搅胀性；反塑性
inverted ram press 逆向柱塞压机
invertible 可逆转的
inverting 转化的
invertomer 反演体
invertose 转化糖
invert point ①转化点；转化温度②(与数高点相对应)最低点；谷
invert solubility 逆向溶解(性)
investigation 调查；研究；勘测；试验
investment 蜡模铸造
investment cost 投资额
inviscid 非黏(性)的
inviscid fluid 非黏性流体
inviscid theory of fluid 非黏性流体理论
invisibility 不可见性
invisible 不可见的；看不见的
invisible chromatogram 不可见色谱
invisible fat 隐(性)油；不可见油
invisible ink 显影油墨；隐色油墨
invisible light 不可见光
invisible loss 看不见的损失；蒸发损失
invisible ray 不可见射线
invisible spectrum 不可见光谱
invisible waste 无形损耗；风耗
in vitro 体外；在活体外〔纤维材料在生物体外进行试验〕
in vitro activation analysis 离体活化分析
in vitro competitive radioassay 体外竞争性放射分析
in vitro culture method 离体培养法
in vitro molecular evolution 体外分子进化
in vitro radio-assay 体外放射测定
in vivo 体内；在活体内〔纤维材料在生物体内进行试验〕
in vivo activation analysis 体内活化分析；活体活化分析
in vivo counting 整体计数；体内计数
in vivo electrochemistry 体内电化学
in vivo electron spin resonance 活体电子自旋共振
in vivo neutron activation analysis 体内中子活化分析*
in vivo nuclear magnetic resonance spectroscopy 活体 NMR 波谱学
involatile 不挥发性
involatile sample 不挥发性试样
involatile substance 不挥发性物质
involatility 不挥发性
involute ①渐开线；渐伸线②渐开的；渐伸的
involute gear 渐开线齿轮
involute gear cutter 牙轮铣刀

inward rectified current 内向整流电流
inyoite 板硼石
iodacetic acid 碘乙酸 CH_2ICOOH
iodacetyl bromide 碘乙酰溴 CH_2ICOBr
iodacetyl chloride 碘乙酰氯 CH_2ICOCl
iodacetyl fluoride 碘乙酰氟 CH_2ICOF
iodacetyl halide 碘乙酰卤 CH_2ICOX
iodacetyl iodide 碘乙酰碘 CH_2ICOI
iodacyl bromide 碘代酰基溴
iodacyl chloride 碘代酰基氯
iodacyl fluoride 碘代酰基氟
iodacyl halide 碘代酰基卤
iodacyl iodide 碘代酰基碘
iodanil 四碘苯醌
iodargyrite 碘银矿
iodate 碘酸盐 MIO_3
iodated 碘化了的
iodate method 碘酸盐法
iodatimetric titration 碘酸盐滴定
iodatimetry 碘酸盐滴定(法)
iodating 碘化的
iodating agent 碘化剂
iodation 碘化作用
iodazide 叠氮碘
iodeosin(=tetraiodo fluorescein) 四碘荧光素
Iodex process 埃奥德克斯过程〔用硝酸洗涤除废气中碘的过程〕
iodic 碘的
iodic acid 碘酸 HIO_3
iodic anhydride 五氧化二碘；碘酸酐 I_2O_5
iodic ether 碘代乙烷 C_2H_5I
iodide 碘化物 MI
iodide extraction system 碘化物萃取体系
iodide in kelp 海草中的碘化物
iodide ion selective electrode 碘离子选择电极
iodide volatility process 碘化物挥发法
iodimetric 碘滴定的
iodimetric analysis ①碘定量(法)②定碘量分析
iodimetric estimation 碘量估测
iodimetric titration 碘滴定法
iodimetry 碘量法*
iodinated 碘化了的
iodinating 碘化的
iodinating agent 碘化剂
iodination 碘化作用
iodine 碘
iodine absorption number 碘吸收值；碘消耗量
iodine absorption value 吸碘值〔炭黑〕

iodine-amylose reagent　碘-直链淀粉试剂
iodine bromide　溴化碘　$IBr;IBr_3;IBr_5$
iodine chloride　氯化碘　$ICl;ICl_3$
iodine colorimetry　碘比色法
iodine complex　碘络合物
iodine coulometer　碘极电量计
iodine cyan(=iodine cyanide)　氰化碘　ICN
iodine cyanide(=iodine cyan)　氰化碘
iodine dioxide　二氧化碘　IO_2
iodine flask　碘瓶*
iodine green　碘绿
iodine in ring　环上的碘
iodine in side chain　侧链上的碘
iodine monobromide　一溴化碘　IBr
iodine monochloride　一氯化碘　ICl
iodine number　碘值*
iodine number flask　碘值(烧)瓶
iodine oxide　氧化碘　$I_2O_4;I_2O_5$
iodine pentafluoride　五氟化碘　IF_5
iodine pentoxide　五氧化碘
iodine plates　碘片
iodine soap　碘皂
iodine solubilization method　碘加溶法
iodine sorption　碘吸着作用
iodine staining　包囊碘液染色法
iodine still　蒸碘器
iodine sulfate　硫酸碘　$I_2(SO_4)_3$
iodine test　碘试法〔用碘试淀粉〕
iodine tetroxide　四氧化二碘　I_2O_4
iodine tincture　碘酊
iodine trichloride　三氯化碘　ICl_3
iodine tungsten lamp　碘钨灯
iodine value　碘值*
iodine water　碘水溶液
iodinin　碘色菌素
iodipin　碘油剂〔兽用〕
iodisan　碘散〔药〕
iodite　亚碘酸盐　MIO_2
iodival(=α-iodoisovaleryl urea)　碘瓦耳；α-碘代异戊酰脲　$(CH_3)_2CHCHICONHCONH_2$
iodizated　碘化了的
iodizating　碘化的
iodization　碘化作用
iodized　碘化了的
iodized fatty acid　碘化脂肪酸
iodized oil　碘化油
iodized starch　碘化淀粉
iodo　碘代　I—

iodoacetal　碘乙缩醛〔俗〕；碘乙醛缩二乙醇　$ICH_2CH(OC_2H_5)_2$
iodoacetaldehyde　碘乙醛　CH_2ICHO
iodoacetamide　碘乙酰胺
iodoacetic acid　碘乙酸　ICH_2CO_2H
iodoacetic bromide　碘乙酰溴　CH_2ICOBr
iodoacetic chloride　碘乙酰氯　CH_2ICOCl
iodoacetic fluoride　碘乙酰氟　CH_2ICOF
iodoacetic halide　碘乙酰卤　CH_2ICOX
iodoacetic iodide　碘乙酰碘　CH_2ICOI
iodoacetone　碘丙酮　ICH_2COCH_3
iodoacetonitrile(=iodomethyl cyanide)　碘甲基氰；碘乙腈　ICH_2CN
iodoacetylene　碘乙炔　$IC \equiv CH$
iodo-acid　碘(代)酸
iodo-acid amide　碘乙酰胺
iodo-aliphatic compound　碘代脂族化合物
iodoalphionic acid　碘阿芬酸
iodo-amino acid　碘代氨基酸　$NH_2RICOOH$
iodo-aurate　碘金酸盐　$M[AuI_4]$
iodo-benzene　碘(代)苯；苯基碘　C_6H_5I
iodobenzene dichloride(=phenyl iodochloride)　二氯化碘代苯　$C_6H_5ICl_2$
iodo-benzoic acid　碘苯甲酸　IC_6H_4COOH
iodobenzoyl chloride　碘苯酰氯
iodobenzyl bromide　碘苄基溴
iodochlorohydroxyquinoline　碘氯代羟基喹啉
iodocyanogen　碘化氰　ICN
iodocyclohexane　碘代环己烷　$C_6H_{11}I$
iodocyclohexanol　碘代环己醇　$HOC_6H_{10}I$
iododecane(=decyl iodide)　癸基碘
iododeoxyuridine　碘(代脱氧尿嘧啶核)苷
iodo-ester　碘代酯
iodoethane(=ethyl iodide)　乙基碘；碘乙烷
iodo-ether　碘代醚
2-iodo-ethyl alcohol　2-碘乙醇
iodoethylene　碘乙烯
iodo-ethyl ester　碘乙酯　$RCOOC_2H_4I$
iodoform(=triiodomethane)　碘仿；三碘甲烷；黄碘　HCI_3
iodoformin　碘仿明
iodoform reaction　碘仿反应
iodoform test　碘仿试验*
iodofuran　碘代呋喃　C_4H_3OI
iodoglycerol　碘代甘油
iodogorgoic acid　碘代珊氨酸；3,5-二碘酪氨酸
1-iodohexane(=hexyl iodide)　1-碘己烷；己基碘　$CH_3(CH_2)_4CH_2I$

iodo-hippurate sodium　碘马尿酸钠
o-iodohippuric acid　邻碘马尿酸
　　　$IC_6H_4CONHCH_2COOH$
iodohydrin　碘醇
7-iodo-8-hydroxyquinoline-5-sulfonic acid(=ferron)
　　7-碘-8-羟基喹啉-5-磺酸；试铁灵
iodo-isopropyl alcohol　碘代异丙醇；1-碘-2-丙醇
　　　$CH_2ICHOHCH_3$
α-iodoisovaleryl urea(=iodival)　碘瓦耳；α-碘代异戊酰脲　$(CH_3)_2CHCHICONHCONH_2$
iodolactonization　碘内酯化
iodole(=tetraiodo pyrrole)　碘咯；四碘代吡咯　I_4C_4NH
iodomercurate　碘汞酸盐
iodomercurate potassium　碘汞酸钾
iodomercuriphenol　碘汞基苯酚　$HgIC_6H_4OH$
iodo-mercury-benzene　碘汞基苯　$HgIC_6H_5$
iodomethane(=methyl iodide)　碘代甲烷；甲基碘　CH_3I
iodomethylate(=methiodide)　甲碘化物；甲碘
iodomethylation　碘甲基化
iodomethyl cyanide(=iodoacetonitrile)　碘甲基氰；碘乙腈　ICH_2CN
iodometric　碘量的
iodometric acid value　碘量法得的酸值
iodometric titration　碘量滴定；碘量法
iodometry　碘量法*
iodonation　碘化*
iodonitrobenzene　碘硝基苯
iodonium　碘鎓；三价碘
iodonium compound　碘鎓化合物
1-iodononane(=nonyl iodide)　1-碘壬烷
　　　$CH_3(CH_2)_7CH_2I$
1-iodooctane　1-碘辛烷
2-iodopentane　2-碘代戊烷
iodophenesic acid　二碘苯酚　$I_2C_6H_3OH$
iodophenic acid　一碘苯酚　IC_6H_4OH
iodophenisic acid　三碘苯酚　$I_3C_6H_2OH$
iodophenol　碘苯酚
iodophenyl acetate　乙酸碘苯酯
p-iodophenylsulfonyl amino acid　对碘苯磺酰氨基酸
p-iodophenyl urethane　对碘苯基尿烷
iodophor　碘递体；碘伏
iodo-phosphonium　碘化鏻　R_4PI
iodophthalein　碘酞酞
iodophthalein sodium(=tetraiodophenolphthalein sodium)
　　碘酞钠；四碘酞酞钠
　　　$NaOOCC_6H_4C(=C_6H_2I_2=O)C_6H_2I_2$
3-iodopropandiol(=glycerin iodohydrin)　碘甘油；3-碘-1,2-丙二醇　$CH_2ICHOHCH_2OH$

2-iodopropane(=isopropyl iodide)　2-碘丙烷；异丙基碘　CH_3CHICH_3
iodopropionic acid　碘丙酸
iodopropylene　碘丙烯
3-iodo-2-propynyl butyl carbamate　氨基甲酸-3-碘-2-丙炔丁酯
iodopyrine　碘(安替)匹林
iodoso　亚碘酰；氧碘基　OI—
iodosobenzene　亚碘酰苯　C_6H_5IO
iodosobenzoic acid　亚碘酰基苯甲酸
iodoso compound　亚碘酰化合物
iodosol　碘(代)百里香酚
iodostarch paper　碘淀粉试纸
iodostarch reaction　碘淀粉反应
iodosylbenzene electrode　亚碘酰苯电极
iodothymol　碘代百里酚　$C_{10}H_{13}OI$
iodotoluene　碘甲苯
iodotrinitromethane　碘代三硝基甲烷；三硝基甲碘　$IC(NO_2)_3$
3-iodotyrosine　3-碘酪氨酸
iodouracil　碘(代)尿嘧啶
iodous　亚碘的
iodous acid　亚碘酸
iodoxy　碘酰基；二氧碘基　O_2I—
iodoxybenzene　碘酰苯　$C_6H_5IO_2$
iodoxy compound　碘酰化合物
iodyl-benzene electrode　碘酰苯电极
iodyl ion　碘酰离子
iodyrite　碘银矿
iolite　堇青石
ion　离子
ion-accelerating voltage　离子加速电压
ion acceleration　离子加速
ion activity　离子活度
ion activity electrode　离子活性电极
ion adsorption　离子吸附(作用)
ionamine blue　离胺蓝
ionamines　离胺染料
ion association　离子缔合
ion-association complex　离子缔合配合物
ion association extraction system　离子缔合萃取体系
ion-atmosphere　离子氛
ion atmosphere effect　离子氛作用；离子氛效应
ion atmosphere radius　离子氛半径
ion backscattering spectrometer　离子反散射谱仪
ion beam　离子束*
ion beam analysis　离子束分析*
ion beam chopper　离子束调制器；离子束切割器

ion beam collimation 离子束准直
ion beam etching 离子束刻蚀
ion beam lithography 离子束印刷术
ion beam scanning 离子束扫描
ion beam spectrochemical analyzer 离子束光谱(化学)分析仪
ion beam sputtering 离子束溅射
ion beam surface probe 离子束表面探针
ion beam width 离子束宽度
ion binding 离子束缚
ion bombardment 离子轰击
ion bombardment ion source 离子轰击离子源*
ion-bombardment mass analyzer 离子轰击质量分析器
ion chamber 离子室
ion channel 离子通道
ion channel switching immunosensor 离子通道免疫传感器
ion chromatogram(IC) 离子色谱图
ion chromatograph 离子色谱仪*
ion chromatography 离子色谱法*
ion cloud 离子云
ion cluster 离子簇；离子群
ion cluster technique 离子簇法；离子群法
ion cluster theory 离子簇理论〔辐射化学〕
ion collection 离子收集
ion collection system 离子收集系统
ion collector 离子收集器
ion collector electrode 离子接收器电极
ion concentration 离子浓度
ion-containing polymer 含离子聚合物
ion cooling 离子冷却
ion core 离子芯*
ion counter 离子计数器
ion counting system 离子计数系统
ion current 离子(电)流
ion current amplifier 离子流放大器
ion current monitor 离子(电)流监控器
ion cyclotron resonance(ICR) 离子回旋共振
ion cyclotron resonance mass analyzer 离子回旋共振质量分析器
ion cyclotron resonance mass spectrometer 离子回旋共振质谱仪*
ion cyclotron resonance mass spectrometry 离子回旋共振质谱法
ion cyclotron resonance spectrometer 离子回旋共振谱仪
iondanthren cloth 阴丹士林布
ion deflection 离子(束)偏转
ion density ratio 离子密度比

ion detector 离子检测器
ion-dipolar interaction 离子偶极性相互作用
ion-dipole bond 离子偶极键
ion-dipole interaction 离子偶极相互作用
ion-dipole physisorption 离子偶极物理吸着
ion displacement 离子位移
ion draw out field 离子拉出场
ion drift technique 离子漂移法
ion-electric focusing 离子电聚焦
ion-electron method 离子电子法
ion emitting surface 离子发射表面
ionene 紫罗烯
ionene polymer 紫罗烯聚合物
ion energy 离子能量
ion energy spread 离子能量发散
ion entrance 离子入口
ion-etching 离子蚀刻(法)；离子浸蚀(法)
ion exchange 离子交换(作用)
ion exchange adsorption 离子交换吸附(法)
ion exchange analysis 离子交换分析
ion exchange bed 离子交换床
ion exchange capacity 离子交换容量；离子交换能力
ion exchange capillary electrokinetic chromatography 离子交换毛细管电动色谱
ion exchange catalysis 离子交换(树脂)催化作用
ion exchange catalyst 离子交换树脂催化剂
ion exchange cellulose 离子交换纤维素
ion-exchange cellulose chromatography 离子交换纤维素色谱法
ion exchange cellulose paper 离子交换纤维素纸
ion exchange chromatography(IEC) 离子交换色谱法
ion exchange column 离子交换柱
ion exchange concentration 离子交换浓缩
ion exchange demineralization 离子交换除盐
ion exchange electrode 离子交换电极
ion exchange electrodialysis 离子交换电渗析
ion exchange electrokinetic chromatography 离子交换电动色谱(法)
ion exchange electrophoresis 离子交换电泳(法)
ion exchange equilibrium 离子交换平衡
ion exchange equilibrium constant 离子交换平衡系数
ion exchange fiber 离子交换纤维
ion exchange filter 离子交换过滤器
ion exchange foam chromatography 离子交换泡沫色谱法
ion exchange kinetics 离子交换动力学
ion exchange layer 离子交换层
ion exchange liquid 离子交换液

ion exchange macromolecule	离子交换大分子
ion exchange material	离子交换材料
ion exchange membrane	离子交换膜*
ion exchange membrane electrode	离子交换膜电极
ion exchange membrane electrodialyser	离子交换膜电渗析器
ion exchange membrane eletrodialysis	离子交换膜电渗析
ion exchange method	离子交换法
ion exchange operation	离子交换操作
ion exchange paper	离子交换纸
ion exchange paper chromatography	离子交换纸色谱(法)
ion exchange phase	离子交换相
ion-exchange pigment	离子交换(型)颜料
ion-exchange polymer	离子交换聚合物
ion exchange potential	离子交换电位
ion exchange process	离子交换法
ion exchange purification	离子交换提纯；离子交换净化
ion exchanger	离子交换剂*
ion-exchange radiochromatography(IERC)	离子交换放射色谱法
ion exchange reaction	离子交换反应
ion exchange regenerant	离子交换(剂)再生剂
ion exchange resin	离子交换树脂*
ion exchange resin column	离子交换树脂柱
ion exchange resin electrode	离子交换树脂电极
ion exchange resin hygrometer	离子交换树脂湿度计
ion exchange resin membrane	离子交换树脂膜；离子交换膜
ion exchange resin membrane electrode	离子交换树脂膜电极
ion exchange resin spectrophotometry	离子交换树脂分光光度法
ion exchange separation	离子交换分离法
ion exchange sephadex	离子交换交联葡聚糖
ion exchange sheet	离子交换片
ion exchange site	离子交换位置
ion exchange softening	离子交换硬水软化
ion exchange surface	离子交换表面
ion exchange system	离子交换系统
ion exchange technique	离子交换技术
ion exchange thin-layer chromatography	离子交换薄层色谱(法)
ion exchange treatment	离子交换处理；离子交换法
ion exchange unit	离子交换装置
ion exchange water	离子交换水
ion exclusion chromatography	离子排斥色谱法*
ion exclusion column	离子排斥柱
ion exclusion packing	离子排阻填料
ion exclusion partition chromatography	离子排斥分配色谱(法)
ion exclusion separation	离子排斥分离(法)
ion exit	离子出口
ion exponent	离子指数
ion floatation	离子浮选*
ion floatation method	离子浮选法
ion flow	离子流
ion fluorescence	离子荧光
ion focusing electrophoresis	离子聚焦电泳
ion fragment	离子碎片
ion fragmentation	离子碎裂
ion-free water	无离子水
ion gas laser	离子气体激光器
ion gettering pump	离子吸气泵
ion gun	离子枪
ionic	离子的
ionic acidity	离子酸度
ionic activity	离子活度
ionic activity coefficient	离子活度系数*
ionic addition reaction	离子加成反应*
ionic adsorption	离子性吸附(作用)
ionically bonded	离子键合的
ionically stabilized product	离子稳定(化)产品
ionic amphoterism	离子两性(现象)
ionic associate	离子缔合物
ionic association	离子缔合*
ionic atmosphere	离子氛*
ionic bond	离子键*
ionic catalyst	离子催化剂
ionic character of bonds	键的离子特性
ionic charge	离子电荷*
ionic chloroprene rubber latex	离子型氯丁胶乳
ionic complex	离子配合物
ionic compound	离子化合物
ionic condensation	离子(引发)缩合(反应)
ionic conductance	离子电导*
ionic conduction	离子传导；离子导电
ionic conductivity	①离子电导性②离子电导率
ionic copolymerization	离子共聚合*
ionic crystal	离子晶体*
ionic crystal hygrometer	离子晶体湿度计
ionic current	离子电流
ionic depositing method	离子沉积法
ionic deposition	离子沉积作用
ionic detergent	离子型去垢剂
ionic dissociation	离子解离*
ionic dissociation energy	电离能；离解能

ionic double layer 离子双(电)层
ionic emulsifier 离子型乳化剂
ionic emulsifying agent 离子(型)乳化剂
ionic environment 离子环境
ionic equation 离子方程式
ionic equilibrium 离子平衡*
ionic exchange membrane 离子交换膜
ionic fluid 离子(射)流；离子束
ionic fluorescence spectrometry 离子荧光光谱法
ionic formula 离子式
ionic fragmentation reactions 离子碎化反应
ionic group 离子基(团)
ionic hydration 离子水合*
ionic hydrophilic group 离子型亲水基
ionic imcompatibility 离子不相容性
ionic initiator 离子型引发剂
ionic interaction 离子相互作用
ionic interaction energy 离子作用能
ionicity parameter 离子性参数*
ionic latex 离子性胶乳
ionic lattice 离子晶格
ionic layer-by-layer self-assembly 层状静电自组装
ionic line 离子线
ionic link(=ionic linkage) 离子键
ionic linkage(=ionic link) 离子键
ionic liquid 离子液体
ionic magnetic susceptibility 离子性磁化率
ionic micellar core 离子型胶束
ionic micelle 离子胶束*
ionic migration 离子迁移
ionic migration number 离子迁移数
ionic mobility 离子迁移率*；离子淌度
ionic molecule 离子(型)分子
ionic nature 离子性质
ionic oligomerics 离子型低聚物
ionic osmotic pressure 离子渗透压
ionic plating 离子镀膜
ionic polarization 离子极化
ionic polymer 离子聚合物*
ionic polymerization 离子(型)聚合*
ionic polymerization catalyst 离子型聚合催化剂
ionic potential 电离(电)势
ionic product 离子积*
ionic radius 离子半径*
ionic reaction 离子反应*
ionic replacement 离子取代*
ionic rubber 离子橡胶
ionics(=ionic surface-active agent) 离子型表面活性剂
ionic sieve 离子筛
ionic solute 离子溶质
ionic solvation 离子溶剂化*
ionic source 离子源
ionic species 离子(存在)形式；离子型态
ionic stabilization processes 离子稳定化过程
ionic-stabilized system 离子稳定体系
ionic strength 离子强度*
ionic strength adjustment buffer 离子强度缓冲剂
ionic strength adjustor(ISA) 离子强度调节剂
ionic strength effect 离子强度效应
ionic surface-active agent(=ionics) 离子型表面活性剂
ionic surfactant 离子型表面活性剂
ionic surfactant intermediate 离子型表面活性剂中间体
ionic theory 离子学说
ionic valancy 离子价
ionic weight 离子量
ionidine 花菱草碱
ionigraphic mobility 离子淌度
ion impact 离子碰撞
ion impact ionization 离子碰撞电离
ion implantation 离子注入*
ion implanted region 离子注入区
ion inclusion 离子存留
ion injection technique 离子注入技术
ion intensity ratio 离子强度比
ion interaction chromatography 离子相互作用色谱法
ion interaction model 离子相互作用模型
ionioic initiator 负离子引发剂
ion-ion emission 离子-离子发射
ion-ion solvent interaction 离子-离子溶剂相互作用
ionisation(=ionization) 电离*；离子化
ionisation potential 电离电位
ionising electrode 电离(化)电极
ionite 离子型结构聚合物
ionium 锾；钍-230〔符号 Io〕
ionium dating method 钍-230 定年代法
ionizable contaminant 离子污染
ionizatin method 电离法
ionization 电离*；离子化
ionization amplifier 电离放大器
ionization buffer 电离缓冲剂；消电离剂
ionization buffer for flame AAS 火焰原子光谱法的电离缓冲剂
ionization by collision 碰撞电离
ionization by light 光致电离
ionization by sputtering 溅射电离
ionization cell 电离(测定)池

ionization chamber 电离室
ionization constant 电离常数*
ionization cross section 电离截面
ionization current 电离电流
ionization degree 电离度
ionization dosimeter 电离剂量计
ionization efficiency 电离效率*
ionization efficiency curve 电离效率曲线
ionization energy 电离能*
ionization equilibrium 电离平衡
ionization event 电离结果
ionization gage detector 电离计检测器
ionization gap method 电离窄缝法
ionization gauge 电离压力计；电离规
ionization heat 电离热
ionization interference 电离干扰*
ionization isomerism 电离异构*
ionization level 电离能级
ionization limit 电离限
ionization loss 电离耗损
ionization manometer 电离作用量计
ionization method 电离法
ionization polymerization 电离聚合
ionization potential 电离电位*
ionization power 电离能力
ionization probability 电离概率
ionization process 电离过程
ionization pulse 电离脉冲
ionization radiation 电离辐射
ionization rate 电离速率
ionization region 电离区域
ionization series 电离序；离子化序列
ionization source 电离源
ionization suppressor 电离抑制剂
ionization tendency 电离倾向
ionization theory 电离学说
ionization vacuum gauge 电离真空规；电离真空计
ionization vacuum meter 电离真空计
ionization voltage 电离电压
ionization yield 电离率
ionized 电离的；离子化的
ionized catalyst 电离催化剂
ionized fraction 电离分数
ionized hydrogen 离子化氢
ionized layer 电离层
ionized state 电离(状)态
ionizer 电离源
ionizing 电离化

ionizing cell 电离室
ionizing electrode 电离(化)电极
ionizing particle 致电离粒子
ionizing potential 电离(电)势
ionizing power 电离本领
ionizing radiation 电离辐射；离子辐射
ionizing solvent 离子化溶剂*
ion kinetic energy 离子动能
ion kinetic energy spectroscopy(IKES) 离子动能谱法*
ion kinetic energy spectrum 离子动能谱
ion laser 离子激光器
ion lens 离子透镜
ion line 离子(谱)线
ion meter 离子计*
ion meter method 离子计法
ion microanalyzer 离子微分析器
ion microprobe 离子微探针
ion microprobe mass analyzer(IMMA) 离子微探针质量分析器*
ion microprobe mass spectrometer 离子微探针质谱计
ion microscope 离子显微镜
ion migration 离子迁移
ion migration spectrograph(IMS) 离子迁移谱仪
ion mobility 离子迁移率；离子淌度
ion mobility detector 离子淌度检测器
ion mobility spectrometry 离子淌度谱法
ion-molecule collision 离子分子碰撞
ion-molecule reaction 离子-分子反应*
ion momentum 离子动量
ion neutralization spectroscopy 离子中和谱法
ionocolorimeter 氢离子比色计
ionodialysis 离子渗析
ionogen(=potential electrolyte) 势电解质*
ionogenic 离子源的〔形成或供应离子的〕·
ionogenic group (潜)离子基团
ionogenic linkage 离化(合)键
ionogenic polar group 离子型极性基团
ionogenic polymer 潜离子(化)聚合物
ionogenic surfactant 离子型表面活性剂
ionogenic system 潜离子体系
ionography 离子(放射)照相法；离子谱法
ionol 紫罗兰醇
ionomer ①离聚物②离子交联聚合物*③离子键聚合物
ionomer coating 离聚物涂料；离子键聚合物涂料
ionomeric 离聚物的；含离子键聚合物的
ionomer resin 离子键树脂
ionometer 离子计
ionone(=irisone) 芷香酮；紫罗酮

φ-ionone(=pseudo ionone) 假性紫罗兰酮
ionone semicarbazone 紫罗酮缩氨基脲
ionophore 实电解质*
ionophoresis （离子）电泳
ionophortic separation （离子）电泳分离
ion optics 离子光学*
ionosphere 电离层
ionotron 静电消除器
ionotropic change 离子移变(作用)
ionotropy 离子移变(作用)
ion pair 离子对*
ion pair chromatography(=paired ion chromatography) 离子对色谱法*
ion pair extraction 离子对萃取
ion pair formation model 离子对形成模型
ion-pairing probe detection 离子对探针检测
ion pair ionization 离子对电离
ion pair polymerization 离子对聚合*
ion pair reagent 离子对试剂
ion pair yield 离子对产率
ion path 离子轨道
ion permeability 离子透过性
ion photodissociation 离子光解
ion photographic plate 离子感光板*
ion probe micro analysis 离子探针显微分析
ion probe microanalyzer 离子探针微区分析器
ion product 离子积；溶解度乘积
ion product constant 离子积常数
ion pump 离子泵
ion quadruplet 离子四重态
ion-radical(=radical ion) 离子基
ion radical polymerization 离子自由基聚合
ion reflection coefficient 离子反射系数
ion repeller 离子推斥极
ion resonance mass spectrometer 离子共振质谱仪
ion retardation 离子阻滞(作用)
ion retardation purification 离子阻滞纯化(作用)
ion retardation resin 离子阻滞型树脂；离子延滞型树脂
ion retardation separation 离子阻滞分离(法)
ion retarding lens 离子减速透镜
ion scanning 离子扫描
ion scattering 离子散射
ion scattering analysis(ISA) 离子散射分析
ion scattering spectrometer(ISS) 离子散射能谱仪
ion scattering spectrometry 离子散射光谱法
ion scattering spectroscopy(ISS) 离子散射能谱(学)*
ion selective disc electrode 离子选择圆盘电极
ion selective electrode 离子选择(性)电极*
ion selective electrode analysis 离子选择(性)电极分析
ion selective electrode detector 离子选择电极检测器
ion selective electrodialysis 离子选择电渗析
ion selective field effect potentiometric sensor 离子选择场效应电位测量传感器
ion selective field effect transistor(ISFET) 离子选择场效应晶体管*
ion selective membrane 离子选择膜
ion selective microelectrode 离子选择微电极
ion selective semicoelectrode 离子选择半导电极
ion selective semiconducting electrode 离子选择性半导电极
ion sensitive emulsion 离子灵敏乳胶
ion sensitive field effect transistor 离子敏感场效应管
ion sensor 离子传感器
ion sequestration 离子螯合作用
ion sieve 离子筛
ion sieve effect 离子筛效应
ion source 离子源*
ion source life 离子源寿命
ion specific electrode 离子选择性(特效性)电极
ionspray 离子喷雾
ion spray junction 离子喷雾接口
ionspray mass spectrometry 离子喷雾质谱法
ion sputtering 离子溅射
ion suppression chromatography 离子抑制色谱法
ion thinner 离子减薄
ion to electron converter 离子-电子转换器
iontophoresis 离子电渗疗法
ion trajectory 离子轨道
ion transmission 离子传输；离子透射；离子透过率
ion transmission efficiency 离子传输率
ion transport 离子迁移
ion trap 离子阱
ion trap detector 离子阱检测器*
ion trap electron mirror 离子(陷)阱电子镜
ion trap mass spectrography(ITMS) 离子阱质谱法
ion trap mass spectrometer 离子阱质谱仪
ion triplet(=triple ion) 三重离子；离子三重态；离子三聚体
ion yield 离子产额
ionyl acetate 乙酸紫罗兰酯
ionylideneacetic acid 芷香亚基乙酸；紫罗兰亚基乙酸
ioxynil 4-羟(基)-3,5-二碘苯甲腈；碘苯腈
ipecac(=ipecacuanha) 吐根
ipecacuanha(=ipecac) 吐根
ipecacuanhic acid 吐根酸 $C_{14}H_{18}O_7$
ipecacuanhin 吐根(素)苷
ipecamine 吐根碱

Ipin method 安品法
IP lamp 燃性试验灯〔试验煤油燃烧性质用灯〕
ipoctal 胍乙啶
ipomeamarone 甘薯黑疤霉酮
ipomeanine 甘薯黑疤霉二酮
ipomoea ①红薯泻根②药薯
ipomoein 野药喇叭苷
ipratropine bromide and salbutamol 异丙托溴铵沙丁胺醇〔药〕
ipratropium bromide 异丙托溴铵〔药〕
ipso position 本位*
ipuranol 紫薯苷
ipurganol 药喇叭醇 $C_{21}H_{32}O_2(OH)_2$
ipurolic acid 3,11-二羟基十四(烷)酸；番红醇酸
IR absorption spectrographic detector 红外吸收光谱检测器
iraldeine(=methyl ionone) 甲基紫罗兰酮
iralia(=isomethyl ionone) 异甲基紫罗兰酮
Irany glass capillary viscometer 艾拉奈玻璃毛细管黏度计
irbesartan 厄贝沙坦〔药〕
irbesartan and hydrochlorothiazide 厄贝沙坦氢氯噻嗪〔药〕
irene 鸢尾烯
iresane 血苋烷
iresenin 血苋宁
iresin 血苋素
iretol 甲氧苯三酚
IR fluorescence 红外荧光*
irgafen 衣妥钠〔药〕；磺胺二甲苯甲酰胺
iridescent glass 虹彩玻璃
iridic (四价)铱的
iridic chloride 四氯化铱 $IrCl_4$
iridic compound 铱化合物
iridic oxide 二氧化铱 IrO_2
iridin 鸢尾定
iridium 铱〔77号元素, 化学符号 Ir〕
iridium anomalies 铱异常*
iridium chloride 氯化铱〔总名: 计有一氯化铱、二氯化铱、三氯化铱和四氯化铱〕
iridium oxide 氧化铱〔指二氧化铱或三氧化二铱〕
iridium sodium chloride 氯化铱钠；氯铱酸钠 $Na_3[IrCl_6]$
iridochloride 二氯化铱 $IrCl_2$
irido-compound 含二价铱化合物
iridodial 虹彩二醛
iridoid 环烯醚萜
iridolactone 虹彩内酯
iridomyrmecin 蚁素

iridosmine(=osmiridium) 铱锇矿
iridous 亚铱(的)〔指三价铱〕
iridous chloride 三氯化铱 $IrCl_3$
iridous compound 三价铱的化合物
iridous oxide 三氧化二铱 Ir_2O_3
irigenin 鸢尾配基
irigenol 鸢尾精酚
iris ①鸢尾属②虹膜③虹状；晕彩
iris absolute 鸢尾净油
irisaldehyde 鸢尾醛
iris diaphram 虹彩光阑〔显微镜的零件〕；可变光阑
irisoid 鸢尾粹
iris oil 鸢尾油
irisone(=ionone) 芷香酮；紫罗酮
IR microcell 红外微型池
IR microscopy 红外显微镜
iron ①熨斗②铁③铁器
ironability 耐熨烫性
iron acetate 乙酸亚铁
iron alum(=ferric alum) 铁明矾
iron angle (三)角铁
iron arc lamp 铁弧灯
iron arsenate coating process 砷酸铁膜处理法
iron bacteria corrosion 铁菌腐蚀
iron-base alloy 铁基合金
iron black 黑锑粉〔是黑色的细锑粉, 不宜译作: 铁黑〕
iron blue 铁蓝
iron blue pigment 铁蓝颜料
iron body valve 铁体阀
iron carbide 一碳化三铁 Fe_3C
iron carbonyl 羰基铁 $Fe(CO)_4$; $Fe(CO)_5$; $Fe_2(CO)_9$
iron carbonyl hydride 氢化四羰基铁 $Fe(CO)_4H_2$
iron casting 生铁铸造
iron catalyst 铁催化剂
iron cell 铁电池
iron cement 铁水泥
iron chain 铁链
iron chloride 氯化铁
iron conduit 铁导管
iron copperas(=green copperas) 绿矾；七水(合)硫酸亚铁
iron copper catalyst 铁铜催化剂
iron crucible 铁坩埚；铁熔埚
iron cyanide 氰化亚铁
iron dichloride 氯化亚铁 $FeCl_2$
iron disulfide 二硫化铁 FeS_2
iron drier 铁催干剂；铁干料
irone 鸢尾酮；甲基芷香酮
iron family element 铁族元素

iron filings 铁屑
iron filler 硬质腻子；铁腻子〔硬质油性腻子的别名〕
iron founding 生铁铸造
iron foundry 铸铁厂
iron garnet 铁榴石；硅酸钙铁矿石
iron gauze 铁丝网
iron glass 铁玻璃
iron gray 铁灰色
iron grit 铁粒；铁砂
iron group 铁系元素*
iron hat 铁帽
iron hinge 铁铰链
ironic 铁的
ironic acetate 乙酸铁 $Fe_2(C_2H_3O_2)_3$
ironic chloride 氯化铁 $FeCl_3$
ironic citrate 柠檬酸铁 $FeC_6H_5O_7$
ironic formate(=ironic formiate) 甲酸铁 $Fe(HCO_2)_3$
ironic formiate(=ironic formate) 甲酸铁
ironic hydroxide 氢氧化铁 $Fe(OH)_3$
ironic lactate 乳酸铁 $Fe_2(C_3H_5O_3)_3$
ironic malate 苹果酸铁 $Fe_2(C_4H_4O_5)_3$
ironic oxalate 草酸铁 $Fe_2(C_2O_4)_3$
ironic oxide 三氧化二铁 Fe_2O_3
ironic phosphate 磷酸铁 $FePO_4$
ironic succinate 丁二酸铁 $Fe_2(C_4H_4O_4)_3$
ironic sulfate 硫酸铁 $Fe_2(SO_4)_3$
ironic sulfide 三硫化二铁 Fe_2S_3
ironing 烫呢；整烫；熨烫
ironing board 烫模
ironing cylinder 熨皮滚筒〔皮革〕
ironing effect 熨烫效应；熨烫现象
ironing fastness 熨烫色牢度
ironing machine 熨平机
iron-in-lubricating-oil test(=iron-in-oil analysis) 润滑油中铁测定法
iron-in-oil analysis(=iron-in-lubricating-oil test) 润滑油中铁测定法
iron liquor 乙酸亚铁溶液 $Fe(C_2H_3O_2)_2 \cdot 4H_2O$
iron-manganese phosphate coating 磷酸铁锰(处理)膜
iron mica 铁云母〔矿〕；镜铁矿；云母氧化铁〔矿〕
iron mineral 铁矿物
iron minium 赭土
iron modulus 铝铁系数
iron monosulfide 一硫化铁；硫化亚铁 FeS
iron mordant 铁质媒染料
iron mould 铁斑
iron nail 铁钉
iron naphthenate 环烷酸铁

iron-nickel accumulator(=Edison storage battery; iron-nickel storage battery) 铁镍蓄电池；爱迪生蓄电池
iron-nickel storage battery(=iron-nickel accumulator) 铁镍蓄电池
iron nitride 一氮化二铁 $Fe_2N;Fe_4N$
iron ocher 赭石；黄土
iron octoate 辛酸铁
iron ore 铁矿
iron organism 铁微生物
iron out 消除；排除
iron oxide 氧化铁
iron oxide black 铁黑；氧化铁黑；四氧化三铁 Fe_3O_4
iron oxide brown 铁棕；氧化铁棕〔是铁红、铁黑、铁黄的混合颜料〕
iron oxide layer 氧化铁层
iron oxide pigment 氧化铁颜料
iron oxide process 氧化铁法〔用氧化铁脱除(石油)气体中硫化氢的过程〕
iron oxide purifier 氧化铁净化器
iron oxide red 铁红；氧化铁红；三氧化二铁 Fe_2O_3
iron oxide yellow 铁黄；氧化铁黄 $Fe_2O_3 \cdot H_2O$
iron paint mill 铁粉漆磨坊
iron pan 铁盘；铁锅
iron pentacarbonyl 五羰铁 $Fe(CO)_5$
iron period 铁器时代
iron phosphate 磷酸铁
iron phosphate coating process 磷酸铁膜处理法
iron phosphide 磷化铁
iron picrate 苦味酸铁
iron pills 含铁药丸
iron pipe 铁管
iron powder 铁粉
iron printing process 印铁法；印铁像；印铁画
iron protocarbonate 碳酸亚铁 $FeCO_3$
iron protochloride 氯化亚铁 $FeCl_2$
iron protosulfide 硫化亚铁 FeS
iron protoxide 氧化亚铁 FeO
iron pyrite 黄铁矿
iron red 铁红
iron resinate 树脂酸铁
iron ring 铁圈；铁环
iron rod 铁棍；铁棒
iron rubber 铁质橡胶
iron scale 铁屑
iron sesquioxide 三氧化二铁 Fe_2O_3
iron sesquisulfide 三硫化二铁 Fe_2S_3
iron shackle ①铁钩环②铁(制)桎梏
iron sheet 铁片；铁板

iron-silicate gel	硅酸铁胶
iron slag	铁熔渣
iron soap	铁皂
iron speck	铁斑
iron stain	铁斑
iron stand	铁架
iron stearate	硬脂酸铁
iron still	铁制蒸馏锅
ironstone	褐铁矿
iron storage battery	铁极蓄电池
iron sulfide	硫化铁
iron sulfide film	硫化铁膜
iron tallate	松浆油酸铁；树脂酸铁
iron tannage	铁鞣
iron tanned leather	铁鞣革
iron tanning	铁鞣
iron tetracarbonyl	四羰(合)铁 $Fe(CO)_4$
iron titanate	①钛酸铁②钛铁矿
iron titanium oxide	氧化钛铁
iron trichloride	三氯化铁 $FeCl_3$
iron trioxide	三氧化二铁 Fe_2O_3
iron vitriol	青矾；铁矾；七水硫酸亚铁 $FeSO_4 \cdot 7H_2O$
iron wax	铁蜡
iron wire	铁丝；铁铉
iron work	铁工厂；炼铁厂
iron yellow	铁黄；氧化铁黄
IR polarized spectrum	红外偏振光谱
IR pyrolysis spectrum	红外热解光谱
IR quenching phosphor	红外猝灭磷光体
irradiance	辐照度；照射度
irradiated coating	(射线)照射固化涂层
irradiated rubber	辐射硫化胶
irradiation	①光渗②照光；照射
irradiation capsule	照射盒；照射罐
irradiation chamber	辐照室
irradiation crosslinking	辐射交联
irradiation damage	辐射损伤
irradiation facility	辐照装置*
irradiation grafting	辐射接枝
γ-irradiation-initiated oxidation	γ辐照引发(的)氧化反应
irradiation polymerization	辐照聚合
irradiation source	辐照源
γ-irradiation technique	γ辐照技术
irradiation time	辐照时间；照射时间
irradiator	辐照器*；辐射体
irrational	不合理的
irrational synthesis	非示构合成；非有理合成
irrecoverable creep	不可复原蠕变
irrecoverable deformation	不可回复形变
irregular block	非规整嵌段*
irregular checking	不规则细裂
irregular close packing	不规则密堆积
irregular convection	不规则对流
irregular cure	不规则硫化
irregular folding	不规则折叠
irregular grain	不规则纹
irregular helical conformation	不规则螺旋形构象
irregularity index	不规则性指数
irregularly shaped	不规则形状的
irregular movement	不规则运动
irregular network	不规则网络
irregular pattern type cracking	不规则形开裂
irregular peak	不规则峰
irregular polymer	非规整聚合物
irregular porosity	不规则孔率
irregular-shaped insert	异型嵌件
irregular-shaped void	不规则状孔隙
irregular-type	不规则型(结晶)
irregular wear	不均匀磨耗
irreproducibility	非重现性
irreproducible	不能再得的；不能再制的
irreversibility	不可逆性
irreversible	不可逆的
irreversible adsorption	不可逆吸附(作用)
irreversible cell	不可逆电池
irreversible chemisorption	不可逆化学吸附
irreversible coagulation	不可逆凝固
irreversible collapse	不可逆皱缩；不可逆塌陷
irreversible colloid	不可逆胶体
irreversible cycle	不可逆循环
irreversible deformation	不可逆变形；永久变形
irreversible denaturation	不可逆变性(作用)
irreversible diffusion	不可逆扩散
irreversible electrode	不可逆电极*
irreversible electrode reaction	不可逆电极反应
irreversible fixation	不可逆结合(固定)
irreversible flow	不可逆流(动)
irreversible gel	不可逆凝胶
irreversible indicator	不可逆指示剂
irreversible inhibitor	不可逆抑制剂
irreversible oxidation reduction	不可逆氧化还原(反应)
irreversible polarographic wave	不可逆极谱波
irreversible potential	不可逆电位
irreversible process	不可逆过程*
irreversible reaction	不可逆反应*
irreversible shrinkage	不可逆收缩

irreversible swelling 不可逆溶胀
irreversible temperature-indicating coating 不可逆型示温涂料
irreversible temperature-indicating paint 不可逆性示温颜料
irreversible temperature -indicating pigment 不可逆性示温颜料
irreversible thermodynamics 不可逆热力学
irrigation ①冲洗法②灌溉；水利
irrigation hose 排灌胶管；灌溉胶管
irrigator ①灌溉器；灌溉车②冲洗器
irritability 过敏性
irritant gas 刺激性毒气
irritant refrigerant 刺激冷却；刺激冷冻
irrotational flow 无旋流
irrotational motion 非旋转运动
IR spectroscopy 红外光谱学
IRT method(=isotope ratio tracer) 同位素比示踪剂法
Irvine phosphorus process 伊尔文制磷法
Irving-Williams series 欧文-威廉斯序
isabnormal 等异常线
isaconitic acid(=isoaconitic acid) 异乌头三酸
isamic acid 衣氨酸 $C_{16}H_{13}N_3O_4$
isanic acid 生红酸；17-十八碳烯-9,11-二炔酸
isanolic acid 赤原酸；生红酸；羟基衣散酸
isano oil 衣散油〔树脂用阻燃剂〕
isano oil coating 衣散油涂料
isaphenic acid 衣酚酸 $C_{16}H_{11}NO_3$
isarol(=ichthammol) 鱼石脂
isatan 靛红烷 $C_{16}H_{12}O_3N_2$
isatic acid(=isatinic acid) 靛红酸；邻氨苯基乙酮酸 $NH_2C_6H_4COCOOH$
isatide 靛红偶 $C_{16}H_{12}N_2O_4$
isatidine 靛红定
isatin 靛红 $C_6H_4COCONH$
isatin anil 靛红缩苯胺 $C_6H_4COC=NC_6H$
└─NH─┘
isatin chloride 靛红化氯
isatinecic acid 靛红裂酸
isatinecine 靛红裂素
isatinic acid(=isatic acid) 靛红酸；邻氨苯基乙酮酸 $NH_2C_6H_4COCOOH$
isatinoxime 靛红肟*
isatogenic acid 靛红原酸 $C_9H_5O_4N$
isatoic acid N-羧氨基苯甲酸 $C_8H_7O_4N$
isatoic anhydride 衣萨酸酐；靛红酸酐
isatophan 甲氧阿托方

isatoxime 衣托肟 $C_8H_6O_2N_2$
isatropic acid 衣卓酸 $C_{18}H_{16}O_4$
I-section steel(=H-section steel) 工字钢
isemycin 伊势霉素
isenthalpic 等焓的
isenthalpic process 等焓过程*
isentropic 等熵的
isentropic change 等熵变化
isentropic compression(=constant entropy compression) 等熵压缩
isentropic flow 等熵流动
isentropic index 等熵指数
isentropic process 等熵过程*
isentropic static continuation 等熵静态延拓
isepamicin 异帕米星〔药〕
iserite 钛铁矿；钛铁砂
isethionate 羟乙(基)磺酸盐
isethionate surfactant 羟乙基磺酸盐型表面活性剂
isethionic acid 羟乙磺酸
isethionic acid ester 羟乙(基)磺酸酯盐
ishkyldite 绢蛇纹石
ishwarene 印马烯
ishwarol 印马醇
ishwarone 印马酮
isiganeite 硬锰矿
isinglass 鱼胶；明胶；云母
island in-sea bicomponent fibre 海岛型双组分纤维
island of superheavy nuclei 超重(核)岛*
islands-in-sea bicomponent fibre 天星型复合纤维；海岛型复合纤维
islands-in-sea type fibre 天星型纤维；海岛型纤维
iso- ①同；等②异
isoabietic acid 异松香酸
isoabsorption point 等吸光点
isoabsorption point-spectrophotometry 等吸收点分光光度法
isoabsorptive point(=isosbestic point) 等吸收点；等光点
isoabsorptive point spetrophotometry 等吸收点分光光度法
isoacetylene 异乙炔
isoaconitic acid(=isaconitic acid) 异乌头三酸
isoacorone 异菖蒲酮
isoadenine 异腺嘌呤
isoalantolactone 异阿兰内酯 $C_{15}H_{20}O_2$
isoalkane 异烷烃
isoalkene 异烯烃
isoalkyl 异烷基

17-isoallopregnane-3β,17β-diol　17-异别孕(甾)-3β,17β-二醇
isoalloxazine　异咯嗪
isoalloxazine adenine dinucleotide　异咯嗪腺苷二核苷酸
isoalloxazine mononucleotide　异咯嗪一核苷酸
isoallyl(=propenyl)　丙烯基
isoambrettolide　异黄葵内酯
isoammodendrine　异沙树碱
isoamoxy(=isopentyloxy)　异戊氧基
isoamyl(=isopentyl)　异戊基
isoamyl acetate　乙酸异戊酯；香蕉油〔俗〕
isoamyl acetic acid　异戊基乙酸；异庚酸　(C_5H_{11})CH_2COOH
isoamyl acetoacetate　乙酰乙酸异戊酯
isoamyl acetone　异戊基丙酮　$(CH_3)_2CHCH_2CH_2COCH_3$
isoamyl alcohol(=3-methyl butanol)　异戊醇；3-甲基丁醇
isoamyl amine　异戊胺　$(CH_3)_2CH(CH_2)_2NH_2$
isoamyl angelate　当归酸异戊酯
isoamyl aniline　异戊基苯胺　$C_6H_5NHCH_2CH(CH_3)_2$
isoamyl anisatate　大茴香酸异戊酯
isoamylase　异淀粉酶
isoamyl benzene　异戊基苯　$C_6H_5CH_2CH_2CH(CH_3)_2$
isoamyl benzoate　苯甲酸异戊酯
isoamyl benzyl ether　异戊基苄基醚　$C_6H_5CH_2OC_5H_{11}$
isoamyl benzyl ketone　异戊基苄基(甲)酮　$C_6H_5CH_2CO(CH_2)_2CH(CH_3)_2$
isoamyl bromide　异戊基溴　$(CH_3)_2CH(CH_2)_2Br$
isoamyl butyrate　丁酸异戊酯　$C_3H_7CO_2C_5H_{11}$
isoamyl caproate　己酸异戊酯　$C_5H_{11}CO_2(CH_2)_2CH(CH_3)_2$
isoamyl caprylate　辛酸异戊酯　$C_7H_{15}CO_2(CH_2)_2CH(CH_3)_2$
isoamyl carbamate　氨基甲酸异戊酯　$H_2NCO_2C_5H_{11}$
isoamyl chloride　异戊基氯　$(CH_3)_2CH(CH_2)_2Cl$
isoamyl chloroacetate　氯乙酸异戊酯　$ClCH_2COOC_5H_{11}$
isoamyl chlorocarbonate　氯甲酸异戊酯　$(CH_3)_2CH(CH_2)_2OCOCl$
isoamyl cinnamate　肉桂酸异戊酯
isoamyl cyanide(=isocapronitrile)　异戊基氰；异己腈　$(CH_3)_2CH(CH_2)_2CN$
isoamyl cyclopenenone　异戊基环戊烯酮
isoamyl dichlorarsine　二氯化异戊胂
isoamyl-4-dimethylamino benzoate　4-二甲氨基安息香酸异戊酯
isoamyl dodecanoate　十二烷酸异戊酯
isoamyl dodecylate　十二烷酸异戊酯
isoamylene　异戊烯

isoamyl ether　(二)异戊醚　$[(CH_3)_2CH(CH_2)_2]_2O$
isoamyl ethyl barbituric acid　异戊基乙基巴比妥酸
isoamyl formate　甲酸异戊酯　$HCO_2C_5H_{11}$
isoamyl 2-furanbutyrate　2-呋喃丁酸异戊酯
isoamyl 2-furanpropionate　2-呋喃丙酸异戊酯
isoamyl α-furfurylacetate　α-糠基乙酸异戊酯
isoamyl furfuryl propionate　糠基丙酸异戊酯
isoamyl heptin carbonate　辛炔酸异戊酯
isoamyl heptylate　庚酸异戊酯
isoamyl hexanoate　己酸异戊酯
isoamyl hexylate　己酸异戊酯
isoamyl 2-hydroxybenzoate　2-羟基苯甲酸异戊酯
isoamylidene(=isopentylidene)　亚戊基
isoamyl iodide　异戊基碘；1-碘-4-甲基丁烷　$(CH_3)_2CHCH_2CH_2I$
isoamyl isobutyrate　异丁酸异戊酯　$(CH_3)_2CHCO_2C_5H_{11}$
isoamyl isocyanide　异戊基胩
isoamyl isonitrile　异戊胩　$(CH_3)_2CH(CH_2)_2NC$
isoamyl isothiocyanate　异硫氰酸异戊酯　$(CH_3)_2CH(CH_2)_2N=CS$
isoamyl isovalerate　异戊酸异戊酯　$C_4H_9CO_2C_5H_{11}$
isoamyl β-ketobutyrate　β-丁酮酸异戊酯
isoamyl ketone　二异戊基(甲)酮　$(C_5H_{11})_2CO$
isoamyl α-ketopropionate　α-氧代丙酸异戊酯
isoamyl lactate　乳酸异戊酯
isoamyl laurate　月桂酸异戊酯；十二酸异戊酯
isoamyl-magnesium-bromide　异戊基溴化镁　$MgBrC_5H_{11}$
isoamyl-malonic acid　异戊基丙二酸　$C_5H_{11}CH(CO_2H)_2$
isoamyl mercaptan　异戊硫醇　$(CH_3)_2CHCH_2CH_2SH$
2-isoamyl-3-methyl-Δ²-cyclopentenone　2-异戊基-3-甲基环戊烯-2-酮
isoamyl naphthalene　异戊基萘　$C_5H_{11}C_{10}H_7$
isoamyl nitrate　硝酸异戊酯　$(CH_3)_2CHCH_2CH_2ONO_2$
isoamyl nitrite　亚硝酸异戊酯　$(CH_3)_2CHCH_2CH_2ONO$
isoamyl nonanoate　壬酸异戊酯
isoamyl nonylate　壬酸异戊酯
isoamyl octanoate　辛酸异戊酯
isoamyl octylate　辛酸异戊酯
isoamylol　异戊醇
isoamyl oleate　油酸异戊酯　$C_8H_{17}CH=CHC_7H_{14}CO_2C_5H_{11}$
isoamyl oxide　异戊醚　$C_5H_{11}OC_5H_{11}$
isoamyl 3-oxobutanoate　3-氧代丁酸异戊酯
isoamyl 2-oxopropanoate　2-氧代丙酸异戊酯
isoamyl oxyacetaldehyde　异戊基含氧乙醛
isoamyl pelargonate　壬酸异戊酯
isoamyl phenylacetate　苯乙酸异戊酯

isoamyl β-phenylacrylate　β-苯基丙烯酸异戊酯
isoamyl phenyl carbinol　异戊基苯基甲醇
　　$(CH_3)_2CH(CH_2)_2CHOHC_6H_5$
isoamyl phenyl ether　异戊基苯基醚；异戊氧基苯
　　$C_5H_{11}OC_6H_5$
isoamyl 3-phenylpropenoate　3-苯基丙烯酸异戊酯
isoamyl propionate　丙酸异戊酯　$C_2H_5CO_2C_5H_{11}$
isoamyl pyruvate　丙酮酸异戊酯
isoamyl salicylate　水杨酸异戊酯　$HOC_6H_4CO_2C_5H_{11}$
isoamyl stearate　硬脂酸异戊酯　$CH_3(CH_2)_{16}CO_2C_5H_{11}$
isoamyl sulfhydrate　异戊硫醇
isoamyl sulfide　异戊基硫醚；硫化异戊基　$(C_5H_{11})_2S$
isoamyl sulfone　(二)异戊基砜　$(C_5H_{11})_2SO_2$
isoamyl thioalcohol　异戊硫醇
isoamyl thiocyanate　硫氰酸异戊酯
　　$(CH_3)_2CH(CH_2)_2SCN$
isoamyl thioglycolate　巯基乙酸异戊酯
isoamyl α-toluate　苯乙酸异戊酯
isoamyl-triethoxy-silicane　异戊代三乙氧基硅
　　$C_5H_{11}Si(OC_2H_5)_3$
isoamyl-triethyl-silicane　异戊代三乙基硅
　　$(C_5H_{11})Si(C_2H_5)_3$
isoamyl-trimethyl-silicane　异戊代三甲基硅
　　$C_5H_{11}Si(CH_3)_3$
isoamyl ω-undecylenate　ω-十一烯酸异戊酯
isoamyl urea　异戊脲　$C_5H_{11}NHCONH_2$
isoamyranol　异白檀烷醇
isoamyrenonol　异白檀烯酮醇
isoandrosterone　异雄甾酮
isoanethole　异大茴香脑
isoangelic acid　异当归酸；顺-2-甲基丁烯酸
　　$CH_3CH=C(CH_3)COOH$
isoanthracene　异蒽　$C_{14}H_{10}$
isoanthraflavic acid　异蒽黄酸；2,7-二羟蒽醌
　　$C_{14}H_6O_2(OH)_2$
isoanthricin　异峨参内酯　$C_{22}H_{22}O_7 \cdot H_2O$
isoantibody　同种抗体；同族抗体
isoantigen　同种抗原；同族抗原
isoapiole　异芹菜脑　$C_{12}H_{14}O_4$
isoaristolochic acid　异马兜铃酸　$C_{17}H_{11}O_7N$
isoaristolone　异马兜铃酮　$C_{15}H_{24}O$
isoartemesia ketone　异蒿属(甲)酮
isoarticulatin　草棉苷
isoartocarpin　异桂木黄素
isoascorbic acid　异抗坏血酸；阿拉伯糖型抗坏血酸
isoatropic acid　异阿托酸
isoaxial　同轴的
isobar　同量异位素*

isobarbaloin　异芦荟苷
isobaric　①同(原子)量异序的②等压的
isobaric activation energy　等压活化能
isobaric atom　同量异序原子
isobaric chromatography　等压色谱法
isobaric heterope　同(原子)量异序元素
isobaric interference　同量异序干扰
isobaric ion　同质异构离子
isobaric mass-change determination　等压质量变化测定
isobaric process　等压过程*
isobaric spin　同位旋
isobarometric　等气压的
isobase　等基线
isobath　等深线
isobebeerine(=isochondodendrine)　异谷树碱
isobehenic acid　异山嵛酸；异二十二(碳)酸
isobergaptene　异香柠檬脑
isobestic point　等消点
isobetanidin　异甜菜素
isobiotol　异侧柏萜醇
isobisabolene　异红没药烯；异防风根烯
isoborneol　异冰片　$C_{10}H_{17}OH$
isobornyl　异冰片基　$C_{10}H_{17}-$
isobornyl acetate　乙酸异冰片酯　$CH_3CO_2C_{10}H_{17}$
isobornyl bromide　异冰片基溴　$C_{10}H_{17}Br$
isobornyl chloride　异冰片基氯　$C_{10}H_{17}Cl$
isobornylene　异冰片烯　$C_{10}H_{16}$
isobornyl formate　甲酸异冰片酯　$HCO_2C_{10}H_{17}$
isobornyl halide　异冰片基卤　$C_{10}H_{17}X$
isobornyl isovalerate　异戊酸异冰片酯
isobornyl methacrylate　甲基丙烯酸异冰片酯
isobornyl propionate　丙酸异冰片酯
isobornyl thiocyanoacetate　氰硫基乙酸异冰片酯
isobornyl triethanolamine maleate　异冰片基三乙醇胺马来酸盐
isobornyl valerate　戊酸异冰片酯
　　$CH_3(CH_2)_3CO_2C_{10}H_{17}$
isobutane　异丁烷　$(CH_3)_2CHCH_3$
isobutanol　异丁醇
isobutene　异丁烯　$(CH_3)_2C=CH_2$
isobutene-diene rubber　丁基橡胶
isobutene rubber　异丁烯橡胶；(聚)丁烯(合成)橡胶
isobutenyl(=2-methylpropenyl)　异丁烯基；2-甲基丙烯基
isobutenyl succinate　琥珀酸异丁烯酯
isobutoxy　异丁氧基　$(CH_3)_2CHCH_2O-$
isobutyl　异丁基　$(CH_3)_2CHCH_2-$
isobutyl acetate　乙酸异丁酯　$CH_3CO_2CH_2CH(CH_3)_2$
isobutyl acetophenone　异丁基苯乙酮

isobutyl acetylene 异丁基乙炔；异己炔 $(CH_3)_2CHCH_2C\equiv CH$
isobutyl acrylate 丙烯酸异丁酯
isobutyl alcohol 异丁醇 $(CH_3)_2CHCH_2OH$
isobutyl aldehyde 异丁醛
isobutyl allyl barbituric acid 异丁基烯丙基巴比妥酸 $C_4H_9(C_3H_2)=C_4H_2O_3N_2$
isobutyl allyl carbinol 异丁基烯丙基原醇
isobutyl amine 异丁胺 $(CH_3)_2CHCH_2NH_2$
isobutyl aminobenzoate 氨基苯甲酸异丁酯 $H_2NC_6H_4CO_2C_4H_9$
isobutyl 2-aminobenzoate 2-氨基苯甲酸异丁酯
isobutyl-p-aminophenol 异丁基对氨基苯酚〔石油防胶剂〕
isobutyl angelate 当归酸异丁酯
isobutyl-aniline N-异丁基苯胺 $C_6H_5NHC_4H_9$
isobutyl anthranilate 邻氨基苯甲酸异丁酯
isobutyl-benzene 异丁基苯 $C_6H_5C_4H_9$
isobutyl benzoate 苯甲酸异丁酯 $C_6H_5CO_2C_4H_9$
isobutyl benzyl carbinol 异丁基苄基原醇
isobutyl benzyl ether 异丁基苄醚
isobutyl benzyl ketone 异丁基苄基(甲)酮；异戊酰甲苯 $C_{12}H_{16}O$
isobutyl bromide 异丁基溴 $(CH_3)_2CHCH_2Br$
isobutyl 2-butenoate 2-丁烯酸异丁酯
isobutyl butyrate 丁酸异丁酯 $C_3H_7CO_2C_4H_9$
isobutylcapramide 异丁基癸酰胺
isobutyl caproate 己酸异丁酯
isobutyl capronate 己酸异丁酯
isobutyl carbamate 氨基甲酸异丁酯 $H_2NCO_2C_4H_9$
isobutyl carbinol 异丁基甲醇；异戊醇
isobutyl carbylamine(=isobutyl isonitrile) 异丁胩 $(CH_3)_2CHCH_2NC$
isobutyl chloride 异丁基氯 $(CH_3)_2CHCH_2Cl$
isobutyl chlorocarbonate 氯甲酸异丁酯 $ClCO_2C_4H_9$
isobutyl cinnamate 肉桂酸异丁酯
isobutyl coenzyme 异丁基辅酶
isobutyl crotonate 巴豆酸异丁酯
isobutyl cyanide(=isovaleronitrile) 异丁基氰；异戊腈 $(CH_3)_2CHCH_2CN$
isobutyl cis-α,β-dimethylacrylate 顺-α,β-二甲基丙烯酸异丁酯
1-isobutyl-3,5-dimethylhexylphosphoric acid(DDPA) 1-异丁基-3,5-二甲基己基磷酸；十二烷基磷酸酯 $(C_{12}H_{25}O)PO(OH)_2$
isobutylene ①异丁烯②异亚丁基
isobutylene-glycol 异亚丁基二醇；2-甲基-1,2-丙二醇 $(CH_3)_2COHCH_2OH$
isobutylene-isoprene copolymer 异丁烯-异戊二烯共聚物；丁基橡胶；异丁橡胶
isobutylene-isoprene latex 丁基胶乳
isobutylene-maleic anhydride copolymer 异丁烯-顺式丁烯二酸酐共聚物
isobutylene-oxide 1,1-二甲基环氧乙烷 $(CH_3)_2CCH_2O$
isobutyl ethyl malonate 丙二酸异丁基乙酯 $CH_2(CO_2C_4H_9)(CO_2C_2H_5)$
isobutyl fluoride 异丁基氟 $(CH_3)_2CHCH_2F$
isobutyl formate 甲酸异丁酯 $HCO_2CH_2CH(CH_3)_2$
isobutyl 2-furanpropionate 2-呋喃丙酸异丁酯
isobutyl furfuryl acetate 糠基乙酸异丁酯
isobutyl furoate 糠酸异丁酯
isobutyl furylpropionate 呋喃基丙酸异丁酯
isobutyl heptanoate 庚酸异丁酯
isobutyl heptylate 庚酸异丁酯
isobutyl hexanoate 己酸异丁酯
isobutyl o-hydroxybenzoate 邻羟基苯甲酸异丁酯
isobutylidene 亚异丁基 $(CH_3)_2CHCH=$
isobutylidene-acetone 亚异丁基丙酮 $CH_3COCH=CHCH(CH_3)_2$
isobutylidyne 次异丁基 $(CH_3)_2CHC\equiv$
isobutyl indole 异丁基吲哚
isobutyl iodide 异丁基碘 $(CH_3)_2CHCH_2I$
isobutyl isobutyrate 异丁酸异丁酯 $(CH_3)_2CHCO_2C_4H_9$
isobutyl isonitrile(=isobutyl carbylamine) 异丁胩 $(CH_3)_2CHCH_2NC$
isobutyl isothiocyanate(=isobutyl mustard oil) 异硫氰酸异丁酯；异丁芥子油 $(CH_3)_2CHCH_2N=CS$
isobutyl isovalerate 异戊酸异丁酯 $C_4H_9CO_2C_4H_9$
isobutyl β-ketobutyrate β-丁酮酸异丁酯
isobutyl ketone 异丁酮
isobutyl lactate 乳酸异丁酯 $CH_3CHOHCO_2C_4H_9$
isobutyl linalool 异丁基芳樟醇
isobutyl mercaptan 异丁硫醇 $(CH_3)_2CHCH_2SH$
isobutyl methacrylate 甲基丙烯酸异丁酯
2-isobutyl-3-methoxypyrazine 2-异丁基-3-甲氧吡嗪
isobutyl cis-2-methyl-2-butenoate 顺-2-甲基-2-丁烯酸异丁酯
isobutyl methyl glyoxal 异丁基甲基乙二醛
2-isobutyl-3-methylpyrazine 2-异丁基-3-甲吡嗪
isobutyl mustard oil(=isobutyl isothiocyanate) 异硫氰酸异丁酯；异丁芥子油 $(CH_3)_2CHCH_2N=CS$
isobutyl nitrate 硝酸异丁酯 $C_4H_9ONO_2$
isobutyl nitrite 亚硝酸异丁酯 $(CH_3)_2CHCH_2ONO$
isobutyl palmitate 棕榈酸异丁酯
isobutyl phenylacetate 苯乙酸异丁酯

$C_6H_5CH_2CO_2C_4H_9$
isobutyl phenyl ether   异丁基苯醚；异丁氧基苯 $C_6H_5OCH_2CH(CH_3)_2$
α-isobutyl phenylethyl alcohol   α-异丁基苯乙醇
isobutyl phenyl ketone   异丁基苯基(甲)酮；异戊酰苯 $C_4H_9COC_6H_5$
isobutyl propionate   丙酸异丁酯 $C_2H_5CO_2C_4H_9$
2-isobutyl pyridine   2-异丁基吡啶
isobutyl quinoline   异丁基喹啉
isobutyl ricinoleate   蓖酸异丁酯 $HOC_{17}H_{32}CO_2C_4H_9$
isobutyl salicylate(=isonefolia)   水杨酸异丁酯
isobutyl stearate   硬脂酸异丁酯 $C_{17}H_{35}CO_2C_4H_9$
2-isobutyl thiazole   2-异丁基噻唑
isobutyl thiocyanate   硫氰酸异丁酯 $(CH_3)_2CHCH_2SCN$
isobutyl-triethyl-silicane   异丁基三乙基硅 $C_4H_9Si(C_2H_5)_3$
isobutyl-trimethyl-silicane   异丁基三甲基硅 $C_4H_9Si(CH_3)_3$
N-isobutylundecylenamide   N-异丁基十一烯酰胺
isobutyl urea   异丁脲 $C_4H_9NHCONH_2$
isobutyl urethane   异丁基尿烷〔俗〕；异丁氨基甲酸乙酯 $C_4H_9NHCO_2C_2H_5$
isobutyl vinyl ether   异丁基乙烯醚
isobutyraldehyde   异丁醛 $(CH_3)_2CHCHO$
isobutyraldehyde diethyl acetal   异丁醛二乙缩醛
isobutyraldehyde diisoamyl acetal   异丁醛二异戊缩醛
isobutyraldehyde semicarbazone   异丁醛缩氨基脲 $C_3H_7CH=N_3CH_3O$
isobutyramide   异丁酰胺 $(CH_3)_2CHCONH_2$
isobutyrate   异丁酸盐(或酯)
isobutyric acid   异丁酸
isobutyric aldehyde   异丁醛
isobutyric anhydride   异丁酸酐
isobutyroin   异丁偶姻；2,5-二甲基-4-羟基-3-己酮 $(CH_3)_2CHCHOHCOCH(CH_3)_2$
isobutyrone(=diisopropyl ketone)   二异丙基(甲)酮 $[(CH_3)_2CH]_2CO$
isobutyronitrile   异丁腈 $(CH_3)_2CHCN$
isobutyrophenone(=isopropyl phenyl ketone)   异丁酰苯；异丙苯基(甲)酮 $(CH_3)_2CHCOC_6H_5$
isobutyropyrrothine   异丁二硫吡咯素
isobutyryl   异丁酰 $(CH_3)_2CHCO—$
isobutyryl chloride   异丁酰氯 $(CH_3)_2CHCOCl$
iso-cadinene   异荜澄烯；异杜松烯
isocamphane   异莰烷 $C_{10}H_{18}$
iso-camphenilol   异莰尼醇
iso-camphenilone   异莰尼酮
isocamphol   异莰醇；异龙脑

iso-campholic acid   异龙脑酸
isocamphor   异樟脑 $C_{10}H_{16}O$
isocamphoric acid   异樟脑酸 $C_8H_{14}(CO_2H)_2$
isocamphoronic acid   异樟脑酮酸〔异樟脑三酸〕 $C_9H_{14}O_6$
isocapric aldehyde   异癸醛 $(CH_3)_2CH(CH_2)_6CHO$
isocaproaldehyde   异己醛
isocaproic acid   异己酸 $(CH_3)_2CH(CH_2)_2CO_2H$
isocaproic aldehyde diethyl acetal   异己醛二乙缩醛
isocapronitrile(=isoamyl cyanide)   异戊基氰；异己腈 $(CH_3)_2CH(CH_2)_2CN$
isocaprylic acid   异辛酸
isocarthamidin   异红花素
isocarvestrene   异香芹萜烯
isocarvone   异香芹酮
isocaryophyllene   异丁子香烯
isocatalysis   异构催化(作用)
isocellobiose   异纤维二糖
isoceryl alcohol   异蜡醇；异二十六醇
isocetic acid   异鲸脑酸
isocetyl   异十六烷基
isochavibetol   异萎叶酚
isochemical indication   同化学性质指示
isocholesterin(=isocholesterol)   异胆甾醇 $C_{27}H_{45}OH$
isocholesterol(=isocholesterin)   异胆甾醇
isocholestryl benzoate   苯甲酸异胆甾醇酯 $C_6H_5CO_2C_{27}H_{45}$
isochondodendrine(=isobebeerine)   异谷树碱
isochore   等容线
isochoric activation energy   等容活化能
isochoric extension   等容延伸
isochoric flow   等容流动
isochoric motion   等容运动
isochoric process   等容过程*
isochroman musk   异色满麝香
isochromatic   等色的
isochromatic curve   等色曲线
isochronal viscoelastic function   等频(变温)黏弹性函数
isochrone   等时线
isochronism   等时性
isochronous   等时的
isochronous cyclotron   等时性回旋加速器
isocinnamalamide   异肉桂酰胺 $C_6H_5C_2H_2CONH_2$
isocinnamic acid   异肉桂酸 $C_6H_5CH=CHCOOH$
isocinnamic anhydride   异肉桂(酸)酐 $(C_6H_5CH=CHCO)_2O$
isocinnamyl chloride   异肉桂酰氯 $C_6H_5CH=CHCOCl$
isocitric acid   异柠檬酸；2-羟基-3-羧基戊二酸

HOOCCHOHCH(CO$_2$H)CH$_2$COOH

isoclinic parameter　等斜参数

isocolloid　同质异性胶体

isocomponent　异构化组分

isocompound　异构化合物

isoconjugate reaction　等共轭反应

isocorydine(=artabotrine;luteanine)　异紫堇定

l-isocorypalmine (=casealutine)　*l*-异紫杷明

isocoumarin　异香豆素　C$_6$H$_4$CH=CHOCO

isocrackate　异构裂化物

isocracking　异构裂化

isocratic　①恒溶剂成分的;恒溶剂组成的②恒溶剂强度的③等浓度的④溶剂成分不变的

isocratic chromatography　①恒溶剂成分色谱法②等度色谱法

isocratic condition　溶剂成分不变的(洗脱)条件

isocratic elution　等度洗脱*;恒溶剂洗脱;恒溶剂成分洗脱(法)

isocratic mode　恒溶剂模式

isocratic operation　恒溶剂成分操作

isocrotonate　异巴豆酸盐(或酯);异丁烯酸盐(酯)

iso-crotonic acid　异巴豆酸

isocrotyl bromide　异丁烯基溴　(CH$_3$)$_2$C=CHBr

isocrotyl chloride　异丁烯基氯　(CH$_3$)$_2$C=CHCl

isocrotyl halide　异丁烯基卤　(CH$_3$)$_2$C=CHX

isocrotyl iodide　异丁烯基碘　(CH$_3$)$_2$C=CHI

isocryptomerin　异柳杉素

isoctane　异辛烷

isocyan　异氰　(CN)

isocyanate　异氰酸盐(酯)

isocyanate adduct　异氰酸酯加合物

isocyanate adhesive　异氰酸酯黏合剂

isocyanate content　异氰酸酯含量

isocyanate equivalent　异氰酸酯当量

isocyanate foam　异氰酸酯泡沫塑料

isocyanate generator　异氰酸酯发生器

isocyanate oil　异氰酸酯改性油

isocyanate plastic(s)　异氰酸酯塑料

isocyanate prepolymer　异氰酸酯预聚物

isocyanate rubber　异氰酸酯橡胶

isocyanate rubber foam　异氰酸酯泡沫橡胶

isocyanate varnish　异氰酸酯清漆

isocyanato　异氰酸基　O=C=N—

isocyanato cyclohexane　异氰酸基环己烷

3-isocyanatomethyl-3,5,5-trimethyl-cyclohexyl isocyanate
　3-异氰酸甲酯基-3,5,5-三甲基环己烷异氰酸酯

isocyanic acid　异氰酸　O=C=NH

isocyanic ester　异氰酸酯

isocyanide　胩;异氰化物　RNC

isocyanilic acid　怪氰酸　NCC(NOH)C(NO$_2$H)CHNOH

isocyano　异氰基　C=N—

isocyanobenzene　异氰苯

isocyanurate　①异氰脲酸酯;三聚异氰酸酯②异氰脲酸盐;三聚异氰酸盐

isocyanuric acid(=fulminuric acid)　异氰脲酸;三聚异氰酸　NCCH(NO$_2$)CONH$_2$

isocyanuric acid ester　三聚异氰酸酯;异氰脲酸酯

isocyanuric acid methyl ester　三聚异氰酸甲酯;异氰脲酸甲酯

isocyanuric acid monoimide　三聚异氰酸一亚胺;异氰脲酸单亚胺

isocyanuric acid triimide　三聚异氰酸三亚胺;异氰脲酸三亚胺

isocyanurimide(=melanuric acid)　异氰尿酰亚胺;黑尿酸

isocyclene　异构环烯

isocyclenone　异环烯酮

isocyclic　①等节环(状)的②同素的③碳环的

isocyclic compound　①碳环化合物②等(原子数)环化合物

isocyclic stem-nucleus　等环母核

isocyclocitral　异环柠檬醛

isocycloheximide　异环己酰亚胺

isodecane(=2-methylnonane)　异癸烷;2-甲基壬烷　(CH$_3$)$_2$CH(CH$_2$)$_6$CH$_3$

isodecyl acrylate　丙烯酸异癸酯

isodecyl alcohol　异癸醇

isodecyl aldehyde　异癸醛

isodecylmethacrylate　甲基丙烯酸异癸酯

isodecyl octyl phthalate　邻苯二甲酸异癸辛酯〔增塑剂〕

isodecyl pelargonate　壬酸异癸酯〔增塑剂〕

isodecyl terephthalate　对苯二甲酸异癸酯〔增塑剂〕

isodensity centrifugation　同密度离心(法)

isodesmic　等键(的)

isodesmic structure　等链结构

isodesmosine　异锁链(赖氨)素

isodextropimaric acid　异(右旋)海松酸

isodextropimarinal　异海松醛

isodextropimarinol(=isopimarinol)　异海松醇

isodialuric acid　异径尿酸

isodiametric　等直径的

isodiapher　同差素〔过剩中子数相同的核素,*A*−2*Z* 相同〕

isodiaphere　等超额中子核素;同差素〔过剩中子数相同的核素,*A*−2*Z* 相同〕

isodiazotate　反重氮酸盐

isodibromosuccinic acid　异二溴丁二酸

COOHC$_2$H$_2$Br$_2$COOH
isodielectric solvent 等介电溶剂
isodihydro caryophyllene epoxide 异二氢环氧石竹烯
isodihydro lavandulal 可可醛；5-甲基-2-(1-甲基乙基)-2-己烯醛
isodihydroxybehenic acid 异二羟基山萮酸
isodimorphism 同二晶(现象)
isodisperse 等分数
isodispersity 等(OH 值)分散性
isododecyl phenol EO 异十二烷基酚聚氧乙烯醚
isodromic controller 等速控制器
isodulcite(=rhamnose) 鼠李糖
isodurene(=1,2,3,5-tetramethylbenzene) 异杜烯；1,2,3,5-四甲基苯；偏四甲苯　(CH$_3$)$_4$C$_6$H$_2$
isodynamic ①等力的②放出等能的③等磁力的
isodynamic law 等力(定)律
isoelectric focusing electrophoresis(IFE) 等电聚焦电泳
isoelectric fractionation 等电点分离
isoelectric heating 等电加热
isoelectric pH value 等电点 pH
isoelectric precipitation 等电沉淀(作用)
isoelectric zone 等电区
isoelectrofocusing 等电聚焦
isoelectronic 等电子的
isoelectronic atom 等电子原子
isoelutropic solvent 同洗脱效果的溶剂
17-isoemicymarin 17-异厄米磁麻苷
isoemodin 异大黄素
isoenantic differentiating temperature 等对映辨识温度
isoenergetic 等能的
isoenthalpic reaction series 等焓反应系列
isoenthalpy 等焓
isoenthalpy change 等焓变化
isoentropic change 等熵变化
isoentropic process 等熵过程
isoentropic reaction series 等熵反应系列
isoentropic tangential modulus 等熵切向模量
isoephedrine (=d-pseudoephedrine) d-假麻黄碱
iso-epitaxy 同质外延
isoequilenin 异马萘雌酮　C$_{18}$H$_{18}$O$_2$
isoequilibrium relationship 等平衡关系
isoerucic acid 异芥酸
isoerythrolaccin 异红紫胶素
isoestragol(=anethol) 大茴香脑
isoestragole 异草蒿脑；茴香脑
isoeugenol 异丁子香酚；对丙烯基邻甲氧基苯酚　C$_6$H$_3$(C$_3$H$_5$)(OCH$_3$)OH
isoeugenol acetate 乙酸异丁子香酚酯
isoeugenol benzoate 苯甲酸丁子香酚酯
isoeugenol benzyl ether 异丁子香酚苄醚
isoeugenol formate 甲酸异丁子香酚酯
isoeugenol methyl ether 异丁子香酚甲醚
isoeugenol phenylacetate 苯乙酸异丁子香酚酯
isofebrifugine 异黄常山碱
isofenchone 异葑酮；异小茴香酮
isofenchyl alcohol 异葑醇　C$_{10}$H$_{18}$O
isoferulic acid 异阿魏酸　CH$_3$OC$_6$H$_3$(OH)CH=CHCOOH
isoflavone 异黄酮*
isoflow type heater 立式圆筒型加热器
isoflurane 异氟烷〔药〕
isofoam contour 等泡线
isoformate 异构重整产物
isoforming 异构重整
isoforming process 异构重整过程
iso-free volume state 等自由体积状态
isogalloflarin 异栝黄素
isogel 等凝胶；同构胃量质凝胶
isogeneic 同基因的
isogeraniol 异牻牛儿醇　C$_{10}$H$_{18}$O
isoglutamine 异谷氨酰胺
isogram 等值线(图)；等高线(图)；等吸光度谱
isoguanine 异鸟嘌呤
isohelenalin 异菊内酯
isohemipinic acid 异半蒎酸
isoheptane(=2-methylhexane) 异庚烷；2-甲基己烷　(CH$_3$)$_2$CHC$_4$H$_9$
isoheptenoic acid 异庚烯酸；3-异丁基丙烯酸　(CH$_3$)$_2$CHCH$_2$CH=CHCOOH
isohexacosane 异二十六(碳)烷　(CH$_3$)$_2$CH(CH$_2$)$_{22}$CH$_3$
isohexane 异己烷　(CH$_3$)$_2$CH(CH$_2$)$_2$CH$_3$
isohexenal 异己烯醛
isohexenoic acid(=3-isopropylacrylic acid) 异己烯酸；3-异丙基丙烯酸　(CH$_3$)$_2$CHCH=CHCOOH
isohexyl 异己基　(CH$_3$)$_2$CH(CH$_2$)$_2$CH$_2$—
isohexyl acetate 乙酸异己酯　CH$_3$CO$_2$C$_4$H$_7$(CH$_3$)$_2$
isohexyl bromide(=5-bromo-2-methylpentane) 异己基溴；5-溴-2-甲基戊烷　(CH$_3$)$_2$CH(CH$_2$)$_2$CH$_2$Br
isohexylidene 亚异己基　(CH$_3$)$_2$CH(CH$_2$)$_2$CH=
isohexylidyne 次异己基　(CH$_3$)$_2$CH(CH$_2$)$_2$C≡
isohexyl naphthalene sulfonate 异己基萘磺酸盐
isohormone 同功激素
isohumulone 异葎草酮
isohydric 等氢离子的
isohydric concentration 等氢离子浓度
isohydric indicator solution 等氢(离子)指示(剂溶)液

isohydric shift　等氢离子转移
isohydric solution　等氢离子溶液
isohydric solvent　等水溶剂
isohydric technique　等氢(离子)技术
isohydrobenzoin　异氢化苯偶姻；偕二苯基乙二醇 $C_{14}H_{12}(OH)_2$
isohydrocarbon　异构烃
isohydrosorbic acid　异氢化山梨酸；异氢化己二烯酸
isohydroxystearic acid　异羟基硬脂酸
isohygrens　等水分线〔皂块失水试验〕
isoimmunization　同种免疫
isoimperatonin　异王草因
isoimperatorin　异前胡醚
isoindigo　异靛
isoindole　异吲哚
3-isoindoline　3-异吲哚啉；二氢异吲哚
1-isoindolinone　异吲哚啉-1-酮
isoindolinyl　异二氢吲哚基；异二氢氮(杂)茚基 $NC_8H_8—$
isoindolyl　异吲哚基；异氮(杂)茚基 $NC_8H_6—$
isoinversion　等反转*
isoionic point　等离子点
isoionic protein　等离子蛋白质
isoionic state　等离子状态
isoionic value　等离子值
isoionone　异紫罗兰酮
isojasmone　异茉莉酮
Iso-Kel process　埃索-凯尔法〔固定床气相异构化法〕
iso-ketopinic acid　异酮蒎酸
isokinetic　等动力学的；等动力的
iso kinetic relationship　等动力学关系
isokinetic sampling　等动力取样；等速采样；等流态取样
isokinetic temperature　等动力学温度
isokinetin　异激动素；2-呋喃甲氨基嘌呤
isokit　同位素药盒；同位素药箱
isokom　等黏线
isolactose　异乳糖
isolated　①孤立的；隔离的②离析出(来)的
isolated-cluster crystal　隔离簇晶体
isolated coil　孤立线团
isolated double bond　孤立双键
isolated electrode(=retarded electrode)　隔离电极
isolated hydrocarbon ring　孤立烃环
isolated mask　绝缘防毒器
isolated polymer molecule　孤立高分子
isolated reaction　孤立反应
isolated system　隔离系统*
isolating　离析的；分离的

isolating valve　隔离阀
isolation　分离；离析；绝缘；隔绝
isolation medium　分离培养基
isolation of pure hydrocarbons　纯烃的离析
isolation valve　隔离阀
isolenic acid(=isolinolenic acid)　异亚麻酸；异-9,12,15-十八碳三烯酸
isoleucine(Ile)　异亮氨酸
isoleucyl　异亮氨酰(基)　$C_2H_5CH(CH_3)CH(NH_2)CO—$
isolimonene　异柠檬烯
isolinderene　异钓樟烯　$C_8H_{14}O_2$
isolinolenic acid(=isolenic acid)　异亚麻酸；异-9,12,15-十八碳三烯酸
isoliquiritigenin　异甘草素；异甘草黄酮
isolog(=isologue)　同构(异素)体
isologous　相同的；同基因的
isologous series　同构(异素)系
isologue(=isolog)　同构(异素)体
isolongifanone　异长叶烷酮
isolongifolenyl acetate　乙酸异长叶烯酯
isolongifolic acid　异长叶酸
isolycopodine　异石松碱
isomaltol　异麦芽酚
isomaltose　异麦芽糖；6-葡萄糖-α-葡萄糖苷
isomaltose-like　类异麦芽糖
isomannide laurate　月桂酸异二缩甘露醇酯
isomannide stearate　硬脂酸异二缩甘露醇酯
isomannite　异甘露醇　$C_6H_{10}O_4$
isomanoene　异泪柏烯
isomate　异构产品；异构化的石油产品
isomate process　(低辛烷值烃)异构过程
Isomax process　埃索麦克斯加氢裂化法
iso-melamine(=imino-melamine)　异三聚氰胺；亚氨型三聚氰胺；异氰脲酸三亚胺
isomenthol　异薄荷醇　$C_{10}H_{20}O$
isomenthone　异薄荷酮　$C_{10}H_{18}O$
isomer　①(同分)异构体；(同分)异构物②同质异能素
isomerate process　固定床异构法
isomer distribution　异构体分布(比例)
isomer generator　同质异能素发生器
isomeric(=isomerical)　(同分)异构的
isomerical(=isomeric)　(同分)异构的
isomeric catalysis　异构催化(作用)
isomeric change　异构化
isomeric compound　异构(化合)物
isomeric hydrocarbon　异构烃
isomeric polymer　异构聚合物
isomeric pseudo-component　假异构体

isomeric state 同分异构态
isomeric three-dimensional projection 等角三维投影图
isomeric transition(IT) 同质异能跃迁
isomeric yield ratio 同质异能素产额比
isomeride 异构体
DL-isomerism DL 异构现象〔一般指光学异构现象，D 指右旋性；L 指左旋性〕
isomerism 异构(现象)
isomerization 异构化
isomerization equilibrium 异构化平衡
isomerization heat 异构热
isomerization method 异构化方法
isomerization of paraffins 烷烃的异构化
isomerization polymerization 异构化聚合*
isomerization process 异构化过程
isomerization reaction 异构化反应
isomerization rubber 异构化橡胶
isomerized 异构化了的
isomerized natural rubber 异构化天然橡胶
isomerized oil 异构化油
isomerized oligomer 异构化低聚物
isomerized rubber 异构化橡校
isomerizing agent 异构化剂
isomerose 异构糖
isomerous 同分异构的；同质异能的
isomer ratio 异构体比率
isomer separation ①(同分)异构体②同质异能素分离
isomer shift 异构体位移
isomery (同分)异构(现象)
isomethyl ionone(=iralia) 异甲基紫罗兰酮
isomethyl retroionone 异甲基逆紫罗兰酮
isometric 等容的
isometric annealing ①等容热处理②定长热处理③等容退火(作用)
isometric heating curve 等容加热曲线
isometric structure (同质)异能结构
isometric system(=cubic system) 等轴晶系
isometry ①等轴(现象)②同分异构(现象)
isomorph 同形体
isomorphic 同形的
isomorphism 类质同晶*
isomorphous 同晶的；同形的
isomorphous compound 同晶化合物
isomorphous elements 同晶(型)元素
isomorphous mixture 同晶混合物
isomorphous polymer 同晶型聚合物
isomorphous replacement (method) 同晶置换(法)*
isomorphous substitution 同晶取代；同晶置换

isomorphous system 同晶系
isomycomycin 异菌霉素
isomyristicin(=isomyristin) 异肉豆蔻素；异肉豆蔻醚
isomyristin(=isomyristicin) 异肉豆蔻素；异肉豆蔻醚
isomyrtanol 异桃金娘烷醇
isonaphthazarin 异萘茜；2,3-二羟基-1,4-萘醌 $(HO)_2C_{10}H_4O_2$
isonefolia(=isobutyl salicylate) 水杨酸异丁酯
isoniazide 异烟肼
isonicotine 异烟碱
isonicotinic acid(=γ-picolinic acid) 异烟酸；γ-吡啶甲酸
isonicotinyl hydrazine 异烟(酰)肼
isonitrile(=carbylamine) 异腈；胩
isonitro(=aci-nitro) 异硝基 $(HO)ON=$
isonitroso(=hydroxyimino) 肟基；异亚硝基 $HON=$
isonitrosoacetone 肟基丙酮 $CH_3COCH=NOH$
isonitrosocamphor 异亚硝基樟脑；肟基樟脑
isonitrosoethyl amyl ketone 肟乙基戊基(甲)酮；1-肟基-3-辛酮 $C_8H_{15}O_2N$
isonitrosomethyl ethyl ketone(=diacetylmonoxime) 肟甲基乙基(甲)酮 $CH_3COC(=NOH)CH_3$
isonitrosomethyl n-hexyl ketone 肟甲基正己基(甲)酮；3-肟基-2-辛酮 $CH_3COC(=NOH)(CH_2)_4CH_3$
isonitrosomethyl n-propyl ketone 肟甲基正丙基甲酮；3-肟基-2-戊酮 $CH_3COC(=NOH)C_2H_5$
isonitroso-propiophenone 肟甲苯基(甲)酮 $C_6H_5COC(=NOH)CH_3$
isonitrosothiocamphor 肟基硫代樟脑
isononane(=2-methyl-octane) 异壬烷；2-甲基辛烷 $(CH_3)_2CHC_6H_{13}$
isononyl acetate 乙酸异壬酯
isooctadecane 异十八(碳)烷
isooctane 异辛烷
isooctyl adipate 己二酸异辛酯
isooctyl alcohol 异辛醇
isooctyl isodecyl adipate 己二酸异辛异癸酯〔增塑剂〕
isooctyl palmitate 棕榈酸异辛酯
isooctyl phenol 异辛基苯酚
isooctyl phenol nonaethylene glycol ether 异辛基酚九乙二醇醚
isooctyl tallate 脂肪酸异辛酯〔增塑剂〕
isoolefine 异烯烃
isooleic acid 异油酸
isoosmotic 等渗的
isoosmoticity 等渗(透压)
isoosmotic pressure 等渗压
isoosmotic solution 等渗(压)溶液
isopan emulsion 等全色乳剂

isoparaffic solvent 异链烷烃溶剂
isoparaffin 异链烷烃
isoparaffin synthesis 异链烷烃合成；异构合成
l-α-isoparteine (=genisteine) l-α-异鹰爪豆碱
isopatchoulane 异广藿香烷
dl-isopelletierine dl-异石榴碱
isopenicillin 异青霉素
isopentane(=2-methylbutane) 异戊烷；2-甲基丁烷 $(CH_3)_2CHC_2H_5$
isopentane separation 异戊烷分离
isopentanethiol 异戊硫醇
isopentanized 含异戊烷的
isopentanizing 加入异戊烷
isopentanoate 异戊酸酯
isopentanoic acid 异戊酸
isopentene 异戊烯 C_5H_{10}
isopentene group 异戊烯基
isopentenyl(=3-methyl-1-butenyl) 异戊烯基
N^6-isopentenyladenine N^6-异戊烯腺嘌呤；玉米素
isopentenylputrescine 异戊烯基腐肉胺
isopentenylpyrophosphate 异戊烯焦磷酸
isopentyl(=isoamyl) 异戊基 $(CH_3)_2CHCH_2CH_2$—
isopentyl 2-furanpropionate 2-呋喃丙酸异戊酯
isopentyl hexanoate 己酸异戊酯
isopentylidene(=isoamylidene) 亚异戊基 $(CH_3)_2CHCH_2CH=$
isopentylidyne 次异戊基 $(CH_3)_2CHCH_2C\equiv$
isopentyl laurate 月桂酸异戊酯
isopentyl mercaptan 异戊硫醇
isopentyl nonanoate 壬酸异戊酯
isopentyloxy(=isoamoxy) 异戊氧基 $(CH_3)_2CHCH_2CH_2O$—
isopentyl phenylacetate 苯乙酸异戊酯
isopentyl propionate 丙酸异戊酯
isopentyl pyruvate 丙酮酸异戊酯
isoperibolic calorimeter 等环境热量计
isoperiplocymarin 异杠柳磁麻苷
isoperiplogenin 异杠柳配基
isophan(e)insulin 鱼精蛋白锌胰岛素
isophorone 异佛尔酮；3,5,5-三甲基-2-环己烯-1-酮
isophorone diamine 异佛尔酮二胺
isophorone diisocyanate(IPDI) 异佛尔酮二异氰酸酯
isophthalal(=isophthalylidene;m-phenylenedimethylidyne) 间苯二亚甲基
isophthalamic acid 间苯二甲氨酸
isophthalate 间苯二酸酯
isophthalic acid(=m-phthalic acid) 间苯二酸 $C_6H_4(CO_2H)_2$
isophthalic aldehyde 间苯二醛 $C_6H_4(CHO)_2$
isophthalic alkyd resin 间苯二(甲)酸醇酸树脂
isophthalic dihydrazide 间苯二甲酰肼
isophthalic ester 间苯二酸酯；异酞酸酯
isophthalonitrile 间苯二氰 $C_6H_4(CN)_2$
isophthaloyl 间苯二酰 —COC_6H_4CO—(m)
isophthaloyl diacid chloride 间苯二酰氯
isophthalyl chloride 间苯二酰氯 $C_6H_4(COCl)_2$
isophthalylidene(=isophthalal;m-phenylenedimethylidyne) 间苯二亚甲基
isophyllocladene 异扁枝烯
isophytol 异植醇；异叶绿醇
isopiestic distillation 等压蒸馏
isopiestic line 等压线
isopiestic method 等压法
isopiestic pressure 等压；恒压
isopilocarpine 异毛果芸香碱
isopimaric acid 异海松酸
isopimarinol(=isodextropimarinol) 异海松醇
isopimelic acid 异庚二酸；2-甲基-2-乙基丁二酸 $C_2H_5(CH_3)CCH_2(CO_2H)_2$
isopimpinellin 异茴芹灵
isopinocampheol 异松涨醇
isopinocamphone 异松涨酮
isopiperitenone 异胡椒烯酮
isopleric line 不等体积线
isopleth ①等值线②等浓度线
isoploid 同倍体；偶倍体
Iso-plus Houdryforming 配套重整〔石油〕
isopolyacid 同多酸
isopolybase 同多碱
isopolymorphism 同多形性
isopolynuclear coordination compound 同多核配合物
isopolyoxometallate 同多金属氧酸盐
isopotential point 等电位点
isoprenaline 异丙肾上腺素〔药〕
isoprene 异戊二烯；2-甲基-1,3-丁二烯 $CH_2=CHC(CH_3)=CH_2$
isoprene-acrylonitrile latex 异戊二烯-丙烯腈胶乳
isoprene alcohol 异戊二烯醇 $CH_2=C(CH_3)CH=CHOH$
isoprene-isobutylene rubber(=butyl rubber) 丁基橡胶
isoprene rubber 异戊二烯橡胶；异戊橡胶
isoprene rubber latex 异戊胶乳
isoprene rule 异戊二烯规则
isoprenoid 类异戊二烯
isoprobability 等概率
isopropanol 异丙醇

isopropanol amide 异丙醇酰胺
isopropanolamine(=α-amino isopropyl alcohol) 异丙醇胺；α-氨基异丙醇 $NH_2CH_2CHOHCH_3$
isopropanol-water mixture 异丙醇-水混合物
isopropenyl(=1-methylvinyl) 1-甲代乙烯基；异丙烯基 $CH_2=C(CH_3)-$
isopropenyl acetate 异丙烯基乙酸酯；乙酸异丙烯酯
isopropenylbenzene 异丙烯基苯
isopropenylbenzene sulphonic acid 异丙烯基苯磺酸
isopropenyl carborane 异丙烯(基)碳硼烷
isopropenyl carborane carboxylic acid 异丙烯(基)碳硼烷羧酸
isopropenyl cyclo-2-hexenyl ketone 异丙烯基环-2-己烯酮
5-isopropenyl-2-norbornene 5-异丙烯基-2-降冰片烯
2-isopropenylpyrazine 2-异丙烯基吡嗪
isopropoxy 异丙氧基 $(CH_3)_2CHO-$
p-isopropoxy-diphenylamine 对异丙氧基二苯胺
isopropoxytitanium triisostearate 三异硬脂酸三异丙氧基钛
isopropyl 异丙基 $(CH_3)_2CH-$
isopropyl acetate 乙酸异丙酯
isopropyl acetone 异丙酮；甲基异丁酮
isopropyl-acetylene 异丙基乙炔；异戊炔 $(CH_3)_2CHC\equiv CH$
3-isopropylacrylic acid(=isohexenoic acid) 异己烯酸；3-异丙基丙烯酸 $(CH_3)_2CHCH=CHCOOH$
isopropyl alcohol 异丙醇；2-丙醇 $(CH_3)_2CHOH$
isopropyl alcohol process for acetone 异丙醇法生产丙酮的工艺
isopropyl allyl-barbituric acid 异丙基烯丙基巴比妥酸 $(C_3H_5)(C_3H_7)C_4H_2O_3N_2$
isopropyl amine(=2-aminopropane) 异丙胺；2-丙胺 $(CH_3)_2CHNH_2$
4-isopropylamine-diphenylamine 4-异丙胺二苯胺〔防老剂〕
isopropylaminoethanol 异丙基氨基乙醇〔乳化剂〕
isopropyl n-amyl ketone 异丙基正戊基(甲)酮；2-甲基-3-辛酮 $CH_3(CH_2)_4COCH(CH_3)_2$
isopropyl aniline 异丙基苯胺 $(CH_3)_2CHC_6H_4NH_2$
isopropylation 异丙基化(作用)
isopropyl benzene(=cumene) 异丙苯；枯烯 $C_6H_5CH(CH_3)_2$
isopropyl benzene hydroperoxide 异丙苯过氧化氢
isopropyl benzoate 苯甲酸异丙酯 $C_6H_5CO_2CH(CH_3)_2$
p-isopropylbenzoyl 对异丙苯甲酰 $p-(CH_3)_2CHC_6H_4CO-$
ar-isopropylbenzyl 异丙苄基 $(CH_3)_2CHC_6H_4CH_2-$
isopropyl benzyl carbinol 异丙基苄基原醇
isopropyl benzyl ether 异丙基苄醚

ar-isopropylbenzylidene 异丙亚苄基 $(CH_3)_2CHC_6H_4CH=$
p-isopropylbenzylidene 对异丙亚苄基
isopropyl benzyl ketone 异丙基苄基(甲)酮 $C_6H_5CH_2COCH(CH_3)_2$
isopropyl bishexadecyl borate 双十六烷基硼酸异丙酯
isopropyl bromide(=2-bromopropane) 异丙基溴；2-溴丙烷 $CH_3CHBrCH_3$
isopropyl n-butyl ketone 异丙基正丁基(甲)酮；2-甲基-3-庚酮 $(CH_3)_2CHCO(CH_2)_3CH_3$
isopropyl tert-butyl ketone 异丙基叔丁基(甲)酮；2,2,4-三甲基-3-戊酮 $(CH_3)_2CHCOC(CH_3)_3$
isopropyl butyrate 丁酸异丙酯 $C_3H_7CO_2C_3H_7$
isopropyl carbylamine(=isopropyl isenitrile) 异丙肼 $(CH_3)_2CHN=C$
isopropyl cellosolve 异丙基溶纤剂
isopropyl chloride(=2-chloropropane) 2-氯丙烷 $CH_3CHClCH_3$
isopropyl chlorocarbonate 氯甲酸异丙酯 $ClCO_2CH(CH_3)_2$
α-isopropyl cinnamic aldehyde α-异丙基(肉)桂醛
isopropylcyclohexane(=hexahydrocumene) 六氢化枯烯；异丙基环己烷 $C_3H_7C_6H_{11}$
isopropyldithiobenzothiazole 异丙基二硫代苯并噻唑〔促进剂〕
isopropyl ether 异丙醚 $ROCH(CH_3)_2$; $(CH_3)_2CHOCH(CH_3)_2$
isopropyl-ethylene(=2-methyl-3-butene) 异丙基乙烯 $(CH_3)_2CHCH=CH_2$
isopropyl fluoride(=2-fluoropropane) 异丙基氟；2-氟丙烷 CH_3CHFCH_3
isopropyl formate 甲酸异丙酯 $HCO_2CH(CH_3)_2$
isopropyl furoate 糠酸异丙酯
isopropyl halide 异丙基卤 $(CH_3)_2CHX$
isopropyl hexanoate 己酸异丙酯
isopropyl hexylate 己酸异丙酯
isopropyl n-hexyl ketone 异丙基正己基(甲)酮 $(CH_3)_2CHCOC_6H_{13}$
p-isopropylhydrocinnamaldehyde 对异丙基氢化肉桂醛
4-isopropyl-2-hydroxy-2-methyl-propiophenone 4-异丙基-2-羟基-2-甲基-苯丙酮
isopropylidene 异亚丙基 $(CH_3)_2C=$
isopropylidene acetone 异亚丙基丙酮
isopropylidene bis-o-cresol 亚异丙基邻甲酚〔防老剂〕
N-isopropylidinemethylamine N-亚异丙基甲胺
2-iso-propylimidazole 2-异丙基咪唑
isopropyl iodide(=2-iodopropane) 2-碘丙烷；异丙基碘 CH_3CHICH_3

isopropyl isenitrile(=isopropyl carbylamine)　异丙胩　$(CH_3)_2CHN=C$

isopropyl isobutyrate　异丁酸异丙酯　$(CH_3)_2CHCO_2CH(CH_3)_2$

isopropyl isonitrile(=isopropyl carbylamine)　异丙胩　$(CH_3)_2CHN=C$

isopropyl isostearoyl diacryl titanate　异硬脂酰二丙烯酰钛酸异丙酯

isopropyl isovalerate　异戊酸异丙酯　$C_4H_9CO_2CH(CH_3)_2$

isopropyl lactate　乳酸异丙酯　$C_2H_4(OH)CO_2C_3H_7$

isopropyl lauryl-myristyl dimethacryl titanate　十二烷基-肉豆蔻(酰)基二甲基丙烯酰钛酸异丙酯

isopropyl-malonic acid　异丙基丙二酸　$C_3H_7CH(CO_2H)_2$

isopropyl mercaptan　异丙硫醇　$(CH_3)_2CHSH$

isopropyl mercaptan copper　异丙基硫醇铜

isopropyl methacrylate　甲基丙烯酸异丙酯

isopropyl methacryl titanate　甲基丙烯酰钛酸异丙酯

isopropyl α-methyl crotonate　α-甲基巴豆酸异丙酯

2-isopropyl-5-methyl-cyclohexanone　2-异丙基-5-甲基环己酮

2-isopropyl-5-methyl-2-hexenal　2-异丙基-5-甲基-2-己烯醛

isopropyl monomethyl-p-aminophenol　异丙基甲基对氨基苯酚〔汽油防胶剂〕

isopropyl mustard oil　异丙基芥子油　$(CH_3)_2CHNCS$

isopropyl myristate　肉豆蔻酸异丙酯

isopropyl naphthalene　异丙基萘

isopropyl α-naphthyl ketone　异丙基α-萘基(甲)酮　$(CH_3)_2CHCOC_{10}H_7$

isopropyl nitrate　硝酸异丙酯

isopropyl percarbonate　过碳酸异丙酯

isopropyl-phenol　异丙基苯酚

4-isopropyl phenyl acetaldehyde　4-异丙基苯乙醛

isopropyl β-phenylacrylate　β-苯基丙烯酸异丙酯

isopropylphenylbiphenyl oxadiazole　异丙苯基联苯基噁二唑　$C_6H_5C_6H_4C_2N_2OC_6H_4CH(CH_3)_2$

isopropyl phenyl carbinol　异丙基苯基甲醇；1-苯代异丁醇　$C_3H_7CHOHC_6H_5$

isopropyl phenyl ketone(=isobutyrophenone)　异丁酰苯；异丙基苯基(甲)酮　$(CH_3)_2CHCOC_6H_5$

3-(p-isopropylphenyl)-propionaldehyde　3-(对异丙苯基)-丙醛

isopropyl phenyl salicylate　水杨酸异丙苯酯

isopropyl propionate　丙酸异丙酯　$C_2H_5CO_2CH(CH_3)_2$

isopropyl pyridine　异丙基吡啶

6-isopropyl quinoline(=lichenol)　6-异丙基喹啉

isopropyl salicylate　水杨酸异丙酯　$HOC_6H_4CO_2C_3H_7$

isopropyl thiocyanate　硫氰酸异丙酯　$(CH_3)_2CHSCN$

isopropyl-β-D-thiogalactoside(IPTG)　异丙基-β-D-硫代半乳糖苷

isopropyl titanate　钛酸异丙酯；钛酸四异丙酯

isopropyl α-toluate　α-苯乙酸异丙酯

isopropyl m-tolyl ketone　异丙基间甲苯基(甲)酮　$CH_3C_6H_4COCH(CH_3)_2$

isopropyl triacryl titanate　三丙烯酰钛酸异丙酯

isopropyl tri(N-aminoethyl-aminoethyl) titanate　异丙基三(N-氨乙基-氨乙基)钛酸酯

isopropyl trianthranil titanate　三氨茴酰钛酸异丙酯

isopropyl tricumylphenyltitanate　三枯基苯钛酸异丙酯

isopropyl tri(dibutyl pyrophosphate) titanate　三(焦磷酸二丁酯)钛酸异丙酯

isopropyl tri(N,N-dimethyl ethylamino) titanate　三(N,N-二甲基乙氨基)钛酸异丙酯

isopropyl tri(dioctyl phosphate) titanate　三(磷酸二辛酯)钛酸异丙酯

isopropyl tri (dioctylphosphato) titanate　三(二辛基磷酰氧基)钛酸异丙酯〔偶联剂〕

isopropyl tri (dioctylpyrophosphato) titanate　三(二辛基焦磷酰氧基)钛酸异丙酯〔偶联剂〕

isopropyl tridodecyl benzesulfonyl titanate　三(十二烷基苯磺酰基)钛酸异丙酯〔偶联剂〕

isopropyl tri(N-ethylamino-ethylamino) titanate　三(N-乙氨基-乙氨基)钛酸异丙酯

isopropyl tri(2-formylphenyl) titanate　三(2-甲酰苯基)钛酸异丙酯

isopropyl triisostearoyl titanate　三异硬脂酰基钛酸异丙酯〔偶联剂〕

isopropyl tri(lauryl-myristyl) titanate　三(十二烷基-肉豆蔻基)钛酸异丙酯

isopropyl tri (methoxyphenyl) titanate　三(甲氧苯基)钛酸异丙酯

isopropyl tri(methyl-butyl pyrophosphate) titanate　三(焦磷酸甲丁酯)钛酸异丙酯

isopropyl triricinoyl titanate　三蓖麻醇酰钛酸异丙酯

isopropyl tris(dioctyl pyrophosphate) titanate　三(焦磷酸二辛酯)钛酸异丙酯

β-isopropyltropolone(IPT)　β-异丙基䓬酚酮；β-异丙基芳庚三烯酚酮　$C_6H_4(OH)(CO)(C_3H_7)$

isopropyl xanthate disulfide　二硫化黄原酸异丙酯〔促进剂〕

isopropy tristearoyl titanate　三硬脂酰钛酸异丙酯

isoproterenol　异丙基肾上腺素

isoproterenol hydrochloride　盐酸异丙肾上腺素

isopulegol　异蒲勒醇；异胡薄荷醇

isopulegol acetate 乙酸异蒲勒醇酯 $CH_3CO_2C_{10}H_{17}$
isopulegone 异蒲勒酮；异胡薄荷酮 $C_{10}H_{16}O$
isopulegyl 异胡薄荷基
isopulegyl acetate 乙酸异胡薄荷酯
isopulegyl formate 甲酸异胡薄荷酯
isopurone 异嘌酮 $C_5H_8N_4O_2$
isopycnic 等偏微比容的*
isopycnic centrifugation 等密度离心(法)
isopycnic gradient centrifugation 等密度梯度离心(法)
isopygmaaein 异矮柏醚
isopyknic 等体积的；等容积的
isopyroine 人字果碱
isoquercitrin 异栎素；异槲皮素
isoquinaldinic acid 异喹哪啶酸
isoquinocycline 异醌环素
isoquinoline 异喹啉*
isoquinolyl 异喹啉基 C_9H_6N-
isoracemization 等消旋*
isorauhimibine 异萝亨宾
isoreactive dyeing system 等活性染色体系
isoreactivity 等活性
isorefractive mixture 等折光指数混合物
isoreserpiline 异利血平灵
isoreserpine 异利血平
isorhamnetin 异鼠李亭
isorhamnose 异鼠李糖
isorhamnoside 异鼠李糖苷
isorheic ①等黏液②等黏的③恒流量的
isorheic elution 恒流量洗脱*
isorheic liquid 等黏度液体
isorheic separation 等黏度分离
isorhodanate 异硫氰酸酯 RNCS
isorhodanic acid 异硫氰酸 HNCS
isorhodanic ester(=isorhodanide) 异硫氰酸酯 RNCS
isorhodanide(=isorhodanic ester) 异硫氰酸酯
isorhodeose 异鼠李糖
isorhodomycin 异紫红霉素
isorhynchophylline 异尖叶(钩藤)碱
isoriboflavin 异核黄素
isoricinoleic acid 异蓖麻酸；羟基十八碳烯酸
isorosindonic acid 异绕森酮酸
isorotation 等旋光度
isorotenone 异鱼藤酮
iso-rubber 异构化橡胶
isosaccharinic acid 异己糖酸
　　$CH_2OHCHOHCH_2COH(CH_2OH)COOH$
isosafroeugenol 异黄樟丁子香酚
isosafrole 异黄樟脑；异黄樟素；4-丙烯基-1,2-亚甲二氧基苯 $CH_2O_2=C_6H_3CH=CHCH_3$
isosantenic acid 异檀烯酸
isosbestic point(=isoabsorptive point) 等吸收点；等吸光点
isosbestic point method 等吸收点法；等吸光点法
isoscope 同位素探伤仪
isosebacic acid 异癸二酸
isoselective polymerization 异构选择聚合
isosepiapterin 异墨蝶呤
isoserine 异丝氨酸 $NH_2CH_2CHOHCO_2H$
isosesamin 异芝麻素
isoshehkangenin 异射干配基
isoshehkanin 异射干苷
isosinomenine 异汉防己碱
Iso-Siv process 异构筛法正烷烃分离
isosmotic 等渗(压)的
isosmoticity 等渗(透)压
isosol 等溶胶；同构异量质溶胶
isosorbide(=1,4,3,6-dianhydrosorbitol) 异山梨醇
isosorbide dinitrate 硝酸异山梨酯〔药〕
isosorbide mononitrate 单硝酸异山梨酯〔药〕
isospecific polymerization 全同立构聚合
isospin triplet 同位旋三重态
isostatic 等压的；均衡的
isostatical anomaly 均衡失常
isostatics 等压线
isostearate 异硬脂酸盐；异硬脂酸酯
isostearic acid 异硬脂酸
isostemonidine 异百部定
isostere (电子)等排物
isosteric (电子)等排的
isosteric heat of adsorption 等量吸附热
isosteric property (电子)等排性
isosterism 等排性；电子等排同物理性(现象)
isostilbene 顺-1,2-二苯基乙烯；异芪；顺式对称二苯代乙烯 $C_6H_5CH=CHC_6H_5$
isostoichiometric quantities 等化学计量数
isostructural 同(结)构的；等结构的；同型的
isostructure 异构造
isosuccinic acid(=methylmalonic acid) 异丁二酸；甲基丙二酸 $CH_3CH(CO_2H)_2$
isosulf 异构硫
isosulfocyanate 异硫氰酸酯 RNCS
isosulfocyanic acid 异硫氰酸 HNCS
isosulfocyanic ester(=isosulfocyanide) 异硫氰酸酯 RNCS
isosulfocyanide(=isosulfocyanic ester) 异硫氰酸酯
isosynthesis 异构(烷烃)合成法

isotachiol 氟硅酸银
isotachoelectrophoresis 等速电泳
isotachophoresis 等速电泳
isotachophoresis injection-coupled with capillary zone electrophoresis 等速电泳-毛细管区带电泳耦合进样
isotachophoresis on paper 纸上等速电泳
isotactic 全同立构(的)；等规立构(的)
isotactic addition 全同(立构)加成；等规加成
isotactic block 全同(立构)嵌段；等规嵌段
isotactic configuration 全同立构构型
isotactic content 全同立构含量
isotactic crystallinity 全同(立构)结晶性；等规结晶性
isotactic diad 全同立构二单元组
isotactic index 全同(立构)指数；等规指数〔指全同立构聚合物的含量〕
isotacticity 等规度；全同立构(规整)度
isotacticity index 全同(立构)规整度指数；等规度指数
isotactic moulding 等压成型〔在加工聚四氟乙烯等材料时〕
isotactic placement 全同(立构)键接
isotactic polymer 全同立构聚合物*；等规聚合物
isotactic polypropylene 全同立构聚丙烯；等规聚丙烯
isotactic polypropylene fiber 全同聚丙烯纤维
isotactic propagation 全同(立构)增长；等规增长
isotactic resin 等规树脂
isotactic sequence 全同(立构)序列
isotactic-syndiotactic co-polymer 全同-间同(立构)共聚物；等规-间规(立构)共聚物
isotactic triad 全同(立构)三单元组；等规三单元组
isotactic unit 全同(立构)单元
isotaxy 全同聚合；等规聚合
isotazettine 异水仙花碱
isoteniscope 等面仪
isotensile head contour 等张力封头曲面
isoteresantalic acid 异对檀香酸
isotetracosane 异二十四(碳)烷 $(CH_3)_2CH(CH_2)_{20}CH_3$
isothebaine 异蒂巴因；异二甲(基)吗啡 $C_{19}H_{21}O_3N$
isotherm 等温线
isothermal(=isothermic) 等温的
isothermal absorption 等温吸收
isothermal analysis 等温分析
isothermal annealing 等温熟炼；等温退火〔热原子化学〕
isothermal bulk relaxation module 等温体积松弛模量
isothermal calorimeter 等温量热计
isothermal change 等温变化
isothermal compressibility 等温压缩性
isothermal compression 等温压缩
isothermal compressor 等温压缩机
isothermal condition 等温条件
isothermal crystallization 等温结晶(作用)
isothermal deformation 等温形变
isothermal distillation 等温蒸馏
isothermal efficiency 等温效率
isothermal electron capture chromatography 等温电子俘获色谱(法)
isothermal evaporation 等温蒸发
isothermal flame ionization chromatography 等温火焰电离色谱(法)
isothermal gas chromatography 等温气相色谱法
isothermal gravimetric analysis 等温重量分析
isothermal heating 等温加热
isothermal heat of adsorption 等温吸附热
isothermal-isobaric chromatography 等温等压色谱法
isothermal-isobaric gas chromatography 等温等压气相色谱(法)
isothermal line 等温线
isothermal linearity 等温线性
isothermal operation of electrothermal atomizer 电热原子化器的等温操作；电热原子化器等温原子化
isothermal parabolic thickening 等温抛物加厚(镀膜)
isothermal period 等温周期
isothermal polymerization 等温聚合
isothermal process 等温过程*
isothermal reactor 等温反应器
isothermal run 等温运转；等温操作
isothermal shear rate 等温剪切速率
isothermal shear relaxation module 等温剪切松弛模量
isothermal spinning threading 等温纺丝线
isothermal static continuation 等温静态延拓
isothermal stress functional 等温应力泛函
isothermal tangential modulus 等温切向模量
isothermal theory 等温理论
isothermal titration calorimetry 等温滴定量热计
isothermal transformation 等温转化
isothermal transformation diagram 等温转变图
isothermal vulcanization 等温硫化
isothermal wall 等温壁
isothermal works 等温机件
isothermic(=isothermal) 等温的
isotherms intersection point 等温线交点
isothianaphthene 异硫茚
isothiazole 异噻唑；1,2-硫氮茂
isothiazolinone 异噻唑啉酮
isothiocyanate 异硫氰酸盐
isothiocyanato(=isothiocyano) 异硫氰基 S＝C＝N—
isothiocyanic acid 异硫氰酸 HNCS

isothiocyanic ester(=isothiocyanide) 异硫氰酸酯 RNCS
isothiocyanide(=isothiocyanic ester) 异硫氰酸酯
isothiocyano(=isothiocyanato) 异硫氰基 S＝C＝N—
isothiourea 异硫脲 RCS(NH$_2$)＝NH
isothreonine 异苏氨酸
isothujapliceane 异大侧柏烷
isothujone 异侧柏酮
isothymol(=carvacrol) 异百里香酚；香芹酚
isotone 同中素；同中子异核素〔中子数相同〕
isotonic ①等渗(压)的②等中子(异位)的
isotonic concentration 等渗(压)浓度
isotonic diluent 等渗稀释
isotonicity 等渗(压)性
isotonic saline solution 等渗压盐溶液
isotonic solution 等渗溶液
isotope 同位素*
isotope abundance 同位素丰度
isotope addition method 同位素加入法；直接同位素稀释法
isotope analysis 同位素分析
isotope assay 同位素分析
isotope broadening 同位素加宽
isotope buildup 同位素累积
isotope capsule 同位素源盒
isotope carrier 同位素载体
isotope cask(=isotope container; isotope flask) 同位素(储运)容器
isotope chart 同位素图表
isotope chemistry 同位素化学
isotope cisternography 同位素脑池造影法
isotope container(=isotope cask; isotope flask) 同位素(储运)容器
isotope cow 同位素母牛发生器
isotope dating 同位素年代测定*
isotope derivative method 同位素衍生物法
isotope dilution 同位素稀释(法)
isotope dilution analysis(IDA) 同位素稀释分析(法)
isotope dilution mass spectrometry(IDM) 同位素稀释质谱法
isotope dilution method 同位素稀释法
isotope dilution-neutron activation analysis 同位素稀释中子活化分析
isotope dilution spark source mass spectrometry 同位素稀释火花源质谱法
isotope effect 同位素效应
isotope enrichment 同位素富集；同位素浓集
isotope exchange method 同位素交换法

isotope exchange reaction 同位素交换反应
isotope excited X-ray fluorescence spectrometry 同位素激发 X 射线荧光法
isotope farming 同位素农作；同位素饲养
isotope flask(=isotope cask) 同位素(储运)容器
isotope fractionation 同位素分离；同位素分凝；同位素分馏
isotope gauge 同位素仪表
isotope generator 同位素电源
isotope geochemistry 同位素地球化学
isotope geochronology 同位素地质年代学
isotope geology 同位素地质学
isotope handling calculator 同位素操作计算尺
isotope hydrology 同位素水文学
isotope ion 同位素离子
isotope irradiator 同位素辐照器
isotope isomer 同位素异构体
isotope-isomeric 同位素异构的
isotope-isomeric mixture 同位素异构混合物
isotope-isomerism 同位素异构(现象)
isotope-isomerization 同位素异构作用
isotope labeling 同位素标记
isotope level detector 同位素液面检测器
isotope mass ratio 同位素质量比
isotope mass spectrometer 同位素质谱仪
isotope mass spectrometry 同位素质谱法
isotope milker ①子体同位素发生器②子体同位素产生物
isotope milking 从母体中分离子体同位素
isotope mixture 同位素混合物
isotope neutralizer 同位素电中和器
isotope peak 同位素峰
isotope pool 同位素库
isotope production reactor 同位素生产(反应)堆
isotope ratio mass spectrometer 同位素比质谱计
isotope ratio tracer method 同位素比示踪物法
isotope replacement method 同位素取代法
isotopes broadening 同位素变宽
isotope separation 同位素分离
isotope separator 同位素分离器
isotope-separator-on-line(ISOL) 在线同位素分离器
isotope shift 同位素位移
isotope side band 同位素边峰
isotope spectrum line 同位素谱线
isotope tagging detection 同位素示踪探测
isotope tracer 同位素示踪剂
isotope tracer technique 同位素示踪技术
isotopic 同位素的

isotopic abundance   同位素丰度
isotopic abundance measurement   同位素丰度测量；同位素丰度测定
isotopically labelled ion   同位素标记离子
isotopic analysis   同位素分析
isotopic carrier   同位素载体
isotopic cluster   同位素簇离子
isotopic competition   同位素竞争
isotopic composition   同位素组分
isotopic correlation safeguards technique   同位素相关核保障监督技术
isotopic dating   同位素测定年龄
isotopic depletion   同位素贫化；同位素耗损
isotopic differentiation   同位素差异
isotopic dilution mass spectrometry   同位素稀释质谱法
isotopic dilution method   同位素稀释法
isotopic dispersion curve   同位素散布曲线
isotopic distribution   同位素分布
isotopic effect   同位素效应
isotopic element   同位(元)素
isotopic enrichment   同位素富集
isotopic exchange   同位素交换
isotopic exchange reaction   同位素交换反应
isotopic foil   同位素箔
isotopic geochronology   同位素地质年代学；同位素地球纪年学
isotopic geology   同位素地质学
isotopic mass   同位素(质)量
isotopic mass spectrometry   同位素质谱法
isotopic measurement   同位素测定法
isotopic molecule   同位素分子
isotopic multiplicity   同位素多重性
isotopic peak   同位素峰
isotopics   同位素组成
isotopic scrambling   同位素攀移
isotopic snow gauge   同位素雪量计
isotopic spin conservation   同位旋守恒
isotopic standard   同位素标准
isotopic target   同位素靶
isotopic thermometer   同位素温度计
isotopic tracer   同位素指示剂；示踪同位素
isotopic tracer technique   同位素示踪技术
isotopic tracing   同位素示踪
isotopic weight   同位素量
isotopology   同位素学
isotopy   同位素学；同位素性质
iso-trans-tactic   反式全同立构
isotriacontane(=melissane)   异三十(碳)烷；蜂花烷 $C_{30}H_{62}$
isotricyclene   异三环萜
isotrilobine(=homotrilobine)   异三裂碱
isotrimorphism   同三晶形(现象)
isotron   同位素分析器
isotrope   各向同性晶体
isotropic   各向同性的
isotropic body   各向同性体
isotropic component of tensor   各向同性张量分量
isotropic etching   各向同性腐蚀
isotropic failure   各向同性衰坏
isotropic fiber   各向同性纤维
isotropic fluid   各向同性流体
isotropic hyperfine coupling constant   各向同性超精细偶合常数
isotropic laminate   ①各向同性层合材料②各向同性层压制件
isotropic material   各向同性材料
isotropic material with memory   各向同性记忆材料
isotropic pitch   各向同性沥青
isotropic plate   各向同性板
isotropic polymer   各向同性聚合物
isotropic pressure   各向同性压力
isotropic stress(=hydrostatic stress)   各向同性应力；静水应力；流体静应力
isotropic substance   同位素
isotropic temperature factor   各向同性温度因子
isotropic tensor(=spherical tensor)   各向同性张量
isotropic thermal parameter   各向异性热参数
isotropic turbostratic carbon   各向同性的湍流碳
isotropic turbulence   各向同性湍动
isotropic turbulent flow   各向同性湍流
isotropy   各向同性现象
isotruxillic acid(=truxinic acid)   吐星酸；异吐昔酸 $(C_6H_5)_2C_4H_4(COOH)_2$
isotype   同种型
isoundecylenyl acetate   乙酸异十一烯酯
isourea   异脲 $NH_2C(OH)=NH$
isouzarigenin   异乌扎配基
isovalent hyperconjugation   等价超共轭
isovaleraldehyde(=3-methyl-butyraldehyde)   异戊醛；3-甲基丁醛 $(CH_3)_2CHCH_2CHO$
isovaleraldehyde diethyl acetal   异戊醛二乙缩醛
isovaleraldehyde diethyl mercaptal   异戊醛二乙缩硫醛
isovaleraldehyde oxime   异戊醛肟 $C_4H_9CH=NOH$
isovaleramide   异戊酰胺 $C_4H_9CONH_2$
isovaleranilide(=isovaleryl aniline)   N-异戊酰苯胺 $C_4H_9CONHC_6H_5$

isovalerate ①异戊酸②异戊酸盐(酯或根)
isovaleric acid(=delphinic acid) 异戊酸
isovaleric aldehyde 异戊醛
isovaleric anhydride 异戊酸酐 $(C_4H_9CO)_2O$
isovaleronitrile(=isobutyl cyanide) 异丁基氰；异戊腈 $(CH_3)_2CHCH_2CN$
isovaleryl 异戊酰 $(CH_3)_2CHCH_2CO-$
isovalerylacetylphenolphthalein 异戊酰乙酰基酚酞
isovaleryl aniline(=isovaleranilide) N-异戊酰苯胺 $C_4H_9CONHC_6H_5$
isovaleryl diethylamine 异戊酰二乙胺 $(CH_3)_2CHCH_2CON(C_2H_5)_2$
isovaline 异缬氨酸
isovalthine 异缬硫氨酸
isovanillic acid 异香草酸 $C_6H_3(CO_2H)(OH)(OCH_3)$
isovanillin(=3-hydroxy-4-methoxybenzaldehyde) 异香兰素；异香草醛；3-羟基-4-甲氧基苯甲醛 $(CH_3O)C_6H_3(OH)CHO$
isovel 等速线
isoversion 催化异构化
isoveryl chloride 异戊酰氯 $(CH_3)_2CHCH_2COCl$
isovincamine 异长春花胺
isoviolanthrene 异紫蒽
isoviolanthrone 异紫蒽酮
isoviolanthrone violet 异紫蒽酮紫
iso-viscosity state 等黏态
isoviscous state 等黏态
isoviscous temperature 等黏温度
isovitamin C 异维生素 C；异抗坏血酸
isovolumic process 等容过程
isoxanthopterin 异黄蝶呤
isoxazole 异噁唑
isoxazolyl 异噁唑基；1,2-氧氮(杂)环戊二烯基 C_3H_2NO-
isoxime 异肟 RCH_2CH_2NO
isoxylitone 异木酮
isoyohimbine(=α-yohimbine) α-育亨宾
isozonide 异臭氧化物
isozyme 同工酶；同功酶
I strain 内张力
itaconate ①衣康酸；亚甲基丁二酸②衣康酸盐(酯或根)
itaconic acid(=methylene-succinic acid) 衣康酸；亚甲基丁二酸 $CH_2=C(CO_2H)CH_2CO_2H$
itaconic anhydride 衣康酸酐；亚甲基丁二酸酐
itaconic ester 衣康酸酯
itai-itai disease 骨痛病〔一种公害病〕
itamalic acid 衣苹酸〔俗〕；羟甲基丁二酸 $CH_2OHCHCOOHCH_2COOH$
item 项目
iterating technique 反复(展开)技术
iteration 迭代(法)；重复
iteration chromatography 循环色谱法
iteratively reweighted least square 重新加权迭代最小二乘算法
iterative method 迭代法*
iterative procedures 迭代过程
iterative target transfer factor analysis 迭代目标转换因子分析法
iterative target transform factor analysis method 迭代目标转换因子分析法
iterative technique 迭代法
-itol〔词尾〕 糖醇
it-plate 石棉橡胶板
itraconazole 伊曲康唑〔药〕
iturin 伊枯草菌素
-ium compound 有机阳离子化合物；镓化合物
ivain 蓍草素；依瓦因 $C_{24}H_{42}O_5$
iva oil 蓍花油
ivory 象牙
ivory black 象牙黑
ivory board 象牙纸板；白卡纸
ivory buff 象牙白色〔浅黄白色〕
ivory nut 象牙果；象牙椰子
ivory paper 象牙纸
ivory soap 象牙皂〔商〕
ivy leaf oil 活血丹叶油
ixiolite 锰钽矿
ixolyte 红蜡石
ixometer 油汁流度计
Izod impact machine 悬臂梁式冲击机
Izod impact strength 伊佐德冲击强度；悬臂梁式冲击强度
Izod impact test(=cantilever beam impact test) 悬臂梁式冲击试验
Izod test 悬臂梁式试验

J

Jablochkoff cell 杰布洛霍夫电池
Jablonski diagram 雅布隆斯基能级图
Jablonski scheme 雅布隆斯基能级图
jaborandi 美洲毛果芸香
jaborandi leaves oil 毛果芸香叶油
jaboridine 美洲毛果芸香碱
jaborine 美洲毛果芸香混碱
J acid J酸；2-氨基-5-苯酚-7-磺酸
jacinth 红锆石
jacinthal(=phenyl acetaldehyde dimethyl acetal) 苯乙醛二甲缩醛
jacinthe(=hyacinth) 风信子
jack 底脚片；挺针片
jacket ①套②套管③夹套
jacket cooling 套管冷却
jacketed （备有）套层的
jacketed crystallizer 套层结晶器
jacketed evaporator 套层蒸发器
jacketed kettle 套锅
jacketed multichamber vessel 带夹套的多室容器
jacketed reactor 夹套式反应器
jacketed seamless kettle 无缝套锅
jacketed still 套层蒸馏器
jacketed syringe 套层注射器
jacketed trough 夹套(冷却)槽(池)
jacketed wall 套墙；套壁
jacket heating 汽套加热
jacket tube 套管
Jackhammer air drill 杰克哈默压气钻
jacking 套料
jackknifing 摺刀切割
Jackob's ladder 杰科布绳梯〔木或铁制有梯级的绳梯〕
jack-pump 油矿泵
jack screw 千斤顶螺旋
jack shaft 起重轴
Jackson candle turbidimeter 杰克逊蜡烛浊度计
Jacobi method 雅克比法
jacobine 夹可宾；千里光碱
Jacobsen rearrangement 雅可布森重排
jacodine(=seneciphylline) 千里光菲灵碱；夹可定
jacoline 吡咯里西啶生物碱；夹可灵
jaconecic acid 夹可酸
jaconine 夹可宁
jacozine 夹可嗪

jacquard board 茶版纸
jacquard openwork structure 贾卡网孔组织
jacquard shedding 提花开口
jacquard straight ties 通丝穿吊法
Jacquet indicator 杰奎特示速仪
jacupirangite 钛铁霞辉岩
jacutinga 富赤铁矿
jade 玉(石)
jadeite 硬玉；翡翠
Jaffe reaction 雅费氏反应〔检肌酸苷〕
jag resistance 抗切口性能
jaipurite 块硫钴矿
jalap 球根牵牛；紫茉莉；药喇叭
jalapic acid 球根牵牛酸 $C_{17}H_{30}O_9$
jalapin 球根牵牛苷；紫茉莉苷
jalapinolic acid 11-羟(基)十六(烷)酸 $CH_3(CH_2)_4CHOH(CH_2)_9COOH$
jalapoid 墨西哥旋花素
jalaric acid 壳脑醛酸 $C_{15}H_{20}O_5$
jalaric ester 紫草茸醇酸脂
Jamaica pepper oil 牙买加胡椒油〔取自 Pimenta officinalis〕
jamaicine 牙买加菜树苦素
Jamba seed oil 芝麻菜子油〔取自芝麻菜 Eruca sativa〕
jamb brush 掸尘刷
jamb duster(=dusting brush) 扫尘刷；掸尘刷
jambosine 蒲桃碱
jambul 蒲桃
jambulol 蒲桃酸 $C_{16}H_8O_9$
James' powder 含锑磷酸钙
James-Martin gas density balance 詹姆斯-马丁气体密度天平
jamesonite 脆硫锑铅矿
Jamin effect 杰明效应
Jamin's interferometer 雅满干涉仪
jamming 人为干扰
jam nut(=lock nut) 锁紧螺母；压紧螺盖〔商〕
Janak's method 耶纳克法
Janecke coordinate 简克坐标〔液-液萃取用〕
janiemycin 贾尼霉素（一种多肽抗生素）
Janssen's equation 詹森方程〔垂直圆筒内粉体粒子间的压力关系式〕
janthinellin 微紫青霉素
Janus green B 烟鲁绿B；詹纳斯绿B

japaconine 日乌头宁
japaconitine 日乌头碱
jap-A-lac 日本大漆 A
japan 日本漆；亮漆
Japan black ①黑沥青(清)漆②黑漆
Japan camphor 日本樟脑 $C_{10}H_{16}O$
Japan clay 蒙脱石
Japan color 亮漆色浆
Japan drier(=liquid drier) 液体催干剂；日本催干剂
Japanese lacquer 日本漆；亮漆
Japanese mint oil 日本薄荷油
Japanese sardine oil 日本沙丁鱼油
Japanese star anise oil 草芥油；日本大茴香油
Japanese tung oil 日本桐油
Japanese wood oil 日本桐油
Japanic acid 二十一烷双酸；日本蜡酸
Japan lacquer 日本漆；亮漆
japanned 漆过的；油过的
japanned leather 漆皮；漆革
japanner (油)漆工
japanners' brown 漆工棕
japanning 涂漆
japanning kettle 漆锅
japanning oven 涂漆用炉；上漆炉
japanning room 漆房；油漆间
Japan sardine oil 日本沙丁鱼油
japan stopper(=japan gold size) ①短油钙脂清漆②调钙脂铅油填孔料用漆
Japan tallow 野漆树蜡
Japan wax 野漆树蜡〔取自 *Rhus succedanea*〕
japopinic acid 日本松节油酸
Japp-Klingemann reaction 杰甫-克林曼反应
jaracanda 玫瑰木
jara jara β-萘基甲基醚
jargon 黄锆石
jargonia 氧化锆
jar mill 瓷制球磨罐
jarosite 黄钾铁矾
jarosite rock 黄钾铁矾岩
jar ring lathe 大口瓶盖垫旋切车床
jasmal 乙酸苄酯
jasminal 茉莉醛
jasmine 茉莉
jasmine absolute 茉莉净油
jasmine aldehyde 茉莉醛；α-戊基(肉)桂醛
jasmine bouquet 茉莉百花型
jasmine concrete 茉莉浸膏
jasmine flower wax 茉莉花蜡

jasmine lactone(=jasmolactone) 茉莉内酯
jasmin oil 茉莉油
jasminolene(=jasmonal H) 茉莉醛 H；2-己基(肉)桂醛
jasmin oxide 茉莉醚；对甲酚苄醚
jasmolactone(=jasmine lactone) 茉莉内酯；顺-2-戊烯基环戊内酯
jasmonal α-戊基肉桂醛；茉莉醛
jasmonal H(=jasminolene) 茉莉醛 H；2-己基(肉)桂醛
jasmonate 茉莉酮酸酯
jasmone 茉莉酮
jasmonic acid 茉莉酮酸
jasmonoid 茉莉酮类化合物
jasmonyl 茉莉酯〔商〕；壬二醇-1,3-单乙酸酯的混合物
jasper 碧玉
jaspilite 碧玉铁质岩
jatamansi 宽叶甘松
jatamansone(=valeranone) 缬草烷酮
jateorhizine(=jatrorhizine) 药根碱
jatex 浓(缩橡)浆
jatrophine 麻风树碱
jatrorrhizine(=jateorhizine) 药根碱
Jaumann tensor 耀曼张量
Java almond oil 爪哇杏仁油
Java basil oil 爪哇罗勒油
Java citronella oil 爪哇香茅油
Java para 爪哇白拉胶
java wax 榕树蜡
javelle 漂白水
jaw breaker 颚式碎裂机
jaw crusher 颚式破碎机
jaw crushing 颚式破碎
jaw oil (海豚)颌油
jaw speed 狭口流速
J cross-polarization technique J 交叉极化法
jecoleic acid 介考裂酸；十九碳烯酸
jecolein 介考裂精
jecoric acid 介考日酸；十八碳三烯酸；肝酸
jecorin 肝糖磷脂
jel 凝胶、冻(胶)
jelletite 绿铁榴石
jellied fuel 凝固燃料；(冻)胶状燃料
jellied gasoline 凝固汽油；胶状(汽)油
jelling agent 胶凝剂
jelly 冻(胶)
jelly detergent (冻)胶状洗涤剂
jelly fungi 胶质菌；木耳
jelly membrane 胶膜
jelly paint 凝胶漆

jelly strength tester　冻胶强度试验仪
jelutong　节路顿胶
Jena glass　耶那玻璃
Jenag strainer　自清洗过滤机(器)
Jena ware　耶那玻璃器
Jenkine cracking　詹金裂化(法)
jenkolic acid　甲烯胱氨酸
Jenner stain　詹纳尔白血球着色剂
Jentsch ignition tester　坚赤着火试验器〔评价燃料着火性质〕
jeppel oil(=bone oil)　骨油
jequiritin(=abrin)　红豆素
jeremejewite　硼酸铝石
jerking table　震淘台
Jersey process　泽西法〔丁烯脱氢制丁二烯法〕
jervine　蒜藜芦碱
jesaconitine　结乌头根碱
jet　①(喷)嘴②喷射③射流
jet aircraft fuel　喷气式飞机用燃料
jet air pump　喷射泵
jet-anvil type fluid energy mill　单喷式气流粉碎机；靶板式气流粉碎机
jet black　烟黑；炭黑
jet blower　喷射嘴
jet chamber　喷雾室
jet cleaning　喷射清洗
jet coal　长焰煤
jet compression　喷射压缩
jet condenser　喷射式冷凝器
jet condenser pump　喷水凝气式泵
jetcrete　喷枪喷射水泥浆；喷浆
jet drying　喷射干燥；喷雾干燥
jet dust counter　(空气)吸附尘埃计数器
jet ejector　喷射器
jet electrode　流汞电极
jet exit　①喷嘴出口②射流出口
jet expansion　射流胀大
jet flow　射流
jet fuel　喷气式发动机燃料
jet looping　(空气)喷射变形加工
jet mixer　喷射混合器
jet mixing system　喷注混合系统
jet molding　①注压(硫化)法②注(射)模(塑)法〔塑料〕
jet mould　喷模；喷射塑模
jet moulding　喷模法；喷射模塑法
jetness　黑度
jetometer　润滑油腐蚀性测定仪
jet orifice　喷射口

jet orifice separator　喷嘴式分离器
jet overflow dyeing machine　喷射溢流绳状染色机
jet propulsion　喷气推进(器)
jet propulsion fuel　喷气式飞机燃料
jet pulverizer　喷射粉磨机
jet pump　喷射泵
jet purifier　喷洗器
jet refuelling vehicle　喷汽式飞机加油车
jet scrubber　喷射洗涤器
jet spinning　喷射纺丝
jet spread　喷射分散
jet spread angle　喷射分散角
jet spun yarn　喷气纱
jet stream　喷流混合法〔胶乳炭黑共沉法之一〕
jet swelling effect　射流胀大效应
jet thrust　射流推力
jetting　①喷射②注射③漩纹〔注塑制品缺陷〕
jetting sump transfer　喷射法输送
jet transfer　射流输送
jet tray　喷射塔板；舌形塔板
jet type dust collector　喷射型(脉动)除尘器；脉冲反吹袋式除尘器
jet velocity　喷射速度
jet wall　喷射壁
jewel　宝石
jewelry　宝石
jib arm　旋臂；摇臂
jig　①簸析机②钻模③样板④夹具
jig boring　座标镗削
jig bush　钻套
jigged-bed　跳汰床
jigger　①簸析机②盘车；辘轳车
jiggering(=jollying)　盘车拉坯
jigger wheel　拉坯盘车
jigging　簸析法
jigging conveyor　振动式斜槽
jigging machine　簸析机
jigging motion　摆动；摇晃运动；往复运动
jigging screen　簸动筛
jigging stenter　簸动拉条机
jig machine(=jig washer)　簸析机
jig stripping　锯齿形吊具除漆
jig washer(=jig machine)　簸析机
jinjili oil　芝麻油
j-j coupling　j-j 偶合
J-modulation method　J 调制法
job　①工作②零件；部件
jockey　薄膜

jockey pulley 托滚；托轮〔运输带用〕
joggle 榫接；啮合扣接榫
joggled lap joint 啮合搭接
johannite 铀铜矾
johimbine 育亨宾碱
Johnson noise 约翰逊噪声
joiner 细木工人；接合者
joiner curtain 连接幕
joint 胶接接头；拼接
joint aging time 黏合期；接合期
joint cathodic protection system 接头阴极接地防腐蚀系统
joint caulk 填缝料；填缝剂
joint cement 接缝胶泥
joint conditioning time(=joint aging time) 接合期
joint condition time 黏合期；接头期
joint coupling 活节连节器；管接头
jointed fibre 接续光纤
joint factor 接合系数
joint filler 接缝填充料
joint flange 连接法兰
joint glue 接合胶
joint grease 接头润滑剂
jointing 接合
jointing compound 密封剂；黏合剂
jointing material ①嵌缝材料②封填胶〔如石棉胶〕
jointing mortar 接缝灰浆
jointing paste 接缝浆
jointing tape 黏接胶带；胶黏带；接缝胶带
jointless 无法兰连接；无缝的
jointless flooring 无接缝铺地材料
joint packing 填充垫圈
joint position 拐点
joint ring 垫圈；垫板
joint roller （壁纸)接缝压平辊
joint sealant 黏(结密)封胶；灌封胶
joint sealer 接缝封闭剂
joint strip 嵌缝胶条；密封胶条
joint tape （壁纸)接缝胶纸带
joint transduction 连锁转导
joint transformation 连锁转化
joint washer 密封垫圈
joist steel 梁钢；工字钢
jojoba butter 霍霍巴脂
jojoba oil 霍霍巴油
jojoba wax 霍霍巴木蜡
Jolly balance 比重天平
jollying(=jiggering) 盘车拉坯

Jolly's spring balance 弹簧秤
Jones furnace 琼斯炉
Jones' method 琼斯法
Jones oxidation 琼斯氧化
Jones reagent 琼斯试剂
Jones reducer(=Jones reductor) 琼斯还原管
Jones reductor(=Jones reducer) 琼斯还原管
Jones reductor method 琼斯还原器法〔测钛白的总钛量〕
jonquil 长寿花
jonquil absolute 长寿花净油
jonquil concrete 长寿花浸膏
jonquille 长寿花
jonquil oil 长寿花油
jordan 锥形精磨机；低速磨浆机
Jordan chest 约旦(储浆)柜
Jordan engine 约旦(打浆)机
jordanite 锑硫砷铅矿
Jordan refiner 约旦精制机
Jordan's double focusing mass spectrograph 乔丹双聚焦质谱仪
Jorissen test 焦瑞生试验〔试甲醛〕
josamycin 交沙霉素
joseite 硫碲铋矿
josenite 晶蜡石
josephinite 镍铁矿
Josephson effect 约瑟夫森效应；约氏效应；超导电子对的超导隧道效应
joule 焦耳〔能量单位〕
Joule effect 焦耳效应
Joule equivalent 焦耳当量
Joule-Kelvin effect 焦耳-开尔文效应
Joule's law 焦耳定律
Joule-Thomson coefficient 焦耳-汤姆孙系数*
Joule-Thomson effect 焦耳-汤姆孙效应*
Joule unit 焦耳当量
journal bearing 轴颈轴承
journal box 轴颈箱
journal compound 轴颈配合物
journal oil 轴颈油
joystick system ①远距离操纵系统②操纵杆操纵系统③十字显示线操作手柄
JP fuel(=jet propulsion fuel) 喷气推进燃料
J phenomenon J现象〔随X射线波长而改变的不连续性吸收现象〕
Judd chart 贾德氏黑白格纸；遮盖力试验纸〔美国标准局专用〕
Judd color difference unit 贾德氏色差单位
Judd graph 贾德氏黑白格纸；遮盖力试验纸

Judd hiding power chart 贾德氏遮盖力试验纸
judge analysis 仲裁分析
judgement criteria 判断标准；评价标准
judgement schematics 判断图式；评价图表；评价公式
judicial chemistry 法医化学
Juerst ebullioscope 尤尔斯特(测醇)沸点计
jug handle 操作用管线〔从蒸馏锅或塔引出〕
jugladin 胡桃素
juglandin 胡桃定
juglandoid 胡桃素
juglanin 胡桃宁
juglans ①胡桃②胡桃皮
juglone(=nucin) 核桃酮
jujube 枣
jujuboside 酸枣仁皂苷
julocrotine 柔黄巴豆碱
julolidine 久洛尼定
jumble beads 红豆
jumbo mixer 巨型混合机
jumbo tire 超低压轮胎
jump ①跳动②(滴定)突跃；跃迁；跃变
jump continuation 跳变延拓
jumper 跨接线
jump frequency 跳变频率
jumping probability 跳变概率
jump in presure 压力突变
jump mashing 急升温打浆法
jump-over connection 越管连接
jump polymerization 突变聚合〔聚合时聚合度突然增加数倍的反应〕；跃升聚合
jump reaction 突变反应〔聚合时聚合度突然增加数倍的反应〕
jump scanning 跳跃扫描
Junbert cell 荣伯特电池
junction ①接头；接界②交点；会合点
junction box header 换热器管束箱
junction house 分类部门〔石油厂的接收原料或产品〕
junction motion 交联点运动
junction potential 接界电位

junction transistor 结式晶体管
junene 刺柏烯；桧烯
junenol 刺柏烯醇；桧醇
junior cave 次级屏蔽室
junipene(=longifolene) 长叶烯
juniper berry oil 刺柏(子)油
juniperene 刺柏二烯；圆柏二烯
juniperic acid 圆柏酸；刺柏酸；桧酸；16-羟(基)十六(烷)酸 $HOCH_2(CH_2)_{13}CH_2COOH$
juniperin 刺柏苦素
juniper oil 刺柏油；桧油
juniperol(=macrocarpol) 刺柏醇；圆柏醇
juniper tar 刺柏焦油；桧焦油
juniper tar oil 刺柏焦油；杜松油〔俗〕
juniper wood oil 刺柏木油
junk 废料堆
Junker's gas calorimeter 荣克气体量热计
Junker's calorimeter 荣克量热器
junket 带乳浆的凝乳
junket ring 压盖衬环
junk ring 压盖衬环
jury pump 备用泵
just-critical system 恰临界系统；刚临界系统
just in-time system 适时系统
jute 黄麻；麻
jute degumming 黄麻脱胶
jute fabric 黄麻织物
jute packing ①(铅油)黄麻填料(密封)②麻丝嵌缝
jute paper 黄麻纸
jute scrim 斯克林黄麻布；平纹网眼黄麻布
jute seed oil 黄麻子油
jute yarn 黄麻纱
juvenescence ①复原现象；还原现象②复壮现象
juvenile 岩浆生的；岩浆源的；新生的
juvenile water 岩浆水
juverimicin 幼霉素
juxtaposition 并置；邻近；接近
juxtaposition metamorphose 接触变形
juxtaposition-twin 并置双晶

K

Kaadener green　绿土；硅酸盐绿
kabaite　陨地蜡（一种存在于陨石中的天然烃）
kachi grass　青香茅
kachi grass oil　青香茅油
kadamba absolute　团花净油
kadsura oil　南五味子油
kaempferide　莰非素
kaempferol　山奈酚
kaerferol　四羟基黄酮
kafir　高粱
kafiroic acid　高粱酸
Kahle's solution　卡耳溶液（一种固定剂）
kahweol　咖啡豆醇
kainic acid(=digenic acid)　红藻氨酸；2-羧甲基-3-异丙烯基脯氨酸；海人草酸
kainite　钾盐镁矾（矿）
kairine　克灵（解热碱）
kairoline (=methyl tetrahydroquinoline)　N-甲基四氢化喹啉　$C_9H_{10}NCH_3$
kaiser roll　开氏辊
kaiser roll machine　开氏辊机
Kaiser transformation　凯撒换值
kakodyl　二甲胂基
kakoxene　磷铁矿
kalaite　含铜绿松石
kalamycin　卡拉霉素
kale　①蔬菜②菜汤③羽衣甘蓝
kali　氧化钾；苛性钾　K_2O
kali ammonsalpeter　氯化钾硝酸铵混合肥料
kalicrete　含铁波特兰水泥
kalimagnesia　含水硫酸钾镁　$K_2SO_4 \cdot MgSO_4 \cdot 6H_2O$
kalimeter(=alkalimeter; Shrolter's apparatus)　碳酸定量器；施罗特碳酸定量器
kalinite　纤维钾明矾
kaliophylite　钾霞石
kali salt　钾盐
kalium-o-phenyl phenate　邻苯基苯酚钾
kallaite　绿松石
Kalle acid　卡耳酸；1-萘胺-2,7-二磺酸
kallitype　铁银（印画）法
Kalman filter　卡尔曼滤波
Kalman filtering method　卡尔曼滤波法
Kalman filter spectrophotometry　卡尔曼滤波分光光度法
Kalman gain　卡尔曼增益

kalsilite　六方钾霞石
kalsomine　刷墙水粉；水性涂料
kalsomine paint　刷墙水粉漆；刷墙水粉涂料
kalzium metal　铝钙合金
kamacite　铁纹石
kamala　咖马粉；粗糠粉
kamala tree　粗糠柴；菲岛桐
kamalin　咖马林；粗糠柴苦素
kamarezite　碱式硫酸铜矿
kambi　栀子胶
Kaminsky　卡明斯基催化剂
α-kamlolenic acid　18-羟基十八碳三烯酸
kammogenin　莰摸配基
kammonin　莰摸宁
kampometer　热辐射计
kanasatine　乌克兰大麻素
kanendomycin(=kanamycin B)　卡那霉素 B
kanerol　卡尼醇
kangaroo leather　袋鼠革
kanirin　氧化三甲胺
kankrinite　钙霞石
kanokonol　坎缬酮醇（取自缬草根挥发油）
kanokonyl acetate　乙酸异坎缬酮酯
kantlex(=kanamycin)　卡那霉素
kanugin　水黄精；甘石黄素
kanyin oil　古香油
kanyl alcohol　卡尼尔醇
kanzuiol(=tirucallol)　甘遂醇　$C_{30}H_{50}O$
kaoliang color　高粱色素
kaoliang oil　高粱油
kaolin(e)(=china clay)　高岭土；瓷土；白陶土
kaolinite　高岭石；高岭土
kaolinite clay　高岭土
kaolinization　高岭土化（作用）
kaolinized granite　高岭土化花岗石
kapillar analyse(=capillary analysis)　毛细(管)分析
kapillar-analysis　毛细分析(法)
kapillary(=capillary)　毛细管
kapnometer　烟密度计
kapok　爪哇木棉〔取自 Ceiba pentandra〕
kapok oil　爪哇木棉油
kapok wax　爪哇木棉蜡
Kappa number　卡伯值
kapron　卡普纶

karabin 夹竹桃树脂
karakin 狗角藤苷
karakul 波斯羔皮
karat(=carat) 开〔量金单位〕
karaya 刺梧桐〔取自 Sterculia urens〕
karaya gum 刺梧桐树胶
karbogel 碳胶〔脱水剂商品名〕
karite butter 烛果油〔取自 Butyrospermum Parkii 的种子〕
karitene 西非牛油树烃
karite oil 烛果油
Karl Fischer moisture determination 卡尔·费歇水分测定法
Karl Fischer reagent 卡尔·费歇尔试剂
Karl-Fischer's method 卡尔·费歇尔水分测定法〔溶剂中的水分定量分析法〕
Karl Fischer titration 卡尔·费歇尔滴定(法)
Karl Fischer titration apparatus 卡尔·费歇尔水分测定仪
Karl Fischer titration method 卡尔·费歇尔滴定法
Kármán equation 卡曼方程
Kármán number 卡曼数
Kármán vortex 卡曼涡
karo-karo-unde absolute 卡罗花净油
karo-karo-unde concrete 卡罗花浸膏
karpholite 硅酸铁锰矿
karstenite 硬石膏 $CaSO_4$
karton 厚纸
karyotyping 核型分析
kaspine leather 开斯宾革
Kassler green 钡绿〔锰酸钡〕
kata-(=cata-) 〔希腊字头〕①渺位②萘状环的1,7位②向下；在下
kata-condensed rings 渺位缩合环
katalase 过氧化氢酶
katalysis 催化作用
katalytic nickel generation(KANIGEN) ①催化制镍法 ②催化镀镍法〔非电解镀镍法之一〕
katamorphism 破碎变质现象
kataphoresis 电泳
katastatic stress 弛存应力
kata thermometer 低温温度计；空调温度计
katergol 火箭燃料
kath(a)emoglobin 变性高铁血红蛋白
katharometer(=catharometer) 热导计；热导气体分析仪；热导池
katharometer supply unit 热导计电源
kathode(=cathode) 阴极；负极
katholysis 阴极电解法
katine 阿拉伯茶碱
kation(=cation) 阳离子

kationic laked pigment 阳离子型色淀颜料
katio oil 烛果油〔取自 Bassia Mottleyana 的种子〕
Katki 卡特基〔印度10～11月收的崟基尼品系紫胶〕
katsu-toxin 蝎毒
Katz funnel 卡茨漏斗
Kaufmann iodine value 考夫曼碘值
kaurene 贝壳杉烯
kaurenic acid 贝壳杉烯酸
kaurenoic acid 异贝壳杉烯酸
kauri 贝壳杉脂；栲树脂
kauri butanol number(=kauri butanol solvency value; kauri butanol value) 贝壳杉脂丁醇(溶解)值
kauri butanol solvency test 贝壳杉脂丁醇(溶液)测定〔某一特殊汽油的芳香度〕
kauri butanol solvency value(=kauri butanol number; kauri butanol value) 贝壳杉脂丁醇(溶解)值
kauri butanol test 贝壳杉脂丁醇试验
kauri butanol value(=kauri butanol solvency value; kauri butanol number) 贝壳杉脂丁醇(溶解)值
kauri copal 贝壳杉珀把
kauri gum 新西兰琥珀
kauri gum solution 贝壳杉树胶溶液
kaurinic acid 贝壳杉酸 $C_{10}H_{16}O_2$
kaurinolic acid 贝壳杉脑酸 $C_{17}H_{34}O_2$
kauri oil 贝壳杉油
kauri pine 贝壳杉
kauri reduction value 贝壳杉脂稀释值
kauri resin 贝壳杉脂；栲树脂
kauri tolerance value test 贝壳杉脂(稀释)容许值试验
kauri varnish 贝壳杉漆
kaurolic acid 贝壳杉油酸 $C_{12}H_{20}O_2$
kauronolic acid 贝壳杉让酸 $C_{12}H_{34}O_2$
kauroresene 贝壳杉脂素
kautchin 萜烯
kautschin 二聚戊烯；双戊烯；杉油精；苎烯
kava(=kawa) ①醉椒②醉椒根
kavaic acid 醉椒酸 $C_{13}H_{12}O_3$
kavain 醉椒素
kavaine 卡瓦根素
kava resin 醉椒树脂
kavatel oil 副大风子油
kawain 醉椒素
kaya oil 椰子油
kayserite 片铝
Kayser's wavel
K-band K
kbar (=kilo
K cap

Keene's cement　基恩水泥；干固水泥
keep　盒底
keesom force　定向力
KE factor　动能因数
keilhauite　钇铈榍石
Keith process (for lead refining)　基思(炼铅)法
Kekulé compound　凯库勒化合物
Kekulé formula　凯库勒式〔指苯结构式〕
Kekulé ring　凯库勒环
K electrons　K层电子
kelley box　盛(压制成块的)润滑脂油杯
kellin　牙签草素；开林
Kellner cell (for hypochlorite)　凯耳讷(次氯酸盐制造)电池
Kellogg cracking process　凯洛格裂化过程
Kellogg hydrocracking process　凯洛格加氢裂化法
Kellogg millisecond furnace　凯洛格毫秒炉
Kellogg process　凯洛格法〔烃类两段重整法〕
Kelly filter　凯利过滤机
kelp　①海草②海草灰
kelp ashes　海草灰
kelp oil　海草油
kelp salt　海草灰盐
Kelvin　开(尔文)〔热力学温度单位〕
Kelvin body (=Voigt body)　开尔文体
Kelvin bridge　开尔文电(阻)桥
Kelvin degree　开尔文温度
Kelvin effect　开尔文效应
Kelvin equation　开尔文公式
Kelvin galvanometer　开尔文式检流计
Kelvin model (=Voigt model)　开尔文模型
Kelvin scale　开尔文温标
Kelvin solid (=delayed elastic solid)　开尔文固体
Kelvin's thermodynamic scale　开氏温标；绝对温标
Kelvin temperature　开尔文温度；开氏温度；绝对温度〔符号 K〕
⋯fert　氯化钾矿
⋯al　开米他
⋯um sublimation apparatus　坎氏真空升华仪
⋯氢化可的松；17-羟-11-脱氢皮质酮
⋯德层〔即电

kentledge　压重料；压舱用的生铁
kepayang oil　克帕荞油
keracyanin　花青素鼠李葡萄糖苷
kerargyrite　角银矿
kerasin　①角苷脂②角铅矿
kerasol　四碘酚酞
keratin　角蛋白
keratin fiber　角蛋白纤维
keratinization　角蛋白化(作用)
kerenes　煤油烯
kerf　切口
kerites　煤油沥青
kerma (词源:kinetic energy released in material)　比释动能；柯玛〔单位：焦耳/千克, 尔格/克〕
kerma rate　比释动能率〔单位:尔格/(克·秒)〕
kermes　①虫胭脂；胭脂虫粉②橘红硫锑矿
kermesic acid　胭脂酮酸
kermesite　橘红硫锑矿
kermes mineral　红锑
kernel oil　①核油②橄榄油
kernite　四水硼砂
kerocain　盐酸普鲁卡因
kerokaine (=procaine)　普鲁卡因
kerole　开罗油〔油母中溶于吡啶和不溶于氯仿的部分〕
kerosene (=kerosene oil)　煤油
kerosene burner　煤油灯
kerosene burning quality　煤油燃烧性质
kerosene burning test　煤油燃烧试验
kerosene color test　煤油颜色测定；煤油色度实验
kerosene degreasing　煤油脱脂
kerosene distillate　煤油馏分
kerosene flash point　煤油闪点
kerosene fraction　煤油馏分
kerosene oil (=kerosene)　煤油
kerosene oil engine　煤油机
kerosene oil-gas　煤油气；重质可燃混合物；重质燃料
kerosene-oil mixture　煤油混合油
kerosene raffinate　精制煤油
kerosene sediment test　煤油中沉积物的测定
kerosene shale (=oil shale)　煤油(母)页岩
kerosene stock　煤油燃料
kerosene stripper　煤油用汽提塔
kerosene sulfur test　煤油中硫的测定
kerotenes　焦化沥青质
Kcrr-ccll　克尔盒
Kerr constant　克尔常数
kerrolic acid　开醇酸
keryl　煤油基

keryl amine 煤油基胺
keryl anisole sulfonate 煤油基甲氧基苯磺酸盐(酯)
kerylbenzene 煤油烷基苯
kerylbenzene sulfonate 煤油基苯磺酸盐
keryl cresol sulfonate 煤油基甲酚磺酸盐
keryl diethanolamine 煤油基二乙醇胺
keryl ether 煤油基醚
kessane 日缬草素
kessanol 日缬草素醇
kessanyl 缬草素基
kessanyl acetate 乙酸日缬草素酯
kessazulene S-愈创木薁
Kessler concentration plant 凯斯勒蒸浓装置
kesso oil 缬草油
kesso root oil 日本缬草油
kessyl acetate 乙酸缬草酯
kessyl alcohol 缬草油醇 $C_{14}H_{24}O_2$
kessyl glycol 日缬草二醇
Kestner acid elevator 凯斯特勒升酸器
Kestner evaporator 凯斯特勒蒸发器
kestose 蔗果三糖
ketal 酮缩醇；缩酮
Ketalar 盐酸氯胺酮
ketaldonyl 酮醛基
ketal formation 酮缩醇的形成
ketamine 氯胺酮〔药〕
ketazine ①〔类名〕(某)酮连氮 $R_2C=NN=CR_2$②(专指 dimethylketazine)四甲基(甲)酮连氮 $[(CH_3)_2C=N]_2$
ketazine process 酮连氮法
ketene(=ethenone) 乙烯酮*
ketene acetal 乙烯酮缩二乙醇 $CH_2=C(OC_2H_5)_2$
ketene acetylation 乙烯酮乙酰化(作用)
ketene diacetal 乙烯酮缩二乙醇
ketene dimer 乙烯酮二聚物；双乙烯酮
ketene generator 乙烯酮发生器
ketene-imine 烯酮亚胺
ketimide 酮酰亚胺
ketimine 酮亚胺
ketimine/ketazine process 酮亚胺/酮连氮法
ketimine process 酮亚胺法
ketine 2,5-二甲基吡嗪 $C_6H_8N_2$
ketipic acid 草酰二乙酸；β,β-己二酮二酸 COOHCH₂COCOCH₂COOH
Ketjen black 科琴黑新增
keto ①(=oxo)氧代 O=②酮基 =CO
keto-acetic acid 丙酮酸 $CH_3COCOOH$
keto acid 酮酸

ketoadipic acid 酮己二酸
ketoadip(o)yl-CoA 酮己二酸单酰 CoA
keto-alcohol 酮醇；氧基醇 $RCOCH_2OH$
keto-aldehyde 酮醛 RCOCHO
ketoalkylation 酮烷基化(作用)
keto-amide 酮酰胺 $RCOCONH_2$
keto-amine 酮胺；氨基酮 RNH_2COR
2-keto-4-butanethiol 2-酮基-4-丁基硫醇
3-ketobutyraldehyde dimethyl acetal 3-酮基丁醛二甲基缩醛
ketobutyric acid 氧代丁酸；丁(邻)酮酸 $CH_3CH_2COCO_2H$
ketocaproic acid 己酮酸
5-keto-carane 蒈酮
keto-carboxylic acid 酮羧酸
ketoconazole 酮康唑〔药〕
ketocoumaran 苯并二氢呋喃酮
γ-ketoctanol γ-酮基辛醇
keto-cyclol prototropy 酮-环醇质子移变
2-keto-3-deoxy galactonic acid 2-酮基-3-脱氧半乳糖酸
2-keto-3-deoxy-7-phosphoglucoheptonic acid 2-酮基-3-脱氧-7-磷酸葡庚糖酸
2-keto-3-deoxy-6-phosphogluconic acid 2-酮基-3-脱氧-6-磷酸葡萄糖酸
ketodestrin 雌酮
keto-dibasic acid 酮二(甲)酸
keto-dicarboxylic acid 酮二(羧)酸
keto-enol 酮-烯醇
keto-enol system 酮-烯醇系
keto-enol tautomerism 酮-烯醇互变异构*
keto ester 酮酸酯*
ketoestrone 酮雌(甾)酮
keto fatty acid 脂肪酮酸
keto fatty alcohol 脂肪酮醇
keto-form 酮式
ketogenesis 生酮(作用)
ketogenic-antiketogenic ratio 生酮抗酮比值
ketogluconate ①酮葡萄糖酸②酮葡萄糖酸盐(酯或根)
ketogluconic acid 酮葡萄糖酸；葡萄糖酮酸
ketoglutaramate 酮戊二酸单酰胺
ketoglutaramic acid 酮戊酰胺酸；酮戊二酸单酰胺
ketoglutarate ①酮戊二酸②酮戊二酸盐(酯或根)
keto-glutaric acid 氧代戊二酸 COOHCH₂CH₂COCOOH
ketoglutaric dehydrogenase 酮戊二酸脱氢酶
ketogulonate ①酮古洛糖酸②酮古洛糖酸盐(酯或根)
ketogulonic acid 酮古洛糖酸
2-ketogulonolactone 2-酮古洛糖酸内酯

ketoheptose 庚酮糖
ketohexonate 己酮糖酸
ketohexonic acid 己酮糖酸
ketohexose 己酮糖
ketohoxonic acid 己酮糖酸
ketohydroxyestrin 雌甾酮
ketoimine 酮亚胺
ketoindole(=oxindole) 羟吲哚
α-ketoisocapric acid α-异癸酮酸
ketoisocaproate 酮异己酸
ketoisocaproic acid 酮异己酸
ketoisovalerate ①酮异戊酸②酮异戊酸盐(酯或根)
ketoisovaleric acid 酮异戊酸
ketoketene 酮式烯酮 $R_2C=CO$
ketol〔类名〕乙酮醇 RCOCHOHR
keto-lactol 内缩酮
ketolase 酮酶
ketole(=indole) 吲哚
ketol-lactol prototropy 酮醇-内酯质子移变
α-ketol rearrangement α-酮醇重排*
ketolysis 酮解(作用)
ketolytic (分)解酮的
ketolytic reaction 酮解反应
keto-manoyl oxide 泪杉醚酮
keto-monobasic acid(=keto-monocarboxylic acid) 一价酮酸 RCOCOOH
keto-monocarboxylic acid(=keto-monobasic acid) 一价酮酸
keto-morpholine 酮基吗啉
ketomycin 酮霉素
keto-myo-inositol 酮肌醇
ketonaldehydmutase 酮醛变位酶
ketone ①〔类名〕酮②〔在个别化合物中的译名〕甲酮
ketone acetal 酮缩醛；缩酮
ketone acid 酮酸
ketone alcohol 酮醇
ketone base 双二甲氨基苯基酮
ketone-benzol-dewaxing process 酮苯脱蜡法
ketone body(=acetone body) 酮体
ketone carbonyl moiety 酮式羰基(部分)
ketone color 含酮色料
ketone condenser 丙酮冷凝器
ketone ester 酮酯 RCOR'COOR"
ketone ether 酮醚；烷氧基酮 RCOCH$_2$OR'
ketone form 酮式
ketone fractionator 丙酮精馏塔
ketone group 酮基〔二价基 $C=O$〕

ketone hydrate 酮水合物
ketone musk 麝香酮
ketone oil 丙酮油
ketone peroxide 酮过氧化物；过氧化某酮
ketone rancidity 酮(臭)败
ketone resin 酮树脂；酮(缩)醛树脂
ketonic 酮的
ketonic acid 酮酸 RCOCOOH
ketonic bond 羰基键
ketonic cleavage(=ketonic hydrolysis) 成酮分解；成酮水解
ketonic cross-linking 酮式交联
ketonic ester 酮酯 RCOR'COOR"
ketonic ester type 酮酯型
ketonic ether 酮醚；烷氧基酮 RCOCH$_2$OR'
ketonic form 酮式
ketonic hydrolysis(=ketonic cleavage) 成酮分解；成酮水解
ketonic link(=ketonic linkage) 羰基键
ketonic linkage(=ketonic link) 羰基键
ketonic oxygen 羰基(中的)氧
ketonic type 酮式
ketonitrile group 氧化腈基；酮腈基
ketonization 酮基化作用
ketonuria(=acetonuria) 酮尿；丙酮尿
ketopalmitate ①酮棕榈酸；酮软脂酸②酮棕榈酸盐(酯或根)
ketopalmitic acid 酮棕榈酸；酮软脂酸
2-keto-6-phosphogluconic acid 2-酮基-6-磷酸葡萄糖酸
2-ketophosphohexonate 2-酮磷酸葡萄糖酸
2-ketophosphohexonic acid 2-酮磷酸己糖酸
β-ketopropane β-酮基丙烷
α-ketopropionic acid α-酮基丙酸
α-ketopropionic aldehyde α-酮基丙醛
9-ketoprostenoate 9-酮前列腺烯酸
ketopurine 次黄嘌呤
ketopyrrolidine 吡咯烷酮；α,γ-丁内酰胺
ketose 酮糖
ketoside 酮苷
ketostearate ①酮硬脂酸②酮硬脂酸盐(酯或根)
ketostearic acid 酮硬脂酸 $C_{18}H_{34}O_3$
17-ketosteroid 17-酮甾类
ketosteroid compound 甾酮类化合物
ketosuccinic acid 草酰乙酸；酮丁二酸
ketotifen 酮替芬〔药〕
ketotriazole 三唑酮
ketotriose 酮丙糖
γ-ketovaleric acid γ-酮基戊酸

ketovinylation 酮乙烯化作用
ketoxime 酮肟
ketoxylose 木酮糖
kettle bodied oil 热炼聚合油；聚合油
kettle cure 罐中硫化
kettle-type reboiler 釜式重沸器；釜式再沸器
kettle wax 锅蜡
kettle wax phase 锅蜡相
keturonate ①糖酮酸②糖酮酸盐(酯或根)
keturonic acid 糖酮酸
ketyl 羰游基
ketyl radical 羰自由基*
keuna saltpetre 混酸铵〔肥料〕
kewda absolute 露兜花净油
key-atom 钥原子
key band 特征谱带
key bond 特征键合；特殊键合
key coat 结合层；打底胶浆；初层
key component 关键组分；主要组分
keyhole shaped orifice 钥匙孔形喷丝孔
keying action ①关键作用②结合作用
keying agent 增黏剂
keying coat ①黏结(固)层②(=primer)底漆
keying strength 咬合强度
keying through of (resin) (树脂)渗透黏牢；渗透性附着
key intermediate 关键中间体
key part 关键部件
key stone 拱石顶
keyway 键槽
key wrench 套筒扳手
khaki color ①土黄色②黄土颜料
khaki drills 卡其布
khari salt 卡里蓝〔印度用于生皮防腐的含硫酸钠等的土盐〕
kharophen 乙酰氨基羟基苯胂酸
khat 阿拉伯茶
khellin 呋喃并色酮
khelloglucoside 开洛醇葡萄糖苷
khusene 客烯
khusenic acid 客烯酸
khusic acid 客酸
khusimol 客烯醇
khusimone 客烯酮
khusinol 客素醇
khus-khus oil 岩兰草油；香根油
khusol 深谷醇；客醇
khusone 客酮
kibble (木)桶

kibbled 碎块的；破碎成块的〔直径约为 0.64cm〕
kibbled glue 碎块胶
kibdelopane(=kibdelophan) 钛铁矿
kick ①冲击②跳动〔仪表指针〕③石油产品的初馏点④汽油的发动性〔在汽车中〕
kicker 揉皮器〔皮革〕
kicker baffle 导向隔板
kick-off point (促进剂)生效温度
kick off temperature (鼓泡剂)分解开始温度；引发温度；生效温度
kick out 析出；分层〔用稀释剂等稀释时〕
kick starter rubber 起动踏脚橡皮
kick up ①骚动②急剧提高汽油辛烷值
kickxia rubber 绢丝橡胶
kid 小山羊革
kidamycin 贵田霉素
kidney oil 肾色油〔松香裂化油, 沸点 250～270℃〕
kidney ore 肾状铁矿；肾铁矿
kidneys 肾形矿脉
kidney shaped fan 肾状扇；肾状(形)喷束〔重的油漆分布在上下两路〕
kidney shape pattern 肾形喷束〔喷束断面图形之一〕
Kienle's functionality theory 金勒氏官能度理论
kier 漂煮锅
kier-boiling 漂煮；煮炼
kiering 漂煮
kies 黄铁矿
kiese gel 硅胶
kieselguhr 硅藻土*
kieselguhr diatomite 硅藻土
kieserite 硫镁矾〔矿〕
kiganene 日柳杉烯
kiganol 日柳杉醇
Kikuchi lines 晶体表面散射的电子流线
kikumycin 菊霉素
kilbrickenite 块辉锑铅矿
Kilburn-Cott furnace 基耳伯-科特电炉
Kiliani aluminum process 基连尼炼铝法
kilkenny coal 无烟煤
killed rubber 过炼胶
killed spirit 焊酸；焊接用的药水
killed steel 镇静钢
killing (裘皮染色)碱处理
kiln ash potash fertilizer 窑灰钾肥
kiln bottom 窑底
kiln burned pine tar(=country tar) 堆积干馏松焦油；窑干馏松焦油
kiln-burning 炭窑制炭法；堆积炭化法

kiln crown　窑顶
kiln dried　(窑中)烘干的
kiln dried lumber　炉烘(干)的木材；窑烘(干)的木材
kiln-dried wood　窑干材
kiln dry　(窑中)烘干
kiln-dry shellac　漂白紫胶
kilned　(窑中)烘干的
kiln gas　窑气；炉烟气
kiln hole　炉门；窑洞
kiln hood　窑罩
kilning　窑烧；窑烘
kiln lining　窑里(层)；里窑层
kiln mill　(一种边磨边送空气的)窑炉转磨
kiln placing　装窑
kiln process　窑烧(炭)法
kiln roasting　窑炉烤烘
kiln shell　窑壳
kiln-site　制炭场
kiln structure　窑的砖结构
kiln tar　窑焦油
kiln white　白霜；吐渣
kilo〔希腊字头〕　①千②(=kilogram)千克
kilobecquerel(kBq)　千贝可勒尔〔=10^3Bq〕
kilocalorie　千卡
kilocurie　千居里
kilodyne　千达因
kilogamma　千微克
kilogauss　千高斯
kilogram　千克
kilogram calorie　大卡；千(克)卡
kilogram-meter　千克·米
kilogray(kGy)　千戈瑞〔=10^3Gy〕
kilojoule　千焦耳
kilolambda　千微升
kiloliter　千升
kilometer　千米；公里
kilorad　千拉德
kilorod facility　千棒设施〔钍元件试验用〕
kilorod program　千棒计划
kilosievert(kSv)　千希沃特
kilostere　千立方公尺
kilowatt　千瓦(特)
kilowatt hour　千瓦(特)小时
kimberlite　角砾云橄岩；金伯利岩
Kimmeridge coal　启莫里煤〔一种可燃页岩〕
kinamycin　醌那霉素
kinase　激酶；致活酶
kind　品种；类；型式

Kindler modification　金德勒变型
kindling point　燃点；着火点
kindling wood　引火柴
kind of oil　油品种
kind of petroleum product　石油产品品种；石油产品类别
kinegraphic control panel　远距离控制板；活动图示的控制板
kinematic　运动学的；动力学的
kinematical boundary condition　运动边界条件
kinematic coefficient　运动系数；动力系数
kinematic coefficient of viscosity　动态黏度系数
kinematic energy correction　运动能修正；运动能校正
kinematic matrix　运动矩阵
kinematic pair　运动副
kinematics　运动学
kinematic similarity　运动相似
kinematic tensor　(运)动张量
kinematic type of capillary viscometer　运动型毛细管黏度计
kinematic viscosimeter　运动黏度计；测定运动黏度用黏度计
kinematic wave　运动波
kinemometer　流速计；感应式转速表
kinetic　①动力(学)的②(运)动的
kinetic acidity　动力学酸度*
kinetic analysis　动力学分析(法)
kinetic balance　动力学平衡；动态平衡
kinetic chain　动力学链
kinetic chain length　动力学链长
kinetic coefficient of friction　动摩擦系数
kinetic colorimetry　动力学比色法*
kinetic control　动力学控制*
kinetic crystallizability　动力学结晶性；动态结晶性
kinetic current　动力电流
kinetic dispersion mill　动力分散磨〔利用漆料的高速旋流和颜料粒子之间冲撞和摩擦原理进行分散〕
kinetic energy　动能
kinetic energy correction　动能改正
kinetic energy effect　动能效应
kinetic equation　动力学方程
kinetic equilibrium　动态平衡
kinetic friction　动摩擦
kinetic friction coefficient　动摩擦系数
kinetic function　动力学函数
kinetic heat effect　动热效应
kinetic hypothesis　分子运动假说
kinetic isotope effect　动力学同位素效应*
kinetic limiting current　动力学极限电流

kinetic masking　动力学掩蔽
kinetic micelle　动力学胶束
kinetic molecular theory　分子运动(理)论
kinetic neutralization test　中和反应速度测定
kinetic parameter　动力学参数
kinetic penultimate effect　前末端效应
kinetic photometry　动力学光度学
kinetic pressure　动(力)压
kinetic property　①运动性质②动力(学)性质
kinetics　动力学
kinetic salt effect　动力学盐效应
kinetic segment　动力学链段
kinetic severity function　动力裂解深度函数
kinetics methodology　动力学方法
kinetics of carburising　渗碳动力学
kinetics of combustion　燃烧动力学
kinetics of crystallization　结晶(化)动力学
kinetics of diffusion　扩散动力学
kinetics of electrode process　电极过程动力学
kinetics of flame　火焰动力学
kinetics of ion formation　离子形成动力学
kinetics of photochemistry　光化学动力学
kinetics of polymerization　聚合动力学
kinetics of reaction　反应动力学
kinetic solvent effect　动力学溶剂效应*
kinetic solvent isotope effect(KSIE)　动态溶剂同位素效应
kinetic spectrophotometry　动力学分光光度法
kinetic spectroscopy　动力学光谱法
kinetics salt effect　动力学盐效应
kinetic stability　动力学稳定性
kinetic template effect　动力学模板效应
kinetic theory　分子运动学说
kinetic theory of elasticity　弹性分子运动论
kinetic theory of fluid　流体分子运动论
kinetic theory of gas　气体的分子运动(理)论
kinetic theory of solution　溶液的分子运动(理)论
kinetic unit　动力单位
kinetic viscosity　动力黏(滞)度
kinetic viscosity index　运动黏度指数
kinetic wave　动力波
kinetic Young's modulus　动态杨氏模量
kineurine　甘油磷酸奎宁
King'blue(=cobalt aluminate)　钴蓝　$Co(AlO_2)_2$
king bolt　大螺丝；大螺旋
King furnace　金氏高温炉
King's desiccator　金氏干燥器
king-size-tanker　大型油船（3～4万吨以上）

King's manometer　金氏压力计
king'yellow　雌黄；三硫化二砷
kinic acid　金鸡纳酸；奎尼酸
kink　扭接
kinking　纵向弯曲；扭转；打扣；打结
kink in surge line　喘振线上的转折点
kink site　扭折位
kink test　纵向弯曲试验
Kinney pump　金尼泵
kinoin　奇诺树脂；热带非洲紫檀树胶
kinomycin(=quinomycin)　醌霉素
kinovin　金鸡纳树皮苷
Kinsel process　金西尔法
kinsho oil　日本金松油
Kipp gas generator　基普气体发生器
Kipp's generator　基普(气体)发生器
kips　①印度驼牛皮；瘤牛皮②印度植鞣驼牛皮坯革
kir　岩沥青
Kirchhoff's law　基尔霍夫定律
kiree(=kiri)　袋渣(紫)胶
kirromycin　黄色霉素
-kis(=times)　〔词尾〕接数词之后表示若干个；若干倍
kish　结晶石墨
kiss-coating　贴胶；挂胶
kiss-roll coater　辊式舐涂机；轻触辊涂机
kiss-roll(er)(=lick roll(er))　①舐(吻)涂辊②给漆辊；舐漆辊③挂胶辊
kiss-roll padding　单面给液法；单面上胶法
Kistiakowsky's equation　基斯梯可斯基公式〔蒸发潜热计算式〕
Kitahara's water bottle　北原式采水器
kitasamycin　吉他霉素〔药〕
kitchen fat　厨用脂
kitchen salt　食盐
kitchen wall paint　厨房墙壁漆
kitol　鲸醇
kiton blue　奇通蓝
kiton colors　奇通染料；三苯甲烷染料
kiton fast green　奇通坚牢绿
kiton fast orange　奇通坚牢橙
kits　漆桶；小桶
Kitson's ebulliometer　基特森沸点计
Kittel centrifugal tray　离心式基特尔塔板
Kittel overflow tray　溢流型基特尔塔板
Kittel plate　基特尔条孔网形塔板
Kittel polygonal tray　多角型基特尔塔板
Kittel standard tray　标准基特尔塔板
Kittel tray　基特尔塔板；斜孔网状塔板

kittool fiber　假桄榔纤维
kitty cracker　小型裂化器
kiwi flavor　猕猴桃香精
Kjeldahl analysis　凯耶达分析法〔测定含氮量的方法〕
Kjeldahl determination　凯氏定氮法
Kjeldahl flask　长颈烧瓶；凯氏烧瓶
Kjeldahl-Gunning method (for nitrogen determination)　凯耶达-冈宁定氮法
Kjeldahl method　凯氏定氮法
Kjeldahl method of nitrogen determination　凯氏定氮法
Kjeldahl's autosystem for nitrogen determination　自动凯氏定氮仪
klaprotholite　脆硫铜铋矿
Klason's lignin　硫酸木素；克拉松木素
Kleinenberg mixture　克莱能柏格混剂〔可可脂、鲸蜡与蓖麻油的混合剂〕
kleinite　氯胺汞矿
Klein's liquid　克莱因试液〔硼钨酸镉的饱和溶液〕
Klett color　克莱特色度
Klett fluorometer　克莱特荧光计
Klett-Summerson colorimeter　克莱特-萨默森比色计
Klett value　克莱特值〔克莱特比色计读数〕
K-line　K 线
Klocknerwerke process　克洛克纳公司方法
kloof buchu　长叶布枯；齿状布枯
klydonograph　(电花)录过量电压器
klystron　速(度)调(制)电子管；速调管
klystron generator　速调管振荡器
klystron oscillator　速调管振荡器
knallgas　氢氧混合气
knar　木节
kneaded　捏和了的
kneaded structure　捏和结构
kneader　捏和机；揉面机
kneading　捏和
kneading machine　捏和机
kneading pulper　纸浆捏和机
kneading trough　捏和桶
K-nearest neighbor method　K 最近邻域法
knee　①弯管；肘管；曲管②膝②曲线上的弯曲处
knee brace　膝形拉条
knee pipe atomizer　弯管喷雾器
knee staker　手工铲皮刀
knickerbocker yarn　彩点纱
knife　①刀②小刀③刀(口)
knife application　刮涂法(涂漆)
knife barker　刀式去皮机
knife block(=knife box)　刀框；刀架子

knife box(=knife block)　刀框；刀架子
knife coat　刮涂法
knife coater　刮刀式涂胶机
knife coating　刮刀式涂胶
knife-curl test　漆膜刮卷试验
knife edge　①支棱②刀口
knife edge of balance　天平的支棱
knife-edge profile　刀刃(形)断面
knife edge test　刀口检验
knife fill(ing)　刮刀填孔(操作)；刮填
knife line(=knife mark)　刮痕；刮纹；刀痕；刀纹〔以刀划线〕
knife line attack　刀线腐蚀
knife line corrosion　刀蚀；线状腐蚀；划线腐蚀
knife mark(=doctor mark)　刮(刀)痕；刮(刀)纹〔刀口有伤刮涂时留下的条状痕迹〕
knife-on-blanket coater　传送带式垫衬刮涂机
knife-on-blanket coating　垫衬刮涂法
knife-over-roll coater　辊衬刮刀涂装机；辊衬刮刀辊涂机
knife-over-roll coating　辊衬刮刀辊涂法
knife switch　闸刀开关
knife tackle　刀组
knife test　刀割试验法〔样板上的漆膜用刀以窄条状割下，检验漆膜的脆性和韧性〕
knife test for plywood　胶合板刀齿试验
knifing filler　刮涂(用)填孔剂；刮涂腻子
knit-de-knit crinkle yarn　假编式变形纱
knit mark　①包滚时间〔开炼〕②复炼时间；回炼时间〔硅橡胶〕③汇合痕〔注塑品〕
knitmesh　织网
knitmesh packing　网卷填料
knitted fabric　针织物
knitted mat　缝编毡
knit time　①复炼时间②热炼时间③包滚时间(开炼)
knitting　①针织；编织②接合③融合
knitting machine oil　针织机油
knitting wool ply machine　绒线合股机
knitting yarn　针织(用)纱
knob-and-tube model　节-管模型
knobbling　熔锤过的铁疙瘩〔炼铁〕
knob knuckle　①关节②转向节③万向接头；铰接；肘接④屋脊
knock characteristic of gasoline　汽油的抗震性
knock-compound　抗震剂
knock die-out　爆震消除
knock die-out curve　爆震消除曲线
knock die-out point　爆震消除点

knock-down 拆卸；拆开；解体；分解
knocker 爆震剂
knock-free 非爆震的
knock-free fuel 非爆震燃料
knock-free operation 无爆震操作
knock indication instrument(=knock indicator) 爆震指示器；爆震器
knock indicator(=knock indication instrument) 爆震指示器；爆震器
knock inducer 爆震诱导物
knocking ①震性②敲
knocking behaviour 爆震行为
knocking characteristic of fuel 燃料的抗震性能
knocking combustion 爆震燃烧
knocking explosion 爆炸
knocking fuel 爆炸燃料
knocking-out pin 推出销；推顶销
knocking zone 爆震区域〔燃烧室〕
knock inhibiting essence 抗震剂；抗爆剂
knock intensity 爆震强度
knock intensity indicator 爆震强度指示器；爆震器
knock intensity method 爆震强度法
knock-limited density index 爆震限制的密度指数
knock-limited performance 爆震限制行为
knock meter 爆震计
knock meter reading 爆震计读数
knock off electrode 敲落电极
knock-off joint(=knock-off post) 可连接的接合器
knock-off post(=knock-off joint) 可连接的接合器
knock on effect 撞击效应
knock-on process 对撞过程；迎头碰撞过程
knockout （塑物)脱模
knockout box 气体分离箱
knockout coil 分离旋管
knockout drum ①分离鼓②分离罐
knockout plate 脱模板；甩板〔塑料〕
knockout press 脱模力；甩力〔塑料〕
knockout tower 分离塔
knock producer 爆震剂〔产生爆震的物质〕
knock property 爆震性
knock rating 爆震率；震率
knock rating of small sample 小量样品的爆震率
knock-reducer 抗爆剂
knock-sedative dope 抗爆混合物；抗爆剂
knock suppressor 抗爆剂
knock tendency 爆震性
knock test 爆震性试验
knock-test engine 爆震试验机

knock testing 爆震作用研究
knock test of gasoline 汽油爆震性试验
knock value 爆震值
Knoevenagel reaction 脑文格反应
Knöfler type extraction apparatus 克涅夫勒式萃取装置
Knoop hardness number 努普硬度值；努氏硬度值
Knoop hardness test 努氏硬度试验
Knoop indentation hardness 努普压痕硬度
Knoop indenter 努普压痕硬度计；努氏压痕硬度计
Knoop microhardness test 努普微硬度试验
Knoop value 努氏(硬度)值
knopite 铈钙钛矿〔钛酸钙矿夹杂氧化铈及氧化铁〕
knopper 鞣伤
knoppern 栎鞣汁
knoppers 鞣伤
Knorr alkalimeter 克诺尔式碳酸定量计
Knorr syntheses 克诺尔合成法
knot ①绳结②结
knot borer 去节机；去节器
knot hole 节孔
knothol mixer 隔膜混合器
knoting strength 打结强度
knot net making 结网机
knot sealer 木节封闭剂
knots sealing 木节封闭
knot strength(=knot tenacity) 打结强度
knot strength ratio 结强度比〔单丝的结强度与拉伸强度之比〕
knot tenacity(=knot strength) 打结强度
knotter （纸浆)节筛
knotter screen 节筛
knotting 去节疤
knotting jar 木节封闭剂专用罐
knotting strength 结强度；打结强度
knotting varnish 木节封闭漆
Knott's colour rules 诺特颜色规律
knotty gneiss 瘤状片麻岩
knotty ore 瘤状矿石
knotty tear(=stick slip tear) 不连续撕裂
knotty wood 多节木料
Knoweles delayed coking process 奴尔斯(外热)延迟焦化过程
Knoweles oven 奴尔斯炉
know how 专门技术；窍门
known ①已知的②已知物
known addition and dilution method 已知添加稀释法
known addition method 已知增量法
known component 已知组分

known sample 已知试样
known solution 已知溶液
known substance 已知物
known subtraction method 已知减量法
Knox cracking 诺克斯裂化法〔气相热裂化法〕
Knox true vapour phase process 诺克斯气相裂化法
Knox unit 诺克斯装置
knoxvillite 叶绿矾
knuckle joint 铰链接合
knuckle thread 圆螺纹
Knudsen burette 海水滴定管；克努森滴定管
Knudsen cell 克努森池
Knudsen diffusion 克努森扩散
Knudsen effect 克努森效应
Knudsen effusion method 克努森隙透法
Knudsen flow 克努森流动
Knudsen force 克努森力〔分子间第四种力〕
Knudsen gauge 克努森真空计
Knudsen ion source 克努森离子源
Knudsen pipette 海水吸移管；克努森吸移管
Knudsen's law 克努森定律
knurl ①滚花②瘤
knurled deflector 凸形导向器
knurling 滚花
Kobel nephelometer 柯贝尔(散射)浊度计
kobenomycin 神户霉素
Kober reagent 科伯试剂
Koch's acid 科赫酸；α-萘胺-3,6,8-三磺酸 $NH_2C_{10}H_4(SO_3H)_3$
Koch's sterilizer 科赫杀菌器
koechlinite 钼铋矿
Koettstorfer number(=saponification value; Koettstorfer value) 皂化值
Koettstorfer value(=saponification value; Koettstorfer number) 皂化值
Kohlrausch bridge 科尔劳许式电桥
Kohlrausch coulometer 科尔劳许电量计
Kohlrausch's dilution law 科尔劳许稀释定律
Kohlrausch's square root law 科尔劳许平方根定律
Kohorn continuous spinner 科霍恩式连续纺丝后处理(联合)机
kojibiose 曲二糖；2-葡萄糖-α-葡萄糖苷
kojic acid 曲酸 $C_6H_6O_4$
Koka flow tester 高化型流变仪
koksaghyz(=koksagyz) 青橡胶草
koksagyz(=koksaghyz) 青橡胶草
koksagyz rubber 青胶蒲公英橡胶
kokubumycin 国分霉素

kokum butter 烛果油〔取自 *Garcinia indica* 的种子〕
kokusagine 香草木碱
kola 可拉(籽)
kolanin 可乐果苷
kolared 可拉红
Kolbe reaction 科尔伯反应
Kolbe-Schmitt carbonation 科尔伯-施密特羧化法
Kolbe synthesis 科尔伯合成
Kolene salt 科伦盐〔熔媒由硝酸钠和亚硝酸盐等组成〕
kollagraph 焊接载量测定器
Kolle flask 克氏扁瓶；甲鱼形瓶
kolm 柯姆煤〔Ronnuma 地方产的富氢煤〕
Kolthoff's buffer solution 柯耳蜀夫缓冲溶液
koluophthisin 考洛杀痨素
Komagome pipette(=safety pipette) 安全移液管
komamycin 古满霉素
komanic acid 吡喃酮-α-甲酸
kondang wax 榕树蜡〔参看 gondang wax〕
kondurangin(=decahydrodihydroxymethoxy fluorenone) 康杜然精；十氢二羟基甲氧基芴酮 $C_{14}H_{22}O_4$
kondurango 南美牛奶藤
konel 铁钛钴镍合金
konesin(=conessine) 地麻素
Konig dye 康尼染料
Konig method 康尼方法
Konig's pyrometer 康尼高温计
konimeter 计尘器；空气尘量计
konkoniogravimeter (空气)尘量计
koniscope 空气尘量计；计尘器
konjaku flour 魔芋粉，蒟蒻粉〔膏化剂〕
konoscopic observation 锥光观察
Konrich method 康里奇氏抗酸细菌染色法
kontrastin (用作比较介质的)氧化锆
Koopmans' theorem 库普曼斯定理
koosmie 库斯米〔印度紫胶品系〕
Koppers-Hasche furnace 科佩斯-哈舍炉
Koppeschaar solution 科别沙尔试液〔即 0.1mol/L 溴液〕
Koppeschaar's titration 科别沙尔滴定
koppite 重烧绿石
Kopp's law 科普定律
Kopp's law of additive volume 科普容积加和定律
koprosterol 粪甾醇
kopsidine 柯蒲定
kopsine 柯蒲碱
kopsingarine 柯蒲加碱
K-orbit K 轨道
kordofan gum 金合欢胶；阿拉伯树胶
korenyl 烷基苯〔别名〕

Korner method　科纳法
kornerupine　钠柱晶石
koronit　柯罗炸药；柯罗那特
Kosanch process (for nickel)　科散奇(制镍)法
kosin　苦苏素
koso　苦苏
kosotoxin　苦苏素
Kossel press　科泽耳式压钠丝管
Kossuth cell (for bromine)　科修思(制溴)电池
Kostanecki acylation　科斯塔尼基酰化作用
Köster's interferometer　科斯特干涉仪
kotoin(=cotoin)　柯桃因
kotomycin　古藤霉素
Köttstorfer number　皂化值
koumine　阔胺
kouminidine　阔胺定
koumiss　马乳酒
kounidine　钩吻碱
Kourbatoff's reagents　考尔巴托夫蚀刻液
koussein　苦苏花素
koussin　苦苏树脂
kousso　苦苏
Kovar(=covar)　科伐(共膨胀)合金
Kovar seal　科伐封接
Kováts index　科瓦茨指数
Kováts retention index　科瓦茨保留指数
Kováts retention index system　科瓦茨保留指数系统
koyamaki oil　日本金松油
Kozeny-Carman equation　康采尼-卡曼方程
Kozeny equation　康采尼方程
K-radiation　K 辐射
Kraemer-Sarnow method　克雷默-萨诺法〔测熔点〕
Krafft point　克拉夫特点*
Krafft temperature　克拉夫特温度〔离子型表面活性剂在水中溶解度陡增的温度〕
kraft(=kraft paper)　牛皮纸；包皮纸
kraft bag paper　牛皮袋纸
kraft liquor　牛皮液；硫酸化液〔造纸〕
kraft mill　牛皮纸工厂
kraft paper(=brown packing paper;kraft)　包皮纸；牛皮纸
Kraft point　临界溶解温度；克拉夫特点〔加热1%的表面活性剂乳浊分散液至溶液立即变成透明时的温度〕
kraft process　牛皮纸浆制法；硫酸盐法制浆
kraft pulp　牛皮纸浆；硫酸盐纸浆
Krames-Kroning's relation　克雷梅斯-克罗尼格关系(式)
krantzitc　树脂石
K-ratio method　系数倍率法
Kratky-porod wormlike chain　克拉特基-波罗蠕虫状链

krausen　涡纹
Kraut's reagent　克劳特微量化学试剂
Krebs unit(s)　克雷布斯黏度单位〔测定高稠度的单位；特别用于颜料浆〕
K-region　K 区
Kreis test　克赖斯三碳不饱和醛试验
kremersite　氯钾铵铁矿
Kremnitz white　克雷姆尼茨白
krennerite　针碲金银矿；斜方碲金矿；白碲金银矿
kresatin　乙酸间甲酚基酯
krigenine　克瑞精　$C_{18}H_{21}O_6N$
krilium　水解聚丙烯腈的钠盐
krinsin　脑氨脂
krith　克瑞〔气体密度的单位名〕
krocodylite　青石棉；虎睛石
kröhnkite　柱钠铜矾
kromycin　克洛霉素〔苦霉素 picromycin 的降解产物〕
Kronecker's delta　克朗内克符号〔数〕
Kronstein's number　克隆斯坦值〔植物油减压蒸馏馏出物量与原油之比，以%表示〕
krügite　镁钾钙矾
kryogen blue　冰精蓝
kryogenin　①冷却剂②3-脲氨基苯甲酰胺
kryometer(=cryometer)　低温计
kryptic acid　氪酸
kryptidine　2,4-二甲基喹啉
krypto〔希腊词头〕　隐匿；潜在
kryptocurine　隐箭毒素
kryptocyanin　隐花青色素
kryptol　石墨、碳化硅、黏土的混合物
krypton　氪〔36号元素，化学符号 Kr〕
krypton adsorption　氪吸附
kryptonate　氪酸盐
kryptonated　氪化的
kryptonated silica　氪化硅胶
kryptonation　氪化(过程)
krypton ionization detector　氪电离检测器
krypton lamp　氪气灯
krypton washout technique　氪洗出术
krypton-water　氪水（包合物）
kryptopyrrole　隐吡咯；2,4-二甲基-3-乙基吡咯
kryptosterol　隐甾醇；羊毛甾醇
kryptoxanthin　隐黄质；玉米黄质
Krystal crystallizer　克里斯塔尔结晶器
krystalglass　①富铅玻璃②富铅玻璃器
Krystal-Oslo type crystallizer　克里斯塔尔-奥斯罗型结晶器
kryton oxide　氧化氪

K-series　K(线)系
kuchersite　油页岩；波罗的海油页岩
kueaf　洋麻
Kugelrohr apparatus　库格尔若蒸馏器
kuhlmannin　6-羟(基)-7,8-二甲氧-4-苯香豆素
Kuhlmann's microbalance　库尔曼微量天平
Kuhn column　库恩塔；多管填料塔
Kuhn-Kuhn equivalent segment length　库恩-库恩等效链长
Kuhn segment length　库恩链长
kuh-seng　苦参
kujimycin　久慈霉素
kukersite　库克油页岩
kukoline　木防己碱
kullgren acid　低磺酸
kullgren lignin　低磺酸化木素
kumquat　金橘
kundrymycin　昆追霉素
Kundt effect　孔脱效应
Kundt's constant　孔脱常数
Kundt's rule　孔脱法则
Kundt's tube　孔脱管
kunhi　滤渣胶
Kunkel's zinc sulfate test　孔克耳硫酸锌检验法
kunomycin　久野霉素
kunzite　紫锂辉石
kupfernickel　红砷镍矿
kupferron　(试)铜铁灵
kupholite　苦伏石
kupramite　吸氨剂
kupramite canister　胆矾吸氨剂滤毒罐
kurchi　止泻木根
kurchine　枯尔钦；止泻木碱
Kurie plot(=Fermi plot)　库里线图〔同费米线图；求β射线最大能量用〕
kuromatsuol　刺柏醇
kuromoji oil　大叶钓樟油
kurrajong oil　枯拉仲油
Kurrol's salt　四聚偏磷酸钾
Kurtz method　库尔兹法〔结构族组成分析法〕
kusamba　玫瑰鸦片浆
kusmi lac　库斯米(品系)紫胶
kussin　苦苏树脂
kuteera gum　刺槐树胶
k value　k 值；黏度(值)
kvar　千伏安与千瓦的比例
kwells　天仙子碱氢溴酸盐
kyanite　蓝晶石
kyanizing process　氯化汞浸注法
kymograph　血压改变记录器
kynurenic acid　犬尿喹啉酸；4-羟基-2-喹啉酸
kynurenine　犬尿氨酸
kynuric acid　犬尿酸；N-草酰邻氨基苯甲酸　$C_9H_7O_5N$
kynurine　犬尿碱；4-羟基喹啉
kyroscopy　①凝固点测定分子量法②产生低温法③合金冷却时现象的研究
kyrtometer　曲度计；曲面测量计
kytomitome　胞质网丝
K_γ factor　K_γ 常数；K_γ 因子〔γ射线剂量率单位，一毫居里点源在距离一厘米处一小时的伦琴数〕

L

l. 左型；左旋
Labarraque solution 拉巴腊克液〔次氯酸钠的水溶液〕
labdan-8,15-diol 赖百当烯-8,15-二醇
labdanolic acid 赖百当酸
labdanol(=isobutyl cinnamate) （肉）桂酸异丁酯
labdanone 岩蔷薇酮；赖百当酮
labdan-8,15,19-triol 赖百当-8,15,19-三醇
labdanum(=rockrose) 岩蔷薇；赖百当〔俗〕
labdanum absolute 岩蔷薇净油；赖百当净油〔俗〕
labdanum concrete 岩蔷薇浸膏；赖百当浸膏〔俗〕
labdanum (gum) 岩蔷薇胶；赖百当胶
labdanum leaf oil 岩蔷薇叶油；赖百当叶油〔俗〕
labdanum oil 岩蔷薇油；赖百当油〔俗〕
labdanum resin 岩蔷薇树脂；赖百当树脂〔俗〕
labdanum resinoid 岩蔷薇香树脂；赖百当香树脂〔俗〕
label ①贴商标②标签
labeled atom 标记原子
labeled compound 标记化合物
labeled molecule 标记分子
labeled soil 标记污垢
labeled substrate 标记底物；标记(酶)作用物；标记基体
label gummer 标签涂胶机
labeling kit (放射性同位素)标记药(物)盒
labeling of monoclonal antibodies 单克隆抗体标记
labeling of radioactive substance 放射性标记
labeling pattern 标记位型；标记式样；标记类型
labeling technique 标记技术；示踪技术
label lacquer 标签漆
label(l)ed atom 标记原子；示踪原子
labelled compound 标记化合物
labelled formaldehyde 标记甲醛；示踪甲醛
labelled pesticide 标记农药
labelled size 标注尺寸；标码
labelled substrate 标记底物；标记(酶)作用物
label(l)er 贴标签机
label(l)ing ①加标记②贴商标
labelling and measuring template 标记与测量模板
label(l)ing machine 贴标签机
labetalol 拉贝洛尔〔药〕
labile 不稳定的；易变的
labile acid 不稳定酸
labile complex 易变配位化合物
labile coordination compound 活性配合物
labile cross-linking 不稳定交联

labile equilibrium 不稳平衡
labile form 不稳形
labile hydrogen 不稳定氢
labile hydrogen atom 不稳定的氢原子
labile ionic bond 不稳定离子键
labile pigment surface 不稳定颜料表面
labile region 不稳定区；易变区
labile state 不稳定态；易变态
lability 不稳定(性)
labilization 使不稳定；使易变
labilized hydrogen atom 活化的氢原子
labilomycin 易毁霉素；抗分枝杆菌抗生素
lab mill 实验室用塑炼机
lab-on-chip 芯片实验室
laboratory ①实验室；试验室②试验所
laboratory apparatus 实验仪器
laboratory apron 实验围裙
laboratory automation 实验室自动化
laboratory coat 实验外衣
laboratory corrosion test 实验室腐蚀试验
laboratory engine test 实验室内燃机试验
laboratory evaluation 实验室(质量)值
laboratory hood 实验室通风橱
laboratory information management system(LIMS) 实验室信息管理系统
laboratory manipulation 实验室管理
laboratory networking 实验室联网化
laboratory octane number 实验辛烷值
laboratory procedure 实验室研究方法
laboratory reagent 实验室试剂
laboratory sample 实验室样(品)
laboratory simulation 实验室模拟
laboratory size extruder 试验用压出机
laboratory size reactor 实验室型反应器
laboratory test 实验室试验
laboratory test-engine method 实验室试验内燃机法
laboratory timer 实验室自动定时仪
laboratory washing machine 实验室洗涤机
laborious test method 繁复的试验方法
Labour pump 拉博尔泵
labradite 钙钠斜长石
labradorite 拉长石；曹灰长石
Labrador tea oil 喇叭茶油；矶蹋躅油〔取自 *Ledum palustre*〕

lab-size equipment　试验用设备
laburnine(=d-trachelanthamidine)　毒豆碱
labware　实验室器皿
labyrinth box　迷宫(式密封)箱
labyrinth-grease seals　迷宫式润滑脂封闭器
labyrinth oil retainer　迷宫式集油器
labyrinth packing　迷宫式填充物
labyrinth piston compressor　迷宫活塞式压缩机
labyrinth pressure ratio　迷宫式密封压力比
labyrinth seal　迷宫式密封
labyrinth sealing　迷宫式封闭
labyrinth structure　迷宫式结构〔熔体过滤器的多孔结构〕
labyrinth teeth　迷宫齿
labyrinth type mud excluder　迷宫式防泥器
labyrinth viewing device　迷宫式窥视装置
labyrinth with sharp-edged strip　锐边带形迷宫密封
lac　①紫胶；虫胶②乳
lac acid　紫胶酸
laca coerula(=lacmus)　石蕊
lacca　紫胶虫
lacca coerula(=litmus)　石蕊
laccaic acid　虫胶红酸〔从紫胶所得〕；紫胶色酸；虫膝酸
laccal　虫漆酚
laccerane　紫胶烷；三十二烷
lacceric acid(=lacceroic acid)　紫胶蜡酸；三十二烷酸；虫胶蜡酸
lacceroic acid(=lacceric acid)　紫胶蜡酸；三十二(烷)酸
laccerol　紫胶蜡醇；三十二(烷)醇；虫胶蜡醇
laccifer lacca　紫胶虫
laccijalaric acid　紫胶壳脑醛酸；虫胶紫茉莉酸；紫草茸酸
laccijalaric ester　紫草茸酸脂
laccishellolic acid　紫胶壳脑酸
laccol　葛漆酚；(虫)漆酚
laccolith　岩盖
lac dye(=lac lake)　紫胶染料
lace curtaining　网帘〔漆膜病态〕
lac encrusted twig　紫梗
lacerate　划破
L acid　L酸；1-萘酚-5-磺酸
lacidipine　拉西地平〔药〕
lacing　编丝；导线
lacinilene　青榆烯
lac insect　紫胶虫
lacker　①漆②上漆③漆器
lacking elasticity　缺乏弹性

lacking toughness　缺乏韧性
lack of coating　漏涂
lack of drying　干性不足；不够干
lack of fill-out　浸润不足〔增强塑料缺陷〕
lack of flexibility　缺乏柔韧性
lack of fuel　燃料缺乏
lack of hiding　遮盖力弱；遮盖不足
lac-la(c)ke(=lac dye)　紫胶染料
lacmoid(=resorcin blue)　间苯二酚蓝
lacmosol　石蕊萃
lacmus　石蕊
lacquer　①天然漆；大漆②挥发性漆③硝基纤维素；喷漆
lacquer article(s)　漆器〔大漆〕
lacquer blue　漆蓝
lacquer coat　漆涂层
lacquer deposit　漆沉积
lacquer diluent　漆冲淡剂
lacquered surface　喷漆表面
lacquer emulsion　硝基漆乳液；硝基乳化液
lacquer enamel　①挥发性瓷漆②硝基瓷漆
lacquerer　(油)漆工
lacquer film　漆膜〔一般指挥发性漆〕
lacquer finish　①(硝基)面漆②涂末道喷漆
lacquer for fabric　纺织品用(硝基)漆
lacquer for laser disc　激光唱片漆
lacquer formation　漆生成
lacquer for metal　金属用硝基漆〔挥发性漆〕
lacquer hanging panel　大漆挂屏〔大漆〕
lacquer head　喷漆分配头
lacquering　①上漆②漆涂层
lacquering machine　喷漆机
lacquer lifting　漆咬底；漆起皱
lacquer-like protecting layer　类漆保护层〔防腐蚀〕
lacquer mar　漆伤；漆咬〔PVC 软质薄膜涂漆后因增塑剂迁移而产生的表面损伤〕
lacquer mar(ring) resistance　抗漆伤性；抗漆咬性
lacquer oil　喷漆(用)油〔可用作硝基喷漆组分的油〕
lacquer petroleum　漆用汽油
lacquer primer　①硝基底漆②挥发性漆底漆
lacquer print　漆印
lacquer putty　①硝基漆腻子(油灰)②灰地〔漆器〕
lacquer sanding sealer　可打磨的硝基封闭剂
lacquer sealer　硝基纤维封闭底漆〔不加填料和颜料〕
lacquer sheathing　漆覆盖层；漆保护皮
lacquer softener　漆用软化剂；硝基增韧剂
lacquer solvent　漆用溶剂
lacquer surfacer　①硝基(纤维)二道浆；硝基纤维整面漆②二道底漆

lacquer test 漆(沉积)试验
lacquer thinner 挥发性漆稀释剂；喷漆稀料
lacquer wares 漆器〔大漆〕
lacquer wax 漆蜡
lac resin 紫胶树脂
lacrimation 催泪(作用)
lacrimator(=lacrymator) 催泪剂
lacrimatory(=lacrymatory) 催泪的
lacrimatory agent 催泪剂
lacrimatory candle 催泪(烟)罐；催泪烛
lacrimatory gas 催泪(性毒)气
lacry- 泪
lacrymator(=lacrimator) 催泪剂
lacrymatory(=lacrimatory) 催泪的
lactacidogen 乳酸精；半乳糖—磷酸 $C_6H_{11}O_5OPO_3H_2$
lactal 呋喃葡烯糖-5-半乳糖苷
lactam 内酰胺*；乳胺
lactam form 内酰胺式
lactamic acid 丙氨酸；2-氨基丙酸 CH_3CHNH_2COOH
lactamide(=lactic amide) 乳酰胺 $CH_3CHOHCONH_2$
lactamidine 乳脒 $CH_3CH(OH)C(NH)NH_2$
lactamine 丙氨酸
lactamization 内酰胺化作用
lactam-lactim tautomerism 肽链(中的)酮醇互变(异构)(现象)
lactan 内酰胺
lactanilide N-乳酰苯胺
lactarazulene 乳菇薁素
lactarinic acid 十八碳-6-酮酸
lactaroviolin 乳菇紫素
lactasin 乳酶生〔药〕
lactate ①乳酸②乳酸盐(酯或根)
lactate condensation compound 乳酸缩合物
lactazam 内酰联胺
lactazone 酯肟
lactcbionic acid 乳糖酸
lactenin 乳烃素
lactescent ①乳的②乳状的
lactescent plant 产胶植物
lactic acid(=milk acid) 乳酸；2-羟基丙酸；丙醇酸 $CH_3CHOHCO_2H$
lactic acid amide 乳酰胺 $CH_3CH(OH)CONH_2$
lactic acid anhydride 乳酸酐 $(CH_3CHOHCO)_2O$
lactic acid series 乳酸系；邻位羟基酸系 $C_nH_{2n}OHCOOH$
lactic aldehyde(=2-hydroxypropion-aldehyde) 乳醛；α-羟基丙醛；丙醇醛 $CH_3CHOHCHO$
lactic amide(=lactamide) 乳酰胺

lactic anhydride 乳酐 $C_6H_{10}O_5$
lactic nitrile 乳腈；丙醇腈；2-羟基丙腈
lactide(s) ①丙交酯②交酯③减水乳酸
lactiferous plant 产胶植物
lactim 内酰亚胺；乳亚胺
lactim form 内酰亚胺式
lactin 内酰亚胺
lactitol(=lactobiosit) 乳糖醇
lacto- 〔词头〕 乳
lactobionic acid 乳糖酸
lactobiose 乳二糖
lactobiosit(= lactitol) 乳糖醇
lactochrome 乳黄素；核黄素；维生素 B_2
lacto-enoic tautomerism 内酯-烯酸互变现象
lactoflavine 乳黄素；核黄素；维生素 B_2 $C_{17}H_{20}O_6N_4$
lactoglyceride 甘油乳酸脂肪酸酯
lactoglycerol 甘油、乳酸和水为 1∶1∶1 的混合物
lactol ①乳醇②邻位羟基内醚〔羟基醛或羟基酮的内醚式〕③内半缩醛*
lacto-lactic acid 乳乳酸 $CH_3CHOHCOOCH(CH_3)COOH$
lactone 内酯*
lactone colouring matters 内酯染料
lactone isomerism 内酯异构现象
lactone process 内酯法〔由美国研制的以环己烷为原料；通过己内酯制取己内酰胺的工艺名称〕
lactone ring 内酯环
lactone ring five membered 五元内酯环
lactone rule 内酯规则
lactonic acid 内酯酸
lactonic leuco-dye 内酯型无色染料；内酯基隐色染料
lactonic ring 内酯环
lactonitrile 乳腈；2-羟基丙腈；丙醇腈 $CH_3CHOHCN$
lactonization 内酯化作用
lactophane 玻璃纸
lactophenine 乳吩吁〔俗〕；N-乳酰乙氧苯胺 $C_2H_5OC_6H_4NHCOCHOHCH_3$
lactoprene 聚酯橡胶
lactoscope 乳酪计
lactose(=milk sugar) 乳糖 $C_{12}H_{22}O_{11}\cdot H_2O$
lactose alkanoate 链烷羧酸乳糖酯
lactoside 乳糖苷
lactosum 乳糖
lactoxime 酯肟
lactoyl(=lactyl) 乳酰；丙醇酰；α-羟丙酰 $CH_3CH(OH)CO-$
lactoyltetrahydropterin 乳酰四氢蝶呤
lactron thread 生橡胶纱线；生橡胶线

lactucarium　山莴苣膏
lactucerol　山莴苣醇
lactulose　乳果糖；二蔗酮糖
lactyl(=lactoyl)　乳酰；丙醇酰；α-羟丙酰　$CH_3CHOHCO$—
lactylic anhydride　乳酸酐　$(CH_3 \cdot 2CHOH \cdot 2CO)_2O$
lactyl lactate　缩二乳酸
lactyl-lactic acid　乳酰乳酸　$CH_3CHOHCOOCH(CH_3)COOH$
lactyl urea　α-乳酰环脲；5-甲代乙内酰胺
lacuna　①空斑②裂孔③气泡
lac varnish　①紫胶清漆；光漆②(=polish)凡立水
lac wax　紫胶蜡
lacy　花边状的；网状的
ladanum(=laudanum)　岩蔷薇；赖百当〔俗〕
ladanum oil　岩蔷薇油；赖百当油〔取自 Cistus creticus〕
ladanum resin　岩蔷薇树脂；赖百当树脂〔俗〕
ladder chain　梯形链
ladder macromolecule　梯形聚合物
ladder -polyimidazopyrrolone　梯形聚苯并咪唑吡咯酮
ladder polymer　梯形聚合物
ladder proofing　防抽丝处理；防脱散处理
ladders　波纹；梯形皱纹〔漆病〕
ladders in painting　梯形纹；梯形痕
ladder structure　梯形结构
ladder study　(用量)递增试验(法)
ladder type network　梯形网络
Ladenburg correction　拉登堡校正
Ladenburg flask　拉登堡式烧瓶
Ladenburg form (of benzene)　(苯的)拉登堡式
Ladenburg formula　拉登堡式
Ladenburg law　拉登堡定律
lading door　装料门
ladle　(长柄)杓
lady's slipper　杓兰根
laevo-(=levo-)　〔拉丁字头〕①左的②左旋的
laevo-configuration　左旋构型
laevoglucose　左旋葡萄糖；果糖
laevoisomer　左旋异构体
laevopimaric acid　左旋海松酸
laevorotary(=laevorotatory)　左旋的
laevo-rotation　左旋(现象)
laevorotatory(=laevorotatory)　左旋的
laevorotatory oil　左旋油
laevulic acid　乙酰丙酸；左旋酸　$CH_3COCH_2CH_2COOH$
l(a)evulinic acid　乙酰丙酸
l(a)evulose　果糖；左旋糖
lag　①落后；滞后②加保温套

lagam balsam　柯拜巴脂
lag coefficient　滞后系数
lagged　①护热的②落后
lagged piping　已保温的管道
lagged pulley　带套鼓轮〔运输带用〕
lagged surface　护热面
lagging　①落后；迟缓②外包〔保温层〕
lagging material　保温材料
lagging shadow　(色谱斑的)落后阴影
lag knock　延迟爆击
lagoon　储留池；氧化塘；生物塘
lagoriolite　硅酸铝钠
lagos silk rubber　绢丝橡胶
lag phase　①停滞阶段②延滞期
Lagrange interpolation　拉格朗日内插
Lagrangian multiplier　拉格朗日乘子
lag time　落后时间；时间差距
laid ledger paper　直纹账簿纸
laid length　敷管长度
laidlomycin　莱特洛霉素；来洛霉素
laid paper　直纹纸；夫士纸
laid-up　拆卸修理
laitance　(水泥)翻沫；浮浆〔混凝土表面的沫状物〕
laitier　浮渣
lake　①色淀；沉淀色料②湖
lake asphalt　湖沥青
lake bed　湖泊沉积矿
lake brine　湖泊盐水
lake colors　色淀染料
laked　漆的
lake dyes　色淀染料
lake pigment(=lake)　色淀颜料；色淀
lake pitch　湖沥青
lake sediment　湖底沉积物
lake toner(=toner)　色淀色原
lake white　色淀白〔矾土白的别名〕
lakh　紫胶
laking agent　固色剂
laksholic acid　壳脑酸
Lalande cell　拉兰德电池
lallemantia oil　拉曼油〔取自 Latlemantia iberica 的籽〕
lambda factor　λ因子
Lambert's law of reflection　朗伯反射定律
Lambiotte retort　兰姆氏釜〔一种早期连续内热干馏釜〕
Lamb-Recht's polymer　湿度温度联测计
lamb reverse(=shearling)　剪毛绵羊革
lamb skin　绵羊皮
lambswool　羔毛织物

lamella 薄片；薄板
lamellar 薄片状的；层状的
lamellar aluminosilicate 层状硅铝酸盐
lamellar boundary slip 层流界面滑移(动)；层流边界滑移
lamellar buff(=flap wheel) 叠层抛光轮
lamellar compound 层状化合物
lamellar crystal 片晶
lamellar deformation 层状形变
lamellar emulsion 层状(液晶)乳液
lamellar-extrusion technique 层状挤压成形技术
lamellar lattice 片晶晶格
lamellar micelle 层状胶束
lamellar model 层状模型
lamellar motion 层状运动
lamellar phase 层状相
lamellar pigment 片状颜料
lamellar rotation 层状旋转
lamellar spacing 层间隙
lamellar structure ①层状(片状)组织(结构)②薄片结构③片晶结构
lamellar twinning during rolling 滚动时层状成对
lamellar type 层型
lamellated 成层的；层状的
lamellated fracture 成层断裂
lamelliform particle (薄)片状粒子
Lamés constant(=Lamés modulus) 拉梅常数
Lamés modulus(=Lamés constant) 拉梅常数
Lamex process 拉米克斯过程〔美国洛斯·阿拉莫斯研究所研究的从钚合金中用电解法提取钚的过程〕
lamina ①薄片；薄板②层状体
laminagraphy 分层照相；层层显影
laminal filler 片状填料
laminar 薄片状的；层状的
laminar boundary layer 层流边界层
laminar composite 层状复合材料；层压复合材料
laminar displacement 层状位移
laminar distortion 层状畸变
laminar film 层流薄膜
laminar flame 层流焰
laminar flame stability 层流火焰稳定性
laminar flow 层流
laminar flow burner 层流燃烧器*
laminar flow cell 层流池
laminar flow extent 层流区
laminar flow zone 层流区
laminar fracture 成层断裂
laminaria 昆布
laminaribiose 昆布二糖

laminarin 昆布多糖；海带多糖
laminariose 昆布糖；海带糖
laminaritol 海带醇；昆布醇
laminar layer 薄片层
laminar liquid film 层状液膜
laminar scale ①层状锈片(皮)；厚锈片②层状氧化皮；片状铁鳞
laminar separation 层流分离
laminar shear 层流剪切
laminar shear flow 层流剪切流
laminar steady flow 层状稳流；稳层流
laminar structure 片状结构
laminar sublayer 层流底层；近壁层流层
laminar-turbulent flow 层流湍流；紊层流
laminar-turbulent transition 层流-湍流转变；层流-湍流过渡
laminar wake 层流尾涡
laminate 层压材料*；层压制品
laminated 成层的；薄片的；层压的
laminated coal 叠层煤
laminated cover 叠层覆盖
laminated covering 薄片覆盖
laminated fabric 胶合织物；叠层织物
laminated film ①层压膜②层叠式漆膜〔抗张强度大，比一层厚膜好〕
laminated glass 层压玻璃
laminated insulation 层状绝缘
laminated magnet 叠层式磁铁；叠片磁铁
laminated material 层压材料
laminated mo(u)lded tube 层状模塑管材
laminated moulding 层压模制品
laminated panel 层压板
laminated phenolic resin 层压酚醛树脂
laminated plastics 层压塑料(制品)
laminated polyethylene film 分层聚乙烯膜
laminated product ①胶合制品②层压制品
laminated section 层压型材
laminated slate 层状页岩；薄片状页岩
laminated spinneret 叠层型复合喷丝头
laminated structure 层状结构；层纹构造
laminated thermosetting plastics 层压热固塑料
laminated vessel 多层式容器
laminated wood 层压板
laminate molded method 层压模压法
laminate molding 层塑法；层压模塑法
laminate plate theory 层合板理论
laminating ①分成薄层②卷成薄片③包以薄片④层压
laminating adhesive 层压黏合剂

laminating agent ①层压剂②层压用树脂；层压用黏合剂
laminating material 层压材料；成层材料
laminating press 层压压机
laminating resin 层压树脂
laminating varnish ①层压树脂液②层合树脂液③层压用清漆
laminating wax 粘贴用蜡
laminating web 层合基料
lamination 层压；层合；分层；夹层
lamination coating 多层贴合〔薄膜等〕
laminator 压膜机
lamine 野芝麻花碱
laming 薄层；薄板
laminine 昆布氨酸；N^6-三甲基赖氨酸内盐
laminography 分层放射摄影图
lamioside 野芝麻苷
lamivudine 拉米夫定〔药〕
Lamm scale (displacement) method 拉姆标度位移法
lamp-base cement 灯泡黏合剂〔电灯泡与其铜头的黏合剂〕
lamp black(=vegetable black) 灯黑；油烟
lamp black and oil blacking 灯黑及油烟涂黑法
lamp burning test 燃灯试验
lamp glass 灯玻璃
lamp method 灯法〔试硫用〕
lamp oil 灯油
lamprey liver oil 八目鳗鱼肝油
lamprobolite 玄武角闪石
lamprophyllite 闪叶石
lamprophyre 煌斑岩
lamp sulfur test 含硫量燃灯试验
lampterin 月夜蕈素
lampterol(=lunamycin) 月夜蕈醇〔即：月亮霉素〕
lamp test 灯试法〔试硫用〕
lana 羊毛
lanadoxin 毛花洋地毒苷
lanafolein 毛花叶英；毛花洋地黄富林苷
lanafuchsine 兰纳品红
lanain 潮湿的羊脂
lanara 酪素纤维
lanarkite 硫酸铅矿
lanasol brown 兰纳盐棕
lanasol colours〔复〕 兰纳盐染料
lanasol orange 兰纳盐橙
lanatoside 毛花(洋地黄)苷
lanatoside C 毛花苷丙〔药〕
lance ①喷枪；喷水器②长矛；双刃小刀
lanced fin 锯齿形翅片

lanceol 澳白檀醇〔取自柳叶檀香 Santalum lancealatum 的油〕；澳檀醇
land 合模面；焊垫
land area 水平承压面积〔压模〕
land carrying capacity 土地承载能力
land disposal 埋于地下
landed mould 凸缘(半溢式)塑模
landed plunger 带平台的柱塞
Lande equation 朗德方程
Lande g factor 朗德g因子
Lande splitting factor 朗德劈裂因子；朗德g因子
land ice 陆(原)冰
landing head （悬挂油罐用)油管头
Landmark process 兰德马克二氮化钛制氨法
land of mold 水平承压面；承压边缘〔压模〕
land-pebbles〔复〕 磷灰土块
land phosphate ①磷灰土②纤核磷灰石
land plaster 石膏
land reclamation 土地开垦
land restoration 土地恢复
landscape ecology 景观生态学
Landskrona band filter 兰次克龙奈带式过滤机
land storage tank 地上储罐
land surface 水平承压面
land width 螺纹顶宽〔螺杆〕
lanestenyl 羊毛甾醇基
langbeinite 无水钾镁矾 $K_2SO_4 \cdot 2MgSO_4$
Lange solution 兰格溶液〔一种胶态金溶液〕
langley 曝光单位
Langmuir adsorption equation 朗缪尔吸附方程式〔固体吸附剂的吸附率方程式〕
Langmuir adsorption isotherm 朗缪尔吸附等温式*
Langmuir balance 朗缪尔表面压力计；朗缪尔表面膜秤
Langmuir-Blodgett film L-B 膜*
Langmuir equation 朗缪尔方程式
Langmuir film balance 朗缪尔膜天平
Langmuir-Hinshelwood mechanism 朗缪尔-欣谢尔伍德机理*
Langmuir isotherm 朗缪尔等温线
Langmuir-Rideal mechanism 朗缪尔-里迪尔机理*
Langmuir's adsorption isotherm 朗缪尔吸附等温线
Langmuir's surface balance 朗缪尔表面压力计；朗缪尔表面膜测定秤
Langmuir's vacuum gauge 朗缪尔真空表(计)
Langmuir theory 朗缪尔理论
Langmuir theory of adsorption 朗缪尔吸附理论
Langmuir type adsorption 朗缪尔型吸附
lanital 人造羊毛

lankamycin 兰卡霉素
lankavacidin(=lankacidin) 兰卡杀菌素
lankavamycin(=lankamycin) 兰卡霉素
lanoceric acid 羊毛蜡酸；二羟三十(烷)酸
lanocerin 羊毛蜡；甘油三羟蜡酸酯
lanolin(=agnin) 羊毛脂
lanolin acetate 羊毛脂乙酸酯；乙酰化羊毛脂
lanolin alcohol 羊毛脂醇；十二(碳)烯醇
lanolin fatty acid 羊毛脂脂肪酸
lanolinum 羊毛脂
lanopalmitic acid 羊毛棕榈酸；一羟(基)棕榈酸；羊毛软脂酸
lanostadiene 羊毛甾二烯
lanostadienol 羊毛甾二烯醇
lanostane 羊毛甾烷
lanostene 羊毛甾烯
lanostenol(=dihydrolanosterol) 羊毛甾烯醇；二氢羊毛甾醇
lanostenone 羊毛甾烯酮
lanosterine 羊毛甾醇
lanosterol 羊毛甾醇
lansene 椰色木烯〔取自越马椰色木 Lansium annamalayannum 木材的精油〕
lansol 椰色木醇
lansoprazole 兰索拉唑〔药〕
lantadene 岩茨烯
lantana flower absolute 马缨丹花净油
lantana oil 马缨丹油
lantanine 马缨丹碱
lanthana 氧化镧 La_2O_3
lanthanide 镧系元素*
lanthanide contraction 镧系收缩
lanthanide-exchanged zeolite 镧系元素置换的沸石
lanthanide hydride 镧系元素氢化物
lanthanide metallide 镧系元素金属化物
lanthanide series 镧系
lanthanide sesquioxide 三氧化二镧系元素
lanthanide shift reagent 镧系位移试剂
lanthanite 镧石
lanthanoid 镧系元素*
lanthanon ①南山酮；马缨酮②镧系元素〔J.K.Marse 于 1947 年建议此项英文名〕
lanthanum 镧
lanthanum acetate 乙酸镧 $La(C_2H_3O_2)_3$
lanthanum alizarin complexon method 茜素镧络合酮法
lanthanum bromate 溴酸镧 $La(BrO_3)_3$
lanthanum bromide 溴化镧 $LaBr_3$
lanthanum carbonate 碳酸镧 $La_2(CO_3)_3$

lanthanum chloranilate 氯冉酸镧
lanthanum chloride 氯化镧 $LaCl_3$
lanthanum dioxysulfate 硫酸二氧二镧 $(LaO)_2SO_4$
lanthanum fluoride electrode 氟化镧电极
lanthanum hexaboride emitter 六硼化镧发射体
lanthanum hydride 三氢化镧 LaH_3
lanthanum hydropyrophosphate 焦磷酸氢镧 $LaHP_2O_7$
lanthanum hydrosulfate 硫酸氢镧 $La(HSO_4)_3$
lanthanum hydroxide 氢氧化镧 $La(OH)_3$
lanthanum iodate 碘酸镧 $La(IO_3)_3$
lanthanum iodide 碘化镧 LaI_3
lanthanum metaphosphate 偏磷酸镧 $La(PO_3)_3$
lanthanum molybdate 钼酸镧 $La_2(MoO_4)_3$
lanthanum nitrate 硝酸镧 $La(NO_3)_3$
lanthanum orthophosphate (正)磷酸镧 $LaPO_4$
lanthanum oxalate 草酸镧 $La_2(C_2O_4)_3$
lanthanum oxide 氧化镧 La_2O_3
lanthanum oxybromide 溴氧化镧 $LaOBr$
lanthanum sesquioxide 三氧化二镧 La_2O_3
lanthanum sulfate 硫酸镧 $La_2(SO_4)_3$
lanthanum sulfide 硫化镧 La_2S_3
lanthanum tartrate 酒石酸镧 $La_2(C_4H_4O_6)_3$
lanthionine 羊毛硫氨酸
lanthopine 兰索品；兰梭平〔一种鸦片生物碱〕
lantol 兰妥〔即兰索品〕
lanum(=lanolin) 羊毛脂
lap ①余面②盖片③边④搭接⑤涂刷漆模时局部增厚
lapachenole 拉帕车脑；牛蒡脑
lapacho bark 拉帕皮
lapachoic acid 拉帕酸〔即:拉帕醇〕
lapachol 黄钟花醌；2-羟-3-异戊烯基萘醌；拉帕醇
lapacho wood 拉帕木；海榄雌
lapidolite 锂云母
lapis albus 天然硅氟化钙
lapis amiridis 刚玉粉
lapis cal aminaris 异极矿
lapis causticus 熔融氢氧化钠(钾)
lapis crucifer 十字石
lapis divinus 铜矾
lapis fluoris 紫石英
lapis lazuli 天青石
lapis lunaris 熔融硝酸银
lapis-ollaris 壶石；皂石〔滑石的变种〕
lap joint 搭接；叠接
lap joint flange 搭接凸缘
lap-joint strength 拉伸剪切强度
Laplace equation 拉普拉斯方程
Laplace's operation 拉普拉斯运算

Laplace transform 拉普拉斯变换
lap of coil 旋管卷；盘管卷
Laporte rule 拉普特规则
lappa 牛蒡
lappaconitine 高乌甲素；拉杷乌头碱
lapped barrel syringe 磨砂吸管注射器
lapped pulp 稀薄纸浆
lapper 成卷机；成网机；研磨机
lappet 浮纹组织
lapping ①压榨②磨光③重叠④研磨
lapping cloth 衬布
lapping in unison 同向垫纱
lapping machine 铺网机
laps （刷路)搭接处局部过厚〔漆病〕
lap-welded pipe 搭焊管
lapwelding 搭焊；搭头焊接
larane 二十(碳)烷
larch agaric 落叶松蕈
larch bark 落叶松皮；落叶松拷胶
larch extract 落叶松拷胶
larch leaf oil 落叶松叶油
larch turpentine 威尼斯松脂
large amplitude oscillatory shear flow 大幅振荡剪切流
large angle strain 大角张力*
large areas of pitting 大面积点蚀
large calcium carbonate 重质碳酸钙〔填料〕
large centaury 海红矢车菊
large chemical complex 大型化学联合企业
large deflection 大挠度
large deformation 大形变
large deformation theory 大形变理论
large-diameter pipe 大口径管
large-duty 高生产率的
large elastic deformation 大弹性形变
large galanga oil 高良姜油
large-lot producer 大型企业
large outside diameter pipe 大外径管
large part 大型模制品；大规格模制品
large-particle-size ①大颗粒②大粒度
large patches 大片斑纹状(泡)
large plate process 大板作业法〔用固定的圆盘及转动的灯和刮板生产炭黑法〕
large pore gel 大孔凝胶
large pore microparticulate column 大孔微粒填料柱
large pore open tube column 大孔径开管柱
large pore separation gel 大孔分离凝胶
large pore spacer gel 大孔间隔凝胶
large production control die LPC 口型
large ring 大环
large ring compound chromatography 大环化合物色谱
large ring lactone 大环内酯
larger particle size dispersion 粒子较大的分散体(液)
large sample 大样本
large scale chromatography 大型色谱法
large-scale experiment 大规模试验
large-scale field test 大规模野外试验
large-scale filtration 工业规模过滤
large shape of brick 大砖模
largest peak 最大峰
largest peak current 最大峰电流；顶点电流；峰电流
largest peak potential 最大峰电位；顶点电位；峰电位
large unilamellar vesicles(LUV) 大单层脂质体
large volume item 大量生产制品
lariat molecule 套索型分子
laricic acid 松蕈酸；落叶松蕈酸
laricinoleic acid 落叶松脑酸
lariciresinol 落叶松树脂醇 $C_{20}H_{24}O_6$
larixine （落叶)松皮素
larixinic acid 落叶松酸
larkspur 飞燕草
larkspur seeds 飞燕草籽
Larmor precession of nuclear magnetic moment 核磁矩的拉莫尔进动
Larmor theorem 拉莫尔定理
larnite 甲型硅灰石 Ca_2SiO_4
larocaine 拉罗卡因〔药，局部麻醉剂〕
larodon 拉罗东〔镇痛退热剂〕
lasalocid 拉沙里菌素；拉沙洛西
lasecon 激光转换器
laser ablation 激光烧蚀；激光消融
laser ablation-resonance ionization spectroscopy(LARIS) 激光烧蚀-共振电离光谱法
laser ablation system 激光烧蚀系统
laser ablative resonance ionization spectroscopy 激光烧蚀共振电离光谱法
laserable material 能发激光的材料；激光材料
laser annealing 激光退火技术
laser-assisted chemical etching 激光辅助化学腐蚀
laser atomizer 激光原子化器
laser beam densitometer 激光束密度计
laser beam excitation 激光束激发
laser beam ionization 激光束电离
laser chromatography 激光色谱
laser confocal fluorescence microscopy 激光共聚焦荧光显微镜
laser desorption(LP) 激光解吸

laser desorption ionization   激光解吸电离
laser desorption ionization source   激光解吸电离源
laser desorption mass spectrometry(LDMS)   激光解吸质谱法
laser desorption technique   激光解吸法
laser detector   激光探测器
laser diffractometer   激光衍射仪
laser diffractometry   激光衍射法
laser diode   激光二极管
laser drilled pinhole   激光钻孔限流器
laser drilling   激光穿孔
laser driven deposition   激光驱动淀积
laser dyes   激光染料
laser enhanced ionization(LEI)   激光增强电离
laser enhanced ionization spectrometry   激光增强电离光谱(法)
laser enhanced ionization spectroscopy   激光增强电离光谱
laser etching   (高分子)光分解蚀刻
laser excited atomic fluorescence spectrometry   激光激发原子荧光光谱法
laser extraction chemistry   激光萃取化学
laser fibre   激光(玻璃)纤维；激光学纤维
laser fibre fineness distribution analyzer   激光式纤维细度分布分析仪
laser flash photolysis   激光闪光光解
laser fluorescence mass spectrometry   激光荧光质谱法
laser fragmentation source   激光碎裂源
laser fusion   激光核聚变
laser hardening   激光相变硬化
laser induced absorption   激光诱导吸收
laser induced atomic fluorescence(LIAF)   激光诱导原子荧光法
laser induced atomic fluorescence spectra   激光诱导原子荧光光谱
laser induced atomic fluorescence spectrometry   激光诱导原子荧光光谱法(法)
laser induced atomic fluorescence spectroscopy   激光诱导原子荧光光谱
laser induced breakdown spectrometry(LIBS)   激光诱导击穿光谱法
laser induced capillary vibration   激光诱导毛细管振动
laser induced chemical reaction   激光诱导化学反应
laser induced fluorescence(LIF)   激光诱导荧光
laser induced fluorescence detector   激光诱导荧光检测器
laser induced fluorescence spectroscopy   激光诱导荧光光谱法
laser induced fluorimetry   激光诱导荧光(测定)法
laser induced molecular fluorescence(LIMF)   激光诱导分子荧光
laser induced molecular fluorescence spectrometry   激光诱导分子荧光光谱法
laser induced photoacoustic detection   激光诱导光声式检测
laser induced photoacoustic spectroscopic detection   激光诱导光声光谱检测
laser induced photocaustic spectroscopy   激光诱导光声光谱法
laser induced photothermal effect   激光诱导光热效应
laser induced predissociation   激光诱导预离解
laser induced Raman scattering   激光诱导拉曼散射
laser interferometry   激光干涉量度法
laser ionization   激光电离
laser ionization mass spectrometry   激光电离质谱法
laser ionization source   激光电离源
laser ionization spectroscopy(LIS)   激光电离光谱(法)
laser ion source   激光离子源
laser isotope separation   激光同位素分离
laser light scatter   激光散射器
laser light scattering   激光散射
laser light scattering detector   激光光散射检测器
laser low temperature fluorescence spectroscopy   激光低温荧光光谱法
laser macro-spectrochemical analysis   激光常量光谱分析
laser mass spectrometry   激光质谱法
laser material   激光材料
laser microanalysis   激光显微分析；激光微区分析
laser micro-emission spectroscopy   激光微区发射光谱法；激光显微发射光谱法
laser microprobe   激光微探针
laser microprobe atomic emission spectrometry   激光显微原子发射光谱分析法
laser microprobe emission spectrometry   激光显微发射光谱分析法
laser microprobe mass analyzer(LAMMA)   激光微探针质量分析器
laser microprobe mass spectrometer(LMMS)   激光微探针质谱仪
laser microprobe mass spectrometry   激光微探针质谱分析
laser microprobe source   激光显微光源
laser microprobe spectrochemical analysis   激光微探针光谱化学分析
laser microscope   激光显微镜
laser microspectral analyzer   激光显微光谱分析仪
laser micro-spectrochemical analysis   激光显微光谱分析

laser modulated electron capture detection　激光调制电子捕获检测
laser multiphoton ion source　激光多光子离子源
laser nephelometer　激光浊度计
laser nephelometric immunoassay　激光浊度免疫分析
laser on-column detection　激光在柱检测
laser optoacoustic detector　激光光声探测器
laser optoacoustic spectroscopy　激光光声谱法
laser optogalvanic spectroscopy　激光电流光谱法
laser paramagnetic resonance absorption　激光顺磁共振吸收
laser photoacoustic spectrometry　激光光声光谱法
laser photoacoustic spectroscopy　激光光声光谱(法)
laser photochemistry　激光光化学
laser photothermal beam deflection effect　激光光热光束偏转效应
laser photothermal deflection effect　激光光热偏转效应
laser photothermal deflection spectroscopy　激光光热偏转光谱
laser photothermal diffraction spectrometry　激光光热衍射光谱法
laser photothermal interference spectrometry　激光光热干涉光谱法
laser photothermal refraction spectrometry　激光光热折射光谱法
laser photothermal spectroscopy　激光光热光谱(法)
laser plasma　激光等离子体(区)
laser plasma spray coating　激光等离子(体)喷涂(镀)
laser probe　激光探针
laser probe mass spectrometry　激光探针质谱法
laser processing　激光加工
laser-produced plasma　激光等离子体
laser pulse desorption　激光脉冲解吸
laser pumping　激光抽运
laser pyrolysis　激光热解
laser pyrolysis gas chromatography　激光热解气相色谱(法)
laser pyrolyzer　激光热解器
laser Raman microprobe　激光拉曼微区探针
laser Raman photocaustic spectroscopy　激光拉曼光声光谱
laser Raman spectrometry　激光拉曼光谱法
laser-reduced capillary vibration detection　激光诱导毛细管振动测量
laser reference　激光内参
laser resonance ionization(LRI)　激光共振电离
laser resonance ionization spectroscopy　激光共振电离光谱法
laser resonant ionization spectrum　激光共振电离光谱
laser single atom detection　激光单原子探测
laser slit-sealing　激光分切热合
laser spectroelectrochemistry　激光光谱电化学
laser spectroscopy　激光光谱法
laser spectrum　激光光谱
laser spectrum analysis　激光光谱分析
laser stimulated Raman scattering　激光增强拉曼散射
laser strainmeter　激光应变计
laser telemetering　激光遥测
laser thermal deflection spectroscopy　激光热偏转光谱法
laser thermal lens calorimetry　激光热透镜量热法
laser thermal lens effect　激光热透镜效应
laser thermal lens microscopy　激光热透镜显微术
laser thermal lens spectroscopy　激光热透镜光谱法
laser thermal lens spectrum　激光热透镜光谱
laser time-resolved fluorescence method　激光时间分辨荧光法
laser time-resolved fluorescence spectroscopy　激光时间分辨荧光光谱法
laser transition　激光跃迁
laser-triggered fusion　激光引发(热核)聚变
laser two-photon ionization　激光双光子电离
laser velocimeter　激光测速计
laser velocimetry　激光测速法
laser welding　激光焊接
lasing　产生激光
lasiocarpine　毛果天芥菜碱
lasserane　白龙胆烷
last　①靴模②最后的
last cut　①最后馏分②重馏分
lastex yarn　胶乳(浸渍)线
lastics　①塑料②弹塑体〔弹性体与塑性体的总称〕
lasting　①钳(鞋)帮②留香
lasting machine　钳帮机
lastometer　崩裂试验机
lastrile　丙烯腈-二烯类共聚纤维
last runnings〔复〕　尾馏分
Laszcynski zinc process　拉斯辛斯基炼锌法
latamoxef　拉氧头孢〔药〕
late barrier　晚势垒
latence(=latency)　潜伏状态
latency(=latence)　潜伏状态
latensification　潜影相强化
latent　潜(伏)的
latent catalyst　潜催化剂
latent converter　潜转化剂
latent curing agent　潜固化剂
latent curing resin　潜在固化树脂

latent defect 潜在缺陷
latent elastic deformation 潜弹性形变
latent energy 潜能
latent factor 本征因子；潜因子
latent force 潜力
latent heat 潜热
latent heat load 潜热负荷
(latent) heat of evaporation 蒸发(潜)热
latent heat of fusion 熔化潜热
latent heat of liquefaction 液化潜热
latent heat of vaporization 汽化潜热
latent image 潜像；潜影
latent period 潜伏期
latent polarity 潜极性
latent solvent 助溶剂；潜溶剂
latent valency 潜(化合)价
latent variable 本征变量
lateral ①横的②侧(面)的③侧生的
lateral adhesion 侧向附着
lateral analysis 横向分析
lateral branching 侧向分支；侧向支化
lateral chain 侧链
lateral clearance 侧隙
lateral concentration profile 横向浓度分布
lateral conjugation 横向共轭作用
lateral deflection 横向偏转
lateral deformation 侧向变形
lateral diffusion 径向扩散
lateral direction 侧向；横向
lateral distribution 横向分布；侧向分布
lateral electrode 侧电极
lateral expansion 侧向膨胀
lateral flow 侧向流动
lateral force microscopy 侧向力显微镜
lateral group 侧基
lateral interaction 侧向相互作用
lateral localization 横向定位
lateral molecular diffusion 侧向分子扩散
lateral order 侧序
lateral order distribution 侧序分布
lateral packing (分子的)侧向堆砌
lateral pressure 侧向压力
lateral refractive index 侧向折射率
lateral register (链分子排列的)横向定位；侧向定位
lateral resistance 侧向阻力
lateral resolution 横向分辨率
lateral section 横断面
lateral slide mould 旁滑式(注射)塑模

lateral sliding 侧向滑动；侧滑
lateral spacing 侧向间距；横向间距
lateral strain 侧向应变
lateral tyre run-out 轮胎横向摆差
late replication 迟复制
lateriomycin 砖红霉素
laterite 红土；砖红壤
laterite soil 砖红壤；红土
lateritic clay 铝红黏土
late slope detection 迟斜检测法
late transition metal catalyst 后过渡金属催化剂
late wood 晚材
latex 胶乳*
latex accelerator 胶乳(用)促进剂
latex adhesive 胶乳黏合剂
latex-agglomerated ion exchanger 乳胶附聚型离子交换剂
latex agglutination test 乳胶凝集试验
latex backing 胶乳衬背；胶乳衬里
latex binder 胶乳黏合剂
latex binder index 乳胶基料指数
latex blend tank 胶乳混合槽
latex bonded fiber 浸胶乳纤维
latex cement 胶乳接合剂
latex coagulum 胶乳凝块
latex-coated fabric 胶乳布
latex coating 乳胶涂料
latex composition(=latex compound) 胶乳配合物
latex compound(=latex composition) 胶乳配合物〔加入拼料〕
latex compounding ①胶乳配合工艺②配料胶乳的制备
latex concentrator 胶乳浓缩机
latex content 含胶乳量
latex cream 胶乳膏
latex cylinder 胶乳罐
latex-deposited article 胶乳沉积制品
latexed 浸了胶乳的；渍浆的
latex enamel 乳胶瓷漆
latex extruding jets 胶乳喷丝头
latex film 胶乳膜
latex foam 泡沫胶；泡沫胶乳
latex froth 胶乳沫
latex froth building machine 胶乳生沫机
latex gelation 胶乳胶凝
latex gelling 胶乳胶凝
latex house paint 建筑用乳胶漆
latexing 胶乳整理；上胶乳
latex ingredient 胶乳拼料

latex masterbatch 胶乳母胶
latex mechanical stability 乳胶机械稳定性
latex mix(=latex mixing) 配合胶乳
latex mixing(=latex mix) 配合胶乳
latex molding 胶乳铸型法
latex mould 浸胶塑模
latexometer 胶乳比重计
latex paint(=rubber-emulsion paint) 乳胶漆
latex particle 胶乳粒子
latex polymerization 乳液聚合
latex proofing 涂胶乳
latex-resorcinoformaldehyde dispersion 胶乳酚醛分散体
latex rubber 胶乳橡胶
latex sponge 胶乳海绵
latex stabilizer 胶乳稳定剂
latex technology 胶乳加工工艺
latex thickener 乳胶增稠剂
latex thickening 胶乳增稠
latex thread 胶乳橡胶线；胶乳胶丝
latex tube(=latex vessel) 胶乳管；橡浆受器
latex vehicle 乳胶漆料
latex vessel(=latex tube) 胶乳管；橡浆受器
latex vulcanizate 胶乳硫化
lathane 384 丁氧乙氧基乙硫腈；L-384；丁氧硫氰醚
lathe 旋床；车床
lathe carrier 车床鸡心夹头
lathe-cut specialties 旋切制品
lathe operation 车床操作
lather ①泡沫②起泡沫③涂以泡沫
latherability 起泡性；起泡能力
lather booster 泡沫促进剂
lather collapse 泡沫破裂；泡沫消失
lather end point 泡沫终点
lathering （使)起泡；(使)发泡
lathering power 起(泡)沫(能)力
lathering property 起泡性(能)
lather modifier 泡沫改进剂
lather oil 纺织用组合油
latherometer 泡沫仪
lather quickness 起泡速度
lather stabilizer 泡沫稳定剂
lather value 泡沫值
lather volume 泡沫体积
lathe stand 小型脚踏车床
lathe tool 车刀(头)
lathosterol 7-烯胆(甾)烷醇
lathytine α-吡啶丙氨酸
laticiferous vessel 胶乳受器
laticometer 胶乳比重计
latin square 拉丁方
Latin square design 拉丁方设计*
latin square method 拉丁方格法
latitude ①纬度②宽容度(感觉)
latitude of emulsion 展度
lattice 晶格*
lattice amplification 晶格放大
lattice arrangement 晶格排列
lattice constants 晶胞参数*
lattice coordination number 晶格配位数；点阵配位数
latticed drum 格子转鼓
lattice defect(=lattice imperfection) 晶格缺陷
lattice deformation 晶格畸变
lattice dislocation 晶格位错
lattice disorder 晶格无序
lattice distance 点阵间距
lattice distortion 点阵畸变
lattice energy 晶格能*
lattice force 晶格力
lattice hole model 点阵空穴模型
lattice image 晶格像
lattice imperfection(=lattice defect) 晶格缺陷
lattice match 晶格匹配
lattice misfit 晶格错配
lattice mismatch 晶格失配
lattice model 点阵模型
lattice of corundum type 刚玉型晶格
lattice of diamond type 金刚石型晶格
lattice of graphite type 石墨型晶格
lattice of rock-salt type 岩盐型晶格
lattice of rutile type 金红石型晶格
lattice of wurtzite type 纤锌矿型晶格
lattice of zincblende type 闪锌矿型晶格
lattice order 晶格序
lattice orientation (晶体)点阵取向
lattice parameters 晶胞参数*
lattice plane (晶)格面*
lattice point (晶)格点*
lattice relaxation 晶格弛豫
lattice resolution 晶格分辨能力
lattice search 格点搜索法；格点寻优法
lattice site 晶格格位
lattice spacing 点阵间距
lattice structure 点阵结构
lattice table 栅条台
lattice theory of liquid 液体点阵理论
lattice truss 格构桁架

lattice-type network   网格形线路；格架形网络；桥形网络
lattice vacancy   晶格空间
lattice vector   (晶)格矢
lattice vibration   晶格振动*；点阵振动
lattice vibrational spectrum   点阵振动光谱
lattice water   晶格水
laudanidine   劳丹定；半日花定
laudanin   降甲劳丹碱；降甲劳丹素
laudanine   劳丹碱；半日花碱
laudanosine   劳丹素；半日花素；N-甲基四氢罂粟碱
laudanum   鸦片酊
Laue diagram   劳厄图
Laue diffraction camera   劳厄衍射照相机
Laue diffraction geometric type   劳厄衍射几何类型
Laue diffraction pattern   劳厄衍射图样
Laue equation   劳厄方程
Laue method   劳厄法
Laue pattern   劳厄图
Laue photography   劳厄照相法
Laue spectrometer   劳厄光谱仪
Laue symmetry group   劳厄群
laughing gas   笑气；一氧化二氮 N_2O
Laughlin continous centrifuge   劳克林连续离心机
laumontite   浊沸石
launching grease(=launching oil)   船用润滑脂
launching oil(=launching grease)   船用润滑脂
launder   ①洗涤槽；流槽②洗涤
launder classifier   洗涤分级器
laundering   洗涤剂
launderometer   耐洗试验器
laundry   ①洗衣作；洗衣店②洗衣房
laundry detergent   洗涤剂
laundry-resistant   耐水洗的
laundry soap   洗衣皂；洗涤用皂
laundry soda   洗衣碱；洗涤用碱
laundry wax   洗衣蜡；洗作石蜡
lauraldehyde(=lauric aldehyde)   月桂醛；十二醛 $C_{11}H_{23}CHO$
lauramide   月桂酰胺；十二酰胺 $C_{11}H_{23}CONH_2$
laurane   月桂烷（一种异十五烷）
laurate   月桂酸酯；十二酸酯；月桂酸盐
laurel berries oil   月桂果油
laurel butter(=laurel fat)   月桂油
laurel fat(=laurel butter)   月桂油
laureline   月桂碱；劳瑞灵
laurel leaves oil   月桂叶油
laurel nut oil   月桂实油
laurel oil   月桂油

laurel tallow   月桂脂
laurel wax   月桂蜡
laurene   月桂烯
lauric acid   月桂酸；十二(烷)酸 $CH_3(CH_2)_{10}CO_2H$
lauric acid glycerol ester   月桂酸甘油酯
lauric aldehyde(=lauraldehyde)   月桂醛；十二醛 $CH_3(CH_2)_{10}CHO$
lauric alkyd resin   月桂酸醇酸树脂
lauric anhydride   月桂酸酐；十二(烷)酸酐 $(C_{11}H_{23}CO)_2O$
lauric lactam   月桂内酰胺；过氧化月桂酰〔聚乙烯聚合用催化剂〕
laurin   月桂精；(三)月桂酸甘油酯
laurine extra   优质羟基香茅醛
laurinol   月桂醇；十二醇
laurite   硫钌锇矿
lauro-dimyristin   甘油-月桂酸二豆蔻酸酯
lauro-distearin   甘油-月桂酸二硬脂酸酯
laurohydroxamic acid   月桂基异羟肟；十二碳异羟肟酸 $C_{11}H_{23}CONHOH$
laurolactam   月桂内酰胺；十二内酰胺
laurolanic acid   樟烷酸
lauroleic acid   月桂烯酸；十二烯酸 $C_{11}H_{21}COOH$
laurolene   樟烯
laurolenic acid   樟烯酸
laurolitsine   新木姜子碱；月桂木姜碱
laurone   月桂酮〔俗〕；12-二十三(烷)酮 $(C_{11}H_{23})_2CO$
lauronic acid   月桂酮酸〔俗〕 $C_9H_{16}O_2$
lauronitrile   月桂腈；十二(烷)腈 $C_{11}H_{23}CN$
lauronolic acid   樟烯酸
laurophenone   月桂苯酮；十二酰基苯 $C_6H_5CO(CH_2)_{10}CH_3$
laurostearin   甘油三桂酸酯
laurotetanine   六驳碱；山鸡椒痉孪碱
lauroyl   月桂酰；十二(烷)酰 $CH_3(CH_2)_{10}CO—$
lauroyl amino acid   月桂酰氨基酸
N-lauroyl-p-aminophenol N-月桂酰对氨基苯酚〔防老剂〕
lauroyl carbamyl guanidine   月桂酰氨基甲酰胍
lauroyl diethanolamine   月桂酰二乙醇胺〔湿润剂〕
lauroyl glutamate   月桂酰谷氨酸盐
lauroyl isethionate   月桂酰基羟基乙磺酸盐
lauroyl peroxide   月桂酰过氧化物；过氧化月桂酰(化合物)〔氯乙烯聚合用催化剂〕
N-lauroyl sarcosinate N-月桂酰基肌氨酸盐(或酯)
laurusan   脱氢间型霉素
laurusin   月桂菌素
lauryl(=dodecyl)   月桂基；十二(烷)基
lauryl acetate   乙酸月桂酯

lauryl acrylate 丙烯酸月桂酯
lauryl alcohol 月桂醇；十二(烷)醇
lauryl aldehyde 月桂醛；十二醛
laurylamide 月桂酰胺；十二烷基酰胺
laurylamine 月桂胺〔促进剂〕
laurylamine acetate 乙酸月桂胺〔促进剂〕
3-laurylamino-propionic acid 3-月桂氨基丙酸
lauryl ammonium acetate 月桂基乙酸铵
lauryl bromide(=dodecyl bromide) 月桂基溴；十二烷基溴 $CH_3(CH_2)_{10}CH_2Br$
lauryl chloride(=dodecyl chloride) 月桂基氯；十二烷基氯 $CH_3(CH_2)_{10}CH_2Cl$
lauryl dimethyl amino oxide 月桂基二甲基氨基氧化物
lauryl dimethyl benzyl ammonium chloride 月桂基二甲基苄基氯化铵
laurylene ①月桂萜烯②十二烯
lauryl ether sulfate 月桂基醚硫酸盐
lauryl gallate(=dodecyl gallate) 没食子酸十二烷酯；没食子酸月桂酯
N-lauryl-γ-hydroxybutylamide N-十二烷基-γ-羟基丁酰胺
lauryl mercaptan 十二烷基硫醇；月桂硫醇
laurylmercaptobutyraldehyde 月桂基硫醇基丁醛〔防老剂〕
laurylmercaptopropionaldehyde 月桂基硫醇基丙醛〔防老剂〕
lauryl methacrylate(LMA) 甲基丙烯酸月桂酯；甲基丙烯酸十二烷基酯
lauryl methacrylate copolymer 甲基丙烯酸月桂酯共聚物
lauryl peroxide 月桂基过氧化物〔硫化剂〕
lauryl phenol 月桂基酚
lauryl phosphate 磷酸月桂酯
lauryl polyglucoside 月桂基聚葡萄糖苷
lauryl propionate 丙酸月桂酯
lauryl pyridinium chloride 氯化月桂基吡啶鎓
lauryl pyridinium sulfate 硫酸月桂基吡啶鎓
lauryl pyrrolidone carboxylate 吡咯烷酮羧酸月桂酯
lauryl sodium sulfate 十二烷基硫酸钠
lauryl sulfate 硫酸月桂酯
lauryl sulfoacetate 磺基乙酸月桂酯
lauryl sulfonic acid 月桂基磺酸
lauryl sulfonyl amide 月桂基磺酰胺
lauryl thiocyanate 硫氰酸十二酯
N-lauryltrialkylmethylamine N-月桂基三烷代甲基胺 $(C_{12}H_{25})NH(CR_3)$
lauryl triethanol ammonium sulfate 月桂基三乙醇硫酸铵
lauryl trimethyl ammonium chloride 月桂基三甲基氯化铵

lauryl-trimethyl-silicane 月桂基三甲基硅烷 $(C_{12}H_{25})Si(CH_3)_3$
Lauson engine 劳森(单缸)内燃机
laustoneu 新灭虱剂
lautal 劳塔尔铜硅铝合金
lautering 过滤
lauter tank(=lauter tub) 滤桶
lauter tub(=lauter tank; lauter tun) 滤桶
lauter tun(=lauter tub) 滤桶
Lauth's violet 劳氏紫
lava 熔岩；火山石
Laval nozzle 拉伐尔喷嘴
lavandin absolute 杂薰衣草净油；拉文定净油〔俗〕
lavandin concrete 杂薰衣草浸膏；拉文定浸膏〔俗〕
lavandin oil 杂薰衣草油；拉文定油〔俗〕
lavandulal 薰衣草醛
lavandulane 薰衣草烷
lavandulic acid 薰衣草酸
lavandulol 薰衣草醇
lavandulyl acetate 乙酸薰衣草酯
lavatory cleaner 盥洗室用洗净剂
lavender 薰衣草(花)
lavender absolute 薰衣草净油
lavender concrete 薰衣草浸膏
lavender flower oil 薰衣草花油
lavender oil 薰衣草油
lavender spike oil(=spike lavender oil) 穗薰衣草油
lavender spirit 薰衣草酒
lavendulin 淡紫灰菌素
lavenite 锆钽矿
laveur 洗气器
lavic 熔岩的
law 定律；规律
lawang bark oil 拉王桂皮油
lawang oil 类豆蔻油
Law cell 劳式电池
law of active mass 活性质量定律
law of causality 因果律
law of chance 机会定律
law of chemical change 化学变化定律
law of combining proportions 化合比例定律
law of combining weight 化合量定律
law of conservation of energy 能量守恒定律
law of conservation of mass 质量守恒定律
law of conservation of massenergy 质能守恒定律
law of conservation of matter 物质守恒定律；质量不灭定律
law of conservation of momentum 动量守恒定律

law of constancy of interfacial angle　(晶)面(交)角不变定律
law of constant composition　成分不变定律
law of constant heat summation　总热量不变定律
law of constant proportion　定比定律
law of definite composition　成分不变定律；固定成分(定)律
law of definite interfacial angles　固定(晶)面角定律
law of definite proportions　定比定律
law of distribution　分配定律
law of equivalent proportions　当量比例定律
law of fixed proportion　定比定律
law of fluorescence　荧光定律
law of freezing-point depression　冰点降低定律
law of gas combining volume　气体化合体积定律
law of gas diffusion　气体扩散定律
law of gas reaction　气体反应定律
law of geometrical crystallography　几何结晶构造定律
law of independent migration　独立移动定律
law of isomorphism　类质同晶定律
Law of Lambert-Beer(=Lambert-Beer law)　朗伯-比尔定律
law of large number　大数定律
law of mass action　质量作用定律
law of mechanical equivalent of heat　热功当量定律
law of mobile equilibrium　移动平衡定律
law of molecular concentration　分子浓度定律
law of multiple proportions　倍比定律
law of octaves　八行周期律
law of parity　宇称定律
law of partial pressure　分压(力)定律
law of partition　分配定律
law of perdurability of matter　物质守恒定律
law of periodicity　周期律
law of perpetual motion　永久运动定律
law of persistance of energy　能量持久定律
law of photochemical equivalence　光化当量定律
law of photochemistry　光化学定律
law of radiation　辐射定律
law of radioactive decay　放射衰变律
law of radioactive displacement　放射位移定律
law of rational indices　有理指数定律
law of reciprocal proportions　互比定律
law of rectilinear diameters　中线定律
law of related composition　成分相关定律
law of size distribution　粒度分布定律
law of solid friction　固体摩擦定律
law of thermal equilibrium　热平衡定律
law of thermoneutrality　热中和定律
law of vapour pressure relations　蒸气压关系定律
Lawrance hydrogen embrittlement tester　劳伦斯式氢脆试验机
lawrencium　铹〔103号元素，化学符号Lr〕
lawsone　2-羟(基)-1,4-萘醌
lawsonite　硬柱石
laxative　轻泻剂
lay　层
layaways　腌鞣池
lay back　刃角磨损
lay-decorating　敷饰
laydown　①沉积作用②搁置③沉淀作用
layer　①腌鞣池②层③焊层
layer analysis　分层分析
layer built cell　积层电池
layer-by-layer self-assembly　交替沉积构筑
layer cake　层饼
layer chromatography　层色谱(法)
layer coagulating by wet method　湿法凝固成膜
layer crystal　层晶
layere　叠鞣池
layered column chromatography　层叠式柱色谱(法)
layered double hydroxide　层状双羟基复合金属氧化物
layered vessel　多层式容器
layering process　层状法
layer lattice　层状晶格；层形点阵
layer lattice type　层格型
layer lay-up design　铺层设计
layer line　层线
layer of carbon　炭渣层
layer of oil molecules　油分子层
layer of oxide　氧化层；氧化膜
layer plating method　夹层培养法
layer structure　层型结构*
layer thickness　层厚度
layer thickness adjuster　层厚度调节器
layer thickness meter　①涂膜测厚仪②薄膜厚度计
layer vat　腌鞣池
lay-flat　(薄膜)平折；平折性
lay-flat adhesive　平服黏合剂〔不卷边，不膨胀的黏合剂〕
lay-flat extrusion　平折挤塑
lay-flat film　平折吹塑薄膜
lay-flat width　折径
laying　敷设〔输送管〕
laying-away　腌鞣
laying down order　铺绵顺序

laying gang　敷设管道的人
laying-off　①停工②下料
laying of (pipe) line　输送管的敷设
laying-out　①敷设线路〔输送管〕②划线
lay-in top　衬垫假罗口
lay off　①停工②下料
lay-off undercoat　涂中间层
lay-on-air dryer　热风气垫式烘干机
layout　①工厂设备布置②敷设线路③石油加工过程流程
layout diagram　布置图；框图；示意图
layout drawing　布置图
layout for drilling　钻孔划线
layout plan　平面图
lay separation　分层；脱层
lay up　①组合叠板②铺贴
lay-up deformation　铺敷变形
lay-up method　叠层法
lay-up molding　铺叠成型
lay wire　模网
lazulite　天蓝石
lazurite　天青石
lazy board　木制支架〔安装输送管用〕
lazy element　惰性元素
L beam　L形梁
LCAO-MO method　原子轨道的线性组合-分子轨道法
L/D ratio(=length/diameter ratio)　长径比
lea　缕纱
lea breaking strain　缕纱断裂应变
leach　①沥滤器②沥滤；浸提
leachability　可浸出性
leach-casting machine　浸提出渣机
leach column　浸沥塔；浸提塔
leached　浸滤了的
leached cossettes　沥滤甜菜丝；沥滤过的甜菜切片
leached hull　浸取过的(元件)壳
leacher　浸取器
leach house　浸提间
leaching　沥滤(法)；沥取(法)；浸提
leaching liquor(=leaching solution)　沥滤液；浸提液
leaching of pesticide in soil　(土壤内)农药淋溶作用
leaching-out　沥滤出；洗出；浸出
leaching process　沥滤法
leaching rate　渗出率；放毒率
leaching solution(=leaching liquor)　沥滤液；浸提液
leaching tank　沥滤槽；浸提桶
leaching tub　沥滤盆
leaching value　浸出值；浸取值

leaching vat　沥滤瓮
leach liquor　沥滤液；浸提液
leach recrystallization process　沥滤重结晶法；浸沥重结晶法
leach residue　浸取残渣
leach tub　沥滤盆；沥取盆
lead　①铅〔82号元素，化学符号Pb〕②(含)铅的③移前④导线；引线
lead accumulator　铅蓄电池
lead acetate　乙酸铅
lead acetate method　乙酸铅法
lead acetate test　乙酸铅试验
lead acetate test paper　乙酸铅试纸
lead acetotartrate　乙酰酒石酸铅
lead aerosols　铅烟雾剂
lead alkyl(=lead alkylide)　烷基铅
lead alkylide(=lead alkyl)　烷基铅
lead alloys　铅合金
lead aluminate　铝酸铅
lead and cadmium release from ceramics　陶瓷铅镉溶出量
lead antimoniate　锑酸铅
lead appreciator　加铅效果改进剂
lead arsenata　①砷酸铅②砷酸氢铅
lead aryl(=leadarylide)　芳基铅
leadarylide(=lead aryl)　芳基铅
lead azide　叠氮化铅
lead azide detonator　叠氮化铅雷管
lead azoimide　叠氮化铅
lead bar　铅条
lead-base grease(=lead grease)　铅皂润滑脂
lead bath　铅浴
lead battery　铅蓄电池(组)
lead billet　铅粒
lead bleach　用乙酸铅和硫酸漂白
lead block test　炸铅试验
lead borate　硼酸铅　$Pb(BO_2)_2$
lead borosilicate　硼硅酸铅
lead brick　铅砖*
lead bromate　溴酸铅　$Pb(BrO_3)_2$
lead bromide　溴化铅
lead button　铅粒
lead cable　铅包电缆
lead carbonate　碳酸铅　$PbCO_3$
lead castle　铅室
lead cerotate　蜡酸铅；二十六(烷)酸铅
lead chamber　铅室
lead chamber crystals〔复〕　铅室结晶

lead chamber gases〔复〕 铅室气
lead chamber pan 铅室底盘
lead chamber process 铅室法
lead chamber space 铅室容积
lead chlorate 氯酸铅 $Pb(ClO_3)_2$
lead chloride 氯化铅
lead chlorite 亚氯酸铅 $Pb(ClO_2)_2$
lead chlorosilicate complex 氯硅酸铅配合物
lead chromate 铬酸铅
lead chromate-coated pigment 铬酸铅包核颜料
lead chrome pigment 铬酸铅颜料
lead-chrome yellow 铅铬黄〔着色剂〕
lead citrate 柠檬酸铅 $Pb(C_3H_6O_7)_2$
lead coated 包铅的
lead coating ①包铅②铅皮
lead coil 铅旋管；铅盘管
lead conductometric value of hop 酒花铅电导值
lead contamination 铅污染
lead coulometer 铅极电量计
lead-covered coil 铅包盘管；铅包蛇管
lead curtain （铅室的）铅墙
lead cyanamide 氰氨化铅
lead deposit 铅沉积
lead dialkyl 二烷基铅
lead dibromide 二溴化铅 $PbBr_2$
lead dichloride 二氯化铅 $PbCl_2$
lead dichromate 重铬酸铅
lead diethide(=lead diethyl) 二乙基铅
lead diethyl(=lead diethide) 二乙基铅 $Pb(C_2H_5)_4$
lead diethyldithiocarbamate 二乙基二硫代氨基甲酸铅〔促进剂〕
lead dihydroarsenate 砷酸二氢铅 $Pb(H_2AsO_4)_2$
lead dimethide(=lead dimethyl) 二甲基铅
lead dimethyl(=lead dimethide) 二甲基铅 $Pb(CH_3)_4$
lead dioxide 二氧化铅 PbO_2
lead distribution （四乙）铅分布
lead dithiocarbamate 二硫代氨基甲酸铅〔促进剂〕
lead dithiofuroate 二硫代呋喃甲酸铅
lead dithionate 连二硫酸铅 PbS_2O_6
lead drier 铅催干剂〔促进油漆干燥〕
lead dust 铅末
leaded ①加铅的②镀铅的③填铅的④包铅的⑤含铅的
leaded fuel 加铅燃料；乙基化汽油
leaded gasoline(=leaded up gasoline) 加铅汽油；乙基化汽油
leaded glove 含铅手套
leaded number 加铅数
leaded petrol 加铅汽油

leaded up gasoline(=leaded gasoline) 加铅汽油；乙基化汽油
leaded zinc 含铅氧化锌；含铅锌白
leaded zinc oxide 含铅锌白；铅化锌白〔ZnO_2+碱式硫酸铅〕
lead electrode 铅电极
leaden ①铅的②铅包的
lead encasing press 套铅机；包铅机
lead-epoxy mix 铅粉环氧树脂混合物
leader 引布；引料法
lead eraser 擦(铅笔)字橡皮
lead ethide(=lead ethyl) 二乙铅；(二)乙基铅
lead ethyl(=lead ethide) (二)乙基铅 $Pb(C_2H_5)_4$
lead ethylsulfate 乙(代)硫酸铅 $Pb(C_2H_5SO_4)_2$
lead extruding press 压铅机
lead extrusion press 压铅(硫化)机
lead ferrocyanide 亚铁氰化铅
lead flake 铅片；碱式碳酸铅
lead fluoride 氟化铅 PbF_2
lead flux 铅(助)熔剂
lead foil 铅箔
lead formate(=lead formiate) 甲酸铅 $Pb(HCO_2)_2$
lead formiate(=lead formate) 甲酸铅
lead-free 无铅的
lead-free drier 无铅催干剂；无铅干料〔俗〕
lead-free fuel 无铅燃料
lead-free gasoline 无铅汽油
lead-free paint 无铅漆
lead-free toy enamel 无铅玩具瓷漆
lead-free zinc oxide 无铅氧化锌
lead fuse wire 保险铅丝
lead glance 方铅矿
lead glass 铅玻璃
lead glaze 铅釉
lead glazing ①铅釉②上铅釉
lead grease 铅皂润滑脂
lead grey 铅灰色
lead halide 卤化铅
lead heat stabilizer 铅系热稳定剂
leadhillite 硫碳铅矿
lead hydrate 氢氧化铅 $Pb(OH)_2$
lead hydrogen arsenate 砷酸氢铅 $PbHAsO_4$
lead hydrogen phosphite 亚磷酸氢铅 $PbHPO_3$
lead hydroxide 氢氧化铅 $Pb(OH)_2$
lead hydroxycarbonate 碱式碳酸铅
lead hydroxynitrate 碱式硝酸铅
lead hypophosphite 次磷酸铅
lead-impregnated apron 衬铅围裙

leading ①领先②主导的；定向的③加铅（汽油加铅则指加四乙铅）
leading edge 前沿
leading electrolyte 先行电解质；前导电解质
leading-in socket 进线插座〔电〕
leading ion 前导离子；先行离子
leading of gasoline 汽油加铅
leading peak 前伸峰*；前延峰
lead in oil 铅油
lead intensification 铅盐加厚法
lead-in wire ①引药线②导入线
lead iodate 碘酸铅 $Pb(IO_3)_2$
lead iodide 碘化铅 PbI_2
lead ion selective electrode 铅离子选择电极
lead isotope dating method 铅同位素(定)年代法
lead joint 铅接头
lead lactate 乳酸铅
leadless color 无铅色料
leadless enamel ①无铅搪瓷②无铅瓷漆
leadless glaze 无铅釉
leadless paint 无铅漆
leadless wood primer 无铅木材底漆
lead line 出油管线
lead lined 铅衬的
lead lined hoods 挂铅管帽
lead lined tank 铅衬里储槽
lead lining 铅衬
lead linoleate 亚油酸铅
lead malate 苹果酸铅
lead matte 粗铅；冰铅
lead mercaptide 烃硫基铅；硫醇铅 $(RS)_2Pb$
lead metaborate 偏硼酸铅 $Pb(BO_2)_2$
lead metaplumbate 偏高铅酸铅 $Pb(PbO_3)$
lead metasilicate 硅酸铅 $PbSiO_3$
lead methide(=lead methyl) (二)甲基铅
lead method of age determination 铅法(地质)年龄测定
lead methyl(=lead methide) (二)甲基铅 $Pb(CH_3)_2$
lead minerals 铅矿(物)
lead molybdate 钼酸铅 $PbMoO_4$
lead monoxide 氧化铅；密陀僧〔俗〕
lead mordant 铅媒染剂
lead myristate 十四烷酸铅
lead naphthenate 环烷酸铅(皂)
lead nitrate 硝酸铅
lead octoate 辛酸铅
lead oleate 油酸铅
lead ore 铅矿
lead ore hearth 溶铅矿敞炉

lead orthophosphate (正)磷酸铅 $Pb(PO_4)_2$
lead orthoplumbate 原高铅酸铅 $Pb_2(PbO_4)$
lead oxalate 草酸铅
lead oxide ①(=lead monoxide)氧化铅；密陀僧〔俗〕②铅的氧化物
lead oxide red 四氧化三铅；红铅；铅丹；光明铅 Pb_3O_4
lead oxychloride 氯氧化铅 $PbCl_2 \cdot PbO; PbCl_2 \cdot 2PbO$
lead palmitate 软脂酸铅
lead pan ①铅盘②铅锅
lead peak 前延峰
lead pentamethylene dithiocarbamate 五亚甲基二硫代氨基甲酸铅
lead peroxide 过氧化铅；二氧化铅
lead peroxide red 红色过氧化铅；红色二氧化铅
lead phenate 苯酚铅 $Pb(OH)OC_6H_5$
lead phenolate 苯酚铅
lead phenolsulfonate 苯酚磺酸铅 $Pb(OHC_6H_4SO_3)_2$
lead phosphate 磷酸铅 $Pb_3(PO_4)_2$
lead phosphite 亚磷酸铅
lead phosphosilicate 磷硅酸铅
lead picrate 苦味酸铅
lead pig 铅块
lead plaster 铅膏
lead plated 镀铅的
lead plating 镀铅
lead plumbate 铅酸铅
lead poisoning 铅中毒
lead powder 金属铅粉；铅末
lead protoxide 一氧化铅
lead pyroarsenate 焦砷酸铅 $Pb_2As_2O_7$
lead pyrophosphate 焦磷酸铅
lead red 红丹；红铅；铅丹；光明铅
lead resinate 树脂酸铅
lead response(=lead responsibility) 受铅性
lead responsibility(=lead response) 受铅性
lead-restricted paint 铅限量油漆；限铅油漆
lead rhodanide 硫氰酸铅 $Pb(SCN)_2$
lead rubber sheet 夹铅胶片〔防X射线用〕
lead salicylate 水杨酸铅〔活性剂〕
lead salt 铅盐
lead screw ①导螺杆②铅螺旋
lead seal 铅封
lead selenate 硒酸铅 $PbSeO_4$
lead selenide 硒化铅 $PbSe$
lead sensitivity 受铅性
lead sesquioxide 三氧化二铅；偏高铅酸铅 $Pb(PbO_3)$
lead sheath 铅护套
lead sheet 铅片

lead silicate 硅酸铅〔稳定剂〕
lead silicate white 硅酸铅白；碱式硅酸铅
lead silicochromate 硅铬酸铅
lead silicofluoride 硅氟化铅
lead silicotitanate 硅钛酸铅
lead smelting 铅的熔炼
lead smelting furnace 熔铅炉；炼铅炉；化铅炉
lead soap 铅皂
lead solder 铅焊料
lead stabilizer 铅稳定剂
lead stearate 硬脂酸铅〔稳定剂〕
lead storage battery 铅蓄电池
lead styphnate 2,4,6-三硝基间苯二酚铅
lead subacetate 碱式乙酸铅
lead subcarbonate 碱式碳酸铅
lead suboxide 一氧化二铅 Pb_2O
lead suboxide anti-corrosive paint 一氧化二铅防锈漆；黑铅粉防锈漆
lead sugar 铅糖*
lead sulfate 硫酸铅
lead sulfide 硫化铅 PbS
lead sulfide sweetening process 硫化铅脱硫醇过程〔或脱臭过程〕
lead sulfite 亚硫酸铅
lead sulfocyanide 硫氰酸铅 $Pb(SCN)_2$
lead superoxide 二氧化铅 PbO_2
lead susceptibility 感铅性
lead susceptibility improver 加铅效果改善剂
lead tallate 松浆油酸铅
lead tartrate 酒石酸铅 $PbC_4H_4O_6$
lead telluride 碲化铅 PbTe
lead tetraacetate 乙酸高铅 $Pb(C_2H_3O_2)_4$
lead tetraalkyl 四烷基铅 PbR_4
lead tetrabromide 四溴化铅；溴化高铅 $PbBr_4$
lead tetrachloride 四氯化铅；氯化高铅 $PbCl_4$
lead tetraethide(=lead tetraethyl) 四乙基铅
lead tetraethyl(=lead tetraethide) 四乙基铅 $Pb(C_2H_5)_4$
lead tetramethide(=lead tetramethyl) 四甲基铅
lead tetramethyl(=lead tetramethide) 四甲基铅 $Pb(CH_3)_4$
lead tetroxide 四氧化三铅；红铅；铅丹；光明铅 Pb_3O_4
lead thiocyanate(=lead thiocyanide) 硫氰酸铅 $Pb(SCN)_2$
lead thiocyanide(=lead thiocyanate) 硫氰酸铅
lead thiosulfate 硫代硫酸铅 $PbSO_3S; PbS_2O_3$
lead titanate 钛酸铅〔可作白色颜料〕 $PbTiO_3$
lead titanate ceramic 钛酸铅陶瓷
lead titration 铅滴定
lead tolerance 容许含铅量；(汽油中)四乙铅容许含量

lead triethyl hydroxide 氢氧化三乙铅 $Pb(OH)(C_2H_5)_3$
lead trinitro-cresylate 三硝基甲酚铅
lead trinitro-resorcinate 三硝基苯间二酚铅
lead trithionate 连三硫酸铅 PbS_3O_6
lead tungstate 钨酸铅 $PbWO_4$
lead vanadate 钒酸铅 $Pb(VO_3)_2$
lead vitriol 铅矾；硫酸铅矿
lead water 铅水；1%碱式乙酸铅溶液
lead white 铅白
lead white paper 铅白纸；碱式碳酸铅纸
lead wire 铅丝
lead wolframate 钨酸铅
leadwork 铅衬
lead yellow 铅黄
lead-zinc accumulator(=lead-zinc storage battery) 铅锌蓄电池
lead-zinc storage battery(=lead-zinc accumulator) 铅锌蓄电池
lead zirconate ceramic 锆钛酸铅陶瓷
leaf alcohol(=cis-3-hexenol) 叶醇；顺式己烯-3-醇
leaf aldehyde 叶醛；顺式己烯-3-醛
leaf fibre 叶纤维
leaf filter 叶(状)滤机
leaf gneiss 片麻岩
leaf gold 金箔
leafing 飘浮；叶浮；浮起
leafing agent 漂浮剂
leafing aluminium 漂浮型铝粉；叶展型铝粉
leafing spatula value (铝粉浆)漂浮值
leafing stabilizer 漂浮稳定剂；叶展稳定剂
leaf spring 薄片弹簧
leaf spring model 板簧模型；片簧模型
leaf stability 漂浮稳定性；叶展稳定性
leaf stamping 彩箔烫印
leaf sumac 漆叶
leaf-type vacuum filter 真空叶滤机
Leahy process 拉埃法〔用硫氨从石油气体脱硫化氢〕
leakage ①泄漏②漏失量
leakage condenser 漏泄冷凝器
leakage current 漏泄电流
leakage detector 探漏仪
leakage factor 漏失系数
leakage flow(=clearance flow) 隙流
leakage resistance 漏电阻
leakage steam 漏泄蒸汽
leakage test 漏泄试验；紧密性试验
leak check 检查漏气
leak clamp 修理夹〔防止输送管漏失的管箍〕

leak detection　检(查)漏(失)；紧密性检查
leak detector　检漏器
leak finding　检漏
leak free　不漏的；密封的
leak gas　漏气
leak hunting　检漏
leak-indicator　漏泄指示器
leakiness　漏泄程度
leaking　漏；漏泄
leakless　不漏的
leak loss　漏泄损失
leak off　漏泄
leak-proof　防漏的；密封的
leakproof fuel tank　防漏的燃料储器
leakproof seal　防漏密封；防漏封口
leak resistance　防止漏失
leak test　漏气试验
leak testing　渗漏试验
leak through　渗漏；滴漏
leaky　漏的；松的；不密的
leaky mold technique　泄漏式模塑技术；溢出式模塑技术
leaky roof　漏顶；不紧密的顶盖
lean　瘠；瘦；贫
lean clay　瘦黏土；贫黏土
lean coal　贫煤
lean concrete　贫混凝土
lean flammability　可燃性下限
lean fuel mixture　贫燃料混合物〔内燃机用空气与燃料的混合物〕
lean gas　贫气
leaning-out effect　贫乏效应
lean lime　贫石灰
lean limit　贫乏限度
lean material　废料；矸石；贫矿
lean mixture　贫燃料混合物
lean mixture maximum power　贫燃料混合物最大功率
lean-mixture rating method　贫混合物评定法〔航空汽油〕
lean mortar　贫灰浆
lean oil　贫油；解吸油
lean oil shale　贫油页岩
lean phase　疏相；烯相
lean solution　废溶液
lean solvent　贫溶剂
learning sample　训练样本
lease　分经
lease tank　油矿油罐
lease tankage　油矿油罐

lease tank vapour conservation　捕集油矿油罐中蒸气
leasing　分绞；分经
least chain length　最小链长度
least median of square regression　最小中位数平方回归
lea strength　缕纱强力
least resistance　最小阻力
least significant difference　最小显著性差异
least square approximation　最小二乘方近似
least square estimation　最小二乘估计
least square fitting　最小二乘法拟合*
least square method　最小二乘法
least squares line　最小二乘曲线
least squares method　最小二乘法
leather　革；皮
leather adhesive　皮革黏合剂
leather bellow　皮风箱
leather board　皮纸板
leather chemist　皮革化学工作者
leather chemists association　皮革化学工作者协会
leather cloth　人造革
leather coating color　皮革涂饰用色料
leather color　①皮革色②染革用染料
leather dope　皮革漆
leather dressing　皮革修整
leather dye(=leather dyestuff)　皮革染料
leather dyestuff(=leather dye)　皮革染料
leatherette　人造革
leather facing　皮衬布；皮面
leather fair　皮革展览会
leather gauge　皮革测厚计
leather goods　皮件
leather grain　革纹理
leather greasing　皮革加脂
leather industry　皮革工业
leather lace　皮条
leather lacing　皮带丝
leather machine　制革机器
leather machine belting　机用皮带；皮带
leather meal　革屑；皮屑〔俗〕
leather measuring machine　量革机
leathern　①似革的②革制的
leatheroid　人造革；纸皮；薄钢纸
leather oil　皮革油
leather packing　①皮革填充②垫革
leather permeability value(LPV)　①皮革渗透值②人造革渗透值
leather picker　皮结
leather press　皮革挤水机

leather research institute 皮革研究所
leather roller 压革机
leather strap 皮带
leather substance 皮革材料
leather substitute 皮革代用品
leather tankage 革屑肥
leather tanning 皮革鞣制
leather technology 皮革工艺学
leather trade 皮革贸易
leather wax 皮革用蜡
leatherwear 皮革服装
leather whip 皮鞭
leathery 似革的
leathery odor 皮革气息
leathery state 似革态
leaven ①发酵②曲子
leavening ①发酵②使发酵
leaving group 离去基团
leavings 残渣
leaving water temperature(LWT) 出水温度
Lebedev butadiene synthesis 列别捷夫丁二烯合成
LeBel-van't Hoff theory 勒贝尔-范特霍夫理论
Leblanc process 吕布兰法〔19世纪工业制碳酸钠的方法〕
lecanoric acid 红粉苔酸；瘤网地衣酸；扁枝衣二酸
Le Chatelier-Brown's principle 勒夏特列-布朗原理
Le Chatelier's principle 勒夏特列原理*
Le Chatelier's thermoelement 热电元件；铂-铂铑合金热电偶
lecithine 卵磷脂
Leclanche cell 勒克兰瑟电池
lectin 植物凝聚素
lectin probes 外源凝集素探针
lectotype 选模标本
lecture experiment 演示实验
leddiene 喇叭二烯
ledene 喇叭烯
Lederer-Manasse reaction 莱德勒-曼讷斯反应
ledermycin(=demethylchlortetracycline) 去甲基金霉素
ledger paper 帐薄纸
leditannic acid 喇叭茶单宁酸
ledixanthin 喇叭茶色素
ledol 喇叭茶醇
ledum camphor 喇叭茶脑；喇叭茶醇
leechee 荔枝
leek oil 韭葱油
Leendertse method for structural group analysis 林得司分析烃类结构族组成方法

leer(=lehr) （玻璃)退火炉
leer pan 退火盘
lees 沉积物；淀积物
Leffler-Grumwald operator 列菲-克勒瓦德算符
leflunomide 来氟米特〔药〕
left ①左(边)②左的
left Cauchy-Green tensor 左柯西-格林张量
left handed crystal ①左晶②左旋结晶
left hand rule 左手定则
left-hand screw thread 左旋螺纹
left-hand thread 左旋螺纹
left-hand worm 逆转螺杆
leftover shear stress 剩余剪切应力
leg ①支架②腿
legal chemistry 法庭化学
Legal reaction 列尬尔反应
legal unit of measurement 法定计量单位
Legendre polynomial expansion 勒让德多项式
legging 拉丝；起黏丝
legibility 清晰度
leg pipe 冷凝器气压管；大气腿
lehr(=leer) （玻璃)退火炉
lehrbachite 硒铅汞矿
lehuntite 钠沸石
leiocom(=dextrin) 糊精
Leipsic yellow 莱比锡黄；柠檬铬黄
Leipzig yellow(=lead chromate) 铬酸铅
Leithner's blue 钴蓝；铝酸钴
L electrons L层电子
lemery salt 硫酸钾 K_2SO_4
lemon chrome 柠檬(铬)黄〔颜料,铅铬黄的一种〕
lemon chrome yellow 铬黄；柠檬黄；铬酸钡黄颜料
lemon-grass oil 柠檬草油；枫茅油
lemon leaf oil 柠檬叶油
lemon mint oil 柠檬香蜂草油
lemon oil 柠檬油
lemon petitgrain oil 柠檬叶油
lemon pips oil 柠檬子油
lemon salt 草酸氢钾；柠檬盐 $KHC_2O_4 \cdot H_2O$
lemon-scented gum tree 柠檬桉
lemon seed oil 柠檬子油
lemon soap 柠檬皂
lemon verbena 防臭木
lemon verbena oil 防臭木油
lemon yellow 柠檬黄〔铬酸铅颜料〕
Leneta drawdown leveling blade 莱内塔流平性试验刮刀
Leneta drawdown leveling value 莱内塔刮样流平值

lengebachite 叶硫砷银铅矿
length 长度；管段
length/diameter ratio(=L/D ratio) 长/径比；L/D比
lengthiness 拉丝性
length mean diameter (粒子)长轴平均直径；长轴平均粒径
length of flow 流动长度
length of intervals 间隔时间
length of life 寿命
length of life test 寿命试验
length of normal 法线的长度；法距
length of oil 油的延展长度
length of panel 板长
length of particle 粒子长径〔指粒子的长轴〕
length of run ①运转时间②展开长度〔专指色谱〕
length of string 拔丝长度；拉丝长度；看丝长〔俗〕
length of useful life 有效寿命
length of weld 焊缝长度
length-to-diameter ratio 长径比
length-to-length cure 逐段硫化
length tolerance 长度公差
length variation 长度偏差
lengthwise(=lengthways) 纵向的；纵长的
lenirobin(＝chrysarobin tetra-acetate) 四乙酸柯桠素
Lennard-Jones fluid 林纳德-琼斯流体
leno breaker 稀缓冲布层〔运输带〕
lens ①(透)镜②眼镜
lens maker's equation 透镜组方程
lenthionine 香菇精
lentinacin 香菇素
lentinan 香菇多糖
lentine 间苯二胺
lentinus edodes 香菇
lentisc 乳香黄连木
lentiscus 乳香黄连木
lentisk 乳香黄连木
lenzitin(=lonzitin) 革裥菌素
leometer 势能增益
leonardite 风化褐煤
leonite 钾镁矾
leonticine 狮足草碱
leonuridine 益母草定
leonurine 益母草碱
leonurinine 益母宁
lepamin 勒帕胺 $C_{20}H_{32}N_2$
lepargylic acid 壬二酸
lepathinic acid 酸模根酸
lepidene 四苯呋喃

lepidine(=4-methyl quinoline) 4-甲基喹啉 $C_{10}H_9N$
lepidocrocite(=lepidocrokile) 纤铁矿
lepidocrokile(=lepidocrocite) 纤铁矿
lepidolite 鳞云母；红云母
lepidometer 测起鳞度器
lepidone 勒皮酮；4-甲基-2-羟基喹啉
leprocidal 医治麻风的
leprotene 麻风菌烯 $C_{40}H_{54}$
leptactinine 细茜碱
leptaflorine 细茜花碱
leptandra 黑根
leptandrin 黑根素〔从北美草本威灵仙根部或者黑根中提取的一种带有苦味的配糖〕
leptandroid 黑根膏
leptazol(＝pentylenetetrazol) 戊四唑〔一种苏醒剂〕
leptin 肥胖荷尔蒙瘦素
leptine I 莱普亭 I
leptodactyline 细指蟾碱；来普达林〔抗高血压药〕
leptogenesis 可纺性
leptokurtosis 凸峰态
leptology 物质细微结构学
leptometer 比黏计
lepton 轻子
leptonema 细线
leptospermone 纤精酮
leptynol 氢氧化钯悬浮液
lercanidipine 乐卡地平〔药〕
lerp 矮桉
Le Seur cell (for caustic soda) 勒苏(制苛性钠)电池
Lesmok powder 莱斯莫克(火)药
less-chalk-resistant 不耐粉化的
less common exchanger 不常见的交换剂
less draft on intermediate drawing frame 倒牵伸
less elastic lay-in warp knitted fabric 小延伸衬纬经编组织
lesser galangal 良姜；红豆蔻
Lessing ring 勒辛环
less-persistent pesticide 低残留农药
less-polluting 低污染的
letdown solvent 调稀溶剂；兑稀溶剂
letdown tank ①调漆槽②兑稀罐
letdown trouble 兑(调)稀故障
letdown vehicle ①兑稀用漆料②调稀料
let-down vessel 排出储罐
let-go 脱胶
lethal agent 致死剂
lethal concentration 致死浓度
lethal dosage 致死剂量

lethal dose 致死(剂)量
lethal equivalent 致死当量
lethal index(=mortality-product) 致死指数
lethal mutation 致死突变
lethal toxicity 致死毒性
lethane 混合杀虫剂
lethane 60 脂肪酸乙硫腈；L-60
lethane 384 special 混合乙硫腈
let-in piece ①垫片②嵌入物③楔
let-off gear 导出装置
let-off pipe 出水管；排水管；排气管
let-off stand 导出装置
letrozole 来曲唑〔药〕
letter balance 邮件天平
lettering compound 刻字橡皮
lettering enamel 书写用(瓷)漆
letter paper 信笺
letterpress 活版印刷
letters patent 专利权证书；专利证；特许证书
lettuce 莴苣
LET value 线能量转移值；传能线密度值
leucacene 苊热裂烃 $C_{54}H_{32}$
leucaenine(=mimosine; leucaenol) 含羞草碱
leucaenol(=mimosine;leucaenine) 含羞草碱
leucaniline 三氨基苯基甲烷
leucanol 萘磺酸钠甲醛缩合物
leucaurine 隐金红科
leucenol 含羞草氨酸
Leuchs's base 刘琪氏碱
Leuchs's rearrangement 刘琪氏重排作用
leucic acid 亮氨酸；白氨酸；2-羟基-4-甲基戊酸
    $C_6H_{12}O_3$
leucinate 亮氨酸酯；白氨酸酯；α-羟基己酸酯
leucine 亮氨酸；白氨酸；异己氨酸
leucinimide 环缩二白氨酸 $C_{12}H_{22}O_2N_2$
leucite(=amphigene) 白榴石
leucitite 白榴石岩
leucitoeder 白榴石面体〔晶形〕
leucitophyre 白榴斑岩
Leuckart reaction 刘卡特反应
leuco(=leuko) ①无色②褪色③白
leuco acid 隐色酸
leucoaniline 白苯胺；甲基三氨基三苯甲烷
    $CH_3C_6H_3(NH_2)CH = (C_6H_4NH_2)_3$
leucoanthocyanidin 无色花色素
leucoanthocyanin 无色花色苷
leucoaurin 白金精；4,4,4-三羟基三苯甲烷；三(对羟苯基)甲烷 $(HOC_6H_4)_3CH$

leucobase 无色母体
leucobase of malachite green 孔雀绿无色母体
    $[(CH_3)_2NC_6H_4]_2CHC_6H_5$
leucobase of pararosaniline 副品红无色母体
    $[(CH_3)_2NC_6H_4]_3CH$
leuco-berberine blue 隐色小檗碱蓝
leuco compound 无色化合物
leucocratic(=light-colored) 浅色的；淡色的
leucocyan 藻兰素
leucocyanidin 无色氰定；无色花青素
leucodelphinidin 无色翠雀定
leucodendron 山龙眼科
leucoderma 白斑病；白癜风
leuco-2,6-dichlorophenolindophenol 无色-2,6-二氯酚靛酚
leucodrin 银树苦素
leuco dye ①无色染料②染料隐色基
leucoflavin 无色核黄素
leucoglycodrin 银树苷
leucoguibourtinidin 无色盖波亭宁
leucoindigo 靛白
leucoline 喹啉
leucomaclurin glycol 儿茶酸母质
leucomalachite green 隐色孔雀绿 $C_{23}H_{26}N_2$
leucomethylene blue 无色美蓝；无色亚甲蓝
leucomidin 白菊定
leucomycin 柱晶白霉素；吉他霉素
leucon 原甲硅酸 $HSi(OH)_3$
leucone 原甲硅酸
leuconic acid 白酮酸；环戊五酮
leucophite 白环蛇纹岩
leucophyre 糟化辉绿岩
leuco-pigment 无色颜料；隐色颜料；白色颜料
leucopterin 无色蝶呤
leucopyrite 低砷铁矿
leucoriboflavin 无色核黄素
leucorobinetinidin 无色刺槐亭定
leucorosolic acid 隐色玫红酸 $C_{20}H_{19}O_3$
leucoscope ①光学高温计②色光光度计
leucosphenite 淡钡钛石
leucothionine 白硫堇
leucotrope 转白剂；拔白剂
leuco vat dye(=leuco vat dyestuff) 无色瓮染料；无色还原染料
leuco vat dyestuff(=leuco vat dye) 无色瓮染料；无色还原染料
leucovorin 甲酰四氢叶酸；亚叶酸
leucoxene 白钛石

leucyl 亮氨酰 (CH₃)₂CHCH₂CH(NH₂)CO—
leucylglycine 亮氨酰甘氨酸
leucyl leucine 亮氨酰亮氨酸
leukanol 短碳链磺酸物〔尤指配制鞣革剂用的磺酸物〕
leuko-(=leuco-) ①无色②褪色③白
leukonic acid 白酮酸；环戊五酮
leukonin 锑酸钠
leukotine 巴拉柯土(树皮)素
leukotriene 白细胞三烯
leuprorelin 亮丙瑞林〔药〕
levamisole 左旋咪唑〔药〕
levamlodipine 左旋氨氯地平〔药〕
levant leather 皱纹革；摩洛哥革
levant storax oil 苏合香油
levant wormseed oil 山道年油；美洲土荆芥油
levarterenol bitartrate 二酒石酸氢去甲肾上腺素
level ①水平②水平面③平坦的④级位
levelability 流平性
level alarm 液位警报；水位警报
level alarm high low 液面上下限警报(器)
level bottle 平液水准瓶
level controller(=level regulator) 液面调节器；控平器；自动调整水平器
level cross mechanism 能级交叉机理
level detection 水平检测
level dyeing 匀染
level dyeing agent 匀染剂
level-dyeing property 匀染性
leveler 匀平剂
level filler (油)液面控制孔
level flask 平液水准瓶
level flesh 平肉〔铲平兽皮近肉面〕
level ga(u)ge ①液面(指示)计②料面(指示)计③水准线
level gauging 物位测量；液面测量；料位测量
level indication controller 液面指示控制器
level indicator 液面指示器
leveling ①均化；拉平②均染
leveling bulb 水准球管；平液球管
leveling characteristics 流平性
leveling effect 拉平效应
leveling end point 流平终点
leveling equation 流平方程式
leveling solvent 拉平溶剂
leveling viscosity 流平黏度
level instrument 测量水平仪
leveller ①调平器②涂布器
level line 水平线；液面线
level(l)ing ①均化②均涂③调平；弄平④流平性

levelling agent 均化剂；匀染剂
levelling arm 矫平机
levelling bottle 平液水准瓶
levelling bulb 平液球管
levelling bulb reservoir 校平储液球
levelling color(=levelling dye; levelling dyestuff) 均染染料
levelling colour 均染染料
levelling dye(=levelling color) 均染染料
levelling dyestuff(=levelling color) 均染染料
levelling effect 拉平效应
levelling instrument 水准器；水平尺；测平仪
levelling machine 平整机
levelling off ①趋于平衡；趋于稳定②均化；均染
levelling off degree of polymerization 极限聚合度
levelling property 均涂性〔涂料〕
levelling red 匀染红
levelling roll 校平辊
levelling screw 校平螺旋；水准螺旋
levelling support 校平架；水准架
levelling tube 平液管；水准管
levelling vessel 平液容器
level meter ①水平仪；水位(液位)指示器②电平表
levelness 匀染性；匀染度
levelness drawdown standard 刮样流度标准板
levelness standard 流平性标准板
level of factor 因素水平*
level of fuel 燃料液面
level of oil 油面
level of significance 显著性水平
level of signification ①显著水准；显著水平②有效级；有效指标
level regulator(=level controller) 液面调节器；控平器；自动调整水平器
level sensor 液位探测器；水位传感器
level tank 液位槽
level transfer nozzle 液位传送器喷嘴
level tube 校平管
level vessel 平液容器
lever 杠杆；杆
lever action air pump 杠杆抽气泵
lever arm swinging model 杠杆臂摇模型
lever press 杠杆式平板机
leverrierite 晶蛭石
levextrel resin 萃淋树脂
leviathan washer 大洗涤机
levigated ①澄出的②磨细的
levigating ①磨细②澄出

levigation ①研磨②水磨
levigation residue ①淘析(淘洗)残余物②水选残余物
levilling of degree of polymerization 极限聚合度；平衡聚合度
levitation melting 悬浮熔融(法)；无坩埚熔融
levo(=kaevi;laevo) 〔拉丁字头〕 左旋的；左的
levo-compound 左旋化合物
levodopa 左旋多巴〔药〕
levodopa and benserazide hydrochloride 多巴丝肼〔药〕
levofloxacin 左氧氟沙星〔药〕
levoglucos(a)e 左旋葡萄糖；果糖
levogyration 左旋
levogyric(=levorotory) 左旋的
levoisomer 左旋异构体
Levol's alloy 勒伏耳银铜合金
levomenthone 臭松油
levopimaric acid 左旋海松酸
levorotary 左旋的
levorotation 左旋
levorotory(=levogyric) 左旋的
levorotatory crystal 左旋(结)晶体
levo-rotatory substance 左旋物质
levorphanol 左啡诺
levothyroxine 左旋甲状腺素〔药〕
levulan 聚左旋糖；聚果糖
levulic acid(=levulinic acid) 乙酰丙酸
levulic aldehyde(=levulinic aldehyde) 乙酰丙醛
$CH_3CO(CH_2)_2CHO$
levulin 菊芋糖
levulinamide 4-氧(代)戊酰胺
levulinate ①乙酰丙酸②乙酰丙酸盐(酯或根)
levulinic acid(=levulic acid) 乙酰丙酸
$CH_3CO(CH_2)_2COOH$
levulinic aldehyde(=levulic aldehyde) 乙酰丙醛
$CH_3CO(CH_2)_2CHO$
levulinic hydrazide 乙酰基丙酰肼
levulinic oxime 乙酰丙酸肟
levulinie acid 乙酰基丙酸
levulosans 聚果糖类
levulosazone 左旋糖脎；果糖脎
levulose 左旋糖；果糖
levyne 插晶菱沸石
Lewis acid 路易斯酸*
Lewis base 路易斯碱*
Lewis color theory 路易斯颜色理论
Lewis complex 路易斯络合物
Lewis filter equation 路易斯过滤公式
Lewisite 路易斯(毒)气

Lewis-Langmuir theory 路易斯-朗缪尔理论
Lewis relation 路易斯关系
Lewis's atom 路易斯型原子
Lewis-Sorel method 路易斯-苏里尔法
Lewis structure 路易斯结构*
Lewis symbols 路易斯符号
Lewis theory 路易斯理论
Lewis white lead 路易斯铅白；碱式硫酸铅白
Lexan 热塑聚碳酸酯
ley 废皂碱水
Leyden blue 莱顿蓝；钴蓝的别名；铝酸钴
Leyden jar 莱顿瓶
L glass L玻璃；含铅玻璃；防辐射玻璃
liana rubber 藤胶
liatris 鹿舌草
liatris oleoresin 蛇鞭菊油树脂
libanotus 乳香
liberation 发出；释出；放出
liberation method 析出法
libocedrol 翠柏酚
libration 实际的(表观的)摆动运动
libriform fibre 韧型纤维
licanic acid 十八碳三烯-4-酮酸
licarcol 芳樟醇
licence ①许可；特许②许可证；特许证；执照
licenced paint store ①专用油漆仓库②有许可证的油漆仓库；执照油漆库
licensed technology 特许转让技术
lichenol(=6-isopropylquinoline) 6-异丙基喹啉
lick roll(er)(=kiss roll(er)) ①舐(吻)涂辊②给漆辊③舐漆辊④挂胶辊
licopersicin(=tomatin) 番茄素
licopin 番茄色素
licorice(=liquorice) 甘草
licorice elixir 甘草酊
licorice extract 甘草浸液
licorice products 甘草产品
licorice saponin 甘草皂角苷
lidenic bond 李登键 [RCH=C(CN)COR]
lidocaine 盐酸利多卡因〔局部麻醉药〕
Lieben reaction 李本反应
Lieben solution 碘的碘化钾溶液
Liebermann test 利伯曼试验
Lieberman-Storch test 利伯曼-斯托奇试验〔检验松香〕
Liebig-Graham condenser 蛇形冷凝管
Liebig method 李比希法
liensinine 莲心碱
life 储存期；寿命

life-cycle ①生活周期；生命周期②(研磨)寿命
life cycle analysis(LCA)　产品生命周期分析
life-cycle cost　寿命周期成本
life element　生物元素
life expectancy　耐用期间
lifeless rubber　无弹力橡皮
life period　寿命；存在时期
life root　千里光〔药〕
life span　使用期限；有效期限
life span of oil　油使用期限
life test　使用期限试验；寿命试验
lifetime　生存期；寿命
lifetime broadening　寿命增宽
life time of chain　链式反应寿命
life time of chain reaction　链反应时间；链反应期限
lifetime of excited state　激发态寿命
lifetime service　长期使用；使用寿命；有效寿命
lift ①升液器②升降机；电梯③举(起)④(上)升⑤升扬高度
lift (air lift)　升扬高度
lifted flame　破碎熄灭了的火焰
lifted land　连接盘起翘
lift engager　提升装置接合器；升降机衔接器
lifter ①升降机②举扬器
lifter roof　储罐浮顶
lifter roof tank　升降式浮顶罐
lift-force flow sensor　提升力流量传感器
lift gas　提升用气体
lifting ①提升；举起；升起②咬底；咬起
lifting effect　腾升效果
lifting fitting　起重用部件
lifting magnet　磁选器
lift of pump　泵的扬程
lift of valve plate　阀片升程；阀片升起高度
lift out　提升；举起
lift pipe　提升管
lift pot　提升斗
lift pump　抽扬泵；升液泵
lift technique ①提升技术②提升设备
lift valve　提升阀；升举阀
lift velocity　上升速度
ligancy　配位数
ligand　配体*
ligand atom　配位原子
ligand chromatography　配位体色谱法
ligand exchange　配体交换*
ligand exchange chromatography　配体交换色谱法
ligand exchanger　配体交换剂

ligand field absorption band　配位(体)场吸收带
ligand field activation energy　配位(体)场活化能
ligand-field effect　配位场效应
ligand field splitting　配体场分裂
ligand field stabilization energy(LFSE)　配位场稳定能
ligand field theory　配体场论*
ligand isomerism　配位基异构(现象)；内层配位体异构(性)现象
ligand-ligand interaction　配体间相互作用*
ligandolysis　配位体解离
ligand polymer　配位(体)聚合物
ligand-transfer reaction　配体转移反应
ligand vapour gas chromatography　配位体蒸气气相色谱法
ligarine　石油醚
ligasoid　液气悬胶
ligating atom　配位原子
ligation　络合物形成(作用)
lig atom　配位原子
light ①光线②发光体③浅(色)的④轻的
light absorber　光吸收剂
light absorbing　吸收光线的
light absorbing agent　紫外线吸收剂
light absorbing chromophore　吸光发色团
light absorbing pigment ①光吸收颜料；吸光颜料②紫外线吸收颜料
light absorption　光吸收
light absorption analysis　光吸收分析
light absorption spectrometry　光吸收光谱法
light actinide　轻锕系(元素)
light adaptation　(对)光适应
light-addressable potentiometric sensor　光可寻址电势测量传感器
light aging(=light ageing)　光(致)老化
light alkylate　轻质烷基化物
light alloy　轻合金
light alumina hydrate　轻体水合氧化铝；轻体矾土白
light amplification by stimulated emission of radiation (=laser) ①激光②激光器
light and dark field image　衍射衬度明场和暗场像
light and water exposure apparatus　光水暴露试验装置
light aromatics　轻质芳烃
light ash　轻苏打
light barrier　光栅
light beam focalizing accessories　光束聚焦附件
light beam oscillograph　光束示波器
light beam welding　光束焊接
light bodied　低黏稠的

light brown 浅棕色
light burned magnesia 轻质煅烧氧化镁
light calcium carbonate 轻质碳酸钙
light camphor oil 樟脑白油〔樟脑油低馏分〕
light carrying fibre bundle 光导纤维束
light-catalysed 光催的
light checking 日晒龟裂
light coal 轻煤；气煤；瓦斯煤
light-colored(=leucocratic) 浅色的；淡色的
light-coloured 浅色的
light colour reclaim 浅色再生橡胶
light concrete 轻混凝土
light-conducting fibre 光导纤维
light consumer 减光器
light corrosion 轻度腐蚀
light creosote (oil) 轻杂酚油
light cross-section method 光切(断)法
light crude 轻原油；石蜡基石油
light cut 轻馏分
light cycle oil 轻循环油
light degradable 可光解的
light degradation 光降解
light density 光密度
light diesel fuel 轻质柴油燃料
light distillate 轻馏分
light distributor 光分布器
light-duty detergent 轻役型洗涤剂；轻垢型洗涤剂
light-duty power manipulator 轻型电动机械手
light effect 光效应
light emission 光发射
light end fission product 轻(量)端裂变产物
light-end product 轻质产品
light ends(=tall oil heads) 松浆油轻馏分〔沸点低的馏分〕
light ends removal column 轻质烃蒸除塔；低沸点馏分蒸除塔
light ends tower 轻馏分(分离)塔；气体分馏塔
light engine oil 轻质发动机油
lightening ①减轻②照亮
lighter ①发光器②点火器③更轻的
lighter bar 轻杆
lighter body 轻质基〔石油〕
lighter tone ①浅色调②(照相)高调
light exposure properties 耐光性；耐晒性
light exposure test 耐光性试验；暴晒试验
light fading 日晒褪色；光照褪色
light-fast 耐光
light fastness 耐晒性；耐光性

lightfastness in tint 色彩的耐光性
light-fastness rating 耐光牢度等级
light fastness standard 耐晒度标准
light fastness test 耐光性试验；耐光度试验
light fastness tester 日晒牢度仪
light filler 轻质填料
light filtered cylinder oil 轻质发动机油
light floral odor 淡花香
light-focusing plastic fibre 光聚焦塑料纤维
light force fit 轻压配合
light fraction 轻馏分
light fruity odor 淡果香
light fuel 轻燃料；易挥发性燃料
light fuel oil 轻燃料油
light fugitive 不耐光的
light gage wire 细钢丝
light gap method 光隙法
light gas oil 轻粗柴油
light gauge steel 轻量型钢
light gray 浅灰色
light grease 不稠的润滑脂
light green ①浅绿色②酸性绿(色淀)〔别名〕
light green note 淡青香
light grey 浅灰色
light-heat aging 光热老化
lighthouse kerosene 灯塔用煤油
lighthouse oil 灯塔用油〔照明信号灯用重质灯油〕
light hydrocarbon 轻质烃
light hydrogen 轻氢
lighting gas 照明气
light inhibitor 光阻聚剂
light initiated polymerization 光引发聚合
light-intensity method 光(强)度法〔测玻璃化温度〕
light interference phenomena 光干涉现象
light leather 轻革
light leather liming 薄革浸灰法
light line 轻质输送管
light liquid isoparaffin 轻质液体异链烷烃
light lubricating oil 轻质润滑油
light magnesia 轻镁土
light magnesium carbonate 碱式碳酸镁
$4MgCO_3 \cdot Mg(OH)_2 \cdot 5H_2O$
light metal 轻金属
light microscope 光学显微镜
light microscopy 光学显微术；光学显微镜检查法
light mixing 光学混频
light naphtha 石脑油；轻溶剂
lightness 光亮度；明度

lightness correction　明度校正；亮度校正
lightness dimension (value)　亮度坐标(值)；明度坐标(值)
lightness of colour　彩色亮度(明度)
lightning　(电)闪
lightning rod　避雷针
light oil　①轻油②优(级松节)油
light oil constituents　轻油组分
light-oil cracking　轻油裂化
light oil distillate　轻油馏分
light oil heater　轻油加热器
light pass length　(吸收)池厚度；光程
light pen control system　光笔控制系统
light penetration　光透射
light permanency　耐晒性；耐光性
light petroleum　石油醚
light phase　轻相
light pink cottage rose　百叶玫瑰；五月玫瑰
light pipe　光管
light platinum metal　铂类轻金属
light polymerization　光聚合(作用)
light press fit　轻压配合
light pressure separator　低压分离器
light pressure steam　低压蒸汽
light process oil　轻(质)操作油
light proof　不透光(的)
light-proofness　①耐光性②遮光性
light protecting agent　防日光老化剂
light quantum　光量子
light rare earths　轻稀土(元素)
light ray　光线
light recoil oil　轻反冲油；反冲系统用轻质油
light recycle stock　轻质循环进料
light red　淡红色的
light red silver ore　硫砷银矿；淡红银矿
light resistance　①耐光性②耐光度
light resistant　耐光的
light resistant container　避光容器
light rocket　照明火箭
light running fit　轻转动配合
light scattering　光散射
light scattering background interference　光散射背景干扰
light scattering detector　光散射检测器
light scattering effect　光散射效应
light scattering measurement　光散射测量(法)
light scattering method　光散射法
light scattering photometer　光散射光度计
light screen　遮光板
light screener　光屏蔽剂

light-screening agent　遮光剂
light selective absorption　光选择(性)吸收
light sensation　①光觉②光敏；感光
light sense　光觉
light sensitive　感光的
light-sensitive cell　光敏电池
light-sensitive compound　光敏化合物
light-sensitive detector　光敏检测器
light sensitiveness(=light sensitivity)　①感光性；光敏性②感光度
light sensitivity(=light sensitiveness)　①感光性；光敏性②感光度
light sensitizer　光敏剂
light sensor　光敏元件；光敏感器
light shade　不饱色；浅色
light shell　照明弹
light sizing　轻浆
light slit method　光隙法
light slushing oil　铸模用润滑油
light soda ash　轻苏打
light solvent naphtha　轻溶剂石脑油
light source　光源
light splitter　分光镜；分束器
light spot　光微斑点
light stability　光稳定性；耐光性
light stability agent　抗光剂
light stability test　光稳定试验
light stabilizer　光稳定剂
light-stable polyethylene　光稳定性聚乙烯
light-stable PU paint　耐光性聚氨酯漆；光稳定性聚氨酯漆
light standard　光标准
light-stick　光棒
light stock　缺料
light tar　轻焦油
light tint　浅色；淡色
light tint paint　浅色漆；淡色漆
light tire　薄壁轮胎；少层(帘布)轮胎
light top roller　轻质辊
light transmission　光传输；光透射；透光
light transmission coefficient　透光系数
light transmittance　透光率
light transmittance pattern　光透射图像
light transmitting fiber　光导纤维
light trap　光阱
light volatile　易挥发的
light volatile fuel　易挥发燃料
light washing　轻度洗涤

light water 轻水
light-water method 水光法〔水光催化磺氧化法〕
light water-moderated nuclear reactor 轻水慢化核反应堆
light-water reactor(LWR) 轻水(冷却和慢化反应)堆
light-weight ①轻量(级)的②轻质的③轻型的；轻便的
light weight coatings 轻被膜；轻(量级)处理膜(膜重250～350毫克/平方英尺)
lightweight coverstock 轻薄覆盖材料；薄型覆盖材料
light weight hide 轻磅皮
light weight pipe 轻质管子
light window 光窗
light wood ①轻木；轻质木材；易燃木②充脂材；油材③明子
light wood oil 轻木油；汽馏松节油
light year 光年
light yellowish brown 浅黄棕色；浅黄相棕色
lignaloe oil 沉香油
lignan 木酚素；木脂素
lignasan 乙基氯汞
ligneous 木材的；木质的；木制的；木的
lignic acid 木质酸
lignification 木质化(作用)
lignified fiber 木化纤维
lignin 木素
lignin extract 亚硫酸纸浆废液(浸)膏
lignin phenol formaldehyde resin 木(质)素酚醛树脂
lignin plastic 木素塑料
lignin resin 木(质)素树脂
lignin sulfonate 木素磺酸盐
lignin sulfonic acid 木质磺酸
lignin sulphite 亚硫酸化木素
lignite 褐煤
lignite benzine 褐煤汽油
lignite bitumen 褐性烟煤
lignite oil 褐煤油
lignite paraffine 褐煤石蜡
lignite tar 褐煤焦油
lignite tar oil 褐煤焦油
lignite tar pitch 褐煤沥青
lignitic coal(=lignitous coal) 褐煤
lignitiferous 褐煤化的
lignitous coal(=lignitic coal) 褐煤
lignocaine hydrochloride 盐酸利多卡因〔局部麻醉药〕
ligno-cellulose 木素纤维素
ligno-cellulose sulfite extract 亚硫酸纸浆废液(浸)膏
lignocellulosic anthracite 木质纤维无烟煤
lignocerane 二十四烷 $C_{24}H_{50}$

lignoceric acid(=tetracosanoic acid) 二十四(烷)酸；木蜡酸 $C_{24}H_{48}O_2$
lignoceryl alcohol 二十四(烷)醇
lignone 木纤维质
lignose 木纤维素
lignosulfonate 磺化油；木素磺化盐
lignosulfonic acid 木质素磺酸
lignosulphonate 木素磺酸盐
lignum 木(材)
ligroin(e) 挥发油；石油醚；粗汽油；溶剂轻油
liguation 熔融；熔解
ligusticumic acid 藁本酸
ligusticum lactone(=n-butylidene phthalide) 藁木内酯；正亚丁基酞内酯
ligustrazine 川芎嗪〔药〕
ligustrin 女贞素
ligustrol 女贞醇
ligustron 女贞酮
likelihood ratio test 似然比检验
like pole 同名极
liker-in 刺辊
lilac absolute 丁香花净油
lilac alcohol 丁香花醇
lilac compound 丁香香精
lilac concrete 丁香花浸膏
lilacin 丁香配基
lilial 铃兰醛；α-甲基对叔丁基苯丙醛
Lilienfeld viscose rayon 里氏黏胶
liliquoid 乳状胶体
Lillie evaporator 李利蒸发器
lilolidine 四氢吡咯并[1,2,3-i,j]四氢喹啉
lily 百合
lily aldehyde 百合醛；铃兰醛
lily concrete 白百合浸膏；铃兰浸膏
lily of the valley 铃兰
lily white 百合白；纯白〔透明石油产品〕
limanol 盐泽泥制剂〔药〕
lima-wood 菩提木
lime acetate 乙酸钙 $Ca(C_2H_3O_2)_2$
lime base grease(=lime grease) 钙基润滑脂
lime blast 石灰斑〔革〕
lime blooming 白华现象
lime blue 石灰蓝（群青和石膏粉的混合物）
lime boil 石灰煮沸〔纤〕
lime borate 硼酸钙
lime burner 石灰窑
lime burning 锻烧石灰
lime carbonate 碳酸钙 $CaCO_3$

lime chloride 氯化钙
lime concrete 石灰混凝土
lime cream 石灰乳
limed ①石灰处理过的②刷了石灰的③用石灰中和的
　　　　④石灰熔融的
lime decontamination 石灰净化；石灰去杂(质)
limed hide 石灰鞣革
limed juice 澄清汁〔制糖〕
limed linseed oil 钙脂亚麻仁油；石灰(精制)亚麻油
limed oil ①钙脂松浆油②钙脂油
limed paint oil ①钙脂清漆〔泛指〕②钙脂松浆油清漆
　　　　〔精制松浆油用石灰处理后加干料〕
limed rosin 石灰松香
limed rosin varnish 钙脂清漆；石灰松香清漆
limed tall oil 钙脂松浆油；石灰松浆油
lime factor 石灰因素
lime feldspar 石灰长石；钙长石
lime flavour 白柠檬香精
lime-free plaster 无石灰(的)灰浆；不含石灰的灰浆
lime gas purification 石灰气纯化
lime glass 石灰玻璃；钙玻璃
lime glaze 石灰釉
lime grease(=lime base grease) 钙基润滑脂
lime green 石灰绿〔孔雀绿和绿土的混合物〕
lime hydrate 消石灰；熟石灰 $Ca(OH)_2$
lime hydration 石灰的消化作用；石灰水化
lime hydroxide 消(熟)石灰；氢氧化钙
lime kiln 石灰窑
lime-kiln gas 石灰窑气
lime kilning 煅烧石灰
lime-kiln with gasogene 有气体发生器的石灰窑
lime lead glass 钙铅玻璃
lime light ①石灰光②著名；注意点
lime magnesia ratio 石灰镁氧比例
lime milk 石灰乳
lime modulus 石灰模数
lime mortar 石灰砂浆
lime mud 苛化泥
limene 橙油倍半萜；红没药烯 $C_{12}H_{24}$
lime nitrate 硝酸钙 $Ca(NO_3)_2$
lime oil 梨莓油；白柠檬油
lime pit 石灰坑
lime plaster 含石灰(的)灰浆；白(石)灰浆
lime purification(=lime purifying) 石灰净化
lime purifying(=lime purification) 石灰净化
lime putty 石灰胶结料；石灰膏
lime red 石灰红〔品红和黏土的混合物〕
lime reel 灰池运皮卷轴〔皮革〕

lime requirement 石灰需要量
lime-resistance 耐钙性
lime-resistant 耐钙的
lime-resistant detergent 耐钙洗涤剂
lime-resistant surfactant 耐钙表面活性剂
lime rock 石灰石
lime rock tower 石灰石塔
lime-sand brick 硅石砖；火砖
lime scum 压滤残渣〔糖〕
lime silicate glass 石灰玻璃
lime slag cement 石灰矿渣水泥
lime slaker 石灰熟化器；化灰器；消石器
lime slaking 石灰消化作用；石灰熟化
lime sludge ①压滤残渣〔糖〕②石灰黏泥
lime soap 石灰皂；钙皂
lime soap dispersant 钙皂分散剂
lime soap disperser 钙皂分散剂
lime soap dispersing action 钙皂分散作用
lime soap dispersing agent 钙皂分散剂
lime soap dispersing power 钙皂分散能力
lime soap dispersing value 钙皂分散值
lime soap streak 钙皂垢痕
lime-soda process 石灰-苏打(软水)法
limespar 方解石
limes reacting dose 反应限量
lime stain 石灰斑(疤)；石灰污迹
lime standard 标准石灰量
lime still 石灰蒸氨器
limes tod 致死限量
limestone 石灰石*
limestone drop test 石灰石落降试验；石灰石冲击试验
limestone elevator 石灰石升降机
limestone flour 石灰石粉；重质碳酸钙
limestone flux 石灰石助熔剂
limestone kiln 石灰窑；石灰炉
limestone oil 石灰石油；石灰石沉积中石油
limestone powder (石灰)石粉；白垩
limestone whiting 白石灰岩
lime sulfur mixture 石(灰)硫(黄)合剂
lime superphosphate 过磷酸钙〔肥料〕
lime treatment (用)石灰处理
limettal 白柠檬醛；白柠檬香基
limette oil 白柠檬油
limettin 白柠檬素
lime wash 刷白；涂白
lime water 石灰水
lime white 石灰白；石灰浆
lime yard 石灰间

liminal ①易觉的②最低量
liminal value 极限值
liming 加石灰；浸灰
liming apparatus ①制石灰乳器②石灰熟化器
liming chamber 石灰乳室〔石灰乳与氨水混合部位〕
liming material 石灰浸物质
liming of resin 树脂的石灰处理
liming out 灰析
liming process 灰浸法；浸灰法
liming stains〔复〕 石灰斑
liming still （制）石灰乳槽
liming tank （加）石灰槽
liming tub ①石灰乳槽②灰鞣槽
liming vat 石灰乳槽
limit 限度；极限
limitation 限度
limit crystallization temperature 极限结晶温度
limit current density 极限电流密度
limit cycle 极限环
limited apparent flow 有限表观流动
limited cathode potential electrolysis 控制阴极电位电解
limited cathode potential method 阴极限压法
limited configuration interaction(LCI) 有限构型相互作用
limited current 极限电流
limited deformation 有限形变
limited expansion of diatomic overlap(LEDO) 有限扩展双原子重叠法
limited flow 有限流动
limited fluid flow 有限流体流动
limited lye 极限碱析液〔三相图上的E点〕
limited miscibility 极限溶混性
limited plastic flow 有限塑性流动
limited pot life 有限活化期；有限活化寿命
limited solubilizing capacity 极限增溶能力
limited swelling 有限溶胀
limited vacuum 极限真空
limit efficiency 极限效率
limiter ①限制器②限幅器
limit formula 极限公式
limit gauge ①极限量规；限制量计②口径限度
limiting adsorption current 极限吸附电流
limiting amplitude of current 电流的极限幅度
limiting association concentration 极限缔合浓度〔发生加溶作用的最低浓度〕
limiting catalytic current 极限催化电流
limiting compliance 极限柔量
limiting concentration 极限浓度
limiting conductivity 极限电导率；当量电导率

limiting current 极限电流
limiting curve 极限曲线
limiting density 极限密度
limiting diffusion coefficient 极限扩散系数
limiting electrode potential 极限电极电位
limiting equivalent ionic conductance 极限等当量离子电导
limiting factor 限制因素
limiting gas velocity 气体极限速度
limiting inherent viscosity(=intrinsic viscosity) 特性黏数；固有黏度
limiting kinetic current 极限动力电流
limiting law 极限定律
limiting load 极限载荷
limiting logarithmic viscosity number 极限对数黏度值
limiting material 浸灰材料〔皮革〕
limiting melting point 极限熔点
limiting migration current 极限迁移电流
limiting modulus 极限模量
limiting orifice 极限孔
limiting oxygen index(LOI) 极限氧指数
limiting proportion 极限比例
limiting retention volume 极限保留体积
limiting rigidity number 极限刚度值
limiting screen 极限筛目
limiting shearing stress 极限剪切应力
limiting sizing 极限筛分
limiting stress 极限应力
limiting stretch rate 极限伸长速率
limiting surface 界面
limiting threshold value 极限阈值
limiting value 极限值
limiting viscosity 特性黏度
limiting viscosity number 特性黏数*
limit of accuracy 准确性极限
limit of detection of impurities 杂质检测极限
limit of determination 测定(下)限
limit of elasticity 弹性极限
limit of fatigue 疲劳限度；疲劳极限
limit of identification 鉴定极限
limit of inaccuracy 误差限度
limit of inflammability 着火极限
limit of miscibility 溶混性极限
limit of proportionality 比例极限
limit of quantification 测定(极)限；定量限
limit of reliability 可靠性极限
limit of saturation 饱和极限
limit of sensitivity 极限灵敏度

limit size 极限尺寸
limits of explosion 爆炸极限
limits of flammability(=limits of inflammability) 可燃性极限
limits of inflammability(=limits of flammability) 可燃性极限
limits of tolerance 容许极限；容许量
limit strength 极限强度
limit test 限度检查
limit test for arsenic 砷的极限试验
limit test for heavy metals 重金属(限度)试验法
limit test for sulfate 硫酸盐极限试验法
limmer 沥青石灰石
limnology 湖沼学
limocitrin 柠檬素；柠檬黄素
limocitrin-3, 7-diglucoside 柠檬素-3,7-二葡萄糖苷
limocitrin-3-glucoside 柠檬素-3-葡萄糖苷
limonane 苧烷
limonene 苧烯；1,8-萜二烯 $C_{10}H_{16}$
dl-limonene dl-苧烯；dl-萜二烯；外消旋苧烯
limonene dioxide 二氧化苧烯
limonene monoepoxide (单)环氧苧；(单)环氧柠檬烯
limonin 柠檬苦素
limonine 柠檬苦素
limonite 褐铁矿
limp leather goods 软皮件
limp state 无弹(性状)态
limy ①胶黏的②石灰质的；含石灰的
linaethol 亚麻酯
linaloe oil 里哪油；沉香油；伽罗木油
linaloe seed oil 伽罗木籽油
linaloe wood oil 伽罗木油；香榄木油
linalol 里哪醇
linalool 里哪木醇；沉香醇；芳樟醇 $C_{10}H_8O$
linalool acetate 里哪醇乙酸酯 $CH_3CO_2C_{10}H_{17}$
linalool diepoxide 二环氧芳樟醇
linalool epoxide 环氧芳樟醇
linalool monoepoxide (单)环氧芳樟醇
linalool monoxide(=epoxylinalool) 环氧芳樟醇
linalo(o)l oil 芳樟醇油
linalool oxide 氧化芳樟醇；环氧芳樟醇
linalyl 里哪基 $C_{10}H_{17}-$
linalyl acetate 乙酸里哪(醇)酯 $CH_3CO_2C_{10}H_{17}$
linalyl 2-aminobenzoate 2-氨基苯甲酸芳樟酯
linalyl anthranilate 邻氨基苯甲酸芳樟酯
linalyl benzoate 苯甲酸芳樟酯
linalyl butyrate 丁酸芳樟酯
linalyl caproate 己酸芳樟酯

linalyl capronate 己酸芳樟酯
linalyl caprylate 辛酸芳樟酯
linalyl cinnamate (肉)桂酸芳樟酯
linalyl ethoxyacetate 乙氧基乙酸芳樟酯
linalyl formate 甲酸芳樟酯
linalyl heptylate 庚酸芳樟酯
linalyl hexanoate 己酸芳樟酯
linalyl hexylate 己酸芳樟酯
linalyl isobutyrate 异丁酸芳樟酯
linalyl isopentanoate 异戊酸芳樟酯
linalyl isovalerate 异戊酸芳樟酯
linalyl methoxyacetate 甲氧基乙酸芳樟酯
linalyl methyl ether 芳樟甲醚
linalyl octanoate 辛酸芳樟酯
linalyl octoate 辛酸芳樟酯
linalyl octylate 辛酸芳樟酯
linalyl phenylacetate 苯乙酸芳樟酯
linalyl 3-phenylpropenoate 3-苯丙烯酸芳樟酯
linalyl propionate 丙酸芳樟酯
linamarin 亚麻苦苷
linarin 里哪苷
linarine (=dl-vasicine) dl-鸭嘴花碱
linarite 青铅矿
lincocin(=lincomycin) 林可霉素
lincolnensin(=lincomycin) 林可霉素
lincomycin(=lincocin; lincolnensin) 林可霉素
lincrusta 油毡纸；彩色拷花墙纸(商标名)
lindane〔俗名〕高丙体六六六；林丹〔一种农用杀虫剂、除草剂〕
lindazulene 母菊薁
Linde copper sweetening 林德铜脱硫法
lindelofidine 长柱琉璃草定
lindelofine 长柱琉璃草分
Lindemann's mechanism 林德曼机理
linden absolute 椴树花净油；菩提花净油〔商〕
linden concrete 椴树花浸膏；菩提花浸膏〔商〕
linden flower oil 椴树花油；菩提花油
linden oil 椴树油
Linde process 林德法〔一种气体液化法〕
lindera lactone 钓樟内酯
lindera leaf oil 乌药叶油
linderane 钓樟烷；乌药醚内酯
lindera oil 乌药油
lindera root oil 乌药根油
linderazulene 钓樟薁
linderene 钓樟烯
linderic acid 5-十二(碳)烯酸；乌药酸
linderol 钓樟醇 $C_{11}H_{22}O$

linderone 钓樟酮
lindestrenolide 乌药根内酯
Linde valve trap 林德阀板
lindgrenite 钼铜矿
lineal density(=linear density) 线密度
line analysis (X射线)线分析
linear alkali consumption rate 线性碱消费率
linear alkene 直链烯烃
linear alkylbenzene 直链烷基苯
linear amorphous polymer 线型无定形聚合物
linear attenuation coefficient 线性消减系数；线性衰减系数
linear calibration 线性校正
linear calibration curve 线性校准曲线
linear capacity 线性容量
linear chain 直链
linear chain polymer 直链聚合物；线型聚合物
linear chelatropic reaction 线性螯合反应
linear chromatography 线性色谱(法)
linear colloid 线性胶体
linear combination 线性组合
linear combination of atomic orbitals(LCAO) 原子轨函数的线性组合(法)
linear condensation polymerization 线型缩聚作用；线型缩聚反应
linear control theory 线性控制论
linear copolymer 线型共聚物
linear correlation 线性相关
linear coupled dynamic theory 线性偶合动态论
linear creep 线性蠕变
linear crystal growth rate 线型结晶生长速率
linear crystallizer 线式结晶器
linear deformation 线性形变
linear density(=lineal density) 线密度
linear dependence 线性相关
linear differential equation 线性微分方程式
linear dilatation(=linear extension) 线膨胀
linear dimension 晶边长度
linear dimer 线型二聚物
linear discriminant analysis 线性鉴别分析
linear discriminant function 线性判别函数
linear dispersion 线色散*
linear distortion 线性畸变
linear drift correction 线性漂移校正
linear dynamic range 线性动态范围
linear elastic fracture mechanics 线弹性断裂力学
linear elastoplastic fracture mechanic 线(性)弹塑性断裂力学

linear electric field effect(LEFE) 线性电场效应
linear elongation ①线性伸长②线性伸长率
linear eluant strength gradient 线性洗脱液强度梯度
linear elution chromatography 线性洗脱色谱法
linear ester 线型酯；直链酯
linear expansibility 线膨胀性
linear expansion 线膨胀
linear expansion coefficient 线膨胀系数
linear expansivity 线(膨)胀系数
linear extinction coefficient 线性消光系数
linear extrapolation method 直线外插法；直线外推法
linear extrusion speed 挤出线速度
linear flow 线性流动；直线流动
linear fluid 线性流体〔即牛顿流体〕
linear free energy 线性自由能*
linear free energy relationship(LFER) 线性自由能关系
linear gas velocity 气体线速
linear gel chromatographic column 线性凝胶色谱柱
linear Gibbs free energy relation 线性吉布斯自由能关系
linear gradient 线性梯度
linear graph 线形图
linear growth rate 线性生长率
linear high polymer 线型高聚物
linear hydrocarbon 直链烃；线型烃
linear hydrocarbon polymer 线型碳氢聚合物
linear ideal chromatography 线性理想色谱(法)
linear independence 线性无关
linear inner interpolation 线性内插法
linear interpolation 线性内插值
linear interpolation method 直线内插法
linear interpolation titration method 线性内插滴定法
linear isotherm 线性等温线
linearity 线性；直线性
linearized flow 线性化流动
linearizer ①线性补偿器；线形补偿器②线性化电路
linear lag 线式取样；线性度〔指仪器响应〕
linear laser spectroscopy 线性激光光谱学
linear law 线性定律
linear learning machine 线性学习机
linear least square fitting 线性最小二乘拟合
linear low density polyethylene(LLDPE) 线型低密度聚乙烯
linearly polarized 线式偏振的
linearly polarized light 线式偏振光
linear macromolecule 线型高分子
linear magnetic field gradient 线性磁场梯度
linear mean diameter 线性平均直径
linear measure 长度测量

linear medium-density polyethylene(LMDPE)　线性中密聚乙烯
linear model　线性模型
linear modulation　线性调制
linear molecule　线型分子*
linear non-ideal chromatography　线性非理想色谱(法)
linear oligoester　线型低聚酯
linear oligomer　线型低聚物；直链低聚物
linear partition isotherm　线性分配等温线
linear path time-of-flight mass spectrometer　直线轨道飞行时间质谱仪
linear phenomena　线性现象
linear polarization method　线性极化法
linear polycondensate　线型缩聚物
linear polycondensation　线性缩聚
linear polyketone　线型聚酮
linear polymer　线型聚合物；线型高聚物
linear polymerization　线型聚合
linear polymer molecule　线型聚合物分子
linear potential sweep　线性电位扫描
linear potential sweep voltammetry　线性电位扫描伏安法
linear prediction(LP)　线性预测
linear prediction method　线性预测法
linear prediction with singular value decomposition (LPSVD)　单值分解的线性预测
linear programming method　线性规划法
linear pulsed time-of-flight mass spectrometer　直线脉冲飞行时间质谱仪
linear pyrolysis rate　线性热解(速)率
linear rate of flow　线性流速
linear regression through the centroid　通过矩心的线性回归
linear regression through the origin　通过原点的线性回归
linear resonance mass spectrometer　直线共振质谱仪
linear restriction　线性约束
linear scale　直线标度
linear scan　线性扫描
linear scanning　线性扫描
linear shrinkage　线向收缩
linear sorption isotherm　线性吸着等温线
linear spectral model　线性谱模型
linear speed　线速度
linear split system　线性分流系统
linear stream　线性流(动)
linear stretching　线性伸长；线性伸缩变化
linear structure　直线(型)结构
linear superpolymer　线型超聚(合)物

linear superposition　线性叠加
linear superposition spectra　谱图的线性叠加
linear sweep anodic stripping voltammetry　线性扫描阳极溶出伏安法
linear sweep chronoamperometry　线性扫描计时电流法
linear sweep differential pulse voltammetry(LSDPV)　线性扫描微分脉冲伏安法
linear sweep oscillopolarography　线性扫描示波极谱法
linear sweep potential　线性扫描电位
linear sweep voltammetry　线性扫描伏安法
linear synthesis　线性合成*
linear temperature programmer　线性程序升温器
linear temperature programming　线性程序升温
linear tempering　线性回火；线性调温
linear theory of viscoelasticity　线性黏弹性理论
linear thermal expansion　线性热膨胀
linear thermodilatometry　线性热膨胀法
linear titration graphy　线性滴定曲线
linear titration plot　线性滴定图解法
linear transformation　线性变换
linear-*trans*-quinacridone　线型反式喹吖啶酮
linear traverse　线性横向绕程
linear trimer　线型三聚体
linear unsaturated polyester　线型不饱和聚酯
linear velocity　线性速度
linear viscoelastic behaviour　线性黏弹特性
linear viscoelasticity　线性黏弹性
linear viscoelasticity with couple-stress　具偶合应力的线性黏弹性
linear-viscoelastic material　线性黏弹材料
linear viscoelastic micropolar material　线性黏弹微极性材料
linear winding　线性缠绕
linear zigzag-shaped polymer　线型锯齿状聚合物
line-blending　在管道中掺和；管道混合
line booster pump station　泵站〔输送管线〕
line break　输送管线断裂
line broadening　谱线增宽*
line checking　线形细裂
line clogging　输送管线阻塞
line current　输送管线口电流
line cut(=line engraving)　线条雕刻；线条镌刻
line drawing　线拉伸
lined ring　衬环
line engraving(=line cut)　线条雕刻；线条镌刻
line-enhanced method　显线法
line factor　谱线因子
line filling　充满足限

line half-width 谱线半宽度
line header section 输送管线的泵站
line-lattice 线型点阵；单向有序
line leather 马具革
line light source 线光源；锐线光源
linen ①亚麻布②亚麻线③亚麻的
linen-backed lining(=paper-faced scrim) 纸面(衬)纱布里
linen finish 布纹(面筛)
linen line system 亚麻长麻纺纱
linen paper 布纹纸
linen yarn 亚麻纱
line of force 力线
line of oil 油组
line of pipes 输送管线
line out 按装置的制度操作
line oven 线炉
line pack storage 输送管中储存产品
line-pairs (光)谱线对
line pattern ①线条式样②线状光谱
line pipe (输送管)线管
line pressure 输送管压力
line production 流水作业法
line profile 谱线轮廓
line pump (向)管线(中)泵送
liner ①衬垫②衬圈〔机工〕
line recorder mass spectrometer 连续记录质谱计
liner energy transfer(LET) 传能线密度
line reversal technique 反转辉线法
liner motor selection 线性电动机选针
liner plate 衬板
line sampling 线式取样
line selection 谱线选择
line shaft 天轴；总轴
line shape analysis 线形分析
line shape of ESR ESR谱线型
line shape of NMR NMR谱线型
line size 管道尺寸
line source 锐线光源；线光源
line spectrum 线状光谱
Lineta leveling test blade method 莱内塔流平试验刮刀法
line-type cracking 线型开裂
line walker 管线巡逻人
line width 谱线宽度
linewidth alternation 线宽交替；交替线宽
line width method for quantitative spectrographic analysis 谱线宽度光谱定量分析法
line width method of quantitative analysis 谱线宽度定量分析法
line with lead 衬铅
line with rubber 衬胶
lingering period 受激期；逗留期
ling liver oil 鳘鱼肝油；鳕鱼肝油
lingo-cellulose 木素纤维素
lingo-cellulose sulfite extract 亚硫酸纸浆废液(浸)膏
linhay 运瓷土的水泥站台
liniment 搽剂；涂沫油；涂沫剂
linin 亚麻素
lining 衬里；里子
lining Babbit 巴氏合金衬层
lining brick 衬砖
lining leather 衬里革
lining shearling 剪毛绵羊衬里革
lining tile 衬瓦
lining tube 衬管
lining wall 衬砌壁
link ①键；键合②环节；连系物
linkage ①键；键合②连锁
linkage force 键合力
linkage heat 键合热；结合热
linkage isomerism 键合异构
linkage valance 键合价
link arm 连杆臂
link-coupled 偶联的
linked 连接的
linked scan 联动扫描
linking 连接
linking agent ①交联剂；键合剂②连接剂；结合剂
linking group 连接基(团)〔分子中连接亲水基和亲油基的部分〕
link mechanism 连杆机构
link molecule 连接分子〔非相邻的晶粒或晶片间的长程连接分子或由于晶片裂开而形成的晶块间的长程连接分子〕
link pin 连杆销
link rod 连杆
linkrusta ①油毡纸②(压)浮雕花纹裱墙(油毡)纸
link-suspended batch centrifuge (间歇式)三足离心机
link together 连接
linnaeite 硫钴矿
linogen cement(=linoleum cement) 油毡胶黏剂
linoleamide 亚麻(酸)酰胺
linoleate ①亚油酸；9,12-十八碳二烯酸②亚油酸盐(酯或根)
linoleate drier 亚油酸盐催干剂
linoleic acid(=linolic acid) 亚油酸；9,12-十八(碳)二烯

酸；罂酸 $C_{18}H_{32}O_2$
linoleic acid dimer 亚油酸二聚体；二聚亚油酸
linoleic acid series 亚油酸系列 $C_nH_{2n-4}O_2$
linoleic-rich oil 富亚油酸油
linolein 亚油精；甘油三亚油酸酯
linolelaidic acid 反亚油酸
linolen(=trilinolein) 亚油精；甘油三亚油酸酯
linolenate ①亚麻酸盐②亚麻酸酯
linolenate driers 亚麻酸(金属)盐催干剂
linolenic acid 亚麻酸；9,12,15-十八(碳)三烯酸 $C_{18}H_{30}O_2$
linolenin(=trilinolenin) 亚麻精；甘油三亚麻酸酯
linoleodistearin 甘油一亚油酸二硬脂酸酯
linoleum 油毡；亚麻油毡
linoleum cement(=linogen cement) 油毡胶黏剂
linolic acid(=linoleic acid) 亚油酸；9,12-十八(碳)二烯酸；罂酸
linolin(=linolein；trilinolin) 亚油精；甘油三亚油酸酯
linoxyn 氧化亚麻油
linseed 亚麻子
linseed fatty acid 亚麻仁脂肪酸
linseed meal 亚麻子饼粉
linseed mucilage 亚麻子黏质
linseed oil 亚麻子油
linseed (oil) monoglyceride 亚麻油(酸)单甘油酯
linseed oil soap 亚麻子油皂
linseed oil varnish 亚麻子油漆
linseed stand oil 亚麻(油)厚油；亚麻聚合油；聚合亚麻仁油
linseed substitute 亚麻子油代用品
linseed varnish oil 亚麻清油
linseys〔复〕 半毛破布
lint(=charpie) 麻布
lint doctor 棉绒刮刀〔纺〕；除尘刮刀
linter ①棉绒②棉绒除去器
linter pulp 棉绒浆
lint-free filter paper 无棉绒滤纸
lint paper 棉绒纸
linurmass 质量流量计
linusic acid(=hexahydroxystearic acid) 六羟基硬脂酸
liothyronine 碘塞罗宁〔药〕
liparite 流纹岩
lipidal 脂醛
lipiodol 碘化罂粟油
Lipkin method 利普金法〔结构族组成分析法〕
lipoic acid 硫辛酸
Li-polymer 高分子锂电池；锂聚合物电池
lipophilic gel chromatography 亲脂凝胶色谱(法)
lipophilic group 亲油基
lipoprotein 脂蛋白
liposoluble 脂溶的
liposome 脂质体
liposome immunoassay(LIA) 脂质体免疫分析法
liposome luminescence 脂质体发光
lippia 立比草
lippianol 过江藤醇
lippia oil 过江藤油
lippiaphenol 过江藤酚；地奥酚
lippione 过江藤酮
Lippmann electrode 李普曼毛细管电极
Lippmann electrometer 李普曼毛细管静电计
lip screen 分级筛
lip seal 唇边式密封
lipsitol 脂肌醇
Lipsky separator 利普斯基分离器
lipstick moulding machine 口红模制机
liquation furnace 熔化炉
liquefactant 熔解物；解凝剂
liquefaction(=liquification) 液化(作用)
liquefaction of coal 煤的液化
liquefaction point 液化点
liquefaction ratio 液化比率
liquefied 液化的
liquefied bituminous material 液态沥青
liquefied butane 液态丁烷
liquefied carbolic acid 液态石炭酸；液态苯酚
liquefied fluid chromatography 液化流体色谱法
liquefied gas 液化气
liquefied methane gas 液化甲烷气
liquefied natural gas(LNG) 液化天然气
liquefied petrolatum 液化石蜡
liquefied petroleum gas(LPG) 液化石油气
liquefied petroleum gas high-pressure holder 液化石油气高压气罐
liquefied phenol 液态酚
liquefied propellant 液化推进剂
liquefier ①稀释剂液化装置②液化剂
liquefying 液化
liquefying plant 液化装置
liquefying-point 液化点；液化温度
liquescence ①可液化性②可冲淡性
liquescency (可)液化性
liquescent 可液化的
liquid ①液体②液态的
liquid abration cleaning 喷湿砂清洗
liquid absolute 液体净油

liquid absorption 吸液量〔颜料对油以外液体的吸收量〕
liquid accelerator 液体促进剂
liquid acetylene 液态乙炔
liquid acrylonitrile butadiene rubber 液体丁腈橡胶
liquid adsorption chromatography 液相吸收色谱法
liquid air 液态空气
liquid air container 液态空气容器
liquid air pollutants 液体大气污染物
liquid amalgam method 液汞齐法
liquidambar resinoid 枫香树脂
liquid ammonia 液态氨
liquid ammonia fabric treating system 织物液氨处理装置
liquid anion exchange electrode 液体阴离子交换电极
liquid anion exchanger 液态阴离子交换剂
liquid application 液施
liquid asphalt 液态沥青
liquid asphaltic product 液态铺路沥青
liquidation ①液化②溶(化分)离法
liquidation of coal 煤的液化
liquid backup 降液管内液层高度
liquid-based formulation ①(颜、填料表面处理的)浆液配方②(以)液量计配方
liquid bed(=fluidized bed) 流体床；沸腾床
liquid bitumen 液态沥青
liquid blasting 液吹法
liquid blending system 液体掺合系统
liquid calcium chloride 氯化钙液
liquid calorimeter 液体量热计
liquid capacity 液体容量
liquid catalyst 液态催化剂
liquid cation exchange electrode 液体阳离子交换电极
liquid cation exchanger 液态阳离子交换剂
liquid caustic 苛性碱液
liquid cell 液体池
liquid cement ①胶液②液体掺碳剂
liquid cement colourant 水泥色浆
liquid charge-transfer chromatography 液体传荷色谱法；液体电荷转移色谱法
liquid charging stock 液体原料
liquid chlorinated wax 液态氯化石蜡
liquid chlorine 液态氯
liquid chloroprene rubber 液体氯丁橡胶
liquid chromatogram 液相色谱(图)
liquid chromatograph ①液相色谱仪②用液相色谱法分析
liquid chromatograph-mass spectrometer 液相色谱-质谱联用仪

liquid chromatography 液相色谱(法)
liquid chromatography-electrocapture detector 液相色谱-电子捕获检测器
liquid chromatography electrochemistry 液相色谱电化学
liquid chromatography-mass spectrography(LCMS) 液相色谱-质谱联用法
liquid chromatography-mass spectrometer 液相色谱-质谱仪；液-质联用仪
liquid chromatography-mass spectrometry 液相色谱-质谱分析法
liquid coating 液体涂料
liquid collector 湿滤器；液体集尘器
liquid colloidal dispersion 液态胶体分散液
liquid colloidal suspension 液态胶体悬浮液
liquid column chromatography 液相柱色谱法
liquid column gauge(=liquid column manometer) 液柱压力计
liquid column manometer(=liquid column gauge) 液柱压力计
liquid compensating yarn tension regulator 液态阻力式张力器
liquid concentrate 液态母料
liquid condensed film 液态凝聚膜
liquid condition 液态；液体状态
liquid conjunction 液连接口
liquid coolant 液体冷冻剂
liquid cooling 液体冷却
liquid core optical fiber 液芯光纤
liquid corrosion 液蚀(作用)
liquid crystal cell 液晶单元
liquid crystal coating 液晶涂料
liquid crystalline phase 液晶相
liquid-crystalline solution 液-晶态溶液
liquid-crystalline state 液-晶态
liquid crystal panel 液晶面板
liquid crystal polymer(LCP) 液晶聚合物；液晶高分子
liquid crystal polymer fibre 液晶聚合物纤维
liquid crystal resin 液晶树脂
liquid crystal spinning 液晶纺丝
liquid crystal state 液晶态
liquid crystal stationary phase 液晶固定相
liquid crystal structure 液晶结构
liquid crystal tunable filter(LCTF) 液晶可调滤光器
liquid curing 液体硫化
liquid cyclone ①液体旋流分离器②液体除尘器
liquid deformation 液体形变
liquid densimeter 液体密度计
liquid desiccant 液体干燥剂

liquid desiccation 液体干燥剂脱水
liquid detergent 液体洗涤剂
liquid diffusion factor 液体扩散因子
liquid diffusion film 液体扩散薄膜
liquid diffusivity 液体扩散率
liquid dispersing 液相分散剂
liquid displacement 液体置换
liquid displacement technique 液体置换法
liquid disposal 废液
liquid dividing head 液体取样头〔从蒸馏柱中〕
liquid drier 液体催干剂；燥液
liquid dye 液体染料
liquid effluent ①排出液②废液；污水
liquid elastic deformation 液体弹性形变
liquid element 液态元素
liquid elevating valve 液面阀
liquid elution chromatography 液相洗脱色谱(法)
liquid encapsulation 液态封装
liquidene 枫香烯
liquid entrainment 液体雾沫
liquid envelope 液体包封
liquid epoxy 液体环氧树脂
liquid epoxy coating 液体环氧涂料
liquid ethylene 液态乙烯
liquid exclusion chromatography 液体排阻色谱法
liquid expanded film 液态扩张膜
liquid-expansion thermometer 液体温度计
liquid extract 液体提出物
liquid extractor 液相萃取器
liquid feed pump 液体进料泵
liquid filled thermometer 液体温度计
liquid filler ①液态腻子②液态填料③加液器
liquid filling machine 浇铸机
liquid film 液膜
liquid film coefficient 液膜系数
liquid film controlling 液膜控制
liquid film hygrometry 液膜测湿法
liquid film lubrication 液膜润滑
liquid film seal 液膜密封
liquid film separation 液膜分离
liquid finish application roll 液体整理剂涂敷辊
liquid Fischer-Tropsch hydrocarbons 费托液体烃〔费托法从一氧化碳和氢合成得到的液体烃〕
liquid flooding 液体溢流
liquid floor soap 液状洗地板皂
liquid flow 液体流
liquid-flow hank dyeing machine 液流式绞纱染色机
liquid flow meter 液体流速计

liquid flow zone 流动态区
liquid fluorescence analysis 液体荧光分析
liquid friction 流体摩擦；液体摩擦
liquid fuel 液体燃料
liquid fuel oil 液体燃料油
liquid gas 液化的气体
liquid gas distributor 液气分布器
liquid-gas fuel 液化气体燃料
liquid glue 液体胶
liquid gold(=bright gold) (陶瓷描金用)装饰金；亮金；金水
liquid gradient 液体梯度
liquid grease 液体润滑脂
liquid gum 液态胶质
liquid hammer 液击；液锤
liquid head 液体压头
liquid header 液体收集器
liquid heater 液体加热器
liquid helium 液态氦
liquid hold-up 液体滞留
liquid honing 液体研磨；水磨；水砂纸打磨；液助打磨
liquid hydrocarbon 液(态)烃
liquid hydrocarbonceous fuel 液烃燃料
liquid hydrogen 液态氢
liquid hydrogen sulfide 液态硫化氢
liquid inelastic deformation 液体非弹性形变
liquid infiltration 液体渗入；液体浸渍；液体渗透
liquid-in-glass thermometer 液体温度计
liquid injection molding(LIM) 液体注射模压成型
liquid ion evaporation 液体离子蒸发
liquid ion exchange 液体离子交换
liquid ion exchange electrode 液体离子交换电极
liquid ion exchange membrane electrode 液体离子交换膜电极
liquid ion exchanger 液态离子交换剂
liquid isoparaffin 液体异链烷烃
liquidity ①液态②流动性
liquid junction cell 液接电池
liquid junction potential 液体接界电势
liquid lamella 液体薄膜
liquid laser 液体激光器
liquid latex (非浓缩的)液态胶乳
liquid layer 液层
liquid-layer column 液层柱
liquid lens 液体透镜
liquid level 液面
liquid-level alarm 液面警报器
liquid-level controller (柱中)液面调节器

liquid leveler　液面计
liquid level gauge　液体水平压力计
liquid level indicator　液面指示器
liquid level meter　液面计
liquid-like phase　类液相
liquid lime chloride　漂白溶液
liquid limit(=sticky point；upper plastic limit)　液态极限
liquid line　(冷冻机)液体管线
liquid-liquid chromatography　液-液色谱法
liquid-liquid equilibrium　液-液平衡
liquid-liquid extraction　液-液萃取
liquid-liquid extraction system　液-液萃取系统
liquid-liquid extractor　液-液萃取器
liquid-liquid interface　液-液界面
liquid-liquid interfacial adsorption　液-液界面吸附
liquid-liquid interfacial polycondensation　液-液(相)界面缩聚(作用)
liquid-liquid miscibility　液-液互混性
liquid-liquid partition　液-液分配
liquid-liquid partition chromatography　液-液分配色谱(法)
liquid-liquid partition system　液-液分配系统
liquid-liquid separation　液-液分离
liquid-liquid solubility　液-液溶解性；液-液溶解度
liquid-liquid spreading phenomenon　液-液展开现象
liquid-liquid system　液-液系统
liquid lubricant　液体润滑剂
liquid lubrication　液体润滑
liquid manometer　液体压力计
liquid medium　液体介质
liquid melt　熔融物
liquid membrane　液膜
liquid membrane electrode　液膜电极
liquid membrane pH electrode　液膜 pH 电极
liquid mercury lift　液汞提升；汞升(催化剂)法
liquid metal extraction　液体金属萃取
liquid metal infiltration　液态金属渗透法
liquid metal ion gun　液态金属离子枪
liquid metal ion source　液态金属离子源
liquid metal soap　液体金属皂
liquid meter　液体流量计；液体计量器
liquid mill　液体磨机
liquid motor fuel　液体动力燃料
liquid nail polish　液体擦指甲剂
liquidness　液态；液状
liquid nitrogen cold trap　液氮冷阱
liquid nitrous oxide　液态氧化二氮
liquid oil drier　油性漆用液体催干剂
liquidometer　液面测量计

liquid-operated　液动的
liquid oxygen　液态氧
liquid packing　水力压紧
liquid paraffin　液体石蜡；石蜡油
liquid penetrate examination　液体渗透试验
liquid peroxide　液体过氧化物
liquid petrolatum　液体石蜡
liquid petrolatum quality test　液体矿脂质量试验
liquid petroleum gas cut　液化石油气馏分
liquid petroleum oil　煤油
liquid phase chemluminescence　液相化学发光
liquid phase chromatography　液相色谱(法)
liquid-phase coal hydrogenation　煤的液相加氢
liquid-phase cracking　液相裂化
liquid-phase cracking process　液相裂化过程
liquid phase epitaxy(LPE)　液相外延
liquid phase fluorination　液相氟化
liquid phase isomerization process　液相异构化过程
liquid phase loading　液相载荷量*
liquid phase operation　液相操作
liquid phase oxidation　液相氧化
liquid phase polymerization　液相聚合
liquid phase reaction　液相反应
liquid-phase refining　液相精制
liquid-phase suspension process　液相悬浮法〔催化剂上的一种合成法〕
liquid-phase thermal cracking　液相热裂化
liquid-piston compressor　液环式压缩机
liquid polishing agent　液体抛光剂
liquid polybutadiene rubber　液体聚丁二烯橡胶
liquid polyisoprene rubber　液体聚异戊二烯橡胶
liquid polymer　液体聚合物；液态聚合物
liquid polymerization　液相聚合法
liquid polysulfide elastomer　液体多硫化物弹性体；液体聚硫化物弹性体
liquid power　液体燃料；汽油
liquid prepolymer　液体预聚物
liquid pressure gauge　液体压力计
liquid prism　液体棱镜
liquid propellant　液体火箭燃料
liquid-pump seal　液泵密封
liquid purification process　液体精制法
liquid quench bath　液体骤冷浴
liquid reaction mo(u)lding　液态反应注塑
liquid reagent　液体试剂
liquid-repellency　抗液性；防液性
liquid-repellent finish　抗液整理
liquid reservoir　储液器

liquid resistance　液态(时)阻力
liquid retention　液体保持(作用)
liquid rigidity　液体刚性
liquid-ring compressor　液环式压缩机
liquid rocket fuel　液体火箭燃料
liquid rosin(=tall oil)　妥尔油
liquid rubber　液态橡胶
liquid scintillation counting(LSC)　液体闪烁计数
liquid-scintillation radioassay　液体闪烁放射性(化验)分析
liquid-scintillation spectrometry　(纤维细度的)液体闪烁光谱测定(法)
liquid-scintillation spectrum　液体闪烁谱
liquid scintillator　①液体闪烁体②液体闪烁器
liquid scrubbing　液体洗涤；液体涤气
liquid seal　液封
liquid-sealed discharge　液封放卸；液封排出
liquid secondary ion mass spectrometry　液体二次离子质谱法
liquid second ion mass spectrometry　液体二次离子质谱法
liquid separator　液体分离器
liquid shampoo　洗发液
liquid sheath　液体护套
liquid shell　液体壳层；液壳
liquid silver　银粉和氯化铅混合物的芳香油悬浮液
liquid slag producer　液态熔渣发生炉
liquid smoke　①分馏木材的首馏分②液烟；木醋酸浸剂
liquid soap　液体皂
liquid soap base　液体皂基
liquid-solid adsorption chromatography　液固吸附色谱法
liquid-solid chromatography　液固色谱(法)
liquid-solid extraction　液固萃取；固液萃取
liquid-solid-gas system　液-固-气体系
liquid-solid interface　液固界面
liquid-solid system　液固系统
liquid solution　液体溶液
liquid solvent wax　溶剂型液蜡
liquid space velocity　液体空间速度
liquid stage　液(体状)态
liquid state　液态；液状
liquid storax　安息香脂；苏合香〔取自 *Liquidambar orientalis*〕
liquid stream　液流
liquid strength　液体强度
liquid styrene butadiene rubber　液体丁苯橡胶
liquid sulfur dioxide　液态二氧化硫
liquid surface adsorption　液面吸附(作用)

liquid surface film　液体表面膜
liquid thermometer　液体温度计
liquid-tight　不透液的；液密的
liquid toilet soap　液体盥洗皂
liquid trap　液体分离器；液阱
liquidus　液相线；液化线
liquidus line　液相线；液(化)线
liquidus temperature　液相线温度
liquid-vapo(u)r interface　液-汽界面
liquid-vapour transition　液汽转变
liquid vehicle　①液体漆料②液体连接料
liquid viscosimeter　液体黏度计
liquid vortex contactor　液体涡动接触器
liquid washing agent　液体洗涤剂
liquid waste　废液；液状污物
liquid waste disposal　废液处理
liquid waste incinerator　废液焚烧炉
liquid water emulsion wax　水乳化液蜡
liquid yield　液体产品产率
liquification(=liquefaction)　液化(作用)
liquified petroleum gas　液化石油气体
liquifier　液化器
liquiritigenin　甘草素
liquiritin　甘草根亭
liquiritoside　甘草根糖苷
liquogel　液状凝胶
liquor　①(水)溶液②液③(蒸馏)酒
liquor ammoniae　氨液
liquor calcis　石灰液
liquor condensate　冷凝液
liquorice(=licorice)　甘草
liquor length(=bath ratio)　浴比
liquor-level regulator　液面调节器
liquor of flints　二氧化硅的硅酸钾溶液
liquor pump　液体泵
liquor ratio　浴比
liquor room　配液室
liroconite　水砷铝铜矿
Lisbon copal(=benguela copal)　里斯本玷柏脂
lisinopril　赖诺普利〔药〕
lisoloid　(内)液(外)固胶体
lissamine green V　丽丝胺绿 V
lissephen　甲酚甘油醚〔松弛肌肉剂〕
liter(=litre)　升
liter flask　(一)升量瓶
liter weight　(一)升重量
litharge　①铅黄；黄丹；密陀僧；一氧化铅②正方铅矿
litharge cement　一氧化铅-甘油黏合剂

litharge-glycerin cement 密陀僧甘油胶黏剂
litharge stock 一氧化铅混合剂
lithargite 密陀僧；黄丹；一氧化铅
lithargysm 铅中毒
lithargysmus 一氧化铅中毒；黄丹中毒
lithergol 单组分火箭燃料
lithia 氧化锂；锂氧 Li_2O
lithia mica 锂云母
lithiated montmorillonite 锂氧蒙脱石
lithiation 锂化˙
lithia water 碳酸氢锂溶液
lithii 锂(的)
lithin 彩色水泥(砂浆)涂料
lithin coating 彩色水泥砂浆(墙壁)涂装
lithionite 锂云母
lithiophyllite 磷锂矿
lithium 锂〔3号元素,化学符号Li〕
lithium acetate 乙酸锂 $LiC_2H_3O_2$
lithium acetylsalicylate 乙酰水杨酸锂
lithium agaricinate 松蕈酸锂
lithium alkoxide 醇锂；烃氧基锂 LiOR
lithium alkyl 烷基锂
lithium alkylide 烷基锂 LiR
lithium aluminate 铝酸锂
lithium aluminium hydride 氢化铝锂
lithium aluminium hydride reduction 氢化铝锂还原〔用 $LiAlH_4$ 还原〕
lithium aluminium tetraethyl 四乙基铝锂
lithium amide 氨基化锂
lithium ammonium sulfate 硫酸锂铵 $Li(NH_4)SO_4$
lithium arsenide 砷化锂 Li_3As
lithium aryl 芳基锂
lithium base grease(=lithium grease) 锂基润滑脂
lithium benzosalicylate 苯并水杨酸锂
lithium bicarbonate 碳酸氢锂 $LiHCO_3$
lithium bichromate 重铬酸锂 $Li_2Cr_2O_7$
lithium bisulfate 硫酸氢锂 $LiHSO_4$
lithium borate 硼酸锂
lithium bromate 溴酸锂 $LiBrO_3$
lithium bromide 溴化锂 LiBr
lithium cacodylate 二甲基胂酸锂
lithium caffeine sulfonate 磺酸咖啡碱锂
lithium carbide 碳化锂 Li_2C_2
lithium carbonate 碳酸锂 Li_2CO_3
lithium cell 锂电池
lithium chlorate 氯酸锂 $LiClO_3$
lithium chloraurate 氯金酸锂 $Li[AuCl_4]$
lithium chloride 氯化锂 LiCl

lithium chloride dew point hygrometer 氯化锂露点湿度计
lithium chloride hygrometer 氯化锂湿度计
lithium chromate 铬酸锂 Li_2CrO_4
lithium citrate 柠檬酸锂
lithium cyanide 氰化锂 LiCN
lithium cyanoplatinate 氰铂酸锂 $Li[Pt(CN)_6]$
lithium cyanoplatinite 氰亚铂酸锂 $Li[Pt(CN)_4]$
lithium deuteride 氘化锂 LiD
lithium deuteriotritide 氘氚化锂
lithium dichromate 重铬酸锂
lithium dimethylcuprate 二甲基铜锂
lithium disilicate 焦硅酸锂 $(Li_2SiO_5)_n$
lithium dithionate 连二硫酸锂 $Li_2S_2O_6$
lithium dithiosalicylate 二硫代水杨酸锂
lithium dodecyl sulfate 十二烷基硫酸锂
lithium-drifted germanium detector 锂漂移锗检测器
lithium-drifted semiconductor detector 锂漂移半导体探测器
lithium-drifted silicon detector 锂漂移硅检测器
lithium ethide 乙基锂 LiC_2H_5
lithium ethoxide 乙醇锂 $LiOC_2H_5$
lithium ethyl aniline N-乙苯胺锂 $LiN(C_2H_5)C_6H_5$
lithium fluophosphate 氟磷酸锂
lithium fluoride 氟化锂 LiF
lithium fluoride analyzing crystal for X-ray optics 氟化锂(X射线)分光结晶
lithium fluoride whisker 氟化锂须晶
lithium fluoroborate 氟硼酸锂 $LiBF_4$
lithium fluosilicate 氟硅酸锂 $Li_2[SiF_6]$
lithium germanate 锗酸锂
lithium glycerophosphate 甘油磷酸锂
lithium grease(=lithium base grease;lithium soap grease) 锂基润滑脂
lithium hydrogen orthophosphate 磷酸二氢锂 LiH_2PO_4
lithium hydrogen sulfate 硫酸氢锂 $LiHSO_4$
lithium hydroxide 氢氧化锂 LiOH
lithium hydroxy-stearate grease 羟基硬脂酸锂润滑脂
lithium hypochlorite 次氯酸锂
lithium hypophosphate 连二磷酸锂 $Li_4P_2O_6$
lithium iodate 碘酸锂 $LiIO_3$
lithium iodide 碘化锂 LiI
lithium isoprene rubber 异戊锂橡胶
lithium lactate 乳酸锂
lithium laurate 月桂酸锂
lithium lubricating grease 锂基润滑脂
lithium manganate 锰酸锂 Li_2MnO_4
lithium metaaluminate 偏铝酸锂 $LiAlO_2$

lithium metaborate 偏硼酸锂 $LiBO_2$
lithium metaborate dihydrate 二水合偏硼酸锂
lithium metaniobate(=lithium niobate) 偏铌酸锂
lithium metaphosphate 偏磷酸锂 $LiPO_3$
lithium metasilicate 硅酸锂 $(Li_2SiO_3)_n$
lithium methide 甲基锂 $LiCH_3$
lithium methoxide 甲醇锂 CH_3OLi
lithium molybdate 钼酸锂 Li_2MoO_4
lithium myristate 肉豆蔻酸锂
lithium niobate(=lithium metaniobate) (偏)铌酸锂〔无机光纤材料〕
lithium nitrate 硝酸锂 $LiNO_3$
lithium nitride 一氮化三锂 Li_3N
lithium nitrite 亚硝酸锂 $LiNO_2$
lithium orthosilicate 原硅酸锂 Li_4SiO_4
lithium oxalate 草酸锂 $Li_2C_2O_4$
lithium oxide 氧化锂 Li_2O
lithium palmitate 软脂酸锂
lithium perchlorate 高氯酸锂 Li_2ClO_4
lithium peroxide 过氧化锂 $Li_2[O_2]$
lithium phenol sulfonate (苯)酚磺酸锂
lithium phosphate 磷酸锂 Li_3PO_4
lithium platinocyanide 氰亚铂酸锂 $Li_2[Pt(CN)_4]$
lithium (poly)butadiene rubber 丁锂橡胶
lithium potassium sulfate 硫酸锂钾 $LiKSO_4$
lithium propoxide 丙醇锂 C_3H_7OLi
lithium pyroborate 焦硼酸锂 $Li_2B_4O_7$
lithium rhodanate 硫氰酸锂
lithium selenate 硒酸锂 Li_2SeO_4
lithium selenide 硒化锂 Li_2Se
lithium selenite 亚硒酸锂 Li_2SeO_3
lithium silicate 硅酸锂 Li_2SiO_3; Li_4SiO_4
lithium soap 锂皂
lithium soap grease(=lithium grease) 锂基润滑脂
lithium stannate 锡酸锂 Li_2SnO_3
lithium stearate 硬脂酸锂；十八(烷)酸锂
lithium sulfantimonate 全硫锑酸锂；四硫锑酸锂 Li_3SbS_4
lithium sulfate 硫酸锂 Li_2SO_4
lithium sulfide 硫化锂 Li_2S
lithium sulfite 亚硫酸锂 Li_2SO_3
lithium sulfocarbolate (苯)酚磺酸锂
lithium sulfocyanide 硫氰酸锂 $LiSCN$
lithium tartrate 酒石酸锂 $Li_2C_4H_4O_6$
lithium tetraborate 焦硼酸锂 $Li_2B_4O_7$
lithium tetracyanoquinodimethane 四氰基对醌二甲烷锂
lithium thallium tartrate 酒石酸铊锂
lithium thiocyanate 硫氰酸锂 $LiSCN$
lithium thiosulfate 硫代硫酸锂 $Li_2S_2O_3$
lithium tritide 氚化锂 LiT
lithium tritoxide 氚氧化锂 $LiOT$
lithium vanadate 钒酸锂 $LiVO_3$; Li_3VO_4
litho- 〔词头〕石
litho felt(=printing blanket) 石印毡；印刷用毡
lithoform 磷酸锌外层
lithogenesis(=lithogenesy) 石的形成
lithogenesy(=lithogenesis) 石的形成
lithographed coating 石(版)印(刷)涂层；平版印刷涂层
lithographic 石印的；平版的;平版印刷术的
lithographic chalk 石印石
lithographic ink 石印墨
lithographic oil 石印油
lithographic varnish 石印清漆
lithography 石印术
lithography roll 石印胶滚
lithoium hydride 一氢化锂 LiH
litholine 石油；原油
lithology 岩石学
lithomarge 密高岭土
litho oil 石印油
litho-paper 石印纸
lithophile element 亲岩元素
lithophylic property 亲硅酸盐性；亲岩石性
lithopone(=zinc baryta white;rubberlith) 锌钡白；立德粉
lithopone-type cadmium pigment 镉钡颜料
lithosphere 岩石圈
litmopyrine 乙酰水杨酸锂
litmus 石蕊
litmus blue 石蕊蓝
litmus milk 石蕊汁
litmus neutral test paper 中性石蕊试纸
litmus paper 石蕊试纸
litmus red test paper 红色石芯试纸
litmus silk 石蕊丝
litmus test 石蕊试验
litmus test paper 石蕊试纸
litmus tincture 石蕊酊
Litol process 固定床催化脱烷基制芳烃过程
litre(=liter) 升
litre flask 升瓶
litsea cubeba oil 山苍子油；木姜子油
Littrow mounting 里特罗装置
Littrow prism 里特罗棱镜
litudinium molybdophosphate(LMP) 二甲基吡啶鎓钼磷酸盐
Litzler computreator 利茨勒自动处理机〔合成纤维帘子

线浸胶机〕
live catalyst　新鲜催化剂
live center　(辊子)有效顶端
live crude　充气原油
live gas　新鲜气体
liveliness　滑爽性；鲜明感性
lively coal　易碎(成块的)煤
lively-twist　绉缩捻
live material　工作物料
Livens bomb　李文氏(投射)弹
live oil　新鲜石油；(含有烃类气体的)新采出的石油
live pit　新灰坑
liver function test　肝功能检查法
livering　肝化〔涂料〕
liver of sulfur　硫肝〔碱金属碳酸盐与硫黄熔融的产物，包括硫化物、多硫化物、碳酸盐、硫酸盐、硫代硫酸盐等〕
live roller conveyer　运动着的滚输送机
liver ore　赤铜矿
Liverpool test　滴定商品氢氧化钠法
liverstone　肝色石〔石膏与石灰石混合物〕
live rubber　高弹性橡胶
liverwort　地钱〔植物〕
live steam　直接蒸汽；活蒸汽
live steam pipe　热蒸汽管
live steam reclaiming process　直接蒸汽(再生)法
living anionic polymerization　活性负离子聚合
living cationic polymerization　活性正离子聚合
living dimer　活性二聚物
living end　(链的)活性端
living insecticide　微生物杀虫剂
living oligomer　活性低聚物
living polymer　活(性)高分子*
living polymerization　活(性)聚合*
living radical polymerization　活性自由基聚合；可控自由基聚合
living ring opening polymerization　活性开环聚合
living sewage　生活污水
living solution　活性溶液
livingstoneite　硫汞锑矿
livitation　浮置技术
lixator　①溶剂②溶解槽
lixivial　①浸滤了的②去了碱的
lixiviant　浸滤剂
lixiviating　①浸滤；浸提②去碱作用
lixiviating tank　浸滤桶
lixiviation　浸滤作用
lixivious　①浸滤了的②去了碱的

lixivium　①浸滤液②灰汁；碱汁
loach oil　泥鳅油
load　①负重②载荷；负荷③载；装填
load area　受压面积
load bearing capacity　载荷容量
load capacity　载荷量
load-carrying ability　(润滑脂)载荷能力
load-carrying additive　载荷能力添加剂
load carrying capacity　(润滑脂)载荷能力
load carrying capacity testing of gear lubricant　齿轮润滑剂承载能力试验
load-deformation curve　载荷-形变曲线
load density　装载密度
loaded　①载荷的；负荷的②阻塞的〔如过滤器〕③填料的
loaded resin　吸附了的树脂；载荷树脂
loaded rubber　填料橡胶
loaded sheets　填料纸
loaded solvent　萃取了的溶剂；负荷溶剂
loaded stock　填料
loaded tyre section width　负荷轮胎断面宽
loaded-up condition　载荷状态
loaded weight　装载重量
load-elongation diagram　荷重伸长图
loader　①承载器②装填器
load-extension curve　荷重伸长曲线
load face　受压面
load factor　①载荷系数②填充系数
load inflation table　承载膨胀表〔轮胎内压〕
loading　①填充剂；填料②装料；填充；填充量③装模④装载
loading agent　填充剂
loading airlock　设备装卸用气闸室
loading and unloading　加载和去载；加载与卸载
loading area　装料区
loading board　承载板
loading capacity　①载荷能力；负荷容量；充填容量②萃取容量③离子交换容量；吸附容量
loading cavity　装料腔
loading density　装填密度
loading depot　装载站；装灌站
loading device　装料设备
loading dock　装载支架
loading end　承载面
loading hopper　加料斗
loading in bulk　散装
loading level　填充量
loading limit　装料容许量

loading location 装载站；装灌站
loading machine 装料机
loading material 填料
loading of coated abrasives (砂布、砂纸打磨中的)砂粒黏滞
loading of stock 填充配合剂
loading pigment 颜料填充剂
loading table 容积表
loading tray 装料盘
loading weight 装载重量；散装重量
loading well 圆筒状装料器
load input tensile tester 输入负荷强力试验仪；输入载荷强力试验仪
load lever (测试)载荷杆
load limit 载荷限度
load metamorphism 承载变性
load per unit of length 单位长度负荷
load pressure 载荷强度
load redundancy ①载荷多余②装料过多
loadstone 四氧化三铁；磁铁矿；磁性氧化铁
load-strain curve 载荷-应变曲线
load test 载荷试验
load voltage 工作电压
loaf sugar 块糖
loam 壤土
loamy 壤质的
loamy soil 壤质土
loba 珞钯(树脂)〔一种马尼剌树脂〕
loban 安息香胶
lobar ①低压②低压的
lobar column 低压柱
lobaric acid 肺衣酸
lobate 分裂的；有裂片的
lobed filament 叶片状异形长丝
lobelane 山梗烷
lobelanidine 山梗烷定；山梗烷醇；山梗醇碱
lobelanine 山梗酮碱
lobelia 北美山梗菜
lobeline 山梗碱；山梗烷醇酮；洛贝林〔用于各种原因引起的中枢性呼吸抑制。临床上常用于新生儿窒息、一氧化碳、鸦片中毒等〕
lobeline hydrochloride 盐酸山梗碱；盐酸洛贝林
lobeline sulfate 硫酸山梗碱；硫酸洛贝林
lobe pump 凸轮泵；多叶回转泵
lobinine 山梗菜次碱
loblolly pine 火炬松
lobster back 曲折管
local 局部的；现场的

local analysis 局部分析
local anesthetic 局部麻醉剂
local annealing 局部退火
local cell 局部电池
local cell corrosion 局部电池腐蚀
local concentration 局部浓度
local conformation 局部构象
local control 局部控制
local corrosion 局部腐蚀
local current 局部电流
local curve fitting method 局部曲线拟合法
local deformation 局部形变
local diamagnetic shielding 局部抗磁屏蔽
local distortion free energy 局部畸变自由能
local element 局部电池；微电池
local field 局部场
local flow 局部流动
local fraction desolvated 局部去溶度
local fraction of atomization 局部原子化度
local fraction volatilized 局部挥发度
local hot spots 局部过热点
localised chemisorption bond 定域化学吸着键
localised orbital(LO) 定域轨道
local isoentropic stress-strain law 局部等熵应力-应变律
localization 定位化
localization energy 定域能
localization method 定位法
localization of spot (色谱)斑定位(法)
localized 定位的；定域的
localized bond 定域键
localized bond model 定域键模型
localized energy 定域能
localized molecular orbital(LMO) 定域分子轨道
localized nucleation 定域成核
localized vibration 局部振动
locally excited state 局部激发态
locally-thin areas of paint (film) 局部漆膜偏薄的部位
local medium effect 局部介质效应
local mode 局部模式
local mode motion 局部模式运动
local optimization 局部优化*
local order 局部有序
local overheating 局部过热
local paramagnetic shielding 局部顺磁屏蔽
local physical deformation 局部自然形变
local plating 局部槽镀
local rate of deformation 局部形变速率
local relaxation mode 局部松弛模式

local repair　局部修理
local stress　局部应力
local stretching　局部拉伸
local superheating　局部过热
local surface tension　局部表面张力
local thermal equilibrium　局部热平衡
local total efficiency atomization　局部总原子化效率
locamphen　樟脑酚碘合剂
locant　位次；位标
Locap gasoline sweetening　洛卡普法汽油脱臭
located field of spin system　自旋系统局部场
locating dowel　定位销
locating hole　定位孔
locating pin　定位销
locating spider　定位芯座
locating spot　①定斑位②定位斑
locating surface　定位面
location reagent　定位试剂；斑点(或谱带)显色剂
lochnericine　洛柯因；洛柯辛碱
lock　①锁②水闸③锁峰〔核磁〕
locked-in air space　(颜料附聚物或干漆膜中)封住的空穴；封闭的气泡；干遮盖
locked-in-air voids　①(颜料附聚物或干漆膜中)封住的空穴；封闭的气泡(空隙)②干遮盖
locked-in end plate　(涂膜器)锁定的端板
locked-in loop structure　闭口线圈结构；毛圈封闭结构
locked-in stress　内应力
locked-seam drum　潜缝桶
locked-seam roller hoop drum　具有滚钩的潜缝桶
Locke's solution　无蛋白质人工血清；乐氏液
lock-in amplifier　锁相放大器
locking　关闭；连接；堵塞
locking-in amplifier　锁相放大器
locking the magnetic field to the radiofrequency　场频连锁
locking wire thread inserts　锁紧型钢丝螺套
lock-nit fabric　纬编锁编织物
lock nut(=jam nut)　锁紧螺母；压紧螺盖〔商〕
lock ring　密封圈
lock sleeve　锁紧套
lock washer　锁紧垫圈
locomotive grease　机车润滑脂
locomotive oil　汽缸油
locust　角豆
locust bean gum　刺槐豆胶
locust beans　角豆
locust tree　刺槐；洋槐
lode　①矿脉②水路；沟
lodermycin(=demethylchlortetracycline)　去甲基金霉素

lodestone　极磁铁矿
lodgment of stock　窝浆；淀浆
Loeffler's methylene blue stain　吕氏美蓝染色液
loess　黄土
Loffler methylene blue　莱夫勒亚甲蓝
loft-dried paper　风干纸
loft drier　箱式干燥器；干燥箱
loft drying　风干
loftiness　丰满手感；蓬松度
lofty handle　弹性手感
loganin　马钱子苷；番木鳖苷
logarithmic decrement　对数衰减；对数减量
logarithmic diluter　对数稀释器
logarithmic dilution method　对数稀释法
logarithmic display　对数显示
logarithmic distribution law　对数分配定律
logarithmic homologue　对数同系物
logarithmic interpolation　对数内插法
logarithmic mean　对数平均
logarithmic mean partial pressure difference　对数平均分压差
logarithmic mean radius　对数平均半径
logarithmic mean temperature　对数平均温度
logarithmic mean temperature difference　对数平均温差
logarithmic normal distribution　对数正态分布*
logarithmic relative retention value　相对保留对数值
logarithmic response　对数响应
logarithmic scale　对数标尺
logarithmic strain(=natural strain; Hencky strain)　固有应变
logarithmic time law　对数时间定律
logarithmic time scale　对数时间标度
logarithmic titration　对数滴定(法)
logarithmic titration graph　对数滴定曲线
logarithmic viscosity　比浓对数黏度
logarithmic viscosity number　比浓对数黏度*；对数黏数
log haul-up　曳木机
logical loss　合理丢失
logical simulation　逻辑模拟
logic gate circuit　逻辑门电路
logic operation　逻辑运算；逻辑操作
logic simulation　逻辑模拟
logic system　逻辑系统
log-log plot　对数标绘
log normal　对数正态
log normal distribution　对数正态分布
log scale　①对数标尺；计算尺②原木检尺③原木材积表
log sector　对数扇形板

log sector disk 对数扇形盘；对数遮光板
log transformation 对数变换*
log washer 分级槽
logwood 苏木
logwood black lake 苏木黑色淀
logwood crystal 苏木精
logwood extract 苏木提出物；苏木萃
logwood paste 洋苏木膏
logy 疲软的；无弹回力的
loiponic acid 哌啶-3,4-二羧酸；洛滂酸 $C_7H_{11}NO_4$
lokav(=Chinese blue) 中国蓝
loline 黑麦草碱
lolin-idine 黑麦草定
lolinine 黑麦草宁
Lomax catalyst 洛马克斯催化剂
Lomax process 洛马克斯法〔加氢裂化、加氢脱硫法〕
lomefloxacin 洛美沙星〔药〕
lomontite 浊沸石
lomustine 洛莫司汀〔药〕
lomycin(=griseomycin) 灰色霉素
lond 采掘段
London clay 伦敦土
London-Eyring-Polanyi PES LEP势能面*
London-Eyring-Polanyi-Sato PES LEPS势能面*
London force(=dispersion force) 色散力；伦敦力
London paste 伦敦灰胶
London purple 伦敦红紫〔杀菌剂〕
London shrinking 伦敦预缩法
London smog event 伦敦烟雾事件
London-van der Waals constant 伦敦-范德华常数
London-van der Waals interaction 伦敦-范德华相互作用
lone electron pair 孤电子对
lone-pair electron 孤对电子*
lone-pair ionization 孤对电离
long bedded furnace 长底炉
long buchu 齿状布枯；长叶布枯
long burner 固定燃烧炉
long center working 大跨距运行
long-chain 长链
long-chain acyltitanate 长链酰基钛酸酯
long-chain alcohol 长链醇
long-chain alkyl derivative 长链烷基衍生物
long-chain branch 长支链*
long chain branched polyethylene 长支链聚乙烯
long chain branching 长链支化
long-chain carboxylic acid 长链羧酸
long chain diacid 长链二酸
long chain dicarboxylic acid 长链二羧酸

long chain fatty acid 长链脂肪酸
long-chain ion 长链离子
long-chain macromolecule 长链大分子
long chain molecule 长链分子
long chain polymer 长链聚合物
long-chain quaternary ammonium ion 长链季铵离子
long-chain sulfoamido acetic acid 长链磺酰氨基乙酸
long cure 长时间硫化
long-distance pipe line 长距离导管(线)
long distillate 宽馏分
long drain oil 长期油
long duration test 耐久试验
longer-chain ammonium salt 长链铵盐
longer-term repainting 长效维修涂装
long-established American usage 美国约定俗成惯例
longevity 使用寿命；耐久性
long exposure type wash primer 长曝型磷化底漆
long-fiber grease 长纤维润滑脂
long-fibred asbestos 长绒石棉
long flame coal 长焰煤
long-flowing 长流动(现象)
long grained plywood 顺纹胶合板
long haul 远程运送
longicatenamycin 长脂链霉素
longifolane 长叶烷
longifolene(=junipene) 长叶烯
longifolic acid 长叶酸
longifolyl acetate 乙酸长叶酯
longifolyl formate 甲酸长叶酯
long-induction-period gasoline 诱导期长的汽油
longipinene 长叶松烷
longitudinal dispersion 纵向渗透
longitudinal mixing 纵向混合
longitudinal modulus 纵向模量
longitudinal relaxation 纵向松弛
longitude 经度
longitudinal 经的；纵的
longitudinal baffle 纵挡板
longitudinal bending 纵向弯曲
longitudinal chain mobilization 纵向链流动(作用)
longitudinal conjugation 纵向共轭作用
longitudinal crimp plate 纵向波纹板
longitudinal deformation 纵向形变
longitudinal detected electron spin echo envelope modulation 纵向检测电子自旋回波包络线调制
longitudinal detection-pulsed ESR 纵向检测脉冲电子自旋共振
longitudinal diffusion 轴向扩散；纵向扩散

longitudinal diffusivity 纵向扩散系数
longitudinal direction 纵向
longitudinal elasticity 纵向弹性
longitudinal feed 纵进(给)
longitudinal flow 纵向流动
longitudinal magnetization 纵向磁化
longitudinal mode 纵模
longitudinal modulus 纵向模量
longitudinal modulus of elasticity 纵向弹性模量
longitudinal order 纵次序
longitudinal ply 长中
longitudinal refractive index 轴向折射率
longitudinal register (分子链排列的)纵向定位;纵向序
longitudinal relaxation 纵向弛豫
longitudinal shearing flow 纵向剪切流
longitudinal shear strength 拉伸剪切强度
longitudinal strain 纵向应变
longitudinal strength 纵向强度
longitudinal stress 纵向应力
longitudinal-transverse shear modulus of elasticity 纵横剪切弹性模量
longitudinal-transverse shear strength 纵横剪切强度
longitudinal tread pattern 纵向胎面花纹
longitudinal void 纵向空穴;纵向空隙
longitudinal wall 纵壁;纵挡板
longitudinal weld 纵向焊缝
longitudinal winding 纵向缠绕
longitudnal seam 纵缝
long lasting 长效的
long-lasting film integrity 长效漆膜完整性
long-lasting protection coating 长效保护涂料;长效防腐蚀涂料
long lasting quality 留香性能
longleaf pine 长叶松
longleaf pine oil 长叶松油
long-life antifouling paint 长效防污漆
long-life grease 长使用期润滑脂〔化学安定的润滑脂〕
long life topcoat 长效面漆;耐久面漆
long-lived 使用期长的;长寿的
long-lived complex 长寿命络合物*
long lived nuclide 长寿命核素
long loaf moulder 长块模
long oil alkyd 长油醇酸树脂
long oil alkyd finish 长油醇酸(面)漆
long-oil-length 长油度
long oil tall oil 长油度松浆油
long oil varnish 长油清漆
longos 纵向小角缠绕

longoze 黄姜花
long period 长周期
long persistence screen 长余辉荧光屏
long production run 大量生产;大批生产
long range 远程的;长距离的
long-range coupling 远程偶合
long-range deformation 长程形变
long range elasticity 广范围弹性;高弹性
long-range electron transfer 长程电子传递
long-range force 长程力
long-range homo nuclear two dimensional chemical shift correlation spectrum 二维长(远)程耦合同核化学位移相关谱
long-range interaction 远程相互作用
long-range interchannel effect 远程通际效应
long-range interference 长程干扰作用
long-range intramolecular interaction 远程分子内相互作用*
long-range order 长程有序
long-range perturbation theory 长程微扰论
long-range rocket fuel 远距离火箭燃料
long-range spin-spin coupling 远程自旋偶合
long-range structure 远程结构*
long residuum 久沸残渣〔石油在常压下,久蒸后所得残渣〕
long-run test 长期试验
long settling of emulsion 乳状液长期澄清
long shaper 长成型装置
long spacing 长空间*
long stapled cotton 长绒棉
long-stroke press 长冲程水压机
long-sweep elbow 远拂肘管
long-term aging 长期老化
long-term aging test 长期老化试验
long-term change 长期变化
long-term compression deflection properties ①长期压缩特性②长期挠曲特性
long-term continuous outdoor exposure 长期连续户外暴晒
long-term corrosion 长期腐蚀
long-term degradation 长期降解
long-term effect 长期效应;长时效应;长时后效
long term efficiency 长期效能;长期收益
long-term exposure 长期暴晒
long-term heat resistance 长时间耐热性;长时间耐高温性
long-term hydrostatic strength characteristics 长期静水压强特性
long term maintenance-free service 长期无检修运转

long-term mechanical property 长期力学性能
long-term modulus 长期模量
long-term nontoxicity test 长期无毒试验
long-term outdoor exposure performance 长期户外暴晒性能
long-term outdoor weatherability 长期户外耐候性
long-term performance test 长期性能试验
long-term poisoning 长期中毒
long-term pressure test 长期耐压试验
long-term research 长远研究
long-term rust preventive painting 长效防锈涂装
long-term stability 长期稳定性
long-term static fatigue failure 长期静态疲劳破损
long-term storage 长期储存
long-term stress resistance 耐长期应力破坏强度
long-term stress-rupture performance 耐长期应力断裂性能
long-term tension 长期张力
long-term thermal stability 长期热稳定性
long-term weathering test 长期耐候性试验；长期暴晒试验
long-throat granulator 长进料口造粒机
long-time burning oil 久燃煤油〔信号灯用〕
long-time burning test 长期燃烧试验
long-time cycle 长(工作)周期
long-time effect 长期效应；长时效应
longtitudinal mixing 纵向混和
long ton 长吨〔1 长吨=1.016 吨〕
long tube absorption cell method 长吸收管法
long tube device 长管装置
long-tube evaporator 长管蒸发器
longulite 长联雏晶；联珠晶子
lonicera oil 忍冬油；金银花油
lonicerin 忍冬苷
lonzitin(=lenzitin) 革祠菌素
looking-glass ore 有光泽赤铁矿
loom ①织机②圆织机
loomite 短纤维滑石
loom nominal reed width 织机公称筘幅
loom oil 织机油；重锭子油
loomstate fabric 坯布；本色布
loop ①周线；线(圈)②(波)腹
loop ager 悬环式蒸化机
loop-breaking strength 钩接断裂强度；环扣断裂强度
loop drying 垂环式干燥；悬环式干燥
looped fabric 针织布
loop expansion pipe 补偿器；涨力弯；膨胀管圈
loopfull 全环的

loop-gap resonator 环路间隙共振腔
looping angle 抱合角；接触角；包角
loop injection system 环式注射系统
loop process reactor 环形管道式连续反应器
loop reactor 回路反应器
loop receiving 接圈
loop strength 互扣强度
loop structure 环状结构
loose 疏松的
loose bar 松配芯棒；活芯棒
loose black 粉末炭黑
loose coal 疏松煤；易破碎煤
loose combination 疏松结合
loose core 松配模芯；活型芯
loose density 散装密度；松密度
loose deposits 疏松沉淀
loose end 裂丝
loose filler 疏松填料
loose fit 松配合
loose flange 活套法兰
loose grain 松面
loose ion pair 松散离子对
loose loop model 松散链圈型
loose machine 撕松机
loose material ①散装材料；散粒料②疏松材料③(化学)不稳定材料
loose measure 粗测
loose membrane 疏松膜；疏松层滤器
loose micelle 松胶束
loose mold 可卸式压模
loosened 分散的；松的
loosened cake 松(蜡)饼
loosened carbon 松的炭渣；分散的炭粒
looseness 不牢固性
loose oxidation products 不稳定氧化产物
loose packing 疏松填充
loose paint flake(s) 疏松的脱落漆片
loose-plate transfer mold 活板式传递塑模
loose punch 活络阳模
loose rocks〔复〕 疏松岩石
loose running fit 松转配合
loose rust ①锈鳞②疏松的锈
loose socket 平滑离合情况
loose structure 松散结构
loose transition state 松散过渡态
loose type flange 松式法兰；活套法兰
lop 砍伐
loperamide 洛哌丁胺〔药〕

lopezite 铬钾矿
lophenol 4-甲基-7-烯胆(甾)烷醇
lophine 洛粉碱；2,4,5-三苯基咪唑
$(C_6H_5)_3C=CNCNH$
lophocerine 冠影掌碱
loping 脉动〔输送石油产品〕
lorandite 红铊矿
loranthyl alcohol 桑寄生醇
loratadine 氯雷他定〔药〕
lorazepam 劳拉西泮〔药〕
L-orbit L轨道
Lorentzen furnace 洛伦兹电炉
Lorentz factor 洛伦兹因子
Lorentzian curve 洛伦兹线型
Lorentzian line shape 洛伦兹线型
Lorentz-Lorentz relationship 洛伦兹-洛伦兹关系式〔遮盖力与颜料-基料折光率之间的近似关系式〕
Lorentz polarization effect 洛伦兹极化效应
Lorentz polarization factor 洛伦兹偏振因子
loretin(=7-iodo-8-hydroxyquinoline 5-sulfonic acid) 试铁灵；7-碘-8-羟基喹啉-5-磺酸
loretinate 7-碘-8-羟基喹啉-5-磺酸金属盐
loricated pipe 内部涂有沥青的管子
loriodendrin 鹅掌楸苦素
lornoxicam 氯诺昔康〔药〕
lorry loader 自动装载机
lorry tyre 载重轮胎
losartan potassium 氯沙坦钾〔药〕
Loschmidt's number 洛斯密德数
losophan 三碘(代)间甲酚
loss angle 损耗角
loss compliance 损耗柔量
loss due to dressing 选矿损失
loss due to leakage 漏失
loss elastic modulus 损失弹性模数；损失弹性模量
Lossen rearrangement 洛森重排作用
loss factor 损耗因数
loss in head 压头损失
loss in octane number 辛烷值降低
loss in weight on drying 干燥失重
loss in weight on heating 加热重量损失；加热失重
loss in weight on prolonged heating 长时(间)加热失重
loss modulus 损耗模量
loss of adhesion 附着力下降(损失)
loss of clarity 丧失透明度；透明度损失
loss of coating integrity 涂层丧失完整性
loss of color 褪色；变色

loss of cycle ①周期损失〔指延长了生产周期损失了有效工时〕②工时损失
loss of drier ①催干剂损失②干料失效
loss of energy 能量损失
loss of flexibility 丧失柔韧性
loss of gasoline by permeation 汽油的渗透损失
loss of gloss 失光；倒光〔漆病〕
loss of heat 热损失
loss of pressure 压力降低
loss of tensile strength 抗张强度损失
loss of transmission 透射率降低
loss of wear 磨耗量
loss of weight 失重
loss of weight in baking 烘减量
loss on heating 加热损失
loss on ignition 烧失量
loss peak 损耗峰
loss shear modulus 损耗剪切模量
loss tangent 损耗角正切值
loss through breathing 石油产品的蒸发损失
loss through standing 储存损耗
loss through volatilization 挥发损失；挥发损耗
loss Young's modulus 损耗杨氏模量；损耗弹性模量
lost cooling time 损失冷却时间
lost of gloss 失光
lost wax process 失蜡(精密)铸造法；熔模铸造法
lot 批；堆
LOTES(low temperature solidification) process 低温固化过程
lot identification mark (产品)批号
lotion 洗剂
lotion soap 香液皂
lot number 批号；批数；签号
lotoflavin 百脉根黄素；牛角花黄素
lot production 大批生产
lot size 批量
lot tolerance 批间允差
lot to lot uniformity 逐批质量一致
loturine 牛角花碱；哈尔满
lotusin 牛角花素
lotus metal 铅锑锡轴承合金
lotus-root-like fibre 藕截面型中空纤维；复孔复合纤维
loup(e) 搅炼铁块
Louver separator 洛弗分离器
lovage 圆叶当归
lovage oil 圆叶当归油〔取自 *Levisticum officinale*〕
lovage root oil 圆叶当归根油
lovastatin 洛伐他汀〔药〕

Lovelock detector(=argon ionization detector)　氩电离检测器；氩离子化检测器；拉夫洛克检测器
Lovibond colour system　罗维邦德色系
Lovibond comparator　罗维邦德比色计
Lovibond tintometer　罗维邦德色调计
low　低的；弱的；贫的
low activity process　低活性法
low added acid concentration　低的酸添加浓度
low alloyed steel　低合金钢
low angle diagram　小角衍射图
low-angle laser light scattering photometer　小角激光散射光度计
low-angle light scattering　小角光散射
low angle scattering　小角散射
low-angle scattering of X-ray　X射线小角散射
low angular sheen　低角度光泽
low ash　少灰的
low atomization pressure　低喷雾压力；低雾化压力
low background scaler　低本底计数器；低背景计数器
low-bake finish　低温烘烤面漆
low-bake furniture finish　低温烘烤家具漆
low baking　低温烘烤
low beam(=dipped beam)　弱光
low binding　弱键联
low bleed liquid phase　低流失液相
low-boiler　低沸点
low boiler cut　低沸点馏分
low-boiling　低沸的；沸点低的
low boiling impurity　低沸点杂质
low boiling point solvent　低沸点溶剂
low bracket gasoline　低辛烷值车用汽油
low bulk　①低松密度②低膨胀；低胀量③小体积
low capacity column　低容量柱
low carbon　①低碳(钢)②低碳的
low-carbon-content graphite　低碳石墨
low carbon residue oil　低焦值油
low-carbon steel　低碳钢
low-cetane　低十六烷的
low cetane standard　低十六烷值标准
low-cetene　低十六烯的
low cold-test distillate　低凝固点馏出物
low-condensed phenolic resin　低缩合度酚醛树脂
low conductivity　低电导率
low-cost water-borne flame retardant　低成本水性阻燃剂
low crosslink density　低交联密度
low-cure water-reducible coating　低温固化水稀释性涂料
low cycle fatigue　低循环疲劳；低周疲劳
low dead volume union　小死体积接头

low-density bleaching　低浓度漂白
low-density branched polyethylene　低密度支化聚乙烯
low density filler　低密度填料
low density lipoprotein(LDL)　低密度脂蛋白
low density polyethylene(LDPE)　低密度聚乙烯
low-density polyethylene plastics(LDPE)　低密度聚乙烯塑料
low-density polymer　低密度聚合物
low dimensional material　低维材料
low distillation thermometer　低温蒸馏温度计
low durometer stock　软质胶料；软胶
low duty tire　轻型载重轮胎
loweite　钠镁矾　$2MgSO_4 \cdot 2Na_2SO_4 \cdot 5H_2O$
low elasticity　低弹性
low-energy electron diffraction(LEED)　低能电子衍射
low-energy electron impact resonance　低能电子冲击共振
low energy electron scattering　低能电子散射
low energy eletron loss spectrometer　低能电子损失(能)谱仪
low energy eletron loss spectroscopy　低能电子损失(能)谱法
low-energy fuel　低能燃料
low energy hydrophobic surface　低(表面)能憎水表面
low energy ion scattering(LEIS)　低能离子散射
low energy ion scattering spectroscopy　低能离子散射谱
low-energy liquid　低(表面)能液体
low energy loss scanning electron microprobe　低能损失扫描电子探针
low energy molecular scattering　低能分子散射
low energy photon detector(LEPD)　低能光子探测器
low-energy solid　低(表面)能固体
low enough viscosity　足够低的黏度
low enriched uranium(LEU)　低浓(缩度)铀
low enrichment leach　低加浓燃料浸取器
lower acetylene　低级乙炔
lower acid　低级酸
lower acidity sulfuric acid　低酸度硫酸
lower agitation propeller　下搅螺旋桨
lower alarm limit(LAL)　下警告限
lower alcohol　低级醇
lower alkyl fatty ester　低碳烷基脂肪酯
lower carrier　底部承载板
lower chord member　下弦构件
lower cloud point　下浊点
lower control limit(LCL)　下控制限
lower critical solution temperature(LCST)　最低临界共溶(溶解)温度
lower explosion limit　爆炸下限

lower guide roll 下导辊
lower heating value 低(发)热值
lower homologue 低级同系物
lower hydrate 低水合物
lowering of freezing point 冰点下降〔测分子量用〕
lower limit of detectability (可)检测下限
lower member 低级物
lower member of the series 同系列中的低级物
lower paraffin hydrocarbons 低级烷烃
lower plastic limit(=plastic limit) 低塑(性)极限
lower platen 下压台；下压板
lower polymer 低聚物
lower ring 下部温带板〔用于球罐〕
lower sample 下层试样〔从液体储备器下层取出〕
lower tank 下部收集器〔容器〕
lower titanium oxide 低钛氧化物；氧化低钛
lower toxic limit 毒性下限
lower warning limit(LWL) 下警戒限
lower yield value 低屈服值
lowest free molecular orbital(LFMO) 最低自由分子轨道
lowest haze 极低雾度
lowest unoccupied molecular orbital(LUMO) 最低未占(分子)轨道
low explosive 低级炸药
low explosive limit 爆炸下限
low-fired pitch fibre 弱火处理的沥青纤维
low flash 低温闪蒸
low-flash oil 低闪点油
low flow 低流动性
low-foamers(=low-foaming detergent) 低泡洗涤剂
low foaming 低泡的
low-foaming detergent(=low-foamers) 低泡洗涤剂
low-foaming surfactant 低泡表面活性剂
low-formaldehyde melamine resin 低甲醛三聚氰胺树脂
low-freezing 低凝固点的
low freezing dynamite 低冻炸药
low-friction film 低摩擦膜
low gloss coating 低光泽涂层
low-grade deposit 低品位矿床
low-grade fuel 低级燃料
low-grade gas 低级煤气；贫煤气
low-grade ore 低级矿；贫矿石
low-gravity fuel 低比重燃料
low head pump 低压头泵；低扬程泵
low-head screen 低头筛；振动筛
low-hearth 精炼炉床
low heating value(=net heating value; net heat of combustion) 净热值；低热值

low-heating value gas 低卡煤气
low hiding strontium chromate 低遮盖力铬酸锶〔锶铬黄〕
low hold-up (塔的)低滞留量
low humidity 低湿度
low hydrogen type electrode 低氢型焊条
low interfacial adhesion 低界面黏合
lowland white fir 低地白枞
low-lead paint 低铅漆
low-lead solvent-thinned priming paint 溶剂稀释型低铅底漆
low level 低平面；低面式；低含量；低水平
low level (condenser) 低面式
low level counting 低(放射性)水平计数
low-level cracking 轻度裂化
low-level jet condenser 低水平注水冷凝器
low level waste(LLW) 低(放射性)水平废物
low-lift pump 低压泵
low-limed cement 低石灰水泥
low limit of size 下限尺寸
low limit of tolerance 下限公差
low-loaded column 低载荷柱
low-loss ferrite 低损耗铁氧体
low-loss glass 低(介电)损失玻璃
low-loss insulation 低(介电)损耗绝缘材料
low-loss optical fiber 低损耗光导纤维
low loss polymer 低介电损耗高聚物
low-mass cut-off 低质量截止点
low-melt-index 低熔体指数
low melting 低熔的；熔点低的
low melting alloy 低熔合金
low-melting ink vehicle 低熔度(点)油墨；联结料；易熔性油墨联结料
low modification ratio 低改性比
low molecular 低分子的
low-molecular plasticizer 低分子(量)增塑剂
low-molecular weight compound 低分子量化合物
low-molecular-weight fraction 低分子量馏分
low-molecular-weight polyester 低分子量聚酯
low-molecular (weight) polymer 低分子(量)聚合物
low-molecular-weight surface active species 低分子量表面活性物质
low-octane 低辛烷值的
low-oil-absorption pigment 低吸油量颜料
low-oil alarm 低油位警报
low overpotential electrode 低过电位电极
low pass filter 低通滤波器；低通滤光片
low-pilling variant 低起毛起球改性纤维

low pitch cone roof　低倾度锥形顶盖（储罐）
low plasticity　低可塑性
low-platinum reforming　低铂重整
low polar region　弱极性区
low-pole　弱极性
low-pollution fuel　低污染燃料
low polymer　低聚物
low polymerization catalyst　低聚合催化剂
low-polymerized precondensate　低聚合度的预缩合物
low porosity urethane gel coat　低孔隙率聚氨酯凝胶涂料
low pour point(=low pour)　低倾点
low pour-point oil(=low pour-test oil)　低倾点油
low pour-test oil(=low pour-point oil)　低倾点油
low power He-Ne laser　低功率氦氖激光器
low preorientation　低预取向(作用)
low pressure　①低压(力)②低(气)压
low pressure aerosol　低压气溶胶
low-pressure distillation　低压蒸馏
low pressure float valve　低压浮球阀
low-pressure gage(＝low-pressure gauge)　低压气压表；低压计
low-pressure gun　低压油枪
low-pressure hot spray　低压热喷涂
low-pressure laminate　低压层压品
low pressure laminating resin　低压层压用树脂
low-pressure lamination　低压层合制品；低压层合(法)
low-pressure liquid chromatography　低压液相色谱
low-pressure moulding　低压模塑(法)
low-pressure oil　低压输送油类
low pressure plasma spraying　低压等离子喷涂
low pressure polyethylene　低压聚乙烯
low-pressure polyethylene fibre　低压聚乙烯纤维
low-pressure polymerization　低压聚合(作用)
low-pressure resin　低压(加工)树脂
low-pressure safety cut-out　低压安全切断器
low-pressure side　低压旋管；冷冻机的蒸发旋管
low-pressure stage　低压段
low-pressure warm-air sprayer　低压温风喷涂机；低压暖风喷涂机
low profile　微缩型
low pulse sedimentation　低脉动沉淀技术
low-range pressure gage　低量程气压表
low rank anthracite　低级无烟煤
low rank bitumen　低级烟煤
low-rate-of-shear viscosimeter　低剪速黏度计；低切变速率黏度计
low reactivity　低活化性；低反应性
low-reading thermometer　低温温度计

low reflux ratio　低回流比
low-resilience　低回弹性(力)
low resolution accurate mass measurement　低分辨准确质量测定
low resolution mass spectrometer　低分辨质谱仪
low resolution mass spectrometry　低分辨质谱法
low resolution NMR　低分辨核磁共振；宽谱线核磁共振
low-rubber compound　低橡胶混合物
Lowry-Brönsted definition of acids and bases　劳瑞-布朗斯特酸碱定义
low severity hydrocracking　低深度加氢裂化
low-shear dispersion　低切变分散
low shear force　低剪切力；低切力；低剪力
low shear rate　低剪切速率；低切速
low-shear viscosity　低剪切黏度；低切变黏度
low shrink(LS)　低收缩性(的)
low-shrinkage fiber　低收缩纤维
low shrink resin　低收缩树脂
low-side roller mill　低面滚磨机
low solid　低固体分
low-solids-content vehicle　低固体分漆料
low-soluble component　低溶性组分
low-solvating plasticizer　低溶剂化增塑剂
low-spin complex　低自旋配合物
low spot　凹陷；明痕
low state of cure　低硫化(程)度
low steel　低碳钢
low-suction controller　抽吸过程中的最低压力调节器
low-sudsers(=low-sudsing detergent)　低泡洗涤剂
low-sudsing detergent(=low-sudsers)　低泡洗涤剂
low sulfur curing system　低硫化系统
low-sulfur fuel　低硫燃料
low sulfur oil　低硫油
low temperature　低温(度)
low-temperature adhesion　①低温附着性②低温黏结
low-temperature annealing　低温退火
low temperature ashing method　低温灰化法
low temperature atomization　低温原子化(法)
low temperature bake　①低温烘烤②低温焙烧
low temperature baking (stoving) enamel　低温烘干涂料
low-temperature bending test　低温弯曲试验；冷弯曲试验
low temperature brittleness　低温脆性
low-temperature brittle point　低温脆化点
low-temperature carbonization　低温干馏
low temperature cell　低温池；低温槽
low-temperature cetane number　低温十六烷值；起动十六烷值

low temperature chromatography 低温色谱(法)
low temperature cinefaction 低温灰化法
low temperature coke 低温焦炭
low-temperature cracking test 低温龟裂试验
low temperature crystallography 低温晶体学
low-temperature curable electrophoretic coating 低温固化电泳漆
low-temperature curing 低温固化
low-temperature disproportionation 低温歧化(作用)
low-temperature distillation 低温蒸馏
low temperature endothermic effect 低温吸热效应
low-temperature extension 低温伸长率
low temperature flexibility ①低温柔韧性②低温(屈)挠度
low-temperature fluidity 低温流动性
low temperature fluorescence spectrometry 低温荧光(光谱)法
low-temperature fluorimetry 低温荧光法*
low-temperature Fourier transform infrared spectroscopy 傅里叶变换低温红外光谱
low temperature fractionation method 低温蒸馏法；低温分馏法
low-temperature gasoline plant 生产汽油的低温装置
low-temperature grease 低温润滑剂
low-temperature heat treatment 低温热处理
low-temperature impact resistance 抗低温冲击性
low temperature infrared spectra 低温红外光谱
low-temperature iron plating 低温镀铁
low-temperature lacquer 低温漆
low-temperature lubrication 低温润滑
low-temperature maturation 低温陈化
low temperature modification 低温型
low-temperature oxidation 低温氧化
low-temperature performance 低温性能
low temperature phosphorescence spectrometry 低温磷光(光谱)法
low temperature phosphorimetry 低温磷光法
low temperature plasma 低温等离子体
low-temperature polymer ①低温聚合物②冷聚橡胶
low temperature polymerization 低温聚合(作用)
low-temperature producer tar 低温发生炉焦油
low temperature recovery process 低温(气体)回收过程
low temperature resistance 耐寒性
low-temperature rubber 低温(聚合)橡胶
low-temperature sensitivity 感低温性；无耐寒性
low-temperature separation 低温分离
low temperature separation process 低温分离；深冷分离
low-temperature setting ①低温固化的②低温固化作用

low-temperature sludge 低温沉淀
low temperature solution polycondensation 低温溶液缩聚(作用)
low-temperature spin bath 低温纺丝浴
low-temperature stability 低温稳定性
low-temperature stiffening 低温硬化
low-temperature sulfuric acid treatment 低温硫酸处理
low-temperature tar 低温焦油
low-temperature tempering 低温回火
low temperature torque test (润滑油)低温扭矩试验
low-temperature toughness 低温韧性；低温韧度
low-temperature treatment 低温处理
low-temperature varnish 低温油漆
low temperature viscose ripening 黏胶低温熟成法
low tempering 低温回火
low tension arc 低压电弧
low-test gasoline 低级汽油；第三类汽油〔前苏联〕
low-tint material 低着色力材料〔着色力为60%～70%〕
low-to-moderate polarity 低中极性
low-viscosity 低黏度
low-viscosity medium 低黏度介质
low-viscosity nitrocellulose 低黏度硝酸纤维素
low-viscosity nitro-cotton 低黏度硝化棉
low-viscosity oil 低黏度润滑油
low-volatile coal 低挥发分煤
low-volatility fuel 低挥发分燃料；重质燃料
low voltage alternating current arc 低压交流电弧
low voltage arc discharge ion source 低压电弧放电离子源
low voltage arc ion source 低压电弧离子源
low voltage capacitor discharge 低压电容放电
low voltage discharge 低压放电
low voltage electrodialysis 低压电渗析
low voltage electrophoresis 低(电)压电泳(法)
low voltage scanning electron microscope 低压扫描电镜
low voltage spark 低压火花
low voltage spark photosource 低压火花光源
low voltage zone electrophoresis 低压区带电泳
low-water season 枯水期
low wool 粗支毛；低支毛
low work factor oil 低工作系数油类
lox 开矿炸药
loxa bark 金鸡纳树皮
loxoprofen 洛索洛芬〔药〕
lozenge 菱形纹
lozenge magnetic stirring bar 菱形磁性搅拌棒
LPF process 浸出沉淀浮选联合法
L-S coupling L-S偶合
lubanol(=coniferyl alcohol) 松柏醇

lubarometer 一种测大气压用仪器
lube 润滑油
lube cut(=lube distillate; lube fraction) 润滑油馏分
lubed-for-life bearing 永久润滑轴承〔即橡胶轴承〕
lube distillate(=lube cut) 润滑油馏分
lube fraction(=lube cut) 润滑油馏分
lube oil(=lubricating oil) 润滑油
lube oil additive 润滑油添加剂
lube oil blending 润滑油调和
lube oil packaging 润滑油包装
lube oil sludge 润滑油淤渣
lube oil technology 润滑油制造工艺
lube oil warehousing 润滑油仓库
lube plant 润滑油工厂
lube stock 润滑油(原)料
lubex (自润滑油中抽出)芳香族物
luboil(=lubricating oil) 润滑油
lubricant ①润滑剂②润滑的
lubricant additive 润滑剂添加剂
lubricant ashless dispersant 润滑油无灰分散剂
lubricant base 润滑油基础油
lubricant bloom(=grease mark) 油迹；油斑
lubricant capillarity 润滑剂毛细管现象
lubricant coating 润滑涂层
lubricant compatibility 润滑剂相容性
lubricant container 润滑剂容器
lubricant demulsifier 润滑油破乳剂
lubricant deterioration 润滑剂变质
lubricant film 润滑膜；润滑层
lubricant for drawing of patterns 注塑成形润滑脱模剂
lubricant performance 润滑剂工作性能
lubricant plasticizer 润滑性增塑剂；软化剂
lubricant separator 润滑剂分离器
lubricant starvation 润滑剂不足
lubricant thickening 润滑脂稠化
lubricated friction 润滑摩擦
lubricated gasoline 加润滑油的汽油
lubricating 润滑的
lubricating agent 润滑剂；隔离剂；脱模剂
lubricating capacity 润滑能力；润滑本领
lubricating coupler 润滑器
lubricating cream 润滑膏
lubricating detergent additive 润滑清净添加剂
lubricating efficiency 润滑效率
lubricating film 润滑油膜
lubricating finish ①润滑整理②润滑(油)剂
lubricating fluid 润滑液体
lubricating gasoline 润滑油汽油混合物
lubricating graphite 润滑石墨
lubricating grease 润滑脂
lubricating gun 润滑油枪
lubricating jelly 凝胶润滑剂
lubricating layer effect 润滑层效应
lubricating oil 润滑油
lubricating oil breakdown 润滑油供应中断
lubricating oil dispersant 润滑油分散剂
lubricating oil distillate 润滑油馏出物
lubricating oil distillation 润滑油蒸馏
lubricating oil emulsion test 润滑油乳化试验
lubricating oil evaporation loss 润滑油蒸发损失
lubricating oil family (石油中)润滑油成分
lubricating oil filter 润滑油过滤器
lubricating oil fine filter 润滑油精滤器
lubricating oil metal test 润滑油金属含量试验
lubricating oil organic acidity 润滑油的有机酸值
lubricating oil processing 润滑油生产
lubricating oil still 润滑油釜
lubricating oil wedge 润滑油楔
lubricating pipe 润滑油管道
lubricating press 润滑油压入器
lubricating property 润滑性质
lubricating quality 润滑质量
lubricating screw 黄油枪
lubricating syringe 润滑剂注射器
lubricating system 润滑系统
lubricating utensils 润滑设备
lubricating value 润滑值
lubricating wick 润滑蕊
lubrication 润滑(作用)
lubrication approximation 润滑近似法
lubrication chart 润滑图表
lubrication equipment 润滑设备
lubrication failure 润滑失效
lubrication guide ①润滑指南②润滑制导
lubrication interval 润滑间隔期
lubrication order 润滑程序
lubrication piping 润滑管道
lubrication point 润滑点
lubrication system 润滑系统
lubrication tank 润滑油杯
lubrication with solid film 固体膜润滑
lubricator ①润滑剂②润滑器③隔离剂
lubricity ①润滑能力②润滑性
lubricity carrier 润滑性质载体
lubricity index 润滑性指数
lucanthone 硫蒽酮〔药〕

Lucas method 卢卡斯法
lucency 透明；发亮
lucensomycin(=etruscomycin) 意北霉素；鲁斯霉素〔多烯大环内酯抗生素〕
lucent 发亮的；半透明的
lucerne 紫花苜蓿
lucidus 光泽的
luciferase 荧光素酶
luciferin 荧光素
luciferinase 荧光素生物酶
lucigenin 光泽精
lucinite 磷铝石
lucium 稀土元素混合物
Luck's indicator(=phenolphthalein) 酚酞
lucoxanthin 番茄黄质
lucumin 路枯马木苷
ludwigite 硼镁铁矿
ludyl 鲁狄尔〔药〕
luetin 黄体素；叶黄素〔药〕
luffa 丝瓜筋
luffanine 丝瓜碱
luffa-seed oil 丝瓜子油
lug ①手柄；耳状柄②突耳；呆耳
lug-cover pail 有盖桶
lug fillet 花纹块圆角
luggage leather 箱包革
Luggin capillary 卢金毛细管*
Lugol's solution 路戈尔碘液
lug tread pattern 横向胎面花纹
lukewarm 微温的；有点温热的
lumazine 2,4-二氧四氢蝶啶
lumbang oil 石栗仁油〔取自 Aleurites moluccana〕
lumber 锯材；成材；制材
lumber recovery 成材出材率；成材得率
lumbritin 蚯蚓亭；蚯蚓素
lumbrofebrin 蚯蚓退热碱
lumen staple fibre 空心(人造)短纤维；中空(人造)短纤维
lumequeic acid 三十(碳)烯酸
lumeter 照度计
lumiandrosterone 光雄甾酮
lumicaeruleic acid 光暗酸；青萤光酸〔黄柏的提取物〕
lumichrome 光色素；二甲(基)异咯嗪
γ-lumicolchicine(=lumicolchicine Ⅱ) γ-光化秋水仙碱
lumiflavin 光黄素；三甲(基)异咯嗪
lumiflavin method 光黄素法
luminal 鲁米那；5-乙基-5-苯基巴比妥酸
  $(C_2H_5)(C_6H_5) = C_4H_2O_3N_2$

luminance 发光度
luminance adaptation 亮度适应过程；亮度适应性
luminance density 发光密度
luminance factor ①亮度系数(因数)；亮度比②规约反射率；反射率比③特定辉度比
luminance level 亮(度)级
luminance purity 亮度纯度
luminance temperature 亮度温度
luminescence 发光*；冷光
luminescence analysis 发光分析法
luminescence center 发光中心*；冷光中心
luminescence co-factor immunoassay(LUCIA) 发光辅助因子免疫分析法
luminescence color 发光颜料
luminescence immunoassay 发光免疫分析
luminescence labeling reagent 发光标记试剂
luminescence method 发光法；光照法；电子激光法
luminescence quantum yield 发光量子产率
luminescence quenching 发光猝灭
luminescence spectrometer 发光光谱仪
luminescent coating 夜光涂料(短期的)
luminescent dyes 荧光染料
luminescent effect 发光效应
luminescent finish 夜(发)光涂料；夜(发)光面漆
luminescent ink 夜光(发光)油墨
luminescent labeling reagent 发光标记试剂
luminescent layer 发光层
luminescent paint 发光涂料
luminescent pigment 发光颜料
luminescent spectrum 发光光谱
luminizing 用放射物质使发光
luminoflavin 发光黄素
luminol 鲁米诺；氨基苯二酰一肼
luminol reaction 3-氨基邻苯二甲酰环肼反应
luminometer 发光计；照明计
luminophor ①发光团②发光体
luminosity 发光度
luminosity coefficient 发光度系数
luminosity curve (芒塞尔色系统)亮度曲线
luminous 发光的
luminous color 发光颜料
luminous densitometer 流明计
luminous density 光密度
luminous directional reflectance ①光定向反射②光定向反射比
luminous efficiency 发光效率
luminous filament pyrometer 发光丝高温计
luminous intensity 发光强度

luminous paint(=radioactive paint) 发光涂料；放射性涂料
luminous pigment 发光颜料；发光涂料
luminous printing ink 发光油墨
luminous quantity 光度量；发光量
luminous ray 光线
luminous reaction 发光反应
luminous reaction zone 发光反应区
luminous reflectance 光反射
luminous reflectance factor 光反射率
luminous sensitivity (感)光(灵)敏度
luminous test 发光试验
luminous transmittance 光透过率；光透射率
lumi-rhodopsin 光视紫红(质)
lumisterol 光甾醇 $C_{28}H_{43}OH$
Lummus cracking process 〔美国〕鲁玛斯公司选择裂化过程
lumoautoradiographic detection 发光放射自显影探测
lumogallion 4-氯-6-(2,4-二羟苯偶氮基)-1-羟苯-2-磺酸；荧光镓试剂
lumophore ①生光团②室温发光物
lump ①块；团②起瘤③起疙瘩④胶团；絮凝物形成
lump absorbent 块状吸收剂
lump breaker 碎矿机；破碎机
lump burner 块矿炉
lump coal 块煤
lumping 结块
lump lime 块石灰；生石灰
lump ore 块矿
lump pyrite 黄铁矿块
lump pyrite burner 黄铁矿块炉
lumps〔复〕 ①块煤②矿块③浆块④团块
lump salt 粗晶盐
lump scrap 胶团
lumpy 块的；成块的
lumpy screener deposit 成块屏蔽剂沉积；块状屏蔽剂沉积物
lumpy sludge 块形沉淀物
lumpy solid samples 块形固体试样
lumulon(=lupulic acid) 忽布酸
lunamarine 月芸碱
lunamycin 月亮霉素；月光菌素
lunar caustic 硝酸银
Lunge nitrometer 朗格定氮计
Lunge's method 朗格方法
Lunge's test 朗格试验
lung injurant agent 伤肺剂
lung irritant 肺部刺激剂
lungwort 兜苔
lup- 羽扇
lupane 羽扇(多环)烷
lupanine 羽扇烷宁；白金雀花碱金雀花碱羽扁豆碱
lupanol 羽扇烷醇
lupanone 羽扇烷酮
luparenol 卢杷烯醇〔取自蛇麻〕
luparol 卢杷醇
luparone 卢杷酮
lupene 羽扇烯
lupenone 羽扇烯酮
lupeol 羽扇醇；羽扇豆醇
lupeose 羽扇糖
lupetazin 二甲基哌嗪
lupetidine 2,6-二甲基哌啶
luphol 卢夫醇；大戟乳脂醇
lupin 羽扇豆
lupinal 羽扇豆醛
lupin alkaloid 羽扇豆碱
lupinane 羽扇豆烷
lupine 羽扇豆
lupine flower oil 羽扇花油
lupinic acid 羽扇豆酸
lupinidine sulfate 硫酸(左旋)鹰爪豆；硫酸金雀花碱
lupinin 羽扇豆苷 $C_{19}H_{32}O_{16}$
lupinine 羽扇豆宁
lupinotoxin 羽扇豆毒素
Lupke resiliometer 卢氏弹性试验机
lupulic acid(=lumulon) 苦味酸；蛇麻酸；忽布酸
lupulin 蛇麻素；忽布素；啤酒花苦味素
lupulinic acid 蛇麻腺酸
lupulone 蛇麻酮
lupulus 蛇麻
Lurgi catalyst 鲁奇催化剂
Lurgi extractor 鲁奇萃取器
Lurgilager metal 鲁奇拉吉合金
Lurgi pressure gasification process 鲁奇加压气化过程
Lurgi process 鲁奇煤气化法
Lurgi-Westfalia extractor 鲁奇-韦斯特法里亚萃取器
luridine 胆碱
lusec 流秒；流塞克〔真空渗漏单位〕
lusomycin 琉沙霉素；光霉素
lussatite 正绿方石英
luster 光泽
luster coating 闪光涂料；闪金属光泽涂料
lustering 上光
lustering agent 上光剂
luster pigment ①闪光颜料②珠光颜料③虹彩颜料

lustre 光泽
lustreless fracture 无泽断面
lustreless paint 无光漆
lustre twill 羽纱
lustrex 多聚苯乙烯薄膜
lustrous 光辉的；灿烂的；闪光的
lustrous rayon 有光嫘萦
lute ①封泥②封闭器③涂油
lutea factor 藤黄因子
luteanine(=artabotrine; isocorydine) 异紫堇定
lutecia 氧化镥 Lu_2O_3
lutecium(=lutetium) 镥〔71号元素，化学符号Lu〕
lutecium chloride 氯化镥 $LuCl_3$
lutecium oxide 氧化镥 Lu_2O_3
luteic acid 淡黄青霉糖酸（一种具高黏性的多糖酸由黄青霉(penicillium luteum) 在含糖液体培养基上生长而得；似为淡黄青霉多糖之丙二酰酯经酸水解可产生葡萄糖）
lutein ①叶黄素〔植〕②黄体制剂〔动物〕
luteinester 叶黄素酯
luteocobaltichloride 三氯化六氨钴 $[Co(NH_3)_6]Cl_3$
luteo-compound 黄(色钴)配盐
luteolin 木犀草素
luteolinidin 3′,4′,5,7-四羟花(色)锌
luteomycin 藤黄霉素
luteo-salt (橘黄色)钴氨盐
luteoskyrin 藤黄醌茜素
luteostal 孕甾酮
lutetium(=lutecium) 镥〔71号元素，化学符号Lu〕
lutetium-diphthalocyanine 镥-二酞菁；镥-双酞菁〔变色有机物〕
lutetium oxide 氧化镥
lutidine(=dimethyl pyridine) 卢剔啶；二甲基吡啶 $(CH_3)_2C_5H_3N$
lutidinic acid(=2,4-pyridine dicarboxylic acid) 卢剔啶酸；2,4-吡啶二甲酸 $C_5H_3N(CO_2H)_2$
lutidinium tungstophosphate(LWP) 二甲基吡啶镕钨磷酸盐
lutidone 卢剔酮 C_7H_9ON
luting ①封闭②塑入
lutoid 黄色体〔存在于胶乳中的形状不规则而带有黄色的粒子主要由蛋白质和类脂物组成黏性很大〕
lutol 硼鞣酸铝
luvangetin 藤橘亭
luvitherm 聚氯乙烯热塑塑料膜
luvungene 藤橘烯
luvungol 藤橘醇
luvungone 藤橘酮

Luwa evaporator 卢瓦蒸发器
Luwesta extractor 卢伟斯塔萃取器〔离心萃取器〕
lux mass 活性黄土
luxurious hand 丰满手感
L-value L值
L'vov furnace 里沃夫炉
L'vov platform 里沃夫平台
lyate 溶剂阴离子
lyate ion 溶剂阴离子
lychee 荔枝
lychee flavor 荔枝香精；荔枝风味
lychnin 剪秋罗苷
lychnose 剪秋罗(四)糖
lycine 甜菜碱；枸杞碱
lycoctonine 牛扁碱
lycomarasmin 番茄菌肽；N-乳酸基甘氨酰天冬酰胺
lycopene 番茄红素 $C_{40}H_{56}$
lycopersene 十氢番茄红素
lycopersicin(=tomatine) 番茄素
lycopodic acid 石松酸
lycopodine 石松碱
lycoramine 石蒜胺
lycoremine(=galanthamine) 雪花胺；加兰他敏
lycorine(=galanthidine) 石蒜碱
lycorisin 石蒜素
lycotetraose 石蒜四糖
lycoxanthin 番茄黄素
lyddite 立德炸药；赖戴特
lydian stone 试金石玉髓
lydimycin 利迪霉素
lye 碱液
lye boil aids in ramie degumming 苎麻碱煮助剂
lye change 废碱液
lye dissolving tank 溶碱槽
lye graduating tank 稠碱槽
lye limit concentration 碱液极限浓度
lye pump 碱液泵
lye ratio 碱液比
lye ring (过量)碱液圈
lye tank(=lye vat) 碱液槽；碱液桶
lye vat(=lye tank) 碱液槽；碱液桶
lygosin 立沽辛；2,2′-二羟基二苯乙烯基甲酮
lying charcoal kiln 卧式炭窑
lying meiler 卧式炭窑
lymecycline 赖甲环素
lynx fat 山猫脂
lyo- ①溶②离
lyocell 绿赛尔

lyogel　液凝胶
lyoluminescence　晶溶发光*；水合发光
lyoluminescence dosimetry　水合发光剂量学
lyoluminescent　水合发光的；晶溶发光的
lyolysis　液解(作用)；溶剂解(作用)
lyometallurgy　水法冶金(学)；萃取冶金(学)；非水溶剂冶金
Lyon furnace　来温电炉
lyonium　溶剂阳离子；溶剂化质子
lyonium ion　溶剂阳离子；溶剂化质子
lyonium salt　溶剂鎓盐〔酸溶于碱性溶剂中生成的盐〕
lyophilic colloid　亲液胶体*
lyophilic gel　亲液凝胶
lyophilic sol　亲液溶胶
lyophilization　冷冻干燥；升华干燥
lyophobe　①疏液体②疏液物
lyophobic　疏液的*
lyophobic colloid　疏液胶体
lyophobic colloidal solution　疏液胶体溶液
lyophobic dispersion　疏液分散(作用)
lyophobicity　疏液性；憎液性
lyophobic sol　憎液溶胶
lyophylization　低压升华干燥法
lyopilization　冷冻干燥
lyosol　水悬胶体
lyosorption　溶剂吸附作用
lyotrope　①感胶离子②易溶物
lyotropic　①感胶离子的②离液的③易溶的
lyotropic liquid crystal　溶致性液晶
lyotropic liquid crystalline polymer　溶致性液晶聚合物；亲液性液晶聚合物
lyotropic mesomorphism　亲液性介晶现象；亲液性介晶态
lyotropic series　感胶离子序*
lyotropic swelling　离子增溶膨胀
lyotropy　增溶性

lyovac process　冰冻干燥水中制品法
lypohydrophilic character　亲水亲脂特性
lyral　新铃兰醛
lysatine　赖酪碱
lysergic acid　麦角酸
lysergic acid diethylamide(LSD)　麦角酸二乙基酰胺；麦角酰二乙胺〔一种迷幻药〕
lysergic hydrazide　麦角酰肼
lysidine　赖西丁；2-甲基-4,5-二氢咪唑　$C_4H_8N_2$
lysidine solution　甲基二氢咪唑溶液
lysidine tartrate　酒石酸甲基二氢咪唑
lysimachia foenum-graecum tincture　灵香草酊
lysimeter　液度(估定)计
lysinal　赖氨醛
lysine(=2,6-diaminocaproic acid)　赖氨酸；2,6-二氨基己酸　$NH_2(CH_2)_4CHNH_2CO_2H$
lysine hydrochloride　盐酸赖氨酸　$C_6H_{14}O_2N_2HCl$
lysitol　酒石酸二甲基哌嗪〔药〕
lysivane　10-(2-二乙氨基丙基)吩噻嗪
lysogeny　溶原性
lysokinase　溶纤维蛋白激酶
lysozyme　溶解酵素
lysyl　赖氨酰　$H_2N(CH_2)_4CH(NH_2)CO—$
lysyl-lysine　赖氨酰赖氨酸
lysyme　溶菌酶
lytic　①溶解的②松解的
lytic agent　溶胞试剂
lytic effect　溶解效应
lytic response　溶解效应
lytropic liquid crystals　溶致液晶
lyxoflavin　来苏黄素
lyxonate　①来苏糖酸②来苏糖酸盐(或酯)
lyxonic acid　来苏糖酸　$CH_2OH(CHOH)_3COOH$
lyxose　来苏糖　$CH_2(CHOH)_4O$
lyxoside　来苏糖苷

M

maaliene　马榄烯
maaliol　马榄醇
mabula panza oil　山柳油
MacAdam color difference　麦克亚当色差
MacAdam color difference equation　麦克亚当色差公式
macaloid adhesive method　复合硅酸盐吸附法
macarbomycin　大炭霉素；马加霉素〔是 1964 年梅泽等由瑞士土壤中的 streptomyces phaeochromogenes 菌培养所得的一种多磷多糖类抗生素,此种抗生素日本已列为饲料添加物,但我国尚未应用,它是动物单用的抗生素〕
macaroni fibre　中空纤维
macaroni press　制通心粉机
macaroni yarn　空心丝；中空丝
MacArthur-Forest process　麦克阿瑟-福莱斯特(炼金)法
macassar oil　马卡发油；望加锡头(发)油
Macbeth color meter　麦克贝斯色彩仪〔比色计〕
Macbeth densitometer　麦克贝斯密度计
Macbeth illuminometer　麦克贝斯照度计
MacCaull corrosion tester　麦考尔腐蚀试验器
macdougallin　仙人掌甾醇；甲(基)胆甾烯二醇
mace　①肉豆蔻②肉豆蔻(树)
mace butter　肉豆蔻油
macene　肉豆蔻烯　$C_{10}H_{18}$
mace oil　肉豆蔻油
macerating　浸渍；冷浸
maceration　浸渍(作用)
maceration extract　浸渍液
maceration method　浸渍法
maceration tank　浸渍槽
maceration water　浸渍水
Mache unit　马谢(放射活性)单位
machilol　桢楠醇　$C_{15}H_{26}O$
machilus leaf oil　楠叶油
machilus oil　楠木油
machinability　切削性；机械加工性
machine　①机(器)；机械②机制
machine building　机械制造；机器制造
machine cut　机械加工；机械切削
machine direction　①轴向②加工方向
machine direction tear　纵向撕裂强度
machine direction tensile strength　纵向抗张强度
machined surface　机器加工面
machine dyeing　机器染色

machine finish　机械光滑
machine glazed finish　机械抛光
machine-glazed paper　机制蜡光纸
machine-gun oil　机枪油
machine made　机制
machine manufacturing　机械制造；机器制造
machine oil　机油〔机器润滑油〕
machine process　机械法
machinery　①机(器)；机(械)②工具；手段；方法
machinery coating　机械用涂料
machinery enamel　机械用漆
machinery lay-out　机械布置
machinery oil　机油
machinery paint　机械(机器)用漆
machinery room　机房；车间
machine scrub　刷磨清洁法
machine shop　机(械)工厂；金工厂；机器房
machine side　机面
machine stability　机上稳定性
machine steel　件钢
machine test　机床试验
machine tool cutting oil　切削工具用的润滑油；乳胶化润滑油
machine tool lubrication　切削工具润滑
machine washable property　可机洗性能
machine washing　机械洗涤；机洗
machine water　网下白水〔纸〕
machine water tray〔复〕　网下水槽
machining allowance　机加工余量
machining mark　机械加工痕迹
machining of metal　金属切削加工
machining oil　切削油
machining property　机械加工性能
Mach number　马赫数
macht metal　铜、锌、铁锻造合金
Mach unit　马赫(速度)单位
M acid　M 酸；1-氨基-5-萘酚-7-磺酸
mackenite metals　耐热 Ni-Cr 合金；耐热 Ni-Cr-Fe 合金
Mackenzie amalgam　麦克汞齐〔Hg-Bi 与 Pb-Hg 合金〕
mackerel oil　鲭鱼油
Mackey test　麦克试验
mackintosh　①防水胶布②胶布雨衣
mackintosh blanket cloth　胶布
mackintosh cloth　橡胶防雨布

Macklow-Smith extrusion viscosity test 米克劳-史密斯挤出黏度试验
Macklow-Smith flow pressure 米克劳-史密斯流动压力
Macklow-Smith plastometer 米克劳-史密斯塑度仪
maclayine 山榄碱
macle ①双晶②短空晶石③矿物中暗斑
Mac-Leod gauge 麦克-劳德真空规
macleyine 原鸦片碱
maclurin 桑酮；桑黄酚；桑橙素
MacMahon packing 麦克马洪填充物
MacMichael viscosimeter 麦克米契尔扭力黏度计
Macquer's salt 砷酸二氢钾 KH_2AsO_4
MacReynold's constant 麦克雷诺常数
macro- 〔希腊字头〕①大的②常量的；大量的③粗视的
macroaggregate 大颗粒
macroanion 大阴离子
macroaxis 长轴
macroazonitrile 大分子偶氮腈
macrobicyclic diamine cryptand 大双环二胺穴醚
macro Brownian motion 宏观布朗运动
macrocarpine 唐松草碱
macrocarpol(=juniperol) 刺柏醇；圆柏醇
macrocation 大阳离子
macrocell 大电池
macrochain 大分子链
macro-climate 宏观气候；大气候
macrocoaervation 常量凝聚
macroconcentration 常量浓度
macroconfiguration 宏观结构
macroconformation 大尺寸构象*
macroconstituent 常量成分
macro-crack 宏观裂缝
macro-creep 宏观蠕变
macrocrimp 大卷曲
macrocrystalline 粗晶的；大块结晶的
macrocrystallinity ①宏观结晶性②宏观结晶度
macrocycle (含15个原子以上的)大环
macrocyclic 大环的
macrocyclic compound 大环化合物
macrocyclic crown 大环冠(醚)
macrocyclic ester 大环酯
macrocyclic ketone 大环酮
macrocyclic lactone 大环内酯
macrocyclic ligand 大环配体
macrocyclic musk 大环麝香
macrocyclic polyether 大环多醚
macrocyclic polymer 大环聚合物
macrocyclic polysiloxane 大环聚硅氧烷
macrocyclic polythiaether 大环多硫醚
macrocyclic siloxane 大环硅氧烷
macrocyclic stereochemistry 大环立体化学
macrocylc compound 大环化合物
macrodeformation 宏观形变
macro-dispersion 粗粒分散体
macrodispersoid 促粒分散胶体
macrodome 长轴坡面
macrodroplet(s) 粗滴
macroelement 常量元素
macro emulsification 粗粒乳化(作用)
macro emulsion 粗(滴)乳状液
macroergic 高能(量)的
macroetching 宏观浸蚀；直观浸蚀
macro-etch test 宏观腐蚀试验
macrofiber 粗视纤维；长纤维
macro-fibril 巨原纤
macrofibrillen structure 大原纤结构
macrofiltration 粗滤(作用)
macro-flow 宏观流动
macrofluid 宏观流体
macrofoam 大泡沫
macroform anisotropy 宏观各向异性
macrogel 大粒凝胶
macrogol 聚乙二醇
macrography ①肉眼检查②大形书写
macroheterogeneity 宏观不均匀性
macroindication 常量指示
macroiniferter 大分子引发转移终止剂
macroinitiator 大分子引发剂
macroion 高分子离子*
macro ionization cross section detector 大电离截面检测器
macrokinetics 宏观动力学
macrolattice 大晶格
macrolide 大环内酯
macro-mechanics 粗观力学
macromer 大分子单体
macro method 常量法
macro-micro crimp 母子型卷曲；宏微复合型卷曲
macro-mixing 大量混合
macromole 大分子
macromolecular 大分子的
macromolecular alignment 大分子排列
macromolecular brush 高分子刷
macromolecular carrier embedment 高分子载体包埋法
macromolecular chain 大分子链；高分子链
macromolecular chain long-range structure 高分子链远

程结构
macromolecular chain short-range structure　高分子链近程结构
macromolecular chemistry　大分子(化合物)化学
macromolecular compound　大分子化合物
macromolecular coupling agent　大分子偶联剂
macromolecular crystallography　大分子晶体学
macromolecular dispersion　大分子分散体
macromolecular entanglement　大分子缠结
macromolecular film adsorption　高分子膜吸附法
macromolecular fluid　大分子流体
macromolecular hemicolloid　大分子半胶体
macromolecular isomorphism　高分子(异质)同晶现象
macromolecular ligand　大分子配体
macromolecular material　高分子材料
macromolecular order　高分子序态
macromolecular organization　大分子组织
macromolecular orientation　大分子取向；高分子取向
macromolecular polymer　大分子聚合物
macromolecular resin network　大分子树脂网络
macromolecular structure　大分子结构；高分子结构
macromolecule　高分子；大分子
macromonomer　大分子单体
macronecine　大叶(千里光)裂碱
macro-nonuniformity　宏观不均匀性；宏观多相性
macronucleus　大核
macroocclusion　最大吸着；大量吸留
macrophylline　大叶千里光碱
macropinacoid　长轴面
macropore　大孔
macropore distribution　粗孔隙分布；大孔隙分布
macroporosity　大孔性；宏观孔隙率
macroporous　大孔(性)的
macroporous polymer　大孔聚合物
macroporous resin　大孔树脂
macroporous silica gel　大孔硅胶
macroporous structure　大孔结构；宏观多孔结构
macroprecipitation　常量沉淀
macro-radical　大分子基团
macroreticular　粗网眼的
macro-reticular packing material　大孔填料；全多孔型填料
macroreticular resin　大网络树脂
macro-reticular silica gel　全多孔硅胶
macro-reticulate aliphatic polymer　大网状脂肪族聚合物
macro-rheology　宏观流变学
macro ring　大环
macroring-chain equilibrium　大环-链间平衡

macroroughness　宏观粗糙度
macror-reticular resin　大网络树脂；大孔树脂
macro sample　常量试样
macroscopic　①粗量的②宏观的③粗视的
macroscopic acidity constant　宏观酸度常数
macroscopic compressibility approximation(MCA)　宏观可压缩性近似
macroscopic constant　宏观常数
macroscopic defect　宏观缺陷
macroscopic elongation　宏观伸长
macroscopic examination　粗量检定；粗视检定
macroscopic irregularity　外观缺陷
macroscopic magnetization vector　宏观磁化强度矢量
macroscopic quantum tunneling effect(MQT)　宏观量子隧道效应
macroscopic reflectance　宏观折射率〔染污研究用语，由照片污垢粒数目估算白度〕
macroscopic surface configuration　宏观表面构型
macroscopic symmetry　宏观对称性
macrose　葡聚糖
macrosome　粗粒体
macrostrain　常量应变；宏应变
macrostress　常量应力；宏应力
macro system　粗(滴)体系
macrotin　升麻素
macrotoid　升麻混素
macrotricyclic diamine cryptand　大三环二胺穴醚
macrovoid　大孔；粗孔；大孔隙
macro-X-ray diffraction　粗聚焦 X 射线衍射
maculanin　钾淀粉
madder　茜草
madder color　茜草色素
madder lake　茜草红色淀
Maddrell salt　长链(的)高分子量偏磷酸钠
Madelung constant　马德隆常数
Madelung synthesis　马德隆合成法
made-to-measure　特制；定制；按定户要求制作
made-to-order　特制；定制；按定户要求制作
made-up article　坯品
made-up fuel oil　补充燃料油
madia oil　紫菀油；麻迪菊油
mafic　镁铁质
mafic minerals〔复〕　镁铁质矿物
mafurite　石盐镁矾与辉石的复合矿
maganese dioxide scavenging　二氧化锰沉淀清除法
magazine creel　复式筒子架
magazine grinder　水浆研磨机；水浆碎木机
magdala red(=naphthalene red)　萘红；麦塔喇红；苏丹红

magenta(=fuchsin)　(碱性)品红；洋红
magenta acid　品红酸
magenta red(=fuchsin)　品红；洋红
maggie　不纯煤
maggot pierced cocoon　穿头茧
maghemite　磁赤铁矿　$\gamma\text{-}Fe_2O_3$
magic(al) angle　魔角
magic angle rotation(MAR)　魔角旋转
magic angle sample spinning　试样魔角自旋
magic angle spinning　魔角自旋
magic eye　调谐指示管
magic hand　机械手
magic integer　幻数
magic integer permutation　幻数排列法
magic integer phase angle　幻数相角
magic integer space　幻数空间
magic island　幻岛
magic mountain　幻山
magic-N nucleus　幻中子核
magic nucleus　幻核*
magic number　幻数
magic-ridge　幻脊
magic tape　万能胶带
magic tee　T形波导支路
magic-Z nucleus　幻质子核
magistery of bismuth　硝酸二羟铋和二硝酸羟铋混合物
　〔$Bi(OH)_2NO_3$ 和 $Bi(OH)\cdot(NO_3)_2$〕
magma　①(稀)糊；稠液②糖糊；洗炼糖膏③岩浆
magma density　稠液密度
magma pump　糊泵；稠液唧筒
Magna forming process　马格纳重整
magnalite　活塞用的铝合金
magnalium　镁铝合金
magnamycin(=carbomycin)　碳霉素
magnane　镁烷
magnarcine　巨水仙碱
magnefen　僵烧白云石
magnesedin　镁菌素
magnesia　镁氧〔矿〕；氧化镁
magnesia alba　白镁氧　$4MgCO_3\cdot Mg(OH)_2\cdot 5H_2O$
magnesia borosilicate glass　硼硅酸镁玻璃
magnesia brick　镁砖
magnesia cement　镁氧水泥
magnesia ceramics　氧化镁瓷；镁氧陶瓷
magnesia-chrome　镁铬合金；铬镁
magnesia citrate　柠檬酸镁　$Mg_3(C_6H_5O_7)_2$
magnesia cordierite　镁堇青石
magnesia fertilizer　镁肥

magnesia glass　含氧化镁(的)玻璃
magnesia goslarite　镁皓矾
magnesia gralmandite　镁钙铁榴石
magnesia lime(=magnesia limestone)　镁氧石灰
magnesia limestone(=magnesia lime)　镁氧石灰
magnesia magma　氧化镁悬浮液
magnesia mica　金云母
magnesia mixture　镁氧混合剂；镁剂
magnesian alum　镁明矾
magnesian calcite　镁方解石；高镁方解石
magnesian chalk　镁白垩
magnesian chromite　镁铬铁矿
magnesian lime(=magnesian limestone)　含镁石灰岩
magnesian limestone(=magnesian lime)　含镁石灰岩
magnesian marble　镁质大理岩；菱镁矿
magnesian nontronite　镁绿脱石
magnesian quicklime　含镁生石灰
magnesian riebeckite　镁钠闪石
magnesian spar　白云石
magnesia saltpetre　镁硝石
magnesia spinel　镁氧尖晶石
magnesia tooth paste　镁氧牙膏
magnesia whisker　氧化镁须晶
magnesioferrite　镁铁矿
magnesio halide derivative　镁卤衍生物
magnesite　菱镁矿*
magnesite brick　镁砖
magnesite refractories　〔复〕　镁氧耐火制件
magnesium　镁〔12号元素，化学符号 Mg〕
magnesium acetate　乙酸镁　$Mg(C_2H_3O_2)_2$
magnesium acid citrate　柠檬酸氢镁
magnesium alkoxide　醇镁；烃氧基镁　$Mg(OR)_2$
magnesium alkyl compound　烷基镁化合物
magnesium alkyl condensation　烷基镁缩合
magnesium alloy die casting　镁合金压(模)铸件
magnesium aluminate　铝酸镁　$Mg(AlO_2)_2$; $Al_2[MgO_4]$
magnesium aluminum silicate　硅酸铝镁〔增稠剂〕
magnesium ammonium arsenate　砷酸铵镁
magnesium ammonium carbonate　碳酸铵镁
　$MgCO_3\cdot(NH_4)_2CO_3\cdot 4H_2O$
magnesium ammonium chloride　氯化铵镁
magnesium ammonium phosphate　磷酸铵镁
　$Mg(NH_4)PO_4$
magnesium amphibole　镁闪石
magnesium anode　镁阳极
magnesium arsenate　砷酸镁　$Mg_3(AsO_4)_2$
magnesium arsenide　砷化镁　Mg_3As_2
magnesium aryl compound　芳基镁化合物

magnesium ascorbyl phosphate 磷酸抗坏血酸酯镁盐
magnesium base grease 镁基润滑脂
magnesium basic carbonate 碱式碳酸镁
magnesium bicarbonate 碳酸氢镁 $Mg(HCO_3)_2$
magnesium bichromate 重铬酸镁 $MgCr_2O_7$
magnesium biphosphate 磷酸二氢镁 $MgH_4(PO_4)_2$
magnesium bisulfate 硫酸氢镁 $Mg(HSO_4)_2$
magnesium borate 硼酸镁 $Mg(BO_2)_2$
magnesium borocitrate 硼硅酸镁
magnesium bromate 溴酸镁 $Mg(BrO_3)_2$
magnesium bromide 溴化镁 $MgBr_2$
magnesium bromoplatinate 溴铂酸镁 $Mg[PtBr_6]$
magnesium butyrate 丁酸镁
magnesium cacodylate 二甲胂酸镁
magnesium calcium carbonate 碳酸镁钙 $MgCO_3 \cdot CaCO_3$
magnesium carbide 碳化镁 MgC_2
magnesium carbonate 碳酸镁
magnesium carbonate (basic) 碱式碳酸镁 $4MgCO_3 \cdot Mg(OH)_2 \cdot 5H_2O$
magnesium cerotate 蜡酸镁；二十六(烷)酸镁
magnesium chlorate 氯酸镁 $Mg(ClO_3)_2$
magnesium chloride 氯化镁 $MgCl_2$
magnesium chlorite 亚氯酸镁
magnesium chloroplatinate 氯铂酸镁 $Mg[PtCl_6]$
magnesium chlorostannate 氯锡酸镁 $Mg[SnCl_6]$
magnesium chromate 铬酸镁 $MgCrO_4$
magnesium chromium ferrite 镁铬铁酸盐
magnesium citrate 柠檬酸镁 $Mg_3(C_6H_5O_7)_2 \cdot 14H_2O$
magnesium copper alloy 镁铜合金
magnesium cyanide 氰化镁 $Mg(CN)_2$
magnesium cyanoplatinate 氰铂酸镁 $Mg[Pt(CN)_4]$
magnesium cyanoplatinite 氰亚铂酸镁 $Mg[Pt(CN)_4]$
magnesium dimethacrylate 二甲基丙烯酸镁
magnesium dioxide 二氧化镁
magnesium dithionate 连二硫酸镁 MgS_2O_6
magnesium dodecyl benzene sulfonate 十二烷基苯磺酸镁
magnesium dodecyl ether sulfate 十二烷基醚硫酸镁
magnesium ethide (二)乙基镁 $Mg(C_2H_5)_2$
magnesium ethoxide(=magnesium ethylate) 乙醇镁；乙氧基镁 $Mg(OC_2H_5)_2$
magnesium ethylate(=magnesium ethoxide) 乙醇镁；乙氧基镁
magnesium ferrite 铁酸镁；镁铁棕
magnesium ferrocyanide 氰亚铁酸镁 $Mg_2[Fe(CN)_6]$
magnesium finishing 镁涂装
magnesium fluoride 氟化镁 MgF_2
magnesium fluosilicate 氟硅酸镁 $Mg[SiF_6]$
magnesium formate(=magnesium formiate) 甲酸镁 $Mg(HCO_2)_2$
magnesium formiate(=magnesium formate) 甲酸镁
magnesium glycerophosphate 甘油磷酸镁
magnesium hydrate 氢氧化镁 $Mg(OH)_2$
magnesium hydride 二氢化镁 MgH_2
magnesium hydrogen arsenate 砷酸氢镁 $MgHAsO_4$
magnesium hydrogen phosphate 磷酸氢镁 $MgHPO_4$
magnesium hydrogen sulfate 硫酸氢镁 $Mg(HSO_4)_2$
magnesium hydroxide 氢氧化镁 $Mg(OH)_2$
magnesium hydroxide method 氢氧化镁精制法
magnesium hydroxyphosphate 氢氧化镁磷酸镁复盐
magnesium hypophosphite 次磷酸镁 $Mg(H_2PO_2)_2$
magnesium hyposulfite 连二亚硫酸镁 MgS_2O_4
magnesium iodate 碘酸镁 $Mg(IO_3)_2$
magnesium iodide 碘化镁 MgI_2
magnesium isopropoxide 异丙醇镁 $Mg(OC_3H_7)_2$
magnesium lactate 乳酸镁 $Mg(C_3H_5O_3)_2$
magnesium lactophosphate 乳酸磷酸镁
magnesium limestone 碳酸钙镁
magnesium magma 氢氧化镁悬浮液
magnesium malate 苹果酸镁 $MgC_4H_4O_5$
magnesium metaborate 偏硼酸镁 $Mg(BO_2)_2$
magnesium metal 镁金属
magnesium metasilicate 硅酸镁 $MgSiO_3$
magnesium methide 二甲镁 $Mg(CH_3)_2$
magnesium methoxide(=magnesium methylate) 甲醇镁；甲氧基镁 $Mg(OCH_3)_2$
magnesium methylate(=magnesium methoxide) 甲醇镁；甲氧基镁
magnesium mineral 镁矿
magnesium molybdate 钼酸镁 $MgMoO_4$
magnesium myristate 肉豆蔻酸镁
magnesium nitrate 硝酸镁 $Mg(NO_3)_2$
magnesium nitride 二氮化三镁 Mg_3N_2
magnesium nitrite 亚硝酸镁 $Mg(NO_2)_2$
magnesium oleate 油酸镁
magnesium orthoborate 正硼酸镁 $Mg_3(BO_3)_2$
magnesium orthophosphate (正)磷酸镁 $Mg_3(PO_4)_2$
magnesium orthosilicate 正硅酸镁 Mg_2SiO_4
magnesium oxalate 草酸镁 MgC_2O_4
magnesium oxide 氧化镁
magnesium palmitate 棕榈酸镁
magnesium peptonate 镁胨〔药〕
magnesium perchlorate 高氯酸镁 $Mg(ClO_4)_2$
magnesium permanganate 高锰酸镁 $Mg(MnO_4)_2$
magnesium peroxide 过氧化镁 $Mg[O_2]$

magnesium phosphate　磷酸镁　$Mg_3(PO_4)_2$
magnesium phosphide　二磷化三镁　Mg_3P_2
magnesium phospholactate　磷酸乳酸镁
magnesium platinocyanide　氰亚铂酸镁　$Mg[Pt(CN)_4]$
magnesium propoxide(=magnesium propylate)　丙醇镁
magnesium propylate(=magnesium propoxide)　丙醇镁
magnesium pyrophosphate　焦磷酸镁　$Mg_2P_2O_7$
magnesium rhodanate　硫氰酸镁　$Mg(SCN)_2$
magnesium ribbon　镁带；镁条
magnesium selenate　硒酸镁　$MgSeO_4$
magnesium silicate　硅酸镁　$MgSiO_3; Mg_2SiO_4$
magnesium silicide　硅化镁　$Mg_2Si; MgSi$
magnesium silicofluoride　氟硅酸镁　$Mg[SiF_6]$
magnesium soap　镁皂
magnesium soap lubricating grease　镁皂润滑脂
magnesium stearate　硬脂酸镁〔润滑剂，稳定剂〕
magnesium sulfate　硫酸镁　$MgSO_4$
magnesium sulfide　硫化镁　MgS
magnesium sulfite　亚硫酸镁　$MgSO_3$
magnesium tartrate　酒石酸镁　$MgC_4H_4O_6$
magnesium thiosulfate　硫代硫酸镁　$MgSO_3S; MgS_2O_3$
magnesium titanate　钛酸镁　$MgTiO_3$
magnesium trisilicate　三硅酸镁
magnesium tungstate　钨酸镁　$MgWO_4$
magnesium wolframate　钨酸镁
magnesol　酸式硅酸镁
magneson(=p-nitrobenzene azoresorcinol)　对硝基苯偶氮间苯二酚；试镁灵　$NO_2C_6H_4N=NC_6H_3(OH)_2$
magnesyl　卤镁基　MgX
magnet　磁铁；磁石；磁体〔物〕
magnet contactor　磁接触器
magnet current　磁铁电流
magnet current trimming circuit　磁铁电流微调电路
magnet for nuclear magnetic resonance　核磁共振用磁体
magnet hydrodynamic micropump　磁流体动力微泵
magnetic　磁(性)的
magnetic absorption　磁吸收
magnetically anisotropic group　磁各向异性基团
magnetically controlled reactive ion etching　磁控反应离子腐蚀
magnetically equivalent　磁全量
magnetic analysis　磁分析
magnetic analyzer　磁分析器
magnetic anisotropy　磁学各向异性
magnetic balance　磁力天平
magnetic behavior　磁行为
magnetic bubble memory　磁泡存储器
magnetic circular dichroic absorption　磁圆二色吸收

magnetic circular dichroism(MCD)　磁性圆二色散
magnetic coil　磁线圈
magnetic concentrator　磁性选矿机
magnetic conductivity permeability　导磁率
magnetic cooling　磁冷却
magnetic coupling　电磁联轴节
magnetic-damped balance　磁阻尼天平；磁抑天平
magnetic declination　磁偏角
magnetic decontamination　磁性纯化〔催化剂〕
magnetic deflection mass spectrometer　磁偏转质谱计
magnetic deflection system　磁偏转系统
magnetic diffusivity　磁扩散率；导磁性；导磁系数
magnetic dipole　磁偶极子
magnetic dipole moment　磁偶极矩
magnetic dipole radiation　磁偶极子辐射
magnetic disturbance　磁性干扰
magnetic drum　磁选鼓
magnetic electron multiplier　磁电子倍增器
magnetic elements　①地磁要素②磁性元件
magnetic equivalence　磁全同
magnetic equivalent nuclei　磁等同核
magnetic examination　磁性探伤检查
magnetic exchange　磁交换
magnetic extracting device　磁选器
magnetic false-twist spindle　磁性假捻锭子
magnetic field　磁场
magnetic field analyzer　磁场分析器
magnetic field automatic scanning　磁场自动扫描
magnetic field deflection　磁场偏转
magnetic field scan　磁场扫描
magnetic field scanning　磁场扫描
magnetic field stability　磁场稳定度
magnetic field strength　磁场强度
magnetic field system of NMR　NMR 谱仪磁场系统
magnetic field welding　磁场焊接
magnetic filter　磁性过滤器
magnetic flaw detecting　磁力探伤
magnetic flow meter　电磁流量计
magnetic fluid control device　磁性流体控制仪
magnetic flux　磁通量
magnetic flux density　磁通密度
magnetic force microscope　磁力显微镜
magnetic force microscopy　磁力显微镜
magnetic gage　磁力测厚计
magnetic gap　磁隙
magnetic gauge　磁力测微计
magnetic guard　磁性钢丝面罩
magnetic hysteresis　磁滞

magnetic hysteresis loop 磁滞回线
magnetic induction 磁感应强度
magnetic inspection 磁力探伤
magnetic iron 磁铁
magnetic iron oxide(=magnetic oxide) 磁性氧化铁；四氧化三铁 Fe_3O_4
magnetic lens 磁透镜
magnetic medium 磁性介质
magnetic meridian 地磁子午线
magnetic mine 磁性水雷
magnetic mirror effect "磁镜"效应
magnetic mixer 磁力混合器
magnetic moment 磁矩
magnetic needle 磁针
magnetic optic 磁光的
magnetic optical rotation 磁致旋光
magnetic oscillograph 电磁示波器；电磁示波管
magnetic oxide(=magnetic iron oxide) 四氧化三铁；磁性氧化铁
magnetic particle examination 磁粉检验
magnetic particle test 磁粉试验
magnetic peak-switching 磁式峰切换
magnetic permeability 磁导率*
magnetic polarizability 磁性极化度
magnetic polarization 磁性极化
magnetic pole 磁极
magnetic polymer 磁性聚合物
magnetic polymer microsphere 磁性高分子微球
magnetic potential 磁势；磁位
magnetic powder 磁性粉末；磁(性)铁粉
magnetic prism 磁棱镜
magnetic pyrite 磁性黄铁矿
magnetic quantum number 磁量子数
magnetic recording pigment 磁性录音颜料
magnetic relaxation 磁松弛
magnetic reluctance 磁阻
magnetic repulsion 磁(排)斥力；磁推斥
magnetic resolution 磁性离解；磁性分离
magnetic resonance force microscopy 磁共振力显微镜
magnetic resonance microscopy 磁共振显微镜
magnetic roll 磁力辊
magnetic roll squeegee 磁性刮浆辊
magnetic rotation ①磁(致)旋光②磁(场)旋光度
magnetic rotatory power 磁旋光本领；磁致偏振转本领
magnetic rubber 磁性橡胶
magnetic sand 磁铁(矿)砂
magnetic saturation 磁性饱和
magnetic scanning 磁扫描
magnetic screening 磁筛选
magnetic sector 扇形磁场
magnetic sector analyzer 磁扇形分析器
magnetic sector spectrometer 扇形磁式质谱仪
magnetic selector 磁选器
magnetic semiconductor 磁性半导体
magnetic sensor 磁传感器
magnetic separation 磁力分离
magnetic separator 磁选器
magnetic shielding 磁屏蔽
magnetic specific heat 磁比热容
magnetic spectrograph 磁谱仪
magnetic spectrum 磁谱
magnetic stirring bar (电)磁搅拌棒
magnetic susceptibility 磁化率
magnetic switch 磁力开关；磁开关
magnetic tape cassette recorder 盒式磁带记录器
magnetic testing 磁力探伤
magnetic texturing spindle 磁性变形转子
magnetic thermobalance 磁(性)热天平
magnetic thickness gauge 磁力测厚仪
magnetic-type gage 磁性测厚仪
magnetic units 磁单位〔真空中磁导率为1〕
magnetic valve 电磁阀
magnetic water 磁(化)水
magnetism ①磁学②磁(性)〔物〕③吸引力；磁力
magnetism shield 磁屏蔽
magnetite(=aimant; aimantine) 磁铁矿
magnetite black 磁性铁黑 Fe_3O_4
magnetization 磁化强度*
magnetization recovery rate 磁化强度的恢复率
magnetization transfer 磁化转移
magnetization vector 磁化矢量
magnetizing current 起磁电流；激磁电流
magnetizing force 磁化力
magneto 久磁电机
magneto-chemistry 磁化学
magneto-controlled sputter coating 磁控溅射镀膜
magnetodiode 磁二极管
magnetoelasticity 磁致弹性
magnetoelectrolysis 磁电解
magnetogyric ratio 磁旋比
magnetohydrodynamics 磁流体动力学
magnetom 磁共振成像
magnetometric analysis 磁性分析
magnetometric titration 磁量滴定法
magnetomotive force 磁通势
magneton 磁子

magneton theory　磁子学说
magneto-optical phenomenon　磁光现象〔物〕
magneto-optical rotation　磁致旋光；磁性光学旋转
magneto-optic effect　磁光效应
magnetophone　磁铁膜录音器
magnetoplasma dynamics(MPD)　磁等离子体动力学
magnetoplumbite　磁性铅酸盐
magnetoresistance　磁致电阻；磁阻效应
magnetoresistive effect　磁阻效应
magneto-resistor　磁电阻
magnetorotation　磁(致)旋光
magnetoscope　验磁器
magnetostriction　磁致伸缩(现象)
magnetostriction apparatus　磁致伸缩仪器
magnetostrictive　磁致伸缩的
magnetothermodynamics　磁热力学
magnetotransistor　磁晶体管
magnet plug monitor technique　磁塞监测技术
magnet pole face　磁极面
magnetron　磁控管；磁(控)电(子)管
magnetron motion　磁控运动
magnet yoke　磁轭
magni-〔词头〕　大
magniferous　含镁的
magnification　①放大②放大率；放大倍数
magnification factor　放大因子
magnification of lens　透镜的放大率
magnification ratio　放大比率
magnifying power　①放大能力②强化③扩大
magnitude　①大小；尺寸②数值③量(度)
magnochromite　镁铬铁矿
l-magnocurarine　l-木兰箭毒碱
d-magnocurarine　d-木兰箭毒碱
magnoflorine　木兰花碱
magnolamine　木兰胺
magnolia coco concrete　夜合花浸膏
magnolia compound　玉兰香精
magnolia flower oil　玉兰花油
magnolia leaf oil　玉兰叶油
magnolin　木兰苷；木兰脂素
magnoline　木兰碱
magnolite　碲酸汞
magnolium　锑铅合金
magnolol　木兰醇　$C_{18}H_{18}O_2$
magnon　磁子
Magnox　①镁诺克斯合金〔含少量铍、铝的抗氧化镁合金〕②镁诺克斯型元件③镁诺克斯型气冷堆
mahareb　苏丹香茅

Mahler calorimeter　马勒量热器
mahogany　桃花心木
mahogany acid　石油磺酸
mahogany petroleum sulfonate　石油磺酸盐
mahogany soap　石油磺酸皂
mahogany sulfonate　石油磺酸盐
mahogany wood oil　红楝子油；红椿油
mahua butter　玛华油
maidenhair　掌叶铁线蕨；铁线蕨
maiden hair tree　银杏；白果
mail handling pigment　邮件分捡颜料
Maillard reaction　美拉德反应
main　①总线②主管③主要的④总的
main air　一次压缩空气；主压缩空气
main amplifier　主放大器
main anode　主阳极
main band　主带
main bearing　主轴承
main cam　主三角
main cell　主电池
main chain　主链
main chain liquid crystalline polymer　主链型液晶聚合物
main-chain scission　主链断裂
main coating oven　主涂层烘干炉；主涂层烘箱
main delivery pipe　主压力管道
main distillate fraction　主馏分
main effect　主效应*
main electrode　主电极
main flue　总烟道
main frame　底盘；主机
main group　主族*
main group element　主族元素
main hydrolysis(=principal hydrolysis)　主水解
mainlaying cost　管道装设费用
main note　主香
main pipe　主管
main pipe line　主管道
main plunger　主活塞
main polymer chain　聚合物主链
main-process stream　(工艺)主流程线；总流程
main product　主要产品
main quantum number　主量子数
main reaction　主(要)反应
main shaft　主轴
main specification of NMR spectrometer　NMR 谱仪主要技术指标
Mainstee evaporator　梅因斯提蒸发器
main stop valve　主停气阀

main switch　总开关
maintained treatment　维修处理；保养处理
main tank　主油箱
maintenance　维持；保养
maintenance area　维修区
maintenance cost　维修费
maintenance downtime　停(机检)修时间
maintenance finish　维修涂料；维修漆
maintenance fitter　维修钳工
maintenance free　不需维修
maintenance-free paint　长效漆；无维修漆
maintenance measure　维修检测
maintenance of soil fertility　土壤肥力维持
maintenance overhaul　经常修理
maintenance paint　维护涂料
maintenance painting treatment　维修涂装处理
maintenance primer　维修(用)底漆
maintenance procedure　保养程序
maintenance supply　维修器材
maintenance work　维修工作
main trunk line　主要干线管路
main unit　主单元；主机；主件
main valve　主阀
main valve disc　主阀盘
mainwash　主洗
main wash heating phase　主洗加热期
main water　自来水
maiolica　陶器〔涂有不透明釉的一种意大利陶器〕
maize　玉米；玉蜀黍；包谷
maize meal　玉米粉；玉米面
maize oil(=corn oil)　玉米油
maize starch　玉米淀粉
maizolith　玉米轴和玉米秆压成的板
Majac jet pulverizer　马亚克对喷式气流粉碎机
majolica　陶器〔涂有不透明釉的一种意大利陶器〕
majolica colour　粗陶器的彩色釉
majolica enamel　粗陶器瓷釉
major　(较)大的；较多的；(较)长的；较重要的
major accident　重大事故
majoran oil　甘牛至油
major axis　长轴〔数〕
major chemical company　大型化学公司
major constituent　主要成分；主要组分
major diameter　大直径
major groove　大沟
major overhaul　大修
major repair　大修
major service　大修

make-up air　补充空气；新气
make-up ammonia　补充氨
make-up carrier (gas)　配载气；补充载气
make-up catalyst　补充催化剂
make-up gas　补充气体；新气；尾吹气
make-up line　(物料)补充管线
make-up lubricating system　装配润滑系统
make-up machine　成型机
make-up machinery　装配机械
make-up oil　补足油〔石油产品的新添加部分〕
make-up pump　补充泵
make-up room(=making-up shop)　装配室；制作室
make-up table　成型操作台
make-up tank　调配槽
make-up unit　装配组
make-up water　补偿给水
making good　①找补②成功；发达③补(赔)偿④矫正；修复
making up　①装配②补足
making-up room(=making-up shop)　装配室
making-up shop(=making-up room)　装配室
malabaricol　岭南臭椿醇
malaccol　马来鱼藤酮
malachite　孔雀石
malachite green　孔雀(石)绿*
malachite green actinometer　孔雀(石)绿感光计
malachite green o-carboxylic acid lactone　孔雀绿-邻羧酸内酯
malachite green toner　孔雀绿色料(色原)〔磷钨酸色淀颜料〕
malacolac　紫胶树脂
malacolite　白透辉石
malacon　水锆石
malakin　马拉锦〔药〕
malakograph　软化率计
malamic acid　苹果酰胺酸　$CONH_2CHOHCH_2COOH$
malamide　苹果酰胺　$CONH_2CHOHH_2CONH_2$
Malaprade reaction(=periodic acid oxidation)　高碘酸氧化反应；马拉破瑞德反应
malaria　疟(疾)
malaridine　咯奈啶〔药〕
malarin　玛拉林(药)
malate　①苹果酸②苹果酸盐(酯或根)
malathion　马拉硫磷；马拉息昂；马拉松〔一种有机磷杀虫剂〕
malayan camphor(=borneol)　冰片；龙脑；莰醇　$C_{10}H_{17}OH$
malay camphor　马来樟脑；龙脑
malchite　微闪长岩

mal-condition 恶劣条件
malcrystalline structure 残晶结构；过渡形结晶构造
maldistribution 分布不良现象
maldonite 黑铋金矿
male-(=malein-)〔词头〕顺丁烯二酸(酰)；马来酸(酰)
malealdehyde 马来醛；顺丁烯二醛 CHOCH=CHCHO
malealdehydic acid 顺丁烯一醛酸；马来酸一醛；甲醛基丙烯酸
maleamic acid 马来酰胺酸 CONH$_2$CH=CHCOOH
maleamide(=maleinamide) 马来酰胺；顺丁烯二酰胺；马来酸二胺
maleamidic acid(=maleamic acid) 马来酰胺酸
male and female 阴阳面；凸凹面
male and female face 阴阳面；凸凹面
male-and-female flange 凸凹面法兰；阴阳面法兰；榫槽法兰
maleanilic acid 马来酸一酰苯胺；顺丁烯一酰苯胺酸
maleanilide 马来酰苯胺
maleate ①顺丁烯二酸；马来酸②马来酸盐(酯或根)
maleated butadiene polymer 马来酸丁二烯聚合物
maleated rosin 马来松香
maleate resin 顺丁烯二酸(酯)树脂；马来(酸)树脂
maleation 马来化(作用)
male die 阳模
male fern 绵马
male fern oil 绵马油
male fitting 阳模配合
male flange 凸面法兰
male force 模塞；阳模
maleic acid 马来酸；顺式丁烯二酸；(E)-丁烯二酸 (=CHCO$_2$H)$_2$
maleic acid chelate 顺丁烯二酸螯合物
maleic acid diamide 顺丁烯(二酸)二酰胺
maleic acid glycol ester 马来酸乙二醇酯
maleic acid monosodium salt 马来酸单钠盐
maleic acid resin 顺丁烯二酸树脂
maleic acid resin varnish 顺丁烯二酸树脂清漆
maleic anhydride 马来酐；顺式丁烯二(酸)酐 (=CHCO)$_2$O
maleic anhydride cyclopentadiene adduct 顺酐环戊二烯加合物
maleic anhydride-polyester 顺丁烯二酸聚酯
maleic anhydride terpinene adduct α-萜品烯-顺丁烯二酸酐加成物
maleic anhydride value 马来酐值；顺丁烯二酐值
maleic dialdehyde 马来醛；顺式丁烯二醛 CHOCH=CHCHO
maleic ester resin 顺丁烯二酸酯树脂

maleic functionality 马来酸官能度；顺丁烯二酸官能度
maleic hydrazide 顺丁烯二酰肼；马来酰肼；抑芽丹
maleic hydrazine 马来酰肼
maleic modified alkyd rosin ester 马来酸酐改性醇酸松香酯
maleic-modified rosin 顺丁烯二酸酐改性松香
maleic oil 马来酸化油；顺丁烯二酸改性油
maleic resin 马来树脂；顺丁烯二酸-丙三醇树脂
maleic semialdehyde 顺丁烯醛酸；马来醛酸
maleic trialkyltin methacrylate 顺丁烯二酸三烷基锡甲基丙烯酸盐
maleic value 二烯值
maleimide 马来酰亚胺；顺丁烯二酰亚胺
maleimycin 马来酰亚胺霉素
male(in)〔词头〕顺丁烯二酸；马来酸
maleinamic acid 马来酰胺酸；顺丁烯酰胺酸
maleinanil 马来酸酐缩苯胺；N-苯基马来酰亚胺
maleinanilic acid 顺丁烯一酰苯胺酸；顺丁烯二酸一酰苯胺
maleinanilide 马来酰苯胺；顺丁烯二酰二苯胺
male(in)imide 顺丁烯二酰亚胺
maleinization 顺丁烯二酸化；马来(酸)化
maleinized oil 顺丁烯二酸酐化油
maleinized tung oil 马来化桐油〔桐油酸与马来酸酐化合〕
maleinoid 顺式异构化合物
maleinoid form 顺式
male mold(=patrix) 阳模；上半模
male mould 阳模
maleopimaric acid 马来海松酸
maleopimaric (acid) anhydride 马来海松酐
maleoyl- 马来酰(基)；顺式丁烯二酰(基)
maleoyl acetoacetate 顺丁烯二酸单酰乙酰乙酸
male punch 阳模冲头
male rotor 阳转子
male screw 阳螺纹
male section 阳模；上半模
male thread 阳螺纹
maleyl- 马来酰(基)；顺丁烯二酰(基)
malfunction 故障；失灵〔机器〕
maliane 橄榄烷
malic acid 苹果酸；羟基丁二酸
malic (acid) diamide 苹果酸二酰胺
malic acid ether 苹果酸醚
malic amide 苹果酰胺 HOC$_2$H$_3$(CONH$_2$)$_2$
malignin 毒曲菌素
mallardite 白锰矾
malleability 展性

malleable (有)展(延)性的；可锤展的
malleable casting 展性铸件
malleable cast iron 可锻铸铁；展性铸铁
malleable detachable chain 展性活动链
malleable iron 可锻铸铁；展性铸铁
malleable pig iron 可锻铸铁；展性生铁
malleable pintle chain 展性锁链
mallee bark 桉树皮
mallet ①短槌②桉树
mallet bark 桉树皮
Mallier chart 马利尔图
Mallinckrodt process 马林克罗特过程〔从铀矿石浸出液用醚类萃取法生产硝酸铀酰的过程〕
mallotoxin 菲律宾野桐毒素；粗糠柴毒卡马拉素
mallow 锦葵
malm 白垩土；钙质砂土〔地质〕
malodor 臭气
malodorous 恶臭的
malol(=ursolic acid) 熊果醇酸
malonamic acid 丙酰胺酸 $CONH_2CH_2COOH$
malonamide 丙二酰胺 $CH_2=(CONH_2)_2$
malonamide nitrile(=cyanoacetamide) 氰基乙酰胺
malonate ①丙二酸②丙二酸盐(酯或根)
malonic acid 丙二酸 $COOHCH_2COOH$
malonic amide 丙二酰胺
malonic anhydride 丙二酸酐 $CH_2(CO)_2$
malonic ester ①丙二酸酯 $COORCH_2COOR$②〔专指〕丙二酸乙酯 $C_2H_5OCOCH_2COOC_2H_5$
malonic semialdehyde 丙二酸半醛
malon oil 黑鲸油
malononitrile 丙二腈 $CH_2(CN)_2$
malonurea(=malonyl urea) 丙二酰脲；巴比妥酸 $C_4H_4O_3N_2$
malonuric acid 马龙尿酸；脲羰基乙酸
malonyl- ①丙二酰(基)②丙二酸单酰(基)
malonyl CoA 丙二酸单酰 CoA
malonyl-CoA-ACP transacylase 丙二酸单酰CoA-ACP 转酰基酶
malonyl urea(=malonurea) 丙二酰脲；巴比妥酸 $C_4H_4O_3N_2$
maloperation 误操作
maloyl- 苹果酰(基)；羟基丁二酰(基) $COCH(OH)CH_2CO-$
Malpica formula 马尔毕卡公式
malt ①麦芽②制麦芽
maltha 软沥青
malthenes〔复〕马青烯；石油脂；石油质〔沥青中溶于石油醚的组分〕

maltol (=2-methyl-3-hydroxy-4-pyrone) 2-甲基-3-羟基-4-吡喃酮；麦芽酚
maltonic acid D-葡萄糖酸
maltosazone 麦芽糖脎
maltose 麦芽糖 $C_{12}H_{22}O_{11}$
malt sugar 麦芽糖 $C_{12}H_{22}O_{11}$
malvalic acid 锦葵酸 $C_{18}H_{32}O_2$
malvidin 锦葵色素；二甲花翠素
malvin 锦葵色素苷；二甲花翠苷
malvon 氧化锦葵苷
mammoth tanker 大型油船〔4～5万吨〕
manaca 番茉莉
manacine 番茉莉碱 $C_{15}H_{22}O_5N_4$
manageable gel 可调凝胶
manageable viscosity ①易于控制的黏度②便于施工的黏度
man conveyor belt 载客运输带；自动人行道
mancophalic acid 马尼剌珀琶树脂
mandarin 橘；柑
mandarin G 酸性橙
mandarin oil 橘子油
mandarin petitgrain oil 橘叶油
mandelate 扁桃酸盐(或酯) $C_6H_5CHOHCOOM$
mandelic acid(=hydroxy-phenylacetic acid) 扁桃酸；苯乙醇酸 $C_6H_5CHOHCO_2H$
Mandelin's reagent 曼得灵试剂
mandelonitrile 扁桃腈；苯乙醇腈 $C_6H_5CHOHCN$
mandeloyl 扁桃酰；α-羟基苯乙酰 $C_6H_5CH(OH)CO-$
mandioc 木薯淀粉
mandragorine 曼陀茄碱
mandrake 欧伤牛草根
mandrel 管芯
mandrel bend 轴棒弯曲；轴棒挠曲
mandrel flexibility 棒轴柔韧性
mandrel forming (管材)型芯缠绕成型
mandrel holder 芯座；芯型支座
mandrelling 拉延
mandrel test 心轴试验法〔试验涂料的柔韧性〕
mandrel test for flexibility 轴棒弯曲柔韧性试验
mandril(=mandrel) 心轴
mangal 锰铝合金
mangan 锰
manganate 锰酸盐 M_2MnO_4
mangancese blende 硫锰矿
manganese 锰〔25 号元素，化学符号 Mn〕
manganese acetate 乙酸锰 $Mn(C_2H_3O_2)_2$
manganese bioxide(=manganese dioxide) 二氧化锰
manganese black 锰黑；软锰矿 MnO_2

manganese BON red　BON系锰红颜料；2-羟基萘-3-甲酸锰红

manganese borate　硼酸锰

manganese boride　硼化锰　$MnB;MnB_2$

manganese bronze　锰青铜

manganese brown　锰棕

manganese carbide　一碳化三锰　Mn_3C

manganese carbonate　碳酸锰

manganese carbonyl　羰基锰

manganese chloride　二氯化锰

manganese copper　锰铜

manganese dioxide　二氧化锰

manganese disilicide　二硅化锰

manganese disulfide　二硫化锰　MnS_2

manganese drier　锰催干剂

manganese family element　锰族元素

manganese fertilizer　锰肥

manganese fluosilicate　氟硅酸锰　$Mn[SiF_6]$

manganese glass　锰质玻璃

manganese green　锰绿

manganese heptoxide　七氧化二锰

manganese horon　锰青铜

manganese hydrogen phosphate　磷酸氢锰

manganese linoleate　亚油酸锰

manganese mica　锰金云母

manganese minerals　锰矿

manganese (Ⅳ) molybdate　钼酸锰

manganese monoxide　一氧化锰；氧化亚锰

manganese mordant　锰媒染剂〔染〕

manganese mud　锰矿渣

manganese naphthenate　环烷酸锰

manganese nitride　二氮化三锰　Mn_3N_2

manganese octoate　辛酸锰

manganese oleate　油酸锰

manganese ore　锰矿

manganese oxide　氧化锰

manganese oxide hydrate　水合氧化锰

manganese oxyhydroxide　碱式氢氧化锰　$MnO(OH)$

manganese peroxide(=manganese dioxide)　二氧化锰

manganese phosphate coatings　磷酸锰镀膜；磷酸锰处理膜

manganese plating　镀锰

manganese powder　①锰粉②二氧化锰

manganese protoxide　一氧化锰　MnO

manganese resinate　树脂酸锰

manganese rosinate　松香酸锰

manganese selenic dihydrate　二水合亚硒酸锰

manganese sesquioxide　三氧化二锰；氧化锰

manganese silicate　原硅酸锰

manganese soap　锰皂

manganese spar　菱锰矿

manganese steel　锰钢

manganese sulfate　硫酸锰

manganese tallate　松浆酸锰

manganese tetrachloride　四氯化锰　$MnCl_4$

manganese-titanium　锰钛合金

manganese trioxide　锰酐；三氧化锰　MnO_3

manganesium　锰〔旧名〕

manganic　①锰的②三价锰的③六价锰的（三价者在盐中注明为三价，六价的酸视为正）

manganic acid　锰酸　H_2MnO_4

manganic anhydride　锰酐；三氧化锰　MnO_3

manganic chloride　三氯化锰　$MnCl_3$

manganic compound　锰化合物；三价锰化合物

manganic hydroxide　三氢氧化锰；三羟化锰　$Mn(OH)_3$

manganic manganous oxide　四氧化三锰

manganic metaphosphate　偏磷酸锰

manganic oxide　三氧化二锰　Mn_2O_3

manganic phosphate　磷酸三价锰

manganic sulfate　硫酸三价锰　$Mn_2(SO_4)_3$

manganiferous　含锰的

manganimetric titration　锰滴定

manganimetry　锰滴定法

manganin　锰镍铜合金

manganite　①亚锰酸盐　$M_4MnO_4;M_2MnO_3$②水锰矿

mangano-compound　二价锰化合物

mangano-manganic oxide　四氧化三锰　Mn_3O_4

manganosite　方锰矿　MnO

manganostilbite　锑砷锰矿

manganous　二价锰的；锰〔二价者在盐中不再注明二价字样〕

manganous acetate　乙酸锰　$Mn(C_2H_3O_2)_2$

manganous acid　亚锰酸　$H_4MnO_4;H_2MnO_3$

manganous ammonium phosphate　磷酸亚锰铵

manganous ammonium sulfate　硫酸亚锰铵

manganous arsenate　砷酸锰　$Mn_3(AsO_4)_2$

manganous butyrate　丁酸锰　$Mn(C_4H_7O_2)_2$

manganous carbonate　碳酸锰　$MnCO_3$

manganous chloride　二氯化锰　$MnCl_2$

manganous chromate　铬酸锰　$MnCrO_4$

manganous compound　二价锰化合物

manganous dihydrogen phosphate　酸性磷酸锰；马日夫盐　$Mn(H_2PO_4)_2$

manganous dithionate　连二硫酸锰　MnS_2O_6

manganous ferrocyanide　亚铁氰化亚锰

manganous fluosilicate　氟硅酸亚锰

manganous glycerinophosphate 甘油磷酸亚锰
manganous hydroxide 二氢氧化锰；二羟化锰 $Mn(OH)_2$
manganous hypophosphite 次磷酸锰
manganous lactate 乳酸锰 $Mn(C_3H_5O_3)_2$
manganous lead resinate 树脂酸铅与树脂酸亚锰的混合物
manganous linoleate 亚油酸锰 $Mn(C_{18}H_{31}O_2)_2$
manganous metaphosphate 偏磷酸亚锰
manganous nitrate 硝酸锰 $Mn(NO_3)_2$
manganous oxalate 草酸锰 MnC_2O_4
manganous oxide 一氧化锰 MnO
manganous phenolsulfonate 苯酚磺酸锰 $Mn(OHC_6H_4SO_3)_2$
manganous phosphate 磷酸锰 $Mn(PO_3)_2$; $Mn_3(PO_4)_2$; $Mn_2P_2O_7$
manganous pyrophosphate 焦磷酸亚锰
manganous silicate 硅酸锰 $MnSiO_3$; Mn_2SiO_4
manganous sulfate 硫酸锰 $MnSO_4$
manganous sulfide 一硫化锰 MnS
manganous sulfite 亚硫酸锰 $MnSO_3$
manganous sulfophenate 苯酚磺酸锰 $Mn(OHC_6H_4SO_3)_2$
manganous tartrate 酒石酸锰 $MnC_4H_4O_6$
mangcorn 混合粒
mangiferin 芒果苷 $C_{19}H_{18}O_{11}$
mangle 轧液机；砑光机
mangler ①砑光机〔造纸及纺织〕②压延机〔橡胶〕
mangnotantalite 锰钽铁矿
mango(=common mango) 芒果
mango butter 芒果脂
mango gum 芒果胶
mangosa oil 楝子油
mangosteen oil 山竹果油
mangosteen tree 倒捻子
mangostin 楝素；倒捻子素
mangrove 红树；栲树
mangrove bark 栲树皮
mangrove bark extract 栲树皮浸膏
mangrove extract 栲树(皮)栲胶
manhole (进)入孔；人孔
man-hole opening 人孔口
man hour 工时
manicure preparations 指甲化妆品
manifold 歧管；多支管；复式接头；管线
manifold clamp 歧管夹；管托架
manifold depression 歧管抽空；支管抽空
manifold exhaust 歧管排气

manifold heat control 歧管加热控制
manifold ignition test 歧管点火试验
manifolding 歧管装置
manifold paper 复印纸；打字纸〔俗〕
manifold pressure 歧管压力；排出管道的压力
manifold vacuum 歧管真空；排出管道的真空
manifold valves 歧管阀；汇流阀
manihoc(=manihot) 木薯
manihot(=manihoc) 木薯
manihot oil 木薯油
manihot rubber 木薯橡胶
manila 蕉麻纸；马尼拉(麻)纸
manila board 蕉麻纸板
manila copal 马尼拉砧柏
manila elemi oil 马尼拉榄香脂油
Manila gum 马尼拉树胶
manila hemp ①麻蕉〔植〕②蕉麻〔指纤维〕
manila kopal 马尼拉砧柏
manila paper 蕉麻纸；马尼拉纸
manila resin 马尼拉树脂
manila rope 蕉麻绳；(白)棕绳
manila wrapping 蕉麻包装纸
manila writing 蕉麻写字纸
manioc(=mariocca) ①木薯②木薯淀粉
maniocca(=manioc) ①木薯②木薯淀粉
manipulation (用手)操作；手术
manipulator 机械手*；操纵者；控制器
manjak 硬化沥青
manjakite 黑云紫苏岩
man-made 人造的
man-made cellulose fibre 人造纤维素纤维
man-made element 人造放射性元素*
man-made fiber 人造纤维
man-made fibre 人造纤维
man-made flock 化学纤维屑
man-made pollutant 人为污染物
man-made rubber 合成橡胶
manna ash 花白蜡树
manna croup 碎麦米
mannan 多缩甘露糖；甘露聚糖
mannase 甘露聚糖酶
Mannheim absorption system 曼海模吸收装置
Mannich base 曼尼希碱*
Mannich base complex 曼尼希基配合物
Mannich condensate 曼尼希缩合物
Mannich reaction 曼尼希反应
mannide 二缩甘露醇
manninotriose 三聚甘露糖；甘露三糖

mannitan 一缩甘露醇
mannitan hexadecanoate 十六(烷)酸失水甘露糖醇酯
mannitan oleate 油酸失水甘露糖醇酯
mannite(=mannitol) 甘露糖醇 $HOCH_2(CHOH)_4CH_2OH$
mannite fatty acid ester 甘露醇脂肪酸酯
mannitic acid 甘露糖酸 $CH_2OH(CHOH)_4COOH$
mannitoboric acid 甘露醇合硼酸
mannitol 甘露糖醇 $CH_2OH(CHOH)_4CH_2OH$
mannitolboric acid 甘露醇合硼酸
mannitol ester 甘露糖醇酯
mannitol hexaacetate 甘露糖醇六乙酸酯;六乙酰化甘露糖醇 $C_6H_8O_6(COCH_3)_6$
mannitol hexanitrate(=hexanitro-mannitol) 六硝基甘露醇;甘露醇六硝酸酯 $(O_2NOCH_2)_2[CH(ONO_2)]_4$
mannitol laurate 月桂酸甘露糖醇酯
mannitose 甘露糖
mannoheptitol 甘露庚糖醇
mannoheptonic acid 甘露庚糖酸
mannoheptose 甘露庚糖 $C_7H_{14}O_7$
mannoheptulose 甘露(型)庚酮糖
mannohydrazone 甘露糖腙
mannonic acid 甘露糖酸 $CH_2OH(CHOH)_4COOH$
mannopeptin 甘露糖肽素;甘露聚糖肽
mannopyranose 吡喃甘露糖
mannosamine 甘露糖胺
mannosaminic acid 甘露糖氨酸
mannosans 甘露聚糖
mannose 甘露糖 $C_5H_{11}O_5CHO$
mannose phenylhydrazone 甘露糖苯腙 $C_6H_{12}O_5=NNHC_6H_5$
D-mannose-6-phosphate D-甘露糖-6-磷酸
α-mannose-1-phosphate α-甘露糖-1-磷酸
mannosidase 甘露糖苷酶
mannoside 甘露糖苷
mannosidosis 甘露糖苷过多症
mannosidostreptomycin 甘露糖链霉素
mannotriose 甘露三糖
mannuronic acid 甘露糖醛酸
manocryometer 加压熔点仪;加压熔点计
manograph 压力记录器
manometer U 型管测压计*
manometer flask 测压计反应瓶
manometer pressure 压力计压力
manometer relay 压力继电器
manometer tube 压力管
manometer tube gland 压力管封闭器
manometric analysis 测压分析(法)

manometric assay 测压法
manometric bomb 测压弹
manometric efficiency 压缩效率;测压效率
manometric fluid(=manometric liquid) 测压液
manometric liquid(=manometric fluid) 测压液
manometric method 测压法
manometric thermometer 压差温度计
manometry 测压法
manool 泪柏醇;泪杉醇
manoscope 流(体)压(力)计;气体密度计
manoscopy 流压术;气体容量分析
manostat 流(体)压(力)器;恒压器;恒压装置
manoyl oxide 泪柏醚
man-rem 人体-雷姆〔辐射剂量单位〕
Mansfeld copper process 曼斯菲尔德炼铜法
Mansfeld's electrode 曼斯菲尔德电极
man-Sv 人体-希沃特〔辐射剂量单位〕
mantissa 尾数
mantle ①罩②(白光)纱罩
mantle carrier 纱罩托
mantle fiber 灯罩纤维
mantle heater 加热用罩;罩式电热器
Manton-Gaulin mill 蒙顿胶体磨
manual ①手动的②手工的
manual airblast 手动空气(鼓风)喷砂;人工空气(鼓风)喷砂
manual assembly 手工成型;手工贴合
manual control 手调;手动控制
manual controller 手动控制器
manual control mode 手动控制模式
manual electric arc welding 手工弧焊
manual finishing 手工清理
manual gas sampling valve 手动气体进样阀
manual injector 手动进样器
manual lubricator 手工润滑
manual mixture control 混合物(组成的)手工调管
manual operation 手工操作
manual plotting 手工做图
manual polarograph 手动式极谱仪
manual powder spray gun 手动粉末喷枪
manual priming (泵的)手工注水
manual proportioning 手工称量
manual range selector 人工量程选择器
manual scan 手工扫描
manual setting 手工装配
manual spray gun 手动喷枪
manual truck 手推车
manual welding 手工焊

manufacture ①制造②制造品③制造厂；工厂
manufactured graphite 人造石墨
manufactured yarn 人造丝
manufacturer's mark 工厂商标
manufacturer's trials 工厂试验
manufacturing assembly drawing 制造装配图
manufacturing automatic system(MAS) 加工自动系统
manufacturing method 生产方法
manuka oil 纤精澳洲茶树油
manumycin 手霉素
manure ①肥料②粪③施肥
manure bate 粪便软化剂
manure salts〔复〕 肥料盐
manure spreader 施肥机
manurial value 肥料价值
manuring 施肥
manus tester 石油闪点测定器
manway 人孔
mapharsen(=oxophenarsine hydrochloride) 马法胂；盐酸2-氨基-4-亚胂酰苯酚
mapharside 2-氨基-4-亚胂酰苯酚盐酸盐
maple ①槭(树)属②槭树
maple sugar 槭糖
maple syrup 槭糖浆；槭树汁
maple tannin 枫树单宁；槭树单宁
mappine(=bufotenine) 蟾蜍色胺
maprotiline 马普替林〔药〕
mar-aging 改变钢的马丁体结构法
Marangoni effect 马兰各尼效应〔因界面张力梯度引起的液流动〕
Maranham balsam 马栾汉香脂
maraniol(=4-methyl-7-ethoxy coumarin) 核桃素；4-甲基-7-乙氧基香豆素
marasin 小皮伞菌素
marasmic acid 小皮伞菌酸
Marasse synthesis(=Kolbe-Schmitt carbonation) 科尔比-施密特羧化法；马拉斯合成法
marasuric acid 马拉酸
marble 大理石；云石
marbled glass 大理石纹玻璃
marbled glazed paper 斑纹蜡光纸〔俗〕
marbled glazed sulfite paper 斑纹蜡光亚硫酸盐纸
marbled grain 粗粒面
marbled lacquer 大理石花纹漆器；云纹漆器〔大漆〕
marbled soap 斑纹皂
marble dust 大理石粉；云石粉
marble figure coating 仿大理石纹涂装法；石纹涂装法
marble flour 大理石粉

marbling ①施加大理石纹②似大理石纹色彩③皮上染花纹
marbling print 云纹印涂；大理石纹印涂
marbling varnish 云纹清漆；大理石纹饰清漆
marcasite 白铁矿〔矿〕
marcescin(=marcesin) 黏赛菌素
marcesin(=marcescin) 黏赛菌素
Marchand tube 马尔昌管〔一种 U 形氯化钙管〕
Marchese copper process 马尔契斯炼铜法
marcitine 腐肤碱
marcomycin 马可霉素
Marcy mill 格子排料式球磨机
marfacing 用改变马丁体结构法产生硬面
margarate ①十七(烷)酸盐 $C_{16}H_{33}COOM$②十七(烷)酸酯 $C_{16}H_{33}COOR$
margaric acid(=heptadecanoic acid) 十七(烷)酸；珠光脂酸 $CH_3(CH_2)_{15}COOH$
margaric aldehyde(=heptadecylaldehyde) 十七(烷)醛 $CH_3(CH_2)_{15}CHO$
margarin 十七烷酸三甘油酯；人造奶油
margarite ①〔旧名，指〕珍珠②珍珠云母〔矿〕
margaron 十六醚 $(C_{16}H_{33})_2O$
margaronitrile 十七腈 $C_{17}H_{33}N$
margarosanite 针硅酸钙铅矿
margin ①边缘②界限③储备(量)④安全系数
marginal effect 边缘效应；边界效应
marginal utility 边际效用
margin of error 误差限度；误差界限
margin of safety 安全限度
margin of stability 稳定限度
margin phenomena(=edge phenomena) 边沿现象
margosa 楝树
margosa oil 楝树油
margosic acid 楝树酸
margosine 楝树碱
Margules's equation 马古利斯方程〔气-液平衡关系〕
marialite 钠柱石
maridomycin 麦里多霉素
Marie Davy cell 玛利戴维电池
Marignac salt 马里纳克盐〔硫酸锡钾〕
marigold ①金盏花②万寿菊
marigold absolute 金盏花净油
marihuana 印度大麻
marijuana 大麻
marina water 海水
marine ①海生的②苦参素(苦参总碱)〔药〕
marine animal oil 海生动物油
marine antifoulant 海洋防污剂；海洋防污毒料

marine anti-fouling paint　海洋防污漆
marine atmosphere　海洋大气
marine bunker facilities　海港油库设备
marine corrosion　海洋腐蚀
marine corrosion rate　海洋腐蚀速度
marine ecosystem　海洋生态系统
marine exposure facilities　海洋暴晒站
marine finish　船舶涂料
marine grey　灰漆
marine growth preventing system　防海洋附着生物涂料体系
marine oil　船用照明油
marine oiler　海洋滑油器
marine paint　船舶漆
marine pollution　海洋污染
marine product　①海产②航海用具
marine sample　海洋样品
marine situation　海洋环境
marine soap　海水皂
marine tankage　油船载重量
marine terminal　海运油库；航海站
marine varnishes　①海上构筑物用漆②船舶用清漆
marinobufagin　海蟾蜍毒精
maripa fat(=maripa oil)　棕榈油
maripa oil(=maripa fat)　棕榈油
maritime climate　海洋性气候
maritimein　金鸡菊苷
maritime pine　海岸松；南欧海松
maritimetin　金鸡菊素；海生菊苷
marjoram　甘牛至草
marjoram oil　甘牛至油
mark　①标记；记号②特性③限度；界线④痕；迹
marked compound　标记化合物
marked dropping bottle　有标记的滴瓶
marker　标记器
marker chromosome　标记染色体
marker enzyme　标志酶
marker-free immunoassay　无标记免疫分析
marker method(=internal standard method)　标示法；内标法
marker peak　标志峰
marketable pollution permits　排污权交易
marketable yarn　商品纱
marketed coating　商品涂料；应市涂料
marketing station　商业基地
mark graduated to deliver　排量标线；排出量刻度线
marking　①作记号②条纹
marking apparatus　(显微镜)记号器

marking felt　打印毡
marking-in　划样
marking ink　打印油墨
marking nut　如树果
marking nut oil　印果油
marking of containers　容器商标
marking-off　划线
marking out　划线
marking paint　①标志漆②路标涂料
Markovitz equation　马科维茨方程
Markovnikov's rule　马尔科夫尼科夫规则*
marks of parasitic plants　寄生植物痕迹
marlaceous　泥灰(岩)的
marl loam　泥灰壤土
marlstone　泥灰石
marly　泥灰(岩)的
marly limestone　泥灰石灰石
maroon lake　紫红色淀
maroon pigment　紫红颜料；栗红颜料
maroon toner　紫红调色剂；紫红色原；栗红色料
maroti fat　海南大风子油
maroti oil　海南大风子油
mar proof　耐磨损性；耐擦伤性
marproofing additive　抗擦伤添加剂；耐磨损添加剂
mar-proofing agent　防划伤助剂
Marquardt index　玛夸特指数〔用红外线吸收法测定酚醛树脂固化程度的指标〕
marquench　分级淬火；淬火时效；马氏体淬火
marquetry　①镶嵌细工②镶嵌工艺品③用玻璃镶嵌的图案
Marquis test　马奎斯试验
mar resistance　耐擦伤性
mar-resistant surface　抗划伤的表面；抗擦伤的表面
marrianolic acid　马连酸；马冉醇酸
married fibre clump　(纤维条内)切口黏结纤维块
marri kino　美叶胺树胶
marring　①刮伤②刮伤性
marrubenol　夏至草醇
marrubiic acid　夏至草酸
marrubiin　夏至草素
marrubium oil　夏至草油
Marseilles soap　马赛皂；橄榄油皂
Mars green　橄榄色铁绿〔铁蓝与天然铁黄的混合物〕
Marshall refiner　马歇尔精制机
marsh buggy tyre　湿地轮胎
marsh gas　沼气；甲烷
marsh mallow root　药用蜀葵根
marsh mallows　药蜀葵

Marsh tea oil　喇叭茶油
Marsh test　马什试验*
marsh trefoil　睡菜
mars pigment　合成氧化铁颜料
Martens densitometer　马滕斯显像密度计
Martens hardness tester　马滕斯硬度计
Martens illuminator　马滕斯照明器
Martens photometer　马滕斯光度计
Martens polarization photometer　马滕斯偏(振)光光度计
Martens sclerometer　马滕斯硬度计
Martens spectroscope　马滕斯分光镜
martial ethiops　磁性氧化铁
martindate abrasion tester　耐磨损测试机
Martin density meter　马丁密度计；马丁比重计
Martin distribution　马丁分配；马丁分布
Martin distribution apparatus　马丁分配装置
Martinet isatin synthesis　马提内特靛红合成法
Martin gas density balance　马丁气体密度天平
Martin's centrifuge　马丁离心器
Martin's filter　马丁过滤器
Martin's test　马丁(热稳定性)试验
martite　假象赤铁矿
Martius yellow　马休黄；2,4-二硝基萘酚钠 $C_{10}H_5O_5N_2Na$
martonite　马当炸药；溴丙酮
marver　乳光玻璃板
mary bush　金盏花
mascagnine　硫铵石
Maser　①微波量子放大器②微波激射器
mashing　①捣碎②浆末③软块
mashing off　煮浆
mashing tub(=mashing tun)　芽浆桶；糖化桶
mashing tun(=mashing tub)　芽浆桶；糖化桶
mash kettle　芽浆锅
mash tub(=mash tun)　芽浆桶；糖化桶
mash tun(=mash tub)　芽浆桶；糖化桶
mashy　压碎的
mask　①面具②掩蔽
maskant　①遮蔽剂；掩蔽剂②遮蔽涂层〔蚀刻时，涂于不需腐蚀的部位，保护原形〕
masked　掩蔽的；隐藏的
masked carbanion　掩蔽碳负离子*
masked element　掩蔽元素
masked group　掩蔽(原子)团
masked isocyanate(=blocked isocyanate)　封闭(型)异氰酸酯
masked radical　掩蔽基
masking　掩蔽
masking action　掩蔽作用

masking agent　掩蔽剂〔分析〕
masking and demasking　隐蔽和解蔽
masking coating　遮蔽涂料；掩蔽涂料
masking compound　掩蔽化合物
masking effect　掩蔽效应
masking index　掩蔽指数
masking material　遮蔽(掩蔽)材料
masking method　掩蔽法
masking paste　防渗碳涂料
masking power　遮盖力〔瓷釉遮盖坯体颜色的能力〕
masking ratio　掩蔽比〔金属配合物总浓度对游离金属离子浓度的比〕
masking reagent　掩蔽(试)剂
masking shield　遮蔽板
Masland single filament flex life tester　梅斯兰德单丝疲劳度测试仪
masonite　汽压处理的木纤维
masonry　①圬工②砖石建筑；(砖、石)砌筑工程
masonry finish　砖石建筑漆；圬工涂料
masonry paint　圬工漆；砖石建筑漆
masonry panel　圬工灰泥样板
masonry primer　圬工底漆；砖石建筑底漆
masonry-stone　石圬土；石砌工程
masonwork　砖石工程；圬工砌筑物
masout　重油
mass　①块；堆②大量③质量
mass ablation rate　重量烧蚀速率；重量消融速率
mass absorption coefficient　质量吸收系数〔单位:厘米^2/克〕
mass action　质量作用；浓度作用
mass action law　质量作用定律；浓度作用定律
mass action principle　质量作用原理；浓度作用原理
mass analyzed ion energy spectrometer　质量分析离子能谱仪
mass analyzed ion energy spectrometry　质量分析离子能谱法
mass analyzed ion kinetic energy spectrometer　质量分析离子动能谱仪
mass analyzed ion kinetic energy spectrometry　质量分析离子动能谱法
mass analyzed ion kinetic energy spectrum　质量分析离子动能谱
mass analyzer　质量分析器
mass analyzer detector　质量分析检测器
Mass and van Krevelen mechanism　马斯-范克里弗伦机理*
mass assignment　质量数确定
mass attenuation coefficient　质量衰减系数
mass balance　物料衡算

mass-basis response   质量(基础)响应
mass-basis sampling   质量取样；质基取样
mass change in reactions   反应的质量变化
mass charge ratio   质荷比
mass chromatogram   质量色谱图
mass chromatography   质量色谱(法)
mass coloration   纺前染色；本体染色
mass colour(=masstone)   ①主色；主色相；表光；表观色相〔单一颜料色浆涂布至刚好完全盖住背景颜色的厚度时，借助反射光所观察到的颜色〕②原色；浓色③墨色〔油墨〕
mass colouring   本体着色
mass concentration   质量浓度
mass conservation   质量守恒；质量不变
mass crystallization   大量结晶
mass defect   质量亏损
mass deficit   质量亏损
mass detection   质量检测
mass detection limit   质量检测限
mass detector   质量检测器
mass difference   质量差
mass diffusion   质量扩散
mass discrimination   质量歧视效应*
mass dispersion   质量色散
mass distribution function   质量分布函数
mass distribution of fission products   裂变产物质量分布
mass distribution ratio   质量分配比
mass doublet   质量双线
mass-dyed   本体染色的；纺前染色的
mass dyeing by injection   注射法整体染色；注射法纺前染色
massecuite(=fillmass;fullmasse)   糖膏
mass effect   质量效应
mass effect of electron multiplier   电子倍增器的质量效应
mass energy cycle   质能循环
mass energy relation   质能关系式
mass finishing   大量整面；大量抛光
mass flow   质量流量
mass flow control   质流控制
mass flow lifting   (固体粒子)密相提升
mass flowmeter   质量流量计
mass flowrate   质量流速
mass flow rate sensitive detector   质量流量敏感型检测器
mass flux   质(量)通量
mass flux ratio   质量通量比；质量流量比
mass flux vector   质(量)通量向量
mass fragmentogram   质量碎片图
mass fragmentography   质量碎片谱法；碎片质谱法

massicot   铅黄；黄丹   PbO
mass indicator   质量指示器
massing   块化
mass integral detector   质量累积检测器
massive article   大型制品；大规格制品
massive-cluster impact ionization   簇轰击电离
massive deposit   块状矿床；块状沉积(物)
mass lamination   大型压板(层压)
mass law   质量定律
mass-law effect   质量定律效应
mass loading   质量负载；重量负载
Massmann furnace   马斯曼高温炉
Massmann high-temperature furnace   马斯曼高温炉
Massmann type heated graphite atomizer   马斯曼型加热石墨(炉)原子化器
mass marker   质量数指示器
mass marking   质量定标
mass measurement   质量测量
mass number   (原子)质量数
massoia bark oil   香厚壳桂皮油
massoia lactone   马索亚内酯
massoia leaf oil   香厚壳桂叶油
mass output rate   出料率；物料输出速度
mass peak   质量峰
mass percent   质量百分数
mass permanence   整体特性；质量特性
mass per unit area   单位面积质量
mass pigmentation   本体着色
mass pigmented fiber   本体着色(的)纤维；纺前着色纤维
mass polymer   本体聚合物
mass polymerization   本体聚合(作用)
mass production   大量生产
mass range   质量范围
mass rate of burning   燃烧的质量速度
mass ratio   质量比
mass resistivity   块体电阻率
mass resolution   质量分辨
mass scale   质量数标度
mass scale assignment   质量标尺指认
mass scan line   质量扫描线
mass scanning   质量扫描
mass selective detector(MSD)   质量选择(性)检测器
mass sensitive quantity   质量敏感值〔随质量而变的值〕
mass separation   质量分离
mass separation/mass identification mass spectrometer   质量分离-质量鉴定质谱仪
mass separation/mass identification spectrometry   质量分离-质量鉴定法*

mass separation-mass spectra characterization 质量分离的质谱检测技术
mass spectra 质谱
mass spectral 质谱的
mass spectral search system 质谱检索系统
mass spectrogram 质谱(图)
mass spectrograph 质谱仪*
mass spectrography 质谱法
mass spectrometer 质谱仪
mass spectrometric analysis 质谱分析
mass spectrometric computation 质谱分析计算
mass spectrometric data 质谱数据
mass spectrometric detection 质谱检测
mass spectrometric detector 质谱检测器
mass spectrometric measurement 质谱测量
mass spectrometric thermal analysis 质谱热分析
mass spectrometry ①质谱法*②质谱分析法
mass spectrometry data centre 质谱分析数据中心
mass spectrometry-mass spectrometry(MS-MS) 质谱-质谱法；质谱-质谱联用法
mass spectrophotometer 质谱分光光度计
mass spectroscope 质谱仪
mass spectroscopic detector 质谱检测器
mass spectroscopy ①质谱学②质谱分析③质谱(分析)法
mass spectrum digitizer 质谱数字转换器
mass spectrum line 质谱线
mass splitting 质谱分裂
mass stability 质谱稳定度
mass standard 质量标样
mass stopping power 质量阻止本领*
mass synchrometer 同步质谱计
mass thickness image 质厚衬度像
mass-to-charge ratio 质荷比（m/e）
mass tone ①主色；主色相；表光；表观色相②原色；浓色③墨色
masstone color 主色；浓色；原色〔未经冲淡的颜色〕
masstone paint 原色漆；单色漆；浓色漆〔单一彩色不含白色颜料的漆〕
masstone pigment 主色颜料〔未用白色颜料冲淡的颜料〕
mass transfer 传质；质量传递
mass-transfer by convection 对流传质
mass-transfer by diffusion 扩散传质
mass-transfer by migration 电迁移传质
mass transfer coefficient 传质系数
mass transfer effect 传质效应
mass transfer factor 传质因子
mass transfer process 传质过程
mass transfer resistance 传质阻力
mass transfer term 传质项
mass transfer theory 质量传递理论；传质理论
mass-transfer valve plate 传质浮阀塔盘
mass transfer velocity 传质速度
mass transport mechanism 质量传递机理；传质机理
mass unit 质量单位
mass velocity 质量速度
mass vibrometer 质量振动计；质量测振计
mass-yield distribution 质量-产额分布
masterbatch ①母炼胶〔橡胶〕②母粒；浓色体
masterbatch color 母炼胶色料
master batching 母炼胶的制备
masterbatch technique 浓色体技术
master color paint 梘色漆〔黄铜色油漆〕
master creep curve 叠合蠕变曲线
master gear 大齿轮
master grind 原色浆；研磨(色)浆；主色浆
master meter 主表；标准仪表；检验表
master mix 高低速双轴混合分散机
master model 原型；标准模型
master plan ①总平面图②总计划；总体规划
master plate 标准干板
master sample 标准样品；标准试样
master-slave manipulator 随动式机械手*
master valve 主阀；总阀
master variable 主变量
master viscosimeter 毛细管型主黏度计；大型(毛细管)黏度计
masterwort oil 欧前胡油
mastic ①黏合剂；胶黏水泥②厚浆涂料〔涂料〕③腻子④乳香
mastic acid 乳香酸
mastic adhesive 腻子胶黏剂
masticadienonic acid 乳香二烯酮酸
masticated copal 塑炼珀把脂；素炼珀把(树)脂
mastication 素炼*
mastication of natural resin 天然树脂的素炼(塑炼)
mastication of resin 树脂塑炼；树脂捏炼
masticator ①撕捏机②素炼机
mastic coating 厚浆涂料
mastic gum ①乳香；玛琦脂②黏合剂
mastiche oil(=mastix oil) 乳香油
mastic-impregnated tape 浸胶胶带
masticinic acid 乳香酸 $C_{23}H_{36}O_4$
masticolic acid 乳香醇酸 $C_{23}H_{36}O_4$
masticonic acid 乳香酮酸 $C_{23}H_{48}O_4$

mastic resinoid 乳香树脂
mastics ①乳香②厚涂涂料；厚浆涂料
mastic sound deadener 玛琍胶(胶泥)型隔音涂料〔涂于车身底部兼有隔音、防水、防尘效果的涂料〕
mastic tree nut oil 乳香油
mastic type caulking material 胶泥型填缝剂
mastix 乳香
mastix oil(=mastiche oil) 乳香油
mast paint 桅杆漆
masut(=mazout; mazut) 重油
mat 毡(片)
matabie 金银粉印花
matabi lactone 马塔内酯
matai resinol 罗汉松树脂酚 $C_{20}H_{22}O_6$
matamycin 马塔霉素
matastable ion 亚稳离子
mat base 垫层
mat binder (玻璃纤维)毡片黏结剂
mat black 无光黑
mat blue 无光蓝
matchamycin 绿茶霉素
matched cam beating up mechanism 共轭凸轮打纬机构
matched column 匹配柱
matched-die mo(u)lding (热成型)对模成型；配套硬模造型；配对硬模压塑法
matched die press-forming 阴阳模成型
matched-mold thermoforming 对模热成型
match factor 匹配因子
matching 摹仿
matching edge 叠合边
matching of stages (压缩)级的配合
matching parts 配件；配合件
matching solubility performance 匹配溶解特性
matching stain 羊毛染色剂；酸性染料
matching surface 配合面
matching test 对比试验
match label paper 火柴标签纸
matchless 不匹配
match-making machinery 火柴制造机(器)
match-mo(u)ld thermoforming 合模热成型；对模热成型
match paper 火柴纸
match wax 火柴蜡
mat coating(=matte coating) 无光涂料
mated tyres confectioning 装配成对轮胎
material ①(材)料；物料；质料②资料③物资④军用器材；军用物资
material balance 物料平衡*
material balance area(MBA) 物料衡算区；物料平衡区；物质收支区
material balance data 物料平衡数据
material balance method 物料平衡法
material constant 材料常数
material control 原材料管理
material control area(MCA) 物料控制区
material derivative 实质导数
material failure 材料失效
material frame indifference principle 实质标架无异原理
material handling equipment 材料装运设备
materialization 物质化(作用)
material needle valve (喷枪)涂料针形阀
material particle (物)质粒(子)
material principal direction 材料主方向
materials accounting method 物料统计法；物料计量法；物料衡算法
material scrap 边角料；残料
material symmetry 实质对称性
material testing machine 材料试验机
material transfer 物质传递
material unaccounted for(MUF) 物料不明损失量；去路不明损失量；物料不衡算量
material use efficiency 材料利用率；材料使用效率
material utilization 原材料利用(率)
mat finish(=matte finish) ①消光处理；无光处理②磨褪〔大漆〕
mat foundation 席形基础；底板基础
mat gloss 平光
mathematical correction method 数学校正法
mathematical expectation 数学期望值
mathematical idealization 数学理想化
mathematical modeling technique 数学模型技术
mathematical model of water quality 水质数学模型
mathematical treatment 数学处理
mathematic model 数学模型
Matheson manifold system 马西森多支管系统
matico 狭叶胡椒
matico camphor 胡椒叶脑
matico oil 狭叶胡椒油
matildite 硫银铋矿
mat lay-up 毡片叠铺
matlockite 氟氯铅矿
matraite 丙硫锌矿；锥锌矿
matrass 蒸馏瓶
matricaria camphor 母菊脑 $C_{10}H_{16}O$
matricaria ester 母菊酯
matricarianol 母菊醇
matricarin 母菊苷

matricin 母菊素
matrine 苦参碱
matrinidine 苦参尼定
matrix ①矩阵②模型③杂矿石；脉石④基块
matrix absorption 基体吸收
matrix assisted laser desorption ionization mass spectroscopy (MALDIMS) 基体辅助激光解吸电离质谱法
matrix-assisted laser desorption-ionization time of flight (MALDI-TOF) 基体辅助激光解吸-离子飞行时间质谱
matrix assisted laser desorption ionization time-of-flight mass spectrography(MALDI-TOFMS) 基体辅助激光解吸电离-飞行时间质谱法
matrix assisted laser desorption ionization-time of flight mass spectrometer 基质辅助激光解吸电离-飞行时间质谱仪
matrix assisted laser desorption mass spectrometry (MALD-MS) 基体辅助激光解吸质谱法
matrix assisted laser desorption- time-of-flight mass spectrography (MALDTO-FMS) 基体辅助激光解吸-飞行时间质谱法
matrix board 字型纸板
matrix correction 基体校正
matrix-dilution method 基体稀释法
matrix effect 基体效应
matrix element 矩阵元
matrixes for embossing 压花板
matrix fiber 基质纤维
matrix interference 基体干扰
matrix inversion 矩阵求逆
matrix inversion procedure 矩阵求逆程序
matrix ion-selective electrode 基体离子选择电极
matrix isolation 基体隔离(法)
matrix isolation spectroscopy 基体分离谱法
matrix isolation technique 基质隔离技术
matrix matching 基体匹配
matrix match method 基体匹配法
matrix material 复合材料〔由树脂与增强材料构成〕
matrix mechanics 矩阵力学
matrix modification 基体改性；基体改进
matrix modifier 基体改进剂
matrix of polymer chain 聚合物链模板；聚合物链基体
matrix polymer(=replica polymer) 复制聚合物；母体移植聚合物
matrix polymerization(=replica polymerization) 复制聚合；母体移植聚合；模拟聚合
matrix reference material 基体标准物质
matrix resin 基体树脂；母体树脂

matrix solid phase dispersion method 基质固相分散提取法
matrix-spun yarn 基质型复合丝；双成分复合丝
matrix suppression 基体抑制
matromycin(=oleandomycin) 竹桃霉素
matsudakeol 松蕈醇
matsu-take alcohol(=3-octenol) 松蕈醇；3-辛烯醇
matt 无(光)泽的；无光的
Mattauch-Herzog double focusing mass spectrometer 马-赫双聚焦质谱仪
Mattauch-Herzog type mass spectrometer 马-赫型质谱仪
mattblack finish 无光黑膜〔粉末渗锌膜〕
matte ①冰铜②无(光)泽的；乌泽的
matte bath 冰铜浴
matte coating(=mat coating) 无光涂料
matte effect 闷光效应
matte finish ①磨褪〔大漆〕②打毛
matte glaze 乌泽釉；乌光釉
matt enamel 无光瓷漆
matte polycaprolactam 无光聚己内酰胺
matter 物质
matter transport 传质
matter waves 物质波
matte side 粗糙面
matte surface ①无光表面②磨褪表面〔大漆〕
matte varnish 无光清漆
matt fibre 无光纤维
matt finishing ①涂无光漆②面漆磨褪(光泽)
Mattiello viscosity 马蒂耶罗黏度〔用于测定印刷油墨〕
matting 消光
matting agent 消光剂
matting effect 消光效应
matting pigment 消光颜料
mattle gum 澳大利亚金合欢胶
matt paint 无泽涂料；无光涂料
matt salt 氟化氢铵
maturation ①老化；陈化②熟化
maturation factor 成熟因子
maturation period 熟化期
mature 成熟的
matured state 熟态
mature wood 成熟材
maturimeter 成熟度测试仪
maturin balsam 马嫩香脂
maturing ①成熟作用②成熟的③熟化〔胶乳〕④陈化〔可发聚苯乙烯珠粒〕
maturing of petroleum 石油的熟化；石油的演化
maturity 熟化度
mauguinite 毛勾炸药；毛勾那特；氯化氰

Maumené number　莫默内数
Maumené test　莫默内试验〔区别干性油与不干性油的试验〕
mauve　①淡紫的②苯胺紫(染料)
mauvein(=aniline purple)　苯胺紫
mauvein(e)　(碱性)木槿紫；苯胺紫
maw oil　罂粟油
maxima　畸峰
maximal　达最大值的
maximal draw ratio　最大拉伸比
maximal work　最大功
maximum　①最大的；最高的；最多的②最大量；最高量；最多量；最大值；最多数；最高数③最高点④畸峰〔极谱〕
maximum air concentration　最大空气浓度
maximum allowable concentration　最高容许浓度
maximum allowable error　最大容许误差
maximum allowable operating temperature　最大允许操作温度
maximum allowable pressure drop　最大容许压力降
maximum allowable sample size　最大允许试样量
maximum allowable working pressure　最高允许工作压力
maximum and minimum thermometer　最高最低温度计
maximum apparent error　最大表观误差
maximum attenuation ratio　最大拉细比(率)
maximum boiling point　最高沸点
maximum boiling-point mixture　最高沸点混合物
maximum bubble method　最大气泡压力法
maximum bubble pressure method　最大泡压法*
maximum clearance　最大间隙
maximum critical value　极大临界值
maximum delivery pressure　最高排出压力
maximum detection limit　最大检测限
maximum effective range　①最大有效范围②最大有效射程
maximum efficiency　最大效率
maximum elongation　扯断伸长率
maximum entropy method(MEM)　最大熵法
maximum error　最大误差
maximum heat resistance　最高耐热性；最大耐热性
maximum humidity of compressed air　压缩空气的最大许可含水量
maximum interval before overcoating　(涂下一道漆的)最长间隔时间
maximum likelihood estimation　最大似然估计
maximum likelihood estimator　极大似然估计量
maximum likelihood method　极大似然法
maximum nonseizure load　最大无卡咬负荷

maximum of the first kind　第一类极大
maximum of the second kind　第二类极大
maximum oil content of compressed air　(压缩气的)最大许可含油量
maximum operating temperature　最高操作温度
maximum output　①最高产率②最高输出
maximum permeability　最大磁导率
maximum permissible concentration(MPC)　最大允许浓度
maximum permissible dose(MPD)　最大允许剂量
maximum permissible exposure(MPE)　最大允许曝射量；最大允许照射量
maximum permissible intake(MPI)　最大允许摄入量
maximum permitted concentration　最大容许浓度
maximum phenomenon　极大现象
maximum pitch time　瓶颈工序；难度工序
maximum power　最高功率
maximum power temperature program　最大功率升温
maximum principal stress　最大主应力
maximum productivity　最大生产率
maximum range　(粒子的)最大行程
maximum reinforcement　最大补强
maximum relaxation time　极大松弛时间
maximum residue limit　最大残留允许量
maximum sampling frequency　最大采样频率
maximum selectivity　最大选择性
maximum soil field capacity　土壤田间最大持水量
maximum specific growth rate　最大比增殖速度
maximum-speed mixture　最高燃速混合物〔具有最大燃烧速度的混合物〕
maximum stress　最大应力
maximum suppressor　极大抑制剂
maximum temperature　最高温度
maximum thermometer　最高温度计
maximum throughput　最大排出量；最大处理量
maximum tolerance　①(毒剂)最大容许量②最大允差
maximum urinary concentration　尿中最大浓度
maximum valence(=maximum valency)　最高价
maximum valency(=maximum valence)　最高价
maximum work　最大功*
maxioffoxacin　莫西沙星〔药〕
maxipen　苯氧甲基青霉素
maxite　硫碳铅矿
maxivalence(=maximum valence)　最高价
Maxton screen　旋转筛
maxwell　麦克斯韦〔磁量单位〕
Maxwell and Harrington impact test　麦克斯韦-哈林顿冲击试验
Maxwell body　麦克斯韦体〔弹黏体〕

Maxwell-Boltzmann distribution 麦克斯韦-玻耳兹曼分布
Maxwell constant 麦克斯韦常数
Maxwell disc (麦克斯韦)配色盘
Maxwell disk 麦克斯韦混色盘
Maxwell distribution law 麦克斯韦分布定律
Maxwell equation of state 麦克斯韦状态方程
Maxwell fluid 麦克斯韦流体
Maxwell isotropic viscous medium 麦克斯韦各向同性黏性介质
Maxwell liquid 麦克斯韦液体
Maxwell model 麦克斯韦模型
Maxwell orthogonal rheometer 麦克斯韦正交流变仪
Maxwell relation 麦克斯韦关系
Maxwell relaxation equation 麦克斯韦松弛方程
Maxwell's disk 麦克斯韦混色盘
maxwhite 白革涂饰剂
max-width 最大宽度
may apple 鬼臼根〔药〕
may blossom 山楂花
may-chang oil 山苍果油
mayer 迈尔〔热容量单位〕
Mayer's hemalum 迈尔染色剂
may flower 桂竹香
may weed 淡甘菊
mazout(=masut) 重油
mazut(=masut) 重油
Mazzoni process 马佐尼(连续制皂)法
McBain-Baker balance 麦克贝恩-贝克秤
McBain centrifuge 麦克贝恩离心机
McCabe-Thiele method 麦凯布-蒂尔法
McCall's equation 麦考尔方程〔介电常数与比重的线性关系式〕
McCance reagent 麦克坎斯试硫剂
McFadyen and Stevens aldehyde synthesis 麦克法迪恩和斯蒂文斯醛合成法
McGill metals 铝青铜合金
McLafferty rearrangement 麦克拉佛特重排
McLaurin process 麦克罗林煤的低温碳化法
McLeod gauge 麦克劳德真空规
McLeod pressure gauge 麦克劳德低压计〔测量1毫米汞柱以下〕
McLeod vacuum gauge 麦克劳德真空规;麦氏真空规;麦氏真空计
McMahon packing 麦克马洪填料;金属网鞍形填料
McReynolds constant 麦克雷诺兹常数
McReynolds phase constant 麦克雷诺兹相常数
Me〔符号〕 甲基

meadow ore 沼铁矿
meadow saffron 秋水仙
meadow sage roots 牧场藿香根
meadow sweet 绣线菊
meadowsweet oil 欧合叶子油;绣线菊油
mean activity 平均活度
mean activity coefficient of ions 离子平均活度系数
mean activity product 平均活度积
mean agitation velocity ①平均激发速度②平均搅拌速度
mean calorie 平均卡
mean camber line 平均中高线
mean carrier velocity 平均载气线速
mean channel depth 平均螺槽深度
mean column pressure 平均柱压
mean cut-off 平均终断;平均填充
mean degree of polymerization 平均聚合度
meander chain configuration 曲折型链
meander configuration 扭曲型;曲折型
meander-shape model 曲棍状模型
mean deviation 平均偏差;平均差异率
mean down time 平均停机时间
mean durability year of coating 涂料平均耐用年限
mean effective diameter 平均有效直径
mean effective pressure 平均有效压力
mean effective viscosity 平均有效黏度
mean error 平均误差
mean field lattice model 平均场晶格模型
mean flaw concentration 平均瑕疵密(集)度
mean flaw-free length 平均无瑕疵长度
mean free path 平均自由程
mean heat capacity 平均热容量
mean hydraulic radius 平均水力半径
mean indicated pressure 平均指示压力
mean length(=length average) 中均长度;平均长度
mean lethal dose 半(数)致死量
mean life 平均寿命
mean linear velocity of carrier gas 载气平均线速*
mean linear velocity of mobile phase 流动相平均线速
mean molar quantity 平均摩尔数量
mean noon sunlight 平正午阳光
mean normal deformation 平均法向形变
mean normal stress 平均正应力
mean of population 总体(平)均值
mean of sample 样本(平)均值
mean pore size 平均孔径
mean radius 平均半径
mean refractive index 平均折射率

mean rigidity　①平均刚性②平均刚度
means　①工具②手段；方法
mean sizes　平均大小；平均尺寸
means of production　生产手段；生产工具
mean spherical model(MSM)　平均球状模型
mean square　均方
mean square chain length　均方链长
mean square deviation　均方偏差
mean square distance　均方距
mean square end-to-end distance　均方末端距
mean square error　均方误差
mean square radius　均方半径
mean square radius of gyration　均方旋转半径
mean square value　均方值
mean standard deviation　平均标准偏差
mean stress　平均应力
mean surface diameter　中均表面直径
mean temperature difference　平均温差
mean value　平均值
mean work absorption　平均功吸收
measling　白斑
measurable peak　可测定峰
measurand　被测量
measure　①量器②量；测③尺寸；分量④方法；步骤
measure analysis　容量分析(法)
measured response　测量响应
measured value　测定值
measured variable　被测变量
measure electrode　测量电极
measurement and test　测试
measurement apparatus for solidification point　凝固点测定装置
measurement cargo　体积货物
measurement error　测量误差
measurement meter　计量表
measurement of friction　摩擦测量
measurement of odor　气味测定
measurement of peak area　峰面积测量法
measurement of surface tension　表面张力测定
measurement of water pollution　水质污染测定
measurement process　测量过程
measurement pump　计量泵
measurement range　测定范围
measuring　测量
measuring and stop motion　测长自停装置；定长自停装置
measuring apparatus of vapor pressure　蒸气压测定装置
measuring appliance　测量仪器；检测仪器(仪表)

measuring arm　(电桥的)测量臂
measuring basis　测量基准
measuring bridge　量电阻桥
measuring buret(te)　量液滴定管
measuring cell　测量池
measuring column　测定柱
measuring cylinder　量筒
measuring device in autoleveller　自调匀整检测机构
measuring element　检测元件
measuring engine　测量机
measuring equipment　控制测量仪器
measuring error　测量误差
measuring flask(=volumetric flask)　(容)量瓶
measuring flask for content　注入式(容)量瓶
measuring flask for delivery　分液(容)量瓶
measuring gage　测厚计
measuring glass　量(液)杯
measuring graduates　刻度量筒
measuring head　①测量头②探头
measuring instrument　测量仪表
measuring machine　量皮机
measuring method of blackening　黑度测定法
measuring microscope　度量显微镜
measuring pipet(te)　刻度吸移管；吸量管
measuring pump　量液泵；计量泵
measuring range　测量限度
measuring spectroscope　波长分光镜
measuring stick　量尺
measuring system　测量系统
meat products analysis　肉制品分析
mebendazol　甲苯咪唑〔药〕
mecamylamine hydrochloride　3-甲氨基异莰烷盐酸盐
mecaprine　麦卡卜林〔治疟药〕
Mecca balsam(=Gilead balsam)　麦加香脂
mechan-〔词头〕　机械
mechanic　①机械(工)匠②机械的
mechanical　①机械的②机(械)工(程)的③机动的
mechanical activation　机械活化
mechanical adhesion　机械黏合
mechanical admittance　力学导纳
mechanical advantage　机械利益
mechanical agitation　机械搅拌
mechanical agitator　机械搅拌器
mechanical air separation　机械空气分离
mechanical analysis　机械分析；粒度分析
mechanical analytical balance　机械分析天平
mechanical anchoring effect　机械黏固效应；机械锚固效应

mechanical antidote   机械解毒
mechanical ashing   机械除灰
mechanical assembly   机械装配；机械组合
mechanical belt   传动带
mechanical blowing   机械发泡
mechanical buffing   机械抛光
mechanical burner   机械炉
mechanical cell   机拌池
mechanical classifier   机动选粒器
mechanical cleaning   机械清洁法
mechanical colorant dispenser   自动调色机；自动混色机
mechanical composition of soil   土壤的机械组成；土性
mechanical conditioning   机械调节
mechanical constitutive relation   力学本构关系式
mechanical cooling tower   机械冷却塔
mechanical counter   机械计数器
mechanical coupling   机械轴节；机械接手
mechanical damage   机械损伤
mechanical degradation   ①机械降解；物理降解②机械损伤；机械性能的降低
mechanical degradation of polymers   聚合物的机械力降解
mechanical denaturation spinning   力学变性纺丝；索引纺丝
mechanical depolymerization   机械解聚
mechanical dewaxing   机械脱蜡
mechanical dimensional stability   力学形稳性；机械尺寸稳定性
mechanical dipping process   机械浸泡法〔硝化纤维素制造〕
mechanical disintegration   机械粉碎
mechanical disruption   机械破裂
mechanical downtime   事故停工时间
mechanical draught   机械通风
mechanical draught cooling tower   机械通风凉水塔；强制通风凉水塔
mechanical efficiency   机械效率
mechanical energy   机械能
mechanical energy balance   机械能平衡
mechanical entrapment   机械夹带
mechanical equivalent   功当量
mechanical equivalent of heat   热功当量
mechanical error   机械误差
mechanical exhaust   机械排汽
mechanical failure   力学破坏
mechanical feed   自动送料器
mechanical filter   ①机械滤器②自动滤器
mechanical flex   机械性屈挠
mechanical flow digram   工程流程图

mechanical foam   机械泡沫
mechanical foaming   机械发泡
mechanical force   机械力
mechanical force feed lubrication   高压下自动润滑
mechanical furnace   机械炉
mechanical gas dispersoid   机械气溶胶
mechanical glazed cap paper   机光厚纸
mechanical glazed poster paper   机光广告纸
mechanical goods   机制品
mechanical goods reclaim   杂胶再生橡胶
mechanical grate   自动栅
mechanical handling   机械操作
mechanical hardening   机械硬化
mechanical impedance   力(学)阻抗
mechanical impurities〔复〕   机械杂质
mechanical independence   机械独立性
mechanical integrator   机械积分器
mechanical kiln   机械炉
mechanical loss   机械损失
mechanical loss factor   机械损失因素
mechanical lubrication   机械润滑；强制润滑
mechanical lubricator   机械润滑器
mechanically ashed producer   机械除灰发生炉
mechanically chargeable zinc-air battery   机械再充式锌-空气电池(组)
mechanically controlled valve   机械控制阀
mechanically rabbled furnace(=mechanically raked furnace)   机械搅拌炉
mechanically raked furnace(=mechanically rabbled furnace)   机械搅拌炉
mechanically-slit yarn   机械切割丝；机械切膜丝
mechanical manometer   自动测压计
mechanical mixture   机械混合物
mechanical model   力学模型
mechanical model with mass   计及质量的力学模型
mechanical model without mass   不计质量的力学模型
mechanical mold   模制品用压模
mechanical muller   机动研磨机
mechanical needle-chipping   机针錾平；机针铲锈
mechanical needling   机械针刺工艺
mechanical octane number   机械辛烷值
mechanical passivity   机械钝性
mechanical performance   机械性能
mechanical piler   机械砌砖机〔设〕
mechanical poking   机械拨火
mechanical polishing   机械抛光
mechanical power house   机动室；动力站〔指水、汽、压缩空气〕

mechanical preparation 机械法预处理
mechanical pressure gauge 刻字压力计
mechanical process 机械法
mechanical producer 机械化煤气发生炉
mechanical programmer 机械程序设计(控制)器(者);程序机构
mechanical property 机械性能
mechanical protection 机械保护
mechanical pulp 机械纸浆;木(制纸)浆
mechanical pulp-free printing paper 道林纸
mechanical pump 机械泵
mechanical pyrite burner 机动黄铁矿炉
mechanical quantity 力学量
mechanical rabble ①机械搅拌②机械搅拌器
mechanical rabbling(=mechanical raking) 机械搅拌
mechanical raking(=mechanical rabbling) 机械搅拌
mechanical reactance 力抗
mechanical refining process (石油产品的)机械精制法
mechanical relaxation 力学松弛
mechanical resistance 力阻
mechanical resonance 机工共振
mechanical response 力学响应
mechanical rubber goods 橡胶工业制品
mechanicals 工业制品
mechanical salt-cake furnace 机动盐饼炉
mechanical sampling 机械取样
mechanical scum dredger 机械化废漆渣清除机
mechanical seal 机械密封;端面密封
mechanical seal water cooler 机械密封(的)水冷却器
mechanical separation 机械分离
mechanical separator 机械分离器
mechanical severing of chemical bond 化学键的机械断裂
mechanical shaker 机械摇动器
mechanical shearforce 机械切变力
mechanical shearing 机械剪切
mechanical sifter 机动筛
mechanical skill 机械技术
mechanical sludge dewatering 污泥机械脱水
mechanical softening 机械软化
mechanical spectrometer 力学频谱计;力学波谱计
mechanical spectrometry ①力学波谱学②力学谱法
mechanical spectroscopy 力学波谱学;力学谱
mechanical spectrum 力学谱
mechanical splitting process 机械法裂离工艺
mechanical stability 机械安定性
mechanical stirrer 机械搅拌器
mechanical stirring 机械搅拌
mechanical stitch ①机械缝合②机械滚压
mechanical stoker (机动)加煤机
mechanical stretcher 机械式张网机
mechanical (superphosphate) den 机械室(放送过磷酸钙)
mechanical system ①力学(体)系②机械系统
mechanical test 机械性能试验
mechanical testing 机械试验
mechanical texturing (纤维)机械变形
mechanical theory of friction 摩擦的机械理论〔凹凸理论〕
mechanical transport interface 机械转移接口
mechanical treatment 机械加工
mechanical ventilation 机械通风
mechanical vibration 机械振动
mechanical viscosity 机械黏度
mechanical washer(=mechanical washer scrubber) ①机械涤气器②机械洗涤器
mechanical washer scrubber(=mechanical washer) ①机械涤气器②机械洗涤器
mechanical water 物理水;非结合水
mechanical wear 机械磨损
mechanical wood pulp 机械木纸浆
mechanical work 机械功
mechanical workout 机械加工
mechanico-chemical reaction 机械化学反应
mechanics 力学
mechanism 机理*
mechanism of addition polymerization 加成聚合的机理(机制)
mechanism of coagulation 凝固机理
mechanism of corrosion 腐蚀机理
mechanism of crosslinking 交联机理
mechanism of degradation 降解机理
mechanism of hydrogen embrittlement 氢脆机理
mechanism of ion fragmentation 离子碎裂机理
mechanism of light fading 日晒褪色机理
mechanism of modification 改性机理
mechanism of polymerization 聚合历程
mechanism of reaction 反应历程
mechanism of soil removal 去污机理(制)
mechanism of styrenation 苯乙烯化机理
mechanism of toxication 致死机理
mechanization 机械化
mechanized 机械化的
mechanized feed 机械化供料
mechanized lorry 机械化货车;具有装卸设备的载重汽车
mechanochemical 机械化学的

mechanochemical breakdown　机械化学塑炼
mechanochemical contraction　机致化学收缩；动力化学收缩
mechanochemical degradation　机致化学降解；动力化学降解
mechano-chemical engine　化学动力机械；化学动力引擎
mechanochemical grafting　机械化学接枝
mechanochemical polycondensation　力化学缩聚
mechanochemical polymerization　机械化学聚合作用
mechano-chemical system　化学动力体系
mechano-chemical wear　机械化学磨损
mechanochemistry　机械化学
mechano-degradation　机械降解
mecholyl　乙酰-β-甲基胆碱
mecilenic acid　十四(碳)烯酸
meclofenoxate　甲氯芬酯〔药〕
meclozine hydrochloride　美其敏；敏克静；盐酸氯苯苄嗪；盐酸氯苯甲嗪；美克洛嗪；美其敏〔抗组胺药〕
mecocyanin　袂康花色苷；龙胆二糖花青苷
meconate　袂康酸盐　$C_5H_2O_3(COOM)_2$
meconic acid　袂康酸　$C_7H_4O_7$
meconidin　袂康定
meconin　袂康宁
mecouinic acid　袂康宁酸　$C_{10}H_{12}O_5$
meconium　①胎便〔新生儿第一次排出的粪便〕②(=opium; papaver)袂康；阿片；罂粟
med fibre　中髓腔毛纤维
media　介质
media density　(研磨)介质密度
median　中位值*
median diameter　中(位)数直径；中值粒径；中径
median effective concentration　有效中浓度
median effective dose　有效中量
median-energy ion scattering(MEIS)　中能离子散射
median gutter　中沟
median immunizing dose　半数免疫剂量
median infection dose　半数感染剂量
median lethal concentration　半致死浓度
median lethal dose(LD_{50})　半数致死剂量
median size　中值粒径；中径
media retention screen　介质阻挡筛网
media-sample boundary　介质样品间界面
media size　(研磨)介质大小
mediator　(分子)介体
medical　医学的；医药的
medical appliance　医疗器材
medical chemistry　医药化学
medical cyclotron　医用回旋加速器

medical glass　医用玻璃
medical grade ^{238}Pu　医学级钚-238〔含 ^{236}Pu 在 0.3ppm 以下〕
medical grade rubber　医用橡胶
medical jurisprudence　法医学
medical polymer　医用高分子
medical reactor　医用反应堆
medical thin cross-section cutter　医用切片机
medicament　药；药剂
medicated oil　药用油
medicated soap　药皂
medication　药疗法
medicinal　医学的；医药的
medicinal carbon　药用(活性)炭
medicinal cosmetic　医用化妆品
medicinal flavour　药用香料
medicinal oil　药用油
medicinal paraffin　药用石蜡
medicinal paraffin oil　药用石蜡油
medicinal perfume　药用香料
medicinal soap　药皂
medicinal wax(=medicinal paraffin)　药用石蜡
medicine　①医学；医药②药物
medicine dropper　医药用滴管
medinal　巴比妥钠
mediocidin　中杀菌素
medium　①介质；介体；媒质②媒介物；培养基③中间的；中级的
medium accelerator　中速促进剂
medium accelerator dosage　中等量促进剂
medium-alkali glass fibre　中碱玻璃纤维
medium boiler　中沸点溶剂
medium boiling solvent　中沸点溶剂
medium-breaking　中度裂化
medium carbon steel　中碳钢
medium chrome yellow　中铬黄〔即铬酸铅〕
medium colour furnace (oxidated) black　中色素炉法(氧化)炭黑
medium crosslink density　中(等)交联密度
medium crushing　中级压碎
medium curing asphalt　中级处理沥青
medium curing cut back (asphalt)　(沥青)中级处理回收段
medium-density polyethylene(MDPE)　中密(度)聚乙烯
medium effect　介质效应
medium filtered cylinder oil　中级过滤气缸油
medium flow rate paper　中流速纸
medium force fit　中级压配合
medium grained　中级粒度的

medium grinding 中级压碎
medium hair grease 中级毛填料润滑脂
medium lift pump 中级升液泵
medium-lot producer 中型企业
medium maintenance 中修
medium molecular weight 中分子量
medium oil 中油
medium oil alkyd 中油度醇酸树脂
medium-oil-length 中油度
medium oil varnish 中油清漆
medium phase suspension 中间相悬浮体
medium pitch 中沥青
medium portable 中型移动式(的)
medium pressure 中等压力；中压
medium pressure accumulator 中压蓄力器
medium pressure Fischer-Tropsch process 中压费-托法
medium-pressure polyethylene 中压聚乙烯
medium processing channel black 中等加工槽黑
medium rank anthracite 中级无烟煤
medium rank bituminite 中级烟煤
medium ring 中环
medium ring compound 中环化合物
medium screening(=medium sizing) 中级筛选
medium-setting emulsified asphalt 中裂乳化沥青
medium shade 中等色〔不深不浅的色相〕
medium shear rate(=mid(dle) shear rate) 中剪切速率；中切速
medium size ①中等尺寸②中等粒度
medium sizing(=medium screening) 中级筛选
medium-soft pitch 半软沥青
medium square 中材；中方材
medium sweep elbow 中级扫拂弯管
medium term corrosion prevention 中效防锈
medium thermal black 中等热裂黑
medium titer 中(等)纤度
medium tone 中间色
medium vinyl polybutadiene rubber 中乙烯基聚丁二烯橡胶
medium-viscosity 中等黏度
medium-viscosity vehicle 中等黏度漆料
medium volume swelling 中(程度)体积溶胀
medium weight hide 中磅皮
medium yellow 中铬黄〔颜料〕
medley 杂拼物；混合物
medmontite ①铜蒙脱石②铜皂石③杂硅孔雀云母
medroxypogesterone 甲羟孕酮〔药〕
medroxyprogesterone acetate 6α-甲基-17α-乙酸基孕甾酮

medullameter 毛髓测定器
medullation 髓(质)化
meerschaum 海泡石
Meerwein-Ponndorf-Vealey reduction(=aluminium alkoxide reduction) 烃氧基铝还原作用；米尔文-庞道夫-沃莱还原作用
megabar 兆巴〔压力单位〕
megabarye 兆巴〔压力单位〕
megacarpidine(=soladulcidine) 茄甜定
megacidin 灭巨杆菌素
megacurie 兆居里
megacycle 兆周
megacyclic musk 大环麝香
megadyne 兆达因
mega-electron volt 兆电子伏特〔符号 MeV〕
megaerg 兆尔格
megafarat 兆法拉
megalomycin 巨大霉素
megameter 兆米
megarad 兆拉德
megaroentgen 兆伦琴〔10^6 伦琴〕
megarrhizin 美瓜根皂苷
megaton 兆吨
megatonnage 兆吨数；兆吨位；兆吨级
megavolt 兆伏特
megeresin 白瑞香脂
megestrol 甲地孕酮
megestrol acetate 乙酸(基)孕甾酮
meglumine iothalamate 麦格鲁明阻 X 射线剂；异泛影葡胺；碘肽葡胺〔水溶性造影剂，静注后从尿中排出，常用于尿路造影，也可用于肾盂、心血管、脑血管等的造影〕
megohm 兆欧
meiler 炭窑
meiler method (process) 堆积制炭法
meionite 钙柱石
Meisenheimer complex 迈森海默配合物
meisoindigotin 甲异靛〔药〕
mekemycin 梅克霉素
Meker burner 麦克灯；网口灯
mekonine 袄康宁
melacacidin 黑木金合欢素；3,3′,4,4′,7,8-六羟基黄烷
melaconite 土黑铜矿
melaleuca oil 白千层油
melaleucic acid 白千层酸
melalinol 白千层醇
melam 蜜白胺
melamac 蜜醛塑料；蜜白胺甲醛塑料

melamine(=cyanuramide)　蜜胺；三聚氰胺　$C_3N_3(NH_2)_3$
melamine-benzoguanamine resin　三聚氰胺-胍胺树脂
melamine coating(=melamine resin coating)　三聚氰胺树脂涂料
melamine decorative laminate　三聚氰胺(塑料)贴面板
melamine flame retardant　三聚氰胺阻燃剂
melamine-formaldehyde resin　三聚氰胺-甲醛树脂*
melamine molding material　三聚氰胺树脂成型材料
melamine-phenol-formaldehyde resin(MPF)　三聚氰胺-酚醛树脂
melamine plastic　蜜胺塑料；三聚氰胺(甲醛)塑料
melamine pyrophosphate　三聚氰胺焦磷酸盐
melamine resin　蜜胺树脂；三聚氰胺-甲醛树脂*
melamine resin adhesive　三聚氰胺树脂黏合剂
melamine resin coating(=melamine coating)　三聚氰胺树脂涂料
melamine resin paint　三聚氰胺树脂漆；氨基树脂漆
melamine resin precondensate　蜜胺树脂预缩物
melamine resin varnish　三聚氰胺(树脂)清漆
melamine-urea resin　三聚氰胺-脲树脂
melamino-formaldehyde resin　蜜胺甲醛树脂
melaminoplast　蜜胺塑料
melaminylphenylarsonic acid　蜜胺基苯胂酸
melampsporin　没食子苷色素
melampyrine(=dulcitol)　卫茅(己六)醇
melampyrite(=dulcitol)　卫茅(己六)醇
melan　草木犀浆
melanerite(=melanterite)　水绿矾
melaniline(=diphenyl guanidine)　蜜苯胺；对称二苯胍
melanin　黑素
melanite　黑榴石
melanosome　黑素
melanothallite　羟氯铜矿
melanterite(=melanterite)　水绿矾
melanuric acid(=isocyanurimide)　异氰尿酰亚胺；黑尿酸
melaphyre　暗玢岩
melarsen　蜜胺基苯胂酸
melarsoprol　美拉胂醇〔治锥虫病药〕
melatonine　N-乙酰基-5-甲氧基色胺；松果素；褪黑素；美拉托宁
Meldola's blue　麦尔多拉蓝
M electron　M 层电子
melee　划玻璃金刚石
melene(=triacontylene)　蜂花烯；三十碳烯　$C_{30}H_{60}$
meletin　橡黄素；楝精
melezitose(=melizitose)　松三糖
melibiose　蜜二糖　$C_{12}H_{22}O_{11}$
melicitose　松三糖

melilite　黄长石
melilitite　黄长石
melilotic acid　黄木樨酸；邻羟苯丙酸
melilotin　黄木樨苷
melilot oil　草木犀油
melilotol　二氢化香豆精
melilotoside　草木犀苷
melilotus　草木犀流浸液〔药〕
melinite　麦宁炸药〔即苦味酸〕
melinonine A　梅侬宁 A
melissane(=triacontane)　蜂花烷；三十烷　$C_{30}H_{62}$
melissa oil　蜂花油
melissate　蜂花酸盐(或酯)；三十(烷)酸盐
melissenol　蜂花烯醇
melissic acid　蜂花酸；三十烷酸　$C_{29}H_{59}COOH$
melissin　蜂花精；甘油三蜂花酸酯
melissyl　蜂花基；三十烷基
melissyl acetate　乙酸蜂花(醇)酯
melissyl alcohol　蜂花醇；三十烷醇　$C_{30}H_{62}O$
melissyl melissate　蜂花酸蜂花酯
melitic　蜜石的
melitose　蜜里糖　$C_{12}H_{22}O_{11}$
melitriose(=raffinose)　蜜里三糖；棉子糖
melittin　蜂毒素
melizitose(=melezitose)　松三糖
mellate　①苯六甲酸盐　$C_6(COOM)_6$②苯六甲酸酯　$C_6(COOR)_6$
mellimide　蜜亚胺　$C_{12}H_3O_6N_6$
mellitate　苯六甲酸酯(或盐)
mellitene　六甲苯　$C_6(CH_3)_6$
mellitic acid(=benzene-hexacarboxylic acid)　苯六甲酸　$C_6(COOH)_6$
mellityl alcohol　五甲基苯甲醇　$(CH_3)_5C_6CH_2OH$
mellon(e)　蜜弄；三聚二氰胺
mellophanic acid　苯偏四甲酸　$C_6H_2(COOH)_4$
mellorite　钙铁硅酸盐岩
Mellot's metal　梅洛特(铋铅锡)合金
mellow　排泄石灰阱〔制革〕
mellowing　柔软处理
mellow lime　中灰
mellow lime liquid　淡石灰水
mellozing　坩埚法金属喷涂
melon　氰尿酰胺
melonal(=2,6-dimethyl-5-hepten-1-al)　甜瓜醛；2,6-二甲基-5-庚烯醛
melon essence　甜瓜香精
melon flavour　甜瓜香精
melonic acid　甜瓜酸

melonite 碲镍矿
melon oil 甜瓜(子)油
melon seed 甜瓜子
melotoxin 甜瓜毒
meloxicam 美洛昔康〔药〕
melphalan 苯丙氨酸氮芥；美法仑〔药〕
melt ①熔化；熔融②熔化物③软化④(渐渐)消失
meltability ①熔性②熔度
meltable 易熔的；可熔化的
melt adhesive 熔融黏结剂；热熔胶
melt blend 熔体混合物
melt-blending 熔体混合；混合熔融
melt blowing technique 熔体喷射技术
melt blown filter 熔喷滤料
melt brown 熔喷法
melt coating 熔融涂装；热喷涂
melt coloration technique 熔体着色技术
melt compounding 热熔混合
melt condensation 熔融缩合
melt cooler (挤出料的)冷却带；熔融物冷却带
melt crystallization 熔体结晶
melt deformation history 熔体形变历程
melt dipping 浸渍熔融物
melt dispersion 熔体分散
melt down 熔化
melt-draw ratio 熔体拉伸比
melt-dyed fibre 熔体染色纤维
melt elasticity 熔体弹性
melt embossing process 熔膜轧纹成型法
melten applied metal 熔敷金属
melten gel 熔胶
melt-extruded film 熔融拉伸膜
melt extrusion 熔体挤出
melt extrusion method 熔体挤出法〔测定聚合物分子量用〕
melt film 熔膜
melt flow 熔体流动
melt (flow) index 熔体流动指数
melt flow instability 熔体流动不稳定性
melt flow rate(MFR) 熔体流动速率
melt fracture 熔体破裂
melt grid type spinning machine 熔栅式纺丝机
melt grown polymer 熔体生长聚合物
melt homogeneity ①熔体均匀性②熔体均匀度
melt hopper 溶体料斗
melt index 熔融指数；熔体指数
melt index apparatus 熔体指数仪〔测定熔体黏度用〕
melt indexer(=melt index plastometer;extrusion type capillary viscometer) 熔体(流动)指数测定仪
melt index plastometer 熔指数测定仪；熔体指数仪
melting 熔化；熔融物
melting analysis 熔炼分析
melting and fusing centrifuge 熔化离心机
melting behavior 熔融性状
melting cone (示温)熔锥
melting curve 熔化曲线；熔融曲线
melting decladding 熔融去壳
melting diagram 熔化图
melting dilation 熔化膨胀
melting effect 熔融效应
melting endotherm 熔融吸热
melting enthalpy 熔融焓
melting entropy 熔化熵
melting expansion 熔融膨胀
melting heat 熔化热
melting interface 熔融界面
melting peak 熔融峰
melting point 熔点
melting point apparatus 熔点测定装置
melting-point determination on varnish resin 清漆树脂的熔点测定
melting point diagram 熔点图
melting point lowering 熔点降低
melting point method 熔点法
melting point tube 熔点测定管
melting pot 熔化锅
melting range (聚合物)熔程；熔点范围
melting rate 熔融速率
melting tank 熔化器；熔化锅
melting temperature 解链温度；熔解温度
melting transition 熔化转变
melting vessel 熔锅
melting viscosity 熔融黏度
melting welding technique 熔结技术
melt loading 熔融装料
melt lump 熔块
meltoryzine 麦芽米曲霉素
melt phase polycondensation 熔融缩聚
melt pigmentation 熔体着色法
melt point 熔点
melt polycondensation 熔融缩聚(作用)；熔体缩聚(作用)
melt polymerization 熔融聚合；熔体聚合
melt processability 熔体加工性能
melt proportioning pump 熔体比例泵；熔体配料泵
melt refining 熔融精制
melt residence time 熔体逗留时间

melt resistance　抗熔性
melt rheology　熔体流变学
melt rheometer　熔体流变仪
melt ring-open polymerization　熔融开环聚合
melts　熔融物
melt section　熔化区
melt-shaping　熔融成形
melt slippage　熔体滑移
melt snagging　熔体阻滞
melt-spinnable polymer　可熔纺聚合物
melt spinning　熔纺*
melt-spinning process　熔纺工艺
melt-spun　熔纺(的)
melt-spun dyed　熔体纺染色的
melt spun fiber　熔纺纤维
melt stagnation　熔体滞流
melt strength　熔体强度
melt strength enhancer　熔体强度增强
melt sump　熔体潭；熔体槽
melt swell(=Barus effect)　巴勒斯效应；熔体挤出胀大
melt throughput　熔体输出量
melt traffic paint　熔融型路线漆；热喷涂路线漆
melt uniformity　熔体均匀性
melt viscometer　熔体黏度计
melt viscosity　熔体黏度
melt zone　①熔体区〔相对于固体粒料区而言〕②熔融段
member　①构件〔机工〕②节；(环中)原子数
membership　隶属度
membership function　隶属函数
membrane antibody　膜抗体
membrane antigen　膜抗原
membrane antigen receptor　膜抗原受体
membrane capacitance　膜电容
membrane capacitance measurement　膜电容测定法
membrane charge　膜电荷
membrane chromatography　膜色谱(法)
membrane disk　膜片
membrane elasticity　膜弹性
membrane electrochemistry　膜电化学
membrane electrode　膜电极*
membrane electrophoresis　膜电泳
membrane emf　膜电动势
membrane equilibrium　膜(渗)平衡
membrane extraction　膜萃取
membrane filter　膜滤器
membrane filter method　膜滤器法
membrane filtration　膜式过滤
membrane filtration culture method　薄滤膜培养法；膜滤培养法
membrane-frit separator(=Grayson separator)　多孔薄膜分离器
membrane immunofluorescence technique　膜免疫荧光技术
membrane inlet mass spectrometry　膜进样质谱法
membrane introduction mass spectrometry　膜进样质谱法
membrane junction　(薄)膜接界
membrane osmometry　隔膜渗透压力测定法
membrane permeability　膜渗透性
membrane potential　膜电位
membrane receptor　细胞膜受体
membrane resistance　膜电阻
membrane selective flow　膜选择性流动
membrane semipermeability　膜半渗透性
membrane separation process　薄膜分离法
membrane separation technique　膜分离技术
membrane separator　隔膜分离器
membrane stress　薄膜应力
membrane system　隔膜法(电泳)
membrane valve　薄膜阀
membrane vesicles　膜囊
membranometer　薄膜式压力计
memel　压花靴面革
memory　①记忆性②存储③弹性复原性
memory circuit　存储电路
memory-delay device in autoleveller　自调匀整记忆延时机构
memory device　存储装置；存储设备
memory effect　①记忆效应*②存储效应③复原效应
memory factor　记忆因子
memory fluid　黏弹性液体
memory function　记忆函数
memory functional　记忆泛函
memory peak　记忆峰
memory phenomenon　①记忆现象②复原现象③存储现象
memory standardization system　存储标准化系统
menachanite　钛铁砂
menadiol　甲萘氢醌〔药〕
menadiol diacetate　二氢萘醌二乙酸酯；维生素 K_4 乙酸酯
menadiol sodium phosphate　2-甲基-1,4-萘氢醌二磷酸酯四钠
menadione　甲萘醌；维生素 K_3；2-甲基-1,4-萘醌
menadione sodium bisulfite　甲基萘醌亚硫酸氢钠；水溶性维生素 K
menakanite　钛铁砂
menaphthone　维生素 K_3

menaphthyl 萘甲基
menaquinone 甲基萘醌类；维生素 K_2 类
Mendeleev(=Mendelyeev) 门捷列夫
Mendeleev's periodic system 门捷列夫周期系
mendelevium 钔〔101 号元素, 符号 Md〕
Mendelyeev(=Mendeleev) 门捷列夫
Mendel(y)eev chart 门捷列夫周期表
Mendel(y)eev group 门捷列夫族〔周期表直行〕
Mendel(y)eev law 门捷列夫周期律
Mendius reaction 门迭斯反应
mendozite 钠明矾
meneghinite 斜辉锑铅矿
mengite 独居石
menhaden oil 鲱油
menilite 硅乳石
meniscus differential thermometer 简易示差温度计
meniscus point 半月点
meniscus reader （弯)液面读镜
menisine 门尼新碱
menisperine 蝙蝠葛碱
menispermoid 蝙蝠葛素
menogene 蒁闹烯；蒁闹二烯
menogerene 蒁闹葛烯；蒁闹三烯
menotropins 尿促性素〔药〕
Menschutkin reaction(=quaternarization) 季铵化作用；门秀金反应
menstruum 溶剂
menstruum universale 万能溶剂
mentha camphor 薄荷醇；薄荷脑
menthadiene 蒁二烯
p-mentha-15-diene 对蒁-15-二烯
dl-1,8(9)-p-menthadiene dl-1,8(9)-对-蒁二烯
menthadione 百里醌
menthandiol 蒁二醇
menthane 蒁烷；薄荷烷 $CH_3C_6H_{10}C_3H_7$
menthanediamine 蒁烷二胺；萜烷二胺
p-menthane hydroperoxide 萜烷过氧化氢
menthanol 蒁烷醇 $C_{10}H_{20}O$
menthanone 蒁烷酮 $C_{10}H_{18}C$
menthanyl 蒁烷基；二氢松油基
menthanyl acetate(=dihydroterpinyl acetate) 乙酸蒁烷酯；乙酸二氢松油酯
menthe 薄荷液
menthene 蒁烯；薄荷烯 $C_{10}H_{18}$
menthenol 蒁烯醇
menthenone 蒁烯酮
menthenyl(=terpinyl) 蒁烯基；松油基；萜品基
menthenyl acetate 乙酸蒁烯酯；乙酸松油酯

menthofuran 薄荷呋喃
menthol 薄荷醇；蒁醇 $C_{10}H_{19}OH$
menthol ester 薄荷醇酯
menthol salicylate 水杨酸蒁醇酯
menthol valerate 戊酸蒁醇酯
menthone 薄荷酮；蒁酮 $C_{10}H_{18}O$
menthone borate 硼酸蒁基酯
menthopinacol 蒁频哪醇
menthyl 蒁基
menthyl acetate 乙酸蒁醇酯；3-萜醇乙酸酯 $CH_3CO_2C_{10}H_{19}$
menthyl amine 蒁胺 $C_{10}H_{19}NH_2$
menthyl anthranilate 邻氨基苯甲酸薄荷酯
sec-menthyl chloride 仲蒁基氯
$tert$-menthyl chloride 叔蒁基氯

$$CH_3CH(CH_2)_2CClCH_2CH_2$$
$$CH(CH_3)_2$$

menthyl cyclohexanol 蒁基环己醇
menthyl ethoxyacetate 乙氧基乙酸薄荷酯
menthyl ethyl ether 蒁基乙基醚
menthyl formate 甲酸蒁酯；甲酸薄荷醇酯 $HCO_2C_{10}H_{19}$
menthyl isovalerate 异戊酸蒁酯；异戊酸薄荷醇酯 $C_4H_9CO_2C_{10}H_{19}$
menthyl lactate 乳酸薄荷酯
menthyl propionate 丙酸薄荷酯
menthyl valerate 戊酸薄荷酯
menyanthes 睡菜
menyanthin 睡菜质
menyanthol 睡菜醇
mepacrine 米帕林；阿的平；疟涤平；甲氯苯氯喹
meperidine hydrochloride 盐酸哌替啶；度冷丁〔镇痛药〕
mephenesin 3-(2-甲苯氧基)-1,2-丙二醇；甲苯丙醇〔松肌药〕
mephenytoin 美芬妥英
mephitic 有臭气的
mephobarbital N-甲基-5-苯基-5-乙基巴妥酸；甲基苯巴比妥
meprobamate 2-甲基-2-丙基-1,3-丙二醇二(氨基甲酸酯)；安宁；眠尔通〔安定药〕
meps(megacycles per second) 兆赫(兹)；每秒兆周
mepyramine maleate 美吡拉敏；顺丁烯二酸美吡胺〔抗组胺药〕
mer ①链节；单体单元②基体
mer-(mere-, meri-) 〔词头〕部分
meralluride 美拉鲁利〔利尿药〕
meraneine 乙酸香叶酯

merbaphen 美巴芬〔药〕
mercadium 汞镉橙〔颜料〕
mercadium orange （硫化）汞镉橙
mercadium pigment 汞镉颜料
mercapfining 脱硫醇精制(法)
mercapsol treating 梅开普索处理
mercaptal 缩硫醛 $RCH(SR)_2$
mercaptalated 缩硫醛化了的
mercaptalation 缩硫醛化作用
mercaptan 硫醇*
mercaptan absorber 硫醇吸收塔
mercaptan acid 巯基酸 $SHRCOOH$
mercaptan separator 硫醇分离器
mercaptan sulfur 硫醇式硫
mercaptan sulfur content 硫醇硫含量
mercaptan-telechelic oligomer 硫醇远螯齐聚物
mercaptide 硫醇盐 RSM
mercaptide ion 烃硫离子 RS—
mercaptining 脱硫醇精制法
mercapto 巯基；氢硫基 HS—
mercaptoacetic acid 巯基乙酸 $HSCH_2CO_2H$
mercaptoamino acid 巯基氨基酸
mercapto arylene thiazole 巯基亚芳基噻唑
2-mercaptobenzimidazole 2-巯基苯并咪唑
2-mercaptobenzimidazole zinc salt 2-硫醇基苯并咪唑锌盐〔防老剂〕
mercaptobenzothiazole 巯基苯并噻唑*
2-mercaptobenzothiazole 2-巯基苯并噻唑；快热粉〔俗〕
mercaptobenzothiazole polysulfide 多硫化巯基苯并噻唑〔促进剂〕
mercapto benzothiazole sulfonic acid 巯基苯并噻唑磺酸
mercaptobenzothiazyl disulfide 二硫化硫醇基苯并噻唑〔促进剂〕
3-mercapto-2-butanone 3-巯基-2-丁酮
mercapto-cinnamic acid 巯基肉桂酸 $HSC_6H_4CH=CHCOOH$
2-mercapto-4,6-diamino-pyrimidine 2-巯基-4,6-二氨基嘧啶
mercaptoethanol 巯基乙醇
mercaptoethylamine 巯基乙胺；半胱胺
mercaptoethylation 巯(基)乙基化(作用)
mercaptoethyliminodiacetic acid(MEIDA) 巯乙基亚氨二乙酸 $HSCH_2CH_2N(CH_2COOH)_2$
β-mercaptoethyltriethoxysilane β-巯乙基三乙氧基硅烷
mercapto group 硫醇基；巯基
2-mercaptoimidazoline 2-巯基咪唑啉〔促进剂〕
mercaptoinosine 巯基肌苷
mercaptol 缩硫醇

mercaptolysis 硫醇解
8-mercapto-p-menthane-3-one 8-巯基对薄烷-3-酮
mercaptomerin sodium 巯汞钠〔药〕
2-mercaptomethylpyrazine 2-巯甲基吡嗪
2-mercapto-5-methyl-tetrahydrotriazine 2-硫醇基-5-甲基四氢三嗪〔促进剂〕
2-mercaptonaphthalene 2-巯基萘
3-mercapto-2-pentanone 3-巯基-2-戊酮
mercaptophenyl 巯苯基
mercaptophenyl dithiodiazolone 巯苯基二硫代二唑酮；试铋哼
mercapto propionate 巯基丙酸盐(或酯)
2-mercaptopropionic acid 2-巯基丙酸
β-mercaptopropionic acid β-巯基丙酸
mercaptopropylsilane 巯基丙基硅烷
γ-mercaptopropyltrimethoxy silane γ-巯醇基丙基三甲氧基硅烷〔偶联剂〕
mercaptopurine 巯嘌呤〔药〕
6-mercaptopurine(6MP) 6-巯基嘌呤
3-mercaptopyruvate 3-巯基丙酮酸
8-mercaptoquinoline 8-巯基喹啉
mercaptosuccinic acid 巯基琥珀酸
mercaptothiazole 巯噻唑
2-mercaptothiazoline 2-巯基噻唑啉〔促进剂〕
mercaptothiazoline 巯基噻唑啉 $C_3H_5NS_2$
mercaptothiocarbamyl sulfide 巯基硫代氨基甲酰硫化物〔促进剂〕
mercapto thiophene 巯基噻吩；噻吩硫醇
mercaptotropolone 巯基草酚酮
mercaptotropone 巯基草酮
mercapto-type-accelerator 硫醇类促进剂
mercaptoundecahydrododecaborate 巯基十一氢十二硼酸盐 $Na_2B_{12}H_{11}SH$
mercapturic acid 硫醚氨酸；巯基尿酸；硫醇尿酸
mercerisation(=mercerization) ①丝光处理②碱化；浸碱作用
mercerization(=mercerisation) ①丝光处理②碱化；浸碱作用
mercerized cotton 府绸；丝光棉
mercerizer 丝光器
mercerizing 丝光处理
mercerizing assistant 丝光助剂
mercerizing bath 丝光处理浴；浸碱处理浴
mercerizing composition 丝光(碱)液成分；碱化液成分
mercerizing machine 丝光处理机
mercerizing resistance 碱化阻力
mercerizing silo 碱液浸渍桶
mercerizing strength (of alkali) 碱化浓度

mercerizing tange　丝光机〔纺织〕
Mercer process　默塞尔丝光处理法
merchantable fuels　商品燃料
merchantable product　商品；标准产品
merchantable timber　商品材
merchant bar　商品条铁
merchant iron　商品铁
merchant steel　商品钢
merchrome　异色异构结晶
mercoid switch　水银(转换)开关
mercuic chloride　氯化汞
mercurammonium-〔词头〕　二汞代氨基
mercurate　①汞化②汞化产物
mercurated　汞化了的；用汞处理了的
mercurating　汞化
mercurating agent　汞化剂
mercuration　汞化*
mercurdiammonium　汞代二氨基
mercurdiammonium chloride　氯化二氨亚汞　$[Hg(NH_3)_2]Cl$
mercuri-　(正)汞基；(二价)汞基　—Hg—
mercuriacetate　乙酸汞化物　$CH_3COOHg·R$
mercurial　①汞制剂②汞的③汞化合物
mercurial antiseptic　汞防腐剂
mercurial barometer　水银气压计
mercurial fahlore　汞黝铜矿
mercurialine　山靛碱
mercurialism　汞中毒；汞毒症
mercurial manometer　水银压力计
mercurial soot　汞炱；冷凝汞〔蒸馏汞矿时所生成〕
mercurial thermometer　水银温度计
mercuriammonium　汞代铵
mercuriated　汞化了的；用汞处理了的
mercuriating　汞化
mercuriating agent　汞化剂
mercuri-bis-compound　亚汞基化合物〔汞的两个价键均与碳原子相连的有机化合物〕
mercuric　(正)汞的；二价汞的
mercuric acetate　乙酸汞　$Hg(C_2H_3O_2)_2$
mercuric acetylide　乙炔汞　HgC_2
mercuric amidosuccinate　酰氨琥珀酸汞　$Hg(C_4H_6N)_2$
mercuric aminophenylarsenate　氨基苯胂酸汞　$HgO_2AsO·C_6H_4NH_2$
mercuric ammonium chloride　氯化氨基汞　$Hg(NH_2)Cl$；$[NH_2Hg]Cl$
mercuric atoxylate　氨基苯胂酸汞　$HgO_2AsO·C_6H_4NH_2$
mercuric benzoate　苯甲酸汞
mercuric bichromate　重铬酸汞　$HgCr_2O_7$
mercuric bromide　溴化汞　$HgBr_2$
mercuric cacodylate　卡可基酸汞；二甲胂酸汞　$Hg[OAsO(CH_3)_2]_2$
mercuric chlorate　氯酸汞　$Hg(ClO_3)_2$
mercuric chloride　氯化汞；升汞　$HgCl_2$
mercuric chloroiodide　氯碘化汞
mercuric compound　正汞化合物；二价汞化合物
mercuric cyanide　氰化汞　$Hg(CN)_2$
mercuric ethyl chloride　氯(化)乙基汞
mercuric ferrocyanide　氰亚铁酸汞　$Hg_2[Fe(CN)_6]$
mercuric fluoride　氟化汞　HgF_2
mercuric fulminate　雷酸汞　$Hg(ON=C)_2$
mercuric iodide　碘化汞　HgI_2
mercuric lactate　乳酸汞　$Hg(C_3H_5O_3)_2$
mercuric lithium iodide　碘化汞锂
mercuric naphtholate　萘酚汞；萘氧基汞　$(C_{10}H_7O)_2Hg$
mercuric nitrate　硝酸汞　$Hg(NO_3)_2$
mercuri-compound　正汞化合物；二价汞化合物
mercuric oxalate　草酸汞　HgC_2O_4
mercuric oxide　氧化汞；一氧化汞；三仙丹　HgO
mercuric oxycyanide　氧氰化汞
mercuric oxyiodide　氧碘化汞
mercuric perchlorate　高氯酸汞　$Hg(ClO_4)_2$
mercuric phenate　苯酚汞
mercuric phenyl acetate　乙酸苯(基)汞
mercuric phosphate　磷酸汞　$Hg_3(PO_4)_2$
mercuric polysulfide　多硫化汞
mercuric potassium cyanide　氰化汞钾　$K[Hg(CN)_3]$
mercuric potassium iodide　碘化汞钾；碘汞酸钾　$K[HgI_3]$; $K_2[HgI_4]$
mercuric pyroborate　焦硼酸汞　HgB_4O_7
mercuric rhodanate(=mercuric rhodanide)　硫氰酸汞　$Hg(SCN)_2$
mercuric rhodanide(=mercuric rhodanate)　硫氰酸汞
mercuric selenide　一硒化汞　$HgSe$
mercuric selenite　亚硒酸汞　$HgSeO_3$
mercuric sulfate　硫酸汞　$HgSO_4$
mercuric sulfide(=red cinnabar)　银朱；朱砂；硫化汞
mercuric sulfocyanate(=mercuric thiocyanate)　硫氰酸汞
mercuric sulfocyanide(=mercuric thiocyanate)　硫氰酸汞
mercuric thiocyanate(=mercuric sulfocyanate; mercuric sulfocyanide; mercuric thiocyanide)　硫氰酸汞　$Hg(SCN)_2$
mercuric thiocyanide(=mercuric thiocyanate)　硫氰酸汞
mercuride　汞化物
mercurification　汞合金化
mercurimetric determination　汞量滴定法
mercurimetric estimation　汞液滴定法

mercurimetric titration 汞量滴定法
mercurimetry 汞量法*
mercuriovegetal 番茉莉
mercurius 汞
mercurius praecipitatus 氧化汞
mercurizate ①汞化②汞化产物
mercurizated 汞化了的；用汞处理了的
mercurizing 汞化
mercurizing agent 汞化剂
mercurization 加汞(作用)；汞化(作用)
mercurization test 汞化试验；乙酸汞试验
mercuroammonium 亚汞代铵(基)
mercurochrome 红汞；红药水〔俗〕；羰汞基荧光黄钠盐
mercurometric titration(=mercurometry) 亚汞量滴定
mercurophen 汞酚盐
mercurophylline 汞茶碱
mercurous 亚汞的；一价汞的
mercurous acetate 乙酸亚汞 $HgC_2H_3O_2$
mercurous azide 叠氮化亚汞
mercurous bitartrate 酒石酸氢亚汞 $HC_4H_4O_6Hg$
mercurous bromate 溴酸亚汞
mercurous bromide 溴化亚汞 $HgBr$
mercurous carbonate 碳酸亚汞 Hg_2CO_3
mercurous chloride 氯化亚汞；甘汞
mercurous citrate 柠檬酸亚汞
mercurous compound 亚汞化合物；一价汞化合物
mercurous cyanide 氰化亚汞 $HgCN$
mercurousdiammonium acetate 乙酸亚汞代二铵
mercurous iodide 碘化亚汞 HgI
mercurous iodobenzene-p-sulfonate 碘苯对磺酸亚汞
mercurous nitrate 硝酸亚汞 $HgNO_3$
mercurous oxide 氧化亚汞 Hg_2O
mercurous phosphate 磷酸亚汞 Hg_3PO_4
mercurous potassium tartrate 酒石酸亚汞钾
mercurous sulfate 硫酸亚汞 Hg_2SO_4
mercurous sulfide 硫化亚汞 Hg_2S
mercurous tartrate 酒石酸亚汞 $Hg_2C_4H_4O_6$
mercuroxyammonium 亚汞氧铵
mercury 汞；水银
mercury accumulation 汞蓄积
mercury air pump(=mercury pump) 汞(蒸气)泵
mercury alkyl(=mercury alkylide) 烷基汞
mercury alkylide(=mercury alkyl) 烷基汞 HgR_2
mercury alloys 汞合金
mercury amalgams (液态)汞合金
mercury arc 汞弧
mercury arc lamp 水银弧光灯
mercury arylide (二)芳基汞 HgR_2

mercury azide 叠氮化汞
mercury balance manostat 汞平衡衡压器
mercury bath ①汞浴②汞浴器
mercury bichloride 氯化汞；升汞 $HgCl_2$
mercury biphenyl 联苯汞
mercury bulb 汞球管
mercury cadmium telluride detector(=MCT detector) 碲化汞-碲化镉复合半导体检测器；MCT 检测器
mercury cathode cell 汞阴极电池
mercury cathode electrolysis 汞阴极电解法
mercury cathode separation 汞阴极分离法
mercury chloride 氯化汞
mercury column 汞柱；水银柱
mercury commutator 汞换向器
mercury compound 汞化合物；汞剂
mercury contact 汞电钥
mercury (contact) cup 汞(接)电盅
mercury-containing sludge 含汞污泥
mercury contamination 汞污染
mercury coulombmeter 汞极电量计；汞极库仑计；水银电量计
mercury density of catalyst 催化剂的汞密度
mercury deposition 汞沉积
mercury dialkyl (二)烷基汞 HgR_2
mercury p,p-dianiline 对二氨基苯基汞 $C_{12}H_{12}N_2Hg$
mercury dibenzyl 二苄汞
mercury diethide(=mercury diethyl) 二乙基汞 $Hg(C_2H_5)_2$
mercury diethyl(=mercury diethide) 二乙基汞
mercury diffusion pump 水银扩散泵
mercury dimethide(=mercury dimethyl) 二甲基汞 $Hg(CH_3)_2$
mercury dimethyl(=mercury dimethide) 二甲基汞
mercury diphenide(=mercury diphenyl) 二苯基汞 $Hg(C_6H_5)_2$
mercury diphenyl(=mercury diphenide) 二苯基汞
mercury displacement (immersion) method 水银置换(浸入)法
mercury distant-reading thermometer 远距离水银温度计
mercury drop electrode 滴汞电极
mercury drop microelectrode 汞滴微电极
mercury dropping 汞滴(电极)
mercury electrode 汞电极；水银电极
mercury ethide(=mercury ethyl) 二乙基汞 $Hg(C_2H_5)_2$
mercury ethyl(=mercury ethide) 二乙基汞
mercury ethyl mercaptide 乙硫醇汞；二乙硫基汞 $(C_2H_5S)_2Hg$
mercury film electrode 汞膜电极

mercury freezing test　水银冷却试验
mercury fulminate　雷酸汞；雷汞　$Hg(ONC)_2$
mercury fume　含汞烟气
mercury furnace　炼汞炉
mercury gatherer　集汞器
mercury gauge　①汞测压计②汞柱③汞表
mercury helide　氦化汞
mercury immersion method　水银浸入法
mercury-in-glass thermometer　汞柱玻璃温度计
mercury interrupter　水银断续器
mercury intrusion method　压汞法
mercury jet electrode　汞流电极
mercury lamp　汞灯
mercury manometer　水银压力计；U形压力计
mercury mercaptide　①硫醇汞　$Hg(SR)_2$②〔专指〕乙硫醇汞　$Hg(SC_2H_5)_2$
mercury-mercurous chloride electrode　汞-甘汞电极
mercury methide(=mercury methyl)　二甲基汞　$Hg(CH_3)_2$
mercury method　水银法〔树脂熔点测定法〕
mercury methyl(=mercury methide)　二甲基汞
mercury minerals　汞矿
mercury naphthide　二萘基汞
mercury nitrate　硝酸汞
mercury ointment　汞软膏
mercury ore　汞矿；水银矿石
mercury oxide　汞氧化物〔包括氧化汞和氧化亚汞〕　HgO；$[Hg_2]O$
mercury penetration method　汞透入法；压汞法
mercury periodide　过碘化汞　$Hg[I_6]$
mercury peroxide　过氧化(亚)汞　$Hg[O_2]$
mercury phenide(=mercury phenyl)　二苯基汞　$Hg(C_6H_5)_2$
mercury phenyl(=mercury phenide)　二苯基汞
mercury pipette　汞吸移管
mercury plated graphite electrode　镀汞石墨电极
mercury plug method　汞塞法
mercury poisoning　汞中毒
mercury pollution　汞污染
mercury pool cathode　汞池阴极
mercury pool electrode　汞池电极*
mercury porosimetry　压汞仪法
mercury porosity-meter　水银孔率测定计；水银空隙率检测计
mercury potassium iodide solution　碘化汞钾溶液
mercury (pressure) gauge　汞压力计
mercury pressure porosimeter　压汞仪
mercury process　用水银作热载体的过程

mercury protochloride　氯化亚汞；一氯化汞　$HgCl$
mercury protoiodide　碘化亚汞　HgI
mercury protoxide　氧化亚汞　$[Hg_2]O$
mercury pump　汞泵；水银泵
mercury purifying tunnel　纯化汞用漏斗
mercury rectifier　水银整流器
mercury-refined lubricant　水银(蒸气蒸馏法)精制的润滑剂
mercury reservoir　水银储存器；存汞池
mercury saccharate　蔗糖汞
mercury scoop　水银捕集器；捕汞器；汞阱
mercury seal　汞封(口)
mercury-sealed　汞封(口)的
mercury-sealed stirrer　水银密封搅拌器；汞封搅拌器
mercury seal flask　汞封(烧)瓶
mercury seal pot　水银密封釜
mercury seal stop-cock　汞封栓
mercury subchloride　一氯化汞；氯化亚汞　$HgCl$
mercury sulfide red　硫化汞红；银朱〔俗〕
mercury-surface lamp　汞(反射)面灯
mercury thermometer　水银温度计
mercury thermoregulators　水银温度调节器
mercury thiocyanate reagent　硫氰酸汞试剂
mercury tongs　汞钳
mercury trap　汞阱；汞捕集器
mercury-type instrument　水银式仪表
mercury vacuum gauge　水银真空计
mercury vacuum pump　水银真空泵
mercury valve　汞(安全)阀
mercury vapor　汞蒸气
mercury vapor (air) pump　汞气泵；汞气(扩散)抽机
mercury vapor jet pump　汞汽喷射泵
mercury (vapor) lamp　汞(汽)灯；水银灯
mergal　胆酸汞鞣酸白蛋白复剂〔药〕
merged peak　合并峰
merging zones technique　合并带技术
meridian　经线；子午线
meridianal isomer　经式异构体
meridian reflection　子午线反射
meridional long spacing reflection　子午线长间距反射
meridional stress　径向应力
meridional velocity　平均速度〔叶片中流体〕
Merimeés yellow　锑酸铅〔黄颜料〕
merino　细纺羊毛纤维
merino wool　美利奴羊毛
meriquinone　1-甲基-4-苯基哌啶-4-羧酸乙酯；维生素K_2
merit number　钢质量数
Merkel chart diagram　默克尔线图〔浓度-熵线图〕

mer mole 摩尔基体量
mero- 〔词头〕部分；局部
merochrome 色异构晶体
merocyanine 部花青
merocyanine-type compound 部花青型化合物
merodesmosine 开链赖氨素
merohedric 晶体缺面的
meropenem 美罗培南〔药〕
meropolar form 半极式
meroquinene 部奎宁 $C_9H_{15}O_2N$
merotropism(=desmotropism) 稳变异构(现象)
merotropy 稳定异构(现象)
Merox process 梅洛克斯硫醇氧化法
merron 质子
merry-go-round photoreactor (立式)旋转光(化学)反应试验仪
mersalyl 汞撒利〔药〕
mersolates 石油磺酸(钠)盐；烷基磺酸(钠)盐
mertenyl 墨烯基
mer weight 基体量
merwinite 镁硅钙石
Merz-Colwell extrusion rheometer 梅尔茨-科威尔挤压式流变仪
mes-(=meso-) 〔希腊字头〕①内消旋②中(间)的③中(位)④介⑤新
mesaconate 中康酸盐(或酯)
mesaconic acid(=methylfumaric acid) 中康酸；甲基富马酸；甲基反式丁烯二酸 $HOOCC(CH_3)=CHCOOH$
mesaconoyl 中康酰 $—COC(CH_3)=CHCO—$
mesalazine 美沙拉嗪〔药〕
mesantoin 美芬妥英；美山妥因〔抗癫痫药〕
mescal buttons 仙人掌芽
mescaline 墨斯卡林；仙人球毒碱(从仙人球中提取的致幻剂)
mesembrene 松叶菊烯
mesembrine 番杏碱；松叶菊碱
mesh (筛)目*
mesh analysis 筛析
mesh cathode 网状阴极
mesh count 网目数
mesh current 网络电流；槽路电流
mesh fabric 网眼织物
mesh filter 筛网过滤器
mesh gauze filter 筛网过滤器
mesh number 筛目；网号；网目号
mesh-of-grind 磨细筛目；研磨细度；研磨粒度〔以通过筛孔的物料百分数计〕
mesh point 网点；网格点；啮合点

mesh range 筛目范围
mesh screen 筛目
mesh sieve 网(状)筛
mesh size 筛目大小
mesic atom 介原子
mesic atom chemistry(=mesoatom chemistry) 介原子化学
mesic chemistry 介子化学*
mesicerin 莱基甘油
mesic molecule 介分子
mesidine(=2,4,6-trimethyl aniline) 莱胺；2,4,6-三甲基苯胺 $(CH_3)_3C_6H_2NH_2$
mesidino(=2,4,6-trimethylanilino) 莱氨基；间三甲苯氨基
mesite 中性岩〔pH 2~2.5〕
mesitene lactone 二甲基香豆灵
mesitic acid 莱酸；5-甲基邻苯二甲酸 $CH_3C_6H_3(COOH)_2$
mesitilol(=mesitylene) 1,3,5-三甲基苯
mesitine spar 菱铁镁矿
mesitite 菱铁镁矿
mesitol(=2,4,6-trimethyl phenol) 莱酚；2,4,6-三甲苯酚 $(CH_3)_3C_6H_2OH$
mesiton 3-二甲氨(基)甲酰氧基-1-甲基吡啶
mesitonic acid 莱酮酸 $C_7H_{12}O_3$
mesitoyl 莱酰
mesityl 莱基；2,4,6-三甲基苯基 $(CH_3)_3C_6H_2—$
mesityl alcohol 莱酚；2,4,6-三甲苯酚 $(CH_3)_3C_6H_2OH$
mesityl carbinol 莱基甲醇；2,4,6-三甲基苄醇
mesitylene(=1,3,5-trimethylbenzene) 莱；1,3,5-三甲基苯 $(CH_3)_3C_6H_3$
mesitylene alcohol 2,4,6-三甲基酚
mesitylene carboxylic acid 2,4,6-三甲基苯甲酸
mesitylene glycerol 莱甘油
mesitylene glycol 莱二醇；甲基苯二甲醇；3,5-二羟甲基甲苯
mesitylene lactone 二甲基香豆灵
mesitylene sulfonyl chloride 均三甲(基)苯磺酰氯
mesitylenic acid(=mesitylinic acid) 莱林酸；3,5-二甲基苯甲酸 $(CH_3)_2C_6H_3COOH$
mesitylenic aldehyde 莱醛；3,5-二甲基苯甲醛
mesityl heptadecyl ketone 莱基十七基(甲)酮；1,3,5-三苄基十七烷基(甲)酮
mesitylic acid 2,4,4-三甲基-5-氧代脯氨酸
mesitylinic acid 莱林酸；3,5-二甲基苯甲酸
mesitylol(=mesitylene) 莱；1,3,5-三甲基苯 C_9H_{12}
mesityl oxide 异亚丙基丙酮；莱基化氧
mesityloxy 莱氧基；2,4,6-三甲苯氧基 $(CH_3)_3C_6H_2O—$

mesna 美司钠〔药〕
meso-(=mes-)〔希腊字头〕 ①内消旋②中(间)的③中(位)④介⑤新
mesoatom chemistry(=mesic atom chemistry) 介原子化学
mesoboric acid 中硼酸 $H_4B_2O_7$; $HBO \cdot H_3BO_3$; $B_2O_3 \cdot 2H_2O$
mesochemistry 介子化学
mesocolloid 介胶体；近胶体
meso compound 内消旋化合物
meso-cortex 间皮质
mesodibromosuccinic acid 中(均)二溴丁二酸；中(均)二溴琥珀酸
meso-dihydroguaiaretic acid 内消旋二氢愈创(木树脂)酸
meso-form 内消旋型
mesogen 介晶
mesogenic group 内消旋配合基
Mesogen-Jacketed polymer 甲壳型液晶高分子
mesohydric tautomerism 氢原子振动互变异构现象
mesohydro tautomerism 氢原子互变异构现象
mesohydry 氢原子振动异构(现象)
meso-inositol 内消旋肌醇；内消旋环己六醇
meso-ionic compound 介-离子化合物
mesoisomer 内消旋异构体
mesomer 内消旋体
mesomeric ①内消旋的②中介的
mesomeric effect 中介效应*
mesomeric ion 中介离子
mesomeric polarization 中介极化作用
mesomeric state ①中介态②稳态
mesomeride 内消旋体
mesomerism ①中介(现象)②稳变异构(现象)；缓变异构(现象)
meso method 半微量法
mesomethylene carbon 亚甲基(桥型)碳〔形成樟脑环内桥的碳原子〕
mesomorphic 介晶的
mesomorphic liquid-crystalline 介晶液晶；变异液晶
mesomorphic phase 介晶相；液晶相
mesomorphic solvent 介晶溶剂
mesomorphic state 介晶态
mesomorphic structure 介晶结构
mesomorphism 介晶性
mesomorphous 平行原子层杂乱交错的
mesomorphous phase 介晶相
mesomorphous region 介晶区；准晶区
meson(=heavy electron) 重电子；介子
meso-naphthadianthrene 中萘二蒽；萘并二蒽
meso-naphthadianthrone 中萘二蒽酮

mesonic atom 介子原子
mesonium 介子素
mesonium chemistry 介子素化学
mesoparaffins 含两个次异丙基的烷烃
meso-periodic acid 中高碘酸；二缩原高碘酸 $I_2O_7 \cdot 3H_2O$; H_3IO_5
mesoperrhenate 中高铼酸盐 M_3ReO_5
mesophase 中间相
mesophasic structure 中间相结构
mesophilic 亲中介态的
mesopore 中孔
mesoporous 中孔(性)的
mesoporphyrin 中卟啉
mesoporphyrinogen 中卟啉原
meso-position ①(杂环异原子)中位②(蒽的)9,10位；中位
mesopyridine 消旋吡啶
mesorcin 苯间二酚
mesorcinol 苯间二酚；均三甲苯二酚 $(CH_3)_3C_6H(OH)_3$
mesoscale 中等尺度；中等规模
mesosilicic acid 中硅酸
mesosome 间体
mesotartaric acid 中酒石酸；内消旋酒石酸 $(CHOHCO_2H)_2$
mesotartrate 中酒石酸盐
meso-thorium 新钍
mesothorium I 新钍 I ^{228}Ra
mesothorium II 新钍 II ^{228}Ac
mesothorium mud 含碳酸新钍和碳酸镭的碳酸钡
mesotomy 内消旋体离析
mesotron(=meson) 介子；重电子
mesoxalate 二羟丙二酸盐(或酯)
mesoxalic acid 中草酸；丙酮二酸 COOHCOCOOH
mesoxalic acid oxime (中)肟基丙二酸
mesoxalyl 中草酰；丙酮二酰 —COCOCO—
mesoxalyl urea 中草酰脲；丙酮二酰脲 $C_4H_2O_4N_2$
mesquite gum 牧豆树胶
messanomycin(=mezzanomycin) 梅扎诺霉素
messenger RNA(mRNA) 信使 RNA
mestinon(=pyridostigmine bromide) 溴化 3-二甲氨基甲酰氧基-1-甲基吡啶；溴吡斯的明；美斯地浓〔重症肌无力用药〕
mesulphen 二甲基噻蒽；甲硫酚〔主要用于杀疥虫，有显著的止痒与消炎的效果〕
mesyl(=methylsulfonyl) 甲磺酰基
mesylation 甲磺酰化(作用)
met-(=meta-)〔希腊字头〕 ①间(位)〔有机系统名用〕②介〔术语用〕③偏〔无机酸用〕④介〔俗名用,

有时指一定数目的聚合物,有时泛指聚合物〕
meta-(=met-)〔希腊字头〕 ①间(位)〔有机系统名用〕 ②介〔术语用〕③偏〔无机酸用〕④介〔俗名用,有时指一定数目的聚合物,有时泛指聚合物〕
metaacetaldehyde 三聚乙醛
metaacetone 3-戊酮
meta-acid 偏酸〔无机酸〕
meta-aluminate 偏铝酸盐 $MAlO_2$
meta-aluminic acid 偏铝酸 $HAlO_2$
meta-antimonate 偏锑酸盐 $MSbO_3$
meta-antimonic acid 偏锑酸 $HSbO_3$
meta-antimonite 偏亚锑酸盐 $MSbO_2$
meta-antimonous acid 偏亚锑酸 $HSbO_2$
meta-arsenate 偏砷酸盐 $MAsO_3$
meta-arsenic acid 偏砷酸 $HAsO_3$
meta-arsenite 偏亚砷酸盐 $MAsO_2$
meta-arsenous acid 偏亚砷酸 $HAsO_2$
metabisulfite 偏亚硫酸氢盐
metabituminous coal 肥煤;中烟煤
metabolic 代谢的
metabolic antagonism 代谢拮抗
metabolic balance 代谢平衡
metabolic inhibition test 代谢抑制试验
metabolic rate 代谢率
metabolic rate analyzer 代谢率分析仪
metabolimeter 基础代谢计
metabolism sensor 代谢型传感器
metabolite 代谢物
metabonomics 代谢物组学
metaborate 偏硼酸盐 MBO_2
metaboric acid 偏硼酸 HBO_2
metacarbonic acid(=carbonic acid) 碳酸 $H_2CO_3;CO(OH)_2$
metacellulose 异纤维素
metacenter 定倾中心
metacetaldehyde 变乙醛;聚乙醛
metacetin N-乙酰对甲氧基苯胺
metacetone 二乙基(甲)酮 $(C_2H_5)_2CO$
metacetonic acid 丙酸 CH_3CH_2COOH
metachemistry 原子结构(化)学
metachloral 介氯醛
metachromasia 因光异色现象;光致变色现象
metachromasy 异染性;变色反应性
metachromatic 因光异色的
metachromic acid 偏铬酸 H_2CrO_4
metacinnabarite 黑辰砂矿
meta-compound 间位化合物
meta cresol 间甲酚
meta cresol purple 间甲酚红紫
meta cresol sulfonphthalein 间甲酚磺酚
metacrolein 介丙烯醛;三聚丙烯醛
metacrylic acid ester 甲基丙烯酸酯
metacryloxy 甲基丙烯酰氧
metacryotic 冻析的
metacrystallic rocks〔复〕 变晶岩
metacrystalline 亚晶的
meta-derivative 间位衍生物
metadiazine(=pyrimidine) 嘧啶
meta directing group 间位定位基*
metadurain 变质暗煤
metadyne 加速-制动联合机
meta-element 母体元素
metaeollagen 变胶原〔高度交联的胶原〕
metafilter 层滤机;层滤器
metafiltration 层滤
metaformaldehyde 变甲醛;聚甲醛
metafuel 三聚乙醛片〔酒精灯燃料〕
metagelatin 变性明胶
metagneiss 变片麻岩
metahemipinic acid 4,5-二甲氧基邻苯二甲酸
metaindic acid 偏铟酸 $HInO_2$
metaiodate (偏)碘酸
meta isomeride 位变异构体
metaisomerism (双键)位变异构现象
meta-isopropyl biphenyl 间异丙基联苯 $C_6H_5C_6H_4CH(CH_3)_2$
metakaolin 偏高岭石
metakaolinite 二水高岭石
metakliny 基团位变
metal 金属
metal accumulation 金属累积作用
metal acetylide 乙炔基金属
metal-adsorbent char 金属吸附炭
metal-air battery 金属-空气电池(组)
metal alcoholate (金属)醇化物;烃氧基金属(盐)
metal alkanoate 金属链烷酸盐
metal alkoxide 金属醇盐;烷氧基金属(盐)
metal alkyl(=metal alkylide) 烷基金属
metal alkylide(=metal alkyl) 烷基金属 MR_n
metal alloy 金属合金
metal alloy coating (金属)合金涂料;(金属)合金涂层
metalammine (金属)氨合物
metalammonia (金属)氨合物
metal arc welding 金属电弧焊
metal arene 金属芳烃
metal aryl(=metal arylide) 芳基金属
metal arylide(=metal aryl) 芳基金属 MR_n

metalation　金属化作用
metal autoclave　金属压热器
metal band conveyor　钢条运输带
metal barometer　金属气压计
metal bath　①金属浴②金属浴器
metal binding　金属键联
metal binding site　金属结合部位
metal block calorimeter　金属热量计
metal block injector　金属块注射器
metal-body lacquer　金属胎漆器〔大漆〕
metal carbonyl　羰基金属
metal carboxylate　金属羧酸盐
metal catalyst　金属催化剂
metal cation　金属阳离子
metal ceramic　金属陶瓷
metal-ceramic coating　金属陶瓷涂层；金属陶瓷涂料
metal ceramic filter　金属陶瓷过滤器
metalceramics　金属陶瓷
metal chelate compound　金属螯合(化合)物
metal-chelated　金属螯合了的
metal chelate extraction　金属螯合物萃取
metal chelate pigment　金属螯合型颜料
metal chelating dyes　金属螯合染料
metalclad　(金属)铠装的；金属包盖的
metal-clad laminate thickness　覆箔板厚度
metal cleaner　金属洗净剂；金属清洗剂
metal cleaner soap　金属洗净皂；金属清洗皂
metal cleaning　金属净洗
metal cleaning soap　金属洗净皂；金属清洗皂
metal cluster　金属簇
metal-coated fibre　镀金属纤维；金属镀覆纤维
metal coated glass fibre　涂金属玻璃纤维
metal coating　①电镀金属法；金属镀膜②金属被膜
metal complex　金属配合物
metal complex catalyst　金属络合催化剂
metal complex dye　金属配位染料
metal complex ion chromatography　金属配合物离子色谱法
metal complex toner　金属配位(型)色料
metal compound　金属间化合物
metal conditioner(=wash primer)　磷化底漆；洗涤底漆
metal-conditioning treatment　金属磷化处理
metal container　金属容器
metal-containing polymer　含金属聚合物
metal contamination on catalyst　催化剂金属中毒
metal corrosion inhibitor　金属腐蚀抑制剂
metal cross linking agent　金属交联剂
metal cutting machine　金属切削机床

metal cutting oil　金属切削油
metal deactivator　金属钝化剂；金属减活化剂
metal decorating finish　金属彩板漆
metal decoration　①金属涂装②金属装潢③彩板印刷
metaldehyde　①聚乙醛②〔专指〕四聚乙醛
metal deposition　金属沉积法
metal disc fuel filter　金属盘燃料过滤器
metal dispersion　金属分散剂
metal-effect vinyl plastic wall paper　有金属效果的乙烯系塑料墙纸
metal electrode　金属电极
metalepsis　取代(作用)
metalepsy　取代作用
metal exchange　金属交换*
metal-faced plywood(=plymax)　金属贴面胶合板
metal fiber　金属纤维；金属丝
metal filler　金属填充剂
metal filter elements　金属过滤器的元件〔指金属托板〕
metal finish　金属涂装
metal finishing　①金属表面处理②金属涂装
metalfluorescent indicator　金属荧光指示剂
metal fog　金属雾
metal foil　金属箔
metal-foil packing　金属箔垫圈
metal founding　金属铸件
metal-free　不含金属的
metal-free phthalocyanine　无金属酞菁颜料
metal-free phthalocyanine blue　无金属酞菁蓝
metal-free porphyrin　无金属卟啉
metal gate　金属栅；金属控制极
metal gauze　金属(丝)网
metal-glazing putty　金属件镶玻璃腻子
metal hose　金属软管；柔性管
metal hydride　金属氢化物；氢化金属
metal hydroxide　金属氢氧化物
metal-hydroxy group　金属-羟基
metalignitious　不结块的
metalimmunoassay(MIA)　金属免疫分析
metal indicator(=metallochromic indicator)　金属指示剂
metal ink　金属油墨
metal inlaying　镶金属
metal insert　金属配件
metal interaction chromatography　金属作用色谱
metal ion buffer　金属离子缓冲液
metal ion buffer solution　金属离子缓冲液
metal ion chelation agent　金属离子螯合剂
metal ion concentration exponent　金属离子浓度指数
metal ion indicator　金属离子指示剂

metal ion stability　(对)金属离子稳定性
metalized azo yellow　闪金光偶氮黄
metal ketyl radical　金属羰游离基　R₂C—ONa
metal lacquer　金属用(喷)漆
metallation　金属化*
metal lead powder　金属铅粉
metal lead primer　金属铅粉底漆
Metallex process　迈塔莱克斯法〔用汞齐将四氯化铀还原成金属铀的方法〕
metallic　金属的
metallic activator(=metallic promoter)　金属性活化剂
metallic alkide　烃基金属
metallic arc　金属电弧
metallic automotive paint　闪(金属)光(泽)汽车漆
metallic bead　金属珠
metallic bond　金属键
metallic brown　锻铁棕
metallic brush　金属刷；钢丝刷
metallic carbonyls　羰基金属
metallic chain backbone polymer　金属主链聚合物
metallic character　金属特性
metallic-coated specimen　镀金属样板
metallic coating　闪(金属)光(泽)涂料；金属涂层
metallic color　金属色
metallic complex salt　金属络盐
metallic conductance　①金属传导②金属电导
metallic conductor　①金属导电体②金属传导体
metallic creep　金属型蠕变
metallic crystal　金属结晶
metallic diaphragm　金属膜层；金属膜片
metallic dip-spin coating　离心浸涂闪光涂料
metallic electrode　金属电极
metallic element　金属元素
metallic fiber　金属纤维
metallic filament　金属丝
metallic film　金属薄膜
metallic filter media　金属过滤介质
metallic finishes　闪光面漆；闪(金属)光(泽)面漆
metallic flake pigment　片状金属颜料
metallic flexible tubing　金属软管
metallic glycerinate　甘油酸金属盐
metallic grit　金属(砂)粒；钢砂〔喷砂处理用的不规则钢砂〕
metallic hydrogen　金属氢
metallic inclusions　内含金属物
metallicity　金属性
metallic lacquer　闪(金属)光漆
metallic lead　金属铅
metallic lead paint　铅粉漆

metallic lubricant　金属润滑剂
metallic luster　金属光泽
metallic lustre　金属光泽
metallic manometer　金属压力计
metallic matrix structural composite　金属基质结构复合材料
metallic nano material　金属纳米材料
metallic naphthenate　环烷酸金属盐
metallic nitroxyls　硝酰金属
metallic packing　金属填料
metallic paint　金属涂料
metallic paper　金属纸
metallic pigment　金属(性)颜料
metallic poison　金属毒
metallic potassim　金属钾
metal(lic) powder　①金属粉(末)②金银粉；铜粉；铝粉
metallic promoter(=metallic activator)　金属性活化剂
metallic radius　金属半径*
metallic resinate　树脂酸金属盐
metallic rosinate　松香酸金属盐
metallics〔复〕　金属粒子
metal(lic) salt　金属盐
metallic sheen　金属光泽
metallic shot　(形状规则的)金属球；钢丸；钢珠〔喷丸处理用〕
metallic soap　金属皂
metallic soap drier　金属皂催干剂
metallic soda reducing process　金属钠还原法
metallic sodium　金属钠
metallic spatula　金属刮刀；钢刮刀
metal(lic) stabilizer　金属稳定剂
metallic stearate　硬脂酸金属盐
metallic stuffing　金属填料
metallic thermometer　金属温度计
metallic titanium-made photoreaction tank　金属钛制光反应槽
metallic travel　(金属)闪光效果；(金属)闪光度
metallic valence　金属价
metallic wallcovering　①墙壁裱糊用金属材料②墙壁装潢用金属材料
metallic yarn　金属丝；金属纤维纱
metalliferous　含金属的
metalliferous dye　(含)金属染料
metalligating　金属结扎
metallikon　金属熔融喷涂
metalline　像金属的
metalling　盖以金属
metallised plastic　镀金属塑料

metallising process 镀金属工艺(过程)；金属涂敷工艺
metallization 敷金属(法)
metallized acid dye 金属络合酸性染料
metallized azo yellow 闪金光偶氮黄
metallized ceramic coating(=ceramic-metal coating) 金属陶瓷涂料
metallized chamber (镀有)金属(薄膜)反应器
metallized coating 金属闪光涂料
metallized dye 金属配位染料
metallized fabric 金属化的网布
metallized fibre 镀金属纤维〔通过喷镀、真空淀积以及化学或电解淀积法，使镍、铜等金属敷镀于纤维表面〕
metallized finish 金属闪光面漆
metallized plastic 镀金属塑料
metallizing 敷镀金属；喷镀金属
metallizing gun 金属喷涂枪
metallo- 〔词头〕金属
metallo acetate metallic salt 金属取代的乙酸金属盐〔如 $Na \cdot CH_2COONa$〕
metal-loaded paint 金属粉涂料
metallocarborane 金属碳硼烷
metallocene 二茂金属配合物〔烯烃定向聚合催化剂〕
metallocene catalyst 茂金属催化剂
metallocene polymer 茂金属聚合物
metallochemistry 金属化学
metallochrome 金属着色剂
metallochromic indicator(=metal indicator) 金属指示剂
metallochromy 金属(表面)染色
metalloenzyme 金属酶
metallofluorescent indicator 金属荧光指示剂
metallogenic map 金属矿物分布图
metallographic 金相(学)的
metallographic hot cell 金相热室
metallographic phase 金相
metallographic test 金相试验
metallography 金相学
metalloid ①准金属②非金属
metalloidal crystal 准金属晶体
metalloid crystal 准金属晶体
metallomethyl siliconate 甲基硅酸金属盐
metallo-organic compound 有机金属化合物
metallo-organic curing catalyst 金属有机固化催化剂
metallo-organic polymer 金属有机聚合物
metalloporphyrin 金属卟啉
metalloprotein 金属蛋白
metallorganic 有机金属的
metalloscope 金相显微镜
metallothionein 金属硫蛋白

metallotrophy 金属移变作用
metal lubricant 金属(添加剂的)润滑材料
metallurgical ①冶金(学)的②冶金的
metallurgical analysis 金相分析；冶金分析
metallurgical coal 冶金煤；炼焦煤
metallurgical coke 冶金焦炭
metallurgical microscope 金相显微镜
metallurgist 冶金师；冶金学家
metallurgy 冶金学
metallyl alcohol(=2-methylallyl alcohol) 2-甲基丙烯醇 $CH_2=C(CH_3)CH_2OH$
metal machining liquid 金属切削液
metal marking (卷钢涂层上的)划痕；刮痕〔漆膜缺陷〕
metal matrix 金属基体；金属基质
metal-matrix composite 金属基质复合材料
metal matrix solid lubricant 金属基质固体润滑剂
metal membrane electrode 金属膜电极
metal-modified 金属改性的
metal nitride 金属氮化物
metalock 金属扣合法
metal organic sulfonate (有机)磺酸金属盐
metal oxide electrode 金属氧化物电极
metal oxide induced polymerization 金属氧化物诱导聚合作用
metal oxides 金属氧化物
metal oxides stationary phase 金属氧化物固定相
metal passivator 金属减活剂
metal phenates 金属酚盐
metal phenyl stearates 苯基硬脂酸金属盐
metal phthalein(=metalphthelein) 金属酞
metal phthalocyanine complex 金属酞菁配合物
metalphthelein(=metal phthalein) 金属酞
metal physics 金属物理(学)
metal pickling 金属的酸浸
metal pigment 金属颜料
metal-plastic adhesive 金属-塑料黏接剂
metal plating 金属镀盖；电镀
metal plating solution 金属镀液
metal polish 金属擦亮剂
metal powder printing 金属粉末印染〔纺〕
metal preparation 金属预处理
metal pre-treating primer 金属预处理底漆；磷化底漆
metal primer 金属打底剂；金属(用)底漆
metal print 印铁
metal processing section 金工工段
metal protective coating 金属保护涂料
metal-protective paint 金属保护漆
metal protector 金属防护剂

metal pulverization 金属粉末化
metal recovery pilot plant 金属回收中间工厂
metal reinforcement 金属补强
metal sanding surfacer 金属砂磨二道漆〔可打磨的二道浆〕
metals comparator 金属材料性能比测器
metal screen filter 金属网滤器
metal-sensitive indicator 金属敏感指示剂
metal sheet 金属薄板(铁皮)
metal shielding 铠装；金属加强层
metal sintered 烧结金属
metal soap 金属皂
metal-soap complexes 金属皂复合物
metal spectroscope 金属分光仪
metal spraying 金属喷镀
metal sulfides 金属硫化物
metal-support interaction 金属载体相互作用*
metal test for lubricating oil 润滑油金属含量测定〔以测定添加剂含量〕
metal toggle frame 金属绷皮架
metal-to-metal adhesive 金属黏合剂；金属胶
metal-to-metal peel strength 金属间黏结离强度；胶接金属剥离强度
metal toner 金属调色剂
metal transport process 金属输运过程
metal tube 金属管
metal vaporizing technique 金属蒸发技术；金属气化技术
metal vapors 金属蒸气
metal vapour discharge lamp 金属蒸气放电灯
metal vinylate 乙烯基金属化物
metal working fluid 金属加工液
metal working lubricant 金属加工润滑材料
metal working soap 金属加工皂
metamagnetism 变磁性
metamer 位变异构体
metamere ①体节②位变异构体
metameric 位变异构的
metameric match 条件等色；条件配色
metameric mismatch(=metamerism) 位变异构(现象)
metameric pair ①位变异构对；同分异构对②条件配色对；条件等色对
metameric state 位变异构态
metameride 位变异构体
metamerism(=metameric mismatch) 位变异构(现象)
metamerism index 条件等色指数
metamers 位变异构体
Metamic 梅塔米(克)金属陶瓷
metamict 原形的异构体位变异构体；介崩体〔一种由于射线作用引起晶格变化而在外形上仍保持原形的化合物〕
metamict mineral 乱晶矿物
metamizol(e) 安乃近
metamolybdic acid 偏钼酸 H_2MoO_4
metamorphic rock 变质(成)岩
metamorphism (岩石)变形现象
metamorphosis ①变形(作用)②变态③变质
metamycin 间霉素
metanephrine 去甲肾上腺素；变肾上腺素；3-O-甲基肾上腺素
metanethole 介茴香脑
metanilamido- 间氨基苯磺酰氨基 m-$H_2NC_6H_4SO_2NH$—
metanilic acid 间胺酸〔俗〕；间氨基苯磺酸 $H_2NC_6H_4SO_3H$
metaniline yellow (酸性)间胺黄
metanil yellow (酸性)间胺黄
metanilyl 间胺酰〔俗〕；间氨基苯磺酰 m—$H_2NC_6H_4SO_2$—
meta-nitroaniline 间-硝基苯胺
meta-nitrophenol 间硝基苯酚
meta-nitrophenoxide anion 间-硝基苯酚盐负离子
metanolic KOH solution 苛性钾甲醇溶液
metantimonate 偏锑酸盐
metantimonic acid 偏锑酸
meta-orientating group 间位定位基
meta-orientation 间位定向
meta-oriented aromatic ring 间位(取代的)芳环
meta-orienting group(=meta-orientating group) 间位定位基
metapeptone 胨的消化产物
metaperiodic acid 偏高碘酸 $I_2O_7 \cdot H_2O;HIO_4$
metaphen(=4-nitro-5-oxymercuro-o-cresol) 4-硝基-5-亚汞氧基邻甲酚；袂塔酚 $CH_3C_6H_2ONO_2Hg$
metaphenylene 间亚苯基
meta-phenylene bismaleimide 间亚苯基双马来酰亚胺〔硫化剂〕
metaphenylene diamine 间苯二胺
metaphenylene diamine hydrochloride 间苯二胺盐酸盐
N,N'-metaphenylene-dimaleimide N,N'-间亚苯基双马来酰亚胺〔硫化剂〕
metaphos 对硫磷
metaphosphate 偏磷酸盐
metaphosphite 偏亚磷酸盐 MPO_2
metaphosphoric acid 偏磷酸；二缩原磷酸 HPO_3
metaphosphoric ester 偏磷酸酯 $RPO_3;ROPO_2$
metaphosphorous acid 偏亚磷酸 HPO_2

metaphosphoryl chloride　偏磷酰氯　PO_2Cl
metaplumbate　偏高铅酸盐　M_2PbO_3
metaplumbic acid　偏高铅酸　H_2PbO_3
meta-position　间位
metapropionaldehyde　多聚丙醛
metapyrocatechase　变儿茶酚酶
metaraminol　间羟胺〔药〕
metaraminol (bi)tartrate　重酒石酸间羟胺；阿拉明；间酚胺〔药〕
metarchon　掩味物
metargon　仲氩
meta rheology　亚流变学
metarsenic acid　偏砷酸
metasilicate　硅酸盐　M_2SiO_3
metasilicate glass　硅酸盐玻璃
metasilicic acid　硅酸〔不叫偏硅酸〕　H_2SiO_3
metasomatic　交代的
metasomatism　交代作用
metasomatose　交代作用
metasomatosis　交代作用
metasome　交代矿物；代替矿物
metastability　①亚稳度②亚稳性
metastable　亚稳(态)的；介稳(态)的
metastable atom　亚稳原子
metastable compound　亚稳化合物
metastable condition　亚稳定状态
metastable crystallite　亚稳微晶；亚稳晶粒
metastable decomposition　亚稳分解
metastable equilibrium　亚稳平衡
metastable fibre　亚稳态纤维
metastable intermediate　亚稳中间体
metastable ion　亚稳离子*
metastable ion decay　亚稳离子裂解
metastable ion peak　亚稳离子峰
metastable limit　亚稳极限
metastable mapping　亚稳作图法
metastable peak　亚稳峰值
metastable phase　亚稳相
metastable scanning　亚稳扫描
metastable solid solution　亚稳态固溶体
metastable state　亚稳态
metastable supersaturation　亚稳过饱和
metastable transition　亚稳过渡；亚稳跃迁
metastannic acid　偏锡酸
metastasis　移位变化；失α微粒变化；转移
metastatic electron　移位电子；移层电子
metastatic thermometer　易位温度计
metastyrene　介苯乙烯　$(C_6H_5CH=CH_2)_n$

metastyrolene　介苯乙烯
meta-subsitution　间位取代(作用)
metasulfite　焦亚硫酸盐
metataxis　机械变化〔岩〕
metatenomeric change　(含氮物)趋稳重排(作用)
metatenomery　(含氮物)趋稳重排(作用)
meta-terphenyl　间三联苯　$C_6H_5C_6H_4C_6H_5$
metathesis　①复分解(作用)；置换(作用)②易位(作用)
metathesis polymerization　易位聚合
metathesis ring-open polymerization　易位开环聚合
metathetical　复分解的；置换的
metathetical reaction　复分解反应；置换反应
metathetic reaction　复分解反应
metathiazole　间噻唑
metatitanate　偏钛酸盐　M_2TiO_3
metatitanic acid　偏钛酸　H_2TiO_3
metatorbernite　偏铜铀云母
metatungstate　偏钨酸盐　$M_2O \cdot 4WO_3 \cdot H_2O$
metatungstic acid　偏钨酸
metavanadate　偏钒酸盐　MVO_3
metavanadic acid　偏钒酸　HVO_3
meta-xylene　间二甲苯
meta-xylylene adipamide　己二酰间苯二甲胺
metazonic acid　中棕酸
meteloidine　曼陀罗碱　$C_{13}H_{21}NO_4$
meteoric　陨降的
meteoric iron　陨铁
meteoric stone　陨石
meteoric water　降水
meteorite　陨石；陨铁
meteorograph　气象记录器；气象计
meteorology　气象学
metepa(=tris(methyl-1-aziridinyl)) phosphine oxide　三(甲基-1-吖丙环基)膦化氧；甲基涕巴
-meter〔词尾〕　计；器；仪
meter(=metre)　①米；公尺②计；表
meter board　仪表板
meter constant　仪表常数
meter diagram　仪表图
metered flow　计量流量
metered lubrication　计量润滑
metered type finish applicator　计量式上油装置
meter error　仪表误差
metering device　计量装置
metering pump　计量泵
metering tank　计量罐
metering valve　计量阀
metering volumetric pump　容积计量泵

metering zone 计量区
meter oil 计器油；计量器用油
meter pump 计量泵
meter reading 计器读数
meter scale 米；公尺
meter stick 米；公尺
metformin 二甲双胍〔药〕
meth- 〔词头〕甲
methacetin N-乙酰对甲氧基苯胺；美沙西丁 $C_9H_{11}O_2N$
methacholine chloride 氯化乙酰甲基胆碱〔药〕
methacrolein(=2-methylacrolein) 异丁烯醛；2-甲基丙烯醛 $CH_2=C(CH_3)CHO$
methacrylaldehyde 异丁烯醛；2-甲基丙烯醛
methacrylamide 甲基丙烯酰胺
2-methacrylamide-2-methylpropanesulfonic acid homopolymer 2-甲基丙烯酰胺-2-甲基丙烷-磺酸均聚物
methacrylate ①异丁烯酸②异丁烯酸盐(酯或根)
methacrylate-based adhesive 甲基丙烯酸酯基黏合剂
methacrylate-butadiene rubber 甲基丙烯酸酯丁二烯橡胶
methacrylate-butadiene-styrene(MBS) 甲基丙烯酸-丁二烯-苯乙烯共聚物
methacrylate ester 甲基丙烯酸酯
methacrylate plastic 甲基丙烯酸塑料
methacrylate-trimethoxysilyl propylmethacrylate copolymer 甲基丙烯酸酯-甲基丙烯酸三甲氧基甲硅(烷)基丙酯共聚物
methacrylato chromic (Ⅲ) chloride(=Volan) 甲基丙烯酸合四氯化铬(Ⅲ)；甲基丙烯酸根合(四)氯化(二)铬
methacryl glycidyl ether 甲基丙烯酰基缩水甘油醚〔稀释剂〕
methacrylic acid(=methylacrylic acid) 异丁烯酸；甲基丙烯酸 $CH_2=C(CH_3)CO_2H$
methacrylic acid ester 异丁烯酸酯；甲基丙烯酸酯
methacrylic chloride 甲基丙烯酰氯
methacrylic ester 甲基丙烯酸酯
methacrylic resin 甲基丙烯酸树脂
methacrylonitrile 甲基丙烯腈
methacryloxy ethyl phosphate 磷酸甲基丙烯酰氧乙基酯
γ-(methacryloxy) propyltrimethoxy silane γ-(甲基丙烯酰氧基)丙基三甲氧基硅烷〔偶联剂〕
methacryl(o)yl- 异丁烯酰(基)；2-甲(基)丙烯酰(基)
methadone 美沙酮〔药〕
methadone hydrochloride 盐酸美沙酮
methal 肉豆蔻醇
methaldehyde 四聚乙醛 $(C_2H_4O)_4$
methallyl(=2-methylallyl) 甲代烯丙基；2-甲代-1-烯基 $CH_2=C(CH_3)CH_2-$

methallyl acetate(=methylallyl acetate) 乙酸甲代烯丙酯
methallyl alcohol(=methylallyl alcohol) 甲代烯丙醇；2-甲基-2-丙烯-1-醇
methallyl amine 甲代烯丙基胺
methallyl bromide 甲代烯丙基溴 $CH_2=C(CH_3)CH_2Br$
methallyl butyrate 丁酸甲基烯丙酯
methallyl chloride 甲代烯丙基氯 $CH_2=C(CH_3)_2CH_2Cl$
methallyl halide 甲代烯丙基卤 $CH_2=C(CH_3)CH_2X$
methallyl iodide 甲代烯丙基碘 $CH_2=C(CH_3)CH_2I$
methallyl sulfonate 甲代烯丙基磺酸盐
methallyl sulfonic acid 甲代烯丙基磺酸
methamphetamine hydrochloride 脱氧麻黄碱盐酸盐〔药〕
methanal 甲醛
methanamide 甲酰胺
methanation 甲烷化作用
methanator 甲烷转化器
methane 甲烷；沼气 CH_4
methane acid 甲酸
methane alcohol 甲醇
methane amide 甲酰胺
methane analyzer 甲烷分析仪
methane arsonic acid 甲胂酸
methane base 双二甲氨基苯基甲烷
methane chloride 一氯甲烷
methanediamine 甲(基)二胺〔环氧固化剂〕
methane dicarboxylic acid 丙二酸
methanedisulfonic acid(=methionic acid) 甲二磺酸 $CH_2=(SO_2OH)_2$
methane fermentation (产)甲烷发酵(作用)
methane hydrocarbon 甲烷系烃
p-methane hydroperoxide 对甲烷过氧化氢〔催化剂〕
methane lean gas 甲烷贫气
methane phosphonic acid 甲膦酸
methane-rich gas 甲烷富气
methane-rich liquid 富甲烷液
methane-seleninic acid 甲基亚硒酸 CH_3SeOOH
methane-selenonic acid 甲基硒酸 CH_3SeO_2OH
methane series 甲烷系
methane-siliconic acid 甲基硅酸
methane stannonic acid 甲基锡酸
methane-steam process 甲烷-水蒸气法；甲烷与蒸汽产生氢气的过程
methane-steam reaction 甲烷-水蒸气反应
methane-sulfonic acid 甲磺酸 CH_3SO_3H
methane sulfonyl chloride 甲磺酰氯
methane sulfonyl fluoride 甲基磺酰氟
methane-tellurinic acid 甲基亚碲酸 CH_3TeOOH

methane-telluronic acid 甲基碲酸 CH_3TeO_2OH
methane thial 甲硫醛
methane thiol 甲硫醇
methanethiol *n*-butyrate 正丁酸甲硫醇酯
methane triacetic acid 甲三乙酸
methano- 亚甲基 —CH_2—
1,6-methano(10)annulene 1,6-亚甲基十碳轮烯
methanoate 甲酸盐(酯)
methanoic acid 甲酸 HCOOH
methanoindene 亚甲基茚
methanol 甲醇 CH_3OH
methanol column 甲醇塔
methanol fuel gel 甲醇胶体燃料
methanolic potassium 甲醇钾
methanolic PVA solution 聚乙烯醇甲醇溶液
methanolizing 甲醇化
methanol synthesis 甲醇合成(法)
methanolysis 甲醇分解(作用)
methanoyl 甲酰氧(基) HCOO—
methantheline bromide 溴甲胺太林；溴本辛〔药〕
methanthiol(=methyl mercaptan) 甲硫醇 CH_3SH
methapyrilene hydrochloride 盐酸美沙吡林〔药〕
methazolamide 醋甲唑胺；甲氮酰胺；甲醋唑胺；美舍唑咪〔药〕
methazonic acid 硝基乙醛肟
methedrine 脱氧麻黄碱
methedrine mandelate 苯乙醇酸脱氧麻黄碱
methemoglobin 高铁血红蛋白
methenamine 六亚甲基四胺；乌洛托品
methene(=methylene) 亚甲基
methene-disulfonic acid 甲二磺酸 $CH_2(SO_3H)_2$
methenyl 次甲基 CH≡
methenyl bromide 三溴甲烷；溴仿
methenyl chloride 三氯甲烷；氯仿
methenyl iodide 三碘甲烷；碘仿
N^5,N^{10}-methenyltetrahydrofolate N^5,N^{10}-次甲基四氢叶酸
methethyl 氯甲烷与氯乙烷混合物
methicillin〔商〕(=2,6-dimethyloxybenzo penicillin) 2,6-二甲氧基苯青霉素
methicillin sodium 2,6-二甲氧基苯基青霉素钠
methide 甲基化物〔如 zinc methide (二)甲基锌〕 $M(CH_3)_n$
methidium 甲锭〔DNA 嵌入剂〕
methimazol 甲巯咪唑；他巴唑〔药〕
methimazole 甲巯咪唑；他巴唑〔药〕
methine and polymethine colouring matter 次甲基和多次甲基染料〔次甲基指 HC≡〕
methiodal 甲碘磺酸钠

methiodal sodium 碘甲磺酸钠
methiodide(=iodomethylate) 甲碘化物；甲碘
methionate 甲二磺酸盐 $M_1O_3SCH_2SO_3M_1$
methionic acid 甲二磺酸 $CH_2(SO_2OH)_2$
methionine 甲硫氨酸；蛋氨酸 $CH_3S(CH_2)_2CHNH_2CO_2H$
methionine synthetase 甲硫氨酸合成酶
methionine transadenosylase 甲硫氨酸转腺苷基酶；甲硫氨酸激酶
methionyl ①甲二磺酰(基) $CH_2(SO_2—)_2$②甲硫氨酰(基) $CH_3SCH_2CH_2CH(NH_2)CO$
methionyl-tRNA transformylase 甲硫氨酰 tRNA 转甲酰基酶
metho- N-甲〔一般指代在 N 上，有时指连在侧链上〕
methobottromycin 甲基波卓霉素
methocel solution 甲基纤维素溶液
methochloride 甲氯化物〔CH_3Cl 的化合物, metho 指甲代在 N 上〕
method by inserting steel sheet on guide way 导轨镶钢法
method for repair or replace freezing water pipe 冷冻法修换水管
methodic error 方法误差
method interference 方法干扰
method of addition 增量法
method of approximation 近似法
method of chamber electrode 室电极法
method of cold vapor 冷蒸气法
method of comparison 比较法
method of continuous variation 连续变化法
method of control sample 控制试样法
method of conversion factor 换算因数法
method of determination of vapor density 蒸气密度测定法
method of direct arc combustion 直接弧烧法
method of direct comparison 直接比较法
method of energy divergence 能量色散法
method of enrichment 浓缩法；富集法
method of evaporation 蒸发法
method of external standard 外标法
method of general analytical curve 持久曲线法
method of Hartman interpolation 哈特曼内插法
method of identification 鉴定法
method of internal normalization 内标归一法
method of internal standard reagent(MISR) 内部标记法
method of lacquering ①涂漆方法②髹饰法；髹漆法〔大漆〕
method of least square 最小二乘法
method of measurement 测量方法
method of non-linear interpolation 非线性内插法

method of operation 操作法；作业法
method of particle size measurement 粒度测定法
method of peak absorption measurement 峰值吸收测量法
method of pendant drop 悬滴法
method of pH control pH 控制法
method of photometric interpolation 光度插入法
method of reference sample 参考试样法
method of rotating sector disk 旋转扇形板法
method of sampling 取样法
method of sampling hydraulic cement 水硬水泥取样法
method of sedimentation equilibrium 沉积平衡法
method of sedimentation velocity 沉降速度法
method of spectral line density comparison 谱线黑度比较法
method of steady graph 固定曲线法
method of substitution 取代法
method of successive approximation 逐次逼近法
method of swelling cut fibre end 切断纤维端溶胀法〔测定纤维定型度〕
method of upright electrode 直立电极法
method of water sampling 水样采集法；采水法
method of weight coefficient 权重系数法
method of zero suppression 零点抑制法
methods of environmental analysis 环境分析方法
method standard 方法标准
methohexitone (sodium) 美索比妥；甲己炔巴比妥；5-烯丙基-1-甲基-5-(1-甲基-2-戊炔基)巴比妥酸〔短效巴比妥类静脉麻醉药〕
methohydroxide 甲羟化物
methoin 5-乙基-3-甲基-5-苯基咪唑-2,4-二酮；美芬妥英
methomethylsulfate 甲基硫酸甲酯化物
methonitrate 硝酸甲酯化物〔MeNO$_3$ 的化合物〕
methopterin-A 甲蝶呤
methose 甲糖 $C_6H_{12}O_6$
methosulfate(=dimethyl sulfate) 硫酸二甲酯
methothrin 甲醚菊酯
methotrexate 氨甲蝶呤
methoxalate 甲草酸盐(或酯) CO(COOM)$_2$
methoxalic acid 甲草酸；丙酮二酸 CO(COOH)$_2$
methoxalyl 甲草酰；乙二酸一甲酯一酰 CH$_3$OOCCO—
methoxamine hydrochloride 盐酸甲氧明；盐酸甲胺〔药〕
methoxide 甲醇盐；甲氧基金属 CH$_3$OM
methoxy 甲氧(基) CH$_3$O—
methoxyacetic acid 甲氧基乙酸 CH$_3$OCH$_2$CO$_2$H
p-methoxyacetophenone 对甲氧苯乙酮
N-methoxyacetyl-phenetidine N-甲氧乙酰对乙氧基苯胺 C$_2$H$_5$OC$_6$H$_4$NHCOCH$_2$OCH$_3$

methoxy-t-amylbenzene 甲氧基叔戊苯 CH$_3$OC$_6$H$_4$C$_5$H$_{11}$
p-methoxy-α-amyl cinnamic aldehyde 对甲氧基-α-戊基(肉)桂醛
4-methoxy-2-amyloquinoline 4-甲氧基-2-戊基喹啉
o-methoxyaniline 邻甲氧基苯胺
ar-methoxyanilino-(=anisidino-) 甲氧苯氨基
methoxybenzaldehyde 甲氧(基)苯甲醛
2-(4-methoxy-benzene diazomercapto)naphthalene 2-(4-甲氧基苯重氮硫醇基)萘
methoxybenzoic acid 甲氧(基)苯甲酸
methoxybenzoyl(=anisoyl) 甲氧苯甲酰；茴香酰
methoxybenzyl 甲氧苄基；甲氧苯甲基 CH$_3$OC$_6$H$_4$CH$_2$—
p-methoxy benzyl acetate 乙酸邻甲氧基苄酯
p-methoxybenzyl alcohol 对甲氧基苄醇；对甲氧基苯甲醇
ar-methoxybenzylidene 甲氧亚苄基；甲氧苯亚甲基 CH$_3$OC$_6$H$_4$CH=
3-methoxybutanol 3-甲氧基丁醇
methoxy butyl acetate 乙酸甲氧基丁酯；丁二醇甲醚乙酸酯
3-methoxybutyl acrylate 丙烯酸 3-甲氧基丁酯
3-methoxybutyl aldehyde 3-甲氧基丁醛
methoxycarbonyl 甲酯基；甲氧羰基 CH$_3$OOC—
methoxychlor(=2,2-bis(p-methoxyphenyl)-1,1,1-trichloroethane) 甲氧氯；甲氧滴滴涕
6-methoxycinchoninic acid(=quininic acid) 6-甲氧基-4-喹啉甲酸；奎宁酸
methoxy cinnamic acid 甲氧基肉桂酸 CH$_3$OC$_6$H$_4$CH=CHCOOH
p-methoxycinnamic acid 对甲氧基(肉)桂酸
p-methoxycinnamic aldehyde 对甲氧基(肉)桂醛
methoxy coumarin 甲氧基香豆素
methoxycresol(=creosol) 甲氧甲酚；4-甲基-2-甲氧苯酚 CH$_3$OC$_6$H$_3$(CH$_3$)OH
methoxy dialkyl acetal(=alkyl formal;formaldehyde dialkyl acetal) 甲氧基型二烷基(醇)缩(甲)醛；烷基醇缩甲醛
2-methoxy-1,4-diazine 2-甲氧基-1,4-二嗪
methoxy dihydrocitronellal 甲氧基二氢香茅醛
methoxydimethylbenzylammonium chloride 氯化甲氧基二甲基苄基铵
methoxyestriol 甲氧雌(甾)三醇
methoxyestrone 甲氧(基)雌(甾)酮
methoxyethyl acctoxystearate 乙酰硬脂酸甲氧乙酯；乙二醇-1-乙酰氧硬脂酸酯-2-甲基醚
methoxyethyl acetyl ricinoleate 乙酰蓖麻酸甲氧乙酯〔增塑剂〕

2-methoxyethyl acrylate 丙烯酸(2-甲氧乙基)酯
methoxy-ethylene maleic anhydride copolymer 甲氧基乙烯马来酐共聚物
methoxyethyl oleate 油酸甲氧(基)乙酯；乙二醇-1-油酸酯-2-甲醚
methoxyethyl ricinoleate 蓖麻酸甲氧基乙酯〔增塑剂〕
methoxyethyl stearate 硬脂酸甲氧基乙酯
methoxy group 甲氧基
3-methoxy-4-hydroxy benzaldehyde 3-甲氧基-4-羟基苯甲醛〔增香剂〕
3-methoxy-4-hydroxy mandelic acid 3-甲氧基-4-羟基扁桃酸；3-甲氧基-4-羟基苯乙醇酸
2-methoxy-3-isobutyl-pyrazine 2-甲氧基-3-异丁基吡嗪
methoxyl 甲氧(基)
methoxylation 甲氧基化(作用)
methoxylglycol acetate 乙酸甲氧基乙醇酯 $CH_3CO_2CH_2CH_2OCH_3$
methoxy-mercuration 甲氧汞化(作用)
methoxymethylamide 甲氧基甲酰胺
methoxymethylated polyamide 甲氧(基)甲基化聚酰胺
methoxy-methylation 甲氧甲基化(作用)
methoxymethyl-ethoxymethyl benzoguanamine 甲氧甲基乙氧甲基苯代胍胺
methoxymethyl isocyanate 异氰酸甲氧基甲酯
N-methyl furfurylamine N-甲基糠胺〔促进剂〕
N-methyl-N,4-dinitrosoaniline N-甲基-N,4-二亚硝基苯胺〔硫化剂〕
N-methyl-2-pentyl-N'-pheny-p-phenylene diamine N-甲基-2-戊基-N'-苯基对苯二胺〔防老剂〕
N-methyl tetrahydrofurfurylamine N-甲基四氢糠胺〔促进剂〕
methoxymethyl melamine 甲氧基甲基蜜胺；甲醚化羟甲基三聚氰
N-methoxymethyl nylon N-甲氧基甲基尼龙〔溶于甲醇作涂料〕
4-methoxy-4-methyl pentan-2-one 4-甲氧基-4-甲基-2-戊酮
2-methoxy-3-methyl pyrazine 2-甲氧基-3-甲基吡嗪
methoxymethyl salicylate 水杨酸甲氧基甲酯 $HOC_6H_4CO_2CH_2OCH_3$
5-methoxy-N-methyl tryptamine 5-甲氧基-N-甲基色胺
methoxyphenol 甲氧基苯酚
methoxyphenyl 甲氧苯基 $CH_3OC_6H_4-$
p-methoxy phenylacetaldehyde 对甲氧苯乙醛
methoxyphenyl acetic acid ①甲氧苯基乙酸②甲氧苯基乙酸二聚体
methoxyphenyl acetone 甲氧苯基丙酮
p-methoxyphenyl alanine 对甲氧基苯丙氨酸

4-methoxy phenyl-2,6-diphenyl pyrylium perchlorate 4-甲氧苯基-2,6-二苯吡喃鎓过氯酸盐
β-p-methoxyphenyl-α-methyl propionaldehyde β-对甲氧苯基-α-甲基丙醛
methoxypolyethylene glycol 甲氧基聚乙二醇；聚乙二醇甲醚
methoxy polyglycol 甲氧基聚乙二醇
methoxy polyoxyethylene alkanoate 甲氧基聚乙氧烯烷酸酯
2-methoxypyrazine 2-甲氧基吡嗪
3-methoxypyridine 3-甲氧基吡啶
4-methoxypyridine N-oxide N-氧化 4-甲氧基吡啶
2-methoxy quinoline 2-甲氧基喹啉
p-methoxystyrene 对-甲氧基苯乙烯
p-methoxystyryl ethyl ketone 对甲氧基苯乙烯基乙酮
6-methoxy-tetrahydro-quinoline 6-甲氧基四氢喹啉
methoxy toluene 甲氧基甲苯
methoxytrimethylsilane 甲氧基三甲基(甲)硅烷
methoxy triphenol 香蒲酚；甲氧基三酚
5-methoxy tryptamine 5-甲氧色胺
3-methoxy tyramine 3-甲氧酪胺
methoxy value 甲氧值
methyl abietate 枞酸甲酯；松香酸甲酯
N-methyl-acetamide N-甲基乙酰胺 $CH_3CONHCH_3$
N-methyl-acetanilide N-甲基乙酰苯胺 $CH_3CON(CH_3)C_6H_5$
methyl acetate 乙酸甲酯 $CH_3CO_2CH_3$
methyl acetic acid 丙酸
methyl acetoacetate 乙酰乙酸甲酯 $CH_3COCH_2CO_2CH_3$
methyl-acetoacetic acid 甲基乙酰乙酸 $CH_3COCH(CH_3)COOH$
methyl-acetoacetic ester ①甲基乙酰乙酸酯 $CH_3COCH(CH_3)COOR$ ②〔专指〕甲基乙酰乙酸乙酯 $CH_3COCH(CH_3)COOC_2H_5$
methylacetoacetyl 甲基乙酰乙酰基
methylacetone 甲基丙酮
methyl acetophenone 对甲基苯乙酮
methyl acetyl 丙酮
methyl acetyl carbinol 甲基乙酰原醇
1-methyl-4-acetyl-Δ^1-cyclohexene 1-甲基-4-乙酰基-1-环己烯
methyl-acetylene(=allylene) 甲基乙炔；丙炔 $CH_3C\equiv CH$
1-methyl-2-acetyl pyrrole 1-甲基-2-乙酰基吡咯
methyl acetyl ricinoleate 乙酰蓖麻酸甲酯〔增塑剂〕
methyl acetyl-salicylate(=methyl aspirin) 乙酰水杨酸甲酯；甲基阿司匹林 $CH_3CO_2C_6H_4CH_3$

sym-methyl acetyl urea 对称甲基乙酰脲 CH₃NHCONHCOCH₃

methylacridine 甲基吖啶

methylacrolein 甲基丙烯醛

N-methyl acrylamide *N*-甲基丙烯酰胺

methyl acrylamidoglycolate methyl ether 2-丙烯酰胺(基)乙醇酸-2-甲醚-1-甲酯

methylacrylate 异丁烯酸盐；甲基丙烯酸盐(或酯) CH₂=C(CH₃)COOM

methylacrylic acid(=methacrylic acid) 异丁烯酸；甲基丙烯酸 CH₂=C(CH₃)CO₂H

methylacryloyl 异丁烯酰(基)；甲基丙烯酰(基)

N-methyladrenaline *N*-甲基肾上腺素

methylaffinity 甲基亲和势；甲基亲和能；甲基亲和性

methylal 甲缩醛(俗)；甲醛缩二甲醇；二甲氧基甲烷 CH₂(OCH₃)₂

methyl alcohol 甲醇 CH₃OH

methyl aldehyde 甲醛

methyl alloxan 甲基阿脲

methylallyl 甲代烯丙基 CH₂=C(CH₃)CH₂—

methylallyl acetate(=methallyl acetate) 乙酸甲代烯丙酯

2-methylallyl alcohol(=metallyl alcohol) 2-甲基丙烯醇 CH₂=C(CH₃)CH₂OH

methylallyl alcohol(=methallyl alcohol) 甲代烯丙醇；2-甲基-2-丙烯-1-醇

methylallyl amine 甲基烯丙基胺 CH₃NHC₃H₅

2-methylallyl butyrate 丁酸 2-甲烯丙酯

methylallyl-carbinol 甲基烯丙基甲醇 (CH₃)(C₃H₅)=CHOH

methylallyl ether 甲基烯丙基醚 CH₃OCH₂CH=CH₂

methylallyl ketone 甲基烯丙基(甲)酮；4-戊烯-2-酮 CH₃COCH₂CH=CH₂

methyl allylphenol 茴香脑

methylal resin 缩甲醛树脂

methylaluminoxane(MAO) 甲基铝氧烷

methyl amidophenol 茴香胺

methylamine 甲胺 CH₃NH₂

methylamine hydrochloride 盐酸甲胺 CH₃NH₂HCl

methylamino 甲氨基

methyl aminoacetate 氨基乙酸甲酯 NH₂CH₂CO₂CH₃

methyl *o*-aminobenzoate(=methyl anthranilate) 邻氨基苯甲酸甲酯 NH₂C₆H₄CO₂CH₃

methyl *p*-aminophenol 甲基对氨基苯酚

methylammonium methyldithiocarbamate 甲基二硫代氨基甲酸甲铵（促进剂）

methyl amphetamine hydrochloride 甲基苯异丙胺盐酸盐

methyl amyl acetate 乙酸甲戊酯；乙酸异己酯

methyl amyl-acetylene 甲基戊基乙炔；2-辛炔 C₈H₁₄

methyl amyl alcohol 甲基戊醇

methyl-*n*-amyl carbinol(=methyl hexyl alcohol) 甲基正戊基甲醇；甲基己醇

methyl amyl ether 甲基戊基醚；甲氧戊烷 CH₃OC₅H₁₁

methyl amyl ketone(=2-heptanone) 甲基戊基(甲)酮；2-庚酮 CH₃(CH₂)₄COCH₃

methyl amyl ketone peroxide 过氧化甲基戊基(甲)酮；过氧 2-庚酮

N-methylaniline(=methyl-phenylamine) *N*-甲基苯胺；甲基苯基胺 C₆H₅NHCH₃

methyl anisate 大茴香酸甲酯

methyl *p*-anisyl ketone 甲基对茴香基(甲)酮 CH₃OC₆H₄CH₂COCH₃

methyl anthranilate(=methyl *o*-aminobenzoate) 邻氨基苯甲酸甲酯 NH₂C₆H₄CO₂CH₃

β-methyl anthraquinone β-甲基蒽醌

methylarene 甲基芳烃

methyl-arsenic acid 甲胂酸 CH₃AsO(OH)₂

methyl arsenious oxide 甲基亚胂酸酐

methylarsine 甲胂 CH₃AsH₂

methyl-arsine dibromide 二溴化甲胂

methyl-arsine dichloride 二氯化甲胂

methyl-arsine-disulfide 二硫化甲胂 CH₃AsS₂

methyl arsine oxide 氧化甲胂 CH₃As=O

methyl-arsine-sulfide 硫化甲胂 CH₃AsS

methyl-arsine-tetrachloride 四氯化甲胂 CH₃AsCl₄

methyl-arsonate 甲胂酸盐 CH₃AsO(OM)₂

methylarsonic acid 甲胂酸；甲基胂酸 CH₃AsO(OH)₂

methylaspirin(=methyl acetyl-salicylate) 乙酰水杨酸甲酯；甲基阿司匹林 CH₃CO₂C₆H₄CO₃

methylate ①甲基化②甲基化产物③甲醇金属④加入甲醇

methylated 甲基化了的

methylated albumin 甲基化清蛋白

methylated albumin kieselguhr(MAK) 甲基清蛋白硅藻土

methylated alcohol 变性酒精

methylated benzene 甲基化(了的)苯

methylated ether 甲基化醚〔用甲醇变性的酒精制成的乙醚〕

methylated melamine 甲醇醚化三聚氰胺

methylated melamine resin 甲醚化三聚氰胺树脂

methylated methylol melamine 甲氧甲基三聚氰胺；甲醇醚化羟甲基三聚氰胺

methylated naphthalene 甲基化萘

methylated pentaethylene hexamine 甲基化五亚乙基六胺〔硫化剂〕

methylated spirit 甲基化酒精〔用甲醇使变性的酒精〕

methylated triethylene tetramine 甲基化三亚乙基四胺〔硫化剂〕
methylating 甲基化
methylating agent 甲基化剂
methylation 甲基化作用〔指加甲基作用和加甲醇使酒精变性的作用〕
methylation in the cold 冷时甲基化
methyl-auramine 甲基金胺 [(CH$_3$)$_2$NC$_6$H$_4$]$_2$CNCH$_3$
methyl azide(=triazo-methane) 叠氮基甲烷 CH$_3$N(N)$_2$
methylbarbituric acid 甲基巴比妥酸；甲基丙二酰脲
methyl behenate 山萮酸甲酯；二十二(烷)酸甲酯 C$_{21}$H$_{43}$CO$_2$CH$_3$
N-methyl-benzamide N-甲基苯甲酰胺 C$_6$H$_5$CONHCH$_3$
methylbenzene(=toluene) 甲苯
methyl benzenesulfonate 苯磺酸甲酯 C$_6$H$_5$SO$_3$CH$_3$
methyl benzethonium chloride 甲基苄索氯铵
methyl 2-benzimidazole carbamate 2-苯并咪唑氨基甲酸甲酯
methyl benzoate 苯甲酸甲酯 C$_6$H$_5$CO$_2$CH$_3$
methyl benzoic acid 甲基苯甲酸
2-methyl-benzoquinone 2-甲基苯醌
2-methylbenzoquinone dioxime 2-甲基苯醌二肟〔硫化剂〕
p-methyl-benzothiazole 对甲基苯并噻唑；2-甲基-1,3-硫氮杂茚 C$_6$H$_4$N═C(CH$_3$)S
methyl benzoylacetate 苯甲酰乙酸甲酯 C$_6$H$_5$COCH$_2$CO$_2$CH$_3$
methyl o-benzoylbenzoate 邻苯甲酰苯甲酸甲酯 C$_6$H$_5$COC$_6$H$_4$CO$_2$CH$_3$
methyl benzoylsalicylate 苯甲酰水杨酸甲酯 C$_6$H$_5$CO$_2$C$_6$H$_4$CO$_2$CH$_3$
methylbenzyl 甲苄基；甲苯甲基 CH$_3$C$_6$H$_4$CH$_2$—
α-methylbenzyl α-甲苄基 C$_6$H$_5$CH(CH$_3$)—
methyl benzylacetate 苄乙酸甲酯；苯丙酸甲酯
N-methylbenzyl amine N-甲苄胺 CH$_3$NHCH$_2$C$_6$H$_5$
N-methylbenzyl-aniline N-甲苄苯胺 C$_6$H$_5$CH$_2$N(CH$_3$)C$_6$H$_5$
methylbenzyl-carbinol 甲基苄基甲醇 CH$_3$CHOHCH$_2$C$_6$H$_5$
methylbenzyl carbinyl propionate 丙酸甲基苄基原酯
methylbenzyl dicoconut fatty acid ammonium chloride 甲苄基二椰子油脂肪酸氯化铵
α-methyl benzyl dimethyl amine α-甲苄基二甲胺〔环氧固化剂〕C$_6$H$_5$CH(CH$_3$)NCH$_3$
methylbenzyl ether 甲基苄基醚；苄氧基甲烷 CH$_3$OCH$_2$C$_6$H$_5$
4-methyl benzylidene camphor 4-甲基亚苄基樟脑

methyl biphenyl 甲基联苯 C$_6$H$_5$C$_6$H$_4$CH$_3$
methyl biphenyl ketone(=p-phenylacetophenone) 甲基联苯基(甲)酮
methyl bismuthine 一甲铋 CH$_3$BiH$_2$
methyl bixin 甲基胭脂树精
methyl blue 甲基蓝
methyl-borate 硼酸甲酯 B(OCH$_3$)$_3$
methyl-boric acid(=methyl-boron-dihydroxide) 甲基硼酸 CH$_3$B(OH)$_2$
methyl-boron-dihydroxide(=methyl-boric acid) 甲基硼酸
methyl branching silicone oil 甲基支链硅油〔脱模剂〕
methyl bromide(=bromomethane) 溴代甲烷 CH$_3$Br
methyl bromoacetate 溴乙酸甲酯 BrCH$_2$CO$_2$CH$_3$
methyl-bromobenzoate 溴苯甲酸酯
methyl bromoethyl ketone 甲基溴乙基(甲)酮；溴丁酮
methyl bromoisovalerate 溴异戊酸甲酯 (CH$_3$)$_2$CHCHBrCO$_2$CH$_3$
methyl p-bromophenyl sulfide 甲基对溴苯基硫醚
methyl bromopropyl ketone 甲基溴丙基(甲)酮
methyl-butadiene 甲基丁二烯；2-甲基丁二烯
3-methylbutanal 3-甲基丁醛
methyl butane 甲基丁烷
2-methylbutane(=isopentane) 异戊烷；2-甲基丁烷 (CH$_3$)$_2$CHC$_2$H$_5$
2-methyl-1-butanethiol 2-甲基-1-丁硫醇
2-methylbutanoic acid 2-甲基丁酸
methylbutanol 甲基丁醇
3-methyl butanol(=isoamyl alcohol) 异戊醇；3-甲基丁醇
3-methyl-2-butanone(=methyl isopropyl-ketone) 甲基异丙基(甲)酮；3-甲基-2-丁酮 CH$_3$COCH(CH$_3$)$_2$
2-methyl-2-butenal 2-甲基-2-丁烯醛
2-methyl-1-butene 2-甲基-1-丁烯 CH$_3$CH$_2$C(Me)═CH$_2$
3-methyl-1-butenyl(=isopentenyl) 异戊烯基
methyl butenyl norbornene 甲基丁烯基降冰片烯
methyl butex 羟苯甲酸甲酯
methyl-butyl acetic acid 甲基丁基乙酸 CH$_3$(C$_4$H$_9$)CHCOOH
N-methylbutylamine N-甲基丁胺
methyl butylbenzene 甲基丁苯
methyl butyl-carbinol 甲基丁基甲醇；2-己醇 CH$_3$(CH$_2$)$_3$CH(OH)CH$_3$
methyl-n-butyl carbinyl acetate 乙酸甲基丁基原酯
methyl butyl ether 甲基丁基醚；甲氧基丁烷 CH$_3$OC$_4$H$_9$
methyl butyl ketone(=2-hexanone) 甲基丁基(甲)酮；2-己酮 CH$_3$COC$_4$H$_9$
methyl butyl ketone process 甲基丁基(甲)酮法

2-methylbutylmercaptan　2-甲基丁硫醇
α-methyl-*p*-*t*-butyl phenylpropionaldehyde　α-甲基对叔丁基苯丙醛；铃兰醛
methyl butyl sulfide　甲基丁基硫醚；甲硫基丁烷　$CH_3SC_4H_9$
methylbutynol　甲基丁炔醇〔尼龙的溶剂及黏度稳定剂〕
3-methyl-butyraldehyde(=isovaleraldehyde)　异戊醛；3-甲基丁醛　$(CH_3)_2CHCH_2CHO$
methyl-butyrate　丁酸甲酯　$C_3H_7COCH_3$
α-methyl butyric acid　α-甲基丁酸；2-甲基丁酸
methylbutyryl-CoA　甲基丁酰辅酶 A
methyl cadmium　甲基镉　$Cd(CH_3)_2$
methyl calcein　甲基钙黄绿素
methyl calcein blue　甲基钙黄绿素蓝
methyl caprate　癸酸甲酯　$C_9H_{19}CO_2CH_3$
2-methyl caproic acid　2-甲基己酸
N-methyl-ε-caprolactam(NMC)　*N*-甲基-ε-己内酰胺
methyl caprylate　辛酸甲酯　$CH_3(CH_2)_6CO_2CH_3$
methyl carbamate　氨基甲酸甲酯　$H_2NCO_2CH_3$
methyl carbic acid(=methyl nadic acid)　甲基降冰片烯二酸
methyl carbic anhydride(=methyl norbornene dioic anhydride)　甲基降冰片烯二酸酐
methyl carbimide　异氰酸甲酯
methyl-carbithionic acid　二硫代乙酸　$CH_3C(SH)=S$
methyl carbitol　二乙二醇一甲醚；甲基卡必醇〔俗〕
methyl carbitol acetate　二乙二醇一甲醚乙酸酯；甲基卡必醇乙酸酯〔俗〕
methyl-carbonate　碳酸二甲酯　$CO(OCH_3)_2$
methyl carbonic acid　甲基碳酸　CH_3COOH
N-methyl-carbostyrile　*N*-甲喹诺酮
methyl carboxylic cellulose　甲基羧基纤维素
methyl carbylamine　甲基胼
methylcarvacryl ketone　甲基香芹基(甲)酮　$CH_3COC_{10}H_{13}$
methyl catechol　愈创木酚
methyl cedryl ether　柏木甲醚
methyl cellosolve　甲基溶纤剂；甲氧基乙醇；乙二醇一甲醚
methylcellosolve acetate(=methoxylglycol acetate)　甲基溶纤剂乙酸酯；甲氧基乙醇乙酸脂　$CH_3CO_2CH_2CH_2OCH_3$
methyl cellulose　甲基纤维素
methyl chavicol　甲基黑椒酚
methyl chloride(=chloromethane)　氯代甲烷　CH_3Cl
methyl chloroacetate　氯乙酸甲酯　$ClCH_2CO_2CH_3$
methyl α-chloroacrylate　α-氯代丙烯酸甲酯
methyl-chlorocarbonate　氯碳酸甲酯；氯甲酸甲酯　$ClCOOCH_3$
methyl chlorofluoride　二氟二氯甲烷
methyl chloroform　甲基氯仿；三氯乙烷
methyl chloroformate　氯甲酸甲酯　$ClCOOCH_3$
2-methylchlorophenoxy acetic acid　2-甲基氯苯氧基乙酸
methyl chlorosilane　甲基氯硅烷
methyl chlorosulfonate　氯磺酸甲酯
methyl-cholanthrene　甲基胆蒽　$C_{20}H_{13}CH_3$
methyl-chromone　甲基色酮　$C_{10}H_8O_2$
methyl cinnamate　肉桂酸甲酯　$C_6H_5C_2H_2CO_2CH_3$
methyl cinnamate type basil　(肉)桂酸甲酯型罗勒
methyl cinnamic acid　甲基肉桂酸　$CH_3C_6H_4CH=CHCOOH$
α-methyl cinnamic aldehyde　α-甲基(肉)桂醛
methyl cinnamyl ether　甲基(肉)桂醚
methyl citrate　柠檬酸甲酯；枸橼酸甲酯
methyl cobalamin　甲基钴胺素
methyl cocaine　甲基可卡因
methyl coniine　甲基毒芹碱
6-methyl coumarin　6-甲基香豆素
methyl-*m*-cresol　间甲酚甲醚
methyl-*p*-cresol　对甲酚甲醚
methyl cresyl phenylacetate　苯乙酸甲基甲酚酯
2-methyl crotonaldehyde　2-甲基巴豆醛
methyl crotonate　巴豆酸甲酯　$C_3H_5CO_2CH_3$
methylcrotonic acid　甲基巴豆酸；甲基丁烯酸
β-methylcrotonyl CoA　β-甲基巴豆酰辅酶
methyl-*o*-cumyl carbinyl acetate　乙酸甲基邻枯茗基原酯
methyl cyanacetate　氰基乙酸甲酯　$CNCH_2COOCH_3$
methyl cyanate　氰酸甲酯　$NCOCH_3$
methyl cyanide　甲基氰；乙腈　CH_3CN
methyl cyanoacetate　氰基乙酸甲酯　$NCCH_2CO_2CH_3$
methyl α-cyanoacrylate(=methyl 2-cyanoacrylate)　α-氰基丙烯酸甲酯〔强黏合剂原料〕
methyl 2-cyanoacrylate(=methyl α-cyanoacrylate)　α-氰基丙烯酸甲酯〔强黏合剂原料〕
methyl cyanoformate　氰甲酸甲酯
methyl cyclobutane　甲基环丁烷　$CH_3C_4H_7$
methyl cycloheptanol　甲基环庚醇　$C_8H_{16}O$
α-methyl cycloheptanone　α-甲基环庚酮
methyl cyclohexadiene　甲基环己二烯　$CH_3C_6H_7$
methyl cyclohexane　甲基环己烷　$CH_3C_6H_{11}$
methyl cyclohexane diisocyanate　甲基环己烷二异氰酸酯〔硫化剂〕
methyl-cyclohexanol　甲基环己醇　$CH_3C_6H_{10}OH$
methyl cyclohexanone　甲基环己酮
methyl cyclohexanone glyceryl acetal　甲基环己酮缩水甘油醚(螺环)〔增塑剂〕

methylcyclohexanone oxime　甲基环己酮肟
methyl cyclohexanone peroxide　甲基环己酮过氧化物〔不饱和聚酯树脂固化剂〕
methylcyclohexene　甲基环己烯　$CH_3C_6H_9$
6-methyl-3-cyclohexene-1-aldehyde　6-甲基-3-环己烯-1-醛
3-methyl-1-cyclohexene-1-aldehyde　3-甲基-1-环己烯-1-醛
3-methyl-Δ^2-cyclohexenone　3-甲基环己烯-2-酮
methyl cyclohexenyl methyl ketone　甲基环己烯基甲酮
methylcyclohexyl acetate　甲基环己基乙酸酯；乙酸甲基环己酯
N-methylcyclohexylamine　N-甲基环己胺　$CH_3NHC_6H_{11}$
3-methyl cyclohexyl carbinol　3-甲基环己基甲醇
methylcyclohexyl carbonate　碳酸甲基环己酯；甲基环己基碳酸酯
4-(2-methylcyclohexyl)-cyclohexan-1-one　4-(2-甲基环己基)环己烷-1-酮
2-(β-methylcyclohexyl)-4,6-dimethyl phenol　2-(β甲基环己基)-4,6-二甲基苯酚〔防老剂〕
methylcyclohexyl isobutyl phthalate　邻苯二甲酸异丁基甲基环己酯
methylcyclohexyl lactate　乳酸甲基环己酯
methylcyclohexyl stearate　硬脂酸甲基环己酯
3-methylcyclopentadecanol(=muscol)　麝香醇；3-甲基环十五醇
3-methylcyclopentadecanone(=muskone)　3-甲基环十五(烷)酮；麝香酮　$CH_3C_{15}H_{27}$=O
3-methyl cyclopentadecylene　麝香烯；3-甲基环十五烯
methylcyclopentadiene dimer　甲基戊二烯二聚物；二聚甲基环戊二烯
methylcyclopentane　甲基环戊烷
methylcyclopentanol　甲基环戊醇
methylcyclopentanone oxime　甲基环戊酮肟
methyl cyclopentene　甲基环戊烯
methyl cyclopentenolone　甲基环戊烯醇酮
γ-methyl cyclotetradecanone　γ-甲基环十四酮
5-methylcytidylic acid　5-甲基胞苷酸
methylcytisine(=caulophylline)　甲基金雀花碱
5-methylcytosine　5-甲基胞嘧啶
methyl 2-decenoate　2-癸烯酸甲酯
methyl decine-carbonate　癸炔羧酸甲酯
methyl dehydroabietate　脱氢枞酸甲酯；脱氢松香酸甲酯
methyl demethyl dehydro abietate　脱甲基脱氢枞酸甲酯
5-methyldeoxycytidine　5-甲基脱氧胞苷
2-methyl-1,4-diazine　2-甲基-1,4-二嗪；2-甲基吡嗪
methyl dibromoanthranilate　二溴氨茴酸甲酯；二溴邻氨基苯甲酸甲酯　$Br_2C_6H_2(NH_2)CO_2CH_3$
methyl dichloroarsine　甲胂化二氯；甲基二氯胂　CH_3AsCl_2
N-methyldiethanol amine　N-甲基二乙醇胺
methyl diethyl amine　甲基二乙胺　$CH_3N(C_2H_5)_2$
methyl dihydroabietate　二氢枞酸甲酯
methyl dihydrojasmonate　二氢茉莉酮酸甲酯
methyldiiodoarsine　甲基二碘胂　CH_3AsI_2
methy1-4-dimethylaminobenzoate　4-二甲氨基安息香酸甲酯
N-methyl-N,4-dinitrosoaniline　N-甲基-N,4-二亚硝基苯胺〔硫化剂〕
2-methyl-4,6-dinonylphenol　2-甲基-4,6-二壬基苯酚〔防老剂〕
2-methyl-1,3-dioxolane　2-甲基-1,3-二氧戊环
methyl diphenyl　甲基联苯
N-methyl-diphenylamine　N-甲基二苯胺　$CH_3N(C_6H_5)_2$
methyl diphenyl carbinol　甲基二苯基甲醇；1,1-二苯基乙醇　$(C_6H_5)_2C(OH)CH_3$
methyl diphenyl ether　甲基二苯醚
methyl-di-n-propylamine　甲基二丙胺　$(C_3H_7)_2NCH_3$
methyl disulfide　二甲二硫；甲基化二硫　CH_3SSCH_3
methyl dithioacetate　二硫代乙酸甲酯
methyl dodecanoate　十二烷酸甲酯；月桂酸甲酯
methyl 10-dodecenoate　10-十二烯酸甲酯
β-methyl dodecyl aldehyde　β-甲基十二醛
methyl donor　甲基供体
methyldopa　甲基多巴〔药〕
methylenation　亚甲基化作用
methylene(=methene)　亚甲基　CH_2=
methylene amino-acetonitrile　亚甲基氨基乙腈
methylene-aniline　亚甲基苯胺　$CH_2NC_6H_5$
N,N'-methylene-bis acrylamide　N,N'亚甲双丙烯酰胺
4,4-methylene-bis(6-t-butyl-o-cresol)　4,4-亚甲基双(6-叔丁基邻甲苯酚)〔防老剂〕
4,4'-methylene-bis(2-chloroaniline)　4,4'-亚甲基双(2-氯苯胺)〔聚氨酯固化剂〕
methylene-bis-o-chloroaniline　亚甲基双邻氯苯胺〔硫化剂〕
4,4'-methylene-bis-cyclohexyl diisocyanate　4,4'-亚甲双环己基二异氰酸酯〔硫化剂〕
methylene bis(4-cyclohexyl isocyanate)　亚甲基二(4-环己基异氰酸酯)
methylene-bis(dialkylphosphine oxide)　亚甲基双(二烷基氧膦)　$R_2POCH_2POR_2$
methylene bis-dialkylphosphonate　亚甲基双(膦酸二烷基)酯　$(RO)_2POCH_2PO(OR)_2$
4,4'-methylene-bis(2,6-di-t-butyl phenol)　4,4'-亚甲双(2,6-二叔丁基苯酚)〔防老剂〕

methylene bis(di-2-ethylhexylphosphine oxide)(MEHDPO) 亚甲基双(二-2-乙基己基氧膦) $(C_8H_{17})_2PO(CH_2)PO(C_8H_{17})_2$

2,2′-methylene-bis(4,6-dimethyl phenol) 2,2′-亚甲基双(4,6-二甲基苯酚)〔防老剂〕

4,4′-methylene-bis(n-ethylaniline) 4,4′-亚甲基双(正乙基苯胺)〔硫化剂〕

2,2′-methylene-bis(4-ethyl-6-t-butyl phenol) 2,2′-亚甲基双(4-乙基-6-叔丁基苯酚)〔防老剂〕

methylene-bis(2-methoxyaniline) 亚甲基双(2-甲氧基苯胺)〔硫化剂〕

2,2′-methylene-bis(4-methyl-6-cyclohexyl phenol) 2,2′-亚甲基双(4-甲基-6-环己基苯酚)〔防老剂〕

2,2′-methylene-bis(4-methyl-6-nonyl phenol) 2,2′-亚甲基双(4-甲基-6-壬基苯酚)〔防老剂〕

1,1′-methylene-bis(2-naphthol) 1,1′-亚甲基双(2-萘酚)〔防老剂〕

methylene bisoxaloacetic ester 亚甲基双草酰基乙酸酯

methylene-bis(4-phenylamine) 亚甲基双(4-苯胺)〔硫化剂〕

methylene-bis(4-phenyl isocyanate) 亚甲基双(4-异氰酸苯酯)〔硫化剂〕

methylene blue 亚甲蓝*

methylene blue absorption 亚甲蓝吸收

methylene blue active substance 亚甲蓝活性物质

methylene blue method 亚甲蓝法

methylene blue milk method 牛奶亚甲蓝(还原)法

methylene blue polymerization inhibitor 亚甲基蓝阻聚剂

methylene blue reaction 亚甲蓝反应

methylene blue stabilizer 亚甲蓝稳定剂

methylene blue test 亚甲蓝试验

methylene bridge 亚甲桥

methylene bromide(=dibromomethane) 二溴甲烷 $CH_2=Br_2$

methylenebutanedioic acid 衣康酸

methylene chloride(=dichloromethane) 二氯甲烷 $CH_2=Cl_2$

methylene cyanide 丙二腈

3-methylene cyclohexene 3-亚甲基环己烯

methylene cyclopropyl glycine 亚甲基环丙基甘氨酸

methylene dianiline 二苯氨基甲烷 $(C_6H_5NH)_2CH_2$

methylene dichloride 二氯甲烷

methylene dicotoin 亚甲醛二可土树皮素

methylene diethyl ether 亚甲基二乙醚；二乙氧基甲烷 $CH_2(OC_2H_5)_2$

methylene diiodide 二碘甲烷

methylene dimalonic acid 亚甲基双丙二酸

methylene dioxy 亚甲二氧基 —OCH$_2$O—

3,4-methylenedioxyphenethyl(=homopiperonyl) 高胡椒基；3,4-亚甲二氧苯乙基

methylene diphenylamino phenyl formate 亚甲基双苯基氨基甲酸苯酯〔黏合剂〕

methylene diphenyl diamine 亚甲基二苯二胺〔促进剂〕

methylene diphenyl diisocyanate 二苯基甲烷-4,4′-二异氰酸酯

methylenediphosphonate(MDP) 亚甲基二膦酸 $CH_2(PO_3H_2)_2$

methylene disalicylic acid 二水杨酸亚甲酯

methylene disulfonyl 亚甲基二磺酰 $CH_2(SO_2)_2=$

methylene ditannin 亚甲醛二单宁酸

methylene fluoride(=difluoromethane) 二氟甲烷 CH_2F_2

methylene glycol 亚甲基乙二醇

methylene green 亚甲绿

methylene group 亚甲基；甲撑 —CH_2—

methylene halide 二卤甲烷 CH_2X_2

methylene imine 亚甲基亚胺

methylene iodide(=diiodomethane) 二碘甲烷 $CH_2=I_2$

5-methylene-2-norbornene 5-亚甲基-2-降冰片烯

methylene-p-toluidine 亚甲基对甲苯胺〔促进剂〕

methylene radical 亚甲基 $=CH_2$

methylene-succinic acid 亚甲基丁二酸 $COOHCH_2C(=CH_2)COOH$

methylene-sulfonic acid 甲二磺酸 $CH_2(SO_3H)_2$

N^5,N^{10}-methylenetetrahydrofolate N^5,N^{10}-亚甲四氢叶酸

methylene-p-toluidine 亚甲基对甲苯胺〔促进剂〕

methylene urea 亚甲基脲；甲撑脲

methylene-violet 亚甲紫

methylenimine(=methylene imine) (亚)甲基亚胺

methylenomycin 次甲霉素

1-methylenpyrrolizidine 1-亚甲基吡咯双烷；亚甲基双稠吡咯啶

methyl eosin 甲基曙红

l-N-methylephedrine l-N-甲基麻黄碱

N-methylepinephrine N-甲基肾上腺素

methyl 9,10-epoxystearate 9,10-环氧基硬脂酸甲酯〔耐寒性增塑剂〕

methyl epoxystearate 环氧硬脂酸甲酯

methyl ergometrine (malete) (顺丁烯二酸)甲基新麦角碱〔药〕

methyl ester 甲酯

methyl ester of rosin 松香甲酯〔增黏剂〕

methylethanolamine 甲基乙醇胺

methyl ether ①甲基醚 $ROCH_3$②甲醚 CH_3OCH_3

4-methyl-7-ethoxycoumarin 4-甲基-7-乙氧香豆素

methyl-ethyl-acetic acid 邻甲基丁酸；甲基乙基酢酸 $CH_3(C_2H_5)CHCOOH$

methyl ethyl-allylamine 甲基乙基烯丙胺
　　$CH_3(C_2H_5)NCH_2CH=CH_2$
methyl ethyl-amine 甲基乙胺 $CH_3NHC_2H_5$
methyl ethyl-t-amyl-carbinol 甲基乙基叔戊基甲醇
methyl ethyl-aniline N-甲基-N-乙基苯胺
　　$C_6H_5N(CH_3)C_2H_5$
methyl ethyl benzyl carbinol 甲基乙基苄基原醇
methyl ethyl carbonate 碳酸甲乙酯 $CH_3OCOOC_2H_5$
methyl ethyl cellulose 甲基乙基纤维素
2-methylethylenediamine tetraacetic acid(MEDTA) 2-甲基乙二胺四乙酸
methylethylene phosphate(MEP) 甲基亚乙基磷酸酯
methyl ethyl ether 甲基乙基醚；甲氧基乙烷
　　$CH_3OC_2H_5$
sym-methyl ethyl-ethylene(=2-pentene) 对称甲基乙基烯；2-戊烯 $C_2H_5CH=CHCH_3$
$unsym$-methyl ethyl-ethylene(=2-methyl-1-butene) 偏甲基乙基乙烯；2-甲基-1-丁烯
methyl ethyl-isoamyl-carbinol 甲基乙基异戊基甲醇
　　$(CH_3)(C_2H_5)(iso\text{-}C_5H_{11})C(OH)$
methyl ethyl-isobutyl-carbinol 甲基乙基异丁基甲醇
　　$(CH_3)(C_2H_5)(iso\text{-}C_4H_9)C(OH)$
methyl ethyl-isohexyl-carbinol 甲基乙基异己基甲醇
　　$C_{10}H_{22}O$
methyl ethyl ketone(=butanone) 甲基乙基(甲)酮；丁酮 $CH_3COC_2H_5$
methyl-ethyl-ketone dewaxing 甲乙酮脱蜡；丁酮脱蜡
methyl ethyl ketone disulfone(=trional) 丁酮缩二乙砜
methylethyl ketone peroxide 甲乙酮过氧化物；过氧化甲乙酮；过氧化丁酮
methyl-ethyl-ketone-water azeotrope 甲基乙基(甲)酮-水共沸混合物
methyl ethyl ketoxime 甲基乙基甲酮肟；甲乙酮肟
methyl ethyl-malonic acid 甲基乙基丙二酸
　　$(CH_3)C_2H_5C=(CO_2H)_2$
methyl ethyl oxalate 草酸甲乙酯 $CH_3O(CO)_2OC_2H_5$
2-methyl-3-ethylpentane 2-甲基-3-乙基戊烷
　　$(C_2H_5)_2CHCH(CH_3)_2$
2-methyl-5-ethylpyrazine 2-甲基-5-乙基吡嗪
methyl ethyl pyridine 甲基乙基吡啶
methylethylthetin 甲基乙基噻亭；甲基乙基硫代羧酸内盐
methyl eugenol 丁子香酚甲醚
methyl exaltone(=muscone) 甲基环十五酮；麝香酮
methylferase 转甲基酶
methylfluorenone 甲基芴酮
methyl fluoride(=fluoromethane) 甲基氟；氟代甲烷
　　CH_3F

methylfluorone 甲基荧光酮
methylfluorosilicone rubber 甲基氟硅橡胶
methyl formate(=methyl formiate) 甲酸甲酯 HCO_2CH_3
methyl formiate(=methyl formate) 甲酸甲酯
methyl formylanilide N-甲基-N-甲酰苯胺
　　$C_6H_5(CH_3)NCHO$
methylfructofuranoside 甲基呋喃果糖苷
6-methylfulvene 亚甲基环戊二烯
methyl fumaric acid(=mesaconic acid) 甲基富马酸；反式甲基丁烯二酸 $CH_3C(CO_2H)=CHCO_2H$
5-methyl-2-furaldehyde 5-甲基-2-呋喃甲醛
2-methyl-3-furanthiol 2-甲基-3-呋喃硫醇；5-甲基糠醛
5-methyl furfural 5-甲基糠醛
N-methyl furfurylamine N-甲基糠胺〔促进剂〕
methyl furfuryl disulfide 甲基糠基二硫
β-methyl-γ-furfurylidene butyraldehyde β-甲基-γ-糠叉丁醛
2-methyl-3,5-furfuryl thiopyrazine 2-甲基-3,5-糠基硫吡嗪
methyl 2-furoate 2-呋喃甲酸甲酯
methyl-furoate 糠酸甲酯 $C_4H_3OCOOCH_3$
α-methyl-β-furylacrolein α-甲基-β-呋喃基丙烯醛
2-methyl-3-furyldisulfide 2-甲基-3-呋喃基二硫
2-methyl-3-furyl tetrasulfide 2-甲基-3-呋喃基四硫
methyl gallate(=gallicin) 棓酸甲酯 $(HO)_3C_6H_2CO_2CH_3$
methyl gallium dichloride 二氯化甲镓
methyl geraniol 甲基香叶醇
methyl glucoside ①甲基葡萄糖苷②甲基苷
methyl glucoside laurate 甲基葡萄糖苷月桂酸酯
methyl glucoside oleate 甲基葡萄糖苷油酸酯
methyl glucoside palmitate 甲基葡萄糖苷棕榈酸酯
methyl glucoside stearate 甲基葡萄糖苷硬脂酸酯
β-methylglutaconyl CoA β-甲基戊烯二酸单酰 CoA
methyl glycerate 甘油酸甲酯 $CH_2OHCHOHCO_2CH_3$
methylglycerin(e) 甲基甘油 $CH_3(CHOH)_2CH_2OH$
methylglycine 甲基甘氨酸
methylglycine oxidase 甲基甘氨酸氧化酶
methyl glycol 乙二醇一甲醚；甲基溶纤剂
methyl glycol acetate 乙二醇一甲醚乙酸酯；甲基溶纤剂醋酸酯〔俗〕
methyl glycollate 乙醇酸甲酯 $HOCH_2CO_2CH_3$
methyl glycosine 肌氨酸；N-甲基甘氨酸
methyl-glyoxal 甲基乙二醛；丙酮醛 CH_3COCHO
methyl glyoxalidine 甲基二氢咪唑
methyl-glyoxime 甲基乙二肟 $C_3H_6N_2O_2$
methyl green 甲基绿
methyl-Grignard-reagent 甲基镁化卤；甲基卤化镁 $XMgCH_3$

methylguanidine 甲胍　$CH_3NHC=NH(NH_2)$
methylguanidine nitrate　硝酸甲胍　$C_2H_7N_3HNO_3$
methylguanidine sulfate　硫酸甲胍　$(C_2H_7N_3)_2H_2SO_4$
methylguanidinoacetic acid　甲基胍(基)乙酸
methyl-guanidyl-acetic acid　N-甲基胍基乙酸
　　$NH_2(NH=)CN(CH_3)CH_2COOH$
methyl halide　甲基卤；卤代甲烷　CH_3X
methyl heptadecyl ketone(=2-nonadecanone)　甲基十七基
　　(甲)酮；2-十九烷酮　$CH_3(CH_2)_{16}COCH_3$
5-methyl-3-heptanone oxime　5-甲基-3-庚酮肟〔防老剂〕
methyl heptenol　甲基庚烯醇
methyl heptenone　甲基庚烯酮
methyl heptine carbonate　2-辛炔酸甲酯
methyl heptoate　庚酸甲酯
methyl heptylate　庚酸甲酯　$CH_3(CH_2)_5CO_2CH_3$
methyl n-heptyl ketone(=2-nonanone)　甲基庚基(甲)酮；
　　2-壬酮　$CH_3CO(CH_2)_6CH_3$
methyl hexabital　海索比妥
methyl hexadecyl ketone　甲基十六基(甲)酮；2-十八(烷)
　　酮　$CH_3(CH_2)_{15}COCH_3$
methyl hexahydro phthalic anhydride　甲基六氢化邻苯二
　　甲酸酐；甲基环己烷邻二甲酸酐
methyl hexalin acetate　乙酸甲基环己酯
2-methylhexane(=isoheptane)　2-甲基己烷；异庚烷
　　$(CH_3)_2CHC_4H_9$
5-methyl-2-hexanone(=methyl isoamyl ketone)　5-甲基-2-
　　己酮；甲基异戊基甲酮
5-methyl-2-hexanone oxime　5-甲基-2-己酮肟〔防老剂〕
methyl hexine carbonate　己炔羧酸甲酯
3-methyl hexyl acetate　乙酸-3-甲基己酯
methyl hexyl alcohol(=methyl-n-amyl carbinol)　甲基正
　　戊基甲醇；甲基己醇
2-methyl hexyl aldehyde　2-甲基己醛
methyl n-hexyl ketone(=2-octanone)　甲基己基(甲)酮；2-
　　辛酮　$CH_3CO(CH_2)_5CH_3$
methylhistamine　甲基组胺
methylhistidine　甲基组氨酸
N^{20}-methylholarrhimine　N^{20}-甲基止泻木明
methyl hydrate　甲醇
N-methylhydrazide　N-甲基酰肼
methylhydrazine　甲肼　CH_3NHNH_2
methylhydrazine sulfate　硫酸甲肼　$CH_3NHNH_2H_2SO_4$
methyl-hydrazine-sulfonic acid　磺酸甲肼
　　$CH_3NHNHSO_3H$
methyl hydrazone　甲腙
2-methyl-1-hydrindone　2-甲基-1-茚满酮
p-methyl hydrocinnamaldehyde　对甲基苯丙醛
methyl-hydrogen-sulfate　硫酸氢甲酯　$CH_3·SO_4H$

methylhydroquinone　甲基氢醌　$CH_3C_6H_3(OH)_2$
methylhydroquinone diacetate　甲基氢醌二乙酸酯；甲基
　　苯对二酚二乙酸酯　$CH_3C_6H_3=(O_2CCH_3)_2$
methyl-hydroselenide　甲硒醇　CH_3SeH
methyl-hydrosulfide　甲硫醇　CH_3SH
methyl hydroxybenzene　甲(基)酚
methyl hydroxybenzoate　羟基苯甲酸甲酯
　　$HOC_6H_4CO_2CH_3$
4-methyl-5-hydroxyethylthiazole　4-甲基-5-羟乙基噻唑
methyl hydroxylamine　甲胺
methylhydroxypyridine　N-甲基羟吡啶
2-methyl-3-hydroxy-4-pyrone(=maltol)　2-甲基-3-羟基-4-
　　吡喃酮；麦芽酚
methylhydroxyquinoline　N-甲基羟喹啉
N-methyl-5-hydroxytryptamine　N-甲基-5-羟色胺
methyl hypochlorite　次氯酸甲酯　CH_3OCl
methyl hypoxanthine　甲基次黄嘌呤
methylic　甲基
methylic alcohol　甲醇
methylidyne　次甲基　$CH\equiv$
N^5,N^{10}-methylidynel-tetrahydrofolate　N^5,N^{10}-次甲基四
　　氢叶酸
1-methylimidazole　1-甲基咪唑
2-methyl imidazole(=glyoxal ethylene)　2-甲基咪唑
1-methylimidazolyl acetic acid　1-甲基咪唑基乙酸
methylimino diacetic acid(MIMDA)　甲基亚氨(基)二
　　乙酸
methylin　甲灵〔指1.多元醇甲醚；2.甘油甲醚；3.甲木质〕
methyl-indenone　甲基二氢化茚酮
methyl-indigo　甲基靛蓝
α-methyl indole　α-甲基吲哚
β-methylindole(=3-methylindole)　3-甲基吲哚
methyl-indone　甲基二氢化茚酮
methyl-inositol　甲基肌醇　$(HO)_5C_6H_6OCH_3$
methyl-iodide(=iodomethane)　甲基碘；碘代甲烷　CH_3I
6-methyl ionone　鸢尾酮
methyl ionone(=iraldeine)　甲基紫罗兰酮
methyl β-irone　甲基-β-鸢尾酮
N-methyl-isatin　N-甲基靛红　$C_6H_4N(CH_3)COCO$

methylisoamyl alcohol　甲基异戊基醇
methyl isoamylamine　甲基异戊基胺
　　$(CH_3)_2CH(CH_2)_2NHCH_3$
methyl isoamyl ether　甲基异戊基醚
methyl isoamyl ketone(=5-methyl-2-hexanone)　5-甲基-2-
　　己酮；甲基异戊基甲酮
methyl isoamyl ketoxime　甲基异戊基(甲)酮肟
　　$CH_3C(=NOH)C_5H_{11}$

β-methyl-β-isobutyl acrolein　β-甲基-β-异丁基丙烯醛
methyl-isobutylamine　甲基异丁基胺　$CH_3NHC_4H_9$
methyl isobutyl carbinol(MIBC)　甲基异丁基甲醇
methyl isobutyl carbinyl acetate　乙酸甲基异丁基原酯
methyl isobutyl ether　甲基异丁基醚　$CH_3OC_4H_9$
methyl isobutyl ketone(MIBK)　甲基异丁基(甲)酮　$CH_3COC_4H_9$
methyl isobutyl ketone peroxide　过氧化甲基异丁基甲酮
methyl isobutyl ketoxime　甲基异丁基酮肟〔防老剂〕
2-methyl-3-isobutyl pyrazine　2-甲基-3-异丁基吡嗪
methyl isobutyrate　异丁酸甲酯　$(CH_3)_2CHCO_2CH_3$
methyl isocyanate　异氰酸甲酯　CH_3NCO
methyl-isocyanide　甲脒　CH_3NC
methyl isocyanurate　异氰尿酸甲酯〔俗〕；三聚异氰酸(三)甲酯　$(CH_3)_3(CON)_3$
methyl isoeugenol　异丁子香酚甲醚
methyl-isohexylcarbinol　甲基异己基甲醇；6-甲基-2-庚醇
methyl isohexyl ketone　甲基异己基(甲)酮；6-甲基-2-庚酮　$(CH_3)_2CH(CH_2)_3COCH_3$
methyl isonitrile　甲脒　CH_3NC
methyl isophthalic acid　甲基-1,3-苯二甲酸
methyl-isoprene　甲基异戊间二烯　$CH_2=C(CH_3)C(CH_3)=CH_2$
methyl isopropenyl ketone　甲基异丙烯基(甲)酮
methyl-isopropylacetylene　甲基异丙基乙炔；4-甲基-2-戊炔　$(CH_3)_2CHC≡CCH_3$
methyl isopropyl ether　甲基异丙基醚　$CH_3PCH(CH_3)_2$
methyl isopropyl-ketone(=3-methyl-2-butanone)　甲基异丙基(甲)酮；3-甲基-2-丁酮　$CH_3COCH(CH_3)_2$
methyl isopropyl ketone semicarbazone　甲基异丙基甲酮缩氨基脲　$C_3H_7(CH_3)C=NNHCONH_2$
methyl isopropyl ketoxime　甲基异丙基(甲)酮肟；3-甲基-2-丁酮肟　$CH_3C(NOH)C_3H_7$
1-methyl-7-isopropyl-phenanthrene　1-甲基-7-异丙基菲
2-methyl-5-isopropyl-phenol　2-甲基-5-异丙基苯酚
2-methyl-5-isopropyl phenylacetaldehyde　2-甲基-5-异丙基苯乙醛
α-methyl-p-isopropyl phenylpropyl aldehyde(=cyclamen aldehyde)　α-甲基对异丙基苯丙醛；仙客来醛
methylisoquinolinium　甲基异喹啉鎓
methyl-isorhodanate(=methylisothiocyanate)　异硫氰酸甲酯
methyl-isorhodanide(=methylisothiocyanate)　异硫氰酸甲酯
methyl-isosulfocyanate(=methylisothiocyanate)　异硫氰酸甲酯
methyl-isosulfocyanide(=methylisothiocyanate)　异硫氰酸甲酯

methyl isothiocyanate(=methyl-isothiocyanide)　异硫氰酸甲酯　$CH_3N=C=S$
methyl-isothiocyanide(=methylisothiocyanate)　异硫氰酸甲酯
methyl-isothiourea　甲基异硫脲　$C_2H_6N_2S$
methyl-isourea　甲基异脲　$NH_2C(OCH_3)NH$
methyl isovalerate　异戊酸甲酯　$(CH_3)_2C_2H_3CO_2CH_3$
methyl jasmonate　茉莉酮酸甲酯
methyl lactate　乳酸甲酯　$CH_3CHOHCO_2CH_3$
2-methyllactic acid　2-甲基乳酸；α-羟基异丁酸　$CH_3C(OH)(CH_3)COOH$
methyl laurate　十二(烷)酸甲酯　$CH_3(CH_2)_{10}CO_2CH_3$
3-methyl laurylaldehyde　3-甲基月桂醛；甲基壬基丙醛
methyl levulinate　γ-戊酮酸甲酯　$CH_3CO(CH_2)_2CO_2CH_3$
methyl linoleate　亚油酸甲酯
methyllycaconitine　甲基牛扁亭
methyllysine　甲基赖氨酸
methyl-magnesium-bromide　甲基溴化镁　$BrMgCH_3$
methyl-magnesium-chloride　甲基氯化镁　$ClMgCH_3$
methyl-magnesium-halide　甲基卤化镁　$XMgCH_3$
methyl-malachite　甲基孔雀绿
methyl-maleic acid　甲基马来酸；甲基顺丁烯二酸；(E)-甲基丁烯二酸　$COOHCH=C(CH_3)COOH$
methyl maleic anhydride　甲基马来酸酐
methyl malic acid　甲基苹果酸　$CH_3C(OH)(CO_2H)CH_2CO_2H$
methyl-malonate　丙二酸二甲酯　$CH_3OCOCH_2COOCH_3$
methylmalonic acid(=isosuccinic acid)　异丁二酸；甲基丙二酸　$CH_3CH(CO_2H)_2$
methylmalonic acid excretion test　甲基丙二酸排泄试验
methylmalonic acid uria　甲基丙二酸尿
methyl-malonic ester　①甲基丙二酸酯　$ROCOCH(CH_3)COOR$　②〔专指〕甲基丙二酸二乙酯　$C_2H_5OCOCH(CH_3)COOC_2H_5$
methylmalonic semialdehyde　甲基丙二酸单醛
methylmalonyl　①甲基丙二酰(基)②甲基丙二单酰(基)
methylmalonyl-CoA isomerase　甲基丙二酸单酰 CoA 异构酶
methyl mandelate　扁桃酸甲酯；苯乙醇酸甲酯　$C_6H_5CHOHCO_2CH_3$
methyl-mandelic acid　甲基扁桃酸　$C_6H_5C(OH)(CH_3)COOH$
methyl mannoside　甲基甘露糖苷　$CH_3OC_6H_{11}O_5$
methyl margarate　十七(烷)酸甲酯　$CH_3(CH_2)_{15}CO_2CH_3$
methyl mercaptan(=methanthiol)　甲硫醇　CH_3SH
methylmercapte　甲硫基
methylmercaptobenzimidazole　甲基硫醇基苯并咪唑〔促

进剂)

4-methyl mercaptobutanal 4-甲硫基丁醛
4-methyl mercapto-2-butanone 4-甲硫基-2-丁酮
3-methyl mercapto-1-hexanol 3-甲硫基-1-己醇
methyl mercaptomethyl propionate 甲硫基丙酸甲酯
β-methylmercapto-propionaldehyde β-甲基硫基丙醛
γ-methyl mercapto propyl alcohol γ-甲硫基丙醇
3-methyl mercapto propyl isothiocyanate 异硫氰酸-3-甲硫丙酯
methyl-mercuric bromide 溴化甲基汞 CH_3HgBr
methyl-mercuric chloride 氯化甲基汞 CH_3HgCl
methyl-mercuric halide 卤化甲基汞 CH_3HgX
methyl-mercuric iodide 碘化甲基汞 CH_3HgI
methyl mercury 甲基汞
methyl methacrylate 异丁烯酸甲酯;甲基丙烯酸甲酯 $C_3H_5CO_2CH_3$
methyl methacrylate-butadiene latex 甲基丙烯酸甲酯丁二烯胶乳
methyl methacrylate polymer 甲基丙烯酸甲酯聚合物;异丁烯酸甲酯聚合物
methyl-methacrylate resin 甲基丙烯酸甲酯树脂
2-methyl-5-methoxy thiazole 2-甲基-5-甲氧基噻唑
methyl methylanthranilate 甲基氨茴酸甲酯;邻甲氨基苯甲酸甲酯 $CH_3NHC_6H_4CO_2CH_3$
methyl 2-methylbutyrate 2-甲基丁酸甲酯
2-methyl-3-(2-methylpropyl) pyrazine 2-甲基-3-(2-甲基丙基)吡嗪
methyl β-methyl-thiopropionate β-甲基硫代丙酸甲酯
2-methyl-3-methylthiopyrazine 2-甲基-3-甲硫基吡嗪
methyl-monosilane 甲基硅烷 CH_3SiH_3
methylmorphine(=codeine) 甲基吗啡碱;可待因
N-methyl morpholine N-甲基吗啉
methyl-mustard oil 甲基芥子油;异硫氰酸甲酯 CH_3NCS
methyl myristate 肉豆蔻酸甲酯;十四酸甲酯 $CH_3(CH_2)_{12}CO_2CH_3$
methyl nadic acid(=methyl carbic acid) 甲基降冰片烯二酸
methyl nadic anhydride 甲基纳迪克酸酐;甲基内亚甲四氢苯酐;甲基降冰片烯二酸酐
methylnaphthalene 甲基萘
methylnaphthohydroquinone 甲(基)萘氢醌;维生素 K_4
methyl naphthoquinone 甲基萘醌
methyl naphthylacetaldehyde 甲基萘乙醛
methyl-β-naphthyl carbinol 甲基-β-萘原醇
methyl naphthyl ether 甲基萘基醚
methyl naphthyl ketone 甲基萘基(甲)酮 $CH_3COC_{10}H_7$
methylnicotinamide 甲基烟酰胺

methyl-nitramine 甲硝胺 CH_3NHNO_2
methyl nitrate 硝酸甲酯 CH_3ONO_2
methyl nitrite 亚硝酸甲酯 CH_3ONO
methyl nitrobenzene 硝基甲苯
methyl nitrolic acid 甲硝肟酸 $HC(NOH)NO_2$
2-methyl-2-nitropropyl acrylate 2-甲基-2-硝基丙基丙烯酸酯
2-methyl-2-nitropropyl methacrylate 2-甲基-2-硝基丙基甲基丙烯酸酯
N-methyl-N-nitro-p-toluenesulfonamide(MNSA) N-甲基-N-亚硝基对甲苯磺酰胺
2-methylnonane(=isodecane) 2-甲基壬烷;异癸烷 $(CH_3)_2CH(CH_2)_6CH_3$
methyl 2-nonenoate 2-壬烯酸甲酯
methyl nonyl acetaldehyde 甲基壬(基)乙醛
methyl nonyl acetaldehyde dimethyl aceta 甲基壬乙醛二甲缩醛
methyl nonyl carbinol 甲基壬基甲醇
methyl n-nonyl ketone(=undecanone) 甲基正壬基(甲)酮;2-十一烷酮 $CH_3(CH_2)_8COCH_3$
methyl nonyl propionaldehyde 甲基壬丙醛
methyl 2-nonynoate 2-壬炔酸甲酯;辛炔羧酸甲酯 $CH_3(CH_2)_5C≡CCOOCH_3$
methyl norbixinate 原胭脂树酸甲酯 $C_{25}H_{30}O_4$
6-methyl-5-norbornene-2,3-dicarboxylic anhydride 6-甲基-5-降冰片烯-2,3 二酸酐;甲基降冰片烯二酸酐
methyl norbornene dioic anhydride(=methyl carbic anhydride) 甲基降冰片烯二酸酐
2-methyl-octane(=isononane) 2-甲基辛烷;异壬烷 $(CH_3)_2CHC_6H_{13}$
6-methyl octanoic acid 6-甲基辛酸
methyl octine carbonate 辛炔羧酸甲酯
methyl octyl acetaldehyde 甲基辛基乙醛;2-甲基癸醛
2-methyl octylaldehyde 2-甲基辛醛
methyl n-octyl ketone(=decanone) 甲基正辛基(甲)酮;癸酮 $CH_3COC_8H_{17}$
methyl 2-octynoate 2-壬炔酸甲酯
methyloic 含有羧基的(—COOH)
methylolacetone 羟甲基丙酮;4-羟基-2-丁酮
N-methylolacrylamide N-羟甲基丙烯酰胺
N-methylol allyl carbamate N-羟甲基氨基甲酸烯丙(基)酯
methylol amide 羟甲基酰胺
methylol-aminoplast 羟甲基氨基塑料
methylolated 羟甲基化的
methylolating 羟甲基化
methylolation 羟甲基化
methylol cellulose 羟甲基纤维素

methylol chloroacetamide 羟甲基氯代乙酰胺
N-methylol compound N-羟甲基化合物
methylol crosslinking agent 羟甲基交联剂
methylol dicyandiamide 羟甲基双氰胺
methylol dimethyl hydantion 羟甲基二甲基海因；羟甲基二甲基乙内酰脲
methyl oleate 油酸甲酯
methyl oleate peroxide 过氧化油酸甲酯
methylol fatty acid 羟甲基脂肪酸
methylol melamine 羟甲基蜜胺；羟甲基三聚氰胺
methylol melamine butyl ether 丁基醚化羟甲基三聚氰胺
N-methylol methacrylamide N-羟甲基甲基丙烯酰胺
methylolphenol 羟甲基苯酚
methylolphenol formaldehyde resin 羟甲基苯酚甲醛树脂〔硫化剂〕
methylolphosphonium 羟甲基磷(根)
methylol riboflavin 羟甲基核黄素
methylol urea 羟甲基脲；脲基甲醇 $NH_2CONHCH_2OH$
N-methylolurethane N-羟甲基氨基甲酸乙酯
methyl orange(=tropeolin-D) 甲基橙；金莲橙-D $(CH_3)_2NC_6H_4N=NC_6H_4SO_3Na$
methyl orange paper 甲基橙试纸
methyl-orange test paper 甲基橙试纸
methyl ortho-silicate 原硅酸甲酯
methyl-oxalacetate 草酰乙酸二甲酯
methyl-oxalacetic ester ①草酰乙酸酯 $ROCOCOCH(CH_3)COOR$②〔专指〕草酰乙酸二乙酯 $C_2H_5OCOCOCH(CH_3)COOC_2H_5$
methyl oxide 甲醚
2-methyloxine 2-甲基喔星；8-羟基喹哪啶
methyloxyaniline 甲氧苯胺；茴香胺
methyl palmitate 十六(烷)酸甲酯 $C_{15}H_{31}CO_2CH_3$
methyl paraben 羟苯甲酸甲酯；尼泊金甲酯；羟甲苯
methyl pelargonate 壬酸甲酯 $CH_3(CH_2)_7CO_2CH_3$
methylpelletierine 甲基石榴碱
methyl pentachlorostearate 五氯硬脂酸甲酯〔增塑剂〕
methyl pentadecyl ketone 甲基十五基(甲)酮；2-十七烷酮 $CH_3(CH_2)_{14}COCH_3$
methylpentadiene 甲基戊二烯
2-methyl pentanal 2-甲基戊醛
methyl pentanediol 甲基戊二醇
3-methyl-2,4-pentanediol 3-甲基-2,4-戊(烷)二醇
4-methyl-2,3-pentanedione 4-甲基-2,3-戊二酮
4-methyl-2-pentanone(=methyl isoamyl ketone) 甲基异戊基(甲)酮；4-甲基-2-己酮 $CH_3COC_5H_{11}$
2-methyl-2-pentenal 2-甲基-2-戊烯醛
methyl pentenyl acetate 乙酸甲基戊烯酯

methyl pentenyl ketone 甲基戊烯基酮
methyl pentenyl salicylate 水杨酸甲基戊烯酯
methyl-pentosan 甲基戊聚糖
methylpentose 甲基戊糖
3-methyl pentyl acetate 乙酸-3-甲基戊酯
N-methyl-2-pentyl-N'-pheny-p-phenylene diamine N-甲基-2-戊基-N'-苯基对苯二胺〔防老剂〕
methylpentynol 3-甲基-1-戊炔-3-醇
methylphenacyl alcohol 甲基苯甲酰甲醇 $CH_3C_6H_4COCH_2OH$
methyl-phenate 苯甲醚
2-methyl phenazine 2-甲基吩嗪 $CH_3C_6H_3=N_2C_6H_4$
methylphenidate 哌醋甲酯〔药〕
methyl phenidine 甲基-N-乙酰乙氧苯胺〔药〕
p-methyl phenol 对甲基苯酚
methyl-phenoxide 苯甲醚 $C_6H_5OCH_3$
methyl phenylacetaldehyde 甲基苯乙醛
methyl phenylacetamide 甲基N-乙酰苯胺〔药〕
methyl phenylacetate 苯乙酸甲酯 $C_6H_5CH_2CO_2CH_3$
methyl-phenylamine(=N-methylaniline) N-甲基苯胺；甲基苯基胺 $C_6H_5NHCH_3$
methyl-phenyl-arsine oxide 甲苯胂化氧 $CH_3C_6H_4As=O$
p-methyl-γ-phenylbutyraldehyde 对甲基-γ-苯丁醛
methyl phenyl carbinol 甲基苯基原醇；α-苯乙醇
methyl phenyl carbinyl acetate 乙酸甲基苯基原酯；乙酸苏合香酯
methyl phenyl diketone 甲基苯基二(甲)酮 $CH_3COCOC_6H_5$
methylphenylene 甲基亚苯基 $CH_3C_6H_3=$
N-methyl-p-phenylenediamine N-甲基对苯二胺 $CH_3NHC_6H_4NH_2$
methyl phenyl ether 甲基苯基醚；茴香醚
N-methyl-β-phenyl-ethylamine N-甲基-β-苯基乙胺
methyl-phenyl-ethyl-carbinol 甲基苯基乙基甲醇 $C_{10}H_{14}O$
methyl-β-phenylethyl ether 甲基-β-苯乙醚
methyl phenyl ethylglycidate 甲基苯基缩水甘油酸乙酯；草莓醛〔商〕；十六醛〔商〕
5-methyl-2-phenyl-2-hexenal 5-甲基-2-苯基-2-己烯醛
sym-methylphenylhydrazine 对称甲基苯基肼 $C_6H_5NHNHCH_3$
unsym-methylphenylhydrazine 不对称甲基苯基肼
1-methyl-3-phenylindone 1-甲基-3-苯基二氢茚酮
methyl phenyl ketone(=acetophenone) 乙酰苯；苯乙酮；甲基苯基(甲)酮 $CH_3COC_6H_5$
4-methyl-1-phenyl-2-pentanol 4-甲基-1-苯基-2-戊醇
4-methyl-1-phenyl-2-pentanone 4-甲基-1-苯基-2-戊酮

methyl phenylpropionate 苯丙酸甲酯
γ-methyl-γ-phenylpropyl alcohol γ-甲基-γ-苯丙醇；3-苯基丁醇
methylphenylsilicone oil 甲基苯基硅酮油
methylphenyl siloxane 甲基苯基硅氧烷
sym-methyl phenylurea 对称甲基苯基脲 $C_8H_{10}N_2O$
β-methyl-δ-phenylvaleraldehyde β-甲基-δ-苯戊醛
methyl-phenyl-vinyl silicone rubber 甲基-苯基-乙烯基硅橡胶
methyl-phosphate 磷酸甲酯
methylphosphine 甲膦 CH_3PH_2
methyl phosphine borine 甲膦合甲硼烷 $CH_3PH_2BH_3$
methyl-phosphinic acid 甲基膦酸 $CH_3PO(OH)_2$
methyl-phosphite 亚磷酸一甲酯 $CH_3OP(OH)_2$
methyl-phospho-acid(=methylphosphonic acid) 甲基膦酸
methylphosphonic diamide 甲基膦二酰胺
methyl-phosphoric acid 磷酸一甲酯
methyl-phosphorous acid ①甲基膦酸 $CH_3PO(OH)_2$ ②亚磷酸一甲酯 $CH_3O(OH)_2$
methyl phthalyl ethyl glycolate 甲基邻苯二甲酰基甘醇酸乙酯〔增塑剂〕
2-methyl-3-phytyl-1,4-naphthoquinone 2-甲基-3-植基-1,4-萘醌；维生素 K_1
N-methyl piperazine N-甲基哌嗪
N-methyl piperidine N-甲(基)哌啶 $C_5H_{10}NCH_3$
methyl piperidine methyl pentamethylene dithiocarbamate 甲基五亚甲基二硫代氨基甲酸甲基哌啶〔促进剂〕
methyl podocarpatrienate(=deisopropyl dehydro abietate) 罗汉松三烯酸甲酯；脱异丙基脱氢枞酸酯
methyl polymethacrylate 聚甲基丙烯酸甲酯〔增塑剂〕
methyl polysiloxane 聚甲基硅氧烷
methyl prednisolone 甲基脱氢皮质(甾)醇〔药〕
methyl propamine 甲基普鲁巴明
2-methyl propanal 2-甲基丙醛
2-methylpropanoic acid 2-甲基丙酸
methyl propargyl ether 甲基炔丙基醚 $HC\equiv CCH_2OCH_3$
2-methylpropenyl(=isobutenyl) 异丁烯基；2-甲基丙烯基
methyl propenyl ketone(=ethylidene acetone) 亚乙基丙酮；甲基丙烯基(甲)酮 $CH_3CH=CHCOCH_3$
methyl propionate 丙酸甲酯 $C_2H_5CO_2CH_3$
methyl-propyl-acetic acid(=2-methyl valeric acid) 甲基丙基乙酸；2-甲基戊酸 $C_2H_5CH_2CH(CH_3)CO_2H$
methyl propyl-acetylene 甲基丙基乙炔；2-己炔
methyl propyl carbinyl acetate 乙酸甲基丙基原酯
3-methyl-5-propyl-2-cyclohexenone 3-甲基-5-丙基-2-环己烯酮
methyl propyl disulfide 甲基丙基二硫醚

methyl propyl-isobutylcarbinol 甲基丙基异丁基甲醇；2,4-二甲基-4-庚醇 $(CH_3)(C_3H_7)(iso-C_4H_9)C(OH)$
methyl n-propyl ketone(=2-pentanone) 甲基丙基(甲)酮；2-戊酮 $CH_3COCH_2C_2H_5$
methyl n-propyl ketoxime 甲基丙基甲肟
2-(2-methylpropyl)-pyridine 2-(2-甲基丙基)吡啶
2-(1-methylpropyl)-thiazole 2-(1-甲基丙基)噻唑
methyl propyl trisulfide 甲基丙基三硫
d-N-methyl-pseudoephedrine d-N-甲基假麻黄碱
methyl pseudoionone 甲基假性紫罗兰酮
methyl psychotrine 甲基九节碱
2-methylpyrazine 2-甲基吡嗪
methyl pyrazinyl ketone 甲基吡嗪基酮
methyl pyridine 甲基吡啶
4-methyl pyridine N-oxide N-氧化-4-甲基吡啶
methylpyridone-carboxylamide 甲基吡啶酮甲酰胺
methyl-2-pyridyl ketone 甲基-2-吡啶基酮
methyl pyrocatechol 甲基焦儿茶酚；甲基邻苯二酚 $CH_3C_6H_3(OH)_2$
methyl-pyrogallol 甲基焦棓酚；甲基连苯三酚 $CH_3C_6H_2(OH)_3$
methyl pyromucate 焦黏酸甲酯；糠酸甲酯
N-methyl-2-pyrrolidinone(NMP) N-甲基-2-吡咯烷二酮
N-methyl pyrrolidone N-甲基吡咯烷酮
β-methylpyrroline β-甲基吡咯啉
methyl 2-pyrrolyl ketone 2-乙酰基吡咯；甲基-2-吡咯基酮
methyl pyruvate 丙酮酸甲酯
4-methyl quinoline(=lepidine) 4-甲基喹啉 $C_{10}H_9N$
methylquinolinium 甲基喹啉鎓
methylquinolizidine 甲基喹嗪；甲基氢化-9-氮(杂)萘
methyl radical 甲基
methyl red 甲基红
methyl red indicator 甲基红指示剂
methyl red test 甲基红试验
5-methylresorcinol 5-甲基间苯二酚
methyl-rhodanate(=methyl-rhodanide) 硫氰酸甲酯 CH_3SCN
methyl-rhodanide(=methyl rhodanate) 硫氰酸甲酯
methyl ricinoleate 蓖麻酸甲酯〔增塑剂〕
methyl rosaniline chloride 氯化甲基蔷薇苯胺；氯化甲基品红(碱)；氯化甲基玫瑰胺
methyl rosaniline hydrochloride 结晶紫〔药〕
methyl rubber 甲基橡胶；二甲基丁二烯橡胶
methyl salicylate 水杨酸甲酯 $HOC_6H_4CO_2CH_3$
methyl-selenide 二甲基硒 $(CH_3)_2Se$
methyl-seleninic acid 甲基亚硒酸 CH_3SeOOH
methyl-silicane 甲基甲硅烷 SiH_3CH_3

methyl-silicone 聚甲基硅氧烷
methyl silicone gum 甲基硅橡胶
methyl silicone oil 甲基硅(氧烷)油；甲基硅油
methyl silicone rubber 甲基硅橡胶
methyl siliconic acid 甲基硅酸 $CH_3Si(OH)_3$
methyl siloxane 甲基硅氧烷
methyl sodium 甲基钠
methylspinazarin 甲基菠菜茜素
3-methyl-spiro-dinaphthopyran(e) 3-甲基-螺-二萘吡喃
methyl-stannic acid(=methyl-stannonic acid) 甲基锡酸
methyl-stannonic acid(=methyl-stannic acid) 甲基锡酸 CH_3SnOOH
methyl stearate 硬脂酸甲脂 $C_{17}H_{35}CO_2CH_3$
10-methylstearic acid 10-甲基硬脂酸
methyl styrene 甲基苯乙烯；乙烯基甲苯
α-methylstyrene-derived product α-甲基苯乙烯-衍生物
methylstyrene rubber 甲基苯乙烯橡胶
methyl styryl carbinyl acetate 乙酸甲基苯乙烯基原酯
methyl styryl ketone 甲基苯乙烯基(甲)酮
methyl-suberate 辛二酸二甲酯 $CH_3OCO(CH_2)_6COOCH_3$
methyl-substituted hydrazine 甲代肼
methyl-succinaldehyde 甲基丁二醛 $CHOCH_2CH(CH_3)CHO$
methyl-succinate 丁二酸二甲酯 $CH_3OCO(CH_2)_2COOCH_3$
methylsuccinic acid(=pyrotartaric acid) 焦酒石酸；甲基丁二酸 $COOHCH_2CH(CH_3)COOH$
N-methyl succinimide N-甲基琥珀酰亚胺；N-甲基丁二酰亚胺
methyl-sulfate 硫酸二甲酯 $(CH_3)_2SO_4$
methyl-sulfhydrate(=methyl-sulfhydryl) 甲硫醇 CH_3SH
methyl-sulfhydryl(=methyl-sulfhydrate) 甲硫醇
methyl-sulfide 二甲硫 $(CH_3)_2S$
methyl-sulfinic acid 甲亚磺酸
ω-(methylsulfinyl)-acetophenone ω-甲基亚磺酰基乙酰苯 $C_6H_5COCH_2SOCH_3$
methyl-sulfinyl-ethane 甲亚磺酰乙烷；甲基乙基亚砜 $(CH_3C_3H_5)S=O$
methyl-sulfocyanate(=methyl-sulfocyanide) 硫氰酸甲酯 CH_3CNS
methyl-sulfocyanide(=methyl-sulfocyanate) 硫氰酸甲酯
methyl sulfoethyl amine 甲基磺基乙胺
methyl sulfonal 乙丁二砜 O=SH
                                           ‖
                                           O
methylsulfonic acid 甲磺酸 CH_3SO_3H

methylsulfonyl(=mesyl) 甲磺酰基 $CH_3SO_2—$
methylsulfonyl chloride 甲磺酰氯 CH_3SO_2Cl
3-methyl-1-(4-sulfophenyl)-2-pyrazolin-5-one 3-甲基-1-(4-磺代苯基)-2-吡唑啉-5-酮
methyl sulfosuccinate ester salt 甲基磺基琥珀酸酯盐
methyl-sulfoxide 二甲亚砜 $(CH_3)_2S=O$
methyl-sulfuric acid O-甲(基)硫酸 CH_3OSO_2OH
methyl sulphonyl 甲磺酰(基) $CH_3SO_2—$
methyl-sulphoxide 二甲亚砜 $(CH_3)_2S=O$
methyl-tartrate 酒石酸二甲酯 $CH_3OCO(CHOH)_2COOCH_3$
methyltartronic acid 甲基丙醇二酸 $CH_3C(OH)(CO_2H)_2$
methyl taurate 牛磺酸甲酯；2-氨基乙磺酸甲酯
N-methyl taurine N-甲基牛磺酸 $CH_3NHCH_2CH_2SO_3H$
methyl-telluride 二甲碲 $(CH_3)_2Te$
N-methyl terephthalhydrazide N-甲基对苯二甲酰肼
methyltestosterone 甲基睾(甾)酮
13-methyltetradecanoic acid 13-甲基十四(烷)酸
N^5-methyl-tetrahydrofolate N^5-甲基四氢叶酸
2-methyltetrahydrofuran-3-one 2-甲基四氢呋喃-3-酮
N-methyl tetrahydrofurfurylamine N-甲基四氢糠胺〔促进剂〕
methyl tetrahydrophthalic anhydride 甲基四氢邻苯二酸酐；4-甲基-1,2,3,6-四氢邻苯二甲酸酐
4-methyl-1,2,3,4-tetrahydroquinoline 4-甲基-1,2,3,4-四氢喹啉
methyl tetrahydroquinoline(=kairoline) N-甲基四氢化喹啉 $C_9H_{10}NCH_3$
1-methyltetrahydro-1,3,5-triazine-4-thione 1-甲基四氢化-1,3,5-三嗪-4-硫酮〔促进剂〕
methyl tetramethylglucoside 四甲基葡萄糖甲苷
methyl-tetrazene 甲基四氮烯 $CH_3NHN=NNH_2$
O-methylthalicberine O-甲基白蓬草碱；唐松草棼碱甲醚
5-methyl-2-thenaldehyde 5-甲基-2-噻吩基甲醛
methyl theobromine 咖啡碱
methyl-thetine 甲基噻亭 $(CH_3)_2SCH_2COO$
4-methyl-5-thiazole ethanol 4-甲基-5-噻唑基乙醇
3-methyl-thiazolidine-2-thione 3-甲基四氢噻唑-2-硫酮〔硫化剂〕
methyl-2-thiazolyl ketone 甲基-2-噻唑基酮
methylthio 甲硫基 $CH_3S—$
2-methyl thioacetaldehyde 2-甲硫基乙醛
5-methylthioadenosine 5-甲硫腺苷
p-methyl-thiobenzophenone 对甲基二苯甲硫酮
3-methylthio butanal 3-甲硫基丁醛
4-methylthio-2-butanone 4-甲硫基-2-丁酮
methyl thiobutyrate 硫代丁酸甲酯

methyl thiocarbamate 硫代氨基甲酸甲酯 NH_2CSOCH_3
methyl thiocyanate(=methyl thiocyanide) 硫氰酸甲酯 CH_3SCN
methyl thiocyanide(=methyl thiocyanate) 硫氰酸甲酯
methyl thiofuroate 硫代糠酸甲酯
methylthio glycolic acid 甲硫乙酸
methylthio group 甲硫基
3-methylthio-1-hexanol 3-甲硫基-1-己醇
3-methylthiol propyl alcohol 3-甲硫基丙醇
2-methyl-5-thiomethylfuran 2-甲基-5-甲硫基呋喃
4-methylthio-4-methyl-2-pentanone 4-甲硫基-4-甲基-2-戊酮
methyl-thionamic acid 甲氨基磺酸 CH_3NHSO_3H
methyl thionine chloride 亚甲蓝
methylthiophene 甲基噻吩
5-methyl-2-thiophene carboxaldehyde 5-甲基-2-噻吩基甲醛
2-methylthiophenol 2-甲硫基苯酚
3-methylthiopropanol 3-甲硫基丙醇
3-methylthiopropionaldehyde 3-甲硫基丙醛
methylthiopyrazine 甲硫基吡嗪
methyl thiouracil 甲基硫尿嘧啶
methyl-thiourea 甲硫脲 $NH_2CSNHCH_3$
2-methylthioxanthone(=2-methyl-9H-thioxanthene-9-one) 2-甲基噻吨酮；2-甲基-9H-噻吨-9-酮
methylthymol blue 甲基百里酚蓝
methyl-tin 二甲锡 $Sn(CH_3)_2$
methyl tin bromide 三溴化甲基锡
methyl tin chloride 三氯化甲基锡
methyl tin iodide 三碘化甲基锡
methyltitanium 甲基钛
methyl-toluate 甲基苯酸甲酯 $CH_3C_6H_4COOCH_3$
methyl toluidine 二甲基苯胺
methyl tolyl ether 甲基甲苯基醚
methyl tolyl ketone 甲基甲苯基(甲)酮 $CH_3COC_6H_4CH_3$
methyl tolyl sulfide 甲基甲苯基硫 $CH_3SC_6H_4CH_3$
methyl transferase 转甲基酶
methyl triacetoxysilane 甲基三乙酰氧基硅烷〔硫化剂〕
methyl tri-t-butylperoxy silane 甲基三叔丁基过氧基硅烷〔偶联剂〕
methyl tricaprylammonium chloride(TCNACl) 氯化甲基三辛基铵 $(CH_3)(C_8H_{17})_3NCl$
methyl trichloroacetate 三氯乙酸甲酯 $Cl_3CCO_2CH_3$
2-methyl-3,3,3-trichloro-1,2-propanediol 2-甲基-3,3,3-三氯-1,2-丙(烷)二醇〔阻燃剂〕
methyl-trichloro-silicane 甲代三氯基硅 CH_3SiCl_3
methyl-triethoxysilicane 甲基三乙氧基硅 $CH_3Si(OC_2H_5)_3$
methyl-triethylsilicane 甲基三乙基硅 $CH_3Si(C_2H_5)_3$

methyl triglycol acrylate 三甘醇一甲醚一丙烯酸酯 $CH_2=CHCOO(C_2H_4O)_3CH_3$
methyl trihydroxyfluorene 甲基三羟基芴
methyl trilaurylammonium nitrate(TLMAN) 硝酸甲基三月桂基铵
methyl-trimethoxysilicane 甲基三甲氧基硅 $CH_3Si(OCH_3)_3$
methyl trimethylacetate 三甲基乙酸甲酯；叔戊酸甲酯 $(CH_3)_3CCO_2CH_3$
methyl tris (ethylmethylketoxime) silane 甲基三乙酮肟基硅烷
methyl trisulfide 甲基三硫
methyl trithiocarbonate 三硫代碳酸甲酯
N-methyltyramine N-甲基酪胺
methyl-tyrosine N-甲(基)酪氨酸 $CH_3NHC_9H_9O_3$
β-methylumbelliferone β-甲基伞形酮〔防日光龟裂剂〕
methyl undecine carbonate 十一炔羧酸甲酯
methyl undecylate 十一酸甲酯
methyl undecylenate 十一烯酸甲酯
methyl n-undecyl ketone(=2-tridecanone) 甲基十一基(甲)酮；2-十三烷酮 $CH_3CO(CH_2)_{10}CH_3$
methyl n-undecyl ketone semicarbazone 甲基十一基(甲)酮缩氨基脲 $C_{11}H_{23}(CH_3)C=NNHCONH_2$
5-methyluracil 5-甲基尿嘧啶
methylurea 甲脲 $CH_3NHCONH_2$
methylurea nitrate 硝酸甲脲 $CH_3NHCONH_2HNO_3$
methyl urethane 甲基尿烷；甲氨基甲酸乙酯 $CH_3NHCO_2C_2H_5$
methyl-uric acid 甲基尿酸 $C_5H_3O_3N_4 \cdot CH_3$
2'-O-methyluridine 2'-O-甲基尿苷
2-methylvaleraldehyde 2-甲基戊醛
methyl valerate 戊酸甲酯 $CH_3(CH_2)_3CO_2CH_3$
α-methylvaleric acid(=2-methyl-valeric acid) α-甲基戊酸；2-甲基戊酸
2-methyl-valeric acid(=α-methylvaleric acid) 2-甲基戊酸；α-甲基戊酸 $C_2H_5CH_2CH(CH_3)CO_2H$
methyl vanillin 甲基香兰素
1-methylvinyl(=isopropenyl) 1-甲代乙烯基；异丙烯基
methylvinyl acetate 乙酸甲基乙烯酯；乙酸异丙烯酯
methyl vinyl ether 甲基乙烯基醚
methyl vinyl ketone 甲基乙烯基(甲)酮；丁烯酮 $CH_2=CHCOCH_3$
2-methyl-5-vinylpyrazine 2-甲基-5-乙烯基吡嗪
methylvinylpyridine-grafted cotton 甲基乙烯基吡啶接枝(的)棉纤维
methylvinylpyridine rubber 甲基乙烯基吡啶橡胶
methyl vinyl silicone rubber 甲基-乙烯基硅橡胶
4-methyl-5-vinylthiazole 4-甲基-5-乙烯基噻唑

methyl violet 甲基紫*
methylviolet base(=pentamethyl pararosaniline) 甲基紫碱 $C_{24}H_{29}ON_3$
methyl violet paper 甲基紫试纸
methyl viologen 甲基紫精；联二-N-甲基吡啶
methyl viologin 甲基紫精
methyl xanthate (乙)黄原酸甲酯 $C_2H_5OCSSCH_3$
methyl xylenol blue(MXB) 甲基二甲酚蓝
methylxyloside 甲基木糖苷
methyl yellow(=dimethyl yellow) 甲基黄
methymycin 酒霉素
methyne 次甲基
methyprylon 甲普龙；甲乙哌酮〔镇静催眠抗惊厥药〕
methystic acid 醉人酸
methysticin 麻醉椒苦素；醉人素〔另有醉椒素〕
metier 多部位干式纺丝机
metier twist 干纺捻度
metlyl branching silicone oil 甲基支链硅油〔脱模剂〕
metoclopramide 甲氧氯普胺〔药〕
metol poisoning 米吐尔中毒
metopon 美托酮；甲基二氢吗啡酮〔镇静药〕
metoprolol 美托洛尔〔药〕
metralac hydrometer 橡胶比重计
metrazol(=pentamethylene tetrazole) 亚戊基四唑
metre(=meter) ①米；公尺②计；表
metric ①米制的②度量的
metrication 换成米制
metric carat 米制(的)开(单位)
metric count 公支；公制支数
metric fine thread 公制细牙螺纹
metric screw thread 公制螺纹
metric system 米制；公制
metric tensor 度量张量〔数〕
metric ton 公吨；千公斤
metric unit 米制单位
metrolac 胶乳比重器〔测干胶含量用〕
metrolac hydrometer 橡浆比重计
metrological accreditation 计量认证
metrological management 计量管理
metrological supervision 计量监督
metrological verification 计量检定
metrological verification regulation 计量检定规程
metrology ①计量学；度量衡学②计量制；度量衡制
metronidazole 2-甲基-5-硝基-1-咪唑基-乙醇；甲硝唑〔药〕
metronom 节拍器
metrotonin 异戊巴比妥
metycaine 苯甲酸γ-(2-甲基哌啶)丙基酯盐酸盐；盐酸美替卡因

metyrapone 甲吡酮；美替拉酮
Mev 百万电子伏
mevaldic acid 3-羟-3-甲戊醛酸
mevalonate ①甲羟戊酸；3-甲基-3,5-二羟(基)戊酸②甲羟戊酸盐(酯或根)
mevalonate-5-pyrophosphate 甲羟戊酸-5-焦磷酸
mevalonic acid(=hiochic acid) 甲瓦龙酸〔俗〕；3,5-二羟基-3-甲戊酸；袂瓦龙酸
Mexican flux 墨西哥(石油中提出的)高硫稀释剂
Mexican onyx 墨西哥缟玛瑙
Mexican poppy seed oil 墨西哥罂粟子油
mexiletine 美西律〔药〕
mexobalamin 甲钴胺〔药〕
Meyer bar(=wire-wound rod) 迈耶绕线棒涂布器
Meyer hardness 迈耶硬度；迈氏硬度
meyerhoffite 三斜硼钙石
Meyer's law 迈耶酯化定律
Meyer's tube 迈耶二氧化碳吸收管
Meyer synthesis 迈耶合成法
meymacite 水钨华
mezanomycin(=mezzanomycin) 梅扎诺霉素
mezcal 龙舌兰酒
mezcaline 仙人球毒碱
mezereon 瑞香
mezereum 瑞香
mezlocillin 美洛西林〔药〕
mezzanine structure 夹层结构
mezzanomycin 梅扎诺霉素
MF resin 蜜胺-甲醛树脂
mgal 毫伽〔=10^{-3} gal=10^{-3} 厘米/秒^2〕
M glass fibre M 玻璃纤维；高模量玻璃纤维
mho 姆欧
mhometer 姆欧计；电导计〔物〕
miamine 氯胺-T
miamycin 迈亚密霉素
miargyrite 辉锑银矿
miarolitic texture 晶洞结构
miascite(=miaskite) 云霞正长岩
miaskite(=miascite) 云霞正长岩
miasm(a) 沼泽地瘴气
miazines 嘧啶类
MIBK wax deoiling 甲基异丁基酮法蜡脱油
mica 云母
mica board 云母板
micaceous 云母的；含云母的；云母状的
micaceous hematite 云母赤铁矿
micaceous iron oxide 云母铁矿
micacite 云母片岩

mica condenser 云母电容器
micado yellow 米卡多黄
mica extender pigment 云母体质颜料
mica flake 云母片
micanite 层合云母板；绝缘石
mica pig 云母晶块
mica powder 云母粉
micas 云母类
mica scale 云母片
mica schist(=mica slate) 云母片岩
mica sheet 云母片
mica slate(=mica schist) 云母片岩
mica specks 云母状斑点
mica splittings 剥制云母；云母片层〔劈裂层〕
mica talc 云母滑石；块滑石
mica-titanium dioxide 云母-二氧化钛
micelianamide 微孢酰胺 $C_{21}H_{28}O_5N_2$
micella 微胶粒；胶束
micellae(=micelle) ①细胞束②胶粒③分子组缨④巢〔橡胶纤维及其他复杂物质单位结构〕
micellar adsorption 胶束吸附
micellar aggregate 微胞凝集体；胶束凝集体
micellar aggregation 胶束集聚(作用)
micellar catalysis 胶束催化
micellar charge 胶束电荷
micellar chromatography 胶束色谱法
micellar colloid 胶束状胶体
micellar copolymerization 胶束聚合
micellar core 胶束内芯
micellar dispersion 胶束分散剂；胶束分散体
micellar electrokinetic capillary chromatography(MECC; MEKC) 胶束电动毛细管色谱法
micellar electrokinetic chromatograph 胶束电动色谱仪
micellar electrokinetic chromatography 胶束电动色谱法
micellar emulsion 胶束乳状液
micellar enhanced spectrofluorometric method 胶束增敏荧光分光法
micellar heterogeneous 胶束多相的；微胞多相的
micellar hydration 胶束水化作用
micellar liquid chromatography 胶束液相色谱法
micellar (molecular) weight 胶束(分子)量
micellar net structure 胶束网状结构微胞网状结构
micellar phase 胶束相
micellar size 胶束大小
micellar solubility(=solubilization capacity) 加溶(容)量
micellar solubilization 胶束增溶作用
micellar (solubilization) spectrophotometry 胶束(增溶)分光光度法

micellar solution 胶束溶液
micellar strand 微胞束；胶束串
micellar structure 胶束结构
micellar surface 胶束表面
micellar sweeping 胶束扫掠
micellar system 胶束体系
micellar theory 胶束理论
micellar thin layer chromatography 胶束薄层色谱法
micelle(=supermolecule) 胶束；微胞
micelle-cluster 胶束簇
micelle core 胶束内芯
micelle critical concentration 胶束临界浓度
micelle enhanced ultrafiltration 胶束增效超滤
micelle formation 胶束形成；胶束化(作用)
micelle-forming concentration 胶束形成浓度；临界胶束浓度
micelle-forming ion 成胶束离子
micelle-forming solute 成胶束溶质
micelle-like aggregate 胶束状聚集体
micelle model 胶束模型
micelle molecular lattice 胶束分子晶格
micelle relaxation 胶束松弛作用
micelles agglomerate 胶束凝聚物
micelle-sensitized derivative variable angle synchronous fluorescence 胶束增敏可变角同步荧光
micelle-sensitized flow injection spectrophotometry 胶束增敏流动注射分光光度法
micelle-sensitized fluorimetry 胶束增敏荧光法
micelle-sensitized kinetic photometry 胶束增敏动力学光度法
micelle shape 胶束形状
micelle stabilization 胶束稳定作用
micelle stabilized room temperature phosphorimetry 胶束增稳室温磷光法
micelle-string 微胞束；胶束串
micelle surface 胶束表面
micelle theory 胶束理论
micelle volume 胶束体积
micetina(=mycetin) 头孢菌亭素
Michael addition 迈克尔加成
Michaelis' buffer solution 米氏缓冲溶液
Michael reaction 迈克尔加成反应
Michelia leat oil 白兰叶油
Michelson interferometer 迈克尔逊干涉仪
Michler's carbinol 米蚩醇 $[(CH_3)_2NC_6H_4]_2$=CHOH
Michler's hydrol 双(二甲基氨基)苯甲醇
Michler's ketone 米蚩酮；4,4'-双(二甲氨基)二苯酮 $[(CH_3)_2NC_6H_4]_2CO$

micoheptin(=mycoheptin) 七烯枝菌素
miconazole 咪康唑〔药〕
miconidin 野牡丹素
miconomycin(=mikonomycin) 米科诺霉素
micra 微米的复数
micranol 小花樟醇
micrinite （煤岩）微粒体
microadding 微量填加
microadsorption column 微吸附柱
microadsorption detector 微吸附检测器
microaerophilic test 微需氧试验
microagglomerate 微(型)附聚体；微凝聚体
microaggregate 小颗粒(聚集体)
microaggregate grain 微聚集粒子
microammeter 微安(培)计
microamount flame atomic absorption spectroscopy 微火焰原子吸收光谱法
microampere 微安(培)
microamperometric titration 微安培滴定；微电流滴定
micro analysis 微量分析
micro (analytical) balance 微量(分析)天平
microanalytical reagent 微量分析试剂
microanalyzer mass spectrometer 微量分析质谱仪
micro and semimicro viscometer 微量和半微量黏度计
microanion 微阴离子
micro-area chemical analysis 微区化学分析
micro-area diffraction 微区衍射
micro-argon detector 微氩检测器
microarray 微型阵列
microautoradiography 显微放射自显影法；微射线自动照相术
microbalance 微量天平
microballoon 微球
microballoon sphere 微球〔把油罐中油面与空气隔绝以减少蒸发损耗〕
microbar 微巴〔压力单位，等于1达因/厘米^2〕
microbarograph 自记式微气压计
microbarometer 微气压计
micro-battery 微电池(组)
micro beaker 微烧杯
microbeam plasma welding 微束等离子焊
microbe electrode sensor 微生物电极传感器
microbe fuel cell 微生物燃料电池
microbe polyester 微生物聚酯
microbe sensor 微生物传感器
microbial degradation 微生物降解
microbial film 微生物膜
microbial membrane electrode 微生物膜电极

microbial sensor 微生物传感器
microbic resolvability 微生物降解性
microbiological activity 微生物活性(活动)
microbiological analysis 微生物学分析法
microbiological assay 微生物学测定法；微生物测定法；微生物测定
microbiological corrosion 微生物腐蚀
microbiological degradation 微生物降解；微生物老化
microbiological detection 微生物检测
microbiological deterioration 微生物降解
microbiological determination(=microbiological assay) 微生物定量法；微生物测定
microbiological fuel cell 微生物燃料电池
microbiological method 微生物测定法
microbiological sensor 微生物传感器
microbiological test 微生物试验
microbiological treatment 微生物处理
microbioscope 微生物显微镜
microbiosensor 微型生物传感器
microbiotic(=antibiotic) 抗菌素
micro blast burner 微型喷灯
microbody 微体
microbore column 微径柱；微孔柱
micro bromination method 微量溴化法
micro-Brownian diffusion 微观布朗扩散
micro-Brownian jump 微布朗跳变；微布朗跃迁
micro-Brownian motion 微观布朗运动
micro-Brownian movement 微观布朗运动
microbubble 微泡(沫)；细小泡沫
microbulb pipette 微球型点样管
microbulking 微膨化
microburette 微量滴定管
micro-burner 微(焰)灯
microcal(l)ipers 千分尺；测微计
microcalorie 小卡；微卡
micro calorimeter 微型量热器；微热量计
microcalorimetric method 微量热法
microcalorimetry 微量热法
microcanomical ensemble 微正则系综
microcap 微升毛细管
microcapillary 微毛细管
microcapillary system 微毛细管体系
microcapsulary 微胶囊技术
microcapsule 微胶囊
microcapsule type adhesive 微胶囊型黏合剂
microcarrier 微载体
micro catalytic reactor 微型催化反应器
microcationic titration 微量阳离子滴定法

micro cell ①微极化池②微液槽③微型比色槽④微电解池
microcellular elastomer 微孔弹体
microcellular filler 微孔填料
micro-cellular polyurethane 微孔状聚氨基甲酸酯
microcellular polyurethane elastomer 微孔聚氨酯弹性体
microcellular rubber 微孔橡胶
microcellular structure 微孔结构；微泡(沫)结构
micro chamber stationary thermocyling PCR chip 微室静态温度循环 PCR 芯片
microchannel detector 微通道检测器
microchannel multiplier 微通道倍增器
microchannel plate 微通道板
microchannel plate detector 微通道板检测器
microchemical analysis 微量化学分析
microchemical apparatus 微量化学仪器
microchemical balance 微量化学天平
microchemical method 微量化学法
microchemical pollution 微量化学污染
microchemistry 微量化学
microchip electrophoresis 芯片电泳
microchromatographic column 微型色谱柱
microchromatography 微量色谱(法)
microchromatoplate 微型色谱板
microcide 抗微生物药
microcidin 苯酚钠和 β 萘酚钠消毒剂
microcin 微生物
microcinematography 显微电影术；显微镜活动摄影术
microcircular technique 微量圆环技术
microcline 微斜长石
microcoacervation 微凝聚
microcolloidal polysilicate 微胶态聚硅酸盐
microcolorimeter 微量比色计
micro-column 微型柱
microcombustion method 微量燃烧法
microcombustion tube analysis method 燃烧管微量分析法
microcomponent 微量组分
microcomputer-controlled recording spectrophotometer 微型计算机控制自记分光光度计
microconcentrator 微型浓缩器
microcone penetrometer 微型针入度计
microconformation 微构象
micro constituent 微量成分
microcontact printing 微接触印刷
micro-continuum approach 微观连续介质法
micro-continuum model 微观连续介质模型
microcosm 小天地盐〔四水磷酸氢铵钠〕

microcosmic bead 磷酸盐珠
microcosmic salt 磷酸氢铵钠 NaNH$_4$HPO$_4$
microcoulometer 微库仑计；微电量计
microcoulometric detector(MCD) 微库仑检测器
microcoulometry 微库仑法；微电量法
microcrack 微裂纹；微裂缝
microcreep 微观蠕变
microcrith 氢原子重量
micro cross section detector 微截面检测器
micro crystal 微晶
microcrystalline 微晶的
microcrystalline cellulose 微晶纤维素
microcrystalline cellulose powder 微晶(型)纤维素粉
microcrystalline cellulose triacetate 微晶三乙酰纤维素
microcrystalline domain 微晶畴
microcrystalline structure 微晶结构
microcrystalline wax 微晶(石)蜡
microcrystallinity ①微结晶性②微结晶度
microcrystallite 微晶(粒)
microcrystallization analysis 显微结晶分析
microcrystalloscopic 微晶学的
microcrystal powder 微晶粉末
microcurie 微居里
micro cuvette 微量比色槽
microdeformable fluid 微观变形流体
micro denier fibre 微细纤维
microdensitometer 微型光密度计；测微光度计
microdensitometry 微密度测定(法)
microdesulfonation 微量脱硫作用
microdetermination 微量测定
microdiagnostics 微诊断法
microdial 微量刻度
microdialysis sampling 微量渗析取样法
microdialysis sampling technique 微量渗析取样技术
microdielectrometer 微介电计
micro diffraction 微衍射
microdiffusion 微量扩散
microdiffusion analysis 微量扩散分析法
microdispersed 微分散的；微弥散的
micro-dispersed filler 微粒分散填充剂
micro dispersion 微弥散；微扩散
microdispersoid 微粒分散胶体
micro distillation apparatus 微量蒸馏装置
microdomain 微区
microdosis 微(小剂)量
microdroplet(s) 微滴
micro-durometer 微型硬度计
microdust trash analyzer 微量杂质分析仪

microdyn tester 微型强力试验仪
micro electroanalyzer 微量电解仪
microelectrode 微电极
microelectrode voltammetric titration 微电极伏安滴定
microelectrolytic determination 微量电解测定
microelectrolytic pump 微量电解泵
microelectrophoresis 微量电泳(法)
microelectrophoretic cell 微型电泳池
micro electro thermofluidic valve 微电热流体阀
microelement 微量元素
microemulsification 微(滴)乳化(作用)
micro emulsion 微乳(状液)*
micro emulsion electrokinetic chromatography 微乳液电动色谱法
micro emulsion polymerization 微乳液聚合
microemulsion stabilized room temperature phosphorimetry (MES-RTP) 微乳状液增稳室温磷光法
microencapsulated isocyanate 微囊化异氰酸酯
microencapsulation 微囊包封
micro environment effect 微环境效应
microequivalent 微(克)当量
microetching 微蚀
microexamination 显微检验;微观检验;微观研究
microexplosion 微爆
micro farad 微法(拉)〔$=10^{-6}$ 法(拉)〕
microfiber 超细纤维
microfiber grease 微纤维(构造)润滑脂
microfibre complex 微细纤维复合体
microfibre-generating conjugate fibre 微细纤维母体复合纤维
microfibril 微纤维
microfibrillarity 微原纤性
microfilm 缩微胶片
microfilter ①微纤维过滤器②微型过滤器
microfiltration 微量过滤(法)
microfiltration manifolds 微量过滤多支管
microfiltration membrane 微滤膜
micro flame ionization detector 微火焰电离检测器
micro flame photometry 微火焰光度法
micro flask 微量烧瓶
micro-flocculate 微絮凝粒
microflow 微流;微观流动
micro flow cell 微流量池
microfluid 微观流体
microfluidic channel 微液管道
microfluidic chips 微流控芯片
microfluidic system integration 微流量系统集成
microfoam rubber 微孔橡胶

micro focusing X-ray tube 微聚焦 X 射线管
microfossil analysis 微量化石分析
micro-Fourier transform infrared spectrometry(M-FTIR) 显微傅里叶变换红外光谱法
microfractography 断面显微术
microfusion method 微量熔化法;微熔鉴定法
microgasburner 微型煤气灯
microgel 微凝胶*
microgel formation 微凝胶形成
microgel particle 凝胶微粒
micro glass beads 细玻璃珠;微粒玻璃珠
microgram titrimetry 超微量滴定法
micrographic method 显微照相法
micrography ①显微摄影②显微(镜)测物理常数法
microgravimetric analysis 微量重量分析
microhardness 显微硬度
micro-hardness tester 微型硬度计
microhardness value 显微硬度(值)
micro-heat of adsorption detector 微量吸附热检测器
microheterogeneity 微观不均一性
micro-high performance liquid chromatography 微柱高效液相色谱法
microhm 微欧姆
microhole bowl 微孔轧辊
microhole elasticity bowl 微孔弹性轧水辊
microhomogeneity 微观(结构)均匀性
micro horizontal buret 微量水平滴定管
micro hot plate 微量(加)热板
micro H-type cell 微量 H 型池
micro-imaging 微成像
microimmunoelectrophoresis 微量免疫电泳
microincineration 显微灰化法;显微燃烧法
micro-indentation tester 微型硬度计
microindication 微量指示
micro-ingredient(=small powder) 小量药品;小药;小料
microinhomogeneity 微不均匀性
micro-injection technique 微注射法
microinstrumentation 仪器微型化
microion 微(分子)离子
micro ion drag pump 微离子牵引泵
micro-IR cell 微型红外池
micro jet 超微气流粉碎机
micro-jet roll coater 微型喷射辊涂机;微喷辊涂机
micro kinetics 微观动力学
micro Kjeldahl method 微量凯耶达(测氮)法;微量凯氏定氮法
microl 小花樟酚
microline 微斜长石

microlite(=microlith)　微晶
microliter　微升
microliter pipet(te)　①微升点样管②微升移液管
microliter syringe　微升注射器
microlite zeolite　微晶(泡)沸石
microlith(=microlite)　微晶
microlitic　微晶结构
microlitic structure　微晶结构
micromanipulator　显微操纵器；微型机械手
micromanometer　微压计
micromasking　微掩模
micromatic setting　微调装置
micromechanics　微观力学；细观力学
micromechanism　微观结构
micro melting point apparatus　微熔点测定仪
micro-membrane suppressor　微膜抑制器
micromerigraph　(空气粉尘)粒径测定仪
micromeritics　粉粒学；粉流学
micro-mesh　微孔筛
micrometer buret　测微滴定管
micrometer eyepiece(=micrometer ocular)　测微目镜
micrometer microscope　微规显微镜
micrometer ocular(=micrometer eyepiece)　测微目镜
micrometer-screw-driven syringe　微米螺旋推动式注射器
micrometer syringe　微量注射器；微型注射器
micrometer thickness gage　微厚计；精密测厚计
micromethod　微量法
micrometry　微测法
micro mica　超微云母粉
micro-micro　微微；皮(可)
micromicrofarad　沙法(拉)；微微法(拉)〔=10^{-12}法(拉)〕
microminiaturization　超小型化
micro-mixing　微量混合；微混
micromoisture meter　微量水分计
micromolar　微摩尔的
micromolar concentrations　微摩尔级(的)浓度；10^{-6}mol/L(数量)级浓度
micromolding in capillary　毛细管内
micromorphic continua　微观形态连续体
micromorphological study　微观形态学研究
micromorphology　微观形态学
micron　微米
micronaire　纤维细度气流测定器
micronaire value　纤维细度值
micron dryer　气流干燥器；微粉干燥机
microne　微胶粒
microneck　微颈

micro-necking　微颈缩
micro-needles　微针状体
micronephelometer　微量散射浊度计
microneutronography　显微中子放射照相法
micronex　气炭黑
micronised wax　微粒蜡；微粉蜡
microniser　粉碎机
micronization　①超微粉碎②超微细化作用；微粉化(作用)
micronized dolomite　微粒白云石
micronized hydrated silica　水合二氧化硅微粉；含水白炭黑微粉
micronized mica　超微(粉碎的)云母粉
micronizer　粉碎机
micronizer mill　气流粉碎机
micron separator　风力分级器；超微风选机
micro-ohm　微欧姆〔=10^{-6}欧姆〕
micro-packed column　微填充柱
microparacrystallite　超微次晶；微次晶籽
microparticle column　微粒柱
microparticle support　微粒状载体
micropenetration　微针入度
micropenetrometer　微针入度计
microphase　微相
microphase separation　微相分离
microphase separation structure of polyurethane block copolymer　聚氨酯嵌段共聚物微相分离结构
microphone　传声器；微音器
microphoto sizer　显微(显微)投影仪
microphysics　微观物理学
micropigment　微细颜料
micropipet(te)　微量吸移管
micropipette aspiration technique　毛细管吸入技术
micro-plant　雏型工厂；台式工厂
microplasma　微等离子体
microplasma zone　微等离子区
microplastometer　微量塑性计
microplex　微粉分粒机
micropoise　微泊〔黏度单位〕
micropolar boundary layer flow　微极边界层流
micropolar fluid　微极性流体
micropolarimeter　微偏振计；微旋光计
micropolariscope　测微偏振镜；偏光显微镜
micropollution　微量污染
micropore　微孔
micropore method　微孔法
micropore specific area　微孔比表面(积)
microporosity　微孔率
microporous　微孔性的

microporous coat 微孔涂层
micro-porous ebonite 微孔硬质胶
microporous film 微孔(薄)膜
microporous hollow synthetic fibre 微孔(性)中空合成纤维
microporous membrane 微孔隔膜；微孔膜
microporous membrane technology 大分子膜技术；微孔膜技术
microporous polymer 微孔聚合物
micro-porous rubber 微孔泡沫胶
microporous sheet 多微孔网片；微孔性片状物
micropowder 超细粉
micropreparative electrophoresis system 微量制备电泳系统
micropreparative TLC 微量制备薄层色谱法
microprobe 微探针
microprobe analysis 微探针分析
microprobe analyzer 微探针分析仪
microprobe spectrometer 微探针光谱仪
microprocessor 微(信息)处理机
microprofile 微表面
microprojection apparatus 显微投影器
microprojector 显微投影仪；显微映像器
micropulse resistance welding 微脉冲电阻焊
micropulverizer 微粉磨机
micropulverizing 精微粉碎；精碎
micropump 微型泵
micropycnometer 微量比重瓶
micropyromerides〔复〕 细粒英石
microradiogram 显微射线照片
microradiographic method 显微放射作图法
microradiography 显微放射照相法；显微放射显影法
microradiography unit 显微射线照相机
microradiometer(=radio-micrometer) 辐射微热计；显微辐射计
micro rapid mill 微型快速磨碎机
microray 微波
micro-reaction 微量反应
micro-reaction technique 微量反应技术
microreactor 微反应器
microresonator 微共鸣器
microreticular 微网状的
microreticular resin 微孔径树脂；小孔树脂
micro-rheology 微观流变学
micro-rotation 微观转动
micro-safety burner 安全微灯
micro sample 微量试样
microsample-injection 微量进样(法)
microsampling 微量取样

micro-scale testing 微量试验
microscope 显微镜
microscope camera 显微照相机
microscope interferometer 显微干涉计；显微干涉仪
microscope stage 显微镜载物台
microscopic acidity constant 微观酸度常数
microscopical analysis ①显微镜分析②显微结晶分析③镜检分析
microscopical identification 显微鉴定
microscopical technique 显微技术
microscopic analysis 显微镜分析
microscopic cloth counter 显微镜式织物密度计
microscopic constant 微观常数
microscopic counting 显微镜计数
microscopic deformation 微观形变
microscopic dissolution time 微观溶解时间
microscopic electrophoresis 显微电泳(法)
microscopic examination 显微观察；金相试验〔冶金〕
microscopic examination of animal feeds 饲料显微镜检法
microscopic fusion analysis 显微镜(下)熔融分析
microscopic inspection 显微检测
microscopic measurement 显微测量；微观测量
microscopic Raman spectroscopy 显微拉曼光谱
microscopic reversibility 微观可逆性*
microscopic structure of coal 煤的显微组织
microscopic test 显微试验；金相试验〔冶金〕
microscopic void 微观孔隙
microsectioning 微切片法
micro-size particles 微米级粒子；超细粒子
microsolubility 微溶性
microsomal antibody 微粒体抗体
microsonic grinder 微型超声研磨机
micro spatula 微量刮勺；微量刮铲
microspectrofluorometry 微量荧光光度法
microspectrometer 显微分光计
microspectrophotometer 显微分光光度计
microspectrophotometric analysis 显微分光光度分析法
microspectrophotometry 显微分光光度法
microspectroscope 显微分光镜
microspectrum 显微光谱
microsphere 微球体
microspheroidal cracking catalyst 裂化粉球催化剂
microspheroidization 微球化处理；微球化作用
microspherulitic 微球晶的
microspike 微端丝
micro-sprayer 微型喷雾器
micro-spray gun 微型喷枪

microstage 显微镜载片台
microstandard 微(量)标准
micro-stopcock 微形活栓
microstrain 微应变
microstress 微应力
micro stress region 微应力区
microstructure 微观结构
microstructure analysis 微结构分析
microstructure of coal 煤的微组织
micro structure of coating 涂层显微结构
microsublimation 微量升华(法)
microsuspension polymerization 微悬浮聚合
microswelling 微溶胀；微膨胀
micro-switch 微型开关
micro-syringe 微型注射器
micro system 微(滴)体系
microtacticity 微观规整性
micro test 微量试验
microtest tube 微量试管
microthermogravimetry 微热重量分析法
micro-thermometer 精密温度计
microthin film 超薄薄膜
micro-thin-layer chromatography 微量薄层色谱法
micro tip 微型尖端
microtitration 微量滴定
microtome 切片机；超薄切片机
micro topography analysis 显微形貌分析
micro-torsion balance 微量扭力天平
microtransfer molding 微转移模塑
microtubule 微管
microturbulence 微湍流
micro unit 微(量)单位
micro Vickers 维氏显微硬度计；维氏微观硬度计
microviscosimeter 微黏度计
microviscosity 微黏性
microvoid 微孔
microvoid coalescence 显微空穴聚结
microvoid coating 微空泡涂料；微孔涂料
microvoid content 微孔含量；微孔容量
microvoid hiding 隐含微孔；微孔型遮盖技术
microvoid structure 微孔结构
microvolt 微伏(特)
microvoltameter 微电量计
microvoltammetric electrode 微伏安电极
microvolumetric analysis 微量容量分析
microwave absorbing paint 微波吸收漆
microwave absorption moisture meter 微波吸收式水分计
microwave-assisted extraction 微波辅助萃取；微波辅助抽提法
microwave bomb 微波弹
microwave bridge 微波桥
microwave cavity 微波(空)腔
micro wave curing 微波硫化
microwave detector 微波检测器
microwave diathermy 微波透热法
microwave digestion(MWD) 微波消解法；微波消化法；微波消解
microwave discharge 微波放电
microwave discharge detector 微波放电检测器
microwave discharge method 微波放电法
microwave drying 微波干燥
microwave electrodeless discharge tube 微波无极放电管
microwave emission detector 微波发射检测器
microwave energy 微波能
microwave excited electrodeless discharge lamp 微波激发无极放电灯
microwave extraction separation 微波萃取分离法
microwave frequency 微波频率
microwave generator 微波发生器
microwave heating 微波加热
microwave hygrometer 微波湿度计
microwave induced optical nuclear polarization 微波诱导光核极化
microwave induced plasma(MIP) 微波诱导等离子体
microwave induced plasma atomic absorption spectrometry 微波诱导等离子体原子吸收光谱法
microwave induced plasma atomic emission spectrometer detector(MIP-AED) 微波诱导等离子体原子发射光谱检测器
microwave induced plasma emission spectrometry (MIP-ES) 微波电感等离子体发射光谱法
microwave induced plasma mass spectrometry 微波诱导等离子体质谱(法)
microwave induced thermoemission spectroscopy 微波诱导热发射光谱(法)
microwave inductive plasma(MIP) 微波诱导等离子体
microwave labeling 微波标记
microwave magnetron 微波磁控(电子)管
microwave moisture measurement 微波水分测量
microwave moisture meter 微波水分计
microwave oven 微波炉
microwave oxygen plasma 微波氧等离子体
microwave plasma(MWP) 微波等离子体
microwave plasma atomic absorption spectrometry 微波等离子体原子吸收光谱法
microwave plasma detector 微波等离子体检测器

microwave plasma emission spectrometric detector 微波等离子体发射光谱检测器
microwave power 微波功率
microwave powered source 微波动力源
microwave radiation 微波辐射
microwave radiation polymerization 微波辐射聚合
microwave region 微波区
microwave resonance procedure 微波共振法
microwave sample digestion system 微波样品消解系统
microwave sample preparation system 微波样品制备系统
microwave spectroscopy 微波光谱
microwave spectrum 微波波谱
microwave system 微波系统
microwave vulcanization 微波硫化
microweighing 微量称量
micro weighing tube 微量称(量)管
micro wire board 微封线(包)线板
micro-X-ray diffraction 微焦X射线衍射
microyield 微屈服
microzone refining 微区精炼
microzyme 微胶粒
midazolam 咪达唑仑〔药〕
mid-boiling point 中沸点
midchain 中链(的)
middle ①中间；中部②中间的；居中的
middle cut 中间馏分
middle density polyethylene(MDPE) 中密度聚乙烯
middle diaphragm 中间光阑
middle distillate 柴油；照明灯油
middle fraction 中间馏分
middle infrared beamsplitter 中红外分束器
middle infrared detector 中红外检测器
middle infrared region 中红外区
middle infrared source 中红外光源
middle infrared spectroscopy 中程红外光谱法
middle lens 中间透镜
middle note 中段香韵
middle oil distillate 中间馏分〔煤油与润滑油之间的馏分〕
middle phase 中间相
middle-pressure liquid chromatography 中压液相色谱
middle-pressure reactor 中压反应器
middle ring 中等环
middle runnings〔复〕 中间馏分
middle shade 中间色
middle soap 中间皂
middle stand oil 中级熟油
middle tank air 中槽空气〔容器中层的蒸气空气空间〕
middlings ①中级品②中砂
middlings flour 粗面粉
midew-proof finish ①防霉涂料(涂装)；防霉面漆②防霉加工(整理)
midew resistance 防霉性
midew-retarding agent 防霉剂
midfeather 中间(墙)壁；中间间隔
midfellow 中间(墙)壁
midget ①(极)小型的②(小)侏儒
midget polyunit 小型叠合装置
Midgley bouncing pin 辛烷值测定机中指示爆震的传感器
midium-viscosity vehicle 中等黏度漆料
midnight ethyl 气体汽油
midodrine 米多君〔药〕
midpoint 中点
mid-point boiling range 平均沸腾温度
mid-position 中点；中间位置
midprobability range 中等概率范围
mid-range 中点值；中列数
midrol 碘甲基苯基吡唑啉酮
midsection ①(塔)中段②中间截面③半节的
mid-value 组中值
mid-way 中路；中间位置
miemite 碳酸钙镁
Miers supersolubility curve 迈尔过溶解度曲线
migma 混合岩浆
migmatite 混合岩
mignonette compound 木犀草香精
mignonette oil 木犀草油
migrant plasticizer 栖移性增塑剂
migrating 迁移；移动
migrating group 迁移基团
migrating plasticizer 迁移增塑器
migration 迁移*
migration aptitude 迁移倾向
migration constant 迁移常数
migration current 迁移电流
migration distance 迁移距离
migration electrography 迁移电谱(分析)法
migration fastness (颜色)渗移牢度
migration index 迁移指数
migration inhibition factor 迁移抑制因子
migration loss (增塑剂)渗移损失
migration of double bond 双键移位
migration of petroleum 石油运移
migration of plasticizer 增塑剂的迁移
migration of the double bond 双键移位

migration origin 迁移起点
migration plasticizer 栖移性增塑剂
migration potential (胶粒)迁移电位；(胶粒)移动电势
migration rate 迁移率
migration-resistant 阻泳移的；抗移位的
migration terminus 迁移终点
migration test 迁移试验
migration time 迁移时间
migration time window 迁移时间窗口
migration transfer printing 泳移转移印花；湿态转移印花
migration tube (离子)迁移管
migration value 迁移值
migration velocity 迁动速度
migratory 迁移的；移动的
migratory aptitude 迁移倾向*
migratory insertion 迁移插入
mikanecine(=platynecine) 阔叶千里光裂碱
Mikrotest dry-film thickness gage 米克洛泰斯特干膜测厚仪
miktoarm polymer 杂臂聚合物
mil ①毫升②密耳；千分之一英寸
Milano rib 罗纹空气层组织
milarite 整柱石
Milas reagent 迈拉斯试剂〔H_2O_2 的叔丁醇溶剂，用 OsO_4 作催化剂〕
mild base 弱碱
mild clay 柔软黏土
mild corrosion 轻度腐蚀
mild cracking 轻度裂化
mildewcide 防霉剂
mildewproof agent 防霉剂
mildew-proof finish ①防霉涂料(涂装)；防霉面漆②防霉加工(整理)
mildew proofing finish 防霉整理
mildew resistance 防霉性；抗霉性
mildew resistant paint 防霉漆
mildew-retarding agent 防霉剂
mild extreme pressure lubricant 中级极压润滑剂
mildly acidic 弱酸性的
mildly detergent oil 含中等量清净分散添加剂的润滑油
mild mercurous chloride 稀氯化亚汞
mild odor 温和香气
mild oxidation 轻度氧化
mild steel 低碳钢；软钢
mild tobacco 淡烟草
mild wear 轻微磨损
mile 英里

mileometer 里程记录器；里程表
milfoil oil 蓍草油
miliolite 风成细粒夹岩
military gas-mask 军用防毒面具
military jet fuel 军用喷气燃料
military lubricant 军用润滑剂
military pipe line 军用管线
military type coating 军用(型)涂料
milk acid(=lactic acid) 乳酸；2-羟基丙酸
milk casein fibre 乳酪(蛋白)纤维
milk glass 乳白玻璃
milkiness ①乳状②乳白色
milking cell 末端电池；补充电池
milk of almonds 杏仁油乳浊液
milk of asafoetida 阿魏乳浊液
milk of barium 氢氧化钡乳浊液
milk of lime 石灰乳
milk-of-lime carbonation process 石灰乳碳酸盐法〔沉淀碳酸钙制法〕
milk of magnesia 氧化镁乳剂
milk of sulfur 硫黄乳；乳(粒)硫
milk products analysis 乳制品分析
milk scale buret 乳白刻度滴管
milk sugar(=lactose) 乳糖 $C_{12}H_{22}O_{11} \cdot H_2O$
milk tester 验乳器
mill ①磨(机)、碾机②压碎③铣④(工)厂⑤铣刀
millability 可轧性
mill apron 返料带
mill bagging 脱滚(现象)
mill banding 包辊(现象)
mill-base 粉碎物料
mill base charge 研磨料装入量
mill base premix 研磨料预混物
mill base slurry ①研磨料浆②漆浆
mill base stream line 研磨料流动线
mill base viscosity ①研磨料黏度②漆浆黏度
mill bed plate 底盘
mill board 书皮纸板
mill bristol 卡(片)纸板
mill clearance 滚距
mill control calculation 压榨管理计算
milled ①磨碎了的②压紧了的
milled clay 漂白土
milled cloth 毡合织物
milled fibre 磨碎纤维
milled rubber 捏炼了的橡胶
milled soap 研制皂
mill engine 压榨机

Miller duplex beater   米勒双效打浆机
millerite   针镍矿
Miller number   米勒数
millet oil   小米油
milletol   鸡血藤醇 $C_{29}H_{50}O_2$
mill fluxing   辊炼塑化
mill grain   磨粒
mill grease   磨机用润滑脂
mill groove   碾槽
milli-〔拉丁字头〕   毫；千分之一
milliammeter   毫安(培)计
milliampere   毫安(培)〔符号 mA〕
milliard   十亿；十万万〔10^9〕
milliatom   毫克原子(量)
millibar   毫巴〔压力单位〕
millibarn   毫靶(恩)〔核反应截面单位〕
millicoulometry   毫库仑分析法；毫电量分析法
millicron   毫微米〔等于 10^{-9} 米,符号 nm〕
millicurie   毫居里〔$37×10^7$ 次衰变/秒, mCi〕
millidarcy   毫达西
milli-equivalent   ①毫(克)当量②毫当量
milliformula weight   毫分子量
milligamma   毫微克
milligram   毫克
milligramage   (放射性)照射量〔每小时每毫克镭〕
milligram-atom   毫克原子
milligramequivalent   毫克当量
milligram-hour   (放射性)照射量〔每小时每毫克镭〕
milligram-ion   毫克离子
milligram method   毫克法〔微量法〕
milligram-molecule   毫克分子(毫摩尔)
milligram per cent   每 100 份中的毫克数
milligray   毫戈瑞
Millikan's rays   宇宙射线
millilambda   毫微升
milliliter   毫升
millimeter   毫米
millimicron   毫微(米)
millimole   ①毫摩尔②毫分子(量)
milling   ①研磨；粉碎②混炼；压炼；轧炼③塑炼④混炼⑤热炼
milling contraction   缩绒收缩
milling cycle   ①辊炼周期②研磨周期
milling green   磨绿
milling head   铣头刀
milling machine   缩呢机
milling machine operation   铣削操作
milling machine operator   铣工
milling off   砑光
milling orange   磨橙；酸性铬橙
milling shrinkage   辊炼收缩
milling soap   ①研制皂②缩绒皂
milling zone   研磨区
millinormal   毫规(度)的；毫克当量的
million   百万；兆
million becquerel   兆贝可勒尔
milliosmol   毫渗透分子
milliphot   毫辐透〔照度单位〕
MILLI plant   米利试验工厂〔西德卡尔斯鲁厄核研究中心制备超铀元素的试验工厂〕
millipore   微孔
millipore filter   微孔过滤器
millipore filtration   微孔过滤
millipore membrane filtration concentration method   微孔薄膜过滤浓集法
millirad   毫拉德
milliroentgen   毫伦琴
mill iron   生铁
millirutherford   毫卢瑟福
millisecond   毫秒〔等于 10^{-3}s〕
millisievert   毫希沃特
millitorr   毫托；微米汞柱
millivolt   毫伏(特)〔符号 mV〕
millivoltmeter   毫伏(特)计
mill knife   磨刀
mill line   轮碾机
mill loading   ①研磨机的加料量②颜料浆中的颜料浓度
mill loss   压榨损失
millman   ①轧钢工②滚轧工人
mill-mixed stock   辊炼料
Millon reaction   米隆反应
Millon's base   米隆碱；氢合氧化汞
mill operator   ①轧钢工②滚轧工人
mill pinlon   碾轮；韶轮
mill release agent   开轧物；启碾物
mill retting   工业浸渍
mill roll opening   滚隙
mill room   压榨室
mill room crane   压榨室起重机
mill run   薄通
mill scale   轧制铁鳞
mill-size chips   小木片；工业用木片
millstone   磨石
millstone grit   磨石；白石
mill tailings   矿山尾矿；矿山矿渣
mill to death   重复碾磨

milo grit 粗碾芦粟粉〔用 Sorghum vulgare 碾制〕
Milontin N-甲基-2-苯基琥珀酰亚胺；苯琥胺
milorganite 活性淤泥肥料
milori blue 米洛丽蓝〔染〕
Milori green(=chrome green) 铬绿〔旧称米洛丽绿〕
milrinone 米力农〔药〕
mimeograph ①油印机②复写板；誊写器③油印
mimeograph machine 油印机
mimeograph paper 蜡纸；油印纸
mimeograph (printing) ink 油墨
mimeograph stencil (sheet) （油印用)蜡纸
mimetite 砷铅矿
mimosa ①含羞草(属)②荆树
mimosa bark ①荆树皮②含羞草皮
min ①(=minimum)最小(值)；最低(值)；最小的；最低的②(=minute)分
mimosa bark extract 荆树皮浸膏
mimosa extract ①荆树皮浸膏②含羞草萃③荆树皮烤胶
mimosine(=leucaenol;leucaenine) 含羞草碱
minaline 吡咯甲酸
Minamata disease event 水俣病事件
minasragrite 钒矾
mine ①采(矿)②矿
mine filling 矿山回填
mine machine oil 矿山机油
mine-prop 矿柱
mineragraphic microscope 矿石显微镜
mineral ①矿(物)；矿石②矿物的③无机的
mineral acid 无机酸*
mineral acid content 无机酸含量
mineral acid neutralization number 无机酸中和数
mineral additive 矿物添加剂
mineral adhesive 硅酸钠
mineral aggregate 矿物集料；无机骨料
mineral alkali 无机碱
mineral black 石墨
mineral blue 矿蓝
mineral burning oil 煤油
mineral butter 凡士林；矿脂
mineral caoutchouc 弹性沥青
mineral carbon 石墨
mineral chameleon 高锰酸钾
mineral charcoal 丝炭；天然木炭
mineral chemistry 矿物化学
mineral coal 丝煤
mineral color 矿物颜料
mineral colza (oil) 重质灯油
mineral compound 无机化合物

mineral cotton 矿棉
mineral cutting oil 矿物切削油〔切削工具用矿质润滑-冷却液〕
mineral dispersion 矿物分散体
mineral dye 无机颜料
mineral ether 石油醚
mineral fat 矿脂
mineral fertilizer 无机肥料〔矿物肥料〕
mineral fiber 矿物纤维
mineral filler 矿物填料
mineral filter 矿物滤器
mineral fuel 矿物燃料
mineral-gelled grease 无机胶凝润脂；矿物质胶态润滑脂
mineral glue 矿物胶〔硅酸钠的水溶液〕
mineral green 矿物绿
mineralization 矿化*
mineralized froth 矿化泡沫
mineralized methylated spirit （甲基紫着色的)甲醇和石脑油混合溶剂
mineralizer ①矿化物②矿化剂
mineralizing 矿化
mineral jelly 矿物冻
mineral kermes 硫氧锑矿
mineral-lard (cutting) oil 矿物-猪油切削油〔矿物润滑油与猪油润滑冷却液〕
mineral lattice 无机物晶体点阵；矿物晶格
mineral linseed oil 矿质亚麻油〔亚麻油的矿质代用品〕
mineral lubricating oil 矿物润滑油
mineral manure 无机肥料；矿物肥料
mineral matter 矿物(质)；矿质
mineral membrane electrode 矿物膜电极
mineralogical 矿物(学)的
mineralogical analysis 矿物组成分析
mineralogical chemistry 矿物化学
mineralography 矿相学
mineralogy 矿物学
mineral oil 矿物油〔指石油、页岩油等矿物来源的油〕
mineral oil spring 矿物油泉；石油喷流
mineral orange(=red lead) 铅丹；红丹
mineral paint 矿物性涂料
mineral paint thinner 矿质油漆稀释剂；油漆的矿物质石油稀释剂
mineral pigment 矿物性颜料
mineral pitch 地沥青；柏油
mineral-polymer composite 矿物-聚合物组分
mineral product 矿产
mineral purple 赭石

mineral quartz　矿物石英；石英石
mineral resin　矿物树脂
mineral resources　矿物资源
mineral rubber　矿物胶；弹性沥青
mineral salt　天然盐；矿盐；岩盐
mineral seal oil　矿质海豹油；重质灯油
mineral separating fluid　离矿液
mineral silicate emulsion　无机硅酸盐乳液
mineral slags　矿渣(磨料)
mineral spirit　一种溶剂油〔约150～200℃馏分〕
mineral spots〔复〕　矿物包体
mineral springs　矿泉
mineral streak　矿脉
mineral substance　矿物质
mineral sulfonate　石油磺酸盐；矿物(油)磺酸盐
mineral sulfur　矿质硫黄
mineral syrup　液体石蜡；石蜡油
mineral tallow　矿蜡
mineral tannage　矿物鞣(法)
mineral tanning agent　矿物鞣剂
mineral tar　①矿质焦油②风化石油
mineral theory　(石油起源的)无机理论
mineral thinner　矿物稀释剂
mineral turpentine　矿质松节油；石油溶剂
mineral violet　矿物紫
mineral water　矿质水
mineral wax　地蜡
mineral white　矿物白；石膏
mineral white primer　碳酸钙油性填孔料
mineral wool　石纤维
mineral yellow　氯氧化铅
miner's inch　矿水标
miner's lamp　矿灯；采矿灯
miner's oil　矿工油〔坑道灯用灯油〕
miner's sunshine(=miner's wax)　矿工蜡〔矿工灯用软石蜡〕
miner's wax(=miner's sunshine)　矿工蜡〔矿工灯用软石蜡〕
mine run　原矿
mine run coal　原煤
minesite　敏勒炸药；敏勒煞特
Minet aluminum process　迈尼特制铝法
minetic　拟晶
mine-timber　坑木
minetisite　砷铅矿
minette　鲕褐铁矿；云煌岩
mine ventilating fan　矿用通风风扇；矿用通风机
mini-〔词头〕　缩；微小(型)
miniature analogue computer　小型模拟计算机
miniature end plate current　微终板电流
miniature mixer-settler　微型混合澄清槽
miniature nuclear battery　小型核电池
miniature sand mill　小(微)型砂磨机
miniaturization　小型化；微型化
miniaturization of instruments　仪器微型化
miniaturized total analysis systems　微全分析系统
minicell　①小型细胞②微型电池
minicomputer　小型计算机；微型计算机
minielectrode　小电极
miniemulsion　微粒乳状液
miniemulsion polymerization　细乳液聚合
miniextruder　微型挤塑机
minifer　单引发-转移剂*
minification　缩小率；缩小尺寸；减少；削减
minifil yarn　粗旦单纤复丝
minim　液量单位
minimal　最小量
minimal flow pump　最小流量泵
minimal inhibitory concentration　最小抑菌浓度
mini-Massmann furnace　小型马斯曼高温炉
mini-mixer-settler　微型混合澄清槽
minimization　最终化；极小化
minimized spangles　①极细的闪光颜料②极小的闪光金属片；极小的闪光塑料片
minimum　①最少的；最低的；最小的②最小(量)；最低(量)；最少(量)；最小数；最少数；最低数③最低点
minimum annular ring　最小环宽
minimum azeotropic mixture　最低共沸物
minimum boiling point azeotrope　最低共沸混合物
minimum clamp stroke　最小合模行程
minimum criterion　最小判据
minimum detectability　最小检出度
minimum detectable activity(MDA)　最小(可)检出(放射性)量；最低(可)检出(放射性)量
minimum detectable amount　最小检出量
minimum detectable concentration　最小检出浓度
minimum detectable quantity　最小检出量
minimum deviation angle　最小偏向角
minimum energy path(MEP)　最低能量途径*
minimum error potential method　最小误差电位法
minimum fatal dose　最小致死量
minimum fill　最低装量
minimum film forming temperature　最低成膜温度
minimum filming temperature　最低成膜温度
minimum flow　最小流量
minimum fluidization　临界流态化；最低流态化

minimum fluidization velocity 最小流(态)化速度
minimum ignition energy 最小点火能
minimum irrigation rate 最小灌注速率
minimum liquid-gas ratio 最小液气比
minimum liquid rate 最小液流量
minimum octane rating 最低辛烷值
minimum overvoltage 最低过电压
minimum peak width 最小峰宽
minimum phenomenon 极小现象
minimum plate height 最低塔板高度
minimum pour point 最低倾点；最低流动点
minimum reflux 最小回流
minimum reflux ratio 最小回流比
minimum residual method 最小残差法*
minimum separation time 最短分离时间
minimum staining reclaim 低污染再生橡胶
minimum theoretical number of plate 最小理论板数
minimum thermometer 最低温度计
minimum unit cost 最低单位成本
minimum void space ①最低的孔隙空间②最低的气孔率
minimum voltage value 最低电压值
minimum wetting rate 最小润湿速度；最小润湿速率〔填料塔〕
minimum work 最小功
mining ①开矿；采矿②采矿(学)③开矿的；采矿的④矿用的
mining engineering 采矿工程
mining explosive 矿用炸药；矿山炸药
mining hose 矿用胶管
mining machinery 采矿机(器)
mininuke 小型核武器
minioluteic acid 微黄青霉酸
mini-plant 中间工厂
minium(=red lead) 铅丹；红铅；四氧化三铅
minivalence 最低化合价
miniviscometer 微型黏度计
mink 貂
mink fat 貂脂
Minkovski distance 明柯夫斯基距离
minocycline(=7-dimethlamino-6-demethyl-6-deoxytetracycline) 二甲胺基四环素；米诺环素
minomycin 美浓霉素；米诺环素
minor (较)小的；(较)少的；(较)短的；不重要的
minor additions 次要配合剂
minor agglomeration 较小的团块
minor base 稀有碱基
minor component 微量组分
minor constituent 少量成分*

minor diameter 小直径
minor element 微量元素
minor groove 小沟
minorine(=vincamine) 长春蔓胺
minor isotopes safeguard technique(MIST) 次要(小量)同位素核监督技术
minoxidil 米诺地尔〔药〕
minseed oil 矿质亚麻油
mint 薄荷
mint bouquet 清凉花香型
mint oil 薄荷油
minty 薄荷香的；凉香的
minty note 薄荷香；凉香
minus ①减②负的
minus deviation 负偏差
minus material 次品
minus sign ①减号②负号
minute ①极少的②微小的③分(钟)④精密的
minute air space(s) 小气泡；小空隙
minute blemish 小瑕疵；小污点；小斑痕
minute bubble 小气泡
minute crack 细裂纹；发状裂纹
minute drop(s) 微小的液滴；极细的液滴(珠)
minute quantity 极少量
min-width 最小宽度
minyulite 水磷钾铝石
miokinine 鸟氨酸三甲基内盐
miotic ①缩瞳药②缩瞳的
miotic effect 瞳孔收缩效应
miotics 缩瞳药〔医〕
MIOX-based paint incorporating aluminium pigment 云母氧化铁铝粉拼用漆；云母铁-铝粉漆
miox intermediate coat 云母氧化铁中涂涂层
mipor 微孔的
mipor rubber 微孔橡胶
mipor scheider 微孔橡胶隔膜
mirabilite 芒硝；硫酸钠矿
miramint 钨钼合金
mirbane oil 硝基苯；密斑油
mire 淤渣；淤泥；矿泥
mire-drum 天然盐水
mirene 米尔烯〔一种双萜烯〕
miropinic acid 米绕品酸
mirror coating 反射镜涂料(膜)
mirror finish 光泽整理；磨光整理
mirror galvanometer(=reflecting galvanometer) 反射镜电流计
mirror image 镜像

mirror image isomerism 镜像异构*
mirror image peak 镜像峰
mirror inspection machine 反光镜式检查机
mirror-like surface (像)镜面一样的表面
mirror nuclei 镜像(原子)核
mirrorstone ①白云母②云母
mirror symmetry 镜面对称*
miry 淤泥的
miscalibration 校准误差；校准失误
miscandenin 薇甘菊素
miscella 溶剂(混合)油；油与溶剂混合物〔萃取〕
miscellaneous 混杂的；各种(各样)的；多方面的；其他的
miscellaneous ether ester 杂醚的酯
miscellaneous paint additive 多用漆添加剂；通用漆添加剂
mischmetal(l) ①稀土金属混合物②商品镧
mischzinn 锡铅锑合金；焊锡
miscibility (溶)混性；可混性；掺混性
miscibility gap 混溶性区
miscible 可(溶)混的
miscible fluid 可混液；混相流体
miscible liquid 可(溶)混液体
miscible solvent 可混溶溶剂
miscilla 合油溶剂
Mises-Hencky flow condition 米赛斯-亨基流动条件
misfit dislocation (晶格)失配位错
misorientation 错排(取向)
mispairing 错配
mispick 缺纬〔纺〕
mispickel(=mispickel arsenopyrite) 砷黄铁矿；毒砂
mispickel arsenopyrite(=mispickel) 砷黄铁矿；毒砂
misplaced atoms 错位原子*
misregistration 记录失真
miss feeding end 无效添绪；失添；空添
missing value 漏失值；漏测值
mist ①(烟)雾②雾
mistabron 巯乙磺酸钠
mistake error 过失误差
mist blower 喷雾器
mist coat ①(湿)罩光薄涂层；雾罩涂层②虚枪膜；虚枪飞雾层
mist collector 集雾器
mist cooling 喷雾冷却；喷雾降温
mistletoe 槲寄生
mist spinning 喷雾纺丝
mist spray 喷雾
mistuning 失调
miter(=mitre) 斜接；45°角接口

miter valve 锥形阀
miticide(=acaricide) 杀螨剂；杀蛆药
mitigal 灭疥〔药〕
Mitis green 巴黎绿〔乙酸铜偏亚砷酸铜复盐〕
mitochondrial DNA(MTDNA) 线粒体DNA
mitochondrial electrode 线粒体电极
mitochondrial RNA(mt RNA) 线粒体RNA
mitomycin 丝裂霉素〔药〕
mitomycin C 丝裂霉素C
mitotic activity 核分裂能力
mitoxantrone 米托蒽醌〔药〕
mitragynine 帽柱碱
mitraversine 异叶帽柱碱
mitre(=miter) 斜线
Mitscherlich desiccator 密切里希保干器
Mitscherlich pulp 密切里希纸浆
mitsubaene 鸭儿芹倍半萜烯
mix ①混合②混合物
mixable (可)溶混的
mix-and-filter system 混合-过滤系统
Mixco extractor 米克斯科萃取塔
mixed 混合的
mixed accelerators〔复〕 ①混合促进剂②联合加速器
mixed acid 混酸〔$HNO_3 + H_2SO_4$〕
mixed acid-base constant 混合酸碱(平衡)常数；(活度-浓度)混示酸碱(平衡)常数
mixed acid ester 混合酸酯
mixed adhesive 混合黏合剂
mixed admission (混合物或燃料的)混合输入
mixed adsorption 混合吸附
mixed agglutination reaction 混合凝集反应
mixed air-interlaced yarn 混合网络丝
mixed aliphatic-aromatic ketone 脂基芳基混合酮
mixed alkyl ether 混合烷基醚
mixed amine 混合胺〔氮上取代基不同〕RR'NH
mixed anhydride 混合酐
mixed aniline point 混合苯胺点
mixed aniline test 混合苯胺试验
mixed antioxidant(=blended antioxidant) 混合防老剂
mixed base ①混合底子②混合碱
mixed base crude (oil) 混合基原油
mixed base crude petroleum 混合基原油
mixed base grease 混合基润滑脂
mixed base oil 混基石油
mixed bed 混合床
mixed bed column (离子交换)混合柱
mixed bed deionization 混合床式去离子
mixed bed ion exchange 混(合)床离子交换

mixed bed ion exchange stationary phase　混合床离子交换固定相
mixed bed system　混合床式
mixed butane　混合丁烷
mixed catalyst　混合催化剂
mixed cement　①联合增碳器②混合水泥
mixed chromatogram　混合色谱(图)
mixed chromatographic test　混合色谱试验
mixed-color method　混色法
mixed column　混合柱
mixed complex(=mixed ligand complex)　混合配合物；杂配物
mixed constant　混合常数*
mixed crystal　混晶*
mixed crystal coprecipitation　混晶共沉淀
mixed crystal membrane electrode　混晶膜电极
mixed double bond　混双键
mixed drum　搅拌式转鼓
mixed dyes〔复〕　混合染料
mixed electrode　混合电极
mixed electrode potential　混合电极电位
mixed ester　混合酯　RCOOR′
mixed ether　混合醚
mixed examination　混熔试验
mixed fatty acid　混合脂肪酸
mixed feed　混合进料法；错流进料法
mixed feeding　混合进料
mixed feed kiln　混合进料窑
mixed feed process　混合进料法
mixed fertilizer　混合肥料
mixed flow　混(合)流(动)
mixed flow compressor　混流式压缩机
mixed flow impeller　混流叶轮
mixed flow pump　混流泵
mixed friction　混合摩擦
mixed gas shield welding　混合气体保护焊
mixed gel column　混合凝胶柱
mixed glyceride　混酸甘油酯
mixed glyceride soap　混合甘油酯皂
mixed goods　混合织物
mixed grades　异级混合物；(石油产品)不同类混合物
mixed immunity　混合免疫；后天免疫
mixed immunofluorescence　混合荧光免疫法
mixed indicator　混合指示剂*
mixed isomers　混合(同分)异构物
mixed ketone　混合酮　RCOR′
mixed lead alkyls　混合烷基铅
mixed leucocyte reaction　混合白细胞反应
mixed ligand complex(=mixed complex)　混合配合物；杂配物
mixed ligand coordination compound　混(合)配(体)配合物*
mixed lubricant　混合润滑物；调合润滑物
mixed melting point　混合熔点
mixed melting point method　混合熔点法
mixed metal cluster　混合金属簇合物*
mixed micelle　混合胶束
mixed monolayer　混合单层
mixed odd-even triglycerides　奇碳数-偶碳数混合甘油三酸酯
mixed organic and inorganic anhydride　混合有机无机酸酐
mixed packing　混合填充(物)
mixed packing column　混合填充柱
mixed packing gas chromatography　混合填充物气相色谱(法)
mixed paint　调和漆
mixed paper stock　混合纸料
mixed phase　混合(固定)相
mixed-phase cracking process　混合相裂化过程
mixed phase hydrocarbon　多相烃
mixed plasticizer　混合增塑剂
mixed polyamide　混合聚酰胺
mixed polyester　混合聚酯
mixed polymer　混合聚合物
mixed polymerization　混合聚合
mixed population non-random heterogeneity　非随机混合总体的不均匀性
mixed potential　混合电位
mixed powder　①混合(火)药②混合(炸)药
mixed rags　①成型②装配③组合
mixed regime of flow　混合流动方式
mixed rock　混合岩石；混成岩石
mixed salt　混盐
mixed sample　混合试样
mixed sols　混合溶胶
mixed solution　混合溶液
mixed solution method　混合溶液法
mixed solvent　混合溶剂
mixed solvent extraction　混合溶剂抽提
mixed stationary phase　混合固定相
mixed stock　混炼胶；胶料
mixed sulfide　混合硫醚　RSR′
mixed superphosphate　①混合酸性磷酸盐②不纯过磷酸钙
mixed surface film　混合表面膜

mixed tannage 混合鞣(法)
mixed triglyceride 混合甘油三酸酯
mixed type 混合型
mixed vaccine 混合疫苗
mixed valence 混合价*
mixed valence compound 混合价化合物*
mixed whites 混合白碎布
mixer ①混合器；混料箱②混频管③混合者④溶解机〔黏胶〕
mixer extruder 挤出混料机
mixer mill 搅拌磨碎机
mixer-settler 混合澄清槽*
mixer-settler bank 混合澄清槽组
mixer-settler extractor 混合沉清萃取器；混合沉降萃取器
mixer slag 混合炉渣
mix-flow process 混流式工艺
mixing 混合；调制
mixing agitator 混合用搅拌器
mixing aid 混炼助剂；操作助剂
mixing and grinding machine 混合研磨机
mixing apron 混合磨护床
mixing arm 搅拌(机)桨叶
mixing battery 气水混合管
mixing chamber 混合室；混炼室；密炼室
mixing column 混合柱；混合塔
mixing condenser 混合冷凝器
mixing current 混合股流
mixing device 混合器；混合设备
mixing drum 混合鼓；混合筒
mixing equipment 混合设备
mixing formula 混合配方
mixing heat 混合热
mixing hollander 混浆机
mixing kettle 混合锅
mixing length 混合长度
mixing length theory 混合长度理论
mixing machine 混合机
mixing mills 混合辊
mixing nozzle 混合喷嘴；混色喷嘴
mixing order 加料顺序
mixing paddle 混合(机)桨叶
mixing pan 混合盘
mixing pan mill 混合盘磨
mixing period 混合期*
mixing point 混合点
mixing pool model 混合池模型
mixing ratio 混合比
mixing rheometer 混合式流变仪
mixing roll 混炼机；混合辊
mixing screw 螺杆混合器；混合螺杆
mixing section 混合段
mixing stirrer arm 搅拌桨叶
mixing tank 混合槽
mixing time 混合期
mixing varnish ①清漆料〔可与各种颜料或颜料浆混合制漆的清漆总称〕②油性调漆料〔包括亮油和桐油清漆用的不同油度的清油〕
mixing vessel 混合容器
mixing water 混合水
mixing white 调色用白墨
mixite 砷铋铜矿
mixotropic series 混溶序〔溶剂按极性增高次序排列，在序列中两种溶剂相距越远，混溶性越差〕
mixtion 金胶；金脚〔大漆〕
mixture ①混合物②混合体③(混)合剂
dl-mixture 外消旋混合体
mixture additivity 混合可加性
mixture colors 混合色
mixture control 混合控制〔混合物组成或性质的控制〕
mixture copaiba 苦配香脂和金合散胶的混合物
mixture design statistic technique 混合物设计统计技术
mixture fabric 混纺织物
mixture heat 混合热
mixture indicator 混合指示剂
mixture law 混合物定律
mixture-method lubrication 混合法润滑〔添加润滑油于燃料内的发动机润滑方法〕
mixture ratio 混合比率〔混合物组成比例〕
mixture strength 混合强度〔燃料对空气比例〕
miyaconitine 宫部乌头碱
mizolastine 咪唑斯汀〔药〕
mizoribine 咪唑立宾〔药〕
MLEV-16 pulse sequence MLEV-16 脉冲系列
mmole(=millimole) 毫摩尔
M.O. 近似值
mobile ①流动的②机动的
mobile adsorption layer 流动吸附层
mobile electron 流动电子
mobile equilibrium 流动平衡；动态平衡
mobile H-tautomerism 流动氢互变异构
mobile hydroen 流动氢原子
mobile hydroxyl 流动羟基；迁移性羟基
mobile impurities 流动杂质
mobile laboratory 活动试验室
mobile liquid 流性液体；低黏度液体

mobile lubrication plant　活动润滑设备
mobile oil　流性油；机油
mobile oily liquid　低黏度油质液体
mobile phase　流动相*
mobile phase front　流动相前沿
mobile phase migration distance　流动相迁移距离
mobile phase presaturation　流动相预饱和
mobile phase sensor　流动相传感器
mobile proton tautomerism　流动质子互变异构
mobility　①淌度②迁移率*③流动性
mobility of ion　离子迁移率
mobilization　活动化*
mobilometer　流变计；淌度计
Mobius system　默比乌斯体系*
mocaya oil　顶束毛榈油〔取自 Acrocomia sclerocarpa〕
moccasin　鹿皮软革
mocha　(铝鞣或醛鞣)绵羊正绒面手套革
mocha stone　苔纹玛瑙
mochyl alcohol　日本雀胶醇
mock gold　金黄铁矿
mock iodine-131　模拟碘-131
mock lead　闪锌矿
mock leno weave　假纱组织
mock ore　闪锌矿
mock silver　银色锡锑合金
mock standard　模拟标准
mockup　样本；模型；模板
mockup shop　模式车间
mockup test　模型试验
mock vermilion　铬酸铅
moclobemide　吗氯贝胺〔药〕
modacrylic　变性聚丙烯腈纤维
modacrylic fiber　改良聚丙烯腈纤维
modal　木代尔
modal fibre　高湿模量黏胶纤维；木代尔纤维
modal length　主体长度
model compound　典型化合物
model engine fuel　典型发动机燃料〔试验性发动机用的标准燃料〕
model fiber　模型纤维
model-fitting　模型拟合
model fringed micelle　缨状微束模型
model hemiester　模型半酯
modeling and parameter estimation　模型化与参数估计
model law　相似定律；模型定律
modelled effect　模塑效应；造型效应
model method　模型法
modelon　毛涤纶

model reaction　模型反应；典型反应
model soil　模拟污垢
mode of appearance　外观(表观)形态
mode of texturing　变形方式
moderate　中等的；适度的
moderate accelerator　中速促进剂
moderate casualty agent　中级伤害剂
moderate cracking　中度裂化
moderate cracking severity　中等裂化深度
moderate duty　中等负荷
moderate-duty compressor　中型压缩机
moderate-duty service　中等负荷操作；一般条件下操作
moderately active　中等活动的
moderately hard water　中(度)硬水
moderately persistent gas　半持久性毒气
moderately polar　中(等)极性的
moderately polar phase　中等极性固定相
moderately soluble　中等溶度的
moderately toxic　中等毒性的
moderately volatile fuel　中等挥发性燃料
moderate operating conditions　中等操作条件
moderate oxidation　中度氧化
moderate polarity　中(等)极性
moderate-speed lubrication　中等速度润滑
moderate-temperature destructive hydrogenation　中温度破坏性加氢
moderate temperature hydrogenation　中温加氢
moderator　①缓和剂②减速剂
modern liquid chromatography　现代液相色谱法
modern refinery　现代化炼油
modification　改性
modification enzyme　改性酶；修饰酶
modification ratio　改性(比)率
modified　改良的；修改的；变了型的
modified alkyd resin　改性醇酸树脂
modified allicylic polyamine　改性脂环族多元胺
modified alliphatic polyamine　改性脂肪族多元胺
modified alloy　改良合金
modified aromatic polyamine　改性芳香族多元胺
modified asphalt　改性沥青
modified Auger parameter　修正型俄歇参数
modified Bauer-Vogel process　改良型鲍尔-沃盖尔法〔铝表面处理法之一，用碱金属铬酸盐处理铝表面〕
modified bilayer lipid membrane　修正双层脂膜
modified bisphenol A epoxies　改性双酚 A 环氧(树脂)类
modified branched polyethylene　改性支化聚乙烯
modified carbon paste electrode　修饰碳糊电极
modified casein　改性乳酪

modified-cellulose 改性纤维素
modified chlorinated rubber 改性氯化橡胶
modified collagen fiber 变形胶原纤维
modified drying oil 改性干性油
modified electrode 修饰电极[*]；改性电极
modified factor analysis 改进的因子分析
modified fringed-micelle structure 改进的缨状微束结构
modified knock intensity method 修正爆震强度法〔按最大爆震强度进行的燃料抗爆性行车试验法〕
modified Lambourn abrasion tester 改良型蓝伯恩磨损试验机
modified latex 改性胶乳
modified oil 改性油
modified olefine polymer 改性烯烃聚合物
modified oleoresinous paint 改性油性树脂漆
modified Ostwald viscometer 改良型奥斯瓦尔德黏度计
modified paper 改性纸
modified phenolic resin 改性酚醛树脂
modified phthalic resin 改性邻苯二甲酸树脂
modified plastisol 改性增塑糊
modified polyamine fibre 改性聚胺纤维
modified polyvinyl alcohol 改性聚乙烯醇
modified resin 改性树脂
modified Reynolds number 修正雷诺数
modified rosin 改性松香
modified rubber 改性橡胶
modified simplex method(MSM) 改进单纯形法
modified soda 改性苏打；碳酸氢三钠 $Na_2CO_3 \cdot NaHCO_3$
modified support 改性载体
modified version 改性型
modified (wax) based anticorrosion coating 改性蜡防腐蚀涂料
modified wax coating 改性蜡涂料
modified wood 改性(木)材
modified zeolite 改性沸石
modified zinc phosphate 改性磷酸锌
modifier 改性剂[*]；改进剂；调节剂；修饰因子
modifying adduct 改性用加成物
modifying agent ①改性剂②调节剂③改良剂④改进剂
modifying asphalt 改性沥青
modifying factor 修正因素；修正因子
modifying resin 改性用树脂
modifying term 修正项
modular 积木式的；组合的
modular instrument system 组合式仪器系统；仪器调制系统；模块化仪器系统
modularity 组合性

modular pilot plant 模式中间工厂
modulated current 调制电流
modulated differential scanning calorimeter 调幅式差示扫描量热仪
modulating amplifier 调制放大器
modulation amplitude 调制幅度
modulation frequency 调制频率
modulation of accelerating voltage 加速电压调制
modulation polarography 调制极谱法[*]
modulation sideband 调制边峰
modulator 调谐剂；调制器
modulator beam 调制束
module ①模件；组件②单元；舱③计量单位
modulus 模数；模量
modulus at rapid deformation 速形变模量；动模量
modulus at slower deformation 缓形变模量
modulus enhancement 模量增强
modulus gradient 模量梯度
modulus in compression 压缩模量
modulus in flexure 挠曲模量；弯曲模量
modulus in shear 剪切模量
modulus in tension 拉伸模量
modulus in torsion 扭曲模量
modulus of compression 压缩模量
modulus of cross elasticity 横向弹性模量
modulus of direct elasticity 直接弹性模数；直接弹性模量
modulus of elasticity 弹性模量；弹性模数
modulus of elasticity in shear 剪切弹性模量
modulus of hydrostatic compression 本体压缩模量；静水压模量
modulus of longitudinal elasticity 纵向弹性模数
modulus of resilience 回弹模量
modulus of rigidity 刚性模量
modulus of rupture(=flexural strength) 弹性模量
modulus of rupture in torsion 扭曲破坏模数；扭曲断裂模量
modulus of shearing 剪切模量
modulus of shearing elasticity 剪切弹性模数；剪切弹性模量
modulus of stretch 拉伸模量；伸长模量
modulus of transverse elasticity 横向弹性模量
modulus of volume expansion(=bulk modulus) 体积模量；本体模量
modulus plateau 模量平坦区
modulus-temperature curve 模量-温度曲线
modulus tester 模量测定仪
modulus to density ratio 模量-密度比

modulus-weight ratio 模量-重量比
Moebius silver-refining process 莫毕士炼银法
Moebius trip type 莫毕士往复型
moellon 油鞣废油〔氧化鱼油混合物〕
mohair 马海毛
mohawkite 砷铜矿
moho 地壳与地幔间的界限
Mohr balance 莫尔比重天平
Mohr condenser 莫尔冷凝器
Mohr liter 莫尔升
Mohr method 莫尔法*
Mohr pinchcock clamp 莫尔弹簧夹
Mohr pipet 莫尔吸量管
Mohr's circle(=circle of stress) 莫尔圆
Mohr's clamp 莫尔夹；弹簧夹
Mohr's clip 莫尔夹；弹簧夹
Mohr's figure 莫尔图
Mohr's method 莫尔法
Mohr's salt 莫尔盐*
Mohs' hardness 莫氏硬度
Mohs' hardness scale 莫氏硬度计
Mohs' scale 莫氏硬度
Mohs'scale of hardness ①莫氏硬度标②莫氏硬度分级法
Mohs value 莫氏值(硬度)
moiety 部分；半个
moire 波纹；云纹
moire pattern 水纹图样
Moissan furnace 莫桑电炉
moissanite 碳硅石
moist (潮)湿的
moist basis 湿基
moist cabinet 湿箱
moist catalysis 湿催化剂
moist chamber ①培养皿②湿室
moist curing 润湿处理〔保持在润湿介质中〕
moistener 湿润器
moistening agent 增湿剂
moistening capacity 润湿能力；可润湿性；增湿能力
moistening of mixture 混合物的润湿〔向燃料混合物内加水〕
moistness 湿度；水分；湿气
moistograph 自计湿度计
moisture absorption 吸湿；水分吸收
moisture absorption characteristics 吸湿特性
moisture absorption test 湿度吸收试验
moisture adsorbability 吸湿性
moisture adsorbent 吸湿剂

moisture analyzer 湿度分析仪
moisture apparatus 测湿器；水分测定仪
moisture balance 水分天平
moisture barrier 防潮层
moisture-barrier property 防潮性
moisture blister resistance 耐湿气起泡性
moisture blow 湿起瘤；水泡(鼓)囊
moisture capacity 湿度；潮度
moisture chamber (控制)湿度展开槽；调湿展开槽
moisture combined ratio 水分结合比
moisture content 含湿量*
moisture content measuring device 水分含量测定装置
moisture content of wood 木材含水率；木材的湿量
moisture content tester 测湿仪；水分测定仪
moisture corrosion test 湿度腐蚀试验
moisture-controlled thin-layer chromatography 控(制)湿(度)薄层色谱(法)
moisture-cured polyurethane coating 湿固化聚氨酯涂料
moisture cure urethane 湿固化聚氨酯
moisture curing 湿固化；潮气固化
moisture detector 湿度检测器
moisture determination 水分测定
moisture determination apparatus 水分检测仪
moisture equilibrium 水分平衡；湿度平衡；吸湿平衡
moisture equivalent 含水当量；水分当量数
moisture factor ①水分因素②湿润系数；含水系数
moisture-free weight 烘干重量；干重
moisture gage 湿度计
moisture index 含水指数；湿润指数
moisture indicating device 水分指示仪
moisture indicator 水分指示器
moisture level ①湿度②水分含量
moisture meter 水分计；湿度计
moisture penetration 透湿性；潮湿渗透性(作用)
moisture percentage 含水率；含湿率
moisture permeability 透湿性
moisture permeation mechanism 透湿机理；透湿机制
moisture pick-up 吸湿
moisture proof 防潮的
moisture proof adhesive tape 防潮胶黏带
moisture proof agent 防潮剂
moisture-proof coating 防潮涂料
moisture-proofness 防潮性
moisture recording controller 水分记录控制器
moisture regain 回潮；吸湿；吸湿性；回湿性
moisture-repellant(=moisture proof) 防潮的；抗湿的
moisture resistance 防潮性；抗湿性
moisture-resistant(=moisture proof) 防潮的；抗湿的

moisture retention ①保湿②保湿性③保湿量
moisture seepage 潮湿渗透
moisture-sensitive material 湿(气)敏(感)材料
moisture sensitive resistance 湿敏电阻
moisture sorption capacity 吸湿能力；吸湿量
moisture sorption isotherm 吸湿等温线
moisture teller 水分(快速)测定仪
moisture tester 水分测量器
moisture test of ice machine oil 冷冻机油的湿度试验
moisture transfer coefficient 水分传递系数；水分移动系数
moisture uptake ①吸湿②吸湿性③吸湿量
moisture vapor permeability 透湿气性；透湿气度；渗潮气性；湿气渗透率
moisture vapor transmission ①透湿气性②透湿气量
moisture vapor transmission cup 透湿杯；潮气渗透杯
moisture vapor transmission rate 透潮率；湿气渗透率
moisturing ①回潮②增湿
moity wool 含杂羊毛
mol(=mole) ①摩尔②克分子
molal ①(重量)摩尔的②摩尔的③重模的；(重量)摩尔(浓度)的
molal concentration 重模浓度；(重量)摩尔浓度
molal conductivity 摩尔电导率
molal depression constant (重量)摩尔冰点下降常数
molal elevation constant (重量)摩尔沸点上升常数
molal free energy 摩尔自由能
molal heat capacity 摩尔热容(量)
molal heat content 摩尔热函(数)
molality 重量摩尔浓度；重模
molal latent heat (质量)摩尔潜热；重模潜热
molal latent heat of vaporization 摩尔蒸发潜热
molal lowering 摩尔下降
molal quantity 摩尔(数)量
molal sensible heat content of gas 摩尔气体显热
molal solution 摩尔溶液；重模溶液；质量摩尔浓度溶液
molal surface 摩尔表面积
molal surface energy 摩尔表面能
molal unit 摩尔单位
molal volume(=molar volume) 摩尔体积
molal weight(=molar weight) 摩尔量
molamma 粒渣(紫)胶
molar ①(体积)摩尔的②摩尔的③容模的；(体积)摩尔(浓度)的
molar absorbance 摩尔吸光度
molar absorbancy index(=molar absorptivity) 摩尔吸收系数；摩尔吸光系数
molar absorption coefficient(=molar absorptivity) 摩尔吸收系数；摩尔吸光系数
molar absorptivity(=molar absorbancy index; molar absorption coefficient) 摩尔吸收系数；摩尔吸光系数
molar average boiling point 摩尔平均沸点
molar cohesion 摩尔内聚力
molar cohesion energy 摩尔内聚能
molar concentration 容模浓度；(体积)摩尔浓度
molar conductance 摩尔电导
molar conductivity 摩尔电导率*
molar depression (of freezing point) 摩尔凝固点降低
molar (electrical) conductivity 摩尔电导率
molar elevation (of boiling point) 摩尔沸点升高
molar energy of vaporization 摩尔蒸发能
molar entropy 摩尔熵*
molar extinction coefficient 摩尔吸收系数；摩尔吸光系数
molar fraction (体积)摩尔分数
molar free energy 摩尔自由能
molar gas constant 摩尔气体常数*
molar heat capacity 摩尔热容*
molar heat content 摩尔量热函(数)；摩尔热函
molar heat of fusion 摩尔熔化热
molar heat of solution 摩尔溶解热
molar heat of vaporization 摩尔蒸发热；摩尔汽化热
molar humidity 摩尔湿度
molar internal energy 摩尔内能*
molarity (体积)摩尔浓度；容模
molar latent heat 摩尔潜热
molar mass 摩尔(质)量；分子量
molar mass average 摩尔质量平均*；分子量平均
molar mass exclusion limit 摩尔质量排除极限*
molar melting entropy 摩尔熔化熵
molar parachor 摩尔等张比容
molar percentage 摩尔百分率；克分子百分率
molar polarization 摩尔极化(度)
molar property 摩尔(溶液)性质
molar ratio 摩尔比率
molar refraction 摩尔折射(度)
molar rotation 摩尔旋光*
molar rotatory power 摩尔旋光度
molar solubility 摩尔溶解度
molar solution 摩尔溶液
molar (surface) area 摩尔(表)面积
molar susceptibility 摩尔磁化率
molar volume(=molal volume) 摩尔体积
molar weight(=molal weight) 摩尔量
molasse 磨砾层(相)
mol-chloric compound 分子氯化合物

mol concentration　摩尔浓度
mold(=mould)　①塑模；(阴)模；模(型)②造型；翻砂③霉菌
moldability　①模压加工性；模压性能②模型最大生产率
moldable　可模压的；适于模压的
moldable synthetic detergent　可模制(的)合成洗涤剂
moldavium　元素87〔未证实〕
mold box　模型箱
mold breaker　揭模起子；撬棒
mold breaking jack　揭模起子；撬棒
mold breathing　模型除气
mold bumping　模型除气
mold cavity　模巢；模腔
mold clamping force　合模力
mold clamping pressure　合模力
mold cleaner　①模型清洁剂②洗模器
mold cleaning　模型清洗操作；洗模
mold cleaning gun　洗模喷枪
mold clearing jack　开模千斤顶
mold clearing temperature　出模温度
mold coating　模具涂料
mold cracker　揭模起子；撬棒
mold deposit　模垢
mold discharging agent　脱模剂
mold dope　脱模剂
mold duplicator　模型复制机
molded diaphragm　模制膜片
molded goods　模制品
molded in place　现场模制
molded mechanicals　模制配件
molded part　模(型)制品
molded piece　模(型)制品
molded resin　模制树脂
mold equalizer　模型平衡器
molder　铸工
mold feeding　填模；装模
mold filling　填模；装模
mold finish　压模表面处理
mold flow property　模内流动性
mold form　模巢；模腔
molding　模塑*
molding box　翻砂箱
molding capacity　模压量；模压额
molding cycle　①成型周期②模塑周期③成型时间④模塑时间
molding machine　①铸模机；制模机②切模机
molding machinery　制模机

molding material　模塑材料；铸模材料
molding powder　塑(料)粉
molding press　压模机
molding resin　成型用树脂
moldings　模(型)制品
molding seam　卸开线〔压模〕
molding shrinkage　①模塑收缩②模塑收缩率③成型收缩④成型收缩率
molding strain　模塑应变
molding temperature　模压温度
molding tolerance　模制公差
mold insert　模型插入物；模制品骨架
mold life　模具使用年限
mold loading　①填模；装模②按发泡胀大率装模
mold locking force　模型闭合力
mold lubricant　脱模剂
mold marks　①模痕；模缝线〔模塑缺陷〕②模具标志
mold materials　模材料
mold oil　脱模油
mold parting agent　脱模剂
mold pressing　模压；模制；用模用制
mold pressure　合模压力
mold prevention agent　防霉剂
mold prevention paint　防霉涂料；防霉漆
mold preventive　防霉剂；杀菌剂
moldproofing coating　防霉涂料
mold release　脱模剂
mold-release agent　脱模剂
mold release spray　脱模喷(雾)剂；喷雾型脱模剂
mold resistance　抗霉性；抗霉力
mold shrinkage　(托)模后收缩
mold split ester linkage　霉菌分解酯链
mold sticking　黏模(现象)
mold stripping　托模；出模；下模
mold test　霉菌试验；防霉试验
mold turn-over　模型周转(率)
mold unloading　脱模；开模；下模
mold vulcanization　模型硫化
mold wash　①铸型涂料②模型用润滑剂；脱模剂
mole　摩尔*
mole-basis response　摩尔响应(值)
molecular　①分子的②摩尔的
molecular abacus　分子算盘
molecular absorption coefficient　分子吸收系数
molecular absorption spectroscopy　分子吸收光谱
molecular acidity　分子酸度
molecular action　分子作用
molecular activation analysis　分子活化分析

molecular addition compound　分子加成化合物
molecular adduct ion source　分子加成离子源
molecular adhesion　分子的附着
molecular adsorption　分子吸附
molecular agglomerate　分子凝聚体
molecular aggregate　分子聚集体
molecular allotropy　分子同素异性(现象)
molecular amplitude　分子幅度
molecular anisotropy　分子各向异性
molecular architecture　①分子构型;分子结构(形式)
　　②分子设计
molecular area　分子面积
molecular arrangement　分子排列
molecular assembly　分子组装;分子组合
molecular association　分子缔合(现象)
molecular asymmetry　分子不对称(性)
molecular attraction　分子吸引
molecular axis　分子轴
molecular backbone　分子主链
molecular backbone chain　分子主链
molecular balance　分子平衡;反应平衡
molecular beacon　分子信标;分子标志
molecular beam　分子束*
molecular beam ion source　分子束离子源
molecular beam-mass analysis　分子束质谱分析
molecular beam sampling　分子束进样
molecular beam technique　分子束技术
molecular biochemistry　分子生物学
molecular biochromatography　分子生物色谱
molecular (boiling point) elevation　分子沸点升高
molecular bond　分子键
molecular breakdown　分子断裂
molecular brush　分子刷
molecular capsule　分子胶囊
molecular chain　分子链
molecular chain length　分子链长
molecular characterization　分子结构特征
molecular chemical skeleton　分子的化学骨架
molecular cleft　分子豁口
molecular cloning　分子扩增;分子纯种化
molecular cluster　分子簇
molecular clustering　分子簇
molecular cohesion　分子内聚(作用)
molecular coil　分子线团
molecular collision　分子碰撞
molecular colloid　分子胶体
molecular combining heat　分子化合热
molecular complex　分子络合物*

molecular composite　分子复合材料
molecular composition　分子组成
molecular compound　分子化合物*
molecular concentration　分子浓度
molecular conductivity　①摩尔电导率;分子导电系数
　　②摩尔传导率
molecular configuration　分子构型
molecular conformation　分子构象*
molecular conjugation　分子共轭(作用)
molecular connectivity　分子连接性指数
molecular constitution　分子结构;分子构成
molecular control agent　分子量调节剂
molecular crowding effect　分子畸变效应
molecular crystal　分子晶体*
molecular deformation　分子形变
molecular deposition film modified electrode　分子沉积膜化学修饰电极
molecular depression　分子凝固点降低
molecular depression constant　分子(凝固点)降低常数
molecular depression constant of freezing point　分子(凝固点)降低常数
molecular depression of freezing point　分子凝固点降低
molecular design　分子设计
molecular diagram　比例分子模型图
molecular diameter　分子直径
molecular diameter method　分子直径法
molecular diffusion　分子扩散*
molecular diffusion term　分子扩散项
molecular diffusivity　分子扩散系数
molecular dimension　分子尺寸;分子尺度
molecular disorientation　分子解取向;分子消向
molecular dispersion　分子分散
molecular dispersivity　分子分散性
molecular distillation　分子蒸馏;高真空蒸馏
molecular distillation apparatus　分子蒸馏器
molecular dosimetry　分子剂量
molecular dynamics　分子动力学
molecular dynamics method　分子动力学方法
molecular dynamics simulation　分子动力学模拟
molecular effusion　分子泻流
molecular electric conductivity　分子电导率
molecular elevation　分子沸点升高
molecular elevation of boiling point　分子沸点升高
molecular emission cavity analysis(MECA)　分子发射空腔分析
molecular emission spectrometry　分子发射光谱法
molecular energy level　分子能级
molecular engram　分子印迹

molecular entanglement 分子缠结
molecular equation 分子反应式
molecular equilibrium 分子平衡
molecular exclusion chromatography 分子排阻色谱法
molecular expansion factor 分子膨胀因素
molecular extinction(=molar absorptivity) 分子消光
molecular extinction coefficient 分子消光系数
molecular field 分子场
molecular film 分子膜；分子层
molecular filter 分子滤器
molecular flexibility 分子柔顺性；分子挠性
molecular flow of gas 气体分子流
molecular fluorescent method 分子荧光分析法
molecular folding 分子折叠
molecular formula 分子式
molecular fraction 摩尔分数；摩尔百分数
molecular fragment 分子碎片
molecular fragment method 分子碎片法*
molecular freedom 分子自由度
molecular free path 分子自由程；分子动距
molecular frequency 分子频率
molecular friction 分子摩擦
molecular gas constant 摩尔气体常数
molecular gas laser 分子气体激光器
molecular geometry 分子几何(结构)*
molecular group(=molecular grouping) 分子团
molecular grouping(=molecular group) 分子团
molecular heat 分子热
molecular heat capacity 摩尔热函；分子热函
molecular hinge 分子关节
molecular hybrid 分子杂交物；分子态掺合体；分子态混合体
molecular hybridization 分子杂交
molecular hypothesis 分子假说
molecular immunology 分子免疫学
molecular imprinting method 分子印记法；分子印渍法
molecular imprinting polymer 分子印迹聚合物
molecular imprinting technique 分子印渍技术；分子印迹技术
molecular inductive capacity 摩尔感应额
molecular integral 分子积分*
molecular interaction 分子间相互作用
molecular interlocking 分子缠结
molecular ion 分子离子*
molecular-ion cluster 分子-离子簇
molecular ionization 分子电离
molecular ionization potential 分子电离电位
molecular ion peak 分子离子峰

molecularity 反应分子数*
molecularity of reaction 反应的分子性
molecular jet 分子喷射
molecular jostling 分子冲撞
molecular lattice 分子晶格
molecular lattice theory 分子晶格理论
molecular leak 分子漏孔
molecular light switch 分子光开关
molecular link(=molecular linkage) 分子键
molecular linkage(=molecular link) 分子键
molecular lowering 分子降低
molecular luminescence 分子发光
molecularly imprinted polymer 分子印迹聚合物
molecular machine 分子机器
molecular magnetic rotation 分子磁(致)旋光
molecular magnetism 分子磁力
molecular magnetorotation 分子磁力旋光
molecular mass 分子量
molecular melting 分子熔解
molecular memory 分子记忆
molecular mobility 分子流动性
molecular model 分子模型
molecular moderator 分子量调节剂
molecular moisture capacity 分子含水量；分子吸湿量
molecular motion 分子运动
molecular multiplier 分子倍增器；分子放大器
molecular multiplier detector 分子倍增检测器
molecular necklace 分子项链
molecular network 分子网络
molecular nucleation 分子成核作用*
molecular number ①分子序(数)②(分子内)原子序数和
molecular orbital(MO) 分子轨道*
molecular orbital energy level 分子轨道能级
molecular orbital method 分子轨道法
molecular orbital theory 分子轨道理论*
molecular order 分子序
molecular organized assembly 分子有序组合体
molecular orientation 分子取向
molecular orientation initiation temperature 分子取向引发温度
molecular oscillation 分子振动；分子振荡
molecular oxygen-containing gas 含分子氧气体
molecular packing density 分子堆砌密度
molecular packing fraction 分子敛集率
molecular packing frequency 分子敛集频率
molecular partition function 分子配分函数*
molecular peak 分子峰
molecular phase hypothesis 分子相假说

molecular polarizability   分子极化度
molecular polarization   分子极化(作用)
molecular preorientation   分子预取向
molecular probe   分子探针
molecular program   分子程序
molecular property   分子性质
molecular pump   分子泵；分子抽机；高真空泵
molecular radius   分子半径
molecular ratio   分子比率
molecular ray   分子射线
molecular reaction   分子反应*
molecular reaction dynamics   分子反应动力学*
molecular rearrangement   分子重排*
molecular recognition   分子识别*
molecular recognition analysis   分子识别分析
molecular refraction   分子折射(度)
molecular refractivity   分子折射率；分子折射度
molecular-related ion   相关分子离子
molecular replacement   分子置换法
molecular replacement technique   分子置换法*
molecular retraction   分子回缩
molecular rheology   分子流变学
molecular rotation   ①摩尔旋光(度)②分子转动
molecular rotatory power   分子旋光度
molecular rupture   分子(链)断裂
molecular scale   分子尺度
molecular scattering   分子散射*
molecular scission   分子断键；分子(键)断裂
molecular selective electrode   分子选择电极
molecular self-assembling   分子自组装
molecular self-assembling immobilization   分子自组装固定法
molecular self-assembly   分子自组装
molecular separator   分子分离器*
molecular shape   分子形状
molecular shuttle   分子梭
molecular sieve   分子筛*
molecular sieve (based) catalyst   分子筛(基)催化剂
molecular sieve chromatography   分子筛色谱(法)
molecular sieve column   分子筛柱
molecular sieve effect   分子筛效应
molecular sieve filter   分子筛过滤器
molecular sieve filtration   分子筛过滤(法)
molecular sieve foreline trap   分子筛前级冷阱
molecular sieves   分子筛
molecular sieve separation   分子筛分离法
molecular sieve sweetening   分子筛脱硫
molecular sieve zeolite   分子筛沸石

molecular sieving   分子筛分离
molecular slip   分子滑动
molecular slippage   分子滑移
molecular solution   分子溶液
molecular solution volume   摩尔溶液体积
molecular sorption   分子吸着
molecular species   分子种(类)
molecular spectroscopy   分子光谱学
molecular spectrum   分子光谱*
molecular stability   ①分子安定性②分子安定度
molecular state   分子状态
molecular still   分子蒸馏器〔高真空蒸馏器〕
molecular stretching   分子伸展
molecular structure   分子结构
molecular surface energy   分子表面能
molecular swarms   分子群集
molecular symmetry   分子对称(性)
molecular template   分子模板
molecular theory   分子(理)论
molecular theory of flow   流动分子论
molecular theory of friction   摩擦的分子理论
molecular thermodynamics   分子热力学*
molecular thermometer   分子温度计〔测分子量〕
molecular tie   分子(间)联结；分子(间)连接
molecular transposition   分子重排
molecular tumbling   分子离散
molecular tweezer   分子钳
molecular type analysis   分子类型分析
molecular vehicle   分子载体
molecular velocity   分子速度
molecular vibration   分子振动
molecular vibrational energy   分子振动能
molecular vibrational frequency   分子振动频率
molecular volume   分子体积*；摩尔体积
molecular weight   分子量
molecular weight average   平均分子量
molecular weight benzene   分子量苯〔冰点降低法用〕
molecular weight chromatography   分子量色谱(法)
molecular weight control   分子量控制
molecular weight control agent   分子量调节剂
molecular weight detector   分子量检测器
molecular weight determination   分子量测定
molecular weight distribution   分子量分布
molecular weight distribution curve   分子量分布曲线
molecular weight distribution differential curve   分子量分布微分曲线
molecular weight distribution of polymer   高聚物分子量分布

molecular weight exclusion limit 分子量排除极限*
molecular weight modifier 分子量调节剂
molecular weight of polymer 高聚物分子量
molecular weight regulator 分子量调节剂；分子量控制剂
molecular weight thermometer 分子量温度计；定分子量用的温度计
molecular weight-viscosity constant 分子量-黏度常数
molecular yarn 分子纱
moleculary homogeneous entity 分子同质体〔不同聚合度的同系聚合物〕
moleculary homogeneous polymer homologs 分子同质聚合物系列
molecular yield 分子产量；分子收率
molecule ①分子②摩尔
molecule-ion(=molecular ion) 分子离子
molecule-mechanical theory of friction 摩擦的分子-机械理论
molecule segment 分子链段
molecule separator 分子分离器
mole fraction 摩尔分数
mole fraction distribution function 摩尔分数分布函数；克分子分数分布函数
mole neutron 摩尔中子
mole number 摩尔数
mole per cent 摩尔百分数
mole ratio 摩尔比；克分子比
mole ratio method 摩尔比法*
mole weight and SI mass units 摩尔质量和 SI 质量单位
Molex process 莫莱克斯法〔分子筛脱蜡制取正构烷烃法之一〕
mol fraction 摩尔分数
molion 分子离子
molions 带负电荷的惰性气体原子群
Molische's reaction 莫里斯克氏反应〔碳水化合物检查〕
Molisch test 莫利希试验
molliclavine 软麦角碱
mollielast 软弹性体
mollient ①使柔软②缓和剂
mollifer ①软化剂②软化器③缓和药
mollification 软化(作用)
mollifier ①软化剂②软化器③缓和药
molliplast 软塑性体
mollisacacidin 柔毛金合欢素；黑荆素；7,3′,4′-三羟基黄烷-3,4-二醇
molochite 孔雀石〔填料〕
moloxide 分子氧化物
molozonide 分子臭氧化物

mol per cent 摩尔百分数
mol percentage 摩尔百分数
mol ratio 摩尔比
molsep(=molecular separation) 分子分离
molten 熔化的；熔融的
molten-bath 熔融浴；熔融釜
molten caustic (soda) 熔融烧碱
molten electrolyte 熔化电解质
molten heat transfer salt 熔盐传热剂；传热熔融盐
molten iron 熔融铁
molten-lead lift (利用)熔铅升举
molten metal 熔融金属
molten metal applicator 熔态金属镀层装置
molten metal decladding 熔融金属去壳
molten particles 熔融粒子
molten polymer 熔融聚合物；聚合物熔体
molten salt breeder reactor(MSBR) 熔盐增殖堆
molten salt chronopotentiometry 熔盐计时电位滴定法
molten salt electrochemistry 熔盐电化学
molten salt electrocutting 熔盐电切割
molten salt electromigration 熔盐电迁移
molten salt mixture 熔盐混合物
molten salt reactor(MSR) 熔盐(反应)堆
molten salt reactor experiment(MSRE) 熔盐反应堆实验
molten salt voltammetry 熔盐伏安法
molten silver extraction process 熔融银萃取法
molten slag 熔渣
molten solvent 熔化溶剂
molten state 熔融状态
molten state viscosity 熔态黏度
molten sulfur 熔融硫
molten zinc decladding 熔融锌去壳
moltep pig 熔融铁
mol volume(=molar volume) 摩尔容积
molybdate 钼酸盐
molybdate orange 钼铬红；钼橘红；钼铬橙；钼橙；钼橙红〔钼酸铅、铬酸铅和硫酸铅的混合颜料〕
molybdate yellow 钼橘黄
molybdena-alumina catalyst 钼铝催化剂
molybdenic ①钼的②三价钼的
molybdenite 辉钼矿*
molybdenous 二价钼的
molybdenum 钼
molybdenum blue 钼蓝
molybdenum bromide 溴化钼
molybdenum chloride 氯化钼〔总名,计有 $MoCl_2$,$MoCl_3$,$MoCl_4$ 和 $MoCl_5$ 等〕
molybdenum coating 钼涂层

molybdenum dichromate　重铬酸(六价)钼　$Mo(Cr_2O_7)_3$
molybdenum dioxide　二氧化钼　MoO_2
molybdenum dioxydichloride　二氯二氧化钼　$(MoO_2)Cl_2$
molybdenum dioxysulfate　硫酸双氧钼　$(MoO_2)SO_4$
molybdenum dioxysulfide　一硫二氧化钼　MoO_2S
molybdenum disulfide　二硫化钼　MoS_2
molybdenum disulfide lubricant　二硫化钼润滑剂
molybdenum etching　钼腐蚀
molybdenum filament　钼丝
molybdenum fluoride　氟化钼〔计有：MoF_3，MoF_4，MoF_6〕
molybdenum hemipentoxide　五氧化二钼　Mo_2O_5
molybdenum hemitrioxide　三氧化二钼　Mo_2O_3
molybdenum hemitrisulfide　三硫化二钼　Mo_2S_3
molybdenum hexacarbonyl　六羰钼　$Mo(CO)_6$
molybdenum hexafluoride　六氟化钼　MoF_6
molybdenum hydroxide　氢氧化钼
molybdenum hydroxypentachloride　五氯氢氧化钼〔双二氯化钼氯氢氧化钼复盐〕　$2MoCl_2 \cdot Mo(OH)Cl$
molybdenum iodide　碘化钼〔计有：MoI_2 和 MoI_4〕
molybdenum lake　钼色淀
molybdenum minerals　钼矿
molybdenum monoxide　一氧化钼　MoO
molybdenum naphthenate　环烷酸钼
molybdenum orange　钼橙
molybdenum oxalate bromide　溴化草酸钼；八溴二草酸六钼
molybdenum oxide　氧化钼〔计有：MoO，Mo_2O_3，MoO_2，Mo_2O_5 和 MoO_3 等〕
molybdenum oxide fibre　氧化钼纤维
molybdenum oxybromide　溴氧化钼〔计有：$MoOBr_2$ 和 $(MoO_2)Br_2$〕
molybdenum oxyfluoride　氟氧化钼〔计有：$MoOF$，$(MoO_2)F_2$ 和 MoO_2F_4〕
molybdenum oxyhydroxydibromide　二溴化一羟一氧合钼　$[(MoO)(OH)]Br_2$
molybdenum oxyhydroxytrichloride　三氯化一羟一氧合钼　$[(MoO)(OH)]Cl_3$
molybdenum pentachloride　五氯化钼　$MoCl_5$
molybdenum pentasulfide　五硫化二钼　Mo_2S_5
molybdenum red　钼(铬)红〔钼酸铅、铬酸铅和硫酸铅的复合颜料〕
molybdenum resistance heater　钼丝电阻加热器
molybdenum sesquioxide　三氧化二钼　Mo_2O_3
molybdenum steel　钼钢
molybdenum sulfide　硫化钼〔计有：Mo_2S_3，MoS_2 和 MoS_3〕
molybdenum tetrabromide　四溴化钼　$MoBr_4$
molybdenum tetrachloride　四氯化钼　$MoCl_4$
molybdenum tetrasulfide　四硫化钼　$MoS_2[S_2]$
molybdenum trichloride　三氯化钼　$MoCl_3$
molybdenum trihydroxide　氢氧化(正)钼；三羟化钼　$Mo(OH)_3$
molybdenum trioxide　三氧化钼　MoO_3
molybdenum trisulfate　硫酸钼　$Mo(SO_4)_3$
molybdenum trisulfide　三硫化钼　MoS_3
molybdenum wire　钼丝
molybdenyl　氧钼基〔通常指 $MoO—$，也指双氧钼基 $MoO_2=$ 和三价氧钼基 $MoO\equiv$〕
molybdenyl bromide　溴化氧钼　$(MoO)Br$
molybdenyl chloride　一氯化氧钼　$(MoO)Cl$
molybdenyl dichloride　二氯化双氧钼　$(MoO_2)Cl_2$
molybdenyl oxalate　草酸氧钼
molybdenyl phosphate　磷酸三氧钼　$(MoO_3)PO_4$
molybdenyl sulfate　硫酸双氧钼　$(MoO_2)SO_4$
molybdenyl tribromide　三溴化氧钼　$(MoO)Br_3$
molybdenyl trichloride　三氯化氧钼　$(MoO)Cl_3$
molybdic　(正)钼〔1. 指三价钼盐；2. 指六价钼酸〕
molybdic acid　钼酸
molybdic anhydride　钼酐　MoO_3
molybdic bromide　三溴化钼　$MoBr_3$
molybdic chloride　三氯化钼　$MoCl_3$
molybdic compound　(正)钼化合物〔三价钼化合物和六价钼化合物〕
molybdic hydroxide　氢氧化(正)钼；三羟化钼　$Mo(OH)_3$
molybdic ocher　钼华
molybdic oxide　三氧化钼　MoO_3
molybdic sulfide　三硫化二钼　Mo_2S_3
molybdite　钼华
molybdoferredoxin(=azofermo)　固氮铁钼(氧还)蛋白
molybdophosphate(=phosphomolybd)　钼磷酸盐；磷钼酸盐
molybdophosphoric acid　磷钼酸
molybdous　二价钼的；亚钼的
molybdous bromide　二溴化钼　$MoBr_2$
molybdous chloride　二氯化钼　$MoCl_2$
molybdous compound　二价钼化合物
molybdous hydroxide　氢氧化亚钼；二羟化钼　$Mo(OH)_2$
molybdous iodide　二碘化钼　MoI_2
molybdous oxide　一氧化钼　MoO
molybdyl　①羟氧钼根〔指:二价的$[MoO(OH)]=$ 和三价的$[MoO(OH)]\equiv$〕②钼酰根
molybdyl dibromide　二溴化羟氧钼　$[MoO(OH)]Br_2$
molysite　铁盐　$FeCl_3$

moly-sulfide 硫化钼
moment ①矩②片刻
momentary 暂时的；霎时的
momentary acidity 暂时酸度
momentary state 暂时状态
momentary value 瞬时值
moment ion matrix effect 瞬间离子基体效应
moment of couple 力偶矩
moment of dipole(=dipole moment) 偶极矩
moment of distribution 分布矩数
moment of flexure 弯曲矩
moment of force 力矩
moment of friction 摩擦力矩
moment of inertia ①惯性矩②转动惯量
moment of molecular anisotropy 分子各向异性矩
moment of momentum 动量矩；角动量
moment of resistance 阻力矩
moment of rotation 转(动)矩
momentum 动量
momentum conservation equation 动量守恒方程
momentum equation 动量方程
momentum flux 动量通量
momentum operator 动量算符
momentum principle 动量原则
momentum relaxation time 动量松弛时间；动量弛豫时间
momentum separator 动量分离器
momentum spectrum 动量谱*
momentum transfer 动量传递
momordicin 苦瓜叶素
momordin 苦瓜定；木鳖子皂苷；苦瓜毒蛋白
momordin Ic 地肤子皂苷 Ic
monacetin(=glycerin monoacetate) 甘油一乙酸酯；一醋精 $CH_3CO_2CH(CH_2OH)_2$
monacid(=monoacid) ①一(酸)价的；一元的；一酸的②一元酸；一碱(价)酸；一价酸
monacid base 一(酸)价碱
monacid salt 酸式盐
monacrin(=acridin-9-amine) 3-氨基邻苯二甲酸酐
monad ①一价物②一价基
monamide 一酰胺
monamine 一元胺
monamycin 摩那霉素
monarda 香蜂草
monardaein 朱唇花葵苷
monardaein chloride 氯化香蜂草因 $C_{44}H_{45}O_{23}Cl$
monarda oil 香蜂草油〔取自 Monurda punctata〕
monardin 香蜂草定

monardin chloride 氯化香蜂草定 $C_{27}H_{31}O_{15}Cl$
monarkite 莽那卡特〔硝铵、硝酸甘油、硝酸钠、食盐炸药〕
Mon-Arsone(=monarsone) 乙肿酸钠 $C_2H_5AsO_3Na_2$
monarsone(=Mon-Arsone) 乙肿酸钠 $C_2H_5AsO_3Na_2$
monastral blue 单星蓝；颜料酞菁
monastral fast blue 单星坚牢蓝
monatomic 单原子的
monatomic acid 一元酸；一(碱)价酸
monatomic base 单价碱；一(酸)价碱
monatomic gas 单原子气体
monatomic molecule 单原子分子
mona wax 泥煤蜡
monazite 独居石*
monchiquite 沸煌岩
Mond fuel gas 蒙德燃料气；作为燃料用的烟煤发生炉气
Mond gas 蒙德煤气
Mond-olefin 蒙德烯
Mond process 蒙德(收回废硫)法
Monel 蒙乃尔合金
Monel metal 蒙乃尔合金；蒙乃尔高强度耐蚀镍铜合金
monensin 金叶树苷
monergol 单一组成喷气燃料
monesia bark 金叶树皮
monesin 金叶树苷；莫能菌素
monethenoid fatty acids 单烯型脂肪酸
Monex process 蒙乃克斯过程〔从浸煮过的未澄清的巴西独居石矿浆中用溶剂萃取法回收钍、铀的过程〕
monite 黄胶磷矿；碳磷灰石
monitor 监测器
monitored control system 监控系统
monitoring car for atmospheric pollution 大气污染监测车
monitoring device 监控装置
monitoring of process variables 工艺过程变数的调节
monitron 放射监视器
monium 稀土元素名〔未证实〕
monkey skin 猴皮
monkey spanner 活动扳手
monkey wrench(=adjustable spanner) 活动扳手
Monk-Gillison mounting 蒙古-吉里逊装置
monk's-hood 乌头
Monlz tray 蒙兹塔板；长条 U 形泡罩塔板
mono- 〔希腊字头〕 一；单
monoacetate 一乙酸酯
monoacetin(=glycerin monoacetate) 一醋精；甘油一乙酸酯 $CH_3CO_2CH(CH_2OH)_2$
monoacetylaniline N-乙酰苯胺 $C_6H_5NHCOCH_3$

monoacetylated 单乙酰代的
monoacetylated cotton 单乙酰(基)化棉
monoacid(=monacid) ①一(酸)价的；一元的；一酸的 ②一元酸；一碱(价)酸；一价酸
monoacid arsenate 砷酸一氢盐 M_2HAsO_4
monoacid base 一价碱；一酸价碱
monoacidic 一(酸)价的；一元的；一酸的
monoacidic base 一元碱；一(酸)价碱
monoacid salt 一酸价盐
monoacylated 单乙酰代的
monoacyloxytitanate-treated mineral 单酰氧基钛酸酯处理的无机物
monoadduct 单(基团)加成物；一元加成物
monoagglomerate latex 单附聚胶乳
monoalkanolamine 单烷醇胺
monoalkoxy 单烷氧基
monoalkyl arsine 一烷基胂 $RAsH_2$
monoalkylated 单烷基取代了的
monoalkylated benzene 单烷基苯
monoalkyl benzene 单烷基苯
monoalkylol amide 单烷醇酰胺
monoalkyl-phosphinous acid 单烷基亚膦酸 $RP(OH)_2$
monoalkyl-phosphonic acid 单烷基膦酸 $RPO(OH)_2$
monoalkyls 单烷基化物
monoalkyl sodium maleate 马来酸单烷基酯钠
monoalkyl sucrose urethane 单烷基蔗糖氨基甲酸乙酯
monoalkyl sulfide 单烷基硫化物
monoamide 单酰胺
monoamide phase 单酰胺类固定相
monoamine 一元胺 RNH_2
monoamine oxidase 单胺氧化酶
monoamino-acid 一氨基酸
monoamino-dicarboxylic acid 一氨基二羧酸 $NH_2R(COOH)_2$
monoamino-monocarboxylic acid 一氨基一羧酸 NH_2RCOOH
monoaminophosphatide 单氨磷脂
monoaminopolyacetic acid 一氨基多乙酸
monoamino-tricarboxylic acid 一氨基三羧酸 $NH_2R(COOH)_3$
monoammonium phosphate 磷酸二氢铵；磷酸一铵 $(NH_4)H_2PO_4$
monoammonium sulfate 硫酸氢铵；硫酸一铵 $(NH_4)HSO_4$
monoarchin 甘油一花生酸酯
monoarylamine 单芳基胺
monoarylated 单芳基取代了的
monoatomic 单原子的

monoatomic acid 一元酸；一(碱)酸
monoatomic base 一元碱；一(酸)价碱
monoatomic gas 单原子气体
monoatomic layer 单原子层
monoaxial stretching 单轴(向)拉伸
monoazo dyes 单偶氮染料
monoazo dyestuff 单偶氮染料
monoazo group 单偶氮基
monoazo pigment 单偶氮颜料
monobasic ①一(碱)价的；一元的②一代的
monobasic acid(=monoprotic acid) 一元酸
monobasic acid ester 一元酸酯；一价酸酯
monobasic alcohol 一元醇；一羟(基)醇
monobasic aluminum nitrate(MONOBAN) 单碱式硝酸铝 $Al(NO_3)_2(OH)$
monobasic calcium phosphate 磷酸二氢钙；一代磷酸钙
monobasic carboxylic acid 一元羧酸
monobasic dihydroxy-acid 二羟一元(羧)酸 $(OH)_2RCOOH$
monobasic ester 一碱价酯
monobasic hydroxy-acid 羟基一元羧酸
monobasic lead chromate-coated silica 包核碱式铬酸铅〔碱式铬酸铅包二氧化硅核〕
monobasic lead sulfate 碱式硫酸铅；一代硫酸铅；硫酸铅白〔俗〕
monobasic magnesium phosphate 一代磷酸镁；磷酸四氢镁
monobasic monohydroxy-acid 一羟一元酸 $OHRCOOH$
monobasic pentahydroxy-acid 五羟一元酸 $(OH)_5RCOOH$
monobasic phenol acid 一元酚酸 HOC_6H_4RCOOH
monobasic potassium citrate 柠檬酸一钾；一碱价柠檬酸钾 $KH_2C_6H_5O_7$
monobasic potassium oxalate 草酸氢钾；一碱价草酸钾 KHC_2O_4
monobasic potassium tartrate 酒石酸氢钾；一碱价酒石酸钾 $KHC_4H_4O_6$
monobasic tetrahydroxy-acid 四羟一元酸 $(OH)_4RCOOH$
monobasic trihydroxy-acid 三羟一元酸 $(OH)_3RCOOH$
monobed 单(一)床
monobehenolin 甘油一山萮酸酯
monobel 单贝尔〔硝铵、硝酸甘油、锯屑、食盐炸药〕
monobenzene 单基取代苯
monobenzyl-para-aminophenol 单苄基对氨基苯酚〔防胶剂〕
monobloc(k) type 整体式；单块式
monobranched hydrocarbons 单支链烃
monobromated 一溴化的

monobromated camphor 一溴化樟脑 $C_{10}H_{15}BrO$
monobromethane ①溴乙烷②溴甲烷
monobromide 一溴化物
monobrominated 一溴化的
monobromination 一溴化作用
monobromo acetanilide 一溴乙酰苯胺
monobromo-acetic acid （一）溴代乙酸 $CH_2BrCOOH$
monobromo-benzene 一溴代苯；溴苯 C_6H_5Br
monobromo-ester 一溴代酯
monobromo-ether 一溴醚
monobromo isovaleryl urea 一溴异戊脲
mono-sec-butoxyaluminium diisopropylate 单仲丁氧基二异丙醇铝
monobutyl-p-aminophenol 一丁基对氨基苯酚
monobutyl itaconate 衣康酸一丁酯
monobutyl phenyl phenol sodium monosulfonate 单丁苯基酚基单磺酸钠
monobutyl phosphate(MBP) 磷酸一丁酯 $(C_4H_9O)PO(OH)_2$
monobutyl phthalate 邻苯二甲酸一丁酯 $HOOCC_6H_4COOC_4H_9$
monobutyl thiourea 一丁基硫脲〔促进剂〕
monobutyrin 甘油一丁酸酯
mono-calcium phosphate 磷酸一钙
monocapryl naphthalene sulfonate 单辛基萘磺酸盐
monocarboxyl cellulose 单羧基纤维素
monocarboxylic acid 一元羧酸 RCOOH
monocarboxylic ester 一元羧酸酯 RCOOR
monocerotin 甘油一蜡酸酯
monochain polyamide 单链聚酰胺
monochain polymer 单链聚合物
monochalcogenide 一硫族元素化物
mono-channel column 单(流)路柱
monochloracetic acid （一）氯代乙酸 $CH_2ClCOOH$
monochlorated 一氯化的
monochlorethane 一氯乙烷；乙基氯 C_2H_5Cl
monochloride 一氯化物
monochlorinated 一氯化的
monochlorination 一氯化作用
monochlorizated 一氯化的
monochloro-acetic acid 一氯代乙酸 $CH_2ClCOOH$
monochloro-acetyl chloride 一氯乙酰氯 $CH_2Cl·COCl$
monochloro-acetyl halide 一氯乙酰卤 $CH_2Cl·COX$
monochloro amine 一氯(代)胺
monochloro-benzene 一氯(代)苯；氯苯 C_6H_5Cl
monochloro copper phthalocyanine 一氯铜酞菁
monochloroderivatives 一氯衍生物
monochloro-diethyl ether 一氯二乙醚 $C_2H_5OCH_2CH_2Cl$
monochloro-dimethyl ether 一氯二甲醚 $ClCH_2OCH_3$
monochloro-ester 一氯酯
monochloro ethane 一氯(代)乙烷
monochloro-ether 一氯醚
monochloro methane 一氯(代)甲烷
monochloro-methyl-ether 氯甲基醚 CH_2ClOR
monochloro-naphthalene 一氯代萘
monochloroselenodiphenylamine 一氯代联苯胺硒
monochlorotoluene 一氯甲苯
monochloro-urea 一氯代脲 $NH_2CONHCl$
monochrom 单色
monochromic (=monochromatic) 单色的
monochromatic(=monochroic) 单色的
monochromatic absorptivity 单色吸收率
monochromatic analysis 单色分析
monochromatic burner 单色灯
monochromatic color scheme 单色系
monochromatic emission 单色发射
monochromatic emissivity 单色发射
monochromatic filter 单色滤色片(器)
monochromatic ion beam 单色离子束
monochromaticity 单色性*
monochromatic lamp 单色灯
monochromatic light 单色光
monochromatic radiation 单色辐射
monochromatic radiography 单色放射线照相术
monochromatic ray 单色光
monochromatic refractometer 单色光折射计
monochromatic X-ray 单色 X 射线
monochromatism 单色感〔不正常的视觉〕
monochromatizing 使单色化；成单色
monochromator 单色仪*
monochrome ①单色的②单铬(染料)〔商〕
monochrome colour 单色；一色〔利用一种颜色的明暗浓淡组成的和谐色〕
monochrometer 单色光镜；单色器；单色仪
monochrome yellow MG 单铬黄 MG
monoclinic 单斜(晶)的
monoclinic crystal 单斜晶
monoclinic crystalline system 单斜晶体系
monoclinic form 单斜晶形
monoclinic prisms 单斜棱晶
monoclinic sulfur 单斜(晶)硫；针形硫
monoclinic system 单斜晶系*
monoclonal antibody 单克隆抗体
monoclonal antibody labelling 单克隆抗体标记
monocolor 单色

monocolor method 单色法
monocomponent 单组分
monocon 聚焦阴极射线管
monocresyl glyceryl ether diacetate 甲苯醚甘油二乙酸酯；甘油一甲苯醚二乙酸酯
monocrystal 单晶
mono-crystalline 单晶的
mono-crystalline silicon 单晶硅
monocyclic 单环的；一环的
monocyclic aromatic 单环芳烃
monocyclic hydrocarbon 单环烃
monocyclic monoterpene 单环单萜(烯)
monocyclic ring 单环
monocyclic sulfide 单环硫化物
monocyclic terpene 单环萜(烯)
monocyclopentadienyl metal compound 单环戊二烯基金属化合物；一茂金属化合物
monodentate ligand 单齿配体
monodiglyceride 单甘油二酸酯
monodisperse 单分散(性)
monodisperse aerosol generation interface(MAGIC) 单分散气溶胶形成接口
monodisperse aerosol generator 单分散气溶胶发生器
monodisperse colloid 单分散(性)胶体
mono-dispersed latex 单分散胶乳
monodisperse medium 单分散介质
monodisperse polymer 单分散(性)聚合物
monodisperse polystyrene 单分散聚苯乙烯
monodisperse suspension 单分散悬浮体
monodispersity 单分散性
monodomain 单相畴
monoene 单烯
monoenoic acid 单烯酸
monoepoxidation 单环氧化(作用)
monoester 单酯
monoethanol 单乙醇的
monoethanolamine 单乙醇胺
monoethanolamine alkyl sulfate 烷基硫酸单乙醇胺盐
monoethanolamine dodecyl sulfate 十二烷基硫酸单乙醇胺盐
monoethanolamine fatty acid 脂肪酸单乙醇胺盐
monoethanolamine lauric ester 月桂酸单乙醇胺酯
monoethanolamine oleate 油酸单乙醇胺
monoethanolamine sulfite 亚硫酸单乙醇胺酯
monoethanol glycol dodecanol ether 单乙醇乙二醇十二醇醚
monoethanol sulfonamide 单乙醇磺酰胺
monoethenoid 单烯型

monoether 单醚
monoethoxymethyl diphenyloxide 一乙氧甲基二苯基氧〔促进剂〕
monoethyl adipate 己二酸一乙酯 $C_2H_5O_2C(CH_2)_4CO_2H$
monoethyl amine 一乙胺
monoethylated 一乙基化的
monoethyl choline （一）乙基胆碱
monoethylenically unsaturated monomer 单烯类不饱和单体
monoethyl ester 一乙基酯
monoethyl ether 一乙基醚
mono-2-ethylhexylphosphoric acid(MEHP;M2-EHPA;H_2MEHP) 单(2-乙基己基)磷酸 $(2-C_2H_5C_6H_{12}O)PO(OH)_2$
monoethylin 一乙灵；甘油一乙醚 $C_3H_5(OH)_2OC_2H_5$
monoethyl malonate 丙二酸一乙酯 $CH_2(CO_2H)(CO_2C_2H_5)$
monoethyl oxalate 草酸一乙酯；乙二酸一乙酯 $C_2H_5O_2CCO_2H$
monoethyl-phosphate 磷酸一乙酯 $C_2H_5O \cdot PO(OH)_2$
monoethyl-phosphoric acid(=monoethyl-phosphate) 一乙磷酸
monoethyl phthalate 邻苯二甲酸一乙酯 $C_6H_4(CO_2C_2H_5)CO_2H$
monoethyl succinate 琥珀酸一乙酯
monoethyl-sulfate 硫酸一乙酯 $CH_3CH_2OSO_3H;CH_3CH_2SO_4H$
monoethyl-sulfite 亚硫酸一乙酯 $C_2H_5O \cdot SO_2H$
monoethyl tartrate 酒石酸一乙酯 $HO_2C(CHOH)_2CO_2C_2H_5$
monoethyl tetrachloro-phthalate 四氯邻苯二甲酸一乙酯 $HO_2CC_6Cl_4CO_2C_2H_5$
mono-fibre yarn 纯纺纱；单一纤维纱
monofil(=monofilament) 单丝；鬃丝；单缕；单纤丝
monofilament(=monofil) 单丝；鬃丝；单缕；单纤丝
monofilament yarn 单丝纱
monofil-coner 单丝锥形筒子络筒机
monofilm 单分子膜
monofluorated 一氟化的
monofluoride 一氟化物
monofluorinated 一氟化的
monofluorination 一氟化作用
monofluorizated 一氟化的
monofluoro-acetic acid 一氟代乙酸 CH_2FCOOH
monofluoro-benzene 一氟代苯；氟苯 C_6H_5F
monofluoro-ester 一氟酯
monofluoro-ether 一氟醚

monofluoro-phosphoric acid 氟基磷酸 H_2PO_3F
monoformal 单缩甲醛
monoformin(=glycerin monoformate) 甘油一甲酸酯；一甲精 $HCO_2CH_2CHOHCH_2OH$
monofrequent 单频率的
monofuel propellant(=monopropellant) 单一组分的喷气燃料；单元喷气燃料
monofunctional 单官能(团)的
monofunctional compound 单官能(基)化合物
monofunctional initiator 单功能起始剂〔高分子〕
monofunctional ion exchanger 单官能离子交换剂
monofunctional molecule 单官能分子
monogen 单价元素
monogenetic ①单色的②单性的
monoglyceride 单酸甘油酯
monoglyceride process 单甘油酯法
monoglyceride stage 醇解阶段
monoglyceride sulfate 甘油一硫酸酯盐
monoglyceryl ester 一甘油酯
monoglycol ester 一甘醇酯
monoglyme 单甘醇二甲醚
monograph ①专论②记录③图
monographic study 专题研究
monohalide 一卤化物
monohalogen-acetone 一卤丙酮
monohalogen-acetylenes〔复〕 一卤炔
monohalogenated 一卤代的
monohalogenated acetone 一卤代丙酮 CH_2XCOCH_3
monohalogenated benzene 一卤代苯 C_6H_5X
monohalogenation 一卤化作用
monohalogen-benzene 一卤代苯 C_6H_5X
monohalogen ester 一卤酯
monohalogen ether 一卤醚
monohalogeno-benzene 一卤代苯 C_6H_5X
monohapto 单配位点
monohydrate 一水合物；一水化物
monohydrate crystals 含一分子结晶水(的)晶体
monohydrating phenol(=monohydric phenol) 一元酚
monohydric 一羟(基)的
monohydric acid 一元酸；一羟基酸
monohydric alcohol 一元醇；一羟基醇
monohydric orthoarsenate 砷酸一氢盐 M_2HAsO_4
monohydric phenol(=monohydrating phenol) 一元酚
monohydric pyrophosphate 焦磷酸一氢盐 $M_3HP_2O_7$
monohydric salt 一氢盐〔仍然含有一个酸式氢的酸式盐〕
monohydrobromide 一氢溴化物 $RHBr$
monohydrochloride 一氢氯化物 $RHCl$
monohydrofluoride 一氢氟化物 RHF
monohydroiodide 一氢碘化物 RHI
mono-hydrolyzable 可单水解的
monohydrolyzable organotitanium treatment 可单水解的有机钛处理
monohydroxy-acid 一羟基羧酸 $OHRCOOH$
monohydroxy-alcohol 一元醇；一羟基醇
monohydroxy-dibasic acid 一羟基二羧酸 $OHR(COOH)_2$
monohydroxyethylamine 一羟基乙胺
monohydroxylated acid 一羟基酸
monohydroxyl derivative 一羟基衍生物
monohydroxy-monobasic acid 一羟基羧酸 $OHRCOOH$
monohydroxy-pentabasic acid 一羟基五羧酸 $OHR(COOH)_5$
monohydroxy-terminated 单羟基端接的
monohydroxy-tetrabasic acid 一羟基四羧酸 $OHR(COOH)_4$
monohydroxy-tribasic acid 一羟基三羧酸 $OHR(COOH)_3$
monoiodated 一碘化的
monoiodide 一碘化物
monoiodinated 一碘化的
monoiodination 一碘化作用
monoiodized 一碘化的
monoiodo-acetic acid 一碘乙酸 CH_2ICOOH
monoiodo-acetic chloride 碘乙酰氯 CH_2ICOCl
monoiodo-benzene 一碘代苯；碘苯 C_6H_5I
monoiodo-ester 一碘酯
monoiodo-ether 一碘醚
mono-isoamyl ethyl malonate 乙基丙二酸一异戊酯 $HOOCCH(C_2H_5)CO_2C_5H_{11}$
monoisobutyl-p-aminophenol 一异丁基对氨基苯酚
monoisopropanolamine 单异丙醇胺
monoisopropylamine 一异丙胺〔活性剂〕
monoisotopic mass 单同位素质量
monoisotopic peak 单种同位素峰
monolaurin 月桂酸甘油单酯；甘油一月桂酸酯
monolayer 单分子层
monolayer adsorption 单分子层吸附
monolayer coverage 单分子层覆盖(率)
monolayer polymer 单分子层聚合物
monolayer polymerization 单分子层聚合(作用)
monoleate 含油酸基化合物
monolith 整体；整料；单块
monolithic column 整体柱
monolithic construction 整体结构

monolithic molding 整体成形
monolupine 臭豆碱
monomelissin 甘油一蜂花酸酯
monomer 单体*
monomer casting 单体浇铸；单体铸塑
monomer cast nylon 单体铸塑尼龙
monomer concentration 单体浓度
monomer content 单体含量
monomer conversion 单体转化(率)
monomer dispersant 单体分散剂
monomer droplet 单体液滴〔乳液聚合时分散于水中的单体液滴〕
monomer electronegativity 单体负电性；单体阴电性；单体负电度
monomer emulsion 单体乳液
monomer fluorescence 单体荧光
monomeric 单体(的)
monomeric acceptor 单体受体
monomeric additive 单体添加剂
monomeric aromatic isocyanate 单体芳族异氰酸酯
monomeric cement 单体黏合剂
monomeric donor 单体给予体
monomeric monolayer 单体单分子层
monomeric reactivity ratio 单体竞聚率
monomeric unit 单体单元*
monomer orientation 单体取向
monomer-plasticizer 单体型增塑剂
monomer-polymer particle 单体-聚合物胶粒
monomer ratio 单体配比
monomer reactivity 单体(聚合)活性
monomer reactivity ratio 单体竞聚率
monomer strainer 单体(过)滤器
monomer transfer 单体转移
monomer unit 单体单元；链节
monomethallic salt 一金属盐
monomethines 一甲碱类
monomethiodide 单甲碘化物
-monomethoxime 〔词尾〕 单甲氧肟
monomethyl adipate 己二酸一甲酯 $CH_3O_2C(CH_2)_4CO_2H$
monomethyl amine 一甲胺
monomethyl arsine 一甲基胂；甲胂 CH_3AsH_2
monomethylated 一甲基化的
monomethylation 一甲基化作用
monomethyl ester 一甲基酯
monomethyl ether 一甲基醚
monomethyl ether of hydroquinone 对苯二酚一甲醚〔防老剂〕
monomethylin(=glycerin α-monomethyl ether) 甘油α-一甲醚；单甲甘油醚；一甲灵 $CH_3OC_3H_5(OH)_2$
monomethylol dimethyl 单羟甲基二甲基
monomethylol dimethyl hydantion 单羟甲基二甲基海因〔俗〕；单羟甲基二甲基乙内酰脲
monomethylolurea 一羟甲基脲；脲基甲醇
monomethyl oxalate 草酸一甲酯；乙二酸一甲酯 $CH_3O_2CCO_2H$
monomethyl-sulfate 硫酸氢甲酯 CH_3OSO_3H
monomethylthionine 单甲基硫堇
monomethyl thiourea 一甲基硫脲〔促进剂〕
monomethyltrichlorosilane (一)甲基三氯(甲)硅烷
monomethyl-uric acid 一甲基尿酸 $C_5H_3NO_3CH_3$
monometric 等轴的
monomolecular ①单(个)分子的；一分子的②单(层)分子的
monomolecular adsorption 单分子吸附*
monomolecular chemical reaction 单分子化学反应
monomolecular decomposition 单分子分解
monomolecular elimination reaction 单分子消除反应
monomolecular film 单分子膜*
monomolecular layer 单分子层
monomolecular membrane 单分子膜
monomolecular reaction 单分子反应
monomolecular zone 单分子层
monomorph 单晶(形)物
monomorphic(=monomorphous) ①单形的②单形体
monomorphous(=monomorphic) ①单形的②单形体
monomyristin 甘油单肉豆蔻酸酯
mononatrium glutamate 谷氨酸钠；味精
mononitraniline 一硝基苯胺
mononitrate ①一硝酸盐②一硝酸酯
mononitration 一硝化
mononitro-benzene 一硝基苯 $C_6H_5NO_2$
mononitro-compound 一硝基化合物
mononitro-derivative 一硝基衍生物
mononitroglycerine 甘油一硝酸酯
mononitromethane 一硝基甲烷〔喷气燃料〕
mononitronaphthalene 一硝基萘
mononitrophenol 一硝酚
mononitrotoluene 一硝基甲苯
mononuclear 单核的
mononuclear aromatics 单环芳香烃
mononuclear complex 单核络合物*
mononuclear coordination compound 单核配合物*
mononuclear wall 单核限界
mononuclidic element 单核素的元素；无同位素的元素
N-mono-sec-octylacetamide(MOAA) N-单仲辛基乙酰胺 $CH_3CONHC_8H_{17}$

monooctyl ester of α-anilinobenzylphosphonic acid (MOABP) α-苯胺苄基膦酸一辛酯 $C_6H_5NHCH(C_6H_5)PO(OH)(OC_8H_{17})$
monooctyl phthalate 邻苯二甲酸一辛酯〔增塑剂〕
monooleate 一油酸
mono-olefin 单烯(属)烃
mono-olefinic compound 单烯属化合物
mono-olefin polymer 单烯类聚合物
monoolein 一油精；甘油一油酸酯 $C_{17}H_{33}COOCH_2CHOHCH_2OH$
mono-oxazines 单噁嗪
monooxide 一氧化物
monooxygenase 单体加氧酶
monopalmitate 甘油一棕榈酸酯
monopalmitin 甘油一棕榈酸酯
monoperacid 单过酸
monoperchromic acid 过(一)铬酸 $HCrO_5$
monoperphthalic acid 单过氧邻苯二甲酸
monophase 单相的
monophase current 单相电流
monophase equilibrium 单相平衡
monophase transformer 单相变压器
monophasic 单相的
monophenol oxidase 一元酚氧化酶
monophenol type antioxidant 单苯酚类防老剂
monophenyl-p-phenylenediamine 一苯基对苯二胺〔防老剂〕
monophosphate 一磷酸盐
monopnictide 一磷族元素化物
monopolar cell 单极电解槽
monopole mass spectrometer 单极质谱仪
monopole radiation 单极辐射
monopole soap 高度磺酸化的脂酸皂
monopol soap 蓖麻油皂；莫诺皂
monopolymer 均聚物
monopotassium phosphate 磷酸二氢钾；磷酸一钾 KH_2PO_4
monopropellant 单组分推进剂；单组分火箭燃料
monoprotic acid(=monobasic acid) 一元酸
monoradical 单价基(团)
monoreactive glyceride 单活性基甘油酯
monosaccharide 单糖
monosaccharose 单糖
monose 单糖
monosilane 甲硅烷
monoskeleton electrode 单骨架电极
mono-soap 一酸皂
monosodium acetylide 乙炔一钠 $HC\equiv CNa$

monosodium arsenate 砷酸二氢钠；砷酸一钠 NaH_2AsO_4
monosodium phosphate 磷酸二氢钠；磷酸一钠
monosodium succinate 琥珀酸钠
monostearate 一硬脂酸
monostearin(=glyceryl monostearate; glycerol monostearate) 甘油单硬脂酰酯
monosubstituted 单基取代了的
monosubstituted carbinol 伯醇；一烷基代甲醇 RCH_2OH
monosubstitution 单基取代
monosucrose alkenyl succinate 蔗糖单链烯基琥珀酸酯
monosulfide 一硫化物
monosulfidic bond 单硫键
monosulfonate 单磺酸盐
monosulfonic acid 一磺酸 RSO_3H
monotactic (构型的)单中心规整性
monotactic polymer 单中心规整(性)聚合物〔包括全同立构、间同立构、反式、顺式等聚合物〕
monotectic 偏晶的
monotectic reaction 偏晶反应
monoterpene 单萜*
monoterpenoid 类单萜
monothioacetal 单缩硫醛
monothiocarbamic acid 一硫代氨基甲酸 $CSNH_2OH;NH_2CSOH;NH_2COSH$
monothiocarbonate 一硫代碳酸盐
monothiocarbonic acid 一硫代碳酸 $CO(SH)OH;CS(OH)_2$
monothioester (一)硫代酯
monotrityl glycerol ether 单三苯甲醇甘油醚
monotropein 水晶兰苷
monotropic 单向转变的
monotropitoside 水晶兰苷；冬绿苷
monotropy ①单变性②单变现象
monotype metal 铸字合金
monounsaturated 单不饱和的
monounsaturated fatty acid 单不饱和脂肪酸
monovalent 一价的；单价的
monovalent alcohol 一价醇；一羟基醇
monovalent base 一价碱；一酸价碱
monovalent cation electrode 一价阳离子电极
monovalent hydrocarbon radical 一价烃基
monovariant 单变(度)的
monovariant system 单变体系
monovinylacetylene 1-丁烯-3-炔
monox 氧化硅
monoxide 一氧化物
Monroe crucible 门罗坩埚

Monsanto resonance elastometer 蒙桑托共振弹性计
Monsel salt 蒙塞尔盐；低硫酸铁
montanate 褐煤酸酯
montanic acid 褐煤酸；二十九(烷)酸 $C_{27}H_{55}COOH$
montanin wax 褐煤蜡；蒙旦蜡〔皮革〕
montanite 碲铋华
montan wax 褐煤蜡
montanyl alcohol(=nonacosanyl alcohol) 褐煤醇；二十九(烷)醇 $C_{29}H_{59}OH$
Monte Carlo simulation 蒙特卡洛模拟
montelukast 孟鲁司特〔药〕
monterey pine 辐射松
monticellite 钙镁橄榄石
montmartrite 石膏
montmorillonite 蒙脱石；蒙脱土
montmorillonite-acrylic ester 蒙脱石-丙烯酸酯
montmorillonite-acrylonitrile system 蒙脱石-丙烯腈体系
montmorillonite clay 蒙脱土
montmorillonite grease 蒙脱土润滑脂
montroydite 橙红石
monzonite 二长岩
Mooney index 门尼黏度
Mooney plastometer 门尼塑度计
Mooney point 门尼值
Mooney shearing disk plastometer 门尼剪切圆盘式塑度计
Mooney shearing-disk viscometer 门尼剪切圆盘黏度计
Mooney viscosity 门尼黏度
moon-knife 月牙刀
moonlight gasoline 月光汽油；自机器油箱内漏出的汽油
moonlight unit(=cesium unit) 铯单位
moonstone 月长石
moor 沼
moor coal 沼煤；松散褐煤(泥煤)
Moore filter 模尔滤机
mooring 系留
mope pole 支撑管道用杆
mop polishing 毡轮抛光；毛刷擦亮
moquette 绒头织物；割绒织物
mora 油拔树籽
moracizine 莫雷西嗪〔药〕
moradiol 桑二醇
M orbit M 轨道
mordant ①媒染剂②媒染的
mordant color(=mordant dye) 媒染料
mordant dye(=mordant color) 媒染料
mordant dyeing 媒染(法)
mordant-fixed dye 固着媒染染料
mordanting ①媒染②酸洗〔冶〕③浸色④浸冶
mordanting bath 媒染液
mordanting dye 媒染料
mordanting process 媒染法
mordant rouge 媒染红
mordant toning 媒染调色法
mordant yellow 媒染黄〔染〕
mordenite 发光沸石
Moreau index 莫氏指数
more easily reducible substance 前放电物质；前还原物质
morelandite 钡砷磷灰石；氯砷钡石
morellin 藤黄素
morenosite 碧矾
morindone 桑酮 $C_{15}H_{10}O_5$
moringatannic acid 桑橙素；桑鞣酸
moringine(=benzylamine) 辣木碱
morisuccus 桑葚汁
Morocco leather 摩洛哥革；搓花革
moroctic acid 十八碳-4,8,12,15-四烯酸
morolic acid 摸绕酸〔三萜系化合物〕
morpankhi oil 崖柏油
morphan 吗吩烷
morphanthridine 吗吩烷啶
morphenol 吗吩醇 $C_{14}H_8O_2$
morphia(=morphine) 吗啡
morphina(=morphine) 吗啡
morphinan 吗啡喃
morphine(=morphia; morphina) 吗啡 $C_{17}H_{19}O_3N$
morphine acetate 乙酸吗啡
morphine benzoate 苯甲酸吗啡
morphine borate 硼酸吗啡
morphine bromide 氢溴酸吗啡
morphine chloride 氢氯酸吗啡
morphine hydrobromide 氢溴酸吗啡
morphine hydrochloride 氢氯酸吗啡
morphine meconate 袂康酸吗啡
morphine methyl bromide 甲基溴吗啡
morphine salicylate 水杨酸吗啡
morphine tartrate 酒石酸吗啡
morphinometry 吗啡含量测定法
morphinum 吗啡
morphol 吗啡酚 $C_{14}H_{10}O_2$
N-morpholide N-酰基吗啉
morpholine 吗啉
morpholine disulfide 二硫化吗啉
morpholine tetrasulfide 四硫化二吗啉〔硫化剂〕

morpholinium compound　吗啉化合物
morpholino　吗啉代〔限指4-位〕
2-morpholinobenzothiazole　2-吗啉基苯并噻唑〔硫化剂〕
N-(morpholinothio)phthalimide　N-吗啉基硫代邻苯二甲酰亚胺〔防焦剂〕
N-morpholinyl benzothiaolesulfenamide　N-吗啉基苯并噻唑次磺酰胺〔促进剂〕
2-(4-morpholinodithio) benzothiazole　2-(4-吗啉二硫代)苯并噻唑〔促进剂〕
2-(4-morpholinothio) benzothiazole sulfenamide　2-(4-吗啉代)苯并噻唑次磺酰胺〔促进剂〕
N-(morpholinothio) phthalimide　N-吗啉硫代邻苯二甲酰亚胺〔防焦剂〕
morpholinyl　吗啉基
N-morpholinyl benzothiaole sulfenamide　N-吗啉基苯并噻唑次磺酰胺〔促进剂〕
morpholinyl methyl phthalimide　邻苯二酰亚胺甲基吗啉〔促进剂〕
morphological repeatability　形态(学)重复性
morphological stability　形态稳定性
morphological structure　形态结构
morphologic model　形态学模型
morphology　形态学
morphology of particles　粒子表面形态(学)；粒子结构
morphology of polymer　聚合物形态学*
morpholone　吗啉酮
morphosane　甲基溴吗啡
morphotropy　变晶影响
morrenine　阿根廷乳草碱
morrhuic acid　鳖肝油酸
morrhuin　鳖肝油碱
morrhuol　鳖肝油素
Morse buret　莫尔斯量管
Morse function　莫尔斯函数
Morse potential function　莫尔斯势函数*
Morse's cone　莫氏圆锥
Morse standard　莫尔斯标准
Morse's taper　莫氏锥度
mortality-product(=lethal index)　致死指数
mortar　研钵*
mortar box　研钵
mortar cement　灰砂水泥
mortar mill　臼研机
mortar mixer　灰浆搅拌机
mortar mixing machine　灰泥混合机
mortar pestle　研杵
mortar resistance　耐砂浆性
mortar sand　灰泥用砂
mortar setting　灰浆凝结
mortar trough　灰浆槽
mortejus　扬液器
mortise　榫眼
mortise joint　榫头接合〔垫圈〕
morus　桑
moryl　氯化氨基甲酰胆碱
mosaic　①斑纹状(的)②镶嵌(的)
mosaic block　镶嵌块
mosaic gold　①黄色铜锌合金②二硫化锡颜料
mosaic liquid crystal　镶嵌液晶
mosaic structure　镶嵌结构*
mosaic structure of crystal　结晶镶嵌构造；晶体镶嵌结构
mosandrite　层硅铈钛矿；褐硅铈矿
mosandrium　钶钇矿中稀土金属钐钇混合物
mosapride　莫沙必利〔药〕
Moscicki furnace　摩西基电炉
Moseley formula　摩斯利公式
Moseley law　摩斯利定律
Moseley number　摩斯利数
Moseley series　摩斯利系列
Moseley spectrum　摩斯利X射线谱
moslene　松油烯
mosquito-net　①蚊帐(纱)②网眼纱
moss agate　苔纹玛瑙
Mössbauer absorber　穆斯堡尔吸收体
Mössbauer-beam experiment　穆斯堡尔束实验
Mössbauer effect　穆斯堡尔效应
Mössbauer isotope　穆斯堡尔同位素
Mössbauer nuclide　穆斯堡尔(效应)核素
Mössbauerography　穆斯堡尔形貌谱法
Mössbauer probe technique　穆斯堡尔探针技术
Mössbauer source　穆斯堡尔源*
Mössbauer spectrometer　穆斯堡尔谱仪*
Mossbauer spectrometry　穆斯堡尔谱法
Mössbauer spectroscopy　穆斯堡尔谱(法)
Mössbauer spectrum　穆斯堡尔(能)谱
mossbunker oil　鲱油
moss copper　苔纹铜
moss gold　苔状金
moss rubber　微孔泡沫胶
moss silver　苔状银
moss starch　苔淀粉
mossy cotton　多绒棉
mossy lead　海绵状铅
most probable distribution　最概然分布；最可几分布
motanic acid　二十九(烷)酸

mother alloy　母合金
mother concentrated granule　色母粒
mother liquor　母液
mother lode　母脉；主矿脉
mother nuclide　母核
mother of coal　煤母
mother of pearl　珍珠母；贝壳
mother oil　(石油)原油
mother solution　母液
mother stock　母分；第一批；主料
mother substance　母体
mother vat　母液瓮；母液桶
mother water　母液
motherwort　益母草
mothicide　杀蛀药
mothproof　防蛀的
mothproof finish(=mothproofing)　防蛀加工
mothproofing(=mothproof finish)　防蛀加工
mothproofing agent　防蛀剂；防蠹剂
motional narrowing　运动变窄
motion law　运动定律
motionless mixer　静态混和器
motion with constant stretch history(MCSH)　恒拉伸史运动
Moto　月吨产量
motometer　转速表；转速计
motor　①电动机；马达②发动机③摩托车
motor benzene(=motor benzol)　动力苯；苯
motor benzol(=motor benzene)　动力苯；苯
motor cycle casing　摩托车外胎
motor cycle oil　摩托车油
motor-driven control valve　电动调节阀
motor-driven pump　电动泵
motor dynamo　电动发电机
motor fuel　发动机燃料；马达用燃料
motor fuel additive　发动机燃料添加剂
motor fuel constituent　发动机燃料组成；车用汽油组成
motor gasoline　动力汽油
motor generator (set)　电动发电机(组)
motor grease　电动机润滑膏
motor hoist　电葫芦
motorised　电动的；机动的
motorised valve　电动阀
motorized nip adjustment　电动调距(装置)
motorized rossing tool　机动刮皮器〔采脂〕
motor method　马达法〔评价燃料抗爆性辛烷值的方法〕
motor-method octane number　马达法辛烷值
motor mix　乙基液〔车用汽油辛烷值添加剂〕

motor-mount pump(=motor pump)　马达泵；电动泵
motor naphtha　马达石脑油；石油英；里格罗因
motor oil　马达油
motor petrol　车用汽油
motor pump(=motor-mount pump)　马达泵；电动泵
motor spirit　车用汽油
motor stirrer　电力搅拌器
motor switch oil　马达开关油；油开关用润滑油
motor valve　马达阀；电动机阀
mottle　斑点；混色斑纹
mottled　花斑
mottled effect　斑点效应
mottled paper　暗斑纸
mottled pig (iron)　斑驳生铁；杂色生铁
mottled silk　斑点丝线；异色合股丝线
mottled soap(=Eschweger soap)　爱氏肥皂；斑纹皂；兰花皂〔俗〕
mottling　麻点
mottling lustre of coal　煤的斑点闪光
mould　①塑模；(阴)模〔物〕；模(型)②造形；翻砂③霉菌
mouldability　模塑性；成形性
mouldable　可塑的；可以模塑的
mo(u)ldable fabric　可(模塑)变形织物
mouldable state　可塑态
mould cavity　模槽
mould clamping　合模
mo(u)ld clamping stroke　合模行程
mould clearing　开模
mo(u)ld close(=mo(u)ld clamp)　闭模
mould cure　模塑硫化〔橡胶〕
mould device　模具
moulded　模制的；制型的
moulded article　模制件
moulded brick　模制砖
moulded casting　模铸
moulded chime lap joint　塑口互搭接头
moulded coal　模制煤
moulded glass　模压玻璃
mo(u)lded-in colour　模塑色
mo(u)lded-in insert　埋塑嵌件；模塑嵌件
mo(u)lded item　模塑件
moulded lacquer　脱胎漆器；压模漆器〔大漆〕
moulded piece　模制件
moulded plastics　模压塑料
mo(u)lded section　模塑型材；模塑异型制品
moulded solid catalyst　模制固体催化剂
moulded steel　模铸钢
mouldenpress　自动压机

moulder ①模②造型物③翻砂物
mo(u)ld hang-up 黏模；挂模
moulding(=molding) ①模制；造型②翻砂
moulding box(=moulding flask) 砂箱
moulding compound 模塑料
moulding compound flowability 模塑料流动性
moulding curing 模压硫化
moulding cycle 模塑(操作)周期；成型(操作)周期
mo(u)lding flash 模塑溢料；模塑飞边
moulding flask(=moulding box) 砂箱
moulding machine 制模机
mo(u)lding machinery trouble shooting 模塑机械故障排除
moulding material 模塑材料；成形材料
moulding method 赋型法；模塑法；成形法
moulding of catalyst 催化剂的塑性；催化剂成型
moulding oil 模油
moulding powder 压型粉
moulding press 模压机
moulding pressure 模塑压力
moulding process 成型过程
moulding scraper 造型刮板
moulding shrinkage 模塑收缩率
moulding temperature 模塑温度
mo(u)lding texturing 造型法变形(工艺)
moulding time 模塑时间
moulding variable ①模塑变量②成型变量
moulding viscosity ①模塑黏度②成型黏度
moulding wax 滑模蜡
mould inhibitor 防霉剂
mould line 合模线
mould loading temperature 装料温度
mould made paper 模制纸
mould oil 模型油
mo(u)ld parting 合模线
mould pressing 模压
mouldproof 防霉的
mould proofing 防霉
mo(u)ldproofing wash 脱模；脱模剂
mould release agent 下模剂
mould-repellant coating 防霉涂料
mould shrinkage 模压收缩；成形收缩
mould splits 模组分
mo(u)ld taper 模具锥度
mould unloading 脱模；下模
mould vulcanization 模塑硫化〔橡〕
mo(u)ld wiper 脱模器
mouldy state 霉状；霉性
mountain-ash 野花楸树

mountain balsam 圣草
mountain beech 山地假山毛榉
mountain blue 石青〔蓝铜矿的俗称；成分为碱式碳酸铜〕
mountain butter 含水纤维状硫酸铝
mountain cork ①石棉②山软木
mountain crystal 水晶
mountain flax ①丝状石棉②泻麻
mountain green 孔雀石；矿山绿
mountain laurel 山月桂
mountain leather 石棉
mountain milk 软方解石
mountain soap 皂石
mountain tobacco 千紫菀花
mounting 安装；装配
mounting compound 安装润滑剂
mounting height 安装高度
mounting hole 安装孔
mounting of grating 光栅装置法
mounting of Rowland circle 罗兰圆装置
mounting plate 固定台；上托板
mouse bite 锯齿；蚀刻缺口
mouthpiece 口模
mouth wash 漱口剂
movable barrier 可移动阻片
movable fit 动配合
movable furnace 活动炉
movable ion 活动离子；移动离子
movable jaw 活动颚
movable member 可动部件
movable partition cell 可动隔板沉淀池
movable platen 动型板；动模板
movable rake 移动耙式导丝器；耙形移动式导丝器
movable spreader 移动式涂布器
movable tank support 活动油罐支架
movable tube-sheet 活动管栅
movement of pesticides in soil 农药移动作用
moving along pin 梳理沿针运动
moving ball type viscometer 动球式黏度计
moving band interface 传送带接口
moving-bar grizzly 动柱栅筛
moving bed 移动床
moving bed adsorption 移动床吸附
moving-bed catalytic reforming 移动床催化重整
moving-bed contact coking apparatus 移动床接触焦化装置
moving-bed contacting 移动床接触
moving bed filter 移动床过滤器
moving-bed ion exchange 移动床离子交换

moving-bed of catalyst　催化剂移动床
moving-bed process　移动床(催化)过程
moving bed reactor　移动床反应器*
moving-bed recycle catalytic cracking　移动床循环催化裂化
moving bed system　移动床式
moving belt　传送带
moving belt interface　传送带接口
moving boundary　移动界面
moving boundary analysis　界面移动分析
moving boundary electrophoresis　移动界面电泳*
moving boundary method　界面移动法*
moving-burden bed reactor　移动床反应器
moving catalyst bed　移动催化剂床
moving coil ammeter　磁电式安培计；动圈式安培计
moving coil galvanometer　圈转电流计〔物〕
moving coordinate system　移动坐标系
moving cylinder press　活动气缸压机
moving film　移动膜
moving grizzly　活动栅筛
moving mercury cathode cell　活动汞阴极电解池
moving of solid　固体的移动
moving parts　活动部分；动件*
moving phase　活动相；流动相
moving plate method　活动干板法
moving plate study　动板试验
moving powder　移动的粉末
moving range　移动极差
moving slot　移动键槽
moving solid bed　移动固体床
moving spreader　活动涂布器
moving state　移动状态
moving vane　动叶片
moving wire detector　移动丝检测器
moving wire hydrogen flame ionization detector　移动丝氢焰电离检测器
moving wire solute transfer detector　移动丝溶质传递检测器
mowragenic acid　雾冰藜原酸
mowrageninic acid　雾冰藜吉宁酸
mowrah　长叶雾冰藜
mowrah fat　雾冰藜脂
mowrah meal　雾冰藜粉
mowric acid　雾冰藜酸
mowrin　雾冰藜碱
m-peak　亚稳峰值
M pipe　M 管；双拆管
M radiation　M 辐射

M-rider　M 形游码
MR type resin　大网络树脂；大孔树脂
ms(=millisecond)　毫秒〔等于 10^{-3} 秒〕
M series　M 系
M shell　M 层
mtric count　公制计数
mu〔希腊字头〕　①μ②千分之一毫米③百万分之一
mucilage　①胶水②黏质③(胶)浆
mucilage dressing　黏液整饰
mucilage season　黏液上光
mucilaginous　①黏液的②分泌黏液的
mucilaginous gum　植物黏胶
mucilaginous solution　黏液
mucinase(=hyaluronic acid)　透明质酸
muck　①腐殖土②湿粪
muck bar(=muck iron)　压条；碾条
mucker　掘壕机；掘沟机
muck iron(=muck bar)　压条；碾条
mucobromic acid　黏溴酸；二溴代丁烯醛酸　$HO_2CCBr=CBrCHO$
mucocellulose　黏纤维素
mucochloric acid　黏氯酸；二氯代丁烯醛酸　$HO_2CCCl=CClCHO$
muconic acid　黏康酸；己二烯二酸　$(CH=CHCO_2H)_2$
muconolactone　黏康酸内酯
mucor　毛霉(属)
mucosa　黏液
mucosa compatibility　黏液相容性
mucosity　黏性
mucous　①黏液的②分泌黏液的
mucous membrane　黏膜；黏液膜
mucous tunic　黏膜
mucuna oil　黎豆油
mudaric acid　牛角瓜酸　$C_{30}H_{46}O_3$
mudarin(=mudarol)　牛角瓜醇
mudarol(=mudarin)　牛角瓜醇
mud-body lacquer　泥胎漆器〔大漆〕
mud box(=mud hole)　澄泥箱
mud cake(=press cake)　滤饼
mud collecting hopper　聚渣斗
mud-cracked　(胶泥状)大龟裂的；大(干)裂缝(的)
mud-cracking　大龟裂
muddy　泥浆状
mud-guard　泥浆板
mud hole(=mud box)　澄泥箱
mud-jacking　容器底部(使底与座边平齐的)水泥涂盖物
mud oil　钻井采油
mud press　滤泥机

mud pump 泥浆泵〔糖〕
mud settler(=mud still) 泥浆沉降器
mud still(=mud settler) 泥浆沉降器
mudstone 泥石；泥岩
mud tank 泥浆桶
Muencke pump 玻璃过滤泵〔连在水龙头上用〕
mufangchine 广防己精
muff 保温套；衬套；套筒
muffle ①隔焰甑；闭〔式烤〕炉②玻璃〔灯〕罩③包裹④灭声器
muffle carrier 玻璃罩架
muffle coat 消音涂装
muffle colors 搪瓷着色料
muffle floor 隔焰炉底
muffle furnace 马弗炉*；高温焚火炉
muffle kiln 隔焰窑
muffle painting ①面釉②釉面③釉彩
muffler 消音器
muffle roaster(=close roaster) 闭式烤炉
muguet 铃兰
muguet compound 铃兰香精
muguol 别罗勒烯醇；3,7-二甲基-4,6-辛二烯-3-醇
mugwort 蒿叶
mugwort oil 艾蒿油
muhuhu oil 木胡油
muirapauma 铁青树
muirapuamine 铁青树碱
mulberry fibre 纸构纤维
mulberry silk 桑蚕丝
mulberry tree 桑(属)
mulch ①覆盖②覆盖料
mulch paper 圆圈纸
mule 走锭细纱机
mule hide 骡皮
mull 研糊
mullein 毛蕊花
mullen 纸强度测试器
muller 研磨机
Müller bath 穆勒(凝固)浴
Müller coulometer 穆勒电量计
Mullerite powder 穆勒赖特火药
Müller's glass(=hyalite) 玻璃蛋白石
mullite 多铝红柱石；莫来石
mull oil 研糊油
mulls (高熔点)膏药
mull technique 研糊技术
mulser 乳化机
mulsh paper 圆圈纸

multen pool 熔池
multhiomycin 多硫霉素
multi- 〔拉丁字头〕多
multiad 多单元组〔高分子链中立体规整性的表示方法〕
multiaddition method 多次标准添加法
multianalyte ①多组分分析②多组分分析物
multianalyte immunoassay 多重免疫分析；多组分免疫分析
multiangle gloss meter 多角光泽计；万向光泽计
multi-angle laser light scattering photometer 多角度激光光散射光度计
multi-angular measurement 多角测量
multi-atomic ion 多原子离子*
multiatom three-dimensional crystal 多原子三维晶体
multiaxial drawing 多轴拉伸
multi-axial stitch bonding machine 多轴向缝编机
multiaxial stress field 多轴应力场
multibladed fan 复叶风扇
multibranched paraffin 多支链烷烃
multi-branched polymer 多支链聚合物
multi-buffered paper chromatography 多缓冲化纸色谱(法)
multi capillary column 多毛细管柱
multicavity mold 多槽塑模
multi-cavity mould 多模腔模具
multi-cavity press 多槽模平板机
multicellar pump 多室泵
multicell heater 多室加热器
multicellular 多室的
multicellular pump 多室泵
multicenter bond 多中心键*
multicenter fragmentation 多中心裂解
multi-center zero differential overlap 多中心零微分重叠法
multicentric integral 多中心积分*
multichain condensation polymer 星形缩聚物；多链缩聚物
multi-chain condensed state 多链凝聚态
multichain polymer 多链聚合物
multi-channel analysis 多通道分析法
multi-channel analyzer(MCA) 多道分析器
multi-channel atomic absorption spectrophotometer 多道原子吸收分光光度计*
multichannel column 多(流)路柱
multi-channel detector 多道检测器
multi-channel imaging spectrophotometer 多道成像分光光度计
multi-channel plate 多道板
multi-channel pulse height analyzer 多道脉冲高度分析器

multi-channel recorder 多路记录器
multi-channel scaling 多道定标
multi-channel scan 多道扫描
multi-channel spectrometer 多通道谱仪*
multi-channel X-ray fluorescence spectrometer 多道X射线荧光光谱仪*
multi-charge fat liquor 多电荷加油(脂)乳液
multichip 超大IC型(多芯片模块)
multiclone 复式旋风分离器
multi-coat lacquer 多道喷漆；多道喷涂用漆
multicollection 多(离子)接收
multi collector mass spectrometer 多接收器质谱仪
multicollinearity 多重共线性
multicolor 多色的；多彩的
multicolor cloth 花布
multicolor coating 多彩花纹饰面涂料
multicolor finish 多色(罩面)漆
multi-color lacquer 多色涂装用漆；多彩漆
multicolor novelty finish 多彩美术漆
multi-color operation 多色喷涂操作
multicolor paint 多色漆
multicolor printing machine 彩色印刷机
multicolor system 多色系统
multicolour dyeing 多色染色
multicolour effect 多色效应
multi-colo(u)r profile extrusion 多色型材挤塑
multicolumn(=multilayer column) 多层柱
multicolumn instrument 多柱仪器
multicolumn operation 多柱操作
multi-compartment reactor 多室反应器；多间隔反应器
multicomponent 多组分的；多元的
multicomponent catalyst 多组分催化剂
multicomponent copolymer 多组分共聚物
multicomponent copolymerization 多组分共聚(作用)
multicomponent distillation 多组分混合物蒸馏
multicomponent fractionation 多组分混合物分馏
multi-component liquid foam processing 多组分液体发泡工艺
multicomponent mixture 多组分混合物
multi-component polymer 多组分聚合物
multi-component polymer fibre 多组分共聚物纤维；共聚物纤维
multicomponent polymerization 多组分共聚(作用)
multicomponent spectrophotometry 多组分分光光度法
multicomponent system 多组分系统；多元体系
multicomputer 多道计算机
multiconfiguration self-consistent field theory(MC-SCF) 多组态自洽场理论*

multiconstituent fiber 多组分纤维
multicopolymerization 多组分共聚合(作用)
multi-core cable 多芯电缆
multi-curie processing 居里级(放射性物质)处理；多居里量处理
multicurie source 多居里源；居里级源
multicusp negative ion source 多交点负离子源
multicycle batch technology 多循环间歇工艺
multicyclone 多管旋风分离器
multicylinder machine 多气缸机器
multicylinder pump 多缸泵
multi-daylight press(=multiplaten press) 多层压机
multideck classifier 多层分粒机
multidentate ligand 多齿配位体；多合配位体
multi-detector chromatograph 多检测器色谱仪
multidimensional 多维(的)
multidimensional chromatograph 多维色谱仪
multidimensional chromatography 多维色谱法
multidimensional detection 多维检测
multi-dimensional detection technique of fluorescence 多维荧光检测技术
multidimensional fluorescence spectrum 多维荧光光谱
multidimensional gas chromatograph 多维气相色谱仪*
multidimensional gas chromatography 多维气相色谱法*
multidimensional micro-total analysis systems 多维微全分析系统
multidimensional NMR spectra 多维NMR谱
multidimensional protein identification technology 多维蛋白质鉴定技术
multi-dip automatic machine 多道自动流(动床)浸涂机
multi-dip product 多次浸渍制品
multidirectional fabric hybrid composites 混杂多向织物复合材料
multidisciplinary approach 多学科研究；综合研究
multidisperse polymer 多分散聚合物
multidose glass vial 多次剂量玻璃瓶
multidose vial 多次剂量瓶
multidraw fractionating tower 多取样点分馏塔
multi-drill head machine 多轴钻床；排钻床
multiduty 多功能
multi-echo imaging 多重回波成像
multi-effect evaporator 多效蒸发器
multi-electrode system 多电极体系
multielement 多元素
multielement activation analysis 多元素活化分析
multielement electrodeless discharge lamp 多元素无极放电灯
multielement hollow cathode lamp 多元素空心阴极灯

multielement lamp 多元素灯
multielement spark emission detector 多元素火花发射检测器
multielement spectroscopy 多元素谱学
multielement substoichiometric separation 多元素亚化学计量分离法；多元素不足化学计量分离法
multielement trace analyses 多元素痕量分析
multi-end reeling machine 多条缫丝机
multienzyme 多酶
multienzyme system 多酶系统
multi-event multichannel analyzer 多种多道分析器
multifactor 多因素
multifeed lubricator(=multifeed oiler) 多点润滑器
multifeed oiler(=multifeed lubricator) 多点润滑器
multifibre cable 多纤光缆
multifilament 长纤维束；复丝
multifilament braid 复丝扭辫
multi-flighted screw 多螺纹螺杆
multiflorin 多花苷
multiform batch ticket 多种形式的批量单
multiform unit vulcanizer 多份硫化器
multifrequent 多频率的
multifuel burner 万能燃烧器
multifunctional 多官能的；多功能的
multifunctional acrylate 多官能团丙烯酸酯
multifunctional enzyme 多功能酶
multi-functional epoxy 多官能(度)环氧
multifunctional epoxy powder coating(s) 多功能环氧粉末涂料
multifunctional exchanger 多官能交换剂
multifunctional exchange resin 多功能基离子交换树脂
multifunctional group ion exchanger 多功能基离子交换剂
multifunctional isocyanate 多官能基异氰酸酯
multifunctionality(=polyfunctionality) 多官能度
multifunctional material 多功能材料
multifunctional molecule 多官能团分子
multifunctional silane 多官能基硅烷
multifunction amine 多功能胺
multifunction catalyst 多功能催化剂
multigap technique 多隙技术
multi-gating 多浇口
multigrade 多品位的；多等级的
multigrade oil 稠化润滑油
multigram-scale 克数量级规模；多克规模
multigraph ①油印机②(旋转)排字印刷机
multigraph paper 蜡纸；油印纸
multigreaser 多点润滑器

multi-guests molecular recognition analysis 多元客体分子识别分析
multi guide bar warp knitted fabric 多梳经编组织
multi-head extruder 多头压出机；多口型压出机
multiheteromacrocycles 杂多大环化合物
multihundred watt radioisotope thermoelectric generator (MHW-RTG) (多)百瓦级放射性同位素热电换能器(能源)
multi-impression mold 多巢(压)模
multiinjector 多用注样器
multi ion detection(MID) 多离子检测
multi ion detector 多离子检测器
multi ionic potential 多离子间膜电位
multi-ion selection(MIS) 多离子选择*
multiisotopic comparator 多同位素比较核素
multikilogram 千克数量；千克规模
multikilowatt 多千瓦的；千瓦级
multilamellar vesicles(MLV) 多层脂质体
multilayer 多(分子)层
multilayer adaline 多层自适应线性神经网络
multilayer adsorption 多(分子)层吸附
multilayer aggregate 多层聚集(体)*
multi-layer blow molding 多层吹塑
multilayer column(=multicolumn) 多层柱
multilayer copolymer 多层共聚物*
multi-layer cylinder 多层圆筒
multilayer electrophoretic coating ①多道电泳涂漆法②多道电泳漆
multi-layer extrusion 多层挤塑
multi-layer fabric 多层织物
multilayer method 多层法
multilayer plating 多层电镀
multilayer polymer 多层聚合物
multilayer sorption 多层吸着
multilayer systems ①多层(成膜)系统②多层成膜涂料
multilayer vessel 多层容器
multilayer welding 多层焊
multilevel 多水平
multilevel ionization 多级电离
multilinear fitting 多元线性拟合
multilinear programmer 多阶线性程序器
multilinear temperature programmer 多级线性程序升温器
multiline lubricating system 多线式润滑系统
multiload treatment 多次(离子交换)吸附处理
multimegacurie 兆居里量级(的)
multimembrane electrodecantation 多膜电渗倾泻
multimeter 万用计；万用表；多量程测量仪表

multimodel chiral liquid chromatographic column 多模式手性液相色谱柱
multimolecular 多分子的
multimolecular adsorption 多分子吸附*
multimolecular film 多分子膜
multimolecular reaction 多分子反应
multinecked flask 多颈烧瓶
multinotch blade 多口板刮板法
multinotch draw-down blade 多口刮刀；多槽刮刀
multi-nozzle 多口喷枪
multinuclear ①多核的②多环的
multi-nuclear magnetic resonance 多核磁共振*
multinuclear probe 多核探头
multinucleate (具)多核的
multinuclei spectrometer 多核分光仪
multiobjective decision optimum 多目标决策优化
multi-object simplex method 多目标单纯形优化法
multi-parameter process monitoring 多参量过程控制
multiparametric analysis 多参数分析
multiparticle spectrometer(MPS) 多粒子谱仪
multipass 多程(测光法)
multipassage mass spectrometer 多通质谱计
multipass and baffled heat exchanger 多程及折流换热器
multipass condenser 多程冷凝器
multipass-counter current system 多(行)程逆流装置
multipass dryer 多程干燥机
multipass exchanger 多程换热器
multipass heater 多程加热器
multi-pass spreading 多遍涂胶
multipass tubular heater 多程列管加热器
multipass welding 多道焊
multipath effect 多路效应
multiphase 多相的
multiphase converter 多相换流器
multiphase flow 多相流动
multiphase polymer 多相聚合物*
multiphase polymer system 多相聚合物体系
multiphase polynomial regression 多段多项式回归
multiphase system 多相系统
multiphasic chemical screening 多相化学筛选法
multiphasic polymer 多相聚合物
multiphoton absorption(MPA) 多光子吸收
multiphoton dissociation(MPD) 多光子解离*
multiphoton excitation(MPE) 多光子激发*
multiphoton ionization 多光子电离
multiphoton non-resonant ionization(MPNRI) 多光子非共振电离
multiphoton transition 多光子跃迁

multi-pigmented system 多种颜料体系
multi-pin shear 多根元件剪切机
multiplane tomographic scanner 多(平)面断层扫描机
multi-platen press(=multi-daylight press) 多层压机；多层平板硫化机
multiple absorption 重复吸收
multiple antibody 多种抗体
multiple antigen 多种抗原
multiple atomic ion 对原子离子
multiple-batch extraction 多级间断提取
multiple bead-rod model 多串珠杆式模型
multiple bedded furnace 多层炉
multiple bed system 多床式
multiple-binder system 多元基料体系；多组分基料体系；多组分成膜物体系
multiple bond (多)重键*
multiple cage mill 多笼粉碎机
multiple-carrier braider 多锭编织机〔8锭以上〕
multiple catalyst 多成分催化剂
multiple cavity mould 多巢压模
multiple-chamber reactor 多室反应器
multiple-charged ion 多电荷离子*
multiple chimneys 复式烟囱；集合烟囱
multiple cleavage 复杂裂开；重复裂开
multiple coated fuel particle 多层涂敷燃料颗粒
multiple-coil (polymerization) process 多反应器或多级(聚合)过程
multiple collector 多接收器*
multiple-column valve 多柱阀
multiple comparator method 多同位素比较法
multiple comparison 多重比较*
multiple-continuous development 多级连续展开
multiple cord tyre 多层外胎
multiple correlation 复合相关；多重相关
multiple daylight press 多层压机〔橡胶〕
multiple decay 多重衰变；分支衰变
multiple deck classifier 多层分级器
multiple deck press 多层压机
multiple detector system 多检测器系统
multiple development 多次展开(法)*
multiple die tubing machine 多模管压机
multiple diffraction 多重衍射
multiple disintegration 倍速蜕变(作用)
multiple dispersion 多重散；复分散
multiple-downcomer sieve tray 多降液管筛板
multiple drilling machine 多轴钻床
multiple effect ①多效②多效性；多方面效应
multiple effect evaporator 多效蒸发器

multiple effect system	多效系统
multiple electrode	多重电极
multiple element lamp	多元素灯
multiple emission	多重发射
multiple emulsion	多重乳状液
multiple enzyme system	多酶体系
multiple excitation	多次激发
multiple extraction	多次萃取
multiple Faraday collector	多法拉第(筒)接收器
multiple feed	多口进料
multiple filament ion source	多灯丝离子源
multiple flighted screw	多螺线螺杆
multiple fractionation	多次分馏；多次分级
multiple frequency excitation	多重频率激发
multiple function	①复官能②复官能(化合)物
multiple headspace extraction	多次顶空提取
multiple-hearth burner(=multiple-hearth furnace)	多床炉
multiple-hearth furnace(=multiple-hearth burner)	多床炉
multiple hearth incinerator	多炉膛焚烧炉
multiple hearth reactor	多室反应器
multiple IDA	多次同位素稀释分析法
multiple immunoassay	多组分免疫分析
multiple impeller pump	多叶泵
multiple internal reflectance	多重内反射
multiple internal reflectance accessories	多次内反射附件
multiple ion detection	多离子检测
multiple ion detector	多离子检测器
multiple ion monitoring	多离子监测*
multiple ion selection(MIS)	多离子选择
multiple irradiation	多重照射*
multiple isotope dilution method	多同位素稀释法
multiple labeling	多标记
multiple lagoon	多段氧化池
multiple layer adhesive	复合膜胶黏剂
multiple linear regression	多元线性回归
multiple linear regression spectrophotometry	多元线性回归分光光度法
multiple line (of spectrum)	多重线(光谱)
multiple link	重键
multiple linkage	重键
multiple magnet mass spectrometer	多磁铁质谱计
multiple-metal multiple-ligand system	多金属多配体系统*
multiple mixer column	多级混合(萃取)柱
multiple mould	多巢压模
multiple muffle furnace	多套炉
multiple nip technology	多级压榨式工艺
multiple opening air inlet	多孔式空气入口
multiple orifice meter	多孔测量计
multiple-orifice viscometer	多孔黏度计
multiple partition	多效分配
multiple pass coat	多道涂布；多层涂装
multiple passette	多效泡罩板塔
multiple pass monochromator	多光路型分光器
multiple path absorption cell	多光路吸收池
multiple phase	多元相
multiple phase emulsion	多重(相)乳状液
multiple photon ionization	多光子电离
multiple piston pump	多缸泵
multiple polar group	多极性基团
multiple pulse experiments	多脉冲实验
multiple pulse line narrowing sequence	多脉冲线窄化序列
multiple-quantum coherence	多量子相干
multiple quantum transition(MQT)	多量子跃迁*
multiple range indicator	多程指示剂
multiple range test	多重极差检验
multiple recording device	多能记录装置
multiple record recorder	多记记录器
multiple recrystallization	多级重结晶
multiple reflection	多次反射
multiple reflection type gas cell	多次反射型气体池
multiple regression analysis	多元回归分析
multiple relaxation	重复松弛；多次松弛
multiple resonance technique	多重共振技术
multiple roll crusher	多辊破碎机
multiple roll mill	多辊磨机
multiple rotary screen	多旋转筛
multiple sampling inspection	多次抽样检验法
multiple scanning	多重扫描
multiple scattering	多重散射*
multiple screw extruder	多轴挤出机；多螺杆挤出机
multiple sea sampler	多筒采水器
multiple side-by-side composite fibre	多层并列型复合纤维
multiple solution method	多重解法
multiple solvent front	多溶剂前沿
multiple sorption column	多级吸着柱
multiple spindle drill	多轴钻床
multiple splitting	多重裂分
multiple spot	复斑
multiple stage compressor	多级压气机
multiple stage drier	多级干燥器
multiple-stage mass spectrometry	多级质谱法
multiple stage model	多级模型
multiple stage press	多级压机

multiple standard addition method　多次标准添加法
multiple-strand test　多根纱线断裂试验
multiple support　多支柱；多架；多座
multiple sweep booth　多刮刀(除垢)喷橱
multiplet　①多重峰*②多重态*
multiple tab gate　复式耳型浇口
multiple-tank storage　多储罐储藏
multiple thin-layer chromatography　多级薄层色谱(法)
multiplet isotope dilution method　多重同位素稀释(分析)法
multiple tooling　多刀切削
multiple transition　多次转变；多重转换
multiplet splitting　多重态分裂*
multiplet theory of catalysis　催化多位理论*
multiple tube burner　多管式燃烧器
multiple-tube column　多管塔
multiple tube method　多管法
multiple tubular reactor　多管式反应器
multiple unit of measurement　倍数计量单位
multiple valence ion　多价离子
multiple vee belt drive　三角带成组传动
multiple wave　重复波
multiple-wick oiler　多点润滑器
multiple wire-drum opener　多齿辊开棉机
multiplex　多路
multiplexer　①多路调制器②转换开关
multiplication　倍增
multiplication effect　放大效应
multiplication of system　体系放大；体系倍增
multiplicative interference　倍增干扰
multiplicity　①多样性；多重性②峰裂数
multiplicity factor　多重性因子*
multiplier　①乘积倍数②倍增器③扩程器
multiplier discrimination　倍增器甄别效应
multiplier gain　倍增器放大因数；倍增器增益
multiplier sensitivity　倍增灵敏度
multi-ply　多层胶合板
multiply-charged ion　多电荷离子
multiply deprotonated molecule　脱质子化分子
multiply finish　复层饰面；多层饰面
multiplying ga(u)ge　倍示(压力)规；放大压力计
multiplying lever balance　倍数杠杆天平
multiplying manometer　倍示压力计
multiply-labelled compound　多重标记化合物
multiply protonated molecule　多质子化分子
multi point adsorption　多点吸附〔炭黑〕
multipoint method　多点法
multipoint recognition　多点识别
multipoint recorder　多点记录器
multipoint rotational viscometer　多点旋转黏度计
multi point thermocouple　多头热电偶
multipolar　多极的
multipolar continuum mechanics　多极连续介质力学
multipole expansion　多极展开*
multi-pole moment　多极矩
multipolymer　共聚物；多元聚合物
multiporous　多(细)孔的
multiport injection reactor　多孔喷射反应器
multiport selection valve　多点选择阀
multiport switching valve　多通转换阀
multiport valve　多点阀
multi-program　多道程序
multipulse(MP)　多脉冲技术
multi-purpose　万能的；多功能的；多效的；多用途的
multipurpose additive　多效添加剂
multipurpose chemical processing plant　多用途化学处理工厂
multi-purpose compact spinning system　多用途短程纺丝系统；多功能紧凑纺丝系统
multipurpose gauge　万能规；多用途规
multipurpose grease　万能润滑脂；多效润滑脂
multipurpose instrument　多性能仪器
multipurpose lubricant　万能润滑油；多用润滑油
multipurpose meter　多用途测量表
multipurpose millbase　多用途研磨料(颜料浆)
multipurpose plant　通用装置；多用途装置
multi-reflect pool　多次反射池
multi-reservoir system　多储器系统
multi-residue analysis　多残留分析
multiresolution analysis(MRA)　多分辨率分析
multi-ring　多环
multi-ring dye　多环染料
multiring hydrocarbon　多环芳香烃
multirod shear　多棒剪切机
multiroll crusher　多辊破碎机
multiroll mill　多滚磨机
multirotation(=mutarotation)　变旋(作用)
multiscrew extrusion machine　多螺杆挤压机
multi-section hopper　多格式储料斗
multisegment composite fibre　多股裂离异形复合纤维
multi-shed weft insertion　多梭口引纬
multislice imaging　多断层成像
multi-solvent system　多溶剂系统
multisource　多性能光源
multi-span option　多间隔选择
multi-speed option　多速选择

multisphere tank 多弧罐；多弧水滴形油罐
multi-spindle drilling machine 多头钻孔机
multi spot 复斑；重斑
multistage 多级
multistage absorption 多级吸收
multistage agitated bed ion exchange 多级搅拌床离子交换
multistage bleaching 多级漂白法
multistage blower 多级鼓风机
multistage column 多级柱
multistage compressor 多级压气机
multistage distillation 多级蒸馏
multistage distribution procedure 多级分配操作
multistage drawing 多级拉伸
multistage efficiency 多级效率
multistage expansion 多级膨胀
multistage extraction 多级抽提
multistage filter 多级过滤器
multistage liquid-liquid extraction 多级液-液萃取
multistage mass spectrometry 多级质谱法
multistage mixer column 多级混合式塔
multistage press 多级压机〔橡〕
multistage pulsed-bed contactor 多级脉冲床萃取器
multistage pump 多级泵
multistage reactor 多级反应器
multistage reforming 多段重整
multistage sampling 多级取样
multistage stabilization of crude oil 原油的多级稳定
multistage stretching 多级拉伸；多阶段伸长
multistage sulfite pulping 多段亚硫酸盐制浆法
multi-stage system continuous scouring and bleaching range 翻板式连续练漂联合机
multistage water-cooled compressor 多级水冷压缩机
multistep attenuator 阶梯减光板
multistep cone pulley 多级塔轮
multistep excitation 多步激发*
multistep pressure reduction 多级减压
multistorey catalytic reactor 多层催化反应器
multi-strand polymer 多股聚合物
multistrand roving 多股玻璃粗纱
multistream heater 多路加热炉
multistream sampling 多流股取样
multi-sulfur linkage 多硫键
multi-sweep method 多重扫描法
multi-sweep polarography 多扫描极谱法*；多重扫描极谱法
multitower fractionation train 多塔分馏系统
multitube furnace 多管炉

multitube knock back type condenser 多管反冲式冷凝器
multitube sulfonator 多管磺化器
multitubular boiler(=fire tube boiler;tubular boiler) 火管锅炉
multitubular condenser 多管冷凝器
multitubular gas condenser 多管气体冷凝器
multitubular heat exchanger 多管式换热器
multitubular high efficiency tower 多管高效塔
multitubular reactor 多管反应器
multitubular steam generator 多管锅炉
multivalence 多价
multivalency 多价
multivalent 多价的
multivalent cation 多价阳离子
multivalent cyclic hydrocarbon radical 多价环烃基
multivalent repression 多价阻遏
multivane air separation 偏转器空气分离
multivariable 多元变量
multivariable data 多元数据
multivariant system 多变体系
multivariate analysis 多元分析
multivariate calibration 多元校正；多变量校正
multivariate data 多元数据
multivariate linear regression 多元线性回归
multivariate normal distribution 多元正态分布
multivariate statistical analysis 多元统计分析
multivariate statistical inference 多元统计推断
multivee filter 皱摺式过滤器；V形皱摺过滤器；瓦棱板式过滤器
multi-vessel system 多容器系
multiviscosity oil 稠化油料
multiwall bag 多层麻袋
multiwalled 多层的
multiwalled vessel 多层式容器
multi-wash test 多次洗涤试验〔测试洗涤剂的抗再沉积能力〕
multi wavelength detector 多波长检测器
multi wavelength linear regression 多波长线性回归
multi wavelength linear regression method 多波长线性回归法
multi wavelength spectrophotometry 多波长分光光度法
multi wavelength UV-Vis detector 多波长紫外-可见光检测器
multiway 多路的；多方面的；多位(加工)形式
multiway cock 多通活栓
multiway trap 多路阱
multi-welding pressure vessel 电渣重熔式压力容器
multi-zone 复区带

multizone cooler　多区域冷却器
mumac　①漆树②漆树属植物的木材③苏模鞣料；野葛鞣料
mu-mesic atom　μ(介)子原子
mu-mesic molecule　μ(介)子分子
mu-metal　导磁合金；μ金属
mummy　普鲁士红；褐色氧化铁粉
muncher　轧碎机
mundic　磁性硫化铁；黄铁矿
mungo　呢绒下脚料次品
Munich lake　胭脂红色淀
municipal refuse　城市垃圾
municipal water distribution system　城市配水系统
munjack　硬化沥青
munjistin　茜草色素
Munktell paper　蒙克特牌滤纸
Munsell chroma　孟塞尔系色品(彩度)；孟塞尔系(色)饱和度
Munsell chroma of neutrals　孟塞尔中性色品(彩度)
Munsell color sphere　孟塞尔色球(仪)
Munsell color system　孟塞尔颜色体系
Munsell hue　孟塞尔系色相
Munsell system　孟塞尔标色系统
Munsell value　孟塞尔系(色)亮度
Munsell value function　孟塞尔系亮度函数
mu oil(=tung oil)　桐油
muon　μ子
muonic atom　μ子原子
muonium　μ子素
muonium chemistry　μ子素化学
muon X-ray analysis　μ介子 X 射线分析
muramic acid　胞壁酸；2-葡萄糖胺-3-乳酸醚
murex　6,6'-二溴靛蓝；骨螺
murexan　骨螺烷〔俗〕；氨基丙二酰脲
murexide　紫脲酸铵
murexine　螺碱
muriacite　硬石膏
muriate of ammonia　氯化铵　NH_4Cl
muriate of potash(=potassium chloride)　氯化钾
muriatic　盐酸化的
Murphree plate efficiency　默佛里塔板效率
murram　红土
murumuru oil　星实桐油
musa textile neonee　蕉麻
muscaridine　蕈毒定；白僵菌素
muscarine　①蝇蕈碱；蕈毒碱；毒菌碱②腐鱼毒
muscicide　杀蝇剂
muscimol　蝇蕈醇

muscimole　蝇蕈醇；氨甲基羟异噁唑
muscol(=3-methylcyclopentadecanol)　麝香醇；3-甲基环十五醇
muscolactone　十五内酯
muscolide(=exaltolide)　环十五内酯
muscone(=methyl exaltone)　甲基环十五酮；麝香酮
muscopyridine　麝香吡啶
muscovite　白云母
musenine　驱虫合欢碱
musennin　驱虫合欢(酸性)树脂
mush　泥煤
mushroom　蕈；蘑菇
mushroom deposits　蕈状沉积〔在灯芯上的〕
mushroom distilling column　蕈状蒸馏塔
mushrooming effect　迅速扩大的影响
mushroom tone　蘑菇香
mushroom trap　蕈状阱
mushroom valve　菌形阀
musk　麝香
musk ambrette　葵子麝香
muskatel　肉豆蔻
musk baur(=1-methyl-5-tert-butyl-2,4,6-trinitrobenzene)　1-甲基-5-叔丁基-2,4,6-三硝基苯；拜氏麝香
musk cumene　3,5,6-三甲基-2,4-二硝基叔丁基苯
musk deer　麝；香麝
musk DTI　萨利麝香；麝香 DTI
muskeg　泥岩沼泽；厚苔泽；苔藓
musk extract　麝香浸剂
muskine　麝香酮
musk ketone　麝香酮
musk lactone　麝香内酯
musk melon　甜瓜
musk odor　麝香香气
musk odorant　麝香香料
muskone(=3-methylcyclopentadecanone)　麝香酮；3-甲基环十五(烷)酮　$CH_3C_{15}H_{27}=O$
musk orchid　角盘兰
muskrat　麝香鼠
musk root　香阿魏根
musk root oil　麝香根油
musk rose　麝香蔷薇
musk seed　木槿籽
musk seed oil　黄葵油；麝葵子油
musk T　麝香 T〔商〕；十三烷二羧酸环乙撑酯
musk tibetene　西藏麝香
musk tincture　麝香酊
musk xylene　二甲苯麝香
musk xylol　二甲苯麝香

musk yarrow oil　麝香蓍油
musky note　麝香香韵
musky odor　麝香香气
muslin　薄纱织物；平布；平纹细布
musquash　麝香鼠
mussanine　驱虫金合欢碱
mussing resistance　防捏皱性
mussite　透灰石；氟碳钙铈矿
mustakone　莫斯德酮
mustard　①芥菜②芥末③芥子气
mustard gas　芥子气
mustard-gas cloud　芥子气云
mustard-gas detector　芥子气探测器
mustard gas safety clothing　芥子气保安衣
mustard leaf(=mustard paper)　芥末硬膏
mustard oil　芥子油
mustard paper(=mustard leaf; mustard plaster)　芥末硬膏
mustard plaster(=mustard paper)　芥末硬膏
mustard seed oil　芥子油
mustine hydrochloride　盐酸氮芥；甲基二氯乙基胺盐酸盐
mustmeter　①浊度计②比浊计
musty　生霉的；霉烂的
musty skein　霉丝
mutagen　诱变剂；诱变因素
mutagenesis　诱变(法)；引起突变
mutagenic radiation　致突变辐射
mutamer　变构物；旋光异构物
mutamerism　变构现象；变旋光现象
mutarotase　变旋酶
mutarotation(=multirotation)　变旋(作用)
mutase　变位酶
mutation　①变更②突变
mutatochrome(=5,8-epoxy-β-carotene oxide)　β-胡萝卜素氧化物
Muthman liquid　穆曼液；四溴化乙炔　$CHBr_2CHBr_2$
mutmeg　肉豆蔻
mutterlauge　母液
muttoin tallow　羊脂
mutton fat　羊脂
mutual　①(相)互的②公共的
mutual action　相互作用
mutual adhesion　相互黏合
mutual attraction　相互吸引
mutual coagulation　①互凝结(作用)②互凝聚(作用)
mutual conductance　互导性
mutual destruction　相互破坏(作用)
mutual diffusion　互扩散*

mutual exchange reaction　互换反应
mutual exclusion　互斥现象
mutual inductance　互感系数
mutual induction　互感应
mutual information　互信息
mutually soluble liquid　互溶液体
mutual polarizability　互极化度；互极化性
mutual polarization　相互极化作用
mutual precipitation　互沉淀
mutual solubility　①互溶度②互溶性
mutual solvation　互为溶剂化作用
mutual solvent　互溶剂
mutual termination　相互终止
mutural polarizability　互极化度；互极化性
mybasan　异烟肼；4-吡啶甲酰肼
mycobacterium phlei　乌体林斯〔药〕
mycocerosic acid　2,4,6-三甲基硬脂酸；结核蜡酸
mycoheptin(=micoheptin)　七烯枝菌素
mycophenolate mofetil　吗替麦考酚酯〔药〕
mycotoxin analysis　真菌毒素分析
mylonite　糜棱岩
myoarsenobenzole　硫胂凡纳明
myoarsphenamine　硫胂凡纳明
myochrysine　金硫基代丁二酸钠　$NaO_2CCH_2CH(SAu)CO_2Na$
myocide　乙酸苯汞
myocrisin　硫代苯酸金钠
myodil　碘苯基十一酸乙酯
myofibril　肌原纤维；肌细胞质纤维
myoglobin test　肌红蛋白试验
myotic　①缩瞳药②收缩瞳孔的
myrabolam　诃子〔植〕
myrbane oil　硝基苯
myrcane　月桂烷
myrceane　月桂烷
myrcene　月桂烯；3-亚甲基-7-甲基-1,6-辛二烯　$C_{10}H_{16}$
myrcenol　月桂烯醇　$C_{10}H_{18}O$
myrcenone　月桂烯酮
myrcenyl　月桂烯基
myrcenyl acetate　乙酸月桂烯酯
myrcia　月桂
myrcia acris　香叶月桂
myrcia oil　月桂油
myria-　〔词头〕万
myrialiter　万升
myriameter　万米
myrica oil　杨梅油
myricetin　杨梅黄酮；3,5,7,3′,4′,5′-六羟黄酮

myricetrin 杨梅苷
myricin 杨梅酯 $C_{30}H_{61}O_2C_{16}H_{31}$
myricinic acid 杨梅酸；三十一(烷)酸 $CH_3(CH_2)_{29}COOH$
myricoid 蜂花素
myricyl 蜂花基；三十烷基 $C_{30}H_{61}-$
myricyl acetate 乙酸蜂花酯 $CH_3CO_2C_{30}H_{61}$
myricyl acid 蜂花酸；三十(烷)酸
myricyl alcohol 蜂花醇 $C_{30}H_{61}OH$
myricyl aldehyde 蜂花醛；三十(烷)醛
myricyl cerotate 二十六酸蜂花酯
myricyl melissate 蜂花酸蜂酯；三十(烷)醇三十一(烷)酸酯
myricyl palmitate 十六酸蜂花酯 $C_{16}H_{31}O_2C_{30}H_{61}$
myristamide(=myristic amide) 肉豆蔻酰胺；十四(烷)酰胺 $C_{13}H_{27}CONH_2$
myristanilide(=myristyl anilide) 肉豆蔻酰苯胺；N-十四(烷)酰苯胺 $C_{13}H_{27}CONHC_6H_5$
myristate 肉豆蔻酸酯；十四(烷)酸酯
myristic acid(=tetradecanoic acid) 肉豆蔻酸；十四(烷)酸 $CH_3(CH_2)_{12}CO_2H$
myristica fragrans leaves oil 肉豆蔻叶油
myristic aldehyde 肉豆蔻醛；十四(烷)醛 $CH_3(CH_2)_{12}CHO$
myristic amide(=myristamide) 肉豆蔻酰胺；十四(烷)酰胺 $C_{13}H_{27}CONH_2$
myristic anhydride 肉豆蔻酸酐；十四(烷)酸酐 $(C_{13}H_{27}CO)_2O$
myristica oil 肉豆蔻油
myristicene 肉豆蔻萜
myristicic acid 肉豆蔻醚酸；3-甲氧基-4,5-甲二氧基苯甲酸 $H_2C(O_2)C_6H_2(OCH_3)(COOH)$
myristicin 肉豆蔻醚 $C_{11}H_{12}O_3$
myristic ketone(=myristone) 肉豆蔻酮；4-二十七烷酮；对称二十七烷酮 $(C_{13}H_{27})_2CO$
myristicol 肉豆蔻脑 $C_{10}H_{16}O$
myristin 甘油三肉豆蔻酸酯
myristinate 肉豆蔻酸酯；十四酸酯
myristodilaurin 甘油一肉豆蔻酸二月桂酸酯
myristodistearin 甘油一肉豆蔻酸二硬酸酯
myristoleic acid(=9-tetradecenoic acid) 肉豆蔻脑酸；9-十四烯酸 $C_{14}H_{26}O_2$
myristone(=myristic ketone) 肉豆蔻酮；4-二十七烷酮；对称二十七烷酮 $(C_{13}H_{27})_2CO$
myristonitrile(=tridecyl cyanide) 肉豆蔻腈；十四(烷)腈 $C_{13}H_{27}CN$
myristostearin 甘油肉豆蔻酸硬脂酸酯
myristoyl 肉豆蔻酰；十四酰 $CH_3(CH_2)_{12}CO-$
myristoyl methyl taurine 肉豆蔻酰甲基牛磺酸盐
myristoyl peroxide 过氧化肉豆蔻酰
myristyl 肉豆蔻基；十四烷基 $C_{14}H_{29}-$
myristyl alcohol 肉豆蔻醇；十四(烷)醇
myristyl anilide(=myristanilide) 肉豆蔻酰苯胺；N-十四(烷)酰苯胺 $C_{13}H_{27}CONHC_6H_5$
myristyl benzyl dimethyl ammonium chloride 肉豆蔻基苄基二甲基氯化铵
myristyl chloride 肉豆蔻酰(基)氯；十四(烷)酰氯 $C_{13}H_{27}COCl$
myristyl laurate 月桂酸肉豆蔻酯
myristyl myristate 肉豆蔻酸十四(醇)酯；肉豆蔻酸肉豆蔻酯
myristyl salicylate 水杨酸肉豆蔻酯
myristyl-trimethylsilicane 肉豆蔻基三甲基硅 $(C_{14}H_{39})Si(CH_3)_3$
myrobalam 诃子〔植〕
myrobalam extract(=myrobalam extract) 诃子栲胶
myronate 黑芥子硫苷酸盐(或酯)
myronic acid 黑芥子硫苷酸
myrosin 黑芥子硫苷酶
myroxin 芥子素 $C_{23}H_{36}O$
myrrh ①没药②没药树③没药树脂
myrrhin 没药脂
myrrh oil 没药油
myrrholic acid 没药酸 $C_{17}H_{22}O_5$
myrrh resin 没药树脂
myrtanal 桃金娘醛
myrtan extract 桉树栲胶
myrtanol 桃金娘烷醇 $C_{10}H_{17}OH$
myrtenal 桃金娘烯醛 $C_{10}H_{14}O$
myrtenic acid 桃金娘烯酸 $C_{10}H_{14}O_2$
myrtenol 桃金娘烯醇 $C_{10}H_{16}O$
myrtenyl 桃金娘烯基
myrtenyl acetate 乙酸桃金娘烯酯
myrtenyl chloride 桃金娘烯基氯 $C_{10}H_{15}Cl$
myrtilin 桃金娘萃
myrtillidin 桃金娘配基
myrtillin 桃金娘灵 $C_{22}H_{22}O_{12}$
myrtillin chloride ①飞燕草素葡萄糖苷②乌饭树提取物
myrtle 桃金娘；香桃木
myrtle oil 桃金娘油；香桃木油
myrtle wax 桃金娘蜡
myrtol 香桃木油；桃金娘油
mystin 硝酸钠和甲醛的混合物〔防腐剂〕
mytilite(=mytilitol) 贻贝酚
mytilitol(=mytilite) 贻贝酚
myvizone 硫乙腙

N

naal grass oil 脉香茅油
naal oil 苏丹香茅油
nabam 代森钠；1,2-亚乙基二(二硫代氨基甲酸钠) NaSSCNHCH$_2$CH$_2$NHCSSNa
nabumetone 萘丁美酮〔药〕
nacarat 胭脂红；洋红
nachrsalz 磷酸钠与磷酸铵的混合物
nacre(=mother of pearl) 珍珠母
nacreous 珠光的；有珍珠光泽的；有虹彩光的
nacreous luster 珠光；虹彩色(泽)
nacreous pigment 珠光颜料
nacrite 珍珠陶土；珍珠石；珠白云母
nadic acid(=5-norbornene-2,3-dicarboxylic acid; carbic acid) 5-降冰片烯-2,3-二酸；降冰片烯二酸
nadic anhydride(=endomethylene tetrahydrophtalic anhydride) 桥亚甲基四氢化邻苯二甲酸酐〔硫化剂〕
nadorite 氯锑铅矿
naematolin 沿丝伞菌素
Nafion 高氟化(离子交换)树脂
naftolens 焦油中不饱和烃
nagelschmidtite (炼钢)平炉炉渣磷酸体
naginata ketone 白苏酮
nagyagite 叶碲矿
nahcolite 天然小苏打
Nahnsen zinc process 南森炼锌法
nailability 受钉性；可钉性
nail enamel 指甲油
nailhead rusting 钉头锈蚀
nail polish(=nail varnish) 指甲漆；指甲油
nail polish paste 指甲膏
nail polish powder 指甲粉
nail polish remover 指甲膏清洗剂
nail varnish(=nail polish) 指甲漆；指甲油
nairomycin 新露霉素
naked fire 明火头；明火
naked flame 明火
naked light 开放光线；无遮盖光线
naked radiator 开放辐射器
naked source 裸源
nakrite 珍珠陶土
nalecin 丁二酸左菌素酯
nalectin(=chloramphenicol) 氯霉素
nalgon tubing 塑料管
nalidixic acid 萘啶酮酸

nalorphine hydrochloride 纳洛芬；N-烯丙基原吗啡盐酸盐
naloxone 纳洛酮
name plate 名牌；号牌
nanaomycin 七尾霉素
nancic acid 乳酸
nancimycin 南锡霉素；利福霉素
nandinine 南天竹碱
nandrolone phenylpropionate 苯丙酸诺龙；19-去甲-17-苯丙酸睾酮
nanmu oil 楠木油
nano- 〔词头〕 纳；毫微〔符号 n；10^{-9}〕
nanoamorphous material 纳米非晶材料
nano and micro relief 微观粗糙度
nano bio material 生物医用纳米材料
nanobulk 纳米块体
nanocage 纳米笼
nanochannel 纳米通道
nanocluster 纳米簇
nanocomposite 纳米复合材料
nanocrystal 纳米晶
nano crystalline material 纳米晶材料
nanodevice 纳米器件
nanoelectrode 纳米电极
nanoelectrospray 微电喷雾
nano farad 纳法(拉)〔10^{-9}法(拉)〕
nano-fiber 纳米纤维
nano-fibril 基原纤
nanofiller 纳米填充剂
nanofilm 纳米薄膜
nanogram 纳克〔10^{-9}克〕
nanohybrid 纳杂化物
nanoindentation 纳米缺口试验；纳米压痕技术
nanolatex 纳米胶乳
nanolithography 纳米加工技术
nanomaterial 纳米材料
nanometer 纳米
nano particle 纳米颗粒
nano powder 纳米粉末
nanoscale 纳米尺度
nano-scale hybrid 纳米级杂化
nanosecond 纳秒；纤秒
nano-second time resolution ultrasonic microscope 纳秒时间分辨超声显微镜

nanosphere 球形纳米聚集体
nano structural material 结构纳米材料
nanostructure assembling system 纳米组装体系
nanostructuresystem 纳米结构体系
nanostructureunit 纳米结构单元
nanotechnology 纳米技术
nanotube 纳米管
nanotube array 纳米管列阵
nanowire 纳米线
nantenine(=domestine) 南天竹宁
nantokite 铜盐；氯化亚铜矿
nap ①毛；绒②起毛
napalin 月桂酸和环烷酸铝
napalite 蜡状烃类
napalm ①纳旁〔一种铝皂〕②铝皂型胶状油③凝固汽油〔通称〕
napelline 乌头碱
napellonine(=songorine) 华北乌头碱
nap fabric 拉绒织物；起毛织物；起绒织物；绒面织物
naphazoline 萘唑啉；2-(1-萘甲基)-咪唑啉
naphazoline nitrate 硝酸萘甲唑啉〔药〕
naphsultam acid 萘-1,8-磺内酰胺
naphsultone 萘-1,8-磺内酯
naphtha 石脑油；(粗)挥发油；粗汽油；轻油
naphtha aceti 乙酸乙酯
naphtha bottoms 粗汽油底油
naphthacene 并四苯
naphthacetol 4-乙酰氨基-1-萘酚
naphtha column 石脑油塔；石脑油吸收塔
naphthacridine 萘并吖啶 $C_{21}H_{13}N$
naphtha cut 石脑油馏分
naphtha distillation 石脑油蒸馏
naphtha fraction 石脑油馏分；粗汽油馏分
naphtha furnace 石脑油炉
naphtha gas 石脑油气
naphtha gas reversion 石脑油气回馏过程
naphtha insolubles 不溶于石脑油部分
naphthal(=naphthylmethylene) 萘亚甲基；萘(甲)醛缩
naphthaldehyde 萘(甲)醛 $C_{10}H_7CHO$
Naphthal dye 纳夫妥染料
naphthalene 萘 $C_{10}H_8$
naphthalene acid black 萘酸黑
naphthalene ball 萘球；萘丸；卫生球〔俗〕
naphthalene carbonal 萘亚甲基 $C_{10}H_7CH=$
naphthalene carboxylic acid 萘甲酸 $C_{10}H_7COOH$
naphthalene chloride 氯代萘 $C_{10}H_7Cl$
naphthalene derivatives 萘的衍生物
naphthalene-1,8-dialdehyde 萘-1,8-二醛

naphthalene diamine 萘二胺
naphthalene 2,1-diazo oxide 氧化 2,1-重氮萘
naphthalene dicarboxylic acid 萘二甲酸 $C_{10}H_6(COOH)_2$
naphthalene dichloride 二氯代萘 $C_{10}H_6Cl_2$
naphthalene-1,5-diisocyanate 亚萘基-1,5-二异氰酸酯〔硫化剂〕
naphthalene diol 萘二酚
naphthalenedione 萘二酮
naphthalene disulfonic acid 萘二磺酸 $C_{10}H_6(SO_3H)_2$
naphthalene fluoride 氟代萘 $C_{10}H_7F$
naphthalene green 萘绿
naphthalene halide 卤代萘
naphthalene heptachloride 七氯代萘 $C_{10}HCl_7$
naphthalene hexachloride 六氯代萘 $C_{10}H_2Cl_6$
naphthalene iodide 碘代萘；萘基碘 $C_{10}H_7I$
naphthalene monocarboxylic acid 萘羧酸；萘甲酸 $C_{10}H_7COOH$
naphthalene monochloride 一氯代萘 $C_{10}H_7Cl$
naphthalene monosulfonic acid 萘磺酸 $C_{10}H_7SO_3H$
naphthalene nucleus 萘环
naphthalene octochloride 八氯代萘 $C_{10}Cl_8$
naphthalene oil 萘油
naphthalene pentachloride 五氯代萘 $C_{10}H_3Cl_5$
naphthalene picrate 苦味酸萘
naphthalene polychloride 多氯代萘
naphthalene polyhalide 多卤代萘
naphthalene red(=magdala red) 萘红；麦塔喇红；苏丹红
naphthalene resin 萘醛树脂
naphthalene ring 萘环
naphthalene rose 萘玫红
naphthalene salts〔复〕 萘盐
naphthalene sulfinic acid 萘亚磺酸
naphthalenesulfonate 萘磺酸盐 $C_{10}H_7SO_3M$
naphthalenesulfonic acid 萘磺酸 $C_{10}H_7SO_3H$
naphthalene sulfonic acid formaldehyde condensation product 萘磺酸甲醛缩合物
naphthalene-sulfonyl chloride 萘磺酰氯
naphthalene-1,4,5,8-tetracarboxylic acid 萘-1,4,5,8-四羧酸
naphthalene tetrachloride 四氯代萘 $C_{10}H_4Cl_4$
naphthalene tetrasulfonic acid 萘四磺酸 $C_{10}H_4(SO_3H)_4$
naphthalene thiol 萘硫酚
naphthalene trichloride 三氯代萘 $C_{10}H_5Cl_3$
naphthalene trisulfonic acid 萘三磺酸 $C_{10}H_5(SO_3H)_3$
naphthalene yellow 萘黄
naphthalic acid 萘二甲酸〔通常指1,8-位，或叫作α-〕 $C_{10}H_6(COOH)_2$

naphthalic acid lactone 萘二酸内酯
naphthalic anhydride 萘二甲酸酐 $C_{10}H_6=(CO)_2O$
naphthalic sulfonic chloride 萘磺酰氯 $C_{10}H_7SO_2Cl$
naphthalide ①萘基金属 $C_{10}H_7M$②N-酰基萘胺 $RCONHC_{10}H_7$
naphthalidine 1-萘胺
naphthalimido 萘二甲酰氨基 $C_{10}H_6(CO)_2N-$
naphthalmine direct black 萘胺直接黑
naphthalocyanine 萘酞菁
naphthalol 水杨酸萘酯
naphthamide 萘甲酰胺
naphthamine blue 萘胺蓝
naphthamine brilliant bordeaux 萘胺亮枣红
naphthamine direct black 萘胺直接黑
naphthamine fast black 萘胺坚牢黑
naphthamine light blue 萘胺浅蓝
naphthamine pure blue 萘胺纯蓝
naphthane(=decalin) 十氢化萘；萘烷
naphthanilide 萘甲酰苯胺
naphthanol 萘烷醇
naphthanthracene 萘并蒽
naphthaometer (石油产品)闪点测定仪
naphtha polyforming 石脑油聚合重整
naphthaquinol 萘醌醇
naphthaquinone 萘醌 $C_{10}H_6O_2$
naphtha reformer 石脑油重整器
naphtha reformer stripper 石脑油重整的汽提塔
naphtha residue 石脑油脚
naphtha scrubber 石脑油吸收塔〔石油气体内〕
naphtha sludge 石脑油酸渣〔透明馏出油精制时的酸渣〕
naphtha soap 石脑油皂
naphtha stock 石脑油料
naphthathiourea α-萘硫脲
naphtha-treating plant 石脑油处理厂
naphthazarine(=5,8-dihydroxy naphthoquinone) 萘茜；5,8-二羟基萘醌 $(HO)_2C_{10}H_4O_2$
naphthazin(e) 氮杂蒽
naphthazole 萘并吡咯
naphthein 石脑油英
naphthenate 环烷酸盐
naphthenate drier 环烷酸盐干燥剂
naphthene ①环烷；环烷属烃 C_nH_{2n}②(旧用名，指)萘
naphthene acid 环烷酸
naphthene alkylation 环烷烃烷基化(作用)
naphthene base(=naphthenic base) 环烷基
naphthene-base crude petroleum 环烷基原油
naphthene base oil 环烷基原油
naphthene content 环烷烃含量

naphthene series 环烷系 C_nH_{2n}
naphthene soap 环烷皂
naphthene sulfonate 环烷磺酸盐
naphthene sulfonic acid 环烷磺酸
naphthenic 环烷的；(脂)环烃的
naphthenic acid ①环烷酸；环酸②环己烷甲酸
naphthenic acid soap 环烷酸皂
naphthenic base(=naphthene base) 环烷基
naphthenic base crude 环烷基原油
naphthenic base crude (oil) 环烷基石油
naphthenic hydrocarbon 环烷烃 C_nH_{2n}
naphthenic lube oil 环烷基润滑油
naphthenic oil(=naphthenoid oil) 环烷油
naphthenic residual oil 环烷基渣油
naphthenics 润滑油中黏度随温度急变的组分
naphthenic soap 环烷皂
naphthenic solvent 环烷溶剂
naphthenic type gasoline 环烷基汽油
naphtheno-aromatic ring 环烷-芳香环
naphthenoid oil(=naphthenic oil) 环烷油
naphthenone 环烷酮
naphthenyl(=naphthylmethylidyne) 萘次甲基 $C_{10}H_7C\equiv$
naphthieno 萘噻基 $-C_{10}H_6S-$
peri-naphthindandione 苯并二氢-1,3-萘二酮
α:β-naphthindandione α:β-苯并-1,3-茚二酮
peri-naphthindane 苯并-2,3-二氢(1H)-萘
α-naphthindane 苯并[e]二氢茚
α-naphthindazole 苯并[g]吲唑
β-naphthindazole 苯并[e]吲唑
naphthindene 萘并环戊烯
naphthinduline 萘对氮蒽兰
α-naphthiondanone 苯并[e]二氢-2-茚酮
naphthionic acid 对氨基萘磺酸；4-氨基-1-萘磺酸 $NH_2C_{10}H_6SO_3H$
naphthisodiazine 二氮杂菲
naphthisotetrazine 四氮杂菲
naphthisotriazine 三氮杂菲
naphthlene bromide 溴代萘 $C_{10}H_7Br$
naphtho- 萘并
naphthoamide 萘甲酰胺
naphthoate 萘甲酸盐(酯) $C_{10}H_7COOM$
naphthochrome colors (复) 萘铬染料
naphthodiazine ①吩嗪②萘并二氮苯
naphthoflavene(=7,8-benzoflavone) 萘黄酮；7,8-苯并黄酮 $C_{19}H_{12}O_2$
naphthofluorene 萘芴
naphthoic acid 萘甲酸 $C_{10}H_7COOH$

naphthoic aldehyde 萘(甲)醛 $C_{10}H_7CHO$
naphthol 萘酚 $C_{10}H_7OH$
naphthol antipyrine 萘酚安替比林
naphthol aristol 二碘-β-萘酚
naphtholate 萘酚盐
α-naphtholbenzene α-萘酚苯
naphthol benzoate 苯甲酸萘酯
naphthol blue(=new blue) 萘酚蓝；亚甲基型染料
naphthol blue black 萘酚蓝黑
β-naphthol butyl ether β-萘丁醚
naphthol disulfonic acid 萘酚二磺酸 $OHC_{10}H_5(SO_3H)_2$
naphthol ether 萘酚醚
naphthol ethyl ether 萘乙醚
naphthol green 萘酚绿；颜料绿
naphthol green B 萘酚绿 B
naphthol methyl ether 萘甲醚
naphthol monosulfonic acid 萘酚一磺酸 $OHC_{10}H_6SO_3H$
naphthology 石油科学
naphthol orange 萘酚橙
naphtholphthalein 萘酚酞 $C_8H_4O_2(C_{10}H_6OH)_2$
naphthol red 萘酚红
naphthol red toner B 萘酚红色原 B
naphthol soap 萘酚皂
naphthol sulfonate 萘酚磺酸盐；羟萘磺酸盐 $C_{10}H_7O_4SM$
naphthol sulfonic acid 萘酚磺酸 $OHC_{10}H_6SO_3H$
naphthol trisulfonic acid 萘酚三磺酸 $OHC_{10}H_4(SO_3H)_3$
naphthol yellow (酸性)萘酚黄
naphthomycin 萘霉素
naphthonitrile 萘甲腈 $C_{10}H_7CN$
naphthophenanthrene 二苯并蒽
naphthophenazine 萘吩嗪 $C_{16}H_{10}N_2$
naphthopicric acid 萘苦酸；2,4,5-三硝基萘酚 $(NO_2)_3C_{10}H_4OH$
naphthoquinaldine 萘喹哪啶；甲基氮菲 $C_{13}H_8NCH_3$
naphthoquinoline 萘喹啉 $C_{13}H_9N$
amphi-naphthoquinone 2,6-萘醌 $C_{10}H_6O_2$
α-naphthoquinone(=1,4-naphthoquinone) α-萘醌；1,4-萘醌
1,4-naphthoquinone(=α-naphthoquinone) α-萘醌；1,4-萘醌
1,2-naphthoquinone(=β-naphthoquinone) 1,2-萘醌；β-萘醌
β-naphthoquinone(=1,2-naphthoquinone) 1,2-萘醌；β-萘醌
naphthoquinoxaline 1,4-二氮杂蒽
naphthoresorcinol 间萘二酚
naphthosalol 水杨酸β-萘基酯
naphthosultone 萘-1,8-磺内酯

naphthotetrazines 四氮杂蒽
naphthothiazoles 萘噻唑
naphthotriazines 三氮杂蒽
naphthoxanthene 苯并夹氧杂蒽
naphthoxazine 吩噁嗪
naphthoxy(=naphthyloxy) 萘氧基 $C_{10}H_7O$—
2-naphthoxypropionamide 2-萘氧基丙酰胺〔晶纹剂〕
naphthoyl 萘酰；萘甲酰基 $C_{10}H_7CO$—
naphthoyl acetonitrile 萘甲酰基乙腈 $C_{10}H_7COCH_2CN$
naphthoyl chloride 萘甲酰氯 $C_{10}H_7COCl$
naphthoyloxy 萘甲酸基 $C_{10}H_7COO$—
naphthoyltrifluoroacetone(NTA; HNTA) 萘甲酰三氟丙酮 $C_{10}H_7COCH_2COCF_3$
naphthyl 萘基 $C_{10}H_7$—
naphthyl acetate 乙酸萘酯 $CH_3CO_2C_{10}H_7$
naphthylacetic acid 萘乙酸 $C_{10}H_7CH_2CO_2H$
naphthyl acetylene 萘基乙炔
naphthyl alcohol α-萘酚
naphthyl aldehyde 萘(基)甲醛
naphthylamine 萘胺 $C_{10}H_7NH_2$
naphthylamine antioxidant 萘胺类防老剂
naphthylamine brown 萘胺棕
naphthylamine disulfonate 萘胺二磺酸盐 $C_{10}H_8O_6NS_2M$
naphthylamine disulfonic acid 萘胺二磺酸 $NH_2C_{10}H_5(SO_3H)_2$
naphthylamine hydrochloride 萘胺化盐酸 $C_{10}H_7NH_2HCl$
naphthylamine monosulfonic acid 萘胺一磺酸 $NH_2C_{10}H_6SO_3H$
2-naphthylamine-6-sulfonic acid 2-萘胺-6-磺酸
α-naphthylamino-azo-benzene α-萘胺基偶氮苯；α-萘红
α-naphthylamino-azo-p-benzene sulfonic acid α-萘胺基偶氮苯对磺酸
naphthyl anthranilate 邻氨基苯甲酸萘酯
α-naphthyl benzoate 苯甲酸α-萘酯 $C_6H_5CO_2C_{10}H_7$
naphthyl carbinol 萘基甲醇
naphthyl cyanide 萘甲腈
2,2'-naphthyl disulfide 2,2'-萘基二硫化物〔阻燃剂〕
naphthylene 亚萘基 —$C_{10}H_6$—
naphthylenediamine 萘二胺
naphthylene-1,5-diisocyanate(NDI) 亚萘基-1,5-二异氰酸酯；萘撑 1,5-二异氰酸酯
β-naphthyl ethyl alcohol β-萘乙醇
naphthyl ethyl ether 萘乙醚
naphthylhydrazine 萘肼 $C_{10}H_7NHNH_2$
naphthylhydrazine hydrochloride 盐酸萘肼 $C_{10}H_7NHNH_2HCl$

naphthyl hydroxide　萘酚
naphthyl hydroxylamine　萘胲　$C_{10}H_7NHOH$
naphthylidene　亚萘基　$CH=CHCH_2C_6H_4C=$
naphthyl isocyanate　异氰酸萘酯　$C_{10}H_7N=CO$
naphthyl mercaptan　萘硫酚
naphthylmercuric　萘汞(基)
naphthylmercuric acetate　乙酸萘汞酯
naphthylmercuric chloride　氯化萘汞
naphthylmethylene(=naphthal)　萘亚甲基；萘(甲)醛缩
naphthyl methyl ether　萘基甲基醚
naphthylmethylidyne(=naphthenyl)　萘次甲基　$C_{10}H_7C\equiv$
naphthyl methyl ketone　萘基甲基(甲)酮
naphthyl naphthalene　联二萘　$C_{10}H_7\cdot C_{10}H_7$
naphthyloxy(=naphthoxy)　萘氧基　$C_{10}H_7O-$
α-naphthyl salicylate(=alphol)　水杨酸α-萘酯　$HOC_6H_4CO_2C_{10}H_7$
α-naphthylthiourea(=antu)　α-萘硫脲；安妥〔毒鼠药〕
α-naphthyl-triethyoxy-silicane　α-萘基三乙氧基硅　$C_{10}H_7Si(OC_2H_5)_3$
naphthyridine　1,5-二氮杂萘　$C_8H_6N_2$
naphthyridinomycin　萘啶霉素
naphtol　萘酚
Napierian logarithms　纳皮耳对数〔自然对数〕
Napier's equation　纳皮耳公式
napiform　芜菁状(的)孔
napkin rings　"餐巾圈"形靶〔指此形状的钚-铝合金靶，用于反应堆辐照制造超钚元素〕
napless　无绒(头)的
Naples yellow　拿浦黄；锑酸铅
napoleonite　含石英、斜长石等矿物的岩石
nappe　①外层②表面
napped fabric　起绒织物
napped jersey　起绒针织物
napper　拉毛机；拉绒机；起绒机
nappiness　起毛性；起绒性
nappy　起绒的；起毛的
naproxen　萘普生〔药〕
napthalyne　萘炔
napthylamine sulfonic acid　萘胺磺酸；氨基萘磺酸　$NH_2C_{10}H_6SO_3H$
naramycin　放线菌酮
narangomycin　纳朗果霉素
narbomycin　冥菌素；那波霉素
narceine　罂粟碱
narceine hydrochloride　盐酸罂粟碱
narceine meconate　袂康酸罂粟碱

narceine sodium salicylate　水杨酸罂粟碱钠
narceine sulfate　硫酸罂粟碱
narceine valerate　戊酸罂粟碱
narceol(=p-cresyl acetate)　乙酸对甲酚酯
narceonic acid　那碎酮酸
narciclasine　水仙环素
narcipoetine(=homolycorine)　高石蒜碱
narcissamine　水仙胺
narcissine　水仙碱
narcissus absolute　水仙花净油
narcissus alkaloid　水仙生物碱
narcissus compound　水仙香精
narcissus concrete　水仙花浸膏
narcissus oil　水仙花油
narcophine　袂康酸吗啡和袂康酸那可汀的复盐
narcosan　环己烯巴比妥
narcosis　麻醉
narcotic　①麻醉药②麻醉的
narcotic poison　致昏迷毒剂
narcotine　那可汀
narcotoline　那可托灵
narcyl　乙基那碎因
narcylene　麻醉用(的)乙炔
naregamine　果阿吐根碱
nargol　核酸银
naringenin　柚(苷)配基；4′,5,7-三羟黄烷酮
naringin(=naringoside)　柚(皮)苷
naringoside(=naringin)　柚(皮)苷
narki metal　耐酸铁硅合金
narobomycin　那波霉素
narrow-boiling range fraction　窄沸程馏分
narrow-boiling range product　窄沸程产品
narrow-bore column　窄孔(径)柱
narrow cut　①窄馏分②窄区(域)切割
narrow fabric without selvage　无织边的窄幅织物
narrow fabric with selvages　有织边的窄幅织物
narrow fraction　窄馏分〔在狭窄温度范围内馏出的馏分〕
narrow gap clearance concentric cylinder　窄隙式同心圆筒(黏度计)
narrowing　收针
narrow-meshed　窄筛分的
narrow molecular weight distribution　窄分子量分布
narrow-mouthed　细口的
narrow-mouthed bottle　细口瓶
narrow-mouthed bottle with ground stopper　细口磨塞瓶
narrow-mouth flask　细口烧瓶
narrow-necked bottle　细颈瓶
narrow particle size distribution　窄粒度分布

narrow ribbon　窄条；窄带
narrow slot　①窄缝②缝隙式流漆口
narrow tube　小直径管
nasal analysis　嗅觉分析；按气味分析
nasal rating　按气味评价
nascent atomic oxygen　新生原子氧
nascent fibre　初生纤维
nascent hydrogen　初生(态)氢
nascent oxygen　初生(态)氧
nascent polymer　初生聚合物
nascent red　初生红
nascent soap　初生皂
nascent soap method　初生皂法
nascent state(=status nascendi)　初生态；新生态
Nash hytor　纳希抽压机
Nash pump　纳希泵
nasopharyngeal applicator　鼻咽敷贴器
nasrol　咖啡磺酸钠
naszogen process　氯酸盐放热化合物；钝性物产氧法
natalite　乙醇乙醚混合物
nataloin　纳塔尔芦荟素
natamycin　纳他霉素；游霉素
nateglinide　那格列奈〔药〕
National Air Sampling Network　国家大气取样网
National Air Surveillance Network　国家大气监视网
National Bureau of Standards　(美国)国家标准局
national depository　国家(废物)储藏库
national standard　国家标准
native　天然的
native asphalt(=native bitumen)　天然沥青
native bitumen(=native asphalt)　天然沥青
native calomel　天然甘汞
native cellulose　天然纤维素
native coke　天然焦
native compound　天然化合物
native defect　本征缺陷*
native element　天然元素
native estate　天然农圃
native immunity　天赋免疫性
native lipid　天然类脂(物)
native metal　自然金属
native mineral wax　天然地蜡
native paraffin　天然石蜡
native pitch　天然沥青
native protein　天然蛋白质
native prussian blue　蓝铁矿
native rubber　天然橡胶
native soda　泡碱
native state of metals　金属的天然状态
native sulfate of barium　天然硫酸钡；重晶石　$BaSO_4$
natriated ion　钠加成离子
natrine　纳春〔生物碱〕
natrium　钠（11号元素，化学符号 Na）
natrium dimethyldithiocarbamate　二甲基二硫代氨基甲酸钠（促进剂）
natrium silicozirconate　硅锆酸钠
natriuretic hormone　钠尿激素
natrocalcite　钠方解石
natrolite　钠沸石
natron　泡碱〔1. 氧化钠；2. 含水苏打 $Na_2CO_3 \cdot 10H_2O$〕
natroncalk　碱石灰〔NaOH 和 CaO 的混合物〕
natronkalk　烧碱石灰
natron lake　泡碱湖
nat-rubber　天然橡胶
natrum　天然碱
Natta catalyst　纳塔催化剂
Natta distribution equation　纳塔分布方程
natural　天然的；自然的
natural abundance　天然丰度；自然丰度
natural abundance of isotopes　同位素的天然丰度
natural accelerator　天然促进剂
natural acid　天然酸
natural adhesive　天然胶黏剂
natural aging　自然老化
natural aging test　天然老化试验
natural air circulation　空气自然循环
natural antibody　自然抗体
natural appearance　天然外观
natural aquatic system　天然水体系
natural background　天然本底
natural base　生物碱
natural black oxide　天然氧化铁黑
natural bleaching　天然漂白
natural blue vitriol　天然蓝矾；天然胆矾；天然五水合硫酸铜
natural breadth　固有幅度；固有幅面
natural brine　天然盐水
natural calcium carbonate　重质碳酸钙
natural catalyst　天然催化剂
natural cellulose　天然纤维素
natural cement　天然水泥
natural changes　天然变化
natural circulation　自然循环
natural-circulation evaporator　自然循环蒸发器
natural circulation water cooling　自然循环的水冷却
natural clay　天然白土

natural coke 天然焦(炭)
natural collagen(ous) fibre 天然胶原纤维
natural colloid 天然胶体
natural color(=natural colour) 本色
natural colored 原色的；本色的；天然色的
natural coloring mattes 天然(着)色料；天然色素
natural colour(=natural color) 本色
natural coloured cotton 天然彩色棉
natural condition test 自然条件下试验
natural contamination 天然污染
natural convection 自然对流
natural cooling 自然冷却
natural copal resin 天然珀把树脂
natural corrosion test 自然(条件)腐蚀试验
natural count 自然计数
natural crack 自然开裂
natural curing 自然发酵；应时发酵〔烟草〕
natural damping 固有阻尼；自然阻尼
natural deterioration 天然劣化(变质)；天然老化
natural draft 自然通风
natural draft cooling tower 自然通风冷却塔
natural draft gas burner 自然通风气体燃烧器
natural draft ventilation 自然通风
natural draught 自然通风
natural draught cooling tower 自然通风凉水塔
natural draw ratio 自然拉伸比
natural durability test 自然耐久性试验
natural dye 天然染料
natural dyestuff 天然染料
natural earth pigment 天然色料
natural ebonite 天然硬质胶
natural evaporation 自然蒸发
natural exposure test 天然暴晒试验
natural fats and oils 天然油脂
natural fertilizer 天然肥料
natural fiber 天然纤维
natural fibre spinning machinery 天然纤维纺织机
natural flavour(ing) 天然食用香料；天然食用调料
natural flaw 自然缺陷
natural floral note 天然花香
natural flower oil 天然花油
natural frequency 固有频率
natural gas line 天然气管道
natural gasoline plant 天然汽油工厂
natural glue 天然胶；天然黏合剂；天然高分子胶黏剂
natural grade 本色级〔树脂品级〕
natural grain 天然粒面
natural graphite 天然石墨

natural gray yarn 天然本色毛纱
natural ground calcium carbonate 天然粉状碳酸钙
natural gum 天然树胶
natural gum resin 天然树胶基树脂
natural gum rubber 天然橡胶
natural half-width 自然半宽度
natural hectorite 天然锂蒙脱石；天然锂皂石
natural high molecular compound 天然高分子化合物
natural high polymer 天然高聚物
natural impurity 天然杂质
natural inhibitor(=natural oxidation inhibitor) 天然抗氧剂
natural iron oxide 天然氧化铁
natural iron oxide pigment 天然氧化铁颜料
natural isolate 天然单离香料
natural isoprene rubber 天然异戊橡胶
natural isotopic abundance 天然同位素丰度
natural latex 天然胶乳
natural latex-based adhesive 天然胶乳型黏结剂
natural leak 自然漏泄
natural leakage 自漏
natural length 自然长度〔羊毛〕
natural length (of wool) 自然长度〔羊毛〕
natural line width 自然线宽*
natural logarithm 自然对数
naturally hard water 天然硬水
naturally occurring emulsifying agent 天然乳化剂
natural mica 天然云母
natural moistening 自然湿润
natural moisturizing factor 天然保湿因子
natural orbital 自然轨道
natural oscillation 固有振荡
natural outdoor weathering test 天然户外耐候性试验
natural oxidation 自然氧化
natural oxidation inhibitor 天然抗氧剂
natural pebble 天然卵石；天然砾石
natural perfume 天然香料
natural plastics ①本色塑料②天然塑料
natural cis-polyisoprene 天然顺式-聚异戊二烯(橡胶)
natural polymer 天然高分子*
natural polymeric material 天然聚合物材料
natural polymeric surface active agent 天然高分子表面活性剂
natural pour point depressant 天然倾点降低剂
natural radioactivity 天然放射性
natural radioelement 天然放射性元素
natural regain 自然吸湿率；正常吸湿率
natural resin 天然树脂*

natural resin characteristic 天然树脂的特性
natural resin for lacquer 硝基漆用的天然树脂
natural resin varnish 天然树脂清漆
natural resonance 固有共振
natural retting 天然浸渍
natural rock 天然沥青
natural rubber 天然橡胶
natural rubber adhesive 天然胶黏合剂；天然胶浆
natural rubber hydrocarbon 天然橡胶烃
natural rubber latex 天然胶乳
natural rubber sheet 天然橡胶片
natural rubber vulcanizate 天然(胶)硫化胶
natural rust 自然锈蚀；自然锈
natural scale ①天然尺寸②固有(自然)量；固有容积③自然比例尺
natural science 自然科学
natural seasoning 天然干燥
natural self-purification 天然自净化作用
natural silk 天然丝
natural soil 天然污垢
natural solid bitumen 天然固体(石油)沥青
natural source 天然源
natural strain(=logarithmic strain；Hencky strain) 固有应变
natural surface active agent 天然表面活性剂
natural synthesis 自然合成
natural synthetic rubber 合成天然橡胶
natural time of viscoelasticity 黏弹性特征时间
natural-tinted fabric 天然本色织物
natural *trans*-polyisoprene 天然反式聚异戊二烯
natural uranium 天然铀
natural ventilation 自然通风
natural vermilion 朱砂〔俗〕；天然银朱；天然硫化汞
natural vibration 固有振动
natural wear 自然损耗
natural weathering 天然老化
natural whiting 天然白垩；天然碳酸钙
natural width 自然宽度
natural wood filler 木材本色填孔料
nature 无色
nature-identical 天然等同物
nature of defect ①缺陷的性质②漆病的性质
nature of friction 摩擦性质
nature of petroleum 石油性质；石油起源
nature rubber 天然橡胶
nature time(=relaxation time) 松弛时间；弛豫时间
nature weathering 天然气候老化
naumannite 硒银矿

Naumene number 诺门数
nauseant 呕吐剂；致呕剂
Nautamixer 诺塔混合器
nautical mile 海里；浬
naval architecture 造船工程
naval brass 船用黄铜
navel 阻捻盘；假捻盘
Navier-Stokes equation 纳维-司托克斯方程
navvy 掘凿器；挖掘器
navy blue 海军蓝
navy heat test 海军生热试验
navy oiler 海军油船
navy pitch 松香基松焦油沥青；松香改性松焦油
navy specification 海军技术规格
NBS color difference equation NBS 色差方程
NC mixture NC 混合剂；烟幕剂〔锌粉、六氯乙烷、高氯酸铵、氯化铵、碳酸钙的混合剂〕
n-dimensional space n 维空间
neamine 新霉胺
near beer 淡啤酒
near-direct proportion 接近直接比例
nearest neighbo(u)r energy 最近邻能
nearest neighbo(u)r frequency 最近邻频率
nearest neighbo(u)r sequence analysis 最近邻顺序分析
near-field optical sensor 近场光学传感器
near-field scanning optical microscope 近场扫描光学显微镜
near-field spectroscopy 近场光谱学
near infrared 近红外(的)
near infrared detector 近红外检测器
near infrared diffusion reflectance spectroscopy 近红外漫反射光谱法
near infrared Fourier transform surface-enhanced Raman spectroscopy 近红外傅里叶变换表面增强拉曼光谱
near infrared ray 近红外线
near infrared region 近红外区
near infrared source 近红外光源
near-infrared spectroscopy 近红外光谱法
near infrared spectrum 近红外线光谱
near IR interferometer 近红外干涉仪
nearly viscometric flow 近测黏流动
nearness(=neartude) 贴近度
near-neutral chroma (接)近中性色(品)
near-solvent action 类溶剂作用
neartude(=nearness) 贴近度
near-ultraviolet light 近紫外光
near-ultraviolet radiation 近紫外辐射
near-ultraviolet ray 近紫外线

near-ultraviolet region　近紫外区
near-ultraviolet spectrum　近紫外光谱
near-zero contact angle　(接)近零(的)接触角
near zero release　近零排放；几乎不排放(废物)
neatening　修补；修整
neatness standard photograph　净度标准样照
neat phase　净相
neat's foot oil　牛蹄油
neat soap　牛油皂；皂粒；皂核；净皂；液晶皂；皂基
neat soap pellet　粒状纯皂
neat soap phase　液晶皂相
neat solvent　纯溶剂；净溶剂
neat viscosity　(混合溶剂中某个组分的)自身黏度
nebenagglutinin　副凝集素
Neber reaction　内布反应；内布重排
nebramycin　硫酸安普霉素；硫酸阿布拉霉素
nebularine　水粉蕈素
nebulite　云状球雏晶体
nebulium　星云素
nebulization　雾化
nebulization chamber　雾化室
nebulization efficiency　雾化效率*
nebulizer　雾化器*
neburon　防莠剂
necessary element　必需元素
neck　颈缩；细颈
neck bush　底箱
neck-down　颈缩
necked-down section　颈缩截面
necked electrode　颈部充填电极
necked-in　向内弯曲(边缘)
necked-out　向外弯曲(边缘)
neck-in　(边缘)向内弯曲
neck in cold drawing　冷拉颈
necking　颈缩现象*
neck migration　颈部迁移
neck-out　(边缘)向外弯曲
necrocitin　坏死菌素
necrocryptoxanthol　黄玉米原维生素 A
nectar　花蜜
needle　①针；针状体②针状结晶
needle antimony　粗锑
needle crystal　针状结晶
needled mat　针刺毡(片)
needle electrometer　指针电位计
needle form of wax crystal　针状蜡结晶
needle guide　针导
needle hub　针头接口

needlelike　针状的
needle-like pigment　针状颜料
needle loom　针刺机
needle lubricator　针形润滑器
needle mesh(=needle number)　针孔筛目
needle nozzle　针形喷嘴
needle number(=needle mesh)　针孔筛目
needle oil　针润滑油〔纺织机器〕
needle ore　针硫铋铅矿
needle paper　包针纸
needle penetration　针头刺入度；针入度
needle penetration of petroleum bitumens　石油沥青针入度
needle penetration test　针入法硬度试验
needle-punching　针刺法
needle shaft　针体；针轴
needle spar　霰石
needle stone　纳沸石
needle stretch(ing) machine　拉针机
needle valve　针(孔)阀
needle valve flow controller　针形阀流量控制器
needle zeolite　钠沸石
needle zinc white　针状锌白
needling　针刺
needling density　针刺密度
needling machine　针刺机
Nef and Spengler method　内夫和斯彭格勒法
nefluorophotometer　比浊荧光光度计
negamycin　负霉素
negapillin　负青霉素
negative adsorption　负吸附*
negative azeotrope　下降性共沸混合物
negative catalysis　负催化(作用)
negative catalyst　负催化剂
negative charge　负电荷
negative colloid　阴性胶体
negative component　阴(性)组分
negative concentration quenching　负浓度猝灭
negative constituent　阴(电)性构分
negative contact agent　负催化剂
negative corpuscle　负电(性)微粒
negative correlation　负相关*
negative cotton　低度硝化(的)纤维素
negative Cotton effect　负考顿效应
negative crystal　负晶体〔物〕；空晶〔矿〕
negative deviation　负偏差
negative die　阴模；下半模
negative difference effect　负差效应

negative differential　负差率
negative doctor test　负含硫试验
negative electric charge　负电荷
negative electricity　阴电；负电
negative electrode　阴极；负极
negative electron　负电子
negative electron affinity photocathode　负电子亲和势光电阴极
negative element　阴性元素
negative feedback　负反馈
negative filter type　负滤光片型
negative group(=negative grouping)　①阴根②负(性)基
negative grouping(=negative group)　①阴根②负(性)基
negative interfacial tension　负界面张力
negative ion　阴离子；负(电性)离子
negative ion chemical ionization　负离子化学电离
negative ion detection　负离子检测；阴离子检测
negative ion field ionization　负离子场电离
negative ion formation　负离子形成
negative ion mass spectrum　负离子质谱
negative let-off　消极式送经
negatively charged　带阴电(荷)的；阴(电)(性)的
negatively charged ion　阴离子；负(电性)离子
negatively charged liquid film (membrane) electrode　阴性液膜电极
negatively charged (material) particle　(带)阴电(荷)微粒；阴(电)(性)微粒
negatively skewed　负偏的；向右倾斜的
negative maximum　负极大
negative meniscus　负液面；(液面)凹形弯曲面
negative normal stress effect　负正应力效应
negative number　负数
negative peak　反峰；负峰；倒峰
negative pion radiotherapy　负π介子放射治疗
negative plate　①阴极板②底片
negative polarity　负极性；负电性
negative pole　阴极；负(电)极
negative positive rule　正负定律〔即中和定律〕
negative potential　阴电势；电位；负电势
negative pressure　负压(力)；真空
negative proton　负质子；阴质子
negative radical　①(电)(性)根②负(性)基
negative ray　阴(极)射线；负电射线
negative reaction　阴性反应
negative replica　复制阴模
negative rotation　负转动
negative skew distribution　负歪斜(偏斜)分布〔粒径分布方式之一〕

negative sol　阴电(荷)溶胶；阴电(性)溶胶；负电(性)溶胶
negative solvatochromism　负的溶剂色变；负的溶剂化显色(现象)
negative spot test bitumen　无残迹沥青
negative staining　负染色
negative stencil　负性感光膜
negative stretch　负拉伸
negative structural viscosity　负结构黏度
negative surface tension　负表面张力
negative synergism　反协同效应
negative temperature coefficient　负温度系数
negative temperature gradient　负温度梯度
negative test　①负结果②反试验
negative thixotropy　负触变性
negative thread　阴螺纹
negative valency　负价
negative work　负功
negativity　(电)负性
negaton　负电子
negatron radiation　负电子辐射
negentropy　负熵
negion　阳离子
negligible over(-)spray loss　过量涂漆(喷涂)损失极少
negligible porosity　微乎其微的孔隙度
negligible residue　可忽视残留
negode　阳极
Negro powder　尼格罗炸药
neighboring cation　相邻阳离子
neighboring electron participation　邻电子参与效应
neighboring line　邻近(谱)线
neighboring paramagnetic shielding　邻区顺磁屏蔽
neighboring valency　邻价
neiopax　列埃欧钯〔阻X射线剂〕
nekal　二丁基萘磺酸钠
N electron　N电子
nelfinavir　奈非那韦〔药〕
Nellen tubing machine　尼林挤出型可塑计
Nellen tubing mechanical plastimeter　尼林管状机械可塑计
Nelson cell　纳尔逊电池
Nelson process　纳尔逊连续纺丝法
Nelson's method　砷钼酸法
Nelson-Somogyi's method　砷钼酸法
nelumbine　莲碱
nemacides　杀线虫剂
nemalite(=nemaline)　纤维水镁石；纤水滑石
nematic compound　向列化合物
nematic copolymer　向列共聚物

nematic crystal structure 向列型晶体结构
nematic liquid crystal 向列型液晶
nematic phase 向列相
nematic state 向列(状)态
nematocides 杀线虫剂
nembutal(=pentobarbital) 戊巴比妥
nemotin 草居蕈素
nemotinic acid 草居蕈酸
Nencki reaction 楞琪反应
neo-〔希腊字头〕 新
neoantergan maleate 顺丁烯二酸吡明安〔药〕
neoantimycin 新抗霉素
neoarsaminol 硫脲嘧啶；新606；新阿斯凡纳明；新苯拉胂；新苯纳胂
neoarsphenamine 新胂凡钠明 $C_{13}H_{13}O_4N_2SAs_2Na$
neoaspergillic acid 新曲霉酸
neoaureothin 新金丝菌素
neobaicalein 新黄芩素
neobornyral(=bornyl isovalerylglycolate) 异戊酸基乙酸冰片酯
neobotogenin 薯蓣配基；Δ^5-22b-螺甾烯-3β-醇-12-酮〔取自 dioscorea mexicana〕
neobrucidine 异不鲁西定
neobrucine 异番木鳖碱；二甲马钱子碱
neocaine 纽卡因；盐酸普鲁卡因
neocalamine 氧化锌、红氧化铁、黄氧化铁混合物
neocarborane 新碳硼烷
neocarcinostatin 新制癌菌素
neocardin 新卡氏菌素
neocarzinostatin(=neocarcinostatin) 新制癌菌素
neocerotic acid 新蜡酸；二十五(烷)酸
neochebulinic acid 新诃黎勒鞣花酸
neocianite 蓝硅酸铜矿石
neocidin 新杀菌素
neocinchophen 新辛可芬〔药〕
neocupferron 新铜铁灵；新铜铁试剂；N-亚硝基α-萘胺 $C_{10}H_7N(NO)ONH_4$
neocupferronate 新铜铁试剂盐
neocuproine(=2,9-dimethyl-1,10-phenanthroline) 新亚铜试剂；2,9-二甲基-1,10-菲咯啉
neocyanine 新喹啉蓝
neodecanoate 新癸酸
neodecanoic acid 新癸酸
neodiarsenol 新洒尔佛散
neodrenal 异丙基肾上腺素〔药〕
neodymia 氧化钕 Nd_2O_3
neodymium 钕〔60号元素，化学符号Nd〕
neodymium acetate 乙酸钕 $Nd(C_2H_3O_2)_3$
neodymium acetyl acetonate 乙酰乙酸钕
neodymium bromate 溴酸钕 $Nd(BrO_3)_3$
neodymium bromide 溴化钕 $NdBr_2$
neodymium carbonate 碳酸钕 $Nd_2(CO_3)_3$
neodymium chloride 氯化钕 $NdCl_3$
neodymium dioxide 二氧化钕 NdO_2
neodymium fluoride 氟化钕 NdF_3
neodymium glass 钕玻璃
neodymium glass laser 钕玻璃激光器
neodymium hydride 氢化钕 NdH_3
neodymium hydrosulfate 硫酸氢钕 $Nd(HSO_4)_3$
neodymium hydroxide 氢氧化钕 $Nd(OH)_3$
neodymium iodide 碘化钕 NdI_3
neodymium molybdate 钼酸钕 $Nd_2(MoO_4)_3$
neodymium nitrate 硝酸钕 $Nd(NO_3)_3$
neodymium oxalate 草酸钕 $Nd_2(C_2O_4)_3$
neodymium oxide 氧化钕 Nd_2O_3
neodymium oxychloride 氯氧化钕 $NdOCl$
neodymium sesquioxide 三氧化二钕 Nd_2O_3
neodymium sulfide 硫化钕 Nd_2S_3
neoergosterol 新麦角甾醇
neofumidin(=neohumidin) 新潮菌素
neogen 银色合金
neogermidine 新杰米定
neohalarsine 马法胂
neoheptane 新七烯
neoherculine 新核枯灵
neohesperidin 新橙皮苷
neohexane 新己烷 $(CH_3)_3CC_2H_5$
neohexane fuel 新己烷汽油
neohexanized 含新己烷的
neohumidin(=neofumidin) 新潮菌素
neohydriol 碘化油注射剂
neohydroxyaspergillic acid 新羟基曲霉酸
neo-α-irone 新α-鸢尾酮
neoisomenthol 新异薄荷醇
neokharsivan 新洒尔佛散；新阿斯凡纳明
neolactose 异乳糖
neolan blue 宜和兰蓝
neolignan 新木脂素
neolites〔复〕 新石
neomenthol 新薄荷醇；新薄荷醇 $C_{10}H_{20}O$
neomethymycin 新酒霉素
neomycin 新霉素〔药〕
neon 氖〔10号元素，化学符号Ne〕
Neonal (powder) 内昂纳耳炸药
neonicotine(=anabasine) 新烟碱；假木贼碱
neon lamp(=neon light) 氖灯；霓虹灯

neonocardin 新诺卡菌素
neon tube 氖管；霓虹管
neoobutin 新奥枯草菌素
neooxine 新羟基喹啉
neopelline 尼喔配林
neopentane 新戊烷；季戊烷 $(CH_3)_4C$
neopentanoate 新戊酸盐(酯)
neopentanoic acid(=pivalic acid) 新戊酸；三甲基乙酸
neo-pentyl 新戊基 $(CH_3)_3CCH_2-$
neopentyl alcohol 新戊醇 $(CH_3)_3CCH_2OH$
neopentylglycol 新戊二醇
neopentylglycol diacrylate 新戊二醇二丙烯酸酯
neopentylglycol diglycidyl ether 新戊二醇二缩水甘油醚
neopentylglycol dimethacrylate 新戊二醇二甲基丙烯酸酯
neopentyl glycol monosuccinate 新戊二醇一琥珀酸酯
neopentyl glycol succinate(NGS) 丁二酸新戊二醇酯
neopentyl halide 新戊基卤 $(CH_3)_3CCH_2X$
neopentyl iodide 新戊基碘 $(CH_3)_3CCH_2I$
neopentyl rearrangement 新戊基重排作用
neophyl 新苯基；2-甲基-2-苯丙基
neophyl chloride 1-氯-2-甲基-2-苯丙烷；新苯基氯
neophytadiene 新植二烯
neopine 尼奥品；β可待因
neopiperitol 新胡椒醇
neoplatyphylline 新阔叶(千里光)碱
neopluramycin 新多色霉素
neopolarograph 新极谱仪
neopolarography 新极谱法
neoprene cement 氯丁胶浆
neoprene-coated sling 包氯丁橡胶的吊索(链钩)
neoprene-coated strap 包氯丁橡胶的曳索
neoprene coating 氯丁橡胶涂料
neoprene flexible tank 氯丁橡胶制韧性容器
neoprene fuel cell 氯丁橡胶燃料容器
neoprene glove 氯丁橡胶手套
neoprene paint 氯丁(二烯)橡胶涂料
neoprene phenolic adhesive 氯丁橡胶-苯酚类黏合剂
neoprene stopper 氯丁橡胶塞
neoprene tubing 氯丁橡胶管
neoprontosil(=azosulfamide; prontosil S; streptozon S) 偶氮磺酰胺；新白浪多息〔药〕
neoprotoveratrine 新原藜芦碱
neosalvarsan 新洒尔佛散；914
neosine 肌碱
neostigmine 新斯的明〔药〕
neostigmine bromide 溴化新斯的明〔药〕
neostrychnine 异马钱子碱
neosynephrin 新生乃复林；苄基麻黄碱；新福林

neotelomycin 新远霉素
neothelomycin 新乳霉素
neothricin 新丝菌素
neothujane 新苧烷；新侧柏烷
neotigogenin 新替告皂苷元
neotridecanohydroximic acid 新十三碳异羟肟酸
neotype 碳酸钙钡矿
neou oil 杷茬油〔取自杷茬 Parinarium macrophyllum 种子，产于西非〕
neovaleraldehyde 新戊醛；三甲基乙醛 $(CH_3)_3CCHO$
neovaricaine 乙醇胺
neoytterbium(=ytterbium) 镱
neozone 苯基苯基胺衍生物
nep 棉结
neper 奈培
nepetalactone 荆芥内酯
nepetalic acid 荆芥酸
nepetalic anhydride 荆芥酸酐
nepeta oil 荆芥油
nepetol 荆芥醇
nephelauxetic effect 电子云重排效应
nephelauxetic parameter 电子云重排参数
nephelauxetic ratio 电子云重排效应比
nephelauxetic series 电子云重排系列
nepheline 霞石
nepheline syenine 霞石正长岩
nephelometer 浊度计；比浊计
nephelometric analysis 浊度分析
nephelometric method 浊度测定法
nephelometric titration 浊度滴定法
nephelometric turbidimeter 散射浊度计
nephelometry 浊度法*
nepho-colorimeter 比浊比色计
nephrite 软玉
nepodin 尼泊定 $C_{18}H_{16}O_4$
neppy cotton 多粒结棉
Neptex process 镎普特克斯过程〔从辐照铀燃料中用磷酸三丁酯萃取法回收纯化镎-237 的过程〕
neptunate 镎酸盐
neptunia 二氧化镎
neptunic acid 镎酸 H_5NpO_6; H_6NpO_6
neptunium 镎
neptunium acetate 乙酸镎
neptunium alkoxide 烷基醇镎 $Np(OR)_4$
neptunium amalgam 镎汞齐
neptunium carbide 碳化镎 NpC; Np_2C_3; NpC_2
neptunium carbonate 碳酸镎
neptunium chloride 氯化镎 $NpCl_3$; $NpCl_4$

neptunium decay series 镎放射系
neptunium diarsenide 二砷化镎 $NpAs_2$
neptunium dioxide 二氧化镎 NpO_2
neptunium ethoxide 乙氧基合镎 $Np(OC_2H_5)_n$
neptunium fluoride 氟化镎 NpF_3; NpF_4; NpF_6
neptunium hexafluoride 六氟化镎 NpF_6
neptunium hydride 氢化镎 NpH_2; NpH_3
neptunium hydroxide 氢氧化镎
neptunium nitrate 硝酸镎 $Np(NO_3)_4$; $NpO(NO_3)_3$; $NpO_2(NO_3)_2$
neptunium oxalate 草酸镎 $Np(C_2O_4)_2$
neptunium oxide 氧化镎 NpO_2; Np_2O_5; Np_3O_8; NpO_3
neptunium oxide-sulfide 氧硫化镎 $NpOS$; Np_2O_2S
neptunium oxyfluoride 氟氧化镎；氟化镎酰 NpO_2F_2
neptunium peroxide 过氧化镎
neptunium phosphate 磷酸镎
neptunium phosphide 磷化镎
neptunium protoxide-oxide 五氧化二镎 Np_2O_5
neptunium pyrophosphate 焦磷酸镎 NpP_2O_7
neptunium selenide 硒化镎
neptunium series 镎系(列)
neptunium silicide 硅化镎 $NpSi_2$
neptunium sulfide 硫化镎 NpS; Np_2S_3; Np_3S_5; Np_2S_5; NpS_3
neptunium tetraethoxide 四乙醇镎 $Np(OC_2H_5)_4$
neptunium tetramethoxide 四甲醇镎 $Np(OCH_3)_4$
neptunoyl (五价)镎酰 NpO_2^+
neptunyl 镎氧离子 NpO_2^{2-}；镎酰
neptunyl acetate 乙酸镎酰；乙酸双氧镎 $NpO_2(CH_3COO)_2$
neptunyl fluoride 氟化镎酰；氟化双氧镎 NpO_2F_2
neptunyl nitrate 硝酸镎酰；硝酸双氧镎 $NpO_2(NO_3)_2$
neptunyl potassium carbonate 碳酸镎酰钾 $KNpO_2CO_3$
neptunyl potassium fluoride 氟化镎酰钾 $KNpO_2F_2$; $K_3NpO_2F_5$
neral 橙花醛；β-柠檬醛 $C_{10}H_{16}O$
neriantin 夹竹桃叶苷
neriin 夹竹桃叶苷
neriodorin 夹竹桃皮苷
nerium D 夹竹桃 D
nerium E 夹竹桃 E
nerium F 夹竹桃 F
Nernst diffusion layer 能斯特扩散层
Nernst effect 能斯特效应
Nernst-Einstein relation 能斯特-爱因斯坦关系式〔扩散定律〕
Nernst equation 能斯特方程*
Nernst factor 能斯特因子

Nernst fuel cell 能斯特燃料电池
Nernst glower 能斯特灯
Nernst-Hartley relation 能斯特-哈特利关系式
Nernst heat theorem 能斯特热定理
Nernstian limit 能斯特(响应)下限
Nernstian potential 能斯特电位
Nernstian slope 能斯特斜率
Nernst lamp 能斯特灯
Nernst partition constant 能斯特分配常数
Nernst-Planck flux equation 能斯特-普朗克通量方程式
Nernst's distribution law 能斯特分配定律
Nernst's torsion balance 能斯特扭力天平
Nernst unit 能斯特流量单位
nerol 橙花醇；3,7-二甲基-2,6-辛二烯-1-醇 $(CH_3)_2C=CHCHCH_2C(CH_3)CH=CH_2OH$
neroli bigarade oil 苦橙油
nerolic ketone 橙花酮
nerolidol 橙花叔醇 $C_{15}H_{26}O$
nerolidyl 橙花叔基
nerolidyl acetate 乙酸橙花叔酯
nerolidyl pyrophosphate 橙花叔醇焦磷酸
neroli oil 苦橙花油
nerol oxide 橙花醚
nerone 橙花酮
nerve ①回缩性；(弹性)复原性②抗变形性
nerve gas 神经错乱性毒气
nerve sedative 神经镇静药
nerve stimulant 神经刺激药
nerviness 回缩性；弹性复原
nervonic acid 神经酸；二十四(碳)烯酸
neryl 橙花基 $C_{10}H_{17}—$
neryl acetate 乙酸橙花酯
neryl butyrate 丁酸橙花酯
neryl formate 甲酸橙花酯
neryl isobutyrate 异丁酸橙花酯
neryl isovalerate 异戊酸橙花酯
neryl phenylacetate 苯乙酸橙花酯
neryl propionate 丙酸橙花酯
nescent silicic acid 初生硅酸
nesosilicates 岛硅酸盐〔各 SiO_4 间无共用氧原子〕
nesquehonite 水碳镁石
nesslerization 奈斯勒比色法；等浓比色法
Nessler jar(=Nessler tube) 奈斯勒比色管
Nessler's colorimetric method 奈氏比色法
Nessler's reagent 奈斯勒试剂
Nessler's solution 奈斯勒溶液
Nessler's test 奈斯勒试验
Nessler tube(=Nessler jar) 奈斯勒比色管

nested mold　多巢(压)模
nesting　嵌置
net absorption　净吸收
net atomic population　净原子布居*
net caloric power　净热值
net caloric value(=net caloric power)　净热值
net calorific power(=net caloric power)　净热值
net calorific value(=net caloric power)　净热值
net charge　净电荷
net cooling　实际冷却；净冷却
net density　净密度
net dipole moment　净偶极矩
net effect　最后效果；综合效果
net efficiency　净效率〔实际的利用效率系数〕
net energy gain　净能量增益
net Faradaic current　净法拉第电流
net gain　净得；净增益
net gas　净气；干气
net gasoline　净汽油；无铅汽油
net heating power　净热值
net heating value(=low heating value; net thermal value)　净热值；低热值
net heating value of gas　气体的净热值
net heat of combustion(=net heating value)　净热值；低热值
nether　天然碱
net (high) polymer　网状高分子；网状高聚物
Nethix process　镎西克斯过程〔离子交换-乙醚溶剂萃取纯化镎的过程，用过氧化氢稳定五价镎，TBP 萃取〕
netilmicin　奈替米星〔药〕
net interaction　净相互作用
net-like structure　网状结构
net load　净载荷
net loss　净损；净亏
net loss of heat　热的净损失
net measure　净测量值
net mobility　净淌度
net negative charge　净负电荷
netoric acid　鱼藤酸
net plane space　网面间隔
net positive suction head(NPSH)　净正吸入压头
net pressure head　可利用压头
net price　净价；实价
net quantity of heat　全部热量
net rate　净速率
net repulsion　净斥力
net repulsive force　净(排)斥力
net residual oil　净残石油

net retention time　净保留时间*
net retention volume　净保留体积*
net-shaped structure　网状结构
net signal　净信号
net solubilization power　净增溶力
net structure　网状结构
nett calender　平板纸压光机〔纸〕
netted　网状的
net temperature drop　净温度降
net thermal value(=net heating value)　净热值；低热值
netting　①网；(金属)丝布②结网
netting analysis　网络分析
netting index(=netting number)　网数
netting number(=netting index)　网数
nettle fiber　荨麻纤维
net ton　净吨〔＝2000 磅〕
net weight　净重
network　网络
network architecture　网络体系结构
network bonding　网状结合；网键合
network chain　网链
network density　网络密度*
network formation　网形成
network function　网络函数
network model　网络模型
network of pipe lines　管道网
network polymer　网状聚合物
networks of atoms　原子网络
network structure　架型结构*
network theory　网络理论
net yield　实际得率；净得率
Neubauer crucible　诺勃氏坩埚
Neuberg ester　诺勃氏酯
Neuberg's scheme　诺勃氏发酵途径
neuchebulagic acid　新诃黎勒鞣花酸
Neuflex process　纽弗莱克斯过程〔冷锆基燃料元件溶解过程，用来制备后处理用料液〕
neumandin　异烟肼
neuralgic point　中枢(控制)点；中枢监测点
neural interface　中性界面
neuraminidase　神经氨(糖)酸苷酶；唾液酸苷酶
neuronal　二乙基溴乙酰胺
neurosin　甘油磷酸钙
neurosporene　链孢类胡卜素；四氢番茄红素；链孢霉素
neurotransmitter　神经(传)递质
neusilber　铜、镍锌合金
neut number　中和值
neuton　零元素〔原子序为 0，原子量为 1〕

neutral ①中和的②中性的
neutral addition compound 中性加成物
neutral analysis 中和分析法
neutral atom 中性原子
neutral atom gun 中性原子枪
neutral axis 中性轴
neutral background 中性色背景
neutral blue 中性蓝
neutral carrier 中性载体
neutral carrier electrode 中性载体电极
neutral carrier membrane 中性载体膜
neutral carrier membrane electrode 中性载体膜电极
neutral catalyst 中性催化剂
neutral clay 中性白土
neutral cleaner 中性清洗剂
neutral colloid 中性胶体
neutral colorant 中性染料；中性颜料
neutral complex 电中性配合物
neutral compound 中性化合物
neutral conjugation 中性共轭(结合)
neutral density filter 中性(密度)滤光片
neutral electrode 中性(电)级
neutral element 中性元素；零族元素
neutral equilibrium 随遇平衡
neutral ester 中性酯
neutral fat 中性脂肪
neutral filter 中性滤光片*
neutral flame 中性焰
neutral flux 中性助熔剂
neutral fragment 中性碎片
neutral fragment loss scanning 中性碎片丢失扫描
neutral grease 天然脂
neutral ionophore 中性离子载体
neutrality ①中性②中和
neutralization 中和*
neutralization agent 中和剂；碱剂
neutralization curve 中和曲线
neutralization equivalent 中和当量
neutralization heat 中和热
neutralization indicator 中和指示剂；酸碱指示剂
neutralization number 中和值；酸值
neutralization of acid with base 酸碱中和作用
neutralization ratio 中和比率；中和比例
neutralization self-solidification process(NSSP) 中和自固化法
neutralization test 中和值测定
neutralization titration(=acid-base titration) 中和滴定；酸碱滴定

neutralization value 中和值
neutralized product 中和产品
neutralized system 中性系(统)；中和系(统)
neutralizer ①中和器②中和剂
neutralizing 中和
neutralizing agent 中和剂
neutralizing chemicals 中和化学药品〔为精制石油产品使用〕
neutralizing tank 中和槽
neutralizing titration 中和滴定
neutralizing treatment 中和处理；碱洗涤
neutralizing well 中和槽
neutralizing with clay 白土中和
neutral leach 中性浸出
neutral ligand 中性配位体
neutral line 中线；中性线；中和线；中立线
neutral lining 中性炉衬
neutral loss 中性产物丢失
neutral loss scan 中性(丢)失扫描
neutral micelle 中性胶束
neutral molecule 中性分子
neutral number 中和值
neutral oil 中性油
neutral oxidation 中性氧化作用〔在中性介质中氧化〕
neutral phosphate extraction 中性磷酸酯萃取
neutral pigment 中性颜料
neutral point 中性点*
neutral polyol 中性多羟基化合物
neutral premetallized dye 中性金属络合染料
neutral pressure process 中和压力法〔回收酒石〕
neutral principle 中性素
neutral product stability 中性产物稳定性
neutral reaction 中性反应
neutral red 中性红*
neutral reduction 中性还原
neutral resin 中性树脂；碱不溶树脂
neutrals 中性物
neutral salt 中性盐；中式盐
neutral salt effect 中性盐效应
neutral slag 中性炉渣
neutral soap 中性皂
neutral soda 中性苏打
neutral softening agent 中性柔软剂
neutral solution 中性溶液
neutral solvent 中性溶剂
neutral solvent mixture 中性混合溶剂
neutral stability 中(性)稳定性；随遇稳定(性)
neutral stability curve 中稳态曲线

neutral step filter 中性阶梯减光板
neutral to litmus 石蕊(测试)中性
neutral violet 中性紫
neutral water glass 中性水玻璃
neutral wedge 消光楔；中性滤光片
neutral wedge photometer 消光楔光度计
neutramycin 中性霉素
neutrapen 青霉素酶
neutrel 流秒〔Sayholt 黏度计在 100°F 下流出所需秒数〕
neutretto 中(性)子
neutrino 中微子；微中子
neutrography(=neutron radiography) 中子射线照相法
neutron 中子
neutron absorber 中子吸收器
neutron-absorbing material 中子吸收剂
neutron absorption 中子吸收作用
neutron absorption analysis 中子吸收分析
neutron activation 中子活化(法)
neutron activation analysis(NAA) 中子活化分析*
neutron-activation kit 中子活化箱〔测中子场特性用〕
neutron beam 中子束
neutron bomb 中子弹
neutron bombardment 中子轰击
neutron capture 中子捕获；中子俘获
neutron capture gamma-ray analysis(NCGA) 中子俘获γ射线分析
neutron capture therapy 中子俘获治疗(法)
neutron counter 中子计数器
neutron crystallography 中子结晶学
neutron curing 中子固化
neutron curtain 中子(防护)幕
neutron-deficient 缺中子的
neutron-deficient nuclide 缺中子核素
neutron diffraction 中子衍射
neutron diffraction analysis 中子衍射分析
neutron flashtube 中子闪光管
neutron flux 中子通量
neutron generator 中子发生器
neutron generator tube 中子发生管
neutron gun 中子枪
neutronics 中子学
neutron-irradiated bromobenzene 中子激化了的溴苯
neutron irradiation 中子照射；中子辐照
neutron moisture gauge 中子湿度计
neutron moisture meter 中子水分计
neutron multiplier assembly 中子倍增器
neutron number 中子数
neutronography 中子照相法

neutron-poor isotope 缺中子同位素
neutron radiography 中子射线摄像法
neutron resonance spectroscopy 中子共振谱学
neutron-rich nuclei〔复〕 富中子核；丰中子核
neutron-rich nuclide 丰中子核素*
neutron-rich side 富中子侧；丰中子侧
neutron scattering 中子散射
neutron shield paint 中子屏蔽(防护)用涂料
neutrons in nuclei 核中子
neutron source 中子源*
neutron spectrometer 中子谱仪
neutrons per second(NPS) 每秒钟(放出)中子数
neutron topography 中子形貌学
neutrophil 中性粒细胞
Neuwieder green 纽维德尔绿
Neville acid 1-萘酚-4-磺酸
Neville and Winther acid 1-萘酚-4-磺酸；奈温酸 $OHC_{10}H_6SO_3H$
nevirapine 奈韦拉平（药）
nevyanskite 天然铱锇合金
Newage vibrating screen 内韦季摆(动)筛
new atom chemistry 新奇原子化学；奇异原子化学
new blue(=naphthol blue) 萘酚蓝；亚甲基型染料
new cacodyl(=arrhenal) 甲基胂酸二钠
new ceramics 新型陶瓷
new feel fibre 新触觉纤维；新感觉纤维
new frontier fibre 新型尖端纤维；新领域纤维
new fuchsin 可溶性品红
newing 酒花酵母
new lime liquor 新灰水
new magenta 新品红
Newman's rule of six 纽曼氏六位规则
Newman projection 纽曼投影式*
new modified simplex 新改进单纯形
new mown hay 新刈草(香)
newness retention 保持永新
new oil 新油；新润滑油
new polymer 新型聚合物
new production reactor(NPR) 新生产堆
new rubber 新胶
news(=newsprint) 新闻纸
newsboard 新闻纸板
news grade groundwood 新闻纸用磨木浆
news grade sulphite pulp 新闻纸用亚硫酸盐木浆
newsprint(=news; newsprinting paper; newsprint paper) 新闻纸
newsprinting ink 油墨
newsprinting paper(=newsprint) 新闻纸

newsprint paper(=newsprint)　新闻纸
new synthetic fibre　新合纤
newton　牛顿〔力的单位 MKS 制〕
Newton's viscosity law　牛顿黏度定律
Newtonian behaviour　牛顿特性
Newtonian body　牛顿稠度
Newtonian flow　牛顿流动*
Newtonian flow equation for tackmeter　黏性计的牛顿流动方程
Newtonian flow model　牛顿流动模型
Newtonian fluid　牛顿流体*
Newtonianism　牛顿性
Newtonian limiting viscosity　牛顿极限黏度
Newtonian liquid　牛顿液体
Newtonian lubricating layer　牛顿润滑层
Newtonian material　牛顿材料
Newtonian melt viscosity　牛顿熔体黏度
Newtonian region　牛顿区
Newtonian shear rate　牛顿剪切速率
Newtonian shear viscosity　牛顿剪切黏度*
Newtonian vehicle　牛顿型漆料
Newtonian viscosity　牛顿黏度*
newtonian viscosity behavior　牛顿黏度特性
newtonite　斜方岭石
newtonium　原元素
Newton refraction　牛顿折射
Newton's alloy　牛顿合金
Newton's concept　牛顿概念
Newton's equation　牛顿方程
Newton's hypothesis　牛顿假说
Newton's law　牛顿定律
Newton's law for resistance　牛顿阻力定律
Newton's law of cooling　牛顿冷却定律
Newton's law of viscosity　牛顿黏性定律
Newton's rings　牛顿环
Newton's viscosity law　牛顿黏度定律
next process unit　下一道工序设备
Neyman-Pearson criterion　奈曼-皮尔逊准则
nezukone　尼楚酮；4-异丙基䓬酮
n-fold axis　n 重轴
ng.(=nanogram)　纳克（$=10^{-9}$ 克）
Ngai camphor　恩盖樟脑；冰片　$C_{10}H_{18}O$
ngaione　恩盖酮；艾纳酮
ngaio oil　乳苦楝蓝油
NH_3 gas-sensing electrode　氨气敏（感）电极
niacin　烟酸；尼克酸；抗糙皮病维生素；维生素 P
niacinamide　烟酰胺；尼克酰胺；抗糙皮病维生素
nialamide　丙酰苄胺异烟肼；尼阿拉米；2-苄基氨甲酰

乙基异烟肼〔抗抑郁药〕
niam fat　非洲栎树油〔取自 *Lophira alata* 的种子〕
niaouli oil　绿花白千层油；袅莉油
nib　碎粒；突出部分
nibyl　铌氧基
nicamide　N,N-二乙基烟酰胺；尼可刹米
nicardipine　尼卡地平〔药〕
niccolite(=arsenical nickel)　红砷镍矿
nicergoline　尼麦角林〔药〕
nicholsonite　含锌霰石
nichrome　镍铬合金
nichrome bead　镍铬合金珠
nichrome coil　镍铬线圈
nichrome source　镍铬丝光源
nichrome triangle　镍铬(合金)三角
nichrome wire　镍铬丝；电热丝
nick　裂纹；裂口；裂缝；隙
nickel　镍〔28 号元素，化学符号 Ni〕
nickel acetate　乙酸镍〔$1.Ni(C_2H_3O_2)_2$; $2.Ni(C_2H_3O_2)_3$〕
nickel alloy　镍合金
nickel alloy steel　镍合金钢
nickel alumide　镍衣铝粉粒
nickel ammine　镍的氨化物
nickel ammonium chloride　氯化镍铵
nickel ammonium sulfate　硫酸镍铵
nickel antimonide　锑化镍　$NiSb$; Ni_2Sb; Ni_5Sb_2; Ni_4Sb_5
nickel antimony titanate　钛酸锑镍；钛镍黄
nickelate　镍酸盐
nickel bead　镍珠
nickel benzoate　苯甲酸镍
nickel benzoylacetonate　苯酰丙酮镍
nickel bis (ethylenediamine) dithiocarbamate　双(乙二胺)二硫代氨基甲酸镍〔防老剂〕
nickel black　镍黑　NiO_2
nickel boat　镍舟
nickel borate　硼酸镍
nickel boron alloy　硼镍合金
nickel brass　镍铜锌合金；镍黄铜
nickel carbonyl(=nickle carbonoxide)　羰基镍　$Ni(CO)_4$
nickel carbonyl catalyst　羰基镍催化剂
nickel cast iron　镍铸铁
nickel catalyzator　镍催化剂
nickel chelate　镍螯合物
nickel chloride　氯化镍　$NiCl_2 \cdot 6H_2O$
nickel-chrome steel　铬镍钢
nickel-chromium alloy　镍铬合金
nickel chromium steel　铬镍钢
nickel chromium triangle　镍铬(合金丝)三角

nickel coatings 镍镀层；镀镍层
nickel-containing heat stabilizer 含镍热稳定剂
nickel crabide 一碳化三镍 Ni_3C
nickel cyanide 氰化镍
nickel-diamond coating 镍-金刚砂涂层
nickel dibutyl dithiocarbamate 二丁基二氨基甲酸镍
nickel diethyldithiocarbamate 二乙基二硫代氨基甲酸镍〔防老剂〕
nickel diisobutyldithiocarbamate 二异丁基二硫代氨基甲酸镍〔防老剂〕
nickel diisopropyl xanthate 二异丙基黄原酸镍〔促进剂〕
nickel dimethyldithiocarbamate 二甲基二硫代氨基甲酸镍〔促进剂〕
nickel dimethyl glyoxime 镍二甲基乙二(醛)肟
nickel etching 镍腐蚀
nickel ethyl xanthate 乙基黄原酸镍〔防老剂〕
nickel-free zircalloy-2 无镍锆-2 合金
nickel glance 辉砷镍矿；硫化砷镍 $NiAsS$
nickel gymnite 硅镍石；含镍水蛇纹石
nickel halogenide 卤化镍
nickelic 高镍；三价镍
nickelic acetate 乙酸高镍 $Ni(C_2H_3O_2)_3$
nickelic compound 高镍化合物
nickelic hydroxide 氢氧化高镍 $Ni(OH)_3$
nickelic nickelous sulfide 辉铁镍矿
nickelic oxide 氧化高镍 Ni_2O_3
nickelic sulfide 硫酸高镍 Ni_2S_3
nickeliferous 含镍的
nickel iron storage battery 镍铁蓄电池；镍极蓄电池
nickel-isomerization 镍(催化)异构化法
nickel isopropyl xanthate 异丙基黄原酸镍〔防老剂〕
nickel-ligand system 镍-配合基体系
nickel matte 镍冰铜；镍锍〔冶〕
nickel mercaptobenzimidazole 硫醇基苯并咪唑镍〔防老剂〕
nickel mercaptobenzothiazole 硫醇基苯并噻唑镍〔促进剂〕
nickel minerals 镍矿(类)
nickel modified polypropylene fibre 镍改性(的)聚丙烯纤维
nickel monoxide 一氧化镍 NiO
nickel nitrate 硝酸镍 $Ni(NO_3)_2 \cdot 6H_2O$
nickel nitride 二氮化三镍 Ni_3N_2
nickelocene 二茂镍
nickel ocher 镍华
nickel octoate 辛酸镍
nickel octyldithiocarbamate 辛基二硫代氨基甲酸镍〔防老剂〕
nickel on carbon catalyst 碳载镍催化剂
nickelous （正)镍；二价镍
nickelous acetate 乙酸镍 $Ni(C_2H_3O_2)_2$
nickelous arsenate 砷酸镍 $Ni_3(AsO_4)_2$
nickelous borate 硼酸镍 $Ni(BO_2)_2$
nickelous bromide 溴化镍 $NiBr_2$
nickelous carbonate 碳酸镍 $NiCO_3$
nickelous chloride 氯化镍 $NiCl_2$
nickelous compound 正镍化合物
nickelous cyanide 氰化镍 $Ni(CN)_2$
nickelous dithionate 连二硫酸镍 NiS_2O_6
nickelous fluoride 氟化镍 NiF_2
nickelous fluosilicate 氟硅酸镍 $Ni[SiF_6]$
nickelous hydrogen fluoride 氟化氢镍 $NiF_2 \cdot 5HF \cdot 6H_2O$
nickelous hydroxide 氢氧化镍 $Ni(OH)_2$
nickelous hypophosphite 次磷酸镍 $Ni(H_2PO_2)_2$
nickelous iodate 碘酸镍 $Ni(IO_3)_2$
nickelous iodide 碘化镍 NiI_2
nickelous-nickelic oxide 四氧化三镍 $NiO \cdot Ni_2O_3$
nickelous nitrate 硝酸镍 $Ni(NO_3)_2$
nickelous nitrite 亚硝酸镍 $Ni(NO_2)_2$
nickelous oxalate 草酸镍 NiC_2O_4
nickelous oxide 氧化镍 NiO
nickelous perchlorate 高氯酸镍 $Ni(ClO_4)_2$
nickelous phosphate 磷酸镍 $Ni_3(PO_4)_2$
nickelous rubidium sulfate 硫酸铷镍 $Rb_2SO_4 \cdot NiSO_4 \cdot 6H_2O$
nickelous selenate 硒酸镍 $NiSeO_4$
nickelous selenide 硒化镍 $NiSe$
nickelous sulfate 硫酸镍 $NiSO_4$
nickelous sulfide 硫化镍 NiS
nickel oxalate 草酸镍 NiC_2O_4
nickel oxide 氧化镍 $NiO; Ni_2O_3; NiO_2; NiO \cdot Ni_2O_3$
nickel pentamethylene dithiocarbamate 五亚甲基二硫代氨基甲酸镍〔防老剂〕
nickel plated 镀(了)镍的
nickel plating 镀镍
nickel plating bath 镀镍浴
nickel powder 镍粉
nickel protoxide 一氧化镍 NiO
nickel pyrite 黄镍矿；硫化镍 NiS
nickel reagent 试镍剂
nickel salipyrine 镍水杨酸和安替比林络盐
nickel salt ①镍盐②硫酸镍
nickel scratch 镍条刮擦；镍刮
nickel sesquioxide 三氧化二镍 Ni_2O_3
nickel silver 镍银
nickel steel 镍钢

nickel sulfate 硫酸镍 $NiSO_4·6H_2O$
nickel superoxide 四氧化镍 NiO_4
nickel tetracarbonyl 四羰基镍 $Ni(CO)_4$
nickel tetrathionate 连四硫酸镍 NiS_4O_6
nickel 2,2′-thiobis(4-octyl phenolate) 2,2′-硫撑-双(4-辛基苯酚)镍〔紫外光吸收剂〕
nickel thiocyanate(=nickel thiocyanide) 硫氰酸镍 $Ni(SCN)_2$
nickel thiocyanide(=nickel thiocyanate) 硫氰酸镍
nickel thiosulfate 硫代硫酸镍 $NiSO_3S$; NiS_2O_3
nickel titanate 钛酸镍；钛镍黄
nickel titanate yellow 钛酸镍黄；钛镍黄
nickel trimethylcyclohexyl xanthate 三甲基环己基黄原酸镍〔防老剂〕
nickel undercoat 镀镍底层
nickel vitriol 硫酸镍 $NiSO_4·7H_2O$
nickel ware 镍制器皿；镍制仪器
nickel yellow 镍黄
nicking 刻痕
nickle carbonoxide(=nickel carbonyl) 羰基镍
nick translation 缺口平移法
niclosamide 氯硝柳胺〔药〕
nicofer 尼可滗
Nicol 尼科耳棱镜
Nicol crossed 正交尼科耳棱镜
Nicol crossed prism 正交尼科耳棱镜
Nicol prism 尼科耳棱晶
Nicolsky-Eisenman equation 尼可尔斯基-艾森曼方程式
Nicol tube 尼科耳管
nicopyrite 镍黄铁矿
nicoteine 降烟碱和毒藜碱混合物；烟草碱
nicotelline 尼古丁
nicotia 烟碱；尼古丁
nicotianine 烟草香素
nicotimine 烟酰亚胺
nicotinaldehyde thiosemicarbazone 烟碱醛缩氨基硫脲〔作为结核抑制剂〕
nicotinamide 烟酰胺〔药〕
nicotinate 烟酸盐(或酯)
nicotine 烟碱；尼古丁
nicotine group 吡啶和吡咯烷类生物碱
nicotine remover 尼古丁除去液
nicotine stain 烟碱污染；尼古丁污染
nicotine sulfate 硫酸烟碱
nicotine tannate 鞣酸烟碱
nicotinic acid 烟酸；尼克酸
nicotinic acid amide 烟酰胺
nicotinic acid hydrochloride 烟酸盐酸盐 $C_5H_4NCO_2HHCl$
nicotinic acid picrate 烟酸化苦味酸 $C_6H_5O_2NC_6H_3O_7N_3$
nicotinic amide 烟酰胺 $C_5H_4NCONH_2$
nicotinohydroxamic acid 烟碱异羟肟酸
nicotinoylglycine 烟酰甘氨酸；烟尿酸
nicotinuric acid 烟酰甘氨酸；烟尿酸
nicotyrine 烟碱烯；二烯烟碱
nicouic acid 鱼藤酮酸
nicouline 鱼藤酮
niddamycin 尼达霉素
nidulin 构巢曲菌素
niello silver 乌银
Niemann's triangle 尼曼三角〔液体在表面上保持平衡时,各有关界面张力间的关系〕
Nierenstein synthesis 尼伦斯坦合成法
Nier-Johnson double focusing mass spectrometer 尼尔-约翰孙型双聚焦质谱仪
Nier-60 mass spectrometer 60°偏转尼尔型质谱仪
Nier's mass spectrometer 尼尔质谱仪
Nier-type mass spectrometer 尼尔型质谱仪
Nieuland catalyst 纽兰德催化剂
nifedipine 硝苯地平〔药〕
Niflex process 尼弗莱克斯过程〔首端过程,用HNO_3-HF溶解不锈钢包壳燃料元件的过程〕
nifurazolidone 呋喃唑酮〔药〕
nigella oil 黑种草子油
nigelline 黑种草碱
niger 皂脚
niger factor 抗黑曲菌素
nigericin 尼日利亚菌素
nigerine(=N, N-dimethyltryptamine) N,N-二甲基色胺
nigerite 锡铝矿
niger oil(=niger seed oil) 皂厂杂油
niger seed oil(=niger oil) 皂厂杂油
nigger 长把手〔从管子伸出的〕
night blindness 夜盲(症)
night-vision equipment 夜视设备
nigraniline 苯胺黑
nigrate 黑沥青〔美国的一种地沥青〕
nigre 皂脚
nigrite 沥青
nigrometer 黑度计
nigrosin 苯胺黑
nigrosine ①尼格(洛辛)②(=aniline black)苯胺黑 $C_{38}H_{27}N_3$
nigrosine base 尼格色基；油溶苯胺黑
nigrosine fat soluble 油溶尼格(洛辛)〔染〕

nigrosine spirit-soluble 醇溶尼格(洛辛)〔染〕
nigrosine water-soluble 水溶尼格(洛辛)〔染〕
nigrotic acid 黑酸；3,6-二羟基-2-磺基-7-萘甲酸
Ni-Hard 含镍耐磨铸铁
nihil album 氧化锌
nikethamide N,N-二乙基烟酰胺；尼克刹米
niketharol N,N-二乙基烟酰胺
Nikiforoff stain 尼奇佛罗夫染色剂
nil 无；零(点)
nil contact angle 无接触角
nil ductility temperature(NDT) 无延性转变温度
Nile blue A 尼罗蓝 A
nilestriol 尼尔雌醇〔药〕
nilic acid 尼里酸；裂叶牵牛子酸
nill 铁屑
nilodin 尼鲁定；尼罗丁〔药〕
nilvar 镍铁合金
nim 印楝
nimbic acid 印楝酸
nimbicitin 楝醇
nimbin 印楝素
nimbinic acid 印楝素酸
nimbiol 酮式楝酚菲
nimesulide 尼美舒利〔药〕
nimodipine 尼莫地平〔药〕
nimonic alloys 耐氧化镍合金
nimustine 尼莫司汀〔药〕
-nin 宁〔生物碱〕
nine-atomic ring(=nine-membered ring) 九元环
nine-membered ring(=nine-atomic ring; nine-ring) 九元环
nine-ring(=nine-membered ring) 九元环
nineteen-membered ring(=nineteen-ring) 十九元环
nineteen-ring(=nineteen-membered ring) 十九元环
Ningpo lac 金漆
Ningpo varnish 金漆
ninhydrin (水合)茚三酮
ninhydrin reaction (水合)茚三酮反应
ninhydrin test (水合)茚三酮试验
ninidrine 水合茚三酮
niob-anatase(=nioboanatase) 铌锐钛矿
niobate 铌酸盐 $MNbO_3; M_3NbO_2; M_4Nb_2O_7$
niobate system piezoelectric ceramics 铌酸盐系陶瓷
niobe oil 尼哦油；苯甲酸甲酯 $C_6H_5COOCH_3$
niobic 铌的
niobic acid 铌酸 $HNbO_3; H_3NbO_4; H_4Nb_2O_7$
niobic anhydride 铌酐 Nb_2O_5
niobite 铌铁矿

niobium 铌〔41号元素，化学符号 Nb〕
niobium alkoxide polymer 铌醇盐聚合物
niobium chloride 氯化铌
niobium minerals 铌矿
niobium oxides 氧化铌
niobium oxychloride 氯氧化铌
niobium pentafluoride 五氟化铌 NbF_5
niobium perovskite 铌钙钛矿
niobium rutile 铌金红石
niobobrookite 铌板钛矿
nioboxy 铌氧基
niobus 三价铌的
nioxime (=1,2-cyclohexanedionedioxime) 1,2-环己二酮二肟 $C_6H_8(NOH)_2$
nip ①夹；摘取②霜害
nipagin 对羟苯甲酸甲酯〔防腐剂〕；尼泊金 HOC_6H_4COOMe
nip angle 捏挤角度
nipasol 对羟苯甲酸丙酯 $C_{10}H_{12}O_3$
nip clearance 辊隙间距；缝间距离
nipecotic acid 3-哌啶甲酸 $C_5H_{10}NCOOH$
niperit 尼帕炸药；尼帕锐特
nip flooding shower 压轧溢流式喷淋器
niphimycin(=nyphimycin) 尼菲霉素
nip opening 滚距
nipple 螺纹接口(管)
nipple joint 螺纹接管；管接头接合
nip roll 压料辊
niranium 含镍镶牙合金
ni-resist 耐热耐蚀镍合金
niribine oil(=black currant flower oil) 黑茶藨子花油
Ni-rich intermetallics 富镍金属互化物；富镍金属间化合物
niromycin(=nairomycin) 新露霉素
nirvanol(=5-phenyl-5-ethylhydantoin) 尼凡诺；5-苯基-5-乙基海因
Nishimoto-Mataga relationship 西本-又贺关系式
nisinic acid 尼生酸；4,8,12,15,18,21-二十四碳六烯酸
nital 硝酸酒精溶液〔蚀刻剂〕
niter(=nitre) 硝石
niter cake 硝饼
niter-cake furnace 硝饼炉
niter oven 硝石炉
niter pot 硝石罐
nitinol 镍钛金属互化物
niton 氡〔86号元素，化学符号 Rn〕；镭射气
nitracetanilide 硝基乙酰苯胺
nitracidium ion 硝酸合氢离子 $H_2NO_3^+$

nitraffine 硝基苯精制物
nitralising (熔融)硝酸钾处理
nitralloy 渗氮钢
nitramide 硝酰胺 $O_2N \cdot NH_2$
nitramine 硝胺
nitramine picrate 硝胺合苦味酸
nitramine rearrangement 硝胺重排作用
nitramino 硝氨基 $O_2NNH—$
nitramon 硝铵火药
nitranilic acid 硝冉酸
nitranilide 苯基重氮酸
nitraniline 硝基苯胺
nitraniline orange(=nitroaniline orange) 硝基苯胺橙
nitratase 硝酸还原酶
nitrate ①硝酸盐②硝酸酯③硝化④硝酸(根)
nitrate additive 硝酸酯添加剂
nitrate asphalt 硝化沥青
nitrate bed 硝石矿床
nitrated 硝化了的
nitrated alcohol 硝化酒精
nitrated cellulose 硝酸纤维素；硝化纤维；硝化棉
nitrate dope 硝基航空涂料
nitrated polyglycerin 硝化聚甘油
nitrate fusion 硝酸盐熔融
nitrate green 硝酸绿〔硝酸铅铬+亚铁氰化铁〕
nitrate group 硝酸根
nitrate ion 硝酸离子
nitrate ion selective electrode 硝酸根离子选择电极
nitrate nitrogen 硝态氮；硝酸盐氮
nitrate of baryta 硝酸钡 $Ba(NO_3)_2$
nitrate of ethylene bromohydrin 溴代乙醇的硝酸酯 $CH_2BrCH_2ONO_2$
nitrate of lime 硝酸钙 $Ca(NO_3)_2$
nitrate of soda 硝酸钠
nitrate radical 硝酸根
nitrate superphosphate 硝酸化过磷酸
nitratine 钠硝石；智利硝石；硝酸钠 $NaNO_3$
nitrating 硝化
nitrating acid 硝化酸
nitrating agent 硝化剂
nitrating centrifuge 硝化离心机
nitrating pot 硝化釜；硝化罐
nitrating separator 硝化离析器
nitrating test 硝化试验
nitration 硝化*
nitration acid heat test 硝酸加热试验
nitration benzol 硝化用苯
nitration grade 硝化级〔适于硝化的规格〕
nitration grade benzene 硝化级苯
nitration grade products 硝化级产品
nitration grade toluene 硝化级甲苯
nitration mixture 硝酸和硫酸的混酸
nitration toluol 硝化用甲苯
nitration xylol 硝化用二甲苯
nitrato- 硝酸基
nitrator 硝化器
nitrator-separator 硝化分离器
nitrazepam 硝西泮〔药〕
nitrazine yellow 硝氮黄；3-[(2,4-二硝基苯基)偶氮基]-4-羟基-2,7-萘二磺酸
nitrazine yellow paper 硝嗪黄试纸
nitre(=niter) 硝石
nitre bed 硝床
nitre cake 硝饼；硫酸氢钠
nitre field 硝床
nitrendipine 尼群地平〔药〕
nitrene 氮宾*；氮烯
nitre oven 硝石炉
nitre pot 硝石罐
nitre-pot system 硝石罐系统
nitriacidium ion 硝酸合氢离子 $H_2NO_3^+$
nitric acid 硝酸
nitric acid absorption tower 硝酸吸收塔
nitric acid bleacher 硝酸漂白塔
nitric acid concentration tower 硝酸浓缩塔
nitric acid ester 硝酸酯 $RONO_2$
nitric acid still 硝酸蒸馏器
nitric acid tail-gas expension turbine 硝酸尾气膨胀透平
nitric anhydride 硝(酸)酐；五氧化二氮
nitric ether 硝酸酯 $RONO_2$
nitric hydrate 水合硝酸〔水 32%〕
nitric nitrogen 硝态氮；硝酸盐中的氮
nitric oxide （一）氧化一氮；氧化氮
nitric plant 硝酸厂
nitridation ①氮化(作用)②渗氮
nitride 氮化物
nitrided catalyst 氮化催化剂
nitrided film 氮化膜
nitride layer 氮化(物)层
nitriding ①氮化②渗氮
nitriding steel 氮化钢
nitriding treatment 氮化处理；渗氮
nitridizing agent 氮化剂
nitridotrisulfuric acid 次氮基三硫酸 $N(SO_3H)_3$
nitrifiable 可硝化的
nitrification 硝酸化作用

nitrification of polymers 聚合物硝化反应
nitrifier 硝化(细)菌
nitrifying 硝化的
nitrifying bacteria 硝化(细)菌
nitrilase 腈水解酶
nitrilation 腈化
nitrilation catalyst 腈化催化剂
nitrile* 腈
nitrile base 叔胺 R_3N
nitrile bonded phase 腈型键合(固定)相
nitrile-butadiene rubber 丁腈橡胶
nitrile ether 腈醚
nitrile grouping 氰基
nitrile latex 丁腈胶乳
nitrile of phenylglycine 苯基氨基酸乙腈
nitrile oxide* 氧化腈
nitrile phenolic adhesive 丁腈酚醛树脂黏合剂
nitrile resin 腈类树脂
nitrile rubber 丁腈橡胶
nitrile-rubber PVC blend 丁腈橡胶-聚氯乙烯掺和物
nitrile silicone 氰硅油
nitrile silicone rubber 腈硅橡胶
nitrile synthesis 腈合成法
nitrilo 次氮基;(三价)氮基 $N\equiv$
nitrilo butadiene rubber 丁腈橡胶
nitrilodiacetic monopropionic acid(NDAP) 次氮基二乙酸一丙酸 $N(CH_2COOH)_2(C_2H_4COOH)$
nitrilotriacetate 次氮基三乙酸酯
nitrilotriacetic acid(NTA) 氨三乙酸*;次氮基三乙酸
nitrilotriacetic acid chelate 氨三乙酸螯合物;三乙酸胺螯合物
nitrilotriethanol 氮川三乙醇;次氮基三乙醇;三乙醇胺
nitrilotrisilane 氮川三硅烷;次氮基三硅烷;三硅烷基胺
nitrilotrisulfonic acid(=nitridotrisulfuric acid) 次氮基三硫酸
nitrine 叠氮 N_3
nitrite 亚硝酸盐(酯或根)
nitrite ion 亚硝酸离子 NO_3^{2-}
nitrite liquor(=nitrite lye) 亚硝碱液
nitrite lye(=nitrite liquor) 亚硝碱液
nitrite nitrogen 亚硝态氮;亚硝酸盐中的氮
nitrite number 亚硝(酸盐)值
nitrite selective electrode 亚硝酸根选择电极
nitrite solution 亚硝碱液
nitrite titration 亚硝酸盐滴定
nitrito- 亚硝酸(基)
nitritocobalamin (亚)硝(酸)钴胺素
nitrizing 渗氮

nitro- 硝基 O_2N-
5-nitro acenaphthene 5-硝基苊
nitroacetic acid 硝基乙酸 $NO_2CH_2CO_2H$
nitroacetyl cellulose 硝酸乙酸纤维素;硝酸醋酸纤维素
nitroacid 硝基酸
nitro-acinitro system 硝基-酸硝基系
nitro-acinitro tautomerism 硝基-酸硝基互变异构
nitroalizarin 硝基茜素
nitroalkane 硝基烷
nitroalkyde paint 硝基纤维改性醇酸树脂漆
nitro-amine 硝胺 $RNHNO_2;RRNNO_2$
nitro-analogues 硝基同系物
nitroaniline 硝基苯胺
nitroaniline orange(=nitraniline orange) 硝基苯胺橙
p-nitroaniline red(=para red) 对位红;对硝基苯胺红
nitro-anions 硝基负离子
nitroanisidine A 硝基茴香胺A
nitroanisole 硝基茴香醚
nitroanthracene 硝基蒽 $C_{14}H_9NO_2$
nitroanthraquinone 硝基蒽醌
nitroanthraquinone sulfonic acid 硝基蒽醌磺酸
nitro-aromatics 硝基芳族化合物
nitroarsenazo I 硝基偶氮胂 I
nitroazobenzene 硝基偶氮苯
4-nitro-trans-azobenzene 4-硝基反式偶氮苯
nitrobarbituric acid 硝基巴比妥酸 $NO_2C_4H_3O_3N_2$
nitrobarite 钡硝石
nitrobenzal-acetophenone 硝基亚苄基乙酰苯 $NO_2C_6H_4CH=CHCOC_6H_5$
nitrobenzaldehyde 硝基苯甲醛
nitrobenzamide 硝基苯甲酰胺
nitrobenzanilide 硝基苯甲酰苯胺
nitrobenzene 硝基苯 $C_6H_5NO_2$
nitrobenzene azonaphthol 硝基苯偶氮萘酚 $NO_2C_6H_4N=NC_{10}H_6(OH)$
p-nitrobenzene azoresorcinol(=magneson) 对硝基苯偶氮间苯二酚;试镁灵 $NO_2C_6H_4N=NC_6H_3(OH)_2$
nitrobenzene azosalicylic acid(=alizarin yellow G) 硝基偶氮苯水杨酸;茜素黄G $NO_2C_6H_4N=NC_6H_3(OH)CO_2H$
p-nitrobenzene diazo-amino-azobenzene(=cadion) 试镉灵
nitrobenzene nitroso naphthol 硝基苯亚硝基萘酚
nitrobenzene process (润滑油的)硝基苯提取过程
nitrobenzene raffinate 硝基苯提纯物
nitrobenzene reduction 硝基苯还原
nitrobenzene sulfonyl chloride 硝基苯磺酰氯
nitrobenzene-sulfuric acid process (润滑油精制的)硝基苯硫酸过程

nitrobenzene thiocyanate 硫氰酸硝基苯酯
nitrobenzidine 硝基联苯胺　$NO_2(NH_2)C_6H_3C_6H_4NH_2$
nitrobenzimidazole 硝基苯并咪唑
nitrobenzoate 硝基苯甲酸盐
nitrobenzoic acid 硝基苯甲酸
nitrobenzol 硝基苯
nitrobenzonitrile 硝基苯甲腈
nitrobenzophenone 硝基苯基苯基(甲)酮
nitrobenzoquinone 硝基苯醌
nitrobenzoylazochromotropic acid 硝基苯偶氮-1,8-二羟基-3,6-二磺酸
nitrobenzoyl chloride 硝基苯甲酰氯
nitrobenzyl 硝基苄基
nitrobenzyl acetate 乙酸硝基苄酯
nitrobenzyl alcohol 硝基苄醇
nitrobenzyl chloride 硝基苄基氯
p-nitrobenzyl-leucomethylene blue 对硝基苄-无色亚甲基蓝
nitrobiphenyl 2-硝基联苯
nitroblue tetrazolium(NBT) 氮蓝四唑
nitroblue tetrazolium day test for neutrophil 中性粒细胞NBT实验
nitrobromoarsenazo 溴硝基偶氮肼
nitrobromoform(=bromopicrin) 硝基溴仿；溴化苦；三溴硝基甲烷　NO_2CBr_3
nitrobruciquinone hydrate(=cacotheline) 卡可西灵；硝基马钱子碱
nitrobutandiol 硝基丁二醇　$CH_3CNO_2=(CH_2OH)_2$
nitrobutane 硝基丁烷
nitro-t-butane 硝基叔丁烷　$(CH_3)_3CNO_2$
nitrocamphor 硝基樟脑　$C_8H_{14}COCHNO_2$

nitrocaphane 硝卡芥〔药〕
6-nitrocaproic acid 6-硝基己酸
nitrocaptax 6-硝基-2-巯基苯并噻唑
nitrocarbamate 硝氨基甲酸酯
nitrocarbol 硝基甲烷
nitrocarbonate 硝基碳酸盐
nitrocarbonitrate 硝基碳酸硝酸复合物
nitrocellulose 硝酸纤维素；硝化纤维(素)
nitrocellulose block 块状硝化纤维素
nitrocellulose chip 硝化纤维(素)漆片；硝化纤维(素)色片
nitrocellulose coating 硝化纤维涂料
nitrocellulose dope 硝化纤维涂料
nitrocellulose emulsion 硝化纤维素乳液；硝化乳液
nitrocellulose film 硝酸纤维素涂膜
nitrocellulose finish 硝化纤维素涂饰剂；硝纤涂饰剂
nitrocellulose isocyanate 硝酸(硝化)纤维素异氰酸酯

nitrocellulose lacquer 硝基漆；硝基纤维漆；硝酸纤维漆；喷漆
nitrocellulose paint 硝基漆；硝化纤维漆
nitrocellulose percentage elongation of film 硝酸纤维素涂膜的伸长率
nitrocellulose powder 硝化纤维素火药
nitrocellulose putty 硝基腻子
nitrocellulose rayon 硝化纤维素嫘萦
nitrocellulose silk 硝化纤维(人造)丝
nitrocellulose viscosity specification 硝酸纤维素黏度指标
nitro-chalk 钾铵硝石；白垩硝〔商用专名〕
nitrochlorobenzene 硝基氯(代)苯
nitrochlorobenzol 硝基氯(代)苯
nitrochloroform(=chloropicrin) 硝基氯仿；氯化苦
nitrochlorophenol 硝基氯苯酚　$HOC_6H_3(Cl)NO_2$
nitrocinnamic acid 硝基肉桂酸　$NO_2C_6H_4CH=CHCOOH$
nitro-color 硝基色料
nitro colouring matters 硝基染料
nitro compound 硝基化合物*
nitro-cotton 硝化棉；硝化纤维素
nitrocresol 硝基甲(苯)酚
nitrocumene 硝基枯烯；3-硝基-1-异丙苯；硝基异丙苯　$NO_2C_6H_4CH(CH_3)_2$
nitrocyclohexane 硝基环己烷　$C_6H_{11}NO_2$
2-nitrocyclohexanone 2-硝基环己酮
nitrocymene 硝基伞花烃；硝基异丙甲苯　$NO_2C_6H_3(CH)_3CH(CH_3)_2$
nitro-derivative 硝基衍生物
nitro-dextrin 硝化糊精
nitro-diazobenzene-perchlorate 过氯酸硝基重氮苯
nitrodiethylaniline 硝基二乙基苯胺
nitrodimethylamine 硝基二甲胺
nitrodimethylaniline 硝基二甲基苯胺
nitrodiphenylmethane 硝基二苯基甲烷　$C_{13}H_{11}NO_2$
nitrodope 硝化涂料
nitrodracylic acid 对硝基苯甲酸
nitro-dye 硝基染料
nitroerythrite 硝化赤藓糖醇
nitro-erythritol 硝化赤藓糖醇；丁四醇四硝酸酯
nitroethane 硝基乙烷　$CH_3CH_2NO_2$
p-nitro-ethylacetanilide N-乙基-N-乙酰基对硝基苯胺　$NO_2C_6H_4N(C_2H_5)COCH_3$
nitroethyl alcohol 硝基乙醇　$HOCH_2CH_2NO_2$
nitroethylene 硝基乙烯　$CH_2=CHNO_2$
nitroferron(=nitro-o-phenanthroline) 硝基亚铁灵
nitrofication process 硝化法

nitrofluorene 硝基芴 $NO_2C_{12}H_7=CH_2$
Nitrofluor process 硝基氟过程〔美国布鲁克海文实验室研究的氟化挥发核燃料后处理方法〕
nitroform(=trinitromethane) 硝仿；三硝基甲烷 $CH(NO_2)_3$
nitroformaldehyde phenylhydrazone 硝基甲醛苯腙 $C_6H_5NHN=CHNO_2$
nitro-2-furancarboxylic acid(=nitropyromucic acid) 硝基-2-呋喃羧酸；硝基焦黏酸 $NO_2C_4H_2OCOOH$
nitrofurantoin 呋喃旦啶；呋喃妥因；呋喃妥英；泌尿康；硝呋妥因〔药〕
nitrofurazone(=5-nitro-2-furaldehyde semicarbazone) 硝基糠腙；5-硝基糠醛缩氨基脲 $NO_2C_4H_3=NNH=CONH_2$
nitrogelatin 硝化甘油炸药
nitrogen ①氮 ②氮气
nitrogen absorption apparatus 氮吸收仪〔炭黑〕
nitrogen adsorption method 氮吸附法〔炭黑〕
nitrogen adsorption surface area 氮吸附法表面积〔炭黑〕
nitrogen analysis 氮量分析
nitrogen apparatus 定氮装置
nitrogenase 固氮酶*
nitrogenated 氮化(了的)
nitrogenated oil 氮化油
nitrogen atmosphere 氮气氛；氮气层
nitrogen balance 氮平衡
nitrogen base 氮碱；含氮的有机碱
nitrogen benzide 偶氮苯
nitrogen blanket 氮气层
nitrogen blowing agent 氮发泡剂
nitrogen bridge 氮桥
nitrogen bulb 定氮球管
nitrogen chain 氮链
nitrogen compound 氮化合物
nitrogen-containing accelerator 含氮促进剂
nitrogen-containing copolymer 含氮共聚物
nitrogen-containing impurity 含氮杂质
nitrogen content 氮含量；总氮量
nitrogen cycle 氮循环
nitrogen degradation （土壤中）氮降解
nitrogen determination 氮测定；凯氏定氮法
nitrogen diluent 氮稀释气
nitrogen dilution of synthesis gas 合成气的氮稀释
nitrogen dioxide 二氧化(一)氮 NO_2
nitrogen equilibrium 氮平衡
nitrogen equivalent 氮当量
nitrogen family 氮族
nitrogen fertilizer 氮肥(料)

nitrogen fixation 固氮(作用)
nitrogen fixation process 固氮法
nitrogen-fixing bacteria 固氮细菌
nitrogen flame detector 氮火焰检测器
nitrogen free extract 可溶性无氮物
nitrogen gas 氮气
nitrogen gas thermometer 氮气温度计
nitrogen heterocyclic ring 含氮杂环
nitrogen iodide 碘化氮
nitrogenization 氮化(作用)
nitrogen lag 氮素留滞
nitrogen lime 氮化石灰
nitrogen manure 氮肥
nitrogen monoxide 一氧化二氮；氧化亚氮
nitrogen mustard 氮芥
nitrogen mustard-N-oxides 含氮芥子-N-氧化物
nitrogenous 含氮的
nitrogenous base 含氮碱
nitrogenous effluent 含氮废水
nitrogenous equilibrium 等氮平衡
nitrogenous fertilizer 氮肥
nitrogenous guano 富氮海鸟粪
nitrogenous manure(=nitrogenous fertilizer) 氮肥
nitrogenous metabolism 氮的代谢作用
nitrogenous-phosphatic fertilizer(=nitrogenous-phosphatic manure) 氮磷肥料
nitrogenous-phosphatic manure(=nitrogenous-phosphatic fertilizer) 氮磷肥料
nitrogenous tankage 含氮槽肥
nitrogen oxide ①氧化氮〔通常指NO 一氧化氮〕②氮的氧化物
nitrogen oxide gas compressor 氧化氮压缩机
nitrogen oxide probe 氮氧化物探头
nitrogen oxides absorption tube 氮氧化物吸收管
nitrogen oxychloride 亚硝酰氯；氯氧化氮 NOCl
nitrogen partition 氮分配
nitrogen pentoxide 五氧化二氮 N_2O_5
nitrogen peroxide ①三氧化(一)氮；过氧化氮 ②二氧化(一)氮的误称
nitrogen-phosphorus detector 氮-磷检测器
nitrogen purging of tank 用氮气扫罐
nitrogen rule 氮规律
nitrogen-selective detector 氮选择性检测器
nitrogen selenide 一氮化硒
nitrogen source 氮源
nitrogen substituted 氮代的
nitrogen sulfide 硫化氮 N_4S_4; N_2S_5
nitrogen sulfochloride 硫氯化氮；一氯四硫化三氮

N_3S_4Cl
nitrogen tetroxide 四氧化二氮 N_2O_4
nitrogen tetroxide-resistant coating 耐四氧化二氮涂料
nitrogen trichloride 三氯化氮 Cl_3N
nitrogen trioxide 三氧化二氮 N_2O_3
nitrogen ylide 氮叶立德*
nitroglycerin(=glycerin trinitrate) 甘油三硝酸酯；硝化甘油 $(O_2NO)_3C_3H_5$
nitroglycerine 硝化甘油；甘油三硝酸酯；硝酸甘油酮 $(O_2NO)_3C_3H_5$
nitroglycerine-amide powder 硝化甘油酰胺炸药
nitroglycerine explosive 硝化甘油炸药
nitroglycerine powder 硝化甘油火药
nitroglycerine substitute 硝化甘油代用品
nitro-glycerol 硝化甘油 $C_3H_5(NO_3)_3$
nitroglycol 硝化甘醇；乙二醇二硝酸酯
nitro-group 硝基
nitroguaiacol 硝基愈创木酚 $NO_2C_6H_3(OCH_3)OH$
nitroguanidine 硝基胍 $NO_2NHC(NH)NH_2$
nitrohalogen compound 卤代硝基化合物
nitro-hydrocarbon 硝基烃
nitro-hydrocellulose 硝基水化纤维素
nitro-hydrochloric acid 王水
nitro-hydrocinnamic acid 硝基氢化肉桂酸 $NO_2C_6H_4CH_2CH_2COOH$
2-nitro-3-hydroxybenzoic acid 2-硝基-3-羟基苯甲酸 $NO_2C_6H_3(OH)CO_2H$
nitrohydroxylamic acid(=N-nitrohydroxylamine) 硝基羟氨酸；N-硝基胲 NO_2NHOH
N-nitrohydroxylamine(=nitrohydroxylamic acid) 硝基羟氨酸；N-硝基胲 NO_2NHOH
nitroic acid 水合硝基酸
nitroil(=nitro oil) 硝化油
nitro indicator 硝基(系)指示剂
nitroisophthalic acid 硝基间苯二甲酸 $NO_2C_6H_3(CO_2H)_2$
nitroisoquinoline 硝基异喹啉 $NO_2C_6H_3=C_3H_3N$
nitrol 硝脑；硝基亚硝基烃
nitrolacquer 硝基漆；硝化纤维漆
nitroleum(=nitroglycerin) 硝化甘油
nitrolic acid 硝肟酸 $RC(=NOH)NO_2$
nitrolignin 硝化木素
nitrolim(e) 氰氨(基)化钙 $CaCN_2; CaN·CN$
nitromagnesite 镁硝石
nitromagnite(=dynamagnite) 硝化甘露醇
nitromesitylene 硝基-1,3,5-三甲苯 $(CH_3)_3C_6H_2NO_2$
nitrometal 硝基金属
nitrometer 测氮管

nitrometer detector 氮量计检测器
nitromethane 硝基甲烷 CH_3NO_2
nitro-methylaniline 硝基甲苯胺
nitro-muriatic acid 王水
nitro musk 硝基麝香
nitron 硝酸试剂*
nitro-naphthalene 硝基萘
nitro-naphthalene-disulfonic acid 硝基萘二磺酸 $NO_2C_{10}H_5(SO_3H)_2$
nitro-naphthalene-monosulfonic acid 硝基萘(一)磺酸 $NO_2C_{10}H_6SO_3H$
nitro-naphthalene-sulfonic acid 硝基萘磺酸 $NO_2C_{10}H_6SO_3H$
nitronaphthoic acid 硝基萘甲酸
nitronaphthol 硝基萘酚
nitronaphthylamine 硝基萘胺 $C_{10}H_6(NO_2)NH_2$
nitronate 氮酸酯
nitrone 硝酮*
nitronic acid〔类名〕 氮酸；氮羧酸 NOOR
nitro-nitrogen 硝态氮
nitro-nitroso- 硝基-亚硝基
nitronium ion 硝镓离子 NO_2^+
nitronium salt 硝镓盐
nitronium tetrafluoroborate 四氟硼酸硝镓；四氟硼酸硝
nitron nitrate 硝酸灵的硝酸盐
nitrononane 硝基壬烷
nitro oil(=nitroil) 硝化油
nitroolefin 硝基烯烃 $RCH=CR'(NO_2)$
nitroorthanilic S 硝基邻氨基苯磺 S
4-nitro-5-oxymercuro-o-cresol (=metaphen) 4-硝基-5-亚汞氧基邻甲苯酚；袂塔酚 $CH_3C_6H_2ONO_2Hg$
nitro-paper 硝化纸
nitro-paraffin 硝基烷
nitro-pentaerythrite 季戊四醇(四)硝酸酯
nitro-phenamine 硝基非那明；毛皮硝基黄
nitro-o-phenanthroline(=nitroferron) 硝基亚铁灵
nitrophenetol 硝基苯乙醚
nitrophenol 硝基苯酚
nitrophenolate 硝基苯酚盐 $NO_2C_6H_4OM$
nitrophenolic dye 硝基酚染料
nitrophenyl-acetic acid 硝基苯乙酸 $NO_2C_6H_4CH_2COOH$
nitrophenylamine 硝基苯胺
nitrophenylarsonic acid 硝基苯胂酸 $NO_2C_6H_4AsO(OH)_2$
nitrophenylation 硝苯基化(作用)
nitrophenyl diethyl phosphate 磷酸硝基苯基二乙酯
nitrophenylene diamine 硝基苯二胺 $NO_2C_6H_3(NH_2)_2$

nitrophenylfluorone 硝基苯荧光酮
nitrophenyl hydrazine hydrochloride 硝基苯肼盐酸盐〔比色测定脂肪酸用试剂〕
nitro-phenyl isocyanate 异氰酸硝基苯酯 $NO_2C_6H_4N=CO$
nitrophoska 硝酸磷酸钾〔磷酸铵、硝酸铵与氯化钾的混合物〕
nitrophthalide 硝基苯酞 $NO_2C_6H_3CH_2OCO$
nitropropane 硝基丙烷
nitroprusside 硝普盐；一亚硝基五氰合(三价)铁酸盐〔含离子的盐〕 $[Fe(NO)(CN)_5]^{2-}$
nitropseudocumene 硝基假枯烯；硝基-1,2,4-三甲苯 $NO_2C_6H_2(CH_3)_3$
1-nitropyrene 1-硝基芘
nitropyrocatechol 硝基邻苯二酚
nitropyromucic acid(=nitro-2-furancarboxylic acid) 硝基-2-呋喃羧酸；硝基焦黏酸 $NO_2C_4H_2OCOOH$
5-nitroquinaldic acid 5-硝基喹哪啶酸
nitroquinaldine 硝基喹哪啶 $NO_2C_9H_5NCH_3$
nitroquinoline 硝基喹啉
nitrosalicylic acid 硝基水杨酸
nitrosalol 硝基水杨酸苯酯 $NO_2C_6H_3(OH)CO_2C_6H_5$
nitrosamine 亚硝胺
nitrosamine rearrangement 亚硝胺重排作用
nitrosamino 亚硝氨基 ONNH—
nitrosates〔类名〕 硝酯肟酸
nitrosating agent(=nitrosation agent) 亚硝基化剂
nitrosation 亚硝化*
nitrosation agent(=nitrosating agent) 亚硝基化剂
nitrose 硫酸硝气溶液
nitrose pigment 亚硝基系颜料
nitrosilk 硝酸丝
nitrosimino 亚硝亚氨基 ONN=
nitrosite ①亚硝酯肟酸②亚氮氧化物
nitroso 亚硝基 ON—
nitrosoamines 亚硝胺
nitrosoaniline 亚硝基苯胺
nitrosobenzene 亚硝基苯 C_6H_5NO
nitrosobenzoic acid 亚硝基苯甲酸
nitroso-t-butane 亚硝基叔丁烷
nitrosocamphor 亚硝基樟脑
nitrosocobalamin 维生素 B_{12}
nitroso color 亚硝基色料
nitroso colouring matter 亚硝基染料
nitroso compound 亚硝基化合物*
nitrosocyclohexane 亚硝基环己烷
nitroso-decarboxylation 亚硝化脱羧作用

nitroso-derivative 亚硝基衍生物
N-nitroso dialkanol amine 亚硝基二烷醇胺
nitrosodiethylamine 亚硝基二乙胺
nitrosodiethylaniline 亚硝基二乙基苯胺
nitrosodimethylamine 亚硝基二甲胺
nitrosodimethylaniline 亚硝基二甲基苯胺
nitrosodiphenylamine 亚硝基二苯胺
N-nitrosodiphenylamine N-亚硝基二苯胺〔防焦剂〕
N-nitrosonaphthyl hydroxylamine N-亚硝基-α-萘胺铵；新铜铁试剂
N-nitrosophenylhydroxylamine ammonium(=cupferron) N-亚硝基-β-胺铵；铜铁试剂
nitroso-dye 亚硝基染料
nitrosoethane 亚硝基乙烷
N-nitroso-ethylaniline(=ethylphenyl-nitrosamine) N-亚硝基-N-乙苯胺 $C_6H_5N(NO)C_2H_5$
nitroso-fluoro-elastomer 亚硝基氟橡胶
nitrosofluoro rubber 亚硝基氟橡胶
nitrosofurfuramide 亚硝基糠醛胺〔促进剂〕
nitroso-group 亚硝基
nitrosoguanidine 亚硝基胍
nitrosohydroxylamines 亚硝基胺
nitrosoketone 亚硝基酮
nitrosolic acid 亚硝肟酸
nitroso-N-methyl-urethane N-亚硝基-N-甲基氨基甲酸乙酯 $(CH_3N(NO)CO_2C_2H_5$
nitroso-β-naphthol 亚硝基β萘酚；毛皮绿
α-nitroso-β-naphthol iron complex α-亚硝基-β-萘酚铁配合物
nitrosonaphthylamine 亚硝基萘胺
nitrosonitric acid 发烟硝酸
nitrosonium 亚硝鎓离子
nitroso oxyquinoline 亚硝基羟基喹啉
nitrosophenol 亚硝基苯酚
o-nitrosophenol 邻亚硝基苯酚
nitrosophenyldimethylpyrazole 亚硝基苯基二甲基吡唑
nitrosophenyl hydroxylamine 亚硝基苯胺
N-nitroso-piperidine N-亚硝基哌啶 $C_5H_{10}NNO$
nitroso-resorcinol 亚硝基间苯二酚 $ONC_6H_3(OH)_2;O=C_6H_3(=NOH)OH$
2-nitrosoresorcinol monomethyl ether 2-亚硝基间苯二酚一甲醚
nitroso-R-salt 亚硝基 R 盐；亚硝基-2-羟基-3,6-萘二磺酸钠；试钴铁灵 $ONC_{10}H_4(SO_3Na)_2OH$
nitroso rubber 亚硝基橡胶
3-nitroso salicylic acid 3-亚硝基水杨酸
nitroso-sulfuric acid 亚硝基硫酸 $NO·HSO_4$; $NO·OSO_2OH$

nitrosothymol(=thymoquinoneoxime) 亚硝基百里酚；百里醌肟 $ONC_{10}H_{12}OH; O=C_{10}H_{12}=NOH$
nitrosotoluene 亚硝基甲苯
N-nitroso-triacetonamine N-亚硝三乙酮胺
nitroso-2,2,4-trimethyl-1,2-dihydroquinoline polymer 亚硝基-2,2,4-三甲基-1,2-二氢化喹啉聚合物〔防焦剂〕
nitrosourea 亚硝基脲
nitro-styrene 硝基苯乙烯 $NO_2C_6H_4CH=CH_2$
nitrostyrolene 硝基苯乙烯
nitro-substitution 硝基取代
nitro-sugar 硝化糖
nitrosulfamide 亚硝基硫酰胺 H_2NSO_2NHNO
nitro-sulfonic acid 硝基磺酸
nitro-sulfuric acid 混酸〔浓硝酸和浓硫酸的混合物〕
nitrosulphonazo III 硝(基)磺偶氮III
nitrosulphophenol AE 硝(基)磺酚 AE
nitrosyl 亚硝酰(基) NO—
nitrosylation 亚硝基化(作用)
nitrosyl bromide 亚硝酰溴；溴化亚硝酰 NOBr
nitrosyl chloride 亚硝酰氯；氯化亚硝酰 NOCl
nitrosyl fluoride 亚硝酰氟 NOF
nitrosyl halide 卤化亚硝酰；亚硝酰卤
nitrosyllium hexafluorouranate 六氟铀酸亚硝酰；六氟铀合亚硝酰 $NOUF_6$
nitrosyl perchlorate 高氯酸亚硝酰
nitrosyl radical 亚硝酰基
nitrosyl ruthenium 亚硝酰钌
nitrosyl sodium 亚硝酰钠 NaNO
nitrosyl-specific detector 亚硝酰特效检测器
nitrosyl sulfate ①硫酸亚硝酰酯 $NO·HSO_4$ ②亚硝基硫酸盐(酯) $NOSO_4M$
nitrosyl sulfuric acid 亚硝基硫酸
nitrosyl sulfuryl chloride 亚硝酰磺酰氯
nitrotartaric acid 硝基酒石酸 $(NO_2COHCO_2H)_2$
nitroterephthalic acid 硝基对苯二酸 $NO_2C_6H_3(COOH)_2$
nitrothiophene 硝基噻吩
nitrothymol 硝基百里酚
nitrotoluene 硝基甲苯
nitrotoluidine 硝基甲苯胺
p-nitro-o-toluidine 对硝基邻甲苯胺
nitrotriaceton amine 硝基三丙酮胺
nitro-trichloromethane 硝基三氯代甲烷
nitrotrimethylolmethane 硝基三(羟甲基甲)烷 $NO_2C(CH_2OH)_3$
nitrotyl(=nitrotyl) 肟基
nitrotyrosine 硝基酪氨酸 $NO_2C_6H_3(OH)CH_2CH(NH_2)CO_2H$
nitroundecane 硝基十一(碳)烷

nitrourea 硝基脲 $NO_2NHCONH_2$
nitrourethane 硝氨基甲酸乙酯 $NO_2NHCO_2C_2H_5$
nitrous 亚硝的
nitrous acid 亚硝酸
nitrous acid ester 亚硝酸酯 RONO
nitrous anhydride 亚硝(酸)酐；三氧化二氮
nitrous compound 亚硝酸化合物
nitrous ether 亚硝基醚
nitrous fumes〔复〕亚硝烟
nitrous gases〔复〕亚硝气
nitrous oxide 一氧化二氮；氧化亚氮
nitrous oxide-acetylene flame 氧化亚氮-乙炔火焰；一氧化二氮-乙炔焰
nitrous oxide dosimeter 氧化亚氮剂量计
nitrous oxide-hydrogen flame 一氧化二氮-氢焰
nitrous oxide sensor 一氧化二氮传感器
nitrous vapours〔复〕亚硝蒸气
nitrous vitriol 硫酸硝气溶液
nitro-varnish 硝基漆；硝化纤维漆
nitrovinylation 硝基乙烯化(作用)
nitroxanthic acid 苦味酸
nitroxide mediated polymerization 氮氧自由基调控聚合
nitroxide radical 氮氧自由基
nitroxime 硝基肟
Nitrox process 一种氮同位素分离过程
nitroxyl 硝酰(基)
nitroxyl chloride 硝酰氯 NO_2Cl
nitro-xylene 硝基二甲苯 $(CH_3)_2C_6H_3NO_2$
nitroxyl fluoride 硝酰氟 NO_2F
nitroxylic acid(=hydronitrous acid) 亚硝酸 HNO_2
nitroxyl radical 硝酰游离基
nitroyl(=nitrotyl) 肟基
nitrum 天然碱
nitryl 硝酰；硝基
nitryl chloride(=nitroxyl chloride) 硝酰氯 NO_2Cl
nivalenol 瓜萎镰菌醇
nivalic acid 地衣酸；松萝酸
nivaquine 硫酸氯喹
nivenite 黑富铀矿
nivitin 山梨糖醇
nizin 尼锌；对氨基苯磺酸锌
nm.(=nanometer) 纳米〔10^{-9}米〕
n-mer n 聚物
NMR absolute sensitivity for detected of nuclei 被测核素的 NMR 绝对灵敏度
NMR apparatus 核磁共振仪
NMR computer tomography 核磁共振计算机断层显像法
NMR control 核磁共振控制

NMR imaging 核磁共振成像
NMR integral (analog) 核磁共振积分(模拟)
NMR line width 核磁共振谱线宽度
NMR microscopy 核磁共振显微学
NMR moisture meter 核磁共振水分计
NMR spectrometer 核磁共振(波谱)仪*
NMR spectrometer with superconducting magnet 超导核磁共振(波谱)仪*
NMR spectroscopy 核磁共振波谱法
NMR stabilizer 核磁共振稳定器
NMR time scale NMR 时标
NMR tomography 核磁共振断层显像法
nobelium 锘〔102 号元素,化学符号 No〕
Nobel oil 诺贝尔油;硝化甘油
nobiletin 川皮苷;川陈皮素
nobiline 贵石槲碱;金石槲碱
noble comber 圆梳机
noble electrode 惰性电极
noble gas 稀有气体
noble gas analysis 稀有气体分析
noble gas mass spectrometer 稀有气体质谱仪
noble laurel 月桂
noble liverwort 地钱
noble metal 贵金属
noble-metal thermocouple 贵金属型热电偶
noble potential 高电势
noble rubber 贵重橡胶〔指硅橡胶、氟橡胶〕
no-bond resonance 无键共振
no bond structure 无键结构
nocarbon recording paper 无碳复写纸
nocardamin 诺卡胺素
nocardianin 诺卡地素
nocardin 诺卡菌素
nocardorubin 诺卡菌红素
no carrier added(NCA) 不加载体*
no constant boiling mixture 非恒沸液
nodakenetin 闹达柯裂亭;紫花前胡苷元
nodakenin 闹达柯宁;紫花前胡苷元
nodal plane 节面
nodal point 节点
node ①(波)节*②结;结点
noded hemispheroid(=noded Horton spheroid) 多弧水滴形油罐
noded Horton spheroid(=noded hemispheroid) 多弧水滴形油罐
no-drip nozzle 无滴口喷嘴〔无排出凝液设备的喷嘴〕
nodular 结节的
nodular cast iron 球墨铸铁

nodular ore 肾状赤铁矿
nodular polymer 球状聚合物
nodular structure 结节结构
nodule ①根瘤②(不规则的)球;粒③小结
nodule point 节点
nodules 瘤状体
nodulizer 球化剂
nodulizing ①附聚作用②烧结作用
no-flow condition 无流动条件;无流动情况
noformicin 诺卡型霉素
nogalamycin 诺加霉素
noise 噪声
noise abatement 消除噪声;噪声减轻
noise abatement coating 消声涂料;消音涂料
noise abatement device 消声器
noise background 噪声背景;噪声本底
noise control 噪声控制
noise deadener ①消声器②隔音(涂)层
noise decoupling 噪声去偶
noise decoupling ^{13}CNMR 碳-13 噪声去偶核磁共振
noise energy 噪声能量
noise factor 噪声因数
noise killer 噪声限制器
noiseless 无噪声
noise level 噪声水平
noise margin 噪声容限
noise meter 噪声测试器;噪声计
noise modulated technique 噪声调制法
noise monitoring 噪声监测
noiseness 噪声(特性)
noise peak 噪声峰
noise pollution monitoring 噪声污染监测
noiseproof feature 防噪声特性
noise ratio 噪声比率
noise silencer 噪声消除器;减噪器
noise snubber 消声器
noise source 噪声源
noise spot ①噪声点;噪声(引起的)斑点②干扰点
noise stimulation 噪声激励
noise suppressor 噪声抑制器
noisy (有)噪声的
noisy base line 噪声基线
noisy chromatogram (有)噪声(的)色谱图
noisy peak 噪声峰
nojirimycin 野尻霉素
no-load 无负荷;空载
no-load characteristic 无载荷特性
no-load friction 无载荷摩擦

no-load nozzle 无载荷喷嘴
no-load run 空载运行；空负荷运行
no-load voltage 无载电压；空载电压
nologenin 延令草皂苷配基
no marking 无划痕
nomenclature 命名法；命名原则
nominal 标称的；额定的；公称的；名义的
nominal area of contact 名义接触面积
nominal aspect ratio （轮胎）公称扁平率
nominal count 公称支数
nominal depth of shade 公称色度；公称色泽深度
nominal diameter 公称直径
nominal diameter of filament 单丝公称直径
nominal dimension 公称尺寸
nominal horse-power 额定马力
nominally achromatic color 标称无彩色
nominal pipe size 管子公称直径；公称管径
nominal size 公称尺寸
nominal strain 标称应变
nominal stress 公标应力
nominal tex 公称特克斯
nominal twist 公称捻度
nominal volume 面值体积；标称体积
nominal wave length 标称波长
nominal weight 面值重量；标称重量
nomogram(=nomograph) 列线图；计算图
nomogram of peak width 峰宽列线图
nomograph(=nomogram) 列线图；计算图
nomography 列线图法
no moving part valve 无活动部件阀微泵
non- 〔拉丁字头〕 不；非；无
nona- 〔拉丁字头〕 九；壬
non-ab initio 非初始
nonablative material 非烧蚀材料
nonabrasive 非磨蚀性
non-abrasive quality 抗磨性
non-absorbency 不吸水性
non-absorbent 非吸收性的
non-absorbent finish 非吸湿性处理
non-absorbent oil 不被吸收油
non-absorbing medium 不吸收性介质
non-absorption line 非吸收线
non-abstractive 非抽提性
non-acid gas 非酸性气体
nonacidics 非酸性物
nonacid oil 非酸性油
non-acid wash 非酸洗
nonacosane 二十九（碳）烷 $C_{29}H_{60}$

nonacosanol 二十九(烷)醇
nonacosanyl alcohol(=montanyl alcohol) 褐煤醇；二十九(烷)醇 $C_{29}H_{59}OH$
nonacosyl 二十九(烷)基 $CH_3(CH_2)_{27}CH_2-$
nonactinic light 无光化性光；非光化性光
non-activated 未活化的
non-activated adsorption 非活化吸附
non-activated stock 未活化的胶料
nonactivating light 非激活化光；非活化光
nonacyclic 九元的〔指九元环的或九个环的〕
nonacyclic compound 九元环化合物
nonadditivity 非加和性的
nonadecadienoic acid 十九碳二烯酸 $C_{18}H_{33}COOH$
nonadecandioic acid(=nonadecane diacid) 十九烷二酸
nonadecane 十九(碳)烷 $CH_3(CH_2)_{17}CH_3$
nonadecane diacid(=nonadecandioic acid) 十九烷二酸 $C_{17}H_{34}(COOH)_2$
nonadecane dicarboxylic acid 十九烷二甲酸；二十一烷二酸 $C_{19}H_{38}(COOH)_2$
nonadecanoic acid 十九(烷)酸 $CH_3(CH_2)_{17}CO_2H$
nonadecanol 十九(烷)醇
2-nonadecanone(=methyl heptadecyl ketone) 甲基十七基(甲)酮；2-十九烷酮 $CH_3(CH_2)_{16}COCH_3$
nonadecendioic acid(=nonadecene diacid) 十九碳烯二酸 $C_{17}H_{32}(COOH)_2$
nonadecene diacid(=nonadecendioic acid) 十九碳烯二酸 $C_{17}H_{32}(COOH)_2$
nonadecene dicarboxylic acid 十九碳烯二甲酸
nonadecenoic acid(=nonadecylenic acid) 十九碳烯酸 $C_{18}H_{35}COOH$
nonadecyl 十九(烷)基 $CH_3(CH_2)_{17}CH_2-$
nonadecyl alcohol 十九(烷)醇
nonadecylenic acid(=nonadecenoic acid) 十九碳烯酸
nonadecylic acid 十九烷酸 $C_{18}H_{37}COOH$
non-adherent material 防黏剂
nonadherent protective layer 防黏保护层〔塑料〕
non-adhesiveness 无黏着性；非附着性
nonadiabatic 非绝热的
non-adiabatic cooling 非绝热冷却
non-adiabatic humidifier 非绝热增湿器
non-adiabatic rectification 非绝热精馏
nonadienal(=violet leaf aldehyde) 壬二烯醛；紫罗兰叶醛
2,6-nonadienal diethyl acetal 2,6-壬二烯醛二乙缩醛
nonadiendioic acid 壬二烯二酸 $C_7H_{10}(COOH)_2$
nonadienoic acid 壬二烯酸 $C_8H_{13}COOH$
nonadienol(=cucumber alcohol) 壬二烯醇；黄瓜醇
non-adsorptive support 非吸附性载体
non-aerated flame 无空气火焰

nonaffine deformation　非仿射形变
non-ageing　未老化的
nonagglomerating　非结块的
non-aggressive　无腐蚀性的
non-air blasting process　非空气喷砂(丸)处理法；无空气喷砂除锈法
γ-nonalactone　γ-壬内酯；椰子醛〔商〕；十八醛〔商〕
non-alcoholic beverage　非酒饮料；不含酒精的饮料
nonaldehyde　壬醛　$C_8H_{17}CHO$
nonalkaili metal sonsor　非碱金属离子传感器
non-alkali glass　无碱玻璃
nonalkali metal sensor　非碱金属离子传感器
non-allele　非等位基因
non-allergenie　非过敏症反应的；无过敏性反应的
non-alternant hydrocarbon　非交替烃*
nonalternant polyene chromogens　非交替的多烯类发色体
nonamer　九聚物
nonamethylene　1,9-亚壬基　$—(CH_2)_9—$
nonamethylene diamine　壬二胺
nonanal　壬醛
nonandioic acid　壬二酸　$C_7H_{14}(COOH)_2$
nonane　壬烷　$CH_3(CH_2)_7CH_3$
nonane diacid　壬二酸　$C_7H_{14}(COOH)_2$
1,9-nonanediamine　1,9-壬二胺　$H_2N(CH_2)_9NH_2$
nonane dicarboxylic acid　壬烷二甲酸；十一烷二酸　$C_9H_{18}(COOH)_2$
nonanediol　壬二醇
nonanediol-1,3-acetate　1,3-壬二醇乙酸酯
nonanedioyl　壬二酰　$—CO(CH_2)_7CO—$
nonanenitrile　壬腈
nonanoate　壬酸酯
nonanoic acid(=nonoic acid)　壬酸　$C_8H_{17}COOH$
nonanol　壬醇〔通常指1-壬醇〕
nonanone(=methyl n-heptyl ketone)　甲基庚基(甲)酮；2-壬酮　$CH_3CO(CH_2)_6CH_3$
2-nonanone(=methyl n-heptyl ketone)　2-壬酮；甲基庚基(甲)酮
nonanoyl　壬酰　$CH_3(CH_2)_7CO—$
nonanoyl vanillylamide　壬酰基香兰胺
nonapeptide　九氨酸肽
non-aqua-system gel column　非水系凝胶色谱柱
non-aqueous　非水的
non-aqueous adhesive　非水溶液(型)黏合剂
non-aqueous catalytic thermometric titration　非水催化热滴定
non-aqueous colloid　非水胶体
non-aqueous cooling medium　非水冷却介质

non-aqueous dispersion　①非水分散(作用)②非水分散体
nonaqueous electrochemistry　非水溶液电化学
nonaqueous finishing agent　非水整理剂
nonaqueous finishing composition　非水整理剂
non-aqueous gel permeation　非水凝胶渗透
non-aqueous ion exchange　非水离子交换
nonaqueous leaching　非水溶液浸出
non-aqueous media　非水介质
nonaqueous mill base　非水漆浆(颜料浆)
non-aqueous phase chromatography　非水相色谱
non-aqueous phase partition chromatography　无水相分配色谱法
non-aqueous phosphating　非水性磷化处理
non-aqueous phosphating primer　非水磷化底漆
non-aqueous polymerization　非水相聚合
non-aqueous processing　非水处理法
non-aqueous reprocessing　非水法后处理*
non-aqueous reversed phase chromatography　无水(相)反相色谱法
non-aqueous solubilization　非水增溶(作用)
non-aqueous solution　非水溶液
non-aqueous solvent　非水溶剂*
non-aqueous titrimetry　非水滴定(分析)法
non-aromatic　非芳香性的
non-aromatic hydrocarbon　非芳香烃
non-asphaltic base oil　非沥青基石油
non-asphaltic petroleum　非沥青质石油
non-asphaltic pyrobitumen　非沥青质热解残渣
non-asphaltic road oil　非沥青质铺路油
non-associated liquid　非缔合液体
non-associating　不缔合的
nonate　壬酸酯
nonatomic ring　九元环
non-azeotrope forming hydrocarbon　不生成共沸液的烃类
nonbasal slip　①非基本滑移②非底层滑移
non-base wash　非碱洗
nonbasic oxide　非碱性氧化物
non-benzenoid aromatic heterocycle　非苯型芳族杂环
non-benzenoid aromatics　非苯型芳族化合物
nonbenzenoid ether　非苯型醚
non-benzenoid hydrocarbon　非苯型烃
non-biodegradable　不能生物降解的
non-BKZ simple liquid(=non Bernstein-Kearsley-Zapus simple liquid)　非BKZ简单液体
non-black pigment　非黑色颜料〔指白色颜料和彩色颜料〕
nonbleeding　①不分流的②不分压的③不渗色

nonbleeding contrast surface　不渗色对照表面
non-bleeding grease　不分油的润滑脂
nonbleeding pigment　不渗色颜料
non-blocking　①不发黏②不黏结③无封闭(作用)
non-blooming　不起霜
non-blooming ingredient　不起霜的拼料
non-blooming stock　不起霜料
non-blooming sulfur　不喷霜硫黄
non-blushing thinner　防潮剂；防白药水〔俗〕
non-bonded interaction　非键相互作用*
nonbonding　非键(合)
nonbonding atomic orbital　非键原子轨道
nonbonding electron　非键电子
nonbonding MO(NBMO)　非键分子轨道
nonbonding (molecular) orbital　非键(分子)轨道
nonbonding orbital　非键轨道
nonbranched　未支化的
non-break oil(=break-free oil)　未裂解(的)油〔用机械法精制油〕
non-brittle　非脆性的
non-burning　不燃的
non-caking black　不结块的炭黑
non-caking coal　不黏结煤
noncalcined clay　未煅烧的瓷土
noncapacitive intermittent DC arc　非电容式间歇直流电弧
non-carbonate hardness　非碳酸盐硬度
non-carbon based polymer　非炭聚合物
non-carbon black filler　非炭黑填料
non-carbon oil　无炭油；不含悬浮炭的润滑油
noncarboxylic anionics　非羧酸盐阴离子表面活性剂
non-catalytic　非催化的
non-catalytic hydrogenation　非催化氢化
non-catalytic polymerization　非催化聚合
non-catalytic process　非催化过程
non-cellulosic fibre　非纤维素纤维；合成纤维
noncentral distribution　非中心分布；无中心分布
non-centrosymmetrical　非中心对称的
noncentrosymmetric structure　非中心对称结构
non-chalking　不粉化的
non-chalking type pigment　不泛白颜料
nonchelated　非螯合的
non-chemical admixture　非化学混合(物)；非化学掺合(物)
non-chlorine-retentive　不吸氯的
non-circular fibre　异形纤维；非圆形截面纤维
non-circulating lubrication　非循环润滑
non-classical　非经典的
nonclassical carbocation　非经典碳正离子*

non-cleanable part　不能清洗部件
non-clearance sleeve roller　无套差胶辊
non-clogging　不堵塞的
non coagulant　非凝固剂
noncoagulating medium　非凝固介质
non-coherent material　散粒状材料
non-coherent scattering　不相干散射
non-cohesive　不附着的
non-coking coal　不结焦煤
non-coking thermal cracking　非结焦热裂化
non-colligative property　非依数性
non-combustibility　不燃性
non-combustible　不燃的
noncombustible coating　非易燃涂层；不燃性涂层
non-combustible material　不燃性材料；不燃物质
non-compatible media　不混溶介质
non-competitive inhibition　非竞争性抑制
non-complete decoupling　不完全去耦
noncompliant coating　不合格的涂料
non-compounded petroleum products　未混合的石油产品
nonconcentric(al)　①非同心的；非同轴的；非同心圆的②非集中的
non-condensable gas　不(可)凝气体
non-condensable gas purger　放气器；不凝气体排除器
non-condensable hydrocarbon　不凝烃类
non-condensing　不凝的；不能冷凝的
nonconditional match(=nonmetameric match)　无条件等色配色；非条件等色配色〔一对颜色在任何照明条件下，由任一观察者观测，其色均相等〕
non-conducting　不(传)导的
non-conducting material　绝缘器材；不导器材
non-conductor　①非导体②非电导体；不导电体
non-congealable oil　不冻结的油类
noncongruent melting point　非相合熔点*
non-conjugated diene　非共轭二烯
non-conjugated fatty acid　非共轭脂肪酸
non-conjugated monomer　非共轭单体*
non-conjugated unsaturation　非共轭不饱和(结构)
non-conjugative monomer　非共轭单体
non-conservative force　非保守力
non-contact atomic force microscope　非接触式原子力显微镜
non-contacting measurement　非接触式测定
noncontamination fuel(=pollution-free fuel)　无污染燃料
non-conterminous　不相连的；不相邻的
non-continuous film　不连续(漆)膜
non-controllable extensional experiment　非可控延伸实验
non-convergency　(波长的)非收敛性

non-convertible 非转化型(的)
nonconvertible coating 非转化型涂料；非转化型涂层
non-convertible film 非转化性漆膜
non-convertible paint 非转化型漆
non-copolymerizable monomeric additive 非共聚(合)用单体添加剂
non-copolymer paint 非共聚物型漆
non-corrodibility 防腐性
non-corrodible 防腐的
non-corroding 不(腐)蚀的
non-corrosibility 非腐蚀性；防腐蚀性
non-corrosive 不锈的；无腐蚀性的
non-corrosive alloy 不锈合金；抗蚀合金
non corrosive lining 防蚀衬里
non-corrosive material 不锈材料
non-corrosiveness 无腐蚀性
non-counter ion 非抗衡离子
non-covalently connected micelles 非共价键合胶束
noncreasable 耐(揉)皱的
non-crease rayon 抗皱嫘萦
non-creasing fabric 不皱布
non-creasing finish 防皱整理
noncrossing rule 不相交规则
non-crushable 耐绉的
non-crystalline 非晶性的
non-crystalline chain orientation 非晶性链取向
non-crystalline copolymer 非晶性共聚物
noncrystalline domain 非结晶区；非晶畴
non-crystalline electrode 非晶体电极
non-crystalline graphite 非晶性石墨
non-crystalline polyolefin 非晶性聚烯烃
non-crystalline state 非晶态
non-crystalline wax 非晶性石蜡
non-crystallizing rubber 不结晶橡胶
non-crystallographic symmetry 非晶体学对称性
noncyclical 非环构的
noncyclic crown ether 非环状冠醚
non-cyclic phosphorylation 非循环磷酸化；非环式磷酸化
non-Daltonian compound 非道尔顿(式)化合物
nondecanoic acid(=nondecylic acid) 十九烷酸
nondecylic acid(=nondecanoic acid) 十九烷酸 $C_{18}H_{37}COOH$
non-degenerate 非简并的；非退化的
nondegraded surfactant 未降解的表面活性剂
non-deliquescent 不潮解的
non-destructive 非破坏性；无损的
non-destructive assay 非破坏性分析

non-destructive detector 非破坏性检测器
non-destructive distillation 非破坏蒸馏
nondestructive estimation 非破坏(性)检验
non-destructive examination 非破坏性检验
non-destructive hydrogenation 非破坏加氢
non-destructive inspection 非破坏性检查
non-destructive reagent 非破坏性试剂
non-destructive test 非破坏性试验；无损探伤
non-destructive testing(NDT) 非破坏性试验；无损探伤；无损试验
non-detachable mold 整体模
non-detectable 不能检测的
nondetergent fatty matter 非洗涤剂脂肪物；不皂化物
non-detergent oil 无去垢性油
nondetergent organic matter 非洗涤剂有机物；不皂化物
non-detonating fuel 不爆震燃料
nondiagram line 非图解线
nondialyzed 未渗析的
non-diffusible calcium 不扩散钙
nondiffusible ion 固定离子；非扩散离子
nondilatant mill base 非膨胀性研磨料
non-dimensional coefficient 无因次系数
non-dimensional parameter 无量纲参数
non-dimming eyepiece 保明片
non-direct current electrolysis 非直流电解
non-dischargeability 不可拔染性
non-discoloration 不褪色；不变色；不脱色
non-discoloring 不变色的
non-discoloring plasticizer 不褪色的增塑剂
non-dispersant 非分散剂的；不含分散剂的
non-dispersion atomic fluorometer 非色散原子荧光分析仪
nondispersive 非色散的
nondispersive analysis 非色散分析*
nondispersive atomic fluorescence spectrometry 非色散原子荧光光谱法
nondispersive spectrometry 非色散光谱法
nondispersive X-ray absorption analysis 非色散X射线吸收分析法
nondispersive X-ray absorption meter 非色散X射线吸收光度计
nondispersive X-ray analyzer 非色散X射线分析器
nondispersive X-ray fluorescence 非色散X射线荧光
nondispersive X-ray spectrometer 非色散X射线光谱仪
non-dissociated 未离解的
non-distillate oil 渣油
non-draining 无穿流
non-draining polymer 无穿流聚合物

non-drip spout 不喷溅溢流嘴
nondrying coating(=grease paint) 不干性油膏
non-drying finish 不干性油膏
non-drying oil 非干性油
non-ductile 无延性的
non-dusting 无粉尘的
non-dusting property 无粉尘性
none ①无②缺（石油产品规格用语）
none gloss varnish 平光清漆；无光清漆
non-elastic 非弹性的
nonelastic deformation 无弹性形变
non-elastic elongation 无弹性延伸
non-elasticity 非弹性
nonelectroactive ion 非电活性离子
non-electrolyte 非电解质；不电离质
non-electrolyte impurity 非电解质杂质
nonelectrolyte solution 非电解质溶液*
non-elution chromatography 非洗脱色谱(法)
nonempirical 非经验的
nonenal 壬烯醛
nonendioic acid(=nonene diacid) 壬烯二酸
nonene 壬烯 C_9H_{18}
nonene diacid(=nonendioic acid) 壬烯二酸 $C_7H_{12}(COOH)_2$
nonene dicarboxylic acid 壬烯二甲酸；十一烯二酸 $C_9H_{16}(COOH)_2$
nonenoate 壬烯酸盐（或酯）
nonenoic acid 壬烯酸 $C_8H_{15}COOH$
nonenol 壬烯醇
α-nonenyl aldehyde α-壬烯醛
nonenyl nitrile 壬烯腈
nonequilibrium 不平衡(的)
nonequilibrium elution 非平衡洗脱
non-equilibrium isotopic exchange method 非平衡同位素交换法
nonequilibrium plasma reactor 不平衡等离子反应器
non-equilibrium polycondensation 非平衡缩聚
non-equilibrium reaction kinetics 非平衡态反应动力学
nonequilibrium situations 非平衡状态
nonequilibrium state 非平衡态
nonequilibrium system 非平衡系统
nonequilibrium theory 非平衡理论
nonequilibrium thermodynamics 非平衡态热力学
non-essential amino acid 非必需氨基酸
non-essential element 非必需元素
non-essential fatty acid 非必需脂肪酸
non-essential water 非组成水(分)
non-etching degreaser 非侵蚀性脱脂剂

non-ex-area 防爆区
non-exchangeable ion 非交换离子
non-extractable 不可萃取的
non-extractable stabilizer system 非萃取性稳定剂体系
non-extractibility 不可抽提性；不可萃取性
nonfading 不褪色(的)
non-fading properties of gasoline 汽油的不变性质
nonfading stain 不褪色的染色剂
non-faradaic admittance 非法拉第导纳
non-faradaic current 非法拉第电流
nonfelting 不毡合的
non-ferrous 非铁的
non-ferrous alloy 非铁(金属)合金；有色(金属)合金
non-ferrous metal 有色金属*
non-ferrous metallurgy 非铁金属冶炼术；有色金属冶炼术
non-ferrous metal working lubricant 非铁金属加工润滑剂
nonfibrillated fibre 非原纤化纤维
nonfibrous filler 非纤维填料
non-fill 花纹缺陷；欠胶；填模不足
non film-forming dressing （木材表面用）不成膜的防腐处理剂
non-filterable 不可过滤的
nonfissionable material 不可裂变物质
non flam contact adhesive 不燃性压合式黏合剂；不燃性触压黏合剂
nonflame 非火焰；无焰
nonflame atomization 非火焰原子化法
nonflame atomizer 非火焰原子化器；无焰原子化器
nonflame atomizing device 非火焰原子化装置
nonflame cell 无焰吸收池
non-flame conveyor belt 防燃运输带
non-flame properties 抗燃性
nonflame sampling 无焰取样法
non-flammable 不易燃的
nonflammable coating 不燃性涂料
non-flammable gas 不燃性气体
nonflammable plasticizer 不燃性增塑剂
non-flowing material 非流动性材料；不流动材料
nonfluid gel 非流体凝胶
non-fluidized column 非流态化柱
non-fluid oil(=non-liquid oil) 不流动润滑油
non-foaming 不起泡的；无泡的
non-foaming detergent 无泡洗涤剂
non-foaming oil 不起泡沫润滑油
non-foaming test 不起泡沫试验
non-fogging plasticizer 不喷霜增塑剂；不栖移增塑剂

non-fouling　不污的
non-fractionating distillation　不分割的蒸馏
non-free-flowing powder　非散粒状粉剂
non-freezing　不冻的
non-freezing lubricating oil　不冻润滑油
non-freezing mixture　不冻混合物
non-freezing solution　不冻液
non-fuel uses of coal　煤的非燃料利用
non-functional compound　无官能化合物〔不含有官能团的化合物〕
non-fusible　不熔的；不能熔化的
nongassing zinc dust　无气泡锌粉〔不产生气体的锌粉〕；稳定型锌粉
non-Gaussian chain　非高斯链
non-Gaussian distribution　非高斯分布
non-Gaussian peak　非高斯峰
non-gel capillary electrophoresis　无胶筛分毛细管电泳
non-gel polymer　无凝胶聚合物
non-gel sieving　无胶筛分
nonglyceride　非甘油酯
nonglyceride component　非甘油酯组分
nongraphitizing carbon　非石墨化碳
non-gumming constituents in gasoline　汽油内不生胶成分
non-gumming fuel　不生胶染料
non-gumming properties　不生胶性质
non-halogen flame retardant cable jacketing materials　无卤阻燃电绝缘材料
non-hazardous　安全的；无危险的
non-hazardous area　安全区；无事故区
non-heat-set material　非热定形材料
nonheat-treated　未热处理的
nonhierarchical-cluster analysis　非系统聚类分析，非谱系聚类分析
nonhomogeneity　不均匀性
nonhomogeneous　①不齐的；非齐性的②不均匀的
non-homogeneous deformation　非均匀形变
nonhomogeneous film　非同质膜
non-homogeneous linear differential equation　非齐线性微分方程(式)
nonhomogeneous system　不均匀体系；非均相体系
non-Hookean elasticity　非胡克弹性
non-hydraulic　非水力的
non-hydrocarbon　非烃
non-hydrocarbon constituents　非烃成分
non-hydrocarbon motor fuel　非烃发动机燃料
nonhydrogen acid　非氢酸
nonhydrolyzable alkyl group　不水解的烷基；非水解性烷基
non-hydrolyzable tannin　非水解鞣质
nonhygroscopic　不吸湿的
non-icing　不结冰的
non-ideal gas　非理想气体
non-ideality parameter　非理想性参数
nonideal solution　非理想溶液*
nonideal solvent blend　非理想溶剂掺混物；非理想的混合溶剂
non-ignitable　不可燃的；不着火的
non-ignitibility　不可燃性
nonimaging macroprobe　非成像宏观探针
non-impingement method　无空气气体燃烧法
non-impinging jet injector　无交叉注流的喷射器
nonimpregnated　未浸渍(过)的
non-inflammability　不燃性
non-inflammable hydraulic fluid　不可燃传动液体
noninteracting　不相互作用的；不相互影响的
noninteracting solvent blend　不相互作用的混合溶剂
noninvasive detection　非破坏性检测
non-ionic　非离子的
non-ionic addition reaction　非离子的加成反应
non-ionic catalysis　非离子催化(作用)
non-ionic compound　非离子化合物
non-ionic detergent　非离子洗涤剂
non-ionic elution　非离子洗脱
nonionic emulsifier　非离子乳化剂
nonionic emulsifying agent　非离子乳化剂
non-ionic fat liquor　非离子(型)加油(脂)乳液
non-ionic hydrophilic component　非离子型亲水成分
non-ionic polyethylene oxide chain　非离子型聚氧化乙烯链
non-ionic reaction　非离子反应
nonionics(=non-ionic surface active agent)　非离子表面活性剂
non-ionic sorbent　非离子吸着剂
nonionic stabilization　非离子稳定作用
non-ionic surface active agent(=nonionics)　非离子表面活性剂
nonionic surfactant　非离子型表面活性剂*
nonionic thickener　非离子型增稠剂
nonionic thioether surfactant　非离子硫醚表面活性剂
nonionized polar group　非离子化极性基团
non-ionizing　不电离的
non-ionizing solvent　不电离溶剂
non-ionogenic linkage　非离子化键(合)
non-ionogenic surface-active agent　非离子表面活性剂
non-ionogenic tenside　非离子洗涤剂；非离子表面活

性剂
nonirradiated 未经照射的
non-irritant ①无刺激性的②非刺激剂(物)
nonirritating 无刺激性的
non-isoentropic flow 非等熵流
non-isothermal calorimeter 非等温量热器
non-isothermal condition 非等温条件
non-isothermal crystallization 非等温结晶
non-isothermal differential scanning calorimetry 非等温示差热扫描量热法
non-isothermal finite linear theory 非等温有限线性理论
non-isothermal flow 非等温流
non-isothermal growth 非等温生长法
non-isothermal operation of electrothermal atomizer 电热原子化器的非等温操作
non-isothermal theory 非等温理论
non-isothermal thermorheologically simple material 非等温热流变性简单材料
non-isothermal viscoelastic fluid 非等温黏弹流体
non-isotopic carrier 非同位素载体*
non-isotropic 非各向同性的
non-isotropic propagation of combustion waves 燃烧波的非同相延伸
non Kekule compound 非凯库勒化合物
nonkinetic analysis 非动态分析(法);非动力学分析(法)
non-knocking condition 不爆震条件
non-knocking fuel 不爆震燃料
non-labelled compound 未标记化合物
non-labelled immunoassay 非标记免疫分析法〔传统免疫分析〕
non-laminar flow 非层流
non-leaded gasoline 工业汽油〔不含四乙基铅的汽油〕
non-leafing aluminium 不漂浮型铝粉;非叶展型铝粉
non-leak detector 非泄漏检测器
non-leaving ligand 非脱离性配位体
non-linear ①非直线(型)的②非线性的
non-linear adsorption isotherm 非线性吸附等温线
non-linear body 非线性体
non-linear calibration 非线性校正
nonlinear chelatropic reaction 非线性螯合反应
non-linear chromatography 非线性色谱法
non-linear continuum mechanics 非线性连续介质力学
non-linear differential equation 非线性微分方程(式)
non-linear effect 非线性效应
non-linear elasticity 非线性弹性
non-linear elastic relation 非线性弹性关系式
non-linear equation 非线性方程
non-linear error 非线性误差
non-linear filtration 非线性过滤
non-linear flow 非线性流
non-linear heterogeneous fluid 非线性多相流体
non-linear ideal chromatography 非线性理想色谱(法)
non-linear isotherm 非线性等温线
non-linear iterative partial least square 非线性迭代偏最小二乘
non-linear iterative partial least square method 非线性迭代偏最小二乘法
non-linearity 非线性
non-linearity split stream 非线性分流
nonlinear large-deformation theory 非线性大形变理论
non-linear laser spectroscopy 非线性激光光谱学
non-linear law 非线性定律
non-linear least square fitting 非线性最小二乘拟合
non-linear mapping 非线性映射
non-linear material with memory 非线性记忆材料
non-linear molecule 非直线(型)分子
non-linear non-ideal chromatography 非线性非理想色谱(法)
non-linear optical effect 非线性光学效应
non-linear optic crystal 非线性光学晶体*
nonlinear optics 非线性光学
nonlinear partial least square method 非线性偏最小二乘法
non-linear phenomena 非线性现象
non-linear photoexcitation 非线性光激发*
non-linear polymer 非线型高聚物
non-linear Raman spectroscopy 非线性拉曼光谱
non-linear regression 非线性回归*
non-linear spectroscopy 非线型光谱法
non-linear spring 非线型弹簧
non-linear structure 非直线(型)结构
non-linear theory of viscoelasticity 非线性黏弹性理论
non-linear thermoviscoelastic solid 非线性热黏弹(性)固体
non-linear viscoelasticity 非线性黏弹性*
non-linear winding 非线性缠绕
non-liquid oil (=non-fluid oil) 不流动润滑油
nonlivering 不肝(稠)化的
non-local coupling 非定域偶合
non-locality 非定域性
non-localized 非定域的
nonlocalized adsorption 非定域吸附
non-localized bond 非定域键*
non-localized molecular orbital 非定域分子轨道
non-lubricated compressor 无(油)润滑压缩机
non-luminous 无光的;不发光的
non-luminous discharge 无光放电

non-luminous flame 无光焰
non-luminous gas 无焰气体；不发光气体
non-magnetic isomer 无磁性异构体
nonmagnetic mass spectrometer 非磁质谱计
nonmelting 不熔融(的)；不熔化(的)
nonmercurial fungicide 无汞防霉剂
non-metal 非金属
non-metal ion sensor 非金属离子传感器
non-metallic 非金属的
non-metallic additive 非金属添加剂
non-metallic coating 非金属涂料；非金属涂层
non-metallic element 非金属元素
non-metallic enamel 无金属光瓷漆
non-metallic inclusion 非金属夹杂物
non-metallic inclusion analysis 非金属夹杂物分析
non-metallic luster 非金属光泽
non-metallic minerals〔复〕 非金属矿物
non-metallic whisker 非金属须晶
nonmetal sublattice 非金属亚晶格
nonmetameric match(=nonconditional match) 无条件等色配色；非条件等色配色〔一对颜色在任何照明条件下，由任一观察者观测，其色均相等〕
non-metameric matching 等色配色
nonmicellar solution 非胶束溶液
non-migrating plasticizer 非迁移型增塑剂
non-migratory 不迁移的
non-mineral constituents 非矿质组分
non-miscible 不(可)溶混的
nonmonotonic error 非单调性误差
nonneppy 无粒结的；无纤维结构
non-Nernstian behavior 非能斯特行为
non-Newtonian coating 非牛顿型涂料
non-Newtonian flow 非牛顿流动*
non-Newtonian fluid 非牛顿流体*
non-Newtonianism 非牛顿性
non-Newtonian liquid 非牛顿液体
non-Newtonian material 非牛顿材料
non-Newtonian parameter 非牛顿参数
non-Newtonian power-law fluid 非牛顿幂律流体
non-Newtonian suspension 非牛顿悬浮体
non-Newtonian viscosity 非牛顿黏度
non-noble metal 非贵金属
nonnormal distribution 非正态分布*
non-normalized relaxation 非标准化松弛
non-nucleated structure 非成核结构
nonoic acid(=nonanoic acid) 壬酸
non oil base 非油基
non-oil degummed ramie 无油精干麻
non oil lubrication 非油润滑
non-oleaginous 非油质的；非油性的
non-oleaginous lubricant 非油性润滑剂
non-orthogonal 非正交的
nonorthogonality 非正交性
nonose 壬糖
non-oxidative deamination 非氧化性脱氨(作用)
non-oxidizable 非氧化性的；不(可)氧化的
non-oxidizing 不氧化的
non-oxidizing atmosphere 非氧化(性)气氛；非氧化环境
non-oxidizing furnace 非氧化炉
non-paired spatial orbital(NPSO) method 非成对空间轨道法
nonparameter test 非参数检验*
nonpenetrating coagulant 非渗透性凝固剂
nonpenetrating stain(=oil stain) 非渗透性着色剂；油性着色剂
non-permanent flow 暂流
non-persistent 不持久的；不稳定的
non-persistent accelerator 非持久性促进剂
non-persistent gas 非持久性毒剂；暂时性毒气
non-petroleum base 非石油基
non-petroleum fluid 非石油质液体
non-petroleum lubricant 非石油润滑剂
non-petroleum origin 非石油起源的
non-petroleum sources 非石油来源
non-phenolic tanning agent 非酚类鞣剂
nonphosphate builder 非磷酸盐助洗剂
nonphosphate detergent 无磷酸盐洗涤剂
non-pigmentary type 非颜料型
non-pigmented dispersion 无颜料的分散液；未着色的分散液
non-pilling 不起球的
non-pinking spirit 不爆震汽油
nonplanarity 非平面性
nonplanar surface 不同平面的表面
non-plastic material 非塑性材料
nonpoint source 非点(污染)源
non-polar 非极性的
non-polar action 非极性作用
non-polar adsorption 非极性吸附
non-polar bond 非极性键*
non-polar bonded phase 非极性键合相
non-polar compound 非极性化合物
non-polar dissociation 非极性离解(作用)
non-polar double bond 非极性双键
non-polarizable 不可极化的；不可偏振的
non-polarizable electrode 不极化电极

non-polarized light 非偏振光；无偏振光
non-polar link 非极性键
non-polar linkage 非极性键(合)
non-polar liquid 非极性液体
non-polar molecule 非极性分子
non polar monomer 非极性单体
non-polar organic molecule 非极性有机分子
non-polar organic reaction 非极性有机反应
non-polar phase 非极性相
non-polar pigment 非极性颜料
non-polar polymer 非极性聚合物*
non-polar resin 非极性树脂
non-polar rubber 非极性橡胶
nonpolar sample 非极性试样
non-polar side chain 非极性侧链
nonpolar solubilization 非极性增溶作用
non-polar solvent 非极性溶剂*
non-polar stationary liquid 非极性固定液
non-polar stationary phase 非极性固定相
non-polar support 非极性载体
non-polar thin film column 非极性薄膜柱
nonpolar Van der Waals force 非极性范德华力
non-polluting coatings 无污染涂料
nonpolluting method 无污染的方法
non-pollution adhesives 无公害黏合剂
non-pored wood 无孔木材
nonporous 无孔的
non-porous catalyst 无孔隙催化剂
non-porous material 无孔材料
non-porous monodisperse packing 无孔单分散填料
non-porous paper 非多孔性纸张
non-porous rubber 密实胶料
non-potable water 非饮用水
non-premium grade gasoline 普通汽油
non-pressure cure 无压硫化
non-pressurized 不加压的；常压的
nonprimitive lattice 复格子
nonpropagating carbonium 非(链)增长型碳鎓离子
non-protective coating 非保护性涂料
non-protein nitrogen(NPN) 非蛋白氮
non-protein-nitrogenous compound 非蛋白氮化合物
non-protonic solvent 非质子性溶剂；无质子溶剂
non-pyrogenic generator 无热原(同位素)发生器
non-pyrogenic kit 无热原药箱
nonquaternary 非季(铵)盐的
nonquaternary cationics 非季铵盐型阳离子表面活性剂
nonquenchable pigment 非猝灭性颜料
nonradiative process 无辐射过程

non-radiative relaxation 无辐射弛豫
non-radiative transition 非辐射跃迁
nonradical cleavage 非游离基裂解
non-radioactive component 非放射性组分
non-radioactive electron capture detector 非放射性电子俘获检测器
nonradiogenic 非放射成因的
nonrandom cross-linked polymer fibre 非无规交联聚合物纤维；梯形聚合物纤维
nonreactant 非反应物质；惰性物质
non-reactive black 非活性炭黑
non-reactive group 非活性基
non-reactive phase 不反应相
non-recording instrument 非记录式仪器
non-recoverable creep 不可恢复蠕变
non-recoverable deformation 不可恢复形变
non-recoverable flow 非返原性流动；塑性流动
non-recoverable shear 不可恢复剪切
non-recoverable storage 不能再回收的储存；永久储存
non-reducing sugar 非还原糖
non-refinable crude 重石油；不加工的原油
non-reflective surface 非反射性表面；无反射的表面
non-refueling duration 不加燃料时间
non-regenerative oven 非再生炉
non-regenerative process 非再生过程
non-reinforced stock 未加固的胶料
nonrelativistic 非相对论(性)的
non-renewable energy resources 不可再生能源
non-reproducible sample 非重复性试样
non-residuum cracking 无残油裂化
non-resinous oil 无树脂质润滑油
non-resolved peak 未分开的峰
non-resonance atomic fluorescence 非共振原子荧光
non-resonance line 非共振线
non-resonance method 非共振法
non-resonant vibrometer 非共振型振动器
non-resonant vibroscope 非共振示振仪
non-retained reference 不被保留的参比物；无保留值的参比物
non-retained solute 无保留溶质
non-return flap valve 止回阀
non-return-to-zero method 不回零法
non-return trap 不可复汽阱
non-return valve(=check valve) 止逆阀；止回阀
non-reversibility 不可逆性
non-reversible 不可逆的；不能反转的
non-reversible coating 不可逆涂层
non-reversible crimp 不可逆卷曲

non-reversible deformation　不可逆形变
non-reversible process　不可逆过程
non-reversible reaction　不可逆反应
non-reversible work hardening　不可逆加工硬化
non-reversible work softening　不可逆加工软化
non-rigid molecule　非刚性分子
non-rigid plastics　非刚性塑料
nonrigid ro(ta)tor　非刚性转子*
non-rising stem　不升杆
non-rising-stem valve　不升杆式阀
non-rosinated type toner　非松香酸盐型色料
non-rusting　不生锈的
non-rusting immersion heater　不生锈的浸没式加热器
non-rust steel　不锈钢
non-sacrificial graphite anode　非牺牲(性)石墨阳极
non-saponifying　不皂化的
non-saponifying oil　不皂化润滑油
non-saturated colour　非饱和色；不饱和色
non scattering material　非散射材料
non-scheduled maintenance　非正规维修；不定期维修；计划外维修
nonscum compound　防脏剂
non-seasonal goods　非应季品；非季节性产品
non-selecting cracking　非选择性裂化
non-selective analyzer　非选择性分析仪
non-selective cracking　非选择裂化
non-selective detector　非选择性检测器
non-selective entrainer　非选择溶剂
non-selective herbicidal oil　无选择性除草油
non-selective oxidation　非选择性氧化(作用)
non-selective phase　非选择性相
non-selective pulse for excitation　非选择性脉冲激发
nonselective reduction process for nitric acid　非选择性还原法制硝酸
nonselective (thermal) excitation　非选择性(热)激发*
nonself　非己的
non-selfigniting bipropellant　不自燃的二组成喷气机燃料
non-sensitized emulsion　非增感乳剂；非增敏乳剂
non-serviceable　不能使用的
nonsetting red lead　不沉降性红丹
non-settling　不沉降的
non-settling slurry　不沉降性料浆
non-shatterable glass　抗震玻璃
non-shock chilling　静止冷却；无振动冷却
non-shrink　防缩*
non-silanized support　非硅烷化载体
non-simple material　非简单材料

nonsingular　非奇异的
non-sinusoidal deformation　非正弦形变
non-sinusoidal oscillation　非正弦振动
non-skid coating　防滑涂料
nonskid deck covering　①防滑甲板革②甲板防滑敷面
non-skid deck paint　防滑甲板漆
nonskid paint　防滑漆
non-sludging oil　不沉淀油；无胶质油；氧化稳定油
non-slurry pelletizing　干法造粒
nonsoap-forming vehicle　不成皂基料；非成皂性漆料
non-soap grease　无皂润滑脂
nonsoap(y) detergent　非皂洗涤剂；合成洗涤剂
non-soluble　不溶解的
non-solvent　非溶剂
non-solvent adhesive(=solventless adhesive)　无溶剂型黏合剂；无溶剂型胶黏剂
non-solvent organosol　非溶剂型有机溶胶
non-solvent type plasticizer　非溶剂型增塑剂
nonsolvent water　不起溶剂作用的水〔指对聚合物起溶胀作用的水〕
non-specific absorption　非特征吸收
non-specific interference　非特征干扰
non-spectral colour　非光谱色
nonspherical particle　非球形粒子
nonspherulitic　非球粒状的；非球晶状的
nonspontaneous process　非自发过程*
nonspray drying process　非喷雾干燥成型工艺
non-spreading　不铺展的；不流散的
non-staining　不污染的
non-staining antioxidant　无污染的抗氧化剂
non-standard equipment　非标设备
non-stationary phase of chain reaction　连锁反应的不稳定相
non-steady flow　非恒稳流动；非稳流
non-steady state　非稳定态
nonstereospecific polymer　非定向聚合物；无规立构聚合物
non-sticking wax　无黏性石蜡
non-sticky　不黏结的；不粘连的
non-sticky sludge　①不黏废漆渣②不黏的泥浆(淤渣)；不黏的沉积物
nonstoichiometric　非化学计量的；非整比的
nonstoichiometric composition　非化学计量组成；非整比组成
nonstoichiometric compound　非整比化合物*
nonstoichiometry　非化学计量性；非定比性
non-stop continuous flow screen changer　网式(熔体)连续过滤器；连续过滤网切换器

non-stop operation 连续操作
non-stop run 连续工作；不间断工作
non-stream lined body 非流线型物体
non-stretch bulked yarn 非伸缩性(膨化)变形纱
non-structural adhesive 非结构黏合剂〔黏合非受力部件用〕
non-structure black 非结构性炭黑
non-sucrose 非糖物
non-sugar 非糖物
non-sulfur pulping 无硫制浆
non-sulfur vulcanization 非硫硫化*
non-superconducting 不超导的；非超导性的
non-superimposable mirror image 不能重叠的镜像
non-suppressed conductance detection 非抑制型电导检测
non-suppressed ion chromatography 非抑制型离子色谱法
non-surface-active 非表面活性的
non-surface-active agent 非表面活性剂
nonsurfactant 非表面活性剂
nonsurfactant material 非表面活性物
non-sweating wax 未发汗石蜡
non-swelling 不溶胀的
non-swelling acid 非膨胀性酸
non-synchronous 非同步的
non-systematic error 偶然误差；非系统误差
non-tacky 非黏性的；不黏的
nontacky mill base 非黏着性研磨料
non-tan(=non-tanning substance) 非鞣质
non-tannin(=non-tanning substance) 非鞣质
non-tanning substance(=non-tan) 非鞣质
non-tans(=non-tanning substance) 非鞣质
non-tapered conduct 非锥形导像管
N-terminal N 末端；氨基末端
nonterminal double bond 非末端双键
nonterminal olefin 非末端烯烃；内烯烃
nonterminated polymer 未封端聚合物
nontermination 非终止；无终止
nonthermal type atomizer 非热原子化器
nonthixotropic 非触变的
nontolerance 不允许残留
non-torque textured filament 不捻回变形长丝；非捻回变形长丝
non-torque type coil like crimping 无扭矩螺旋状卷曲
non-toxic 无毒的
non-toxic catalyst 无毒催化剂
non-toxic coatings 无毒涂料
non-toxic formulation 无毒配方
non-toxic fungicide 无毒防霉剂

nontoxic fungistat 无毒制霉剂
nontoxicity 无毒性
non-toxic paint 无毒漆
non-toxic pigment 无毒颜料
non-toxic plasticizer 无毒增塑剂
non-toxic smoke candle 无毒烟幕罐
non-toxic stabilizer 无毒稳定剂
nontracer isotopic method 非示踪物的同位素法
non-tracking 无(电花)径迹
non-transparency 不透明性
nontransparent 不透明的
non-trivial solution 非平凡解
nontronite 囊脱石
non-tunable 不可调的
non-turbulent 非湍流的
non-turbulent flow 非湍流
non-uniform adsorption 非均匀吸附
non-uniform combustion 不均匀燃烧
non-uniform flow 非匀流
non-uniformity 不均匀性
non-uniform pipe 不匀导管
non-uniform polymer 多分散性聚合物
non-uniform scale 不均匀刻度
nonuploid 九倍体
nonuploidy 九倍性
nonurea adduct-forming material 尿素不加合物
non-UV absorbing 不吸收紫外光的
non-valent 无价的；零价的
non-variant system 无变量系
nonvinylic 非乙烯的
non-viscous distillate 低黏度馏出物
non-viscous flow 非黏性流(动)
non-viscous Hookean material 非黏性胡克材料
non-viscous lubrication 非黏滞润滑
non-viscous neutral (oil) 不黏中性油
non-volatile 不挥发的
non-volatile bonding agent 非挥发性键合剂
non-volatile content 不挥发分；不挥发物含量
non-volatile matter 不挥发物质
non-volatile swelling agent 不挥发的溶胀剂
non-volatility 不挥发性
non-volatilizable residue 不挥发性残余物
non washable distemper 不耐擦洗的水浆涂料
non-washing plate and frame filter press 不可洗式板框式压滤机
non-Weissenberg fluid 非韦森堡流体
non-weldable steel 不可焊接的钢
non-wettable 不可润湿的

non-wetting 不润湿的
nonwinterized oil 未冬化油
non-wiping 不移位的〔黏合剂〕
nonwoven card 非织造梳理机
non-woven fabrics 无纺布*
non-woven scrim 非织造稀松布
nonwovens insulation material 非织造隔声隔热材料
nonwover-like fibre 非织造布状制法薄膜纤维
non-yellowing 不泛黄性；不泛黄的
nonyellowing coating 不泛黄涂料；不变黄涂料
non-yellowing film 不泛黄漆膜
non-yellowing polyurethane resin 不泛黄聚氨酯树脂
non-yellowing properties 不泛黄性
non-yield point behaviour 无屈服值特性
nonyl 壬基 $CH_3(CH_2)_7CH_2-$
nonyl acetate 乙酸壬酯
nonyl acid(=pelargonic acid) 壬酸
nonyl alcohol 壬醇 $CH_3(CH_2)_7CH_2OH$
nonyl aldehyde 壬醛
nonyl aldehyde dimethyl acetal 壬醛二甲缩醛
nonyl amine 壬胺 $CH_3(CH_2)_7CH_2NH_2$
nonyl anthranilate 邻氨基苯甲酸壬酯
nonylate 壬酸酯
nonylated diphenylamine 壬基化二苯胺〔防老剂〕
nonyl benzene 壬基苯
nonyl caproate 己酸壬酯 $C_5H_{11}CO_2C_9H_{19}$
nonyl caprylate 辛酸壬酯
nonyl carbinol 癸醇
nonyl dimethyl benzyl ammonium halide 壬基二甲基苄基卤化铵
nonylene 壬烯
nonylene-2,4-xylenol 壬烯基-2,4-二甲基苯酚〔防老剂〕
nonyl formate 甲酸壬酯 $HCO_2C_9H_{19}$
nonylic acid 壬酸 $C_8H_{17}COOH$
nonyl iodide(=1-iodononane) 1-碘壬烷 $CH_3(CH_2)_7CH_2I$
nonyl isobutyrate 异丁酸壬酯
γ-nonyl lactone γ-壬内酯；椰子醛〔商〕；十八醛〔商〕
nonyl methacrylate 甲基丙烯酸壬酯
nonylone 二辛基(甲)酮
nonyl phenol 壬基酚
nonyl phenol EO 壬基酚环氧乙烷；壬基酚聚氧乙烯醚
nonyl phenol ether phosphate 壬基酚醚磷酸盐
nonyl phenol ether sulphate 壬基酚醚硫酸盐
nonyl phenol ethoxylate 壬基苯酚乙氧基化物
nonyl phenol formaldehyde resin 壬基酚甲醛树脂
nonylphenol polyethylene glycol 壬基苯酚聚乙二醇〔湿润剂〕
p-nonylphenylphenylamine 对壬基苯基苯胺〔防老剂〕

α-n-nonylpridine-N-oxide α-正壬基吡啶氧化物 $C_9H_{19} \cdot C_5H_4NO$
nonyl propionate 丙酸壬酯
nonyne 壬炔
nonzeolite 非沸石
non-zero 无零
nonzero spin 非零自旋
non-zinc epoxy-primer 无锌环氧底漆
noodle-cutting machine 空心面切制机
nook 角落；转角处；隐蔽处
nootkatane 诺卡烷
nootkatene 诺卡烯
nootkatin 香柏素
nootkatone 诺卡酮
nopinane 诺品烷；诺蒎烷 C_9H_{16}
nopinene 诺品烯 $C_{10}H_{16}$
nopinic acid 诺品酸
nopinol 诺蒎醇 $C_9H_{16}O$
nopinone 诺蒎酮
nopol 诺卜醇
nopyl 诺卜(醇)基
nopyl acetate 乙酸诺卜酯
nor- ①降；去甲②正〔指 normal〕
noradrenaline bitartrate 去甲肾上腺素〔药〕
noragathenic acid 原贝壳松烯酸
noralpinone 降良姜酮
noratropine 降阿托品；降颠茄碱；去甲阿托品
N orbit N 轨道
norbixin 降胭脂树素 $C_{24}H_{28}O_4$
norbornadiene 降冰片二烯
norbornane 降冰片烷 C_7H_{12}
norbornanol 降冰片烷醇
norbornene 降冰片烯
5-norbornene-2,3-dicarboxylic acid (=carbic acid; nadic acid) 5-降冰片烯-2,3-二酸；降冰片烯二酸
5-norbornene-2-methanol(=2-hydroxy methyl-5- norbornene) 5-降冰片烯-2-甲醇；2-羟甲基-5-降冰片烯〔涂料用聚合物改性剂〕
5-norbornene-2-methyacrylate 5-降冰片烯-2-甲醇丙烯酸酯〔醋酸乙烯乳液聚合物交联剂〕
norborneol 降冰片
norbornylene 降冰片烯
norcamphane 降莰烷 C_7H_{12}
norcamphanyl 降莰基 $C_7H_{11}-$
norcamphene 降莰烯 C_8H_{12}
norcamphor 降樟脑
π-norcamphor(=santenone) π-降樟脑；檀烯酮
norcamphotenic acid 降龙脑烯酸

norcarane 降莰(烷) C_7H_{12}
norcarene 降莰烯 C_7H_{10}
norcarenone 降莰酮
norcaryophllenic acid 降石竹烯酸
norcholane 降胆烷
norcholesterol 降胆甾醇
norconidendrin 降铁杉内酯
nordenskiöldine 硼钙锡矿
Nordhausen acid 发烟硫酸
nordihydroguaiaretic acid 去甲二氢愈创木酸
nordin 诺定素
norecsantalal(=nortricycloekasantalal) 降檀香醛
norecsantalic acid 降(外)檀香酸
norecsantalol 降(外)檀香醇
norephedrine 降麻黄碱;去甲麻黄碱
norepol 多羟醇-聚植物油酸缩合物
no-residuum cracking 非残油式裂化
norethandrolone 乙诺酮〔药〕
norethindrone 炔诺酮
norethisterone 炔诺酮〔药〕
norethynodrel 异炔诺酮-炔雌醇甲醚
norfloxacin 诺氟沙星〔药〕
norfolk island pine 南洋杉
norhomocamphoric acid 降高樟脑酸
norhyoscine(=norscopolamine) 降莨菪胺
nor-hyoscyamine 降天仙子胺
no-rinse-system 无冲洗系统
norisocampholic acid 降异龙脑酸
norite 苏长岩
norium 混合稀土金属
norlaudanosoline 全去甲劳丹碱;去甲基罂粟碱
nor-leucine 正亮氨酸;正白氨酸
norleucyl 正亮氨酰(基);己氨酰;正白氨酰(基) $CH_3(CH_2)_3CH(NH_2)CO—$
norm ①范数;模方〔数〕②规范
Norma-Hoffman stability test 挪玛-哈夫门安定性试验
normal ①垂直(的);正交(的);法向(的)②正规;正常;标准③正态;简正④当量(的);规度(的)
normal acid 正酸
normal aliphatic primary mercaptan 正构脂肪(族)伯硫醇
normal alloy 正常合金
normal analytical zone 正常分析区
normal atmosphere 正常大气;标准大气
normal atom 常态原子
normal benzene 标准汽油
normal benzine 标准轻质汽油
normal boiling point 标准沸点
normal burette 标准滴定管

normal butane 正丁烷
normal butanol 正丁醇
normal butyl alcohol 正丁醇〔溶剂〕
normal calomel electrode 当量甘汞电极
normal camphor oil 樟脑原油
normal carbon chain 正碳链;直碳链
normal cell 标准电池
normal cellulose 正常纤维素
normal chain 正链;直链
normal combustion (混合物的)完全燃烧
normal complex 正常配合物
normal component 法向分量
normal compound 正构化合物
normal concentration 规(定浓)度;标准浓度;(克)当量浓度
normal condition(=normal state) 标准(状)态
normal cooling 正常冷却
normal covalency 正常共价
normal curve 正态曲线
normal deformation 法向形变
normal density 正常密度
normal deviate 正态偏差
normal direct-line fluorescence 正常直跃线荧光
normal dispersion 正常分散
normal distribution 正态分布*
normal dull 普通消光;正规消光
normal duty conveyor 中型运输带;中负荷运输带
normal elastic modulus 正弹性模量
normal electrode 标准电极
normal electrode potential 标准电极势
normal element 标准电池
normal emissivity 标准发射率
normal equation 正规方程
normal extension 法向拉伸
normal fluid 标准液体
normal force 法向力;垂直力
normal freezing 正常凝固
normal gas solution 标准气体吸收液
normal Gaussian distribution function 正态高斯分布函数
normal glass 标准玻璃
normal glow discharge 正常辉光放电
normal glucose tolerance 正常糖耐量
normal ground joint 标准磨口接头
normal growth 正常生长
normal heptane (正)庚烷
normal hexane 正己烷
normal homologue 正同系物

normal hydrocarbon　直链烃
normality　规度；规定浓度；当量浓度
normality test　正态性检验
normalization　①归一法 ②正态化 ③正常化
normalization condition　归一化条件
normalization factor　归一化因子
normalization mass　归一化质量
normalization method　归一化法
normalization of multi-chromatogram of GC　气相色谱多谱图归一定量法
normalization of ratios　同位素比归一化
normalization of spectra　谱归一化
normalize　①规度化 ②正常化 ③正火〔冶金〕
normalized　正规化的
normalized intensity　归一化强度*
normalized structure factor　归一化结构因子
normalizing　①规度化 ②正常化 ③正火〔冶金〕
normal juice　原汁
normal latex　普通胶乳
normal lead silico chromate　正硅铬酸铅〔包核颜料〕
normal line　图表线
normal liquid　正常液体
normal load　法向载荷
normal mode　简正模式*
normal mode of vibration　简正振动
normal modulus　中定伸强力
normal octane　正辛烷
normal olefine　正(构)烯烃；直链烯烃
normal operating losses(NOL)　正常运行损失
normal operation　正常操作
normal operation loss(NOL)　正常运行损失
normal paraffin　正链烷
normal paraffin hydrocarbon(NPH)　正烷烃 C_nH_{2n+2}
normal paraffin sulfate　直链烷烃硫酸盐
normal phase　正相
normal phase bonded plate　正相键合板
normal phase chromatography　正相色谱法
normal phase ion-pair chromatography　正相离子对色谱法
normal phase method　正相(展开)法
normal portland cement　正常波特兰水泥〔即普通水泥〕
normal potential　标准电位(势)
normal pressure　(正)常压(力)〔760 毫米汞柱压力〕
normal pressure and temperature　标准状态
normal pressure chemiionization method　常压化学电离法
normal pressure chemiionization source　常压化学电离源
normal pressure effect　法向压力效应
normal pressure reactor　常压反应器

normal pressure synthesis　常压合成
normal probability paper　正态概率纸
normal pulse　常规脉冲法
normal pulse polarography(NPP)　常规脉冲极谱法
normal pulse voltammetry　常规脉冲伏安法
normal range　正常范围
normal rayon twist　正常人造丝捻度
normal reduction potential　标准还原电势
normal running temperature　正常操作温度
normal saline　生理盐水
normal salt　中式盐；正盐
normal saturation　(展开槽)常规饱和
normal saturation chamber　常规饱和式展开槽
normal sea water　标准海水
normal shut-down　定期检修；正常停工
normal silver-silver chloride electrode　标准银-氯化银电极
normal solution　规度溶液；当量溶液
normal state(=normal condition)　标准(状)态；基态
normal stepwise-line fluorescence　正常阶跃线荧光
normal stress　法向应力*；正应力
normal stress coefficient　法向应力系数
normal stress difference　法向应力差
normal stress effect　法向应力效应
normal stress phenomena　法向应力现象
normal stress pump　法向应力泵
normal structure black　中结构炭黑
normal substitution　正常取代(作用)
normal sulfur cure(=elemental sulfur cure)　(元素)硫硫化
normal temperature　①(正)常温(度) ②标准温度
normal temperature oscillating dyeing machine　常温振荡试样机
normal temperature-pressure(NTP)　标准温(度)压(力)；标准状态
normal test sieve　标准试验筛
normal trichromatism　正常(三)色视觉
normal uranium　标准铀〔天然铀 ^{235}U 含量为 0.7115%〕
normal valency　(正)常价
normal variate　正态变量
normal vibration　简正振动*
normal viscosity　正常黏度；标准黏度
normal volume　正常体积
normal wear　正常磨损
normal weight　①规定量 ②正常重量
normal Zeeman effect　正常塞曼效应
normenthane　降蓋烷；异丙基环己烷
normetanephrine　去甲变肾上腺素；去甲-3-O-甲基肾上腺素

normocytin 维生素 B_{12} 和维生素 B_{12b} 制剂
norm system 岩石分类法〔按理论的化学成分〕
normuscol 环十五醇
normuscone 环十五酮
nornicotine 降烟碱；去甲烟碱 $C_9H_{12}N_2$
nornidulin 降巢曲菌素
noropianic acid 降鸦片酸 $C_8H_6O_5$
nor paraf(=normal paraffin) 正链烷烃
norpinane 降蒎烷 C_7H_{12}
norpinic acid 降蒎酸 $(CH_3)_2C_4H_4(COOH)_2$
norpluviine 降雨石蒜碱；去甲雨石蒜碱
norprodigiosin 去甲灵菌红素
d-norpseudoephedrine(=cathine) d-降假麻黄碱
norsabite 洛沙巴特
norsantonin 降山道年；去甲山道年
norsclareol oxide 降香紫苏醚
norscopine 降莨菪品〔取自茄科 Datura metel〕
norscopolamine(=norhyoscine) 降莨菪胺
norsolanellic acid 氧化胆汁酸
norsterane 降甾烷
norsteroid 降甾族化合物；去甲甾类
norstictic acid 降斑点酸
19-nortestosterone 19-去甲睾(甾)酮；诺龙
nortricycloekasantalal(=norecsantalal) 降檀香醛
nortropine 降托品；盐酸去甲托品醇
nortropinone 降托品酮；降颠茄酮；去甲托品酮
norvancomycin 去甲万古霉素〔药〕
norwegian saltpeter 挪威硝石；挪威硝酸钙肥料
no sag 无流挂〔流挂等级之一〕
noscapine 那可汀〔药〕
nose 刀尖
nosean 黝方石
noseanit 黝方岩
nose bar 凸杆；撑杆
nose cup 鼻罩
nose gas 喷嚏(性毒)气；致嚏(性毒)气
noselite 黝方石
nosepiece 换镜旋座；目镜转换盘
nose primer 信嘴
no smear time 不黏尘干时间
nosophen 四碘酚酞
no-swell finish 抗溶胀整理
not-acidified 未酸化的；未加酸的
notalin 桔霉素
notalysin 青霉溶素；帚菌溶素
notatin ①点霉素②葡萄糖氧化酶
notation 符号
notation and classification of nuclear spin system 核自旋体系的表示法与分类
notation of crystal-face 晶面标记法
notch 凹口；缺口；切口
notch bar 凹口试杆
notch bar impact test 凹口试杆冲击试验
notch bending test 切口挠曲试验
notch brittleness 缺口脆性
notch ductility 缺口塑性
notched-bar impact test 切口试条冲击试验
notched edge of blade 带槽的刮刀刃
notched gage 齿状测厚仪
notched impact strength 切口冲击强度
notch effect 切口效应
notch sample 有刻痕试片
notch sensitivity 切口敏感性
notch toughness 缺口韧性
notch (weir) 凹口
not detected 未检测出的
note 香(韵)；香调；韵调；香型
not-go gauge 不过端量规
not good merchantable 不良商品；深茶色〔石油产品颜色标号〕
nothosmyrnium oil 紫茎芹油
nothosmyrnol 紫茎芹脑
noticeable wear 明显磨损；显著磨损
notomycin 脊霉素
no-touch production 不触摸生产(法)
not-reversible 不可逆的；不能反转的
not-saturated 不饱和的
not specified 未说明技术条件的；未标明的
no-twist roving (for over-unwinding) 绝对无捻粗纱
noumeite 硅镁镍矿
nourishing cream 有营养的乳油
nourishing paste 羔鞣剂〔革〕
nourrice tank 燃料桶
nourseothricin 诺尔丝菌素
novaculite 磨石
novain 肉毒素
novarsenobenzene(=novarsenobenzol) 新洒尔佛散；914；新胂凡钠明 $C_{13}H_{13}O_4As_2N_2Na$
novarsenobenzol(=novarsenobenzene) 新洒尔佛散；914；新胂凡钠明
novaspirin 新阿司匹林
novel 新(奇)的
novic acid 诺维酸
novobiocic acid 新生酸
novobiocin(=albamycin;biotexin;cardelmycin) 新生霉素
novocain(e)(=procaine) 普鲁卡因；奴佛卡因

novocain oxide 氧化普鲁卡因
novolac(=novolak) 线型酚醛清漆
novolac adhesive 酚醛树脂胶黏剂
novolac epoxy 酚醛型环氧树脂
novolac resin 酚醛清漆树脂
novomycin 新新霉素
novursane 诺乌烷
nowak process 湿式纤维板制法
nowetting 不润湿
n-oxide of aminophosphine oxide 氨氧化物氧膦(类化合物)
noxious 有毒的；毒害的
noxious effect 有害作用；有毒作用
noxious gas 有害气体；有毒气体
noxious solvent 有毒溶剂
noxiversin 解毒菌素
nozzle ①(喷)嘴②喷油嘴
nozzle chip (喷枪的)喷头
nozzle clamp 喷嘴夹
nozzle erosion 喷嘴腐蚀(烧蚀)
nozzle filter 喷丝头滤器
nozzle flame 喷嘴(上部)火焰
nozzle flow meter 喷嘴流量计
nozzle manifold 支管注嘴
nozzle mixer 喷嘴混合器
nozzle of retort 烧甄喷嘴
nozzle opening reinforcement 接管开孔补强
nozzle process 喷嘴法
nozzle separation 喷嘴分离(法)*
nozzle-skimmer 喷嘴分离器
nozzle valve 喷嘴阀
^{239}Np cow 镎239的"母牛"〔指^{243}Am, 可从"挤牛奶"中提取^{239}Np〕
npsh(=net positive suction head) 净正吸入压头
N shell N层
nsutite 六方锰矿, 恩苏塔锰矿
nuance 香调；香气近似；(色调、音调等的)细微差别
nuancing 变调
nub ①小块(尤指粘粑脂硬块)②小瘤；小疖
nubecula 混浊(症)
nubuck 正绒面革
nub yarn 结子纱
Nuchar 纽恰尔牌活性炭
nucidine 努斯定
nuciferal 日楂醛
nuciferine 莲碱
nuciferol 日楂醇
nuciform 核桃状的

nucin(=juglone) 核桃酮
nucite 环己六醇
nuclear ①核的②(原子)核的③有核的；含核的
nuclear activation analysis 核活化分析
nuclear analytical method 核分析(法)
nuclear artillery (原子)核炮队
nuclear atom 含核原子；有核原子；核型原子
nuclear atom theory 含核原子学说
nuclear battery 核电池*
nuclear boiling(=nucleate boiling) 泡核沸腾
nuclear-binding energy 核结合能
nuclear bombardment 核子轰击
nuclear bond 核键
nuclear bromination 环上溴代反应
nuclear carbon 环中的碳
nuclear-centered wave function 核中心波函数
nuclear charge 核电荷*
nuclear chemical engineering 核化学工程
nuclear chemistry 核化学*
nuclear chemistry and technology 核化学化工*
nuclear chlorination 环上氯代反应
nuclear clock 核时钟
nuclear compound 含环化合物
nuclear cratering 核(爆破)开洞；核开坑
nuclear decay constant 核衰变常数
nuclear decomposition 核衰变；核分裂
nuclear dipole-dipole interaction 核偶极-偶极作用
nuclear disintegration energy 核蜕变能
nuclear division 核分裂
nuclear ecology 核生态学
nuclear electric quadrupole interaction(NEQI) 核电四极相互作用
nuclear electric quadrupole moment 核电四极矩
nuclear electron 核电子
nuclear electronics 核电子学
nuclear energy 核能
nuclear envelope 核膜
nuclear equation (原子)核方程式
nuclear excitation by electron transition(NEET) 电子跃迁致核激(效应)；尼特(效应)
nuclear facilities 核设施
nuclear fission (原子)核分裂(作用)
nuclear fluorination 环上氟代反应
nuclear fragmentation 核断裂
nuclear fuel 核燃料*
nuclear fuel cycle 核燃料循环*
nuclear fuel reprocessing 核燃料后处理*
nuclear fusion 核融合

nuclear gamma-ray resonance adsorption(NGR)　核γ射线
　　共振吸收〔即穆斯堡尔效应〕
nuclear geochemistry　核地球化学*
nuclear g factor　核 g 因子
nuclear halogen　环上卤素
nuclear halogenation　环上卤代反应
nuclear homologue　核同系物
nuclear homology　核同系性
nuclear induction　核子感应
nuclear iodination　环上碘代反应
nuclear isobar　核同量异位素
nuclear isomer　(核)同质异能素*
nuclear isomerism　同核异构(现象)
nuclearity　核性
nuclear know-how　核知识；核技能
nuclear Lande factor　核朗德因子
nuclear laser　核激光(器)
nuclear logging　核测井*
nuclear magnetic double resonance(NMDR)　核磁双共振
nuclear magnetic moment　核磁矩*
nuclear magnetic multiple resonance spectroscopy　多重核磁共振波谱法
nuclear magnetic-nuclear quadrupole double resonance　核磁-核四极双共振
nuclear magnetic relaxation　(原子)核磁(性)松弛
nuclear magnetic resonance(NMR)　核磁共振*
nuclear magnetic resonance computerized tomography (NMRCT)　核磁共振计算机化断层显像*
nuclear magnetic resonance condition　核磁共振条件
nuclear magnetic resonance imaging　核磁共振成像
nuclear magnetic resonance spectrometer　核磁共振波谱仪
nuclear magnetic resonance spectroscopy　核磁共振波谱法*
nuclear magneton　核磁子〔磁矩单位〕
nuclear mass　核质量
nuclear medicine　核医学*
nuclear membrane　核膜
nuclear microanalysis　核微量分析
nuclear microprobe　核微探针*
nuclear mining　核(爆炸)采矿
nuclear nitrogen　环中的氮
nuclear Overhauser effect(NOE)　核欧沃豪斯效应*
nuclear packing　核(量)紧束(作用)
nuclear paramagnetism　核顺磁性
nuclear particle　①(原子)核粒②(子)核子
nuclear partition function　核配分函数*
nuclear pharmaceuticals　核药物*

nuclear pharmacy　核药学
nuclear pore film　核孔膜
nuclear power plants　核电站
nuclear precession　核进动
nuclear pressure vessel　核压力容器
nuclear proton　核质子
nuclear pumped laser　核(泵浦)激光
nuclear pure　核纯(的)
nuclear quadruple resonance(NQR)　核四极共振
nuclear quadrupole coupling　核四极耦合
nuclear quadrupole double resonance spectrometer　核四极双共振波谱仪
nuclear quadrupole interaction　核四极矩相互作用
nuclear quadrupole moment　核四极矩*
nuclear quadrupole resonance　核四极共振
nuclear radiation　核辐射
nuclear radiation detector　核辐射探测器
nuclear reaction　(原子)核反应
nuclear reaction analysis　核反应分析
nuclear reaction cross section　核反应(横)截面
nuclear reactor　核子反应器
nuclear relaxation　核弛豫
nuclear resonance　核共振
nuclear resonance fluorescence　核共振荧光
nuclear resonance level　核共振能级
nuclear resonance spectrograph　核共振摄谱仪
nuclear resonance spectroscopy　核共振光谱学
nuclear safeguards technique　核保障监督技术*
nuclear sap　核液
nuclear satellite signal　核伴线信号
nuclear screening constant　核屏蔽常数
nuclear separation　核间距离
nuclear single side band method　核单边带法
nuclear soling　微孔大底
nuclear spin　核自旋*
nuclear spin echo　核自旋回波
nuclear-spin quantum number　核自旋量子数
nuclear spin resonance　核旋共振
nuclear spin resonance spectroscopy　核自旋共振波谱法
nuclear stimulation　核刺激(法)
nuclear stockpile　核储备
nuclear structure　①核结构②(原子)核结构
nuclear substituted　环上取代的
nuclear substitution　环上取代(作用)
nuclear symbol　(原子)核符号
nuclear theory　(原子)核学说
nuclear track technique　核径迹技术
nuclear transformation　(原子)核转变

nuclear transmutation 核蜕变
nuclear Zeemann correlated electron-nuclear double resonance 核塞曼相关电子-核双共振
nucleate boiling(=nuclear boiling) 泡核沸腾
nucleated 核的
nucleated nylon 核化耐纶
nucleated polypropylene 核化聚丙烯
nucleated resin 核化树脂
nucleating 核化
nucleating agent 成核剂
nucleating glass 成核玻璃
nucleating point 晶核形成点；成(晶)核点
nucleation 成核作用
nucleation theory 核晶形成理论；成核理论
nucleator 成核剂；核化剂
nuclei〔复〕 ①核②晶核
nucleic acid 核酸
nucleicacidase 核酸酶
nucleic acid blot technique 核酸吸印技术
nucleic acid probe 核酸探针
nucleic acid purification system 核酸提纯系统
nucleidic mass 原子量〔据占多数的天然同位素计算〕
nuclei mode 核模；艾肯粒子
nuclein ①核素②核质
nucleinate 核酸盐
nucleo-cosmochronology 核子宇宙年代学
nucleolus（复 nuclei) ①核②环核③核心
nucleon 核子；单子
nucleonic density meter 核子密度计
nucleonics 核子学〔物〕
nucleonic(s)gage 测厚规
nucleonic specific gravity meter 核子比重计
nucleophile 亲核体*
nucleophile-assisted unimolecular electrophilic substitution 亲核体协助单分子亲电取代*
nucleophilic 亲核的
nucleophilic addition 亲核加成
nucleophilic atom localization energy 亲核原子定域能
nucleophilic displacement 亲核置换
nucleophilic group 亲核(性)基团
nucleophilicity 亲核性*
nucleophilic reaction 亲核反应*
nucleophilic reagent 亲核试剂
nucleophilic substitution 亲核取代
nucleophobic 疏核的；受电子的；亲水的
nucleose 核蛋白
nucleosidase 核苷酶
nucleoside 核苷

nucleoside diphosphate 二磷酸核苷
nucleoside monophosphate 单磷酸核苷
nucleotidase 核苷酸酶
nucleotide 核苷酸
nucleus ①环②(原子)核；(晶)核③细胞核 ④核心
nucleus formation ①(晶)核生成(作用)②(原子)核生成(作用)
nucleus formation speed 晶核形成速度
nucleus of crystal 晶核；晶籽；结晶中心
nucleus of flame （引入燃烧混合物的)火花
nucleus substituted 环上取代的
nucleus substitution 环上取代反应
nuclide 核素*
nuclide chart 核素图*
nuclide generator 核素发生器
nuclide mass 核素质量
nuclophilic reactivity 亲核反应性
Nuffield weatherometer 纽费尔德老化试验机
nugget 熔核
nuisance 公害
nuisance analysis 公害分析
nuisanceless technology 无公害工艺
nujol ①医药用润滑油②液体石蜡
nujol mull 石蜡糊；医用石蜡研糊
nujol mull method 石蜡糊法
null ①零位②零
null adjustment 零位调节
null comparison method 示零补偿法
null-current potentiometric titration 零电流电位滴定
null-current potentiometry 零电流电位法
null detector 零点检测器
null electrode 零(位)电极
null hypothesis 零假设*；原假设
null indicator 示零器
null instrument 平衡点测定器
null method(=zero method) 零点法；衡消法
null method of measurement 零差测量法
null point 零点
null-point pH 零点 pH
null-potential pH 零电位 pH
null potential pM 零电位 pM
null seeking recorder 找零记录仪
null type 零型
null valence 零价
null valency 零价
number ①数(目)；值②数字③(计)数
number-average degree of polymerization 数均聚合度
number-average kinetic chain length 数均动力学链长度

number-average micellar weight　数均胶束量
number-average molar mass　数均分子量*
number-average molecular weight　数均分子量*
number distribution function　数量分布函数*
numbering　位次编排
number mean molecular weight　数均分子量
number of components　组分数*
number of effective plates　有效塔板数*
number of effective theoretical plates　有效理论塔板数
number of ethylene oxide units　环氧乙烷单位数
number of flexing　挠曲次数
number of flexing to break　挠曲寿命；挠曲断裂次数
number of gill faller　针梳机针板号数
number of level　水平数
number of observations　观测值数目；测定值数目
number of oxyethylene unit　氧乙烯单位数；环氧乙烷单位数
number of passes　程数〔热变换器〕
number of plates　塔板数
number of product molecules　产品分子数
number of replications　重复(测定)次数
number of theoretical plates　理论塔板数*
number of transfer unit　传质单元数
numeral　①数字②数字的
numerate　数；计算
numerator　分子〔数〕
numeric　①数值的；数字的②分数
numerical　数字的
numerical analysis　数值分析
numerical aperture　数值孔径
numerical characteristic of distribution　分布的数字特征
numerical control　数字控制
numerical data processing　数据处理
numerical equation　数字方程(式)
numerical estimate　数值估计
numerical genetic algorithm　数值遗传算法
numerical integration　数字积分
numerical mean length　数值平均长度
numerical method　数值计算法
numerical taxonomy　数值分类
numerical value　数值
numerical variable　①数字变量②数变词；数变项
numeric character　数字符号
numeroscope　数字记录器；示数器
numoquin　纽莫昆；乙基氢化叩卜林〔药〕
nungu oil　牛油树油〔取自 Butyrospermum Parkii 的种子〕
nupercaine(=dibucaine)　地布卡因；奴白卡因
nupharidine　萍蓬汀
nupharine　萍蓬碱
Nurex process　纳雷克斯法〔尿素脱蜡法〕
nursimycin　诺尔斯霉素
nurture　①营养物②培体
Nusselt number　努塞尔特数
nut　①螺母；螺(旋)套（物）②铁螺母③硬果；坚果
nutans type barley　曲穗大麦
nutch　上下真空滤器
nutgall　梧子；五倍子；没食子
nutmeg　肉豆蔻
nutmeg butter　肉豆蔻脂
nutmeg oil　肉豆蔻油
nut mill　脱壳磨机
nut oil　胡桃油
nutrient　营养素；营养品；养分
nutrient allowance　营养允许量；营养需要量
nutrient matter　营养物质
nutrilite　营养料
nutriment　营养物
nutrimental　营养的
nutrition　①营养②营养学
nutritional　营养的
nutritional significance　营养意义
nutrition analysis of fruits and vegetables　果蔬营养分析
nutritive value　营养(价)值
nutritive value index　营养价指数
Nutsche funnel　布氏漏斗
nutsch filter　(真空吸滤)过滤器
Nutting-Scott Blair equation　纳丁-斯科特布莱尔方程
Nutting's equation　纳丁方程
Nutting's spectrophotometer　纳丁分光光度计
nutty　坚果香
nux moschata　肉豆蔻
nux vomica　马钱子；番木鳖
N→V transition　N→V 跃迁
NW acid　NW酸；萘-1-酚-4-磺酸
n-well　n 阱
nybomycin　尼博霉素
nyctalopia　夜盲(症)
nyctanthin　感夜树素
Nylander reagent　酒石酸钾钠、氢氧化钠和硝酸氧铋的溶液
Nylander test　尿中糖检验
nylon　锦纶(尼龙)
nylon base laminate　尼龙布基层合塑料
nylon blended fabric　尼龙混纺织物
nylon chiffon　尼龙薄绸
nylon paint　聚酰胺(树脂)涂料；尼龙涂料〔俗〕

nylon paste 尼龙浆
nylon powder coating(s) 尼龙粉末涂料
nylon reinforcing mesh 尼龙增强(筛)网
nylon tow sewing thread 尼龙复丝缝纫线
nyphimycin 尼菲霉素

nysfungin 制霉素〔药〕
nystagmus 眼球震颤
nystatin(=fungicidin) 制霉菌素
Nytril fibre 奈特里尔纤维；偏氰乙烯共聚物纤维

O

oak 栎(树)；槲(树)；橡(树)
oak apple 栎子
oak bark 栎树皮；槲树皮
oak bark extract 槲树皮浸膏〔皮革〕
oak bark tannage 槲树皮鞣法〔皮革〕
oak extract 栎栲胶
oak moss 橡苔；栎扁枝衣
oak moss absolute 橡苔净油
oak-moss resin 栎藓树脂；橡苔树脂
oakum 麻根；麻穰子；填絮
oak varnish 栎木清漆；橡木清漆
oak-wood extract 槲木浸膏
oar leather 桨革
oat 燕麦；雀麦
oaten 燕麦的
oat malt 燕麦芽
oatmeal ①燕麦粉②燕麦粥
oatmeal paper 粉纸
obakulactone 黄柏内酯
obakunone 黄柏酮 $C_{26}H_{30}O_7$
obamegine 黄皮树碱
oba oil 喔巴油；加蓬依苦木油
O-belt(=round belt) 圆型带；圆断面形带
Obermüller's test 奥伯米勒胆甾醇试验
Oberphos 过磷酸钙肥料粒
obital pairing 轨道成对
obital symmetry 轨道对称性
object ①事物②目的；宗旨③反对
object angle 物角
object color ①客观色彩②物体色彩(颜色)
object distance 物距
object glass 物镜；载物片；载玻片
objectionable constituent 有害成分
objectionable impurities 有害杂质
objectionable intermingling 互不相容；不易黏合，互不交融
objective ①物镜②目的；目标③客观的
objective aperture (显微镜的)物镜孔(径)
objective astigmatic corrector 物镜像散校正器
objective driving 目标驱动
objective function 目标函数
objective lens 物镜
objective program 目标程序
objective size 目的纤度

object lens 物镜
object mode of appearance 物体外观(表观)形态
oblate ellipsoid 扁(圆)椭球体
oblate ellipsoid head contour 扁椭球封头曲面
oblate symmetric top molecule 扁对称陀螺分子
oblatum 糯米纸；淀粉纸
oblique angle 斜角〔包括锐角和钝角〕
oblique bore stopcock 斜孔活栓；斜孔活塞
oblique crystal 单斜晶
oblique joint 斜接接头
oblique lattice 斜点阵
oblique ray 斜射(光)线
oblique reflection 斜反射
obliterating power ①遮盖力②涂盖力；埋(湮)没力
obovoid 倒卵形
observability 可观测性
observation ①观察；观测②按语
observation angle 观测角
observation door 看火门
observation error 观测误差
observation height 观测高度
observation port 看火门
observation sites in flame 火焰观察位置
observatory 天文台；观象台
observed difference 观测(值)差
observed temperature 观测温度
observed value 观测值；测量值
obsidian 黑曜岩
obstruction 阻碍
obstruction factor 阻碍因子
obturator ring 闭塞环；活塞环
obtusatic acid(=ramalic acid) 树花地衣酸
obtusilic acid 4-十碳烯酸；4-癸烯酸
obtusin 纯菇菌素；决明素；1,7-二羟基-2,3,8-三甲氧基-6-甲基蒽-9,10-二酮
obtutin 奥枯草菌素
occasional lubrication 非正规润滑
occasional stirring 间歇搅拌
occidane 金钟烷
occidentalol 金钟柏醇
occluded 封留的；夹杂的；包藏的
occluded gas 包藏气
occluded oil 吸着油
occluded resin 吸着树脂

occluded water 吸留水；包藏水
occludent 吸留剂
occluder 遮光板
occlusion 包藏*
occlusion cellulose 包藏纤维素；夹杂纤维素
occlusion coprecipitation 吸留共沉淀
occlusion polymerization 包藏聚合
occult 隐蔽的；看不见的
occulting light 明灭灯；明灭相间灯；连闭灯
occult peak 隐蔽峰
occupational exposure 职业曝射；职业暴露
occupied surface state 占据表面态
occurrence ①存在②遭遇；事件
ocean depot 海洋燃料库
ocean development 海洋开发
ocean-going tanker 海洋油轮
oceanic sediment 海洋沉积
oceanium 海洋群落
ocean pollution 海洋污染
ocean terminal 海洋基地
ocher(=ochre) ①赭石②赭色
ocherous deposit 赭色沉积物
ocher red 豆沙红
ochraceous 赭土的；赭色的
ochran 黄铝土
ochre(=ocher) ①赭石②赭色
ochre mutation(=UAA mutation) 赭石突变
ochreous 赭土的；赭色的
ocimene 罗勒烯 $C_{10}H_{16}$
ocimenone 罗勒烯酮
O'connel's relation 奥康内尔关系〔塔效率与液体黏度、挥发度的关系式〕
ocotilla wax(=ocotillo wax) 墨西哥刺木蜡
ocotillo wax(=ocotilla wax) 墨西哥刺木蜡
octa-(=octo-) 〔希腊字头〕八；辛
octabromide test 八溴化物试验〔鉴别油脂中鱼油含量的方法〕
octabromo diphenyl 八溴联苯〔阻燃剂〕
octachlor 氯丹
octachloro-camphene(=toxaphene) 八氯莰烯；毒莰烯
octachlorocyclopentene 八氯环戊烯〔硫化剂〕
octachloropropane(=perchloropropane) 八氯丙烷，全氯丙烷 C_3Cl_8
octacosane 二十八(碳)烷 $C_{28}H_{58}$
octacosyl 二十八(烷)基 $CH_3(CH_2)_{26}CH_2—$
octad ①八价物②八价的③八素组
octadeca- 〔希腊字头〕十八
octadecadienoic acid 十八碳二烯酸 $C_{17}H_{31}COOH$

octadecamethylene diamine 十八烷二胺
octadecandioic acid(=octadecane diacid) 十八烷二酸；十八双酸 $C_{16}H_{32}(COOH)_2$
octadecandiol 十八烷二醇
octadecane(=octodecane) ①(正)十八(碳)烷②十八(碳)(级)烷
octadecane diacid(=octadecandioic acid) 十八烷二酸；十八双酸 $C_{16}H_{32}(COOH)_2$
octadecane dicarboxylic acid 十八烷二甲酸；二十烷二酸 $C_{18}H_{36}(COOH)_2$
octadecanoate 十八(烷)酸盐；硬脂酸盐
octadecanoic acid 十八(烷)酸；硬脂酸 $C_{17}H_{35}COOH$
octadecanol 十八(烷)醇
octadecanonaene 十八碳九烯
octadecanoyl 十八(烷)酰基 $CH_3(CH_2)_{16}CO—$
octadecaoctaene 十八碳八烯
octadecendioic acid(=octadecene diacid) 十八(碳)烯二酸
1-octadecene 1-十八(碳)烯 $CH_3(CH_2)_{15}CH=CH_2$
octadecene diacid(=octadecendioic acid) 十八(碳)烯二酸 $C_{16}H_{30}(COOH)_2$
octadecene dicarboxyli acid 十八(碳)烯二羧酸；二十(碳)烯二酸 $C_{18}H_{34}(COOH)_2$
octadecene nitrile 十八(碳)烯腈
octadecenic acid 十八(碳)烯酸 $C_{17}H_{33}COOH$
octadecenyl alcohol 十八烯醇
octadecenyl aldehyde 十八烯醛
octadecenyl amine 十八烯胺
octadecenyl dimethyl ethyl ammonium chloride 十八烯基二甲基乙基氯化铵
octadecoic acid(=octadecylic acid) 十八烷酸；硬脂酸
octadecyl(=octodecyl) 十八(碳)烷基 $CH_3(CH_2)_{16}CH_2—$
octadecyl acetate 乙酸十八(烷)醇酯
octadecyl alcohol 十八(烷)醇 $CH_3(CH_2)_{16}CH_2OH$
octadecylamine 十八胺〔防焦剂〕
octadecyl ammonitm bentonite 十八烷基铵膨润土
octadecyl ammonium caprylate 辛酸十八(烷)铵
octadecyl ammonium carboxylate 羧酸十八(烷)铵
octadecyl bromide 十八(烷)基溴
octadecyl carbamate 氨基甲酸十八烷基酯
octadecyl chloromethyl ether 十八烷基氯代甲基醚
N-octadecyldiethanolamine N-十八烷基二乙醇胺〔防老剂〕
octadecyl dimethyl benzyl ammonium chloride 氯化十八烷基二甲基苄基铵〔稳定剂〕
octadecyl dimethyl benzyl ammonium halide 十八烷基二甲基苄基卤化铵
octadecylic acid(=octadecoic acid) 十八(烷)酸；硬脂酸 $C_{17}H_{35}COOH$

octadecylic iodide 十八(烷)基碘；1-碘代十八碳烷 $CH_3(CH_2)_{16}CH_2I$	octa-klor(=chlordane) 氯丹；八氯化甲桥茚
octadecyl isocyanate 异氰酸十八醇酯〔硫化剂〕	octal ①八进制的②八面的；八边的
n-octadecyl mercaptan 正十八烷硫醇	γ-octalactone γ-辛内酯
octadecyl methacrylate 甲基丙烯酸十八酯	δ-octalactone δ-辛内酯
octadecyl palmitate 棕榈酸十八(烷)醇酯	n-octaldehyde 正辛醛
octadecyl silane 十八烷基键合硅胶	octaldehyde dimethyl acetal 辛醛二甲缩醛
octadecyl trimethyl ammonium bromide 溴化十八烷基三甲基铵〔硫化剂〕	octalin 八氢化萘
	octalobal cross section 八叶形横截面
octadecyl trimethyl ammonium chloride 氯化十八烷基三甲基铵〔阻燃剂〕	octamer 八聚物
	octamethylcyclotetrasiloxane 八甲基环(化)四硅氧烷；八甲基化环四硅氧烷
octadecyne 十八(碳)炔	
octadecynoic acid 十八(碳)炔酸	octamethylene 1,8-亚辛基 —$(CH_2)_8$—
octadentate 八齿(的)；八合(的)	1,4-octamethylenebenzene 1,4-亚辛基苯
octadiene 辛二烯 C_8H_{14}	octamethylenediamine 辛二胺
octadienoic acid 辛二烯酸 $C_7H_{11}COOH$	octamethyleneglycol(=1,8-octandiol) 1,8-亚辛基二醇；1,8-辛二醇 $HO(CH_2)_8OH$
octadienol 辛二烯醇	
2,3-octadione(=acetyl-caproyl) 2,3-辛二酮 $CH_3(CH_2)_4COCOCH_3$	octamethylpyrophosphoramide(OMPA) 八甲基焦磷酰胺
	octamethyltrisiloxane 八甲基三硅氧烷
octaethyl cyclotetrasilazane 八乙基环丁硅氮烷；八乙基环四硅氮烷	octamylose ①八糖体②八聚淀粉糖
	octanal 辛醛
octaethylporphyrin(OEP) 八乙基卟啉	octanal dimethyl acetal 辛醛二甲缩醛
Octafining process Octafining 工艺〔邻、间二甲苯催化异构化制对二甲苯〕	octandioic acid(=octane diacid) 辛二酸
	1,8-octandiol(=octamethyleneglycol) 1,8-辛二醇；1,8-亚辛基二醇 $HO(CH_2)_8OH$
octafluorohexane diol 八氟己二醇	
octafluoro-3,5-hexanedione(OFHD) 八氟-3,5-己二酮 $C_2F_5COCH_2COCF_3$	octane ①(正)辛烷②辛(级)烷
	octane carboxylic acid 壬酸
octa-fluorozirconate 八氟锆酸盐 $M_n(ZrF_8)$	octane curve 辛烷曲线
	octane diacid(=octandioic acid) 辛二酸 $C_6H_{12}(COOH)_2$
octaforming 八碳重整	
octafunctional initiator 八官能起始剂	octane dicarboxylic acid 辛烷二甲酸；癸二酸 $C_8H_{16}(COOH)_2$
octafunctional stabilizer 八官能稳定剂	
octagonal cross-section 八角形横截面	octanedioic acid 辛二酸
octahedra 八面体	octanedioyl 辛二酰 —$CO(CH_2)_6CO$—
octahedral 八面的	octane fade 辛烷值下降
octahedral arrangement 八面体排列(结构)	octane improvement 改善辛烷值；改善爆炸稳定性
octahedral compound 八面体化合物*	octane level 辛烷值；辛烷特性
octahedral hybrid orbital 八面体杂化轨道	octane number(=octane rating) 辛烷值
octahedral hydroxyl group 八面体排列的羟基基团	octane number nomograph 辛烷值列线图
octahedral radius 八面体半径	octane number response (发动机对汽油)爆震稳定性的灵敏度
octahedral stress 八面体应力	
octahedral structure 八面体结构	octane promoter 抗爆剂〔改善爆震性能的添加剂〕
octahedrite 八面石	octane rating(= octane number) 辛烷值
octahedronlike molecule 八面体状分子	octane ratio 辛烷值
octahydrocoumarin 八氢香豆素	octane requirements 辛烷要求；对燃料爆震稳定性的要求；对燃料辛烷值的要求
octahydroestrone 八氢雌(甾)酮	
octahydro-naphthalene 八氢化萘	octane scale 辛烷值尺度
octahydro retene 八氢惹烯	octane selector 辛烷值选择器
octakis-〔词头〕 八个	octane unit 辛烷单位〔石油〕
	octane upgrading 辛烷值提级

octane value　辛烷值
octanoate　辛酸酯
octanohydroxamic acid　辛基异羟肟酸；八碳异羟肟酸　$C_7H_{15}CONHOH$
octanoic acid　辛酸　$C_7H_{15}COOH$
octanol　辛醇〔通常指:1-辛醇〕
1-octanol　1-辛醇　$CH_3(CH_2)_6CH_2OH$
2-octanol　2-辛醇　$CH_3(CH_2)_5CHOHCH_3$
2-octanone(=methyl *n*-hexyl ketone)　甲基己基(甲)酮；2-辛酮　$CH_3CO(CH_2)_5CH_3$
octanoyl　辛酰(基)　$CH_3(CH_2)_6CO—$
octant rule　八区规则*
octaoxacyclotetracosane　八氧杂环二十四烷；24-冠(醚)-8　$C_{16}O_8H_{32}$
octapeptide　八肽；八缩氨酸肽
octaphenyluranocene　八苯基环辛四烯铀　$U[C_8H_4(C_6H_5)_4]_2$
octatetraene　辛四烯
octathiacyclooctacosane(OTO)　八硫杂环二十八烷
octatomic ring　八元环
octatriacontanoic acid　三十八烷酸
2,4,6-octatriene　2,4,6-辛三烯
octavalent　八价的
octaverine　奥他维林；苯基异喹啉衍生物〔骨骼肌松弛药〕
octazone　辛腙
2-octenal　辛烯-2-醛
octendioic acid(=octene diacid)　辛烯二酸
octene　辛烯　C_8H_{16}
octene diacid(=octendioic acid)　辛烯二酸　$C_6H_{10}(COOH)_2$
octene dicarboxylic acid　辛烯二甲酸；癸烯二酸　$C_8H_{14}(COOH)_2$
octenoic acid　辛烯酸　$C_7H_{13}COOH$
octenol　辛烯醇
octet　①八隅体；八角(体)②八重态；八重线③八重峰
octet formula　八隅式；八角式
octet rule　八隅规则；八角定则
octet stability　八隅体安定性
octet theory　八隅(电子)学说
α-octine-α-carboxylic acid　α-辛炔羧酸
octivalence　八价
octivalent　八价的
octo-(=octa-)〔希腊字头〕　八
octoacetyl sucrose　八乙酰基蔗糖
octoate　①辛酸盐②(=salt of 2-ethylhexoic acid)2-乙基己酸盐〔油漆催干剂〕
octoate drier　辛酸盐催干剂

octocosoic acid　二十八烷酸　$C_{27}H_{55}COOH$
octocrylene　氰双苯丙烯酸辛酯
octocyclic compound　八环化合物
octodeca-〔词头〕　十八
octodecane(=octadecane)　①(正)十八(碳)烷②十八(碳)(级)烷
octodecyl(=octadecyl)　十八(碳)烷基
octodecyl acetate　乙酸十八(烷)醇酯
octodecyl alcohol　十八(烷)醇
octodecyl palmitate　棕榈酸十八(烷)醇酯
octofollin(=benzestrol)　辛叶素；异辛雌酚
octohydroxy-arachidic acid　八羟基花生酸
octohydroxylated acid　八羟基酸
octohydroxy-stearic acid　八羟基硬脂酸
octoic acid　辛酸　$C_7H_{15}COOH$
octopamine　真蛸胺；章鱼胺
octopin　真蛸晶；章肉晶
octopine　真蛸碱；章鱼(肉)碱
octoploid　八倍体
octosan　聚辛糖
octovalence　八价
octovalent　八价的
octreotide　奥曲肽〔8肽〕〔药〕
octulose　辛酮糖
octulosonic acid　辛酮糖酸
octupole　八极
octupole moment　八极矩
octyl　辛基　$CH_3(CH_2)_6CH_2—$
octyl acetate　乙酸辛酯
n-octyl alcohol　辛醇　$CH_3(CH_2)_6CH_2OH$
octyl aldehyde　辛醛
octyl aldehyde diethyl acetal　辛醛二乙缩醛
octyl aldehyde dimethyl acetal　辛醛二甲缩醛
n-octyl amine　辛胺　$CH_3(CH_2)_7NH_2$
octylate　辛酸盐(酯)
octylated diphenylamine　辛基化二苯胺〔防老剂〕
n-octyl benzoate　苯甲酸辛酯
octyl benzyl phthalate　邻苯二甲酸辛苄酯〔增塑剂〕
n-octyl bromide　辛基溴；1-溴辛烷　$CH_3(CH_2)_6CH_2Br$
n-octyl butyrate　丁酸辛酯
γ-octyl-γ-butyrolactone　γ-辛基-γ-丁内酯
octyl carbinol　辛基甲醇；壬醇
n-octyl chloride(=1-chloro-octane)　辛基氯；1-氯辛烷　$CH_3(CH_2)_6CH_2Cl$
octyl cresol　辛基甲酚
octyl cresyl phthalate　邻苯二甲酸辛基甲苯酯〔增塑剂〕
octyl decyl adipate　己二酸辛癸酯〔增塑剂〕
n-octyl-*n*-decyl phthalate　邻苯二甲酸正辛正癸酯〔增

octyl-decyl phthalate 邻苯二甲酸辛癸酯
n-octyl-*n*-decyl trimellitate 偏苯三酸正辛正癸酯〔增塑剂〕
octyl dimethyl aminobenzoic acid 二甲基对氨基苯甲酸辛酯
octyl dimethyl benzyl ammonium halide 辛基二甲基苄基卤化铵
octyl diphenyl phosphate 磷酸辛基二苯酯〔增塑剂〕
octyl dodecanol 辛基十二醇
octyl dodecyl myristate 肉豆蔻酸辛基十二醇酯
octylene 辛烯
1,3-octylene glycol 1,3-辛二醇
octylene glycol titanate 钛酸(二个)辛二醇酯〔偶联剂〕
octylene oxide 环氧辛烷
octylenic acid 辛烯酸 $C_7H_{13}COOH$
octyl epoxy stearate 环氧硬脂酸辛酯〔增塑剂〕
n-octyl fluoride(=1-fluoro-octane) 1-氟辛烷 $CH_3(CH_2)_6CH_2F$
octyl formate 甲酸辛酯 $HCO_2C_8H_{17}$
octyl gallate 没食子酸辛酯
octyl heptoate 庚酸辛酯
n-octyl heptylate 庚酸辛酯
n-octyl hydrogen *n*-octylphosphonate (正)辛基膦酸单(正)辛酯；(正)辛基膦酸正辛-氢酯 $(C_8H_{17}O)(C_8H_{17})PO(OH)$
octylic acid 辛酸 $C_7H_{15}COOH$
octyl isobutyrate 异丁酸辛酯
octyl-isodecyl phthalate 邻苯二甲酸辛基异癸酯〔增塑剂〕
n-octyl isovalerate 异戊酸辛酯
n-octyl mercaptan 正辛硫醇
octyl methoxycinnamate 甲氧基肉桂酸辛酯
octyl octanoate 辛酸辛酯
n-octyloxy propionitrile 正辛氧基丙腈
octyl phenol 辛基酚
octyl phenol polyglycol ether 辛基酚聚乙二醇醚
octyl phenol (tertiary) (叔)辛(基)酚
octyl phenoxyethanol 辛基苯氧基乙醇〔表面活性剂〕
n-octyl phenyl acetate 苯乙酸正辛酯
octyl phenyl phosphate 磷酸辛基苯酯〔增塑剂〕
octylphenylphosphinic acid 辛基苯基次膦酸 $(C_8H_{17})(C_6H_5)PO(OH)$
p-octylphenyl salicylate 水杨酸对辛基苯酯
octyl phosphate 磷酸辛酯〔增塑剂〕
octyl phosphate triethanolamine salt 磷酸辛酯三乙醇胺盐
octyl phosphonic acid 辛基膦酸

octyl polyoxyethylene glycol 辛基聚氧乙烯乙二醇
n-octyl propionate 丙酸辛酯
octyl pyridinium bromide 辛基溴化吡啶鎓
octyl resorcinol 辛基间苯二酚
octyl salicylate 水杨酸辛酯
n-octyl terephthalate 对苯二甲酸正辛酯〔增塑剂〕
octyl toluate 甲苯甲酸辛酯
octyl trimethyl ammonium decane sulfonate 癸烷磺酸辛基三甲基铵
octyl-trimethyl-silicane 辛基三甲基硅 $C_8H_{17}Si(CH_3)_3$
octyndioic acid(=octyne diacid) 辛炔二酸
octyne 辛炔 C_8H_{14}
1-octyne(=*n*-hexyl acetylene) 正己基乙炔；1-辛炔 $CH_3(CH_2)_5C\equiv CH$
octyne diacid(=octyndioic acid) 辛炔二酸 $C_6H_8(COOH)_2$
octyne dicarboxylic acid 辛炔二羧酸；癸炔二酸 $C_8H_{11}(COOH)_2$
octynic acid 辛炔酸 $C_7H_{11}COOH$
ocular ①目镜②眼见的
ocular estimation 目估；目测
ocular eye piece 目镜
ocular lens 接目镜
ocular micrometer 目镜测微计
ocular network micrometer 目镜网络测微计
odallin 海芒果苷
OD color(=optical density color) 光密度色度
odd ①额外的②零星的③奇(数)的④单(数)的
odd alternants anion 奇数交替烃负离子
odd-electron 奇(数)电子
odd electron bond 奇(数)电子键
odd electron compound 奇(数)电子化合物
odd-electron ion 奇电子离子·
odd-electron molecule 奇(数)电子分子
odd-even alternation 奇偶交替；奇偶循环
odd-even effect 奇偶效应
odd-even element 奇偶元素
odd-even nuclei 奇-偶核〔核内质子为奇数，中子为偶数〕
odd function 奇函数
odd mass 奇质量
oddment 余料
odd molecule 奇(数)电子分子
odd-numbered 奇数的；奇碳原子数的
odd-odd element 奇奇元素
odds 余料
odd-shaped 异形的；畸形的
odd-shaped cross section 异形截面
odd-shaped orifice 异形孔

odd sheets　不合规格的纸张
odd soap　奇碳原子数皂
Oderberger colloid mill　奥特尔伯格胶体磨
odometer　测距器
odor　气味〔指香味和臭味〕
odor additive　添味剂；香味添加剂
odorant　①添味剂；增味剂②香料
odorator　喷香器
odor characteristic　香气特征
odor classification　香气分类
odor concentration　臭味浓度
odor control additive　调味添加剂
odor eliminator　气息净化剂
odor emission rate　①排臭等级②臭味散发率；臭味放出速率
odor engineering　香化工程〔橡胶〕
odor free　无气味；无恶臭
odoriferous　有气味的
odoriferous body　含香物
odoriferous material　香料
odorimeter　气味计〔测定气体中添味剂浓度的仪器〕
odorimetry(=olfactometry)　气味测定(法)
odorin　臭葱素
odor index　恶臭(强度)指数
odor inhibitor　恶臭抑制剂；臭气清除剂
odor intensity　气味强度
odor intensity index　臭气强度指数
odoriphor group　发香团
odorization　添味(作用)
odorizer　添味剂加入器
odorizer tank　添味剂储槽
odorless　无气味的；无臭的
odorless alkyd flat　无臭醇酸平光(乳胶)漆
odorless bitumen　无臭沥青
odorless coating　无臭涂料
odorless drier　无臭催干剂
odorless kerosene(OK)　无臭煤油〔组成约为：正烃60%、异烃10%、萘30%、芳烃1%〕
odorless mineral spirit　无(臭)味石油溶剂油
odorlessness　无气味
odorless paint　无(臭)味漆
odorless solvent　无臭溶剂
odor nuance　香气近似；香调
odorometer　气味计
odorous　有香气的；香的
odorousness　气味浓度
odorous principle　香气主要成分
odorous substance　有气味物质；恶臭物质

odor permanency　气味持久性
odor pollution　恶臭污染
odor-producing oxidation drying　发臭的氧化干燥(固化)
odor prolongation　香气延持
odor recognizer　气味识别
odor sampling　气味采样
odor source　恶臭源
odor strength　臭味强度；臭度
odor test　香气试验
odor theory　气味学说
odor threshold　恶臭极限值；恶臭阈值
odor type　香型
odour　气味；香气；气息
odourless　无气味的；无臭的
odourless fixatives　无香气定香剂
odour-producing component　气味组分；产生气味的组分
odour test　气味试验
od-ray　生物冷光；生物荧光
odyssic acid　奥台草酸
odyssin　奥台蕈素
Oechsle　奥氏糖度表
oeillet　香石竹；康乃馨
oenanthal　庚醛
oenanthic　酒香；酿香
oenanthic acid　庚酸　$C_6H_{13}COOH$
oenanthic aldehyde　庚醛　$C_6H_{13}CHO$
oenanthic ether　酒醚；庚酸乙酯；人造康酿克油
oenanthol　庚醇
oenantholactam　庚内酰胺
oenantholactone　庚内酯
oenanthotoxin　水芹毒素
oenanthyl(=heptanoyl)　庚酰　$CH_3(CH_2)_5CO—$
oenanthylic acid　庚酸　$C_6H_{13}COOH$
oenanthylic aldehyde　庚醛　$C_6H_{13}CHO$
oenanthylidene acetone　庚叉丙酮
oenidin　锦葵色素
oenin　锦葵色素-3-β-葡萄糖苷
oenology　酒学
oenometer　酒度计〔测定酒的浓度〕
oenotannin　酒单宁
oeolotropic　各向异性的
Oerliken cell　欧厄利肯电池
oerstedite　含氧化钛的水合锆石
oestriol(=estriol)　雌(甾)三醇　$C_{18}H_{24}O_3$
oetiocholanone　初胆烷酮
Oettel chlorate process　奥伊特耳氯酸盐法
Oettel coulometer　奥伊特耳电量计
off-air　废空气

off-air scrubber 废空气洗涤器
offal 废料；下脚料
off-axis 偏轴
off-axis tensile 偏心张力拉伸
off center illusion 视觉中心偏离错视
off-center load 偏心荷重
off-color 变色；不标准颜色；不正常颜色
off-color gasoline 变色汽油
off-color industry 炭黑工业
off-color material 变色材料
off-color product 变色产品〔比要求颜色标准为深的产品〕
off-column 离柱
off-contact 架空
off-cut 切余纸板
off-design behaviour 非设计工况；非设计特性
off-design conditions 超出设计(规定的)条件
off-diagonal 对角线旁的
off-directional tensile test 偏轴拉伸测试
offensive odour 恶臭
off flavour 臭气；臭味
off-gas 废气；尾气；出口气
off-gas benzene 废气(中含的)苯
off-gas line 废气管
off-grade 等外品
official (acceptance) test 法定试验；国家认可的试验；正式认可的试验
official compound 法定化合物
official method 法定(分析)法
official sample 法定试样
official test 法定试验
officinal 药典的；法定的；药用的
off-line ①脱机②离线
off-line analysis 离线分析
off-line capillary electrophoresis-matrix-assisted laser desorption/ ionization-mass spectrometry 毛细管电泳基质辅助激光解吸电离质谱离线检测
off-line computer 断机计算机
off-line data handling 断机数据处理
off-loading 卸下；卸料
off-loading point 卸荷点
off-load period 卸荷期；非载荷期
off-loom contraction 下机缩率
off-loom weft density 下机纬密
off-machine sizing (纸)机外施胶
off odour 臭气；臭味
off oil 次等油
off peak load 非高峰负荷

off-port flame 喷嘴外部火焰
off-position 停位；关位
off resonance 偏共振*
off resonance decoupling 偏共振去耦
off resonance proton spin decoupling 质子偏共振自旋去耦
offscouring 废物
offscum 废渣
off season 非生产季节〔糖〕；淡季
off-set 偏置*
off-set bend 迂回管
off-set die head 直角机头〔压出机〕
off-set facility 辅助设施；非生产设施
offset gravure coater 转移式槽辊涂布机
offset modulus 补偿模量
off-set oil 印刷油
off-set paper 胶版印纸
offset variable 位移变量
offset yield strength 补偿屈服强度
off shade 败色〔染〕
offshore discharge 近海排放
offshore oil delivery 海底油管输送
offshore oil delivery pipe 海底输油管道
offshore oil tank 海上油罐
offshore tanker terminal 海上中转油罐
offshore unloading 海底管道卸载
off-side of bearing 轴承的(润滑油)出口边
off-side tank 石油箱
off-site ①厂区外；工地外；单位外②装置外
off-site surveillane 场外监视；非现场监视
off-size 非规定大小；非规定尺寸
off sorts 等外品
off specification 不合格的
off specification material 不合规格的物料；等外品
off-stream 侧馏分〔自蒸馏塔侧取出的馏分〕
off-stream case 停用反应器
off-stream pipe line 停用管线
off-stream unit 停用设备
offtake pipe 排水管
off-test product 不合格产品
off water 废水
off-white (近于)纯白的；灰白色的；米色的
O-fiber 小纤维
ofloxacin 氧氟沙星〔药〕
Oglialoro reaction 奥格里阿洛若反应
ohba-kusu oil 大叶樟油
ohm 欧(姆)
ohmammeter 欧(姆)安(培)计

ohmic curve　欧姆曲线
ohmic polarization　欧姆极化*
ohmic potential drop　欧姆电势降*
ohmmeter　欧姆计
ohmoil　加醇醛α-萘胺的矿物油
Ohm's law　欧姆定律
OH value　羟基值
ohyamycin(=formycin B)　间型霉素 B
oiazines　邻二氮苯
-oic acid〔词尾〕　酸
oil　①油②石油
oil abrasion　润滑油的磨蚀作用
oil absorbent　吸油剂
oil absorbent adhesive　吸油性黏合剂
oil absorber　油吸收剂；吸油剂
oil-absorbing polymer　吸油性高分子
oil absorption　吸油量
oil absorption end point　吸油量终点
oil-absorption (number)　吸油值
oil absorption of pigment　颜料吸油量
oil absorption point　吸油点
oil-absorption power　吸油塔
oil absorption test　吸油试验
oil absorption value　吸油值
oil addition　加油；添加润滑油
oil aeration　油的通(空)气(处理)
oil affinity　亲油性
oil ageing　润滑油的老化
oil agent　(上)油剂；润滑剂
oil (air) pump　油(抽气)泵
oil analysis　油分析；原油分析
oil and fat　油脂
oil and gas separator　油气分离器
oil-and-water trap　油-水收集器
oil annealing　油淬火
oil antifoaming agent　润滑油的抗沫剂
oil atomization　油雾化
oil atomizer　喷油器
oil back　装油设备〔石油加工厂用以往铁路油槽车装油的设备〕
oil baffle　(润滑)油挡板(导板)；挡油板
oil baffle plate　挡油板
oil bag　油袋
oil barge　油驳船
oil base　油基
oil base coating　油基漆；油性漆
oil-based eggshell paint　蛋壳光油基漆；油基半光漆
oil-based masonry paint　油基圬工漆；油性(基)砖石砌体(用)漆
oil-based mastic sealant　①厚浆型油基填缝料②油基砂胶封闭剂
oil-based media　油基介质
oil base drilling fluid　油基钻孔液体
oil-base paint　油性漆；油基漆
oil base polish　油基光亮剂
oil-bath　①油浴②油浴器；油锅
oil-bath filter　油浴过滤器〔利用油作为过滤介质〕
oil-bath lubrication　油浴润滑
oil-bearing　载油的；含油的
oil bearing layer　含油层
oil-bearing shale　含油页岩
oil binder　①油性基料②油性黏合剂
oil black carbon　油(充)炭黑
oil black nigrosine　油溶苯胺黑；油溶对氮蒽黑
oil-black plant　石油炭黑厂
oil bleaching　油漂白；油脱色
oil bleaching apparatus　油漂白仪器
oil bloom　油霜〔油的荧光〕
oil body　油的体度〔润滑油的黏度或稠度〕
oil bodying　油的聚合
oil boiling pot　煮油锅
oil boom　油栅
oil-bound distemper　油基水浆涂料；油基墙粉涂料
oil-bound film　(润滑)油膜
oil bound water paint　水乳化漆；油基水乳化漆
oil box　油盒；润滑油盒；油箱
oil brake　油刹车
oil brand　油品牌号；油品商标
oil breaking　油裂解
oil break switch　油开关
oil brush　润滑油刷
oil buffer　油类避震器
oil burner　(烧)油炉；燃油器
oil-burning　烧油；用油做燃料的
oil-burning cooker　烧油烹饪器
oil-burning stove　烧油火炉
oil-can　油壶；油罐
oil-can lid　油壶盖
oil-can lubrication　油壶润滑
oil-can ratchet autolock　油壶的自动盖头
oil cap　油杯；油壶
oil capacity　油容量；容油量
oil car　油槽车
oil-carbon sludge　碳质残渣
oil-carrier(=oil-carrying ship)　油船
oil-carrying ship(=oil-carrier)　油船

oil-catalyst slurry 催化剂油浆；油与催化剂的悬浮液
oil catcher 润滑油收集器；润滑油分离器
oil catch ring 润滑油收集环
oil cavity 润滑油沟；润滑孔
oil-cell bottom 油箱沉积物
oil chamber 润滑油室；储油器
oil change 换油；润滑油的更换
oil change period 换油期
oil channel 油沟
oil charge 充油
oil circuit breaker 油开关
oil circulating reactor 油循环反应器
oil circulating system 润滑油循环系统
oil circulation 油循环；润滑油循环
oil circulation gauge 油循环压差计
oil-circulation process 油循环过程
oil clarifier 润滑油澄清器
oil classification 油分类
oil-clay mixture 油-白土混合物
oil cleaner 油净化器
oil cleaning system 油净化系统；润滑油清洁系统
oil cloth 油布
oil cloth drier 油布干燥器
oil-coated 涂油的
oil cock wrench 润滑油开关扳子
oil coil 油旋管；油盘管
oil coke 油焦；石油焦炭
oil cold test 油品混浊点(凝固点)测定
oil collecting system 集油系统
oil collection boom 集油栅
oil-collector 润滑油收集器
oil color(=oil colour) 油溶性染料
oil colour(=oil color) 油溶性染料
oil column 油柱
oil combustion 石油燃烧
oil conduit 油管
oil connecting pipe 油连接管
oil conservation 石油资源保护
oil consumption 油消耗
oil container 油容器
oil-contaminated water 含油污水
oil contamination 油的沾污；油污染
oil content 含油量
oil-content gauge (油箱中)存油指针
oil control ability of ring 活塞环润滑油控制性能
oil control characteristic (润滑)油的控制特性
oil control ring 护油圈；护油环
oil control room 石油系统控制室
oil control thermostat 油控恒温器
oil conversion process 石油转化过程
oil-cooled 油冷却的
oil cooler 油冷却器〔用油作冷却介质的冷却器〕
oil cooler position indicator 油冷却器位置指示器
oil cooler unit 油冷却器单元；润滑油辐射器
oil cooling 油冷却
oil cooling rib 油冷却肋管
oil corrosion 油腐蚀〔油中酸性物质的腐蚀〕
oil corrosion test 油腐蚀试验〔石油产品的金属腐蚀试验〕
oil country tabular goods 石油工业用管材
oil crust 油壳；固体油层
oil cup 油杯；油储存器；仪器上进油杯
oildag 石墨滑油；石墨润滑剂；胶体石墨
oil damping 油类减震
oil darkening 油变黑
oil decolorization 油脱色
oil decomposition 油分解〔油的氧化或老化〕
oil deflector 油折流器
oil degumming 油脱胶
oil dehydrating 油品脱水；润滑乳胶脱水
oil dehydrating plant 油品脱水装置
oil dehydrator 油品脱水器
oil delivery 石油输送
oil delivery truck 运送石油产品的车辆；油槽车
oil delustering 油消光
oil demand 石油需要量；油品需要量
oil demulsibility 油品的破乳化性
oil denuding （吸着)油的脱吸
oil deposits 油沉渣
oil-depth gauge 储油测量器
oil desulfurization 原油脱硫
oil deterioration 油品变质；油老化〔油品质量恶化〕
oil development 石油开采
oil dewaxing 油脱蜡
oil diffusion pump 油扩散泵
oil-diluent gasoline 润滑油稀释用汽油
oil dilution （润滑)油稀释
oil dilution tank （润滑)油稀释槽
oil dipper 油杓
oil discharge 卸油
oil dispense 油料分配；油料输送
oil dispenser 油料分配设备
oil dispensing equipment 油料分配设备
oil-dispersible 可分散于油内的；在油内生成悬浮体的
oil-dispersing property 油的分散性质
oil-displacing agent 驱油剂

oil-dissolving solvent　油溶性溶剂；溶解于油的溶剂
oil distillate　油的馏分〔石油或润滑油馏出油〕
oil distributing box　润滑油分配盒
oil distributor pipe　油分配器的管道
oil dock　①油船坞②油船码头
oil dope　油品添加剂；润滑油添加剂
oil drag　油摩擦
oil drag loss　油摩擦损失
oil drain　油类流出孔
oil drain drum　排油罐
oil drain interval(=oil drain period)　换(润滑)油期
oil drain period(=oil drain interval)　换(润滑)油期
oil drain pump　吸油泵
oil drain valve　放油阀
oil dregs　油料沉淀
oil drier　油催干剂；油干料〔俗〕
oil drilling　石油钻井
oil drilling pipe　石油钻井管；石油钻井杆
oil dripping　油料滴漏
oil droplet　小油滴；油珠
oil-drowned　油浸的
oil drum　油桶
oil drum stand　油桶架
oil dual thermometer　油用温度计
oil duct　油沟
oiled　上油的；油过的；涂油的；油化的
oiled cloth　油布
oiled leather　涂油革
oiled ore　油化矿石
oiled paper　油纸
oil ejector　油喷射器
oil-electric drive　柴油发电驱动
oil electric engine　柴油发电机〔与石油发动机相联结的电动机〕
oil eliminator　除油器；油分离器
oil emulsion　油乳胶
oil emulsion adjuvant　油乳化佐剂
oil emulsion composition　油乳胶组成
oil emulsion mud　油乳胶浆
oil emulsion paint　油乳胶漆
oil engine　柴油机
oil engineer　石油工程师
oil enriched rubber　充油橡胶
oil equipment　石油设备
oiler　加油器〔商〕
oil-extended polybutadiene rubber　充油聚丁二烯橡胶
oil-extended rubber　油充橡胶；油增塑橡胶
oil-extended thermoplastic styrene butadiene rubber　充油热塑丁苯橡胶
oil-extender　石油软化剂
oil-extraction plant　炼油厂；提油厂
oil extractor　油提取器
oil exudation　渗油
oil face　油面
oil factor　①吸油值②油类质量因素
oil failure　供油中断
oil faucet　油支管
oil feed　进油；给油
oil feeder　进油器；自动润滑设备；滴油器
oil-feeding device　进油设备
oil feed injector　油料注入器；进油注射器
oil-feed pump　进油泵
oil-feed tank　进油槽
oil felt pad　油毡垫；油毡
oil fence　油栅；油拦
oil field　(石)油田
oil field emulsion　油田乳胶〔石油开采时〕
oil field gas　油田气
oil field tank　油田油罐
oil-filled cable　封油绝缘电缆
oil filled cable oil　油浸电缆油
oil filler　油料加入器；加油口
oil filler cap　油料加入器盖；加油口盖
oil filler hole　油料加入器孔；加油口
oil filler pipe　油料加入器管；加油管
oil filling　装油
oil film　油膜
oil film breakdown(=oil film failure)　油膜破坏
oil film failure(=oil film breakdown)　油膜破坏
oil film stiffness　油膜刚度
oil film strength　油膜强度
oil filter　滤油器
oil filter body　滤油器本体
oil filter cartridge　滤油器芯子
oil-filter deposits　滤油器沉积
oil-filter element　滤油器滤心
oil filter housing　滤油器本体
oil-filtering unit　滤油设备
oil-filter pipe　滤油管
oil-filter screen　滤油器网
oil-fired　烧油的
oil-fired furnace　烧油炉
oil firing　烧油
oil-firing burner　油燃烧器
oil-fixative pigment　固油颜料
oil flash　油的闪蒸；油的闪点

oil flowmeter 油量计
oil foaming 油生成泡沫
oil fog generator 油雾发生器
oil fogger 油雾化器
oil fog lubrication 油雾润滑
oil foot 油脚；油渣
oil footings 油脚
oil fraction 油的馏分
oil free compressor 无油压缩机
oil free labyrinth compressor 无油润滑迷宫式压缩机
oil-free lubrication （汽缸》无油润滑
oil-free petrolatum 无油石油膏
oil-free washable distemper ①无油耐擦洗水浆涂料(刷墙粉)②无油可洗性绘画颜料
oil fuel 油燃料
oil-fuel depot 油燃料仓库
oil-fuel pump 油燃料泵
oil-fuel tank 油燃料罐；油槽〔船上〕
oil-fuel transfer pump 油燃料输送泵
oil funnel 油漏斗
oil gallery 油沟
oil gas 油气
oil gas contact 油气接触面
oil gas generator 油气发生器
oil gasification 油的气化；液体燃料的气化
oil gas interface 油气界面
oil gas ratio 油气比
oil gas tar 油气焦油
oil gas tar pitch 油气焦油的沥青
oil gathering system 集油系统
oil gauge ①油品比重计②油规；量油计；油位表
oil-gauge cock 油面管开关
oil-gauge glass 油面玻璃〔观察油面的玻窗或玻管〕
oil-gauge pipe 油面指示管
oil-gauge rod 油面测杆
oil-gauge stick 油面测杆
oil gear 油传动装置
oil genesis 石油起源；石油生成
oil gloss coating 有光油性漆
oil gold size(=gold size) ①油性金胶；金胶油②金漆；金胶；贴金用漆
oil grain 含油种子
oil groove （润滑）油槽
oil guard 油防护；油保护
oil gun 油枪；注油器
oil-gun adapter 油枪连接器
oil-gun lubrication 油枪润滑；高压下润滑
oil-gun nipple 油枪接头

oil handling 油处理〔油的净化、加工、制取、注入、提出或运送〕
oil hand pump 手压油泵
oil hardening ①油淬火②油硬化
oil hazard 油料事故；油料火险
oil header 油料分布器；集油管
oil heated 油热的
oil heated brick chamber 砖砌燃油燃烧室
oil heater 油热器
oil heating 油热法
oil heating installation 油热设备
oil hold 油舱
oil-holder 盛油器
oil hole 油孔
oil hose 运油软管
oil hulk 石油驳船
oil-hydrogen mixture 油与氢气的混合物
oil hydrometer 油比重计
oil hydrosol 油水溶胶；油水乳液
oil identifying 油的鉴定
oil immersion 油浸法
oil immersion test 油浸试验
oil-impregnated 油浸的
oil-in 油入口
oil indentifying 油鉴定
oil indicator 示油器
oil industry 石油工业
oil industry wastewater 石油工业废水
oiliness 油性
oiliness additive 油性添加剂〔提高润滑性能的添加剂〕
oiliness carrier 油性载体
oiliness compound （润滑油中)油性化合物
oiliness degree 油性度
oiliness film 油性膜
oiliness improver 油性改进剂
oiliness index 油性指标
oiliness test 油性试验
oiling ①润滑②加油；油化
oiling agent 加油剂
oiling chart 润滑图表
oiling off 揩油
oiling off machine 揩油机
oiling pad 油毡；油垫衬
oiling ring 润滑环
oiling wick 润滑芯子
oil injection 油喷射
oil inlet pipe 油引入管；油支管
oil in reserve 储存油〔经常存在于管道内与油罐内的

油类〕
oil in service　使用的油；在使用中的油料
oil in sight　可察的石油储量
oil-insoluble sludge　不溶于油类的残渣
oil inspection kit　成套检查油料仪器
oil in storage　管道储油；储存油料
oil insulator　油类绝缘体
oil intercepter　油料捕集器
oil in water　水包油(型)；油在水中(型)
oil in water dispersion　水包油分散体(液)
oil in water emulsion　水包油乳状液
oil in water type　水包油型；水中油型
oil-in-water type emulsion　水包油型乳状液；油水相乳状液
oil jacket　油套；油夹套
oil jetty　石油码头
oil joint　抗油垫衬
oil-laden powder　含油粉末；油污粉末
oil lamp　油灯
oil lantern　油灯
oil layer　油层；油膜；(油类溶剂精制时的)精制层
oil-leaching property　油萃取性；油浸提性
oil lead　引油管线
oil leak　漏油
oil leak detector　漏油口测定器
oil length　①含油率②油的聚合程度
oil length of alkyd　醇酸树脂的油长(度)
oil length of varnish　油漆的含油率
oilless bearing　无油轴承
oil-level　油面
oil-level cock　油面计开关
oil-level dipstick　量油尺；油面测量杆
oil-level gauge　油面指示器；油量计
oil-level indicator　油面指示器
oil-level mark　油面标志
oil-level pipe　油面管；油类平衡管
oil-level plug　油面测量计栓
oil-level tell-tale　油面信号灯
oil life　油的寿命；润滑油的使用期
oil light　煤油灯
oil-line　油管
oil-line plugging　油管堵塞
oil-line pump　油管泵；石油管道用泵
oil-line scavenge　油管的吹洗
oil liver　油肝〔润滑油内形成的胶状物或固体物质〕
oil-loading facilities　装卸石油产品的设备
oil-loading rack　装卸石油产品的栈桥
oil loving　油溶的；亲油性的

oil lubrication　油类润滑
oil manifold　油管
oil manometer　油压力计
oil manufacture　油品生产
oil-marred surface　油损表面
oil mass　油量
oil mastic　①油性腻子②油性砂胶
oil mat　①沥青面层②沥青垫层
oil meal　油粕；油粉
oil measurement　油计量
oil-measuring pump　计量油泵
oil-measuring rod　量油尺
oil menstruum　(气体收集设备内的)石油溶剂
oil meter　油量计
oil metering system　油计量系统
oil-micro filter　微型滤油器
oil migration　石油运移
oil-mill technique　油磨法
oil mist　油雾
oil misting　油雾化
oil-mist lubrication　油雾润滑
oil-modified alkyd　油改性醇酸
oil modified latex paint　油改性乳胶漆
oil modified maleic alkyd　油改性顺(丁烯二酸)酐醇酸树脂
oil modified polyurethane varnish　油改性聚氨酯清漆
oil modified urethanes　油改性聚氨酯
oil mordant　石油媒染剂
oil-mull technique　油磨法
oil-naphtha solution　油的石脑油溶液
oil number　①(按美国汽车工程师协会分类的发动机)润滑油牌号②吸油量
oil of ants　糠醛
oil of apple　戊酸戊酯
oil of arachis　花生油
oil of (artificial) almond　人造杏仁油；苯甲醛
oil of badian　八角茴香油
oil of bananas　香蕉水〔俗〕；乙酸戊酯
oil of bay　桂油
oil of bayberry　月桂果油
oil of benne　(芝)麻油
oil of caoutchouc　生橡胶油
oil of checkerberry　白珠树油
oil of china wood　桐油
oil of cognac　①科涅克油②庚酸乙酯
oil of cuscus　香根草油；岩兰草油
oil of dogfish　鲛油
oil of dolphin　海豚油

oil of fir 冷杉油
oil of flaxseed 亚麻子油
oil of florence 橄榄油
oil of garlic 蒜油；烯丙基硫醚
oil of gingelly 芝麻油
oil of glonoin 硝化甘油
oil of goosefoot 藜油
oil of gourd 黄瓜油
oil of groundnut 花生油
oil of lemon 柠檬油
oil of maize 玉蜀黍油
oil of melissa 柠檬草油
oil of mignonette 木犀草油
oil of mirbane 硝基苯；密斑油
oil of mustard 芥子油
oil of paraffin 石蜡油
oil of patchouly 广藿香油
oil of pears 乙酸戊酯
oil of sperm 鲸蜡油
oil of spike 穗薰衣草油
oil of tar 焦油
oil of turpentine 松节油
oil of verbena 马鞭草油
oil of vitriol 浓硫酸
oil of wintergreen 冬青油；水杨酸甲酯
oil origin 石油起源
oil outlet temperature 油出口温度
oil-out line 油流出管道
oil-overflow valve 油溢流阀
oil oxidation stability 油氧化稳定性
oil package 石油产品容器
oil packaging 油的包装〔往小桶装油〕
oil packing paper 包装油纸
oil pad 油垫；润滑垫衬；润滑填料
oil pan 油盘；油底壳
oil pan tray 油盘槽
oil paper 油纸
oil passageway 油管；油沟
oil paste of color 油性颜料浆；油性色浆
oil pastes ①色浆②油性色浆
oil patch 油斑
oil phase 油相
oil pick-up 油杓
oil pipe 油管
oil pipe line 石油管路
oil piping 油管安装
oil piping layout 油管布置
oil pistol 油类分布开关

oil pocket 油包；储油容器
oil pointer 油标；油面指示器
oil polish 擦光油〔擦光家具用低黏度油〕
oil pollution 油污染；石油污染
Oil Pollution Act 油污法案〔指美国 1924 年制定的禁止石油污染沿岸海水法案〕
oil polymer 充油聚合物
oil pool 油池
oil pot 润滑油器；储油容器
oil potential 估计最高储油量
oil power 石油动力；石油机
oil press 榨油机
oil pressure 油压
oil-pressure adjusting screw 油压调节螺丝
oil pressure cut-out 油压切断器
oil-pressure damper 油压减震器
oil-pressure gauge 油压表
oil-pressure gauge connection 油压表连接管
oil-pressure gauge (feed) pipe 油压表管路
oil-pressure indicator 油压指示计
oil-pressure pipe 油压管
oil-pressure pump 油压泵
oil pressure regulator 油压调整器
oil-pressure relief valve 油压减压阀
oil-pressure shut-off switch 油压遮断器
oil-pressure stabilizer 油压稳定器
oil-pressure valve 油压阀
oil primer 油性底漆
oil primer coat 油性底漆层
oil primer coating ①油性底漆②涂油性底漆
oil process 油(再生)法
oil-processing unit 石油加工装置
oil product 石油产品
oil production 石油生产
oil-proof 耐油的；防油的
oil proofing ①防油②防油的
oil-proofing agent 防油剂；耐油剂
oil-proofness 耐油性；油稳定性
oil-proof paint 耐油漆
oil pudding(=oil pulp) 油悬浮皂〔在制造润滑脂中用以稠化润滑油〕
oil pulp(=oil pudding) 油悬浮皂〔在制造润滑脂中用以稠化润滑油〕
oil pump ①油(抽气)泵②(送)油泵
oil pump body(=oil pump case) 油泵本体
oil pump bypass valve 油泵支管阀
oil pump capacity 油泵能力
oil pump case(=oil pump body) 油泵本体

oil pumper ①油泵工作员②引导润滑油至气缸表面部分的活塞环
oil pump gear 油泵齿轮
oil pumping 油品泵送
oil pumping room(=oil pump room) 油泵室
oil pumping system 油品泵送系统
oil pump room(=oil pumping room) 油泵室
oil pump screen 油泵过滤器；油泵滤网
oil pump shaft 油泵轴
oil purification 油品净化；油品精制
oil purifier 油品净化器；净油器
oil putty 油性腻子；油灰
oil quenching ①油急冷②油淬；油中淬火
oil radiator 油辐射器
oil-reactive phenolic resin 油变性酚醛树脂
oil reactivity 油反应性
oil receiver 油接受器
oil reclaimer 油类精制器；废油再生装置
oil reclaiming(=oil reclamation) 废油的再生
oil reclamation(=oil reclaiming) 废油的再生
oil recovery 油的回收
oil rectifier 油类精馏器
oil recycle process 油循环过程
oil red 油红
oil refiner pack 油类精制的过滤器
oil refinery 石油加工厂
oil refinery plant 石油加工厂
oil refining 石油炼制；炼油
oil-refining capacity 炼油能力
oil-refining properties 石油加工特性
oil regenerator 石油再生器
oil relief valve 润滑油系统减压阀
oil removal filter 除油过滤器
oil removal process 除油法
oil removing 脱油；除油
oil-repellent ①拒油②抗油剂
oil reserves 石油藏量
oil reservoir 油箱；油罐；沉降槽
oil residue(=oil residuum) 油渣；蒸馏石油的残渣；残油
oil resinification 油结胶；油固化
oil resistance 抗油性；耐油性
oil resistant 抗油的；耐油的
oil resistant coating 耐油涂料
oil resistant primer 耐油底漆
oil-resistant rubber 抗油橡皮
oil-resistant synthetic rubber 耐油性合成橡胶
oil-resisting 抗油的
oil-resisting sheath 抗油外壳

oil resisting test 耐油性试验
oil resources 石油资源
oil retainer (曲轴的)润滑油保持环
oil-retaining plug 润滑管线系统栓
oil-retaining property 润滑油保持性能
oil-retaining rim 润滑油保持凸边
oil-retaining ring(=oil ring) 油环；润滑油保持环
oil return passage 油回流管道
oil return pipe 油回流管
oil rig 抽油装置
oil ring(=oil-retaining ring) 油环；润滑油保持环
oil-ring bearing 油环轴承；有油圈的轴承
oil-ring groove (在活塞上)油环槽
oil-ring plugging 油环卡住
oil-ring sticking 油环黏结
oil rotary pump 油转泵
oil salvage 废油再生
oil sand 油砂；石油砂
oil saponification value 油皂化值
oil scavenge pump 吸油泵
oil scoop 油类收集器
oil scraper piston ring 除油活塞环
oil screen 油筛；精制石油的滤网
oil-screen clogging 油筛堵塞；油过滤器的堵塞
oil scupper 油槽
oil seal 油封(口)
oil sealed stuffer 油封填料
oil seal housing 油封箱
oil seal leather 油封革
oil seal ring 油封圈
oil sediments 油中沉淀物
oil seed 含油种子
oil separating 油分离
oil separating barge 分离油的驳船
oil separator 油分离器
oil servicing 油供应
oil-servicing facilities 供油设备
oil shale(=kerosene shale) 煤油(母)页岩
oil shale distillation 油页岩干馏
oil-shale fuel 页岩油
oil-shale kerogen 页岩油母
oil-shale retort 油页岩干馏釜
oil-shale retorting 油页岩干馏
oil-shale retorting plant 油页岩干馏厂
oil ship 油船；油轮
oil shock absorption 油减震器
oil shortage 石油不足
oil shrinkage loss 石油蒸发损失

oil sight glass 油面观察玻璃
oil singler 分油器
oil skimmer 撇油器；油撇取器
oil skimmer barge 撇油船
oil skimmer vessel 撇油器
oil skimming tank 撇油罐
oilskin 油布
oil skinning 石油皮囊装运
oil slick 浮油；油花
oil slick boom 水面油栅；浮油栅
oil slinger 抛油环；甩油环
oil sludge 油泥
oil smoke 油烟
oil-soaked 浸油的
oil-softener 油类软化剂
oil solubility 在油内的溶解度
oil-soluble 油溶的；可溶于油的
oil-soluble dyes 油溶染料
oil soluble emulsifying agent 油溶性乳化剂
oil soluble group 油溶性基团
oil-soluble inhibitor ①油溶性抗氧化剂②油溶性缓蚀剂
oil-soluble metal salt 油溶性金属皂(金属盐)
oil soluble non-ionic active agent 油溶性非离子活性剂
oil-soluble perfume 油溶性香精
oil-soluble phenolic resin 油溶性酚醛树脂
oil-soluble resin 油溶性树脂
oil soluble surfactant 油溶性表面活性剂
oil-solvent blend(=oil-solvent solution) 油与溶剂的混合物
oil-solvent solution(=oil-solvent blend) 油与溶剂的混合物
oil sources 油源
oil specifications 油料技术规格
oil spill 漏油；浮油
oil spill boom 漏油栅
oil spill bubble harrier 漏油泡栅
oil splash gear 喷溅油用的齿轮
oil splitting 油脂裂解
oil splitting and re-esterification 油的分解和再酯化
oil splitting plant 油脂水解车间
oil spot 油点
oil spout 喷油
oil spray 油喷淋
oil sprayer 油喷淋器
oil spring 油泉；石油泉
oil stability 油品安定性
oil stain(=nonpenetrating stain) 非渗透性着色剂；油性着色剂

oil stainer 油性着色剂
oil starvation 油缺乏
oil-steam mixture 石油-水蒸气混合物
oil-steel reaction 油钢反应
oil still 油精馏器
oil-still tube support 石油蒸馏炉的炉管支架
oil stock 油储藏
oil stock loss 油储藏损失
oil stock loss control 油储藏损失控制
oil stone 油石；油砥石
oil stopper 油性填孔料
oil storage 油储藏
oil storage barge 石油驳船
oil storage tank (储)油槽
oil storage vessel (储)油罐
oil store 油库；石油储藏所；储油仓库
oil storing 石油储存
oil strainer 油料过滤器
oil stream 油流〔在管道内流动的油流〕
oil stripper 刮油器
oil stripping 油纹
oil substitute 油代用品
oil suction pipe 吸油管
oil suction pipe adapter 吸油管连接器
oil suction pump 吸油泵
oil sump 油沉淀池
oil sump capacity 油沉淀池容量
oil sump strainer 油沉淀池过滤器
oil sump tank 油收集槽
oil supply 油供应
oil supplying 油供应
oil supply pipe 供油管道
oil supply port 给油口
oil supply system 供油系统
oil supply tank 供油槽
oil supply volume 供油容量
oil surfacer 油性二道浆
oil surface temperature 油的表面温度
oil surface tension 油的表面张力
oil swell 油溶胀；油泡胀
oil syphon 油虹吸管
oil tagging system 油示踪法；石油标识法
oil tank 油罐
oil tank car 油槽车
oil tank cooler 油槽冷却器
oil tanker 油槽
oil tank heater 油槽加热器
oil tank hopper 油槽入口接受过滤器

oil tank truck 油槽车
oil tank vent 油槽排气管
oil tannage 油鞣(法)
oil-tanned deerskin 油鞣鹿革
oil-tanned leather 油鞣革
oil tanning 油鞣
oil tar 焦油
oil-tar sludge 焦油油泥
oil temperature regulator 油温调节器
oil tempering 油回火
oil terminal 油料仓库
oil test 油密试验
oil test engine 油料试验机
oil thief 油料取样器
oil thinning 油的稀释
oil thixotropic behavior 油的触变行为
oil throw 抛油环
oil thrower 喷油嘴；挡油圈
oil throwing 油喷淋；油雾化
oil-tight 不透油的
oil-tight bulkhead 不透油隔膜；不透油间壁
oil tightening 油封填充；不透油垫圈
oil tightness 不透油(性)
oil tintometer 油色调计
oil transferring 石油输送
oil trap 集油槽；油收集器
oil tray 油盘；油盆
oil treater 油净化器；油处理器
oil-treatment of coal 煤的油处理
oil trough 集油盘
oil true color 油本色
oil tube 油管
oil type 油型
oil-type air cleaner 油浴空气过滤器
oil type analysis 油型分析
oil vaporizer 油类蒸发器
oil vapour 油(蒸)气
oil vapour pump 油蒸气泵
oil vermilion 油朱红
oil viscosity 油黏度
oil volatility 油的挥发性
oil wash 油洗
oil washing 油洗
oil waste 废油
oil-water boundary 石油水煤气
oil-water emulsion ink 油基水乳化油墨
oil-water gas 石油水煤气
oil-water interface 油与水之间的交界面

oil-water ratio 油水比
oil-water separator 油水分离器
oil-water sludge 油水沉渣
oil-wax mixture 油-石蜡混合物
oil way 润滑油沟
oil wearing quality 油的防磨损性质〔润滑油降低润滑表面机械的磨损性能〕
oil wedge 油楔
oil well (石)油井
oil-well drilling machinery 凿油井机；钻油井机
oil wet 油湿润的
oil-wet particle 油润湿的粒子
oil wetted viscous filter 油黏附过滤器；油黏滞过滤器
oil wheel 加脂转鼓
oil wheeling 转鼓加油；重革填充
oil white 油白
oil wiper ring 刮油环
oily ①油的；含油的；油性；油质的②多油的；油腻的③油滑的
oil yard 石油厂场地
oily bitumen 油沥青
oily discharge 带油废水
oil yellow 油黄
oily flame 油焰；油脂火焰
oily flavor 油溶性食用香精
oily layer 油层
oily liquid 油状液体
oily lubricant 油性润滑剂
oily materials 油性材料
oily matter 油状物质
oily moisture 油潮气
oily perfume 油溶性香精
oily phase 油相
oily pollutant 含油污物；油性污物
oily polymer 油状聚合物
oily sewer 含油污水排水管
oily skin lotion 油肤膏
oily soil 油质污垢
oily sol 油状溶胶
oily waste liquor 含油废液
oily (waste) water 含油废水；含油污水
-oin 〔词尾〕 偶姻〔作俗名用。两个醛分子聚合成羟基酮；如 benzoin 苯偶姻,$C_6H_5CH \cdot OHCOC_6H_5$; butyroin 丁偶姻等〕
ointment 软膏；药膏
ointment base 软膏基料
ointment for gas protection 防毒软膏
ointment of zinc oxide 氧化锌软膏

oiticica oil 奥蒂树油
O'Keefee distribution 奥基夫分配
okenite 硅钙石
okinalein 白头翁英；翁因〔白头翁的一种提取物〕
okinalin 白头翁灵〔白头翁的一种提取物〕
Oklo phenomena 奥克洛现象*
okra-seed oil 秋葵子油
-ol 〔词尾〕醇；酚〔芳香族化合物〕
^{18}O labeling 氧-18 标记
olamide 油酰胺〔脱模剂〕
olanzapine 奥氮平（药）
olation 羟连作用*；羟桥合
ol bridge 羟桥*
old age theory 地龄增加重元素理论
old corrugated container(OCC) 废瓦楞箱(纸板)
old fustic 染色桑膏
old iron 废铁
old paper 废纸；旧纸
old paper stock 废纸纸浆
Oldroyd model 奥尔德罗伊特模型
Oldroyd's convected derivative 奥尔德罗伊特对流导数；奥尔德罗伊特迁移导数
Oldshye-Rushton column 奥尔德休-拉什顿塔
oleaginous 含油的
oleaginous fluid 油质液体
oleaginous material 含油物质
oleaginousness 含油量
oleaginous seed 含油种子
oleamide(=oleylamide) 油酰胺
oleanane 齐墩果烷
oleander oil 欧夹竹桃油
oleandomycin(=amimycin) 竹桃霉素
oleandrigenin 夹竹桃配基；夹竹桃苷元
oleandrin 夹竹桃苷 $C_{30}H_{48}O_3$
oleandrose 齐墩果糖；夹竹桃糖
oleanene 土当归烯；齐墩果烯；高根二醇
oleanol 石竹酸
oleanolic acid(=caryophyllin) 石竹素 $C_{30}H_{48}O_3$
olease 油酸酯酶
oleastene 橄榄油烃
oleasterol 橄榄油甾醇
oleate 油酸盐 $C_{18}H_{33}O_2M$
oleate ester 油酸酯
oleate soap 油酸皂
olefiant 成油的；生油的
olefiant gas 成油气
olefin 烯烃
olefination 烯化作用；成烯作用

olefin-carbon monoxide copolymer 烯烃-一氧化碳共聚物
olefin conversion process 烯烃转化法
olefin disproportionation 烯烃歧化(作用)
olefine 链烯
olefine acid 烯酸 $C_nH_{2n-1}COOH$
olefin(e) alcohol 烯醇
olefine aldehyde 烯醛 $C_nH_{2n-1}CHO$
olefin(e) complex 烯烃配合物；烯烃配位化合物
olefine copolymer(OCP) 烯烃共聚物
olefin(e) ketone 烯酮
olefine polymer oil 烯烃聚合油〔自烯类叠合所得到的合成润滑油〕
olefine polymer oil from Fischer-Tropsch gasoline 费托法烯烃聚合油〔费托法汽油馏分中烯烃聚合所制成的合成润滑油〕
olefine polymer oil from Fischer-Tropsch waxes 费托石蜡法烯烃聚合油〔以费托法烃类合成过程中所生成的石蜡经裂解所生成的高级烯聚合而成的合成润滑油〕
olefine-substituted aromatic 烯烃取代芳香烃
olefin fibre 烯烃纤维
olefin hydrocarbon 烯烃
olefinic 烯属的
olefinic acid(=olefinic carboxylic acid) 烯羧酸；烯属酸 $C_nH_{2n-1}COOH$
olefinic alcohol 烯醇 $C_nH_{2n-1}CH_2OH$
olefinic amino amide 烯氨基酰胺
olefinic bond 烯键
olefinic carbohydrate 烯属碳水化合物
olefinic carbon 烯碳
olefinic carboxylic acid(=olefinic acid) 烯羧酸；烯属酸
olefinic constituent 烯组分
olefinic fuel 烯烃燃料；含不饱和烃的燃料
olefinic hydrate 烯族水化物
olefinic hydrocarbon 烯烃
olefinic link(=olefinic linkage) 烯键
olefinic linkage(=olefinic link) 烯键
olefinic polymerization 烯烃聚合
olefinic rubber 烯烃橡胶
olefinic silica 烯属二氧化硅
olefinic terpene 烯属萜烯
olefinic unsaturation 烯烃不饱和度
olefin metathesis 烯烃复分解；烯烃歧化
olefin matathesis polymerization 烯烃易位聚合
olefin oligomerization 烯烃齐聚
olefin oxidation 烯烃氧(作用)
olefin oxide 氧化烯烃
olefin plastic(s) 烯烃塑料

olefin polymerization 烯烃聚合
olefin polymer oil 烯烃聚合油；聚烯烃油
olefin polysulfide 多硫化烯烃
olefins 烯烃
olefin silver complex 烯烃合银；烯烃银配盐
olefin sulfate 硫酸烯烃酯
olefin sulfonate 烯属磺酸酯
oleic 油酸的
oleic acid 油酸；顺 9-十八碳烯酸
oleic acid chloride 油酰氯
oleic acid diethanolamide 二羟乙基油酰胺
oleic acid ester 油酸酯
oleic acid glyceride 甘油油酸酯
oleic acid series 油酸同系物 $C_nH_{2n-2}O_2$
oleic alcohol 油醇
oleic series 丙烯酸系列
oleic soap 油甘皂；三油酸甘油酯皂
oleiferous 油性的
olein(=triolein) 油精；三油精；三油酸甘油酯 $(C_{17}H_{33}COO)_3C_3H_5$
oleinic acid 油酸 $CH_3(CH_2)_7CH=CH(CH_2)_7COOH$
olein soap 油甘皂；三油酸甘油酯皂
oleite 硫代蓖麻酸钠
oleo- 〔拉丁字头〕①油的②油酰(基)
oleo-creosote 杂酚油油酸酯
oleodipalmitin 甘油二软脂酸油酸酯
oleodisaturated glyceride 甘油二饱和甘油酯
oleodistearin 甘油二硬脂酸油酸酯
oleo-gear 油减震器
oleogel 油凝胶〔以油为分散剂的〕
oleo-guaiacol 愈创木酚油酸酯
oleo-leg 减震器支柱；油减震器底盘支柱
oleomargarine(=margarine) 人造奶油；真珠油
oleometer 油比重计
oleo nitrile 油基腈
oleo oil ①含油酸的固定油②牛奶脂；液牛油
oleopalmitate 油酸软脂酸盐；油酸棕榈酸盐
oleophilic 亲油的
oleophilic group 亲油基
oleophilic iron oxide 亲油氧化铁
oleophilicity 亲油性
oleophilic moiety 亲油部分
oleophilic resin 亲油树脂
oleophobic 疏油的
oleophobic colloid 疏油胶体
oleophylic 亲油的
oleophylic colloid 亲油胶体
oleophylic gellant 亲油胶凝剂

oleophylic graphite 亲油石墨
oleo-pneumatic brake 油传动刹车
oleoptene(=stearoptene) 玫瑰蜡
oleo red 油红
oleo refractometer 油折射计
oleoresin 含油树脂；精油树脂；松脂
oleoresin anise 茴香油树脂
oleoresin aspidium 绵马油树脂
oleoresin basil 罗勒油树脂
oleoresin capsicum 辣椒油树脂
oleoresin caraway 香芹油树脂
oleoresin cubeb 荜澄茄油树脂
oleoresin ginger 姜油树脂
oleoresin lupulin 啤酒花油树脂
oleoresin marjoram 甘牛至油树脂
oleoresin origanum 牛至油树脂
oleo-resinous 油树脂
oleoresinous coating(s) 油基树脂涂料
oleoresinous flat 油性树脂平光漆
oleoresinous paint 油性树脂漆
oleo-resinous undercoat 油改性树脂中间层
oleoresinous varnish 油性树脂清漆
oleoresin parsley 芹菜油树脂
oleosacchara 油糖
oleo shock absorber 油减震器
oleosol 固体润滑油
oleo-soluble 油溶性的
oleostasis property 油滞性
oleostearine 油硬脂；压制脂
oleo stock ①原浆；原浆②工业牛脂
oleostrut 减震支柱；油减震器支柱
oleo undercarriage 油减震器底盘
oleovitamin A 维生素 A 油
oleoxy chloride 油酰氯 $CH_3(CH_2)_7CH=CH(CH_2)_7COCl$
oleoxy methyl taurate 油酰甲基牛磺酸盐
oleoyl 油酰 $CH_3(CH_2)_7CH=CH(CH_2)_7CO-$
oleoyl anisidine sulfonate 油酰甲氧基苯胺磺酸盐
oleoyl chloride 油酰氯
oleoyl ethyl anilide 油酰乙(基)苯胺
oleum ①发烟硫酸②〔拉丁文〕油
oleum calcis 潮化氯化钙液
oleum spirit 油精〔沸程范围在 300~400°F 之间的石油馏分〕
oleuropeen 油橄榄苦素〔一种苦苷，在加工油橄榄时应予除去〕
oleuropeic acid 油酸
Olex 烷烯分离法

oleyl 油基；十八碳-9-烯基
oleyl alcohol 油醇
oleylalcohol disulfate 油醇二硫酸盐
oleylamide(=oleamide) 油酰胺
oleyl amine 油胺；十八(碳)烯胺
oleylcetyl alcohol 油基鲸蜡醇
oleyl ether sulphate 油基醚硫酸盐
oleyl nitrile 油酰腈
oleyl sulfate 油基硫酸盐
oleyl triethanol amine 油酰三乙醇胺
olfactie 悟〔气味强度单位〕
olfaction 嗅感(作用)
olfactometry(=odorimetry) 气味测定(法)；嗅觉测定法
olfactory 嗅感；香感
olfacty 臭味单位
ol group 桥羟基*
olibanol 乳香醇
olibanoresin 乳香脂
olibanum 乳香
olibanum oil 乳香油
olibanum resin 乳香树脂
olibanum resinoid 乳香树脂
-olic acid 脑酸〔在俗名中用〕；炔酸；油酸
-olide〔词尾〕 交酯
Oliensis spot test 奥林萨斯试验〔测定沥青中裂化烃〕
olifiant gas 成油气；乙烯
oligo-〔希腊字头〕 寡；少
oligoacrylonitrile 丙烯腈低聚物
oligoamide 酰胺低聚物
oligoamino acid 氨基酸低聚物
oligoclase 奥长石
oligoclasite 奥长岩
oligocyclocarbonate 低环碳酸酯
oligoester 酯类低聚物
oligo-1,6-glucosidase 低聚-1,6-葡萄糖苷酶
oligomer 低聚物*；低聚体
oligomer distribution 低聚物分布
oligomer extraction 低聚物萃取
oligomeric 低聚的
oligomeric condensation 低聚缩合(作用)
oligomeric dicarboxylic acid 低聚二羧酸
oligomeric dimethylolurea 低聚二羟甲基脲
oligomeric diol 低聚二醇；二醇低聚物
oligomeric material 低聚物材料
oligomeric vinyl phosphonate 低聚膦酸乙烯酯
oligomerization 低聚反应*
oligomolecular supermolecule 低聚分子超分子
oligomycin 寡霉素
oligonite 菱锰铁矿
oligonucleotide 低(聚)核苷酸
oligonucleotide array 寡核苷酸阵列
oligoolefine 寡烯〔指含 3,4 个烯键〕
oligophrenia phenylpyruvica 智力发育不全性苯丙酮尿症
oligopolymer 低聚物
oligosaccharidase 低聚糖酶；寡糖酶
oligosaccharide 寡糖*；低聚糖
oligosaccharide sulfate 低聚糖硫酸盐
oligose 低聚糖；寡糖
oligosilicic acid 寡硅酸
oligoterephthalic acid glycol ester 对苯二甲酸乙二醇酯低聚物
oligotrophy 营养不足；贫养；寡养
oligourethane 氨基甲酸酯低聚物；低聚氨酯
oliguria 少尿(症)
olimycin(=orymycin) 稻霉素
Olin claritymeter 奥林透明度仪
olistomerism 残途同归现象
olivaceous 橄榄色的
olivacin 橄色菌素
olive castile soap 马赛皂
olive-green 橄榄绿
olive-kernel oil 橄榄仁油
olivenite 橄榄铜矿
olive oil 橄榄油
olive oil soap 橄榄油皂
Oliver filter 奥利弗过滤器
olivetoric acid 戊基地衣缩酚酸；油地衣酸
olive-yellow 橄榄黄
olivil 橄榄树脂素
olivine 橄榄石
olivine-apatite rock 橄榄磷灰岩
olivine diabase 橄榄辉绿岩
olivomycin 橄榄霉素
Olmstead clay (美国)奥姆司梯黏土
Olsen flow tester 奥尔森流动试验机
Olsen's durometer 奥氏硬度计
Olsen's testing machine 水泥抗张强度测定器
Olsen type hardness tester 奥尔森型硬度试验机
omal 三氯苯酚
ombuine 商陆精
omega- ①〔希腊字母〕ω ②末位的
omega-lauric lactam ω-十二内酰胺；ω-月桂内酰胺
omega sulfonation 磺甲基化（ω-磺化）
omegatron 回旋质谱仪*
omeprazole 奥美拉唑〔药〕
OM-ESR spectroscopy 定向调制-电子自旋共振波谱(法)

omni-　全(部)；总；遍
omni-directional pressure　各向均匀压力
omni-factor　多因素
on a large scale　大规模的
on-and-off controller　双位式控制器
on-and-off service　利用双位阀门的调节过程
on-axle　正轴
once-run distillate　一次馏液
once-run kerosene　直馏灯油
once-run oil　页岩原油蒸馏时得到的最后馏分
once-through　①单程的；一次的②单循环的；一次操作(完成)的
once-through conversion(=one-pass operation)　非循环过程；单程转化
once-through cracking　单程裂化；非循环裂化
once-through fuel cycle　一次通过式燃料循环*
once-through material balance　一次通过物料平衡；单程物料平衡
once-through method　一次法〔非循环或非再生的技术过程，溶剂或吸附剂的不经再生而只用一次的方法〕
once-through operation　单程操作
once-through pipe still　一次蒸发管式炉；常压管式炉
once-through process　非循环过程；单程过程
onchocerciasis　肉丝虫病
oncogen　癌基因
oncogene　致癌基因
oncology　肿瘤学
on-column　在柱
on-column derivatization　柱中衍生化；柱上衍生
on-column detection　柱上检测
on-column direct laser detection　直接激光在柱检测
on-column electrical conductivity detection　在柱电导率检测
on-column enrichment　柱上富集
on-column injection　柱上注射；柱头注射；柱头进样
on-column injection system　柱头注射系统；柱头进样系统
oncolysis　肿瘤消除；溶癌作用
on-contact printing　密贴式印刷
oncostatin　制瘤素；抑瘤素
oncotic agent　增渗剂
oncovin(=vinblastine)　长春新碱
ondansetron　昂丹司琼〔药〕
on-detector system　直联检测器系统
ondometer　波长计；测波器；波形测量仪
one　①—②—(个)的；单(独)的；单(个)的
-one〔词尾〕　酮
one-and-one-half-screw machine　双螺杆喂料式单螺杆挤塑机
one-bath process　一浴法；单浴法
one-brush bottle washer　单刷洗瓶器
one-button filter plate　单纽滤板
one-button plate　单纽板
one-can urethane coating　一罐装聚氨酯涂料；单组分聚氨酯涂料
one coat　一道漆；一道涂层
one-color indicator　单色指示剂
one-component system　单组分系统*
one-cylinder motor　单缸发动机
one-deck classifier　单板分级机
one-deck drawplate oven　单屉烤炉
one-deck oven　单层炉
one-degree benzene　一度苯
one-dimensional　单向的
one-dimensional chromatography　单向色谱(法)
one-dimensional comparative chromatography　单向比较色谱法
one-dimensional development　单向展开(法)
one-dimensional extension　一维拉伸
one-dimensional failure　一维衰坏；单向破坏
one-dimensional flame propagation　火焰前缘的单向分布
one dimensional lattice　一维点阵
one-dimensional NMR spectra　一维核磁共振谱
one-dimensional order　单向有序；线型有序
one-dimensional structure　一维结构*
one drop liquid chromatography　一滴液相色谱法
one-electron approximation　单电子近似*
one-electron bond　单电子键
one-ended initiator　单端引发剂
one-ended living polymer　单端活聚合物
one-ended oligomer　单端(活)低聚物
one-factor experiment　单因子试验；单因素试验
one-fluid cell　单液电池
one-fluid theory　单流体(学)说
one-glaze ware　单色釉
one-molecule-deep layer　一分子厚的(润滑剂)层
one package polyurethane enamel　单组分聚氨酯磁漆
one package stabilizer　复合稳定剂；综合稳定剂
one pack coatings　一罐装涂料；单组分涂料
one-pack stabilizer　单组分稳定剂
one pack system　单组分
one-pass operation(=once-through conversion)　非循环过程；单程转化
one-period　一(个)周期
one-phase current　单相电流

one-piece 一体(的);整体(的)
one-pit system 一槽法
one-point adjustment 一点调节
one point method 单点法
one-pot coating 单罐装涂料;单组分涂料
one-pot method 一锅法
one pot system 单罐装系统
one pour method 一次注型法
one-quarter-filled shell effect 四分之一充满(电子)壳层效应
oneset 整体的〔非组成部分的〕
one-shot lubrication 一次润滑系统
one-shot molding (聚氨酯泡沫)一步成型
one-shot molding process 一次铸型法
one-shot pump 单塞泵
one-shot rigid urethane foam 一步法硬聚氨酯泡沫
one-shot urethane foam 一步法聚氨酯泡沫胶
one-sided confidence interval 单侧置信区间
one-sided criterion of significance 单侧显著性检验
one-sided etching 单侧腐蚀
one-sided precoat 单面预涂
one-side test 单侧检验
one side welding 单面焊
one-stage alkyd process 一步醇酸(树脂)合成法
one-stage conversion of gases 气体的一段转化过程
one-stage disproportionation 一步歧化(法)
one-stage hydrolysis 一段水解;一级水解
one-stage nitration 一段硝化(法)
one-stage process ①一段(硝化)法②一段法;一步法③(酚醛树脂)单级制备法
one-stage resin ①一步加热树脂②单级(酚醛)树脂
one-stage thermal liquefaction 一级热液化
one-step conversion 一步转化(法)
one-step curing 一次硫化法
one-step elution 一步洗脱(法)
one-step high-speed spin-draw-winding process 一步法高速纺丝拉伸卷绕工艺
one-step high-speed spinning-stretching-crimping-cutting process 一步法高速纺丝、拉伸、卷曲和切断工艺
one-step polymerization 一段聚合;一段叠合
one-step process of acrylic fibre 聚丙烯腈纺丝原液一步法工艺
one-step synthesis 一步合成
one-stop station 各种技术服务站
one-tailed test 单侧检验*
one-tank model 单槽模型
one thread sizing 单丝上浆
one-throw yarn 一次加捻多股线

one-touch gelation 触发胶化
one-trip bottle 一次性使用瓶;不回收瓶
one-turn cap 一转盖头
one-way analysis of covariance 一种方式协方差分析
one-way analysis of variances 一种方式方差分析
one-way chromatogram 单向色谱(图)
one-way classification 一种方式分组
one-way cock 单向阀门;单通活栓
one-way drawing 单向牵伸
one-way layout ①一元配置;一元构型②一元分类
one-way orientation 单向取向
one-way strip chromatogram 单向条色谱图
one-way valve 单向阀
one-way vision mirror 半透明镜
one-zone drawframe 单区拉伸机
onflow 流入;进气
onion oil 洋葱油
onion-shaped discharge 葱头形流出〔指自黏度计喷嘴流出〕
onit(=cyclotrimethylene trinitramine) 渥尼脱;翁尼特;环三亚甲基三硝胺
onium compound 鎓类化合物;(电负性元素)最高正价化合物
onium group 鎓基
onium ion 鎓离子
onium salt 鎓盐
onium surfactant 鎓类表面活性剂
on-line 联机;线上〔计算机〕
on-line analysis 线上分析;流线分析;在线分析
on-line analyzer 在线分析仪;(生产)线上分析仪
on-line colorimetric analysis 流线比色分析;在线比色分析
on-line computer 联机计算机
on-line connection of chromatograph and electronic digital computer 色谱仪和电子数字计算机联用
on-line coulometer 流线电量计;在线库仑仪
on-line data handling 联机数据处理
on-line degasser 在线脱气装置
on-line detection 柱上检测
on-line electrical stacking 在线电堆集
on-line electrochemistry 在线电化学
on-line instrumentation ①在线仪表装置②在线仪表检控〔生产线上的仪表测试和控制〕
on-line Kalman filter 在线卡尔曼滤波器
on-line LC-MS 在线液相色谱-质谱联用
on-line mass spectrometry 线上质谱分析
on-line moisture meter 在线水分计
online monitoring of water quality 水质在线监测

on-line oven　流水线炉；联机炉
on-line preconcentration　在线预浓集
on-line real time detection　在线实时检测
on-line reduction　在线还原
on-line separation　线上分离；流线分离；在线分离
on-line unit　联机设备；流线设备；线上设备
only-coat fibre　全皮纤维
onocerin(=onocol)　芒柄花素；芒柄花萜
onocol(=onocerin)　芒柄花素；芒柄花萜
on-off　开关(式)
on-off control　开关控制
on-off controller　开关控制器
on-off reaction　(链的)增减反应
on-off state　断续状态
on-off switch　换向开关
on-off time proportional controller　开关时间比例控制器
on/off valve　开关阀
onofrite　硒汞矿
onomycin　小野霉素
ononid　中性芒柄花根素
ononin　芒柄花苷
-onose〔词尾〕　酮糖
-onosic acid　酮糖酸
on-port flame　喷嘴部分火焰
on position　插入位置；开关的工作位置
on rust paint　带锈涂料
on-schedule delivery　按时(准时)交货
onset　开始反应
onset of turbulence　涡流运动的开始
onset potential　起始电位；呈现电位
on-side of bearing　轴承油入口边
on-site　在工地上；场(厂)区内；单位内；装置内
on-site analysis　现场分析
on-site inspection　现场检验(检查)
on-site reprocessing　现场后处理；就地后处理；就堆建后处理工厂
on-site reprocessing plant　(反应堆或核电站)现场后处理工厂；就堆所建后处理工厂
on-site training　现场训练
on-stream　在操作中；在工作过程中；在运转中；开车
on-stream analysis　流程分析；线上分析
on-stream analyzer　流程分析器
on-stream chromatograph　流程色谱仪
on-stream chromatography　流程色谱法
on-stream efficiency　开工率(即设备的有效使用期与修理期的比例)
on-stream inspection　不停车检查；运转中检查
on-stream maintenance　开工期间维修
on-stream period(=on-stream time)　连续开工期限〔石油加工设备的〕
on-stream pressure　操作压力
on-stream time(=on-stream period)　连续开工期限〔石油加工设备的〕
on tap　上栓
on-the-button control　按钮控制
on-the-job dependability　现场使用可靠性
on-the-spot disposal　就地处理
onychograph　指甲毛细管搏动描记器
onyx　缟玛瑙
oocyan(=oocyanin)　蛋壳青素
oocyan(in)　胆绿素；蛋壳青素
oölite　鲕状岩
oölitic limestone　鲕状灰岩
oölitic texture　鲕状结构
oometer　量蛋器
ooze　淤泥
ooze corrosion　海泥腐蚀
ooze leather　植鞣(铬鞣)软牛革；植鞣绒面革
oozy　淤泥的
opacification　乳浊化；乳浊状；不透明
opacifier　遮光剂；(使)不透明剂
opacifying agent　不透明剂
opacifying effect　遮盖效应
opacifying pigment　不透明颜料
opacifying power　遮盖力
opacifying property　遮盖性；遮光性
opacifying strength　遮盖强度
opacimeter　暗度计；不透明度仪
opacin　碘酞
opacity　①不透(明)性②不透(明)度；暗度③浑浊度
opacity pigment　不透明颜料
opal　①乳白的②蛋白石〔矿〕
opalescence　乳光；乳色
opalescent　乳光的；乳色的
opalescent glass　乳光玻璃；不透明玻璃
opalescent lacquers　珠光漆
opal flashes　蛋白石采色
opal glass　乳白玻璃；玻璃瓷
opaline　①蛋白石的②乳白的
opalized wood　蛋白石化木
opalizer　消光剂；遮光剂；不透明剂；乳浊剂
opal jusper　蛋白石化木
opal oil　乳白油〔与亚麻仁油相混合的硫酸精制石油馏分〕
opalwax　乳白蜡〔一种氢化植物蜡〕
OPA process　二异辛基磷酸萃取法

opaque 不透明的
opaque background 不透明背景
opaque body 不透明体
opaque coating 不透明涂料；不透明涂层
opaque color 不透明色
opaque copolymer 不透明共聚物
opaque enamel 乳白搪瓷
opaque finish 不透明涂饰剂
opaque fused silica 不透明熔融硅石
opaque glass ①乳白玻璃②不透明玻璃
opaque glaze 乳浊釉；乳白釉
opaqueness 不透明性②不透明度③不透性④不透度
opaque paint 不透明的涂料；遮光涂料
opaque pigment 不透明颜料
opaquer 遮光剂
opaque stain 不透明着色剂
opaque white 乳白(色)
open ①开②断③(敞)开的；敞口的
open air 空气；户外
open-air atmosphere 露天环境〔测试〕
open air drying 露天干燥
open-air exposure ①露天曝置②露天暴晒
open-air retting 露天沤麻
open-air weathering 露天气候老化〔测试〕
open-and-shut controller 开关控制器
open and shut valve 双位开关；双效开关
open assembly 晾胶
open assembly time 晾置时间
open-band twist Z捻；反手捻
open bearing 开启轴承
open-bed 开(放式)床
open-bed chromatography 开床色谱法
open black ash furnace 敞式黑灰炉
open bubble 表面气泡〔塑料缺陷〕
open-bull gear drive 大齿轮传动；间接传动
open burning coal 非结焦性煤
open casting 敞模铸塑
open cast molding 无压铸型
open cell 开口电池
open-celled structure 开孔结构
open cell foamed plastic 开孔泡沫塑料
open-cell product 开孔泡沫制品
open chain 开链
open chain compound 开链化合物
open chain hydrocarbon 开链烃
open chamber 开口室〔柴油机直接喷射的燃烧室〕
open channel 明渠
open channel blocking 阻断通道开放

open circuit 断路；开路
open circuit battery 暂流电池组
open circuit cell 暂流电池〔物〕
open-circuit crushing 开线路压碎
open-circuit grinding 开路研磨
open-circuit operation 断路操作
open circuit potential 开路电势*
open circuit voltage 开路电压；空载电压
open coated capillary 空心涂渍毛细管
open column ①空心柱〔指毛细管柱〕②开柱〔指薄层〕
open container ①敞口容器②开顶集装箱
open-cup flash-and-fire test 开杯闪点与燃点试验
open-cup flash-point 开杯闪点
open cup flash test 开口闪点试验；开杯闪点试验
open cup test 开杯试验
open-cup tester 开杯试验器
open cure 无模硫化
open-cycle gas turbine 开放循环气轮机
open delivery 明流式
open-discharge filter-press 开口卸料压滤机
open drive 空车运转
open-edge gutter 向外卷边槽
opened lap 开棉球
open end (管子)开口端
open-end(ed) steam pipe 开口的蒸汽管
open-end hole 开口端孔
open end perform 开口型坯
open end shaving machine 敞开式削匀机
opener 开棉机
open evaporator 开敞式蒸发器
open-faced drum 空心轴颈转鼓
open filter 敞式沙滤器
open fire 明火；活火
open fire tester 开杯燃点测定器
open flame 明火；活火
open-flame furnace ①敞炉；平炉；有焰炉②开炉
open flash point 开杯闪点
open flash point test 敞口闪点试验
open front 前(面)开式；前开门式
open-fronted cabinet 正面(前面)开式橱；(柜、箱、壳)体前开门式橱
open furnace 敞炉；平炉
open gauge 开口压力计
open-gear lubrication 开口齿轮润滑
open grain ①大孔隙性；大孔隙率②(木材)粗纹理
open heap 敞放土堆
open-hearth 平炉
open-hearth furnace 平炉

open-hearth process　平炉法
open-hearth refining　平炉炼钢法
open-hearth steel　平炉钢
open-heater　敞口炉；敞口蒸发器
open hole　空心
open hole capillary column　空心毛细管柱
open hole column　空心柱
open impeller　开式叶轮
open-impeller pump　开式叶轮泵
opening　①洞，孔；缝；隙②开孔③通路
opening crack　大龟裂；露底龟裂；大开口裂缝
opening for feed(=feed nozzle)　进料口
opening of bond(=opening of link)　键的断开
opening of link(=opening of bond; opening of linkage)　键的断开
opening of linkage(=opening of link)　键的断开
opening of ring　环的裂开
opening of sieve　筛孔
opening type　襟型
opening up　打开
opening wedge　揭模(用)铁楔
open installation　露天装置
open ion pair　开敞离子对
open joint　接头空隙
open manometer　开口压力计
open mixer　开炼机
open pan mixer　敞口盘式混合器
open-pit method　(矿物填料)露天开采法
open pore　开孔
openpore coating　①白茬涂漆；白坯涂漆；本色涂漆②打底子油
open-pore finishing　①白茬涂装；白坯涂装②(木器)白茬涂装法；白坯涂装法
open pot stand oil　开釜聚合油
open return bend　U形管
open roaster　开式烤炉
open sand filter　敞式砂滤器
open-seas skimmer　海面撇油器
open shell　开壳层*
open shell ion　开壳层电子
open shell system　开壳(层)体系
open side of blade　刮刀的开口边
open soaper　开口皂洗器〔纤维〕
open splice　开缝
open split　开口分流
open split type GC-MS interface　开口分流型气相色谱-质谱接口
open steam　直接水蒸气
open steam coils　直接水蒸气旋管
open steam cure　①直接水蒸气硫化②直接水蒸气处治
open structure　松散结构
open-syphon experiment　敞口虹吸实验
open system　敞开系统*
open tank filter　敞滤槽
open-textured　(质地)疏松的；松散的；不致密的
open to atmosphere　与大气通连的
open-top mold　敞口模
open-top pail　开口桶
open trench　明沟
open tube　开口管；空心管
open tube mercury manometer　开口管水银压力计
open tube test　开口管试验
open tubular capillary packed column　空心毛细管填充柱
open tubular chromatography　开口管柱色谱法；空心柱色谱法
open tubular column　空心柱*；开口管柱
open type　开口式；敞开式
open-type heat exchanger　开口式热交换器
open (type) impeller　开式叶轮
open-vat　①敞口瓮②制纸瓮；无盖浆槽
open vulcanization　无模硫化*
open water-cooling tower　敞式水冷塔
open-width impregnator　平幅浸轧机
operability　适用性；可操作性
operability of process　操作的可能性
operand　运算对象，运算数；操作数
operating　操作的；工作的
operating air　操作气体；操纵空气
operating aisle　操作走廊
operating area　操作区
operating characteristic　操作特性
operating characteristic curve　操作特性曲线
operating charges　操作消耗量
operating condition　操作条件；工作条件
operating cost　操作费用；生产费用
operating cycle　操作周期；运转周期
operating duty　①工作负载(状态)②操作规程
operating efficiency　操作效率
operating failure　操作故障
operating features　操作特点
operating flexibility of tray　塔板操作弹性
operating floor　操作台
operating fluid　工作液体
operating handle　操作手柄
operating head　工作压头
operating inferiority　①操作拙劣②生产亏损

operating latitude 运转幅度；操作范围
operating life 工作寿命；使用寿命
operating line 工作线；操作线
operating manual 运行手册；操作手册
operating overhead expense 非操作费用
operating panel 操作面板；操纵台
operating parameter ①操作参数②运行参数
operating performance 操作手续
operating point 操作点
operating post 工作岗位
operating pressure 操作压力
operating procedure 工作原理图；操作步骤；操作程序
operating program 操作程序
operating radius 作用半径；运行半径
operating refinery （现在生产的）石油加工厂
operating repair 日常维修
operating sequence 操作程序
operating speed 操作速度
operating system management 操作系统管理
operating temperature 操作温度
operating temperature range 操作温度范围
operating tension 工作张力
operating threshold 工作限；工作阈
operating time 运行时间；操作时间
operating variable 操作变量
operating volatility 操作挥发度
operating voltage 操作电压；动作电压；工作电压
operation 操作；运转手续；处理
operational amplifier for conductance measurement 电导测定中的有效(运算)放大器
operational calculus 运算微积分〔数〕
operational character ①运算字符②控制字符
operational condition 操作条件
operation(al) cost 操作费；运转费；开车费用
operational definition of the pH scale pH 的操作定义
operational error 操作误差
operational function simulating 运算函数模拟
operational procedure 操作规程
operational reliability ①操作(运行)可靠性②运算可靠性
operational use time 有效利用时间；有效工作时间
operation analysis 操作分析
operation code 工艺卡
operation control 操作控制
operation error 操作误差
operation factor(=operation ratio) ①运行(转)率；开工率②运算率
operation flow sheet 操作顺序图；操作流程图
operation guide 制导；导向
operation handle arm 操作手柄
operation hours 操作小时；设备运行时间
operation in tandem 联立操作
operation of expectation 期望值运算
operation plan 施工方案(计划)；作业方案(计划)
operation procedure 操作程序；操作规程
operation rate 运行率；开工率
operation ratio(=operation factor) ①运行(转)率；开工率②运算率
operation recorder 操作记录器
operation research 运筹学
operation reserve ①开车准备；运转准备②业务准备金
operation sequence 工序；操作工序
operation sequence control 操作(运算)程序控制
operation sheet 操作卡片；操作记录表；工序卡
operation standard 操作标准
operation variable 操作变数
operator ①操作者②(运)算符〔数〕③操纵基因
operator equation 算符方程
operator error 操作者误差
operator gene 操纵基因
operon 操纵子
ophelic acid 樟牙菜苷酸
ophicalcite 蛇纹大理石
ophiobollin 蛇孢菌素
ophiolite 蛇纹石
ophiotoxin 眼镜蛇毒素
ophioxylin 蛇根木苷；蛇藤苷
ophite （闪化）辉绿岩
ophitic texture 辉绿结构
ophitron 小型微波发生器
ophthalmin 维生素 A
ophthalmo- 〔词头〕眼
ophthalmoleukoscope 偏振光色感计
ophthalmoscope 检眼镜
ophtocillin(=xanthocillin) 黄青霉素
opial 盐酸阿片碱
opianine(=narcotine) 那可汀
opianyl 袂康酸
opiate 鸦片制剂
opium 鸦片；罂粟
opium alkaloid hydrochloride 盐酸鸦片碱
opium alkaloids 罂粟碱
opium vinegar 罂粟制品
opium wax 鸦片蜡
opopanax 红没药；防风药
opopanax gum 红没药树脂；防风根树脂

opopanax oil　红没药油；防风根油
opoponax　红没药；防风根
Oppenauer oxidation　奥本奥尔氧化作用
Opperman source　奥泼曼光源
opponent-colors color space　对立色色空间
opposed　①相对立(抗)的②不同的
opposed conformation(=eclipsed conformation)　重叠构象
opposing reaction　对峙反应*
opposite direction　反向；对向；相反的方向
opposite electricity　相反电荷
opposite poles　异名极
opposite pressure　反压力
opposition　①对抗；抵抗②反对
oppositional allele　对立等位(基因)
oppression　压迫；抑制
opsopyrrole　3-甲基-4-乙基吡咯
optic　①旋光的②视觉的；视力的
optical　①光(学)的②旋光的
optical aberration　光学像差
optical absorption spectroscopy　光学吸收光谱法
optical absorption transition　光学吸收跃迁
optical active polymer　光活性聚合物*
optical activity　光学活性*
optical alignment　光学调整；光学校直
optical ammeter　光学电流计
optical analysis　光学分析
optical anisotropy　光学各向异性
optical antimer(=optical antipode)　旋光对映体
optical antipode(=optical antimer)　旋光对映体
optical attenuation coefficient　光衰减系数
optical axis　光轴
optical axis of crystal　晶体光轴*
optical beam scanning　光束扫描
optical beat　光学拍频
optical bench　光具座
optical bleaches　荧光增白剂
optical bleaching　光漂白
optical bleaching agent　荧光增白剂；光学增白剂
optical brightener　荧光增白剂
optical brightening agent　荧光增白剂
optical calcite　光学方解石；冰洲石
optical cavity　光孔穴
optical character　光学特性
optical chemical sensor　光化学传感器
optical chopper　斩光器；光调制器
optical chromatography　光色谱法
optical clarity　光学透明度；光学清晰度
optical clear　光学透明(的)

optical comb　光梳
optical comparator　光学比较器
optical conductivity　光导性；光导率
optical constant　光学常数
optical contact　无界面干扰线的接触
optical cycle　旋光循环
optical densitometer　光密度计
optical densitometric method　光密度(分析)法
optical density(OD)　光密度
optical density color　光密度色度
optical depolarization　去偏振(作用)
optical depression　旋光性降低
optical detection of magnetic resonance(ODMR)　光检测磁共振
optical diffraction　光衍射
optical dispersion　光色散
optical distortion　光畸变
optical distortion value　光学畸变值
optical dye　荧光染料
optical efficiency　光效率
optical electrobalance　光电天平
optical electron　光电子
optical electronic regulation　光电(子)调节
optical emission spectrography　发射光谱法
optical emission spectrometer　光发射光谱仪
optical emission spectroscopy　发射光谱
optical enantiomorph　旋光对映体；光学对映体
optical exaltation　旋光性增强；超加折射
optical excitation　光激发
optical excited atom　光激发原子
optical extinction coefficient　消光系数
optical fiber　光学纤维；光导纤维
optical fiber chemical sensor　光纤化学传感器
optical fiber flow cell　光导纤维流动池
optical fiber for remote sensing　遥感光导纤维
optical fiber resonance Raman spectroscopy　光纤共振拉曼光谱
optical fiber sensor　光纤传感器
optical fibre　光导纤维
optical-fibre acoustic sensor　光纤声传感器
optical-fibre active connector　光纤有源连接器
optical-fibre jacket　光纤套层
optical filter　滤光片；滤光器
optical forward scattering　光学前向散射
optical gain　光增益
optical glass　光学玻璃
optical grating　光栅
optical guided wave fibre　光波导纤维

optical hairiness counter 光学毛羽计数器
optical holography 光全息照相；全息记录材料
optical index 光学指数〔表示黏度与折光率之间的关系〕
optical induction 光学诱导*
optical instrument 光学仪器
optical isomer(=optical isomeride) 旋光异构体；旋光异构物
optical isomeride(=optical isomer) 旋光异构体；旋光异构物
optical isomerism 旋光异构*
optical isotropy 光学各向同性(现象)
optical lever 旋转杠杆
optical limit sensor 光敏限位传感器
optically active 旋光的；起偏振转(作用)的
optically active absorption band 光学活性吸收谱带
optically active carbon 旋光碳(原子)
optically active covalent compound 旋光共价化合物
optically active enantiomorph 旋光对映体
optically active exchanger 旋光(离子)交换剂
optically active form 旋光体
optically active isomer 旋光异构体；旋光异构物
optically active polymer 光学活性高分子；旋光聚合物
optically active resin 旋光树脂
optically active stationary phase 旋光固定相
optically active substance 旋光物
optically anisotropic melt 光学各向异性熔体
optically anisotropic mesophase pitch 光学各向异性中间相沥青
optically detected electron-electron spin double resonance 光检测电子-电子自旋双共振
optically detected electron-nuclear double resonance 光检测电子-核双共振
optically detected electron-nuclear-nuclear triple resonance 光检测电子-核-核三重共振
optically detected magnetic resonance 光检测磁共振
optically empty 光真空的
optically gated electrophoresis 光门电泳
optically inactive 不旋光的；非旋光的
optically inactive substance 不旋光物
optically isotropic 光学各向同性的
optically pure 旋光纯的
optically stimulated exo-electron emission 光激发外逸电子发射
optically transparent electrode 光透电极
optically transparent vitreous carbon electrode 光透玻璃碳电极*
optical microanalyzer 光学微分析器

optical microscope 光学显微镜
optical microscopy 光学显微术
optical microwave double resonance(OMDR) 光学微波双共振
optical modulator 光调制器
optical multichannel analyzer 光学多道分析器
optical null method 光平衡法
optical null principle 光学零位原理
optical path 光路；光程
optical path difference 光程差
optical path length 光程长度；光路长度；光程
optical perturbation-electron paramagnetic resonance 光微扰-电子顺磁共振
optical plastics 光学塑料
optical polish 光学抛光
optical property 光学性质
optical purity 光学纯度*
optical pyrometer(=ardometer) 光学高温计*；光测高温计
optical quenching 光猝灭
optical receiver 光接收器
optical resolution 旋光拆开；光学离析；光学分辨
optical resonator 光学共振腔
optical retardation 光滞后
optical rotary dispersion 旋光色散
optical rotation 旋光性*；旋光度
optical rotatory dispersion 旋光色散*
optical rotatory dispersion analysis 旋光色散分析
optical rotatory power 旋光本领
optical rotatory substance 旋光物质
optical scanning 光学扫描
optical scattering coefficient 光散射系数
optical sensitization 光学敏化
optical sensor 光学传感器
optical signal detection 光信号检测
optical spectral analyzer 光学光谱分析器
optical spectrometer 光学光谱仪
optical spectroscopy 分光法
optical spectrum 光谱
optical spectrum analyzer 光谱分析器
optical superposition 旋光叠合现象
optical system 光(学)系(统)
optical thermal analysis 光学热分析
optical thickness 光学厚度
optical time-domain reflectometer 光时域反射计
optical transmission factor 透光系数
optical transmittance 光透射比；光透射率；光透射系数
optical tube length 光管长度

optical waveguide 光波导
optical wedge 光学楔;消光楔
optical whitener 光学增白剂
optical whitening agent 荧光增白剂
optic fibre diameter analyzer 光学纤维直径分析仪
opticity 旋光性;旋偏振性
optics 光学
optimal 最适(宜)的
optimal block design 最优区组设计*
optimal control 最优控制
optimal design 最优设计
optimal design of experiment 最优试验设计
optimal estimate 最优估计*
optimal estimation 最优估计
optimal molar ratio 最佳克分子比
optimal performance 最佳特性
optimal value 最优值*
optimeter 光电比色计
optiminimeter 光学测微计
optimization ①最优化②优选法
optimization of flow 流速优化
optimization point 优化点
optimized method 优选法
optimizing 最佳化
optimum ①最佳的;最佳值;最适(宜)的②最适宜点;最适宜情形
optimum allocation 最优分配
optimum angle 最佳角
optimum bridge resistance 最佳电桥电阻
optimum capacity 最佳能力;最佳容量;最佳处理量
optimum column efficiency 最佳柱效
optimum column length 最佳柱长
optimum column temperature 最佳柱温
optimum concentration 最佳浓度;最适浓度
optimum condition 最佳条件
optimum configuration 最优构型
optimum cure ①最适硫化②最适处治
optimum design 最佳设计
optimum efficiency 最佳效率
optimum eluant velocity 最佳洗脱液速度
optimum feed location 最宜进料位置;最佳进料点
optimum film thickness 最佳漆膜厚度
optimum flip angle 最佳倾倒角
optimum flow 最佳流量
optimum flow rate 最佳流速
optimum fusion temperature 最佳熔化温度
optimum gas pressure 最佳气压
optimum gas velocity 最佳气体流速

optimum operating temperature 最佳操作温度
optimum operation 最佳操作(法);调优(技术)
optimum performance 最佳性能;最佳操作特性
optimum pH 最适 pH;最佳 pH
optimum pigment-vehicle ratio 最佳颜基比
optimum pitch 最佳螺距
optimum practical flow rate 最佳实际流速
optimum practical gas velocity 最佳实际气速
optimum procedure 最优化程序
optimum pulse angle 最佳倾倒角
optimum range 最佳量程
optimum reflux 最适回流
optimum reflux ratio 最佳回流比
optimum resolution 最佳分辨率
optimum sampling fraction 最优抽样比
optimum size ①最佳尺寸②最佳粒度
optimum successive overrelaxation 最优逐次超弛豫
optimum temperature 最适温度
optimum temperature difference 最适温差
optimum temperature of solubilization 最佳增溶温度〔非离子表面活性剂的浊点〕
optimum temperature sequence 最优温度序列
optimum transmission 最佳透光度
optimum transmittance 最佳透光度
optional 任意的
optional equipment 附加设备
optional programmability 可选择的程序可编性
optional verification 非强制(性)检定
optoacoustic cell 光声池
optoacoustic detector 光声探测器
optoacoustic spectroscopy 光声光谱法
optoacoustic spectroscopy detector 光声光谱检测器
optoacoustic spectrum 光声光谱
optochine 乙氢去甲奎宁;纽莫昆〔药〕
optode 光极
opto-electronical wrap detector 光电式绕辊检测器
optoelectronics 光电子学
optogalvanic effect 光电流效应
optogalvanic spectroscopy 光电流光谱法
opto-isolator 光隔离
optoporation 光透入法
optrode 光极
oral contraceptive 口服避孕药
oral lethal dose 口服致死量〔半数致死量试验〕
oral toxicity 口服毒性
oraluton 脱水羟基孕甾酮
orange ①橙(子);橘子②橙色的
orange Ⅲ (=methyl orange) (酸性)三号橙;甲基橙;金

莲橙 D
orange base 橙碱
orange blossom 橙花
orange blossom odor 橙花气味
orange blossom oil 橙花油
orange brown 橙棕
orange chrome 铬橙
orange chrome yellow 橙铬黄〔碱式铬酸铅〕
orange dextran 橙色葡聚糖
orange dye 橙色染料
orange flower absolute 橙花净油
orange flower oil 橙花油
orange flower water 橙花水
orange juice 橙汁；橘子汁
orange lac 橙色紫胶
orange lake 橙色色淀
orange lead(=orange mineral) 铅橙
orange lead chromate 铅铬橙；铬酸铅橙
orange leaf oil 橙叶油
orange mineral(=orange lead) 铅橙
orange oil 橙油
orange oxide 橙色氧化物
orange pale 淡橙黄色的〔润滑油标准色〕
orange peel ①橙皮②橘皮皱纹
orange-peel finish 橘皮状皱纹漆
orange-peel oil 橘皮油
orange pigment 橙色颜料
oranger crystals β-萘甲酮
orange red 橙红
orange rind 陈皮
oranger liquid 萘甲酮混合物
orange root 毛茛根
orange-seed oil 橙子油
orange shellac 橙色虫胶；橙色紫胶；精制虫胶片
orange skin 橘皮
orange sweet oil 甜橙油
orange syrup 橙皮糖浆
orange tincture 橙皮酊
orange toner 橙色色原(调色料)
orange Ⅳ(=tropeolin OO) (酸性)四号橙；金莲橙 OO；二苯氨基偶氮对苯磺酸
orange vat pigment 橙瓮颜料
orange yellow 橙黄
orangite 橙黄石
Oranienburg grained soap 奥堡皂
orarsan 乙酰氨基羟苯基胂酸
oraviron 甲基睾丸甾酮
orbenin 甲氯苯异唑青霉素

orbit ①轨道②眼窠
orbital ①轨道的②轨函数；轨道
orbital angular momentum 轨道角动量
orbital arrangement 轨道排列
orbital degeneracy 轨道简并度
orbital effect 轨道效应
orbital electron 核外电子
orbital electron capture 轨道电子捕获；核外电子捕获
orbital energy 轨道能量
orbital motion 轨道运行
orbital overlap population 轨道重叠布居
orbital pad sander (便携)轨道式砂垫磨光机
orbital phase conservation 轨道位相守恒
orbital quantum number 轨道量子数
orbital sander 轨道式喷砂机；轨道式打磨机(磨光机)
orbital symmetry 轨道对称(性)
orbital symmetry rule 轨道对称定则
orbital theory 轨函数(学)说
orbital valence force field(OVFF) 轨道价力场
orbital wave function 轨道波函数
orbitread process 缠贴胎面胶翻胎法
orcein 苔红素；地衣红
orcein dye 苔红素染料
orchard heating oil 果园取暖用燃料
orchidae 水杨酸戊酯〔增塑剂〕
orchid composition 兰花香精
orchid compound 兰花香精
orchil(=tournesol;orselle) 苔藓红素；苔藓色素；石蕊素
orchinol 红门兰醇
orcin(=orcinol) 苔黑素；苔黑酚；5-甲基-1,3-苯二酚
orcine test 5-甲基-1,3-苯二酚试验
orcinol(=5-methyl-resorcinol; orcin) 苔黑酚；苔黑素；5-甲基-1,3-苯二酚 $CH_3C_6H_3(OH)_2$
orcinol phthalein 苔黑酚酞 $C_{22}H_{16}O_5$
β-orcylic aldehyde 甲基苔黑醛；2,5-二甲基-4,6-二羟基苯甲醛
ordeal bark 毒树皮
ordeal bean 毒豆
order ①秩序②级；次；次序；等级
order-disorder distribution 有序-无序分布(状态)
order-disorder (phase) transition 有序-无序(相)转变
order-disorder transformation 有序-无序转变
order-disorder transition 有序-无序转变
order distribution 序态分布
ordered alloy 有序合金
ordered aromatic copolyamide 有序芳族共聚酰胺
ordered arrangement 有序排列
ordered copolymer 有序共聚物

ordered copolymerization　有序共聚(作用)
ordered domain　有序(微)区
ordered-form　有序型
ordered heterocyclic copolymer　有序杂环共聚物
ordered heterocyclic-imide copolymer　有序(的)杂环酰亚胺共聚物
ordered lattice　有序晶格
ordered linear copolymer　有规线型共聚物
ordered point defect　有序点缺陷
ordered polymer　有序聚合物
ordered region　有序区
ordered sequence　有序序列
ordered solid solution　有序固溶体
ordered structure　有序结构
ordering　①调整②有序化(转变)
ordering beat treatment　有序化热处理
ordering parameter　排列次序参数；调整参数；有序化参数
ordering reaction　有序化反应
orderly shutdown　正常停车
orderly three-dimensional arrangement　有序三维(度)排列
order number　原子序(数)
order of addition　加料顺序；加药顺序
order of compound　①化合物次序②化合物目数
order of diffraction　衍射级
order of excellence　优劣顺序
order of magnitude　数量级
order of multiple-quantum coherence　多量子相干的阶
order of prefix　词头的次序；字头的次序
order of radical　基名的次序
order of reaction　反应级数
order of resonance　共振级
order of spectrum　光谱级
order of volatility　挥发顺序
order parameter　序参数
order sorter　(光谱)级数分类器
order statistics　顺序统计量
ordinal　①序数②按序的③属于某目的
ordinal number　序数
ordinal test　①定序试验②定目试验
ordinary　(寻)常的；普通的
ordinary bond(=ordinary link)　单价键
ordinary bright cone　普通有光丝筒子；普通有光筒子丝
ordinary differential equation　常微分方程(式)
ordinary elastic deformation　常弹性形变；普(通)弹(性)形变
ordinary elasticity　普通弹性

ordinary extract　普通浸膏〔未经亚硫酸盐处理的〕
ordinary iron　铸铁
ordinary least square　普通最小二乘
ordinary least square fitting　普通最小二乘拟合
ordinary link(=ordinary linkage)　单价键
ordinary linkage(=ordinary link)　单价键
ordinary loop knot　筒子结
ordinary portland cement　普通硅酸盐水泥
ordinary pressure　(寻)常压(力)
ordinary ray　寻常光线；寻常射线
ordinary superphosphate　过磷酸钙
ordinary temperature　(寻)常温(度)
ordinary valence　主(要化合)价；常见价
ordinary water　普通水；轻水
ordinate　纵(坐)标(数)
ordination number　原子序(数)
ordnance　兵工(制)品；军需品
ordosite　河套岩；鄂尔多斯岩
ore　矿；矿石
ore assay mill　试矿磨机
ore band　矿带；矿脉
ore-bearing rocks　〔复〕(含)矿岩
ore bed　矿层
ore bedding　矿床
ore body　矿体
ore cleaning　选矿
ore crusher　碎矿机
ore deposit　矿床
ore dressing　选矿
ore furnace　煅矿炉
oregonensin　奥地灵芝素
oregon maple　大叶槭
ore grinder　矿石研磨机
ore hand picking(=ore hand sorting)　人工选矿
ore hand sorting(=ore hand picking)　人工选矿
ore hearth　熔矿炉；膛式炉
ore microscope　矿石显微镜
ore mill　磨矿机
ore milling　选矿
oreodaphne　月桂油
oreodaphnol　月桂油醇
ore of phosphorus　磷矿石
ore pocket　矿袋
ore process　生铁矿石法
ore roaster　煅矿炉
ore roasting　煅烧矿石
ore roasting chamber　煅矿室
ore separator　矿石分离器

ore sorting　(拣)选矿(石)
ore stamp　捣矿石锤
organ　①器官②工具
organelle　细胞(小)器
orgrange chrome yellow　橙铬黄〔碱式铬酸铅〕
organic　有机的
organic accelerator　有机促进剂
organic acid　有机酸
organic acid content　有机酸含量
organic acid dizirconyl　有机酸三氧化二锆
organic acidity　①有机酸性②有机酸度
organic acid peroxide　有机酸过氧化物；过氧化有机酸
organic adhesive　有机胶黏剂
organic analysis　有机分析
organic anion　有机阴离子
organic barrier coat　有机隔离涂膜；有机屏蔽涂层
organic base　有机碱
organic bentonite　有机(改性)膨润土
organic bentonite gel　有机膨润土凝胶
organic binder　有机漆基；有机成膜物
organic biosensor　有机相生物传感器
organic blowing agent　有机发泡剂
organic bound chlorine　有机结合氯
organic cation　有机阳离子
organic cation-organic anion complex　有机阳离子-有机阴离子络合物
organic chemical pollutant　有机化学污染物
organic chemicals　有机药品
organic chemist　有机化学家；有机化学工作者
organic chemistry　有机化学
organic chromatography　有机色谱法
organic chromogenic reagent　有机显色剂
organic clay　有机瓷土；有机白土
organic coating　有机涂层
organic colorant　①有机着色剂②有机染料③有机颜料
organic color pigment　有机彩色颜料
organic color reagent　有机显色剂
organic combustion　有机(物)燃烧
organic compound　有机化合物
organic condensation agent　有机缩合剂
organic contaminant　有机污染物
organic contamination　有机污染
organic content　有机质含量
organic coprecipitant　有机共沉淀剂
organic corrosion inhibitor　有机阻蚀剂；有机防锈剂
organic degradation　①有机降解②发霉
organic derivative　有机衍生物
organic derivative of alumino-silicate　铝硅酸盐的有机衍生物

生物
organic detritus　有机残渣
organic diluent　有机稀释剂
organic dip-spin coatings　浸渍离心(法)有机涂料
organic disulfide　有机二硫化物　R_2S_2
organic dust　有机灰尘；有机粉尘
organic electrochemistry　有机电化学
organic electrode process　有机电极过程
organic electrolyte　有机电解质
organic elemental analysis　有机元素分析
organic element polymer　有机元素聚合物
organic enhancement　有机增强
organic ester　有机酯
organic ether　有机醚
organic exchanger　有机离子交换剂
organic fiber　有机纤维
organic fiber optic material　有机光(学)纤(维)材料
organic fiber reinforced plastic　有机纤维增强塑料
organic filler　有机填料
organic film deposition　有机膜沉积
organic film etching　有机膜腐蚀
organic flocculant　有机絮凝剂
organic fluorescent substance　有机荧光物质
organic functional group　有机官能团
organic grease　有机润滑脂
organic group　有机基(团)
organic halide　有机卤化物；卤有机化合物
organic hydroperoxide　有机过氧化氢；过氧化氢有机(化合)物
organic inhibitor　有机抑制剂
organic inorganic hybrid material　有机-无机杂化材料
organic ion　有机离子
organic ion exchanger　有机离子交换剂
organic isocyanate　有机异氰酸酯
organic laser　有机激光器
organic layered silicate　有机层状硅酸盐
organic ligand　有机配体
organic liquid　有机液体
organic liquid gel　有机液体胶体
organic luminophor　有机发光体
organic mass spectrometer　有机质谱仪
organic mass spectrometry　有机质谱法
organic matrix　有机(聚合物)基体；有机基质
organic matter　有机物(质)
organic media　有机介质
organic mercurials　有机汞制剂；有机汞化物
organic mercury determination　有机汞测定
organic metal　有机(化)金属

organic mineral　有机矿物
organic modifier　有机改进剂
organic nitrogenous　有机氮
organic nitrogenous fertilizer　有机氮肥
organic peracid　有机过酸　RCOOOH
organic peroxide　有机过氧化物
organic phase　有机相
organic phase biosensor　有机相生物传感器
organic phosphate　磷酸酯
organic phosphite　有机亚磷酸酯
organic phosphor determination　有机磷测定
organic phosphorous insecticide　有机磷杀虫剂
organic photoconductive film　有机光导薄膜
organic photoconductor　有机光导体
organic pigment　有机颜料
organic pigment Florida exposure test　有机颜料佛罗里达暴晒试验
organic pigment permanence to heat　有机颜料的耐热性
organic pigment permanence to light　有机颜料的耐光性
organic polar solvent　有机极性溶剂
organic pollutant　有机污染物
organic pollutant analysis　有机污染物分析
organic polymer　有机高分子
organic polysulfide　有机多硫化物
organic precipitant　有机沉淀剂
organic promoter　有机助催化剂；有机促进剂；有机助聚剂
organic protective coating　有机保护涂料
organic qualitative analysis　有机定性分析
organic radical　①有机基②有机根
organic raw material　有机原料
organic reaction　有机反应
organic reagent　有机试剂
organic reagents for precipitation　有机沉淀剂
organic reinforcing agent　有机补强剂
organics　有机物
organic salt　有机盐
organic semiconductor　有机半导体*
organic silicon compound　有机硅化合物
organic slime　有机泥煤〔一种藻类物质生成的煤〕
organic sludge　有机淤泥
organic solvent　有机溶剂
organic solvent degreasing　有机溶剂脱脂(法)
organic solvent effect　有机溶剂效应
organic stabilizer　有机稳定剂
organic substance　有机物(质)
organic substrate　①有机基质②有机物底材
organic sulfide　有机硫化物

organic sulfur　有机硫
organic sulfur compound　有机硫化合物
organic sulfur determination　有机硫测定
organic suspending agent　有机悬浮剂
organic symbol　有机符号
organic synthesis　有机合成
organic terminology　有机术语
organic theory　有机理论
organic thinner　有机稀释剂
organic titanate　①有机钛酸酯②有机钛酸盐
organic titanium　有机钛
organic toner　有机色料；有机色原
organic ultraviolet-radiation absorber　有机紫外线辐射吸收剂
organic volatile compound　有机挥发(性化合)物
organic waste　有机废弃物
organic waste water　有机废水
organic yellow　①有机黄(颜料)②(=Hansa yellow)耐晒黄；汉撒黄
organic zeolite　有机沸石
organic zinc-rich primer　有机富锌底漆
organidin　碘化甘油
organism　有机体；生物体
organisol(=organosol)　有机溶胶；稀释增塑糊
organization　组织
organization of gas protection　防毒组织
organized assembles　有序组合体
organized ferment　活体酶
organized polymer　组织化聚合物
organized polymerization　有序聚合；组织聚合
organized polymer zone　有机(化)聚合物层；有机(化)聚合物区
organo-〔词头〕　有机(的)
organoacetoxysilane　(有机)乙酰氧基硅烷
organoactinide　有机锕系元素化合物
organo-alkali compound　有机碱化合物
organoalkoxysilane　烷氧基硅烷
organo-aluminium　有机铝
organo-aluminium compound　有机铝化合物
organoaluminium sesquichloride　有机铝倍半氯化物
organo-alumino-silicate　有机铝硅酸盐
organobentonite　有机膨润土
organo-borane　有机硼烷
organoboration　有机硼化(作用)
organoboron catalyst　有机硼催化剂
organoboron compound　有机硼化合物
organoboron polyamide　有机硼聚酰胺
organoboron siloxane　有机硼硅氧烷

organocationic 有机阳离子(的)
organochlorine compound 有机氯化物
organochlorine residue 有机氯残留量
organochlorosilane 有机氯硅烷
organo-chromium compound 有机铬化合物
organo-chromium reagent 有机铬试剂
organoclay(=organopolysilicate) 有机黏土；聚有机硅酸盐
organoclay-thickened latex 有机瓷土增稠胶乳
organoclay-thickener 有机瓷(白)土增稠剂
organodisilane 有机乙硅烷；有机二硅烷
organoferric 有机铁的
organo-functional group 有机官能(基)团
organogel 有机凝胶
organogenous sediment 有机沉淀物
organo-germanium oxide 有机锗氧化物
organo-halogen-silane 有机卤代硅烷
organoiron compound 有机铁化合物
organolanthanide 有机镧系元素化合物
organo-lead compound 有机铅化合物
organoleptic 器官感觉的〔可以引起器官感觉的〕
organoleptic test 器官感觉试验
organoleptic threshold 风味阈值
organolite 有机碱交换料；离子交换树脂
organolithium compound 有机锂化合物
organo-magnesium compound 有机镁化合物
organo-magnesium halide 有机镁卤化合物 RMgX
organomercurial 有机汞制剂
organo-mercuric compound 有机汞化物
organo-mercuric halide(=organo-mercuric salt) 有机汞卤化合物 RHgX
organo-mercuric salt(=organomercuric halide) 有机汞卤化合物
organomercurous fungicide 有机汞防霉剂；有机汞杀菌剂
organo-metal(=organo-metallic compound) 有机金属化合物
organo-metallic antiknock 有机金属抗爆剂
organo-metallic catalyst 有机金属催化剂
organo-metallic chemistry 有机金属化学
organometallic complex 有机金属配合(络)物
organo-metallic compound(=organo-metal) 有机金属化合物
organo-metallic condensation polymer 有机金属缩聚物
organometallic monomer 有机金属单体
organometallic polymer 金属有机聚合物*
organometallic reagent 有机金属试剂
organo-metallics 有机金属化合物

organometallic stabilizer 有机金属稳定剂
organo-metalloidal compound 有机准金属化合物
organo-metal substitution 有机金属取代作用
organo montmorillonite 有机蒙脱土；有机高岭土
organo-peroxide 有机过氧化物
organophilic ①亲有机物质的②亲表官的
organophilic bentonite 亲有机物质的膨润土；有机质膨润黏土
organophilic carrier 亲有机载体
organophilic gel 疏水凝胶
organophobic 疏有机性的；疏有机物质的
organophosphate 有机磷酸盐；有机磷酸酯
organophosphite 有机亚磷酸酯
organo phosphorous compound 有机磷化合物
organophosphorous insecticide 有机含磷杀虫剂
organophosphorus polyamide 有机磷聚酰胺
organophosphorus polymer 有机磷聚合物
organopolysilane 有机聚硅烷
organopolysilicate(=organoclay) 有机黏土；聚有机硅酸盐
organoradiomercurial 有机放射性汞制剂
organosilan 有机硅烷
organosilanediol 有机硅烷二醇
organosilicane resin 硅树脂
organosilicate 有机硅酸盐
organo-silicic oil 有机硅润滑剂；硅质润滑油
organosilicon 有机硅(化合物)
organo-silicon compound 有机硅化合物
organo-silicone 有机硅氧烷
organo-silicone rubber 有机硅橡胶
organosilicon heatresiant paint 有机硅耐高温漆
organosilicon oxide 有机硅氧化物
organosilicon polyamide 有机硅聚酰胺
organosilicon polyester 有机硅聚酯
organosiliconpolymer 有机硅聚合物
organosilmethylene 亚甲联二硅基〔结构为—Si—CH$_2$—Si—〕
organo-siloxane 有机硅氧烷
organosiloxane polymer 有机硅氧烷聚合物
organosilyl ①有机硅的②甲硅烷基
organosmectite 有机绿土
organosol 有机溶胶；稀释增塑糊
organosol paper coating 有机溶胶纸张涂料
organosoluble cellulose ester 有机(溶剂)溶解性纤维素酯
organosoluble cellulose ether 有机(溶剂)溶解性纤维素醚
organo tin 有机锡

organo-tin antifouling paint　有机锡防污漆
organotin-carboxylate　有机锡羧酸盐
organo-tin compound　有机锡化合物
organotin (containing) paint　(含)有机锡漆
organo-tin mercaptide　有机锡硫醇盐
organotin stabilizer　有机锡稳定剂
organotitanate coupling agent　有机钛酸酯偶联剂
organotrialkoxysilane　有机三烷氧基硅烷
organouranium compound　有机铀化合物
organ-pipe leather　风琴管革
organza　透明硬纱
orgeat　杏仁糖浆
orientable　可取向的；可定位的
oriental agate　(东方)玛瑙
oriental amethyst　紫刚玉
oriental aromatic　东方型香料
oriental emerald　绿刚玉
oriental garnet　柘榴石
oriental hyacinth　红锆石
oriental lacquer　中国大漆；中国天然漆
oriental lapis　青金石〔天然群青〕
oriental note　东方香韵
oriental odor　东方型香
oriental powder　藤黄与硝酸钾混合物
oriental ruby　红宝石
oriental sapphire　真蓝宝石
oriental sweetgum　安息香香脂
oriental topaz　黄宝石；黄刚玉
orientated polymer　取向聚合物
orientating group　定向基；定位基
orientation*　取向*；定位；定向
orientational disorder　取向无序
orientational entropy　取向熵
orientational strengthening　取向增强
orientation angle　①取向角②方位角
orientation birefringence　取向双折射
orientation blow mo(u)lding(=stretch blow mo(u)lding)　取向吹塑
orientation complex　定向配合物
orientation contrast　取向对比
orientation crystallization　取向结晶
orientation degree　取向度
orientation distribution function　取向分布函数
orientation effect　取向效应
orientation factor　取向因素
orientation force　定向力
orientation function　取向函数
orientation half-angle　取向半角

orientation half-width　取向半值幅；取向半宽度
orientation hysteresis　定向滞后
orientation index　取向指数；晶向指数
orientation law　定向(定)律
orientation matrix　取向矩阵
orientation modulation-electron spin resonance　定向调制-电子自旋共振
orientation modulation-electron spin resonance spectroscopy　定向调制-电子自旋共振波谱法
orientation (molar) polarization　(摩尔)定向极化(度)
orientation of micelle　胶束取向
orientation parameter　取向参数
orientation pattern　取向式样；取向图样
orientation polarizability　定向极化率
orientation polarization　偶极子转向极化；定向极化
orientation ratio　取向比
orientation release stress　解取向应力；消(除)取向应力
orientation rule　定向法则
orientation theory of thixotropy　触变(性)取向论
orientation uniformity　取向均一性
orientative experiment　定向试验
orientator　定向器
orient-blow system　①取向吹塑法②取向吹塑装置
oriented　定向的
oriented adsorption　定向吸附
oriented chain　取向链
oriented-crystalline state　取向结晶态
oriented crystallization　取向结晶(作用)
oriented double refraction　取向双折射
oriented fibre　取向纤维；拉伸纤维
oriented film　定向膜
oriented gel　取向凝胶
oriented growth　取向生长
oriented layer　定向层
oriented phase　定向相
oriented polarization　取向极化
oriented polymerization　取向聚合
oriented region　取向区
oriented slit tape system　取向切膜法制条装置
oriented structure　定向结构
oriented wedge　定向楔
orienting　定向
orienting effect　定向效应
orienting effect of group　基团的定向效应
orienting group　定向取代基
orienting roll　导向滚；导辊
orientomycin　东霉素；环丝氨酸
orient yellow　天然硫化镉；镉黄

orietz 蔷薇辉石
orifice ①小孔②口型③喷丝孔④孔板
orifice baffle 锐孔挡板
orifice check valve 小孔止回阀
orifice coefficient 锐孔系数；流量系数〔液体流经锐孔的〕
orifice column 筛板塔
orifice column mixer 筛板塔混合器
orifice control valve 锐孔调节阀；孔板节流阀
orifice differential 孔流分压计
orifice discharge 锐孔流量
orifice flowmeter 孔板流量计
orifice gas scrubber 锐孔气体洗涤器
orifice holder 锐孔固定装置
orifice land 口模成型面
orifice meter 孔板流量计；细孔流速仪
orifice meter coefficient 锐孔流量计系数
orifice metering 锐孔测量法
orifice method （计量的)锐孔法
orifice mixer 孔流混合器
orifice plate 孔板
orifice plate flowmeter 孔板流量计
orifice plate pulse column 孔板脉冲柱
orifice relief 口型槽沉
orifice restriction 锐孔收缩
orifice rheometer 孔式流变仪
orifice support tube 锐孔支管
orifice valve 阻尼阀
orifice viscometer 锐孔黏度计
origanol 牛至醇
origanum oil 牛至油
origin ①来源；起源②原点
original 最初的；原来的；开始的
original adoption 最初实施；最初采用(通过)
original brightness 原始亮度
original colour 原色；最初的颜色；原来的颜色
original design 原设计
original grating 原形光栅
original initiator concentration 原始引发剂浓度
original line 起点线
original mold 原型；样模
original plasticity 原始可塑性
original point 起点
original position statistic distribution analysis 原位统计分布分析
original shape 原来的形状；初始形状
original stock 原始混合物
original twist 初捻

original value 原始数据；起初值
origoester acrylate 丙烯酸酯低聚物
O-ring O 形环；O 形垫圈〔密封圈〕
Orion ion membrane electrode 奥龙离子膜电极
oripavine 东罂粟碱；3-氧去甲蒂巴因
orixine 和常山碱
orizabin 球根牵牛苷；喇叭脂
ormolu 锌青铜；铜锌锡合金
ormolu varnish 镀金漆；仿金漆
ormosanine(=piptamine) 红豆树宁；苦豆碱
ormosin 红豆树素
ormosine 红豆树碱
ormosinine 红豆树新宁
ornamental ①装饰用的；增光的②装饰品
ornamental brick 釉面砖；装饰性花砖
ornamental enamel 装饰性搪瓷品
ornamental glass (ware) 装饰性玻璃(器皿)
ornamental moulding 艺术造型
ornamental plaster 装饰性粉饰；华丽粉饰
ornamental plating 装饰性电镀层
ornamycin 装饰霉素
ornidazole 奥硝唑〔药〕
ornithine and aspartate 鸟氨酸天冬氨酸〔药〕
ornithine decarboxylase 鸟氨酸脱羧酶
ornithuric acid(=dibenzoyl ornithine) 鸟尿酸；N,N-二苯甲酰鸟氨酸
orobol 二羟四氢黄酮
oroboside 香豌豆苷
orogen 造山带
orogenic 切向压缩力的
oropon 阿鲁朋（人造脱灰剂）
orosomycin 山霉素
orotic acid 乳清酸；4-羧基尿嘧啶 $C_5H_4N_2O_4$
oroxylin 木蝴蝶素
orpiment(=arsenic orange) 雌黄
orris 香菖蒲〔调酒味用药草〕
orris absolute 鸢尾净油
orris butter 鸢尾脂
orris concrete 鸢尾浸膏
orris oil 鸢尾油
orris resinoid 鸢尾树脂
orris rhizome 鸢尾根
orris root 鸢尾根
Orr's (zinc) white(=lithopone) 奥尔白；立德粉；锌钡白
Orr white 锌钡白
Orsat analysis 奥萨特(气体)分析(法)
orseille(=orselle) 苔色素
orse(i)llic acid 苔色酸

orselle(=orchil;orseille)　苔色素
orsellin　苔色灵
orsellinic acid　苔色酸；4,6-二羟-2-甲苯甲酸
orsomycin(=orosomycin)　山霉素
orsudan　奥尔苏丹〔药〕
orthanilamide　邻氨基苯磺酰胺　$NH_2C_6H_4SO_2NH_2$
orthanilic acid　邻氨基苯磺酸　$NH_2C_6H_4SO_3H$
orthanilic K　邻磺素 K
orthanilic S　邻氨基苯磺 S
orthene　乙酰甲胺磷〔缓效性有机磷杀虫剂〕
orthin　4-羟基-2-肼基苯甲酸；奥丁
orthite　褐帘石
ortho-〔希腊字头〕　①正②原③邻(位)
ortho-acetate　原乙酸酯　$CH_3C(OR)_3$
ortho-acetic acid　原乙酸　$CH_3C(OH)_3$
orthoacid　原酸*
ortho-alkylation　邻位烷基化作用
ortho-aluminate　原铝酸盐
ortho-aluminic acid　原铝酸　H_3AlO_3
ortho-aminophenyl glyoxalic lactim(=isatin)　邻氨基苯乙醛(酸)内酰亚胺；靛红
ortho-amyl phenol　邻戊基苯酚
ortho-and peri-fused　单边互稠
ortho-antimonate　正锑酸盐　M_3SbO_4
ortho-antimonic acid　正锑酸　H_3SbO_4
ortho-antimonite　原亚锑酸盐　M_3SbO_3
ortho-antimonous acid　原亚锑酸　H_3SbO_3
ortho-arsenate　正砷酸盐　M_3AsO_4
ortho-arsenic acid　正砷酸　H_3AsO_4
ortho-arsenite　原亚砷酸盐　M_3AsO_3
ortho-arsenous acid　原亚砷酸　H_3AsO_3
orthobaric density　本压密度；标准密度
orthobaric volume　标准容积
ortho-borate　原(正)硼酸盐　M_3BO_3
ortho-boric acid　原(正)硼酸　H_3BO_3
ortho-carbonate　原碳酸盐(酯)
orthocarbonic acid　原碳酸　$C(OH)_4$
ortho-chlor-para nitraniline　邻氯对硝基苯胺
orthochromatic emulsion　正色乳剂
orthochromatic film　正色片
orthochrome　原色母　$C_{23}H_{22}N_2(C_2H_5I)$
ortho-chromic acid　原铬酸　H_6CrO_6
orthoclase　正长石
ortho-compound　邻位化合物
ortho-cresol　邻甲酚
orthodane　荠苧烷
orthodene　荠苧登
ortho-derivative　邻位衍生物
orthodiazine　邻二嗪；1,2-二氮杂苯
ortho-dichlorobenzene　邻二氯苯
ortho-directing group　代用品；取代基
orthodox paint　传统油漆；旧式油漆
orthodox six-stage phosphating plant　传统的六步法磷酸盐化处理装置
ortho ester　原酸酯
orthoferulic acid　邻阿魏酸；2-羟(基)-3-甲氧(基)肉桂酸
orthoflow catalytic cracking　正流型流化催化裂化
orthoflow fluid catalytic cracker　正流流体催化裂化装置
orthoflow fluid cracking unit　正流流体裂化装置
orthoflow process　正流过程
orthoform　3-氨基-4-羟基苯甲酸甲酯；原仿；俄妥仿〔药〕　$C_6H_3(COOCH_3)OH(NH)_2$
ortho-formate(=ortho-formiate)　原甲酸酯　$OR_3(HC)$
ortho-formiate(=ortho-formate)　原甲酸酯
orthoformic acid　原甲酸　$HC(OH)_3$
orthoforming process　正流重整过程
orthoforming unit　正流重整装置
orthoform new　新俄妥仿〔药〕
orthoform old　老俄妥仿〔药〕
ortho-fused　单边稠
orthogneiss　正片麻岩
orthogonal　①正交的②矩形的
orthogonal array　正交数组；正交阵列
orthogonal atomic orbital　正交原子轨道
orthogonal basis　正交基〔数〕
orthogonal coordinate　垂直坐标；正交坐标
orthogonal coordinate system　正交坐标系〔数〕
orthogonal design　正交设计
orthogonal design of experiment　正交试验设计
orthogonal experimental design　正交试验设计
orthogonal function　正交函数
orthogonal function spectrophotometry　正交函数分光光度法
orthogonal injection　正交注入
orthogonality　正交性
orthogonalization　正交化*
orthogonalization recurrence selection method　正交递归选择法
orthogonal jump weighted centroid method　正交跳跃加权形心单纯形法
orthogonal layout(=orthogonal table)　正交表
orthogonal matrix　正交矩阵
orthogonal ply laminate　正交层合板
orthogonal polynomial　正交多项式
orthogonal polynomial regression　正交多项式回归
orthogonal pressure　垂直压力

orthogonal regression design 正交回归设计法
orthogonal rheometer 正交流变仪
orthogonal system 正交系
orthogonal table(=orthogonal layout) 正交表
orthogonal tensor 正交张量〔数〕
orthogonal transformation 正交变换
orthogonal viscoelastic body 正交黏弹体
orthogranite 正花岗岩
ortho-helium 正氦
orthohydrogen 正氢*
ortho-hydroxy-phenylazo-naphthol sulphonic acid 邻羟基苯偶氮-β-萘酚磺酸
orthoiodohippurate(OIH) 邻碘马尿酸盐
ortho-isomer(=ortho-isomeride) 邻位异构物
ortho-isomeride(=ortho-isomer) 邻位异构物
orthokinetic 同向移东
orthokinetic coagulation 同向凝结(作用)
orthokinetic flocculation 同向絮凝(作用)
ortho-localization energy 正交定域能
ortho-molybdic acid 原钼酸 H_6MoO_6
orthonitraniline orange 邻硝基苯胺橙
orthonitric acid 原硝酸
ortho-nitrogen 正氮
orthonormal 正交归一的
orthonormal basis 标准正交基
orthonormal orbital 正交归一轨道
orthonormal system 标准正交系；正交归一系
ortho orange 邻硝基苯胺橙
ortho-orientating group(=ortho-orienting group) 邻位定向基团
orthopanchromatic 全色
ortho-para directing groups 邻对位定向基团
ortho-para orientation 邻对位定向
ortho-para ratio 邻位-对位比率
orthopedic plaster 金氏白粉〔一种掺明矾的刷墙石膏粉〕
orthopedic polymer 矫形用聚合物
ortho-periodic acid 原(正)高碘酸
ortho-phenylene diamine 邻苯二胺
ortho-phosphate 正磷酸盐
orthophosphate acetyltransferase (正)磷酸乙酰基转移酶
ortho-phosphite (原)亚磷酸盐 M_2HPO_3
orthophosphoric acid (正)磷酸
ortho-phosphoric acid ester 正磷酸酯
orthophosphoric monoester phosphohydrase 磷酸(酯)酶；(正)磷酸酯水解酶
ortho-phosphorous acid (原)亚磷酸 H_3PO_3
orthophyre 正长斑岩
ortho-plumbate 原高铅酸盐 M_4PbO_4
ortho-plumbic acid 原高铅酸 H_4PbO_4
ortho-plumbic oxide 原高铅氧化物；二氧化铅
ortho-position 邻位
ortho-positronium 正-正电子素；正电子偶素
orthoquartzite 火成石英岩
orthorhombic 正交的
orthorhombic form 正交形
orthorhombic system 正交(晶)系
orthorhombic unit cell 正交晶胞
orthoselection 直向选择；定向选择
ortho-silicate ①原硅酸盐 M_4SiO_4 ②原硅酸酯 $Si(OR)_4$
ortho-silicic acid 原硅酸
ortho-siliformic acid 原甲硅酸；三羟基硅烷 $HSi(OH)_3$
ortho-silinic acid 原硅酸 H_4SiO_4;$Si(OH)_4$
orthosiphonin 直管草苷
ortho states α态；原态
ortho-sulfate 原硫酸盐
ortho-sulfuric acid 原硫酸
orthotaxy 级状聚形
ortho-terphenyl 邻三联苯 $C_6H_5C_6H_4C_6H_5$
orthothiazine 邻噻嗪；1,2-硫氮杂苯
ortho-thiocarbonic acid 四硫代原碳酸 $C(SH)_4$
ortho-titanate 原钛酸盐 M_4TiO_4
ortho-titanic acid 原钛酸 H_4TiO_4
orthotolidine arsenite method 邻联甲苯胺亚砷酸盐法
ortho tolidine method 邻联甲苯胺法
orthotropic 正交(各向)异性(的)
orthotropic elastic material 正交异性弹性材料
orthotropic hardening 正交异性硬化
orthotropic plate 正交异性板
orthotropy 正交各向异性
ortho-tungstic acid 原钨酸
ortho-vanadate 正钒酸盐 M_3VO_4
ortho-vanadic acid 正钒酸 H_3VO_4
ortho-water 普通水
orthoxazine 邻嗪；1,2-氧氮杂苯
ortho-zirconic acid 原锆酸 H_4ZrO_4
orticant ①发痒的②发痒剂
orticant action 发痒作用
orticant agent 发痒剂
ortizon 过氧化氢合尿素
Orton rearrangement 奥顿重排作用
orymycin 稻霉素
oryzachlorin 稻氯曲菌素
oryzacidin 杀稻菌素
oryzacidine 米曲杀菌素

oryzamycin 稻病霉素
oryzanal 米糠素
oryzanol 谷维素〔药〕
oryzasizine(=oryzacidine) 米曲杀菌素
oryzoxymycin 白叶枯霉素
orzata 杏仁糖浆
osage orange 桑橙
osamine 糖胺
-osan 聚糖
osazone 脎
osazone reaction 成脎反应
osazone test 糖脎试验
oscillant elastometer 振荡弹性计
oscillating 摆动的；振动的
oscillating agitator 振荡搅拌器
oscillating blade microtome 振动式超薄切片机
oscillating bob viscometer 振动摆球黏度计
oscillating bond 振动键
oscillating buffing machine 振荡式磨革机
oscillating chute 振动式斜槽
oscillating-crystal method 振荡晶体法
oscillating-cylinder viscometer 振荡圆筒式黏度计
oscillating damper 减振器
oscillating discharge 振动放电
oscillating discharge source 振荡放电离子源
oscillating disk rheometer 振动圆盘式流变仪；振荡盘式流变计
oscillating-disk viscometer 振荡圆盘式黏度计
oscillating double bond 摆动双键
oscillating drop 振荡滴
oscillating feeder 摆动进料器
oscillating flow 振荡流(动)
oscillating function 摆动函数；振动函数
oscillating jet technique 振荡射流法〔测动表面张力〕
oscillating link(=oscillating linkage) 振动键
oscillating linkage(=oscillating link) 振动键
oscillating machine 摆动机
oscillating motion 摆动运动
oscillating paint shaker 振荡式油漆搅拌机
oscillating plunger pump 摆动活塞泵
oscillating reaction 振荡反应
oscillating sander (往复式)砂带磨床；振动式磨光机
oscillating screen 振动筛
oscillating thermionic valve 振荡热离子真空管
oscillation 摆动；振动；振荡
oscillational quantum number 振动量子数
oscillation frequency 振动频率
oscillation lubricator 摆动润滑器

oscillation of gas 气体摆动
oscillation photograph 回摆图
oscillation theory 振动(理)论
oscillator 振荡器
oscillator circuit 振荡线路；振荡(回)路
oscillator strength 振子强度
oscillator type apparatus for high frequency titration 高频滴定振荡器
oscillatory convection 振荡对流
oscillatory discharge 振动放电
oscillatory motion 振动
oscillatory normal stress 振荡正应力
oscillatory response of mechanical model 力学模型的振荡响应
oscillatory rheometer 振动式流变仪
oscillatory rotational viscometer 振荡旋转黏度计
oscillatory shear 振荡剪切
oscillatory shear flow 振荡剪切流
oscillatory viscoelastometer 振荡黏弹计
oscillogram 示波图
oscillograph(=oscillometer) 示波器
oscillographic analysis 示波分析
oscillographic conductometric titrimetric method 示波电导滴定法
oscillographic coulometric titration 示波库仑滴定
oscillographic polarograph 示波极谱仪
oscillographic polarography 示波极谱法
oscillographic potentiometric titration 示波电位滴定(法)*
oscillographic potentiometric titrimetric method 示波电位滴定法
oscillographic titration 示波滴定(法)*
oscillography 示波法
oscillometer(=oscillograph) 示波器
oscillometric titration(=high frequency titration) 高频滴定
oscillometry 示波量法；高频分析
oscillopolarographic titration 示波极谱滴定(法)*
oscillopolarography 示波极谱法
oscilloscope 示波器
oscilloscopic chromatography 示波色谱(法)
oscilloscopic polarography 示波极谱法
oscillosynchroscope 同步示波器
oscine 莨菪灵
-ose 〔词尾〕 糖
Oseen-Goldstein correction 奥西恩-戈尔茨坦改正
Oseen's approximation 奥奥恩近似式
oseltamivir 奥司他韦〔药〕
oshaic acid 藁本酸

-oside〔词尾〕 糖苷
Oslo crystallizer 奥斯陆结晶器
osmane 木樨烷
osmanthus concrete 桂花浸膏；木樨浸膏
osmate 锇酸盐
osmate radical 锇酸根
osmic 锇的
osmic acid 锇酸
osmic acid anhydride 锇酸酐
osmic anhydride 四氧化锇 OsO_4
osmic compound 四价锇化合物
osmic hydroxide 四氢氧化锇；四羟化锇 $Os(OH)_4$
osmics 臭味学
osmiophilic 亲锇的
osmiridium(=iridosmine) 铱锇矿
osmium 锇
osmium chloride 氯化锇〔总名，计有：二氯化锇 $OsCl_2$、三氯化锇 $OsCl_3$ 和四氯化锇 $OsCl_4$〕
osmium dating method 锇定年代法
osmium dichloride 二氯化锇 $OsCl_2$
osmium dioxide 二氧化锇 OsO_2
osmium disulfide 二硫化锇 OsS_2
osmium fluoride 八氟化锇 OsF_8
osmium monoxide 一氧化锇 OsO
osmium oxide 氧化锇 $OsO; Os_2O_3; OsO_2; OsO_4$
osmium potassium chloride 氯锇酸钾
osmium sesquioxide 三氧化二锇 Os_2O_3
osmium sodium chloride 氯锇酸钠
osmium sulfide 硫化锇 $OsS_2; OsS_4$
osmium tetrachloride 四氯化锇 $OsCl_4$
osmium tetrasulfide 四硫化锇 OsS_4
osmium tetroxide 四氧化锇 OsO_4
osmium trichloride 三氯化锇 $OsCl_3$
osmocene 二茂锇
osmoceptor 发香团受体
osmolality 克分子渗透压重量浓度
osmolity 渗透度；渗透性
osmology 嗅觉；渗透学
osmometer ①渗透计②香度计
osmometry 渗透压(力)测定法
Osmond iron 奥斯孟铁
osmondite 碳化铁在 α-铁里的固态溶液；奥氏体
osmophore 发香团
osmophore group 生臭团〔产生臭味的基团〕
osmophor group 发香团
osmoscope 渗透试验器
osmosis 渗透(作用)*；渗析
osmotaxis 趋渗性

osmotic 渗透的
osmotic agent 渗透剂
osmotic balance ①渗透天平②渗透平衡
osmotic bleaching 渗透漂白
osmotic cell 渗透池
osmotic coefficient 渗透系数
osmotic concentration 渗透浓度
osmotic dehydration 渗透脱水(作用)
osmotic effect 渗透效应
osmotic equilibrium 渗透平衡
osmotic equivalent 渗透当量
osmotic exchange 渗透交换
osmotic force 渗透力
osmotic imbibition 渗透性吸液
osmotic membrane 渗透膜
osmotic potential 渗透势
osmotic pressure 渗透压*
osmotic pressure gradient 渗透压梯度
osmotic pressure of polymer solution 高聚物溶液的渗透压
osmotic shrinkage 渗透收缩
osmotic solvation 渗透溶合作用
osmotic stress 渗透应力
osmoticum 渗压剂
osmotropism 向渗性
Osmund furnace 高风箱炉；奥斯孟炉
Osmund iron 奥斯孟铁
osmyl 气味
-osone〔词尾〕 邻酮醛糖
-osonic acid〔词尾〕 酮糖酸
osotetrazine(=v-tetrazine) 接四嗪；1,2,3,4-连四嗪
osotriazolazimine 接三唑并叠氮；1,5-二氢-1,2,3-三唑并[4,5-d]-1,2,3-三唑
osotriazole(=2H-1,2,3-triazole) 接三唑；2H-1,2,3-三唑
osram 锇钨合金
ossalin 狐骨髓油
ossamycin 奥萨霉素
ossein ①骨胶原②生胶质
ossein silk 骨胶原人造丝；生胶质人造丝
osteolite 土磷灰石
osteolith 土(骨)磷灰石
Osterstrome liquid phase process 奥斯托斯脱罗莫液相过程〔利用吸附剂加工裂化馏出物〕
osthenol 欧前胡酚
osthol 蛇床子素 $C_{15}H_{16}O_3$
osthole 王草素；蛇床子素
ostranite 锆石
ostreasterol 牡蛎甾醇

ostreogrycin(=ostreogricin)　蛎灰菌素
ostrich fat　驼鸟脂
ostrich skin　驼鸟皮
ostrich yarn　卷曲绒毛线
Ostromyslenskii butadiene synthesis　奥斯卓米斯伦斯基丁二烯合成法
ostruthin　欧前胡精；王草黄〔从王草中取出的黄色结晶物〕
ostruthol　欧前胡脑
Ost's method for cleaning the electrode　奥斯特电极清洁法
Ostwald absorption coefficient　奥斯特瓦尔德吸收系数
Ostwald color system　奥斯特瓦尔德色系(列)
Ostwald curve　奥斯特瓦尔德曲线
Ostwald decade rheostat　奥斯特瓦尔德十进变阻器；奥氏十进变阻器
Ostwald equation　奥斯特瓦尔德电离度方程式
Ostwald process　奥斯特瓦尔德(制硝酸)法
Ostwald ripening　奥斯特瓦尔德熟化；奥氏熟化
Ostwald rule　奥斯特瓦尔德稳定态规律
Ostwald's absorption coefficient　奥斯特瓦尔德吸收系数
Ostwald's colorimeter　奥氏比色计
Ostwald's dilution law　奥斯特瓦尔德稀释定律
Ostwald's pycnometer　奥斯特瓦尔德比重计
Ostwald's viscometer　奥斯特瓦尔德黏度计
Ostwald U-tube　奥斯特瓦尔德 U 形管〔测黏度〕
Ostwald visco(si)meter　奥斯特瓦尔德黏度计
osyris oil　沙针油
osyritin(=osyritrin)　沙针苷；芸香苷
otan　二羟基丙酮
otavite　菱镉矿
otaylite　膨润土；斑脱岩
othonecine　奥索(千里光)裂碱
othosenine(=tomentosine)　奥索(千里光)碱
otoba butter　肉豆蔻脂
otobain　肉豆蔻脂素
otoba wax　肉豆蔻脂
otobite　肉豆蔻蜡
otolith　耳石
otter　①水獭②水獭皮
otto　玫瑰油；香精油
Otto-cycle engine　奥图-循环发动机
otto of rose　玫瑰油
O type sealing washer　O 型密封圈
ouabagenin　乌巴配基
ouabain　乌本(箭毒)苷；G 毒毛旋花苷
oudenone　小奥德酮；长根菇素
ounce　盎司；英两

ounce strength　盎司强度〔一种测定氨水浓度的单位，即中和一加仑氨水所需要硫酸的英两数〕
ouricuri oil　小冠椰子油
ouricury wax　小冠巴西棕蜡
ouroboros　含尾蛇〔希腊点金术符号〕
-ous〔词尾〕　亚〔中文译作词头〕
outage　预留容量〔蓄水池为了水膨胀〕
outage table　预留容器空间表
outboard bearing　外装轴承
outbreak　破裂；断裂；中断
outburn　烧完
outburst　爆发
outcome　结果；结论
outcrop　露头
outcross　异型杂交
outdoor aging　室外老化
outdoor aging characteristics　耐室外老化特性
outdoor durability　户外耐久性
outdoor equipment　室外设备
outdoor exposure test　室外暴露试验
outdoor exposure test stand　室外暴晒试验架；露天暴晒试验架
outdoor life　室外寿命
outdoor paint　户外用漆
outdoor station　户外暴晒站
outdoor storage　室外储存；露天储存
out(door) stress-crack life　室外应力龟裂寿命
outdoor weatherability　户外耐候性
outdoor weather resistance　室外耐候性
out-draw textured yarn　二步法拉伸变形丝；外拉伸变形丝
outer　外(面)的；(在)外的
outer bearing　外轴承
outer bed　外床
outer bed volume(=outer volume)　外床体积
outer casing　外壳；外罩
outer coagulant　外部凝固剂
outer coating　外层；外皮
outer complex　外络合物
outer cone　外锥
outer coordination sphere　外配位层
outer cover　外罩
outer diameter　外径
outer flame　外(层火)焰
outer Helmholtz plane　外部亥姆霍兹面
outer housing　外壳；外罩
outer loop　远核圈
outer mantle of flame　火焰外光轮

outermost electron 最外层电子
outermost monolayer crosslink 最外单分子层交联
outer of roundness 椭圆度；不圆度
outer-orbital complex 外轨配合物
outer-orbital configuration 外轨构型
outer orbital coordination compound 外轨配合物
outer orbital mechanism 外轨机理
outer phase 外相
outer ply 表层
outer race 轴承外环
outer reamer 外扩孔锥；外扩孔器
outer reference electrode 外参比电极
outer sleeve 外套筒
outer space 外层空间；宇宙空间；太空
outer sphere 外层
outer sphere complex 外层配合物
outer sphere complexation 外层配位；外球配位
outer sphere mechanism 外层机理
outer stack 外砌体
outer tube 外管
outer volume(=outer bed volume) 外床体积
outfall 泄出口；泄出管；泄出
outfit ①装置②配备；装备
outflow 流出
outflow resistance 流出阻力；排放阻力
outgas 除气
outgassed ionexchange 脱气离子交换
outgassing 除气作用
outgassing cooler 除气冷却器；脱气冷却器
outgassing rate 脱气速率；除气速率
outgoing beam 发出光束；射出光束
outgoing gas 出气
outgrowth 悬出；横出；侧出
outlet ①出口管②出口
outlet at bottom 底部流出口；排除污物管
outlet connection 出口联结管
outlet diagram 出口图；喷嘴出口速度三角图
outlet end detection of electrical conductivity 柱端电导率检测
outlet of still 炉出口〔管状炉的〕
outlet pipe 出口管；排出管；放出管；流出管
outlet pressure 出口压力
outlet seam 缝份
outlet side of pump 泵的排出端
outlet sleeve 加压套管；出口套管
outlet stool 出口支管
outlet temperature 出口温度
outlet transfer line 出口传送管线

outlet valve 排出阀
outlet velocity 出口速度；流出速度
outlet water 废水
outlier ①异常值*；离群值②分离物
outline ①大纲；概要②轮廓；略图
out-line analysis 线外分析；离线分析
outline-blanking die 外形冲切模
outline drawing 略图；草图；外形图；轮廓图
outline of process 生产过(流)程简图
outlying observation 反常(观测)值
out of date 过期的；过时的
out of door 室外的；露天的
out of gas 缺油；燃料用完
out of gear 切断的；不工作的；脱离开的
out of mesh 切断的；脱离开的
out of operation(=out of work) 不能工作的；工作有妨碍的；失效的
out-of-order 无次序(的)；出故障(的)；不正常(的)；混乱(的)
out-of phase ①异相②失相；失真；不同相③不在位相上的
out-of-phase component 异相分量
out-of-phase modulus 损耗模量；异相模量
out-of-pile inventory (反应)堆外(燃料)投入量；堆外库存周转量
out-of-pile test loop 堆外试验回路
out-of-plane bending 面外偏移
out-of-plane bending vibration 面外弯曲振动；面外形变振动
out-of-plane deformation vibration 面外形变振动；面外弯曲振动
out of plane vibration (分子)面外振动
out of register 不对齐；不配对
out of repair 失修的；处于不正常状态的
out of roundness 不圆
out-of-run 失效
out-of-sequence operation 违反操作程序
out of service 失效
out of shape 失去正常形状的
out of size 非正常大小的
out of use 不能使用的
out of work(=out of operation) 不能工作的；工作有妨碍的；失效的
outperformed standard 超(过)标(准)；高于标准(的)
output ①出口②产量③输出④排(出)量
output amplifier 输出放大器
output attenuation 输出衰减
output beam 输出束

output efficiency 输出效率
output hum level 输出交流声电平
output network 输出网络
output of column 塔的生产能力；塔产物物料平衡
output resistor 输出阻抗器
out-shell electron 外层电子
outshot ①废品②等外品(指原料)
outside boundary limit 界区外的
outside calipers 外卡钳
outside circulation 外环流
outside diameter 外直径
outside durability 户(室)外耐久性
outside finish 外面粉饰；外面修饰
outside gradient 外梯度
outside house paint 外用建筑漆
outside indicator (液)外指示剂
outside-in filter 外缘-中心型过滤器
outside stator cup 外定子圆筒；外定子套筒
outside storage tank 外部储料槽
outside stripping section 外部汽提段
outsize 非标准尺寸制品
outsize tire 大型轮胎
outsole 脚；基底
outspent 废的；用过了的
outsqueezing 压出；榨出
out stroke 排气冲程
out-to-out 总尺寸；总长度；总宽度
outward flange 向外凸缘
outward rectified current 外向整流电流
outward unit normal vector 外向单位法向向量
ouvarovite 钙铬榴石
oval 卵圆形
ovalbumin 卵清蛋白
ovalene 卵苯；卵烯
oval flowmeter 椭圆流量计
ovalicin 卵假散囊菌素
ovality 椭圆率
ovallized column 椭圆形柱
oven ①炉②烘箱
oven ageing 炉内老化；热致老化；热力老化
oven aging test 炉内老化试验；热力老化试验
oven-air flow rate 烘箱空气流量(流速)
oven atmosphere 炉气
oven bottom 炉底；炉床
oven bottom bread 底火面包
oven cavity atmosphere 烘炉内气氛(体)；炉膛气氛
oven chamber 炉室
oven coke 炉焦(炭)

oven cool-down circuit 加热炉冷却电路
oven cure 烘炉硫化
oven drawing (恒温)加热箱拉伸
oven-dried 烘干
oven-dried wood 绝干材；全干材
oven dry 烘干；烤干
oven drying 烘(炉式)干(燥)
oven-dry weight 炉式干燥重量；烘干重量
oven flue 炉灶烟道
oven gas 焦(炭)炉气
oven hearth 炉底；炉床
oven heating indicator circuit 炉加热指示器电路
oven heat stability (聚氯乙烯)烘箱法热稳定度
oven inner liner 加热炉内衬
oven make up air 空气补充室
oven method 烘箱干燥法〔测定水分〕
oven port 炉门；窑洞
oven process 炉法
ovenstone 耐火石
oven test 耐热试验
over/under tolerance 正负公差；容许公差
overabundance of lines 谱线过剩；谱线过多
over-acceleration 促进剂过量
over-acetylation 过度乙酰化(作用)
overaging 过(度)老化
overall 总(括)的
overall absorption coefficient 总吸收系数
overall absorptivity 总吸收系数
overall accuracy 总准确度
overall attenuation level 总衰减电平
overall chain length 总链长
overall coefficient 总系数
overall coefficient of heat transfer 总传热系数
overall coefficient of heat transmission 总传热系数
overall coefficient of oxygen transfer 氧传递总系数
overall combined feed ratio 总的综合进料比〔循环原料与新原料量的比例〕
overall composition 总成分
overall crystallinity 总结晶度
overall dimension 总尺寸；外形尺寸
overall dissociation constant 总离解常数
overall efficiency 总效率；总有效利用系数
overall efficiency of atomization 原子化总效率
overall film porosity 漆膜总孔隙率
overall formation constant 总形成常数
overall heat balance 总热平衡
overall material balance 总物料平衡
overall noise level 总噪声电平

overall performance rating　综合性能等级
overall polymerization speed　总聚合速度
overall porosity　总孔隙率
overall properties　全部性能；综合性能
overall rate　总率
over-all rate equation of chemical reaction　总反应速度方程(式)
overall reaction　总反应
overall reduction　总还原作用
overall resolution efficiency　总分离效能指标；总分离效率；总分辨效率
overall size　总尺寸
overall stability constant　总稳定常数
overall thermoelastic modulus　总热弹模量；总热弹性模量
overall uncertainty　总不确定度；扩展不确定度
overall volume shrinkage　总体积收缩量
overall wear characteristic　总磨损特性
overall yield　总收率
over-and-under controller　自动控制器
overbake　烘烤过度；焙(灼)烧过度；烧损
overbake yellowing　过度烘烤泛黄(现象)〔漆病〕
overbaking　烘烤过度
overbased　高碱性的
overbased lube oil　高碱性润滑油
overbate　过度软化
over-beating　打浆过度
overbleaching　过度漂白
overbrushing　①多刷；多涂〔漆膜增厚〕②刷漆过多；涂漆过量
over-burden　①(放射性)过量积存②覆盖层
overburden stripping　覆盖层剥离
overburning　烧损；烧毁
overburnt　过烧的
overcarbonated juice　过饱充汁
overcarbonation　充碳酸气过饱和
over center type coiler　大圈条
overcharge　充电过度
overchlorinated　过氯化的
over-churning　(碱纤维素的)过酯化；过黄原酸化
over-cleaning　浸洗过度；酸洗过度
overcoat　罩面层
overcoatability　面漆配套性；再涂性
overcoating　①外敷层；保护涂层②外套料；大衣料
over-compression　预缩过度
over condensed resin　过度缩聚树脂
overcook　蒸煮过度
overcooking　(油)热炼过度

overcool　过(度)冷(却)
overcooled　过冷却的
over-cooled engine　过分冷却的发动机
overcooling　过冷
overcount　计数过度
overcracking　过度裂化
over cure　过硫*；过固化
overcuring　①过固化②过熟化③硫化过度
overcurrent　过载电流
over damping　过(度)阻尼
overdesign of cracking unit　有余量的裂化装置设计
overdimensioned　超尺寸的
overdosage　过量；剂量过度
over-drastic　过度的；过于激烈
overdried　过干的
overdrive　超速行驶；加速移动
over-driven　上动的
over-driven buhrstone mill　上动细砂质石磨
over-driven centrifuge　上悬式离心机
over-driving　超传动；超负荷
over dry　过(分)干(燥)的
overdrying　过干
over-dry weight　绝对干重；烘干至不变重量
over dyeing　套染
overetched　过蚀刻的；蚀刻过度的
over-exposure　过曝射；超照射
overfall　①溢流；外溢；溢出②溢出口③回浆
overfeeding　过量进料
overfill　过量填注
overfire　过度燃烧；过热
overfiring　烧损；烧毁
overflocculation　①过度絮凝②结大块
overflow　①溢流②溢出③溢水路
overflow alarm　溢流警报
overflow annular lip　溢流环状嘴
overflow cavity　溢(出)式模巢
overflow chamber　溢流室
overflow cock　溢流栓
overflow dam　(蒸馏塔塔盘的)溢流堰
overflow gallery pipe　溢流收集管
overflow gate　溢流门
overflow indicator　溢流指示器
overflowing　溢流
overflow line　溢流线
overflow liquor　溢流(鞣)液
overflow mill　溢流磨
overflow mould　溢流模；挤压模
overflow nipples　(蒸馏塔塔盘的)溢流短管

overflow pipe(=overflow tube)　溢流管
overflow plate　溢流板
overflow port　溢流口
overflow reactor　溢流反应器
overflow tank　溢流罐
overflow tap　溢流栓
overflow trap　溢流阱
overflow trough　溢流(斜)槽
overflow tube(=overflow pipe)　溢流管
overflow (type) mold　溢(出)式压模
overflow valve　溢流阀
overflow vat　溢流鞣池
overflow vessel　溢流容器
overflow viscometer　溢流黏度计
overflow weir　溢流堰
overfoaming　溢泡
overfueling of engine　发动机燃料供应过量
over gassed spun silk yarn　焦黄丝
over-gassed yarn　焦纱
overglaze　①釉面的②面釉
overglaze color　釉面颜料
overglaze decoration　釉面(彩)
overglaze pigment　釉上颜料；面釉颜料
overgraining　过度粒化；过度析皂
over grind　过度粉碎
overground　过度粉碎的；过细的
overgrowth　①附晶生长②生长过度③蔓延
overhang　总浮空
overhanging applicator blade　涂布器外伸刀片
overhanging edge　①外伸的刀刃②松边
overhang roll　悬辊〔设〕
overhauling　大修；拆修；卸修；翻修
overhaul instruction　大修指导
overhaul life　大修周期
overhaul stand　大修台
Overhauser effect　欧沃豪斯效应；核极化效应
overhead　①塔顶馏出物②高出地面的；架设的；架空的
overhead bright stock　馏出的光亮油〔塔顶馏出的高质量油〕
overhead caulking　顶部边盖
overhead condenser　塔顶(产物)冷凝器
overhead convection type of pipe still　向上对流式管子炉
overhead convection type of reactor　向上对流式反应器
overhead cost　间接费用；总开支
overhead cylinder stock　气缸油头馏分
overhead distillate　头馏分
overhead drum　塔顶(冷凝)鼓
overhead expenses　杂费
overhead fraction　塔顶馏分
overhead line　架空电缆；高架线
overhead mixture　塔顶混合物
overhead monorail conveyor　架空单轨传送带
overhead naphtha(=overhead-taken naphtha)　塔顶汽油馏分
overhead oil　塔顶馏出油
overhead pipe　架空管道
overhead position welding　仰焊
overhead product　塔顶馏出物
overhead stream　顶流；顶部物流；塔顶物流
overhead-taken naphtha(=overhead naphtha)　塔顶汽油馏分
overhead tank　压力槽；压力罐
overhead travelling crane　桥式起重机
overhead valve cylinder　架设开关筒
overhead vapor condenser　塔顶馏出蒸气冷凝器
overhead welding　仰焊
overheat　过热
overheated　过热的
overheated spot　过热点
overheated zone　过热区
overheater　过热器
overheating　过热
overheat pan　过热锅
overheat prevention　防止过热
overhung-type centrifugal compressor　外悬式离心压缩机
overhydration　水合过度；水中毒
overhydrocracking　深度加氢裂化
over-ignited　过度灼烧的；热解温度过高的
overindation　唧送
overlacquer　罩光喷漆
overladen　超载；过负荷
overlaid plywood　贴面胶合板
overlap　①重叠②焊瘤〔焊接〕
overlap coefficient　重叠系数
overlap integral　重叠积分
overlap joint　重叠接合
overlap of elution curve　洗脱曲线重叠
overlap of peaks　峰重叠
overlapping blade　交叉的叶片
overlapping folded chain model　重叠状折叠链模型
overlapping peaks　重叠峰
overlapping resolution mapping(ORM)　重叠分辨图形法
overlapping run　搭头焊道
overlapping wave　重叠波
overlap population　重叠分布
overlay coating　贴面层；表面镀层(涂层)

overlay mat 覆盖毡片
overlay painting interval 涂装间隔〔二道涂漆之间的间隔时间〕
overlay painting test 重涂试验
overlay welding 堆焊
overlength separator 超长(切片)分离器；分选机
overlimed 加灰过量的
overload 超负荷；过载
overload tank 过载槽
overload valve 超载阀；过载活门
overload wear 过载磨损
overlubricate 过量润滑
over-lubrication 润滑过量
over-mastication ①素炼过度②过度研磨
over mature fibre 过成熟纤维
overmilling 过度捏和
overmixing 过度混合
overmo(u)lding 重叠注塑
over oiling 用油过度；加油过度
over oxidation 过度氧化
overpacking ①过充模〔注塑〕②超灌料〔使泡沫塑料增大密度〕
overpainting ①涂面漆；涂罩光漆②涂末道漆
over-pass 溢流挡板
over-pigmentation ①颜料过量②颜料分过多
overplumped 过肥的〔革〕
overplumping 过度膨胀
over point 初馏点
overpotential 超电势；过电位
overpotential resistance 过电压电阻
overpotential term 超电位项
overpowering odor 强烈气味
overpressure 过压；超压
over pressured liquid chromatography 过压液相色谱法
over pressured thin layer chromatography 过压薄层色谱法
overpressure valve 过压阀
overpressurized TLC 过压薄层色谱法
overprime 燃料过量注入
overprint coating ①罩光涂料②复色叠印涂料〔在印好的版面上再印另一种颜色和谐的涂膜〕
overprint finish 罩印清漆
overprinting 罩印；套印
overprint varnish 罩印清漆；复印清漆
overproof 超标准的
overproof spirit 过浓的酒；超标准酒
overreaction 过度反应；反作用过强
overreduction 过度还原

overrefining 过度精制
overrich mixture 过富混合物
overrun ①溢流②膨胀量
overrunning 溢流的
oversalting 过(度)盐析
oversampling 过速采样
oversaturated 过饱和的
over-saturation 过饱和
over-shooting 过头射击；照射过头
overshoot line (extra line) 过击线；外来线
over-shot buffer 绒面革磨里机
overshot tank filling 自顶部注入油罐内
oversize 筛上料
oversize tyre 过大轮胎
oversoaking ①过度浸湿；浸水(渗)过度②(裱糊时)刷胶浆时间过长纸被泡烂；浸泡过度
oversource heater (电热)逸热辐射加热器
overspeed test 超速试验
overspill 溢出物
overspray ①喷溅性②过(度)喷(涂)
overspray collection hopper 过剩粉末收集箱；过喷粉末收集器
oversprayed material 过喷的材料〔粉末或油漆〕
over-spray loss 过喷涂的损失
overspray sludge 过喷的废漆〔残漆〕
overstability 过稳定性
overstable 超稳定；很稳定
overstep of end point 超过终点
overstock 过度储备
overstrained 应变过度的
overstretching 过拉伸
over stripe bleeding 条格状渗色
overstuffed 多油的；涂油过多的
oversulfonate 过磺化
oversulfuring of gasoline 汽油加硫过量〔在亚铅酸钠净化法中〕
oversupply 过度供应
overtanned 过鞣的；鞣过了度的
overtemperature warning 超温警报
over-the-counter 非处方药物
over-the-top vapour 自塔顶排出蒸汽
overtime 加班时间
overtitration 滴过了头
overtone 浓色；色调过深；泛频峰；泛韵〔指多香气〕
overtone band 倍频谱带；泛频谱带
overtone frequency 泛频；倍频
over/under tolerance 正负公差；容许公差
over-up dipping process 叠层浸渍法

overventilation 换气过度
overvoltage(=supertension) 超(额)电压；过电压
overvoltage electrode 超(电)压电极
overvulcanization 过度硫化；硫化过度
overweight coating 涂漆过量；涂层过厚
ovo 蛋；卵
Ovoco classifier 奥俄克分矿运送器
ovoflavin 核黄素；维生素 B_2
ovostab 苯甲酸雌二醇酯
ovothricin 卵丝菌素
Owen process 桉油浮选法
owner indicator 自显指示剂
own viscosity 自身黏度
oxa- 噁；氧杂
7-oxabicyclo[2.2.1] heptane 7-氧杂二环[2.2.1]庚烷
oxacid(=oxy-acid) 含氧酸
oxacillin 噁西林；5-甲基-3-苯基-4-异噁唑青霉素；青霉素 p-12
oxacyclobutane(=oxetane) 氧杂环丁烷
oxacyclopropane 环氧乙烷；氧杂环丙烷
oxadiazine 噁二嗪；氧二氮(杂)苯；氧二氮苴
oxadiazole 噁二唑；氧二氮茂
oxadiazolinethione 噁二唑啉硫酮
1,2,4-oxadiazolone-5 1,2,4-噁二唑啉-5-酮
11-oxahexadecanolide 11-氧杂十六内酯
oxalacetate 草乙酸盐(酯) $ROCOCOCH_2COOR'$
oxalacetic acid 草乙酸；丁酮二酸 $HOOCCOCH_2COOH$
oxalacetic carboxylase 草乙酸羧酶
oxalacetic ester 草乙酸酯〔有时专指：草乙酸乙酯〕 $ROCOCOCH_2COOR'$
oxalactone 氧杂内酯
oxalaldehyde 乙二醛；草醛 CHOCHO
oxalamide 草酰胺；乙二酰胺 $NH_2COCONH_2$
oxalanilide N,N'-二苯基乙二酰胺
oxalate ①草酸②草酸盐(酯或根)
oxalate coating 草酸盐(处理)被膜
oxalate method 草酸盐法
oxalate turbidity 草酸盐混浊(度)
oxaldehyde(=glyoxal) 乙二醛；草醛 $(CHO)_2$
oxaldiureide dioxime 草酰二脲二肟
$[H_2NCONHC(=NOH)]_2$
oxalene 草烯 —N=C—C=N—
oxalethyline 联甲胺羰 $C_2H_5NHCOCONHC_2H_5$
oxalic acid 草酸；乙二酸 HO_2CCO_2H
oxalic acid diethyl ester 草酸二乙酯
oxalic acid dinitrile 草腈；乙二腈

oxalic acid series （乙)二酸系
oxalic aldehyde 乙二醛；草醛 $(CHO)_2$
oxalic amide 草酰胺
oxalic dialdehyde 乙二醛
oxalic dianilide N,N'-二苯基乙二酰胺
oxalic monoamide 草单酰胺
oxalic monoanilide 草单苯胺
oxalic monoureide 草脲酸
oxalimide 草酰亚胺
oxaliplatin 奥沙利铂〔药〕
oxalium 草酸氢钾
oxalmethylin N,N-二甲基乙二酰胺
oxalmethyline 联甲胺羰 $CH_3NHCOCONHCH_3$
oxalo- 草〔俗〕；乙二酸一酰基
oxaloacetamide 草酰乙酰胺
oxaloacetate decarboxylase 草酰乙酸脱羧酶
oxaloacetic acid 草酰乙酸；丁酮二酸
oxaloacetic acid-enolphosphate 磷酸烯醇草酰乙酸；磷酸烯醇丁酮酸
oxalodiacetic acid 草二乙酸；草酰 3,4-己二酮二酸
$COOHCH_2COCOCH_2COOH$
oxalopropionamide 草酰丙酰胺
oxalosuccinic acid 草酰琥珀酸
oxaluramide 草尿酰胺；氨基草酰脲 $C_3H_5O_3N_3$
oxaluria 草酸尿
oxaluric acid 草尿酸；脲基乙酮酸
$NH_2CONHCOCO_2H$
oxaluric amide 草尿酰胺；脲基乙酮酸胺 $C_3H_5O_3N_3$
oxalyl 草酰；乙二酰 —COCO—
oxalyl chloride 草酰氯；乙二酰氯 ClCOCOCl
oxalysine 溶菌素
oxam 一氧化一氰 CNO
oxamate ①草氨酸盐②草氨酸酯
oxamethane 草氨酸乙酯 $C_2O_2NH_2(C_2H_5O)$
oxamic acid(=oxaminic acid) 草氨酸；氨羰基甲酸 NH_2COCO_2H
oxamicetin 氧反菌素
oxamic hydrazide(=amino-oxamide) 草氨酰肼；氨基草酰胺；氨羰基甲酰肼 $NH_2COCONHNH_2$
oxamide 草酰胺；乙二酰二胺 $(CONH_2)_2$
oxamidine 氨肟 $RC(=NOH)NH_2$
oxamido- 草酰氨基；乙二酰氨基 $H_2NCOCONH—$
oxaminic acid(=oxamic acid) 草氨酸；氨羰基甲酸 $H_2NCOCOOH$
oxammonium hydrochloride 盐酸羟胺 $NH_2OH·HCl$
oxammonium sulfate 硫酸羟氨 $(NH_2OH)_2·H_2SO_4$
oxamoyl(=oxamyl) 草氨酰；氨基乙二酰 $H_2NCOCO—$
oxamycin 草霉素；环丝氨酸；氧霉素

oxamyl(=oxamoyl) 草氨酰；氨基乙二酰 H₂NCOCO—
oxane 噁烷〔俗〕；氧丙环；环氧乙烷 CH₂CH₂O
oxanilic acid(=phenyloxamic acid) 苯胺羰酸 C₆H₅NHCOCOOH
oxanilide N,N′-草酰二苯胺 (CONHC₆H₅)₂
oxanthranol 蒽二酚 C₁₄H₁₀O₂
oxanthrol(=oxanthrone) 蒽酚酮 C₆H₄(COCHOH)C₆H₄
oxanthrone(=oxanthrol) 蒽酚酮
oxapicene 䓛并吡喃
oxathiane 氧硫杂环己烷
oxathiene 氧硫杂环己烯
oxathietane 噁噻烷；邻氧硫杂环丁烷
oxathiin 氧硫杂环己二烯
oxatollic acid 草陶酸〔俗〕；二苄基乙醇酸 HOC(CH₂Ph)₂COOH
1-oxa-4,7,10-triazacyclododecane 1-氧杂-4,7,10-三氮杂环十二烷
oxatriazole 噁三唑；氧三氮(杂)戊
oxatyl(=carboxyl) 羧基
1,2-oxazetidine 1,2-氧氮杂环丁烷
oxazine 噁嗪*
oxazine colouring matters 噁嗪染料
oxazine dye 噁嗪染料
oxazinomycin 噁嗪霉素
oxazinone 噁嗪酮
oxazinyl 噁嗪基 ONC₄H₄—
oxazole 噁唑*
oxazolidine 噁唑烷；1,3-氧氮杂环戊烷 OCH₂NHCH₂CH₂
oxazolidinedione 噁唑烷二酮
oxazolidinone 噁唑烷酮
oxazolidinyl 噁唑烷基 ONC₃H₆—
oxazolidinylethyl methacrylate 甲基丙烯酸噁唑烷基乙酯；甲基丙烯酸氧氮戊环基乙酯
oxazolidone 噁唑烷酮
oxazoline 噁唑啉 OCH=NCH₂CH₂
oxazoline wax 噁唑啉蜡（其结构类似唑啉）
oxazolinyl 噁唑啉基 C₃H₄NO—
oxazolone 噁唑酮
oxazolyl 噁唑基 C₃H₂NO—
oxazones 噁嗪酮；羟噁嗪
ox bile extract 牛胆萃
ox blood 牛血；霁红
oxdiazine 噁二嗪；一氧二氮杂苯 C₃H₄ON₂
oxdiazole 噁二唑 C₂H₂ON₂

oxene(=oxirene) 环氧乙烯
oxetane(=oxacyclobutane) 氧杂环丁烷
oxetane polymer 氧杂环丁烷聚合物
oxetane resin(=chlorinated polyether resin) 氧杂环丁烷树脂；氯化聚醚树脂
oxethyl〔词头〕 乙氧基 C₂H₅O—
oxethyl toluene sulfonamide 乙氧基甲苯磺酰胺
oxetone 螺[4.4]二氧己烷
Oxford process 用硫化钠从铜中分离镍法
Oxford unit 青霉素强度单位；牛津单位
oxgall 牛胆汁
ox-hide collagen 牛皮胶元
oxhydrile ion 羟氧离子 HOO—
oxiacalcite 草酸钙石
oxicracking 氧化裂解
oxidability 可氧化(性)
oxidable 可氧化的
oxidant 氧化剂
oxidant gas 助燃气
oxidant recorder 氧化剂记录器
oxidant smog 氧化剂烟雾
oxidapatite 氧磷灰石
oxidase(=oxydase) 氧化酶
oxidate ①氧化物②氧化
oxidated 氧化了的
oxidated asphalt 氧化沥青
oxidating 氧化
oxidation 氧化*
oxidation accelerator 氧化促进剂
oxidation air 氧化空气
oxidation base 氧化显色碱〔染〕
oxidation bleaching 氧化漂白
oxidation catalysis 氧化催化
oxidation catalyst 氧化催化剂
oxidation channels treatment 氧化沟法处理
oxidation characteristic of lubricating oil 润滑油的氧化性能
oxidation cleavage 氧化分裂；氧化裂解
oxidation column 氧化塔
oxidation current 氧化电流
oxidation discharge 氧化拔染
oxidation ditch 氧化沟
oxidation drying 氧化干燥
oxidation dye 氧化染料
oxidation film 氧化膜
oxidation flame 氧化焰
oxidation gaseous effluent 氧化废气〔氧化时气态流出物〕

oxidation-inhibited grease   氧化抑制润滑脂
oxidation-inhibited oil   氧化抑制油
oxidation inhibitor(=oxidation retarder)   氧化抑制剂
oxidation number   氧化值
oxidation of lubricating oil   润滑油的氧化〔由于氧化而在油内生成沉淀〕
oxidation of odor by ozone   臭气臭氧氧化法
oxidation of polymer   聚合物的氧化反应
oxidation oven   氧化炉
oxidation panel   氧化段；氧化组
oxidation polymerization   氧化聚合
oxidation pond   氧化池
oxidation potential   氧化电位*
oxidation preventive   抗氧化剂
oxidation processes of ammonia   氨氧化法
oxidation process   氧化法
oxidation products   氧化产物
oxidation promotor   氧化促进剂
oxidation rate   氧化速度
oxidation reaction   氧化反应
oxidation (reclaiming) process   氧化(再生)法
oxidation-reduction   氧化还原(作用)*
oxidation-reduction catalyst   氧化还原催化剂
oxidation-reduction cell   氧化还原电池
oxidation-reduction electrode   氧化还原电极
oxidation-reduction indicator(=redox indicator)   氧化还原指示剂
oxidation-reduction initiator   氧化还原引发剂
oxidation-reduction method   氧化还原(滴定)法
oxidation-reduction polymerization   氧化还原聚合
oxidation-reduction potential   氧化还原电位*
oxidation-reduction potential determination   氧化还原电位测定
oxidation-reduction reaction   氧化还原反应
oxidation-reduction system   氧化还原体系
oxidation-reduction titration(=redox titration)   氧化还原滴定
oxidation-reduction wave   氧化还原波
oxidation resistance   抗氧化性
oxidation-resistant oil   氧化保持润滑油
oxidation retarder(=oxidation inhibitor)   氧化抑制剂
oxidation rinsing   氧化冲洗
oxidation rule   氧化法则
oxidation scission   氧化断裂
oxidation sludge   氧化渣
oxidation stability   氧化稳定性
oxidation state   氧化态；氧化级
oxidation step   氧化步骤

oxidation susceptibility   氧化性能
oxidation sweetening   氧化脱硫
oxidation tendency   氧化倾向
oxidation test   氧化试验〔石油产品的〕
oxidation test in bomb   氧弹氧化试验
oxidation test of lubricant oil   润滑油氧化试验
oxidation tower   氧化塔
oxidation transformation   氧化转变
oxidation treatment of waste water   废水氧化处理法
oxidation type alkyd resin   氧化(固化)型醇酸树脂
oxidation unit   氧化装置
oxidation value   氧化值
oxidation wave   氧化波
oxidation zone   氧化带
oxidative   氧化的
oxidative aging   氧(化性)老化
oxidative attack   氧化侵蚀
oxidative capacity   氧化能力
oxidative carbamylation   氧化氨甲酰化(作用)
oxidative catalyst   氧化催化剂
oxidative cationic polymerization   阳离子型氧化(催化)聚合
oxidative chlorination(=oxychlorination)   氧氯化作用
oxidative chlorophosphination   氧促氯磷化作用
oxidative coupling polymerization   氧化偶联聚合
oxidative crosslinking   氧化交联
oxidative cyclization   氧化环化(作用)
oxidative damage   氧化性损伤*
oxidative deaminization   氧化脱氨(作用)
oxidative decanning   氧化去壳
oxidative decarboxylation   氧化脱羧*
oxidative decladding   氧化去壳
oxidative decomposition   氧化分解作用
oxidative decomposition of oil   油的氧化分解作用
oxidative degradation   氧化降解(作用)
oxidative degradation of polymer   聚合物氧化降解反应
oxidative dehydrogenation   氧化脱氢
oxidative destruction   氧化分解
oxidative fluorination   氧化氟化作用
oxidative nitration   氧化硝化作用
oxidative nitrosation   氧化亚硝化(作用)
oxidative phosphorylation   氧化磷酸化(作用)
oxidative photodimerization   氧化光致二聚合作用
oxidative polymer   氧化性聚合物
oxidative polymerization   氧化聚合*
oxidative rancidity   氧化臭败
oxidative scission   氧化断裂
oxidative slagging   氧化造渣；氧化渣化

oxidative sparging 氧化性鼓泡
oxidative stability 氧化稳定度；氧化稳定性
oxidative stress-cracking 氧化应力龟裂
oxidative thermostability 热氧化稳定性
oxidative unhairing 氧化脱毛
oxide 氧化物
oxide catalyst 氧化物催化剂*
oxide coating 氧化物层
oxide compound 氧化物
oxide cracking catalyst 氧化裂化催化剂
oxide debris 氧化皮；氧化碎片
oxide-drossing furnace 氧化撇渣炉；氧化造渣炉
oxide film 氧化膜
oxide film on anode 阳极氧化膜
oxide gel 氧化物凝胶体
oxide of chromium 氧化铬
oxide of iron 氧化铁；铁丹
oxide pigment 氧化物颜料
oxide purification 氧化精制〔即氧化铁精制〕
oxide salt 氧化物盐类
oxide skin 氧化物层
oxide wax 氧化石蜡
oxidic 氧化的
oxidic film 氧化膜
oxidic potentiometric stripping analysis 氧化电势溶出法
oxidiferous 含氧化物的
oxidimetry 氧化(还原)测定(法)；氧化(还原)滴定(法)
oxidizability ①氧化性②氧化度
oxidizable 可氧化的
oxidizable hydrocarbons 可氧化的烃
oxidization 氧化作用
oxidized (被)氧化的
oxidized asphalt 氧化沥青
oxidized benzidine 氧化联苯胺
oxidized bitumen 氧化沥青；吹制沥青
oxidized black 氧化炭黑
oxidized carbon filament (预)氧化碳(纤维)长丝
oxidized metal 氧化金属
oxidized oil 氧化油
oxidized paraffin wax(=oxidized petroleum wax) 氧化石蜡
oxidized petrolatum 氧化凡士林
oxidized petroleum wax(=oxidized paraffin wax) 氧化石蜡
oxidized rubber 氧化橡胶
oxidized sludge process 氧化污泥法
oxidized still 氧化炉
oxidizer 氧化剂

oxidizer discharge 氧化剂输出(释放)
oxidizing 氧化的
oxidizing acid 氧化酸
oxidizing agent 氧化剂
oxidizing air-drying varnish 氧化气干型清漆
oxidizing atmosphere 氧化氛
oxidizing baking varnish 氧化烘烤型清漆
oxidizing chamber 氧化室
oxidizing coating 氧化(固化)型涂料
oxidizing column 氧化塔
oxidizing dodecyl dimethyl amine 十二烷基二甲胺氧化物
oxidizing electrode 氧化电极
oxidizing enzyme 氧化酶
oxidizing flame 氧化焰
oxidizing flux 氧化(助)熔剂
oxidizing fusion 氧化熔融
oxidizing phthalic alkyd resin 氧化(固化)邻苯二甲酸醇酸树脂
oxidizing power 氧化本领；氧化力
oxidizing property 氧化性(质)
oxidizing reaction 氧化反应
oxidizing roasting 氧化煅烧
oxidizing solution 氧化性溶液
oxidizing tower 氧化塔
oxidizing type alkyd resin 氧化型醇酸树脂
oxidizing zone 氧化层；氧化带
oxido- ①(环)氧；氧桥②氧化(还原)
oxido-indicator 氧化还原指示剂
oxido-reductase 氧化还原酶
oxido-reduction 氧化还原(作用)
oxidosome 氧化体
oxidotransformation 氧化转变
oximase 肟酶
oximate 肟盐
oximation 肟化(作用)
oximation reaction 肟化反应
oxime 肟*
oxime anhydride(=lactazone) 内肟；丁肟酸酐；酯肟
oxime-blocked 2-methylphenyl isocyanates 肟封闭的2-甲(基)苯(单)异氰酸酯
oxime rearrangement 肟重排
oximide 草酰亚胺 COCONH
oximido(=hydroxyimino) 肟基 HON=
oximidobenzotetronic acid 肟基苯并特窗酸
oximido compound 羟亚氨基化合物
oximino(=oximido) 肟基；羟氨基 HON=

oximino acid 肟基酸
oximinoglutaric acid 肟基戊二酸
oximinoketone 肟(基)酮
oxinate 8-羟基喹啉盐
oxindigo 氧靛蓝
oxindole(=ketoindole) 羟吲哚
oxine(=8-hydroxyquinoline) 8-羟基喹啉
oxine chelate 8-羟基喹啉螯合物
oxine derivatives 8-羟基喹啉衍生物
oxine extraction 8-羟基喹啉萃取
oxiracetam 奥拉西坦〔药〕
oxirane 环氧乙烷 CH_2CH_2O
oxirane formation(=epoxidation) 环氧化作用
oxirane polymer 聚环氧乙烷；环氧乙烷聚合物
oxirane ring 环氧环
oxirane value 环氧值
oxirene 环氧乙烯 $CH=CHO$
oxitol(=phenyl cellosolve) 苯基溶纤剂
oxo ①氧代②〔在无机化合物中通常指〕氧合 O=
oxo acid 含氧酸
oxo alcohol 羰基合成醇
oxoallobetulin 氧代别桦木脑
oxo-bottom base grease 氧底基润滑脂
oxo bridge 氧桥
oxo-bridging 氧桥键合
2-oxo-1-buanol 2-氧代-1-丁醇
3-oxobutanal dimethyl acetal 3-氧代丁醛二甲缩醛
oxobutanol(=hydroxy butanone) 氧代丁醇；羟基丁酮
oxocarbonium ion 氧碳鎓离子
oxo catalyst 羰基合成催化剂
oxo chemicals 羰基类化学品
oxocol 氧化(法)合成醇类
oxo-compound 氧基化合物
2-oxo-4,5-diaminoparabanic acid 2-氧-4,5-二氨基环乙二酰脲
oxodioxalato titanate(Ⅳ) 氧(代)二草酸根合钛(Ⅳ)酸盐
oxoepistephamiersine 氧代表千金藤默星碱
9-oxoferruginol(=sugiol) 9-氧代铁锈酚
α-oxoglutarate α-氧代戊二酸盐(酯)
2-oxoglutarate amiontransferase 2-氧代戊二酸氨基转移酶
oxo group 桥氧基
oxoheptyl acetate 乙酸氧代庚酯；酮庚基乙酸酯
2-oxohexamethylen(e) imine(=caprolactam) 2-氧代六亚甲亚胺；己内酰胺
oxohexane(=hexanone) 氧代己烷；己酮

oxohexyl acetate 乙酸氧代己酯；酮己基乙酸酯
oxolation 氧连作用*；氧桥合(作用)
oxolation polymer 氧配位聚合物〔金属连接在氧基上的聚合物〕
oxolin 1,2,3,4-萘四酮
oxomalonic acid 二羟丙二酸
oxometallate 金属氧酸盐
oxometallic acid 金属氧酸
oxomonocyanogen 一氧化一氰 CNO
oxonation(=oxo reaction) 羰化反应
oxonic acid 1,4,5,6-四氢-4,6-二氧-1,3,5-三嗪-2-羧酸
oxonine 氧䓬；氧杂环壬四烯
oxonio 氧鎓(离子) H_3O^+
oxonite 苦味酸的硝酸溶液
oxonium 氧鎓；烊
oxonium compound 氧鎓化合物*
oxonium form 氧鎓型；烊型
oxonium ion 氧鎓离子*
oxonium salt extraction 烊盐萃取
oxonols 氧杂菁
3-oxononane 3-氧代壬烷
oxopalmitic acid 8-酮基十六烷酸
oxopentabromotungstate(V) 一氧五溴化钨(V)酸盐
2-oxopentanedioic acid(=α-ketoglutaric acid) 2-氧代戊二酸；α-酮基戊二酸
4-oxopentanoic acid(=levulinic acid) 4-氧代戊酸；γ-酮基戊酸；乙酰丙酸
oxophenarsine hydrochloride(=mapharsen) 盐酸氧苯胂；盐酸 3-氨基-4-羟苯基胂化氧；马法胂
oxophenic acid 邻苯二酚
oxophile ①亲氧的②极性的
oxo process 羰基合成*
oxoproline 羟脯氨酸
16-oxoprometaphanine 16-氧代原间千金藤碱
2-oxopropanal 2-氧代丙醛
3-(2-oxopropyl) coronaridine 3-(2-羰基-丙基)冠狗牙花碱
oxo-reaction(=oxonation) 羰基合成；含氧化合物合成〔含氧化合物合成反应〕
oxosilane(=siloxane) 硅氧烷
7-oxo-β-sitosterol 7-酮基-β-谷甾醇
oxo solvent 含氧溶剂
17-oxosparteine(=d-hydroxysparteine) d-羟基鹰爪豆碱
oxosuccinic acid 2-氧代丁二酸；丁酮二酸
oxo sulfate 羰基醇硫酸盐(酯)
oxo synthesis 羰基合成(法)
oxo synthesis gas 羰基合成气
6-oxotestosterone 6-氧睾(甾)酮

oxotomycin 氧玉米霉素
oxoxylin 木蝴蝶苦素
oxozone 四原子氧 O_4
oxozonide 氧臭氧化合物
ox-red indicator 氧化还原指示剂
oxy- 〔希腊字头〕 ①氧化②氧(代)③含氧的
oxyacanthine 尖刺碱
oxyacetaldehyde 含氧乙醛
oxyacetate 羟(基)乙酸;乙醇酸
oxyacetic acid(=hydroxyacetic acid) 羟基乙酸;乙醇酸
oxyacetone 羟丙酮
oxy-acetylene 氧-乙炔
oxy-acetylene blowpipe 氧-乙炔吹管
oxy-acetylene cutting 气割;氧-乙炔切割
oxy-acetylene flame 氧炔焰*;氧乙炔焰
oxy-acetylene torch 氧-乙炔炬
oxy-acetylene welding 气焊;氧-乙炔焊
oxy-acid(=oxacid;oxygen acid) 含氧酸
oxyalizarin 红紫素
oxyalkylatable 可烷氧化的
oxyalkylation 烷氧基化(作用)
oxyalkyl chain 烷氧基链
oxyalkylene 氧化烯
oxyalkylene group 氧化烯基
oxyalkylene unit(s) 氧化烯单位
oxyaluminum 羟基铝;氢氧化铝
oxyaluminum cation 氧化铝阳离子
oxyalyl urea 草酰脲 $C_3H_2N_2O_3$
oxyamide 〔类名〕 羟基酰胺 R(OH)CONH$_2$
oxyamination 羟氨基化*
oxyamine 〔类名〕 羟基胺
oxyammonia ①羟氨;氢氧化胺②胺 NH$_2$OH
oxyanthracene 蒽酚
oxyasiaticoside 氧积雪草苷
oxyaspergillic acid 氧曲霉酸
oxyaustenite 氧化奥氏体;氧化γ铁固熔体
oxyazide 羟叠氮化物
oxyazo 〔词头〕 羟偶氮
oxybenzalization 羟基缩苯醛化;羟基苯亚甲基化
oxybenzene 羟苯;苯酚
oxybenzoic 羟苯甲酸的
2-oxybenzothiazole 2-羟基苯并噻唑〔促进剂〕
oxybenzoxazole 羟苯并噁唑
oxybenzoyl homopolyester 氧基苯甲酰均聚酯
oxybicillin N,N'-二氧代乙二胺青霉素
oxybiotin 氧代生物素
p-oxybis-benzenesulfonyl hydrazide 对氧双苯磺酰肼〔发泡剂〕

oxybromide 溴氧化物
oxybutylene 氧化丁烯
oxybutynin 奥昔布宁〔药〕
oxycalorimeter (用)氧量热器
oxycarbide 碳氧化物
oxycationic 含氧阳离子的
oxycellulose 氧化纤维素
oxychalcogenide 氧硫族元素化物〔氧硫、氧硒、氧碲化物〕
oxychloride 氯氧化物
oxychloride cement 氯氧水泥
oxychloride lead-refining 氯氧化物炼铅法
oxychlorination(=oxidative chlorination) 氧氯化作用
oxychlororaphine 氧绿菌素
oxycholesterol 羟胆甾醇
oxycoal gas 氧煤气;灯用煤气;照明气
oxycodone hydrochloride 盐酸羟考酮;盐酸14-羟基二氢可待因酮
oxy-compound ①氧基化合物②羟基化合物
oxydase(=oxidase) 氧化酶
oxydation(=oxidation) 氧化(作用)
oxydative degradation 氧化降解(作用)
oxydehydrogenation 氧化脱氢
oxydic 氧化的
oxydic film 氧化膜
oxydiethanoic acid 氧联二乙酸 HOOCCH$_2$OCH$_2$COOH
oxydiethylene- 氧联二乙基 O(C$_2$H$_4$—)$_2$
N-oxydiethlene-2-benzothiazole-sulfenamide N-氧联二亚乙基-2-苯并噻唑次磺酰胺〔促进剂〕
N-oxydiethylenethiuram disulfide 二硫化 N-氧联二亚乙基秋兰姆〔促进剂〕
2,2'-oxydipropionitrile (ODPN) 2,2'-二氰基乙醚
oxydizing flame 氧化焰
oxydol 过氧化氢;双氧水
oxydone 氧化酮
oxydoreduction reaction 氧化还原反应
oxyethane 环氧乙烷
oxyethylated castor oil fatty amide 氧乙烯(化)蓖麻油脂肪酰胺
oxyethylated fatty amide 氧乙烯(化)脂肪酰胺
oxyethylation 乙氧基化(作用)
oxyethyl chain 乙氧基链
oxyethylene group 氧(化)乙烯基
oxyethyl group 乙氧基
oxyfluoride 氟氧化物
oxy-flux cutting 氧熔剂切割
oxyful 双氧水
oxygen ①氧②氧气

oxygen absorbent 吸氧剂
oxygen absorber 氧吸收剂
oxygen absorption 氧吸收
oxygen absorption rate 吸氧速度
oxygen absorption rate constant 氧吸收速率常数
oxygen absorption test 吸氧试验；(石油产品的)诱导期测定
oxygen acceptor 氧(接)受体
oxygen-acetylene flame 氧(气)-乙炔焰
oxygen acid(=oxy-acid) 含氧酸
oxygen ag(e)ing 氧致老化
oxygen analyzer 氧(气)分析计
oxygenase (加)氧酶
oxygenate analyzer 含氧化合物分析仪
oxygenated ①充了氧的②氧化了的③用氧饱和了的
oxygenated black 氧化炭黑
oxygenated oil 氧化油
oxygenated products 氧化产品；含氧产品
oxygenated rubber 氧化橡胶
oxygenated water 充氧水
oxygenation 氧合作用
oxygenator 充氧器
oxygen balance 氧平衡
oxygen bleaching 氧漂白
oxygen bomb 氧弹；氧气瓶
oxygen-bomb aging 氧弹老化
oxygen bomb aging test 氧弹老化试验
oxygen bomb calorimeter 氧弹量热器
oxygen bomb method 氧弹法
oxygen bomb test 氧弹老化试验；加压气热老化试验
oxygen bridge 氧桥
oxygen bubbles 氧气泡
oxygen capacity 氧容量
oxygen carrier 氧载体*
oxygen cell(=aeration cell) 充气电池；氧气电池
oxygen compound 氧化合物
oxygen consumption 氧消耗
oxygen consumption rate 耗氧速率
oxygen-containing gas 含氧气体
oxygen content 氧含量
oxygen-convertible material 氧转化型材料
oxygen convertible phthalic resin 氧转化型邻苯二甲酸树脂
oxygen cutting 氧化切割；气割
oxygen cycle 氧循环
oxygen cylinder 储氧钢筒；氧气瓶
oxygen debt 氧债
oxygen deficiency 缺氧；亏氧

oxygen deficit 氧亏
oxygen demand 需氧量
oxygen-denuded air 贫氧空气
oxygen depletion ①缺氧；亏氧②氧耗尽；氧用尽
oxygen diffusion type corrosion 氧扩散型腐蚀
oxygen dissociation curve 氧解离曲线
oxygen effect 氧效应
oxygen electrode 氧电极*
oxygen enhancement ratio(OER) 氧效应增强比
oxygen-enriched air 富氧空气
oxygen-enriched atmosphere 富氧气氛
oxygen-enriched environment 富氧环境
oxygen evolution 释氧(作用)；放出氧
oxygen excess 氧过剩
oxygen exchange reaction 氧交换反应
oxygen family element 氧族元素
oxygen flask combustion 氧瓶燃烧
oxygen flask method 氧瓶(燃烧)法
oxygen flux cutting 氧熔剂切割
oxygen-free 不含氧的；无氧的
oxygen-free atmosphere 无氧大气；无氧气氛
oxygen-free carrier gas 无氧载气
oxygen-functionality-grouping 氧官能(度)基(团)
oxygen gas 氧气
oxygen heater 氧气加热器；氧气预热炉
oxygen heterocycle 含氧杂环
oxygen heterocyclic ring 含氧杂环
oxygen-hydrogen flame 氧(气)-氢(气)焰
oxygen-hydrogen ratio 氧氢比(率)
oxygen index 需氧指数；氧指数
oxygen index value 氧指数值
oxygenizement ①氧化(作用)②充氧(作用)
oxygen lance cutting 氧矛切割
oxygen line 氧气供应管道
oxygen machining 氧气切割
oxygen mask 氧(气)面具
oxygen mass transfer coefficient 氧传质系数
oxygenmeter 氧气计；氧气表
oxygen-oil rocked fuel 氧-油火箭燃料
oxygenolysis 氧化分解(作用)
oxygenous 氧的；氧气的
oxygen overvoltage (阳极放)氧过电压
oxygen plasma 氧等离子体
oxygen polarography 氧极谱法
oxygen pressure ag(e)ing test 加压氧气热老化试验
oxygen protection apparatus 氧(气)防护器
oxygen protection room 氧气避毒室
oxygen quenching 氧抑制；氧猝熄；抑制氧化

oxygen release 释氧；氧释放
oxygen-releasing compound 释氧化合物
oxygen-rich flame 富氧焰
oxygen sag curve (水流)溶氧下垂线
oxygen-saturated 氧饱和的
oxygen saturation curve 氧饱和曲线
oxygen saturation value 氧饱和值
oxygen scavenger 氧清除剂；去氧剂
oxygen sensor 氧传感器；透气膜电极
oxygen-shielded flame 氧屏蔽火焰
oxygen specific response of the flame ionization detection system 氧特效响应火焰电离检测器
oxygen supply 供氧器
oxygen transfer 氧传递
oxygen-transfer agent 氧传送剂；氧转移剂
oxygen transfer coefficient 传氧系数
oxygen transfer rate 传氧速率
oxygen treatment 氧(气)(治)疗(法)
oxygen treatment apparatus 氧(气)(治)疗器
oxygen tree 氧树
oxygen turbo-compressor 氧气透平压缩机
oxygen uptake 摄氧
oxygen uptake rate 摄氧速率
oxygen uptake test 摄氧试验
oxygen vacancies 氧空穴
oxygen value 氧值〔汽油〕
oxygen vapor pressure thermometer 氧蒸气压温度计
oxygen wave 氧波
oxyhalide(=oxyhalogenide) 卤氧化物
oxyhalide ion 卤氧(根)离子
oxyhalogen 卤氧
oxyhalogen-acid 卤氧酸
oxyhalogenation 卤氧化(作用)
oxyhalogen compound 卤氧化合物
oxyhalogenide(=oxyhalide) 卤氧化物
oxyhalogen ion 卤氧(根)离子
oxyhydrase 解水酶
oxyhydrazides 羟(某)酰肼
oxyhydrogen 氢氧(气)；氢氧的
oxyhydrogen blowpipe 氢氧吹管
oxyhydrogen cell 氢氧电池
oxyhydrogen flame 氢氧焰
oxyhydrogen light 氢氧(碳)光
oxyhydrogen reaction 氢氧反应
oxyhydrogen voltameter ①氢氧电量计②气电量计
oxyhydrogen welding 氢氧焊
oxyhydroquinone 1,2,4-苯三酚
oxyhydroxide 羟基氧化物*

oxyindole(=hydroxy-indole) 羟(基)吲哚
oxyiodide 氧碘化物
oxyjavanicin 氧爪哇镰菌素；镰红菌素
oxyketone dye 醌型染料
-oxyl〔词尾〕 烃氧基〔如: methoxyl 甲氧基, phenoxyl 苯氧基〕
oxylactone 羟(某)内酯
oxylan 对苄(基)苯基甲酰胺
oxylepidine 联苯甲酰苯乙烯 $(C_6H_5CHCOC_6H_5)_2$
oxyleucotin 巴拉木素
oxyliquit 液氧炸药
oxylith 过氧化钠和漂白粉的混合物
oxyluciferin 氧化型荧光素
oxy-luminescence 氧化发光
oxyluminescence thermal analysis 氧化发光热分析
oxymalonic acid 羟基丙二酸
oxymatrine 氧化苦参碱
oxymel 药蜂蜜
oxymercuration 羟汞化*
oxymethoxyallylbenzene 丁子香酚
oxymethylconiferin 顶香花配基
oxymethylcresoltannin 羟甲基甲酚单宁
oxymethylene 甲醛
oxymethylene diphosphoric acid 氧亚甲焦磷酸
oxymorphine 脱氢吗啡
oxymuriate 氯酸盐
oxymuriatic acid 氯酸
oxyn 氧化干性油〔干性油的双键加上氧〕
oxynaphthoic acid α-萘酚酸
oxynaphthoxazime(=mukogen) 氯化二甲基苯基对铵-β-氧萘噁嗪
oxynarcotine 羟基那可汀
oxynervone 羟基神经苷酯；羟基烯脑苷脂
oxynervonic acid 羟基神经酸
oxyneurine 甜菜碱
oxynicotine 氧化烟碱
oxynicotinic acid 2-羟烟碱酸
oxynitrate 含氧硝酸盐
oxynitration 羟硝化作用〔同时加上羟基和硝基〕
oxynitride 氧氮化物
oxynitriles 羟某腈
oxynitroso 亚硝酸基
oxypathy 酸中毒
oxyphenarisine hydrochloride 奥盼胂盐酸盐〔药〕
oxyphenazone 羟苯腙
oxyphencyclimine hydrochloride 氧苯环亚胺盐酸盐
oxyphenoxazone 羟苯腙
oxyphenyl 羟苯基

oxyphilic element 亲氧元素
oxyphor 羟樟脑
oxyphosphorane ①邻苯二酚正膦酸三烷基酯②邻二酚正膦酸三烷基酯
oxyphthalic acid 羟苯二甲酸
oxyphyres〔复〕 淡色斑岩
oxypicolinic acid 4-羟吡啶-2-甲酸
oxypnictide 氧磷族元素化物〔氧磷、氧胂、氧锑、氧铋化物〕
oxypolymerization 氧化聚合
oxypressin 催产加压素
oxy-propane flame 氧-丙烷火焰
oxypropylation 丙氧基化作用
oxypropylene 氧化丙烯
oxypropylene group 氧(化)丙烯基
oxypropyl group 丙氧基
oxyproteinic acid 羟基蛋白酸
oxyprotosulfonic acid 氧蛋白磺酸
oxypyridne 羟吡啶
oxypyroracemic acid 羟基丙酮酸 $CH_2OHCOCOOH$
oxypyrrolnitrin 氧硝吡咯菌素
oxyquercetin 羟栎精
oxyquinaseptol 羟喹邻酚磺酸
oxyquinazoline 羟喹唑啉
8-oxyquinolin copper 8-羟基喹啉铜
oxyquinoline 羟喹啉
oxyquinolinic acid 2-羟吡啶-5,6-二羧酸
oxysalt 含氧盐
oxy sebacic acid 羟(基)皮脂酸
oxysorb 吸氧(剂)
oxysparteine 氧化鹰爪豆碱
oxysporin 尖孢菌素
oxystearic acid 羟基硬脂酸
oxystearin 氧化硬脂精
11-oxysteroid 11-氧甾类
oxysuccinic acid 羟丁二酸；苹果酸
oxysulfate 含氧硫酸盐
oxy-sulfonation 羟化磺化(作用)
oxytetracycline 土霉素；氧四环素
oxythiamine 羟基硫胺
oxythiazole 羟噻唑
oxytiglic acid 环氧惕各酸；α,β二甲基甘油内醚酸
oxy-Tobias acid 2-萘酚-1-磺酸
oxytocic ①催产的②催产药
oxytocic hormone 催产素
oxytocin 催产素
oxytol acetate 乙酸溶纤剂；乙二醇乙醚乙酸酯
oxytolerant 耐氧的

oxytoluol 甲酚
oxytoxin 氧化毒素(产物)
oxytrisulfotungstate 三硫氧钨酸盐
oxytropism 向氧性；趋氧性
oxyurushic acid 羟漆酸
oxyvanadium compound 钒氧化合物
oyamycin 大谷霉素
oyster shell whiting 贝壳粉
oyster white support 灰白色载体
oz 盎司；英两
ozamin(=benzopurpurin) 苯并红紫
oz ap 药量盎司
oz av 常衡盎司
oz fl 液体盎司
ozobenzene 臭氧苯
ozocerite(=ozokerite) 地蜡
ozogen 过氧化氢
ozokerite(=ozocerite) 地蜡
ozokerite yellow 地蜡黄
ozonation 臭氧化(作用)
ozonation treatment of waste water 废水臭氧化处理法
ozonator(=ozonizer) 臭氧化发生器
ozone 臭氧 O_3
ozone ageing 臭氧老化
ozone attack 臭氧侵袭
ozone box 臭氧老化试验箱
ozone chemiluminescence detector 臭氧化学发光检测器
ozone concentration 臭氧浓度
ozone-containing air 含臭氧的空气
ozone crack 臭氧裂解
ozone cracking 臭氧龟裂
ozone decomposition reaction 臭氧分解反应
ozone degradability 臭氧降解性
ozone degradation 臭氧降解(作用)
ozone fading 臭氧褪色
ozone generator 臭氧发生器
ozone hole 臭氧洞
ozone-induced chemiluminescence 臭氧诱导化学发光
ozone-induced cracking 臭氧龟裂
ozone layer 臭氧层
ozone layer depletion 臭氧层损耗
ozone monitor 臭氧监测仪
ozone monitor analysis method 臭氧监测分析方法
ozone paper 臭氧试纸
ozone proof 耐臭氧化；抗臭氧的
ozone puncture 臭氧蚀穿
ozone resistance 抗臭氧性
ozone resistants 抗臭氧剂

ozone scavenger 臭氧清除剂
ozone spectrophotometer 臭氧分光光度计
ozone value 臭氧值；臭氧价
ozone weather meter 臭氧老化试验机；臭氧天候老化试验机
ozonidate 臭氧剂
ozonidation 臭氧化(作用)
ozonide 臭氧化物
ozonization 臭氧化
ozonized air 臭氧化空气
ozonized ether 过氧化氢乙醚溶液
ozonizer(=ozonator) 臭氧化发生器
ozonizing tower 臭氧化塔
ozonolysed natural rubber 臭氧分解的天然胶
ozonolysis 臭氧解*；臭氧化分析法
ozonolytic aging 臭氧(分解)老化
ozonometer 臭氧计
ozonometry 臭氧定量法
ozonoscope 臭氧检验器
ozonosphere 臭氧层
ozonosphere damage monitoring 臭氧层破坏监测
ozotetrazole 连四嗪

P

Paal-Knorr synthesis(=Knorr synthesis)　帕耳-诺尔合成法
PAB-cellulose　对氨(基)苄基纤维素
pachimeter　测重机
pachnolite　霜晶石
pachometer　测厚计
pachyman　茯苓聚糖　$(C_6H_{10}O_5)_n$
pachymic acid　茯苓酸
pachyrhizid　豆薯苷
pachyrhizus　豆薯；凉薯；葛；地瓜
pachysamine　富贵草胺；花粉杨胺碱
pachysandienol　粉蕊黄杨二烯醇
pachysonol　富贵草酮醇；粉蕊黄杨酮醇
pachytene stage　粗线期〔指细胞减数分裂前期的第三期期间配对的染色体变粗并分裂成染色单体〕
pacite　毒砂
pack　装填；填充
package　①包装②小批件数③组合件④晶体管壳
package boiler　快装锅炉
package build　卷装成型
package collection device　组装式空气调节器
packaged adhesive　（小)包装的黏合剂
packaged plant(=packaged unit)　小型装置；可移动装置
packaged production　小批生产
packaged unit(=packaged plant)　小型装置；可移动装置
packaged vacuum unit　小型真空设备
package elevator　包装品升运机
package performance analyzer　筒子性能分析仪；卷装性能分析仪
packager　打包机；包装机
package stability　储存稳定性
packaging　包装
packaging container　包装容器
packaging film　包装薄膜
packaging machine　包装机
packaging machinery　包装机械
packaging material　包装材料
packaging operation　包装操作
packaging plant(=packaging unit)　包装工厂(设备)
packaging-shaping defect　筒子成形疵点
packaging twine　打包绳
packaging unit(=packaging plant)　包装工厂(设备)
pack carburizing　装箱渗碳；固体渗碳
pack chromizing　装箱渗铬；固体渗铬
packed　包装了的

packed array　合并数组
packed bed　填充床
packed bed filter　填充层过滤器；滤清器
packed capillary column　填充毛细管柱
packed column　填充柱*
packed column chromatography　填充柱色谱法
packed column reactor　填充柱反应器
packed density　填充密度
packed distillation column(=packed column)　填充塔；填充蒸馏塔
packed dryer　填充式干燥机
packed extraction column　填充(式)萃取塔
packed extractor　填充提取塔
packed factor　填充因素
packed height　填充高度
packed layer　填充层
packed micropore column　微径填充柱
packed pigment bed　堆积颜料层
packed purging column　填充吹洗塔
packed reaction column　填充反应柱
packed reaction tower　填充反应塔
packed space　填充容积
packed spray tower　填充淋塔
packed tower　①填充塔②填料塔
packed weight　填充量
packer　包装机；打包机
packer's oil　包装喷射油〔喷射水果或蛋类用的低黏度石油产品〕
pack hardening　装箱(渗碳)硬化
packing　①填充物；填料②填充；装(色谱)柱③垫革
packing arrangement　堆积排列
packing box　填料函；填料盒
packing case　填料函
packing chromatography　填充色谱(法)
packing coefficient　堆积系数
packing defect　堆砌缺陷；排列缺陷
packing density　填充密度；堆积密度
packing drier　填充物干燥器
packing dryer　填充物干燥器
packing effect　敛集效应〔物〕
packing factor　填充因子
packing fraction　紧束分数；敛集率〔物〕
packing gas chromatography　填充气相色谱(法)
packing gland　填料格兰；填料压盖

packing grease 密封润滑脂
packing height 填充高度
packing house by-product 肉类加工厂副产物
packing layer 填充层
packing leather 皮碗
packing lubrication 填充润滑
packing machine 包装机
packing material 填充材料；填充物
packing non-uniforming factor 填充物不均匀因数
packing operation 填充操作
packing paper 包装纸
packing pattern (颜料粒子的)堆积方式
packing plate 垫板
packing press ①填料压机②包装机
packing procedure 填充手续
packing ring 密封环；轴封环；垫圈
packing size 填充物尺寸
packing space 填充空间
packing specification 包装规格
packing spiral 螺旋形垫料
packing structure 填充物结构
packing support 填充物支架
packing tightening 填充物紧密化；紧密填充
packing timber 包装材
packing tower 填料塔
packing up 包装
packing value 堆积值
packing void fraction 填料(层)空隙度
packing volume 填充体积
packless ①未包装的②无填充的
packless seal 无包扎密封
packless valve 无填料阀
packsand 细(粒)砂岩
pack screen (活)钢丝滤板
paclitaxel 紫杉醇〔药〕
Pacol dehydrogenation process(=Pacol process) 帕科尔脱氢法〔直链烷烃脱氢制直链烯烃的方法；UOP 法〕
Pacol dehydrogenation unit 帕科尔脱氢装置〔UOP 脱氢装置〕
Pacol-Olex combination process 帕科尔-奥来克斯联合法〔直链烷烃脱氢-烷、烯分离生产直链单烯烃〕
Pacol process 帕科尔(脱氢)法〔直链烷烃催化脱氢；UOP 法〕
pactacin(=pactamycin) 密旋霉素
pactamycin(=pactacin) 密旋霉素
pad ①垫板②填塞
padcure 浸轧热处理；浸轧固化
padded cloth 填塞用布

padding ①填料；填塞物②棉絮③轧染
padding machine 压染机
padding mangle 初染机
padding table 拭浆台
paddle ①踏板②桨(叶)；搅棒；划槽
paddle agitator 桨式搅动器
paddle door 划槽门
paddle drum bleacher 桨鼓漂白机
paddle dyeing 划槽染色
paddle method of dyeing 搅槽染法
paddle mixer 桨式混合机
paddle mixer with intermeshing fingers 桨板混合机
paddle pump 桨式泵
paddle stirrer(=anchor stirrer) 桨式搅拌器
paddle-type Stomer viscometer 桨叶式斯托默黏度计
paddle vat 划槽
paddle viscometer 桨式黏度计
paddle wheel 桨轮
paddling 搅拌
paddling process 搅拌法
paddock stool 牛肝菌
pad-jig dyeing 浸轧卷染法
pad lubricator 垫状润滑器
pad sander 布团砂光(打磨)机
pad stain 布团搓涂用着色料；搓涂用着色剂
pad-steam dyeing 轧蒸染色
paecilomycerol 拟青霉菌素
p(a)eonidin 甲基花青素；芍药色素
p(a)eonin 甲基花青苷；芍药色素苷
paeonine 芍药根碱
p(a)eonol 芍药酮；2-羟基-4-甲氧基苯乙酮
paeonolide 鼠李芍药苷；芍药交酯
paeonoside 芍药糖苷；丹皮苷
pagodite 寿山石；宝塔石
pail ①提桶②桶〔容量〕
pain phosphorus 块状磷
paint 油漆*
paintability 可涂漆性；可涂饰性
paintable finish 油漆涂饰
paintable type of parting agent 油漆型脱模剂
paint additive 涂料添加剂
paint-and-batten method 涂油漆和钉板法〔防海生钻孔动物蛀蚀〕
paint and varnish 色漆和清漆
paint application 油漆施工(应用)
paint bake 油漆烘烤；油漆烘干
paint baking over 油漆烘烤过度；过度烘烤
paint base ①漆料；漆基②底漆③油漆打底

paint bath 漆槽
paint blower ①压缩空气喷涂机②喷漆枪；喷漆器
paint brush 漆刷；漆帚
paint-brush cleaner 漆刷清洗剂；漆刷洗涤剂
paint build-up 油漆积累；油漆堆积；油漆积厚
paint-burner 火焰除漆器
paint can 漆罐；漆桶
paintcast 铸型涂料
paint channel 油漆管道(路)
paint circulating pump 油漆循环泵
paint coat 涂层；漆膜
paint coat cushion 油漆缓冲层
paint-coated ①涂漆的②画(油)画的；彩绘的
paint container 油漆储罐
paint delivery 涂料供应；供漆量
paint deterioration 油漆变质
paint dipping 浸涂(漆)
paint-drainage holes 油漆排放口；排漆口〔指废漆排放〕
paint draw-down bar 油漆刮涂棒
paint dump tank 油漆接受罐；进漆罐
painted ①涂(了)漆的②染(了)色的
painted article 涂漆的工件(制品)
painted hides 漆皮〔革〕
painted interchangeability 涂装的可交替性
painted softwood finish 软(质)木壁板涂饰
painted surface 涂漆面
painter (油)漆工；漆画工
painter's caulk 油漆填缝料；漆工填缝剂
painter's naphtha 白节油〔溶剂,漆用石脑油〕
painter's apron 油漆工围裙
painter's colic 铅中毒性绞痛
paint filler 油漆填料
paint film 涂料薄膜
paint film tester 漆膜试验机
paint formulation 油漆配方
paint grade titanium dioxide 涂料级二氧化钛；钛白粉
paint grinder 油漆研磨机
paint grinder mill ①炼漆机②颜料磨
paint guide (喷枪)涂料导管
paint harling 岩石饰面法；花岗石粒(片)溅射饰面涂法
paint heater 油漆加热器
paint holding capacity (漆刷的)蘸漆量
paint hose 输漆软管
painting ①涂(油)漆②上色③涂抹
painting defect 涂装缺陷
painting for depilation 涂灰浆脱毛
paint(ing) gun 喷漆枪；喷涂枪
painting line 标色水线〔轻载和重载水线之间部分〕

painting line (boot topping) 标色水线〔轻载和重载水线之间部分〕
painting machine 上色机
painting method 涂抹法
painting nozzle 涂料喷头
painting out 用漆涂盖
painting period (重新)涂装周期
painting plaster 灰泥(墙面)涂漆；灰泥涂饰
painting porous coating 涂(渗透性)多孔涂料
paint ingredient 油漆成分
painting technique 涂装技术
paint latex 漆用乳胶
paint-like consistency 油漆状稠度
paint media 漆料
paint mill 涂料磨机
paint mixer 涂料混合器
paint naphtha 调漆油；油漆溶剂油
paint oil 调漆油
paint paste ①漆浆②厚漆
paint patching (油漆)补修；补漆
paint patch panel 涂料比较试验板
paint pigment 漆用颜料
paint primer 底漆
paint remover 脱漆剂
paint rock 铁赭土
paint roller mill 涂料磨机
paint scrubber 洗漆刷
paint-skin 漆皮〔罐内油漆表面结皮〕
paint sprayer 油漆喷枪；喷漆器
paint spray nozzle 油漆喷嘴
paint strainer 滤漆筛
paint stripping 脱漆；漆膜的剥离
paint thinner 涂料稀释剂
paint tumbler 滚筒式浸涂机
paint vehicle 漆料
paipunine 百部碱
pair ①对；偶②一副；一套③配对
paired 配对的；成对的
paired-bond dissociation 双键离解(作用)
paired burette 配对滴定管
paired column 成对柱
paired comparison 成对比较*
paired comparison experiment 成对比较试验
paired electrons 成对电子
paired eyepiece 双目镜
paired ion chromatography(=ion pair chromatography) 离子对色谱法*
pairing 对合；成对

pairing energy 配对能；成对能
pairing energy effect 配对能量效应
pairing faulty 配对缺陷
pairing twist factor 对偶捻系数
pair production （电子）偶形成
pair test 二点试验法；配对试验
palatine chrome black S 宫殿铬黑 S；酸性茜素黑 SN
palatine fast blue GGN 宫殿坚牢蓝 GGN；C.I.酸性蓝 158
palatine red 宫殿红〔染〕
palau 金钯合金〔约含 80%Au〕
pale 苍(色)；浅
pale crepe 苍皱(橡)胶
pale fawn 淡黄褐色；淡糙米色
pale flag 香根鸢尾
pale-green bottle 浅绿瓶
pale milky cast on coating film 漆膜上泛浅乳白色荧光
pale oil 苍色油；浅色润滑油
paleotemperature determination 古代温度测定
pale pink 浅桃红
pale red 浅红
pale rosin 浅色松香
palest varnish 最浅色清漆
palette 调色板；托板
pale yellow 浅黄
palicourine 巴柯碱
paligorskite 坡缕石
palisade layer 栅状层
palitantin 徘徊青霉素；5-庚基-1,3-二烯-2,3-二羟基-6-(羟甲基)环己烷酮
palladiazo 钯偶氮
palladic ①(正)钯的；四价钯的②钯制的③含钯的
palladic compound 正钯化合物
palladichloride 氯钯酸盐
palladic oxide 二氧化钯 PdO_2
palladic sulfide 二硫化钯 PdS_2
palladinized asbestos 披钯石棉
palladious 亚钯的
palladised charcoal 披钯木炭
palladium 钯〔46 号元素，化学符号 Pd〕
palladium asbestos 披钯石棉
palladium black 钯黑
palladium carbon 载钯(活性)炭
palladium catalyst 钯催化剂
palladium charcoal 披钯木炭
palladium chloride 氯化钯〔总名，计有：二氯化钯 $PdCl_2$ 和四氯化钯 $PdCl_4$〕
palladium dioxide 二氧化钯 PdO_2
palladium generator 钯发生器
palladium hydride 氢化钯 Pd_2H
palladium hydroxide 氢氧化钯 $Pd(OH)_2$; $Pd(OH)_4$
palladium monoxide 一氧化钯 PdO
palladium oxide 氧化钯 $PdO; PdO_2$
palladium separator 钯分离器
palladium sponge 海绵(状)钯
palladium sulfide 硫化钯
palladium transmodulator 钯调制器
palladium tube 钯管
pallado 亚钯的
pallador 铂钯热电偶
palladous compound 亚钯化合物
palladous hydroxide 氢氧化亚钯；二羟化钯 $Pd(OH)_2$
palladous oxide 一氧化钯 PdO
palladous sulfate 硫酸钯 $PdSO_4$
pallamine 胶态钯
pallasite 石铁陨石；橄榄陨铁
pallet ①垫衬②码垛盘③底板
palletizing 码垛堆积
pallet truck 码垛车
Pall ring 鲍尔环；带孔环形填料
palm ①棕榈(树)②棕榈香
palmae 棕榈油
Palmaer phosphate 帕麦尔磷肥
palmarosa 玫瑰草
palmarosa oil(=pamorusa oil) 香茅油；掌玫油
palmatine 巴马亭；非洲防己碱
palmatrubine 巴马土宾
palm butter(=palm grease) 棕榈油
palm compound 棕榈香精
palmellin (淡水)藻红素
palmering 柔软加光整理
palmetto (北美)棕榈果
palm grease(=palm butter) 棕榈油
palmic 棕榈的
palmic acid 棕榈酸；十六(烷)酸；软脂酸 $CH_3(CH_2)_{14}COOH$
palmierite 硫钾钠铅矿
palmin 纯可可脂；棕榈酯；软脂酸酯
palmital 棕榈醛；软脂醛
palmitaldehyde 棕榈醛；软脂醛
palmitamide(=palmitic amide) 棕榈酸酰胺；十六(烷)酰胺 $C_{15}H_{31}CONH_2$
palmitate ①棕榈酸；软脂酸；十六(烷)酸②棕榈酸盐(酯或根)
palmitic acid(=palmitinic acid) 棕榈酸；十六(烷)酸；软脂酸 $CH_3(CH_2)_{14}CO_2H$
palmitic acid amide 棕榈酸酰胺；十六(烷)酸酰胺

$C_{15}H_{31}CONH_2$
palmitic acid glyceride 甘油棕榈酸酯
palmitic aldehyde 棕榈醛 $CH_3(CH_2)_{14}CHO$
palmitic aldehyde oxime 棕榈醛肟 $C_{15}H_{31}CH=NOH$
palmitic amide(=palmitamide) 棕榈酸酰胺；十六(烷)酰胺 $C_{15}H_{31}CONH_2$
palmitic anhydride 棕榈(酸)酐 $(C_{15}H_{31}CO)_2O$
palmitic monoethanolamide 棕榈酰单乙醇胺；棕榈酸单乙醇酰胺
palmitin(=tripalmitin) 棕榈精；三棕榈精；(三)棕榈酸甘油酯 $(C_{15}H_{31}COO)_3C_3H_5$
palmitinic acid(=palmitic acid) 棕榈酸；十六(烷)酸；软脂酸 $CH_3(CH_2)_{14}COOH$
palmitodistearin 甘油二硬脂酸一棕榈酸酯
palmitoleic acid(=physetoleic acid) 抹香鲸烯酸；棕榈烯酸；9-十六(碳)烯酸
palmitoleostearin 棕榈油(酰)硬脂(酰)甘油酯
palmitoleoyl 棕榈油酰(基)
palmitolic acid(=7-hexadecynoic acid) 棕榈炔酸；7-十六(碳)炔酸 $CH_3(CH_2)_7C\equiv C(CH_2)_5COOH$
palmitone(=dipentadecyl ketone) 棕榈酮；16-三十一(烷)酮；对称三十一酮 $(C_{15}H_{31})_2CO$
palmitonitrile 棕榈腈；十六(烷)腈 $C_{15}H_{31}CN$
palmitoyl 棕榈酰；十六(烷)酰 $CH_3(CH_2)_{14}CO-$
palmitoyl chloride 棕榈酰氯 $C_{15}H_{31}COCl$
palmitoyl glucose 棕榈酰基葡萄糖；葡萄糖棕榈酸酯
palmitoyl methyltaurate 棕榈酰基牛磺酸甲酯
palmityl ①棕榈基；十六(烷)基②棕榈酰；十六(烷)酰
palmityl alcohol 棕榈醇
palmityl amine 棕榈基胺；十六(烷)基胺
palmityl chloride 棕榈酰氯
palmityl trimethyl ammonium chloride 十六烷基三甲基氯化铵
palm-kernel oil 棕榈仁油
palm-nut oil 棕榈坚果油
palm oil(=palm pulp oil) 棕榈油
palm oil fatty acid 棕榈油脂肪酸
palm oil grease 棕榈油脂
palm oil pitch ①棕榈油沥青②脂肪酸沥青
palm oil soap 棕榈油皂
palm olein 棕榈油精
palmolive 棕榄(香型)
palmolive soap 棕榄皂
palm pulp oil(=palm oil) 棕榈油
Palmquist apparatus 帕姆奎斯特二氧化碳器
palm test 手掌法(汽油挥发性试验)
palm wax 棕榈蜡
paludrine 白疟特灵

palustric acid 长叶松酸
palustrine(=equisetine) 沼泽(木贼)碱
palustrol 喇叭茶(萸)醇
palygorskite 坡缕石
pamaquine 帕马奎宁〔药〕
Pamela process 一种高放废液玻璃固化过程〔德国〕
pamidronate disodium 帕米膦酸二钠〔药〕
pamorusa oil(=palmarosa oil) 香茅油；掌玫油
pan ①盘(天平)；(称量)盘②锅③〔词头〕全；总
panabase 黝铜矿
panacea 万能药
panacene 人参萜
panacon 人参酮
panama bark 皂树皮
Panapak packing 帕纳帕克填料
panaquilon 人参奎酮
pan arrest 盘托
panaxcoside 人参糖苷
Panax ginseng 人参
panaxin 人参英；人参辛苷
panaxsapogenol 人参皂草精醇
PAN-based carbon fibre 聚丙烯腈基碳纤维；PAN基碳纤维
pan bench 蒸发盘组
pan boiling 锅煮
pan bread 盘式面包；模制面包
pan burner 盘炉
panchromatic emulsion 全色乳剂
panchromatic film 全色胶片
panchromatograph 多用色谱仪
pancock coil 旋管；蛇管
pan control apparatus 盘式控制仪
pan conveyer 盘式运输机〔设〕
pancratine(=haemanthidine) 血安生
pancreatic enzyme 胰酶
pancreatic islet cell 胰岛细胞
pancreatin ①胰酶②胰酶制剂
pancreatitis 胰腺炎
pan cure 盘蒸硫化；粉蒸硫化
pancuronium bromide 泮库溴铵〔药〕
pandermite 白硼钙石
Pandia continuous digester 潘迪亚(管式)连续蒸煮器
pandiene value 全二烯值
pan drier 盘式烘干器
pane ①(门窗的)一块玻璃②锤顶③镶玻璃(板)
panel ①板条②翼片③面板④表盘
panel (board) 配电盘
panel clamp 样板夹

panel coker 成漆板焦化器〔炼油〕
panel coking test 成漆板焦化试验〔评价油的炭化情况〕
panel cooler 平板冷却器；板式冷却器
panel length 节间长度
panel size 片坯尺寸
panel spalling test （耐火砖的）格子散裂试验
panel test 评审检验
Paneth-Fajans-Hahn adsorption rule 潘-法-罕吸附规律*
Paneth's rule 潘纳思放射性元素吸附规律
Paneth technique 潘纳思技术
pan feeder 盘式给料器；盘式给漆器
pan-furnace 锅与炉
pangamic acid(=vitamin B_{15}) 潘氨酸，维生素 B_{15}
pan gas 锅气〔指硫酸盐锅冒出的气体〕
pan head rivet 圆头铆钉
pan holder 镫芯〔天平〕
pan house 蒸发盘室
paniculatine 甜茅碱
panipenem and betamipron 帕尼培南倍他米隆〔药〕
pan mill 碾磨盘
pan milling 碾磨
pan mixer 锅式混合机
panmycin(=tetracycline) 四环素
pannic acid 绵马酸
panning 选淘
pannonit 硝铵、硝酸甘油、食盐炸药
panogen(=methylmercuric dicyandiamide) 双氰胍甲汞
panosialin 泛涎菌素
pan pipe 锅管
pan process 直接蒸气(再生)法；油(再生)法
pan (regeneration) process 盘型再生法〔橡胶〕
pan residue 馏渣；蒸馏余液
pan roll 碾机辊
pans 中间罐；接受罐
pan salt 锅盐
pan scale 锅垢〔硬水的沉淀锅垢〕
pansman 蒸煮工
pan straddle 容器支架
pansupari 槟榔、槟榔子和石灰的混合物
pansy 三色堇；蝴蝶花
pantachromatic 多色的；全消色差的
pantachromism(=pantachromism) 多色(现象)
pantal 铜、锰、镁、铝合金
pantaxanthin 五黄质；海胆黄质
pantellerite 碱流岩
pantetheine 泛酰巯基乙胺
pantethin 潘特生；双泛酰硫乙胺 $C_{18}H_{32}O_5N_2S$
panthenol 泛醇；泛酰醇

panthomycin(=pantomycin) 双氢链霉素泛酸盐
pantile 波形瓦
pantocaine 潘妥卡因；对卡因
pantochromic 多色的
pantochromism(=pantachromism) 多色(现象)
pantocrine 鹿茸精
pantograph ①比例绘图仪②胸廓描记器
pantoic acid 泛解酸
pantomorphism 全对称现象
pantomycin 双氢链霉素泛酸盐
pantonine 泛氨酸；α-氨基-β,β-二甲基-γ-羟基丁酸
pantopaque 碘苯基十一酸乙酯
pantopon 鸦片总碱
pantoprazole 泮托拉唑〔药〕
pantothenamide 泛酸酰胺
pantothenate ①泛酸②泛酸盐(酯或根)
pantothenic acid 泛酸
$HOCH_2C(CH_3)_2CH(OH)CONH(CH_2)_2COOH$
pantothenol 羟泛酸；维生素 B_5
pantothenyl alcohol 泛醇；本多生醇
N-(pantothenyl)-β-aminoethanethiol 泛酰巯基乙胺
pantoyl 泛酰基
pantoyltaurine(=thiopanic acid) 泛磺酸
pan vulcanization 罐中硫化法〔橡〕
papain 木瓜蛋白酶
papaver 罂粟；鸦片
papaveraldine 罂粟啶
papaveramine 罂粟胺
papaveric acid 罂粟酸 $C_{16}H_{13}NO_7$
papaverine 罂粟碱
papaverinol 罂粟醇 $C_{20}H_{21}O_5N$
papaveroline 罂粟灵〔周围血管扩张药〕$C_{16}H_{13}O_4N$
paper ①纸②论文
paper adhesive 纸用黏合剂
paper adsorption chromatography 纸吸附色谱(法)
paper analysis 纸上分析
paper blotter press 框式压滤机
paper board 纸板
paper bridge 纸桥
paper chromatoelectrophoresis 纸色谱电泳
paper chromatogram 纸色谱(图)
paper chromatographic distribution 纸色谱分布
paper chromatographic method 纸色谱分析法
paper chromatographic scanner 纸色谱扫描器
paper chromatography 纸色谱(法)
paper clay 浆土
paper cloth 纸织物；纸布
paper coal 薄层褐煤

paper coating　纸张涂层
paper colors　染纸颜料
paper condenser　纸介(质)电容器
paper constant　(色谱)纸常数
papercore plywood　纸心胶合板
paper disc　纸盘
paper-disk chromatography　(多层)圆纸色谱(法)〔将多片面积相同的圆形色谱纸叠层在管内，进行展开操作〕
paper-disk method　圆纸法；纸盘法
paper electrophoresis　纸电泳(法)
paper electrophoresis method　纸电泳法
paper electrophoretic separation　纸电泳分离
paper-faced scrim(=linen-backed lining)　纸面(衬)纱布里
paper-face overlay(=paper plastic overlay)　纸基塑料贴面
paper filter　纸滤器
paper hangings〔复〕　糊墙纸
paper laminate　纸层薄板
paper machine　造纸机
paper maker　造纸工
paper-maker's alum　造纸明矾　$Al_2(SO_4)_3 \cdot 18H_2O$
paper-making machinery　造纸机(器)
paper matrix　纸模
paper mill　造纸厂
paper mill foam oil　造纸厂泡沫油
paper partition chromatography(=papyrography)　纸分配色谱(法)
paper-pile(=chromatopile)　(色谱)纸堆
paper plastic overlays(=paper-face overlays)　纸基塑料贴面
paper pulp　纸浆
paper pulp filter　纸浆过滤器
paper radiochromatography(PRC)　纸放射色谱法
paper retention analysis　纸保留分析(法)
paper roll　纸辊
paper scale thermometer　纸刻度温度计
paper shredder　切废纸机
paper size　纸的尺码
paper sizing　施胶
paper-solvent equilibrium　纸溶剂平衡
paper spar　薄片方解石
paper speck　纸斑
paper speed　纸带速度
paper stock　纸料
paper strip　条形纸
paper strip chromatography　条纸色谱(法)
paper strip method　条纸法；条纸色谱法
paper tape　纸带
paper thickness gauge　纸厚度计

paper thimble　纸壳筒
paper towel　纸巾
paper transport　记录纸传递
paper twine　纸绳；纸纱；纸线
paper wax　纸蜡
paper web　纸幅
paper wood　造纸材
paper yarn　纸绳
paper yellow　①纸黄色②纸黄〔染〕
papery shale　纸板
papilla　乳头状突起
pappe　纸板
papra-amidophenol　对氨基苯酚〔毛皮染料〕
papyrography(=paper partition chromatography)　纸分配色谱(法)
par　同等；平等
para-　①〔位次〕对②〔作词根用〕仲〔如仲碘酸、仲氢、仲班酸、仲蒽〕③副〔俗名〕
para-acid　①仲酸〔无机〕②对位酸〔有机〕
para-amidophenol　对氨基苯酚〔毛皮染料〕
para-aminoacetophenone　对氨基苯乙酮
para-aminobenzoic acid　对氨基苯(甲)酸
para-aminophenol　对氨基苯酚
para-anthracene　仲蒽；聚二蒽
para arrowroot　木薯淀粉
para balsam　帕拉香脂
parabanic acid　仲班酸；乙二酰脲　$C_3H_2O_3N_2$
parabasalt　普通玄武岩
paraben(=para-hydroxybenzoate)　对羟基苯甲酸酯
parabituminous　易结烟煤的
parabola　抛(物)线
parabola mass spectra　抛物线质谱
parabola mass spectrograph　抛物线质谱仪
parabolic　抛(物)线的；抛物的
parabolic mirror　抛物面反射镜
parabolic path　抛物线轨道
parabolic velocity distribution　抛物线速度分布
paraboloid　抛物面
paraboloid condenser　抛物面聚光器
paraboloid of pressure　压力抛物面体
paraboloid of revolution　旋转抛物
parabuxin　副黄杨碱
paracellulose　薄壁组织的纤维素
paracetaldehyde　三聚乙醛
paracetamol　4-乙酰氨基酚
parachemistry　仲化学〔无机和有机化学以外的化学〕
parachloromercuribenzene sulfonate(PCM-BS)　对氯汞苯磺酸　$ClHgC_6H_4SO_3H$

parachloromercuribenzoate(PCMB)　对氯汞苯甲酸 $ClHgC_6H_4COOH$
para-chlorometacresol　对氯代间甲酚
para-chlorometaxylenol　对氯代间二甲苯酚
para-chlorophenol　对氯苯酚
parachor　①等张比(较)容(积);等张体积②(摩尔)等张比容;摩尔等张体积
parachor value　等张比容值
paracme　衰退期;缓解期
para-compound　对位化合物
paraconic acid　仲康酸;5-氧代-3-四氢呋喃酸 $C_4H_5O_2COOH$
paraconiine　仲康因　$C_8H_{15}N$
paracoto　副可土树皮
paracotoin　副可土因
paracoumarone　聚氧茚
paracoumarone-indene resin(=coumarone-indene resin)　苯并呋喃-茚树脂
paracresol　对甲酚
para-crystal　次晶;酝晶
paracrystalline　次晶的
paracrystalline phase　次晶相
paracrystalline state　次晶态
paracrystallinity　次晶度
paracyanogen　仲氰;多聚氰　$(CN)_n$
paracyclophane　二聚二甲苯一羧酸;二聚(对)二甲苯邻甲酸
para-dehydrochlorination　脱共轭氯化氢(作用)
para-derivative　对位衍生物
para-diazine(=pyrazine)　对二嗪;吡嗪;1,4-二氮杂苯
para-dichlorobenzene　对二氯苯
para-diethylsuccinic acid　内消旋二乙基丁二酸
para-dihydroxybehenic acid　聚二羟基山嵛酸
para-dihydroxystearic acid　聚二羟基硬脂酸
para-dimethylaminobenzaldehyde　对二甲氨基苯甲醛
paradioxybenzene　对苯二酚
para-directing group　对位定向基
paradise nut oil　乐园坚果油〔取自 Leicythis zabucajo〕
para-disubstitution derivative　对位二取代衍生物;对位双取代衍生物
para-dye　偶合染料〔染〕
paraelectric phase　顺磁电相
paraffin　石蜡
paraffinaceous　石蜡的
paraffinaceous petroleum　含石蜡石油
paraffin acid　石蜡酸
paraffin aldehyde　石蜡醛
paraffin-asphalt petroleum　石蜡沥青石油

paraffin base　石蜡基;石蜡底子;链烷(属)烃基
paraffin-base crude(=paraffin-base oil)　石蜡基石油
paraffin-base oil(=paraffin-base crude;paraffin-base petroleum)　石蜡基石油
paraffin-base petroleum(=paraffin-base oil)　石蜡基石油
paraffin bath　石蜡浴
paraffin bottle　石蜡瓶〔灯油的容器,英国名〕
paraffin butter　石蜡脂
paraffin coal　石蜡煤
paraffin coating　石蜡封盖;石蜡涂层
paraffin content　石蜡含量
paraffin crude oil　石蜡基原油
paraffin degreasing　石蜡脱脂
paraffin distillate　石蜡馏分
paraffine(=paraffin)　①石蜡;硬石蜡②链烷(属)烃
paraffin embedding　石蜡打底
paraffin flux　石蜡渣油
paraffin gas　烷烃气体;石蜡气体
paraffin grade wax　商品石蜡
paraffin grinding method　石蜡油研磨法
paraffin hydrocarbon　烷属烃;链烷烃;饱和(开)链烃
paraffinic acid　链烷酸;石蜡族酸
paraffinic base crude　石蜡基原油
paraffinic hydrate　链烷水化物
paraffinic hydrocarbon　链烷烃;烷属烃
paraffinicity　链烷烃含量;石蜡含量
paraffinic oil(=paraffin oil)　石蜡油
paraffinic solvent　石蜡族溶剂
paraffin intermediate crude　石蜡中间基石油
paraffin jelly　(药用)石蜡油膏
paraffin mass　石蜡块
paraffin method　石蜡法
paraffin oil(=paraffinic oil)　石蜡油
paraffin oxidation　石蜡氧化
paraffin oxidation style　石蜡氧化法
paraffin paper　(石)蜡纸
paraffin press　石蜡过滤器
paraffin pressed distillate　石蜡馏分
paraffin refined wax　纯石蜡;精制石蜡
paraffin removal　清蜡;除蜡
paraffin-rich　富石蜡的;含大量石蜡的
paraffins　烷属烃
paraffin scale(=paraffin scale wax)　粗石蜡;未精制石蜡
paraffin scale wax(=paraffin scale)　粗石蜡;未精制石蜡
paraffin-sealed stirrer　石蜡液封搅拌器
paraffin series　石蜡(物)系
paraffin slack wax　疏松石蜡
paraffin slop　不合格石蜡

paraffin soap 石蜡皂
paraffin sulfonate 石蜡烃磺酸盐；链烷磺酸盐
paraffin sweated wax 粗石蜡
paraffin sweating 石蜡发汗作用
paraffinum 石蜡
paraffinum durum 硬石蜡；固体石蜡
paraffinum liquidum 矿脂；凡士林油
paraffinum liquidum leve 矿脂；凡士林油
paraffinum molle 矿脂；凡士林
paraffin wax 石蜡*
paraffin wax emulsion 石蜡乳状液
paraffin wax melting point test 石蜡熔点试验
paraffin wax quality test 石蜡质量试验
paraffin xylol 石蜡二甲苯溶液
paraflow 巴拉弗洛〔一种抗凝剂〕
paraform 聚甲醛
paraformaldehyde 低聚甲醛*；多聚甲醛
parafuchsin 副品红
paragel 硅凝胶；巴拉盖尔〔一种抗凝剂〕
paragenesis 共生
paragneiss ①副片麻岩②水成片麻岩
paragonite 钠云母
Paraguay tea 巴拉圭茶
parahelium 仲氦
parahemophilia 副血友病
parahydrogen 仲氢*
para-hydroxybenzoate(=paraben) 对羟基苯甲酸酯
para-isomer(=para-isomeride) 对位异构体
para-isomeride(=para-isomer) 对位异构体
paralactic acid 副乳酸 $CH_3CHOHCOOH$
paraldehyde 仲(乙)醛；三聚乙醛 $(C_2H_4O)_3$
paraldol 仲醛醇〔俗〕；二聚 3-羟丁醛；二聚间羟丁醛 $(C_4H_8O_2)_2$
paraleucaniline(=triaminotriphenyl-methane) 三氨基三苯甲烷；副白苯胺 $(NH_2C_6H_4)_3CH$
para-linkage 对(位)键合；对键(结)构
parallax 视差
parallel ①平行的②并联的③同方向的④相似的；符合的
parallel acquisition 并行获得
parallel analysis 平行分析
parallel band 平行带
parallel beam 平行光束
parallel-beam glossmeter 平行光束光泽计
parallel centrifugal analyzer 平行式离心分析仪
parallel-chain crystal 平行链晶体*
parallel circuit 并联电路
parallel column 并联柱
parallel component of magnetization 磁化矢量平行分量

parallel connection 并联接法
parallel current ①并联电流②并流
parallel-current jet condenser 并流注水凝气器
parallel determination 平行测定
parallel direction 平行方向
paralleled plate plastometer 平行板式可塑计
parallel feed 并流送料法
parallel feeding 并行送料；混合进料
parallel FIA 平行流动注射分析
parallel fill-out 平行进料〔注塑〕
parallel flow 并流；平行流
parallel-flow drier 并流干燥器
parallel-flow evaporator 并流蒸发器
parallel grain plies 平行组合
parallel groove adhesion 平行沟线附着力
parallelism ①平行②平行度
parallelization tester (纤维)平行度试验仪
parallel-jawed clamp 平行咬夹
parallel laminate 顺纹层压制品
parallel laminated 平行层压的；同向(纹理)层压的
parallel laying 平行铺网
parallelosterism 同晶型组与化学成分或物理性质的关系
parallel plate capacitor 平行板电容器
parallel plate electrode 平行板电极
parallel plate interceptor(PPI) 平行板拦截器
parallel plate method 平行板法〔测定表面张力〕
parallel plate oil separator 平行板油分离器
parallel plate plastometer 平行板塑性计
parallel plate rheogoniometer 平行板流变性测定仪
parallel plate viscometer 平行板黏度计
parallel pyrolysis 并联热解
parallel rays 平行光线
parallel reaction 平行反应
parallel rods screen 平行杆筛
parallel strip 平行条；平行板条
parallel test 平行试验
parallel triple filament method 平行三灯丝法
parallel type DC amplifier 并列型直流放大器
parallel type spray drying 并流式喷雾干燥
parallel vibration 平行振动
para-localization energy 仲定域能；对值定域能
paraluminite 丝铝矾
paralysant gas 麻痹性毒气
paralyser ①催化毒物②麻痹药
paralysis 麻痹；瘫痪
paralysol 钾来苏儿〔甲酚与甲酚钾的混合物〕
paralyst 麻痹药
paralyzant 麻痹剂

paralyzer(=paralyser) 麻痹药
para magenta 偶合品红
paramagnet 顺磁体
paramagnetic 顺磁(性)的
paramagnetic body 顺磁体
paramagnetic broadening 顺磁加宽；顺磁增宽
paramagnetic complex 顺磁性配合物
paramagnetic compound 顺磁化合物
paramagnetic effect 顺磁效应*
paramagnetic oxygen analyzer 顺磁氧分析器
paramagnetic property 顺磁性
paramagnetic property of oxygen 氧的顺磁性
paramagnetic reagent 顺磁试剂
paramagnetic relaxation reagent 顺磁性弛豫试剂
paramagnetic resonance(PMR) 顺磁共振
paramagnetic resonance absorption 顺磁共振吸收
paramagnetic resonance method 顺磁共振法
paramagnetic ring current 顺磁环电流*
paramagnetic shielding 顺磁屏蔽*
paramagnetic shift 顺磁位移*
paramagnetic shift reagent 顺磁性位移试剂*
paramagnetic substance 顺磁物质*
paramagnetic susceptibility 顺磁磁化率；顺磁性系数
paramagnetism 顺磁性
para-mandelic acid 仲扁桃酸；苯乙醇酸 $C_6H_5CHOHCOOH$
paramecin 草履虫素
χ-parameter χ参数
parameter estimation 参数估计*
parameter matching 参数契合法
parameter method 参数法
parameter of specular gloss 镜面光泽参数
parameter test 参数检验*
paramethadione 3,5-二甲基-5-乙基-噁唑-2,4-二酮；甲乙双酮；对甲双酮
paramethyl red 副甲基红；对甲基红
parametral ratio (晶标)轴率
parametrical nonlinearity 参数非线性
parametric devices 参量器件
parametric model 参数模型
parametric oscillator 参量振荡器
parametric oscillatory amplifier 参量振荡放大器
parametrization 参量化
paramide 蜜亚胺
paramisan sodium 氨基水杨酸钠
para-molybdic acid 仲钼酸
paramorph 同质异晶体；同质异形体
paramorphine(=thebaine) 蒂巴因；二甲基吗啡；副吗啡

paramorphism 同质异晶(现象)
paramount 最高的；头等的
paramucic acid 仲黏酸
paranaphthalene 蒽
paranephrine 肾上腺素
paranitraniline red 对硝基苯胺红
paranitraniline S 对硝苯胺S〔染〕
para-nitrophenoxide anion 对硝基苯酚负离子
para-nitroso-N,N-dimethylaniline 对亚硝基-N,N-二甲基苯胺
para-nitrosotoluene 对亚硝基甲苯
paranox 巴拉诺克斯〔一种润滑油多效添加剂〕
paraoctyl phenol 对辛基苯酚
para-orientating group(=para-orienting group) 对位取代基
para-orientation 对位取向；对位取代
para-oriented ring 对位取向环；对位取代环
para-orienting group(=para-orientating group) 对位取代基
para-ortho-hydrogen conversion 对-邻位氢转换
parapectic 果胶糖酸
para-periodic acid 仲高碘酸；一缩原高碘酸 $I_2O_7 \cdot 5H_2O;H_5IO_6$
paraphenylene diamine salt 对苯二胺盐
para-phenylene diisocyanate 对-亚苯基二异氰酸酯
para phenyl phenol 对苯基苯酚
para position 对位
para-positronium 仲正电子素；仲电子偶素
paraquat 对草快；百草枯〔农药〕
paraquinanisol 对喹啉甲醚
paraquinoid 对醌型
paraquinoid structure 对醌结构
paraquinonedioxime 对醌二肟
para red 对位红
pararosaniline(=triaminotriphenyl-carbinol) 三氨基三苯甲醇；碱性副品红；副蔷薇苯胺 $HOC(C_6H_4NH_2)_3$
pararosaniline base 副品红碱
pararosaniline chloride 副品红
pararosaniline dyes 副品红染料
pararosaniline hydrochloride 盐酸副品红
pararosolic acid(=rosolic acid) 玫红酸
parartrose 麦际
Para rubber seed oil 帕拉橡子油
parasitic error 粗差；粗大误差；寄生误差；疏失误差
parasiticidal 杀寄生虫的
parasiticide 杀寄生物药
parasiticin 苄青霉素
parasitic reaction 寄生反应

parasorbic acid 仲山梨酸；花楸酸
parasporin 巴拉波任〔细菌杀虫剂〕
para states β态
parasympathetic blocking agent 副交感神经麻醉剂
paratartarics 外消旋式
para-terphenyl 对三联苯 $C_6H_5 \cdot C_6H_4 \cdot C_6H_5$
para-tertiary butyl phenol 对特丁基苯酚；对叔丁酚〔俗〕
para-tertiary butyl phenol mercaptan 对叔丁基苯酚硫醇
parathene 烷基化环烃
parathesin 对氨基苯甲酸乙酯
parathiazine 对噻嗪；1,4-硫氮杂苯
parathion(=thiophos) 对硫磷；一六○五〔农药〕
paratone 巴拉东〔一种增黏剂〕
paratonere 颜料红〔染〕
paratose 泊雷糖
paratropic plane 经向面
para-tungstate 仲钨酸盐 $M_2O \cdot 12WO_3 \cdot nH_2O$；$3M_2O \cdot 7WO_3$；$M_2O \cdot 2WO_3$；$M_2O \cdot 3WO_3$
para-tungstic acid 仲钨酸
paraxin(=chloramphenicol) 氯霉素
paraxylene 对二甲苯
parazon 对羟基联(二)苯
parchment ①羊皮纸②(假)羊皮纸；植物羊皮纸
parchment imitation 仿羊皮纸
parchmentized 羊皮纸化的〔纸〕
parchmentized fibre 硬化纸板纤维
parchmentizing 浓硫酸处理(使皮纸化)；羊皮化
parchmentizing paper 羊皮化纸
parchment leather dressing 鼓皮修整
parchment paper (假)羊皮纸
parchmoid 仿羊皮纸
parchmyn 仿羊皮纸
parcolite method 蒸气喷涂磷酸盐处理法
PAR derivatization PAR 衍生化法
pared bark surface 割皮面积〔橡〕
paregoric 樟脑鸦片酊
pareira 帕雷(亦拉)〔生药〕
pareirine 软齿花根碱；杷瑞任〔生物碱〕
parenamine 帕雷胺；雷帕明
parent ①根源②母(体)的
parental DNA 亲代 DNA
parental retention 母体(形式)保留
parent compound 母体化合物
parent element 母体元素
parent HC's chromatography 母体烃色谱法
parent ion 母离子
parent ion stability 母离子稳定性
parent latex 原料胶乳

parent material ①原材料②母材
parent metal 母材；基层金属
parent molecule 母体分子
parent-molecule ion 母分子离子
parent name 母体名
parent nuclide 母体核素*
parent oil 本源油；母体油
parent peak 母体峰
parent polymer 母体聚合物
parent radioisotope 母体放射性同位素
parent resin 母体树脂；本源树脂
parent stock 母料；母体原料
parent substance 母体；母物
parent vinyl ester 乙烯酯母体；起始乙烯酯
Parex process 帕来克斯法〔固定床液相连续分离对二甲苯法〕
pargasite 韭角闪石
parget ①灰泥；涂灰泥②石膏③墁饰花纹
parhelium 仲氦
parianite 帕里尼达沥青
paricine 杷日素
paridine 杷日定
parietal ①周壁的；周缘的②膜壁的
parietic acid 大黄酸
parietin 蜈蚣苔素；大黄素-7-甲醚
pariglin 菝葜苷
parillic acid(=parillin) 副菝葜酸；副菝葜皂苷
parillin(=parillic acid) 副菝葜皂苷；副菝葜酸
parinaric acid 十八碳四烯酸
paring machine 剥皮机
paring reaction 修边反应
Paris black(=lamp black) 灯黑；烟炱
Paris blue 巴黎蓝；碱性甲基蓝
Paris green(=urania green) 巴黎绿
Parisian blue(=Prussian blue) 铁蓝
Parisier-Parr method 帕里色-帕尔法
Parisier-Parr-Pople method PPP 法*
parisite 氟碳钙铈矿
parison 型坯；料坯
parison die swell(=parison swell) 型坯离模膨胀
parison hang time 型坯悬垂时间
parison maker 型坯工〔玻〕
parison mo(u)ld 雏型模；初型模
Paris violet(=methyl violet) 甲基紫
Paris white 巴黎白；碳酸钙粉
parity 宇称*
parity conservation 宇称守恒*
parity of wave function 波函数的宇称性

parity selection rule　宇称选择定则
parkerization(=parkerisation)　磷酸盐化处理膜(法)
parkerized　磷酸盐处理的；磷化处理的
parkerizing process　磷化；磷酸盐处理
Parker's cement　罗马水泥
parkesine　硝化纤维素塑料
Parkes process (for lead refining)　帕克斯(炼铅法)
parkine　派克木碱
Parlin cup viscosity　巴林杯黏度
parma blue　巴马兰；三苯基玫苯胺染料
Parma violet　巴马紫罗兰
parmelin　梅衣素
parmone　异紫罗兰酮
parol　石蜡燃料
paroline　液体石油膏
paromomycetin　巴龙菌素
paromomycin(=farmiglucin;gabbromycin;humatin; humycin)　巴龙霉素
paromycin　巴龙霉素
paronite　石棉橡胶板
paronite mill　石棉橡胶辊
paroxazine　对噁嗪；1,4-氧氮(杂)苯
paroxetine　帕罗西汀〔药〕
Parr bomb fusion method　帕尔弹式熔融法
Parr calorimeter　帕尔量热器
Parr surface turbidimeter　帕尔表面浊度计
parsec　秒差距
parsley　欧芹
parsley camphor　欧芹樟脑；芹菜脑
parsley fruit　欧芹子
parsley leaves oil　欧芹叶油
parsley oil　欧芹油
parsley root　欧芹根
parsley-seed oil　欧芹子油
parsnip oil　蔓荆油；欧防风油
Parsons sulfur recovery process　帕森斯硫回收法
part　①部(分)②本分③脚色④分开；分离
part cargoes　部分装载〔未装满的铁路油槽车，指定在不同地点卸出石油产品的铁路油槽车〕
part dimension　部件尺寸；零件大小
parthenclide　银胶菊内酯；小白菊内酯
parthenicine　银胶菊碱
parthenin　银胶菊宁；银胶菊素
parthenine　银胶菊碱
partheniol　银胶菊醇
partial　部分的；局部的
partial acetolysis　部分乙(酸水)解
partial acetylation　部分乙酰化

partial analysis　部分分析
partial argentation resin chromatography　部分银化树脂色谱(法)
partial automation　半自动化
partial balance　部分结算
partial bond fixation　键(的)部分固定化
partial coefficient(=distribution coefficient; partition coefficient)　分配系数
partial combustion　部分燃烧
partial combustion burner　部分燃烧(裂解)炉
partial combustion continuous process　连续不完全燃烧法〔炭黑〕
partial condensation　部分冷凝
partial condenser　分凝器
partial confounding　部分混区
partial correlation coefficient　偏相关系数*
partial cracking　部分裂化
partial decay constant　部分衰变常数
partial derivative　偏导(函)数
partial desulfurization　部分脱硫(法)
partial differential equation　偏微分方程(式)
partial dislocation　局部位错〔晶体〕
partial dismantling(=partial dismounting)　部分拆卸
partial dismounting(=partial dismantling)　部分拆卸
partial draining　局部穿流*
partial epimerization　部分差向(立体)异构化(作用)
partial equation　部分方程式
partial ester　偏酯
partial excitation　部分激发
partial flow filter　支管过滤器
partial glyceride　偏甘油酯
partial half-life　部分半衰期
partial heat(=differential heat)　微分热；定浓热
partial hydrogenation　部分氢化
partial hydrolysis　部分水解
partial hydrolysate　部分水解物
partial-immersion thermometer　部分浸没式温度计
partial ionization　局部电离
partial ionization theory　局部电离(理)论
partial ladder polymer　部分梯形聚合物
partial least square method(PLS)　偏最小二乘法
partial least squares regression spectrophotometry　偏最小二乘分光光度法
partial lubrication　局部润滑
partially balanced incomplete block design　部分平衡不完全区组设计
partially branched polymer　部分支化聚合物
partially controllable flow　部分可控流动；局部可控流

partially crystalline polymer　部分结晶聚合物
partially developed fibre　半熟纤维
partially drawn yarn　低拉伸丝；预拉伸丝
partially hydrogenated　部分氢化的
partially intergrown knots　半活节
partially miscible　部分溶混的
partially miscible liquid　部分溶混液
partially racemic compound　部分外消旋化合物
partially relaxed Fourier transform　部分弛豫傅里叶变换
partially spent　部分报废；局部耗损
partial miscibility　部分(溶)混性
partial miscible liquid　部分(溶)混的液体
partial molal capacity　偏摩尔容量
partial molal content　偏摩尔含量
partial molal density　偏摩尔密度
partial molal entropy　偏摩尔熵
partial molal free energy　偏摩尔自由能
partial molal heat content　偏摩尔热函
partial molal quantity　偏摩尔(数)量
partial molal specific heat　偏摩尔比热
partial molal volume　偏摩尔体积
partial molal work content　偏摩尔功函
partial molar enthalpy　偏摩尔焓*
partial molar free energy　偏摩尔自由能*
partial molar quantity　偏摩尔量*
partial molar volume　偏摩尔体积*
partial oxidation　部分氧化
partial oxidation cracking　部分氧化裂化
partial oxidation process　部分氧化法
partial oxidation reactor　部分氧化反应器；乙炔炉
partial phosphate ester salt　偏磷酸酯盐
partial polymer　部分聚合物；预聚物
partial potential(=chemical potential)　化学势
partial pressure　分压(力)；部分压力
partial pressure analysis　分压强分析
partial pressure analyzer　分压强分析器
partial pressure gauges　分压计
partial pressure gradient　分压梯度
partial pressure mass spectrometer　分压强质谱计
partial racemic mixture　部分消旋混合物；局部消旋混合物
partial racemization　部分外消旋(作用)
partial rate factor　分速度系数*
partial reaction　部分反应
partial reduction　部分还原
partial reflector　部分反射器；部分反射镜
partial reflux　部分回流
partial reflux operation　部分回流操作
partial regeneration　部分再生
partial regression coefficient　偏回归系数
partial relaxation　部分松弛；局部松弛
partial saponification number　部分皂化值
partial solvent　部分溶剂
partial specific volume　微分比容
partial stabilized gasoline　部分稳定的汽油
partial stabilizer　部分稳定器
partial stripping agent　部分退色剂
partial structure expansion　部分结构扩展
partial synthesis　部分合成
partial thixotropy　局部触变性
partial vacuum　部分真空；不完全真空
partial valence(=partial valency)　余价；部分价
partial valency(=partial valence)　余价；部分价
partial vaporization　局部汽化
partial vapor pressure　蒸气分压
partial volume　分容
partial vulcanization　部分硫化
α-particle　α粒(子)
particle　①粒子②(颗)粒③质点
particle adhesion　粒子黏着
particle agglomerate　粒子附聚物
particle beam　微粒束；粒子束
particle beam interface　粒子束接口
particle board　碎料板；刨花板
particle charge　点电荷
particle cloud　粒子云
particle cluster　粒子团
particle collector　集尘装置；颗粒收集器
particle concentration　粒子浓度；尘含量
particle counting immunoassay　颗粒计数免疫分析
particle detector　粒子检测器
particle diameter　粒(子直)径
particle diameter effect　粒径效应
particle distribution　粒子分布
particle distribution pattern　粒径分布图形(模式)
particle eletrophoresis(=microscopic eletrophoresis)　粒子电泳
particle fineness　颗粒细度
particle fluence　粒(子)流(量)密度
particle-image velocimetry　粒子成像流速测定法
particle induced ionization　粒子诱导离子化
particle induced X-ray emission　粒子诱导 X 射线发射
particle induced X-ray emission analysis　粒子激发 X 射线发射分析
particle induced X-ray emission microanalysis　粒子激发 X 射线发射微分析
particle induced X-ray emission spectrum(PIXE)　粒子诱

导 X 射线发射光谱
particle interaction　粒子相互作用
particle matter　微粒物质
particle migration　粒子移动
particle morphology　粒子形态
particle orientation　粒子取向
particle packing pattern　粒子堆积模式
particle-particle boundaries　粒子-粒子(间)界面
particle-particle interaction energy　颗粒间的界面作用能
particle-particle packing　(催化剂)紧密包装
particle path　粒子路途；粒子轨道
particle radius　颗粒半径
particle scattering factor　粒子散射因子*
particle scattering function　粒子散射函数*
particle separator　颗粒分离器
particle shape　颗粒形状
particle size　①粒子大小②颗粒大小③细度
particle size analysis　粒度分析
particle size analyzer　粒度分析仪
particle size control agent　粒度控制剂
particle size determination　(粉末)粒度测定(鉴定)
particle size dispersion　粒度分布；颗粒大小分布
particle size distribution　粒子大小分布
particle size distribution determination　粒径分布测定
particle size effect　粒度效应
particle size measurement　粒度测定
particle sizer　粒度分级器
particle size reduction　①粒度减小②粉碎；研磨
particle technology　粒子工艺学；粉体工艺学
particle theory of radiation　光辐射的微粒说
particle-to-particle collision　粒子对粒子的撞击
particle weight　粒子重量
particle weight scale　粒子重量标尺；粒子(重量)秤
particular sampling　颗粒物采样
particulate　①粒子；微粒②粒状；颗粒的
particulate contamination　颗粒物沾污
particulated bed　粒子床
particulate-filled composite　填料粒子的组分；散填料组成
particulate filler　粒状填料
particulate fluidization　散式流化
particulately-fluidized phase　流化颗粒相
particulate matter(PM)　颗粒物
particulate matter sampler　大气颗粒物采样器
particulate pollutant　微粒污染物
particulate solids　(催化剂)粉碎固体粒子
parting　分金〔用热浓硫酸将金银分开〕
parting acids　分银的各种强度硝酸

parting agent　脱模剂；模型润滑剂；隔离剂
parting compound　脱模剂；隔离剂
parting furnace　分离炉
parting plane　界面；际面
parting slip　隔片
parting surface　分离(界)面
partinium　钨铝合金
part integration　分部积分
partition　①分配；分布②分开③间壁④配分
partition capacity　分配容量
partition chromatography　分配色谱法*
partition coefficient(=distribution coefficient)　分配系数
partition coefficient of pesticide　农药分配系数
partition column　分配柱
partition effect　分配效应
partition equilibrium　分配平衡
partition equilibrium constant　分配(平衡)指数
partition function　配分函数*
partition gas chromatography　分配气相色谱法
partitioning agent　分配剂
partitioning cycle　分离循环
partitioning device　分配设备
partition isotherm　分配等温线
partition law(=distribution law)　①分配(定)律②分布(定)律
partition line(=parting line)　分模线
partition liquid　分配液
partition method　隔离方法
partition polarography　分配极谱法
partition rate　分配速率
partition ratio　分配比；分配率
partition ring　十字环
partition wall　隔墙；间壁；隔开墙
partly miscible liquid　部分混溶液体
partography(=paper partition chromatography)　纸分配色谱(法)
part per billion(ppb)　十亿分之一〔10^{-9}〕
part per million(ppm)　百万分之一〔10^{-6}〕
part per trillion(ppt)　万亿分之一〔10^{-12}〕
part run　未满管筒子
parts　零件
parts drawing　零件图
part sectioned view　局部剖视图
parts list　零件目录表
parts per billion　十亿分之(几)
parvoline　二乙基吡啶　$C_9H_{13}N$
parvule　小粒；小球；小丸
parvuline　小小菌素

parylene(=poly-para xylylene) 聚对亚苯基二甲基
Pascalian fluid(=inviscid fluid) 帕斯卡流体;非黏性流体
Paschen-Back effect 帕邢-贝克效应
Paschen-Runge mounting 帕邢-龙格装置
pasomycin 步霉素
pasque flower 白头翁
pass ①传递②通过③过(去)④焊道
passage ①通道②行程
passage type 过渡形式
pass band 通带
pass energy 通能
passenger tyre 轻轮胎
passette 泡罩
passimeter 内径指示规
passing machine 筛滤机
pass into solution 溶入溶液
passion-flower 粉色西番莲
passion-flower absolute 西番莲净油
passion fruit 鸡蛋果
passivating 被动化
passivating agent 钝化剂
passivating film 钝化膜
passivating layer 钝化层
passivating treatment 钝化处理
passivation 钝化*
passivation layer 钝化层
passivation potential 钝化电势*
passivator ①减活剂②钝化剂
passive ①钝态的②被动的
passive absorption 被动吸收
passive activation analysis 无源活化分析
passive agglutination test 被动凝集试验
passive condition 钝态
passive film 钝化膜*
passive gamma-ray assay 无源γ射线分析;固有γ射线分析
passive gamma ray method 无源γ射线法;固有γ射线法
passive gas protection 消极防毒
passive interrogation 无源探询*
passive metal 钝态金属
passive neutron assay 无源中子分析;固有中子分析
passive neutron method 无源中子法;固有中子法
passive resistance 钝态抵抗
passive state 钝态
passive state of metal 金属(的)钝态
passive transfer (敏化)被动传递
passive transport 非主动运输;被动运转
passive valve 被动阀

passivity ①钝性②钝态
pass over ①通入②过渡
pass sequence 焊道顺序
past column reaction 柱后反应
paste ①糊②浆糊
paste adhesive 膏状黏合剂;糊状胶黏剂
paste blue 普鲁士蓝
paste board 胶纸板
paste color 色浆
paste cutting compound 糊状切削化合物;冷却切削工具用的糊状物
pasted plate 涂浆板
paste driers(=patent driers) 糊状催干剂;膏状催干剂
paste drying 贴板干燥
paste dye 浆状染料
paste filler 膏状填孔剂
paste filling machine 装膏机
paste for printing 印刷油膏
paste-like product 糊状产品
pastel shade 浅色调
paste lubricant(=pasty lubricant) 糊状润滑剂
paste-making resin 制糊树脂
paste method 浆糊法
paste mill 磨浆机
paste mixer 调浆机;调糊机
paste molding ①糊塑②糊塑法
paste of soap 皂膏;皂浆;皂胶〔制肥皂〕
paste paint 厚漆
paste point machine 浆点机
paste red lead 红铅油膏
paste roller mill 磨浆机
paste rouge 胭脂(膏)
paste shampoo 洗发膏
paste sheeting (生产固体喷气燃烧的)糊状片
paste soap 膏状皂
pasteurising varnish 耐温水清漆〔耐巴氏杀菌时70~80℃温水〕
pastil(=pastille) 锭(剂)
pastille(=pastil) 锭(剂)
pastiness 浆糊状态
pasting ①裱糊②贴板
pasting brush 裱糊刷子;涂浆刷子
pasting dryer unit 贴板干燥组
pasting frame 贴板架
pasting machine 裱糊机
pasting plate washing unit 贴板冲洗组
pastometer 巴氏温度计
pasty ①(浆)糊状的②(肉)馅饼

pasty agent　糊料；增稠剂
pasty lubricant(=paste lubricant)　糊状润滑剂
pasty mass　糊状物质
pasty sludge　糊状残渣
pat　①小块；饼②合宜的
patch　①补丁②金属补片③小块地
patch belt　救急带；修理带
patch fuel　块状燃料
patching　修补
patching rubber material　修补胶料〔橡〕
patch of carbon　局部积炭
patch of grey　小块灰色补片
patchoulene　绿叶烯　$C_{15}H_{24}$
patchoulene oxide　氧化广藿香烯；氧化绿叶烯
patchoulenol　广藿香烯醇；绿叶醇
patchoulenone　广藿香烯酮；派超力烯酮
patchouli alcohol　绿叶醇；广藿香醇
patchoulin　绿叶灵　$C_{15}H_{26}O$
patchouli oil　绿叶油；广藿香油
patchoulion　广藿香酮；绿叶酮
patch plug　补片
patch test　贴附性试验
patch thermocouple　接触热电偶
patchy anisotropism　不规则各向异性(现象)
pate　①头部；颈部〔皮革〕②前额
patent　①专利(特许证)；专利权②明显的③专利件
patentability　可专利(性)
patent blue　专利蓝〔属于三苯甲烷类染料〕
patent coated paper　特制涂布纸
patent damp-resisting lining　特制防潮(糊墙纸)夹衬(层)
patent driers(=paste driers)　糊状催干剂；膏状催干剂
patent finish　漆革涂饰剂
patent flour　特级面粉
patent fuel　专利燃料
patent knotting primer　专用木节封闭底漆
patent leather　漆皮；人造革
patent vermillion(=mercuric sulfide)　硫化汞
patent white board　特别涂布白纸板
Patera process　帕特拉冶银法
path　路径
path difference　(光)程差
path length　光程
pathochemistry　病理化学
pathocidin　祛病菌素
path of migration　迁移途径；迁移程
path of steepest ascent　最速上升路线
pathogen　病原(体)
pathogenic　致病的

pathogenicity　致病性；病原性
pathogeny　病因
pathologic　①病理(学)的②(有)病的
pathological　①病理(学)的②(有)病的
pathological chemistry　病理化学
pathological crystal　变态晶体
pathologic reaction　病理反应
pathology　①病理学；病害(学)②病理
pathophysiology　病理生理学
patina　带色薄锈
patination　①生铜锈；布满铜绿②生锈
patio process　混汞法〔析金银〕
pat of cement　(水泥)饼
patrix(=male mold)　阳模；上半模
patronite　绿硫钒矿
pattern　①图案；花样；型式②模型③类型④摹制⑤图形
pattern analysis　图形分析
pattern coefficient　图形系数
pattern depth　花纹深度
pattern embossing　压花
pattern enamel　①皱纹瓷漆②(有花纹的)美术漆
pattern inspection　样板检验
pattern recognition　模式识别
pattern recognition optimization method for multiple target
　　多目标模式识别优化法
pattern sheet　印花膜；印花纸
pattern stability test　模型稳定性试验
pattern staining　粉末电沉积图案
Patterson search technique　帕特森寻峰法*
Patterson synthesis　帕特森合成
Pattinson process　帕廷森法〔从铅中分离银〕
patuletin　万寿菊素
patulin　展开青霉素；展青霉素；棒曲霉素
paucidisperse　少量分散
paucidisperse system　少量分散系统
paucine　巴柯碱
paucity　少数；少量
pauco nuts　五壳豆果
Pauli (exclusion) principle　泡利(不相容)原理
Pauling electronegativity scale　鲍林电负性标度*
Pauling process　鲍林法
Pauling structure　鲍林结构
Pauli principle　泡利原理
paullinia (paullinio) tannin　巴西可可单宁
paullinio tannin　泡林藤单宁
Pauly silk　铜氨丝
Pauly's reaction　泡利反应
pavement structure　铺砌结构

paving asphalt 铺路沥青
paving asphalt cement 铺路沥青水泥
paving brick 铺地砖
paving brick clay 铺地砖土
Pavy's solution 佩维测糖溶液
pawl ①爪②棘爪
pawpaw(=papaw) 木瓜
payload capacity 有效载重量
Payne's process 潘恩木材防火法
pay-off 送料；送料装置
pay ore 值得加工的矿石
p-block element p区元素*
P-bomb 飞弹；P式炸弹
P-cellulose 磷酸纤维素
P-collagen P胶原
PDF powder diffraction file cards PDF粉末衍射卡片
peach aldehyde 桃醛；γ-十一烷酸内酯
Peachey process 皮契(硫化)法
Peachey process of vulcanization 皮契硫化法
peach-face micro-powder fabric 桃皮绒微粉感织物
peach flavour 桃子香精
peach-kernel oil 桃仁油
peach leaves 桃树叶
peach skin fabric 桃皮绒
peach wood 洋苏木
pea coal 颗粒煤
peacock coal 闪光煤
peacock copper 斑铜矿
peak ①峰②峰值；最大值
peak absorption 峰值吸收
peak absorption method 峰值吸收法*
peak and valley structure 峰谷结构
peak area 峰面积*；谱峰面积
peak area integration 峰面积积分
peak area measurement 峰面积测量法
peak area ratio 峰面积比
peak area response 峰面积响应(值)
peak asymmetry 峰不对称性
peak base 峰底*
peak base width 峰底宽度
peak broadening 峰加宽；峰展宽
peak capacity 峰容量；最高容量
peak center 峰中心点
peak centroid assignment 峰质心测定法
peak coefficient 峰形系数
peak compensation 峰校正
peak concentration 峰收缩
peak content of chromatography 色谱峰容量

peak contraction ①峰收缩②峰密集
peak current 峰电流*
peak cutting 截峰
peak detection 峰检测
peak dispersion 峰分散
peak distortion 峰畸变
peaked casing 胎体褶皱
peak efficiency of unit 设备的最大生产率
peak factor 峰值因数；振幅因数
peak flame temperature 火焰最高温度
peak flatness 峰顶平坦度
peak form 峰形
peak half-width 半峰宽
peak height 峰高*
peak height digitizer 峰高数字转换器
peak height measurement 峰高测量(法)
peak height method 峰高(定量)法
peak height response 峰高响应(值)
peak hopping measurement 峰跳跃测定
peak hopping mode 峰跳跃方式
peak identification 峰鉴别；峰鉴定
peak identifier 峰鉴别器；峰鉴定器
peaking 暗沟；凸顶
peak intensity 峰强度
peak intensity method 峰值强度法；顶点强度法
peak inversion 峰(形)转化
peak jumping technique 峰跳扫法
peak kurtosis 峰峭度；峰突化
peak load 高峰负荷
peak load operation 最高负荷操作
peak matching 峰匹配
peak matching circuit for the mass difference measurement 测量质量差的峰匹配电路
peak matching method 峰匹配法*
peak matching technique 峰匹配技术
peak matching unit 峰匹配单元
peak maximum ①峰顶点②峰最大值
peak melting point 最高熔点
peakness 峰度
peak notation in three dimensional spectra 三维NMR谱峰分类
peak number 峰数
peak-odor 顶香
peak overlapping 峰重叠
peak-peak matching 峰-峰匹配
peak performance 最大生产率
peak pickling 谱线读取
peak point 峰值点

peak polymerization rate	高峰聚合速率；最大聚合速率
peak position	峰位
peak potential	峰电位
peak pressure	最大压力
peak pressure indicator	最大压力指示器
peak profile	峰轮廓
peak rate of conversion	最高转化率
peak ratio	峰比率
peak resolution	峰分辨度
peak response	最大灵敏度；峰值响应
peak search	谱线读取
peak selection	峰选择
peak sensor	峰传感器
peak shape	峰形
peak shear strength	最大剪切强度
peak shift	峰位移
peak shift technique	峰位移技术
peak size	峰大小
peak skew	峰扭曲
peak slope tangent	峰斜切线
peak spreading	峰扩展
peak stability	峰稳定度
peak stress	峰值应力
peak switching	峰切换
peak symmetry	峰对称性
peak synthesis	峰合成
peak tail	峰(拖)尾
peak temperature	峰值温度；出峰温度
peak temperature rise	最高的温度上升
peak time	出峰时间
peak-to-average ratio	峰值与平均值之比
peak to background ratio	峰背之比
peak-to-peak	(正负)峰间的；由最大值到最小值
peak-to-peak amplitude	正反峰间隔值
peak-to-peak current	峰-峰电流
peak to peak linewidth	峰-峰宽度
peak-to-peak noise	(正反)峰间噪声(值)
peak-to-peak potential	峰峰电位
peak-to-trough ratio	峰谷比
peak-to-valley ratio	峰谷比
peak valley	峰谷
peak value	峰值
peak wavelength	峰值波长
peak width	峰宽*
peak width at base	基线处峰宽
peak width at half	半峰宽
peak width at half height	半高峰宽
peak width at inflection point	拐点处峰宽
peak with heading	前延峰
peaky curve	高峰曲线；巅值曲线
pea limestone	豆石；豆石灰石
peanut fiber	花生纤维
peanut hull meal	(落)花生壳粉
peanut oil	花生油
peanut oil fatty acid	花生油脂肪酸
peanut ore	黑钨矿
peanut phosphatide	花生磷脂
pea protein fiber	大豆蛋白纤维
pear essence	梨(子)香精
pear flavour	梨(子)香精
pearl	①灰白色②珍珠；真珠
pearl alum	珠矾；硫酸铝
pearl ash	珍珠灰；粗碳酸钾
pearl black	珠状炭黑
pearlescence	珠光粉
pearlescent effect	珠光效应
pearlescent lacquer	珠光漆
pearlescent pigment	珍珠颜料
pearlescing agent	珠光剂
pearl essence	珠光粉；鱼鳞粉
pearleverlasting	山萩
pearl filler	碳酸钙填料
pearl finish	珠光漆
pearl glue	颗粒胶
pearl grain	珠格令〔重量单位〕
pearl hardening	硫酸钙块
pearliness	珠光；珍珠光泽
pearling agent	珠光剂
pearlite	①珍珠岩②珠光体
pearlitic	珠层(铁)的；珠粒(体)的
pearl mill	珠磨；瓷球球磨机
pearl-necklace	珠链
pearl-necklace model	珠链模型
pearl opal	蓝白色；蛋白石
pearl polymerization(=bead polymerization)	珠状聚合；成珠聚合(法)
pearl spar	白云石
pearl spring model	珠簧模型
pearlstone(=(expanded)pearlite)	珍珠岩；(膨胀)珍珠岩
pearl string	珠串
pearl string model	珠链模型
pearl string reactor	珠串反应器
pearl white	珍珠白〔指:氯氧化铋〕 BiOCl
pearly	①珠状的②珍珠色的
pearly pigment	珠光颜料
pear oil	梨油〔指：乙酸戊酯 $CH_3CO_2C_5H_{11}$〕

pear-seed oil 梨子油
pear-shaped tube 梨形管
Pearson symbol 皮尔松符号
pea stone 豆石;豆石灰石
peat 泥煤;泥炭
peatbog 泥煤沼
peat brick 泥煤砖
peat charcoal 泥煤焦炭
peat coal 泥煤;泥炭
peat coke 泥煤焦炭
peat dust 泥煤尘;泥煤粉
peat fibre 泥炭纤维
peat gas 泥煤气体
peat paraffin 泥煤石蜡
peat tar 泥煤焦油
peat-tar acid 泥煤焦油酸
peat-tar pitch 泥煤沥青
peat wax 泥煤蜡
peaty 泥煤的
pebble 卵石;砾(石);石子
pebble bed 卵石床;卵石热载体床
pebble finish 疙疸漆;斑纹漆
pebble grain 碎石粒纹
pebble graining 搓碎石花
pebble heater 卵石加热器
pebble heater process 卵石加热器过程
pebble mill 卵石球磨机
pebble-oil ratio 卵石-油比〔卵石热载体与原料油比例〕
pebble packing 卵石填料
pebble phosphate 亚磷酸盐团
pebble powder 砾状(火)药
pebbling ①磨成小球②使成珠皮的过程〔革〕
pebbling machine 粒面机〔革〕
pecan oil 胡桃油;佩甘油
peccary ①野猪②野猪手套革
peck 佩克;容量单位〔等于二加仑〕
Peclet number 皮克里特(准)数
Peclet number based on flame speed 火焰速度的皮克里特准数
Peclet number based on jet velocity 喷射速度的皮克里特准数
pectase(=pectolase) 果胶酶
pectate 果胶酸盐
pectenine 梳碱
pectenoxanthin 梳黄质;蛤黄质
pectic acid 果胶酸
pectin 果胶
pectine 果胶

pectin grade 果胶的胶凝度
pectin haze (葡萄酒)果胶混浊
pectinic acid 果胶酯酸
pectin jelly 果胶冻
pectin substance 果胶物质
pectisation 凝结;胶凝作用
pecto-cellulose 果胶纤维素
pectograph 胶干图形
pectography 胶干图形学
pectolase(=pectase) 果胶酶
pectolinarin 果胶里哪苷;大蓟苷;柳穿鱼苷
pectolysis 果胶溶解(作用)
pectose 果胶糖;果蔬胶
pedal ①踏板②脚(的)
Pederson process (for alumina) 彼得森(制氧化铝)法
pedesis 布朗运动
pedestal bearing 支承轴承
pedestal pad 底座
pedestal-type centrifugal compressor 支承式离心压缩机
pedicel 小梗;柄
pedigree 谱系
pedigree probability 系谱概率
peel ①皮②去皮
peelable mask 可剥胶
peel adhesive test 剥离试验
peel angle 剥离角
peel bond strength 剥离强度
peeler 去皮机;剥皮机
peeling 剥离;去皮;剥皮
peeling rate 剥离速度
peeling resistance ①抗剥离力②耐剥离性
peeling strength 剥离强度*
peeling tester 剥离试验仪
peel oil 果皮油
peel ply 剥离层
peel reclaim 胎面再生橡胶
peel strength 剥离强度
peel test 剥离试验
peener 喷丸装置
peening (用锤尖)锤击
peening shot 喷丸
peep door 观察孔
peephole 窥孔;观测孔
Peet-Grady method 皮特-格腊迪法〔测定石油杀虫剂杀虫效果的方法〕
pefloxacin 培氟沙星(药)
peg ①(木)钉;木栓②标柱
peganine(=vasicine) 骆驼蓬碱

peganite 磷矾土
pegmatite 伟晶岩
pegmatitic structure 伟晶(岩)结构
peg switch 栓转电闸
pegu catechu 儿茶
peiminane 贝母烷 $C_{27}H_{45}N$
peimine 贝母素甲；浙贝母碱
peiminine 贝母素乙；去氢浙贝母碱
peiminone 贝母酮 $C_{27}H_{41}O_3N$
peitunitze 白爪〔景德镇所产白土〕
pekea-nut oil 白毫石油
pelagite 海底锰结核
pelargonaldehyde 壬醛 $CH_3(CH_2)_7CHO$
pelargonamide 壬酰胺 $CH_3(CH_2)_7CONH_2$
pelargonane 天竺葵烷
pelargonate 壬酸盐
pelargone 9-十七(烷)酮；天竺葵酮
pelargonic acid(=nonyl acid) 壬酸 $CH_3(CH_2)_7CO_2H$
pelargonic alcohol 壬醇
pelargonic aldehyde 壬醛 $CH_3(CH_2)_7CHO$
pelargonic aldehyde oxime 壬醛肟 $C_8H_{17}CH=NOH$
pelargonic ester 壬酸酯
pelargonidin 天竺葵色素；花葵素 $C_{15}H_{10}O_5HCl$
pelargonidin chloride 氯化天竺葵色素
pelargonin 天竺葵色素苷
pelargonitrile 壬腈 $CH_3(CH_2)_7CN$
pelargonyl(=nonanoyl) 壬酰 $CH_3(CH_2)_7CO—$
N-pelargonyl p-aminophenol N-壬酰基对氨基苯酚
pelargonyl chloride 壬酰氯 $CH_3(CH_2)_7COCl$
pelargonyl vanillylamide 壬酰基香兰酰胺
peliconite rock 水磨土岩
Péligot blue 佩利果特蓝〔一种水合氧化铜〕
Péligot salt 佩利果特盐〔氯铬酸钾〕
Péligot tube 佩利果特 U 形管
peliomycin 佩里霉素
pelite 泥质岩
pelitic 泥质的
pelitic structure 泥岩状结构
pelitization 泥化(作用)
pellagra 糙皮病；玉米红斑病
pellagramin 烟酸
pellagra-preventive factor 抗糙皮病因子；维生素 PP
pellected catalyst 丸状催化剂
pellet ①丸；片②弹丸〔药〕③压(成)丸
pellet bed 丸片床
pellet density 颗粒密度
pellet drier 丸粒(催化剂)干燥器
pellet dryer 丸状(催化剂)干燥器

pelleted catalyst 丸状催化剂
pellet fuel 丸片燃料；球丸燃料
pellet head 造粒机头
pelletierine 石榴碱
pelletierine hydrochloride 盐酸石榴碱
pelletierine sulfate 硫酸石榴碱
pelleting 压丸；压片；造粒
pelleting cup 接丸杯；接片器
pelleting machine 造粒机
pelleting of carbon black 炭黑制片
pelleting press 压丸器；压片器
pelletization 造粒；粒化
pelletized black(=beaded black) 粒状炭黑
pelletizer 造粒机；压片机
pelletizing 造粒；制成丸子
pelletizing drum 制丸鼓
pelletizing process 制丸过程
pellet method 压片法
pellet press 压丸机
pellet technique 压片技术
pellicle ①膜；薄膜②(照相)软片③菌膜；浮膜
pellicular 薄膜的
pellicular adsorbent 薄膜型吸附剂
pellicular exchanger 涂敷薄膜交换剂
pellicular ion-exchanger 薄壳型离子交换剂
pellicular ion exchange resin 薄膜型离子交换树脂
pellicular microbead 薄壳型微珠
pellicular microbead support 薄壳型微珠载体
pellicular packing 薄壳型填充剂*
pellicular resin 涂敷薄膜树脂
pellicular support 薄壳载体
Pellin-Broca prism 佩林-布洛卡棱镜
pellitorine 墙草碱；派立托胺
pellitory root 除虫菊根
pellotine 佩落碱；(墨西哥)仙人掌己碱
pellucidness ①透明(性)②透明度③透光度〔糖〕
pelopium 不纯的铌
peloponium 铌
pelosine 箭毒素
pelotherapy 自然物外用疗法
pelrosilane 芥烷
pelt 毛皮；生皮；裸皮
peltatin 盾叶鬼白素；足叶草脂素
Peltier effect 珀尔帖效应；电-热效应
Peltier heat pump 珀尔帖热泵
peltogynol 盾母醇
peltry ①皮囊②皮货③风箱
pemosors(=multilayer copolymer) 多层共聚物

pemphigic acid　绵虫蜡酸
pemphigus　天疱疮
pemphigus alcohol　绵虫蜡醇
pempidine　1,2,2,6,6-五甲基哌啶；潘必啶〔降压药〕
penagar　含青霉素的5%琼脂溶液
penak　琥珀色(树)脂
penaldic acid　喷醛酸；青霉醛酸
penams　青霉烷类
pen arm　(记录)笔杆
penatin　点青霉素
penbritin　氨苄青霉素
pencil　条；光线锥
pencil hardness test　铅笔硬度试验
pencil lacquer　铅笔漆
pencilled　辐射的
pencil-line flow　射束线流
pencillium corylophilum　顶青霉素
pencil of rays　光流；光束
pencil scratching tester　铅笔硬度试验机
pencil slate　笔板岩
pencil stone　叶蜡石；滑石
pencil test　铅笔试验
pendant chain　侧链
pendant drop method　悬滴法（测定表面张力）
pendant group(=side group; leteral group)　侧基
pendent drop method　悬滴法
pendulum fibre bundle tensile tester　摆式束纤维强力仪
pendulum hammer　摆锤打桩机
pendulum hardness　摆测硬度
pendulum impact test　摆锤式冲击试验
pendulum jigger　摆锤打光机〔革〕
pendulum machine　摆锤打桩机
pendulum oiler　摆锤润滑器
pendulum plastometer　摆锤式塑性计
pendulum press　摆锤压榨机
pendulum rebound　摆锤回弹
pendulum-rocker hardness　摆杆硬度
pendulum roller　摆锤辊子
pendulum rolling machine　摆式压光机
pendulum tester　摆锤式试验仪
pendulum type friction machine　摆锤式摩擦试验机
pendulum viscosimeter　摆锤黏度计
penems　青霉烯类
penetrability　贯穿性；穿透性
penetrable　①可穿透的②(可)渗透的
penetrameter(=penetrometer)　针入式硬度计；透度计
penetrameter sensitivity　透度计灵敏度
penetrant　①渗透剂②穿入物
penetrant test　渗透试验
penetrating agent　渗透剂
penetrating finish　渗透性封闭剂；(木材用)低黏度油(清漆)
penetrating inspection　渗透探伤
penetrating odor　刺激气味；尖锐气味
penetrating oil　渗透(润滑)油
penetrating orbit　贯穿轨道
penetrating power　①贯穿本领〔物〕；穿透力②渗透能力
penetrating power of anion into complex　阴离子的络合能力
penetrating property　渗透性
penetrating quality　渗透性能
penetrating stain(=oil stain)　油性着色剂
penetration　①贯穿；穿透；刺入②针入度③熔深④透入；渗入
penetration asphalt　渗透用沥青
penetration complex　透入配合物
penetration degree　针入度
penetration distance　渗透距离
penetration dyeing　渗透染色
penetration effect　穿透效应*
penetration index　渗透指数
penetration membrane　渗透膜
penetration method　(铺路或建筑用的)沥青渗透法
penetration mode　穿透模式
penetration number　渗透值
penetration of bitumens　沥青的针入度
penetration of electric field　电场渗透
penetration of grease　润滑脂针入度
penetration of heat　热的穿透
penetration rate　焊透率
penetration test　针入度试验
penetration theory　渗透理论
penetration twin　穿插孪晶
penetrative index　针入(度)指数
penetrator　过烧〔焊〕
penetrometer　针入度计；贯穿计；穿透计；刺度计
penetrometer cone　针入度计圆锥体
penetrometer number　针入度测定值
penetron　γ射线穿透仪
penetro-viscometer　针入式黏度计
Penex process　佩内克斯戊烷-己烷异构化法
penfluridol　五氟利多〔药〕
penichromin　色青霉素
penicidin　青霉杀菌素；棒曲霉素
penicillamine　青霉胺
penicillanic acid　青霉烷酸
penicillic acid　青霉酸
penicillin　青霉素

penicillin A 青霉素 A；青霉氧化酶
penicillin dihydro F 戊青霉素
penicillin F 青霉素 F；2-戊烯青霉素
penicillin G 青霉素 G；苄青霉素
penicillin K 青霉素 K；庚青霉素
penicillin sodium 青霉素钠
penicillin V 青霉素 V；苯氧甲基青霉素
penicillin X 青霉素 X；对羟苄青霉素
penicin 6-氨基青霉烷酸
penillonic acid 青霉酮酸
penitaline 意大利青霉素
penner oil(=pennyroyal oil) 胡薄荷油
penniclavine 羽麦角碱
pennine 叶绿泥石
Penning gauge 彭宁规
Penning ionization 彭宁电离
Penning ion trap 彭宁离子阱
penninite 叶绿泥石
pennogenin 喷闹配基
pennolene white oil 白色药用润滑油
pennone 四甲基-2-戊酮
pennyroyal mint 胡薄荷
pennyroyal oil(=penner oil) 胡薄荷油
pennyweight 重量单位
pen recorder 笔式记录器
pen recording polarograph 笔录极谱仪
Penrhyn powder 彭赖恩(炸)药
Pensky-Martens apparatus 彭斯克-马丁测闪点器
Pensky-Martens closed tester 彭斯克-马丁密闭式(石油产品闪点)试验器
Pensky-Martens flash point test 彭斯克-马丁闪点试验
pen speed (记录)笔速
penstock 压头管线
penta〔希腊字头〕 ①五；戊②季戊炸药③五氯酚
penta-acetylated 五乙酰代的
pentaamino〔词头〕 五氨基
pentaaminobenzene 五氨基苯；苯五胺 $(NH_2)_5C_6H$
pentaatomic 五原子的
pentabasic 五元碱性的
pentabasic acid 五元(碱)酸
pentabasic alcohol 五元醇
pentabasic carboxylic acid 五元碱羧酸
pentabasic ester 五元碱酸酯
pentaborane 戊硼烷
pentabromated 五溴化的
pentabromide 五溴化物
pentabrominated 五溴化的
pentabromination 五溴化反应

pentabromizated 五溴化的
pentabromization 五溴化反应
pentabromo- 五溴(代)
pentabromoacetone 五溴丙酮 $Br_3CCOCHBr_2$
pentabromoaniline 五溴苯胺 $C_6Br_5NH_2$
pentabromobenzene 五溴苯 Br_5C_6H
pentabromochlorocyclohexane 五溴氯环己烷〔阻燃剂〕
pentabromocyclopentadienyl 五溴环戊二烯(基)
pentabromodiphenyl ether 五溴二苯醚
pentabromodiphenyl oxide 五溴二苯醚〔阻燃剂〕
pentabromoethane 五溴乙烷 $CHBr_2CBr_3$
pentabromoethyl benzene 五溴乙基苯〔阻燃剂〕
pentabromophenol 五溴苯酚 Br_5C_6OH
pentabromotoluene 五溴甲苯〔阻燃剂〕
pentabutyryl glucose 五丁酰葡萄糖 $C_6H_7O_6(COC_3H_7)_5$
pentacarboxylic acid 五羧酸
pentacene 并五苯
pentacetate 五乙酸盐
pentacetylglucose 五乙酰葡萄糖 $C_6H_7(OCOCH_3)_5$
pentachlorated 五氯化的
pentachloride 五氯化物
pentachlorinated 五氯化的
pentachlorination 五氯化作用
pentachlorizated 五氯化的
pentachlorization 五氯化作用
pentachloro- 五氯(代)
pentachloro-ammine-platinate 五氯氨铂盐
pentachloroaniline 五氯苯胺 $Cl_5C_6NH_2$
pentachlorobenzene 五氯苯 Cl_5C_6H
pentachlorodiamine 五氯二胺〔防老剂〕
pentachloroethane 五氯乙烷 $CHCl_2CCl_3$
pentachloroethylbenzene 五氯乙基苯 $Cl_5C_6C_2H_5$
pentachlorophenol 五氯苯酚〔农药〕
pentachlorophenolate 五氯酚盐
pentachlorophenol dodecyl amine salt 五氯苯酚十二烷基胺盐
pentachlorophenyl ethylene diamine 五氯苯亚乙基二胺
pentachlorophenyl laurate 月桂酸五氯苯酯〔杀菌剂〕
pentachlorophenylmercaptan 五氯硫酚〔塑解剂〕
pentachlorothiophenol 五氯硫酚
pentacid base 五价碱
pentacite 季戊四醇醇酸树脂
pentacontane 五十烷
pentacontyl 五十(烷)基 $CH_3(CH_2)_{48}CH_2-$
pentacoordinate siloxane 五配位硅氧烷
pentacosamic acid 脑酮酸
pentacosandioic acid(=pentacosane diacid) 二十五烷二酸

pentacosane 二十五(碳)烷 $C_{25}H_{52}$
pentacosane diacid(=pentacosandioic acid) 二十五烷二酸 $C_{23}H_{46}(COOH)_2$
pentacosane dicarboxylic acid 二十五烷二羧酸 $C_{25}H_{50}(COOH)_2$
pentacosanoic acid(=pentacosoic acid) 二十五烷酸 $C_{24}H_{49}COOH$
pentacosoic acid(=pentacosanoic acid) 二十五烷酸
pentacosyl 二十五(烷)基 $CH_3(CH_2)_{23}CH_2-$
penta-cyanoaquoferrate 五氰一水合铁酸盐
pentacyanocarbonyl cobaltate(Ⅱ) 五氰一羰合钴(Ⅱ)酸盐
pentacyanocarbonyl ferrate(Ⅱ) 五氰一羰合铁(Ⅱ)酸盐
pentacyanoiron complex 五氰铁复(配)盐
pentacyclic 五环的
pentacyclic compound 五环化合物
pentacyclic ring 五核环
pentad ①五价物②五价元素③五价基
pentadecadienoic acid 十五碳二烯酸 $C_{14}H_{25}COOH$
pentadecanaloxime 十五(烷)醛肟 $C_{14}H_{29}CH=NOH$
pentadecandioic acid(=pentadecane diacid) 十五烷二酸
n-pentadecane 正十五(碳)烷 $CH_3(CH_2)_{13}CH_3$
pentadecane diacid(=pentadecandioic acid) 十五烷二酸 $C_{13}H_{26}(COOH)_2$
pentadecane dicarboxylic acid 十五烷二羧酸；十七二酸 $C_{15}H_{30}(COOH)_2$
pentadecane sulfonate 十五烷基磺酸盐
pentadecanoic acid 十五烷酸 $C_{14}H_{29}COOH$
1-pentadecanol(=pentadecyl alcohol) 十五(烷)醇 $CH_3(CH_2)_{13}CH_2OH$
pentadecanolide 环十五内酯
pentadecanone 十五烷酮
pentadecanoyl 十五(烷)酰 $CH_3(CH_2)_{13}CO-$
pentadecendioic acid(=pentadecene diacid) 十五(碳)烯二酸
pentadecene diacid(=pentadecendioic acid) 十五(碳)烯二酸 $C_{13}H_{24}(COOH)_2$
pentadecene dicarboxylic acid 十五(碳)烯二羧酸；十七(碳)烯二酸
pentadecenic acid(=pentadecenoic acid) 十五(碳)烯酸 $C_{14}H_{27}COOH$
pentadecenoic acid(=pentadecenic acid) 十五(碳)烯酸
pentadecenyl phenol 十五(碳)烯基酚
ω-pentadeclactone ω-十五内酯
pentadecoic acid 十五烷酸 $C_{14}H_{29}COOH$
pentadecyl 十五(烷)基 $CH_3(CH_2)_{13}CH_2-$
pentadecyl acetate 乙酸十五(烷)酯 $CH_3CO_2C_{15}H_{31}$
pentadecyl alcohol(=1-pentadecanol) 十五(烷)醇 $CH_3(CH_2)_{13}CH_2OH$
pentadecyl amine 十五(烷)胺 $C_{15}H_{31}NH_2$

2-pentadecylbenzoquinone dioxime 2-十五烷基苯醌二肟〔硫化剂〕
pentadecyl bromide(=1-bromopentadecane) 十五(烷)基溴；1-溴十五烷 $CH_3(CH_2)_{13}CH_2Br$
pentadecyl caproate 己酸十五(烷)酯 $C_5H_{11}CO_2C_{15}H_{31}$
pentadecyl dimethyl benzyl ammonium 十五烷基二甲基苄基铵
pentadecylendioic acid(=pentadecylene diacid) 十五(碳)烯二酸
pentadecylene diacid(=pentadecylendioic acid) 十五(碳)烯二酸 $C_{13}H_{24}(COOH)_2$
pentadecylene dicarboxylic acid 十五(碳)烯二羧酸；十七(碳)烯二酸 $C_{15}H_{28}(COOH)_2$
pentadecylenic acid 十五(碳)烯酸 $C_{14}H_{27}COOH$
pentadecylic acid 十五(烷)酸 $C_{14}H_{29}COOH$
pentadecylic alcohol 十五烷醇
3-pentadecyl-4-nitrosophenol 3-十五烷基-4-亚硝基苯酚〔防老剂〕
pentadecynic acid 十五(碳)炔酸 $C_{14}H_{25}COOH$
2,4-pentadienal 2,4-戊二烯醛
pentadienal 戊二烯醛
2,3-pentadiene(=dimethyl allene) 2,3-戊二烯；二甲基丙烯
pentadiene(=pentylene) 戊二烯亚戊基
1,3-pentadiene(=piperylene) 1,3-戊二烯；戊间二烯 $CH_3CH=CHCH=CH_2$
pentadiene carboxylic acid 山梨酸
pentadienide anion 戊二烯负离子
pentadienoic acid 戊二烯酸 C_4H_5COOH
pentadigalloylglucose 双棓酰单宁；五双鞣酰基葡萄糖
pentadiine 戊二炔
1,2-pentadiol 1,2-戊二醇 $C_3H_7CHOHCH_2OH$
1,5-pentadiol(=pentamethylene glycol) 1,5-亚戊基二醇；1,5-戊二醇 $CH_2(CH_2CH_2OH)_2$
pentaene 五烯
pentaerythrite 季戊四醇 $C(CH_2OH)_4$
pentaerythrite tetra-acetate 季戊四醇四乙酸酯 $C(CH_2O_2CCH_3)_4$
pentaerythrite tetranitrate 季戊四醇四硝酸酯
pentaerythritol 季戊四醇 $C(CH_2OH)_4$
pentaerythritol abietate 季戊四醇松香酸脂
pentaerythritol acrylate polyfunctional monomer 季戊四醇丙烯酸酯多官能单体
pentaerythritol alkyl ether 季戊四醇烷基醚
pentaerythritol diester 季戊四醇二酯
pentaerythritol diphosphate 二磷酸季戊四醇酯
pentaerythritol ester 季戊四醇酯
pentaerythritol fatty ester 季戊四醇脂肪酸酯

pentaerythritol maleic resin　马来酸-季戊四醇树脂
pentaerythritol monolaurate　季戊四醇单月桂酸酯
pentaerythritol monooleate　季戊四醇单油酸酯
pentaerythritol monopalmitate　季戊四醇单棕榈酸酯
pentaerythritol phosphite　季戊四醇亚磷酸酯
pentaerythritol phthalic resin　邻苯二甲酸-季戊四醇树脂
pentaerythritol resin　季戊四醇树脂〔增黏剂〕
pentaerythritol stearate　季戊四醇硬脂酸酯
pentaerythritol tallate　松浆(油)酸季戊四醇酯
pentaerythritol tetracrylate　四丙烯酸季戊四醇酯〔硫化剂〕
pentaerythritol tetrakis(diphenyl phosphite)　季戊四醇四(亚磷酸二苯酯)
pentaerythritol tetramyristate　四肉豆蔻酸季戊四醇酯
pentaerythritol tetrastearate　四硬脂酸季戊四醇酯〔硫化剂〕
pentaerythritol tetrathioglycollate　四巯基乙醇酸季戊四醇酯
pentaerythritol triacrylate　三丙烯酸季戊四醇酯〔硫化剂〕
pentaether　五醚；四甘醇二丁醚　$C_4H_9O(C_2H_4O)_4C_4H_9$
pentaethyl-antimony　五乙锑　$Sb(C_2H_5)_5$
pentaethylated　五乙基化的
pentaethyl benzene　五乙基苯　$(C_2H_5)_5C_6H$
pentaethylenehexamine(=penten)　五亚乙基六胺　$[(NH_2CH_2CH_2)_2NCH_2—]_2$
pentafluorated　五氟化的
pentafluoride　五氟化物
pentafluorinated　五氟化的
pentafluorination　五氟化反应
pentafluorizated　五氟化的
pentafluorization　五氟化反应
pentafluorodimethylheptanedione(PFD-MHD)　五氟二甲基庚二酮　$CF_3CF_2COCH_2COC(CH_3)_2CH_3$
pentafluoroelastomer　五氟类橡胶
pentafunctional initiator　五功能起始剂
pentafungin　戊抗真菌素
pentagalloyl glucose　黄棓酰单宁；五没食子酰葡萄糖
pentaglucose　戊糖
pentaglycine　五甘氨酸
pentaglycol　2,2-二甲基丙二醇；五甘醇
pentagonal bipyramid　五角双锥
pentagraph　缩放仪；比例绘图器
pentahalide　五卤化物
pentahalogenated benzene(=pentahalogeno-benzene)　五卤代苯
pentahalogeno-benzene(=pentahalogenated benzene)　五卤代苯　C_6HX_5
pentahapto　五合〔例：二茂铁 ferrocene 类〕
pentahedron　五面体

pentahydrate　五水合物
pentahydric acid　五(碱)价酸
pentahydric alcohol　五元醇
pentahydro-〔词头〕五氢
pentahydroxy-〔词头〕五羟(代)
pentahydroxy-acid　五羟基酸　$(OH)_5RCOOH$
pentahydroxy-alcohol　五元醇
pentahydroxy-benzene　苯五酚　$C_6H(OH)_5$
pentahydroxybenzophenone　五羟基二苯(甲)酮　$(HO)_2C_6H_3COC_6H_2(OH)_3$
pentahydroxy dibasic acid　五羟二(碱)价酸　$(OH)_5R(COOH)_2$
pentahydroxyl compound　五羟基化合物
pentahydroxy monobasic acid　五羟一(碱)价酸　$(OH)_5RCOOH$
pentaiodated　五碘化的
pentaiodide　五碘化物
pentaiodinated　五碘化的
pentaiodination　五碘化反应
pentaiodizated　五碘化的
pentaiodization　五碘化反应
pentaiodo〔词头〕五碘
pentaiodobenzene　五碘苯　C_6HI_5
pentaiodoethane　五碘乙烷　CHI_2CI_3
pentakis〔词头〕五个
pental　三甲基乙烯
pentalandite(=pentlandite)　镍黄铁矿〔矿〕；硫镍铁矿
pentalane　并环戊烷
pentalarm　戊硫醇混合物
pentaldol　戊醛醇〔俗〕；2,2-二甲基-3-羟基丙醛　$(CH_3)_2C(CHO)CH_2OH$
pentalene　并环戊二烯
pentalin　五氯乙烷
pentaline　双级旋片泵
pentalkylated　五烃基代的
pentalobal cross section　五叶形(横)截面
pentamer　五聚物；五节聚合物
pentamethine cyanine　五甲炔花青；五甲川菁
pentamethoxyl red　五甲氧基红
pentamethyl-〔词头〕五甲基
pentamethylaminobenzene　五甲基苯胺　$(CH_3)_5C_6NH_2$
pentamethylated　五甲基代的
pentamethylbenzene　五甲基苯　$(CH_3)_5C_6H$
pentamethylbenzoic acid　五甲代苯酸　$(CH_3)_5C_6CO_2H$
pentamethylene　1,5-亚戊基　$—CH_2(CH_2)_3CH_2—$
pentamethylene bromide　1,5-二溴戊烷
pentamethylene diamine　戊二胺；尸胺
pentamethylene glycol(=1,5-pentadiol)　1,5-亚戊基二醇；

1,5-戊二醇　$CH_2(CH_2CH_2OH)_2$
pentamethylene oxide　氧己环　$(CH_2)_5O$
pentamethylene sulfide　五甲撑硫；硫杂己环
pentamethylene tetramine　亚戊基四胺；五甲撑四胺
pentamethylene tetrazole　亚戊基四唑
　　　$N(CH_2)_5C=(N_3)$
pentamethyloctadecyl-1, 3-propane-diammonium dichloride　二氯化五甲基十八烷基-1,3-丙二铵
pentamethylol melamine　五羟甲基三聚氰胺
pentamethylpararosaniline(=methyl violet)　五甲基副品红；甲基紫
pentamethylphenol　五甲基苯酚　$(CH_3)_5C_6OH$
pentamidine　喷他脒〔药〕
pentamidine isothionate　异硫代羟酸五脒
pentammine　五氨合物
pentamycin　戊霉素
pentanal　戊醛
pentandioic acid　戊二酸
1,5-pentandiol diacrylate　1, 5-戊二醇二丙烯酸酯
pentane　戊烷
pentane carboxylic acid　己酸
pentane diacid　戊二酸　$C_3H_6(COOH)_2$
pentane dicarboxylic acid　戊二羧酸；庚二酸　$C_5H_{10}(COOH)_2$
pentanedioic acid　戊二酸
1,5-pentanediol　1,5-戊二醇〔硫化剂〕
pentanediol　戊二醇
2,4-pentanedione　2,4-戊二酮；戊间二酮；乙酰丙酮
2,3-pentanedione　2, 3-戊二酮
pentane lamp　戊烷灯
pentanes plus　戊烷以上的烃
pentane thermometer　戊烷温度计
pentanethiol　戊硫醇　$Me(CH_2)_4SH$
pentanitrophenol　五硝基苯酚　$C_6(NO_2)_5OH$
pentanitrophenol ether　五硝基苯酚醚
pentanizer　戊烷馏除器
pentanizing　戊烷化
pentanizing column(=pentanizing tower)　戊烷馏除塔
pentanizing tower(=pentanizing column)　戊烷馏除塔
pentanoate　戊酸盐(酯或根)
pentanoic acid　戊酸　C_4H_9COOH
pentanol　戊醇
2-pentanone(=methyl n-propyl ketone)　2-戊酮；甲基丙基(甲)酮
2-pentanone oxime　2-戊酮肟
pentan-thiol　戊硫醇
pentanuclear(=pentanucleate; pentanucleated)　五环的；五核的
pentanucleate(=pentanuclear)　五环的；五核的
pentanucleated(=pentanuclear)　五环的；五核的
pentaose　戊糖
pentaoxacyclooctadecane　五氧杂环十八烷　$C_{13}O_5H_{26}$
pentaphene　戊芬；二苯并[b,h]菲
pentaphenylethane　五苯基乙烷　$(C_6H_5)_5C_2H$
pentapropyl glucose　五丙基葡萄糖
penta resin　季戊四醇酯胶树脂〔由松香和季戊四醇制成〕
pentaricinoleic acid　五聚蓖酸
pentarylated　五芳基代的
pentasodium triphosphate　三聚磷酸(五)钠
pentasol　工业戊醇
penta-substitution product　五取代产物
pentasulfide　五硫化物
pentathionic acid　连五硫酸　$H_2S_5O_6$
pentathiotriphosphoric acid　五硫代三磷酸
pentatomic　五原子的；五元的
pentatomic acid　五(碱)价酸
pentatomic alcohol　五元醇
pentatomic base　五价碱
pentatomic phenol　五元酚
pentatomic ring　五节环；五原子环
n-pentatriacontane　(正)三十五(碳)烷　$CH_3(CH_2)_{33}CH_3$
18-pentatriacontanone(=stearone)　硬脂酮；18-三十五(烷)酮　$C_{35}H_{70}O$
pentavalence　五价
pentavalent　五价的
pentavalent alcohol　五元醇
pentazane　吡咯烷
penten(=pentaethylenehexamine)　五亚乙基六胺　$[NH_2CH_2CH_2)_2NCH_2—]_2$
pentazdiene　五氮二烯　$HN=NNHN=NH$
pentazocine　镇痛新
pentazole　五唑　$N=NN=NNH$
pentazolyl　五唑基　$N=NN=NN—$
pentazyl　五唑基
pentelide　第五族元素化物
pentels　第五族元素
2-pentenal　2-戊烯醛
pentenate　戊烯酸盐(酯)
pentendioic acid(=pentene diacid)　戊烯二酸
1-pentene(=α-amylene)　1-戊烯　$C_2H_5CH_2CH=CH_2$
2-pentene(=β-amylene)　2-戊烯
pentene diacid(=pentendioic acid; pentenedioic acid)　戊烯二酸　$C_3H_4(COOH)_2$

pentene dicarboxylic acid　戊烯二羧酸；庚烯二酸　$C_5H_8(COOH)_2$
pentenedioic acid(=pentene diacid)　戊烯二酸
3-pentene-1-yne(=pirylene)　3-戊烯-1-炔
pentenic acid　戊烯酸　C_4H_7COOH
pentenoic acid　戊烯酸
pentenol　戊烯醇
pentenomycin　环戊烯霉素
pentenyl　戊烯基；2-戊烯基　$CH_3CH_2CH=CHCH_2-$
pentethide　五乙基化物
pentevalent　五价的
penthrite(=tetranitro-pentaerythrite)　季戊四醇四硝酸酯；季戊炸药
pentinic acid(=α-ethyltetronic acid)　喷亭酸；α-乙基季酮酸　$C_2H_5C_4H_3O_3$
pentinoic acid　戊炔酸　$CH_3CH_2C≡CCOOH$
pentite　戊五醇
pentitol　戊五醇；戊糖醇　$CH_2OH(CHOH)_3CH_2OH$
pentlandite(=pentalandite)　镍黄铁矿〔矿〕；硫镍铁矿
pentobarbital　戊巴比妥
pentobarbitone　戊巴比通〔药〕
pentoic acid　戊酸　C_4H_9COOH
-pentol（词尾）　五醇
pentolinium tartrate　酒石酸喷托铵〔药〕
pentonic acid　戊糖酸　$CH_2OH(CHOH)_3COOH$
pentopyranose　吡喃戊糖；吡喃糖
pentosamine　戊糖胺
pentosan　多缩戊糖；戊聚糖
pentosanase　戊聚糖酶
pentosan hydralysis　戊聚糖水解
pentosazone　戊糖脎
pentose　戊糖
pentoseenes　戊糖烯
pentostam(=sodium stibogluconate)　葡萄糖酸锑钠
pentoxazone　环戊噁草酮；噁嗪酮〔除草剂〕
pentoxide　五氧化物
pentoxifylline　己酮可可碱〔药〕
pentoxime　戊肟〔俗〕；环戊五酮肟
pentoxy　戊氧基
pentoxyverine　喷托维林〔药〕
pentravel　指针移动
pentriacontane　(正)五十(碳)烷
pentrite　四硝基赤藓醇
penturonic acid　戊糖醛酸
pentyl　①戊(烷)基　$CH_3(CH_2)_3CH_2-$②季戊炸药；奔斯乃特；奔梯耳
iso-pentyl　异戊基　$(CH_3)_2CHCH_2CH_2-$
sec-pentyl　仲戊基　$CH_3CH_2CH_2CH(CH_3)-$

tert-pentyl　叔戊基　$CH_3CH_2C(CH_3)_2-$
pentyl acetate　乙酸戊酯
2-pentyl acrolein　2-戊基丙烯醛
pentyl alcohol　戊醇
pentyl amine　戊胺
pentyl butyrate　丁酸戊酯
α-pentylcinnamaldehyde　α-戊基肉桂醛
α-pentylcinnamyl acetate　乙酸-α-戊基肉桂酯
α-pentylcinnamyl alcohol　α-戊基肉桂醇
α-pentylcinnamyl formate　甲酸-α-肉桂酯
pentylene(=pentadiene)　戊二烯亚戊基
pentyl formate　甲酸戊酯
2-pentyl furan　2-戊基呋喃
pentyl-2-furyl ketone　戊基-2-呋喃基酮
pentylhexamethyleneimine　戊代(环)六亚甲基亚胺
pentyl hexanoate　己酸戊酯
pentylidene　亚戊基　$CH_3(CH_2)_3CH=$
pentylidyne　次戊基　$CH_3(CH_2)_3C≡$
pentyloxy　戊氧基　$CH_3(CH_2)_3CH_2O-$
2-pentyl pyridine　2-戊基吡啶
n-pentyl vinyl carbinol　正戊基乙烯基甲醇
pentyne　戊炔　C_5H_8
pentynoic acid　戊炔酸　$C_2H_5C≡CCOOH$
penultimate effect　前末端基效应
penultimate unit　前末端单元
Penzold's reagent　彭若德糖试剂
peonin　甲基花青苷；芍药色素苷
peonine　芍药碱
peonol　芍药醇　$C_9H_{10}O_3$
people's gas mask　民用防毒面具
peperite　红褐色火山凝灰岩
pepper box　有虻眼的皮革
pepper cress oil　独行菜油
pepper leaf oil　胡椒叶油
peppermint　薄荷
peppermint camphor　薄荷醇；3-萜醇　$C_{10}H_{19}OH$
peppermint oil　薄荷油
peppermint water　薄荷水
pepper oil　胡椒油
pepper oleoresin　胡椒油树脂
pepper-poison gas　胡椒毒气
peppery(=bitty)　涂膜上凸起的小块或颗粒
pepsase　胃蛋白酶
pepsin　胃蛋白酶
pepthiomycin　肽硫霉素
peptic　①胃的②胃蛋白酶的
peptidase　肽酶
peptide　肽*

peptide bond 肽键
peptide chain 肽链
peptide sequencer 肽序列(分析)仪
peptides mapping 肽图谱
peptidoglycan 肽葡聚糖；黏肽
peptimycin 肽霉素
peptisation 解胶
peptizate 胶溶体
peptization 胶溶*
peptization by resins (树脂)胶溶作用
peptizator 胶溶剂；胶化剂
peptized fuel 胶溶燃料
peptized rubber 增塑橡胶
peptizer 塑解剂*；胶溶剂
peptizing 胶溶
peptizing action 胶溶作用
peptizing agent 胶溶剂；胶化剂
peptizing power 胶溶能力
peptizing property 胶溶性质
peptonization 陈化(作用)
peptonizing tube 渗透析胨管
per- 〔拉丁字头〕 ①高②过③全
perabrodil 碘司特〔2,2′-二羟基二乙胺和 3,5-二碘-4-吡啶酮-N-乙酸的盐〕
peracetic acid 过乙酸 CH_3CO_3H
peracetylated rubber 过乙酰化橡胶
peracid 过酸*
peracid decomposition 过酸分解
peracid ester 过氧酸酯
peracidity 过酸性
peracid salt 过(氧)酸盐
perbasic 高碱性的
perbenzoic acid (=peroxybenzoic acid) 过苯甲酸 $C_6H_5CO_2OH$
perborate 过硼酸盐 $MBO[O_2]; M_2B_4O_6[O_2]$
perborate bleaching 过硼酸盐漂白
perboric acid 过硼酸 $HBO[O_2]$
perboric acid soda 过硼酸钠
perbromate 高溴酸盐 $MBrO_4$
perbrome-acetone 全溴丙酮；六溴丙酮 CBr_3COCBr_3
perbromide 过溴化物
perbromo- 〔词头〕 全溴(代)〔表示所有 C 上的 H 全被 Br 取代了，但保持官能团特性所必须的 H 除外〕
perbromo-carbon 全溴化碳
perbromo-ethane 全溴乙烷；六溴乙烷 CBr_3CBr_3
perbromo-ether 全溴乙醚 $(C_2Br_5)_2O$
perbromo-ethylene 全溴乙烯；四溴代乙烯 $CBr_2=CBr_2$

perbromo-hydrocarbon 全溴代烃
perbromo-propionaldoxime 全溴丙醛肟 $CBr_3CBr_2CH=NOH$（注意：$CH=NOH$ 中的 H 并未被取代）
percaine 辛可卡因；拍卡因〔一种局部麻醉剂〕
percale 高密度薄纱类织物
percamin 拍卡民〔局部麻醉药〕
percarbide 过碳化物
percarbonate 过碳酸盐 $M_2C_2O_6$
percarbonic acid 过碳酸 $H_2C_2O_6$
percarboxylate 过(氧)羧酸酯(盐)
per cent 百分之(几)；每一百；百分中
percent absorption 吸光百分率
percent abundance 百分率(相对)分布量
percentage 百分数；百分率
percentage absolute humidity 绝对湿度百分数
percentage by volume 体积百分数
percentage by weight 重量百分数
percentage composition 百分(数)组成
percentage concentration 百分(比)浓度
percentage conversion 转化百分数；转化率
percentage crystallinity 结晶百分率
percentage elongation 百分伸长
percentage error 百分(数)误差；误差百分数
percentage excess reactant 过剩反应物百分数
percentage extraction (百分)萃取率
percentage humidity 百分(数)湿度；湿度百分数；相对湿度
percentage liquid phase 液相百分数
percentage loss of weight 减量百分率；失重百分率
percentage moisture content 含水率
percentage of blister 起泡的百分率〔漆病〕
percentage of carbon residue 残碳(含量)百分数
percentage of d-character d 特征百分率
percentage of elongation 伸长率
percentage of free water 游离水含率
percentage of grafting 接枝率
percentage of permissible deviation 容许偏差(百分)率
percentage of variation 不匀率
percentage of water hold 含水率
percentage point 分位点；百分点
percentage reduction in area 断面收缩率
percentage relative humidity 百分相对湿度；相对湿度百分数
percentage standard deviation 百分标准(偏)差
percentage test 百分比测定
percentage transmission 透射百分率
percent alcoholysis 醇解率
percent by volume 体积百分数；体积百分比；容量百

分比
percent by weight 重量百分数
percent conversion 转化率
percent crystallinity 结晶度
percent error 百分误差
percent extraction 萃取率
percent hydrolysis 水解率
percent of grade 斜度百分数
percent pigment volume 颜料体积百分比
percent polymerization 聚合率
percent recovery 回收率；总馏出率
percent solid 固体含量
percent test 挑选试验
percent thermal shrinkage 热收缩率
percent transmission(=percent transmittance) 透光百分率
percent transmittance(=percent transmission) 透光百分率
percent transmittancy 透光百分率
percent unsaturation 不饱和率
percent volume solids(=solids by volume) 固体体积百分率
perceptron 感知机
perch ①皮革的致柔②皮革破绽的弥补
perching 铲软（皮革）
perching knife 铲刀
perching machine 铲软机
perchlorate 高氯酸盐 $MClO_4$
perchlorate explosive 高氯酸盐炸药
perchlorate extraction system 高氯酸盐萃取体系
perchlorate lead-refining process 高氯酸盐炼铅法
perchloratocerate 高氯酸根合铈酸盐
perchlorethane 全氯乙烷
perchlorether 全氯乙醚
perchlorethylene 全氯乙烯
perchloric acid 高氯酸 $HClO_4$
perchloric acid anhydride 七氧化二氯；高氯酸酐
perchloric acid dihydrate 二水合高氯酸
perchloric acid ether 高氯酸乙酯
perchloric acid hydrate 水合高氯酸
perchloride 高氯化物〔含氯最多者〕
perchlorinated polyvinyl chloride 全氯代聚氯乙烯
perchlorination 全氯化
perchlorizing 过氯化；全氯化
perchloro 全氯(代)〔代替所有 C 上的氢〕
perchloro-butadiene 全氯丁二烯 C_4Cl_6
perchloro-carbon 全氯化碳
perchloro-ethane 全氯乙烷 C_2Cl_6
perchloro ether 全氯乙醚 $Cl_5C_2OC_2Cl_5$
perchloro-ethylene 全氯乙烯 $CCl_2=CCl_2$
perchloro-hydrocarbon 全氯代烃
perchloro methyl-mercaptan 全氯甲硫醇 Cl_3CSCl〔注意：SH 中的 H 也被取代了〕
perchloroparaffin 全氯化石蜡
perchloropentacyclodecane 全氯五环癸烷〔阻燃剂常与 Sb_2O_3 并用〕；灭蚁灵
perchloropropane(=octachloropropane) 八氯丙烷；全氯丙烷 C_3Cl_8
perchloryl 全氯酸基
perchloryl fluoride 过氯酰氟
perchromate 过铬酸盐
perchromic acid 过铬酸
perclene 四氯乙烯；全氯乙烯
Perco catalytic disulphurization process 汽油催化脱硫的培柯过程
Perco copper sweetening 培柯铜脱硫(法)
Perco cycloversion catalytic reforming 培柯循环转化的催化重整
Perco dehydrogenation process 培柯催化脱氢过程
Perco HF alkylation 培柯氟化氢(催化)烷基化过程
percolate ①渗出液②滤出液
percolater 渗滤器
percolating ①渗透②滤过几层的吸着物
percolating clay(=percolation clay) 渗滤白土
percolating filter 渗滤器
percolation 渗滤
percolation clay(=percolating clay) 渗滤白土
percolation deasphalting method 渗滤脱沥青法
percolation extractor 滤过式浸出器
percolation filter 渗滤器
percolation filtration 渗滤
percolation method 渗滤法
percolation process 渗滤法
percolation tank 渗滤槽
percolation test 渗透试验
percolation threshold （炭黑导电塑料）突增界限(值)；渗透阈值
percolation treatment 渗滤处理
percolation vat 渗滤桶
percolation yield 渗滤产率
percolator 过滤器；渗滤器
percolumbic acid 过铌酸 $HNbO_2[O_2]$
percrystallization 透析结晶(作用)
percussion 碰(撞)；撞击；打击
percussion cap 碰炸帽；撞击帽
percussion figure 撞击(裂纹)图像
percussion fuse 碰炸引信；击发引信
percussion mortar 桩臼

percussion powder(=fulminating powder) 雷爆火药
percussion primer 碰炸起爆管；撞击起爆管
percussion welder 储能焊机
percyanoolefine 全氰烯烃
perdeuterated 全氘化的
perdeuterated fatty acid 全氘化脂肪酸
perdeuterated organic compound 全氘化有机化合物
perdeuterioalkane 全氘烷
perdeuteriomethane 全氘甲烷 CD_4
perdeuteroalkane 全氘烷
perdeuteromethane 全氘甲烷
perdistillation 透析蒸馏(作用)
perdurability ①时间；延续时间②耐久性；持久性
peredimycin(=peresimycin) 贝勒霉素
pereirine(=geissoschizoline) 缝籽木早灵；缝籽碱
pereiro bark 缝籽木皮
peresimycin(=peredimycin) 贝勒霉素
perester 过酸酯*
perester radical (过氧)酸酯基
perezinone 墨西哥菊酮
perezol 墨西哥金菊酸酒精溶液
perezon 墨西哥金菊酸
perezone 三褶菊酮
perfect black body 理想黑体；完全黑体；绝对黑体
perfect combustion 完全燃烧
perfect condition 理想状态
perfect crystal 全整晶体
perfect discharge 完全出料法
perfect elasticity 理想弹性；完全弹性
perfect fluid 理想流体；完全流体
perfect focusing mass spectrometer 完全聚焦质谱计
perfect gas 理想气体
perfection dust collector 十全聚尘机
perfection of crystal 晶体的完整性
perfection of lattice 晶格的完整性
perfect lubrication 完全润滑
perfectly elastic 完全弹性的
perfectly elastic body 完全弹性(物)体
perfectly elastic collision 完全弹性碰撞
perfectly oriented structure 完全取向结构
perfect mixing 完全混合
perfect plasticity 理想塑性；完全塑性
perfect plate 理想板
perfect radiator 完全辐射体〔物〕
perfect simple distillation 理想简单蒸馏
perfect solution 完美溶液*
perfect vacuum 完全真空；绝对真空
perfect white body 理想白体；完全白体；绝对白体

perfect white diffuser 全白漫射体
perferrate 高铁酸盐 M_2FeO_4
perferrite 亚铁酸盐
Perflex process 派尔弗莱克斯法
perfluocarbon 全氟化碳
perfluoride 过氟化盐；全氟化盐
perfluorinated aliphatic acid 全氟脂肪酸
perfluorinated ethylene-propylene copolymer 四氟乙烯-六氟丙烯共聚物
perfluorination 全氟化(作用)
perfluorine oil 全氟烷烃油
perfluoro 全氟(代)
perfluoroacryloyl fluoride 全氟丙烯酰氟
perfluoroaliphatic group 全氟脂族基
perfluoroalkane oil 全氟烷油
perfluoro alkene 全氟链烯烃
perfluoro(alkoxy alkane)(PFA) 全氟烷氧基树脂
perfluoroalkyl 全氟烃基
perfluoro alkyl alkylene mercapto group 全氟烷基亚烷基巯基
perfluoro alkyl amidine 全氟(代)烷基脒
perfluoroalkylation 全氟烷基化(作用)
perfluoroalkylene triazine 全氟亚烷基三嗪；全氟撑三嗪
perfluoro alkyl sulfonic acid 全氟(代)烷基磺酸
perfluoro alkyl sulfonyl amido alkanol 全氟烷基磺酰氨基(链烷)醇
perfluoro alkyl triazine polymer 全氟烷基三嗪聚合物〔橡胶〕
perfluoroallene 全氟丙二烯
perfluorobutyramidine 全氟丁脒
perfluoro caprylic acid 全氟辛酸
perfluoro-carbon 全氟化碳
perfluorocarbon fibre 全氟(化)碳纤维；碳氟纤维
perfluoro carboxylic acid 全氟(烷基)羧酸
perfluoro-compound 全氟化物；过氟化物
perfluorodecalin 全氟萘烷 $C_{10}F_{18}$
perfluoro decanyl ether 全氟癸基醚
perfluorodimethylcyclohexane 全氟二甲基环己烷 C_8F_{16}
perfluoro-ether 全氟乙醚 $(C_2F_5)_2O$
perfluoro ethylene 全氟乙烯；四氟乙烯
perfluoro (ethylene-propylene) 全氟(乙烯-丙烯)塑料
perfluoroglutarodiamidine 全氟戊二脒
perfluoro-hydrocarbon 全氟代烃
perfluorokerosene(PFK) 全氟煤油
perfluoro-2-methylbicyclo[4.4.0] decane(PFMD) 全氟-2-甲基双环[4.4.0]癸烷
perfluoromethyl cyclohexane(PFMCH) 全氟甲基环己烷 C_7F_{14}

perfluoro octane sulfonate 全氟辛烷磺酸盐
perfluoro octane sulfonyl fluoride 全氟辛烷磺酰氟
perfluoro octane sulphonate(PFOS) 全氟辛烷磺酸酯
perfluoro octyl phosphate salt 全氟辛基磷酸盐
perfluoroorganometallic compound 全氟有机金属化合物
perfluoroparaffin 全氟(代)烃
perfluoropolymer 全氟聚合物；过氟化聚合物
perfluoropropane 全氟丙烷
perfluoro propylene 全氟丙烯
perfluorotributylamine 全氟三丁胺 $(C_4F_9)_3N$
perforated 多孔的；冲孔的
perforated bottom 多孔底
perforated cage filter 多孔笼式过滤器
perforated drum heat-setter 圆网式热定型机；多孔辊筒式热定型机
perforated grain powder 空心条药
perforated pipe 多孔管
perforated plate 多孔板；滤板
perforated plate column (多层)孔板(蒸馏)塔
perforated plate distillation column 孔板蒸馏塔
perforated plate tower 孔板塔
perforated plywood 有孔胶合板
perforated roller 多孔卷轴；网眼辊
perforated rotating disc column(PRDC) 多孔转盘(萃取)柱
perforated screen 多孔板筛
perforated steam spray 喷汽器
perforated tape 穿孔带
perforated tape controlled 穿孔纸带控制的
perforated tray 多孔板塔盘
perforated wall centrifuge 多孔式离心机
perforated washer 多孔水洗器
perforated water spray 喷水器
perforating action 成孔作用
perforating machine 穿孔机；打眼机
perforation 穿孔
perforator 穿孔机；打眼机
performance ①性能；特性②演绎
performance additive 改性助剂
performance characteristic 工作特性；操作特性
performance chart 操作性能图；特性图
performance curve 特性曲线
performance enhancers for the polymer industry 聚合物增性剂
performance index 特性指数；性能指数
performance number 功率值
performance parameter 性能参数
performance plastic 功能塑料
performance-proved product 使用合格的制品
performance test ①性能试验②运转试验
performed polymer 预聚体
performic acid 过甲酸
performing 预成型；压片
Perform tray 泊尔佛姆塔板〔具有碎流挡板的斜孔推液式塔板〕
perfringens 气性坏疽抗毒素
perfume ①香料②香水③香味
perfume base 香基
perfume-blotter 闻香(纸)条
perfume chemistry 香料化学
perfume composition ①香精②香料组分
perfume compound 香精
perfume concrete 香料吸着剂
perfumed article 加香制品
perfumed base 香料底子
perfumed card 香(纸)片
perfumed incense 香熏剂
perfumed ribbon 香带
perfumed soap 香皂
perfume emulsion 乳液香精
perfume fixative 香料固定剂
perfume formulation 香精配方
perfume ingredient 香精成分；香精组分
perfume material 香料
perfume oil 芳香油
perfume preserving agent 香精保藏剂
perfumer ①香料制造者②香剂
perfumery ①香料；香水；香物②香料制造厂
perfumery compound 香精
perfumery oil 香料油
perfumery tincture 香料酊剂
perfuming 加香
perfuming cosmetic 香妆品
perfuming detergent 加香洗涤剂
perfusion 灌注(法)
perfusion chromatography 灌注色谱法；灌流色谱法；灌注层析
perfusion chromatography packing 灌注色谱填料
pergamyn 羊皮纸
pergamyn paper 耐油纸
pergenol 过硼酸钠和酒石酸氢钠混合物
pergolide 培高利特〔药〕
perhalide 全卤化物
perhalocarbon(=perhalohydrocarbon) 全卤化碳
perhalogenation 全卤化(作用)
perhalogeno 全卤(代)〔代替所有 C 上的氢〕

perhalogenomethane　全卤甲烷
perhalohydrocarbon(=perhalocarbon)　全卤化碳
perhydrate　过氧化氢合物*
perhydridase　过氢酶
perhydride　过氢化物
perhydro-〔词头〕　全氢化
perhydroanthracene　蒽烷；全氢化蒽
perhydroaromatic system　全氢化芳香系
perhydrocarotene　全氢胡萝卜烯
perhydrocrocetin　全氢藏花酸
perhydrocyclopentanophenanthrene　环戊烷多氢菲
perhydrogenated rosin　全氢化松香
perhydrol　强双氧水〔含30%过氧化氢〕
perhydronorbixin　全氢降红木素
perhydrophenanthrene　菲烷；全氢化菲
perhydrosqualene　全氢化角鲨烯
perhydrotriazole thione derivative　全氢三唑硫酮衍生物
perhydrotriphenylene　全氢三苯撑
perhydrovitamin A　全氢(化)维生素A
peri-〔希腊字头〕　①迫；(邻)近；环绕②萘环的1,8-或4,5-位
peri-acid　迫位酸（俗）；8-氨基-1-萘磺酸
perianthopodin　坏安坡定；花被足定
peri-bridge　迫位桥
pericarp　果皮
periclase　方镁石
pericline　钠长石
peri-compound　迫位化合物；萘环1,8-或4,5-位的化合物
pericondensed hydrocarbon　迫位缩合烃
peri-condensed rings　周环并合
pericyclic reaction　周环反应*
pericyclivine　环佩日文碱
pericyclo-〔词头〕　架环的
pericyclo-compound　架环化合物
peri-derivative　迫位衍生物；萘环1,8-或4,5-位的衍生物
perido-steatite　橄榄蛇纹岩
peridot　橄榄石
peridotites　橄榄岩
peri effect　近位效应*
perigee　近地点
perihelion　近日点
perikinetic　异向运动的
perikinetic coagulation　异向凝结(作用)
perilla alcohol　紫苏醇　$C_{10}H_{16}O$
perilla aldehyde　紫苏醛；紫苏糖〔俗〕
perilla citriodora　柠檬紫苏
perilla herb oil　紫苏油
perilla ketone(=perillone)　紫苏酮

perillalcohol　紫苏醇
perillaldehyde　紫苏醛　$C_{10}H_{14}O$
perilla oil　紫苏子油；荏油
perillartine　紫苏亭　$C_{10}H_{15}ON$
perilla sugar　紫苏糖；紫苏醛
perillene　紫苏烯
perillic acid　紫苏酸
perillone(=perilla ketone)　紫苏酮
perillyl　紫苏基
perillyl acetate　乙酸紫苏酯
perillyl alcohol　紫苏子醇
perimeter(=scotometer)　目场计〔物〕
perimidine　咱啶；萘嵌间二氮杂苯　$C_{11}H_8N_2$
perimidinyl(=perimidyl)　咱啶基；萘嵌间二氮(杂)苯基　$N_2C_{11}H_7-$
perimidyl(=perimidinyl)　咱啶基；萘嵌间二氮(杂)苯基
perimorph　包被矿物
perimycin　表霉素；培里霉素；真菌霉素
perinaphthene　周萘
perinaphthenone　周萘酮；萘嵌苯酮
peri-naphthindene　蒽嵌环己烯
perinaphthindenylium cation　周萘茚阳离子
perinaphthodiazine　周萘二嗪
perinaphthotriazole　周萘三嗪
perindopril　培哚普利〔药〕
perinone　紫环酮
period　①(时)期②周期
periodate　①高碘酸②高碘酸盐
periodate oxidized cellulose　高碘酸盐氧化纤维素
periodate titration　高碘酸钾(滴定)法*
period element　(同)周期元素
periodic　①周期的；定时的②间歇的③高碘的
periodic acid　高碘酸
periodic acid oxidation(=Malaprade reaction)　高碘酸氧化反应；马拉破瑞德反应
periodical　①周期的；定时的②定期刊行的③间歇的
periodical operation　间歇操作；周期性操作
periodical tapping　间隔采脂
periodic chain　周期链
periodic chemical reaction　间歇化学反应
periodic classification　周期分类
periodic coagulation　间歇凝固(作用)；间歇凝聚(作用)
periodic concentration polarization　周期浓差极化
periodic copolymer　周期共聚物*
periodic crystallization　反复结晶；间歇结晶
periodic dipole fluctuation　周期性偶极起伏
periodic family　周期族
periodic flow　周期性流动

periodic function 周期函数
periodic group 周期类
periodic heterogeneity 周期不均匀性
periodic inspection 定期检修；小修
periodicity 周期性
periodic kiln 间歇窑
periodic law 周期律
periodic law of elements 元素周期律*
periodic motion 周期运动
periodic oscillation 周期振荡
periodic phenomena 周期现象
periodic precipitation 间歇沉淀
periodic proof test 定期试用试验
periodic property 周期性(质)
periodic reverse current 周期反向电流
periodic sampling 周期性进样
periodic spiral 周期螺列
periodic system 周期系(统)
periodic systematic sampling 周期性系统取样
periodic system of elements 元素周期系*
periodic table 周期表
periodic table of elements 元素周期表*
periodic test 周期试验
periodide 高碘化物
periodo- 全碘(代)〔代替所有 C 上的氢〕
periodo-carbon 全碘化碳
periodo-ethane 全碘乙烷；六碘乙烷 C_2I_6
periodo-ether 全碘乙醚 $(C_2I_5)_2O$
periodo-ethylene 全碘乙烯；四碘代乙烯 $CI_2=CI_2$
period of decay 衰变期
period of half change ①半变期②半寿期
periodo-hydrocarbon 全碘代烃
peripheral ①周围的；外围的②外面的；外部的
peripheral device 外围设备
peripheral discharge mill 周缘出料磨
peripheral equipment 外围设备；外部设备
peripheral jet 圆周喷射
peripheral radius 外周半径
peripheral ratio 缘速比
peripheral reaction 圆周反应〔即表皮反应〕
peripheral speed 圆周速度
peripheral twist extent 捻度传递长度
peripheral velocity 圆周线速度
peripherine(=tolazoline) 陶拉唑啉
periphery ①周围；周边②圆周③圆体的外面
periphery turbine pump 有周缘叶片的透平泵
periplogenin 杠柳配基；萝藦苷辅基
peri position 近位*

perished metal 过烧金属
perished staple 脆弱棉纤维
perished surface 受损表面
perishing 漆膜老化破坏〔漆膜因老化引起的失光、开裂、脱落等的统称〕
peristalsis 蠕动
peristaltic pump 蠕动泵
peristaltin 药鼠李苷
peristaltin(e) 药鼠李树皮
periston 聚乙烯基吡咯烷酮
peritectic reaction 转熔作用；转熔反应
peritectics ①转熔②包晶
peritectic temperature 转熔温度
Perkin's mauve 佩金苯胺紫
Perkin's reaction 佩金芳醛脂酸缩合反应
Perkin's violet 佩金苯胺紫
perlate salt 磷酸氢二钠盐 Na_2HPO_4
perlatolic acid 珠光酸
Perlin viscosity cup 柏林黏度杯
perlite 珍珠岩
perlitic structure 珠光结构
Perloksystem Perlok 丝束直接成条机
perloline 黑麦草碱
perlon-1(=caprone) 贝纶-1；卡普隆
perlon roller 贝纶辊
permafrost 常年地冻的；永久冻土
permalloy 坡莫合金；镍铁导磁合金
permanence 永久性；持久性
permanence of color 保色性
permanence of ink 墨水之持久性
permanence to light 耐光性
permanency 永久性；持久性
permanent 永久的；持久的
permanent artificial standard solution 代用标准液
permanent bend 永久弯曲
permanent character 永久特性；恒定特性
permanent chemical modifier 持久化学改进剂
permanent compression set (永久)压缩变形
permanent deformation 永久形变
permanent dipole 永久(性)偶极
permanent dipole force 永久偶极力
permanent dipole moment 永久偶极矩*
permanent distortion 永久畸变
permanent dye 持久染料；不变色染料
permanent elongation 持久伸长；余留伸长
permanent expansion 持久膨胀；余留膨胀
permanent filmforming material 固有成膜物质
permanent fire-retardant rayon 永久性阻燃黏胶丝

permanent flame 持久火焰；恒定火焰
permanent friction loss 永久摩擦损失
permanent gas 永久气体
permanent green 永久绿〔铬黄或锌黄加氧化铬〕
permanent hardness 永久硬度
permanent hardness of water 水的永久硬度
permanent hard water 永久硬水
permanent lubrication 永久润滑；终身润滑；一次润滑
permanent luminescent pigment 恒效发光颜料
permanently tacky adhesive 永黏性黏合剂
permanent magnet 永久磁铁
permanent magnet alloy 永磁合金
permanent-memory norm 永恒记忆范数
permanent modification technique 持久化学改进技术
permanent output 长期生产率
permanent plasticizer 长效增塑剂
permanent precipitation 永久沉淀
permanent preparation 永久制备
permanent-press fabrics 永久性压烫织物
permanent retarder 永久性阻聚剂
permanent set ①固定伸张②永久应变；永久变定③最后凝结
permanent sizing 耐洗上浆整理
permanent starch 永久上浆剂
permanent strain 永久应变
permanent tack 永久性回黏
permanent twist 固定扭转；永久扭转
permanent-type antifreeze 永久型抗冻液
permanent white 钡白；硫酸钡
permanent yellow 永固黄〔着色剂〕
permanganate 高锰酸盐
permanganate bleach (用)高锰酸盐漂白
permanganate bleaching (用)高锰酸盐漂白
permanganate method 高锰酸盐法
permanganate number 高锰酸盐值
permanganate of potash 高锰酸钾
permanganate oxidation (用)高锰酸盐氧化
permanganate titration 高锰酸钾(滴定)法*
permanganate value 高锰酸(盐)值
permanganic acid 高锰酸　$HMnO_4$
permanganic anhydride 高锰酸酐；七氧化二锰　Mn_2O_7
permanganimetric method 高锰酸钾(滴定)法
permanganometric titration 高锰酸钾滴定法
permanganometry 高锰酸盐滴定法
permanganyl 高锰酰　MnO_3—
permanganyl chloride 高锰酰氯　MnO_3Cl
permanganyl fluoride 高锰酰氟　MnO_3F
permeability ①渗透性*②渗透率

permeability coefficient 渗透系数
permeability cup 渗透(性试验)杯
permeability factor 渗透因子
permeability limit 渗透限度；渗透极限
permeability measurement 渗透测定法
permeability of gas 毒气渗透性
permeability of water 透水性
permeability test ①渗透率试验②穿透性试验
permeability to heat 热的可透度
permeability-weepage 渗漏
permeable 渗透性的；可渗透的
permeable-lining furnace 可透衬层炉
permeable membrane 可透性膜
permeable membrane interface 渗透膜连接装置
permeable plastic 可透塑料
permeable soil 渗透性土壤
permeameter 渗透计
permeant 渗透剂
permeaselective membrane 选择性渗透膜
permeaselectivity 选择渗透性
permeate 渗透
permeate agent 渗透剂
permeation 渗透
permeation barrier ①阻渗层②阻渗器
permeation coefficient (肥皂的)透水系数
permeation constant 渗透常数
permeation flaw detection 渗透探伤法
permeation limit 渗透极限
permeation limit molecular weight 渗透极限分子量
permeation range 渗透范围
permeation rate 浸透速度；浸透速率
permeation separation 渗透分离
permeation separation unit 渗透分离装置
permeation total limit 渗透总限度
permeation tube 渗透管
permeator 渗透器；渗透装置
permethrin 苄氯菊酯
per mille 千分之(几分)；千分率
permissible 容许的
permissible deviation 容许偏差
permissible dose 允许剂量
permissible error 容许误差
permissible explosive 安全炸药；合格炸药
permissible limit 容许极限；容许限度
permissible line 可容许线
permissible load 允许负荷
permissible out of roundness 允许不圆度
permissible sample size 允许试样量

permissible stress 许用应力
permissible sulfur 允许硫含量
permissible tolerance 容许公差
permissible variation 容许偏差
permittivity 介电常数
permivar 镍钴铁合金
permolybdate 过钼酸盐 $xR_2O \cdot yMoO_2[O_2]$
permolybdic acid 过钼酸 $H_2MoO_3[O_2]; H_2MoO_2[O_2]$
permonosulfuric acid(=peroxy-monosulfuric acid) 过一硫酸 $H_2SO_3[O_2]$
permselective cationic membrane 选择性阳离子渗透膜
permselective diaphragm 选择性渗透隔膜
permselective ion-exchange membrane 选择透过性离子交换膜
perm-selective membrane 选择性渗透膜
permselectivity 选择渗透性
permutation ①蜕变②取代；置换
permutational isomer 置换异构体；排列异构体
permutation symbol 排列符号
permutite 滤砂；软水砂；人造沸石
permutite base exchange 滤砂离子交换
permutoid 交换体
permutoid reaction 交换反应；交换体沉淀反应
permutoid swelling 交换溶胀
per-needling 预针刺
pernitric acid 过硝酸
pernitroso-camphor 二亚硝基樟脑 $C_{16}H_{16}O_2N_2$
pernoston β-溴烯丙基仲丁基丙二酰脲
pernot furnace 反射搅炼炉
peroctoate ①过(氧)辛酸盐②过(氧)辛酸酯
perofskite 钙钛矿
perogram 纸色谱图
perolene 载热体〔联苯-联苯醚混合物〕
peronine 苄基吗啡盐酸盐
peropyrene 靴二蒽
perortho ester 过原酸酯
perosmic 高锇的
perosmic acid (高)锇酸 H_2OsO_4
perosmic anhydride 四氧化锇 OsO_4
Perot lamp 波洛特式(汞气)灯
perovskite 钙钛矿*
perovskite structure 钙钛结构矿
peroxamine 过氧化胺
peroxidase 过氧化酶
peroxidase-anti-peroxidase technique 过氧化酶-抗过氧化酶技术
peroxidate 过氧化物
peroxidating 过氧化

peroxidation(=peroxidization) 过氧化作用
peroxide 过氧化物*
peroxide accelerator 过氧化物促进剂(磷化处理)
peroxide bleaches 过氧化物漂白剂
peroxide bomb calorimeter 过氧化量热器
peroxide bond 过氧化物键
peroxide breakdown 过氧化物破坏
peroxide bridge 过氧桥
peroxide catalyst 过氧化物催化剂
peroxide crosslinking 过氧化物交联*
peroxide cure 过氧化物硫化
peroxide decomposer 过氧化物分解剂
peroxide effect 过氧化物效应
peroxide-forming substance 生成过氧化物的物质
peroxide fusion 过氧化物熔融
peroxide initiator 过氧化物引发剂
peroxide number 过氧化值
peroxide of barium 过氧化钡 BaO_2
peroxide of benzoyl 苯酰化过氧
peroxide of hydrogen 过氧化氢 H_2O_2
peroxide of iron 三氧化二铁
peroxide radical 过氧化物游离基
peroxide theory of gum formation 胶质生成的过氧化理论〔石油〕
peroxide value 过氧化值
peroxide vulcanization 过氧化物硫化
peroxide vulcanizing agent 过氧化物硫化剂
peroxidization 过氧化作用*
peroxidizing property 过氧化性
peroxo- 〔词头〕过氧化
peroxo bridge 过氧桥*
peroxocarbonate 过碳酸盐(酯)
peroxochromate(=peroxychromate) 过铬酸盐
peroxochromic acid(=peroxychromic acid) 过铬酸
peroxo-compound 过氧化合物
peroxodicarbonate(=peroxydicarbonate) 过(氧)二碳酸盐
peroxodisulfate 过二硫酸盐
peroxomonocarbonate 过(氧)一碳酸盐
peroxophosphate 过磷酸盐
peroxy- 过氧
peroxyacetic acid 过氧乙酸
peroxyacetyl nitrate(PAN) 硝酸过氧化乙酰
peroxy acid(=peracid) 过氧酸〔含—O—O—基的酸〕
peroxybenzoic acid(=perbenzoic acid) 过苯甲酸
peroxybenzoyl nitrate(PBN) 过苯酰基硝酸酯
peroxyboric acid 过硼酸 HBO_3
peroxycarbonate 过氧碳酸盐
peroxychromate(=peroxochromate) 过铬酸盐

peroxychromic acid(=peroxochromic acid) 过铬酸
peroxy compound 过氧化合物
peroxydase 过氧化物酶
peroxydicarbonate(=peroxodicarbonate) 过(氧)二碳酸盐
peroxydicarbonates 过氧化二碳酸酯类
peroxy dicumyl 过氧化二异丙苯
peroxy-disulfate(=persulfate) 过(二)硫酸盐
peroxy-disulfuric acid(=persulfuric acid) 过(二)硫酸 $H_2S_2O_8$
peroxydol 过硼酸钠
peroxyesters 过氧化酯类
peroxyformic acid 过甲酸 HCOOOH
peroxy ketal 过氧缩酮;过氧酮缩醇
peroxyketal-cured styrene-polyester 过氧酮缩醇固化的苯乙烯-聚酯
peroxyl(=hydrogen peroxide) 过氧化氢
peroxy-monosulfate 过一硫酸盐
peroxy-monosulfuric acid(=permonosulfuric acid) 过一硫酸
peroxy-nitrate 过硝酸盐
peroxy-nitric acid 过硝酸
peroxyoxalate 过氧化草酸盐(酯);过氧草酰类化合物
peroxy oxygen 过氧(化物)的氧;过氧态氧
peroxypropionyl nitrate 过氧丙酰基硝酸酯
peroxy radical 过氧自由基
peroxysuccinic acid 过氧丁二酸;二(丁二酸一酰化)过氧 $(HOOCC_2H_4CO)_2O_2$
peroxysulfuric acid 过硫酸
peroxytitanic acid(=pertitanic acid) 过(氧)钛酸
peroxy (type) catalyst 过氧(化物)型催化剂
peroxyuranic acid 过氧铀酸
perparaldehyde 过仲醛 $C_6H_{12}O_4$
per pass conversion 单程转化率
perpendicular ①垂直;正交②垂(直)线
perpendicular cut shape 直剪印〔绒类织物疵点〕
perpendicular direction 垂直方向
perpendicular drop method (作)垂线法
perpendicularity 正交;垂直
perpendicular vibration 垂直振动
perpetual 永恒的;永久的
perpetual extension 持恒拉伸
perpetual motion 永恒运动
perpetual rest 持恒静态
perpetual rheological history 持恒流变(历)史
perpex 甲基丙烯酸甲酯
perphenazine 奋乃静;羟哌氯丙嗪〔抗精神病药〕
perphosphoric acid 过(二)磷酸 $H_4P_2O_6[O_2]$
perpivalate ①过新戊酸盐②过新戊酸酯

perpyrophosphate 过焦磷酸盐
perrhenate 高铼酸盐 $M·ReO_4$
perrhenic acid 高铼酸 $HReO_4$
perroleum asphalt 石油(地)沥青
perrotine 波若丁印花机
perrutenate 过钌酸盐 $MRuO_4$
perry 梨汁酒
persalt 过酸盐
perseite(=mannoheptitol; perseitol) 鳄梨糖醇;甘露庚糖醇
perseitol(=mannoheptitol; perseite) 鳄梨糖醇;甘露庚糖醇 $C_7H_{16}O_7$
perseleno- 过硒亚基 Se=Se=
perseulose 鳄梨酮糖;半乳庚酮糖
Pershbecker furnace 佩希伯克尔(转式)炉
per shift 每班
persian 印度植鞣绵羊坯革
Persian gulf red 铁红
Persian red 铬红;波斯红
persic oil 桃仁油
persilicic 富硅的
persimmon oil 柿子油
persistence ①持久(性)②持久度
persistence length 相关长度*
persistence of energy 能量常住;能量守恒
persistency ①持久性②持久度
persistent 持久的
persistent acceleration 持久性促进剂;迟延性促进剂
persistent agent 持久(毒)剂;长效剂
persistent bioaccumulative and toxic chemical 持久性的生物可积累性毒性化合物
persistent characteristic 持久特性
persistent gas 持久(的)毒气
persistent line(=rays ultimate) 住留谱线;最后线
persistent odor 持久香气
persistent organic pollutants(POPs) 持久性有机污染物
persistent radical 持续自由基
persistent strength 持久强度
persistent toxic substances 持久性有毒污染物
persisting elongation 持久伸长;余留伸长
persitol 鳄梨糖醇;甘露庚糖醇
personal equation 个人公式
personal error 个人误差
personal error in dipping 个人计量误差
Persooz hardness 帕萨兹硬度〔单位:秒〕
persorption ①(气孔)渗入吸附②吸混(作用)
perspective view (立体)透视图
perspiration 出汗;发汗

perspiration deposit 出汗沉积物
perspiration fastness 耐汗坚牢度
perspiration fastness tester 耐汗渍(色)牢度试验仪
perspiration resistance 耐汗性
perspirometer 耐汗渍牢度试验器
perstearic acid 过硬脂酸
perstoff 双光气
persulfate(=peroxy-disulfate) 过(二)硫酸盐
persulfate initiator 过(二)硫酸盐引发剂
persulfate ion 过(二)硫酸根离子
persulfate method 过(二)硫酸盐法
persulfate oxidation 过(二)硫酸盐氧化(作用)
persulfate salt 过(二)硫酸盐
persulfide 过硫化物
persulfuric acid(=peroxy-disulfuric acid) 过(二)硫酸 $H_2S_2O_8$
persulphate initiator 过硫酸盐引发剂
per-swollen 预溶胀的
pertantalic acid 过二钽酸 $HTaO_2[O_2]$
pertechnetate 高锝酸盐
pertechnyl fluoride 氟化高锝酰 TcO_3F
perthane 二氯双(乙基苯基)乙烷；乙滴滴
perthio- 〔词头〕过硫〔仅用于取代 O 时〕 S=S=
perthiocarbonate 过硫碳酸盐 $M_2CS_3[S_2];M_2CS_4$
perthiocarbonic acid 过硫碳酸 $H_2CS_3[S_2];H_2CS_4$
pertitanate(=peroxytitanate) 过钛酸盐
pertitanic acid(=peroxytitanic acid) 过钛酸 H_4TiO_5
pertite 柏载特
pertucin 穿孔假胞菌素
pertungstate 过钨酸盐 $M_2WO_5[O_2];M_2WO_2[O_2]$
pertungstic acid 过钨酸 $H_2WO_5[O_2];H_2WO_2[O_2]$
perturbation ①微扰；扰动②摄动
perturbational molecular orbital(PMO) 微扰分子轨道
perturbation calculation 微扰计算
perturbation function 微扰函数
perturbation method 微扰法；摄动法
perturbation molecular orbital 微扰分子轨道
perturbation theory 微扰理论*
perturbed angular correlation(PAC) 受扰角关联
perturbed dimension 扰动尺寸*
perturbed structure 微扰结构
perturbing variables 干扰变量
pertussis vaccine 百日咳疫苗
Peru balsam(=Peru balsam oil；Peruvian balsam) 秘鲁香脂
Peru balsam oil(=Peru balsam) 秘鲁香脂
per unit 每单位量的；每一设备的
per unit area 每一单位面积

per unit length 单位长度
per unit volume 每一单位体积
per unit weight 每一单位重量
peruol 苯甲酸苄酯
peruscabin 苯甲酸苄酯
Peruvian balsam(=Peru balsam) 秘鲁香脂
Peruvian bark 金鸡纳树皮
Peruvian guano 秘鲁鸟粪
peruvin 肉桂醇
peruviol 橙花油醇
pervanadic acid 过钒酸 $HVO_2[O_2]$
pervaporation 全蒸发(过程)
pervasiveness 能嗅度
pervesterol 藻脂甾醇
pervious ①可透的②可渗透的
perviousness ①可透性②渗透性
pervitine 脱氧麻黄碱
perxenate 过氙酸盐(根)
perylene 花
perylene maroon 花紫红
perylene pigments 花系颜料
perylene radical cation 花基阳离子
perylene scarlet 花猩红
peryllartine 紫苏糖
perzirconic acid 过锆酸 $H_4ZrO_5[O_2]$
PES(polyester) 聚酯
PE spectrum(=photoelectron spectrum) 光电子能谱
pesticide analysis 农药分析
pesticide chemistry 农药化学
pesticide hazard 农药公害
pesticide identification 农药确证
pesticide pollution 农药公害
pesticide residue 农药残留(量)
pesticide residue analysis 农药残留分析
pesticide residue field trial 农药残留田间试验
pesticide residues analysis 农药残留物分析
pesticides 农药
pesticide specification 农药规格
pesticin 鼠疫巴氏菌素
pestle (研)杵
pestle mill 捣锤
peta 拍它；千兆兆
petabecquerel 千万亿贝可勒尔
petalite 透锂长石
petal-like conjugate fibre 花瓣状(截面)复合纤维
petalon 盘形氢核
petasin 蜂斗精
petasinolide 蜂斗交酯

petasol 蜂斗醇
petcock(=pet valve) 小型旋塞
Peter-Gries' method 彼得-格里斯方法
Peter's apparatus 彼得装置〔测定真沸点的精馏装置〕
pethidine 哌替啶；度冷丁〔麻醉镇痛药〕
pethidine-hydrochloride 盐酸哌替啶；度冷丁
petiole 叶柄
petitgrain bergamot oil 香柠檬叶油
petitgrain bigarade oil(=bitter orange leaf oil) 苦橙叶油
petitgrain lemon oil 柠檬叶油
petitgrain mandarin oil 橘叶油
petitgrain oil 橙叶油
petitgrain pomelo oil 香泡叶油
petitgrain sweet orange oil 甜橙叶油
Petreco electrical desalting process 皮特莱柯电法脱盐过程
petrel wax 海燕蜡
petrichor(=petrochor) 潮土油
Petri dish 陪替氏培养皿〔医〕
petrification 石化(作用)
petrified wood 木化石
petrifying 石化
petrifying liquid 防潮液；石化液〔喷于墙上以不透潮湿〕
petrin 岩菌素；蝶呤；四氮奈
petrl ointment 石油软膏
petroacetylene 石油乙炔
petrobenzene 石油苯
petrocene 石油省〔蒽的异构物〕
petrochemical 石油化学的
petrochemical industry 石油化学工业
petrochemical intermediate 石油化学中间产品
petrochemical manufacture 石油化学产品的生产
petrochemical materials 石油化工原料
petrochemical plant 石油化学工厂
petrochemical processing 石油化学加工
petrochemical reactions 石油化学反应
petrochemicals 石油化学制品
petrochemical storage 石油化学产品的储藏
petrochemical storage vessel 石油化学产品的储藏器
petrochemical waste 石油化学废弃物
petrochemical works 石油化工厂
petrochemistry 石油化学
petrochor(=petrichor) 潮土油
petro-ether 石油醚
Petroff equation 佩特罗夫摩擦系数公式
Petroff reagent 佩特罗夫试剂
petrogas 液体丙烷
petrograd standard 软木材体积标准

petrographical 岩石学的
petrographic investigation 岩相研究
petrography 岩石(分类)学；岩石组织学
petrol 汽油
petrolat(=petrolatum) 矿脂；凡士林
petrol atomizer 汽油雾化器
petrolatum(=petrolat) 矿脂；凡士林
petrolatum album 白矿脂；白凡士林
petrolatum jelly 矿脂
petrolatum liquidum(=petrolatum oil) 矿脂；凡士林油
petrolatum melting point 矿脂熔点
petrolatum oil(=petrolatum liquidum) 矿脂；凡士林油
petrolatum ointment 矿脂；凡士林
petrolatum stock 矿脂原料
petrolatum wax 矿脂蜡
petrolax 液体矿脂
petrol balance 汽油秤
petrol barge 汽油驳船
petrol capacity 汽油装载容量
petrol carrier 油船；汽油载运船
petrol cock 汽油阀门
petrol coke 石油焦
petrol consumption 汽油消耗
petrol content 汽油馏分含量
petrol-content gauge 汽油量表
petrol delivery 汽油供应(输送)
petrol-depth gauge 汽油表
petrol dripping ①汽油流出②流进汽油
petrolene 石油烯；软沥青〔沥青中溶于己烷的部分〕
petroleum 石油
petroleum acids〔复〕 石油酸类
petroleum additive 石油添加剂
petroleum analysis 石油分析
petroleum aromatics 石油芳烃
petroleum asphalt 石(油)(地)沥青
petroleum base 石油基
petroleum-based 石油基的
petroleum benzene 石油苯；焦苯
petroleum benzine 石油醚
petroleum bitumen 石油沥青
petroleum black 石油炭黑
petroleum bloom 石油起霜作用
petroleum butter 凡士林；矿脂；石油膏
petroleum can 小油桶
petroleum car (铁路)油槽车
petroleum ceresin 石油地蜡
petroleum chemical plant 石油化工厂
petroleum chemicals 石油化学产品

petroleum chromatography 石油色层
petroleum coal 固体石油
petroleum coke 石油焦炭；石油焦
petroleum composition 石油组成
petroleum cracking 石油裂解
petroleum cracking process 石油裂化过程
petroleum crude 原油
petroleum crude oil 石油原油
petroleum cut(=petroleum distillate) 石油馏分
petroleum demulsification 石油破乳作用
petroleum derivative 石油衍生物
petroleum derived 石油衍生的
petroleum distillate(=petroleum cut) 石油馏分
petroleum distillation 石油蒸馏
petroleum drying oil 石油系(合成)干性油
petroleum emulsifying agent 石油乳化剂
petroleum engineering 石油工程
petroleum ester 石油酯
petroleum ether(=petrolic ether) 石油醚
petroleum ether extract 石油醚萃取物
petroleum ether insoluble 石油醚不溶物
petroleum ether soluble matter 石油醚可溶物
petroleum expansion 石油加热膨胀
petroleum extract 石油萃取物；石油抽提物
petroleum feeder 石油喷射器
petroleum fermentation process 石油发酵过程
petroleum formation 石油生成
petroleum fraction 石油馏分
petroleum fuel oil 石油燃料油
petroleum gas 石油气体
petroleum gas oil 石油气体油
petroleum genesis 石油形成
petroleum geology 石油地质
petroleum grease 石油润滑脂
petroleum hydrocarbon 石油烃
petroleum industry 石油工业
petroleum jelly(=vaseline) 矿脂；凡士林；石油冻；石油膏
petroleum legislation 石油法案
petroleum leve 石油醚〔沸点 40～60℃〕
petroleum lubricant 石油润滑剂
petroleum lubricating grease 石油润滑脂
petroleum metering instruments 石油控制测量仪器
petroleum microbiology 石油微生物学
petroleum motor oil 石油车用润滑油
petroleum naphtha(=ligroin) 石脑油；石油英
petroleum nitrogen bases 石油中的含氮碱
petroleum odour 石油气味

petroleum oil 石油润滑油
petroleum oil flux 重油
petroleum ointment 矿脂；凡士林
petroleum origin 石油起源
petroleum paraffin 石油石蜡
petroleum pipe line 石油管道
petroleum pitch 石油沥青
petroleum pitch fibre 石油沥青纤维
petroleum pollution 石油污染
petroleum processing 石油加工
petroleum processing wastes 石油加工废弃物
petroleum producer 石油企业家
petroleum product 石油产品
petroleum production 采油
petroleum refiner 石油炼制者
petroleum refinery 炼油厂；石油炼厂
petroleum refinery capacity 石油炼制能力
petroleum refinery waste 石油炼厂废弃物；炼油厂废弃物
petroleum refining 石油加工
petroleum refining furnace 石油炼制炉
petroleum refining industry 石油炼制工业；炼油工业
petroleum reforming 石油重整
petroleum reserves 石油储量
petroleum residual oil 石油渣油
petroleum residue 石油脚
petroleum resin 石油树脂*
petroleum resin emulsion 石油树脂乳液
petroleum science 石油科学
petroleum ship 油船
petroleum soap 石油皂
petroleum sodium sulfonate 石油磺酸钠
petroleum solid 石油固体
petroleum solvent 石油溶剂
petroleum specialties〔复〕 特殊石油产品
petroleum spirit ①石油溶剂油〔美〕②闪点在 0℃ 以下的软质烃类混合物〔英〕
petroleum still 石油蒸馏釜
petroleum storage depot 油库
petroleum substitutes(=petroleum supplements) 石油代用品
petroleum sulfonate 石油磺酸盐
petroleum supplements(=petroleum substitutes) 石油代用品
petroleum sweetening 石油脱硫(法)
petroleum tailings〔复〕 石油脚〔石油蒸馏残余物〕
petroleum tank 油槽；油罐
petroleum tank wagon 铁路油槽车

petroleum tar　石油沥青；石油焦油
petroleum tester　石油测定器
petroleum testing methods　石油试验法
petroleum thinner　石油稀释剂
petroleum transport　石油输送
petroleum vapour　石油蒸气
petroleum wax　石油蜡
petrol-feed pump　加油泵；汽油泵
petrol-filling station　加油站
petrol filter　汽油过滤器
petrol gauge　汽油表
petrol-gauge dial　汽油表刻度盘
petrol hose　汽油胶管
petrolic ether(=petroleum ether)　石油醚
petroliferous　含石油的
petroliferous shale　含石油页岩
petrolift　燃料泵
petroline　①(缅甸仰光石油中提出的)固体石蜡②(英国苏格兰页岩中提出的)煤油
petrol lamp　汽油灯
petrol level　汽油液面
petrol-level gauge　汽油表
petrology　岩石学
petrol-oil-lubricants　燃料-润滑剂
petrol overflow tube　汽油溢流管
petrol-pressure gauge　汽油压力计
petrol pump　汽油泵
petrol sediment bulb　汽油沉降器
petrol tank　汽油箱；汽油储存器
petrol tanker　汽油油槽车
petronaphthalene　石油萘
petronol　液体石蜡；液体石油脂
petrosapol　石油软膏
petroselic acid　岩芹酸；洋芫荽子酸
petroselinic acid　岩芹酸；6-十八(碳)烯酸
petroselinolic acid　岩芹炔酸
petroselinum　岩芹；洋芫荽〔即 parsley 石芹，因各有衍生名，故从外文订两个名字〕
petrosilane　岩芹烷；二十(碳)烷　$C_{20}H_{42}$
petrosilex　燧石
petrosio　液体矿脂
petrous　石质的；化石的；固体的
petrox　油酸铵皂化的石蜡油
petroxolin　油酸铵皂化的石蜡油
petsalit　柏塞里特〔炸药〕
petticoat　有圆锥口的软管
petunidin　矮牵牛(苷)配基；3'-甲花翠素
petunin　矮牵牛苷

pet valve(=petcock)　小型旋塞
petzite　碲金银矿
peucedanine　前胡精
peucedanum oil　前胡油
peucenin　前胡宁
pewter　锡铅合金
pexitropy　冷却结晶作用
pez　地沥青
pfeilring reagent　范木裂脂试剂
pferidekraft　马力米制单位〔75千克/秒〕
Pfeufer's green　绿菌染料
P.F. process(=piston-flow process)　脉动流工艺；间歇流工艺〔聚醋酸乙烯酯水解工艺〕
PFT(=pulsed Fourier transform)　脉冲傅里叶变换
PFT-NMR spectrometer　脉冲傅里叶变换核磁共振仪
PFT-NMR spectroscopy　脉冲傅里叶变换核磁共振谱
pfund arc　普芬德电弧
Pfund cryptometer　普芬德遮盖力计
Pfund film gauge　普芬德涂膜厚度计
Pfund gauge(=Pfund film gauge)　普芬德(漆膜)厚度计
Pfund hardness number　普芬德硬度值
Pfund indentation hardness　普芬德压痕硬度
PGC(=pyrolysis gas chromatography)　热解气相色谱(法)
Ph〔符号〕苯基
phacolite　扁豆菱沸石
pH adjustment tank　pH 调节槽；酸碱度调节槽
phaeanthine(=1-tetrandrine)　亮花木碱；汉防己甲素；粉防己碱
phaeo-〔词头〕深色的；微黑的
phaeochromin(=feochromin)　暗色菌素
phaeofacin(=phaeophacin)　生暗菌素
phaeohemins〔复〕棕腐质
phaeomelanin　棕黑素
phaeophacin　生暗菌素
ph(a)eophorbide　脱镁叶绿甲酯酸
phagocytic luminescence　吞噬细胞发光
phalamycin　亮霉素
phallin　溶血菌毒
phalloidin　毒伞素；鬼笔环肽
phanerites〔复〕显晶岩
phanodorn　非罗多；环巴比妥；乙基环己烯基巴比妥酸　$(C_2H_5)(C_6H_9)$=$C_4H_2O_3N_2$
phanquone　4,7-菲咯啉-5,6-二醌；安痢平
phantastica　致幻觉药
phantom atom　虚拟原子
phantom network　虚幻网络
phantom polymer　虚幻聚合物
phantom view　经过透明壁的内视图

pharaoh's serpents 硫氰酸汞
pharbitic acid 牵牛脂酸
pharbitin 牵牛脂苷 $C_{54}H_{96}O_{27}$
pharmaceutical ①药物的②药物
pharmaceutical chemistry 药物化学；制药化学
pharmaceutical emulsion 药用乳液
pharmaceutical material 药
pharmaceutical ointment 药用软膏
pharmaceutical preparations 药物制剂
pharmaceutical purity 药物纯
pharmaceuticals 药物
pharmaceutical soap 药皂
pharmaceutics 制药学
pharmacist 药剂师
pharmacodynamics 药效学
pharmacognosy 生药学
pharmacokinetic parameter 药物动力学参数
pharmacokinetics 药物动力学；药代动力学；药动学
pharmacolite 毒石
pharmacologic activity 药物学活性
pharmacological 药理学的
pharmacological action 药理作用
pharmacological effect 药理效应
pharmacological property 药理学性质
pharmacologist 药理学家
pharmacology 药理学；药物学
pharmacometrics 药物计量学
pharmacopeia 药典
pharmacophore 药效团
pharmacopoeia 药典
pharmacosiderite 毒铁矿
pharmacotherapy 药疗法
pharmacy ①药学②药房
pharmagel 高纯明胶；药用明胶
pharmiglucin 药胶菌素
phase ①位相②物相③相
phase allotropy 同素异形相
phase analysis 物相分析；相分析
phase analysis by X-ray diffraction X射线衍射物相分析
phase angle 相角
phase angle of radio-frequency pulse 射频脉冲相位角
3-phase auto-transformer 三相自耦变压器
phase behaviour 相特性
phase behaviour of greases 润滑脂的相特性
phase bleed 相流失
phase boundary 相界
phase boundary potential 相界电位
phase change 相变*

phase chemistry 相化学
phase contact area 相界面积
phase contrast(PC) 相位对比
phase contrast microscope 相衬显微镜；位相显微镜
phase contrast microscopy 相衬显微术；位相显微术
phase converter （电流）换相器
phase correction 相位校正
phase current 相电流
phase cycle 相位循环
phase detection microscope 相位探知显微镜
phase detection microscopy 相检测显微镜
phase diagram 相图*
phase difference 相位差
phase disentrainment 相夹带剔除
phase distortion 相位畸变
phase drift 相位漂移
phased-yarn 异相位(自捻)纱
phase encoding 相位编码
phase encoding gradient 相位编码梯度
phase equilibrium 相平衡*
phase error on interferogram 干涉图的相位误差
phase fault phase hologram 相位型全息图
phase holography 相全息术
phaseic acid （红花）菜豆酸
phase interface 相界面
phase inversion 倒相；转相；换相〔连续相和分散相倒换〕
phase-inversion membrane 相变形成膜
phase-inversion polymerization 相转化聚合*
phase inversion temperature （乳状液的）相变型温度；转相温度
phase lag 相位滞后
phase locked loop(PLL) 锁相环
phase map 相图
phase memory time 相位记忆时间
phase microscope 相衬显微镜；位相显微镜
phaseolic acid 菜豆酸；5,8,12-三羟十二(烷)酮酸
phaseoline 菜豆碱
phaseolunatin 菜豆亭
phaseolunatinic acid 菜豆亭酸
phaseomannite(=inositol) 肌醇
phase plate 相移片
phase problem 相角问题
phase ratio 相比(率)*
phase region 相域
phase relation 位相关系*
phase relationship 相态关系
phase-resolved fluorescence 相分辨荧光
phase reversal 相的反转

phase reversal of emulsion 乳液转相；乳液的相逆转
phase rule 相律*
phase selective alternating current polarography 相敏交流极谱法*
phase selective anodic stripping voltammetry 相敏阳极溶出伏安法
phase sensitive AC adsorption stripping voltammetry 相敏交流吸附溶出伏安法
phase sensitive AC polarography 相敏交流极谱法
phase sensitive AC voltammetry 相敏交流伏安法
phase sensitive detector 相敏检测器；位相敏感性检测器
phase sensitive polarograph 相敏极谱仪
phase sensitive polarography 相敏极谱法
phase-separating paper 相分离纸
phase separation 相分离；析相作用
phase separation fluoroimmunoassay 相分离荧光免疫分析
phase separation model 析相模型
phase-separation model of micelle formation 胶束形成的析相模型
phase separation temperature 析相温度
phase separator 分相器
phase shift 相转变
3-phase slip-ring induction motor 三相滑环式感应电动机
phase solubility analysis 相溶(解)度分析
phase space 相空间
phase species analysis 物相形态分析
phase specific constant 相特征常数
3-phase squirrel cage induction motor 三相鼠笼式感应电动机
phase surface 相表面
phase titration (分)相滴定
phase-transfer catalysis 相转移催化*
phase transfer catalyst 相转移催化剂
phase transformation 相(转)变
phase transformation detector 相转化检测器
phase transition 相变*
phase transition humidity sensor 相转湿度传感器
phase transition temperature 相变温度
phase variation 位相变化；相变异
phase velocity 相位速度；相速度〔物〕
phase volume 相体积
phase volume fraction 相体积分数
phase volume ratio 相体积比
phase volume theory 相体积理论
phasic property 相性
phasor 移相器

pH buffer solution pH 缓冲溶液
pH electrode pH 电极
phellandral 水芹醛
phellandrene 水芹烯
phellandric acid 水芹酸
phellandrol 水芹醇
phellem 木栓
phellonic acid 软木醇酸；二十二烷羟酸
phellopterin 珊瑚菜质；珊瑚菜内酯
-phen- ①〔意译〕苯②〔音译〕芬，吩〔与苯无关〕
phenacaine(=holocaine) 非那卡因；芬那卡因（即：哈洛卡因）〔一种局部麻醉药，尤用于眼科〕
phenacemide 苯乙酰脲
phenacethydrazine 乙酰基苯肼
phenacetin(e) 非那西汀；N-乙酰基对乙氧苯胺 $CH_3CONHC_6H_4OC_2H_5$
phenacetol 苯氧基丙酮
phenacetolin 迪吉讷(Degener)指示剂
phenacetylaniline(=phenyl acetanilide) N-苯乙酰胺；N-苯乙酰苯胺 $C_6H_5CH_2CONHC_6H_5$
phenacite 硅铍石 $BeSiO_4$
phenacyl 苯甲酰甲基 $C_6H_5COCH_2-$
phenacyl-acetone 苯甲酰甲基丙酮 $C_6H_5COCH_2CH_2COCH_3$
phenacyl alcohol 苯甲酰甲醇 $C_6H_5COCH_2OH$
phenacyl bromide 苯甲酰甲基溴 $C_6H_5COCH_2Br$
phenacyl butyrate 丁酸苯甲酰原酯
phenacyl chloride 苯甲酰甲基氯 $C_6H_5COCH_2Cl$
phenacyl ester 苯乙酮酯 $RCOOCH_2COC_6H_5$
phenacyl halide 苯甲酰甲基卤 $C_6H_5COCH_2X$
phenacylidene 苯甲酰亚甲基 $C_6H_5COCH=$
phenalene 非那烯
phenalgin 氨苯 $C_6H_5NH_3$
phenamine 非那明
phenamine hydrochloride(=phenocoll hydrochloride) 盐酸非那明 $C_{10}H_{14}O_2N_2HCl$
phenamine salicylate 水杨酸非那明 $C_{10}H_{14}O_2N_2C_7H_6O_3$
phenanthrahydroquinone 菲氢醌；9,10-菲二酚 $C_{14}H_8(OH)_2$
phenanthraquinone(=phenanthrenequinone) 菲醌 $C_6H_4(CO)_2C_6H_4$
phenanthrene 菲 $(C_6H_4CH)_2$
phenanthrene carboxylic acid 菲羧酸 $C_{14}H_9COOH$
phenanthrene dione 菲二酮；菲醌
phenanthrene hydroquinone 菲氢醌
phenanthrene nucleus 菲核；菲环

phenanthrenequinone(=phenanthraquinone) 菲醌 $C_{14}H_8O_2$
phenanthrenequinone dioxime 菲醌二肟 $C_{14}H_{10}O_2N_2$
phenanthrenequinone dioxime anhydride 菲醌二肟酐 $C_{14}H_8ON_2$
phenanthrenequinone monoxime 菲醌一肟 $C_{14}H_9O_2N$
phenanthrene ring 菲环
phenanthrenol 菲酚
phenanthrenone 菲酮 $C_6H_4COCH_2C_6H_4$
phenanthridine 菲啶 $C_{13}H_9N$
phenanthridinone 菲啶酮
phenanthridinyl 菲啶基；5-氮菲基 $NC_{13}H_8$—
phenanthridone 菲啶酮
phenanthrine 菲
phenanthro 菲并
phenanthrol 菲酚 $C_{14}H_9OH$
phenanthroline 菲咯啉
phenanthroline ion(=ferroin) 试亚铁灵
phenanthrone 菲酮
phenanthrophenazine 二苯吩嗪
phenanthryl 菲基〔有5种异构物〕 $C_{14}H_9$—
phenanthrylene 亚菲基 —$C_{14}H_8$—
phenanthryne 菲炔
phenaphthacridone 苯并萘吖啶酮
phenarsazine 吩吡嗪；砷氮杂蒽 $C_{12}H_8AsN$
phenarsazine chloride(=adamsite) 氯化吩吡嗪；亚当氏毒气 $C_{12}H_9AsClN$
phenarsazine fluoride 氟化吩吡嗪
phenarsen 4,5-二氯苯-1,3-二磺酰胺盐酸盐
phenasic acid 五取代苯酚
phenate(=phenolate) ①酚盐②(苯)酚盐；石炭酸盐
phenate stabilizer 酚盐稳定剂
phenazine 吩嗪 $C_6H_4=N_2=C_6H_4$
phenazinone 吩嗪酮
phenazinyl 吩嗪基；夹二氮蒽基 $N_2C_{12}H_7$—
phenazo 4,4′-双(4-羟基偶氮苯)-3,3′-二硝基联苯 $[HOC_6H_2N_2(NO_2)C_6H_3]_2$
phenazone 非那宗；二甲基苯基吡唑酮；安替比林〔药〕 $C_{11}H_{12}ON_2$
phenazonium 二甲基苯基吡唑酮鎓
phendic acid 吩(茚)满酸
phendioxin 苯并-1,4-二氧六环
phene(=benzene) 苯
phenedin 非那西汀〔药〕
phenegol 对硝基磺酸钾基酚汞
phenelyl(=ethoxyphenyl) 乙氧苯基
phenelzine 苯乙肼
phenelzine sulfate 硫酸苯乙基肼

phenenyl 三价苯基〔均、偏或连〕 $C_6H_3\equiv$
phenergan 非那根；异丙嗪
phenesic acid 二取代酚
phenesterin(e) 胆甾醇对苯乙酸氮芥
phenetetrol 1,2,3,4-苯四酚
phenethicillin 苯氧乙基青霉素
phenethyl 苯乙基 $C_6H_5CH_2CH_2$—
phenethyl acetate 乙酸苯乙酯
phenethyl alcohol 苯乙醇
phenethylamine 苯乙胺
phenethyl benzyl ketone 苯乙基苄基(甲)酮
phenethylene 苯乙烯
phenethyl pyridinium 苯乙基吡啶(鎓)盐
phenethyl serine 苯乙基丝氨酸
phenetide〔类名〕 N-酰乙氧基苯胺 $C_2H_5OC_6H_4NHCOR$
p-phenetidine 对氨基苯乙醚
phenetidine 氨基苯乙醚；乙氧基苯胺；苯乙定〔俗〕 $C_2H_5OC_6H_4NH_2$
phenetidine camphorate 樟脑酸乙氧基苯胺
p-phenetidine citrate 柠檬酸化对氨基苯乙醚
phenetidines 氨基苯乙醚；乙氧基苯胺〔合成药物类名〕
phenetidine salicylate 水杨酸乙氧基苯胺
phenetidino 乙氧苯氨基〔邻、间或对〕 $C_2H_5OC_6H_4NH$—
phenetole(=ethyl phenolate) 苯乙醚；乙氧基苯 $C_2H_5OC_6H_5$
phenetyl 乙氧苯基 $(C_2H_5O)C_6H_4$—
phenetyl-urea 乙氧基苯脲 $C_2H_5OC_6H_4NHCONH_2$
phengite 多硅白云母
phenglutarimide hydrochloride 芬格鲁胺盐酸盐〔抗胆碱药〕
-phenic ①〔意译〕苯②〔音译〕芬，吩〔与苯无关〕
phenic acid 苯酚
phenicate ①酚盐②用酚消毒
phenicin 芬尼菌素；猩红青霉菌素；盐酸苯异丙肼
phenide 苯基金属 MC_6H_5
phenidine 芬尼定〔药〕
phenidone 菲尼酮
phenil 苯基 C_6H_5—
phenin 芬宁〔药〕
phenindamine tartrate 酒石酸苯茚胺〔药〕
phenindione 苯茚二酮〔药〕
pheniodol 碘阿芬酸；碘苯丙酸
phenisic acid 三代苯酚
phenixin 四氯化碳 CCl_4
phenkapton 芬硫磷〔药〕
phenmethyl 苯甲基；苄基 $C_6H_5CH_2$—
phenmethyltriazine 苯并甲基三嗪
phenmetrazine hydrochloride 芬美曲嗪盐酸盐〔药〕

phenmiazine(=quinazoline) 间二氮杂萘；1,3-二氮杂萘；喹唑啉
phenobarbital 苯巴比妥 $C_{12}H_{12}O_3N_2$
phenobarbital sodium 苯巴比妥钠；鲁米那钠 $C_{12}H_{11}O_3N_2Na$
phenobarbitone 苯巴比通；鲁米那〔药〕 $C_{12}H_{12}O_3N_2$
phenocoll 非诺可
phenocoll hydrochloride(=phenamine hydrochloride) 盐酸非那明 $C_{10}H_{14}O_2N_2HCl$
phenodiazine 二氮（杂）萘〔α为12,β为13〕
phenodin(=hematine) 苏木精
phenogen 同型体
phenol 酚*
phenol absorber tower 苯酚吸收塔
phenol acetaldehyde resin 苯酚乙醛树脂
phenol-acetone unit 苯酚-丙酮装置；异丙基苯氧化装置
phenol acid 羟基芳酸
phenol adsorption capacity 酚吸附容量；酚吸附能力
phenol aldehyde ①酚醛②苯酚醛
phenol antioxidant 酚类防老剂
phenol-arsonic acid 羟基苯胂酸 $HOC_6H_4AsO(OH)_2$
phenolate(=phenate) ①酚盐②(苯)酚盐 C_6H_5OM
phenolated alkyd 桐油酚醛树脂改性醇酸树脂〔简称〕
phenolate process （气体）酚盐精制过程
phenol benzoate 苯甲酸苯酯 $C_6H_5COOC_6H_5$
phenol bismuth 苯氧二羟铋 $Bi(OH)_2OC_6H_5$
phenol-blocked isocyanate 酚类封端异氰酸酯
phenol blue (苯)酚蓝
phenol camphor (苯)酚-樟脑混合物
phenol coefficient 苯酚系数
phenol derivatives 苯酚衍生物
phenol disulfonate method 苯酚二磺酸酯法
phenol disulfonic acid 苯酚二磺酸 $C_6H_3(OH)(SO_3H)_2$
phenol-2,4-disulphonic acid colorimetry 酚二磺酸比色法
phenol ester 苯酚酯
phenol ether 苯酚醚
phenol ether derivative 酚醚衍生物
phenol ether resin 苯酚醚树脂
phenol extraction 苯酚提取
phenol extraction process(=phenol treating) 苯酚精制过程
phenol-formaldehyde condensation 苯酚甲醛缩合反应
phenol formaldehyde ratio 苯酚/甲醛比；酚/醛比
phenol-formaldehyde resin 酚醛树脂*；苯酚-甲醛树脂
phenol-formaldehyde sulfonate (苯)酚(甲)醛磺酸盐
phenol-furfural resin(PFF) 酚呋喃树脂
phenol glucuronic acid(=phenyl glucuronide) 葡萄糖苯苷酸

phenol glue 酚醛胶
phenol glycerin 酚的甘油溶液〔16%〕
phenol haloalkyl ether 苯基卤代烷基醚
phenolic ①酚的②(苯)酚的
phenolic acid 酚酸
phenolic adhesive 酚醛树脂黏合剂
phenolic alcohol 酚醇
phenolic aldehyde 酚醛
phenolic antioxidant 酚类防老剂
phenolic cement 酚醛树脂胶接剂
phenolic composition 酚醛(树脂)压塑料
phenolic dimer 酚类二聚物
phenolic effluents 含酚废水
phenolic epoxy resin 酚醛环氧树脂
phenolic ester 酚酯
phenolic ether 酚醚
phenolic foam 酚醛泡沫体
phenolic glue 酚醛胶
phenolic group ①酚基②(苯)酚基
phenolic hydroxyl 酚式羟基
phenolic ion exchange resin 酚醛离子交换树脂
phenolic ketone 酚酮
phenolic lacquer 酚醛(真)漆
phenolic laminate 酚醛塑料层合板；酚醛层板
phenolic laminated fiber 酚醛(树脂)层压纤维板
phenolic-modified alkyd 酚醛改性醇酸树脂
phenolic note 酚气息
phenolic novolac 线型酚醛(清漆)树脂
phenolic odor 酚气息
phenolic plastic 酚醛塑料
phenolic resin 酚醛树脂*
phenolic resin balls 酚醛树脂球
phenolic resin fibre 酚醛树脂纤维
phenolic resin varnish 酚醛(树脂)清漆
phenolics 酚醛塑料
phenolic stabilizer 酚型稳定剂
phenolic steroid 酚类甾族化合物
phenolic sulfide 苯酚硫醚〔防老剂〕
phenolic sulfonamide 苯酚磺(酰)胺
phenolic tanning agent 酚类鞣剂
phenolic type antioxidant 酚类防老剂
phenolic varnish 酚醛(树脂)清漆
phenolic waste liquor 含酚废液
phenolic waste water 苯酚废水
phenolic water drum 苯酚水槽
phenol-indene resin 酚醛-茚树脂
phenol-ketone isomerism 酚酮异构
phenol-keto tautomerism 酚-酮互变异构*

phenol lignin 苯酚木素
phenol-modified alkyd resin 酚醛改性醇酸树脂
phenoloid (藻类植物的)酚性化合物
phenol oil 苯酚润滑油
phenol oxidase 酚氧化酶
phenolphthalein indicator 酚酞指示剂
phenolphthalein test paper 酚酞试纸
phenolphthalexon 酚酞氨络剂
phenolphthalin 酚酞啉；还原酚酞 $C_{20}H_{16}O_4$
phenolplast 酚醛塑料
phenolquinine 苯酚奎宁
phenol reagent 酚试剂
phenol red (苯)酚红*
phenol removal process 除酚法
phenol resin 酚树脂
phenols 酚类
phenol storage 苯酚罐
phenolsulfonate 苯酚磺酸盐；羟基苯磺酸盐 $HOC_6H_4SO_3M$
phenol sulfonic acid 苯酚磺酸 $OHC_6H_4SO_3H$
phenolsulfonic acid resin 苯酚磺酸树脂〔阳离子交换树脂〕
phenolsulfonphthalein 酚磺酞；酚红 $C_{19}H_{14}O_5S$
phenol sulfuric acid 苯酚硫酸酯
phenol terpene 酚萜(缩合物)
phenol tetraiodophthalein sodium 酚四碘酞钠〔药〕 $NaOOCC_6I_4C(=C_6H_4=O)C_6H_4ONa$
phenol tower 苯酚塔
phenol treating(=phenol extraction process) 苯酚精制过程
phenol tricarboxylic acid 苯酚三羧酸 $HOC_6H_2(CO_2H)_3$
phenol water disposal 酚水处理
phenolysis 酚解
phenomenological analysis 唯象分析
phenomenological coefficient 唯象系数
phenomenological equation 唯象方程
phenomenological model theory 唯象模型理论
phenomenological rheology 唯象流变学
phenomenological theory 唯象论
phenomenological theory of elasticity 弹性唯象论
phenomenological treatment 唯象处理
phenomenon 现象
phenomenon of detonation 爆震现象
phenomenon of osmosis 渗透现象
phenomycin 酚霉素
phenon ①同型种②表观群
phenonaphthazine 苯并吩嗪
phenones 苯某酮
phenonium ion 苯氧离子 $C_6H_5O—$

phenophenanthrazine 苯并菲嗪
phenopiazine 对二氮(杂)萘
phenoplast 酚醛树脂
phenopyrine 二甲基苯基吡唑酮
phenoquinone 苯酚合苯醌
phenoresorcin 苯酚-间苯二酚混合物
phenosafranine 酚藏花红*
phenosal 吩那沙〔药〕
phenosalyl 苯酚、水杨酸、薄荷醇、乳酸混合物
phenose 酚糖 $C_6H_6(OH)_6$
phenoselenazine 吩硒嗪 $C_{12}H_9NSe$
phenosic acid 四代苯酚
phenostal 草酸二苯酯 $C_{14}H_{10}O_4$
phenosuccin 对乙氧苯基琥珀亚胺
phenothiazine(=thiodiphenylamine) 吩噻嗪
$C_6H_4NHC_6H_4S$
phenothiazine-type antioxidant 吩噻嗪型抗氧化剂〔润滑油用〕
phenothiazinyl 吩噻嗪基
phenothiazone 吩噻嗪酮 $C_{12}H_7NOS$
phenothioxin 吩噻噁 $C_{12}H_8OS$
phenotype 表型
phenoxarsine 吩噁吡 $C_{12}H_9AsO$
phenoxaselenin 吩噁硒 $C_{12}H_8OSe$
phenoxatellurin 吩噁碲 $C_{12}H_8OTe$
phenoxazine 吩噁嗪 $C_{12}H_9ON$
phenoxetol 苯氧基乙醇
phenoxide(=phenolate) ①酚盐②(苯)酚盐
phenoxthine(=phenothioxin; dibenzo thioxine) 吩噻噁
$C_6H_4OC_6H_4S$
phenoxy 苯氧基 $C_6H_5O—$
phenoxy acetaldehyde 苯氧基乙醛
phenoxy-acetamide 苯氧基乙酰胺 $C_6H_5OCH_2CONH_2$
phenoxy acetic acid 苯氧基乙酸 $C_6H_5OCH_2CO_2H$
phenoxy acetone 苯氧基丙酮 $C_6H_5OCH_2COCH_3$
phenoxy alkyl ammonium halide 苯氧基烷基卤化铵
phenoxybenzamine 苯氧苄胺
phenoxy benzamine hydrochloride 盐酸酚苄明
phenoxy benzoic acid 苯氧基苯甲酸 $C_6H_5OC_6H_4CO_2H$
phenoxy-butyronitrile 苯氧基丁腈 $C_6H_5O(CH_2)_3CN$
phenoxy caffeine 苯氧基咖啡碱
phenoxydinaphth-fuchsonedicarboxylic acid 苯氧基二萘并品红酮二羧酸；萘铬绿 G
phenoxy ethanol 苯氧基乙醇
phenoxyethyl acetate 乙酸苯氧乙酯
phenoxyethyl alcohol 苯氧基乙醇 $C_6H_5OCH_2CH_2OH$

phenoxy ethyl amide 苯氧基乙酰胺
phenoxy ethyl amine 苯氧乙基胺
phenoxyethyl isobutanoate 异丁酸苯氧基乙酯
phenoxymethylpenicillin 青霉素V〔药〕
phenoxy methylpenicillin potassium 苯氧基甲基青霉素钾
phenoxy propandiol 苯氧丙二醇
phenoxypropionitrile 苯氧基丙腈〔晶纹剂〕
phenoxy resin 苯氧基树脂
phenoxy silicate 苯氧基硅酸酯(盐)
phenpiazine 对二氮(杂)萘
phensuximide N-甲基-α-苯基琥珀酰亚胺
phenthiazine(=phenothiazine) 吩噻嗪
phenthiazone 吩噻嗪酮
phenthiol(=thiophenol) 苯硫酚 C_6H_5SH
phenthylcarbamide 苯乙基脲
phentolamine 酚妥拉明〔药〕
phentolamine hydrochloride 酚妥拉明〔药〕
phentriazine 苯并三嗪
phenyl 苯基 C_6H_5-
phenylacetaldehyde 苯乙醛 $C_6H_5CH_2CHO$
phenylacetaldehyde dibenzyl acetal 苯乙醛二苄缩醛
phenylacetaldehyde diethyl acetal 苯乙醛二乙缩醛
phenylacetaldehyde diisoamyl acetal 苯乙醛二异戊缩醛
phenyl acetaldehyde dimethyl acetal(=jacinthal) 苯乙醛二甲缩醛
phenylacetaldehyde glycerinacetal 苯乙醛甘油缩醛
phenylacetamide 苯乙酰胺 $C_6H_5CH_2CONH_2$
phenyl acetanilide(=phenacetylaniline) N-苯乙酰胺；N-苯乙酰苯胺 $C_6H_5CH_2CONHC_6H_5$
phenylacetate ①乙酸苯酯 $CH_3COOC_6H_5$ ②苯乙酸盐 $C_6H_5CH_2COOM$
phenylacetic acid 苯乙酸 $C_6H_5CH_2CO_2H$
phenylacetic aldehyde 苯乙醛
phenylacetic anhydride 苯乙(酸)酐 $(C_6H_5CH_2CO)_2O$
phenyl-acetone(=methyl benzylketone) 苯基丙酮；甲苄基(甲)酮
p-phenylacetophenone(=methyl biphenyl ketone) 对苯乙酮；甲基二苯(甲)酮 $C_{12}H_9COCH_3$
phenylacetyl 苯乙酰 $C_6H_5CH_2CO-$
phenylacetyl chloride 苯乙酰氯 $C_6H_5CH_2COCl$
phenylacetylene 苯(基)乙炔 $C_6H_5C\equiv CH$
phenylacetylglutamine 苯乙酰谷氨酰胺
phenylacetylglycine 苯乙酰甘氨酸；苯乙尿酸
phenylacridine 苯基吖啶 $C_6H_5C_{13}H_8N$
phenyl acrylate 丙烯酸苯酯
α-phenylacrylic acid α-苯基丙烯酸
phenylacrylic acid 苯基丙烯酸

2-phenylacrylic acid(=atropic acid) 2-苯基丙烯酸；阿托酸 $C_6H_5C(=CH_2)COOH$
2-phenylacryloyl(=atropoyl) 2-苯基丙烯酰；阿托酰 $C_6H_5C(=CH_2)CO-$
3-phenylacryloyl(=cinnamoyl) 3-苯基丙烯酰；肉桂酰 $C_6H_5CH=CHCO-$
phenylalanine 苯基丙氨酸 $C_6H_5CH_2CH(NH_2)CO_2H$
phenylalanine-3,4-quinone 苯丙氨酸-3,4-醌
phenylalanyl 苯丙氨酰(基)
phenyl aldehyde 苯甲醛
phenyl alkyl ether 苯烷基醚
phenyl alkylsulfonate 石油磺酸苯酯；石油酯〔增塑剂〕
phenylallene 苯丙烯
phenyl-allylene 苯丙炔
phenylamine 苯胺
2-phenylamino-3-methyl-6-(N-ethyl-p-toly amino) fluoran 2-苯基氨-3-甲基-6-(N-乙基-对-甲苯氨基)荧烷
2-phenyl-4-amylene aldehyde 2-苯基-4-戊烯醛
phenylaniline(=aminobiphenyl) 苯基苯胺；氨基联苯 $C_6H_5C_6H_4NH_2$
phenylanilineurea 苯基苯胺脲；二苯氨基脲
phenylanthracene 苯基蒽 $C_6H_5C_{14}H_9$
N-phenylanthranilic acid N-苯氨茴酸；邻苯氨基苯甲酸
phenyl-arsenimide 苯砷亚胺 $C_6H_5As=NH$
phenyl-arsenite 苯亚胂酸盐 $C_6H_5As(OM)_2$
phenyl-arsenoxide(=phenyl-arsine oxide) 苯亚胂氧化物
phenylarsine 苯胂 $C_6H_5AsH_2$
phenyl-arsine oxide 苯胂化氧 $C_6H_5As=O$
phenyl-arsine oxychloride 苯胂氧化二氯 $C_6H_5AsOCl_2$
phenyl-arsine sesquisulfide 双苯胂化三硫 $(C_6H_5)_2As_2S_3$
phenyl-arsine sulfide 苯胂化硫 $C_6H_5As=S$
phenyl-arsine tetrachloride 苯胂化四氯 $C_6H_5AsCl_4$
phenyl-arsonate 苯胂酸盐 $C_6H_5AsO(OM)_2$
phenylarsonic acid 苯胂酸；苯砷酸 $C_6H_5AsO(OH)_2$
phenyl-arsonous acid 苯亚胂酸 $C_6H_5As(OH)_2$
phenylate 苯醚 ROC_6H_5
phenylated 苯代的
phenylathranilic acid 苯基氨茴酸 $C_6H_5NHC_6H_4CO_2H$
phenylating 苯基化；引入苯基
phenylating agent 苯基化剂；加苯剂
phenylation 苯基化(作用)
phenylazide(=triazobenzene) 叠氮基苯 $C_6H_5N_3$
phenylazo 苯偶氮基 $C_6H_5N=N-$
1-phenylazo-2-anthrol 1-苯基偶氮-2-蒽酚
phenylazobenzoyl chloride 苯偶氮基苄酰氯
phenylazobenzthiazole 苯偶氮基苯并噻唑
phenylazojulolidine dyes 苯基偶氮久洛尼定啶染料
phenylazomethane 苯偶氮基甲烷

phenylazonaphthalene 苯偶氮基萘
phenylazo β-naphthylamine 苯基偶氮-β-萘胺
phenylazonaphthylamine 苯偶氮基萘胺
phenylazothiazole 苯偶氮基噻唑
phenylazothiophene 苯偶氮基噻吩
phenylbenzamide N-苯基苯酰胺
phenylbenzene(=biphenyl) 联(二)苯
phenylbenzhydryl 苯基羟苄基
phenylbenzoate ①苯甲酸苯酯 $C_6H_5COOC_6H_5$ ②苯基
    苯酸盐 $C_6H_5C_6H_4COOM$
N-phenylbenzohydoxamic acid N-苯基苄羟肟酸
2-phenyl-1,4-benzopyrone 2-苯基-1,4-苯并吡喃酮;黄酮
2-phenylbenzopyrylium ion 2-苯基苯并吡(喃)䓬;花
    (色)䓬
phenyl-benzoylene-urea 苯基亚苯酰基脲
    $C_6H_4NHCON(C_6H_5)CO$
phenyl-benzyl-carbinol 苯基苄基甲醇;1,2-二苯乙醇;
    均二苯代乙醇 $C_6H_5CHOHCH_2C_6H_5$
phenyl benzyl carbinyl acetate 乙酸苯基苄基原酯
phenylbenzyl ketone(=desoxybenzoin) 脱氧苯偶姻;二苯
    乙酮;苯苄基(甲)酮 $C_6H_5COCH_2C_6H_5$
phenyl benzyl phthalate 邻苯二甲酸苯基苄酯〔增塑剂〕
phenylbenzyl tin chloride 苯基苄基氯化锡
phenyl-benzylurethane 苯基苄基尿烷〔俗〕;N-苯苄氨
    基甲酸乙酯 $C_6H_5CH_2(C_6H_5)NCO_2C_2H_5$
phenylbiphenylyl oxadiazole(PBD) 苯基联苯基㗁二唑
phenyl-bismaleimide 苯基双马来酰亚胺
phenyl-borate 硼酸苯酯 $B(OC_6H_5)_3$
phenylboric acid(=phenyl-boron dihydroxide; phenyl
    boronic acid) 苯基硼酸 $C_6H_5B(OH)_2$
phenylborine 苯基甲硼烷
phenylboronate 苯基硼酸盐
phenyl boron dichloride 苯基二氯硼烷
phenyl-boron dihydroxide(=phenylboric acid) 苯基硼酸
phenyl boronic acid(=phenylboric acid) 苯基硼酸
    $C_6H_5B(OH)_2$
phenyl-bromide(=brombenzene) 苯基溴;溴苯 C_6H_5Br
phenyl bromoacetate 溴乙酸苯酯 $BrCH_2CO_2C_6H_5$
phenylbutazone 保泰松〔药〕
2-phenyl-2-butenal 2-苯基-2-丁烯醛
α-phenyl butyraldehyde α-苯基丁醛
phenyl butyrate 丁酸苯酯 $C_3H_7CO_2C_6H_5$
phenylbutyric acid 苯基丁酸 $C_6H_5(CH_2)_3COOH$
phenyl-cacodyl 苯基卡可基;四苯连砷
    $(C_6H_5)_2AsAs(C_6H_5)_2$
phenyl carbamate 氨基甲酸苯酯 $NH_2CO_2C_6H_5$
3-phenylcarbamido(=3-phenylureido) 苯氨基甲酰氨基;

苯脲基 $C_6H_5NHCONH—$
phenylcarbamoyl 苯氨羰基;苯氨基甲酰 $C_6H_5NHCO—$
phenylcarbinol 苯甲醇
phenyl carbitol 二乙二醇一苯醚
phenyl-carbonate O-苯基碳酸盐(酯) C_6H_5OCOOM;
    C_6H_5OCOOR
phenyl-carbonic acid O-苯基碳酸;碳酸一苯酯
    C_6H_5OCOOH
phenyl carbylamine dichloride 苯胩化二氯
    $C_6H_5N=CCl_2$
phenyl cellosolve(=oxitol) 苯基溶纤剂
phenylchinaldine 苯基甲基喹啉
phenylchinoline 苯基喹啉
phenyl-chloride 苯基氯;氯苯 C_6H_5Cl
phenyl chloroacetate 氯乙酸苯酯 $ClCH_2CO_2C_6H_5$
p-phenyl-ω-chloroacetophenone 对苯基-ω-氯乙酰苯
    $C_6H_5C_6H_4COCH_2Cl$
phenyl-chloroform 苯基氯仿;α,α,α-三氯甲苯
    $C_6H_5CCl_3$
phenyl cinnamate (肉)桂酸苯(酚)酯
phenyl-cinnamic acid 苯基肉桂酸;2-苯代肉桂酸
    $C_6H_5CH=C(C_6H_5)COOH$
phenyl citronellyl ether 苯基香茅醚
phenyl cresyl oxide 苯基甲酚醚
phenyl cresyl phthalate 邻苯二甲酸苯基甲苯酯〔增塑剂〕
phenylcrotonic acid 苯基巴豆酸
    $C_6H_5CH_2CH=CHCO_2H$
phenyl crotonylene 苯基巴豆炔 $C_6H_5CH_2C≡CCH_3$
phenylcumalin 苯(基)吡喃酮
phenyl-cyanamide 苯氨腈 C_6H_5NHCN
phenyl-cyanate 氰酸苯酯 C_6H_5OCN
phenyl cyanide 苯基氰
phenylcyclohexane(=cyclohexylbenzene) 苯基环己烷;环
    己基苯 $C_6H_5CH(CH_2)_4CH_2$
phenylcyclohexanol 苯基环己醇
phenyl cyclohexyl ether 苯基环己醚
N-phenyldiacetamide N-苯基二乙酰胺
phenyl-diaryl-oxyarsine 苯亚胂酸二芳酯
    $C_6H_5As(OR)_2$
phenyl-dibromide 二溴(代)苯 $C_6H_4Br_2$
phenyldibromoarsine 二溴化苯胂
phenyldicarbinol 苯二甲醇
phenyldichlorarsine 二氯化苯胂
phenyl-dichloride 二氯(代)苯 $C_6H_4Cl_2$
phenyldichloroarsine 苯基二氯胂;苯胂化二氯
    $C_6H_5AsCl_2$
phenyldichlorophosphine 苯基二氯膦;二氯苯膦

$C_6H_5PCl_2$
phenyldiethanol-amine　苯基二乙醇胺；苯基双羟乙基胺　$(HOCH_2CH_2)_2NC_6H_5$
phenyl-difluoride　二氟代苯　$C_6H_4F_2$
phenyl-dihalide　二卤代苯　$C_6H_4X_2$
phenyldihydronaphthalene　苯基二氢萘
phenyldihydroquinazoline　苯基二氢喹唑啉　$C_6H_4CH_2N(C_6H_5)CH=N$
phenyl-dihydroxy-arsine　苯亚胂酸　$C_6H_5As(OH)_2$
2-phenyl-4, 5-dihydroxymethyl imidazole　2-苯基-4,5-二羟甲基咪唑
phenyl-diiodide　二碘(代)苯　$C_6H_4I_2$
phenyl-dimethyl-arsine cyanobromide　溴氰化二甲苯砷　$(CH_3)_2C_6H_3AsBrCN$
phenyl-dimethyl-arsine dichloride　二氯化二甲苯砷　$(CH_3)_2C_6H_3AsCl_2$
phenyl-dimethyl-arsine-dihydroxide　二氢氧化二甲苯砷　$(CH_3)_2C_6H_3As(OH)_2$
N-phenyl-N'-(1,3-dimethylbutyl)-p-phenylenediamine　N-苯基-N'-(1,3-二甲基丁基)对苯二胺〔防老剂〕
phenyl dimethyl carbinyl acetate　乙酸苯基二甲基原酯
phenyldimethylpyrazolone　苯基二甲吡唑啉酮
phenyldiphenylcarbinol　苯基二苯基甲醇
phenyl disulfide　二硫苯
phenylene　亚苯基　—C_6H_4—
phenylenebisazo　苯双偶氮基　—N=NC_6H_4N=N—
phenylene blue　吲达胺
phenylene-diacetic acid　苯二乙酸　$C_6H_4(CH_2COOH)_2$
phenylene diamine　苯二胺
p-phenylenediamine derivative antioxidant　对苯二胺衍生物防老剂
phenylene-diarsonic acid　苯二砷酸　$C_6H_4(AsH_2O_3)_2$
phenylene diazo　苯双偶氮基　—N=NC_6H_4N=N—
phenylene diazo sulfide　苯双偶氮硫
m-phenylene diisocyanate　间-苯二异氰酸酯
p-phenylene diisocyanate　对-亚苯基二异氰酸酯
N,N'-m-phenylene dimaleimide　N,N'-间亚苯基双马来酰亚胺〔硫化剂〕
phenylene dimercaptan　苯二硫醇
phenylenedimethylene　苯二甲基〔邻、间或对〕—$H_2CC_6H_4CH_2$—
phenylenedimethylidyne　苯二亚甲基〔邻、间或对〕=HCC_6H_4CH=
phenylene-1,3-disulfonyl hydrazide　亚苯基-1,3-二磺酰肼〔发泡剂〕
m-phenyleneguanidime　间亚苯胍
phenylene-sulfourea(=phenylene thiourea)　亚苯基硫脲
phenylene thiourea(=phenylene-sulfourea)　亚苯基硫脲
phenylene urea　亚苯基脲
phenylephedrine hydrochloride　苯肾上腺素；去氧肾上腺素〔药〕
phenylephrine　去氧肾上腺素〔药〕
phenyl ester　苯酯
phenylethane　苯乙烷
phenylethanol　苯基乙醇
phenylethanol amine　苯乙醇胺
phenyl ether　①苯醚　RO$C_6H_5$②二苯醚　$C_6H_5OC_6H_5$
phenylethyl　苯乙基
phenylethyl acetate　乙酸苯乙酯　$C_6H_5C_2H_4O_2CCH_3$
phenyl ethyl alcohol　苯乙醇
β-phenyl-ethylamine　β-苯基乙胺
phenylethyl anthranilate　邻氨基苯甲酸苯乙酯
phenylethyl barbituric acid　苯巴比妥
phenylethyl benzoate　苯甲酸苯乙酯　$C_6H_5CO_2C_2H_4C_6H_5$
phenylethyl butyrate　丁酸苯乙酯
phenylethyl cinnamate　(肉)桂酸苯乙酯
phenylethyl dimethyl carbinol　苯乙基二甲基原醇
phenylethyl dimethyl carbinyl acetate　乙酸苯乙基二甲基原酯
phenylethyl dimethyl carbinyl isovalerate　异戊酸苯乙基二甲基原酯
phenylethylene　苯基-1, 2-亚乙基　$C_6H_5CHCH_2$—
phenylethylene oxide(=styrene oxide)　氧化苯乙烯　$C_6H_5CHOCH_2$
phenylethyl ethoxyacetate　乙氧基乙酸苯乙酯
phenylethyl formate　甲酸苯乙酯
phenylethyl hydantoin　5, 5-苯基乙基乙内酰脲
phenylethyl isobutyrate　异丁酸苯乙酯
phenylethyl isovalerate　异戊酸苯乙酯
phenylethyl methoxyacetate　甲氧基乙酸苯乙酯
phenylethyl methyl ethyl carbinyl acetate　乙酸苯乙基甲基乙基原酯
phenyl-ethyl-mustard oil　苯乙基芥子油；异硫氰酸苯乙酯　$C_6H_5CH_2CH_2NCS$
phenylethyl phenylacetate　苯乙酸苯乙酯
phenylethyl propionate　丙酸苯乙酯
phenylethyl salicylate　水杨酸苯乙酯；柳酸苯乙酯
phenylethyl valerate　戊酸苯乙酯
ω-phenyl fatty acid　端苯基脂肪酸
phenyl-fluoride　苯基氟；氟苯　C_6H_5F
phenyl fluorone　苯基荧光酮
phenylformamide　N-甲酰基苯胺
phenyl formate　甲酸苯酯　$HCO_2C_6H_5$

phenylformic acid 苯甲酸
phenyl geranyl ether 苯基香叶醚
phenyl glucosazone 苯基葡萄糖脎〔即：葡萄糖脎〕
  $(C_6H_5NHN)_2C_6H_{10}O_4$
phenylglucosidase 苯基葡萄糖苷酶
phenyl glucuronide(=phenol glucuronic acid) 葡萄糖苯苷酸
phenyl glycidyl ether 苯基缩水甘油醚〔稀释剂〕
phenylglycine 苯基甘氨酸 $C_6H_5NHCH_2CO_2H$
phenylglycine-o-carboxylic acid 苯基甘氨酸邻羧酸
  $HO_2CC_6H_4NHCH_2CO_2H$
phenylglycine ethyl ester 苯基甘氨酸乙酯
  $C_6H_5NHCH_2CO_2C_2H_5$
phenylglycocoll 苯基甘氨酸
phenyl-glycol diacetate 苯乙二醇二乙酸酯
phenyl-glycolethylene ether 苯乙二醇次乙醚
phenylglycollic acid 苯乙醇酸
phenyl-glycolmethylene ether 苯乙二醇次甲醚
phenyl-glycolmonoacetate 苯乙二醇单乙酸酯
phenylglyoxal 苯甲酰甲醛
phenylglyoxylic acid 苯甲酰甲酸
phenyl group 苯基
phenyl-halide 苯基卤 C_6H_5X
phenyl heteric polyols 苯基混嵌多元醇〔表面活性剂〕
phenyl-hexabromide 六溴(代)苯 C_6Br_6
phenyl-hexachloride 六氯(代)苯；六六六 C_6Cl_6
phenyl-hexafluoride 六氟(代)苯 C_6F_6
phenyl-hexahalide 六卤(代)苯 C_6X_6
phenyl-hexaiodide 六碘(代)苯 C_6I_6
phenyl hexyl ketone 苯基己基(甲)酮 $C_6H_5COC_6H_{11}$
N-phenyl-N'-hexyl-p-phenylene-diamine N-苯基-N'-己基对苯二胺〔防老剂〕
phenyl hydrate 苯酚
phenyl hydrazine 苯肼 $C_6H_5NHNH_2$
phenylhydrazine hydrochloride 盐酸(化)苯肼
phenylhydrazine levulinic 苯肼乙酰丙酸
phenyl hydrazine-p-sulfonic acid 苯肼对磺酸；对肼基苯磺酸 $H_2NNHC_6H_4SO_3H$
phenylhydrazine urea 苯肼脲
phenylhydrazone 苯腙 (=NNHPh)
phenylhydrazoneacetal 乙醛苯腙
phenylhydrazoquinoline 苯肼基喹啉 $C_6H_5(NH_2C_9H_6N)$
phenyl-hydrogen-sulfate 苯硫酸；硫酸苯氢酯 $C_6H_5HSO_4$
phenyl-hydrosulfide 苯硫酚 C_6H_5SH
phenyl hydroxide 苯酚
phenylhydroxylamine 苯胲 C_6H_5NHOH
phenylic 苯基的

phenylic acid ①(苯)酚；石炭酸②酚 C_6H_5OH
phenylid 苯胺
phenylidene(=cyclohexadienylidene) 环己二烯亚基
2-phenylimidazole 2-苯基咪唑
phenylimino 苯亚氨基 $C_6H_5N=$
phenyl indane dicarboxylic acid 苯基茚满二羧酸
phenylindanedione 苯基二氢化茚-1,3-二酮
phenyl-iodide 碘代苯；苯基碘 C_6H_5I
phenyl iodochloride(=iodobenzene dichloride) 二氯化碘代苯 $C_6H_5ICl_2$
phenyl ionone 苯基紫罗兰酮
phenyl isoamyl carbinyl acetate 乙酸苯基异戊基原酯
phenyl isoamyl ketone 苯基异戊基(甲)酮；异己酰苯 $C_5H_{11}COC_6H_5$
phenyl isobutyl carbinyl acetate 乙酸苯基异丁基原酯
phenyl isobutyl ketone 苯基异丁酮
phenyl isocyanate 异氰酸苯酯 $C_6H_5N=CO$
phenyl-isocyanide 苯胼 C_6H_5NC
phenylisonitramine N-异硝基苯胺
phenyl isopropyl ketone 苯基异丙基(甲)酮；异丁酰苯 $C_6H_5COC_3H_7$
N-phenyl-N'-isopropyl-p-phenylenediamine N-苯基-N'-异丙基对苯二胺〔防老剂〕
phenyl-isorhodanate(=phenyl isothiocyanate) 异硫氰酸苯酯
phenyl-isorhodanide(=phenyl isothiocyanate) 异硫氰酸苯酯
phenyl-isosulfocyanate(=phenyl isothiocyanate) 异硫氰酸苯酯
phenyl-isosulfocyanide(=phenyl isothiocyanate) 异硫氰酸苯酯
phenyl isothiocyanate(=phenyl-isorhodanate；phenyl-isorhodanide；phenyl-isosulfocyanate；phenyl-isosulfocyanide；phenyl-isothiocyanide) 异硫氰酸苯酯 $C_6H_5N=C=S$
phenyl-isothiocyanate method 异硫氰酸苯酯法
phenyl-isothiocyanide(=phenyl isothiocyanate) 异硫氰酸苯酯
phenylium 苯氧乙酸
phenyl ketone 二苯(甲)酮
phenylketonuria 苯丙酮酸尿(症)
phenyllactazam 苯基内酰联胺
phenyl-lactic acid 苯基乳酸 $CH_3C(C_6H_5)OHCOOH$
phenyl linalyl ether 苯基芳樟醚
phenyl lithium 苯基锂 C_6H_5Li
phenyl-magnesium-bromide 苯基溴化镁 $MgBrC_6H_5$
phenyl-magnesium-chloride 苯基氯化镁 $MgClC_6H_5$
phenyl-magnesium-halide 苯基卤化镁 C_6H_5MgX

N-phenyl melamine N-苯基三聚氰胺
phenyl menthyl ether 苯基薄荷醚
phenylmercaptan(=thiophenol) 苯硫酚 C_6H_5SH
β-phenylmercaptobutylaldehyde β-苯基硫醇基丁醛〔防老剂〕
3-phenyl-5-mercapto-1,3,4-thiazole-2-thione 3-苯基-5-硫醇基-1,3,4-噻二唑-2-硫酮〔促进剂〕
phenyl mercurial(=mercury phenyl) 二苯基汞
phenyl mercuric acetate 乙酸苯汞
phenyl mercuric acrylate 丙烯酸苯汞
phenylmercuric bromide 苯基溴化汞 C_6H_5HgBr
phenylmercuric chloride 苯基氯化汞 C_6H_5HgCl
phenyl mercuric dioctyl sulfosuccinate 琥珀酸二辛酯磺酸苯汞
phenylmercuric iodide 苯基碘化汞 C_6H_5HgI
phenyl mercuric nitrate 硝酸苯汞
phenyl mercuric propionate 丙酸苯汞
phenyl mercuric salicylate 水杨酸苯汞
phenyl-mercuric-salt 苯基汞化卤 $HgXC_6H_5$
phenyl mercury borate 硼酸苯汞
phenyl mercury dodecenyl succinate 丁二酸十二烯酯苯汞
phenyl mercury oleate 油酸苯汞
phenyl mercury propionate 丙酸苯汞
phenyl mercury succinate 琥珀酸苯汞
phenylmethane 甲苯
phenyl-methyl-arsinic acid 苯基甲基次胂酸 $C_6H_5(CH_3)AsOH$
1-phenyl-3-methyl-6-t-butyl phenol 1-苯基-3-甲基-6-叔丁基苯酚〔防老剂〕
phenyl-methyl-carbinol 苯基甲基甲醇 $C_6H_5CHOHCH_3$
phenyl methyl carbinyl acetate 乙酸苯基甲基原酯；乙酸苏合香酯
phenyl methyl carbinyl formate 甲酸苯基甲基原酯
phenyl methyl carbinyl propionate 丙酸苯基甲基原酯
phenylmethylether 茴香醚
N-phenyl-(N'-methylheptyl)-p-phenylenediamine N-苯基-(N'-甲基庚基)对苯二胺〔防老剂〕
2-phenyl-4-methyl-5-hydroxymethyl imidazole 2-苯基-4-甲基-5-羟甲基咪唑
2-phenyl-4-methyl imidazole 2-苯基-4-甲基咪唑
phenyl methyl ketone 乙酰苯；甲基苯基(甲)酮 $C_4H_5ON_2C_6H_5$
phenyl-methyl silicone rubber 苯基甲基硅橡胶
phenyl-monobromide 一溴(代)苯 C_6H_5Br
phenyl-monochloride 一氯(代)苯 C_6H_5Cl
phenyl-monofluoride 一氟(代)苯 C_6H_5F
phenyl-monohalide 一卤(代)苯 C_6H_5X

phenyl-monoiodide 一碘代苯 C_6H_5I
N-phenyl-morpholine N-苯基吗啉
  $(CH_2)_2O(CH_2)_2NC_6H_5$
phenyl-mustard oil 苯基芥子油；异硫氰酸苯酯 C_6H_5NCS
phenylnaphthalene 苯基萘 $C_6H_5C_{10}H_7$
phenyl-1-naphthylamine 苯基-1-萘胺 $C_6H_5NHC_{10}H_7$
N-phenyl-α-naphthylamine N-苯基-α-萘胺〔防老剂〕
phenyl-β-naphthylamine 苯基-β-萘胺
phenylnaphthyl carbazole 苯基-α-萘基咔唑
phenylnaphthyl ketone 苯萘基(甲)酮
N-phenyl-N'-β-naphthyl-p-phenylenediamine N-苯基-N'-β-萘基对苯二胺〔防老剂〕
phenylnitramine 苯硝胺；N-硝基苯胺
phenyl-nitro-cinnamic acid 硝基苯基肉桂酸 $NO_2C_6H_4CH=CHCOOH$
phenyl nitromethane 苯基硝基甲烷
phenylnitrone 苯基硝酸灵
phenylo 苯基
phenylo boric acid 苯硼酸 $C_6H_5B(OH)_2$
N-phenyl-N'-2-octyl-p-phenylenediamine N-苯基-N'-2-辛基对苯二胺〔防老剂〕
phenylog 联苯物
phenylogic series 联苯物系列〔如 $C_6H_6, C_{12}H_{10}, C_{18}H_{14}$ 等〕
phenylon 安替比林〔药〕
phenylor 苯酚基；羟苯基
phenylo salicylic acid 苯基水杨酸
phenylosazone 苯脎
phenyl-oxalacetic ester 苯基草乙酸酯 $ROCOCOCH(C_6H_5)COOR; C_2H_5OCOCOCH(C_6H_5)COOC_2H_5$
phenyloxalate 草酸二苯酯
phenyloxamic acid(=oxanilic acid) 苯胺羰酸 $C_6H_5NHCOCOOH$
phenyloxydisulfide 二硫化二苯氧
phenylparaconic acid 苯基仲康酸；苯基丁内酯-β-甲酸
phenyl-paraffin alcohols 苯基脂醇
phenyl-pentabromide 五溴(代)苯 C_6HBr_5
phenyl-pentachloride 五氯(代)苯 C_6HCl_5
phenyl-pentafluoride 五氟(代)苯 C_6HF_5
phenyl-pentahalide 五卤(代)苯 C_6HX_5
phenyl-pentaiodide 五碘代苯 C_6HI_5
1-phenylpentane 1-苯基戊烷
3-phenyl-4-pentenal 3-苯基-4-戊烯醛
phenyl-peri acid 苯基-1-萘胺-8-磺酸
α-phenylphenacyl(=desyl) 二苯乙酮基；α-苯甲酰苯甲基
phenylphenol 苯基苯酚

o-phenyl phenol　邻苯基苯酚〔防霉剂〕
phenyl-phosphate　磷酸苯酯　$C_6H_5OPO(OH)_2$
phenyl-phosphenylic acid(=phenylphosphinic acid)　①苯膦酸②苯次膦酸
phenylphosphine(=phosphaniline)　苯膦　$C_6H_5PH_2$
phenylphosphinic acid(=phenyl-phosphenylic acid)　①苯膦酸　$C_6H_5OP(OH)_2$②苯次膦酸　C_6H_5HPOOH
phenyl-phosphite　亚磷酸苯酯　$C_6H_5OP(OH)_2$
phenyl-phospho-acid(=phenyl-phosphonic acid)　苯(基)膦酸　$C_6H_5PO(OH)_2$
phenyl-phosphonic acid(=phenyl-phospho-acid)　苯(基)膦酸
phenyl phosphonyl chloride　苯基膦酰二氯　$C_6H_5POCl_2$
phenylphosphonyl dichloride(=phosphenyl oxychloride)　苯膦酰二氯　$C_6H_5POCl_2$
phenyl phosphonyl-difluoride　苯基膦酰二氟
phenyl-phosphorous acid　苯膦酸　$C_6H_5PO(OH)_2$；$C_6H_5OP(OH)_2$
phenyl-phthalamic acid　苯基邻氨甲酰基苯甲酸　$C_6H_5C_6H_3(CONH_2)COOH$
N-phenylphthalimide(=phthalanil)　N-苯基邻氨甲酰亚胺；N-邻氨甲酰苯胺　$C_6H_5N=C_8H_4O_2$
phenyl polychloride　多氯代苯
phenyl polyfluoride　多氟代苯
phenyl polyhalide　多卤代苯
phenyl polyiodide　多碘代苯
phenyl potassium　苯基钾　KC_6H_5
2-phenyl propane　异丙苯；枯烯
1-phenyl-1, 2-propanedione-2-(O-ethoxycarbonyl) oxime　1-苯基-12-丙二酮-2-(O-乙氧甲酰基)肟
phenylpropanol　苯丙醇〔药〕
phenylpropiolic acid　苯丙炔酸　$C_6H_5C\equiv CCO_2H$
β-phenylpropionaldehyde　β-苯丙醛；氢化(肉)桂醛
phenylpropionaldehyde cyclic ethylene acetal　苯丙醛环乙二醇缩醛
phenylpropionaldehyde dimethyl acetal　苯丙醛二甲缩醛
phenylpropionate　丙酸苯酯　$C_2H_5CO_2C_6H_5$
phenylpropionic acid　苯基丙酸
phenylpropiophenone　苯基苄基(甲)酮
phenylpropyl acetate　乙酸苯丙酯
phenylpropyl alcohol　苯丙醇
phenylpropyl aldehyde　苯丙醛
phenylpropyl cinnamate　(肉)桂酸苯丙酯
phenylpropyl formate　甲酸苯丙酯
phenyl propyl ketone　苯丁酮
α-(3-phenylpropyl)-tetrahydrofuran　α-(3-苯丙基)四氢呋喃
3-phenyl-1-propyne　3-苯基-1-丙炔
phenylpyrazolone　苯基吡唑酮　$C_9H_8N_2$
phenylpyridine　苯基吡啶

phenyl-pyruvic acid　苯丙酮酸　$C_6H_5CH_2COCOOH$
phenylpyruvic oligophrenia　苯丙酮酸尿性智力发育不全
phenylquinaldine　苯基喹哪啶；苯基甲喹啉　$C_{16}H_{13}N$
phenylquinoline　苯基喹啉　$C_{15}H_{11}N$
phenyl rhodanate(=phenyl rhodanide)　硫代氰酸苯酯　C_6H_5SCN
phenyl rhodanide(=phenyl rhodanate)　硫代氰酸苯酯
phenyl salicylate(=salol)　水杨酸苯酯；萨罗　$HOC_6H_4CO_2C_6H_5$
phenyl salicylic acid　苯基水杨酸
phenyl-selenide　二苯硒　$(C_6H_5)_2Se$
phenyl silane　苯基硅烷
phenylsilane resin　苯硅烷树脂
phenyl silicone　聚苯基硅氧烷
phenyl sodium　苯基钠　NaC_6H_5
phenyl stearic acid　苯基硬脂酸
phenyl stearic amine　苯基硬脂基胺
N-phenyl succimide(=succinanil)　琥珀酰苯胺　$(CH_2CO)_2NC_6H_5$
phenylsulfinyl(=benzene sulfinyl)　苯亚磺酰　C_6H_5SO-
N-phenylthiophthalimide　N-苯基硫代邻苯二甲酰亚胺〔防焦剂〕
phenyl-thiourea　苯硫脲　$C_6H_5NHCSNH_2$
phenyl-tiglic acid　苯基惕各酸　$C_6H_5CH=C(C_2H_5)CO_2H$
phenyltin　①二苯锡②苯锡基　$PhSn$
phenyl-tin-chloride　三氯化苯锡　$C_6H_5SnCl_3$
phenyltin tribenzyl　苯基三苄基锡
phenyltoin　二苯基乙内酰脲钠
phenyltoluene　苯基(代)甲苯
N-phenyl-N'-(p-toluenesulfonyl)-p-phenylenediamine　N-苯基-N'-(对甲苯磺酰基)对苯二胺
phenyl toluene-sulphonate　甲苯磺酸苯酯
phenyltolyl　苯基甲苯
phenyl-o-tolyl guanidine　苯基邻甲苯胍〔促进剂〕
phenyl tolyl ketone　苯基甲苯基(甲)酮
phenyl triazine derivative　苯基三嗪衍生物
phenyl tribromide　三溴(代)苯　$C_6H_3Br_3$
phenyl-trichloro-silane　苯基三氯硅烷　$C_6H_5SiCl_3$
phenyl-trichloro-silicane　苯基三氯硅烷　$C_6H_5SiCl_3$
phenyl triethoxysilane　苯基三乙氧基硅烷〔交联剂〕
phenyl-triethyl-silane　苯基三乙基硅烷　$C_6H_5Si(C_2H_5)_3$
phenyl-triethyl-silicane　苯基三乙基硅烷　$C_6H_5Si(C_2H_5)_3$
phenyl trifluoride　三氟(代)苯　$C_6H_3F_3$
phenyl trihalide　三卤(代)苯　$C_6H_3X_3$
phenyl triiodide　三碘(代)苯　$C_6H_3I_3$
phenyl trimethicone　苯基三甲基硅氧烷
phenyltrimethyl ammonium hydroxide　氢氧化-N-苯基三

甲铵
phenyl-trimethylammonium iodide 碘化-N-苯基三甲铵
6-phenyl-2,2,4-trimethyl-1,2-dihydroquinoline 6-苯基-2,2,4-三甲基-1,2-二氢化喹啉〔防老剂〕
phenyl-trimethylsilane 苯基三甲基硅烷 $C_6H_5Si(CH_3)_3$
phenyl undecyl ketone 苯基十一(烷)酮
phenyl-urea 苯脲 $C_6H_5NHCONH_2$
3-phenylureido(=3-phenylcarbamido) 苯氨基甲酰氨基；苯脲基 $C_6H_5NHCONH—$
N-phenylurethane 苯氨基甲酸乙酯 $C_6H_5NHCO_2C_2H_5$
phenylvaleric acid 苯戊酸
phenyl vinyl ketone 苯基乙烯基(甲)酮
phenyl-xanthate 苯基黄原酸盐 C_6H_5OCSSM
phenyl-xanthogenate 苯基黄原酸盐 C_6H_5OCSSM
phenyl-xanthogenic acid 苯基黄原酸 C_6H_5OCSSH
phenyl-xanthonate 苯基黄原酸盐 C_6H_5OCSSM
phenyl-xanthonic acid 苯基黄原酸 C_6H_5OCSSH
phenyl-xanthydrol 苯基呫吨氢醇 $C_6H_5C_{13}H_8O(OH)$
phenytoin sodium 苯妥英钠〔药〕
phenzoline 苯基二氢喹唑啉
pheohemin 黑氯血红素；氯铁黑卟啉
pheophorbide 脱镁叶绿甲酯一酸
pheophytin 脱镁叶绿素
pheoretin 大黄胶素
pheromone 信息素*
pheron 酶蛋白；脱辅基酶
phesin 非生〔药〕；磺酸钠非那西汀
pH glass electrode pH 玻璃电极
pH gradient elution pH 梯度洗脱
phial(=vial) 管(形)瓶
-philic〔希腊字尾〕 亲〔在中文中为词头〕
phillipite 天然铜铁矾
Phillips beaker 菲利普烧杯
Phillips catalytic isomerization 菲利普催化异构化(法)
Phillips gauge 菲利普真空计
phillipsite 钙十字石
Phillips process 菲利普法〔丁烷脱氢制丁二烯法〕
phillygenin 非丽配基 $C_{21}H_{24}O_6$
phillyrigenin 非丽属配基 $(C_{30}H_{48}O_4)$
phillyrin 非丽苷；连翘苷
philosopher's stone 哲人石；点金石
philosopher's wool 氧化锌
philosophical chemistry 哲理化学
phinatas oil 含酚杀虫油
pH indicator pH 指示剂
pH indicator absorbance ratio method pH 指示剂吸光度比值测定法
phlean 梯牧草果聚糖

phlegma 冷凝液
phleocidin 杀草菌素
phleomycin 腐草霉素
phlobaphene 栎鞣红；红粉
phlobatannin 红粉鞣质
phloem fibre 韧皮纤维
phlogistication 除氧(作用)
phlogistic theory 燃素(学)说
phlogiston 燃素
phlogiston theory 燃素(学)说
phlogopite 金云母
phlolaphenes 脂酸焦儿茶酚酯
phloracylophenone 根皮酰苯
phloretic acid 根皮酸 $C_9H_{10}O_3$
phloretin 根皮素；根皮苷配基
phlorhizin 根皮苷
phloridzin(=phlorizin) 根皮苷
phlorite 红块高岭土
phlorizein 氧化根皮苷
phlorizin(=phloridzin) 根皮苷
phloroacetophenone 根皮乙酰苯；乙酰间苯三酚 $C_6H_2(OH)_3COCH_3$
phlorobenzophenone 苯根皮酚
phloroglucin(=phloroglucinol) 间苯三酚；均苯三酚；藤黄酚；根皮酚 $(HO)_3C_6H_3$
phloroglucinol(=phloroglucin) 间苯三酚；均苯三酚；藤黄酚；根皮酚 $(HO)_3C_6H_3$
phloroglucinol phthalein 间苯三酚酞；梧子色素
phloroglucinol reaction 间苯三酚反应
phloroglucinol triethyl ether 间苯三酚三乙醚 $C_6H_3(OC_2H_5)_3$
phloroglucinol trimethyl ether 间苯三酚三甲醚 $C_6H_3(OCH_3)_3$
phloroglucinol trioxime 间苯三酚肟；环己间三肟 $C_6H_6(=NOH)_3$
phloroglucinol triphenyl ether 间苯三酚三苯醚 $C_6H_3(OC_6H_5)_3$
phloroglucite 1,3,5-环己三醇
phloroglucitol 间环己烷三醇；1,3,5-环己烷三醇 $C_6H_{12}O_3$
phlorol 邻乙基苯酚 $C_8H_{10}O$
phlorone 对二甲基苯对醌 $C_8H_8O_2$
phlorose(=α-glucose) 根皮糖；α-葡萄糖
phloxin 根皮红；四溴二氯荧光黄
phloxine 焰红染料；酸性红色染料
phloxine toner 焰红色原(色料)
pH meter pH 计*
pH monitor pH 监控器

pH number　pH 值
-phobic〔希腊词尾〕　疏；憎〔在中文中为词头〕
phocenic acid(=valeric acid)　戊酸
phocenin　甘油三戊酸酯　$C_{18}H_{32}O_6$
phoenicin(=phenicin)　芬尼菌素
phoenicine　绯红素
phoenicochroite　红铬铅矿
phoeophorbide　脱镁叶绿酸
phoeophorbin　脱镁叶绿二酸
phoeophytin　脱镁叶绿素
pholcodine　福尔可定；啉吗啡；福尔定〔药〕
phomalactone　基点霉内酯
phomine　基点菌素
phonene　同声组
phonic　声的
phono-〔希腊字头〕　声；音
phonochemical　声化学的
phonochemical reaction　声化(学)反应
phonochemistry　声化学
phonolite　响岩
phonometer　声强计
phonon　声子
phonosensitive　感声的
phorbide　脱镁叶绿环类
phorbin　①脱镁叶绿母环类②脱镁叶绿环类
phorbol　佛波醇
phorene　佛尔烯
phoretic　原子间致导电性的
phormium fiber　新西兰麻纤维〔取自 phormium tenax〕
phormium tenax　新西兰麻
phorone(=sym-diisopropylidene acetone)　佛尔酮；对称亚异丙基丙酮　$[(CH_3)_2C=CH]_2CO$
phoronomy　(运)动、时(间)与相对性学
phorylresin　磷酰酚树脂
phosgenation　光气化(作用)
phosgene　光气*
phosgenite　角铅矿
phosistor　光敏晶体管
phospha-　磷杂
phosphagen(=phosphocreatine)　磷酸肌酸　$H_2PO_3NHC(=NH)N(CH_3)CH_2COOH$
phospham　磷胺　PN_2H
phosphaniline(=phenylphosphine)　苯膦
phosphaphenanthrene oxides　氧化磷杂菲类
phospharseno　偶磷砷基　—P=As—
phosphatase　磷酸(酯)酶
phosphate　磷酸盐(酯)
phosphate acetyltransferase　磷酸乙酰基转移酶

phosphate bead　磷酸盐熔珠(试验)
phosphate bead reaction　磷酸盐熔珠反应
phosphate bond energy　磷酸键能
phosphate buffer　磷酸盐缓冲剂
phosphate buffered saline(PBS)　磷酸盐缓冲溶液
phosphate buffered saline tween(PBS-T)　加吐温的磷酸盐缓冲溶液
phosphate-buffer solution(PBS)　磷酸盐缓冲溶液
phosphate builder　磷酸盐助剂
phosphate coating process　磷酸盐被膜处理法
phosphate-containing detergent　含磷酸盐洗涤剂
phosphate crown　磷铬黄
phosphated　磷酸盐化的
phosphated alcohol　磷酸化(脂肪)醇；(脂肪)醇磷酸酯盐
phosphate desulphurization(=phosphate process)　磷酸脱硫过程
phosphated fabric　磷酸盐化织物
phosphated glyceride　磷酸化甘油酯
phosphated metal　磷酸盐化金属
phosphated polyoxyethylenated alcohol　磷酸化聚氧乙烯脂肪醇；脂肪醇聚氧乙烯醚磷酸(酯)盐
phosphated polyoxyethylenated alkylphenol　磷酸化聚氧乙烯烷基酚；烷基酚聚氧乙烯醚磷酸(酯)盐
phosphate ester　磷酸酯
phosphate fertilizer　磷肥(料)
phosphate-free detergent　无磷酸盐洗涤剂
phosphate glass solidification　磷酸盐玻璃固化(法)
phosphate ion　磷酸根离子
phosphate method　磷酸盐法
phosphate of ammonia　磷酸铵　$(NH_4)_3PO_4$
phosphate of lime　磷酸钙　$Ca_3(PO_4)_2$
phosphate plasticizer　磷酸酯(类)增塑剂
phosphate potential　磷酸势
phosphate process(=phosphate desulphurization)　磷酸脱硫过程
phosphate rock　磷酸盐岩
phosphate solution pump　磷酸盐溶液泵
phosphate tannage　磷酸盐鞣(法)
phosphate tanning　磷酸盐鞣
phosphate treatment　磷化处理；磷酸基处理
phosphate-type surfactant　磷酸盐(或酯)型表面活性剂
phosphate weigher　磷酸盐量计
phosphatic　磷酸盐的
phosphatic feed　磷酸盐饲料
phosphatic fertilizer　磷肥(料)
phosphatic guano　磷酸鸟粪石
phosphatic manure　磷肥(料)
phosphatic rock　磷灰石

phosphatic slag 含磷熔渣
phosphatidalcholine 缩醛磷脂酰胆碱
phosphatidalserine 缩醛磷脂酰丝氨酸
phosphatidate ①磷脂酸②磷脂酸盐(酯或根)
phosphatide 磷脂
phosphatidic acid 磷脂酸
phosphatidylcholinase 磷脂酰胆碱酶
phosphatidylcholine 磷脂酰胆碱；卵磷脂
phosphatidyl ethanolamine 磷脂酰乙醇胺
phosphatidyl glycerol 磷脂酰甘油
phosphatidylinositol 磷脂酰肌醇
phosphatidyl serine 磷脂酰丝氨酸
phosphating 磷酸盐化；磷化〔金属的表面处理〕
phosphating classification 磷化处理分类；磷化处理种类
phosphating coat 磷化膜
phosphating ferrous metal 黑色金属的磷化(处理)
phosphating plant 磷(酸盐)化装置
phosphating process 磷(酸盐)化处理法
phosphatins (动物组织中)有机磷酸化物
phosphation 磷酸化
phosphatization 磷化
phosphatize ①磷(酸盐)化②(=phosphate coat)磷化膜
phosphatizing 磷化
phosphato-dimolybdic acid 磷二钼酸 $H_3PO_2(MoO_4)_2$
phosphato-molybdic acid 磷钼酸 $H_3PO_2 \cdot MoO_4$
phosphato-titanate 磷酸酯合钛酸酯
phosphato-tungstate 磷钨酸盐
phosphato-tungstic acid 磷钨酸 $P_2O_5 \cdot mWO_3 \cdot nH_2O$
phosphaturia 磷酸盐尿
phosphazene 磷腈
phosphazene polymer 磷腈(类)聚合物
phosphazide 叠氮膦
phosphazine 膦嗪
phosphazo 偶磷氮基 —P═N—
phosphene(=phosphurane) 磷杂环戊二烯
phosphenic acid 氧次膦酸 $O═P(O)(OH)$
phosphenous acid 氧卑膦酸 $O═POH$
phosphenyl 苯膦基
phosphenylic acid 苯(基)膦酸 $C_6H_5PO(OH)_2$
phosphenylic oxychloride 苯膦氧化二氯 $C_6H_5POCl_2$
phosphenyl oxychloride(=phenylphosphonyl dichloride) 苯膦酰二氯 $C_6H_5POCl_2$
phosphide 磷化物
phosphinate polymer 次膦酸盐聚合物
phosphine 膦*
phosphine borine 膦硼烷
phosphine-bridged polymer 膦桥聚合物
phosphine imide 亚胺膦 H_3PNH

phosphine oxide 氧(化)膦 $H_3PO; R_3PO$
phosphine sulfide 硫(化)膦；三烃硫膦 $R_3PS; H_3PS$
phosphinic acid ①次膦酸 $RHPOOH; R_2POOH$②〔混用时，指 phosphonic acid〕膦酸 $RP(O)(OH)_2$
phosphinico 磷酸亚基 $(HO)OP═$
phosphinidene 亚膦基 $HP═$
phosphinidyne 次膦基 $P≡$
phosphinimine 膦亚胺
phosphinimyl 亚氨膦基 $H_2(HN)P—$
phosphinimylidene 亚氨亚膦基 $H(HN)P═$
phosphinimylidyne 亚氨次膦基 $(HN)P≡$
phosphinium 膦鎓 PH_4^+
phosphino 膦基 $H_2P—$
phosphino aryl boranes 膦芳基硼烷类
phosphinodifluorophosphine 偏二氟双膦 H_2PPF_2
phosphinoso 羟亚膦基 $HOP═$
phosphinothioyl 硫膦基 $H_2(S)P—$
phosphinothioylidene 硫亚膦基 $H(S)P═$
phosphinous acid 三价膦酸 $R_2P(OH)$
phosphinoxides 氧化膦类
phosphinyl 氧膦基 $H_2(O)P—$
phosphinylidene 氧亚膦基 $H(O)P═$
phosphinylidyne 次氧膦基；膦酰 $P(O)≡$
phosphite 亚磷酸盐 $MPO_3; M_2HPO_3; M_4P_2O_5$
phosphite chelator 亚磷酸酯螯合剂
phosphite-coordinated titanate 亚磷酸配位的钛酸盐(酯)
phosphite ester 亚磷酸酯
phospho- 二氧磷基 $O_2P—$
phospho acid ①膦酸 $RPO(OH)_2$②次磷酸 $R_2PO(OH)$
phosphobenzene 偶磷苯
phosphocalcite 轻斜磷铜矿
phosphocerite 磷镧铈矿
phosphochalcite 轻斜磷铜矿
phospho-cozymase 磷酸辅酶；二氧磷基辅酶
phosphocreatine(=phosphagen) 磷酸肌酸 $H_2PO_3NHC(═NH)N(CH_3)CH_2COOH$
phosphodiesterase 磷酸二酯酶
phosphoferrite 铁磷锰矿
phosphofluoric acid 六氟磷酸
phosphogypsum 磷石膏
phospholipase 磷脂酶
phospho-lipid(=phospho-lipin) 磷脂
phospho-lipin(=phospho-lipid) 磷脂
phosphomolybd(=molybdophosphate) 钼磷酸盐；磷钼酸盐
phosphomolybdate 磷钼酸盐
phosphomolybdate method 磷钼酸盐法
phosphomolybdic acid 磷钼酸 $H_3PO_3MoO_4$

phosphomolybdic pigments 磷钼酸盐系(色淀)颜料
phosphonate polymer 膦酸盐聚合物
phosphonate treatment 膦酸酯处理；膦酸盐处理
phosphonation 磷酸化作用
phosphonazo I 偶氮膦 I
 $(HSO_3)_2C_{10}H_3(OH)_2N_2C_6H_4PO_3H_2$
phosphonazo Ⅲ 偶氮膦Ⅲ
 $(HSO_3)_2C_{10}H_2(OH)_2[N_2C_6H_4PO_3H_2]_2$
phosphonazo R 偶氮膦 R
phosphonia 磷鎓杂
phosphonic 膦(酸)的
phosphonic acid 膦酸 $RPO(OH)_2$
phosphonic (acid) amide 膦(酸)酰胺
phosphonic acid ester 膦酸酯
phosphonic acid type resin 膦酸型树脂
phosphonic chloride 膦酰氯
phosphonio 磷鎓基 H_3P^+—
phosphonite 亚膦酸盐(酯)
phosphonitrile 磷腈
phosphonitrile chloride cyclic trimer 三聚氯化磷腈
phosphonitrile fluoroelastomer 膦腈氟橡胶
phosphonitrilic chloride 氯化磷腈
phosphonitrilic polymer 磷氮聚合物；磷腈聚合物〔指含—P=N—结构的聚合物〕
phosphonitrogen 磷氮肥
phosphonitryl 磷氮基 =PN
phosphonium 磷鎓；鏻 PR_4^+
phosphonium bromide 溴化鏻 $PH_4Br;R_4PBr$
phosphonium chloride 氯化鏻
phosphonium halide 卤化鏻 $PH_4X;R_4PX$
phosphonium hydroxide 氢氧化鏻 $PH_4^+OH^-;PR_4^+OH^-$
phosphonium iodide 碘化鏻
phosphonium ion 鏻离子*
phosphonium salt 鏻盐*
phosphonium stabilizer 鏻化(合)物稳定剂
phosphono- 膦酰基 $(HO)_2OP$—
phosphonoacetone 二氧磷基丙酮 $O_2PCH_2COCH_3$
phosphonoamidino 膦羧脒基
phosphonodithioic acid 二硫代膦酸 H_3POS_2
phosphonodithious acid 三硫代亚膦酸 $(HS)_2PH$
phosphonomycin(=fosfomycin) 磷霉素
phosphononitridic acid 氮代膦酸 $(HO)P(N)H$
phosphononitridothioic acid 硫代氮代膦酸 $(HS)P(N)H$
phosphonoso ①羟氧亚磷基 $(HO)OP=$ ②羟氧膦基 $(HO)OPH$—
phosphonosuccinic acid ester sodium 膦酸丁二酸酯钠(盐)
phosphonothiolic acid 硫代膦酸 $(HO)(HS)P(O)H$
phosphonothiolothionic acid 硫代硫膦酸

(HO)(HS)P(S)H
phosphonothionic acid 硫代膦酸；膦硫羰酸
 $(HO)_2P(S)H$
phosphonothious acid 硫代亚膦酸 $(HO)(HS)PH$
phosphonotrithioic acid 三硫代膦酸；二硫代硫膦酸
 $(HS)_2P(S)H$
phosphonous acid 亚膦酸 $RP(OH)_2$
phosphonylation 膦羧化作用
phosphonyl chloride 膦酰氯
phosphonyl dichloride 膦酰二氯
phosphopantetheine 磷酸泛酰巯基乙胺
phosphophyllite 磷叶石
phosphor 无机发光材料*；磷光物质
phosphoralkylation 膦酸烷基酯化反应〔导入—$RP(O)(OR)_2$基团的反应〕
phosphoramidate 氨基磷酸酯
phosphoramide 膦酰胺
phosphoramidic acid 氨基磷酸 $H_2N(HO)_2PO$
phosphoramidimidic acid 亚氨代氨基磷酸
 $(HO)_2(H_2N)P(NH)$
phosphorane 正膦 PH_5
phosphoranedioic acid 正膦二酸 $H_3P(OH)_2$
phosphoranediyl 正膦亚基 H_3P
phosphoranepentayl 正膦五基
phosphoranepentoic acid 正膦五酸；原膦酸 $P(OH)_5$
phosphoranetetrayl 正膦四基
phosphoranetetroic acid 正膦四酸 $HP(OH)_4$
phosphoranetrioic acid 正膦三酸 $H_2P(OH)_3$
phosphoranetriyl 正膦次基 H_2P=
phosphoranoic acid 正膦(单)酸 H_4POH
phosphoranyl 正膦基 H_4P—
phosphorate 膦酸化物〔含=P_2O_4基的化合物〕
phosphorated 含磷的
phosphorated oil 磷酸化油
phosphor bronze 磷青铜
phosphorescence 磷光*
phosphorescence analysis 磷光分析*
phosphorescence excitation spectrum 磷光激发光谱
phosphorescence intensity 磷光强度*
phosphorescence microwave double resonance(PMDR) 磷光微波双共振
phosphorescence peak 磷光峰
phosphorescence spectrum 磷光光谱*
phosphorescent coating 磷光漆；夜光漆(涂料)
phosphorescent decay 磷光体余辉
phosphorescent paint 磷光涂料
phosphorescent pigment 磷光颜料；夜光颜料

phosphorescent substance 磷光体
phosphorescent yarn 磷光纱
phosphoret(t)ed hydrogen ①磷的氢化物②膦
phosphor hydrogen compounds 磷氢化合物
phosphoric 磷的；五价磷的
phosphoric acid 磷酸
phosphoric acid-diatomite catalyst 磷酸-硅藻土催化剂
phosphoric acid tanning agent 磷酸鞣剂
phosphoric acid tris-dialkylamide 六烷基替磷酰(三)胺 $[R_2N]_3PO$
phosphoric amide chloride 氯化磷酰胺
phosphoric anhydride 磷(酸)酐；五氧化二磷
phosphoric chloride 五氯化磷 PCl_5
phosphoric ester 磷酸酯
phosphoric ether(=triethyl phosphate) 磷酸三乙酯
phosphoric triamide 磷酰三胺
phosphorimeter 磷光计*
phosphorimetric analysis 磷光分析
phosphorimetry 磷光光度法
phosphorimidic acid 亚氨代磷酸〔HN 代 O〕 $(HO)_3P(NH)$
phosphor incendiary bomb 黄磷纵火弹
phosphorite ①亚磷酸盐(酯)〔含 P_2O_3〕②磷钙石；磷灰岩
phosphorization 磷化*
phosphoro 偶磷 —P=P—
phosphoro-amidate 氨基磷酸盐
phosphorobenzene 偶磷苯 $C_6H_5P=PC_6H_5$
phosphoroclastic reaction 磷酸裂解反应
phosphorodithioate 二硫代磷酸酯
phosphorodithioic acid 二硫代磷酸 $H_3PO_2S_2$
phosphoro-imidate 亚氨膦酸盐
phosphorolysis 磷酸解(作用)
phosphorometer 磷光计
phosphoroscope 磷光计
phosphoroso 氧磷基 OP—
phosphorous 亚磷的；三价磷的
phosphorous acid 亚磷酸 H_3PO_3
phosphorous acid ester 亚磷酸酯
phosphorous anhydride 三氧化二磷；亚磷酐 $P_2O_3;P_4O_6$
phosphorous compound 磷化合物
phosphorous-containing polyester 含磷聚酯
phosphorous hemiselenide 三硒化二磷
phosphorous nitride dichloride 二氯化氮化磷
phosphorous nitride dichloride tetramer 二氯化氮化磷四聚物
phosphorous nitride dichloride trimer (=dichlorophosphazine) 二氯化氮化磷三聚物；二氯膦嗪

phosphorous pentoxide 五氧化二磷
phosphorous test for lubricating oil 测定润滑油(中磷含量的)亚磷酸法
phosphor tin 磷锡
phosphor-tungstic green 磷钨酸绿
phosphor-tungstomolybdic lake 磷钨钼酸色淀
phosphorus 磷〔15 号元素，化学符号 P〕
phosphorus-based function group 含磷功能基
phosphorus bromonitride 二溴氮化磷 $PNBr_2$
phosphorus bronze 磷青铜
phosphorus chloride 氯化磷〔总名，计有：二氯化磷 PCl_2、三氯化磷 PCl_3 和五氯化磷 PCl_5〕
phosphorus dichloride 二氯化磷 $PCl_2;P_2Cl_4$
phosphorus diiodide 二碘化磷 $PI_2;P_2I_4$
phosphorus family 磷族
phosphorus family element 磷族元素
phosphorus flame detector 磷火焰检测器
phosphorus hydrides〔复〕 磷氢化合物 $PH_3;P_2H_4;P_{12}H_6;P_5H_2;P_9H_2$
phosphorus in lubricating oils 润滑油含磷量
phosphorus nitride 氮化磷 $P_3N_5;P_4N_6;PN$
phosphorus-nitrogen backboned polymer 磷-氮主链聚合物
phosphorus ore 磷矿石
phosphorus oxybromide 三溴氧化磷 $POBr_3$
phosphorus oxychloride 磷酰氯；三氯氧化磷 $POCl_3$
phosphorus oxyfluoride 三氟氧化磷；磷酰氟 POF_3
phosphorus pentabromide 五溴化磷 PBr_5
phosphorus pentachloride 五氯化磷 PCl_5
phosphorus pentafluoride 五氟化磷 PF_5
phosphorus pentasulfide 五硫化二磷 P_2S_5
phosphorus pentoxide 五氧化二磷 P_2O_5
phosphorus printing 磷印试验
phosphorus rock 磷矿石
phosphorus sesquisulfide 三硫化四磷
phosphorus smoke blasting apparatus 磷烟喷雾器
phosphorus suboxide 一氧化四磷 P_4O
phosphorus sulfochloride 三氯硫化磷 $SPCl_3$
phosphorus sulfofluoride 三氟硫化磷 SPF_3
phosphorus tetroxide 四氧化二磷 P_2O_4
phosphorus thiochloride 三氯硫磷 $PSCl_3$
phosphorus tribromide 三溴化磷 PBr_3
phosphorus trichloride 三氯化磷 PCl_3
phosphorus trifluoride 三氟化磷 PF_3
phosphorus triiodide 三碘化磷 PI_3
phosphorus trioxide 三氧化二磷 P_2O_3
phosphorus ylide 磷叶立德*
phosphoryl 磷酰基 OP≡

phosphorylate 磷酸(酰)化
phosphorylated cellulose 磷酸纤维素；磷酰化纤维素
phosphorylated cotton(=cellulose orthophosphate) 磷酸纤维素
phosphorylated monoglyceride 磷酰化甘油单(酸)酯
phosphorylating 磷酸(酰)化
phosphorylating agent 磷酰化剂
phosphorylation 磷酸(酰)化(作用)
phosphoryl bromide chloride 氯溴化磷酰；磷酰氯溴
phosphoryl chloride 磷酰氯
phosphorylcholine 磷酸胆碱
phosphorylethanolamine 磷酸乙醇胺
phosphorylglyceric acid 磷酸甘油酸
phosphoryl nitride 磷酰基化氮 PON
phosphoryl triamide 磷酰三胺 PO(NH$_2$)$_3$
phospho-serine 磷酸丝氨酸
　　　PO$_3$H$_2$OCH$_2$CHNH$_2$COOH
phosphotaurocyamine 磷酸脒基牛磺酸
phosphotransacetylase 磷酸转乙酰酶
phosphotungstate 磷钨酸盐
phospho-tungstic acid(=phospho-wolframic acid) 磷钨酸
　　　P$_2$O$_5$·24WO$_3$·nH$_2$O;H$_2$[P(W$_2$O$_7$)$_6$]$_7$·nH$_2$O
phospho-tungstic green 磷钨酸绿
phosphotungstomolibdic pigment 磷钨钼酸颜料
phosphotungstomolybdic acid 磷钨钼酸
phospho-tungstomolybdic lake 磷钨钼酸色淀
phosphowolframate 磷钨酸盐
phospho-wolframic acid(=phospho-tungstic acid) 磷钨酸
phosphurane(=phosphene) 磷杂环戊二烯
phosphuretted 含低磷的
phosphuret(t)ed hydrogen ①磷化氢②膦
phosteam 蒸气磷化法
phostonic acid 烷基亚膦酸
phot 辐透〔照度单位：厘米烛光〕
photic 辐射的
photics 光学
photistor 光敏晶体管
photo- 〔希腊字头〕①光②感光的
photoabsorption 光吸收*
photoabsorption coefficient 光吸收系数
photoaccoustic detection 光声检测*
photo-acid generating system 光生酸体系
photo-acid-generation reaction 光致产酸反应
photoacoustic cell 光声池
photoacoustic detection 光声检测法
photoacoustic detector 光声检测器
photoacoustic imaging(PAI) 光声成像
photoacoustic immunoassay 光声免疫分析法
photoacoustic infrared spectroscopy 光声红外光谱法
photoacoustic sound 光学声
photoacoustic spectrometer 光声光谱仪
photoacoustic spectrometry(PAS)(=optosonic spectrometry) 光声(光)谱法*
photoacoustic spectroscopy 光声光谱学
photoactinic (发)光化射线的；能产生光化作用的
photo-actinic action 光化作用
photoactivation 光活化*
photoactive coating 光敏涂料
photoactive polymer 光活性聚合物
photoactivity 光敏性；光活化性；感光性
photo-addition 光化加成反应
photoadduct 光(致)加合物
photoadsorption 光致吸附
photoag(e)ing 光老化
photoallergen 光敏变应原
photoallergic mechanism 光变应性机理
photoallergy 光变应性
photoanalytical chemistry 光分析化学
photoanilide rearrangement 光致苯胺重排
photo-annealing 光退火；光熟炼
photoassisted decomposition 光助分解
photoassisted oxidation 光助氧化(作用)
photoassociation 光缔合
photobacteriomycin 光菌霉素
photobleaching 光漂白*
photocatalysis 光催化*
photocatalyst 光催化剂
photocatalytic activity 光催化活性
photocatalytic degradation 光催化降解
photocatalyzed gel 光催化凝胶
photocathode 光阴极
photocell(=photoelectric cell) 光电池；光电管
photochemical 光化学的
photochemical absorption 光化(学)吸收
photochemical absorption law 光化吸收定律
photochemical activation 光化活化
photochemical activity 光化活性
photochemical addition 光化加成(作用)
photochemical aerosol 光化学气溶胶
photochemical after effect 光化学后效
photochemical air pollution 光化学空气污染
photochemical alkylation 光化烷基化(作用)
photochemical catalysis 光催化(作用)
photochemical catalyst 光(化学)催化剂
photochemical cell 光化学电池
photochemical chlorination 光化学氯化

photochemical crosslinking 光化学交联
photochemical decomposition 光化分解
photochemical degradation 光化降解
photochemical degradation of pollutant 污染物的光化学降解
photochemical dissociation 光化离解
photochemical effect 光化学效应
photochemical efficiency 光化效率
photochemical enrichment 光化学浓集〔浓缩〕
photochemical equilibrium 光化平衡
photochemical equivalent 光化当量
photochemical etching 光化学腐蚀
photochemical excitation 光化激发
photochemical fog 光化学雾*
photochemical formation 光化形成
photochemical free-radical formation 光化学自由基形成
photochemical induction 光化诱导
photochemical initiation 光化引发
photochemical inversion 光化转化
photochemical kinetics 光化动力学
photochemically-induced 光化学诱导的
photochemically induced oxidation 光化学诱导氧化
photochemical oxidant 光化学氧化剂
photochemical ozonization 光化臭氧化(作用)
photochemical polymerization 光化聚合
photochemical primary process 光化学初始过程
photochemical process 光化学过程*
photochemical ratio 光化比
photochemical reaction 光化学反应*
photochemical reactivity 光化反应性
photochemical rearrangement 光化学重排*
photochemical scavenging 光化学清除
photochemical secondary process 光化学后续过程
photochemical sensitization 光化(学)敏化作用
photochemical separation 光化学分离法
photochemical smog 光化学烟雾
photochemical temperature effect 光化温度效应
photochemical threshold 光化门槛；光化阈；光化限
photochemical transformation 光化学转化
photochemical vapor deposition 光化学蒸气沉积
photochemical yield 光化产量
photochemistry 光化学*
photo-chlorination 感光氯化(作用)
photochromic 光致变色的
photochromic fibre 光致变色纤维
photochromic pigment 光敏颜料
photochromic polymer 光致变色高聚物
photochromism 光致变色*

photocleavage 光致开裂；光致断裂
photocolorimeter 光比色计
photocolorimeter for flowing sample 流体试样光电光度计
photocolorimetry 光比色法
photocombustion 光燃烧
photoconducting 光电导的
photoconduction 光电导
photoconductive cell 光导管；光敏电阻；光电池
photoconductive coating 光电导性涂料
photoconductive detector 光导检测器
photoconductive effect 光电导(性)效应
photoconductive fiber 光导纤维*
photoconductive pigment 光电导性颜料
photoconductive polymer 光(电)导聚合物*
photoconductive sensor small-signal model 光电传感器小信号模型
photoconductivity 光电导性*
photoconductor 光电导体*
photocontraction 光致收缩
photoconversion 光转换
photocopolymerization 光致共聚合(作用)
photocopying 照相复印；影印
photocopy toner 照相复印色料(色原)
photo-creep method 光测蠕变法
photocrosslinkable 可光致交联的
photocrosslinkable polymer 光交联聚合物
photocrosslinking 光致交联
photocurable 可光致固化的；光熟化的
photocurable polymer 光固化聚合物
photo-cure 光固化
photocuring 光固化
photo-curing material 光固化材料
photocuring reaction 光(致)固化反应
photocurrent 光电流
photocyclized 光致环化的
photodecomposition 光解*
photodegradable 可光降解的
photodegradable polymer 光降解聚合物*
photodegradation 光降解*
photodegradation of polymer 聚合物的光降解
photodegradative weathering reaction 光降解老化反应
photodensitometer 光密度计；光密度摄影仪
photodensitometric quantitation 光密度定量
photodensitometry 光密度分析法
photodepolymerization 光解聚(作用)
photodesorption 光解吸(作用)
photodetachment 光致分离

photodetector 光检测器
photo-deterioration 光致蜕化；光致变质
photodichroism 光二向色性
photo-dimerization 光二聚合反应
photodiode 光电二极管；光敏二极管
photodiode array 光电二极管阵列
photodiode array detector 光二极管阵列检测器
photodiode small-signal model 光电二极管小信号模型
photodissociation 光解离*
photodynamic 在光中发荧光的
photodynamic therapy 光动态疗法
photo effect 光电效应
photoejection of electron 电子的光致发射
photoelastic analysis 光弹性分析
photoelastic effect 光弹性效应
photo-elasticity 光弹性
photoelastic method 光测弹性法
photoelastic polymer 光弹性聚合物
photoelastic test 光弹性试验
photoelectret 光致驻极体
photoelectric 光电的
photoelectric absorptive coefficient 光电吸收系数
photoelectrical effect 光电效应
photoelectric amplifier 光电放大器
photoelectric cell(=photocell;photronic cell) 光电池；光电管
photoelectric colorimeter 光电比色计*
photoelectric colorimetry 光电比色法
photoelectric color method 光电比色测定法；测定(石油产品)颜色的光电法
photoelectric cross-section 光电离截面
photoelectric detection system 光电检测装置
photoelectric detector 光电检测器
photoelectric device 光电器件
photoelectric dew point hygrometer 光电露点湿度计
photoelectric direct reading spectrometer 光电直读光谱计*；光电直读光谱仪；光量计
photoelectric dust detector 光电烟尘浓度(测定)计
photoelectric effect 光电效应
photoelectric emission 光电发射
photoelectric emission microscope 光电发射(电子)显微镜
photoelectric Fourier transformation 光电傅里叶变换
photoelectric glossmeter 光电光泽计
photoelectric illuminometer 光电照度计
photoelectric infrared detector 光电红外检测器
photoelectricity ①光电②光电学
photoelectric multiplier 光电倍增管

photoelectric multiplying tube(PMT) 光电倍增管
photoelectric peak 光电峰
photoelectric phenomenon 光电现象
photoelectric photometer 光电光度计
photoelectric photometry 光电测光法
photoelectric pyrometer 光电高温计
photoelectric reactor 光电反应器
photo-electric scanning cotton length tester 纤维长度照影仪
photoelectric spectrocolo(u)rimeter 光电分光测色计；光电分光光度计；光电分光色度计
photoelectric spectrophotometer 光电分光光度计
photoelectric spectrophotometry 光电分光光度法
photoelectric spectropolarimeter 光电旋光分光光度计
photoelectric spectroscopy 光电能谱法
photoelectric steelometer 光电析钢仪
photoelectric titration 光电滴定
photoelectric transducer 光电换能器
photoelectric tristimulus colorimeter 光电三刺激值比色计
photoelectric tube 光电管
photoelectric turbidometer 光电浊度计
photoelectrocatalysis 光电催化*
photoelectrochemical cell 光电化学电池
photoelectrochemical effect 光电化学效应
photoelectrochemical etching 光电化学腐蚀
photoelectrochemical process 光电化学过程
photoelectrochemistry 光电化学*
photoelectroconductive polymer 光电导高聚物
photoelectrometer 光电计
photoelectromotive force 光电动势
photoelectron 光电子
photoelectron spectroscopy(PES) 光电子能谱法
photoelectron spectroscopy of inner shell electron(PESIS) 内层电子光电子能谱法
photoelectron spectrum 光电子能谱
photoelectro-spectrophotometer 光电分光光度计
photoelimination 光消去反应*
photoemission 光电(子)发射
photoemissive sensor 光电发射传感器
photoemissive tube 光电管
photoemitted electron 光(发射)电子
photoenergy 光能
photoengraving 照相制版；照相凸版
photoengraving method 照相雕刻法
photoengraving process 照相制版法；光刻法；光蚀刻法
photo-equilibrium 光平衡
photo-etching 光刻(法)；光蚀(技)术
photo-excitation method 光激发法

photoexcited 光激的
photoextinction 消光(作用)
photoextinction method 比浊分析法
photofading 光褪色
photo-fission 光分裂(作用)
photofission product 光(致)裂变产物
photoflavins 9-烷基咯嗪
photo-flood lamp 照相泛光灯
photofluorimeter 荧光计
photofluorometer 荧光计
photofluoroscope 荧光屏
photofragmentation 光碎片化*
photofragment spectroscopy 光致(分子)碎片光谱法
photogalvanic cell 光化学电池
photogalvanometer 微量光电流计
photogelatin 感光底片胶
photogen 发光体；发光源
photogene 页岩煤油
photogeneration 光生作用
photogenic 发光的；发磷光的
photografted 光致接枝的
photograph ①相片②照相；摄影
photographic 照相的
photographic brightness 照相亮度
photographic cell 感光元
photographic chemicals 摄影用化学品
photographic chemistry 照相化学
photographic densitometer 照相黑度计
photographic densitometric analysis 摄影光密度分析(法)
photographic developer ①显像剂②显影剂
photographic emulsion 照相乳剂；感光乳剂
photographic emulsion detector 照相乳剂检测器
photographic evaluation 摄影测定
photographic film 照相软片
photographic filter 照相滤光器
photogaphic intensifier 照相强化剂
photographic negative 照相底片
photographic parer 相纸
photographic photometry 照相光度学；照相分光测定法
photographic plate 照相底片；干片
photographic positive 照相正片
photographic process 照相法
photographic recording polarograph 摄影极谱仪
photographic reducer ①照相减薄剂②照相还原剂
photographic screen 滤色屏
photographic sensitivity 照相感光度
photographic sensitizer 照相敏化剂；照相增感剂
photographic spectrophotometry 照相分光(光度)测定法
photographic spectroscope 照相分光镜
photographic spectrum 照相感光范围
photographic study of combustion 燃烧过程的照相研究法
photography 照相术〔物〕
photogravure 照相凹版印刷
photohalide 感光性卤化物
photohalogenation 光卤化*
photo-hardening 光硬化(作用)
photohole 光穴
photohyalography 照光蚀刻术
photo-induced crosslinking reaction 光诱导交联反应
photoinduced electron transfer 光诱导电子转移
photo-induced polymerization 光(致)聚合
photoiniferter 光引发转移终止剂
photo-initiated oxidation 光引发氧化反应
photoinitiated polymerization 光引发聚合
photoinitiation 光引发(作用)
photoinitiator 光敏引发剂*
photoion 光离子
photoionization 光离子化*；光致电离
photoionization detector(PID) 光致电离检测器*；光离子化检测器
photoionization ion source 光致电离离子源
photoionization process 光电离过程
photoion-photoelectron coincidence mass spectrometer 光离子-光电子符合质谱仪
photoirradiation 光辐照
photoisomeric change 感光异构变化
photoisomerism 感光异构(现象)
photoisomerization 光异构化*
photokinesis 光动现象；光激运动；趋光性
photolabile 对光不安的
photolithographic off-set process 照相平版胶印印刷
photolithography 光刻(蚀)法
photoluminescence 光致发光
photoluminescence polymer 光致发光聚合物
photolysis 光(分)解(作用)
photolysis gas chromatography 光解气相色谱(法)
photolyte 光解质
photolytic 光解的
photolytic chain scission 光致链断裂；光解性链断裂
photomacrograph 低倍放大摄影照片
photomagnetic capture 光磁俘获
photomagnetism 光磁性
photometer 光度计
photometric 光测的；计光的
photometrical accuracy and reproducibility 光度精度和光度重现性

photometric analysis 光度分析
photometric curve 测光曲线
photometric detector 光电检测器
photometric error 光测误差；光度读数误差
photometric measurement 光电测定法
photometric scales 光泽(光度)特性标度
photometric standard 测光标准
photometric tin detector 光度计式锡检测器
photometric titration 光度滴定(法)*
photometric titration apparatus 光度滴定装置
photometry 光度学；计光术
photomicrograph 显微照相；显微照片
photomicrography 显微摄影术
photomicroscope 照相显微镜
photomicroscopy 显微照相术
photomultiplier 光电倍增管*
photomultiplier detector 光电倍增检测器
photomultiplier tube 光电倍增管
photon 光子
photon activation analysis(PAA) 光子活化分析*
photon beam 光子束
photon bombardment 光子轰击
photon bunching 光子集束效应
photon capture 光子捕获
photon correlation spectroscopy 光子相关光谱；光子相关谱法
photon counting spectrophotometer 光子计数分光光度计
photon counting technique 光子计数法
photon detector 光子检测器
photon effect 光子效应
photon energy 光子能量
photonephelometer 光电浊度计
photoneutron 光中子
photoneutron logging 光中子测井
photoneutron source 光中子源
photon excited atom 光子激发原子
photon excited electron 光子激发电子
photon excited X-ray fluorescence analysis 光子激发的X射线荧光分析
photon fluence 光子流量
photonic bandgap 光子带隙
photonic chip 光子芯片
photonics 光子学
photon impact(PI) 光子冲击
photon induced action 光子激发(诱导)作用
photon induced field ionization 光子诱导场电离
photon induced scanning Auger microscope 光子激发扫描俄歇显微镜

photonitrosation 光亚硝化(作用)
photo-nitrosation of cyclohexane process(=PNC process) (己内酰胺的)环己烷光(致)亚硝化法
photon-pumped electrochemical etching 光子注入电化学腐蚀
photonuclear activation analysis 光核活化分析
photonuclear reaction 光核反应
photooptical sensitizer 光学照相敏感剂
photo-oxidant stabilizer 防光老化剂
photooxidation 光氧化*
photo-oxidation chain-cleavage 光氧化致链断裂
photooxidation treatment of waste water 废水光氧化处理法
photo-oxidative degradation 光氧化降解*
photo-oxide 光氧化物
photooxydation 感光氧化作用；光致氧化作用
photopeak 光峰
photopen recorder 光电笔记录器
photoperiod 光周期
photoperiodism 光周期现象
photophora 发光器
photophoresis 光泳现象
photophysical process 光物理过程*
photopic 光适应
photoplasticity 光塑性
photopolymer ①光(致)聚合物②感光聚合物；光敏聚合物
photopolymerisable 光聚合的
photo polymerization 光(致)聚合*
photopolymerizer 光聚合剂
photopone 光音器
photopredissociation 光预解离*
photoprocess ①光(化学)过程②光学处理；光学加工
photoproduct 光化产品
photo-protection ①光防护②光防护装置
photoradiochromatogram 放射色谱光显影图；光放射色谱图
photoradiochromatography 光放射色谱法
photo rapid dyestuff 感光显色染料
photoreaction 光反应
photoreactivation 光复活(作用)
photorearrangement 光重排*
photoreceptor 光(感)受器
photo-recording polarograph 照相式极谱仪；摄影极谱仪
photoredox behavior 光氧化还原行为
photoredox reaction 光致氧化还原*
photoreduction 光还原*
photoreduction indicator 光致还原指示剂

photo-refractive 光折变
photoresist 光致抗蚀剂；光刻胶
photoresistance 光敏电阻
photoresistance effect 光电效应
photoresponse 光感应；光响应
photoresponsive polymer 光响应高分子
photo responsive system 易感光体系
photoreversibility 光致可逆性
photo-rheologic method 光测流变法
photosensitised autoxidation 感光自动氧化
photosensitiser 感光剂
photosensitive 光敏的
photosensitive adhesive 光敏胶黏剂；光敏性黏合剂
photosensitive area 光敏区
photosensitive cell 光电池；光电管
photosensitive control film 光敏控制胶带
photosensitive emulsion 感光乳液
photosensitive glass 光敏玻璃
photosensitive plastic 光敏(性)塑料；感光性塑料
photosensitive polymer 光敏聚合物*；光敏高分子
photosensitive resin 感光性树脂
photosensitive surface 感光表面
photosensitive zero detector 光敏零点检测器
photosensitivity 光敏性〔指1.感光性；2.感光灵敏度〕
photosensitivity of iron carbonyl 羰基化铁的光敏性
photosensitization 光敏化*
photo sensitized initiation 光敏引发(作用)
photosensitized oxidation of paraffin 石蜡的光敏氧化
photosensitized polymerization 光敏聚合(作用)
photosensitized reaction 光敏反应
photosensitizer 光敏剂*
photosensitizing test 光敏试验
photosensor 光敏器(件)
photosetting 光固化
photo solid-state polymerization 光引发固态聚合(作用)
photosource 光源
photosphere 光层
photostability 耐光性
photostabilization 耐光性
photostabilizer 耐光剂；光稳定剂
photostable 不感光的
photostationary state 光稳(定)态
photostimulated ionization 光激电离*
photosulfoxidation 光(致)磺氧化(作用)
photosynthesis 光合(作用)
photosynthesis analyzer 光合作用分析仪
photosynthetic cycle 光合循环
photosyntometer 光合计

phototendering 光脆裂效应
photothermal beam deflection effect 光热光束偏转效应
photothermal deflection densitometer 光热偏转密度计
photothermal deflection spectrometer 光热偏转光谱仪
photothermal diffraction spectrometer 光热衍射光谱仪
photothermal spectrometry(PTS) 光热光谱法
photothermography 光温度记录法
photothermoluminescence 光热致发光
phototransistor 光敏晶体管
phototrode 光极
phototronic cell 光电池
phototropy 光致变色
phototube 光电管
phototurbidometry 光电比浊度测定法
photovoltaic cell 光电池*
photovoltaic detector 光电池检测器
photovoltaic effect 光生伏打效应
photovolt densitometer 光电密度计
photovulcanization 光硫化(作用)
photoxide 光氧化物
photronic cell(=photoelectric cell) 光电池
pH paper pH 试纸*
phreatic water 地下水
pH-recorder pH 记录器；氢离子浓度记录器
pH-regulator pH 调节器
phrenazole(=metrazole) 环戊四唑
phrenosin 羟脑苷脂
phrenosinic acid 二十四醇酸；脑羟酸；α-羟二十四酸 $CH_3(CH_2)_{21}CHOHCOOH$
phrenosterol 脑甾醇
pH reversal method pH 反馈法
pH sensor pH 传感器
phsophopyridoxamine 磷酸吡哆胺
pH-stabilizer pH 稳定剂
pH test thread pH 试验纤维
phthaiocyanine dye 酞菁染料
phthalal 邻苯二亚甲基
phthalaldehyde 苯二醛（通常指邻苯二醛）
phthalaldehydic acid 苯醛酸；邻甲酰苯甲酸
phthalamic acid 邻氨甲酰苯甲酸；邻氨羰基苯甲酸 $C_8H_7O_3N$
phthalamide 邻苯二酰胺 $C_6H_4(CONH_2)_2$
phthalamidic acid 邻羧基苯甲酰胺
α-phthalamidoglutarimide α-邻苯二亚甲基戊亚胺
phthalamoyl 邻氨羰苯甲酰；邻苯二甲酸一酰胺一酰 $H_2NCOC_6H_4CO—$
phthalanil(=N-phenylphthalimide) N-邻苯二酰苯胺 $C_6H_5N=C_8H_4O_2$

phthalanone 邻羟甲基苯甲酸内酯
phthalate 邻苯二甲酸盐(酯) $C_6H_4(COOM)_2$
phthalate ester 邻苯二甲酸酯
phthalation 邻苯二甲酰化(作用)
phthalazine(=2, 3-benzodiazine) 2,3-二氮杂萘 $C_8H_6N_2$
phthalazinium 酞嗪(镐)盐
phthalazinyl 酞嗪基；2,3-二氮杂萘基 $C_8H_5N_2$—
phthalazone 2,3-二氮杂萘酮
phthaldiamide(=phthalamide) 邻苯二甲酰胺
phthalein complexone 酞氨羧络合剂
phthalein indicator 酞类指示剂
phthaleins 酞类〔深色有机化合物及染料的类名〕
phthalein violet 酞紫
phthalexone S 酞腙 S
phthalhydrazide 邻苯二甲酰肼 $C_8H_6N_2O_2$
m-phthalic acid 间苯二甲酸
o-phthalic acid 邻苯二甲酸
p-phthalic acid 对苯二甲酸
phthalic acid 苯二甲酸〔一般指邻苯二甲酸，旧用俗名酞酸〕 $C_6H_4(CO_2H)_2$
o-phthalic acid diglycidyl ester 邻苯二甲酸二缩水甘油酯
phthalic acid series 邻苯二甲酸系列
phthalic aldehyde 邻苯二醛 $C_6H_4(CHO)_2$
phthalic alkyd resin 邻苯二甲酸醇酸树脂；苯酐醇酸树脂
phthalic amide 邻苯二甲酰胺
phthalic anhydride 邻苯二甲酸酐 $C_6H_4(CO)_2O$
phthalic diamide 邻苯二甲酰胺 $C_6H_4(CONH_2)_2$
phthalic ester 邻苯二甲酸酯 $C_6H_4(COOR)_2$
phthalic imidine 邻苯二甲酰亚胺
phthalic monoester 邻苯二甲酸单酯
phthalic nitrile 邻苯二腈
phthalide 2-苯并[c]呋喃酮 $C_6H_4CH_2OCO$
phthalidene 2-苯并[c]呋喃酮亚基
phthalide resin 邻苯二甲酸树脂；酞酸树脂〔俗〕
phthalidyl 2-苯并[c]呋喃酮基
phthalidylidene 苯并[c]呋喃酮亚基
phthalidylideneacetic acid 苯并[c]呋喃酮亚基乙酸
phthalimide 苯邻二甲酰亚胺 $C_6H_4(CO)_2NH$
phthalimidine 苯并[c]吡咯酮
phthalimido 苯二(甲)酰亚氨基 COC_6H_4CON—
phthalimidoacrylic acid 邻苯二(甲)酰亚氨基丙烯酸
phthalimidomethyl group 邻苯二甲酰亚氨(基)甲基
phthalimidoxime 苯二甲亚胺肟
phthaline 二 R 甲基苯甲酸 $R_2CHC_6H_4COOH$
phthalizine 酞嗪

phthalocyanine(=phthalocyanin) 酞菁〔染料〕
phthalocyanine blue 酞菁蓝
phthalocyanine dyes and pigments 酞菁染料和颜料
phthalocyanine grease(=phthalocyanine pigment grease) 酞菁润滑脂
phthalocyanine intermediate 酞菁素
phthalocyanine lubricant(=phthalocyanine pigment grease) 酞菁润滑脂
phthalocyanine pigment 酞菁颜料
phthalocyanine pigment grease(=phthalocyanine grease; phthalocyanine lubricant) 酞菁润滑脂
phthalocyanin(=phthalocyanine) 酞菁〔染料〕
phthalone 酞酮
phthalonic acid 邻羧基苯乙酮酸
phthalonitrile 邻苯二甲腈 $C_8H_4N_2$
phthaloperine 酞吡啉
phthalophenone 二苯代酚酞 $(C_5H_5)_2CC_6H_4COO$
phthaloyl(=phthalyl) 邻苯二甲酰 —COC_6H_4CO—
phthaloylamino acid 邻苯二甲酰氨基酸
phthalthrin 胺菊酯
phthaluric acid 邻苯二甲酸一酰脲
phthalyl(=phthaloyl) 邻苯二甲酰 —COC_6H_4CO—
phthalyl alcohol 邻苯二醇
phthalyl chloride 邻苯二甲酰氯 $C_6H_4(COCl)_2$
phthalyl content (邻)苯二(甲)酰含量
phthalylglutamic acid 邻苯二甲酰谷氨酸
phthalyl glycine 邻苯二甲酰甘氨酸
phthalyl hydrazide 邻苯二(甲)酰肼 $C_6H_4CONHNHCO$
phthalyl hydroxamic acid 邻苯二甲酰(基)羟肟酸
phthalylidene 苯邻二亚甲基 $(C_6H_4)(CH=)_2$
phthalyl phenylhydrazide 邻苯二甲酰基苯酰肼
phthalyl sulfathiazole 邻苯二甲酰基磺胺噻唑
phthalyl synthesis 邻苯二甲酰基合成(法)
phthienoate 结核菌烯酸盐
phthienoic acid 结核菌烯酸
phthiocerol 结核菌醇
phthiocol 结核菌萘醌；2-甲基-3-羟基-1,4-萘醌 $C_{11}H_8O_3$
phthioic acid 结核菌酸
phthiomycin 痨霉素
phthoric acid〔旧名〕 氢氟酸
phtotropism 光色互变
pH unit pH 单位
phunki(=rhooki) 成熟紫胶
pH value pH 值*

pH value comparator　pH 值比较器
pH viscosity　pH 黏度
phycite　赤藓醇
phycocolloid　藻胶
phyllanthin　叶下珠脂素
phyllanthol　叶下珠醇
phyllite　千枚岩；硬绿泥石
phylloaetioporphyrin　叶本卟啉
phyllocladene　扁枝烯
physa〔希腊文〕　气
physalin　酸浆果红素；桐酸玉黍黄素
physalite　硅酸铝矿
physeptone　6-二甲氨基-4,4-二苯基庚子酮盐酸盐
physeteric acid　抹香鲸酸；5-十四(碳)烯酸
physetoleic acid(=palmitoleic acid)　抹香鲸烯酸；棕榈烯酸；9-十六(碳)烯酸
physic　①药物②泻药
physical　①物理的②物理(学)的
physical absorption　物理吸收
physical abuse　机械损伤
physical adsorption　物理吸附
physical agent　物理因数
physical aging　物理老化
physical analysis　物理分析
physical and chemical vapour deposition　物理、化学气相沉积
physical antioxidant　物理性防老剂
physical appearance　物理性质
physical atomic weight　物理原子量
physical behavior　物理行为
physical blend(s)　物理掺和物；冷拼体
physical blowing　物理发泡
physical catalyst　物理催化剂〔能够促进或改变化学反应的辐射能〕
physical change　物理变化
physical-chemical process　物理化学法
physical chemistry　物理化学
physical classification　物理分类(法)
physical component of tensor　张量(的)固有分量
physical condition　①物理状态②物质状态
physical constant　物理常数
physical crosslinking　物理交联
physical crushing　物理粉碎法
physical deactivation　物理钝化法
physical dimension　外形尺寸；外缘尺寸
physical entanglement　物理缠结
physical equation　物理公式
physical equilibrium　物理平衡

physical foam　物理发泡
physical foaming agent　物理发泡剂
physical form　外形；外观
physical hygrometer　干湿球湿度计
physical incompatibility　物理不相容性；物理的互斥性
physical interference　物理干扰
physical intumescent flame retardancy　物理膨胀阻燃
physical inventory　实测库存量；实测物料量
physical isomer　物理异构物
physical isomerism　物理(性)异构(现象)；物理异性(现象)
physical-mechanical property　物理-机械性能
physical mixture　物理状态混合物
physical modification　物理改性；物理变性
physical operation　物理操作
physical organic chemistry　理论有机化学
physical phenomenon　物理现象
physical property　物理性质；物理性能
physical pseudomorphy　物理性假同晶(现象)
physical quantity　物理量值
physical refining process　(石油的)物理加工方法
physicals　物理性能
physical science　物理科学
physical softener　物理(性)软化剂
physical solution　物理溶液
physical solvent　物理(性)溶剂
physical state　物(理状)态
physical testing　物理(性)试验
physical unit　物理单位
physical vapour deposition(PVD)　物理蒸气沉积法(镀膜)
physicist　物理学家
physic-nut oil　麻风子油〔取自 *Jatropha curcas* 的种子〕
physico-chemical　物理化学的
physico-chemical analysis　物理化学分析
physico-chemical constant　物(理)化(学)常数
physico-chemical process　物理化学法
physico-chemical (sewage) treatment　(污水)物理化学处理
physicomechanical property　物理机械性质
physics　物理(学)
physiochemical(=biochemical)　生(物)化(学)的；生(理)化(学)的
physiochemistry　生理化学
physiography　自然地理学
physiological action　生理作用
physiological activity　生理活动(性)
physiological chemistry　生理化学
physiological classification　生理分类(法)
physiological equilibrium　生理平衡
physiological oxidation　生理氧化(作用)

physiological reduction 生理还原(作用)
physiological salt 生理盐分
physiological salt solution 生理食盐溶液
physiological solution 生理溶液
physiological unity 生理统一体
physisorption 物理吸着
physisorption film 物理吸附膜
physite 赤醇
physodallic acid 囊状地衣酸 $C_{26}H_{30}O_8$
physodic acid 囊(状)地衣酸
physostigma 毒扁豆
physostigmine 毒扁豆碱
physostigmine salicylate 水杨酸毒扁豆碱
phytadiene 植二烯 $C_{20}H_{38}$
phytane 植烷 $C_{20}H_{42}$
phytanic acid 植烷酸
phytanol(=dihydrophytol) 植烷醇 $C_{20}H_{41}OH$
phytanolate 植烷醇盐
phytene 植烯 $C_{20}H_{40}$
phytic acid(=inositol hexaphosphoric acid; phytinic acid) 肌醇六磷酸
phytin ①肌醇六磷酸钙镁；非丁②白木耳
phytinic acid(=phytic acid) 肌醇六磷酸
phytoactin 植病活菌素；多肽霉素
phytoalexin 植物抗毒素
phytobacteriomycin 植菌霉素
phytochelation 植物螯合作用
phytochrome (植物)光敏色素
phytoene 八氢番茄红素
phytoextraction 植物萃取作用
phytohormone 植物激素*
phytol 植醇；叶绿醇 $C_{20}H_{39}OH$
phytol ketone 植物酮
phytomelin 芸香素
phytomenadione 维生素 K_1
phytonadione 维生素 K_1
phytopharmacy 植物药物学
phytopyrrole 植物色素吡咯
phytostabilization 植物固化作用
phytosterin(=phytosterol) 植物甾醇
phytosterol(=phytosterin) 植物甾醇
phytosynthesis 植物合成(作用)
phytosynthetic production 植物合成生产
phytotoxicant 植物毒素
phytotoxicity 植物毒性
phyto-toxicity test of pesticide 农药药害试验
phytotoxic metabolite 植物毒性代谢物
phytovolatilization 植物挥发作用

phytyl 植基；叶绿基 $C_{20}H_{39}$—
piaselenole 苯并[c]硒二唑
piazines(=para-diazines) 对二嗪类
piazthiole 苯并[c]噻二唑
pi-bond π键
picacic acid 鹊鬼伞酸
picamar 丙基焦棓酚二甲醚 $C_{13}H_{16}O_3$
picene 苉；二萘品(并)苯 $C_{22}H_{14}$
piceneketone 苉(甲)酮 $C_{21}H_{12}O$
piceneperhydride 苉化过氢；过氢苉 $C_{22}H_{36}$
picenic acid 苉酸 $C_{21}H_{14}O_2$
picenoqinone 苉醌 $C_{22}H_{12}O_2$
piceoside 云杉苷
pichurim beans 黄樟子
pick 纬纱；投梭
pick count 织物纬密
picked ore 选出矿
picker displacement mechanism 皮结移位机构
pickeringite 镁明矾
pick glass 织物分析镜
picking 浸酸
picking band 打梭皮带
picking belt 选矿带
picking draft 排绵
picking table 摇床；选矿床
picking-up 咬底；咬起
pickle ①腌菜②泡黄瓜③腌鱼④浸酸〔皮革〕
pickle alum 硫酸铝；(钾)明矾；(铝)钾矾
pickle bath 浸酸浴
pickle change 加盐析变；(肥皂生产的)第二次盐析
pickled 浸酸的
pickled pelt 浸酸裸皮
pickled skin 浸酸皮
pickle patch 酸洗斑点
pickles tyre cord 无纬轮胎帘子布
pickle wash lye (肥皂生产中的)第二次盐析(的含甘油)碱液
pickling ①酸浸②浸渍
pickling agent ①浸渍剂②酸浸剂
pickling bath(=pickling tub) 酸洗池(槽)
pickling brittleness 酸浸脆性
pickling embrittlement 酸洗致脆
pickling inhibitor 酸蚀抑制剂；防酸蚀剂
pickling liquor 酸浸液；浸渍液
pickling oil 防锈油
pickling process ①酸浸法②浸渍法
pickling tank 酸浸槽
pickling tub(=pickling bath) 酸洗池(槽)

pickling waste water 酸洗废水
picknometer(=picnometer) ①比重瓶②比重管
pick test 抽样检验
pick up ①传感器②拾音器
pick-up fraction of gasoline 汽油的启动馏分
pick-up metal(=tramp metal) 混入金属
pick-up pump 真空泵
pick-up reaction 拾取反应
pick-up roll 引辊；取胶辊
picnometer(=picknometer; pycnometer) ①比重瓶②比重管
pico-〔词头〕皮(可)；沙；微微〔符号 p〕
picoammeter 皮(可)安(培)计；微微安培计
pico farad 皮(可)法(拉)（10^{-12} 法(拉)）
picogram 皮(可)克；微微克〔10^{-12} 克〕
picolinamide 吡啶酰胺；氮苯酰胺
picoline 甲基吡啶
picolinealdehyde-2-quinolylhydrazone 皮考啉醛-2-喹啉腙
picolinic acid(=pyridine carboxylic acid) 吡啶甲酸 C_5H_4NCOOH
picolinic acid ester 吡啶甲酸酯；皮考林酸酯
picolinium 甲基吡啶鎓；皮考啉鎓
picolinium molybdophosphate(PIMP) 甲基吡啶鎓钼磷酸盐
picolinium tungstosilicate(PIWSI) 甲基吡啶鎓钨硅酸盐
picolinyl 皮考啉基；甲代吡啶基
picolyl 吡啶甲基；皮考基
2-picolyliminodiacetic acid(H_2PIDA) 2-甲基吡啶基亚氨基二乙酸 $(CH_3C_5H_3N)N(CH_2COOH)_2$
picometer 皮(可)米〔10^{-12} 米〕
picomole 皮(可)摩尔〔10^{-12} 摩尔〕
picomole quantities 皮(可)摩尔(数量)级
picosecond 皮(可)秒〔10^{-12} 秒〕
picotite 铬尖晶石
picraconitine 苦乌头碱
picradonidin 黑藜芦苷
picral 苦醇〔苦味酸的酒精溶液〕
picramate 苦氨酸盐；4,6-二硝基-2-氨基苯酚盐 $(NO_2)_2(NH_2)C_6H_2OM$
picramic acid(=4,6-dinitro-2-aminophenol) 苦氨酸；4,6-二硝基-2-氨基苯酚 $HOC_6H_2(NH_2)(NO_2)_2$
picramide(=picryl amine) 苦酰胺；苦基胺 $(NO_2)_3C_6H_2NH_2$
picraminazo 苦胺偶氮；偶氮苦胺
picraminazo chrome 苦胺偶氮铬
picramnine 苦姆宁；苦姆碱
picranisic acid(=picric acid; 2, 4, 6-trinitrophenol) 苦味酸；2, 4, 6-三硝基苯酚 $(NO_2)_3C_6H_2OH$

picrasmine 苦树苷 $C_{35}H_{46}O_9$
picrate 苦味酸盐
picrate powder 苦酸盐(炸)药
picratol 苦味酸根
picric acid(=picranisic acid; 2,4,6-trinitrophenol; picronitric acid) 苦味酸；2,4,6-三硝基苯酚 $(NO_2)_3C_6H_2OH$
picric acid method 苦味酸法
picrin 苦味碱
picrinite 苦味酸；苦酸(炸)药
picrite 橄苦岩
picro-〔词头〕苦
picroaconitine 苦乌头碱
picroadonidine 黑藜芦苦素
picrocarmine 苦胭(脂)红
picrocrocin 苦藏花素
picroerythrin 苦红素
picrol 苦醇
picrolichenic acid 苦地衣酸
picrolite 硬蛇纹石
picrolonic acid 苦酮酸 $C_{10}H_8O_5N_4$
picromerite 软钾镁矾
picronigrosin 苦苯胺黑
picronitric acid(=picric acid) 苦硝酸〔即：苦味酸〕
picropodophyllin 鬼柏苦 $C_{22}H_{22}O_8$
picroroccelin 海石蕊苦素 $C_{20}H_{22}O_4N_2$
picrosclerotine 麦角苦碱
picrotin 苦亭 $C_{15}H_{18}O_7$
picrotone 苦酮 $C_{14}H_{16}O_3$
picrotonol 木防己毒乙酮醇
picrotoxin 木防己苦毒素
picrotoxinin 木防己苦毒宁
picryl 苦基；间三硝苯基；2,4,6-三硝基苯基 2,4,6-$(NO_2)_3C_6H_2$—
picryl acetate 乙酸苦味基酯
picryl amine(=picramide) 苦酰胺；苦基胺 $(NO_2)_3C_6H_2NH_2$
picryl chloride 苦基氯；2,4,6-三硝基氯苯 $ClC_6H_2(NO_2)_3$
2-picrylhydrazyl 2-苦基偕腙肼
picryl sulfide 苦基硫；六硝基二苯
Pictet crystals 皮克特晶体〔水合二氧化硫〕
pictorial display 图像显示
pictorialization （用)图表(显)示；图表化
Pictor liquid 皮克托液体〔SO_2 和 CO_2 的混合物〕
picture framing 厚边
picul 担；石〔中国重量单位；合百市斤〕
picylene 苊芴
pi(π)-donation π-给电子现象
piece ①件；块②修补③接合

piece cutter　冲切机
piece dyeing　匹染
piece-goods　零件货物
piece-up　接头
piecewise fitting　逐段拟合
piecewise linear regression　逐段线性回归
piedmontite　红帘石
pie filling　饼馅
pierce tap　钢丝带〔橡〕
piercing test　贯穿试验
piericidin　杀粉蝶菌素；虫螨霉素
pier process　水煤气发生过程
pie-shaped piece　盘形片
pieso-　〔词头〕压力
pie-still heater　管式炉加热器
pietraverdite　阿尔卑斯浅绿凝灰岩
pi(π)-excessive heteroatoms　π-额外杂原子
piezo-　〔希腊字头〕压；压力
piezochemistry　高压化学
piezocontrol　压电整
piezocrystal　压晶
piezocrystallization　加压结晶
piezodialysis　加压渗析
piezo effect　压电效应
piezo-electric apparatus　压电仪(器)
piezoelectric ceramics　压电陶瓷
piezo-electric chemistry　压电化学
piezoelectric composite　压电复合材料
piezo-electric crystal　压电晶体
piezo-electric crystal hygrometer　压电晶体湿度计
piezo-electric detector　压电检测器
piezo-electric DNA sensor　压电DNA传感器
piezo-electric effect　压电效应
piezo-electric enzyme sensor　压电酶传感器
piezo-electric immunoassay　压电免疫测定
piezo-electric immunosensor　压电免疫传感器
piezoelectricity　①压电(现象)②压电学
piezo-electric material　压电材料
piezo-electric microbe sensor　压电微生物传感器
piezo-electric moisture analyzer　压电水分分析器
piezoelectric polymer　压电高分子；压电性聚合物
piezo-electric powder　压电粉末
piezo-electric quartz crystal resonator　压电石英晶体共振器
piezo-electric quartz detector　压电石英检测器
piezo-electric quartz mass sensor　压电石英质量传感器
piezo-electric quartz sensor　压电石英晶体传感器
piezo-electric sorption detector　压电吸着检测器
piezo-electric spectroelectrochemistry　压电光谱电化学
piezoelectric transducer　压电换能器
piezoelectron　压电电子
piezoeletric quartz sensor　压电晶体传感器
piezo-gauge　压力计
piezojunction effect　压结效应
piezoluminescence　压电发光
piezometer　测压计；流压计；水压计
piezometer ring　匀压计
piezometer tube　匀压计管
piezometric hole　量压孔
piezometric ring　压力计环
piezoquartz manometer　加压水晶压差计
piezoresistor　压敏电阻
piezo-thermoluminescence　压力热致发光
pig　①生铁(块)②猪
pig-and-ore process　生铁矿石法
pig bed　铸床；浇铸场；出铁场
pig boiling　生铁沸腾
pig boiling process　生铁沸腾法
pig chasing(=pig tracing)　沿管道随清管器运动试探
pigeon's manure bate　鸽粪软化
pigeon dung　鸽粪；鸽肥
pigeon dung bate　鸽粪软化
pigeon fat　鸽脂
pigeon-hole　①鸽笼②鸽笼式架
pigeon-holed arch　多孔火拱
pigeon manure　鸽粪
pig gauge　栓规
pigging　(管道内部的)清管器清理
pig hide　猪皮
pig iron　铁锭；生铁；铣铁
pig lead　铅锭；生铅
pig liver fat　猪肝脂
pigment　①色素〔生物〕②色料③颜料
pigmentability　①颜料捏合性；颜料掺和性②颜料着色性
pigmentary(=pigmental)　①(含)颜料的；(含)色素的②颜料级的
pigmentation　①颜料淀积(作用)②色素形成(作用)
pigment binder ratio　颜料漆料比
pigment binding power index　颜料结合力指数
pigment bleeding　颜料渗色
pigment bordeaux　颜料枣红
pigment brown　颜料棕
pigment carrier　颜料载体
pigment characteristics　颜料特性
pigment chrome yellow　颜料铬黄

pigment crowding 颜料密集〔颜料过于密集〕
pigment dispersing agent 颜料分散剂
pigment dispersion 颜料分散体
pigment dispersion micelle 颜料分散胶束
pigment dyeing 涂料染色
pigment dyes(=dry pigments) 有机颜料
pigment dyestuff 颜料性染料；有机颜料〔指不含金属离子的〕
pigmented 着色的
pigmented coating ①着色涂料②色漆
pigmented compound 填料混炼胶〔橡〕
pigmented filler 着色填孔剂
pigmented glass 有色玻璃
pigmented layer 着色层
pigmented preservation 着色防腐剂
pigmented rubber emulsion 橡胶乳化色漆；着色橡胶乳液
pigmented shellac 着色虫胶；紫胶色漆
pigmented yarn 纺(前)染(色)丝；无光化纤纱；无光丝
pigment extender(=extender pigments) 体质颜料
pigment fast red 快红颜料
pigment fast violet R 颜料坚牢紫 R；二噁嗪紫
pigment finish 涂颜料
pigment flocculate 颜料絮凝体
pigment flooding 涂料变色(作用)〔干燥或加热时的〕
pigment-grade calcium silicate 着色用硅酸钙
pigment grind 色浆；漆浆
pigment identification 颜料鉴别(定)；颜料识别
pigmenting(=pigmentation) ①着色②轧颜料浆；压色浆③颜料化
pigmenting power (颜料的)着色力
pigmenting property 着色性能
pigment lake 色淀颜料〔含金属离子的有机颜料〕
pigment matrix 颜料基质；颜料母体
pigmentolysis 解除色素(作用)
pigmentophilic anchor 亲颜料性锚固；亲颜料性锚碇
pigment orange 颜料橙
pigment packing factor 颜料堆积因数；颜料密集因数
pigment padding 颜料轧染(法)
pigment paper ①色纸；颜料纸②(=carbon tissue)炭素纸
pigment paste 颜料膏
pigment pulp 颜料浆〔专指颜料的水分散液〕
pigment ratio 颜料比(率)
pigment reduction 颜料冲淡
pigment reduction method 色素还原法
pigment sediment 颜料沉积物
pigment settling gage 颜料沉降测定仪
pigment slurry 颜料浆；水浆颜料
pigment-to-binder ratio 颜(料)基(料)比

pigment volume ①颜料体积②(=pigment volume concentration)颜料体积浓度
pigment volume concentration(PVC) 颜料体积浓度
pigment volume ratio 颜料比容；颜料体积比
pigment wettability 颜料润湿性
pigment wiping filler stain (木器用)颜料填孔着色剂
pigment yellow 颜料黄
pig metal 生炼金属
pig mould 生浇〔金属〕
pig's foot grease 猪脚脂
pigtail 盘管
pig tracing(=pig chasing) 沿管道随清管器运动试探
pikrococin 藏花醛苷
pikromycin(=picromycin) 苦霉素
Pila-fractionation (使用低级烃为溶剂的)润滑油分馏
pilchardine 鲱油
pilchard oil 沙丁(鱼)油
pile ①堆②堆(叠)③木桩④电堆⑤铀堆
pile burning(=bin cure) 仓储硫化；早期硫化〔橡胶〕
piled beets〔复〕 垎(壕堆)藏甜菜〔糖〕
pile fabric 起毛织物
pile-irradiated 堆照(过)的
piliganine 石松碱
piling process 堆烧(炭)法
pill ①丸；片②药九；丸剂
pillar ①(显微镜)基柱②柱(子)
pillar inlay net 衬纬编链网孔
pillaromycin 基柱霉素
pilled-catalyst 丸形催化剂
pilling 起球
pilling effect 起球现象
pilling tester 起毛起球试验仪
pillow container 枕形容器
pilocarpidine 毛果芸香定
pilocarpine 毛果芸香碱
pilocereine 毛仙影掌碱
pilocerine 毛仙影掌碱
piloquinone 毛醌素
pilosine 毛果芸香素
pilosinine 毛果芸香素宁
pilosomycin 毛发霉素
pilot ①指示②排障器
pilot burner 引火灯
pilot cracking unit 中间裂化装置
pilot ion(=reference ion) 比较离子；参比离子
pilot ion method 指示离子法
pilot lamp 指示灯
pilot (light) burner 引火灯

pilot-operated controller　自动控制器
pilot plant　①中试装置②试验工厂
pilot process　中间过程
pilot-scale　中间工厂规模
pilot-scale production　中间工厂规模生产；中间试制
pilot study　试验性研究；中间规模研究
pilot test　中试；典型试验
pilot torch　点火灯
pilot unit　中间试验工厂
pilot valve　导向阀
pimanthrene　海松烯；1,7-二甲基菲
pimaradiene　海松二烯
pimarate　海松酸(甲)酯
pimaric acid(=dextropimaric acid)　(右旋)海松酸　$C_{20}H_{30}O_2$
pimaricin　匹马菌素
pimaric-type acid　海松酸型酸
pimarinal　海松醛
pimarinol　海松醇
pimelate　庚二酸盐(酯)　$COOM(CH_2)_5COOM$；$COOR(CH_2)_5COOR$
pimelic acid　庚二酸　$(CH_2)_5(CO_2H)_2$
pimelic dinitrile　庚二腈　$(CH_2)_5(CN)_2$
pimelinketone　环己酮
pimelite　镍皂石；脂光蛇纹石
pimelonitrile　庚二腈
pimeloyl　庚二酰　—$CO(CH_2)_5CO$—
pimeloyl chloride　庚二酰氯
pimenta berry oil　众香子油
pimenta leaf oil　众香叶油
pimentic acid　多香果酸
pimento oil　多香果油〔取自 *Pimenta officinalis*〕
pimento seed oil　甘椒子油
pimpinella　茴芹
pimpinellin　茴芹素
pimple　(表面上的)小突起；疙瘩〔塑料〕
pimpling　起丘疹
pin　①针；销②栓
-pin　频〔生物碱〕
pinachrome　松色素
pinacoid　轴面体
pinacol　①四甲基乙二醇；2,3-二甲基-2,3-丁二醇　$[(CH_3)_2COH]_2$②邻二叔醇类　$[R_2COH]_2$③频哪醇
pinacol condensation　邻二叔醇重排作用
pinacol conversion(=pinacolic rearrangement)　邻二叔醇重排作用
pinacolic rearrangement(=pinacol conversion)　邻二叔醇重排作用

pinacolone　频哪酮
pinacolone　①(=pinacoline；*t*-butyl methyl ketone)　叔己酮；3,3-二甲基-2-丁酮　$CH_3COC(CH_3)_3$②叔酮　$RCOCR_3$
pinacolone rearrangement(=pinacon-pinacolin transformation)　频哪酮重排作用
pinacolone semicarbazone　叔己酮缩氨基脲　$C_5H_{12}C=NNHCONH_2$
pinacol rearrangement　频哪醇重排
pinacolyl　频哪基；2-叔己基
pinacolyl acetate　频哪基乙酸酯　$CH_3CO_2C_6H_{13}$
pinacolyl alcohol　频哪基醇；2-叔己醇　$(CH_3)_3CCHOHCH_3$
pinacone(=tetramethyl ethylene ketone)　2,3-二甲基-2,3-丁二醇；频哪醇
pinacon-pinacolin transformation(=pinacolone rearrangement)　频哪酮重排作用
pinacyanol　频哪氰
pinacyanol chloride　氯化频哪氰醇
pinakoid　轴面(体)
pinakryptol　频哪隐绿
pinakryptol green　频哪隐绿
pinakryptol yellow　频哪隐黄
pinalic acid　戊酸
pinane　蒎烷　$C_{10}H_{18}$
pinane hydroperoxide　氢过氧化蒎烷　$C_{10}H_{18}O_2$
pinang　槟榔
pinanol　蒎烷醇；马鞭草醇
pinanone　蒎烷酮；马鞭草酮
pinaster seed oil　松果油
pinaverdol　频哪绿
pinaverium bromide　匹维溴铵〔药〕
pin bundle　针形元件束
pincelin(=pinselin)　青霉抗菌素
pincers　(咬口)钳(子)
pincette　镊子；小钳子
pinch　①夹紧；捻(紧)②(一)撮；微量
pinchbeck　金色铜；铜锌合金
pinchcock　弹簧夹
pinchcock clamp　弹簧(节流)夹；节流夹
pinched composition　夹紧组成
pinched cord　乱线；线绳错位
pinched tube　压制限流器
pinch effect　收缩效应；收聚效应
pinching　收聚〔接近平衡线和操作线交点处〕
pinch roll　夹紧辊
pinckneyin　品克呢亚苷
pineapple essence　凤梨香精；波萝香精

pineapple fiber 凤梨麻；波萝麻
pineapple flavour 凤梨香精；波萝香精
pine bouquet 松林百花香；松林百花型
pine camphor 松醇；松脑 $C_{10}H_{16}O$
pine cone oil 松节油
pine gum 松脂
pine leaf oil(=pine needle oil) 松针油
pine marten fat 松皮脂
pinene 蒎烯 $C_{10}H_{16}$
pinene dichloride 蒎烯化二氯 $C_{10}H_{16}Cl_2$
pine needle oil(=pine leaf oil) 松针油
α-pinene epoxide 环氧-α-蒎烯
pinene epoxide 环氧蒎烯
pineneglycol 蒎烯二醇
pinene hydrate(=homopinol) 水合蒎烯；高蒎醇 $C_{10}H_{18}O$
pinene hydrochloride 盐酸蒎烯；蒎烯化氢氯 $C_{10}H_{16}HCl$
pinenol 蒎烯醇；马鞭草烯醇
pinenone 蒎烯酮；马鞭草烯酮
pine oil 松油
pine oleoresin 松脂
pine pitch 松焦油沥青
pine resin ①松香②松脂
pine root crude oil 松根原油
pine-seed oil 松子油
pine-shaving reaction 松木(材)显色反应
pine tar 松焦油
pine tar oil 松焦油
pine tar pitch 松焦油沥青
pine-tree oil 松树油
pine wood oil 松木油
pine wood tar 松木焦油
piney 松林香气
piney tallow 松木硬脂
pin gate 针形浇口
pin ga(u)ge 栓规〔商〕
pinging 震性
ping-pong mechanism 乒乓机制
pin grain 海豹革粒纹
pinguin 土木香油
pinhole 针孔
pinhole leak 针孔裂缝
pin-hole plotter 打孔器
pinhole porosity 针眼；小气眼；砂眼
pinholing 针孔
pin-holing corrosion 针孔形腐蚀；针孔形锈蚀
pinic 松的

pinic acid 蒎酸 $C_9H_{14}O_4$
pinic acid diester 松脂酸双酯
pinicortannic acid 赤松单宁酸 $(C_{16}H_{18}O_{11})_2 \cdot H_2O$
pinidine 松里汀
pinifolic acid 松叶酸
pinine glycol 松香芹二醇；松香苄二醇；蒎烷二醇 $C_{10}H_{18}O_2$
pinion 副齿轮；小齿轮
pinion (gear) 小齿轮；传动齿轮
pinion grease 传动齿轮用润滑脂
pinion ratio 传动齿轮比
pinipicrin 松叶苦素
pinitannic acid 松单宁酸
pinite 蒎立醇
pinitol 右旋肌醇甲醚
pink color ①铬锡红(颜料)②桃红色〔俗〕
pink damask rose 大马士革玫瑰
pink discolouration 发红〔皮革〕
pink group (粉)红色类(载体)
pinking(=knocking) 震性
pink noise 粉红色噪声
pin knots 针节
pink salt 锡盐〔通常指氯锡酸铵或二氯化锡〕
pink support (粉)红色载体
pin load (十字头)销负荷
pinnatifidin 羽菊定
pinned mixing section 销钉式混炼段
pinned stock 漆刷箍紧毛根；锁紧毛根
pinnoite 柱硼镁石
pinoacetaldehyde 松乙醛
pinoalkanal 松烷醛
pinobanksin 短叶松素
pinocamphane 松莰烷 $C_{10}H_{18}$
pinocampheol 松莰醇
pinocamphone 蒎莰酮
pinocarveol 松香芹醇
pinocarvone 松香芹酮 $C_{10}H_{14}O$
pinocembrin 松属素
pinocytosis 胞饮(作用)
pinoglycol 蒎烯二醇 $C_{10}H_{18}O_3$
pinol 蒎脑〔不叫松醇；松醇是 coniferyl alcobol〕
pinol glycol 蒎脑二醇
pinol hydrate 水合蒎脑 $C_{10}H_{18}O_2$
pinoline 松香烃
pinomyricetin 松杨梅精；松树杨梅色素
pinonate 蒎酮酸盐(酯)
pinone 蒎酮 $C_{10}H_{16}O$
pinonic acid 蒎酮酸 $C_{10}H_{16}O_3$

pinononic acid 低漈酮酸
pinoquercetin 西黄松黄酮；6-甲基槲皮酮
pinoresinol 松脂酚
pinostrobin 乔松酮；5-羟(基)-7-甲氧(基)黄烷酮
pinosylvin 赤松素；3,5-二羟(基)芪
pin-point 鉴别；挑选
pin-roller fibrillator 针辊式裂膜装置；针辊式原纤化装置
pin-roller snagging tester 刺辊式钩丝仪
pinselin 青霉抗菌素
pint 品脱〔容量单位〕
pint cup 一品脱(的)杯(罐)
Pintsch gas (粗柴油)高温裂解气体
Pintsch gas tar 高温裂解焦油
Pintsch process 高温裂解
pin-type 插入式的；针型的
pinwheel leather measuring machine 针轮量革机
Piobert effect 多晶形铁和钢的表面标志；派奥伯特效应
pioglitazone 吡格列酮〔药〕
pioklastin 片绿脓菌素
Piola-Finger tensor 皮奥拉-芬格张量
Piola-Kirchhoff stress 皮奥拉-基尔霍夫应力
piomycin 多效霉素；多氧霉素
pipage 管道；管子；管道系统
pipanol hydrochloride 苯海索〔药〕
pipe ①管(子)②装管(子)
pipe bend 管肘
pipe bender 弯管机
pipe bending machine 弯管机
pipe bundles 管束
pipe capacity 管线容量
pipe carrier 管架
pipe centering apparatus 管子找中装置；管子对中仪
pipe clamp 管夹
pipe classifier 管式分粒器
pipe clay 管土
pipe cleaning pig 清管器
pipe clip 管钩
pipe closer 管闸；管塞
pipe coating process 管道涂敷法
pipe coil(=spiral pipe) 蛇管
pipecolic acid(=pipecolinic acid) 2-哌啶酸
pipecoline(=2-methyl piperidine) 2-甲基哌啶 $CH_3C_5H_{10}N$
pipecoline methylpentamethylenedithiocarbamate 甲基五亚甲基二硫代氨基甲酸甲基哌啶〔促进剂〕
pipecolinic acid(=pipecolic acid) 2-哌啶酸
pipe collapsing device 压平管子工具

pipe column 管束
pipe connecting device 接管装置
pipe connection 管道连接；管子接头
pipe cooler 管式冷却器
pipe coupling 管连接
pipecuronium bromide 哌库溴铵〔药〕
pipe cutting machine 切管机
piped 管的；管状的
pipe dope 管子涂料
pipe duct 管道；输送管；导管
pipe elevator 水压升降机
pipe end aligner 管端对准器
pipe expander 扩管器
pipe extractor 抽出管子用设备
pipe filter 管式过滤器
pipe fitting 接管零件
pipe flange 管法兰；管子凸缘
pipe-flange packing 管法兰垫
pipe flexibility 管子韧性
pipe flow 管流；导管流动
pipe furnace(=pipe heater) 管式炉
pipe gang 管道班；管道队
pipe head 管道进口端
pipe heater(=pipe furnace) 管式炉
pipe holder 管支架
pipe installation 管线铺设
pipe joint 管接合
pipe joint composition 管接头处垫料
pipe-layer 管道安装工
pipe laying 管道安装；铺设管道
pipe laying system 管道铺设系统
pipe leak 管裂缝
pipe line 管道
pipe line agitator 液体连续搅拌机
pipe-line anchor (海底)管道固定锚
pipe-line blending 管道内(石油产品的)混合
pipeline center 管道找中器
pipe-line cleaner 管道清洁器；清管器
pipe-line communication 管线通信
pipe line compressor 管道压缩机；增压压缩机
pipe-line construction 管道建设
pipe-line control valves 管道控制阀
pipeline drying pig 管道干燥清管器
pipe-line electronics 管道电子通信设备
pipe-line equipment 管道设备
pipe-line fluid network calculator 管道流量网计算器
pipe-line fracture test 管道破裂试验
pipeline gas 管道气

pipeline gauge　管道流速计
pipeline gauger　管线测量员
pipeline inspection pig　管道清洁器；清管器
pipeline interface　管道接口；管道交界
pipeline interface detector　管道接口探测器
pipeline laying　管道铺设
pipe-line layout　管道设计
pipe-line maintenance　管道维护
pipe-line mileage　管道长度英里数
pipe-line oil　管道油
pipeline pig　清管器
pipeline plugging pig　管道堵塞器；堵管器
pipeline positioning　管道定位
pipe-line pumping station　管道泵站
pipe-line purging　管道清扫
pipeliner　管道工
pipeline reactor　管道式反应器
pipe-line reconditioning　管道修理
pipeliner's cradle　管道摇架
pipe-line run　(石油产品)管道输送量
pipe-line scraper　管道刮刀
pipe-line spans　管道跨距
pipeline stiffener　管道支肋
pipe-line stock　管道储存油
pipeline stopcock　管道旋塞
pipe-line storage　管道储存量
pipe-line tank　管道槽
pipe-line transmission capacity　管道输送量
pipe-line transportation　管道输送
pipeline valve　管道阀
pipeline valve actuator　管道阀起动器
pipe-line walker　管道巡查员
pipe-line wrapping　管道绝缘
pipelining　管道安装；管道敷设
pipe locator　地下管道探测仪
pipe-man　管道安装工
pipe manifold　管道汇集器
pipe manifold valves　管道汇集器阀门
pipemidic acid　吡哌酸〔药〕
pipe network　管网；管道系统
pipe plug　管塞；塞子；塞头
pipe position indicator　管位指示器
pipe precipitator　管式沉降器
pipe protection　管道保护
pipe prover(=pipe tester)　管道检验仪
piperacillin　哌拉西林〔药〕
pipe rack　管道支架
piperamide　胡椒酰胺

pipe range　管道网
piperaquine　哌喹〔药〕
piperazine(=diethylenediamine)　哌嗪
$$NHC_2H_4NHC_2H_4$$
piperazine adipate　己二酸哌嗪
piperazine bis-dithiocarbamate　哌嗪双氨荒酸盐
piperazine citrate　柠檬酸哌嗪
piperazinedione　哌嗪二酮　$C_4H_6O_2N_2$
piperazine ferulate　阿魏酸哌嗪〔药〕
piperazine phosphate　磷酸哌嗪
piperazine quinate　奎尼酸哌嗪
piperazinium　哌嗪鎓
piper betel oil　蒌叶油
pipe reducer　渐缩管；(不同管道连结的)大小头
piperic　胡椒的
piperic acid　胡椒酸；β-(3,4-亚甲二氧基苯)基-2,4-戊二烯酸　$CH_2O_2=C_6H_3C_5H_5O_2$
piperidic acid　γ-氨基丁酸
piperidide　胡椒脂
piperidine　哌啶
piperidine methyl phthalimide　邻苯二酰亚胺甲基哌啶〔促进剂〕
piperidine pentamethylene dithiocarbamate　亚戊基二硫代氨基甲酸哌啶
piperidinium　哌啶鎓
piperidinium salt　哌啶鎓盐
piperidino　哌啶子基(指1位基而言)　$C_5H_{10}N—$
piperidone　哌啶酮　C_5H_9NO
piperidyl(2-, 3-, 或 4-)　哌啶基；氮杂环己基　$C_5H_{10}N—$
piperidylidene　亚哌啶基　$C_5H_9N=$
piperidyl urea　哌啶基脲；哌啶甲酰胺
piperidyl urethane　哌啶基尿烷；哌啶(基)N-甲酸乙酯
piperine　胡椒碱
pipe ring　管道环
piperinic acid　胡椒酸
pipe riser　升管器
piperitenol　薄荷烯醇
piperitenone　薄荷烯酮
piperitenone oxide　薄荷烯酮醚
piperitenoxide　氧化胡椒烯
piperitol　薄荷醇
piperitone　薄荷酮　$C_{10}H_{16}O$
piperitone oxide　氧化薄荷酮；薄荷酮醚
piper kadsura　细心青蒌藤
piperocaine hydrochloride　哌罗卡因盐酸盐
piperolidine　δ-毒芹侧碱
pipe roller　滚管机

piperonal ①(=heliotropin)胡椒醛；3,4-亚甲二氧基苯甲醛 $CH_2O_2;C_6H_3CHO$ ②(=peperonylidene)亚胡椒基 $CH_2O_2=C_6H_3CH=$
piperonal-acetophenone 亚胡椒基乙酰苯；胡椒醛缩乙酰苯 $C_8H_6O_2=CHCOC_6H_5$
piperonaldehyde 胡椒醛
piperonyl 胡椒基；3,4-亚甲二氧苄基；3,4-亚甲二氧苯甲基 $(CH_2O_2)C_6H_3CH_2—$
α-piperonyl acetaldehyde α-胡椒基乙醛
piperonyl alcohol 胡椒基醇；3,4-亚甲二氧(基)苯甲醇 $CH_2O_2=C_6H_3CH_2OH$
piperonyl butoxide 胡椒基丁醚
piperonyl cyclohexenone(=piperonyl cyclonene) 胡椒基环己烯酮
piperonyl cyclonene(=piperonyl cyclohexenone) 胡椒基环己烯酮
2-piperonyl ethanol 2-胡椒基乙醇
piperonylic acid 胡椒基酸；3,4-亚甲二氧基苯甲酸 $CH_2O_2C_6H_3CO_2H$
piperonylidene 亚胡椒基；3,4-甲二氧苯亚甲基 $3,4-(CH_2O_2)C_6H_3CH=$
piperonylidene butyraldehyde 亚胡椒基丁醛
α-piperonylidene propionaldehyde α-亚胡椒基丙醛
piperonyloyl 胡椒基酰；3,4-甲二氧苯(甲)酰 $3,4-(CH_2O_2)C_6H_3CO—$
piperonyl phloroglucinol dimethyl ether(=protocotoin) 间苯三酚胡椒基二甲基醚；原可土因；原可土树皮素 $(CH_3O)_2C_{14}H_8O_4$
piperovatine 卵形椒碱〔取自卵形椒 Piper ovatum 的叶、枝、根〕 $C_{16}H_{21}O_2N$
piperyl 胡椒酰
piperylene(=1,3-pentadiene) 1,3-戊二烯；戊间二烯 $CH_3CH=CHCH=CH_2$
piperylene rubber 戊二烯橡胶
piperylhydrazine 哌啶胺
pipe saddle 管道鞍；鞍形管道修理夹头
pipe sampling 管式取样器
pipe scale 管垢
pipe scraper(=pipe-line scraper) 管道刮刀
pipe sealing 封管的；封接管口的
pipe sealing compound 封管化合物
pipe sheet 管板
pipe socket 管套口
pipe spacer 定距管
pipe still 管式炉；管式蒸馏釜
pipe still distillation 管式炉蒸馏
pipestone(=catlinite) 烟斗泥
pipe stool 管道扇形座架

pipe straightener 直管器
pipe string 管道支线
pipe string hanger 管道吊架
pipe structure preparation 管道结构准备工作〔管道表面涂保护层前的准备工作〕
pipe support 管道支架
pipe system 管道系统
pipet(=pipette) 吸移管；吸量管；球管；移液吸管
pipe tap 锥管螺纹
pipe tester(=pipe prover) 管道检验仪
pipe-thread 管螺纹
pipe train 管材挤塑机组
pipet safety filling attachment 安全移液管
pipet stand(=pipet support) 移液管架
pipet support(=pipet stand) 移液管架
pipette(=pipet) 吸移管；吸量管；球管；移液吸管
pipette rack 吸移管架
pipette stand 吸移管架
pipette viscometer 滴管黏度计
pipettor 吸移管管理器
pipet washer 移液管洗涤器
pipet with single bench mark 单标线吸量管
pipewalker(=pipe-line walker) 管道巡查员
piping ①管(子)②管(子)系(统)③管道④管连接
piping design 配管设计
piping diagram 管道分布图
piping hanger 管道挂钩
piping installation 管道施工
piping insulation 管道保温
piping manifold 管道网
piping system 管道系统
piping time 配管时间
pipi root 商陆根
pipitzahoac 金菜根
pipitzahoic acid 金菜酸
pipitzahoin 金菜根色素
pipoint technique 精密技术
pipotiazine 哌泊塞嗪〔药〕
pipsissewa extract 梅笠草浸液
pipsylamino acid 对碘苯磺酰氨基酸
piptamine(=ormosanine) 红豆树宁；苦豆碱
pique crepe 凹凸绉
piquia fat 白胡桃油
piquiarana oil 柯雅卡油
piracetam 吡拉西坦〔药〕
Pirani gauge 皮拉尼真空计
pirarubicin 吡柔比星〔药〕
piricularin 稻瘟菌素；梨形孢菌素

piririma oil 椰子油〔取自 Cocos syagres 的种子〕
pirn 纬管纱；纡子
piroglycerina 硝化甘油
pirolatin 鹿蹄草亭
piroxicam 吡罗昔康〔药〕
pirssonite 钙水碱
pirylene(=3-pentene-1-yne) 3-戊烯-1-炔
pisangceric acid(=pisangceryl acid) 香蕉蜡酸
pisangceryl acid(=pisangceric acid) 香蕉蜡酸
pisangceryl alcohol 香蕉蜡醇 $C_{13}H_{22}OH$
pisangcerylic acid 香蕉蜡酸 $C_{24}H_{48}O_2$
pisanite 铜绿矾
pisatin(=phytoalexin) 豌豆素
piscidic acid 番石榴酸 $C_{11}H_{10}O_7$
piscidin 番石榴素
piscose 阿洛酮糖
pisiform 豌豆形的
pisolite 豆石〔矿〕
pissasphalt 天然沥青
pistachio nut oil(=pistachio oil) 乳香黄连木油
pistachio oil(=pistachio nut oil) 乳香黄连木油
pistacite 绿帘石
pistol powder 手枪火药
pistomesite 菱镁铁矿
piston 活塞
piston blower 活塞式鼓风机
piston body 活塞体
piston buffer change valve 活塞式缓变阀
piston buret 活塞滴定管
piston bush 活塞衬套
piston clearance 活塞间隙
piston compressor 活塞式压缩机
piston core sampler 活塞岩芯采样器；活塞取样管
piston cup 活塞皮碗
piston displacement 活塞位移
piston flow 活塞流
piston force 活塞力
piston gauge 活塞式压力计
piston groove 活塞槽
piston-in-cylinder capillary viscometer 筒内活塞式毛细管黏度计
piston leather 活塞(涨圈用)革
piston meter 活塞流量计
piston oil 活塞油
piston packing 活塞密封圈
piston packing leather 活塞填密皮
piston pin 活塞销
piston pneumatic pump 活塞气动泵；气动活塞泵
piston pump 活塞泵
piston ring 活塞密封圈
piston rod 活塞杆
piston rod thrust 活塞杆推力
piston speed 活塞速度
piston stroke 活塞冲程；活塞行程
piston-type area meter 活塞式面积流量计
piston-type pressure controller 活塞式压力控制器
piston valve 活塞阀
piston wrench 活塞扳手
pit ①坑；槽②矿井③纹孔
pit aperture 纹孔口
pit asphalt 软沥青
pitayamine 毕达金鸡纳碱
pit border 纹孔缘
pit car oil 黑色润滑油
pitch ①沥青；木沥青②螺距③齿距④间距
pitch bay 沥青坑
pitch binder 沥青黏结剂
pitch blende 沥青铀矿
pitch coal 沥青煤
pitch coke 沥青焦炭
pitch coking 沥青焦化
pitch cooler 沥青冷却器
pitched-base carbon fibre 沥青基碳纤维
pitch-epoxy resin 沥青-环氧树脂
pitcher 柄盂
pitch error 节径误差
pitchers 瓷渣
pitch filler 沥青填料
pitching 前后震动
pitch line 齿距线；节线
pitch oil 冷杉油〔取自 Abies pectinata 的种子〕
pitch oxidation 沥青氧化
pitch peat 沥青泥煤
pitch pocket 油眼
pitch sample 沥青样
pitch softening point 沥青软化点
pitch stone 松脂岩
pitch-tree oil 柏树油
pitch troubles〔复〕 沥青障
pitchy 沥青的
pitchy iron 沥青铁；砷铁浮渣
pit coal 沥青煤
pit corrosion 点蚀
pith ①(木)髓〔物〕；树心②生气；精力
pithecolobine 猴耳环碱
pith flecks 健斑

pit hole 石洞〔储存气体用〕
pith ray 髓射线
pit kiln 炼焦炉
pit kiln process 坑烧(炭)法
pit leak 小裂孔
pitman 连接杆；摇杆
pitocin 催产素
pitometer(=Pitot gauge) 毕托压差计
Pitot gauge(=pitometer) 毕托压差计
Pitot meter 毕托压差计
Pitot static tube 静压管
Pitot tube 毕托(流速测定)管
pit-pair 纹孔对
pit patches 小裂孔封补
pitressin 加压素；抗利尿激素
Pitrowsky test 毕氏缩二脲试验
pits〔复〕 凹点(塑料)
pittacol 六甲氧基玫红酸
pitted tank 坑内储槽
pitticite 土砷铁矾
pitting ①小孔②纹孔式③麻点；点腐蚀④锈斑
pitting corrosion 小孔腐蚀
pitting corrosion depth 点蚀深度
pitting corrosion speed 点蚀速度
pitting corrosion test 点(腐)蚀试验
pitting factor 孔蚀率
pitting point 节点
Pittsburgh flux 皮特斯堡沥青
pituri 澳洲薛杜蕾；皮特尤里树(一种类似烟草的树)
piturine 薛杜蕾碱
pit water 矿井水
pit wood 矿材；坑道木
Pitzer strain 皮策张力*
piuri 印度黄
pivalaldehyde 新戊醛；三甲基乙醛 $C_5H_{10}O$
pivalate 新戊酸酯
pivalic acid(=neopentanoic acid) 新戊酸；三甲基乙酸 $C_5H_{10}O_2$
pivalic aldehyde 新戊醛；三甲基乙醛 $(CH_3)_3CCHO$
pivalic ketone(=hexamethylacetone;pivalone) 六甲基丙酮 $(CH_3)_3CCOC(CH_3)_3$
pivalolactone 新戊内酯
pivalone(=pivalic ketone) 六甲基丙酮 $(CH_3)_3CCOC(CH_3)_3$
pivaloyl(=pivalyl) 新戊酰；三甲基乙酰 $(CH_3)_3CCO—$
pivalyl(=pivaloyl) 新戊酰；三甲基乙酰 $(CH_3)_3CCO—$
pivalyl aldehyde 新戊醛；三甲基乙醛
pivalyl bromide 新戊酰溴 $(CH_3)_3CCOBr$

pivalyl chloride 新戊酰氯；三甲基乙酰氯 $(CH_3)_3CCCl$
pivalyl halide 新戊酰卤 $(CH_3)CCOX$
pivoted float 球形浮标
pivot tank 中央储槽
pizotifen 苯噻啶〔药〕
pK method pK法
place ①地方；地位②位置③位(数)④次序
placebo ①空白对照剂②安慰剂
placed in-line 串联的
place isomerism 部位同分异构(现象)
placement 键接
placer 砂矿
placer deposit 砂积矿床
placer gold 砂金
placer mining 淘金
placing 放；置；装
plagioclase 斜长石
plagioclasite 斜长岩
plagionite 斜硫锑铅矿
plague vaccine 鼠疫疫苗
plain 素色
plain agar 普通琼脂
plain bearing 普通轴承；滑动轴承
plain carbon steel 普通碳钢
plain colour 单色；素色〔指单一的颜色〕
plain core pin 平面成孔销
plain curve 平坦曲线*
plain draw mold 简易压模
plain dyeing 染素色〔染单一颜色〕
plain end 平管口
plain end (of pipe) (管子的)平端
plain face 平面
plain-face corrugated flange 平面皱纹边缘
plain flange 平面法兰
plain flash line 平合模线
plain gear 正齿轮
plain goods 平纹织物〔纤〕
plain insert 光面嵌件
plain linoleum 单色油毡
plain mineral oil 普通矿物润滑油
plain oil 普通润滑油
plain oil cup 普通润滑器
plain pipette 未刻吸移管
plain roll 光滚筒
plain sedimentation 简单沉降
plain-straight-face flange 平面法兰；平面凸缘
plain tube 光(滑)管
plain-type Horton spheroid 无内部加固的水滴形油罐

plain weave　①平织②布纹纸；平织纸
plait　编织的绳子；褶边
plaited　①折(褶)的②编成的
plaited paper filter　褶纸滤器
plaited surface　鳞面
plaiter　(浆板自动)折褶机
plait point　褶点
plan　①计划；设计②平面图
planar　①平面的②平的
planar bilayer lipid membrane　平板双层磷脂膜
planar chirality　平面手性
planar chromatography　平面色谱法*
planar coordination compound　平面配合物
planar defect　面缺陷
planar inversion　平面反演
planar molecule　平面分子*
planar orientation　沿面取向；平面取向
planar point　平点
planar segment　平面链段
planar solution　平面解
planar winding　平面缠绕
planar zigzag conformation　平面锯齿构象
planar zigzag form　平面曲折形；平面锯齿形
planar zigzag structure　平面锯齿(形)结构
planchet　(小)金属片
Planck constant　普朗克常数
Planckian radiator　普朗克辐射体；黑体
Planck law of radiation　普朗克辐射定律
Planck radiation law　普朗克辐射定律
Planck's potential　普朗克函数
plane　①平面②刨(子)③刨削④悬铃木
plane analysis　面分析
plane concentrating monochromator　平面汇集单色仪
plane Couette flow　二维库爱特流动
plane detonation front　爆炸波前平面
plane diffraction grating　平面衍射光栅
plane flow　平面流
plane formula　平面式
plane fracture　平面破裂
plane grating　平面光栅
plane grating spectrograph　平面光栅光谱仪
plane lattice　平面晶格；平面点阵
plane mirror　平面反射镜
plane mirror analyzer　平面镜分析器
plane of incidence　入射(平)面
plane of polarization　偏振(平)面；偏振面
plane of reflection　反射面
plane of rupture　裂断面

plane of slip　滑动面
plane of symmetry　对称面
plane of vaporization　汽化平面
plane orientation number　平面取向数
plane Poiseuille flow　二维泊肃叶流动
plane polarized light　平面偏振光
plane polarizer　平面偏光镜
planer　刨床
planer operator　刨工
planer tool　刨刀
plane shear strength　平面剪切强度
plane slanted bottom　平面斜底
plane strain　平面应变
plane stress　平面应力
plane symmetric isomerism　平面对称异构(现象)
planet agitator　回绕式搅拌器
planetary　行星式的
planetary atmosphere　行星雾围
planetary electron　行星式电子
planetary stirrer　行星式搅拌器
planetary stirring machine　行星式搅拌机
plane tomography(=planigraphy)　平面断层照相法
planet stirrer　行星式搅拌器
plane-type grinding mill　行星式粉碎机
plane tyre　飞机轮胎
plane wave　平面波
plangi　单染图案染法；绞缬
planigraphy(=plane tomography)　平面断层照相法
planimeter　面积仪；求积仪
planimetry　面积法；求积法
planing　刨削
planing machine　刨床
planing operation　刨削操作
plank　板条
planking　缩绒；毡合
planning　计划
planocaine　盐酸普鲁卡因
plano-concave　平凹的
plano-convex　平凸的
planoform　对氨基苯甲酸丁酯
planography　平板印刷；石印
planomycin　平霉素
plant　①(制造)厂；工厂；车间②植物③装置；设备④种植
plant acid　植物酸
plant aerosol　植物用气溶胶
plantagin　车前苷
plantainlily　玉簪

plant air   工艺空气
plant alkaloid   植物碱
plant analysis   植物分析
plantation rubber   栽培橡胶
plant cellulose   植物纤维素
plant chromatograph   工业色谱仪
plant chromatography   工业色谱(法)
plant effluent   工厂废水；工厂排放液
plantenolic acid   车前烯醇酸
plant fiber   植物纤维
plant filter   工厂过滤设备
plant hormone(=auxin)   茁长素；植物生长素
plant inlet   工厂入口
plant layout   工厂布置；车间布置；工艺布置
plant load factor   设备负荷系数
plant outlet   工厂出口
plant pigments   植物色素
plant piping   工厂管道系统；工厂管道安设
plant practice   工厂实习
plant rubber   天然橡胶
plant safety rules   工厂安全条例
plant-scale   ①工业规模的；大规模的②工厂规模
plant-scale equipment   工厂规模设备
plant screen efficiency   工厂掩护效率
plant-size equipment   生产型设备
plant's operating loss   工厂的操作损失
plant sterol   植物甾醇
plant trial   工厂条件下试用；生产上试用
plant unit   设备；装置
plant use   生产(上)使用
plant water   生产用水
plan view   平面图；俯视图
plaque   瓷花金属板
plasma   ①等离子体②等离子区③血浆④深绿玉髓
plasma arc   ①等离子(电)弧②电弧等离子体装置
plasma arc furnace   等离子(电弧)炉
plasma arc technology   等离子弧技术
plasma arc welding   等离子弧焊
plasma atomic fluorescence spectrometry   等离子体原子荧光光谱法
plasma body   等离子体
plasma carburising   等离子渗碳(法)
plasma carburizing   等离子渗碳(法)
plasma chemical vapor deposition   等离子化学蒸气沉积
plasma chemistry   等离子体化学
plasma chromatograph   等离子体色谱仪
plasma chromatography   等离子体色谱法*
plasma chromatography-mass spectrometry   等离子体色谱-质谱法
plasma confinement   ①等离子约束②等离子吸持
plasma cutting   等离子切割
plasma damage   等离子体损坏
plasma decay   等离子体衰变(减)
plasma desorption   等离子体解吸
plasma desorption mass spectrometry(PDMS)   等离子体解吸质谱法
plasma diagnostics   等离子体诊断学
plasma differential arc(PDA)   等离子差动电弧
plasma discharge   等离子体放电
plasma discharge light sources   等离子体放电光源
plasma display   等离子体显示(器)
plasma drug concentration   血药浓度
plasma-enhanced chemical vapor deposition   等离子体增强化学气相沉积
plasma flame   等离子火焰
plasma gas chromatography   等离子体气相色谱法
plasma generator   等离子体发生器
plasmagram   等离子体色谱图
plasma gun   等离子枪
plasma heating   等离子体加热
plasma-induced polymerization   等离子诱导聚合
plasma jet   等离子射流
plasma jet flame   等离子流火焰
plasma jet process   等离子流工艺
plasma jet source   等离子激发光源
plasma light source   等离子体光源
plasma loss peak   等离子体损失峰
plasma nitriding   等离子渗氮
plasma photosource   等离子体光源
plasma polymer   等离子聚合物
plasma polymerization   等离子聚合*
plasma power control(PPC)   等离子体功率控制
plasma processing   等离子体加工
plasma source   等离子体(光)源*
plasma spray(ed) coating   等离子喷镀层
plasma spraying   等离子喷涂
plasmaspray interface   等离子体喷雾式接口
plasma state   等离子态
plasma stream reactor   等离子流反应器
plasma substitute blood   代浆全血
plasmatorch   等离子体炬
plasmatorch tube   等离子体炬管
plasma transferred arc(PTA)   等离子传递电弧
plasma transferred arc fuse   等离子传递电弧熔融
plasmatron   等离子管；等离子流发生器
plasma welding   等离子焊

plasma zone　等离子区
plasmid　质粒；等离子粒团
plasmin determination　纤溶酶测定
plasminogen determination　纤溶酶原测定
plasmolysis　质壁分离；胞质皱缩
plasmon　①质粒基因组②细胞质基因组
plasmoquine　扑疟母星；扑疟喹
plast-agglomerator　冷相废丝再造粒机
plastainer　塑料容器
plastelast　塑弹性物；弹性塑料
plaster　①烧石膏；熟石膏②墙粉；灰泥粉刷料③硬膏；膏药
plaster block　石膏块
plaster board　糊墙纸板
plaster casting　石膏模注型法
plaster-cellulose type filler　石膏灰泥-纤维素型填孔料
plaster concrete　石膏混凝土
plastering　①灰泥；灰浆②抹灰；用石灰粉饰③石膏细工
plaster kiln　石膏窑
plaster mixer　石膏浆混合器
plaster model　石膏模型
plaster molding　用石膏模注型法
plaster mould　石膏模
plaster of colophony　松香(合铅)凝膏
plaster of lead　铅膏
plaster of Paris　烧石膏；熟石膏
plaster primer　灰泥底漆；灰泥墙耐碱底漆
plaster stone　石膏(石)
plastic　塑料*
plastic after effect　塑性后效
plastic alloy(=polyblend)　塑料合金；共混塑料
plastic and glass fiber pipes　塑料与玻璃纤维管
plastic anisotropy　塑性各向异性
plastic article　塑料制品
plasticating capacity　①塑炼能力〔每小时塑炼的千克数〕②挤出能力
plasticating cycle　塑炼周期
plastication　①塑化；增塑作用；软化②塑炼(作用)
plasticator　塑化器；塑炼机
plastic autoclave　塑料压热器
plastic binder　塑性黏合剂；混合增塑剂
plastic-bonded　塑性黏合
plastic cement　塑胶；塑料黏结剂
plastic clay　塑性黏土；陶土
plastic coal　塑性煤
plastic-coated fabric　涂塑布
plastic-coated fiber　塑料涂层纤维；涂塑料纤维
plastic-coated screen (net)　涂塑网布

plastic coating　塑料涂层；塑料涂膜
plastic coefficient　塑性系数
plastic composite　塑料复合材料
plastic consistency　塑性稠度
plastic container　塑料容器
plastic content　含塑量
plastic crystal　塑晶
plastic deformation　塑性形变
plastic disperse system　塑性分散系
plastic dosimeter　塑料剂量计
plastic fat　胶性脂肪；塑性脂肪
plastic film　塑料薄膜
plastic flexibility　可塑柔韧性
plastic floating deck　(储罐的)塑料浮动顶盖
plastic flow　塑性流动
plastic flow curve　塑性流动曲线
plastic flow equation　塑性流动方程
plastic flow model　塑性流(动)模型
plastic fluid　塑性流体
plastic fluidity　塑性流动性
plastic foam(s)　泡沫塑料
plastic fracture　塑性破坏
plastic friction　塑性摩擦
plastic friction welding　塑料摩擦焊接
plastic fuel-binder　燃料黏合剂
plastic hinge　塑性铰
plastic hysteresis　塑性滞后(现象)
plasticification zone(=melting zone)　塑化区；塑炼区
plastic incinerator　塑料焚烧炉
plastic inelasticity　塑性非弹性
plasticised fabric　塑性化纺织物
plasticising co-monomer　增塑性共聚单体
plasticising monomer　增塑性单体
plasticising oil　增塑油
plasticity　可塑性*
plasticity agent　增塑剂
plasticity index　可塑性指数；范性指数
plasticity number　可塑值；可塑度
plasticity of asphalt　沥青的可塑性
plasticity of greases　润滑脂的塑性
plasticity range　塑性(温度)范围
plasticity-recovery number　可塑性恢复值；塑性恢复值
plasticity retention (percentage)　塑性保持率
plasticity stability　塑性稳定性
plasticity test　可塑性试验
plasticity value　塑性值；可塑值
plasticization　增塑作用*
plasticization process　增塑(再生)法

plasticized powder	增塑粉末
plasticized taut relaxation	塑化张力弛豫
plasticizer	增塑剂*
plasticizer absorption	①增塑剂吸收②增塑剂吸收量
plasticizer-adhesive	增塑-黏合剂
plasticizer extender(=co-plasticizer)	增塑增容剂
plasticizer extraction	增塑剂析出；增塑剂萃取
plasticizer extraction test	增塑剂的萃取试验
plasticizer exudation	增塑剂渗出
plasticizer fogging	增塑剂雾化(成雾)
plasticizer loading	增塑剂用量；增塑剂含量
plasticizer loss	增塑剂损耗
plasticizer migration	增塑剂迁移；增塑剂渗移
plasticizer oil	油类增塑剂
plasticizer permanence	增塑剂的耐久性
plasticizer-stabilizer	增塑-稳定剂
plasticizer volatility	增塑剂的挥发性
plasticizing	塑化*
plasticizing agent	增塑剂
plasticizing-bath	塑化浴
plasticizing efficiency	增塑效率
plasticizing rate	①塑化程度②增塑剂含率〔增塑剂在树脂(涂料)中的百分率〕
plasticizing stabilizer	增塑稳定剂
plasticizing wood	塑化木材
plasticizing zone	塑化区
plastic layer	可塑层
plastic limit	塑性极限
plastic material	①塑料②塑性物质
plastic matrix	①塑料母料；塑料基料②塑料字模〔印刷〕
plastic melt	塑料熔体
plastic memory	塑性记忆
plastic microball	塑料粉球
plastic mobility	塑性淌度
plastic-nylon boom	塑料-尼龙水栅
plastico-dynamics	塑性动力学
plasticorder	塑度计；塑性计
Plasti-Corder rheometer	普拉斯蒂-科德流变仪
plastico-statics	塑性静力学
plastico-visco-elasticity	塑性黏弹性
plastic packing	塑料垫圈；塑料轴封
plastic paint	①塑性漆②干酪素拉毛涂料
plastic paste	塑料糊
plastic phase	塑化相；塑化阶段
plastic propellant	塑性推进剂
plastic property	塑性
plastic pump	塑料泵
plastic range	塑性范围
plastic range test	塑性范围试验
plastic resilience	塑性回弹
plastic resistance(=internal friction)	塑性阻力
plastics additive	塑料添加剂
plastic scintillator	塑料闪烁器
plastics coating	塑料涂层
plastic solid	塑性固体
plastic spraying technique	塑料喷涂技术
plastics solidification	塑料固化
plastic stage	可塑态；可塑阶段
plastic state	塑(性状)态
plastic strain	塑性应变
plastic strength	塑性强度
plastic sulfur(=elastic sulfur)	弹性硫
plastic surgery	整形手术；整形外科
plastic suspension	塑性悬浮体
plastic tank	塑性罐
plastic tanker	塑性槽车
plastic tape	①塑性带；塑性复印带；塑性复制带②塑料带
plastic torsion	塑性扭力
plastic viscosity	塑性黏度
plastic viscosity equation on tackmeter	黏性计的塑性黏度方程
plastic viscosity limitation on rotational viscometer	旋转式黏度计的塑性黏度极限
plastic ware	塑料器具
plastic waste	塑性废物；废塑料
plastic wax	塑性石蜡
plastic wood	塑用木；硝酸纤维素(木粉)浆
plastic yield(=yield value)	塑变值；塑流点
plastic yielding	塑性屈服
plastic yield-point	塑(料)流点
plastic yield test	塑流试验
plastic yield value	塑流值；塑性屈服值
plastic zone	塑性区；塑性范围
plastid	质体
plastification	①增塑作用；软化②塑炼
plastificator	①(螺杆)塑炼机②增塑剂
plastifier	增塑剂
plastifying	增塑的；塑化的
plastilock	(用合成橡胶改性的)酚醛树脂黏合剂
plastimeter	塑性计；塑度计；可塑计
plastimetry	可塑度测定法；测塑法；塑性测定法
plastipaste	增塑糊
plastisol	增塑溶胶*
plastisol coating	塑溶胶涂料
plastisol ink	塑溶胶型油墨；热熔油墨

plastisol wrinkle finish　塑溶胶皱纹漆
plastispray　塑料粉末喷涂；喷塑
plastocyanin　质体蓝素
plastoelastic body　塑弹性物体
plastoelastic deformation　塑弹形变
plastoelasticity　塑弹性
plastogel　塑性凝胶
plastogram　塑性变形图
plastograph　塑性形变记录仪；塑性仪
plastom　质体基因组
plastomer　塑性体；塑料
plastometer　塑性计；塑度计
plastometer constant　塑性常数
plastometer index　胶质层指数
plastometric set　塑性变形
plastometry　塑性测定法
plastoponic　塑料种植法〔用泡沫塑料代土壤〕
plastoquinone　塑体醌
plasto rubber　增塑橡胶
platable resin　可电镀树脂
plate　①板；版；片②极板③色谱板④照相(底)片
plate amalgamation　混汞析金(法)
plate-and-cone viscometer　板锥式黏度计
plate and frame (type) filter press　板框式压滤机
plateau　①金属①②平顶〔往往指曲线〕③平台
plateau atomization　平台原子化
Plateau border　普拉特奥边界〔气泡间三薄膜相遇处，形成微小三角形液柱，该处的曲率半径是负值〕
plateau cure　正硫化
plateau effect　①(硫化)平顶曲线效应②平顶曲线效应
plateau effect of viscosity　黏度的平稳效应
plateau phenomenon　平坦现象〔指曲线〕
plateau region　平坦区〔指曲线〕
plateau value　平稳值
plate battery　板状电池(组)
plate bending machine　卷板机
plate calender　平板纸砑光机
plate-chopping machine　板式切割机
plate chromatography　板色谱(法)〔薄层色谱(法)〕
plate coater　(色谱)板涂布器
plate column　①多层(蒸馏)柱②板式塔
plate composition　(蒸馏塔)塔板上液体组成
plate cooler　板式冷却器
plate crystal　片状结晶
plate crystal monochromater　平面晶体单色器
plate culture　平皿培养(法)
plated balls　金属覆盖球
plate determination　塔板数的计算与确定

plate development　(色谱)板展开
plated finish　熨平涂饰剂
plate distillation column　塔板蒸馏柱
plated item　电镀件；电镀产(制)品
plate edge planing machine　刨边机
plate efficiency　(塔)板效率
plate efficiency factor　塔板效率因子
plate electrode　极板
plate electrophoresis　板电泳
plate evaporator　片式蒸发器
plate exchanger　板式换热器
plate fan　钢板风扇
plate feeder　平板供料器
plate filter　板式过滤器
plate-fin heat exchanger　散热片式换热器；板翅式换热器
plate form of wax crystals　石蜡结晶的片状结构
plate glass　平板玻璃；玻璃砖
plate goods　片状商品
plate heater　板式加热器
plate heat exchanger　板式换热器
plate height　塔板高度
plate iron　铁板；铁皮
platelets　小片状体；片晶
plate level gauge　板式液面计
plate-like　片状
plate-like particle　片状粒子
plate-like structure　片状构造
plate method　板法
platen　(压机)压板
platen area　平板面积
platenomycin　普拉特霉素
platen press　压板压蒸机
plate number　塔板数(目)
plate orifice　锐孔板
plate out　一次性起霜；沉析；沉析贴层
plate-out agent　防积垢剂
plate-out resistance　防积垢(沉积)性
plate precipitator　板状沉降器
plate processing　照片底板处理
plate release　贴板剥离
plate roller　滚板机
plates and bubble caps　塔板和泡罩
plate shale　板页岩
plate shearing machine　剪板机
plate singeing machine　板式燎毛机
plates of similar polarity　同极板
plate spacing　(塔)板距

plate storage rack	薄层板储箱
plate sulfate	板状硫酸盐；硫酸盐首次结晶部分
plate supply	板极(屏极)电源；阳极电源
plate surface	镀盖表面；电镀表面
plate theory	塔板理论
plate theory equation	塔板理论方程
plate thickness	板厚度
plate-to-plate calculation	逐板计算
plate tower	板式塔
plate-type condenser	板式冷凝器
plate-type evaporator	板式蒸发器
plate-type heat exchanger	板式换热器
plate valve	片状阀
plate wax crystals	片状石蜡结晶
plate weight	片状砝码
plate work	板金工加工
plate working	板金工加工
platform	平台；站台；天平托盘；天平称量盘
platformate	铂重整产品；铂重整汽油
platform atomization	平台原子化
platform balance	台秤
platform conveyer	平台运输器
Platformer	铂重整装置〔环球油品公司，即UOP〕
platform furnace	平面石墨炉
platform furnace graphite	平台石墨炉
platforming(UOP)	铂重整
platforming process(UOP)	铂重整过程
platforming reactions	铂重整反应
platform scale(=platform weighing scale)	台秤
platform truck	平台式起重车
platform weighing scale(=platform scale)	台秤
platina	粗铂
platinammine	铂氨化物
platinammonium	氨合铂离子
platinate	铂酸盐 M_2PtO_4
plating	①(电)镀②镀(敷)③熨平；印纹
plating action	①镀敷作用②电镀作用
plating bath	(电)镀浴
plating lines	电镀(流水)线
plating machine	①印纹机②拆布机③压光机
plating paper	印纹纸
plating pen	镀笔
plating press	压光机
plating tank	电镀槽
plating vat	电镀瓮；电镀箱
plating yarn carrier	添纱导纱器
platinibromide	溴酸盐
platinic	(正)铂的；四价铂的
platinic bromide	溴化铂 $PtBr_4$
platinic chloride	氯化铂 $PtCl_4$
platinic compound	(正)铂化合物
platinichloride	氯铂酸盐 $M_2[PtCl_6]$
platinic hydroxide	氢氧化铂；四羟化铂 $Pt(OH)_4$
platinic iodide	碘化铂 PtI_4
platinic oxide	氧化铂 PtO_2
platinic sulfate	硫酸铂 $Pt(SO_4)_2$
platinic sulfide	硫化铂 PtS_2
platiniferous	含铂的
platiniridium	天然铂铱合金
platinization	镀铂(作用)
platinized	镀铂的；披铂的
platinized asbestos	披铂石棉；(载)铂石棉
platinized carbon electrode	镀铂碳电极
platinized carbon granule	载铂碳粒；镀铂碳粒
platinized charcoal	披铂炭
platinized platinum electrode	镀铂铂电极
platinized silica gel	披铂硅胶；载铂硅胶
platinizing	镀铂
platinizing bath	镀铂浴
platinochloride	氯亚铂酸盐 $M_2[PtCl_4]$
platinocyanide	氰亚铂酸盐 $M_2[Pt(CN)_4]$
platinoid	铜、锌、镍、钨合金；假白金
platinous bromide	溴化亚铂 $PtBr_2$
platinous chloride	氯化亚铂 $PtCl_2$
platinous compound	亚铂化合物
platinous cyanide	氰化亚铂 $Pt(CN)_2$
platinous hydroxide	氢氧化亚铂 $Pt(OH)_2$
platinous iodide	碘化亚铂 PtI_2
platinous sodium chloride	氯亚铂酸钠 $Na_2[PtCl_4]$
platinous sulfide	硫化亚铂
platinum	铂；白金
platinum asbestos	披铂石棉；载铂石棉
platinum based catalyst	铂基催化剂
platinum black	铂黑
platinum black electrode	铂黑电极
platinum boat	铂舟
platinum catalyst	铂催化剂
platinum charcoal	披铂炭
platinum chloride	氯化铂 $PtCl$；$PtCl_2$；$PtCl_3$；$PtCl_4$
platinum cladding	用铂包盖其他金属
platinum-cobalt color standard	铂钴色标准
platinum cone	白金锥
platinum contact	铂催化剂
platinum contact process	铂催化法
platinum crucible	白金坩埚；铂坩埚
platinum dichloride	二氯化铂 $PtCl_2$

platinum dicyanide 二氰化铂 Pt(CN)$_2$
platinum dioxide 二氧化铂 PtO$_2$
platinum dish 白金杯
platinum electrode 铂电极
platinum family element 铂族元素
platinum filament 铂丝
platinum foil 铂箔；白金箔
platinum group 铂系元素
platinum guaze 铂网
platinum loop 白金圈
platinum microelectrode 铂微电极
platinum mineral 铂矿
platinum monochloride 一氯化铂 PtCl
platinum monoxide 一氧化铂 PtO
platinum nozzle 铂喷嘴
platinum oxide 氧化铂 PtO; Pt$_2$O$_3$; PtO$_2$
platinum-oxide catalyst 氧化铂催化剂
platinum platinum-nickel thermocouple 铂-铂镍温差电偶
platinum platinum-rhodium thermocouple 铂-铂铑温差电偶
platinum-platinum thermoelement 铂-铂热敏元件
platinum pumice 披铂浮石
platinum reforming 铂重整
platinum resistance sensor 铂电阻敏感器
platinum resistance temperature sensor 铂电阻测温计；铂电阻感温计
platinum resistance thermometer 铂阻温度计
platinum-rhenium reforming 铂-铼重整
platinum sesquioxide 三氧化二铂 Pt$_2$O$_3$
platinum sheet 铂片；白金片
platinum sponge 铂绵
platinum still 白金蒸馏釜
platinum sulfide 硫化铂 PtS; PtS$_2$
platinum tetrachloride 四氯化铂
platinum thermometer 铂温度计
platinum tube 白金管
platinum ware 铂皿
platinum wire 铂丝；白金丝
platinum wire electrode 铂丝电极
platinum wire loop 铂丝圈
platinum wire temperature sensor 铂丝温度敏感器
platinum yellow 铂黄；氯铂酸碱性金属盐
Plato degree 柏拉图度
platosammine 二氨合亚铂化物 PtX$_2$·2NH$_3$
Plato unit 100克麦芽汁中总固体重量
platreater 芳香烃浓缩物加氢精制设备
platreating (环球油品公司)铂精制(过程)〔芳香烃浓缩物加氢精制法〕
platten ①弄平②制箔
plattnerite 块黑铅矿
plattner mortar 冲击钵
platynecine(=mikanecine) 阔叶千里光裂碱
platy particle 片(扁平)状粒子
platyphylline 阔叶千里光碱；狗舌草碱
Plauson colloid mill 普劳逊胶体磨
plazolite 杂水榴石
pL buffer solution pL 缓冲溶液
pleasant effect 舒适效应；舒适感
pleasant note 愉快香韵
pleasing sweetness 愉快甜香
pleated sheet 折叠片
pleated sheet structure 折叠结构
pleating 对折
pleiad 同位素群
pleiadiene 偕双烯型化合物 ═C═
pleio- 多
pleionomer 准聚合物；均低聚物
pleionomerization 亚聚合(作用)
pleiotropic gene 多效性基因
pleiotropy 多效性
plenum ①正压(风)室②压力通风系统；强制通风
plenum chamber ①送气室；送风室②喷混室
plenum system 压力通风系统
plenum ventilation 压力通风
pleochroic 多色的
pleochroic halo 多色晕(环)
pleochroism(=polychroism) 多色(现象)
pleochromatic 多色的
pleochromatism 多色(现象)
pleocidin 满杀菌素
pleomorphic ①多晶的②多形态的；群落局变
pleomorphism(=polymorphism) (同质)多晶(现象)
pleomycin 满霉素
pleonast(e) 镁铁尖晶石
plesiocurietherapy 贴近放射治疗法
plesiotherapy 贴近治疗法
plessite 合纹石
Plessy's green 磷酸铬
pleurin 侧耳菌素
pleurisy foot 马利筋
pleuromutilin 截短侧耳素
pleurotin 灰侧耳菌素
plevacol 甲醛和氨基苯甲酰丁子香酚三甲酚合剂
plexifilament (超细纤维)丛丝
plexus (超细纤维)丛；(超细纤维)网状组织

pliability 揉曲性
pliability test(=bend test)　①弯曲试验②韧性试验〔沥青〕
pliancy　挠曲性；柔韧性；可塑性
pliant flow　缓慢流动
plicacetin　折皱菌素
plicatic acid　大侧柏酸
pliers　手钳
plinol　鸢醇；1,2-二甲基-3-异丙基环戊烷的羟基衍生物
pliowax　橡脂石蜡
plioweld　橡胶(与)金属结合法
plisse crepe　泡泡纱
plodder　①蜗压机②压条机
plodding　模压
ploidy　倍性；倍数性
plot　标绘；制图
Plotnikow effect　射线纵散射效应
plot plan　平面布置图
plot polarograph　手动式极谱仪；打点式极谱仪；绘图极谱仪
plotted　标绘的；绘制的
plotting　标绘；制图
plotting accuracy　制图精度
plotting paper　标绘纸；制图纸〔商〕
plough　窄式刮刀〔离心机〕
plough blade mixer　犁片混合器
plucked wool　拔的羊毛〔不是剪的〕
Plucker tube　普氏二极管
plucking of metals　金属掘出
plug　插脚；塞柱
plug-and-ring forming　塞-圈成型法
plug-assist air-press forming　柱塞助压压气成型
plug brake　反接制动；逆转制动
plug cock　旋塞
plug connection　①插塞接触②插接法
plug contact　①插塞接触②插接法
plug flow　(活)塞式流(动)
plug flow reactor　活塞流反应器
plug gang　(管道与炉内的)焦炭清除队
plug gauge　塞规
plugged　塞紧的
plugged nozzle　堵塞的喷嘴
plugged-off tube　焦炭堵塞管
plugged screen　堵塞的过滤器
plugging　堵塞
plugging agent　封堵剂；堵水剂〔油田〕
plugging inhibitor　防堵剂
plugging of catalyst system　催化剂系统的堵塞
plug hole　塞孔

plug-in　①插入②插入式的
plug-in board　插件板
plug-in isotemperature controller　插入式等温控制器
plug of clay　泥塞〔冶金〕
plug resistance　插塞电阻
plug-type pycnometer　塞形比重计
plug up　塞紧
plug valve　旋塞阀
plug wedge　楔形塞
plug welding　塞焊
plumb　①铅锤；直锤〔商〕②测锤
plumb acetate　乙酸铅
plumbage crucible　石墨坩埚
plumbagin　石苈蓉萘醌
plumbaginous　石墨的
plumbago　石墨
plumbagol　石苈蓉醇
plumbago pot(=plumbago crucible)　石墨坩埚
plumbane　烃基铅
plumbate　高铅酸盐
plumbate pigment　铅酸盐颜料
plumb bob　铅锤；直锤；测锤
plumbean(=plumbeous)　①(正)铅的②铅的
plumbeous(=plumbean)　①(正)铅的②铅的
plumber　铅匠；水管工人
plumbery　铅器
plumbic　高铅的；四价铅的
plumbic acetate　乙酸铅
plumbic acid　高铅酸
plumbic chloride　氯化高铅；四氯化铅　$PbCl_4$
plumbic compound　高铅化合物
plumbic glaze　铅釉
plumbichloride　氯高铅酸盐　$M_2[PbCl_6]$
plumbic ocher　棕色氧化铅
plumbic oxide　氧化高铅；二氧化铅　PbO_2
plumbic sulfate (basic)　碱式硫酸高铅　$Pb(OH)_2SO_4$
plumbiferous　含铅的
plumbiferous glaze　铅釉
plumbifuoride　氟高铅酸盐　$M_2[PbF_6]$
plumbism　铅中毒
plumbite　铅酸盐
plumbite of soda　铅酸钠
plumbite process　铅酸钠精制过程
plumbite sweetener　铅酸盐香化器；铅酸盐脱臭装置
plumbite sweetening　铅酸盐香化
plumbite treatment　铅酸盐处理
plumb line　铅垂线；直垂线〔物〕
plum bob　(天平)铅锤；测锤；直锤；挂锤

plumbocalcite 铅方解石
plumbogummite 水磷铝铅矿
plumbomycin 铅色霉素
plumbo-plumbic oxide(=red lead oxide) 四氧化三铅；氧化铅十三氧化二铅〔红丹的化学成分〕
$Pb_3O_4; PbO \cdot Pb_2O_3$
plumbosolvency 铅在液体中的溶解程度
plumbous 铅的；二价铅的
plumbous acetate (basic) 碱式乙酸铅
$Pb(C_2H_3O_2)_2 \cdot Pb(OH)_2$
plumbous acid 铅酸 H_2PbO_2〔二价铅，中文视作正铅，四价为高铅〕
plumbous antimoniate 偏锑酸铅 $Pb_3(SbO_4)_2$
plumbous bichromate 重铬酸铅 $PbCr_2O_7$
plumbous chromate 铬酸铅 $PbCrO_4$
plumbous compound （正）铅化合物
plumbous cyanide 氰化铅 $Pb(CN)_2$
plumbous ferrocyanide 氰亚铁酸铅 $Pb_2[Fe(CN)_6]$
plumbous hyposulfate 连二硫酸铅 PbS_2O_6
plumbous hyposulfite 连二亚硫酸铅 PbS_2O_4
plumbous metaplumbate 偏铅酸铅 $Pb(PbO_3)$
plumbous nitrite 亚硝酸铅 $Pb(NO_2)_2$
plumbous orthoplumbate 原高铅酸铅 $Pb_2(PbO_4)$
plumbous oxide 氧化铅；一氧化铅 PbO
plumbous pyrophosphate 焦磷酸铅 $Pb_2P_2O_7$
plumbous rhodanate 硫氰酸铅 $Pb(SCN)_2$
plumbous silicofluoride 氟硅酸铅 $Pb[SiF_6]$
plumbous subacetate 碱式乙酸铅
$Pb(C_2H_3O_2)_2 \cdot Pb(OH)_2$
plumbous subcarbonate 次碳酸铅；碱式碳酸铅
$2PbCO_3 \cdot Pb(OH)_2$
plumbous sulfate (basic) 碱式硫酸铅 $PbSO_4 \cdot PbO$
plumbous sulfite 亚硫酸铅 $PbSO_3$
plumbous tungstate 钨酸铅 $PbWO_4$
plumbyl 铅烷基 $H_3Pb—$
plume 烟流；烟条；卷流
plumer block 轴承箱；轴承座
plumeria oil 鸡蛋花油
plumericin 鸡蛋花素
plumerin 缅栀子苷
plum flavour 李子香精
plumiera 鸡蛋花皮
plumieric acid 喜蛋花酸 $C_{20}H_{24}O_{12}$
plumierin 鸡蛋花苷
plum-kernel oil 梅仁油
plummet ①测锤②准绳
plumose 羽毛状的
plumose growth 羽毛状生长

plumose mica 石棉状白云母
plumping 膨胀
plumping agent 膨胀剂
plumping power （灰)膨胀本领；灰膨胀度〔皮革〕
plumpness 丰满弹性
Plunck's law of radiation 普朗克辐射定律
plunge battery 浸液电池
plunger 柱塞；塞子
plunger forward time 柱塞推进时间
plunger mold 柱压式模
plunger oil pump 柱塞式油泵
plunger preplasticating injection mo(u)lding machine 柱塞预塑化式注塑机
plunger pump 柱塞泵
plunger-type transfer molding 柱塞式传递模制法
plunging pole 铁棒〔拨火用的〕
plural component spray equipment 多组分喷涂设备；多组分喷枪
plural gel 复合凝胶
plurallin 多色菌素
plural-vortex turbulent zone 多涡流湍动区
pluramycin 多色霉素
plurimolecular 多分子的
Plurix process 普路里克斯过程〔离子交换回收钚过程。一种从辐照钚靶中提取超钚元素的前处理过程，现已不用〕
Pluronic grid representation 普卢兰尼克方格图
plus ①加②正的
plush （长）毛绒
plus material 面料；筛面料
plus mesh 筛上料；筛上物
plus-pressure 正压；压力；加压(力)
plus-pressure furnace 加压炉
plus sign ①加号②正号
plus strand 正链；正股
pluto 放射性检查计
plutonate 钚酸盐
plutonia 二氧化钚
plutonia molybdenum cermet(PMC) 氧化钚-钼金属陶瓷
plutonia sol 氧化钚溶胶
plutonic 火成的
plutonic acid 钚酸 $H_4PuO_5; H_6PuO_6$
plutonium 钚〔94号元素，化学符号 Pu〕
plutonium-bearing scrap 含钚碎屑(片)
plutonium carbide 碳化钚
plutonium carbonitride 碳氮化钚 $PuCN$
plutonium fuel 钚燃料
plutonium hexafluoride 六氟化钚 PuF_6

plutonium hydride　氢化钚
plutonium hydroxide　氢氧化钚
plutonium load-out vessel　钚卸料容器
plutonium metaphosphate　偏磷酸钚　Pu(PO$_3$)$_4$
plutonium nitrate　硝酸钚　Pu(NO$_3$)$_3$; Pu(NO$_3$)$_4$; PuO$_2$(NO$_3$)$_2$
plutonium oxalate　草酸钚　Pu(C$_2$O$_4$)$_2$
plutonium oxide　氧化钚　PuO; PuO$_2$
plutonium peroxide　过氧化钚
plutonium phosphate　磷酸钚　PuPO$_4$
plutonium pyrophosphate　焦磷酸钚　PuP$_2$O$_7$
plutonium reclamation facility　钚回收设备
plutonium recycle test reactor(PRTR)　钚再循环试验堆
plutonium sesquioxide　三氧化二钚　Pu$_2$O$_3$
plutonium sesquitelluride　三碲化二钚　Pu$_2$Te$_3$
plutonium sulfate　硫酸钚　Pu(SO$_4$)$_2$
plutonium tetrafluoride　四氟化钚　PuF$_4$
plutonium tetraisopropoxide　四异丙氧基钚　Pu[OCH(CH$_3$)$_2$]$_4$
plutonium tricyclopentadienide　三环戊二烯合钚　Pu(C$_5$H$_5$)$_3$
plutonium trifluordie　三氟化钚　PuF$_3$
plutonyl　钚酰*
plutonyl hydroxide　氢氧化钚酰　PuO$_2$(OH)$_2$
plutonyl nitrate　硝酸钚酰　PuO$_2$(NO$_3$)$_2$
pluviometer　雨量计
ply　①合叠②板片；层；木材薄片
plyability　①层压性②重叠性；复合性
ply adhesion　层间附着力；剥离强度
ply adhesion strength　剥纸强度；层间附着(黏结)强度
ply angle　铺层角
ply drop　铺层递减
ply gage　①测厚计②布层厚度
ply group　增强纤维层组；铺层组
plying　并线；合股
plymax(=metal-faced plywood)　金属贴面胶合板
plymetal　夹金属胶合板
ply separation machine　层离机
ply stacking sequence　铺层顺序
ply strain　铺层应变
ply stress　铺层应力
plywood　层(压木)板；黏板；胶合板
ply yarn　合股线
PMA pigment　①磷钼酸颜料②(=precipitated basic dye blue)(磷钼酸)碱性蓝色淀
P-matrix method　P矩阵法
pM buffer solution　pM缓冲溶液
PMR spectrum　质子核磁共振谱

pNa glass electrode　pNa玻璃电极
pneumacator　气动发送器
pneumatic　①气体的；有气体的②空气的；有空气的③气体(力)学的
pneumatical-assisted electrospray　气相电子喷射
pneumatically assisted electrospray ionization　气动辅助电喷雾离子化
pneumatically controlled sample leak　气动控制进样漏孔
pneumatically membrane-operated gas sampling valve　气动膜式气体进样阀
pneumatic bridge hygrometer　气桥湿度计
pneumatic cell　气体测试器；气动吸收池
pneumatic chromatograph　气动色谱仪
pneumatic classifier　气力分级器
pneumatic clutch brake　气动离合式刹车
pneumatic control　气动控制
pneumatic conveyer　气动运输机
pneumatic conveyer dryer　气流干燥器
pneumatic conveying　气动运送
pneumatic conveying dryer　气流干燥设备
pneumatic conveying plate　气动吹送装置
pneumatic conveyor　气动运输机；气流输送器
pneumatic cross-flow nebulizer　气动交叉式雾化器
pneumatic cushioning　空气弹簧
pneumatic detector　气动检测器
pneumatic diaphragm control valve　气动薄膜调节阀
pneumatic double-pen recorder　气动双针记录器
pncumatic dricr　气流干燥机；气流干燥器
pneumatic drill　风钻
pneumatic dryer　气流干燥器
pneumatic elutriator　风力分级器
pneumatic fiber-fineness indicator　气流式纤维细度仪
pneumatic flotation　泡沫浮选
pneumatic hose　空气(压力)胶管；耐压胶管
pneumatic intensifier pump　气动放大泵；气动增强泵
pneumatic irregularity tester　气流式不匀率试验仪
pneumatic jet amplifier detector　气动喷嘴放大检测器
pneumatic jig　风筛
pneumatic machinery　风动机械
pneumatic nebulization　气动雾化
pneumatic nebulizer　气动雾化器
pneumatic odorizer　气体添味剂加入器
pneumatic outlet　气体出口
pneumatic pipeline pig　气动清管器
pneumatic plant　气动装置
pneumatic positioner valve　气动阀门定位器
pneumatic pump　气动泵*
pneumatic rabbit　气动兔〔快速输送放射源用的气动小盒〕

pneumatic restriction 气流限制
pneumatics 气体力学
pneumatic separation 气流分离
pneumatic stirring process （压）气搅(拌)法
pneumatic stretcher 气动拉伸器
pneumatic system 气动系统
pneumatic test 气压试验
pneumatic thickener 气压稠料器；气压脱水机
pneumatic tool 风动工具；气动工具
pneumatic-tool oil 气体输送工具润滑油
pneumatic transmission 压缩空气传动
pneumatic transmitter 压缩空气传动器
pneumatic transport 气动输送
pneumatic trough 集气槽
pneumatic tyre 气胎
pneumatic valve 气动阀
pneumatogen ①物理发泡剂②气体发生器
pneumatolysis 气化(作用)
pneumeractor 测量石油产品量的记录仪
pneumin 亚甲杂酚油
pneumokoniosis 肺尘病
pneumo(no)coniosis 肺尘病
pneumotexturing 气流变形；喷气变形
pneutronic level controller 气动电子位面控制器
pneu-vac drier 真空气流干燥器
pneu-vac dryer 真空气流干燥器
pnicogen 磷属元素*
pnictide 磷属元素化物*
poaching 漂洗
poaching engine 漂白机
pockeling 溶液表面发展园道
Pockers' cell 普克尔盒
pocket ①窝矿矿袋②(口)袋
pocket accumulator 袋插蓄电池
pocket builder ①布筒机②制带机
pocket chamber 小型电离室
pocket dosimeter 袖珍剂量计
pocketed heat 蓄热；内部发热
pocket grinder 压榨碎木机
pocket ionization chamber 小型电离室
pocket-like recess 袋形夹套
pocket magnifier 袖珍放大镜
pockets 孔穴〔塑料制品缺陷〕
pockets in tank 油罐(底部)的凹陷部分
pocking 水泡〔漆病〕
pocking mark ①麻点②橘皮〔漆病〕
pockmarking 麻点
poco oil 水薄荷油

Podbielniak analysis 波氏法精密分馏
Podbielniak centrifugal extractor 波氏离心萃取器
Podbielniak extractor 波氏萃取器〔膜式离心萃取器〕
Podbielniak precise distillation apparatus 波氏精密分馏装置
podocarpane 罗汉松烷
podocarpic acid 罗汉松酸
podocarpinol 罗汉松醇
podocarprene 罗汉松烯
podolite 碳磷灰石
podomycin 足霉素
podophyllic acid 鬼臼酸
podophyllin 鬼臼脂
podophyllotoxin 鬼臼毒
podophyllum 鬼臼根
podopimardiene 坡刀海松二烯
podzolisation 三层化(作用)
poeticine 诗水仙碱
Poggendoriff bichromate cell 波根多里弗重铬酸盐电池
Poggendoriff compensation method 波根多里弗补偿法
pogostone 广藿香酮
Pohle and Mehlenbacher method 波利-梅伦巴克法〔测甘油单酸酯〕
Pohl's commutator 坡耳整流器
poidometer 重量计
poikilothermic 不定温的
poikilothermism 适应温度能力
poine 草甸菌素
point ①点②尖端③要点
point analysis 点分析
point-blank ①直射的；正射的②明白的
point by point 点到点(的聚合)
point-by-point test 逐步试验
point charge 点电荷
point condition 点态
point defect 点缺陷
point discharge 尖端放电
point dislocation 点位错
point efficiency 点效率
pointer 指针
pointer gauge 指针表
pointer scale 刻度板
point estimate 点估计
point estimation 点估计*
point function 点函数
point group 点群
pointilistic effect 点子花纹效应；点画效应
pointillism ①点画法②点彩(画)派

pointing ①标定②灰浆勾缝
pointing noise 噪声
point lattice 点阵
point load 点载荷
point of application 应用点；使用点
point of chain rupture 链破裂点
point of coagulation 凝结点；凝固点；絮(胶)凝点
point of contact 接触点
point of detonation 爆震点
point of discharge 放电点
point of discontinuity 曲线折点
point of draw 取出点〔汽油塔〕
point of equivalent 等当点
point of flammability 闪点；燃点
point of fluidization 流化点
point of fusion 熔点
point of ignition 着火点；燃烧点
point of inflection 回折点；折点
point of instability 不稳定点
point of inversion 转化点
point of neutralization 中和点
point of reversal in density 密度(曲线)回折点；密度(曲线)倒转点
point of sample 点样点
point of solidification 固化点
point of take-off (馏出物自蒸馏塔的)取出点
point of wave inception 起波点
point-of-zero charge(PZC) 零点电荷
point of zero electric charge 零电荷点
pointolite lamp 点(光)源灯
point pollution source 点污染源
point source 点源*
point to plane method 点对面法
point to point method 点对点法
point value 点值
Poirrier's blue 泡依蓝
poise ①泊〔黏度单位〕②平衡
poised 抗氧化还原的
Poiseuille equation 泊肃叶方程
Poiseuille flow 泊肃叶流动
Poiseuille's law 泊肃叶定律
poising action 平衡作用
poising agent (氧化还原作用)平衡剂
poison ①毒②毒物；毒药③毒(害)；毒死④布毒
poison effect 毒物效应
poison hemlock oil 钩吻叶芹油
poisoning ①中毒②布毒
poisoning of catalyst 催化剂中毒

poisonous ①(有)毒的②中毒的
poisonous dose 中毒量
poisonous effect 毒效(应)
poisonousness 毒性
poisonous property 毒性
poison tower 除毒塔
poison vapors 毒蒸气
poison vine 番木鳖
Poisson distribution equation 泊松分布方程
Poisson's distribution 泊松分布*
Poisson's equation 泊松公式
Poisson's law 泊松定律
Poisson's number 泊松数
Poisson's ratio 泊松比
Poisson tailing 泊松形峰尾
Poisson type 泊松型〔重均分子量与数均分子量近乎相等的分子量分布型式〕
poivrette 碎橄核
poke 拨火；添火 ；透炉
poke hole 拨火孔
poker 拨火棒；火钩
poker mechanism 拨火装置
poking 拨火
poking hole 拨火孔；搅孔〔冶〕
Polaises reaction 波氏反应
polar ①极性的②极化的③磁极的④地极的
polar absorbent 极性吸收剂
polar absorption 极性吸收
polar activation 极性活化(作用)
polar activator 极性活化剂
polar adsorbent 极性吸附剂
polar adsorption 极性吸附(作用)
polar axis 极性轴
polar binding 极性键联
polar bond 极性键
polar bonded phase 极性键合相
polar catalyst 极性催化剂
polar chain molecule 极链分子；有极链分子
polar colors(=polar dies)〔复〕 极性染料〔染〕
polar compound 极性化合物
polar contribution 极性分布
polar coordinate 极坐标〔数〕
polar covalence 极性共价
polar crystal 极性晶体
polar cycloaliphatic solvent 极性环(化)脂(肪)族溶剂
polar decomposition principle 极分解原理
polar dies(=polar colors)〔复〕 极性染料
polar effect 极性效应

polar fluid 极性流体
polar formula 极性式
polar grey 极性灰
polar group 极性基团
polarimeter 旋光计；偏光计
polarimetric analysis 旋光分析
polarimetric test 极化试验
polarimetry(=polariscopy) 旋光测定(法)；偏振测定(法)；测偏振术
polarine 发动机润滑油
polarisability 极化性；极化率
polarisation(=polarization) ①极化②偏振化
polarisation curve 极化曲线
polariscope 旋光镜；偏振(光)镜；旋光计
polariscope tube 旋光计管；偏振(光)镜管
polariscopy(=polarimetry) 旋光测定(法)；偏振测定(法)；测偏振术
polarite 坡拉炸药；卜拉赖特；铋铅钯矿
polarity 极性
polarity formula 电子式
polarity hypothesis 极性假说
polarity index ①极性指数②偏光性指数
polarity indicator 示极器
polarity of chain (分子)链的极性
polarity reversal 极性反向
polarity reverse 极性反转
polarity switch 极性开关
polarizability 可极化性*；极化率
polarizability component 极化率分量
polarizability difference 极化率差值
polarizability effect 极化效应
polarizable 可极化的
polarizable electrode 可极化电极
polarizable group 可极化基
polarization ①极化(作用)②偏振化(作用)③极化(度)；(电)极化强度
polarization analysis 偏振(光)分析法
polarization analyzer 检偏片；检偏振(光)镜
polarization battery 极化电池组
polarization capacity 极化量
polarization cell 极化电池
polarization colorimeter 偏光比色计
polarization current 极化电流
polarization curve 极化曲线
polarization effect 极化效应
polarization electrode 极化电极
polarization factor 偏振化因子*
polarization fluorometer 偏光荧光计
polarization function 极化函数
polarization increase factor 极化增强因子
polarization infrared technique 偏振红外光技术
polarization interference microscope 偏光干涉显微镜
polarization interferometer 偏振干涉仪
polarization microscope 偏光显微镜
polarization-modulated electron-nuclear double resonance 极化调制电子-核双共振
polarization of electrode 电极极化
polarization of light 光偏振
polarization overpotential 极化超电势
polarization photometer 旋光光度计
polarization plane 偏振面
polarization potential(=polarization voltage) 极化电位；极化电压
polarization resistance 极化电阻
polarization resistance method 极化电阻法〔新的评价漆膜耐腐蚀性的方法〕
polarization rotation 偏振旋转
polarization spectrometer (偏振)光分光计
polarization spectrum 偏振光谱
polarization tensor 极化张量
polarization titration 极化滴定
polarization transfer 极化转移
polarization tube 起偏振管；旋(偏)光管
polarization voltage(=polarization potential) 极化电位；极化电压
polarized ①偏振化了的②极化了的
polarized electrode 极化电极
polarized electron 极化电子
polarized fluorescence 极化荧光；偏振荧光
polarized indicator electrode 极化指示电极
polarized light 偏振光*；极化光
polarized light microscope 偏光显微镜
polarized optical microscopy 偏振光显微术
polarized relay 极化继电器
polarized target 极化靶
polarized X-ray spectroscopy 偏振光X射线光谱法
polarizer 起偏振器
polarizing angle 偏振角
polarizing battery 极化蓄电池
polarizing current 极化电流
polarizing disc(=polaroid) (人造)偏振片；起偏振片
polarizing effect 极化效应
polarizing electrode 极化电极
polarizing filter 极化滤波器
polarizing fluorescence 偏振荧光
polarizing interferometer 偏振光干涉仪

polarizing microscope 偏(振)光显微镜
polarizing power 极化力
polarizing prism 起偏振棱镜
polarizing spectrophotometer 偏振分光光度计
polarizing voltage 极化电压
polar link 极性键
polar linkage 极性键(合)
polar liquid 极性液体
polar liquid phase 极性液相
polar lyophilic group 极性亲液基团
polar mobility 极性迁移率
polar molecule 极性分子；有极分子
polar monomer 极性单体*
polar-nonpolar solubilization 极性物-非极性物之间的增溶作用
polar number 极(性)价数；电价
polarogram 极谱图
polarograph 极谱仪
polarographic analysis 极谱分析(法)；极化分析(法)
polarographic assay 极谱测定
polarographic catalytic wave 极谱催化波
polarographic chronoamperometry 极谱计时电流法
polarographic coulometry 极谱库仑法
polarographic current 极谱电流
polarographic detector 极谱检测器
polarographic dissolved oxygen sensor 极谱溶氧传感器
polarographic flow-through detector 极谱流通检测器
polarographic immunoassay 极谱免疫分析法
polarographic indicator 极谱指示剂
polarographic ion constant 极谱离子常数
polarographic maximum 极谱极大
polarographic maximum suppressor method 极谱最大抑制法〔测临界胶束浓度〕
polarographic method 极谱测定法
polarographic micelle point 极谱法胶束点；极谱法临界胶束浓度
polarographic scanner 极谱扫描器
polarographic spectrum 极谱波谱
polarographic titration 极谱滴定；极化滴定
polarographic wave 极谱波
polarography 极谱法
polarography at constant potential 恒电位极谱法
polaroid-photography 偏振光摄影法
polarometric titration 极谱滴定(法)；极化滴定(法)
polaron 极化子
polar organic media 极性有机介质
polar pattern overboot 安全导电胶靴
polar phase 极性相
polar pigment 极性颜料
polar pleated sheet 极性折叠片
polar polymer 极性聚合物*
polar reaction 极性反应
polar rubber(=active rubber) 活性橡胶；极性橡胶
polar sample 极性试样
polar side chain 极性侧链
polar solute 极性溶质
polar solvent 极性溶剂
polar solvent liquid 极化溶剂液体
polar stationary phase 极性固定相
polar substitution 极性取代
polar support 极性载体
polar swelling agent 极性溶胀剂
polar symmetry axis 极性对称轴
polar theory 极化理论
polar winding 长丝缠绕；极向缠绕
polar zone 定向偶极区
poldine methylsulfate 甲硫泊尔定；甲基硫酸波定〔药〕
pole ①极②顶点③(磁)极④(电)极⑤竿
pole boundary 磁极边界
pole cup 电极杯
pole distance 极间距离
pole drying 电极干燥
pole effect 电极效应
pole face (磁)极面
pole figure 极图
pole marks 极标
Polenske value 波伦斯基值；不溶解挥发脂肪酸值
pole paper 极谱纸
pole piece 极靴
pole piece of magnet 极靴
pole shoe 极蹄
pole stock 杆柱材
pole tension 极电压
pole terminal 极端
polianite 黝锰矿
policeman 淀帚*
policing ①控制②清扫
poling (木)竿炼(法)〔冶金〕；穿管；套棒
polish ①擦亮剂；擦光物②擦亮；磨光③虫胶清漆④抛光⑤抛光剂
polishability 抛光性
polishable 可以擦的
polish abrasive 抛光磨料
polished 光面
polished face 磨光面
polished finish 抛光处理

polished integral restrictor 磨口限流器
polished lacquer 推光漆〔大漆〕
polished lacquer wares 推光漆器
polished painting 推光漆画〔大漆〕
polished surface 抛光(的)表面
polisher 磨光器
polishes ①醇溶性清漆；金油〔俗〕②虫胶清漆
polishing filter 精炼过滤器；净化过滤器
polishing machine ①磨光机②轧光机③抛光机
polishing marks 抛光磨痕
polishing material 磨光物；擦亮剂
polishing medium 研磨剂；抛光剂
polishing oil 擦亮油〔革〕
polishing paste 研磨膏剂
polishing slate 磨石片
polishing tool 抛光工具
polishing varnish 擦亮清漆；擦光(清)漆
polishing wax 上光蜡
polishing wheel ①抛光轮②细砂轮
polka dot paint 多彩涂料
pollen 花粉
pollen analysis 花粉分析
Pollent and Cross plastometer 波兰特-克劳斯塑性计
pollucite 铯榴石
pollutant 污染物；沾污物
pollutant analyzer 污染物分析器
pollutant burden 污染物负荷
pollutant characterization 污染物表征
pollutant concentration 污染物浓度
pollutant discharging license 排污许可证
polluted water 污水
polluting strength 污染强度
pollution 污染
pollution by radionuclide 放射线核素污染
pollution charge 排污收费
pollution control 污染控制
pollution degree 污染度
pollution-free 无污染的
pollution-free energy(=energy without pollution) 无污染能源
pollution-free fuel(=noncontamination fuel) 无污染燃料
pollution-free technology 无污染工艺
pollution index 污染指数
pollution indicating organism 污染指示生物
pollution intensity 污染强度
pollution monitoring 污染监测
pollution nuisance 污染公害
pollution of automobile exhaust gas 汽车尾气污染
pollution of heavy metal 重金属污染
pollution source 污染源
pollution speciation 污染物形态
pollution treatment of potable water 饮用水除污染处理
pollux 铯榴石
polmarosa oil(=rusa oil) 玫瑰草油
polonide 钋化物
polonite 钋酸盐
polonium 钋〔84号元素，化学符号Po〕
poly- 〔希腊字头〕多；聚
polyacenaphthylene 聚苊
polyacene ①聚烯烃②多并苯
polyacetaldehyde 聚乙醛
polyacetal resin 聚缩醛树脂
polyacetal(s) 聚缩醛(类)
polyacetamidoacrylic acid 聚乙酰胺丙烯酸
polyacetylene 聚乙炔
polyacetylenic ester 聚乙炔酯
polyacetylvinylamine 聚乙酰氨基乙烯
polyacid 多酸*
polyacid base 多(酸)价碱；多元碱
polyacrylamide 聚丙烯酰胺
polyacrylamide bed 聚丙烯酰胺柱床
polyacrylamide exchanger 聚丙烯酰胺交换剂
polyacrylamide gel 聚丙烯酰胺凝胶
polyacrylamide gel electrophoresis 聚丙烯酰胺凝胶电泳
polyacrylate 聚丙烯酸酯
polyacrylate binder 聚丙烯酸酯黏合剂
polyacrylate dispersion 聚丙烯酸酯分散体；聚丙烯酸酯乳液
polyacrylate emulsion 聚丙烯酸酯乳液
polyacrylate rubber 聚丙烯酸酯橡胶
polyacrylate rubber latex 聚丙烯酸酯胶乳
polyacrylate sodium 聚丙烯酸钠
polyacrylic 丙烯酸类聚合物
polyacrylic acid(PAA) 聚丙烯酸
polyacrylic acid zirconium salt 聚丙烯酸锆盐
polyacrylic coating 聚丙烯酸系涂层
polyacrylic ester 聚丙烯酸酯
polyacrylic ester latex (聚)丙烯酸酯胶乳
polyacrylic fiber 聚丙烯酸(酯)纤维；腈纶
polyacrylic plastics 聚丙烯酸塑料
polyacrylic rubber 聚丙烯酸酯橡胶
polyacrylonitrile 聚丙烯腈
poly(acrylonitrile-covinylpyrrolidone) 丙烯腈-乙烯基吡咯烷酮共聚物
polyacrylonitrile fiber 聚丙烯腈纤维
polyactivation 多物活化

polyacyloxalamidrazone fibre　聚酰基草酸脒腙纤维
polyad　①多价物②多价的
polyaddition　聚加成反应*；逐步加成聚合
polyaddition reaction　加(成)聚(合)反应
polyaddition reactor　加聚反应器
polyadduct　聚加合物
polyadelphite　锰铁榴石
polyadenylic acid(poly(A))　多聚腺苷酸
poly-β-alanine(=nylon-3)　聚β氨基酸；锦纶-3；尼龙-3
polyalcohol(=polyol)　多元醇
polyalkane　聚链烷
polyalkemer　聚烯烃
polyalkenamer　开环聚烯烃；环烯烃开环聚合物〔对于环烯烃的聚合产物，由于聚合机理的不同，一种是开环聚合，一种是保留环打开双键的聚合〕
polyalkene　聚烯烃
polyalkoxide　聚烷氧化物
polyalkoxylated polyol　聚烷氧基多元醇
polyalkoxysilane　聚烷氧基硅烷
polyalkyl acrylate　聚烷基丙烯酸酯；聚丙烯酸烷基酯
polyalkylated　多烷基化的
polyalkylation　多烷基化作用
polyalkylbenzene　多烷基苯
polyalkylene amide　①聚亚烷基酰胺②聚酰胺树脂
polyalkylene glycol　聚(亚烷基)二醇
polyalkylene glycol diester　聚烷撑二(醇)酯
polyalkylene glycol ether　聚烷撑二醇醚
polyalkylene glycol lubricant　聚(亚烷基)二醇润滑剂
polyalkylene glycol-modified polyester　聚烷撑二醇改性聚酯
polyalkylene glycol monoalkyl ether　聚烷撑二醇单烷基醚
polyalkylene glycol oil　聚(亚烷基)二醇(润滑)油
polyalkylene oxide　聚环氧烷；聚醚
polyalkylene sulfide　聚亚烃化硫
polyalkylene sulfone　聚亚烃砜
polyalkylene synthetic lubricant　聚烯合成润滑剂
polyalkylene terephthalamide　聚对苯二甲酰亚烷基二胺
polyalkyl oxide rubber　聚烷基环氧橡胶
polyalkyl polyphenylamine　聚烷基聚苯胺〔硫化剂〕
polyalkylsulfone　聚烷基砜
polyalkyltitanate　钛酸多烷基酯(催化剂)
polyallomer　异质同晶聚合物*；聚异质同晶体
polyallomer resin　异质同晶共聚树脂
polyalloy　高分子合金
polyallyl cyclohexene　聚烯丙基环己烯
poly(allyl diglycol carbonate)　碳酸-二乙二醇酯-烯丙醇酯树脂

polyallylmethacrylate　聚甲基丙烯酸烯丙酯
poly(alpha-methylstyrene)(PMS)　聚α-甲基苯乙烯
poly alpha olefin　聚α-烯烃
polyaluminoorganosiloxane　聚有机铝硅氧烷
polyaluminosiloxane　聚铝硅氧烷
polyamic acid　聚酰胺(基)酸〔制造聚酰亚胺的中间体〕
polyamic amide　聚酰胺酰胺
polyamidation　聚酰胺化
polyamide　聚酰胺*
polyamide-acid polymer　聚酰胺酸聚合物
polyamide binder　聚酰胺黏合剂
polyamide-ester　聚酰胺酯
polyamide ether　聚酰胺醚
polyamide fiber　聚酰胺纤维；锦纶；尼龙
polyamide film　聚酰胺薄膜
polyamide-forming　生成聚酰胺的
polyamide-imide　聚酰胺-酰亚胺
polyamide intermingled yarn　锦纶混纤丝
polyamide-pheno(-formaldehyole)resin　聚酰胺-酚醛树脂
polyamide plastic　聚酰胺塑料
polyamide-polyester-polyether blend　聚酰胺-聚酯-聚醚混合物
polyamide resin　聚酰胺树脂
polyamide-thickened gel　聚酰胺稠化润滑脂
polyamide thin-layer plate　聚酰胺薄层板
polyamide transparent sewing thread　聚酰胺透明缝纫线
polyamide-type hardener　聚酰胺型固化剂
polyamidine　聚脒
polyamidoamines　聚酰胺型胺类
polyamidoarylate　芳基化聚酰胺
polyamidoester fibre　聚酰胺酯纤维
polyamidohygrostrepton(=phytoactin)　植病活菌素
polyamidoimide　聚酰胺酰亚胺
polyamine　多胺；聚胺
polyamine-methylene resin　聚亚甲胺树脂
polyamine salt　聚胺盐
polyamine treatment　聚胺处理；多胺处理
polyamino acid　聚氨基酸
poly(aminoamic acid)　聚氨基酰胺酸
polyaminoamide　聚氨基酰胺
poly-p-aminobenzoic acid　聚对氨基苯甲酸
poly-10-amino capric acid　聚-10-氨基癸酸；聚ω-氨基癸酸
poly(ω-amino caproic acid)　聚(ω-氨基己酸)；尼龙 6；聚酰胺 6
poly(8-amino caprylic acid)　聚(8-氨基辛酸)；尼龙 8；聚酰胺 8

polyaminoester(s) 聚酰胺酯(类)
polyaminohygrostreptin(=phytostreptin) 植病链菌素
polyaminoresin 聚胺树脂
polyaminotriazole 聚氨基三唑
polyaminotriazole fibre 聚氨基三唑纤维；聚氨基三氮杂茂纤维
polyaminoundecanoamide 聚氨基十一酰胺
poly-11-amino undecanoic acid 聚-11-氨基十一酸；聚ω-氨基十一酸
polyammonium 聚铵
polyampholyte 两性聚电解质
polyamphoteric electrolyte 两性聚电解质
polyanhydride(s) 聚酐(类)
polyaniline 聚苯胺
polyanion 聚阴离子
polyanthin 多花素
polyanthinin 多花素宁
polyanthraquinone-bisbenzazole fibre 聚蒽醌-双吲哚纤维
polyanthroline 聚蒽咯啉
poly (A) polymerase 多(聚)腺苷酸聚合酶
polyaramide 聚芳酰胺
polyargyrite 方辉锑银矿
polyaromatic heterocycles 芳烃杂环聚合物
polyaryl amide(PARA) 聚芳酰胺
polyarylate 聚芳酯
polyarylated 多芳基化的
polyarylation 多芳基化反应
polyarylene 聚芳撑
polyaryleneimide 聚芳撑酰亚胺
polyarylene sulfide fibre 聚芳基硫醚纤维
polyarylester 聚芳酯
polyaryl ether 聚芳醚；芳香族聚醚
polyaryletherketone(PAEK) 聚芳醚酮
polyaryloxy silane 聚芳氧基硅烷
polyarylsulfone(PASU) 聚芳砜
polyarylsulphone 聚芳砜
polyase 聚合酶；多糖酶
polyaspartic acid 聚天冬氨酸
polyatomic 多原子的
polyatomic acid 多元酸
polyatomic alcohol 多元醇
polyatomic ion 多原子离子
polyatomic molecule 多原子分子
polyatomic phenol 多元酚
polyatomic ring 多节环
polyazanaphthalene 多氮杂萘
polyazelaic polyanhydride 聚壬二酸聚酐〔固化剂〕
polyaziridine 聚氮丙啶

polyaziridine crosslinking agent 聚氮丙啶交联剂
polyazo dye 多偶氮染料
polybase 多碱*
polybase crude oil 混合基石油
polybasic ①多碱(价)的；多元的②多代的
polybasic acid(=polyprotic acid) 多(碱)价酸〔无机〕；多元酸〔有机〕
polybasic alcohol 多元醇
polybasic anion 多碱(元)阴离子
polybasic carboxylic acid 多元羧酸
polybasic complex 多元配合物
polybasic ester 多元酸的酯
polybasic weak acid 多元弱酸
polybasite 硫锑铜银矿
poly(p-benzamide) 聚对苯甲酰胺〔耐高温纤维〕
polybenzene 聚苯
polybenzimidazole 聚苯并咪唑
polybenzimidazole membrane 聚苯并咪唑膜
polybenzimidazole resin 聚苯并咪唑树脂
polybenzimidazonium salt 聚苯并咪唑鎓盐
polybenzimidazopyrrolone 聚苯并咪唑并吡咯酮
polybenzine(=polymer gasoline) 聚合汽油
polybenzobistrizole phenanthroline fibre 聚苯并双三唑氮杂菲纤维
polybenzothiadiazine dioxide 聚二氧化苯并噻二嗪
polybenzothiazole 聚苯并噻唑
polybenzothiazole amide 聚苯并噻唑酰胺
polybenzothiazole imide 聚苯并噻唑酰亚胺
polybenzotriazole 聚苯并三唑
polybenzoxadiazole 聚苯并二唑
polybenzoxazinone 聚苯并嗪酮
polybenzoxazole 聚苯并唑
polybenzthiazole imide 聚苯并噻唑酰亚胺
poly-β-benzyl-L-aspartate 聚-L-天冬氨酸-β-苄酯
polybenzyl methacrylate 聚甲基丙烯酸苄酯
polyblend 高分子共混物；掺混料
polyblend fibre 聚合物混(合)纺纤维；聚合物混合体纤维
polyblock 嵌段共聚物
polybond 聚硫橡胶黏合剂
polybromated 多溴化的
polybromide 多溴化合物
polybrominated 多溴化的
polybrominated diphenyl ethers(PBDEs) 多溴联苯醚
polybromination 多溴化反应
polybromizated 多溴化的
polybromization 多溴化反应
polybromocarbon(=polybromohydrocarbon) 多溴烃

polybromohydrocarbon(=polybromocarbon)　多溴烃
polybromoprene　聚溴丁二烯
1,4-polybutadiene　1,4-聚丁二烯
polybutadiene　聚丁二烯*
polybutadiene-acrylic copolymer　聚丁二烯-丙烯酸共聚物
polybutadiene-acrylonitrile(PBAN)　聚丁二烯-丙烯腈
polybutadiene fibre　聚丁二烯纤维
polybutadiene latex　(聚)丁二烯胶乳
polybutadiene rubber　聚丁二烯橡胶
polybutadiene rubber latex　聚丁二烯胶乳
polybutadiene-styrene(PBS)　聚丁二烯-苯乙烯
polybutenamer　环丁烯开环聚合物
polybutene　聚丁烯*
poly(butyl acrylate)(PBA)　聚丙烯酸丁酯
polybutylene adipate　聚己二酸亚丁酯〔增塑剂〕
polybutylene glycol　聚亚丁基二醇
polybutyleneglycol-bis-3-aziridinopropionate　聚丁二醇双-3-氮丙啶丙酸酯
polybutylenes　聚丁烯
poly(butylene terephthalate)(PBTP)　聚对苯二甲酸丁二醇酯
polybutyl isocyanate　聚异氰酸丁酯〔硫化剂〕
polybutyl methacrylate　聚甲基丙烯酸丁酯
polybutyl methacrylate emulsion　聚甲基丙烯酸丁酯乳液
poly(butyl vinyl ether)　聚丁基乙烯基醚
polybutyrolactam　聚丁内酰胺
polycaproamide　聚己酰胺〔尼龙6〕
polycaprolactam　聚己内酰胺
polycaprolactam powder　聚卡普纶粉
polycaprolactone　聚己酸内酯
polycaprolactone glycol　聚己内酯二醇
polycaprolactonetriol　聚己内酯三醇
poly capryllactam　聚辛内酰胺〔尼龙8〕
polycarbam(in)ate　聚氨基甲酸酯
polycarbodiimide　聚碳化二亚胺〔防老剂〕
polycarboimide(PCD)　聚碳酰亚胺
polycarbonamide　聚碳酰胺
polycarbonate　聚碳酸酯
polycarbonate diol　聚碳酸酯二醇
polycarbonate fibre　聚碳酸酯纤维
polycarbonate film　聚碳酸酯薄膜
polycarbonate polyol　聚碳酸酯(型)多元醇
polycarborane methylacrylate　聚甲基丙烯酸碳硼烷(代甲)酯
polycarborane siloxane　聚碳硼烷基硅氧烷
polycarboxylate　聚羧酸酯
polycarboxylate-coated titania dispersion　聚羧酸盐(酯)包膜二氧化钛(的)分散体
polycarboxylic acid　①聚羧酸②多元羧酸
polycarbylamine(=polyisonitrile)　聚胩；聚异腈
polycation　聚阳离子
polychloral　吡啶和三氯乙醛产物
polychlorated　多氯化的
polychloride　多氯化物
polychlorinated　多氯化的
polychlorinated biphenyls(PCBs)　多氯联苯
polychlorinated dibenzo-p-dioxin(s)(PCDD(s))　氯代二苯并二噁英
polychlorination　多氯化反应
polychlorizated　多氯化的
polychlorization　多氯化反应
polychloroalkane　多氯烷烃
polychlorobiphenyl　多氯联苯
poly-2-chlorobutadiene-1,3-chloroprene rubber　氯丁橡胶
polychlorocarbon(=polychlorohydrocarbon)　多氯代烃
polychloro copper phthalocyanine　多氯酞菁铜
polychloroether　氯化聚醚；聚氯醚
polychlorohydrocarbon(=polychlorocarbon)　多氯代烃
polychloroparaffin　多氯化石蜡；多氯代石蜡
polychlorophosphonate　聚氯膦酸酯
polychloroprene　聚氯丁二烯
polychloroprene latex　氯丁胶乳
polychloroprene rubber　氯丁橡胶
polychlorostyrene　聚氯苯乙烯
polychlorosubtilin　多氯枯草菌素
poly(chlorotrifluoroethylene)(PCTFE)　聚三氟氯乙烯
polychlorovinyl　聚氯乙烯
polychlorparaffin　多氯化石蜡
polychroism(=pleochroism)　多色(现象)
polychrom(=esculin)　多色素；七叶灵
polychromate　①多铬酸盐②多色性物体
polychromatic　多色的
polychromatic enamel　多彩瓷漆
polychromatic fibre　热敏变色纤维〔能随温度的变化而改变色泽的纤维〕
polychromatic finish　①多色面漆；多彩涂层②闪光涂层；闪光漆
polychromatic light　多色光
polychromatic radiation　多色辐射；杂色辐射
polychromatism　多色性；多色现象
polychromatophile　亲不同色的
polychromator　多色仪
polychrome　多色的
polychrome finish　①多色面漆②多色面饰
polychrome lacquer　彩漆

polycistron 多顺反子
polyclonal antibody 多克隆抗体
polycoagulant 凝聚剂
Polyco catalytic polymerization process (美)叠合过程公司焦磷酸铜催化叠合过程
polycomplex 多重络合物；配位聚剂
polycomplexation 配位聚(作用)
polycomponent 多组分
polycondensate 缩聚物；多聚物
polycondensation 缩聚作用
polycondensation catalyst 缩聚催化剂
polycondensation equilibrium 缩聚平衡
polycondensation reactor 缩聚反应器
polycondensation vessel 缩聚釜
polycondensing 缩聚
polycoordination 配位缩聚；配位聚合
polycoumarone 聚香豆酮；聚氧杂茚
polycrase 复稀金矿
polycrystal 多晶*
polycrystalline 多晶的
polycrystalline fibre 多晶纤维
polycrystalline film capacity 多晶膜容量；多晶胶片容量
polycrystalline graphite 多晶石墨
polycrystalline material 多晶物质；多晶材料
polycrystalline metal whisker 多晶金属须晶
polycrystalline pigment 多晶形颜料
polycrystalline polymer 多晶形聚合物
polycrystalline silicon 多晶硅
polycrystalline state 多晶态
polycrystalline structure 多晶结构
polycrystalline zirconium dioxide fibre 多晶二氧化锆纤维
polycyclamide 聚环酰胺
polycyclic 多环的
polycyclic aromatic hydrocarbon(PAHs) 多环芳烃
polycyclic benzenoid compound 多环的苯型化合物
polycyclic compound 多环化合物
polycyclic hydrocarbon 多环烃
polycyclic musk 多环麝香
polycyclic naphthene 多环环烷
polycyclic polyene polymer 多环多烯聚合物
polycyclic quinones 多环的醌类化合物
polycyclic ring 多核环
polycyclic saturated hydrocarbon 多环饱和烃
polycyclic system 多环体系
polycyclization 聚环作用
polycyclobutene 聚环丁烯
polycyclohexadiene 聚环己二烯

poly-cyclohexanedimethylene dibromoterephthalate 聚二溴对苯二甲酸环己二甲酯
poly-1,4-cyclohexylene adipamide 聚1,4-己二酰环己二胺
poly-1,4-cyclohexylene dithiocarbamide 聚硫脲环己二胺
polycyclohexyl methacrylate 聚甲基丙烯酸环己酯
poly-p-cyclohexylphenyl methacrylate 聚甲基丙烯酸对环己基苯酯
polycyclopentadiene 聚环戊二烯
polycyclo-rubber 环化橡胶；多环化橡胶；多环橡胶
polycyclotrimerization 多环三聚(作用)
polycysteine 聚胱氨酸
polydecamethylene adipamide 聚亚己基癸二酰胺
polydecamethylene adipate 聚己二酸亚癸基酯
polydecamethylene formal 聚亚癸基甲醛
polydecamethylene glutarate 聚戊二酸亚癸基酯
polydecamethylene glycol carbonate 聚碳酸十二烷二醇酯
polydecamethylene hydroquinone diglycollate 聚对苯二酚二缩羟乙酸亚癸基酯
polydecamethylene oxalate 聚乙二酸亚癸基酯
polydecamethylene oxamide 聚亚癸基乙二酰胺
polydecamethylene phenylene diacetamide 聚亚癸基亚苯基二乙酰胺
polydecamethylene phthalate 聚邻苯二甲酸亚癸基酯
polydecamethylene resorcinol diglycollate 聚间苯二酚二缩羟乙酸亚癸基酯
polydecamethylene sebacate 聚癸二酸亚癸基酯
polydecamethylene succinate 聚丁二酸亚癸基酯
polydecamethylene undecandicarboxylate 聚十三烷二酸亚癸基酯
polydecandicarboxylic anhydride 聚十二烷二酸酐
polydentate compound 多齿化合物
polydentate effect 多配位基团效应
polydentate ligand 多齿配体
polydeter 多行列式
polydextran gel 葡聚糖凝胶
polydiacetylene 聚二乙炔〔有机光纤材料的原料〕
polydialkysiloxane 聚二烷基硅氧烷
poly(diallyl isophthalate) 聚间苯二甲酸二烯丙酯
polydiallyl phthalate 聚邻苯二甲酸二烯丙酯〔增塑剂〕
polydiarylamine 聚二芳胺〔防老剂〕
polydibenzyl malonyl urea 聚二苄基丙二酰脲
polydichlorophosphazene 聚氯化膦腈
polydichlorostyrene 聚二氯苯乙烯
polydicyclopentadiene resins 聚二环戊二烯树脂
polydideuteroethylene 聚二氘乙烯

polydiene 聚二烯烃；二烯类橡胶
polydiene rubber （聚）二烯橡胶
polydiethylene adipate 聚己二酸二亚乙酯〔增塑剂〕
polydiethyleneglycolmethacrylate 聚甲基丙烯酸缩二(乙二醇)酯
polydihydronaphthalene 聚二氢萘
polydiisopropylcarbinol methacrylate 聚甲基丙烯酸二异丙基甲醇酯
polydiketopiperazine 聚哌嗪二酮
polydimethyl butadiene 聚二甲(基)丁二烯
polydimethylsiloxane 聚二甲基硅氧烷〔脱模剂〕
polydimethylsiloxane rubber 聚二甲基硅氧烷橡胶
polydimethylsilylene 聚二甲基甲硅撑
polydiolefin 聚二烯烃
polydioxanone 聚对二氧环己酮
poly(diphenyl ether sulfone) 聚二苯醚砜
polydisperse 多分散
polydisperse aerosol 多分散气溶胶
poly-dispersed particles 多分散(性)微粒
polydisperse latex 多分散胶乳
polydisperse polymer 多分散性聚合物
polydisperse polystyrene 多分散聚苯乙烯
polydisperse system 多分散系统
polydispersion 多分散性
polydispersity 多分散性
polydispersity index(PID) 多分散性指数
polydivinyl acetylene 聚二乙烯(基)乙炔
polydroxyl kanoate 微生物聚酯
polydymite 辉镍矿
polyelectrode 混合电极
polyelectrolyte 高分子电解质；聚电解质
polyelectrolyte complex 聚电解质复合物
polyelectrolyte flocculation 聚合电解质絮凝(作用)
polyelectrolyte molecule 聚电解质分子
polyenanthoamide 聚庚酰胺
polyene 多烯烃；聚烯烃
polyene chromogens 多烯类发色体
polyene fatty acid 多烯脂肪酸；多双键脂肪酸
polyene group-containing fibre 含聚烯基纤维
polyene phosphatidyl choline 多烯磷脂酰胆碱〔药〕
polyene pigment 多烯色素
polyenergic 多能的；非单色的
polyene structure 多烯结构；多烯结构
polyenic 多烯的
polyenic compound 多烯化合物
polyenized 多烯化的
polyenoic acid 多烯酸
polyenoid 多烯的

polyenoid fatty acid 多(乙)烯型脂肪酸；多双键脂肪酸
polyenoid system 多烯系统；共轭双键系统
polyepichlorohydrin 聚环氧氯丙烷
polyepihalohydrin 聚表卤代醇
polyepoxide 聚环氧化物
polyepoxyether 聚环氧醚
polyepoxypropylaniline 聚环氧丙基苯胺
polyester 聚酯；涤纶
polyester/isocyanate paint 聚酯/异氰酸酯漆
polyester adipate 聚己二酸酯〔增塑剂〕
polyesteramide 聚酰胺酯
polyesteramine 聚酯胺
polyester based polyurethane rubber 聚酯型聚氨酯橡胶
polyester capillaries 聚酯丝
polyester cashmere 聚酯开士米(织物)
polyester chain 聚酯链
polyester coating 聚酯涂料
polyester cord 聚酯帘线
polyester diisocyanate 聚酯(型)二异氰酸酯
polyester-diol 聚酯(型)二醇
polyester ether 聚酯醚
polyester fiber 聚酯纤维；涤纶
polyester film 聚酯薄膜
polyester foam 聚酯泡沫塑料
polyester gauze 涤纶薄纱；涤纶纱
polyester gel coat resins 聚酯凝胶涂料树脂
polyester glutarate 聚戊二酸酯〔增塑剂〕
polyester glycol 聚酯二醇
polyester habotai 涤丝纺
polyesterification 聚酯(化)(作用)
polyesterimide 聚酯酰亚胺
polyester isocyanate 聚酯异氰酸酯
polyester laminate 聚酯层压板
polyester phosphate 聚酯磷酸酯
polyester plasticizer 聚酯增塑剂
polyester polyol 聚酯型多元醇
polyester resin 聚酯类树脂
polyester rubber 聚酯橡胶
polyester synthetic lubricant 聚酯合成润滑剂
polyester terephthalate 聚对苯二甲酸酯
polyester-type polyacrylate 聚酯型聚丙烯酸酯
polyester types of polyurethane rubber 聚酯型聚氨酯橡胶
polyester-urea 聚酯脲
poly(ester urethane)(PAUR) 聚酯型聚氨酯；聚酯型氨基甲酸酯
polyester wire enamel 聚酯漆包线漆
polyetair 微蜂窝状聚醚弹性体
polyethenoxy alkanolamide 聚氧乙烯烷醇酰胺

polyethenoxy alkylphenol ether 聚氧乙烯烷基苯酚醚
polyethenoxy alkylphenols 聚氧乙烯烷基苯酚醚〔表面活性剂〕
polyethenoxy ether 聚氧乙烯醚
polyether 聚醚*
poly(ether amide) 聚醚酰胺
polyetheramine 聚醚胺
polyether block 聚醚嵌段
polyether block amide(PEBA) 聚醚嵌段酰胺
polyether di-isocyanate 聚醚(型)二异氰酸酯
polyether diol 聚醚二醇
polyether ester fibre 聚醚-酯纤维
poly(ether-ether-ketone)(PEEK) 聚醚醚酮*
polyether foam 聚醚泡沫塑料
polyether glycol ①聚醚(型)二醇②(=polyethyleneglycol)聚乙二醇
poly(etherimide) 聚醚酰亚胺
polyetherin 多醚菌素
poly(ether-ketone)(PEK) 聚醚酮*
poly(ether-ketone-ketone)(PEKK) 聚醚酮酮
polyether oil 聚醚(润滑)油
polyether plasticizer 聚醚增塑剂
polyether-polyketone 聚醚醚酮
polyether polyol 聚醚(型)多(元)醇
polyether polythioether 聚醚聚硫醚〔乳化剂〕
polyether (poly) urethane 聚醚型(聚)氨基甲酯酯；聚醚型聚氨酯
polyether rubber 聚醚橡胶
polyethersulfides 聚醚硫化物〔硫化剂〕
poly(ether sulfone) 聚醚砜*
polyethersulphone fibre 聚醚砜纤维
polyether surfactant 聚醚类表面活性剂
polyether triol 聚醚三醇
polyether types of polyurethane rubber 聚醚型聚氨酯橡胶
poly(ether-urethane)(PEUR) 聚醚氨酯*
polyethoxy alkylamine 聚乙氧基烷基胺
polyethoxy alkylaryl ether 聚乙氧基烷基芳基醚
polyethoxy alkylphenol 聚乙氧基烷基酚
polyethoxy amine surfactant 聚乙氧基胺(型)表面活性剂
polyethoxy ether 聚乙氧基醚
polyethoxy glyceride 聚乙氧基甘油酯
polyethoxylate 聚乙氧基化物
polyethoxylated castor oil 聚乙氧基脂肪酸
polyethoxylated cetyl stearyl alcohol 聚乙氧基化鲸蜡基代硬脂醇
polyethoxylated nonylphenol 聚乙氧基化壬基酚；壬基酚聚氧乙烯醚
polyethyl acrylate 聚丙烯酸乙酯

polyethylated 多乙基(代)的
polyethylation 多乙基(代)反应
polyethylenamine 聚乙烯胺
polyethylene 聚乙烯
polyethylene adipate 聚己二酸亚乙基酯
γ-(polyethyleneamino) propyltrimethoxy silane γ-(多亚乙基氨基)丙基三甲氧基硅烷〔偶联剂〕
polyethylene bag 聚乙烯袋
polyethylene carbonate 聚碳酸亚乙基酯
polyethylenediaminecellulose 聚乙二胺纤维素
polyethylene-4,4′-diphenylene dicarboxylate 聚 4,4′-联苯二甲酸亚乙基酯
polyethylene diphenyl methane-4,4′-dicarboxylate 聚二苯甲烷-4,4′-二甲酸亚乙基酯
polyethylene disulfide 聚二硫化乙烯
polyethylene ether nonionics 聚乙烯醚非离子(表面活性剂)
polyethylene fiber 聚乙烯纤维
polyethylene foams 聚乙烯泡沫塑料
polyethylene glycol 聚乙二醇；聚氧乙烯
polyethylene glycol adipate 聚己二酸乙二醇酯
polyethylene glycol azeleate 聚壬二酸乙二醇酯
polyethylene glycol decanedicarboxylate 聚十二烷二酸乙二醇酯
polyethyleneglycol diacrylate 聚乙二醇二丙烯酸酯
polyethylene glycol dibenzoate 聚乙二醇二苯甲酸酯〔增塑剂〕
polyethylene glycol di(2-ethylhexoate) 聚乙二醇二(2-乙基己酸)酯
polyethyleneglycol diglycidylether 聚乙二醇二缩水甘油醚
polyethylene glycol dilaurate 聚乙二醇二月桂酸酯
polyethylene glycol dimethacrylate 聚乙二醇二甲基丙烯酸酯〔硫化剂〕
polyethylene glycol diphenate 聚 2,2′-联苯二甲酸乙二醇酯
polyethylene glycol-3,3′-diphenylene dicarboxylate 聚 3,3′-联苯二甲酸乙二醇酯
polyethylene glycol diphenylsulfide dicarboxylate 聚二苯硫醚-4,4′-二甲酸乙二醇酯
polyethylene glycol distearate 聚乙二醇二硬脂酸酯
polyethylene glycol dodecandicarboxylate 聚十四烷二酸乙二醇酯
polyethylene glycol ether 聚乙二醇醚
polyethylene glycol ether phosphate 聚乙二醇醚磷酸盐(或酯)
polyethylene glycol ether sulfate 聚乙二醇醚硫酸盐(或酯)
polyethylene glycol fumarate 聚反丁烯二酸乙二醇酯
polyethylene glycol hydroquioxyl diacetate 聚对苯二酚

二缩羟乙酸乙二醇酯
polyethylene glycol isophthalate 聚间苯二甲酸乙二醇酯
polyethylene glycol lauryl ether 聚乙二醇月桂基醚
polyethylene glycol maleate 聚顺丁烯二酸乙二醇酯
polyethylene glycol malonate 聚丙二酸乙二醇酯
polyethylene glycol monalkanoate 聚乙二醇单烷酸酯
polyethylene glycol monolaurate 聚乙二醇单月桂酸酯〔分散剂〕
polyethyleneglycol monomethacrylate 聚乙二醇单甲基丙烯酸酯
polyethylene glycol monooleate 聚乙二醇单油酸酯
polyethylene glycol 1,4-naphthalene dicarboxylate 聚1,4-萘二甲酸乙二醇酯
polyethylene glycol 2,6-naphthalene dicarboxylate 聚2,6-萘二甲酸乙二醇酯
polyethylene glycol nonandicarboxylate 聚十一烷二酸乙二醇酯
polyethylene glycol nonylphenyl ether 聚乙二醇壬基苯基醚；壬基酚聚氧乙烯醚
polyethylene glycol octylphenyl ether sulfate 聚乙二醇辛基苯基醚硫酸盐
polyethylene glycol oxalate 聚乙二酸乙二醇酯
polyethylene glycol oxide 聚氧化乙烯
polyethylene glycol perfluorodecene ether 聚乙二醇全氟癸烯醚
polyethylene glycol phenyl ether ester 聚乙二醇苯基醚酯
polyethylene glycol phthalate 聚邻苯二甲酸乙二醇酯
polyethylene glycol quaternized inner salt 聚乙二醇季铵内盐
polyethylene glycol resorcinol diglycolate 聚间苯二酚二缩羟乙酸乙二醇酯
polyethylene glycol sebacamide 聚癸二酰乙二胺
polyethylene glycol sebacate 聚癸二酸乙二醇酯
polyethylene glycol stearate 聚乙二醇硬脂酸酯
polyethylene glycol suberate 聚辛二酸乙二醇酯
polyethylene glycol succinate 聚丁二酸乙二醇酯
polyethylene glycol sulfophenyl ether 聚乙二醇磺基苯基醚
polyethylene glycol terephthalate 聚对苯二甲酸乙二醇酯
polyethylene glycol terpene ether 聚乙二醇萜烯醚
polyethylene glycol tetrasulfide 聚四硫化乙烯
polyethylene glycol undecandicarboxylate 聚十三烷二酸乙二酯
polyethylene imine 聚乙烯亚胺
polyethylene isophthalate 聚间苯二甲酸亚乙酯〔增塑剂〕
polyethylene lozenge 聚乙烯菱晶
polyethylene malonate 聚丙二酸亚乙基酯

polyethylene monofilament 聚乙烯单丝
polyethylene oil 聚乙烯(润滑)油
poly(ethylene oxide) 聚环氧乙烷；聚氧化乙烯
polyethylene powder coating 聚乙烯粉末涂料
polyethylene powder disk technique 聚乙烯粉末压片法
polyethylene sebacate 聚癸二酸亚乙酯〔增塑剂〕
polyethylene (squeeze type) wash bottle 聚乙烯(挤压型)洗瓶
polyethylene succinate 聚丁二酸亚乙酯〔增塑剂〕
polyethylene tape 聚乙烯胶带〔绝缘带〕
poly(ethylene terephthalate)(PET) 聚对苯二甲酸乙二酯*
polyethylene wax 聚乙烯蜡
polyethylenimine 聚氮丙啶〔硫化剂〕
polyethylidene 聚亚乙基
polyethyl methacrylate 聚甲基丙烯酸乙酯〔黏合剂〕
polyethyltriethoxysilane 聚乙基三乙氧基硅烷
polyferrocene 聚二茂铁
polyferrocenylene 聚亚二茂铁基
polyfluocarbon(=polyfluohydrocarbon) 多氟烃
polyfluohydrocarbon(=polyfluocarbon) 多氟烃
polyfluorated 多氟化的
polyfluoride 多氟化物
polyfluorinated 多氟化的
polyfluorination 多氟化反应
polyfluorizated 多氟化的
polyfluorization 多氟化反应
polyfluoroacrylate 聚氟代丙烯酸酯
polyfluoroacylphosphinate 多氟乙酰次膦酸酯
polyfluoroacylphosphonate 多氟乙酰膦酸酯
polyfluoroalkanesulfonate salt 多氟烷磺酸盐
polyfluoroalkylthio alcohol 多氟烷硫醇
polyfluorocarbon(=polyfluorohydrocarbon) 多氟烃
polyfluorocarbon fibre 聚氟碳(化合物)纤维；聚氟烃纤维
polyfluoroethylene 聚氟乙烯
polyfluorohydrocarbon(=polyfluorocarbon) 多氟烃
polyfluoro kerosene 全氟煤油
polyfluoroprene 聚氟丁二烯
polyform 聚合重整
polyformal 聚甲醛
poly formaldehyde(=polyoxy methylene) 聚甲醛
polyformate(=polyform distillate) 聚合重整馏分
polyform distillate(=polyformate) 聚合重整馏分
polyforming(=polyform process) 聚合重整
polyform process(=polyforming) 聚合重整
polyfructofuranoside 聚呋喃果糖苷
polyfunctional 多官能的；多功能的
polyfunctional acid 多官能酸

polyfunctional alcohol 多官能醇
polyfunctional aliphatic isocyanate 多官能脂肪族异氰酸酯
polyfunctional amine 多官能胺
polyfunctional aromatic isocyanate 多官能芳(香)族异氰酸酯
polyfunctional aziridine 多官能氮丙啶〔交联剂〕
polyfunctional base 多官能碱
polyfunctional carbamate 多官能(性)氨基甲酸酯
polyfunctional catalysis 多功能催化作用
polyfunctional catalyst 多功能催化剂
polyfunctional compound 多官能化合物
polyfunctional condensation 多官能缩合
polyfunctional epoxy resin 多官能环氧树脂
polyfunctional extractant 多官能团萃取剂；多功能基萃取剂
polyfunctional glycidylether 多官能缩水甘油醚
polyfunctional group 多官能团
polyfunctional initiator 多官能印花剂
polyfunctional isocyanate 多官能异氰酸酯
polyfunctionality(=multifunctionality) 多官能度
polyfunctional monomer 多官能单体
polyfunctional structure 多官能结构
polyfungin 多真菌素
polygalacturonase 聚半乳糖醛酸酶
polygalacturonic acid 聚半乳糖醛酸
polygalic acid 远志酸
polygalin 远志灵 $C_{32}H_{54}O_{18}$
polygalite(=polygalitol) 远志糖醇
polygalitol(=polygalite) 远志糖醇
polygamarin 远志苦苷
polygarskite 水合硅酸铝镁矿
polygas 聚合汽油
polygen 多种价元素〔有两种以上的化合价〕
polygenetic ①多色的②多性的
polygenetic dye 多色染料；多色性染料
polygermanosiloxane 聚有机锗硅氧烷
polyglucosan 聚葡萄糖胶；葡聚糖
polyglutamic acid 聚谷氨酸
polyglycerol 聚甘油
polyglycerol ester 聚甘油酯
polyglyceryl phthalate 聚邻苯二甲酸甘油酯
polyglycidyl ether 多缩水甘油醚；聚缩水甘油醚
polyglycine 聚甘氨酸
polyglycol 聚乙二醇
polyglycol distearate 聚乙二醇二硬脂酸酯
polyglycol ether 聚乙二醇醚
polyglycolic acid 聚羟基乙酸

polyglycollide fibre 聚乙交酯纤维
polygodial 蓼二醛
polygon 多边形；多角形
polygonal 多边的；多角的
polygonal cell 多角形热室
polygonin 虎杖苷
polygonization 多边形化
polyguanidine 多胍
polyhalide 多卤化物
polyhalite 杂卤石
polyhalocarbon 多卤烃
polyhalogenated 多卤化的
polyhalogenated dibenzofuran 多卤化氧芴；多卤化二苯并呋喃
polyhalogenation 多卤化反应
polyhalogen-benzoic acid 多卤苯酸
polyhalogenohydrocarbon 多卤烃
polyhalohydrocarbon 多卤烃
polyheavy water 反常水；聚重水〔现已否定其存在〕
polyhedral 多面的
polyhedral borders liquid cell 多面边界液胞
polyhedral capsid 多面体衣壳
polyhedral flame 多面体火焰
polyhedral oligomeric silsesquioxane 低聚倍半硅氧烷
polyhedral rearrangement 多面体重排
polyhedron 多面体
polyheptamethylene adipinamide 聚亚庚基己二酰胺
polyheptamethylene carbonate 聚碳酸亚庚基酯
polyheptamethylene pimelamide 聚亚庚基庚二酰胺
polyheptamethylene sebacamide 聚亚庚基癸二酰胺
polyheptamethylene succinate 聚丁二酸亚庚基酯
poly-1-heptene 聚1-庚烯
polyheterocycles 聚杂环类
polyheterocyclic amide fibre 聚杂环酰胺纤维
polyhexadecane dicarboxylic anhydride 聚十八烷二酐
polyhexafluoropentylene adipate 聚己二酸六氟戊二(醇)酯
polyhexafluoropentylene isophthalate 聚间苯二酸六氟戊二(醇)酯
polyhexafluoropropylene 聚六氟丙烯；聚全氟丙烯
polyhexafluoropropylene oxide 聚六氟氧丙化烯
poly(hexamethylene adipamide) 聚己二酰己二胺
polyhexamethylene adipate 聚己二酸亚己基酯
polyhexamethylene azelamide fibre 聚壬二酰己二胺纤维
polyhexamethylene azeleate 聚壬二酸亚己基酯
polyhexamethylene carbonate 聚碳酸亚己基酯
polyhexamethylene diphenyl dicarboxylate 聚联苯二酸亚己基酯
polyhexamethylene dodecan dicarboxylate 聚十四烷二

酸亚己基酯
polyhexamethylene formal 聚亚己基缩甲醛
polyhexamethylene hydroquinonediglycollate 聚对苯二酚二缩羟乙酸亚己基酯
polyhexamethylene laurylamide fibre 聚十二酰己二胺纤维
polyhexamethylene oxalate 聚乙二酸亚己基酯
polyhexamethylene phenylene dipropionamide 聚亚己基亚苯基二丙酰胺
polyhexamethylene pimelate 聚庚二酸亚己基酯
polyhexamethylene resorcinol diglycollate 聚间苯二酚二缩羟乙酸亚己基酯
polyhexamethylene sebacamide 聚亚己基癸二酰胺
polyhexamethylene sebacate 聚癸二酸亚己基酯
polyhexamethylene succinate 聚丁二酸亚己基酯
polyhexamethylene sulfide 聚六亚甲基硫；聚己硫醚
polyhexamethylene terephthalamide 聚亚己基对苯二甲酰胺
polyhindered phenol 聚位阻酚〔抗氧化剂、热稳定剂〕
polyhydrate 多水合物
polyhydrazide 聚酰肼
polyhydrazone 聚腙
polyhydric 多羟(基)的
polyhydric acid 多元酸；多(碱)价酸
polyhydric alcohol 多元醇
polyhydric alcohol fatty acid ester 多元醇脂肪酸酯
polyhydric monoester 多元醇单(脂肪酸)酯
polyhydric phenol 多元酚
polyhydric salt 多酸式盐；多氢盐
polyhydrocyanic acid 聚氢氰酸
polyhydrogensiloxane 聚氢硅氧烷
polyhydronaphthalene 聚二氢化萘
polyhydrone 多聚水
polyhydroxy-acid 多羟基酸
polyhydroxy-alcohol 多羟基醇
polyhydroxy-benzene 多羟基苯
polyhydroxybenzophenone 多羟基二苯甲酮
polyhydroxy-compound 多羟基化合物
polyhydroxy-dibasic acid 多羟基二元酸
polyhydroxy ether 聚羟基醚
polyhydroxyether resin 聚羟基醚树脂；聚苯氧树脂
polyhydroxyethyl glycolate 聚羟乙酸乙二醇酯
polyhydroxylated 多羟基化的
polyhydroxylated compound 多羟基化合物
polyhydroxylated polyether oligomer 多羟基聚醚低聚物
polyhydroxylation 多羟基化反应
polyhydroxy-monobasic acid 多羟基一元酸
polyhydroxy phenol 多羟基苯酚

polyhydroxy reactant 多羟基反应物；多元醇成分
polyhydroxy-tribasic acid 多羟基三元酸
poly-5-hydroxy-2,3,4-trimethoxy valeric acid 聚5-羟基-2,3,4-三甲氧基戊酸
polyimidazopyrrolone 聚咪唑并吡咯酮
polyimidazoquinazoline 聚咪唑并喹唑啉
polyimide 聚酰亚胺
polyimidesulfone(PISU) 聚酰亚胺砜
polyimidoylamidine 聚亚氨基脒
polyindene 聚茚
polyindigo 聚靛蓝
polyinosinic acid-polycytidylic acid 聚肌苷酸多聚胞苷酸
polyiodated 多碘化的
polyiodide 多碘化合物
polyiodinated 多碘化的
polyiodination 多碘化反应
polyiodized 多碘化的
polyiodization 多碘化反应
polyiodocarbon(=polyiodohydrocarbon) 多碘烃
polyiodohydrocarbon(=polyiodocarbon) 多碘烃
polyion 聚离子
polyion complex 聚离子复合物；聚离子络合物
polyion radical 聚离子自由基
polyisobutene 聚异丁烯
polyisobutyl acrylate 聚丙烯酸异丁酯
polyisobutylene 聚异丁烯
polyisobutylene rubber 聚异丁烯橡胶
polyisobutyl methacrylate 聚甲基丙烯酸异丁酯
polyisocyanate ①聚异氰酸酯②多异氰酸酯
polyisocyanate monomer 多异氰酸酯单体
polyisoprene 聚异戊二烯*
polyisoprene latex (聚)异戊(二烯)胶乳
polyisoprene rubber 聚异戊二烯橡胶；异戊橡胶
polyisoprene rubber latex 聚异戊二烯胶乳
polyisopropenyl methyl ketone 聚异丙烯甲基酮
polyisopropyl methacrylate 聚甲基丙烯酸异丙酯
polyisothianaphthene 聚异硫茚
polyisotopic peak 多同位素峰
polyketide 聚酮化合物
polyketoacidomycin 多酮酸霉素
polyketone 聚酮
poly(lactic acid) 聚乳酸
polylactide 聚丙交酯；聚交酯
polylactone 聚内酯
polylaminate structure 多层结构
polylauryl methacrylate 聚甲基丙烯酸月桂酯
polyligand complex 多配基配合物

polyliner　多孔分流梭
polylithiation　聚锂化(作用)
polylithionite　多硅锂云母
polylol　多元醇
polylysine　聚赖氨酸
polymer　聚合物*
ω-polymer　ω聚合物
polymer abrasion　高聚物磨耗
polymer accumulator　聚合产品收集器
polymer aggregated state　高聚物的聚集态
polymer alloy　聚合物合金
polymer architecture　聚合物结构
polymerase　聚合酶
polymerase chain reaction(PCR)　聚合酶链式反应；PCR扩增法
polymer backbone　聚合物骨架
polymer binder　聚合物黏合剂
polymer blend　高分子共混物；共混聚合物
polymer-blend fibre　共混纤维
polymer blending　聚合物共混
polymer blend spinning method　聚合物混合纺丝法
polymer bonding agent　聚合物黏合剂
polymer brushes　聚合物刷
polymer bubble tower　聚合物泡罩塔
polymer buildup　聚合物结垢(现象)
polymercaptan　聚硫醇
polymer catalyst　高分子催化剂*
polymer-cement-based composite　聚合物-水泥基复合材料
polymer chain　聚合物链
polymer chain effect　高分子链效应
polymer characterization　聚合物(特征)鉴定
polymer chemistry　高分子化学*
polymer chips　聚合物切片
polymer chips heater　聚合物切片加热器
polymer coagulant　高分子凝结剂；高分子凝集剂
polymer coated electrode　聚合物涂覆电极
polymer colloid　高分子胶体
polymer complex　聚合物配合物
polymer concrete　聚合物混凝土
polymer containing metal group　含金属(基团)聚合物
polymer content　聚合物含量
polymer creep　高聚物蠕变
polymer crystal　高分子晶体
polymer crystalline structure　高聚物的晶态结构
polymer crystallite　高分子微晶
polymer crystallography　高分子晶体学
polymer degradation　聚合物降解
polymer deposit　聚合体沉积物

polymer dielectric humidity sensor　聚合物电介(质)湿度传感器
polymer dispersion(=emulsion polymer)　乳聚橡胶
polymer dope　聚合物纺丝液
polymer drain line　(由白土塔引出)聚合物凝液的管道
polymer drug　高分子药物
polymer-drug conjugate　高分子-药物结合体
polymer dust　聚合物粉尘
polymer effect　高分子效应
polymer elastomer state　高聚物的高弹态
polymer electret　聚合物驻极体*；高聚物驻极体
polymer emulsion　合成胶乳
polymer fatigue　高聚物疲劳
polymer film　聚合物膜
polymer finisher　(聚合物的)后缩聚器
polymer finishing　聚合物后缩聚
polymer-finishing rate　聚合物后缩聚速率
polymer flocculant　聚合物絮凝剂
polymer formation　聚合物形成
polymer-forming　形成聚合物的
polymer-forming tendency　形成聚合物趋势
polymer formulator　聚合物配方
polymer fragment　聚合物的碎片
polymer gasoline(=polybenzine)　聚合汽油
polymer gel　聚合物凝胶
polymer gelation　聚合物胶凝化
polymer gel slough　聚合物凝胶状垢坍落物
polymer glass state　高聚物的玻璃态
polymer grade　聚合级；聚合用规格
polymer-homologue　聚合同系物；同系聚合物
polymeric　聚合的
polymeric acceptor　聚合(接)受体
polymeric additive　高分子添加剂*
polymeric adsorbate　高分子吸附物
polymeric adsorbent　吸附树脂
polymeric aggregate　聚合物粒料
polymeric alcohol　聚合醇
polymeric amide　聚酰胺
polymeric antioxidant　聚合型防老剂
polymeric associative type thickener　缔合型高分子增稠剂；缔合聚合物型增稠剂
polymeric bilayer　聚合双分子层；双分子层聚合物
polymeric carrier　高分子载体
polymeric chain　聚合链
polymeric chelates　聚合物螯合物
polymeric 2-chlorobutadiene(=polychloro-prene; neoprene)　聚氯丁二烯；聚氯丁烯；氯丁橡胶
polymeric coalescent agent　聚合性聚结剂〔能参加聚合

反应的）
polymeric compatibilizer 聚合物型相容剂
polymeric compound 聚合物；高分子化合物
polymeric condensing reagent 高分子缩合试剂
polymeric constituent 聚合物组分；聚合物成分
polymeric crystallinity 高聚物的结晶度
polymeric derivatization reagent 聚合衍生化试剂
polymeric dispersant 聚合物分散剂
polymeric dispersion stabilizer 聚合物分散稳定剂
polymeric donor 聚合物给体
polymeric drug 高分子药物
polymeric electrolyte 聚合物电解质
polymeric encapsulate 聚合物包封
polymeric fiber 聚合物纤维
polymeric flocculant 高分子絮凝剂
polymeric form 多聚形
polymeric halogenation reagent 高分子卤代试剂
polymeric homologue 同系聚合物
polymeric hydrocarbon 聚合烃
polymeric isocyanate 聚异氰酸酯
polymeric latex 聚合胶乳；聚合乳胶
polymeric ligand 聚合配合基
polymeric liquid 聚合液体
polymeric matrix 聚合物基体；聚合物母体
polymeric membrane 高分子膜
polymeric modification 聚合改性
polymeric modifier 聚合改性剂
polymeric multilayer 聚合多分子层；多分子层聚合物
polymeric nucleating agent 聚合晶核剂
polymeric peroxide(s) 聚过氧化物(类)
polymeric phosphine reagent 含膦高分子试剂
polymeric plasticizer 高分子型增塑剂
polymeric polyisocyanate 聚异氰酸酯；高分子多异氰酸酯
polymeric precursor 预聚物
polymeric quaternary ammonium compound 高分子季铵化合物
polymeric quinone 聚合(苯)醌
polymeric reagent 聚合试剂
polymeric semiconductor 高分子半导体
polymeric separator 聚合分离器
polymeric stiffening agent 聚合型增硬剂
polymeric substance 聚合物
polymeric sulfobetaine 聚合型磺基铵乙内酯
polymeric surface active agent 高分子表面活性剂
polymeric surfactant 聚合物表面活性剂*
polymeric system 聚合(物)体系
polymeric thickener 聚合增稠剂
polymeric unit 聚合物单元
polymeride(=polymer) 聚合物
polymer impregnated concrete 聚合物(树脂)浸渍混凝土
polymer-impregnated concrete plate 聚合物浸渍水泥板
polymer impregnated tritiated concrete(PITC) 聚合物浸渍含氚混凝土
polymerisable 可聚合的
polymerisable contaminant 可聚污染物
polymerisate 聚合产物
polymerisation conversion 聚合转化率
polymerisation equilibrium 聚合平衡
polymerisation in solution 溶液聚合
polymerisation power 聚合能力
polymerisation reactivity 聚合活性；聚合反应性
polymerisation stopper 阻聚剂
polymeriser ①聚合物②聚合器
polymerism 聚合现象*
polymer isomer 聚合物同分异构体
polymerizable compound 可聚合化合物
polymerizate 聚合产物
polymerization 聚合(反应)*
polymerization accelerator 聚合加速剂；聚合促进剂
polymerization activator 聚合活化剂
polymerization activity 聚合活性
polymerization agent 聚合剂
polymerization apparatus 聚合装置；聚合设备
polymerization autoclave 聚合高压釜；高压聚合釜
polymerization catalyst 聚合催化剂
polymerization chemicals 聚合剂
polymerization controller 聚合调节剂
polymerization-coupling reactant 聚合偶联剂
polymerization cycle 聚合周期
polymerization degree 聚合度
polymerization-depolymerization equilibrium 聚合-解聚平衡
polymerization equilibrium 聚合平衡
polymerization exotherm 聚合反应温升；聚合放热(量)
polymerization extruder 螺杆聚合机
polymerization finishing 后聚合(作用)
polymerization floor temperature 聚合下限温度
polymerization furnace 聚合炉
polymerization gasoline 聚合汽油
polymerization-grade 聚合级；聚合用规格
polymerization-grade propylene 聚合级丙烯；适于聚合的丙烯
polymerization index of polyphenol 多酚物质聚合指数
polymerization inhibitor 阻聚剂
polymerization in homogeneous phase 均相聚合

polymerization initiating zone 聚合引发(反应)区
polymerization initiator 聚合引发剂
polymerization in situ 就地聚合
polymerization kinetics 聚合动力学*
polymerization losses 聚合损失
polymerization mechanism 聚合机理
polymerization modifier 聚合调节剂
polymerization of isomerized monomer 异构化单体聚合(作用)
polymerization plant 聚合装置
polymerization procedure 聚合工艺规程；聚合程序
polymerization process 聚合过程
polymerization product 聚合产品
polymerization promotor 聚合促进剂
polymerization rate 聚合速率
polymerization reaction engineering 聚合反应工程
polymerization reaction mechanism 聚合反应机理
polymerization regulator 聚合调节剂
polymerization retardation 聚合抑制
polymerization retarder 聚合抑制剂；阻聚剂
polymerization speed 聚合速度
polymerization stabilizer 聚合稳定剂
polymerization starter (聚合)引发剂
polymerization system 聚合体系
polymerization technique 聚合技术；聚合工艺
polymerization temperature 聚合温度
polymerization terminator 聚合终止剂
polymerization thermodynamics 聚合热力学
polymerization time 聚合时间
polymerization unit 聚合设备；聚合工厂
polymerization velocity 聚合速度
polymerized antioxidant 聚合型防老剂
polymerized castor oil 聚合蓖麻油
polymerized fatty acid 聚合脂肪酸
polymerized linseed oil 聚合亚麻仁油
polymerized marine oil ①聚合海洋动物油②聚合鱼油
polymerized oil(=stand oil) 聚合油；定油；厚油
polymerized plasticizer 高分子型增塑剂
polymerized rosin 聚合松香
polymerized solute 聚合溶质
polymerized substance 聚合物
polymerized 2,2,4-trimethyl-1,2-dihydroquinoline 聚2,2,4-三甲基-1,2-二氢化喹啉〔防老剂〕
polymerized tung oil 聚合桐油
polymerizer 聚合釜；聚合器；聚合物
polymerizing 聚合
polymerizing agent 聚合剂
polymerizing catalyst 聚合催化剂
polymerizing condition 聚合条件
polymerizing factor 聚合系数；聚合因子
polymerizing pipe 聚合管
polymerizing power 聚合能力
polymer layered silicates nanocomposites 高聚物-黏土纳米复合材料
polymer-making autoclave 高压聚合釜；压热聚合釜
polymer matrix 聚合物母体
polymer melt 聚合物熔体
polymer-melt temperature 聚合物熔体温度
polymer-metal complex 聚合物-金属配合物
polymer-metal interface 聚合物-金属界面
polymer mixture 聚合物混合体
polymer modification 聚合物改性
polymer modified asphalt 聚合物改性沥青
polymer modified-concrete 聚合物改性混凝土
polymer and modified concrete and mortar 聚合物改性混凝土和改性水泥砂浆
polymer modifier 聚合物改性剂
polymer modulus 高聚物模量
polymer moisture content monitor 聚合物含水率监控器
polymer molecular weight 高聚物分子量
polymer molecule 聚合物分子
polymer monolayer 单分子层聚合物；聚合(物)单分子层
polymer morphology 聚合物形态学
polymer nano material 高分子纳米材料
polymer network 聚合物网络
polymer nomenclature 聚合物命名法
polymer non-crystalline structure 高聚物的非晶态结构
polymer oil 聚合润滑油
polymer orientation 聚合物取向
polymerous 聚合状的
polymer pair 聚合物对
polymer phase 聚合物相
polymer physical chemistry 高分子物理化学
polymer physics 高分子物理；聚合物物理
polymer plant 聚合装置
polymer-polymer complex 聚合物-聚合物配合物
polymer-poor phase 贫聚合物相*
polymer processing 聚合物加工
polymer pump 聚合泵
polymer radical 聚合物游离基
polymer radical concentration 聚合物游离基浓度
polymer reactant 高分子试剂
polymer reaction rate 聚合反应速率
polymer reagent 高分子试剂
polymer rheology 聚合物流变学
polymer ribbon 带状聚合物

polymer-rich phase 富聚合物相*
polymers 各种原子百分(数)成分相同的化合物
polymer science 聚合物科学
polymer segment 聚合物链段；高分子链段
polymer series 聚合物组系
polymer single crystal 高聚物单晶
polymer slurry 聚合物稀浆；聚合物悬浮液
polymer-soluble dye 能溶于聚合物的染料
polymer solution 高分子溶液；聚合物溶液
polymer solvent 聚合物溶剂
polymer-solvent interaction 聚合物-溶剂相互作用*
polymer-solvent interaction parameter 聚合物-溶剂相互作用参数
polymersome 聚质体
polymer spherulite 高聚物球晶
polymer stabilizer 聚合物稳定剂
polymer stand 带状聚合物
polymer strength 高聚物强度
polymer stress relaxation 高聚物的应力弛豫
polymer-substrate interaction 聚合物-底物相互作用
polymer substrate ion exchanger 聚合物基质离子交换剂
polymer support 高分子载体
polymer surfactant 高分子表面活性剂
polymer swelling 高聚物溶胀
polymer thick film 厚膜糊
polymer-through-out rate 聚合物通过速率；聚合物生产率
polymer type antioxidant 聚合物型抗氧剂；聚合型防老剂
polymer unit 聚合物单元
polymer viscous flow state 高聚物的黏流态
polymer waste 高聚物废料
polymer whisker 高分子须晶
polymer yield 高聚物屈服
polymetallic carbonyls 多金属羰基化物
polymetalloorganosiloxane 聚有机金属硅氧烷
polymetaphosphate 聚偏磷酸盐(酯)
polymeter 复式物性计
polymethacrylate 聚甲基丙烯酸酯
polymethacrylic 聚甲基丙烯酸
polymethacrylic acid 聚甲基丙烯酸
polymethacrylimide(PMI) 聚甲基丙烯酰亚胺
polymethacrylonitrile 聚甲基丙烯腈
polymethin dyes 聚甲炔染料；多次甲基染料；甲川染料
polymethine 聚次甲基；聚甲炔
polymethoxy acetal 聚甲氧基缩醛；聚甲氧基(型)二甲醇缩甲醛
polymethoxy bicyclic oxazolidine 聚甲氧基双环噁唑烷

polymethoxy dimethyl acetal(=polymethoxy acetal) 聚甲氧基二甲醇缩(甲)醛
polymethyl acrylate 聚丙烯酸甲酯
polymethyl acrylic acid 聚甲基丙烯酸
polymethyl aromatics 聚甲基芳香烃
polymethylated 聚甲基的
polymethylation 聚甲基化(作用)
polymethyl-benzene 聚甲基苯
poly-2-methylbutadiene(=polyisoprene) 聚-2-甲基丁二烯
polymethyl chloroacrylate 聚氯丙烯酸甲酯
poly(methyl-α-chloroacrylate)(PMCA) 聚α-氯代丙烯酸甲酯
polymethylene 聚亚甲基
polymethylene glycols 聚亚甲基二醇类
polymethylene polyphenylamine 聚亚甲基聚苯胺〔防老剂〕
polymethylene polyphenylene polyisocyanate 多亚甲基多亚苯基多异氰酸酯
polymethylene polyphenyl isocyanate 聚亚甲基聚苯基异氰酸酯〔硫化剂〕
polymethylene sebacamide 聚癸二酰甲二胺
polymethylene tetra sulfide 四硫化聚亚甲基
polymethyl galacturonase 聚果基半乳糖醛酶
polymethyl-α-halogenoacrylate(s) 聚α-卤(代)丙烯酸甲酯(类)
polymethyl hydrogen siloxane 聚甲基氢硅氧烷
polym(ethyl methacrylate)(PMMA) 聚甲基丙烯酸甲酯
polymethylol aminotriazine 多羟甲基氨基三嗪
polymethylolated urea 多羟甲基化脲
poly(4-methyl-1-pentene) 聚 4-甲基-1-戊烯*
polymethylphenylsiloxane 聚甲基苯基硅醚
polymethyl polyphenylamine 聚甲基聚苯胺
polymethyl siloxane 聚甲基硅氧烷
poly-α-methylstyrene 聚α-甲基苯乙烯
polymignite 铌铈钇矿
polymine(=polyethylene imine) 聚乙烯亚胺
polymixin 多黏菌素
polymolecular 多分子的
polymolecular assembly 多分子聚集体
polymolecular facial film 高分子面膜；高分子美容膜
polymolecularity ①多分子性②多分散性
polymolecularity correction 多分散改正
polymolecular layer 多分子层
polymolecular non-reversible reaction 多分子不可逆反应
polymolecular reaction 多分子反应
polymolecular reactions of cracking 裂化的多分子反应
polymonochlorotrifluoroethylene 聚一氯三氟乙烯
polymorph 多晶型物
polymorphic 多晶型的

polymorphic configuration 多形性构型；多晶型构型
polymorphic form 多晶型物
polymorphic inversion 多晶型转化
polymorphic substance (同质)多晶型物
polymorphic transformation 多晶型变换；多晶型转变
polymorphic transition 多晶型变换
polymorphism(=pleomorphism) (同质)多晶型(现象)
polymorphous 多晶型的
polymorphy 多晶型现象
polymycin 多链丝霉素
polymyxin 多黏菌素
polymyxin B 多黏菌素B
polynactin 浏阳菌素
polynaphthenic acid 聚环烷酸；生成焦质酸类
poly-3,7-naphthylene terephthalate 聚对苯二甲酸萘-3,7-二酯
polynaphthyridine ring 多萘啶环
polynitrobenzene 聚硝基苯
polynitroethylene 聚硝基乙烯
polynitrogen system 多氮系
polynomial equation 多项式方程
polynomial fitting 多项式拟合
polynomial regression 多项式回归
polynonamethylene adipamide 聚亚壬基己二酰胺
polynonamethylene adipate 聚己二酸亚壬基酯
polynonamethylene azeleate 聚壬二酸亚壬基酯
polynonamethylene carbonate 聚碳酸亚壬基酯
polynonamethylene formal 聚亚壬基缩甲醛
polynonamethylene resorcinaldiglycolate 聚间苯二酚缩二羟基乙酸亚壬基酯
polynonamethylene sebacamide 聚亚壬基癸二酰胺
polynonamethylene succinate 聚丁二酸亚壬基酯
polynorbornene 聚降冰片烯
polynortricyclene 聚三环烯
polynosic 富纤；富强纤维；黏胶；黏液丝
polynosic fibre 波里诺西克纤维；富强纤维
polynuclear 多核的；多环的
polynuclear aromatics 多核芳香烃
polynuclear complex 多核络合物*
polynuclear complex salt 多核络盐
polynuclear compound 多环化合物
polynuclear coordination compound 多核配合物
polynuclear hydrocarbon 多环烃
polynuclear metal complex 多核金属配合物
polynuclear plane(=honeycomb structure) 多环结构；蜂窝状结构
polynucleated 多核的；多环的
polynucleotidase 多核苷酸酶

polynucleotide 多核苷酸
polynucleotide ligase 多核苷酸连接酶
polynucleotide nucleotidyltransferase 多核苷酸转核苷酰酶；多核苷酸磷酸化酶
polynucleotide phosphorylase 多核苷酸磷酸化酶；多核苷酸转核苷酰酶
polyoctadecamethylene formal 聚亚十八烷基缩甲醛
polyoctamethylene adipamide 聚亚辛基己二酰胺
polyoctamethylene carbonate 聚碳酸亚辛基酯
polyoctamethylene malonate 聚丙二酸亚辛基酯
polyoctamethylene sebacamide 聚亚辛基癸二酰胺
polyoctamethylene sebacate 聚癸二酸亚辛基酯
polyoctamethylene suberate 聚亚辛基二酰胺
polyoctanoyllactam 聚辛内酰胺；尼龙-8
poly(1-octene) 聚(1-辛烯)
polyoctylene elastomer 聚辛烯弹性体
polyol 多羟基化合物；多元醇
polyol/diisocyanate ratio 多元醇/二异氰酸酯比率
polyolefin 聚烯烃
polyolefin acid 聚烯酸
polyolefine 聚烯烃
polyolefin elastomer 聚烯烃橡胶
polyolefin ionomer 聚烯烃离聚物；离子键型聚烯烃
polyolefin plastic 聚烯烃塑料
polyolefin plastomer 聚烯烃塑弹体
polyol (poly) urethane 多元醇型聚氨酯
polyol surfactant 多元醇类表面活性剂
polyoma virus 多瘤病毒
polyorganometallosiloxane 聚有机金属硅氧烷
polyorganosilicate 聚有机硅酸盐；有机黏土
polyorganosilicate graft polymer 聚有机硅酸盐接枝聚合物
polyorganosiloxane 聚有机硅氧烷
polyorganosiloxane compound 聚合有机硅氧烷化合物
polyorganostannosiloxane 聚有机锡硅氧烷
polyorganotitanosiloxane 聚有机钛硅氧烷
polyose 多糖类；聚糖
polyose unlike sugar 非糖性聚糖
polyoxamide 聚乙二酰胺
polyoxazole 聚唑
poly(2-oxetanone) 聚(氧杂环丁-2-酮)
polyoxide 多氧化物
polyoxin 多氧菌素；保丽安；多抗霉素
polyoxometallate 多金属氧酸盐
polyoxometallic acid 多金属氧酸
polyoxy- 〔词头〕多氧
polyoxyacid 多缩含氧酸
polyoxyalkylenated 聚氧烷撑化的；聚氧烯化的

polyoxyalkylenated triglyceride 聚氧烯化甘油三酸酯
polyoxyalkylene 聚氧化烯；聚乙二醇
polyoxyalkylene bis-thiourea 聚亚氧烷基双硫脲
polyoxyalkylene ether 聚氧化烯醚
polyoxyalkylene fatty acid ester 脂肪酸聚氧烷撑酯
polyoxyalkylene fluoroalkyl ether 聚氧化烯氟代烷基醚
polyoxyalkylene glycol 聚亚氧烷基乙二醇
polyoxyalkylene glycol alkyl ether 聚氧化烯二醇烷基醚
polyoxyalkylene segment 聚氧烷撑链段；聚氧烯链段
polyoxyalkylene siloxane copolymer 聚氧化烯-硅氧烷共聚物
polyoxyalkylene sulfide 聚氧化烯基硫醚
polyoxyalkylene sulfonate 聚氧化烯磺酸盐
polyoxyamylene 聚氧化戊烯
polyoxybutylene 聚氧化丁烯
polyoxybutylene glycol 聚氧化丁二醇
polyoxybutylene polyoxyethylene block polyols 聚氧化丁烯-聚氧化乙烯嵌段多元醇〔表面活性剂〕
poly-ω-oxycaprylic acid 聚(端)羟辛酸；聚ω-羟辛酸
polyoxycyclohexene 聚氧化环己烯
polyoxyethylated 聚氧乙撑化的；聚氧乙烯的
polyoxyethylated alkylphenol 聚氧乙基化烷基酚；烷基酚聚氧乙烯醚
polyoxyethylated amine 聚氧乙基化胺
polyoxyethylated cardanol 聚氧乙基化腰果酚
polyoxyethylated castor oil 聚氧乙基化蓖麻油
polyoxyethylated coconut oil alcohol 聚氧乙基化椰(子)油醇；椰(子)油醇聚氧乙烯醚
polyoxyethylated lauryl itaconate 聚氧乙基化月桂基衣康酸酯
polyoxyethylated propylene glycol 聚氧乙烯化丙二醇
polyoxyethylation 聚氧乙烯化作用；环氧乙烷加成作用
polyoxyethylenated acetylenic glycol 聚氧乙烯化炔二醇
polyoxyethylenated alcohol 聚氧乙烯化(脂肪)醇；(脂肪)醇聚氧乙烯醚
polyoxyethylenated alkylnaphthol 聚氧乙烯化烷基萘酚；烷基萘酚聚氧乙烯醚
polyoxyethylenated alkylphenol 聚氧乙烯化烷基酚；烷基酚聚氧乙烯醚
polyoxyethylenated alkyl phosphate 聚氧乙烯化烷基磷酸酯
polyoxyethylenated amine 聚氧乙烯化(脂肪)胺
polyoxyethylenated castor oil 聚氧乙烯化蓖麻油
polyoxyethylenated cycloalkylphenol 聚氧乙烯化环烷基酚；环烷基酚聚氧乙烯醚
polyoxyethylenated glyceride 聚氧乙烯化甘油酯
polyoxyethylenated glyceryl tristearate 聚氧乙烯化甘油三硬脂酸酯
polyoxyethylenated mercaptan 聚氧乙烯化硫醇
polyoxyethylenated nonylphenol 聚氧乙烯化壬基酚；壬基酚聚氧乙烯醚
polyoxyethylenated phosphate 聚氧乙烯化磷酸酯
polyoxyethylenated polyoxypropylene glycol 聚氧乙烯化聚氧丙二醇
polyoxyethylenated sorbitan ester 聚氧乙烯化失水山梨糖醇酯
polyoxyethylene 聚氧乙烯
polyoxyethylene alcohol 聚氧乙撑醇；聚乙烯醇
polyoxyethylene alkylamide 聚氧乙烯烷基酰胺
polyoxyethylene alkyl amine 聚氧乙烯烷基胺
polyoxyethylene alkyl ether 聚氧乙烯烷基醚
polyoxyethylene alkylphenol 聚氧乙烯烷基酚
polyoxyethylene alkyl phenyl ether 聚氧乙烯烷基苯基醚
polyoxyethylene alkyl phenyl ether sulfate 聚氧乙烷基苯基硫酸盐
polyoxyethylene alkyl thioether 聚氧乙烯烷基硫醚
polyoxyethylene aryl phenol 聚氧乙烯芳基酚
polyoxyethylene carboxylic ester 聚氧乙烯羧酸酯
polyoxyethylene ester surfactant 聚氧乙烯酯类表面活性剂
polyoxyethylene ether 聚氧乙烯醚
polyoxyethylene fatty acid 聚氧乙烯脂肪酸
polyoxyethylene fatty acid ester 聚氧乙烯脂肪酸酯
polyoxyethylene fibre 聚氧乙烯纤维；聚氧乙撑纤维
polyoxyethylene glucoside ester 聚氧乙烯葡萄糖苷酯
polyoxyethylene glyceride 聚氧乙烯甘油酯
polyoxyethylene glycerol ether 聚氧乙烯甘油醚
polyoxyethylene glycerol laurate ester 聚氧乙烯甘油月桂酸酯
polyoxyethylene glycerol stearate 聚氧乙烯甘油硬脂酸酯
poly(oxyethylene glycol) 聚乙二醇
polyoxyethylene glycol alkyl ether 聚氧乙烯二醇烷基醚
polyoxyethylene glycol ester 聚氧乙烯二醇酯
polyoxyethylene hexadecanol 聚氧乙烯十六(烷)醇
polyoxyethylene hydroabietyl alcohol 聚氧乙烯氢化松香醇
polyoxyethylene isooctylphenol 聚氧乙烯异辛基酚
polyoxyethylene lanolin alcohol 聚氧乙烯羊毛脂醇
polyoxyethylene lauric acid ester 聚氧乙烯月桂酸酯
polyoxyethylene lauryl alcohol 聚氧乙烯月桂醇
polyoxyethylene lauryl ether 聚氧化乙烯十二烷基醚〔稳定剂〕
polyoxyethylene mercaptan 聚氧乙烯硫醇
polyoxyethylene nonylphenol 聚氧乙烯壬基酚；壬基酚聚氧乙烯醚

polyoxyethylene nonylphenyl ether 聚氧化乙烯壬基苯基醚〔湿润剂〕
polyoxyethylene octadecanol 聚氧乙烯十八醇；十八醇聚氧乙烯醚
polyoxyethylene octadecylamine 聚氧化乙烯十八烷基胺〔稳定剂〕
polyoxyethylene octylphenol ether 聚氧乙烯辛基酚醚；辛基酚聚氧乙烯醚
polyoxyethylene octyl phenyl ether 聚氧乙烯辛基苯基醚；聚乙二醇辛基苯基醚
polyoxyethylene oleate 聚氧乙烯油酸酯
polyoxyethylene oxypropylene(POEOP) 聚氧乙烯氧丙烯；聚乙丙二醇
polyoxyethylene oxypropylene glycol 聚氧乙烯氧丙二醇
polyoxyethylene pentaerythritol 聚氧乙烯季戊四醇；季戊四醇聚氧乙烯醚
polyoxyethylene phosphate 聚氧乙烯磷酸酯
polyoxyethylene phosphonate 聚氧乙烯膦酸酯
polyoxyethylene-polyoxypropylene block co-polymer 聚氧乙烯-聚氧丙烯嵌段共聚物
polyoxyethylene polyoxypropylene ether 聚氧乙烯-聚氧丙烯醚
polyoxyethylene polyoxypropylene glycol 聚氧乙烯-聚氧丙烯二醇
polyoxyethylene polyoxypropylene phosphate 聚氧乙烯-聚氧丙烯醚磷酸酯盐
polyoxyethylene ricinoleate 聚氧乙烯蓖麻酸酯
polyoxyethylene ricinoleic acid 聚氧乙烯蓖麻醇酸；蓖麻醇酸聚氧乙烯酯
polyoxyethylene sorbitan carboxylic ester 聚氧乙烯山梨糖醇酐羧酸酯
polyoxyethylene sorbitan monooleate 聚氧乙烯山梨糖醇酐单油酸酯
polyoxyethylene sorbitan monopalmitate 聚氧乙烯山梨糖醇酐单棕榈酸酯
polyoxyethylene sorbitan monostearate 聚氧乙烯山梨糖醇酐单硬脂酸酯
polyoxyethylene sorbitan tall oil ester 聚氧乙烯失水山梨糖醇妥尔油(酸)酯
polyoxyethylene sorbitan trioleate 聚氧乙烯山梨糖醇酐三油酸酯
polyoxyethylene sorbitan tristearate 聚氧乙烯山梨糖醇酐三硬脂酸酯
polyoxyethylene sorbitol ester 聚氧乙烯山梨糖醇酯
polyoxyethylene stearate 聚氧乙烯硬脂酸酯
polyoxyethylene stearic acid ester 聚氧乙烯硬脂酸酯
polyoxyethylene succinate 聚氧乙烯丁二酸酯
polyoxyethylene sucrose ester 聚氧乙烯蔗糖酯
polyoxyethylene sucrose oleate 聚氧乙烯蔗糖油酸酯
polyoxyethylene sucrose palmitate 聚氧乙烯蔗糖棕榈酸酯
polyoxyethylene sucrose stearate 聚氧乙烯蔗糖硬脂酸酯
polyoxyethylene sulfonamide 聚氧乙烯磺酰胺
polyoxyethylene tallate 聚氧乙烯妥尔油酸酯
polyoxyethylene tall oil ester 聚氧乙烯妥尔油酯
polyoxyethylene tallow sucroglyceride 聚氧乙烯牛油蔗糖甘油酯
polyoxyethylene tetradecanol 聚氧乙烯十四烷醇
polyoxyethylene tetradecylphenol 聚氧乙烯十四烷基酚
polyoxyethylene thioalcohol 聚氧乙烯硫醇；硫醇聚氧乙烯醚
polyoxyethylene thioester 聚氧乙烯硫酯
polyoxyethylene thioether 聚氧乙烯硫醚
polyoxyethylene tridecanol 聚氧乙烯十三醇；十三醇聚氧乙烯醚
polyoxyethylene tridecyl alcohol 聚氧乙烯十三醇；十三醇聚氧乙烯醚
polyoxyethylene tridecyl ether phosphate 聚氧乙烯十三烷基醚磷酸酯
polyoxyethylene-type surfactant 聚氧乙烯型表面活性剂
polyoxyl-40-stearate 硬脂酸-40-聚烃氧基酯
polyoxy methylene(=poly formaldehyde) 聚甲醛
polyoxymethylene resin 聚甲醛树脂
poly-ω-oxynonane carboxylate 聚(端)羟癸酸；聚ω-羟癸酸
poly-ω-oxyoctadecane carboxylate 聚(端)羟十九酸；聚ω-羟十九酸
polyoxypropylene glycol 聚氧化丙二醇〔热敏剂〕
polyoxytetramethylene 聚四氢呋喃
poly-ω-oxyundecane carboxylate 聚(端)羟十二酸；聚ω-羟十二酸
poly-ω-oxyundecanoic acid 聚(端)羟十一酸；聚ω-羟十一酸
polyoxyvalerate 聚(端)羟戊酸；聚ω-羟戊酸
poly-paper 压塑(封面)纸；塑料涂层纸
polyparadinitrosobenzene 聚对二亚硝基苯〔活性剂〕
polyparaphenylene 聚对苯撑
polypectate gel 聚果胶酸盐凝胶
polypentaerythritol 聚季戊四醇
polypentamethylene adipamide 聚亚戊基己二酰胺
polypentamethylene adipate 聚己二酸亚戊基酯
polypentamethylene azelamide 聚亚戊基壬二酰胺
polypentamethylene dedecanedicarboxylamide 聚亚戊基十四烷二酰胺
polypentamethylene formal 聚亚戊基缩甲醛

polypentamethylene glutaramide 聚亚戊基戊二酰胺
polypentamethylene nonandicarboxylamide 聚亚戊基十一烷二酰胺
polypentamethylene oxalate 聚乙二酸亚戊基酯
polypentamethylene sebacamide 聚亚戊基癸二酰胺
polypentamethylene sebacate 聚癸二酸亚戊基酯
polypentamethylene suberamide 聚亚戊基辛二酰胺
polypentamethylene succinate 聚丁二酸亚戊基酯
polypentamethylene terephthalate 聚对苯二甲酸亚戊基酯
polypentamethylene tetradecanedicarboxylamide 聚亚戊基十六烷二酰胺
polypentamethylene undecanedicarboxylate 聚十三烷二酸亚戊基酯
polypentanamer 聚戊烯(橡胶)
polypeptide 多肽
polypeptide of fibroin 丝素肽
poly(perfluoropropene) 聚全氟丙烯
polyperoxide 聚过氧化物
polyphase 多相的
polyphase current 多相电流
poly-phase emulsion 多重(相)乳状液
polyphase equilibrium 多相平衡
polyphenanthridine 聚菲啶
polyphenol 多酚
polyphenol type antioxidant 多酚防老剂
polyphenylene 聚亚苯基
poly(p-phenylene) 聚对亚苯
poly-m-phenylene adipamide 聚己二酰间苯二胺
poly(phenylene ether) 聚苯醚
polyphenylene ethyl 聚乙基苯；聚对二甲苯
polyphenylene methyl 聚甲基苯
poly(phenylene oxide) 聚苯醚
poly (phenylene oxide) deprecated(PPO) 聚苯醚
poly-m-phenylene sebacamide 聚癸二酰间苯二胺
poly(p-phenylene sulfide) 聚对亚苯硫醚；聚苯硫醚
polyphenylene sulfide fibre 聚苯硫醚纤维
poly(phenylene sulfone)(PPSU) 聚苯砜
poly(p-phenylene terephthalate) 聚对苯二甲酸对苯二酯*
polyphenylene triazole 聚苯撑苯基三唑〔耐高温涂料、黏合剂用聚合物〕
polyphenylene vinylene 聚苯乙炔
polyphenylether 聚苯醚
polyphenylethyl 聚乙基苯
polyphenylmethyl 聚甲基苯
polyphenyloprene 聚苯基丁二烯
polyphenylsiloxanol 聚苯基硅氧烷醇

polyphenyl thioether 聚苯硫醚
polyphospharic acid 多磷酸
polyphospharic acid chelate 多磷酸螯合物
polyphosphate 多磷酸盐
polyphosphazene 聚磷腈
polyphosphides 多磷化物
polyphosphinate 聚次膦酸酯；聚次膦酸盐
polyphosphonitrile 聚磷腈；磷腈聚合物
polyphosphoric acid 多磷酸
polyphthalamide(PPA) 聚邻苯二甲酰胺
polyphthaloyl urea 聚酞酰脲
polypimelic anhydride 聚庚二酐
polypivalolactone 聚新戊内酯
polyplant 聚合装置
polyploidy 多倍性
polyporenic acid 多孔蕈酸
polyporin 多孔蕈素
polypren bridge 聚戊烯桥
polyprene 聚戊二烯
polyprenol 多萜醇
polypropene 聚丙烯
polypropene twine 聚丙烯细绳
polypropylene 聚丙烯
polypropylene adipate(PPA) 聚己二酸丙二醇酯〔增塑剂〕
polypropylene fiber 聚丙烯纤维；丙纶
polypropylene glycol 聚丙二醇
polypropyleneglycol diacrylate 聚(异)丙二醇二丙烯酸酯
polypropyleneglycol diglycidyl ether 聚丙二醇二缩水甘油醚
polypropyleneglycol dimethacrylate 聚丙二醇二甲基丙烯酸酯
polypropyleneglycol monomethacrylate 聚丙二醇单甲基丙烯酸酯
polypropylen(e)imine 聚甲基吖丙啶
polypropylene laurate 聚月桂酸亚丙酯〔增塑剂〕
polypropylene oxalate 聚草酸亚丙基酯
poly(propylene oxide) 聚环氧丙烷；聚氧化丙烯
polypropylene phthalate 聚邻苯二甲酸亚丙基酯
polypropylene powder coating(s) 聚丙烯粉末涂料
polypropylene sebacate 聚癸二酸亚丙酯〔增塑剂〕
polypropylene sulfide 聚硫化丙烯；聚环硫丙烷；聚丙硫醚
polypropylene tetramer 四聚丙烯
poly-n-propyl methacrylate 聚甲基丙烯酸正丙酯
polypropyrene 聚丙烯
polyprotic acid(=polybasic acid) 多元酸
polyprotonic acid 多元酸

polypyknotic 双异重组分的
polypyrazine 聚吡嗪
polypyrimidone quinazolone 聚嘧啶酮-喹唑啉酮
polypyromellitimide 聚均苯四酰亚胺
poly(pyromellitimido-1,4-phenylene) 聚均苯四酰亚胺-1,4-亚苯*
polypyrrolidone 聚吡咯烷酮
polyquinazoline dione 聚喹唑啉二酮
polyquinoxaline 聚喹喔啉
polyquinoyl 多醌基
polyradical 多基
polyreaction 聚合反应
polyrecombination 再化合聚合(作用)
polyric oil α-甲基苯乙烯和蓖麻油混合物
polyrubber 聚合橡胶〔用聚酯树脂溶液制成〕
polysaccharase(=polyase) 多糖酶；聚合酶
polysaccharide(=polysaccharose) 多糖
polysaccharose(=polysaccharide) 多糖
polysalt 聚合盐；高分子盐
polysebacic anhydride 聚癸二酐
poly sebacic polyanhydride 聚癸二酸酐
polyset process 多步定形
polysilicate 硅酸盐聚合物；聚硅酸盐
polysilicic acid 多硅酸
polysilicic acid sol 聚硅酸溶胶
polysilicone rubber (聚)硅橡胶
polysilicon etching 多晶硅腐蚀
polysiloxane 聚硅氧烷
polysiloxane-aluminium soap grease 聚硅醚铝皂润滑脂
polysiloxanediol 聚硅氧烷二醇
polysiloxane grease 聚硅醚润滑脂
polysiloxane rubber 聚硅氧烷橡胶；硅橡胶
polysilsesquioxane 聚倍半硅氧烷
polysoap 聚皂〔高分子表面活性剂〕
polysorb 药用有机聚合物载体
polysorbate-80 多乙氧基醚
polystep reaction 多步反应
polysterol(=polystyrene) 聚苯乙烯
polystichalbin 耳蕨白素
polystichin 耳蕨素
polystichinin 耳蕨宁
polystichinol 耳蕨醇
polystichocitrin 耳蕨柠檬素
polystichoflavin 耳蕨黄素
polystictin 木云芝素
polystyrene 聚苯乙烯
polystyrene burette 聚苯乙烯滴定管
polystyrene-divinylbenzene resin 聚苯乙烯-二乙烯苯树脂
polystyrene film 聚苯乙烯薄膜
polystyrene gel 聚苯乙烯凝胶
polystyrene glycol 聚苯基乙二醇
polystyrene latex 聚苯乙烯胶乳
polystyrene peroxide 聚过氧化苯乙烯
polystyrene plastic 聚苯乙烯塑料
polystyrenesulfonic acid type resin 聚苯乙烯磺酸型树脂
polystyrol 聚苯乙烯
polystyrolsulfon acid 聚苯乙烯磺酸
polystyryl cation 聚苯乙烯阳离子
polystyryl radical 聚苯乙烯自由基
polystyryl sodium 聚苯乙烯钠
polysuberic anhydride 聚辛二酸酐
polysubstituted 多取代的
polysubstitution 多取代(作用)
polysubstitution compound 多取代化合物
polysuccinonitrile 聚丁二腈；聚琥珀腈
polysulfide 聚硫化物*
polysulfide crosslink 多硫键
polysulfide crosslinked structure 多硫(化合物的)交联结构
polysulfide elastomer 聚硫弹性体
polysulfide rubber 聚硫橡胶
polysulfide rubber latex 聚硫胶乳
polysulfidic bond 多硫键
polysulfonamide fibre 聚磺酰胺纤维
polysulfonate copolymer 聚磺酸酯共聚物
polysulfone 聚砜
polysulfone hollow fibre 聚砜中空纤维
polysulfur nitride 聚氮化硫 $(SN)_x$
polysulphide polymer 多硫聚合物
polysulphonamide 聚磺酰胺
polysuspensoid 多相悬(浮)胶体
polysynthetic twins 聚片双晶
polytactic polymer 多有规立构聚合物
polytene ①聚乙烯②聚乙烯纤维
polytene chromosome 多线染色体
polyterephthalamide 聚对苯二甲酰胺
polyterephthalate 聚对苯二甲酸酯
poly(terephthalic anhydride) 聚对苯二甲酸酐
poly(terephthalic hydrazide) 聚对苯二甲酰肼
polyterpene 多萜(烯)
polyterpene resin 多萜树脂；萜烯树脂
polytetradecamethylene carbonate 聚碳酸十四烷基酯
polytetradecamethylene formal 聚十四烷基缩甲醛
polytetradecanoic anhydride 聚十四烷酸酐
poly(tetrafluoroethylene) 聚四氟乙烯

polytetrafluoroethylene gasket 聚四氟乙烯垫圈
polytetrahydrofuran 聚四氢呋喃〔溶剂〕
polytetramethylene adipamide 聚亚丁基己二酰胺
polytetramethylene azelamide 聚亚丁基壬二酰胺
polytetramethylene azeleate 聚壬二酸亚丁基酯
polytetramethylene carbonate 聚碳酸亚丁基酯
polytetramethylene ether glycol 聚四亚甲基醚乙二醇〔增塑剂〕
polytetramethylene glycol 聚丁二醇
polytetramethylene hydroquinone diglycollate 聚对苯二酚二缩羟基乙酸亚丁基酯
polytetramethylene nonandicarboxylic amide 聚亚丁基十一烷二酰胺
polytetramethylene pimelamide 聚亚丁基庚二酰胺
polytetramethylene resorcinol diglycollate 聚间苯二酚二缩羟基乙酸亚丁基酯
polytetramethylene sebacamide 聚亚丁基癸二酰胺
polytetramethylene sebacate 聚癸二酸亚丁基酯
polytetramethylene suberamide 聚亚丁基辛二酰胺
polytetramethylene succinate 聚丁二酸亚丁基酯
polytetramethylene terephthalate 聚对苯二甲酸四亚甲酯〔增塑剂〕
polythene(=polyethylene) 聚乙烯
polythene cup method 聚乙烯杯法
polythiaether 聚硫醚
polythiazole 聚噻唑
polythiazoline 聚噻唑啉
polythioamide 聚硫酰胺
N-polythiocyclohexylamine-N-cyclohexyl benzothiazolesulfenamide N-多硫代环己胺-N-环己基苯并噻唑次磺酰胺〔促进剂〕
polythioester 聚硫酯
polythioether 聚硫醚*
polythiol 聚硫醇；多硫醇
polythionate 连多硫酸盐 $M_2S_xO_6$
polythionic acid 连多硫酸 $H_2S_xO_6$
polythiophen gel 聚噻吩凝胶
polythiourea 聚硫脲
polythiourethane 聚硫氨酯；聚硫代氨基甲酸乙酯
poly(titanium acetylacetonate) 聚乙酰丙酮钛
polytitanoorganosiloxane 聚有机钛硅氧烷
polytopal isomerism 多面体异构
poly(tributoxy titanium) 聚钛酸丁酯；聚三丁氧基钛
polytrichloropropylene 聚三氯丙烯
polytridecamethylene succinate 聚丁二酸十三烷基酯
polytridecanoyllactam 聚十三内酰胺
polytriethylene adipate 聚己二酸三亚乙酯〔增塑剂〕
polytrifluorobutadiene rubber 聚三氟丁二烯橡胶
polytrifluorochloroethylene(PTFCE) 聚三氟氯乙烯
poly-p-trifluoromethylstyrene 聚对三氟甲基苯乙烯
polytrifluorostyrene 聚三氟苯乙烯
polytrimethylene adipate 聚己二酸亚丙基酯
polytrimethylene carbonate 聚碳酸亚丙基酯
polytrimethylene ether glycol 聚三亚甲基醚乙二醇〔增塑剂〕
polytrimethylene formal 聚亚丙基缩甲醛
polytrimethylene hexadecane dicarboxylate 聚十八烷二酸亚丙基酯
polytrimethylene hydroquinoxyl diacetate 聚对苯二酚二缩羟基乙酸亚丙基酯
polytrimethylene oxalate 聚乙二酸亚丙基酯
polytrimethylene phthalate 聚邻苯二甲酸亚丙基酯
polytrimethylene resorcinoxyl diacetate 聚间苯二酸二缩羟乙酸亚丙基酯
polytrimethylene sebacate 聚癸二酸亚丙基酯
polytrimethylene succinate 聚丁二酸亚丙基酯
poly(tripropoxy titanium) 聚钛酸丙酯；聚三丙氧基钛
polytropic change 多向变化；体积压缩变化
polytropic compression 多变压缩
polytropic cycle 多变循环
polytropic efficiency 多变效率
polytropic exponent 多变指数
polytropic extruder 多热源挤塑机
polytropic extrusion 多热源挤塑
polytropic head 多变压头
polytropic process 多方过程*
polytropism （同质)多晶(现象)
polytropy 多变性
polytypism 多型性
polytyrosine 聚酪氨酸
polyundecamethylene alcohol 聚十一醇
polyundecamethylene diamine sebacate 聚十一烷二胺癸二酸
polyundecamethylene glycol 聚十一烷二醇
polyundecamethylene glycol oxalate 聚乙二酸亚十一烷基二醇酯
polyundecandiol carbonate 聚碳酸十一烷二醇酯
poly-unit(=polymerization plant) 聚合装置
polyunsaturated fatty acid 多不饱和脂肪酸
polyunsaturated oil 多不饱和油
polyunsaturation 多不饱和(反应)
polyuranate 多铀酸盐
polyurea 聚脲
polyurea fiber 聚脲纤维
polyurethane 聚氨基甲酸酯*
polyurethane adhesive 聚氨酯胶黏剂；聚氨酯黏合剂

polyurethane elastic fiber 聚氨酯弹性纤维；氨纶
polyurethane elastomer 聚氨酯弹性体
polyurethane elastomeric fiber 聚氨酯弹性纤维
polyurethane fiber 聚氨基甲酸乙酯纤维
polyurethane finish 聚氨酯面漆
polyurethane foam 聚氨酯泡沫塑料
polyurethane pitch coating 聚氨酯沥青涂料
polyurethane powder coating 聚氨酯粉末涂料
polyurethane primer 聚氨酯底漆
polyurethane reaction-injection molding 聚氨酯反应性-注射成形法
polyurethane resin adhesive 聚氨酯树脂黏合剂
polyurethane rubber 聚氨酯橡胶
polyurethane rubber latex 聚氨酯胶乳
polyurethane sealant 聚氨酯密封胶
polyurethane-urea 聚氨酯-脲
polyurethane zinc rich primer 聚氨酯富锌底漆
polyuridylic acid 聚尿苷酸；多尿苷酸
polyuronide 多糖醛酸苷
polyvalency 多价
polyvalent 多价的
polyvalent alcohol 多元醇
polyvalent cation 多价阳离子
polyvalent compensating cation 多价补偿阳离子
polyvalent counter immunoelectrophoresis 多价逆流免疫电泳
polyvalent metal 多价金属
polyvalent retreatment plant 多用途(核燃料)后处理工厂
polyvinyl 乙烯基聚合物
polyvinyl acetaldehyde(=polyvinyl acetal) ①聚乙烯醇缩乙醛②聚乙烯醇缩醛(类)
polyvinyl acetal(s) 聚乙烯醇缩乙醛；聚乙烯醇缩醛(类)
poly(vinyl acetate) 聚乙酸乙烯酯
polyvinylacetate-acrylic copolymer 醋酸乙烯-丙烯酸共聚物
polyvinyl acetate-chloride copolymers 醋酸乙烯-氯乙烯共聚物
polyvinyl acetate emulsion 聚醋酸乙烯乳液
polyvinyl acetate emulsion paint 聚乙酸乙烯酯乳液涂料
polyvinyl acetate latex 聚乙酸乙烯胶乳
poly(vinyl alcohol) 聚乙烯醇
polyvinyl alcohol fiber 聚乙烯醇纤维
polyvinyl alcohol film 聚乙烯醇薄膜
polyvinyl alcohol-vinylamine copolymer 聚乙烯醇-乙烯胺共聚物
polyvinyl amine 聚乙烯胺
polyvinyl benzene sulfonate 聚乙烯苯磺酸盐
polyvinylbenzene sulfonic acid 聚乙烯基苯磺酸

polyvinyl bromide 聚溴乙烯
poly(vinyl butyral) 聚乙烯醇缩丁醛
polyvinyl butyral acetal 聚乙烯醇缩丁醛
poly(vinyl carbazole) 聚乙烯基咔唑
poly(vinyl chloride) 聚氯乙烯
poly(vinyl chloride-acetate) 氯乙烯-乙酸乙烯酯共聚物
polyvinyl chloride coating 聚氯乙烯树脂涂料
polyvinyl chloride fiber 聚氯乙烯纤维；氯纶
polyvinylchloride latex 聚氯乙烯胶乳
(poly)vinyl chloride organosol 聚氯乙烯有机溶胶
(poly)vinyl chloride plastisol (聚)氯乙烯塑溶胶
polyvinyl chloride-vinyl acetate 聚氯乙烯-乙酸乙烯酯；氯-醋共聚物
poly(vinyl chloride vinyl methyl ether)(PVM) 聚(氯乙烯-甲基乙烯基醚)
polyvinyl chloroacetate 聚氯代乙酸乙烯酯
polyvinyl cyclohexene 聚乙烯环己烯
polyvinyl dichloride 聚二氯乙烯
polyvinyl dimethylamine 聚乙烯二甲胺
polyvinyldine chloride 聚偏氯乙烯
polyvinylene 聚亚乙烯基
poly(vinylene chloride) 聚1,2-二氯亚乙烯*
polyvinyl ester 聚乙烯酯
polyvinyl ether 聚乙烯醚
polyvinyl ethyl ether 聚乙烯(基)乙醚
poly(vinyl fluoride) 聚氟乙烯
polyvinyl fluoride fibre 聚氟乙烯纤维
poly(vinyl formal) 聚乙烯醇缩甲醛
polyvinyl formal fibre 聚乙烯醇缩甲醛纤维
polyvinyl formate 聚甲酸乙烯酯
polyvinylhalide 聚卤(代)乙烯
poly(vinylidene chloride) 聚偏1,1-二氯乙烯*
polyvinylidene dicyanide 聚偏氰乙烯
poly(vinylidene fluoride) 聚偏1,1-二氟乙烯*
polyvinyl iodide 聚碘乙烯
polyvinyl isobutyl ether 聚乙烯基异丁基醚〔增塑剂〕
polyvinyl mercaptan 聚乙烯硫醇
polyvinyl methylamine 聚乙烯甲胺
polyvinyl methyl ether 聚乙烯基甲基醚
polyvinylmethylketone 聚乙烯基甲基(甲)酮
poly(vinyl methyl terephthalate) 聚(对苯二甲酸乙烯甲酯)
polyvinyl phthalimide 聚乙烯苯二酰亚胺
poly vinylpyrene 聚乙烯芘；聚乙烯嵌二萘
polyvinyl pyridine N-oxide 聚乙烯吡啶-N-氧化物
polyvinyl pyridine soap 聚乙烯吡啶皂
polyvinylpyrrolidone(PVP) 聚乙烯吡咯烷酮
polyvinyl resin 聚乙烯(基类)树脂
polyvinyl stearate 聚硬脂酸乙烯酯

poly(vinyl sulfate)　聚乙烯磺酸盐
polyvinyltoluene　聚甲基苯乙烯
polyvinyl triethoxysilane　聚乙烯基三乙氧基硅烷〔硫化剂〕
polyvinyl vinylidene chloride　氯乙烯-偏二氯乙烯共聚物
polyvitaminosis　缺多种维生素症
polywater　反常水；聚水；不冻水〔现在已否定其存在〕
poly-wire tyre　合成纤维-钢丝轮胎
polyxylenecarbonate　聚碳酸对苯二亚甲基酯
polyxylene ether　聚二甲苯醚
poly-*p*-xylene phenylene diacetate　聚亚苯基二乙酸对苯二亚甲基酯
polyxylene sebacamide　聚对苯二亚甲基癸二酰胺
polyxylol　聚二甲苯
polyxylose　聚木糖
poly-*p*-xylylene　聚对苯二亚甲基
polyyne　聚炔烃
polyzonal　多区带的；多区域的
polyzonal development　多区域展开
pomade　花香膏；香脂〔吸收天然花油后的脂肪油〕
pomade absolute　香脂净油
pomegranate extract　石榴浸液
pomegranate rind　石榴皮
pomelo(=pompelmuse)　香泡；柚〔植物〕
pomiferin　橙桑黄酮
pommel(=pommel of cork)　前鞍〔皮革〕
pommel of cork(=pommel)　前鞍〔皮革〕
Pomolio-Celdecor process　碱氯法(制浆)；氯化法(制浆)
pompelmuse(=pomelo)　香泡；柚〔植物〕
Pompey red　铁红
pompion oil　南瓜子油
PONA analysis　烷烃-烯烃-环烷烃-芳香烃定量分析；族组成分析
ponceau　①丽春花②丽春红〔颜色〕
Ponchon-Savarit graphical method　邦昌-萨洼里特图解法
pondage　蓄水量
pond retting　浸沤法(麻)
poney mixer　容器替换式立式搅拌机
pongam oil　水黄皮油〔取自 *Pongamia glabra*〕
ponianic(=gelutong)　节路顿(树)脂
ponsol colors　滂梭燃料〔染〕
ponsol golden orange　滂梭金橙〔染〕
pontachrome blue　滂铬蓝〔染〕
pontachrome blue-black　滂铬蓝黑〔染〕
pontachrome colors　滂铬色
pontacyl fast violet　滂酰快紫〔一种三苯甲基酸染料〕
pontacyl light green　滂酰浅绿
pontamine blue　滂胺蓝〔染〕

pontamine colors　滂胺燃料〔染〕
pontamine diazo scarlet　滂胺偶氮猩红
pontamine fast red　滂胺坚牢红
pontamine sky-blue　滂胺天蓝
pontanin　梧子色素
pontianak copal　坤甸；坤甸树胶
pontianak gum　朋地安脂
pontil　(取熔融玻璃用的)铁杆
pontol　仲醇和叔醇混合物
pontoon roof　浮顶
pontoon storage tank　浮式储罐
pony mixer　容器可换式搅拌机；立式搅拌机；打浆筒〔俗〕
pony roll　①筒子；卷线管②盘卷③卷轴
pony roll cutter　切盘纸机
pool　沉淀池；污水池
pool condenser　液池电容器
pooled sample variance　合并样本方差
pooled standard deviation　合并标准(偏)差*
pooled variance　合并方差*
poona fat　浦那脂〔取自 *Cototropis gigantea*〕
poonahlite　钙沸石
poor　①不良的；劣(质)的②瘦弱的③贫瘠的④贫穷的
poor combustion　不良燃烧；不完全燃烧
poor concrete　贫混凝土
poor conductor　①不良导体②不良(电)导体
poor cracking stock　不易裂化的原料
poor diesel fuel　劣质柴油机燃料；低十六烷值柴油机燃料
poor efficiency　低生产率；低利用系数
poor gas　贫燃气
poor ignition quality fuel　劣质柴油；低十六烷值(柴油机)燃料
poor mixture　劣质混合物
poor oil　劣质油
poor ore　贫矿石
poor placement　错位
poor pressing wax　不过滤石蜡
poor quality oil　劣质油
poor register　合模不充分
poor rock　贫矿石
poor separation　不良分离；不清晰分离
poor solvent　不良溶剂；劣溶剂
popcorn　爆玉米香
popcorn plastics　米花状塑料；爆花粒状塑料
popcorn polymer(=ω-polymer)　端聚物；ω-聚合物；米花状聚合物
popcorn polymerization　米花状聚合(作用)
poplar buds oil　杨树芽油
Pople's SCF method　玻扑尔自洽场法

poplin　府绸
poplox　热胀性硅酸钠
poponax resin　没药树脂
poppet valve　盘阀；提升阀
popping　爆裂
poppy　罂粟
poppy oil　罂粟子油
poppy seed　罂粟子
poppy seed oil　罂粟子油
pop safety valve　快泄安全阀
pop strength　耐破强度
pop test　（织物）破裂强度试验
popular name　俗名；通用名
population　总体*
population deviation　总体偏差
population inversion　布居反转*；粒子数反转
population mean　总体(平)均值*
population on nuclear spin energy levels　核自旋能级上的布居数
population parameter　总体参数
population variance　总体方差*
populene　杨烯
poplin　杨属灵
populoid　山杨皮素
pop-up indicator　机械指示器
P/O ratio　磷氧比
porcelain　瓷(器)
porcelain ball　瓷球
porcelain ball mill　瓷(制)球磨机
porcelain boat　瓷舟
porcelain body　瓷骨；瓷体
porcelain burner　瓷燃烧炉
porcelain casserole　瓷勺皿
porcelain clay　瓷土〔矿〕
porcelain cleat　瓷夹板
porcelain clip　瓷夹
porcelain color　瓷色料
porcelain crucible　瓷坩埚
porcelain cup　瓷杯
porcelain dish　瓷皿
porcelain dish gum　瓷皿(测定)的胶质
porcelain enamel　搪瓷
porcelain filter(ing) crucible　瓷过滤坩埚
porcelain filter stick　瓷滤棒
porcelain fracture　瓷断面
porcelain glaze　瓷釉
porcelain insulator　瓷绝缘物
porcelain mill　瓷磨

porcelain nozzle　瓷质纺丝头
porcelain plaque　瓷渣
porcelain spot plate　瓷点滴板
porcelain utensils　①瓷仪器②瓷器皿
porcelain ware　瓷器
porcellanite　陶瓷状变岩
porcellanous　陶瓷的
porch paint(=deck paint)　甲板漆
pore　①(微)孔；气孔；毛孔②管孔
pore diameter　孔径
pore distribution　①孔分布②孔径分布
pore foaming agent　微孔发泡剂
pore-free surface　无孔表面
pore fungus　多孔菌
pore limit electrophoresis　限孔电泳
pore nucleating agent　微孔形成剂；微孔成核剂〔用于制高吸水性纤维〕
pore opening　气孔的大小
pores　孔隙
pore size　孔径大小
pore size analyzer　孔径分析仪
pore size distribution　孔径分布
pore structure　孔结构
pore volume　孔体积
pore volume of catalyst　催化剂孔隙度
porfiromycin　紫菜霉素；甲基丝裂霉素
poricin　卧孔菌素
poriferasterol　多孔甾醇
porocel　高度粉碎的高活性白土
porochrom　白色硅藻土载体
porocrepe　泡沫胶片
poromeric material　微孔性合成材料；透气性(人造革)材料
poromerics　微孔材料
porometer　微孔测径仪
porosimeter　孔度计
porosity　孔隙率*；多孔性
porosity apparatus　孔率仪
porosity factor　多孔度系数；细孔系数
porosity function　孔隙率函数
porosity index　孔隙率指数
porosity meter　孔率计；孔隙率测定仪
porosity tester　孔率检验器；气孔度测验器
porous　①多孔的；疏松的②素烧(瓷)的
porous alumina　多孔氧化铝
porous barrier　多孔膜；多孔隔板
porous bed　多孔床
porous binder　多孔基料；多孔性黏结剂

porous body ①多孔物体②素烧瓷体
porous carrier 多孔载体
porous cell 素烧筒
porous cellular structure 渗透性微孔结构
porous ceramic junction 多孔陶瓷连接
porous ceramic separator 多孔陶瓷分离器
porous coating 渗透性涂料
porous collagen fiber matrix 多孔胶原纤维基底
porous crystalline structure 多孔结晶结构
porous cup 素烧杯
porous-cup method 素烧杯法
porous diaphragm 多孔隔膜
porous diaphragm cell 多孔隔膜电解池
porous diffusion 疏松扩散；多孔扩散
porous diffusion electrode 多孔扩散电极
porous electrode 多孔电极
porous filling paste 填料浆糊
porous film 多孔膜
porous filter cylinder 素烧滤筒
porous flint 多孔质火石；多孔燧石〔磨盘材料〕
porous glass 多孔玻璃
porous glass separator 多孔玻璃分离器
porous glass tube 多孔玻璃管
porous granules 多孔(催化剂)粒子
porous graphite 多孔石墨
porous graphitic carbon 多孔石墨碳
porous ion exchange resin 多孔性离子交换树脂
porous layer 多孔层
porous layer adsorbent 多孔层吸附剂
porous layer bead 多孔层珠
porous layer glass bead 多孔层玻璃珠
porous layer open column 多孔层空心柱
porous layer open tubular column 多孔层开口管柱
porous layer support 多孔层载体
porous medium 疏松介质
porous membrane 多孔膜〔分离气体的多孔过滤器〕
porous metal electrode 多孔金属电极
porous metal filter 多孔金属过滤器
porous metal separator 多孔金属分离器
porous microbead support 多孔微珠载体
porous nature 多孔性
porousness 多孔性
porous network 多孔网络
porous packing 多孔填充物
porous plastic 多孔塑料；泡沫塑料
porous plate 素烧(瓷)板
porous plug 多孔塞
porous polymer 多孔聚合物
porous polymer beads 高分子多孔微球
porous polymer beads GLS column 多孔聚合物气液固色谱柱
porous polymer monolithic column 多孔聚合(物)整体柱
porous polymer packing 多孔聚合物填料
porous porcelain cup 素烧(瓷)杯
porous reticular polymer 多孔网状聚合物
porous rubber 多孔橡胶
porous sheet (=clay plate) 素烧(瓷)板
porous shell 多孔外壳
porous silica bead 多孔硅珠
porous silica gel 多孔硅胶
porous silicon 多孔硅
porous silver separator 多孔银分离器
porous solid bed 疏松固体床
porous sorbent 多孔性吸着剂
porous stainless steel separator 多孔不锈钢分离器
porous structure 多孔结构
porous support 多孔载体
porous surface 多孔表面
porous swarm model 疏松群集模型
porous tile 素烧(瓷)板
porous wood 有孔材；阔叶材
porous xerogel 多孔干凝胶
porpezite 钯金
porphin(e) 卟吩
porphin ring 卟吩环
porphobilinogen 胆色素原；3-丙酸基-4-乙酸基-5-氨甲基吡咯
porphrazine (=tetrazaporphin) 四氮卟吩；紫菜碱
porphyrazine (=tetraazoporphine) 四氮卟吩；紫菜嗪
porphyric acid 壳苔酸
porphyrilic acid 紫菜酸 $C_{16}H_{10}O_7$
porphyrin 卟啉*
porphyrine 紫菜碱 $C_{21}H_{25}O_2N_3$
porphyrin-ferrochelatase 卟啉亚铁螯合酶
porphyrin macrocycle 卟啉大环
porphyrinogen 卟啉原；还原卟啉
porphyrinogenic steroid 生卟啉甾类；生卟啉类固醇
porphyrinuria 卟啉尿
porphyrite 玢岩
porphyrization 粉碎(作用)
porphyromycin (=porfiromycin) 紫菜霉素；甲基丝裂霉素
porphyropsin 视紫(质)
porphyroxine 紫鸦片碱
porphyry 斑岩
porpoise oil 海豚油

porpoise skin 海豚皮
porporino ①黄粉金②血卟啉
porret(=leek) 韭葱
port ①汽门；水门②葡萄酒
portable 手提的；轻便的
portable blower 轻便鼓风器
portable chromatograph 便携式色谱仪
portable compressor 移动式压缩机；小型压缩机
portable conveyor 活动运输机；移动式运输带
portable cord 移动线；拉线〔电〕
portable disk viscometer 便携盘式黏度计
portable electrostatic spraying unit 移动式静电喷涂装置
portable instrument 手提式仪器
portable ion meter 携带式离子计
portable irradiation facility 移动式辐照装置
portable irradiator 轻便辐照器
portable light 手提灯；便携灯
portable lubrication 轻便润滑〔润滑器与油枪润滑〕
portable magnetic thickness gauge 轻便磁性测厚仪
portable mould 手提模
portable neutron generator 携带式中子发生器
portable nozzle blast system 活动喷嘴喷砂系统
portable outfit 移动式装备；轻便式装备
portable pH meter 便携式 pH 计
portable pipeline system 可移动的管道系统
portable pumping unit 可移动泵设备
portable retort 手提式曲颈甑
portable sampler 手提取样器
portable steam generator 移动式蒸气发生器
portable storage tank 可移动储存器；油桶
portable yarn evenness tester 便携式电容均匀度试验仪
portamycin(=streptolydigin) 门霉素〔即：利迪链菌素〕
portative power 负荷能力；载重量
portion 部分
portion-wise addition 分批添加
portland blast-furnace slag cement 高炉矿渣(硅酸盐)水泥
portland cement 波特兰水泥；硅酸盐水泥
portland cement concrete (普通)水泥混凝土
portlandite 氢氧钙石
Portland stone 波特兰石
portmanteau leather 旅行皮包革
Portugal petitgrain oil 葡萄牙橙叶油
port wine 葡萄(汁)酒
poryzamycin 治稻瘟霉素
positional isomer 位置异构物
position correlation 正相关
positioned weld 定位焊

positioner ①胎具②定位器
position finder 测位器
positioning stop 定位制动器；定位器
position isomer 位置异构物
position isomerism 位置异构
position of absorption peak 吸收峰的位置
position sensitive detector 位(置)灵敏检测器
position sensitive photodetector 位(置)敏(感)光检测器
position sensitive proportional counter 位敏正比计数器
position sensitivity detector 位置灵敏检测器
position welding 定位焊
positive ①正的②阳(电)性的③(照相)正片④正像
positive adsorption 正吸附(作用)
positive azeotrope 正共沸混合物
positive capillary electrokinetic chromatography 正相毛细管电色谱
positive catalysis 正催化反应
positive catalyst 正催化剂
positive charge 正(阳)电荷
positive column 阳极(辅助)塔
positive component 阳性组分
positive constituent 阳(电)性构分
positive cooling 补助冷却
positive correlation 正相关
positive correlation coefficient 正相关系数
positive Cotton effect 正考顿效应
positive crystal 正晶体
positive definite 正定的
positive die 阳模；上半模
positive difference effect 正差效应
positive discharge 脉动流出
positive displacement blower 正位移鼓风机
positive-displacement compressor 正位移压缩机；容积式压缩机
positive displacement flow meter 容积式流量计
positive displacement meter 正压移动计
positive displacement pump 正位移泵；容积式泵
positive earth 阳(正)极接地
positive electricity 阳电；正电
positive electrode 阳(电)极；正(电)极
positive electron 阳电子；正(电)子
positive element 阳性元素
positive feed 强制进料；机械进料
positive feedback 正反馈
positive filter type 正滤光器型
positive group ①阳根②正(性)基
positive hole current(=hole current) 空穴电流
positive infinitely variable gear 无级变速器

positive ion　阳离子；正(电性)离子
positive ion bombardment source　正离子轰击源
positive ion chemical ionization(PICI)　正离子化学电离
positive ion mass spectrum　正离子质谱
positive latex　阳电荷胶乳
positively charged　带阳电(荷)的；正(电)性的
positively charged electron　阳电子；正(电)子
positively charged ion　阳离子；正(电性)离子
positively charged liquid film (membrane) electrode　阳性液膜电极
positively charged (material) particle　带阳电(荷)粒子；正(电)(性)微粒
positively charged sol　(带)阳电(荷)溶胶
positive maximum　正极大(值)
positive mould　阳压模；不溢式压模
positive-negative ion mass spectrometry　正负离子质谱法
positive nucleus　阳性核
positive ore　实矿石
positive peak　正峰
positive plate　①阳极板②正片
positive polarity　正极性；正电性
positive pole　阳极；正极
positive potential　阳电势；正电势；正电位
positive pressure　正压(力)；压力
positive prime　自然回水；离心吸入
positive radical　①阳根；正(电)(性)根②正(性)基；正(电)(性)基
positive ray　正电射线；阳(极)射线
positive reaction　阳性反应；正反应
positive replica　复制阳模
positive rotation　正旋(光)
positive seal　正压密封
positive semidefinite　半正定的
positive skew(ed) distribution　正偏(歪)斜分布〔粒径分布方式之一〕
positive sol　阳电(荷)溶胶；阳(电)性溶胶
positive spot test bitumen　斑点试验阳性的沥青
positive substituent　阳性取代基
positive temperature coefficient device　正温度系数端子
positive terminal　阳极端
positive uniaxial crystal　正单轴晶体
positive valency　正价
positron　正子；阳电子；正电子
positron annihilation　正电子湮没
positron camera　正电子照相机
positron decay　正电子衰变
positron emission computerized tomography(PET; PECT)　正电子发射断层扫描
positron emission tomography　正电子发射断层显像法
positronium　正电子素
positronium chemistry　正电子素化学
positronium chloride　氯化正电子素
positronium formation　正电子素形成(现象)
positronium reaction　正电子素反应
positron lifetime measurement　正电子寿命测量(法)
positron scintigraphy　正电子闪烁照相法
positron tomography　正电子断层显像法
posode　阳极
posologic　剂量的
posology　剂量学
possible ore　待证矿石
post　①支柱②接线柱；端子③黏合剂废料
post-　〔拉丁字头〕后；继
post-absorption　后吸收
post acceleration detector　后加速检测器
postactivated　后(段)活化(的)
postadhesion　后附着；后黏合；后黏结
post-alkaline cleaner wash　后碱清洗
post-bleach　补充漂白；后漂白
postbulkable　可后膨化变形的
post-calcination treatment　后煅烧处理
post card paper　明信片纸
post chlorization　后氯化
post-chromatographic derivatization　柱后衍生光度法
post coat　后涂漆〔成品涂漆〕；现场涂装；现场涂漆
post-coating cooling　涂后冷却
post column　①柱后②后置柱
post column derivatization　柱后衍生化(法)
post column reaction　柱后反应
post column reactor　柱后反应器
post-combustion　后燃
post condensation　后缩合(作用)
post-condenser　后(冷)凝器；后凝缩器
post-consumer waste　消费者用过的废料
postcooling　后冷却
post crystallization　后结晶
post cure　①后硫化；二次硫化②后硬化〔塑料〕③后熟化
post-cure reaction　后固化反应
post curing treatment　后固化处理；后熟化处理
post-defecation　二次澄清
post-defecation juice　澄清汁〔糖〕
post-defecation lime　二次澄清(糖)汁用石灰
postdeposition heat treatment　沉积后热处理
post discharging substance　后放电物质
post drop　后期降温

posteffect polymerization 后效应聚合
posterior distribution 后验分布；验后分布
posterior probability 后验概率；验后概率
poster paper 广告纸
post expansion 后发泡
postextension swell 拉伸后胀大
post-extractive operation 萃取后操作
post-form die 后成型模
post forming 热后成形；二次成形
post-handling 后处理
post-hardening 后硬化
postheat 后热
post-heating 后加热
post-heating operation 后加热操作
post-heat treatment 后热处理；二次热处理
posthydrolysis 后水解
postincubation-capture method 温浴后-捕获方法
post ionization 后电离
post-irradiation examination 辐照后(照射后)检验
postirradiation oxidation 辐照后氧化
post-metallocene initiator 后茂金属引发剂
postmicellar region 胶束化后(浓度)区
post-molding operation 出模后工序
post-molding stability 成形后稳定性
post-mold shrinkage 出模收缩
post neutralization 后中和
post pallet 柱式托板；带柱的托板〔运搬用具〕
postpeak 后峰*
post plasticity 后增塑性
post-plasticization 后增塑(作用)
post polymerization 后聚合*
postprecipitation 继沉淀*；后沉淀
post processing 后处理
post pulse 后脉冲
post-radioactivation 后放射活化(法)；可活化示踪(法)
post-reacted 后反应的
post-reaction 补充反应
postreaction heat treatment 反应后热处理
post-reactor 补充反应器
post-reactor polymer finishing 出釜后聚合物处理
post rinse 再漂洗；后漂洗
post-shaping treatment 后成形处理
postshot drilling 爆后钻探
post shrinkage 后收缩
post softening 后软化
post solvent ignition technique 溶剂后点火法
post-source decay 源后裂解；源后衰减技术
post-source decay matrix-assisted laser desorption 源后衰减基质辅助的激光解吸
post-stripper linac 剥离器后级直线加速器
post-synthetic gap 合成后间隙
post-transition element 过渡后元素*
post-treatment 后处理；后加工
post-type mixer 立式搅拌机；打浆筒〔俗〕
postulate 假说
postulated mechanism 假设机理
postural hypertension 姿势性高血压(症)；体位性高血压
post vulcanization 后硫化(作用)
postwave 后波*
postweld heat treatment 焊后热处理
post yield 后屈服；继续屈服
pot 罐；釜；锅；壶；钵；盆
potale 酒糟
pot-and-muffle furnace 坩埚-马弗炉
pot and pan process 罐与锅法
pot arch 加温炉
potarite 天然钯汞齐
potash 钾碱*
potash alum (=potassium alum) (铝)钾矾；(钾)明矾
potash basic slag 钾碱炉渣
potash black-ash 粗碳酸钾；粗钾碱
potash blue 铁蓝；华蓝；钾碱蓝
potash bulb 钾(碱)球(管)
potash cartridge 钾碱筒
potash chlorate 氯酸钾 $KClO_3$
potash chloride 氯化钾 KCl
potash feldspar 钾长石
potash fertilizer 钾肥(料)
potash glass 钾玻璃
potash-lead glass 钾铅玻璃
potash-lime glass 钾钙玻璃
potash lye 钾碱液
potash manure 钾肥(料)
potash mica 钾云母
potash muriate 氯化钾 KCl
potash prussiate 赤血盐
potash soap 钾皂
potash tank 苛性钾槽
potash water 含钾泉
potash water glass 钾水玻璃；硅酸钾
potassa 氢氧化钾；苛性钾 KOH
potassamide 氨基钾 KNH_2
potassa sulfurata 硫化钾 K_2S
potassic 含钾的
potassic basic slag 钾碱炉渣
potassic fertilizer 钾肥(料)

potassic glass   钾玻璃
potassic manure   钾肥(料)
potassic super(phosphate)   过磷酸钾
potassii nitras   硝酸钾 KNO_3
potassil〔拉丁文〕   碳酸钾
potassio-   钾代
potassium   钾〔19号元素,化学符号K〕
potassium acetate   乙酸钾 $KC_2H_3O_2$
potassium acid iodate   碘酸氢钾 $KH(IO_3)_2$
potassium acid phthalate   酞酸氢钾;邻苯二甲酸氢钾
potassium acid tartrate   酒石酸氢钾
potassium alcoholate(=potassium alkoxide)   ①醇钾②烃氧基钾 ROK
potassium alkoxide(=potassium alcoholate)   ①醇钾②烃氧基钾 ROK
potassium alum   (钾)明矾;(铝)钾矾
  $K_2SO_4 \cdot Al_2(SO_4)_3 \cdot 24H_2O; KAl(SO_4)_2 \cdot 12H_2O$
potassium aluminate   铝酸钾 K_3AlO_3
potassium aluminium sulfate   硫酸铝钾;(铝)钾矾;明矾
  $K_2SO_4 \cdot Al_2(SO_4)_3 \cdot 24H_2O; KAl(SO_4)_2 \cdot 12H_2O$
potassium aluminum silicate   硅酸铝钾
potassium americyl carbonate   碳酸镅酰钾 $KAmO_2CO_3$
potassium amide   氨基钾
potassium aminochromate   氨基铬酸钾
potassium ammonium tartrate   酒石酸铵钾
  $KNH_4C_4H_4O_6$
potassium amyl sulfate   硫酸戊酯钾
potassium antimonate   锑酸钾〔通常指 $KSbO_3$〕 $KSbO_3$;
  K_3SbO_4; $K_4Sb_2O_7$
potassium antimonyl tartrate   酒石酸氧锑钾;吐酒石
potassium argentocyanide   氰银酸钾;氰化银钾
  $K[Ag(CN)_2]$
potassium-argon age   钾-氩(法测得的)年龄
potassium-argon dating method   钾氩定年代法
potassium-argon method   钾-氩法
potassium-argon method of age determination   钾氩法(地质)年代测定
potassium arsenate   砷酸钾 $KAsO_3$; $K_3As_2O_7$; $K_2As_2O_7$
potassium arsenite   亚砷酸钾 $KAsO_2$; K_3AsO_3; $K_4As_2O_5$
potassium aurate   金酸钾 $KAuO_2$
potassium auric bromide   溴化金钾;溴金酸钾
  $K[AuBr_4]$
potassium auric chloride   氯化金钾;氯金酸钾
  $K[AuCl_4]$
potassium auric cyanide   氰化金钾;氰金酸钾
  $K[Au(CN)_4]$
potassium auric iodide   碘化金钾;碘金酸钾 $K[AuI_4]$
potassium aurocyanide(=potassium aurous cyanide)   氰亚金酸钾;氰化亚金钾 $K[Au(CN)_2]$
potassium aurous cyanide(=potassium aurocyanide)   氰化亚金钾;氰亚金酸钾
potassium azide   叠氮酸钾 $K[N_3]$
potassium benzene-diazotate   苯重氮酸钾
  $C_6H_5N=NOK$
potassium benzilate   二苯基乙醇酸钾
  $(C_6H_5)_2C(OH)COOK$
potassium benzodisulfonate   苯二磺酸钾 $C_6H_4(SO_3K)_2$
potassium beryllate   铍酸钾 K_2BeO_2
potassium bicarbonate   碳酸氢钾 $KHCO_3$
potassium bichromate(=potassium dichromate)   重铬酸钾
potassium binoxalate(=potassium bioxalate)   草酸氢钾
  KHC_2O_4
potassium bioxalate(=potassium binoxalate)   草酸氢钾
potassium bisulfate   硫酸氢钾 $KHSO_4$
potassium bisulfite   亚硫酸氢钾 $KHSO_3$
potassium bitartrate   酒石酸氢钾 $KHC_4H_4O_6$
potassium borate   硼酸钾 KBO_2; K_3BO_3; $K_4B_2O_7$
potassium borofluoride   氟硼酸钾 $K[BF_4]$
potassium borohydride   硼氢化钾 KHB_4
potassium bromate   溴酸钾 $KBrO_3$
potassium bromate method   溴酸钾(滴定)法
potassium bromaurate   溴金酸钾 $K[AuBr_4]$
potassium bromide   溴化钾 KBr
potassium bromide disc method   溴化钾片法
potassium bromide tablet method   溴化钾(压)片法
potassium bromo chlorobenzene sulfonate   溴氯苯磺酸钾
potassium bromoplatinate   溴铂酸钾 $K_2[PtBr_6]$
potassium bromoplatinite   溴亚铂酸钾 $K_2[PtBr_4]$
potassium tert-butoxide   叔丁醇钾 $(CH_3)_3COK$
potassium butyl xanthate   丁基黄原酸钾
potassium butyrate   丁酸钾 $KC_4H_7O_2$
potassium cacodylate   二甲胂酸钾 $(CH_3)_2AsOOK$
potassium carbonate   碳酸钾
potassium ceric sulfate   硫酸铈钾;四硫酸根合铈酸钾
  $K_4[Ce_2(SO_4)_4]$
potassium cerous nitrate   硝酸亚铈钾 $K_2[Ce(NO_3)_4]$
potassium chlorate(=potassium oxymuriate)   氯酸钾
  $KClO_3$
potassium chloraurate   氯金酸钾 $K[AuCl_4]$
potassium chloride   氯化钾 KCl
potassium chloride bridge   氯化钾(盐)电桥
potassium chlorite   亚氯酸钾 $KClO_2$
potassium chloropalladate   氯钯酸钾 $K_2[PdCl_6]$
potassium chloropalladite   氯亚钯酸钾 $K_2[PdCl_4]$
potassium chloroplatinate   氯铂酸钾
potassium chloroplatinite   氯亚铂酸钾 $K_2[PtCl_4]$

potassium chlorostannate 氯锡酸钾 $K_2[SnCl_6]$
potassium chromate 铬酸钾 K_2CrO_4
potassium chrome alum 钾铬矾；硫酸铬钾 $K_2SO_4 \cdot Cr_2(SO_4)_3 \cdot 24H_2O$；$KCr(SO_4)_2 \cdot 12H_2O$
potassium chromicyanide 氰铬酸钾 $K_3[Cr(CN)_6]$
potassium chromium oxalate 草酸铬钾；三草酸根合铬酸三钾
potassium chromium sulfate 硫酸铬钾；钾铬矾 $KCr(SO_4)_2 \cdot 12H_2O$
potassium cinnamate 肉桂酸钾 $KC_9H_7O_2$
potassium citrate 柠檬酸钾 $K_3C_6H_5O_7$
potassium cobalticyanide 氰高钴酸钾；氰化高钴钾 $K_3[Co(CN)_6]$
potassium cobaltinitrite 亚硝高钴酸钾 $K_3[Co(NO_2)_6]$
potassium cobaltocyanate 氰酸钴钾；四氰酸根钴酸钾 $K_2[Co(OCN)_4]$
potassium copper lead nitrite 亚硝酸铅铜钾
potassium copper oxalate 草酸铜钾
potassium cuprocyanide 氰亚酮酸钾 $K_3[Cu(CN)_4]$
potassium cyanate 氰酸钾 KOCN
potassium cyanaurite 氰亚金酸钾；氰化亚金钾 $K[Au(CN)_2]$
potassium cyanide 氰化钾
potassium decyl sulfate 癸烷基硫酸钾
potassium diborane 乙硼烷(合)钾 $K_2B_2H_6$
potassium dibutyldithiocarbamate 二丁基二硫代氨基甲酸钾〔促进剂〕
potassium dichlorocuprite 氯亚铜酸钾 $K[CuCl_2]$
potassium dichloroisocyanurate 二氯异氰尿酸钾
potassium dichromate(=potassium bichromate) 重铬酸钾
potassium dichromate method 重铬酸钾法
potassium dichromate process 重铬酸钾法
potassium dihydrogen arsenate 砷酸二氢钾 KH_2AsO_4
potassium dihydrogen arsenite 亚砷酸二氢钾 KH_2AsO_3
potassium dihydrogen hypophosphate 连二磷酸二氢二钾 $K_2H_2P_2O_6$
potassium dihydrogen phosphate 磷酸二氢钾 KH_2PO_4
potassium dihydrogen phosphite 亚磷酸二氢钾 KH_2PO_3
potassium diisopropyl naphthalene sulfonate 二异丙基萘磺酸钾
potassium dimethyldithiocarbamate 二甲基二硫代氨基甲酸钾〔促进剂〕
potassium disilicate 焦硅酸钾 $(K_2Si_2O_5)_n$
potassium disulfate 焦硫酸钾 $K_2S_2O_7$
potassium disulfatoindate 二硫酸根合铟酸钾 $K_2SO_4 \cdot In_2(SO_4)_3 \cdot 24H_2O$；$KIn(SO_4)_2 \cdot 12H_2O$

potassium electrode based on valinomycin 缬氨霉素钾电极
potassium ester 钾酯
potassium ferrate 高铁酸钾 K_2FeO_4
potassium ferric oxalate 草酸铁钾 $K_3Fe(C_2O_4)_3$
potassium ferricyanide 铁氰化钾；赤血盐
potassium ferrite 铁酸钾 $KFeO_2$
potassium ferrocyanide 亚铁氰化钾；黄血盐
potassium fluocolumbate 七氟铌酸钾 $K_2[NbF_7]$
potassium fluooxycolumbate 五氟一氧铌酸钾 $K_2[NbOF_5]$
potassium fluoplatinate 氟铂酸钾 $K_2[PtF_6]$
potassium fluoprotactinate 氟镤酸钾；七氟镤酸钾 $K_2[PaF_7]$
potassium fluoride 氟化钾 KF
potassium fluoscandate 氟钪酸钾 $K_3[ScF_6]$
potassium fluosilicate 氟硅酸钾 $K_2[SiF_6]$
potassium fluosilicate method 氟硅酸钾法〔测定玻璃中二氧化硅含量的方法〕
potassium fluotantalate 氟钽酸钾；七氟钽酸钾 $K_2[TaF_7]$
potassium fluotitanate 氟钛酸钾
potassium fluozirconate 氟锆酸钾 $K_2[ZrF_6]$
potassium hexachlorindate 六氯铟酸钾 $K_3[InCl_6]$
potassium hexachlorothallate 六氯铊酸钾；六氯(正)铊钾 $K_3[TlCl_6]$
potassium hydrate 氢氧化钾 KOH
potassium hydride 氢化钾 KH
potassium hydrogen arsenite 亚砷酸氢钾 K_2HAsO_3
potassium hydrogen oxalate 草酸氢钾
potassium hydrogen phosphate 磷酸氢钾 K_2HPO_4
potassium hydrogen phosphite 亚磷酸氢钾 K_2HPO_3
potassium hydrogen sulfate 硫酸氢钾 $KHSO_4$
potassium hydrosulfide 氢硫化钾 KHS
potassium hydrotartrate 酒石酸氢钾 $KHC_4H_4O_6$
potassium hydroxide 氢氧化钾；苛性钾
potassium hydroxide adsorption method 氢氧化钾吸着法
potassium-hydroxide number 氢氧化钾值；KOH 值〔皂化值〕
potassium hypermanganate 高锰酸钾；灰锰养〔俗〕 $KMnO_4$
potassium hypobromite 次溴酸钾 KOBr
potassium hypochlorite 次氯酸钾 KOCl
potassium hypophosphate 连二磷酸钾 $K_4P_2O_6$
potassium hypophosphite 次磷酸钾 KH_2PO_2
potassium hyposulfate 连二硫酸钾 $K_2S_2O_6$
potassium hyposulfite 连二亚硫酸钾 $K_2S_2O_4$
potassium indium alum 铟钾矾

$K_2SO_4 \cdot In_2(SO_4)_3 \cdot 24H_2O; KIn(SO_4)_2 \cdot 12H_2O$
potassium iodate 碘酸钾 KIO_3
potassium iodate starch paper 碘酸钾淀粉试纸
potassium iodaurate 碘金酸钾 $K[AuI_4]$
potassium iodide 碘化钾
potassium iodide (starch) test paper 碘化钾(淀粉)试纸
potassium iodohydrargyrate(=mercuric potassium iodide) 碘化汞钾；碘汞酸钾 $K_2[HgI_4]$
potassium ion selective electrode 碘离子选择电极
potassium iron(Ⅲ) sulfate 硫酸铁钾
potassium isopropyl xanthate 异丙基黄原酸钾〔促进剂〕
potassium lanthanum nitrate 硝酸镧钾；五硝镧酸钾 $K_2[La(NO_3)_5]$
potassium lanthanum orthophosphate 磷酸镧钾 $K_3[La_2(PO_4)_3]$
potassium laurate 月桂酸钾〔稳定剂〕
potassium manganate 锰酸钾 K_2MnO_4
potassium 2-mercaptobenzothiazole 2-硫醇基苯并噻唑钾〔促进剂〕
potassium mercuric iodide 碘化汞钾；碘汞酸钾 $K_2[HgI_4]$
potassium metaarsenate 偏砷酸钾 $KAsO_3$
potassium metaarsenite 偏亚砷酸钾 $KAsO_2$
potassium metabisulfite 焦亚硫酸钾 $K_2S_2O_5$
potassium metaborate 偏硼酸钾 KBO_2
potassium metaperiodate 偏高碘酸钾
potassium metaphosphate 偏磷酸钾 KPO_3
potassium metasilicate 硅酸钾 $(K_2SiO_3)_n$
potassium methide 甲基钾 KCH_3
potassium methoxide(=potassium methylate) 甲醇钾 $KOCH_3$
potassium methylate(=potassium methoxide) 甲醇钾
potassium methylene diisopropyl naphthalene sulphonate 亚甲基二异丙基萘磺酸钾〔分散剂〕
potassium mineral 钾矿
potassium molybdate 钼酸钾 K_2MoO_4
potassium monoethyldithiocarbamate 一乙基二硫代氨基甲酸钾〔促进剂〕
potassium nitrate 硝酸钾；钾硝；火硝；硝石
potassium nitride 氮化钾
potassium nitrite 亚硝酸钾 KNO_2
potassium orthoarsenite 亚原砷酸钾 K_3AsO_3
potassium orthophosphate (正)磷酸钾
potassium osmate(=potassium perosmate) (高)锇酸钾 K_2OsO_4
potassium oxide 氧化钾 K_2O
potassium oxymuriate(=potassium chlorate) 氯酸钾 $KClO_3$

potassium palladichloride 氯钯酸钾 $K_2[PdCl_6]$
potassium palladochloride 氯亚钯酸钾 $K_2[PdCl_4]$
potassium pentamethylenedithiocarbamate 五亚甲基二硫代氨基甲酸钾〔促进剂〕
potassium pentasulfide 五硫化二钾 K_2S_5
potassium perborate 过硼酸钾 $KBO[O_2]$
potassium percarbonate 过(二)碳酸钾 $K_2C_2O_6$
potassium perchlorate 高氯酸钾
potassium perchlorate explosive 高氯酸钾炸药
potassium perchromate 过铬酸钾 $K_2[Cr_2O_4(O_2)_4]; K_3[Cr(O_2)_4]$
potassium periodate 高碘酸钾 $KIO_4; K_3IO_5; K_5IO_6$
potassium periodate titration 高碘酸钾滴定
potassium permanganate 高锰酸钾；灰锰养〔俗〕 $KMnO_4$
potassium permanganate method 高锰酸钾法
potassium permanganate process 高锰酸钾法
potassium perosmate(=potassium osmate) (高)锇酸钾 K_2OsO_4
potassium peroxide 过氧化钾 KO_2
potassium peroxodisulphate(=potassium persulphate) 过(氧化二)硫酸钾
potassium per(oxy) borate 过硼酸钾
potassium per(oxy) nitrate 过(氧)硝酸钾
potassium per(oxy) titanate 过钛酸钾
potassium perruthenate 高钌酸钾 $KRuO_4$
potassium persulfate 过(二)硫酸钾 $K_2S_2O_6[O_2]; K_2S_2O_8$
potassium persulphate(=potassium peroxodisulphate) 过(氧化二)硫酸钾
potassium phosphate 磷酸钾 $KPO_3; K_3PO_4; K_4P_2O_7$
potassium phosphite 亚磷酸钾 K_3PO_3
potassium platinichloride 氯铂酸钾 $K_2[PtCl_6]$
potassium platinocyanide 氰亚铂酸钾 $K_2[Pt(CN)_4]$
potassium plumbate 高铅酸钾 K_2PbO_3
potassium plumbite 铅酸钾 K_2PbO_2
potassium pyroantimonate(=potassium pyroantimoniate) 焦锑酸(二氢二)钾 $K_2H_2Sb_2O_7$
potassium pyroantimoniate(=potassium pyroantimonate) 焦锑酸钾
potassium pyroarsenate 焦砷酸钾 $K_4As_2O_7$
potassium pyroborate 焦硼酸钾 $K_2B_4O_7$
potassium pyrophosphate 焦磷酸钾 $K_4P_2O_7$
potassium pyrosulfate 焦硫酸钾 $K_2S_2O_7$
potassium pyrosulfite 焦亚硫酸钾 $K_2S_2O_5$
potassium rhodanate 硫氰酸钾；硫氰化钾 $KSCN$
potassium ruthenate 钌酸钾 K_2RuO_4
potassium salt of 2-mercaptobenzothiazole 2-硫醇基苯并噻唑钾盐〔促进剂〕

potassium scandium sulfate 硫酸钪钾；三硫酸根合钪酸钾 $K_3[Sc(SO_4)_3]$
potassium selenate 硒酸钾 K_2SeO_4
potassium selenide 硒化钾 K_2Se
potassium selenite 亚硒酸钾 K_2SeO_3
potassium silicate 硅酸钾 $K_2SiO_3; K_4SiO_4$
potassium silicofluoride 氟硅酸钾 $K_2[SiF_6]$
potassium silver cobaltinitrite 六硝高钴酸银二钾 $K_2Ag[Co(NO_2)_6]$
potassium sodium aluminum silicate 铝硅酸钠钾
potassium sodium carbonate 碳酸钠钾 $KNaCO_3$
potassium sodium cobaltinitrite 六硝高钴酸钠二钾 $K_2Na[Co(NO_2)_6]$
potassium sodium tartrate 酒石酸钾钠
potassium stannate 锡酸钾 K_2SnO_3
potassium sulfantimonate 全硫锑酸钾；四硫代锑酸钾 K_3SbS_4
potassium sulfate 硫酸钾 K_2SO_4
potassium sulfatothorate 硫代钍酸钾
potassium sulfide 硫化钾 $K_2S; K_2[S_2]; K_2[S_3]; K_2[S_4]; K_2[S_5]$
potassium sulfite 亚硫酸钾 K_2SO_3
potassium sulfocarbonate 全硫碳酸钾 K_2CS_3
potassium sulfocyanate 硫氰酸钾；硫氰化钾 KSCN
potassium sulfocyanide 硫氰酸钾；硫氰化钾
potassium superoxide 超氧化钾
potassium tantalifluoride 氟钽酸钾；七氟钽酸钾 $K_2[TaF_7]$
potassium tellurate 碲酸钾 K_2TeO_4
potassium tellurite 亚碲酸钾 K_2TeO_3
potassium test for lubricating oil 润滑油钾含量试验
potassium tetraborate 四硼酸钾 $K_2B_4O_7$
potassium tetraphenylborate(=potassium tetraphenylboron) 四苯基硼(酸)钾
potassium tetraphenylboron(=potassium tetraphenylborate) 四苯基硼(酸)钾
potassium tetrathionate 连四硫酸钾 $K_2S_4O_6$
potassium tetroxalate 二草酸三氢钾 $KH_3(C_2O_4)_2 \cdot 12H_2O$
potassium thiocyanate 硫氰酸钾；硫氰化钾
potassium thiosulfate 硫代硫酸钾 $K_2SO_3S; K_2S_2O_3$
potassium titanate 钛酸钾
potassium titanate fiber 钛酸钾纤维
potassium titanium oxalate 草酸钛钾
potassium trichromate 三铬酸钾；(二缩)三铬酸钾二钾
potassium triiodide 三碘化钾
potassium tripolyphosphate 三聚磷酸钾
potassium tungstate(=potassium wolframate) 钨酸钾 K_2WO_4
potassium vanadate 钒酸钾 $KVO_3; K_3VO_4$
potassium wolframate(=potassium tungstate) 钨酸钾
potassium xanthate(=potassium xanthogenate; potassium xanthonate) 黄原酸钾 ROCSSK; C_2H_5OCSSK
potassium xanthogenate(=potassium xanthate) 黄原酸钾
potassium xanthonate(=potassium xanthate) 黄原酸钾
potassium zincate 锌酸钾 K_2ZnO_2
potassium zinc sulfate 硫酸锌钾 $K_2SO_4 \cdot ZnSO_4 \cdot 6H_2O$
potassium zirconium fluoride 氟化锆钾 K_2ZrF_8
potato flavor 土豆香精
potato gum 大戟胶
potato spirit 杂醇油
potato starch 马铃薯淀粉
potcher 漂洗槽
pot crusher 罐式压碎机
pot dissolver 罐式溶解器
potency ①效(力)；效能；效验 ②力量
potency of accelerator 催速剂效力
potent ①有效的 ②有力的
potent green odor 强烈青香
potential ①势；位 ②电势；电位 ③潜(在)的；可能的
ζ-potential ζ电势*
potential alcohol 实得酒精产量
potential analysis 电位分析法
potential associative thickener 电位(势)缔合增稠剂
potential barrier 势垒；位垒
potential buffer 电位缓冲液
potential buffer solution(=potential mediator) 电位介体；电位缓冲溶液
potential build-up curve 电位升起曲线
potential concept 势差概念；位差概念
potential-current curve 电位-电流曲线
potential decay curve 电位衰减(降落)曲线
potential delustering power 潜在消(退、除)光能力
potential-dependent chronoamperometry 定电位计时电流法
potential difference ①势差；位差 ②电势差；电位差
potential difference radiofrequency detector 电位差射频检测器
potential drop 电势降
potential dynamic method 动电位法
potential electrolyte(=ionogen) 势电解质*
potential energy 势能；位能
potential energy barrier 势垒；位垒
potential energy function 势能函数
potential energy profile 势能剖面
potential energy surface(PES) 势能面

potential fertility 潜肥力
potential field 势场；位场
potential flow 势流
potential function 电势函数
potential gradient 电势梯度
potential gradient detector(PGD) 电位梯度检测器
potential gum(=existing gum) 原在胶
potential gum in gasoline 汽油中的潜在胶
potential gum value 潜在胶值
potential head(=potential pressure) 位(置高)差
potential hill 势垒；位垒
potentially-reactivity 潜反应性
potential measurement 电位测定；电势测量
potential mediator(=potential buffer solution) 电位介体；电位缓冲溶液
potential method 电势法〔测定表面张力〕
potential metric drum 分压轮
potential of zero charge 零点荷电位；零电荷电势
potential-pH diagram 电势-pH 图*
potential pressure(=potential head) 位(置高)差
potential scanning voltammetry 电位扫描伏安法
potential step chronocoulometry 电位阶跃计时库仑法
potential step method 电位阶跃法*
potential theory 势论；位论
potential-time curve 电位-时间曲线
potential trough 势坑
potential twist turn 潜在捻数
potential value 电势值；电位值
potential well 位阱
potentiating agent 增效剂
potentiation 势差现象
potentiodynamic method 动电位法
potentiometer ①电势计；电位计②分压器
potentiometer oil 电势差计油
potentiometer pyrometer 热电温度计；高温电位差计
potentiometer recorder 电势计记录器
potentiometer titration 电势滴定；电位滴定
potentiometric analysis 电势分析；电位分析
potentiometric curve 电势滴定曲线
potentiometric detector 电位检测器
potentiometric determination 电势测定(法)
potentiometric determination of pH of solution 溶液 pH 的电位测定(法)
potentiometric differential titration 电势差示滴定(法)
potentiometric gas sensor 电位式气体传感器
potentiometric method 电势滴定法；电位滴定法
potentiometric microanalysis 电势差法微分分析
potentiometric microdetermination 电势微量测定

potentiometric optical pyrometer 电位计式光测高温计
potentiometric recorder 电势记录器；电位记录器
potentiometric stripping analysis 电位溶出分析*
potentiometric titration 电势滴定(法)；电位滴定(法)
potentiometric titration at zero current 零点电流电位滴定；无电流电位差滴定
potentiometric titration with attackable electrode 异种金属电极电位差滴定
potentiometric titration with bimetallic electrode system 双金属电极电位差滴定
potentiometric titrator 电位滴定仪*
potentiometric wheel 电势计轮
potentiometry 电势分析法；电位分析法
potentiostat 恒电位仪*
potentiostatic coulometry 恒电位库仑法
potentiostatic electrolysis 恒电位电解
potentiostatic method 恒电势法*
potentiostatic polarograph 恒电位极谱仪
pot eye 瓷圈
pot filter 筒式过滤器
pot flushing (颜料)釜挤(水)法
pot furnace 罐炉
pot gas 烧硫炉气体
potheater 压热釜
pothole 地面深穴
pot lead 石墨
pot life 储存期*
pot liquor 可蒸馏的液体
pot mill 球形磨
pot (of) spinning machine ①纺丝罐②离心罐
pot process 罐法〔制造硝化纤维素的一种工业方法〕
pot retainer 料槽托板
pot roasting 锅焙烧〔冶〕
pot spinning machine 罐纺机
pot stability 活化期稳定性
pot steel 坩埚钢
pot still 罐(式蒸)馏器
potstone 粗皂石
potten wood 腐朽材
potter's earth 陶土
potter's lead 粗粒方铅矿
potter's ore 粗粒方铅矿
potter's wheel 陶轮；拉坯轮车
pottery ①陶器②陶器制造术③陶器制造所
pottery casting 陶器铸坯
pottery clay 陶土
pottery kiln 陶瓷窑
pottery lathe 陶器镟床

pottery moulding 陶器模制
pottery turning 陶器镟坯
pottery ware 陶器
potting ①陶器制造②装缸；缸封；埋嵌
potting and embedding varnish 浇注型密封清漆
potting process 罐烧法〔灼烧硝酸盐制取氧化氮的方法〕
potting syrup 浇注液
pottle 罐儿〔容量名，等于4品脱，约2.3升〕
pot-type transfer molding 罐式传递模制法
pot valve 罐阀
pot vitrification process 罐中玻璃化法
pound ①磅②捣；重打
poundage ①磅数②(以磅计的)产量
poundal 磅达〔英制力的单位〕
pound atom 磅原子；磅(衡)原子
pound calorie 磅卡
pound-centigrade unit 磅-摄氏单位
pounding out of lubricant 润滑油(自润滑点)挤出
pound mole 磅分子；磅衡分子
pound-molecule 磅分子
pound-mole of hydrocarbon 烃的磅分子数
pourability 倾倒性
pourable 可浇注的；可灌入的
pourable liquid ①易浇注液体②易流动液体
Pourbaix diagram 甫尔拜图；电位pH图
pour depressor(=pour point depressant) 倾点下降剂
poured asphalt 浇铸沥青
pour foam test 倾注起泡试验
pouring 灌塑；灌料；铸封
pouring attachment ①喷头②浇注器
pouring drum 倾出鼓
pouring-in hole 倾入孔
pouring mold 铸型用模；铸型
pouring pail 浇铸桶
pour inhibitor(=pour point depressant) 倾点下降剂
pour instability(=pour reversion) 凝固温度的不稳定性
pour point(=flow point) 倾点；流点；流动点；倾倒点；浇铸点
pour point additive(=pour point depressant) 倾点下降剂
pour point depressant(=pour depressor; pour inhibitor; pour point additive; pour point reducer) 倾点下降剂
pour point depressor 倾点下降物
pour point improver 倾点改进剂
pour point reducer(=pour point depressant) 倾点下降剂
pour point temperature 倾点温度；流(动)点温度；倾倒点温度；浇铸点温度
pour point test(=pour test) 倾点试验
pour reversion 倾点的不稳定性；倾点回升
pour stability 倾点的稳定性
pour test(=pour point test) 倾点试验
povidon iodine 聚乙烯吡咯烷酮碘
powder ①粉；粉末②(化装用的)粉；脂粉③粉剂；药粉④磨成粉(状)；研末
powder adhesive 粉末黏结剂；粉状胶黏剂
powder agglomerate 粉(末结)块
powder air mix 粉末空气混合物
powder arc method 粉末电弧法*
powder B B火药
powder bed coating 流动床粉末涂料
powder binder ①粉末基料②粉末连接料(油墨)③粉末黏合剂
powder blower 吹粉器
powder blue 氧化钴
powder bottle 粉瓶
powder building 粉末结块
powder camera 粉末照相机；粉末(衍射)摄影箱
powder cask 火药桶；炸药桶
powder chamber 粉末舱；粉末室
powder characteristics tester 粉体特性测定器
powder cloud (喷出的)雾化粉末云团
powder clump 粉末结块
powder coal(=powdered coal) 粉煤；煤粉
powder coating ①粉末涂料；粉末涂层②粉末涂装
powder-core composite 粉芯复合料
powder crystal 粉晶
powder cutting 氧熔剂切割
powder delivery air 粉末输送气流
powder delivery pump 供粉泵；粉末输送泵
powder density 视密度；表观密度
powder diagram 粉末(衍射)图
powder (diffraction) camera 粉末(衍射)摄影箱
powder diffraction pattern 粉末衍射图
powder dip coating 粉末浸涂法
powdered (磨成)粉状的
powdered activated carbon 粉状活性炭
powdered adhesive 粉末(粉状)黏合剂
powdered aluminum 铝粉
powdered anthracite 粉状无烟煤
powdered catalyst 粉状催化剂
powdered caustic (soda) 粉状苛性钠
powdered coal(=powder coal) 粉煤；煤粉
powdered coke 粉焦
powdered custic (soda) 粉状苛性钠
powdered detergent 粉状洗涤剂
powdered distemper(=calcimine) 刷墙粉
powdered extract 粉状栲胶

powdered flavor	粉状香料
powdered fuel	粉状燃料
powdered glue	粉状胶合剂
powdered graphite	粉状石墨
powdered lubricant	润滑粉；粉质润滑材料
powdered metallic lead	粉状金属铅；金属铅粉
powdered opium	鸦片粉(末)
powdered ore	矿粉；粉状矿
powdered plastic	粉末塑料
powdered pulp	粉状纸浆
powdered quartz	石英粉；结晶型二氧化硅
powdered reaction	粉末反应
powdered resin coating	粉末树脂涂料
powdered rubber	粉末橡胶
powdered scrap	粉末废料；粉屑
powdered soap	皂粉
powdered solid fuel	粉状固体燃料
powdered solids	粉状固体粒子
powdered steatile(=talc)	滑石粉
powdered sugar	糖粉
powdered vulcanized rubber	粉状硫化橡胶
powdered whiting	重质碳酸钙
powder electrostatic coating	粉末静电涂装
powder feeder	供粉器
powder filler	填充粉
powder finish line	粉末涂装线
powder flame spraying	(金属)粉末火焰喷镀法
powder flow measurement	粉料流动性测定
powder flowmeter	粉末流量计
powder funnel	漏粉斗；填粉料
powder glazing	火药的打光作用
powder grain	火药粒
powder grain density	火药粒密度
powder grain size	火药粒大小
powder hand gun	手提式粉末(涂料)喷枪
powdering	粉化
powdering agent	隔离剂；打粉剂
powder jet	粉末喷嘴(枪)
powder-loaded polymer	粉末填充聚合物
powder mass	粉末块
powder metallurgy	粉末冶金学
powder method	粉末法
powder mixer	粉末混合机
powder molding	①粉料成型②粉料成型件
powder morphology	粉末形态
powder pattern simulation	粉末图拟合
powder pockets	粉块；粉瘤
powder point machine	粉点机
powder premix	粉状预混料
powder processing	粉末加工
powder recycle pump	粉末循环泵
powder rubber	粉状橡胶
powder sample	粉末试样
powder series	幂级数
powder shifter	筛粉机
powder sinter molding	粉末烧结成形
powder spray booth	喷粉橱；粉末喷涂间
powder sprayer	喷粉器
powder sprinkling technique	撒料法
powder structure determination	粉晶法结构测定
powder to air ratio	粉气比；粉末与空气之比
powder train	火药串(列)；导火线
powder trier	火药试验机
powder X-ray diffraction	粉末 X 射线衍射；多晶 X 射线衍射(法)
powder X-ray diffractometry	粉末 X 射线衍射法；多晶 X 射线衍射法
powder X-ray photograph	粉末 X 射线照相
powdery	粉状的；粉末化了的
powdery emulsion paint	粉末状乳胶漆；乳胶粉末涂料
powellamine	鲍威胺
powellane	鲍威烷
powelline	鲍威灵
powellite	钼钨钙矿
powellizing	浸硬
power	①率②功率③动力④(乘)幂〔数〕；方
power amplifier	功率放大器
power belt	传动带
power booster fuel	功率提高燃料〔含提高发动机功率添加剂的燃料〕
power broadening	功率展宽
power cable	动力电缆
power cleaning	①(=blast cleaning)喷砂清理；喷砂除锈②动力(机械)清理
power-compensation differential scanning calorimetry	功率补偿型差示扫描量热法
power consumption	功率消耗；动力消耗
power control circuit	功率控制电路
power current	动力电(流)
power density	功率密度
power distillate	动力馏分；未精制的拖拉机煤油〔介于煤油与粗柴油之间的馏分〕
power-driven pump	机械带动泵；马达泵
power efficiency	功率系数；利用有效系数
power equation	功率方程；幂方程
power equipment	动力设备

power factor   功率因数
power feed   自动进料；机械进料
power fluid(=power liquid)   动力液体
powerforming   强化重整
power formula   功率式；乘方公式
power fraction of gasoline   汽油动力馏分
power fuel   动力燃料
powerful animal   强烈动物(香料)香〔指麝香、灵猫香等〕
powerful fruity odor   强烈果香
powerful microscope   高倍显微镜
power function   ①幂函数②功效函数
power fuse   电力保险丝
power gas   动力煤气
power house   ①动力厂②发电厂
power insulator   强流绝缘体
power kerosene   动力煤油
power law   幂律；幂函数式
power law equation   幂律方程
power law fluid power law fluid model   幂定律流体模型
power law index   幂律指数
power law model   幂律模型
power liquid(=power fluid)   动力液体
power loader   机器装载器
power loss   功率损失
power mains   输电线
power method   乘幂法
power model   幂模型
power of absorption   吸收本领
power of test   检测(能)力
power pack   电源部分
power panel   电力闸板
power plant   ①动力厂②发电厂
power press   强压机
power pump   动力泵
power rectifier   电力整流器
power series   幂级数
power source   电源
power spectrum   功率谱
powerstat   调压变压器
power stretch fabric   弹力织物；高弹织物
power supply   电源
power surge   动力高峰
power transformer   电力变压器
power transmission   ①传动②电力输送
power-transmission fluid   动力传递液
power transmission line   电力输送线；递能线
power truck   电动车
power unit   动力单位
power utilization index(PUI)   功率利用指数
power wheel   动力轮
pox   ①水痘②(=small-pox)天花
poxvirus   痘病病毒
Poynting effect   坡印廷效应
Poynting-Thomson body(=anelastic body)   坡印廷-汤姆森体
Poynting-Thomson solid(=anelastic solid)   坡印廷-汤姆森固体
Po-Yoak oil(=Po-Yok oil)   坡唷油
Po-Yok oil(=Po-Yoak oil)   坡唷油〔取自西非的 *Afrolicania elaeosperma* 和 *Parinarium sherbroense*〕
pozzolana   火山灰
pozzolanic action   凝硬作用
pozzuolana   白榴火山灰
PPI rheometer(=pressure products industries rheometer)   高压产品工业流变仪
practical   实用的；应用的
practical chemistry   实用化学
practical column temperature   实用柱温
practical formulation   实用配方；生产配方
practical spreading rate   实际涂布面积
practical unit   实用单位
practical viscosity measurement   实际黏度测定
practice   ①实验②实践③实用
practomycin   实霉素
praequine   帕马奎宁〔药〕
Prager-Jacobson classification   普腊格-雅各布逊分类法
pragilbert   瓦特与磁流强度比例
pragmoline   溴化乙酰胆碱
prairie-god plant   (二等设备的)小规模炼油厂
pralina   6-甲基香豆素
Prandtl body(=elasticoplastic body)   普兰托体
Prandtl number(Pr)   普兰托数〔无因次数群，等于运动黏度×密度×比热/热导率〕
Prandtl's universal law of friction   普兰托一般摩擦定律
prangenin   黄原醇丁醚
prase   ①葱绿色②葱绿玉髓
praseocobaltichloride   一氯化二氯四氨络钴 $[Co(NH_3)_4Cl_2]Cl$
praseodymia   三氧化二镨；氧化镨 Pr_2O_3
praseodymium   镨〔59号元素，化学符号Pr〕
praseodymium bromate   溴酸镨 $Pr(BrO_3)_3$
praseodymium bromide   溴化镨 $PrBr_3$
praseodymium carbonate   碳酸镨 $Pr_2(CO_3)_3$
praseodymium chloride   氯化镨 $PrCl_3$
praseodymium disulfide   二硫化镨 PrS_2
praseodymium fluoride   氟化镨 PrF_3

praseodymium hydride 氢化镨 PrH_3
praseodymium hydrosulfate 硫酸氢镨 $Pr(HSO_4)_3$
praseodymium hydroxide 氢氧化镨 $Pr(OH)_3$
praseodymium iodide 碘化镨 PrI_3
praseodymium molybdate 钼酸镨 $Pr_2(MoO_4)_3$
praseodymium nitrate 硝酸镨 $Pr(NO_3)_3$
praseodymium oxalate 草酸镨 $Pr_2(C_2O_4)_3$
praseodymium oxychloride 氯氧化镨 PrOCl
praseodymium peroxide 过三氧化二镨 $Pr_2O[O_2]$
praseodymium phosphate 磷酸镨 $PrPO_4$
praseodymium sesquioxide 三氧化二镨；氧化镨 Pr_2O_3
praseodymium sulfate 硫酸镨 $Pr_2(SO_4)_3$
praseodymium sulfide 硫化镨 Pr_2S_3
praseolite 堇云石；绿石英
prasinomycin 葱绿霉素
prasterone 普拉睾酮〔药〕
Pratt process 普来特(气相裂化)过程
Pratt unit 普来特(气相裂化)系统装置
pravastatin 普伐他汀〔药〕
Prayon continuous filter 泼锐翁连续过滤机
praziquantel 吡喹酮〔药〕
prazosin 哌唑嗪〔药〕
pre- 〔拉丁字头〕预；前；先；在上
preabsorption 预吸收
pre-accelerated polyester 预催化固化聚酯
preadduct 预加成物
preadsorption 预吸附
pre-aging 预停放
pre-amp(=preamplifier) 前置放大器
preamplification 前置放大作用
preamplifier(=pre-amp) 前置放大器
preannihilative emission 预湮灭发射
prearcing or presparking effect （电弧或电火化)预燃效应
pre-assembled 预装配的；预组装的
preblend 预混
pre-blowing technique 预吹胀法
pre-boarder 预定形机
preboarding 预定形；织袜纤维热压处理
prebodying 预热炼
prebreaker 预破碎机
prebrightening 预增白
pre-burn 空白灼烧；灼烧恒重
preburning time 预燃时间
precalciferol 前钙化醇
precalcined 初步煅烧的
precalibration 预校正
precancerous 能发展成癌的；尚未成癌的
precarcinogen 前致癌物

precast concrete 预制混凝土
precast hole 预铸孔
precautions and procedures 注意事项及步骤
preceding reaction 前行反应
preceramic polymer 陶瓷前驱聚合物
precess 旋进
precession 旋进；进动
precessional orbit 旋进轨道
precession camera 旋进相机
precession photography 旋进照相法
pre-chill 预冷却；预冷冻
prechiller 预冷却器
prechlorination 预氯化
precholecalciferol 预胆钙化醇
precious 宝贵的；贵(重)的
precious garnet 贵榴石
precious metal 贵金属
precious stone 宝石
precipitability(=precipitation threshold) 沉淀度；沉淀性；临界沉淀点
precipitable 可沉淀的
precipitant 沉淀剂
precipitate ①沉淀(物)②沉(淀)出③使沉淀
precipitated 沉淀的；沉(淀)出的
precipitated alumina hydrate 沉淀矾土白
precipitated asphalt 沉淀的沥青
precipitated barium carbonate 沉淀碳酸钡
precipitated barium sulfate 沉淀硫酸钡〔填料〕
precipitated baryte 沉淀硫酸钡
precipitated basic dye blues ①磷钨酸碱性蓝色淀②磷钼酸碱性蓝色淀
precipitated basic dye violets ①磷钨酸碱性紫色淀②磷钼酸碱性紫色淀
precipitated bone 沉淀骨质
precipitated calcium carbonate 沉淀法碳酸钙
precipitated calcium superphosphate 沉淀过磷酸钙
precipitated catalyst 沉淀的催化剂
precipitated chalk 沉淀碳酸钙
precipitated magnesium carbonate 沉淀碳酸镁
precipitated phosphate(=precipitated superphosphate) 沉淀磷酸钙〔一种钙肥料〕
precipitated pigment 沉淀颜料
precipitated resinate 沉淀树脂酸盐
precipitated silica 沉淀二氧化硅
precipitated superphosphate(=precipitated phosphate) 沉淀磷酸钙
precipitated vapor 淀析的蒸气
precipitated whiting 沉淀白垩

precipitate-impregnated membrane electrode　沉淀填嵌型膜电极
precipitating　沉淀的
precipitating agent　沉淀剂
precipitating bath　沉淀浴；凝固浴
precipitating inhibitor　沉淀抑制剂；阻沉剂
precipitating tank　沉淀槽
precipitation　沉淀(作用)；析出
precipitation acidity　降水酸度
precipitation analysis　沉淀分析(法)；沉淀滴定(法)
precipitation bath　沉淀浴；凝固浴
precipitation chromatography　沉淀色谱(法)
precipitation dewaxing process　沉淀脱蜡过程
precipitation-exchange resin　沉淀交换树脂
precipitation figure　(铬盐溶液的)浑浊度
precipitation form　沉淀形(式)
precipitation fractionation　沉淀分级*
precipitation from homogeneous solution　均匀沉淀*
precipitation gelation　沉淀凝胶化(作用)
precipitation gravimetry　沉淀重量(分析)法
precipitation hardening　沉淀硬化
precipitation heat　沉淀热
precipitation immunoassay　沉淀免疫分析
precipitation indicator　沉淀指示剂
precipitation method　沉淀法*
precipitation naphtha　沉淀石脑油〔测定润滑油沉淀值的汽油溶剂〕
precipitation number　沉淀值
precipitation-peptization　沉淀-胶溶(作用)
precipitation point value　沉淀点值
precipitation polymerization　沉淀聚合*
precipitation process　沉淀过程
precipitation reaction　沉淀反应
precipitation separation　沉淀分离(法)
precipitation tank　沉淀器
precipitate temperature　沉淀温度
precipitation test　沉淀试验
precipitation threshold(=precipitability)　沉淀度；沉淀性；临界沉淀点
precipitation titration　沉淀滴定(法)*
precipitation titration indicator　沉淀滴定指示剂
precipitation transformation　脱溶相变〔晶体〕
precipitation value　沉淀值
precipitation vat　沉淀瓮；澄清瓮
precipitation volume　沉淀体积
precipitator　沉淀器；聚尘器
precipitator tower　沉淀塔
precipitimetric indicator　沉淀滴定指示剂

precipitin　沉淀素
precipitinogen　沉淀素原
precipitin reaction　沉淀素反应
precipitometer　沉淀计
precise distillation　精密蒸馏
precise fractionation　精密分馏
precise mass determination　精密质量测定
precise package build　精密卷绕成形
precision　精密度*
precision balance　精密天平
precision burette　精密滴定管
precision chemical analysis　精密化学分析
precision colorimetry　精密比色分析法
precision fractional distillation　精密分馏
precision injection machine　精密注塑机
precision instrument　精密仪器
precision machining　精密机械加工
precision molding　精密模塑法；精密造型法
precision null-point potentiometry　精确零点电位法
precision of regression equation　回归方程的精密度
precision of spectrophotometry　分光光度法的精密度
precision photometer detector　精密光度计检测器
precision polymerization　精密聚合
precision sliding valve　精密滑动阀
precision thermometer　精密温度计
precision work　精加工
preclean　预清洗
pre-coagulation　早期凝固
pre-coagulum　早期凝块；早期凝固物
precoat　①预涂渍②预涂布③预涂层④底漆；打底子；上底
precoated aluminum　预涂铝材
precoated layer　预涂层
precoated material　预涂材料
precoated metal sheet　预涂金属板
precoated plate　预涂布(色谱)板
precoater　预涂机
precoat filter　预涂(复)层过滤机
precoating　①底漆；上底〔油漆〕②预涂渍
precoat metal plate　预涂金属板
precoat-type filter　预涂型过滤器
precocious　①早熟的②早慧的；发育过早的
precocity　早熟(性)；早发育
precocity of accelerator　促进剂的早期作用
precoking　预焦化
precollagen　前胶原
precolored bead　预着色料珠
precolored granule　预着色粒料

pre-column ①前置柱*②柱前
pre-column chiral derivatization 柱前手性衍生
pre-column derivatization 柱前衍生
pre-column flow control 柱前流量控制
pre-column-inlet splitter 柱前入口分流器
pre-column reaction 柱前反应
precombustion 预燃烧
precombustion chamber 预燃烧室
precombustion reaction 预燃烧反应
precommisioning test 投料前试车
precompounding 预混
precompressed air 预压缩的空气
precompression 预压缩
precomputed 预算的
pre-concentration 预浓缩；预富集
preconcentrator 预浓缩器
precondensate ①预缩合②预缩合物
precondensation 预缩合
precondensation vessel 预缩合釜
precondensator ①预缩合器②预冷凝器
precondenser 预冷凝器
precondition ①预老化②预稳定③预处理
preconditioner 预调节器
preconditioning ①预老化②预稳定③预处理
precontamination 初期污染
precool 预冷
precooler 预冷却器
precooling 预冷却
precoverage 预覆盖度
precritical 临界前的；亚临界的
precrushing 预碎
precure ①预硫化；预固化②预熟化
precuring ①预硫化；预固化②预熟化
precursor 母体*；前身；产物母体
precursor ion 先驱离子；母离子
precursor nuclide 前驱核素
precursor polymer 前体聚合物
precursor-product relation 前身-产物关系；母体-子体关系
precut 预切割
precut column 预切割柱
precut device 预切割装置
pre-deaerator 预脱泡器；预除气器
predefecation 预澄清；初步澄清〔糖〕
predefecation juice 预澄清汁；预灰汁
pre-defined physical property 预定物理性能
pre-degasser 预脱气塔；预脱气器
predetermination 预定性

pre-determined characteristics 预定性能
predetermined level 预定高度
predetermined nucleation 预成核作用
predetermined pressure 预定压力
prediagnostic 预先诊断
predicted response 预报响应值
prediction interval 预期区间；预测区间
prediction set 预示集
prediction value 预报值
predictive coulometry 预测库仑法
predictive residual error sum of squares 预期残差平方和
prediluted 预稀释的
prediluted oil 预先稀释的(润滑)油
pre-dip 预浸
predispersed pigment preparations 预分散颜料制剂
pre-dispersed sulfur(=coated sulfur) 包得硫黄；预分散硫黄
predisperser 前置分光器；光谱分级器
predispersion device 前置色散装置
predissociation 预离解*
pre-dissolving tank 预溶解罐(槽)
predistillation 初步蒸馏
prednisolone 脱氢皮(质甾)醇；氢化泼尼松；泼尼松龙
prednisone 强的松；脱氢可的松；泼尼松
predominance-region diagram 优势区域图
predominant colour 主色
predominant hue 主色相
predominant quantity 优势成分(重)量；主要成分的重量
predominant wavelength 主波长
predose 预剂量；初始剂量；本底剂量
predry 预干
predryer 预烘机；预干机
predrying 预干燥；预烘燥
pre-dyed 预染色的
preelectrolysis method 前电解法
pre-electrophoresis 预电泳
preembedded 预埋的；预嵌入的
pre-emulsified 预乳化的
preemulsion 预制乳状液〔未经匀化的粗乳状液〕
pre-enzyme 前体酶；酶原
pre-equilibrated 预平衡过的
pre-equilibration 预平衡；前置平衡
pre-equilibration column 前置平衡柱
pre-evaporation 初步蒸发
pre-evaporator 初步蒸发器
pre-evolution test 简化初步评价法
preexisting mineral 先成矿物
pre-expanded bead 预发颗粒*

pre-expansion 预发泡
preexponential factor 指数前因子*
pre-exposure 预曝光
preextraction 预提取；预浸出
pre-extraction step 萃取操作前的步骤；预萃取步骤
prefab ①预制的②预制品；预制件
prefab primer 预涂底漆
prefabricated paste 预制色浆
prefabrication 预制；预加工
preference 宁择；特选
preferential adsorption 选择性吸附
preferential combustion 优先燃烧(作用)
preferential evaporation 选择性蒸发；优先蒸发
preferential oxidation of hydrogen 氢的优先氧化
preferential selectivity(=preterential solubility) 有择溶解度
preferential solubility(=preferential selectivity) 有择溶解度
preferential solvent 有(选)择(的)溶剂
preferential spectrophotometry 优先分光光度法
preferential wetting 优先湿润
preferment ①酶原；前酶②提升；升级
preferred conformation 优化的构象
preferred numbers 优先数系
preferred orientation 择优取向(定位)；优先配位
prefilling press 成形预压机
prefilter 预滤机；前滤器
prefinish 预涂
prefinished metal plate 预涂金属板
prefinished plywood 预饰面胶合板
prefinisher 预缩聚器；前缩聚器
prefinishing 预整理；前整理
prefix 字头；词头
preflame oxidation 预燃氧化
preflame reaction 预燃反应
preflash 预闪蒸
prefloc 预絮凝粒
preflooding 预液泛
pre-fog 先期灰化
preform ①压片②雏形；塑坯预塑
pre-formed adduct 预制加合物
preformed catalyst 成型催化剂
preformed gum 燃用树脂；烧火的树脂
preformed packing (分馏装置的)可调节的填充物
preformed polymer 预聚物
preformed thin layer 预涂薄层
pre-former 预成形机
preforming 压片

preforming machine 压片机
preform molding 塑坯模制法
preformylation 预甲酰化
prefractionation 初步分馏
prefractionator 初步分馏塔
pre-frothing 预发泡
pre-fusion 预熔融
pregeijeren 前吉烯
pregeijerene 前吉(枝)烯
pre-gel ①预凝胶②预胶凝；预成胶③预胶化层〔塑料〕
pregnandiol 孕(甾)二醇
pregnane 孕(甾)烷
pregnanediol 孕(甾)二醇
pregnanolone 孕(甾)烷醇酮
pregnant organic phase 萃取了溶质的有机相；富金属有机相；含有金属的有机相；饱和有机相
pregnene 孕(甾)烯
pregneninolone 17-乙炔睾(甾)酮
pregnenolone 孕(甾)烯醇酮
pregraphitic structure 前石墨化结构
pregrattite 西奥白云母
pregrounding 粗压碎
pregwood 浸胶木材
preheat 预热
preheated injector 预热的注射器
preheater 预热器
preheater furnace 预热炉
preheating 预热
preheating by dielectric losses(=HF heating) 高频预热
preheating chamber 预热室
preheating device 预热装置
preheating evaporator 预热蒸发器
preheating of air 空气的预热
preheating of fuel 燃料的预热
preheating of fuel oil 燃料油的预热
preheating section 预热段
preheating section of kiln 窑的预热段
preheating temperature 预热温度
preheating time 预热时间
preheating tube 预热管
preheating zone 预热区
prehnite 葡萄石
prehnitene 连四甲苯；1,2,3,4-四甲基苯 $(CH_3)_4C_6H_2$
prehnite rock 葡萄石岩
prehnitic acid 连苯四酸；1,2,3,4-苯四甲酸 $C_6H_2(COOH)_4$
prehnitilic acid 连三甲基苯甲酸；2,3,4-三甲基-1-苯甲酸 $(CH_3)_3C_6H_2COOH$

prehnitylic acid 2,3,4-三甲基苯甲酸 $(CH_3)_3C_6H_2COOH$
prehydrolysis 预水解
preignition 预燃(作用)
preignition chamber 预燃室
preignition period 预燃期
preimpregnate (混合成分)保混剂
preimpregnated bonding sheer 预浸黏结片
pre-impregnated fabric 预浸渍织物
pre-impregnated mat 预浸渍毡
preimpregnated reinforced plastic 预浸增强塑料
pre-impregnated roving 预浸渍无捻粗纱
preimpregnation 预浸
prein 压平布面
preinitiation complex 起始前复合物
pre-irradiation ①辐照前；照射前②预辐照法
pre-irradiation treatment 辐照前处理
prelimer 预加灰器；预澄清器
preliminary ①初步的②初级的③预备的
preliminary aging 预先老化〔橡〕
preliminary assay 预测定；预试验
preliminary breaker 预先压碎机
preliminary breaking 预先压碎
preliminary calibration curve 预备校正曲线
preliminary crusher 预先压碎机；初级压碎机
preliminary crushing 预先压碎
preliminary dip 预浸
preliminary distillation (石油的)初步蒸馏
preliminary electrolysis 预电解
preliminary examination 初步检定；初步试验
preliminary filter 初步过滤器
preliminary finish 预整理；前处理
preliminary fuel filter 燃料粗过滤
preliminary heating 预热
preliminary heat treatment in oxidizing 初步氧化热处理
preliminary orientation 初步取向
preliminary pre-stressing 初次预加应力
preliminary roasting 初步煅烧
preliminary roller 前辊
preliminary sizing 粗筛选
preliminary step ①初步②预备步骤
preliminary study 初步研究
preliminary test 初步试验
preliminary treatment 初步处理；预处理
preliminary washing 预洗；初洗
preliming 预加灰（糖）
preliming tank 预灰槽
pre-load ①预载；预加负荷②最初轴向负荷〔法兰垫〕
Prelog's rule 普雷洛格规则*

premary amyl alcohol 伯戊醇
premasticated rubber 塑炼胶；预塑炼胶
premastication 预先捏和作用
premature ageing 前期老化；早期老化
premature brittle fracture 早期脆化破坏
premature catalysis 早期催化(作用)
premature coagulation 早期凝固
premature condensation 早期缩合
premature cure 早期固化；早期硬化
premature curing 早期固化
premature expansion 早期发泡
premature failure of paint 漆膜的过早破损(失效)
premature failure of the coating 涂层的早期破坏(失效)
premature firing 过早点火
premature gelation 早期胶凝
premature oxidation 过早氧化
premature polymerization 早期聚合
premature setting 过早固化
premature sintering 早期烧结
premature vulcanization(=bin cure) (橡胶)早期硫化；仓储硫化
premelt 预熔化
premesophase pitch 预中间相沥青
premetabolite 前代谢物
pre-metallised dye 预金属络合的染料
premicellar association 胶束化前缔合
premicellar region 胶束化前(浓度)区
premier ①最前的；首要的②最早的
premier alloy 镍铁铬锰合金
Premier colloid mill 普雷迈尔胶体磨
premier jus 牛肾外围组织油
Premier mill 普雷迈尔磨
premium engine oil(=premium motor oil) 高级车用汽油
premium grade 优等品
premium grade gasoline(=premium motor fuel) 高级汽油
premium motor fuel(=premium grade gasoline) 高级汽油
premium motor oil(=premium engine oil) 高级车用汽油
premium-priced fuel 高价燃料
premium properties 优良性能
premium-type paint 高质量油漆
premix 预混料
premix burner 预混合型燃烧器
premixed burner 预混燃烧器
premixed flame 预混焰
premixer 预混合器
premixing 预混合
premnine 腐婢碱
premolding 预压型

premoulding 预先铸模
prenderol(=2,2-diethyl-1,3-propanediol) 甫任德醇〔俗〕；2,2-二乙基-1,3-丙二醇
pre-neutralization 预中和
prenitic acid 苯四酸 $C_6H_2(COOH)_4$
prenitol 连四甲苯
prenol 异戊烯醇
prenophorin(=pyrenophorin) 核球壳菌素
prenyl 含异戊(间)二烯基的
prenyl acetate 乙酸异戊烯酯
preoiler 预先加油器
pre-oiling 预充油；预加油
preoperational test 试运转；预运转试验
pre-op test 运行前试验
preorientation 预取向
pre-orientation degree 预取向度
preoxidized 预先氧化的
prepacked 预先装入的
prepacked column 预填充柱
prepacked support 预填充载体
preparation ①制备②制剂；制品③调制；调合〔香精生产中〕
preparation facility ①制备装置②调制设备
preparation jar 陈列样品玻缸
preparation of analytical sample 分析样品制备
preparation of greases 润滑脂的制备
preparation of indicator solution 指示剂溶液配制
preparation of sample 样品制备
preparation period 准备期
preparation room ①预备室②制备室
preparation scale 制备级(的)
preparation vessel 制备槽；调制槽
preparative chromatograph 制备色谱仪
preparative chromatography 制备色谱(法)
preparative column 制备柱
preparative column electrophoresis 制备柱电泳法
preparative countercurrent chromatography 制备反流色谱法
preparative electrophoresis 制备电泳
preparative gas chromatograph 制备气相色谱仪
preparative gas chromatography 制备气相色谱法
preparative layer 制层
preparative layer chromatography 制层色谱(法)
preparative liquid chromatograph 制备液相色谱仪*
preparative partition chromatography 制备分配色谱(法)
preparative photosynthesis 制备性光化合成
preparative plate number 制备塔板数
preparative radiation chemistry 制备辐射化学

preparative scale plate number 制备级塔板数
preparative scale sample 制备级试样
preparative thin layer chromatography 制备性薄层色谱法
preparative zone electrophoresis 制备区带电泳(法)
preparator 选矿机；精选机
prepared alkyd paint 醇酸调和漆
prepared chalk 研细白垩；沉降白垩
prepared degras 植物油和动物油的混合物
prepared lard 精制猪油
prepared lithopone-zinc white paint 锌钡白-锌白调和漆
prepared paint 调和漆
prepared suet 精制羊油
prepared tar 精制焦油
preparing agent(=pre-processing agent) 预加工剂；预处理剂
preparing gill 理条针梳
prepeak 前峰*
prepeak drift rate 前峰漂移速率
prephenic acid 预苯酸
pre-phosphated sheet 磷化预处理的(金属)板
prepilot plant 雏形工厂；台式工厂
preplasticator 预塑化器
preplasticizer 预增塑剂
preplasticizing 预塑化
preplastificator 预塑化器
pre-plodder 预压条机
prepolyamide salt 聚酰胺原盐
prepolycondensate 预缩聚物
prepolymer 预聚物
prepolymer gel 预聚物凝胶
prepolymerization 预聚合
pre-polymerizer 预聚合器；前聚合器
pre-preg ①含有化学增稠剂的模(型)垫②预浸渍
prepreg cured thickness 预浸材料固化厚度
prepreg mat 预浸(玻璃纤维)网片
prepreg mo(u)lding 预浸成形(法)
prepressing 预压榨
preprocess 预处理；预(先)加工
preprocessing aid 预操作助剂
preproduction 试制；试生产；小批(量)生产
pre-production check 生产前检验
pre-production trial 试生产
prepurification 预纯化(作用)
prereacted material 预反应物料
prereacted-urethane coating 预反应型聚氨酯涂料〔单组分聚氨酯涂料〕
pre-reduction 预(先)还原

prerefining 预先精炼
prerequisite 先决条件；必要条件
preresonance Raman spectroscopy 预共振拉曼光谱法
prerigidized yarn inserting method 刚性棒插入法
pre-ripening 预熟成
preroasting 初步煅烧
prerotation 预旋
prerupture flow 破坏前流动
prerupture response 破坏前响应
pre-saponification 预皂化(作用)
presaturation column 预饱和柱
presaturator 预饱和器
prescouring 预洗炼
prescription ①药方②处方；开方
prescription balance 药剂天平
pre-selected masses 预选质量数
presence of unsaturates 不饱和化合物的存在；不饱和烃的含量〔石油〕
present gum in gasoline 汽油中的显胶
preseparation column 预分离柱
preservation ①防腐(作用)②保存
preservation test (储藏)防腐性试验
preservative ①保存剂②防腐剂；杀菌剂③防老剂
preservative agent 保存剂
preservative coat 防腐层
preservative engine oil 防护机器油
preservative lubricating oil 防护润滑油
preservative oil 防护油
preservative treated plywood 防腐胶合板
preserved food 罐头食品
preserved latex 保存胶乳
preserved plywood 防腐胶合板
preserving ①保藏②做成罐头
preserving agent 防腐剂
preset ①预调②预置
preset control 程序控制；预先给定控制
preset count and time 预置时间与预置计数
pre-setting ①预定形②预调整
pre-settling tank 初步沉淀箱
preset value 预调值；初调值
preshrinking 预收缩
pre-shrunk 已预缩的；预收缩的
presintering 初步熔结
pre slit optics 外光路系统
presoak 预浸渍
presoil 预染污
prespark curve 预燃曲线
presparking period 预燃时间

press ①压机；压力机②压榨机③压④印刷机⑤印刷物⑥压床；冲床
pressability of wax distillate 石蜡馏出物的压榨性能
pressable 可压榨的
pressable wax 可压榨的石蜡
press atomization 加压喷雾
pressboard 压板
press button 按钮
press button control 按钮控制
press cake(=mud cake) 滤饼
press casting 压力铸造
press cloth 滤布
press cylinder 压榨筒
press dewaxing 压滤脱蜡(法)
press distillate(=pressure distillate) 加压馏出物
press drip 压滤机油滴
press drop 压(力下)降
press dumper 压滤机卸料装置
pressed 压榨
pressed active carbon 成型活性炭；定型颗粒活性炭
pressed barium sulfate 压制硫酸钡
pressed beet pulp(=pressed cossettes) 榨过的甜菜渣〔糖〕
pressed charge 压制燃料〔固体火箭燃料〕
pressed cossettes(=pressed beet pulp) 榨过的甜菜渣〔糖〕
pressed disc method 压片法
pressed disc technique 压片技术〔红外光谱〕
pressed distillate 石蜡馏分
pressed-fibre board 纤维板；壁板
pressed film 压膜
pressed finish 压光面饰
pressed fuel 压制煤块
pressed halide disk technique 卤化物压片技术
pressed oil 脱蜡油
pressed pellet membrane electrode 压片膜电极；混晶膜电极
pressed plutonium oxide(PPO) 压制过的氧化钚
pressed pulp 榨过的甜菜渣〔糖〕
presser 压榨器
presser filter 压滤机
press factor 挤压因数
press felt 压毡
press felt roll 压毡辊
press-felt stretcher 压毡延伸(纤维)机
press filter 压滤机
press fit 压配合
press-flow extrusion rheometer 压流型挤出流变仪
press heater 热压(硫化)锅
press house 压榨间

press-in 压入
pressing 挤压；加压
pressing damaged filament 压伤丝
pressing in 压进；压入；压合
pressing machine ①压力机②压(榨)机
pressing paper 粗面滤纸
pressing quality 压榨本领
pressing roll(er) 压辊
pressing speed 压合速度；合模速度
pressing tank 压榨槽
pressing time 压制时间
pression 压力
press juice ①压出汁②滤汁
press knife 冲刀
Pressley tester for fibre bundle strength 卜氏束纤维强力试验机
press lye 黑液；压榨碱液
press mo(u)ld 压模
press mud 压滤泥浆
press oil 压型油
pressostat 恒压器
press over system （鞣液）溢流法
presspahn 压制板
press part 压榨部
press plate ①压板②加热板③印刷(铅)板
press polish 压光
press pump 压榨泵
press ratio ①压缩度(比)；挤压率②压(榨)出力
press roll 压辊
press roll lever 压辊杠杆
press roll weight 压辊砝码
presstite 普列斯塑料
presstite joint 普列斯塑料接头〔用"普列斯塑料"塑胶封闭的管子接头〕
pressure 压力；压强
pressure above bubble point 高于饱和点压力
pressure-actuated 压力起动的
pressure aging 加压老化
pressure alarm system 压力警报信号系统
pressure-and-vacuum release valve(=pressure-and-vacuum vent valve) （油罐的）呼吸阀
pressure-and-vacuum vent valve(=pressure-and-vacuum release valve) （油罐的）呼吸阀
pressure atomization 加压喷雾
pressure at right angles 垂直压力
pressure bag 压力包；压力袋
pressure-bag mo(u)lding 加压袋压法成型
pressure balance 压力平衡

pressure balancing 压力均衡法
pressure bell 压力钟；调压钟
pressure blower 压力吹气机；压力鼓风器
pressure boost 增加压力
pressure booster 增压器
pressure bottle 耐压瓶
pressure bottoms 蒸馏釜残渣
pressure-break （层压塑料内的）压裂缝
pressure broadening 压力展宽；压致(谱线)增宽
pressure bubble-plug assist vacuum forming 气胀模塞助压真空成型
pressure bubble vacuum snap-back forming 气胀阳模真空反吸成型
pressure circulation system 压力循环系统
pressure coat 压力沉附物〔在蒸馏器或容器壁上沉积的石油产品〕
pressure coefficient 压力系数
pressure coefficient of viscosity 黏度的压力系数
pressure combustion 加压燃烧
pressure compensation 压力补偿
pressure container 喷涂压料罐
pressure control equipment 压力控制设备
pressure controller 压力控制器
pressure control valve 压力控制阀
pressure conversion factor 压力转换因子
pressure conversion table 压力转换表
pressure cooker 加压蒸煮器
pressure cooling 加压冷却
pressure correction factor 压力校正因子
pressure cracking process 加压裂化过程
pressure curtain coating 压力幕涂法
pressure curve 压力曲线
pressured column 加压柱
pressure dependence parameter 压力相关参数
pressure difference 压(力)差
pressure differential 压(力)差
pressure diffusion 加压扩散
pressure displacement cell 汞极排压电解槽
pressure distillate(=press distillate) 加压馏出物
pressure distillate bottoms 加压蒸馏残渣；裂化汽油再蒸馏残渣
pressure distillation 加压蒸馏
pressure distribution 压力分配
pressure drag 压阻；压致曳力〔物〕
pressure drop 压(力)降
pressure drop coefficient 压力损失系数
pressure dwell 加压期间
pressure effect 压力效应

pressure efficiency　压力效率
pressure energy　压能；压头
pressure-enthalpy chart　压焓图
pressure equalization　压力均匀化；均压化
pressure equalizer　均压器
pressure evaporation　分压蒸馏法
pressure expulsion system　加压卸料系统
pressure extraction　加压萃取；压力萃取
pressure face　压力面；管压面
pressure factor　压力系数
pressure-feed　压力送料；加压装料
pressure-feed filler　压力送料设备
pressure-feed gasoline　压力输送汽油
pressure-feed gun　压料式喷枪
pressure-feed lubrication　压力送料润滑
pressure-feed tank　压力送料槽
pressure figure　压像
pressure filter　压滤器
pressure-filter-bulb method　球压滤法
pressure-filter-tube method　管压滤法
pressure filtration　加压过滤
pressure flash tower　加压闪蒸塔
pressure flask　耐压烧瓶
pressure float　压力浮标
pressure flow　有压流；回流；逆流
pressure fluid　施压流体
pressure fractionation　加压分馏
pressure fuel　裂化重油
pressure fuel system　压力输送燃料系统
pressure gas　压缩气体
pressure gasoline　裂化汽油
pressure gas plant　稳压煤气工厂
pressure ga(u)ge　压力表
pressure gauge thermometer　示压温度计
pressure governor　节压器
pressure gradient　压力梯度
pressure gradient correction factor　压力梯度校正因子
pressure grease cup　压力输送润滑脂的润滑器
pressure gun　黄油枪
pressure gun grease　黄油枪用润滑脂
pressure head　压头
pressure hose　压力胶管；耐压胶管
pressure indicating controller　压力指示控制器
pressure indicator　气压表
pressure intensity　压力强度
pressure jump　压力跃变*
pressure lifting　压力升高
pressure line　耐压管线

pressure loss　压力损失
pressure loss conversion chart　压力损失换算表
pressure lubricating system　加压润滑系统
pressure lubrication　加压润滑
pressure lubricator　加压润滑器
pressure maintenance process　油层压力保持法
pressure manometer　差示压力计
pressure mark　压印
pressure marking　①压痕②光泽斑驳；明暗失调〔漆病〕
pressure measurement　压力测量
pressure melter type spinning machine　压熔式纺丝机
pressure meter　压力计
pressure modulus　压力模数
pressure molding　加压模塑
pressure mottling(=pressure marking)　光泽斑驳；明暗失调〔漆病〕
pressure mottling resistance　抗压痕性
pressure moulding　加压模塑
pressure naphtha　裂化石脑油
pressure nozzle　压力喷嘴
pressure nozzle atomization　加压喷雾
pressure oil　压送油
pressure oil pipe　压送油管
pressure oil pump　压油泵
pressure opportunity　加压时机
pressure packing　加压填充
pressure packing method　加压填充法
pressure pad　传压垫
pressure pan　高压罐；高压釜
pressure pipe　耐压管
pressure piping　耐压管
pressure polish　加压磨光
pressure polymerization reactor　高压聚合反应器
pressure programmed gas chromatography　程序变压气相色谱(法)
pressure programmer　程序变压(力)器
pressure programming　程序变压(力洗脱)*
pressure proof inspection　加压检查法
pressure propellant　①加压推进器②加压推进剂
pressure pulling　加压提拉
pressure pump　压气泵
pressure range　压力(变化)范围
pressure rating　压力等级
pressure ratio　压力比；压缩比
pressure recorder　压力记录器
pressure recording controller　压力记录控制器
pressure reducer　减压器
pressure reducing station　减压站

pressure reducing transformer 减压(变压)器
pressure reducing valve 减压阀
pressure reduction 减压
pressure reforming 加压重整
pressure register 记压器
pressure regulating valve 调压阀
pressure regulator 压力调节器
pressure regulator valve 压力调节阀
pressure release valve 放压阀；卸压阀
pressure relief 解除压力
pressure relief valve 卸压阀；安全阀
pressure relieving system 卸压系统
pressure rendering 压力炼油
pressure reservoir 储压器
pressure resistance 加压阻力
pressure roll coater(=gravure coater) 压力辊涂机；触压辊涂机
pressure rubber tubing 耐压橡皮管
pressure sand filter 加压砂滤器
pressure saponification 加压皂化
pressure schedule 压力表
pressure seal 加压密封
pressure-sensitive adhesion 压敏黏合
pressure sensitive adhesive 压敏型黏合剂；压敏胶黏剂
pressure-sensitive adhesive tape 压敏胶黏带
pressure-sensitive cement 压敏黏合剂
pressure sensitive color 压敏色素
pressure-sensitive film 压敏(性)薄膜
pressure-sensitive polymer 压敏聚合物
pressure-sensitive tape 压敏胶黏黏带
pressure sensor 压力传感器
pressure side 加压侧
pressure sintering 加压烧结
pressure still 裂化炉
pressure-still distillate 裂化炉馏出物
pressure-still tar 裂化炉焦油
pressure storage tank 高压储罐
pressure surge 压力
pressure switch 压力开关
pressure tank 压力槽
pressure tap 压力计接口
pressure tar 裂化焦油
pressure technique 高压技术
pressure-temperature coefficient 压力-温度系数
pressure test 压力试验；试压
pressure tester 压力试验器
pressure testing with product 产品压力试验〔石油〕
pressure texture 压缩结构

pressure thermit welding 加压热剂焊
pressure-tight 密闭的；受压不漏气的
pressure tight lock system 压力封闭系统
pressure tightness 气密性；不渗透性
pressure trapping 人工增压
pressure treader 压力上胎面机
pressure tubing 耐压(橡皮)管
pressure-type thermometer 压差式温度计
pressure unit 压力单位
pressure unloading 压力降低
pressure vacuum gauge 真空压力计
pressure vacuum vent valve 真空呼吸阀
pressure valve 压力阀
pressure vapour lamp 煤油蒸气灯
pressure vent valves 通风阀；呼吸阀
pressure vessel 高压容器；压力容器
pressure volume diagram 压力-体积图表
pressure-volume-temperature 压力-体积-温度
pressure-volume-temperature relations 压力容积温度关系
pressure-volume-temperature superposition 压力-体积-温度叠加
pressure warning unit 压力计
pressure washing 压洗
pressure water ①压力水；加压水②压出汁〔糖〕
pressure wave 压力波
pressure weight gallon cup 压力重量加仑杯
pressure welding 压焊
pressure zone 压力区
pressurization 高压密封法；加压法
pressurized air 压缩空气
pressurized aqueous combustion(PAC) 压水燃烧法；湿空气氧化法
pressurized container ①加压容器；压力容器②压缩空气箱
pressurized crystallization apparatus 加压结晶装置
pressurized depolymerization tower 压力解聚塔
pressurized sample 加压试样
pressurized still 受压蒸馏釜
pressurized system 受压系统
pressurized water nuclear reactor 压水型原子反应器
pressurized water reactor(PWR) (加)压水(反应)堆
presssre reducer 减压器
press vulcanization 加压硫化
press vulcanizer 平压硫化机
press water ①压力水；加压水②压出汁〔糖〕
press welding 压焊
prestone 普列斯通〔低凝固点液体乙二醇系防冻剂〕

Prestone cooling 普列斯通冷却
prestrain 预应变
prestress 预应力
prestressed pull bar method 预应力拉杆法
prestretched film 预拉伸薄膜
pre-stretching 预伸张
pre-stripper linac 剥离器前级直线加速器
presulfiding reaction 预黄化反应
presumptive test 推定试验；假定试验
preswollen 预溶胀的；预膨胀的
pre-swollen material 预溶胀材料
pretanning 预鞣
pretanning agent 预鞣剂
pretested packing 预制填充物
prethickening 预浓缩；预增稠
pretreated filler 预处理填料
pretreating agent 预处理剂
pretreating finish ①预处理②预处理剂
pretreatment 预处理；前处理
pretreatment of paper 滤纸预处理；滤纸前处理
pretreatment primer(=wash primer) 前处理底漆；磷化底漆
pretreatment streaks 预处理条痕状污迹
pretreatment unit 预处理装置
pretzel 双圈饼干
prevalution test 石油产品的模型试验
preven(ta)tive maintenance 预防性检修
prevention ①防止②预防
preventive ①预防的②预防药；预防剂
preventive coating 防护涂层
preventive maintenance 预防性保养
preventive maintenance measure 预防性保养检测；维护保养措施
preventive solution 预防溶液
preventol 防腐粉
previously vaporized charge 预先汽化进料
previously varnished 预涂(清)漆的
prevulcanization 预(先)硫化(作用)；过早硫化
prevulcanized 预硫化的；预硬化的
pre-warming 预热
prewashing 预洗
prewashing column 预洗涤塔
prewash phase 预洗期
prewash tank 预洗槽；预洗桶
prewave 前波*
prewood 浸脂(胶)木材
pribramite ①波西米亚闪锌矿②针铁矿；镉闪锌矿
price 价(格)；价值

price-density profile (产品)价值密度分布；适用性密度分布
priceite 白硼钙石
pricker ①通针〔商〕②刺
prickly pear-seed oil 棘梨子油
prill 金属小球
prilling granulator 造粒塔；造粒装置
prilling tower 造粒塔
primaquine 伯氨喹〔药〕
primaquine phosphate 磷酸伯氨喹〔药〕
primary ①第一的；(最)初的；原(来)的②初级的③伯的；连上一个碳原子的④原线圈；原线蟠〔物〕⑤一代的〔无机盐〕
primary aberration 第一级像差
primary acid 伯酸
primary activity (催化剂)初期活性；基本活性
primary adsorption 第一类吸附；化学吸附
primary aggregate(=primary structure) 原结构；一次结构〔炭黑〕
primary air ①初级空气②一次空气；一次风
primary air-fuel ratio 初级空气与燃料量之比
primary alcohol 伯醇
primary aliphatic amine 伯脂肪胺
primary alkylammonium derivative 伯烷基铵衍生物
primary alkyl peroxide 伯烷基过氧化物
primary amine 伯胺
primary amine value 伯胺值
primary ammonium 伯铵
primary ammonium phosphate 磷酸二氢铵
primary aromatic amine 芳香族伯胺
primary arsenate 一代砷酸盐 MH_2AsO_4
primary bath 主浴；第一浴
primary battery 原电池*
primary bond ①主价键②主价键合
primary bonding force ①主价键(合)力②伯级键合力
primary calcium phosphate 一代磷酸钙 $Ca(H_2PO_4)_2$
primary carbon 伯碳(原子)
primary carbon atom 伯碳原子
primary carbon black particle 原结构炭黑粒子
primary cell 原电池；初级(反应)电池；次电池
primary cellulose acetate 初级醋酸纤维素
primary characteristic 原有特性
primary circuit 初级电路；原电路
primary cleaner 初级清洗剂
primary coil 原线圈
primary color 原色
primary colorant 主着色剂
primary combustion 初级燃烧

primary combustion zone　初级燃烧区
primary coolant　一次冷却剂；一次制冷剂
primary coulometric analysis　直接电量分析；定电位电量分析
primary coulometric titration　初级库仑滴定；第一类库仑滴定
primary coulometry　初级库仑法
primary cracking　初级裂化
primary creep　初级蠕变
primary crusher　粗碎机；初碎机
primary crystal　初凝晶
primary crystallinity　主结晶度
primary crystallization　主结晶
primary degradation　初级降解(作用)
primary dispersion　①主色散区②初级分散
primary distillation　初级蒸馏
primary driers　主催干剂
primary electrode　原电极；基本电极
primary electrolysis　初级(反应)电解(作用)
primary element　原电池；初级(反应)电池
primary emission　一次发射；一次排放
primary excitation　初级激发
primary filter　初级过滤；基色滤色器
primary filtrate　初级过滤液
primary flash distillate　轻(的)石油馏分
primary fractionator　一级分馏塔；初(分)馏塔
primary gas　主气
primary gluing　一次胶合；初步胶合
primary heater　第一加热器；初加热器
primary high polymer　一次性高聚物
primary homothallism　初级同宗配合(现象)
primary hydroxyl　①伯羟基②伯醇
primary ion　主要离子
primary ion beam　原离子束；一次离子束
primary ionization　初始电离；一次电离
primary ionization current　初始电离电流
primary ion mass spectrometer　一次离子质谱仪
primary ion penetration　初级离子穿透深度
primary ion source　初级离子源
primary iron phosphate　一代磷酸铁；磷酸二氢铁
primary isotope effect　一级同位素效应*
primary linkage(=strong lingkage)　一级键合；强键
primary material　原料
primary nitro-compound　伯硝基化合物
primary nitroparaffin　伯硝基烷
primary normal stress difference　第一应力差
primary nucleus　原核
primary oil　初级油；原(煤)油

primary oxide　原氧化物
primary particle(=direct particle)　原始粒子
primary particulate　一次颗粒物
primary petroleum　初级石油
primary phosphate　一代磷酸盐
primary phosphine　伯䏲
primary plasticizer　主增塑剂
primary pollutant　首要污染物
primary precipitate　主要沉淀；起始沉淀
primary process　初级过程
primary product　①初(步)产物；初次产物②主要产物
primary propylene chlorohydrin(=2-chloro-1-propanol)　2-氯-1-丙醇　$CH_3CHClCH_2OH$
primary protein derivatives　初步衍生
primary proteose　初䏡
primary purification　初步精制
primary radiation　初级辐射
primary radical termination　初级自由基终止*
primary reaction　①初级反应②主要反应
primary reaction zone　初级反应区
primary reduction　初级还原
primary reference fuel(=primary standard fuel)　第一参比燃料；正标准燃料
primary reference material　一级标准物质
primary resin　原树脂
primary rocks〔复〕　原生岩类
primary salt　一代盐〔一个 H 被取代〕
primary salt effect　原盐效应
primary sample　初次试样；初次样品
primary sampling unit　初次抽样单位
primary seal　初级封液；初级封闭
primary-secondary alcohol　伯仲醇　$RCHOHCH_2OH$
primary sewage treatment　污水一次处理；污水一级处理
primary sludge　原始污泥；原污泥
primary solvent　一级溶剂
primary sorption　一次吸附
primary spectroscopic standard　原光谱标准
primary standard　一级标准*；基准物
primary standard fuel(=primary reference fuel)　第一参比燃料；正标准燃料
primary standard of wavelength　原始波长标准
primary standard substance　基准物(质)
primary state　始态
primary stress　一次应力
primary structure　原结构；一次结构
primary subtractive colo(u)r　减原色
primary sulfur(=elemental sulfur)　元素硫
primary tar　原焦油

primary-tertiary alcohol 伯叔醇 (OH)CR$_2$CH$_2$OH
primary tower 初级塔；初馏塔
primary training gasoline 初级教练用航空汽油
primary treatment 一次处理
primary truss 主桁架
primary unit cell 初级晶胞
primary valence(=primary valency) 主价
primary valence bond 主价键
primary valency(=primary valence) 主价
primary voltage 初级电压
primary vulcanization 第一次硫化；定型硫化
primary water 原生水
primary wave(=P-wave) 初波
primary white reference 第一参比标准白色；原始基准白色
primary X-ray 原级 X 射线
primary zinc phosphate 一次磷酸锌；磷酸二氢锌
prime ①第一的；原始的②最重要的③置火药备放④发火
prime cement 打底胶浆；底浆；初层
prime city naphtha(=prime-cut naphtha) 首城石脑油〔密度为 0.7093～0.6919 的石油溶剂〕
prime coat 底漆
prime-coating ①底层涂布②底层涂料；底漆
prime cost 成本
prime crude oil 上等原油
prime-cut naphtha(=prime city naphtha) 首城石脑油〔密度为 0.7093～0.6919 的石油溶剂〕
primed ①供准备的②灌引动水入唧筒
prime filter 初级过滤器
prime mover 原动机
prime pigment 主颜料；着色颜料〔具有着色力和遮盖力的颜料〕
primer 底漆*
primer absorption 底漆吸收
primer coating 涂底剂
primer curing oven 底漆固化炉
primer-detonator ①始爆剂；起爆剂②始发爆管；始发爆器
primer dip tank 底漆浸涂槽
primer extension 引物延伸(技术)
primer fluid 起动液
primer for bridge 桥梁底漆
primer for rusty surface 带锈底漆
primer jus 上等脂
primer knife 底漆刮刀
primer line 启动管路
primer oil 桐油底漆料；二道浆用熟桐油
primer pump 起动泵
primer sealer 封闭底漆〔兼具底漆和封闭剂两种功能〕
primer seat lubricating grease 垫表面的润滑脂
primer strand 元股
primer surfacer 底漆二道浆；头二道混合底漆
primer-topcoat coating 底面合一涂料
prime steam 湿蒸汽
prime steam lard 蒸制猪油
primeval isotope 原始(时代)的同位素
primeval lead 原始铅〔指同位素组成为原始组成〕
primeverin 樱草苷
primeverose 樱草糖
prime white kerosene(=prime white oil) 上等白色煤油
prime white oil(=prime white kerosene) 上等白色煤油
primidone 扑米酮〔药〕
primin 樱草素
priming ①蒸溅；飞沫②引动；起爆；发火③初给④引火药
priming apparatus 引火器
priming brush 涂底色用画笔
priming can 注油器
priming chamber 起动室
priming charge 引爆药
priming coat 底漆；底施；底层
priming composition 最初的组成
priming funnel 引动水的漏斗
priming inductor 起动泵用的喷射设备
priming line 灌注管路
priming oil 打底用油
priming paint 底漆
priming pigment 底漆(用)颜料
priming powder 起爆粉
priming pump 引液泵
priming reaction 活化反应
priming sugar 加啤酒用的葡萄糖
priming (tower) 泡沫充塞
priming tube 起爆管
priming valve 初级阀
primitive colour 基本色〔指光谱颜色〕
primitive fiber 微纤维
primitive lattice 初基点阵
primitive reaction 基本反应；本源反应
primitive rocks〔复〕 原生岩类
primitive soap 基皂
primordial radioelement 原生放射性元素
primordial radionuclide 原生放射性核素
primordial sediment 原生沉积物
primordium 原基
primrose ①樱草②淡黄(色)的

primrose chrome 铬樱红
primrose oil 樱草油
primulagenin 报春花素
primulaverin 樱草根苷
primulin bases 硫化染料
primulin(e) 樱草灵
primulin(e) dye 樱草灵(类)染料
primuline (printing) process 樱草素印像法；重氮盐印像法
primulite 樱草粒糖
primverin 樱草苷
primverose 玉米糖
primycin 伯霉素
Prince rupert drops 梨状熔融玻璃滴
Prince's metal 黄铜
principal 主要的
principal axis 主轴
principal axis of strain 应变主轴
principal axis of strain tensor 应变张量主轴
principal axis of stress 应力主轴
principal axis of stress tensor 应力张量主轴
principal bond 主键
principal chain 主链
principal component analysis 主成分分析
principal component regression method 主成分回归法
principal constituent 主成分
principal coordinate analysis 主坐标分析
principal defecation 最后澄清〔糖〕
principal deformation 主形变
principal direction 主方向
principal element 主要元素
principal elongation 主伸长
principal extension ratio 主拉伸比
principal fermentation 主要发酵作用
principal focal point 主(要)焦点
principal hue 主色相
principal hydrolysis(=main hydrolysis) 主水解
principal invariant of tensor 张量(的)主不变量
principal length 主体长度
principal link(=principal linkage) 主键
principal linkage(=principal link) 主键
principal monomer 主要单体
principal orientation direction 主取向方向
principal quantum number 主量子数
principal reaction 主要反应
principal series 主线系
principal splitter 主分割塔
principal strain 主应变
principal stress 主应力

principal valence 主(要化合)价
principal value 主值
principal value of strain tensor 应变张量主值
principle ①原理；原则②(要)素
principle of Clausius 克劳修斯原理
principle of coordinate invariance 坐标不变性原理
principle of corresponding state 对应状态原理
principle of detailed balance 精细平衡原理
principle of determinism 定数论原理
principle of dimensional invariance 量纲不变性原理
principle of entropy increase 熵增原理
principle of entropy production 熵原理
principle of equipresence 等存原理
principle of internal standard 内标原理
principle of local action 局部作用原理
principle of material frame indifference 物质标架无差异原理
principle of material indifference 物质无差异原理
principle of material objectivity 物质客观性原理
principle of maximum multiplicity 最大多重性原理
principle of maximum overlapping 最大交盖原则
principle of microreversibility 微观可逆性原理
principle of mobile equilibrium 动态平衡原理
principle of similarity 相似定理；相似原理；相似法则
principle of superposition 叠加原理
principle of time-temperature superposition 时温等效
print ①印刷②印模③(痕)迹④印刷物⑤印画〔照片等〕
printability 可印性
printable polypropylene film 可印刷的聚丙烯膜
printed board thickness 印制板厚度
printed calico 印花布
printed circuit 印刷电路
printed circuit board 印刷电路板
printed cloth 印花布
printed coating ①印刷涂层②印铁涂层
printed linoleum 印花油毡
printer ①印花工②打印器
printer-plotter 打字绘图机
printer's ink 印刷墨；油墨
printer's liquor 乙酸亚铁溶液
print-grade 印刷级
printing ①印刷②印相③印染
printing blanket(=litho felt) 石印毡；印刷用毡
printing blue 印染蓝；印刷蓝
printing color 印染染料
printing dye 印染染料
printing induline 印染引杜林染料
printing ink 印刷墨；油墨

printing integrator 打印积分器
printing machine 印刷机
printing paper 印书纸;道林纸
printing paste 印染浆
printing roller 印染辊
printing table 印染台
printing without grey 无灰印色
printout 印出(数据);打印
print out recorder 印出记录器
print-out unit 打印输出单元
print resistance 抗黏污性;抗印痕性
print test(=print resistance test) 防黏污试验;防转印性试验;压痕试验
print through 压透,过度挤压
print works〔复〕 印花机
priori distribution 先验分布;验前分布
priori probability 先验概率;验前概率
priority pollutant 优先污染物
prism ①棱镜②棱柱体;棱晶③三棱形
prismatic （三）棱形的
prismatical ①棱柱的②棱镜的
prismatic form (of benzene) （苯的)棱柱式〔即拉登堡式〕
prismatic formula 棱柱式
prismatic sulfur 三斜硫黄;（三)棱形硫黄
prismatine 柱晶石
prism for infrared 红外用棱镜
prism formula 棱柱式
prism infrared spectrometer 棱镜红外分光计
prism material 棱镜材料
prism powder 棱柱火药
prism spectrograph 棱镜光谱仪
prism spectrometer 棱镜分光计
pristane 姥鲛烷;朴日斯烷
pristanic acid 降植烷酸
pristimerin 抗革兰氏阳性球菌素
pristinamycin 原始霉素
probability 概率*
probability curve 概率曲线
probability density 概率密度
probability density function 概率密度函数
probability distribution 概率分布
probability factor 概率因子
probability of ionization 电离概率
probability paper 概率(坐标)纸
probability sampling 概率抽样
probable 可几的;可能的;近真的
probable error 概率误差
probable ore 可能的矿藏

probartital sodium 普罗比妥钠;(异)丙巴比妥钠;5-乙基-5-异丙基巴比妥酸钠
probation 检验;验证;鉴定
probation report 鉴定报告;检验报告
probe ①探头*;探针②探测;探测器③直接进样器
probe atomization 探针原子化
probe beam 探测束
probe insertion 探头插孔
probenecid 丙磺舒;羧苯磺丙胺
proberberine 原小檗碱
probe rod 探针杆;探杆
probe technique 探针技术
probing step 探测阶段
probolog 电测定器〔检验热交换器管路缺陷〕
probucol 普罗布考〔药〕
procainamide 普鲁卡因胺〔药〕
procainamide hydrochloride 普鲁卡因酰胺盐酸盐
procaine 普鲁卡因〔药〕
procaine base 普鲁卡因碱
procaine benzylpenicillin 普鲁卡因青霉素〔药〕
procaine hydrochloride 盐酸普鲁卡因
    $NH_2C_6H_4COOCH_2CH_2N(C_2H_5)_2 \cdot HCl$
procaine nitrate 硝酸普鲁卡因
procaine penicillin 普鲁卡因青霉素
procarbazine 丙卡巴肼〔药〕
procaterol 丙卡特罗〔药〕
procedural variable 工艺过程变量
procedure ①步骤②手续③方法④程序
procedure error 操作误差
procedure for analysis 分析操作规程
procedure of test 分析手续;试验手续
proceomycin 高霉素
process 过程*
processability 加工性能;操作性能
process additive 操作助剂
process aid 操作助剂
process analysis 过程分析
process analysis chart 工序分析表
process analytical chemistry 过程分析化学
process analyzer 过程分析仪
process and instrument diagram 工艺及仪表流程图
process calculation 工艺计算
process capability chart 工序精度点图;工序能力图表
process capacity 加工能力
process cell 工艺热室
process chart 工艺流程图
process chemical 操作助剂
process chemistry 过程化学;工艺化学

process chromatograph 流程色谱仪
process condition 工艺操作条件
process connection 工艺连接件
process control 过程控制
process controller 过程控制器
process-control viscometer 程序控制黏度计
process control word 过程控制代码
process cost 加工费；操作费用
process design 工艺设计
process development 过程开发
process diode array analyzer 过程二极管阵列分析器
process drawing 工艺流程图
process dynamics 过程动力学
processed gas 加工过的气体；精制过的气体；脱硫气体
processed isotope 化学加工精制放射性同位素
processed product 加工产物
processed rubber 再炼胶
process evaluation 工艺过程评价
process explanation 工艺说明；过程说明
process flow diagram 工艺流程图
process flow sheet 工艺流程图
process fluid test 工作流体试车
process for gun running 树胶熔融法(过程)
process furnace 管式炉；加热炉
process gas chromatograph 流程气相色谱仪
process gas chromatography (工艺)流程气相色谱(法)；工程管理气相色谱(法)
process gas scrubber 气体回收洗涤塔
process history 历程
process hold-up time 操作中断时间
processibility 加工性能
process industry 加工工业
processing 加工*
processing agent 加工助剂
processing aid 操作助剂
processing and molding of polymer 高分子加工成型
processing behavior 加工性能
process(ing) consideration 工艺(加工)条件；工艺过程概要(原理)
processing ease 操作容易；便于加工
processing equipment 工艺设备
processing facility 工艺设备
processing frequency 回旋频率
processing know-how (工艺)操作秘诀(诀窍)
processing loss 加工损耗
processing material 操作助剂
processing method 加工方法
processing of adhesion 胶黏的工艺过程
processing of crude oil 原油的加工
processing oil 加工油
processing parameter 工艺参数
processing paste 加工糊
processing plant 石油加工厂；炼油厂
processing property 操作性能；工艺性能
processing quality 工艺质量；加工质量
processing route 加工路线；加工方法
processing safety 操作安全性
processing set-up 工艺布置
processing temperature 加工温度；作业温度
processing units 加工设备
processional frequency 回旋频率
process ion meter 流程离子计
process lag 加工滞后；过程滞后
process line 工艺管道
process mass spectrometer 流程质谱仪
process monitoring ①操作过程的检调；加工过程的监视②程序控制
process monitor system (工艺)过程监控系统
process oil 操作油
processor ①信息处理机②加工程序；加工机械
process parameter 工艺参数
process period 加工周期
process pH meter 流程pH计
process piping 加工管线
process piping flowsheet 工艺管道流程图
process pump 工艺过程用泵
process residue 工艺残渣；工艺废料
process return gasket 介质回路(端盖)垫圈
process return header 介质回路端盖
process shrinkage 加工收缩(率)
process steam 生产用水蒸气
process-stream analyzer 工艺流程分析器
process stub 介质回路端盖固定板
process tankage 生物废料制的肥料
process unit 工艺设备
process utilities 工艺过程公用事业
process waste 工艺废物
process water 生产用水
process white 沉淀硫酸钡；钡白
procetane 柴油的添加剂
prochamazulene(=matricin) 母菊素
prochamazulenogen 原地甘菊精
prochiral 前手性；原手性
prochirality 前手性的；原手性的
prochlorite 蠕绿泥石；扇石
prochlorperazine 普鲁氯嗪〔药〕

procured retreading 预硫化翻胎
procyanine 原花色素
procyclidine hydrochloride 开马君；卡马特灵〔药〕
prodegradant 降解助剂；助老化剂
prodigiosin 灵菌红素
prodorit 普罗多里特〔一种石英与沥青捏炼产品〕
Prodorit tower 普罗多里特(吸收)塔
produce ①生产②产品
produced crude oil 开采到的原油
produced quantity 开采量
produce in situ 原地生产；就地生产
producer ①发生器②(炉煤气)发生炉；制气炉③生产者
producer-coloured 原液染色的；纺前着色的
producer furnace (煤气)发生炉
producer gas (发生)炉煤气
producer gas coal tar 发生炉煤焦油
producer gas coal tar pitch 发生炉煤焦油沥青
producer gas generator ①气体发生器②煤气发生炉
producer gas plant 煤气厂
producer gas process 煤气发生过程
producer gas tar 发生炉焦油
producer gas tar pitch 发生炉焦油沥青
producers stock ①原料②商品
producible 可产生的
producing capacity 生产能力；产量
producing depth(=producing horizons) ①(石油)床层的深度②生产水平
producing horizons(=producing depth) ①(石油)床层的深度②生产水平
product ①产物；(生)产品；生成物②(乘)积
product assurance 产品保证
product cavity 模巢
product concentration 产品浓度
product cut 产品馏分
product distribution ①产品分配②产品组成
product-emitter flame 产物发射焰
product improvement test 产品改进试验
product innovation ①产品革新(改进)②新产品；新制品
product ion 子离子
production ①生产；制造②生产物③生产
production batch 生产批次
production bottle neck 生产薄弱环节
production control 生产管理
production flow 生产流程
production formula 生产配方
production halts 生产中断；停产
production in series 顺序生产；流水作业的生产
production life 生产设备有效使用寿命
production line 生产流水线
production practice 生产实践
production rate 生产率
production run 大量生产；成批生产
production run(s) equipment 生产流水线用设备
production schedule 生产计划；生产进度
production scheme 生产流程图
production sequence 生产程序
production test ①生产(条件下)试验②生产上试用
production unit 生产设备
productive capacity 生产量；生产能力
productive maintenance 生产保全
productivity 产率
productivity function 产率函数
productivity rate ①生产②产率
product molecules 反应产物分子
product of mortality(=lethal index) 致死积
producton run 大量生产；成批生产
product property 制品性能
product reflux 产品回流
product rheology 产品流变性
product separation 产品分离
products-line (石油)产品的管线
products pipe-line operation (石油)产品管线的操作
products spotter 石油产品混合相的指示计
product stabilizer (石油)产品的稳定器
product stock 产品储存
product stream (石油)产品管流
product take-off (从分馏柱)收取石油馏出物的连接管
product tank 成品油罐
product terminal (石油)产品输送终点
product testing 产品试验
product-use 制品用途
product variables 制品差异
product yield 成品收率
proenzyme 酶原
pro-eutectoid 先共晶的
proferment 酶原
profile ①外形；轮廓②剖视图③分布型
profile cutter 成型刀具
profiled acrylic fibre 聚丙烯腈异形纤维
profile depth 涂层深度；涂层厚度〔涂层表面由峰到谷的平均值〕
profiled fibre 异表纤维；异形截面纤维
profile die 型材挤塑口模
profiled material 异型材
profile drag 轮廓曳力
profiled strip 异型钢带

profile extrudate 异型挤出件
profile extrusion 型材挤塑
profile fiber 异形纤维
profile height (涂膜)剖面高度；断面高度
profile length 型材长度
profile of diffraction line 衍射线分布图
profile pattern ①剖(断)面图②表面粗糙度；表面平整性
profile scanner 一维扫描机；线扫描机
profile shapes 异型材
profile shell 压型辊
profiling 压型
profiling calender 胎面机
profiling machine 靠模工具机；仿形工具机
profiling milling machine 靠模铣床；仿形铣床
profiling roll 压型辊
profilogram 性质改变轮廓图
profilometer 外形仪
profilometry 表面光度测定法
proflavin 二氨基吖啶；原黄素
proflavine(=3, 6-diaminoacridine sulfate) 硫酸原黄素；硫酸-3, 6-二氨基吖啶 $C_{13}H_{11}N_3 \cdot H_2SO_4$
proflavine cream 硫酸原黄素膏
profoamer 泡沫促进剂
progallin-p 棓酸丙酯
progenitor ①原粒子②原本
progeny test 后代测验；系谱测验
progesterone 黄体酮〔药〕
prognosis 预后
program(=programme) 程序
program board 程序控制台(盘)
program control 程序控制
program-controlled dyeing machine 程控染色机
program controller(=programming controller) 程序控制器；程序检验器
program counter 程序计数器
program end 程序(终)端；程序末尾
program evaluation and review technique 程序评审法；计划估评法
program installation 程序控制设备；程序装置(计)
program interrupt 程序中断
programmability 程序可编性
programmable computer 程(序)控(制)计算机
programmable oven 程(序)控(制)加热炉
programmable power supply 程(序)控(制)电源
programmable wavelength detector 程序波长检测器
programme(=program) 程序
programmed acquisition 程序采集
programmed current chronopotentiometry 程序电流计时电位法*
programmed flow gas chromatography 程序变流气相色谱(法)
programmed heating 程序加热
programmed model 程序方式
programmed multiple development 程序多次展开(法)*
programmed pressure 程序变压(力)
programmed pressure gas chromatography 程序变压气相色谱(法)
programmed run 程序控制；程序操作
programmed scan 程序扫描
programmed solvent 程序变溶剂
programmed temperature gas chromatography 程序升温气相色谱(法)
programmed temperature vaporizer 程序升温蒸发器
programmed temperature works 程序升温机件
programmed titration method 程控滴定法
programme peak selector 程序峰选择器
programmer 程序器
programming 程序(控制)的
programming controller(=program controller) 程序控制器；程序检验器
programming cool 程序冷却
programming flow gas chromatography 程序变流气相色谱法
programming language 程序设计语言
programming pressure 程序变压力；程序升压
programming start 程序起动
programming system 程序(控制)系统
program parameter 程序参数
program pulse 程序脉冲
program read-in 程序读入
program register 程序寄存器
program run 程序运行
program segment 程序段
program selection 程序选择
program simulator 程序模拟器
program speed 程序速度
program stop 程序停止(指令)；程序停机
program tape 程序带
program temperature controller unit 温度程序控制单元
program timer 程序计时器
program word 程序字码；程序语
progression ①级数②前进；增进③进行
progressive ①进步的②累进的③前进的；发展的
progressive defecation 逐步澄清
progressive development(=ultra-development; continuous running development) 流出展开

progressive drier 逐步干燥器
progressive error 累积误差；累计误差
progressive failure 逐步损坏
progressive freezing 逐步冷冻法
progressive gluing 连续胶合
progressive hydrolysis 逐步水解
progressive plastic yield 渐进塑性屈服
progressive poisoning （催化剂）递增中毒；累积中毒
progressive shear thickening 递增剪切变稠化
progressive shear thinning 递增剪切稀化
progress variable 进展变量
proguanil(=chloroguanide) 氯胍
proguanil hydrochloride N-对氯苯基-N-异丙基双胍盐酸盐
progynon B 保女荣 B；苯甲酸雌二醇
prohibitive amount 限制使用量；限额
prohibitive consumption 最高的消耗量
proidonite 氟硅石
proiprietary formula 专用配方
project ①计划；设计②投射；射出③投影
project area 投影面积
projected area molding 柱塞注压机
projected-scale 投影标尺
projected-scale balance 投影标尺天平
projectile ①弹丸；射弹②(以力)射出的③轰击粒子；入射粒子
projectile impingement 弹丸撞击式分散设备
projecting apparatus 幻灯装置；投影装置器
projection ①投射②射影③突出物；支头
projection comparator 投影比长仪
projection formula 投影式
projection lantern 幻灯
projection nucleus 投影环
projection operator 投影算符
projection reconstruction 投影重建
projection scale 投影标尺
projection velocity 喷射速度；抛射速度
projection welding 凸焊
projective algorithm 投影算法
projective pursuit 投影寻踪法
projective pursuit regression 投影寻踪回归
projective rotation factor analysis 投影旋转因子分析
projector ①投射器②投射炮③放映机④投影仪
project salt vault 盐库计划〔放射性废物处置用〕
prokaryote 原核生物
prokeyvit 维生素 K_3
proknock 诱震剂
proknock properties of gasoline 汽油的助爆震性能

prolamine 麸蛋白；醇溶谷蛋白
prolate 扁长的
prolate drop 扁长液滴
prolate ellipsoid 扁长椭圆球体
prolate symmetric top molecule 长对称陀螺形分子
prolate trochoidal mass spectrometer 长摆线质谱仪
proliferating polymer 增殖聚合物
proliferation 增殖
proliferin 多育曲菌素；增殖蛋白
proliferous polymerization 增殖聚合*
proline(Pro) 脯氨酸
prolon 蛋白质纤维
prolong 冷凝管
prolonged antifouling power 长效防污能力
prolonged can stability 长期罐藏稳定性
prolonged downward 下伸；下延
prolonged treatment 延伸处理
promacetin(=sodium 4,4-diaminodiphenylsulfone-2-acetyl sulfonamide) 4,4-二氨二苯砜-2-乙酰基磺酰胺钠
promazine hydrochloride 10-(3-二甲氨基丙基)吩噻嗪盐酸盐；盐酸丙嗪；盐酸普马嗪
prometaphanine 原间千金藤碱
promethazine 异丙嗪
promethazine chlorotheophyllinate 氯茶叶碱异丙嗪〔药〕
promethazine hydrochloride 盐酸异丙嗪；非那根
promethium 钷
promin(=sodium 4,4′-diamino-diphenyl-sulfone-N,N'-didextrose sulfonate) 普乐民；4,4′-二氨基-二苯砜-N,N'-二葡萄糖磺酸钠〔治疗肺结核和麻风药物〕
prominal(=methylphenobarbital) 普罗米那；甲基苯巴比妥
prominent peak 主峰
promoter 助催化剂*
promoter action 促进作用
promoter of vulcanization 硫化活性(促进)剂
promoting agent 增进剂；活性剂
promotor 促进剂；助催化剂
promotor for alkylation 烃化促进剂
prompt 瞬时
prompt gamma-ray analysis 瞬发γ射线分析
prompt neutron activation analysis(PNAA) 瞬发中子活化分析；中子俘获γ射线分析
promutagen 预诱变剂；前诱变剂
proneness 倾向性
pronestyl 4-氨基-N-2-二乙氨乙基苯甲酰胺；百浪斯迪普鲁卡因酰胺
pronethalol hydrochloride 盐酸丙萘洛尔〔药〕
prontosil 百浪多息〔治疗化脓性细菌引起的疾病的特效

药〕；偶氮磺胺；2,4-二氨基偶氮苯-4-磺酰胺
prontosil rubrum(=protosil)　百浪多息；偶氮磺胺
prontosil soluble(=neoprontosil)　可溶性百浪多息；新百浪多息
proof　①耐；防②证明③实验④(酒类的)标准强度⑤规定的；合乎标准的
proof box　保险箱
proofed cloth　胶布；防水布
proofed sleeve　橡皮管
proofer　检验者
proof fabric　胶布
proof gallon　标准加仑〔酒精浓度〕
proofing　①涂胶；刮胶；上胶②证明
proofing cabinet　(烘面)试验橱
proofings〔复〕　胶布
proof mortar　试粉研钵
proof resilience　①弹性极限应力②(=resilient energy)弹性能
proof sample　样品
proof spirit　规定酒精
proof stick　试验棒；塞尺
proof strength　保证强度；允许强度
proof stress　弹性极限应力；耐力；试验应力
proof test　安全试验
pro-oxidant　助氧(化)剂
pro-oxygenic agent　助氧(化)剂
pro-oxygens　助氧(化)剂
propadiene　丙二烯
propadrine(=norephedrine)　普鲁巴准；去甲麻黄碱
propaesin(=para-aminobenzoic acid propyl ester)　普罗帕辛；对氨基苯酸丙酯
propafenone　普罗帕酮〔药〕
propagating chain end　增长链端
propagating radical　(链)增长自由基
propagation　①传播②繁殖③增长
propagation coefficient　增殖系数；增长系数
propagation constant　传播常数
propagation crosslink　交联增长
propagation energy　(裂纹)扩散能量
propagation of crystalline entity　晶体生长
propagation of error　误差传递
propagation of heat　热的传布；传热
propagation process　(链)增长过程
propagation rate constant　(链)增长速率常数
propagation reaction　(链)增长反应
propagation tear resistance　(薄膜)抗蔓延撕裂性
propagation velocity　传播速度
propalanine　氨基丁酸

propaldehyde　丙醛　CH_3CH_2CHO
propamidine(=4, 4-diamidinodiphen-oxy-propane)　二脒二苯氧基丙烷
propamine　普鲁巴明
propamine sulfate　硫酸苯丙胺；硫酸普鲁巴明
propanal　丙醛　CH_3CH_2CHO
propanamide　丙酰胺
propandioic acid(=propane diacid)　丙二酸
propane　丙烷　$CH_3CH_2CH_3$
propane-acid process　丙烷-酸法〔用丙烷和酸的润滑油精制〕
propane-air mixture　丙烷和空气的混合物
propane burner　丙烷喷灯；丙烷燃烧器
propane diacid(=propandioic acid)　丙二酸
propane dinitrile　丙二腈
propanedioic acid　丙二酸
propanediol-1,2-carbonate　丙二醇-1,2-二碳酸酯
1,2-propanediol dibenzoate　二苯甲酸-1, 2-丙二醇酯
propane-propene cut(=P-P cut)　丙烷-丙烯馏分
propane-propene fraction(=P-P fraction)　丙烷-丙烯馏分
propane selanol(=propyl selenomercaptan)　丙硒醇　$CH_3CH_2CH_2SeH$
propane sultone　丙磺酸内酯
propane tetracarboxylic acid　丙烷四羧酸　$(COOH)_2CHCH_2CH(COOH)_2$
n-propane thiol　正丙硫醇
propanetricarbonyl　丙三羰基
propane tricarboxylic acid　丙三羧酸　$COOHCH_2CH(COOH)CH_2COOH$
propanetriol　丙三醇
propanetriol tribenzoate　三苯甲酸丙三醇酯
1,2,3-propanetriol tribenzoate　三苯甲酸-1, 2, 3-丙三醇酯
propanoic acid　丙酸　CH_3CH_2COOH
propanol　丙醇
n-propanol(=n-propyl alcohol)　正丙醇
1-propanol　1-丙醇；(正)丙醇
2-propanol　2-丙醇；异丙醇
propanolate(=propylate)　丙醇盐；丙氧基金属
propanolato(=propoxy)　①丙醇化物②丙氧基金属
propanolato-titanium　①丙氧基钛②钛酸丙酯
propanoldiacid　丙醇二酸　$COOHCHOHCOOH$
propanolon acid　丙醇酮酸　$CH_2OHCOCOOH$
propanone　丙酮
propantheline　丙胺太林
propantheline bromide　溴丙胺西林；丙胺太林；普鲁本辛〔药〕
propanthiol　丙硫醇
propargyl　炔丙基　$HC{\equiv}CCH_2-$

propargyl acetate 乙酸炔丙酯
propargyl alcohol 炔丙醇 HC≡CCH₂OH
propargyl aldehyde 炔丙醛
propargyl bromide(=3-bromo-1-propyne) 炔丙基溴；3-溴-1-丙炔 HC≡CCH₂Br
propargyl chloride(=3-chloro-1-propyne) 炔丙基氯；3-氯-1-丙炔 HC≡CCH₂Cl
propargyl derivative 炔丙基衍生物
propargyl halide(=propargylic halide) 炔丙基卤
propargylic acid(=propiolic acid) 丙炔酸
propargylic halide(=propargyl halide) 炔丙基卤
propargylic rearrangement 炔丙基重排作用
propargyl iodide(=3-iodo-1-propyne) 炔丙基碘；3-碘-1-丙炔 HC≡CCH₂I
propargyl isocyanate 异硫氰酸炔丙酯
propasin 对氨苯甲酸丙酯
propazone 普鲁帕宗；5,5-二丙基-2,4-二氧嘧唑啉
propeimine 前贝母素 C₂₄H₄₀O₃
propellane 螺桨烷
propellant 推进剂；发射药；火箭燃料
propeller 螺旋桨；推进器
propeller agitator 叶片式搅拌器；螺旋桨式搅拌器
propeller antiicing liquid 螺旋推动机的抗冻液
propeller blade 螺旋桨叶
propeller fan 轴流通风机；螺旋桨风扇
propeller mixer 推进式混合器
propeller pump 螺旋泵
propeller shaft 螺旋轴
propeller stirrer 螺旋桨搅拌机
propeller turbine 螺旋桨式透平(涡轮机)；轴流定桨式水轮机
propeller type fan 螺旋桨式风扇
propeller type mixer 螺旋桨式混合器
propeller type stirrer 螺旋桨式搅拌器
propenal 丙烯醛
propene 丙烯
propene dicarboxylic acid 丙烯二羧酸；戊烯二酸 C₃H₄(COOH)₂
propenenitrile 丙烯腈
propene thiol 丙烯基硫醇
2-propene-1-thiol 2-丙烯-1-硫醇
propenoate 丙烯酸酯
propenoic acid 丙烯酸 C₂H₃COOH
propenol 丙烯醇
propenoxylated 丙氧基化的
propensity (自然)倾向
propenyl(=isoallyl) 丙烯基 CH₃CH=CH—
p-propenyl anisol 对丙烯基茴香醚

propenylbenzene 丙烯基苯 C₆H₅CH=CHCH₃
propenyl carbonate 碳酸丙烯酯
propenyl chloride 氯丙烯；丙烯基氯
propenyl cyanide 丙烯基腈
propenylene 1,3-亚丙烯基 —CH₂CH=CH—
propenylguaiacol 丙烯基愈创木酚 CH₃CH=CHC₆H₃(OH)OCH₃
propenyl hydrate(=glycerol) 甘油
propenylidene 亚丙烯基 CH₃CH=C=
propenyl isothiocyanate(=propenyl mustard oil) 丙烯基芥子油；异硫氰酸丙烯酯
propenyl mustard oil(=propenyl isothiocyanate) 丙烯基芥子油；异硫氰酸丙烯酯
propenyl phenol 丙烯基酚
propenyl propyl disulfide 丙烯基丙基二硫
propenyl propyl ketone 丙烯基丙基酮
propenyl-2,4,5-trimethoxy benzene(=asarone) 丙烯基-2,4,5-三甲氧苯；细辛脑
propenyl trinitrate 硝化甘油
4-propenyl veratrole 4-丙烯基藜芦醚
propeptone 半胨
proper fuel 合格的燃料
proper function(=eigen function) 本征函数
proper inflation 适度内压
proper lubrication 合适的润滑
proper mass(=rest mass) 静质量
proper mixture ratio 合适的混合比
proper orthogonal group 正常正交群
proper property 固有性能；适当性能
proper rotation 真转动
property 性(质)；本性；特性
property modification 改性
property-modifying additive (聚合物)改性添加剂
property selective detector 特性选择检测器
property-test 性能试验
proper value(=eigen value) 本征值
propethylene(=isopropenyl vinyl ether) 异丙烯乙烯醚
prophetin 喷瓜苷
prophylactic ①预防的②预防药
prophylaxis 预防
propicillin 苯氧乙基青霉素
propilidene 亚丙基
propine(=propyne) 丙炔
propinol 炔丙醇
propinyl 炔丙基
β-propiolactone β-丙醇酸内酯
propiolaldehyde(=propynal；propargyl aldehyde) 丙炔醛 HC≡CCHO

propiolate 丙炔酸盐(酯)
propiolic acid(=propargylic acid) 丙炔酸 HC≡CCOOH
propiolic acid series 丙炔酸系
propiolic alcohol 丙炔醇 HC≡CCH$_2$OH
propiolic halide 卤丙炔 HC≡CCH$_2$X
propioloyl(=propiolyl) 丙炔酰 HC≡CCO—
propiolyl(=propioloyl) 丙炔酰
propion 3-戊酮
propionaldehyde 丙醛 CH$_3$CH$_2$CHO
propionaldehyde oxime 丙醛肟
propionaldehyde semicarbazone 丙醛缩氨基脲；亚丙基氨基脲 C$_2$H$_5$CH=NNHCONH$_2$
propionaldoxime 丙醛肟 C$_2$H$_5$CH=NOH
propionamide 丙酰胺 C$_2$H$_5$CONH$_2$
propionamido 丙酰氨基 CH$_3$CH$_2$CONH—
propionanilide N-丙酰苯胺 C$_2$H$_5$CONHC$_6$H$_5$
propionate ①丙酸②丙酸盐(酯或根)
propione 3-戊酮
propionic acid 丙酸 CH$_3$CH$_2$CO$_2$H
propionic aldehyde 丙醛 CH$_3$CH$_2$CHO
propionic anhydride 丙酸酐 (C$_2$H$_5$CO)$_2$O
propionin 丙酸菌素
propionitrile 丙腈 CH$_3$CH$_2$CN
propiono 丙酰基
propionthioaldehyde 丙硫醛
propionyl 丙酰 CH$_3$CH$_2$CO—
propionyl bromide 丙酰溴 C$_2$H$_5$COBr
propionyl chloride 丙酰氯 C$_2$H$_5$COCl
propionylcholine 丙酰胆碱
propionyl fluoride 丙酰氟 C$_2$H$_5$COF
propionyl iodide 丙酰碘 C$_2$H$_5$COI
propionyloxy 丙酸基；丙酰氧基 CH$_3$CH$_2$COO—
propionyl salicylic acid 丙酰基水杨酸
propiophenone 苯基乙基(甲)酮
propiopyrrothin(=aureothricin) 丙二硫吡咯素〔即：金丝菌素〕
β-propiosultone β-丙磺酸内酯
propipette 吸移管；吸液管；吸量管
propofol 丙泊酚〔药〕
propolis 蜂胶
proponal 二异丙巴比妥酸〔药〕
proportion ①比例②配合
proportional ①比例(的)②比例量
proportional bag sampling system 比例包囊取样系统；比例取样法
proportional band (控制仪器的)比例范围
proportional (by) parts 按份比例
proportional controller 比例控制器
proportional counter 正比计数器*
proportional counting region 比例计数区域
proportional draw-off device 按比例选取石油产品的设备
proportional error 比例误差
proportional gas meter 比例气体计量器
proportionality 比例(性)
proportional(ity) limit 比例限度；配比限度
proportional limit in shear 剪切比例极限
proportional meter 比例计
proportional mixing (按)比例混合
proportional positioning control 比例定位控制
proportional pump 比例泵；定量泵
proportional reset controller 比例重定调节器；等速调节器
proportional sampling 比例抽样*
proportional sampling device 按比例的取样器；定量取样器
proportional scale 比例尺
proportional spreading rate 涂布比率
proportional temperature control 比例温度控制
proportional timer 比例计时器
proportional valve 比例阀；定量阀
proportional volume gradient 比例体积梯度
proportioned streams 配合管流〔按需要量配合的管流〕
proportioner ①比例调节器；比例装置②定量给料器
proportioning (按比例定量)配合
proportioning by volume 按体积配合〔按体积比例定量配合〕
proportioning by weight 按重量配合〔按重量比例定量配合〕
proportioning controller 比例控制器；配比控制器
proportioning device (比例定量)配合设备
proportioning hopper 配比加料斗
proportioning machine 配料机；比例投料机
proportioning pigment-binder ratio 按颜-基比配料
proportioning plant (比例定量)配合装置
proportioning pump 配比泵；比例泵
proportioning valve 配合阀
proportion limit 比(例)限(度)
proportionlity 比例(性)；比例关系
proportion meter 比例计；配料计
proportionner 配合加料(漏)斗；测量加料(漏)斗
proportion of ingredients 拼份的比例
propoxide 丙氧化物；丙醇盐 CH$_3$CH$_2$CH$_2$OM
propoxy 丙氧基 CH$_3$CH$_2$CH$_2$O—
propoxylate 丙氧基化物
propoxylated ether 丙氧基化醚

propoxylation 丙氧基化(作用)
propoxyphene hydrochloride 盐酸右丙氧芬〔药〕
propranolol 普萘洛尔〔药〕
proprietary 专利的
proprietary agent 专利品
proprietary material 专用材料
proprietary phosphating solution 专卖(利)磷酸盐处理液；专卖(利)磷化处理液
proprietary term ①专利名②专用术语
prop-shaft 螺旋桨轴
propulsion 推进
propulsive 推进的；促进的
propulsive cathode (氢)气推(液)阴极
n-propyl 正丙基 $CH_3CH_2CH_2—$
sec-propyl(=isopropyl) 异丙基 $(CH_3)_2CH—$
N-propylacetanilide N-丙基-N-乙酰苯胺 $C_6H_5N(COCH_3)CH_2C_2H_5$
propyl acetate 乙酸丙酯 $CH_3CO_2CH_2C_2H_5$
propyl-acetic acid 丙基乙酸；戊酸 C_4H_9COOH
propyl acetoacetate 乙酰乙酸丙酯 $CH_3COCH_2CO_3H_7$
propyl acetone 丙基丙酮；甲基丁基甲酮
propylacetylene(=1-pentyne) 戊炔；丙基乙炔；1-戊炔 $C_2H_5CH_2C{\equiv}CH$
propylal 丙缩醛；二丙醇缩甲醛；二丙氧基甲烷 $CH_2(OCH_2C_2H_5)_2$
n-propyl alcohol(=n-propanol) 正丙醇
propylaldehyde 丙醛 CH_3CH_2CHO
n-propylamine(=1-aminopropane) 正丙胺；1-丙胺 $CH_3CH_2CH_2NH_2$
propyl aminobenzoate 氨基苯甲酸丙酯 $NH_2C_6H_4CO_2C_3H_7$
propyl-ammonium smectite 丙基铵绿土
N-propyl aniline N-丙基苯胺 $C_6H_5NHCH_2C_2H_5$
propyl anisole 丙基苯甲醚
propyl arsonic acid 丙(基)胂酸 $C_3H_7AsO(OH)_2$
propylate(=propanolate) 丙醇盐；丙氧基金属
propylben 对羟基苯甲酸丙酯；尼泊金丙酯
propyl benzene 丙苯 $C_6H_5CH_2C_2H_5$
propyl benzoate 苯甲酸丙酯 $C_6H_5CO_2C_3H_7$
o-propylbenzoic acid 邻丙基苯甲酸 $C_3H_7C_6H_4CO_2H$
propyl benzoylbenzoate 苯酰基苯甲酸丙酯 $C_6H_5COC_6H_4CO_2C_3H_7$
propyl benzyl cellulose 丙基苄基纤维素
propyl benzyl ketone 丙基苄基(甲)酮 $C_3H_7COCH_2C_6H_5$
propyl boric acid(=propyl boron dihydroxide) 丙基硼酸 $C_3H_7B(OH)_2$
propyl boron dihydroxide(=propyl boric acid) 丙基硼酸

propyl bromide(=1-bromopropane) 丙基溴；1-溴丙烷 $CH_3CH_2CH_2Br$
propyl-bromo-propenyl-barbituric acid 丙基溴丙烯基巴比妥酸 $(C_3H_7)(CH_3CBr{=}CH){=}C_4H_2O_3N_2$
propyl butyl-carbinol 丙基丁基甲醇；4-辛醇 $C_3H_7CHOHC_4H_9$
propylbutyl ether 丙基丁基醚；丙氧基丁烷 $C_3H_7OC_4H_9$
propyl butyrate 丁酸丙酯 $C_3H_7CO_2C_3H_7$
propyl caproate 己酸丙酯 $C_5H_{11}CO_2CH_2C_2H_5$
propyl caprylate 辛酸丙酯 $C_7H_{15}CO_2CH_2C_2H_5$
propyl carbinol(=n-butyl alcohol) 丙基甲醇；正丁醇
n-propyl carbylamine(=propyl isonitrile) 丙胩 $C_2H_5CH_2N{\equiv}C$
propyl cellosolve 丙基溶纤剂；乙二醇一丙醚 $CH_3CH_2CH_2OCH_2CH_2OH$
propyl cellulose 丙基纤维素
propyl chloride(=1-chloropropane) 丙基氯；1-氯丙烷 $CH_3CH_2CH_2Cl$
propyl chlorocarbonate 氯甲酸丙酯 $ClCO_2C_3H_7$
propyl cinnamate 肉桂酸丙酯 $C_8H_7CO_2C_3H_7$
propyl cyanate 氰酸丙酯 C_3H_7OCN
propyl cyanide 丙基氰；丁腈 C_3H_7CN
propyl cyclohexane 丙基环己烷 $C_3H_7C_6H_{11}$
propyldioctylamine 丙基二辛基胺 $C_3H_7(C_8H_{17})_2N$
propyl disulfide 丙基二硫
propyldithiopropane 丙基二硫丙烷
propylene ①(=propene)丙烯 $CH_3CH{=}CH_2$②1,2-亚丙基 $CH_3CHCH_2—$
propylene alcohol 丙二醇
propylene bromide 二溴(化)丙烯 $CH_3CHBrCH_2Br$
propylene bromohydrin 丙溴醇 $CH_3CHBrCH_2OH$
propylene carbonate(=4-methyl-2,5-dioxolone) 碳酸丙二醇酯；1,2-丙二醇碳酸酯；4-甲基-2,5-二氧杂环戊-1-酮
propylene chloride 氯(化)丙烯 $CH_3CHClCH_2Cl$
sec-propylene chlorohydrin(=1-chloro-2-propanol) 丙氯仲醇；1-氯-2-丙醇 $CH_3CHOHCH_2Cl$
propylene cyanide(=1,2-dicyanopropane) 丙邻二腈；1,2-二氰基丙烷 $CH_3CH(CN)CH_2CN$
propylene diamine(=1,2-diaminopropane) 丙邻二胺；1,2-二氨基丙烷 $CH_3CH(NH_2)CH_2NH_2$
1,3-propylenediamine-2-ol tetraacetic acid(DPTA) 1,3-丙二胺-2-醇-四乙酸
propylene dichloride 二氯(化)丙烯 $CH_3CHClCH_2Cl$
propylene diester 丙二酯
propylene dihalide 二卤(化)丙烯 CH_3CHXCH_2X

propylene diiodide 二碘(化)丙烯 CH_3CHICH_2I
propylene dimer 丙烯二聚物
propylene dimerization 丙烯二聚(法)
propylene epoxide 氧化丙烯；甲基氧丙环 C_3H_6O
propylene-ethylene polyallomer 丙烯-乙烯异质同晶共聚物
α-propylene-glycol α-丙二醇 $CH_3CHOHCH_2OH$
β-propylene-glycol β-丙二醇 $CH_2OHCH_2CH_2OH$
propylene glycol alginate 海藻酸丙二酯
propylene glycol n-butyl ether 丙二醇正丁醚〔新型溶纤剂系助溶剂〕
propylene-glycol diacetate 丙二醇二乙酸酯 $(CH_3CO_2)_2CH_2CH(CH_3)$
propylene glycol dibenzoate 二苯甲酸丙二醇酯
propylene glycol dicaprate 丙二醇二癸酸酯
propylene glycol diglycidylether 丙二醇二缩水甘油醚
propylene glycol dipelargonate 丙二醇二壬酸酯
propylene glycol dipropionate 丙二醇二丙酸酯
propylene glycol methyl ether acetate 丙二醇(单)甲醚乙酸酯
propylene-glycol monoacetate 丙二醇一乙酸酯 $CH_3CO_2CH_2CHOHCH_3$
propylene glycol monoethyl ether 丙二醇一乙醚
propylene glycol monoethyl ether acetate 丙二醇单乙醚乙酸酯
1,2-propylene glycol monolaurate 1,2-丙二醇单月桂酸酯
propylene glycol monomethyl ether 丙二醇单甲醚
propylene glycol monomethyl ether acetate 丙二醇单甲醚乙酸酯
propylene glycol monooleate 丙二醇单油酸酯
propylene glycol monopalmitate 丙二醇单棕榈酸酯
propylene glycol monostearate 丙二醇单硬脂酸酯
propylene halide 二卤(化)丙烯 CH_3CHXCH_2X
propylene imine(=2-methyl aziridine) 2-甲基吖丙啶
propylene iodide 二碘(化)丙烯 CH_3CHICH_2I
propylene liquid 液态丙烯
propylene oligomer 丙烯低聚物
propylene oxide 氧化丙烯；1,2-环氧丙烷
propylene oxide adduct 环氧丙烷加成物；氧化丙烯加成物
propylene product 丙烯成品
propylene resin 丙烯类树脂
propylene steam 气态丙烯
propylene sulfide 硫化丙烯；硫丁环 $(CH_2)_3S$
propylene tetramer(=tetrapropylene) 四聚丙烯
propylene trimer 丙烯三聚物
propyl ester 丙酯 $RCOOC_3H_7$
propyl ether ①丙醚 $C_3H_7OC_3H_7$ ②丙基醚 C_3H_7OR

propyl ethylene(=1-pentene) 丙基乙烯；1-戊烯 $C_2H_5CH_2CH=CH_2$
propylfluorenone 丙基芴酮
propyl fluoride(=1-fluoropropane) 丙基氟；1-氟丙烷 $CH_3CH_2CH_2F$
propyl formate 甲酸丙酯 $HCO_2CH_2C_2H_5$
propyl-2-furanacrylate 2-呋喃基丙烯酸丙酯
propyl α-furancarboxylate α-呋喃羧酸丙酯 $C_4H_3OCO_2C_3H_7$
α-propyl-β-(α-furyl)acrolein α-丙基-β-(α-呋喃基)丙烯醛
propyl-β-furylacrylate β-呋喃丙烯酸丙酯 $C_4H_3OCH=CHCO_2C_3H_7$
propyl gallate 棓酸丙酯 $(HO)_3C_6H_2CO_2C_3H_7$
propyl glycol 丙基乙二醇
propyl glycolate 乙醇酸丙酯 $HOCH_2CO_2C_3H_7$
propyl halide 丙基卤 C_3H_7X
4-propyl-4-heptanol 4-丙基-4-庚醇；三丙基甲醇 $(C_2H_5CH_2)_3COH$
n-propyl heptine carbonate 庚炔碳酸丙酯
propyl heptylate 庚酸丙酯
propyl hexanoate 己酸丙酯
propylhexedrine 甲基(苄代甲代)甲基胺
n-propyl hexylate 己酸正丙酯
propyl hexyl ketone 丙基己基(甲)酮；4-癸酮 $C_3H_7COC_6H_{13}$
propyl hydrogen-sulfate 硫酸氢丙酯 $C_3H_7HSO_4$
propyl hydrosulfide 丙硫醇 C_3H_7SH
propyl p-hydroxybenzoate 对羟基苯甲酸丙酯 $HOC_6H_4COOC_3H_7$
propyl hydroxylamine 丙胲 $C_2H_5CH_2NHOH$
propylic alcohol 丙醇
propylidene 亚丙基 $CH_3CH_2CH=$
propylidene bromide 1,1-二溴丙烷
n-propylidene butyraldehyde 正亚丙基丁醛
propylidene chloride 1,1-二氯丙烷；亚丙基二氯
α-propylidene enanthaldehyde α-亚丙基庚醛
propylidene halide 亚丙基二卤；二卤(代)丙烷 $CH_3CH_2CHX_2$
3-propylidene phthalide 3-亚丙基-2-苯并[c]呋喃酮
propylidyne 次丙基 $CH_3CH_2C\equiv$
propyl iodide(=1-iodopropane) 丙基碘；1-碘丙烷 $CH_3CH_2CH_2I$
propyliodone 丙碘酮〔药〕
propyl isobutyl-carbinol 丙基异丁基甲醇；2-甲基-4-庚醇 $C_3H_7CHOHC_4H_9$
propyl isobutyl ketone 丙基异丁基(甲)酮；2-甲基-4-庚酮 $C_2H_5CH_2COC_4H_9$
propyl isobutyrate 异丁酸丙酯 $(CH_3)_2CHCO_2C_3H_7$

propyl isocyanate 异氰酸丙酯 C_3H_7NCO
propyl isocyanide 丙胩 $C_2H_5CH_2N{\equiv}C$
propylisome 丙基增效剂
propyl isonitrile(=n-propyl carbylamine) 丙胩 $C_2H_5CH_2N{\equiv}C$
propyl isopropyl ether 丙基异丙基醚 $C_2H_5CH_2OCH(CH_3)_2$
propyl isorhodanate(=propyl isothiocyanate) 异硫氰酸丙酯
propyl isorhodanide(=propyl isothiocyanate) 异硫氰酸丙酯
propyl isosulfocyanate(=propyl isothiocyanate) 异硫氰酸丙酯
propyl isosulfocyanide(=propyl isothiocyanate) 异硫氰酸丙酯
propyl isothiocyanate(=propyl isorhodanate; propyl isorhodanide; propyl isosulfocyanate; propyl isosulfocyanide; propyl isothiocyanide) 异硫氰酸丙脂 $CH_3CH_2CH_2NCS$
propyl isothiocyanide(=propyl isothiocyanate) 异硫氰酸丙酯
propyl isovalerate 异戊酸丙酯 $C_4H_9CO_2CH_2C_2H_5$
propylite 青磐岩
propyl lactate 乳酸丙酯 $C_2H_4(OH)CO_2C_3H_7$
propyl levulinate 4-氧戊酸丙酯 $CH_3CO(CH_2)_2COOC_3H_7$
propyl malonic acid 丙基丙二酸 $C_3H_7CH(CO_2H)_2$
propylmercaptan 丙硫醇 $CH_3CH_2CH_2SH$
propylmercuric bromide 溴化丙基汞 $CH_3CH_2CH_2HgBr$
propylmercuric chloride 氯化丙基汞 $CH_3CH_2CH_2HgCl$
propylmercuric iodide 碘化丙基汞 $CH_3CH_2CH_2HgI$
propyl methacrylate 甲基丙烯酸丙酯
propyl methyl heptenyl acetate 乙酸丙基甲基庚烯酯
propyl methyl trisulfide 丙基甲基三硫
n-propyl mustard oil 丙基芥子油
propyl naphthalene sulfonate 丙基萘磺酸盐
N-propyl naphthylamine N-丙基萘胺 $C_{10}H_7NHCH_2C_2H_5$
propylnitramine 丙基硝胺 $C_2H_5CH_2NHNO_2$
propyl nitrate 硝酸丙酯 $CH_3CH_2CH_2ONO_2$
propyl nitrite 亚硝酸丙酯 $CH_3CH_2CH_2ONO$
propyl nitrolic acid 丙基硝肟酸
propyloic 羧乙基 —CH_2CH_2COOH
propylparaben 对羟苯甲酸丙酯
propyl pelargonate 壬酸丙酯 $C_8H_{17}CO_2C_3H_7$
o-propylphenol 邻丙基苯酚 $C_3H_5CH_2C_6H_4OH$
n-propyl phenylacetate 苯乙酸丙酯

propyl phenyl-carbinol 丙基苯基甲醇；1-苯基-1-丁醇 $C_3H_7CHOHC_6H_5$
propyl phenyl ether 丙基苯基醚；苯丙醚 $C_2H_5CH_2COC_6H_5$
propyl phenyl ketone(=n-butyrophenone) 丙基苯基(甲)酮 $C_2H_5CH_2COC_6H_5$
propyl-phosphine 丙膦 $C_3H_7PH_2$
propyl propionate 丙酸丙酯 $C_2H_5CO_2CH_2C_2H_5$
2-propyl pyridine 2-丙基吡啶 $C_3H_7C_5H_4N$
propylpyridonium salt N-丙基吡啶酮盐
N-propylpyrrolidine N-丙基吡咯烷
propyl red 丙基红
propyl rhodanate(=propyl rhodanide) 硫氰酸丙酯 C_3H_7SCN
propyl rhodanide(=propyl rhodanate) 硫氰酸丙酯
propyl ricinoleate 蓖麻醇酸丙酯
propyl salicylate 水杨酸丙酯 $HOC_6H_4CO_2C_3H_7$
propyl selenomercaptan(=propane selanol) 丙硒醇 $CH_3CH_2CH_2SeH$
3-propyl-spiro-dibenzopyran 3-丙基-螺-二苯并吡喃
propyl succinic acid 丙基丁二酸 $C_3H_7C_2H_3(CO_2H)_2$
propyl sulfate 硫酸二丙酯 $(C_3H_7)_2SO_4$
propyl sulfhydrate 丙硫醇 C_3H_7SH
propyl sulfide 二丙硫醚 $C_3H_7SC_3H_7$
propyl sulfocyanate(=propyl thiocyanate) 硫氰酸丙酯
propyl sulfocyanide(=propyl thiocyanate) 硫氰酸丙酯
propyl thioacetate 硫代乙酸丙酯
propyl thiocyanate(=thiocyanide) 硫氰酸丙酯
propyl thiocyanide(=propyl thiocyanate) 硫氰酸丙酯
propyl thioether 二丙硫醚 $C_3H_7SC_3H_7$
n-propyl thiol 正丙硫醇
propylthiouracil 丙硫氧嘧啶；丙基硫尿嘧啶
propyl thiourea 丙硫脲 $C_3H_7NHCSNH_2$
p-propyl-toluene(=p-methylpropyl-benzene) 对丙基甲苯；对甲基丙苯 $CH_3C_6H_4CH_2C_2H_5$
propyl p-toluenesulfonate 对甲苯磺酸丙酯 $CH_3C_6H_4SO_3C_3H_7$
propyl-trichlorosilicane 丙基三氯硅 $C_3H_7SiCl_3$
propyl-triethoxysilicane 丙基三乙氧基硅 $C_3H_7Si(OC_2H_5)_3$
propyl triethyl ammonium halide 丙基三乙基卤化铵
propyl-triethylsilicane 丙基三乙基硅 $C_3H_7Si(C_2H_5)_3$
propyl-trimethylsilicane 丙基三甲基硅 $C_3H_7Si(CH_3)_3$
propyl trisulfide 丙基三硫
propyl-urea 丙脲 $C_3H_7NHCONH_2$
propyl urethane 丙氨基甲酸乙酯；丙基尿烷〔俗〕 $C_3H_7NHCO_2C_2H_5$
propyl valerate 戊酸丙酯 $C_4H_9CO_2C_3H_7$

propyl-vinyl-carbinol(=1-hexene-3-ol) 丙基乙烯基甲醇；1-己烯-3-醇 $C_6H_{11}OH$
propyl xanthate 丙黄原酸盐 C_3H_7OCSSM
propyl xanthic acid 丙黄原酸 C_3H_7OCSSH
propyl xanthogenate 丙黄原酸盐 C_3H_7OCSSM
propyl xanthogenic acid 丙黄原酸 C_3H_7OCSSH
propyl xanthonate 丙黄原酸盐 C_3H_7OCSSM
propyl xanthonic acid 丙黄原酸 C_3H_7OCSSH
propynal(=propiolaldehyde; propargyl aldehyde) 丙炔醛 HC≡CCHO
propyne(=propine) 丙炔 $CH_3C≡CH$
propynoic acid 丙炔酸 HC≡CCOOH
propynol 炔丙醇
2-propynyl 2-丙炔基
pro rata 按比例
proration ①石油(和气体)开采的集中控制②按比例的分配
prorennin 凝乳酶原
pro-R group 前R基团
pros- 〔希腊字头〕前
prosapogenin 前皂配基
proseptazine(=benzyl sulfanilamide) 苄基磺胺
pro-S group 前S基团
proshesis 取代；置换
prosize 大豆蛋白松香胶料
prosol 粟醇
prosopite 水铝氟石
prospecting 探矿
prospective oil 石油储量
prospective ore 可采矿
prosposition (萘环的)平位；(萘环的)2,3位
prostaglandin(PG) 前列腺素
prostanoic acid 前列腺烷酸
prostanoid 前列腺素类(化合物)
prostaphin 5-甲基-3-苯基-4-异噁唑青霉素钠
prostat 普适泰〔药〕
prostephanaberrine 原千金藤那布任碱
prosthetic group 辅基
prostigmin(e) 新斯的明
protactinide 镤化物
protactinium(=protoactinium) 镤
protactinium oxychloride 氯氧化镤
protactinium pentachloride 五氯化镤 $PaCl_5$
protactinium pentoxide 五氧化二镤 Pa_2O_5
protactinium peroxide 过氧化镤
protactinium sulfate 硫酸镤
protamine 鱼精蛋白〔药〕
protaptin 变形菌肽素

protargyl 元素原
protectant ①防护剂；保护剂②防老(化)剂
protected group 被护基
protected reversing thermometer 防压倒转温度计
protected sol 被护溶胶
protecting ①保护；防护②保护的；防护的
protecting agent 保护剂
protecting band 垫带
protecting cap 保险帽
protecting casing 保险罩
protecting colloid(=protective colloid) 保护胶体
protecting crust 防护壳
protecting film 防护膜
protecting group 保护基
protecting tube 防护管
protection 保护；防护
protection against corrosion 防腐蚀
protection against rust 防锈
protection by metallic coating 金属涂层保护法
protection by oxide film 氧化物膜保护法
protection by paints and lacquer 涂漆保护法
protection class (静电发生器的)保护等级
protection effect 保护效应
protection helmet 防护头兜
protection method 保护法；防护法
protection suit 防护服
protection test 防护试验
protective 保护的；防护的
protective action 保护作用
protective agent ①防护剂；保护剂②防老(化)剂
protective antigen 保护性抗原
protective barrier 保护屏蔽；保护层
protective cap ①防护帽②防护罩
protective clothing 防护衣
protective coating 保护涂层
protective colloid(=protecting colloid) 保护胶体
protective colloid action 保护胶体作用
protective color 保护(颜)色
protective cover ①覆盖层；保护罩②保护涂料
protective cream 防护膏
protective effect 保护效应
protective enzyme 防护酶
protective film 保护膜；保护层
protective foil 保护膜
protective gas 保护气体
protective group 保护基
protective layer 保护层
protective oxide layer 氧化保护膜

protective powder coating ①保护性粉末涂料②防腐蚀粉末涂料
protective property 保护性能
protective screen 防护屏
protective shell 保护层；保护壳
protective storage 防护储存〔按照消防范围的储存〕
protective strip coating 可剥性保护涂料
protective system ①保护系统②防腐体系；防腐涂料
protective value (润滑脂的)防护性能
protector ①防护器②保护者
protector of respiration 呼吸防护器
prot(e)id(e) 蛋白质
protein adhesive 蛋白质黏合剂
protein-aldehyde plastic 蛋白醛塑料
protein-aldehyde resin 蛋白醛树脂
proteinase 蛋白(水解)酶
protein binding assay 蛋白结合测定(法)
protein blotting 蛋白质印迹
protein carrier 载体蛋白
protein chip 蛋白质芯片
protein-doped matrix crystals 蛋白质掺杂基质结晶
protein error 蛋白质误差
protein fiber 蛋白质纤维
protein finishes 蛋白质漆；蛋白质涂饰剂
protein gel 蛋白凝胶
protein hydrolysis 蛋白(质)水解
protein index 蛋白质指数
protein kinase 蛋白激酶
protein nitrogen 蛋白质氮
proteinoid 类蛋白(质)
protein-peptide sequencer 蛋白质-肽序列分析仪
protein plastic 蛋白质塑料
protein-polyamide condensation product 蛋白质-聚酰胺缩合物
protein powder 蛋白质粉末
protein refractometer 蛋白计
protein regenerated fibre 再生蛋白质纤维
protein resin 蛋白质树脂
protein sequenator 蛋白质序列分析仪
protein synthesizer 蛋白质合成器
protein wave 蛋白质波
proteolysis 蛋白水解(作用)；蛋白酶解
proteolytic activity 解蛋白活度
proteolytic enzyme 蛋白(水解)酶
proteome 蛋白质组
proteomics 蛋白质组学
protheite 钙铁辉石
prothrombin 凝血酶原；凝血因子Ⅱ
prothrombinase 凝血酶原酶；促凝血球蛋白；凝血因子Va
prothromboplastic factor 促凝血酶原激酶因子
prothromboplastin 促凝血酶原激酶原
protic 质子的
protic solvent 质子溶剂*；质子传递溶剂
protiodide 碘化物最小碘量的
protionamide 丙硫异烟胺〔药〕
protium 氕*
proto- 〔希腊字头〕原
protoactinium 镤〔91号元素，化学符号Pa〕
protoactinorhodin 原放线紫红素
proto(a)etioporphyrin 原本卟啉
proto-albumose 原朊
protoanemonin 原白头翁素
protobastite 顽火辉石
protobitumen 原沥青〔有机物转化为石油的最初阶段〕
protobromide 溴化亚(某)；低溴化物
protocalcium 原钙；初钙
protocatechuic acid 原儿茶酸；3,4-二羟苯甲酸 (HO)$_2$C$_6$H$_3$COOH
protocatechuoyl 原儿茶酰；3,4-二羟苯甲酰 (HO)$_2$C$_6$H$_3$CO—
protocatechuyl 原儿茶基；3,4-二羟苄基 (HO)$_2$C$_6$H$_3$CH$_2$—
protocatechuyl alcohol 原儿茶醇；3,4-二羟苄醇 (HO)$_2$C$_6$H$_3$CH$_2$OH
protocetraric acid 原冰岛衣酸；梅木地衣酸
protochloride 氯化亚(某)；低氯化物
protocidin 杀原虫菌素
protocollagen 本胶原(蛋白)
protocotoin (=piperonyl phloroglucinol dimethyl ether) 间苯三酚胡椒基二甲基醚；原可土因；原可土树皮素 (CH$_3$O)$_2$C$_{14}$H$_8$O$_4$
protocrocin 原藏花素
protocurarine 原箭毒碱
protocurine 马钱子箭毒碱
protodestruxin 原腐败菌素
protoemetine 原吐根碱
protoenzyme 原酶
protoferriheme 高铁血红素
protofibre 原(生)纤维
protofibril 原纤维
protofluorine 原氟；初氟
protogen 硫辛酸
protogenic 生质子的；原生的
protogenic impurity 供质子的杂质
protogenic solvent 给质子溶剂；酸性溶剂
protohematin 正铁血红素

protoheme(=heme) 正铁血红素
protohemin 氯化血红素
protohydrogen 原氢；初氢
protokaolin 原高岭土
protokaolin coacervate 原高岭土凝聚层
protolichesterinic acid 原苔甾酸
protolignin 原(本)木素
protolysis 质子传递作用*；质子迁移
protolysis reaction 质子迁移反应
protolyte 质子传递物
protolytic solvent(=protonic solvent) 质子溶剂
protolytic stability constant 金属酸式配合物的稳定常数
protomycin 原虫霉素
proton ①质子；(正)质子②氢核；气核
proton acceptor 质子接受体
proton acceptor solvent 质子受体溶剂；受质子溶剂
proton-activated ester bond 质子活化的酯键
proton activation analysis(PAA) 质子活化分析
proton activity 质子活度
proton affinity 质子亲和力*
protonated 质子化了的
protonated ligand 质子化配体
protonated molecular ion 质子化分子离子
protonating agent 质子化(试)剂
protonation 质子化*
protonation constant 质子化常数
proton balance 质子平衡
proton beam 质子束
proton-binding energy 质子结合能
proton bombardment 质子轰击
proton-bridged ion 质子桥接离子
proton chemical shift 质子化学位移
proton condition 质子条件
proton donor 质子给(予)体
proton donor solute 给质子溶质
proton enhanced NMR 增强质子核磁共振
proton equilibrium 质子平衡
proton exchange 质子交换
proton exchange reaction 质子交换反应
proton excited X-ray spectrometry 质子激发 X 射线光谱法*
proton hydrate 水合质子；水化质子
protonic acid 质子酸〔放出质子的化合物〕
protonic solvent 质子溶剂
proton induced X-ray emission analysis(PIXE) 质子激发 X 射线发射分析*
proton inert solvent 质子惰性溶剂
protonium ion 氢鎓离子

proton magnetic resonance(PMR) 质子磁共振
proton magnetic resonance spectroscopy 质子磁共振光谱法
proton microbeam 质子微束；质子聚焦束
proton microprobe 质子微探针
proton microscope 质子显微镜
proton NMR spectrometry 质子核磁共振光谱法
proton NMR spectrum 质子核磁共振谱
proton noise decoupling 质子噪声去偶
proton nuclear magnetic resonance 质子核磁共振
proton nuclear magnetic resonance spectroscopy 质子核磁共振谱法；质子核磁共振谱学
proton number 质子数
protonolysis 质子分解
proton- proton coupling 质子-质子耦合
proton- proton dipolar interaction 质子-质子偶极相互作用
proton pump 质子泵
proton reference level 质子参考水平
proton relaxation enhancement 质子弛豫增强
proton-rich side 富质子侧；丰质子侧
proton shifting 质子移动(作用)
proton theory of acid 酸的质子理论
proton theory of acid and base 酸碱质子理论
proton transfer 质子传递*
proton transfer polymerization 质子转移聚合(作用)
proton transfer reaction 质子转移反应
protoparaffin 原石蜡
protoparaffin wax 原石蜡〔粗石油内所含有的石蜡，无定形的石蜡〕
protopectin 原果胶
protopetroleum 原石油；初级石油
protophile 亲质子物
protophilia 亲质子性
protophilic 亲质子的
protophilic solvent 亲质子溶剂
protophobe 质子供体
protophobic solvent 疏质子溶剂
protophyllin 原叶素
protopine 前鸦片碱
protopolymer 原生聚合物
protoporphyrin 原卟啉
protoproduct 木材废料制的内燃机燃料
protoproteose 原胨；水溶性初胨
protosal 甘油水杨基甲酸酯
protosalt 低价金属盐
protostephanine 原千金藤碱
protosubstance 矿脂胶凝质
protosulfate 低硫酸盐；含硫酸基最少的硫酸盐〔类同

低卤化物,不能叫作原硫酸盐 ortho-sulfate〕
prototropic 质子异变的
prototropic change 质子异变
prototropic rearrangement 质子转移重排*
prototropy 质子转移*
prototype 原型;主型
prototype equipment 定型设备
prototype kilogram 千克原器
prototype plant 原型工厂
prototype reactor 原型(反应)堆
protoveratrin 原藜芦素
protoveratridine 原藜芦定
protoveratrine 原藜芦碱
protoxide 氧化亚某;低氧化物
protoxydic 氧化低价的;低氧化的
protracted break 长时受力损坏
protracted test 疲劳试验
protracted test machine 疲劳试验机
protractor 测角器;分度规
protruded packing 多孔填料;冲压填料
protruding edge 露出的棱角(边缘)
protruding end ①露出的尾端②断纤维
protuberance ①红焰②突出部分
protyle theory (物质)单元(学)说
proustite 淡红银矿
provamycin(=spiramycin) 螺旋霉素
proved 探明的〔资源〕;肯定的
provenience study (古物)来源和出处研究
proven system 验证系统
prover 校准仪
proving ①校对②发酵
proving room 发酵室
proviridomycin 原绿霉素
provitamin 前维生素;维生素原
provitamin A 前维生素 A;维生素 A 原
provitamin D 前维生素 D;二氢麦角甾醇
provitamin D₂ 前维生素 D₂;麦角甾醇
provocation 激发(作用)
proxam 黄原酸异丙酯
proximate analysis 近似分析;组分分析;构分分析;实用分析
proximate calculation 近似计算
proximate composition 近似组成
proximation 迫近;近似
proximity 接近;邻近
proximity effect 接近效应
proximity exploder 遥控引信;近爆信管;近爆引信
proximity exposure 邻近曝光
proximity fuse 接近引信;近炸引管;低空爆炸信管
proxylin apparatus 生氧防毒器
prozane 三氮烷 H_2NNHNH_2
prulaurasin 桂樱苷;dl-扁桃腈葡萄糖苷
prumycin 普鲁霉素
prunasin 野黑樱苷;d-扁桃腈葡萄糖苷
prunol(=ursolic acid) 乌索酸
Prussian black 铁黑
Prussian blue 普鲁士蓝
Prussian blue test 普鲁士蓝试验
Prussian brown 铁棕
Prussian green 普鲁士绿
prussiate 氰化物 MCN
prussiate of iron(=iron blue) 铁蓝
prussic acid 氢氰酸 HCN
prussi complex (六)氰合(高)铁(Ⅲ)盐
prussine 氰
prusso salt 普鲁索盐;五氰合铁酸盐
psalliotin 黑伞蕈素
psammite 砂屑岩
psephite(=psephyte) 砾质岩
psephyte(=psephite) 砾质岩
pseudacetic acid 丙酸
pseudaconitine 假乌头碱
pseudo-〔希腊字头〕 假;似
pseudoacacia concrete 刺槐花浸膏;洋槐花浸膏
pseudo-acid 假酸
pseudo-acidity 假酸性
pseudoaconitine 假乌头碱
pseudoadsorption 假吸附(作用)
pseudo-affine deformation 准仿射形变
pseudo-alloy coating 假合金涂层
pseudoallyl(=isopropenyl) 异丙烯基 $CH_2=C(CH_3)-$
pseudoalum 假矾
pseudo aromaticity 拟芳香性;假芳香性
pseudoasymmetric 假不对称的
pseudoasymmetric carbon 假不对称碳
pseudoasymmetry 假不对称
pseudoatom 假原子
pseudoazimide(=indodiazole) 茚并二唑
pseudobase 假碱
pseudo-basicity ①假碱性②假碱度
pseudobinary system 假二元体系
pseudobochmite 假勃姆石
pseudobrookite 铁板钛矿
pseudobrucine 假番木鳖碱;假二甲马钱子碱
pseudobutylene 2-丁烯
pseudocannel coal 假烛煤

pseudo-capacity 假电容
pseudocatalysis 假催化(作用)
pseudo cationic living polymerization 假正离子活聚合
pseudo cationic polymerization 假正离子聚合
pseudocellulose 半纤维素
pseudochalcedony 假玉髓
pseudochirality 假手性
pseudo-chlorogenin 假鸡骨常山碱
pseudocholestene 假胆甾烯
pseudo cobalamin 假钴胺(素)
pseudocomponent 假组分
pseudo-composition method 代用组成法
pseudo-compound 假化合物
pseudoconhydrine 假羟基毒芹碱
pseudo-contact shift(=dipolar shift) 偶极位移
pseudo countercurrent process 假逆流过程
pseudo critical constant 假临界常数
pseudo crossing 赝交叉
pseudocrystal 假晶；赝晶
pseudocrystalline aluminosilicate 假晶硅酸铝
pseudocumene(=1,2,4-trimethylbenzene) 假枯烯；1,2,4-三甲基苯 $C_6H_3(CH_3)_3$
pseudocumidine(=2, 4, 5-trimethylaniline) 假枯胺；2, 4, 5-三甲基苯胺 $(CH_3)_3C_6H_2NH_2$
pseudocumidino(=2, 4, 5-trimethylanilino) 假枯氨基；2, 4, 5-三甲苯氨基
pseudocuminol 假枯茗醇；三甲基苯酚 $HOC_6H_2(CH_3)_3$
as-pseudocumyl(=2, 3, 5-trimethylphenyl) 偏枯基；2, 3, 5-三甲苯基
s-pseudocumyl(=2, 4, 5-trimethylphenyl) 均枯基；2, 4, 5-三甲苯基
v-pseudocumyl(=2, 3, 6-trimethylphenyl) 连枯基；2, 3, 6-三甲苯基
pseudocurarine 夹竹桃碱
pseudocyanate 假氰酸盐
pseudocyclocitrylidene acetone 假性亚环柠檬基丙酮；假性环柠檬叉丙酮
pseudo-deconvolution 准去卷积法
pseudo-derivative IDA 假衍生物同位素稀释分析法
pseudo diaminocyclohexane 假二氨基环己烷
pseudo-diffusion coefficient 假扩散系数
pseudo-echo transformation filter 伪回波变换滤波
pseudo-eigen value 假本征值
pseudoelaeostearic acid 假桐酸
pseudoelastic property 假弹性
pseudo-emulsion 假乳状液〔极不稳定的乳状液〕
pseudo end point 假终点；终点不清
pseudo-ephedrine 假麻黄碱

d-pseudoephedrine(=isoephedrine) d-假麻黄碱
pseudo-equilibrium compliance 假平衡柔量
pseudo-equilibrium modulus 假平衡模量
pseudoequilibrium state 假平衡态
pseudo-equilibrium zone 假平衡区
pseudo ester 拟酯
pseudo-factor 拟因素
pseudofinite strain 假有限应变
pseudo first order 准一级
pseudo first order reaction 准一级反应*
pseudogas state 假气态
pseudo gel 假凝胶
pseudo-ginsheng 三七
pseudoglucosazone 假葡萄糖脎
pseudo-gums〔复〕 假胶；潜胶〔石油〕
pseudohalgen 拟卤素
pseudohalide 拟卤化物
pseudohalogen 拟卤素*
pseudohardness 假硬
pseudoheavy media 假重介质；假重液
pseudohexagonal crystalline system 准六方晶系
pseudo-high-dilution 拟高稀释
pseudohomolycorine 假高石蒜碱
pseudohyoscyamine 降天仙子胺
pseudo-indole(=indolenine) 假吲哚 C_8H_7N
pseudoindolyl 假吲哚基；异氮杂茚基〔有7种异构体〕 C_8H_6N-
pseudo in situ 准原位
pseudo-ionone 假紫罗酮 $C_{13}H_{20}O$
pseudoirone(=3-methyl pseudoionone) 假鸢尾酮；3-甲基假紫罗兰酮
pseudoisochromatic plate(=PIC plate) 伪等色图；色盲图
pseudoisomerism 假(同分)异构(现象)
pseudoisotope 假同位素
pseudoisotopic dilution analysis 假同位素稀释分析法
pseudojervine 假杰尔碱；拟藜芦碱
pseudolanthanide 准镧系元素
pseudolattice state 假晶格态
pseudolayer 假层
pseudo level 拟水平*
pseudo-liminar flow 拟层流
pseudolinear 假线性的
pseudoliving polymerization 假活性聚合
pseudolycorine 假石蒜碱
pseudo-matrix isolation(PMI) 伪矩阵隔离
pseudomer 假异构体
pseudomerism 假(同分)异构(现象)
pseudomonomolecular reaction 假单分子反应

pseudomorph(=pseudomorphous) 假同晶的
pseudomorphine 假吗啡
pseudomorphosis 假同晶
pseudomorphous(=pseudomorph) 假同晶的
pseudomorphy 假同晶
pseudonitrol 假硝醇 $R_2C(NH_2)NO$
pseudonorephedrine 去甲伪麻黄碱
pseudo-orthorhombic 准正交(晶)的
pseudopelletierine 假石榴碱
pseudoperiodicity 假周期
pseudo-peroxide 假过氧化物
pseudophase 假相
pseudo-pin hole 假针孔〔由空气中的尘埃粒子引起的针孔〕〔漆病〕
pseudoplastic 假塑性*
pseudoplastic behavior 假塑性性能
pseudoplastic consistency 假塑性稠度
pseudoplastic flow 假塑性流动
pseudoplastic flow curve 假塑性流动曲线
pseudoplastic fluid 假塑性流体
pseudoplastic hysteresis loop 假塑性滞后回线
pseudo-plasticity 假塑性；非宾厄姆塑性
pseudoplastic material 假塑性材料
pseudoplastic pigment 假塑性颜料
pseudoplastic Reynolds number 假塑性雷诺数
pseudoplastic rheological 假塑性流变
pseudoplastic viscosity 假塑性黏度
pseudo potential 赝势
pseudopotential method 假位势法
pseudopurpurin 假红紫素；三羟蒽醌羧酸
pseudoraceme 假外消旋体
pseudoracemic mixed crystal 假外消旋混(合结)晶
pseudoracemic mixture 假外消旋混合物
pseudorandom number 伪随机数
pseudo-reference electrode 准参比电极
pseudo-reinforcement 假补强；填充(作用)
pseudoreserpine 假利血平；假蛇根碱
pseudo-resistance 假抗病力〔胶树〕
pseudorotation 假旋转
pseudo saccharin chloride 氯代异糖精
pseudo-salt 假盐
pseudoschistosity 假片理性
pseudo-selectivity 准选择性*
pseudosmaragdite 假绿闪石
pseudo-solid 假固体(态)
pseudo-solid condition 假固态
pseudosolution 假溶液；胶体溶液
pseudo-stable 假稳定的

pseudo-stable flow 拟稳态流
pseudo-stationary phase 准固定相；假固定相
pseudostreptomycin 假链霉素
pseudostrychnine 假马钱子碱
pseudosymmetry 假对称(现象)
pseudotanning 假鞣
pseudotautomerism 假互变异构(现象)
pseudo termination 假终止
pseudoternary system 假三元体系
pseudo-thiohydantoic acid 假硫脲基乙酸
pseudothiourea 假硫脲
pseudotropine 假托品
pseudo-unimolecular reaction 准单分子反应*
pseudo-urea 假脲〔即:异脲〕 $OHC(=NH)NH_2$
pseudo-uric acid 假尿酸 $C_4H_3O_3N_2NHCONH_2$
pseuouridine 假尿(嘧啶核)苷
pseudo viscosity 假黏度；非"牛顿"黏度
pseudoviscous fluid 假黏性流体
pseudovitamin B_{12} 假维生素 B_{12}
pseudo-vulcanizate 假硫化橡胶〔橡〕
pseudowax 假石蜡
pseudo wet-bulb temperature 假湿球温度
psia(=pounds per square inch, absolute) 磅/平方英寸〔绝对压力〕
psicofuranine 狭霉素 C；阿洛酮糖腺苷
psi factor ψ因子
psig(=pounds per square inch, gauge) 磅/平方英寸〔表压〕
psilocin 二甲-4-羟色胺
psilocybin 二甲-4-羟色胺磷酸
psilopeganum sinense oil 裸芸油；山麻黄油
psoralea corylifolia 补骨脂
psoralene 补骨脂素
psoraline(=caffeine) 补骨脂灵
psoromic acid 荼㿉衣酸〔取自一种地衣 Psoroma crassum〕
psuedo-cell 准晶胞*
psuedo-periodicity 准周期性*
psuedo-symmetry 准对称性*
psychomimetic 引起幻觉的〔指药物〕
psychosine(=sphingosyl galactoside) 吐根素；(神经)鞘氨醇半乳糖苷
psychotrine 九节碱；吐根微碱
psychotropic drug 精神治疗药物
psychrometer 湿度计；干湿球湿度计
psychrometric chart 湿度图
psychrometric difference 湿度差
psychrometric method 湿度测量法
psychrometric ratio 干湿比；湿度比
psychrometry 湿度测定法

psylla alcohol 叶虱醇
psyllaic acid 叶虱酸 $CH_3(CH_2)_{31}COOH$
psylla wax 叶虱蜡
psyllic acid(=psyllostearylic acid) 叶虱酸；三十三(烷)酸 $C_{32}H_{65}COOH$
psyllic alcohol(=psyllostearyl alcohol) 叶虱醇；三十三(烷)醇 $C_{33}H_{67}OH$
psyllostearyl acetate 叶虱醇乙酸酯；乙酸三十三(烷)醇酯
psyllostearyl alcohol(=psyllic alcohol) 叶虱醇；三十三(烷)醇
psyllostearyl benzoate 叶虱醇苯甲酸酯
psyllostearylic acid(=psyllic acid) 叶虱酸；三十三(烷)酸
pteridine 蝶啶
pteridyl 蝶啶基 $C_6H_3N_4-$
pterigospermin 辣木素
pterin 蝶呤
pterobilin 蝶蓝素
pterocarpine 紫檀素
pteroic acid 蝶酸
pterostilbene 蝶芪
pteroyl 蝶酰 $C_{14}H_{11}N_6O_2-$
pteroylglutamic acid(=folic acid) 蝶酰谷氨酸；叶酸；维生素 Bc
pteroylmonoglutamic acid 蝶酰谷氨酸；叶酸；维生素 Bc
pteroylmonoylutamic acid 维生素 Bc
pteroyltriglutamic acid 蝶酰三谷氨酸
pterygospermim 辣木素
ptomaine 尸碱；尸毒
Pt wire NMR thermometer 铂丝核磁共振温度计
puberonic acid 短毛酮酸〔取自短毛青霉菌 Penicillium puberulum〕
puberulic acid 软毛青霉酸
puberulonic acid 软毛青霉二酸酐
Pubex process 〔词源: plutonium batch extraction〕 普贝克斯过程〔从提取超钚元素用的辐照过钚靶中用 D2EHPA/二乙苯从酸性溶液中回收和纯化未烧尽钚的间歇萃取过程〕
public hazard 公害
public nuisance 公害
public nuisance analysis 公害分析
public water 公用水
puccin 血根素
pucherite 钒铋矿
puchiin 荸荠英
puckered fabric 皱纹织物
puckered ring 折叠环*
puckering vibration 褶皱振动

pudding furnace(=puddler) 搅炼炉；炼铁炉
puddler(=puddling furnace) 搅炼炉
puddling 搅炼(作用)
puddling furnace(=puddler) 搅炼炉；炼铁炉
puddling process 搅炼过程
puddling slag 搅炼炉渣
puerarin 葛根素〔药〕
puff 疏松；膨突；泡起
puffing 晶胀现象
puffing agent 膨胀剂；增稠剂
puff-paste 层叠面团
puffy body(=false body) ①假稠性②假稠度
pug ①捏土②捏土机
pugging 捏和
pugging mill 捏和碾磨机
pug mill 捏和碾磨机
puk ①溅跑②发酵液跑锅
pukateine 蒲卡特因〔由樟科植物 Laurelia novae zelandae 得到的一种生物碱〕
puking 溅跑现象；翻跑现象〔发酵〕
pulanomycin(=planomycin) 平霉素
pulchellin 天人菊灵
pulcherrimin 普切明
pulcherriminic acid 美好菌素酸
pulegene 蒲勒烯
pulegenone 蒲勒烯酮
pulegium 长叶薄荷
pulegol 长叶薄荷醇
pulegone 长叶薄荷酮
pulenene 普楞烯
pulespenone 胡薄荷烯酮；长叶薄荷烯酮
pulforming 拉引成型
Pulfrich photometer 普耳弗里奇光度计
Pulfrich refractometer 普耳弗里奇折射计
pull air velocity 排(抽)气风速
pull-away (漆膜)脱落
pull back 拉回〔泵〕
pull back device 抗弯装置；滚筒定位装置
pull down 软化〔革〕
pulled bead 胎圈绷紧
pulled surface ①传动表面②粗糙表面；起皱表面
puller ①拎板机②拉者
puller belting 传动带
pulley ①滑轮；滑车②皮带轮
pulley belt 传动带
pull-in effect 回缩效应
pulling motion 拉动；牵(引运)动
pulling up (涂料)拉起

pull-in torque 启动转矩；输入转矩
pullman bread 长面包
pull off strength 拉出强度
pull-out coat and composite 塑料织物挤出涂布复合
pullout strength 拉离强度
pull-out torque 输出转矩
pull-push extraction system 推拉萃取体系
pull rate 提拉速度
pull resistance 抗拉性
pullulan 可溶性支链淀粉
pull-up （层压板）层脱；脱层
pulmoform 二愈创木酚基甲烷
pulp ①纸浆；纸粕②(水)浆③果肉；骨浆
pulp assay 纸浆试验
pulp board 纸浆板
pulp catcher 受浆器〔纸〕
pulp density 纸浆密度
pulp digester 蒸煮器
pulp disc method 纸圆盘法〔抗菌素的定性检测〕
pulp drier 甜菜废丝干燥器
pulp elevator 甜菜废丝提升器
pulper 碎浆机
pulp filter 纸浆过滤器
pulp flume 甜菜废丝烟道〔糖〕
pulpifying 打成浆；使化为纸浆
pulpiness 浆状
pulping 打(成)浆；碎浆；制浆
pulping engine 碎浆机
pulping machine 研磨机；捣碎机
pulp kneader 碎浆机
pulp mill 纸浆厂
pulp mill digester 蒸煮锅
pulp mill screenings 纸浆筛余物
pulpous ①浆状的②果肉的③柔软的
pulpous state 浆状
pulp press 甜菜废丝压榨器〔糖〕
pulp press water 甜菜废丝压榨水〔糖〕
pulp process paper 纸浆法纸；打浆法纸
pulp pump 废丝泵〔糖〕
pulp screen 筛浆机；(纸)浆筛
pulp sheet making machine 抄浆机
pulp sluice 纸浆堰
pulp slurry 液体浆
pulp stone 磨浆石
pulp strainer 纸浆筛滤器
pulp twine 纸捻纱；纸拈纱〔纺〕
pulp washer 洗浆池
pulp water 废丝水〔糖〕

pulp web 浆幅
pulp wood 造纸木材；制浆木材
pulpy ①浆状的②果肉的③柔软的
pulque 龙舌兰发酵汁
pulsaire collector ①脉冲式微粉捕集器②脉冲式除尘器
pulsating ①脉动(的)；脉冲的②片断的
pulsating current 脉动电流
pulsating equipment 脉动装置
pulsating flow 脉动流
pulsating light 脉动光
pulsating load 脉动负载；脉动载荷
pulsating output 脉动输出
pulsating plunger pump 脉动柱塞泵
pulsating pressure 脉冲压力
pulsating pump 脉动泵
pulsating screen 脉动筛
pulsating voltage 脉动电压
pulsation 脉动
pulsation damper 脉冲消除装置；脉冲阻尼器
pulsation phenomenon 脉动现象
pulsation point 喘振点；脉动点
pulsator classifier 脉动分级器
pulsator washer 脉动式洗衣机
pulse 脉冲*；脉动
pulse amplifier 脉冲放大器
pulse amplitude 脉冲幅度
pulse calorimetry 脉冲量热法
pulse-chase 脉冲追踪(术)
pulse chronopotentiometry 脉冲计时电位法
pulse column 脉冲柱
pulse coulometry 脉冲库仑分析法；脉冲库仑滴定法
pulse counting 脉冲计数
pulse current 脉冲电流
pulse damper 脉冲阻尼器
pulsed beam 脉冲束
pulsed chamber 脉冲室
pulsed clone contactor 脉冲旋流萃取器
pulsed column 脉动塔
pulsed compression wave 脉冲压缩波
pulsed current 脉冲电流
pulsed cyclone contactor 脉冲旋流萃取器
pulsed discharge detector(PDD) 脉冲放电检测器
pulsed discharge electron capture detector 脉冲放电电子捕获检测器
pulse delay 脉冲延迟
pulsed electron capture detector 脉冲电子俘获检测器
pulsed electron impact ion source 脉冲电子碰撞离子源
pulsed electron paramagnetic resonance 脉冲电子顺磁

共振
pulsed extraction column　脉冲萃取塔
pulsed extractor　脉冲萃取器
pulsed field gel electrophoresis　脉冲电场凝胶电泳
pulsed field gradient technique　脉冲梯度场技术
pulsed filter　脉动过滤器
pulsed flame photometer detector(PFPD)　脉冲火焰光度检测器
pulsed Fourier transform(PFT)　脉冲傅里叶变换
pulsed Fourier transform NMR spectrometer　脉冲傅里叶变换核磁共振(波谱)仪*
pulsed-gas fluidation　脉冲气体流态化
pulsed glow discharge　脉冲辉光放电
pulsed glow discharge ion source　脉冲辉光放电离子源
pulsed ion extraction　脉冲引出
pulsed ionizing beam　脉冲离子束
pulsed laser　脉冲激光器
pulsed laser beam　脉冲激光光束
pulsed laser photoacoustic calorimetry　脉冲激光光声(微量)量热法
pulsed matrix isolation　脉冲矩阵隔离
pulsed neutron activation　脉冲中子活化
pulsed neutron interrogation(PNI)　脉冲中子探询(法)
pulsed neutron reactor　脉冲中子(反应)堆
pulsed neutron source　脉冲中子源
pulsed packed tower　脉冲填料塔
pulsed photothermal deflection spectroscopy　脉冲光热偏转光谱法
pulsed positive ion-negative ion chemical ionization　脉冲正-负离子化学电离
pulsed pyrolysis unit　脉冲热解单元
pulsed sieve-plate column　脉冲筛板柱
pulse duration　脉冲持续时间
pulsed voltage　脉冲电压
pulse excitation　脉冲激发
pulse extraction column　脉动抽提柱
pulse field gel electrophoresis　脉冲电场凝胶电泳
pulse Fourier transform nuclear magnetic resonance spectrometer　脉冲傅里叶变换核磁共振谱仪
pulse Fourier transform nuclear magnetic resonance spectroscopy　脉冲傅里叶变换核磁共振谱
pulse-free chromatographic pump　无脉冲色谱泵
pulse-free flow　无脉冲流
pulse-free pump　无脉冲泵
pulse-free signal　无脉冲信号
pulse frequency　脉冲频率
pulse frip angle　脉冲倾倒角
pulse generator　脉冲发生器
pulse height analyzer　脉冲振幅分析器；脉冲高度分析器
pulse height discriminator　脉冲高度鉴别器
pulse height distribution analysis(PDA)　脉冲高度分布分析
pulse height distribution curve　脉冲高度分布曲线
pulse height selector　脉冲高度选择器
pulse homonuclear decoupling　脉冲同核去耦
pulse injection　脉动注射
pulse interval　脉冲间隔
pulse ionization chamber　脉冲电离室
pulse labelling　脉冲标记(术)
pulse laser　脉冲激光器
pulseless pump　无脉动泵
pulse mode pyrolyzer　脉冲式热解器
pulse nebulization　脉冲雾化法
pulse period　脉冲周期
pulse photothermal deflection spectrum　脉冲光热偏转光谱
pulse propagation　脉冲(波)传播
pulse propagation method　脉冲传播法
pulse pump　脉冲泵
pulse pyrolysis　脉冲热解；脉冲高温分解
pulse pyrolysis gas chromatography method　脉冲热解气相色谱(法)
pulser　脉冲发生器
pulse radiolysis　脉冲辐解*
pulse radiopolarography　脉冲放射极谱法
pulse scheme about two dimensional spectra　二维NMR谱脉冲序列的时序图
pulse sequence and time-order scheme　脉冲序列与时序图
pulse spectrometry　脉冲能谱学
pulse technique　脉冲技术
pulse type reactor　脉冲(型)反应器
pulse voltammetry　脉冲伏安法
pulse wave　脉冲波
pulse welding　脉冲焊
pulse width　脉冲宽度
pulse width modulation　脉冲宽度调制
pulsing rate　脉冲速率
pulsometer　①气压扬水机②自动抽酸机③脉搏计
pultruded section　拉挤型材
pultruder　拉挤成型机
pultrusion　管材挤出；拉挤成型
pultrusion machine　拉挤成型机
pulverability　(可)粉化性
pulverator　粉碎器
pulverin(=barilla)　草灰；苏打灰；海草灰苏打
pulverising mill　微粉碎机；粉碎磨
pulverizable　可以粉化的；可成粉末的

pulverization 粉碎(作用);研末(作用)
pulverized ①(磨成)粉状的②喷雾的
pulverized chalk(=dry processed chalk) 干磨碳酸钙〔填料〕
pulverized coal 粉煤;煤粉;煤末
pulverized dirt 粉状屑
pulverized fuel 粉状燃料
pulverized fuel pipe 粉状燃料的管线
pulverized rubber 橡胶粉
pulverized soap stone 滑石粉
pulverized specimen 粉状样品
pulverizer ①粉磨机②喷雾器
pulverizing ①精细粉碎②喷雾;雾化
pulverizing element 研磨机件
pulverizing mill 微粉碎机;粉碎磨
pulverizing sample 粉碎样品
pulverous 粉状的
pulverulence 粉末状态
pulverulent 粉状的
pulvic acid 普耳文酸;枕酸
pulvilloric acid 粉青霉酸;青霉酸
pulvinate 普耳文酸盐;枕酸盐
pulvinic acid 普耳文酸;枕酸
pulvomycin 粉霉素
pumacite(=pumice) 浮石
pumice(=pumacite; pumice stone) 浮石;轻石
pumice concrete 浮石水泥
pumice in lumps 块状浮石
pumiceous 浮石(质)的
pumiceous texture 浮石结构;泡沫状结构
pumice sand 浮石粉〔填料〕
pumice soap 浮石皂
pumice stone(=pumice) 浮石;轻石
pumice stone concrete 浮石水泥
pumicite ①(=pumice)浮石②火山尘埃
pumiliol 山松醇
pummeling (油鞣革)揉软
pump 泵;唧筒;抽机
pumpability 可泵性
pumpability test 泵抽试验
pumpable 可泵抽的
pump around circuit 循环回流
pumpback 回抽;反流
pump beam 泵浦光束
pump block 泵套
pump case 泵壳
pump check valve 泵止逆阀
pump circulation 泵唧循环;强制循环
pump cooling 泵唧冷却
pump delivery 泵排量
pump disc 泵盘
pump displacement 泵送本领
pump duty 泵排量
pumped laser 泵激激光
pump exhausting 泵抽
pump feed evaporator 泵供液蒸器;强制循环蒸发器
pump gun 泵唧黄油枪
pump head 泵压头;泵的压力
pump house 泵站;泵房
pumping 泵送
pumping action ①泵的动作②泵的作用
pumping capacity 泵唧效能
pumping equipment 泵设备
pumping fluid 泵抽液体
pumping installation 泵的装置
pumping limit 喘振限;泵(唧)送极限
pumping losses 泵唧损失
pumping main 泵唧主管
pumping out 泵唧排出;泵出
pumping-out line 泵唧排出(石油产品)管线
pumping over 泵送;唧送
pumping plant 泵站
pumping point 喘振点;泵出点;唧动点;抽动点
pumping power ①泵唧功率②泵送能力
pumping rate 泵唧速度;增压速度
pumping speed 抽气速率
pumping station 泵站
pumping unit 泵的装置
pumping value 泵送值;泵汲值〔指泵送规定压力而言〕
pump inlet valve 泵吸入阀
pump installation 泵的装配
pumpkin-seed oil 南瓜子油
pump laser 泵激激光
pumpless 无泵术
pump lift 泵的扬程
pump log 木制溢流导管
pump lubrication 泵唧润滑;加压;润滑
pump main 泵唧主管;泵的导出管
pump-mixing system 泵搅拌系统
pump-mix type 泵混合型;泵搅拌型
pump off 泵唧排出;泵出
pump outlet valve 泵(排)放出阀
pump output(=pumping capacity) 泵排量;泵出量
pump out valve 抽空阀
pumpover 泵送;唧送
pump overcapacity 泵的过容量
pump pipette 泵吸移管

pump piping 泵唧管线
pump plant 泵站
pump priming 泵的充溢
pump ram 泵柱
pump runner 泵输
pump station 泵站
pump tang 泵用榫轴
pump turbine ①涡轮泵②水泵-水轮机
pump-type circulation system lubrication 泵式循环润滑系统
pump valve 泵阀
punch ①冲床；冲压机②穿孔器；穿孔③果汁饮料
punch card control system 穿孔卡控制系统
punched card 穿孔卡片
punched-card devices 冲孔卡片法；在打孔卡片上计算分馏过程的方法
punched card method 穿孔卡片法
punched paper tape 穿孔纸带
punched-plate screen 冲孔板筛
punched tape 穿孔纸带
puncher 穿孔器；冲孔器
punching 穿孔；冲孔
punching die 冲模；冲刀
punching machine ①轧切机；冲压机②冲孔机；打孔机；穿孔机
punching pad 冲压胶垫
punch pin 冲头
punch press 冲床；冲压机
punch tape controlled programming system 穿孔纸带控制程序系统
punch tool 打孔器
punch work 冲孔工作
punctuation (密码)标点法
puncture 刺扎；刺孔
puncture voltage 击穿电压
pungenin 松针苷；二羟苯乙酮葡萄糖苷
pungent ①刺激的②刺鼻的③刺舌的
pungent odor 刺激臭味；尖刺气息〔香料〕
pungent taste 刺激味
punicic acid 石榴油酸；十八碳三烯酸
punicine 石榴素
punizin 紫螺紫素
punjum waste 黏并废丝
punt(=punty) 铁杆
puntee(=punty) 铁杆
punty(=punt; puntee) (取熔融玻璃用的)铁杆
punty feeder 铁杆送料器〔玻〕
pupal acid 蛹酸

pupal fat 蛹油；蛹脂
pure 纯(净)的；清洁的
pure air 洁净空气
pure asphalt 纯沥青
pure chemistry 纯化学
pure compound 纯化合物
pure copper 红铜
pure deformation 纯形变
pure flow 直线流动
pure gas ①纯煤气②纯净气体
pure gum stock 纯胶料；纯生橡胶
pure isooctane 纯异辛烷；2,2,4-三甲基戊烷
pure kraft paper(=unglazed kraft paper) 棕色平面包皮纸
pure line 纯系
pure liquid lubrication 纯液体润滑
purely elastic substance 纯弹性物质
purely plastic flow 纯塑性流动
purely viscous fluid 纯黏性流体
purely viscous substance 纯黏性物质
pure mineral oil 纯矿油
pureness ①纯度②纯洁
pure Newtonian flow 纯牛顿流动
pure organic liquid 纯有机液体
pure oxygen 纯氧
pure purples track 纯紫(色)径迹
pure reclaim 纯再生胶〔橡胶〕
pure rubber 纯(净)生胶
pure sample 纯样
pure shear 纯剪切
pure soap content 纯皂含量
pure soluble blue 纯溶蓝〔商〕；酸性水溶青
pure strain 纯应变
pure stress 单向应力
pure substance 高纯度物质；纯物质
pure turning spectrum of gas molecule 气体分子的纯转动光谱
pure viscosity 纯黏性
pure viscous flow 纯黏性流
pure water 纯水；净水
Purex process (词源:pluotnum-uranium refining by extraction) 钚雷克斯流程〔从辐照铀燃料用磷酸三丁酯萃取法回收纯化铀钚〕
Purex waste 钚雷克斯过程废液
purgatin 轻泻素；二乙酸三羟蒽醌
purgative ①(泄)泻剂；泻药②泄泻的③清除的；有清除力的
purgative gas 清洗气
purgatol 泻脑 $C_{18}H_{12}O_7$

purge ①清洗；清除②使清洁
purge-and-trap methods 吹扫和捕集方法
purge cock 放水旋塞
purge flow 清洗(气)流
purge gas 吹扫气体；排气
purge oil 冲洗油
purge pipe 排气管
purge valve 放(空)气阀
purging ①清洗；清除②换气③使泻清
purging by alternating current electrolysis 交流电解清洗
purging method 冲洗方法；清洗方法
purging nut oil 麻风子油
purification 提纯；净化；纯化；精制
purification efficiency 纯化效能；净化效率
purification of gas 气体纯化
purification of water 水的净化；净水(法)
purification tank 净化池
purification technique 提纯技术；净化技术；精制技术
purification tower 提纯塔；净化塔
purified 精制的；精炼的；纯净的
purified clay 精制瓷土
purified gas 净化气
purified grade 精制级
purified petroleum benzin 纯苯
purified product 精制产品
purified water 净化水
purifier 净化器
purifyer 提纯器
purifying 精制；纯化；提纯；净化
purifying agent 提纯器；净化器；纯净器
purifying box 净化箱
purifying column 净化塔
purifying material 纯净物料
purine base 嘌呤碱
puring vat 脱灰桶；鞣瓮
purity 纯度*
purity analysis 纯度分析
purity check 纯度检验
purity checking 检定纯度
purity coefficient 纯度
purity determination 纯度测定
purity determined by differential scanning calorimetry 差示扫描量热法测定纯度
purity drop 纯度降低
purity of color 色纯度
purity of hone 色相纯度
purity quotient 纯度
purity test 纯度试验

Purlex process 珀莱克斯过程；"纯化提取"法〔改进的 Amex 法，从铀矿浸出液中用胺类萃取法制得高纯铀化合物〕
purl lace fabric 双反面纱罗织物
purple ①红紫；绀(色)；青莲(色)②变成红紫色
purple carmine 红紫酸铵
purple copper 斑铜矿
purple lake 红紫色淀
purple lustrous powder 红紫色闪光粉末
purple of cassius 金锡紫
purple pigment 红紫色颜料
purple red 紫红
purple rugose rose 紫玫瑰
purplish-black 红紫黑
purpure 紫色
purpureal 红紫色的；绀色的；青莲色的
purpureo 红紫配盐
purpureo-cobaltichloride 二氯化一氯五氨合钴；红紫氯钴盐 $[Co(NH_3)_5Cl]Cl_2$
purpuric 红紫(酸)的
purpuric acid 红紫酸〔紫尿酸是 violuric acid〕 $C_8H_5N_5O_6$
purpurin 红紫素；1,2,4-三羟基蒽醌 $C_{14}H_8O_5$
purpurogallin 红棓酚
purpurogenone 红紫精酮
purpuromycin 绛红霉素
purpuroxanthene(=xanthopurpurin) 1,3-二羟基蒽醌
purpuroxanthic acid 红紫黄原酸 $C_{15}H_3O_6$
purpurum bromocresolis 溴甲酚紫
purree 印度黄
purreic acid ①印度黄酸②优黄酸
purrenone 优呫吨酮
purrone 优呫吨酮
Pusey-Jones hardness tester 普西-琼斯硬度试验器
pushback plunger 回程柱塞
push bar 推杆
push-bar conveyor 刮板式运输机
push-bottom oiler 按底润滑器；薄膜润滑器
push-button 按钮
push button actuator 按钮起动器
push-button box 按钮箱
push-button control 按钮控制
push-button pipe line station 按钮泵站
push-button switch 按钮开关
push-button syringe 按钮注射器
push-button tank gauging 按钮油罐测量法
push-button valve 按钮阀
pusher 推杆

pusher-bar conveyer 推杆运输机
pusher machine 推焦车
pusher mechanism 推料机构装置
pusher pump 压气泵
pusher side 推焦面
pusher-type 推送(料)式的；(强)压式的
pusher-type fan 压风机；压力通风机
push fit 推入配合
pushing ①推②推焦
pushing machine 推焦车
push-on starter 按钮起动器
push-out 推出；排出
pushover bar 推杆；推出器
push pole 推杆
push-pull effect 推拉效应*
push-pull propeller stirrer 推拉螺桨搅拌器
push-pull rod 推拉杆
push-pull test 推拉试验；推挽试验
push-pull type 推拉型
push rod 推杆
push-type centrifuge 推送式离心机
push-type conveyer 推式输送机
push-type lubricating fitting 加压型的润滑填料
pustulan 石脐素
pustulant 起脓疱剂
put into gear 起动；开动
put into operation 实行；实施；投入运行；投入生产
put into service 交付使用，投入运行；启用
put on stream 投入生产；开动
put on trial 试验
put out 关；停止
put out of action(=put out of service) 关；停止
put out of service(=put out of action) 关；停止
putrefaction test 腐败试验〔水性粉末涂料的检验项目之一〕
putrefactive ①腐败的②引起腐败的
putrefactive alkaloids 腐败生物碱
putrescine 腐胺；1,4-丁二胺 $NH_2(CH_2)_4NH_2$
putrid soak 发酵浸水
putting-out 平展；伸展
putting-out cylinder 伸展机刀轴
putty chaser 油灰磨
putty up 刮底漆腻子
puzzolana 白榴火山灰
puzzolana cement 火山灰水泥
puzzolane 白榴火山灰
puzzolanic admixture (白榴)火山灰加料
PVA adhesive 聚醋酸乙烯黏结剂

PVC coating 聚氯乙烯涂料
PVC dispersion 聚氯乙烯分散液
PVC gravure ink 聚氯乙烯凹版油墨
PVC membrane electrode 聚氯乙烯膜电极
PVC stabilizer 聚氯乙烯稳定剂
PVD-electron beam 物理蒸气沉积电子束
pX unit pX单位
pX value pX值
pycnometer 比重瓶
pycnometer method 比重瓶测定比重法
pycnometric 测比重的
pycnometric density 比重瓶测量密度法
pycnosis 固缩现象
pyelectan 肾影碘；1-甲基-3,5-二碘-4-吡啶酮-2,6-二羧酸钠
pygmaein 矮柏醚
pyknometer 比重瓶；比重计
pyknometry 比重瓶测定法
pylon ①柱台②定向塔；定向起重机
pylumin process 扑铝明法〔铝表面化学处理法之一；其专用处理液牌号为 Pylumin,化学成分为碱性铬酸盐加碳酸钛或碳酸铬，或磷酸氢二钠〕
pyo (绿)脓菌素
pyoceram 玻璃瓷
pyocin 脓菌素
pyoclastin 片绿脓菌素
pyo compounds 浓菌素类化合物
pyoctanin(=methyl violet) 甲基紫；脓单宁
pyocyanase 绿脓菌酶
pyocyanic acid 绿脓菌酸
pyocyanin 脓青素
pyocyanine 绿脓菌素
pyocyanolysin 绿浓菌溶素
pyod 热电偶；温差电偶
pyoflurorescein 脓荧光黄
pyoktanin(=methyl violet) 甲基紫；脓单宁
pyoktanin blue(=methyl violet) 甲基紫；脓单宁蓝
pyoktanin yellow(=auramine) 金胺；脓单宁黄
pyolipic acid 绿脓脂酸
pyoluene 脓硫烯 $C_5H_9O_2SN$
pyoluteorin 藤黄绿脓菌素
pyorubis 绿脓菌红素
pyosin 脓胞素
pyostacin(=pristinomycin) 原始霉素
pyoxanthin 脓黄质
pyoxanthose 半绿脓青素
pyperomyl butoxide 胡椒基丁氧醚
pyr- 〔词头〕 焦

pyracetic acid 焦木酸
pyraconitine 焦乌头碱
pyracrimycin 吡丙烯霉素
pyrahexy(=hexa-hydrocannabinol) 六氢大麻醇
pyraloxin(=pyrogallol oxide) 氧化焦没食子酚
pyramid 棱锥形；角锥〔数〕；金字塔(形)
pyramidal 棱锥(形)的；角锥(形)的
pyramidal inversion 棱锥形倒反
pyramidal pit 棱台形坑
pyramidal point 棱锥尖
pyramidal rotating screen 棱锥旋转网
pyramidal structure 棱锥结构
pyramid hardness 棱锥体硬度
pyramid indenter （硬度计)角锥形压头
pyramidon(=aminopyrine) 匹拉米董；氨基比林〔药〕
pyramine 嘧胺；2-甲-4-氨-5-羟甲基嘧啶
pyran(=pyrane) 吡喃 C_5H_6O
pyrane(=pyran) 吡喃
pyranil black 一种黑色硫化染料
pyranium salt 吡喃盐
pyranofructose 吡喃果糖
pyranoglucose 吡喃葡萄糖
pyranohexose 吡喃己糖
pyranoid 吡喃型的
pyranoid form 吡喃型
pyranoid ring 吡喃环
pyranone(=pyrone) 吡喃酮
pyranopentose 吡喃戊糖
pyranose 吡喃糖
pyranose form 吡喃型
pyranoside 吡喃糖苷
pyrantel pamoate 双羟萘酸噻嘧啶〔药〕
pyranthrene 皮蒽 $C_{30}H_{16}$
pyranthrone(=indanthrene golden orange G) 皮蒽酮染料；阴丹士林金黄 G
pyranthrone orange 皮蒽酮橙；阴丹士林橙
pyrantin(=p-ethoxyphenyl-succinimide) 吡喃亭 $C_2H_5OC_6H_4N(COCH_2)_2$
pyranyl 吡喃基 OC_5H_5—
pyranylation 吡喃基化
pyrargyrite(=acrosite) 硫锑银矿；深红银矿
pyrathiazine 吡咯拉坐；匹拉噻嗪〔药〕
pyrazinamide(=pyrazinoic acid amide) 吡嗪酰胺
pyrazine(=1, 4-diazine) 吡嗪 N=CHCH=NCH=CH
pyrazine-2,3-dicarboxylic acid 吡嗪-2,3-二羧酸
pyrazine ethanethiol 吡嗪基乙硫醇
pyrazine methanethiol 吡嗪基甲硫醇

pyrazinodiisoindole 吡嗪并二异吲哚
pyrazinoic acid amide(=pyrazinamide) 吡嗪酰胺
pyrazinyl 吡嗪基 $N_2C_4H_3$—
2-pyrazinyl ethyl mercaptan 2-吡嗪基乙硫醇
pyrazinyl methyl mercaptan 吡嗪基甲硫醇
pyrazinyl methyl sulfide 吡嗪基甲基硫醚
pyrazofurin 吡唑呋喃菌素
β-pyrazol-1-alanine β-吡唑丙氨酸
pyrazole 吡唑*
pyrazole blue 吡唑蓝
pyrazolidine 吡唑烷 $HNCH_2CH_2CH_2NH$
pyrazolidinone(=pyrazolidone) 吡唑烷酮
pyrazolidinyl(=pyrazolidyl) 吡唑烷基 $N_2C_5H_7$—
pyrazolidone 吡唑烷酮 $C_3H_6ON_2$
pyrazolidyl(=pyrazolidinyl) 吡唑烷基 $N_2C_5H_7$—
pyrazoline 吡唑啉；二氢化吡唑 $C_3H_6N_2$
pyrazoline dithiocarbamate 吡唑啉氨荒酸盐
pyrazolinium 二氢化吡唑鎓
pyrazolinyl 吡唑啉基 $N_2C_3H_5$—
pyrazolone(=1, 2-pentadiazenone) 吡唑啉酮 $COCH_2CH=NNH$
pyrazolone maroon 吡唑啉酮紫红
pyrazolone red 吡唑啉酮红
pyrazolone ring 吡唑啉酮环
pyrazolone yellow 吡唑啉酮黄
pyrazoloquinazolone pigment 吡唑喹唑(啉)酮颜料
pyrazolyl 吡唑基 $N_2C_3H_3$—
pyrazomycin 吡唑霉素
pyrazotol yellow 吡唑酚黄
pyrene 芘 $C_{16}H_{10}$
pyreneite 黑钙铁榴石
pyrenite 甲基-三硝基苯基硝胺
pyrenol 芘醇（百里酚等的混合物）
pyrenophorin 核球壳菌素
pyrenum 百里香酚、苯甲酸钠和水杨酸钠的混合物
pyrenyl 芘基 $C_{16}H_9$—
pyrethrin 除虫菊酯
pyrethroid 拟除虫菊酯
pyrethrol 除虫菊醇 $C_{21}H_{34}O$
pyrethrolone 除虫菊醇酮
pyrethrone 除虫菊酮
pyrethrosin 除虫菊精
pyrethrum camphor 除虫菊脑
pyrethrum extract 除虫菊萃
pyretol 除虫菊醇
Pyrex ①派热克斯牌②派热克斯(牌硬)玻璃

Pyrex glass 派热克斯(牌硬)玻璃
pyrgeometer 地面(大气)辐射强度计
pyrheliograph 直接日照(射)强度计;日光直射强度计
pyrheliometer 日射强度计
pyridazine(=1,2-diazine) 哒嗪 N=CHCHCHCH=N
pyridazinone 哒嗪酮 $C_4H_6ON_2$
pyridazinyl 哒嗪基 $N_2C_4H_3-$
pyridazone 哒酮 $C_4H_4ON_2$
pyridination 使成吡啶盐
pyridindole 吡啶哚;吡啶并氮茚 $C_{11}H_8N_2$
pyridine 吡啶*
pyridine acid 吡啶酸
pyridine bases 吡啶碱类
pyridine-butadiene latex 丁吡胶乳
pyridine butadiene rubber 丁吡橡胶
pyridine carboxylic acid(=picolinic acid) 吡啶甲酸 C_5H_4NCOOH
pyridine-(Ca-saponite) complex 吡啶-(钙-皂石)复合物
pyridine colorimetric method 吡啶比色法
pyridine derivatives 吡啶衍生物
pyridine dicarboxylic acid 吡啶二羧酸 $C_5H_3N(COOH)_2$
pyridine-2,3-dicarboxylic acid(=quinolinic acid) 喹啉酸;2,3-吡啶二羧酸
pyridine dihydrochloride 二盐酸吡啶 $C_5H_5N \cdot 2HCl$
pyridine diisocyanate 吡啶二异氰酸酯
pyridine disulfonic acid 吡啶二磺酸 $C_5H_3N(SO_3H)_2$
pyridine dye 吡啶染料;氮苯染料
pyridine hydrochloride 盐酸吡啶 C_5H_5NHCl
4-pyridine methanethiol 4-吡啶基甲硫醇
pyridine monocarboxylic acid 吡啶一羧酸 C_5H_4NCOOH
pyridine monosulfonic acid 吡啶一磺酸 $C_5H_4NSO_3H$
pyridine oxide 氧化吡啶;氧化氮苯
pyridine N-oxide(=pyridine N-oxyl) N-氧化吡啶
pyridine N-oxyl(=pyridine N-oxide) N-氧化吡啶
pyridine-pentacarboxylic acid 吡啶五羧酸 $C_5N(COOH)_5$
pyridine-pyrazolone reagent 吡啶-吡唑啉酮试剂
pyridine ring 氮苯环;吡啶环
pyridine-SO_3 complex 吡啶-SO_3络合物〔硫酸化剂或磺化剂〕
pyridine sulfonamide 吡啶磺酰胺
pyridine-sulfonic acid 吡啶磺酸 $C_5H_4NSO_3H$
pyridine thiocyanate reaction 吡啶-硫氰酸铵沉淀反应
pyridine tricarboxylic acid 吡啶三甲酸
pyridinium 吡啶鎓 $C_5H_5NH^+$

pyridinium cyclopentadienylide N-环戊二烯吡啶鎓
pyridinium molybdophosphate(PMP) 钼磷酸吡啶鎓
pyridinium salt 吡啶盐
pyridinium tungostophosphate(PWP) 钨磷酸吡啶鎓
pyridinium tungostosilicate(PWSi) 钨硅酸吡啶鎓
pyridinium uranyl chloride 氯化铀酰吡啶鎓 $(C_5H_6N)_2UO_2Cl_4$
pyridinium wollframophosphate 钨磷酸吡啶鎓
pyridino 吡啶并
pyridinophane 吡啶并环烷
pyrido 吡啶并
pyrido-carbazole 吡啶并咔唑
pyridol 吡啶酚 C_5H_4NOH
pyridomycin 吡啶霉素
pyridone 吡啶酮;羟基吡啶
pyridone orthophosphate 正磷酸吡啶酮
pyridopyridine 吡啶并吡啶 $C_8H_6N_2$
pyridoquinoline 吡啶并喹啉 $C_{12}H_8N_2$
pyridostigmine 3-二甲氨基甲酰氧基-1-甲基吡啶
pyridostigmine bromide 溴化 3-二甲氨基甲酰氧基-1-甲基吡啶 $C_9H_{13}O_2N_2Br$
pyridotropolone 吡啶并草酚酮
pyridoxal 吡哆醛
pyridoxal phosphate 磷酸吡哆醛
pyridoxamine 吡哆胺
pyridoxamine phosphate 磷酸吡哆胺
pyridoxic acid 吡哆酸
pyridoxine 吡哆素〔类名,指具有维生素 B_6 活性的天然吡啶衍生物,包括吡哆醇、吡哆胺、吡哆醛等〕
pyridoxine diocatanoate 二辛酸吡哆醇酯
pyridoxine methyl ether 吡哆素甲醚 $CH_3OC_8H_{10}O_2N$
pyridoxin(e) phosphate 磷酸吡哆素
pyridoxine tripalmitate 吡哆素三棕榈酸酯
pyridoxin hydrochloride 维生素 B_6 盐酸盐
pyridoxiol 吡哆醇
pyridoxol 吡哆醇
pyridyl 吡啶基 NC_5H_4-
pyridylacetic acid(HPAC) 吡啶乙酸 $C_5H_4NCH_2COOH$
3-(α-pyridyl) acrylic acid 3-(α-吡啶基)丙烯酸
pyridylaldehyde 吡啶甲醛
pyridyl-2-aldoxime 吡啶基-2-醛肟
2-(2-pyridylazo)-5-diethylaminophenol(PADAP) 2-(2-吡啶偶氮)-5-二乙氨基苯酚
1-(2-pyridylazo)-2-naphthol 1-(2-吡啶基偶氮)-2-萘酚
4-(2-pyridylazo) resorcinol(PAR) 4-(2-吡啶基偶氮)间苯二酚
pyridylidene 亚吡啶基
4-pyridyl methyl mercaptan 4-吡啶基甲硫醇

pyridylpyridinium dichloride 二氯化吡啶基吡啶
pyriform 梨状的
pyrilamine maleate 马来酸吡拉明;顺丁烯二酸吡纳明〔药〕
pyrimetamine 嘧啶甲胺
pyrimethamine 乙胺嘧啶〔药〕
pyrimidine(=metadiazine) 嘧啶
pyrimidine bases 嘧啶碱类
pyrimidine cluster 嘧啶团
pyrimidine dimer 嘧啶二(聚)体
pyrimidine dione 嘧啶二酮
pyrimidine nucleoside 嘧啶核苷
pyrimidine tetrone 嘧啶四酮
pyrimidine trione 嘧啶三酮
pyrimidinyl(=pyrimidyl) 嘧啶基;间二氮苯基 $N_2C_4H_3$—
pyrimido 嘧啶并
2H-pyrimido[4,3-α] isoquinoline 2H-嘧啶并[4,3-α]异喹啉
pyrimido[4,5-b] quinoline 嘧啶并[4,5-b]喹啉
pyrimido-isoquinoline 嘧啶并异喹啉
pyrimidone 嘧啶酮 $C_4H_6N_2O$
pyrimidyl(=pyrimidinyl) 嘧啶基;间二氮苯基
pyrindane 4-氮杂全氢茚
pyrindigo 吡啶靛蓝
pyrindine ①4-氮茚②〔有时指〕5-氮茚
pyrindol 吡咯并[1,2-α]吡啶
pyrindole(=indolizine) 中氮茚;吲嗪 C_8H_7N
pyrindoxylic acid 4-氮茚酸
pyrite 黄铁矿*
pyrite burner 黄铁矿炉
pyrite burner gas 黄铁矿炉气
pyrite cinder 黄铁矿烬滓
pyrite copper ore 黄铜矿
pyrite dust 黄铁矿粉
pyrite dust burner 黄铁矿粉燃烧炉
pyrite fines〔复〕 碎黄铁矿
pyrite fines burner(=pyrite fines roaster) 碎黄铁矿燃烧炉
pyrite fines roaster(=pyrite fines burner) 碎黄铁矿燃烧炉
pyrite furnace 黄铁矿炉
pyrite furnace gas(=pyrite gas) 黄铁矿炉气
pyrite gas(=pyrite furnace gas) 黄铁矿炉气
pyrite hand-rabbled burner(=pyrite hand-rabbled furnace; pyrite hand-rabbled roaster; pyrite hand-worked burner) 黄铁矿手工搅拌炉
pyrite hand-rabbled furnace(=pyrite hand-rabbled burner) 黄铁矿手工搅拌炉
pyrite hand-rabbled roaster(=pyrite hand-rabbled burner) 黄铁矿手工搅拌炉
pyrite hand-worked burner(=pyrite hand-rabbled burner) 黄铁矿手工搅拌炉
pyrite kiln 黄铁矿窑
pyrite mechanical burner(=pyrite mechanical furnace; pyrite mechanical roaster) 黄铁矿机械炉
pyrite mechanical furnace(=pyrite mechanical burner) 黄铁矿机械炉
pyrite mechanical roaster(=pyrite mechanical burner) 黄铁矿机械炉
pyrite oven 黄铁矿炉
pyrite oven gas 黄铁矿炉气
pyrite roaster 黄铁矿焙烧炉
pyrite roaster gas 黄铁矿焙烧炉气
pyrites 硫化铁矿类
pyrite shelf burner(=pyrite shelf furnace; pyrite shelf roaster) 黄铁矿屉式炉
pyrite shelf furnace(=pyrite shelf burner) 黄铁矿屉式炉
pyrite shelf roaster(=pyrite shelf burner) 黄铁矿屉式炉
pyrite smalls〔复〕 碎黄铁矿
pyrites sulfur 黄铁矿硫
pyrithiamine 吡啶(代噻唑)硫胺;抗硫胺
pyrithione 2-巯基吡啶氧化物
pyritic ①高温的②黄铁矿的
pyritical ①高温的②黄铁矿的
pyritic process(=pyritic smelting) 高温冶炼
pyritic smelting(=pyritic process) 高温冶炼
pyritic sulfur 黄铁矿硫
pyritiferous rock 黄铁岩
pyritinol 吡硫醇〔药〕
pyritous ①高温的②黄铁矿的
pyro- 〔希腊字头〕 火;热;高温;焦
pyroabietic acid 焦松香酸
pyroacetic acid 焦木酸
pyro-acid 焦酸〔在无机酸中指:一水缩二某酸〕
pyroaersenate 焦砷酸盐 $M_4As_2O_7$
pyro alcohol 甲醇
pyroantimonate 焦锑酸盐 $M_4Sb_2O_7$
pyroantimonic acid 焦锑酸 H_4Sb_2O
pyroantimonite 焦亚锑酸盐 $M_4Sb_2O_5$
pyroantimonous acid 焦亚锑酸 $H_4Sb_2O_5$
pyroarsenic acid 焦砷酸 $H_4As_2O_7$
pyroarsenite 焦亚砷酸盐 $M_4As_2O_5$
pyroarsenous acid 焦亚砷酸 $H_4As_2O_5$
pyroaurite 鳞镁铁矿;碳酸镁铁矿
pyrobain(=guaiacol enzylicether) 派鲁卡因;愈创木酚苄醚
pyrobetulin 焦桦木脑

pyrobitumen 焦沥青
pyrobituminous shale 焦沥青页岩；油页岩
pyroborate 焦硼酸盐 $M_2B_4O_7$
pyroboric acid 焦硼酸；四硼酸 $H_2B_4O_7$
pyrocalciferol 焦钙化醇；焦骨化醇
pyrocarbon coating 热解碳涂层
pyrocatechin 焦儿茶酚；邻苯二酚
pyrocatechol 焦儿茶酚；邻苯二酚 $C_6H_4(OH)_2$
pyrocatechol monomethyl ether 焦儿茶酚单甲醚
pyrocatechol sulfonphthalein(=pyrocatechol violet) 邻苯二酚紫
pyrocatechol violet(=pyrocatechol sulfonphthalein) 邻苯二酚紫
pyrocatechu aldehyde 焦儿茶醛 $(OH)_2C_6H_3CHO$
pyrocatechuic acid （焦）儿茶酸；2,3-二羟(基)苯甲酸
pyrocellulose 焦纤维素；高氮硝化纤维素
pyrocellulose powder 焦纤维素火药；高氮硝化纤维火药
pyrochemical processing 高温化学处理
pyrochemistry 高温化学
pyrochlor 烧绿石
pyrochlore constitution 烧绿石结构
pyrochlore structure 焦绿石结构
pyrochroite 片水锰矿
pyrocinchonic acid 焦辛可酸；二甲代丁烯二酸
　　$COOHC(CH_3)=C(CH_3)COOH$
pyroclastic 有火成和水成料的火山灰
pyroclavine 焦棒碱；焦棒麦角素
pyroclean 热浸脱脂；热浸净化
pyroclor 皮罗克勒〔一种变压器油的商名〕
pyrocoll 焦咯；二羰化二吡咯
　　$C_4H_3N=(CO)_2=NC_4H_3$
pyrocollodion 焦珂罗酊；高硝珂罗酊
pyrocomane(=1,4-pyrone) 1,4-吡喃酮；焦考曼
pyro-condensation 高温缩合；热缩
pyro-conductivity 热电导性
pyrocurzerenone 焦蓬莪术酮；焦莪术呋喃烯酮
pyrodextrin 焦糊精
pyrodigit 数字显示温度指示器
pyrodin 乙酰苯肼
pyroelectric 热电的；热电物质
pyroelectric behavior 热电行为
pyroelectric detector 热电检测器
pyroelectric effect 热电效应
pyroelectricity ①热电②热电学
pyroelectric polymer 热电性高分子
pyrogaelol 焦酚；1,2,3-苯三酚
pyrogallate 焦棓酸盐

pyrogallic acid 焦棓酸〔即：焦棓酚〕；焦性没食子酸
pyrogallol 连苯三酚
pyrogallol-carboxylic acid 焦棓酚羧酸 $(HO)_3C_6H_2CO_2H$
pyrogallol derivatives 焦棓酚衍生物
pyrogallol 1, 3-dimethyl ether(=2, 6-dimethoxyphenol) 焦棓酚-1, 3-二甲醚；2, 6-二甲氧基苯酚 $(CH_3O)_2C_6H_3OH$
pyrogallol monoacetate(=eugallol) （一）乙酸焦棓酚 $CH_3COOC_6H_3(OH)_2$
pyrogallol oxide(=pyraloxin) 氧化焦没食子酚
pyrogallol salicylate 焦棓酚水杨酸酯
pyrogallol sulfonphthalein 焦棓酚磺酞
pyrogallol tannin 焦酚单宁
pyrogallol tanning material 焦棓酚鞣料；没食子类鞣料
pyrogallol triacetate 焦棓酚三乙酸酯 $C_6H_3(O_2CCH_3)_3$
pyrogallol trimethyl ether 焦棓酚三甲醚 $C_6H_3(OCH_3)_3$
pyrogen ①焦精〔染料〕②热原
pyrogene colors〔复〕 焦精染料〔染〕
pyrogene deep black 焦精深黑
pyrogene direct blue 焦精直接蓝
pyrogene dyes 焦精染料
pyrogene grey 焦精灰〔染〕
pyrogeneous 火成的
pyrogenetic decomposition 热解作用；高温分解(作用)
pyrogene yellow O 焦精黄O
pyrogen-free 无热原
pyrogenic ①热解的②生热的③焦化的；火成的
pyrogenic activity 生热活性
pyrogenic decomposition 热解；高温分解
pyrogenic distillation 高温蒸馏；干馏
pyrogenic effect 热原效应；发热效应
pyrogenic reaction ①焦化反应②生热反应
pyrogenic rocks〔复〕 火成岩
pyrogenic silica(=fumed silica) 热解氧化硅
pyrogenous 干馏的；高温蒸馏的
pyrogenous asphalt 干馏沥青；高温分解沥青
pyrogen process 高温法
pyrogen testing 热原试验
pyrogram 热解图
pyrograph 裂解色谱；热谱
pyrographitic oxide 氧化石墨酸
pyrohydrolysis 高温水解
pyrohydrolytic analysis 高温水解分析
pyroil 皮罗依〔一种润滑油多效能添加剂的商名〕
pyrojapaconitine 焦性乌头碱
pyrokomane 4-吡喃酮
pyrolaxon(=gallamine triethiodide) 皮罗拉克松；三乙碘

化没食子铵；三乙碘化三(β-二乙氨正氧基)苯〔药〕
pyrolene 鹿蹄草烯
pyrolic alloy 镍铬合金
pyroligneous 焦木的
pyroligneous acid 焦木酸；木乙酸
pyroligneous distillate 焦木馏出物
pyroligneous vinegar 焦木醋
pyrolignite of lime 木屑；木粉
pyrolitic boron nitride 热解氮化硼
pyrology 热工学
pyrolusite 软锰矿
pyrolutite 火山灰
pyrolysated 热分解的；热解的
pyrolysis 热解*；裂解
pyrolysis apparatus 热解器
pyrolysis chamber 热解室
pyrolysis coils 热解旋管
pyrolysis curve 热解曲线
pyrolysis furnace 热解炉
pyrolysis gas 裂解气
pyrolysis gas chromatography 裂解气相色谱法；裂解色谱仪
pyrolysis gas oil 裂解柴油
pyrolysis gasoline 裂解汽油
pyrolysis GC-MS 热解气相色谱-质谱法
pyrolysis mass spectrometry(PY-MS) 热解质谱法
pyrolysis oven 热解炉
pyrolysis product 热解产物
pyrolysis reactor 热解反应器
pyrolysis temperature 热解温度
pyrolysis thin-layer chromatography 热解薄层色谱法
pyrolysis time 热解时间
pyrolysis tube furnace 裂解管式炉
pyrolysis unit 热解单元；热解组件
pyrolythic acid 三聚氰酸
pyrolytic 热解的
pyrolytically coated graphite tube 热解涂层石墨管
pyrolytic carbon 焦化石墨；热解石墨
pyrolytic carbon-coated fuel 高温热解碳涂敷燃料
pyrolytic-chromatography 裂解色谱法
pyrolytic conversion 高温转化
pyrolytic cracking 热裂(作用)；高温裂化
pyrolytic decomposition 高温分解；高热分解
pyrolytic degradation 热降解
pyrolytic elimination 热解消除*
pyrolytic graphite 热解石墨
pyrolytic infiltration 热解渗滤
pyrolytic polymer 热解聚合物*

pyrolytic reaction 热解反应
pyrolytic silicon carbide 热解碳化硅
pyrolytic spectrum 热解光谱*
pyrolytic surface coating 热解表面涂覆(法)
pyrolytic technique 热解技术
pyrolytic thin layer chromatograph 热解薄层色谱仪
pyrolyzate(=pyrolysis product) 热解产物
pyrolyzate spectra 热解光谱
pyrolyzer 热解器；裂解进样器
pyromagnetic 热磁的
pyromeconic acid(=β-hydroxypyrone) 焦袂康酸；3-羟基对吡喃酮
pyromellitdiimide 苯均四酰二亚胺
pyromellitic acid 1,2,4,5-苯四酸 $C_{10}H_6O_8$
pyromellitic anhydride 1,2,4,5-苯四酸酐
pyromellitic dianhydride 苯均四酸二酐〔催化剂〕
pyromellitic diimide 1,2,4,5-苯四甲酰二亚胺
pyromellitic ester 均苯四酸酯
pyromellitonitrile 1,2,4,5-苯四腈
pyro-met(=pyrometallurgical) 高温冶金(的)
pyrometallurgical process 高温冶金术
pyrometallurgy ①热冶学②热冶术
pyrometamorphism 热力变质；高温变质
pyrometasomatism 热力交代作用
pyrometer 高温计
pyrometer cone 高温计熔锥
pyrometer couple 高温热电偶
pyrometer protecting tube 高温计保护套管
pyrometer sighting tube 高温计窥视管
pyrometric cone (示温)熔锥
pyrometric cone equivalent value 熔锥比值；热锥比值
pyrometric control 用高温计控制(检查)
pyrometric gage 高温规
pyrometry 高温测定(法)；测高温术
pyromorphite 磷氯铅矿
pyromorphous 火晶的
pyromucate 焦黏酸盐(酯) C_3H_3OCOOM; C_3H_3OCOOR
pyromucic acid(=α-furancarboxylic acid) α-呋喃甲酸；焦黏酸
pyromucic aldehyde 焦黏醛；糠醛
pyromucic amide 焦黏酰胺 $C_4H_3OCONH_2$
pyromucic anhydride 焦黏酐 $(C_4H_3OCO)_2O$
pyromucic nitrile 焦黏腈 C_4H_3OCN
pyromucyl 焦黏酰 C_4H_3OCO-
pyromucylchloride 焦黏酰氯 C_4H_3OCOCl
pyromusic acid(=pyromucic acid) 焦黏酸
N-pyromycinone N-吡咯霉素酮
pyronaphtha 焦石脑油；重煤油

pyrone 吡喃酮 $C_5H_4O_2$
1,4-pyrone(=pyrocomane) 1,4-吡喃酮；焦考曼
pyrone carboxylic acid 吡喃酮羧酸
pyronene 吡喃酮烯
pyronine 焦宁〔染料〕
pyronine dye 焦宁染料
pyronine G 焦宁G
pyronone α,γ-吡喃酮
pyro-oxidation-reduction 高温氧化还原(过程)
pyroparaffine 焦石蜡；重质蜡
pyrope 镁铝榴石
pyrophaeophorbide 嗜焦素
pyrophanite 红钛锰矿
pyrophore (撞燃的)引火物
pyrophoric 引火的；生火花的
pyrophoric alloy 引火合金
pyrophoric lead 发火铅
pyrophoric powder 引火粉
pyrophoric reaction 引火反应
pyrophorous 引火的；自燃的
pyrophorus 引火物；自燃物
pyrophosphate ①焦磷酸②焦磷酸盐(酯)
pyrophosphate exchange reaction 焦磷酸交换反应
pyrophosphate method 焦磷酸盐法
pyrophosphato-titanate 焦磷酸酯合钛酸酯
pyrophosphite 焦亚磷酸盐(酯) $M_4P_2O_5$
pyrophosphito-titanate 焦亚磷酸酯合钛酸酯
pyrophosphodiamic acid 二氨基焦磷酸
pyrophosphonate 焦膦酸盐(酯)
pyrophosphoric acid 焦磷酸
pyrophosphorolysis 焦磷酸解作用
pyrophosphorous acid 焦亚磷酸 $H_4P_2O_5$
pyrophosphoryl 焦磷酰 $O=POP=O$
pyrophosphoryl chloride 焦磷酰氯 $P_2O_3Cl_4$
pyrophyllite 叶蜡石〔天然无水硅酸铝〕
pyropolymer 热解聚合物；焦化聚合物
pyroprobe 裂解探针
pyro-probe-ribbon 热解取样带
pyroprocess 高温过程；高温法
pyroprocessing 高温处理
pyroracemamide 丙酮酰胺 $MeCOCONH_2$
pyroracemic acid 丙酮酸 $CH_3COCOOH$
pyroracemic aldehyde 丙酮醛 CH_3COCHO
pyroschist(=pyroshale) 焦页岩；可燃性油页岩
pyroscope 辐射热度计
pyroshale 焦页岩；可燃性油页岩
pyrosilicate 焦硅酸盐
pyrosine 赤藓红

pyrosol 高温溶胶；熔溶胶
pyrostat 高温(保持)器
pyrostilpnite 火色硫锑银矿
pyrosulfate 焦硫酸盐 $M_2S_2O_7$
pyrosulfate fusion 焦硫酸盐熔融
pyrosulfuric acid 焦硫酸；一缩二(正)硫酸 $H_2S_2O_7$
pyrosulfurous acid 焦亚硫酸；一缩二亚硫酸 $H_2S_2O_5$
pyrosulfuryl 焦硫酰 S_2O_5
pyrosulfuryl chloride 焦硫酰氯 $(S_2O_5)Cl_2$
pyrotartaraldehyde 焦酒石醛；甲基丁二醛
　　$CHOCH_2CH(CH_3)CHO$
pyrotartaric acid(=pyrovinic acid) 焦酒石酸；甲基丁二酸
　　$COOHCH_2CH(CH_3)COOH$
pyrotartrate(=pyrovinate) 焦酒石酸盐(酯)；甲基丁二酸盐(酯) $COOMCH_2CH(CH_3)COOM$；
　　$COORCH_2CH(CH_3)COOR$
pyrotechnic ①焰火的②焰火制造的
pyrotechnic composition 烟火药，混合炸药
pyrotechnics(=pyrotechny) 焰火制造术
pyrotechny(=pyrotechnics) 焰火制造术
pyroterebic acid 4-甲基-3-戊烯酸
pyrothioarsenate 焦硫代砷酸盐
pyrotic 腐蚀的
pyrotic smelting 高温冶炼
pyrotol process 加氢脱烷基法
pyrotritaric acid(=uvic acid) 乌韦酸；2,5-二甲基-3-呋喃羧酸 $(CH_3)_2C_4HOCOOH$
pyrouric acid 三聚氰酸
pyrovanadate 焦钒酸盐 $M_4V_2O_7$
pyrovanadic acid 焦钒酸 $H_2V_2O_7$
pyrovinate(=pyrotartrate) 焦酒石酸盐(酯)；甲基丁二酸盐(酯) $COOMCH_2CH(CH_3)COOM$；
　　$COORCH_2CH(CH_3)COOR$
pyrovinic acid(=pyrotartaric acid) 焦酒石酸；甲基丁二酸
　　$COOHCH_2CH(CH_3)COOH$
pyrovoltage 热电压
pyroxene 辉石
pyroxenite 辉石
pyroxilin(=pyroxylin) 火棉；低氮硝化纤维素
pyroxylic spirit 甲醇
pyroxylin 硝酸纤维素；焦木素；火棉；低氮硝化纤维素
pyroxyline 硝化棉；火棉；漆用硝基纤维素
pyroxylin(e) cement 火棉胶
pyroxylin(e) finish 硝基漆
pyroxyline lacquer 焦木素漆；火棉漆
pyroxyline primer 硝化纤维底漆
pyroxyline silk 焦木素丝；硝化(纤维人造)丝
pyroxylin finish 硝化纤维涂饰剂

pyroxylin silk 低氮硝酸纤维素丝；焦木素丝
pyrozinc process 高温锌(萃取)过程
pyrrhite 烧绿石
pyrrhosiderite 针铁矿
pyrrhotine 磁黄铁矿
pyrrhotite 磁黄铁矿
pyrrilium 吡喃鎓化合物；吡喃锌型化合物
pyrroaetioporphyrin 焦初卟啉
pyrrocoline 8-吡咯并吡啶；中氮茚；焦可林
pyrrocolino-steroids 中氮茚并甾类(化合物)
pyrrodiazole 三唑 $C_2H_3N_3$
pyrrole(=imidole) 吡咯
pyrrole-2-aldehyde(=α-pyrryl-aldehyde) 吡咯-2-甲醛；
α-吡咯基甲醛 $CHOC_4H_4N$
pyrrole-antibiotic 吡咯抗菌素
pyrrole-α-carboxylic acid 吡咯-α-羧酸
pyrrole ring 吡咯环
pyrrole test 吡咯试验；氮茂试验
pyrrolidine(=tetrahydropyrrole) 吡咯烷；四氢化吡咯
$(CH_2)_4$=NH
pyrrolidine alkaloids 吡咯烷生物碱类
pyrrolidine carboxylic acid 吡咯烷羧酸 C_4H_8NCOOH
pyrrolidine-2-carboxylic acid 2-吡咯烷羧酸
pyrrolidinedione 琥珀酰亚胺
pyrrolidine dithiocarbamate 吡咯烷氨荒酸
pyrrolidinetrione 吡咯烷三酮
pyrrolidinium 吡咯烷鎓(盐)
pyrrolidinomethyl tetracycline 吡咯烷甲基四环素
pyrrolidinyl(=pyrrolidyl) 吡咯烷基 C_4H_8N-
pyrrolidone 吡咯烷酮
pyrrolidone carboxylic acid 吡咯烷酮羧酸
$C_4H_6ONCOOH$
pyrrolidyl(=pyrrolidinyl) 吡咯烷基
pyrroline(=dihydropyrrole) 吡咯啉；二氢化吡咯
$CH_2CH_2NHCH=CH$
pyrroline carboxylic acid 二氢吡咯羧酸
pyrrolinium compound (五价氮)吡咯啉化合物
pyrrolinium hydrochloride 盐酸吡咯啉
pyrrolinyl 吡咯啉基 NC_4H_6-
pyrrolizidine 吡咯联啶
pyrrolizidine-9-hydrochloride 双稠吡咯啶-9-盐酸盐
pyrrolizine 吡咯烷
pyrrolnitrin 硝吡咯菌素；吡咯尼群
pyrrolo 吡咯并
pyrrolo-indole 吡咯并吲哚
pyrrolopyrrole pigment 吡咯并吡咯颜料
pyrroloquinoline ①吡咯并喹啉 $C_{11}H_9N$②任何含有一
个吡咯环及一个喹啉环的
pyrrolyl(=pyrryl) 吡咯基 NC_4H_4-
pyrrolylcarbonyl(=pyrroyl) 吡咯基甲酰；吡咯羰基
C_5H_3NO
pyrrolylene(=butadiene) 丁(间)二烯
CH_2=$CHCH$=CH_2
pyrromonazole 吡唑
pyrromycin 吡咯霉素
pyrromycinone 吡咯霉素酮
pyrrone 吡酮
pyrroporphyrin 焦卟啉
pyrrotriazole 焦三唑；1,2,3,4-四唑
pyrroyl(=pyrrolylcarbonyl) 吡咯甲酰
pyrryl(=pyrrolyl) 吡咯基
α-pyrryl-aldehyde(=pyrrole-2-aldehyde) 吡咯-2-甲醛；
α-吡咯基甲醛 $CHOC_4H_4N$
pyrryl thiazole 吡咯基噻唑
pyrulic acid(=$trans$-10-heptadecen-8-ynoic acid) 十七
(碳)烯-10-炔-8-酸
pyruric acid 三聚氰酸
pyruvaldehyde 丙酮醛 CH_3COCHO
pyruvate 丙酮酸盐(酯) $CH_3COCOOM$; $CH_3COCOOR$
pyruvate carboxylase 丙酮酸羧化酶
pyruvate dehydrogenase 丙酮酸脱氢酶
pyruvate kinase 丙酮酸激酶
pyruvate oxidase 丙酮酸氧化酶
pyruvate phosphokinase 丙酮酸(磷酸)激酶
pyruvic acid 丙酮酸 $CH_3COCOOH$
pyruvic alcohol 丙酮醇；羟基丙酮 CH_3COCH_2OH
pyruvic aldehyde 丙酮醛 CH_3COCHO
pyruvic dehydrogenase(=pyruvic oxidase) 丙酮酸脱氢
酶；丙酮酸氧化酶
pyruvic ketolase 丙酮酸酮酶
pyruvic nitrile 丙酮腈
pyruvic oxidase(=pyruvic dehydrogenase) 丙酮酸脱氢
酶；丙酮酸氧化酶
pyruvonitrile 丙酮腈 CH_3COCN
pyruvoyl 丙酮酰 CH_3COCO-
pyrvinium panoate 扑蛲灵；扑蛲喹；吡文驱(肠虫)药
pyrvolidine 胡萝卜叶碱
pyrylium 吡(喃)锌；吡喃鎓
pyrylium compound 吡喃鎓化合物
pyrylium salt 吡喃鎓盐
pythonic acid 蟒蛇胆酸
pyx〔拉丁文〕 沥青
pyx liquida 木焦油

Q

Q-branch   Q 分支
Q detector   定性，定量检测器
Q, e scheme   Q,e 概念
Q gas counter   "Q"气体计数管
Q-gasoline(=Q-grade gasoline)   合格的汽油；"Q"级汽油
Q-grade gasoline(=Q-gasoline)   合格的汽油；"Q"级汽油
Q-meter   Q 表；品质因数表；优质计
Q-switch   Q 开关
Q-switch laser   Q 开关激光器
qua-〔拉丁字头〕   ①拟；伪；假；准；似②模拟；仿真
quantitative paper(=quantitative filter paper)   定量滤纸
quad alloy   四元合金〔指含钼、锆、铌、钛的铀合金〕
quadrangle   四角(平面)形
quadrant   圆周的四分之一；象限
quadrant balance   扇形天平
quadrant electrometer   象限静电计
quadrant electrostatic meter   象限静电计
quadrant style Shore durometer   扇形肖氏硬度计
quadrate   ①方材；方钢②方块物；正方形；正方形的③平方
quadratic drift correction   二次漂移校正
quadratic equation   二次方程式
quadratic form   二次型
quadratic interpolation   二次插值法
quadratic mean deviation   均方(偏)差
quadratic system   正方晶系
quadratic term   二次项
quadrature detection   正交检测
quadrature-lagging   后移 90°；滞后 90°
quadrature phase detection   正交检波；正交相位检测
quadri-〔拉丁字头〕   四
quadribasic   ①四碱价的；四元的②四代的
quadribasic acid   四价酸；四元酸
quadricovalent   四配价的
quadridentate   四配位体
quadridentate chelate   四配位体螯合物
quadridentate ligand   四齿配位体
quadrifidin   四担蕈素
quadrilateral   ①四边形的②四边形
quadrilineatin   四线曲菌素
quadrillon   ①亿亿亿〔10^{24}〕②千万亿〔美国用法，10^{15}〕
quadrimolecular   四分子的
quadrimolecular reaction   四分子反应
quadrine α-氨基(正)丁酸
quadripolar   四极的；四端的

quadripole   四极
quadripolymer   四元聚合物
quadrivalence(=quadrivalency)   四价
quadrivalency(=quadrivalence)   四价
quadrivalent   四价的
quadrivalent anion   四价阴离子
quadrivalent atom   四价原子
quadrivalent cation   四价阳离子
quadrivalent derivative   四价衍生物
quadrivalent element   四价元素
quadrol N,N,N',N'-四(2-羟基丙基)乙二胺
quadroxide   四氧化物
quadru-〔词头〕   四
quadruple   ①四倍②四倍的③以四乘的④增加四倍
quadruple bond   四重键
quadruple burner   四头灯
quadruple effect   ①四效②四效(式)的
quadruple effect evaporator   四效蒸发器
quadruple ion   四重离子
quadruple link(=quadruple linkage)   四(价)键
quadruple linkage(=quadruple link)   四(价)键
quadruple metering pump-gear   四出轴式计量泵传动齿轮
quadruple point   四相点
quadruple unit attachment   四元附件
quadrupolar analyzer   四极分析器
quadrupolar axialization   四极轴向化
quadrupolar relaxation   四极弛豫
quadrupole   四极
quadrupole bar   四极杆
quadrupole broadening   四极展宽
quadrupole coupling   四极耦合
quadrupole coupling constant   四极耦合常数
quadrupole coupling tensor   四极耦合张量
quadrupole double resonance   四极双共振
quadrupole field   四极场
quadrupole-hexapole- quadrupole mass spectrometer   四极-六极-四极质谱仪
quadrupole hyperfine structure   四极超精细结构
quadrupole ion   四极离子
quadrupole ion storage   四极离子存储器
quadrupole ion store   四极离子阱
quadrupole lens   四极透镜
quadrupole magnet   四级磁体；四极磁铁

quadrupole mass analyzer 四极杆质量分析器
quadrupole mass filter 四极滤质器
quadrupole mass spectrometer(QMS) 四极质谱仪*
quadrupole mass spectrometry(QMS) 四极质谱法
quadrupole moment 四极矩
quadrupole- quadrupole coupling 四极-四极耦合
quadrupole- quadrupole splitting 四极-四极分裂
quadrupole- quadrupole transition 四极-四极跃迁
quadrupole radiation 四极辐射
quadrupole relaxation 四极弛豫
quadrupole residual gas analyzer 四极残余气体分析器
quadrupole resonance 四极矩共振
quadrupole resonance spectrometer 四极共振波谱仪
quadrupole rod 四极杆
quadrupole rod assembly 四极杆组件
quadrupole spectrometer 四极(质)谱仪
quadrupole splitting 四极分裂
quadrupole thermal ionization mass spectrometer 四极热电离质谱仪
quadrupole thermal-programmed mass detector 四极程序升温质量检测器
quail 凝结
quaker buttons 马钱子
qualification ①资格；条件②赋与资格；合格③规格
qualification program 鉴定程序；鉴定方案
qualification test 合格试验，鉴定试验
qualimeter X射线硬度测定仪
qualimetrics 品质计量学
qualitative ①定性的②描性的；描述的③大概的④质的
qualitative analysis 定性分析
qualitative analytical chemistry 定性分析化学
qualitative application 定性应用
qualitative assay 定性检验
qualitative chemical analysis 化学定性分析
qualitative criteria 定性判据
qualitative determination 定性测定
qualitative electrophoretic technique 定性电泳技术
qualitative elementary analysis 元素定性分析
qualitative filter paper 定性滤纸
qualitative reaction 定性检验；定性反应
qualitative spectral analysis 光谱定性分析
qualitative spectroanalysis 光谱定性分析
qualitative spectrometric analysis 光谱定性分析
qualitative spot test 定性斑点试验
qualitative tensile property 定性拉伸性能
qualitative test 定性试验；定性检验；定性测定
qualities of aspect 外观质量
quality ①质量②特性③纯度④品位

quality arbitration ①质量检定②质量商定
quality assurance 质量保证；保证质量〔可达免检的质量〕；质量保险
quality assurance for environmental monitoring 环境监测质量保证
quality assurance in the environmental monitoring 环境监测中的质量保证
quality booster 质量改善剂
quality characteristic 特性值
quality coefficient 质量系数
quality concrete 优质混凝土
quality control 质量控制*；质量管理
quality control between laboratory 实验室间质量控制
quality control chart 质量管理图；质量控制图
quality control checking 质量控制检查
quality control of blood bank 血库质量控制
quality control of environmental analysis 环境分析质量控制
quality control procedure 质量管理规范
quality control sample 质量控制样
quality control testing 质量控制试验；质量管理试验
quality control time 质量控制时间
quality control within laboratory 实验室内质量控制
quality determination 质量测定
quality determination of fats 油脂质量鉴定
quality determination of plant 植物品质测定
quality evaluation 质量鉴定；质量评价
quality factor 品质因子；品质因数
quality forging 优质锻件
quality grade 质量等级
quality index 质量指标；质量指数
quality inspection 质量检验
quality inspection of cosmetics 化妆品质量检验
quality level 质量标准
quality lubricant 优质润滑剂
quality management 质量管理
quality management sample 管理样
quality monitor 质量监控器
quality number 品质支数
quality of aspect 外观质量
quality of fit 配合等级
quality of fuel 燃料质量
quality of gasoline 汽油质量
quality of tolerance 公差等级
quality part 高级部件；合格品
quality retention 质量保持率
quality specification ①质量标准；质量说明书②技术规格；技术规范

quality standard   质量标准
quality supervision   质量监督
quality surveillance   质量监督
quantasome   光能转化体；量子体
quantification(=quantitation)   定量；量化；以数量计定量
quantile fractile   分位数
quantimeter   X射线剂量计
quantitation   定量
quantitation in situ   原位定量法
quantitative   ①定量的②量的
quantitative analysis   定量分析*
quantitative analysis of end group   端基定量分析
quantitative analytical chemistry   定量分析化学
quantitative assay   定量测定；定量分析
quantitative criteria   定量判据
quantitative determination   定量测定
quantitative differential thermal analysis   定量差热分析
quantitative distribution   数量分配；定量分布(配)
quantitative electron spin resonance   定量电子自旋共振
quantitative evaluation   定量测定；定量评价(估)
quantitative experiment   定量实验
quantitative filter paper   定量滤纸
quantitative image analysis system   定量图像分析系统
quantitatively   定量的；数量上
quantitative mass spectrometry   定量质谱法
quantitative organic elemental analysis   有机元素定量分析
quantitative paper chromatography   定量纸色谱(法)
quantitative parameter   定量参数
quantitative reaction   定量反应
quantitative response   定量响应
quantitative sensitivity   定量灵敏度
quantitative spectral analysis   光谱定量分析
quantitative spectrochemical analysis   定量光谱化学分析
quantitative spectrometric analysis   光谱定量分析
quantitative structure-activity relation   定量构效关系
quantitative test   定量试验
quantitative test for arsenic   砷定量试验
quantitative work   定量操作
quantities delivered   运出量
quantities received   接收量；得到的量
quantity   ①量；值；数；参数②大量
quantity of alkali consumption   碱消耗量
quantity of dry combustion gas   干燃烧(煤)气量
quantity of electricity   电量
quantity of heat   热量
quantity of information   信息量
quantity of light   光量
quantity of radiant energy   辐射能量；辐射能通量
quantity of reflux   回流量
quantity of substance   物质的量
quantity of transition material   过渡原料量；过渡熔体量
quantity of wet combustion gas   湿燃烧(煤)气量
quantity production   大量生产；成批生产
quantity retention   质量保持率
quantivalence(=quantivalency)   化合价；原子价
quantivalency(=quantivalence)   化合价；原子价
quantivalent(=multivalent)   多价的
quantization   量子化(作用)
quantization of energy   能量量子化
quantized   量子化的；量化的
quantized system   量子化体系
quantocorder   光电直读式发射光谱仪
quantometer   光量计；光电光谱仪
quantorecorder   ①光量计；辐射强度测量计②光子计数器
quantovac   光电直读式真空发射光谱仪
quantum   量子；定量；和；量
quantum bit   量子位
quantum chemistry   量子化学
quantum collision   量子碰撞
quantum condition   量子(化)条件
quantum crystal   量子晶体
quantum effect   量子效应
quantum efficiency   量子效率
quantum electrodynamics   量子电动力学
quantum field   量子场
quantum group   量子群
quantum jump   量子(性)跳变
quantum liquid   量子液体
quantum mechanical theory   量子力学理论
quantum mechanics   量子力学
quantum notation   ①量子标志②量子符号表示法
quantum number   量子数
quantum partition function   量子配分函数
quantum pharmacology   量子药理学
quantum relation   量子关系
quantum restriction   量子(条件)限制
quantum size effect   量子尺寸效应
quantum state   量子态
quantum statistical mechanics   量子统计力学
quantum statistics   量子统计
quantum theory   量子理论
quantum transition   量子跃迁
quantum trap   量子阱
quantum unit   量子单位
quantum value   量子值

quantum yield 量子产率*
quarantine 检疫
quarantine inspector 检疫员
quardrature-axis 正交轴线
quarentoxide 四氧化物
quarezite 石英岩；石英砂
quark 夸克
quarry 采石场
quarry-faced 粗面的；毛面的(石料等)
quarry face of stone 粗(凿)石面
quarrying 采石
quarry stone 石块
quarry storage for heating oils 燃料油的石矿储藏
quarry-tile 方砖；缸砖；大瓷砖
quart 夸(脱)
quartation (硝酸)析银法
quart can 夸脱罐；夸脱漆罐
quart cup 一夸脱的杯(罐)
quarter ①四分之一②季；三个月③刻(钟)
quarter bend 直角弯
quartering 四分(法)*
quarter light(=quarter vent) 边窗
quartern 四等分；四分之一
quarternary alloy 四元合金
quarternary structure 四级结构
quarter phase 两相的；双相的；二相的
quarter sponge method 1/4 水发面法〔面包〕
quarter turn drive 直角(转动)传动
quarter vent(=quarter light) 边窗
quarter wave plate 四分之一波(长)片；四分之一波长板
quarter wave potential 四分之一波电位
quartet 四重峰*；四重线
quart-hard annealing 低硬度退火
quartied calcium carbonate 重质碳酸钙
quartile 四分位
quartz 石英
quartz apparatus 石英仪器；石英器皿
quartz boat 石英舟；石英皿
quartz (Bunsen) burner 石英(本生)灯；石英燃烧器
quartz cell 石英液槽；石英比色皿；石英吸收池；石英容器
quartz condensing-lens 石英聚光镜
quartz container 石英皿；石英容器
quartz crucible 石英坩埚
quartz crystal microbalance 石英晶体微量天平
quartz etching 石英腐蚀
quartz exchange column 石英离子交换柱
quartz fiber 石英纤维

quartz fiber manometer 石英丝压力计
quartz filter sand 石英过滤砂
quartz glass 石英玻璃
quartzification 石英化(作用)
quartz infrared lamp 红外线石英灯；石英红外灯
quartz iodine lamp 石英碘灯
quartz iodine tungsten lamp 石英碘钨灯
quartzite 石英岩
quartzitic 石英岩的
quartz lens 石英透镜
quartz mercury arc lamp 石英水银电弧灯
quartz mercury lamp 石英汞灯
quartz nozzle 石英喷嘴
quartz oscillator 石英振子；石英晶体振荡器
quartz pipe 石英管
quartz plate 石英片
quartz prism 石英棱镜
quartz resonator 石英振子；石英晶体振荡器
quartz rock 石英岩
quartz sand 石英砂
quartz sleeve 石英套筒
quartz spectrograph 石英(棱镜)摄谱仪
quartz spectrophotometer 石英分光光度计
quartz spectroscope 石英分光镜
quartz spring 石英弹簧
quartz surface induced luminescence 石英表面诱导发光
quartz thermometer 石英温度计
quartz tube 石英管
quartz tube atom-trapping 石英管原子捕集法
quartz-tube thermometer 石英温度计
quartz wedge analyzer 石英劈检偏器
quartz wedge compensator 石英楔补偿器；石英劈补偿器
quartz wool(=silica wool) 石英棉
quartzy 石英的；水晶的
quartz yarn 石英纱〔石英纤维与废纺纤维混纺而成〕
quary 脱脂乳渣；脱脂酸凝乳
quasar 发强射线的远天体；类星射电源
quasi- 〔拉丁字头〕 似；准；拟
quasi-aromatic compound 似芳族化合物
quasi-atomic model 准原子模型
quasi-bound atom 准结合原子
quasi-bound electron 准束缚电子
quasi-bound state 准结合态
quasi-chemical equilibrium 准化学平衡
quasi-chemical equilibrium of defect 缺陷的类化学平衡*
quasi-chemical method 似化学方法；半化学方法
quasi-conjugation(=hyper conjugation) 超共轭效应；似共轭效应

quasi-continuous process　准连续法
quasi-continuum　准连续区
quasi-crosslink　准交联；似交联
quasi-crystal　准晶体
quasi-crystalline lattice　准晶格
quasi-crystalline state　准结晶状态
quasi-crystalline structure　似晶(体)结构
quasi-cyclic　准循环的
quasi-diffusion　准扩散(的)
quasi-elastic　准弹性的；似弹性的
quasi-elastic light scattering　准弹性光散射
quasi-elastic scattering　似弹性散射
quasi-elastic vibration　准(似)弹性振动
quasi-equilibrium　准平衡
quasi-equilibrium state　准平衡态
quasi-equilibrium theory(QET)　准平衡理论
quasi-equilibrium theory of mass spectrum　质谱准平衡原理
quasi-eutectic　①伪共晶的②伪共晶体
quasi-factorial design　准因子设计
quasi-Fermi level　准费米能级
quasi-fibrous　准纤维状的；似纤维状的
quasi-flow　准流动；半流动
quasi-focusing X-ray diffractometer　准聚焦X射线衍射仪
quasi-free electron　准自由电子*
quasi-free-vortex　准自由旋涡
quasi-homogeneous　准均匀的；准均质的
quasi-homogeneous material　准均匀材料
quasi-hydrodynamic lubrication　似流体动力学的润滑；半流体润滑
quasi-isothermal　似等温的
quasi-isotropic　准各向同性(的)
quasi-isotropy　类无向性；准各向同性
quasi-lattice model　准晶格模型
quasi-lattice theory(QLT)　准晶格理论
quasi-linear　准线性的
quasi-linear elasticity　准线性弹性
quasi-linearization　准线性化
quasi-liquid membrane(QLM)　准液膜
quasi-living polymerization　准活性聚合
quasi-molecular ion　准分子离子
quasi-monochromatic light　准单色光
quasi-optical wave　准光波
quasi-ordered region　准有序区
quasi-para orientation　准对位定位；似对位取向
quasi-particle　准粒子
quasiperiodic crystal　准周期性晶体

quasipermanent deformation　似(永)久形变
quasi-phosphonium compound　似鏻化合物
quasi-plasticity　①似塑性②似塑度
quasi-prepolymer process　准预聚物法
quasi recemate　准外消旋体*
quasi-recemic compound　准外消旋化合物
quasi reversibility　准可逆
quasi-reversible change　准可逆变化
quasi-reversible wave　准可逆光
quasi-single-strand polymer　准单股聚合物
quasi-soft　半软质
quasi-stable element　准稳元素
quasi-stable island　准稳岛
quasi-stable isotope　准稳定同位素〔指碘-129、镎-237等长寿命同位素〕
quasi-stable state　准稳态
quasi-static　准静态(的)
quasi-static change　似静态变化
quasi-static process　似静过程
quasi-stationary　准稳态的
quasi-stationary flow　准稳流
quasi-stationary process　准静态过程
quasi-steady flow　准稳流
quasi-steady state　准稳态
quasi-superheavy element　准超重元素
quasi thixotropy　准触变性
quasi-unimolecular reaction　准单分子反应
quasi-viscous　准黏性的
quasi-viscous creep　准黏性蠕变
quasi viscous effect　似黏性效果
quasi-viscous flow　准黏性流
quasi-viscous liquid(=non-Newtonian fluid)　似黏滞液体；非牛顿液体
quassic acid　苦木酸　$C_{30}H_{38}O_{10}$
quassin　苦木素　$C_{22}H_{28}O_6$
quaternaries　季(铵)盐(类)；四元的
quaternarization(=quaternization)　季铵化反应；季铵化作用
quaternary　①四元的②四价的③季的；连上四个碳原子的
quaternary adduct　四元加合物
quaternary alkanol　季化(链)烷醇
quaternary alkanolammonium type resin　季烷醇铵型树脂
quaternary alkoxypropyl ammonium halide　四(元)烷氧基丙基卤化铵
quaternary alkyl ammonium hydroxide　氢氧化烷基季铵
quaternary alkyl ammonium type resin　烷基季铵型树脂
quaternary alkyl aryl methyl ammonium chloride　四(元)

烷基芳基甲基氯化铵
quaternary alkylphosphonium salt 季磷盐；四烷基膦盐 R_4PX
quaternary alloy 四元合金
quaternary amines 季铵类
quaternary aminoethyl(QAE) (季)铵乙基
quaternary ammonium 季铵
quaternary ammonium ampholytic 季铵型两性表面活性剂
quaternary ammonium base 季铵碱
quaternary ammonium base cellulose 季铵碱纤维素
quaternary ammonium compound 季铵化合物*
quaternary ammonium fluorosilicate 氟硅酸季铵盐
quaternary ammonium halide 季铵卤化物 R_4NX
quaternary ammonium hexafluorosilicate 六氟硅酸季铵盐
quaternary ammonium hydrate 季铵碱 R_4NOH
quaternary ammonium hydroxide 氢氧化季铵；季铵碱
quaternary ammonium montmorillonite 季铵蒙脱石
quaternary ammonium nitrate 硝酸季铵盐
quaternary ammonium pentachlorophenate 五氯苯酚季铵盐
quaternary ammonium perchlorate 高氯酸季铵盐
quaternary ammonium polymetaphosphate 聚偏磷酸季铵盐
quaternary ammonium salt 季铵盐
quaternary ammonium softener 季铵(型)柔软剂
quaternary ammonium sulfate 硫酸季铵盐
quaternary ammonium thiocyanate 硫代氰酸季铵盐
quaternary ammonium type resin 季铵型树脂
quaternary arsenical compound 季砷合物；季砷鎓化合物 R_4AsX
quaternary arsonium compound 季钟化合物；季砷鎓化合物
quaternary base 季碱
quaternary borohydride 硼氢化季盐
quaternary carbon atom 季碳原子
quaternary cationics 季(铵)盐阳离子表面活性剂
quaternary cationic surfactant 季(铵)盐类阳离子表面活性剂
quaternary compound ①季化合物②四元化合物
quaternary copolymer 季铵共聚物；四元共聚物
quaternary detergent 季(铵)盐类洗涤剂
quaternary exchange resin 四组分(离子)交换树脂
quaternary germicidal surfactant 季(铵)型杀菌性表面活性剂
quaternary halide 卤化季盐
quaternary liquid system 四元液体系统
quaternary mixture 四元混合物
quaternary morpholinium alkyl sulfate 季(铵)化吗啉硫酸烷基酯盐
quaternary myristoylcholine chloride 季(铵)化肉豆蔻酰胆碱氯化物
quaternary nicotinium sulfate 季(铵)化烟碱硫酸盐
quaternary nitrogen 季(铵)氮
quaternary phosphine 季膦 R_4PX
quaternary phosphonium compound 季磷化合物；季磷鎓化合物
quaternary phosphonium hydroxide 季磷碱；氢氧化季磷 R_4POH
quaternary pyridinium resin 吡啶(鎓)盐树脂
quaternary pyridinium salt 吡啶(鎓)盐
quaternary salt 季盐
quaternary stibonium hydroxide 季锑碱；氢氧化季锑 R_4SbOH
quaternary structure 四级结构
quaternary surfactant 季(铵)盐表面活性剂
quaternary system 四元系统
quaternization(=quaternarization) 季铵化作用
quaternized 季碱化的；季铵化的
quaternized glucamine 季铵化葡萄糖胺；季化五羟基己胺
quaternized guanidinium salt 季铵化胍盐
quaternized imidazoline 季铵化咪唑啉
quaternized methylol amide 季铵化羟甲基酰胺
quaternized polyvinyl alcohol 季铵化聚乙烯醇
quaternizing agent 季铵化剂
quaternizing reagent 季碱化剂
quaterphenyl 四联苯
quaterpolymer 四元聚合物
quatrimycin 四一霉素(即：差向四环素)
Quebec screen test 魁北克筛分试验〔石棉干法分级试验〕
quebrachamine 白雀木皮胺 $C_{19}H_{26}N_2$
quebrachine(=yohimbine) 育亨宾
quebrachite 白雀木醇
quebrachitol 白雀木醇；白坚木皮醇
quebracho ①白雀树皮②白雀树
quebracho extract 白雀树萃；白雀树皮提取物；坚木栲胶
quebracho gum 白雀木胶
quebrachomine 白雀碱
quebracho tannin 白雀木鞣质；坚木鞣质
quebracho wood 白雀木
queen metal 锡锑焊料
queen's delight 大戟科植物草乌桕
queen substance 蜂王浆信息素

queen's yellow 皇后黄；碱式硫酸汞
quench ①淬火；骤冷②熄灭，灭火③止渴
quench air 骤冷空气
quench condensation 骤冷凝
quench cooler 急冷器
quench crack 淬火裂纹
quench duct 骤冷丝室
quenched and tempered steel 调质钢
quenched blast furnace slag 骤冷高炉矿渣；淬火高炉矿渣
quenched region 猝灭区；暗区
quenched room temperature phosphorimetry(QRTP) 猝灭室温磷光法
quencher 猝灭剂；淬火剂
quench gas 急冷气；骤冷气
quench hot 高温淬火
quenching 猝灭*
quenching agent ①(紫外线)猝灭剂②淬火剂
quenching and tempering (钢的)调质处理
quenching apparatus 淬火器
quenching bath 淬火浴
quenching broadening 猝灭变宽
quenching car 淬火车
quenching chamber 淬火室；骤冷室
quenching circuit 猝灭电路
quenching compound (紫外光)猝灭化合物；激发(态)抑制剂
quenching condition ①骤冷条件②淬火条件
quenching constant 猝灭常数*
quenching crack 淬致裂痕
quenching cross section 猝灭截面*
quench(ing) delay 淬火延迟
quenching diameter 猝灭直径
quenching distortion 淬火变形(畸变)
quenching effect 猝灭效应*
quenching effect of atomic fluorescence 原子荧光猝灭效应
quenching fluorometry 猝灭荧光测定法
quenching furnace 淬火炉
quenching gas 猝灭气体
quenching medium 猝灭剂；猝灭介质
quenching of fluorescence 荧光猝灭
quenching of gases 气体骤冷
quenching oil ①急冷油；骤冷油②淬火油
quenching stress 骤冷应力
quenching temperature 淬火温度
quench liquid 冷却液；骤冷液
quench oil(=quenching oil) 淬火油

quench pump 骤冷泵
quench stock 骤冷馏出物；骤冷原料
quench tank 骤冷槽
quench temperature 骤冷温度
quench tower 急冷塔；骤冷塔
quercetagetin 六羟黄酮；槲皮万寿菊素；栎草亭
quercetagetin-3,6-dimethyl ether 槲皮万寿菊素-3,6-二甲醚
quercetagetin-3,6-dimethyl ether-7-o-D-pyranoglucoside 槲皮万寿菊素-3,6-二甲醚-7-o-D-吡喃葡萄糖苷
quercetin 槲皮素〔广泛分布于许多食用和药用植物比如洋葱、茶叶、苹果、银杏等中的一类黄酮类化合物，用作营养添加剂〕；五羟黄酮；栎精
quercetin-3,3-dimethyl ether 槲皮素-3,3-二甲醚
quercetinic acid 栎精酸
quercetin pentamethyl 五甲基栎精
quercetone(=quercetin chalcone) 槲皮素黄酮；栎酮
quercic acid 栎辛酸
quercimetin(=quercitrin) 栎素；栎皮苷
quercin 槲皮苦素；栎辛
quercinic acid 栎辛酸
quercinin 栎曲菌素
quercitannic acid 橡树皮单宁酸〔无定形略带棕黄色或红色的白色粉末，易溶于水和醇〕 $C_{17}H_{16}O_9$
quercitannin 槲皮鞣酸
quercite(=quercitol;cyclohexanpentol) 栎醇；环己五醇 $CH_2(CHOH)_4CHOH$
quercitin ①豕草花粉苷②栎精
quercitol 栎醇；环己五醇 $C_6H_7(OH)_5$
quercitrinic acid 栎素酸
quercitrin(=quercimetin) 栎素；栎皮苷
quercitron 栎皮粉
quercitron lake 黑栎黄色淀
queretaroic acid 栎焦油酸；南美烟管仙人掌酸
questiomycin 寻霉素
quetenite 赤铁矾
quetiapine 喹硫平〔药〕
quetsch 上浆装置
queue 行列；队
Quevenne lactometer 牛乳比重计
Quevenne's iron 还原铁
quick access memory 快速存取存储器
quick acting accelerator 快速促进剂
quick acting catalyst 快速催化剂
quick aging 快速老化
quick article change system of weaving machine 织机快速品种变换系统

quick burning fuse 速燃引信
quick clay 过敏性黏土；不稳定黏土
quick-cleaning perform extruder 快速清理式型坯挤出机
quick cooking 速煮
quick cooling and grinding 急冷粉碎；骤冷粉碎
quick cure ①快速硫化法②快速腌制
quick-disconnect 快速拆卸活节头
quick-dissolving 速溶的
quick-dissolving soap 速溶皂
quick-dry facing wash 快干饰面涂料
quick-drying ①快干的②快干
quick drying boiled oil 快干性熟油
quick drying enamel 快干瓷漆
quick drying ink 快干油墨
quick drying lacquer 快干漆
quick drying oil 快干油
quick drying paint 快干漆
quick drying primer 快干底漆
quickening liquid 催镀液
quick freezing 快速冻(法)
quick-frozen solution 速冻溶液
quick-hardening 快速硬化；速凝的(水泥)
quick hardening cement 快硬水泥
quick-hardening lime 快硬石灰
quick hardening Portland cement 快硬(波特兰)水泥；快硬硅酸盐水泥
quick heating and cooling test 骤冷骤热试验
quick laboratory ball mill 快速实验室球磨机
quicklime 生石灰*；氧化钙
quicklime grease 钙基润滑脂
quick locking thread (纺丝组件)快速锁紧式螺纹
quick match 速燃引信头
quick-mo(u)ld-changing injection mo(u)lding machine 快速换模注塑机
quick-opening gate valve 速启阀门
quick-opening level 速启杠杆
quick-opening valve 速启阀
quick-operating 快动的；迅速操作的
quick purge valve 快速清洗阀
quick-reading flow sheets 快读流程；简单流程
quick-release handle (涂漆垫的)快松脱手柄
quick releave valve 快泄阀
quick-response control system 快速反应控制系统；高灵敏度控制装置
quick response regulation 快速应变调整；应急调整
quick response transducer 小惯性传感器
quick-sand 流砂
quick setting cement ①快干胶水②快干水泥

quick-setting ink 快干油墨
quicksilver 汞；水银
quick solder 软焊
quickstick 快黏
quickstick test 快黏试验
quick stopping mechanism 紧急停车装置
quick tannage 快鞣法
quick tanning 速鞣
quick test 快速试验
quick-turning steel 快削钢
quick vinegar process 快速酿醋法
quickwash shrinkage tester 快速洗水缩水试验机
quid pro quo 代替物；交换物
quiescence 静止期
quiescent bed 静止床
quiescent fluidized bed 平稳流态化床
quiescent layer 静止层
quiescent load 静止进料；静载荷
quiescent nonisothermal crystallization 静态非等温结晶
quiescent state 静态
quiet colour 素色
quiet steel 软钢(低碳钢)；全脱氧钢；全镇静钢
quill ①导火线②羽毛管
quillaia(=quillaia bark) 皂树皮
quillaia bark(=quillaia) 皂树皮
quillaia extracts 皂树皮萃取物
quillaia saponaria 皂树；肥皂树
quillaic acid 皂皮酸
quillaja saponaria 皂皮树
quillaja-saponin 皂树皂苷
quillaja-sapotoxin 皂树皂毒素
quilted structure fabric 绗缝织物
quin- ①奎〔指奎宁系化合物〕②喹〔指喹啉系化合物〕
quina 奎哪；金鸡纳皮
quinacetine 奎哪亭
quinacetine sulfate 硫酸奎哪亭
quinacridine 喹吖啶
quinacridone 喹吖啶酮
quinacridone gold 喹吖啶酮金黄〔颜料〕
quinacridone quinone 喹吖啶酮醌
quinacridone red 喹吖啶酮红
quinacridone violet 喹吖啶酮紫；二氢喹吖啶二酮
quinacrine 奎吖因〔即阿的平〕
quinaform 奎哪仿
quinalbarbitone sodium 司可巴比妥；速可眠
quinaldic acid 喹哪啶酸；喹啉-2-羧酸
quinaldine 喹哪啶〔即2-甲基喹啉〕 $C_9H_6NCH_3$
quinaldine carboxylic acid 喹哪啶羧酸

quinaldinehydroxamic acid 喹哪啶异羟肟酸
quinaldine red ①甲基氮䓬红②喹哪啶红
quinaldinic acid 喹哪啶酸；2-喹啉羧酸 C_9H_6NCOOH
quinaldinium compound 喹哪啶鎓化合物
γ-quinaldone γ-羟基-2-甲基喹啉
quinaldylamine 喹哪啶胺；2-氨基甲基喹啉
quinalgen(=analgen) 喹哪晶；乙氧基苯甲酰氨基喹啉
quinalizarin 醌茜素
quinalizarinsulphonic acid 醌茜素磺酸
quinamicine 奎米素；奎纳米辛
quinamidine 奎脒 $C_{19}H_{24}N_2O_2$
quinamine 奎胺 $C_{19}H_{24}N_2O_2$
quinane 奎烷 $C_{20}H_{24}N_2$
quinaphthol 奎萘酚；磺酸β-萘酚喹啉
quinardic acid(=quinardinic acid) 喹哪啶酸
quinardinic acid(=quinardic acid) 喹哪啶酸
quinary ①五个(一组)的②五进位制；五进制
quinary alloy 五元合金
quinary steel 五元合金钢
quinaseptol(=diaphthol) 迪阿索耳；间磺酸邻氧喹啉〔尿道消毒剂〕；奎色醇
quinaseptol silver(=argentol) 奎色醇银
quinasitinic acid 奎亭酸 $C_9H_7NO_4$
quinate 奎尼酸盐 $(HO)_4C_6H_7COOM$
quinate of urotropine 乌洛托品的奎尼酸盐
quinazerin(=quinizarin) 醌茜；奎札因
quinazine(=quinoxaline) 喹嗪；喹噁啉
quinazo 喹唑啉并；间二氮萘并
quinazoline(=phenmiazine) 喹唑啉；间二氮杂萘；1,3-二氮杂萘 $C_6H_4CH=NCH=N$
quinazoline-based polyamide 喹唑啉基的聚酰胺
quinazolinyl(=quinazolyl) 喹唑啉基；间二氮(杂)萘基 $C_8H_5N_2-$
2-quinazolone 喹唑(啉)酮 $C_8H_6N_2O$
quinazolyl(=quinazolinyl) 喹唑啉基；间二氮(杂)萘基
quinazo [4,3-b] quinazol-8-one 喹唑啉并[4,3-b]喹唑酮
quince 榅桲；榅桲
quince oil 榅桲油
quince oil acid 榅桲油酸
quindoline(=10H-indolo[3,2-b] quinoline]) 10H-吲哚[3,2-b]喹啉；喹叨啉 $C_{15}H_{10}N_2$
quinene 奎烯 $C_{20}H_{22}N_2O$
quinetum 奎宁母
quinhydrone 醌氢醌；对苯醌合对苯二酚 $C_6H_4O_2·C_6H_4(OH)_2$
quinhydrone electrode 醌氢醌电极
quinia(=quinine) 奎宁

quinic 奎宁的
quinic acid 奎尼酸；金鸡纳酸；1,3,4,5-四羟(基)-1-环己烷羧酸 $(HO)_4C_6H_7COOH$
quinicine(=quinotoxine) 奎尼辛
quinidamine 奎尼胺 $C_{19}H_{24}N_2O_2$
quinide 奎尼内酯
quinidine 奎尼丁〔药〕
quinidine gluconate 葡萄糖酸奎尼丁
quinidine hydrochloride 盐酸奎尼丁 $C_{20}H_{24}O_2N_2·HCl$
quinidine sulfate 硫酸奎尼丁
quinindene 奎茚；苯并异吲哚；苯并4-氮茚 $C_4H_4C_8H_4N$
quinindole 奎吲哚 $C_{11}H_3N_2$
quinine(=chinine) 奎宁；金鸡纳碱；金鸡纳霜
quinine acetate 乙酸奎宁
quinine acetyl salicylate 乙酰水杨酸奎宁
quinine albuminate 白蛋白合奎宁
quinine alkaloids 奎宁生物碱类
quinine antimonate 锑酸奎宁
quinine arrhenate 甲基胂酸奎宁
quinine aspirin 乙酰水杨酸奎宁
quinine bisulfate 酸式硫酸奎宁；金鸡纳霜；重硫酸奎宁
quinine bromate 溴酸奎宁
quinine camphorate 樟脑酸奎宁
quinine carbolate 石炭酸奎宁
quinine dihydrochloride 二盐酸奎宁
quinine ethyl carbonate(=euquinine) 碳酸乙酯奎宁；无味奎宁；优奎宁
quinine ferricyanide 氰铁酸奎宁
quinine glycerophosphate 甘油磷酸奎宁
quinine hydrochloride 盐酸奎宁
quinine iodosulfate(=herapathite) 碘硫酸奎宁
quinine phenol sulfonate 苯酚磺酸奎宁
quinine salt 奎宁盐
quinine sulfate 硫酸奎宁
quinine sulfate periodide 硫酸奎宁合高碘化物
quinine sulfoguaiacolate 硫愈创木酚奎宁
quinine trisulfoacetyl creosote 三硫乙酰杂酚油奎宁
quininic acid(=6-methoxycinchoninic acid) 奎宁酸；6-甲氧基喹啉-4-羧酸 $CH_3OC_9H_5NCO_2H$
quininium ion 奎宁鎓离子
quininone 奎宁酮 $C_{20}H_{22}N_2O_2$
quinisatin 奎靛红
quinisatinic acid 奎靛红酸
quinisatinoxime 喹靛红肟
quinisoamyline 奎异戊灵
quinite(=p-cyclohexandiol) 对环己二醇 $C_6H_{10}(OH)_2$
quinitol 对环己二醇 $C_6H_{10}(OH)_2$

quinium(=mefloquine)　甲氟喹；奎宁母
quinizarin(=quinazerin)　醌茜；奎札因；1,4-二羟基蒽醌 $C_{14}H_8O_4$
quinizine(=antipyrine)　安替比林
Quinlin and Weiser method　奎林-韦塞法〔硅胶凝胶柱色谱定量分离甘油单、双、三酸酯的方法〕
quino　①喹啉并②〔词头〕奎诺
quinoa　奎藜籽
quinocarbonium　醌碳鎓
quinochromes(=thiamine)　醌色素；硫胺；维生素 B_1
quinocycline　醌环素
quinoform(=chinoform)　奎诺仿；碘氯羟基喹啉
quinogen　醌精
quinoid　醌型
quinoidal moiety　醌型部分；醌式基部分
quinoid compound　醌型化合物
quinoid form　醌型
quinoidine(=chinioidina)　奎诺酊
quinoid structure　醌型结构
quinol　对苯二酚；醌醇〔1. (二)氢醌 $O=C_6H_4=HOH$；2. 甲代二氢醌 $O=C_6H_4(OH)CH_3$〕
quinol imide　醌醇亚胺
quinolinamine(=quinolyl amine)　喹啉(基)胺；2-氨基喹啉
quinolinate transphosphoribosylase　喹啉酸转磷酸核糖酶
quinolinazo　喹啉偶氮
quinolinazo R　喹啉偶氮 R
quinoline(=chinoline)　喹啉　C_9H_7N
quinoline acids　喹啉酸类
quinoline aldehyde　喹啉甲醛
quinoline-2-aldehyde-2-quinolylhydrazone(QAQH)　喹啉-2-醛-2-喹啉腙
quinoline blue　喹啉蓝；氮萘蓝
quinoline-2-carboxylic acid　喹啉-2-羧酸；喹哪啶酸
quinoline-8-carboxylic acid　喹啉-8-羧酸
quinoline carboxylic acid　喹啉羧酸　$C_9H_6N \cdot COOH$
quinoline colouring matters　喹啉染料
quinoline dicarboxylic acid　喹啉二羧酸
quinoline dye　喹啉染料；氮萘染料
quinoline ethiodide　磺化 N-乙基喹啉〔腐蚀抑制剂〕
quinoline methochloride　喹啉甲氯化物；氯化甲基喹啉
quinoline nuclei　喹啉核
quinoline thiocyanate　硫氰酸喹啉
quinoline yellow　(酸性)喹啉黄
quinoline yellow lake　喹啉黄色淀
quinolinic acid(=pyridine-2,3-dicarboxylic acid)　喹啉酸；2,3-吡啶二羧酸
quinolinic anhydride　喹啉酸酐
quinolinium compound　喹啉鎓化合物

quinolino　喹啉并
8-quinolinol　8-羟基喹啉　$C_9H_6N(OH)$
quinolinolate　喹啉醇化物
quinolinoxazole　喹啉并噁唑　$C_{10}H_8ON_2$
quinolizine　喹嗪　C_9H_9N
quinolizino　喹嗪并
$1H$-quinolizino〔1,8-ab〕quinolizine　$1H$-喹嗪并〔1,8-ab〕喹嗪
quinolone(=carbostyril)　喹诺酮；2-羟基喹啉
quinolone-methide　N-甲基-2-亚甲基喹啉
quinol phosphate　磷酸对苯二酚酯
quinolyl　喹啉基〔有 7 种异构体〕　C_9H_6N-
3-(2-quinolyl) acrylic acid　3-(2-喹啉基)丙烯酸
quinolyl amine(=quinolinamine)　喹啉(基)胺；2-氨基喹啉
8-quinolyl mercaptan　8-喹啉基硫醇
quinolyl oxyalkyl ammonium halide　喹啉基氧烷基卤化铵
quinolyl thioether　喹啉硫醚
quinomycin　醌霉素
quinondiimine　醌二亚胺
quinone　醌*
o-quinone　①邻醌②邻(苯)醌
quinone-anilide　醌苯胺；2,5-二苯氨基对苯醌
quinone chlorimide　醌氯亚胺
quinone-N-chlorimide　醌-N-氯亚胺
p-quinonedianil　对苯醌缩二苯胺
quinone dichlorimide　醌二氯亚胺
p-quinone diimine　对醌二亚胺
quinone imide　醌亚胺　$NH=C_6H_4=O$
quinone-imine　醌亚胺
quinone methide　醌的甲基化物
quinone monoimine　醌(单)亚胺
quinone monoxime　醌-肟
quinone oxime dye　(苯)醌肟染料
quinone pigments　醌颜料
quinone polymer　苯醌聚合物
quinones　醌类
quinone tanning　醌鞣
quinonimine　醌亚胺
quinonimine dye　醌亚胺染料
quinonoid　醌型
quinonoid compound　醌型化合物
quinonoid form　醌型构型
quinonoid structure　醌型结构
quinonyl(=benzoquinonyl)　醌基　$C_6H_3O_2-$
quinophan　奎诺芬；阿托方　$C_6H_5C_9H_5NCOOH$
quinophenol(oxine)　奎诺(苯)酚；8-羟基喹啉；喹啉酚
quinopyrine(=chinopyrine)　奎诺比林
quinoquinazoline　喹啉并喹唑啉

quinoquinazolone 喹啉并喹唑(啉)酮
quino[4,3-b] quinoline 喹啉并[4,3-b]喹啉
α-quinoquinoline(=pyrido[2,3-b] quinoline；1,10-naphtho-diazine) 吡啶并[2,3-b]喹啉；1,10-二氮杂蒽
quinoral(=chinoral) 奎诺醛；奎宁合氯醛
quinotannic acid 奎诺单宁酸；奎诺鞣酸 $C_{14}H_{16}O_9$
quinoticine 奎诺剔素
quinotidine 奎诺剔定
quinotine 奎诺亭
quinotoxine(=quinicine) 奎尼辛
quinotropine 奎诺托品；奎尼酸乌洛托品
quinovaic acid 异鼠李酸；奎诺瓦酸
quinovic acid 奎诺酸
quinovin 奎诺温
quinovose 异鼠李糖；6-脱氧葡萄糖
quinoxaline(=quinazine) 喹噁啉 $C_6H_4N=CHCH=N$
quinoxaline-2,3-dithiol 喹噁啉-2,3-二硫酚
quinoxalinyl(=quinoxalyl) 喹噁啉基；对二氮萘基 $C_8H_5N_2—$
quinoxalo 喹噁啉并
quinoxalone 喹噁酮
quinoxalophenazine 喹噁啉并吩嗪
quinoxalyl(=quinoxalinyl) 喹噁啉基；对二氮萘基 $C_8H_5N_2—$
quinoxime 醌肟
quinoxin 奎诺克辛；亚硝基酚
quinoxyl 奎诺昔尔；喹碘方
quinpropyline 奎丙灵
quinquamycin 五烯霉素
quinque- 〔拉丁字头〕 五
quinquedentate ligand 五齿配位体
quinque-molecular 五分子的
quinquemolecular reaction 五分子反应
quinquennial index 五年索引
quinquephenyl 五联苯
quinquevalence(=quinquevalency) 五价
quinquevalent nitrogen 五价氮
quinqui- 五
quinquidentate ligand 五齿配位体
quinquiphenyl 对联五苯
quinquivalent 五价的
quinssopropyline 奎异丙灵
quintavalent 五价
quintessence 浓萃〔浓的萃取液〕
quintet 五重峰
quintillon ①百万亿亿亿〔10^{30}〕②百亿亿〔美国用法，10^{18}〕
quintomycin 五氮霉素
quintone 萜烯树脂〔C_5馏分纯戊二烯树脂〕
quintozene(=PCNB) 五氯硝基苯
quintuple ①五倍②五倍的③以五乘之④增加五倍
quintuple point 五相点
quintuple space 五维空间
quinuclidine 奎宁环
quinuclidinyl 奎宁环基 $NC_7H_{12}—$
quinuclidone 奎宁环酮
quire 刀〔纸张单位，25张为一刀〕
quirk ①深槽沟；凹部②火道③花边花纹④斜角镶条
quitenidine 奎特尼定 $C_{19}H_{22}O_4N_2$
Q unit Q 单位〔热量单位，$1Q=10^{18}$ 英热单位 Btu〕
quotane(=1-β-dimethylaminoethoxy-3-n-butyl-isoquinoline) 喹坦；二甲异喹
quota sampling 定额抽样
quotation ①引证；指引②引用字句③估价单
quotation mark 引(用符)号
quotient ①商②系数
quotient method 商法
quoting ①引用；引证②报价
Q-value Q 值*

R

rabbet 槽口
rabbeted end 槽口
rabbet joint 槽口接合
rabbit brush 一枝黄〔一种多年生橡胶植物〕
rabbit serum albumin 兔血清白蛋白
rabbit serum globulin 兔血清球蛋白
rabble (长柄)耙
rabble arm 耙柄
rabble blade 耙齿
rabble blade holder 耙齿把
rabbling 搅拌〔炼铁〕
rabbling burner(=rabbling furnace; rabbling roaster) 搅拌炉
rabbling door 搅拌孔
rabbling furnace(=rabbling burner) 搅拌炉
rabbling roaster(=rabbling burner) 搅拌炉
rabelaisin 拉伯雷毒苷
rabeprazole 雷贝拉唑〔药〕
R absorption band R吸收带
raC 射碳
racahout 橡实粉
race board 走梭板
racemate ①外消旋物②外消旋酒石酸盐
racemation 外消旋(作用)
raceme ①外消旋体；外消旋物②总状花序〔生物〕
racemic 外消旋的
dl-(DL)-racemic ①外消旋体②外消旋的
racemic acid ①外消旋酸②外消旋酒石酸
 COOH(CHOH)$_2$COOH
racemic body 外消旋体
racemic compound 外消旋化合物
racemic mixture 外消旋混合物
racemic modification 外消旋(变)体
racemic polylactide 外消旋聚丙交酯
racemic solid solution 外消旋固体溶液*
racemic tartaric acid 外消旋酒石酸
 COOH(CHOH)$_2$COOH
racemic tartarte 外消旋酒石酸盐
racemism 外消旋(性)
racemization 外消旋化*
racemization heat 外消旋热
racemized 外消旋的
racemoid ①外消旋物②外消旋的
racephedrine 消旋麻黄碱

race track effect 轨道效应
2R-acid 2R酸；2-氨基-8-萘酚-3,6 二磺酸
R-acid(=2-naphthol-3,6-disulfonic acid) R酸；2-萘酚-3,6-二磺酸
rack ①架②齿条
rack-and-pinion 齿条齿轮传动装置
rackarock 氯酸钾-硝基苯炸药
rack car 棚架车
rack driven 齿条传动
racked stitch 扳花组织；波纹组织
rack gear 齿条；齿条传动
racking configuration 装架方式；框(台)架结构
racking of rubber 橡胶整理
rackman 油桶管理员
rack mounting 架(上安)装的
rack rod 齿条
rack work 齿条加工
raCl 射氯
rad 拉德〔辐射吸收剂量单位，等于 100 尔格/克〕
radappertization 辐射灭菌；彻底的杀菌
radar 雷达；无线电探测器
radar-absorbent coatings 雷达电磁波吸收涂料；反雷达涂料
radar-absorbent finish 雷达电磁波吸收漆(漆层)；反雷达漆(涂层)
radar absorbent material 雷达电磁波吸收材料
radar absorptivity 雷达电磁波吸收能力
radcrete "放射"混凝土；辐射处理混凝土
raddeamine 蕾蒂胺
raddeanine 蕾蒂宁〔取自蕾蒂贝母 Fritillaria raddeana〕
raddle 代赭石；土赤铁矿
radgas 放射性气体
radiac instrumentation 放射性探测仪器
radiacmeter 放射性辐射测定器
radial ①径向的；辐向的②沿视线的③辐射(形)的④光线的⑤半径的
radial acceleration 径向加速度
radial arm 旋臂
radial bearing 径向轴承
radial bladed impeller 径向式叶轮
radial burner 辐形灯
radial carcass 子午线(结构)胎体
radial chromatography(=circular chromatography) 径向(展开)色谱(法)；环形色谱(法)

radial clearance 径向间隙
radial clutch 径向离合器
radial compression column 径向压缩柱
radial compression separation system 径向压缩分离系统
radial compressive stress 径向压应力
radial compressor 离心式压缩机
radial contraction 径向收缩
radial coordinate ①径向配位②极坐标
radial copolymer 星形共聚物
radial cracking 径向裂纹；辐射裂纹
radial cut ring 径向切口环〔压缩机填料〕
radial development 径向展开(法)
radial development chromatography 径向展开色谱
radial diffuser 径向扩压器；径向扩散器
radial diffusion 径向扩散
radial dilation 径向扩容；径向膨胀
radial dilution effect 径向稀释效应
radial drill 旋臂钻床〔机工〕
radial drilling machine 旋臂钻床
radial electronic analyzer 辐射电子分析器
radial electrostatic field analyzer 径向静电场分析器
radial facing 旋面〔商〕
radial flow chromatography 径向流动色谱
radial flow column 径流柱
radial grooved filter plate 辐射凹纹滤板
radial impeller 径向叶轮
radial inflow 向心式
radial inflow compressor 向心式压缩机
radial loading 径向载荷
radially differentiated structure effect 径向异构效应
radial magnetic field analyzer 径向磁场分析器
radial normal stress 径向正应力
radial outflow method of cooling 径向外流冷却法
radial outflow quench 径向外吹骤冷；由内及外环吹骤冷
radial plunger oil pump 旋转往复油泵；旋转柱塞油泵
radial quench unit 径向骤冷装置
radial reactor 径流式反应器
radial refractive-index profile 径向折射率分布
radial run-out 纵向偏心度；径向偏心度
radial section (of coal) (煤的)辐射剖面
radial shrinkage 径向收缩
radial stiffness 径向劲度
radial stress 径向应力
radial swelling 径向溶胀；径向膨胀
radial symmetry(=actinomorphy) 放线对称；辐射对称性
radial vane 径向叶片
radial wear ①径向磨损②径向磨损量

radial wire (-cord) tire 钢丝子午(线)轮胎
radiameter 放射计
radian 弧度
radiance ①辐射②照射(作用)
radian frequency 角频率；圆频率
radiant ①辐射的②发光的③发热的④照耀的
radiant arc furnace 辐射电弧炉
radiant capacity 辐射功率；辐射能力
radiant cooling method 辐射冷却法
radiant dryer 辐射烘箱；辐射干燥器
radiant emittance 辐射率；辐射发射率；辐射能流密度
radiant energy 辐射能
radiant-energy spectrum 辐射能谱
radiant exitance 辐射出射度；辐射度
radiant flux 辐射通量
radiant flux density 辐射通量密度
radiant heat 辐射热
radiant heat absorber 辐射热吸收器
radiant heat baking 辐射热烘烤(烘干)
radiant heat density 辐射热强度
radiant heat drying 辐射热干燥
radiant heater 辐射加热器
radiant heating 辐射加热(法)
radiant heat transfer coefficient 辐射传热系数
radiant heat zone 辐射热区(层)
radiant intensity 辐射强度
radiant jacket 辐射加热夹套
radiant light 辐射光
radiant matter 辐射物(质)
radiant oven 辐射烘干炉
radiant panel burner 辐射板式燃烧器
radiant power ①辐射本领②辐射功率
radiant quantity 辐射量
radiant ratio 辐射率
radiant resistance furnace 辐射电阻炉
radiant roof tubes 有辐射罩的管子
radiant section (管式炉的)辐射部分
radiant section outlet 辐射部分的出口
radiant state 辐射态
radiant tubular heater 辐射型管式炉
radiant-type furnace 辐射型管式炉
radiant-type pipe still 辐射型管式炉
radiant wall tubes 辐射壁管
radiated 辐射的
radiated power 辐射功率
radiated spar 纤晶石
radiating 辐射的
radiating capacity 辐射本领

radiating heat 辐射热
radiating power 辐射功率
radiating surface 辐射面
radiation ①辐射②放射③放射物④辐射线⑤放射线⑥照射(作用)
radiation absorber 辐射吸收剂
radiation absorption 辐射吸收
radiation absorption analysis 放射线吸收分析
radiation-actuate 辐射诱发；辐射激励
radiational deactivation 辐射减活化；辐射去活化；辐射失活
radiational excitation 辐射激发
radiation annealing 辐射退火；辐射熟练
radiation appliance 辐射装置
radiation arc furnace 辐射电弧炉
radiation auto-oxidation 辐射自氧化*
radiation biochemistry 辐射生物化学
radiation breeding 辐射育种
radiation buffer 光谱(辐射)缓冲剂
radiation catalysis 辐射催化
radiation-chemical chromatography 放射化学色谱(法)
radiation chemical engineering 辐射化工*
radiation chemical reaction 放射线(诱发)化学反应
radiation-chemical reduction 辐射化学还原
radiation chemistry 辐射化学*
radiation cleavage 辐射裂解
radiation coefficient 辐射系数
radiation coloration ①射线着色作用②射线显色作用
radiation constant 辐射常数；辐射恒量
radiation crosslinking 辐射交联*
radiation crosslinking of polymer 聚合物辐射交联反应
radiation curable coating 辐射固化涂料
radiation curable polymer 辐射固化聚合物
radiation curable system 辐射线固化涂料(体系)
radiation cure ①辐射熟化②辐射处理
radiation-cured 辐射硬化的；辐射熟化的；辐射处理的
radiation cured coating 辐射固化涂料
radiation curing 辐射固化*
radiation damage 辐射变质；辐射损伤
radiation damage inhibitor 防放射线剂
radiation decomposition 辐射分解
radiation-degradable polymer 辐射降解聚合物
radiation degradation 辐射降解
radiation degradation of polymer 聚合物辐射降解
radiation density 辐射密度
radiation detection 辐射检测
radiation detector 辐射探测器
radiation dose 辐射剂量
radiation dosimeter 射线剂量计；辐射剂量计
radiation dosimetry 辐射剂量学
radiation effect ①辐射效应②照射效应
radiation efficiency 辐射效率
radiation electrochemistry 辐射电化学
radiation energy 辐射能
radiation equilibrium 辐射平衡
radiation error 辐射误差
radiation field 辐射场
radiation fins 辐射翅；波纹套；辐射叶
radiation formula 辐射式
radiation gradient 辐射梯度
radiation-grafted 辐射接枝的
radiation grafting 辐射接枝
radiation hardening 辐射硬化
radiation heat 辐射热
radiation heater 辐射加热器
radiation heating surface 辐射受热面
radiation heat transfer 辐射热传递；辐射传热
radiation hypothesis 辐射假说
radiation immobilization 辐射固定化
radiation-induced aberration 辐射诱变畸形
radiation-induced bulk polymerization 辐射诱导本体聚合(作用)
radiation-induced chlorination 辐射诱导氯化(作用)
radiation induced crosslinking 辐射诱导交联
radiation-induced graft copolymerization 辐射诱导接枝共聚合(作用)
radiation-induced grafting 辐射诱导接枝
radiation-induced ionic polymerization 辐射诱导离子聚合
radiation-induced lethal 辐射诱发致死因子
radiation induced mutation 辐射诱发突变
radiation-induced polymerization 辐射诱导聚合(作用)
radiation-induced radical reaction 辐射诱导自由基反应
radiation induced reaction 放射线(诱发)化学反应
radiation-initiated polymerization 辐射引发聚合*
radiation injuries 辐射损伤
radiation inversion 辐射逆温
radiation ion polymerization 辐射离子聚合*
radiation law 辐射(定)律
radiationless relaxation 无辐射弛豫
radiationless transition 无辐射跃迁
radiation level 辐射能级；辐射强度
radiation loss 辐射损失
radiation measurement 放射测量
radiationmeter 伦琴(辐射)计；X射线计
radiation modification 辐射改性

radiation moisture measurement 辐射水分测定法
radiation pasteurization 辐射巴氏杀菌
radiation polymerization 辐射聚合*
radiation power 放射功率
radiation preservation 辐射保藏
radiation preservation of food 辐射保藏食品
radiation pressure 辐射压(力)；辐射压强
radiation processing 辐射加工*
radiation proof glass fibre 防辐射玻璃纤维
radiation proof paint 防辐射漆
radiation protection 辐射防护
radiation protective paint 防辐射涂料
radiation pyrometer 辐射高温计；辐射测温计
radiation resistance 抗辐射性能
radiation resistance furnace 辐射电阻炉
radiation resistant paint 耐辐射漆
radiation resistant polymer 耐辐射聚合物
radiation safety 放射性安全
radiation scavenger 辐射清除剂
radiation screen 辐射罩；辐射屏
radiation self-decomposition 放射性自分解
radiation-sensitive bond 辐射线敏感键
radiation shielding 辐射屏蔽
radiation shielding coating 放射线屏蔽涂料；辐射防护涂料；防辐射涂料
radiation shielding glass 放射线屏蔽玻璃；辐射防护玻璃；防辐射玻璃
radiation shielding material 射线屏蔽材料
radiation shielding window 辐射屏蔽窗
radiation source 辐射源
radiation source implant 植入辐射源
radiation sterilization 辐射灭菌*
radiation syndrome 辐射并发(症)
radiation synthesis 辐射合成
radiation thermometer 辐射温度计
radiation thermometry ①辐射计温学②辐射计温术
radiation value 辐射值
radiation vulcanization 辐射硫化
radiative lifetime 辐射寿命
radiative transition 辐射跃迁
radiator ①辐射器②辐射体③散热器；暖气片；暖气管
radiator drain 水箱出水管
radiator enamel 散热器瓷漆；暖气片用瓷漆
radiator filler 散热器进料孔
radiator grill 散热器护栅
radiator paint 散热器用漆
radiator strainer 散热器粗滤器
radiator valve 散热器阀；恒温控制器

radiator vent line 散热器通风管
radical ①基；原子团；游离基；自由基②根③基本的
radical acceptor 自由基接受体
radical addition polymerization 自由基加聚反应
radical anion 自由基负离子*
radical-anion initiator 自由基-阴离子引发剂；游离基阴离子引发剂
radical catalyst 游离基催化剂
radical cation 自由基正离子*
radical chain reaction 自由基链反应〔高分子〕
radical chain terminator 自由基链终止剂
radical change 游离基变换
radical copolymerization 自由基共聚合*
radical coupling 自由基偶合
radical crosslinking 自由基交联(作用)
radical-cured 自由基固化的
radical electronegativity 自由基电负性
radical formation 自由基形成
radical former 自由基生成体
radical in-flow 径向内吹
radical-initiated polymerization 自由基引发聚合
radical initiation 自由基引发
radical initiator 自由基引发剂
radical ion 自由基离子*
radical mechanism 游离基机理
radical out-flow 径向外吹
radical pair 自由基对
radical pair mechanism 自由基对机理
radical polymerization 游离基(引发)聚合(反应)；自由基(引发)聚合(反应)
radical reaction 自由基间的反应
radical (reaction)-yield-detected magnetic resonance 自由基(反应)产量检测磁共振
radical ring opening polymerization 自由基开环聚合
radical scavenger 自由基捕获剂*
radical screw clearance 挤压机径向螺杆间隙
radical telomerization 自由基型调聚反应；游离基型调聚反应
radical terminator 游离基终止剂
radical transfer 基团转移
radical transfer reaction 自由基转移反应
radical-trapping agent 游离基俘获剂
radical valence ①根价②基(化合)价
radical weight 原子团重〔原子团的原子量和〕
radicle ①根②基
radicofunctional name 根基官能名称
radio ①无线电；射电〔物〕
radio- 〔拉丁字头〕 放射；(放)射线；辐射

radioactinium 射𨱎 ^{227}Th
radioactivation 辐射激活(作用)
radioactivation analysis 放射化分析；活化分析
radioactive 放射(性)的
radioactive aerosol 放射性气溶胶
radioactive analysis 放射分析法
radioactive analyzer 放射性分析器
radio active anti-fouling 放射性防污
radioactive antigen-binding assay 放射性抗原结合测定
radioactive artificial soil 放射性人造污垢
radioactive background 放射性本底
radioactive bombardment 辐照；辐射
radioactive capture 放射捕获
radioactive chain 放射性衰变链
radioactive change 放射性变化
radioactive clathrate technique 放射性笼形包合物技术
radioactive colloid 放射性胶体
radioactive component 放射性组分
radioactive concentration 放射性浓度
radioactive constant 放射性常数
radioactive contamination 放射性污染*
radioactive contamination monitoring 放射性污染监测
radioactive counter 辐射性计数器
radioactive dating 放射性测定年代
radioactive decay 放射性衰变
radioactive decay chain 放射性衰变链
radioactive decay constant 放射性衰变常数
radioactive decay law 放射性衰变律
radioactive decay scheme 放射性衰变纲图
radioactive decay series 放射性衰变系
radioactive decontamination 放射性去污*
radioactive deposit 放射性淀质
radioactive deposition 放射性沉积物
radioactive detector 放射性检测器
radioactive dirt 放射性污垢
radioactive disintegration 放射性蜕变
radioactive displacement law 放射性位移定律
radioactive dust 放射性灰尘
radioactive earth 放射性土
radioactive electron capture detector 放射性电子俘获检测器
radioactive element 放射(性)元素
radioactive equilibrium 放射性平衡*
radioactive fallout 放射性落尘；放射性沉降物
radioactive family 放射系
radioactive foil electron capture detector 放射性箔电子俘获检测器
radioactive half life 放射半衰期

radioactive immunoassay 放射免疫测定
radioactive indicator (=radiother) 放射指示剂
radioactive indicator method 放射性指示剂法
radioactive ink 放射性油墨
radioactive iodinated human serum albumin 放射性碘标记的人血清白蛋白
radioactive isotope 放射性同位素
radioactive labelling 放射性标记
radioactive level ga(u)ge 放射性液面计
radioactive life 放射性寿命
radioactive luminous paint 放射性夜明漆
radioactively labelled substance 放射性标记物
radioactive marker 放射性标记物
radioactive mineral 放射性矿物
radioactive nitrogen 放射性氮；射氮
radioactive nuclide 放射性核素
radioactive nuleus 放射性核
radioactive paint(=luminous paint) 发光涂料；放射性涂料
radioactive protection 放射性防护
radioactive purity 放射性纯度
radioactive radiation 放射性辐射
radioactive ray 放射线
radioactive sample tray 放射性测定用样品盘
radioactive secular equilibrium 放射性长期平衡
radioactive series 放射系(列)
radioactive soil 放射性污垢
radioactive source 放射源
radioactive standard 放射性标准
radioactive standard solution 放射性标准溶液
radioactive standard source 放射性标准源
radioactive static eliminator 放射性静电消除器
radioactive substance 放射性物质
radioactive tracer 放射性示踪物
radioactive-tracer-fibre 放射性示踪纤维
radioactive tracer method 放射性示踪剂(分析)法
radioactive transformation 放射衰变现象
radioactive transient equilibrium 放射性暂时平衡
radioactive tritium source 放射性氚源
radioactive waste repository 放射性废物处置库*
radioactive waste treatment 放射性废物处理
radioactivity 放射性*
radioactivity detector 放射性检测器
radioactivity logging 放射性测井
radioactivity measurement 放射性测定
radioactivity monitoring 放射性监测
radioaerosol 放射性气溶胶
radio-allergo-sorbent test(RAST) 放射过敏原吸附试验

radioanalysis 放射性分析
radioanalytical chemistry 放射分析化学
radioanalytical diagnosis of nuclear weapons 核武器放射分析法诊断
radioanalytical physics 放射分析物理
radioapplicator 放射性敷贴器
radioassay 放射性检测*
radioassay detector 放射验定检测器
radio-autograph 放射(同位素)显迹图
radioautographic analysis 放射自显影分析
radioautography 放射自显影法
radiobench 放射性工作台
radiobiochemistry 放射生物化学
radiobiogeochemistry 放射生物地球化学
radiobiology 放射生物学
radio bismuth 放射性铋
radiocarbon 放射性碳
radiocarbon chronology 放射性碳测定年代法；放射性碳纪年法
radiocarbon dating 放射性碳测定年代
radiocardiography 心放射图法；放射心动描记法
radiochemical 放射化学试剂
radiochemical analysis 放射化学分析
radiochemical behavior 放射化学行为
radiochemical graft 放(射)化(学)接枝
radiochemical grafting 放射化学接枝
radiochemically pure 放射化学纯
radiochemical modification 放射化学改性
radiochemical neutron activation analysis(RNAA) 放化中子活化分析
radiochemical polarography 放射(化学)极谱法
radiochemical polymerization 辐射化学聚合
radiochemical processing plant 放射化学处理工厂
radiochemical pure 放(射)化(学)纯
radiochemical purity 放(射)化(学)纯度
radiochemical replacement 放射化学排代；放射化学置换
radiochemicals 放射化学试剂(药品)
radiochemical separation 放射化学分离*
radiochemical synthesis 放射化学合成
radiochemistry 放射化学
radiochemotherapeutic 放射化疗的
radiochemotherapy 放射化(学)疗法
radiochlorine 射氯
radio-chromatogram 放射色谱
radiochromatogram scanner 放射色谱扫描器
radiochromatograph 辐射色谱(法)
radiochromatographic separation 辐射色谱分离法

radiochromatographic technique 放射性色谱技术
radiochromatography 放射色谱法
radiochrometer X射线硬度测定仪
radiochromic 放射致色的；辐射致色的
radiochromic dye 放射致色染料；辐射致色染料
radiocobalt 放射性钴
radiocolloid 放射性胶体
radiocontamination 放射(性)污染
radio-controlled pump station 用无线电自动控制的泵站
radiocounting 放射性计数
radiocrystallography 放射性结晶学
radiocytology 放射细胞学
radiodermatitis 放射性皮炎
radio detecting and ranging(=radar) 无线电探测定位(器)；雷达
radiodiagnosis 放射性诊断
radiodiagnostic agent 放射性诊断试剂
radioecology 放射生态学
radioelectret 辐射驻极体
radioelectrochemical analysis 放射电化学分析
radioelectrochemistry 放射电化学
radioelectrophoresis 放射电泳
radioelectrophoretic method 放射电泳法
radioelement 放射性元素
radio-emanation 镭射气
radioencephalography 脑放射照相法
radioenvironmental chemistry 放射环境化学
radioenzymatic assay 放射酶学测定(法)
radio frequency ①无线电频(率)②射频〔物〕
radio frequency cold crucible method 射频感应冷坩埚法
radio frequency curing 射频熟化
radio frequency detector 射频检测器
radio frequency discharge detector 射频放电检测器
radio frequency excited electrodeless discharge lamp 射频激发无极放电灯
radio frequency excited ion laser 射频激发离子激光器
radio frequency field 射频场
radio frequency generators 射频发生器
radio frequency gluing 高频胶合
radio-frequency heater 射频加热器；高频加热器
radio-frequency heating 射频加热；高频加热
radiofrequency heat sealing 射频热合
radio-frequency induction(RFI) 射频感应
radio frequency ion trap mass spectrometer 射频离子阱质谱仪
radio frequency mass spectrometer 射频质谱仪
radio frequency oscillator 射频振荡器
radio frequency oxygen plasma 射频氧等离子体

radio-frequency plasma 射频等离子体
radio frequency polarograph 射频极谱仪
radio frequency polarography 射频极谱法*
radio-frequency preheating 射频预热
radiofrequency sealing 射频热合
radio-frequency spark ion source 射频火花离子源
radio frequency spark source 射频火花(离子)源
radio frequency spectral emission detector 射频光谱发射检测器
radiofrequency spectroscopy 射频波谱法
radiofrequency torch 射频(火)炬
radio frequency transformer 射频变压器；无线电频变压器
radio-frequency voltage 射频电压
radio frequency wave 射频波
radio-frequency welding 射频焊接；高频焊接
radio gas chromatography(RGC) 射频气体色谱(法)
radiogenic heat 辐射热
radiogenic helium 放射产生的氦
radiogenic lead 放射产生的铅
radio-geophysical chemistry 放射地球物理化学
radiogram ①射线照相②无线电报
radiograph (放)线照相
radiographic (放)射线照相的
radiographic effect (放)射线照相效应
radiographic examination 射线照相检查
radiographic inspection 射线照相检查
radiographic test 射线探伤
radiographic testing machine 放射线探伤仪
radiographic thickness gauge 放射线厚度计
radiography 射线照相法；放射显影法
radiography analysis 放射线照相分析
radiogravimetry 放射性重量分析法
radiohalo 放射晕
radiohalogen 放射性卤素
radio heater 射频加热器
radioheating 射频加热
radio heat meter 辐射量热计
radiohistography 放射组织自显术
radiohydrochemical index 放射水文化学指数
radioimmunoanalyzer 放射免疫分析器
radioimmunoassay(RIA) 放射免疫测定(法)
radioimmunoassay instrument 放射免疫测定仪
radioimmunoassay kit(RIA kit) 放射免疫分析药箱
radioimmunoelectrophoresis 放射免疫电泳
radioimmunofluorescent antibody technique 放射免疫荧光抗体技术
radioimmunologic 放射免疫学的
radioimmunology 放射免疫学

radioimmunoprecipitation assay(RIP) 放射免疫沉淀测定(法)
radioimmuno-rocketphoresis autography 放射免疫火箭电泳自显影法
radioimmunosorbent test(RIST) 放射免疫吸附试验
radioinactivation 放射灭活
radioindication method 放射性指示(分析)法
radioindicator 放射性指示剂
radioiodinated 放射性碘(标记)的
radioiodinated serum albumin(RISA) 放射性碘标记血清(色)蛋白
radioiodinated steroid 放射性碘标记甾族化合物
radioiodine 放射性碘
radio ionization detector 放射性电离检测器
radioisomerization 放射异构(现象)
radioisotope 放射性同位素
radioisotope applicator 放射性同位素敷贴剂*
radioisotope arteriography 放射性同位素动脉造影术
radioisotope battery 放射性同位素电池
radioisotope cisternography 放射性同位素脑池造影法
radioisotope customer 放射性同位素用户
radioisotope detector 放射性同位素检测器
radioisotope gauge 放射性同位素计
radioisotope generator 放射性同位素发生器
radioisotope heater unit(RHU) 放射性同位素加热装置
radioisotope instrument 放射性同位素仪表
radio isotope monitor technique 放射性同位素(示踪原子)监测技术
radioisotope on-stream analysis 放射性同位素流线分析(法)
radioisotope-powered cardiac pacemaker 放射性同位素(能源)心脏起搏器
radioisotope renogram 放射性同位素肾图
radio-isotope resistant paint 防放(辐)射性同位素漆
radioisotope scanner 放射性同位素扫描机
radioisotope scanning 放射性同位素扫描
radioisotope smoke alarm(RISA) 放射性同位素火灾烟雾报警器
radioisotope source 放射性同位素(激发)源
radioisotope suit(RI suit) 操作放射性同位素用工作服
radioisotope thermoelectric generator(RTG) 放射性同位素热电发生器；放射性同位素热电源
radioisotope tracer method 放射性同位素示踪法
radioisotope ventriculography 放射性同位素脑室造影法
radioisotopic tracer 放射性同位素示踪物；放射性同位素指示剂
radio-labeled 放射性标记的
radio-labeled compound 放射性标记化合物*
radiolabeled soil 放射性标记污垢

radiolabeled surfactant 放射标记表面活性剂
radio-labeling 放射性标记*
radio-labelled compound 放射性标记化合物
radiolabelling method 放射性标记方法
radiolead 射铅；镭G
radioligand assay 放射配体测(法)
radiolite 钠沸石
radiolocation 雷达定位
radiological chemistry 放射化学
radiological physics 放射物理学
radiology (应用)辐射学；放射学
radiolucent X射线阻碍
radioluminescence 射线发光(现象)
radioluminous material 放射发光材料
radioluminous timepiece 放射性发光钟表
radiolysis 辐解作用
radiolysis of alkanes 石蜡烃的辐解作用
radiolytic cleavage 辐解分裂
radiolytic initiated polymerization 辐射聚合
radiolytic polarography 射解极谱法
radiolytic stability 辐射(分解)稳定性
radiometallography 辐射金相学；放射金相学
radiometallurgy 放射冶金学
radiometer ①放射计②辐射计
radiometer gauge 辐射气压计
radiometric analysis 放射分析；辐射测量分析
radiometric calorimetry 放射量热法
radiometric detector 放射度检测器
radiometric ga(u)ge 放射性计器〔测量仪表〕
radiometric polarography 放射极谱法
radiometric sorter 放射性分选计
radiometric titration 放射性滴定*
radiometrology 放射计量学
radiometry 放射分析法
radiomicrobiological assay 放射微生物分析*
radiomicrometer(=micro-radiometer) 显微辐射计
radiomimetic drug 放射模拟药
radiomimetics 类辐射作用物质
radiomutant 辐射突变体
radiomutation 辐射突变；放射性突变
radion(=radium emanation) 镭射气〔即氡〕
radionickel 放射镍
radionickel exchanged kaolinite 放射镍交换高岭土
radio-nitrogen 射氮
radionuclide 放射性核素
radionuclide battery(RNB) 放射性核电池
radionuclide clearance technique 放射性核素清除技术
radionuclide fluorescence analysis 放射性核素荧光分析

radionuclide kinetics 放射性核素动力学
radionuclide metrology 放射性核素计量学
radionuclide migration 放射性核素迁移
radionuclidic purity 放射性核(素)纯
radiopaper chromatography 放射纸色谱法
radiopaque 不透射线的
radiopaque fibre 辐射屏蔽性纤维；射线阻挡性纤维
radioparent 透射线的
radiopasteurization 辐射杀菌(作用)；辐射消毒(作用)
radiopharmaceutical chemistry 放射药物化学
radiopharmaceuticals 放射性药物
radiopharmaceutical therapy 放射药物治疗*
radiopharmacist 放射性药物学家；放射性药剂师
radiopharmacy 放射药物学*
radiophosphorus 放射(性)磷
radiophotoluminescence 放射光致发光*
radiophotoluminescent dosimeter 放射光致发光剂量计
radiophotoluminescent glass 放射光致发光玻璃
radiophotovoltaic conversion 辐射光电转换
radiopolarogram 放射极谱图
radiopolarographic detector 放射极谱检测器
radiopolarography 放射极谱法
radiopolymerization 辐射聚合(作用)
radiopreservation 辐射保藏
radioprotectant 辐射防护剂*
radioprotective compound 辐射防护化合物；抗辐射化合物
radioprotector 辐射防护剂
radio-quantum number 放射量子数
radioreagent analysis 放射性试剂分析
radioreagent method 放射试剂法
radioreceptor assay 放射性受体测定(法)
radio-release determination 放射性释放测定*
radioresistance 抗辐射性
radioruthenium 射钌；放射性钌
radioscannogram 放射扫描图
radioscintigraphy 放射性闪烁照相法
radioscope 放射镜
radiosensitivity 放射致敏性
radiosensitization 辐射敏化*
radiosensitizer 放射致敏剂
radiosensitizing effect 放射敏化效应
radio-sodium 射钠
radiostereoassay 放射立体化学分析法
radio sterilization 辐射杀菌；辐射致不育
radio sulfur(=isotopic sulfur) 放射硫；同位素硫
radiosynthesis 放射合成
radio-tellurium 射碲；钋

radiother(=radioactive indicator) 放射指示剂
radiotherapeutic 放射治疗的
radiotherapeutics 放射治疗学
radiotherapist 放射性治疗工作者
radiotherapy 放射线疗法；辐射治疗法
radiothermal analysis 放射热分析
radiothermics 射频加热技术
radiothermochromatography 放射热色谱法
radiothermoluminescence 辐射热致发光
radio thin layer chromatography(RTLC) 放射性薄层色谱法
radiothor 放射指示剂
radiothorium 射钍 ^{228}Th
radiotolerance 耐辐照(限)度
radiotoxicity 放射毒性
radiotoxicology 放射毒理学
radiotoxin 放射性毒素
radiotracer ①放射指示剂②(放射)示踪剂
radiotracer leak detector 放射示踪检漏仪
radiotreatment 放射处理；辐射处理
radiotronic 辐射波频操作的
radio valve ①整流器②无线电真空管
radiovision (无线)电视；无线电传真
radiovoltaic conversion 辐射电转换
radiovoltaic generator 辐射电源
radiovulcanization 无线电硫化；高频硫化
radish-seed oil 萝卜子油
radium 镭〔88号元素，化学符号Ra〕
radium A 镭A ^{218}Po
radium appliance 镭装置
radium B 镭B ^{214}Pb
radium bath water 医用含镭水
radium bromide 溴化镭 $RaBr_2$
radium C 镭C ^{214}Bi
radium C′ 镭C′ ^{214}Po
radium C″ 镭C″ ^{210}Tl
radium carbonate 碳酸镭 $RaCO_3$
radium chloride 氯化镭 $RaCl_2$
radium compress 镭敷(剂)
radium D 镭D ^{210}Pb
radium drinking water 含镭饮料水
radium E 镭E ^{210}Bi
radium electroscope 镭验电器
radium emanation(=radion) 镭射气〔即氡〕
radium equivalent 镭当量
radium F 镭F ^{210}Po
radium G 镭G ^{206}Pb
radium needle 镭针

radium plaque 小镭板；镭点源
radium series 镭族
radium sulfate 硫酸镭 $RaSO_4$
radium therapy 镭锭疗法
radium units 镭量单位
radius 半径
radius clearance 径向间隙
radius of action 作用半径
radius of crank 曲柄半径
radius of curvature 曲率半径
radius of gyration 回转半径
radius ratio 半径比
radon 氡〔86号元素，化学符号Rn〕
radon chloride 氯化氡
radon fluoride 氟化氡
radonoscope 氡定量计
radon scrubber 洗氡器
radstone 辐射处理石料
radurization 辐射杀菌；有选择的杀菌〔延长储存期〕
radwaste 放射性废物
radwaste final disposal 放射性废物最终处置
radwaste management 放射性废物管理
radwood 辐射处理木料；木塑料
rafaelite (阿根廷)地沥青；斜羟氯铅矿；钒地沥青
raffia wax(=raphia wax) 酒椰蜡
raffinate 萃余液*
raffinate furnace 残液加热炉
raffinate layer(=raffinate phase) (溶剂精炼润滑油的)残油层
raffinate oil 残油；抽余油
raffinate phase(=raffinate layer) (溶剂精炼润滑油的)残油层
raffinate run tank 残油储罐
raffinate stream 提余液流；残液流
raffinate surge tank 残油缓冲罐
raft test 筏排试验〔船底漆的试验方法之一〕
rag ①毛刺②除去毛刺③压花，滚花④轧槽堆焊
rag boiler 破布蒸煮器
rag bolt 棘螺栓
rag cutter 破布切断机
ragging technique 去毛刺技术
rag knife 切布刀
rag mix 碎布胶料
rag rolled finish(=rag rolling) 碎布滚花涂装
rag rolling ①碎布(辊)滚花涂装法②干膜滚花涂装法
rags 碎布；破布
rags calender 碎布胶料压光机〔橡〕
rag screw 地脚螺栓；棘螺栓

Ragsky test 拉格斯基试氯仿法
rags mixing 碎布胶料
rag sorting room 碎布选择室；选布间
ragweed 豚草
rag wheel 碎布抛光轮
ragwort 千里光；美狗舌草
rag wrapping 灰色包装纸
railroad 铁路；铁道
railroad fuel 机车燃料
railroad writing 铁路写字纸
rail tank car(=rail tanker) 铁路油槽车
rail tanker(=rail tank car) 铁路油槽车
railway 铁路；铁道
railway journal box grease(=railway wagon axle grease) 货车轴用润滑脂
railway rail 钢轨；铁轨；轨铁
railway wagon axle grease(=railway journal box grease) 货车轴用润滑脂
railwlay freight car oil 铁路用粗润滑剂；车箱润滑剂
raindeck extractor 淋降板抽提塔
rain effect 雨滴效应
rain glass 气压表；晴雨表
raining 淋降现象〔塔板〕
rainout 雨除
rain-proof type 防雨型
rain spotting(=water spotting) 水渍；水斑；雨斑〔漆病〕
rain tree 雨树〔泰国紫胶虫寄主树〕
rain water 雨水
raised fabric 起毛织物〔纺〕
raised face 凸面
raised-face flange 凸面法兰
raised grain 木纹隆起
raised head bolt 凸头螺栓
raised pattern 凸纹花样
raising 咬底；咬起；起绒；拉绒
raising agent 膨松剂
raising platform 升降台
raisin seed oil 葡萄子油
Raiskii spark 拉氏火花
Raiskii spark generator 拉氏火花发生器
raisnomycin 雷斯诺霉素
rake ①耙；(长柄)耙②(灰皿)耙③倾料
rake angle 斜度角〔机工〕
rake arms (刮去催化剂用的)长柄耙
rake classifier 耙式选粒器
rake conveyer 耙式运送机
rake mixer 耙式混合器
rake product 粗粒

rake stirrer 耙搅器
raking ①搜集②耙③倾料
RaLa process 镭镧提取过程〔从冷却两天的材料试验堆燃料元件中回收钡-140 的过程〕
raloxifene 雷洛昔芬〔药〕
ralstonite 氟钠镁铝石
ram ①水压机活塞②撞击器③压力扬汲机
ram accumulator 柱塞式蓄力器
ramalic acid 树花地衣酸
ramalinolic acid 乙种构橘苔酸
Raman activity 拉曼活性
Raman difference spectroscopy 差分拉曼光谱
Raman frequency 拉曼频率
Raman inactivity 拉曼非活性
Raman induced Kerry effect spectroscopy 拉曼诱导凯利效应光谱法
Raman scattering 拉曼散射
Raman shift 拉曼位移
Raman spectroelectrochemistry 拉曼光谱电化学法
Raman spectrograph 拉曼光谱仪
Raman spectrometry 拉曼光谱法
Raman spectrophotometer 拉曼分光光度计
Raman spectroscopy 拉曼光谱法
rambutan wax 红毛丹蜡
ram cylinder 水压机气缸
Rame-Hart goniometer 拉姆-哈特测角计
ram extruder 柱塞式压出机
ramie 苎麻；青麻
ramie carding machine 苎麻梳麻机
ram injection machine 柱塞式注压机
ramipril 雷米普利〔药〕
ram-jet fuel 冲压式喷气发动机燃料
rammelsbergite 斜方砷镍矿
ramming 锤击；撞击
ramming machine 出焦车
ram mixer 柱塞式密炼机
ramollescence 软化作用
ramp ①斜面②装料(滑)台
ram packing 阀杆封垫
ramped temperature program 斜坡升温程序
ram pot 液压筒
ram pressure 风筒压力
ram pump 柱塞泵
ramp-up time (往复导丝)增速时间
ram ratio (气体的)动态压缩比
ramsayite 褐硅钠钛矿
Ramsay-Shield equation 拉姆齐-谢尔德公式*
Ramsbottom carbon residue 兰氏残炭

Ramsbottom coke 兰氏残炭
Ramsbottom method 兰氏法〔测定油中残炭含量〕
ram's horn test 落锤试验
ram side 机面
ram travel 柱塞行程
ram-type pump 柱塞泵
ranch 大牧场
rancidity test of fats 油脂酸败试验
randanite 硅藻土
random ①任意②无(一定)目的的③无规的
random access memory(RAM) 随机存取存储器；读写存储器
random and pseudorandom numbers 随机数和伪随机数
random arrangement 无规排列
random chain 无规链
random chain scission 链无规断裂
random charge 无规则电荷
random chemomechanical fibrillation 无规化学机械原纤化
random coil 无规线团
random coiling molecule 无规线团状分子
random coiling polymer 无规卷曲聚合物
random coil model 无规线团模型
random coil molecule 无规线团形(高)分子
random composite fibre 随机复合纤维
random conformation 无规构象
random copolymer 无规共聚物
random copolymerization 无规共聚合*
random crosslinking 无规交联
random crystallization 无规结晶
random degradation 无规降解
random dispersion ①无规则分散作用②不规则分散(度)
random equilibration 无规平衡
random fluctuation 无规涨落
random heterogeneity 随机不均匀性
random inspection 抽查；抽验
randomization of systematic error 系统误差随机化
randomized block design of experiment 随机区组试验设计
randomized complete block design 随机完全区组设计
random labeled 随机标记的；不定位标记的
random laying 杂乱铺网
random linear copolymer 无规线型共聚物
random loose packing 无规则松散堆集
randomly coiled conformation 无规盘旋构象
randomly-oriented 无规取向(的)
random mechanical fibrillation process 无规机械原纤化法

random model 随机模型
random mutagenesis 随机突变
random network 无规网络
random noise 不规则噪声；随机噪声
random orientation 无规则取向
random packing ①无规则填充②不规则填充物
random phase approximation 无规相近似
random phase method 随机相角法
random polyesteramide 无规聚酯酰胺酯
random polymer 无规聚合物
random polymerization 无规聚合
random structure 无规结构
random turbulence 无规则湍动
random walk chain 无规行走链
random walk model 无规行走模型
random wind 无规则卷曲
Rando-webber 兰多成网机
Raney catalyst 拉尼催化剂
Raney nickel 拉尼镍
Raney nickel alloy 拉尼镍合金
Raney nickel catalyst 拉尼镍催化剂
Raney's alloy 拉尼镍铝合金
ranga butter 奶油树油〔取自 *Pentadesma butyracea* 的种子〕
range 极差*；范围
range coal 块煤
range control chart 极差控制图
rangeeni lac 銮基尼紫胶〔印度紫胶品系〕
range finder 测远器；测远仪
range housings 炉罩
range of boiling 沸腾范围
range of chemical shift 化学位移范围
range of definite counting 固定计数范围
range of stability 安定阶段；稳定范围
range of temperature 温度范围
range of thermometer scale 温度计读数范围
range of titration jump 滴定突跃范围
range of use 应用范围；用途种类
range of work 加工范围
range oil 厨房用重煤油
ranitidine 雷尼替丁〔药〕
Rankine cycle 兰金循环
Rankine degree 兰金度数
Rankine scale 兰金温度标
Rankine temperature 兰金温度
ranking 排序法
rankinite 硅钙石
rank of tensor 张量级
Ransburg electro-air gun 兰斯堡电离空气喷枪〔喷雾粒

子带电）
Ransburg electro-hydraulic gun 兰斯堡液压静电喷枪
Ransburg electrostatic blade coater 兰斯堡叶片式静电涂漆机
ranunculine 毛茛碱
Raoult law 拉乌尔定律
rap 潮湿绝缘体；(包装管子用的)绝缘体
rapakivi 奥长环斑岩
rapanone(=2,5-dihydroxy-3-tridecyl-1,4-benzoquinone) 雷帕酮；酸藤子醌
raphanin 莱菔子素
raphia 棕榈叶纤维
raphia alcohol 棕榈叶醇
raphia wax(=raffia wax) 棕榈叶蜡；酒椰蜡
raphides 叶内草酸钙小晶体
rapic acid 菜子酸
rapid accelerator 快速催速剂
rapid aging 快速老化
rapid aging method 快速老化法
rapid analysis 快速分析
rapid analysis method 快速分析方法
rapidazol 快磺素
rapid-breaking emulsified asphalt 快裂乳化沥青
rapid catalyst 快速催化剂
rapid chemical analysis 快速化学分析
rapid coagulation 快速凝结
rapid cure adhesive 速(变)定胶黏剂
rapid-curing 快固的；快凝的
rapid-curing asphalt(=rapid-curing cutback) 快干(铺路)沥青
rapid-curing cutback(=rapid-curing asphalt; rapid-curing paving binder) 快干(铺路)沥青
rapid curing cutback asphalt 快干(铺路)沥青
rapid-curing paving binder(=rapid-curing cutback) 快干(铺路)沥青
rapid degradation 迅速降解
rapid determine of red fuel 红油的快速鉴定
rapid developer 快速显影液
rapid-drying varnish 快干清漆
rapid dry process 快速干燥法
rapid fast blue 快速坚牢蓝
rapid fast colors 快速坚牢颜料
rapid fast orange 快速坚牢橙
rapid fast pulver process 粉状鞣剂转鼓速鞣法
rapid fast scarlet 快速坚牢猩红
rapid filter 快速过滤器
rapid fixer with hardener 快速坚膜定影液
rapid floculation 快速絮凝(作用)

rapid flow (mixing) method 快速流动(混合)法
rapid freezing technique 快速冷冻技术
rapid hardening cement 快硬水泥
rapid heater 快速加热器
rapid liming 快速浸灰(法)
rapid open-access channel electrophoresis 开管快速电泳
rapid paper 快(性印像)纸
rapid polymerization 快速聚合
rapid preoxidation 快速预氧化
rapid process(=r-process) 快过程
rapid radioimmunoassay 快速放射免疫测定
rapid reaction 快速反应
rapid scan Fourier transform NMR correlation spectroscopy 快速扫描傅里叶变换核磁共振相关波谱法
rapid scan monochromator 快速扫描单色仪
rapid scanning interferometer 快速扫描干涉仪
rapid-setting emulsified asphalt 快凝乳化沥青
rapid solidification 快速凝固法
rapid specific surface area analyzer 快速比表面分析仪
rapid steel 高速钢
rapid steel tool 高速钢工具
rapid stock-removal 快速切削；快速磨削
rapid tannage 速鞣法
rapid tanning of sole leather 底革速鞣
rapid test of spirit 白酒快速检验
rapid tool steel 高速工具钢
rapin 芸苔素
rapinic acid 菜子油中异油酸
rare ①稀有的；稀少的②稀疏的；稀薄的③珍贵的④半熟的
rare alkaline earth metal 稀有碱土金属
rare base 稀有碱基
rare earth ①稀土的②稀土(金属)元素
rare-earth additive 稀土(化合物)添加剂
rare earth alloy 稀土合金
rare earth chelate compound 稀土元素螯合物
rare earth drier 稀土(金属)催干剂
rare earth element 稀土元素*
rare earth fission products 稀土裂变产物
rare earth metal 稀土金属*
rare earth minerals 稀土矿物类
rare earth naphthenate 环烷酸稀土金属盐〔催干剂〕
rare earth octoate 辛酸稀土金属盐〔催干剂〕
rare-earth oxide 稀土金属氧化物
rare earths 稀土族
rare earth soap 稀土皂
rare element 稀有元素*
rarefaction 稀疏(作用)

rarefiable 稀化的；稀疏的
rarefied 低于大气压的；稀薄
rarefied gas 稀薄气体
rarefied mixture 稀疏混合物
rare gas 稀有气体*
rare gas clathrate 稀有气体笼形包合物
rare gas removal pilot plant 分离稀有气体中间工厂
rare metal 稀有金属
rare mixture 稀有混合物
Raschig ring 腊希环；腊希圈
Raschig's method 腊希(苯酚合成)法
Raschig tubes〔复〕 腊希圈；填充圈
raschite 若贾特〔铵硝炸药之一〕
rasmosin 升麻树脂
rasorite 斜方硼砂
rasp 打磨齿片；钢丝轮
raspberry 覆盆子；悬钩子
raspberry aldehyde 覆盆子醛；二十醛〔指混合香料〕
raspberry essence 覆盆子香精
raspberry flavour 覆盆子香精
raspberry ketone 覆盆子酮
raspberry seed oil 覆盆子油
rasping machine 磨光机
rasping process 磨锉工序；大抛工序
ratany 南美远志根
ratchet ①棘轮；棘轮机构②棘齿；棘爪
ratchet arrangement 棘轮装置
ratchet driver 棘轮转动装置
ratchet gearing 棘轮转动装置；棘轮机构
ratcheting 棘轮现象
ratchet-type grease cup 棘轮式注油器
ratchet type jack 棘轮式千斤顶；伞齿轮式起重器
rate ①速率②率
rate accelerating material(RAM) 速率加速剂；增速剂
rate constant 速率常数*
rate constant of polymerization 聚合速率常数
rate constant of propagation （链)增长速率常数
rate constant of reaction 反应(的)速率常数
rate constant of termination （链)终止速率常数
rate controlling step 速率控制步骤；决定反应速度的步骤
rated ①额定的；标称的②计算的；设计的
rated capacity 计算容量；额定容量
rated consumption 定额消耗
rate-dependent viscoelastic mode 含速率黏弹性模式
rate-determining stage 反应速率决定性步骤
rate determining step 决速步*；速度决定步骤
rated horsepower 额定功率；额定马力
rated load 额定负荷；额定负载
rated output 额定生产量
rated power 额定功率
rated pressure 额定压力；规定压力
rated speed 额定速度；额定转速
rated voltage 额定电压；标称电压
rated wear 额定磨损量；规定的使用寿命〔指磨损到不能使用时为止的磨损量〕
rate equation 速率方程*
rate-limiting factor 限速因素
rate-limiting reaction 限速反应
rate meter 速率计
rate of adsorption 吸附速率
rate of aging 老成速率；老化速率
rate of application 施工速度
rate of aspiration 提吸速率
rate of catalyst flow 催化剂流速
rate of charring 成炭速率
rate of coagulation 凝固速率
rate of crosslinking 交联速度
rate of cure ①硫化速率②固化速率
rate of decomposition 分解速率；分解率
rate of dispersion 分散速率
rate of drying 干燥速率
rate of efflux 流出速度
rate of energy absorption 能量吸收速率
rate of expansion 膨胀速率；膨胀比
rate of film build 成膜(速)率
rate of flow ①流速；流率②移动率③流量
rate of flow controller 流速控制器
rate of flow in rinse 淋(冲)洗流速
rate-of-flow meter 流量计
rate of growth ①增长率；增长速度②(晶体)长大速率③成长率；生长速度
rate of heat flow 热流速度；热流率
rate of heating 加热速度
rate of heat production 热量生成速率
rate of heat release 散热速率
rate of heat transfer 传热速率
rate of initiation 引发速率
rate of initiation reaction 引发反应速率
rate of interruption 断续比
rate of liquid aspiration 液体吸喷速率
rate of liquid consumption 液体消耗速率
rate of loss 损耗率
rate of mass transfer 传质速率
rate of migration ①色移速度(率)②(增塑剂)迁移速度(率)
rate of moisture absorption 吸湿速度；吸湿率

rate of non-stationary state 非稳态速率
rate of orientation 取向速率
rate of oxidation 氧化速率
rate of oxygenation 充氧速率
rate of permeability 渗透率
rate of permeation 渗透速度；渗透率
rate of photo-tendering 光致脆化速率
rate of polymerization 聚合速率
rate of productivity 生产率
rate of propagation （裂纹）扩展速度；蔓延速度
rate of reaction 反应速率；反应速度
rate of release 渗出率；释放速率
rate of sample aspiration 样品吸喷速度；进样速度
rate of scale formation 生垢率
rate of sedimentation 沉积速率；淀积速率
rate of self-purification 自净速率
rate of settling 沉降速率
rate of shear 剪切速率；剪速；切变速率
rate of shearing strain(=rate of shear) 切变速率；剪切速率
rate of solidification 凝固速度
rate of solvation 溶剂化速率
rate of vaporization 气化速率
rate of volatilization 挥发速率
rate of volume flow 体积流动速率
rate of washing 洗涤速率
rate of wear 磨耗率
rate of weight loss 失重率
rate of withdraw 引出速率
rate process 速率过程
rate process theory(=reaction rate theory) 速率过程理论
rate regulator 速率调节器
rate theory 速率学说
rat fish oil 银鲛油；鼠鱼油
Rathenan-Suter sodium process 腊锡南-苏特制钠法
rathite 斜方砷铅矿；灰砷铅矿
rathole 鼠洞
ratholite 针钠钙石
raticide 杀鼠药
rating ①额定；额定值②评价
rating for light fastness 耐光等级
rating of deposits 储量的评价；储量的评级
rating of fuel 燃料的评价；燃料的评级
rating of merit （发动机、燃烧室等的）性能评价
ratio 比(率)；比(例)
ratio arm 比例臂
ratio estimate 比推定量
ratio flow control 流量比例控制

ratio flow controller 流量比控制器
ratio-frequency welding 高频焊
ratio gear 变速轮
ratio meter 比率计
ratio method 比率法；比例法；比值法
rational （合）理的；有理的
rational analysis 示构分析；状态分析
rational formula 示构式；示性式
rational number 有理数
rational synthesis 示构合成；有理合成
rationing of petroleum products 石油产品的定量分配
ratio of component 组成物比例；组分比
ratio of compression 压缩比
ratio of damping 阻尼比
ratio of enlargement 放大率
ratio of non-threading 空穿率
ratio of run-in 送经比
ratio of specific heat 热容比
ratio of surface to volume(=specific surface area) 表面与体积之比；比表面积
ratio of the CMC 临界胶束浓度比
ratio plug valve 比率调整阀
ratio recording electric-null system 电比率记录式电学零位平衡系统
ratio-recording method 记录比例法
ratio separator 比率分离机；比率分离器；离心分离机
ratio spectrum derivative UV-spectrophotometry 比光谱导数紫外分光光度法
rat poison 鼠毒
rat-tail file 圆锉；鼠尾锉(刀)
rattle 响(亮程)度
rattlesnake fat 响尾蛇脂
rattlesnakeroot 远志根
rat unit 鼠单位
raubasine(=ajmalicine) 阿吗碱
raubasinine(=reserpinine) 利血平宁
rauhimbine(=corynanthine) 柯楠质；萝亨宾碱
raunescine 毛萝芙木碱；萝莱碱
raunormine(=deserpidine) 去甲氧利血平；地舍平
raupine(=sarpagine) 蛇根精
rauwolfia alkaloids 萝芙藤碱类
rauwolfia serpentina alkaloids 萝芙藤碱类
rauwolfine 萝芙碱
rauwolfinine 萝芙碱宁
rauwolscane 萝芙烷
rauwolscine 萝芙素
rauwolscinyl alcohol 萝芙素醇
ravelling(s) ①(织物)拆散的纱；散开的纱②乱纱

raw ①生的；原(状)的②未制(炼)的；不熟的③粗的
raw acid 原酸
raw asbestos 粗石棉
raw belt 带坯
raw blending fluid 粗混合油〔未经分类的粗混合物〕
raw catalyst 新催化剂；未还原催化剂
raw charge(=raw feed) 新进料
raw coal 原煤
raw coke 生焦
raw copper 粗铜；泡铜
raw cotton 原棉
raw crude producer gas 未净化的发生炉煤气
raw data 原始数据
raw distillate 粗馏物；粗馏出油
raw edge plain belt 切边平带
raw edges 裂边；毛边
raw fabric 坯布；原色布
raw feed(=raw charge) 新进料
raw fuel stock 粗燃料油
raw gasoline 粗汽油；不纯汽油
raw glaze 生釉
raw glycerine 粗甘油
raw hide ①生皮②血光皮〔指羊皮纸一类的生皮〕
rawhide lace 血光皮条
raw hide maul 生牛皮槌
raw lacquer （中国）生漆
raw linseed oil 生亚麻油
raw liver oil 鳁鱼肝油
raw lube stock 粗润滑油馏分
raw material 原(材)料
raw material properties 原料的性质
raw meal 生料
raw mill 生料磨(机)
raw Mooney 生胶门尼黏度
raw natural gas 未精制的天然气
raw natural gasoline 未稳定的天然汽油
raw nickel matte 粗制镍
raw oil 原料油；粗制油；未精制的油料
raw oil pump （管路和裂化设备的)原料油泵
raw ore 原矿石
raw paper 原纸；纸坯
raw petroleum coke 生石油焦；粗石油焦
raw pig iron 铸造用生铁
raw polymer 原料聚合物
raw press-distillate 粗裂化馏出物
raw producer gas 粗发生炉煤气
raw pulp 生纸浆；原浆
raw ramie 生苎麻
raw rubber 生橡胶
raw rubber block 生胶块
raw sanitary water 未经处理的生活污水
raw sewage 原污水；未处理的污水
raw sienna 富铁黄土
raw silk 生丝
raw spirit 无水酒精
raw steel 粗钢；原钢
raw stock 原料(生皮)
raw umber 富锰棕土
raw waste water 原污水；未处理的废水
raw yarn 原色纱
ray ①(射)线②(光)线
Raybin's reaction 重氮尿嘧啶试验
ray filter 滤光镜；滤光器
γ-ray-initiated polymerization γ射线引发聚合
ray inspection 射线探伤
β-ray ionization detector β射线电离检测器；β射线离子化检测器
Rayleigh criterion 瑞利判据；瑞利准则
Rayleigh-Jeans law 瑞利-吉恩斯定律
Rayleigh method 瑞利法
Rayleigh ratio 瑞利比
Rayleigh scattering 瑞利散射
Rayleigh scattering light 瑞利散射光
Rayleigh scattering method 瑞利散射法
Raymond mill 雷蒙粉碎机；雷蒙磨
Raymond ring roller mill 雷蒙磨；摆式磨粉机；环辊式磨粉机
ray of light 光线
rayon 嫘萦；人造纤维；人造丝
rayon band 人造扁丝；黏胶扁丝
rayon cord 嫘萦帘线；黏胶帘线
rayon cut staple 人造短纤维；人造切段纤维
rayon fabric 人造丝织物
rayon filament 嫘萦丝
rayon pulp 嫘萦浆；人造丝浆
rayon staple fiber 嫘萦短纤维
rayon top 嫘萦条
rayon tow 嫘萦纤维束
rayon warp size 嫘萦经浆料
rayon yarn 嫘萦丝；人造丝
rayophane 玻璃纸
ray proofing 防辐射的
β-ray standard β射线标准
rays ultimate(=persistent line) 住留谱线；最后线
γ-ray test γ射线探伤
razor blade scraper 薄片刮刀

razor stone 细磨石
R-band R带
R-control chart 极差控制图*
re- 〔拉丁字头〕 ①再；重；复；更②回；向后③相互；回复④(相)反
re-acetylated 重(新)乙酰化的
reacetylation 重乙酰化(反应)；二次乙酰化(反应)
reach rod 拉杆
reactable naphthene 可反应的环烷；可脱氢的环烷
react acid 呈酸性反应
reactance ①电抗②力抗
reactance factor 电抗因数；无功(效)功率
reactant ①反应体②反应剂
reactant gas 反应气体
reactant resin 活性树脂
react basic 呈碱性反应
reacted gas 反应气；裂化气
reacting (起)反应的
reacting capacity(=reaction capacity) 反应本领
reacting furnace 反应炉
reacting mass 反应物料
reacting power (化学)反应本领
reacting substance (起)反应物
reacting weight 反应量
reaction ①反应；作用②反作用③反动力
reaction accelerator ①反应加速器②反应加速剂
reaction arrester 反应阻止剂
reaction bath monitor 反应槽监测器
reaction bonding 反应黏合
reaction box 反应箱
reaction capacity(=reacting capacity) 反应本领
reaction chamber 反应室
reaction channel 反应通道
reaction chromatography 反应色谱法
reaction coil 反应旋管
reaction column ①反应柱②反应塔
reaction condition 反应条件
reaction constant 反应常数
reaction control 反应控制
reaction control agent 反应调节剂
reaction coordinate 反应坐标
reaction coulometric detector 反应库仑检测器
reaction cross-section 反应截面
reaction differential thermal analysis 反应差热分析法
reaction distillation 反应蒸馏
reaction drum 反应室；反应鼓
reaction energy barrier 反应能垒
reaction environment 反应环境

reaction equation 反应方程式
reaction front 反应前沿
reaction gas chromatography 反应气相色谱法*
reaction heat 反应热
reaction impedance 反应阻抗
reaction injection molding(RIM) 反应注射模塑；反应注射成型
reaction in-situ 原位反应
reaction intermediate 反应中间体
reaction isochore 反应等容线
reaction isotherm 反应等温式
reaction kettle 反应釜；反应器
reaction kinetics 反应动力学
reactionless 惰性的；无反应的
reactionlessness 反应(上的)惰性
reaction liquid 反应液体
reaction mass 反应物料
reaction mass spectrometry 反应质谱法
reaction mechanism 反应机理*
reaction medium 反应介质
reaction mixture 反应混合物
reaction network 反应网络
reaction of functional groups on polymer 聚合物官能团反应
reaction of pendant group 侧基反应
reaction of propagation (链)增长反应
reaction of the third order 第三级反应
reaction order 反应级数*
reaction partner 反应参与物
reaction path 反应途径
reaction path degeneracy 反应途径简并
reaction period 反应周期
reaction preference 反应选择性；反应的优选性
reaction probability 反应概率
reaction product 反应产物
reaction promoter 反应促进剂
reaction rate 反应速率
reaction rate constant 反应速率常数
reaction rate equation 反应速率方程式
reaction rate theory(=rate process theory) 速率过程理论
reaction scheme 反应图解
reaction sensitivity 反应灵敏度
reaction soldering 反应焊接；还原焊接
reaction spinning 反应纺丝
reaction stage(=reaction step) 反应阶段
reaction step(=reaction stage) 反应阶段
reaction still 反应锅(釜)；反应塔
reaction stream 反应流

reaction stress 反作用应力
reaction system 反应系统
reaction test for grease 润滑脂反应试验〔测定润滑脂酸碱性的方法〕
reaction time 反应时间*
reaction tower 反应塔
reaction train 反应系列
reaction trajectory 反应轨迹
reaction turbine 反击式涡轮机
reaction velocity 反应速度
reaction vessel 反应(容)器；反应锅
reaction wood 反应材
reaction yield 反应产量
reaction zone 反应区
reactivating 再活化
reactivating temperature 再活化温度
reactivation 再活化(作用)；重激活〔物〕
reactivation bonding 再活化结合
reactivation chain polymerization 再活化链聚合
reactivation cycle 再活化周期
reactivation gas 再活化(催化剂的)气体
reactivation in situ 就地再活化(催化剂)
reactivation line 再活化管路
reactivation of catalyst 催化剂的再活化
reactivation of column 柱的再活化；柱复活
reactivation of the catalyst 催化剂的再活化作用
reactivation of the earth 白土的再活化作用
reactivation process 再活化过程；再生过程
reactivation zone (转子式除湿器)再生区
reactivator 再生器
reactive 反应的
reactive adhesive 活性黏合剂
reactive behavior 反应性能(特性)；反应活性；活性表现(变化)
reactive chlorine 活性氯
reactive coating 活性涂料
reactive colour 活性染料
reactive coupler 活性成色剂；活性偶合剂
reactive curing agent 活性固化剂
reactive diluent 活性稀释剂
reactive dye 活性染料
reactive dyeing 活性染料
reactive dyestuff 活性染料
reactive epoxy coating 活性环氧涂料
reactive extrusion 反应性挤塑
reactive fibre 活性纤维；反应性纤维
reactive force 反应力；作用力
reactive group 反应基

reactive heat-melting adhesive 反应性热熔胶
reactive hydrocarbon 活性烃
reactive hydrogen 活性氢；活泼氢
reactive hydrogen atom 活泼氢原子
reactive hydroxylic species 活性羟基物
reactive intermediate 活泼中间体*
reactive ion enhanced etching 反应离子增强腐蚀
reactive ion etching 反应离子刻蚀
reactive monomer 活性单体
reactiveness(=reactivity) ①反应性②活动性
reactive phenolic resin 活性酚醛树脂
reactive pigment 活性颜料
reactive plasticizer 活性增塑剂
reactive polymer 反应性聚合物*
reactive power 无功功率；无效功率
reactive power factor 无功功率因数
reactive processing 反应性加工
reactive quencher 活性猝灭剂
reactive resin 活性树脂；反应型树脂〔认为可与干性油起反应〕
reactive sealing 反应密封
reactive silane 活性硅烷
reactive site 反应活性部位
reactive species 活性种*
reactive thinner 活性稀释剂(稀料)
reactive-type flame retardant 活性阻燃剂；反应型阻燃剂
reactivity(=reactiveness) ①反应性②活动性
reactivity coefficient 活性系数；反应性系数
reactivity index 反应活性指数
reactivity parameter 活性参数；反应性参数〔反应能力指标〕
reactivity ratio(=monomer reactivity ratio) (单体)反应竞聚率
reactivity ratio of monomer 单体竞聚率
reactor ①反应剂②反应器③反应堆
reactor chemistry 反应堆化学
reactor control 反应器控制
reactor cooling 反应器冷却
reactor design 反应器设计
reactor dilution 反应器稀释
reactor drum 反应鼓；反应器
reactor effluent 反应器流出物
reactor engineering 反应器工程学
reactor feed 反应器进料
reactor fouling 反应器积垢
reactor grid 反应器栅
reactor guard section 反应器防护部分
reactor integrated reprocessing 堆厂一体化后处理；就堆

建厂(核燃料)后处理
reactor pressure vessel　反应堆压力容器
reactor produced radionuclides　反应堆生产的放射性核素
reactor product　反应器产品；反应产品
reactor product cooler　反应产品冷却器；产品冷却器
reactor shield　反应堆防护屏(蔽)
reactor top pressure　反应器顶部的压力
reactor type　反应堆型式
reactor utilization　反应器利用(率)
reacylation　再酰化作用
readability　①可读度；读出能力②清晰度
reader　读数器
read error　读数误差
reader trace　记录仪划线
read in　读入；计入
reading accuracy　读出精度
reading frame displacement　密码位移；移码
reading glass　读数镜
reading-in　穿经
reading lens　读数镜；放大镜
reading microscope　读数显微镜
reading-out integral　显示积分
reading tachometer　读数转速表
Readman phosphorus process　里德曼制磷法
read off　读出；显示
read only memory(ROM)　只读存储器
read-out　读出
read-out integrator　读出积分器
read-out system　读出系统；显示系统
read pulse and phase　读脉冲及相位
read write memory　读写存储器
ready　①有(准)备的②轻便的
ready coating　①快速涂装；快速罩面②快速浇面
ready commodity　成品
ready-formed　现成的；预制的；已制好的
ready-made　现成的；制就的
ready-made goods　现成制品
ready mixed aluminum paint　铝粉调和漆
ready mixed concrete　预拌混凝土
ready mixed paint　调和漆
ready time　(水性压合式黏结剂的)合压时间
ready-to-mix aluminum paint　现用现混铝粉漆
ready-to-mix pack　现用现混(合)包装〔修补灰泥的软砂与水泥的分装包装〕
ready-to-spray　立即可喷涂的；即喷的
ready to use column　现成柱；备用柱
ready to use spray reagent　已配好的喷雾剂；备用喷雾剂

reagent　试剂
reagent blank　试剂空白
reagent bottle　试剂瓶*
reagent bottle with ground stopper　带玻璃磨塞的试剂瓶
reagent color-developing method　试剂显色法
reagent for group precipitation(=group precipitant)　组沉淀试剂
reagent for nickel　试镍剂
reagent grade　试剂级别*
reagent identification　试剂鉴定
reagent paper　试纸
reagent-proof　抗试剂的
reagent resistance　耐试剂性
reagent solution　试剂溶液
reagent sprayer　试剂喷雾器
reagent water　试剂水
reagglomeration　再附聚(作用)
reaggregation　再聚集
reaggregation agent　再聚集剂
real　①实(在)的②真(实)的
real area of contact　实际接触面积
real crystal　实际晶体
real density　真(实)密度
real density of catalyst　催化剂的真密度
real end　①后端②(火箭或发动机)尾部
realgar　雄黄*
real gas　实际气体；真实气体
realizable limit　确认限度；可信限(度)；鉴定极限
real number　实数
real peak　真峰
real potential　真实电位
real solubility　真溶(解)度
real solution　实际溶液
real specific gravity　真比重
real time　实时
real-time analysis　实时分析；快速分析
real-time high resolution mass spectrometry　快速高分辨质谱分析；实时高分辨质谱法
real-time moisture analysis　实时水分分析法
ream　①铰孔②扩孔③令〔纸张计数单位，为480～500张〕
reamed hole　铰孔
reamer　铰刀
ream weight　①令重〔纸〕②垛重〔木；砖的〕
reannealing　重退火
rear　①后部②后(面)的
rear axle　后轴
rear-axle lubricant　后轴润滑剂
rear barrel temperature　机筒后部温度

rear lacquered surface 反面涂漆的表面；背面涂漆的表面
rearrangement 重排*
rearrangement catalyst （分子)重排催化剂
rearrangement ion 重排离子
rearrangement ion peak 重排离子峰
rearrangement of the radical 基团重排
rearrangement peak 重排峰
rearrangement product 重排(产)物
rearrangement reaction 重排反应
rear spring 后弹簧〔机工〕
rear valving 后部排气
Réaumur thermometer 雷默温度计
reblading 换刀
reblending 再次混合；重混合
reblunge 重混合
reboiler 再沸器；重沸器
reboiler coil(=reboiler section) 再煮旋管
reboiler section(=reboiler coil) 再煮旋管
reboil heat 重(煮)热
reboiling 再煮；重热
reboil ratio 再沸比；重沸比
rebound 回弹，反跳；回跳
rebound degree 回弹率
rebound elasticity 回弹性；弹回性
rebound hardness 回弹硬度；反跳硬度
rebound hardness tester 回弹硬度试验仪
rebound model 反弹模型
rebound pendulum 回弹摆锤
rebound pendulum machine 回跳打桩机
rebound resiliometer 回弹性测定计
rebreaker run coal 二级加料煤
rebuilding 全翻新
rebuilt lac(=rubulac) 再生紫胶
reburner 转化炉
reburning 再烧过程
recalescence 复辉(现象)
recalescent point 复辉点
recalibration 再次校准
recapped tyre 胎面翻新轮胎
recapper 轮胎翻新器
recapping 胎面翻新
recarbonation 再碳酸化(作用)
recarburizer 再碳化剂
Recatro process 雷卡脱法制合成气过程
receding colour 似远色
receiver ①接受器②接收机③转化炉
receiver gases 接收气体〔从石油加工设备直接得到的气体〕
receiver pipe 接受管；油罐
receiver pressure control 接受器的压力控制
receiver recorder 二级记录器
receiving 接收；接受
receiving bin 接受仓
receiving box 接受器
receiving drum 接受器
receiving flask 接收瓶；收集瓶
receiving house （石油工厂的)收油车间
receiving opening 装料孔
receiving point 进料给定点；受料给定点
receiving tank 接受罐；收油站
receiving terminal 接受站；收油站
receiving vacuum tube 收讯真空管
recent fossil resin 新生化石树脂
recent resin 新(采的天然)树脂
recentrifuge 再次离心
recentrifuging 再次离心
receptacle ①储器②插坐
receptor ①接受器②受体
receptor model 受体模型
recess ①凹穴②炮眼
recessed (filter press) plate 安框滤板
recessed plate press 凹板式压滤机
rechargeable zinc-air battery 再充(电)式锌-空气电池(组)
recheck 重新检查
rechipper 复切(木片)机；精削机；复研机
rechroming 重新镀铬
recipe ①取药②配方
recipient vessel 接受器
reciprocal cell 倒易晶胞
reciprocal combustion 互燃(作用)；往复燃烧
reciprocal dispersion 色散率倒数；倒易色散率
reciprocal lattice 倒易点阵；倒易晶格
reciprocal lattice vector 倒格矢；倒易点阵向量
reciprocal linear dispersion 倒数线色散*
reciprocal motion 往复运动
reciprocal of apparent density 松密度的倒数
reciprocal ohm(=mho) 倒欧姆；姆(欧)
reciprocal precipitation 相互沉淀(作用)
reciprocal reaction 往复反应；可逆反应
reciprocal recurrent selection 交互反复选择
reciprocal relation 倒数关系；可逆关系
reciprocal salt pair 平衡盐对
reciprocal second 倒(易)秒
reciprocal sensibility （天平的)感量
reciprocal shaker 往复振荡器

reciprocal space 倒易空间
reciprocal specific turbidity 倒数比浓浊度
reciprocal vector 倒易向量*
reciprocating 往复的；来复的
reciprocating blower 往复式鼓风机
reciprocating compressor 活塞压气机；往复压气机
reciprocating-conveyor continuous centrifuge 往复式连续推料离心机
reciprocating diaphragm pump 往复式隔膜泵
reciprocating feeder 往复式加料器
reciprocating feed pump 往复式进料泵
reciprocating mass 往复质量
reciprocating motion 往复运动
reciprocating piston pump 往复式活塞泵*
reciprocating plate extraction column 往复振动板式萃取柱
reciprocating positive displacement pump 往复式正压排液泵
reciprocating pump 往复泵
reciprocating pump station 往复泵站
reciprocating screw injection machine 往复式螺杆注压机
reciprocating sieve 振动筛；往复筛
reciprocating spraying machine 往复式喷涂机
reciprocating (steam) engine 往复(蒸汽)机
reciprocating steam pump 往复蒸汽泵
reciprocator 往复喷涂机
reciprocity ①倒易；反比②交互；相互(关系)
reciprocity law 倒易(定)律；反比定律
reciprocity law failure 互易律失效
recirculated air 循环空气
recirculating air filter 循环空气过滤器
recirculating chromatography 循环色谱(法)
recirculating heater 环流加热器
recirculating main 环流总管〔在船上分配液体燃料的总管线〕
recirculating oil 再循环油
recirculating pump 循环泵
recirculating ratio 循环比；循环系数
recirculating system ①循环系统②(在船上)分配液体燃料的系统
recirculation (再)循环；(复)环流
recirculation furnace(=recirculation still) 循环炉；(裂化装置中)循环气体的炉子
recirculation priming (泵的)循环引动
recirculation pump 再循环泵
recirculation still(=recirculation furnace) 循环炉；(裂化装置中)循环气体的炉子
recirculation unit 再循环装置；循环搅拌装置

reclaimator 螺杆脱硫机；再生机
reclaimed 再生的；翻造的；收复的；回收的
reclaimed oil 再生油
reclaimed rubber(=reconverted rubber) 再生胶
reclaimed water 再生水
reclaimer 回收设备
reclaiming 翻造；收复；回收
reclaiming aid 再生助剂
reclaiming by centrifuge 离心法回收；离心法精制
reclaiming by filtration 过滤法回收；过滤法精制
reclaiming by gravity 澄清法回收；澄清法精制
reclaiming oil 再生用油
reclaiming process 再生工艺
reclaim mix 翻造混炼胶
reclaim recovery efficiency 再生回收(效)率
reclaim sulfuric acid 回收硫酸
reclaim technology 再生工艺
reclaim vulcanizate 翻造硫化胶
reclamation 再生；回收；翻造
reclassification 再次分级
reclining twill 缓斜纹
recoatability 再涂适应性；再涂性
recoat adhesion 再涂附着力
recoating test 再涂试验；罩漆试验
recoating time(=recoat time) 再涂时间；下道漆涂装时间
recognition 特异识别性〔分子印迹技术的特点之一〕
recoil 反冲*；退回；退缩；反冲力
recoil atom 反冲原子
recoil chemistry 反冲化学
recoil electron 反冲电子
recoil energy 反冲能
recoil ion 反冲离子
recoil-labeling 反冲标记
recoilless emission 无反冲能发射
recoil "milking" technique 反冲"挤奶"法〔指从母体放射性同位素中提取反冲子体〕
recoil nuclei 反冲核
recoil oil 反冲油；后座油
recoil product 反冲产物
recoil radiation 反冲辐射
recoil separation 反冲分离
recoil species 反冲粒种
recoil spectrometer 反冲谱仪
recombinant 重组体
recombinant DNA 重组DNA
recombination ①再化合②复合③重组
recombination continuous spectrum 复合连续光谱
recombination fluorescence 重组荧光

recombination of ion and electron 离子电子复合
recombination rate 复合率；再化合速率
recombination reaction layer 再结合反应层
recombiner 复合器；复合剂
recompounding 再次配合〔橡胶〕
recompression 再次压缩
recon 重组子；交换子
reconcentration 再浓集；再浓缩
recondition ①重调节②修理
reconditioning ①重调节；回收②修理
reconditioning system 回收系统
reconnaissance （管路的）探查
reconstituted oil 再造油；翻造油〔用选定的不饱脂肪酸重新酯化的油〕
reconstructed chromatogram 重整色谱图
reconstructed ion chromatogram 重建离子流色谱图
reconstruction total ion current 重建总离子流
reconverted rubber(=reclaimed rubber) 再生胶
recooling tower 二次冷却塔
record ①记录②资料；数据
record chart 记录表；记录纸
recorder 记录器
recorder deflection 记录器偏转度；记录器行程
recorder response 记录器响应
recorder scan 记录器扫描
recording controller 记录控制器
recording flow controller 自记流量控制器
recording flow meter 自记流量计
recording gas meter 记录式气体流量计
recording gear 记录传动装置
recording hygrometer 自记湿度计；湿度记录器
recording infrared spectrophotometer 记录式红外分光光度计
recording instrument 记录式仪器
recording liquid level gauge 自记液面仪；液面记录器
recording meter 自记计数器
recording oscillograph 记录示波器
recording pen lifter 记录笔升降器
recording pen linkage 记录笔尖的传动装置
recording polarograph 记录式极谱仪
recording pressure gauge 自记压力计
recording pyrometer 自记高温计
recording Raman spectrophotometer 记录式拉曼分光光度计
recording spectrometer 记录式光谱仪
recording spectrophotometer 记录式分光光度计
recording thermometer 记录式温度计
recording titrator 记录式滴定计

recoverable ①可复(原)的②可回收的
recoverable deformation gradient history 可恢复形变梯度史
recoverableness ①可复(原)性②可回收性
recoverable shear 可恢复剪切
recoverable strain 可恢复应变
recovered 回收的
recovered acid 回收酸；再生酸
recovered alcohol 回收酒精
recovered carbon 回收活性炭
recovered grease 回收脂膏
recovered manganese mud 回收锰泥
recovered oil 回收油
recovered solvent 回收的溶剂
recovered temperature 回收温度；馏出温度
recovering of vaporized hydrocarbon 回收气态烃；轻质油回收
recovery 回收*
recovery factor 回收系数
recovery gas turbine 回收气体透平；回收气体涡轮
recovery installation （溶剂)回收装置
recovery of heat 热的回收
recovery of oil 油的回收
recovery of solvent 溶剂回收
recovery of storage-battery 蓄电池恢复作用
recovery percent 回收百分数
recovery plant 回收设备；再生工厂
recovery pressure 回收压力；回升压力；压力恢复
recovery ratio 恢复率；回收率；回缩率
recovery system 回收设备
recovery test ①回收试验②采收率试验
recovery total 回收总量
recovery tower 回收塔
recovery unit 回收装置
recovery value 回复值
recovery volume 回收体积
recovery waste heat 回收余热；废热回收
recovery yield of monomer 单体回收率
recovery zone 回收段；回收区
recracking 再裂化
recrement 余渣；废物
recrystal 重结晶；再结晶
recrystallization 再结晶(作用)
recrystallization zone 再结晶区
recrystallizer 重结晶器
rectangle 长方(形)
rectangle form pool 矩形池
rectangular 长方(形)的

rectangular alignment 直角校正
rectangular array 矩(形)阵列；长方阵列
rectangular chamber 矩形展开槽
rectangular coordinate 直角坐标
rectangular crystal 矩形(结)晶体
rectangular distribution 矩形分布〔粒径分布方式之一〕
rectangular groove 矩形槽口
rectangularity 垂直度
rectangular kiln 长方窑
rectangular mesh 长方网格；长方筛孔(目)
rectangular mesh screen 长方筛；矩形筛
rectangular pipe 长方管；矩管
rectangular plate 矩形板
rectangular sedimentation tank 平流式沉淀池
rectangular strip 矩形(色谱)条
rectangular system 长方系〔物〕；直角系；矩形系
rectangular tank 长方形容器；长方形油罐
rectangular wave current 方波电流
rectangular weir 矩形堰
rectifiable 可精馏的
rectification ①精馏②调整；矫正③整流④矫频
rectification column(=rectification tower) 精馏塔
rectification of defects 矫正缺陷；修复缺陷
rectification still 精馏釜
rectification test 精馏试验〔分馏时的酸性试验〕
rectification tower(=rectification column) 精馏塔
rectification tube 精馏管
rectification under vacuum 真空精馏；减压精馏
rectified ①精馏过的②整流的
rectified aniseed oil 精炼茴油
rectified arc 整流电弧
rectified high voltage spark 整流高压火花
rectified oil 精馏精油
rectified spirit 精馏酒精
rectifier ①整流器②精馏器③纠正仪
rectifier with constant current output 恒电流输出整流器
rectifier with constant voltage output 恒电压输出整流器
rectifying 精馏(过程)
rectifying column ①精馏柱②精馏塔
rectifying plate 精馏板；分馏板
rectifying section 精馏段
rectifying still 精馏釜
rectifying tower 精馏塔
rectifying tray 精馏浅盘〔分馏柱中的浅皿〕
rectilinear comb machine 直行精梳机；平梳机
rectilinear coordinate paper 直线坐标纸
rectilinear flow 直线流
rectilinear scanner 直线扫描仪

rectilinear system 直线系
rectilinear transmittance 平行透光率
rectometer 精馏计
rectorite 累托石；钠板石
recuperability 回收可能性
recuperable ①可以同流换热的②可以回收的
recuperated ①同流换热的②回收的
recuperated rubber 再生胶
recuperation ①复原；复元②同流换热(法)；同流节热；继续收热(法)
recuperation of heat 同流换热(法)
recuperative ①同流换热的；复热的②复原的；复元的
recuperative burner(=recuperative furnace; recuperative oven) 同流换热炉
recuperative furnace(=recuperative burner) 同流换热炉
recuperative gas furnace 间壁回热式燃气炉
recuperative oven(=recuperative burner) 同流换热炉
recuperative pot furnace (同流)换热玻璃熔炉
recuperative system 同流换热法；同流节热法
recuperator ①同流换热器②能量回收器
recuperator blocks 同流换热器块
recuperator furnace 同流换热器炉
recuperator lubricant 同流换热器的润滑剂
recuperator tower 同流换热塔
recuperatory ①同流换热的②回收的；复原的
Recuplex process 雷丘普莱克斯法〔从渣和坩埚废料中用磷酸三丁酯萃取法回收钚的过程〕
recurability 复焙烘性
recure 再硫化
recurrence formula 递推公式
recurrence method 递推法*
recurring 再熟化
recurring amide radical 重复酰胺基
recurring group 重复基
recurring segment 重复链段；重复链节
recurring structural unit 重复结构单元
recursion formula 循环公式
recycle (再)循环
recycle acid 循环酸
recycle back 反向循环
recycle chromatography 循环色谱法
recycle circuit 循环线路
recycled fuel 再循环燃料
recycled off-gas 循环尾气；循环废气
recycled powder hose 循环粉末的软管；回收粉末的软管
recycle fraction 循环馏分
recycle fuels plant 再循环燃料工厂〔将硝酸钚转换成二

氧化钍并与二氧化铀混合压片制成燃料〕
recycle gas 循环气体
recycle gas blower 循环(气)鼓风机
recycle gas furnace 循环气体炉
recycle gas oil 循环瓦斯油
recycle gasoline 循环汽油
recycle gas water scrubber 循环气水洗塔
recycle gel permeation chromatography 循环凝胶渗透色谱(法)
recycle liquid pump 循环液泵
recycle pump 循环泵
recycle ratio 循环比；循环系数
recycle ratio of gaseous constituents 气态组分的循环比
recycle stock 再循环物料
recycle stream 再循环液流；返回液流
recycle system 循环系统
recycle unit 循环设备
recycle valve 循环阀
recycling (再)循环
recycling chromatography 循环色谱法
recycling elution 再循环洗脱
recycling procedure 再循环操作
recycling separation 循环分离
red acid 1,5-二萘酚-3,7-二磺酸
red angola copal 红色安哥拉玴脂
red antimony 红锑矿
red arsenic(=arsenic disulfide) 二硫化二砷；硫化砷；雄黄〔俗〕
red arsenic glass 雄黄
red bauxite 铁铝土
red bole 红玄武土
red brass 红色黄铜
red brittlement 热脆；热脆性
red brittleness 热脆性；红脆性
red cabbage extract 甘蓝萃
red cadmium 镉红
red cake 市售钒酸钠
red camphor oil 樟脑红油；樟脑重油
red cedar leaves oil 血柏叶油
red cedar wood oil 红雪松木油；血柏叶油
red cell folate assay 红细胞叶酸测定
red chalk 红垩
red chrome 铬红；铬酸铅
red cinnabar(=mercuric sulfide) 银朱；朱砂；硫化汞
red clay 红黏土；紫砂；红陶土
red clover oil 红三叶油
red cobalt 钴华
red coloration 红色；着红色

red compressed asbestos sheet 石棉红纸板
red copper 紫铜；赤铜
red copper ore 赤铜矿
red copper oxide(=cuprous oxide) 红色氧化铜；氧化亚铜红
reddingite 磷锰矿
reddish 带红相的(色)；带红头的(色)
reddish black 红黑
reddish blue 红蓝色
reddish brown 红棕
reddish orange 红橙色
reddish tint 红(色)光；红相；红色头〔俗〕
reddish tone 红光；红头〔俗〕
reddish violet 红紫
reddish yellow 红黄
reddle 红赭石
red enamel 红瓷漆
red engine oil 红机油
redeposition ①再沉淀②(污垢的)再沉积
red feather 红火焰
red fiber sheet 红色硬化纸板
Redfield efflux type viscometer 雷德菲尔德流出型黏度计
red filter 红色滤光器
redflair(=flairred) (蓝色)红光位移；向红位移〔在日光下是蓝色在白炽光下是红紫色〕
red flowering ironbank 红花桉
red gum 红树胶
red heat 赤热
red hematite 红赤铁矿
red-hot 赤热的
red-hot carbon 赤热炭
redilution 再稀释
red indigo 红靛；紫花
red ink 红墨水
red iron oxide 氧化铁红；三氧化二铁〔着色剂〕
re-disperse (使之)再分散
redispersible latex 再分散胶乳
redispersion 再分散(作用)
redistillation 再蒸馏(作用)；重蒸馏
redistilled 再度蒸馏了的
redistilled oil 再蒸馏精油
redistilled water 重蒸(馏)水
redistribution 再分配
redistribution baffle 再分布挡板
redistribution procedure 再分配法
redistribution reaction 再分布反应
redistributor 再分布器

red lac 紫胶
red lead(=mineral orange; minium) 铅丹；红丹；红铅粉；四氧化三铅
red lead anti-rust paint 红丹防锈漆
red lead-coated pigment 包核红丹颜料〔以二氧化硅为核，上包红丹的颜料〕
red lead ore 铬铅矿
red lead oxide 四氧化三铅；红丹
red lead paint 铅丹漆
red lead primer 红丹底漆
Redlich-Kister equation 雷德利克-基斯特方程
red (light) laser 红光激光器
red limit 红光极限
red liquor 红碱液〔染〕；红液乙酸铝媒染液
red litharge 红相黄丹〔六角形结晶一氧化铅〕
red machine oil 红机油
red manganese 红锰矿
red mercury iodide 碘化汞
red mercury oxide 氧化汞 HgO
red mercury sulfide 红色硫化汞；辰砂 HgS
red metal 红色铜合金
red mud 红泥
redness 红色
red ocher 红赭石；代赭石〔矿〕；氧化铁
red oil 红油
red orpiment 雄黄
redox 氧化还原(作用)
redox agent 氧化还原剂
redox analyzer 氧化还原分析器
redox buffer 氧化还原缓冲剂
redox buffer solution 氧化还原缓冲液
redox catalysis 氧化还原催化
redox catalyst 氧化还原催化剂
redox-catalyst system 氧化还原催化体系
redox catalyzed polymerization 氧化还原催化聚合(反应)
redox chelometric titration 氧化还原螯合滴定
redox chemistry 氧化还原过程化学
redox drain cleaning （具)氧化还原(作用)的排水管清洗剂
redox electrode 氧化还原电极
redox equilibrium 氧化还原平衡
redox grafting process 氧化还原接枝法
red oxide ①红色氧化物②紫红漆
red oxide micropigment 微细氧化铁红颜料
red oxide of iron 三氧化二铁
red oxide of mercury 氧化汞红
red oxide paint 紫红漆
redox indicator(=oxidation-reduction indicator) 氧化还原指示剂
redox-initiated graft polymerization 氧化还原引发的接枝聚合作用
redox initiate polymerization 氧化还原引发聚合(作用)
redox initiation 氧化还原引发(作用)
redox initiator 氧化还原引发剂*
redox ion exchanger 氧化还原离子交换树脂
redox mechanism 氧化还原机理
redoxokinetic effect 可逆电极的整流效应
redox oligomer 氧化还原低聚物
redox polymer 电子交换聚合物
redox polymerization 氧化还原聚合
redox potential 氧化还原电位；氧化还原电势
redox potential detector 氧化还原电位检测器
redox precipitation 氧化还原沉淀
redox process 氧化还原法
REDOX process 雷道克斯流程*
redox resin 氧化还原树脂*
redox rubber 冷聚丁苯橡胶
redox series 氧化还原系列
redox system 氧化还原(引发)系统
redox thermochemistry 氧化还原热化学
redox titrant 氧化还原滴定剂
redox titration(=oxidation-reduction titration) 氧化还原滴定
red phosphorus 红磷；赤磷
red pine seed oil 红松子油
red potassium prussiate 赤血盐钾；铁氰化钾
red precipitate 红色沉淀物；红色的氧化汞 HgO
red prussiate of potash 铁氰化钾；赤血盐
red prussiate of soda 铁氰化钠 $Na_3[Fe(CN)_6]$
red-rapid photographic plate 红快型感光板
redress ①恢复；复元②矫正
red rock 红岩；全晶质霏细斑岩
redruthite 浑铜矿
red schorl 红电气石；金红石
red seal 红印(级)〔氧化锌普通质量级；立德粉的 ZnS 含量为 28%～30%的产品级〕
red sensitive 红敏的
red sensitive phototube 红敏光电管
red sensitive plate 红敏底片
red shade (颜色)带红头的；带红光的
red shift(=bathochromic effect) 向红效应；红移
red short iron 红脆铁
red shortness(=hot shortness) 热脆性；热缩性
red spruce oil 赤云杉油
red stain ①红色着色剂；红色料②红变；赤变③露红铜；红铜锈

red stone　氧化铁
red stortness　热脆性
red support　红色载体
red tan solution　红棕色溶液；红褐色溶液
red thyme oil　红百里油
red tide　赤潮
red toner　标准(偶氮)色基玫瑰红
reduced　①(被)还原的②减低的③折合的④简化的
reduced bath　还原浴
reduced catalyst receiver　再生催化剂的接受器
reduced color　①还原色②还原色料
reduced compliance　对比柔量
reduced concentration　对比浓度；约化浓度
reduced co-ordinates　对比坐标
reduced copper gauze　还原铜网
reduced coupling constant　约化偶合常数
reduced crude　拔顶油；残油〔分馏过汽油和煤油的石油〕
reduced density　对比密度；约化密度
reduced equation of state　对比状态方程*
reduced factor　对比因子；折合因子
reduced flange　异径法兰
reduced flow　简化流动
reduced fuel oil　重质重油
reduced heat　换算热量
reduced iron　还原铁
reduced isothermal　对比等温线
reduced lyes　配比碱
reduced mass　折合质量
reduced mobile phase velocity　折合流动相速度
reduced modulus　对比模量
reduced mordant　还原媒染剂
reduced oil　拔顶油；残油
reduced ore　还原矿
reduced osmotic pressure　比浓渗透压
reduced pigment packing factor　降低的颜料堆积因数
reduced pigment volume concentration　对比颜料体积浓度
reduced plate height　折合塔板高度；简化塔板高度
reduced pressure　①对比压②减压
reduced pressure discharge　低压放电
reduced pressure distillation　减压蒸馏；真空蒸馏
reduced shade　冲淡(的)色(相)；浅色
reduced specific viscosity　折合比浓黏度
reduced specimen　缩尺试片；小型试片
reduced state　对比态
reduced steam　减压水蒸气
reduced stone　碎石
reduced stress　对比应力；约化应力

reduced strychnine method　士的宁还原法；番木鳖碱还原法
reduced tee　异径三通管
reduced temperature　对比温度；还原温度；换算温度
reduced time　对比时间
reduced tint　冲淡色
reduced transparency　对比透明度
reduced turbidity　比浓浊度
reduced variable　折算变量；对比变数
reduced velocity　折合速度；简化速度
reduced viscosity　比浓黏度*
reduced volume　对比体积
reduced white lead　还原铅白
reducer　①还原剂②还原器③大小头；异径管④减压阀⑤退黏剂
reducer tee　异径三通
reduce unit　还原晶胞
reducibility　①换算性②还原性；再现性
reducibility of catalyst　催化剂的可还原性
reducibility test　稀释性能试验；还原性能试验
reducible　可还原的
reducibleness　可还原性
reducible nitrogen　可还原氮
reducing　①还原的②减低的③折合的
reducing ability　还原能力
reducing action　还原作用
reducing agent　还原剂
reducing atmosphere　还原空气
reducing bath　还原浴
reducing bleach　还原漂白剂
reducing catalyst　还原催化剂
reducing coupling　异径联结
reducing cross　异径十字头
reducing-depth thread　不等深螺纹
reducing device　减速器
reducing dye　还原染料
reducing elbow　异径弯头
reducing electrode　还原电极
reducing ell　异径弯头
reducing end group　还原性(末)端基
reducing factor　折合因数；简化因数；对比因子
reducing fittings　异径管件
reducing flame　还原焰
reducing flux　还原性熔剂
reducing furnace　还原炉
reducing fusion　还原熔融
reducing gas　还原气
reducing impurity　还原性杂质

reducing machine 磨碎机；粉碎机
reducing mechanism ①粉碎装置②减速器
reducing medium 还原剂；还原介质
reducing of crude oil 原油蒸去轻质油
reducing pipe 渐缩管
reducing-pitch thread 螺距渐减螺纹
reducing power ①还原本领②消色力
reducing property 还原性(质)
reducing reagent 还原剂
reducing screen 减光屏
reducing solution 还原性溶液
reducing stabilizer 还原性稳定剂
reducing still 轻质油蒸馏釜
reducing tee 渐缩三通管
reducing union 缩口接头
reducing unit (裂化油或)轻质油蒸馏装置
reducing value 还原值
reducing valve 减压阀
reducing zone 还原层；还原带
reductant 还原剂
reductase 还原酶
reductibility 还原性
reductic acid(=reductinic acid) 还原酸；1,2-二羟基-3-酮环戊烯
reductimetry 还原滴定法
reductinic acid(=reductic acid) 还原酸；1,2-二羟基-3-酮环戊烯
reduction ①还原②简化③折合④减少；减低；减小⑤减速
reduction accelerator 还原加速剂；还原促进剂
reduction-aeration 还原气体法
reduction carrier 还原载体
reduction chromatography 还原色谱法
reduction coefficient 换算系数
reduction current 还原电流
reduction dimerization 还原二聚作用
reduction discharge 还原拨染
reduction discharge printing 还原拔染印花法
reduction drive 减速(齿轮)传动；减速机
reduction elution 还原洗脱法
reduction error 缩分误差
reduction factor(=reduction coefficient) 换算系数
reduction furnace 还原炉
reduction gear 减速齿轮
reduction gear box 减速(齿轮)箱
reduction gear ratio 减速(齿轮)比
reduction gear turbine oil 减速齿轮透平油
reduction gyratory 回转压碎机
reduction heat 还原热

reduction inhibitor agent 还原抑制剂
reduction method ①还原法②(氧化)还原(滴定)法
reduction of area 断面收缩率
reduction of dimensionality 降维(处理)
reduction of sample size 样品尺寸的减小；样品量减小
reduction of weighing to vacuo 真空校准(称量)
reduction-oxidation 还原氧化
reduction-oxidation polymerization 氧化还原聚合
reduction paste 冲淡色浆
reduction potential 还原电位*
reduction process 还原过程
reduction rate 减速比
reduction ratio ①减小系数②磨碎比③减速(齿轮)比
reduction reextraction 还原反萃(取)
reduction sleeve 减速套筒
reduction tube 还原管
reduction value 还原值
reduction wave 还原波
reductive 还原的
reductive acylation 还原酰化*
reductive agent 还原剂；脱氧剂
reductive alkylation 还原烷基化*
reductive amination 还原性氨化(作用)
reductive ammonolysis(=hydroammonolysis) 还原性氨解(作用)
reductive cyclization 还原性环化(作用)
reductive deamination 还原性脱氨基(作用)
reductive dehalogenation 还原性脱卤(作用)
reductive desulfuration 还原性脱硫(作用)
reductive methylation 还原性甲基化(作用)
reductive ozonolysis 臭氧还原分解(作用)
reductive potentiometric stripping analysis 还原电势溶出法
reductive smog 还原性烟雾
reductive sulfonation 还原性磺化(作用)
reductometry 还原滴定法
reductone 还原酮；二羟丙烯醛
reductor 还原器
red ultramarine 红群青
redundancy 冗余信息
red vitriol 赤矾
red, white and blue test 红、白、蓝试验〔伯胺、仲胺和叔胺的亚硝酸试验〕
Redwood orifice viscometer 雷德伍德锐孔黏度计
Redwood second 雷德伍德黏度计流出秒数
Redwood viscosimeter 雷德伍德黏度计
Redwood viscosity 雷氏黏度
redye 复染

red zinc ore 红锌矿
Reech's theorem 李奇氏(流体弹性)定理
Reed inkometer 里德油墨计
reed pulp 苇浆
reed valve 簧片阀
reel ①卷轴（机工）②丝管筛③绕线车
reelability 解舒
reel drum 卷筒
reeled paper 捆纸
reeler 络丝机
reelimination 再消去；再删除
reeling ①摇丝；绕丝②卷；绕
reeling machine ①摇丝机；绕丝机②卷取机
reel-off gear 松卷装置
reel oven 转炉
re-enforced ①(被)增强的；(被)加强的②加钢筋的；加钢芯的
reenrichment 再浓缩；再浓集
reentrant face 重入面；再面
re-entrant part 凹槽(角)部件；空腔部件
re-entrant portion 凹陷的部位；空腔部位
reequilibrate 再平衡
reequipment 更新设备
reextraction 反萃取
refabrication 再加工；再制备
referee method 仲裁法
referee test 仲裁试验
reference ①参比；参考②参考文献
reference arm （电桥)参比臂
reference beam 参比光束；参考光束
reference capillary 参比毛细管
reference cell 参比池
reference cell cavity 参比池腔
reference chromatogram 参比色谱(图)
reference color 标准色；参比色
reference column 参比柱
reference compound 参比物
reference dimension 基准尺寸
reference dose 参考剂量
reference electrode 参比电极
reference electrode filling solution 参比电极充填液
reference element 参比元素
reference element technique 参比元素技术
reference file 参比文件
reference gas 参比气体；标准气
reference holder 参比物支持器
reference interval 参考范围
reference ion(=pilot ion) 比较离子；参比离子
reference level 参考水平
reference light source 标准光源
reference line 参比线；起读线
reference liquid 参比液；标准液
reference mark 参比刻度；零点；参比标记
reference material 标准物质
reference material certificate(RM) 标准物质证书
reference method 标准方法
reference peak location 参比峰位
reference photomultiplier 参比光电倍增管
reference point 参比点
reference potential 参比电位
reference reaction 参比反应
reference sample 参比试样；标准样
reference scale 参比刻度
reference solution 参比溶液
reference spectral line 参比谱线
reference standard 参比标准；比照标准
reference state 参比状态
reference stimuli 参考原色
reference substance 参比物；标准物
reference system 参比系统
reference voltage 参比电压
refikite 褐煤树脂
refillable 适于再装的〔油桶〕
refiltered oil 再度过滤的油；回收油；用过滤法回收油
refiltering 再过滤
refinable crude 可精炼的原油
refined 精炼的；精制的
refined abrasive 精细磨料
refined aluminum 精铝；精炼铝
refined asphalt 精制沥青
refined asphaltic bitumen 精制的沥青
refined bleached lac 脱蜡漂白紫胶
refined glycerine 精制甘油
refined grade solvent naphtha 精制级溶剂石脑油〔沸点不超过155℃〕
refined lead 精炼铅；软铅
refined linseed oil 精炼胡麻子油
refined oil 精制油品；精炼油品
refined oil flash tower 精制油闪蒸塔
refined oil mixture 精制油混合物
refined oil stripper 精制油的汽提塔
refined paraffin wax 精石蜡
refined product 精制石油产品
refined rubber 再炼胶
refined salt 精盐
refined salt cake 精制盐饼

refined soda ash　精苏打灰；碳酸钠
refined tar　精制焦油
refined tin　精炼锡
refined turpentine　精制松节油
refined wax　精制石蜡
refined white alkali　精制白碱；纯碱
refined wood rosin　精制木松香
refined wool fat　精制羊毛脂
refinement　提纯；精炼
refiner　①精制机；精研机②匀料机③匀浆机
refinery　①精炼厂②炼糖厂
refinery building program　炼油厂建设计划
refinery calculation chart　炼油厂计算图表
refinery coke　石油焦
refinery compressor　炼油厂压缩机
refinery control　炼油厂生产控制
refinery cooling equipment　炼油厂冷却设备
refinery distillation　精馏作用
refinery dock　炼油厂码头
refinery dross　精炼炉渣
refinery effluents　炼油厂排污
refinery electrification　炼油厂电气化
refinery emulsion　炼油厂乳浊体
refinery engineer　炼油厂工程师
refinery flares　炼油厂废气
refinery furnace　精炼炉
refinery gas　炼油厂气
refinery instrumentation　炼油厂测量仪器设备
refinery loading rack　炼油厂起重机
refinery pit　精制槽
refinery processing　炼油厂加工过程
refinery process units　炼油厂工艺设备
refinery products　石油加工产品
refinery scum　精炼炉渣
refinery sludge　炼油厂酸渣
refinery stock　炼油厂原料
refinery tank　炼油厂油罐
refinery waste water　炼油废水
refinery waste-water separator　炼油厂废水分离器
refinery water　炼油废水
refining　①精炼；精制；提炼②匀料；匀浆
refining depth　精制深度
refining earth　精制用白土
refining equipment　炼油设备
refining fusion　精(制熔)炼
refining heat　精炼热
refining loss　炼耗
refining method　(石油)加工法

refining mill　精研机
refining oil　精炼油
refining operation　精炼
refining plant　净化装置；精制装置
refining solvent　(选择性)精制溶剂
refining steel　精炼钢
refining tallow　精炼牛脂
refining tank　精炼桶
refining value　精糖值
refining yield　精炼得率
refinish enamel　重涂用瓷漆；修补用瓷漆
refinishing paint　重涂涂料；维修漆；修补漆
reflectance　反射度
reflectance factor　反射比；反射系数
reflectance gloss meter　反射光泽计
reflectance measurement　反射度测量
reflectance meter　反射(率)计
reflectance photometer　反射光度计
reflectance spectroscopy　反射光谱(法)
reflectance value　①反射值；反射率②亮度
reflectdensitometer　反射密度仪
reflected beam　反射束
reflected ray　反射线
reflected wave　反射波
reflecting coating　①反光涂料；反光涂层②反光镀层
reflecting galvanometer(=mirror galvanometer)　反射镜电流计；反射镜检流计
reflecting microscope　反射显微镜
reflecting mirror　反射镜
reflecting power　①反射本领②反射比
reflecting prism　反射棱镜
reflection　①反射*②反映*
reflection angle　反射角
reflection densitometer　反射光密度计
reflection density　反射密度；反射浓度
reflection factor　反射系数
reflection goniometer　反射测角器
reflection grating　反射光栅
reflection high energy electron diffraction　反射式高能电子衍射
reflection high energy electron diffraction spectroscopy (RHEED)　反射高能电子衍射能谱
reflection index　反射率；反射指数
reflection law　反射定律
reflection measurement　反射测定(法)
reflection meter　反射计
reflection method　反射法；反射式探伤法
reflection ray　反射(光)线

reflection reduc(t)ing coating   反射衰减涂料(层);减反射涂料
reflection spectrum   反射光谱*
reflection stepped grating   反射式阶梯光栅
reflection symmetry   反射对称
reflection topography   反射形貌术
reflection-type differential refractometer   反射型示差折射计
reflection type phase constant microscope   反射型相位差显微镜
reflective beads   反光玻璃珠〔路标漆用〕
reflective coating   反光涂料;反射涂层
reflective index   反射指数
reflective pigment   反光颜料
reflective property   反射性
reflective spectrophotometry   反射分光光度法
reflective surface   反射面
reflectivity   ①反射系数;反射率;反射比②反射性
reflectometer   反射计
reflector   ①反射望远镜②反射器③(中子)反射器④反射镜
reflectorized paint   反光漆
reflectorizing glass bead   反光玻璃珠
reflectorizing of paint   涂料反光
reflector lamp   反光灯
refleshing machine   灰皮去肉机
reflex blue   碱性蓝
reflex condenser   回流冷凝器
reflex-copying method   反射(摄影)记录法
reflexion coefficient   反射系数
reflexion grating   反射光栅
reflex pattern   对称花纹
reflocculation   再絮凝
reflow   逆流;反流
reflow coating   紊流平涂料
refluence   ①反向电流②反流;逆流
reflux   ①回流;反流②回流的
reflux accumulator(=reflux drum)   回流液储器;回流液接受器
reflux atomizer   回流喷雾器
reflux coil   回流旋管
reflux column   ①回流(蒸馏)柱②回流(蒸馏)塔
reflux condenser(=reflux exchanger)   回流交换器;回流冷凝器
reflux conditions   回流条件
reflux cooler   回流冷却器
reflux cooling   回流冷却(法)
reflux distillation   回流蒸馏

reflux divider   (分馏塔上)采取回流液样品的设备
reflux drum(=reflux accumulator)   回流液储器;回流液接受器
reflux duty   回流热耗
reflux exchanger(=reflux condenser)   回流交换器;回流冷凝器
reflux extraction flowsheet   回流萃取流程
refluxing   回流
refluxing backflow   回流
refluxing coil   回流旋管
reflux line   回流线路;回流管线
reflux pump   回流泵〔抽回流液用〕
reflux rate   回流速率
reflux ratio   回流比(率)
reflux splitter   回流分配器
reflux supply   回流供应
reflux to product ratio   回流系数
reflux tower   回流塔
reflux valve   回流阀
refolding   蜷曲;再折叠
reformate   重整产品
Reformatsky reaction   瑞福马斯基反应
reformed gasoline   重整汽油
reformed rubber   改造橡胶
reformer   ①重整炉;重整装置②烃水蒸气转化装置
reformer pretreating   重整预处理(法)
reforming   ①重整〔指石脑油加工〕②转化〔指烃类与水蒸气转化生成 H_2+CO 的过程〕③改良
reforming catalyst   重整催化剂
reforming catalyst chloration   重整催化(剂)氯化
reforming catalyst reduction   重整催化(剂)还原
reforming furnace   重整炉〔增加汽油辛烷值的炉子〕
reforming gas   转化气
reforming Perco process   贝柯(催化)重整法
reforming plant   重整车间
reforming process(=reforming)   重整
reforming purge gas   重整排出气
reforming reaction   重整反应
reforming reactor   重整反应器
reforming recycle gas   重整循环气
reforming system   重整系统
reforming tube   重整炉管
reforming unit(=reforming plant)   重整工厂
refrachor   等折比容〔化合物的物理常数〕
refracted   折射的
refracted light   折射光
refracted ray   折射线
refracted wave   折射波

refraction ①折射②折射(度)
refraction analysis 折射分析法
refraction analyzer 折射分析仪器
refraction angle 折射角
refraction attenuated glass optical fibre 渐变型玻璃光导纤维
refraction coefficient 折射系数
refraction constant 折射常数
refraction law 折射定律
refraction of X-ray X射线折射
refractive exponent(=refractive index) 折射率；折光指数
refractive index(=refractive exponent) 折射率；折光指数
refractive index detector 折射率检测器；(差示)折光检测器
refractive index gradient 折光指数梯度
refractive index increment 折光指数增量
refractive index of extender pigment 体质颜料折光指数
refractive index unit full scale(RIUFS) 满刻度折射率单位
refractive lens 折射型镜头
refractiveness 折射性
refractive power 折射本领
refractivity ①折射系数；折射率差②折射性；折射本领
refractivity intercept 比折射度
refractometer 折光计*；折射计；折射仪
refractometric analysis 折射分析
refractometric titration 折射滴定
refractometry 折光测定法
refractor 折射器
refractories 耐火材料
refractoriness ①耐熔性②耐熔度③不应性
refractory alloy 耐火合金
refractory brick 耐火砖
refractory cements 耐火水泥；耐火胶黏材料
refractory chamber 耐火室
refractory clay 耐火泥；耐火黏土
refractory concrete 耐火混凝土
refractory fiber 耐火纤维
refractory fibre reinforced plastic 耐火纤维增强塑料
refractory gas oil 耐热瓦斯油；抗热性高的粗柴油
refractory glass 耐火玻璃
refractory glass fiber 耐火玻璃纤维
refractory insulator 耐火绝缘体
refractory iron ore 耐火的铁矿石
refractory-lined ovens 耐火炉
refractory lining 耐火衬
refractory material 耐火材料
refractory metal 高熔点金属
refractory metal fibre reinforced ceramic 耐高温金属纤维增强陶瓷
refractory mortar 耐火泥；耐火砂浆
refractory nitride 耐火氮化物
refractory oxide 高熔点氧化物；耐高温氧化物
refractory paint 耐火涂料；耐火漆
refractory pebbles 耐火石
refractory period 不应期
refractory pigment 难熔颜料；硬质颜料
refractory porcelain 耐火瓷
refractory products〔复〕 耐火产品
refractory reaction product 耐火反应产物
refractory rocks〔复〕 耐火岩石
refractory slag 难熔炉渣
refractory stock 抗热原料油；难于裂化的原料油
refractory surface 耐火面
refractory tube 耐火管
refractoscope 折射检验器
refrangibility 折射率
refrax 金刚砂砖
refrigerant ①制冷剂；冷冻剂②制冷的
refrigerant oil 冷冻机油
refrigerated 冷冻的；冷却的
refrigerated centrifuge 冷冻离心机
refrigerated cooling 冷冻冷却
refrigerated tanker 冷冻油船；冷冻液输送罐
refrigerated truck 冷藏车
refrigerated water 冷冻水
refrigerating 制冷的
refrigerating capacity 冷冻能力；致冷量
refrigerating coil 制冷旋管
refrigerating cycle 制冷循环
refrigerating fluid 冷冻液
refrigerating machine 制冷机
refrigerating plant 制冷装置
refrigeration ①制冷(作用)；(人工)冷冻(作用)②制冷学
refrigeration compressor 冷冻机；制冷压缩机
refrigeration equipment 制冷设备；冷冻机
refrigeration facility 冷冻设施；制冷设备
refrigeration filling 致冷介质
refrigeration machine 制冷机；冷冻机
refrigeration system 冷冻系统
refrigeration ton 冷冻吨〔冷冻设备的冷冻能力单位〕
refrigerative 制冷的；冷却的
refrigerator(=congealer；freezer) ①制冷器；冷冻器；冷冻机②冷藏箱
refrigerator compressor lubricant 冷冻机润滑剂
refrigerator finishes 电冰箱漆
refrigerator oil 冷冻机油

refrigerator tray 致冷盘；冷冻盘
refrigeratory 冷却的；制冷的
refringence 折射度
refueler 加燃料器；加油器
refuelling 更换燃料
refuelling depot 加燃料站；加油站
refuelling machine 装(燃)料机
refuelling point 加燃料站；加油站
refuelling unit 加油设备
refuin(=anthramycin) 回避菌素〔即：氨茴霉素〕
refuse 废物；渣
refuse bag ①返料袋；筛余料袋②废物袋；垃圾袋
refuse destructor furnace 废渣焚烧炉
refuse furnace 垃圾(焚化)炉
refuse incineration 垃圾焚化
refuse incinerator 垃圾焚化厂
refuse lac (残)渣(紫)胶
refuse landfill 垃圾填埋场
refuse oil 废油
refuse-polymer composite 废渣-聚合物复合材料
refuse silk 废丝
refusion 再熔(法)
regain 回潮；回潮率
regasification 再气化；再蒸发
regenerant ①再生剂②再生的
regenerated carbon 再生(活性)炭
regenerated catalyst 再生催化剂
regenerated cellulose 再生纤维素
regenerated cellulose fibre 再生纤维素纤维
regenerated cellulose rayon 再生纤维素嫘萦
regenerated clay 再生陶土
regenerated protein fiber 再生蛋白质纤维
regenerated solvent 再生溶剂
regenerated wool 再生羊毛
regenerating 再生的过程
regenerating column 再生塔；回收塔
regenerating cycle 再生循环
regenerating furnace ①再生炉②交流换热炉
regeneration 再生作用*
regeneration air (用于)再生(的)空气
regeneration efficiency 再生效率
regeneration electrode array 再生电极阵列
regeneration gas 再生气体〔再生催化剂时所产生的〕
regeneration in situ 在原处再生
regeneration level 再生水平；再生水准
regeneration of doctor solution 铅酸钠溶液的再生
regeneration of heat 交流换热(法)；间断收热(法)
regeneration period 再生周期
regeneration rate 再生速率
regeneration system 再生系统
regeneration temperature 再生温度
regeneration zone 再生区域；再生层
regenerative 交流换热的；蓄热的
regenerative chamber 再生内胎
regenerative cycle 再生循环
regenerative device 交流换热装置
regenerative fuel cell 再生式燃料电池
regenerative furnace ①再生炉②交流换热炉
regenerative heat exchanger 交流换热器
regenerative heating 交流换热
regenerative lubricating oil 再生润滑油
regeneratively reforming 再生式重整
regenerative oven ①再生炉②交流换热炉
regenerative pump 涡流泵；再生泵
regenerative system 交流换热法
regenerator ①再生炉②交流换热炉③再生剂
regenerator column 再生塔
regenerator kiln ①交流换热炉②再生炉
regeneratory furnace ①再生炉②交流换热炉
region of disorder 无序区；不规则区
regiospecific 区域专一的；配向性
register ①记录②记录器③定位器
registering chronograph 自记计时器；自动计时器
registration 合模；盖模
Regnault cell 勒纳尔特电池
Regnault value 勒纳尔特值〔一立方米氢的重量〕
regression 回归
regression analysis 回归分析*
regression coefficient 回归系数*
regression curve 回归曲线*
regression equation 回归方程
regression line 回归线
regression sum of squares 回归平方和*
regression surface 回归曲面
regression variance 回归方差
regressive shear thickening 递降剪切稠化
regressive shear thinning 递降剪切稀化
regrinding 再次研磨
regrind supply hopper (表面)抛光喂料斗
regroover 再次刻纹机
regrooving 再次刻纹
Reguir cell 勒奇尔电池
regular ①正规的②照例的③端正的④规则的
regular alternation 规则交变
regular barrel 正规桶
regular basis 定期的制度；经常的制度

regular basket weave 方平组织
regular bleached lac 普通漂白紫胶
regular block 规整嵌段
regular checking 定期检查
regular colo(u)r channel black 普通色素槽黑
regular column 正规柱
regular convention formula 习用配方；普通配方
regular copolyoxamide 规则共聚草酰胺
regular crystal 正方晶
regular-density gel 普通密度硅胶
regular engine oil(=regular motor oil) 普通车用机油
regular grade gasoline(=regular motor gasoline) 普通(级)汽油
regularity 均匀度〔纱或丝的〕；规整性
regular lacquer 传统的硝基漆
regular motor gasoline(=regular grade gasoline) 普通(级)汽油
regular motor oil(=regular engine oil) 普通车用机油
regular-ordered polymer 规则有序聚合物
regular packing 规则填料；规整填料
regular polymer 规整聚合物
regular reflection 正常反射；镜面(单向)反射
regular roll-coater 通用辊涂机
regular solution 正规溶液
regular solution theory 正规溶液理论
regular system 等轴晶系
regular tenacity rayon 标准强力人造丝
regular turning tool 规则车刀
regular twill 正则斜纹
regular wear 均匀磨耗
regular wear curve 磨损规律曲线
regulated polymer 有规聚合物
regulating block 调节装置；调整嵌段
regulating transformer 调压变压器；调节变压器
regulating valve 调节阀
regulation ①调节；调整②调整率
regulation cock 调节栓
regulation range 调节范围
regulator ①调节剂②调节器
regulator for downhill installation 管路中反流的调节器
regulator gene 调节基因
regulator of level 水平调节器
regulator section (在管路上的)调节器部分
regulator solution 调节(溶)液
regulator valve 调压阀
regulatory method of analysis 公定分析方法
regulus ①(熔矿所得的)金属渣；熔块②硫化复盐
regulus antimony 锑块

regulus mirror 锑镜
regulus of antimony 锑块
regurgitation 反流；回流
rehandling 重复劳动；返工
reheat 再(加)热；二次加热
reheat boiler 再热锅炉
reheat crack 再热裂纹
reheater ①再热炉②再热器
reheat factor 再热系数
reheat furnace 再热炉
reheating 再(加)热
reheating furnace 再热炉
reheat shrinkage 再热收缩
reheat turbine 再热式汽轮机
rehmanin 地黄宁
rehybridization 再次杂化
rehydration 再水化；再水合
Reichert-Meissl value 赖克特-迈斯耳值
Reichert number 赖克特数
Reid vapour pressure 雷德蒸气压
Reid vapour pressure bomb 雷德蒸气压力测定弹
Reid vapour pressure test 雷德蒸气压试验
reignition 再燃烧；再发火；再点火
Reimer's reaction 赖默酚-氯仿反应
Reimer-Tiemann reaction 赖默-梯曼反应
reincorporation 复原
reindeer skin 驯鹿皮
reineckate 雷纳克特酸盐
Reineckate salt 雷纳克特盐
Reinecke's acid 雷纳克酸；四硫氰基二氨合铬酸
Reinecke's salt 雷纳克盐；雷氏盐
Reiner-Rivlin fluid 赖讷-里夫林流体
reinforced bar 钢筋；螺纹钢筋
reinforced composite 增强复合材料
reinforced concrete 钢筋混凝土
reinforced concrete bed 钢筋混凝土地基
reinforced concrete construction 钢筋混凝土结构
reinforced concrete storage 钢筋混凝土油罐
reinforced gel coat 增强凝胶涂层
reinforced material 加强(材)料；加固料
reinforced plastic(s) 增强塑料
reinforced plywood 强化胶合板
reinforced polycarbonate 增强聚碳酸酯
reinforced polyester 增强聚酯
reinforced radial car tyre 增强子午线轿车轮胎
reinforced reaction injection molding 增强反应注塑法；增强反应性注射成形法
reinforced reaction molding 增强反应成型

reinforced rib 加强肋
reinforced rubber 补强橡胶〔含有补强剂的橡胶〕
reinforced stock 加固橡胶
reinforced thermoplastic(RTP) 增强热塑性塑料
reinforced thermoplastic resin 热塑性增强树脂
reinforced thermosetting plastic 增强热固性塑料
reinforced tyre 加固外胎
reinforcement 增强材料；增强剂
reinforcement concrete 钢筋混凝土
reinforcement layer 增强层
reinforcer 增强剂
reinforce strip 补强胶条；钢圈外包布
reinforcing 增强；补强
reinforcing action 补强作用
reinforcing agent 增强剂；补强剂
reinforcing aids 增强助剂
reinforcing band 补强带；增强带
reinforcing effect 增强效应；补强效应
reinforcing filler 补强填充剂；活性填充剂
reinforcing material 加强(材)料
reinforcing pigment 增强性颜料；补强颜料
reinforcing plate method 补强板法
reinforcing ring 加强圈；加固环
reinforcing size 增强型浸润剂；偶联浸润剂
reinforcing steel bar 增强钢筋
reinforcing whiting 活性碳酸钙
reinite 方钨铁矿
reinitiate polymerization 再引发聚合
re-initiation 再(次)引发*
rein leather 马具革
Reinsch test 赖因什检小量砷法
reirradiation technique 再照射技术
reject(=rejected log) 等外材；淘汰材；不合规格材
rejected log(=reject) 等外材；淘汰材；不合规格材
rejected material 废料；废品
rejected part 不合格部件；废品
reject filter 废料滤机
rejection 排斥
rejection coefficient 阻透系数
rejection of heat 散热
rejection rate 废品率，舍弃率
rejection region 舍弃域；拒绝域；否定域
reject value 剔除值
rejuvenated rubber 再生胶
rejuvenation 恢复过程；(黏胶的)嫩化
relamination 多层板压合
related substance 有关物质
relational degree 关联度

relation of stoichiometry 化学计量关系
relative abundance 相对丰度
relative accuracy 相对准确度
relative acidity 相对酸度
relative activity 相对活度
relative aperture 相对孔径
relative area response 相对面积响应(值)
relative asymmetry 相对不对称(性)
relative atomic mass 相对原子质量
relative atomic weight 相对原子量
relative bias 相对偏倚
relative biological effectiveness 相对生物学效应
relative blackness 相对黑度；比较黑度
relative brightness 相对亮度
relative burnup 相对燃耗
relative carbon response factor 相对碳(质量)响应因子
relative catalytic activity 相对的催化活性
relative concentration 相对浓度
relative correction factor 相对校正因子
relative crystallinity 相对结晶度；相对结晶性
relative deformation 相对形变
relative deformation gradient 相对形变梯度
relative density 相对密度
relative detection limit 相对检出限
relative deviation 相对偏差*
relative dielectrical constant 相对介电常数
relative dry hiding power 相对干遮盖力
relative efficiency 相对有效系数
relative electric potential difference 相对(电极)电位差
relative equilibrium 相对平衡
relative error 相对误差*
relative external standard deviation 相对外标偏差
relative extractability 相对可萃(取)性
relative flexibility 相对柔软(韧)性
relative flow 相对流动
relative frequency 相对频数
relative front 相对前沿
relative humidity 相对湿度
relative index of syndiotacticity 间同立构相对指数
relative intensity 相对强度
relative intensity of analytical line pair 分析线对相对强度
relative intensity of spectral line 谱线相对强度
relative internal standard deviation 相对内标偏差
relative isotope abundance determination 相对同位素丰度测量
relative isotopic abundance 相对同位素丰度
relative levelling 相对流平性

relative luminosity 相对发光度
relative luminosity curve 相对发光度曲线
relative luminous efficiency 相对发光效率
relative measurement 相对测量
relative method 相对法
relative method (of activation analysis) 相对法(活化分析)
relative modulus 相对模数
relative moisture 相对湿度
relative molal free energy 相对摩尔自由能
relative molal heat content 相对摩尔(量)热含；相对摩尔热焓
relative molar response 相对摩尔响应(值)
relative molecular mass 相对分子质量
relative molecular weight 相对分子量
relative motion 相对运动
relative parameter 相对参数
relative partial free energy 相对微分自由能；相对定浓自由能
relative peak area 相对峰面积
relative peak sharpness 峰相对尖锐度
relative penetration ratio 相对渗透率
relative permittivity 相对介电常数
relative phase 相对相
relative photometer 比较光度计
relative polarity 相对极性
relative polarity of stationary liquid 固定液的相对极性
relative reactivity 相对活性；相对反应性
relative receptivity of magnetic nuclei 磁性核的相对可接受度
relative redundancy 相对冗余信息
relative resolution product 相对分离度积
relative response 相对响应
relative response factor 相对响应因子
relative response value 相对响应值
relative retention 相对保留值*
relative retention ratio 相对保留比
relative retention time 相对保留时间
relative retention time method 相对保留时间法
relative retention value 相对保留值
relative retention volume 相对保留体积
relative R_f value 相对比移值；相对 R_f 值
relative sample sensitivity 相对试样灵敏度
relative selectivity 相对选择性
relative sensitivity 相对灵敏度
relative sensitivity coefficient 相对灵敏度系数
relative sensitivity factor 相对灵敏度系数
relative solvatation 相对溶剂化作用
relative solvation 相对溶剂化作用

relative specular transmittance 相对镜透射比
relative standard deviation(RSD) 相对标准偏差
relative systematic error 相对系统误差
relative tack 相对黏性
relative tensor 相对张量
relative thermal conductivity 相对导热率；相对导热性
relative thermodynamic stability 热力学比稳性；热力学相对稳定度
relative transmittance 透射比；相对透射率
relative vapour density 相对蒸气密度〔与空气或氢气比较〕
relative vapour pressure 相对蒸气压
relative velocity 相对速度
relative viscosity 相对黏度〔条件黏度〕
relative viscosity increment 相对黏度增量
relative volatility 相对挥发度
relative volume 相对体积
relativistic 相对论性的；相对论的
relativistic effect 相对论效应
relativity ①相对性②相对论
relativity theory 相对论
relaxation 弛豫(作用)；松弛
relaxational effect 弛豫效应
relaxational potential 弛豫势能
relaxational potential model 弛豫势能模型
relaxation balance 弛豫平衡
relaxation curve 松弛曲线；弛豫曲线
relaxation effect 松弛效应；弛豫效应
relaxation force 弛豫力*
relaxation function 弛豫函数
relaxation in NMR 核磁共振中的弛豫
relaxation kinetics 弛豫动力学
relaxation mechanism 弛豫机理；弛豫机制；弛豫历程
relaxation method 弛豫法*
relaxation modulus 弛豫模量
relaxation of electron distribution 电子分布弛豫*
relaxation of stress 应力松弛
relaxation parameters 弛豫参量
relaxation phenomena 松弛现象；弛豫现象
relaxation reagent 弛豫试剂*
relaxation solvent 弛豫溶剂
relaxation spectrum 弛豫谱*
relaxation technique 松弛法；弛豫法
relaxation test 松弛试验；弛豫试验
relaxation time 弛豫时间*
relaxation time measurement 弛豫时间测量
relaxation type absorption 松弛型吸收
relaxation type dispersion 松弛型色散
relaxation velocity 松弛速度

relaxed stage 松弛停放时间
relaxin 耻骨松弛激素；松弛肽
relaxometer 应力松弛计；张弛测量器
relay ①替续器②继电器③接管
relay counter 继电器式计数器
relayed coherence transfer spectroscopy 接力相干转移谱；接力同核位移相关谱
relay plunger 继电器插棒
relay pump stations （在管路中的)替续泵站；中间泵站
relay synthesis 接替合成
relay tank 中间油罐
re-leaching 再浸取
releasability 可剥离性；脱模性
releasant ①剥离剂②脱模剂
release agent 隔离剂；脱模制
release coating 防黏涂料；脱模涂料；剥离涂层
release line ①工厂管路②从管式炉到分馏塔或反应器的管路
release mesh 合格粒度
release paper 隔离纸
release property 防黏着性
release treatment 消除处理
release valve 泄气阀
releasing agent ①释放剂②脱模剂
releasing factor 释放因子
releasing hormone 释放激素
releasing liquid 脱模液；离模液
relevant reactor 相关反应器
reliability 可靠性；可行性；可信性
reliability design 可靠性设计
reliability index 可靠性因子
reliability ranking 可靠性顺序
reliability theory 可靠性理论
reliable work horse 可靠工作马力；使用安全马力
relic 残余的；残留的；蚀余的
relic structure 残余结构
relict 残渣
relief ①浮凸②减压
relief coating 浮雕涂料；凸纹涂料
relief condenser 排放冷凝器
relief fitting ①溢流塞；出口②减压润滑器
relief gas 排出气；废气；吹气
relief holder 均衡储气器
relief hole 放水孔
relief lines 切片刀痕
relief liquor 放出液；排放液
relief port 放气口
relief valve 安全阀；减压阀
relief valve connection 安全阀接口
reliever 减压装置
relieving ①减压；排气；放气②解除
relining 换衬
relocation 重新定位
relomycin 雷洛霉素；氢泰乐菌素
reluctance 磁阻
rem(=roentgen equivalent man) 雷姆；人体伦琴当量
remanence 顽磁(感应强度)
remark 备注
remedial measure 修理方法；补救措施
remedy 医药
remedy allowance 公差
remelter 再熔器
remelting 再熔(法)
remelting furnace 再熔炉
remethylation 再次甲基化
remilking 再次提取子体同位素
remilled rubber 再炼胶
remineralization 再矿化
rem ionization chamber(RIC) 雷姆电离室
remix 再混合；复拌
remnant polarization 剩余极化强度
remobility 再迁移性
remoistenable adhesive 再湿性黏合剂
remollescent 软化的
remolten 再度熔化的
remolten scrap 再熔废料
remote connector 远距离连结器
remote control 远程操作；遥控
remote control apparatus 遥控操作装置
remote control equipment 遥控设备
remote-controlled piping station 远距离控制的泵站
remote-control rack 远距离控制起重机
remote control system 远距离控制系统
remote cup （高压喷涂装置的)远距供漆罐
remote handling system 遥动系统；遥控操作系统
remote head pump 遥控泵头
remote indication 远距离指示
remote maintenance 遥控维修；远距间接维修
remote manipulator 遥控机械手
remote manual control 手动遥控；远距离手控
remote measuring device 遥测装置
remote measuring method 遥测法
remote operated balance 遥控操作天平
remote-operated controller 远距离控制器
remote pipe connector 遥控接管器
remote pipetter 遥控移液器

remote pumping unit 远距离控制的泵装置
remote reencapsulation 遥控再封装；遥控分装
remote sampling 遥控取样；远距离取样
remote sensing 遥测
remote sensing and measurement technology in environmental monitoring 环境监测的遥感遥测技术
remote sensing by induction 感应遥测
remote sensing for atmospheric pollution 大气污染遥测
remote sensing for water pollution 水污染遥测
remote sensing Fourier transform infrared spectroscopy 遥感傅里叶变换红外光谱法
remote sensor 遥测传感器
remote setpoint 遥控设定(点)
remote viewing equipment 遥视设备；远距离观察设备
remoto-control apparatus 遥控操作装置
removability 或移动性；可拆性
removable agitator 移动式搅拌器
removable bottom 活底
removable caps (分馏塔板的)活帽(盖)
removable eliminator plate 可拆卸的节流板；活动节流板
removable needle 可拆卸针头
removable pan 活盘〔面包〕
removable varnish kettle (可)移动式(清漆)热炼锅
removal ①除去②移去
removal by filtration 滤除
removal by suction 抽除
removal of carbon monoxide 脱(除)一氧化碳
removal of filtrate by air 吹气脱水；吹气助滤
removal of foam 消除泡沫；消泡
removal of oxygen 脱氧；除氧
removal of water 脱水
removals 木材年(度采)伐量
remove overhead 除去(上头的)轻质油
remover ①脱(涂)膜剂②洗净剂
removing 除去
renamycin(=lenamycin) 羊毛霉素
renardine 肾形(千里光)碱
renastacarcin 雷纳抗癌素
renaturation 变性复原(作用)；复性
render 炼油
render alkaline 碱化
render and set(=renderset) 两层抹灰
rendering 炼油
rendering kettle 炼油锅
rendering lath 抹灰板条
render visible 使可目测的
renewable energy sources 可再生能源
renewals 备件
renewal tire 更新轮胎
Rennerfeld furnace 伦讷菲尔德电炉
Renolds number 雷诺数
renormalization 重正化；重新归一化；再次归一化
renovated rubber 再生胶
renovated tyre 翻新轮胎
rensselaerite 假晶滑石
rent 裂缝；裂口
re-odorant 除臭剂
reoil 再上油
reordering 重有序化；重成序
reorientation 再定向；重取向
reoxidation 再次氧化
rep 伦琴当量
repack 改装
repack with grease 改装润滑脂
repaglinide 瑞格列奈〔药〕
repainted period 重涂周期
re-painting (旧漆打掉)重涂；重新涂漆；维修涂装
repainting hours 重涂时间〔指维修重涂〕
repair 修理
repaired plywood 修补过的胶合板
repairing welding 补焊
repair outfit 修补工具
repair sheets 修补胶
repair welding 维修焊接
repeatability 重复性*
repeatability error 重现性误差
repeat distance (链段)重复距离
repeated addition 叠加；重复相加
repeated analysis 反复分析；重复分析
repeated bending tester 弯曲疲劳试验机
repeated compression tester 反复式压缩试验机
repeated cracking 再裂化；多次裂化
repeated crystallinity 重复结晶性
repeated crystallization 反复结晶(作用)
repeated deformation 反复变形
repeated development 反复展开
repeated flexure 频繁弯曲
repeated hydrogenation 再加氢；多次加氢
repeated impact tester 反复式冲击试验机
repeated load 交变载荷；反复载荷
repeated loading 反复加压
repeated stress 反复应力；交变应力
repeated structural unit 重复结构单元
repeated tension and compression test 拉压疲劳试验
repeated tension tester 反复张力试验机

repeated test 重复试验；重复检验
repeated trial 重复试验
repeating chain unit 重复链(结构)单元
repeating decimal 循环小数
repeating group 重复基
repeating segment 重复链段；重复链节
repeating silicon-carbon atoms 硅-碳原子交替(重复)
repeating structural unit 重复结构单元
repeating unit 重复单元
repeat unit structure 重复单元结构；重复链段结构
repellency 防护性
repellent ①相斥的②拒斥剂
repeller plate 推斥板
repelling effect 排斥效应
repeptization 再胶溶〔凝聚胶体因用纯液或溶液高度稀释而发生的胶溶现象〕
repercolation 再渗滤(作用)
repetition (重)复(再)现；重复
repetition frequency 重复频率
repetition rate 重现率
repetition work 重复工作
repetitive chromatographic apparatus 循环式色谱仪器
repetitive error 重复误差
repetitive process 迭代法
repetitive production 反复(重复、周期性)生产
repetitive scan 重复扫描
repetitive scrubbing 反复洗涤
repetitive sequence 重复序列
rephosphorization 回磷
repipe 置换管子
replaceable 可置换的
replaceable hydrogen 可置换氢
replaceable ion 可置换离子
replaceable plate holder (打浆机)板式可换夹持器
replaceable tread tyre 活胎面轮胎
replaced 重做的；更换的；置换的
replacement ①置换；取代②移位③补给④复置；放回⑤轮换
replacement chromatography 置换色谱法
replacement design 等代设计
replacement metathesis reaction 复分解反应；置换反应
replacement name 置换名称
replacement part 备件
replacement reaction 复分解反应；置换反应
replacement rinse 冷热交替冲洗
replacement series 置换(次)序；取代(次)序
replacement substoichiometry 置换亚化学计量法；置换不足当量法

replacement tanning agent 取代鞣剂
replacement titration 置换滴定(法)*
replacing ①置换；取代②更换
replacing acid 取代酸
replacing of catalyst 催化剂的更换
replastering 重新抹灰；重新用石膏灰浆粉饰
replenisher ①补充剂②补充器③调节器
replenishing solution 补充液
replenishment chemicals 补给化学药品
replica 复制物；复制品；复制试样
replica grating 复制光栅
replica molding 复制模塑法
replica of surface 表面的复制品(复印型)
replica plant 中间工厂
replica plating ①镀法复制②影印培养
replica polymerization(=matrix polymerization) 复制聚合；母体移植聚合；模拟聚合
replicase 复制酶
replicate 平行测定
replicate determination 平行测定
replicate measurement 重复测量
replication 复制；重复
replication cycle 复制周期
replicative enzyme 复制酶
replicator 复制基因
replica-type mechanism 复制式机理；模板式机理
repolymerizability (活聚体)再聚合性
repolymerization 再聚合*
reporter gene 报告基因
reporter group 信息基团
Reppe chemistry 雷帕化学〔用高压乙炔合成各种烯烃衍生物的化学〕
Reppe reaction 雷帕(合成)反应
Reppe synthesis 雷帕合成(法)
reprecipitation 再(度)沉淀
reprecipitation method 再沉淀法〔用以精制高分子物质〕
representation 表示；表象
representative data 典型数据
representative formula 典型配方
representativeness of sample 样品的代表性
representative sample 代表性样品
repress 再压；补行加压
repress brick 再压砖
repressing 再压；补加压力
repression 再压缩；抑制
repress oil 再压油
repressor 阻遏物
repressuring gasoline 加压汽油；加丁烷汽油

reprocess 后处理；再处理
reprocessed 后处理过的
reprocessed plastic(s) 再生塑料
reprocessing 再加工
reprocessing analysis （核燃料)后处理分析
reprocessing plant （核燃料)后处理工厂
reprocessing rate （核燃料)后处理能力(速率)
reprocessor （核燃料)后处理单位〔工厂；研究所〕；后处理工作者
reproducibility 再现性*
reproducible ①可再产的②可再制的；可重现的
reproducible temperature 再现温度
reprogression 组织循环重排
reptation chain 蛇形链
reptation model 蛇形模型
reptation theory 蛇形理论
repulp filter 再压(石蜡)滤器
repulp filter feed tank 再压滤器的进料罐
repulsion ①推斥②斥力
repulsive potential 排斥势
repulsive potential energy surface 推斥型势能面*
repurification 再纯化
repurifier 再(提)纯器
repurity 再纯化
required NPSH 所需的净正吸入压头；额定的净正吸入压头
required power 需要功率
reradiating surface 再辐射面
re-reeling 扬返；复摇
rerefined oil 再生润滑油
rerefining 再精制
rerun bottoms 再蒸馏后的残油
rerun column 再蒸馏塔
rerunning 再(度)蒸馏
rerunning plant （石油产品)再蒸馏设备
rerunning still 再蒸馏锅
rerunning tower 再蒸馏塔
rerunning unit （石油产品)再蒸馏设备
rerun oil 再蒸馏油
rerun plant 再蒸馏设备
rerun still 再蒸馏釜
rerun yield 再蒸馏产率
re-rusting 再生锈；又生锈
resacetophenone(=2,4-dihydroxy acetophenone) 雷琐苯乙酮〔俗〕；2,4-二羟基苯乙酮 $(HO)_2C_6H_3COCH_3$
resaldol 雷琐多
resalgin 二羟基苯甲酸安替比林
resaponifying 重皂化；再皂化

resazoin(=resazurin) 刃天青
resazurin(=resazoin) 刃天青
re-screeded 重新刮平的；重新找平的
rescud 再刮；皮的重行刮光
rescue orange 亮橙；救生橙(颜料)〔颜色醒目，用于救生器具〕
rescuing work 救生工作
research approach 科研途径；科研方法
research microscope 研究用显微镜
research octane (number) 研究法辛烷值
reseda absolute 木犀草净油
reseda oil 木犀油
reseda root oil 木犀草根油
resene 氧化树脂；含氧树脂；碱不溶树脂
reserpine 利血平〔药〕
reserpinine(=raubasinine) 利血平宁
reserve ①储备物②储备(的)③藏量④保留
reserve acidity 储备酸度
reserve alkalinity 储备碱度
reserve battery 储备电池(组)
reserve carbohydrate 储备碳水化合物
reserved pump 备用泵
reserve equipment 备用设备
reserve parts 备件
reserves （石油的)储量；储备
reserve tank 储(油)罐
reservoir ①储液槽②储存器③蓄水池
reservoir burette （带)储瓶滴定管
reservoir capacity 油罐容量
reservoir priming （用泵)增压加油
reservoir sensor 储液式传感器
reservoir-type grease cup 油罐式油壶；油罐式油杯
reservoir vessel 储存容器
reset control 预置控制
reset controller 重调(动作)控制器；重定(动作)控制器
reset time ①重定时间②复位时间
resettling 再度澄清
re-shearing 再剪切
residence time 停留时间*；滞留时间
residfining process 渣油加氢精制法
residual 残差*；残余的
residual acid 残留酸
residual acrylonitrile content 残余丙烯腈含量
residual activity 剩余活度
residual adsorption ①残留吸附；永久吸附②残留吸附量
residual affinity 剩余亲和势
residual and pollution by organochlorine pesticides 有机

氯农药残留
residual and pollution by organophosphorus pesticides　有机磷农药残留
residual ash　残余灰分
residual asphalt　残余沥青
residual basic hydrolysis method　剩余碱水解法
residual bilinearization　残差双线分解法
residual bitumen　残余地沥青
residual capacity　剩余电容
residual charge　残余电荷
residual charge stock　裂化用残油；裂化用渣油
residual chloride analyzer　余氯分析器
residual chlorine　余氯
residual clay　原生黏土；残余黏土
residual coupling constant　剩余偶合常数*
residual current　残余电流*
residual current compensator　残余电流补偿装置
residual cylinder stock　残余汽缸油
residual deformation　残余变形；永久变形
residual degree of freedom　残余自由度
residual effect　(剩)余效(应)
residual elongation　残余伸长；永久伸长
residual entropy　残余熵*
residual extension　剩余伸长
residual film　残留膜
residual flower　残花
residual fraction　尾馏分；残余馏分
residual fuel oil　残余燃料油
residual gas　残余气体
residual gas analysis　剩余气体分析；残留气体分析
residual gas analyzer　残余气体分析器
residual hardness　残留硬度
residual hardness of water　水的残留硬度
residual humidity　残留水分
residual indentation　①残余空穴；残余凹陷；残余陷穴②残余压痕(凹痕)；永久压痕
residual internal strain　残余内应变
residual liquid　残余液体
residual liquid junction potential　残余液接电位
residual load　残余(轴向)负荷〔法兰垫〕
residual moisture　残留水分
residual molding stress　残余成型应力
residual monomer　残余单体
residual nitrogen　非蛋白氮
residual oil　渣油；残油
residual paint film　残留漆膜
residual paraffin　残余石蜡
residual percentage crimp　残余卷曲率；剩余卷曲率

residual photolytic activity　残余光活性
residual pitch　残余沥青
residual plot　残差图
residual pressure drop　(塔板上)剩余压力降低
residual product　残余产物；残油
residual range　剩余区域
residual ray　剩余射线
residual ray filter　残余射线过滤器
residual resistance　残余电阻
residual rocks〔复〕　残余岩石
residual saturation　剩余饱和度
residual sericin　残胶比；残胶量
residual set　残余变形；永久变形
residual shear tress　残余切应力
residual shrinkage　残余收缩
residual silanol　残余硅醇基
residual soil　残留污垢
residual solvent resonance　残余溶剂共振
residual standard deviation　残余标准偏差
residual stock　残余油料
residual streaks　条状残留污迹
residual stress　剩余应力
residual stretch　剩余伸长
residual styrene content　残余苯乙烯含量
residual sulfur content　残硫量
residual sum of squares　残差平方和
residual tack　残余黏性
residual tackiness　残余黏性
residual tar　残余焦油
residual titration　剩余滴定；反滴定；返滴定
residual torque tester　残余扭矩测定仪
residual-type operation(=residuum cracking process)　残余式裂化操作过程
residual valence(=residual valency)　剩余(化合)价
residual valence force　残余价力
residual valence theory　剩余(化合)价学说
residual valency(=residual valence)　剩余(化合)价
residual value　剩余值；残留值
residual variance　残余方差*
residual volatile matter　残余挥发物
residual volume　剩余体积
residual water　残留水
residuary　剩余的
residue　①残余②滤渣③残留物
residue distribution of particle size　残留粒子(粒度)分布
residue gas　残余气；干气
residue on evaporation　蒸发残渣
residue on ignition　灼烧残渣

residue on sieve 筛余(物)
residues 残余
residue sequence 残余链区〔合成树脂及塑料〕
residues of combustion 燃烧残余
residues theorem 留数定理
residue tar 残余焦油；釜焦油
residuum 残油；残渣(油)
residuum coking 残油焦化
residuum cracking process(=residuum process) 残油式裂化操作过程
residuum grease 残油润滑脂
residuum hydroconversion 渣油加氢转化(法)
residuum hydrodesulfurization 渣油加氢脱硫(法)
residuum pressure process 残油式加压裂化过程
residuum process(=residuum cracking process) 残油式裂化操作过程
resilience ①回弹；弹(回)性②回能；弹能
resilience energy 弹(回)能(量)
resilience factor 回弹系数
resilience meter 弹性计
resilience test 弹性试验
resiliency 回弹力；(回)弹性；缓冲性
resilient 弹性的
resilient-elasticity recovery 弹性变形恢复
resilient energy 回弹能(量)
resilient floor 弹性地板
resilient flooring 弹性铺地材料
resilient floor tile 弹性地面砖
resilient floor tile adhesive 弹性地板层(用)黏合剂
resilient leather 韧性革
resilient support(ing) member 弹性支座
resilient tile ①弹性瓦②弹性砖；弹性地面砖
resilin 节枝弹性蛋白
resiliometer 回弹计；弹性(测定)计
resin 树脂*
resin acid(=resinolic acid) 树脂酸
resin adhesive 树脂型胶黏剂；树脂型黏合剂
resin alloy(=polyblend) 树脂合金；树脂共混
resinamines 含氮树脂
resin anion(RA) 阴离子树脂
resin applicator ①树脂涂饰器②树脂流注器
resinate 树脂酸盐或酯
resinated barium lithol red 树脂酸盐化钡立索红；树脂处理钡立索红
resinated pigment 树脂酸(盐)化颜料；树脂酸颜料
resinate drier 树脂酸盐催干剂
resinated type toner 松香酸盐型色料(色原)
resination ①树脂整理②用树脂浸透

resination process 树脂酸盐化方法；树脂酸盐化过程
resin based adhesive 树脂基胶黏剂
resin bed 树脂床
resin blush 树脂致白(现象)
resin blushing 树脂致白(现象)
resin-bonded pigment 彩色树脂型颜料
resin-bound exterior paste filler 外用树脂基浆状填孔剂
resin canal 树脂道
resin capacity(=exchange capacity) 树脂(交换)容量
resin casting body 浇铸体
resin cation(RC) 阳离子树脂
resin cement 树脂胶浆
resin column 离子(交换)柱
resin concentrate 树脂母料
resin concrete 聚合物胶结混凝土；树脂混凝土
resin content 树脂含量
resin density 树脂密度
resin dispersion ①树脂分散(作用)②树脂分散体(液)
resin dispersion micelle 树脂分散胶束
resin duct 树脂导管
resin emulsion 树脂乳液
resin emulsion flat 无光树脂乳液(乳胶漆)
resin emulsion paint 树脂乳化漆；乳胶漆
resinene 中性树脂
resineon 树脂香油〔药用〕
resin ester 树脂酯；酯化树脂
resin extender 树脂增量剂；树脂填料
resin exudation 树脂渗出
resin-filled rubber 树脂配合橡胶
resin filling 树脂填充
resin finishing 树脂整理
resin flexibilizer 树脂软化剂；树脂增韧剂
resin flow 树脂流动率
resin flush removal 去溢胶
resin formation (燃料中)树脂(质的)形成
resin-free 不含树脂的
resin-free pulp 无树脂浆粕
resin guaiac 愈创木树脂
resinic acid 树脂酸
resiniferous 含有树脂的；含脂的
resinification 树脂化(作用)
resinifying 树脂化
resinifying agent 树脂化剂
resin impregnation 树脂填充
resin-in column process 柱式离子交换树脂吸附法
resin injection molding 树脂注塑成型
resin-in-pulp process (矿)浆料离子交换过程；树脂矿浆吸附过程

resin intermediate 树脂中间体
resin kettle 树脂(炼聚)锅
resin-loaded paper 饱和树脂纸
resin-loading-encapsulation technique 树脂吸附封装技术
resin matrix 树脂母体(塑料)
resin matrix composite 树脂基复合材料；聚合基复合材料
resin-membrane electrode (离子交换)树脂膜电极
resin modified 树脂改性的
resin monomer 树脂单体
resin of podophyllium 鬼臼树脂
resinogen 树脂原
resinography 显微树脂学
resinoic acid 树脂型酸
resinoid 热固性树脂
resin oil 树脂油
resinol 树脂醇
resinolic acid(=resin acid) 树脂酸
resinophore 易树脂化
resinophore group 易树脂化基团
resinotannol 树脂单宁醇
resinous 树脂的
resinous acid 树脂酸
resinous area 多脂表面；含脂表面〔如树节疤等〕
resinous exchanger 树脂交换剂
resinous exudation 树脂状渗出物(分泌物)
resinous impurities 树脂状杂质
resinous lustre 树脂状光泽
resinous material 树脂状物(质)；树脂质
resinous matrix 树脂状基体；树脂状基质
resinous matter 树脂状物质
resinousness ①树脂性②树脂度
resinous odour 树脂味
resinous oil 树脂质油
resinous phenol impregnation 用树脂酚浸渗
resinous plasticizer 树脂(型)增塑剂
resinous spue(s) 树脂渗出物
resinous substance(=resinous matter) 树脂状物质
resinous varnish 树脂清漆
resinous wood 明子；充脂材
resinous wood distillation 明子干馏
resinox 酚-甲醛树脂；酚-甲醛塑料
resin package 树脂封装
resin padding 树脂打底；树脂浸轧
resin paste 树脂糊
resin phase 树脂相
resin pickup ①浸树脂；蘸树脂②浸树脂量
resin-pigment ratio(=pigment-binder ratio) 树脂颜料比；颜基比
resin plaster 树脂(铅复)膏〔药〕
resin pocket 树脂囊；(层压品夹层内的)树脂淤积
resin pressure pot 树脂压力罐
resin recession 树脂下陷；树脂凹缩
resin rich area 富胶区；富树脂区
resin rich layer 富树脂层
resin smear 胶糊渣；胶渣
resin-soaked wood 充脂(木)材
resin soap 树脂皂；松香皂
resin softening point 树脂软化点
resin solvency of spirits 汽油溶解树脂质本领
resin spewing 树脂渗出
resin spirit 树脂精
resin spot test (离子交换)树脂点滴试验
resin stain 树脂污染〔木节渗出树脂而污染漆膜〕
resin-starved area 贫树脂区；缺胶区
resin streak 树脂流纹；(层压品表面的)树脂条痕
resin surge pot 树脂搅动罐
resin tannage 树脂鞣法
resin tanning 树脂鞣
resin tapping(=turpentining) 采(割松)脂
resin transfer molding 树脂传递模塑；树脂传递成型
resin treated plywood 树脂处理胶合板
resin treatment 树脂加工(处理)
resin varnish 树脂清漆
resin viscosity 树脂黏度
resiny 树脂的
resistance ①阻力②电阻③抵抗④内阻
resistance box 电阻箱
resistance bridge 电阻桥
resistance-capacitance coupled amplifier 阻容耦合放大器
resistance capacity 电阻容量
resistance cellulose test 强纤维试验
resistance coil 电阻线圈
resistance concept 阻力观念
resistance electrode system 电阻电极系统
resistance element 电阻元件
resistance emitter 电阻加热器
resistance furnace 电阻(电)炉；阻力炉
resistance ga(u)ge 阻力计；电阻规
resistance head 压头
resistance heated atomizer 电阻加热原子化器
resistance hygrometer 电阻湿度计
resistance of friction 摩擦阻力
resistance of gas mass transfer 气相传质阻力
resistance of liquid mass transfer 液相传质阻力

resistance of mass transfer 传质阻力
resistance of medium 介质阻力
resistance paint 电阻漆
resistance pyrometer 电阻高温计
resistance strain gage 电阻应变仪
resistance strain meter 电阻应变仪
resistance test 抗腐蚀试验；抗药性试验
resistance thermometer 电阻温度计
resistance time ①抗毒时间②抵抗时间
resistance to abrasion 磨蚀阻力
resistance to acids and bases 耐酸碱性
resistance to ag(e)ing ①耐(抗)老化性②老化阻力
resistance to air loss 气密性
resistance to alkali 耐碱性
resistance to atmosphere 耐候性
resistance to atmospheric corrosion 耐大气腐蚀性
resistance to biodeterioration 耐生物老化性；生物老化抵抗力
resistance to bleed(ing) 抗渗色性
resistance to blistering 抗起泡性
resistance to blocking 抗粘连性
resistance to blush 抗发白力
resistance to boiling water 耐沸水性
resistance to breakage 耐破损性
resistance to brittleness 抗脆化性
resistance to bronzing ①抗泛铜光性②抗青铜斑性
resistance to caustic soda 耐苛性钠性
resistance to chalking 耐粉化性
resistance to chemical reagents 耐药品性
resistance to chlorine 耐氯性；抗氯性
resistance to cigarette burns 耐香烟灼烧性
resistance to cleaning agent 抗去污剂性
resistance to cold 耐寒强度
resistance to color change by ultraviolet 耐紫外线变色性
resistance to compression 抗压强度
resistance to concentrated acids 耐浓酸性
resistance to continual dampness 长期耐湿性
resistance to corrosion 耐腐蚀性
resistance to corrosion cracking 耐腐蚀龟裂性
resistance to corrosion seawater 耐海水腐蚀性
resistance to corrosive chemicals 耐化学药品腐蚀性
resistance to cracking 抗龟裂性；抗开裂性
resistance to crazing 抗细裂(纹)性
resistance to creep 抗蠕变性(力)
resistance to crocking ①抗(摩擦)掉色性；抗污染性②抗洇色性
resistance to crushing 抗压碎性；抗碾性
resistance to degradation at processing temperature 耐热加工降解性；耐加工温度降解性
resistance to degradation by sunlight 抗日光降解性；日光降解抗力
resistance to detergents 耐洗涤剂性
resistance to deterioration ①抗变质力②抗老化性③抗蜕(退)化力
resistance to deterioration on aging 抗老化变质性；抗老化蜕(退)化力
resistance to deterioration on weathering 耐暴晒老化性；耐气候老化变质性
resistance to dilute acids 耐稀酸性
resistance to dirt pickup 防吸尘性
resistance to dry cleaning solvents 耐干洗溶剂性
resistance to elements 耐候性；耐候化性能
resistance to embrittlement 抗脆化性
resistance to erosion 耐(抗)侵蚀性
resistance to fatigue 耐疲劳性
resistance to flame 耐燃烧性
resistance to flexing 抗挠曲性
resistance to flow (对液体、气体的)流动阻力
resistance to fogging 抗雾性
resistance to heat 耐热性
resistance to heat discolouration 抗热变色性
resistance to heat distorsion 耐热畸变性；耐热变形性
resistance to humid-ag(e)ing exposure 耐湿老化性；抗吸潮老化性
resistance to hydrolysis 抗水解性
resistance to impact 碰撞阻力
resistance to impact shock 抗冲震性；抗冲击强度
resistance to indentation 抗压陷性
resistance to light of coating 涂料的耐光性
resistance to lubricating 耐润滑油性
resistance to marring 抗划伤性；抗擦(损)伤性
resistance to mass transfer 传质阻力
resistance to outflow (对气体、液体的)流出阻力
resistance to ozone cracking 耐臭氧龟裂性；抗臭氧开裂性
resistance to penetrating dirt 抗(耐)污物渗透性
resistance to plastic flow 抗塑流性
resistance to poisoning (催化剂)抗毒本领
resistance to rain 耐雨水性；抗雨水性
resistance to ripping 抗撕裂性
resistance to rupture 抗破坏性
resistance to salt (spray) fog 抗盐雾性
resistance to salt water 耐盐水性；抗盐水性
resistance to scraping 抗刮性
resistance to scuffing 耐磨损性；抗擦伤性
resistance to settling 抗沉淀性；抗沉降性

resistance to shear　剪切阻力
resistance to shock　抗震性；耐冲击性
resistance to sidewall bruising　(轮胎)抗侧壁破裂性
resistance to solvent　耐溶剂性
resistance to spalling　破裂强度〔玻〕
resistance to stone-chipping　抗石击性
resistance to stress-crazing　抗应力细裂性
resistance to sulfide staining　抗硫化物污染性；抗硫污染性
resistance to surface delamination　抗表面剥落性
resistance to swelling　耐溶胀性
resistance to thermooxidation　抗热氧化性
resistance to ultraviolet discoloration　耐紫外线变色性
resistance to ultraviolet (ray) radiation　耐紫外线辐射性
resistance to water soak　抗吸湿性
resistance to wear　耐磨性；耐穿
resistance to weathering　耐候化性能
resistance to yellowing　抗泛黄性
resistance welding　接触焊
resistance (wire) strain gage　电阻(丝)应变仪
resistant metal　耐蚀金属
resistant to alkali　耐碱的
resistant to corrosion　抗腐蚀
resistant to saline solution　耐盐水的
resistant to tarnishing　抗锈蚀
resister(=resistor)　电阻器
resist ink　耐油墨的；防油墨的
resistive coating　电阻漆
resistivity　电阻率；电阻系数
resistivity against fire　耐火性
resistivity against water　耐水性
resistivohmeter　静电(喷涂专用)电阻表
resistor(=resister)　电阻器
resistor furnace　电阻炉
resist pattern　(光致)防蚀图形
resist printing　防染印花
resist sagging　防流挂
resite　酚醛树脂 B；不熔酚醛树脂
resitol　酚醛树脂 C；半熔酚醛树脂
reslurry　重新打浆
resmethrin　灭虫菊；苄呋菊酯
resocyanin　β-甲基伞形酮
resoflavin　间苯二酚黄素
resoiling　再染污
resol　酚醛树脂 A；可熔酚醛树脂
resolidification　再凝固(作用)；重新固化
resol resin(=resole)　甲阶酚醛树脂；可熔性酚醛树脂
resolubilization　再增溶(作用)

resolution　①拆分*②分离度*③分辨力④分辨率
resolution capacity　分辨能力
resolution distance　分辨距离；解析距离
resolution loss　分辨损失
resolution of precipitate　沉淀的溶解
resolution power　分辨本领
resolution product　分离度积
resolutive polymer　分解性高分子
resolvent　①溶剂(媒)；分解物②有溶解力的
resolver　溶剂；溶媒
resolving　解析过程
resolving agent　拆解试剂
resolving limit　分辨极限
resolving power　分辨本领〔解析光学异构的本领〕
resolving time　分辨时间
resolving voltage　分辨电压
resonance　共振*
resonance absorption　共振吸收
resonance absorption transition　共振吸收跃迁
resonance atomic fluorescence　共振原子荧光
resonance broadening　共振(谱线)增宽；共振展宽；共振变宽
resonance capture　共振俘获
resonance cavity　谐振腔
resonance condition　共振条件
resonance corona detector　共振电晕检测器
resonance curve　共振曲线
resonance detector　共振检测器
resonance dispersion　共振色散
resonance effect　共振效应*
resonance energy　共振能*
resonance enhanced multiphoton ionization　共振增强多光子电离
resonance excitation　共振激发*
resonance filter　共振滤器
resonance fluorescence　共振荧光*
resonance frequency　共振频率
resonance hump　共振峰
resonance indication tube　谐振指示管
resonance integral　共振积分
resonance ionization mass spectroscopy　共振电离质谱法
resonance ionization mass spectrum　共振电离质谱
resonance ionization spectroscopy　共振电离光谱法
resonance ion spectrum(RIS)　共振离子谱
resonance laser ablation(RLA)　共振激光烧蚀
resonance light scattering　共振光散射
resonance line　共振线*
resonance method　共振法

resonance monochromator 共振单色器
resonance monochromerer 共振单色计
resonance potential 共振电势
resonance Raman effect 共振拉曼效应
resonance Raman scattering 共振拉曼散射
resonance Raman spectroelectrochemistry 共振拉曼光谱电化学法
resonance Raman spectrometry(RRS) 共振拉曼光谱法
resonance Rayleigh scattering 共振瑞利散射
resonance screen 共振筛
resonance spectrometer 共振光谱仪
resonance spectrum 共振谱
resonance state 共振状态
resonance theory 共振论*
resonance type absorption 共振型吸收
resonance type bending fatigue tester 共振型弯曲疲劳试验机
resonance type dispersion 共振型分散；共振型色散
resonance value 共振值
resonant absorption 共振吸收
resonant frequency 共振频率
resonant humidity sensor 共振湿度传感器
resonant ionization spectrometry 共振电离光谱法
resonant transformer 共振变压器
resonant vibrometer 共振示振仪
resonating electron 共振电子
resonator ①共振腔②共振电子排布
resopyrine 雷琐比林〔药〕
resorcin 间苯二酚；雷琐酚；雷琐辛 $C_6H_4(OH)_2$
resorcinal resin 间苯二酚甲醛树脂
resorcin blue(=lacmoid) 间苯二酚蓝
resorcin brown 间苯二酚棕
resorcin diacetate 间苯二酚二乙酸酯 $C_6H_4(O_2CCH_3)_2$
resorcin dibenzoate 间苯二酚双苯甲酸酯 $C_6H_4(O_2CC_6H_5)_2$
resorcin dimethyl ether 间苯二酚二甲醚 $C_6H_4(OCH_3)_2$
resorcine-azo-benzene-sulfonic acid(=tropeolin O) 金莲橙O；偶氮苯间二酚磺酸
resorcin monoacetate 间苯二酚一乙酸酯
resorcin monomethyl ether 间苯二酚一甲醚 $HOC_6H_4OCH_3$
resorcinol 间苯二酚；雷琐酚；雷琐辛
resorcinol carbolate 间苯二酚石炭酸混合物
resorcinol diglycidyl ether 间苯二酚二缩水甘油醚；间苯酚二环氧甘油醚
resorcinol formaldehyde donor silica 间甲白；间苯二酚-甲醛给予体-白炭黑
resorcinol-formaldehyde resin 间苯二酚-甲醛树脂

resorcinol-formaldehyde resin adhesive 间苯二酚-甲醛树脂黏合剂
resorcinol monoacetate 单乙酸间苯二酚酯
resorcinol monobenzoate 间苯二酚单苯甲酸酯〔防老剂〕
resorcinol phthalein 间苯二酚酞
resorcinol resin 间苯二酚树脂
resorcinol test 间苯二酚试验〔测定对苯二酸酯增塑剂或醇酸树脂中的邻苯二甲酸〕
resorcinol yellow 间苯二酚偶氮苯磺酸
resorcitol 1,3-环己二醇
resorcyl ①间羟苯基②间二羟苯基
resorcylate 间(二)羟苯甲酸盐(酯)
resorcylic acid 二羟基苯甲酸；雷琐酸 $(OH)_2C_6H_3COOH$
resorption 回吸(作用)；吸除(作用)
resorpyrine 雷琐比林〔药〕
resorufin(=9-hydroxy-3-isophenoxazone) 试卤灵；9-羟基-3-异吩噁唑酮 $HOC_{12}H_6ON=O$
resources 储藏；资源
respiration chemiluminescence 呼吸发光
respiration coefficient 呼吸系数
respiratory acidosis 呼吸性酸中毒
respiratory alkalosis 呼吸性碱中毒
respiratory box 呼吸箱；滤毒箱
respiratory quotient 呼吸商；呼吸系数
respirometer 呼吸测定计
response 响应(值)*
response curve 响应曲线
response data 响应数据
response factor 响应因子
response function 响应函数
response limit 灵敏度；响应极限
response speed 响应速度；反应速度
response surface 响应面
response time 响应时间
response to tetraethyl lead 对四乙铅的感受性
response value 响应值
responsivity 响应性；反应性
respropiophenone 异丙基苯基(甲)酮
rest ①座；架；支持物②静止
restaking 再刮软
resteeping 再浸渍；重浸渍
rest energy 静能
resterification 再酯化(作用)
restilling 再蒸馏
resting state 静止状态
restless steel 不锈钢
rest mass(=proper mass) 静质量〔物〕

restoration 复原
restored acid 回收的酸
restored rubber 再生胶
restore method with plastic deformation 塑性变形修复法
restoring 回收；再生
restoring force 回复力；复原力
rest period 放置(时)期
rest point 停点
rest point shifting 停点移动
rest potential 静止电位
rest preventer 防锈剂
restrained end 约束端
restrainer 抑制剂
restraining effect 抑制作用
restricted area molding 窄面模塑
restricted diffusion 限制扩散
restricted diffusion chromatography 被阻扩散色谱法；有限扩散色谱法
restricted flow 限止流；受限流
restricted Hartree-Fock(RHF) method 限制(的)哈特莱-弗克法
restricted non-isothermal theory 狭义非等温理论
restricted oil supply 有限的供油；油的有限供应
restricted rotation 阻碍旋转*
restricted sudsing detergent 控泡洗涤剂
restricted theory of relativity 狭义相对论
restriction 限制
restriction endonuclease 限制性核酸内切酶
restriction flowmeter 节流式流量计
restriction orifice 节流孔板
restrictive gel 排阻性凝胶
restrictive stability 受限稳定性
restrictive valve 节流阀
restrictor ①限制器②限流(量)器
resublimation 再升华(作用)
resublimed 再度升华了的
resublimed iodine 再升华碘
resufacing welding 堆焊
result ①结果②得数
resultant ①生成物；(反应)产物②总和
resultant accuracy 总准确度
resultant count 总支数
resultant draft 实际牵伸；总拉伸
resultant error 综合误差；总误差
resultant of reaction 反应生成物；反应产物
resultant stimulus 刺激和；总刺激
resurfacing 再铺平；再涂层
resuscitation 回生

resuscitation gas 回生气
resuspending 再悬浮
resuspension 再悬浮(作用)
resveratrol 白藜芦醇；3,4,5-三羟(基)芪
resynthesis 再合成
retainable pollutant 保守性污染物
retained concentration 保留浓度
retainer bolt 止动螺钉
retainer plate 模型套板；托模板
retaining board 垫板〔冷压机用〕
retaining ion technique 保留离子法
retaining ring 定位环；扣环；扣套
retaining screen 阻滞筛
retamine 瑞它明；鹰爪豆碱
retanning 复鞣
retanning agent 复鞣剂
retardance 迟缓；减慢
retardancy 阻滞性
retard anionic polymerization 阻滞阴离子聚合
retardant ①阻滞剂；抑制剂②缓染剂
retardation 缓聚作用*
retardation constant 阻滞常数
retardation effect 阻滞效应
retardation factor 阻滞因数；滞留因数
retardation function 推迟作用
retardation kinetics in polymerization 聚合过程的减速动力学
retardation plate 延迟板
retardation quotient 阻滞系数
retardation spectra 推迟时谱
retardation time 推迟时间*；延时
retardation (time) spectrum 推迟(时间)谱*
retarded 减速的；被阻滞的
retarded autoxidation 延迟自氧化(作用)
retarded compliance 延迟柔量
retarded deformation 延迟形变
retarded elastic deformation 推迟弹性形变
retarded elasticity 推迟弹性；延迟弹性
retarded elasticity function 推迟弹性函数
retarded electrode(=isolated electrode) 隔离电极
retarded flow 推迟流；减速流
retarded motion 减速运动
retarded oxidation 延迟氧化；缓慢氧化
retarded reaction 阻滞反应
retarded spontaneous recovery 推迟自发回复
retarded swelling property 缓溶胀性
retarded viscose 延迟(凝固的)黏胶
retarder 缓聚剂*；阻聚剂

retarder grid assembly 抑止栅组件
retarder solvent 挥发迟缓溶剂；挥发缓慢溶剂
retarder thinner ①(硝基漆)防潮剂；防白药水〔俗〕②缓干稀释剂；缓干稀料
retarding agent 阻滞剂；迟延剂；抑制剂
retarding effect (萃取)阻滞效应；抑制效应；推迟效应
retarding electrode ①抑止电极②减速电极
retarding force 减速力〔物〕
retarding grid 减速栅极
retarding potential 抑止电位；减速电位
retarding potential difference technique 减速电位差法
retarding rate constant 阻滞速率常数
retarding reaction 阻滞反应；延迟反应
retarding solvent 缓溶剂
retarding thermal decomposition 延迟热分解
retene(=1-methyl-7-isopropylphenanthrene) 惹烯；1-甲基-7-异丙基菲 $C_{18}H_{18}$
retenequinone 惹醌
retentate 渗余物
retention ①保留*②保留值
retention aid 助留剂
retention analysis ①残留分析②保留分析(法)
retention basin 储留池；储水池
retention behavior 保留行为
retention constant 保留常数
retention data 保留数据
retention distance 保留距离
retention factor 保留因子
retention forecasting 保留(值)预测
retention gap ①滞留带②保留间隙；保留隙口
retention gap column 保留间隙柱
retentiongram 保留分析图；保留分析曲线
retention increment 保留(值)增量
retention index 保留指数*
retention index qualitative method 保留指数定性法
retention mechanism 保留机理
retention of adhesion 保留附着力；黏结力的保持性
retention of clarity 澄清度保持性；澄清度持久性
retention of color 保色性；着色稳定性
retention of configuration 构型保持*
retention of embossing 压花(压纹)保持性(度)
retention of flexibility 挠曲性的保持(力)
retention of hollowness (纤维的)空腔保持性
retention of surface gloss 表面光泽保持性(持久性)
retention of ultimate elongation 极限伸长保持性
retention of viscosity 黏度的保持
retention of whiteness 白度坚牢性；白度保持性；白度稳定性

retention order 保留顺序
retention parameter 保留参数
retention pin 止动梢；定位梢
retention process 保留过程
retention property 保留性能
retention qualitative method 保留值定性法
retention ratio 保留(值)比
retention temperature 保留温度
retention time 保留时间*
retention time counter 保留时间计数器
retention time distribution 保留时间分布
retention value 保留值
retention volume 保留体积*
retention window 保留时间窗口
retentive activity 保留活性
retentivity ①保持性；缓和性②顽磁性
retentivity of pesticides 药剂持留能力；药剂黏着性
Retger's law 雷奇尔混合晶体物性定律
rethickening 再次增稠
reticular 网状的
reticular fiber 网状纤维
reticular layer 网状层
reticular polymer 网状聚合物
reticular structure 网状结构
reticular tissue 网状组织
reticulated 网状的
reticulated urethane foam 网构聚氨酯泡沫塑料
reticulated vitreous carbon electrode 多孔玻碳电极
reticulate structure 网状结构
reticulation 小皱纹；起皱纹
reticule 标(度)线；网袋
reticulocyte production index 网织红细胞生成指数
retinal ①视黄醛；视网膜醛②视网膜的
retinene 视黄醛；视黄素
retinoid 类维生素A
retinol ①视黄醇；维生素A②松香油
retort ①曲颈甑②甑；蒸馏甑；(蒸馏)罐〔例如制煤气用〕③(干馏)釜
retort bulb 曲颈甑的球管
retort carbon 甑碳
retort clay 甑土
retort coke 甑馏焦(炭)；蒸馏焦(炭)
retort condenser 甑馏冷凝器
retort furnace 甑式炉
retort gas 甑中产生的气体
retort gasoline 用甑生产的汽油
retort gas tar 甑中得到的煤焦油
retort graphite 甑馏石墨

retort holder 曲颈甑架
retort house 甑室；甑组
retorting 甑馏；干馏
retorting of shales 页岩干馏
retorting process 甑式干馏过程
retort neck 曲(颈)甑颈
retort oven 甑式炉
retort pine tar 甑馏松焦油；干馏松焦油
retort producer 甑式发生炉
retort residue 甑渣
retort tar 甑馏焦油
retouch 修饰；润色
retouching paint ①修版墨②修补漆
retracted film 回缩膜*
retracting drop 收缩液滴
retraction 收缩；回缩；回复；复原
retraction modulus 回复模数
retreadability 翻新适合性；有翻新价值
retreaded tyre 翻新车胎
retreader 翻新器
retreading 胎面翻新
retreating 再精制；再加工
retreatment 再精制；再加工
retree paper 缺损纸
retrievability 可回收性；可恢复性
retrievable storage 可回收储存
retrievable surface storage facility (RSSF) （放射性废物）可回收地面储存设施
retro- 〔词头〕向后
retroaction ①反力②逆反应
retroalkylation 逆烷基化作用
retro Diels-Alder reaction 逆狄尔斯-阿尔德反应*
retrogradation 退减(作用)
retrograde aldol condensation 逆羟醛缩合*
retrograde condensation 反缩合
retrograde dissociation 退减离解
retrograde reaction 退变反应；退化反应
retrogression ①逆反应②消退③退化④后退
retroirone 逆鸢尾酮
retroisomeric system 反回异构系
retronecanol 倒千里光裂醇
retronecine 倒千里光裂碱
retronecinic acid 倒千里光裂酸 $C_{10}H_{16}O_5$
retropinacolic conversion 反频哪酮重排作用
retropinacol rearrangement 逆频哪醇重排*
retrorsine 倒千里光碱
retrosynthesis 逆合成
retro-system 反回系

retting ①沤麻②浸解
retting pit 纤维浸解坑
rettle 反应釜
returnable 可回收使用的
returnable container(s) ①可回收使用的容器②可反复使用的集装箱
return-bend 回管
return-circuit ring 反向导流器〔离心压缩机的〕；弯道
return conduit 回流管；溢流管
returned acid 回流酸
returned sludge 回流污泥
return flow compressor 回流式压缩机
return flow prevention valve 止回阀
return line 回流管
return passage 回路
return pipe 回管
return pump 抽空泵
return shipping cost 来回输送的成本
return speed 返回速度
return spring 回动弹簧
return stroke 回程；反冲程
return swing arm drip 摆动溢流管
return trap 可复阱
return tube 溢流管
return water 白水；回水〔纸〕
retusine 吊裙草碱；凹猪屎豆碱
retzbanyite 块辉铅铋矿
Reunion basil oil 留尼汪罗勒油；丁香罗勒油
reusability 再用性；重复使用性
reusable 可以再用的
reusable cotton waste 再用棉；回花
reusable spun silk waste 回绵
reuse 再用
reuse water 再生水；回用水
reusinite 类树脂石
reutilization 再用
revaporization 再汽化(作用)；再蒸发(作用)
revaporizer 二次蒸发器
revealing 显色
revealing agent 显色剂
revelation 显色
reverberatory burning 返焰灼烧
reverberatory roasting 返焰灼烧
reversal ①反转；颠倒②撤消
reversal coulometry 逆向库仑法*
reversal current chronopotentiometry 反向电流计时电位法
reversal peak 倒峰

reversal point 转换点
reversal pulse polarography(RPP) 反向脉冲极谱法
reversal valve 可逆阀；换向阀
reverse-acting valve 可逆阀；换向阀
reverse air blast 反向鼓风
reverse alternate length (聚合物)反向交替长度
reverse atom transfer radical polymerization(RATRP) 反向原子转移自由基聚合
reverse capillary electrokinetic chromatography 反向毛细管电色谱
reverse charge 反向充电
reverse combustion 互燃；往复燃烧
reverse coupling 反向偶合
reverse current ①反流；逆流②反向电流
reverse-current drier 回流式干燥器
reverse current plugging system 反转停机装置
reverse cylinder 倒转滚筒
reversed bending 反向弯曲
reversed current drier 逆流干燥器
reversed current relay 反流替续器〔物〕
reversed direct injection burner 逆式直接喷入燃烧器
reverse deionization 反向脱盐；反脱盐
reverse detergent 逆性洗涤剂（防锈）
reversed flow 换向流；逆转流
reversed geometry （双聚焦）倒置结构
reverse dilution method 逆稀释法
reversed least square-derivative spectrophotometry 逆最小二乘-导数分光光度法
reversed line 自蚀(光谱)线
reversed micelle 逆胶束
reversed micelle enhanced fluorescence 逆胶束增强荧光法
reversed micelle enhanced fluorimetry 反向胶束增强荧光法
reversed micelle-stabilized room temperature fluorimetry 逆胶束增稳室温荧光法
reversed octant 反象限
reversed of stress 应力交变
reverse double focusing mass spectrometer 反置双聚焦质谱计*
reversed passive hemagglutination 反向被动血凝试验
reversed phase 反相
reversed phase chromatography 反相色谱法
reversed phase column 反向柱
reversed phase extraction chromatography 反相萃取色谱法
reversed phase foam chromatography 反相泡沫色谱法
reversed phase high performance liquid chromatography 反向高效液相色谱法
reversed phase ion pair chromatography 反向离子对色谱法
reversed-phase liquid-liquid partition chromatography 反相液液分配色谱(法)
reversed phase method 反相法
reversed phase micelle extraction 反向胶束萃取
reversed phase packing 反相填充物
reversed phase paper chromatograph 反相纸色谱法
reversed phase partition 反相分配
reversed phase partition chromatography 反相分配色谱法*
reversed phase sorbent 反相吸着剂
reversed phase suspension polymerization 反相悬浮聚合
reversed phase system 反相系统
reversed phase thin layer chromatography 反相薄层色谱法
reversed phase thin layer plate 反相薄层板
reversed radioimmunoassay 逆放射免疫分析(测定)
reverse draw forming 反向拉伸成型
reversed spin 反向自旋
reversed transcriptive enzyme 逆转录酶；反转录酶
reverse electroendosmotic flow 反转电渗流
reverse electrophoresis 逆电泳(现象)
reverse electrophoretic coating 逆电泳涂漆
reverse emulsion 反相乳液
reverse feedback 负反馈
reverse fit search 逆向拟合检索
reverse-flow baffle 回流挡板
reverse-flow gas process 换向流发生气体过程
reverse flow technique 换向技术；逆流技术
reverse-flow tray 换向流型挡板
reverse flow viscometer 逆流黏度计
reverse IDA 逆同位素稀释分析*
reverse impact test 反冲击试验
reverse isotope dilution analysis 逆同位素稀释分析
reverse isotopic dilution method 逆同位素稀释法
reverse isotopic exchange method 逆同位素交换法
reverse leach 反浸
reverse leach process 反浸法；颠倒浸出法
reverse library search 逆向谱库检索
reverse locknit 经绒平组织
reverse micelle extraction 反胶团萃取
reverse osmosis ①反渗透②反向渗析
reverse osmosis desalination 反渗透脱盐
reverse osmosis membrane 反渗透膜
reverse osmosis unit 反渗透装置
reverse osmosis water treatment 反渗透水处理法

reverse phase chromatography　反相色谱法
reverse phase evaporation vesicles(REV)　反相蒸发脂质体
reverse phase ion pair chromatography　反相离子对色谱法
reverse polarity　反极性
reverse polymerization　逆聚合
reverse pulse polarography　逆向脉冲极谱法
reverse ratio rubber　高苯乙烯(丁苯)橡胶
reverse reaction　逆反应
reverse rotation　反向旋转
reverse search　逆检索
reverse sequence　反顺序；反向序列
reverse side coating　背面涂漆
reverse strike　反序共沉淀分出
reverse-taper nozzle　逆锥度注嘴
reverse transcriptase　逆转录酶；反转录酶
reverse transcription　反转录
reverse translation　逆向翻译
reversibility　可逆性；可逆溯性
reversible　可逆的；可倒的；可回溯的
reversible addition fragmentation chain transfer(RAFT)　可逆加成断裂链转移
reversible adsorption　可逆吸附
reversible cell　可逆电池
reversible change　可逆变化
reversible circulation　反向电流
reversible circulation valve　双向循环阀
reversible cleaning tip　可逆向清洗的喷嘴；双向清洗式喷嘴
reversible colloid　可逆胶体
reversible cycle　可逆循环
reversible deformation　可逆形变
reversible diffusion　可逆扩散
reversible electrode　可逆电极*
reversible electrode potential　可逆电极电势*；可逆电位
reversible electrode reaction　可逆电极反应
reversible electrolysis　可逆电解
reversible engine　可逆(发动)机
reversible filter　可逆加压过滤器
reversible gel　可逆凝胶
reversible hydrolysis　可逆水解
reversible inhibitor　可逆抑制剂
reversible khaki　双面卡
reversible photochemical reaction　可逆光化学反应；光化学可逆反应
reversible polarographic wave　可逆极谱波
reversible polarography　可逆极谱法
reversible polymerization　可逆聚合
reversible precipitation　可逆沉淀(作用)
reversible process　可逆过程*
reversible reaction　可逆反应*
reversible reference electrode　可逆参比电极
reversible setting out machine　可逆平展机
reversible sol　可逆溶胶
reversible susceptibility　可逆磁化率
reversible swelling　可逆溶胀
reversible temperature-indicating coating　可逆性示温涂料
reversible temperature-indicating pigment　可逆性示温颜料
reversible thermocoating　可逆示温涂料
reversible thermo paint　可逆性示温涂料
reversible transition　可逆转变
reversible wave　可逆波*
reversible work　可逆功*
reversible work hardening　可逆加工硬化
reversible work softening　可逆加工软化
reversibly contractile　可以逆收缩的
reversing　反向；回动
reversing gear　回动装置
reversing lever　回动杠杆
reversing machine　翻钢机
reversing mechanism　换向机构
reversing pump　反向泵
reversing switch　换向开关
reversing thermometer　颠倒温度计
reversing valve　回转阀；可逆阀；回动阀
reversing water bottle　颠倒采水器
reversion　返硫*
reversion of gas　气体烃的转化；气体烃的裂化
reversion of kerosene　煤油的储存变质
reversion phase chromatography　反相色谱(法)
reversion spectroscope　倒转分光镜
reversion test　变质试验
reverted calcium phosphate　复原磷酸钙
revertex　浓缩胶乳；蒸浓胶乳
revertible(=revertive)　可逆的；可倒的
revertive(=revertible)　可逆的；可倒的
revision test　重复试验
revive　①复活②还原成金属
revivification　复活(作用)；再生
revivification of catalyst　催化剂的复活
revivification of solution　溶剂的复活；溶剂活性的复原
revivifier　①交流换热器②再生器
revivifying　复活；再生

revolcving feed table　旋转供料台
revoluting drum drier　转筒式干燥机
revolution　①旋转②循环；公转
revolution counter　转速计
revolution indicator　转速计
revolution meter　转速表；旋转计数器
revolution per second(rps)　每秒钟转数；转/秒
revolver　滚筒；旋转器；转换器
revolving black ash furnace　旋转黑灰炉
revolving burner　旋转炉
revolving center　活顶尖
revolving cylinder　①回转滚筒(针筒)②旋转汽缸
revolving drier　转筒干燥器
revolving drum　转筒
revolving drum cooler　转鼓式冷却器
revolving drum drier　转筒式干燥机
revolving drum washer　旋转鼓式洗涤器
revolving filter　旋转过滤器
revolving furnace　旋转炉
revolving grate　旋转炉箅
revolving horizontal drum　卧式(旋)转鼓；卧式转筒
revolving joint　旋转接头
revolving screen　旋筒筛
revolving (superphosphate) den　旋转(过磷酸钙)室
revolving tray oven　旋转浅炉
revolving turret　转盘式换筒装置
revolving wheel　旋转砂轮
revulcanization　再(次)硫化
revultex　硫化蒸发胶乳
re-water　①再浇②纸浆残水
rewelding　重焊
rewetting　再润湿；回潮
rewetting agent　再润湿剂
rewetting test　再润湿试验
rewinder　复卷机
rewinding and edge trimming machine　倒布整边机
rework cost　返工费(用)
reworking　再次加工
reworking of spent catalyst　废催化剂的再次加工〔即再生〕
rework solution　返回(溶)液；返工(溶)液；再加工(溶)液
Rexforming process　列克斯重整法
Reynold's analogy　雷诺模拟
Reynold's criteria　雷诺准则；雷诺判据
Reynold's equation　雷诺公式
Reynold's foil　雷诺箔〔蜡牛皮纸和黏胶膜夹盖的铅箔〕
Reynold's liquid　雷诺液体
Reynold's lubrication equation　雷诺润滑方程式
Reynold's number　雷诺数

Reynold's plastic　雷诺塑性
Reynold's stress　雷诺应力
Reynold-Stanton equation　雷诺-斯丹顿公式
Reynold's transport theorem　雷诺输运定理
Reynold's turbulence　雷诺湍流
R_f　比移值
R_f value　R_f值*
RG acid　RG 酸；1-萘酚-3,6-二磺酸
rhabdophane　磷锶镨矿
rhamnetin　鼠李亭；鼠李醚；栎精-7-甲基醚　$C_{16}H_{12}O_7$
rhamnicogenol　鼠李科精醇
rhamnicoside　鼠李科苷
rhamnin　鼠李宁
rhamninose　鼠李三糖　$C_{18}H_{32}O_{14}$
rhamnite(=rhamnitol)　鼠李糖醇
rhamnitol(=rhamnite)　鼠李糖醇　$C_6H_{14}O_5$
rhamnofluorin　鼠李氟
rhamnogalactoside　鼠李半乳糖苷
rhamnoglucoside　鼠李葡萄糖苷
rhamnoheptonic acid　鼠李庚酮酸
rhamnoic acid　鼠李酸
rhamnol　鼠李醇
rhamnolipid　鼠李糖脂
rhamnomannoside　鼠李甘露糖苷
rhamnose(=isodulcite)　鼠李糖　$C_6H_{12}O_5 \cdot H_2O$
rhamnose phenylhydrazone　鼠李糖苯腙　$C_6H_{12}O_4=NNC_6H_5$
rhamnoside　鼠李糖苷
rhamnosterin　鼠李甾醇　$C_{18}H_{28}O_2$
rhamoxanthin　鼠李黄质
rhapontigenin　祁卢配基；丹叶大黄素
rhatanin　娜檀宁；N-甲基酪氨酸
rhe　流值〔流度单位〕
rheadine　大黄定
rheic acid　大黄根酸　$C_{20}H_{16}O_9$
rhein(=rheinic acid)　大黄酸；4,5-二羟蒽醌-2-羧酸　$C_{15}H_8O_6$
rheinic acid(=rhein)　大黄酸；4,5-二羟蒽醌-2-羧酸　$C_{15}H_8O_6$
rheinolic acid　大黄醇酸
rhenate　铼酸盐　M_2ReO_4
rheniforming　铼重整
rhenite　亚铼酸盐　M_2ReO_3
rhenium　铼　〔75 号元素，化学符号 Re〕
rhenium black　铼黑
rhenium dioxide　二氧化铼　ReO_2
rhenium disulfide　二硫化铼　ReS_2
rhenium filament　铼灯丝

rhenium heptasulfide 七硫化二铼 Re_2S_7
rhenium heptoxide 七氧化二铼 Re_2O_7
rhenium hexafluoride 六氟化铼 ReF_6
rhenium-osmium dating method 铼-锇定年代法
rhenium oxide 氧化铼 $Re_2O_3; ReO_2; ReO_3; Re_2O_7$
rhenium oxychloride 氯氧化铼 $ReOCl_4; ReOCl_5$
rhenium oxyfluoride ①四氟氧化铼②二氟二氧化铼 ③（总称）氟氧化铼 $ReOF_4; Re_2O_2F_2$
rhenium pentachloride 五氯化铼 $ReCl_5$
rhenium peroxide 过氧化铼 Re_2O_8
rhenium sesquioxide 三氧化二铼 Re_2O_3
rhenium tetrachloride 四氯化铼 $ReCl_4$
rhenium trichloride 三氯化铼 $ReCl_3$
rhenium trioxide 三氧化铼 ReO_3
rheo- 〔希腊字头〕 流
rheochor （摩尔）等黏比容
rheochord 滑线变阻器
rheochrysin 大黄苷
rheodestruction 流变破坏
rheodichroism 流变二色性
rheoelastic extrusion(=elastic melt extrusion) 流变弹性挤塑
rheogeniometer 流变性测定仪
rheogoniometry 流变性测定法
rheogram 流变图
rheograph 流变记录器
rheoid 大黄剂
rheological 流变(学的)
rheological agent 流变剂
rheological analysis 流变分析
rheological behaviour 流变特性
rheological body 流变体
rheological characteristics 流变特征
rheological coefficient 流变系数
rheological constant 流变常数
rheological diagram 流变图
rheological equation(=constitutive equation) 流变方程；本构方程
rheological equation of state 流变状态方程
rheological equilibrium 流变平衡
rheological history 流变史
rheological hysteresis 流变滞后
rheologically simple fluid 流变性简单流体
rheologically simple solid 流变性简单固体
rheological model 流变模型
rheological nomenclature 流变学词汇；流变学术语
rheological parameter 流变参数
rheological phenomena 流变现象

rheological property 流变性质
rheological response 流变响应
rheological shear stress 流变切应力
rheological test curve 流变试验曲线
rheological thermal analysis(RTA) 流变热分析
rheological unit 流变单元
rheological variables 流变可变参数
rheological voluminosity(=voluminosity) 容积度
rheological yield condition 流变屈服条件
rheological yield stress 流变屈服应力
rheologist 流变学家
rheology 流变学*
rheology control agent 流变控制剂；流变调节剂
rheology modifier ①流变改性(改进)剂；流变性调节剂 ②黏弹性调节剂
rheology of suspension 悬浮体流变学
rheology remains 流变残余
rheomalaxis(=soften in flow) 失稠性；流动致软性
rheometer ①电流计②血流速度计③流变仪
rheometric test 流变性试验
rheometry 流变测定法
rheo-optical property 流变双折射光学性质
rheo-optics 流变光学
rheopectic 震凝；抗流变的；流凝的
rheopectic flow 震凝流动
rheopectic fluid 震凝流体；抗流变流体；流凝流体
rheopecticity ①震凝性②震凝度
rheopexic 震凝的；抗流变的；流凝的
rheopexy 震凝性*；触变性
rheopurgin 大黄普精
rheoscope 验电器
rheostan 变阻合金
rheostat 变阻器
rheotannic acid 大黄单宁酸；大黄鞣酸 $C_{26}H_{26}O_{14}$
rheotron 通用流变仪
rheovisco-elastometer 流变黏弹计
rheoviscometer 流变黏度计
Rh factor Rh 因子；凝集因子
rhigolene 利可冷〔局部麻醉剂〕
rhinanthin 玄参苷
rhine metal 来因金〔一种铜锡合金〕
Rhine polluted event 莱茵河污染事件
rhinestone 来因石〔一种水晶〕
rhizocarpic acid 地衣黄素酸
rhizocholic acid 根胆酸
rhizome 根茎
rhizonaldehyde 瑞藏醛
rhizonic acid 瑞藏酸 $C_{10}H_{12}O_4$

rhizoplane 粘根土壤粒
rhizosphere 根区
rho ρ〔希腊字母〕
rhodacene 绕达省；玫红省 $C_{30}H_{20}$
rhodalline 烯丙基硫脲
rhodamine 罗丹明〔染〕
rhodamine 3B 罗丹明 3B
rhodamine B 罗丹明 B〔染〕
rhodamine B lake 罗丹明 B 色淀〔染〕
rhodamine 6G 罗丹明 6G*
rhodamine G (lake) 罗丹明 G(色淀)〔染〕
rhodamine reds 罗丹明红
rhodan 双硫氰酸；连二硫氰酸 $(SCN)_2$
rhodanate 硫(代)氰酸盐
rhodanic acid ①硫(代)氰酸 HSCN②罗丹酸〔即：罗丹宁〕
rhodanic ester 硫氰酸酯 RSCN
rhodanide 硫氰酸盐(或酯)；硫氰化物 MSCN
rhodanide reaction 硫氰酸盐反应；硫氰化物反应
rhodanine 罗丹宁〔即：罗丹酸〕 $SCH_2CONHCS$
rhodanine method 罗丹宁法
rhodanizing 镀铑
rhodanometry ①氰值测(油)法②硫氰酸盐定量(法)
rhodan value 硫氰值
rhodate 铑酸盐 M_2RhO_4
rhodeite 万年青糖醇
rhodenin 万年青宁
rhodeol 万年青糖醇 $C_6H_{14}O_5$
rhodeoretin 万年青亭糖
rhodeose 万年青糖 $C_6H_{12}O_5$
rhodexin 万年青素
rhodiene 万年青烯 $(C_{10}H_{16})_x$
rhodime 2-噻唑亚胺
rhodinal 玫瑰醛
Rhodin cell (for caustic soda) 若丁(制苛性钠)电池
rhodinic acid 香茅酸〔主要成分为单萜烯〕
rhodinol 玫瑰醇〔香茅醇和香叶醇的混合物〕
rhodinolic acid 玫瑰酸
rhodinyl 玫瑰基
rhodinyl acetate 乙酸玫瑰酯
rhodinyl benzoate 苯甲酸玫瑰酯
rhodinyl butyrate 丁酸玫瑰酯
rhodinyl formate 甲酸玫瑰酯
rhodinyl isovalerate 异戊酸玫瑰酯
rhodinyl phenylacetate 苯乙酸玫瑰酯
rhodinyl propionate 丙酸玫瑰酯
rhodite 天然铑金合金

rhodium 铑〔45 号元素，化学符号 Rh〕
rhodium black 铑黑
rhodium carbonyl catalyst 羰基铑催化剂
rhodium cesium alum 铑铯矾
  $Cs_2SO_4 \cdot Rh_2(SO_4)_3 \cdot 24H_2O; CsRh(SO_4)_2 \cdot 12H_2O$
rhodium chloride 三氯化铑 $RhCl_3$
rhodium complex 铑配合物
rhodium dioxide 二氧化铑 RhO_2
rhodium hydroxide 氢氧化铑 $Rh(OH)_3$
rhodium monoxide 一氧化铑 RhO
rhodium nitrate 硝酸铑 $Rh(NO_3)_3$
rhodium oil 旋花油〔取自 Convolvulus〕
rhodium oxide 氧化铑 $RhO; Rh_2O_3; RhO_2$
rhodium platinum catalyst 铑铂催化剂
rhodium platinum gauze 铑铂网
rhodium sesquioxide 三氧化二铑 Rh_2O_3
rhodium sulfate 硫酸三价铑 $Rh_2(SO_4)_3$
rhodium sulfide 一硫化铑 RhS
rhodium wood oil 玫瑰木油
rhodizite 硼锂铍矿
rhodizonic acid 玫棕酸；环己烯二醇四酮
rhodo-〔词头〕 玫红
rhodochrosite 菱锰矿
rhodo complex compound 紫配合物
rhododendrin 杜鹃素
rhododendrol 杜鹃醇 $C_{10}H_{14}O_2$
rhodoflavin 红核黄素
rhodol 对甲氨基酚
rhodolite 红榴石；铁镁铝榴石
rhodonite 蔷薇辉石
rhodonite rock 蔷薇辉石岩
rhoduline heliotrope 若杜林天介紫〔苯胺颜料〕
rhoeadic acid 罂粟酸
rhoeadine 丽春花碱
rhoeagenine 丽春花宁
rhoetzite 蓝晶石
rhombic ①正交(晶)的②菱形的
rhombic dodecahedron 菱形十二面体
rhombic mica 金云母
rhombic quartz 长石
rhombic spar 斜方白云石
rhombic sulfur 斜方硫
rhombic system 正交晶系；斜方晶系
rhombifoline 菱叶(野决明)碱
rhomboclase 板铁矿
rhombohedra 菱面体
rhombohedral ①菱形的②菱面体的
rhombohedral iron ore ①赤铁矿②菱铁矿

rhombohedral lattice	三方点格*
rhombohedral system	菱形晶系
rhombohedron	菱面体
rhomboid	①似菱形的②长菱形；平行四边形
rhomboidal	斜方形的
rhombspar	白云石
rhombus	菱形；斜方形
rhometer	电阻率计
rhooki(=phunki)	成熟紫胶
rhotanium	钯铑合金
rhumbatron	高速电子轰击器
rhusin	盐肤木素粉
rhusoid	盐肤木素粉
rhynchophylline(=mitrinermine)	钩藤碱
rhyolite	流纹岩
rhysimeter	流速计
rhythmic deposition	间歇淀积(作用)
rhythmic precipitation	间歇沉淀(作用)
rhythmic reaction	间歇反应
RIA test package	放射免疫分析试验箱
ribavirin	利巴韦林〔药〕
ribbed funnel	线沟漏斗
ribbed glass	柳条玻璃
ribbed roller	肋辊
ribbed smoked sheet	皱纹烟胶
ribbing	条纹〔漆病〕
ribbon	①带；带状物②木桁；板条③打印带④饰带⑤钢卷尺
ribbon blender	螺条混合器
ribbon breaker	防叠装置
ribbon clay	带状黏土
ribbon conveyor	螺条运输机
ribbon cylinder	绕带式筒体
ribbon drier	螺条干燥机
ribbonfil	扁丝
ribbon iron	扁钢；条钢；带钢
ribbonization	成带性
ribbon lap machine	并卷绕机；精并卷机
ribbon mixer	螺条混合机
ribbon (polymer)	条带(聚合体)
ribbon powder	带状(火)药
ribbon pyrolyzer	带状裂解器
ribbon-screw mixer	(单圆锥筒行星式)带状螺旋混合机(器)
ribbon stirrer	螺条搅拌器
ribbon winding vessel	绕带式容器
ribichloric acid	猪殃殃酸
ribodesose(=desoxyribose)	脱氧核糖
riboflavin(=vitamin B_2)	核黄素；维生素 B_2
riboflavin kinase	核黄素激酶
riboflavin tetrabutyrate	四丁酸核黄素
riboflavinylglucoside	核黄素酰葡糖苷
ribonic acid	核糖酸 $CH_2OH(CHOH)_3COOH$
ribonuclease	核糖核酸酶
ribonucleic acid(RNA)	核糖核酸
ribonucleic acid analysis	核糖核酸分析
ribo (nucleo) side	核(糖核)苷
ribo (nucleo) tide	核(糖核)苷酸
3-ribonucleotide phosphohydrolase	3-核苷酸磷酸水解酶；3-核苷酸酶
ribose	核糖 $C_4H_9O_4CHO$
riboside	核(糖核)苷
ribosoe	核糖体
ribosome	核(糖核)蛋白体；核糖体
ribosome RNA(rRNA)	核糖体 RNA
ribotide	核(糖核)苷酸
Rice-Ramsperger-Kassel-Marcus theory	RRKM 理论*〔反应动力学理论〕
Rice-Ramsperger-Kassel theory	RRK 理论*
rice straw pulp	稻秆浆粕
rich	富有的
rich amine	单乙基胺的饱和溶液
Richard coulombmeter	里查德电量计
rich clay	富黏土；沃土；肥土
rich coal	肥煤
rich color	浓色
richellite	土氟磷铁矿
rich gas	富煤气
rich in hydrogen	富氢的
rich iron	富硅铁
rich lead	富铅
rich lime	肥石灰
rich liquid	富液
rich mixture	油脂混合物
richmondite	四水磷铝石
rich naphtha	饱和石脑油
rich oil stabilization	富油的稳定
rich ore	富矿；好矿
rich shade	浓色相
rich solution	浓溶液；富溶液
rich water condenser	富水冷凝器
ricinate	蓖麻油酸盐
ricinelaidate	反蓖麻油酸盐 $C_{17}H_{32}(OH)COOM$
ricinelaidic acid	反蓖麻酸；反-12-羟基-9-十八(碳)烯酸 $C_{18}H_{34}O_3$
ricinene alkyds	蓖麻碱醇酸
ricinic acid	蓖麻酸

ricinine 蓖麻碱 HC=HCC(OCH₃)=CCNN(CN₃)CO
ricinoleate 蓖麻醇酸酯
ricinoleate acid 蓖麻酸
ricinoleate ester 蓖麻油醇酯
ricinoleic acid(=ricinolic acid) 蓖麻油酸；12-羟基-9-十八(碳)烯酸 $C_{18}H_{34}O_3$
ricinoleic acid amide 蓖麻油(酸)酰胺
ricinoleic acid inner ester 蓖麻油酸内酯
ricinoleidin 甘油三蓖麻油酸酯
ricinolein 甘油三蓖麻油酸酯
ricinoleyl glycerine 甘油-蓖麻醇酸酯；甘油-蓖麻油酸酯
ricinolic acid(=ricinoleic acid) 蓖麻油酸；12-羟基-9-十八(碳)烯酸 $C_{18}H_{34}O_3$
ricinstearolic acid 蓖麻硬脂炔酸；12-羟基-9-十八(碳)炔酸 $C_6H_{13}CHOHCH_2CCC_7H_{14}CO_2H$
ricinus oil 蓖麻油
rickardite 碲铜矿
rickshaw tire 人力车轮胎
riddle 粗筛；格筛
riddlings〔复〕 粗筛余料
rider 游码*
rider bar 游码杆
rider carrier 游码钩
rider hook 游码钩
rider notch 游码凹口
rider peak 驼峰
rider pick up 游码移动装置
rider roll 接触胶皮辊
ridge estimate 岭估计
ridgeless variable speed drive 无级变速传动
ridge regression 岭回归
ridge-runner 管道
ridity 干燥度；枯燥
riebeckite 钠闪石
Riedel method 里德尔法〔计算临界压力〕
Riedel-Plank-Miller equation 里德尔-普兰克-米勒方程式〔求蒸汽压及汽化潜热〕
rifamp(ic)in 利福平
rifamycin 利福霉素
rifapentine 利福喷汀〔药〕
riffle board ①去轻馏分用的容器②(在管路上的)缓冲器
riffled table 淘洗台
riffler 沉砂槽；条板；除砂盘
riffling 用金刚砂水磨
rifled pipe ①输送黏油的内螺丝管〔约有3米长的螺丝〕②焙烤管

rifled tube(=rifled pipe) ①输送黏油的内螺丝管②焙烤管
rifle grenade 枪榴弹
rifle gun 来福枪
rifle powder 步枪火药
rifocin(=rifamycin) 利福霉素
rifomycin(=rifamycin) 利福霉素
rift crack(=heart check) 心裂〔木材干裂〕
rig ①装配②装具；装备③试验台④凿井机
rigesity 糙度
rig for testing 试验架
rigging equipment 装配设备
rigging machine 贴合机
rigging up(=rigging out) 安装；装配
right ①右(边)的②直的③合宜的
right angle 直角
right angle atomizer 直角喷雾器
right angle bend 直角弯管
right angle clamp 直角夹
right angle drive 直角传动
right angle stopcock 储水活栓；直角活栓
right circular cylinder coordinate 竖圆柱坐标系
right-handed crystal ①右晶②右旋结晶
right-hand(ed) screw 右旋螺旋
right-hand mill 右传动开炼机
right hand rule 右手规则
right hand thread 右旋螺纹
right hand worm 右旋螺纹
right reading image 正像
rigid axle 刚性轴
rigid bisphenol 刚链双酚类(化合物)
rigid block 刚性嵌段
rigid body 刚(性)体
rigid cellular plastic 硬质微孔塑料；硬质泡沫塑料
rigid chain 刚性链*
rigid chain polymer 刚性链聚合物*
rigid coordinate 刚性坐标
rigid coupling 刚性联轴节
rigid ellipsoid 刚性椭球
rigid-flex printed board 硬软合板
rigid foam 硬质泡沫
rigid-free-draining molecule 刚性自由穿流分子
rigid glass solvent 玻璃状溶剂
rigid granules (催化剂的)固体颗粒
rigidified rubber 硬化橡胶
rigidity ①刚性②稳定性③刚度
rigidity agent 硬化剂
rigidity and flexibility (高分子链的)刚性与柔性
rigidity modulus 刚性模量

rigidity modulus of elasticity　刚度弹性模量
rigidization　刚性化
rigid macromolecule　刚性大分子
rigid matrix electrode　固定基体电极
rigid plastic　硬质塑料
rigid polymer　刚性聚合物；塑料
rigid radical　刚性基团
rigid rod　刚性棒
rigid-rod like polymer　刚性杆形聚合物
rigid rod polymer　刚棒高分子
rigid rolls　刚性压辊
rigid rotator　刚性转子*
rigid rubber　坚韧生胶
rigid sol　①硬质(增塑)溶胶；硬质增塑糊②增硬溶胶
rigid solid(=Euclidean body)　欧几里得体；刚性体
rigid temperature　硬化温度
rigid urethane foam　硬质聚氨酯泡沫塑料
Rillieux evaporator　里留克斯蒸发器
Rilsan coating　尼龙-11 粉末涂料
rim band　边带
rim chafing　轮辋耐磨损性；轮辋磨损
rimifon　雷米封；异烟酰肼；对吡啶酰肼
rimmed steel　沸腾钢
rimose　有裂缝(的)；开裂的
rim plating apparatus　边缘电镀器
rim spead　轮辋速度；圆周速度
rim stress　边缘应力；轮缘应力〔机〕
rim strip　垫带
rim tape　垫带
rimuene(=totarene)　桃柘烯；芮木烯；芮木泪柏烯
rimu pine　芮木泪柏
rimu resin　芮木；泪柏树脂〔取自芮木泪柏 Dacrydium cupressinum〕
rim velocity　周边速度
rim well　轮辋的鞍边
-rin　①扔〔生物碱音译〕②〔意译〕华；苷
rind　果皮
ring　①环；圈②环(形)的③环绕；包围④鸣(铃)；鸣钟
ring and ball apparatus　环球仪；沥青软化点测定器〔一种测定沥青软化点的仪器〕
ring and ball melting point　环球法熔点
ring and ball method　环球法〔测熔点〕
ring-and-ball test　环球法试验；沥青软化点测定试验
ring assembly　集合环〔非稠合的环如(二)联苯、三联苯〕
ring baffle　环形挡板
ring balance(=ring manometer)　环状压差天平
ring bobbin conical building　细纱圆锥形卷绕
ring bolt　带环螺栓〔机工〕

Ringbom's curve　林博曲线
ring breakage　环破裂
ring brominated　环上溴代
ring bromination　环上溴代作用
ring casing　环室
ring-chain tautomerism　环-链互变异构*
ring chlorinated　环上氯代
ring chlorination　环上氯代作用
ring chromatography　环形色谱法
ring cleavage　环破裂
ring-closing reaction　环合反应
ring closure　环合*
ring-closure reaction　环合反应
ring colorimetry　环圈比色法
ring compound　环状化合物
ring-containing diamine　含环二胺(类)
ring content　(分子中的)环含量；环数
ring contraction　环缩小(反应)*
ring conversion　环转化作用
ring crusher　环式压碎辊
ring crushing　用环辊压碎
ring cutter　环形截管器
ring (detachment) method　(拉)环法〔测表面张力〕
ring dyeing　环染
ringed spherulite　环带球晶
ring electrode　环电极
ring electrography　环电(移)谱法
Ringelmann smoke chart　林格曼烟浓度图
ring enlargement　扩环(反应)*
Ringer's solution　林格氏溶液
ring expansion　扩环(反应)*
ring-expansion polymerization　开环聚合
ring filling　环的饱和；环的填充
ring flange　法兰盘；环法兰
ring fluorinated　环上氟代
ring fluorination　环上氟代作用
ring formation　成环(作用)
ring forming addition polymerization　成环加成聚合(作用)
ring furnace　环形炉
ring furnacing　用环形炉加热
ring gage　环规
ring gap atomizer　环隙喷雾器
ring gasket　环形垫料
ring gauge　环规；弹簧压力计
ring grizzly　环筛
ring halogenated　环上卤代
ring halogenation　环上卤代作用
ring heater　环形加热器

ring homology　环状同系(现象)
ring hydrogen　环上的氢
ring iodinated　环上碘代
ring iodination　环上碘代作用
ring isomerism　环异构
ring joint　环接头；环状接合
ring ketone　环酮
ring kiln　环形窑
ring main　闭合干线；环状干线〔电〕
ring manometer(=ring balance)　环状压差天平
ring method　环轮法；拉环法〔测表面张力方法〕
ring modulus　环模数；环状系数
ring nitrogen　环中的氮(原子)
ring nozzle　环形喷嘴
ring-oiled sleeve bearing　油环润滑式滑动轴承；油环式滑动轴承
ring oiler　环形加油器
ring oiling　油环润滑
ring opening　开环(作用)
ring-opening copolymerization　开环共聚合*
ring opening metathesis polymerization(ROMP)　开环易位聚合
ring opening polyaddition　开环加成聚合
ring-opening polymerization　开环聚合*
ring oven　环炉*
ring oven method　环炉法*
ring oven technique　环炉技术
ring packed column　环状填料塔
ring packed tower　环状填料塔
ring packing　环状填充物
ring porous wood　环孔材
ring powder　环状火药
ring reaction　环状反应
ring rider　环状游码
ring roll crusher　环辊压碎机
ring roll mill　环辊研磨机
ring roll pulverizer　环辊粉磨机
ring sample　环状试片；环形试样
ring scission　环破裂
ring scission polymerization　开环聚合(作用)
ring shaped　环形的；环的
ring-shaped hydrocarbon　环烃
ring specimen　环形试样
ring spinning frame　环锭细纱机
ring stand　环架；铁环架
ring stiffener　加强圈
ring strain　环张力
ring structure　环状结构

ring substituted　环上取代
ring substitution　环上取代作用
ring symbol　环状符号
ring system　环状系统
ring tensile　环状试片抗张力
ring test　环(的)试验
ring test piece　环式试验片
ring thread gage　环状牙规
ring tripod　环(形)三脚架
ring twister　环锭捻丝机
ring-type mold　环状模型
ring valve　环形阀；环状阀
rinkite　层硅铈钛矿
Rinmann's green　林曼绿；锌酸钴 $CoZnO_2$
rinneite　钾铁盐
rinnic acid　红松树脂酸
rinsability　可漂洗性；可洗净性
rinse　漂(洗)；轻洗
rinse additive　漂清助剂；调理助剂
rinse aids　漂清助剂
rinse clean　冲洗；漂洗(净)；刷洗(净)
rinse cycle　漂清周期
rinser　冲洗器
rinse solution　冲洗(溶)液
rinse system　漂洗系统；冲洗系统
rinse tank　漂洗槽
rinse water　漂洗水；冲洗水
rinsing　漂洗；冲洗
rinsing bath　冲洗浴；淋洗浴
rinsing cycle　漂清周期
rinsing vat　漂洗瓮
rio arrowroot　木薯
riomycin(=fluvomycin)　河霉素〔即：河流霉素〕
Rio Tinto process　雷廷托提铜法
rip　洗涤器；刮板；刮刀；清管器
rip bottom tank　有多孔底的桶
ripener　催熟剂
ripening　熟成*；成熟
ripening degree　成熟度
ripening index　熟成指数
ripe sludge　熟污泥
ripidolite　铁绿泥石
rip load　致裂载荷
rippite　锐派特〔炸药〕
ripple　涟波；波纹
ripple column　波纹塔
ripple finish(=crinkle finish)　皱纹(罩面)漆；波纹面饰
ripple method　涟波法〔测表面张力〕

ripple tray 波纹塔板
ripple voltage 脉动电压
riptography 沉淀滴定法
risamicin 利萨霉素
riser ①气门②溢流口③(泡罩塔板上的)升气管④立管⑤升井⑥升降机
rise velocity of flame gas 火焰气体上升速度
risic acid(=rissic acid) 日斯酸；2-羧基-4,5-二甲氧基苯氧基乙酸 $C_6H_2(OCH_3)_2COOHOCH_2COOH$
rising bubble viscometer 上升气泡黏度计；升泡法黏度计
rising-film evaporator 升膜式蒸发器
rising solution 提浸溶液〔革〕
rising stem 升杆
risk management system 事故管理系统
risk probability 风险概率
risperidone 利培酮〔药〕
rissic acid(=risic acid) 日斯酸；2-羧基-4,5-二甲氧基苯氧基乙酸 $C_6H_2(OCH_3)_2COOHOCH_2COOH$
RIS-TOF mass spectrometer 激光共振电离飞行时间质谱仪
ritalin 苯哌啶；醋酸甲酯〔商〕
ritonavir 利托那韦〔药〕
Ritz combination principle 里兹并合原理
rivanol 利凡诺〔药〕；乳酸-6,9-二氨基-7-乙氧基吖啶
rivelling 条纹
rivel varnish 皱纹清漆
river basins 河流流域
river pipe 过河管子；(横渡河的)重管子
river retting 流水沤麻
riverside clay 河边白土
river terminal 内河石油库
river water 河水
river (water) die away test 河水衰减试验
rivet ①铆钉②铆接
riveted 铆钉的；帽钉的
riveted flange 铆着凸缘
riveted joint 铆钉接合
riveted pipe 铆接管
riveted tank 铆接油罐
rivet head 铆钉头
rivet hole 铆孔
riveting 铆接
rivit ①铆钉②铆接
rivited bond 铆接
rivotite 碳酸铜锑矿
R lattice R心格子
RNA polymerase RNA聚合酶
RNA splicing RNA剪接
RNA virus RNA病毒
road ①(道)路②途径
road asphalt 筑路沥青
road binders 路油
road dust 路尘〔测去污力用〕
road emulsion 铺路用乳胶体〔沥青在水中的稀乳胶体〕
road knock test 汽油抗爆性的行车试验
road line paint 公路划线漆
road marking paint 路标漆
road metal 筑路金属材料
road mix 路旁混合〔沥青〕；路旁拌合
road octane number 行车辛烷值
road oil 路油；铺路沥青
road tank car 油槽汽车
road tanker(=road tank car) 油槽汽车
road tar 筑路焦油
road test 行车试验
roan 漆叶鞣绵羊革
roast ①焙烧②焙烧生成物
roasted copperas pigment 焙烧硫酸亚铁制的铁红颜料
roasted ore 焙烧过的矿石
roasted pyrite 焙烧黄铁矿
roaster 焙烧炉
roaster ash 焙烧苏打灰〔碳酸钙与碳酸氢钙〕
roaster furnace 焙烧炉
roaster gas 焙烧炉气
roast gas 焙烧发生气
roasting 焙(烧)；炼；烤
roasting-and-reaction 焙烧和反应
roasting-and-reduction 焙烧和还原
roasting furnace 焙烧炉
roasting heap 焙烧堆
roasting heat 焙烧热
roasting kiln 焙烧窑；烤窑
roasting oven 焙烧炉
roasting pile 焙烧堆
roasting space 焙烧空间
roasting stall 焙烧围圈
roasting test 焙烧试验
Roats' blower 鲁氏鼓风机
Robatel extractor 罗巴特尔萃取器
Robert evaporator 罗伯特蒸发器
robin 刺槐毒素
robinetinidol 刺槐亭醇
robinetin(=3,3',4',5',7-pentahydroxyflavone) 刺槐亭；3,3',4',5',7-五羟基黄酮
robinia 刺槐；洋槐
robinin 刺槐素

Robison ester 罗比森酯；葡萄糖-6-磷酸
robtein(＝2′,3,4,4′,5-pentahydroxychalcone) 刺槐因
robust estimate 稳健估计
robust Kalman filter 稳健卡尔曼滤波
robust linear regression 稳健线性回归法
robust regression 稳健回归
robust statistics 稳健统计
roccellic acid 石蕊酸；2-甲基-3-十二烷基琥珀酸
roccelline 酸性红 A〔染〕
Rochelle salt 罗谢尔盐；四水合酒石酸钾钠 $KNaC_4H_4O_6 \cdot 4H_2O$
Rochon prism 罗甸棱镜
rock alum 钾明矾石
rock asphalt 岩沥青
rock breaker 岩石破碎机
rock breaking 岩石破裂
rock cork 石软木
rock crusher 岩石压碎机
rock crushing 压碎岩石
rock crystal 水晶〔无色的石英〕
rock drill 钻石机
rocker conveyor 摆动式运输机
rocker hardness tester 摆杆硬度计
rocker shaft 摇轴
rocker vat 动荡池
rocker yard 动荡鞣池组
rocket ①火箭②火箭(炮弹)③一种焰火
rocket ammunition 火箭(炮弹)
rocket-borne mass spectrometer 火箭携带质谱仪
rocket electrophoresis 火箭电泳*
rocket (engine) fuel 火箭(发动机)燃料；喷气燃料
rocket immunoelectrophoresis 火箭免疫电泳
rocket propellant 火箭推进剂
rock fibre 石绒；岩石纤维
rock-flush 中心无沟槽花纹
rock-forming minerals〔复〕造岩矿物
rock gas 天然气；天然煤气
rocking 摇摆的；摇动的
rocking brass furnace (for zinc refining) 摇摆式(炼锌)电炉
rocking cell 摇摆式电解槽
rocking equipment 振荡装置
rocking sieve 摇摆筛
rocking trough 摇摆槽
rocking type 摇摆型；摇动型〔来回摆动的〕
rocking vibration 平面摇摆振动
rock leather 一种石棉
rock maple 糖槭

rock meal 岩乳
rock milk 岩乳
rockogenin 岩配基；$5\alpha,22\alpha$-螺甾烷-$3\beta,12\beta$-二醇〔取自 Agave gracilipes〕
rock oil 石油
rock packed column 石头填料蒸馏塔
rock phosphate 磷酸岩；磷酸石
rock phosphorite 磷灰岩
rock quartz 石英
rock rheological property 岩石流变性
rockrose 岩蔷薇；赖百当〔俗〕
rock ruby 红柘榴石
rock salt 岩盐*
rock salt prism 岩盐棱镜
rock sand 岩砂
rock silk 细丝石棉
rock smasher 压石机
rock storing of oil 岩石中石油的储藏
rock tallow 伟晶蜡石
rock tar 原油
Rockwell hardness 洛氏硬度
rocuronium bromide 罗库溴铵〔药〕
rod coil block copolymer 刚-柔嵌段共聚物
rod-curtain precipitator 条屏沉降器；静电沉淀器
rodilone 罗地砜；双(对乙酰氨基苯)砜
rodless angle (molding) press 无杆角式压塑机
rodlike chain 棒状链
rodlike chain model 棒形链模型
rodlike polymer 棒状高分子
rodlike polymer molecule 棒状高分子；刚性链高分子
rod mill 棒磨机
rods 棒条体
rod-shaped micelle 棒状胶束
rod-shaped particle 棒状粒子〔颗粒〕
rod shaped polyelectrolyte 棒形聚(合)电解质
rod-shaped porcelain media 棒形瓷质研磨介质；瓷棒介质
rod sulfur 硫棒；棒状硫黄
rod wax 蜡棒；蜡条；棒状蜡
roentgen 伦琴〔射线剂量单位〕
roentgen-equivalent-man(rem) 人体伦琴当量；雷姆
roentgen machine X光机
roentgenogram X射线照相
roentgenography X射线照相法
roentgenology X射线学；伦琴射线学
roentgenometry 伦琴射线测试法
roentgenoscope 伦琴射线透视机；X光机
roentgen photograph 伦琴射线照相

roentgen(r)　伦琴〔射线辐射量单位〕
roentgen rays　伦琴射线；X射线
roentgen tube　伦琴管
roentgen unit　伦琴单位
Roesler's process　罗斯勒法〔从金分离铜、银〕
roe stone　鱼卵石
rofecoxib　罗非昔布〔药〕
rogascope　测金属线裂纹或直径的电磁器
rogitine　吩妥胺〔药〕
rogue peak　畸峰
Roha salt　碳酸氢钠、硫酸钠、碳酸钙、碳酸镁和菖蒲与茴芹粉、苦艾等的混合物
Rohrbach's solution　氯化钡和氯化汞的溶液
Röhrig tube　洛氏提脂管
Rohrschneider constant　罗尔施奈德常数
Rohrschneider phase constant　罗尔施奈德相常数
Rohrschneider's constant　罗尔施奈德常数
roily oil　浊油；混浊了的油
roll　①辊②卷③轧；辗滚；转动
roll adjustment　滚距调整；调距
rollarounds　移动油罐用的辊子
roll arrangement　滚筒配置
roll brown　棕卷(橡)胶
roll camber(=roll crown)　压辊中高度
roll coating　辊涂
roll crusher　滚(压)碎机
roll crushing　滚轧粉碎
rolled　滚压的
rolled angle　角钢
rolled glass　轧制玻璃
rolled gold　金箔；包金
rolled iron　轧制钢
rolled laminated tube　滚压层压管
rolled lead　薄铅板；铅皮
rolled leather　轧制革
rolled metal　轧制金属
rolled roofing　卷材铺顶
rolled section　轧制革
rolled sheet material　轧制钢板
rolled sheet metal　轧制薄板
rolled steel　辊轧钢；轧制钢材
rolled steel wire　盘条；铁丝盘
rolled tin　锌板；锌皮
rolled-to-death rubber　过炼胶
rolled-up stock　压制件
rolled up tank　折叠式油罐
roll-embossed fibre route　纹辊裂膜成纤法
roller　①滚柱②辊③托辊

roller adjusting device　墨辊调节装置
roller analysis　滚柱分析
roller application　辊涂施工
roller bearing　滚柱轴承
roller black　辊筒炭黑
roller box　辊式箱
roller calender　砑光机
roller carbon black　辊筒炭黑
roller cleaner　①辊筒清洗剂〔清洁剂〕②辊涂脱漆剂
roller cleaning knife　洗墨辊刮刀
roller coat　辊涂
roller coatability　辊涂性
roller coater　辊涂机
roller coating　辊涂
roller coating enamel　辊涂瓷漆
roller coating machine　辊涂机
rollercoat line　辊涂流水线
roller conveyor　辊式运输机
roller core　涂漆辊芯
roller drier　辊筒式干燥机
roller feed motion　辊式喂给装置
roller fibre length tester　罗拉式长度分析仪
roller flaker(=roller flaking mill)　制片轧磨
roller flaking mill(= roller flaker)　制片轧磨
roller grizzly　滚动筛
roller hoop drum　加强桶
roller jar mill　活轴球磨
roller leather　机轴革
roller mill　轧制机
roller milling　轧制
roller opening　碾缝
roller pattern paint　辊涂花纹漆〔建筑物内壁装饰用〕
roller sander　砂纸辊打磨机
roller spiral　螺旋滚动运输带
roller stitcher　压合滚
roller-type pump　滚柱式泵
rollformer exit table　滚形机出口滚道；自动卷材机出口滚道
rollforming　滚轧成形；成形轧制；辊锻成形；自动成卷
rollforming-foaming　①滚压成形泡沫②滚压成形发泡
rollforming speed　滚轧成型速度
roll gauge　①轧辊(型)缝②墨辊压力计〔测量仪〕
roll grinder　滚筒研磨机
roll grinding machine　滚筒研磨机
rolling　①捏炼；压炼②滚动③卷边〔浸渍制品〕④滚压
rolling and quartering　滚动四分法；对角滚折四分法
rolling-ball viscometer　滚球式黏度计
rolling beam apparatus　旋转杆(动态)仪

rolling circle amplification 滚环扩增技术
rolling-circle model 滚环模型
rolling friction 滚动摩擦；第二类摩擦
rolling friction method 滚动摩擦法
rolling hardening 滚压强化
rolling jack 打光机
rolling list 荷叶边〔非织造布疵点〕
rolling loss 滚动损失
rolling machine 滚光机
rolling mill ①轧机机②轧钢机
rolling off 轧去；卷离
rolling oil 碾油；辊子油
rolling pin 柱塞
rolling press ①压光机〔造纸及纺织〕②压延机〔橡胶〕
rolling resistance 滚动阻力
rolling sphere viscometer 转动落球黏度计；滚球式黏度计
rolling stock ①(铁道)有轨运输工具②轧制材料
rolling up 缠绕；卷起
rolling-up device 卷纸设备
roll kiss coater 辊式涂舐机
roll mill ①轧制机②轧钢机
roll milling 轧制
roll mixing 辊压捏合
roll nip 滚距
roll oil 轧用油；用于冷轧软金属的油
roll on-roll traffic 拖车
roll opening 滚距
roll out 轧去
roll plate culture 转碟培养法
roll product 轧制产物
roll release agent 防黏滚剂
roll resistance 滚动阻力
roll roofing 卷材铺顶
rolls ①辊子②轧制机
roll scale 轧制铁鳞
roll separation 滚距
roll setting 飞刀辊调节
roll shell 轮缘
roll spew 卷裹渗出
roll surface speed 周速度
roll tube culture 转管培养法；旋管培养法
roll wave 滚波
Roman cement 罗马水泥；天然水泥
romanium 钨铜镍铝合金；罗曼铝合金
Roman ocher 罗马赭石；橙赭石
Roman purple 罗马红紫
roman vitriol 硫酸铜

romeite 锑钙石
rondomycin(=6-methyleneoxytetracycline) 6-亚甲基土霉素
rongalite 雕白粉〔商〕；次硫酸氢钠与甲醛的加成物 $CH_2O \cdot NaHSO_2 \cdot 2H_2O$
röntgen(=roentgen) 伦琴〔单位〕
röntgen equivalent physical 物理伦琴当量
röntgenogram 伦琴照片；X射线照片
röntgen ray(=X-ray) 伦琴射线；X射线
rood 罗特〔面积测量单位〕
roof cladding 屋面材料；屋面覆盖层
roof coating 屋顶漆
roof drain 浮顶油罐的排水系统
roofing asphalt 铺顶沥青
roofing felt 屋顶油毡
roofing flux 屋顶柏油
roofing granules 屋面颗粒材料；屋顶用碎渣(石)块
roofing paint 屋顶漆
roofing paper 屋顶油纸
roofing slate 屋顶岩板
roofing tile 屋瓦
roof manway 油罐顶上的入孔
roof openings 油罐顶上的开口；罐顶入孔
roof radiant tube 炉顶辐射管
roof supports 罐顶桁架〔油罐〕
roof tiles 屋瓦
roof walkway 盖顶踏板〔油罐〕
room 室；房(间)；车间；工段
room temperature 室温；常温
room temperature cure ①室温固化②室温熟化③室温硫化〔橡胶〕
room temperature curing 室温固化
room temperature curing varnish 室温固化清漆；常温固化清漆
room temperature phosphorescence 室温磷光
room temperature phosphorimetry 室温磷光法
room-temperature setting 室温固化
room-temperature setting adhesive 室温固化黏合剂
room temperature stability 室温稳定性
room temperature superconductor 室温超导体
room temperature vulcanization 室温硫化；常温硫化
room temperature vulcanized silicone rubber 室温硫化硅橡胶
root 根部
root crack 根部裂纹
root diameter 螺杆根部直径
root face ①钝边②齿根圆
Roothaan's equation 鲁桑方程
root mean square 均方根

root-mean-square deviation 均方根偏差；标准偏差
root-mean-square end-to-end distance 均方根末端距
root-mean-square error 标准误差；均方根误差
root-means square value 均方根值
root opening(=root gap) (焊缝)根部间隙
root rubber 根橡胶
Roots blower 罗茨(式)鼓风机
root stock 根茎
rope ①绳(子)②串③束；捆
rope dyeing 捆染
rope grease 绳脂
rope impregnating mangle 绳状浸染机
rope-operated ladder 绳(索)梯
rope soap 悬挂皂
rope-type head 绳式喷头
rope washer(=rope washing-machine) 洗绳机
rope washing machine(=rope washer) 洗绳机
ropey(=ropy) ①绳状物②黏的；可拉成丝的
ropiness ①丝纹②黏性
roping 条痕
ropivacaine 罗哌卡因〔药〕
ropy ①黏的②可拉长成线的
ropy finish 绳状刷痕；绳串病〔漆病〕〔湿边时间短促，刷痕未流平，出现成串的绳状厚膜〕
rosaniline 蔷薇苯胺；品红(碱)；玫苯胺 $(CH_3C_6H_3NH_2)COH(C_6H_4NH_2)_2$
rosaniline blue 蔷薇苯胺蓝(碱性蓝)
rosaniline chloride 氯化蔷薇(玫)苯胺；氯化品红
p-rosaniline method 对蔷薇苯胺法
roscherite 碱磷钙锰铁矿
roscoelite 钒云母
rose ①玫瑰红(色)②蔷薇(属)
rose Bengal 玫瑰红*；孟加拉玫瑰红；四碘四氯荧光素
rose burner 蔷薇花形灯头
rose copper 红铜
Rose crucible 罗斯坩埚
roselite 玫瑰砷酸钙石
rose metal 铋铅铅合金
Rose-Miles method 罗斯-迈尔斯法〔倾注起泡试验〕
Rosenblad heat exchanger 罗森布拉德换热器
Rosenhain calorimeter 罗森痕量热器
rosenonlactone 玫瑰酮内酯
Rosenstein process 罗森斯坦法〔从氯与水煤气制盐酸〕
Rosensthiel's green 罗森斯蒂绿；锰酸钡
Rosentiehl green 钡绿〔锰酸钡〕
roseo-cobaltichloride 三氯化一水五氨钴 $[Co(NH_3)_5H_2O]Cl_3$
rose oil(=attar of rose) 玫瑰油

roseolic acid 玫瑰酸
rose oxide 氧化玫瑰；玫瑰醚；2-(2-甲基丙烯-1-基)-4-甲基四氢吡喃
rose oxide ketone 氧化玫瑰酮
rosephenone 结晶玫瑰；三氯甲基苄醇乙酸酯
rose pink ①淡玫瑰红色②玫瑰红〔色淀颜料〕
rose process 熔融锌沉淀提金法
rose quartz 蔷薇石英
rosette copper 花式铜；盘铜
rosette crystal 莲座晶体
rosette graphite 菊花状石墨
rosette of needles 针状体束
rose vitriol 硫酸钴
rose water 蔷薇水；玫瑰水
rosewood 蔷薇木；玫瑰木
rose-wood oil 蔷薇木油；玫瑰木油
rosiglitazone 罗格列酮〔药〕
rosilic acid 罗斯酸
rosin(=colophony) 松香；松脂
rosin acid 松香酸
rosin acid derivative 松香酸衍生物
rosin adduct 松香加成物
rosin alcohol 松香醇
rosinate 松脂酸盐(酯)
rosinated toner 松香酸盐(化)色料(色原)
rosinated type toner 松香酸盐型色料；松香酸盐型色原
rosinate soap 松香皂
rosinate varnish 松脂酸酯制清漆
rosin content 松香含量
rosin derivative 松香衍生物
rosindone(=rosindulone) 蔷薇引杜(林)酮 $C_{22}H_{14}N_2O$
rosindonic acid 绕森酮酸
rosinduline 蔷薇引杜林；玫红对氮蒽
$HN=C_{10}H_5=NC_6H_4NC_6H_5$
rosindulone(=rosindone) 蔷薇引杜(林)酮 $C_{22}H_{14}N_2O$
rosined (grained) soap 松香皂
Rosin equation 罗辛公式〔估算旋风分离的极限粒度〕
rosiness 玫瑰色(红)
rosin essence 松香精
rosin ester 松香酯
rosin ethylene glycol ester 松香乙二醇酯
rosin flux 焊剂
rosing 着玫红颜色
rosin glycerin ester 甘油松香(酯)；酯胶〔俗〕
rosin grease 松脂油膏
rosin jack 闪锌矿
rosin joint 虚焊接；未焊牢的连接

rosin lye (皂胶加)松香(后的含甘油)碱液
rosin-maleic adduct 顺(丁烯二酸)酐松香加成物
rosin milk 松香乳液
rosin modified phenolic resin 松香改性酚醛树脂
rosin oil 松香油
rosin oil adulterant(=rosin oil adultment) 松香油的(石油)代用品
rosin oil adultment(=rosin oil adulterant) 松香油的(石油)代用品
rosinol 松香油
rosin pentaerythritol ester 季戊四醇松香酯
rosin pitch 松香沥青
rosin resin 松香(基)树脂
rosin size 松香胶料
rosin sizing 上松香胶
rosin soap 松香皂
rosin speck 松香斑
rosin spirit 松香精;松脂醇
rosin spot 松香斑
rosin standard 松香标准
rosin test (qualitative) 松香定性试验
rosin type (sample) 松香型(样品)
rosin varnish 松香清漆
rosinweed 松香草
rosiny 松香的
rosinyl 松香基
rosite 分解了的钙长石;硫铜锑矿
rosmarinic acid 迷迭香酸
rosmarinine 迷迭香宁
rosolate(=corallinate) 玫红酸盐
rosolic acid(=aurin; corallin; pararosolic acid) 玫红酸;金精
rosonolactone 玫红内酯
Rossi-Peakes (flow) test 罗西-皮克斯(流动)试验〔测定粉末涂料流动速度的方法〕
Rossi-Peakes flow tester 罗西-皮克斯流动试验机
rossite 水钒钙石
Rossler-Edelmann process (for zinc extraction) 罗斯勒-埃德曼炼锌法
Ross-Miles apparatus 罗斯-迈尔斯仪〔测发泡能力〕
Ross-Miles lather method 罗斯-迈尔斯泡沫测定法
Ross-Miles method 罗斯-迈尔斯(泡沫测定)法
rosterite 铯绿柱石
rosthornite 棕色树脂
rostone 人造石
rosy ①玫瑰色的②玫瑰红的
Rota-film evaporator 转膜蒸发器
Rota-film still 转膜蒸馏器

rotamer 旋转异构体*
rotamerism(=geometric isomerism) 几何异构(现象)
rotameter(=rotermeter) 转子流量计
Rotap test (dry) (石棉)干式转打振动筛分试验
rotary 旋转;回转
rotary agitator 旋转式搅拌器
rotary airlock valve 旋转气塞阀
rotary air pump 旋转空气泵
rotary annular column 旋转环隙塔
rotary atomizer 旋转雾化器
rotary autoclave 旋转式压热器
rotary bending tester 旋转弯曲试验机
rotary blower 旋转鼓风机;回转鼓风机
rotary boiler (旋转)蒸球;旋转锅炉
rotary breaker 旋转压碎机
rotary Brownian motion 转动布朗运动
rotary chopper 转刀切碎机
rotary column 旋转塔
rotary compression molding press 旋转式压缩成型机
rotary compressor 回转式压缩机
rotary condenser 旋转式冷凝器
rotary continuous filter 旋转连续(过)滤机
rotary converter 转动换流机;旋转变流机
rotary crusher 旋转压碎机
rotary cup atomizer 转杯雾化器
rotary cutter 旋转式切断机;转刀切断机
rotary diffusion 转动扩散
rotary diffusion coefficient 旋转扩散系数
rotary digester 旋转浸煮器
rotary disc extractor 转盘萃取器
rotary disc filter 转盘过滤器
rotary disc knife 圆盘刀
rotary disc meter 转盘式流量计
rotary disc valve 转盘阀
rotary-disk pulsed extractor 转盘脉冲抽提器
rotary-displacement compressor 旋转置换压缩机
rotary drier(=rotatory drier) 旋转式干燥器
rotary drum (旋)转鼓
rotary drum drier 转鼓(式)干燥器
rotary drum feeder 转鼓(式)加料器
rotary drum filter 转鼓(式)过滤机
rotary drum mixer 转鼓(式)混合机
rotary drum vacuum filter 转筒真空(式)过滤机
rotary dryer 转鼓式干燥机;转筒式干燥机
rotary dry vacuum pump 旋转真空干燥泵
rotary evaporator 旋转式气化器
rotary filter 旋转滤机
rotary finishing 旋转抛光

rotary flowmeter 转子流量计
rotary furnace 旋转炉；转炉
rotary gauge 转轮式湿膜测厚仪
rotary gear pump 旋转齿轮泵
rotary grate 旋转炉箅
rotary grate gas producer 转箅气体发生炉
rotary grate shaft kiln 转箅筒形窑
rotary hammer mill 旋转锤磨机
rotary hearth incinerator 回转床式焚烧炉
rotary honing machine 旋转式搪磨机
rotary induction system 旋转输送系统〔在旋转发动机上输送燃料的系统〕
rotary inertia 转动惯量
rotary injection machine 回转式注压机
rotary-inversion axis 反轴*
rotary joint 回转接头
rotary kiln 旋转窑；回转炉
rotary knife 圆盘刀
rotary knife feeder 旋转式刮刀加料器
rotary machine 回转机械
rotary mercury pump 旋转汞泵
rotary movement 转动
rotary oil burners 旋转石油喷灯
rotary oiler 旋转加油器
rotary oven 旋转炉
rotary pelleting machine 旋转造粒机
rotary pipe bending machine 旋转式弯管机
rotary piston 回转式活塞
rotary-piston lump 旋转式活塞泵
rotary plating barrel 旋转电镀桶
rotary-plunger pump 旋转式柱塞泵
rotary positive displacement pump 旋转式正压排液泵
rotary preforming press 旋转压片机
rotary pump(=rotatory pump) 旋转泵
rotary rectifier 旋转整流器
rotary retort 旋转甑
rotary rheometer 转动流变仪
rotary sander 砂轮机
rotary screen 旋转筛
rotary scrubber 旋转涤气机
rotary seal 回转轴封；转动密封
rotary short-shot process 回转欠料注塑法
rotary singeing 圆筒烧毛
rotary slaker 旋转消(石)灰器
rotary spraying machine 旋转式喷涂机
rotary squeezer 旋转压榨机
rotary steam joint 旋转式汽管接头
rotary switch 旋钮开关

rotary table 转盘；转台
rotary table feeder 旋转盘式送料器
rotary thermal rectifying column 旋转式精馏柱
rotary vacuum dryer 旋转式真空干燥器
rotary vacuum pump 旋转真空泵
rotary valve 回转阀
rotary-vane feeder 旋叶送料器
rotary vane type pump 旋转式叶片泵
rotary washing nozzle 旋转洗涤喷嘴
rotary wire brush 旋转式钢丝刷〔除锈、脱漆用〕
rotated dropping mercury electrode 旋转滴汞电极
rotated square layout 转角正方形排列
rotated triangular layout 转角三角形排列
rotating 旋转
rotating agitator 旋转式搅拌器
rotating anode X-ray source 旋转阳极X射线光源
rotating anode X-ray tube 旋转阳极X射线管
rotating arc welding 旋转电弧焊
rotating blade 转动叶片
rotating-blade blasting hardening 旋片喷丸强化
rotating bob viscometer 转锤式黏度计
rotating coaxial-cylinder viscometry 旋转同轴圆筒测黏法
rotating combustion bomb calorimeter 旋转燃烧弹式量热器
rotating concentric cylinder viscometer 同心圆筒式旋转黏度计
rotating concentric tube distilling column 旋转同心管分馏柱
rotating-cone distilling column 转锥分馏柱
rotating coordinates 旋转坐标系*
rotating cradle 旋转摇架
rotating crystal method 旋转晶体法
rotating crystal pattern 旋晶衍射图
rotating cup viscometer 转杯式黏度计
rotating cylinder visco(si)meter (旋)转筒式黏度计
rotating disc column 转盘塔
rotating disc contactor(RDC) 转盘(式)接触器(萃取器)
rotating disk electrode 旋转圆盘电极*
rotating disk electrode voltammetry 旋转圆盘电极伏安法
rotating disk viscometer 旋转圆盘式黏度计
rotating drum incinerator 转鼓式焚烧炉
rotating drying drum (旋转式)干燥鼓
rotating electrode 旋转电极*
rotating electrode method 旋转电极法
rotating element 转动元件
rotating extractor 离心萃取器

rotating frame　旋转坐标系*
rotating furnace　旋转(焙烤)炉
rotating interrupter　旋转断续器
rotating magnet　旋转式磁铁
rotating mass　回转质量
rotating mercury electrode　旋转汞电极
rotating mercury film disc-ring electrode　旋转汞膜盘环电极
rotating metal microelectrode　旋转金属微电极
rotating nozzle　回转式喷丝头；旋转式喷丝嘴
rotating pan mixer　转盘混合器
rotating part　旋转部分
rotating plasma device　旋转等离子体装置
rotating plate　旋转台；配药台；剂量分定台
rotating platform electrode　旋转平台电极
rotating platinum disk electrode　旋转铂盘电极
rotating platinum electrode(RPE)　转动铂电极
rotating platinum microelectrode　旋转铂微电极
rotating pump　旋转泵
rotating ring-disk electrode　旋转环-盘电极*
rotating sample magnetometer(RSM)　旋转样品磁强计
rotating screen　旋转筛
rotating seat　旋转座；动环术；动密封座
rotating sector　旋转窗；旋转阑；旋转扇形板
rotating sector disk　旋转扇形板
rotating sector disk method　旋转光阑法
rotating sector method　旋转光阑法
rotating sector mirror　旋转扇形镜
rotating shutter time-of-flight spectrometer　(带)转动快门的飞行时间能谱仪
rotating shuttle　旋梭
rotating spreader　旋转分流梭
rotating thermometer　旋转温度计
rotating thin layer chromatograph　旋转薄层色谱仪*
rotating thin layer chromatography　旋转薄层色谱法*
rotation　旋转*
rotation/oscillation photography　回转/回摆照相法
rotational analysis　旋光分析
rotational casting　离心浇铸
rotational diffusion　转动扩散*
rotational diffusion coefficient　旋转扩散系数
rotational distortion　旋转畸变
rotational echo double resonance　旋转回波双共振
rotational energy　转动能
rotational equilibrium　转动平衡
rotational fine structure　转动精细结构
rotational flow　旋流
rotational isomer　内旋(转)异构体

rotational isomerism　旋光异构现象；旋转异构现象
rotational isomerization　内旋(转)异构化；旋转异构化
rotational laminar displacement　旋转层状位移
rotational level　转动能级
rotational line　旋转线
rotational little-chamber counter-current chromatography　旋转小室逆流色谱
rotationally cooled laser-induced fluorescence　旋转冷却激光诱导荧光
rotational method　旋转法；转筒法〔测定黏度〕
rotational molding　旋转模塑
rotational quantum number　转动量子数
rotational Raman spectroscopy　转动拉曼散射
rotational rheometer　旋转流变仪
rotational spectroscopy　转动光谱法
rotational spectrum　转动光谱*
rotational speed　旋转速度；循环速度
rotational strength　内旋强度；旋光强度
rotational sub-level　旋转亚能级
rotational transition　转动跃迁
rotational viscometer　旋转黏度计
rotation diffusion constant　转动扩散常数
rotation per minute(RPM)　每分钟转数
rotation platinum electrode　旋转铂电极
rotation pump　机械泵
rotation relaxation　(高分子链的)旋转弛豫
rotation side band　旋转边带
rotation spectrum　转动光谱
rotation speed　转速
rotation vibration spectrum　转振光谱
rotation-viscometer　旋转黏度计
rotator　①旋转器②旋转反射炉③转子；转动体
rotatory　①旋转的②旋光的
rotatory Brownian motion　转动布朗运动
rotatory diffusion　转动扩散
rotatory diffusion constant　转动扩散常数
rotatory dispersion　旋光色散(现象)
rotatory drier(=rotary drier)　旋转式干燥器
rotatory evaporator　旋转蒸发器
rotatory-inversion　旋转-倒反*
rotatory power　旋光本领
rotatory pump(=rotary pump)　旋转泵
rotaversion　反顺转变(作用)
rotaxane　轮烷*
rotenic acid　鱼藤酸　$C_{12}H_{12}O_4$
rotenoid　类鱼藤酮
rotenolone Ⅰ　鱼藤醇酮Ⅰ
rotenolone Ⅱ　鱼藤醇酮Ⅱ

rotenone 鱼藤酮 $C_{23}H_{22}O_6$
rotenone hydrochoride 盐酸化鱼藤酮
rotenonone 鱼藤酮酮；鱼藤二酮
rotermeter(=rotameter) 转子流量计
rotex screen 转动筛分机
rothic acid 绕雌酸 $C_{14}H_{12}O_7$
rothoffite 粒榴石
rotoblast(ing) 转筒喷砂(丸)；旋转喷砂(丸)
Rotocel extractor 罗多西尔抽提器
rotoclone collector 旋风集尘(粉)器；旋风收集器
roto-dipping (汽车车身)旋转式浸涂
rotodip plant 旋转式浸涂装置〔汽车车身的磷化处理和涂底漆〕
rotodip process 旋转浸涂法
rotoflex (帘子线的)耐疲劳试验
rotoforming 旋转成型
rotogravure paper 转轮影印纸
rotometer 旋转流量计
rotomill 单转子连续混炼机
rotophore 旋光中心；旋光基
rotopiston pump 旋转活塞泵
rotor ①转(动)子；转动件②转片③旋度
rotor aeration test apparatus 转子通气试验机
rotor assembly 转子组合体；转子组
rotor beater mill 转子锤磨机
rotor blade 动叶片
rotoreflected beam 回射束
rotor evaporator 转子式蒸发器
rotor nog 转子凸棱
rotor plate viscometer 转(动)板黏度计
rotor pump 旋转泵
rotor spinning machine 转杯纺纱机
rotor spinning unit 转杯纺纱单位
rotory feeder 旋转进料器
rotoscope 急动检验器
roto-sifter 回转筛
rotospray process 转喷涂工艺
rotproof(ed) yarn 抗霉纱
rotproofing 防腐处理；防霉处理
rotproofness 耐腐性
rot resistance 防霉
rot-resistant 防腐的；抗腐的
rot resistor 防腐剂
rotten knots 腐朽节
rotten spot 烂斑〔革〕；陷沼斑；沼穴
rottenstone 擦亮石
rotten timber 腐朽材
rottsite 镍叶绿泥石
rottweil powder 罗特韦耳(火)药；硝化棉、硝化甘油、石蜡火药
rotundifolone 圆叶酮；环氧胡薄荷酮
rotundine 罗通定〔药〕
rouge ①红铁粉②胭脂
rouge flambe 含铜红釉
rougemontite 橄钛辉长岩
rough 橘皮；麻点
rough bare suction hose 露线式吸引胶管
rough burning 粗燃烧
roughcast effect 石子黏面效应
rough coal 原煤
rough-cutting oil 粗切削油
rough deposit 粗沉积物
roughening machine 磨毛机；磨光机
roughening tool 磨锉工具
roughening transition 粗糙化转变
roughening wheel 磨锉轮；砂轮
rougher 粗选池
rough etching 粗(浸)蚀
rough finish 粗面
rough forging 粗锻
roughing 粗加工；初步加工
roughing filter 粗滤器
roughing out 粗分离
roughing pocket 粗筛袋
roughing-up machine 磨毛机；磨光机
rough lumber 粗锯制板
rough machined 粗加工的
rough machining 粗加工
roughness 糙度；粗度
roughness axis 糙度轴
roughness curve 糙度曲线
roughness factor 粗糙度系数；粗糙度因子
roughness tester ①表面粗糙度测定仪②光洁度测定器
rough niter 氯化镁 $MgCl_2$
rough purification 初步净化
rough sizing 粗(筛)选
rough spots 粗糙斑点
rough surface 粗糙面；糙面
rough surface colloid mill 糙面胶体磨
rough-tanned leather 粗鞣革
rough textured film 粗糙膜；质地粗糙膜
rough tube 粗糙管道
rough vacuum 低级真空
rough weight 毛重
rough working 粗加工
roundabout process 间接过程

roundabout system　回转式流水作业法
round bar iron　圆铁
round billet　小圆材
round-body gasket　圆型封圈
round-body packing ring　圆型轴封
round-bottomed flask(=round-bottom flask)　圆底烧瓶
round-bottom flask(=round-bottomed flask)　圆底烧瓶
rounded angle　圆角
rounded orifice　圆孔
rounded stepwise gradient　圆阶式梯度
rounded wiping edge　倒圆的防尘圈(边)缘
round-ended cylinder　圆底筒
rounder end　鹅颈管
round factor　舍入因数
round file　圆锉
round flask　圆底烧瓶
round hole sieve　圆孔筛
rounding　①使成圆形②生皮分割〔皮革〕
rounding error　舍入误差
rounding knife　分割刀
rounding of data　数据舍入法
rounding table　分割台
round maximum　圆形极大
round-meshed　有圆孔的
roundness　圆度
round-off error　修约误差*；舍入误差
round shaft　圆手柄；圆把
round-shaped pattern　圆形喷束
round steel　圆钢
round tank　圆形桶；圆桶形油罐
round-the-clock process　连续过程
round-the-clock production　连续生产
round timber　圆材
round top peak　圆顶峰
round-trial　轮流试验
Roussin's salt　鲁辛盐
roustabout　(炼油厂)不熟练的工人
route profile　管子线路的外形(或纵切面图)
routine　例行；常规的
routine analysis　常规分析*
routine cleaning　定期清洗
routine diagnostic analysis　常规诊断分析
routine experiment　例行试验
routine inspection　常规检查，例行检查
routine laboratory test　例行实验室试验
routine method of determination　例行测定方法
routing　挖刨
routing analysis　例行分析

rovalising　金属磷酸膜被覆法
rovibronic spectrum　电子振转光谱*
roving　无捻粗纱
roving denier indicator　张力式纤度感知器
Rowland circle　罗兰圆
rowlandite　硅氟铁钇矿
Rowland mounting　罗兰装置
row line　行线
row-nucleated cylindrite　横列成核织态结构
row of tubes　(在裂化炉中的)管子行列
roxenol　氯二甲苯酚
roxithromycin　罗红霉素〔药〕
royal water　王水
r-process(=rapid process)　快过程
R salt　R酸盐
R-S system of nomenclature　R-S 命名体系*
R titration　R滴定
rub　摩擦
ruban　2-(4-喹啉基甲基)奎宁环
rub bar　橡胶棒
rubbed brickwork　磨砖对缝砌筑(砌体)
rubbed concrete　水磨石
rubbed film　涂的薄膜
rubbed finish　磨光饰面
rubber　①橡胶*②橡皮
rubber accelerator　橡胶(硫化)促进剂
rubber acid　橡皮硫酸〔水中生物氧化橡皮中的硫而生成〕
rubber adhesive　橡胶类黏合剂；橡胶系黏合剂
rubber adhesive plaster　橡皮膏
rubber alloy　并用胶；共混胶〔高聚物和橡胶的掺和物〕；橡塑胶
rubber and cork sheeting　夹胶软木板
rubber antioxidant　橡胶防老剂
rubber-asbestos plate　橡胶石棉板
rubber base paint　橡胶系涂料；橡胶基漆
rubber bead core　橡胶球心轴
rubber bearing　①橡胶轴承②橡胶垫板
rubber bearing plant　橡胶植物
rubber belt　橡皮带；胶带
rubber belting　橡皮带；胶带
rubber bromide　溴化橡胶
rubber buffer　橡胶缓冲装置；缓冲橡胶制品
rubber bulb with reservoir　橡皮二连球；双连球管
rubber bumper　橡胶缓冲装置；缓冲橡胶
rubber bung　橡胶塞
rubber canvas hose　橡胶帆布软管
rubber cap　橡皮罩
rubber cargo hose　运输石油产品的橡皮管

rubber cement　橡胶泥
rubber chemicals　橡胶药品；橡胶助剂
rubber clay　橡胶用陶土
rubber cloth mask　橡皮布面具
rubber-coated　涂了胶的；上了胶的
rubber coating　涂胶
rubber color　橡胶着色剂
rubber composition(=rubber compound)　橡胶混合物
rubber compound(=rubber composition)　橡胶混合物
rubber compounding　橡胶配合*
rubber constituent　橡胶成分
rubber cover(=rubber covering)　涂胶
rubber-covered roll　橡皮套辊
rubber-covered wire　橡皮皮线
rubber covering(=rubber cover)　涂胶
rubber crepe sheet　白皱片
rubber crumb　废胶粉；胶粉；橡胶粒
rubber derivate paint　橡胶衍生物涂料
rubber dibromide　二溴化橡胶〔计算橡胶烃用〕
rubber dispersion　橡胶分散液
rubber dough　橡胶泥
rubber effect　橡胶效应〔伸长时结晶化〕
rubber elasticity　橡皮弹性
rubber-emulsion paint(=latex paint)　乳胶漆
rubber estate　橡胶园
rubber (fabric) drum　橡胶帆布桶；软桶
rubber film　橡胶薄膜；胶膜
rubber finger　橡皮指套
rubber flag　橡皮帚
rubber formulary　橡胶配方手册
rubber friction tape　橡胶摩擦带
rubber gage　橡胶测厚计
rubber gasket　橡胶密封垫；橡胶封圈
rubber gel　橡胶凝胶体
rubber goods　①橡皮制品②橡胶制品
rubber goods making machinery　橡皮用品制造机
rubber grade (carbon) black　橡胶用炭黑
rubber green　橡皮绿；树胶绿
rubber hand pump　橡胶手揿泵
rubber headed hammer　橡胶头锤
rubber hose　橡皮软管
rubber hydrocarbon　橡胶烃
rubber hydrochloride　盐酸橡胶
rubber hydrohalide　卤氢化橡胶
rubberiness　弹性
rubbering　挂胶；覆胶
rubber ingredient　橡胶配合剂；橡胶助剂
rubber insulated　橡皮绝缘的；胶包的

rubber insulated copper wire　胶包线
rubber insulation　①橡皮绝缘；橡胶绝缘②橡皮布
rubber insulation wire　绝缘线；包皮线〔电〕
rubber isomer　橡胶异构体
rubber item　橡胶制品
rubberized　贴的；上胶的；涂胶的；涂了橡胶的
rubberized fabric　胶布
rubberized tricot lining　涂胶编织衬布
rubberizing　贴胶；涂胶；上胶
rubberizing textile　涂(有)橡胶(的)织物
rubber latex　橡胶胶乳*
rubber-like　橡胶状的
rubber-like behaviour　似橡胶特性
rubber-like elasticity　似橡胶弹性
rubber-like liquid　似橡胶液体
rubber-like polyamide　橡胶状聚酰胺；高弹性聚酰胺
rubber-like polymer　橡胶状聚合物
rubber-like product　类似橡胶物；高弹性物
rubber-like solid　似橡胶固体
rubber-like tubing　类橡胶软管
rubber lined　橡皮衬里的
rubber lined bearing　衬胶轴承
rubber lined hose　橡皮衬里管
rubber lining　橡皮衬
rubberlith(=Lithopone)　橡胶用锌钡白；立德粉
rubber lubricant　橡皮润滑剂〔不损伤橡皮的非石油润滑剂〕
rubber lump　橡胶凝块
rubber machine belting　橡皮机器带
rubber machinery　橡皮工业机械
rubber mass　橡胶基质；胶体；胶料
rubber matrix　橡胶基质；胶体；胶料
rubber membrane dialysis　橡胶膜渗析
rubbermeter　橡皮计；测胶硬度计
rubber mill　开放式炼胶机；开炼机
rubber milling　磨橡胶
rubber modified plastics　橡胶改性塑料*
rubber mould　橡胶硫化模；橡胶衬里塑模
rubber mo(u)lding press　平板硫化机
rubber neck　(管子接头的)橡皮套管
rubbernet　橡胶磁铁〔橡胶与钡铁氧体混合制成的磁铁〕
rubber pad　橡胶垫
rubber paste　橡胶糊
rubber phase　橡胶相
rubber-phile　亲橡胶性
rubberphilic　亲橡胶的
rubber-phobe　疏橡胶性
rubber pig　①生胶块②混炼胶卷

rubber plant 橡胶植物
rubber plantation 橡胶园
rubber plastic blend 橡塑共混胶；橡塑并用胶
rubber plasticizing agent 橡胶增塑剂
rubber plating 金属镀胶
rubber plug 橡皮插塞
rubber poison 橡胶毒物〔如铜、锰〕
rubber policeman 橡胶淀帚
rubber polymer 橡胶；合成胶
rubber powder 橡胶粉；粉状橡胶
rubber producting plant 橡胶植物
rubber reclaiming agent 橡胶再生剂
rubber reinforced polystyrene 橡胶补强聚苯乙烯
rubber-reinforcing filler 橡胶补强填料
rubber-reinforcing resin 橡胶补强用树脂
rubber resin 橡胶树脂
rubber-resin alloy 橡胶树脂并用胶；耐冲击性塑料
rubber-resin blend 橡胶树脂并用胶；耐冲击性塑料
rubber roll 橡皮辊
rubber roller 橡皮辊子
rubber scarlet 橡胶大红
rubberscope ①橡胶照射仪②铜锰测定仪〔快速铜锰含量测定仪〕
rubber scraps〔复〕 橡胶碎屑
rubber seal 橡皮填圈；橡皮封口；橡皮密封环
rubber seed oil 橡胶树子油
rubber septum 橡胶隔片
rubber shaft bearing 橡胶轴承
rubber sheathed flexible cord 橡皮线
rubber sheet ①生胶片②胶片
rubber sheet cutter 橡皮切片机
rubber sheeter 压片机〔橡〕
rubber smoked sheet 烟胶片；红橡皮
rubber sole 橡胶(鞋)底
rubber solution 橡胶胶水
rubber solution strainer 橡胶胶水滤液
rubber solvent 橡胶溶剂
rubber sources 橡胶资源
rubber sponge 海绵橡胶
rubber state 橡胶态
rubber stock 橡胶混合物
rubber stopper 橡皮塞(子)
rubber straining machine 滤胶机
rubber styrene graft 苯乙烯接枝橡胶
rubber substitute 橡胶代用品
rubber (suction) ball 橡皮(吸气)球
rubber suction cups 橡胶吸盘
rubber sulfide 硫化橡胶
rubber-sulfur mix 橡胶硫黄混合物
rubber-sundries 橡胶小件杂品
rubber-supporting bearing 橡胶轴承
rubber-surfaced roll 胶滚
rubber tape 橡皮胶带
rubber tester 橡胶试验器
rubber-textile article 橡胶-织物物品
rubber thread 橡胶线
rubber-to-metal adhesive 橡胶结合金属用黏合剂
rubber tooling 橡胶加工设备
rubber transition 橡胶态转变
rubber tree 橡胶树
rubber trimming 毛边（橡）
rubber tube(=rubber tubing) 橡皮管(子)
rubber tube expander 张(橡皮)管器
rubber tubing(=rubber tube) 橡皮管(子)
rubber unvulcanizate 未硫化胶料；混炼胶
rubber uses 橡胶用途
rubber washer 橡皮(垫)圈
rubber washing machine 洗胶机
rubber web 胶片
rubbery 橡胶状的
rubbery character 橡胶类似性质
rubbery consistency 橡胶态稠度
rubbery copolymer 橡胶状共聚物
rubbery deformation 橡胶态形变
rubbery flow 橡胶状流动
rubbery flow zone 橡胶态流动区
rubber yielding plant 产胶植物
rubbery network ①橡胶(态)网络②橡胶交联网
rubbery plateau zone 橡胶态平坦区；橡胶态高弹区
rubbery polymer 橡胶状聚合物
rubbery shear modulus 橡胶态剪切模量
rubbery state 橡胶(状)态
rubbery substance 橡胶性物质
rubbery tensile modulus 橡胶态拉伸模量
rubbing 研磨；打磨(性)
rubbing abrasion 磨损
rubbing fastness 耐摩擦程度
rubbing flat 砂光；打(磨)毛
rubbing leathers 搓条皮圈
rubbing machine 摩擦机
rubbing oil 摩擦用油
rubbing paste 抛光膏；抛光浆
rubbing surface 摩擦面；打磨的表面
rubbing test 摩擦试验；揉搓试验
rubbing tester 摩擦(牢固度试验)机
rubbing twist 搓捻

rubbing varnish 可打磨清漆
rubble(=rubble stone) 卵石；巨砾
rubble stone(=rubble) 卵石；巨砾
rubbone 橡皮酮
rubean 红氨酸
rubeanate 红氨酸盐
rubeane(=rubeanic acid; rubeane hydride) 红氨酸；红氨；二硫代乙二酰胺 $NH_2CS \cdot CSNH_2$
rubeane hydride(=rubeane) 红氨酸；红氨；二硫代乙二酰胺
rubeanic acid(=dithio-oxamide) 二硫代草酰胺；红氨酸 $(C=SNH_2)_2$
rubefacient 引赤药；发红剂
rubellin 鲁白宁毒苷
rubellite 红电气石
rubene 红烯
ruberite 赤铜矿
ruberythric acid(=rubianic acid) 茜根酸；玉红氨酸 $C_{26}H_{28}O_{14}$
rub fastness(=rub proofness) 耐磨程度
rubi- 玉红
rubia 茜根
rubiadin 茜根定
rubianic acid(=ruberythric acid) 茜根酸；玉红氨酸 $C_{26}H_{28}O_{14}$
rubiazol 羧基偶氮磺胺
rubican 茜草苷；茜(草)素
Rubicast process 鲁比卡斯特法〔结皮聚氨酯泡沫塑料发泡法〕
rubicelle 橙尖晶石
rubicene 玉红省 $C_{26}H_{14}$
rubichrome 玉红色素
rubidine 玉红啶 $C_{11}H_{17}N$
rubidium 铷〔37号元素，化学符号Rb〕
rubidium acetate 乙酸铷
rubidium bicarbonate 碳酸氢铷；重碳酸铷 $RbHCO_3$
rubidium bichromate 重铬酸铷 $Rb_2Cr_2O_7$
rubidium bisulfate 硫酸氢铷 $RbHSO_4$
rubidium bisulfite 亚硫酸氢铷 $RbHSO_3$
rubidium bromate 溴酸铷 $RbBrO_3$
rubidium bromide 溴化铷 $RbBr$
rubidium carbonate 碳酸铷 Rb_2CO_3
rubidium chlorate 氯酸铷 $RbClO_3$
rubidium chloride 氯化铷 $RbCl$
rubidium chloroplatinate 氯铂酸铷 $Rb_2[PtCl_6]$
rubidium chlorostannate 氯锡酸铷 $Rb_2[SnCl_6]$
rubidium chromate 铬酸铷 Rb_2CrO_4
rubidium dithionate 连二硫酸铷 $Rb_2S_2O_6$
rubidium fluoride 氟化铷 RbF
rubidium fluosilicate 氟硅酸铷 $Rb_2[SiF_6]$
rubidium hydride 氢化铷 RbH
rubidium hydrogen sulfate 硫酸氢铷 $RbHSO_4$
rubidium hydroxide 一氢氧化铷；一羟化铷 $Rb(OH)$
rubidium indium alum 铟铷矾
rubidium iodate 碘酸铷 $RbIO_3$
rubidium iodide 碘化铷 RbI
rubidium manganate 锰酸铷 Rb_2MnO_4
rubidium nitrate 硝酸铷 $RbNO_3$
rubidium oxide 氧化铷 Rb_2O
rubidium perchlorate 高氯酸铷 $RbClO_4$
rubidium periodate 高碘酸铷 $RbIO_4$
rubidium peroxide 过四氧化二铷 $Rb_2[O_4]$
rubidium persulfate 过(二)硫酸铷 $Rb_2S_2O_6[O_2]$
rubidium platinichloride 氯铂酸铷
rubidium selenate 硒酸铷 Rb_2SeO_4
rubidiumsilicate 硅酸铷 Rb_2SiO_3; Rb_2SiO_4
rubidium-strontium age 铷-锶法测定年龄*
rubidium-strontium dating method 铷-锶定年代法
rubidium-strontium method of age determination 铷-锶(地质)年代测定法
rubidium sulfate 硫酸铷 Rb_2SO_4
rubidium sulfate flame detector 硫酸铷火焰检测器
rubidium sulfide 硫化铷 Rb_2S
rubidium tartrate 酒石酸铷 $Rb_2C_4H_4O_6$
rubidium thiocyanate 硫氰酸铷 $RbSCN$
rubiginous (铁)锈色的；赤褐色的
rubijervine 玉红杰尔碱；红藜芦碱 $C_{27}H_{43}O_2N$
rubine ①玉红 ②红宝石
rubine toner 红宝石色料；红宝石色原
rubitannic acid 玉红单宁酸 $C_{14}H_{22}O_{12}H_2O$
rub-off 揉搓脱色〔薄膜、片材〕
rub-off constant 揉搓常数〔PVC薄膜〕
rub-off resistance 耐揉搓性
rub-on ①擦亮；磨光滑 ②拌和；揉和
rub-out action 研composition作用；研展作用
rub-out method ①刮刀混合法〔吸油量测定法〕②摩擦分散法〔颜料絮凝测定法〕
rub-out oil absorption data 刮刀研合法吸油量数据
rub-out test (颜料吸油量)捏合试验
rub-out time 研合时间
rub proofness(=rub fastness) 耐磨程度
rubrax 矿质橡胶
rubredoxin 红素氧还蛋白
rubrene 红荧烯 $C_{42}H_{28}$
rub resistance ①耐摩擦性；耐打磨性 ②耐揉搓性
rubroskyrin 瑰天精

rub stone　磨石
rub test　摩擦试验
rub tester　研磨试验机
rubulac(=rebuilt lac)　再生紫胶
ruby　红宝石*
ruby alamandine　红尖晶石
ruby arsenic　雄黄
ruby balas　红尖晶石
ruby blende　红色闪锌矿
ruby continuous laser　红宝石连续(波)激光器
ruby copper　赤铜矿
ruby glass　宝石红玻璃；玉红玻璃
ruby lac(=ruby shellac)　宝石(紫)胶
ruby laser　红宝石激光器
ruby mica　针铁矿
ruby red　宝石红
ruby shellac(=ruby lac)　宝石(紫)胶
ruby silver　淡红银矿
ruby spinel　红尖晶石
ruby sulfur　雄黄　As_2S_2
ruby zinc　红闪锌矿
rudaceous　砾状的
Rudolfi's equation　鲁道尔菲电离程度方程式
rue　芸香
rue oil　芸香油〔取自 $Ruta\ graveolens$〕
rufi(=rufo)　绛
rufigallic acid(=rufigallo)　绛酸
rufigallol(=rufigallic acid)　绛酸
rufin　绛脂　$C_{21}H_{20}O_8$
rufinosporin　变红孢菌素
rufiopin　绛雅片素
ruflanic acid　1,4-二羟蒽醌-2-磺酸
rufo(=rufi)　绛
rufochromomycin　绛色霉素
rufol　绛醇　$C_{14}H_{10}O_2$
rufomycin　绛霉素
rug cleaning soap　洗毯皂
rugged catalyst　稳定催化剂；具有机械强度的催化剂
rugged face　粗糙面
ruggedness　坚固性；耐久性；高强度
rugosimeter　糙度计
rugosity　粗糙度
rugosity factor　粗糙度因素
rug shampooing detergent　洗毯剂
Ruhrbenzin process　德国鲁尔汽油公司合成石油过程
Ruhrchemie process　德国鲁尔化学公司合成石油过程
ruhrgasol　鲁尔汽油
Ruhrol process　鲁尔丁辛醇法

ruled counting plate　分格计数板
ruled grating　刻线光栅
rule of absent reflection(=extinction rule of refraction)　(折射)消光规则
rule of hyper-conjugation　超共轭法则
rule of log-log translation　对数-对数平行移动定律
rule of mixture　混合定律
rule of mutual attraction　向心法则*
rule of mutual exclusion　互不相容规则
rule of mutual solubility　相似相溶原则
rule of similarity　相似规则
rule of six　六位规则〔第六位上的取代基阻碍第一位上的官能反应〕
rule of superposition　重叠规律
rule of surface　表面规则
rule of thumb　拇指规则
rule of thumb test　(石油产品的)简单评价方法
rule of valence summation　原子价总和规律
rules for rounding off　修约规则
rumbatron　电子加速器
rumbling(=tumbling)　(小件)转鼓(内)浸涂法；滚筒内浸涂法
rumbling compound　抛光剂
rumex　黄酸模
rumicin　羊蹄根素
rumpometer　水压式漆膜抗张强度计
run　①程②流动③斜槽④试验；运转
runaway condition　失控状态
run-away corrosion　骤增腐蚀
runback　回流(管)
run by gravity　重力流
run coal　软松煤；软烟煤；原煤
run down　馏出
run-down box　①观察匣②探照灯
run-down drum　馏出油接受器
run-down house　(炼油厂的)接见室；联合部门
run-down legs　垂直的溢流管
run-down lines　接受线路
run-down pipe　溢流管
run-down storage(=run-down tank)　馏出油接受罐
run-down tank(=run-down storage)　馏出油接受罐
run dry　①在干燥的情况下操作②干燥
run empty　空车运转
run free　空车运转
rung　①轮辐②扶梯横档
run gum　再熔胶
run idle　空车运转
run in　①注入；流入②试运转；试车

run-kuri gum　再(熔)制烤胶
run length　运转周期
run light　空车运转
runner　①(注射塑模内的)流道②引铁水的地沟③回转石〔磨石上旋转的石〕④转子⑤叶轮
runnerless injection molding　无流道冷料注塑
runner mill　回转磨石
runner milling　碾磨机
running　①流动②流挂〔涂料〕
running balance　动(态)平衡
running buffer　电泳缓冲液
running cost　运行成本
running expense　运转费用
running fit　松动配合；转动配合
running gear　移动装置；行动装置；运行机构
running gel　电泳凝胶
running idle　空转
running-in　①试车；试运转②跑合运转；磨合运转
running-in ability　磨合性
running index　运转指数
running indicator　运转指示器
running-in operation　试运转
running nip　运转辊距
running of gum　树胶的再度熔化
running piping　输送管道
running point　蒸馏点
running repair　验修
running sample　平均样品；运转样品
running speed　运转速度；运行速度
running submergence　工作浸没度
running test　试探性试验；运转试验
running time　运转时间
running water rinse　流动水淋洗(冲洗)
running water wash　流动水洗
runny plaste　流膏；软膏
run off　①流出；溢出；泄出②流失
run-off pipe　泄水管
run-off plate　引出板
run-off tap　泄水龙头；泄水旋塞
run-of mine　①(质量)一般的②未分级的；未筛分(精选)的③原矿
run of mine coal　原煤
run-of-mine ore　原矿
run-on plate　引弧板
run period　流出周期
runs test　游程检验
run tank　馏分收集器；馏出物收集器
runtime　运转时间

Rupert's drops　梨形玻璃滴
rupture(=breakage)　①破裂；断裂②裂缝；裂口③洞穿
rupture coefficient　破裂系数
rupture criterion　断裂判据
rupture cross-section　破裂断面
rupture disc　破裂盘；断裂盘
rupture disk　防爆膜；防爆片
ruptured surface　破裂面；断裂面
rupture efficiency factor　破碎效率因素
rupture elongation　拉伸断脱伸长率
rupture factor　断裂系数
rupture life　持久期限；破裂寿命
rupture of agglomerates　附聚物破碎(率)
rupture point　破裂点
rupture strength　抗裂强度
rupture stress　致裂应力
rupture test　断裂试验；持久试验
rupturing capacity　破裂力值
rupturing of ring　(碳氢化合物)环破裂
rural tyre　农村适用轮胎
rusa oil(=polmarosa oil)　玫瑰草油
ruscogenin　鲁斯可皂苷元
russellite　钨铋矿
Russell-Saunders coupling　鲁塞尔-桑德尔偶合
russet　植鞣坯革
russet brown　黄褐色
Russia(n) leather　俄罗斯革
Russian leather bordered with pebble grain　镶粒俄罗斯革
Russia oil　俄罗斯焦油
rust　①锈；铁锈②生锈
rust blister　锈蚀疱
rust blistering　泡疹；锈泡
rust bloom　锈霜；锈污
rust cement　铁管接合油泥
rust cleaning　除锈
rust color　铁锈色
rust-colored　铁锈色的
rust conversion primer　转化型带锈底漆
rust-converter coating　转化型带锈涂料
rust corrosion　锈蚀
rust creep　锈蚀蔓延；锈迹蔓延
rust creepage　锈蚀蔓延
rust deposit　积锈；锈垢
rust-free surface　无锈表面
rust grade　①锈级〔生锈的等级〕②除锈等级
rust grade scale　锈蚀分级标准
rust grease(=rust inhibiting lubricant)　防锈润滑脂

rustiness 生锈
rusting ①生锈②生锈的
rusting factor 生锈因素
rusting mechanism 生锈机理
rusting rate 锈蚀速率；生锈程度
rustings 锈斑；锈迹
rust inhibiting 防锈
rust inhibiting agent 防锈剂
rust inhibiting coating 防锈涂料
rust inhibiting lubricant(=rust grease) 防锈润滑脂
rust inhibiting paint 防锈漆
rust inhibiting primer 防锈底漆
rust inhibition 防锈作用
rust inhibitive paint 防锈漆
rust inhibitive primer 防锈底漆
rust-inhibitive washes(=chemical conversion coating) 锈转化型底漆；带锈洗涤底漆
rust inhibitor 抗腐蚀添加剂；防锈剂
rustless 不锈的
rustless iron 不锈钢
rustless metal 不锈金属
rustless steel 不锈钢
rust preventing agent 防锈剂
rust preventing coating 防锈涂料
rust preventing grease 防锈脂
rust preventing oil 防锈油
rust prevention 防锈(作用)
rust prevention test 防锈试验
rust preventive 防锈剂
rust preventive additive 防锈添加剂
rust-preventive agent 防锈剂
rust preventive oil 防锈油；防腐(蚀)油
rust-proof 抗锈的；不锈的
rust-proofing (使)抗锈
rustproofing polyurethane 防锈聚氨酯
rust-proof oil 抗腐蚀油；金属表面防锈油
rust remover 除锈剂
rust resistance 耐锈力；防锈性
rust resisting 抗(防)锈的
rust resisting paint 防锈漆
rust resisting steel 不锈钢
rust scale 铁锈屑
rust-stabilizing pigment 防锈颜料
rust stable 防锈的；耐锈蚀的
rust surface 生锈表面
rusty 生锈的；铁锈色的
rusty brown 锈褐色的
rusty degree 生锈等级；锈蚀等级
rusty stain 锈斑；锈迹
rutacultin 异羽叶芸香素
rutaecarpin 吴(茱)黄次碱
rutaevin 吴茱黄苦素
rutamycin 鲁塔霉素；芦他霉素
rutarin 芸香素
ruthenate 钌酸盐 M_2RuO_4
ruthenic (正)钌的；四价钌的
ruthenic acid 钌酸
ruthenic chloride 四氯化钌 $RuCl_4$
ruthenic compound (正)钌化合物
ruthenic oxide 二氧化钌 RuO_2
ruthenious(=ruthenous) 亚钌的；二价钌的
ruthenium 钌〔44号元素，化学符号Ru〕
ruthenium bromide 溴化钌
ruthenium carbonyl 羰合钌
ruthenium catalyst 钌催化剂
ruthenium chloride 氯化钌 $RuCl_2$; $RuCl_3$; $RuCl_4$
ruthenium dioxide 二氧化钌 RuO_2
ruthenium-dipyridine complex 钌-联吡啶配合物
ruthenium hydrochloride 氯钌酸 H_2RuCl_6
ruthenium hydroxide 三氢氧化钌；三羟化钌 $Ru(OH)_3$
ruthenium oxide 氧化钌 RuO; Ru_2O_3; RuO_2
ruthenium sesquioxide 三氧化二钌 Ru_2O_3
ruthenium silicide 硅化钌
ruthenium tetroxide 四氧化钌 RuO_4
Ruthenobery zinc process 腊森伯里炼锌法
ruthenocene 二茂(合)钌 $(C_5H_5)_2Ru$
ruthenous(=ruthenious) 亚钌的；二价钌的
ruthenous chloride 二氯化钌 $RuCl_2$
ruthenous compound 亚钌化物
ruthenous oxide 一氧化钌；氧化亚钌 RuO
rutherford(rd) 卢瑟福〔放射性强度单位，$1rd=10^6$ 衰变/秒〕
Rutherford back scattering(RBS) 卢瑟福背散射
Rutherford back scattering spectroscopy(RBS) 卢瑟福背散射谱法
Rutherford-Bohr atom model 卢瑟福-玻尔原子模型
rutherfordine 菱铀矿
Rutherford prism 卢瑟福棱镜
rutic acid 芸香酸
rutile 金红石
rutile-calcium white (金红石型)钛钙白〔钛钙复合颜料〕
rutile mixed phase pigment 金红石型混相颜料
rutile pigment 金红石型颜料
rutile pigmentary 颜料级金红石型(二氧化钛)
rutile titania 金红石型二氧化钛
rutile titanium dioxide 金红石型二氧化钛；金红石型钛白

rutile-type structure　金红石型结构
rutilization　金红石(晶型)化
rutilization catalyst　金红石化催化剂
rutin(=eldrin)　芸香苷；芦丁
rutinic acid　芸香亭酸〔取自芸香 Ruta graveolens〕$C_{25}H_{28}O_{15}$
rutinose　芸香糖
rutinoside　芸香糖苷
rutoside　芸香苷；芦丁；维生素 P
Rybczynski-Hadamard law　赖勃斯基-哈达马德定律〔关于乳状液中液珠沉降速度的定律〕
Rydberg constant　里德伯常数
Rydberg fundamental constant　里德伯基本常数
Rydberg number　里德伯数
Rydberg's formula　里德伯频率振动公式
Rydberg state　里德伯态
Rydberg transition　里德伯跃迁
ryegrass　麦草
rymer(=reamer)　①铰刀②铰床
ryridomycin　莱里多霉素

S

sabadilla ①沙巴草；喷嚏草；沙巴达②沙巴草子
sabadilla seed oil 南美灯心草子油
sabadillic acid 沙巴酸
sabadilline 沙巴灵
sabadine 沙巴定
sabadinine 沙巴碱；赛藜芦碱
sabal 蓝棕〔植物〕
sabalol 蓝棕醇
sabatrine 沙巴春
sabbatia 美苦草〔植物〕
sabbatin 美苦草苷
sabina glycol 圆柏二醇
sabina hydrate 水合桧烯
sabinaketone(=sabinene ketone) 桧烯酮 $C_9H_{14}O$
sabinane 桧烷 $C_{10}H_{18}$
sabina oil 桧油
sabine 赛(宾)〔吸声单位〕
sabinene 桧萜；桧烯
sabinene hydrate 水合桧烯
sabinene ketone(=sabinaketone) 桧烯酮
sabinic acid 桧酸；12-羟基十二(烷)酸
sabin oil 桧油
sabinol 桧萜醇
sabinyl 桧基
sabinyl acetate 乙酸桧酯
sable ①黑貂②黑貂皮
sabromin 沙波明；二溴山萮酸钙 $Ca(C_{22}H_{41}O_2Br_2)_2$
saccharamide 糖二酰胺 $C_6H_{12}O_6N_2$
saccharan 多聚糖
saccharate ①糖二酸盐②蔗糖盐
saccharetin 糖蔗衣〔取自糖甘蔗〕 $(C_5H_7O_{10})_n$
saccharic acid ①葡萄糖二酸②糖酸
saccharide 糖类*
sacchariferous 含糖的
saccharification 糖化作用
saccharifying 糖化过程
saccharimeter 糖量计〔测旋光〕
saccharimetry 糖量测定(法)
saccharin(=o-benzoic sulfimide; saccharol;garantose; gluside) 糖精；邻磺酰苯甲酰亚胺 $C_7H_5O_3NS$
saccharine 糖精
saccharine sodium 糖精钠
m-saccharinic acid 偏糖精酸
 $CH_2OH(CHOH)_2CH_2CHOHCOOH$
p-saccharinic acid 仲糖精酸
 $CH_2OHCH_2COH(COOH)CHOHCH_2OH$
saccharinic acid 糖精酸；己糖酸 $C_6H_{12}O_6$
saccharinol 糖精
saccharinose(=saccharin) 糖精
saccharin sodium salt(=soluble saccharin) 糖精钠；可溶糖精
saccharinum 糖精
saccharization 糖化作用
saccharobiose(=sucrose) 蔗二糖；蔗糖
saccharogalactorrhea 乳汁多糖
saccharogen 糖原
saccharogenic method 糖化法
saccharogenic power 糖化本领；糖化力
saccharoid ①糖块状的②糖质物③粒状物
saccharol 糖精
saccharolactic acid(=mucic acid) 糖乳酸；黏酸
saccharolactone 葡萄糖二酸单内酯
saccharol(=saccharin) 糖精
saccharometer 糖液比重计
saccharose(=sucrose) 蔗糖 $C_{12}H_{22}O_{11}$
Sachse process 萨克斯法〔甲烷部分燃烧制乙炔〕
S-acid S 酸；1-氨基-8-萘酚-4 磺酸
2S-acid (=1,8-dihydroxynaphthalene-2,4 disulfonic acid) 2S 酸；1,8-二羟基萘-2,4 二磺酸
sack cleaner(=sack cleaning machine) 净袋机
sack cleaning machine(=sack cleaner) 净袋机
sack cloth 麻袋布；粗麻布
sack elevator 袋式升降机
sacking 装液过程
sack packer 装袋机
Saco-Lower tester 机械式均匀度试验仪
sacred bark tree 药鼠李
sacrificial action 牺牲作用
sacrificial anode 牺牲阳极
sacrificial layer 牺牲层
sacrificial metal coating 牺牲金属保护层
sacrificial process 牺牲层工艺
sacrificial protection ①牺牲金属保护(法)②牺牲阳极保护法③锌或铝阳极保护
saddening 媒染(后)固着处理；(色泽)黯淡处理
saddening agent 媒染固着剂；黯淡剂
saddle ①鞍形物；鞍座②座架；座板；滑板③管托；支管架④(谐振曲线的)凹谷

saddle leather 鞍皮
saddle marks 鞍印
saddle packing 马鞍形填料
saddle point 鞍点*
saddle repair clamp 马鞍形修理夹
saddle seat 鞍形座
saddle soap 马鞍形皂
saddle support 鞍式支座
saddle tee 马鞍形三通
safa 聚酰胺纤维
safe accelerator 安全促进剂;迟延性促进剂
safe allowable load 安全允许负载
safe allowable stress 安全允用应力
safe coefficient 安全系数
safe concentration 安全浓度
safe dose 安全剂量
safeguard ①防护;保护②防护物
safeguards analysis 安全保障分析;核监督分析
safeguards material control system 安全监督(核)材料控制系统
safeguards technique 安全保障技术;核监督技术
safe heat transfer 安全传热
safe level ①允许浓度②允许用量
safe light lamp 暗室灯
safe operation 安全操作
safe range of stress 疲劳极限;应力安全范围
safety 安全
safety alarm device 安全报警装置
safety belt 安全带
safety bulb 安全球
safety cage (油罐操作用的)安全室
safety can 安全罐
safety catch 安全制动装置
safety cock 安全栓
safety code 安全规程
safety coefficient 安全系数
safety color 安全色〔安全标志颜色〕
safety cut-off ①安全开关②安全切断
safety cutout 安全切断器;安全切断
safety device 安全防护装置
safety drilling method 安全钻孔法
safety equipment 安全装置
safety explosive 安全炸药
safety factor ①安全率②安全因素
safety flask 安全烧瓶
safety fuel 安全燃料
safety funnel 安全漏斗
safety funnel tube 安全漏斗管

safety fuse ①安全导火线②安全引信;定时引信
safety gate 安全门
safety glass 安全玻璃;不碎玻璃
safety goggles 防护眼镜;护目镜
safety guard 防护装置〔如防护罩、防护栏〕
safety hardener 安全固化剂
safety head 安全盖
safety ink 安全墨
safety lamp 安全灯
safety lighting fitting 安全照明装置
safety margin 安全限度
safety match 安全火柴
safety measure 安全测试
safety mercury pipette 安全汞吸移管
safety nitrogen storage tank 安全(用)氮储槽
safety paper 保险纸
safety pin ①安全销机②(安全)扣针
safety pipette(=Komagome pipette) 安全移液管
safety plug 安全插头
safety powder 保险粉;连二亚硫酸钠
safety precaution 安全防护
safety regulation 安全规程
safety solvent 安全溶剂
safety stop 安全装置;紧急停车装置
safety stud 防滑钉〔轮胎〕
safety switch 安全开关;紧急开关
safety syphon 安全虹吸管
safety throw-out bar 安全关闭器连杆
safety tread 安全轮胎
safety tube 安全管
safety valve 安全阀
safety vent 安全(气)孔
safety wire 紧急开关拉绳
safe working pressure 安全操作压力
safe working strength 安全操作强度
safe yield 安全产量
saffian leather 山羊革;摩洛哥革
safflorite 斜方砷钴矿
safflower 红花
safflower oil 红花油
safflower red 红花油;藏红花色
safflower yellow(=carthamus yellow) 红花黄色素
saffron 藏花;藏红花
saffron extract 藏红花浸剂
saffron glucoside 藏红花苷
saffron oil 藏花油
saffron substitute 藏红花代用品
saflor yellow 红花黄

safran ①番红花色素；藏红花色素②藏红色；橘黄色
safranal 藏花醛 $C_{10}H_{14}O$
safranine (碱性)藏红 $C_{18}H_{14}N_4$
safranine dye (碱性)藏红染料
safranine O (碱性)藏红O $C_{18}H_{14}N_4$
safranine T 藏红T；碱性藏红
safranine test (碱性)藏红试验
safraninol 藏红醇 $C_{18}H_{13}ON_3$
safranol ①藏花醇 $C_{18}H_{12}O_2N_2$②藻类定性素
safrene 黄樟烯 $C_{10}H_{16}$
safro-cugenol 黄樟丁子香酚
safrole(= shikimol) 黄樟(油)素
safron 藏花；藏红花
safrovanillin 黄樟香兰素
safty analysis report 安全分析报告
safty shield 安全防护罩
sag 垂挂*
saga oil (=sage oil) 鼠尾草油；撒尔维亚油
sagapenum 阿魏树脂〔取自阿魏 *Ferula*〕
sag curve 挠度曲线；垂度曲线
sag data 流挂数据；流挂等级
sagebrush 蒿
sagebrush oil 蒿油
sage clary oil 欧丹参油；香紫苏油
sage oil(=saga oil) 鼠尾草油；撒尔维亚油
saggar(=sagger) 烧箱；烧盆
sag ga(u)ge 垂度计；弛度计
sagger(=saggar) 烧箱；烧盆
sagger body 烧箱坯质
sagger clay 烧箱土
sagger grog 烧箱渣
sagger house 烧釉窑
sagger placing 装烧箱
sagger press 烧箱压机
sagging ①沉降②流挂③倾塌
sagging beam method (下)沉梁法
sagging rate 流挂速度
sag index applicator 刮涂法流挂指数试验仪
sag-index blade 流挂指数刮刀
sagittol β-桉叶醇
sag-liner 流挂线规
sag-liner drive 流挂线规
sago 西(谷)米
sag prevention 防流挂；抗流淌试验
sag resistance 抗流挂性；流挂稳定性
sags and runaway 流挂和滴落
sag temperature ①流挂温度②熔垂温度；淌流温度〔塑料〕
sag tester 流挂试验器
sag tie 吊杆
sagy 加过紫芫香料的
sahlinite 黄砷氯铅矿
sahlite 次透辉石
sailcloth 篷帆布；厚篷帆布
sailing ship 帆船
saiodine 一碘二十二酸钙 $Ca(C_{21}H_{42}ICO_2)_2$
sajodin 一碘二十二酸钙
Sakaguchi test 坂口试验
sakuranetin 樱花亭 $C_{16}H_{14}O_5$
sakuranin 樱花苷 $C_{22}H_{24}O_{10}$
sal ①盐②硅铝质
salable product 合格品；正品
salacetol(=salicylacetol) 水杨酸丙酮酯 $C_6H_4(OH)COOCH_2COMe$
sal acetosella 草酸氢钾
salad oil 生菜油；色拉油
sal aerotus 碳酸氢钾
sal alembred(=sal alembroth) 氯化汞胺 $[NH_2Hg]Cl$；$Hg(NH_2)Cl$
sal alembroth(=sal alembred) 氯化汞胺
salamandaridine 蝶螈定
salamander 切顶石墨锥；耐火架
salamanderine 蝶螈碱 $C_{24}H_{60}O_5N_2$
salamandrite 纸柏
sal amarum 硫酸镁
salamide 精磺胺；水杨酰胺 $HOC_6H_4CONH_2$
sal ammoniac 硇砂；氯化铵
salantol 水杨酸丙酮酯 $HOC_6H_4COOCH_2COCH_3$
salazine 水杨嗪；双亚水杨基连氮 $HOC_6H_4CH=NN=CHC_6H_4OH$
salazinic acid 水杨嗪酸
salazolon(=salipyrine) 沙利比林；水杨酸安替比林
salbutamol 沙丁胺醇
sal communis 食盐；氯化钠
saldanine 曼陀罗碱
saleable drawn yarn 正品拉伸丝；可出售的拉伸丝
saleable quality 合格质量；正品级
sale gauge 售量仪
sal enixum 硫酸氢钾 $KHSO_4$
sal epsom 泻盐；硫酸镁
saleratus 碳酸氢钾
saleripol(=gentisyl alcohol) 龙胆醇
sal ethyl (=ethyl salicylate) 水杨酸乙酯
sal fossile 食盐
sal glauberi 芒硝；硫酸钠
salhypnone 水杨安眠酮 $C_6H_4(OOCC_6H_5)COOCH_3$
salic 硅铝质

salicifoline 柳叶木兰碱
salicilobine 柳叶山梗碱
salicin(e) 水杨苷 $C_{13}H_{18}O_7$
salicine bordeaux R 柳醇枣红R
salic minerals 硅铝质矿物
salicoside 水杨苷;山杨苷;柳醇
salicyl 水杨基;邻羟苄基 $HOC_6H_4CH_2-$
salicylacetol(=salacetol) 水杨酸丙酮酯
salicylal 水杨醛
salicylalcohol 水杨醇;邻羟苯甲醇 $C_7H_8O_2$
salicylaldehyde(=o-hydroxy-benzaldehyde) 水杨醛;邻羟基苯醛 HOC_6H_4CHO
salicylaldehyde acetylhydrazone 水杨醛缩乙酰腙
salicylaldehyde glucose 水杨醛葡萄糖
salicylaldimine 水杨醛亚胺
salicylaldoxime 水杨醛肟
salicylamide 水杨酰胺 $HOC_6H_4CONH_2$
salicylamidoxime 水杨胺肟
salicylanilide N-水杨酰基苯胺 $HOC_6H_4CONHC_6H_5$
salicylase 水杨酶
salicylate ①水杨酸盐②水杨酸酯 OHC_6H_4COOM; OHC_6H_4COOR
salicylazochromotropic acid 水杨基偶氮变色酸
salicylfluorenone 水杨芴酮
salicyl fluorone 水杨荧光酮
salicylhydroxamic acid 水杨基羟肟酸 $(HO)C_6H_4CONHOH$
salicylic acid (=o-hydroxybenzoic acid) 水杨酸;邻羟基苯甲酸 $HOC_6H_4CO_2H$
salicylic acid collodion 水杨酸胶棉
salicylic aldazine 水杨醛连氮〔发泡剂〕
salicylic aldehyde 水杨醛;邻羟苯甲醛
salicylic amide 水杨酰胺
salicylic anhydride(=salicylide) 水杨酐;水杨酸内酯
salicylic glycerinformaldehyde 甘油水杨酸甲酸酯
salicylide(=salicylic anhydride) 水杨酐;水杨酸内酯
salicylidene 亚水杨基;亚邻羟苄基;邻羟苯亚甲基 $HOC_6H_4CH=$
salicylidene acetamide 亚水杨基乙酰胺 $HOC_6H_4CH=NCOCH_3$
salicylidene-o-aminophenol 亚水杨基邻氨基酚
salicylidene amino-2-thiophenol 亚水杨基氨基-2-硫酚
salicylidene benzamide 亚水杨基苯酰胺 $HOC_6H_4CH=NCOC_6H_5$
salicylidene p-phenetidine 亚水杨基对乙氧基苯胺 $HOC_6H_4CH=NC_6H_4OC_2H_5$
salicylidene thiohydantoin 亚水杨基海硫因 $HOC_6H_4CH=C_3H_2ON_2S$

salicylimine 水杨亚胺
salicylol 水杨油 $C_7H_8O_2$
salicylonitrile 水杨腈;邻羟基苯甲腈 HOC_6H_4CN
salicyloyl 水杨酰;邻羟苯甲酰 $o\text{-}HOC_6H_4CO-$
salicyl-p-phenetidine 水杨基对氨基苯乙醚
salicylresorcinol 水杨基间苯二酚 $C_{13}H_{10}O_4$
salicyl salicylic acid 水杨基水杨酸
salicyluric acid 水杨尿酸;水杨酸甘氨酸
salicyl yellow 水杨基黄
salicylyl 水杨酰 $C_6H_4(OH)CO-$
saliferous 含盐的
salifiable (能变)成盐的
salification 成盐作用;盐渍化
salifying 成盐
saligallol 水杨棓酚
saligenin ①(=salicoside)水杨苷②(=salicylalcohol)水杨醇
saligenol 水杨醇
salimenthol 水杨蓋醇 $C_6H_4(OH)COOC_{10}H_{19}$
salimeter (=salometer) 盐(液比)重计
salimetry 盐分析法
salina ①海水蒸发槽②盐湖;盐田
salinaphthol (=betol) 水杨萘酚
saline ①盐水②含盐的③咸的
saline matter 盐分
saline media 含盐介质
salineness 含盐度
salines 盐泉;盐地
saline solution 0.6%食盐水
saline taste 咸味
saline water 咸水;盐水
saliniferous 含盐的
salinigrin 柳黑苷
salinimeter 盐量计
salinity 盐浓度;咸度;含盐量
salinity determination 盐含量测定
salinity gradient 含盐量梯度
salinometer ①(电导)调谐器②盐液密度计
salinous ①盐的②咸的③腌的
saliphen(=saliphenin) 水杨汾 $C_{15}H_{15}O_3N$
saliphenin(=saliphen) 水杨汾
salipyrene 沙利比林
salipyrine(=salpyrine) 沙利比林;水杨酸安替比林
saliretin 水杨亭
salirgan 水杨干;沙利尔干
saliseparin 菝葜苷;副菝葜皂角苷
salit(=borneol salicylate) 水杨酸冰片酯
salitannol 水杨酸单宁酸缩合产物 $C_{14}H_{10}O_7$
saliter 硝石;钠硝;硝酸钠 $NaNO_3$

salithymol 水杨百里酚 $C_6H_4(OH)COOC_{10}H_{13}$
salitre 硝石；钠硝；硝酸钠 $NaNO_3$
Salkowski's solution 磷钨酸溶液
sall(=phenyl salicalate) 萨罗；水杨酸苯酯〔芳香剂〕
sall ring 小环*
salmak 氯化铵
sal marinum 海盐；氯化钠
salmiac 氯化铵 NH_4Cl
salmic acid 鲑红酸
sal mirabile 芒硝；硫酸钠
salmon gum 萨门胶〔取自 *Eucalyptus salmonoflora*〕
salmon oil 鲑鱼油
salochinine 萨罗奎宁
salocoll 萨罗考
salol(=phenyl salicylate) 萨罗；水杨酸苯酯 $HOC_6H_4CO_2C_6H_5$
salolphosphinic acid 萨罗次膦酸
salometer (=salimeter) 盐(液比)重计
Salom lead-refining process 萨洛姆炼铅法
salophen 萨罗汾 $C_6H_4(OH)COOC_6H_4NH(COCH_3)$
saloquinine 萨罗奎宁
salosalicylide(=disalicylide) 双水杨内酯
sal perlatum 磷酸钠
sal prunella 盐硝；精制硝酸钠
salpyrine(=salipyrine) 沙利比林；水杨酸安替比林
sal sapientiae 氯化汞铵
sal sedatirum 硼砂
sal sedative 硼酸
sal soda 苏打；十水(合)碳酸钠
sal soda tank 苏打槽
salsolidine 猪毛菜定
salsoline 猪毛菜碱
salt ①盐；盐类②食盐
salt air corrosion ①盐气腐蚀②海风腐蚀
salt air heat exchanger 盐空气热交换器
sal tartari 碳酸钾
saltating velocity 跃动速度；跳跃速度
saltation flow 跃动流
saltation point 沉积点〔流态化〕
salt atmosphere 含盐大气
salt bath 盐浴
salt-bath dip brazing 盐浴钎(浸)焊
salt bed 盐床；盐层
salt binding 盐黏合；盐结合
salt box 提盐器
salt brick 盐砖
salt bridge 盐桥
salt bridge supported bilayer lipid membrane 盐桥支撑双层磷脂膜
salt brine 盐生；盐汁
salt burned 盐渍坏的；盐灼
salt-cake 盐饼；芒硝
salt-cake furnace 盐饼炉
salt cake glass 硫酸盐玻璃
salt cake pan 芒硝锅
salt-cake pot 盐饼锅
salt-cake process 盐饼法
salt cake roaster 芒硝炉
salt catch(=salt catcher; salt chamber) 受盐器
salt catcher(=salt catch) 受盐器
salt cellulose 盐纤维素
salt chamber(=salt catch) 受盐器
salt-coated test 涂盐试验；浸盐试验
salt content 含盐量；盐含量
salt correction 盐效校正
salt cote 盐泉
salt cycle process 盐循环过程
salt deposit 盐层
salt dome 盐丘
salt drum 受盐器；集盐器
salt dye(=direct dye) 直接染料
salt effect 盐效应*
salt efflorescence 盐霜
salt elevator 升盐器
saltern 盐场
salt error 盐(误)差
Saltex process 萨尔特克斯过程；熔盐提取法〔从辐照二氧化铀及二氧化钚燃料元件中用熔盐中沉淀过滤的方法回收钚的高温化学过程〕
salt extractor 提盐器
saltfield 盐田
salt figure 盐值〔黏胶液在标准状态下凝固所需的盐液浓度〕
salt-filled exhaust valve 充盐排气阀；钠盐冷却的排气阀
salt filter 滤盐器
salt-filtering phenomenon 渗盐现象
salt flux 盐熔剂
salt fog 盐雾
salt fog chamber 盐雾箱
salt fog resistance 耐盐雾性
salt fog test 盐(水喷)雾试验
salt formation 盐的形成
salt forming 成盐的；形成盐的
salt forming agent 成盐剂
salt-forming group 成盐基(团)
salt-forming reactor 成盐反应器

salt-free process	无盐过程*
salt furnace	盐釜；煮盐锅
salt garden	盐池
salt gauge	盐浮计；盐水比重计
salt glaze	盐釉
salt glazed	上盐釉的
salt glazed brick	瓷砖
salt glazing	盐釉
salt grinder	食盐粉碎机
salt hardening	盐(浴)淬(火)
salt hydrate	水合盐；盐合水
salt ice	盐冰；冰盐混合物
salt index	盐值〔黏胶溶液〕
salt index method	盐值法〔测定黏胶成熟度的方法〕
saltiness	含盐性；咸度
salting evaporator	析盐蒸发器
salting-in	盐溶*
salting-in effect	盐溶效应
salting liquid	盐溶液
salting liquor	盐水；盐汁
salting-out	盐析*
salting-out agent	盐析剂
salting-out chromatography	盐析色谱法
salting-out effect	盐析效应*；盐析作用
salting-out electrolyte	盐析电解质
salting-out elution chromatography	盐析洗脱色谱(法)
salting-out evaporator	盐析蒸发器
salting-out of solvent	溶剂的盐析
salting-out paper chromatography	盐析纸色谱(法)
salting stain	盐斑
salting strength	盐析强度
salting up	沉出盐粒
saltish	微咸
salt lagoon	咸水湖
salt lake	盐湖；盐池
saltlike	盐状的
salt lime	石膏；硫酸钙
salt linkage	盐键
salt liquor	盐水；盐汁
salt loaded	用盐饱和了的
salt loading	用盐饱和
salt marsh	盐沼
salt meter	盐(液比)浮计；盐液密度计
salt milling	盐(助)研磨法
salt mine	岩盐坑
salt mixture	(熔化)盐(热载体)混合物
saltness	含盐度
salt of lemon	草酸氢钾 $KHC_2O_4 \cdot H_2O$
salt of phosphorus	磷酸氢铵钠 $NH_4NaHPO_4 \cdot 4H_2O$
salt of sorrel	草酸氢钾 $KHC_2O_4 \cdot H_2O$
salt of tartar	酒石酸氢钾
salt of vitriol	皓矾；锌矾
salt of wormwood	碳酸钾
salt out	盐析；加盐分离
salt out by sodium hydroxide	用氢氧化钠盐析
salt oven	盐釜；煮盐锅
salt pan	(煮)盐锅
salt pan scale	(煮)盐锅巴
salt patenting	盐浴淬火；盐浴等温处理
saltpeter(=saltpetre)	硝石*；钾硝
saltpetre(=saltpeter)	硝石；钾硝
saltpetre salt	硝石盐
saltpetrous	①硝石的②含硝的
salt pit	盐坑；盐井
salt point	盐析点；盐值
salt poit test	盐值测定
salt polymer	盐聚体
salt receiver	受盐器；集盐器
salt rejection	脱盐(率)；盐剔除
salt rock	岩盐
salts	盐类
salt screen	荧光屏
salt separator	析盐器
salt settler	盐沉清器
salt settling	盐的沉降
salt settling tank	盐沉清器(槽)
salt shower test	淋盐水试验
salt-soap grease	盐-皂润滑剂
salt-solubilized	盐溶液化的；盐增溶(化)的
salt solution reservoir	盐溶液储槽；盐水储槽
salt splitting capacity	中性盐分解能力；中性盐分解容量
salt spray	盐喷雾；盐雾
salt spray booth	盐雾试验箱
salt-spray cabinet	盐雾箱
salt spray chamber	盐雾试验箱
salt spray condition	盐水喷雾条件
salt spray fog	盐雾
salt spray resistance	耐盐雾(性)
salt spray test	盐喷试验
salt spray testing	盐水喷雾试验
salt spring	盐泉
salt stains	盐斑皮
salt system	盐系统
salt tank	盐桶
salt transport process	盐迁移过程
salt water	海水；咸水

salt water detergent　海水(用)洗涤剂
salt water dip test　浸盐水试验
salt water immersion test　盐水浸渍试验
salt water soap　海水(用)肥皂
salt work　盐厂
salty　盐的
Saltzman method　萨尔茨曼法〔氮氧化物的湿式分析法〕
salufer　氟硅酸钠
salumin　水杨酸铝
salvaged pipes　旧管子
salvage material　回收材料；材料回收
salvage shop　(美国)修理工厂
salvage sump　(炼油厂下水道的)石油捕集器
salvarsan(=arsphenamine)　洒尔佛散；胂凡钠明〔梅毒特效药〕；六〇六　[AsC$_6$H$_3$(OH)NH$_2$]$_2$HCl・2H$_2$O
salve　①药膏；油膏剂②止痛药；治疮药③敷药膏
salve bases　制备药膏用的凡士林(原料)
salvelike(=salvy)　药膏状的
salvene　鼠尾草烯
salve oil　药膏油
salviol　苧酮　C$_{10}$H$_{16}$O
salvis　对氯水杨酸钠
sal volatable　碳酸铵
sal volatile　碳酸铵
salvosal lithia　萨罗亚磷酸锂
salvy(=salvelike)　药膏状的
Salzburg vitriol　胆矾；五水(合)硫酸铜；蓝矾　CuSO$_4$・5H$_2$O
samandaridine　蝾螈定
samandarine　蝾螈碱
samaria　氧化钐　Sm$_2$O$_3$
samaric　三价钐的；(正)钐的
samaric bromate　溴酸钐　Sm(BrO$_3$)$_3$
samaric bromide　溴化钐　SmBr$_3$
samaric carbonate　碳酸钐　Sm$_2$(CO$_3$)$_3$
samaric chloride　氯化钐　SmCl$_3$
samaric compound　正钐化合物
samaric fluoride　氟化钐　SmF$_3$
samaric hydride　三氢化钐　SmH$_3$
samaric hydropyrophosphate　焦磷酸氢钐　SmHP$_2$O$_7$
samaric hydrosulfate　硫酸氢钐　Sm(HSO$_4$)$_3$
samaric hydroxide　氢氧化钐　Sm(OH)$_3$
samaric metaphosphate　偏磷酸钐　Sm(PO$_3$)$_3$
samaric nitrate　硝酸钐　Sm(NO$_3$)$_3$
samaric orthophosphate　(正)磷酸钐　SmPO$_4$
samaric oxalate　草酸钐　Sm$_2$(C$_2$O$_4$)$_3$
samaric oxychloride　氯氧化钐　SmOCl
samaric sulfate　硫酸钐　Sm$_2$(SO$_4$)$_3$
samaric sulfide　硫化钐　Sm$_2$S$_3$
samarium　钐〔62号元素，化学符号Sm〕
samarium dichloride　二氯化钐　SmCl$_2$
samarium diiodide　碘化亚钐　SmI$_2$
samarium oxide　氧化亚钐　Sm$_2$O$_3$
samarium sesquioxide　三氧化二钐　Sm$_2$O$_3$
samarium trichloride　氯化钐　SmCl$_3$
samarium triiodide　三碘化钐　SmI$_3$
samarous　亚钐(化物)；二价钐的
samarous chloride　二氯化钐　SmCl$_2$
samarous iodide　碘化亚钐　SmI$_2$
samarous sulfate　硫酸亚钐
samarskite　铌钇矿
samatine　山马蹄碱
sambucus　接骨木花
sambunigrin　黑接骨木苷；苯乙腈葡萄糖苷
same-size ratio　等比量；一比一
samidin　萨米定
sammie　(木屑)回潮
samming　①均湿(法)②陈化(作用)
samming machine　均湿机
sammy　(木屑)回潮
samol　水杨盖酯
samonin　皂草苷
samphire oil　欧海莲子油
sample　①样本；标本②试样；样品③抽样；取样④进样；送样
sample addition method　样品添加法
sample analyzer　试样分析器
sample and hold amplifier (S/H)　采样保持放大器
sample application　点样
sample applicator　点样器
sample blow in technique　撒样法
sample boat　试样舟皿
sample bottle　样品瓶
sample capacity　样本容量
sample cell　①样品管②样品池
sample chamber　试样室
sample changer　样品转换装置
sample characterization　样品表征
sample cleanup　样本净化
sample collection　试样收集
sample collector　取样器
sample concentration　试样浓缩
sample concentration method　试样浓缩法
sample concentrator　试样浓缩器
sample condensation　试样冷凝
sample connection　取样口

sample contamination 试样污染
sample cutter 样品切割器
sampled-current polarography (=tast polarography) 采流极谱法*
sampled-current voltammetry 采流伏安法*
sample delivery point 进样点
sample deviation 样本偏差*
sample dimensionality 样品维数
sample disposal system 样品处理系统
sample drawing 抽样品
sample drum 打样转鼓〔皮革〕
sample enrichment injection of chromatography 色谱富集进样
sample evaporation chamber 试样气化室
sample evaporator 试样气化器
sample excitation 样品激发
sample extraction 样本提取
sample flash vaporizer 试样闪蒸器
sample gas 样气
sample gasification 试样气化
sample-handling system 处理样品系统；进样系统
sample heater 样品加热炉
sample injection 注射试样；进样
sample injection port 试样注入口；注样口
sample injector 进样器
sample inlet system 试样入口系统
sample in sample out method 含样去样检测法
sample integrity 样品完整性
sample introduction 进样
sample introduction system 进样系统
sample introduction under pressure 压力下进样
sample length 溶液厚度
sample light beam 试样光束
sample line 样品管线
sample load 试样载荷
sample loop 进样环管
sample loss 试样损耗
sample mean 样本(平)均值*
sample nozzle 取样喷嘴
sample of residue analysis 残留分析样本
sample on-line pretreatment 试样在线预处理
sample path 试样路径
sample path length 溶液厚度〔光在溶液中所经过的距离，通常以厘米为单位〕
sample point 取样点；取样口
sample preparation 样品制备
sample preparation system 样品制备系统
sample presentation 送样

sample pretreatment 试样预处理
sample probe 试样探针
sample purity 试样纯度
sampler ①取样器②进样器
sample rate 进样速率
sample recovery 试样回收
sample reduction and division 试样缩分
sample reduction equipment 样品缩分设备
sample rotation rate 样品旋转速率
sample size 试样量；进样量；样本容量；样本大小
sample size loss 试样量损耗
sample solution 样品溶液；试液
sample splitter 试样分流器
sample spot 试样斑
sample spotter 点样器
sample stacking 样品堆积效应
sample standard deviation 样本标准(偏)差
sample streaker 试样划痕器
sample structure 试样结构
sample subtraction method 样品减量法
sample survey 样品鉴定
sample switching 试样转换
sample system miniaturization 取样系统小型化
sample take-off 样品输出口
sample technique 取样法
sample thief 取样器
sample throughput 试样处理量
sample train 取样系(统)
sample transfer operation 试样转移操作(法)
sample turret barrel 六角形试样器
sample value 样本值*
sample valve 进样阀
sample valve septum 进样阀隔片
sample variance 样本方差*
sample zone 样品带
sampling 取样*；采样；抽样
sampling action 进样动作
sampling and analyzing unit 取样和分析单位
sampling bias 抽样偏差
sampling boat 取样舟
sampling boat technique 进样舟技术
sampling condition ①进样条件②取样条件
sampling cone 取样锥
sampling cup 取样杯
sampling depth 取样深度
sampling device ①取样装置②进样装置
sampling distribution 抽样分布
sampling error 取样误差

sampling gun 取样枪
sampling inspection 抽样检验*
sampling inspection plan 抽样检验方案
sampling interval ①进样间隔②取样间隔
sampling jug 取样瓶
sampling nozzle 取样管嘴
sampling plate 抽样板
sampling point interval 采样点间隔
sampling polarograph 采流极谱仪
sampling probe 取样针
sampling pump ①取样泵②进样泵
sampling ratio 取样比
sampling survey 抽样调查
sampling system ①取样系统②进样系统③取样法④进样法
sampling technique 采样技术；取样技术
sampling test 抽样检验
sampling time 取样时间
sampling tube 取样管
sampling unit 取样装置；取样单位
sampling valve indicator 取样指示计
Samrex process 萨姆莱克斯过程〔锕、镉、稀土元素分离纯化过程；接在 Lacex 过程之后〕
samsonite 散锁那特〔炸药〕；硫锑锰银矿
sanatron 窄脉冲多谐振荡器
sanclomycin(＝tetracycline) 四环素
sand ①砂(子)②砂地
sandability 打磨性能
sandalidol 人造檀香
sandal oil 檀香油
sandal soap 檀香皂
sandalwood 檀香木
sandalwood oil 檀香油
sandarac gum 山达脂胶
sandarac(h) ①山达脂；桧树胶②雄黄
sandaracing 涂山脂
sandarac oil 山达脂油；雄花油
sandaracolic acid 山达酸
sandaracopimaric acid 山达海松酸；柏脂海松酸
sand asphalt (地)沥青砂岩
sandatone 檀香酮
sand-bath ①砂浴②砂浴器
sand-bed filter 砂床过滤器
sand belt 砂皮带
sand blast 砂喷；砂吹；砂磨
sand blasted finish 喷砂精整；喷砂打光
sand blaster 喷砂机
sand blast hose 喷砂胶管
sand blasting 喷砂(清洁)处理
sand blasting equipment 喷砂设备
sand blasting machine 喷砂机
sand blower 喷砂器
sand blowing 喷砂打光
sand carrying capacity 流砂量
sand cast finish 砂模铸件处理
sand casting ①砂型铸造②砂型铸件
sand-catcher 捕砂槽；除砂器
sand cement 砂水泥
sand-cement grout 沙泥砂浆
sand cloth (金刚)砂布
sand containing paint 含砂涂料
sand cracking 砂子裂解
sand dry 脱砂干燥
sanded 砂磨(光了)的
sanded plywood 砂光胶合板
sand-eel oil 砂鳝鱼油
Sandell index 桑德尔指数
Sandell's sensitivity 桑德尔灵敏度
sander 研磨机
Sanderson drying time meter 桑德森干燥时间计
sanders wood 檀香木
sand filter 砂滤器
sand filtering 砂滤
sand filtering device 砂(层过)滤器；砂滤装置
sand finish ①砂浆抹面②硅砂防滑漆
sand finish paint 硅砂防滑漆〔以硅砂为骨料的防滑和装饰性涂料〕
sand glass 砂玻璃
sand grind 砂磨
sanding property (可)打磨性
sanding sealer 掺砂火漆
sanding through 砂透
sandish 砂(质)的
sandiver 玻璃沫
sand jet 喷砂器
sand leach 砂滤
sand-lime brick (石)灰砂(粒)砖
sand mill 砂磨机
sand-mill premix 砂磨预混合料
sand mold casting 砂型铸件
sandothrene blue 山道士林蓝
sandothrene colors 山道士林染料
sandothrene dark blue 山道士林深蓝
sandothrene pink 山道士林桃红
sand paper (金刚)砂纸
sandpaper block 砂纸打磨块

sand pit 砂坑
sand pump 砂泵
sand roughness 砂粒粗糙性
sand scale 砂垢
sand seal 砂封
sand slime separator 砂泥分离器
sand soap 砂肥皂
sand spit 砂洲
sandstone 砂岩；砂石
sandstone still 砂石蒸馏器
sand-suction hose 吸砂胶管
sand sump 砂槽
sand trail 落砂试验
sand trap 砂槽；除砂盘
sand trap section 砂槽系统
sand trial 落砂试验
sand-washing 砂洗除锈；水力喷砂除锈
sandwich ①层叠物②夹层结构③复合结构材料
sandwich assay 夹心分析
sandwich chamber(=S-chamber) S(展开)室；S(展开)槽
sandwich compound 夹心化合物*
sandwich coordination compound 夹心配合物
sandwich copolymer 嵌段共聚物；夹心共聚物
sandwiched type 偶合型〔复合半导体纳米微粒的结构型式之一〕
sandwich fiber hybrid composite 夹芯型混杂纤维复合材料
sandwich heating 夹层加热；两面加热
sandwich hybridization 夹心杂交
sandwich inclusion compound 夹层包合物
sandwiching 插入；夹入
sandwich injection 夹层注射
sandwich isolator 剪切型防振器
sandwich laminate 夹芯层压材料
sandwich layer 夹层；芯层
sandwich material 夹层材料；芯材
sandwich molecule 蜂窝状夹层分子
sandwich panel 夹心板
sandwich structure 夹层结构
sandwich technique 夹层技术
sandwich zone technique 夹层带技术
sandy 砂(质)的
sandy clay 砂土
sandy marl 砂质泥灰岩
sandy seal 砂封
S and Z twist combination 顺反捻混合
sanfordization 防缩
sanforizing 防缩(皱)处理；机械预缩整理

sanforzing agent 防缩剂
sangbo crepe brocade 桑波缎
Sanger reagent 山格试剂〔即二硝基氟苯(DNFB)〕
Sanger's method 二硝基苯基化法
sanguilutine 血根黄碱
sanguinarine(e) 血根碱
sanguinarine nitrate 血根碱硝酸盐
sanguinarine sulfate 血根碱硫酸盐
sanguirubine 血根红碱
sanguisorba 地榆
sanguisorbin 地榆英
sanicle 变豆菜
sanicula 变豆菜
sanidine 透长石；玻璃长石
sanidinite 透长(石)岩
sanitary landfills 卫生填埋
sanitation inspection 卫生检验
sankey diagram 流程图
sanoform 二碘水杨酸甲酯 $C_6H_2I_2(OH)COOMe$
sanorin 山罗林
sanshoamide 山椒酰胺
sanshool 山椒醇；花椒麻味素
sanshotoxin 山椒毒
santa herb 山达草
santal 紫檀酸
santalal 檀香醛
santalane 檀香烷
santalbinic acid 白檀子油酸；十八碳烯炔酸
santal camphor 檀香脑
santalene 檀香萜 $C_{15}H_{24}$
santalenic acid (=santalin) 檀香烯酸
santalic acid 紫檀色素
santalin(=santalenic acid) 檀香烯酸
α-santalinenone α-檀香灵烯酮
santal oil 檀香油
santalol 檀香醇
santalone 檀香酮
santal wood oil 檀香木油
santalyl 檀香基
santalyl acetate 乙酸檀香酯
santalyl carbonate 碳酸檀香酯
santalyl chloride 檀香基氯 $C_{15}H_{23}Cl$
santalyl formate 甲酸檀香酯
santalyl isovalerate 异戊酸檀香酯
santalyl phenylacetate 苯乙酸檀香酯
santalyl propionate 丙酸檀香酯
santalyl salicylate(=santyl) 水杨酸檀香酯
santel 水基丙烯酸树脂(乳剂)

santene 檀烯 C_9H_{14}
santenenic acid 檀烯酸
santenic acid 檀酸
santenol 檀烯醇 $C_9H_{16}O$
santenone 檀烯酮
santiganine 山剔干碱
santiganine hydrochloride 山剔干盐酸盐
santol 檀紫素
α-santolinenone α-檀灵烯酮
santomerse 润湿剂；烷化芳基磺酸盐
santonica 山道年花
santonic acid 山道(年)酸 $C_{15}H_{20}O_4$
santonin (=anhydride of santonic acid) 山道年；山道酸酐 $C_{15}H_{18}O_3$
santoninate 山道年酸酯
santoninic acid 山道年酸 $C_{15}H_{20}O_4$
santoninoxime 山道年肟 $C_{15}H_{18}O_2(NOH)$
santonous acid 山道亚酸
santyl(=santalyl salicylate) 水杨酸檀香酯 $HOC_6H_4CO_2C_{15}H_{23}$
sap green(=bladder green) ①暗绿色②树液绿〔一种暗绿色天然染料〕
sapietic acid 杉皮酸
sapinine 乌桕碱
sapling 幼树
sapo 橄榄油肥皂
sapodilla 人心果
sapogenol 皂草精醇
sapo glycerinatus 甘油皂
sapo jalapinus 药喇叭皂
sapo kalinus 亚麻油钾皂
sapo molis 橄榄油钾皂
saponaceous 肥皂的
saponaretin 牡荆色素
saponaria 肥皂草
saponarin 皂草苷
saponated petroleum 皂化石油〔白油或凡士林和油酸，氨水的混合物〕
saponetin 皂草亭
saponide 合成洗涤剂
saponifiability 皂化性
saponifiable 可皂化的
saponifiable content (可)皂化物含量
saponifiable group 可皂化的基
saponifiable matter 可皂化物
saponifiable oil 皂化油
saponification 皂化*
saponification agent (=saponifier) 皂化剂

saponification column 皂化塔〔连续皂化装置〕
saponification degree 皂化程度；皂化率
saponification equivalent 皂化当量
saponification flask 皂化(烧)瓶
saponification glycerin 皂制甘油
saponification number 皂化值*
saponification rate 皂化速率
saponification ratio(=saponification value) 皂化值
saponification reaction 皂化反应
saponification value(=saponification ratio) 皂化值
saponified 皂化(了)的
saponified acetate 皂化醋酸纤维
saponified acetate rayon 皂化醋酸人造丝
saponified acetate yarn 皂化醋酸丝
saponified cellulose acetate 皂化纤维素醋酸酯
saponified oil 皂化油
saponified oleine 皂化了的甘油三油酸酯
saponified polymer 皂化聚合物
saponifier ①皂化剂②皂化器
saponifying 皂化
saponifying agent 皂化剂
saponin 皂草苷；皂角苷
saponin P 皂草苷P
saponite 滑石粉
sapota 山榄果；人心果
sapotalene 山榄烯 $C_{13}H_{14}$
sapote gum 人心果胶；山榄果胶
sapotin 山榄碱
sapotinetin 山榄裂碱
sapotoxin 皂角毒苷
sapphire 蓝宝石*
sapphire whisker 蓝宝石须晶
sapphirine 假蓝宝石
saprine 腐肉碱
saproconite 粒状深海灰岩
saprolite 腐泥土
sapromixite (含)藻煤
sapropel ①腐泥②腐泥煤
sapropelic coal 腐泥煤
sapropelitic 似煤(的)；似沥青(的)
sap rot 边腐
sap stain 边材变色
sap wood 边材
sarafloxacin monohydrochloride monohydrate 盐酸沙氟沙星
saran 偏氯纶；莎纶〔偏二氯乙烯与氯乙烯共聚物〕
saran latex 聚偏氯乙烯胶乳
sarcodactylis oil 佛手油

sarcolactate 肌乳酸盐〔即: 右旋乳酸盐〕
sarcolactic acid (=*p*-lactic acid) 肌乳酸 CH₃CHOHCOOH
sarcolite 肉色柱石
sarcosine (=*N*-methylglycine) 肌氨酸; *N*-甲基甘氨酸 CH₃NHCH₂COOH
sard 肉红玉髓
sardinianite 单斜铅矾
sardonyx 缠丝玛瑙
sargasso 马尾藻
sarpagine(=raupine) 蛇根精
sarracenine 瓶子草碱
sarracine 瓶千里光碱
sarracine-*N*-oxide *N*-氧化瓶千里光碱
sarsaponin(=sarsasapogenin glycoside) 丝兰皂苷; 萨洒皂苷
sarsasapogenin 萨洒皂草配基; 菝葜皂苷元
sarsasapogenin glycoside 萨洒皂草配基配糖体
sarsasaponin 拔葜苷
sassafras artificial KH 合成二氢黄樟(油)素
sassafras oil(=sassafras wood oil) 黄樟油
sassafras pith 黄樟木髓
sassafras wood oil (=sassafras oil) 黄樟油
sassoline 天然硼酸
satavic acid (=tetrahydroxy-stearic acid) 萨它酸; 四羟基硬脂酸
sateen ①纬面缎纹②假缎; 缎纹布〔棉织物〕
satellite band 附属谱带
satellite electron 卫星(状)电子
satellite extruder 卫星型挤塑机
satellite gear 行星齿轮
satellite line 伴线
satellite peak 伴峰; 卫星峰
satin ①纬面缎纹②假缎; 缎纹布
satinage 轧光整理
satine ①缎光整理②轧光
satined paper 釉光纸; 缎光纸
satin emulsion 缎光乳胶漆; 半光乳胶漆
satin film 半透明薄膜
satin finish 半光涂料; 缎面处理
satinite 硫酸钙(石)
satinizing 光泽整理
satinizing lacquer 斑洒金(用)漆; 梨子地漆〔大漆〕
satin lustre 缎面光泽; 缎光
satin paper 蜡光纸
satin spar 纤维石
satin stone 纤维石
satin weave 缎纹组织
satin white 缎光白〔由硫酸钙和铝酸钙共沉淀而得〕

sativic acid 洒剔酸 $C_{18}H_{36}O_6$
sativin 蒜头素
saturability 饱和能力
saturable 可饱和的
saturable core reactor 可饱和铁心扼流圈
saturant 饱和剂
saturated 饱和的
saturated acid 饱和酸
saturated acyclic hydrocarbon 饱和无环烃; 饱和链烃
saturated adsorption band 饱和吸附带
saturated air 饱和空气
saturated air mixture 饱和的(可燃蒸气)空气蒸气混合物
saturated alcohol 饱和醇
saturated aliphatics 饱和脂(肪)族化合物
saturated calomel electrode 饱和甘汞电极
saturated carbon ring 饱和碳环
saturated color 饱和色
saturated compound 饱和化合物
saturated concentration 饱和浓度
saturated cycle 饱和的(环烷)环
saturated dihalide 二卤代烷 $C_nH_{2n}X_2$
saturated dissolved oxygen 饱和溶解氧
saturated fatty acid 饱和脂肪酸
saturated humidity 饱和湿度
saturated hydrocarbon 饱和烃
saturated layer 饱和层
saturated liquid 饱和溶液
saturated long chain fatty acid 饱和长链脂肪酸
saturated mixture 饱和混合物
saturated monomer 饱和单体
saturated naphtha 饱和石脑油
saturated oil 饱和油
saturated oligomer 饱和低聚物
saturated optical nonresonant emission spectroscopy 饱和光学非共振发射光谱法
saturated polybasic acid 饱和多元酸
saturated polyester 饱和聚酯*
saturated polyester plastic 饱和聚酯塑料
saturated rock 饱和岩
saturated rubber 饱和橡胶
saturated silicate 饱和硅酸盐
saturated solution 饱和溶液
saturated state 饱和状态
saturated steam 饱和蒸汽
saturated triglyceride 甘油三饱和酸酯
saturated vapor line 饱和蒸气(曲)线
saturated vapor pressure 饱和蒸气压
saturated vapo(u)r 饱和蒸气

saturated vegetable oil 饱和植物油
saturated volume 饱和容积
saturating capacity 饱和量
saturating capacity of paper 纸的饱和性
saturating composition 浸渗组分
saturating paper 油纸
saturating speed 饱和速度
saturating tank 浸渍槽
saturation ①饱和*②饱和度；章度〔物理色度学〕
saturation activity 饱和放射性
saturation adsorption 饱和吸附
saturation adsorption band 饱和吸附谱带
saturation amount 饱和量
saturation analysis 饱和分析
saturation capacity 饱和量
saturation coefficient 饱和系数
saturation current 饱和电流
saturation current method 饱和电流法
saturation curve 饱和曲线
saturation degree 饱和度
saturation efficiency 饱和度
saturation exponent 饱和指数；饱和率
saturation factor 饱和因子*
saturation hydrogenation 饱和加氢作用
saturation index 饱和指数
saturation isomerism 饱和异构
saturation isotherm 饱和等温线
saturation limit 饱和极限
saturation method 饱和法
saturation of double bonds 双键的饱和
saturation point 饱和点
saturation pressure 饱和压力
saturation ratio 饱和比
saturation recovery 饱和回复法
saturation recovery method 饱和回复法
saturation region 饱和区
saturation spectroscopy 饱和光谱学
saturation tank 饱和桶
saturation temperature 饱和温度
saturation tower 饱和塔
saturation transfer 饱和转移
saturation value 饱和值
saturator 饱和剂；饱和器；浸胶机
saturent 浸渍剂
saturex 饱和器
saturite 饱和溶液沉积物〔在蛇纹岩中的细矿粒构造〕
saturnism （中）铅毒
saturn salt 乙酸铅 $Pb(C_2H_3O_2)_2$

saturnus 铅
sauconite 锌蒙脱石；硅铝锌铅石；羟锌矿
saunders 檀香
Saunderson-Milner zeta space color difference equation 桑德森-米尔纳ζ空间色差方程
Saunders-Treloar plastomer 桑德斯-特雷罗拉塑性计
saunders wood 檀香木
sausage casing 烧瓶套
sausage-shaped micelle 腊肠状胶束
saussureic acid 风毛菊酸
saussurite 槽化石
Sauter mean diameter 沙得平均直径
save 洋苏叶
save-all ①防溅器②白水回收装置〔造纸〕
save-all boxe 防溅箱
save-all tray 白水盘〔造纸〕
Savelsberg-Wanuschaff process 萨维耳斯伯格-万诺恰夫(炼镍)法
savine 桧；圆柏
savine oil 桧油
savings 滤屑〔纤维〕
savorquin 二碘羟基喹啉
savory ①美味的②香薄荷
savory oil 香薄荷油
saw 锯
sawara leaf wax 椹叶蜡〔取自日本花柏的叶〕
saw-blade type impeller 锯齿状搅拌叶轮
sawdust 锯屑；锯末；木屑
sawdust link 锯屑链
sawdust room 锯屑室
sawhorse 锯木架
sawing 锯开；锯断
sawing machine 锯床
sawing mill （鞣料)锯磨
saw machine 锯床
saw plametto berries 盖屋棕榈果
saw tooth box 锯齿箱
sawtooth edge splitting 锯齿形刀口裂膜(法)
saw-tooth polarography 锯齿波极谱法*
saxatilic acid 石地衣酸
saxicoles 岩苔类
saxol 液体石蜡油
saxoline 液体石蜡油
Saxonia powder 色克桑尼亚火药
saxonite 色克桑那特〔炸药〕
Saybolt chromometer (=Saybolt colorimeter) 赛波特比色计
Saybolt color 赛波特色度

Saybolt colorimeter　赛波特比色计
Saybolt ring number of kerosene　煤油的赛波特环值
Saybolt universal viscosity　赛氏通用黏度
Saybolt viscosimeter　赛波特黏度计
Saybolt viscosity test　赛波特黏度试验
S-bend pipe　S型管
s-block element　s区元素*
scabbed rust　瘤状锈
scabiolide　糙叶菊内酯
scaffold　脚手架
scaffold bridge　脚手桥
scaffolding　搭脚手架
scagliola　人工大理石
scalar　标(量)的；无向(量)的；纯量的
scalar coupling　标量偶合*
scalar-density　标密(度)
scalar function　标量函数
scale　①秤②标度；制度③比例尺④刻尺度
scale aerometer　目测比重计
scale beam　秤杆
scale coefficient　污垢系数
scale copper　薄铜片
scaled-down model　缩尺模型；按比例缩小的模型
scale deposit　污垢沉积物
scaled eyepiece　刻度目镜；目镜测微计
scale effect　①标度效应②放大效应
scale expansion　量程扩大；量程扩展；标尺扩展
scale factor　标度因子*
scale fiber (viscose)　有鳞(片)纤维(黏胶)
scale formation　结垢
scale incrustation　水垢；水锈
scale inhibitor　阻垢剂
scale load　标准载荷
scale off　①剥落；片落②脱皮；扒皮
scale of hardness　硬度标度
scale of roughness　粗糙标度
scale operation　大规模生产
scale pan　天平盘
scale pan arrester　天平盘托
scale paraffin　片状石蜡
scaler　定标器*；计数器
scale range　刻度(标度)范围；刻度量程
scale reading　标尺读数
scale removal　清除氧化皮；除鳞
scales　鳞状体
scale stone　硅灰石
scale traps　固体沉降槽
scale-up problems　工艺过程放大问题

scale value　标度值
scale wax　鳞状蜡
scaliness　起鳞程度
scaling　①标度*；比例变换②结垢③剥落；腐蚀
scaling between coated film　漆膜层间剥落
scaling circuit　定标电路；计数电路
scaling down　按比例缩小
scaling furnace　(铁皮)镀锡炉
scaling index　腐蚀指数
scaling off　剥落
scaling rate　结垢速率
scaling resistance　抗片落性；抗鳞爆性；抗起鳞性
scaling theory　标度理论
scaling-up　放大；按比例增加
scallops　扇形
scalp　拔顶〔从残油中取出有价值的组分而降低残油质量的方法〕
scalper　护筛粗网
scalper screen　粗筛
scalping　粗筛
scalping screen　粗粒筛；大眼筛；脱介筛〔为筛除大块粗粒用〕
scalp off　脱屑
scalp pomade　头油
scaly　鳞状
scammonin　番薯苷
scammony　番薯
scammony resin　番薯树脂
scan axis　扫描轴
scan coil　扫描线圈
scandia　氧化钪
scandium　钪〔21号元素，化学符号Sc〕
scandium bromide　溴化钪　$ScBr_3$
scandium carbonate　碳酸钪　$Sc_2(CO_3)_3$
scandium chloride　氯化钪　$ScCl_3$
scandium fluoride　氟化钪　ScF_3
scandium hydroxide　氢氧化钪　$Sc(OH)_3$
scandium hydroxynitrate　碱式硝酸钪　$Sc(OH)(NO_3)_2$
scandium iodide　碘化钪　ScI_3
scandium nitrate　硝酸钪　$Sc(NO_3)_3$
scandium oxide　氧化钪　Sc_2O_3
scandium oxynitrate　硝酸氧化钪　$(Sc_2O)(NO_3)_4$
scandium oxysulfate　硫酸氧化钪　$(Sc_2O)(SO_4)_2$
scandium phosphate　磷酸钪　$ScPO_4$
scandium salts　钪盐
scandium sulfate　硫酸钪　$Sc_2(SO_4)_3$
scandium sulfide　硫化钪　Sc_2S_3
scandium tritide　氚化钪　ScT

scan input 扫描输入
scanner 扫描器
scanner detector 扫描检测器
scanning 扫描
scanning agent 扫描剂
scanning and transmission electron microscopy 电子扫描透射显微镜
scanning Auger electron spectrometer 扫描俄歇电子能谱仪
scanning Auger microprobe 扫描俄歇微探针
scanning calorimetry 扫描量热法
scanning camera 扫描照相机
scanning capacitance microscope 电容扫描显微镜
scanning chronopotentiometry 扫描计时电位法
scanning densitometer 扫描密度计
scanning device 扫描器；扫描装置
scanning electrochemical microscope (SECM) 扫描电化学显微镜
scanning electrochemical microscopy 扫描电化学显微法
scanning electron beam 扫描电子束
scanning electronic microscope 扫描电子显微镜
scanning electron micrograph 扫描电子显微照片
scanning electron microprobe 扫描电子微探针
scanning electron microscope (SEM) 扫描电子显微镜；扫描电镜
scanning electron microscope fractograph 断口组织电子扫描显微照片
scanning electron microscopy 扫描电子显微法
scanning electron photomicrograph 扫描电子显微照片
scanning force microscope (SFM) 扫描力显微镜
scanning frequency 扫描频率
scanning infrared spectrophotometer 扫描红外分光光度计
scanning ion microprobe 扫描离子微探针
scanning ion microscope 扫描离子显微镜
scanning microspectrophotometer 扫描显微分光光度计
scanning mode 扫描方式
scanning near-field optical microscopy 扫描近场光学显微镜
scanning pattern 扫描图形
scanning probe microscope (SPM) 扫描探针显微镜
scanning probe microscopy (SPM) 扫描探针显微法
scanning rate 扫描速率
scanning refractometer 扫描折射计
scanning speed 扫描速率
scanning system 扫描系统
scanning transmission electron microscope (STEM) 扫描透射式电子显微镜
scanning tunneling microprobe 扫描隧道显微镜

scanning tunneling microscope 扫描隧道显微镜
scanning tunneling microscopy (STM) 扫描隧道显微法
scanning tunneling spectroscopy (STS) 扫描隧道谱(法)
scanning tunnel luminescence (STL) 扫描隧道发光技术
scanning tunnel microscope (STM) 扫描隧道显微镜*
scanning type microscopic luster meter 扫描型显微光泽计
scanning velocity modulation 扫描速度调制
scan polarography 扫描极谱法
scan probe microscopy 扫描探针显微镜
scant 次材〔不够规格尺寸的成材或半成材〕
scan transmission electron microscope 扫描透射电子显微镜
scapolite 方柱石
scapolitization 方柱石化作用
scar 创痕；伤痕
scarf ①嵌接〔机工〕②嵌接处③斜面
scarfed plywood 斜接胶合板
scarfing 火焰表面清理
scarfing joint 斜接头
scarf joint 斜接接头；嵌接
scarlet 猩红色
scarlet chrome 铬猩红〔铬酸铅、钼酸铅和硫酸铅的混合颜料〕
scarlet fast G base (=4-nitro-o-toluidine) 猩红坚牢G色基；4-硝基邻甲苯胺
scarlet G base 猩红G色基
scarlet lake 猩红色淀
scarlet phosphorus 猩红磷〔即紫磷〕
scarlet red (=C.I.solvent red 24) 猩红
scarved joint 嵌接接头
scatol (=skatole) 粪臭素；3-甲基吲哚
scatter 散射；分散；扩散
scatter coefficient 散射系数
scatter diagram 散点图(表)
scattered beam 散射束
scattered electrons 散射电子
scattered intensity 散射强度
scattered ion 散射离子
scattered-light 散光的
scattered ray 散射线
scattered wave theory 散射波理论
scatterer 散射体
scattering ①(分)散②散射③漫散；播散；驱散
scattering angle 散射角
scattering coefficient(=coefficient of scatter) (光)散射系数
scattering cross section 散射截面
scattering diluent 散射剂

scattering dissymmetry 散射不对称性
scattering efficiency 散射效率
scattering electron 散射电子
scattering high energy electron diffraction 散射高能电子衍射
scattering intensity 散射强度
scattering light analysis 散射分析
scattering loss 散射损失
scattering particles 散射粒子
scattering pigment 散射颜料
scattering process 散射过程
scattering strength 散射强度
scattering tint(ing) strength 散射着色力
scattering unit 散射单元
scattering vector 散射矢量
scattering volume 散射体积
scavenge area 扫气(孔)面积；排气(换气)面
scavenge filter 清扫过滤器
scavenge port 换气口
scavenger 清除剂
scavenger pipe 排出管
scavenger precipitation 清除沉淀
scavenging 清除(的)；清扫的
scavenging-precipitation ion exchange process 清除沉淀-离子交换法
scavenging process 清除过程
sceleratine 硬壳汀
scene analysis 现场分析
scent ①香味②(=perfume)香料
scented soap 香皂
scenting agent 香料
scentless mayweed 淡甘菊
scent materials 香料
scentometer 气味测定仪
Schaal's indicator(=alizarin) 茜素
Schäffer's acid 2-萘酚-6-磺酸
Schäffer's pressure gauge 薛佛氏压力计
S-chamber (=sandwich chamber) S(展开)室；S(展开)槽；夹层槽
schapbachite 硫铅铋银矿
Schapper testing machine 薛珀检测器
Schardinger dextrin α-糊精
Scharples type ultracentrifuge 夏普列斯超速离心机
scheduled shutdown 计划停工
schedule of price 估价表
scheduling 计划
Scheele's green 亚砷酸氢铜 $CuHAsO_3$
scheelite(=calcioscheelite) 白钨矿

scheererite 板晶蜡
scheibeite 红铬铅矿；夏伯矿
Scheibel column 夏培尔塔〔搅拌填料塔〕
schema ①草图；流程图；设计图②方案；计划；设计③电路
schematic 图解的；示意的；概略的
schematical sideview 侧视示意图
schematic arrangement (分子结构的)示意图
schematic cutaway 剖视图；剖开(内视)图
schematic diagram 示意图
schematic drawing 简(略)图；示意图
schematic flow sheet 流程示意图
schematic illustration 示意图
schematic layout 示意图
schematic representation 略图；简图；示意图
scheme 图像；方案；设计
scheme of colo(u)r ①配色方案(设计)②着色法(绘画)
schemochrome 结构色
Scherring process for triacetate 舍林法三醋酸纤维素生产流程
schieferspar 层方解石
Schiff base 席夫碱
Schiff solution 席夫醛试液
Schiff's reagent 席夫试剂；品红试剂
Schiller and Naumann equation 希勒-瑙曼方程
schiller spar 绢石
Schilling's U-tube 西林U字管
schinus berry oil 加州胡椒油
schinus leaf oil 加州胡椒叶油
schinus oil 加州胡椒油；肖乳香油〔取自加州胡椒树 Schinus molle〕
schisandrin 五味子素
schist 页岩；片岩；板岩
schistic 页岩的
schistoid 片状岩
schistose 片状的；页状的
schistose mica 云母片岩
schistosity 片岩性
schistous rocks 片状岩；页状岩
schizandrin 五味子素
schizophrenic micellization 多重胶束化
schlieren ①不同折射度②条纹
Schlieren optical detector 司列伦光学检测器
schlieren optical system 暗线照相光学系统
schlieren optics 条纹光学
schlieren photograph 纹影照相
schlieren technique 光纹技术
Schlippe's salt 全硫锑酸钠 Na_3SbS_4

Schlotterbeck reaction 什罗特尔贝克合成酮反应
Schmidt number 施密特数
Schmidt process (for hydrogen and oxygen) 施密特(制氢氧)法
Schmidt test 施密特检胶试验
Schmoluchowski's equation 席摩勒绰夫斯基方程式
schneebergite 铁锑钙石
schneiderite 硝铵;二硝基萘;树脂炸药
Schnitzer's green ①磷酸铬绿②氧化铬绿〔别名〕
Schob pendulum tester 肖伯摆锤弹性计
Schoenemann reaction 熊曼反应
Schoenflies symbol 熊弗利符号
schoenite 软钾镁矾
Schold high-speed shot mill 肖尔德高速丸磨机
Schold shot mill 肖尔德高速丸磨机
Schöllkopf's acid 萧尔科夫酸
Schon and Shulz ebulliometer 舍恩-舒尔茨沸点升高计
Schönherr process 舒赫(制硝酸)法
school copybook paper 学校笔记本纸
Schoop hypochlorite process 朔普次氯酸盐法
Schoop process (for hydrogen and oxygen) 朔普(制氢氧)法
Schopper cloth abrasion tester 肖伯尔型织物耐磨试验仪
Schopper folding machine 肖伯尔折叠(验纸)机
Schopper folding tester 肖伯尔折叠试验机
Schopper freeness 肖氏打浆度
Schopper rebound 肖伯尔回弹性
Schopper tensile tester 肖伯尔抗拉试验仪
Schopper testing machine 肖伯尔试验机
schorl 黑电气石
schorlomite 钛榴石
schorl rock 黑电气石岩
Schott-Gen's scrubbing bottle 肖特-根洗气瓶
Schottky defect 肖特基缺陷*
Schottky field emission source 肖特基场发射电子源
schou oil 氧化豆油
schradan 八甲磷;八甲基焦磷酰胺(杀虫剂)
Schramm heat resistance 施拉姆耐热性(耐热度)
Schramm test 施拉姆试验
schraufite 喀尔巴阡山琥珀
schreibersite 磷铁镍陨石
Schreiner calender 施赖讷砑光机
schreinering 织品加密
Schrödinger atom 薛定谔原子
Schrödinger equation 薛定谔方程
Schrödinger representation 薛定谔表象
Schrödinger wave equation 薛定谔波动方程

schroeckingerite 美国铀矿
Schrotter's apparatus (=alkalimeter; kalimeter) 施罗特碳酸定量器
Schrötter carbonate determination apparatus 施罗特碳酸定量器
Schuckert cell (for hypochlorite) 舒克特(制次氯酸盐)电池
Schultz number 舒茨染料分类号
Schulze-Hardy rule 舒尔策-哈代规则*
Schulze's rule 舒尔策定则〔随离子的化合价而变异的沉淀效应〕
Schulz-Zimm distribution 舒尔茨-齐姆分布
Schumann plate 舒曼感光板
Schumann rays 舒曼射线
schungite 次石墨
Schuster gage 磁性测厚计
Schwartzschild relationship 施瓦兹希尔德关系式
Schwarz-Bezemer equation 施瓦兹-贝泽默公式〔计算乳状液的液珠大小分布的公式〕
Schwarzl's approximation formula 施瓦兹尔近似式
schwatzite 汞黝铜矿
Schweinfurth green 乙酸铜、亚砷酸铜复盐 $Cu(C_2H_3O_2)_2 \cdot 3Cu(AsO_2)_2$
Schweitzer automatic spraying equipment 施威茨尔自动喷涂装置
Schweitzeros reagent (=cupriammonium solution) 铜铵溶液
Schweitzer system 施威茨尔系统;施威茨尔自动喷涂装置
sciadinone 金松酮
sciadopitene 金松烯 $C_{20}H_{32}$
sciagraph ①X射线照相②人体辐射暗中照相
science 科学
science of flow 流动科学
science of lubrication 润滑科学
scientific 科学的
scientific research 科学研究
scillabiose 绵枣儿二糖;鼠李糖葡萄糖苷
scillain 海葱因
scillarabiose 海葱二糖 $C_{12}H_{22}O_{11}$
scillaren 海葱苷
scillin 海葱灵
scillipicrin 海葱苦
scilliroside 海葱糖苷
scillitin 海葱亭
-scine 醒〔生物碱〕
scintanalyzed reagent 闪烁分析试剂
scinticamera 闪烁照相机

scinticounting 闪烁计数〔测量放射性〕
scintigram 闪烁图*
scintigraphy 闪烁照相法
scintillant 闪烁体
scintillating agent 闪烁剂
scintillating crystal 闪烁晶体
scintillating liquid 闪烁液体
scintillating solution 闪烁溶液
scintillation 闪烁(现象)
scintillation counter 闪烁计数器
scintillation counting 闪烁计数
scintillation crystal 闪烁晶体
scintillation detector 闪烁探测器*
scintillation fission counter 闪烁裂变计数器
scintillation material 闪烁物
scintillation probe 闪烁探针
scintillation screen 闪烁屏
scintillation solution 闪烁液
scintillation spectrometer 闪烁分光计；闪烁(能)谱仪
scintillation spectrometry 闪烁光谱测定法
scintillation vial 闪烁管
scintillator ①闪烁剂②闪烁器
scintillography 闪烁图法
scintillometer 闪烁计；闪烁计数器
scintilloscope 闪烁镜
scintiphoto 闪烁照相图
scintiphotography 闪烁照相术
scintiscan 闪烁扫描图
scintiscanning 闪烁扫描*
scintitomogram 闪烁断层图*
scintitomography 闪烁断层照相法；闪烁断层显像法
scissile bond 易分裂键
scission ①裂开②切开；剪断③分离
scissionable bond 可断裂键
scission of link 断键
scissoring 剪切
scissoring vibration 剪式振动
scissor vibration 剪形振动；剪振
sclerolac 紫胶树脂
sclerometer 硬度计；肖氏硬度计
scleron 司克勒龙铝合金
scler(o)scope ①金属硬度计②肖氏硬度计
scleroscope hardness (验出的)硬度
scleroscope hardness test 肖氏硬度试验
scleroscope hardness test of Share 肖(尔)氏回跳硬度试验
scolecite 钙沸石
scolecitite 钙沸石岩
scoliodonic acid 斜齿鲨酸〔一种二十四碳五烯酸〕 $C_{24}H_{36}O_2$
scoop 勺；杓
scoop bin 多格式储料罐
scoop feed elevator 离心式提升机
scoop feeder 勺式加料器
scoop wheel 勺轮
scoop wheel cell 勺轮电解池
scoparin 金雀花素 $C_{20}H_{20}O_{10}$
scoparone 滨蒿内酯；扫帚艾酮
scopolamine 莨菪胺
scopolamine butylbromide 丁溴东莨菪碱
scopolamine hydrobromide 氢溴酸莨菪胺
scopoletin(=7-hydroxy-6-methoxycoumarin) 莨菪亭；7-羟基-6-甲氧基香豆素
scopolic acid 莨菪酸
scopolin 莨菪苷
scopoline 莨菪灵
scopomannite 莨菪甘露液〔盐酸莨菪胺的10%甘露醇水溶液，药用〕
scopometer 视测浊度计
scopometry 视测浊度测定法
scopulite 羽雏晶
scorbutanin 维生素C
scorch ①过早硫化②烧焦；焦化③烧痛
scorched earth 焦土
scorched rubber 早期硫化橡胶
scorching 焦烧；过早硫化
scorching quality 抗焦(烧)性能
scorch retarder 防焦剂
scorch safety 焦烧安全性；防焦性
scorch test 过早硫化试验；焦烧试验
scorchy accelerator 易焦烧促进剂
scorchy sensitivity 易焦烧性
score 模面伤痕
score cutter 截纸机
scored 已修整过的
scored hide 有刀伤的生皮
scored skins 修整过的皮革
scoreline 刮痕〔挤塑制品表面纵向沟纹〕
score matrix 得分矩阵
scoria 炉渣；熔渣
scoriaceous 矿渣的
scorification 铅析(金银)(法)；造渣
scorifier 试金坩埚
scoriform 渣状的
scorifying 煅烧
scoring 研伤；刮伤；划伤
scorodite 臭葱石

scotch broom　金雀花
scotch tape adhesion　透明胶带黏附〔测油墨的黏附牢度〕
Scotch tape test　(附着力)斯科奇胶带试验
scotogram　人身辐射像
scotograph　人身辐射像
scotographic　暗室显影的
scotography　暗室显影(法)
scotoma　暗点；盲点
scotometer(=perimeter)　目场计
Scott evaporator　斯科特蒸发器
Scott pendulum machine　斯科特摆锤式试验机
Scott tensilometer　斯科特拉伸试验机
Scott tester　斯科特试验机
Scott viscosimeter　斯科特黏度计
Scott volumeter　斯科特容量计
scoulerine　金黄紫堇碱
scourage　洗余水
scoured silk　熟丝
scourer　①洗刷器②擦洗剂；擦净剂；洗净剂
scouring　①擦净；洗净②侵蚀
scouring agent　擦洗剂；擦净剂
scouring cinder　侵蚀性熔渣
scouring effect　洗净效应
scouring liquor　擦洗液；擦净液
scouring machine　冲洗机
scouring pad　擦洗(纸)片
scouring rush　木贼
scouring soap　擦洗皂；擦净皂
scouring stone　冲洗石
scouring table　冲洗台
scouring trough　擦洗池
scouring water　洗水；洗液
Scove kiln　泥封窑；斯寇夫窑
scrall　螺旋管
scrambling　混乱
scrap brass　碎黄铜
scrap crepe　碎屑绉片〔橡胶〕
scrape　刮
scrape adhesion　划痕附着力
scraped film evaporator　刮(板)膜式蒸发器
scraped plywood　刮光胶台板
scraped surface heat exchanger　刮板式换热器
scrape off　刮去
scrape-off type remover　刮除型脱漆剂
scraper　①刮刀②刮板③铲运机④刮料机
scraper blade　刮板
scraper bucket　铲斗
scraper chasing　跟随刮刀
scraper conveyor　刮板式运输机
scraper flight　刮板带
scraper flight conveyer　刮板(式)运输机
scraper ring　刮油环
scraper trap　刮刀沟槽
scrap film　胶卷断片
scrap handling　废品处理
scraping　①刮削；刮面②刮工
scraping knife　刮刀
scraping method　铲刮除锈法；铲刮清洗(清理)法
scraping ring　刮环
scraping tool　刮刀
scrap iron　废铁
scrap leather　废皮；碎皮〔皮革〕
scrapless forming process　无边角料热成型法
scrappage　废物；报废(率)
scrap process　废钢法
scrap rate　废品率
scrap rubber　废胶
scrap rubber chopping　废胶切碎
scrap rubber selection　废胶分选
scrap rubber washing　废胶洗涤
scraps　金属渣
scrap soap　边角皂
scrap stock　①边角料②废料
scratch　①刮痕②微伤③划记号
scratch hardness　刮痕硬度
scratch hardness tester　划痕硬度试验仪
scratching　刮痕；划痕
scratching test　漆膜划痕试验
scratch lathe　擦光机；磨光镟床
scratch method　划痕法
scratch off　刮去
scratch oil test　油蚀性的测定
scratch polish　擦光
scratch-proofness　耐划性
scratch resistance　抗刮性
scratch resistant　耐划痕的；抗划伤的
scratch test　刮痕(硬度)试验
screen　①屏；幕②荧光屏；投影屏③隔板；屏障④筛子；筛网⑤屏栅板⑥晶格⑦滤光镜
screen analysis　筛析
screen analysis curve　筛析曲线
screen analysis test　颗粒分析试验；筛析试验
screen aperture　筛孔；筛号
screen banks　筛组
screen bottom tray　筛底盘
screen capacity　筛效率

screen centrifuge　筛式离心机
screen classifier　筛选机
screen cloth　筛布；筛网
screen decanter　筛式沉降器；过滤式沉降器
screen deck　筛板
screen dot　网点；网眼点纹
screened　①遮蔽的②筛过的
screened cushion　屏蔽垫层；遮垫
screened indicator　遮蔽指示剂
screened stock　已筛浆料；细浆
screened without pigment　滤除颜料
screen efficiency　筛选效率
screener　屏蔽剂
screen filter　筛滤器；筛滤板
screen floating　筛余物
screen grid　屏栅
screening　①筛(选)；筛分②遮蔽
screening ability　掩蔽能力；屏蔽能力
screening agent　掩蔽剂
screening area　筛孔面积；网眼面积
screening capacity　①遮蔽能力②筛选能力
screening constant　屏蔽常数
screening doublet　屏蔽双线
screening effect　屏蔽效应
screening efficiency　筛选效率；筛分系数
screening experiment　筛选实验
screening formulation　优选配方
screening machine　筛选机
screening material　屏蔽材料
screening method　筛选法
screening number　屏遮数
screening of catalyst　催化剂的筛选
screening operation　筛选
screening rate　筛选率
screenings　筛余物
screening specifications　筛选规格
screenings pulp　木节纸浆
screening surface　筛面
screening test　筛选试验
screening trommel　转筒筛
screening varnish　丝漏清漆
screening wrapping　木节包装纸
screen mat　过滤栅网
screen material　屏蔽材料
screen mesh　①筛眼；(筛)目；(筛)号；筛孔②网眼
screen mill　筛磨
screen number　筛孔数目
screen opening　筛孔

screen pack changer　换网器
screen painting　①丝漏涂漆法②丝网印刷；网版印刷
screen pipe　筛滤管；有筛滤板的管子
screen plate　筛板
screen plugging　筛网堵塞
screen printing　筛网印花；绢网印花；丝漏印刷
screen range　筛级
screen residue　筛余物；筛渣
screen separator mill　筛分磨
screen size　筛号
screen sizing　用筛分级
screen tailings　筛余物
screen test　筛析
screen-type centrifuge　筛筒式离心机
screen wall counter　屏蔽计数器
screw　①螺旋②螺钉③螺丝④螺杆
screw axis　螺旋轴
screw cap　螺丝帽
screw capper　螺丝帽机床
screw channel depth　(螺杆)螺槽深度
screw channel width　(螺杆)螺槽宽度
screw clamp　螺旋夹
screw cock　螺旋(活)栓
screw compressor　螺杆压缩机
screw compressor clamp　螺旋压管夹
screw conveyor　螺旋运输器
screw conveyor flight　螺旋运输器刮板
screw conveyor trough　螺旋运输器槽
screw conveyor type extractor　螺旋输送式萃取器
screw-cutting machine　螺纹切削机床
screw-cutting oil　攻丝润滑油；切螺丝钉用的润滑油
screw die　板牙
screw dislocation　螺旋位错
screw-downs　螺距下调
screw-driven pump　螺旋驱动泵
screw-driven syringe pump　螺旋传动注射泵
screw drum　螺旋型转鼓
screwed　攻丝的；切过的；车丝的
screwed connection　螺旋连接；螺纹接头
screwed fittings　螺旋管件
screwed flange　螺纹法兰
screwed hole　螺纹孔
screwed hose coupling　螺旋式胶管接头
screwed joint　螺纹接头；螺管接头；螺栓接头
screwed pipe　螺纹管
screwed plug　螺纹栓
screwed tube　螺纹管
screwed union　螺旋连接器；螺旋接头

screwed valve　螺旋阀
screw efficiency　螺杆(捏合)效率
screw extruder　螺杆挤出机
screw-extruder-evaporator　螺杆挤压(蒸发)机
screw extrusion　螺旋挤压
screw extrusion machine　螺杆压出机
screw extrusion press　螺杆挤压机；螺杆压榨机
screw feeder　螺杆加料器；搅龙
screw feed grease cup　润滑脂的螺旋输送器
screw flange　螺栓连接法兰
screw flight　螺杆螺纹
screw ga(u)ge　螺纹(量)规
screw hoop　环首螺栓；带环螺栓
screw-impeller pump　旋桨泵
screw-in compression　止动螺钉；紧压螺钉
screw injection machine　螺杆柱塞并用式注压机
screw injection molding machine　螺杆式注塑机
screw-in type　旋入式
screw-kneader　螺旋捏和机
screw lift (elevator)　螺旋提升机
screw machine　螺杆压出机
screw-melter-fed spinning machine　螺杆(式)纺丝机；螺杆式熔(化器喂)料纺丝机
screw micrometer　螺旋千分尺；螺旋测微计
screw mixer　螺杆式混(捏)合机
screw motion　螺旋运动
screw pelletizer　螺杆造粒机
screw pinch-cock　螺旋(节流)夹
screw pitch　螺杆螺距
screw plasticator　螺杆塑炼机
screw plug ga(u)ge　螺纹塞规
screw press　螺杆压榨机
screw-propeller　螺旋桨
screw pump　螺杆泵
screw-ram injection machine　螺杆柱塞并用式注压机
screw return pressure　螺杆回压力
screw root　螺杆中心轴
screw rotation　螺旋转动
screw snap gauge　螺纹卡规
screw speed adjustment　①螺杆速度调节②螺杆调速装置
screw spring　螺旋弹簧
screw stem　螺杆中心轴
screw structure　①螺旋结构②螺杆构造
screw tap　螺丝攻；丝锥
screw terminal　螺丝头
screw thread　螺纹
screw top　(容器的)螺旋塞(盖)

screw-top cup　有螺旋盖的加油器
screw type continuous centrifuge　螺旋式连续离心机
screw type crusher　螺旋式破碎机
screw-type extruder　螺旋(式)压出机
screw-type extrusion machine　螺杆压出机
screw-type mixer　螺旋式混和器
screw worm　螺杆
screw wrench　螺钉板手
scribbling　预梳；粗梳
scribe　①划痕②雕合；合缝
scribe and strip testing　划线剥落试验〔附着力检验法〕
scribed specimens　划线样板
scriber　画线针；画线器
scriber tester　划线法(附着力)试验器
scribing tool　划线工具
scrim　稀松无纺织物
scrion　一种均质离子交换膜
scripton　转录子
scroll　①涡管②涡卷状
scroll-cam traverse　螺旋槽筒式往复导丝装置
scroll centrifuge　螺卷轴离心机
scroll land　螺杆螺纹棱面
scroll output　(螺杆)压出量
scroop　丝鸣
scrooping feel　丝鸣感
scrub　①洗涤；擦洗；刷②洗气；洗气③灌木
scrub agent　擦洗剂；磨洗剂
scrubbability　刷洗性；擦净性
scrubbability tester　擦洗(性)试验机；刷洗(性)试验机
scrubbable　耐擦洗的；可擦洗的
scrubbed　精制的；精炼的；纯净的
scrubbed extract　洗涤过的萃取液
scrubbed gas　①纯气体②纯煤气
scrubbed solvent　洗涤过的溶剂
scrubber　①涤气器②洗涤器③擦洗粉
scrubber collector　洗涤收集器
scrubber tank(=scrubber tower)　洗气罐
scrubber tower(=scrubber tank)　洗气罐
scrubbing　①洗涤；擦洗；刷②洗气；洗气
scrubbing action　擦洗作用；刷洗作用
scrubbing agent　擦洗剂；刷洗剂
scrubbing bottle　洗气瓶
scrubbing dust collection　洗涤除尘
scrubbing effect　洗涤效果
scrubbing filter　涤气过滤器
scrubbing oil　洗涤油；涤气油
scrubbing section　洗涤段；洗涤区
scrubbing soap　擦洗皂

scrubbing stage　洗涤级
scrubbing tower　涤气塔
scrub board(=base board)　基线板；踢脚板
scrub column　洗涤柱
scrub oak　胭脂栎
scrub raffinate　洗余液；洗涤废液
scrub-resistance test　抗湿磨试验
scrub soap　擦洗皂
scrub solution　洗涤液；萃洗剂
scrub test　洗涤试验；擦洗试验
scruff　表面污垢
scruple　①极微之量②20谷〔药剂衡量单位〕
scud　白云岩
scudding　推挤
scudding knife　切纸刀；推挤刀〔皮革〕
scuff　磨损；磨损处
scuffing　碰擦；磨损
scuff mark　擦伤痕迹
scuff resistance　耐擦伤性
scull　桶结壳
scullcap　黄芩
scum　①浮渣②吐渣③浮膜；表面生长
scumble　涂暗色
scumble glaze　①辉映罩光②辉映抛光
scumbling　涂暗色
scum collecting bin　浮渣(泡沫)收集器(槽)
scum dredger　①废漆渣清除机②除沫器
scummer　撇(浮)渣杓
scumming　①起泡沫；形成浮渣②除去泡沫〔浮渣〕
scummy　浮渣
scum pipe　吐渣管；排渣管
scum rubber　泡沫橡胶
scurf　皮垢
scurvy grass　辣根菜
scutcher　打麻机
scutellarin　黄芩素
sea basin　海水池
sea-bed disposal　海床处置(法)；(投入)海底处置
Seabord process of gas purification　西鲍尔特气体纯化法
sea brine　盐海水
sea buchthorn　沙枣
sea buchu　海布枯
sea cabbage　甘蓝
sea coal　海运煤
sea disposal　海洋处置·
seafennel oil　海茴香油
seagoing pipe line　海底油管

sea island cotton　海岛棉
seal　封接；密封
sealability　密封能力
sealable tank　密封罐
seal adaptability　密封适应性
sealant　密封胶；密封腻子；密封材料；防渗漏剂
sealant tape　密封胶带
seal box　①沉箱②气压箱
seal cement　固封接合剂
seal coat　①封闭底漆②封闭层
seal coating　密封涂层
sealed cavity　密封空腔
sealed cooling　闭式冷却法
sealed distillator　密封蒸馏器
sealed-in　封闭的
sealed package　封闭包装
sealed source　密封源·
sealed surface　密封表面
sealed tube　封闭管
sealed tube furnace　封闭管炉
sealed-tube method　封管法
sealer　封闭器
sealer coat　封闭层
sealer for knot　木节封闭剂
sea level　海平面
seal hanger　封液的悬浮体〔浮顶〕
sealing　熔接；封闭；密封；热合
sealing adhesive　密封胶黏剂
sealing bar　熔接棒
sealing compound　密封接合
sealing connection　密封接头
sealing effect　密封效应；封闭效应
sealing element　封闭体；封闭物
sealing filler　封闭填孔剂；封闭填料
sealing flange　密封法兰，密封凸缘
sealing fluid　封闭液体
sealing force　密封力
sealing gasket　密封垫
sealing-in　焊接；熔接
sealing joint　密闭接头
sealing leak　密封泄漏
sealing leg(=seal leg)　(催化剂的)汽压升降器
sealing liquid　封闭液体
sealing machine　封口机；压盖机
sealing medium　密封介质
sealing-off　脱焊；开焊
sealing of metal coating　金属涂层的密封性
sealing ply　密封层

sealing property   密封能力
sealing ring   密封环
sealing run(=backing run)   ①(焊接)焊根②背焊
sealing surface   密封面
sealing test   封闭性实验
sealing treatment   封闭处理
sealing wax   火漆
sea-lion oil   海狮油
seal leg(=sealing leg)   (催化剂的)汽压升降器
seal off   封离;封闭;密封
seal off stain   封闭底色(工序);底色封闭
seal oil   海豹油
seal point   液封点〔塔板〕
seal pot   受液盘;密封釜
seal ring   封口圈
seal skin   海豹皮
sealskin finish   海豹皮面饰
seal tank   气压箱
seam   缝
seam allowance   缝份
seamed pipe   有缝管
seam flammability   接缝耐燃性
seam-free product   无缝制品
seamless   无缝的
seamless article   无缝制品
seamless pipe   无缝管
seamless steel pipe   无缝钢管
seamless tube   无缝管
sea moss(=Irish moss)   爱尔兰苔;海苔
seam sealer   填缝料
seam welding   缝焊
sea of instability   不稳定海
sea-port bulk station   港埠(石油)仓库
Searle viscometer   塞尔(旋转圆筒式)黏度计;旋转圆筒式黏度计
sea salt   海盐
sea-salt fine particle   海盐微粒
seaside atmosphere   海滨大气
seaside atmospheric corrosion   海滨大气腐蚀
sea soap   海水皂
season   ①晒干②调味③季节
seasonal balancing   按季节温度选择汽油组成;季节平衡
seasonal climatic conditions   季节性气候条件
seasonal grades   季节级别
seasonal variation   季节性变化
season cracking   ①老化开裂;陈化开裂②干裂
seasoned   ①已晒干的②调了味的
seasoned wood   风干(木)材;风化材

seasoning   ①涂光②调味③储放④风干;风化
seasoning extract   调味浸剂
seasoning machine   上光机
seasonless   未平整的
seat cushion   座垫
seating time   磨合时间;跑合时间
seat of valve   阀(盘)座
seat stays   立叉
seawater bronze   耐蚀青铜〔铜、镍、锡、锌、铋的合金〕
sea water corrosion   海水腐蚀
sea water desalination plant   海水淡化装置(设备)
sea water detergent   海水洗涤剂
sea water immersion test   海水浸渍试验
sea water immersion test of paint film   漆膜浸海水试验
sea water jet test   海水喷射试验
sea water pump   海水泵
sea water resistance   耐海水性
sea water soap   海水皂
sea water washing   海水洗涤
sea water zone   海水部位
sea wax   海蜡
seaweed   海藻;海草
seaweed char(=seaweed charcoal)   海藻煤
seaweed charcoal(=seaweed char)   海藻煤
sea weed corrosion   海藻腐蚀
seaweed fiber   海藻纤维
seaweed gel   海藻胶
seaweed wax   海藻蜡
sea-wolf liver oil   海狼肝油
sea wormwood   海滨蒿
seaworthy packing   耐航包装
sebacate   癸二酸盐(酯) $MOOC(CH_2)_8COOM$
sebacic acid   癸二酸 $(CH_2)_8(CO_2H)_2$
sebacic acid ester plasticizer   癸二酸酯增塑剂
sebacic dihydrazide   癸二酸二酰肼
sebacic dinitrile   癸二腈 $(CH_2)_8(CN)_2$
sebaconitrile   癸二腈
sebacoyl   癸二酰 $—CO(CH_2)_8CO—$
sebacoyl chloride   癸二酰氯
sebate   ①癸二酸盐 $COOM(CH_2)_8COOM$②癸二酸酯 $ROOC(CH_2)_8COOR$
sebiferic acid   乌桕酸
sebum   皮脂
sec   仲〔1.指 CH_3……$CH(CH_3)$—型支链烃基;2.指二元胺及 R_2CHOH 型的醇〕
secaline   黑麦碱;三甲胺 $(CH_3)_3N$
secalonic acid   黑麦酮酸 $C_{14}H_{14}O_6$
SE-cellulose(=sulfoethyl cellulose)   磺乙基纤维素

sechometer (手摇)感应电机
seco- 〔用作环系命名的系统词头〕断
seco alkylation 断裂烷基化*
secobarbital 司可巴比妥
seco-ebruicolic acid 闭联齿孔酸
second ①第二(的)②秒(钟)
second absorption band 第二吸收(谱)带
secondary ①仲〔指 $CH_3……CH(CH_3)$ 一型支链烃基或指二元胺及 R_2CHOH 型的醇〕②第二的③次级的；次等的④副的；次要的⑤副(线)圈⑥二代〔无机盐，酸中有两个酸式 H 被取代〕
secondary aberration 第二级像差
secondary acetate 二代乙酸盐
secondary acid 二级酸
secondary activation agent 辅助活化剂
secondary additive 辅助配合剂
secondary adsorption 第二类吸附；物理吸附
secondary air 二次空气；二次风
secondary alcohol 仲醇
secondary aliphatic amine 脂肪族仲胺
secondary amine 仲胺
secondary amine value 仲胺值
secondary amino group 仲胺基
secondary amyl 仲戊基
secondary arsenate(=secondary arseniate) 二代砷酸盐 M_2HAsO_4
secondary arseniate(=secondary arsenate) 二代砷酸盐
secondary basic accelerator 碱性促进助剂
secondary bath (第)二浴
secondary battery 二次电池(组)；蓄电池(组)
secondary bond ①二次键合；二次结合②次价键
secondary bonding force 二次键力
secondary breaker 二级分碎机
secondary butyl alcohol 仲丁醇〔溶剂〕
secondary calcium phosphate(=dibasic calcium phosphate) 二代磷酸钙；磷酸氢钙
secondary carbon 仲碳原子
secondary carbon atom 仲碳原子
secondary cell 二次电池；蓄电池
secondary cellulose acetate 二醋酸纤维素
secondary circuits ①次要电路②二次回路
secondary coil 副(线)圈
secondary color 复色；化合色；调和色〔指绯红、黄和浅蓝〕
secondary combustion 次级燃烧
secondary combustion zone 次级燃烧区
secondary contamination 二次污染
secondary coulometric titration 二级库仑滴定

secondary coulometry 次级库仑法
secondary cracking 二次裂化；补充裂化
secondary creep 第二阶段蠕变；蠕变恒速区
secondary crusher 中碎机
secondary crushing 二级压碎
secondary crystallinity 次级结晶度；后期结晶度
secondary crystallization 后期结晶
secondary decomposition 再度分解
secondary diamine 仲二胺
secondary dilution solvent 二次稀释溶剂
secondary discharge 次级放电
secondary drier 次催干剂
secondary effect ①次步效应②副效应；附带效应
secondary electrolysis 次级电解
secondary electron 二次电子
secondary electron detector 次级电子检测器
secondary electron emission 二次电子发射
secondary electron image 次级电子像
secondary electron multiplier 二级电子放大器
secondary emission ①二次排放②二次发射
secondary extinction ①二次消光②二次衰减
secondary extraction oil 二次萃取油
secondary factor 次级因素
secondary filter ①二次过滤；精密过滤②精滤器
secondary flow (第)二次流动；次生流
secondary focusing 二级聚焦
secondary force 次级力；范德华力
secondary fractionator 二级分馏塔
secondary fuel filter 燃料精滤器
secondary gas 辅气
secondary heater 第二加热器；副加热器
secondary heating circuit 辅助热油回路
secondary hexyl alcohol 仲己醇
secondary hydroxyl ①仲羟基②仲醇
secondary inertia force 二次惯性力
secondary ion 次级离子*
secondary ion bombardment 二次离子轰击
secondary ion mass spectrography (SIMS) 二次离子质谱法
secondary ion mass spectrometer 二次离子质谱仪
secondary ion mass spectrometry (SIMS) 次级离子质谱法
secondary ion mass spectroscopy (SIMS) 次级离子质谱(法)
secondary ion mass spectrum 二次离子质谱
secondary ion microanalyzer 二次离子微量分析器
secondary ion optics 二次离子光学系统
secondary isotope effect 二级同位素效应*
secondary linkage(=weak linkage) 二级键；弱键
secondary messenger 第二信使

secondary metal　次生金属
secondary meter　二次仪表
secondary mineral　次生矿物
secondary minimum　次极小
secondary neutral mass spectrometry　二次中性粒子质谱法
secondary nitroparaffin　仲硝基烷
secondary nucleation　二次成核
secondary octyl alcohol　仲辛醇〔消泡剂〕
secondary oil recovery　二次采油
secondary particle　二级粒子；次级粒子
secondary particulate　二次颗粒物，二次气溶胶
secondary phosphine　仲膦
secondary plasticizer　次级增塑剂
secondary pollutant　二次污染物
secondary pollution　二次污染
secondary polymer　二级聚合物
secondary polymerization reaction　二级聚合反应；次级聚合反应；后期聚合反应
secondary process　次级过程*
secondary product　①二步产物；第二产品②次要产物
secondary radiation　次级辐射
secondary ray　次级射线
secondary reaction　①副反应②次级反应
secondary reaction zone　次级反应区
secondary recycle isobutane　再循环异丁烷
secondary reduction　次级还原(作用)
secondary reference fuel　副校准燃料
secondary reference material　二级标准物质
secondary relaxation　次级弛豫*
secondary relaxation temperature　次级弛豫温度
secondary salinization　次生盐渍化
secondary salt　①(=dibasic salt)二代盐②副盐
secondary salt effect　副盐效应
secondary sample　二次样品
secondary sampling unit　二次取样装置
secondary seal　二级封闭
secondary sedimentation　二次沉淀
secondary sedimentation basin　二次沉淀池
secondary sewage treatment　二次污水处理；污水二级处理
secondary solvent　二级溶剂；辅助溶剂
secondary solvent effect　次级溶剂效应
secondary spectroscopic standard　二级光谱标准
secondary spectrum　二级光谱
secondary stabilizer　二级稳定器
secondary standard　二级标准*
secondary standard of measurement　副计量基准

secondary standard of wavelength　次级波长标准
secondary structure(=transient structure)　暂时结构；二次结构〔炭黑〕
secondary sulphide ore　次生硫化矿
secondary surface analyzer　二级表面分析器
secondary surface preparation　二次表面处理
secondary tar　次级焦油
secondary-tertiary alcohol　仲叔醇
secondary transition　次级转变
secondary treatment　二次处理
secondary valence　副价*
secondary valence bond　次价键
secondary valence force　次化合价力；副价力
secondary valency　次(化)合价；副价
secondary wave　次波
secondary white reference　副校准参比标准白色；副比照标准白色
secondary X-ray　次级 X 射线
secondary X-ray fluorescence　次级 X 射线荧光
secondary X-ray fluorescence spectrometry　次级 X 射线荧光光谱法
second axiom of rheology　流变学第二公理
second bath　(第)二浴
second carbonation　二次碳酸饱和作用
second chemical equilibrium　二次化学平衡
second class grade chemical　二级试剂
second CMC　第二临界胶束浓度
second derivative potentiometric titration　二阶导数电位滴定
second derivative potentiometric titration curve　二阶导数电位滴定曲线
second derivative spectroscopy　二阶导数光谱学
second derivative spectrum　二次导数光谱；二次微分谱
second dimension　第二(展开)方向
second evaporator　二次蒸发器
second exitation　二次脱离
second-generation　第二代
second-grade resin　二等树脂
second-growth einsteinium　次生锿
second-growth isotope　次生同位素
second harmonic AC voltammetry　二阶谐波交流伏安法*
second harmonic alternating current polarography　二阶谐波交流极谱法
second harmonic alternating current voltammetry　二阶谐波交流伏安法
second ionization constant　次级电离常数
second ion mass spectroscopy　二次离子质谱法
second kind oxidation-reduction electrode　第二类氧化还

原电极
second law of absorption 光吸收第二定律
second law of thermodynamics 热力学第二定律*
second level ionic line 二级离子线
second level ionization 二级电离
second messenger 第二信使
second moment 二次矩
second Newtonian region 第二牛顿区
second order crystallization 二级结晶(作用)
second-order differential titration 二次微分滴定
second-order effect 二阶效应；次级效应
second-order elasticity 二阶弹性
second-order fluid 二阶流体
second order ionization potential 二级电离电势；二级电离电位
second-order microfluid 二阶微观流体
second order moment 次级矩
second order nucleation 二级成核作用
second order phase change 二级相变
second order phase transition 二级相变*
second order rate constant 二级(反应)速率常数
second order reaction 二级反应*
second order reduced density matrix 二阶约化密度矩阵
second order spectrum 二级图谱*
second-order stress 二阶应力
second-order temperature 二级转变温度；玻璃化温度
second-order tensor 二阶张量
second-order thermomechanical disturbance 二阶热力学扰动
second order transition (第)二级转变
second order transition point 二次转变点；二级转变点
second-order transition temperature 二级转变温度；玻璃化转变温度
second order viscoelastic fluid 二阶黏弹流体
second order viscoelasticity 二阶黏弹性
second order viscoelastic theory 二阶黏弹理论
second-order viscous effect 二阶黏性效应
second painting 二次涂漆
second patent flour 二级专利粉
second press 二榨油
second pyrolysis 次级热解
second-stage treatment 二级处理
second stop-clock 停表；计秒表
second surface 背面
second twist 复捻
second Virial coefficient 第二维里系数
second zone coagulation 第二酸(度)区凝固〔胶乳〕
sectional 分级的；分区的；分段的

sectional fibre populousness 截面纤维稠密度；纤维填充密度
sectional iron 型铁
sectional mold 镶合塑模
sectional repair 局部修补
sectional titration 分级滴定
sectional width (轮胎的)断面宽度
section cutter 切片机
section hopper 分层储料斗
section modulus 断面模量
sections 型铁
section steel 型钢
section topography 截面形貌学
sector 扇形片(板)
sector disc 扇形盘
sector field direction focusing 扇形场方向聚焦
sectorial pipe-coupling method 分段联接管路
sector instrument 扇形场仪器
sector magnetic analyzer 扇形磁分析器
sector mass spectrometer 扇形质谱仪
sector method 分区法
sector mirror 扇形镜
sector-shaped rotor 扇形转子
sectrometer 真空管滴定计
secular 长期的；久期的
secular aberration 长期光行差〔天文〕
secular broadening 久期增宽
secular equation 久期方程*
secular equilibrium 久期平衡
secular motion 特征运动
secular variation 时效变化；长期变化
secunda oil 页岩太阳油
securite 安全炸药
security valve 安全阀
sedamine 镇静胺；景天胺
dl-sedamine dl-景天胺
sedanoic acid 瑟丹酸 $C_{12}H_{20}O_3$
sedanolide 芹菜子交酯；瑟丹内酯 $C_{12}H_{18}O_2$
sedanonic acid 瑟丹酮酸
sedanonic anhydride 瑟丹酸酐 $C_{12}H_{18}O_3$
sedan paint 轿车漆
sedatin N-戊酰对乙氧基苯胺
sedative ①镇静剂②镇定的
sediment 沉降物；沉积物
sedimental(=sedimentary) 沉淀的
sedimentary 沉积的
sedimentary clay 沉积泥
sedimentary deposits 沉积物

sedimentary rock 沉积岩；水成岩
sedimentation 沉降*
sedimentation analysis 沉降分析法〔土壤〕
sedimentation average molecular weight 沉降平均分子量
sedimentation balance 沉降天平
sedimentation basin 沉淀池
sedimentation boundary 沉降界面
sedimentation centrifuge 离心沉降机
sedimentation chamber 沉淀槽
sedimentation coefficient 沉降系数*
sedimentation constant 沉降系数
sedimentation equilibrium 沉降平衡*
sedimentation equilibrium method 沉降平衡法*
sedimentation field flow fractionation 沉降场流分离
sedimentation fractionation 沉降分级
sedimentation glass 沉淀杯
sedimentation method 沉降法
sedimentation number 沉降值
sedimentation of coke 焦炭沉积
sedimentation of fuel 燃料沉积
sedimentation of suspension 乳胶体沉积；悬浮物的沉积
sedimentation potential 沉降势*
sedimentation rate 沉积(速)率
sedimentation (separation) technique 沉降(分离)技术
sedimentation tank 沉积槽；沉淀池
sedimentation-type centrifuge 沉降式离心机
sedimentation value 沉降值
sedimentation velocity 沉降速度
sedimentation velocity method 沉降速度法
sedimentator 沉淀器；离心器
sediment box 沉积箱
sediment bulb 沉降器；沉淀器
sediment chamber 沉积室
sediment collector 沉积收集器
sediment-filled phase 沉积物填充相；沉积相
sedimenting boundary 沉降界面
sedimentograph 沉降测定器
sedimentometer 沉降仪
sedimentometric analysis 沉淀体积分析
sediment pan 沉积盘；沉淀池
sediment sampler 底泥采样器，沉积物采样器
sediment sampling 底泥采样
sediment tank 澄清槽〔橡胶〕
sediment transport 沉积物运移
sediment trap 沉淀池
sediment volume 沉降体积
sedimetry 沉降分析法
sedond cut file 二号粗锉；中锉

sedulene 瑟杜烯
sedulone 瑟杜酮
Seebeck coefficient 塞贝克系数；热电系数
Seebeck effect 塞贝克效应；热电效应
seed ①晶种②(种)子
seed-blanket reactor 点火区-再生区反应堆
seed cotton 子棉
seed crystal 晶种；籽晶
seed disinfectant 种子消毒剂
seed drier 谷物干燥机
seeded crystallization 接种结晶
seeded emulsion polymerization 种子乳液聚合
seeded miniemulsion polymerization 种子细乳液聚合
seed fat 植物油
seed fibre 种子毛纤维
seed grain 结晶母粒
seed hair 种子纤维
seed huller ①去皮机②去谷机
seeding 播种；放入晶种
seeding polymerization ①种子聚合②接种聚合(作用)
seed kernel oil 谷仁油
seed lac (颗)粒(紫)胶
seed mixer ①晶种槽②种子混合器
seed nuclei 晶种核
seed oil 种子油
seed pod fiber 籽皮纤维
seed polymer 种子聚合物
seed polymerization 种子聚合
seed sower 播种机
seeds type 雏型；种子型
seed tank 晶种桶
seed wool 棉绒
seed(y) finish 麻面；麻脸〔漆病〕
seepage 渗漏
seepage basin 渗透池
seepage-proofing 防渗
seepage rate test 渗出速率试验
seepage spring 渗出泉
seepage trench disposal 渗出沟处置(法)
seep in 渗入；漏入
seeping 渗出
seep oil 渗出石油〔在油母页岩矿藏中的石油〕
Segas process 从渣油制造高热量气体的过程
Seger cone 西格(示温熔)锥
seggar ①火泥②火泥箱
segmental ①链段的②扇形的；片段的
segmental belt 活节运输带
segmental Brownian motion 链段布朗运动

segmental bush　拼合式轴衬
segmental flow　链段流动；链段滑移
segmental friction factor　链段摩擦因子
segmental jump　分段跳变
segmental jump frequency　链段跃迁频率
segment(al) mobility　链段流动性
segmental motion　链段运动
segment analysis　分部分析
segment anisotropy　链段各向异性
segment-arc spinneret　弧形组合喷丝板
segmentation　节段法
segment copolymerization　嵌段共聚(作用)；链段共聚(作用)
segmented continuous flow analysis　间隔式连续流动分析
segmented copolymer　多嵌段共聚物*
segmented polyether urethane　医用聚氨酯
segmented polymer　链段聚合物
segment interaction　链段相互作用
segment-interaction parameter　链段间作用参数
segment long spacing　片段长间距
segment motion　链段运动
segmentor　间隔分段器；相分隔器
segment rotation　链段旋转
segmer　链段
segregated oil(=separated oil)　析得干性油
segregation　分离；离析；分聚
segregation gas chromatography　分离气相色谱(法)
sehta　辉钴矿
Seidel density　赛德耳黑度
Seidel density method　赛德耳黑度法
Seidel transformation　赛德耳换值
Seidlitz powder　塞德利茨粉〔缓泻剂的一种〕
seifert solder　铝用焊药
Seignette salt　酒石酸钾钠　$KNaC_4H_4O_6$
seimple elixir　香药酒
seismic intensity　地震烈度
seismic load　地震载荷
seismic wave　地震波
seismograph　地震仪
seismology　地震学
seismometer　地震仪
seismoscope　地震仪；地震器；地震示波仪；地震记录仪
seizure　胶住；咬住
seizure delay method　延迟胶住法〔试验润滑油在轴承中的胶住性能〕
seizure of the threads　螺纹卡住；螺纹咬死
seizuring load　胶住载荷
seizuring pressure　胶住压力
sekikaic acid　石花酸
sekisamin　瑟奇萨明；二氢石蒜碱.
sekisanine　瑟奇萨宁；二氢石蒜碱
sekisanolin　瑟奇萨脑灵；二氢石蒜脑灵
selachoceric acid　鲨蜡酸
selacholeic acid　鲨油酸
selachyl alcohol　鲨油醇
selagite　黑云粗面岩
Selas furnace　西拉斯炉
Selas gas burner　西拉斯煤气炉
Selas process　西拉斯(公司)法〔管式炉裂解生产烯烃法〕
selectance　选择度；选择系数
selected area diffraction　选区电子衍射
selected area electron diffraction　选区电子衍射
selected crude oil　挑选过的原油；特定原油
selected finish　特定面漆；选(择)定的面漆
selected ion monitoring (SIM)　选择离子监测
selected ordinate method　选择坐标法；图解分析法；选择波长法
selected reaction monitoring (SRM)　选择反应监测
selection　选(择)；挑(选)
selection check　抽查
selection criteria　选择准则(判据)
selection reference　选择基准
selection rule　选择定则
selection rule for atomic spectrum　原子光谱选择定则
selection theory　选择学说
selection valve　选择阀
selective absorption　选择性吸收
selective aggregation　选择聚集
selective alkylation　选择性烃化反应
selective anodic process　选择性阳极过程
selective azeotropic distillation　选择性共沸蒸馏
selective bottom-up　自下而上选择；自底向上选择；倒选
selective catalytic conversion　选择催化转化
selective catalytic cracking　选择催化裂化
selective catalytic hydrogenation　选择性催化加氢
selective chlorination　选择氯化(法)
selective combustion　选择燃烧
selective control chart　选择控制图
selective corrosion　选择性腐蚀
selective cracking (process)　选择裂化；多炉裂化；分别裂化
selective crude　选择的原油；特定原油
selective cupping　伐前选采(脂)法
selective decoupling　选择性去耦
selective degradation　选择性降解

selective destructive hydrogenation 选择破坏加氢
selective detection 选择性检测
selective detector 选择性检测器
selective diffusion 选择性扩散
selective displacement 选择性顶替；选择性取代
selective electrode 选择电极
selective elution 选择性洗脱
selective entropy ①选择熵②选择的平均信息量
selective epitaxial growth 选择性外延生长
selective evaporation 分馏；精馏
selective extractant 选择性萃取剂；特效萃取剂
selective extraction 选择性提取
selective finishing 选择精制；选择性修饰
selective flotation 选择性浮选
selective four-coil unit 选择四炉裂化装置
selective hydrocracking 选择加氢裂化
selective hydrogenation 选择加氢
selective hydrolysis 选择性水解；优先水解
selective hydroperoxidation 选择性氢过氧化(作用)
selective hydrotreatment 选择加氢处理
selective indium halide flame emission detector 选择性卤化铟火焰发射检测器
selective interchangeability 选择互换性
selective ion detection 离子选择检测
selective ion detector 离子选择检测器
selective ion exchange resin 选择性离子交换树脂
selective ion permeability 离子选择渗透性
selective ion-sensitive electrode 选择性离子敏感电极
selective iron leaching 选择性浸铁法；优先浸铁(作用)
selective irradiation 选择辐射
selective localization 选择性定位
selective mixed-phase cracking 选择混合相裂化
selective modulation 选择调制
selective optrode membrane 选择光极膜
selective oxidation 选择性氧化；分别氧化
selective permeation 选择性渗透
selective phase 选择(性)相
selective polarization transfer 选择性极化转移
selective polymer 选择聚合物
selective polymerization 选择聚合
selective polymerization process 选择聚合过程
selective population inversion 选择性布居数反转
selective precipitation 选择沉淀
selective reaction 选择反应
selective reagent 选择(性)试剂*
selective rectification 选择分馏
selective reduction 选择性还原作用
selective reduction process for nitric acid 选择性还原法制硝酸
selective reflection 选择性反射
selective rule 选择规则
selective sensitivity 选择性灵敏度
selective solvent 选择性溶剂
selective solvent extracted oil 选择溶剂提取油
selective solvent extraction 选择溶剂萃取
selective solvent process 选择溶提法
selective-solvent-refined oil 选择溶剂精制油
selective sorbent 选择性吸着剂
selective spin decoupling 选择性自旋去偶
selective stationary phase 选择性固定相
selective stripping 选择(性)反萃(取)
selective test 选择性试验
selective transmission 选择性透射
selective treatment 选择加工；选择处理
selective uniaxial orientation 选择性单轴取向
selective uniplanar orientation 选择性单面取向
selective volatilization 选择性挥发
selective wetting 选择润湿
selectivity 选择性*
selectivity coefficient 选择系数
selectivity constant 选择性常数
selectivity control 选择性调节；取代调节
selectivity curve 选择性曲线
selectivity factor 选择性因子
selectivity index 选择性指数
selectivity of catalyst 催化剂的选择性
selectivity ratio 选择性比
selectoforming 选择重整
selector 选择器
selector range 选择器量程
selector valve 选择器阀
selectrode 选择性电极〔一种多孔石墨棒型的多用离子选择性电极〕
selegiline 司来吉兰
selenanthrene 硒士林；9,10-二硒杂蒽
selenate 硒酸盐 M_2SeO_4
selenate radical 硒酸根
selenazole 硒唑
selenazoline 硒唑啉
selenic ①四价硒的；(正)硒的②六价硒的
selenic acid 硒酸
selenic chloride 四氯化硒 $SeCl_4$
selenide ①硒化物②硒醚 M_2Se
selenide of silver 一硒化二银 Ag_2Se
selenine hydrochloride 盐酸瑟勒宁
seleninic acid (有机基)亚硒酸；硒代亚磺酸

selenino 硒宁基；亚硒酸一酰基；亚硒酸基 (HO)OSe—
seleninyl 亚硒酰·OSe=
selenious(=selenous) ①二价硒的；亚硒的②四价硒的
selenious acid 亚硒酸 H_2SeO_3
selenious acid anhydride 亚硒(酸)酐；二氧化硒 SeO_2
selenious chloride 氯化亚硒 Se_2Cl_2
selenious oxide 二氧化硒 SeO_2
selenious yeast 硒酵母
selenite ①亚硒酸盐②透(明)石膏
selenite radical 亚硒酸根
selenitum 透(明)石膏；玄精石
selenium 硒〔34号元素，化学符号 Se〕
selenium cadmium pigment(=cadmium red) 硒镉颜料；镉红
selenium cell 硒电池
selenium chloride 氯化硒〔总名，计有：一氯化硒 Se_2Cl_2 和四氯化硒 $SeCl_4$〕
selenium compound 硒的化合物
selenium dehydrogenation 硒脱氢作用
selenium dialkyl (二)烷基硒 R_2Se
selenium dibutyldithiocarbamate 二丁基二硫代氨基甲酸硒〔促进剂〕
selenium diethyl 二乙硒 $(C_2H_5)_2Se$
selenium diethyl dithiocarbamate 二乙基二硫代氨基甲酸硒
selenium dimethyl 二甲硒 $(CH_3)_2Se$
selenium dioxide 二氧化硒 SeO_2
selenium dipropyl 二丙硒 $(C_3H_7)_2Se$
selenium ethide 二乙硒 $(C_2H_5)_2Se$
selenium hexafluoride 六氟化硒 SeF_6
selenium methide 二甲硒 $(CH_3)_2Se$
selenium nitride 氮化硒 $Se_2N_2; Se_4N_4$
selenium oxide 氧化硒 $SeO_2; SeO_3$
selenium oxychloride 二氯氧化硒
selenium photocell 硒光电池
selenium photoelectric cell 硒光电池
selenium propyl 二丙硒 $(C_3H_7)_2Se$
selenium rectifier 硒整流器
selenium ruby glass 硒红玻璃
selenium specific detector 硒专用检测器
selenium sulfide 一硫化硒 SeS
selenium trioxide 三氧化硒 SeO_3
seleno 硒基 —Se—
seleno-acid ①硒代磺酸②硒代酸
selenocyanic acid 硒代氰酸
selenocyano 氰硒基 N≡CSe—
selenocystathionine 丙氨酸丁氨酸硒醚；胱硒醚
selenocystine 硒代胱氨酸
selenoid 螺线管
selenol 硒醇
selenole 苯并硒二唑
selenomercaptan 硒醇 RSeH
selenomethionine 硒代蛋氨酸
selenomycin 月霉素
selenone 硒(代)砜
selenonic acid (有机基)硒酸；硒代磺酸
selenonium 硒
selenonium rectifier 硒整流器
selenono 硒酸一酰；硒羧基 $HO_3Se—$
selenonyl 硒酰基 $O_2Se=$
selenophen 硒吩
selenophenol 苯硒酚 C_6H_5SeH
selenophosphate additive 硒代磷酸盐添加物
selenophthalide 硒代苯酞
selenopyrimidine 二嘧啶硒
selenopyronine 硒呫吨
selenosemicarbazide 氨基硒脲
selenourea 硒脲 NH_2CSeNH_2
selenous(=selenious) ①二价硒的；亚硒的②四价硒的
selenous acid 亚硒酸
selenoxanthone 硒代呫吨酮
selenuretted 含硒氢的
selenyl ①氢硒基 HSe—②氧硒基 =SeO
selenylation 硒化*
selenyl chloride 二氯氧化硒 $SeOCl_2$
self-absorption 自吸收
self absorption and self reversal 自吸和自蚀*
self absorption broadening 自吸变宽；自吸展宽
self-absorption effect 自吸效应
self-acting lubricator 自动润滑器
self-acting scale 自动秤
self-acting seal 自紧密封
self-acting thermostat(=automatic thermostat) 自调恒温器；自动恒温器
self-acting valve 自动阀
self-actor 自动机
self-adapting 自适应的
self-adhesion 自黏(作用)；自黏力
self-adhesive ①自黏型黏合剂②自黏的
self-adhesive backing 自黏衬里
self-adhesive label 自黏标签；不干胶标签
self-adhesive tape 自黏带
self-adjusting 自动调整
self-adjusting pipette 自动吸移管
self-adjusting weigher 自动秤量器

self aging 自然老化
self-aligning bearing 自校式轴承
self-alignment 自动排列成行
self-alkylation 自烷基化作用
self-assembled film 自组膜
self-assembled monolayer 自组装单分子层
self-assembled monolayer membrane 自组装单分子膜
self assembling membrane 自组膜
self assembling technique 自组装技术
self-assembly 自组装
self-assembly monolayer modified electrode 自组装膜修饰电极
self-association and self-dissociation 自动解离
self-avoiding walk chain 自回避行走链
self avoiding walk model 自避随机行走模型
self-balancing recorder 自平衡记录仪
self-balancing system 自平衡系统
self-basifying 自调碱度(的)
self-beam tray 自身梁式塔板
self-beat spectroscopy 自拍谱法
self-blowing drum filter 自吹鼓式过滤器
self-bodying 自稠化
self-boil 自沸
self-bonded fibre 自黏合纤维
self-bonding coating 自黏结涂层
self-bonding material 自黏结材料
self-calibration 自动校准
self-catalyzed reaction 自催化反应
self-charring plate 自炭化板
self-check 自检；自动检验
self chemical ionization 自(身)化学离子化
self-cleaning 自动清洗
self-cleaning centrifuge 自动清洗离心机；自动排渣离心机
self-cleaning cyclone separator 自洁式旋风分离器
self-cleaning effect 自动净化效应；自净作用；自洁效应
self-cleaning filter 自洁式过滤器〔具自洁作用的滤清器〕
self-cleaning grizzly 自动清洗栅
self-cleaning oil filter 自动清洗的油过滤器
self-cleaning paint 自(清)洁型涂料
self-cleaning twin screw extruder 自净式双螺杆挤出机
self-closing gauge seal 测量表的自动关闭盖
self color 本色；单色〔运用一种颜色的浓淡明、暗组成的和谐色调〕
self-colored material 自着色材料
self colour (=nature colour) 天然色；本色；单色
self-combustible 自燃的
self-combustion 自燃
self-compensating 自动补偿
self-condensation ①自冷凝②自缩合作用
self-condensing vinyl polymerization 自缩合乙烯基聚合
self-consistent charge and configuration (SCCC) 自洽电荷和构型
self-consistent field (SCF) 自洽场
self-consistent field scattered wave(SCFSW) technique 散射波自洽场法
self-consistent field scattered wave $X\alpha$ method (SCF-SW-$X\alpha$) $X\alpha$多重散射波自洽场法
self-consistent field theory 自洽场理论
self-consistent molecular orbital procedure 自洽场分子轨道法
self-consistent units 自相一致的测量单位
self-contained cracking 独立裂化
self-contained drive 机内传动
self-contained propellant 独立推进剂
self-cooling 自冷却
self-crimpable composite filament yarn (可)自卷曲的复合丝
self-crosslinking 自交联(作用)
self-crosslinking acrylic resin 自交联丙烯酸树脂
self-crowding factor 自集因子
self cure ①自干；自固化；室温固化②自硫化；室温硫化
self-curing 自动硫化
self-curing adhesive 自固化黏合剂
self-curing cement 自动硫化胶
self-decomposition 自分解
self-demagnetization 自去磁
self depolarization 自消偏振
self diffusion 自扩散
self-diffusion coefficient 自扩散系数
self-diffusion of nuclear spin 核自旋的自扩散
self discharge 自放电
self-dumping (渗滤器的)自动卸料
self-electrode 自电极〔由试样形成的作为分析用的电极〕
self-emptying (划槽)自动出料
self-energized seal 自紧式密封
self-energy 本征能量；内(禀)能(量)；自具能(量)
self-enhancement 自增强
self-etching paint 自动浸蚀(洗涤)底漆；自动磷化底漆
self-etching primer 磷化底漆；自蚀底漆
self-etch primer(=wash primer) 自磷化底漆；洗涤底漆；自蚀底漆
self-excitation 自激发；自激励
self-excited vibration 自激振动
self extinguishing 自熄性
self-extinguishing polymer 自熄聚合物

self-feeder 自动加料器；自给器
self-feeding 自动加料
self-feeding grease cup 自动加油的脂杯
self-fluxing alloy 自熔融合金
self-fluxing alloy powder 自熔性合金粉末
self-fluxing property 自融合性
self focused laser beam 自聚焦激光束
self-focusing laser 自聚焦激光器
self-fusible ore 自熔矿
self-glazing emulsion 自动上釉乳状液
self-gripping nylon tape fastener （自锁)尼龙搭扣
self-hardening steel 自硬钢
self-igniting bipropellant 自燃双料推进剂
self-ignition 自燃
self-ignition point 自燃(着火)点
self-ignition temperature 自燃温度；自燃点
self-inclusion phenomenon 自包结现象
self-inductance 自感
self-induction 自感(应)
self-inflammable 自燃的
self-inflammable mixture 自燃混合物
self-initiation 自引发
self-integrating method 自积分法
self-ionization 自(身)电离
self-ionization spectroscopy 自电离谱
self-irradiation damage 自辐照损伤
self-leafing glass flake 自漂浮玻璃(磷)片
self-lifting property 自咬底性
self-lubrication coating material 自润滑涂层材料
self-luminous 自发光的
self-luminous paint 自发光涂料
self-modulation 自调制
self-occluding mask 自封闭掩模技术
self-oil feeder 自动加油器
self-oiling 自动加油润滑
self-oligomerization 自低聚合
self-operated controller 自动控制器
self-operated regulator 自动调节器
self-ordering rigid-chain polymer 自(动有)序刚性链聚合物
self-organization 自组合
self organizing artificial neural network 自组织人工神经网络
self-organizing feature map 自组织特征映射
self-oscillation regime 自激振荡态
self-plasticizing action 自增塑作用；内增塑作用
self-polishing antifouling paint 自抛光防污漆
self-polishing wax 自抛光性(乳化)石蜡；(液态)水(性)乳液石蜡；水分散性石蜡

self-polymerization(=bulk polymerization) 本体聚合(反应)
self-primed paint 自成底漆
self-priming pump 自吸泵；自引泵
self-priming tank 自引桶〔有自动调节液面的桶〕
self-priming topcoat 底面两用面漆；底面合一面漆
self-process 自加工
self-propagating 自动传播的
self-propagating exothermic reaction 自动传播的放热反应
self-propagation 自增长*
self-propelled 自动推进(式)的
self-propelling tripper 自动推进(式)倾料器
self-purging cell 自动净化池
self-purging trap 自动净化槽
self-purification 自净(作用)
self-purification process 自净过程
self quenching 自猝灭*
self-quenching counter 自猝灭计数管
self-radiolysis 自辐解*
self-recording 自记的；自动记录的
self-recording barometer 自录气压计；自记气压计
self-recording dilatometer 自(动)记(录)式膨胀计
self-recording instrument 自记式仪器
self-recording measurement 自记测量；自(动)测量
self-refrigeration 自动冷凝
self-registering barometer 记录式气压计
self registering hair hygrometer 毛发式自记湿度计
self-registering instrument 自动记录仪
self-regulating fuel pump 自动调节的燃料泵
self-regulation 自动调节
self-reinforcing polymer 自增强聚合物*
self-renewing surface protectant 自更新表面保护剂
self resonating vibroscope 自谐振示振仪
self-reversal ①自蚀②自倒转
self-scanning diode array 自扫描二极管阵列
self-scanning pyroelectric array 自扫描热电阵列
self scattering 自散射
self-sealing 自封的；自动封闭的
self-sealing coke-oven door 自动封闭焦化炉门
self-sealing fuel cell(=self-sealing fuel tank) 自封闭燃料箱
self-sealing fuel tank(=self-sealing fuel cell) 自封闭燃料箱
self-sealing paint 自封闭漆；自动封闭漆〔一种兼具封闭底漆作用的涂料〕
self-sealing plastics 自动封闭塑胶接合剂
self-sealing ring 自紧式密封环
self-sealing rubber septum 自密封橡胶隔片

self sealing septum	自封闭隔片
self-sealing tank	自动封闭桶
self-serve station	自动加油站
self-service instrument	自动化仪表
self shadowing(=self-shielding)	自屏蔽
selfshield arc welding	自保护电弧焊
self-shielding(=self shadowing)	自屏蔽
self-shielding effect	自屏蔽效应
self-shielding factor	自屏(蔽)因子
self-shrinking cylinder	自紧式圆筒
self-starting	自动启动
self-stratifying formulation	①自动分(成)层配方②自动分(成)层涂料
self-stratifying powder coating	底面自分层粉末涂料
self-sufficient unit	自足的设备
self-supporting stack	自架烟囱
self-supporting umbrella roof	没有梁(的油罐)圆形顶盖
self-suspending	自动悬浮
self-sustaining reaction	自持(链锁)反应
self-sustaining reactor	自持(反应)堆
self-synthesis	自合成
self termination	自终止(作用)
self-texturing paint	自变花纹漆；自成(生)花纹漆
self-texturing yarn	自变形丝
self-thermosetting(=self-curing;ambient cure;self-vulcanizing)	自(热)固化；室温固化
self-tone	①原色调②素色〔无花纹，单一种彩色〕
self-unloading pump	自动卸料泵
self-venting pouring drum	有溢流塞的加料桶
self-vulcanizing	自动硫化*
self-vulcanizing adhesive	自硫化黏合剂
selinane	蛇床烷
selinene	蛇床烯；芹子烯
selinene hydrochloride	盐酸蛇床烯
selinenol(=cudesmol)	蛇床烯醇；瑟灵烯醇
selinum oil	蛇床子油
sellaite	氟镁石
selocide	硒的硫化钾铵复合剂〔杀虫药〕
selwynite	黄赭石
semecarpus	打印果
semetic state	齐端状态
semi-	〔拉丁词头〕①半②部分的；不完全的
semi-acetal	醛缩一醇；半缩醛
semialdehyde	半醛
semi-aniline finish	半苯胺涂饰剂
semi-aniline leather	半苯胺革
semi-anthracite(=semi-anthracitic coal)	半无烟煤
semi-anthracitic coal(=semi-anthracite)	半无烟煤
semiaromatic film	半芳香族膜
semi-asphaltic flux	半石油沥青
semi-asphaltic oil(=semi-asphaltic petroleum)	半沥青油
semi-asphaltic petroleum(=semi-asphaltic oil)	半沥青油
semi-Auger process	半俄歇过程
semi-automatic lubrication	半自动润滑
semi-automatic metal arc welding	半自动金属电弧焊
semi-automatic mold	半自动塑模
semi-automatic operation	半自动化操作
semi-automatic quantitative sampling	半自动定量进样
semi-automatic welding	半自动焊
semi-automation	半自动化
semi-ballon tyre	半低压轮胎
semibatch reactor	半间歇式反应器
semibenzene	半苯；对二烷基亚甲基环己二烯
semi-bituminous	半沥青(的)
semi-bituminous coal	半烟煤
semi-blown asphalt	半吹气沥青
semiboiled process	半沸煮法；热法〔煮皂〕
semiboiled soap	半沸煮皂；热法皂
semi-boiling	半沸煮的
semi-boiling process	半沸煮法
semi-boiling soap	半沸煮皂；热法皂
semicarbazide	①氨基脲②脲氨基 $NH_2NHCONH_2$
semicarbazide hydrochloride	氨基脲化氢氯 $NH_2CON_2H_3HCl$
semicarbazido	脲氨基 $H_2NCONHNH-$
semicarbazino	脲亚氨基 $H_2NCONHN=$
semicarbazone	缩氨基脲*
semicarbazono	脲亚氨基 $H_2NCONHN=$
semi-β-carotenone	β-胡萝卜素单酮
semi-chemical pulp	半化学纸浆
semi-chrome leather	混合铬鞣革
semi-chrome tanning	半铬鞣
semi-clear film	半透明漆膜
semiclosed-type impeller	半开式叶轮
semi-coke	半焦(炭)
semi-coking	半焦化(作用)；低温炼焦
semi-colloid	半胶体
semi-colloidal	半胶体的
semi-commercial production	中间工厂规模生产；中间试制
semi-commercial unit	半工业装置
semi-conducting electrode	半导体(电)电极
semiconducting glass fibre	半导体玻璃纤维
semiconducting insulation	半导体绝缘(层)〔电缆〕
semiconducting organic polymer	有机半导体聚合物
semiconducting polymer	高分子半导体*

semi-conductive thin-film detector 半导体薄膜检测器
semi-conductivity 半导电性
semiconductor 半导体
semiconductor cell 半导体电池
semiconductor crystal 半导体晶体
semiconductor detector 半导体检测器*
semiconductor electrochemistry 半导体电化学
semiconductor electron multiplier 半导体电子倍增器
semiconductor integrated circuit 半导体集成电路
semiconductor laser 半导体激光器
semiconductor oxide 半导体氧化物
semiconductor radiation detector 半导体放射检测器
semiconductor rectifier 半导体整流器
semiconductor solid (radiation) detector (SSD) 半导体固体检测器
semiconductor varnish 半导体漆
semiconservative replication 半保留复制
semi-continuous 半连续的
semi-continuous activated sludge test 半连续活性污泥试验
semi-continuous centrifuge spinning method 半连续离心纺丝法
semi-continuous distillation 半连续蒸馏
semi-continuous grease production 润滑脂的半连续生产
semi-continuous leaching 半连续浸取
semi-continuous polymerization 半连续聚合*
semi-continuous process 半连续过程
semi-crystal 半水晶
semi-crystalline ①半晶体②半晶质(的)
semi-crystalline polymer 半结晶聚合物*
semi-crystalline region 半结晶区
semi-crystalline rocks 半晶质岩
semicure 半固化；半硫化
semi-cured 半硫化了的
semi-cyclic bond(=semi-cyclic link) 半环键
semi-cyclic double bond 半环双键
semi-cyclic double link 半环双键
semi-cyclic double linkage 半环双键
semi-cyclic link(=semi-cyclic bond; semi-cyclic linkage) 半环键
semi-cyclic linkage(=semi-cyclic link) 半环键
semi-diaphanous 半透明的
semi-differential current 半微分电流
semi-differential polarography 半微分极谱法*
semi-diffusion 半扩散
semi-diffusion cell 半扩散池
semidine 半联胺；重苯胺；苯氨基苯胺
　　$RC_6H_4NHC_6H_4NH_2$

semidine change 半联胺变化
semidine rearrangement(=semidine transposition) 半联胺重排作用
semidine transposition(=semidine rearrangement) 半联胺重排作用
semi-direct ammonia recovery 部分直接氨的回收
semidissolution(=acid swelling) 酸溶胀法；半溶解法
semi-drop centre rim 半深槽轮辋
semi-dry blotting 半干印迹(法)
semi-dry friction 半干摩擦
semi-drying oil 半干性油
semi-dry process 半干法
semi-dry pulp 半干(纸)浆
semi-dry surface 半干表面
semi-dry winding 半干法缠绕
semi-dull 近无光
semi-durable adhesive 半耐久性黏合剂
semi-ebonite 半硬质胶〔橡胶〕
semi-empirical method 半经验法
semiempirical molecular orbital method 半经验分子轨道法*
semiempirical quantum chemical 半经验量子化学的
semi-faience 半瓷釉
semi-finished product 半成品
semi-finishing 半精加工
semi-flash condition 半闪光条件
semi-flash mould 半溢料式模具
semiflat 半光
semi-flat band method 半带法〔橡胶〕
semi- flexible chain polymer 半柔性链聚合物
semi-float 无纤维再生胶
semi-fluid ①半流②半流体
semi-fluid friction 半液体摩擦；半流体摩擦
semi-fluid grease 半流润滑脂
semi-fluidized bed 半流化床
semi-fluid lubrication 半流润滑
semifossil resin 半化石树脂
semi-fusion method 半熔法
semi-girder 悬臂梁
semi-gloss 半光泽
semi-gloss acrylic latex paint 半光丙烯酸乳胶漆
semi-gloss coating 半光漆
semi-gloss enamel 半光磁漆
semi-gloss lacquer ①半光天然漆②半光挥发性漆③半光硝基纤维漆
semigloss latex grind 半光乳胶研磨料
semigloss latex mill base 半光乳胶漆浆
semi-gloss material 半光材料
semi-gloss paint 半光漆

semi-gloss white finish　半光白色面漆
semi-grained soap　半粒肥皂
semi-hardboard　半硬质纤维板
semihardened paint film　半固化漆膜
semi-hard rubber　半硬橡胶
semi-homogeneous liquid　半均相液体
semi-homogeneous strain　半均匀应变
semi-homogeneous stress　半均匀应力
semihydrate　半水合物
semihydrogenation　半氢化
semi-ideal solution　半理想溶液
semi-integral current　半积分电流
semi-integral polarography　半积分极谱法*
semi-interpenetrating polymer network(SIPN)　半互穿聚合物网络
semi-isolated fluidized catalyst bed　半封闭流化态催化剂床
semiladder polymer　半梯形聚合物
semilethals　半致死因子
semi liquid　半液体的
semi-liquid phase process　半液相过程
semi-logarithmic coordinate paper　半对数坐标纸
semi-logarithmic plot　半对数图
semi-logarithmic scale　半对数标度
semilog paper　半对数图纸
semi-lubricant　半润滑剂
semi-metal　半金属*
semi-metallic molecular solid　半金属分子固体
semi-metallic packing　半金属填料
semimetallocene　单茂金属化合物
semi-methyl thymol blue　半甲基百里酚蓝
semi micelle　半胶束*
semimicro-　①半微(量)②半微
semimicro analysis (=meso analysis)　半微量分析*
semimicro (analytical) balance　半微量(分析)天平
semi-micro aniline point test　半微量苯胺点试验
semimicro-chemistry　半微量化学
semimicro column　半微柱
semimicro fractionation　半微量分馏
semi-micro hydrogenation apparatus　半微量加氢装置
semimicro method　半微量法
semi-milled finishing　半缩绒整理
semimuffle furnace　半马弗炉
semi-normal solution　半当量溶液
seminose　甘露糖
semi-opaque　半透明的
semi-opaque coloured coating　半透明着色涂层
semi-ordered region　半有序区

semi-organic chemistry　半有机化学
semi-organic pigment　半有机颜料
semioxamazide(=amino-oxamide)　氨基草酰肼 $NH_2COCONHNH_2$
semioxamazone　缩氨基草酰肼
semipaste paint　半厚油漆
semi-pearl polymerization　半悬浮聚合；半成珠聚合(作用)
semi-permeability　半透过性
semi-permeable　半(渗)透性的
semi-permeable layer　半透层
semi-permeable materials　半渗透性材料
semi-permeable membrane　半透膜
semipersistent gas　半持久性毒气
semipervine　茉莉碱
semiphenomenological theory　半唯象理论
semipinacol rearrangement　半频哪醇重排*
semi-plant　中间工厂
semi-plant scale equipment　半工厂设备
semi-plastic state　半塑性状态
semi-polar　半极性的
semi-polar bond　半极性键
semipolar double bond　半极性双键
semi-polar double link　半极性双键
semi-polar double linkage　半极性双键
semi-polarity　中极性
semi-polar link(=semi-polar linkage)　半极性键
semi-polar linkage(=semi-polar link)　半极性键
semi-polar liquid phase　中极性液相
semi-polar stationary liquid　中极性固定液
semi-polar stationary phase　中极性固定相
semipolymerized　半聚合的
semi-porous surface　半多孔表面
semipositive mould　半溢式塑模
semiprecious　次贵重的；次等的
semiprecious stone　次等宝石
semi-preparation column　半制备柱
semi-preparative　半制备的
semi-preparative column　半制备柱
semi-preparative separation　半制备规模分离；小量生产规模分离
semi-prepared　半精制的
semi-prepared oil paint　油性半调和漆
semi-prepared paint　半调和漆
semi-prepolymer technique　半预聚体工艺
semi-product　半成品
semi-production equipment　中间(工厂)生产设备
semi-protic solvent　半质子性溶剂
semiquantitative　半定量的

semiquantitative analysis	半定量分析
semi-quantitative assessment	半定量估值
semiquantitative determination	半定量测定
semiquantitative spectral analysis	光谱半定量分析
semiquantitative spectrometric analysis	光谱半定量分析*
semiquinone	半醌*
semiradial reciprocating compressor	扇形往复式压缩机；星形往复式压缩机
semi-rational formula	半经验公式
semi-refined oil	半精制油料
semi-refined wax	半精制蜡
semi-reinforcing agent	半促进剂；半补强剂（橡胶）
semi-reinforcing furnace black	半补强炉黑
semiremote handling	半远距离操作
semi-reversibility	半可逆性
semi-reversible sorption isotherm	半可逆吸着等温线
semi-rigid	半固的；半硬的
semi-rigid chain	半刚性链
semirigid-chain polymer	半刚性链聚合物
semirigid plastic	半硬质塑料
semi-scale production	半工业化生产
semisilica brick	半硅砖
semiskilled labour	半熟练劳动
semi-solid	半固体
semi-solid bituminous materials	半固体沥青物质
semi-solid lubricant	半固体润滑剂
semistable	半稳定的
semistable dispersion	半稳定分散体
semisteel	半钢
semi-synthetic fibre	半合成纤维〔例如醋酯纤维〕
semi-synthetic perfume	半合成香料
semi-systematic name	半系统名称
semi-transparent	半透明的
semi-transparent microporous coating	半透明微孔涂层
semi-transparent mirror	半明镜
semi-transparent stain	半透明着色料；半透明染色剂
semi-valence(=semi-valency)	半价
semi-valency (=semi-valence)	半价
semivitreous	半玻璃化的
semivitreous china	半玻化瓷
semivitrified	半玻璃化的；半呈玻璃态的
semi-vulcanization	半硫化(作用)
semi-water gas	半水煤气
semi-works	中间工厂；中试车间
semi-work scale plant	半工厂装置
semi-works plant	中间工厂
semi-works production	中间工厂规模生产；中间试验
semi-worsted yarn	半精梳纱
semixylenol orange (SXO)	半二甲(苯)酚橙
semolina	荨苈；独行莱
sempervine	钩吻素丙〔1915年首次从美洲钩吻中分离得到的〕
sempervirine	常绿钩吻碱
semprelin	液体石油膏
semustine	司莫司汀
senaite	铅锰钛铁矿
senarmontite	方锑矿
senecifolic acid	千里光叶酸 $C_{10}H_{16}O_6$
senecifolidine	千里光叶定
senecifoline	千里光叶碱
senecifoline hydrochloride	盐酸千里光叶碱
senecifolinine	千里光叶宁
senecine	千里光因
senecio alkaloids	千里光碱类
senecioate	千里酸酯
senecioic acid	千里光酸；异戊烯酸 $(CH_3)_2C=CHCOOH$
senecioic acid tropine ester	千里光酸托品酯
senecionine	千里光宁
senecioyl	千里光酰；异戊烯酰 $(CH_3)_2C=CHCO-$
senegal	阿拉伯树脂
senegal gum(=gum Arabic)	阿拉伯树胶
senega root oil	远志油
senegenin(=senegeninic acid)	远志配基；远志酸
senegeninic acid(=senegenin)	远志酸；远志配基
senegin	远志精
senescence	衰老
seneski	天然焦炭
senfolomycin	桑福洛霉素
senna	番泻叶
senna leaf wax	山扁豆叶蜡〔取自山扁豆属植物〕
senna leaves	番泻叶
sennatin	番泻叶碱
senomycin	濑野霉素
sensibiligen	过敏原
sensibility	①灵敏度②敏感性
sensibility in practice	实感量；实际灵敏度
sensibilization agent	敏化剂
sensibilized	敏化作用
sensible heat	显热
sensible heat of fluid	流体显热
sensing crystal	敏感晶体
sensing electrode	敏感电极
sensing element	敏感元件
sensitive	①灵敏的；敏感的②感光(性)的
sensitive cell	光电池；光电管

sensitive color 灵敏色
sensitive differential pressure device 灵敏压差计
sensitive drill 精式钻床
sensitive emission line 灵敏发射谱线
sensitive line 灵敏线
sensitive mercury differential manometer 灵敏水银示差压力计
sensitiveness ①灵敏；敏感②灵敏性③灵敏度；敏感度
sensitiveness test 敏感度试验
sensitive plant 含羞草
sensitive reaction 灵敏反应；敏感反应
sensitive volume 有效体积
sensitivity 灵敏度*
sensitivity adjustment screw 灵敏度调节螺旋
sensitivity coefficient 灵敏度系数
sensitivity curve 灵敏度曲线
sensitivity drift 灵敏度漂移
sensitivity factor 灵敏度因子
sensitivity index 灵敏度指标
sensitivity limit 检测极限
sensitivity ratio 灵敏度比率
sensitivity test 敏感性试验
sensitivity to contamination 对污浊的敏感性
sensitization 敏化(作用)；活性化
sensitization test 致敏试验
sensitized atomic fluorescence 敏化原子荧光
sensitized emulsion 增敏乳剂
sensitized fluorescence 敏化荧光*
sensitized fluorescence of atom 原子敏化荧光
sensitized laser material 敏化激光材料
sensitized paper 感光纸
sensitized phosphorescence 敏化磷光*
sensitized polarograph 敏化极谱仪
sensitized reaction 增敏反应
sensitized room temperature phosphorimetry 敏化室温磷光法
sensitizer 敏化剂；增感剂
sensitizing 敏化过程
sensitizing agent 敏化剂；增感剂
sensitometer 曝光表；感光仪
sensitometric 感光(学)的；露光深浅的
sensitometric curve 感光(深浅)曲线
sensitometry 感光测量学；感光度测定
sensor 传感器*；敏感元件
sensory test 感官试验；感官鉴定
sentinel hole ①监察孔②磨损试孔〔催化剂U形管〕
sentinel pyrometer 高温计
sentinel valve 调节阀；操纵阀

separability 可分离性
separable conjugate fiber spinning method 裂离型复合纤维纺丝法
separableness 可分性
separant 隔离剂
separant coating 涂隔离剂
separate analysis 分别分析
separate application ①分别涂布②分别施胶(黏合)
separated application adhesive 组分分涂黏合剂
separated flame 分离火焰
separated flow 分(离)流
separated isotope 分离同位素
separated oil(=segregated oil) 析得干性油
separate oil supply 润滑油分别供应；分别润滑
separate out 分出；离出；析出
separate-phase indicator 分相指示剂
separate-phase titration 分相滴定法
separate pot mo(u)ld 分离料腔式模具
separate venturi 独立的文氏管；分离的文氏管
separating agent 分离剂
separating bath 分开热浴
separating bowl 分离篮
separating box 分离箱
separating by dilution 稀释分离
separating capacity 分离额；分离本领
separating centrifuge 分离机；离心分离机
separating coefficient 分离系数
separating column 分离柱
separating efficiency 分离效率
separating element 分离单元*
separating factor 分离因数
separating flask 分液瓶
separating fluid 分重液体
separating funnel (=separatory funnel) 分液漏斗
separating layer 分离层
separating machine 分离机
separating mechanism ①分离机理②分离装置
separating of grease 润滑脂的分离
separating of oil 油的分离
separating power 分离能力
separating property 分离性能
separating screen 分选筛
separating size 分离度
separating tank 澄清槽
separating unit 分离单元
separating wall ①挡板②隔墙；隔板
separation 分离*
separation analysis 分离分析

separation by development(=separation by displacement) 置换分离
separation by displacement(=separation by development) 置换分离
separation cell 分离池
separation chamber 分离槽；展开槽
separation coal 精选煤
separation coefficient 分离系数
separation column 分馏塔
separation criterion 分离判据
separation development 分离展开
separation efficiency 分离效率
separation equipment 分选设备
separation facility 分离装置
separation factor 分离因子*；分离系数
separation function 分离函数
separation gel 分离胶
separation identification system 分离鉴定系统
separation intensity factor 分离强度因数
separation number 分离数*
separation optimization 分离最佳化
separation parameter 分离参数
separation plant 分离工厂
separation potential 分离势*
separation problem 分离课题
separation process 分离过程
separation quotient 分离商
separation reaction 分离反应
separation-reaction-separation technique 分离-反应-分离技术
separation sieve 分离筛
separation system 分离系统
separation temperature 分离温度
separation test for greases 润滑脂分油试验
separation value 分离值
separation work 分离功*
separative duty 分离本领
separator 分离器
separator box 分离箱〔分离炼油厂废水中水和油〕
separator column 分离柱
separator diaphragm 隔板隔膜
Separator-Nobel dewaxing process 三氯乙烷溶剂润滑油脱蜡过程
separator-scrubber 分离涤气机
separator sediment 分离器沉积物
separator tank 分离槽
separatory funnel(=separating funnel) 分液漏斗
separatory head （注塑机)分流芯头

sephadiamine 千金藤二胺
sepia ①乌贼；墨鱼②乌贼染料
sepiapterin 墨蝶呤
sepiolite 海泡石
septa-〔拉丁字头〕 七
septavalence(=septavalency) 七价
septavalency(=septavalence) 七价
septavalent 七价的
septazine(=benzylsulfanilamide) 苄基磺胺
septet 七重峰；七重线
septic sewage （腐败的)污水
septic tank 化粪池
septivalence(=septivalency) 七价
septivalency(=septivalence) 七价
septivalent 七价的
septometer 空气污度计
septum 中隔；隔膜
septum buffer 隔膜缓冲器
septum injector 隔片注射器
septumless injector 无隔片注射器
septumless sampler 无隔片进样器
septumless syringe 无隔片注射器
septum magnet 切隔磁铁
septum sampling 隔膜进样
septurit 磺胺-己胺复剂
sequanator 序列分析仪
sequence ①序列②(顺)序③链区④结果
sequence analysis 顺序分析
sequence chromatography 序列色谱法
sequence control 程序控制；顺序控制
sequence distribution 序列分布
sequence length 序列长度
sequence-length distribution 序列长度分布*
sequence monomer 单体序列
sequence of volatility 挥发顺序*
sequence rule 顺序法则
sequence-specific artificial nuclease 序列专一性的人工核酸酶
sequencial reaction 连续反应
sequencing 测序
sequencing by hybridization 杂交测序法
sequentex yarn 顺序拉伸变形丝；外拉伸变形丝
sequential analysis 序贯分析*
sequential automatic X-ray spectrometer 顺序型自动X射线谱仪
sequential control 顺序控制
sequential copolymer 序列共聚物*
sequential determination 顺序测定(法)

sequential differential 系列差分
sequential inductively coupled plasma spectrometer 顺序等离子体光谱仪
sequential interpenetrating polymer network 有序互相贯穿聚合物网络
sequential lamination 接续性压合法
sequential mass spectrometry 序贯质谱法
sequential modular approach 序贯模块法
sequential multiple autoanalyzer 连续多路自动分析器
sequential plasma spectrometer 顺序型等离子体光谱仪
sequential polymerization 序列聚合*
sequential regularity 序列规则性
sequential sampling 序贯抽样*
sequential sampling inspection 序贯抽样检验
sequential search 序贯寻优*
sequential test 序贯试验
sequent injection analysis 顺序注射分析
sequester （多价）螯合剂
sequestering action 螯合作用
sequestering activity 螯合活性
sequestering agent (=chelating agent) （多价）螯合剂
sequestering power 螯合能力
sequestrant(=chelating agent) （多价）螯合剂
sequestration(=chelation) （多价）螯合作用
sequestric acid 乙二胺四乙酸
sequiatannic acid(=sequoia tannin) 红杉单宁酸
sequoiatannic acid 红杉鞣质(单宁)酸
sequoia tannin(=sequiatannic acid) 红杉单宁
sequoyitol 红杉醇
serge 哔叽
serge blue （碱性）亚甲兰；哔叽兰
serial dilution 连续稀释
serial number ①连续数值②序号
serial production 系统生产；成批生产
sericin 丝胶
sericite (mica) 丝云母；绢云母
series ①系(列)；组；型②序③串联④组〔生物分类〕⑤串⑥级数
series arc furnace 串电弧炉
series-bypass method 串联旁路方法
series circuit 串联电路
series classification 顺序分类
series dialysis 串联式透析法
series electron 系电子
series flow 串流
series limit 系限
series limit continuous spectrum 系限连续光谱
series of cells 电池组

series of compounds 化合物系
series of dihydroxy-terminated poly (ethylene oxide) 双羟基端接的聚乙二醇系列〔即不同聚合度的聚乙二醇〕
series of lines 光谱(线)系
series of permanent artificial standards 稳定的代用标准系列
series of X-ray lines X射线系
series-parallel operation 串并联操作
series pipe still 管组蒸馏炉
series pyrolysis 系列热解
series spot welding 单面多点点焊；单面点焊；连续点焊
serimeter 验丝计；强力延伸试验器
serine 丝氨酸 $HOCH_2CH(NH_2)COOH$
serine methylester 丝氨酸甲酯 $HOCH_2CH(NH_2)COOCH_3$
serinol 丝氨醇
seriplane test 纱匀度试验
serolipase 血清脂酶
serological pipette 血清学取样器
serological test 血清学检查法
serology 血清学
seronine 血清素
sero-reaction 血清反应
serotherapy 血清疗法
serozyme 凝血酶原
serpentaria 蛇根
serpentary 蛇根马兜铃
serpentine ①蛇根碱；利血平②蛇纹石
serpentine cooler 蛇管冷却器
serpentine superphosphate 苦土过磷酸盐；蛇纹石矿过磷酸盐
serpentinine 蛇根亭宁
serpex 塞佩克斯碱性耐火材料
serpine 蛇根平
serpinine(=tetraphyllicine) 四叶萝芙辛
serratagenic acid 三对节酸
serrated belt 齿形三角带；风扇带
serrated edge （锯）齿形边
serrated ridge 锯齿状隆起物〔粒子表面形状〕
serration 锯齿(形)；裂瓣
serration (of cross section) 裂瓣(截面)
sertraline 舍曲林
serum ①血清②清液③乳清〔橡胶〕
serum albumin 血清清蛋白；血清白蛋白
serum cap 橡胶盖
serum cholesterol 血清胆固醇
serum cholinesterase 血清胆碱酯酶
serum globulin 血清球蛋白
serum potassium 血清钾；血钾

serum protein　血清蛋白
serum protein refractometer　血清蛋白(折射)计
service　服务；维护；保养
serviceability　使用性能；适用性
serviceability limitation　使用性能范围
serviceability temperature　使用温度
serviceable　可用的；正常的；能操作的
serviceable life　使用期
serviceable range　使用范围
service condition　使用条件
service department expenses　服务部门费用
service durability　耐用性；耐久性
service garage　服务站
service gas mask　军用防毒面具
service kit　维修箱
service life　使用寿命
service line　①作业线②动力管道
service parts　备用零件
service pipe　工作管道
service regulation　使用规则；操作规程
service reservoir(=service tank)　工作油罐
service-shock resistance　实用上的耐冲击性
service shop　维修车间；检修厂
service station　服务站；修理站；加油站
service station bulk plant　加油站油罐厂
service tank　工作油罐
service temperature　使用温度
service water　家用水
servicing　①服务；技术服务②维修；保养
servo　①伺服机构②随动系统
servo-drafter　随动牵伸装置；伺服牵伸装置
servo-lubrication　中央润滑
servo mechanism　伺服机构
servomotor　伺服电动机
servoscribe　伺服扫描
servoscribe recorder　伺服记录器
servoscribe scanning　伺服扫描
servo system　伺服系统
seryl　丝氨酰　HOCH$_2$CH(NH$_2$)CO—
sesame oil　芝麻油
sesamol　芝麻酚
sesamolin　芝麻酚林
sesbania　田菁〔一种绿肥〕
seselin　邪蒿素
sesqui-〔拉丁字头〕　一倍半；一又二分之一；二分之三
sesquibenihene　倍半贝尼烯
sesquibenihidiol　倍半贝尼黑二醇
sesquibenihiol　倍半贝尼黑醇

sesquicarbonate　倍半碳酸盐
sesquicarbonate of soda　(二)碳酸氢三钠　NaHCO$_3$·Na$_2$CO$_3$·2H$_2$O
sesquichamene　倍半查烯〔取自钝花柏 Chamaecyparis octusa〕
sesquichloride　倍半氯化物；三氯化二某
sesquicitronellene　倍半香茅烯；倍半香草烯
sesquicryptol　倍半杉醇
sesquifulvalene　倍半富瓦烯
sesquigoyol　倍半告衣醇
sesquioxide　倍半氧化物[*]
sesquiquinone　倍半醌
sesquisalt　倍半盐
sesquisilicate　倍半硅酸盐
sesquisoda　倍半钠碱
sesquistearate　倍半硬脂酸盐(酯)
sesquisulfide　倍半硫化物；三硫化二某
sesquiterpene　倍半萜[*]
sesquiterpenoid　倍半萜(烯)化合物
sessile algae　无柄藻；固着藻类
sessile bubble method　静泡法；固定泡法；停泡法
set　①(一)套；组；叠②凝固③变定；永久变形④固定⑤固定伸张⑥安置
setacyl direct violet　瑟它酰直接紫
Setaflash closed tester　塞塔闭杯闪点试验器
set after break　扯断永久变形
setamine black　毛胺黑
setamine blue　毛胺蓝
setamine colors　毛胺染料
setamine pink　毛胺桃红
set and locking screw　锁紧螺旋
set at break　断裂变定
set free　游离
set gas purification　湿煤气净化(法)
set gauge (=setting gauge)　定位规；校正规
set nut　①锁紧螺母②定位紧母，调整螺母
setocyanine O　瑟陶花青O
set of bands　谱带组
setoflavine　毛黄素
setoglaucine　毛罂蓝
setoglaucine O　毛罂蓝O
set point　①凝固点；凝结点②沉淀点③调定点；给定值
set point control　定点控制；凝结点控制；调整点控制
set screw　①固定螺钉②调节螺钉
set temperature　硬化温度
setter　耐火架子
sett grease　冷煮润滑脂；冷煮钙皂润滑脂
sett grease soap　冷煮润滑脂皂

set time 硬化时间
setting ①凝结；凝固；硬化②变定③装置④位置⑤安装⑥硬化
setting accelerator 促凝剂
setting action 变定作用；凝固作用
setting agent 固化剂；硬化剂；定形剂
setting bath 沉降槽
setting coat ①末道涂层②末道粉刷层
setting control 定位控制
setting device 调节装置
setting dial 仪器标尺
setting medium ①定形介质②凝固介质
setting-out ①压水法②出发③平展〔皮革〕；伸展
setting-out cylinder 平展机刀轴
setting-out knife 平展刀
setting-out machine ①除屑机②平展机；伸展机
setting-out sleeve 平展机滚轴套筒
setting plane 装配平面图
setting point 凝结点；凝固点
setting process 沉降法
setting rate 凝固速率；硬化速率
setting slurry 沉降性浆料
setting temperature 凝固温度；固化温度
setting time 硬化时间
setting treatment 定形处理；硬化处理
setting up 安装
setting-up point ①缩(聚)合反应温度②凝固(结)温度
setting-up screw 调距螺栓
settled bed 澄清床
settled grained soap 澄清颗粒皂
settled grease 澄清油脂
settled process 沉降法；静置法
settled solution 澄清溶液
settled tar 沉淀(木)焦油
settlement ①沉淀；澄清；下沉②沉淀物③解决；固定
settlement on storage 储存沉淀；库存沉淀〔漆病〕
settlement separate 沉降分离
settlement space 沉降空间
settlement test 沉降试验
settle out 沉降；沉积；沉淀出来
settler 澄清器；沉降器
settling ①沉降②沉清
settling area 沉降截面
settling basin 沉降池〔石油炼厂处理污水用〕
settling bath 沉降浴；沉淀浴；澄清浴
settling bowl 沉降池
settling box 沉降箱
settling capacity 沉降力

settling centrifuge 沉降式离心机
settling chamber 沉降室
settling classifier 沉降分级器
settling column 沉降室
settling cone 沉降锥
settling data 沉降数据
settling down 沉下；沉降
settling drum 沉降鼓
settling height 沉降高度；沉降距离
settling out 沉出
settling particle 沉降粒子
settling pit 沉降坑〔石油〕
settling pocket 沉降室
settling pond 澄清池
settling process 沉降法
settling property 沉降性能
settling rate 沉降速率
settling ratio 沉降系数
settlings 沉淀
settling section 澄清区；澄清段；沉降段
settling tank 沉清槽；澄清桶
settling test 沉降试验
settling time 沉降时间
settling tub 沉淀槽
settling vat 沉降瓮
settling velocity 沉降速度
settling velocity equation 沉降速度方程式
settling vessel 沉降器
settling volume 沉降体积
set to touch 指触干燥
set-to-touch time 指触干时间
set up ①变定②装置妥(当)；装置；装备
set-up effect 开始(硫化)效应*
set up of mix 混炼胶的变定
seven-membered ring(=seven ring) 七元环
seven ring(=seven-membered ring) 七元环
seventeen-membered ring(=seventeen-ring) 十七元环
seventeen-ring(=seventeen-membered ring) 十七元环
several lines method 多线法
severe condition 苛刻条件
severe contamination 严重污染
severe heating 急剧加热
severe marking 严重划伤；严重划痕
severe oxidation 严重氧化
severe pitting 严重点蚀
severe radiation 强辐射
severe stress 危险应力
severe test 严格试验；繁重工作条件下试验

severe washing test 强化洗涤试验
severe wear 严重磨损
severity 苛刻度〔加工深度〕
severity factor 强度系数;(测定工艺过程操作强度的)强度因数
severity of quench 急冷度
sevoflurane 七氟烷
sew ①缝合②汁液③下水管道
sewage ①阴沟水;污水②下水道③下水道系统
sewage aeration 污水曝气
sewage discharge standard 污水排放标准
sewage disposal 污水处理
sewage disposal system 污水处理系统
sewage disposal work 污水处理厂
sewage farm 污水灌溉田
sewage gas 沼气〔一般组成为60%～70%甲烷和30%～40%二氧化碳〕
sewage irrigation disposal 污水灌溉处置
sewage lagoon 污水池
sewage loading 污水负荷量
sewage oxidation pond 污水氧化池
sewage pipe 污水管
sewage plant 污水处理厂
sewage purification 污水净化
sewage sludge (污水处理生成的)污泥
sewage sludge disposal 污泥处理
sewage stream 污水系统;泥水流
sewage treatment plants 污水处理厂
sewage treatment structure 污水处理结构
sewage treatment system 污水处理系统
sewage work 污水处理厂
sewer 下水道;阴沟
sewerage 污水
sewer gas 沟道气
sewer pipe 污水管;沟管
sewer pipe clay 沟管土
sewing-machine oil 缝纫机油
sewing-up 缝合
sexa- 〔拉丁字头〕六
sexadentale chelate 六配位螯合物
sexadentate 六配位
sexadentate ligand 六齿配体;六合配体
sexamer 六聚物;六节聚合物
sexavalence(=sexavalency) 六价
sexavalency(=sexavalence) 六价
sexavalent 六价的
sexi-covalent 六价共价的
sexidentate ligand(=sexadentate ligand) 六齿配位体

sexiphenyl 联六苯 $C_{36}H_{26}$
sexivalence(=sexivalency) 六价
sexivalency(=sexivalence) 六价
sexivalent 六价的
sextant 六分仪
sextate 甲基环己醇乙酸酯
sextet 六重峰
sextol 甲基环己醇
sextol adipate 己二酸二甲基环己醇酯
sextuple-effect evaporation 六效蒸发
sextuple-effect evaporator 六效蒸发器
seybertite 褐脆云母
Seymond disc crusher 赛芒德盘压机
sfax 针茅
s glass fibre s玻璃纤维;强玻璃纤维
shackle ①钩环〔机工〕;铁扣②桎梏
shaddock 香泡;柚
shaddock oil 香泡油;柚皮油
shade 明暗调整色;色调
shaded-off (颜色)冲淡
shade of color 浓淡;色调;色彩的微差
shadowgram ①X射线照相②人身辐射相
shadow mask 阴影掩模
shaft ①轴②(矿)井③升降井④柱身⑤旋钮⑥手轮
shaft bearing 轴承
shaft bracket 轴架
shaft bushing 轴瓦;轴衬
shaft coupling 联轴节
shaft furnace 竖式炉
shaft horsepower 轴功率;轴马力
shafting oil 传动油;轴油
shaft kiln 竖(式)窑
shaft neck 轴颈
shaft packing 轴封
shaft run-out 轴(径向)跳动
shaft seal 轴封
shaft sleeve 轴套
shaft work 轴功
shag ①粗毛②切碎的烟草
shagreen 鲨革;绿革
shake 摇动;振动
shakedown load 安定状态载荷
shakeless deckle 稳定框
shake lotion 多层色化妆水
shake-off 震脱;抖落
shake-off effect 携出效应
shake-out machine 摇出机〔测定石油中沉淀物与水分含量的离心机〕

shake-proof 防震的
shaker 摇动器；摇动机
shake rails 记录胎
shaker belting 振动输送带
shaker conveyor 摇动运输器
shaker screen 摇动筛
shaker type collector ①振荡式微粉捕集器②振荡式除尘器
shake tube method 标准管摇动测泡法
shake up 摇动；摇荡
shake-up peak 携出峰
shaking 摇动；摇振
shaking apparatus 摇动器；振摇机
shaking chute 振动斜槽
shaking conveyer 摆动运输机
shaking culture 振荡培养法
shaking feeder 摆动供给器
shaking grizzly 摆动栅
shaking machine 摇动机
shaking mixer 摇动混合器；振荡混合器
shaking screen 摇动筛
shaking shoot 振动斜槽
shaking sieve 摇动筛
shaking trough 摇动斜槽
shale 页岩
shale distillation 页岩蒸馏
shale distillation retort 页岩蒸馏甑
shale gasoline 页岩汽油
shale naphtha(=shale oil naphtha) 页岩石脑油
shale oil 页岩油
shale oil cracking 页岩油裂化(法)
shale oil extraction 页岩油抽提
shale oil naphtha(=shale naphtha) 页岩石脑油
shale oil processing 页岩油加工
shale oil recovery 页岩油生产；页岩油的回收
shale oven 页岩干馏炉
shale paraffin 页岩石蜡
shale pit 页岩坑；页岩槽
shale pyrolysis 页岩热解；页岩热分解
shale retort 页岩干馏甑
shale retorting 页岩干馏
shale retorting process 页岩干馏过程
shale rock 页岩
shale silicate 页岩硅酸盐
shale spirit 页岩汽油〔英国〕
shale tar 页岩焦油
shale tar pitch 页岩沥青
shale tar still 页岩焦油蒸馏釜
shale wax 页岩石蜡
shallot 亚实基隆葱
shallow flight worm 浅螺纹螺杆
shallow fuel bed 薄煤层床
shallow underground burial 浅层埋藏*
shallu 高粱
shaly 页岩的
shammy 麂皮；油鞣革
shamoy 麂皮
shampoo 洗发剂
shampoo powder 洗发粉
shank ①柄；项②胫
Shank's extraction process 香克提取法
Shank's lixiviating system 香克沥滤法
Shank's process 香克法〔一种连续式固体抽提法〕
Shank's washing system 香克洗涤法
shanking cylinder 腿革平展机刀轴
shank setting machine 腿革平展机
Shank's extraction process 香克提取法
Shank's lixiviating system 香克沥滤法
Shank's lixiviating tank 香克沥滤桶
Shannon's theorem 仙农定理
shaped brick 异型砖
shaped excitation waveform 定形激发波形
shaped pulse 特形脉冲；整形脉冲
shaped section 型材
shaped washing agent 固体洗涤剂；块状洗涤剂
shape factor ①形状因数②形状因素
shape index 形状指数
shapeless 无定形的
shape memory alloy 形状记忆合金
shape-memory polymer 形状记忆高分子；形状记忆聚合物
shape memory silk 形状记忆生丝
shape of catalyst 催化剂颗粒形状
shape of molecule 分子形状
shape of spot 斑形状
shaper ①成形机②成形器③牛头刨床
shape retention 外形保持性
shape-retentive finish 保形整理
shape selective catalysis 择形催化*
shape-selective sorbent 择形吸着剂
shaping ①成形；成型②修刨；刨
shaping cycle 成型周期
shaping jig 成型工具；成型夹具
shaping machine 牛头刨床
shaping of catalyst 催化剂成型
shaping operation 刨削操作；成形操作

shaping plate ①样板；规②轨距规
shaping without stock removal 无屑成形
shared-cluster crystal 共簇晶体
shared electron 共享电子；共价电子
shared electron pair 共享电子对；共价电子对
shark liver oil 鲨鱼肝油
shark oil 鲨鱼油
shark ray oil 鲨鱼翅油
shark skin 鲨鱼皮
sharp coat(=sharp colour) ①(砂浆面)铅油封闭层；封闭色浆②钻生(大漆)
sharp-corrugated roll 尖纹辊筒
sharp-cut filter 斩波滤光器
sharp-edge coverage 锐缘覆盖力
sharp-edged 锋利的；尖锐的；刀刃状的
sharp-edged flat throttles 锐边平板节流
sharp-edged orifice 锐(边)孔；锐缘口〔机工〕
sharp-edge orifice meter 锐孔流量计
sharpener ①削(锋)刀②磨快器；磨刀器
sharpening 削尖
sharpening agent 去毛加速剂
sharpening of front 前沿变尖
sharpening stick 磨石
sharp flavor 辛辣味；辛辣调料
sharp instrument 带尖的工具
Sharples acid treating process 夏普勒斯(离心机)酸处理过程
Sharples centrifuge 夏普勒斯离心机；连续式高速离心机
Sharples continuous centrifugal 夏普勒斯连续离心法
Sharples dewaxing process 夏普勒斯(离心机)脱蜡过程
Sharples supercentrifuge 夏普勒斯超速离心机
sharp line 尖锐(谱)线；清晰(谱)线
sharpness ①锐度②敏锐③清晰度
sharpness index 敏锐指数*
sharpness of cuts (精馏塔内)馏分分割精确度
sharp paint 快干漆
sharp peak 尖峰；锐峰
sharp separation 清晰分离；快速分离
sharp series 锐系
shatter-index 震裂系数
shatter-index test(=shatter test) 震裂试验
shatter-proof 不碎的；防震的；防冲的
shatter proof glass 不碎玻璃；耐震玻璃
shatter-resistant 抗碎裂
shatters ①碎片②岩屑③废石
shatter test(=shatter-index test) 震裂试验
shaved weight 削匀重

shave hook ①镍刀钩②心形刮刀③铅锉
shaver ①刮刀②削匀工
shaving ①刮平；修整；剃；削②修理；削匀；刮平
shaving beam 刮皮毛架
shaving cream 剃须膏
shaving cylinder 削匀机刀轴
shaving knife 刮刀；消匀刀
shaving machine 削匀机
shaving mill (鞣料)削磨机
shaving powder 剃须粉
shavings ①削下薄片②纸屑
shaving soap 剃须皂
shaving stick 剃须皂条
shaving wash machine 刮洗机
shaving weight 刮重；刨重
Shawinigan black(=acetylene black) 乙炔炭黑
shea butter 牛油树脂
shea cambium 牛油树形成层
sheaflike structure 束状结构
shea nut fat 牛油树坚果脂
shear ①剪截机②剪(切)；割③大绞剪
shear axis 剪切轴
shear bond strength 剪切黏合强度
shear breakage 剪(切破)裂
shear brittleness ①剪切脆性②剪切脆度
shear compliance 剪切柔量
shear connector 剪刀接合器；受切接合部件
shear crack 剪切龟裂；切变裂缝；剪裂缝
shear creep 剪切蠕变
shear cut machine 铡断机
shear deformation 剪切形变
shear degradation 切变降解
shear displacement 剪切位移
shear elasticity(=shear modulus) 切变模量
shear flow 剪切流
shear force 剪切力；切力
shear fracture 剪切破坏
shear-free flow 无剪切流
shear gradient 剪切梯度
shear hardening 剪切硬化
shear-housing 剪断机室
sheariness 切变性
shearing 剪切*
shearing action 剪切作用
shearing degradation 切变降解
shearing disk viscometer 剪切圆盘式黏度计
shearing fatigue tester 剪切疲劳试验仪
shearing force 剪切力

shearing machine 剪切机；剪床
shearing modulus 剪切模量
shearing modulus of elasticity 剪切弹性模量
shearing rate 剪切速率
shearing resilience 剪切回弹
shearing resistance 剪切阻力
shearing strain 剪切变形
shearing strength 抗切强度；抗剪强度
shearing stress 剪切应力
shearing test 剪切试验；切变试验
shearing test apparatus 剪切试验仪；切力试验仪
shearing tester 剪切试验机
shearing traction 剪切引力
shearing work 剪工工作
shear initiation stress 剪切起始应力
shear-leach process 剪断-浸取法
shearling(=lamb reverse) 剪毛绵羊革
shear load 剪切负荷
shear log pole 起重木(杆)
shear modulus 剪切模量*
shear plane 剪切面
shear plate viscometer 剪切板黏度计
shear-rate hardening 切速硬化
shear rate-shear stress relation 切速-切应力关系式
shear-rate softening 切速软化
shear-rate thickening 切速稠化
shear-rate thinning 切速稀化
shear rate thixotropy 切速触变(现象)
shear relaxation adapter 剪切应力松弛附加器
shear resistance 抗剪强度
shear rigidity 剪切刚性
shear sandwich 剪切式防震器
shear-sensitive resin 剪(切)敏(感)性树脂
shear softening 剪切软化
shear stability 剪切安定性
shear stability test 剪切稳定性试验
shear steel 高速切削钢
shear stiffness 抗剪刚度
shear strain 剪切应变*
shear strength 剪切强度*
shear stress 剪切应力*
shear stress relaxation modulus 切应力松弛模量
shear structure 切变结构*
shear tenacity ①剪切韧性②剪切韧度
shear tension 剪切张力
shear thickening(=dilatancy) 剪切稠化；胀流性
shear thinning 剪切稀化*
shear-thinning effect 剪切稀化效应

shear viscosity 剪切黏度*
sheath ①鞘；套；护套②皮鞘
sheath core fiber 皮芯纤维
sheathed flame 鞘型火焰；屏蔽火焰*
sheathed heater (活络)夹套加热器；护套加热器
sheathed wire 铠装线；金属护皮电线
sheath flow 套液
sheath flow cuvette 鞘流池
sheath flow interface 鞘流接口
sheath flow liquid 鞘流液
sheath flow pool 鞘流池
sheathing bronze 镶嵌用青铜
sheathing paper 绝热纸
sheathless interface 无鞘接口
sheave 带槽滑车轮
shed 梭口
sheen(=angular gloss) 斜向光泽度
sheep's foot oil 羊脚油
sheep-foot roller 羊足压路机
sheep laurel 山月桂
sheep leather (绵)羊革
sheep oil 羊毛脂
sheep pelts 羊皮〔皮革〕
sheep-skin ①(绵)羊皮②羊皮纸
sheepskins cut in shearing 白碴羊皮〔皮革〕
sheep sorrel 酸模
sheep's wool (绵)羊毛
sheerness 透明性；细薄度
sheet asbestos 石棉板
sheet asphalt 片状沥青
sheet bar 铁条片
sheet brass 黄铜板
sheet coloring(=sheet dyeing) 全张着色
sheet crystals 片状晶体
sheet dyeing(=sheet coloring) 全张着色
sheeted 片状的
sheeter 压片机
sheet extruder 压片压出机；螺杆压片机
sheet glass 玻璃片
sheet-glass process 平板玻璃制造法
sheet-glass tank furnace 平板玻璃池窑
sheet high polymer 片状结构高聚物
sheeting 压片
sheeting calender 平板纸压光机
sheeting machine 压片机
sheeting mill 压片机
sheeting process 压片工序
sheeting roll 薄板轧辊

sheetings 阔幅平布；被单布
sheet iron 铁板；铁皮
sheet lead 铅皮
sheet-like crystallite 片状微晶
sheet metal work 板金工作
sheet metal working 板金工加工
sheet mill 薄板轧机
sheet mold 片料吹(气塑)模
sheet molding compound(SMC) 片状模塑料
sheet mould 片料吹(气塑)模
sheet out 成片
sheet packing (成张)包装胶片
sheet paper 纸张
sheet polymer 片型聚(合)物
sheet pulp 浆张
sheet rolls 薄片轧辊
sheets 片材
sheets cutting 裁板
sheet steel 钢板
sheet steel enamel 钢片搪瓷
sheet steel enclosure 薄钢板外壳
sheet steel housing 钢板槽型
sheet steel piling 钢板桩
sheet steeping 浆板浸渍；浆粕浸渍
sheet structure 片状结构
sheety crepe 花边绉片
sheet zinc 锌皮
Sheffield plate 包银铜片
shekanin 射干英
shelf ①架(子)②放上架子③(转鼓)挡板
shelf burner 盘架炉；搁架炉
shelf burning 盘架炉燃烧
shelf curing(=bin curing) 存放自硫化
shelf drier 盘架干燥室
shelf dryer 柜式干燥机
shelf furnace 盘架炉；搁架炉
shelf goods 小(罐)包装漆
shelf kiln 盘架炉；搁架炉
shelf life 储存期；储存寿命；搁置寿命
shelf oven 盘架炉
shelf pyrite burner 架式燃黄铁矿炉
shelf test 储存试验
shelf-time 存放时间
shell 壳；(容器)筒体；壳体
shellac 紫胶(片)；虫胶(片)
shellac bond 紫(虫)胶黏合剂
shellac coating 紫胶涂料
shellac ester 紫胶酯；紫胶片酯

shellac formaldehyde resin 紫(虫)胶甲醛树脂
shellacked 涂紫胶的
shellac plastic 紫胶塑料
shellac polish 紫胶上光剂
shellac substitute 紫胶代用品；虫胶代用品
shellac varnish 紫胶清漆
shellac wax 紫胶蜡
Shell aging test 壳牌(公司)老化试验
shellalic acid 紫胶酸
shell and coil condenser 壳式蛇管冷凝器
shell and drier 列管干燥机
shell-and-plate (heat) exchanger 板壳式换热器
shell and tube condenser 管壳式冷凝器；列管式冷凝器
shell and tube cooler 管壳式冷凝器
shell and tube evaporator 管壳式蒸发器；列管式蒸发器
shell and tube exchanger 管壳式换热器；列管式换热器
shell and tube heat exchanger 列管式换热器
shell and tube type heat exchanger 壳管式热交换器
shell and tube-type reactor 管壳式反应器
shellane 壳烷
Shell benzene recovery process (美国)壳牌(石油公司)油提蒸馏法)苯回收法
shell capacity (薄壳容器)容积
Shell catalytic hydrogenation process (美国)壳牌(石油公司)催化加氢过程
Shell catalytic polymerization process (美国)壳牌(石油公司)催化聚合法
shell construction 薄壳结构
shell cover 端盖；封头
shell-crosslinked knedel 壳交联胶束
shellene 壳烯 $C_{13}H_{20}$
Shell four ball test 壳牌(公司)四球试验
shell fuse 炮弹引信
Shell hot filtration test 壳牌(石油公司)热过滤试验
Shell hydrodesulfurization process 壳牌(石油公司)氢化脱硫过程
Shell hydroformylation 谢尔烯烃醛化法
shell innage 容器充满部分
shell lime 介壳石灰(石)；贝石灰
shell limestone 介壳石灰石
shell of boiler 锅炉套〔化工〕
shell of column 柱段
shell of mill 磨子套
shell of pipe 管壳
shell of roll 轮缘
shell of tank 容器外壳；罐体
shellolic acid 紫胶酸
shell outage 容器未充满部分

shell pass	壳程
Shell phosphate desulfurization	(美国)壳牌(石油公司)磷酸盐脱硫法
Shell phosphate purification process	(美国)壳牌(石油公司)磷酸盐脱硫精制法
Shell process	壳牌(公司)法〔由丁烯制丁二烯法〕
shell reamer	空心绞刀
shell ring	(容器)筒节
shell rock	介壳石灰岩
shell side	壳方；壳程〔换热器〕
shell-still	简单的蒸馏釜
shell-still battery	瓮电池
shell-to-shell distance	(容器外)壳间距离
shelly texture	贝壳状结构
shepherd's purse	芥菜
sherardised coating	粉末渗锌(涂敷)层
sherardising process	粉末渗锌法；粉末渗锌过程
sherardising temperature	粉末渗锌温度
sherardizing	镀(上)锌粉
sherardizing galvanizing	粉末镀锌
Sherwood number	雪伍德数
sherwood oil	石油醚
Shewhart's control chart	休哈特控制图
shibuol	柿涩酚 $C_{14}H_{20}O_9$
shield	①罩②屏③保护
shielded	隔离的
shielded arc welding	气体保护焊
shielded cave	屏蔽室*
shielded-cave facility	屏蔽(地下)室设施
shielded cell	屏蔽箱*
shielded cubicle	屏蔽小室
shielded flame	屏蔽火焰*
shielded flask	屏蔽容器
shielded nuclide	屏蔽核素
shielded plasma	屏蔽式等离子体
shielded track calutron	密封屏蔽轨道式电磁分离器
shield gas	保护气体
shielding	①屏蔽*②隔离
shielding agent	掩蔽剂
shielding constant	屏蔽常数*
shielding effect	屏蔽效应*
shielding efficiency	屏蔽效率
shielding gas	保护气
shielding length	屏蔽长度
shielding low tension insulator	隔离低压绝缘子
shield retaining	隔离器
shield-row tube	屏蔽排管
Shiff base	泄馥基〔含氮基的香料〕
shift	①移动②变速；调挡③变速器④值班
shift converter	变换炉
shift factor	平移因子
shift reagent	位移试剂*
shift relaxation reagent	无位移试剂
shift technique	位移技术
shikimene	莽草素
shikimic acid	莽草酸
shikimin	莽草素
shikimitoxin	莽草毒
shikimol(=safrole)	黄樟(油)素
shikonin(=d-alkannin)	紫草宁
shilajatu	矢拉价土〔印度矿胶〕
shim	补条
shim coil	匀场线圈*
shimming	匀场*
shimoburo base	西母碱类
shimosite	西摩塞特〔炸药〕
shiner	①发光物；发光体②(无光漆)发花现象；吸光(漆病)
shingle	木瓦
shinglelike structure	(片状)颜料在涂料中形成的敷瓦状结构；叠瓦状结构
shingler	挤渣压力机；锻铁机
shingle stain	板墙油色；板墙涂料；屋顶涂料
shingling	挤渣；压挤；锻铁
shingling hammer	锻锤；锻铁锤
shining ore	镜铁矿
shinomycin	志野霉素
shiny side	光面
shionon	紫菀酮 $C_{34}H_{56}O$
ship	①船②(装)运
ship auger	(软膏或半固体石油产品采选试样)舟形棒
shipboard irradiation (of seafood)	(海味水产品)船上照射
shipboard-type	船用型；舰用型
ship bottom anti-fouling paint	船底防污漆
ship-bottom paint	船底涂料；船底漆
ship hull paint	船壳漆；船体漆
ship in bulk	散装船；散装发运
ship oven	船式炉
shipper/receiver difference(SRD)	运输容器/接收容器计量差
shipper enrichment	发方(测得的或标识的)浓缩度
shipper pole	移带
shipper rod	运转杆〔机工〕
shipping	①装运②航业
shipping capsule	运输小盒
shipping case	装运箱；装搬箱
shipping damage	运输损伤

shipping point 航运点；航运站；石油仓库
shipping specifications 航运规程
shipping terminal 航运基地；码头仓库
shipping ticket 运货单；航运文件
shipping weight 航运重量
ship's bottom paint 船底油漆
ship tanks 油船
ship-to-shore pipe line 船岸输油管〔将油从船输到岸上用的输油管〕
ship-to-shore submarine pipe line 船岸海底输油管〔将油从船输到岸上用的海底输油管〕
shirlacrol 酚焦油和氢氧化钠溶液
shirlan 歇利防霉剂
shirlastain 鉴别织物纤维染色剂
shish-kebab 串型多晶结构
shish-kebab structure 串晶结构*
shiu oil 芳油；芳樟油
shive 亚麻皮；亚麻屑
shivering 粉碎
shock ①震动②电震③休克④爆破⑤冲击
shock absorber 减震器
shock-absorber oil 减震油
shock-absorbing capacity 防充能力；缓冲性能
shock-absorbing desk 消震台
shock absorbing pad 防震垫
shock absorption 冲击吸收；缓冲作用
shock chilling 骤冷；激冷
shock coagulation 冲击凝固
shock cooling 骤冷；激冷
shock discharge 冲击卸料；撞击排料；震动排放
shock elasticity 冲击弹性
shock factor 冲击系数
shock-heating 骤热
shock liquid 防震液
shock load 冲击载荷
shock loss 冲击损失
shock molding 冲压模塑
shock motion 冲击运动
shock pendulum 冲击摆锤
shock ply 冲击层
shockproof ①耐震的；抗震的②耐电击的
shock reducer 缓冲器
shock-resistance 耐冲击性；防冲性
shock resistant 抗震物
shocks ①冲击②缓冲装置
shock strength 冲击强度
shock suppressor 吸震器
shock testing machine 震动试验机

shock test of extreme-pressure lubricant 特压润滑剂的震动试验
shock test of filter 过滤器震动试验
shock test of greases 润滑脂震动试验
shock transmissibility 冲击传递率
shock tube 激波管*
shock tube method 冲击波管法
shock wave 冲击波
shoddy ①再生胶②废胶粉③低吸再生胶④回弹毛
shoe ①瓦(状物)②导向板
shoe calender 鞋底压延机
shoe counter 皮鞋支跟
shoe cream 皮鞋油
shoe lining leather 鞋里革；鞋里皮
shoemaker 制鞋工人
shoe of tank 油罐浮顶座板
shoe stiffener 皮鞋硬衬
shoe strings ①皮鞋带②管状石油层
shogaol 生姜酚；姜烯酚 $C_{17}H_{24}O_3$
shogging motion 横移运动
shogyuene 牛樟烯
Shohl's micro buret(te) 肖尔氏微量滴定管
shonamic acid 肖楠酸
shonkinite 等色岩
shoot (植物)地上部分；枝条
shop air 车间气源
shop coat primer 预涂底漆
shop drawing 装配图；工作图
shop fabricated 车间制作的
shop line 车间风管
shop painting 预涂；工厂涂装(施工)
shop primer 工厂底漆
Shore creep 肖氏蠕变
Shore durometer 肖氏硬度计
Shore durometer hardness 肖氏压痕硬度
Shore hardness(=durometer-hardness) 肖氏硬度
Shore hardness tester 肖氏硬度计
shore pipe line 海滨管道
Shore scleroscope 肖尔反弹式硬度计
shore tank 海滨油罐
shorl 黑电(气)石
shorl rock 黑电(气)石岩
short annealing 快速退火
short bed continuous development 短床连续展开
short boiling range product 窄沸程产品
short-chain alcohols 短链醇类
short-chain ammonium 短链铵
short-chain branch 短支链*

short-chain olefin　短链烯烃
short circuit　①泄流②短路
short-circuit amperometric titration　短路安培滴定；短路电流滴定
short-column chromatography　短柱色谱(法)
short-cooled fuel　短期冷却(的核)燃料
shortcut attachment　简便切断装置
short-cut method　简便方法
short cycle curing　短时硫化
short cycle reclaiming process　快速再生法
short damping　快速蒸熏
short ends　①短材②箱盒用材③薪材
shortened phosphating process　简化磷酸盐处理过程
shortening　缩短
shortening property　起酥性
shorter-chain-length polyamide　短链聚酰胺
short fiber　短纤维
short-fibered asbestos　短绒石棉
short-fiber grease　短纤维润滑脂
short filling　注料不足
short-flame coal　短焰煤
short-gauge compression test　短标距压缩试件法
short grained　细粒的
short grained wood　细纹木材
shorthand　速记法
short iron　脆性铁
shortite　碳酸钠钙矿
short-life intermediates　短寿命中间化合物
short-lived activity　短期活度
short lived nuclide　短寿命的核素
short-lived transient　短寿(命)瞬态(产物)
short mix　快速混合〔专指油脂碱炼过程〕
short molding　欠压；缺胶
short-necked flask　短颈瓶
short nipple　短螺纹接套
short oil　短油；聚合程度不大的油
short oil alkyd　短油醇酸树脂
short oil varnish　短油清漆
short out　短路
short pass　短路
short paste　贫油色浆
short-path distillation　短路蒸馏；分子蒸馏；极窄馏分蒸馏
short-path thermal desorption　短程热解析(进样)
short period　①短周期②短(时)期
short production run　小批量生产
short range　短程；近程
short range force　短程力*

short-range interaction　近程相互作用
short-range interchannel effect　短程通际效应
short-range intramolecular interaction　近程分子内相互作用*
short range order　短程有序*
short-range perturbation theory　短程微扰理论
short-range structure　近程结构*
short ream　小令〔每令480张纸〕
short residence time(SRT)　短停留时间
short residuum　脆残油；真空(塔)蒸馏残渣；浓缩残油
short ripening　短期熟成
short rubber　低级橡胶〔生胶〕
short run　小批量
short run production　小批量生产
short run test　短程试验；快速试验
short shot　缺料
short spacing　短间距
short stapled cotton　粗绒棉
short steam　低压蒸汽
short stereosequence　立构短链
short stock　短纤维纸浆
short stopped chain　断链
short stopped polymerization　速止聚合*
short stopped reaction　速止反应；中止反应
shortstop(per)　速止剂
short stopping　速止(作用)；急速中止
short stopping agent　速止剂；链(锁)终止剂
short stopping of chain　链的速止
short stopping of reaction　反应的急速中止
short-stroke press　短程压缩泵
short tandem repeat　短串联重复序列
short term curing　短时硫化
short-term denier uniformity　短程旦数均匀度；短程纤维均匀性
short-term heat resistance　短时耐热性；短时耐高温性
short ton　短吨〔1短吨=0.9072吨〕；美(吨)〔=2000磅〕
short volume ratio　低容积比；低液体系数
short wave　短波
short wavelength cutoff　短波限
short wavelength limit　短波限
short weight　欠重
short wide columns　短粗柱
shot　①发射②细粒；小球
shot blast　喷丸处理
shot-catalyst principle　短催化原则〔利用比催化密实的惰性热载体操作的原则〕
shot-dye　闪色染色
shot effect　①闪光效应②散粒效应

shot-gun powder 鸟枪火药
shot lubrication 油枪润滑
shot mill 钢丸磨；丸磨机
shot noise 散粒(效应)噪声
shot peening ①喷丸加工；喷丸处理②喷丸硬化
shot point 爆(心)点
shot survey 爆炸测量(石)
shot-to-shot turnaround 注射间隔时间
shoulder ①肩②颈皮③钝边④台肩
shoulder angle 肩斜角
shouldered impeller 轮背凸起的叶轮
shoulder of absorption curve 吸收曲线肩(峰)
shoulder peak 肩峰
shoulder seal 轴向密封；轴封
shoulder tool 肩刀
shovel 铲(子)
shovel sampling 用铲取样
shoving hole 推料孔
shoving pan 推料锅
shoving rake 推料耙
showdomycin 焦土霉素
shower ①大量降下物②大量③大量下降
shower coat 喷淋涂装
shower coater 喷淋涂装机；喷淋涂布机
shower cooler 喷淋冷却器
shower cooling 喷射冷却
shower pipe 喷水管
shower water 喷淋水
showing through 不盖底
shows 产量〔石油，煤油〕
shoyualdehyde β-乙酰巴豆醛
shrapnel 榴霰弹
shredded rubber 橡胶屑
shredder 撕碎机；粉碎机；研磨机
shredding 撕碎；粉碎
shredding machine 撕碎机
shrinkage ①(收)缩②缩误
shrinkage allowance 收缩留量
shrinkage cavity 缩孔
shrinkage crack 缩裂；收缩裂缝
shrinkage cracking 缩裂
shrinkage curve 收缩曲线
shrinkage factor 收缩因数
shrinkage in addition polymer 加聚物的皱缩(作用)
shrinkage mark 收缩皱纹
shrinkage ration 收缩率
shrinkage strain 收缩变形
shrinkage stress 收缩应力

shrinkage temperature 收缩温度
shrink bobbin 收缩筒管
shrink-fit vessel 套合式容器；缩合式容器
shrink fixture 防缩器
shrink head 收缩头；冒口
shrinking 收缩
shrinking effect 收缩效应
shrinking temperature 收缩温度
shrink jig (=shrink fixture) ①防缩模②防缩夹具
shrink leather 皱纹革
shrinkproof top 防缩羊毛条
shrink rate 收缩率
shrink resistance 防缩性；抗缩性
shrink stress 收缩应力
shrink tension 收缩张力
shrivel 皱缩
shrivel finish (=ripple finish) 皱纹(罩面)漆
shrivelled 皱织
shrivelling 皱纹；折纹
shrivel varnish 皱纹清漆
shroud ①屏蔽；掩蔽②套筒；罩盖③侧板
shrouded impeller 闭式叶轮
shrunk-and-peened flange 皱合法兰
shrunk-and-rolled flange 皱卷法兰
shuffle box 换气箱
shunt ①支路②分流器
shunt filter 支管过滤器
shunting action 分流作用
shunt meter 并联电流计
shunt trip coil 并联脱扣线圈
shurry pump ①淤浆泵②料浆泵
shut-down 停工；关闭
shut-down inspection 停工检查
shut-down maintenance 停工维修
shut-down period 停工期
shut-down schedule 停工期工作计划
shut-in storage 封井储油
shut-off block 关闭部件
shut-off head 关闭压头
shut-off nozzle 开关式喷嘴；阀控喷嘴
shut-off valve 关闭阀
shutter ①(光)闸②开闭器
shutting-down 关闭；停止；停工
shuttle ①梭(子)②穿梭
shuttle-block oil pump 有导板的旋转油泵
shuttle conveyor 穿梭运输器
shuttle mechanism 穿梭机制；往返机制
shuttle-type feed 摆动式供料

shuttle vector 穿梭载体
sial 硅铝带
sialic acid(=sialinic acid) 唾液酸
sialidase 唾液酸酶
sialinic acid (=sialic acid) 唾液酸
sialon 硅铝氧氮陶瓷
siamese blow mo(u)lding 连体式吹塑；一模多穴吹塑法
siaresinolic acid 斯阿树脂酸〔泰国安息香中的一种树脂酸〕$C_{30}H_{48}O_4$
siaresinotanrol 斯阿树脂鞣醇
sib-analysis 同胞分析
sibucao 苏木
siccation 干燥(作用)
siccative ①(=drier)催干剂；干料②干燥的
siccative oil 干性油
siccity 干燥
sickle pump 镰式泵；具镰刀形分布器的齿轮泵
sidac 玻璃纸
side angle press 角(式)压机
side application 旁施
side arm 侧臂；单面横臂
side band 边带*
side bottom 旁脚；塔下部引出物
side by side 并列；并排
side-by-side composite fibre 并列型复合纤维
side cap 旁盖
side chain 侧链；支链
side chain bromination 侧链溴代作用
side chain carbon 侧链碳(原子)
side chain chlorination 侧链的氯代作用
side chain compound 侧链化合物
side chain fluorination 侧链氟代作用
side chain halogenation 侧链卤代作用
side chain iodination 侧链碘代作用
side chain isomer 侧链异构体
side chain isomerism 侧链异构作用
side chain liquid crystalline polymer 侧链型液晶聚合物
side chain motion 侧链运动
side chain nitrogen 侧链氮(原子)
side chain radical 侧链基
side creel 落地纱架
side-cut 侧馏分；侧取馏出物
side-cut distillate(=side-draw; side-draw naphtha) 侧馏分；侧取馏出物
side discharge 侧卸
side-draw(=side-cut distillate) 侧馏分；侧取馏出物
side-draw control 侧取控制
side-draw naphtha(=side-cut distillate) 侧馏分；侧取馏出物

side draw plate(=side draw tray) 侧取塔板
side draw tray(=side draw plate) 侧取塔板
side effect ①边界效应②副作用
side feed opening 侧面进料口
side-fired recirculating still 侧热循环釜；炉侧加热与气体循环管式炉
side-flue oven 侧面出烟炉
side frame 侧面机架；侧壁；侧板
side group(=pendant group; leteral group) 侧基
side injection molding 侧面注压法
side isomery 侧链异构(现象)
side leather 牛皮半张鞋面革
side-line product 副线产品
side loading 侧加料
side neck 侧筒；支管
side-neck flask 支管烧瓶
side-on complex 侧连配合物
side-on coordination 侧向配位*
side outlet elbow 支流拐管
side outlet tee 支流三通管
side plate of rolls 挡轮板；挡辊板
side pressure 侧压
side product 副产品
side rail 导轨
side ram press 角式水压泵
side reaction 副反应*；支反应
side reaction coefficient(=alpha coefficient) 副反应系数*；α系数
side-reaction-protecting capacity 副反应防止能力
side reflux 侧回流
side relief 侧面放汽(管路)
side ring (多环化合物)侧环
siderite ①菱铁矿②陨铁③蓝石英
siderography ①磷铁矿表面研究②钢板雕刻术
siderolite 铁陨石
side roll 旁辊〔压延机〕
siderology(=siderurgy) 冶铁学
siderophile element 亲铁元素
siderophore 铁载体
sideroplesite 镁菱铁矿
sideropyrite 黄铁矿
siderostat 恒光轴器
siderotilate 五水硫酸亚铁矿
side run-off 侧线馏分；塔测抽出物
siderurgy(=siderology) 冶铁学
side seal pillar 端面气封柱；端面密封柱
side spacing 旁侧间距
side stream 侧线馏分；塔侧抽出物

sidestream distillation 侧流蒸馏
side-stream stripper(=side stripper) 侧线(馏分)汽提塔
side stripper(=side-stream stripper) 侧线(馏分)汽提塔
side-stripping column 侧线汽提塔
side stud bolt 双头螺栓
side substitution 侧链取代(作用)
side sway 横向摆动
side theory 侧链免疫理论
side throw 横向摆动
side thrust 侧压
side-to-side baffles 缺圆折流板
sidetrim unit 两侧裁边装置
side tube 支管；侧管
side-tube flask 支管烧瓶
side valency 副价
sidewall fender 胎侧护胶
side wall filter 侧壁过滤器
sidewall stiffness 胎侧刚性；胎侧刚度
side-window X-ray tube 侧窗式X射线管
sidonal 奎尼酸哌嗪
Sidot's blende 人造硫化锌
Siegbahn notation 西格巴恩标志
siege 炉底；炉床
Siemens furnace 西门子炉
Siemens-Halske copper process 西门子-哈尔斯基炼铜法
Siemens Martin furnace 西门子-马丁(返焰)炉；平炉
Siemens-Martin steel 西门子-马丁钢；平炉钢
Siemens ozonizer 西门子臭氧化器
Siemens process 西门子(炼铁)法
Siemens producer 西门子煤气发生炉
sienna 黄土(颜料)
Sierra Leone copal 塞拉利昂玷玛脂〔颜色浅可制白漆〕
sieve ①筛(子)②筛(分)
sieve analysis 筛析〔粒度分析〕
sieve catalyst 分子筛催化剂
sieve cloth 筛布
sieve diaphragm 筛网隔膜
sieve effect 筛孔效应
sieve grate 筛箅
sieve mesh 筛眼；筛孔
sieve method 筛选法；逐步淘汰法
sieve opening 筛目；筛孔
sieve-plate 筛板
sieve-plate column(=sieve-plate tower) 筛板塔
sieve-plate tower(=sieve-plate column) 筛板塔
sieve ratio 筛比；筛制
sieve residue 筛渣；筛余物
sievert(Sv) 西弗特〔剂量当量单位，1 Sv=100rem〕

sieve scale 筛垢
sieve series 分级筛；系列筛
sieve shaker 摇筛机
sieve sorbent ①筛状吸附剂②分子筛
sieve sorption pump 分子筛吸附泵
sieve test 筛分试验
sieve texture 筛状结构
sieve tray 筛板；筛盘
sieve-tray column(=sieve-tray tower) 筛板塔
sieve-tray tower(=sieve tray column) 筛板塔
sieve vent 筛分装置放空口
sieve with blower 鼓风筛；风选筛
sieving 筛(分)
sieving action 筛分作用
sieving analysis 筛析
sieving department 筛选车间
sieving machine 筛选机；筛分机
sieving mechanism (喷粉柜内过喷粉末的)筛分装置
sieving medium 筛分介质
sieving test 筛分试验
si face si 面*
sift ①筛(分)②精查
sifter 筛(子)
sifting ①筛选；过筛②过滤
sifting machine 筛选机；筛分机
sift-proof bin 防漏式储存罐
sight bubbler 可视起泡器
sight feed 可视进料
sight-feed lubricator(=sight-feed oiler) 可视进料润滑器；控制润滑
sight-feed oiler(=sight-feed lubricator) 可视进料润滑器；控制润滑
sight flow indicator 可视流量指示器
sight gauge (测量煤油灯火焰高度与宽度的)观测尺
sight glass 视镜
sight hole 验火孔
sight indicator 可视指示剂
sighting 涂色；着色
sighting agent 可视指示剂
sigma complex σ复合物
sigma effect σ效应
sigma gas calorimeter σ气体量热计
sigma mixer(=Z-blade mixer) 曲拐式搅拌机
sigmamycin 十八霉素〔四环素与竹桃霉素合剂〕
sigma phenomenon σ现象
sigma reaction σ反应
sigmatropic reaction σ移位反应；单键转移反应
sigmatropic rearrangement σ迁移重排*

sigmatropic rearrangement reaction　σ迁移反应
signal　信号
signal amplification　信号放大
signal amplifier　信号放大器
signal bandwidth　信号带宽
signal carrying line　信号载线
signal center　①征象峰②预兆峰③信号峰
signal intensity　信号强度
signal level　信号电平
signal multiplier method　系数倍率法
signal-noise ratio (S/N)　信噪比*
signal of dispersion　色散信号
signal pattern　信号图像；信号图案
signal peak　信号峰(值)
signal peptide　信号肽
signal recovery system　信号再生系统
signal sequence　信号序列
signal-to-background ratio　信背比
signal-to-noise ratio　(信)(号)噪(声)比
significance　显著性
significance level　显著性水平*
significance test　显著性检验*
significance testing procedure　显著性检验程序；显著性检验步骤
significant difference　显著性差异*
significant figure　有效数字*
significant level　显著性水平
signification test　显著性检验
sign of rotation　旋转标志
sign paint　广告(牌)漆
sign test　符号检验*
sig water　稀碱液〔用于涂饰前揩拭皮面〕
Sikes hydrometer　席克斯比重计〔测醇液浓度〕
sikimin　八角萜；日本八角烯
sila-　硅杂
silage　青储(饲)料
silance finish　硅烷(类)表面涂饰剂
silandiol　硅烷二醇
silane(=silicohydride)　硅烷　Si_nH
silane coupler　硅烷偶联剂
silane coupling agent　硅烷偶联剂
silane finish　硅烷(类)表面涂饰剂；硅烷处理(剂)
silane resin　硅烷树脂
silanion　硅负离子　R_3Si^-
silanization　硅烷化
silanized　硅烷化过的
silanized support　硅烷化载体
silanizing　硅烷化

silanizing agent　硅烷化剂
silanol　硅烷醇
silanol-bromine complex　硅烷醇-溴配合物
silanol group　硅烷醇基
silanophilic interaction　亲硅醇基效应；亲硅羟基作用
silantriol　硅烷三醇
silastic　硅橡胶
silastic paste　硅橡胶浆
silastic stocks　硅橡胶坯料
silastomer(=silicone rubber)　硅橡胶
silathiane　硅硫烷
silavans　含硅碳氮的耐高温聚合物
silazane　硅氮烷　$H_3Si(NHSiH_2)_nNHSiH_3$
silazanecarboxylic ester　硅氮烷羧酸
silazine link　硅氮键
silbamin　氟化银
silberol　苯酸磺酸银；羟基苯磺酸银
silchrome (steel)　硅铬耐热钢
silencer　消声器
silencer-filter　消声过滤器
silencer mounting　消音器〔即防振器〕
silene　硅宾；硅烯；烷基亚甲硅基　$R_2Si=$
silent discharge　无声放电
silent gear　无声齿轮；塑料齿轮
silent paint　消音漆
Silesia explosive　氯酸钾、硝化树脂强炸药
Silesian furnace　蒸锌炉
silex　①石英粉②燧石
silex glass　石英玻璃；硅玻璃
silexite　英石岩
Silex process　西莱克斯过程〔硅胶吸附萃取色谱法从辐照燃料中回收铀钚的过程〕
silhouette　①线条②轮廓③侧面影像
silhydrite　水硅石；水石英
Silibinin　水飞蓟宾
silica　硅石；二氧化硅
silica abrasive　石英粉磨料
silica adsorbent　硅胶吸附剂
silica aerogel　硅补强剂；白炭黑
silica aerogel-thickened grease　硅石-空气凝胶稠化的润滑剂
silica-alumina anionic leg　硅铝阴离子结构
silica-alumina catalyst　硅铝催化剂
silica-alumina hydrogel　硅铝水胶
silica-alumina ratio　(氧化)硅铝比
silica-base catalyst　硅基催化剂
silica black　硅酸黑〔含少量炭黑的氧化硅〕
silica brick　硅砖

silica carbide 碳化硅
silica cement ①硅石水泥；火山灰水泥②硅胶泥
silica chain structure 二氧化硅链结构
silica CMC 硅胶CMC〔硅胶中加羧甲基纤维素〕
silica-coated lead chromate 二氧化硅处理的铬酸铅；二氧化硅包核铬酸铅
silica coke oven 硅石炼焦炉；用硅砖作内衬的焦化炉
silica compound 白炭黑(填充)胶料
silica content 硅酸含量；二氧化硅含量
silica core (包核颜料的)二氧化硅核
silica dish 石英碟〔测定汽油胶质用〕
silica fiber 二氧化硅纤维；石英纤维
silica fibre reinforced metal 硅石纤维增强金属
silica flask 石英烧瓶
silica flatting agent 二氧化硅消光剂
silica flour(=crystalline flour) 晶粉；硅粉〔填料〕
silica gel (氧化)硅胶
silica-gel adsorption 硅胶吸着
silica-gel bead 硅胶珠
silica gel chromatography 硅胶色层(分离)法
silica gel chromatoplate 硅胶色谱板
silica gel desiccator 硅胶式干燥器
silica-gel drier 硅胶干燥剂
silica-gel filler 硅补强剂；白炭黑
silica gel G 硅胶G〔硅胶中含10%～15%的烧石膏〕
silica-gel grease 硅胶填充润滑脂
silica gel H 硅胶H
silica-gel percolation 硅胶渗滤；硅胶渗滤分馏
silica-gel plate 硅胶板
silica-gel powder 硅胶粉
silica gel-sintered glass plate 硅胶烧结玻璃板
silica-gel sphere 硅胶球
silica-gel substrate ion exchanger 硅胶基质离子交换剂
silica gel thin-layer chromatography 硅胶薄层色谱(法)
silica GF 254 硅胶GF254〔硅胶中加烧石膏和254纳米发光的荧光剂〕
silica glass 石英玻璃
silica-hexa-resorcinol 间甲白；白炭黑-六亚甲基四胺-间苯二酚
silica hydrogel 二氧化硅水凝胶〔合成含水二氧化硅〕
silica index 硅氧指数
silicam 二亚氨基硅 $Si(NH)_2$
silica-magnesia catalyst 硅镁催化剂
silica minerals 硅氧矿物
silica modulus 硅氧系数
silican carbide sharpening stone 金刚砂磨石
silicane 硅烷*
silica pigment 硅补强剂；白炭黑

silica refractories 硅土耐火制件
silica removal 除硅；去硅
silica replica 二氧化硅复制品
silica rich water 富硅水
silica-rich zeolite 高硅沸石
silica sand 硅砂；石英砂
silica skeleton 硅石骨架
silica sol 硅溶胶
silica solution 硅溶胶
silica tannage 硅鞣法
silica tanning 硅鞣
silicate 硅酸盐(酯)
silicate adhesive 硅酸盐黏合剂
silicate analysis 硅酸盐分析
silicate cement 硅酸盐水泥
silicate ceramics 硅酸盐陶瓷
silicate coating 硅酸盐涂料
silicate-containing detergent 含硅酸盐洗涤剂
silicate crown 硅酸盐冕玻璃
silicate degree 硅酸度
silicated soap 硅酸盐皂
silicate ester 硅酸酯
silicate glass 硅酸盐玻璃
silicate glass fiber 硅酸盐玻璃纤维
silicate minerals 硅酸盐矿物
silicate of zinc 硅酸锌 Zn_2SiO_4
silicate paint 硅酸盐漆
silicate pigment 硅酸盐颜料
silicate platelets 硅酸盐片晶；硅酸盐小片
silicate polymer 硅酸盐聚合物
silicate rock 硅石
silicate soft soap 硅酸盐软皂
silica test 硅含量测定
silicate wool 矿棉
silication(=silicatization) 硅化作用*
silica-titania ceramic (氧化)硅-钛陶瓷
silicatization(=silication) 硅化作用
silica TLC plate 硅胶薄层色谱板
silica tube 石英管；硅管
silica type filler 硅补强剂；白炭黑
silica wall-coated open tubular column 二氧化硅覆壁开管柱
silica white 硅补强剂；白炭黑
silica wool(=quartz wool) 石英棉
silica-zirconia catalyst 硅锆催化剂
siliceous 硅质的；含硅的
siliceous algae 硅藻
siliceous deposits 硅沉积；石英状沉积物

siliceous earth 硅藻土	silicon boride 硼化硅 SiB_3; SiB_6
siliceous filler 硅质填料	silicon-boron coating 硅硼涂(敷)层
siliceous reinforcing agent 硅补强剂；白炭黑	silicon brass 硅黄铜
siliceous rocks 硅质岩类	silicon bromide chloride fluoride 氟氯溴化硅
siliceous sinter 硅渣	silicon bronze 硅青铜
siliceous slag 硅土矿渣	silicon carbide(=carborundum) 碳化硅；金刚砂
silice-reinforced SBR 白炭黑补强丁苯橡胶	silicon carbide abrasive 碳化硅磨料；金刚砂磨料
silicic acid 硅酸	silicon carbide canals (管式炉的)碳化硅烟道
silicic acid anhydride 硅(酸)酐；二氧化硅 SiO_2	silicon carbide coated fibre 碳化硅涂覆纤维
silicic acid ester 硅酸酯	silicon carbide fiber 碳化硅纤维
silicic acid tanning agent 硅酸鞣剂	silicon carbide heating element 碳化硅加热原件
silicide 硅化物	silicon carbide waterproof paper 碳化硅防水砂纸
silicification(=silification) 硅化作用	silicon-carbon bond 硅-碳键
silicified tuff 硅化凝灰岩	silicon chlorides 氯化硅 Si_nCl_{n+2}
silicified wood 硅化木	silicon-containing copolymer 含硅共聚物
silicifying 硅化作用	silicon-containing polymer 含硅聚合物
silicious 硅质的	silicon controlled rectifier 硅可控整流器
silicious clay 硅土	silicon defoamer 硅质去沫剂
silicoacetic acid 硅乙酸 CH_3SiOOH	silicon detector 硅检测器
silico acetylene 硅乙炔 $HSiSiH$	silicon dioxide 二氧化硅 SiO_2
silicoamino-acid 硅氨基酸	silicone (聚)硅氧烷
silicobenzoic acid 硅苯甲酸；苯基甲硅酸 $(C_6H_5SiOOH)_n$	silicone alkyd 有机硅(改性)醇酸树脂
	silicone-alkyd combination 有机硅-醇酸的化合作用
silicobromoform 硅溴仿；三溴甲硅烷 $HSiBr_3$	silicone alloy 聚硅氧烷
silico butane 丁硅烷 Si_4H_{10}	silicone antifoam agent 有机硅消泡剂
silico-calcium 硅钙	silicone base(=silicone gum) 硅橡胶
silicochloroform 硅氯仿；三氯甲硅烷 $HSiCl_3$	silicone coating 硅酮涂层
silico-chromium 硅铬合金	silicone elastomer 硅氧烷弹性体
silico-decitungstic acid 四水合十钨硅酸	silicone emulsion 乳化硅油〔脱模剂〕
silicoethane 乙硅烷 Si_2H_6	silicone fluid 硅树脂液；硅酮液
silico ethylene 硅乙烯 $H_2Si=SiH_2$	silicone gasket 硅橡胶垫圈
silicofluoric acid 硅氟酸；六氟合硅氢酸 $H_2[SiF_6]$	silicone grease 硅(润滑)脂
silicofluoride 氟硅酸盐；氟硅化物	silicone gum 硅橡胶纯胶料
silicoglaserite α-硅钙石	silicone hydride 硅烷
silicoheptane 三乙基甲硅烷；三乙基甲硅 $SiH(C_2H_5)_3$	silicone insulating coating 有机硅绝缘涂料
silicohydride(=silane) 硅烷	silicone macromolecule 有机硅大分子
silicoiodoform 硅碘仿；三碘甲硅烷 $HSiI_3$	silicone-nitrile rubber 硅腈橡胶
silicol(=hydroxysilane) 羟基硅烷 R_3SiOH	silicone oil 硅(氧烷)油
silicomagnesiofluorite 氟硅钙镁石；硅镁萤石	silicone oil bath 硅油浴
silicomanganese 硅锰合金	silicone oil emulsion 硅油乳液
silico-manganese steel 硅锰钢	silicone paint 有机硅漆；硅酮涂料；聚硅氧烷漆
silicomethane(=monosilane) 硅甲烷；甲硅烷 SiH_4	silicone plastics 有机硅塑料；聚硅氧烷塑料
silicomolybdate 硅钼酸盐	silicone polish 有机硅抛光剂
silicomolybdic acid 硅钼酸	silicone polymers 硅氧烷聚合物
silicomonazite 硅独居石	silicone putty 有机硅腻子
silicon 硅〔14号元素，化学符号 Si〕	silicone resin 有机硅树脂*
silicon alkyl 烃基硅	silicone resin coating 有机硅树脂涂料
silicon alloys 硅合金(类)	silicone rubber(=silastomer) 硅橡胶

silicone rubber gasket 硅(橡)胶垫圈
silicone rubber gun-applied mastic sealant 喷涂型硅橡胶厚浆填缝料
silicone rubber membrane 硅橡胶膜
silicone rubber septum 硅橡胶隔片
silicon ester 硅酸酯
silicon ethyl 四乙基硅〔因为有乙硅烷、甲硅烷,故基字不宜略去〕 $Si(C_2H_5)_4$
silicone type surface active agent 硅氧烷系表面活性剂
silicone varnish 有机硅清漆；聚硅氧烷清漆
silicone-wax polish 有机硅-石蜡抛光剂
silicon hexaboride 六硼化硅 SiB_6
silicon hexachloride 六氯化二硅；氯乙硅烷
silicon hydride 氢化硅；硅烷
silicon hydride chlorid (氢)氯硅烷；氢氯化硅
silicon-hydrogen bond 硅-氢键
silicon hydroxide 氢氧化硅；原硅酸 $Si(OH)_4$
siliconic acid 烃基硅羧基酸 RSiOOH
silicon intensified target 硅增强靶
silicon intensified target vidicon 硅增强靶光导摄像管
silicon iodide chloride 氯碘化硅
silicon iron sheet 硅铁片
siliconising(=siliconizing) (扩散)渗硅；硅化处理
silicon isothiocyanate 异硫氰酸硅
siliconiting 硅烷化
siliconium ion 硅正离子
siliconization 硅氧烷化
siliconized plate 硅钢片
siliconizing(=siliconising) (扩散)渗硅；硅化处理
silicon methyl 四甲基硅 $Si(CH_3)_4$
silicon modified rubber 硅(改性)橡胶
silicon mold release agent 硅脱模剂
silicon nitride 四氮化三硅 Si_3N_4
silicon nitride whisker 氮化硅须晶
silicon-nitrile rubber 硅-腈橡胶
silicono 硅羧基 HOOSi—
silicon oil 硅油
silicon-organic dicarboxylic acid 有机硅二羧酸
silicon oxide 氧化硅 $SiO;SiO_2$
silicon oxide polymer 氧化硅聚合物
silicon oxygen bond 硅氧键
silicon oxysulfide 氧硫化硅 SiOS
silicon-oxy tetrahedron 硅氧四面体
silicon-photodiode detector 硅光二极管检测器
silicon polymer 硅聚合物
silicon rectifier welder 硅整流焊机
silicon rhodanide 硫氰酸硅
silicon rubber 硅橡胶*

silicon selective detector 硅选择性检测器
silicon-silicon bond 硅-硅键
silicon steel 硅钢
silicon steel sheet 硅钢片
silicon steel sheet varnish 硅钢片清漆
silicon sulfide 硫化硅 $SiS;SiS_2$
silicon target vidicon tube 硅靶光导摄像管
silicon tetrabutyl 四丁基硅 $Si(C_4H_9)_4$
silicon tetrachloride 四氯化硅
silicon tetraethyl 四乙基硅 $Si(C_2H_5)_4$
silicon tetrafluoride 四氟化硅
silicon tetraiodide 四碘化硅 SiI_4
silicon tetramethyl 四甲基硅
silicon tetraphenyl 四苯基硅 $Si(C_6H_5)_4$
silicon tetrapropyl 四丙基硅 $Si(C_3H_7)_4$
silicon thiocyanate 硫氰酸硅
silicon-treated paper 用硅处理的纸
silicon valley 硅谷
silicon varnish 硅树脂清漆
silicon wafer 硅片
silicon-zirconium 硅锆合金
silicoorganic compound 有机硅化合物
silicooxalic acid 硅草酸 $(SiOOH)_2$
silicophosphate 硅磷酸盐
silicopropane 丙硅烷 Si_3H_8
silico triboride 三硼化硅
silicotungstate 硅钨酸盐
silicotungstic acid(=silicowolframic acid) 硅钨酸 $SiO_2 \cdot 12WO_3 \cdot 4H_2O; H_8[Si(W_2O_7)_6]$
silico-tungstic acid test 硅钨酸试验〔测定灯油颜色稳定性的试验〕
silicowolframic acid(=silicotungstic acid) 硅钨酸
silicyl 甲硅烷基 H_3Si—
silicylene 亚(甲)硅烷基
silicyl oxide 氧化硅烷
silification(=silicification) 硅化作用
silithiane 硅硫烷 $H_3Si(SSiH_2)_nSSiH_3$
silk 蚕丝
silk board 提花纸板
silk boiling 煮丝
silk boil-off liquor 煮丝废液
silk bolting cloth 丝(绢)筛布
silk coated wire 丝包线
silk-covered wire 丝包线
silk crepe 真丝绉
silk damping 泡丝
silk emulsion 丝光乳胶漆
silkete 丝光(工艺)

silk fabric 丝织物
silk gloss paint 丝光漆
silk glue 丝胶
silkiness 丝光性
silking 丝纹；丝光
silk knitted article 真丝针织品
silk spinning 丝纺
silkweed 海藻丝状纤维
silky fibre 丝纤维
silky luster 丝光光泽
sillimanite ware 硅线石器皿
siloxane(=oxosilane) 硅氧烷
siloxane crosslinking acrylic coatings 硅氧烷交联型丙烯酸涂料
siloxane-glycol copolymer 硅氧烷-乙二醇共聚物
siloxane polymer 硅氧烷聚合物
siloxy 甲硅烷氧基 H_3SiO-
silphenylene （耐热）硅亚苯基树脂
silumin 铝硅合金
silvecarvon 林香芹酮
silver ①银②银制物③银(制)的④似银的⑤镀银⑥变为银白色⑦银白色
silver acetate 乙酸银 $AgC_2H_3O_2$
silver acetylide 乙炔银
silver-alloy brazing 银钎焊
silver alloys 银合金
silver amalgam 银汞合金；银汞齐
silver amminobromide 溴化氨合银 $[Ag(NH_3)]Br$
silver amminochloride 氯化氨合银 $[Ag(NH_3)]Cl$
silver arsenate(=silver arseniate) 砷酸银 Ag_3AsO_4
silver arseniate(=silver arsenate) 砷酸银
silver arsenide 砷化银
silver arsenite 亚砷酸银 Ag_3AsO_3
silver azide 叠氮化银
silver bath 银锅；银浴器
silver beet 银甜菜
silver bicarbonate 碳酸氢银 $AgHCO_3$
silver bichromate 重铬酸银 $Ag_2Cr_2O_7$
silver blister 银泡
silver brazing 银焊
silver bromate 溴酸银 $AgBrO_3$
silver bromide 溴化银
silver bullion 银条
silver carbide 乙炔银 Ag_2C_2
silver carbonate 碳酸银 Ag_2CO_3
silver catalyst 银催化剂
silver chlorate 氯酸银 $AgClO_3$
silver chloride 氯化银

silver chloride electrode 氯化银电极
silver chloroacetate 氯乙酸银 $AgC_2H_2ClO_2$
silver chloroplatinate 氯铂酸银 $Ag_2[PtCl_6]$
silver chromate 铬酸银 Ag_2CrO_4
silver cinnamate(=silver cinnamylate) 肉桂酸银 $AgC_9H_7O_2$
silver cinnamylate (=silver cinnamate) 肉桂酸银
silver citrate 柠檬酸银 $AgC_6H_5O_7$
silver coated glass bead 涂银玻璃珠
silver-coated reaction vessel 镀银反应釜
silver coulometer 银电量计
silver cyanate 氰酸银 $AgOCN$
silver cyanide 氰化银 $AgCN$
silver diamminohydroxide 氢氧化二氨合银 $[Ag(NH_3)_2]OH$
silver diamminonitrate 硝酸二氨合银 $[Ag(NH_3)_2]NO_3$
silver dichromate 重铬酸银 $Ag_2Cr_2O_7$
silver dithionate 连二硫酸银 $Ag_2S_2O_6$
silvered film 镀银层
silvered glance(=argentite) 辉银矿
silvered (window) glass 镀银玻璃板
silver electrode 银电极；裸银电极
silver ferricyanide 氰铁酸银 $Ag_3[Fe(CN)_6]$
silver ferrocyanide 氰亚铁酸银 $Ag_4[Fe(CN)_6]$
silver fir 银枞
silver fir oil 银枞油
silver fluoride 氟化银 AgF
silver fluosilicate 氟硅酸银 $Ag_2[SiF_6]$
silver foil 银箔
silver fulminate 雷酸银
silver gauze 银丝网
silver glance 辉银矿
silver granule 银粒
silver group 银族
silver hydrogen phosphate 磷酸氢银 Ag_2HPO_4
silver hydrogen sulfate 硫酸氢银 $AgHSO_4$
silver hydroxide 氢氧化银 $AgOH$
silver hypochlorite 次氯酸银 $AgOCl$
silver hyponitrite 连二次硝酸银 $Ag_2N_2O_2$
silver hypophosphate 连二磷酸银 $Ag_2PO_3; Ag_4P_2O_6$
silver-impregnated zeolite （浸）银沸石
silveriness 银白(色)
silvering 镀银
silvering bath 镀银浴
silver ink 银粉油墨；铝粉油墨
silver iodate 碘酸银 $AgIO_3$
silver iodide 碘化银
silver ketene 烯酮银

silver lacquer 银器喷漆〔银器皿罩光漆〕
silver lactate 乳酸银 $AgC_3H_5O_3$
silver leaf 银箔
silverly ①银的②白色的③银制的
silver manganate 锰酸银 Ag_2MnO_4
silver-medal 银奖
silver mesh 银网〔做电极〕
silver metaphosphate 偏磷酸银 $AgPO_3$
silver microelectrode 银微电极
silver mirror 银镜
silver mirror reaction 银镜反应
silver mirror test 银镜试验*
silver molybdate 钼酸银 Ag_2MoO_4
silvern 银(制)的
silver necking 银颈
silver net 银网
silver nitrate 硝酸银
silver nitride 一氮化三银 Ag_3N
silver nitrite 亚硝酸银 $AgNO_2$
silver number 银值
silver ore 银矿
silver orthoarsenate(=silver orthoarseniate) 原砷酸银 Ag_3AsO_4
silver orthoarseniate(=silver orthoarsenate) 原砷酸银
silver orthoarsenite 原亚砷酸银 Ag_3AsO_3
silver oxalate 草酸银 $Ag_2C_2O_4$
silver oxide 氧化银 Ag_2O
silver paint 银粉漆；铝粉漆
silver paper 银箔
silver perchlorate 高氯酸银 $AgClO_4$
silver permanganate 高锰酸银 $AgMnO_4$
silver peroxide 过氧化银 $Ag_2[O_2]$
silver phenolsulfonate 苯酚磺酸银 $AgC_6H_5O_4S$
silver phosphate 磷酸银 $AgPO_3$; Ag_3PO_4; $Ag_4P_2O_7$
silver phosphide 二磷化银 AgP_2
silver plate 镀银件
silver plated 镀银的
silver plating 镀银
silver plating bath 镀银浴
silver point 银点
silver potassium cyanide 氰化银钾；氰银酸钾 $K[Ag(CN)_2]$
silver potassium fulminate 雷酸银钾
silver powder ①银粉②铝粉
silver pyrophosphate 焦磷酸银 $Ag_4P_2O_7$
silver reductor 银还原剂
silver seal 银印级〔立德粉的 ZnS 含量为 60%的品级〕
silver selenate 硒酸银 Ag_2SeO_4

silver selenide 一硒化二银 Ag_2Se
silver sesquioxide 三氧化二银 Ag_2O_3
silver silicofluoride 氟硅酸银 $Ag_2[SiF_6]$
silver-silver chloride electrode 银-氯化银电极
silver-silver iodide electrode 银-碘化银电极
silver soap 银光皂；珠光皂
silver sodium chloride 氯化银钠 $Na[AgCl_2]$
silver sodium cyanide 氰化银钠；青银酸钠 $Na[Ag(CN)_2]$
silver sodium thiosulfate 硫代硫酸一银三钠 $Na_3[AgSO_3S)_2]$
silver sol 银溶胶
silver solder 银焊条
silver soldering 银焊
silver spray process 喷银法〔表面处理〕
silver stibide 一锑化三银 Ag_3Sb
silver streak(ing) 银纹〔注塑品缺陷〕；浆糊斑
silver suboxide 一氧化四银 Ag_4O
silver sulfate 硫酸银 Ag_2SO_4
silver sulfide 硫化银 Ag_2S
silver sulfite 亚硫酸银 Ag_2SO_3
silver sulfophenylate 苯酚磺酸银 $AgC_6H_5O_4S$
silver tartrate 酒石酸银 $Ag_2C_4H_4O_6$
silver telluride 碲化银 Ag_2Te
silver tetraborate 四硼酸银 $Ag_2B_4O_7$
silver thiocyanate(=silver thiocyanide) 硫氰酸银 $AgSCN$
silver thiocyanide(=silver thiocyanate) 硫氰酸银
silver thiosulfate 硫代硫酸银 Ag_2SO_3S; $Ag_2S_2O_3$
silver titration coulometer 银滴定电量计
silver tree 银树
silver trinitrophenolate 三硝基苯酚银
silver voltameter 银电量计
silverware 银器
silver wattle 银荆树
silverweed 野艾菊
silver white 银白
silver wire 银线；银丝
silvery ①银的②银色的③银制的
silvery fracture 银白色断面
silver zeolite 银沸石
silver-zinc storage cell 银锌蓄电池*
silvestrene 枞萜；1-甲基-3-异丙烯基-1-环己烯 $C_{10}H_{16}$
silveterpin 林萜
silvichemicals 林(产)化(工)产品
silvinate 松香酸盐
silyl 甲硅烷基 H_3Si-
silylacizing 甲硅烷基化

silylamine　甲硅烷(基)胺
silylamino　甲硅烷氨基　H_3SiNH-
silylanization　甲硅烷化
silylanizing　甲硅烷基化
silylated　甲硅烷基化的
silylating agent　甲硅烷基化剂
silylation　硅烷(基)化*
silyl bromide　溴甲硅烷；甲硅烷基溴
silyl compound　甲硅烷基化合物
silyldisilanyl　异丙硅烷基　$(H_3Si)_2SiH-$
silylene　亚甲硅烷基　$H_2Si=$
silyl fluoride　氟化甲硅烷
silylidyne　次甲硅烷基　$HSi\equiv$
silyl ketene acetal　①乙烯酮缩二甲硅醇②二甲基乙烯酮缩(甲醇)；三甲基硅醇
silyloxy　甲硅氧基　H_3SiO-
silyl phenoxy alcohol　甲硅烷基苯氧基乙醇
silylthio　甲硅烷硫基　H_3SiS-
silymarin　水飞蓟素
sily oxide(=disiloxane)　二硅氧烷；乙硅醚
sima　硅镁圈；硅镁带
simarouba bark　苦楝皮
simaroubidin　苦楝素
simaruba　苦楝
simazine　西玛三嗪；2-氯-4,6-双(乙氨基)-S-三嗪
similar asymmetric carbon atom　相似的不对称碳原子
similar drug　类药品
similarity　相似；类似；相像；同样
similarity coefficient　相似性系数
similarity law　相似定律
similarity number　相似数
similarity principle　相似原理
similarity rule　相似定律
similarity transformation　相似变换
simile (printing) paper　模造纸
similitude theory　相似理论
Simon's test　西蒙氏检验〔检验伯胺与仲胺〕
simoniz　汽车蜡
simple　①单味药②采草药
simple acid dye　普通酸性染料
simple adsorption　单纯吸附
simple anhydride　酸酐；简单酐　$(RCO)_2O$
simple atomic spectrum　简单原子光谱
simple batch distillation　单程蒸馏
simple batch still　单程蒸馏釜
simple beam impact machine(=Charpy impact machine)　单梁(式)冲击(试验)机
simple classifier　普通分粒器

simple collision theory(SCT)　简单碰撞理论*
simple cyanine　单菁
simple derivative IDA　单一衍生物同位素稀释分析法
simple distillation　简单蒸馏(作用)
simple ether　简单醚
simple extension　简单拉伸
simple extensional flow　简单拉伸流动
simple fluid　简单流体
simple fluid with fading memory　衰退记忆简单流体
simple function　单官能
simple gear(ing) extruder　单速压出机
simple glass　普通玻璃；钾钠玻璃
simple glyceride　同酸甘油酯
simple goiter　单纯甲状腺肿；瘿病
simple harmonic motion　简谐运动
simple insert needle　普通插入式针头
simple-integral viscoelastic fluid model　简单积分型黏弹流体模型
simple interaction　二因素交互效应
simple ketone　简单酮；对称酮
simple lachrymator　单纯催泪剂
simple linear polymer　线型聚合物
simple lipid　单纯脂质
simple liquid　单纯液体
simple material of Noll　诺尔简单材料
simple microscope　单式显微镜
simple monoclinic　简单单斜
simple multiple contact operation　简单多级接触操作
simple multiple extraction　并流多级萃取
simple neutral red　简单中性红
simple nonpolar species　简单非极性组分
simple orthorhombic　简单正交
simple partial condensation　简单部分冷凝
simple pass tubular heater　单程列管加热器
simple pipe tee reactor　单管三通反应器
simple plastic flow　简单塑性流
simple plasticizer　低分子增塑剂
simple polymerization　纯聚合〔共聚合的对称语〕
simple pump　单缸泵
simple push　简单推力
simple radical　简(单)基
simple random sampling　简单随机抽样
simple reaction　简单反应
simple rectilinear shearing motion　简单直(线)剪(切)运动
simple salt　简单盐
simple shape　简单形状；外形简单
simple simplex　基本单纯形*

simplest formula　最简(单的化学)式
simple structure　静定结构
simple substance　单质
simple sugar　单糖
simple supported　简支的
simple thermodynamic material　简单热力学材料
simple torsion　简单扭转
simple triclinic　简单三斜
simple trommel　单筒筛
simple turbine type mixer　简单涡轮混合机
simple viscous liquid　简单黏性液体
simplex　①单(纯)②单缸
simplex crusher　单式压碎机
simplex-duplex pump　单效双泵
simplex optimization　单纯形优化*
simplex optimization method　单纯形优化法
simplex pump　单缸泵
simplex reciprocating pump　单缸来复泵
simplex xanthating machine　简式黄化机
simplicity　简单(性)；简易；简明
simplified titration method　简化滴定法
simplified topper　简易拔顶装置
simplifying model　简化模型
Simpson's rule　辛普逊定则
simtex yarn　同时拉伸变形丝
simulate absorption　受激吸收
simulated annealing　模拟退火
simulated annealing algorithm　模拟退火算法
simulated body fluid　人体模拟溶液
simulated conditions　模拟条件；相似条件
simulated environmental test　模拟环境试验
simulated flow　模拟流动
simulated reference material　模拟标准物质
simulated service test of greases　润滑脂的模拟应用试验
simulated spectrum　模拟谱*
simulate emission　受激发射
simulate test　模拟试验
simulate treatment process　模拟处理法
simulation　模拟
simulation method　模拟法
simulation quantity　模拟量
simulation test　模拟试验
simulator　模拟器
simultaneous automatic X-ray spectrometer　同时型自动 X 射线谱仪
simultaneous coupling method　同步偶联方法
simultaneous determination　同时测定
simultaneous electrodeposition　同步电沉积法；同步电(泳)沉积法
simultaneous equation　联立方程(式)
simultaneous equation method　解线性方程组法；联立方程法
simultaneous excitation　同时激发
simultaneous interpenetrating network　同步互穿网络
simultaneous irradiation　同时辐照法
simultaneous multielement analysis　多元素同时分析
simultaneous multielement capability　多元素同时分析能力
simultaneous observation　同时观测
simultaneous plasma spectrometer　同时型等离子体光谱仪
simultaneous reaction　联立反应
simultaneous technique　(同时)联用技术
simultaneous techniques of thermal analysis　热分析联用技术
simultaneous test　同步试验；对比试验
simultaneous titration　联合滴定
simultaneous X-ray spectrometer　同时式 X 射线光谱仪
simvastatin　辛伐他汀
sinactine　华尖碱
sinalbin　白芥子硫苷
sinamine　芥子胺；烯丙氨腈　$C_4H_6N_2$
sinanomycin　西那诺霉素
sinapic acid　芥子酸　$C_{11}H_{12}O_5$
sinapine　芥子
sinapinic acid　芥子酸
sinapis　芥子
sinapolin　芥子灵；双烯丙基脲　$C_{21}H_{24}O_4N_3$
sinapyl alcohol　芥子醇
Sinclair-Baker reforming process　辛克莱-贝克型重整过程
Sinclair hydrotreating process　辛克莱加氢处理法
-sine　〔词尾〕素
sine bar　正弦杆
sinensal　甜橙醛
sine wave　正弦波
sing around method　四周声响法
singeing machine　①点火机②烧毛机
singer　烧毛机
singe stove　烧毛炉
single　①单(一)的；单纯的②一次的
single abrader　单纱耐磨试验仪
single acting　单动
single acting compressor　单作用压缩机
single acting piston　单动活塞
single acting pump　单动泵

single admission　单向入口
single base powder　单料药
single-batch extraction　单釜提取
single-bead string reactor　单珠串联反应器
single beam atomic absorption spectrometer　单光束原子吸收光谱仪
single beam atomic absorption spectrophotometer　单光束原子吸收分光光度计
single beam balance　单臂天平
single-beam crane　单梁起重机
single beam infrared spectrometer　单光束红外光谱仪；单光束红外分光计
single beam non-recording ultraviolet spectrophotometer　单光束非记录式紫外分光光度计
single beam spectrophotometer　单光束分光光度计*
single beam-splitter　单光束分束器
single beater lap machine　单打手成卷机
single bed system　单床式
single-blade kneading　单叶捏和机
single blade mixer　单桨搅拌机
single blind　带单圈的盲板
single bob method　单浮子法
single bond　单键*
single-bridged polymer　单桥聚合物
single cavity　单(模)槽
single cavity mold　单横腔模具；单巢模
single cavity roll　空心滚筒
single-cell furnace　单室炉
single chain compound　单链化合物
single-chain condensed state　单链凝聚态
single channel analyzer (SCA)　单道分析器
single-channel chart pen type　单道记录笔式(记录器)
single channel ICP monochromator　单通道电感耦合等离子体单色仪
single-channel injection valve　单道注入阀
single channel pulse height analyzer　单波道脉冲高度分析器
single chip processor　单片机
single coat　单层
single coating　①单涂层②一次涂饰
single collector　单接收器*
single-color cloth　单色布
single-color printing machine　单色印染机
single color system　单色系统
single column　单柱
single-column ion chromatography　单柱离子色谱法
single-contact controller　自动控制器
single crystal　单晶*

single crystal diffractometer　单晶衍射仪
single crystal electrode　单晶电极
single crystal fibre　单晶纤维；须晶
single crystal filament　单晶丝；须晶
single crystal graphite　单晶石墨
single crystal membrane electrode　单晶膜电极
single crystal X-ray diffraction　单晶 X 射线衍射(法)
single crystal X-ray diffraction system　单晶 X 射线衍射系统
single crystal X-ray diffractometer　单晶 X 射线衍射仪
single cylinder compressor　单缸压缩机
single-cylinder machine　单烘缸机
single cylinder steeping press　单柱式浸(渍)压(榨)机
single-cylinder test stand　单缸试验室装置
single damper　单气闸
single-deck　单层；单板
single-deck classifier　单板分级机
single-decker oven　单层炉
single-degree-of-freedom dynamical system　单自由度动态系
single die tubing machine　单铸拉管机；单镙模压机
single distilled　一次蒸馏的
single drop machine　单流机〔面包〕
single drum drier　单鼓干燥器
single effect　单效
single effect distillation　单效蒸馏(作用)
single-effect evaporation　单效蒸发
single-effect evaporator　单效蒸发器
single electrode　单电极
single-electrode potential　单电极电位
single electron approximation　单电子近似
single electron occupied molecular orbital　单电子已占的分子轨道
single electron transfer　单电子转移*
single-end twisting machine　单纱加捻机
single-entry centrifugal pump　单一入口离心泵
single extraction　一次萃取；一次抽提
single factor　单因素
single fertilizer　单纯肥料
single fibre specific strength tester　单纤维比强度仪
single filament　单灯丝
single-filling duck　单纬帆布
single firing　一烧
single flame　单火焰
single flash　一次闪蒸；一次蒸馏；急骤蒸馏
single-flash pipe still　一次闪蒸管式釜
single-flash system　一次闪蒸系统
single flight conveyor belt　单跨运输带

single-flow centrifugal compressor　单流离心压缩机
single fluid cell　单液电池
single-fluid MSBR　单流体熔盐增殖堆
single-fluid nozzle　单相喷嘴〔压力式喷嘴〕；单流喷嘴
single focus(ing)　单聚焦
single-focusing instrument　单聚焦仪器
single focusing mass spectrometer　单聚焦质谱仪*
single fold　单折叠；折边互咬
single force　集中力
single front type　单列式；单面式
single hydrophilic group type surfactant　单亲水基型表面活性剂
single IDA　(一步)同位素稀释法
single immunodiffusion　单向免疫扩散试验
single impeller　单叶轮
single-impeller pump　单叶泵
single impression mold　单巢模
single-inlet impeller　单吸叶轮
single ion detector　单离子检测器
single ion monitoring　单离子监测*
single-layer coating　①单层涂层②一道涂装
single layer method　单层法
single-layer welding　单层焊
single-line lubricating system　单线式润滑系统
single-line manifold　单道流路
single link(=single linkage)　单键
single linkage(=single link)　单键
single linked　单键的
single mechanical seal　单端面机械密封
single mixing　单段混炼
single-mode　横模
single molecule detection　单分子检测
single-molecule force spectroscopy　单分子力谱仪
single nickel salt　①(简)单镍盐②硫酸镍
single nucleotide polymorphism　单碱基多态性
single open-flame sunshine carbon arc lamp　单开式日光型碳弧灯
single pan balance　单盘天平*
single-pan substitution balance　单盘置换天平
single paper backing sheet　单层背纸(版)；单背纸型
single particle　单一粒子
single particle light scattering　单粒子散射
single pass　单程
single pass condenser　单向容器器
single-pass conversion　单程转化
single-pass dryer　单程干燥机
single-pass exchanger　单程热交换器
single-pass operation　单程操作

single pass system (=single beam system)　单光路(方)式；单光束(方)式
single-pass welding　单道焊
single pass yield　单程产量
single-phase auto-transformer　单相自耦变压器
single-phase α-brass　单相α-黄铜〔含锌达 27%〕
single phase chromatograph　单相色谱仪
single phase current　单相电流
single-phased alloy　单相合金
single-phase nozzle　单相喷嘴〔压力式喷嘴〕
single phase voltage　单相电压
single photon camera　单光子照相机*
single photon counting　单光子记数
single photon emission computerized tomography (SPECT)　单光子发射计算机化断层显像*
single piece work　单件加工
single-pipe system　单管系统
single pique interlock　双罗纹集圈网眼织物
single piston reciprocating pump　单活塞往复泵
single-ply cloth　单层纺织品
single-point cutting tool　单刃刀具
single polymer　单组分聚合物
single pour slush mo(u)lding　一次灌糊搪塑
single product　单一制品
single product plant　单一产品工厂
single pulse shock tube　单脉冲激波管*
single-purpose machine　专用机床
single-quantum transition　单量子跃迁
single radiant section furnace　单一辐射段炉
single-ring naphthene　单环环烷烃
single-roll breaker　单辊压碎机
single roll coater　单辊涂布机
single roll crusher　单辊压碎机
single-roll feeder　单轨进料器
single-run welding　单道焊
singles　单纯锑；90%纯的还原锑；90%Sb
single screw extruder　单螺杆压出机
single-screw oil pump　单螺旋油泵
single section　单断面
single segmental baffle　单弓形折流板
single service rack　单面注油栈桥
single-shell rotary drier　单壳旋转干燥鼓
single slot burner　单缝燃烧器
single solvent　单一溶剂
single-solvent extraction　单一溶剂提取
single-solvent process　单一溶剂(提取)法
single-spiral partition ring　单螺纹隔环
single spread　单面涂(胶)布

single spreading 单面涂布
single-stage axial fan 单级轴流风扇
single-stage compression 单级压缩
single stage crystallizer 单级结晶器
single-stage evaporation 单级蒸发
single-stage evaporator 单级蒸发器
single-stage extraction 单级提取；一段抽提
single-stage process 单级(酚醛树脂)制备法
single stamp mill 单锤碎矿机
single-steeping 一次浸渍
single-step 单线的
single step compressor 单级压缩机
single-step perturbation technique 一步扰动技术
single-step process 一步法(酚醛)
single-step resin 一步法(酚醛)树脂
single-storey tapping 单层采脂法
single strand ①(绳、线等)单股②单丝；单纤维
single strand chain 单排链条
single-strand polymer 单股聚合物*
single-stroke preforming press 单冲程预塑机
single suction 单吸(式)
single-suction volute pump 单吸离心泵
single sweep method 单相扫描法
single-sweep oscillographic polarograph 单扫描示波极谱仪
single-sweep oscillographic polarography 单扫描示波极谱法
single sweep polarography 单扫描极谱法
single sweep tee 三通管
single-sweep voltammetry 单扫描伏安法
singlet ①单(线)态②单峰*③单一
single tank 单一油罐
single-tank storage 单一油罐储存
single test 一次试验
single texturing 单根变形
single thread 单螺纹
single thread tester 单纱强度试验仪
singlet linkage 单电键
single top coater 顶辊单面涂漆机；辊下单面涂漆机
single topping 单面贴胶
single top (roll) coater (单辊)上辊施涂机
singlet state 单线态；单重态
singlet-triplet transition 单态-三态转移
single tube tyre 单管轮胎
single-U groove welding 单面 U 形槽焊
single value decomposition 奇异值分解法
single-vane rotary pump 单翼旋转泵
single vaporization 一次蒸发；简单蒸发

single V-groove V 型槽
single-V groove welding 单面 V 形槽焊
single-voltage-sweep polarography 单电压扫描极谱法
single wash 一次洗涤(法)
single wavelength selector 单波段选择器
single withdrawal procedure 单相除去法
single-wormed screw 单螺纹螺杆；单头螺杆〔压出机〕
single-worm plodder 单螺杆压条机
single yarn 单(股)纱
singlings 初馏物
singly-blocked isomers 单封闭异构体
singly bonded 单键连接的
singly-charged bidentate ligand 单核二齿配位体；单核二合配位体
singly colored 单色的
singly excited configuration 单激发模型
singly linked 单键连接的
singly occupied MO(SOMO) 单电子已占分子轨道
singular 奇异的
singular matrix 奇异矩阵
singular point 奇(异)点
singular solution ①奇(异)解②特殊溶液
sinigrin (=potassium myronate) 黑芥子硫苷酸钾
sinigroside 芥子黑糖
sinine 新宁
sinistrin 海葱糖 $C_6H_{10}O_5$
sinistrose 左旋糖
sink ①水斗；水槽；漏水池②汇
sinkability ①吸湿性②沉降量
sinkage (冷膜)下陷
sinkaline (=choline) 胆碱
sink-and-float separation 沉浮分离法
sinker cam-cap 生克罩；袜机沉降片罩
sink flow ①汇流②水斗流动
sink-hole 泥箱；污水池
sinking pump 浸没泵
sinking time 时间〔表示润湿力的数值〕
sinoacutine 清风藤碱
sinodiosgenin 华地奥配基
sinolprocess 单程法；自一氧化碳与氢合成烃类过程
sinomenine (=cucoline) 汉防己碱
sinopaipunine 华百部碱
sinoreserpine 华利血平
sinostemonine 华百部碱
sinter ①多孔状淀土②溶渣③多孔④熔结⑤泉华
sinterability 可烧结性
sinter coating ①(粉料)烧结涂布②烧结涂层
sinter cone 泉华丘

sinter (-) corundum ①烧结金刚砂；刚玉②矾土陶瓷
sinter deposits 泉华沉积
sintered 烧结的；熔结的；粘结的
sintered alumina cylinder 熔融烧结柱形氧化铝〔一种研磨介质〕
sintered aluminium powder (SAP) 烧结铝粉
sintered catalyst 烧结催化剂
sintered crucible 烧结坩埚
sintered glass (=fritted glass) 烧结玻璃；多孔玻璃
sintered glass bead 烧结玻璃珠
sintered glass crucibles 烧结玻璃坩埚
sintered glass filter 烧结玻璃滤器
sintered-glass filter crucible (烧结)玻璃砂(滤)坩埚*
sintered glass filter funnel 烧结玻璃砂滤漏斗
sintered glass funnel 烧结玻(璃)板漏斗
sintered iron catalyst 烧铁催化剂
sintered material 烧结物料
sintered-metal filter 烧结金属过滤器
sintered particle aggregate 烧结的粒子聚集体
sintered plate 烧结板*
sinter-fused 熔结
sintering 烧结*；熔结；热压结
sintering coal 烧结煤；炼焦煤
sintering furnace 烧结炉
sintering machine 矿石烧结机
sintering metal 烧结金属；金属陶瓷
sintering method 烧结法；熔结法；软化成形法
sintering of catalyst 催化剂的烧结
sintering oven ①烧结炉②(粉末涂料)高温固化炉；高温烘炉
sintering point 烧结点；烧结温度
sintering powder 烧结粉末
sintering process 烧结法；烧结过程
sintering temperature 烧结温度
sinter molding 烧结成型
sinter-roasting 烧结
sinter stones 泉华石
sintetics 合成产品
sinusoidal baseline drift 波状基线漂移
sinusoidal denier variation 正弦曲线状纤度变化
siomycin 盐屋霉素
sipalin 己二酸环己酯-己二酸甲基环己酯混合物
sipeimine 西贝母碱
sipeimol 西贝母醇
sipeimone 西贝母酮
siphon(=syphon) ①虹吸管②虹吸
siphonage 虹吸能力
siphonaxanthin 管藻黄质

siphon barometer 虹吸气压计
siphon bend 虹吸弯管
siphon condenser 虹吸式冷凝器
siphon gauge 虹吸压力计
siphon height 虹吸高度
siphon lubricator 虹吸润滑
siphon neck 虹吸管颈
siphon oiler 虹吸加油器
siphon pipe 虹吸管
siphon suction 虹吸润滑器
siphon trap 虹吸闸门
sipylite 褐钇铌矿
sirenin 雌诱素
sirius 黏胶纤维
Sirocco fan(=squirrel cage type-fan) 西罗克风扇；多叶片式风扇；鼠笼式风扇
sirolimus 西罗莫司
sisal 剑麻；西沙尔麻
sisal fiber 剑麻纤维
sisal hemp 剑麻
sisomycin 紫苏霉素；西梭霉素
sissing(=cissing) 缩边〔漆病〕
site 部位；定域体；地点；位置
site assembly 现场装配
site-directed immobilization 定点固定
site-directed mutagenesis 定点突变
site group analysis 基群分析
site-model theory 位置模型理论；模型布置理论
site selection fluorescence spectrometry 选择性荧光光谱法
site-selective excitation 位置选择激发
siting analysis 选址研究；选址分析
sitostane 谷甾烷
sitosterin(=sitosterol) 谷甾醇
β-sitosterol β-谷甾醇
sitosterol(=sitosterin) 谷甾醇
SI units SI 单位；标准国际单位制
six coordinate complex 六配位络合物；六配位复盐；六配位配合物
six-membered ring(=six-ring) 六元环
six member transition 六元过渡态
six-plan tomoscan 六面断层扫描图谱
six-ring (=six-membered ring) 六元环
six stage washer 六步洗涤装置
sixteen-membered ring(=sixteen-ring) 十六元环
sixteen-ring(=sixteen-membered ring) 十六元环
six-way valve 六通阀
sizability 施胶性能

size ①胶料②上胶③大小；大度；尺寸
size analysis 粒度分析
size applicator(=sizing applicator) 浸胶器；施胶机
size-bound distemper 胶(黏)性墙粉
size circulating squirter 唧筒输浆泵
size coat (彩印)底漆
size coating for cloth 织物用上浆涂料
size composition 尺寸组成；大小组成
size content 浸润剂含量
size control 大小控制；尺寸控制
sized 按大小排好了的；筛过的
size data 筛分数据
size degradation 打小；粉碎；磨碎
size distemper 填充刷墙粉
sized material 筛析品
sized ore 分级矿石
sized paper 施胶纸
sized warp 有浆经纱
sized warp winding regulator 浆轴卷绕机构
size exclusion chromatography (SEC) 尺寸排阻色谱法*；体积排斥色谱；排阻色谱
size exclusion column 筛析柱
size exclusion theory 体积排斥理论
size fastness 上胶
size-fit 尺寸匹配作用
size frequency curve 粒度频率曲线
size migration 浸润剂迁移
size of mesh 筛目大小；筛号
size of paper (记录)纸幅
size of particles 粒子尺寸；粒子大小；粒度
size of sample 试样规格；样板尺寸
size of upper limit 上限线密度
sizepine P 膏状强化松香胶
sizepine W 石蜡松香胶
size pot 浆罐
size proportionate sampling 比例取样
size putty 快干腻子〔用贴金漆调制〕
sizer 填料器；上胶器
size range 粒度范围
size recovery 回浆处理
size reduction 打小；磨碎；粉碎；研碎；捣碎；碎解
size reduction machinery 粉碎机械
size residue 浸润剂残留
size sample 测粒度样品
size sensitivity (凝胶)孔径灵敏度
size separation 粒析；颗粒离析
size separation equipment 分选设备
size sieving capacity 尺寸筛分能力

size silk 小丝；纤度丝
size test 粒度分析
size-weight ratio (颗粒)样重比率
sizing 定型*；定尺寸；定大小
sizing agent ①胶黏剂②浆料③施胶剂④上浆剂
sizing analysis 筛析
sizing applicator(=size applicator) 浸胶器；施胶机
sizing emulsion 浸渍剂；浸渍液
sizing ingredient ①黏合剂②浸渍剂
sizing machine 浆纱机
sizing material 胶料；浆料
sizing of minerals 矿物的分级
sizing of ore 矿石的分级
sizing of refining equipment 加工设备能力计算
sizing of sugar 糖的分级
sizing plot 筛析图表
sizings 胶料
sizing system 定型装置
sizing test 粒度分析
sizing varnish 封闭清漆；上浆清漆；贴金漆
sizing vat 填料瓮；填料桶
sizy 胶黏的
skarn 矽卡岩
skate conveyer 滑道输送器
skate liver oil 鳐鱼肝油
skatole(=3-methylindole) 粪臭素；3-甲基吲哚
skatoxyl 粪臭基
skatoxylsulfuric acid 粪臭基硫酸
skein 绞纱；绞丝
skein booking 绞丝打包
skeining 成绞
skein making 绞丝(工艺)
skeletal aromatic ring 骨架芳(族)环
skeletal atom 骨架原子；基干原子
skeletal catalyst 骨架催化剂*
skeletal composite 骨架增强复合材料
skeletal deformation vibration 骨架变角振动；骨架形变振动
skeletal isomerization 骨架异构化(作用)
skeletal vibration (分子)骨架振动
skeleton ①骨架②骨骼
skeleton catalyst(=skeletal catalyst) 骨架催化剂
skeleton construction ①框架结构；骨架式结构②钢骨建筑；钢筋混凝土结构
skeleton crystal 骸晶
skeleton electrode 骨架电极
skeleton key 万能钥匙
skeleton polymer 骨架聚合物

skeleton structure 骨架结构
skeleton symbol (结构)简式
skeleton view 透视图
skelgas(=pentane) 戊烷
skep 聚乙烯丙烯橡胶
sketch 草图；简图
skew boat conformation 扭船型构象*
skew conformation 邻位交叉构象
skewed distribution ①偏态分布②非对称分布
skewed peak 斜峰
skew factor 斜扭系数
skew form 邻位交叉式
skew front 不整齐前沿；斜前沿
skewing 交叉
skewing mechanism 交叉装置
skewness 偏度
skew ratio 偏斜率
skew symmetric 不对称的
skew-symmetric tensor 斜对称张量
skiadin 含碘油〔射线照相用〕
skiagram 射线照片；X射线照片
skiagraph 射线照片；X射线照片
skiameter X射线线量计
skid 溢流纹〔注塑缺陷〕
skid force 侧滑力
Skidmore crucible 斯琪模坩埚〔残炭测定仪用〕
skid resistance 抗滑性
skid-resistant coating 防滑涂料
skill of application 施工技巧；应用技艺
skill of labour 劳动技巧；劳动熟练(程度)
skill of the formulator 配方设计师的技巧
skim 胶清
skim calender 贴胶压延机
skim coat ①极薄的涂膜②贴胶胶片
skim coat calender 平板纸压光机
skim-coating 贴胶
skim gate 除渣器
skimmed latex 撇皮胶乳
skimmer 撇乳器；锥形分器
skimmianine(=β-fagarine) 茵芋碱
skimmin 茵芋苷
skimming baffle 分离挡板
skimming calender 平板纸压光机
skimming of waste water 自废水表面上撇取石油
skimming plant 撇蒸厂〔自石油中蒸馏出轻质馏分〕
skimming pond 撇油池〔分离石油中污水及残渣的水池〕
skimming processing 撇蒸过程〔蒸馏出石油中轻馏分〕
skimming refinery 轻油炼油厂
skimming resistance 抗剥离力
skimmings 浮渣
skimming tank 撇油槽〔分离石油中污水及残渣〕
skim rubber 胶清橡胶
skim serum 胶清
skim stock 敷涂混合物
skin and core effect 皮心效应
skin coat (泡沫塑料之表面上的)表皮涂层
skin cracking 皮层热裂
skin cream 润肤霜
skin density (泡沫塑料)表皮密度
skin effect 皮层效应*
skin foam(s) 结皮泡沫塑料
skin-gas effect 皮层气泡(附着)效应
skin irritant ①皮肤刺激剂②刺激皮肤的
skin irritating gas 皮肤刺激(毒)气
skin irritation 皮肤刺激性
skin-making machine 造皮机
skin material 蒙皮材料
skinned 剥皮
skinning 削皮；剥皮
skinning bed 剥皮架
skinning test 结皮试验
skin-protecting agent 护肤剂
skin protection 皮肤防护
skin reaction 皮肤反应
skin-reactive factor 皮肤反应因子
skins 碎革
skin sensitization 皮肤敏化(作用)
skin tanker 单皮槽〔石油产品直接与外壳接触〕
skin temperature 表皮温度
skin tension 表皮张力
skin toning lotion 润肤洗液
skip 翻斗车；料车；起重箱
skip hoist 料车升降机
skipping 漏涂
skip welding 跳焊
skirting ①踢脚板；护墙板②边缘
skirting board ①踢脚板；壁脚板②柱础；柱石
skirt leather 植物鞣牛皮坯革
skirt support 裙座
skittle 小艇
skiver 粒面革；剖革；次等羊皮
skiving ①切片②削；刮
skiving machine 切片机；剖革机
sklero- 〔词头〕 硬
skleron 硬合金
sklodwskite 硅镁铀矿

Skoglund condenser 斯寇格兰德冷凝器
Skraup's synthesis 斯克洛浦合成
skull 炉瘤；熔铁上的渣
skullcap 黄芩
Skull melting 射频感应冷坩埚法
skull reclamation process (熔融精炼)坩埚渣壳燃料回收过程；渣皮回收过程
skunk bush 山茱萸
skunk cabbage 臭菘
skunk oil 臭鼬油
skunks (臭)鼬臭石油
skutterudite 方钴矿
sky blue 天蓝；青
skylight 天窗
skyrin 醌茜素
sky-star type composite fibre 天星型复合纤维
slab ①厚块；(厚)片；板②背板③铁块④切片⑤板皮⑥平板
slabber 肥皂压片机
slabbing machine 压片机
slabby coal 层状煤
slab chromatography 平板色谱(法)
slab-cooling machine 肥皂冷压机
slab-cutting machine 切皂机
slab dissolver 平板式溶解器
slab electrophoresis 平板电泳
slab grating 木格选别机；木浆除滓机
slab oil 胶块油；白色矿物油
slab rubber 胶块；板状橡胶
slabs 胶块；板状橡胶
slabstone 石板
slab wax 板状石蜡
slack ①煤屑的②松懈的
slack barrel 渣屑桶；非液体用桶；装石蜡用桶（容量110千克）
slack bin 矿渣料斗
slack coal 煤屑
slacked lime 消(熟)石灰
slack fired 欠火的；欠煅的
slacking test 风化试验
slack melt 半熔(炼)；弱(热)裂解
slack-melt congo resin 初熔；(半熔)刚果(砧坯)树脂
slack melt copal 半熔砧坯
slackmelt resin 半熔树脂〔熔化后仍然是部分可溶〕
slackness 松弛
slack side tension 松弛侧张力〔传动带〕
slack size 薄薄上胶
slack tank 空罐；备用罐
slack washing machine 洗辫组机
slack wax 疏松石蜡
slag 炉渣；矿渣
slag breaker 碎渣机
slag breaking 打碎矿渣；压碎矿渣
slag brick 矿渣砖
slag cement 矿渣水泥
slagceram 炉渣瓷
slag crusher 碎渣机
slag crushing 打碎矿渣；压碎矿渣
slag crust 渣皮
slag deposit 渣沉积
slag dump 堆渣场
slag fiber 矿渣纤维
slag formation 造渣；制造矿渣；渣化
slag forming 造渣的
slagging 成渣
slagging gas producer 放液渣的气体发生炉
slagging medium ①助熔剂②焊剂
slagging process 造渣过程
slaggy 矿渣的
slag hearth 渣炉
slag hole 渣口
slag inclusion 夹渣
slag notch 渣口
slag pot 渣桶
slag resistance 抗渣性
slag roasting 烧渣的
slag tap 出渣
slag-tap gas producer 出渣气体发生炉
slag-type abrasive 矿渣型磨料
slag welding 电渣焊
slag wool 矿渣棉
slaked lime 熟石灰*
slakin 造渣；制造矿渣
slaking quick lime 消化(的)生石灰
slaking tank 消和槽
slant chute 斜溜槽
slant culture 斜面培养
slanting cut 斜割法
slapdash 粗抹面；粗涂；不规则地抹面〔建筑〕
slap dashing 粗涂饰〔建筑〕
slash ①割开②刀痕③林区废料
slasher 断木机
slashing 割裂
slat 板条
slat conveyer 翻板式输送机
slate ①板岩②石板

slate clay　板岩土；黏板岩
slate coal　板岩煤
slate dust(=slate flour)　①石粉；(黏)板岩粉②瓦灰
slate flour(=slate dust)　①石粉；(黏)板岩粉②瓦灰
slate pit　板岩坑〔作季节性储藏石油产品用〕
slate powder　①(黏)板炭粉②瓦灰
Slater determinant　斯莱特行列式*
Slater's theory　斯莱特理论*
Slater-type orbital (STO)　斯莱特型轨道
slate scudding machine　石刀推挤机
slate white　碱式碳酸铅；铅白
slating　铺石板
slat packed column　板条填充塔
slat-packed tower　板条填充塔
slat packing　条片填充
slaty coal　板岩煤
slaty limestone　板状石灰石
slaughter house　屠宰场
slave drive　辅机联动；辅助设备联动
sleaker　刮子
sledge microtome　铲式超薄切片机
sledge mill　槌磨
sleeker　刮刀〔皮革〕
sleeking　磨光
sleeking glass　磨光玻璃
sleeking tub　磨光盆
sleeky　光滑的
sleeper　①铁路枕木②地龙③小搁栅
sleepiness　失光〔漆病〕
sleepy　①倒光②(轻度)失光〔漆病〕
sleeve　套管；套筒
sleeve junction　套管液接
sleeve-type reference electrode　套筒式参比电极*
sleeving　套管
slice　①(薄)片②浆③切片④堰；板
slice bar　搅拨杆；拨火棒；长把火铲
slicer　切片机
slicing machine　切片机
slicker　①刮刀②油布雨衣
slicking　磨光
slicking glass　磨光玻璃
slicking-in roller　压平辊；精压辊
slick joint　滑动接头
slick-surfaced leather　光面皮革
slide　①片②(载)片；滑动片④滑(动)
slide bearing　滑动轴承
slide damper　滑动闸板
slide fastener　拉炼；拉锁
slide fit(=sliding fit)　滑动配合
slide glass　滑片；载片
slide oil　滑油；黑色润滑剂〔闪点为282℃的润滑油〕
slider　滑触头；滑子
slide rule　计算尺
slide sample valve　进样滑阀
slide valve　滑动阀
slide-vane pump　滑动叶板泵
slide wire　滑线
slide wire bridge　滑线电桥
sliding　滑动；滑移
sliding bearing　滑动轴承
sliding block　天平游码
sliding damper　活档
sliding fit(=slide fit)　滑动配合
sliding flue damper　烟道活挡
sliding friction　滑动摩擦
sliding packing　往复式轴封
sliding-vane compressor　滑板式压缩机
sliding-vane pump　滑板泵
sliding-vane vacuum pump　滑板式真空泵
sliding weight balance　游码式天平
Sligh oxidation test　斯拉氧化试验〔斯拉法油氧化稳定性测定〕
slightly enriched uranium　稍加浓铀；低加浓铀
slightly mobile potassium　微动性钾
slightly reddish cast　轻泛红现象
slightly soluble　微溶
slightly viscoelastic fluid　微黏弹性流体
slight sag　轻微流挂；轻度流挂〔流挂等级之一〕
slight yellow tinge　(带)浅黄色相
slime　①(黏)泥；矿泥；淀渣；残渣②黏液
slime pump　矿泥泵
slimer　细粒摇床
slimes　矿泥
slimes concentrator　矿泥浓缩器
slime separator　矿泥分离器
slime spots　浆斑
slimy cake　糊状(石蜡的)饼块
slinger(=slinger ring)　吊环
slinger ring(=slinger)　吊环
sling hygrometer(=sling psychrometer)　摇动湿度计
slinging eye　吊索眼
sling psychrometer　手旋(干湿球)湿度计；摇动湿度计
slink lamb　绵羔羊裘皮
slinning cabinet　纺丝通道
slip　①泥釉；滑尼②滑脱；滑移③错误④逃(逸)⑤(窄)条

slip additive 滑爽添加剂；润滑添加剂
slip agent 增滑剂；滑爽剂；滑润剂〔加在涂料组分之中，漆膜干后它能从漆膜渗出起润滑作用〕
slip agent masterbatch 爽化剂母粒
slip aids 滑动助剂
slip band 滑移带
slip casting 滑移浇铸
slip coating ①润滑涂料；润滑涂层②滑爽涂料；滑爽涂层
slip compound 光滑助剂〔油墨〕
slip cover 滑动盖子
slip direction 滑移方向
slip-fit connection 套式接合
slip glaze 泥釉
slip joint 滑动接头
slip kiln 烘硬窑
slip line 滑移线
slip on flange 平焊法兰
slip-on head 滑动头；移动头
slippage 滑移；滑脱；推(滑粘)合
slippage factor 滑动因子
slippage flow 滑流
slipper 闸瓦；滑块；制动器
slipper dipping(=underbody dipping) （车身)移动浸涂；(车身)下(身)浸涂
slipperiness 光滑性；平滑性
slippery feel 手感滑润
slipping 滑动
slipping agent 增滑剂；润滑剂
slip resistance ①耐滑性②抗剪强度③防侧滑性
slip resistance agent 防滑剂
slip-ring ventilator 滑环通风机
slip sheet 滑爽垫片；防粘连衬纸
slit 狭缝*
slit coil 窄带卷；纵剪带卷〔宽卷材纵切成窄的卷材〕
slit focusing method 缝隙聚焦法
slit function 缝隙函数
slit programme 狭缝程序
slitter 纵断器；切纸机
slitting saw 切割锯
slitting shear machine 切条机
slit ultramicroscope 缝隙超显微镜
slit width 狭缝宽度
slit yarn 纵裂纤维；切膜扁丝
sliver ①裂片；薄片；碎料②裂(成细长条)
slivering 并条
sliver lap machine 精条卷机
sliver screen 碎料筛

Slocomb machine 平臂刮软机
Slocomb staker 平臂刮软机
slop 废油；不合格石油产品
slop cut 废馏分；不合格馏分
slope 斜度；斜率
slope curve 斜度曲线
slope detection 斜率检测
slope detector 斜率检测器
slope magnification effect 斜率放大效应
slope ratio method 斜率比法
slope sensitivity 斜率灵敏度
sloping baffle 倾斜挡板
sloping bench 斜台
sloping bottom 斜底
sloping bottom tank 斜底罐；锥底油罐
sloping cascade surface （球磨时的)倾斜瀑布面
sloping catalyst line 倾斜催化剂管道
sloping chute 斜槽
sloping front boundary 斜前沿边缘
sloping hearth 斜炉底
sloping roof furnace 斜顶炉
slop line 废线；不合格石油产品管线
slop oil 废油；不合格石油产品
slop padding 浸轧(工艺)
slops ①废油②废水；污水③蒸馏废液
slop tank 废罐储存不合格石油产品的油罐
slop tension 倾斜张力〔传动带〕
slop wax(=slop wax cut) 废蜡；未经过滤的石蜡；原料石蜡
slop wax cut(=slop wax) 废蜡；未经过滤的石蜡；原料石蜡
slop wax fraction 废蜡馏分
slosh （石油产品自管道内)漏出
slot ①槽；沟；缝；窄口②开槽③长眼；长方形孔
slot-and-crank drive 曲柄连杆传动
slot area 齿缝面积
slot atomizer 缝隙式雾化器；缝隙式喷雾器；缝隙式喷嘴
slot burner 窄口喷灯；缝式燃烧器
slot coke oven 平行炼焦炉
slot jet 扁孔喷嘴
slot mesh 长眼
slot mesh plate(=slot mesh screen) 长眼筛板
slot mesh screen(=slot mesh plate) 长眼筛板
slot opening 齿缝开度
slot punched plate 长眼板
slot tear 穿孔撕裂
slotted ①(开了)长眼的②有槽的

slotted drum　开槽桶
slotted opening　裂隙开口
slotted plate　长眼筛板
slotted punched plate(=slotted punched screen)　长眼板
slotted punched screen(=slotted punched plate)　长眼板
slotted quartz tube　开槽石英管；开缝石英管
slotted quartz tube atom trapping　缝管石英管原子捕集法
slotted screen　长眼筛
slotted tube atom trap　缝管原子捕集器；缝管原子阱
slotted tube atom trapping atomic absorption spectrometry　缝管原子捕集原子吸收光谱法
slotted tube resonator　开槽式管状谐振器
slotted tube trapping-atomic absorption spectrometry　缝管捕集-原子吸收光谱法
slotting machine　开槽机
slot weld　槽焊
slot width　齿缝宽度
slovikite　菱镁铁矾
slow　①慢的；(徐)缓的②减(低)速(度)
slow-breaking emulsified asphalt　慢裂乳化沥青
slow burning　①缓慢燃烧②(耐燃试验)缓燃(品)
slow-burning smokeless powder　慢燃无烟火药
slow cement　慢干水泥
slow chemisorption　慢速化学吸附
slow coagulation　缓慢凝结
slow combustion　缓燃(烧)
slow combustion pipette (=gas-combustion pipette)　缓燃(烧)吸移管；气体燃烧吸移管
slow-curing asphalt　慢干道路沥青〔用柴油馏分稀释的道路沥青〕
slow curing cut back (asphalt)　慢干道路沥青
slow-curing paving binder　慢干的液体道路沥青
slow curing resin　慢干树脂
slow discharge　迟缓放电
slow-drying　干燥缓慢
slow-evaporating cosolvent　慢(速)蒸发助溶剂
slow-evaporating diluent　慢挥发稀释剂
slow-evaporating solvent　慢挥发溶剂
slow fixation　缓慢结合(固定)
slow flow rate paper　慢流速纸
slowing-down radiation　慢化辐射
slowing-down spectrometer　慢化能谱仪
slow match　慢燃引信头
slow motion　慢动(作)
slow motion clamp　慢动夹
slow neutron　慢中子
slow oxidation　缓缓氧化
slow oxidation of fuel　燃料的缓慢氧化作用
slow passage　慢通过
slow process　慢过程
slow-release fibre　缓慢释放纤维
slow setting cement　慢固水泥
slow-setting emulsified asphalt　慢凝乳化沥青
slow speed mixer　慢速度混合器
slow spontaneous recovery　缓慢自然恢复
slow stock　黏状(纸)浆
slubbing　粗纱卷
slub yarn (rayon)　竹节丝(嫘萦)
sludge　①污泥②淤渣
sludge accumulation　淤渣积累
sludge acid　淤渣(硫)酸〔从含硫沥青中制出的硫酸〕
sludge acid phosphate　淤渣酸磷酸盐
sludge acid separating tank　淤渣硫酸分离槽
sludge age　泥龄
sludge asphalt　淤渣沥青
sludge bulking　污泥膨胀
sludge chamber　淤渣池
sludge cock　淤渣阀
sludge coking plant　淤渣焦化厂
sludge conditioning　淤渣改质；淤渣处理
sludge dehydrating　污泥脱水
sludge deposit(ion)　淤渣沉积
sludge determination　淤渣测定
sludge dewatering　污泥脱水
sludge digestion　污泥消化
sludge diluent　淤渣稀释剂
sludge-dispersant　淤渣分散剂
sludge-dispersing agent　淤渣分散剂
sludge disposal　污泥处理
sludge distillate separator　淤渣蒸馏分离器
sludge emulsion　淤渣乳胶
sludge filtration　淤渣过滤
sludge formation　淤渣生成
sludge-forming constituents　形成淤渣的组分
sludge gas　淤渣气
sludge growth index　污泥增殖指数〔污泥增加重量对进水中 BOD 重量的比〕
sludge impoundment　污泥池
sludge incineration　污泥焚烧
sludge incinerator　污泥焚烧炉
sludge inhibitor(=sludge preventive)　防渣剂；淤渣生成抑制剂
sludge in oil　油中的淤渣
sludge-laden lubricant　含渣润滑剂
sludgeless oil　无渣油
sludge loading　污泥负荷〔活性污泥厂每日投入的 BOD

对活性污泥的重量比〕
sludge number 淤渣值；沉淀值
sludge oil 淤渣油；酸渣抽出油
sludge oil tar 淤渣油焦油
sludge particles 淤渣颗粒
sludge prevention 防止淤渣生成
sludge preventive(=sludge inhibitor) 防渣剂；淤渣生成抑制剂
sludge promoter 淤渣(生成)促进剂
sludge-proof 防渣的〔防止形成润滑油乳胶的，防止生成淤渣的〕
sludge pump 污泥泵；泥浆泵
sludge remover 淤渣清除器
sludge-suspending agent 淤渣悬浮剂
sludge test 淤渣试验〔变压器油的沉淀试验；氧化试验〕
sludge thickening 污泥浓缩
sludge treatment 污泥处理
sludge water 淤渣水〔与水混合的酸渣，含酸渣废水〕
sludging 成渣的
sludging test(=sludge test) 成渣试验
sludging time ①成渣时间②淤渣生成速度；氧化沉淀物生成速度
sludging value 淤渣值〔油内沉淀生成值〕
sludgy deposit 淤渣沉积
slug ①栓②煅屑
slugging 腾涌
sluggish flow(=sluggish flowing) 缓慢流动
sluggish flowing(=sluggish flow) 缓慢流动
sluggish lubricant 黏滞润滑油；流动性低的润滑油
sluggish precipitation method 缓慢沉淀法
slugs 未燃烧着的燃料；未蒸发的燃料液滴
sluice ①水门〔机工〕②(水门)沟③长斜水槽④(自水门)流出的水⑤淘洗⑥以水沟搬运
sluice box 聚金箱
sluice gate 水门；水闸
sluice opener 堰式开松槽
sluice valve 闸式阀；闸门(阀)
sluicing ①洗涤；洗净②精洗
sluicing pipe 洗涤管
slum 润滑油渣
slumping 滑挂〔漆病〕；滑移
slump resistance 抗塌陷性；防滑塌性；抗坍落性
slump test 重陷试验
Slurex process 斯勒雷克斯过程；浆液萃取过程
slurried catalyst 与原料混合的催化剂
slurry ①淤浆②泥(浆)③浆；生料
slurry coating 水浆涂料
slurry counterpart 水浆(二氧化钛)的对应产品(同等产品)

slurry density 水浆密度
slurry drier 淤浆干燥机
slurry exchanger 淤浆交换器〔催化剂悬浮体与原料油热交换器〕
slurry filter-cake 淤浆滤饼
slurry loading 浆液装车
slurry mixer 拌浆器
slurry oil 淤浆油；糊状物稀释油；悬浮催化剂粉末油料
slurry-packed method 淤浆填充法
slurry packing 匀浆填充(法)
slurry packing method 淤泥填充法
slurry packing technique 浆液充填法
slurry-phase reactor 淤浆反应器
slurry polymerization 淤浆聚合
slurry polymerization process 淤浆聚合法〔悬浮催化剂聚合过程〕
slurry pond 淤浆池
slurry process 淤浆法〔由一氧化碳与氢合成烃类的油悬浮过程〕
slurry ring 生料圈
slurry sampling 悬浊液进样
slurry tank 淤浆槽
slurry-type copper chloride treating 淤浆式氯化铜处理〔氯化铜与黏土混合的汽油干精制法〕
slush 烂泥；淤泥
slush box 脂膏盒
slush casting 中空铸型法
slush coating (铁桶内壁)灌浸涂料
slushing ①减水(作用)抗(腐)蚀；抗湿②涂油灰
slushing compound 抗蚀润滑剂
slushing grease 抗蚀润滑脂
slushing oil 抗蚀油
slush molding 中空模塑；涂凝模塑
slush moulding 涂凝模塑(法)
slush pulp 浓浆
slush pump 污水泵
slush stock 加厚浆
slushy 淤泥的
small amplitude shear oscillation 小振幅剪切振荡
small angle 小角的；低角的〔X 射线衍射图〕
small angle elastic scattering 小角弹性散射
small-angle laser scattering detector 小角度激光散射检测器
small angle light scattering 小角光散射
small angle scattering 小角散射*
small angle X-ray diffraction 小角 X 射线衍射
small angle X-ray diffraction method 小角 X 射线衍射法
small angle X-ray scattering X 射线小角散射

small anti-gas apparatus 小型防毒器
small aperture separator system 小隙缝分离系统
small arms 小型武器
small-bore cylindrical tube 小孔圆筒状管
small calorie(=small calory) 小卡
small calory(=small calorie) 小卡
small coal 细煤
small diameter shaft 小直径搅拌轴
small-duty pipe still 小生产率管式炉；小能量管式炉
small fennel oil 小茴香油
small galangal 良姜；红豆蔻
small graphite 粉粒石墨
small-lot production(=small-scale production) 小规模生产
small molding 小(件)模制品；小规格模制品
small molecule crystallography 小分子晶体学
small part 小(件)模制品；小规格模制品
small-particle 小粒(子)；小颗粒
small pipe line 小直径管线
small pore separation gel 小孔分离凝胶
small-porosity gel 细孔凝胶
small-pox 天花
small prunel 夏枯草
small quantity 少量
small sample 小样本
smalls burner 矿末炉
smalls burning 矿末灼烧
small-scale production(=small-lot production) 小规模生产
small-scale test 小型试验
small-shape 小样板
small size effect 小尺寸效应
small spot XPS 小面积 XPS 谱仪
smalls roaster 矿末煅烧炉
smalls roasting 矿末煅烧
small strip 小板条
small surfaces 小面积区〔包括桅杆、甲板室、排水口等〕
small transfer line 小运输线；补充管道；支管道
small unilamellar vesicles (SUV) 小单层脂质体
smalt 大青
smaltine(=smaltite) 砷钴矿
smaltinus 暗蓝色的
smaltite(=smaltine) 砷钴矿
smaragd green 绿闪石绿〔氧化铬绿的一种〕
smaragdine(-nus) 绿玉色；艳绿色
smaragdite 绿闪石
smaragdo-chalchit 氯铜矿
smart material 智能材料
smart polymer 智能高分子材料
smart transmitter 灵敏变送器；灵敏传感器
smartweed 红蓼
smasher-type mill 撞击式研磨机
smashing dispersion equipment 撞击式分散设备
smaze(=smoke+haze) 烟霾；干性烟雾
SM-cellulose 磺甲基纤维素
smear 胶糊渣；胶渣
smearer mill 抹研机；捻式磨；捻磨机
smearer-type mill 捻式磨；抹研式研磨机
smearing culture 涂抹培养
smearing dispersion equipment 捻磨〔抹研式分散设备如三辊磨、胶体磨〕
smearproof paint 耐脏漆；防涂污漆；防涂抹漆
smear test 擦拭(法)检查
smeary 油污的
smectic ①近晶状液晶的②脂状(的)
smectic compound 近晶化合物
smectic liquid crystal 近晶型液晶；层列型液晶
smectic mesophase 近晶中介相
smectic phase 近晶相；层滑型介晶相
smectic state (液晶)近晶态；(液晶分子)碟状结构态
smectic structure 近晶型(液晶)结构；层列(液晶)结构
smectite 绿土；蒙脱石
smectite structure 绿土结构；蒙脱石结构
smell ①嗅②臭(味)；气味
smelling blotter 闻香纸条
smelling salt 鼻盐；碳酸铵 $(NH_4)_2CO_3$
smelter ①熔炉②冶金厂③炼厂
smelter flue 熔炉烟道
smelter hearth 熔炼炉
smeltering 熔炼；冶炼
smeltery 熔炼厂
smelting 熔炼；熔化
smelting charge 熔炉装料
smelting furnace 熔炼炉
smelting furnace hearth 熔炼炉
smelting point 熔点
smelting pot 熔炼坩埚；熔炼罐
smelting process 熔炼法
smelting soda 炼碱
smelting trial 熔炼试验
smelting works 熔炼厂
smilacin 菝葜素；菝葜苷
smilagenin 菝葜配基
smilagenone 菝葜精酮
smilax 菝葜；丝米
smilonin 菝葜宁

smirnovine 没药豆碱
Smith's approximate formula 史密斯近似式
Smith crucible 史密斯坩埚
Smith-Hieftje background correction method 自吸收校正背景法
smithing coal(=smithy coal) 锻冶煤
smithite 斜硫砷银矿
smithsonite 菱锌矿
smithy coal(=smithing coal) 锻冶煤
smog (=smoke+fog) 烟雾
smog aerosol 烟雾气溶胶
smog episode 烟雾中毒事件
smog-forming hydrocarbon 形成烟雾的烃
smoke ①烟②(烟)熏；以烟驱逐
smoke abatement device 消烟除尘设备
smoke agent 烟雾剂；发烟剂
smoke apparatus 放烟器
smoke bell (煤油灯)灯帽；烟罩
smoke black 烟黑
smoke bomb 烟雾炸弹；烟幕弹
smoke candle 烟雾罐；烟幕烛
smoke condenser 聚烟器
smoke cured 烟熏的；熏好了的
smoke curing 烟熏制
smoked 烟熏过的
smoke damage 烟害
smoke density 烟雾密(浓)度
smoke density factor 烟雾密度系数
smoke depressant system 抑烟系统；消烟系统
smoked glass 烟色玻璃
smoked magnesium oxide 气相法氧化镁；煅烧法氧化镁
smoke dried 烟熏干燥；熏干的
smoke dry 烟熏干燥；熏干
smoke drying 烟熏干燥；熏干
smoked sheet 烟片(橡胶)
smoke dust 烟尘
smoke extraction unit 烟雾提取装置〔熄灭时用〕
smoke filter 烟雾过滤器
smoke flue 烟道
smoke funnel 发烟筒
smoke haze 烟霾
smoke house 烟熏室
smokeless 无烟的
smokeless burner ①无烟燃烧器②喷灯
smokeless fuel 无烟燃料
smokeless powder 无烟(火)药
smokeless propellant 无烟发射药；无烟推进剂
smoke point 发烟点

smoke prevention 防烟法
smoke protection 烟雾防护
Smoker method 斯莫克法〔理论塔板计算法〕
smokescope 烟雾密度测定器
smoke screen 烟幕
smoke shell 烟雾(炮)弹；烟幕弹
smoke stack 烟囱
smoke stain resistance 抗烟雾污染性
smoke stone 烟晶
smoke substance 烟雾剂
smoke suppressant 防烟剂
smoke suppression 烟雾抑制(作用)
smoke tanning 烟鞣
smoke volatility index 烟挥发指数
smokiness 发烟性
smoking ①发烟②(烟)熏
smoking burner 发烟灯
smoky ①烟的②有烟的
smoky quartz 墨晶；烟色石英
smolder 发烟燃烧；熏烧
Smoluchowski equation 斯莫路柯维斯基公式
Smoluchowski theory 斯莫路柯维斯基理论
smooth-drying fabric 干(燥自)平织物；免烫织物
smoother roller ①压平辊②匀料辊③轧光辊
smoothers 滑粉〔添加于润滑剂内的微粒固体润滑物〕
smooth fibre 平滑纤维
smooth file 细锉
smoothing of analytical signal 分析信号平滑(处理)
smoothing of cloth 布的熨平
smoothing press 光泽辊
smoothing roll 光泽辊
smooth muscle 平滑肌
smoothness of texture 织构平滑性
smooth note 圆和香韵
smooth outflow 均匀流出
smooth plating 压平
smooth roll 光泽辊
smooth surface colloid mill 光面的胶体磨
smooth-tactility 光滑感
smooth technique 平滑技术
smooth-texture grease 均匀结构润滑脂
smooth tube 光滑管道
smoulder 发烟燃烧；熏烧
smudge coal 天然焦炭
smudge oil 花园用燃料油〔果园防霜冻用燃料〕
smudging 污染；污点〔漆病〕
smuts 片状炭黑
snab sample 定时取样的样品

snail 涡管；涡轮；涡形板
snail wire 虾米螺丝；蜗形导丝器
snake-cage amphoteric ion exchange resin 蛇笼型两性离子交换树脂
snake-cage ion exchanger 蛇笼树脂
snake-cage polyelectrolyte 蛇笼型聚电解质
snake cage resin 蛇笼(状)树脂
snake lily 鸢尾
snake poison 蛇毒
snake root oil 蛇根油；加拿大细辛根油
snake skin 蛇皮
snake venom 蛇毒(液)
snaking (管材)清垢
snapback fiber 松紧纤维；弹性纤维
snap-closed drum 有紧箍的桶
snap-lever oiler 有弹簧盖的润滑器
snap-off 弹回高度
snapper(=grab-bucket sampler) 抓斗式采泥器
snappiness 柔韧性；高弹性
snappy rubber 高弹性橡胶
snap switch 快动开关；弹簧开关
snarl 卷缩
sneezing gas 喷嚏性毒气
sneezing powder 喷嚏粉
snifting 吸入空气
sniol 聚氯乙烯纤维
sniperscope 夜袭镜；红外线瞄准镜
snippet 小片
snoop (=snooper) 探测器
snoop bubble meter 探泡计
snooper(=snoop) 探测器
snooperscope 红外线夜望镜；夜间探测器
snoop leak detector 漏气检查器
snore-hole 鼾孔
snout 嘴；口
snow load 雪载荷
snowout 雪除
snow white 雪白
snuff ①鼻烟②嗅；闻③气味；香气
snuff bean 零陵香；黑香豆
Snyder reagent 4,7-二羟基-1,10-二氮杂菲
Snyder's fractionating column 斯奈德分馏柱
Snyder solvent strength parameter 斯奈德溶剂强度参数
soakage ①浸湿性②吸水量
soak cure 热渗透硫化
soaked bobbin ①浸卷机②浸糖桶
soaked in solvent 浸于溶剂中；用溶剂浸泡
soaker ①裂化反应室〔石油〕②浸渍剂

soaker tubes(=soaking tubes) 裂化反应管
soaking 浸水
soaking chamber 裂化反应室
soaking drum ①裂化反应鼓②浸水转鼓〔皮革〕
soaking pit ①浸水槽〔皮革〕②均热炉
soaking section 裂化反应段
soaking section outlet 裂化反应段出口
soaking time ①裂化反应时间②浸水时间〔皮革〕
soaking tub 浸渍管
soaking tubes(=soaker tubes) 裂化反应管
soaking vat 浸水池
soaking water 浸水
soaking weight 湿重
soak in lime 浸(石)灰(革)
soak liquor 浸渍液
soak test 浸泡试验
soak vat 浸水池
soap (肥)皂
soap ashes 草灰碱
soap bar 皂条
soap bark 皂树皮
soap base 皂坯
soap-based powder 皂坯洗涤粉
soap base grease 皂基润滑脂
soap blank 皂块〔打印前〕
soap blender 捏皂机
soap body 皂体
soap boiler 煮皂工
soap boiling 煮皂；肥皂煮制
soap boiling copper(=soap boiling kettle) 煮皂锅
soap boiling kettle(=soap boiling copper; soap boiling pan) 煮皂锅
soap boiling pan(=soap boiling kettle) 煮皂锅
soap bubble 皂泡
soap bubble flowmeter 皂泡流量计
soap bubble leak detection 皂泡检漏法
soap bubble method (测定火焰基本速度的)皂泡法
soap builder 肥皂助洗剂；肥皂增效助剂
soap cauldron 煮皂锅
soap chip 皂片；皂屑
soap chip drier 皂片干燥机
soap chipper 刨皂机
soap chipping 刨皂；肥皂刨平
soap chromatography 皂色谱法
soap composition 皂用香精
soap compound 皂用香精
soap content 皂含量
soap cooling frame 冷皂框

soap copper　煮皂锅
soap coverage　皂吸附量
soap crude (glycerine)　皂水粗甘油
soap crutcher　搅皂机
soap crutching pan　肥皂混合罐
soap curd　①皂粒②皂垢〔不溶性脂肪酸钙、镁盐〕
soap cutter(=soap cutting machine)　切皂机
soap cutting machine(=soap cutter; soap cutting table)　切皂机
soap cutting table(=soap cutting machine)　切皂机
soap drier　肥皂干燥机
soap drying plant　肥皂干燥室
soap dye　皂模
soaped-filled　灌皂的
soap emulsion　(有)乳化剂(的)乳液
soaper　煮皂工
soap extraction　皂萃取
soap factor　(润滑脂内)肥皂(含量)因数
soap factory　肥皂制造厂；肥皂煮制厂
soap fastness　耐皂性
soap-fast red　耐皂红(萘酚红)
soap fat　制皂用脂
soap fat liquor　肥皂乳化的加油(脂)液
soap film　皂膜
soap film flowmeter　皂膜流量计
soap film gas meter　皂膜气量计
soap film meter　皂膜计
soap film tensor　皂膜张力
soap fitting　肥皂整理
soap flake　(肥)皂片
soap flaking　片皂；皂片的制造
soap flaking rolls　皂片机
soap frame　(凝)皂框
soap framing　框皂；将肥皂注入皂框中
soap free emulsion　无皂乳状液
soap germicide　皂用杀菌剂
soap grain　皂粒；皂核
soap graining　皂的盐析
soap grease　皂基(润滑)脂
soap hydrate　肥皂水合物；水合肥皂
soapiness　皂滑性
soaping　皂洗；皂煮
soap-in-oil system　皂-油系统
soap kettle　煮皂锅
soap lake　肥皂色淀〔碱性染料和肥皂的沉淀物，可用于着色〕
soap lather　肥皂泡沫
soapless　无皂的

soapless detergent　无皂洗涤剂〔指合成洗涤剂〕
soapless shampoo　无皂洗发剂
soapless soap　合成洗涤剂
soaplike　类皂的
soap liniment　软皂的70%酒精溶液〔加香料用作涂敷剂〕
soap lubricant　皂液；肥皂水〔润滑剂〕
soap lye　皂(碱)液
soap lye glycerine　皂液甘油
soap maker　煮皂工
soap making　制皂；肥皂制造
soap making machine　制皂机
soap micelle　皂胶束
soap mill　研皂机
soap milling　肥皂的研压
soap mixing tank　肥皂混合槽
soap mould　皂模
soap moulding　冲压肥皂
soap noodles　细皂条；面条(皂)
soap oil　皂油
soap-oil dispersion　皂油分散体
soap packing machine　肥皂包装机
soap pan　皂化锅；煮皂锅
soap paste　皂糊
soap perfume　皂用香料
soap perfuming　在肥皂中加入香料
soap-phase structure　皂相构造
soap plant　肥皂(制造)厂
soap plates cooling plant　皂板冷却装置
soap plodder　肥皂压条机
soap plodding　肥皂压条
soap powder　皂粉
soap press　压皂机
soap process　皂化处理
soap pump　皂泵
soap resistance　抗皂性
soap-resistance test　耐皂试验
soap ribbon　皂带
soap roller mill　压皂磨辊
soap root　皂根
soap scent　肥皂香料
soap scenting　在肥皂中加入香料
soap-sensitive　肥皂过敏
soap-sensitive individuals　肥皂过敏者
soap shampoo　含皂洗发剂
soap shaving machine　刨皂床
soap sheets　肥皂纸片
soap slab　皂条；条皂
soap slabber(=soap slab cutter)　切皂条机

soap slab cutter(=soap slabber)　切皂条机
soap solution　皂碱液
soap spot　皂渍
soap stamping press　压皂机
soapstock　皂脚〔油脂碱炼〕；皂料〔制皂〕
soapstone　皂石
soapstoning　涂隔离剂
soap structure　肥皂构造
soap substitute　代用皂；肥皂代用品
soap tablet　皂丸
soap test of greases　润滑脂内皂的测定
soap titration method　皂滴定法
soap treatment(=soap process)　皂化处理
soap-tree oil　皂树油
soap-type emulsifier　皂型乳化剂
soap value test　皂化值测定
soapwood　皂树
soap work　肥皂(制造)厂
soap wort　皂根
soap wrapper　肥皂纸；包皂纸
soap wrapping machine　包皂机
soapy flavor　肥皂味
soapy water sensitivity　肥皂水过敏性
sobita　酒石酸铋钠
sobrerol　水合藦醇
sobrerone　松萜
soccerball leather　足球革
socket　①套②承窝〔机工〕③插座
socketed pipe(=socket pipe)　套管
socketed tube(=socket pipe)　套管
socket joint　套筒联接；筒接
socket pipe(=socketed pipe; socketed tube; socket tube)　套管
socket tube(=socket pipe)　套管
Socony oxidation susceptibility　沙孔奈公司氧化感受性
soda　纯碱*
soda alum　钠明矾　$Na_2SO_4 \cdot Al_2(SO_4)_3 \cdot 24H_2O$；$NaAl(SO_4)_2 \cdot 12H_2O$
soda-asbestos(=ascarite)　苏打-石棉；烧碱石棉
soda ash　苏打灰；(钠)碱灰；纯碱
soda ash glass　钠玻璃
soda ash roaster　苏打灰煅烧机
soda ash roasting　煅烧苏打灰
soda block　苏打块；石碱
soda boil　碱煮
soda boiling　碱煮的
soda brine　苏打卤
soda cellulose　碱纤维素

soda circulating pump　苏打循环泵
soda crystal　苏打结晶；钠碱晶
soda feldspar　钠长石
sodafining　碱洗；碱精制
soda glass　钠玻璃
soda grease(=sodium-base grease)　钠基(润滑)脂
soda-lime　碱石灰
soda lime glass(=soda lime silica glass)　钠钙玻璃
soda lime process　苏打石灰法；石灰碱(软水)法
soda lime silica glass (=soda lime glass)　钠钙玻璃
soda lime tube　苏打石灰管
sodalite　方钠石*
sodalite structural unit　方钠石结构单元
soda-lye　氢氧化钠浓溶液
soda lye causticizing　碱水苛化；用碱水碱化
soda lye wash　碱水洗涤
sodamide　氨基(化)钠　NH_2Na
soda mint　碳酸氢钠
soda niter　天然硝石；智利硝(石)
soda nitrate　智利硝(石)
soda nitre　天然硝石
soda oil(=sod oil)　油鞣回收油
soda pearl ash glass(=soda potash glass)　钠钾玻璃
soda potash glass (=soda pearl ash glass)　钠钾玻璃
soda powder　①苏打粉②智利硝石和石墨制的炸药
soda process　苏打法；苛性钠法
soda pulp　碱法纸浆
soda pulping　碱化纸浆
soda salt　苏打；碳酸钠　Na_2CO_3
soda slag　生铁脱硫时产生的炉渣
soda sludge　苏打淤渣
soda soap　钠皂
soda solution　苏打溶液；碱液
soda tar　苏打焦油
sodation　碳酸钠去垢(法)
soda wash　碱洗；苏打洗涤
soda-wash solution　碱洗溶液
soda-wash tower　碱洗塔
soda water　苏打水
sodio-acetoacetic ester　①钠代乙酰乙酸酯　$CH_3COCHNaCOOR$②〔专指〕钠代乙酰乙酸乙酯　$CH_3COCHNaCOOC_2H_5$
sodio-alkylmalonic ester　①钠代烷基丙二酸酯　$ROCOCRNaCOOR$②〔专指〕钠代烷基丙二酸乙酯　$C_2H_5OCOCRNaCOOC_2H_5$
sodio-cyanoacetic ester　①钠代氰基乙酸酯　$CNCHNaCOOR$②〔专指〕钠代氰基乙酸乙酯　$CNCHNaCOOC_2H_5$

sodio-derivative 钠衍生物
sodio-ethylmalonic ester ①钠代乙基丙二酸酯 ROCOCNa(C_2H_5)COOR ②〔专指〕钠代乙基丙二酸乙酯 C_2H_5OCOCNa(C_2H_5)COOC_2H_5
sodio-ketoester 钠代酮酸酯
sodiomalonic ester 钠代丙二酸酯 NaHC(COOR)$_2$
sodiomethylmalonate 钠代甲基丙二酸盐(酯)
sodio-methylmalonic ester ①钠代甲基丙二酸酯 ROCOCNa(CH$_3$)COOR ②〔专指〕钠代甲基丙二酸乙酯 C_2H_5OCOCNa(CH$_3$)COOC_2H_5
sodion 钠离子
5-sodiosulfoisophthalic acid 5-钠磺基间苯二甲酸
sodium 钠〔11号元素, 化学符号 Na〕
sodium 3,5-bis(carbomethoxy) benzenesulfonate 3,5-二甲酯基苯磺酸钠
sodium abietate 枞酸钠; 松香酸钠
sodium acetate 乙酸钠
sodium acetate trihydrate 三水(合)乙酸钠
sodium acetylide ①乙炔(二)钠 ②乙炔钠
sodium acid arsenate(=sodium acid arseniate) 酸式砷酸钠
sodium acid arseniate(=sodium acid arsenate) 酸式砷酸钠
sodium acid phosphate 酸式磷酸钠〔磷酸二氢钠 NaH$_2$PO$_4$ 和磷酸氢二钠 Na$_2$HPO$_4$〕
sodium acid sulfide 酸式硫化钠 NaHS
sodium 2-acrylamido-2-methylpropane sulfonate α-丙烯酰氨基-2-甲基丙烷磺酸钠
sodium acrylate 丙烯酸钠
sodium alcoholate(=sodium alkoxide) 醇钠; 烃氧基钠
sodium alginate 藻酸钠
sodium alizarinsulfonate 茜素磺酸钠; 茜素红
sodium alkanesulfonate 链烷基磺酸钠
sodium alkoxide(=sodium alcoholate) 醇钠; 烃氧基钠 NaOR
sodium alkyl 烷基钠 R—Na
sodium alkyl amino-propionate 烷基氨基丙酸钠
sodium alkylate(=sodium alcoholate) 醇钠; 烃氧基钠
sodium alkyl benzene sulfonate 烷基苯磺酸钠
sodium alkyl naphthalene sulfonate 烷基萘磺酸钠
sodium alkylsiliconate 烷基硅酸钠
sodium alkyl sulfate 烷基硫酸钠
sodium alkyl-sulfinate 烷基亚磺酸钠 RSO$_2$Na
sodium allyl sulfonate 烯丙基磺酸钠
sodium alum 钠矾
sodium aluminate 铝酸钠 NaAlO$_2$; Na$_3$AlO$_3$
sodium aluminium chlorohydroxy lactate 氯代羟基乳酸铝钠(络盐)
sodium aluminium silicate 铝硅酸钠
sodium aluminofluoride 氟化铝钠; 冰晶石
sodium aluminomethylsiloxane 铝甲基硅氧烷
sodium aluminosilicate(=sodium silicoaluminate) 硅酸铝钠
sodium aluminum fluoride 氟铝酸钠; 氟化铝钠; 冰晶石 Na$_3$AlF$_8$
sodium aluminum hydride 氢化铝钠 NaAlH$_4$
sodium aluminum silicate 硅酸铝钠
sodium aluminum tetraethyl 四乙基铝钠
sodium amalgam 钠汞齐
sodium americyl triacetate 乙酸镅酰钠 NaAmO$_2$(C$_2$H$_3$O$_2$)$_3$
sodium amide 氨基(化)钠 NaNH$_2$
sodium aminobenzoate 氨基苯甲酸钠
sodium aminosalicylate 氨基水杨酸钠
sodium ammonium biphosphate 磷酸氢铵钠 Na(NH$_4$)(HPO$_4$)·4H$_2$O
sodium ammonium phosphate 磷酸铵钠
sodium amyl sulfate 戊基硫酸钠
sodium aniline arsonate 对氨基苯胂酸钠
sodium antimonate(=sodium antimoniate) 锑酸钠 Na$_3$SbO$_4$; Na$_4$Sb$_2$O$_7$
sodium antimoniate(=sodium antimonate) 锑酸钠
sodium argentocyanide 氰银酸钠; 氰化银钠 Na[Ag(CN)$_2$]
sodium arsanilate 阿散酸钠; 氨基苯胂酸钠 NH$_2$·C$_6$H$_4$·AsO(ONa)$_2$
sodium arsenate(=sodium arseniate) 砷酸钠 NaAsO$_3$; Na$_3$AsO$_4$; Na$_4$As$_2$O$_7$
sodium arseniate(=sodium arsenate) 砷酸钠
sodium arsenide 砷化钠 Na$_3$As
sodium arsenite 亚砷酸钠 NaAsO$_2$; Na$_3$AsO$_3$; Na$_4$As$_2$O$_5$
sodium aryl 芳基钠 Ar—Na
sodium ascorbate 抗坏血酸钠
sodium azide 叠氮化钠
sodium base grease(=sodium soap grease) 钠皂润滑脂; 钠基润滑脂
sodium base sulfite pulping 碱性亚硫酸钠制浆法
sodium benzene-arsonate 苯胂酸钠 C$_6$H$_5$AsO(ONa)$_2$
sodium benzoate 苯甲酸钠 C$_6$H$_5$COONa
sodium benzyl succinate 苄基琥珀酸钠
sodium benzyl sulfanilate 苄基对氨基苯磺酸钠
sodium beryllate 铍酸钠 Na$_2$BeO$_2$
sodium bicarbonate 碳酸氢钠; 小苏打
sodium bichromate(=sodium dichromate) 重铬酸钠; 红矾钠

sodium bifluoride　氟氢化钠　NaHF$_2$
sodium binoxalate(=sodium bioxalate)　草酸氢钠　NaHC$_2$O$_4$
sodium bioxalate(=sodium binoxalate)　草酸氢钠
sodium biphosphate　磷酸二氢钠　NaH$_2$PO$_4$
sodium bismuth thiosulfate　硫代硫酸铋钠；卡诺氏试剂
sodium bisuccinate　丁二酸氢钠　NaHC$_4$H$_4$O$_4$
sodium bisulfate　硫酸氢钠　NaHSO$_4$
sodium bisulfite　亚硫酸氢钠　NaHSO$_3$
sodium bitartrate　酒石酸氢钠　NaHC$_4$H$_4$O$_6$
sodium-bonded fuel　钠黏合(反应堆)燃料
sodium borate(=sodium tetraborate)　四硼酸钠；硼砂
sodium borobenzoate　硼苯甲酸钠
sodium borohydride　硼氢化钠；氢化硼钠
sodium boron salicylate　水杨酸合硼酸钠
sodium boryl sulfate　硫酸氧硼(根)钠　Na(BO)SO$_4$
sodium bromate　溴酸钠　NaBrO$_3$
sodium bromaurate　溴金酸钠　Na[AuBr$_4$]
sodium bromide　溴化钠　NaBr
sodium bromite　亚溴酸钠　NaBrO$_2$
sodium-butadiene rubber　丁钠橡胶
sodium butoxide(=sodium butylate)　丁醇钠　NaOC$_4$H$_9$
sodium butylate(=sodium butoxide)　丁醇钠
sodium butylnaphthalenesulfonate　丁基萘磺酸钠
sodium butyrate　丁酸钠　C$_3$H$_7$COONa
sodium cacodylate(=arsycodile)　二甲次胂酸钠
sodium caprate　癸酸钠　C$_9$H$_{19}$COONa
sodium caproate　己酸钠　C$_5$H$_{11}$COONa
sodium caprolactam(ate)　己内酰胺钠
sodium caprylate　辛酸钠　C$_7$H$_{15}$COONa
sodium carbide　碳化钠；乙炔钠　Na$_2$C$_2$
sodium carbonate　碳酸钠；纯碱　Na$_2$CO$_3$
sodium-carbon dioxide polymer of butadiene　丁二烯的钠-二氧化碳聚合物
sodium carboxy methyl cellulose　羧基甲基纤维素钠
sodium carboxymethylcellulose fibre　羧甲基纤维素钠纤维
sodium carrier　钠载体
sodium catalyzed polybutadiene　丁钠橡胶
sodium cellulosate　纤维素钠
sodium cellulose glycol(l)ate　羧甲基纤维素钠
sodium cellulose xanthate　纤维素黄酸钠
sodium cellulose xanthogenate　纤维素黄(原)酸钠
sodium cerous nitrate　硝酸亚铈钠　Na$_2$[Ce(NO$_3$)$_4$]
sodium cerous orthophosphate　磷酸亚铈钠　Na$_3$[Ce$_2$(PO$_4$)$_3$]
sodium cetanesulfonate　十六烷基磺酸钠
sodium cetyl alcohol sulfate　十六烷醇硫酸钠

sodium cetyl sulfate　十六烷基硫酸钠
sodium chlorate　氯酸钠
sodium chloraurate　氯金酸钠　Na[AuCl$_4$]
sodium chloride　氯化钠；食盐　NaCl
sodium chloride index　氯化钠值；盐值
sodium chlorite　亚氯酸钠　NaClO$_2$
sodium chloroacetate　氯乙酸钠
sodium 2-chloro-3-nitrotoluene-5-sulfonate　试钾灵；2-氯-3-硝基甲苯-5-磺酸钠
sodium chloropalladite　氯亚钯酸钠　Na$_2$[PdCl$_4$]
sodium chloroplatinate　氯铂酸钠　Na$_2$[PtCl$_6$]
sodium chloroplatinite　氯亚铂酸钠　Na$_2$[PtCl$_4$]
sodium chromate　铬酸钠　Na$_2$CrO$_4$
sodium cinnamate(=sodium cinnamylate)　肉桂酸钠　NaC$_9$H$_7$O$_2$
sodium cinnamylate(=sodium cinnamate)　肉桂酸钠
sodium citrate　柠檬酸钠　Na$_3$C$_6$H$_5$O$_7$
sodium cobaltinitrite　亚硝酸钴钠　Na$_3$[Co(NO$_2$)$_6$]·1/2H$_2$O
sodium columbate　铌酸钠　NaNbO$_3$；Na$_3$NbO$_4$
sodium cooling　(金属)钠冷却
sodium coulometer　钠电量计
sodium cyanamide　氰氨基钠　Na$_2$N·CN；Na$_2$CN$_2$
sodium cyanate　氰酸钠　NaOCN
sodium cyanide　氰化钠　NaCN
sodium cyanoacetic ester　①钠代氰基乙酸酯　CNCHNaCOOR②〔专指〕钠代氰基乙酸乙酯　CNCHNaCOOC$_2$H$_5$
sodium cyanoborohydride　氰基硼氢钠
sodium cyclamate(=sodium cyclohexylsulfamate)　环氨酸钠；环己基氨基磺酸钠〔甜味剂〕　C$_6$H$_{11}$NHSO$_3$Na
sodium cycle　钠循环
sodium cyclohexylethyldithiocarbamate　环己基乙基二硫代氨基甲酸钠〔促进剂〕
sodium N-cyclohexyl-N-palmitoyl taurate　N-环己基-N-棕榈酰牛磺酸钠〔稳定剂〕
sodium cyclohexylsulfamate(=sodium cyclamate)　环氨酸钠；环己基氨基磺酸钠〔甜味剂〕　C$_6$H$_{11}$NHSO$_3$Na
sodium decamolybdate　(九缩)十钼酸二钠
sodium 4,4′-diaminodiphenylsulfone-2-acetyl sulfonamide(=promacetin)　4,4′-二氨二苯砜-2-乙酰基磺酰胺钠
sodium diamino-diphenylsulfone-N,N′-didextrose sulfonate(=promin)　二乙氨基二苯砜-N,N′-二葡萄糖磺酸钠
sodium diamyl sulfosuccinate　二戊基磺化琥珀酸钠〔湿润剂〕
sodium dibutyl dithiocarbamate　二丁基氨(基)荒酸钠〔促进剂〕
sodium dibutyl naphthalene sulfonate　二丁基萘磺酸钠

sodium di-*t*-butyl phosphate 二叔丁基磷酸钠〔增塑剂〕
sodium dichromate(=sodium bichromate) 重铬酸钠
sodium diethylaniline-azobenzene sulfonate(=ethyl orange) 乙基橙；二乙基苯胺偶氮苯磺酸钠
sodium diethylbarbiturate 二乙基巴比妥酸钠
sodium diethyldithiocarbamate (试)铜锌灵；二乙基二硫代氨基甲酸钠
sodium dihydrogen arsenate(=sodium dihydrogen arseniate) 砷酸二氢钠 NaH$_2$AsO$_4$
sodium dihydrogen arseniate(=sodium dihydrogen arsenate) 砷酸二氢钠
sodium dihydrogen arsenite 亚砷酸二氢钠 NaH$_2$AsO$_3$
sodium dihydrogen phosphate 磷酸二氢钠 NaH$_2$PO$_4$
sodium dihydrogen phosphite 亚磷酸二氢钠 NaH$_2$PO$_3$
sodium 2,3-dimercaptopropane sulfonate(=unithiol) 2,3-二巯基丙烷磺酸钠
sodium dimethyldithiocarbamate 二甲基二硫代氨基甲酸钠〔促进剂〕
sodium 4,4-dimethyl-4-silapentane sulfonate(DSS) 4,4-二甲基-4-硅代戊磺酸钠*
sodium dioctyl sulfosuccinate 二辛基磺化琥珀酸钠〔湿润剂〕
sodium dioxide 过氧化钠 Na$_2$[O$_2$]
sodium diphenylaminesulfonate 二苯胺磺酸钠*
sodium diphenyl-ketyl 钠化二苯酮(游)基
sodium disilicate 二硅酸钠 (Na$_2$Si$_2$O$_5$)$_n$
sodium disulfide(=sodium persulfide) 过硫化钠；二硫化(二)钠 Na$_2$S$_2$
sodium dithionate 连二硫酸钠 Na$_2$S$_2$O$_6$
sodium dithionite 连二亚硫酸钠 Na$_2$S$_2$O$_4$
sodium di(tolyldiazo-bis-8-amino-1-naphthol-3,6- disulfonate) 双(甲苯基偶氮氨基萘酚二磺酸钠)
sodium-divinyl rubber 丁钠橡胶
sodium D lines 钠 D 双线
sodium dodecane sulphonate 十二烷基磺酸钠
sodium dodecyl benzene sulfonate (SDBS) 十二烷基苯磺酸钠
sodium error (=alkaline error) 钠(误)差
sodium ester 钠酯〔是钠盐又是酯〕 ROCORCOONa
sodium ethide 乙基钠 NaC$_2$H$_5$
sodium ethoxide 乙醇钠；乙氧钠 NaOC$_2$H$_5$
sodium ethyl 乙基钠 NaC$_2$H$_5$
sodium ethylate 乙醇钠；乙氧钠 NaOC$_2$H$_5$
sodium ethyl carbonate 碳酸乙酯钠
sodium ethylcyclohexyl dithiocarbamate 乙基环己基二硫代氨基甲酸钠〔促进剂〕
sodium ethylene diamine tetracetate 乙二胺四乙酸钠
sodium 2-ethylhexyl sulfate 2-乙基己基硫酸钠〔乳化剂〕

sodium ethylsiliconate 乙基硅酸钠
sodium ethyl-sulfate 乙硫酸钠 C$_2$H$_5$SO$_4$Na
sodium ethyl-xanthate(=sodium ethyl-xanthogenate) 乙基黄原酸钠 C$_2$H$_5$OCSSNa
sodium ethyl-xanthogenate(=sodium ethyl-xanthate) 乙基黄原酸钠
sodium fatty-acylate solution 脂肪酰化钠溶液
sodium ferricyanide(=sodium prussiate) 赤血盐钠；铁氰化钠
sodium ferrite 铁酸钠 NaFeO$_2$ 〔ferrous acid 中文叫(正)铁酸〕
sodium ferrocyanide 亚铁氰化钠
sodium fluoride 氟化钠 NaF
sodium fluoride bead 氟化钠(熔)球
sodium fluoroacetate(=1080) 氟代乙酸钠；1080
sodium fluoroaluminate 氟铝酸钠；冰晶石
sodium fluoroborate 氟硼酸钠
sodium fluorophosphate 氟磷酸钠
sodium fluorosilicate 氟硅酸钠
sodium fluoscandate 氟钪酸钠 Na$_3$[ScF$_6$]
sodium fluosilicate 氟硅酸钠
sodium fluozirconate 氟锆酸钠 Na$_2$ZrF$_6$
sodium formaldehyde sulfoxylate 甲醛合次硫酸氢钠 CH$_2$O·NaHSO$_2$
sodium formate(=sodium formiate) 甲酸钠
sodium formiate(=sodium formate) 甲酸钠
sodium fulminate 雷酸钠 C=NONa
sodium fusion (加)钠熔(化)(作用)
sodium fusion method 钠熔法
sodium fusion test 钠熔试验
sodium germanate 锗酸钠 Na$_2$GeO$_3$
sodium glucoheptonate 葡庚糖酸钠〔稳定剂〕
sodium glutamate 谷氨酸钠；味精
sodium grease(=sodium-base grease) 钠基(润滑)脂
sodium hexachlorothallate 六氯铊酸钠
sodium hexadecyl sulfate 十六烷基硫酸钠〔湿润剂〕
sodium hexafluorophosphate 六氟磷酸钠
sodium hexametaphosphate 六偏磷酸钠
sodium hexaselenide 六硒化钠
sodium hydrate 氢氧化钠 NaOH
sodium hydrate-cellulose xanthate 纤维素黄酸钠
sodium hydride 氢化钠 NaH
sodium hydrogen arsenate(=sodium hydrogen arseniate) 砷酸氢二钠 Na$_2$HAsO$_4$
sodium hydrogen arseniate(=sodium hydrogen arsenate) 砷酸氢二钠
sodium hydrogen arsenite 亚砷酸氢二钠 Na$_2$HAsO$_3$
sodium hydrogen carbonate 碳酸氢钠 NaHCO$_3$

sodium hydrogen fluoride 氟氢化钠；酸性氟化钠
sodium hydrogen hypophosphate 连二磷酸氢钠；酸式连二磷酸钠
sodium hydrogen phosphate 磷酸氢二钠 Na_2HPO_4
sodium hydrogen phosphide 磷化氢钠
sodium hydrogen pyroantimonate 焦锑酸氢钠；酸式焦锑酸钠
sodium hydrogen selenide 硒氢化钠
sodium hydrogen selenite(=sodium hydroselenite) 亚硒酸氢钠
sodium hydrogen sulfate 硫酸氢钠 $NaHSO_4$
sodium hydroselenite(=sodium hydrogen selenite) 亚硒酸氢钠
sodium hydrosulfide 硫氢化钠 NaHS
sodium hydrosulfite 连二亚硫酸钠；保险粉 $Na_2S_2O_4 \cdot 2H_2O$
sodium hydrotartrate 酒石酸氢钠 $NaHC_4H_4O_6$
sodium hydroxide 氢氧化钠；苛性；烧碱
sodium hydroxide method 氢氧化钠法〔测定白土活性的氢氧化钠滴定法〕
sodium-hydroxide treatment 氢氧化钠处理
sodium hydroxylamine sulfonate 胲基磺酸钠 $NHOH \cdot SO_3Na$
sodium hydroxymethane sulfinate 羟甲基亚磺酸钠
sodium hydroxymethane sulfonate 羟甲基磺酸钠
sodium hydroxy methyl glycinate 羟甲基甘氨酸钠
sodium hypoborate 连二硼酸钠
sodium hypobromite 次溴酸钠 NaOBr
sodium hypochlorite 次氯酸钠 NaOCl
sodium hypochlorite process 次氯酸钠法
sodium hypochlorite solution 次氯酸钠液；漂白液
sodium hyponitrite 次硝酸钠 $Na_2N_2O_2$
sodium hypophosphate 连二磷酸钠 $Na_2PO_3;Na_4P_2O_6$
sodium hypophosphite 次磷酸钠 NaH_2PO_2
sodium hyposulfate 连二硫酸钠 $Na_2S_2O_6$
sodium hyposulfite(=hypo；sodium thiosulfate) 硫代硫酸钠；大苏打 $Na_2S_2O_3$
sodium indigodisulfonate 酸性靛蓝；靛二磺酸钠
sodium indium alum 铟钠矾 $Na_2SO_4 \cdot In_2(SO_4)_3 \cdot 24H_2O$; $NaIn(SO_4)_2 \cdot 12H_2O$
sodium iodate 碘酸钠 $NaIO_3$
sodium iodide 碘化钠 NaI
sodium iodomethanesulfonate 碘(代)甲磺酸钠
sodium ion selective electrode 钠离子选择电极
sodium iridichloride 氯铱酸钠；六氯合三价铱酸钠 $Na_3[IrCl_6]$
sodium isobutyl naphthalene sulfonate 异丁基萘磺酸钠〔湿润剂〕

sodium isopropoxide(=sodium isopropylate) 异丙醇钠 $NaOC_3H_7$
sodium isopropylate(=sodium isopropoxide) 异丙醇钠
sodium isopropyl xanthate 异丙基磺酸钠
sodium lactate 乳酸钠 $NaC_3H_5O_3$
sodium lamp 钠灯
sodium lanthanum nitrate 硝酸镧钠；五硝合镧酸钠
sodium laurate 月桂酸钠；十二酸钠〔分散剂〕
sodium laureth sulfate 月桂基(聚氧乙烯)醚硫酸钠
sodium lauroylmethyl taurate 月桂酰甲基牛磺酸钠
sodium lauroyl sarcosine 月桂酰肌氨酸钠
sodium lauryl benzene sulfonate 月桂基苯磺酸钠；十二烷基苯磺酸钠〔分散剂〕
sodium lauryl ether sulfate 月桂基乙醚硫酸钠
sodium lauryl sulfate 十二烷基硫酸钠
sodium laurylsulfonate 月桂基磺酸钠；十二(烷)基磺酸钠
sodium lignosulfonate 木质素磺酸钠〔分散剂〕
sodium lithol tonner 钠立索(尔)色料(色原)
sodium magnesium silicate 硅酸镁钠
sodium malate 苹果酸钠 $Na_2CH_4O_5$
sodium maleate 马来酸钠 $Na_2C_4H_2O_4$
sodium malonate 丙二酸钠 $Na_2C_3H_2O_4$
sodium manganate 锰酸钠 Na_2MnO_4
sodium 2-mercapto benzothiazole 2-硫醇基苯并噻唑钠〔促进剂〕
sodium mercuric thiocyanate 硫氰酸汞钠 $Na[Hg(SCN)_3]$
sodium metaarsenite 偏亚砷酸钠 $NaAsO_2$
sodium metabisulfite 焦亚硫酸钠 $Na_2S_2O_5$
sodium metaborate 偏硼酸钠 $NaBO_2$
sodium metaperiodate 偏高碘酸钠
sodium metaphosphate 偏磷酸钠 $NaPO_3$
sodium metasilicate 偏硅酸钠 Na_2SiO_3
sodium metastannate 偏锡酸钠 $Na_2Sn_5O_{11}$
sodium metavanadate 偏钒酸钠 $NaVO_3$
sodium metavanadate method 偏钒酸钠法
sodium methide 甲基钠 $NaCH_3$
sodium methoxide 甲醇钠 $NaOCH_3$
sodium methyl 甲基钠 $NaCH_3$
sodium methyl-acetoacetic ester ①钠代甲基乙酰乙酸酯 $CH_3COC(CH_3)NaCOOR$ ②〔专指〕钠代甲酰乙酸乙酯 $CH_3COC(CH_2)NaCOOC_2H_5$
sodium methyl-acetylide 丙炔钠；甲基乙炔钠 $CH_3C\equiv CNa$
sodium methyl-arsonate 甲胂酸钠 $CH_3AsO(ONa)_2$
sodium methylate 甲醇钠 $NaOCH_3$
sodium methylene bis-naphthalene sulfonate 亚甲基双萘

磺酸钠
sodium methylene diisopropyl naphthalene sulfonate 亚甲基二异丙基萘磺酸钠〔分散剂〕
sodium methyl mercaptide 甲硫醇钠
sodium N-methyl-N-oleyl taurate N-甲基-N-油酰基牛磺酸钠〔稳定剂〕
sodium molybdate 钼酸钠 $NaMoO_4$
sodium naphthalene 萘基钠
sodium naphthol-azobenzene sulfonate (=tropeolin OOO) 金莲橙 OOO；酸性橙；萘酚-偶氮苯磺酸钠
sodium nitrate 硝酸钠；智利硝
sodium nitre 钠硝石；智利硝(石) $NaNO_3$
sodium nitride 一氮化三钠 Na_3N
sodium nitrilo triacetate 次氮基三乙酸钠；氮川三乙酸钠
sodium nitrite 亚硝酸钠 $NaNO_2$
sodium p-nitrobenzene-azo-salicylate (=alizarin yellow R) 茜素黄 R
sodium nitroferricyanide 亚硝酸铁氰化钠
sodium 2-(p-nitrophenylazo) chromotrope 2-(对硝基苯偶氮基)铬变酸钠；铬变素 2B
sodium nitroprussiate 硝普酸钠 $Na_2[Fe(CN)_5NO]·2H_2O$
sodium nitroprusside 硝普酸钠；亚硝基铁氰化钠
sodium nitroprusside reaction 硝普酸钠反应
sodium octamolybdate 八钼酸钠
sodium octyl phosphate 磷酸辛酯钠
sodium oleate 油酸钠
sodium oleate gel 油酸钠凝胶
sodium oleyl p-anisidine sulfonate 油酰基对甲氧基苯胺磺酸钠〔稳定剂〕
sodium orthoarsenite 原亚砷酸钠 Na_3AsO_3
sodium orthophenylphenate 邻苯基(苯)酚钠
sodium orthophosphate (正)磷酸钠 Na_3PO_4
sodium orthophosphate (dimetallic) 二代磷酸钠 Na_2HPO_4
sodium orthophosphate (monometallic) 一代磷酸钠 NaH_2PO_4
sodium orthophosphate (trimetallic) 磷酸钠 Na_3PO_4
sodium orthosilicate 原硅酸钠 Na_4SiO_4
sodium orthovanadate 原钒酸钠 Na_3VO_4
sodium oxalate 草酸钠
sodium oxaloacetic ester ①钠代草乙酸酯 $ROCOCOCHNaCOOR$②〔专指〕钠代草乙酸乙酯 $C_2H_5OCOCOCHNaCOOC_2H$
sodium oxide 氧化钠 Na_2O
sodium palmitate 棕榈酸钠
sodium paramolybdate 仲钼酸钠；四缩七钼酸六钠
sodium paratungstate 仲钨酸钠 $5Na_2O·12WO_3·28H_2O$

sodium pentaborate 五硼酸钠
sodium pentachlorophenate 五氯酚钠〔农药〕
sodium pentachlorophenol 五氯苯酚钠
sodium pentachloro phenolate 五氯苯酚钠盐
sodium pentamethylene dithiocarbamate 亚戊基二硫代氨基甲酸钠〔促进剂〕
sodium perborate 高硼酸钠
sodium perborate tetrahydrate 过硼酸钠四水合物
sodium percarbonate 过(二)碳酸钠 $Na_2CO_2[O_2]$; $Na_2C_2O_4[O_2]$
sodium perchlorate 高氯酸钠 $NaClO_4$
sodium periodate 高碘酸钠 $NaIO_4$; Na_3IO_5; Na_5IO_6
sodium permanganate 高锰酸钠 $NaMnO_4$
sodium peroxide 过氧化钠
sodium peroxozirconate 过(氧)锆酸钠
sodium persulfate 过(二)硫酸钠 $Na_2S_2O_8$; $Na_2S_2O_6[O_2]$
sodium persulfide (=sodium disulfide) 过硫化钠；二硫化(二)钠 Na_2S_2
sodium perxenate 过氙酸钠
sodium phenate 苯酚钠 $NaOC_6H_5$
sodium phenide 苯基钠 NaC_6H_5
sodium phenolate 苯酚钠 $NaOC_6H_5$
sodium-phenolate process (苯)酚钠脱硫过程
sodium phenolsulfonate 苯酚磺酸钠 $OHC_6H_4SO_3Na$
sodium phenoxide 苯酚钠 $NaOC_6H_5$
sodium phenyl-arsenite 苯亚胂酸钠 $C_6H_5As(ONa)_2$
sodium phenyl-arsonate 苯胂酸钠 $C_6H_5AsO(ONa)_2$
sodium phenylate 苯酚钠 $NaOC_6H_5$
sodium-o-phenyl phenolate 邻苯基苯酚钠〔防霉剂〕
sodium phosphate 磷酸钠 Na_3PO_4
sodium phosphide 一磷化三钠 Na_3P
sodium phospho-12-molybdate 磷-12-钼酸钠
sodium phosphotungstate 磷钨酸钠
sodium picrate 苦味酸钠 $(NO_2)_3C_6H_2ONa$
sodium platinichloride 氯铂酸钠 $Na_2[PtCl_6]$
sodium platinochloride 氯亚铂酸钠 $Na_2[PtCl_4]$
sodium platinocyanide 氰亚铂酸钠 $Na_2[Pt(CN)_4]$
sodium platinous chloride 氯亚铂酸钠 $Na_2[PtCl_4]$
sodium plumbate 高铅酸钠 Na_2PbO_3
sodium plumbite 铅酸钠 Na_2PbO_2
sodium plumbite treatment 铅酸钠处理
sodium plutonyl acetate 乙酸钚酰钠 $NaPuO_2(C_2H_3O_2)_3$
sodium polyacrylate 聚丙烯酸钠〔增稠剂〕
sodium polybutadiene rubber 丁钠橡胶
sodium polymer 钠聚合物
sodium polymerization 钠(引发)聚合(作用)
sodium polymethacrylate 聚甲基丙烯酸钠〔增稠剂〕
sodium potassium carbonate 碳酸钠钾 $KNaCO_3$

sodium potassium silicate 硅酸钾钠；钾钠泡花碱 $(Na_2O \cdot K_2O) \cdot nSi_2O$
sodium-potassium tartrate 酒石酸钠钾 $KNaC_4H_4O_6$
sodium press 压钠(丝)器
sodium primary phosphate 磷酸二氢钠
sodium propoxide(=sodium propylate) 丙醇钠 $NaOC_3H_7$
sodium propylate(=sodium propoxide) 丙醇钠
sodium prussiate(=sodium ferricyanide) 赤血盐钠；铁氰化钠
sodium pump 钠泵
sodium pyroantimonate(=sodium pyroantimoniate) 焦锑酸钠
sodium pyroantimoniate(=sodium pyroantimonate) 焦锑酸钠
sodium pyroarsenate(=sodium pyroarseniate) 焦砷酸钠 $Na_4As_2O_7$
sodium pyroarseniate(=sodium pyroarsenate) 焦砷酸钠
sodium pyroborate 焦硼酸钠 $Na_2B_4O_7$
sodium pyrophosphate 焦磷酸钠 $Na_4P_2O_7$
sodium pyrosulfate 焦硫酸钠 $Na_2S_2O_7$
sodium pyrosulfite 焦亚硫酸钠 $Na_2S_2O_5$
sodium reduction process (金属)钠还原法
sodium resinate 树脂酸钠；树脂皂
sodium rhodanate 硫氰酸钠 NaSCN
sodium ricinoleate 蓖麻油酸钠〔稳定剂〕
sodium rubber 丁钠橡胶
sodium salicylate 水杨酸钠
sodium salt ①钠盐②（专指）氯化钠
sodium salt of 2-azodicarboxylate 偶氮二羧酸钠盐〔发泡剂〕
sodium salt of 2-mercaptobenzothiazole 2-硫醇基苯并噻唑钠盐〔促进剂〕
sodium salt of sulfonated methyloleate 磺化甲基油酸钠盐〔湿润剂〕
sodium sand 钠砂；粒状金属钠
sodium scandium sulfate 硫酸钪钠；三硫酸根合钪酸钠 $Na_3[Sc(SO_4)_3]$
sodium selenate 硒酸钠 Na_2SeO_4
sodium selenide 硒化钠 Na_2Se
sodium selenite 亚硒酸钠 Na_2SeO_3
sodium-sensitized thermoionic detector 钠敏化热离子检测器
sodium sesquicarbonate 碳酸氢三钠 $Na_2CO_3 \cdot NaHCO_3 \cdot 2H_2O$
sodium sesquisilicate 倍半硅酸钠
sodium silicate 硅酸钠；水玻璃 Na_2SiO_3
sodium silicate adhesive 硅酸钠黏合剂

sodium silicoaluminate(=sodium aluminosilicate) 硅酸铝钠
sodium silicofluoride 硅氟化钠
sodium silico-12-molybdate 硅-12-钼酸钠
sodium silico-zirconate 硅锆酸钠 Na_2ZrSiO_5
sodium soap 钠皂
sodium soap grease(=sodium base grease) 钠皂润滑脂；钠基润滑脂
sodium sorbate 山梨酸钠
sodium stannate 锡酸钠 Na_2SnO_3
sodium starch glycollate 羟基乙酸淀粉钠
sodium stearate 硬脂酸钠
sodium succinate 琥珀酸钠；丁二酸钠 $C_4H_4O_4Na_2$
sodium sucrate 蔗糖钠
sodium sulfanilate 氨基苯磺酸钠；磺胺酸钠 $NH_2C_6H_4SO_3Na$
sodium sulfantimonate(=sodium sulfantimoniate) 全硫锑酸钠 Na_3SbS_4
sodium sulfantimoniate(=sodium sulfantimonate) 全硫锑酸钠
sodium sulfate 硫酸钠
sodium sulfate-sensitized flame photometric detector 硫酸钠敏化火焰光度检测器
sodium sulfhydrate 氢硫化钠 NaHS
sodium sulfide 硫化钠 Na_2S
sodium sulfide lime 硫化钠石灰混合剂
sodium sulfide settling tank 硫化钠沉降槽
sodium sulfite 亚硫酸钠 Na_2SO_3
sodium sulfocarbonate 全硫碳酸钠 Na_2CS_3
sodium sulfocyanate 硫氰酸钠 NaSCN
sodium sulfonate 磺酸钠〔分散剂〕
sodium sulfovinate ①硫酸酯钠 RSO_4Na②〔专指〕硫酸乙酯钠 CH_5SO_4Na
sodium sulfoxylate formaldehyde 甲醛次硫酸钠
sodium superoxide 超氧化钠
sodium tallowate soap 牛脂钠皂
sodium tartrate 酒石酸钠 $Na_2C_4H_4O_6$
sodium taurocholate 牛磺胆酸钠 $NaC_{26}H_{44}NSO_7$
sodium tellurate 碲酸钠 Na_2TeO_4
sodium tellurite 亚碲酸钠 Na_2TeO_3
sodium tetraborate(=borax;sodium borate) 四硼酸钠；硼砂
sodium tetradecene sulfonate α-十四烯烃磺酸钠
sodium tetraphenylborate 四苯基硼酸钠
sodium tetraphenyl boron 四苯(基)硼酸钠 $NaB(C_6H_5)_4$
sodium tetraphosphate 四磷酸钠
sodium tetrapolyphosphate 四多磷酸(六)钠
sodium tetrathionate 连四硫酸钠 $Na_2S_4O_6$

sodium tetrazodiphenyl-naphthionate(=congo red) 刚果红
sodium thallide 一铊化钠 NaTl
sodium thermoionic detector 钠热离子检测器
sodium thioacetate 乙硫羟酸钠 CH_3COSNa
sodium thioantimonate(=sodium thioantimoniate) 全硫锑酸钠 Na_3SbS_4
sodium thioantimoniate(=sodium thioantimonate) 全硫锑酸钠
sodium thioarsenate(=sodium thioarseniate) 全硫砷酸钠 Na_3AsS_4
sodium thioarseniate(=sodium thioarsenate) 全硫砷酸钠
sodium thiocarbonate 全硫碳酸钠 Na_2CS_3
sodium thiocyanate 硫氰酸钠 NaSCN
sodium thiosulfate 硫代硫酸钠；大苏打 $Na_2S_2O_3$
sodium thiosulfate softener 硫代硫酸钠软化剂
sodium trichlorophenate 三氯酚钠
sodium tridecyl phosphate 十三烷基磷酸钠〔乳化剂〕
sodium triphenylcyanboron 三苯氰硼钠 $Na(C_6H_5)_3CNB$
sodium tripolyphosphate (STPP) 三磷酸钠
sodium trititanate 三钛酸钠 $Na_2Ti_3O_7$
sodium tungstate 钨酸钠 Na_2WO_4
sodium uranate 铀酸钠 Na_2UO_4
sodium uranyl acetate 乙酸铀酰钠 $NaUO_2(C_2H_3O_2)_3$
sodium vanadate 钒酸钠 $NaVO_3$；Na_3VO_4
sodium vapor lamp 钠蒸气灯；钠光灯
sodium vinyl sulfonate 乙烯基磺酸钠
sodium wire 钠线
sodium wire press 钠线压制机
sodium wolframate 钨酸钠 Na_2WO_4
sodium xanthate(=sodium xanthogenate; sodium xanthonate) 黄原酸钠 ROCSSNa；EtOCSSNa
sodium xanthogenate(=sodium xanthate) 黄原酸钠
sodium xanthonate(=sodium xanthate) 黄原酸钠
sodium zeolite 钠沸石
sodium zincate 锌酸钠 Na_2ZnO_2
sodium zirconate 锆酸钠 Na_2ZrO_3
sodium zirconium sulfate compound 硫酸锆钠复盐 $Zr(SO_4)_2 \cdot 2Na_2SO_4 \cdot 4H_2O$
sod oil(=soda oil) 油鞣回收油
sodos 磷酸二氢钠和碳酸氢钠的混合物
SOD viscosimeter SOD 黏度计〔测量高黏度油类与润滑脂用〕
sodyl 钠氧基
soedomycin 添田霉素
sofnolite(=sofnol soda-lime G) 碱石灰锰酸制剂
soframycin(=neomycin) 新霉素
soft ①(柔)软的②软水的③不含酒精的

soft acid 软酸
soft agent 软化剂
soft alkyl benzene 软性烷基苯〔指易为生物降解的直链烷基苯〕
Softal surface tension kit 索夫塔尔表面张力检测箱
soft asphalt 软沥青
soft base 软碱
soft beta emitter 软β发射体
soft beta radiation 软β辐射
soft black 软质炭黑
soft blast 喷软粒(处理)；喷软砂处理
soft burning 轻烧
soft carbon(=soft black) 软质炭黑〔如热裂法炭黑〕
soft cast iron 软生铁
soft centered structure 柔心结构
soft clay 塑性黏土；陶土
soft cloth 软布
soft coal 烟煤
soft coherent paste 软凝聚浆
soft cook ①蒸煮不足②软浆蒸煮
soft core 软心轴
soft detergent 软性洗涤剂〔易为生物降解者〕
soft distemper 软质刷墙粉〔可打磨的水胶性大白粉刷墙粉〕；低档浅色墙粉
soft drink 软饮料〔不含酒精的饮料〕
softened rubber 软化橡胶
softened water 软化水；软化了的水
softener ①软化剂；柔软剂②软化器③增塑剂
softener paste 软化脂膏
softening 软化
softening agent ①软化剂；柔软剂②软水剂③软化器
softening and antistatic agent 柔软及抗静电剂
softening of water 水的软化
softening point 软化点
softening point apparatus 软化点测定仪
softening point measurement 软化点测定
softening point thermometer 软化点(测定)温度计
softening power 软化能力
softening temperature 软化温度*
softening water 软水
soft fat 软脂
soft-feel coating 柔软感涂料
soft feel(ing) 柔软感；手感柔软
soft fiber 软纤维
soft firing 轻烧
soft gel 软(质)凝胶
soft gel-like matrix 软拟胶骨架
soft glass 软玻璃

soft goods 纺织品
soft grease 软膏
soft grease lubrication 软膏润滑
soft gum 软胶
softifume 烟砂（氰化钠、氯酸钠和砂的混合物）
soft ionization 软电离
soft ionization technique 软电离技术
soft iron 软铁
soft lead 软铅
soft leather 软革
soft lithography 软光刻；软刻蚀
soft lump 软块
soft matter 软物质
soft metal 软金属
soft mud brick machine 软泥造砖机
soft mud process 软泥制坯法
softness (柔)软度
softness index 软度指数
softness number 软度值
softness parameter (碱)软度参数
soft packing 软填料
soft paraffin 软石蜡
soft paste porcelain 软瓷器
soft petroleum ointment 矿脂；凡士林软膏
soft physics 软物理
soft pig iron 软生铁
soft pitch 软沥青
soft plastic 软塑料
soft polymer 软质聚合物
soft porcelain 软瓷器
soft potash soap 软钾皂
soft pulse 软脉冲
soft pulse electron spin echo envelope modulation 软脉冲电子自旋回波包络线调制
soft radiation 软(性)射线
soft ray 软(性)射线
soft resin 软树脂
soft rubber 软橡皮
soft shale 软页岩
soft-shell/hard-core type latex 软壳/硬芯型胶乳
soft silica ①软硅石；软石英②无定形二氧化硅
soft silk 熟丝；脱胶丝
soft-sized paper 吸水纸
soft-slick tactility 软滑感
soft sludge 软淤渣；软沉积物
soft soap 软肥皂
soft solder 软焊条
soft soldering 软焊

soft solid materials 半固体物料；塑性物质
soft steel(=mild steel) 低碳钢；软钢
soft tar pitch 软焦油沥青
soft-textured pigment 软质颜料
soft twist yarn 低捻纱；弱捻纱
soft washing agent 软性洗涤剂
soft water 软水
soft wax 软蜡；易熔石蜡
soft wax tank 软蜡槽
soft white clay 软质陶土
soft white gum 软白胶
soft wood 软材；针叶材
soft wood charcoal 针叶材木炭；软材木炭
soft wood distillation 针叶材干馏；软材干馏；松根干馏
soft wood tar 针叶材焦油；软材焦油；松焦油
soft wood tar pitch 针叶材焦油沥青；软材焦油沥青；松焦油沥青
soft X-ray 软 X 射线
soft X-ray appearance potential spectroscopy 软 X 射线表观位能能谱(法)
soft X-ray source 软 X 射线源
sogasoid 固气溶胶
soggy 欠硫；硫化不足
Sohio process 索亥俄法〔氨氧化制丙烯腈法〕
Sohio process for acrylonitrile 丙烯腈合成法；氨氧化法
soil ①土壤②施肥料③污垢
soilability 染污程度
soil acidity 土壤酸度
soil adsorption coefficient 土壤吸附系数
soil amendment ①土壤改良②土壤改良剂
soil analysis ①污垢分析②土壤分析
soil auger 土样钻取器；土壤取样钻
soil bacteria 土壤细菌
soil bacteriology 土壤细菌学
soil burial test 埋土试验
soil capillary water holding capacity 土壤毛细管持水量
soil carrying capacity 携污容量
soil carrying property 携污性
soil class 土性
soil clay mineral 土壤黏土矿物
soil conditioner 土壤改良剂；土壤调节剂
soil conservation 水土保持
soil content 污垢含量
soil corrosion 土力侵蚀
soil degradation 土壤退化
soiled 污染的
soiled cloth 污布
soiled cotton 污布

soiled plate　染污(的)薄板
soiled swatch　污布样
soil erosion　土壤侵蚀
soil fertility　土壤肥度
soil field capacity　土壤田间持水量
soil horizon　土层
soiling　染污
soiling agent　染污剂〔测去污力用〕
soiling formula　污液配方
soiling solution　污液
soil inoculation method　土壤接种法
soil invader　土壤寄居菌
soil load(ing)　污垢负载
soil maximum hygroscopicity　土壤最小吸湿量
soil maximum moisture capacity　土壤最大分子持水量
soil mineral　土壤矿物
soil moisture　土壤自然含水量
soil moisture monitoring　土壤水分监测
soil organic matter　土壤有机质
soil permeability coefficient　土壤渗透系数
soil pollution monitoring　土壤污染监测
soil profile　土壤剖面
soil reaction　土壤反应
soil redeposition inhibitor　污垢再附着抑制剂
soil removability　去污能力
soil removal　去污
soil removal efficiency　去污率
soil-removing action　去污作用
soil-repellent　防污剂
soil-repellent finishing　拒污整理；防污整理
soil ring　污垢环纹
soil salination　土壤盐碱化
soil sampler　土壤采样器
soil sampling　土壤采样
soil saturated moisture capacity　土壤饱和含水量
soil science　土壤科学
soil stabilizer　土壤稳定剂
soil suspender　污垢悬浮剂
soil-suspending ability　悬污能力
soil-suspending action　悬污作用
soil-suspending agent　污垢悬浮剂
soil-suspending power　悬污能力
soil thermometer　土壤温度计
soil water　土壤水分
soil water constant　土壤水分常数
soja　大豆
sojasterol　大豆甾醇
sojourn time　停留时间

sol　溶胶*
Sola catalytic process　(降低锅垢生成的)索拉催化法
soladulcamarine　茄甜苦碱
soladulcidine(=megacarpidine)　茄甜定；蜀羊泉定
soladulcidintetroside　茄甜定丁糖苷；蜀羊泉定丁糖苷
solamargine　糖苷生物碱边缘茄碱
solancarpidine(=solasodine)　茄解定
solandrine　茄君〔一种茄碱〕
solanellic acid　茄呢酸；胆汁六酸　$C_{23}H_{34}O_{12}$
solanesol　茄呢醇
solanic acid　茄酸
solanidine　茄啶　$C_{27}H_{43}ON$
solanin　茄灵
solanine　茄碱
solanion　茄镰孢菌素
solanocapsidine　茄辣椒定；玉珊瑚啶；茄卡西定
solanocapsine　假椒茄素；毛叶冬珊瑚碱
solanone　茄酮
solanorubin(=licopene)　茄玉红
solanthren dye　搔兰士林染料
solanthrene black　搔兰士林黑
solanthrene blue　搔兰士林蓝
solanthrene brilliant green　搔兰士林亮绿
solanthrene colors　搔兰士林染料
solantine blue　搔兰亭蓝
solantine colors　搔兰亭染料〔商名，一类直接染料〕
solantine pink　搔兰亭桃红
solapalmitine　三裂茄素
solapson　索拉松
solar　太阳的；日的
solar blind detector　日盲型检测器
solar cells　太阳能电池
solar constant　太阳常数
solar energy　太阳能
solar evaporation　暴晒蒸发
solar furnace　太阳(能)炉
solargentum　银-白明胶合剂
solarisation(=solarization)　①暴晒(作用)②光致淀粉减少(作用)
solarization(=solarisation)　①暴晒(作用)②光致淀粉减少(作用)
solar oil　太阳油；索拉油
solar pan　盐池；盐田
solar radiation　太阳辐射
solar radiation intensity　太阳照射强度
solar radiation pressure　太阳辐射压(力)
solar rays　太阳射线
solar salt　太阳盐；晒制盐

solar scarlet 太阳猩红
solar spectrum 太阳光谱
solar year 太阳年；回归年
solasodine 茄解定；水解羟基茄碱〔茄科植物中含有的有毒的糖苷生物碱〕
solasonine 茄解碱；羟基茄碱
solate 液化凝胶
solation 溶胶化(作用)
solbrol 对羟基苯甲酸甲酯
sol coating 溶胶涂装法
solder ①焊(接)②焊剂；焊锡③软钎料
solderability 可焊性；钎焊性
solder acid 焊锡水
solder bath 金属熔化浴(测定石油产品自燃点用)
soldered joint 焊接；点焊接合
soldering 软钎焊
soldering apparatus 焊接器
soldering flame 焊接焰
soldering fluid 焊(接)液
soldering flux 助焊剂；焊液
soldering lamp 喷灯；焊接灯
soldering tin 焊锡
solder iron 焊铁
solder mask 绿漆；防焊膜
solder mask ink 阻焊油
solder paster 焊锡膏
sold resin ①售出的树脂②成品树脂
sole ①唯一的②(脚)底(板)③鲽(鱼)
sole binder 单独基料；单一基料
sole crepe 底绉片
soledon brilliant purple 搔勒董亮红紫
soledon colors 搔勒董染料
soledon yellow 搔勒董黄
sole flue 底烟道；小烟道
sole fracture 断底
soleing plate 底板
sole leather 底革
sole leather brushing and finishing machine 底革刷光机
sole leather finishing 底革涂饰
sole leather roller 底革辊筒机
solemycin(=soulmycin) 苏洛霉素
solene 汽油
solenite 索冷那特〔炸药〕
solenoidal vector ①无散向量②螺线向量
solenoid electric valve 电磁控制阀
solenoid operated valve 电磁阀
solenoid valve 电磁阀
sole pigment 专用颜料；单用颜料

sole plate 底板
Solex oil process 索莱克斯炼油法〔用丙烷分离植物与动物油脂〕
Solex process 索莱克斯过程
solfatara 硫气孔
solferino 品红
solf paste end point 软浆点〔吸油量终点〕
sol-fraction 溶胶部分
solf-textured pigment 软质地颜料〔易于分散的颜料〕
sol-gel method 溶胶-凝胶法*
sol-gel microsphere 溶胶-凝胶(法制备的)微球粒
sol-gel process 溶胶-凝胶过程*
sol-gel processing 溶胶-凝胶处理(过程)
sol-gel technique 溶胶-凝胶技术
sol-gel technology 溶胶-凝胶工艺
sol-gel transformation 溶胶-凝胶转化
sol-gel vibratory compaction method 溶胶-凝胶振动增实法
solid ①固体②固体的③立体④立体的
solid acid catalyst 固体酸催化剂
solid-air interface 固-气界面
solid alcohol 固体酒精
solid analysis 固体分析
solid angle 立体角
solid bearing 固体轴承；整体轴承
solid bed of catalyst 催化剂的固体床；催化剂的紧密床层
solid-bed reactor (催化剂)固体床反应器
solid binder (催化剂的)固体黏合剂
solid bitumen 固体(油)沥青
solid bituminous materials 固体沥青物质
solid blue 固体蓝
solid body 固体
solid bowl 无孔转鼓〔离心机〕；实壁转筒
solid bowl type screw decanter 筒沉降螺旋卸料离心分离机
solid brick 实心砖
solid carbon 固体碳
solid carbon dioxide 固体二氧化碳
solid cement 固体接合剂
solid chunks (白土内)固体结块
solid coal 硬煤
solid coke 固化焦
solid color ①素色；单色②实地色〔纺织〕
solid color enamel 单-色(瓷)漆；实色漆；素色瓷漆
solid concentration 固体浓度
solid condition 固(体状)态
solid content 固体含量

solid copper sweetening process 固体铜甜化法；氯化铜脱硫
solid-core electrolytic cable 固体芯线电缆
solid-core packing 实心填充物
solid-core support 实心载体
solid crystal 固体结晶
solid culture 固体培养
solid cure 高度硫化
solid detergent 固体洗涤剂
solid diffusion 固相扩散
solid dispersion 固态分散体
solid drier 固体干燥剂；燥漆
solid electrolyte 固体电解质
solid epitaxy 固相外延
solid expansion thermometer 固体膨胀温度计
solid explosive 固体炸药；固体喷气燃料
solid extract 固体萃；固体提取物
solid filling 固体充填
solid film 固态膜
solid-film lubricant 固体膜润滑剂
solid-film lubrication 固体膜润滑
solid filter-aids 固体助滤剂
solid Fischer-Tropsch hydrocarbons 一氧化碳与氢合成而得的固体烃
solid fluorescence analysis 固体荧光分析
solid fluorimeter 固体荧光计
solid fluorimetry 固体荧光法
solid foam 固体泡沫体
solid foamed material 固体起沫剂
solid friction 固体摩擦
solid fuel 固体燃料；固态燃料
solid-gas sol 固气溶胶
solid-gas solution 固气溶体
solid-gas system 固气(物)系
solid gel 固态凝胶
solid gel fuel 固体胶燃料
solid geometry 立体几何(学)
solid green 固体绿
solid gum 固体胶
solid heterogeneous catalyst 固体多相催化剂
solid hydrocarbon fuel 固体烃燃料
solidifiability 凝固性
solidifiable 可凝固的；可固化的
solidification 凝固(作用)；固(体)化(作用)
solidification heat 固化热
solidification of radwaste 放射性废物固化
solidification point 凝点；固化点
solidification rate 固化速率；凝固速率
solidification rate parameter 固化速率参数
solidification temperature 固化温度
solidification value 凝固点；固化温度
solidified 固(体)化的
solidified alcohol 固化酒精
solidified carbon dioxide 固体二氧化碳；干冰
solidified gasoline 固化汽油
solidified kerosene 固化灯油
solidified oil ①固化油②氧化油
solidified oleoresin(=scrape) 毛松香
solidified petroleum product 固化石油产品
solidifying 固化；凝固
solidifying agent 固化剂
solidifying characteristics of oil 油料的固化特性
solidifying gasoline 凝固汽油
solidifying medium 凝固介质
solidifying point 凝固点
solidifying pressure 固化压力
solidity 固态
solid koji 固体曲
solid laser 固态激光器
solid level 固体含量；固体浓度
solid-like effect 似固效应
solid line 实线
solid-liquid chromatography 固-液色谱(法)
solid-liquid cyclone 固-液旋流(分离)器
solid-liquid equilibrium 固-液平衡
solid-liquid extraction 固体-液体提取；固液提取
solid-liquid interfacial tension 固液界面张力
solid-liquid lubricant 固液润滑剂
solid-liquid separation 固液分离
solid lubricant 固体润滑剂；固态润滑剂
solid lubrication 固体润滑
solid medium 固体培养基
solid membrane electrode 固膜电极
solid metal soap 固体金属皂
solid microelectrode 固态微电极
solid microsphere 实心微球
solid natural rubber 固体天然橡胶
solidness ①硬度②硬性
solid oil 固体润滑油
solid oligomer 固体低聚物
solid oxidant 固体氧化剂
solid packing ring 无接头的轴封
solid paraffin 固态石蜡
solid particle 固体粒子
solid perfume 固状香精
solid petrolatum 固体矿脂

solid petroleum product　固体石油产品
solid phase　固相
solid phase competitive binding fluoroimmunoassay　固相竞争结合荧光免疫分析
solid phase diffusion　固相扩散
solid phase extraction　固相萃取
solid phase extractor (SPE)　固相萃取器
solid-phase extrusion　固相挤出
solid phase fluorescence immunoassay　固相荧光免疫分析
solid phase microextraction　固相微萃取；固相微量抽提
solid phase polycondensation　固相缩聚*
solid phase polymerization　固相聚合*
solid phase press　固相压力成形
solid phase radioimmunoassay (SPRIA)　固相放射免疫分析
solid phase reaction　固相反应
solid phase reflection spectrophotometry　固相反射分光光度法
solid phase spectrophotometry　固相分光光度法
solid-phase structure of grease　润滑脂固相结构
solid phase two-site fluoroimmunoassay　固相双位点荧光免疫分析法
solid phosphoric acid catalyst　固体磷酸催化剂
solid phosphoric acid slurry　糊状磷酸催化剂
solid phosphor laser　固态磷光体激光器
solid piston　实心活塞
solid point(=solidification point)　凝固点；固化点
solid polymer　固体聚合物
solid polymerization　固相聚合(法)
solid propellant　固体推进剂；固体喷气燃料
solid pulverulent filler　微粉状填料
solid pump　固体吹散泵
solid radical initiator　固态自由基引发剂
solid radwaste　固体放射性废物
solid reagent　固体试剂
solid residue　固体残渣
solid resin　固体树脂
solid rotation blade　固体旋转式叶片
solid rotor　固体转子
solid rubber　硬橡皮；硬质橡胶
solids　固体粒子；硬粒
solid sample　固体样品
solid sample insertion probe　固体进样插入探头
solid sampler　固体进样器
solids by volume(=volume solid)　容积固体分；固体体积含量
solids by weight　固体重量含量
solids constituent　固体分
solids content　固体含量
solids (content) index　固体(量)指数
solids content in volume　固体体积含量
solids flow　固体粒子流动〔催化剂流动〕
solid shade　素色；一色；单色
solid shade paint　单色漆
solid shaft　实心轴
solids handling technique　固体物质输送技术
solids holdup　固体粒子停留时间
solid soap　固体皂
solid-solid equilibrium　固体间平衡
solid-solid reaction　固-固相反应
solid-solid transition　固-固相转移
solid solution　固溶体*；固(态)溶液
solid solution alloy　固溶体合金
solid solution cermet(SSC)　固溶体金属陶瓷
solid spectrophotometry　固相分光光度法
solids pickup point　固相混合点
solids recovery equipment　固体物质回收设备
solid state　固(体状)态
solid state chemistry　固态化学
solid-state circuit　固态电路
solid-state construction　固体构造
solid-state detector　固态探测仪
solid-state diffusion　固态扩散
solid state electrode　固态电极
solid state electrolyte　固态电解质
solid state FET pH probe　固态场效应晶体管 pH 探头
solid state infrared modulator　固态红外调制器
solid state interaction　固态相互作用
solid state polyelectrolyte　固态高分子电解质
solid-state polymerization　固态聚合
solid-state power proportioning controller　固态功率比例控制器
solid-state-proportioning controller　固态比例控制器
solid state reaction　固相反应*
solid state track detector (SSTD)　固态径迹检测器
solid storage engineering test facility (SSETF)　固体放射性废物储存工程试验设备
solid substrates room temperature phosphorimetry　固体基质室温磷光法
solid support　载体
solid supported bilayer lipid membrane　固体支撑双层磷脂膜
solid surface chemiluminescence　固体表面化学发光
solid suspension opals　(固体)粒子悬浮乳白玻璃
solid target　固体靶
solid thermometer　固体温度计

solid titration　固体(滴定剂)滴定
solid-to-solid transition　固-固相转变
solid twist　实捻
solid tyre　实心(轮)胎
solid urethane rubber　固体聚氨酯橡胶
solidus　固相线
solidus isoconcentration curve　固相等浓度曲线
solidus temperature　固相线温度
solid velocity　固体粒子混合速度〔催化剂混合速度〕
solid violet　固体紫
solid-wall bowl centrifuge　实体鼓壁式离心机
solid wall centrifuge　无孔离心机；实壁离心机
solid wall pressure vessel　整体式压力容器；单层式压力容器
solid waste　固体废物
solid waste disposal　固体废物处理
solid waste incinerator　固体废物焚烧炉
solid yellow　固体黄
soling　鞋底胶片
soling plate　底板
Solinox process　索林诺克斯过程〔快堆不锈钢壳核燃料用锑-铜合金液体金属去壳法〕
solion　溶液离子管
soliquoid　悬浮液；悬浮体
soliton　孤子
sol latex　溶液胶乳
Soller collimator　索勒准直器
Soller slit　索勒缝隙
solodization　(土壤)脱碱(作用)
solozone　过氧化氢
sol rubber　溶橡胶
sol solution　溶胶溶液
solubilisation　增溶溶解；增溶(作用)；加溶
solubilised　溶解了的
solubilised sulfur dye　溶性硫化染料
solubilised vat dye　溶性还原染料
solubility　①溶(解)度②溶(解)性；(可)溶性
solubility analysis　溶度分析
solubility behavior　溶解行为
solubility coefficient　溶度系数
solubility constant　溶解度常数
solubility curve　溶度曲线
solubility effect　溶解效应
solubility exponent　溶度指数
solubility fractionation　溶度分级
solubility isotherm　溶度等温线
solubility law　溶解度定律
solubility limit　溶度极限

solubility of residue　残渣溶解度
solubility of soap　肥皂溶解度
solubility parameter　溶度参数*
solubility product　溶(解)度(乘)积
solubility product constant　溶度积常数
solubility promoter　增溶剂；助溶剂
solubility temperature coefficient　溶解度温度系数
solubility-temperature curve　溶解度温度曲线
solubility test　溶解试验
solubility value　溶解值
solubilization　加溶(作用)；增溶
solubilization agent　增溶剂；溶解剂
solubilization capacity(=micellar solubility)　加溶(容)量
solubilization chromatography　加溶色谱法；增溶色谱法
solubilization curve　加溶曲线
solubilization end point　加溶终点
solubilization isotherm　加溶等温线
solubilization limit　加溶极限
solubilization measurement　增溶(作用)测量装置
solubilized sulfur dye　溶性硫化染料
solubilized system　加溶体系
solubilized vat dye　溶性还原染料
solubilizer　加溶剂；增溶剂
solubilizing　增溶能力
solubilizing agent　加溶剂〔指增加溶解性的试剂〕；增溶剂
solubilizing group　加溶基〔指增加溶解性的基团〕；增溶基
solubilizing micelle　增溶胶束
solubilizing power　加溶能力
solubilizing reaction　增溶反应；(增溶)溶解反应
solubity constant　溶解度常数
soluble　①可溶的②可以解决的③可解释的
soluble analysis　可溶性分析
soluble anode　可溶性阳极
soluble barbital(=soluble barbitone)　溶性巴比妥 $(C_2H_5)_2C_4O_3N_2HNa$
soluble barbitone(=soluble barbital)　溶性巴比妥
soluble bitumen　溶性沥青
soluble blue　溶性蓝
soluble colorant　可溶着色剂
soluble concrete　可溶性浸膏
soluble cotton　硝化纤维素
soluble drier　液体干料；可溶性干料
soluble dye　可溶染料
soluble fluorescein　溶性荧光素；荧光素钠 $C_{20}H_{12}O_5Na_2$
soluble glass　水玻璃；可溶性硅酸钠；泡花碱

soluble gluside 糖精钠 $C_7H_4O_3SNNa$
soluble gum 溶性胶
soluble iodophthalein 溶性碘酚酞；四碘酚酞钠 $C_{20}H_8O_4I_4Na_2$
soluble ionic precursor 水溶性离子母体
soluble lead salt 可溶铅盐
soluble matter 可溶物
soluble mercury 溶性汞
soluble metal content 可溶性金属含量
soluble monolayer 可溶性单层
solubleness ①溶(解)度②溶(解)性；(可)溶性
soluble neutron absorber 可溶性中子吸收剂
soluble neutron poison 可溶性中子毒物
soluble nylon resin 可溶性尼龙树脂
soluble oil 可乳化油
soluble oil cutting fluid(=soluble oil emulsion) 可乳化油切削液；可乳化油乳胶
soluble oil emulsion(=soluble oil cutting fluid) 可乳化油切削液；可乳化油乳胶
soluble oil paste 可乳化油膏
soluble perfume 水溶性香精
soluble phenobarbital(=soluble phenobarbitone) 苯巴比妥；鲁米那钠 $C_{12}H_{11}O_3N_2Na$
soluble phenobarbitone(=soluble phenobarbital) 苯巴比妥；鲁米那钠
soluble phosphoric acid (可)溶性磷酸
soluble redwood 可溶性红木
soluble saccharin(=saccharin sodium salt) (可)溶性糖精；糖精钠 $C_7H_4O_3NSNa$
soluble salt 可溶性盐
soluble soil 可溶性污垢
soluble starch 可溶淀粉
soluble sulfathiazole(=sulfathiazole sodium) 溶性磺胺噻唑；磺胺噻唑钠
soluble sulfathiazone(=sulfathiazole sodium; sulfathiazolum soluble) 可溶性磺胺噻唑；磺胺噻唑钠
soluble sulfur 可溶性硫黄
soluble tartar 酒石酸钾铵
soluble tartrate 酒石酸钾
solum 防潮湿料层；土壤表层
soluseptazine 苯丙磺胺苯二磺酸钠
solustibosan 双氧锑根葡萄糖酸钠
solute 溶质；溶解物
solute balance 溶质衡算
solute boundary layer 溶质边界层
solute distribution 溶质分布
solute gas 溶解气
solute migration distance 溶质迁移距离

solute peak 溶质峰
solute property detector 溶质性能检测器
solute-solvent interaction 溶质-溶剂相互作用
solute transfer 溶质传递
solute zone 溶质层
solution ①溶液*；溶体②溶解
solution adhesive 溶液黏合剂
solution adsorption method 溶液吸附法
solution balance 溶液天平
solution birefringence 溶液双折射
solution calorimeter 溶液量热计
solution ceramic coating 溶液型陶瓷涂层
solution coating 浸液涂漆；溶液浸渍
solution coating method 溶液涂渍法
solution complex 溶液配合物
solution conductivity detector 溶液电导检测器
solution cooler 溶液冷却器
solution depression 溶液的冰点降低
solution dilution factor 溶液稀释因子
solution-dyed 溶液染色
solution flash spinning method 溶液闪蒸纺丝法
solution fluorimetry 溶液荧光法
solution gas 溶解气
solution heat 溶解热
solution heater 溶液加热器
solution heat treatment 固溶热处理
solutioning machine 涂胶机；刮浆机
solution method 溶液法
solution mining 溶液采矿
solution mixer 溶液混合器
solution mixing tank 溶液混合槽
solution of ammonium hydroxide 氢氧化铵溶液
solution of cellulose derivative 纤维素衍生物溶液
solution polycondensation 溶液缩聚
solution polymerization 溶液聚合*
solution polymerization process for polyacrylonitrile 聚丙烯腈溶液聚合工艺
solution polymerized butadiene styrene rubber(SSBR) 溶聚丁苯橡胶
solution power 溶解力〔溶解其他物质的本领〕
solution pressure 溶解压(力)；溶解压强
solution property detector 溶液性能检测器
solution reclaiming method 溶液回收法
solution residue technique 溶液残渣技术
solution salt ①溶液盐②苄胺基苯磺酸钠 $C_6H_5CH_2NHC_6H_4SO_3Na$
solution spinning 溶液纺丝
solution spraying 溶液喷涂

solution spraying friction plating 摩擦电喷镀
solution strainer 溶液过滤器
solution strength 溶液浓度
solution tank 溶液槽
solution technique 溶液技术；溶解技术
solution tension 溶解压
solution theory 溶解理论
solutizer 硫醇溶解加速剂
solutizer-air regenerative process 硫醇溶解加速剂-空气再生法
solutizer process 硫醇溶解加速剂过程
solutizer regenerator 硫醇溶解加速剂再生塔
solutizer-steam regenerative process 硫醇溶解加速剂-蒸汽再生法
solutizer-tannin process 硫醇溶解加速剂-单宁法
solvability 溶剂化能力
solvatable cationic component 能溶剂化的阳离子组分
solvate 溶剂合物*
solvated electron 溶剂化电子*
solvated hydrogen ion 溶剂化氢离子
solvated layer 溶剂化层
solvated polymer 溶剂化聚合物
solvated proton 溶剂化质子*；溶剂阳离子
solvated resin 溶剂化树脂
solvate isomerism 溶剂合异构
solvate theory 溶剂化学说
solvating ability 溶剂化能力；溶剂化可能性
solvating action 溶剂化作用
solvating effect 溶剂化效应
solvating plasticizer 溶剂化增塑剂
solvating power 溶剂化本领(能力)
solvation 溶剂化(作用)
solvation accelerator 溶剂化促进剂
solvation effect 溶剂化效应
solvation energy 溶剂化能
solvation rate 溶剂化速率
solvatochromism 溶剂化显色(现象)
solvatochromism effect 溶致变色效应
solvatophobic bonding 疏溶剂化结合
Solvay carbonator 苏尔维碳酸化器
Solvay-Kellner cell (for caustic soda) 苏尔维-凯耳讷(制苛性钠)电池
Solvay process 苏尔维法
Solvay solution 苏尔维液
Solvay tower 苏尔维塔
solvency 溶解本领
solvency action 溶解作用
solvency of petroleum spirit 汽油溶解本领
solvency power 溶解力
solvent 溶剂*
solvent action 溶剂作用
solvent-activated adhesive 溶剂活化胶黏剂
solvent adhesive 溶剂型胶黏剂
solvent-air flush check valve 溶剂冲-空气吹(洗)转换开关
solvent assisted micromolding 溶剂协助微模塑
solvent attack 溶剂浸蚀
solvent balance 溶剂平衡
solvent-based adhesive 溶剂型胶黏剂；溶液型黏合剂
solvent based coating 溶剂型涂料
solvent-based polyisocyanate 溶剂型聚(多)异氰酸酯
solvent-based system 溶剂型体系
solvent-binder ratio 溶剂基料比
solvent binding 溶剂束缚
solvent bleeding 溶剂(性)渗色〔漆病〕
solvent-borne 含溶剂的；溶剂型(的)*
solvent-borne type 溶剂型
solvent borne urethanes 溶剂型氨基甲酸酯
solvent brushing 溶剂刷洗
solvent cage 溶剂笼
solvent cage effect 溶剂的笼蔽效应
solvent-cast film 溶剂注膜
solvent cell 溶剂池
solvent circulation rate 溶剂循环速率
solvent cleaner 溶剂洗净剂
solvent column (=solvent tower) 溶剂抽提塔；抽提蒸馏塔
solvent compartment 溶剂室
solvent compatibility 溶剂相容性
solvent composition 溶剂组成
solvent concentration 溶剂浓度
solvent condenser 溶剂冷凝器
solvent consumption 溶剂消耗量
solvent contamination 溶剂污染
solvent correction 溶剂校正
solvent cracking 溶剂分解；溶剂裂解
solvent-craze resistance 耐溶剂银纹(发白)性；抗溶剂细裂性
solvent-crazing 溶剂致银纹
solvent crystallization 溶剂结晶法〔分离油的不同组分的一种方法〕
solvent cut ①溶剂馏分 ②溶剂稀释
solvent cycle 溶剂循环
solvent deasphalting 溶剂脱沥青
solvent decarbonizing 溶剂脱碳(法)
solvent degreasing 溶剂脱脂

solvent delivery system 溶剂输送系统
solvent deresining 溶剂脱树脂(法)
solvent detergent 溶剂洗涤剂
solvent dewaxing (用)溶剂脱蜡
solvent dewaxing-deoiling 溶剂脱蜡-脱油(法)
solvent dewaxing process 溶剂脱蜡法
solvent dewaxing unit 溶剂脱蜡装置
solvent diluent 溶剂稀释剂
solvent dispersion 溶剂分散
solvent drum dyeing machine 鼓筒式溶剂染色机
solvent dye 溶剂染料
solvent effect 溶剂效应*
solvent efficiency 溶剂效率
solvent elimination technique 溶剂峰消除技术
solvent embrittlement 溶剂脆化(发脆); 溶剂致脆
solvent emulsion 溶剂乳状液
solvent epoxy varnish 溶剂型环氧清漆
solvent etching 溶剂腐蚀; 溶剂蚀刻
solvent ether (二)乙醚 $C_2H_5OC_2H_5$
solvent evaporation 溶剂蒸发
solvent extractable matter 溶剂萃取物; 溶剂提取物
solvent-extracted oil 溶剂萃取油; 溶剂抽出油
solvent extractible 溶剂可抽(取)的; 溶剂可萃取的
solvent extraction 溶剂萃取*
solvent extraction generator 溶剂萃取(同位素)发生器
solvent extraction method 溶剂萃取法
solvent extraction process 溶剂提取法
solvent extraction separation 溶剂萃取分离
solvent extraction test 溶剂萃取试验
solvent extraction tower 溶剂提取塔
solvent fat liquoring 溶剂加油
solvent flashing 挥发溶剂; 溶剂闪蒸; 溶剂晾干
solvent flow blotting 溶剂流印迹
solvent fractionation 溶剂分馏; 溶剂选择分离
solvent-free 无溶剂的
solvent-free alkylammonium 无溶剂烷基铵
solvent-free analysis 无溶剂分析
solvent-free coating 无溶剂涂料
solvent-free encapsulated pigment 无溶剂胶囊颜料
solvent-free injection 无溶剂注射
solvent-free liquid epoxy resin 无溶剂液体环氧树脂
solvent-free melt extrusion 无溶剂熔体挤出
solvent-free paint 无溶剂漆
solvent-free polyisocyanate 无溶剂多异氰酸酯
solvent-free polyurethane adhesive 无溶剂聚氨酯黏合剂
solvent-free shaping 无溶剂成形
solvent-free single layer complex 无溶剂(分子)单层配合物

solvent front 溶剂前沿; 展开剂前沿
solvent fumes 溶剂烟雾
solvent gradient (elution) chromatography 溶剂梯度洗脱色谱法
solvent hold-up 溶剂滞留
solvent imbalance 溶剂不平衡
solvent-impoverished vehicle 贫溶剂漆料
solvent impregnated resin 浸渍树脂
solvent-impregnating resin 溶剂浸渍树脂
solvent inclusion 溶剂包藏作用
solvent index 溶剂值(指数)
solvent-induced 溶剂引致的
solvent-induced crazing 溶剂引致的银纹; 溶剂性银纹
solvent-in-pulp 溶剂矿浆萃取; 溶剂浆液萃取
solvent-in-pulp extraction 矿浆萃取
solvent isotope effect (SIE) 溶剂同位素效应
solventized soap 溶剂化皂
solvent-laden 带溶剂的; 含溶剂的
solvent layer 溶剂层
solventless adhesive(=non-solvent adhesive) 无溶剂型胶黏剂; 无溶剂型黏合剂
solventless coating 无溶剂涂料
solventless epoxy coating 无溶剂环氧涂料
solventless fabric coating 无溶剂(的)织物涂层
solventless polyester varnish 无溶剂聚酯清漆
solventless varnish 无溶剂清漆
solvent load ①(大气中)溶剂载荷溶剂(挥发)量 ②溶剂用量(含量)
solvent loss 溶剂损失
solvent makeup 溶剂补充
solvent medium 溶剂介质
solvent method 溶剂法
solvent migration index 溶剂迁移指数
solvent naphtha 溶剂石脑油
solvent neutral oil 溶剂中性油
solvent-nonsolvent method 溶剂-非溶剂法
solvent-nonsolvent system 溶剂-非溶剂体系
solvent oil 溶剂油
solvent-oil ratio 溶剂和油的比例; 溶剂消耗量
solvent organosol 溶剂型有机溶胶
solvent paste wax 溶剂型石蜡膏
solvent peak 溶剂峰
solvent phase 溶剂相
solvent polarity 溶剂极性
solvent polarity indicator 溶剂极性指示剂
solvent polarity parameter 溶剂极性参数
solvent polish(ing) 溶剂抛光
solvent polymerization 溶解聚合〔在溶液中聚合〕

solvent pop　(溶剂滞留)起泡现象〔漆病〕
solvent popping　溶剂滚沸
solvent power　溶解本领
solvent prechiller　溶剂预冷器
solvent pressing　加溶剂压滤〔油类脱蜡时应用〕
solvent process　溶剂精制法〔润滑油溶剂选择精制过程〕
solvent process oil　溶剂精制油
solvent programmer　溶剂程序变换器
solvent programming　程序变溶剂
solvent programming elution　程序变溶剂洗脱
solvent pump　溶剂泵
solvent radiolysis　溶剂辐(射分)解
solvent raffinate　溶剂精制液
solvent reactivated　(用)溶剂活性化的
solvent reclaim　溶剂法再生胶
solvent reclaimer　溶剂回收设备
solvent reclamation　溶剂回收
solvent recovering system　溶剂回收系统
solvent recovery　溶剂回收
solvent recovery plant　溶剂回收厂
solvent recuperation　溶剂回收
solvent recycler　溶剂循环器
solvent reducible coating　溶剂可稀释涂料；溶剂性涂料
solvent refined　溶剂精制的
solvent-refined oil　溶剂精制油；溶剂选择精制油
solvent refining(=solvent treating; solvent treatment)　溶剂精制；溶剂处理
solvent refining agent　溶剂精制剂
solvent refining process　溶剂精制法
solvent regain　吸溶剂量
solvent regeneration　①溶剂再生②萃取剂再生
solvent regeneration process　溶剂再生法
solvent regenerator　溶剂再生器
solvent release agent　溶剂型脱模剂
solvent removal　(用)溶剂清除(除去)
solvent removal effect　去溶作用
solvent removal system　去溶剂系统
solvent removal technique　溶剂清除法
solvent reservoir　溶剂储器
solvent resistance　耐溶剂性
solvent resistant coating　耐溶剂涂料
solvent resistant column　抗溶剂柱〔凝胶过滤柱〕
solvent resistant grease　抗溶剂润滑脂
solvent resistant hose　耐溶剂胶管；汽油胶管
solvent resonance suppression　溶剂共振抑制
solvent retention　溶剂存留量；溶剂滞留性
solvent-rich letdown vehicle　富溶剂调漆料
solvent-rich vehicle　富溶剂漆料

solvent salt milling　溶剂盐研磨法
solvent scouring　干洗；溶剂擦洗
solvent seal(ing)　溶剂封合
solvent segregation　溶剂分离法〔分离油的不同组分〕
solvent selectivity　溶剂选择性
solvent-separated ion-pair　溶剂分离的离子对
solvent separation　溶剂分离法
solvent separator　溶剂分离器
solvent series (=eluotropic series)　洗脱序(列)；溶剂序
solvent shift　溶剂位移
solvent silicone varnish　溶剂型有机硅清漆
solvent slop　污染溶剂；废溶剂
solvent soap　溶剂皂
solvent solubility　溶剂溶解性
solvent-soluble dye　溶剂溶解染料
solvent-soluble soap　溶剂溶化皂
solvent-solute pair　溶剂-溶质对
solvent solution　溶剂溶液
solvent spun fibre　溶纺纤维；溶液法纤维
solvent stabilization operator　溶剂稳定化算符
solvent-starved vehicle　贫溶剂漆料
solvent strength　溶剂浓度；溶剂强度
solvent strength gradient　溶剂强度梯度
solvent strength parameter　溶剂强度参数
solvent stripper　溶剂型脱漆剂
solvent-support ratio　溶剂载体比
solvent swell　溶剂溶胀
solvent-swollen　溶剂溶胀
solvent tannage　溶剂鞣法
solvent test　溶剂试验
solvent-thinned paint　溶剂稀释漆；溶剂型漆
solvent tolerance　溶剂容忍度；溶剂的稀释极限
solvent tolerance of resin　树脂在溶剂中的(最大)溶解度
solvent tower(=solvent column)　溶剂抽提塔；抽提蒸馏塔
solvent transmission　溶剂透过率
solvent treating(=solvent refining)　溶剂精制；溶剂处理
solvent treating plant　溶剂精制厂
solvent treatment(=solvent refining)　溶剂精制；溶剂处理
solvent trough　溶剂槽
solvent type adhesive　溶剂型黏合剂
solvent-type paint　溶剂型色漆
solvent-type plasticizer　溶剂型增塑剂
solvent-type polish　溶剂型抛光剂(膏)
solvent uptake　溶剂吸取(量)
solvent-vapor gradient　溶剂-蒸气梯度
solvent vapor recovery plant　溶剂蒸气回收装置
solvent vapour degreasing　溶剂蒸气脱脂
solvent volatility　溶剂挥发性

solvent washing effect 溶剂洗提影响
solvent waste disposal 废溶剂排放
solvent water 溶剂水分
solvent wax 溶剂型液体(石)蜡
solvent winterization 溶剂冬化法
solvent wipe-off method 溶剂擦洗法
solvent wiping 溶剂擦拭
solvesso 一种含芳烃量高的溶剂
Solvex cracking furnace 索维克斯裂化炉〔有悬吊挡板的火焰自两边加热的裂化炉〕
solvoacid 溶剂合酸
solvobase 溶剂合碱
solvochromism 溶剂变色性
solvolysis 溶剂解
solvolysis reaction 溶剂分解反应
solvolyte 溶剂化物
solvolytic counterion 溶剂化对离子
solvolytic dissociation 溶剂离解(作用)
solvolytic reaction 溶剂化反应
solvolyze(=solvolysis) 加溶剂分解
solvophilic 亲溶剂的
solvophobic 憎溶剂的
solvophobic chromatography 疏溶剂色谱法
solvophobic interaction 疏溶剂作用
solvophobic interaction principle 疏溶剂作用理论
solvophobic theory 疏溶剂理论
solvosalt 溶剂合盐
solvus 溶解度曲线；溶线
solway colors 搔尔威染料
solway purple 搔尔威红紫
somatochirality 体型手性
Sommelet reaction 萨姆勒特反应〔从苄胺和甲醛制苯甲醛法〕
Sommerfeld notation 索默菲尔德标志法
sommos 氯乙醛酒精合剂
somnifacient 催眠剂
somniferol 催眠醇
somnirol 茄醇
soneryl 5-乙基-5-丁基丙二酰脲
sonic 音速的；声音的
sonic agglomeration 声波凝聚(作用)
sonic agglomerator ①声波凝聚器②声波除(集)尘器
sonic altimeter 声波测高计
sonic anemometer 声波风速计
sonication ①声波振荡②声波破碎③声处理④超声波法
sonic barrier 音障
sonic boom 音爆

sonic-electronic flowmeter 声-电流量计
sonic gas analyzer 声响式气体分析计
sonic hygrometer 声速湿度计
sonic hygrometry 声速测湿法
sonic nucleation 声波核晶过程
sonic precipitation ①声波除尘；声波集尘(作用)②声波沉淀(作用)
sonic sifter 声频筛
sonic sifter analysis 声波筛析法
sonic spray ionization 超声喷雾电离
sonic velocity detector 声速检测器
sonic velocity method 声速法〔测定分子取向〕
sonic washing (超)声波洗涤(法)
sonic washing machine (超)声波洗涤机
sonimetric analysis(=sonimetry) 超声波分析(法)；声波分析(法)
sonimetry (=sonimetric analysis) 超声波分析(法)；声波分析(法)
sonography 声图描记法
sonolator (可见语言)声谱显示仪
sonoluminescence 声致发光
sonora gum (墨西哥)酚油木胶
soot ①烟灰；煤炱；炭黑②煤烟
soot blower 烟灰吹除机
soot blower system 吹灰系统
soot blowing 烟灰吹除
soot-chamber test (静电)烟尘室试验
soot collector 滤烟器；集灰器
soot deposits 烟灰沉积
sootfall 烟灰沉降
sootflake 烟灰薄片
soot formation 烟灰生成；炭黑生成；烟炱生成
soothing cream 润肤膏
soothing oil 润肤油
soot inhibitor 烟灰抑制剂
soot of mercury 汞炱
soot type soil 烟灰类污垢
soot-water suspension 烟灰-水悬浮液
sooty 烟灰的；炭黑的；烟炱的
sooty coal 泥煤；劣质软煤
sooty mold 烟霉；煤灰霉病
sophisticated bottom system 高级(先进)的船底漆体系
sophisticated categories ①精细分类②尖端技术部门
sophisticated fluid system 精密流体泵
sophisticated paint line 高精密涂装线
sophistication 掺杂
sophocarpinol 槐子壳醇
sophoramine 槐胺

sophoranol 槐醇
sophorin 槐苷〔取自槐属植物 Sophora species〕
sophorine 槐碱；金雀花碱
sophorine A (=hwae-hua-menine-A) 槐苷 A；槐花米甲素
sophorine B (=hwae-hua-menine-B) 槐苷 B；槐花米乙素
sophorine C (=hwae-hua-menine-C) 槐苷 C；槐花米丙素
sophorolipid 槐糖脂
sophorose 2-葡糖-β-葡糖苷
sophoroside 槐糖苷
soporific 安眠药
sorbate 山梨酸酯 $CH_2CH=CHCH=CHCOOR$
sorbed water 吸着水
sorbency 吸着力；吸着性
sorbent 吸着剂
sorbent carrier 吸着剂载体
sorbent material 吸附剂；吸着剂
sorbic acid 山梨酸；2,4-己二烯酸
$CH_3CH=CHCH=CHCOOH$
sorbic alcohol 山梨醇；2,4-己二烯醇
sorbic aldehyde 山梨醛；2,4-己二烯醛
sorbic ester 山梨酸酯
sorbide 脱(缩)二水山梨(糖)醇
sorbierite 山梨醇
sorbin 山梨糖
sorbing 吸着
sorbing agent 吸着剂
sorbinose 山梨糖
sorbitan ①脱水山梨(糖)醇
$HOCH_2CHOHCHCHOHCHOHCH_2O$ ②山梨聚糖
sorbitan carboxylic ester 脱水山梨(糖)醇羧酸酯
sorbitan dodecanoate 失水山梨糖醇十二烷酸酯
sorbitan ester 脱水山梨(糖)醇(羧酸)酯；山梨(糖)醇酐酯
sorbitan fatty acid ester 失水山梨糖醇脂肪酸酯
sorbitan fatty ester 失水山梨糖醇脂肪酸酯
sorbitan isostearate 失水山梨糖醇异硬脂酸酯
sorbitan laurate 失水山梨糖醇月桂酸酯
sorbitan monolaurate(=Span 20) 失水山梨糖醇单月桂酸酯；司盘 20
sorbitan oleate 失水山梨糖醇油酸酯
sorbitan palmitate 失水山梨糖醇棕榈酸酯
sorbitan sesquioleate 失水山梨糖醇倍半油酸酯
sorbitan stearate 失水山梨糖醇硬脂酸酯
sorbitan trioleate 失水山梨糖醇三油酸酯
sorbite(=sorbitol) 山梨(糖)醇
sorbitol(=sorbite) 山梨(糖)醇 $HOCH_2(CHOH)_4CH_2OH$
sorbitol ester 山梨(糖)醇酯
sorbitol hexaacetate 山梨(糖)醇六乙酸酯 $C_6H_8O_6(COCH_3)_6$
sorbitol laurate 山梨(糖)醇月桂酸酯
sorbitol polyglycidylether 山梨糖醇多缩水甘油醚
sorbitol stearate 硬脂酸山梨糖醇酯
sorbitol tallate 松浆(油)酸山梨糖醇酯
sorbityl monododecanoate 山梨(糖)醇单月桂酸酯；山梨醇单十二酸酯
sorbonitrile 山梨腈
sorbose 山梨糖 $C_6H_{12}O_6$
sorburonic acid 山梨糖酮酸
sorbyl alcohol(=sorbic alcohol) 山梨醇；2,4-己二烯醇
Sorel cement 索瑞耳胶结料
Sorel's method 索瑞耳法
Sorensen indicator 索楞逊指示剂
Sorensen phosphate 索楞逊磷酸盐；磷酸二氢钠
Sorensen symbols 索楞逊符号
Sorensen value 索楞逊值；氢离子浓度
sores 伤处；疮毒
Soret band 索雷谱带
Soret effect 索雷效应〔液体混合物的热扩散〕
Soret potential 索雷电位；热-浓差电位
sorghum 高粱
sorghum molasses 高粱饴；高粱糖浆
sorghum oil 高粱油
sorghum syrup 高粱饴
sorgo(=sorghum) 高粱
sorgo cane(=sorghum) 高粱
sorigenin 鼠李苷配基
sorinja oil(=behen oil) 山萮油
sorption 吸着(作用)
sorption capacity 吸着能力
sorption detector 吸着检测器
sorption equilibrium 吸着平衡
sorption extraction 吸附萃取(法)
sorption gradient 吸着梯度
sorption hysteresis 吸着滞后现象
sorption isotherm 吸着等温线
sorption isotherm curves 等温吸湿曲线
sorption property 吸着性能
sorption pump 吸着泵
sorption rate 吸着速率
sorption ratio 吸着比
sorption site 吸着点
sorption strength 吸着强度
sorptive capacity 吸着能力
sorptive force 吸着力
sorptive medium 吸着介质
sorptive power 吸着能力

sorptographic analysis 吸附(色层)分析
sorptography 色层法
sorptometer 吸着检测仪
sorrel 酸模；酸模叶
sorrel salt 一水草酸氢钾 $KHC_2O_4 \cdot H_2O$
sort ①类(别)②方法③分类
sorter (纤维长度)分析器
sorting 拣选；分级；分类
sorting bed 拣选床；选粒床
sorting belt 拣选(运输)带
sorting belt conveyer 拣选带式运输机
sorting chute 拣选斜槽
sorting conveyer 拣选运输机
sorting equipment 分级装置；分类设备
sorting floor 拣选台
sorting grizzly 拣选筛
sorting table 拣选台；摇床
sorting test 选别试验
sorting yard 拣选场
sosoloid 固溶体；固态溶液
sound absorbent material 吸音材料
sound absorbing coefficient 吸音系数；吸声率
sound absorbing paint 吸音漆；隔音涂料
sound absorption 吸音作用；消声力
sound absorption coefficient 吸声系数
sound barrier 音障；隔音物
sound cement 固定性水泥
sound damping 隔音的；防噪声的
sound-damping qualities 消音量；隔音量
sound deadening 隔音的；防噪声的
sound deadening sheet 隔音膜；消音薄板
sound diminishing paint 隔音漆；消音漆
sound emission 声发射
sounding ①测深②触探③响
sounding line 测(水)深用绳；测深索
sounding rod (测定石油产品或水高度的)测量杆
sound-insulating materials 隔音材料
sound insulation 隔音；绝音；消音
sound intensity level 声强级
sound knots 健全节
sound-level meter 声级计；噪声计
soundness test 固性试验
sound plaster ①坚固灰浆层(垫层)②硬石膏层
sound pollution 噪声污染
sound pressure 声压
sound pressure level 声压强度；响度
sound-pressure meter 声压计
sound-proofing property 隔音性能

sound-proof material 隔音材料
sound propagation 声传播
sound radiation 声辐射
sound spectrograph 声谱仪
sound spectrum 声谱
sound speed 声速
sound suppressor 消声器
sound textured surface 坚硬(实心)的立体花纹表面
sound transparent rubber 传声橡胶
sound velocity 声速
sound vibration 声振动
sound wave propagation method 声波传播法
sound waves 声波
sound wood 良木；好材
soup 胶浆
soup compound 浆状物；汤状物
soup stick 吸杆；麦杆；蜡纸杆
sour 酸的
sour bark 酸性树皮
sour bath 酸浴；酸液
source (离子)源；放射源
γ-source γ源
source-based nomenclature 以原料为基础的命名法
source capsule 源盒；源管
source-coupler loss 光源-耦合器损耗
source flow 源流
source housing 离子源室
source impedance 电源阻抗
source name 源名
source of light 光源
source of pollution 污染源
source of radiation 辐射源
source parameters 离子源参数
source program 源程序
source pump station (管线上)起点泵站
source recognition 污染源识别
sour cherry 欧洲酸樱桃
sour corrosion 二氧化硫腐蚀；(含硫的)酸性腐蚀
sour crude 酸性原油；含硫原油
sour crude oil 酸性原油；含硫石油；含硫量高的石油
sourdine 消声器；噪声抑制器
sour dip 底革酸浸
sour distillate 酸性馏分；含硫馏分
sour dry gas 酸性干气；含硫量高的干燥石油气
sour gas 酸气
sour nature gas 高硫天然气
sourness ①酸性②酸度
sour odour 酸味

sour oil 含硫油；酸性油
sour orange(=bigarade orange) 酸橙；苦橙
sour quench water 酸性骤冷水
sour sulfur 酸性硫
sour water 酸性水
sourwood 酸叶石楠
Sovafining 索伐精制(法)
Sovaforming 索伐重整(法)
Soxhlet's extractor 索格利特萃取器
Soxhlet apparatus 索格利特抽提器
Soxhlet continuous extractor 索格利特连续萃取器；索氏连续萃取器
Soxhlet extraction 索氏萃取
Soxhlet extraction method 索格利特萃取法；索氏萃取法
Soxhlet extract method 索氏抽提法
Soxhlet method 索格利特法
Soxhlet's extractor 索氏萃取器
Soxhlet solution 索格利特溶液
soya-bean 大豆
soya bean fatty acid 大豆脂肪酸
soya bean oil(=soya oil; soybean oil) 豆油
soy(a)-bean oil fatty acid 豆油脂肪酸
soya fatty acid 大豆(油)脂肪酸
soya oil(=soya bean oil) 豆油
soyasterol 大豆甾醇
soyate 豆油脂肪酸盐
soybean 大豆纤维
soybean alkyd 豆油醇酸树脂
soybean oil(=soya bean oil) 豆油
soybean oil plasticizer 豆油增塑剂
sozal 对酚磺酸铝
soziodol(=sozoiodol) 二碘酚磺酸
sozoiodol(=soziodol) 二碘酚磺酸 $C_6H_4O_4I_2S$
sozoiodolate 二碘酚磺酸盐
sozoiodolic acid 二碘酚磺酸 $C_6H_2I_2OHSO_3H$
sozolic acid(=aseptol) 苯酚-2-磺酸
space ①空间；宇(物)；地位；距离②隔开
space buff 隔层磨光轮
space cavity 空隙
space charge 空间电荷
space charge effect 空间电荷效应
space charge-limited (受)空间电荷限制的
space charge region 空间电荷区
space chemistry 立体化学
spaced bucket elevator 间隔斗提升机
space deodorant 室内芳香剂；室内除味剂
space diagram ①空间(立体)图②矢量图
space-distribution interference 空间分布干扰

spaced peaks 彼此分开的峰
space factor 占空因数；方向性系数；填充系数
space field flow fraction 空间场流分级
space formula 立体式
space frame 空间构架；立体构架
space group(=spacer;space grouping) ①空间群②间隔基；间隔团
space grouping(=space group;spacer) ①空间群②间隔基；间隔团
space group symbol 空间群符号
space isomerism 空间异构现象；立体异构现象
space lattice 立体晶格；空间格子；空间格点
space manifold 空间流形
space mass spectrometry 空间质谱分析
space model 立体模型；空间点阵
space network 立体网状结构
space-network high polymer 立体网形高聚物
space network polymer 立体网状聚合物；立体网状高聚物
space nuclear auxiliary power(SNAP) 空间核辅助能源
space polycyclic compound 体形多环化合物；立体多环化合物
space polymer 立构聚合物；立体聚合物
space quantization 空间量子化
spacer ①空间群②间隔基；间隔团③嵌木④垫板⑤隔片；隔离物
spacer arm effect 间隔臂效应
space rate 空速
spacer block (模具)隔块
space resolution 空间分辨率
spacer gel 成层胶；间隔凝胶
spacer plate 分离板
spacer ring 间隔环
spacer rod (色谱缸)分隔棒
space screen 立体筛
space spray 室内喷洒剂；室内芳香剂
space-time 时空
space-time efficiency 空时效率
space-time yield 时空产率
space unit 晶胞
space velocity 空速*
spachtling(=spackle) 填孔剂；填缝剂；填泥料
spacial distribution 空间分布
spacing ①间隔；间距②复位；定距
spacing collar 限位套筒；限位圈
spacing of lattice 晶格间隔；点阵间距
spacing of particle 颗粒间距；粒子间距
spacing parameter 间隔参数

spacing piece　定距片
spacing tube　定距管
spackle(=spackling compound)　填泥料；腻子料；腻子
spackling　填孔剂；填缝剂；填泥料
spade　铲(子)；锄(头)
SPADNS(=sulfophenylazo-chromotropic acid)　钍锆试剂；磺基苯偶氮变色酸
spaghetti drying rack　(漆布)绝缘(套)管干燥架
spallation　散裂
spallation neutron source　散变中子源
spallation product　散裂产物；散变产物
spallation reaction　散裂反应
spalled masonry　碎裂的砖石建筑
spalling　散裂
spalling resistance　抗散裂强度
spalling test　散裂试验
spallogenic radionuclide　散裂生成放射性核素
spandex　弹力纤维
spandex fiber　弹性纤维；松紧纤维
spangles(=glitter)　闪光颜料
spangolite　氯铜矾
Spanish hop's oil　西班牙酒花油
Spanish red oxide　西班牙氧化铁红〔专指西班牙马拉加地区产的天然氧化铁红〕
Spanish white　西班牙白；天然碳酸钙〔别名〕
span length　跨长；跨距
spanner　扳手；扳紧器
spar　①桅杆②(晶)石
sparassol　重菇醇
spar buoy　浮标
spare colour module　备用彩色粉末舱(储柜)
spare detail　备用零件
spare fan　备用(排)风扇
spare part　备件〔机工〕
spare varnish　维修用漆；备用清漆
sparger　①配器②分布器③喷射器
sparge water　冲洗水
sparging　喷射
sparging (jet) dyeing machine　喷射式染色机
sparingly soluble　微溶的
spark　①火花②电花③闪光④微物⑤发火花；发电花
spark advance　火花提前
spark comparison method　火花比较法
spark cutting　电火花切割
spark discharge　火花放电
spark discharge detector　火花放电检测器
spark discharge type recorder　火花放电型记录仪
spark excitation　火花激发

spark gap width　火花隙宽度
spark generator　火花发生器
sparking alloy　引火合金
sparking electrode　火花电极
spark ionic source　火花离子源
spark ionization　火花电离
spark ionization source　火花离子源
spark line　火花谱线
sparkling　火花
sparkling heat　焊接热；火花热
sparkling luster　耀眼光泽；闪烁光泽
sparkling water　汽水
sparkling wine　汽酒
spark plasma　火花等离子体
sparkproof　不起火花的
sparkproof tools　防火花工具〔青铜铍合金或其他非金属工具〕
spark source　火花源
spark source ionization　火花源电离
spark-source mass spectrograph　火花源质谱仪
spark source mass spectrometer　火花源质谱仪
spark source mass spectrometry　火花源质谱法
spark spectrum　火花光谱*
spark spectrum analysis　火花光谱分析
spark test　火花试验
spark tester　电火花试验机
spark-test tool　火花试验工具
sparry(=spathic)　晶石的
sparsiflorine　散花(巴豆)碱
sparsogenin　稀疏菌素
sparsomycin　稀疏霉素
spartalite　红锌矿
sparteine　鹰爪豆碱；金雀花碱
sparteine bisulfate　鹰爪豆碱硫酸盐
sparteine hydriodide　鹰爪豆碱氢碘酸盐
sparteine hydrochloride　鹰爪豆碱盐酸盐
sparteine sulfate　鹰爪豆碱硫酸盐
spartium　①金雀花②鹰爪豆
spartium alkaloids　鹰爪豆生物碱类
spar varnish　桅杆清漆
spasm　痉挛
spasmolytic　解痉药；镇痉药
spasmotin　痉挛碱
spasmotoxin　破伤风毒素
spatchel type knifing filler　(木料粗纹理用)快干型刮涂填孔料
spathic(=sparry; spathose; spathous)　晶石的
spathic iron ore　菱铁矿

spathose(=spathic) 晶石的
spathous(=spathic) 晶石的
spathulenol 斯巴醇
spatial 空间的；立体的
spatial arrangement 空间排列；空间布置
spatial array detection 空间矩阵检测
spatial chemistry 立体化学
spatial coherence 空间相干性
spatial configuration 立体构型
spatial configuration of molecules 分子的空间结构
spatial conjugation 立体共轭
spatial content 空间容量
spatial coordinate 空间坐标
spatial distribution 空间分布
spatial distribution interference 空间分布干扰
spatial gradient 空间梯度
spatial isomerism 立体异构；空间异构
spatial nuclei 三维核；立体核
spatial relation 空间关系；立体关系
spatial resolution 空间分辨率
spatial structure 立体结构
spatial symmetry 空间对称性
spatial velocity (火焰)空间速度；传播速度
spattening finishing 溅涂法
spatter 溅
spatter coating 溅彩涂装；喷溅涂装
spatter dash 喷洒；喷涂；溅彩
spatter finish ①溅彩涂料(面漆)②溅彩涂装法
spatter gun 溅涂喷枪；溅彩喷枪
spattering 溅
spattering finishing 溅涂法
spatter method ①溅彩涂装②阴极溅射沉积法
spatter paint 溅彩漆
spatter resistance 抗喷溅性
spatter spraying 溅彩喷涂法
spatter stain (木器)喷溅着色法
spatting 喷溅麻点〔漆病〕
spatula(=spatule) 刮勺；刮铲
spatular painting 刮涂
spatula rub-out 刮刀混合；刮刀捏合
spatula rub-out method 刮刀混合法〔吸油量测定法〕
spatula test (铝粉浆)漂浮率试验
spatule (=spatula) 刮勺；刮铲
spavin 煤层下耐火泥
spearmint oil 薄荷油〔取自 mentha〕；留兰香油
spear pyrite 矛黄铁矿
spec-finder 谱线检索
special alluminium bronze 特殊铝青铜(合金)

special asphalt 特种沥青；专用沥青
special boiling point spirits 特定沸点石油溶剂油
special bronze 特殊青铜；无锡青铜
special coating 特种涂料
special corrosion-inhibitive non-drying primer 特种不干性防腐蚀打底涂料
special cross-section 特殊截面；异形截面
special dye 特种染料
special emission standard ①特殊排放标准②污染物特殊扩散标准
special fabric 特形织物
special fissionable materials 特种可裂变物质〔加浓铀和钚〕
special fixture 专用夹具
special grade chemical 特级试剂
special industrial solvents 特种工业用溶剂；工业专用溶剂
special insulation 特制绝缘层；特制隔热层
specialist additive 专用添加剂；特种添加剂
speciality chemicals 特种化学药品；特制药品
speciality coating 特种涂料
speciality plasticizer 特殊增塑剂
speciality polymer 特殊性能高分子；特种高分子
speciality rubber 特种橡胶
speciality standard 专业标准
specialized conventional sulfur dye 特型惯用硫化染料
specialized paint 特种油漆；专用油漆
special nuclear material 特殊核材料
special oil 特殊油料
special parts and components 特别专件及配件
special plasticizer 特种增塑剂；专用增塑(韧)剂
special plywood 特种胶合板
special primer 特种底漆
special purpose ingredient 特种配合剂
special purpose rubber 特种橡胶；专用橡胶
special reaction 特殊反应
special reagent 特效试剂
special routines 专门规定
special steel 特种钢
special surface active agent 特种表面活性剂
special theory of relativity 狭义相对论
specialties 特殊产品
special tool steel 特殊工具钢
specialty chemicals 专用化学品
specialty elastomer 特种橡胶
specialty fiber 特种纤维
specialty monomer 特制单体；专用单体
specialty plasticizer 特制增塑剂；专用增塑剂
specialty polymer 专用聚合物

speciation 形态
speciation analysis 形态分析
species ①种；类②物种
species analysis 形态分析
specific ①特效的②特效药③特殊的④比的
specific absorption 吸收率；吸收比；吸收系数
specific absorption band 特殊吸收(谱)带
specific absorption coefficient 比吸收系数
specific absorption coefficient method 比吸收系数法
specific absorptivity 比吸光系数*
specific activity 比活度*；比放射性(活度)
specific adhesion 特性黏合；比黏合
specific adsorption 特性吸附*
specific area 比面积
specification 规格；技术规范；说明书
specification coating 专用涂料
specification gasoline 汽油规格
specification limit 规格限度；标准界限
specification of ion meter 离子计指标
specification requirements 技术规格要求
specification standard 规格标准
specific breaking work 拉伸断裂比功
specific capacity ①比电容②比容量
specific charge 荷质比
specific condition of exposure 暴露的特殊条件
specific conductance 电导率；电导系数；比电导
specific conductivity 比电导率
specific crossover point 交叉率
specific damping capacity 比阻尼容量
specific density 比重
specific duty 单位生产量
specific dynamic action 特种动力作用
specific electric conductivity 比电导率
specific elongation 延伸率；相对伸长
specific end use (制品)特定用途
specific energy 比能
specific energy loss 比能(量)损失
specific enthalpy 比热含(量)
specific entropy 比熵
specific extinction 比消光(系数)；比吸光(系数)
specific extinction coefficient 比消光系数
specific filtration resistance 过滤比；比阻力；比滤阻
specific flexural modulus 比挠曲模量
specific flexural rigidity 比挠曲刚度
specific fuel consumption 比燃料消耗
specific gas constant 比气体常数
specific gravimeter 比重计
specific gravity 比重

specific gravity balance (=hydrostatic balance) 比重天平
specific gravity bottle 比重瓶
specific gravity cell 比重电池
specific gravity correction factor 比重改正因数
specific gravity fraction 比重分离份
specific gravity hydrometer 比重计
specific gravity of gas 气体的比重
specific gravity of pigment 颜料的比重
specific gravity test 比重测定
specific gravity tester 快速比重测定仪
specific growth rate 比增殖速度
specific heat 比热
specific heat at constant pressure 定压比热
specific heat at constant volume 定体(积)比热；恒容比热
specific heat capacity 比热容
specific heat per unit mass 单位物质的比热
specific heat per unit volume 单位体积的比热
specific heat per unit weight 单位重量的比热
specific heat ratio 比热比；绝热指数
specific Helmhotz free energy 单位亥姆霍兹自由能
specific humidity 比湿度
specific humidity of gas 气体的比湿度
specific impulse 比冲量；比推力
specific inaccuracy 误差率
specific indicator 专一指示剂
specific inductive capacitance 介电常数；电容率
specific inductive capacity(=specific inductivity) 介电常数；电容率
specific inductive constant 介电常数
specific inductivity(=specific inductive capacity) 介电常数；电容率
specific information price 信息比价
specific installation 特殊设备；单独设备
specific insulation resistance 比绝缘电阻
specific interaction of the ions 离子特殊相互作用
specific interfacial area 比界面积
specific interference 特征干扰
specific internal energy 比内能
specific intrinsic viscosity 比特性黏度
specific inventory 比投料量；比装载量
specific investment 比投资
specific ion electrode 离子选择电极
specific ion-exchange membrane electrode 专一性离子交换液膜电极
specific ionization 电离比值
specificity ①特(异)性；特征②专一性
specific loss 比(损)耗
specific magnetization 磁化率

specific mass　密度
specific module　比模量
specific modulus　比模量
specific molding pressure　单位模压力
specific oxygen uptake rate　氧吸收比速度
specific parachor　比等张比容
specific performance index　特性指数
specific permeability　比渗透率
specific polarization　比极化度
specific pore volume　比孔容
specific power　比动力；单位功率
specific power input　单位功率输入
specific pressure　比压；单位压力
specific property　特性
specific-purpose incinerator　特种用途焚烧炉
specific γ-ray constant　比γ射线常数〔单位：伦·米^2·小时$^{-1}$·居里$^{-1}$〕
specific rate　比速度
specific rate constant　比速常数
specific ratio　比率
specific reaction　特异反应；特效反应；专一反应；专属反应
specific reaction rate　反应比速
specific reagent　特效试剂
specific refraction　折射系数
specific refractive power　折射率；比折射力
specific refractivity　折射率系数；折射率差度
specific refractory power　折射率
specific resistance　电阻率；电阻系数
specific resistivity　比电阻
specific retention volume　比保留体积
specific rigidity　比刚度
specific rotary power　旋光率
specific rotation　比旋光*
specific rotatory power　比旋光度
specific speed　比速；比转数〔离心泵或离心风机〕
specific strength　比强度
specific stress power　单位应力功率
specific surface　比表面
specific surface area　比表面积
specific surface area analyzer　比表面积分析仪
specific surface free energy　比表面自由能
specific surface tension　比表面张力
specific surface tension test　比表面张力试验
specific susceptibility　比磁化率；比磁化系数
specific tenacity　比强度；比折断阻力；比抗张力
specific tensile strength　比拉伸强度
specific test　特效试验

specific thermal conductivity　比导热率
specific viscosity　增比黏度
specific volume　比容；体积度
specific volume in situ　现场比容
specific volume resistance　比容电阻率
specific weight　比重
specific yield　个别产率
specified　指定的
specified circuit　指定(电)路
specified criteria　给定的标准(技术条件)；明细规范
specified humidity　规定湿度
specified lubricant　合规格润滑剂
specified temperature　规定温度
specified time　规定时间
specifying of hydraulic fluid　传动流体的(技术)规定
specimen　样品；标本
specimen bottle　标本瓶；试样瓶
specimen current　样品电流；吸收电流
specimen handling　进样
specimen thickness effect　试样厚度效应
speciomycin　特霉素
speciosine　美艳(秋水仙)碱
speck　①斑点②缺点③微物④使有斑点
speckle　①小斑点②加斑点
speckled appearance　图像的斑点性
speckled finish　斑点花纹涂装法
specky　①微粒〔漆病〕〔漆膜表面上的极小粒子〕②有微粒的
specky fabric　斑点织物
specky grind　斑点打磨；磨光微粒
specpure　光谱纯的
spectinomycin　大观霉素
spectral　光谱的
spectral absorption　光谱吸收
spectral analysis　光谱分析(法)
spectral band　光谱带
spectral band intensity　光谱带强度
spectral bandpass　光谱通带
spectral band width　光谱带宽(度)
spectral blank background　光谱空白背景
spectral blank signal　光谱空白信号
spectral buffer　光谱缓冲剂*
spectral calculating board　光谱计算板
spectral chromatography　光谱色谱法
spectral classification　光谱分类
spectral color　光谱颜色
spectral comparator　光谱比长仪
spectral composition　光谱组成

spectral concentration 光谱浓度
spectral density 光谱密度
spectral diffusion 谱扩散
spectral distribution 光谱分布
spectral distribution curve 光谱分布曲线
spectral dye method 光谱染料法〔测临界胶束浓度〕
spectral editing 谱编辑
spectral energy distribution 光谱能(量)分布
spectral filter 滤光器；滤光片
spectral frequency 光谱频率
spectral half width 谱线半峰宽
spectral instrument 光谱仪器
spectral intensity 光谱(谱线)强度
spectral interference 光谱干扰*
spectral line 光谱线；波谱线
spectral line half width 谱线半宽度
spectral line intensity 谱线强度*
spectral line interference 谱线干扰*
spectral line overlap 谱线重叠
spectral line profile 谱线轮廓
spectral line self-absorption 谱线自吸
spectral line self-reversion 谱线自蚀
spectral line series 光谱线系
spectral line width 光谱线宽度
spectral locus(=spectrum locus) 光谱轨迹
spectral luminous efficiency 光谱发光效率
spectral match 光谱匹配
spectral method 光谱(分析)法
spectral order 光谱级
spectral overlap 光谱重叠；谱线重叠
spectral pattern 谱型
spectral perturbation 光谱微扰作用
spectral photographic equipment 摄谱仪
spectral photographic plate 光谱感光板*
spectral photometer 光谱光度计
spectral power distribution curve 光谱功率分布曲线
spectral projector 光谱投影仪
spectral purity 光谱纯度
spectral radiance 光谱辐射
spectral radiant power 光谱辐射功率
spectral range 光谱限度；光谱范围
spectral reflectance 光谱反射率
spectral reflectance curve 光谱反射曲线
spectral region 光谱区
spectral resolution 光谱分辨率
spectral response 光谱响应
spectral search 光谱检索
spectral selectivity 光谱选择性

spectral sensitivity 光谱灵敏度
spectral series 光谱(线)系
spectral-shift reactor 谱移(反应)堆
spectral slit width 光谱狭缝宽度
spectral structure correlation 谱图-结构相关
spectral subtraction method 光谱差减法
spectral term 光谱项
spectral transmittance 光谱透射比
spectral tube 光谱管
spectral types 光谱类型
spectral width 谱宽
spectra method 光谱(分析)法
spectra metrics 光谱计量
spectroanalysis 光谱分析
spectrochemical analysis 光谱化学分析
spectrochemical carrier 光谱化学载体
spectrochemical pure 光谱化学纯
spectrochemical series 光谱化学系列*
spectrochemistry 光谱化学
spectrocolo(u)rimetry 光谱光度测色法；分光比色法；光谱色度学
spectrocomparator 光谱比较仪；光谱比长仪
spectrodensitometer 分光光密度计
spectrodensitometry 分光密度测定法
spectroelectrochemistry 光谱电化学*
spectrofluorimetry 分光荧光法
spectrofluorometer 分光荧光计*
spectrofluorometric detector 荧光光谱检测器
spectrofluorophosphorimeter 荧光磷光分光计
spectroginiophotometer 分光测角光度计
spectrograde 光谱级(的)；光谱纯(的)
spectrogram 光谱图
spectrograph 摄谱仪*；光谱仪
spectrographic 光谱的
spectrographical identification 光谱鉴定(法)
spectrographic analysis 光谱(分析)法
spectrographic camera 光谱照相机
spectrographic detection 光谱侦查；光谱检测
spectrographic grade 光谱级
spectrographic identification 光谱鉴定
spectrographic method 光谱(分析)法
spectrography 摄谱学；光谱学；光谱法
spectroheliograph 日光光谱仪
spectrometer 分光计；光谱仪
spectrometer cliff 分光计片
spectrometric 光谱测定的；度谱的
spectrometry 光谱测定(法)；度(光)谱术
spectrophosphorimetry 磷光分光光度法

spectrophotofluorimetry 荧光分光光度法
spectrophotofluorometer 荧光分光光度计
spectrophoto-fluorometric detector 荧光分光光度检测器
spectrophotometer 分光光度计*
spectrophotometric 分光光度的
spectrophotometric analysis 分光光度分析
spectrophotometric color match 分光光度计法配色
spectrophotometric curve 分光光度曲线；光谱辐射曲线
spectrophotometric determination 分光光度测定(法)
spectrophotometric evaluation 分光光度推定法
spectrophotometric hiding 分光光度(法)遮盖力
spectrophotometric measurement 分光光度测定
spectrophotometric method 分光光度法
spectrophotometric reflectance curve 分光光度反射曲线
spectrophotometric study 分光光度测定研究
spectrophotometric titrating apparatus 分光光度滴定仪
spectrophotometric titration 分光光度滴定
spectrophotometric trichromatic colo(u)rimeter 分光光度三色色度仪
spectrophotometry 分光光度测定(法)
spectrophotophosphorimeter 磷光分光计
spectropolarimeter 旋光分光计
spectropolarimetric method 旋光分光法
spectropolarimetry 旋光分光法
spectroradiometer 辐射光谱仪
spectroradiometric 辐射分光的
spectroradiometry 辐射光谱法
spectroscope 分光镜
spectroscopically pure 光谱纯*
spectroscopic analysis 分光镜分析(法)
spectroscopic analysis method 光谱分析法
spectroscopic buffer 光谱(辐射)缓冲剂
spectroscopic carrier 光谱载体
spectroscopic element detector 分光元素检测器
spectroscopic identification 光谱鉴定
spectroscopic laboratory 光谱实验室
spectroscopic lamp 光谱灯
spectroscopic method 分光镜(分析)法
spectroscopic prism 分光棱镜
spectroscopic pure 光谱纯
spectroscopic sensor 分光传感器
spectroscopic solvent 光谱纯溶剂
spectroscopic splitting factor 谱线裂距因数
spectroscopic standard 分光镜标准
spectroscopic study 分光镜分析
spectroscopic term 光谱项*
spectroscopic term symbol 光谱项符号
spectroscopy 光谱学

spectrum ①光谱②波谱③谱
spectrum analysis 光谱分析
spectrum analyzer 光谱分析仪
spectrum atlas 光谱图表
spectrum burner 光谱灯
spectrum chart 光谱图
spectrum color 光谱色
spectrum lamp 光谱灯
spectrum line 光谱线
spectrum monitor technique 光谱监测技术
spectrum projector 映谱仪*；光谱投影仪
spectrum range (光)谱区
spectrum series 光谱线系
spectrum stripping 谱图扣除
spectrum subtraction 光谱差减法
specular 镜的；镜状的
specular cast iron(=specular pig iron) 镜铁
specular coal 镜煤
specular gloss 镜面光泽
specular hematite(=specularite) 镜铁矿
specular iron(=specularite) 镜铁矿
specularite(=specular hematite; specular iron) 镜铁矿
specular metal(=speculum metal) 制镜合金
specular pig iron(=specular cast iron) 镜铁
specular reflectance 镜反射
specular reflectance beam(=gloss beam) 镜面反射光束
specular reflectance excluded (SCE) 漫反射率〔不包括镜面反射率〕
specular reflectance included (SCI) 总反射率〔包括镜面反射率和漫反射率〕
specular reflection 镜面反射；定向反射；单向反射；规则反射
specular reflectivity 镜反射率
specular transmittance 镜面透射比
speculum metal(=specular metal) 制镜合金
speed ①速率②速度③速动
speed belt 速带
speed change box 变速箱
speed changer 变速器
speed increaser 增速器
speed indicator 速率计；速度计
speed line mill 快速圆盘磨
speed mark 速力计
speed of cooling 冷却速度
speed of flow 流速
speed of propagation 传播速度
speed of reaction 反应速度；反应速率
speed of registration 记录速度

speed of revolution 转速
speed of rotation ①旋光速度②旋转速度
speed of water absorption 吸水率
speedometer 测速计
speed ratio 速比
speed reducer 减速机减速齿轮
speed reducing gear 减速齿轮
speed regulator 速度调节器
speed setter 速度设定器；定速器
speed-up gears 增速齿轮
speed variator 变速装置
speedway-traffic marking paint 高速公路划线漆；高速公路路标漆
spegazzinine 查盾宁
speise ①天然砷化物②硬渣
speiss cobalt 不纯的砷钴矿
speleology 成洞学；洞窟学
speleomycin 岩洞霉素
spelter 商品锌锭
spelter coating 锌涂层
spelter solder 锌焊药
spelter works 炼锌厂
spencerite 硅碳铁锰矿；单斜磷锌矿
Spencer method 斯彭瑟（分级）法〔测定高分子的分子量分布〕
spent ①用完的②余(下)的；废的
spent acid 废酸
spent acid revivification 余酸收复
spent bark 废树皮；废鞣料
spent catalyst 废催化剂
spent caustic 废碱；废烧碱
spent charge ①废炉料②馏锌渣
spent clay 废土
spent doctor 废博士溶液；试硫液
spent extractant 使用过的萃取剂
spent ferric oxide 废氧化铁
spent fuel 乏燃料*
spent fuel processing plant 燃烧过的燃料处理工厂
spent fuel shipping cask 使用过(废)燃料运输容器
spent gas 废气
spent lime 废石灰
spent lime liquor 废灰液
spent liquor 废液
spent lye 废碱液
spent lye change 换去废碱液
spent lye from soap 废皂碱水
spent lye treatment 废(碱)液处理
spent material 废料

spent meal 脱脂粉
spent ore 废矿
spent oxide 废氧化物
spent pickle liquor 酸洗废液
spent process water 工艺废水
spent reactivation gas 废再生气
spent reagent 废试剂
spent residue 废物
spent scrub stream 废洗涤液
spent shale 废油页岩
spent soap lye 废皂碱水
spent soda 废纯碱
spent solvent 用过的溶剂
spent sulfite liquor 亚硫酸盐纸浆废液
spent tan 废鞣液
spent tanning material 废鞣料
spent tan press 废鞣料压机
spermaceti(=spermaceti oil) 鲸蜡
spermaceti oil (=spermaceti) 鲸蜡
spermaceti wax 鲸蜡
spermidine(=N-(3-aminopropyl)-butane-1,4-diamine) 亚精胺；N-(3-氨丙基)-1,4-丁二胺
spermine(=gerontine) 精胺；精素 $NH_2(CH_2)_3NH(CH_2)_4NH(CH_2)_3NH_2$
sperm oil 鲸蜡油
sperm oil fatty acid 鲸油脂肪酸
sperm oil fatty acid EO adduct 鲸油脂肪酸环氧乙烷加成物
spermol 鲸蜡醇
sperrylite 砷铂矿
spessartite 锰铝榴石；斜煌岩
spew 溢料缝；流失胶；溢边；溢胶
spewing ①(=extruding)压出②渗出；流出
spewing machine 压出机
spew pips 毛刺；气孔胶
sphacelic acid 麦角酸
sphacelotoxin 痉挛毒素
sphaerite 球磷铝石
sphalerite(=zinc blende) 闪锌矿
sphene 楣石
sphenic 楔形的
sphenoid 半面晶形
sphenolith 岩楔
spherand 环状冠醚
sphere ①(圆)球②范围
sphere colloid 球状胶体
sphere of instability 不稳定区
sphere-sphere rheometer 双球式流变仪

spherical 球形〔胶束形状〕
spherical aberration 球面像差
spherical bead(s) 圆珠；圆球
spherical body 球体
spherical catalyst 球形催化剂
spherical cavity 球状穴；圆穴(洞)
spherical cluster 球状团簇；球状群集体
spherical coordinate 球面坐标
spherical deflector analyzer 球偏转能量分析器
spherical diameter 球(形)直径
spherical entity 球形实体
spherical foam chromatography 球形泡沫色谱(法)
spherical gas-holder 球形储气罐
spherical Gaussian orbital(SGO) 球型高斯轨道
spherical ground glass joint 球形磨口玻璃接头
spherical heater 球形加热器
spherical holder 球形储罐
spherical ion exchange resin 球形离子交换树脂
spherically-shaped cavity 球形空穴
spherical micelle 球型胶束*
spherical mirror 球面(反射)镜
spherical model calculation 球对称模型计算法
spherical pellet(s) 圆球；圆丸
spherical polar coordinate 球极坐标
spherical polar coordinate system 球极坐标系
spherical pressure tank 球形压力罐〔储藏气体与蒸气压大的石油产品用的能保持较大内部压力的球形储罐〕
spherical radius 球面曲率
spherical roller bearing 鼓形滚柱轴承
spherical shell 球(形)壳
spherical tank 球形罐
spherical tensor 球张量
spherical top 球陀螺
spherical top molecule 球形陀螺分子
spherical valve 球阀；截止阀
spherical vessel 球形容器
spherical void 球形孔隙
sphericity 球形度
sphericity value 球度值
spheriolite 菱磷铝岩
spherocobaltite 球泡酸钴矿
spherocolloid 球形胶体
spheroid ①球状体②球状容器
spheroidal 球体的
spheroidal catalyst particles 球状催化剂粒子
spheroidal crystal 球形结晶
spheroidal-granite cast iron 球墨铸铁
spheroidal particle 球形颗粒

spheroidal state 球体状态
spheroides 球状体
spheroid form 球状体
spheroidization 球化(作用)
spheroid tank 球状油罐
spherometer 球径计
spheromycin 新生霉素
spherulite 球晶*
spherulite radial growth rate 球晶辐向生长速率
spherulite rock 球粒岩
spherulites 球状微结晶；球晶
sphincter 松紧口装置
sphingomyelin (神经)鞘磷脂
sphingosyl galactoside(=psychosine) 吐根素；(神经)鞘氨醇半乳糖苷
sphondin 牛防风定；牛防风素
spice and herb 辛香料
spice bush 西洋蜡梅
spice bush seed oil 香叶树子油
spicular 针状的
spider ①芯型支座②多腿支架③多肢架〔固定晶种用〕
spider legs 云卷花纹；流云花纹
spider-web hydrocarbon 多枝链烃
spiegel (=spiegel iron) 镜铁
spiegeleisen 镜铁
spiegel iron(=spiegel) 镜铁
spigot 插销；塞子；方榫
spigot discharge 卸料孔
spigot joint 接嘴
spigot products 砂粒状产品(由分送器口输出)
spike ①掺料②增量；增敏；示踪③钉④尖峰⑤活性种
spiked sample 加料样品；加料试样
spiked solution 掺料溶液；增敏溶液；示踪溶液
spike isotope 添加同位素；示踪同位素
spike lavender 穗薰衣草
spike lavender oil 穗薰衣草油
spikenard 甘松香
spikenard oil 甘松油
spike oil 穗薰衣草油
spike pulse 窄脉冲
spiking 掺加(示踪剂)*
spiking isotope ①加同位素指示剂②掺料
spiking peak 尖峰
spilanthene 千日菊烯
spilanthol(=affinin) 千日菊酰胺
spilite 细碧岩
spillway 溢流道
spilt cure 穿插硫化〔多层平板机〕

spin 自旋*
spinacene(=squalene) 菠菜烯；角鲨烯；2,6,10,15,19,23-六甲基-2,6,10,14,18,22-二十四碳六烯；三十碳六烯
spinacin (角)鲨素；咪唑并吡啶甲酸
spinacine 菠菜素
spin-adapted configuration 自旋匹配组态*
spin adduct 自旋加合物
spin-allowed transition 自旋容许跃迁*
spin angular momentum 自旋角动量
spin-assembly technique 旋涂-自组装技术
spinasterol 菠菜甾醇
spin bath 沉降槽
spin-coating 旋转涂膜法；甩膜法
spin-coloration (合成纤维)抽丝着色
spin-coloring 抽丝(纺丝)着色
spin conservation 自旋守恒
spin coupling constant 自旋偶合常数
spin decoupling 自旋去偶*
spin degeneracy 自旋简并性
spin delocalization (SD) 自旋离域
spin density 自旋密度
spindle ①心轴②指轴③锭子
spindle collar 轴套
spindle distillate 锭馏物
spindle fibre 纺锤丝
spindle guide 轴导
spindle hander-up 锭子磨圆头机
spindle lubricant 锭子油；锭子润滑油
spindle moulder 塑锭机〔面包〕
spindle of filter 过滤器旋转轴
spindle oil 锭子油
spindle-shaped 锭状的
spindle socket 轴套
spindlette 假捻小转子
spin doublet 自旋双线
spin-draw process 纺拉成形法
spin-draw-take-up system 纺丝拉伸卷绕装置
spin-draw-winding machine 纺丝拉伸卷绕(联合)机
spin echo 自旋回波*
spin-echo correlated spectroscopy (SECSY) 自旋回波相关谱*
spin echo difference (SED) 自旋回波差
spin echo double resonance 自旋回波双共振
spin echo Fourier transform 自旋回波傅里叶变换
spin-echo method 自旋回波法
spin-echo refocusing 自旋回波重聚焦
spin-echo spectroscopy 自旋回波谱法
spin-echo technique 自旋回波技术

spin effect 自旋效应
spinel 尖晶石*
spinel structure (晶体的)针状结构；尖晶石结构
spinel type pigment 尖晶石型颜料
spin flip Raman laser 自旋反转拉曼激光器
spin-forbidden transition 自旋禁阻跃迁*
spin free complex 高自旋络合物
spin-free quantum chemistry 无自旋量子化学
spin friction 自旋摩擦
spin immuno-assay 自旋免疫测定法
spin isomer 自旋异构体
spin label 自旋标记；自旋标记物
spin labeled ①自旋示踪②顺磁标记
spin labeling 顺磁标记；自旋标记
spin labeling EPR (SL-EPR) 自旋标记电子顺磁共振
spin label method 自旋标记法
spin-lattice relaxation 自旋-晶格弛豫
spin-lattice relaxation time in rotation reference 旋转坐标系中的自旋-晶格弛豫时间
spinlay 纺丝成网；纺丝成布
spin locking 自旋锁定
spin matrix 自旋矩阵
spin membrane immuno-assay 自旋膜免疫测定法
spin moment 自旋矩
spin-momentum vector 自旋动量矢量
spin multiplicity 自旋多重性
spinnability 可纺性*
spinner ①旋转器②机头罩
spinner culture 迴转培养
spinneret 纺丝头；喷丝板；喷丝头；喷丝嘴
spinnerette 纺丝头；喷丝板；喷丝头
spinnert ①旋转式喷雾嘴②多孔喷丝头〔白金合金制〕③多孔口型〔压出机〕
spinning ①自旋②纺丝；纺纱
spinning acid 纺丝酸
spinning assistant 纺纱(工艺)助剂；纺丝(工艺)助剂
spinning-band column 旋带精馏塔
spinning band distillation column 自旋带蒸馏柱
spinning bath 纺丝浴
spinning bath stretch 纺丝浴拉伸
spinning box 纺丝罐
spinning cake (纺丝)丝饼
spinning can 纺丝罐
spinning cell 纺丝套筒
spinning detonation 旋焰爆震；有特殊火焰前峰传播的爆震
spinning die (=spinning jet) 喷丝头
spinning disc atomizer 旋转式圆盘雾化器；喷雾转盘

spinning-disk applicator　转盘式涂膜器
spinning drop interfacial tensiometer　旋滴界面张力仪
spinning duration　纺丝期间
spinning electron　自转电子；自旋电子
spinning fly　飞花
spinning funnel　纺丝漏斗
spinning head　喷丝头
spinning-in-coefficient　纺纱系数
spinning jet(=spinning die)　喷丝头
spinning line　纺纱流程；纺丝流程；纺纱路线；纺丝线路
spinning machine　纺丝机
spinning machinery　①旋压机床②离心机③纺丝机
spinning metering pump　纺丝计量泵
spinning nozzle　纺丝头
spinning oil　纺丝油剂
spinning paper　纺织纸
spinning pot　纺丝罐
spinning power　纺丝本领
spinning process　纺丝过程
spinning pump　纺丝泵
spinning quality of viscose　黏胶的纺丝性能
spinning side bands　旋转边峰；旋转边带
spinning solution　纺丝溶液
spinning top　陀螺
spinning way　纺丝程
spin nutation　自旋章动
spinodal decomposition　亚稳相分离
spin-off　副产品；派生产品
spinometer　纺丝测速计
spin orbital　自旋轨道
spin orbital splitting　自旋-轨道分裂
spin-orbit coupling　自旋轨道耦合*
spin-orbit coupling constant　自旋-轨道偶合常数
spin-orbit interaction　自旋与轨道相互作用
spin-orbit orientation　自旋-轨道取向
spin out　纺长
spin packet　自旋包
spin paired complex　自旋成对配合物
spin-paired coordination compound　自旋成对配合物*
spin pairing　自旋成对*
spin polarization (SP)　自旋极化
spinpot spinner　离心罐式纺丝机
spin probe　自旋探针
spin-probe technique　自旋取样技术；自旋探针技术
spin quantum number　自旋量子数*
spin resonance　自旋共振
spin-spin coupling　自旋-自旋偶合*
spin-spin exchange interaction　自旋-自旋交换相互作用

spin-spin interaction　自旋间相互作用
spin-spin relaxation　自旋-自旋弛豫
spin-spin splitting　自旋-自旋裂分*
spin-spin splitting system　自旋-自旋裂分系统
spin susceptibility　自旋磁化率
spin temperature　自旋温度
spin tensor　自旋张量
spinthariscope　闪烁镜
spintharoscope　闪烁镜
spin tickling　自旋微扰
spin trap　自旋捕捉剂
spin trapping　自旋捕捉技术
spin trapping method　自旋捕捉法
spinulosin　小刺青霉素；羟烟曲霉醌
spin-unrestricted Hartree-Fock method(SUHF)　非限定自旋哈特莱-福克法
spin warp　自旋翘曲
spin wave function　自旋波函数*
spin wave spectrum　自旋波谱
spin welding　旋转焊接
spiracin　甲基羧基水杨酸
spiraea oil　绣线菊油；欧合叶子油
spiraeic acid(=salicylic acid)　绣线菊酸；水杨酸
spiral　①旋管②螺旋③螺旋的④蜷线
spiral agitator　螺旋搅拌器
spiral bevel gear　斜锥齿轮
spiral blade　螺旋阻板
spiral brush sifter　螺旋刷
spiral burner　旋管灯头
spiral burr　螺纹
spiral chute　螺旋(式)斜槽
spiral coil　螺纹旋管；旋管
spiral coil cooling tube　蛇形冷却管；冷却盘管
spiral column　螺旋形(精馏)塔
spiral condenser (=coil condenser)　蛇形冷凝器；螺管形冷凝器
spiral conveyer　螺旋运输机
spiral conveyor　螺旋运输机
spiral distortion　①螺旋形畸变②各向异性失真
spiral drill　麻花钻
spiral electrode　螺线状电极
spiral element　(压差计的)螺旋形管
spiral feeder　螺状送料器
spiral flow tank　旋流箱
spiral flow test　螺旋流动试验；螺线流动度试验
spiral grain　螺旋纹理；螺旋木纹
spiral head　分度头
spiral heater　旋管加热器

spiral heat exchanger 螺旋板式热交换器
spirality （棉纤维）转曲度
spiral lay 螺旋线角度
spiral-lobe compressor 螺杆压缩机
spirally traced pipe-line insulation 管道绝缘层的螺旋形安装
spiral method 螺旋(割)法
spiral pipe(=pipe coil) 蛇管
spiral-plate heat exchanger 螺旋形片状热交换器
spiral polymer 螺旋聚合物
spiral ribbon mixer 螺带式搅拌机
spiral ring 螺旋环
spiral riveted pipe 螺纹铆接管
spiral-screen column 旋筛(精馏)塔
spiral separator 螺旋分离器
spiral shaped column 螺旋形柱
spiral stirrer 螺旋搅拌器
spiral structure 螺旋结构
spiral terrace （高分子结构的）螺旋台阶
spiral tube heat exchanger 螺旋管式换热器
spiral tube manometer 螺旋管压力计
spiral unwinding machine 螺旋解开机
spiral valve tray 带螺旋叶片的浮阀塔板
spiral vibratory conveyor 螺旋振动输送器
spiral whirls 螺旋形旋涡
spiral winding 螺旋缠绕
spiral-wound gasket 螺旋形垫衬
spiramycin 螺旋霉素
spirane 螺烷*
spirane structure 螺环结构；螺旋结构
spirit ①酒精；醇②酯剂
spirit acid 浓乙酸
spirit-based poultice 醇溶性膏糊
spirit black 醇溶黑
spirit blue 醇溶青
spirit-borne vinyl resin 醇溶性乙烯树脂
spirit-burning stove （燃）酒精炉
spirit color 醇溶染料
spirit dye(=spirit dyestuff) 醇溶染料
spirit dyestuff(=spirit dye) 醇溶染料
spirit enamelling 挥发性磁漆
spirit gauge 酒精比重计
spirit lamp 酒精灯
spirit of alum 硫酸
spirit of ammonia 氨水
spirit of copper 乙酸铜制的乙酸
spirit of ethyl-nitrite 亚硝酸乙酯的酒精溶液
spirit of hartshorn 鹿角酒
spirit of niter 硝酸乙酯的酒精溶液
spirit of nitrous ether 亚硝酸酯的酒精溶液
spirit of salt 盐酸
spirit of sulphur 硫酸
spirit of turpentine 松节油
spirit of wine 酒精；乙醇；火酒
spirit of wood 甲醇；木醇
spirit orange 醇溶橙
spirit red 醇溶红
spirit-rinsable paint remover 溶剂刷洗型脱漆剂
spirits of rosin(=rosin spirit) 松香烃
spirit soluble dye(=spirit soluble dyestuff) 醇溶性染料
spirit soluble dyestuff(=spirit soluble dye) 醇溶性染料
spirit soluble gum 醇溶性树胶
spirit-soluble resin 醇溶性树脂
spirit stain 酒精着色剂〔含有一点树脂或不含树脂的酒精颜料溶液〕
spirit thinned shellac 醇溶性紫胶
spirit thinned type ①醇溶(性)型②溶剂稀释型；溶剂型
spirit transfer 醇溶性转印
spirit-type paint 醇溶性漆
spirituous ①酒精的②醇的
spirit varnish 醇溶性清漆
spirit yellow G 醇溶黄 G
spiro〔希腊字头〕 螺；螺旋
spiroannulation 螺增环*
spiro atom 螺原子
spirobicyclohexane 螺二环己烷
spirobindene 螺二茚
spiroborate 螺硼酸酯
spirochin 凤尾辣木素
spirocid 醋酰胺脒；乙酰胺羟苯胂酸
spiro-compound 螺环化合物
spirocyclane 螺环烷〔专指：螺戊烷〕
spirocyclic compound 螺环化合物
spiro[4.5] decane 螺[4.5]癸烷
spirodilactone 螺环双内酯
spiroform 螺仿；乙酰水杨酸苯酯
spiroglycol diglycidyl ether 螺环二醇二缩水甘油醚
spiroheptane 螺[3.3]庚烷
spiroheptane dicarboxylic acid 螺庚烷二羧酸
spiroheterocyclic compound 螺杂环化合物
spiro hydrocarbon 螺环烃
spiro(no)lactone 螺甾内酯
spiro[4.4]nonane 螺[4.4]壬烷
spiropentane 螺戊烷
spiro polymer 螺环聚合物；螺旋结构聚合物
spirosal 水杨酸羟乙酯

spirostan 螺甾烷
spirostanone 螺甾烷酮
spiro system 螺环系
spiro union 螺接〔一个原子连接在两个环上,例如
       CH₂   CH₂
      CH₂ C CH₂
       CH₂   CH₂ 〕
spirozid(=acetarsone) 乙酰胺胂
spirrillum 螺旋状
spirt ①冲;溅②喷进
spitting 漆雾;喷溅(物)
spitting spray (喷枪)喷路颤振;喷路波动
spitzkasten 锥形选粒器
splash 喷溅
splash baffle 防喷溅挡板
splash bar 搅棒
splash circulating system 喷射循环润滑系统
splasher 折焰板
splash feed 喷射送料
splash head 喷射头;防溅球管
splashing 喷射
splashing device 喷射设备
splash loading 喷射加料
splash lubricating system 喷射润滑系统
splash lubrication 喷射润滑;飞溅式润滑
splash pocket 喷射囊;喷射时为润滑剂充满的空间
splash (spray) zone ①喷溅区②飞溅带
splash zone corrosion 飞溅带腐蚀
splash zone painting 飞溅带涂装
splay 斜孔
splicing 接头;拼接
splicing machine 粘结机
splicing tape 电线绝缘包布;黑胶布
splindle fitting 鼓形接头
spline function 样条函数
spline smoothing 样条函数平滑法
splint coal 暗硬煤
splinter wood 纵裂木材
split 裂缝
split beam microscope 光束分裂显微(镜)测厚仪
split beam spectrophotometer 分光束分光光度计
split-blip ①双峰②裂峰
split-brick 剖半砖;半厚砖
split-bucket elevator 分段斗形(催化剂)升扬器
split burner 裂口灯
split-casing pump 裂壳泵〔外壳可沿水平面分离的泵〕
split caustic treatment 裂化碱处理〔裂化馏分的苛性钠精制〕
split cavity 分裂槽
split chase mould 对开模子
split clamp 对开夹
split coil 分流盘管
split extraction 分裂提取
split fibre 裂散纤维
split-film 劈裂纤维膜〔由126毫米宽的薄膜拉伸成细纤维〕
split flow 分流;分开流动;平行流动
split-flow heater 分流加热器
split-flow spinning machine 热风分流(型)干式纺丝机
split group 除下的原子团
split index 劈裂指数
split journal box 分裂式轴头箱
splitless capillary column 无分流毛细管柱
splitless injector 无分流注射器
splitless sampling 不分流进样
split-loop injection 分支环注入技术
split mold 对开模子;瓣合塑模
split nut 拼合螺母
split pin 开口销;开尾销
split-plot design 裂区(试验)法
split-plot experiment design 分割试验设计
split point 分流点
split product 分裂产物;裂解产物
split ratio 分流比
split repair clamp 可拆卸的修理夹
split ring mould 裂环模子
split run 交替通蒸气〔水煤气发生器〕
split sampling 分流进样法
split solvent technique 拼和溶剂技术
split stream 分流
split stream injector 分流注射器
split stream sampling 分流进样
splittability 裂变性;易分裂性
split tear 剖层撕裂
splitted 分裂的
splitter 分流器*
splitter injector 分流进样器
splitter inlet 分流器入口
split test 劈裂试验;开裂试验
splitting ①分裂;裂解②剖皮③割开④劈裂
splitting ability 撕裂性能
splitting action 分裂作用;裂解作用
splitting and shaving 起层及削里
splitting by frost 冻裂
splitting constant 分裂常数;裂解常数

splitting efficiency	分束效率
splitting equipment	分割设备
splitting gas	裂解气
splitting gauge	片皮测厚计
splitting horse	片皮架
splitting machine	片皮机；剖层机
splitting mechanism	裂解机理
splitting of chain	链断裂
splitting-off	①裂口②分裂③分离
splitting of heavy emulsion	重乳胶的破坏
splitting ratio	分流比；分割比(率)
splitting resistance	①抗开裂性②开裂阻力
splitting system	分流系统
splitting tallow	裂解牛脂
splitting tank	裂化桶
splitting test	分裂试验
splitting up	分裂；裂开
split-tube thief	分流管取样
split volatile matter ratio	分挥发率
split-volute-case pump	外壳可沿水平面拆开的离心泵
split-xanthation	分段黄化
splotch	斑渍；污点；污迹
spodogram method	灰像法
spodumene	锂辉石
spoilage	变坏
spoil bank	管沟损坏
spoiled casting	残缺铸件
spoke	幅
sponge	海绵
sponge deposit	海绵状镀层
sponge filter	海绵滤器
sponge grease	海绵状润滑脂
sponge iron	海绵(状)铁
spongelike structure	海绵状结构
sponge method	海绵法
sponge paste	海绵状膏
sponge petroleum coke	海绵石油焦
sponge plastic	海绵(状)塑料；多孔塑料
sponge platinum	铂绒
sponge polymer	海绵(状)聚合物
sponge process	海绵法
sponge product	海绵制品
sponge rubber	海绵(状)橡胶
sponges	海绵皂〔制造润滑脂用皂〕
sponge stirring machine	海绵搅拌机
spongilite	海绵岩
spongin(e)	海绵硬蛋白
sponginess	海绵性；海绵现象
sponging	①以海绵揩拭②揩油液
sponging agent	发泡剂
spongiolite	海绵硅灰土
spongy	似海绵状的
spongy cure	海绵状硫化
spongy iron	海绵(状)铁
spongy lead	海绵(状)铅
spongy material	海绵状材料
spongy metal	海绵状金属
spongy platinum	海绵铂
spongy uranium	海绵铀
spontaneous aberration	自发畸变
spontaneous combustion	自发燃烧；自燃
spontaneous combustion coal	自燃煤
spontaneous crystallization	自发结晶
spontaneous disintegration	自发蜕变
spontaneous elastic recovery	自发弹性恢复
spontaneous electrogravimetry	自发电重量法
spontaneous elongation	自发伸长
spontaneous emission	自发发射*
spontaneous emulsification	自发乳化(作用)
spontaneous emulsion	自发乳状液；稳定乳状液
spontaneous evaporation	自然蒸发
spontaneous fire	自然火
spontaneous firing	自然燃烧
spontaneous fission (SF)	自发裂变
spontaneous fission counting	自发裂变计数
spontaneous fission dating method	核裂变定年代法
spontaneous heating	自然加热
spontaneous hydrolysis	自发水解
spontaneous ignition	自发着火
spontaneous ignition point	自燃点
spontaneous ignition temperature	发火点温度；自发着火点温度
spontaneously fissible isomer	自发裂变同质异能态
spontaneously inflammable	可自燃的
spontaneous nucleation	自发成核
spontaneous polarization	自发极化
spontaneous polarization method	自发极化法
spontaneous polymerization	自发聚合
spontaneous process	自发过程*
spontaneous radiation	自发辐射
spontaneous Raman scattering	自发拉曼散射
spontaneous reaction	自发反应*
spontaneous restoration	自发复原
spontaneous termination	自发终止
spontaneous transfer reaction	自转移反应
spontaneous transition	自发跃迁

spontaneous wetting 自发润湿
spool ①卷线筒；线轴②双端法兰管〔两端有法兰的管子〕③短管
spool axis 卷筒；卷轴
spool stage 有边筒管座
spool test 绕轴试验
spool valve 短管阀
spoon ①匙(子)；勺②以匙取出
spoon fed 勺式灌送
spoon test 匙勺试验(法)
spoonwort 辣根菜油〔取自 Cochlearia officinalis〕
spoon wort oil 辣根菜油
sporadic nucleation 自发成核
sporting powder 猎枪(火)药
sporting shot-gun powder 猎(用鸟)枪火药
sports goods 体育用品
spot ①(色谱)斑②斑点③色斑④点滴⑤点样；加样
spot analysis 点滴分析；斑点分析
spot-and-stain repellency 抗污点污渍性
spot application (斑)点状点样
spot asphalt quality test 点滴沥青质量试验；点滴试验
spot chromatography 斑点色谱(法)
spot coater 滴涂机
spot collector 斑收集器
spot colorimetry 点滴比色法
spot contact bearing 滚珠轴承
spot cure 局部硫化
spot detection 斑点检测
spot diameter 斑直径
spot edge 斑边缘
spot facing 刮孔口平面
spot finishing(=spotting in) 找补；打补钉〔小面积修补〕
spot gluing 点胶合；局部胶合
spot height-to-width ratio 斑高宽比
spot indicator 点滴指示剂
spot lube-oil evaluation 润滑油就地评价；润滑油在发动机上直接评价
spot-out method 点滴法
spot paper 点滴试纸
spot plate 点滴板*
spot priming 填补
spot printing ①点涂②点状痕迹③印相过度
spot putting 局部填腻子；找补腻子
spot quality test 点滴定性分析
spot reaction 斑点反应
spot reconcentration 斑再浓缩
spot reconditioning 就地修理；不拆卸修理
spot removal method 斑消除法

spot sequence 出斑顺序
spot size 斑大小
spottedness(=spottiness) 斑点
spotted schist 斑点片岩
spotter 点样器
spot test 斑点试验*；点滴试验
spot test analysis 点滴试验分析
spottiness(=spottedness) 斑点
spotting 花点；花斑；橘皮；麻点
spotting agent 去斑剂〔干洗用〕
spotting detergent 去斑洗涤剂〔干洗用〕
spotting guide 点样模板
spotting in(=spot finishing) 找补；打补钉〔小面积修补〕
spotting soap 去斑皂〔干洗用〕
spotting template 点样模板
spot welding 点焊
spot welding machine 点焊机
spot welding primer 点焊底漆
spot width 斑点宽度
spout ①嘴②槽③出口；喂料口④溜槽
spout brush(=striker brush) 触发式柄芯供漆长柄刷；吊锤式漆刷
spouted bed 喷射床
spouting of liquid 液体喷射
spouting velocity 喷射出口速度
spout of funnel 漏斗嘴
spray ①喷(射)②喷雾③喷淋④喷显剂〔色谱〕
sprayability ①喷涂性②雾化性
sprayable urethane 喷涂型聚氨酯橡胶
spray absorber 喷淋吸收器
spray agent 喷显剂
spray and wipe painting 衍彩法〔大漆〕
spray angle 喷射角度
spray apparatus 喷雾器
spray atomizer 喷淋雾化器
spray base ①喷距〔喷嘴与工件表面的距离〕②喷底漆
spray bonderizing 喷射法磷化处理
spray-bonderizing solution 喷雾磷化溶液
spray booth 喷柜
spray broom 多嘴喷枪；多嘴喷头
spray bulb for powder 吹粉球
spray cap 喷雾罩
spray catcher 捕雾器
spray chamber(=atomizer chamber) 喷雾室；雾化室
spray cistern 喷雾槽
spray cleaner 喷雾清洁(洗)剂；喷雾除垢剂
spray coating 喷着涂装；喷射涂装
spray column ①喷雾柱②喷雾塔

spray condenser 喷雾冷凝器
spray containment area 喷涂安全区；喷涂隔离区
spray controller 喷涂控制器
spray-cooled side 淋冷的侧壁
spray cooler 喷雾冷却器
spray cooling 喷雾冷却(法)
spray cooling tower 喷雾冷却塔
spray crystallization 喷雾结晶
spray cutter 潄边喷嘴
spray degreasing 喷淋脱脂法
spray deoiling 雾化脱油(法)
spray disc 喷雾盘
spray-dried 喷雾干燥的
spray-dried detergent 喷雾干燥(的)洗涤剂
spray dried latex 喷雾(法)胶乳
spray drier 喷雾干燥器
spray dry 喷雾干燥
spray dryer 喷雾干燥器
spray drying 喷雾干燥
spray-drying process 喷雾干燥法
spray drying tower 喷雾干燥塔
spray-dust 喷漆粉尘〔喷漆的起粒现象〕
spray dyeing 喷染
sprayed and fused deposition 喷熔沉积法；熔融喷镀法
sprayed coil cooler 喷淋式蛇管冷却器
sprayed deposit 喷雾沉积
sprayed earth 喷制土；硫酸活化铝矾土
sprayed insulation 喷淋绝缘
sprayed metal coating 金属熔融喷镀层
sprayed rubber 喷制橡胶
sprayed tank 喷淋罐；有淋水设备的油罐
sprayer ①喷雾器；喷洒器②喷头
sprayer nozzle 喷管；喷嘴
spray evaporator 喷雾蒸发器
spray extraction column 喷淋式抽提塔
spray fan 喷束；喷雾扇〔指喷束的锥形断面〕
spray feed 喷送
spray finish 喷涂涂饰剂
spray finishing 喷涂
spray gun 喷枪
spray gun housing 喷枪枪座
spray gun phosphating 喷雾磷化处理法
spray header 喷淋水管；喷淋头
spray impact detector 喷撞检测器
spraying 喷雾法；喷涂
spraying cabin 喷雾匣
spraying cleaner 喷洗器
spraying equipment 喷涂设备

spraying glazing 喷釉
spraying gun 喷雾器；雾化器；喷枪
spraying hose 喷雾胶管
spraying lacquer 喷涂用漆
spraying machine 喷雾机；喷涂机
spraying method 喷涂法
spraying of fuel 燃料雾化
spraying performance 喷雾性能；喷涂效能
spraying pistol 喷枪
spraying position ①喷涂位置②喷镀位置
spraying reagent 喷显剂
spraying shop 喷漆间；喷漆场
spraying soap powder 喷皂粉
spraying station ①喷漆站②喷涂段流水线；喷漆工位
spraying varnish 喷涂清漆
spraying welding 喷焊
spray lacquer 喷漆
spray loss 喷雾损失；喷雾飞逸
spray lubrication 喷淋润滑
spray mask （喷漆工用)面罩；呼吸口罩
spray metal plating 喷镀金属法
spray method 喷雾法
spray mixing 喷雾混合
spray mix process 喷雾混合法
spray mottle 橘皮麻点；斑点〔漆病〕
spray needle 喷针
spray nozzle 喷嘴
spray oil ①喷淋油②杀虫油
spray oiler 喷淋润滑器
sprayometer 喷液比重计
spray paint 喷漆
spray painting 喷涂
spray painting equipment 喷涂机
spray painting plant 喷涂(漆)设备
Spraypak packing 斯普雷帕克填料；金属网交织排列填料
spray particle size 喷淋粒子大小
spray pattern 喷路；喷纹
spray penetration 喷淋深度；喷淋穿透尺度
spray phosphating plant 喷雾磷化装置
spray pipe 喷水管
spray piping 喷淋管道(系统)
spray pond 喷淋池
spray process ①喷涂法②喷淋法
spray producer 喷雾器
spray reagent 喷显试剂
spray regime ①喷射方式②喷射状态③喷射范围
spray rinsing 喷射清洗

spray scrubber 喷液涤气器
spray separator 喷淋分离器
spray solidification 喷淋固化(过程)
spray spinning 喷纺成形
spray tank 喷洗槽
spray test 雾化试验；喷雾试验
spray thrower 喷淋器
spray time 喷涂时间；喷雾时间
spray tip ①(喷枪)喷嘴；喷头②喷嘴梢
spray tower 喷粉塔
spray trap 〔点灯法测定硫含量仪器的〕雾滴分离器
spray-type air cooler 喷淋式空气冷却器
spray-type cement 喷用胶浆
spray-type drier 喷淋式干燥器
spray type evaporator 喷淋式蒸发器
spray-type humidifier 喷雾增湿塔；喷雾增湿装置
spray unit ①喷雾设备②喷涂设备
spray up 喷射成型
spray valve 喷淋阀
spray viscosity 喷涂黏度；施工黏度
spray washer 喷洗器
spray washing machine 喷洗机
spray water 喷淋水；雾状水
spray webbing ①喷丝②被覆包装；喷丝包装
spray welding coating 喷焊层
spray zone 喷漆区
spread ①涂胶②扩展；铺展③布置④安装施工段⑤涂抹食品〔果酱、黄油等〕
spreadability 覆盖性
spreadable life （黏合剂)可涂期；使用期
spread area(=spreading area) 涂布面积
spread-coat 展涂；刮涂
spread coater ①刮漆机②刮胶机
spread coating 刷涂法；刮涂法
spreader ①涂布器；涂铺器②扩张器③铺展剂
spreader bar 平压机
spreader calender 擦胶压延机
spreader frictioning 压延擦胶
spreader roller (涂布)展匀辊
spread film 铺展膜
spreading 铺展；涂布；涂层
spreading agent 铺展剂；涂铺剂；展着剂
spreading area(=spread area) 涂布面积
spreading behaviour of oil 油的溢流性质
spreading blade 展涂刀；刮涂(胶)板
spreading calender 等速混压机；涂胶压延机
spreading capacity 涂布率；遮盖本领〔油漆〕
spreading chest 压延头
spreading coefficient 铺展系数
spreading compound 涂铺胶料
spreading device (=spreader) 涂铺器
spreading factor 扩散因素
spreading force 铺展力
spreading function 加宽函数
spreading kinetics 铺展动力学；展布动力学
spreading knife 刮刀；涂胶刀
spreading machine 刮胶机；涂胶机
spreading parameter 扩展参数
spreading peak 扩展峰
spreading power 涂布面积〔油漆〕
spreading pressure 铺展压力
spreading rate 扩展速率；涂布率
spreading roller 展涂辊；刮胶辊
spreading solvent 铺展溶剂
spreading wetting 铺展润湿
spread-layer chromatography 涂层色谱(法)
spreadometer ①平行板黏度计②涂胶机
spread plate method 平板涂布培养法
spread source 分布源
spread suspension method 悬浮液涂布观察法
spready hide 大张皮；大开张薄皮
Sprengel explosive 斯普伦炸药
Sprengel pump 高真空泵
Sprengel tube 高真空管
spring 弹簧〔模拟固体弹性用〕
spring and dashport model 弹簧和黏壶模型〔指描述弹性和黏性的流变模型〕
spring back 回弹
spring balance 弹簧秤
spring bumper 弹簧减震器
spring caliper 弹簧卡钳
spring chuck 弹簧夹盘
spring clamp 弹簧夹
spring cleaner 弹簧清洁器
spring constant ①弹簧常数②刚性度常数
spring-dashport model 弹簧-黏壶模型
spring grease cup(=spring lubricator) 弹簧润滑器
spring hinge 弹簧铰链
spring hook 弹簧钩
springiness 弹(簧)性
springing 弹性
spring leaf 弹簧(叶)片
spring loaded nozzle 弹簧加压喷嘴
spring-loaded pressure relief valve 弹簧式安全阀
spring-loaded seal 弹簧加压密封
spring-loaded valve 弹簧阀

spring lubricator(=spring grease cup)　弹簧润滑器
spring pinchcock　弹簧夹
spring rate　弹簧振频；刚性度振动频率
spring reducing valve　弹簧减压阀
spring regulator　弹簧调速器
spring roll　弹簧辊
spring safety valve　弹簧安全阀
spring seat　弹簧座
spring steel　弹簧钢
spring tension　弹簧张力
spring tester　弹簧试验机
spring tool　弹簧刀
spring type thermobalance　弹簧式热天平
spring washer　弹簧垫圈
spring water　泉水
spring wood(=early wood)　早材；春材
springy resilience　弹簧状回弹
sprinker　喷洒器；喷水车
sprinkler　泼洒器；洒水器
sprinkler head　喷水头
sprinkling　喷洒；溅洒
sprinkling can　喷淋润滑器
sprinkling filter　滴滤池
sprinkling machine　喷水器
sprinkling system　喷洒系统；洒水系统；喷淋系统
s-process(=slow process)　慢过程
sprocket　链轮
sprocket chain　铰链
sprocket wheel　链轮
spruce　云杉
spruce bark　云杉树皮
spruce extract　①云杉(树皮)栲胶②亚硫酸纸浆废液(浸)膏
spruce oil　云杉油
spruce pulp　鱼鳞松浆；枞木浆；云杉木浆
spruce tannin　云杉鞣酸；云杉单宁
sprue　①浇口；铸口②流道
sprue granulator　注口冷料粉碎机
sprue groove　铸道
sprung joint　①弯关节；弯接头②(墙壁涂层的)拐角搭接处(接缝)
sprung shoe　弓形垫块；弓形滑脚
sprunstron　聚丙烯化纤
spud　剥皮刀
spue　压铸硫化
spueeze bottle　(牙膏袋式)塑料挤瓶
spuing balance　弹簧秤
spum-colored　纺时着色
spum-dyed rayon　色纺嫘萦

spume　①泡②起泡
spumescence　泡沫性
spumescent　起泡的
spumific　发泡
spumous(=spumy)　(多)泡沫的
spumy(=spumous)　(多)泡沫的
spun-colo(u)red　纺时着色
spun cotton　棉纱
spun-dyed　纺时染色
spun-dyeing injector　纺前着色注射器
spun glass(=fiber glass)　玻璃纤维
spun roving　定长毛纱；毛圈粗纱
spun silk yielding　出棉率
spun yarn　短纤纱；精纺纱；细纱
spur feterita　高粱
spur gear　正齿轮
spur-gear lubrication　正齿轮润滑
spurious band　乱真谱带；虚假谱带
spurious count　乱真计数
spur-line　支线
spurred rye　麦角
spurting　①溅散②喷
sputter　溅(散)；溅射
sputter coating　①溅射涂膜②溅射涂装
sputter deposition　溅射沉积
sputtered film　喷镀薄膜
sputtered glass　溅镀玻璃
sputtered ion　溅射离子
sputtered neutral mass spectrometer (SNMS)　溅射中性粒子质谱仪
sputtered neutral mass spectrometry　溅射中性粒子质谱法
sputter equipment　①溅花装置②溅射装置
sputter etching　溅射腐蚀
sputtering　①喷镀②溅射
sputtering chamber　溅射室
sputtering ion pump　溅射离子泵
sputtering neutral mass spectrometry　溅射中性粒子质谱法
sputtering phenomenon　溅射现象
sputtering process　①溅射法②溅射涂装法；喷镀法
sputtering rate　溅射率；溅射速率
sputtering technology　①溅花工艺②溅射工艺
sputtering yield　溅射产额
sputter ion pump　溅射离子泵
sputter pump　溅射泵
squalane　角鲨烷；异三十烷
squalene(=spinacene)　菠菜烯；角鲨烯；2,6,10,15,19,23-六甲基-2,6,10,14,18,22-二十四碳六烯；三十碳六烯
squalid　①不洁的②油封的

squamatic acid 鳞片酸
square ①正方(形)②(正)方的③平方④乘方；自乘⑤枋材⑥方〔各种板材单位〕⑦角尺
square billet 方坯料
square centimeter 平方厘米
square-cut ①裁方；切成方形(直角)②(丝毛纤维)等长切断
square dense fabric 方形密织物
squared-off cascade 方形化级联；直角化级联
square edged orifice 方孔
square elbow 直角弯管
square end ①方尖；方头②(漆刷)平头
square file 方锉
square foot 平方英尺
square grid ①方格网②正方形栅格
square head bolt 方头螺栓
square hybrid orbital 正方形杂化轨函数
square iron 方铁
square lacquer cabinet 正方漆柜〔大漆〕
square log(=balk) 枋材；大木
square mesh 方网眼
square mesh screen 方网眼筛
square meter 平方米
square pitch arrangement 正方形排列
square planar complex 正方平面(形)络合物
square point (刀、铲的)方头
square pulse 矩形脉冲
square-root Kalman filter 平方根卡尔曼滤波器
square root of the variance 方差的平方根值；标准偏差
square shank drill 方柄钻
square steel 方钢
square tank 正方形储罐
square thread 方螺纹
square wave 方波
square wave-form oscillator 方波形振荡器
square wave polarograph 方波极谱仪
square wave polarography 方波极谱法*
square wave voltammetry 方波伏安法*
square woven fabric 平布；平织布
squaric acid(=dihydroxy cyclobutenedione) 方形酸；二羟基环丁烯二酮
squarine 方酸菁
squarylium cyanine 方酸菁
squash oil 南瓜油
squatting 沉下；沉降
squawroot 洋葳严仙
squeegee ①汽车前窗刮水器②隔离胶；油皮胶；夹层胶
squeegee calender 油皮压延机
squeegee coat 揉搓涂层
squeegee paste 滚印色料
squeeze ①挤压力②滚距
squeeze film 挤压膜
squeeze-out 溢胶
squeezer ①压榨机②压铆器；压铆机
squeeze roll 压水辊
squeeze squirt 鹤首式洗瓶
squeezing 压榨
squeezing action (涂料受)挤压作用；挤走〔辊涂时，压力过大漆被挤跑的现象〕
squeezing bowl 压水轴
squeezing machine 压榨机
squeezing out 榨出
squeezing process 挤水
squeezing roll 挤压滚
squeezing test 压扁试验
squib 爆筒
squirming 蠕动(作用)
squirrel cage 松鼠笼
squirrel cage blower 鼠笼式鼓风机
squirrel-cage disintegrator 笼式解磨机
squirrel cage mill 笼式磨机
squirt 注射器；喷枪
squirt-can lubrication 油枪润滑；手提润滑器润滑
squirt feeder 喷射送料器
squirt gun 油枪
squirting 喷射；喷出(净)
squirt pump 喷射泵
squirt sampling pump 进样注射泵
SRL electrolytic dissolver SRL 电解溶解器
SRS technique(=separation-reaction-separation technique) 分离-反应-分离技术
St. Venant body 圣维南体
St. Venant compatibility 圣维南相容性
St. Venant principle 圣维南原理
St. Venant's flow condition 圣维南流动条件
stabbing salve 管道配件润滑剂
stab culture 穿刺培养
stabilisator 离心均质机
stability ①稳定性；安定性；复原性②坚固性；牢固性；耐久性③强度；刚度
stability condition 稳定性条件
stability constant 稳定常数*
stability criterion 稳定性准则
stability diagram 稳定图
stability factor ①稳定系数②刚度系数
stability in hard water 耐硬水性

stability in storage 储存稳定性
stability in use 使用稳定性
stability limit 稳定极限
β-stability line β稳定线
stability meter 稳定性试验仪
stability of emulsion 乳胶稳定性
stability of flow 流动稳定性
stability of gloss 光泽稳定性
stability of ink 墨水稳定性
stability of reference material 标准物质的稳定性
stability of regression equation 回归方程的稳定性
stability of sensitivity 灵敏度稳定性
stability product (=over-all stability constant; cumulative constant) 稳定(常数)积
stability setting (干、湿热法)稳定性定形
stability test 稳定性试验
stability test of reference material 标准物质稳定性检验
stability to aging 老化稳定性；耐老化性
stability to degradation 降解稳定性
stabilization 稳定(作用)
stabilization agent 稳定剂
stabilization tank 缓冲罐
stabilizator 稳定剂
stabilized dicalcium phosphate dihydrate 稳定的磷酸二钙二水合物〔牙膏磨料〕
stabilized gasoline 稳定汽油；去丁烷汽油
stabilized grease 稳定润滑脂
stabilized latex 稳定胶乳
stabilized natural gasoline 稳定天然汽油
stabilized non-torque yarn 稳定的无捻回弹力丝
stabilized radical 稳定自由基
stabilized temperature plateau furnace 稳定温度石墨炉平台技术
stabilized temperature platform (STP) 稳温平台
stabilized temperature platform furnace (STPF) 稳温平台石墨炉(技术)
stabilizer ①稳定器②稳定剂
stabilizer column(=stabilizer tower) 稳定塔
stabilizer feed pump 稳定塔进油泵
stabilizer gas 稳定塔气体
stabilizer overhead 稳定塔顶排出气体
stabilizer packed column 填充稳定塔
stabilizer plant 稳定厂；稳定装置
stabilizer reboiler 稳定塔再沸器
stabilizer tower(=stabilizer column) 稳定塔
stabilizing 稳定；安定
stabilizing agent 稳定剂
stabilizing fin 稳定翅；稳定尾

stabilizing gas 稳定气体
stabilizing layer 稳定层
stabilizing mechanism 稳定化机理
stabilizing plasticizer 稳定增塑剂
stabilizing pressure valve 稳压阀
stabilometer 稳定性试验机
stable burning 稳定燃烧
stable detonation 稳定爆炸
stable distillate fuel 稳定馏出燃料
stable emulsion 稳定乳状液
stable equilibrium 稳定平衡
stable free radical polymerization(FRP) 稳定自由基聚合
stable gasoline 稳定汽油
stable gel structure 稳定凝胶构造
stable hydrocarbon 稳定烃；饱和烃
stable ion 稳定离子
stable island 稳定岛*
stable isotope 稳定同位素*
stable isotope analysis 稳定同位素分析
stable isotope dilution mass spectrometry 稳定同位素稀释质谱法
stable isotope dilution technique 稳定同位素稀释法
stable isotope ratio mass spectrometer 稳定同位素比质谱仪
stable isotope tracer method 稳定同位素示踪法
stable isotopic tracer 稳定同位素示踪
stable nuclide 稳定核素*
stable pair 稳定对
stable pour-point 稳定倾点；稳定凝固点
stable range 稳定范围
stable scopolamine 莨菪胺氢溴酸盐-甘露糖复剂
stable state 稳定态
stachydrine 水苏碱；脯氨酸二甲内盐
stachyose 水苏糖
stack ①堆积②容积单位③组套④叠式储存器
stackability 堆叠性；堆垛性；层叠性
stack area 烟道切面积
stack-clone contactor 堆积旋流接触器
stack cooker 层式蒸缸
stack damper 烟道气闸
stack dilution factor 烟囱稀释因数
stacked detector 堆积式检测器
stacked discs outside friction 叠盘式外摩擦
stacked-foil technique 叠箔技术
stacked-hydroclone contactor 堆积式水力旋流接触器(萃取器)
stacked lamella-like structure 叠层片晶状结构
stacked packing 堆积填充物

stacked plate chromatographic column 叠板色谱柱
stacked plot 堆砌图；堆积图
stacked silica bead 堆积硅珠
stacked-stage extractor 叠级萃取器
stacker 堆垛机；叉式电池车
stack flue （水平）烟道
stack gas 烟道气
stack gas analyzer 烟道气分析器
stack gas purifier 烟道气净化器
stacking ①区带浓缩技术②堆积成垛
stacking density 堆积密度
stacking effect 堆集效应；堆垛效应
stacking factor 堆集因数；填充系数
stacking fault 堆垛层错*
stacking gel 积层凝胶
stacking machine 堆垛机；叉式起重车
stack losses 烟道损失；烟道气带出的热损失
stack molding 叠模压塑
stack of lamellae 片晶叠层；片晶堆垛
stack paint 烟囱涂料
stack sampling 烟道取样
stack wood 堆垛木材
stactometer 滴量计
Staeger test 斯泰格(油料氧化稳定性)试验
staff ①杖；棒；竿②柄
Stafford process 司坦福(干馏)法〔一种连续干馏作业〕
staff-tree oil 南蛇藤油；杆树油
stage ①台②(阶)段③级
stage aeration 阶段曝气；逐步曝气
stage breaking(=stage crushing) 分级压碎
stage cooking 分段蒸煮
stage countercurrent refining method 分段逆流精制法
stage crushing 分级压碎
stage digestion 多段式消化
staged reactor 多级反应器
stage drying 分段干燥
stage efficiency 分段效率；级效率
stage extraction 分段提取
stage filter 分级过滤器
stage flotation 分级浮选；连续浮选
stage grafting 分级接枝
stage heating 逐步加热
stage micrometer 分级测微计
stage number (=number of plate) 塔板数
stage of reaction 反应阶段
stage plank 脚手板
stage pump 多级泵
stage reduction 分级粉碎

stagewise ①阶梯的②分段的
stagewise contactor 多级(式)接触装置
stagewise countercurrent oxidation system 阶段式逆流氧化装置
stagewise gradient 阶梯梯度
stagewise operation 分段操作
stagger ①拐折②蹒跚
staggered 交错的；参错的
staggered air heater 拐折空气加热器
staggered arrangement 交错布置；错排
staggered bond (相互)交错键
staggered conformation 对位交叉构象*
staggered form 参差式
staggered tubes 拐折管排；交错管排
staggering stitch 线迹歪斜
staging ①手脚架②工作台③分段法；分级法
staging life(=shelf life) 储藏寿命；存放期；储藏期限
stagnant 停滞的
stagnant ambient medium 静止周围介质
stagnant boundary layer 静止边界层
stagnant catalyst 固定催化剂
stagnant condition 不流动(静止)状态；停滞状态
stagnant film 滞膜
stagnant gas film 静止气膜
stagnant liquid film 静止液膜
stagnant medium 静止的介质
stagnant mobile phase 停滞流动相
stagnant water 死水
stagnant zone 静止区域
stagnation point ①临界点②驻点；滞留点
stagnation pressure 滞止压力；驻点压力
stagnation temperature ①临界温度②滞止温度；驻点温度
stagonometer(=stalagmometer) (表面张力)滴重计
stain ①着色剂；染(色)剂②污点③着色
stainable 可染色的
stain control agent 防污剂
stained ①涂(了)漆的②染(了)色的
stained cloth 染色布
stained glass ①冰屑玻璃②(烧成)彩画的玻璃
stained paper 花纸
stained wood 着色木
stainer 染匠
stain for wood 木材着色
staining ①着色；染色②刷染法
staining agent 染色剂；着色剂
staining bath 染槽
staining jar (载片)染色缸

staining method 着色法
staining power 染色本领；着色力
staining reagent 着色试剂
staining stand 染架
staining test 着色试验
stainless 不锈的
stainless oil 不锈油；无斑油
stainless steal drying tumbler 不锈钢烘筒烘燥机
stainless steel 不锈钢
stainless steel helices 不锈钢填料
stainless steel syringe needle 不锈钢注射针头
stain proofing 防锈处理
stainproofing agent 防污染剂
stain removal 去污(斑)
stain remover 去污剂
stain resistance 耐污染性〔不指环境污染〕
stain-resistant paint 耐污染漆
stain-resistant topcoat 抗污染面漆；耐污染面漆
stain test 污染试验；斑痕试验
stain varnish 有(着)色清漆
stain white 缎光白〔白色颜料，主要成分为硫酸钙和氧化铝〕
staircase method 梯段法
staircase polarography 阶梯波极谱法
staircase reaction 梯段反应
staircase voltammetry 阶梯波伏安法；阶梯扫描伏安法
stair handrail 楼梯扶手
stake ①桩②加桩③拉软
staking 拉软；刮软
staking machine 拉软机；刮软机
stalactite 钟乳石
stalagmite 石笋
stalagmometer(=stagonometer) (表面张力)滴重计
stalagmometric titration 表面张力滴定(法)
stalagmometry 表面张力测定法
stalagmones 降表面张力质
stale lime 陈石灰；已潮的石灰
stale wastewater 腐臭污水
stalloy 高硅钢片
stall roasting 泥窑焙烧
Stallwood jet device 斯托伍德气室
Stammsche method 碱性高锰酸盐滴定法
stamp ①捣磨；压(碎)②(模)冲③压杆④印模⑤捣击机⑥压滚
stampability 冲压性；模锻性
stamp battery ①连续重击杵②捣矿杵
stamp breaking(=stamp crushing) 捣碎
stamp crushing(=stamp breaking) 捣碎
stamp die 捣矿砧
stamp duty 捣碎量
stamped ①已捣碎的②已盖印的
stamped concrete 捣固混凝土
stamped method 压印法
stamped overshoe 模压套鞋
stamped packing ring 冲切式密封环
stamper ①杵②捣击机③打号机
stamping ①模冲②模冲片③锤击④打印
stamping ink 印盒墨
stamping knife 冲刀
stamping machine ①平压切断机；冲切机②冲压机③打印机
stamping mill 捣磨
stamping mold 冲压模
stamping press 打印模
stamp mill 捣磨机
stamp milling 捣磨
stamp mortar 捣锤研碎机
stamp mortar screen 捣研筛
stamp-pad ink 打印墨
stamp pestle 捣杵
stamp product 捣碎产物
stamp shoe 捣蹄
stamp stem 捣杆
stanch fibre 止血纤维
stanchion 支柱；支架
stand ①架台②载体
standard ①标准②标准物；标准样③标准的
standard accessories 标准附件
standard acid solution 标准酸溶液
standard acid test 酸度的标准测定法
standard addition 标准物加入(法)
standard addition calibrating technique 标准加入校准法
standard addition method 标准加入法*
standard affinity 标准亲和力
standard air 标准空气
standard alkali 标准碱溶液
standard alkali solution 标准碱溶液
standard analysis 标准分析
standard anionics 标准阴离子表面活性剂
standard antitoxin 标准抗毒素
standard atmosphere 标准大气；标准温湿度
standard atmosphere pressure 标准大气压力；正常压力
standard bisphenol A epoxy 标准双酚 A 环氧(树脂)类
standard black ①标准黑②标准黑沥青漆
standard blend 标准(混合)物
standard block of hardness 标准硬度块(片)

standard brass Saybolt test lamp　赛波特标准黄铜灯
standard buffer　标准缓冲剂
standard buffer solution　标准缓冲溶液
standard buret(=standard burette)　校准滴管；标准滴定管
standard burette (=standard buret)　校准滴管；标准滴定管
standard burn test　标准燃烧试验
standard candle　标准烛
standard capacity　标准容量
standard cationics　标准阳离子表面活性剂
standard caustic　标准氢氧化钠溶液
standard cell　标准电池
standard clock　标准钟
standard code　标准规定；法规
standard color scale　标准色等级
standard color solution　标准比色液
standard color system　标准色系
standard column　标准柱
standard condenser　标准电容器
standard conditions　标准状态
standard curve　标准曲线
standard density　标准密度
standard depth of colour　标准色度
standard depth of shade　标准色浓度〔浓淡深度〕
standard deviation (SD)　标准(偏)差*
standard deviation between laboratory　室间标准差
standard deviation of sample　样本标准偏差
standard distillation test　标准蒸馏试验
standard distribution function　标准分布函数
standard drawdown sheet　标准刮样纸〔美国用〕
standard drive　基准传动；齿轮传动
standard electrode　标准电极
standard electrode potential　标准电极电势；标准电极电位
standard electromotive force　标准电动势
standard elution volume　标准洗提体积
standard equilibrium constant　标准平衡常数
standard equipment　定型设备；标准设备
standard error　标准误差*
standard evaporator　标准式蒸发器
standard filter　标准滤光片
standard flat bark crepe　标准树皮绉片
standard free energy change　标准自由能变化*
standard free energy of formation　标准生成自由能*
standard fuel　标准燃料
standard gas　标准气体
standard half-cell　标准半电池
standard heat (enthalpy) of formation　标准生成热(焓)
standard heat of combustion　标准燃烧热
standard heat of formation　标准生成热
standard humidity　标准湿度
standard hydrogen electrode (SHE)　标准氢电极
standard illuminant　标准光源；标准照明
standard impact specimen　标准冲击试样(样板)
standard international atmosphere　国际标准气压
standard interpretation procedure　标准解析程序
standard iodine　标准碘
standardization　①标准化*②标定；定位③校准
standardized bend test　标准化弯曲试验
standardized curve　标准(化)曲线；校准曲线
standardized intensity　标准化强度
standardized liquid　标准液体
standardized regression coefficient　标准回归系数*
standardized solution　标准溶液
standardized value　标准化值
standardized variable　标准化变量
standardizing　标准化
standardizing box　标准化负荷测定机
standard Kittel tray with a packed intermediate　板间充填料的克特尔塔板
standard lamp　标准灯
standard light both　标准光源箱
standard light source　标准光源
standard map　标准图
standard material　标准品
standard method of analysis　标准分析法
standard methods of environmental analysis　环境分析标准方法
standard moisture regain　标准回潮率
standard multipoint recorder　标准多点记录器
standard normal distribution　标准正态分布
standard nozzle　标准喷嘴
standard nylon screw finisher　锦纶的标准型螺杆后缩聚器
standard oil process　美孚法〔烃类部分氧化制合成气法〕
standard opal glass　标准乳浊玻璃
standard operating procedure (SOP)　标准操作步骤
standard oxidation reduction potential　标准氧化还原电位
standard pallette knife　标准型调漆(色)刀
standard pH buffers　标准 pH 缓冲溶液
standard piece glass　织物(标准)分析镜
standard potential　标准电位；标准电势
standard pressure　标准压力〔760 毫米汞柱〕
standard reagent　标准试剂
standard recipe　标准配方
standard reference color　标准参比色；标准色

standard reference material (SRM)　标准参考物质；标准物质
standard regain　标准回潮
standard relative luminosity　标准相对视觉灵敏度
standard resistor　标准电阻(器)
standard sample　标准(试)样
standard sampling container　标准取样容器
standard sand　标准砂
standard scale　①标准温标；绝对温标②标准刻度③标准尺④标准秤
standard screen　标准筛
standard screen scale　标准筛制
standard sea water　标准海水
standard series method　标准系列比色法
standard sieve　标准筛
standard single electrode potential　标准电极电位
standard soil　标准污垢
standard soiled cloth　标准污布
standard solution　标准溶液*
standard source　标准源
standard source of radioactivity　标准放射源
standard specification　标准规格
standard specific gravity　标准比重
standard specimen　标准试样(片)
standard spectra collection　标准光谱集
standard spectral transmittance value　标准光谱透射值
standard spectrum　标准光谱*
standard state　标准(状)态
standard state chemical potential　标准态化学势
standard steam pressure　标准蒸汽压
standard storage tank　标准储罐
standard substance　标准物；基准物
standard subtraction method　标准减量法
standard temperature　标准温度
standard test　标准试验
standard (thermodynamic) scale　标准(热力学)温标；绝对温标
standard thermometer　标准温度计
standard tip　标准喷嘴(喷头)
standard titrimetric substance　标准滴定物
standard tolerance　标准公差
standard uncertainty　标准不确定度
standard unit　标准单位
standard vertical tube evaporator　标准竖管蒸发管；管式蒸发管
standard volume　标准体积
standard wavelength　标准波长
standard weather meter　标准耐气候牢度试验仪

standard weights　标准砝码
standard white kerosene　标准白煤油
standard white plaque　标准白瓷板〔测白度用〕
standard wire thread inserts　标准型钢丝螺套
stand-by column　备用柱
stand heap charing　直堆烧炭(法)
stand-in　模拟；代用物；冷试验代用品
standing　储藏；静置
standing bath　续染浴；连缸；老脚水
standing current　基流
standing (evaporation) losses　储存时的蒸发损失
standing pipet(te)　立式点样管
standing storage　长期储藏
standing tank　固定储罐
standing time　①停留时间②停台时间
standing wave(=wave formation)　驻波
stand linseed oil　①厚(熟)亚麻(仁)油；亚麻聚合油；亚麻厚油②调墨油〔油墨〕
stand oil(=polymerized oil)　聚合油；定油；厚油
stand pipe　竖管；立管
stannanes　锡烷(类)
stannate　锡酸盐
stannate radicle　锡酸根
stannekite　煤中树脂状烃
stannic　(正)锡的；四价锡的
stannic acid　锡酸
stannic bromide　(四)溴化锡　$SnBr_4$
stannic chloride　(四)氯化锡　$SnCl_4$
stannic chromate　铬酸锡　$Sn(CrO_4)_2$
stannic compound　(正)锡化合物；四价锡化合物
stannic disulfide　二硫化锡　SnS_2
stannic ethide　四乙基锡　$Sn(C_2H_5)_4$
stannic ethyl hydroxide　氢氧化三乙基锡　$SnOH(C_2H_5)_3$
stannic fluoride　(四)氟化锡　SnF_4
stannic hydride　氢化锡　H_4Sn
stannic hydroxide　氢氧化锡　$Sn(OH)_4$
stannic iodide　碘化锡　SnI_4
stannic mesotetrapyridylporphyrin　内消旋四吡啶基卟啉锡；四吡啶基中卟啉锡
stannic methide　四甲基锡　$Sn(CH_3)_4$
stannic nitride　四氮化三锡　Sn_3N_4
stannic oxide　氧化锡；二氧化锡　SnO_2
stannic oxychloride　二氯氧化锡　$SnOCl_2$
stannic phenide　四苯基锡　$Sn(C_6H_5)_4$
stannic sulfate　硫酸锡　$Sn(SO_4)_2$
stannic sulfide　硫化锡　SnS_2
stanniferous　含锡的
stannine　黄锡矿

stannising 镀锡
stannite 亚锡酸盐 M_2SnO_2
stannoacetic acid 甲基锡酸 CH_3SnOOH
stannometric titration(=stannometry) (氯化)亚锡(还原)滴定法
stannometry(=stannometric titration) (氯化)亚锡(还原)滴定法
stannonate 亚锡酸盐
stannonic acid (烃基)锡酸 $RSnOOH$
stannonic ester 亚锡酸酯
stannonium 一烃基锡烷
stannous 亚锡的；二价锡的
stannous bromide 溴化亚锡 $SnBr_2$
stannous chloride 氯化亚锡；二氯化锡 $SnCl_2$
stannous chromate 铬酸亚锡 $SnCrO_4$
stannous citrate 柠檬酸亚锡 $Sn_3(C_6H_5O_7)_2$
stannous compound 亚锡化合物
stannous ethide 二乙基锡 $Sn(C_2H_5)_2$
stannous 2-ethylhexoate 2-乙基己酸亚锡
stannous fluoride 氟化亚锡 SnF_2
stannous hydroxide 氢氧化亚锡 $Sn(OH)_2$
stannous iodide 碘化亚锡 SnI_2
stannous malate 苹果酸亚锡 $SnC_4H_4O_5$
stannous maleate 马来酸亚锡 $SnC_4H_2O_4$
stannous methide 二甲基锡 $Sn(CH_3)_2$
stannous octoate 辛酸亚锡
stannous oxalate 草酸亚锡 SnC_2O_4
stannous oxide 氧化亚锡；一氧化锡 SnO
stannous phenide 二苯基锡 $Sn(C_6H_5)_2$
stannous pyrophosphate 焦磷酸亚锡
stannous sulfate 硫酸锡 $SnSO_4$
stannous sulfide 一硫化锡 SnS
stannous tartrate 酒石酸亚锡 $SnC_4H_4O_6$
stannyl 甲锡烷基 H_3Sn-
stannylene 甲锡亚烷基 $H_2Sn=$
stanozolol 司坦唑醇
stantienite 黑琥珀
Stanton number 斯坦顿数
staphisagria 翠雀子
staphisagria alkaloids 翠雀生物碱类
staphisagrine 翠雀碱
staphisagroine 翠雀副碱
staphisaine 翠雀草碱
staphisine 斯塔飞燕草碱
staple diagram (纤维)长度分布图
staple fiber ①常产纤维；定长纤维②人造棉(花)
staple-fibre aftertreatment machine 短纤维后处理机；人造短纤维后处理机
staple-fibre woven fabric 定长纤维织物
staple-garnetting 短纤维扯松工艺
staple glass fiber 常产玻璃纤维；标准玻璃纤维
staple length 纤维长度
staple yarn 定长纤维纱
stapling 纤维长度分级
star anise 八角茴香
star aniseed 八角
star aniseed oil 八角茴香油；大茴香油
star anise oil 八角茴香油
star antimony 精制锑
star bowls 精锑块材
starch ①淀粉 $(C_6H_{10}C_5)_n$②(淀粉)浆③(上)浆
starch adhesive 浆糊；面浆
starch coating 淀粉涂料
starch column 淀粉柱
starch dialdehyde 淀粉二醛
starch ether paste powder 淀粉醚浆糊粉
starch finishing 上浆
starch flour 淀粉(末)
starch gel electrophoresis 淀粉凝胶电泳
starch glue 淀粉胶(水)；淀粉浆糊
starch glycerin 甘油淀粉
starch glycerite 甘油淀粉
starch gum 糊精
starch hexanitrate 淀粉六硝酸酯
starch indicator 淀粉指示剂
starchiness 淀粉性
starching 上浆
starching clay 高岭土；白黏土；膨润土
starching machine 上浆机
starch iodide 淀粉碘化物；碘化淀粉
starch iodide paper 淀粉碘化物试纸
starch iodide reaction 淀粉碘化物反应
starch iodide test 淀粉碘化物试验
starch machine 上浆机
starchness 淀粉度
starch nitrate 硝酸淀粉
starch paper 淀粉(试)纸
starch partition chromatography 淀粉分配色谱法
starch paste 浆糊
starch pasting 淀粉糊化
starch phosphate 磷酸淀粉
starch soluble 溶性淀粉
starch solution 淀粉溶液
starch sugar 淀粉糖
starch sultone 淀粉磺内酯
starch test paper 淀粉试纸

starch value 淀粉值
starchy 淀粉的
star comparator 星号比较仪
star connection 星状连接；星芒接法
starex 浮油松香
star feeder 星状加料器
star formation 星状组合
Stark broadening 斯塔克展宽
Stark effect 斯塔克效应
stark rubber 冷冻(橡)胶
star-like micelle 星形胶束
starlite 蓝锆石
star metal 精制锑
star miktoarm polymer 星形杂臂聚合物
star polymer 星形聚合物*
star-shaped arrangement 星形排列；星形布置
star-shaped carbon cell 星状碳极电池
star-shaped sinter candle 星型烧结烛形滤芯
start ①开始②开动
start button 起动按钮
start expansion temperature 初始膨胀温度
start gradient 起始梯度
starting 起动；开动；起始；启动
starting characteristic of fuel 燃料的起动特性
starting crude 起动原油
starting device 起动装置〔电工〕
starting fluid 起动(用柴油机)燃料
starting fraction 初馏分；起始馏分
starting gas 原料气
starting heater 始热器
starting leak source 始漏点
starting line 起点线
starting material 原材料
starting monomer 原料单体
starting motor 起动马达
starting paraffin 原料石蜡；原来石蜡
starting point 起点
starting point concentration 始点浓度；初始浓度
starting stave (容器)的第一竖板
starting switch 起动开关
starting temperature 起始温度
starting test 起动试验
starting volatility 起动时(汽油)挥发度
starting voltage 启动电压
starting vortex 起始涡流
start spot 起始斑
start-up 开工；启动；开车
start-up cost 试运转费用

start up from cold(=cold start up) 冷态起动
start-up procedure 试运转顺序
start-up speed 初速
start-up test 试运转；试车
starvation 缺胶
starved area 欠胶面
starved feeding 欠喂；供料不足
starved glue line 缺胶线〔层压塑料〕
starved joint 欠胶接头；缺胶接头
starved line 缺胶层
starved surface(=hungry surface) ①(漆膜)丰满度差的表面②干瘪的表面③塌渗的表面
starving 饿料；贫料
starving out 漏涂
Stas pipet 斯塔斯吸量管
stassfurtite 纤硼石
Stassfurt potash salt 斯塔斯弗特钾盐
Stassfurt salt 斯塔斯弗特盐
state (状)态；情况
state analysis 状态分析；示构分析
state coordinate 状态坐标
state diagram 状态图；平衡图
state function 状态函数
state graph 状态图
state hysteresis 状态滞后
state isomerism 状态同分异构(现象)
state of control 控制状态
state of cure 硫化程度；固化程度
state of disarray 混乱状态；无序状态
state of matter 物态；物质状态
state of oxidation 氧化状态
state of strain 应变状态
state of stress 应力状态
state of the art 技术水平；工艺水平
state of vulcanization 硫化程度
state property 态性
state selection 选态
state selectivity 状态选择性
state-to-state reaction dynamics 态-态反应动力学*
state verification 状态测试
static ①天电②静(止)的
statically indeterminate 静不定的
statically indeterminate structure 静不定结构
statical strain indicator 静态应变仪
static and dynamic balance 动静平衡
static atom 静止原子
static balance 静平衡
static bed 固定床

static bed of catalyst　催化剂固定床
static capacity　界面容量；积分容量
static catalytic cracking　固定床催化裂化
static characteristic of reaction system　反应体系的静态特征
static characteristics　静态特性
static charge　静电荷
static charge tester　静电测定仪
static chemiluminescence measurement　静态化学发光测量
static coating method　静态涂渍法
static compliance　静态柔量
static conducting device　导静电装置
static controller　固定控制器
static deformation　静态形变
static dielectric constant　静电介电系数
static discharge head　排出静压头
static electricity　静电
static elimination　消除静电
static eliminator　静电消除器
static energy　静压能
static equilibrium　静态平衡
static error　静态误差
static fatigue　静态疲劳
static fatigue failure　静疲劳失效
static flow　静流；层流
static-free hose　无静电胶管；不带电的胶管
static friction　静摩擦
static friction coefficient　静摩擦系数
static friction of absorbed film　吸附膜的静摩擦
static gas washer　静力煤气净化器
static head　静压高差；静压头
static headspace analysis　静态顶空分析法
static hold-up　静滞留
static inhibitor　静电消除剂；抗静电剂
static interfacial tension　静界面张力
static ion exchange　静态离子交换
static light scattering　静态光散射
static load(ing)　静负载
static lubrication　静态润滑
static mass spectrometer　静态质谱仪*
static mercury drop electrode　静汞滴电极
static method　静态法
static modulus　静态模量
static opening　静压测孔
static phase　静止相
static plate manometer　静压板状压差计
static pressure　静压(力)

static pressure difference　静压差
static pressure level controllers　静压头控制器
static pressure pen　静压头指针
static pressure tube　静压测量管
static problem　静态问题；静力问题
static-prone fibre　易产生静电纤维
static quenching　静态消光；静态猝灭
static resistant polyamide　抗静电聚酰胺
static rigidity　静态刚性
statics　静力学
static seal　静液封；静密封
static secondary ion mass spectrometry　静态次级离子质谱
static situation　静止状态；静止环境
statics of fluid　流体静力学
static sparking　静电打火；静电火花
static split system　静态分流系统
static splitting　静态分流
static stability　静态稳定性
static strain　静应变
static stress　静应力
static suction lift　静止吸入高度
static surface tension　静态表面张力*
static test　静态试验
static tower malting system　静止塔式制麦系统
static tyre test　静态轮胎试验
static vacuum mass spectrometer　静态真空质谱仪
static voltage　静电压
static water drop corrosion test　静态水滴腐蚀试验
static wetting effect　静止润湿作用
static wire　导电丝
station　站
stationary　①静止的②固定的；稳定的③不变的；定立的
stationary bed　固定床
stationary bob　静止浮子
stationary collapsible container　固定可折叠容器（由坚固帆布制成）
stationary cylindrical screen　固定式圆筒筛分机
stationary digester　固定式蒸煮釜
stationary disk　静止盘
stationary electric current　稳定电流
stationary electrode　静止电极
stationary electrode polarography　静止电极极谱法
stationary electrode voltammetry　静止电极伏安法
stationary film　静止薄膜
stationary flow　稳流
stationary furnace　固定炉
stationary heavy-duty type paste mixer　固定式稠度漆浆

混合(搅拌)机
stationary hysteresis　稳态滞后现象
stationary jaw　固定夹板
stationary lip(=fixed lip)　定模唇
stationary liquid　固定液
stationary liquid phase　液体固定相
stationary liquid polarity　固定液极性
stationary mercury electrode　固定汞电极；汞池电极
stationary method　静态法
stationary migration velocity　固定迁移速度
stationary parasitism　停留寄生；固定寄生
stationary partition liquid　固定分配液
stationary phase　固定相
stationary phase bed　固定相床
stationary phase bleeding　固定相流失
stationary phase coating　固定相涂布
stationary phase gradient　固定相梯度
stationary phase index　固定相指数
stationary phase pollution　固定相污染
stationary phase volume　固定相体积
stationary platen　定压板；定模板
stationary platinum electrode　静铂电极；固定铂电极
stationary platinum microelectrode　固定铂微电极；静铂微电极
stationary potential　稳态电势
stationary screen　固定筛
stationary screen condenser　固定式筛网集丝器
stationary seat　静(环)座〔机械密封〕
stationary solid bed　固定固体床
stationary solid phase　固体固定相
stationary solvent phase　静止溶剂相
stationary state　定态；稳态
stationary tripper　固定式倾料器
stationary vane　静叶片
stationary vessel　固定式容器
stationary wave　驻波；定波
station pointer　三杆分度仪
statistic　①统计的②统计量
statistical analysis　统计分析
statistical analysis of analytical data　分析数据的统计处理
statistical assumption　统计假设
statistical bias　统计偏差；统计偏倚
statistical chain　统计链
statistical coil　统计线团
statistical copolymer　统计(结构)共聚物
statistical decomposition　统计分解
statistical distribution　统计分布

statistical distribution theory　统计分配定律
statistical ensemble　统计系综；统计总体
statistical entropy　统计熵
statistical error　统计误差
statistical evaluation　统计分析；统计评价
statistical expectation　统计期望值
statistical fluctuation　统计性涨落
statistical hypothesis　统计假设
statistical inference　统计推断*
statistical isolinear multiple component analysis method (SIMCA method)　统计等线多组分分析法
statistical mean　统计平均值
statistical mechanics　统计力学
statistical method　统计法；平均法
statistical pattern recognition　统计模式识别
statistical quality control　统计(法)质量管理
statistical retention time　统计停留时间
statistical segment　统计链段*
statistical simulation spectrophotometry　统计模拟分光光度法
statistical temperature scale　统计温标
statistical test　统计检验*
statistical thermodynamic method　统计热力学方法
statistical thermodynamics　统计热力学
statistical weight　统计权重*
statistic copolymer　无规嵌段共聚物
statistics　①统计(学)②统计法
statolen　匍枝青霉素
stator　①挡板②定子；定片
stator blades　静叶片；固定叶片
Statron gun　斯塔特朗喷枪〔法国 Sames 公司设计〕
stat unite　厘米-克-秒静电制；CGS 静电制
status nascendi(=nascent state)　初生态；新生态
staubosphere　灰尘层
staurolite　十字石
stave　桶板
stave pipe　条木管
stavesacre seed　翠雀子
stave sheets　储罐壁板；储罐竖立板
stave wood　桶(板)材
stay　①拉撑②撑条；拉条
staybelite　氢化松香
staybolt　拉紧螺钉；拉紧螺栓
Staybright(=Staybrite)　斯特布赖特镍不锈钢
staybwood　未压缩木材
stay in grade　品质稳定
steadily accelerated fluid　匀加速流体；稳恒加速流体
steadiness　①稳定性②均匀性

steady　①稳(固)的②固定的；稳恒的
steady baseline　稳定基线
steady bearing　支撑轴承〔防止长轴摆动〕
steady compliance　稳态柔量
steady flow　稳流
steady flow process　稳流过程
steady seepage　等量渗透
steady shear compliance　稳剪切柔量
steady simple shear flow　简单剪切稳流
steady state　稳态
steady-state electroanalytical method　稳态电分析法
steady state loading　稳态载荷
steady state measurement　稳态测量
steady-state method　稳态(处理)法
steady state response　稳态响应
steady-state signal　稳态信号
steady state vibration　稳态振动
steady vibration　稳态振动
stealthy technique　隐身技术
steam　①蒸汽；水蒸气②蒸③以蒸汽发动
steam accumulator　蒸汽储蓄器
steam activation　蒸汽活化(作用)
steam admission side　进蒸汽侧
steam ager　蒸汽熟化器；蒸箱
steam-air activation　蒸汽空气活化(作用)
steam-air decoking method　蒸汽-空气除焦法
steam and air blown producer　蒸汽及吹气发生器
steam and air mixture　蒸汽空气混合气
steam and gas mixture　蒸汽煤气混合气
steam-and-solvent condenser　蒸汽与溶剂冷藏器
steam asphalt　蒸汽处理沥青
steam atmospheric distillation　常压蒸汽蒸馏
steam atomizer　蒸汽喷油器
steam atomizing　蒸汽雾化
steam autoclave　蒸汽加热加压釜；蒸汽高压釜
steam-bath　①蒸汽浴②蒸汽浴器
steam blown　蒸汽吹制的
steam-blown poke hole　蒸汽吹孔
steamboat coal　航运用煤
steam boiler　①蒸汽锅炉②(蒸)汽锅
steam-bottom still　底部蒸汽加热蒸馏釜
steam boundary curve　蒸汽界面曲线
steam bubble　蒸汽泡
steam calorifier　蒸汽热水器
steam calorimeter　蒸汽量热计
steam can　蒸汽发生器
steam chamber　蒸汽室
steam channel　蒸汽沟

steam chest　蒸汽柜；蒸汽夹板；汽室
steam cleaning　蒸汽清洁法；蒸汽清洗
steam cleaning gun　蒸气清洗枪〔清除旧漆膜用〕
steam cleaning unit　蒸汽清扫装置
steam coal　锅炉用煤
steam coil　蒸汽旋管
steam-coil-heated　蒸汽旋管加热的
steam-coil-heater tank car　蒸汽旋管加热器的油槽车
steam color printing　蒸汽印染
steam condenser　蒸汽冷凝器
steam conditioning　蒸汽给湿；蒸汽调湿
steam conduit　蒸汽管道
steam consumption　蒸汽消耗
steam-cooked grease　蒸汽煮沸润滑脂
steam-cooked reclaim　蒸热回收胶
steam cooking　蒸煮
steam curing　蒸汽硫化
steam cylinder　蒸汽缸
steam-cylinder lubrication　蒸汽缸润滑
steam-cylinder oil　蒸汽缸油
steam-cylinder stock　蒸汽缸油
steam dealkylation　蒸汽脱烷基化(作用)
steam distillation　蒸汽蒸馏
steam distilled　(蒸)汽(蒸)馏的
steam-distilled pine oil　蒸气蒸馏松油
steam-distilled (wood) rosin　木松香
steam-distilled (wood) turpentine　木松节油
steam dome　汽包
steam drier　蒸汽干燥器
steam driven　蒸汽动力的
steam driven pump　蒸汽泵
steamdrum　上汽锅；蒸汽锅筒
steam dryer　蒸汽干燥器
steam drying　蒸汽干燥(法)
steamed　吹蒸汽的
steamed cracking unit　蒸汽裂化装置
steamed mechanical pulp　蒸汽机制纸浆
steamed wood turpentine　汽馏松节油
steam ejector　蒸汽喷射器
steam-end efficiency　汽缸效率
steam engine　蒸汽机
steam-engine lubricant　蒸汽机(汽缸)润滑剂
steamer　蒸汽发生器
steam exhaust　排汽(装置)
steam explosion method　蒸汽爆裂法
steam film　蒸汽膜
steam fittings　蒸汽管件
steam flow meter　蒸汽流量计

steam funnel	蒸汽漏斗
steam gas	过热蒸汽
steam gauge	蒸汽表
steam generator	蒸汽发生器
steam gravity	蒸汽密度
steam gun	蒸汽喷枪
steam gun method	蒸汽喷射法；蒸汽喷枪法
steam hammer	汽锤
steam header	蒸汽汇集器；蒸汽室；汽包
steam heat	蒸汽热
steam heated	蒸汽加热的
steam-heated cal(l)andria	蒸汽加热
steam-heated degreasing plant	蒸汽加热脱脂装置
steam-heated evaporator	蒸汽加热蒸发器
steam-heated pipe line	蒸汽加热管道
steam-heated rotary dryer	蒸汽加热式回转干燥器
steam heated still	蒸汽加热蒸馏釜；汽馏釜
steam heater	蒸汽加热器
steam heating	蒸汽加热；蒸汽供暖
steam hoist	蒸汽升降机
steam hose	蒸汽软管；输送蒸汽用软管
steam hydrocarbon reformer	烃蒸汽转化炉
steaming	蒸热；通入蒸汽；蒸汽加工；汽蒸定形〔纺织〕
steaming in a rotary pressure cooker	回转加压蒸煮
steaming machine	蒸热机
steaming operation	蒸汽通入操作（裂化装置）
steaming out	吹汽
steaming out tank	吹汽槽
steaming power	蒸汽生产额
steaming time	汽蒸时间
steam iron generator	蒸汽铁屑生氢器；氢气发生器
steam iron hydrogen	蒸汽铁屑产生的氢气
steam jacket	①蒸汽套②蒸汽套管
steam jacketed	蒸汽套的
steam jacketed mold	蒸汽夹套模
steam jenny phosphating	移动式蒸汽喷雾磷化处理法
steam jet	①蒸汽喷嘴②蒸汽喷射
steam jet agitator	蒸汽搅动器
steam jet air pump	蒸汽泵；汽轮泵
steam jet blower	蒸汽鼓风机；汽轮鼓风机
steam jet booster	蒸汽喷射增效器
steam-jet compression	蒸汽喷射压缩(法)
steam-jet cooling system	蒸汽喷射冷凝系统
steam-jet refrigerating machine	蒸汽喷射致冷机
steam jet vacuum pump	蒸汽喷射真空泵
steam-laden	水蒸气饱和的；含水蒸气的
steam lift	蒸汽提升器
steam line	蒸汽线路；供汽管
steam lubrication	蒸汽润滑
steam molecule	蒸汽分子
steam nozzle	蒸汽喷嘴
steam out	蒸汽吹出
steam period	通蒸汽期
steam phosphating	蒸汽磷化处理法；热磷化
steam pipe	蒸汽管子
steam pipe coil	蒸汽旋管
steam pipe expansion loop	蒸汽管膨胀圈
steam pipe line	蒸汽管线
steam pipe oven	蒸汽加热炉
steam plant	蒸汽厂
steam plate	蒸汽(加热)板
steam platen press	蒸汽平压
steam pockets	蒸汽包
steam point	沸点
steam poisoning	蒸汽中毒（催化剂）
steam pressure	蒸汽压(力)
steam pressure gauge	汽压表
steam pump	蒸汽泵
steam-purge zone	蒸汽吹换区
steam rate	汽耗；汽耗率
steam-refined	汽炼的；蒸汽精制的
steam-refined asphalt	汽炼沥青；蒸汽精制的沥青
steam-refined cylinder stock	汽炼汽缸油；重(润滑)油
steam-refined oil	汽炼油
steam-refined residuum	汽炼渣油；蒸汽精制油
steam-refined stock	汽炼油料；残留汽缸油；残留润滑油馏分
steam refining	汽炼的；蒸汽精制；蒸汽蒸馏
steam reforming	蒸汽转化
steam return line	蒸汽返回管道；冷凝蒸汽管线
steam roller	蒸汽碾路机
steam run	蒸汽吹炼
steam run gas	蒸汽吹出的气体
steam saturator	蒸汽饱和器
steam sealing	蒸气封闭处理(氧化膜)
steam separator	蒸汽分离器；凝汽罐
steam-size coal	蒸汽级煤；14.3～2.5mm 级无烟煤
steam slot	走汽缝
steam-smothering	蒸汽灭火
steam-smothering line	灭火蒸汽管道
steam sparging device	蒸汽喷射装置
steam spinning	蒸汽纺丝法
steam splitting process	蒸汽裂解法
steam-spraying	蒸汽喷涂
steam-spray process	蒸气喷涂法〔以过热蒸汽代替空气喷涂〕

steam stamp 蒸汽捣矿机
steam stamp cylinder 汽捣活塞
steam stamp mill 汽捣磨
steam sterilization 蒸汽灭菌
steam sterilizer 蒸汽消毒器
steam sterilizing 蒸汽消毒；蒸汽灭菌
steam still 蒸汽蒸馏器
steam strainer 滤汽器
steam stripped (蒸)汽提(馏)的
steam stripper 汽提机
steam stripping (蒸)汽提(馏)
steam stripping tower 蒸汽分离塔；汽提塔
steam stuffing box 蒸汽密封盒
steam superheater 蒸汽过热器
steam superheating (用)蒸汽过热
steam supply 蒸汽供应
steam supply line 蒸汽供应线
steam table 蒸汽表
steam tension 蒸汽压(力)；蒸汽张力
steam tight 不透蒸汽的；汽密的
steam trace (加热)蒸汽管道
steam tracer line 蒸汽伴热管
steam tracing 伴热蒸汽管
steam trap 汽阱；凝汽缸
steam trap assembly 凝汽阀组
steam treatment 水蒸气处理
steam tube dryer 蒸汽管干燥器
steam turbine (蒸)汽涡轮
steam turbine lubricating system 蒸汽叶轮机润滑系统
steam turbine lubrication 蒸汽叶轮机润滑
steam turbine oil 蒸汽
steam-volatile oil 水蒸气蒸馏挥发油
steam vulcanization 蒸汽硫化
stearaldehyde 硬脂醛；十八(烷)醛
stearamide 硬脂酰胺；十八(烷)酰胺 $C_{17}H_{35}CONH_2$
stearanilide 硬脂酰苯胺 $C_{17}H_{35}CONHC_6H_5$
stearate 硬脂酸盐 $C_{17}H_{35}CO_2M$
stearate radical 硬脂酸根
stearic acid 硬脂酸；十八(烷)酸 $CH_3(CH_2)_{16}COOH$
stearic acid amide(=stearic amide) 硬脂酸酰胺；硬脂酰胺
stearic acid derivative 硬脂酸衍生物
stearic acid film 硬脂酸膜
stearic acid glyceride 甘油硬脂酸酯
stearic aldehyde 硬脂醛；十八(烷)醛 $CH_3(CH_2)_{16}CHO$
stearic amide(=stearic acid amide) 硬脂酰胺
stearic anhydride 硬脂(酸)酐；十八(烷)(酸)酐 $(C_{17}H_{35}CO)_2O$

stearic ethanolamide 硬脂乙醇酰胺
stearic glyceride 甘油硬脂酸酯
stearic hydrazide 硬脂酰肼
stearin(=tristearin) 硬脂精；三硬脂精；(三)硬脂酸甘油酯 $C_3H_5(OOCC_{17}H_{35})_3$
stearine oil 硬脂油
stearine pitch 硬脂沥青
stearinery 硬脂制造业
stearo-dilaurin 二桂酸硬脂酸甘油酯
stearodiolein 硬脂酸二油酸甘油酯
stearo-dipalmitin 二棕榈酸硬脂酸甘油酯
stearolactone 硬脂酸内酯；十八(烷)酸内酯
stearo-lauro-myristin 硬脂酸月桂酸肉豆蔻酸甘油酯
stearolic acid (=9-octadecynoic acid) 硬脂炔酸；9-十八(碳)炔酸 $C_8H_{17}C\equiv C(CH_2)_7CO_2H$
stearo-myristin 硬脂酸肉豆蔻酸甘油酯
stearo-myristo-laurin 硬脂酸肉豆蔻酸月桂酸甘油酯
stearone (=18-pentatriacontanone) 硬脂酮；18-三十五(烷)酮 $C_{35}H_{70}O$
stearonitrile 硬脂腈；十八(烷)腈 $C_{17}H_{35}CN$
stearo-palmito-olein 硬脂酸棕榈酸油酸甘油酯
stearophenone 硬脂苯酮；十八碳酰苯
stearoptene(=oleoptene) 玫瑰蜡〔香精油的固体氧化部分〕
stearoxylic acid 硬脂氧酸；二氧代硬脂酸 $CH_3(CH_2)_7COCO(CH_2)_7COOH$
stearoxyl trimethyl silane 硬脂氧基三甲基硅烷
stearoyl 硬脂酰；十八烷酰 $CH_3(CH_2)_{16}CO-$
stearoyl isethionate 硬脂酰基羟基乙磺酸盐
stearyl 硬脂酰；十八烷酰
n-stearyl acrylate 丙烯酸正十八酯
stearyl alcohol 十八烷醇
stearyl aldehyde 硬脂醛
stearylamide (SA) 硬脂酰胺；十八酰胺
stearylamine 硬脂胺；十八胺
stearylamine acetate 乙酸硬脂酰胺〔促进剂〕
stearyl amino propionic acid 硬脂氨基丙酸
stearyl betaine 硬脂基甜菜碱
stearyl cellulose 硬脂酰纤维素
stearyl chloride 硬脂酰氯；十八烷酰氯 $C_{17}H_{35}COCl$
stearyl dimethicone 硬脂基二甲基硅氧烷
stearyl dimethyl benzyl ammonium chloride 十八烷基二甲基苄基氯化铵
stearyl glycyrrhetinate 硬脂基甘草亭酸酯
stearyl imidazol 硬脂基咪唑
stearyl methacrylate 甲基丙烯酸硬脂醇酯；甲基丙烯酸十八酯
stearyl pyridinium bromide 硬脂基溴化吡啶鎓
stearyl stearate 硬脂酸十八酯

stearyl trimethyl ammonium chloride 硬脂酰三甲基氯化铵〔促进剂〕
steatite 滑石；皂石
steatite bobbin 块滑石线圈骨架
steatite ceramics 块滑石陶瓷
steatitic 块滑石的
stebisimine 千金藤比斯碱
stechiometry(=stoichiometry) 化学计算(法)；化学计量学
steclin(=tetracycline) 四环素
Stedman packing 斯特曼填料；金属网锥形规则填料
Stedman packing column 斯特曼填充塔
steel ①钢②钢(制)的③似钢的④钢制品；钢块
steel alloy 合金钢
steel ball bearing 钢球轴承
steel band 钢带；载重带
steel band tape 钢带〔测量储罐用〕
steel bar 钢条
steel bomb 钢弹；钢制反应釜
steel brush 钢刷(子)
steel casting 钢铸件
steel chain 钢链
steel clad 钢片衬
steel coil 钢旋管
steel-cord conveyor belt 钢丝绳运输带
steel-cutting compound 钢材切削化合物
steel cylinder 钢筒
steele acid (可氧化的)松香酸
steel file 钢锉
steel foil 钢箔
steel forging 锻钢
steel framework 钢框；钢架
steel grits 钢砂〔喷洗用〕
steel hammer 钢锤；铁锤
steel hardening oil 钢材硬化油
steel ingot 钢锭；钢块
steel lined 衬钢的
steel magnet 磁钢
steel mortar 钢研钵
steeloscope 看谱镜
steel plate 钢板
steel plate thickness gauge 铁板厚度计
steel plating 镀钢
steel product 钢材
steel roll 钢辊
steel rule 钢尺
steel sections(=steel shapes) 型钢
steel shapes(=steel sections) 型钢
steel sheet pile 钢板桩
steel shot 钢砂；钢丸
steel spatula 钢刮刀
steel strip 钢带；带钢
steel structure coating 钢结构用涂料
steel tape 钢(卷)尺
steel wire 钢丝
steel wire cord 钢丝帘线
steel wire rope 钢丝绳
steel wool 钢(丝)棉
steel works (铸)钢厂
steelyard 大磅秤；弹簧秤；吊秤；杆秤
steely iron 炼钢(用)铁
Steenbock unit 斯廷博克单位〔一种维生素单位〕
steeper ①浸渍器②较陡的
steepest ascent 最速上升法*
steepest-ascent procedure 最陡上升法
steepest descent 最速下降法*
steeping 浸渍；浸碱
steeping and pressing tank 浸碱压榨机
steeping bowl 浸渍碗
steeping cell 浸渍器
steeping fluid 浸渍液
steeping liquor 浸渍(碱)液
steeping lye 浸渍碱液
steeping press 浸压机
steeping tank 浸渍槽
steeping trough 浸渍槽
steeping vat 浸渍槽；浸胶槽
steeple compound stamp 尖柱捣磨
steep-roof asphalt 坡用沥青；高坡度顶盖用沥青
steering 转向；操纵方向
steering shaft 转向轴
steering wheel 操向轮
Stefan-Boltzmann constant 斯蒂芬-玻耳兹曼常数
Stefan-Boltzmann law 斯蒂芬-玻耳兹曼定律
Stefan-Boltzmann law of radiation 斯蒂芬-玻耳兹曼辐射定律
Stefan's law of diffusion 斯蒂芬扩散定律
Steffen's process 斯特芬法
steffimycin(=steffisburgensimycin) 斯堡霉素；司替霉素
steffisburgensimycin(=steffimycin) 斯堡霉素
steigerite 水钒铝矿
Steiner bubble viscometer 斯坦纳气泡黏度计
stelite 钨铬钴合金
stellar evolution 星体演化
stellar nucleosynthesis 星体核合成
stellar r-process 星体快速过程

stellar spectra 星光谱
stellar s-process 星体慢过程
stellasterol 星鱼甾醇
stellate crystals 星状晶体
stellerite 淡红沸石
stellite ①针钠钙石②硅灰石③斯特莱特硬质合金〔钨铬钴合金〕
Stelometer 斯特洛束纤维强力机
stemless funnel 无管漏斗
stem nucleus 主链
stemonidine 次百部碱；百部定；史弟蒙尼定碱
stemonine 百部碱
stench 臭气；恶臭
stencil ①横板②镂花模板③(油印用)蜡纸
stencil finishing 镂花涂装
stenciling 镂花涂装
stencil paint 镂花涂装用漆
stencil sheet (油印用)蜡纸
stender dish (生物)染色皿
stenocarpine 皂荚碱
stenol 石烯醇
stenosine 砷酸甲酯钠
stenter 展幅机
stentering 拉幅(工艺)；伸展的；展幅的
stentering machine 展幅机；拉幅机
step ①步②步骤③阶
step addition polymer 逐步加成聚合物
stepblender 梯形混合机
step by step analysis 分部分析；分步分析
step-by-step design 逐步设计；分段设计
step by step method 按步方法
step-by-step test 逐步试验
step change 阶跃
step chromatography 台阶色谱法*
step colorimeter 分步比色计
step copolymerization 逐步共聚合
step-down gear 减速机
step-down transformer 降压器
step filter 阶式光楔
step function 阶跃函数
4-step godet 四级导丝盘
step-growth polymerization 逐步聚合(作用)；逐步增长聚合
stephamiersine 千金藤默星碱
stephanine 千金藤碱
stephanite 脆银矿
stephanoline 千金藤诺灵
stepholine 千金藤福灵

step index type 步进指数型
stepinonine 氧代千金藤默星碱
step joint Z形接头〔轴封〕
step labyrinth 阶梯式迷宫密封
step ladder polymer 分段梯形聚合物
stepless speed change device 无级变速装置
stepless speed variation 无级变速(调速)
step of reaction 反应阶段
steponine 异千金藤碱
stepped godet 多级导辊；多级导丝盘
stepped temperature program 阶梯升温程序
steppe salt 原草盐
step piston 阶梯形活塞；级差活塞
step polycondensation 逐步缩聚
step polymerization 逐步聚合
step precision winding machine 有级精密卷绕络筒机
step pulse 阶梯脉冲波
step reaction 逐步反应
step reaction polymerization 逐步聚合*
step sector 阶梯光阑
step sector method 阶梯光阑法
step site 台阶位
step size 步长
step sizing 连续筛选
step speed change 有级变速
step speed regulation 有级调速
step-test procedure 逐级测试法；分级试验法
step up cure 分段硫化
step-up transformer 升压器
step weakener 阶梯减光板
step wedge 阶式光楔
step width 步长
stepwise ①逐步的；分段的②阶式的③梯段的
stepwise addition 逐步加入；逐步添加
stepwise approximation method 逐步近似法
stepwise complex formation constant 逐级配合物生成常数
stepwise cracking 逐步龟裂(开裂)
stepwise decomposition 逐级分解*
stepwise degradation 逐级降解
stepwise development 分步展开
stepwise device 阶梯装置；分阶装置
stepwise discriminate analysis 逐步判别分析
stepwise dissociation 逐级解离*
stepwise elution 阶కి洗脱
stepwise-elution analysis 逐次洗脱分析
stepwise excitation 分步激发
stepwise formation constant 逐级形成常数

stepwise gradient　阶式梯度
stepwise gradient device　阶式梯度装置
stepwise hydrolysis　逐级水解
stepwise hydrolysis constant　逐级水解常数
stepwise instability constant　逐级不稳定常数
stepwise leveling process　逐步流平法；逐步流平过程
stepwise line fluorescence　阶跃线荧光
stepwise method　分段(依次计算)法
stepwise method of McCade and Thiele　(决定精馏塔塔板数的)马克开勃-齐利图解分段(依次计算)法
stepwise polymerization　逐步聚合
stepwise regression　逐步回归*
stepwise stability constant(=consecutive stability constant)　逐级稳定常数
stepwise synthesis　分步合成
stepwise titration　分步滴定(法)*
steramide　乙酰磺胺
sterandryl　丙酸睾丸甾酮
sterane　甾烷；11β,17,21-三羟基孕甾-1,4-二烯-3,20-二酮
steranthrene　立蒽
sterate group　硬脂酸基(根)
stercorin　粪甾醇
stercorite　磷钠铵石
sterculia gum　苹婆胶；梧桐胶
sterculia urens　刺苹婆
sterculic acid　苹婆酸；9,10-亚甲基油酸
stere　立方米
stereo-　〔希腊字头〕　立体；固(体)
stereo-analogs　立体类似物
stereo-binocular microscope　双目立体显微镜；双筒立体显微镜
stereoblock　立构嵌段*
stereoblock copolymer　立构规正嵌段共聚物
stereoblock polymer　立(体)构(形)规正嵌段聚合物
stereocenter　立构(规正)中心
stereochemical　立体化学的
stereochemical change　立体化学变化
stereochemical formula(=stereo-formula)　立体化学式
stereochemically　立体化学的
stereochemical orientation　立体(化学)取向
stereochemical specificity　立体化学专一性
stereochemical symmetry　立体化学的对称性
stereochemistry　立体化学*
stereochromy　立体彩饰法
stereocomplex　定向络合物；立体络合物
stereodiagram　立体图
stereo-directed polymer　立体定向聚合物

stereoeffect　立体效应
stereoelective polymerization　立构有择聚合
stereoelectivity　立构有择性；立构规整有择性
stereo-electronic effect　空间电子效应
stereo-electron microscopy　立体电子显微镜检术
stereo-formula(=stereochemical formula;stereochemic formula)　立体化学式
stereograft polymer　立构接枝聚合物
stereo-homo-polymer　立构均聚物
stereohybridization　立构(规正)杂化作用
stereoinversion　①立构倒置②立构倒置体
stereoirregular　立体无规的
stereo-isomer　立体异构体；空间异构体
stereoisomeric　立体异构的
stereoisomeric form　立体(同分)异构型
stereoisomeric monomer　立体异构单体
stereoisomeride　立体异构体
stereoisomerism　立体异构(现象)*
stereo-isometric elastomer　立体异构橡胶
stereoisotactic polymer　立构等规聚合物；立构全同聚合物
stereo-lithography　立体蚀刻
stereomer　立体异构体
stereomeric　立体异构的
stereomeride　立体异构体
stereometer　体积计；立体测量仪
stereometric formula　立体式
stereometry　测体积学
stereo-ordered polymer　立构有序聚合物
stereopairs　立体双像；立体照片对
stereopolybutadiene　有规立构聚丁二烯橡胶
stereo-polymerization　定向聚合
stereopticon(=balopticon)　投影放大器
stereorandom copolymer　立构无规共聚物
stereorandom polymerization　立构无规聚合
stereoregular　有规立构的；定向的
stereoregular fibrous polymer　立构规整的纤维状聚合物
stereoregularity　立体有规性
stereoregular polybutadiene　有规立构聚丁二烯橡胶
stereoregular polymer　有规立构聚合物*
stereoregular polymerization　定向聚合；立构规整聚合
stereo-regular rubber　有规立构橡胶
stereoregular structure　有规立构结构；立构规整结构
stereoregulated　立构规整的
stereoregulated polymerization　立构规整聚合
stereo-regulation　立体调节
stereorepeating unit　立构重复单元*
stereo-rubber　有规立构橡胶

stereos 有规立构橡胶
stereoscan 立体扫描
stereoscan electron microscope 立体扫描电子显微镜
stereoscanning electronmicrograph 立体扫描电子显微照片
stereoscan photograph 扫描电镜照片
stereoscope 体视镜
stereoscopic electron micrograph 体视电子显微照相；立体电子显微照相图
stereoscopic microscope 体视显微镜
stereoscopic photograph 体视照相
stereoscopic picture 体视相片
stereoscopic synthesis 体视合成
stereoscopy 立体观测；体视术
stereoselective polymerization 立构有择聚合(作用)
stereoselective ring A 甾择环A
stereoselective total synthesis 立体有择全合成
stereoselectivity 立体选择性*
stereoskiagraphy 体视X射线照相术
stereospecific 立体有择的；立体定向的
stereospecific adsorbent 立体有择吸附剂
stereospecific catalyst 立体有择催化剂
stereospecific copolymerization 立构有规共聚合
stereospecificity 立体专一性*
stereospecific polymer 立体有择聚合物；立体定向聚合物；有规立构聚合物
stereospecific polymerization 有规立构聚合(作用)；定向聚合
stereospecific reaction 立体有择反应
stereospecific rubber 有规立构橡胶
stereospecific step-growth condensation reaction 立体有择逐步增长缩合反应
stereospecific structure 有规立构结构
stereospecific synthesis 立体有择合成
stereospecific template 有规立构模板；定向模板
stereospecific Ziegler-Natta polymerization 立体有择齐格勒-纳塔型聚合
stereostructural formula 立体结构式
stereostructure 立体结构
stereosymmetric rubber 对称立构橡胶
stereotacticity 立构规整性
stereotactic polymerization 定向聚合
stereotaxis 趋实体性；趋能性
stereotopochemistry 立构局部化学
stereotropism 向实体性；亲实体性
stereotype ①铅版〔印刷工业〕②制铅版③刻板
stereotyping 制铅板
stereo zoom microscope 变焦距体视显微镜

steric 空间(排列)的；位的
sterically defined 空间定位的；立体结构上已确定了的
sterically hindered phenol (空间)位阻酚
steric assistance 空间助效
steric barrier 立体障碍；位阻
steric compatibility 空间相容度
steric compression 空间压缩；挤压
steric configuration 空间构型；立体构型
steric course 空间过程；立体化学过程
steric direction 空间导向
steric effect 位阻效应；空间效应
steric exclusion 位阻排阻；空间排斥
steric exclusion chromatography 空间排阻色谱(法)
steric exclusion model 空间排阻模型
steric factor 空间因子*
steric hindrance 位阻(现象)
steric hindrance effect 立体位阻效应
steric influence 位阻影响
steric interaction 空间相互作用
steric isomerism 立体异构现象
steric isotope effect 立体同位素效应；空间同位素效应
steric linkage 立体键合；体型键合
steric order 立构有序；立体秩序
steric protection 位阻保护作用
steric requirement 空间条件
steric restriction 空间阻碍
steride(=steroid) 甾族化合物
sterigma 小梗
sterile board 无菌箱
sterile culture ①无菌培养②纯菌培养
sterile cupboard 无菌箱
sterile room 无菌室
sterilization ①消毒；灭菌②绝育
sterin 硬脂酸精；甘油硬脂酸酯
sterioside 甾苷
sternbergite 硫铁银矿
Stern diffuse double layer 斯特恩扩散双电层
Stern double layer 斯特恩双层
Stern effect 斯特恩效应
Stern-Gerlach experiment 斯特恩-格拉赫实验
Stern layer 斯特恩扩散双电层
Stern potential 斯特恩电位；斯特恩电势
stern tube 轴管
sternutation 发嚏(作用)
sternutative ①喷嚏剂②发嚏的
sternutatory gas 喷嚏(性)(毒)气
sterny 粗粒的
stero-bile acid 甾族胆汁酸

steroid 甾族化合物*
steroidal 甾族的
steroidal amine 甾族胺
steroid glycoside 甾类糖苷
steroid nucleus 甾核
sterol 甾醇；固醇
sterol acetate 乙酸甾醇酯
sterol ester 甾醇酯
sterone 甾酮
steropton 香精蜡
sterro metal 铁锡锌铜合金
stertite(=talc) 滑石〔一种水合硅酸镁〕
Stetefeldt furnace 斯特费尔特焙烧炉
Stevenson's rule 史蒂文森规则
stevioside 甜菊苷
Stewart-Kirchhoff law 斯图尔特-基尔霍夫定律
sthene 斯亭〔力单位，为1000牛顿或10达因〕
sthenosage 防水处理
stib- 锑
stibacetin 乙酰氨基苯脒酸钠
stibamine 脒胺；对氨基苯脒酸钠
stibarseno 偶锑砷基 —Sb=As—
stibate 锑酸盐〔1.偏 $MSbO_3$; 2.正 M_3SbO_4; 3.焦 $M_4Sb_2O_7$〕
α-stibazole α-芪唑；α-苯乙烯基吡啶 $NC_5H_4CH=CHC_6H_5$
stibazole 芪唑；苯乙烯基吡啶
γ-stibazole γ-芪唑；4-苯乙烯基吡啶
stibial （正）锑的；五价锑的
stibiate 锑酸盐〔1.偏 $MSbO_3$ 2.正 M_3SbO_4; 3.焦 $M_4Sb_2O_7$〕
stibiated 含锑的
stibic(=antimonic) 锑的；五价锑的
stibiconite 黄锑华 $H_2Sb_2O_5$
stibide(=antimonide) 锑化物
stibine ①锑化(三)氢②䏲
stibine hydroxide 氢氧化二烃基锑；氢氧化䏲 $Sb(OH)R_2$
stibinico- 亚䏲羧基 (HO)OSb=
stibino- 䏲基 H_2Sb—
stibinoso ①二羟锑基 $(HO)_2Sb$—②羟锑基 HOSb=
stibious(=antimonous) 三价锑的
stibite 黄锑华
stibium 锑
stibnate 锑酸盐〔1.偏 $MSbO_3$; 2.正 M_3SbO_4; 3.焦 $M_4Sb_2O_7$〕
stibnic(=antimonic) 锑的
stibnide 锑化物
stibnite 辉锑矿*

stibnous(=antimonous) 亚锑的
stibo- 锑酰 O_2Sb—
stibonazo III 偶氮䏲III
stibonic acid 䏲酸；二烃基锑酸 R_2SbOOH
stibonium 锑鎓〔指有机五价锑化合物〕
stibonium compound 锑鎓化合物
stibonium hydroxide 氢氧化四烃基锑 $Sb(OH)R_4$
stibonium iodide 四烃基锑化碘 $SbIR_4$
stibono- 䏲羧基 $(HO)_2OSb$—
stibophen 䏲芬
stibosan 䏲散
stiboso- 亚锑酰 OSb—
stibous(=antimonous) 亚锑的
stibyl(=stibino-) 䏲基
stibylene 亚䏲基 HSb=
stick ①棍；棒；杆②粘(贴)
sticked overshoe 胶鞋
sticker 黏着剂；固着剂
stick fast 黏牢；黏紧
stick gum 黏胶
stick(i)ness ①黏着性②发黏(现象)
sticking agent 黏着剂；固着剂
sticking coefficient 黏附系数
sticking patch 黏垫片
sticking plaster 黏皮膏〔橡胶〕
sticking point 黏附点；黏附温度
sticking tape 胶黏带；胶水纸
stick(ing) temperature 黏附温度；黏着温度
stick lac 原紫胶；紫梗原胶
stickle-back oil 九刺鱼油
stick of sulfur 硫棒
stick-on sole 胶合鞋底
stick phosphorus 棒状磷
stick shellac 紫胶棒〔用于修饰木材〕
stickstofflost(=stickstoff lost) 氮芥毒气
stick sulfur 硫棒；棒状硫黄
stick vat 吊鞣池
sticky 刷涂阻力；胶黏的；黏附的
sticky material 黏着材料；黏合剂
sticky oil 黏石油
sticky paint film 黏性漆膜；发黏的漆膜
sticky paint (particle) 黏性油漆(粒子)〔指排放空气所含的黏性漆雾粒〕
sticky point 黏结点；胶黏点；黏附温度
stiction 黏滞
stiff backbone 刚性主链；硬挺主链
stiff batch 稠厚装料
stiff chain 硬性链；刚性链

stiff clay 硬泥
stiff differential equation 病态微分方程
stiffened aluminium sheet 加强铝板
stiffened panel 加筋板
stiffener ①硬化剂②增稠剂③刚性体④加强件⑤补强条；补强板
stiffener material 增强板材
stiffen flow 黏滞流
stiffening ①上浆②僵化③使硬④加强
stiffening agent 硬化剂；硬挺整理剂
stiffening fluid 硬化流体
stiffening of grease 润滑脂固化
stiffening plastic substance 硬化塑性物质
stiffening plate 补强板
stiffening rib 加强肋
stiffening ring 补强环；补强圈
stiffening temperature 硬化温度
stiff fibre 硬纤维
stiff flow 难流动(性)
stiff mud 硬泥
stiff mud brick 坚硬泥砖
stiff mud brick machine 硬泥造砖机
stiffner 钢圈外包布；补强胶条
stiffness 僵硬性；劲度；挺度
stiffness factor 劲度因子
stiffness index 劲度指数
stiffness modulus 劲度模量
stiffness tester 劲度试验仪
stiffness-toughness balance 刚度与韧性的平衡
stiff paint 厚漆
stiff paste 浓膏
stiff paste form 浓(稠)膏状物；黏糊膏状物
stiff shaft 刚性轴
stigmastane 豆甾烷 $C_{29}H_{52}$
stigmastanol 豆甾烷醇 $C_{29}H_{52}O$
stigmastenol 豆甾烯醇
stigmasterol 豆甾醇 $C_{29}H_{48}O$
stigmatic 消(像)散
stigmator 消像散器
stilba〔词头〕 锑杂
stilbamidine 䏡脒
stilbazo(=stilbene-4,4′-bis-(1-azo-3,4-dihydroxy-benzene) (2,2′-disulfonate) 二苯乙烯-4,4′-双(1-偶氮-3,4-二羟基苯)-2,2′-二磺酸盐；芪偶氮
stilbene 芪；1,2-二苯乙烯；均二苯代乙烯 C_6H_5CH=CHC_6H_5
stilbene α-carboxylic acid 芪-α-羧酸；α-苯基肉桂酸
stilbene diamine 芪二胺 $C_6H_5CH(NH_2)CH(NH_2)C_6H_5$

stilbene-diol 芪二酚 HOC_6H_4CH=CHC_6H_4OH
stilbenedithiolato 二硫醇化芪
stilbene dye 芪染料；二苯乙烯染料
stilbene hydrate 水合芪 $C_{14}H_{14}O$
stilbene-sulphonic acid 1,2-二苯乙烯磺酸
stilbenyl 芪基；均二苯乙烯基
stilbestrol 己烯雌酚
stilbite 辉沸石
stilboestrol 己烯雌酚
stile ①门梃；窗梃②侧柱
still ①蒸馏釜；蒸馏锅②酿酒场③蒸馏④静止的⑤不起泡的
stillage 釜馏物
stillage bottoms 釜脚
stillage gas 釜馏气
stillage residue 釜残渣
stillage return system 釜馏物回流装置
still air 静止的空气
still bottoms 釜脚
still coke 釜馏焦；蒸馏焦
still column 蒸馏柱
still-cooled cylinder 自冷式汽缸
still dome 釜室
still gas 釜馏气
still grease 釜馏润滑脂
still head 蒸馏头
stilling 釜馏
stilling box 釜箱
stillingia oil 梓油
stillingia stand oil 梓油聚合油；梓油厚油
stillingic acid 乌桕酸；2,4-癸二烯酸
stillingine 草乌桕碱；银叶大戟碱
still kettle 蒸馏釜
still liquor 釜馏液
stillman 锅炉工
still-porous coating 静态多孔镀层
still pot 蒸馏釜
still preheater 预热釜
still residue 蒸馏余液；馏渣
still setting 釜衬砖
still shell 釜壳
still steam 蒸馏用蒸汽
still tube 蒸馏管
still wax 釜馏蜡
stilpnomelane 黑硬绿泥石
stilt 承坯架
stimulant 兴奋剂；刺激剂
stimulated absorption 受激吸收

stimulated absorption transition 受激吸收跃迁
stimulated emission 受激发射*
stimulated emission transition 受激发射跃迁
stimulated radiation 受激辐射
stimulated Raman gain 受激拉曼增益
stimulated Raman scattering (SRS) 受激拉曼散射
stimulated Raman scattering effect 受激拉曼散射效应
stimulated restoration 受激回复
stimulation echo 激发回波
stinging nettles 荨麻
stink ①臭气②发臭气
stink bomb 臭气弹
stink cupboard 通风橱
stinkstone 臭石；臭灰岩
stinkweed 曼陀罗
stink wood oil 臭木油
stipitat(on)ic acid 密挤青霉素；密挤青霉酸
stipple(=stippling) ①(湿膜)拂平，荡平②点蘸涂法；蓓蕾漆糅饰法〔大漆〕③拉花涂法④点彩(点刻)法
stippled coating 复色拉毛涂层
stipple finish ①点蘸涂饰法②点蘸面漆
stipple paint ①点蘸漆；拉花漆，点彩漆②蓓蕾漆〔大漆〕
stippler ①点彩笔；点彩刷②拉花刷(笤梳)
stippling(=stipple) ①(湿膜)拂平(荡平)②点蘸涂法；蓓蕾漆糅饰法〔大漆〕③拉花涂法④点彩(点刻)法
stipule 托叶
Stirling's approximation 斯特林近似(法)
stirred flow reactor 流动搅拌反应器
stirred-tank reactor 搅拌釜反应器
stirrer ①搅拌器；搅动器②搅棒
stirrer arm 搅拌(机)桨叶
stirring apparatus 搅拌装置
stirring machine 搅拌器
stirring mill 搅拌机
stirring motion 湍流；涡流
stirring rake 搅拌桨
stirring rod 搅棒
stirring screw conveyer 螺旋搅拌输送机
stirring-type mixer 搅拌式混合器；搅拌式拌合机
stirring-up rake 翻料肥
stirrup (U 形)夹头
stirrup rest 镫座〔天平〕
stitch 缝迹；针迹，压合；滚压
stitch brake lining 闸边皮
stitch chain 线辫
stitcher 滚压机；压合辊
stitching ①滚压②滚压器〔橡胶〕
stitching oil 滚压油

stitch mat 缝编毡
stitch tear 缝合撕裂
S titration S 滴定
stizolobic acid 羧基-γ-吡喃酮丙氨酸
stizolobinic acid 羧基-α-吡喃酮丙氨酸
stizolophine 百金菊碱
stoadite 斯脱代特〔含钨钼镍的硬合金钢〕
Stobie beater 斯托比打浆机
Stobie furnace 斯托比电炉
stochastic 随机的
stochastic Liouville's equation 随机刘维方程
stochastic method 概率法
stochastic model 随机模型
stochastic paper 随机纸；统计分析纸
stochastic process 随机过程
stochastics (=inductive statistics) 归纳统计学
stochastic sampling 随机采样；随机抽样
stochastic variable 随机变数
stock-catalyst ratio 原料-催化剂比
stock chest 纸料柜
stock color paste 原配色浆；原料色浆
stock cutter 切料机
stock distributor 物料分配器
stock dyeing 散纤维染色
stocked items 库存物件
Stockholm tar 松焦油
stock indicator 装料指示器
stockinette ①弹力织物；平针织物②针织品边角料〔作擦洗纱布材〕
stock liquor 浓鞣液
stock oil 纺织厂用除尘油
stock pan 接料盘〔开炼机〕
stock pile 储备品；藏量
stock-piling 储存
stock polymer 原料聚合物
stock room 储料间
stocksaver 捕浆器
stock soap 基皂
stock solution 储(备溶)液
stock tank 储罐；油罐
stock tank barrels 油罐桶数〔换算成标准状态下的数量〕
stock tank oil 库存石油；油罐石油〔换算成标准状态下的数量〕
stock tub 储藏盆
stock vat 染色瓮
stock white 白色浆；白漆浆
Stoddard's naphtha(=Stoddard's solvent) 干洗溶剂汽油
Stoddard's solvent(=Stoddard's naphtha) 干洗溶剂汽油

stoichiometric(=stoichiometrical)　化学计算的；化学数量的
stoichiometrical (=stoichiometric)　化学计算的；化学数量的
stoichiometric amount　化学计算量
stoichiometric calculation　化学计算(法计算)
stoichiometric chemistry　论量化学；化学计量化学
stoichiometric coefficient　化学计量系数
stoichiometric composition　化学计量成分；化学计算成分
stoichiometric compound　整比化合物*
stoichiometric concentration　化学计量浓度*
stoichiometric constant(=concentration constant)　计量常数
stoichiometric equation　化学计算方程式
stoichiometric flame　化学计量火焰
stoichiometric impurity　化学计量杂质；化学计算杂质
stoichiometric number　计量数；反应数
stoichiometric point　计量点*
stoichiometric proportion　化学计量比
stoichiometric ratio　化学计量比
stoichiometric relation　(化学)数量关系
stoichiometric relationship　化学计算关系
stoichiometric substitution　化学计量取代
stoichiometry(=stechiometry)　化学计算(法)；化学计量学
stoke hole　拨火孔
stoker　加煤机
stoker coal　加煤机用煤
stoker grade　加煤机用品级
stoker oil　加煤机灯油(矿灯油)
stoker size coal　加煤机级煤
stokes　斯托克斯〔动力黏度单位，等于1厘米^2/秒〕
Stokes fluorescence　斯托克斯荧光
Stokes' formula　斯托克斯公式
Stokes' law　斯托克斯定律
Stokes line　斯托克斯线
Stokes' number　斯托克斯数
Stokes' radius　斯托克斯半径；分子回旋半径
Stokes-Raman line　斯托克斯-拉曼线
Stokes' reagent　斯托克斯试剂
Stoke's scattering　斯托克斯散射
Stoke's shift　斯托克斯位移
Stokes' theorem　斯托克斯定律
Stoke viscosity　斯托克黏度〔动力黏度之一〕
stoking　①拨火②鼓上磨击③(油鞣革)揉软
stolzite　钨铅矿
stone bolt　地脚螺栓
stone breaker　碎石机

stone breaking　打碎石头
stone catcher　采石器
stone chippings　碎石；碎石子(片)
stone chip resistance　抗石击性
stone cistern　石槽
stone coal　石煤；块状无烟煤；硬煤
stone condensing tower　石制冷凝塔
stone crusher　碎石机
stone crushing　打碎石头
stone dust　石粉
stone fiber　岩石纤维
stone filter　石滤器
stone flax　石棉
stone gap clearance　石磨间隙
stone green　石绿
stone grit size　(石磨的)石面粒度；石料粒度
stone kiln　石(炭)窑
stone muller　石辗机
stone oil　石油
stone pine oil　石松油〔取自 Pinus picea〕
stone plate　石板
stoner　碎石机
stone roll　石辊；石辗
stone roll beater　石辊打浆机
stone roll breaker　石辊打浆机
stoneroot　石薄荷
stone rose oil　杜鹃花油
Stoner quanta　斯同纳量子
stone screen　石子筛
Stone's tension clamp　斯通张力夹
stone still　火砖炉
stoneware　①缸器；缸瓷；粗陶(器)②缸器的；缸瓷的
stoneware clay　缸(瓷)土；粗陶土
stoneware glaze　缸瓷釉
stoneware pipe　陶制管
stoneware pump　陶制泵
stoneware receiver　缸瓷收受器
stoneware shapes　缸器填料
stone wax　地蜡
stone wool fibre　石(毛)纤维；岩石纤维
stone-work　石工
stoning　碎石的过程
stoning machine　碎石机
stonobel　斯通诺贝尔
stony　石(头)的
stony fracture　石裂
stool　模底板
stop　①停(止)②塞住

stop bath　停止浴
stop block　触止块〔机工〕；车辖〔机工〕
Stop bracket gasoline　高辛烷值汽油
stop button　制动按钮
stop-clock　立止钟
stop-cock　（活）栓；转闩；管闩；活塞
stopcock buret(=stopcock burette)　活栓滴定管
stopcock burette(=stopcock buret)　活栓滴定管
stop-cock grease　活栓脂膏
stop-cock remover　拨栓器
stop codon　终止密码子
stop-flow　停流操作
stop flow analysis　截流分析
stop-flow fluorimetry　停流荧光分析法
stop-flow injection　停流进样*
stop-flow spectrophotometry　停流分光光度法
stop-off lacquer　电镀隔绝涂料；电镀屏蔽涂层；防镀漆
stoppage　停止；阻滞
stopped-belt sampling　停车取样
stopped flow analyzer　停流分析器
stopped flow cell　停流流通池
stopped flow gas chromatography　截流气相色谱（法）
stopped flow method　停止流动法*
stopped-flow technique　停流技术
stopped flow voltammetry　停流伏安法
stopped polymer　断链聚合物
stopper　①塞子②阻聚剂
stopper cock　活栓；活塞
stoppered bottle　塞好的瓶子
stop pin　止动销；固定销
stopping　嵌填；填塞
stopping ability　制动能力
stopping potential　遏止电位
stopping power　阻止本领
stop plate　盲板
stopple　塞
stop pod　限位垫板
stop point　停点
stop-start chromatograph　间断式色谱仪
stop-start operation　停车-启动操作
stop-valve　节流阀；截止阀
stop-valve lubricator　节流阀润滑器
stop watch　停表；秒表
storability　耐储性
storage　①储藏②(储藏)库
storage barge　储油驳船（暂时储放油品用）
storage battery　蓄电池(组)
storage battery plate　蓄电池板

storage bin　储料斗
storage bunker　储煤仓
storage capability　储存能力
storage capacity　储存容量；油罐容量
storage cell　蓄电池
storage compliance　储能柔量
storage effect　储能效应
storage efficiency　储存效率
storage elastic(ity) modulus　储存弹性模量(系数)
storage hopper　储料斗
storage life　储存期限；搁置寿命；适用期
storage losses　储存时的损失
storage meter　仓库计量器
storage modulus　储能模量*
storage period　停放时间；放置时间
storage pipe line　仓库管线
storage piping installation　油库管线装设
storage plant　储油场；油罐场
storage polymer　耐储存聚合物
storage pool　储存池
storage property　耐储存性
storage reservoir　储藏仓库
storage stability　储藏稳定性；耐储存性
storage stability of greases　润滑脂的储存安定性
storage stability test　存储稳定性试验
storage tank　储槽；储桶；储罐
storage terminal　码头(港口)仓库；转运基地
storage test　储存试验〔储存90天测定汽油胶质生成速度〕
storage time　储存期限；储存寿命
storage vault　储存库
storage water heater　蓄热器
storax　①苏合香脂②苏合香；安息香
storax oil　苏合香油
Storch-Lieberman test　施托希-利伯曼试验〔检验松香〕
Storch-Morawski reaction　施托希-莫拉夫斯基反应
Storch-Morawski test　施托希-莫拉夫斯基试验〔树脂的定性试验方法〕
store　①商店②堆积；大量
store crane　仓库起重机
stored waveform inverse Fourier transform　储存波形逆傅里叶变换
store holder　储料器
store-room　储藏室
stores　堆积物；备用品
storesin　苏合香树脂
store tank　储槽；储存槽
storey twist machine　层式加捻机；多层加捻机
storing　储藏

storing cistern 储槽
Stormer viscometer 斯氏黏度计
stovaine 斯妥乏因
stovarsol(=acetarsone) 醋酰胺胂；3-乙酰氨基-4-羟基苯胂酸
stove ①窑②炉
stove black ①耐高温黑沥青漆〔烘烤炉及钢烟窗等用〕②石墨〔别名〕
stove coal 火炉用无烟煤〔50～30mm 级的无烟煤〕
stoved acrylic enamel 烘干丙烯酸磁漆
stove distillate 燃炉用的轻油
stove finish 烘干的油漆
stove fuel 火炉燃料油；家用重油
stove gasoline 火炉用汽油；取暖或照明用汽油
stove oil 点炉用的油
stove pipe coupling method 火炉管接合法
stove polish ①擦炉料②烤用擦料
stove small coal 火炉小块煤；4 号无烟煤〔直径 47～25mm〕
stove tile 搪炉砖；面砖；瓷砖
stoving black varnish 黑色烤漆
stoving corrosion-resistant finish 耐腐蚀烘烤面漆
stoving enamel 烘漆；烘瓷漆
stoving finish 罩面烘漆
stoving lacquer 烘烤(烘干)喷漆
stoving of the paint 油漆烘烤
stoving oven ①烘干炉②焙烧炉
stoving resin 烘烤型树脂
stoving temperature 烘烤温度
stoving time 烘烤时间
stoving varnish 清烘漆
stoyer 饲料
strafing 扫射
straggling 歧离；分散；无序；散布
straggling effect 离散效应；歧离效应
straight ①直的②正确的
straight alcohol 纯酒精
straight alkyd resin 未改性醇酸树脂
straight arm 直臂；直桨叶
straight arm mixer 直臂混合机
straight arm stirrer 直臂搅拌机
straight asphalt 直馏沥青
straight burr 直纹
straight capillary 直毛细管
straight chain 直链
straight chain alcohol 直链醇
straight chain alkane 直链烷
straight chain compound 直链化合物
straight chain fatty acid 直链脂肪酸
straight chain higher alcohol 直链高碳醇
straight chain hydrocarbon 直链烃
straight-chain molecule 直链分子
straight chain monohydric aliphatic alcohol 直链一元脂肪醇
straight-chain paraffin 直链烷烃
straight chain paraffin sulfate 直链烷烃硫酸盐
straight chain polymer 直链高分子
straight chain reaction 直链反应
straight chrome tanning 纯铬鞣
straight-cylindrical 直筒形的
straight dipping 单纯浸渍法
straight distillation 直馏
straight drying oil 直链干性油
straight dynamite 正系炸油(爆)药
straight edge ①(直)尺②直边
straight-edge (film) applicator 直线(直刃)式涂膜器
straight-edge scraper blade 直缘刮板；直刀刮板
straight-edge spreader 直刃式涂膜器；直刃式涂布器；直线涂布器
straight ends package 直边筒子；直边卷装
straightener 矫正器
straightening 压直；矫直；调直
straightening press 手压直机
straightening vanes 导直叶片
straight-forward fractional distillation(=straight fractional distillation) 直接分馏
straight fractional distillation(=straight-forward fractional distillation) 直接分馏
straight grain 直纹
straight leaded gasoline 直馏加铅(四乙铅)汽油
straight line 直线；回归直线
straight linear polyethylene 直链聚乙烯；纯线型聚乙烯
straight-lined manufacturing 流水生产
straight-line flow 直线流动
straightline-flow continuous production 流水线连续生产
straight line furnace 直线炉
straight line motion ①直线运动②直线流动
straight-line pad sander 直线(往复)式布团砂光(打磨)机
straight-lode compressor (罗茨型)转子压缩机
straight method 直捏法
straight mineral cutting oil 纯矿物性切削油
straight mineral oil 纯矿物油；直馏矿物油
straight molecular orientation 分子单轴取向；分子单向排列
straight nickel steel 直镍钢
straight petroleum product 纯石油产品

straight phenol-formaldehyde resin 净酚-甲醛树脂
straight polymer 纯聚合物
straight product 纯产品
straight reciprocating motion 直线往复运动
straight resin 净树脂
straight riveted pipe 铆接管
straight run 直馏法；直馏馏分；直馏产品
straight-run asphalt 直馏沥青
straight-run bitumen 直馏沥青
straight-run clear gasoline 直馏纯(无四乙铅)汽油
straight-run cold rubber 无油低温丁苯胶
straight run distillate 直馏(出)物；直馏(出)油
straight run distillation 直馏；直接蒸馏
straight-run gasoline 直馏汽油
straight-run motor fuel 直馏发动机燃料
straight-run naphtha 直馏石脑油
straight-run oil 直馏油
straight-run pitch 直馏沥青
straight-run product 直馏(石油)产品
straight-run spirit 直馏汽油(英)
straight-run stock 直馏油料
straight side long pitch steel chain 直边长节式钢链
straight side tyre 直边外胎
straight silicon polymer 纯有机硅聚合物；单一有机硅聚合物
straight soap 纯皂〔由单一油脂制成〕
straight styrene-butadiene rubber 普通丁苯胶〔无油无炭黑丁苯胶〕
straight synthetics 单纯合成洗涤剂〔无助洗剂者〕
straight through extrusion 直通压出
straight through flow 直通流动
straight-through joint 平直接合
straight-through labyrinth 直通式迷宫密封
straight-through process 直通过程；非循环过程
straight way valve 通行阀
strain ①应变；胁变②张力③系〔生物分类〕④(粗)滤⑤族；种⑥菌株⑦小种
strainability ①过滤性能②过滤率
strain-age cracking 应变时效裂纹
strain aging 应变时效
strain amplitude 极限应变平均差
strain analysis 应变分析
strain birefringence 应变双折射
strain capacity 应变能量
strain centre 应变中心
strain concentration factor 应变集中因子
strained oil 滤过油
strained rubber 应变橡胶

strain energy 应变能
strainer 粗滤器
strainer cartridge 粗滤片筒
strainer chamber 粗滤室；滤清室
strainer filter 粗滤器；线网滤器；滤片滤器
strainer head 粗滤器头
strainer plug 粗滤堵塞
strainer-slabber 滤胶压片机
strain ga(u)ge 应变片；应变仪；应变测试仪
strain gauge technique 应变技术
strain-gauging 应变测量
strain hardening 应变硬化
strain impulse 应变冲量
strain indicating lacquer 应变指示漆；裂纹漆
straining basket 粗滤篮
straining box 粗滤箱
straining frame 绷皮架〔皮革〕
straining funnel 粗滤漏斗
straining installation 过滤设备
strain limiting load 应变极限载荷
strain matrix 应变矩阵
strain(o)meter 应变仪；伸长仪；张力计
strain optical coefficient 应变光学系数
strain roll 松紧辊
strain softening 应变软化
strain tensor 应变张量
strain tester 应变试验仪
strain theory 张力学说；应变学说
strake 铁箍；侧板；底板
stramonium seeds 曼陀罗子
strand 原丝；股
strand coating 线材包塑〔电线、钢丝之类线材用塑料包皮〕
strand displacement 链取代
strand granulator 铸带条切粒机；铸带条切片机
strand integrity 原丝集束性
strand preparation and cutting unit 铸带切片装置
strand system 原丝系列
strange peak 假峰
strangulation effect 钳住效应；法线效应
strap ①狭条②皮带；布带③套板④盖板；垫片
strap butts 身皮带革
strapped joint 盖板接头
strapping table 计量表
strass 假钻石；斯特拉斯假金刚石
strata (单数为 stratum) ①层②岩层③薄片
Stratcold acid treating process 斯特拉柯尔特酸处理法〔低温下用硫酸处理汽油〕

Stratco lead-sulfide process 斯特拉斯科硫化铅法〔精制汽油〕
Stratford centrifuge treating system 斯特拉福特离心处理系统〔精制石油产品〕
Stratford process 斯特拉福特法〔用脱色白土精制裂化馏分〕
stratification ①层叠形成②分层作用③层化；层次
stratifide film 层状膜
stratified fluid 层叠流体；层状流体
stratified mixture 层状混合物
stratified plastic 层压塑料
stratified rocks 成层岩
stratified sampling 分层抽样
stratographic 色谱的
stratographic analysis 色谱分析
stratography 色层法
stratometer 土壤硬度计
straton 层子
stratosphere 平流层；同温层
stratum ①层②岩层；地层
Straubel prism 斯特劳贝尔棱镜
Straubel's atomizer 静电喷雾器
strawberriff 草莓酸；2-甲基-2-戊烯酸
strawberry aldehyde 草莓醛；十六醛；β-甲基-β-苯基缩水甘油酸乙酯
strawberry essence 草莓香精
strawberry flavour 草莓香精
strawberry-guava 草莓番石榴
strawberry oil 草莓油
strawberry tree 莓果树
straw board 草纸板
straw cutter 切草机
straw distillate 草黄油馏分
straw fiber 草纤维
straw mushroom 草菇
straw oil 草黄油
straw pulp 草纸浆
straw wrapping 草料包装纸
stray crystal 杂(散)晶
stray current corrosion 漏泄电流腐蚀
stray electrical field 杂散电场
stray light 杂散光*；漫射光
stray radiation 杂散光
stray signal 寄生信号；杂散信号
streak ①条纹；色线②(一)层③加条纹④割沟；侧沟
streak application 条状点样
streak camera 条纹相机
streak chromatography 条痕色谱(法)

streak culture 划线培养
streaking 开(割)沟
streaking technique 喷显技术
streak inoculation 划线接种
streak line 条纹线〔塑料制品缺陷〕
streak method 划线法；涂抹法；痕色(试验)法
streak on porcelain 素烧瓷上的条痕
streak plate 条痕板
streak reagent 喷显剂
streak test 条痕试验；色线试验
streaky coat 条痕状涂膜〔漆病〕〔喷枪方向倾斜，喷成厚薄交替的带状膜〕
streaky dyeing 染色条痕花
streaky structure 条状组织
stream ①(水)流；(气)流②流
stream breaker 碎流装置；碎流板
stream day 连续开工日〔石油厂车间〕
streamer ①雪崩电子流②流光；射光
stream factor (全负荷)生产定额
stream gravity 流分比重；流重
stream handling 流动式处理；连续进料；连续输送
stream hours 连续工作时数
streaming ①流动②洗矿
streaming birefringence 流动双折射
streaming dichroism 流动二色性
streaming electrode 泉汞电极
streaming fluid 流动流体
streaming light scattering 流动光散射
streaming mercury cathode 流汞阴极
streaming mercury electrode 流汞电极*
streaming potential 泳动电势*
streaming steam method 流动蒸汽法
streamline 流线；流线型
streamline boundary layer 流线边界层
streamlined ①流线型的；层流的②顺流安装的；连成一体的③现代化的
streamlined body 流线型物体
streamlined filter 流线式过滤器〔液体沿滤叶流动的叶片滤过器〕
streamlined flow 直线流动
streamlined motion 流线运动〔液体的流线式或匀整的运动〕
streamlined pressure tank 流线型压力油罐
streamlined process 直线过程
streamlined return-bend header 流线型回弯头；流线型回管
streamline filter 流线式滤器
streamline filtration 流线式过滤

streamline flow 流线形流动；层流
streamline motion 流线运动；直线流动
streamline production technique 流水作业的生产法
stream of electrons 电子流
stream on (=on-stream) 在流程中
stream pollution 河流污染
stream splitter 气(或液)流分流器
stream switching 物流转换
stream time 连续开工时间；工作周期
stream-to-stream time 停工检修期
stream variables 川流参变量
Strecker reaction 斯特雷克尔氨基酸反应
Strecker synthesis 斯特雷克尔合成〔一种从 α-羟基腈制备氨基酸的合成法〕
Stredford process 斯特雷福德法；蒽醌法〔净化水煤气和焦炉气〕
street dirt 街尘〔测去污力用〕
street elbow 长臂肘管
strengite 粉红磷铁矿
strength ①强度②力
strength analysis 强度分析
strengthened coating 强化涂层
strengthener 增强材料；增强剂
strengthening action 补强作用
strengthening agent 补强剂
strength imparting material 加强物料〔橡胶〕
strength property 着色性能
strength retention 保留强度
strength tester 强度试验仪
strength-to-density ratio 强度-密度比
strength-to-weight ratio 强度-重量比
streptase 链激霉
streptogenin 链霉配基；链球菌促长肽
streptomycin 链霉素
streptomycin hydrochloride 盐酸链霉素
streptomycin sulfate 硫酸链霉素
stress 应力；胁强；压力
stress analysis 应力分析
stress-at-break 断裂应力
stress at definite elongation 定伸应力
stress birefringence 应力双折射
stress build up 应力聚集
stress coating 应力龟裂涂料；应力裂纹涂层
stress compensation 应力补偿
stress concentration 应力集中
stress concentration factor 应力集中系数
stress corrosion 应力腐蚀
stress corrosion cracking 应力腐蚀裂开

stress crack 应力龟裂；应力开裂
stress-crack failure 应力龟裂破损；应力开裂破损
stress cracking 应力开裂；应力断裂
stress crazing 应力银纹
stress crystallinity ①应力结晶性②应力结晶度
stress cycle 应力周期
stress decay 应力衰减；应力松弛
stress-deformation curve 应力形变曲线
stress-deviation 应力偏差；应力差
stress-difference 应力差
stress distribution 应力分布
stressed polymer 应力聚合物
stressed shell 预应力外壳
stress gradient 应力梯度
stress intensity 应力强度
stress intensity factor 应力强度因子*
stress optical coefficient 应力光学系数
stress optical constant 应力光学常数
stress relaxation 应力松弛
stress relaxation processability tester 应力弛豫加工性能测试仪
stress relief test 应力消除试验〔热收缩试验〕
stress rupture 应力破裂
stress rupture strength 应力破裂强度
stress-rupture testing of tube 管子的应力破裂试验〔抗内压试验〕
stress singularity 应力奇点
stress softening 应力软化
stress strain 应力应变
stress strain curve 应力-应变曲线
stress-strain loop 应力-应变滞后圈
stress tensor 应力张量
stress whitening 应力致白
stress wrinkle 应力折皱
stress yield 屈服应力
stretchability 拉伸性；抽伸性
stretch-blow mo(u)lding 拉坯吹塑成型
stretch-fit 弹性尺寸的；尺寸广适性的
stretching ①伸缩②拉伸
stretching board 伸张板
stretching force 张(拉)力
stretching frame 伸张架
stretching frequency 拉伸频率
stretching machine 拉幅机
stretching motion 拉伸运动
stretching strain 拉伸应变
stretching vibration 拉伸振动
stretch orientation 拉伸定向

stretch ratio 拉伸比
stretch roll 松紧辊
stretch spinning 拉伸纺丝
stretch tensor 拉伸张量
stretch thermoforming 拉伸热成型
stretch-twister 拉伸加捻机
stretch-type 弹力型
stretch vibrometer 拉伸示振仪；拉伸振动仪
stretchy 有弹性的；能伸长的
stretch yarns 弹力丝
stria 条纹
striated 成纹的
striated structure 条纹结构〔晶体〕
striate gypsum 纤维石膏
striation 辉纹
strike 预镀
striker brush(=spout brush) 触发式柄芯供漆长柄刷；吊锤式漆刷
strike-through 透底
strike-through time 渗透时间〔非织造布性能指标〕
striking 共沉淀分出(作用)；光沉淀捕集
striking current 冲击电流
striking machine 扩展机
striking-out 展平
striking-out machine 平展机；伸展机
string ①线；绳②一行；一列
string discharge (filtration) 绳卸饼法
stringer 吊绳〔沿线路吊管子用〕
stringer bead 新管子上最初的焊缝
string for varnish 清漆(热炼)拉丝(试验)
stringiness 拉丝性
stringing 沿管线敷管
string proof test 纤维试验
string stress 收缩应力
string test 拉丝试验；看丝
stringy 黏稠的；拉丝的
stringy cotton 索丝
strip (滤纸)条；条纹；色条
strip calender 条胶压延机〔橡胶〕
strip chart 长条记录纸
strip chart recorder 长条纸记录器
strip chromatography 带状色谱法
strip coating ①活盖②可剥涂层
strip column 反萃取柱
stripe 条纹；色条
stripe contact laser 带状接触激光器
strip electrode 带形电极
strip electrophoresis 纸条电泳

stripe test 色条试验
strip-extraction 反萃取
strip extraction column 反萃取柱
strip heater ①汽提加热器②带状(电热丝)加热器
strip isotherm 反萃(取)等温线
strip leveller 带钢平整器
strip liquor 反萃液
strip material 条状材料
strip mill 出条辊〔电缆〕
strip of soap 皂片；皂屑
strippable adhesive 可剥性黏合剂
strippable coating ①活盖②可剥涂层
strippable composition ①可剥组分②可剥涂料
strippable film 可剥膜；可剥漆膜
strippable paint 可剥漆
strip packing gland 对开填料压盖
strippant ①洗涤剂②解吸剂③剥色剂
stripped ①汽提过的；已去轻油部分的②已洗涤过的；已解吸了的
stripped atom 剥电子层原子
stripped charcoal 已解吸的活性炭
stripped gas 已除去汽油烃的干气
stripped oil 汽提油；无轻油的石油
stripped soap 皂片；皂屑
stripped solution 贫化液；反萃过的溶液
stripped solvent 反萃过的溶剂
stripper ①反萃取器②汽提塔③剥离器④贫化段(区)；剥淡段(区)⑤剥离剂
stripper apparatus 脱模装置；揭模
stripper bolt 脱模螺栓
stripper efficiency 汽提塔效率；解吸塔效率
stripper operator 汽提设备操作工；气体汽油车间操作工
stripper plant 汽提车间
stripper plate ①分馏柱塔板②挤压板
stripper-plate mold 脱(膜)板塑模；丁字塑模
stripper solution ①剥膜液②汽提溶液
stripper trap 汽提塔阱
stripper zone 汽提段
stripping 反萃取*；溶出；溶脱
stripping action 脱漆作用
stripping agent ①反萃(取)剂②剥色剂
stripping analysis 提溶(极谱)法
stripping coefficient 反萃系数
stripping column ①脱除柱②反萃取柱
stripping column stills 汽提塔蒸馏器
stripping cracking 汽提裂化
stripping device 脱模工具
stripping drum(=stripping column) 汽提塔

stripping effect　（基料）抽出效应
stripping electrography　溶出电谱法
stripping emulsion method　剥离乳胶法
stripping factor　解吸因数
stripping film　可剥漆膜
stripping-film technique　剥离膜技术
stripping gas　解吸气
stripping knife　①剥漆刀②剥皮刀
stripping lacquer　可剥涂料；临时涂层
stripping liquid　液态吸收剂；解吸液
stripping method　脱漆方法
stripping model　夺取模型
stripping molding　脱模模塑
stripping of catalyst with gas　用气体汽提(粉末状)催化剂
stripping of gas　从气体中提取汽油
stripping of oil　油的气提
stripping of soap　肥皂刨平
stripping oil　吸收油；洗涤油
stripping operation　脱漆操作；剥漆操作
stripping paper　条纹纸
stripping plant steam generation　汽提车间水蒸汽发生
stripping polarography　溶出极谱法
stripping process　汽提过程
stripping reaction　(核的)剥裂反应
strippings　①轻油部分②解吸产物
stripping section　剥淡段；贫化区；反萃段
stripping solution　反提(取)剂
stripping stage　①汽提段②解吸段
stripping steam　汽提用水蒸气
stripping still　汽提蒸馏器
stripping strength　剥离强度
stripping technique　分离技术
stripping test　剥离试验
stripping tower　汽提塔
stripping vessel　①汽提釜②解吸釜
stripping voltammetry　溶出伏安法
strip steel　带钢
strip thermocouple　条状热电偶
strip transfer method　纸条转移法
strip winding　绕组
strip winding vessel　绕带式容器
stripy　条纹状的；有条纹的
strobe scanning electron microscope　频闪扫描电镜
strobe tachometer　频闪转速计；闪光测速器
strobophonometer　爆震测声计〔测量汽油在发动机中爆震时声音强度的仪器〕
stroboscope　①频闪观察仪；闪光测速仪；频闪仪②转

速很高的机器
stroboscope photography　频闪观测器照相术
stroboscopic analysis　频闪观测分析
stroboscopic lamp　频闪放电管
stroboscopic pulse radiolysis(SPR)　闪频脉冲辐解
strobotach　频闪转速计
stroke pump　冲击泵
strok method　划线法〔测定施胶度〕
stromeyerite　硫铜银矿
strong acetate　强力醋酯纤维
strong acid　强酸
strong-acid cation exchange fibre　强酸性阳离子交换纤维
strong-acid cation exchange resin　强酸性阳离子交换树脂
strong-acid number　强酸中和值
strong acid type ion exchanger　强酸型离子交换剂
strong aqua　浓氨水
strongback　定位板
strong base　强碱
strong-base anion exchanger　强碱性阴离子交换剂
strong-base anion exchange resin　强碱性阴离子交换树脂
strong-base number　强碱中和值
strong base type ion exchanger　强碱型离子交换剂
strong brine　浓盐水
strong caustic　浓氢氧化钠溶液
strong change　碱化脱盐
strong coal　强煤
strong collision assumption　强碰撞假设
strong coupling system　强耦合系统
strong electrolyte　强电解质
strong fatty note　强烈油脂气息
strong fibre　强力纤维
strong floral odor　强烈花香
strong interfacial bond　强界面黏(结)合
strong interference line　强干扰线
strong lime　新灰；强灰
strong linkage(=primary linkage)　一级键合；强键
strong liquor　强碱水
strongly acidic cation exchanger　强酸性阳离子交换剂
strongly acidic resin　强酸性(阳离子交换)树脂
strongly basic anion exchange resin　强碱性阴离子交换树脂
strongly basic anion ion exchange fiber　强碱性阴离子交换纤维
strong lye　强碱水
strongly phosphoric acid decomposition method　浓磷酸

分解法
strongly polar hydrocarbon 强极性烃
strong magnetic nuclei 磁性强核
strong metal-support interaction (SMSI) 金属载体强相互作用*
strong oxide-oxide interaction(SOOI) 氧化物间强相互作用*
strong plastic deformation 强烈塑性变形法
strong room 保险库
strong salt-cake 强盐饼
strong segregation limit 强分凝极限
strong segregation theory 强分凝理论
strong (sodium) sulfate 强硫酸钠
strong solution 浓溶液
strong solvent 强溶剂；活性溶剂
strong stand oil 稠熟油
strong tack 强黏性
strong vat 强碱瓮
strong viscose rayon 强力黏胶人造丝
strong yellow 老黄(色)
strontia 氧化锶 SrO
strontia hydrate 氢氧化锶 $Sr(OH)_2$
strontianite 菱锶矿
strontia water 氢氧化锶
strontium 锶〔38号元素，化学符号Sr〕
strontium acetate 乙酸锶 $Sr(C_2H_3O_2)_2$
strontium arsenide 砷化锶 Sr_3As_2
strontium arsenite 亚砷酸锶
strontium barium niobate 铌酸锶钡
strontium (base) grease 锶基脂
strontium bicarbonate 碳酸氢锶 $Sr(HCO_3)_2$
strontium bichromate 重铬酸锶 $SrCr_2O_7$
strontium binoxalate(=strontium bioxalate) 草酸氢锶
strontium bioxalate(=strontium binoxalate) 草酸氢锶 $Sr(HC_2O_4)_2$
strontium biphosphate 磷酸二氢锶 $Sr(H_2PO_4)_2$
strontium bisulfate 硫酸氢锶 $Sr(HSO_4)_2$
strontium bisulfite 亚硫酸氢锶 $Sr(HSO_3)_2$
strontium bitartrate 酒石酸氢锶 $Sr(HC_4H_4O_6)_2$
strontium borate 硼酸锶
strontium bromate 溴酸锶 $Sr(BrO_3)_2$
strontium bromide 溴化锶 $SrBr_2$
strontium calcium chromate 铬酸锶钙
strontium calcium yellow 锶钙黄
strontium carbide 二碳化锶 SrC_2
strontium carbonate 碳酸锶 $SrCO_3$
strontium chlorate 氯酸锶 $Sr(ClO_3)_2$
strontium chloride 氯化锶 $SrCl_2$

strontium chromate 铬酸锶 $SrCrO_4$
strontium citrate 柠檬酸锶 $Sr_3(C_6H_5O_7)_2$
strontium cyanide 氰化锶 $Sr(CN)_2$
strontium dating method 锶定年代法
strontium dichromate 重铬酸锶 $SrCrO_7$
strontium dithionate 连二硫酸锶 SrS_2O_6
strontium fluoride 氟化锶 SrF_2
strontium fluosilicate 氟硅酸锶 $Sr[SiF_6]$
strontium-90 foil 锶-90箔
strontium formate 甲酸锶 $Sr(HCO_2)_2$
strontium glycerophosphate 甘油磷酸锶 $SrC_3H_7PO_6 \cdot xH_2O$
strontium hydrogen phosphate 磷酸氢锶 $SrHPO_4$
strontium hydrosulfide 硫氢化锶 $Sr(HS)_2$
strontium hydroxide 氢氧化锶 $Sr(OH)_2$
strontium hyposulfate 连二硫酸锶 SrS_2O_6
strontium hyposulfite 连二亚硫酸锶 SrS_2O_4
strontium iodate 碘酸锶 $Sr(IO_3)_2$
strontium iodide 碘化锶 SrI_2
strontium lactate 乳酸锶 $Sr(C_3H_5O_3)_2$
strontium lithol 锶立索
strontium malate 苹果酸锶 $SrC_4H_4O_5$
strontium maleate 马来酸锶 $SrC_4H_2O_4$
strontium manganate 锰酸锶 $SrMnO_4$
strontium metaborate 偏硼酸锶 $Sr(BO_2)_2$
strontium molybdate 钼酸锶；钼(锶)白
strontium monophosphate 磷酸氢锶 $SrHPO_4$
strontium nitrate 硝酸锶 $Sr(NO_3)_2$
strontium nitride 二氮化三锶 Sr_3N_2
strontium nitrite 亚硝酸锶 $Sr(NO_2)_2$
strontium orthophosphate (正)磷酸锶 $Sr_3(PO_4)_2$
strontium oxalate 草酸锶 SrC_2O_4
strontium oxide 氧化锶 SrO
strontium permanganate 高锰酸锶 $Sr(MnO_4)_2$
strontium peroxide 过氧化锶 $Sr[O_2]$
strontium phosphate 磷酸锶
strontium phosphide 二磷化三锶 Sr_3P_2
strontium platinocyanide 氰亚铂酸锶 $Sr[Pt(CN)_4]$
strontium rhodanate 硫氰酸锶 $Sr(CNS)_2$
strontium saccharate 糖二酸锶 $2SrO \cdot C_{12}H_{22}O_{11}$
strontium selenate 硒酸锶 $SrSeO_4$
strontium silicate 硅酸锶 $SrSiO_3$; Sr_2SiO_4
strontium (soap) grease 锶皂润滑脂
strontium stearate 硬脂酸锶
strontium sulfate 硫酸锶 $SrSO_4$
strontium sulfide 硫化锶 SrS
strontium sulfite 亚硫酸锶 $SrSO_3$
strontium superoxide 过氧化锶 $Sr[O_2]$

strontium tartrate 酒石酸锶 $SrC_4H_4O_6$
strontium tetraborate 四硼酸锶 SrB_4O_7
strontium thiocyanate 硫氰酸锶 $Sr(CNS)_2$
strontium thiosulfate 硫代硫酸锶 $SrSO_3S;SrS_2O_3$
strontium tungstate(=strontium wolframate) 钨酸锶 $SrWO_4$
strontium unit 锶单位
strontium white 锶白
strontium wolframate(=strontium tungstate) 钨酸锶
strontium yellow 锶黄；铬酸锶
strophanthin K 毒毛花苷 K
strop leather 磨刀皮带
Strouhal number 斯特罗哈数
STR process 潜艇热中子堆燃料处理过程〔沸腾氢氟酸溶解锆合金壳高浓铀燃料法〕
struck atom 被击原子
struck green 优质(铅)铬绿
structoset prepolymer 定结构预聚物
structural 结构的；构造的
structural adhesive 结构型胶黏剂
structural agent 结构化剂；立(体结)构形成剂
structural analysis 结构分析
structural asymmetry 结构不对称性
structural behaviour 结构性能
structural bond 结构胶接件
structural chemistry 结构化学
structural composite materials 结构复合材料
structural element 结构元素
structural flaw 结构缺陷
structural foam molding 结构泡沫成型
structural genomics 结构基因组学
structural group 结构族
structural group analysis 结构族分析
structural heterogeneity 结构不均匀性；结构非均性
structural heterozygote 结构杂合体；结构杂合子
structural homogeneity 结构均一性；结构均匀性
structural imperfection 结构不完整性
structural inhomogeneity 结构不均匀性
structural integrity 结构完整性
structural isomer 结构同分异构体
structural isomeric 结构同分异构的
structural isomeride 结构同分异构体
structural isomerism 结构同分异构(现象)
structural matrix 结构基体；结构基质
structural member 结构部件
structural model 结构模型
structural motif 结构基元
structural parachor 结构等张比容
structural parameter 结构参数
structural plywood 结构胶合板
structural purity 结构纯度
structural rearrangement 结构重排
structural representation 结构示意图
structural rheology 结构流变学
structural stability 结构稳定性
structural steel 结构钢
structural steel coating 钢构筑物涂料
structural stone 建筑石料
structural texture analysis 结构纹理分析
structural turbulence 结构湍流
structural viscosity 结构(化)黏度
structural viscosity index 结构黏度指数；结构黏度系数
structure analysis 结构分析；组织分析
structure-based nomenclature 以结构为基础的命名法
structure collapsing treatment 结构致密处理
structure determination 结构测定
structured paint 触变漆
structured pigment 结构性颜料；造型性颜料
structure factor 结构因子
structure formula 结构式
structure index 结构指数
structure isomer(=structure isomeride) 结构同分异构体
structure isomeride(=structure isomer) 结构同分异构体
structure isomerism 结构同分异构(现象)
structure rating (颜)结构分级
structure semivariant 结构半不变量
structure-symbol 结构符号
structure type 结构类型
structure unit 结构单元
structurization 结构化
struvite 鸟粪石
strychnine 马钱子碱；士的宁
strychnine acetate 乙酸马钱子碱
strychnine arsenate 砷酸马钱子碱
strychnine citrate 柠檬酸马钱子碱
strychnine glycerophosphate 甘油磷酸马钱子碱
strychnine hydrochloride 盐酸马钱子碱
strychnine iodate 碘酸马钱子碱
strychnine lactate 乳酸马钱子碱
strychnine nitrate 硝酸士的宁；硝酸番木鳖碱
strychnine phenol-sulfonate 苯酚磺酸马钱子碱
strychnine phosphate 磷酸马钱子碱
strychnine salicylate 水杨酸马钱子碱
strychnine sulfate 硫酸马钱子碱
strychninic acid 马钱子碱酸
strychninium 马钱子镕

strychninolic acid 马钱子碱醇酸
strychninonic acid 马钱子碱酮酸
stucco (塑像)灰泥；粉饰灰泥
stucco coating (水)泥(石)灰砂浆拉毛造型涂料
stud 双头螺栓
stud bolt 双头螺栓
Student's t distribution 学生氏 t 分布
stud welding 电栓焊
stuff ①(材)料②本质；要素③填充
stuff chest ①纸料柜②储料柜
stuff coloring(=stuff dyeing) 布料着色；呢料着色
stuff dyeing(=stuff coloring) 布料着色；呢料着色
stuffed ①已塞满了的②喂饱了的
Stuffer law 斯塔弗二砜皂化定律
stuff goods 毛织物；毛料；毛织品；呢绒
stuffing ①加脂(法)②填充
stuffing box 填料函；填料箱；填料盒
stuffing box cover 填料函上盖；填料箱压盖
stuffing box free design 无填料函式(密封)装置
stuffing box gland 填料函压盖；填料函格兰
stuffing drum 加脂转鼓〔皮革〕
stuffing mill 加脂转鼓
stuffing wheel 加脂转鼓
stuff loading 布的填料
stuff pump 纸料泵
stuff sizing 布的上胶
stump ①伐根②残干③树楂
stumpage 市售木料
stump turpentine 残干松节油
stumpwood ①根株材②明子
stupp 粗汞华〔从矿石中蒸出的粗制汞升华物〕
stupping 脱模；脱芯
sturgeon 鲟鱼
sturgeon liver oil 鲟鱼肝油
sturgeon oil 鲟鱼油
Sturmer viscometer 斯特梅尔型黏度计
sturnutatory 引起喷嚏的
Stutzer's reagent 斯塔策尔试剂〔氢氧化铜的甘油溶液〕
stylolitic structure 缝合结构
stylomycin(=puromycin) 嘌呤霉素
stylopine 刺罂粟碱
stylus 描形针；触针
styphnate 收敛酸盐(酯)；2,4,6-三硝基苯间二酚的盐(酯) $(NO_2)_3C_6H(OM)$；$(NO_2)_3C_6H(OR)_2$
styphnic acid 收敛酸；2,4,6-三硝基苯间二酚
styptic 止血药
stypticine 止血素；盐酸可他宁 $C_{12}H_{15}O_4NHCl$
styptol 止血醇

styracin 苏合香英；苯丙烯酸苯丙烯酯
$C_6H_5CH=CHCOOCH_2CH=CHC_6H_5$
styracine 肉桂酸肉桂酯 $C_{18}H_{16}O_2$
styracitol 苏合香醇
styracol 苏合香脑；肉桂酸愈创木酚酯
$C_6H_5CH=CHCOOC_6H_4OCH_3$
styralyl(=styroyl; styrolyl) 苯乙基
styralyl acetate 乙酸苏合香酯；乙酸甲基苯基原酯
styralyl alcohol 苏合香醇
styralyl butyrate 丁酸苏合香酯
styralyl propionate 丙酸苏合香酯
styrax (=storax) 苏合香；安息香
styrax benzoin 安息香
styrax oil 苏合香油；安息香油
styrax resin 苏合香树脂；安息香树脂
styrax resinoid 苏合香树脂；安息香树脂
styrenated alkyd 苯乙烯改性醇酸树脂
styrenated blown soybean oil 苯乙烯化吹制豆油
styrenated dehydrated castor oil 苯乙烯化脱水蓖麻油酸
styrenated diphenylamine 苯乙烯化二苯胺〔防老剂〕
styrenated linseed oil 苯乙烯改性亚麻(籽)油；苯乙烯化亚麻油
styrenated oil 苯乙烯化油
styrenated phenol 苯乙烯酚
styrenated polyester 苯乙烯化聚酯
styrenated soybean oil 苯乙烯化豆油
styrenated tall oil 苯乙烯化松浆油
styrenation 苯乙烯化
styrene ①(=phenylethylene)苯乙烯；苏合香烯 $C_6H_5CH=CH_2$②(=phenylethylene)苯亚乙基 $PhCHCH_2-$；$PhCH_2CH=$
styrene-acrylonitrile copolymer 苯乙烯-丙烯腈共聚物
styrene-acrylonitrile plastic(SAN) 苯乙烯-丙烯腈塑料
styrene-acrylonitrile resin 苯乙烯-丙烯腈树脂
styrene alkyd resin 苯乙烯改性醇酸树脂
styrene-alpha-methylstyrene plastic(SMS) 苯乙烯-α-甲基苯乙烯塑料
styrene bromohydrin β-溴-α-苯乙醇 $C_6H_5CHOHCH_2Br$
styrene-butadiene block copolymer(SBS) 苯乙烯-丁二烯嵌段共聚物
styrene-butadiene copolymer 苯乙烯-丁二烯共聚物
styrene-butadiene latex 苯乙烯-丁二烯胶乳
styrene-butadiene plastic(SB) 苯乙烯-丁二烯塑料
styrene-butadiene (resin) coating 苯乙烯-丁二烯树脂涂料；丁苯树脂涂料
styrene-butadiene (resin) emulsion 苯乙烯-丁二烯树脂乳化液；丁苯(树脂)乳液

styrene butadiene rubber(SBR) 丁苯橡胶
styrene butadiene rubber latex 丁苯胶乳
styrene-butadiene-styrene 苯乙烯-丁二烯-苯乙烯
styrene butadiene styrene block copolymer(SBS) 苯乙烯-丁二烯-苯乙烯嵌段共聚物
styrene copolymerization 苯乙烯共聚反应
styrene-divinylbenzene copolymer 苯乙烯-乙烯苯共聚物
styrene isoprene butadiene rubber(SIBR) 苯乙烯-异戊二烯-丁二烯橡胶
styrene isoprene styrene block copolymer(SIS) 苯乙烯-丁二烯-苯乙烯嵌段共聚物
styrene maleic anhydride copolymer 苯乙烯-顺丁烯二酸酐共聚物
styrene-maleic anhydride plastic 苯乙烯-马来酐塑料
styrene-methylmethacrylate copolymer 苯乙烯-甲基丙烯酸甲酯共聚物
styrene-α-methylstyrene copolymer 苯乙烯-α-甲基苯乙烯共聚物
styrene monomer 苯乙烯单体
styrene oligomer 苯乙烯低聚物
styrene oxide 氧化苯乙烯 $C_6H_5 \cdot C_2H_3O$
styrene plastic 苯乙烯塑料
styrene resin 苯乙烯类树脂
styrene-rubber plastic(SRP) 聚苯乙烯橡胶改性塑料
styrene solubility 苯乙烯溶解度
styrene suppressant 苯乙烯抑制剂
styrene type ion exchange resin (聚)苯乙烯型离子交换树脂
styrol(=styrene) 苯乙烯；苏合香烯
styrolene ①肉桂塑料〔一种聚苯乙烯塑料〕②(=phenylethylene)苯亚乙基
styrolene acetate 乙酸苏合香酯
styrolene alcohol(=styryl alcohol) 肉桂醇
styrolyl(=styroyl; styralyl) 苯乙基
styrolyl alcohol(=methyl benzyl alcohol) 苏合香醇；甲基苄醇
styrolyl diacetate 双乙酸苯乙二酯
styron ①(=cinnamic alcohol)肉桂醇②肉桂塑料〔一种聚苯乙烯塑料〕
styrone 肉桂醇
styroyl(=styrolyl; styralyl) 苯乙基
styroyl acetate 乙酸苏合香酯；乙酸甲基苯基原酯
styroyl alcohol 苏合香醇；甲基苯基原醇
styryl 苯乙烯基 $C_6H_5CH=CH-$
styryl alcohol 肉桂醇 $C_6H_5CH=CHCH_2OH$
styrylamine 苯乙烯胺 $C_6H_5CH=CHNH_2$
styryl carbinol 苯乙烯基甲醇；肉桂醇

styryl ketone 二苯乙烯(甲)酮
styryl methyl ketone 苯乙烯基甲基(甲)酮
styryl yeast 苯乙烯酵母
stysadin 束梗孢菌素
suaveolent 芳香的
sub- 〔拉丁字头〕①下；低；亚；次②亚〔生物分类〕③副；辅助
subacetate 碱式乙酸盐
subacid 微酸(性)的
subacidity 微酸性
subambient 低于室温的；低温的
subambient chromatography 低温色谱(法)
subambient temperature 低温
subassembly ①组件；部件；预装件②局部装配
subatmospheric pressure 低于大气压(的)
subatomic 亚原子的
subatomic debris 亚原子碎片
subatomic decomposition 亚原子分解
subatomic particle 亚原子微粒
subatomic phenomenon 亚原子现象
subatomic reaction 亚原子反应
subatomics 亚原子学
sub-bituminous coal 次烟煤
subcarbonate 碱式碳酸盐
subcavity gang mould 死模子
subcavity mould 死模子
subchain 子链
sub-chain motion 链段运动
subchloride 低氯化物；氯化低价物〔中文中不沿用这个旧用英文名称，而视其化学式命名〕
subchloride of mercury 一氯化汞；氯化亚汞 $[Hg_2]Cl_2$
subcolloidal micelle 亚胶态胶束
subcolloidal structure 亚胶态结构
subcontractor 承包工厂
subcooled 低温冷却的
subcritical fluid 次临界流体
subcrustal 地壳下的
subcrystalline structure 亚晶态结构
subdivision ①细分；再分②分部；小类
subdivision of air flow 空气气流的分支
subdivision of current 电流分支
subdivision of gross samples 总试料的细分；大样细分
subeno 苏北依〔关节炎药〕；苯·丁二酸一苄酯钙盐 $C_6H_5COOCaOOCCH_2CH_2COOCH_2C_6H_5$
suberamide 辛二酰胺
suberane 环庚烷；软木烷
suberate ①辛二酸②辛二酸盐(酯或根)
suberene(=cycloheptene) 环庚烯

suberenel 栓花椒醇
suberic acid 辛二酸 $(CH_2)_6(CO_2H)_2$
suberic aldehyde 辛二醛 $(CH_2)_6(CHO)_2$
suberol(=cycloheptanol) 环庚醇；软木醇
$$CH_2(CH_2)_5CHOH$$
suberone(=cycloheptanone) 环庚酮；软木酮
$$CH_2(CH_2)_5CO$$
suberonitrile 辛二腈
suberosin 软木花椒素
suberoyl 辛二酰 —$CO(CH_2)_6CO$—
suberyl 环庚基
suberyl alcohol 环庚醇
suberylarginine 辛二酰精氨酸
subfibril 亚原纤维
subfluoride 低氟化物；氟化低价物
subfraction 细(分)馏分
subfractionation ①细分馏②细分级③精馏
sub-frame 机座；底座
subgallate 碱式棓酸盐
sub grain 亚晶粒
subgroup 副族*
subgroup A 主族；A族
subgroup B 副族；B族
subgroup element 副族元素
subhalide 低卤化物；卤化低价物
subhedral minerals 岩状矿物
subindanthrene dye 亚士林染料
subiodide 低碘化物；碘化低价物
sub-laminate 子层合板
sublamine 升胺；乙二胺合硫酸汞
sublate salts 浮选促集盐
sublation 离子浮选促集过程
sublattice 亚点阵；子晶格；亚晶格
sublayer ①下层②次层；亚表层
sub-level 亚水平；次水平
sublimable 可升华的
sublimate ①升华物②升汞
sublimation 升华*
sublimation apparatus 升华器
sublimation drying 升华干燥
sublimation heat 升华热
sublimation ink 升华性油墨
sublimation of carbon 碳的升华
sublimation procedure 升华法
sublimation pump 升华泵
sublimatogram 升华图

sublimatography (混合物)升华谱分离法
sublimator 升华器
sublimatory ①升华器②升华用的
sublimed 升华了的
sublimed blue lead 升华法铅蓝；升华法碱式硫酸铅蓝
sublimed iodine 升华碘
sublimed lead 升华铅；蓝铅；碱式硫酸铅
sublimed sulfur 升华硫；硫华
sublimed white lead 升华白铅〔碱式硫酸铅和氧化锌的混合物〕
sublimer ①升华材料②升华器
subliming 升华
subliming pot 升华皿
submarine ①海底的②潜水艇
submarine illumination 水下照度
submarine line(=submarine pipe line) 海底管线
submarine-line terminal 海底管线终点
submarine mine 水雷
submarine mountain 水下山
submarine photometer 水下光度计
submarine pipe line(=submarine line) 海底管线
submarine ridge 水下脊
submerged 沉没的；浸入的；浸没的；潜没的
submerged arc cladding 埋弧包覆；潜弧镀层
submerged arc cracker 沉没电弧裂化器
submerged arc welding 埋弧焊
submerged cathode cell 沉阴极电池
submerged coil condenser 沉浸式蛇管冷凝器
submerged combustion 浸没燃烧
submerged condenser 潜管冷凝器
submerged culture ①深层培养②浸没培养
submerged electrode 浸埋电极；插入式电极
submerged lubrication 浸入式润滑
submerged orifice 浸液隔膜
submerged pump 液下泵
submerged storage tank 沉没式油罐；水下油罐
submerged structure 水下构筑物
submerged tank 潜没油罐；水下油罐；地下油罐
submerged tube condenser 潜管冷凝器
submerged tube evaporator 潜管蒸发器
submerged tube type evaporator 潜管蒸发器
submergence 沉入；浸入；浸没；潜没
submersed 沉没的；浸入的；浸没的；潜没的
submersible oil boom 沉没式油栅
submersible tank battery barge 潜罐油船；水下油罐组货船
submersion 沉入；浸入；浸没；潜没
submersion section 浸洗段

submetallic 半金属的
submicellar solution 逊胶束溶液〔浓度低于临界胶束浓度的表面活性剂溶液〕
submicelle 逊胶束
submicro analysis 超微量分析
submicro capillary 亚微毛细管
submicrocrack 亚微观裂纹
submicrogram 亚微克(量)；低于微克(量)的
submicrogram quantity 亚微克量
submicromethod(=ultramicromethod) 超微量法
submicron 亚微细粒
sub-micron dispenser 亚微(粒子)分散机
submicron particle 亚微(米)粒子；亚微细粒〔普通显微镜下看不见的粒子〕
sub micro probe 亚微探针
submicrosample 超微量试样
submicroscopic(al) structure 亚微观结构
submicroscopic micelle 亚微观胶束
sub-millimeter wave 亚毫米波
submolecule(=segment) 链段；亚分子
sub-multiple 几分之一
subnanogram 低于毫微克(数量级)；低于纳克(数量级)
subnanomole 亚纳摩尔
subnanosecond 亚毫微秒；亚纳秒
subneat soap 次液晶皂
subnitrate 碱式硝酸盐
subnormal 正常以下的
subnormal density 亚常态密度；准常态密度〔低于正常的密度〕
subnormal structure black 低结构炭黑
subnormal temperature 亚(于)常温；低于正常温度
subordinate quantum number 次量子数
suboxide 低氧化物*
subpermanent set 非永久变形
subphosphate 碱式磷酸盐
sub product 副产品
subsalicylate 碱式水杨酸盐
subsalt 碱式盐
sub-sample ①子样品；子样②二次抽样
sub-sampling(=two-stage sampling) 二次进样(法)
subscript 下标
subsensitivity 亚灵敏度
subsequent coat 下一道漆；下一道涂膜
subsequent deformation 后成变形
subsequent handling 后续工序
subsequent treatment 后处理
subsidence settling 泥渣沉积
subsidence vat 沉降瓮

subsidency 沉降；沉下
subsider 沉降槽
subsidiary air supply 补充的空气供给
subsidiary reaction 副反应
subsidiary stress 附加应力
subsidiary valence 副(化合)价
subsidiary valency 副价
subsieve-size apparatus 亚筛粒度分析仪
subsilicate 碱式硅酸盐(酯)
subsoil 下层土
subsolidus data 亚固线数据
subsolution 面下溶液
subsonic ①亚声(的)②亚声速(的)
subsonic diffuser 亚声速扩压器
subspectrum 亚谱
substance 实物；物质
substance spot 试样斑
substandard 不标准的
substantial ①真实的；实在的②坚实的
substantial six-membered rings 基本的六节环
substantiation 证实
substantive ①直接的②直(接)染的③独立的④真的
substantive azo dye(=direct cotton dye) 直接偶氮染料；直接棉染料
substantive color(=substantive dye) 直接染料
substantive dye(=direct dye;substantive color) 直接染料
substantive dyeing 直接染色
substantivity 亲和性；直接(上染)性
substituent 取代基
substituent constant 取代基常数
substituent effect 取代基效应
substituent group 取代基
substituent uniformity 取代基均匀度
substitute ①代用品②取代③代替者
substituted ①取代的②代替的
substituted acrylontrile 取代丙烯腈
N-substituted amide N-取代酰胺
substituted aromatics 取代的芳香化合物
substituted benzene 取代苯；苯的同系物
α,γ-substituted benzyl hydroperoxide α,γ取代的苄基过氧化氢
substituted compound 取代化合物
substituted ethylene 取代乙烯
substituted organosilicon compound 取代的有机硅化合物
substituted phenol 取代酚
substituted polysiloxane 替代的聚硅氧烷
substituted sulfuryl amide 替代的磺酰胺

substitute lubricant 润滑剂代用品
substitute material 代用品
substitute natural gas 合成天然气；天然气代用品
substituting 取代
substituting agent 取代剂
substituting group 取代基
substitution 取代*
ω-substitution 链端取代作用
substitutional defect 取代缺陷
substitution chelometric titration 置换螯合滴定
substitution derivate 取代衍生物
substitution index 取代系数
substitution in ring 环上取代
substitution in side chain 支链取代
substitution product 取代产物
substitution reaction 取代反应；排代反应
substitution value 取代值；取代度
substitution weighing method 置换称量法
substitutive derivative 取代衍生物
substitutive name 取代名称
substoichiometric 亚化学计量的；不足化学计量的；低于化学计量的
substoichiometric analysis 亚化学计量分析
substoichiometric compound 亚化学计量化合物
substoichiometric extractant 亚化学计量萃取剂
substoichiometric extraction 亚化学计量萃取
substoichiometric isotope-dilution analysis 亚化学计量同位素稀释分析
substoichiometric reverse isotope dilution method 亚化学计量反同位素稀释法
substoichiometry ①亚化学计量学②亚化学计量法
substract 底物
substraction 扣除；减去
substraction chromatography 扣除色谱(法)
substractive color process 减色法
substractive name 减缩名称
substractive process 扣除法
substrate ①底物②基质③载体④固定液
substrate coating 底材涂装
substrate conditions 底材状态；被涂物状态
substrate constant 底物常数；基质常数
substrate extrusion coating(=extrusion coating) 挤压涂装
substrate labeled immunoassay 底物标记免疫分析法
substrate material 基质材料
substrate nature 底材性质
substrate pollution 底质污染
substrate specificity 底物特异性；基质专一性
substrate surface 底材表面；被涂物表面；基材表面

substrate treatment 底材处理
substrate web 基材
substrate wetting agent 底材润湿助剂
substratum 底层；胶层
substruction 基本结构；亚结构
subsulfate 碱式硫酸盐
sub-superheavy elements 亚超重元素
subsurface 表面下的；液面下的；地下的
subsurface corrosion ①表面下的腐蚀②地下腐蚀
subsurface defect 表面下缺陷；内部缺陷
subsurface disposal 地下排放
subsurface drainage 浅地表排水
subsurface incineration 地下焚烧
sub-surface pollution 地下污染
subsurface storage 地下储藏
subterranean 地下的；隐蔽的；掩盖的
subterranean disposal 地下处置*
subterranean granite 地下花岗石(岩)
subterranean leak 地下裂漏〔管道或油罐〕
subterranean storage 地下油罐；地下油库
subterranean tank 地下油罐
subterranean water 地下水
subthreshold 亚阈(能)；低于阈(能)
subthreshold neutron interrogation technique 低于阈能的中子探询技术
sub-tow 分(股)丝束
subtractive colorant mixture theory 减色法混色理论
subtractive color mixing 减色混合
subtractive colour matching 减色法配色
subtractive colour system 减色法系统
subtractive colour triangle 减色三角形
subtractive complementary colour 减色法补色
subtractive correction 减色校正法；减色法配光
subtractive method 减色法；相减法
subtractive mixing of colors 减色法混色
subtractive primaries 减色法三基(原)色；相减合成三基(原)色
subtractive process 脱除(杂质)过程
subtractive synthesis 减色法合成
subtractive system 减色法系统
subtractive-type process 减色法
subtract time 减法时间
subtribe 亚族
subtropical exposure facility ①亚热带暴晒架②亚热带暴晒站
subtropical test facility 亚热带(暴晒)试验站(架)
subtropical (zone) exposure test(ing) 亚热带暴晒试验
subunit 亚单位；亚基

subwater pipe line 水下管线
subwaxy phase 次蜡状相
subwaxy soap 次蜡状皂
subzero 零下；低温
sub-zero engine oil 低温机油；寒区机油〔凝固点低于−53.8℃的机油〕
sub-zero gear oil 寒区齿轮油；低温齿轮油〔凝固点低于−53.8℃的齿轮油〕
sub-zero oil 低温润滑油〔凝固点低于−53.8℃的润滑油〕
sub-zero temperature 零下温度；负温
sub-zone monitoring 分区监测
succeeding coat of paint 下一道漆的涂膜
succeeding vat 连续槽
successive 连续的
successive addition method 逐次(级)添加法
successive analysis 逐级分析；连续分析
successive approximate method 逐次近似法*；逐次逼近法
successive approximation 逐步逼近法
successive contrast 色感递次对比度
successive decay 逐次衰变；逐级衰变；连续衰变
successive flash vaporization 连续闪蒸
successive polymerization 逐次聚合；连续聚合
successive reaction 连续反应；逐级反应
succimide 琥珀酰亚胺；(正)丁二酰亚胺
succinaldehyde 琥珀醛；丁二醛 $CHOCH_2CH_2CHO$
succinamic acid 琥珀酰氨酸；丁二酸一酰胺；丁酰氨酸 $NH_2COCH_2CH_2COOH$
succinamide(=succinic diamide) 琥珀酰胺；丁二酰胺 $NH_2COCH_2CH_2CONH_2$
succinamoyl(=succinamyl) 琥珀酰氨酰；氨羰丙酰；丁二酸一酰氨一酰基 $H_2NCOCH_2CH_2CO—$
succinamyl(=succinamoyl) 琥珀酰氨酰；氨羰丙酰
succinanil(=N-phenyl succimide) 琥珀酰苯胺 $(CH_2CO)_2NC_6H_5$
succinanilide N-琥珀酰二苯胺 $C_6H_5NHCOCH_2CH_2COHNC_6H_5$
succinate ①琥珀酸；丁二酸②琥珀酸盐(酯或根)
succinate-acetoacetate CoA transferase 琥珀酸-乙酰乙酰 CoA 转移酶
succinate method 琥珀酸盐法
succinchloroimide(=succinic chlorimide) 琥珀酰氯亚胺 $(CH_2CO)_2NCl$
succindialdehyde 琥珀醛；丁二醛 $CHOCH_2CH_2CHO$
succinelite 琥珀酸〔从琥珀制得〕
succinic〔词根〕 琥珀(酸或酰)
succinic acid 琥珀酸；丁二酸 $(CH_2CO_2H)_2$
succinic acid peroxide 过氧化琥珀酸

succinic acid semialdehyde 琥珀酸半醛
succinic aldehyde 琥珀醛；丁二醛 $(CH_2CHO)_2$
succinic amide 琥珀酰胺
succinic anhydride 琥珀酐；丁二(酸)酐 $(CH_2CO)_2O$
succinic chloride 琥珀酰氯；丁二酰氯 $(CH_2COCl)_2$
succinic chlorimide(=succinchloroimide) 琥珀酰氯亚胺 $(CH_2CO)_2NCl$
succinic diamide(=succinamide) 琥珀酰胺 $(CH_2CONH_2)_2$
succinic-fumaric equilibrium 琥珀酸-反丁烯二酸平衡
succinic monoamide 琥珀一酰胺 $NH_2COC_2H_4CO_2H$
succinic oxidase(=succinoxidase) 琥珀酸氧化酶
succinic peroxide 琥珀(一)酰化过氧 $(COOHCH_2CH_2CO)_2O_2$
succinimide 琥珀酰亚胺 $(CH_2CO)_2NH$
succinimido 琥珀酰亚氨基 $COCH_2CH_2CON—$
N-succinimidyl 3-(2-pyridyldithio) propion-ate(SPDP) N-琥珀酰亚胺 3-(2-吡啶二硫基)丙酸酯
succinin 琥珀脂
succinite 琥珀
succinoamino 琥珀酰氨基 $(CH_2CO)_2N—$
succinol 琥珀油
succinonitrile 琥珀腈；丁二腈 $(CH_2CN)_2$
succinosuccinic acid 琥珀酰琥珀酸
succinosuccinic ester 琥珀酰琥珀酸酯
succinoxidase(=succinic oxidase) 琥珀酸氧化酶
succinoylation 琥珀一酰化
succinyl 琥珀酰；丁二酰 $—COCH_2CH_2CO—$
succinyl chloride 琥珀酰氯 $ClCOCH_2CH_2COCl$
succinyl dichloride 琥珀酰氯 $COClCH_2CH_2COCl$
succinylosuccinic ester 琥珀酰琥珀酸酯 $ROCOC_6H_6O_2COOR$
succinyl oxide 琥珀酰化氧；琥珀酰酐 $C_4H_4O_3$
succinylsulfathiazole 琥珀酰磺胺噻唑
succnanilic acid 琥珀酰苯胺酸
succus 汁；液
suck-back 倒吸
sucker ①吸管②吸者③吸枝〔植物〕④吸盘
sucker rod wax 吸杆蜡；唧筒杆上分出的污蜡
suck in 吸入；抽入
sucking 吸引；抽吸
sucking fit 推入配合；密配合
sucking jet pump 吸引喷嘴泵
sucralfate 硫糖铝
sucramine 糖精的铵盐
sucrate 蔗糖金属派生物
sucrochemistry 蔗糖化学

sucroglyceride 甘油蔗糖酯
sucrol(=dulcin) 甘素；甜精
sucrose 蔗糖
sucrose acetate isobutyrate 蔗糖乙酸异丁酸酯
sucrose alkoxymethyl ether 蔗糖烷氧基甲基醚
sucrose benzoate 蔗糖苯甲酸酯
sucrose diester 蔗糖二酯
sucrose distearate 蔗糖二硬脂酸酯
sucrose dodecanoate 蔗糖十二(烷)酸酯
sucrose dodecyl ether 蔗糖十二烷基醚
sucrose ester 蔗糖酯
sucrose ether 蔗糖醚
sucrose fatty ester 蔗糖脂肪酸酯
sucrose monoester 蔗糖单酯
sucrose monostearate 蔗糖单硬脂酸酯
sucrose octaacetate 蔗糖八乙酸酯
suction ①空吸；吸(取)②吸力
suction air 抽吸空气
suction air chamber(=suction bell) 抽吸空气室
suction apparatus 抽吸装置
suction basket 吸入口网子
suction bell(=suction air chamber) 抽吸空气室
suction bend 吸引管弯头
suction bottle(=filtering flask; filter flask) 吸滤瓶
suction conveyor 真空输送机
suction couch roll 真空伏辊
suction eye 吸入孔〔叶轮〕
suction fan 吸风机；吸风电扇
suction filter (空)吸滤器
suction flask(=suction bottle) 吸滤瓶
suction fuel line 燃料吸入线
suction funnel 吸滤漏斗；抽滤漏斗
suction gas producer 空吸(炉)煤气发生炉
suction gauge 吸力计；真空计
suction head ①吸升力②吸水头
suction header 吸入集管
suction heater 吸入加热器
suction hose 吸力头；抽吸软管
suction inlet 吸入口
suction lead 吸入导管
suction lift ①吸引高度②吸引提升机
suction line 吸入管线
suction main 吸入总管；吸入干管
suction manifold 吸引支管
suction method of cleaning 抽吸清洗法；真空清洗法〔配气网的〕
suction mill 吸磨机
suction nozzle 吸入嘴

suction opening 吸气口；抽风口
suction passage 吸入孔道
suction performance 抽气率
suction phase 吸引相
suction piece 吸入连接管
suction pipe 吸管
suction pipe losses 吸管损失〔水力〕
suction pipet(=suction pipette) 移液管
suction pipette(=suction pipet) 移液管
suction port 吸入孔
suction potential 吸势
suction press 吸水压榨；真空压榨
suction press roll 吸水压榨辊；吸压辊
suction pressure 吸引压力
suction producer 吸气式发生器
suction pulsation 吸引管中的脉动
suction pump 空吸泵；空吸抽机
suction resistance 吸引阻力
suction ring 泄漏环；抽空密封环
suction roll 吸(水)辊
suction seal 吸力密封
suction side 吸引端；吸引的一边
suction strainer 吸滤
suction stroke 吸气冲程
suction tank 吸入罐；吸入槽〔位置高于泵者〕
suction temperature 吸入温度
suction tube 吸管
suction valve 吸入阀
suction velocity 吸入速度
suction vortex 吸入旋涡
suction water 吸入水
suction well 吸入井〔位置低于泵者〕
suction zone 吸引区
sudan(=benzeneazoresorcinol) 苏丹；苯偶氮间苯二酚 $C_6H_5N_2C_6H_3(OH)_2$
sudanite rock 滑石和菱镁矿的混合物
sudan red 苏丹红
sudan yellow 苏丹黄
sudden change (滴定)突跃
sudden discharge 瞬时放电；骤然放电
sudden shock(=bounce) 突发性震动；冲击；冲撞
sudermo 二甲基噻蒽
sudorific 发汗药
suds ①顽固泡沫〔润滑油、肥皂水或水溶液中的〕②黏稠介质中的空气泡③肥皂水
suds booster 增泡剂
suds depressant 抑泡剂
suds end point 泡沫终点

sudsing 起泡
suds lubrication 肥皂水润滑〔金属切削时用〕
suds-stabilizing agent 泡沫稳定剂；稳泡剂
suds suppressing agent 抑泡剂
suds suppressor 抑泡剂
suede effect 麂皮效应；绒面效应
suede finish ①皱纹面漆②仿麂皮漆③波纹整理；仿麂皮整理
suede leather 绒面革
suede shearling 剪毛绵羊绒面革
suet 板油
suet oil 板油
suet substitute 板油代用品
suety 板油的
sufentanyl 舒芬太尼
sufficient lubrication 充分润滑
suffix 词尾；字尾
suffocating gas 窒息性毒气
suffocation 窒息(作用)
sugar 糖(类)；蔗糖
sugar alcohol 糖醇
sugar-aldehyde reaction 糖-醛反应
sugar apple 番荔枝
sugar assimilation test 糖同化试验
sugar beet 制糖甜菜
sugar beet slices 甜菜片
sugar bin 糖仓
sugar campaign 糖役
sugar-cane 糖甘蔗
sugar cane juice 甘蔗汁
sugar-cane mill 压蔗机
sugar cane molasses 蔗糖蜜
sugar-cane wax 甘蔗蜡
sugar carbonate 碳酸糖(酯)
sugar chemicals 糖的化工产品
sugar cleansing 糖的净化作用
sugar-coated (包有)糖衣的
sugar concentration graduation 糖度线
sugar content 含糖量
sugar degree 糖度；含糖量
sugar derivatives 糖衍生物
sugar detergent 糖(酯类)洗涤剂
sugar determination 糖测定
sugar dish 糖碟
sugar drier 烘糖机器
sugared ①加糖的②糖饯的；糖腌的
sugar ester 糖酯
sugar ether 糖醚
sugar flask 糖瓶
sugar formazan 糖(缩甲)
sugar granulation 糖成粒析出
sugariness ①甜度②甜性
sugaring ①加糖的②糖饯的
sugar juice 糖汁
sugar juice defecation 糖汁的澄清
sugar juice defecation sludge 糖汁澄定渣
sugar juice extraction 糖汁提取法
sugar juice liming 糖汁加(石)灰
sugar juice pump 糖汁泵
sugarlike polyose 类糖多聚糖
sugar lime 葡萄糖钙
sugar maple 糖枫
sugar maple syrup 枫糖浆
sugar maple tree 糖槭树
sugar melter 溶糖器
sugar mercaptal 糖缩硫醇
sugar mill 制糖厂；糖厂；糖作坊
sugar of lead 铅糖；乙酸铅 $Pb(C_2H_3O_2)_2$
sugar phosphate (=sugar phosphoric ester) 糖磷酸酯
sugar phosphoric ester (=sugar phosphate) 糖磷酸酯
sugar pine 砂糖松
sugar refining 糖的精炼
sugar sulfate 糖硫酸酯
sugar sulfuric ester 糖硫酸酯
sugar surfactant 糖(酯类)表面活性剂
sugar tolerance 糖耐量
sugar tolerance curve 耐糖曲线
sugar washing 糖的洗涤
sugarwood 宽果苦榄蓝
sugary 糖样的
suggesting jasmin 似茉莉香
suggestion 香感；香韵
suggestive 香感；香韵
sugi gum 杉脂；杉树胶；松脂
suginene 日柳杉烯
sugiol (=9-oxoferruginol) 9-氧代铁锈酚
sugiresinol 柳杉树脂酚
sugordomycin 库马霉素；苏哥多霉素
suint 羊毛粗脂
suint ash 羊毛粗脂灰
suint salt 羊毛粗脂盐
suitability distillation 评价润滑油潜含量的蒸馏法
suitamycin 吹田霉素
sulf- (词头) 硫(代)；磺基〔表示有硫存在〕
sulfa 磺胺〔磺胺剂俗名词头，指对氨基苯磺酰氨基〕
$NH_2C_6H_4SO_2NH—$

sulfabenzamide 苯酰磺胺
sulface combustion 表面燃烧
sulfacetamide 乙酰磺胺
sulfacetamide sodium 乙酰磺胺钠
sulfacetimide 乙酰磺胺
sulfacid 硫磺酸 RCOSH; RSO$_3$H
sulfactin 硫放线菌素
sulfactol 硫代硫酸钠
sulfadiazine(=2-sulfanilylpyrimidine) 磺胺嘧啶 H$_2$NC$_6$H$_4$SO$_2$NHC$_4$H$_3$N$_2$
sulfadiazine silver 磺胺嘧啶银
sulfadiazine sodium 磺胺嘧啶钠
sulfadiazine zinc 磺胺嘧啶锌
sulfadimethoxine 磺胺二甲氧哒嗪；磺胺二甲氧嗪；磺胺二甲氧基嘧啶
sulfadimethylpyrimidine(=sulfadimidine) 磺胺二甲嘧啶
sulfadimidine(=sulfadimethylpyrimidine; sulfamethazie) 磺胺二甲嘧啶
sulfadimidine sodium 磺胺二甲嘧啶钠
sulfadoxine 磺胺多辛
sulfa drug 磺胺(制)剂
sulfaethylthiadiazole 磺胺乙基噻二唑
sulfaethylthiazolone 磺胺乙基噻唑酮
sulfafurazole 磺胺二甲异噁唑
sulfaguanidine(=sulfanilylguanidine) 磺胺胍；磺胺脒；对氨基苯磺酰胍
sulfahydantoin 磺胺海因；磺胺乙内酰脲
sulfaldehyde 硫醛
sulfamate 氨基磺酸盐(酯) RNHSO$_3$M; RNHSO$_2$OR
sulfamation 磺胺化作用
sulfamerazine(=sulfamethyldiazine) 磺胺甲基嘧啶
sulfamerazine sodium 磺胺甲基嘧啶钠
sulfamethazine(=sulfadimidine) 磺胺二甲嘧啶
sulfamethazole 磺胺二甲噁唑
sulfamethizole 磺胺甲基硫(代)二嗪；磺胺甲二唑；磺胺甲基噻二唑
sulfamethoxydiazine 磺胺甲氧二嗪
sulfamethoxypyridazine 磺胺甲氧哒嗪
sulfamethyldiazine(=sulfamerazine) 磺胺甲基嘧啶
sulfamethylthiazole 磺胺甲基噻唑
sulfamic 氨磺酰基
sulfamic acid 氨基磺酸 RNHSO$_3$H; R$_2$NSO$_3$H
sulfamic acid chloride 氨基磺酰氯
sulfamic acid salt 氨基磺酸盐
sulfamide ①硫酰胺 SO$_2$(NH$_2$)$_2$ ②〔词尾〕磺酰胺 RSO$_2$NH$_2$
sulfamide-formaldehyde resin 磺酰胺-甲醛树脂
sulfamidic acid 氨基磺酸

sulfamidobarbituric acid 磺氨基巴比妥酸
sulfamine ①〔词头〕氨磺酰 NH$_2$SO$_2$— ②〔词尾〕磺酰胺 RSO$_2$NH$_2$
sulfamine-benzoic acid 氨磺酰苯甲酸 NH$_2$SO$_2$C$_6$H$_4$COOH
sulfaminic acid 氨基磺酸 NH$_2$SO$_2$OH
sulfamino(=sulfoamino) 磺氨基
o-sulfaminobenzoic acid 邻氨磺酰苯甲酸 NH$_2$SO$_2$C$_6$H$_4$COOH
sulfamoyl(=sulfamyl) 氨磺酰 H$_2$NSO$_2$—
sulfamoyl benzoic acid 氨磺酰苯甲酸
sulfamyl(=sulfamoyl) 氨磺酰
sulfane 硫烷 S$_x$H$_y$
sulfanilamide(=sulfanilic amide) 磺胺；对氨基苯磺酰胺
sulfanilamido 磺氨基；对氨苯磺酰氨基 H$_2$NC$_6$H$_4$SO$_2$NH—
sulfanilamido radical 磺氨基
sulfanilate 磺胺酸盐；对氨基苯磺酸盐 NH$_2$C$_6$H$_4$SO$_3$M
sulfanilic acid 磺胺酸；对氨基苯磺酸 NH$_2$C$_6$H$_4$SO$_3$H
sulfanilic amide(=sulfanilamide) 磺胺；对氨基苯磺酰胺 NH$_2$C$_6$H$_4$SO$_2$NH$_2$
sulfanilyl 磺胺酰；对氨基苯磺酰 H$_2$NC$_6$H$_4$SO$_2$—
sulfanilylguanidine(=sulfaguanidine) 磺胺胍；对氨基苯磺酰胍；磺胺脒
2-sulfanilylpyridine(=sulfapyridine) 磺胺吡啶；2-磺胺吡啶
sulfanilyl radical 磺胺酰基；对氨基苯磺酰基
N^4-sulfanilyl sulfanilamide(=disulon) 磺胺酰磺胺；双磺胺
sulfanole 磺烷油〔商名，指 N-脂烃磺酰胺〕
"Sulfan" process 三氧化硫连续磺化法〔制烷基磺酸盐〕
sulfantimonate(=sulfantimoniate) 硫代锑酸盐 M$_3$SbS$_4$
sulfantimoniate(=sulfantimonate) 硫代锑酸盐
sulfantimonic acid 硫代锑酸 H$_3$SbS$_4$
sulfantimonide 硫锑化物
sulfantimonite 硫代亚锑酸盐 M$_3$SbS$_3$
sulfapyrazine 磺胺吡嗪
sulfapyridine(=2-sulfanilylpyridine; sulfidin) 磺胺吡啶
sulfaquinoxaline 磺胺喹噁啉
sulfarsenate(=sulfarseniate) 硫代砷酸盐 M$_3$AsS$_4$
sulfarseniate(=sulfarsenate) 硫代砷酸盐
sulfarsenic acid 硫代砷酸 H$_3$AsS$_4$
sulfarsenide 硫砷化物
sulfarsenite 硫代亚砷酸盐 M$_3$AsS$_3$
sulfarsphenamine 硫胂凡纳明；4,4'-二羟基偶胂苯-3,3'-二乙代亚硫酸钠
sulfasalazine 柳氮磺吡啶
sulfasolucin 磺胺溶素

sulfasomidine 磺胺二甲异嘧啶；磺胺异嘧啶；磺胺索米定
sulfasuxidine(=succinyl-sulfathiazole) 磺胺杀克啶；丁二酰磺胺噻唑
sulfatase 硫酸酯酶
sulfate ①硫酸酯 R_2SO_4 ②硫酸盐 M_2SO_4
sulfate adenylyl transferase 硫酸腺苷酰转移酶
sulfate cellulose 硫酸纤维素
sulfated alkanolamide 烷醇酰氨硫酸盐；硫酸化烷醇酰胺
sulfated alkyl ether 烷基醚硫酸盐；硫酸化烷基醚
sulfated alkylolamide 烷醇酰氨硫酸盐；硫酸化烷醇酰胺
sulfated amide 硫酸化酰胺
sulfated amyl oleate 硫酸化油酸戊酯
sulfated ash value 硫酸盐灰分值
sulfated butyl oleate 油酸丁酯硫酸盐；硫酸化油酸丁酯
sulfated castor oil 硫酸化蓖麻油
sulfated cod oil 硫酸化鳕鱼油
sulfate detergent 硫酸盐(型)洗涤剂
sulfated ethoxylated alcohol 乙氧基化醇硫酸盐；硫酸化乙氧基化醇
sulfated ethoxylated alkylphenol 乙氧基化烷基酚硫酸盐；硫酸化乙氧基化烷基酚
sulfated fatty acid monoglyceride 硫酸化甘油单脂肪酸酯
sulfated fatty alcohol 硫酸化脂族醇
sulfated fatty ester 硫酸化脂肪(族)酯
sulfated glyceride 硫酸化甘油酯
sulfate digester 硫酸(纤维)消化器
sulfated lauryl alcohol 月桂醇硫酸盐；硫酸化月桂醇
sulfated monoglyceride 硫酸化甘油单脂肪酸酯
sulfated oil 磺化油
sulfated residue test 残余硫酸试验
sulfated surfactant 硫酸盐型表面活性剂
sulfate hemiester 硫酸半酯
sulfate ion 硫酸根离子
sulfate method 硫酸盐法
sulfate of ammonia 硫酸铵
sulfate of lime 硫酸钙
sulfate of potash 硫酸钾
sulfate of soda glass 硫酸盐玻璃
sulfate pigment 硫酸盐颜料
sulfate process 硫酸盐(制纸浆)法
sulfate process titanium dioxide 硫酸法二氧化钛；硫酸法钛白
sulfate pulp 硫酸盐纸浆
sulfate pulping 硫酸盐纸浆

sulfate radical 硫酸根 SO_4^{2-}
sulfate-reducing bacteria 硫酸盐还原细菌
sulfate resin 磺化树脂
sulfate rosin 磺化松香
sulfate saturator 硫酸盐饱和器
sulfate sulfur 硫酸盐式硫；硫酸盐中的硫
sulfate terpentine oil 硫酸盐纸浆松节油
sulfate transferase 硫酸转移酶
sulfathalidine N-酞酰磺胺噻唑
sulfathiazole 磺胺噻唑
sulfathiazole sodium(=sulfathiazolum soluble) 磺胺噻唑钠；可溶性磺胺噻唑
sulfathiazoline 磺胺噻唑啉
sulfathiazolum soluble(=sulfathiazole sodium) 可溶性磺胺噻唑；磺胺噻唑钠
sulfathiodiazole 磺胺噻二唑
sulfatidase 硫(脑)苷脂酶
sulfatidate 硫酸(脑苷)酯；脑硫酯
sulfatidate sulfatase 硫(脑)苷脂硫酸酯酶
sulfatide 硫(脑)苷脂
sulfating 硫酸化
sulfating roasting 硫酸盐焙烧
sulfation 硫酸盐化作用
sulfato 硫酸根基 $SO_4—$
sulfatobetaine 硫酸基甜菜碱
sulfatocerate 硫酸根合铈酸盐
sulfatoceric acid 硫酸根合铈酸
sulfatocobalamin 硫酸钴胺
sulfatostannate 硫酸根合锡酸盐
sulfatosulfonate 硫酸根合磺酸盐
sulfenamide(=sulphenamide) 次磺酰胺
sulfenamide type accelerator 次磺酰胺类促进剂
sulfenanilide N-亚磺酰苯胺
sulfenazoxine 8-羟基-5-(对磺苯偶氮基)喹啉
sulfenic acid 次磺酸 RSOH
sulfenyl ①〔词头〕亚氧硫基 OS= ②〔词中〕亚磺酰 —S(O)— ③〔单用〕氧硫基 =SO ④(烃)硫基 RS—
sulfenylation 亚磺酰化*
sulfenyl halides 氧硫基卤
sulfetrone 4,4'-双(γ-苯丙胺基)二苯砜-$\alpha,\gamma,\alpha',\gamma$-四磺酸四钠
Sulfex-F process 索尔费克斯-F过程〔用 H_2SO_4-HF 溶解铀锆燃料的首端过程〕
Sulfex process 索尔费克斯过程〔用 H_2SO_4 溶解不锈钢包壳燃料的首端过程〕
sulfhydrate 氢硫化物 MSH
sulfhydryl(=mercapto) 氢硫(基)；巯(基) HS—

sulfhydryl group 氢硫基；巯基 HS—
sulfhydryl reagent 巯基试剂
sulfidal 胶态硫
sulfidation 硫化(作用)
sulfidation-resistance 耐硫化性
sulfide ①硫化物②硫醚
sulfide anti-oxidant buffer (SAOB) 硫化物抗氧化缓冲液
sulfide black-ash 黑色硫化钠；粗硫化钠
sulfide churn 黄原化鼓
sulfide colors 硫化染料
sulfide contactor 硫化物接触器
sulfide drum 黄原化鼓
sulfide dye 硫化钠浴中用的染料
sulfide film 硫化膜
sulfide fusion 硫化(物)熔融
sulfide ion 二价硫离子
sulfide linkage 硫键
sulfide method 硫化物法
sulfide painting machine 涂灰浆机
sulfide staining 硫化物污染斑；硫化污染
sulfide stain resistance 耐硫化污染性
sulfide sulfur 硫化物型硫
sulfide titration 硫化物滴定
sulfidin(=sulfapyridine; 2-sulfanilylpyridine) 磺胺啶；磺胺吡啶
sulfiding 形成硫化物
sulfidion 二价硫离子
sulfidity 硫化度
sulfidizing 形成硫化物
sulfilimine(=sulfimine) (烃基)硫亚胺 R_2SNH
sulfime 硫肟 $RCH(=NSH)$
sulfimide 磺酰亚胺 $(SO_2NH)_2$
sulfimine(=sulfilimine) (烃基)硫亚胺
-sulfinate〔词尾〕 亚磺酸盐 RSO_2M
sulfination 亚磺化(作用)
sulfindigotic acid 硫靛酸
sulfine 锍化物；四价硫的有机化合物 R_3SX
sulfine oxide 氢氧化三烃基硫 R_3SOH
-sulfinic acid〔词尾〕 亚磺酸 RSO_2H
sulfinid 糖精
sulfino 亚磺基；亚硫酸一酰 $(HO)OS—$
Sulfinol process 萨菲努尔法〔除气体中二氧化硫、硫化氢法〕
sulfinpyrazone 1,2-二苯基-4-[2-(苯亚磺酰)乙基]-3,5-吡唑烷二酮
sulfinyl(=thionyl) 亚硫酰基；亚磺酰 —SO—
sulfinyl amine 亚磺酰胺

sulfinyldiacetic acid 亚硫酰二乙酸
sulfion 硫离子 S^{2-}
sulfisooxazole 磺代异噁唑
sulfitating agent 亚硫酸化剂
sulfitation ①亚硫酸化(作用)②亚硫酸处理
sulfite ①亚硫酸盐②亚硫酸酯
sulfite cellulose 亚硫酸(盐)纸浆
sulfite cellulose extract 亚硫酸盐纸浆提出物
sulfite cellulose liquor(=sulfite cellulose waste lye) 亚硫酸盐废碱液
sulfite cellulose waste lye(=sulfite cellulose liquor) 亚硫酸盐废碱液
sulfite cooking acid(=sulfite cooking liquor) 亚硫酸盐蒸煮液
sulfite cooking liquor(=sulfite cooking acid) 亚硫酸盐蒸煮液
sulfited extract 亚硫酸盐提取物
sulfited fish oil 亚硫酸化鱼油
sulfited mimosa extract 亚硫酸化荆树皮浸膏
sulfited oil 亚硫酸化油
sulfite ion 亚硫酸根离子
sulfite liquor 亚硫酸盐废液
sulfite method 亚硫酸盐法
sulfite process 亚硫酸盐(制纸浆)法
sulfite pulp 亚硫酸盐纸浆
sulfite pulping 煮制亚硫酸盐纸浆
sulfite waste liquor(=sulfite waste lye) 亚硫酸盐废液
sulfite waste lye(=sulfite waste liquor) 亚硫酸盐废液
sulfite wrapping (paper) 包装纸
sulfiting ①亚硫酸盐化(作用)②亚硫酸处理
sulfitolysis 亚硫酸盐解
sulfo-(=sulpho-)〔拉丁字头〕 ①硫(代)②磺基 —SO_3H
sulfoacetic acid 磺基乙酸 HO_3SCH_2COOH
sulfoacid ①磺酸②硫代酸 RSO_3H
sulfoacylation 磺基乙酰化作用 $CORSO_3H$
sulfoalkylation 磺烷基化作用 RSO_3H
sulfoalkyl fumarate 磺基富马酸烷基酯；磺基反丁烯二酸烷基酯
sulfoalkyl methacrylate 磺基甲基丙烯酸烷基酯；甲基丙烯酸烷基酯磺酸盐
sulfoamidic acid 氨基磺酸 $RNHSO_3H$
sulfoamino(=sulfamino) 磺氨基
sulfoaminobenzoic acid 磺氨基苯甲酸
sulfoarsenide 磺砷化物
sulfoarylation 磺芳化作用
sulfoaspartate 磺基天冬氨酸
sulfobenzide 二苯砜 $(C_6H_5)_2SO_2$
o-sulfobenzoic acid 邻磺基苯甲酸

$HO_3SC_6H_4CO_2H \cdot 3H_2O$
sulfobetaine 磺基三甲铵乙内酯
sulfobromophthalein sodium 硫溴邻苯二甲酸钠
sulfocarbamide 硫脲
sulfocarbanilide 对称二苯硫脲 $CS(NHC_6H_5)_2$
sulfocarbazone 硫卡巴腙 $H_2NN=CNHNH_2$
　　　　　　　　　　　　　　　　　$|$
　　　　　　　　　　　　　　　　　SH
sulfocarbimide 硫氰酸化物
sulfocarbodiazone 双偶氮硫代酮基化物
sulfocarbolate 酚磺酸盐(酯) $HOC_6H_4SO_3M$
sulfocarbolic acid 苯酚磺酸 $HOC_6H_4SO_3H$
sulfocarbonate 硫代碳酸盐(酯) $M_2CS_3;CS(SR)_2$
sulfocarbonic acid 硫代碳酸 $H_2CS_3;CS(SH)_2$
sulfocarbons 硫碳化合物
sulfochlorides (烃基)磺酰氯 RSO_2Cl
sulfochlorination 氯磺化作用
sulfochlorophenol S 磺氯酚 S
sulfocidin 硫杀菌素
sulfocompound 含硫化合物
sulfocyan 硫氰酸盐
sulfocyanate 硫氰酸盐(酯) MSCN;RSCN
sulfocyanic acid 硫氰酸；硫代氰酸 HSCN
sulfocyanic ester 硫氰酸酯 RSCN
sulfocyanide 硫氰酸盐(或酯) MSCN; RSCN
sulfocyclorubber 磺基环化橡胶
S-sulfocysteine 犀氨酸；S-磺酸半胱氨酸
sulfodiperacid 过二硫酸 $H_2S_2O_8$
sulfoether 硫醚 RSR
sulfoethylcellulose 磺乙基纤维素〔含有 $HSO_3CH_2CH_2—$
　　置换基的纤维素〕
sulfofatty acid salt 磺基脂肪酸盐
sulfofication 硫化(作用)
sulfoform 硫仿；硫化三苯基
sulfogalactosylceramide 硫酸半乳糖基酰基鞘氨醇
sulfogen black 硫精黑
sulfogen green 硫精绿
sulfoglutarate 磺基戊二酸盐(酯)
sulfo-group 磺酸基
sulfoguaiacin 磺基愈创木酚
sulfohalide 卤硫化物
sulfohydrate 硫氢化物
sulfohydrazide 磺酰肼〔发泡剂〕
sulfohydroxystearic acid 磺基羟基硬脂酸
sulfoichthyolic acid 磺基鱼石脂酸
sulfoid 胶态硫
sulfoisophthalate 磺基异酞酸酯；磺基间苯二甲酸酯
sulfoisophthalic acid 磺基间苯二酸

sulfoitaconate diester 磺基衣康酸二酯；磺基亚甲基丁二酸二酯
sulfolane 环丁砜*
sulfolane method 环丁砜法
sulfolane unit 环丁砜(抽提)装置
sulfolauric acid 磺基月桂酸
sulfoleate 磺基油酸盐；油酸酯磺酸盐
sulfoleic acid 磺化油酸
sulfolene 环丁烯砜；1,1-二氧化二氢噻吩
sulfolipide 硫(脑)苷脂；脑硫脂
sulfolipins 硫脂(类)
sulfolite 自然硫岩
sulfomalonamide 磺基丙二酰胺
sulfomalonate 磺基丙二酸盐(酯)
sulfometaboric acid 硫代偏硼酸 HBS_2
p-sulfo-o-methoxybenzene-azo-dimethyl-α-napthylamine
　　对磺基邻甲氧基苯偶氮二甲基-α-萘胺
sulfomethyl amide 磺基甲酰胺
sulfomethylated 磺甲基化的
sulfomethylation 磺甲基化作用〔即引入 $HSO_3CH_2—$〕
sulfomethyl cellulose 磺甲基纤维素
sulfomethyl ester 磺基(脂肪酸)甲酯
sulfomethyl palmitate 棕榈酸甲酯磺酸盐；磺基棕榈酸甲酯
sulfomonomer 磺基化单体
sulfomucin 硫黏蛋白
sulfomycin 硫霉素
sulfomyristic acid 磺基肉豆蔻酸
sulfon 嗍呐〔染料词头〕
sulfon acid blue 嗍呐酸性青
sulfonal 嗍呐哪；眠砜甲烷；索佛那；二乙眠砜
　　$(CH_3)_2C(SO_2C_2H_5)$
sulfonamic 氨磺酰基(的)
sulfonamic acid 氨基磺酸 $RNHSO_3H$
sulfonamide(=sulfamine)〔词头〕 氨磺酰 $NH_2SO_2—$
sulfonamide-aldehyde resin 氨磺酰-(甲)醛树脂
sulfonamide-formaldehyde resin 氨磺酰-甲醛树脂
sulfonamide plasticizer 磺酰胺类增塑剂
sulfonamide resin 磺酰胺树脂
sulfonamide resistant strain 抗磺胺菌株
sulfonamido- 亚磺酰氨基 $—SO_2NH—$
sulfonamido benzoic acid 氨磺酰苯甲酸
sulfonamido-crysoidin(=prontosil) 偶氮磺胺
sulfonaphthaleins 磺酞类
sulfonaphthol 磺基萘酚；萘酚磺酸 $HSO_3C_{10}H_6OH$
sulfonaphthyl stearic acid 磺基萘基硬脂酸
sulfonatability (可)磺化性
sulfonate ①磺化②磺酸盐(酯)

sulfonated 磺化的
sulfonated aliphatic alcohol 磺化脂肪醇
sulfonated aliphatic hydrocarbon 磺化脂肪烃；脂肪烃磺酸盐
sulfonated alkane 链烷磺酸盐；磺化链烷烃
sulfonated alkene 磺化链烯烃；链烯磺酸盐
sulfonated alkyl acrylate 磺化丙烯酸烷基酯
sulfonated alkylbenzene 烷基苯磺酸盐；磺化烷基苯
sulfonated alkylbiphenyl ether 磺化烷基二苯醚
sulfonated alkyl naphathalene 磺化烷基萘；烷基萘磺酸盐
sulfonated alkylphenol 磺化烷基酚
sulfonated amide 磺酰胺
sulfonated amidoester 磺化酰胺基酯
sulfonated amine 磺化胺
sulfonated biphenyl alkyl ether 磺化二苯基烷基醚
sulfonated bodies 磺化物
sulfonated butyl oleate 磺化油酸丁酯
sulfonated castor oil 磺化蓖麻油
sulfonated coal 磺化煤
sulfonated copper phthalocyanine 磺化铜酞菁
sulfonated diamide 磺化二酰胺
sulfonated diphenyl alkane 磺化二苯基烷；二苯烷磺酸盐
sulfonated ester 磺化酯；酯磺酸盐
sulfonate detergent 磺酸盐洗涤剂
sulfonated fatty amide 磺化脂肪(族)酰胺
sulfonated fatty ester 磺化脂肪(族)酯
sulfonated hydrocarbon 磺化烃；烃磺酸盐
sulfonated ketone 磺化酮
sulfonated lignin 磺化木质素；木质素磺酸盐
sulfonated methyl oleate 磺化油酸甲酯〔湿润剂〕
sulfonated monomer 磺化单体
sulfonated naphthol 磺化萘酚
sulfonated neat's foot oil 磺化牛脚油
sulfonated oil 磺化油
sulfonated α-olefin α-烯烃磺酸盐
sulfonated olefinic ester 磺化烯烃酯
sulfonated oleic acid 磺化油酸
sulfonated petroleum 磺化石油；石油磺酸盐
sulfonated phenol 磺化苯酚
sulfonated polyethylene 磺化聚乙烯
sulfonated polystyrene type resin 磺化聚苯乙烯型树脂〔离子交换树脂〕
sulfonated pyrrole 磺化吡咯；磺化氮杂茂
sulfonated rape oil 磺化菜子油
sulfonated red oil 磺化红油〔即土耳其红油〕
sulfonated retene 磺化䓛烯

sulfonated soybean oil 磺化豆油
sulfonated stearic acid 磺化硬脂酸
sulfonated styrol resin 磺化苯乙烯树脂
sulfonated type 磺化型
sulfonated wool oil 磺化羊毛油
sulfonate grease 磺化脂
sulfonate group-containing acrylonitrile co-polymer 含磺酸盐基的丙烯腈共聚物
sulfonate-modified polyester 磺化改性的聚酯
sulfonating 磺化
sulfonating agent 磺化剂
sulfonation 磺化*
sulfonation flask 磺化度测定瓶
sulfonation in acid treating 在酸处理时磺化
sulfonation number 磺化值
sulfonation reaction 磺化反应
sulfonation residue 磺化残渣
sulfonation value 磺化值
sulfonator 磺化器
sulfon cyanine black 蒯花青黑
sulfone 砜*
sulfone bislysine 双赖氨酸砜
sulfone methanes 砜甲烷类〔眠砜〔催眠药物〕 $R_2C(SO_2R)_2$
sulfone phthalein 磺酞
sulfonethylmethane(=trional) 眠砜乙基甲烷；三乙眠砜；2,2-二(乙砜基)丁烷 $CH_3(C_2H_5)C=(SO_2C_2H_5)_2$
sulfonic acid 磺酸*
sulfonic acid amide ①磺酰胺②硫酰胺 RSO_2NH_2
sulfonic acid bromide 磺酰溴 RSO_2Br
sulfonic acid chloride (烃基)磺酰氯 RSO_2Cl
sulfonic acid fluoride (烃基)磺酰氟 RSO_2F
sulfonic acid group 磺(酸)基 HSO_3-
sulfonic acid halide 磺酰卤 RSO_2X
sulfonic acid iodide 磺酰碘 RSO_2I
sulfonic acid type cation ion exchange fibre 磺酸型阳离子交换纤维
sulfonic derivatives 磺基衍生物
sulfonic group 磺基 HSO_3-
sulfonic resin 磺酸型树脂
sulfoniobetaine 磺基甜菜碱
sulfonitric medium 硫酸-硝酸介质
sulfonium 锍
sulfonium compound 有机四价硫化合物；锍化物 R_3SX
sulfonium halide 卤硫锇化物 R_3SX
sulfonium hydroxide 氢氧化三烃基硫 R_3SOH
sulfonium iodide 碘化锍 R_3SI

sulfonium ion 硫鎓离子
sulfonmethane(=sulfonal) 眠砜甲烷；二乙眠砜
sulfonphthalein indicator 磺酞类指示剂
sulfonyl(=sulfuryl) ①磺酰 —SO₂—②硫酰
sulfonylation 磺酰化*
sulfonyl azide 苯磺酰叠氮〔发泡剂〕
4-sulfonylazido benzoic acid 4-磺酰叠氮安息香酸 HOOCC₆H₄SO₂·N₃
sulfonyl chloride 磺酰氯
sulfonyl compound 磺酰化物；硫酰化物
sulfonyldianiline 磺基二苯胺〔固化剂〕
sulfonyl hydrazide 磺酰肼〔发泡剂〕
sulfonyl isocyanate 异氰酸磺酰酯 ROSO₂NCO
sulfonylurea 磺酰脲类
sulfoparaldehyde 仲乙硫醛；三聚乙硫醛 (CH₃·2CHS)₃
sulfophenol 邻磺基苯酚
sulfophenyl 磺苯基 HO₃SC₆H₄—
sulfophenylate ①苯酚磺酸盐(酯) OHC₆H₄SO₃M；OHC₆H₄SO₂OR②硫酸苯酯盐 C₆H₅SO₄
sulfophenylazochromotropic acid 磺基苯偶氮变色酸；钍锆试剂
p-1-sulfophenyl-3-methyl-5- pyrazolone 对-1-磺代苯基-3-甲基-5-吡唑啉酮
sulfo-phenyl succinic acid alkylester 磺基苯丁二酸烷基酯
sulfophenyl triazine 磺苯基三嗪
sulfophilic element 亲硫元素
sulfophthalidine 磺胺酞啶
sulfopropyl 磺丙基
sulforaphen 萝卜硫素
sulforhodanate 硫氰酸盐(酯) MSCN;RSCN
sulforhodanide(=sulforhodanate) 硫氰酸盐(酯)
sulforicinate 磺化蓖麻醇酸盐
sulfo-ricinoleic acid 磺基蓖麻(醇)酸；磺化蓖麻酸
sulfor monochloride 一氯化硫 S₂Cl₂
sulfosalicylate 磺基水杨酸盐
sulfosalicylic acid 磺基水杨酸*
sulfosalicylic acid test 磺基水杨酸试验
sulfosalt ①含硫的酸的盐②磺酸盐 RSO₃M
sulfoselenide 硫硒化物
sulfosemicarbazide (某)氨基硫脲
sulfosol 硫酸溶胶
sulfostannate 硫代锡酸盐 M₂SnS₃
sulfostannic acid 硫代锡酸；全硫锡酸 H₂SnS₃
sulfostannite 硫代亚锡酸盐 M₂SnS₂
sulfostannous acid 硫代亚锡酸；全硫亚锡酸 H₂SnS₂
sulfosuccinamate 磺化琥珀酰胺酸盐(酯)
sulfosuccinate 磺基丁二酸盐(酯)；磺基琥珀酸盐(酯)
sulfosuccinic acid 磺基琥珀酸；磺基丁二酸

sulfosuccinic acid-diiso-butylester-S-sodium salt 丁二酸二异丁酯磺酸钠
sulfosuccinic ester 磺基琥珀酸酯
sulfosuccinimide 磺基琥珀酰亚胺
sulfo-sulfonate 磺基磺酸盐〔二磺酸盐〕
sulfourea 硫脲 CS(NH₂)₂
sulfovinate ①烃基硫酸盐；硫酸烃酯 RSO₄M②〔专指〕乙基(代)硫酸盐 C₂H₅SO₄M
sulfovinic acid ①烃基硫酸 RSO₄H②〔专指〕乙基(代)硫酸 C₂H₅SO₄H
sulfoxidation 磺基氧化作用
sulfoxide 亚砜*
sulfoxone sodium 阿地砜钠；硫福宋钠；索发克松钠
sulfoxonium 氧化锍
Sulfox process 空气氧化法污水处理过程
sulfoxy 硫氧基；亚硫酰
sulfoxylate 次硫酸盐 M₂SO₂; MHSO₂
sulfoxylic acid 次硫酸 H₂SO₂; S(OH)₂
λ-sulfur λ硫
sulfur(=sulphur) ①硫黄；硫(磺)②硫(黄)的③硫化〔燃料〕
sulfur acid 含硫酸
sulfur alcohol ①硫醇②乙硫醇 RSH
sulfur anhydride 三氧化硫
sulfurate 加硫的
sulfurated ①加(了)硫的②硫化的
sulfurated hydrogen 硫化氢 H₂S
sulfurated lime 硫化钙 CaS
sulfurating reagent 硫化剂
sulfuration(=thionation) 硫化作用
sulfuration of mineral oil 矿物油硫化
sulfur-based functional group 含硫功能基
sulfur-bearing crude 含硫原油
sulfur-bearing gases 含硫气体
sulfur-bearing oil 含硫石油
sulfur black 硫化黑
sulfur black paste 膏状硫化黑
sulfur bloom 硫黄华；硫霜
sulfur blooming 喷硫(现象)
sulfur blue 硫化蓝
sulfur bomb 硫弹
sulfur bridge 硫桥
sulfur bromide 溴化硫 S₂Br₂
sulfur brown 硫化棕
sulfur burner 烧硫炉
sulfur-cake hopper 硫块储斗；硫块进料斗
sulfur carbanion 硫碳阴离子
sulfur chloride 氯化硫〔1. 一氯化硫 S₂Cl₂; 2.二氯化硫 SCl₂〕

sulfur chloride vulcanization 氯化硫溶液硫化;冷硫化
sulfur-chlorinated cutting oil 含硫氯化切削油
sulfur colors 硫化染料
sulfur compound 硫化合物
sulfur-containing aroma 硫黄气味
sulfur content 硫含量
sulfur corrosion 硫(化合物)腐蚀作用
sulfur crack 硫蚀裂纹
sulfur cross-linking 硫黄交联;硫黄硫化
sulfur-cured vulcanizate 硫黄硫化胶
sulfur dichloride 二氯化硫 SCl_2
sulfur dioxide 二氧化硫 SO_2
sulfur-dioxide-bensol process 二氧化硫和苯精制法
sulfur dioxide concentration 二氧化硫浓度;亚硫酸气浓度
sulfur dioxide converter 二氧化硫转化器
sulfur dioxide corrosion 二氧化硫腐蚀
sulfur dioxide fumigation 二氧化硫熏蒸法
sulfur dioxide probe 二氧化硫探头
sulfur dioxide reduced chrome 二氧化硫还原的铬鞣液
sulfur dioxide refining 二氧化硫精制
sulfur dioxide sensor 二氧化硫传感器
sulfur direct blue 硫化直接蓝
sulfur dispersion 硫悬浮体
sulfur distribution 硫分布
sulfur-donor 硫给予体;给硫体;硫载体
sulfur donor agent 给硫剂
sulfur donor reagent 给硫(配体)试剂
sulfur dye 硫化染料
sulfur dyeing 硫化染料
sulfur dyestuff 硫化染料
sulfur elimination 脱硫;去硫
sulfureous 硫的
sulfuret 硫化物
sulfur ether 硫醚 RSR
sulfuretted hydrogen 硫化氢
sulfuretted oil 硫化油
sulfur family 硫族
sulfur family element 硫族元素
sulfur feed stock 含硫进料
sulfur flame detector 硫火焰检测器
sulfur flour 硫粉
sulfur flowers 硫华
sulfur fluorescein 硫荧光素
sulfur fluoride 氟化硫 S_2F_2; SF_2; SF_4
sulfur-free accelerator 无硫促进剂
sulfur free fuel 无硫燃料
sulfur-free gasoline 脱硫汽油

sulfur-free vulcanization 无硫硫化;无硫硬化
sulfur fuel 含硫燃料
sulfur green T 硫化绿 T
sulfur group 硫族
sulfur heterocyclic ring 含硫的杂环
sulfur hexafluoride 六氟化硫
sulfur hexaiodide 六碘化硫
sulfur-hydrocarbon slurry 硫烃淤浆
sulfuric (正)硫
sulfuric acid 硫酸
sulfuric acid alkylation(=sulfuric alkylation) 硫酸烃化
sulfuric acid anhydride 硫酸酐;三氧化硫 SO_3
sulfuric acid bath 硫酸浴
sulfuric acid bubble jar 硫酸鼓泡器
sulfuric acid chamber 硫酸室;铅室
sulfuric acid ester 硫酸酯 $ROSO_3H$; $(RO)_2SO_2H$
sulfuric (acid) ester detergent 硫酸酯洗涤剂
sulfuric acid heat test 硫酸发热试验
sulfuric acid hydration process 硫酸水合法
sulfuric acid lignin 硫酸木素
sulfuric acid manostat 硫酸恒温浴
sulfuric acid method ①硫酸(试)法②硫酸法
sulfuric acid mist 硫酸雾
sulfuric acid monohydrate 一水(合)硫酸
sulfuric acid process 硫酸法
sulfuric acid refining(=sulfuric acid treatment) 硫酸精制
sulfuric-acid rubber 硫酸橡胶〔如环化橡胶〕
sulfuric acid test (检验甾醇用的)硫酸试验法
sulfuric acid tower 干燥塔
sulfuric acid treatment(=sulfuric acid refining) 硫酸精制
sulfuric alkylation(=sulfuric acid alkylation) 硫酸烃化
sulfuric anhydride 硫(酸)酐;三氧化硫 SO_3
sulfurication of polymer 高聚物的磺化反应
sulfuric chloride 硫酰氯
sulfuric dioxide 二氧化硫
sulfuric ester 硫酸酯
sulfuric ester of glycerol 甘油硫酸酯
sulfuric ether 乙醚 $C_2H_5OC_2H_5$
sulfuric monohydrate 一水(化)合物
sulfuring up 起硫霜;喷硫
sulfur insulated switch 硫绝缘开关
sulfur iodide 二碘化二硫 S_2I_2
sulfurization 硫化*
sulfurized 硫化的
sulfurized asphalt 加硫沥青
sulfurized base oil 含硫原油
sulfurized boiled oil 硫化熟油
sulfurized-chlorinated mineral oil 硫化氯化矿物油

sulfurized cutting oil 硫化切削油
sulfurized fat 硫化脂
sulfurized grease 硫化润滑脂
sulfurized oil 硫化油
sulfurized polyolefin 硫化聚烯烃
sulfurized tall oil 硫化松脂油
sulfurizing 硫化作用
sulfurizing of asphalt 沥青的硫化
sulfur khaki 硫化茶褐
sulfurless vulcanizing agent 无硫硫化剂
sulfur liver 硫肝〔硫化钾和多硫化钾的混合物〕
sulfur monofluoride 一氟化硫 S_2F_2
sulfur monoxide 一氧化硫 SO
sulfur mould 硫黄模子；硫盘
sulfur mustard 硫芥子气
sulfur number 硫值〔每100毫升中毫克数〕
sulfur oil 硫化油
sulfur oil slurry 含硫油浆
sulfur olive green 硫化橄榄绿
sulfur olive oil 硫化橄榄油
sulfurous 亚硫
sulfurous acid 亚硫酸
sulfurous acid anhydride 亚硫酸酐；二氧化硫 SO_2
sulfurous esters 亚硫酸酯(类)
sulfurous fuel 含硫燃料
sulfurous gas 含硫气体
sulfurous iron ore 硫铁矿
sulfur oven 烧硫炉
sulfur oxide 氧化硫 SO; S_2O_3; SO_2; SO_3
sulfur oxychloride 亚硫酰氯；氯化亚砜；氯氧化硫 $SOCl_2$
sulfur oxyfluoride 二氟二氧化硫
sulfur phosphorus compound 硫磷化合物
sulfur poisoning 硫中毒
sulfur recovery plant 硫回收车间
sulfur recovery process 硫回收过程
sulfur removal 脱硫
sulfur removal plant 脱硫车间
sulfur resistant catalyst 抗硫催化剂
sulfur sensitive catalyst 易为硫中毒的催化剂
sulfur sesquioxide 三氧化二硫
sulfur soap 硫黄皂
sulfur stained yarn 黄斑丝
sulfur station 硫化站
sulfur still 硫化釜
sulfur stove 烧硫炉
sulfur subbromide 二溴化二硫
sulfur subiodide 二碘化二硫

sulfur tank 硫化槽；亚硫酸盐化器
sulfur tannage 硫黄鞣剂
sulfur test 硫黄测定
sulfur tetrafluoride 四氟化硫 SF_4
sulfur trioxide 三氧化硫
sulfur vulcanization 硫黄硫化
sulfur waste 废硫
sulfury ①硫的②硫黄气
sulfur yellow G 硫化黄 G
sulfur yellow S 硫化黄 S
sulfuryl(=sulfonyl) ①磺酰 $-SO_2-$②硫酰
sulfuryl bromide 磺酰溴 SO_2Br_2
sulfuryl chloride 硫酰氯 SO_2Cl_2
sulfuryl diamide 硫酰二胺
sulfuryl fluoride 二氟二氧化硫；硫酰氟 SO_2F_2
sulfur ylide 硫叶立德*
sulindac 舒林酸
sulph 硫(代)
sulphate 硫酸盐(酯)
sulphate cook 硫酸盐蒸煮
sulphated fatty alcohol 硫酸化脂肪醇
sulphate-gluconate electrolyte 硫酸-葡萄糖酸盐电解质
sulphate hardness 硫酸盐硬度
sulphate process 硫酸盐法〔浆粕制法〕
sulphation factor 硫酸化因子
sulphenamide(=sulfenamide) 次磺酰胺
sulphication 硫化作用
sulphication attack 硫化腐蚀
sulphidation 硫化作用
sulphide discoloration 硫化变色；硫化物变色
sulphide dye(=sulphur dye) 硫化染料
sulphide staining 硫化变黑；硫化(物)污染
sulphide stress corrosion 硫化物应力腐蚀
sulphidizing 磺化；硫化
sulphite 亚硫酸盐(酯)
sulphite process 亚硫酸盐法
sulphite staining 亚硫酸盐污染
sulphiting(=sulfiting) ①亚硫酸化②亚硫酸盐处理
sulpho ①硫(代)②磺基
sulpho-acetate 磺基乙酸酯
sulpho-acetic acid 磺基乙酸
sulphoalkyl acrylate 丙烯酸磺基烷酯
sulphobetaine 磺基甜菜碱；磺基三甲铵乙内酯
sulphocarboxyazo Ⅲ 磺羧偶氮Ⅲ
sulpho chloride 磺酰氯
sulphochlorophenol AE 磺氯酚 AE
sulphocresolazo Ⅲ 磺甲酚偶氮Ⅲ
sulphonamide〔词头〕 氨磺酰

sulphonate ①磺化②磺酸盐(酯)
sulphonated butyl ricinoleate 磺化蓖麻醇酸丁酯
sulphonated coal 磺化煤
sulphonated oil 磺化油
sulphonated tallow 磺化(动物)脂
sulphonating 磺化
sulphonating agent 磺化剂
sulphonation 磺化(作用)
sulphonator 磺化器
sulphonazo 偶氮砜
sulphone(=sulfone) 砜 RSO_2R
sulphone amide derivative 磺胺衍生物
sulphonic acid 磺酸
sulphonitrophenol K 磺硝基酚 K
sulphonyl chloride 磺酰氯
sulphopone 锌钙白〔硫酸钙加硫化锌〕
sulphosalicylazo III 磺水杨基偶氮III
sulphosalicylic acid 磺基水杨酸
sulphosalt 磺酸盐
sulphosuccinamate 磺基琥珀酰胺酸盐(酯)
sulphosuccinate 磺基琥珀酸盐(酯); 磺基丁二酸盐(酯)
sulphourea 硫脲
sulphoxy reducing agent 亚硫酰(化合物)还原剂
sulphur(=sulfur) ①硫(黄)②硫(黄)的③硫化〔染料〕
sulphurate(=sulfurate) 硫化
sulphur bloom 硫霜
sulphur bridge 硫桥
sulphur dioxide 二氧化硫
sulphur dioxide emission 二氧化硫排放量(放出量)
sulphur dye(=sulphide dye) 硫化染料
sulphuric acid oxidation coating 硫酸(阳极)氧化膜
sulphuric ether 乙醚
sulphuring 熏硫; 亚硫酸处理; 用二氧化硫保藏
sulphurization ①硫化(作用)②加硫黄(混炼)
sulphurization agent 硫化剂
sulphurized lubricant 硫化润滑剂
sulphurizing 硫化
sulphurous anhydride 亚硫酸酐; 二氧化硫
sulphur print test 硫印试验
sulphur removal 去硫;(黏胶纤维)脱硫
sulpiride 舒必利
sultaine 磺基甜菜碱
sultam ①〔类名〕磺内酰胺②〔专指〕萘-1,8-磺酸内酰胺
sultone ①〔类名〕磺内酯②〔专指〕萘-1,8-磺酸内酯
sulvanite 硫钒铜矿
Sulzer packing 苏采尔填料; 苏尔兹填料; 金属网波纹填料

sum 总数; 和
sumac ①漆树属②漆叶
sumac extract 漆叶浸膏
sumach ①漆树属②漆叶
sumac(h) bark 漆树树皮
sumac(h) berries 漆树果
sumach extract 漆叶栲胶
sumaching 漆叶处理
sumach tanned 苏模鞣制的
sumac tanned 漆叶鞣制的
sumaresinol 安息香胶酸
sumaresinolic acid 苏门树脂脑酸
Sumatra camphor 苏门答腊樟脑〔即: 冰片; 莰醇〕 $C_{10}H_{17}OH$
sumatrol 苏门答腊酚; 苏门答腊鱼藤酚
sumbul ①苏布②苏布根; 麝香根
sumbulene 苏布烯
sumbulic acid 苏布酸
sumbul oil 苏布油; 麝香根油
summarized chromatogram 简化色谱图
summative precipitation 加和法沉淀〔分子量分级名称〕
summator 加法器
summer black oil 夏季黑油; 夏用黑机油
summer gasoline 夏季汽油
summer grade 夏用; 夏季级〔石油产品〕
summer grade gasoline 夏季级汽油
summer oil 夏用油
summer-oil level 夏季加油高度
summer petrol 夏用汽油
summer savory 夏香薄荷
summer white oil 夏用白油
summer wood ①晚材; 夏材②大木; 大材
summer yellow oil 夏用黄油
summit current 峰电流; 顶点电流
summit potential 峰电位; 顶点电位
sum of cross-products 交叉乘积和
sum of square of deviations 偏差平方和
sum of square of residues 残差平方和
sum-over-states (SOS) 状态和
sum-of-states method 状态和法
sump 池; 槽
sump liquor 污油槽(废)液
sump oil 沉积池油; 污油
sump pit 放空坑〔迅速放空炼油设备时用〕
sump pump 污水泵; 污油泵
sump strainer 沉积池滤器
sump tank 废油罐
sump thermostat 储槽控制恒温器

sun atmosphere　太阳大气
sunbleach　暴晒漂白法
sun blue　直接蓝
sunburn　晒斑
sunburn cream　晒斑膏
sunburned(=sunburnt)　晒伤的
sunburn-proof factor　防晒指数
sunburnt　晒伤的
sun care preparations　防晒化妆品
sun-checking　晒裂；日光龟裂
suncrack　晒裂
sun cracking　晒裂
sun-crazing　晒裂
sun dried hide　晒干皮
sunfast　耐晒的
sunfish liver oil　翻车鱼肝油；太阳鱼肝油
sunflower　向日葵
sunflower cake　向日葵油饼
sunflower gearing traversing unit　葵花轮式啮合往复装置
sunflower oil　向日葵油；葵花油
sunflower plaiter　葵花轮式折叠器
sunflower seed　向日葵籽
sunflower seed oil　向日葵籽油
sun green　直接绿
sunitizing　卫生剂
sunken oil storage　地下油库
sunken pipe　地下管道
sunken tank　地下储罐
sunk type switch　埋装式开关
sun light　日光
sunlight checking　日光致龟裂
sunlight-induced degradation　日光诱发的老化作用
sunlight resistance　耐日光性
sunlight screening additive　日光屏蔽添加剂；防晒添加剂
sunn　印度麻纤维；菽麻纤维〔取自 Crotalaria juncea〕
sunning of kerosene　煤油日晒脱色
sun patent leather out of doors　室外用耐晒专利皮
sun-proof　耐晒的
sun-screening agent　防晒剂
sun screening compound　防晒化合物；防晒剂
sun screen preparations　防晒黑化妆品
sunshine arc fadeometer　日照弧光褪色试验器
sunshine unit　锶单位
sunstone　日长石
sunt　（水提）金合欢胶
suntan　晒黑；晒黑剂

suntan preparations　防晒斑化妆品；晒黑化妆品
sun test　日晒试验〔石油产品颜色安定性〕
super　〔拉丁字头〕①过；超；高于②总〔生物分类〕
super absorbent polymer　高吸水性高分子
super accelerator　超促进剂
super acid　超酸*
super acidic catalyst　超强酸催化剂
superacidity　过度酸性
superacid polymer　过(量)酸(的)聚合物
superacid solution　过量乙酸溶液
superacidulated　过酸化的
superacidulation　过酸化作用
superactinide　超锕系元素
superactinide element　后锕系元素
superactinide series　超锕系元素
superactive catalyst　超活性催化剂
superactivity　超活性
superaddition　附加物
super ageing mix　超老化混合胶
superalkali　苛性钠；氢氧化钠
superalkalinity　碱性过度
superalloy　高温合金
superamide　过(氧)酰胺
super-anthracite　超级无烟煤
superatmospheric pressure　超级大气压
superazeotropic distillation process for dilute nitric acid concentration　共沸蒸馏稀硝酸浓缩法
super balloon　超压轮胎
super basic catalyst　超强碱催化剂
superbine　南非海葱苦素
supercage　超笼
supercalender　高度砑光机；超滚压机；多辊压延机
supercarbonate(=bicarbonate)　碳酸氢盐　$MHCO_3$
supercarburize　过度掺碳
super-cell　硅藻土助滤剂
super centrifugation　超速离心
supercentrifuge　超速离心机
supercharge　增加器；增压器
supercharged test method(=supercharge method)　增压进料试验法
supercharge method(=supercharged test method)　增压进料试验法
supercharge octane number　增压辛烷值
supercharger　增压器
supercharging　增压作用
supercharging device　增压装置
superchlorination　过氯化(作用)
super clear laboratory　超净实验室

supercoat 保护涂层
supercoil 超螺旋
supercombat gasoline 超级战斗机用汽油；115/145 航空汽油〔染成暗红色的战斗机用高级汽油〕
superconductance 超导
superconducting alloy 超导合金
superconducting cable 超导电缆
superconducting carbonitride 超导碳氮化物
superconducting characteristic 超导特性
superconducting filament 超导丝
superconducting generator 超导发生器
superconducting helix 超导螺旋线
superconducting motor 超导电机；超导马达
superconducting multifilament 超导复丝
superconducting multifilamentary wire 超导金属复丝(线)
superconducting quantum interference devices 超导量子干涉器件
superconducting state 超导态
superconducting synchrotron 超导同步加速器
superconducting thin film 超导薄膜
superconducting transition temperature 超导转变温度
superconductive path 超导电路
superconductive polymer 超导聚合物
superconductor 超导体*
supercooled liquid 过冷液体
supercooled micelle 过冷胶束；深冷胶束
supercooled state 过冷态
supercooled vapour 过冷蒸气
supercooler 过冷却器
supercooling 过冷
supercritical 超临界的
supercritical fluid 超临界流体
supercritical fluid chromatograph 超临界流体色谱仪
supercritical fluid chromatograph-supercritical fluid chromatograph (SFC-SFC) 二维超临界流体色谱仪
supercritical fluid chromatography 超临界流体色谱法
supercritical fluid extraction (SFE) 超临界流体萃取
supercritical gaseous discharge 超临界排出气
supercritical liquid 超临界液体
supercritical liquid chromatography 超临界液相色谱法
supercritical mobile phase 超临界流动相
supercritical pressure boiler 超临界压力锅炉
supercrust rocks 表成岩
supercurrent 超(导)电流
supercyclone 超旋风器；超分尘器
superdelocalizability 超离域性；超非定域性
super-draw method 超拉伸法
super-dry ①过干燥②过干燥的

superelasticity 超弹性
super-element 超(重)元素
super-equivalent adsorption 超当量吸附
superexchange interaction 超交换相互作用
superfacial velocity 表观速度
superfast loading 快速装料
super-fast polymerization 超速聚合
superfatted soap 富脂(肥)皂
superfatting agent 富脂剂
super fibre 超强力纤维
superficial 表面的
superficial cementation(=surface cementation) 表面渗碳
superficial degradation 表面降解
superficial expansion 表面膨胀
superficial linear vilocity ①表观线速度②空塔线速度
superficially porous particle 表面多孔颗粒
superfiltrol 超滤土〔一种磨细的漂白土〕
superfine cement 超细水泥
superfine fibre 超细纤维
superfine fuel atomizer 超细燃料喷雾器
superfine orange shellac 优等橙色紫胶
superfine precipitated calcium carbonate 超细沉降碳酸钙
superfine silicone dioxide 超微细粒子白炭黑
superfine white 超精白〔煤油颜色品级〕
superfluid 超流体
superfluidity 超流动性
superfluous ends 倒断头
superfractionation 超精馏
superfractionation technique 超精馏工艺
superfractionator 超精馏塔
superfraction column 超精馏柱
super-fuel 超级燃料
superfusion 过熔(融)
super-gasoline 超级汽油；高抗爆性汽油
superglycerinated oil 超甘油化油
supergrip 快凝胶
super grip tire 超耐滑轮胎
superheated steam 过热蒸汽
superheated steam generator 过热蒸汽发生器
superheated vapor 过热蒸气
superheated vapor drier 过热蒸气干燥器
superheater 过热器；过热炉
superheating 过热
super heat resisting material (超级)耐热材料
superheavy bomb 超重(元素制的)弹
superheavy element 超重元素
superheavy hydrogen 超重氢

superheavy nucleus 超重核
superheavy particle 超重粒子
super helix 超螺旋
superhigh pressure boiler 超高压锅炉
superhigh pressure compressor 超高压压缩机
super high shrinkage in boiling water 超高沸水收缩率
superhighspeed 超高速的
super high tenacity rayon 超强力人造丝
super HILCA 超重离子直线加速器
superhybrid composite 超混杂复合材料
superhyperfine structure 超超精细结构
superimpose ①重叠②附加
superimposed peak 重叠峰
superimposed (pre-distillation) retort 重复(再蒸馏)甑
superintendent ①管理者；监督②管理的；指挥的
super ion-conductive polymer 超离子导电聚合物*
superior fuel 超级燃料
superior grease 超级润滑脂
superior layer 上层；高层
superior processing rubber 易操作橡胶
superior rubber 超级橡胶
superlattice 超(结晶)格子；超点阵
superlattice ordering 超晶格有序化
super-light foam 超轻质泡沫塑料
superloy 一种碳钢
supermalloy 镍钼铁锰合金
supermethylation 超甲基化(作用)
supermicro-analysis 超微量分析
super-microcrater fibre 超微凹坑纤维〔纤维表面有无数微坑，可产生深色效应〕
supermicro method 超微量法
super micron mill 超微粉碎机
supermicroscope 超显微镜
super modified controlled weighted centroid simplex method 超改进控制加权形心单纯形法
super modified simplex method 超改进单纯形法
supermol 胶束
supermolecular 超分子的
supermolecular lattice 超分子晶格
supermolecular structure 超分子结构
supermolecular texture 超分子织态结构
supermolecule 超分子
supermolecule approach 超分子方法
supernatant ①上层清液②(浮于)上层的
supernatant (fluid) 上清液
supernatant layer 清液层
supernatant liquid 沉清液体；清液层
superneat soap 超液晶皂

super-normal structure black 高结构炭黑
supernuclei 超(重)核
superoctane fuel(=superoctane number fuel) 超辛烷值燃料〔辛烷值大于100的燃料〕
superoctane number fuel(=superoctane fuel) 超辛烷值燃料〔辛烷值大于100的燃料〕
superol 硫酸邻羟基喹啉
superoxide (=peroxide) 过氧化物；超氧化物
superoxide anion 过氧化物阴离子
superoxide dismutase 超氧化物歧化酶*
superoxide radical 超氧自由基
superoxidized linseed oil 过氧化亚麻油
superoxidized oil 过氧化油
superpalite 氯甲酸三氯甲酯〔毒气〕；双光气
superpan emulsion 超全色乳剂
superparamagnetism 超顺磁性
superpeptized fuel 超胶溶燃料
superphosphate ①过磷酸钙②酸性磷酸盐
superphosphate den 过磷酸钙储藏室
superphosphate excavator 过磷酸钙挖掘机
superphosphate fertilizer 过磷酸钙肥料
superphosphate mixer 过磷酸钙混合机
super pneumatic tyre(=super balloon) 超压轮胎
superpolyamide 超高分子(量)聚酰胺
superpolyester 超聚酯
superpolymer 超高聚物
superpolymerization 超(分子量)聚合
superposability 可叠加性
superposition 叠加
superposition method ①叠加法②多层法
superposition principle 叠置原理
superposition theorem 叠加定理
superpotential 超电位
superprecipitation 超沉淀
superpressure 超压；超计大气压
superradiance 超发光
superrich mixture 过富(燃料-空气)混合物
super rubber 超级橡胶
super-saturability 过饱和度
supersaturated 过饱和的
supersaturated solution 过饱(和)溶液
supersaturated vapor 过饱和蒸气
supersaturation 过饱和*
supersaturation ratio 过饱和比
supersaturation solution 过饱和溶液
superselection rule 超选择定则
supersensitive 超高灵敏度的
supersensitivity 超灵敏度

super separator 超微分级器
superservice station 高级修车加油站
supersiliceous zeolite 高硅沸石
super snail 特慢(法)〔核磁〕
"supersoft" surfactant "超软性"表面活性剂〔极易生物降解的〕
supersolidification 过凝固(现象)
supersolubility 超溶解度；过溶度
supersonic ①超声的②超声速的
supersonic absorption 超音吸收
supersonic aircraft oil 超声航空油
supersonic beam source 超声束源
supersonic detector 超声速探伤器
supersonic dispersed oil 超声分散油
supersonic flaw detector 超声波探伤器
supersonic frequency 超声频(率)
supersonic generator 超声速发生器
supersonic jet fluorimetry 超声喷雾荧光法
supersonic jet synchronous fluorimetry 超声喷雾同步荧光法
supersonic machining 超声波加工
supersonic molecular beam 超声分子束
supersonic sounding 超声探测法
supersonic speed 超声速度
supersonic tannage 超声波鞣法
supersonic velocity 超声速(度)
supersonic wave 超声波
superspruce extract 精制的亚硫酸纸浆废液(浸)膏
supersteel 高速钢
superstoichiometric 过当量的；超化学计量的
superstructure paint (船舶)上层建筑漆
super sublimed white lead 碱式硫酸铅白
super suds 高泡型洗涤剂
supertension(=overvoltage) (超)(额)电压
super thin cut 超薄切片法
superthreshold 超阈(能)
superthreshold neutron interrogation technique 高于阈能的中子探询技术
supertransuranic element 铀后元素
super tyre 超级轮胎
supervarnish 桐油清漆
super viscose fiber 超强黏胶纤维
supervised pattern recognition 有监督模式识别
supervision 监督；管理
supervisory control 监控
supervoltage 超电压
supervoltage radiotherapy 超压放射治疗
super-washable 超耐洗的

super water 反常水；超聚水；不冻水〔现已否定其存在〕
superwaxy phase 超蜡状相
superwaxy soap 超蜡状皂
supper fine grinding 超细粉碎
supper micron mill 超微粉碎机
supplemental equipment 补充设备
supplemental firing 补充烧炉
supplemental grinding 补充打磨；找补打磨
supplementary air intake 补充空气加入
supplementary bond angle 键角补角
supplementary condition 补充条件；附加条件
supplementary fuel 增补燃料
supplementary furnace 补充炉
supplementary product 辅料
supplementary speed reduction means 辅助减速装置
supplementary valency 补价
supplement 2 oil 补2号油〔含添加剂约11%的在繁重条件下应用的润滑油〕
supply ①供应；供给②填补③备有之物
supply capacity lag 供热容量滞后〔指石油加工热控制过程〕
supply column 供应塔
supply container 供应容器
supply contract 供应合同
supply failure 供应中断；储存用罄
supply line 供应管线
supply main 供应总管
supply of fuel 燃料供应
supply of oil 油的供应
supply pack 电源装置
supply ring 供油环；进料圈；油环
supply side 供应方面
supply stream 供应流
supply system 供应系统
supply tank 供应油罐
supply unit 电源装置
supply valve 供给阀
supply voltage 电源电压
support ①架；支座②载体
support clamp 框架
support-coated open tubular column (SCOTC) 载体涂层开口管柱
support coating 载体涂布
support density 载体密度
supported bilayer lipid membrane 支撑双层磷脂膜
supported catalyst 载体上的催化剂
supported liquid membrane (SLM) 液膜支撑
support effect 载体效应

supporter 载体
support induced crystal growth 载体诱导晶体生长
supporting agent 载剂
supporting device 支承装置
supporting electrode 辅助电极；支持电极
supporting electrolyte(=indifferent electrolyte) 支持电解质
supporting equipment 辅助设备
supporting gas 载气
supporting idler 托滚〔运输带〕
supporting liquid 载液
supporting material ①载体②支承材料
supporting medium 载体
supporting membrane 支撑膜
supporting pillar 支持柱
supporting-point roller ①托轮(辊)；支撑轮(辊)②随动辊
supporting pole 中央支持柱
supporting roll 承压滚
supporting structure of catalyst 催化剂载体结构
supporting substrate 基材；底材
supporting yard 支持场地
support leg 支柱；支脚
support of the catalyst 催化剂载体
support plate 支承板；垫板；载板
support prime 载体涂底液
support ring 支承环
support rod 载体棒
support stand 支架
support surface adsorption 载体表面吸附作用
suppository 栓剂；塞药；坐药
suppressed column 抑制柱
suppressed conductance detection 抑制型电导检测
suppressed ion chromatography 抑制型离子色谱法
suppressing agent 抑制剂
suppressing of detonation(=suppressing of knocking; suppression of knock) 爆震的抑制
suppressing of knocking(=suppressing of detonation) 爆震的抑制
suppression 抑制；熄灭；消除
suppression of knock(=suppressing of detonation) 爆震的抑制
suppressor 抑制剂*
suppressor column 抑制柱
suppressor effect 抑制剂效应
suppressor gene 压制基因；校正基因
suppressor grid 抑制栅
suppurative ①化脓剂②化脓的
supra- 超；上

suprafacial ①同侧；同面的②表面上的
supralethal irradiation 超致死剂量照射
supramolecular 超分子的
supramolecular device 超分子器件
supramolecular material 超分子材料
supramolecular polymer 超分子聚合物
supramolecular structure 超分子结构
supramolecule 超分子
supra polymer 超高分子
suprarenalin 肾上腺素
suprarenaline 肾上腺素
suprarenin 肾上腺素
suprasterol 超甾醇；过照甾醇
supravital staining 离体活体染色；体外活体染色
suramin 苏拉明〔一种抗锥虫及丝虫药〕
surface 表面*
surface /volume ratio 表面/体积比；面积/容积比；面容比
surface absorber 表面吸收器
surface acoustic wave 表面声波
surface acoustic wave-impedance sensor 阻抗式表面声波传感器
surface acoustic wave sensor 表面声波传感器
surface action 表面作用
surface-activated clay 表面活化陶土
surface activation 表面激活
surface active 表面活性的
surface active agent 表面活性剂
surface-active anion 表面活性阴离子
surface-active cation 表面活性阳离子
surface-active collector 表面活性促集剂
surface active composition 表面活性组分
surface-active impurity 表面活性杂质
surface-active indicator 表面活性指示剂
surface-active ion 表面活性离子
surface active substance 表面活性物质
surface activity 表面活性；表面活度
surface adhesion 表面黏附
surface aeration equipment 表面曝气装置
surface ag(e)ing 表面老化
surface air cooler 表面空气冷却器
surface alkalinity ①表面碱性②表面含碱量
surface analysis 表面分析
surface analyzer 表面分析仪
surface and microarea analysis technology 表面和微区分析技术
surface antibody 表面抗体
surface antigen 表面抗原

surface applicator　表面敷贴器
surface area　表面积
surface area analyzer　比表面(积)分析仪
surface area apparatus　表面积仪
surface atom concentration　表面原子浓度
surface atomizer　液面喷雾器
surface availability　表面有效度
surface availability of catalyst　催化剂的有效表面
surface balance　表面天平
surface bed　表面床
surface blemish　表面缺陷
surface blister　表面发泡(剂)
surface blush　雾面
surface-bonded water　表面结合水
surface-catalyzed photodegradation reaction　表面-催化的光降解反应
surface-catalyzed reaction　表面催化的反应
surface cementation(=superficial cementation)　表面渗碳
surface charge　表面电荷
surface charge density　表面电荷密度
surface chelate compound　表面螯合物
surface chemical shift　表面化学位移
surface chemistry　表面化学
surface chromatography　表面色谱(法)
surface cloudiness　表面浑浊性(度)
surface coarsening　(模具)表面粗化
surface coat　表面涂层
surface coating　表面涂层
surface coating formulations　表面涂料配方
surface coating technique　表面涂层技术
surface cock　液面控制旋塞；液面控制活栓
surface coefficient　表面系数
surface coefficient of heat transfer　表面散热系数
surface color　表面色
surface complexation reaction　表面络合反应
surface compound technique　表面复合技术
surface concentration　表面浓度
surface condenser　表面冷凝器
surface conductance　表面电导
surface conductivity　表面电导率
surface contact　表面接触
surface container　表面调整剂
surface contamination　表面污染
surface corrosion　表面腐蚀
surface coverage　表面覆盖度
surface covered　①涂敷面积；涂布面积②涂敷效率；涂布速度
surface crack　表面裂纹；表面裂缝

surface crystallography　表面晶体学
surface crystal nucleation　表面晶核化过程
surface current classifier　表流分级器
surface damage　表面损伤
surface decoration　表面修饰
surface density　表面密度
surface diffusion　表面扩散
surface diffusion coefficient　表面扩散系数
surface dipole moment　表面偶极矩
surface disturbance　表面损伤
surface duality　表面二象性
surface effect　表面效应
surface elasticity　表面弹性
surface elastic modulus　表面弹性模量
surface element　表面元素
surface elongation viscosity　表面伸长黏度
surface emission flame photometric detector　表面发射火焰光度检测器
surface emission ion source　表面发射离子源
surface emissivity　表面发射系数
surface energetics　表面能学
surface energy　表面能
surface energy level　表面能级
surface energy of liquid　液体的表面能
surface energy of polymer　高聚物表面能
surface engineering　表面工程
surface enhanced Raman scattering　表面增强拉曼散射
surface enhanced Raman scattering spectrum (SERS)　表面增强拉曼散射谱
surface enhanced Raman spectrometry (SERS)　表面增强拉曼光谱法
surface enhanced Raman spectroscopy　表面增强拉曼光谱法
surface enhanced resonance Raman scattering　表面增强共振拉曼散射
surface enhanced resonant Raman spectroscopy　表面增强共振拉曼光谱法
surface enrichment　表面富集
surface entropy　表面熵
surface evaporation　表面蒸发(作用)
surface excess　表面超额*
surface excess concentration　表面过剩浓度
surface film　表面膜
surface film balance　表面膜天平
surface film potential　表面膜势
surface filtration　表面过滤
surface finish　表面加工；表面抛光
surface fit　曲面拟合

surface flow 表面流动
surface flow classifier 表面流分级器
surface force 表面力
surface fracture 表面破损
surface free energy 表面自由能
surface gauge 划准盘
surface-generating reactor 表面增生型反应器
surface grafting 表面接枝
surface hardener 表面硬化剂
surface hardening 表面硬化
surface imperfection 表面缺陷
surface imprinting 表面印渍法
surface-induced dissociation 表面诱导解离
surface ionic site 表面离子点
surface ionization 表面电离
surface ionization mass spectrometry 表面电离质谱法
surface lacquer 表面喷漆
surface layer 表层；面层
surface microlayer 表面微层
surface mobility 表面迁移率
surface modification 表面改性；表面改质
surface modification of polymer 聚合物表面改性
surface modification technique 表面改性技术
surface modified fibre 表面改性纤维
surface modifier 表面改性剂
surface-modifying agent 表面改性剂
surface orientation effect 表面定向效应
surface parameter 表面参数
surface phase 表面相
surface pH electrode 表面 pH 电极
surface photomodification 表面光致改性
surface photovoltaic spectroscopy 表面光电(压)光谱法
surface pin 平推顶杆
surface plasma resonance 表面等离子体共振技术
surface plasma resonance absorption 表面等离子体共振吸收
surface plasticity 表面塑性
surface plate ①平板②表面板③平台
surface poise 表面泊〔表面黏度单位〕
surface polymerization 界面聚合(作用)
surface porosity 表面孔率
surface potential 表面电位
surface potential detector 表面电位检测器
surface preparation 表面预处理
surface preparation of adhesion 胶黏表面处理
surface printing 表面印染
surface pyrometer 表面高温计
surfacer 二道底漆；二道浆
surface reaction 表面反应
surface relaxation 表面弛豫
surface renewal theory 表面更新理论
surface resistance 表面电阻
surface resistivity 表面电阻系数；表面电阻率
surface roughness 表面粗(糙)度
surface roughness factor 表面粗糙性因子
surface sandblast 表面喷砂(处理)
surface segregation 表面偏析
surface shear viscosity 表面剪切黏性
surface speed 钻针表面(切线)速度
surface state analysis 表面态分析
surface tack 表面发黏
surface tack eliminator 表面防黏剂
surface tackiness 表面黏性
surface tensiometer 表面张力计
surface tension 表面张力*
surface tension apparatus 表面张力测定仪
surface tension balance 表面张力测定计；表面张力天平
surface tension effect 表面张力效应
surface tension gradient 表面张力梯度
surface tension method 表面张力法
surface tension of latex 胶乳表面张力
surface transport 表面迁移
surface treating agent 表面处理剂
surface treatment 表面处理
surface-uniformity gloss 平面均匀光泽
surface unit 表面单位
surface viscometer 表面黏度计
surface viscosimeter 表面黏度计
surface viscosity 表面黏度
surface water 地面水；地表水
surface X-ray absorption spectroscopy 表面X射线吸收谱*
surfacing build-up welding 堆焊
surfacing mat 表面毡(片)
surfacing of roofing material 铺顶表面平整
surfacing process 镀面工艺；贴面工艺
surfactancy 表面活性
surfactant (=surface-active agent) 表面活性剂
surfactant electrode 表面活性剂电极
surfactant micelle 表面活性剂胶束
surfactant partitioning 表面活性剂分配
surfactant property 表面活化性能
surfactant selective electrode 表面活性剂选择性电极
surfactant selective electrode method 表面活性剂(离子)选择电极法〔测临界胶束浓度〕
surfaction 表面改质；表面改性

surfactivity 表面活性
surfatron 同轴表面波激励器
surfused liquid 过冷液(体)
surfusion 过冷(现象)
surge chamber 调节室
surge control 喘振控制；喘振防护
surge damper 减震器
surge drum 收集筒；平衡筒
surge effect 涌浪效应
surge hopper 浪涌(自动)加料斗
surge impedance 涌浪组抗
surge plate 涌浪挡板
surge pot 缓冲罐；平衡罐
surge pump 涌浪泵；膜式泵；隔膜泵
surgery 外科(学)
surge tank 缓冲缶；稳压缶；气室
surgical adhesive 橡皮膏
surgical soap 药(用)皂
surging 涌浪；突波
surging force 涌浪力；脉冲力；冲击力
surging of gas 气体的涌浪
surging shock 涌浪撞震；液捶
surgumycin 苏尔古霉素
surinamine N-甲基酪氨酸；柯丫树碱
surpalite 聚光气
surplus ①多余的；剩余的②残料；余料
surplus air 剩余空气
surplus capacity 剩余容量；过剩容量
surplus factor 剩余因子
surplus material 剩余材料
surplus oil 剩余油
surplus paint 多余油漆
surplus pressure 剩余压力
surplus sludge 剩余污泥
surplus solution 剩余溶液
surplus stock 剩余原料
surplus valve 溢流阀
surrogate reference material 代用标准物质
surrogate standard 代标标样
surrosion 腐蚀增重(作用)
surrounding medium 环境介质
surroundings 环境*
surrounding temperature 周围温度
surrounding water jacket 外层水夹套
survey ①测量②察视
surveying ①测量(术)；测量(学)②测量
survey meter (放射性污染)检测器
survey scan 全谱扫描

survival probability 残存概率；存活率
susceptance 电纳
susceptibility ①(=magnetic susceptibility)磁化率②感受性③敏感度
susceptibility correction 敏感性校正〔校正溶剂对核磁共振谱的影响〕
susceptibility factor 敏感因素
susceptibility of gasoline 汽油的敏感性(对四乙铅而言)
susceptibility to dehydrogenation 对脱氢作用的敏感性
susceptibility to photooxidation 光(致)氧化敏感性
susceptible ①敏感的②易染的
susceptiveness 灵敏度
susceptometry 电纳分析
susceptor 感受器；接受器
suspendability 悬浮力；悬浮性能
suspend conveyor 起吊运输机；悬空传送带
suspended ①悬浮的②暂(停)的
suspended arch 悬拱
suspended carbon 悬浮的碳
suspended centrifuge 上挂式离心机
suspended level viscometer 气承液柱黏度计
suspended matter 悬浮物；悬浮质
suspended oil 悬浮油类
suspended particles 悬浮粒子
suspended particulate 悬浮颗粒物；飘尘
suspended phase 悬浮相
suspended pipe line(=suspension pipe line) 架空管道
suspended refractory arch 架空耐火拱穹
suspended sludge 悬浮淤渣
suspended solid 悬浮固体
suspended state 悬浮(状)态
suspended substance 悬浮物；悬浮体
suspended transformation 悬浮转变(作用)
suspended type of furnace 悬吊式炉〔在钢架上悬吊的裂化炉〕
suspended wall lining 悬吊式壁面铺盖层
suspender 吊鞦池
suspending agent 悬浮剂
suspending capacity 悬浮能力
suspending medium 悬浮介质
suspending phase 悬浮相
suspending power 悬浮力
suspending velocity 悬浮速度
suspension 悬浮液；悬浊液
suspension agent (=suspending agent) 悬浮剂
suspension bed 悬浮床
suspension bunker 悬吊进漏斗
suspension colloid 悬(浮)胶(体)

suspension dispersing agent 悬浮分散剂
suspension dyeing 悬浮染色
suspension effect 悬浮效应
suspension line 悬挂索〔吊管道用〕
suspension medium 悬浮介质
suspension method 悬浮法
suspension packing 悬浮填充
suspension pipe line(=suspended pipe line) 架空管道
suspension polymer 悬浮聚合物
suspension polymerization 悬浮聚合*
suspension rheology 悬浮体流变学
suspensions of powdered catalyst 粉末催化剂悬浮体
suspension stability 悬浮稳定性
suspension stabilizer 悬浮稳定剂
suspension system ①悬浮(体)系②摇动法
suspension time 悬浮时间
suspensoid 悬浮体；悬胶(体)
suspensoid catalytic cracking process 悬胶催化裂化过程
suspensoid cracked fuel 悬胶催化裂化燃料
suspensoid media 悬胶介质
suspensoid state 悬胶态
sussexite 硼锰矿
sustainable development 可持续发展
sustained firing 长射；持续喷射
sustained off-resonance irradiation 持续去共振照射
sustained release dosage 缓释剂
sustained wave 持续波
Sutherland-Einstein equation 萨瑟兰-爱因斯坦方程
suture 缝合
suxamethonium bromide 溴化琥珀酰胆碱
suxamethonium chloride 氯化琥珀胆碱
Svedberg unit 斯韦柏单位〔沉降系数单位〕
swabber 装管工
swabbing 刷浆；刷色
swab-man 管道清洁工
swab method 棉拭检验法
swage 铁模；铁型
swagelok (管子的)接套；接头紧锁螺母
swagelok coupling 套管接头
swaging 锻造
swaging machine 锻造机；锻冶机
Swale powder 斯韦耳(炸)药
swamp bay oil 沼泽油梨油；湿地桂油
swamp ore 沼铁矿
Swan band 斯旺(谱)带
swan neck 鹅颈式管；弯头管
swan neck system 鹅颈管装置；弯(曲)管装置
swap action valve 交换作用阀；快速作用阀

Sward rocker 斯瓦德摆杆硬度计
swarf 细铁屑
swash-plate (罐中)隔板
swatch (小块)布样
sweat band leather (帽中)汗带皮
sweat distillate(=sweat oil) 发汗油〔石蜡发汗时所得的油〕
sweated scale wax 片状发汗石蜡
sweated wax 发汗石蜡
sweater 发汗器
sweating heat 焊接热
sweating metal 焊接金属
sweating of ore 煅烧矿石
sweating of wax 石蜡的热熔
sweating process 发汗法
sweating soldering 热熔焊接
sweating stove (石蜡)发汗炉
sweating time (亚麻子油腻子制造中的)热熔时间
sweating water 表面凝结水；结露水
sweat iron 焊铁
sweat oil(=sweat distillate) 发汗油〔石蜡发汗时所得的油〕
sweat pan 发汗盘〔石蜡工厂发汗间用〕
sweat pit 发汗槽
sweat roll 蒸汽滚筒
sweat wax 发汗石蜡
swedged 减小直径的
Swedish putty 瑞典腻子〔石膏-纤维素型填孔料和油漆或聚醋酸乙烯黏合剂混合制成的稠浆〕
Swedish standard rust grade 瑞典标准锈蚀等级
sweep ①扫描②清扫；吹扫
sweep beam 扫描(电子)束
sweep-blast(ing)(=brush-off blast) 刷(拭)除锈(级)喷砂；刷净级喷砂(处理)
sweep boom 扫油板〔石油厂捕集池水面上的薄板〕
sweep coil 扫描线圈
sweep diffusion method 扫描扩散法
sweep electron microscope 扫描式电子显微镜
sweeping 吹扫；刮除
sweeping gas 吹扫气体
sweeping-in action of lubricant 润滑剂的刮入动作
sweeping purging 冲洗；清除
sweep interval 扫描时间
sweep recurrence rate 扫描复现率；扫描重复率
sweep time 火焰扩展时间
sweep volume ①扫过容积②扫描容积
sweep width 扫描宽度
sweep wood 沉底木
sweet acacia 金合欢
sweet almond oil 甜杏仁油

sweet basil oil （甜）罗勒油
sweet bay(=bay) 月桂
sweet birch 北美甜桦
sweet birch oil （甜)桦油；水杨酸甲酯
sweet cane 白菖蒲
sweet cherry 欧洲甜樱桃
sweet chervil 欧洲没药
sweet chestnut 欧洲栗
sweet citron 甜柠檬
sweet clay 鲜白土；再生漂白土
sweet corrosion 载硫腐蚀；淡水腐蚀〔淡水中的 CO_2 腐蚀〕
sweet crude oil 低硫原油
sweet distillate(=sweetened distillate) 脱硫馏分
sweet dry gas 脱硫干(石油)气
sweetened distillate(=sweet distillate) 脱硫馏分
sweetener ①脱硫设备；香化设备；用博士溶液精制汽油的设备②增甜剂；增香剂
sweetening 脱硫；香化；博士溶液精制；变甜;甘味料
sweetening agent 脱硫剂；香化剂
sweetening of hydrocarbon 烃类脱硫，烃类香化
sweetening of sour distillate 酸性馏分脱硫
sweetening process 脱硫过程；香化过程；香化处理；除去含硫化合物
sweetening still 脱硫蒸馏釜；香化蒸馏器
sweetening treatment 脱硫；香化；用博士溶液精制
sweet extract 柔性栲胶
sweet fennel 甜茴香
sweet flag 白藤
sweet floral 甜的花香
sweet gas 无硫气〔石油〕
sweet gasoline 无硫汽油
sweet golden rod oil 香一枝黄花油；甜秋金草油
sweet grass 白菖蒲
sweet gum tree 胶皮糖香树
Sweetland filter 斯威特兰滤器
Sweetland filter press 斯威特兰压滤机〔滤清石油产品用〕
sweet lemon 甜柠檬
sweet limon oil 甜柠檬油
sweet mandins 香薷草
sweet marjoram oil 甘牛至油；香马郁兰油
sweet melon 甜瓜
sweet milfoil 香薷草
sweet myrrh 红没药
sweet oil 脱硫油
sweet orange blossom oil 甜橙花油
sweet orange oil 甜橙油
sweet orange peel 甜橙皮
sweet pea 香豌豆；香鲨豆
sweet petroleum product 无硫石油产品
sweet roasting 去硫煅烧
sweet scented verbena 防臭木
sweet-smelling gasoline 香汽油〔去硫醇无臭味的汽油〕
sweet sorghum 高粱；蜀黍
sweet spirit of nitre 亚硝酸乙酯的酒精溶液
sweet stock 无硫原料
sweet tan liquor 鲜鞣液
sweet violet 香堇菜；紫罗兰
sweet water pump 甜水泵
sweet wood ruff 香车叶草
sweet yarrow 香薷草
swellability 溶胀性；溶胀度
swellant 溶胀剂，泡胀剂
sweller 膨胀剂；溶胀剂
swell factor 膨胀系数
swell-gelation(=swell-gelatination) 膨润-凝胶化(作用)
swelling 溶胀*
swelling ability （纤维)溶胀性；溶胀本领
swelling acid 膨胀性酸
swelling agent 溶胀剂
swelling anisotropy 溶胀各向异性
swelling capacity ①溶胀度②溶胀性
swelling clay 膨胀瓷土；膨胀黏土
swelling coefficient 膨胀系数
swelling cut fibre ends method 切断纤维端溶胀法〔测定尼龙变形丝定型度的方法〕
swelling degree 膨胀度
swelling effect 溶胀效应
swelling heat 膨胀热
swelling index 溶胀指数
swelling liquid (=solvent) 溶剂
swelling measurement 溶胀测量
swelling of paint coating （油漆)涂层溶胀
swelling power ①溶胀本领②溶胀力
swelling pressure 溶胀压力
swelling property 溶胀性
swelling ratio 溶胀比
Swenson-Walker crystallizer 斯温森-沃克结晶器
swept volume 活塞排量
swerchirin 獐牙菜精
swimming particles strainer 悬浮微粒捕集器
swimming pool coating 游泳池涂料
swimming pool paint 游泳池用漆
Swinburne-Ashcroft process (for zinc chloride) 斯温伯恩-阿舍夫(制氯化锌)法
swinestone 臭石灰；沥青石灰石

swing　①(摇)摆；摇②回转③吊
swing axle　摆动轴
swing check valve　摇板式逆阀
swing gate creel　转门式筒子架
swing hammer mill　摆锤磨
swing hammer pulverizer　摆锤式粉碎机
swinging beam hardness tester　摆杆硬度计
swinging bucket　悬挂吊桶式(转子)
swinging conveyer　摆动运输机
swinging gauge method　吊表法〔测注油高度〕
swinging screen　摆动筛
swinging sieve　摆动筛
swinging strength　摆动力量；抗摆强度
swinging suction　摇头吸料管；摇头虹吸管
swinging table　推旋台
swinging tray　吊盘〔面包〕
swinging tray oven　吊盘炉〔面包〕
swinging-vane oil pump　转叶式油泵
swing joint　旋转接头
swing line(=swing pipe)　吊管；摆管〔油罐〕
swing man　代班人
swing method　摆动法(天平)
swing pipe　吊管；摆管〔油罐〕
swing reactor　摇摆式反应器
swing sieve　摇筛
swing sledge mill　摆槌磨
swing-up inspection window　刮片式视孔
swirl　回荡
swirl combustion chamber　涡流式燃烧室
swirling flow　旋流
swirling fuel injection　涡流燃烧喷射
swirl nozzle　旋流喷嘴
swirls　漩涡式缺陷
Swiss pine oil　瑞士松油〔取自 Pinus cembra〕
switch　①电闸②转辙器
switchable neutron source　(可)开关中子源
switchboard model　接线板模型
switch box　(电)闸盒
switch capsule　开关盒
switcher　调转工〔管理石油进入油罐〕
switch grease　电闸润滑脂
switching column technique　色谱柱切换技术
switching time　切换时间
switch oil　电闸油；变压器油
switch on　接通；接入
switch sampling　切换(阀)进样
switch the feed　转换进料
switch valves　转换阀；转换开关
swivel　①转体；旋转部分②旋转；旋回
swivel bottom vice　转盘虎钳
swivel hook　转动钩；铰链环
swivel joint　旋转接合
swivel pipe　旋转管
swivel replication　转环复制
swivel table　转台；转盘
swivel type Mac-Leod gauge　旋转型麦氏真空计
SWMILL sand grinder　斯沃米尔型砂磨机(间歇式)
swollen　溶胀的
swollen cellulose　溶胀纤维素
swollen rubber　溶胀橡胶
swollen state　溶胀态
sycoceryl　无花基　$C_{17}H_{27}CH_2$—
sycoceryl alcohol　无花醇　$C_{17}H_{27}CH_2OH$
syderolite　陶土
sydnone　悉尼酮*
syenite　黑花岗岩；正长岩
sylva　森林
sylvan　邻甲呋喃；2-甲基呋喃
sylvanite　针碲金矿
sylvate　松香酸盐
sylvecarvone　枞香芹酮
sylvestrene　枞油烯
sylveterpin(=carveterpin)　枞萜二醇；香芹萜二醇
sylveterpineol　枞松油醇
sylvic acid　松香酸
sylvic oil　树脂油
sylviculture　林学
sylvine　钾石盐；钾盐
sylvinite　钾石盐
sylvite　钾盐
sym(=symmetrical)　对称；均
symbasis　联基
symbiotic fixation　共生固氮(法)
symbiotic growth　共生
symbiotic nitrogen fixation　共生固氮作用
symbol　符号；记号
symbolic addition method　符号附加法
symbol of point group　点群符号
symbol weight　符号量
symcenter　对称中心
sym-dichlorotetrafluoroacetone hydrate　均二氯四氟丙酮水合物
sym-di-o-tolyl guanidine　均二邻甲苯基胍
symmetria　对称
symmetric　对称的
symmetrical　对称的

symmetrical chiasma 对称交叉
symmetrical compound 对称化合物
symmetrical diagram 对称图解
symmetrical factor 对称因子;对称因素
symmetrical laminate 对称层合板
symmetrical peak 对称峰
symmetrical powder cloud 对称的粉末雾
symmetrical ring 对称环
symmetrical solute band 对称溶质谱带
symmetrical standard state 对称标准态
symmetric(al) state 对称态
symmetrical stretching vibration 对称伸缩振动
symmetrical titration 对称滴定
symmetrical top 对称陀螺
symmetrical wave 对称波
symmetric(al) wave function 对称波函数
symmetric carbon atom 对称碳原子
symmetric center 对称中心*
symmetric matrix 对称矩阵
symmetric molecule 对称分子
symmetric top molecule 对称陀螺(形)分子
symmetric tread pattern 对称式胎面花纹
symmetric vibration 对称摆动
symmetrization 对称化作用
symmetry 对称(现象);对称性
symmetry-adapted basis 对称性匹配基*
symmetry-adapted configuration 对称性匹配组态*
symmetry-allowed 对称允许的
symmetry axis 对称轴
symmetry center 对称中心
symmetry class 对称类型
symmetry element 对称元素*
symmetry factor 对称因素*
symmetry forbidden 对称性禁阻*
symmetry forbidden reaction 对称禁阻反应*
symmetry operation 对称操作*
symmetry orbital 对称轨道
symmetry plane 对称面
symmetry selection rules 对称选择定则
sympathetic ink 隐显墨水
sympathetic reaction 交感反应
sympeda 交叉对称
symphorol 咖啡碱磺酸盐
symplesite 砷铁矿
symplex structure 对称结构
symproportionation 对称歧化(作用)
sympton ①征兆②症状
sym-tetrachloroacetone 对称四氯丙酮;1,1,3,3-四氯丙酮 $(Cl_2CH)_2CO$
sym-tetrachloroethane 对称四氯乙烷;1,1,2,2-四氯乙烷 $Cl_2CHCHCl_2$
sym-tetraphenylethane 对称四苯乙烷 $[(C_6H_5)_2CH]_2$
syn- 〔希腊及拉丁字头〕 ①顺式②同;共;与
synadelphite 砷铝锰矿
synaeresis(=syneresis) 脱水收缩(作用)
synalbumin 抗胰岛素
synaldoxime 顺式醛肟
synanthrin 菊糖;菊粉
synaptic current 突触电流
synartetic acceleration 邻位加速
syncaine(=procaine) 顺卡因〔即:普鲁卡因〕
syncartesis 离子化物质过渡状态的稳定
synchrocyclotron 同步回旋加速器
synchronal 同步的;同时的
synchroneity 同步性
synchronism 同步性
synchronization 同步(作用)
synchronization control 同步控制;同步调整
synchronized cyclic capillary electrophoresis 同步循环毛细管电泳
synchronized oscillation 同步振动
synchronized timing 同步成圈
synchronizer 整步器;同步器
synchronizing 整步;同步
synchronous 同步的;同时性的
synchronous analysis 同时分析
synchronous converter 同步换流机
synchronous derivative fluorimetry 同步导数荧光法
synchronous fluorescence 同步荧光
synchronous fluorescence quenching method 同步荧光猝灭法
synchronous fluorimetry 同步荧光分析法
synchronous motor 同步电动机
synchronous process 同步过程
synchronous radiation total reflection X-ray fluorescence analysis 同步辐射全反射 X 射线荧光分析
synchrophasotron 同步稳相加速器
synchroscope 同步示波器
synchrotron 同步加速器
synchrotron radiation (SR) 同步辐射*;同步辐射源
synchrotron radiation photoelectron spectroscopy (SRPES) 同步辐射光电子能谱法
synchrotron radiation X-ray fluorescence spectrometry 同步辐射 X 射线荧光法
syncillin 苯氧乙基青霉素
synclinal conformation 顺错构象*

syncurine 碘化癸烷双胺
syncyanin 脓蓝素
syndesine 联赖氨酸；羟赖氨醛醇
syndet(=synthetic detergent) 合成洗涤剂
syndetergent 合成洗涤剂
syndiazo compound 顺式重氮化合物
syndiazotate 顺重氮酸盐
syndioduotactic stereoregularity 间同双立构规整性
syndio form 间同(立构)型
syndiospecificity 间规度
syndiotactic 间同的；间规的
syndiotactic addition 间同(立构)加成
syndiotactic configuration 间同(立构)构型
syndiotactic index 间同(立构)指数
syndiotacticity 间同度；间同立构(规整)度
syndiotactic placement 间同(立构)键接
syndiotactic polymer 间同立构聚合物；间规聚合物
syndiotactic polymerization 间同(立构)聚合
syndiotactic polypropylene 间规聚丙烯
syndiotactic polyvinyl alcohol 间同(立构)聚乙烯醇
syndiotactic propagation 间同(立构)增殖
syndiotactic sequence 间同(立构)序列
syndiotactic structure 间同结构；间规结构
syndiotactic triad 间同(立构)三(单)元组
syndiotactic unit 间同立构单元
syndrome 症候群；综合症
syndyotactic polymer 间规(立构)聚合物；间同(立构)聚合物
synephrine 脱氧肾上腺素
syneresis 脱水收缩(作用)
synergetic coefficient 协同系数
synergetic effect 协同效应；增效作用
synergetic extraction 协同萃取
synergetic luminescence 协同发光
synergia 协同；增效
synergic 协同的；增效的
synergic reagent 协(同)萃(取)剂
synergic reflex 协同反射；协作反射
synergism 协合作用；增效作用
synergist 协萃剂；增效剂
synergistic action 协合作用
synergistic agent 增效剂
synergistic catalyst 增效催化剂；协合催化剂
synergistic coefficient (SC) 协萃系数
synergistic effect 协同效应*
synergistic emulsifier 多效乳化剂
synergistic extractant 协萃剂
synergistic extraction 协同萃取
synergistic flame retardant 增效性阻燃剂
synergistic reaction 协同作用
synergistic reactive surfactant 具有协同作用及反应活性的表面活性剂
synergistic system 协(同)萃(取)体系
synfacial reaction 同面反应*
syn-form 顺式
synfuel 合成燃料
syngas 合成气
syngenite 钾石膏
syn-isomerism 顺式(同分)异构
synol catalyst 辛诺催化剂
synol products 辛诺产品
synol synthesis 辛诺合成〔熔铁催化剂由合成气体制醇、烯、烷液体混合物〕
synonym 异名
synourin oil 脱水蓖麻油
syn-oxime 顺式肟
synperiplanar conformation 顺叠构象*
syn-position 顺位
synroc 合成岩石
syntactic 复合铸塑的
syntactic pattern recognition 句法模式识别
syntan 合成鞣剂
syntexis 同熔作用
synthalin 十烷双胍
synthal process 辛塔法〔由一氧化碳与氢气按辛塔法合成烃类〕
synthermal 同温度的
synthesis 合成*
synthesis ammonia 合成氨
synthesis converter 合成转化器；合成反应器
synthesis cycle 合成循环
synthesis gas 合成气
synthesis of polymer 高分子合成
synthesis reactor 合成反应器
synthesis temperature 合成温度
synthesized fiber 合成纤维
synthesizing of hydrocarbon fuels 烃燃料合成
synthetase 合成酶
synthetic adhesive 合成黏合剂；合成胶黏剂
synthetic alcohol 合成酒精
synthetic ammonia 合成氨
synthetic aviation fuel ①合成航空燃料②辛烷值为100的航空汽油③加氢航空汽油
synthetic balata 合成巴塔树胶
synthetic bar 合成块皂；块状合成洗涤剂
synthetic bees wax 合成蜂蜡

synthetic boundary cell　形成界面池
synthetic calcium carbonate　合成碳酸钙
synthetic calcium silicate　合成硅酸钙
synthetic calibration sample　合成校准样品
synthetic camphor　合成樟脑
synthetic catalyst　合成催化剂
synthetic cement　合成胶浆
synthetic chemistry　合成化学
synthetic chloroprene rubber　合成氯丁(二烯)橡胶
synthetic crude　合成原油
synthetic depolymerized rubber　合成解聚橡胶
synthetic detergent　合成洗涤剂
synthetic drying oil　合成干性油
synthetic dye　合成染料
synthetic dyestuff　合成染料
synthetic earth　合成白土；活化水矾土
synthetic ebonite　合成硬质胶
synthetic elastomer　合成弹性体
synthetic enamel　合成磁漆
synthetic erionite　合成毛沸石
synthetic ester lubricant　合成酯类润滑剂
synthetic fat liquor　合成加油剂
synthetic faujasites　合成八面沸石
synthetic feedstock　合成储备原料
synthetic fiber　合成纤维
synthetic fibre ion exchanger　合成纤维离子交换剂
synthetic fibre laminate　合成纤维层压制品
synthetic fine silica　合成微细二氧化硅
synthetic fixed nitrogen　合成固定氮
synthetic flower oil　人造花香油；人工调制花香油
synthetic fluid　合成液体
synthetic fluid cracking catalyst　合成液体裂化催化剂
synthetic fuel　合成燃料
synthetic gas　①合成气②合成(煤)气
synthetic gasoline　合成汽油
synthetic gem　合成宝石
synthetic glycerine　合成甘油
synthetic grease(=synthetic oil grease)　合成油润滑脂
synthetic gutta-percha　合成古塔波胶；反式(聚)异戊(二烯)橡胶
synthetic hard water　人造硬水
synthetic hectorite　合成锂蒙脱石
synthetic hematite　合成氧化铁红
synthetic high polymer　合成高分子(物)；合成高聚物
synthetic ion-exchange resin　合成离子交换树脂
synthetic iron oxide　合成氧化铁红
synthetic latex　合成胶乳
synthetic leather　合成皮革

synthetic lipid　合成类脂物
synthetic liquid fuel　合成液体燃料
synthetic long chain alcohol　合成长链醇
synthetic lubricant　合成润滑剂
synthetic lubricant fluid　合成润滑油
synthetic lubricating oil　合成润滑油
synthetic macromolecule　合成高分子
synthetic magmetite　(磁性)氧化铁黑；合成氧化铁黑
synthetic method　合成(方)法
synthetic mica　合成云母
synthetic mixture　合成混合物
synthetic mordenite　合成发光沸石
synthetic motor oil　合成机油
synthetic musk　合成麝香；人造麝香
synthetic mustard oil　合成芥子油；合成烯丙基芥子油 $CH_2=CHCH_2NCS$
synthetic natural gas　合成天然气
synthetic oil　合成油
synthetic oil grease　合成油润滑脂
synthetic opals　合成蛋白石
synthetic organic pigment　合成有机颜料
synthetic organics　合成有机药物
synthetic oriental lacquer　合成大漆
synthetic pearl essence　合成珠光粉
synthetic perfume　合成香料
synthetic petrolatum　合成矿脂
synthetic petroleum　合成石油
synthetic plastic(s)　合成塑料
synthetic polyisoprene　异戊(二烯)橡胶
synthetic polymer　合成聚合物*
synthetic polymer bonding　合成聚合物黏合
synthetic polymer dispersion　合成聚合物分散体
synthetic polymer emulsion　合成聚合物乳液
synthetic polymeric flocculant　合成高分子絮凝剂
synthetic product　合成产物
synthetic reference material　合成标准物质
synthetic reference solution　合成参比溶液
synthetic resin　合成树脂
synthetic resin adhesive　合成树脂黏合剂
synthetic resin bond　合成树脂黏合剂
synthetic resin cement　合成树脂胶接剂
synthetic resin emulsion　合成树脂乳液
synthetic resin glue　合成树脂胶
synthetic resin molding compound　塑料粉
synthetic resin varnish　合成树脂清漆
synthetic route　合成路线
synthetic rubber　合成橡胶
synthetic rubber latex　合成胶乳

synthetic rubber mix　合成混胶
synthetic rubber resin　合成橡胶树脂
synthetic rubber seal　合成橡胶密封
synthetics　合成品
synthetic scheme　合成方案；合成流程图
synthetic sea water　人造海水；合成海水
synthetic segmented copolymer　合成嵌段共聚物
synthetic sewage　人造污水
synthetic silica　合成二氧化硅
synthetic size　合成胶料；合成上浆剂
synthetic skewed peak　合成斜峰
synthetic sodium butadiene rubber　合成丁(二烯)钠胶
synthetic soil　人造污垢
synthetic spindle oil　合成锭子油〔增塑剂〕
synthetic standard method　合成标准法
synthetic staple　合成短纤维
synthetic tannin　合成单宁
synthetic tanning agent　合成鞣剂
synthetic tanning material　合成鞣料
synthetic thickening agent　合成增稠剂
synthetic tower　合成塔
synthetic wax　合成蜡
synthetic wetting agent　合成湿润剂
synthetic wood　合成木材
synthetic yellow ocher　合成黄赭石〔氧化铁黄加硅酸铝等制成〕
synthetic zeolite　合成沸石
synthin　①合成烃类②合成元件
synthine process　辛亭过程〔由煤或天然气合成燃料的过程〕
synthol　合成燃料
syntholub　合成润滑剂
synthometrics　合成计量学
synthomycin(=syntomycin)　合霉素；消旋氯霉素
synthon　合成纤维
synthovo　二氢二乙基(化)己烯雌酚
syntomycin(=syntomycin)　合霉素；消旋氯霉素
syntony　谐振；共振；调谐
syntony model　谐振子模型
syntropan　安普洛托品磷酸盐；僧托烷
syn-type　顺式；顺(基)型
syphon(=siphon)　①虹吸管②虹吸
syphon barometer　虹吸气压计
syphon pipe(=syphon tube)　虹吸管
syphon tube(=syphon pipe)　虹吸管
Syrian asphalt　(叙利亚)地沥青
syringa-aldehyde　丁香醛
syringaldazine　丁香醛连氮
syringaldehyde　丁香醛；4-羟基-3,5-二甲氧苯甲醛
syringa oblata concrete　紫丁香花浸膏
syringaresinol　丁香树脂醇
syringe　注射管；注射器
syringe burette　注射滴定管
syringe extractor　注射器形萃取器
syringe graduation　注射器刻度
syringe guide　注射导引器
syringe injection　注射器进样
syringe needle　注射器针头
syringenin　丁香苷元；丁香配基
syringe nozzle　注射器喷嘴
syringe pipette　注射器吸液管
syringe plunger　注射器柱塞
syringe pump　注射泵；活塞泵
syringe reaction　注射器针管反应
syringe reaction chromatography　针管反应色谱(法)
syringe sampling　注射器进样
syringetin　丁香黄素；丁香亭
syringe-type pump　注射泵
syringic acid(=4-hydroxy-3,5-dimethoxybenzoic acid)　丁香酸；4-羟基-3,5-二甲氧基苯甲酸　HO(CH$_3$O)$_2$C$_6$H$_2$CO$_2$H
syringidin　丁香定
syringin　丁香苷
syringomycin　丁香假单胞菌素
syringone　丁香酮
syringyl alcohol(=4-hydroxy-3,5-dimethoxy benzyl alcohol)　丁香醇；4-羟基-3,5-二甲氧基苄醇
syrup　浆；浆状物
syrup acacia　阿拉伯胶糖浆
syrup of orange　橘皮汁
system　①系②体系③装置④(方)法⑤物系
system analysis method　系统分析法
systematic absences　系统消光
systematical error　系统误差
systematic analysis　系统分析
systematic error　系统误差
systematic name　系统名(称)；学名
systematic nomenclature　系统命名法
systematic samples　系统样品
systematic sampling　系统抽样*；系统取样
systematic testing　系统试验
system check　系统检验
system component　(涂料)配方成分；涂料组成(成分)
system factor　系统因子
system for nuclear auxiliary power(SNAP)　核辅助动力系统

systemic ①系统的②全身的
systemic anaphylaxis 全身过敏
systemic error 系统误差
systemic insecticide 内吸(性)杀虫剂
systemic poison 神经系毒剂
systemic toxic ①神经系毒剂②毒害神经系的
systemic toxic agent 神经系毒剂
system matrix combination method 系统矩阵组合法
system model 系统模型
system of absolute units 绝对单位制
system of cgs units cgs单位制
system of compounds 化合物系统
system of elements 元素系统
system of fertilization 施肥方法
system of incomplete mutual solubility 不完全互溶物系
system optimization 系统优化
system peak 系统峰
system program 系统程序
system sampling 系统取样
system simulation 系统模拟
systems measurement 系统测定
system spectral bandpass 系统光谱通带
szaibelyite 硼镁石
Szilard-Chalmers effect 齐拉-切尔曼斯效应*
Szilard-Chalmers method 齐拉-切尔曼斯法
szmikite 锰矾
szomolnokite 水铁矾

T

tabacum 烟草
tabashir 竹黄；竹节石
tabashis 竹黄；竹节石
tabby ①绉绸；波纹绸②绉绸的
Taber abrader 泰(伯尔)氏耐磨试验机；泰氏磨损试验机
Taber abrasion index 泰(伯尔)氏磨耗指数
Taber index 泰氏磨耗指数
Taber machine 泰(伯尔)氏磨损试验机
table ①表②桌(子)③台④(薄)片⑤摇床〔矿〕
table feeder 平板加料器
table-inking 调墨台(板)
table linen 桌布
table of random number 随机数表
table of variance analysis 方差分析表
table oil 台油；陶瓷厂用油
table press 台式压力机
table roll 台辊
table roll section 台辊段
table salt 食盐
table stacker 平台堆皮机
tablet (药)片；片剂
tablet agent 压片剂
tablet compressing machine(=tablet compression machine) 压片机
tablet compression machine(=tablet compressing machine; tablet machine) 压片机
tablet(ing) machine 造粒机
tablet machine(=tablet compression machine) 压片机
tablet method 片剂法
tablet press 压片机
tabletting 压片
tabletting machine 压片机
tabletting press 压片机
tablettized 压成片的
tabling ①制表②摇床精选〔矿〕
tabloid 小药片
tabular ①表的②平片状的
tabular crystal 平片状结晶
tabular value 表值
tabun 塔崩；垂龙 83；二甲氨基氰磷酸乙酯 $(CH_3)_2NP(OC_2H_5)(O)CN$
tacamahac 塔柯胶
tacciometer 表面黏度计
tache 斑点；色斑

tachhydrite 溢晶石
tachia 它虫阿〔南非南美龙胆科植物〕
tachiol 氟化银
tachogram 转速(记录)图
tachometer 转速表；转速计
tachometer drive 旋速计传动；转数表传动
tachycardia 心搏过速
tachylite 玄武玻璃
tachyol 氟化银
tachysterin 速甾醇；环裂甾醇
tachysterol 速甾醇〔麦角固醇的同质异构体是麦角固醇或光甾醇紫外线辐射的情况下生成的，进一步的辐照可产生维生素 D_2〕 $C_{28}H_{43}OH$
tack 黏性
tackability 黏着能力；增黏能力
tack cloth ①黏性布〔除尘用〕②(磨平用)浸胶布
tack eliminator 防黏剂
tack-free ①不黏手②指压干〔涂料〕
tack-free time 消黏时间；表面黏性消失时间
tackifier 黏合剂；增黏剂
tackifying ability 黏着能力；增黏能力
tackifying resin (增)黏性树脂
tackiness 黏性；发黏；黏着性；黏手性
tackiness agent 黏合剂
tack life 黏性消失时间
tackmeter 黏性试验机；黏合性试验机
tackoscope 黏性器；测黏仪
tack producer 增黏剂
tack-producing agent 增黏剂
tack-producing material 增黏剂
tack time 黏性时间
tacky mill base 黏性研磨料
tackyness 黏着性
tacky producer 增黏剂
tacky temperature 发黏温度
taconite 铁燧岩；南美贫铁矿
tacrolimus 他克莫司〔药〕
tactical classification 战术的分类(法)
tactic block 有规立构嵌段*
tactic content 立构规整程度
tacticity 立构规整度*；立构规整性
tacticity effect on dynamic property 动态性质构型规整效应
tactic polyepoxypropane 有规立构聚环氧丙烷

tactic polymer 有规立构聚合物；有规聚合物
tactic polymerization 有规立构聚合*
tactile sensor 触觉传感器
tactility 触感
tactoid 微团聚体
tactosol 有规溶胶〔具有光学异向性的胶体粒子〕
tact system 流水作业(线)
Tadema structural group analysis 塔德玛结构族分析
tadeonal 水蓼醛；塔蓼醛
taeniacide 杀绦虫剂
taeniafuge 驱绦虫剂
taenite 镍纹石；天然铁镍合金
Tafel equation 塔费尔公式*
Tafel slope 塔费尔斜率
tag 标签
tagatone 塔格酮
tagatose 塔格糖
tagaturonic acid (=(2S,3R,4S)-2,3,4,6-tetrahydroxy-5-oxo-hexanoic acid) 塔格糖醛酸；(2S,3R,4S)- 2,3,4,6-四羟基-5-氧代己酸 $C_6H_{10}O_7$
Tag closed tester 泰格密封闪点试验器
tagetes absolute 万寿菊净油
tagetes oil 万寿菊油
tagetone 万寿菊酮
tagetonol 万寿菊酮醇
tagged atom (=tracer atom) 标记原子；加有标记的原子；示迹原子；显迹原子
tagged compound 标记化合物
tagged detergent 标记洗涤剂
tagged element 标记原子
tagged-ligand transfer 标记配位体转移
tagged soil 标记污垢
tagging 标记
tagging reagent (衍生化)标记试剂
tagma 象牙椰子
tagulaway 萨布香脂
tail ①峰尾②谱带尾③拖尾
tail board loaders 尾板装车机
tail board (of the classifier) 分级器的尾板
tail current 尾电流
tailed peak 拖尾峰
tail-end-drive conveyor 后鼓轮驱动的运输带
tail fin 尾翼
tail formation 拖尾(形成)
tail fraction 尾馏分
tail gas 尾气；废气
tail gas analyzer 尾气分析器
tail house 接收调拨间

tailing 谱尾；拖尾；尾料
tailing edge 后沿
tailing effect 拖尾效应
tailing factor 拖尾因子*
tailing flight 尾部螺纹
tailing fraction 尾馏分
tailing ion 拖尾离子
tailing off 拖尾
tailing peak 拖尾峰*
tailing pond 尾矿池
tailing reducer 减尾剂；尾部减缩剂
tailings ①尾渣②尾矿③尾馏分
tailings pile 尾矿堆
tailing wheel 尾轮
tail oil 尾油；最后馏分
tailor ①改装；修整②缝制；剪裁
tailor coating 特制涂料；专用涂料
tailored excitation 裁剪激发；单一能量激发
tailored radiation 单纯辐射；单一(能量)辐射
tailor-made 定做的；特制的；专用的
tailor-made coating 特制涂料；专用涂料
tailor-made column packing 定做的柱填充物
tailor-made oil 订制油；特制油
tailor-made paint 特制漆；专用漆
tailor-made polymer 特制聚合
tailor-made tank trucks 订制油槽车
tailor-make product 订制产品；特制产品；定做产品
tailover 筛余物；筛渣
tail pipe 尾管
tail pipe flame 柴油机尾管焰
tail-producing site 拖尾部位
tail pulley 尾滑轮
tail pump 残液泵
tail rod 尾杆
tail rod catcher 尾杆罩
tails assay 尾料分析；尾气分析
tail(-to)-tail arrangement 尾-尾排列
tail-to-tail linking 尾尾连接
tail-to-tail polymer 尾-尾聚合物
tail-to-tail structure 尾接结构
tail vat 尾池〔鞣池组中最淡的〕
tail water 废水
tail-weight 尾重〔划在尾部的生皮重量标记〕
taimycin (=thaimycin) 泰霉素
taint-free ①无污染的②不腐败的③未变色的
Tainton process (for zinc extraction) 泰恩汤(提锌)法
taitomycin 泰东霉素
taiwanol 台湾杉醇

takatonine 高唐碱
take 所取的量
take-away belt 接取装置；引出装置
take-away conveyor 接取装置；引出装置
take away pump 送料泵
take delivery 提货
take down comb 牵拉梳；穿纱板；起底板
take-off ①取出；移去②输出③输出轴；功率输出端
takeoff apron ①(三辊磨)刮刀板②(=takeoff knife)刮漆刀
take-off equipment 退卷装置；输出装置
take-off gear 接取装置
takeoff knife(=takeoff apron) 刮漆刀；刮漆板
take-off machinery 接取装置
take-off rate 流出速度
take-off time 取出(馏分)时间
take-up gear 导出装置
take-up of vehicle 吸收漆料
take-up stand 导出装置
talang (树上割的)割汁沟
talaron 蓝霉菌素
talc(=talcum;powdered steatile) 滑石粉
talcing 涂隔离剂；撒粉；扑粉
talc-like strand 滑石状纤维；类滑石纤维
talcous 滑石的
talc powder 滑石粉
talcum(=talc) 滑石
talcum powder 滑石粉
talisai oil 榄仁树油〔取自榄仁树 Terminalia catappa 的种子〕
talite 塔罗糖醇 $CH_2OH(CHOH)_4CH_2OH$
talitol 塔罗糖醇
Tallalay process(=hydrogen peroxide freezing process) 过氧化氢冷冻法
tallalkyd 松浆油醇酸树脂
tallate 树脂酸盐
tall can 高(型)罐
tall form beaker 高型烧杯
tall horn beaker 高形烧杯
talloel(=tall oil) 松浆油；妥尔油
tall oil(=tallol; talloel) 妥尔油；松浆油
tall oil fatty acid 松浆油脂肪酸
tall oil heads(=light ends) 松浆油轻馏分〔沸点低的馏分〕
tall oil pitch 松浆油沥青
tall oil rosin (木浆)浮油松香
tall oil soap 松浆油皂
tall oil soap grease 松浆油皂基润滑脂
tallol(=tall oil) 妥尔油；松浆油
talloleic acid 松浆油酸

tallow wood 小帽桉
tall varnish 松浆油清漆；油清漆〔俗〕
talmi 镀金黄铜
Talspesk process〔词源：Trivalent actinidelanthanide separate〕 塔尔斯皮克法〔酸性膦酸酯萃取分离三价锕系和镧系过程〕
talus ①废料②距骨；踝骨；斜面
tama fat 奶油树脂
tamala oil 梓樟油
tamarind 酸荚罗望子树
tamarind extract 罗望子浸剂
tamarind seed gum 罗望子胶；枸子胶
tamarind-seed oil 罗望子油
tamarugite 斜钠明矾 $NaAl(SO_4)\cdot 6H_2O$
Tamele's apparatus 达姆勒装置
Tammann crucible(=Tammann tube) 塔曼坩埚；塔曼管
Tammann temperature 塔曼温度*
Tammann tube (=Tammann crucible) 塔曼管；塔曼坩埚
tammy 布筛
tamoxifen 他莫昔芬〔药〕
tamped volume 夯实体积
tamper ①(中子)反射器②(中子)反射剂③干预④惰层；反射层⑤干扰
tamper-proof 防干扰
tamper-resistant 抗干扰
tampicin 牵牛树脂 $C_{34}H_{54}O_4$
tamping ①夯实；捣固；填塞②管口扩张韧性试验③装填；压型
tamping bar 捣捧
tamponing ①擦涂；揩涂〔用棉团或纱团蘸漆反复多次擦于器物之上以增加厚度〕②搓漆〔大漆〕
tan ①鞣(革)②褐(色)
tanacetene 艾菊萜 $C_{10}H_{16}$
tanacetin 艾菊苦素
tanacetone 艾菊酮
tanacetum oil 艾菊油
tan ball 球状鞣料渣
tan bark 树皮鞣料
tan bark waste 废树皮鞣料
tan-brown shade 暗褐色调；深棕褐相
tan cake 鞣饼
tandem accelerator 串列式加速器
tandem airplane 纵列多翼机
tandem calender 成对压光机
tandem cards 双联梳棉机
tandem coating 串列涂装法
tandem compound 串列复式蒸汽机；单轴(多缸)汽轮机
tandem double focusing mass spectrometer 串联双聚焦

质谱计
tandem drive 串联传动；联动
tandem electron impact chemical ionization source 串级电子轰击化学电离源
tandem fuel cycle 串级燃料循环
tandem immunoelectrophoresis 串联免疫电泳
tandem jigger 串联卷染机
tandem line 串联辊涂线；双辊串联涂装流水线〔卷材涂装用，底漆和面漆一次涂完〕
tandem mass spectrometer 串联质谱计
tandem mass spectrometry 串联质谱法
tandem reaction sequence 连续反应过程*
tandem source 串联光源；级联光源
tan drum 鞣制转鼓
tanekaha 叶枝松皮
tangent error 正切误差
tangent flexural modulus 正切挠曲模量
tangent formula 正切公式
tangent galvanometer 正切电流计
tangential strain 切向应变；剪应变
tangential stress(=shearing stress) 切向应力
tangential traction 切向引力
tangential velocity 切向速度
tangent incision method 切线切割法
tangent modulus 切线模数
tangent of the dielectric loss angle 介电损耗角正切
tangent oxidation potential 正切氧化电位
tangent reduction potential 正切还原电位
tangerine 红橘
tanghinia 毒海果
tang-kui ①当归浸膏②当归
tangle 昆布
tanglelacing 交络；网络
tan house 鞣料库
tank ①桶；罐；池；槽；柜②库
tank accessories 油罐附件
tankage ①装罐储存②罐容量
tankage installation 油罐安装
tank agitator 油罐搅拌器
tank air 油罐空气；油罐中混合气
tank air-mover 油罐空气驱除器；驱除(油罐或油槽中)混合气用的压缩机
tank air temperature 油罐空气温度；(油罐中)蒸气空气温度
tank anchoring 油罐固定
tank balloon 油罐气囊
tank barge 油槽驳船
tank base 油罐底

tank battery 油罐组；油罐场
tank battery barge 油罐组驳船
tank body truck 油槽汽车
tank bottom 罐底；槽底
tank bottom chime 油罐底缘
tank bottom emulsion 罐底乳状液
tank breathing diaphragm 油罐呼吸隔膜
tank calibration 油罐校正
tank cap 油罐盖
tank capacity 油罐容量
tank car 油槽车
tank car blower 油槽车吹扫机
tank car bottom shell 油槽车底壳
tank car dome 油槽车穹室
tank car dome head 油槽车穹顶
tank car head 油槽车顶
tank car ladder 油槽车梯子
tank car lining 油槽衬
tank car loading rack 油槽车装载架
tank car shell sheet 油槽壳皮
tank cell 槽式电池
tank circuit 储能电路
tank classification 油罐分类
tank classifier 分级槽；槽式分级器
tank cleaning 油罐清洗
tank clean-out opening 油罐清扫口
tank coating 油罐涂料层
tank coil 油罐旋管
tank-coil formula 油罐旋管公式
tank compartment 油罐分隔室
tank connections 油罐连接管
tank content gauge 油罐量器
tank cooler 油罐冷却器
tank cross member 油罐横梁
tank crystallization 槽式结晶(作用)
tank crystallizer 槽式结晶器
tank culture 水箱式栽培
tank deck 油台；油罐顶
tank deck float 油台浮子；油罐顶浮子
tank design 油罐设计
tank details 油罐结构零件
tank dilution (油)罐内稀释
tank dome 油罐圆顶
tank drain 油罐放出口
tank elevation 油罐高度
tanker 油槽船
tanker aft 油槽船尾部
tanker beam 油槽船梁

tanker depth 油槽船吃水深度	tank mileage 油罐英里程；油箱英里程
tanker discharging 油槽船排出；油槽船卸油	tank miles 油罐英里数；油箱英里数
tanker displacement 油槽船排水量	tankometer 油罐(含量)计
tanker draft 油槽船吃水	tank operation 油罐操作
tanker length 油槽船长度	tankoscope (油罐)透视灯
tanker loaded draft 油槽船装载吃水	tank outage 油罐中损耗
tanker loading 油槽船装载	tank outlet plugging 塞住油罐出口
tanker loading terminal 油槽船装油码头	tank pad 油罐基础；油罐座
tanker longitudinal bulkhead 油槽船纵向隔框	tank park 油罐场
tanker mooring 油槽船的系船所	tank plate 油罐钢板
tanker oil compartment 油槽船船舱	tank pressure 油罐中压力
tanker ship(=tanker) 油槽船	tank pressure gauge 油罐压力计
tanker stokehold 油槽船火舱	tank rack 油罐架
tanker tonnage 油槽船吨位	tank railing 油罐顶栏杆
tanker transportation 油槽船运输	tank reactor 罐式反应器；釜式反应器
tanker turnround 油槽船周转过程	tank relief valve 油罐放气阀
tank farm 储油站；油库	tank replacement 更换油罐
tank farm pipe line 油罐场管线	tank rim 油罐边缘
tank farm stock 油罐场储存油料	tank riveting 油罐铆合
tank filler 油罐加油口	tank roof cave 油罐顶檐
tank filler cap 油罐加油口盖	tank scale 油罐秤
tank filling system 油罐装油系统	tank seam testing 油罐缝试验
tank filter valve 油罐过滤阀	tank selector valve 油罐选择阀
tank fire 油罐失火	tank setting 油罐安装
tank floating 油罐浮动	tank shell 油罐壳体
tank foundation 油罐基础	tank shell thickness 油罐壳厚度
tank furnace 槽炉	tank ship 油槽船
tank gauge 油罐标尺	tank shop 储(油)罐工场
tank gauging 测量油罐中液体	tanks-in-series model 多槽串联模型
tank grounding 油罐接地	tank site 油罐位置
tank hatch 油罐顶小门	tank sizing 油罐容量计算
tank heating coil 油罐加热旋管	tank sound pipe 油罐探测管
tank heating line 油罐加热管线	tank spacer 油罐隔板；油罐支柱
tank installation 油罐装置	tank spacing 油罐间距
tank insulation 油罐(热)绝缘	tank station 油罐站
tank joint 油罐接缝	tank steamer 油罐汽船
tank life 使用期	tank strap 油桶箍条
tank lighter 油罐驳船；油驳	tank strapper 箍桶工
tank lining 油罐衬里	tank strapping 箍桶
tank liquor 槽液	tank sump 油罐沉积槽
tank lorry 油槽汽车	tank support 油罐支柱；油罐基础
tank lubrication 油箱润滑	tank support system 油罐支架系统
tank lubricator 油箱润滑器	tank surface reconditioning 油罐表面修整
tank maintenance 油罐保养	tank top 油罐顶盖
tank manhole 油罐人孔	tank top ring 油罐顶环
tank manifold 油罐支管	tank-trailer 油槽拖车
tank manifold valves 油罐支管阀	tank truck 油槽汽车
tank microflora 油罐微生物	tank truck loading rack 油槽汽车负载架

tank truck pump　油槽汽车泵
tank truck trailer　油罐汽车拖车
tank trueing　油罐整形
tank type vulcanizer　硫化罐
tank valve　油罐阀；油罐呼吸阀
tank vapour　油罐蒸气
tank vapour recovery　油罐蒸气回收
tank vapour space　油罐蒸气空间
tank ventilation(=tank venting)　油罐换气
tank venting(=tank ventilation)　油罐换气
tank vent pipe　油罐通气管
tank vessel　油船
tank volume charts　油罐容积图表
tank wagon　油槽车
tank wagon price　油槽车上交货价格
tank waste　①废料场②渣
tank water riding　油罐浮水移动
tank welding　油罐焊接
tank white　(外用)油罐白漆〔自洁性外用白色金属表面涂料〕
tank winch　油罐绞车
tank work　油罐装置
tan liquor　鞣液
tan liquor heater and cooler　鞣液加热冷却器
tan liquor pump　鞣液泵
tan mill　鞣料磨；碾树皮机
tannage　①鞣革②鞣法
tannal　鞣酸铝；单宁酸铝
tannalbin　鞣酸蛋白
tannate　单宁酸盐
tannated toner　单宁酸(盐)色料
tannate of lime　单宁酸钙
tanned　鞣过的
tanner　制革工人
tanner oil　鞣革油
tannery　鞣革厂；皮革厂
tannery roll　制革胶滚
tannery sludge　制革厂污泥
tannic acid　单宁酸；鞣酸
tannic acid soap　单宁酸皂；鞣酸皂
tannigen　单宁精乙酰鞣酸
tannin　单宁*
tannin-based wash-coat　单宁基磷化层
tannin extract　单宁萃；单宁提取物
tanning　鞣(革)；鞣制
tanning agent　鞣剂
tanning drum　鞣制转鼓
tanning extract　单宁萃；单宁提取物

tanning industry　制革工业
tanning liquor　鞣液
tanning material　鞣料
tanning matter　鞣料；单宁物质
tanning tumbler　鞣革转筒
tanning vat　鞣瓮
tannin of persimmon(=persimmon tannin)　柿单宁；柿汁单宁
tannin soap　单宁皂
tannoform　鞣仿；亚甲单宁
tannogen　乙酸鞣质
tannometer(=barkometer)　巴克表；鞣液比重计
tannon　六亚甲基四胺单宁
tannopin　六亚甲基四胺单宁
tannyl acetate　单宁乙酸酯
tan pit　鞣瓮
tan pot　鞣槽
tanshinol　丹参醇
tanshinone　丹参酮
tanshinon IIA　丹参酮 IIA〔药〕
tan stove　鞣料炉
tansy oil　艾菊油〔取自 Tanacetum vulgare〕
tantalate　钽酸盐
tantalic　①含钽的②五价钽的；正钽的
tantalic acid　钽酸
tantalic bromide　五溴化钽
tantalic chloride　五氯化钽
tantalic compound　五价钽化合物
tantalic fluoride　五氟化钽
tantalic oxide　五氧化二钽
tantalifluoride　氟钽酸盐
tantalite　钽铁矿
tantalous　三价钽的；亚钽的
tantalous bromide　三溴化钽　$TaBr_3$
tantalous chloride　三氯化钽　$TaCl_3$
tantalum　钽〔73号元素，化学符号 Ta〕
tantalum boat　钽舟
tantalum bromide　溴化钽
tantalum carbide　碳化钽　$TaC; Ta_2C$
tantalum chloride　氯化钽〔总名，计有三氯化钽 $TaCl_3$ 和五氯化钽 $TaCl_5$〕
tantalum dioxide　二氧化钽　TaO_2
tantalum filament　钽丝
tantalum filament atomizer　钽丝原子化器
tantalum fluoride　五氟化钽　TaF_5
tantalum lining　钽衬
tantalum minerals　钽矿
tantalum nitride　一氮化钽　TaN

tantalum oxide 氧化钽 TaO_2;Ta_2O_5
tantalum pentabromide 五溴化钽 $TaBr_5$
tantalum pentachloride 五氯化钽 $TaCl_5$
tantalum pentafluoride 五氟化钽 TaF_5
tantalum pentoxide 五氧化二钽 Ta_2O_5
tantalum plate atomizer 钽片原子化器
tantalum potassium fluoride 氟化钽钾；氟钽酸钾；七氟合钽酸钾 $K_2[TaF_7]$
tantalum sampling boat 钽取样舟
tantalum silicide 二硅化钽 $TaSi_2$
tantalum sulfide 硫化钽 TaS_2
tantalum tribromide 三溴化钽 $TaBr_3$
tantalum trichloride 三氯化钽 $TaCl_3$
tantanum filament 钽灯丝
tantcopper 硅铜
tantiron 硅铁
tantnickel 硅镍
tan vat 鞣槽；鞣池
tan waste 废鞣料
tan waste press 废鞣压滤器
tanyard ①鞣革厂；皮革厂②鞣池间
tap ①龙头②轻击③放(出)液(体)；放水④螺旋公〔机工〕
tapability 成带性
tapazol 他巴唑；甲巯咪唑
tap bolt 有头螺栓
tap cinder 搅炼炉渣
tap cock 水管栓
tape ①带；带材②锭带
tape backing 磁带背面；磁带背衬
tape conductor 带状(超)导体；带状(超)导线
taped hole gauge 锥形孔规
tape extrusion line 带条挤出生产线
tape method 纱带(试验)法〔测润湿力〕
tape molding compound 带状模塑料
taper ①锥销②拔梢③锥形；锥度
taper-bar grizzly 楔形炉箅
tapered bar 楔杆
tapered block copolymer 递变嵌段共聚物*
tapered bottom 锥底
tapered column 锥形柱
tapered copolymer 标记共聚物；示踪共聚物；锥形共聚物
tapered increaser 锥形扩大器
tapered interchangeable ground joint 锥形标准磨口接头
tapered pipe 异径管
tapered plug 锥形塞
tapered reducer 锥形减小器
tapered section method 斜削法

tapered slot 斜沟
tapered strip 锥形条
tapered tension 渐减张力
taper gauge 内径规
tapering lap joint 尖塔接头
tapering shape (截)锥式
taper line grating 织物密度板
taper pipe 异径管
taper-shaped blade 锥形刮刀；油画调色刀
taper thread 圆锥形螺纹
tapestry satin 织锦缎
tape test 胶带试验
tape welding 带焊
tape winding ①带缠②带缠制件
tape without selvage 无织边带
tape with selvage 有织边带
tape yarn 扁丝
tap funnel 滴液漏斗
tap hole ①漏孔②放液口
tapioca 木薯淀粉
tapioca flour 木薯(淀)粉
tapping ①放液②割浆；割胶③采(割松)脂
tapping channel (树上割的)割汁沟
tapping floor 浇铸场
tapping furnace 放液炉；液流电炉
tapping hole 浇铸孔
tapping panel 鼓击框
tapping pipe 泄水管
tapping point ①(管道)分叉点②(液体)放料点③流漆点
tapping response 叩抚反应
tap pipe 泄水管
tap water 自来水
tar 焦油
tar accumulator 焦油沉积器
tarachrome 蒲公英色素
tar acid 焦油酸
tar acid oil 焦油酸油
tar asphalt 焦油沥青
tara tannin 刺云实单宁
taraxacerin 蒲公英华
taraxacin 蒲公英素
taraxacum rubber 蒲公英橡胶
taraxanthin 蒲公英黄质
taraxasterol 蒲公英甾醇
taraxasterone 蒲公英甾酮
taraxerane 蒲公英赛烷
taraxeranol 蒲公英赛烷醇
taraxerene 蒲公英赛烯

taraxerenone 蒲公英赛烯酮
taraxerol 蒲公英赛醇
taraxol 蒲公英醇
tar base 焦油碱
tar-bitumen 焦油沥青
tar boiling plant(=tar distillery) 焦油蒸馏车间
tar bottom 焦油底
tar brush 焦油刷子
tar-building 形成焦油的
tarbuttite 三斜磷锌矿
tar camphor(=naphthalene) 萘
tar catcher 提焦油器；焦油分离器
tarchonyl alcohol 非洲樟醇
tar concrete 焦油混凝土
tar condenser 焦油冷凝器
tar condensing 焦油冷凝
tarconine 他可宁
tar cooler box 焦油冷却箱
tar cracking 焦油裂化
tar cracking zone 焦油裂化区域
tar crudes 焦油粗油
tar cuts〔复〕 焦油馏分
tar dehydrating still 焦油脱水蒸馏釜
tar destruction producer 焦油分解气体发生器
tardin 缓青霉素
tar distillate 焦油馏分
tar distillation 焦油蒸馏
tar distillery(=tar boiling plant) 焦油蒸馏车间
tare ①皮重②定皮重③配衡体④配衡
tare bottle 配衡瓶
tared 配衡的
tared filter 配衡滤器
tared flask 配衡烧瓶
tarelaidic acid 6-十八(碳)烯酸
tar emulsion 焦油乳液
tare shot 配衡小球；减重小球
tare weight 皮重；毛重
tar extraction 焦油提取
tar extractor 焦油提取器
tar filling 焦油填充
tar fog 焦油雾
tar formation 焦油的形成
tar-forming 形成焦油的
tar from lignite 褐煤焦油
tar from lignite schist 页岩焦油
tar gas 焦油气
target ①靶(子)②目标③对阳极
target chamber 靶室
target chemistry 靶化学*
target compound 靶子化合物
target element in X-ray source X 射线管靶材
target factor analysis 目标因子分析法
target foil 靶箔
target gas 碰撞气
target holder 靶托*
target hole 靶孔
target mass 工件；被涂物
target nucleus 靶核*
targetry 制靶法*
target solvent 目标溶剂
target transformation factor analysis 目标转换因子分析
target-type 靶型〔氯氧化铋的晶形〕
targusic acid(=lapachol) 黄钟花醌；2-羟基-3-戊烯基-α-萘醌
tar heavy oil 重焦油
taring 配衡；定皮重
tariric acid 塔日酸；5-十八(碳)炔酸
tar light oil 轻焦油
tar-mac(=tar-macadam) 柏油碎石铺路
tar-macadam(=tar-mac) 柏油碎石铺路
tar main 焦油总管
tar-making process 制焦油过程
tarnish ①失(去)光泽②无光泽
tarnish film 失泽膜
tarnishing 锈蚀*
tarnish inhibitor 晦暗抑制剂
tarnish resistant ①防(金属)变色的②防(金属)失光的
tar number(=tar value) 焦油值
taro 芋头
tar oil 焦油
taroxylic acid 塔氧酸；6,7-二氧代十八酸
tarp(=tarpaulin) 焦油帆布
tar paper 焦油纸；沥青油纸
tarpaulin 焦油帆布
tar pitch 焦油沥青
tar plug 焦油塞；放焦油口的塞头
tar pool 焦油池
tar product 焦油产品
tarragon 龙蒿
tarragon oil 龙蒿油
tarragon vinegar 龙蒿醋
tarred 涂焦油的
tarred felt 焦油毡
tarry 焦油状的
tarry cut(=tarry distillate) 焦油馏分
tarry distillate(=tarry cut) 焦油馏分

tarry material 焦油状物
tarry matter 焦油状物质
tarry oil 焦油状原油
tarry residue 焦油状残渣
tarry vapour 焦油蒸气
tar sand 焦油砂
tar-saturated 浸焦油的；焦油浸制的
tar saturator 焦油浸制机
tar seam 焦油缝
tar separation 焦油分离
tar separator 焦油分离器
tar soap 焦油皂
tar softening point 焦油软化点
tar spirit 煤焦系溶剂
tar spray(ing) hose 喷浇沥青胶管
tar spraying tank 焦油喷布车；喷焦油槽车
tar still 焦油蒸馏釜
tar stock 焦油原料
tar stripper 焦油汽提器
tar stripper naphtha vapour 焦油汽提器中的石脑油蒸气
tar stripping unit 焦油汽提装置
tartan 格子呢
tar tape 焦油胶布
tartar 酒石；酒石酸氢钾 $KHC_4H_4O_6$
tartar emetic 吐酒石；酒石酸氧锑钾
tartaric 酒石的
tartaric acid(=2,3-dihydroxy succinic acid) 酒石酸；2,3-二羟基丁二酸 HOOCCHOHCHOHCOOH
tartaric acid monoamide 酒石酸一酰胺 $NH_2CO(CHOH)_2CO_2H$
tartaric emetic 吐酒石；酒石酸氧锑钾 $K(SbO)C_4H_4O_6 \cdot 1/2H_2O$
tartarlithine 酒石酸氢锂
tartar salt 钾碱；碳酸钾 K_2CO_3
tartarus 酒石酸氢钾
tar test 焦油试验
tartrate ①酒石酸②酒石酸盐(酯或根)
tartrated antimony 吐酒石；酒石酸氧锑钾 $K(SbO)C_4H_4O_6 \cdot 1/2H_2O$
tartrate precipitation 酒石酸盐沉淀
tartrate stability 酒石酸盐稳定性
tartratolead 酒石酸根合铅
tartrazine 酒石黄
tartrazine lake 酒石黄色淀
tartrazine yellow 酒石黄
tartronate ①丙醇二酸②丙醇二酸盐(酯) COOMCHOHCOOM;COORCHOHCOOR
tartronic acid(=hydroxymalonic acid) 丙醇二酸；羟基丙二酸 $HOCH(CO_2H)_2$
tartronyl urea(=dialuric acid) 丙醇二酰脲
tar value(=tar number) 焦油值
tar vapor 焦油蒸气
tar vaporizer 焦油蒸发器
tar viscosimeter 焦油黏度计
tar well 焦油收集器
tar-yielding 形成焦油的；产焦油的
taseometer 应力计
tasimeter 微压计
tasmanone 塔马桉酮
taspine 塔斯品碱
taspinic acid 塔斯品酸
taste ①味②尝(味)
taste sense 味觉
taste sensor 味觉传感器
taste threshold 味阈值
taste threshold value 味阈值
tast polarograph 间滴极谱仪
tast polarography 采流极谱(法)
tast voltammetry 取样直流伏安法
Tatoray process 塔特莱法〔由甲苯歧化制苯及二甲苯法〕
Tatoray unit 塔特莱(歧化)装置
tatralite(=tetryl) 特刷拉特〔炸药〕
tattooing 刺花
taurate 牛磺酸盐；氨基乙磺酸盐
tauremycin 牛蒡素
tauride 氨基乙磺酸盐
taurine 牛磺酸；氨基乙磺酸；牛胆碱 $NH_2CH_2CH_2SO_3H$
taurine-N,N-diacetic acid(TDA;H₃TDA) 氨基乙磺酸-N,N-二乙酸
taurocarbamic acid 牛磺脲酸；脲基乙磺酸 $H_2NCONHCH_2CH_2SO_3H$
taurocholate ①牛磺胆酸②牛磺胆酸盐(酯或根)
taurocholic acid 牛磺胆酸 $C_{26}H_{45}O_7NS$
taurocyamine 脒基牛磺酸
tauryl 牛磺酰；氨基乙磺酰 $H_2NCH_2CH_2SO_2-$
taut 拉紧的；绷紧的
tautness 拉紧；紧固度；张紧度
tauto- 〔词头〕互变〔异构〕
tautochrone 等时曲线
tautocyanate 互变(异构)氰酸酯 RN=CO
tautomer(=tautomeride) 互变(异构)体
tautomerase 互变异构酶
tautomeric 互变(异构)的
tautomeric effect 互变(异构)效应
tautomeric equilibrium 互变(异构)平衡

tautomeric ratio 互变异构比
tautomeric structure 互变(异构)结构
tautomeride(=tautomer) 互变(异构)体
tautomerism 互变异构*
tautomerization 互变异构化*
tautomerizing 互变(异构)现象
tautourea 互变脲 $NH_2C=NHOH$
taut-wire printing 紧丝印刷(法)
taw 硝皮
tawer 盐硝皮工人；生鞣皮工人
tawing (明矾)硝制
tawing paste (明矾)硝制剂
taxicatin 红豆杉苷；3,5-二甲氧苯酚葡萄糖苷
taxifolin 紫杉叶素
taxine A 紫杉碱 A
taxis ①构型规整性②向性；趋性③排列；次序
taxisterol 毒甾醇
taxites 斑杂岩
taxocline 杂交差型
taxometer 量片器
taxus 紫杉
Taylor bag filter 泰勒袋滤器
Taylor diffusion 泰勒扩散
Taylor four-roll mill 泰勒四滚筒磨
Taylor-Prandtl modification of Reynolds analogy 泰勒-普兰特对雷诺相似性的修改
Taylor's equation 泰勒方程式
Taylor vortex 泰勒旋涡
Taylor vortex flow 泰勒涡流
Taylor-White process 泰勒-怀特炼钢法
Taylor wire method 泰勒线法
tazettadiol 水仙花二醇
tazettamide 水仙花酰胺
tazettamine 水仙花胺
tazettine 水仙花碱 $C_{18}H_{21}O_5N$
tazettinemethine 水仙花甲碱
tazettineneomethine 水仙花新甲碱
tazettinol 水仙花碱醇
tazettinone 水仙花碱酮
TBA number 硫代巴比妥酸值
TBA test 硫代巴比妥酸试验〔测油脂酸败〕
t-bend 丁字弯头
tbilimycin 第比利霉素
TBP-Hexone process 磷酸三丁酯-异己酮过程〔溶剂萃取处理辐照浓缩铀燃料过程〕
TBP-23 process 磷酸三丁酯-23 过程〔从辐照钍中提取铀-233，与 Thorex 流程类似，但不回收钍〕
TBP-25 process 磷酸三丁酯-25 过程；高浓铀钚雷克斯流程〔从辐照高浓缩铀燃料回收纯化铀-235〕
t-criterion t 检验
t-distribution t 分布*
tea ①茶②茶叶③茶树
tea catechin 茶叶儿茶酸
tea cup 茶杯量
TEAE-cellulose(=triethylaminoethyl cellulose) 三乙氨基乙基纤维素
teak 柚木
tea lead 包茶的铅箔
tea leaf oil 茶叶油
teal oil 芝麻油
team design 成套设计
teamwork 联合动作；协同动作
tea oil 茶(子)油
tear ①撕裂；撕破；裂开②磨损；破坏；裂缝
tearability 易撕裂性；撕裂度
tear-and-wear 磨损；磨耗
tear drop ①滴珠；珠状流挂〔漆病〕②亮点〔树脂〕③泪滴〔塑料制品缺陷〕④滴料
tear gas 催泪瓦斯；催泪(性)毒气
tearing 撕裂；扯裂
tearing energy 撕裂能
tearing resistance 撕破强度
tearing strength 抗撕强度*
tearing tester 撕裂检验器
tearing toughness 抗撕(裂)强度
tea root 美卫茅根
tea rose 香水月季
tea rose concrete 香水月季花浸膏
tear propagation 撕破延展
tear resistance 抗撕裂性；撕裂性能
tear resistance tester 抗撕试验机
tear speed 撕裂速率
Tears sulfur recovery process 蒂尔斯硫回收法
tear strength 撕裂强度
tease 起毛；起(拉)绒；梳理
tea-seed oil 茶(子)油
teasel 刺果起毛机；起绒刺果
teasing 亚麻整梳
tea spoon 茶匙量
teat pipette 橡皮头移液管
tebelon(=isobutyl oleate) 油酸异丁酯
technetium 锝
technic ①技巧；技艺②工艺的③工(艺)学
technical ①技术的；工艺的②学术的③专门的；专业的
technical accident 技术性事故
technical analysis 工业分析

technical application 工业应用；工程应用
technical cloth 工业用布
technical economical index 技术经济指标
technical fabric 工业(用)织物
technical factor 技术因数(因素)
technical gelatine 工业明胶
technical grade 工业级；工业用
technical grading 工艺分级；技术等级
technical isooctane(=technical octane) 工业异辛烷
technicality ①专门②细微的区别
technical know-how 技术秘密；技术诀窍(秘诀)技术知识
technical leather 工业用革
technical license 技术执照；技术许可证
technically bound sulfur(=true combined sulfur) 有机结合硫
technically pure 工业纯(净)
technically specified rubber 工艺分类橡胶
technical manual 技术规范；技术手册
technical octane(=technical isooctane) 工业异辛烷
technical property 工艺性能
technical pure 工业纯(净)
technical rubber goods 橡胶工业制品
technical scale 工业规模
technical specification 技术规格
technical term 术语；专门名词
technical term of chemistry 化学术语
technical white oil 工业用白油
technical word 术语
technician 技术员
technicolo(u)r 天然色；彩色
technics 艺术(学)；艺术论
technique ①技术②技巧
technique of manufacture 制造技术
technological 工(艺)学的
technological property 工艺性能
technology 工(艺)学；工(业)技(术)学
tecleine 柚木芸香碱
Teclu burner 特克卢燃烧器；双层转筒燃烧器
tecomin(=lapachol) 拉帕醇；黄钟花醌
tectioite ①轧成杆状材料②构造岩
tectite 风化硅铝石
tectochrysin 柚木柯因
tectonic axis 构造轴
tectonics 构造地质学
tectoquinone 鸢尾醌
tectoridin 鸢尾苷
tectorigenin 鸢尾黄素

tedion 2,4,5,4-四氯二苯砜
tee ①T(形)管②丁字铁；T字铁③T(字)形④三通
tee fluid flow meter 三通液体流量计
tee joint 三通；T形接管；T形接头(焊接)
teel oil 芝麻油
teeming ①铸造②铸造物；铸件
tee piece T(形)管；丁字管
tee pipe ①丁字管；T(形)管②丁(字)形③三晶
tee reducing on outlet 支路异径丁字管
tee reducing on run 直路异径丁字管
tee-slot 丁字(形)槽
tees powder 梯司炸药
teeter chamber(=teeter column) 搅拌室
teeter column(=teeter chamber) 搅拌室
teetering 簸震
teeth spacing 齿距
tee welding T形焊
tegafur 替加氟〔药〕
tegaserod 替加色罗〔药〕
teglam fat 南洋同翼香脂〔取自 Isoptera bornensis 的种子〕
tegument ①皮；外被②天然壳
teicoplanin 替考拉宁〔药〕
tektite 熔融石；玻陨石
telangiectasia 毛细管扩张
telcomer 嵌聚物
telecesium machine 远距离铯治疗机
telechelic 链端有官能基的；遥爪的
telechelic butadiene (rubber) 遥爪合丁二烯〔橡胶〕
telechelic oligomer 遥爪齐聚物
telechelic polymer 遥爪聚合物*；远螯聚合物
telechelic polyolefin 链端官能基聚烯烃；远螯聚烯烃
telechelic rubber 遥爪橡胶；远螯橡胶
telecinobufagin 远华蟾蜍精〔参考华蟾蜍精 cinobufagin〕
telecobalt machine 远距离钴治疗机
telecobalt therapy 远距离钴治疗法
telecopolymerization 共调聚反应
telecurie therapy 远距照射疗法；远距镭疗法
telegauge 遥测仪；远距离测量仪表
telegraphic paper 电报用纸
telegraphing resistance 抗污渍(迹)潜移性
teleidoscope 液层扫描镜
telekinesy 远距离作用
telemeter 遥测计
telemetering 远距离测定；遥测
telemetering system of pipe line 管线遥测系统
telemetry 遥测术
teleoperator 遥控机械手；遥控操作员
telepancurie therapy 全身性照射疗法

telephotography ①远距离照相术②传真电报
teleradium therapy 远距镭疗法
teleroentgentherapy 远距X射线疗法
telescope 望远镜
telescope flint 望远镜用燧石
telescopic ①望远镜的②远视的③套筒的；套管的；伸缩的
telescopic concentric cylinder viscometer 伸缩式同心圆筒黏度计
telescopic distortion 套筒式畸变；层流畸变
telescopic effect 肩颈效应；颈缩效应
telescopic flow 套筒式流动；层流
telescopic gas-holder 套筒储气柜
telescopic joint 套管连接
telescopic laminar displacement 套筒式层状位移
telescopic laminar flow 套筒式层流
telescopic lap splice 套接接头
telescopic movement 套筒式运动；层流运动
telescopic oiler 套管加油器
telescopic optical pyrometer 套筒式光测高温计
telescopic screw 套筒螺旋〔机工〕
telescopic working platform 套筒工作台
telestimulator 遥控刺激器
teletherapy 远程(放射)治疗*
telethermometer 遥测温度计
teletype control 远距离控制
television 电视
telfairic acid 亚油酸；顺式9,12-十八碳二烯酸
tellerium acid 碲代酸
Teller rosette 特勒花环填料
tellerwort 血根
telling test 辨别试验；鉴别试验
telltale ①信号装置②警报器③计数器；寄存器
tellurate 碲酸盐(酯)
tellurate radical 碲酸根
telluretted hydrogen 碲化氢
telluric (正)碲〔指四价碲化物或六价碲酸〕
telluric acid 碲酸
telluric acid anhydride 三氧化碲 TeO_3
telluric bismuth 碲铋矿
telluric bromide 四溴化碲 $TeBr_4$
telluric chloride 四氯化碲 $TeCl_4$
telluric iodide 四碘化碲 TeI_4
telluric lead 碲铅矿
telluric lines 大气吸收线
telluric ocher 黄碲矿
telluric oxide 三氧化碲 TeO_3
telluric silver 碲银矿

telluride ①碲化物②碲醚③碲(根)
tellurinic acid （烃基)亚碲酸；碲代亚磺酸
tellurious 亚碲(的)
tellurite ①亚碲酸盐 $M_2TeO_3$②黄碲矿 TeO_2
tellurium 碲
tellurium bichloride 二氯化碲
tellurium bromide 四溴化碲 $TeBr_4$
tellurium chloride 氯化碲〔总名，计有二氯化碲 $TeCl_2$ 和四氯化碲 $TeCl_4$〕
tellurium dichloride 二氯化碲 $TeCl_2$
tellurium diethyl 二乙基碲 $Te(C_2H_5)_2$
tellurium diethyl dithiocarbamate 二乙基二硫代氨基甲酸碲
tellurium diiodide 二碘化碲 TeI_2
tellurium dimethyldithiocarbamate 二甲基二硫代氨基甲酸碲〔促进剂〕
tellurium dioxide 二氧化碲 TeO_2
tellurium fluoride 氟化碲 $TeF_4;TeF_6$
tellurium hexafluoride 六氟化碲 TeF_6
tellurium iodide 碘化碲 $TeI_2;TeI_4$
tellurium nitrate 硝酸碲
tellurium ochre 黄碲矿
tellurium oxide 氧化碲 $TeO_2;TeO_3$
tellurium oxychloride 二氢一氧化碲 $TeOCl_2$
tellurium tetrabromide 四溴化碲 $TeBr_4$
tellurium tetrachloride 四氯化碲 $TeCl_4$
tellurium tetrafluoride 四氟化碲 TeF_4
tellurium tetraiodide 四碘化碲 TeI_4
tellurium trioxide 三氧化碲 TeO_3
telluro- 碲基 —Te—
tellur(o)bismuth 辉碲铋矿
tellurometer 微波测距仪；雷达测距仪
telluronic acid （烃基)碲酸；碲代磺酸
tellur(o) nickel 碲镍矿
telluronium 碲(鎓) R_3Te^+
telluronium iodide 碘化碲(鎓) R_3TeI
tellur(o) ocher 二氧化碲
tellurous acid 亚碲酸 H_2TeO_3
tellurous acid anhydride 亚碲酸酐；二氧化碲 TeO_2
tellurous bromide 溴化亚碲
tellurous chloride 二氯化碲；氯化亚碲 $TeCl_2$
tellurous oxide 二氧化碲；亚碲酸酐 TeO_2
telluryl 氧碲基 OTe=
telluryl chloride 二氯一氧化碲 $TeOCl_2$
telmisartan 替米沙坦〔药〕
telochrome 电视阴极射线管
telocopolymerization(=intertelomerization) 共调聚反应
telogen 调聚体；远控聚合反应的连锁反应链载体

teloidine 特洛碱；三羟莨菪烷
telomer 调聚物*
telomeric alcohol 调聚醇
telomeric reaction 调聚反应
telomerization 调聚反应*
telomerization reactor 调聚反应器
telomerized polymer 调节聚合物
telomerizing 调聚
telopsis 电视
teloschistin 远裂亭
telsit 敌息特〔炸药〕
temparin 双香豆素
temper ①回火②回火度；钢的硬度；强度
tempera paint 壁画漆；壁画色(料)
temperary repair 临时修理
temperature 温度
temperature amplifier 温度放大器
temperature balance 温度平衡
temperature band 温度带；温度范围
temperature boundary layer 温度边界层
temperature buffer 温度缓冲
temperature buzzer 温度警报
temperature change 温度变动
temperature coefficient 温度系数
temperature coefficient of density 密度的温度系数
temperature-compensating device 温度补偿装置
temperature compensation 温度补偿
temperature-composition diagram 温度-组成图
temperature contrast 温度差；差度不均匀分布
temperature control 温度控制
temperature controlled phase-separation immunoassay (TCPSIA) 控温相分离免疫分析法
temperature controller 温度控制器
temperature correction 温度校准
temperature deformation curve 温度形变曲线
temperature detector 探温器；测温检测器
temperature difference 温(度)差
temperature diffuse scattering 温度漫散射
temperature diffusivity 温度扩散系数；热传导
temperature-distillation chart 蒸馏温度曲线图
temperature drop 温度(下)降
temperature drop correction 温度(下)降校正
temperature effect 温度效应
temperature enthalpy curve 温度-热函曲线
temperature-entropy diagram 温熵图
temperature equalization 温度平衡；温度匀等
temperature equilibrator 温度补偿器
temperature equilibrium 温度平衡

temperature error 温度误差
temperature expansion coefficient 温度膨胀系数
temperature extremes 温度极限
temperature factor ①温度因素②温度因数
temperature field 温度场
temperature fluctuation 温度波动
temperature-fluidity characteristic 温度-流度关系特性
temperature gauge 温度计
temperature gauge unit 计温单位
temperature gradient 温度梯度
temperature gradient bar 温度梯度板
temperature gradient chromatography 温度梯度色谱法
temperature homogeneity 温度均匀性
temperature-humidity graph 温湿曲线
temperature hyperbola 温度双曲线
temperature index 温度指数〔表征聚合物的燃烧性能〕
temperature indicating coating 示温涂料
temperature indicating crayon 示温蜡笔；示温(色)铅笔；示温棒
temperature-indicating paint(=heat-sensitive paint) 示温漆
temperature indicating pigment(=heat indicating pigment) 示温颜料
temperature indicator 温度指示器
temperature inversion 逆温现象
temperature level 温度位
temperature limitation 温度极限
temperature measuring system 温度测量系统
temperature meter 温度计
temperature of charge 物料温度；反应物温度
temperature of combustion 燃烧温度
temperature of deflection under load 载荷挠曲温度
temperature of explosion 爆炸温度
temperature of solidification 凝固温度；凝固点
temperature plug 测量温度用的热电偶
temperature-pressure curve 温度-压力曲线
temperature probe 测温探头(探针)
temperature profile 温度分布型；温度线图
temperature program 升温程序；程序升温
temperature programmed chromatography 程序升温色谱法
temperature programmed desorption (TPD) 程序升温脱附*
temperature programmed gas chromatography (TPGC) 程序升温气相色谱法
temperature programmed oxidation (TPO) 程序升温氧化*
temperature programmed reaction spectrum (TPRS) 程序升温反应谱*
temperature programmed works 程序升温机件

temperature programmer 程序升温器
temperature programming 程序升温
temperature quenching 温度消光；温度猝灭
temperature radiation 温度辐射
temperature ranges 温度范围
temperature rate 升温速率
temperature ratio 温度比
temperature recorder 温度记录器
temperature recorder-controller 温度记录控制器；温度记录调节器
temperature regulator(=thermo-regulator) 温度调节器
temperature resistance 耐温性
temperature-resistant grease 耐温润滑脂
temperature retention 保温性
temperature rise 温度上升
temperature-sensing element 温度敏感元件
temperature sensing feed-back 温度敏感反馈
temperature sensing probe 温度敏感探针
temperature-sensitive material 热敏性材料；温度敏感性材料
temperature sensitive paint 示温漆；变色漆；温度指示漆
temperature sensitive paper 热敏纸
temperature sensitive pellet 热敏球
temperature sensitivity 温度敏感性；温度灵敏度
temperature sensor 热敏元件；温度传感器
temperature-setting dial 温度调节度盘
temperature shock 温度震扰；热震
temperature-slope adjustment 温度-斜率调节
temperature-slope compensation 温度-斜率补偿
temperature stability 温度稳定性
temperature strain 温度应变
temperature stress 温度应力；温差应力
temperature-stress-strain diagram 温度-应力-应变图
temperature switch 热动开关
temperature-time curve 温度-时间曲线
temperature-time equivalence 温度-时间等效
temperature-time factor 温度-时间因数
temperature-time limit 温时极限
temperature transmitter 温度变送器；温度传感器
temperature upper limit 温度上限
temperature valve 温度阀
temperature variation 温度变化
temperature warping 温度翘曲；温度扭曲(弯曲)
temper brittleness 回火脆性
temper carbon 回火碳
tempered oil 调和油
tempered steel 回火钢
tempering ①回火②人工老化③混合；调和④退鞣〔皮革〕
tempering bath 回火浴
tempering coil 调温旋管
tempering color 回火颜色
tempering furnace 回火炉
tempering oil 回火油
tempering stove 回火炉；硬化炉
tempering tank 混合桶；混合槽
tempering tower 缓冲塔
templat(=template) 模板；样板
template(=templat; templet) 模板；样板
template chromatography 模板色谱法
template effect 模板效应
template guest molecule 模板客体分子
template host molecule 模板主体分子
template polymerization(=matrix polymerization) 模板聚合*
template reaction 模板反应*
template replication 模板复制
template strand (模板)股；(模板)链
template synthesis 模板合成*
templet(=template) 模板
temple tree 鸡蛋花
temporal coding 时间编码
temporal match 时间匹配
temporal modulation 时间调制
temporal resolution 时间分辨
temporal variation 瞬时变化；瞬间变化
temporary acceptable daily intake 暂定每日允许摄入量
temporary cross-link 瞬时交联(键)
temporary fuel 暂时燃料；临时燃料；代用品燃料
temporary hardness ①暂时硬度②暂时硬性
temporary hard water 暂硬水
temporary maximum residue limit 暂定最高残留限量
temporary method 暂用方法
temporary protective coat 临时性保护涂层
temporary rust prevention agent 临时性防锈剂
temporary set 暂时变定
temporary storage 暂时储存
temulentine 黑麦草子碱；毒草碱
tenacious adsorption 黏滞吸附
tenacious adsorption site 黏滞吸附点
tenacious site 黏滞点
tenacity 韧性；韧度
tenaculum 钩针
tendency of wet film to bubble 湿膜嗜起(鼓)泡性；湿膜易起(鼓)泡性
tendency to bubble 易起(鼓)泡性；嗜鼓泡性
tender ①柔软的②煤水车③在管中输送的部分油品

tender pulp 软纸浆
tender side 操作面；前侧
tenebrimycin(=nebramycin) 妥布霉素；暗霉素
tenemycin(=nebramycin) 妥布霉素；暗霉素
teniacide 杀绦虫剂
teniafuge 驱绦虫剂
teniposide 替尼泊苷〔药〕
ten-membered ring(=ten-ring) 十元环
tennantite 砷黝铜矿
tenorite 黑铜矿
ten-pounds gasoline 十磅汽油
ten-pounds gauge pressure 十磅表压力
ten-ring(=ten-membered ring) 十元环
tensammetric curve 张力曲线*
tensammetric peak 张力峰*
tensammetry 张力法*
tensibility 可伸长性；伸长率
tenside 表面活性剂
tensile ①张力的；抗张的②拉伸的
tensile breaking force 拉伸断裂强力
tensile breaking strength 拉伸断裂强度
tensile breaking tenacity 拉伸断裂强度
tensile break strength 拉伸断裂强度
tensile compliance 拉伸柔量
tensile creep 拉伸蠕变
tensile creep compliance 拉伸蠕变柔量
tensile crimp modulus 拉伸卷曲模量
tensile curve 拉伸曲线
tensile elasticity 拉伸弹性
tensile elongation 拉伸长度
tensile energy 抗张本领
tensile figure 拉伸指数
tensile force 张力；拉力
tensile force-extension relation 张力-拉伸关系式
tensile heat distortion 热变形张力
tensile load 拉伸载荷；张力
tensile machine 拉力试验机
tensile modules of elasticity 拉伸弹性模量
tensile modulus 拉伸模量；抗张模量
tensile-permanent set 拉伸变定；永久伸长
tensile product 抗张积
tensile property 拉伸性能
tensile rigidity 拉伸刚度
tensile shear strength 拉伸剪切强度
tensile strain 拉伸应变；张应变
tensile strength 拉伸强度*
tensile strength test 抗张强度试验
tensile strength tester 拉伸强度测定仪

tensile stress 拉伸应力；张应力
tensile stress relaxation 拉伸应力弛豫；拉伸应力松弛
tensile test 张力试验；拉伸试验
tensile testing apparatus 张力试验器
tensile testing machine 张力试验机
tensile test specimen 拉伸试片
tensile ultimate strength 拉伸极限强度
tensile viscosity 拉伸黏性；拉伸黏度
tensile yield ①扯断伸长率②张力屈服值；张力屈服点
tensility 可拉伸性
tensilometer 拉伸仪
tensimeter （流体）压强计；压力计
tensioactive 表面活性(的)
tensiometer 张力计；拉力计
tensiometric property 表面活性
tensiometric titration 表面张力滴定法
tensional modulus 张力模量
tension band 张力带
tension bar 拉杆
tension control braking system 张力调节式制动装置
tension cracking 拉伸龟裂
tension gauge 张力计
tension impact 拉伸冲击
tensioning load 张力载荷
tension knock-off device 消除张力装置
tension levelling 张力平整
tension link 拉杆
tension meter 张力计
tension modulus 张力模量
tension pulley 张力滑轮
tension set 永久变形
tension side 紧边；主动边〔皮带〕
tension spring 拉簧；牵力弹簧
tension stiffening 拉伸硬化
tension strain 拉伸应变
tension test 张力试验
tension tester 拉力试验机；张力试验机
tensometer 强力试验机；张力仪
tensor analysis 张量分析
tensoscan 张力监控仪
tent 浮盖
tentacle supports 触角载体
tentative 暂行的；试行的
tentative experiment 初步试验
tentative method 暂行方法
tentative method of analysis 分析试行方法；分析暂行方法
tentative specifications 暂行规格

tentative standard 暂行标准
Tentelew process 滕提柳法
tenter 拉幅机
tentering machine 伸幅机
tentering of weft 经纬架
tenth meter 10^{-10} 米；埃
tenth-normal 分规的；十分之一当量浓度的
tenth-normal solution 分规溶液
tenuity 稀薄度
tenulin 薄菊灵 $C_{17}H_{22}O_5$
tephigram 熵温图
tephrite 碱玄岩
tephroite 锰橄榄石
tephrosal 灰叶醛 $C_{10}H_{16}O$
tephrosin 灰叶素
tepid 微热的；有点温热的
tepyl acetate 3-丁基-5-甲基四氢化-2H-吡喃-4-醇乙酸酯

ter- [拉丁字头] 三
tera- [词头] 万亿；兆兆；垓 [符号 T；10^{12}]
teracalorie 兆兆卡；万亿卡 (=10^{12} 卡)
teracidic base 三价碱
teraconic acid 芸康酸；异亚丙基丁二酸
  $(CH_3)_2C=C(CO_2H)CH_2COOH$
teracrylic acid 2,3-二甲基-2-戊烯酸
  $(CH_3)_2C=C(CH_3)CH_2COOH$
terahydroxyflavone 四羟基黄酮
teratogenesis 致畸作用
teratogenicity 致畸作用
teratogenicity test 致畸试验
teratolite 银髓
teratoma 畸胎瘤
terazosin 特拉唑嗪 [药]
terbia 氧化铽 Tb_2O_3
terbinafine 特比萘芬 [药]
terbium 铽 [65 号元素，化学符号 Tb]
terbium bromide 溴化铽 $TbBr_3$
terbium carbonate 碳酸铽 $Tb_2(CO_3)_3$
terbium chloride 氯化铽 $TbCl_3$
terbium hydroxide 氢氧化铽 $Tb(OH)_3$
terbium nitrate 硝酸铽 $Tb(NO_3)_3$
terbium oxide 氧化铽 Tb_2O_3
terbium oxychloride 氯氧化铽 TbOCl
terbium sesquioxide 氧化铽 Tb_2O_3
terbromide 三溴化合物
terbutaline 特布他林 [药]

terchebin 诃子素
terchloride 三氯化合物
tercopolymer 三元共聚物
terdentate ligand 三齿配位体；三合配位体
terebene 芸香烯；单萜烯混合物
terebenthene(=turpentine) 松节油
terebentylic acid 松节酸 $C_8H_{10}O_2$
terebic acid(=terebinic acid) 芸香酸；4-甲(基)-3-羧酸基-1,4-戊内酯 $(CH_3)_2CCH(CO_2H)CH_2CO$
terebine 涂料催干剂
terebine drier 液体锰铅催干剂
terebinthina 松节油
terephthalal(=terephthalylidene) 对苯二亚甲基 $=CHC_6H_4CH=$
terephthalaldehyde resin 对苯二甲醛树脂
terephthalate 对苯二酸盐(酯)
terephthalic acid(=p-phthalic acid) 对苯二酸 $C_6H_4(CO_2H)_2$
terephthalic acid diglycidylester 对苯二甲酸二缩水甘油酯
terephthalic acid ester 对苯二(甲)酸酯
terephthalic aldehyde 对苯二醛 $C_6H_4(CHO)_2$
terephthalic-group content 对苯二(甲)酸基团含量
terephthalonic acid 对羧甲酰基苯甲酸
terephthalonitrile 对苯二腈 $C_6H_4(CN)_2$
terephthaloyl 对苯二酰 —COC_6H_4CO—
terephthalyl alcohol 对苯二甲醇
terephthalyl chloride 对苯二酰氯 $C_6H_4(COCl)_2$
terephthalylidene(=terephthalal) 对苯二亚甲基
teresantalic acid 对檀香酸；檀油酸
teresantalol 对檀香醇
teresantalyl 对檀香基
teresantalyl acetate 乙酸对檀香酯
terfenadine 特非那定 [药]
Terg-O-Tometer (振荡式)涤垢仪 [测去污力]
terhalide 三卤化合物
teriodide 三碘化合物
terlinguarite 黄氯汞矿
term ①项 [数] ②期(限)③术语④名(称)
term analysis 光谱项分析
term energy 谱项能
terminal acetylene link 末端炔键
terminal addition 末端加成
terminal aldehyde group (链)端醛基
terminal amino acid residue 末端氨基酸残基
terminal amino group 末端氨基
terminal amino group content (链)端氨基含量

terminal and subterminal hydroxylation 端碳及亚端碳羟化
terminal bond 端键
terminal carbon 末端碳原子
terminal carboxyl 末端羧基
terminal carboxyl group 末端羧基
terminal delivery 码头交货
terminal dispensing station 码头配油站
terminal disposal of radwaste 放射性废物最终处置
terminal double bond(=terminal double link) 末端双键
terminal double link(=terminal double bond) 末端双键
terminal effect 端基效应
terminal facilities 油库设备；码头设备
terminal falling velocity 极限降落速度
terminal functionality 端基官能度；链端官能度
terminal group(=terminal grouping) 端基
terminal grouping(=terminal group) 端基
terminal hydrophilic group 末端亲水基；亲水端基
terminal hydroxy group 羟端基；(链)端羟基
terminal hydroxyl 末端羟基
terminal incorporation 末端参入
terminal initiation 链端引发作用
terminal inversion 末端倒位
terminal isothiocyanate group (链)端异硫氰酸基—NCS
terminal link(=terminal linkage) 端键
terminal linkage(=terminal link) 端键
terminal olefin (末)端烯烃；α-烯烃
terminal olefine 端烯烃
terminal olefinic bond(=terminal olefinic link) 末端烯键
terminal olefinic link(=terminal olefinic bond) 末端烯键
terminal plug 末端堵塞
terminal polar group (末)端极性基；极性端基
terminal pulleys 前后鼓轮
terminal pump station 末端泵站；终点泵站
terminal radical 终端游离基
terminal speed 极限速度；临界速度
terminal station 终点站
terminal storage 最终储存
terminal triple bond(=terminal triple link) 末端三键
terminal triple link(=terminal triple bond) 末端三键
terminal unsaturation 末端不饱和
terminal velocity 终点；末端速度
terminal voltage (路)端电压
terminated 封端的；链端的；终止的
terminated polymer 封端聚合物
terminated radical 端基；链端基团
terminating electrolyte (TE) 终末电解质
termination 终止(作用)
termination agent (链的)终止剂
termination by cyclization 环化终止(反应)
termination codon 终止密码子
termination coefficient 链终止系数
termination of chain 链的终止
terminator 终止剂
terminolic acid 榄仁树脑酸；终油酸 $C_{30}H_{48}O_6$
term of spectrum 光谱项
termolecular 三分子的
termolecular mechanism 三分子机理
termolecular reaction 三分子反应
termostate oil cooler 恒温油冷却器
term overlapping 谱项重叠
term splitting 谱项分裂*
ternary 三元的
ternary acid 三元酸
ternary collision 三元碰撞
ternary complex 三元络合物*
ternary composition diagram 三元组成图
ternary compound 三元化合物
ternary copolymer 三元共聚物
ternary copolymerization 三元共聚合
ternary diagram 三元图解
ternary electrolyte 三元电解质
ternary eutectic 三元共晶体；三元低共熔点混合物
ternary fission 三分裂；三裂变
ternary phase diagrams 三元相图
ternary polymerization 三元共聚(作用)
ternary solution 三元溶液
ternary steel 三元钢
ternary system 三元体系
ternary system ceramic 三元系陶瓷
ternary vapour-liquid equilibrium 三元蒸气-液体平衡
terne plate 镀铅铁板
ternitrate 三硝酸盐(酯)
terone 试钛灵
terotechnology 维修工艺学
teroxide 三氧化物
terpadiene 萜二烯
terpadienone 萜二烯酮
terpane(=menthane) 萜烷；蓋烷
terpene 萜*
terpene acids 萜酸类
terpene alcohol 萜(烯)醇
terpene aldehyde 萜(烯)醛
terpene dihydrochloride 二氢氯酸萜
terpene diphenyl sulfonate 萜烯二苯磺酸盐

terpene hydrate 水合萜烯
terpene hydriodide 氢碘酸萜
terpene hydrocarbon 萜烯烃〔再生剂〕
terpene hydrochloride 氢氯酸萜
terpeneless oil 无萜油
terpene maleic anhydride condensation product 萜烯-顺丁烯二酸酐缩合物
terpene peroxide 过氧化萜烯
terpene phenol 萜(化)酚
terpene resin 萜烯树脂
terpene-resinic acid blend 萜烯-树脂酸共混料
terpene solvent 萜烯溶剂
terpenic 萜烯(类)的
terpenic acid 萜烯酸
terpenic alcohol 萜烯醇
terpenic jalaric acid 萜烯壳脑醛酸
terpenic series 萜烯系
terpenoid 萜类化合物；类萜
terpenol 萜烯醇
terpenol phosphate 磷酸萜烯醇
terpenone 萜烯酮
terpenyl 萜烯基
terpenylic acid 萜烯基酸
terphenyl 三联苯 $C_6H_5C_6H_4C_6H_5$
terphenyl compound 三苯化合物
terpilene(=terpinylene) 萜基烯 $C_{10}H_{16}$
terpilene dihydrochloride 桉脑
terpilenol(=terpineol) 萜品醇；松油醇
terpine 萜品；1,8-萜二醇 $C_{10}H_{20}O_2$
terpine hydrate 水合萜品；水合1,6-萜二醇 $C_{10}H_{20}O_2·H_2O$
terpine laurate 月桂酸萜品〔湿润剂〕
terpinene 萜品烯；松油烯 $C_{10}H_{16}$
α-terpinene α-萜品烯；1,3-萜二烯 $C_{10}H_{16}$
terpinene maleic anhydride adduct 萜品烯-顺丁烯二酸酐加成物
terpineol(=terpilenol) 萜品醇；松油醇
terpine resin 萜烯树脂
terpinol 萜品油〔混合物〕
terpinolene 萜品油烯；异松油烯 $C_{10}H_{16}$
terpinyl 萜品基；松油基
terpinyl acetate 乙酸萜品酯；1-萜烯-8-醇乙酸酯；乙酸松油酯 $CH_3CO_2C_{10}H_{17}$
α-terpinyl anthranilate 邻氨基苯甲酸-α-松油酯
α-terpinyl butyrate 丁酸-α-松油酯
terpinyl cinnamate (肉)桂酸松油酯
terpinylene(=terpilene) 萜基烯
α-terpinyl ethoxyacetate 乙氧基乙酸-α-松油酯

terpinyl formate 甲酸萜品酯 $HCO_2C_{10}H_{17}$
terpinylic acid 萜品基酸
α-terpinyl isovalerate 异戊酸-α-松油酯
α-terpinyl methoxyacetate 甲氧基乙酸-α-松油酯
terpinyl phenylacetate 苯乙酸松油酯
terpinyl propionate 丙酸松油酯
terpolyamide 三元共聚酰胺
terpolycyantoamino-formaldehyde resin 三聚氰胺甲醛树脂〔硫化剂〕
terpolymer 三元共聚物
terpolymer EP rubber 三元乙丙橡胶
terpolymerization 三聚作用
terposol 溁醚
terpyridyl 三联吡啶
terra alba 白土；硬石膏
terra cariosa 硅藻土
terrace 平台
terrace furnace 梯台式炉
terracinoic acid 土霉酸；地霉酸
terra-cotta ①空心砖②琉璃瓦③混合陶器
terra-cotta clay 混合陶土
terrain ①地面；地形②岩层
terra japonica 棕儿茶
terramycin(=oxytetracycline) 土霉素
terra rossa 红土；化石性红土
terrazzo concrete(=terrazzo) 水磨石(混凝土)
terrazzo seal 水磨石封闭剂；水磨石封闭蜡〔耐碱性树脂或蜡的有机溶剂或水的溶液〕
terrecin 土曲菌素
terreic acid 土曲霉酸
terrein 土曲霉酮；二羟基丙烯基环戊烯酮
terrestrial (大)地的；陆地的
terrestrial gravitation 地球引力
terrestrial magnetism ①地磁②地磁学
terre verte(=green earth) 绿土
terrific speed 极大(高)速率
terry swatch 毛圈布布样
tersulfate 三硫酸盐
tert- 〔词头〕叔〔1.指 $CH_3≡C(CH_3)_2$ 型支链烃基；2.指三元胺及 R_3COH 型的醇〕
tert-aliphatic monoglycidylether 叔脂肪族单缩水甘油醚
tert-amyl 叔戊基 $CH_3CH_2C(CH_3)_2$—
tert-butanol 叔丁醇
tert-butoxy 叔丁氧基 $(CH_3)_3CO$—
tert-butyl 叔丁基 $CH_3C(CH_3)_2$—
tert-butyl alcohol 叔丁醇
tert-butylamine 叔丁胺
tert-butyl chloride 特丁基氯

tert-butyl hydroperoxide 叔丁基氢过氧化物
2-tert-butyl-malachite green 2-叔丁基孔雀绿
tert-butyl perbenzoate 过苯酸特丁酯〔聚合催化剂〕
tert-butyl permaleic acid （酸式）过马来酸叔丁酯
tert-butyl peroxide 过氧化叔丁基；叔丁基过氧化物
tert-butyl per(oxy) acetate 过(氧)乙酸叔丁酯
tert-butyl per(oxy) isobutyrate 过氧异丁酸叔丁酯
tert-butyl peroxy isopropyl carbonate 碳酸叔丁基过氧异丙酯
tert-butyl per(oxy) pivalate 过氧新戊酸叔丁酯
tert-butyl perphthalic acid 叔丁基过氧邻苯二甲酸
p-tert-butylphenol 对叔丁基苯酚
tert-butyl vinyl ether 特丁基乙烯基醚
tertdodecyl benzoic acid 叔十二烷基苯甲酸
tertiary(=tert-) 叔（1.指 $CH_3 \equiv C(CH_3)_2$ 型支链烃基；2.指三元胺及 R_3COH 型的醇）
tertiary acetylenic glycol 叔炔二醇
tertiary alcohol 叔醇 R_3COH
tertiary alkyl 叔烷基 $CH_3 \equiv C(CH_3)_2-$
tertiary alkyl peroxide 叔烷基过氧化物
tertiary amine 叔胺 R_3N
tertiary amine catalyst 叔胺催化剂
tertiary amine surfactant 叔胺表面活性剂
tertiary amine value 叔胺值
tertiary ammonium 叔铵
tertiary amyl 叔戊基
tertiary amyl alcohol 叔戊醇 $C_2H_5(CH_3)_2COH$
tertiary arsenate 三代砷酸盐 M_3AsO_4
tertiary arsine cyanohalide 氰卤化叔钾 R_3AsXCN
tertiary arsine dichloride 二氯化叔钾 R_3AsCl_2
tertiary arsine oxide 氧化叔钾 $R_3As=O$
tertiary arsine oxyhalide 氧卤化叔钾化
tertiary base 叔碱
tertiary butanol 叔丁醇
tertiary butyl 叔丁基 $(CH_3)_3C-$
tertiary butylated hydroquinone 叔丁基氢醌
tertiary butyl hydroquinone 叔丁基氢醌
tertiary calcium phosphate(=tribasic calcium phosphate) 三代磷酸钙；正磷酸钙
tertiary carbon 叔碳(原子)
tertiary carbon atom 叔碳原子
tertiary creep 三重蠕变
tertiary hydrocarbon 叔烃
tertiary hydrogen 叔氢
tertiary hydroxyl group 三级羟基
tertiary iron phosphate 三代磷酸铁；磷酸铁
tertiary lipid 无磷或氮的脂类
tertiary mixture 三元混合物
tertiary monocarboxylic acid 叔碳一羧酸；叔碳酸
tertiary nitroparaffines 叔硝基烷
tertiary oil recovery 三次采油
tertiary pentyl 叔戊基
tertiary phosphate 三代磷酸盐
tertiary phosphine 叔膦 R_3P
tertiary sample 三级样品
tertiary solid 三元固体
tertiary standards of wavelength 三级波长标准
tertiary structure 三级结构
tertiary treatment （污水)三级处理
tertiomycin 叔霉素
tertnonyl benzoic acid 叔壬苯甲酸
tertnonyl toluene 叔壬基甲苯
teruchiomycin 照内霉素
tervalence 三价
tervalency 三价
tervalent 三价的
Tesla coil 特斯拉线圈
tessellated 镶嵌细工的
tesseral system 等轴晶系
test ①试验；试法②检验(法)
testane 睾丸烷
test bar 试验杆
test batch 试验的一批
test beard 排须状纤维试样
test bed(=test bench) 试验台；试验床
test bench(=test bed) 试验台；试验床
test block 试验台
test blocking 封闭试验
test board 试验板；仪表板
test body 试样；样品
test burner 试验灯
test cock 试栓
test condition 试验条件
test control 对比用试验标准；(对照)标样
test data 试验数据
test desk 试验台；试验桌
test duration 试验时间
test dye 试验用染料
tested recipe 检定的配方；成熟的配方
test engine data 试验机数据
test environment 测试环境；试验环境
test equipment 试验设备
tester ①检验器②测定器；试验器；试验计；试验机
test-fence 暴晒架；试验架
test figure 试验数字
test for chloride limit 氯化物(限度)试验法

test for contamination　污染试验
test formulation　试验配方
test gauge　试验压力计
test glass　试管
test hole　试验孔
test house　试验站
test indicator　试验指示器
testing　试验；检验；化验
testing battery　检验电池组
testing campaign　试验循环
testing certificate　试验检定证书
testing column　试验柱
testing equipment　试验设备
testing field(=testing ground)　试验场
testing gallery　试验隧道
testing ground(=testing field)　试验场
testing laboratory　检验室
testing machine　试验机
testing machine of mechanical strength　机械强度试验机；力学强度试验机
testing method　试验法；检验方法
testing procedure　试验程序；试验步骤
testing screen　试验用筛
testing sieve　实验(室用)筛
testing sieve shaker　实验筛摇撼机
testing station　试验站
test jar　试验缸
test load　试验负荷
test marketed　试销
test-meter　试验仪表
test method　试验方法
test-mixer　试验混合器
test mixture　试验混合物
test model　试验模型
test of calorific value　热值试验；卡值试验
test of distribution type　分布类型检验
test of goodness of fit　适合度检定；拟合优度检验
test of hypothesis　假设检验
test of outlier　异常值检验
test of significance　显著性检验
test oil　试验油
test oscillator　试验振荡器
testosterone propionate　丙酸睾酮〔药〕
testosterone undecanoate　十一酸睾酮〔药〕
test paper　试纸*
test period　试验期
test-piece　试样；试件
test plant　试验车间；试验装置
test portion　试验部分
test pressure　试验压力
test pressure gauge　试验压力计
test procedure　试验方法；试验程序；试验操作程序（规程）
test program　试验计划
test pump　试验泵
test recipe　试验配方
test rig　试验设备；试验台
test room　试验室
test run(s)　①试生产；小生产②试样的制备③运行试验；试运转；试车
test sample　试样
test set　试验集
test shaft　试验井；探井
test sheet　试验取样
test sieve　试验(室用)筛
test sieve analysis　(试验)筛析
test skein　试验绞丝；绞纱试样
test solution　试液*
test specification　试验规格
test specimen　试件
test specimen cutting press　试样切割机
test stand　试验台
test statistic　检验统计量*
test substance　试样；样品
test tank　试验罐
test-tube　试管；试验管
test-tube brush　试管刷
test tube centrifuge　试管离心机
test tube clamp　试管夹
test-tube holder　试管夹
test tube method ag(e)ing tester　试管式老化试验机
test-tube rack(=test-tube stand)　试管架
test tube stand (=test-tube rack;test tube support)　试管架
test tube support(=test tube stand)　试管架
test tube with ground stopper　①带塞试管②玻璃磨塞试管
test unit　试验装置；试验单元
test vehicle　试验车辆
test work　试验工作
tetan　四硝基甲烷
tetanics　抽风毒药；致抽风药
tetanthrene　四氢化菲　$C_{14}H_{14}$
tetanus　破伤风(症)
tetanus toxoid　破伤风菌疫苗
tetany　痉挛；搐搦
tetartohedral　四半面的；具有四分之一结晶体对称面的
tetra-〔希腊字头〕　四

tetra-acetylated 四乙酰化了的
tetra acetyl ethylene diamine 四乙酰基乙二胺
tetraacetyl-glucose 四乙酰葡萄糖 $C_6H_8O_2(OCOCH_3)_4$
tetraacetyl-hydrazine 四乙酰肼 $[(CH_3CO)_2N]_2$
tetra acetyl methylene diamine 四乙酰亚甲二胺
tetra-acylated 四酰基化了的
tetraalkoxyphosphonium ion 四烷氧基磷离子
tetraalkyl ammonium base 四烷基铵碱
tetra-alkyl ammonium hydroxide 氢氧化四烃基铵 R_4NOH
tetraalkyl ammonium perfluoro-alkane sulfonate 四烷基全氟链烷磺酸铵
tetraalkyl ammonium salt 四烷基季铵盐
tetra-alkyl arsonium chloride 氯化四烷基钾 R_4AsCl
tetra-alkylated 四烷基的
tetra-alkyl lead 四烷基铅
tetraalkylsilane 四烷基甲硅烷
tetraalkyl thiourea 四烷基硫脲〔促进剂〕
tetraalkylthiuram disulfide 二硫化四烷基秋兰姆〔促进剂〕
tetra-allyloxy-silicane 四烯丙氧基硅 $(C_3H_5O)_4Si$
tetraallyl phosphonosuccinate 膦酸丁二酸四烯丙酯
tetraallyl thorium 四烯丙基钍 $Th(C_3H_5)_4$
tetraallyl uranium 四烯丙基铀 $U(C_3H_5)_4$
tetraaminobiphenyl sulfone 四氨基联苯砜
tetra-amino-3,3-dimethyl diphenyl methane 四氨基-3,3-二甲基二苯甲烷 $[CH_3(NH_2)_2C_6H_2]_2CH_2$
tetra amino nickel 四氨基镍
tetraamylbenzene 四戊基苯 $(C_5H_{11})_4C_6H_2$
tetraamyl-silicane 四戊基硅 $(C_5H_{11})_4Si$
tetra-arylated 四芳基(化了)的
tetra-atomic 四原子的
tetra-atomic acid 四(碱)价酸
tetra-atomic alcohol 四元醇〔含有四个羟基的醇〕
tetra-atomic base 四价碱
tetra-atomic phenol 四元酚
tetra-atomic ring 四元环
tetraazaphenalene 四氮杂萘并苯
tetraazoporphine(=porphyrazine) 四氮卟吩；紫菜嗪
tetrabase 四甲基二氨基二苯甲烷
tetra base paper(=tetra paper) 臭氧试纸
tetrabasic ①四碱价的；四元的②四代的
tetrabasic acid(=tetraprotic acid) 四元酸
tetrabasic alcohol 四元醇〔含有四个羟基的醇〕
tetrabasic carboxylic acid 四羧酸 $R(COOH)_4$
tetrabasic ester 四羧酸酯
tetrabasic hydroxy acid 四价羟基酸
tetrabasic lead sulfate 四(碱)价硫酸铅；四代硫酸铅
tetrabenazine 四苯喹嗪 $C_{19}H_{27}NO_3$

tetrabenzoporphyrazine(=phthalocyanine) 四苯并紫菜嗪；酞菁
tetrabenzyl- 〔词头〕 四苄基
tetrabenzyl-silicane 四苄基硅 $(C_6H_5CH_2)_4Si$
N,N,N',N'-tetrabis(2-hydroxypropyl) ethylenediamine N,N,N',N'-四双(2-羟丙基)乙二胺〔交联剂〕
tetraborane 四硼烷 B_4H_{10}
tetraborate 四硼酸盐 $M_2B_4O_7$
tetraboric acid 四硼酸 $H_2B_4O_7$
tetrabromated 四溴化的
tetrabromide 四溴化物
tetrabrominated(=tetrabromizated) 四溴化的
tetrabromizated(=tetrabrominated) 四溴化的
tetrabromo- 〔词头〕 四溴
tetrabromoaniline 四溴苯胺
tetrabromo-benzene 四溴苯 $C_6H_2Br_4$
tetrabromo-bisphenol A 四溴双酚A；4,4'-亚异丙基双(2,6-二溴苯酚)〔阻燃剂〕
tetrabromo-m-cresolsulfonphthalein 四溴间甲苯酚磺酞 $C_{21}H_{14}O_5Br_4S$
tetrabromoethane 四溴乙烷
tetrabromo-ethylene 四溴代乙烯 $Br_2C=CBr_2$
tetrabromofluorescein 四溴荧光素；酸性曙红〔染〕
tetrabromomethane 四溴甲烷 CBr_4
tetrabromophenolphthalein 四溴苯酚酞 $C_{20}H_{10}O_4Br_4$
tetrabromophenol sulfonphthalein(=bromophenol blue) 四溴苯酚磺酞；溴酚蓝
tetrabromophthalic anhydride 四溴邻苯二甲酸酐 C_6Br_4COOCO
tetrabromopyrrol 四溴吡咯 C_4HBr_4N
tetrabromoquinone(=bromanil) 四溴醌；四溴代(对)苯醌 $O=C_6Br_4=O$
tetrabromothiophene 四溴噻吩 $BrC=BrCCBr=CBr-S$
tetrabutylammonium chloride 氯化四丁基铵
tetrabutylammonium iodide 碘化四丁铵 $(C_4H_9)_4NI$
tetrabutylammonium salt 四丁铵盐 $(C_4H_9)_4NX$
tetrabutyl lead 四丁基铅 $Pb(C_4H_9)_4$
tetrabutyl orthotitanate 原钛酸四丁酯
tetrabutyl pyrophosphate 焦磷酸四丁酯
tetrabutyl silicane 四丁基硅 $(C_4H_9)_4Si$
tetrabutylthiuram disulfide 二硫化四丁基秋兰姆
tetrabutyl tin 四丁基锡 $Sn(CH_2CH_2CH_3)_4$
tetrabutyl titanate(TBT) 钛酸四丁酯
tetrabutyl zirconate 锆酸(四)丁酯；四丁氧基锆
tetracaine 丁卡因〔药〕
tetracaine hydrochloride 盐酸丁卡因

tetracarboxylic acid　四羧酸
tetracarboxylic ester　四羧酸酯
tetracarp　四氯乙烯
tetracene　并四苯
tetracetate　四乙酸盐(酯)
tetrachlorated　四氯化的
tetrachloride　四氯化物
tetrachlorinated(=tetrachlorizated)　四氯化的
tetrachlorizated(=tetrachlorinated)　四氯化的
tetrachloroalkane　四氯(化)链烷
tetrachloroamericate　四氯镅酸盐　$MAmCl_4$
tetrachlor(o)-aniline　四氯苯胺
tetrachloroanthracene　四氯蒽　$C_{14}H_6Cl_4$
tetrachloroanthraquinone　四氯蒽醌　$C_{14}H_4Cl_4O_2$
tetrachloro-benzene　四氯(代)苯　$C_6H_2Cl_4$
tetrachlorobenzoquinone　四氯苯醌
tetrachlorobisphenol　四氯双酚
tetrachloro-compound　四氯化合物
tetrachlorocyclopentadiene　四氯环戊二烯
tetrachloroethane　四氯乙烷
tetrachloroethylene　四氯乙烯；全氯乙烯　$Cl_2C=CCl_2$
tetrachloroethylene carbonate　碳酸四氯乙烯酯
N-tetrachloroethylidenephosphoramidic dichloride　二氯化N-四氯亚乙基氨基磷酰　$CCl_3CCl=NPOCl_2$
tetrachlorohydroquinone　四氯代氢醌；四氯苯对二酚　$(HO)_2C_6Cl_4$
tetrachloroisoindolinone　四氯异吲哚啉酮
tetrachloro-methane　四氯代甲烷；四氯化碳　CCl_4
tetrachlorophenolphthalein　四氯酚酞　$C_{20}H_{10}O_4Cl_4$
tetrachlorophthalic acid　四氯邻苯二甲酸　$Cl_4C_6(CO_2H)_2$
tetrachloro-phthalic anhydride　四氯代苯二甲酸酐；四氯酞酸酐
tetrachloroquinone(=chloranil)　四氯代(苯对)醌　$O=C_6Cl_4=O$
tetrachlorothiophene　四氯噻吩　C_4Cl_4S
tetrachromate　四铬酸盐　$M_2Cr_4O_{13}$
tetrachromic acid　四铬酸　$H_2Cr_4O_{13}$
tetracid　四酸
tetracid base　四(酸)价碱
tetracoccus　四联球菌
tetracontane　四十烷　$CH_3(CH_2)_{38}CH_3$
tetracontyl　四十(烷)基　$CH_3(CH_2)_{38}CH_2-$
tetracosandienoic acid　二十四碳二烯酸　$C_{23}H_{43}COOH$
tetracosandioic acid(=tetracosane diacid)　二十四烷二酸
tetracosane　①(正)二十四(碳)烷②二十四(碳)(级)烷
tetracosane diacid(=tetracosandioic acid)　二十四烷二酸　$C_{22}H_{44}(COOH)_2$
tetracosane dicarboxylic acid　二十四烷二羧酸；二十六烷二酸　$C_{24}H_{48}(COOH)_2$
tetracosanic acid　二十四(烷)酸　$CH_3(CH_2)_{22}COOH$
tetracosanoic acid　二十四(烷)酸
tetracosanol　二十四(烷)醇
tetracosendioic acid(=tetracosene diacid)　二十四碳烯二酸
tetracosene diacid(=tetracosendioic acid)　二十四碳烯二酸　$C_{22}H_{42}(COOH)_2$
tetracosene dicarboxylic acid　二十四碳烯二羧酸　$C_{24}H_{46}(COOH)_2$
tetracosenic acid(=tetracosenoic acid)　二十四碳烯酸　$C_{23}H_{45}COOH$
tetracosenoic acid(=tetracosenic acid)　二十四碳烯酸
tetracosyl　二十四(烷)基　$CH_3(CH_2)_{22}CH_2-$
tetra-p-cresoxy-silicane　四对甲苯氧基硅　$(CH_3C_6H_4O)_4Si$
tetracyanoethylene　四氰乙烯　$(NC)_2C=C(CN)_2$
tetracyanopropane　四氰丙烷
tetracyano-p-quinodimethane　四氰基对醌二甲烷
tetracyclic　四环的
tetracyclic compound　四环化合物
tetracyclic hydrocarbon　四环烃
tetracyclic ring　四核环
tetracycline(=ambramycin;cyclomycin)　琥珀霉素；四环素
tetracycline antibiotics　四环系抗菌素
tetracyclinonitrile　四环素腈
tetracyclohexyltin　四环己基锡　$(C_6H_{11})_4Sn$
tetracyclone　四环酮
tetracyclopentadienyl neptunium　四环戊二烯合镎　$Np(C_5H_5)_4$
tetracyclopentadienyl uranium　四环戊二烯合铀　$U(C_5H_5)_4$
tetracylated　四酰基化了的
tetrad　①四价的②四素组；四分组
tetradecadienoic acid　十四碳二烯酸　$C_{13}H_{23}COOH$
tetradecaethylene glycol　十三缩(水)十四(个)乙二醇；十四甘醇
tetradecaheptaene　十四碳七烯
tetradecahexaene　十四碳六烯
δ-tetradecalactone　δ-十四内酯
tetradecandioic acid(=tetradecane diacid)　十四烷二酸
tetradecane　①(正)十四(碳)烷②十四(碳)(级)烷　$CH_3(CH_2)_{12}CH_3$
tetradecane diacid(=tetradecandioic acid)　十四烷二酸　$C_{12}H_{24}(COOH)_2$
tetradecane dicarboxylic acid　十四烷二羧酸；十六双酸

$C_{14}H_{28}(COOH)_2$
tetradecanoate 十四烷酸酯
tetradecanoic acid 十四(烷)酸 $C_{13}H_{27}COOH$
n-tetradecanol 正十四烷醇 $C_{14}H_{29}OH$
tetradecanoyl 十四(烷)酰 $CH_3(CH_2)_{12}CO-$
tetradecendioic acid(=tetradecene diacid) 十四烯二酸
2-tetradecene 2-十四(碳)烯 $CH_3(CH_2)_{10}CH=CHCH_3$
tetradecene 十四(碳)烯
tetradecene diacid(=tetradecendioic acid) 十四(碳)烯二酸 $C_{12}H_{22}(COOH)_2$
tetradecene dicarboxylic acid 十四(碳)(一)烯二羧酸 $C_{14}H_{26}(COOH)_2$
tetradecenic acid(=tetradecenoic acid) 十四(碳)烯酸 $C_{13}H_{25}COOH$
tetradecenoic acid(=tetradecenic acid) 十四(碳)烯酸
tetradecoic acid 十四(烷)酸 $C_{13}H_{27}COOH$
tetradecyl 十四(烷)基 $CH_3(CH_2)_{12}CH_2-$
tetradecyl acetate 乙酸十四(烷)酯 $CH_3CO_2C_{14}H_{29}$
tetradecyl alcohol 十四醇 $CH_3(CH_2)_{13}CH_2OH$
tetradecyl aldehyde 十四醛
tetradecylamidopropyl dimethyl benzyl ammonium chloride 氯化十四烷酰氨丙基二甲基苄基铵〔胶凝剂〕
tetradecylamine 十四(烷)胺 $C_{13}H_{27}CH_2NH_2$
tetradecyl ammonium chloride 十四烷基氯化铵
tetradecyl caproate 己酸十四(烷)酯 $C_5H_{11}CO_2C_{14}H_{29}$
tetradecylendioic acid(=tetradecylene diacid) 十四(碳)烯二酸
α-tetradecylene(=l-tetradecene) α-十四(碳)烯；1-十四(碳)烯 $CH_3(CH_2)_{11}CH=CH_2$
tetradecylene diacid(=tetradecylendioic acid) 十四(碳)烯二酸 $C_{12}H_{22}(COOH)_2$
tetradecylene dicarboxylic acid 十四(碳)烯二羧酸 $C_{14}H_{26}(COOH)_2$
tetradecylenic acid 十四(碳)烯酸
tetradecyl mercaptan 十四烷基硫醇
tetradecyl methyl pyridinium chloride 十四烷基甲基氯化吡啶鎓
tetradecyl naphthalene sulfonate 十四烷基萘磺酸盐
tetradecyl oxirane 十四烷基环氧乙烷
tetradecyl polyoxyethylene glycol 十四烷基聚氧乙烯二醇
tetradecyl pyridinium bromide 十四烷基溴化吡啶鎓
tetradecyl trimethyl ammonium chloride 十四烷基三甲基氯化铵
tetradecynic acid(=tetradecynoic acid) 十四(碳)炔酸 $C_{13}H_{23}COOH$
tetradecynoic acid(=tetradecynic acid) 十四(碳)炔酸
tetrad effect 四素组效应
tetradentate 四配位基(的)
tetradentate ligand 四齿配位体；四啮配位体
tetrad grouping 四素组分组
tetra(2,2-diallyloxymethyl-1-butoxy) titanium di(di-tridecyl phosphite) 四(2,2-二烯丙氧甲基-1-丁氧基)钛合二[亚磷酸二(十三酯)]
tetradymite 辉碲铋矿
tetraene 四烯
tetraethanolammonium hydroxide 氢氧化四(羟乙基)铵 $(HOCH_2CH_2)_4NOH$
tetraethel lead 四乙铅
tetraethide 四乙基金属 $M(C_2H_5)_4$
tetraethoxy-silicane 四乙氧基硅 $(C_2H_5O)_4Si$
tetraethyl〔词头〕 四乙基
tetraethylamine hydroxide 氢氧化四乙胺〔活性剂〕
tetraethylammonium bromide 溴化四乙铵 $(C_2H_5)_4NBr$
tetraethylammonium chloride 氯化四乙铵 $(C_2H_5)_4NCl$
tetraethylammonium hydroxide 氢氧化四乙铵 $(C_2H_5)_4NOH$
tetraethylammonium iodide 碘化四乙铵 $(C_2H_5)_4NI$
tetraethylammonium sulfate 硫酸四乙铵 $[(C_2H_5)_4N]_2SO_4$
tetraethylated 四乙基化的
tetraethylbenzene 四乙基苯
tetraethyl compound 四乙基化合物
tetraethyldiaminobenzophenone 二(N-二乙氨基苯基)(甲)酮
tetraethyldiaminodiphenylmethane 二(N-二乙氨基苯基)甲烷 $[(C_2H_5)_2NC_6H_4]_2CH_2$
tetraethyldiaminotriphenylcarbinol 二(N-二乙氨基苯基)苯甲醇 $[(C_2H_5)_2NC_6H_4]_2C(OH)C_6H_5$
tetraethyl-diarsine 四乙化二砷；双二乙胂 $(C_2H_5)_2AsAs(C_2H_5)_2$
tetraethylene-glycol(=tetraglycol) 三水缩四乙二醇；四甘醇 $(CH_2OCH_2)_3=(CH_2OH)_2$
tetraethylene glycol diacrylate 四甘醇二丙烯酸酯
tetraethylene glycol dimethacrylate 四甘醇二甲基丙烯酸酯〔硫化剂〕
tetraethylene pentamine 四亚乙基五胺；三缩四乙二胺 $NH_2(CH_2CH_2NH)_3CH_2CH_2NH_2$
tetraethyl ethanetetracarboxylate 乙四羧酸四乙酯 $[(C_2H_5O_2C)_2CH]_2$
tetraethyl gas 四乙铅汽油
tetraethyl germanium 四乙锗
tetraethyl lead(=tetraethyl plumbate) 四乙铅 $Pb(C_2H_5)_4$
tetraethyl orthocarbonate 原碳酸四乙酯 $C(OC_2H_5)_4$
tetraethyl orthosilicate 原硅酸四乙酯
tetraethyl plumbate(=tetraethyl lead) 四乙铅

tetraethyl propane-tetracarboxylate 丙四羧酸四乙酯 $(CH_2CHCH)(CO_2C_2H_5)_4$
tetraethylpyrophosphate(TEPP) 焦磷酸四乙酯；特普
tetraethyl radiolead 四乙基放射铅
tetraethyl silicane 四乙基硅 $Si(C_2H_5)_4$
tetraethyl-succinic acid 四乙代丁二酸 $COOHC(C_2H_5)_2C(C_2H_5)_2COOH$
tetraethyl tetrazene 四乙基四氮烯
tetraethylthiuram disulfide 二硫化四乙基秋兰姆
tetraethyl tin 四乙基锡 $Sn(C_2H_5)_4$
tetraethyl urea 四乙基脲 $(C_2H_5)_2NCON(C_2H_5)_2$
tetrafluorated 四氟化的
tetrafluoride 四氟化物
tetrafluorinated(=tetrafluorizated) 四氟化的
tetrafluorizated(=tetrafluorinated) 四氟化的
tetrafluoro-*trans*-azobenzene 四氟代反式偶氮苯
tetrafluoroberyllate 四氟铍酸(根或盐) BeF_4^{2-}; M_2BeF_4
tetrafluoroborate 四氟硼酸根(盐) BF_4^-; MBF_4
tetrafluoroborate ion-selective electrode 四氟硼酸根离子选择性电极
tetrafluoro compound 四氟化合物
tetrafluoroethylene 四氟乙烯
tetrafluoroethylene-hexafluoropropylene copolymer fibre 四氟乙烯-六氟丙烯共聚纤维；全氟乙丙烯共聚纤维
tetrafluoro-methane 四氟代甲烷；四氟化碳 CF_4
tetra-functional epoxy 四官能度环氧树脂；四功能环氧树脂
tetrafunctional initiator 四功能起始剂
tetrafunctional stabilizer 四官能(基)稳定剂
tetragalloyl erythrite 四棓酰赤藓醇
tetraglycol 四甘醇；三水缩四(乙二醇) $(HOCH_2CH_2OCH_2CH_2)_2O$
tetragonal 四方形的
tetragonal cell 四方晶胞；正方晶胞
tetragonal crystal 四方晶体
tetragonal phase 正方晶相
tetragonal pyramid 正方锥
tetragonal spheroidal 正方晶系
tetragonal system 四方晶系*
tetrahalide 四卤化物
tetrahalogenated 四卤化的
tetrahalogenated benzene(=tetrahalogeno-benzene) 四卤代苯 $C_6H_2X_4$
tetrahalogeno-benzene(=tetrahalogenated benzene) 四卤代苯
tetrahalogeno-compound 四卤化合物
tetrahalophthalic acid ester 四卤代邻苯二(甲)酸酯
tetrahedral(=tetrahedronal) 四面体的
tetrahedral complex 四面体配合物
tetrahedral configuration 四面体构型*
tetrahedral hybrid orbital 正四面体杂化轨函数
tetrahedral interstice 四面体间隙
tetrahedral isomorphous substitution 四面体同晶系取代(作用)
tetrahedral layer 四面(体)层
tetrahedrally-symmetrical molecule 四面体对称分子
tetrahedral packing pattern (粒子的)四面体堆积方式〔堆积模型〕
tetrahedral radius 正四面体半径
tetrahedral silica network 四面体二氧化硅网络
tetrahedral silicate 四面体硅酸盐
tetrahedral structure 四面体结构
tetrahedrite 黝铜矿
tetrahedron 四面体
tetrahedronal(=tetrahedral) 四面体的
tetrahedronal atom 四面体的原子
tetrahedronal carbon 四面体状碳
tetrahedron group 四面体群；四面体基团
tetraheptoxy-silicane 四庚氧基硅 $(C_7H_{11}O)_4Si$
tetraheptylammonium nitrate 硝酸四庚铵 $(C_7H_{15})_4NNO_3$
tetrahexahedron 二十四面体
tetrahexine 四六烯菌素
tetrahydrate 四水合物
tetrahydric 四氢化的
tetrahydric acid 四元酸
tetrahydric alcohol 四元醇〔含有四个羟基的醇〕
tetrahydric phenol 四元酚
tetrahydric salt 四酸式盐
tetrahydro〔词头〕 四氢化
tetrahydroabietic acid 四氢枞酸
tetrahydroabietyl alcohol 四氢枞醇
tetrahydroaldosterone 四氢醛甾酮
tetrahydroalstonine 四氢鸭脚木碱
tetrahydrobenzene(=cyclohexene) 四氢化苯；环己烯
tetrahydrobenzoic acid 四氢化苯甲酸
tetrahydrobiopterin 四氢生物蝶呤
tetrahydrobisabolene 四氢红没药烯
tetrahydro-butene 四氢丁烯
tetrahydrocaryophyllene 四氢石竹烯
tetrahydrocitral 四氢柠檬醛
tetrahydro-compound 四氢化合物
dl-tetrahydrocoptisine(=corydalis E) 延胡索素 E
tetrahydrocorticosterone 四氢皮(质甾)酮
tetrahydrocortisol 四氢皮(质甾)醇
tetrahydrocortisone 四氢可的松
tetrahydro-α-cyperone 四氢-α-香附酮

tetrahydroeucarvone 四氢优香芹酮；四氢优葛缕酮
tetrahydrofolate(THF) 四氢叶酸
tetrahydrofolate dehydrogenase 四氢叶酸脱氢酶
tetrahydrofolic acid 四氢叶酸
tetrahydroform 三亚甲基亚胺；丙亚胺
tetrahydrofuran(=tetramethylene oxide) 四氢呋喃
tetrahydrofuran polymer 四氢呋喃聚合物
tetrahydrofuran polymer oil 四氢呋喃聚合油
tetrahydrofurfuryl 氢糠基；四氢化糠基
$CH_2CH_2CH_2OCHCH_2-$
tetrahydrofurfuryl acetate 乙酸四氢糠酯
$CH_3CO_2CH_2C_4H_7O$
tetrahydrofurfuryl acrylate 丙烯酸四氢糠酯
tetrahydrofurfuryl alcohol 四氢糠醇；四氢化呋喃甲醇
$C_4H_7OCH_2OH$
tetrahydrofurfuryl amine 四氢糠胺（促进剂）
tetrahydrofurfuryl benzoate 苯甲酸四氢糠酯
$C_6H_5CO_2CH_2C_4H_7O$
tetrahydrofurfuryl butyrate 丁酸四氢糠酯
$C_3H_7CO_2CH_2C_4H_7O$
tetrahydrofurfuryl caproate 己酸四氢糠酯
$C_5H_{11}CO_2CH_2C_4H_7O$
tetrahydrofurfuryl cinnamate 肉桂酸四氢糠酯
tetrahydrofurfuryl ester 四氢化糠酯
tetrahydrofurfuryl lactate 乳酸四氢糠酯
$CH_3CHOHCO_2CH_2C_4H_7O$
tetrahydrofurfuryl laurate 十二(烷)酸四氢糠酯
$C_{11}H_{23}CO_2CH_2C_4H_7O$
tetrahydrofurfuryl methacrylate 甲基丙烯酸四氢糠酯
tetrahydrofurfuryl oleate 油酸四氢糠酯
tetrahydrofurfuryl palmitate 十六(烷)酸四氢糠酯
$C_{15}H_{31}CO_2CH_2C_4H_7O$
tetrahydrofurfuryl phthalate 邻苯二甲酸四氢糠酯（增塑剂）
tetrahydrofurfuryl propionate 丙酸四氢糠酯
$C_2H_5CO_2CH_2C_4H_7O$
tetrahydrofurfuryl salicylate 水杨酸四氢糠酯
$HOC_6H_4CO_2CH_2C_4H_7O$
tetrahydro-2-furylmethyl acetate 乙酸四氢-2-呋喃甲酯
tetrahydro-2-furylmethyl propionate 丙酸四氢-2-呋喃甲酯
tetrahydrogen sulfato boric acid 四(硫酸氢根合)硼酸
tetrahydrogeraniol 四氢香叶醇
tetrahydrogeranyl acetate(=dihydrocitronellyl acetate) 乙酸四氢香叶酯；乙酸二氢香茅酯
tetrahydroglyoxaline 四氢化甘噁啉；咪唑啉
tetrahydroirone 四氢鸢尾酮
tetrahydroisopimaric acid 四氢异海松酸

tetrahydrojasmone 四氢茉莉酮；2-戊基-3-甲基环戊酮
tetrahydrolinalool 四氢芳樟醇
tetrahydrolinalyl acetate 乙酸四氢芳樟酯
tetrahydromanool 四氢泪杉醇
tetrahydro-p-methyl quinoline 四氢对甲基喹啉
tetrahydromuguol 四氢别罗勒烯醇
tetrahydromuguol acetate 乙酸四氢别罗勒烯酯
tetrahydronaphthalene(=tetralin) 四氢化萘
tetrahydronaphthalene musk 四氢萘麝香
tetrahydronaphthalic anhydride 四氢化邻苯二甲酸酐（硫化剂）
tetrahydropalmatine 四氢帕马丁（药）
tetrahydrophthalic acid 四氢化邻苯二甲酸
tetrahydrophthalic anhydride 四氢化邻苯二甲酸酐
tetrahydropimaric acid 四氢海松酸
tetrahydropseudoionone 四氢假性紫罗兰酮
tetrahydropteridine 四氢蝶啶
tetrahydropyrane 四氢吡喃
tetrahydropyrrole(=pyrrolidine) 四氢吡咯 $(CH_2)_4=NH$
tetrahydroquinoline 四氢喹啉
tetrahydroquinone 四氢化醌 $OC=(CH_2)_4=CO$
tetrahydroquinoxaline 四氢喹噁啉
tetrahydro retene 四氢惹烯
tetrahydrorotenone 四氢化鱼藤酮
tetrahydrosylvane 四氢邻甲呋喃 $CH_3C_4H_7O$
tetrahydro-thiazole 四氢噻唑
tetrahydrothiazole-2-thion 四氢噻唑-2-硫酮（促进剂）
tetrahydrothiophene 四氢噻吩
3-tetrahydrothiophenone 3-四氢噻吩酮
tetrahydrotoluene(=methylcyclohexene) 四氢化甲苯；甲基环己烯 $CH_3C_6H_9$
tetrahydro-β-vetivol 四氢-β-岩兰草醇
tetrahydroxy acid 四羟基酸
tetrahydroxy adipic acid 四羟基己二酸
$COOH(CHOH)_4COOH$
tetrahydroxy alcohol 四元醇（含有四个羟基的醇）
tetrahydroxy benzene 四羟基苯；苯四酚 $C_6H_2(OH)_4$
tetrahydroxybenzophenone 四羟基二苯甲酮（紫外线吸收剂）
tetrahydroxy compound 四羟基化合物
tetrahydroxy dibasic acid 四羟基二元酸 $(OH)_4R(COOH)_2$
tetrahydroxyethylethylenediamine 四羟基乙基乙二胺（交联剂）
tetrahydroxyflavonol 四羟基黄酮醇 $C_{15}H_{10}O_7$
tetrahydroxylated 四羟基化的
tetrahydroxyl compound 四羟基化合物
tetrahydroxy monobasic acid 四羟基一元酸 $(OH)_4RCOOH$

tetrahydroxy phenol 四羟基苯；苯四酚 $(OH)_4C_6H_2$
tetrahydroxy-quinone 四羟基醌；四羟基苯对醌 $(HO)_4C_6O_2$
tetrahydroxy-stearic acid 四羟基硬脂酸
tetraiodated 四碘化的
tetraiodide 四碘化物
tetraiodinated(=tetraiodizated) 四碘化的
tetraiodizated(=tetraiodinated) 四碘化的
tetraiodo-benzene 四碘(代)苯 $C_6H_2I_4$
tetraiodo compound 四碘化合物
tetraiodo-ethylene 四碘(代)乙烯 $I_2C=CI_2$
tetraiodofluorescein(=iodeosin) 四碘荧光素
tetraiodo-methane 四碘代甲烷；四碘化碳 CI_4
tetraiodophenolphthalein 四碘酚酞 $C_{20}H_{10}O_4I_4$
tetraiodophenolphthalein sodium salt 四碘酚酞钠 $C_{20}H_8O_4I_4Na_2$
tetraiodophenolsulfonphthalein 四碘苯酚磺酞 $C_{19}H_{10}O_5I_4S$
tetraiodophthalic anhydride 四碘酞酐〔俗〕；四碘代邻苯二(酸)酐 $I_4C_6(CO)_2O$
tetraiodopyrrole(=iodole) 四碘吡咯；碘咯 I_4C_4NH
tetraiodotetrachlorfluorescein 四碘四氯荧光素；玫瑰红 $C_{20}H_4O_5I_4Cl_4$
tetraiodothyroacetic acid 四碘甲腺乙酸
tetraisoamoxy-silicane 四异戊氧基硅 $(C_5H_{11}O)_4Si$
tetraisoamyl lead 四异戊基铅 $Pb[CH_2CH_2CH(CH_3)_2]_4$
tetraisoamyl silicane 四异戊基硅 $(C_5H_{11})_4Si$
tetraisoamyl tin 四异戊基锡 $[(CH_3)_2CHCH_2CH_2]_4Sn$
tetraisobutoxy-silicane 四异丁氧基硅 $(C_4H_9O)_4Si$
tetraisobutyl-lead 四异丁基铅 $Pb[CH_2CH(CH_3)_2]_4$
tetraisopropoxy titanium 四异丙氧基钛；钛酸四异丙酯
tetraisopropylbenzene 四异丙苯 $[(CH_3)_2CH]_4C_6H_2$
tetraisopropyl di(dilaurylphosphito) titanate 二(亚磷酸二月桂酯)钛酸四异丙酯
tetraisopropyl-lead 四异丙基铅 $Pb[CH(CH_3)_2]_4$
tetraisopropyl titanate(=titanium isopropylate) 异丙醇钛；钛酸(四)异丙酯
-tetraketone 〔词尾〕 四(甲)酮
tetrakis 〔词头〕 四个
tetrakis(ethoxymethyl) benzoguanamine 四乙氧甲基苯代三聚氰二胺
tetrakis(2-ethyl-1,3-hexane diolato) titanium 钛酸四异辛二(醇)酯
tetrakis hexahedron 四重六面体
tetralin(=tetrahydronaphthalene) 四氢化萘 $C_6H_4CH_2(CH_2)_2CH_2$
tetraline(=tetralin) 四氢化萘

tetralol 四氢萘酚
tetralon acid 乙二胺四乙酸
tetralone 四氢萘酮
tetralute 甲状腺素(柱)测定药箱
tetralyl 四氢萘基
tetramer 四聚物；四聚体
tetramethide 四甲基金属 $M(CH_3)_4$
tetramethoxy-compound 四甲氧基化合物
tetramethoxy-silicane 四甲氧基硅 $(CH_3O)_4Si$
tetramethyl 〔词头〕 四甲基
tetramethylammonium 四甲基铵
tetramethylammonium azide 叠氮化四甲基铵
tetramethylammonium bromide 溴化四甲铵 $(CH_3)_4NBr$
tetramethylammonium chloride 氯化四甲铵 $(CH_3)_4NCl$
tetramethylammonium formate 甲酸四甲铵
tetramethylammonium hydroxide 氢氧化四甲铵 $(CH_3)_4NOH$
tetramethylammonium iodide 碘化四甲铵 $(CH_3)_4NI$
tetramethylammonium saponite 四甲铵皂石
tetramethyl-arsonium hydroxide 氢氧化四甲钟 $(CH_3)_4AsOH$
tetramethylated 四甲基化的
tetramethyl-benzene 四甲基苯 $(CH_3)_4C_6H_2$
1,2,4,5-tetramethylbenzene(=durene) 1,2,4,5-四甲基苯
tetramethyl-benzidine N-四甲联苯胺 $[(CH_3)_2NC_6H_4]_2$
tetramethyl-benzoquinone 四甲基苯醌 $(CH_3)_4C_6O_2$
tetramethyl-bis(3,5-dimethyl-4-hydroxyphenyl) sulfone 四甲基双(3,5-二甲基-4-羟基苯基)砜〔防焦剂〕
tetramethyl-1,4-butanediamine 四甲基-1,4-丁二胺〔硫化剂〕
1,1,3,3-tetra-methylbutyl hydroperoxide 1,1,3,3-四甲基丁基过氧化氢
(1,1,3,3-tetra-methylbutyl peroxy)-2-ethylhexanoate (1,1,3,3-四甲基丁基过氧化)-2-乙基己酸酯
tetramethyl compound 四甲基化合物
tetramethylcyclobutanediol 四甲基环丁二醇
tetramethyl-diaminobenzhydrol 四甲基二氨基二苯甲醇
tetramethyl-diaminobenzophenone 四甲基二氨基二苯(甲)酮
N-tetramethyl diaminodiphenylcarbinol(=Michler's hydrol) N-四甲基二氨基二苯基甲醇；米希勒醇
tetramethyl-diaminotriphenylmethane 四甲基二氨基三苯基甲烷
tetramethyl diphosphine 四甲基偶膦 $(CH_3)_2P·P(CH_3)_2$
tetramethylene ①四亚甲基；1,4-亚丁基 —$CH_2(CH_2)_2CH_2$— ②(=cyclobutane)环丁烷
tetramethylene adipate 己二酸四亚甲基酯；己二酸丁二

醇酯
tetramethylene amine 四亚甲基胺
tetramethylene-diamine 四亚甲基二胺；1,4-丁二胺 $H_2N(CH_2)_4NH_2$
tetramethylene diammonium acetate 乙酸丁二铵
tetramethylene diguanidine 四亚甲基二胍
tetramethylene diisocyanate 二异氰酸四亚甲酯
tetramethylene diisothiocyanate 四甲撑二异硫氰酸酯
tetramethylene dimethacrylate 二甲基丙烯酸丁二(醇)酯
tetramethylene glycol 1,4-丁二醇
tetramethylene imine 环丁亚胺；四亚甲基亚胺
tetramethylene maleic anhydride 四亚甲基马来酸酐
tetramethylene oxide(=tetrahydrofuran) 四氢呋喃
tetramethylene pentamine 四亚甲基五胺〔硫化剂〕
tetramethylene phosphate 磷酸四亚甲酯
tetramethylene sulfide 四氢噻吩 $(CH_2)_4S$
tetramethylene sulfone(=sulfolane) 四氢噻吩砜；环丁砜
tetramethylene sulfoxide 四亚甲基亚砜
tetramethyl-ethylene 四甲基乙烯 $(CH_3)_2C=C(CH_3)_2$
tetramethyl ethylene glycol dimethyl silicon ether 四甲基乙二醇二甲基硅醚〔结构控制剂〕
tetramethyl ethylene ketone(=pinacone) 2,3-二甲基-2,3-丁二醇；频哪醇
tetramethyl glucopyranose 四甲基吡喃葡萄糖
tetramethyl-glucose 四甲基葡萄糖
tetramethyl glucoside 四甲基葡萄糖苷
tetramethyl guanidine 四甲基胍〔促进剂〕
tetramethyl-hexamethylene diamine 四甲基代六甲(撑)二胺
tetramethyl-lead(=tetramethyl plumbane) 四甲基铅 $Pb(CH_3)_4$
tetramethyl leucaniline 二(二甲氨基苯基)苯甲烷
tetramethylmethane 四甲基甲烷
tetramethylol benzoguanamine 四羟甲基苯代三聚氰二胺
tetramethylol cyclohexanol 四羟甲基环己醇
tetramethylol glycoluril 四羟甲基甘脲
tetramethylol methane tetraacrylate 季戊四醇四丙烯酸酯
tetramethylol urea 四羟甲基脲
tetramethyl-oxamide 四甲基草酰胺
2,3,5,6-tetramethylphenyl(=duryl) 2,3,5,6-四甲(基)苯基；杜基
tetramethyl phenylene 2,3,5,6-四甲(基)亚苯基 $(CH_3)_4C_6H-$
tetramethyl-p-phenylene(=durylene) 四甲(基)代对亚苯基
tetramethyl phenylene diamine 二(二甲氨基)苯二胺 $C_6H_4[N(CH_3)_2]_2$
tetramethyl phosphonium 四甲基磷；四甲基磷鎓 $(CH_3)_4P^+$
tetramethyl phosphonium chloride 氯化四甲基磷
tetramethyl phosphonium iodide 碘化四甲基磷 $(CH_3)_4PI$
tetramethylpiperidone(=triacetonamine) 三丙酮胺；四甲基哌啶酮 $C_9H_{17}ON$
tetramethyl plumbane(=tetramethyl lead) 四甲基铅
2,3,5,6-tetramethylpyrazine 2,3,5,6-四甲基吡嗪
2,2,6,6-tetramethyl pyridine N-oxyl N-氧化-2,2,6,6-四甲基吡啶
tetramethyl-p-quaterphenyl 四甲基对四联苯 $H[C_6H_3(CH_3)]_4H$
tetramethylsilane 四甲基硅(烷) $Si(CH_3)_4$
tetramethyl silicane 四甲基硅 $Si(CH_3)_4$
tetramethyl stannane(=tetramethyl tin) 四甲基锡 $(CH_3)_4Sn$
tetramethyl-succinic acid 四甲基丁二酸 $[(CH_3)_2CCO_2H]_2$
tetramethyltetraethylthiuram disulfide 二硫化四甲基四乙基秋兰姆〔促进剂〕
tetramethylthiazine 四甲基噻嗪
tetramethyl thiourea 四甲基硫脲〔促进剂〕
tetramethyl-thiuram disulfide 二硫化四甲基秋兰姆 $[(CH_3)_2NCSS]_2$
tetramethyl tin(=tetramethyl stannane) 四甲基锡 $(CH_3)_4Sn$
tetramethyl triaminotriphenyl methane 二(二甲氨基苯基)氨基苯基甲烷 $[(CH_3)_2NC_6H_4]_2CHC_6H_4$
tetramethyltrioxypurine 四甲基三氧代嘌呤
tetramethyl-urea 四甲基脲 $[(CH_3)_2N]_2CO$
tetramethyl-uric acid 四甲基尿酸 $C_5H_4O_3(CH_3)_4$
tetramethyl xylylene diisocyanate 四甲代二甲撑二异氰酸酯；四甲代苯二亚甲基二异氰酸酯
tetramido 四酰氨基
tetramine 四胺
tetramino 四氨基
tetrammine 四氨合物
tetrammine cobaltrichloride 三氯化四氨钴 $[Co(NH_3)_4]Cl_3$
tetrammine platinous chloride 二氯化四氯铂 $[Pt(NH_3)_4]Cl_2$
tetramolecular 四分子的
tetramolecular reaction 四分子反应
tetramorphism 四晶(现象)
tetramycin 四霉素；四环素；四环素碱
tetrandrine 汉防己甲素〔药〕
tetrane 丁烷
tetranitrate 四硝酸酯

tetranitrated 四硝基化的
tetranitrated compound(=tetranitro-compound) 四硝基化合物
tetranitro-aniline 四硝基苯胺
tetranitro-anisole 四硝基苯甲醚
tetranitro-anthraquinone 四硝基蒽醌
tetranitrocarbazol 四硝基咔唑
tetranitrochrysazin 四硝基柯嗪；四硝基-1,8-二羟基蒽醌 $(NO_2)_4C_{14}H_2(OH)_2O$
tetranitro-compound(=tetranitrated compound) 四硝基化合物
tetranitrodiglycerine 四硝基二甘油；二缩甘油四硝酸酯 $C_6H_{10}N_4O_{13}$
tetranitro-dihydroxydiphenyl 四硝基二羟基联苯 $[(NO_2)_2C_6H_2OH]_2$
tetranitro-diphenyl 四硝基联苯
tetranitro-diphenylamine 四硝基二苯胺
tetranitrodiphenyl disulfide 四硝基二苯二硫 $[(NO_2)_2C_6H_3S]_2$
tetranitrodiphenyl ether 四硝基二苯醚 $[C_6H_3(NO_2)_2]_2O$
tetranitrodiphenyl-methane 四硝基二苯甲烷 $C_{13}H_8(NO_2)_4$
tetranitro-ethylaniline 四硝基乙基苯胺
tetranitro-glycerine 甘油四硝酸酯
tetranitrol 季戊四醇四硝酸酯 $C_4H_6(ONO_2)_4$
tetranitromethane 四硝基甲烷 $C(NO_2)_4$
tetranitro-methylaniline(=tetryl) 四硝基甲基苯胺；特屈儿
tetranitro-naphthalene 四硝基萘
tetranitro-pentaerythrite(=penthrite) 季戊四醇四硝酸酯；季戊炸药
tetranitro-phenol 四硝基苯酚
tetranthera 柠樟树皮
tetranuclear(=tetranucleate; tetranucleated) 四环的；四核的
tetranucleate(=tetranuclear) 四环的；四核的
tetranucleated(=tetranuclear) 四环的；四核的
tetranucleotide 四核苷酸
tetraoctyl bis(ditridecyl phosphite) titanate 二[亚磷酸双(十三烷基酯)]钛酸四辛酯
tetraoctyleneglycol titanium 钛酸四异辛二(醇)酯
tetraoctyloxy titanium di(dilaurylphosphite) 四辛氧基钛二(亚磷酸二月桂酯)
tetraoxypyrimidine 四氧嘧啶
tetra paper(=tetra base paper) 臭氧试纸
tetrapartite chromosome 四分染色体
tetrapeptide 四肽
tetraphene 苯并蒽

tetraphenothio germanium 四苯硫基锗
tetraphenoxy germanium 四苯氧基锗
tetraphenoxy-silicane 四苯氧基硅 $(C_6H_5O)_4Si$
tetraphenylarsonium chloride 氯化四苯钾
tetraphenylated 四苯基化的；四苯基代的
tetraphenylboron 四苯硼
tetraphenylene 亚四苯基 ($=C_6H_4)_4$
tetraphenyl ethylene 四苯乙烯 $(C_6H_5)_2C=C(C_6H_5)_2$
tetraphenyl furan 四苯基呋喃
tetraphenyl guanidine 四苯胍 $HN=C[N(C_6H_5)_2]_2$
tetraphenyl-hexaoxacyclooctadecadiene 四苯基六氧环十八醚二烯
tetraphenyl hydrazine 四苯肼 $(C_6H_5)_2NN(C_6H_5)_2$
tetraphenyl lead 四苯基铅 $Pb(C_6H_5)_4$
tetraphenyl methane 四苯基甲烷 $(C_6H_5)_4C$
tetraphenylphosphonium chloride 氯化四苯基磷；四苯磷氯 $(C_6H_5)_4PCl$
tetraphenylporphin dianion 四苯卟吩的二价负离子
tetraphenyl pyrazine 四苯基吡嗪
tetraphenyl silicane 四苯基硅 $(C_6H_5)_4Si$
tetraphenyl-silicon 四苯基硅 $Si(C_6H_5)_4$
tetraphenyl stannum 四苯基锡〔交联剂〕
tetraphenylstibonium bromide 溴化四苯(基)锑
tetraphenyl succinic acid 四苯基琥珀酸
tetraphenyl tin 四苯基锡 $(C_6H_5)_4Sn$
tetraphenyl urea 四苯基脲 $[(C_6H_5)_2N]_2CO$
tetraphosphine 四膦 $H_2PPH \cdot PHPH_2$
tetraphosphine oxide 氧化四膦
tetraphosphorus monoselenide 一硒化四磷
tetraphyllicine(=serpinine) 四叶萝芙辛
tetraploid 四倍体
tetrapolymer 四元共聚物
tetrapotassium peroxy diphosphate 过二磷酸四钾〔引发剂〕
tetrapotassium pyrophosphate 焦磷酸四钾
tetrapropenyl succinic anhydride 四丙烯基琥珀酸酐〔硫化剂〕
tetrapropoxy-silicane 四丙氧基硅 $(C_3H_7O)_4Si$
tetrapropylammonium hydroxide 氢氧化四丙基铵 $(C_3H_7)_4NOH$
tetrapropylammonium iodide 碘化四丙铵 $(C_2H_5CH_2)_4NI$
tetrapropylene(=propylene tetramer) 四聚丙烯
tetrapropylene alkylphenol 四聚丙烯烷基酚
tetrapropylene benzene 四聚丙烯基苯
tetrapropylene benzene sulfonate 四聚丙烯基苯磺酸盐
tetrapropyl-lead 四丙基铅 $Pb(C_3H_7)_4$
tetrapropyl orthocarbonate 原碳酸四丙酯 $C(OCH_2C_2H_5)_4$
tetrapropyl-silicane 四丙基硅 $(C_3H_7)_4Si$

tetrapropylthiuram disulfide 二硫化四丙基秋兰姆〔促进剂〕
tetrapropyltin 四丙基锡 $(CH_3CH_2CH_2)_4Sn$
tetraprotic acid(=tetrabasic acid) 四元酸
tetraricinoleic acid 四聚蓖麻醇酸
tetrasaccharide 四糖
tetrasilane 四硅烷；丁硅烷
tetrasodium N-(1,2-dicarboxyethyl)-N-octadecylsulfosuccinate N-(1,2-二羧乙基)-N-十八烷基磺化琥珀酸四钠〔防凝胶剂〕
tetrasodium edetate hydrate 乙二胺四乙酸四钠水合物
tetrasodium ethylenediamine tetraacetate 乙二胺四乙酸四钠〔稳定剂〕
tetrasodium pyrophosphate 焦磷酸四钠
tetrasodium salt of ethylene diamine tetraacetic acid 乙二胺四乙酸的四钠盐
tetrasorb 甲状腺素(树脂)测定药箱
tetrastearoyl titanate 钛酸四硬脂酰酯
tetrastyrene 四聚苯乙烯 $C_{32}H_{32}$
tetrasubstituted 四元取代
tetra substitution product 四取代产物
tetrasulfide 四硫化物
tetraterpene 四萜(烯)
tetrathiacyclotetradecane 四硫代环十四烷
tetrathiafulvalene 四硫富瓦烯
3,3′-tetrathio-bis(2-methylfuran) 3,3′-四硫代双(2-甲基呋喃)
tetrathiodimorpholine 四硫代二吗啉〔硫化剂〕
tetrathionate 连四硫酸盐 $M_2S_4O_6$
tetrathionic acid 连四硫酸 $H_2S_4O_6$
tetratolyl lead 四甲苯基铅
tetratolyl silicane 四甲苯基硅
tetratolyl tin 四甲苯基锡
tetratomic ①四原子的②四羟基的
tetratomic acid 四(碱)价酸
tetratomic alcohol 四元醇〔含有四个羟基的醇〕
tetratomic base 四价碱
tetratomic ring 四元环
tetratriacontane 三十四(碳)烷 $CH_3(CH_2)_{32}CH_3$
tetravalence(=tetravalency) 四价
tetravalency(=tetravalence) 四价
tetravalent 四价的
tetravalent alcohol 四元醇〔含有四个羟基的醇〕
tetravalent base 四价碱
tetrazanaphthalene 四氮杂萘
tetrazane 四氮烷 $H_2NNHNHNH_2$
tetrazaporphin(=porphrazine) 四氮吡吩；紫棕碱
tetrazene〔类名〕 四氮烯 $NH_2NHN=NH$
v-tetrazine 连四嗪
tetrazine 四嗪；四氮杂苯 $C_2H_2N_4$

s-tetrazine 对称四嗪
a-tetrazine 不对称四嗪
tetrazo(=bisazo;bisdiazo) 双偶氮 $(-N=N-)_2$
tetrazo compound 双偶氮化合物
tetrazole 四唑
tetrazoline colorimetry 四唑啉比色法
tetrazolium 四唑(鎓)
tetrazolium bioautography 四唑鎓生物自显影法
tetrazolium blue 四唑鎓蓝
tetrazolo 四唑并
tetrazolyl 四唑基
tetrazone 四氮腙；偶二氮化合物 $R_2NN=NNR_2$
tetrazotic acid 四氮酸
tetrazotization 双偶氮化(作用)；四氮化(作用)
tetrazotized 四氮化的；双偶氮化了的
tetrazyl 四唑基
tetrazyl hydrazine 四唑肼
tetrelide 第四族元素化物
tetrels 第四族元素
tetren(=tetraethylenepentamine) 四(亚)乙(基)五胺
tetrenolin 四烯醇素
tetrin 四烯菌素
tetrinic acid 特春酸；α-甲基特窗酸
tetritol 丁糖醇
tetrodonine 河豚卵毒素
tetrodotoxin 河豚毒
tetrol 四醇〔化合物结构上有四个羟基〕
tetrolaldehyde(=tetrolic aldehyde) 丁炔醛
tetrole 呋喃
tetrolic acid 2-丁炔酸
tetrolic aldehyde(=tetrolaldehyde) 丁炔醛
tetronal 特妥那；四乙眠砜；3-戊酮缩二乙砜 $(C_2H_5)_2C(SO_2C_2H_5)_2$
tetronate 4-羟(基)乙酰乙酸内酯
-tetrone〔词尾〕 四酮
tetronerythrin 羽赤素
tetronic acid 特窗酸；季酮酸；4-羟乙酰乙酸内酯
tetronimide 特窗酸亚胺
tetrose 四糖 $C_4H_8O_4$
tetroxide 四氧化物
tetruronic acid 丁糖酮酸
tetryl(=2,4,6-trinitrophenylmethylnitramine) 特屈儿；2,4,6-三硝基苯甲硝胺 $(NO_2)_3C_6H_2N(CH_3)NO_2$
tetryl formate 甲酸异丁酯
teucrin 石蚕苷
tex 特〔纤度单位,1000米纤维重1克为1特〕
Texaco hydrofinishing 德士古加氢精制(法)
Texas cedar wood oil 得克萨斯柏木油

texibond 聚乙酸乙烯酯类黏合剂
texilscope 混纺(成分)分析仪
texogenin 丝兰肖配基〔取自 Yucca schottii〕
texrope 三角皮带
textile ①织物②纺织的
textile assistant 纺织(用)助剂
textile auxiliary 纺织助剂
textile bleaching agent 纺织品漂白剂
textile board 织板；压板（纺）
textile coating 织物涂料；织物涂胶
textile composite material 纺织复合材料
textile degumming detergent 织物脱胶用洗涤剂
textile end product 纺织复制品
textile fabric 织物
textile finishing agent 纺织品整理剂
textile glass 纺织玻璃纤维
textile glass multifilament product 连续玻璃纤维制品
textile glass staple fibre product 定长玻璃纤维制品
textile lacquer 织品用漆；织物用漆
textile lubrication 纺织机的润滑
textile machinery leather 纺织机用革
textile material ①纺织材料②纺织品
textile mill effluent 纺织厂排出(的)废(物)
textile oil 纺织用油
textile processing 纺织加工
textile size 纺织型浸润剂
textile soap 丝光皂
textile softener 纺织柔软剂
textile spirit 纺织品用溶剂汽油
textile spool paper 卷线(筒)纸
textile treating agent 纺织品处理剂；纺织原料加工助剂
textile washing 纺织品洗涤
textile wetting 纺织品润湿
textolite 层压胶布板；织物酚醛树脂夹板
textryl 板片结构
texture ①结构；构造②组织③网纹④质地⑤纹理
texture analysis 织构分析
textured coating ①立体花纹美术漆；浮雕涂料②斑纹漆
textured masonry paint 浮雕型圬工漆；立体花纹砖石建筑漆
textured paint(=plastic paint) ①立体花纹漆②低浮雕漆③(碎)石粒黏面漆；塑性漆
textured surface glass bead 网纹表面玻璃珠
textured yarn 膨松纱；结构纱；花色纱
texture-finished paint 立体花纹美术漆；立体花纹饰面漆；塑性漆
texture finishing 整理
texture of cloth 织物组织
texture of color pigment 彩色颜料的质地
texture of loaf 面包组织
texture of wood 木材组织；木理
texture rating (颜料)质地分级；质地等级
texturing ①纹饰②刻花；压花；滚花
texturing agent 结构改进剂
texturing surface coating 浮雕花纹表面涂料；立体花纹表面涂料
texturization 变形
texturized yarn 变形纱；膨体纱
texturizing agent 花纹造型剂
tex universal yarn numbering system 纱线通用的特制
thalassemia 地中海贫血(症)
thalazole 邻苯二甲酰磺胺酰噻唑
thalenite 红钇石
thalictrine 白蓬草碱；唐松草碱
thalidomide 沙利度胺（药）
thallation 铊化
Thalleoquin reaction 拓奎反应；绿奎宁反应
thallic ①(正)铊的；三价铊的②含(正)铊的
thallic bromide 三溴化铊 $TlBr_3$
thallic chloride 三氯化铊 $TlCl_3$
thallic fluodichloride 一氟二氯化铊 $TlFCl_2$
thallic fluoride 三氟化铊 TlF_3
thallic hydroxide 氢氧化铊 $Tl(OH)_3$
thallic iodide 三碘化铊 TlI_3
thallic nitrate 硝酸铊 $Tl(NO_3)_3$
thallic oxide 三氧化二铊 Tl_2O_3
thallic oxychloride 氯氧化铊 $TlOCl$
thallic oxyfluoride 氟氧化铊 $TlOF$
thallic peroxide 过五氧化三铊 $Tl_3O_3[O_2]$
thallic sulfate 硫酸铊 $Tl_2(SO_4)_3$
thallic sulfide 三硫化二铊 Tl_2S_3
thalline(=6-methoxytetrahydroquinoline) 沙啉；6-甲氧基四氢化喹啉〔药用〕；扁长草碱 $CH_3OC_9H_{10}N$
thalline salicylate 水杨酸四氢对甲氧基喹啉
thalline sulfate 硫酸四氢对甲氧基喹啉
thalline tartrate 酒石酸四氢对甲氧基喹啉
thallium 铊〔81 号元素，化学符号 Tl〕
thallium acetate 乙酸亚铊 $TlC_2H_3O_2$
thallium alcoholate 醇亚铊；烃氧基亚铊 $TlOR$
thallium amalgam-thallous chloride electrode 铊汞齐-氯化亚铊电极
thallium bromide 溴化铊 $TlBr$; $TlBr_2$; $TlBr_3$
thallium bromodichloride 二氯一溴化铊 $TlBrCl_2$
thallium chloride 氯化铊 $TlCl$; $TlCl_2$; $TlCl_3$
thallium chlorodibromide 一氯二溴化铊 $TlClBr_2$
thallium dibromide 二溴化铊 $TlBr_2$

thallium dichloride 二氯化铊 TlCl$_2$	thallous cyanide 氰化亚铊；一氰化铊 TlCN
thallium diethyl chloride 氯化二乙铊 TlCl(C$_2$H$_5$)$_2$	thallous fluoride 氟化亚铊；一氟化铊 TlF
thallium diethyl hydroxide 氢氧化二乙铊 Tl(OH)(C$_2$H$_5$)$_2$	thallous formate(=thallous formiate) 甲酸亚铊 TlHCO$_2$
thallium difluoride 二氟化铊 TlF$_2$	thallous formiate(=thallous formate) 甲酸亚铊
thallium diiodide 二碘化铊 TlI$_2$	thallous hydrofluoride 氢氟化亚铊 Tl(HF$_2$)
thallium ethide(=thallium ethyl) (三)乙基铊 Tl(C$_2$H$_5$)$_3$	thallous hydroxide 氢氧化亚铊 TlOH
thallium ethyl(=thallium ethide) (三)乙基铊	thallous iodate 碘酸亚铊 TlIO$_3$
thallium fluoride 氟化铊 TlF; TlF$_2$; TlF$_3$	thallous iodide 碘化亚铊；一碘化铊 TlI
thallium formate(=thallium formiate) 甲酸亚铊 TlHCO$_2$	thallous nitrate 硝酸亚铊 TlNO$_3$
thallium formiate(=thallium formate) 甲酸亚铊	thallous oxide 氧化亚铊；一氧化二铊 Tl$_2$O
thallium iodide 碘化铊 TlI; TlI$_2$; TlI$_3$	thallous perchlorate 高氯酸亚铊 TlClO$_4$
thallium methide(=thallium methyl) 三甲铊 Tl(CH$_3$)$_3$	thallous phosphate 磷酸亚铊 Tl$_3$PO$_4$
thallium methyl(=thallium methide) 三甲铊	thallous rhodanate 硫氰酸亚铊 TlSCN
thallium monobromide 溴化亚铊；一溴化铊 TlBr	thallous selenate 硒酸亚铊 Tl$_2$SeO$_4$
thallium monochloride 氯化亚铊；一氯化铊 TlCl	thallous sulfate 硫酸亚铊 Tl$_2$SO$_4$
thallium monofluoride 氟化亚铊；一氟化铊 TlF	thallous sulfide 硫化亚铊；一硫化二铊 Tl$_2$S
thallium monoiodide 碘化亚铊；一碘化铊 TlI	thalmine 小白蓬草碱；小唐松草碱
thallium monoxide 一氧化二铊 Tl$_2$O	thalviol(=thujone) 崖柏酮；艾菊酮
thallium oxide 氧化铊 Tl$_2$O; Tl$_2$O$_3$	thamnolic acid 地茶酸
thallium ozone test paper 铊臭氧试纸	thanatol 乙氧基苯酚
thallium pentasulfide 五硫化二铊 Tl$_2$S$_5$	thanomin 乙醇胺
thallium sesquioxide 三氧化二铊 Tl$_2$O$_3$	thapsic acid 十六碳二酸 HOOC(CH$_2$)$_{14}$COOH
thallium sesquisulfide 三硫化二铊 Tl$_2$S$_3$	thawing ①熔化；融化②变软
thallium spectrum line 铊光谱线	thawing and freezing test 融(化及冰)冻试验
thallium sulfide 硫化铊 Tl$_2$S; Tl$_2$S$_3$; Tl$_2$S$_5$	thaw point 露点
thallium tribromide 三溴化铊 TlBr$_3$	theaflavin 茶黄素〔茶中多元酚氧化产物〕
thallium trichloride 三氯化铊 TlCl$_3$	theamin 茶胺 C$_7$H$_8$N$_4$O$_4$NHC$_2$H$_4$OH
thallium trifluoride 三氟化铊 TlF$_3$	theanine 茶氨酸；N-乙基-γ-谷氨酰胺
thallium trifluoroacetate 三氟乙酸铊	thearubigin 茶玉红精
thallium triiodide 三碘化铊 TlI$_3$	theaspirane 茶螺烷
thallium trinitrate(TTN) 硝酸铊	theaspirone 茶香螺酮；茶螺烯酮
thallium trioxide 三氧化二铊 Tl$_2$O$_3$	thebaine(=paramorphine) 蒂巴因；二甲基吗啡；副吗啡 C$_{19}$H$_{21}$O$_3$N
thallium trisulfide 三硫化二铊 Tl$_2$S$_3$	thebaine hydrochloride 盐酸蒂巴因
thallium yellow 铊黄；铬酸铊	thebaine tartrate 酒石酸蒂巴因
thallosic 含一价和三价铊的	thebainol 蒂巴因酚
thallosic choride 四氯化二铊 TlCl·TlCl$_3$;Tl[TlCl$_4$]	thebainone 蒂巴因酮
thallosic oxide 一氧化铊 Tl$_2$O·Tl$_2$O$_3$	thebaol 蒂巴酚；3,6-二甲氧基-4-羟基菲 C$_{16}$H$_{14}$O$_3$
thallous 亚铊的；一价铊的	thebenedine 萘并异喹啉
thallous acetate 乙酸亚铊 TlC$_2$H$_3$O$_2$	thebenol 蒂奔酚
thallous alum 铊矾 Tl$_2$SO$_4$·Al$_2$(SO$_4$)$_3$·24H$_2$O; TlAl(SO$_4$)$_2$·12H$_2$O	thebenone 蒂奔酮
	the 14 Bravais lattices 14 种布拉威点格
thallous bromate 溴酸亚铊 TlBrO$_3$	thecodine 蒂可定；二羟可待因
thallous bromide 溴化亚铊；一溴化铊 TlBr	the 32 crystallographic point groups 32 种晶体学点群
thallous carbonate 碳酸亚铊 Tl$_2$CO$_3$	the 230 crystallographic space groups 230 种晶体学空间群
thallous chlorate 氯酸亚铊 TlClO$_3$	
thallous chloraurate 氯金酸亚铊 Tl[AuCl$_4$]	theelin(=estrone) 雌酮
thallous chloride 氯化亚铊；一氯化铊 TlCl	theelol(=estriol) 雌三醇；16,17-二羟甾酚

theetsee　马来黑漆
theine(=caffeine)　茶碱；咖啡因
Thelen evaporator　锡伦蒸发器
Thelen pan　锡伦盘
thelephoric acid　革菌酸　$C_{20}H_{12}O_9$
ThEm　钍射气
Thenard blue　特纳德蓝；铝酸亚钴　$Co(AlO_2)_2$; Al_2CoO_4
thenardite　天然无水芒硝
Thenard's blue test　特纳德蓝试验；铝酸亚钴试验
thenoyl　噻吩甲酰　SCH=CHCH=CCO—
thenoylacetone　噻吩甲酰丙酮
thenoyltrifluoroacetone　噻吩甲酰三氟丙酮
thenyl　噻吩甲基　$C_4H_3SCH_2$—
thenyl alcohol　噻吩甲醇
thenyl diamine hydrochloride　噻吩甲基二氨基盐酸盐
thenylidene　噻吩亚甲基　C_4H_3SCH=
theobroma oil　可可油；可可脂
theobromine　可可碱；3,7-二甲基黄嘌呤
theobromine acetyl salicylate　乙酰水杨酸可可碱
theobromine barium sodium salicylate　水杨酸钠合可可碱钡
theobromine calcium salicylate　水杨酸可可碱钙
theobromine hydrochloride　盐酸可可碱　$C_7H_8N_4O_2 \cdot HCl$
theobromine lithium　可可碱锂
theobromine lithium salicylate　水杨酸锂可可碱
theobromine salicylate　水杨酸可可碱
theobromine sodium acetate　乙酸钠可可碱
theobromine sodium formate　甲酸钠可可碱
theobromine sodium salicylate　水杨酸可可碱钠
theobromose　可可碱锂
theocin(=theophylline)　茶碱；1,3-二甲基黄嘌呤
theodolite(=transit)　经纬仪
theogallin　3-邻没食子酰奎尼酸
theoline　一种石油芳烃
theonacete　乙酸钠可可碱
theophorine　甲酸钠可可碱
theophylline(=theocin)　茶叶碱；1,3-二甲基黄嘌呤
theophylline ethylenediamine　茶碱乙二胺
theophylline hydrate　水合茶叶碱
theophylline sodium　茶叶碱钠
theophylline sodium formate　甲酸钠茶叶碱
theorem　定理
theoretical　理论的
theoretical air　理论空气(量)
theoretical chemistry　理论化学
theoretical coverage　理论涂布面积
theoretical crystallography　理论晶体学
theoretical cycle　理论循环
theoretical decomposition voltage　理论分解电压
theoretical diffusion coefficient　理论扩散系数
theoretical distribution　理论分布
theoretical distribution curve　理论分配曲线
theoretical efficiency　理论效率
theoretical efficiency of screen　网目理论功率
theoretical end point　理论终点
theoretical equation　理论公式
theoretical equivalent number of effects　理论有效当量数
theoretical error　理论误差
theoretical extraction curve　理论萃取曲线
theoretical horsepower　理论功率
theoretical indicator card　理论指示图
theoretically complete combustion　理论完全燃烧
theoretically dry　绝对干燥的
theoretically perfect plate(=theoretically perfect tray)　理想板
theoretically perfect tray(=theoretically perfect plate)　理想板
theoretically total conversion　理论完全转化
theoretical peak area　理论峰面积
theoretical physics　理论物理
theoretical plate(=theoretical tray)　理论(塔)板
theoretical plate number　理论塔板数
theoretical power　理论功率
theoretical resolution　理论分辨率
theoretical resolving power　理论分辨本领
theoretical spreading rate　理论涂布面积
theoretical tray(=theoretical plate)　理论(塔)板
theoretical value　理论值
theoretical volumetric efficiency　理论容积效率
theoretical yield　理论(上)产量
theory　理论；学说
theory of acid and base　酸碱理论
theory of acid-base electrolytic dissociation　酸碱电离理论
theory of analogy　相似理论
theory of chemical equilibrium　化学平衡理论
theory of combustion　燃烧学说
theory of conduction　①传导理论②电导理论
theory of continuous media　连续介质理论
theory of dimensions　因次理论
theory of electrolytic dissociation　电离理论
theory of fluctuation　波动理论；涨落理论
theory of free radicals　游离基理论

theory of friction 摩擦理论
theory of hard and soft acid and base 软硬酸碱理论
theory of interionic attraction 离子互吸理论
theory of ionization 电离理论；离子化学说
theory of random walk model 随机模型理论
theory of rate process 速率过程理论
theory of reaction rate 反应速率理论
theory of relativity 相对论
theotannin 茶叶鞣质
theralite 企猎岩
therapeusis 治疗；疗法
therapeutical 治疗的
therapeutical agent 治疗剂
therapeutical chemistry 治疗化学
therapeutic index 治疗指数
therapeutics 治疗学
therapic acid 治疗酸；十八碳四烯酸
therapy ①治疗学②疗法
therm ①克卡〔物〕②英国(煤气)热(量)单位
therma-analysis instrumentation 热分析仪器
thermae 温泉
thermal ①热的②温(热)的
thermal activation 热活化*
thermal activation energy 热激活能
thermal addition 热力加成作用
thermal adhesive 热敏性黏合剂；热熔性黏合剂
thermal adsorption detector 吸附热检测器
thermal adsorption modulator 吸附热调制器
thermal adsorption stripping chromatography 热解吸色谱法
thermal aging 热陈化*；热老化
thermal aging test 热老化测试；热老化实验
thermal agitation 热搅动
thermal alkylation 热烃化
thermal analysis 热分析
thermal analysis curve 热分析曲线
thermal analyzer 热分析仪
thermal anomaly 热反常
thermal baffle 隔热板
thermal balance 热平衡；热天平
thermal barrier 绝热层；保温层
thermal behavior 热行为；热性能
thermal black 热裂炭黑
thermal bleaching 热漂白
thermal bonding 热黏合
thermal bounce 热反弹
thermal boundary layer 热边界层
thermal breakdown 热断裂；热分解

thermal broadening 热变宽
thermal bulb-type thermometer 热球式温度计
thermal calendaring 热轧法
thermal capacity 热容量
thermal capillary pumping 热毛细作用微泵
thermal chlorination 热氯化
thermal chromatography 热色谱法
thermal circulation 热循环
thermal cleavage 热裂解；热分裂
thermal coalescence 热聚结
thermal coefficient 导热系数；导热率
thermal coefficient of expansion 热膨胀系数
thermal color ①示温涂料②热变色
thermal column 热柱
thermal compensation 热补偿
thermal compensator 热补偿器
thermal compressor 喷射器压缩机；热力压缩机
thermal condition 热状况
thermal conduction ①导热性②热导率③热传导
thermal conductivity ①导热系数；热导率②导热性
thermal conductivity analyzer 热导分析器
thermal conductivity bridge 热导率桥
thermal conductivity cell 热导池
thermal conductivity cell detector 热导池检测器
thermal conductivity detector (TCD) 热导式检测器
thermal conductivity gauge 热导真空规
thermal conductivity method 热导法
thermal conductivity tester by absolute plate method 直接平板法导热仪
thermal constant 温度常数
thermal content 热含量
thermal contraction 热收缩
thermal control paint 调温漆；温控涂料
thermal control surface coating 调温(控温)表面涂料(涂层)
thermal convection 热对流
thermal conversion of gases 气体的热转化
thermal conversion process 热转化过程
thermal couple 热电偶
thermal cracker 热裂化装置
thermal cracking 热(裂)解
thermal cracking gas chromatography 热解气相色谱(法)
thermal creep 热力蠕变
thermal critical point 热临界点
thermal current 热流
thermal curtain 热屏；热障
thermal cycle 热循环
thermal cyclization 热环化(作用)；热成环(作用)

thermal cyclization reaction 热环化反应
thermal decomposer 热分解器
thermal decomposition 热(力分)解(作用)
thermal decomposition test 热解试验
thermal decomposition type initiator 热分解型引发剂
thermal deformation 热变形(作用)
thermal deformation under load 加负荷热变形
thermal degradation 热降解
thermal degradation of polymer 聚合物热降解
thermal degradation profile 热降解分布
thermal degradation reaction 热降解反应
thermal dehydration 热力脱氢作用
thermal denaturation 热变性
thermal depolarized light intensity 热消偏振光强度法
thermal depolymerization 热解聚(作用)(法)
thermal desorption 热脱附；热解吸
thermal desorption method 热解吸法
thermal desorption spectroscopy 热脱附谱
thermal desorption spectrum (TDS) 热脱附谱
thermal destruction 热分解；热降解；热断裂
thermal destruction curve 热力破坏曲线；热致死曲线
thermal detector 热检测器
thermal diffusion 热扩散
thermal diffusion column 热扩散柱
thermal diffusion fractionation 热扩散分离
thermal diffusion ratio 热扩散速率
thermal diffusion zone 热扩散区
thermal diffusivity ①热扩散系数②热扩散性
thermal dilation 热膨胀
thermal dilatometry 热膨胀测定法
thermal discharge ①放热②放热量③热显示
thermal dispersion 热散逸；热弥散；热分散
thermal dissociation 热力离解作用
thermal distillation 热蒸馏
thermal distortion 热畸变
thermal drift 热漂移
thermal drop feed oiler 热差点滴加油器
thermal effect 热效应
thermal efficiency 热效率
thermal effluent 热废液
thermal electromotive force 温差电动势
thermal electron 热(发射)电子
thermal electron emission 热电子发射
thermal element 热敏元件
thermal ellipsoid 热椭圆体
thermal emission electron gun 热电子发射枪
thermal emission electron source 热发射电子源
thermal endothermic decomposition 吸热分解

thermal energy 热能
thermal energy analyzer 热能分析器
thermal engine 热机
thermal entropy 温度熵
thermal environment 热环境
thermal equilibrium 热平衡
thermal equivalent of work 热功当量
thermal etching 热腐蚀；热蚀刻
thermal excitation 热激发
thermal expansion 热膨胀
thermal expansion coefficient 热膨胀系数
thermal expansivity 热膨胀系数
thermal explosion 热爆炸
thermal feedback 热反馈
thermal field 热场
thermal field flow fraction 热场流分级
thermal flux 热流
thermal forming 热成型
thermal fractionation 热分级
thermal fractionation thin layer chromatograph 热分离薄层色谱仪
thermal fragment 热裂
thermal fragmentation 热裂(作用)
thermal gradient 热梯度
thermal gradient method 热梯度法
thermal gravimetric analysis 热重分析
thermal gravimetric analysis instrument 热失重分析仪
thermal gravimetric analysis method 热重分析法；热重量分析法
thermal head 热位差
thermal history 热史；受热历程
thermal Hofmann degradation 霍夫曼热降解
thermal hydrolysis 热水解
thermal hysteresis 热滞后现象
thermal incineration 热焚烧
thermal incinerator 热焚烧炉
thermal-induced fragment 热致裂解(分子)碎片
thermal infrared multispectral scanner 热红外多光谱扫描仪
thermal-initiated polymerization 热引发聚合(作用)
thermal initiation 热引发*
thermal instability 热不稳定性
thermal insulating coating 隔热涂料
thermal insulating material 绝热材料
thermal intake length 热量引入区
thermal ionization 热电离
thermal ionization energy 热电离能
thermal ionization ion source 热电离离子源

thermal ionization mass spectrometer 热电离质谱仪
thermal isomerization 热异构化(作用)
thermalized recoil atom 热能化的反冲原子
thermal lamination ①热复合②热层压成型
thermal laser lens effect 激光热透镜效应
thermal lens effect 热透镜效应
thermal lensing spectroscopy 热透镜光谱学
thermal lens spectrometry 热透镜光谱法
thermal lens spectrophotometry 热透镜分光光度法
thermal life 热时效
thermal limit 热限界
thermal linear expansion coefficient 线性热膨胀系数
thermal load 热负荷
thermal losses 热损失；热损耗；热消耗量；耗热量
thermal lubricator(=thermal oiler) 热润滑器；热加油器
thermally assisted fluorescence 热助荧光
thermally assisted resonance atomic fluorescence 热助共振原子荧光
thermally assisted resonance fluorescence 热助共振荧光
thermally assisted stepwise line fluorescence 热助阶跃线荧光
thermally equivalent 均匀受热
thermally foamed plastic 热发泡沫塑料
thermally indicating material 热指示材料；示温材料
thermally oxidized black(=channel black) 热氧化炭黑；槽法炭黑
thermally reversible gel 热逆变凝胶
thermally sensitive resin 热敏性树脂
thermally sensitized 热敏(感)的
thermally sensitized polymerization 热敏聚合(作用)
thermally splitting 热裂化
thermally stable 热稳定的
thermally stable polymer 耐热性聚合物
thermally stable sulfone group 热稳定性砜基
thermally stimulated current 热释electric流法
thermally stimulated discharge current 热刺激放电流法
thermally unstable fibre 非耐热纤维；热不稳定纤维
thermal mass flowmeter 热质流量计
thermal measurement 热(量)测量
thermal medium boiler 热载体加热炉
thermal meter 测热仪表
thermal motion 热运动
thermal naphtha 热裂化石脑油
thermal neutron 热能中子；慢中子
thermal neutron laminagraphy 热中子薄层照相法
thermal noise 热噪声
thermal oil 导热油；热媒
thermal oil discharge 导热油排出

thermal oiler(=thermal lubricator) 热加油器
thermal oil return 导热油回流管
thermal oil supply 导热油供给管
thermal oil vent 导热油放空管
thermal oxidation 热氧化
thermal-oxidative aging 热氧化老化
thermal oxidative degradation 热氧化降解
thermal-oxidative plasticization 热氧化塑炼；热氧化软化
thermal oxidative treatment 热氧化处理
thermal parameter 热参数
thermal pericyclic reaction 热周环反应
thermal plasticization 热塑炼
thermal plug 热塞子
thermal pollution 热污染
thermal polycondensation 热缩聚(作用)
thermal polymerization 热聚合*
thermal potential 热势；热位
thermal power 热功率；发热量
thermal precipitation 温差沉淀
thermal precipitator 热沉淀器
thermal pressure 热压力
thermal property 热性质
thermal radiation 热辐射
thermal radius 热半径
thermal randomness 热混乱度
thermal rectification 热精馏
thermal reflectance spectroscopy 热反射光谱法
thermal reforming 热重整
thermal relaxation 热弛豫
thermal relaxation time 热松弛时间
thermal residuum cracking 热裂残油
thermal resistance ①热阻②热电阻
thermal resistance coefficient 隔热系数；热阻系数
thermal resistor 热敏电阻
thermal ripening 加热成熟
thermal rupture 热分解
thermals 热裂法炭黑
thermal sensibility 热灵敏性
thermal sensitizer 热敏剂
thermal separation 热分离
thermal-setting 热变定；热定形；热固化
thermal shock 热(冲)击
thermal shock cracking 热振动龟裂；热冲击断裂
thermal shocking 热冲击；热振荡；温度急变
thermal shock test 热冲击试验
thermal shrinkage 热收缩
thermal shrinkage behaviour 热收缩性状

thermal-shrinkage differential 热收缩差异
thermal shrinkage stress 热收缩应力
thermal shrinkage stress curve 热收缩应力曲线
thermal softening 热软化；热塑炼
thermal spike 热峰；热尖；热钉
thermalspray 热喷雾
thermal spraying 热喷镀；热喷涂
thermal spraying coating 热喷涂涂层
thermal spraying deposit 喷涂层
thermal spraying gun 热喷涂枪
thermal spraying ionization (TSI) 热喷雾电离
thermal stability ①热稳定性②热稳定度〔可耐最高温度〕
thermal stability index 热稳定性指数
thermal step chromatography 热分级谱法
thermal still 热蒸馏釜
thermal storage 蓄热
thermal strain 热应变
thermal stress 热应力
thermal stress-cracking(TSC) 热应力开裂
thermal stress fatigue 热应力疲劳
thermal stretch 热拉伸
thermal-swing cycle 温度转换循环
thermal telomerization 热调聚反应
thermal test 耐热性试验
thermal titration 热滴定
thermal trace 热示踪
thermal transient analysis 热瞬变分析
thermal transmission coefficient 传热系数
thermal transmittance 透热度；透热率；传热系数
thermal treatment 热处理
thermal type of resonance detector 热共振检测器
thermal unit 热(量)单位
thermal value 发热量；热值
thermal value of fuel oil 燃料油的发热量
thermal velocity 热速度
thermal vibration 热振动
thermal volatilization analysis 热挥发分析
thermal voltage 热电压
thermal voltaic transducer 热电转换器
thermal wear 热磨损
thermal wedge 测温楔；测温锥
thermal yellow 受热泛黄
thermal zone 感热区
thermatomic black(=thermax) 契马黑；热原子炭黑
thermatomic process 热原子过程
thermax(=thermatomic black) 契马黑；热原子炭黑
thermel 温差电偶；热电偶
thermic 热的

thermie 兆卡
thermil 瑟密〔热量单位〕
thermion 热离子
thermionic 热离子的
thermionic activity 热离子活度
thermionic cell 热离子电池
thermionic converter 热离子换能器；热离子转换器
thermionic current 热离子电流
thermionic detector 热离子检测器
thermionic effect 热离子效应
thermionic emission 热离子发射
thermionic emission detector 热离子发射检测器
thermionic fuel element(TFE) 热离子堆燃料元件
thermionic ionization detector 热离子化检测器
thermionic ionization gauge 热离子电离计
thermionic ion source 热离子源
thermionic rubidium silicate detector 硅酸铷热离子检测器
thermionics 热离子学
thermionic source 热离子源
thermionic vacuum tube 热离子真空管
thermisistor 热敏电阻器
thermistor 热敏电阻；热子
thermistor detector 热敏电阻检测器
thermistor electrode 热敏电阻电极
thermistor thermometer 热敏电阻温度计
thermistor vacuum gauge 热敏电阻真空计
thermite 铝热剂
thermite bomb 铝热剂(炸)弹
thermite method 铝热(剂)法
thermite process 铝热(剂)法
Thermit reaction 西尔米特反应；铝热反应
thermit welding 热剂焊
thermo- 〔希腊字头〕 热
thermoacoustimetry 热传声法
thermoanalysis 热分析
thermoanalytical microscopy 热分析显微术
thermobalance 热天平
thermobarometer 温度压力计
thermobattery 热电池(组)
thermo-bonding powder 热黏合粉
thermobulb 热球
thermocatalytic 热催化的
thermochemical 热化学的
thermo-chemical breakdown 热化学塑炼
thermochemical calculation 热化学计算
thermochemical cycle 热化学循环
thermochemical equation 热化学方程式

thermochemical equilibrium 热化学平衡
thermochemical kinetics 热化学动力学
thermochemical measurement 热化学量度
thermochemical method 热化学(方)法
thermochemical standard 热化学标准
thermochemical titration 热化学滴定
thermochemical treatment 化学热处理
thermochemistry 热化学
thermochor 分子体积与温度关系
thermochromism 热色现象；热致变色
thermochrose 选吸热线(作用)
thermo-color ①示温涂料②热变色
thermocompression evaporation 热压蒸发
thermocompressor 热压机
thermo-controller (泡沫)热敏控制剂
thermocooling 温差环流冷却
thermocouple 热电偶；温差电偶
thermocouple ammeter 温差电偶安培计
thermocouple detector 热电偶检测器
thermocouple gauge 温差电偶计
thermocouple needle 热电偶针
thermocouple pyrometer 热电偶高温计
thermocouple thermometer 热电偶温度计
thermocouple vacuum gauge 温差电偶(式)真空计
thermocouple well 热电偶管
thermocross 热电偶
thermocurrent 热电流
thermo detector 热探测器；热检波器；温差探测器
thermodielectric analysis 热介电法
thermodiffusion 热扩散
thermo-diffusion fractionation 热扩散分级
thermodiffusiophoresis 热扩散泳
thermodilatometric analysis 热膨胀分析
thermodilatometry 热膨胀分析法
thermodin 乙酰基对乙氧苯基乌拉坦〔解热镇痛剂〕
thermo-dissociation 加热离解(作用)
thermodynamic 热力学的
thermodynamic acidity 热力学酸度
thermodynamical coordinate 热力学坐标；热力学参数
thermodynamical criterion of equilibrium 平衡的热力学判据
thermodynamic(al) efficiency 热力学效率
thermodynamical equilibrium 热力学平衡
thermodynamical function 热力学函数
thermodynamically admissible process 热力学容许过程
thermodynamically equivalent sphere 热力学等效球
thermodynamical probability 热力学概率
thermodynamical restriction 热力学约束；热力学限制

thermodynamical system 热力学系统
thermodynamic analysis 热力学分析
thermodynamic change 热力学转化
thermodynamic concentration 热力有效浓度
thermodynamic constant 热力学常数
thermodynamic control 热力学控制*
thermodynamic data 热力学数据
thermodynamic duct ①热力学管②液体喷气发动机
thermodynamic efficiency 热力学效率
thermodynamic energy 热力学能
thermodynamic equilibrium 热力学平衡*
thermodynamic equilibrium constant 热力学平衡常数
thermodynamic exchange equilibrium constant 热力学交换平衡常数
thermodynamic function 热力学函数
thermodynamic liquid crystalline polymer 热力学液晶聚合物；热致性液晶聚合物
thermodynamic of polymer solution 高分子溶液的热力学
thermodynamic parameter 热力学参数*
thermodynamic partition coefficient 热力学分配系数
thermodynamic potential 热力学势
thermodynamic pressure 热力学压力
thermodynamic probability 热力学概率
thermodynamic property 热力学性质*
thermodynamic quality of solvent 溶剂热力学性质
thermodynamic relation 热力学关系
thermodynamics 热力学*
thermodynamics of constrained material 受限材料热力学
thermodynamics of elasticity 弹性热力学
thermodynamics of material with memory 记忆材料热力学
thermodynamic stability 热力学稳定性
thermodynamic system 热力学体系
thermodynamic temperature 热力学温度
thermodynamic temperature scale 热力学温标
thermodynamic variable 热力学变量
thermoelastic deformation 热弹性形变
thermoelastic effect 热弹效应
thermoelastic inversion 热弹性逆转
thermoelastic inversion point 热弹性逆转点
thermo-elasticity 热弹性
thermoelastic theory 热弹性理论
thermoelectric 温差电的；热电的
thermoelectrical sorting method 热电材料鉴定法
thermoelectric battery 热电堆
thermoelectric cell 热电元件
thermoelectric coefficient 温差电系数
thermo-electric conversion 热电转换

thermoelectric cooling　热电致冷
thermoelectric couple(=thermo (electric) element)　热电偶；温差电偶
thermoelectric current　温差电流；热电流
thermoelectric effect　温差电效应
thermo (electric) element (=thermoelectric couple)　温差电偶；热电偶
thermoelectric force　温差电势；热电势
thermoelectric galvanometer　热电电流计
thermoelectricity　①温差电；热电(现象)②温差电学；热电学
thermoelectric material　热电材料
thermoelectric pile　热电偶
thermoelectric power　温差电势率
thermoelectric pyrometer　热电偶高温计
thermoelectric refrigeration　热电致冷
thermoelectric series　热电序；温差电序
thermoelectric thermometer　热电温度计
thermoelectrochemistry　热电化学
thermo-electrometer　热电计
thermoelectrometry　热电学法
thermoelectromotive force　温差电动势；热电动势
thermoelectron　热电子
thermoelectronic cooler　热电冷却器
thermoelectronic effect　热电子效应
thermo-element(=thermo-couple)　温差电偶；热电偶
thermo-energy analyzer　热能分析器
thermofixation　热固化
thermofix process　热定形(工艺)；热固着(工艺)
thermoflux test　热通量试验
Thermofor　塞摩福(型)流动床
Thermofor (catalytic) cracking　塞摩福(型)流动床催化裂化
Thermofor catalytic cracking unit　塞摩福(型)流动床催化裂化装置
Thermofor (catalytic) reforming　塞摩福(型)流动床催化重整
Thermofor clay burning process　塞摩福(型)流动床白土灼烧再生过程
Thermofor continuous cracking plant　塞摩福(型)流动床连续裂化车间
Thermofor continuous percolation process　塞摩福(型)流动床连续渗滤过程
Thermofor cracking process(=Thermofor catalytic cracking)　塞摩福(型)流动床催化裂化法
Thermofor kiln　塞摩福(型)流动床催化剂再生炉
thermoforming　热成型
thermofractography　热裂谱法；热分级谱法

thermofusible　热熔的
thermofusion device　热熔装置
thermo-galvanometer　热电电流计；温差电偶电流计
thermogram(=differential thermal analysis curve)　差示热分析图
thermograph(=thermometrograph)　温度记录器；自记温度计
thermographic compound　热熔化合物；热敏性化合物
thermographic phosphor　热敏性磷光体〔指热猝灭性磷光体〕
thermographic type pigment　热敏型颜料
thermography　①自动记录测温法②热熔凸印法；热压凸印术
thermo-gravimetric analysis　热解重量分析法
thermogravimetric analyzer　热重分析器
thermogravimetric curve　热重曲线
thermogravimetry　热重分析法；热(解)重(量)分析法；热重法
thermohardening resin　热固性树脂
thermo-hydrodealkylation　热加氢脱烷基(作用)
thermohydrograph　温湿计；温度湿度记录器
thermohydrometer　热比重计
thermoindicator paint　示温漆；变色漆；温度指示漆
thermoiniferter　热引发转移终止剂
thermoionization mass spectrograph　热电离质谱仪
thermo-junction　热电偶；温差电偶
thermokalite　热碱
thermolator　温度调节器；调温器
thermolize　表面热处理
thermolized　表面热处理的
thermoluminescence　热释发光*；热发光(现象)
thermoluminescence thermal analysis　热发光热分析
thermolysis　①散热(作用)②热分解
thermolysis curve　热解曲线
thermolysis gas chromatography　热解气相色谱
thermolysis-resistant　耐热解的
thermolytic dissociation　热解
thermolytic reactivity　热解反应性
thermomagetometry　热磁学法
thermomagnetic analysis　热磁分析
thermomagnetic converter　热磁换能器
thermomagnetic curve　热磁曲线
thermomagnetic effect　热磁效应
thermomagnetometry　热磁法
thermomechanical analysis　热机械分析(法)；温度-形变分析(法)
thermomechanical analyzer　热机械分析仪
thermomechanical coupling behaviour　热力学偶合特性；

热变形偶合特性
thermomechanical curve 热-力学曲线；热变形曲线
thermomechanical effect 热力学效应；热变形效应
thermomechanical erosion 热力学侵蚀
thermomechanical generated radical 热力学产生的基团
thermomechanical property 热机械性能；热力学性能
thermo-mechanical pulp(TMP) 预热机械(磨木)浆
thermomechanical stability 热-力学稳定性；热变形稳定性
thermomechanics ①热力学；热变形学②热机械学
thermo-melt material 热溶材料
thermometal 双金属
thermo-metamorphism 热同素异形(现象)；热力变质
thermometer 温度计
thermometer adjustment 温度计调整
thermometer bulb 温度计测温包
thermometer calibration 温度计校正
thermometer collar 温度计套圈
thermometer conversion 温度计换算
thermometer correction 温度计改正数
thermometer error 温度计误差
thermometer reader 温度计读数机
thermometer reading 温度计读数
thermometer scale 温度计标度
thermometer stem 温度计枢轴
thermometer well 温度计插池
thermometric ①温度的②测温的
thermometric conductivity 热(传)导率
thermometric fluid 测温液体；温度计用液体
thermometric fixed point 温度固定点
thermometric gas 充温度计的气体
thermometric hydrometer 测温比重计
thermometric method 热计量法
thermometric titration 热滴定(法)*
thermometric titration calorimetry 测温滴定量热法
thermometric titrimetry 温度滴定法
thermometry 计温学*
thermo micro-application separation 热微量转移
thermomicropolar theory 热微极性理论
thermomicroscopy 热显微术
thermomigration 热迁移
thermonatrite 水碱；一水碳酸钠 $Na_2CO_3 \cdot H_2O$
thermonegative 吸热的
thermo-negative reaction 吸热反应
thermo-neutrality 热中和(性)
thermonuclear chemical engineering 热核化学工程
thermonuclear explosion 热核爆炸
thermonuclear neutron source 热核中子源
thermonuclear reaction 热核反应
thermo-oxidation 热氧化
thermo-oxidative 热氧化的
thermo-oxidative aging 热氧老化
thermooxidative crosslinking 热氧化交联
thermooxidative degradation 热氧化降解
thermo-oxidative pyrolysis 热氧化裂解
thermo-oxidative stability 热氧化稳定性
thermopaint 示温涂料
thermopair 热电偶
thermoparticle analysis 热粒子分析
thermoparticulate analysis 热微粒法；热微粒分析
thermophilic 嗜热的；耐热的
thermophone 传声温度计；传声温度器
thermophore 蓄热器
thermophoresis 热泳法
thermophotometry 热光学法
thermo pigment 示温颜料
thermopile 温差电堆；热电堆
thermoplast 热塑(性)塑料
thermoplastic 热塑性塑料
thermoplastic acrylic resin 热塑性丙烯酸树脂
thermoplastic adhesive 热塑(性)黏结剂
thermoplastication 热塑炼
thermoplastic bitumen product 热塑性沥青制品
thermoplastic coating 热塑性涂料
thermoplastic composite 热塑性复合材料
thermoplastic elastomer(TPE) 热塑性弹性体
thermoplastic elastomer polyether(PEBA) 聚酯热塑弹性体
thermoplastic fiber 热塑性纤维
thermoplastic film ①热塑性薄膜②热塑性漆膜
thermoplastic flow 热塑流
thermoplasticity 热(熔)塑性
thermoplastic material 热塑塑料
thermoplastic matrix 热塑性基体
thermoplastic paint 热塑性漆
thermoplastic plastic 热塑性塑料
thermoplastic polyester 热塑性聚酯
thermoplastic polyester powder 热塑性聚酯粉末
thermoplastic polyester powder coating 热塑性聚酯粉末涂料
thermoplastic polymer 热塑性聚合物
thermoplastic polyurethane 热塑性聚氨酯
thermoplastic powder coating 热塑性粉末涂料
thermoplastic resin 热塑性树脂
thermoplastic rubber 热塑性橡胶
thermoplastic styrene butadiene rubber 热塑丁苯橡胶
thermoplastic textured yarn 热塑性变形丝

thermoplastic tile 热塑性瓦
thermoplastic urethane(TPU) 热塑性聚氨酯
thermo-plastic varnish 热塑性清漆
thermoplastification 热塑化(作用)
thermoplastified polymer 热塑化聚合物
thermopolymer 热聚物
thermopolymerization 热聚合
thermopositive 放热的
thermopositive reaction 放热反应
thermopotential 热力势
thermopren 环化橡胶
thermopren cement 环化橡胶黏合剂
thermoprene 环化橡胶
thermoprene cement 环化橡胶胶浆
thermopsine 野决明碱
thermoradiation 热辐射〔联合消毒法〕
thermoradiography(TRG) 放射热谱法
thermoradiometric titration 放射量热滴定
thermoradiometry(TRM) 放射量热法
thermo-reduction(=thermite process) 铝热(剂)法
thermoreflectometry 热反射光法
thermoregulated 温度调节的；调温的
thermoregulation 温度调节
thermo-regulator(=temperature regulator) 温度调节器
thermorelay 热动(温差)继电器
thermoretractile 热缩性的
thermoreversible 热致可逆的
thermoreversible complexation 热致可逆络合
thermorheologically dissipation function 热流变(性)耗散函数
thermorheologically simple material 热流变性简单材料
thermorheological simplicity 热流变(性)简化；热流变单一性
thermos bottle 保温瓶
thermoscope 验温器
thermo-sensibility 热敏性
thermosensitive luminescence 热敏发光
thermosensitive luminescence polymer 热敏发光聚合物
thermo-sensitivity 热敏性
thermosensitizing agent 热敏剂
thermoset ①热固性②热变定的
thermoset acrylic 热固性丙烯酸
thermoset extrusion 热挤塑法
thermoset plastic 热固塑料
thermoset polymer 热变定聚合物；热固性聚合物
thermoset polyurethane 热固聚氨酯
thermoset resin 热固性树脂
thermosetting 热固性

thermosetting acrylate 热固性丙烯酸酯
thermosetting acrylic resin 热固性丙烯酸树脂
thermosetting adhesive 热固性黏合剂
thermosetting cement 热固性胶泥
thermosetting coating 热固性涂料
thermosetting composite 热固性复合材料
thermosetting film 热固性漆膜
thermosetting lacquer 热固性喷漆〔挥发性漆〕
thermosetting material 热固性物质
thermosetting matrix 热固性基体
thermosetting molding material 热固性模塑料
thermosetting paint 热固性漆
thermosetting plastic 热固塑料
thermosetting polyester resin 热固性聚酯树脂
thermosetting polymer 热固(性)聚合物
thermosetting powder coating(s) 热固性粉末涂料
thermosetting resin 热固性树脂
thermosetting resin adhesive 热固性树脂黏合剂
thermosetting varnish 热固油漆
thermoset varnish 热固(性)清漆
thermoshrinking 热收缩
thermoshrinking behaviour 热收缩性状
thermosiphon 热虹吸管
thermosiphon cooling 热虹吸管冷却
thermosiphon reboiler 热虹吸式重沸器
thermosistor 调温器
thermosol 热溶胶
thermosol dying process 热溶染色法
thermosoling 热溶胶化；热溶(法)
thermosonimetry 热发声法；热声波分析法
thermospectrometry 热光谱法
thermosphere 热层
thermospray ①热喷涂②火焰喷涂
thermospray interface 热喷雾接口
thermospray ionization 热喷雾电离
thermostabile(=thermostable) 耐热的
thermostability 热稳定性
thermostabilization 热稳定化(作用)
thermo-stabilizer 热稳定剂
thermostable 耐热的
thermostat ①恒温箱②恒温槽③温度继电器
thermostat body 恒温器体
thermostat body top cap 恒温器体顶盖
thermostat by-pass pipe elbow 恒温器旁路管肘管(弯头)
thermostat container 恒温箱
thermostatic(al) ①恒温(器)的②热静力(学)的
thermostatical control 恒温控制
thermostatically controlled 恒温控制的

thermostatic bath	恒温槽
thermostatic carburettor	恒温化油器
thermostatic circulation bath	恒温循环水浴
thermostatic control	恒温控制；恒温调节
thermostatic control valve	恒温控制阀
thermostatic oven	恒温加热炉
thermostatic regulator	恒温调节器
thermostat layer	恒温层
thermostat(t)ed trough	恒温槽
thermosteresis	热损失
thermo stress	热应力
thermosyphon	热虹吸管；温差对流系统
thermosyphon horizontal reboiler	热虹吸卧式再沸器
thermosyphon reboiler	热虹吸式再沸器
thermotaxy	①热排聚形 ②热排性
thermotension	热张力
thermo tester	升华色牢度仪；耐干热色牢度仪
thermotolerant	耐热的
thermotropic(=caloritropic)	亲热的；向热的；趋热的
thermotropic liquid crystal	热致性液晶
thermotropic liquid crystalline polymer	热致性液晶聚合物；热致液晶高分子
thermotropic mesomorphism	热致性介晶；向热型介晶
thermotropism	向热性
thermovent	散热口
thermoviscoelasticity	①热黏弹性 ②热黏弹度
thermoviscosimeter	热黏度计
thermoviscosity	热黏度
thermovolumetric analysis	热容量分析
thermovulcanizate	热硫化胶
thermowell	温度计插孔；温度计套管；热电偶(温度计)套管
thermsilid	硅铁耐酸合金
theta solvent	θ溶剂
theta state	θ态
theta temperature	θ温度
thetin(=thetine)	噻亭；硫代三烃基内酯
thetine(=thetin)	噻亭；硫代三烃基内酯 R_2SCH_2COO
theveresin	黄夹竹桃素
thevetin	黄夹竹桃苷
thevetose	黄夹竹桃糖
the window of migration time	迁移时间窗口
thia	①噻 硫杂 ②硫代
thiabendazole	噻唑苯并咪唑；噻苯哒唑
thiacetamide	硫代乙酰胺 CH_3CSNH_2
thiacetate	硫代乙酸盐(酯) CH_3COSM; CH_3COSR
thiacetic	硫代乙酸的
thiactin(=bryamycin)	硫活素；藓霉素
thiadiazine	噻二嗪
thiaindan	苯并二氢噻吩
thial	硫醛
thialdine	噻啶 $(CH_3)_3C_3H_4S_2N$
thiamazole	甲巯咪唑〔药〕
thiambutosine	N-对二甲氨基苯基-N'-对丁氧苯基硫脲
thiamer	硫聚物
thiamide	硫羰胺 $RCSNH_2$
thiamine	硫胺素；维生素 B_1
thiamine carboxylic acid	硫胺羧酸
thiamine chloride	氯化硫胺
thiamine disulfide	二硫胺
thiamine hydrochloride(=vitamin B$_1$)	盐酸硫胺素；维生素 B_1
thiamine methyl disulfide	甲基二硫胺素
thiamine mononitrate	硝酸硫胺
thiamine monophosphate	一磷酸硫胺素
thiamine propyl disulfide	丙基二硫胺素
thiamine pyrophosphate	硫胺焦磷酸素
thiamin S(=phenothiazine)	吩噻嗪；夹硫氮杂蒽
thianaphthene(=thionaphthene)	硫茚
thianaphtheno-	硫茚并
thianaphthenol	硫茚酚
2(3H)-thianaphthenone	2(3H)-硫茚酮
thianaphtheno[6,5-b] thianaphthene	硫茚并[6,5-b]硫茚
thianaphthenyl	硫茚基 C_8H_5S-
thianthrene(=diphenylene disulfide)	噻蒽
thiapyran(=thiopyran)	噻喃
thiapyrones	噻喃酮
thiapyrylium	噻喃䓬
thiapyrylium compound	噻喃䓬化合物
thiation	硫杂化
thiatriazole	噻三唑
thiatropylidene	噻䓬
thiaxanthene	噻吨
thiazamide	噻吖胺
thiazan	噻嗪烷 C_4H_9NS
thiazine	噻嗪 C_4H_5NS
thiazine colouring matters	噻嗪染料
thiazine dye	噻嗪染料
thiazinyl	噻嗪基；硫氮苯基 C_4H_4NS-
thiazole	噻唑*
thiazole colouring matter	噻唑染料
thiazole dye	噻唑染料
thiazole purple	噻唑紫
thiazoles	噻唑类
thiazole type accelerator	噻唑型促进剂

thiazolidine 噻唑烷；四氢噻唑 C_3H_7NS
thiazolidinethion-2 四氢噻唑-2-硫酮〔促进剂〕
thiazolidino- 噻唑烷并
thiazolidinyl(=thiazolidyl) 噻唑烷基 C_3H_6NS-
thiazolidomycin 噻唑霉素
thiazolidone 噻唑烷酮
thiazolidonyl 噻唑烷酮基
thiazolidyl(=thiazolidinyl) 噻唑烷基
thiazoline 噻唑啉；二氢噻唑 C_3H_5NS
thiazolinium compound 噻唑啉(鎓)化合物
thiazolinone 噻唑啉酮
thiazolinyl 噻唑啉基 C_3H_4NS-
thiazolium compounds 噻唑(鎓)化合物
thiazoltriazol 噻唑并三唑 $C_4H_3N_3S$
thiazolyl 噻唑基 C_3H_2NS-
2-(4-thiazolyl)-4-azabenzimidazole 2-(4-噻唑基)-4-氮杂苯并咪唑〔高效低毒防霉剂〕
4-(2-thiazolylazo)-resorcinol 4-(2-噻唑基偶氮)间苯二酚
thiazyl 噻唑基
thick ①厚的②稠密的③浓厚的
thick article 厚壁制品；厚型制品
thick beaded edge 厚瘤边；串珠厚边
thick coat 厚涂层
thick coating 厚涂层
thick cylinder 厚壁圆筒
thickened 增稠的；稠化了的
thickened fuel 稠化燃料
thickened gasoline 增稠汽油
thickened grease 增稠润滑脂
thickened hydrocarbon 增稠烃
thickened lubricant 增稠润滑剂
thickened mordant 增稠媒染剂
thickened mud 稠滤渣
thickened oil 吹制油；氧化油
thickened printing color 增稠颜料
thickened pulp 浓纸浆
thickened sludge 浓缩污泥
thickener ①增稠器②增稠剂
thickener response 增稠剂响应性；增稠剂反应能力〔加入增稠剂效应〕
thickening 稠化；稠化过程；增稠过程
thickening agent(=thickening material) 增稠剂
thickening effect 增稠效应
thickening efficiency 增稠率
thickening material(=thickening agent) 增稠剂
thickening of paint film 漆膜增厚
thickening power 增稠本领
thickening rate 增稠速率

thick film column 厚膜柱
thick-film epoxy powder coating(s) 厚膜环氧粉末涂料
thick-film lubrication 厚膜润滑
thick-film rust preventive 厚膜防锈的
thick juice 稠汁；浓汁(糖)
thick juice body 稠汁壳(体)
thick juice press 稠汁压滤机
thick juice saturation 稠汁饱和
thick juice sulfitation 稠汁亚硫酸化
thick layer 厚涂层；厚膜
thick layer chromatography 厚层色谱法
thick liquid 黏稠液体
thick mash method 浓醪法
thick meniscus growth 厚弯液面生长(法)
thick molding compound (TMC) 厚片模塑料
thick needle 粗针头
thickness ①厚度；厚薄②稠度
thickness agent 稠化剂
thickness gauge 厚度计；厚度规
thickness loss 厚度损耗
thickness measurement 厚度测定
thickness of a fabric 织物厚度
thickness of coating system 涂层的厚度
thickness of liquid film 液膜厚度
thickness of panel 板厚
thickness of printing paste 印油的稠度
thickness of root face 钝边高度
thickness of slime 矿泥稠度
thickness of wall 壁厚
thickness piece 厚薄规；厚隙规
thickness tester 测厚仪；厚度计
thick oil 黏稠油品
thickol coating 聚硫橡胶涂料
thickol-epoxy combination 聚硫橡胶-环氧复合(物)
thickol polymer 聚硫橡胶聚合物
thick paste 稠膏；稠浆
thick place 粗段；粗点；粗位
thick sludge 黏稠淤渣
thick slurry process 浓浆法
thick target 厚靶
thick target yield 厚靶产额
thick-walled 厚壁的
thick-walled needle 厚壁针头
thick-walled tube 厚壁管
thief 取样
thief hatch(=thief hole) 取样口；取样孔
thief hole(=thief hatch) 取样口；取样孔
thief sample 容器内指定部位取的试样

thief sampling	容器内指定部位取样
Thiel-Stoll solution	梯耳-斯图耳溶液；饱和高氯酸铅溶液
thieno-	噻吩并
thienofuran	噻吩并呋喃
thienoisothiazole	噻吩并异噻唑
thienone	噻吩酮
thienopyridine	噻吩并吡啶
thienyl	噻吩基　C_4H_3S-
thienyl acrylic acid	噻吩基丙烯酸；3-硫茂基丙烯酸
thienylalanine	噻吩丙氨酸
thienyl diphenylmethane	噻吩二苯基甲烷
2-thienyl disulfide	2-噻吩基二硫
thienyl ketone	二噻吩基甲酮
thienyl mercaptan	噻吩硫醇
thienyl methyl ketone	噻吩基甲基酮
thienyl phenylketone	噻吩基苯基甲酮
thienyl pyrimidine	噻吩基嘧啶
2-thienylthiol	2-噻吩基硫醇
Thierry zinc furnace	塞利炼锌炉
Thies process	梯斯提金法
thieving paper	感水纸
thiirane	硫杂丙环　CH_2CH_2S
thiirene	硫杂丙烯环　$CH=CHS$
thilane	四氢噻吩　C_4H_8S
thimble	①壳筒②套管
thimerosal (=sodium ethylmercurithiosalicylate)	乙基汞硫代水杨酸钠
thin	①薄的②稀(薄)的③使稀(薄)
thin alloy zone crystallization (TAZC)	薄合金区域结晶
thin core	薄基板；内层板
thin cylinder	薄壁圆筒
thin end	细经
thin film	薄膜
thin-film chromatography	薄膜色谱(法)
thin-film column	薄膜柱
thin-film column packing	薄膜柱填充物
thin-film deaerator	薄膜脱泡器；薄膜脱气塔
thin-film distillation	薄膜蒸馏
thin film epoxy powder coating(s)	薄膜环氧粉末涂料
thin-film evaporation unit	薄膜蒸发装置
thin-film evaporator	薄膜式蒸发器
thin film evaporometer	薄膜蒸发计
thin film lining	薄膜衬里
thin-film lubrication	薄膜润滑
thin film mercury electrode	薄膜汞电极
thin-film psychrometer	薄膜式湿度计
thin film rectifier	薄膜精馏器
thin-film rust preventive	薄膜防锈的
thin-film scanner	薄膜扫描器
thin film technology	薄膜技术
thin film vacuum still	薄膜真空蒸馏釜
thin flaky pigment	薄片状颜料
thin fluid	稀薄流体
thin gas mixture	稀薄气体混合物
thin-ga(u)ge goods	薄壁制品；薄型制品
thin-gauge sheet	薄型片材
thin juice	稀汁
thin juice body	稀汁
thin juice sulfitation	稀汁亚硫酸化
thin layer	薄层
thin layer bed	薄层床
thin layer chromatogram scanner	薄层色谱扫描仪
thin layer chromatograph	薄层色谱仪
thin-layer chromatographic analysis	薄层色谱分析
thin-layer chromatographic densitometry	薄层色谱光密度(定量)法
thin-layer chromatographic fractionation	薄层色谱分离(法)
thin-layer chromatographic plate	薄层色谱板
thin layer chromatography	薄层色谱法
thin layer chromatography scanner	薄层色谱扫描仪
thin layer chromatography scanning	薄层色谱扫描法
thin layer chronopotentiometry	薄层计时电位法
thin layer detector cell	薄层检测器池
thin-layer electrophoresis	薄层电泳(法)
thin layer gel chromatography	薄层凝胶色谱法
thin-layer gel filtration	薄层凝胶过滤(法)
thin layer gel permeation chromatography	薄层凝胶渗透色谱法
thin layer partition chromatography	薄层分区色谱法
thin layer plate	薄层板*
thin layer potentiometry	薄层电位法
thin layer radiochromatography	薄层放射色谱法
thin layer rod	薄层棒
thin layer rod chromatography	薄层棒色谱法
thin-layer scanner	薄层扫描仪
thin-layer thermochromatography	薄层热色谱法
thin layer voltammetry	薄层伏安法
thinly fluid	液状的；液态的；液体的
thinly liquid	易流动；黏度小的液体
thinned lubricant	稀释的润滑剂
thinned oil	稀释石油；充气石油
thin negative	浅底片
thinner	①稀料②稀释剂；冲淡剂

thinness 单薄；手感单薄
thinning (=reducing) 稀释
thinning drow ①稀释；兑稀；调稀②变细；变薄
thinning limits 稀释限度
thinning of liquid film 液膜变薄
thinning of oil 油的稀释
thinning ratio 稀释比；稀释率
thinning tank 稀释罐；调漆罐
thinning vessel 兑稀釜；调稀罐
thin oil 稀油
thinolite 薄水石
thin out 稀释
thin paste 稀浆
thin plate orifice 薄孔板
thin projections 尖锐突出物
thin-shell 薄壳
thin stand oil 稀熟油
thin target 薄靶
thin-target yield 薄靶产额
thin-wall(ed) container 薄壁容器
thin-walled tube 薄壁(卷绕)筒管
thin window 薄窗
thio 硫代 S=
thioacetal 硫缩醛
thioacetaldehyde 乙硫醛
thioacetamide 硫代乙酰胺 CH_3CSNH_2
thioacetanilide N-硫代乙酰苯胺 $C_6H_5NHCSCH_3$
thioacetate 硫代乙酸盐(酯) $CH_3COSM; CH_3COSR$
thioacet-dimethylamide N-硫代乙酰二甲胺 $CH_3CSN(CH_3)_2$
thioacetic acid 硫代乙酸；乙硫羟酸 CH_3COSH
thioacetin 硫代乙酸甘油酯
thioaceto-acetic ester 硫乙酰乙酸酯
thioacetone 丙硫酮
thioacetophenone 硫代苯乙酮
thioacetyl 硫代乙酰
thio acid 硫羰酸*
thioacid amide 硫代酰胺 $RCSNH_2$
thioacylation 硫代酰化作用
thio-alcohol 硫醇
thio-aldehyde ①硫醛②乙硫醛
thio-allyl ether(=diallyl sulfide) 硫代烯丙醚；烯丙基硫醚 $(CH_2=CHCH_2)_2S$
thioameline 烷基硫代三聚氰二胺；烷基硫代胍胺
thioamide 硫代酰胺
thioamidization process 硫(代)酰胺化过程
thioamyl alcohol 硫代戊醇
thioanhydride 硫代酸酐

thioanilide N-硫代(某)酰苯胺
thioaniline(=diaminodiphenyl sulfide) 硫苯胺〔俗〕；二氨基苯基硫 $(NH_2C_6H_4)_2S$
thioanisole 苯硫基甲烷；茴香硫醚 $C_6H_5SCH_3$
thioantimonate(=thioantimoniate) 硫代锑酸盐 M_3SbS_4
thioantimoniate(=thioantimonate) 硫代锑酸盐
thioantimonic acid 硫代锑酸 H_3SbS_4
thioarsenate(=thioarseniate) 硫代砷酸盐 M_3AsS_4
thioarseniate(=thioarsenate) 硫代砷酸盐
thioarsenic acid 硫代砷酸 H_3AsS_4
thioarsenious acid(=thioarsenous acid) 硫代亚砷酸
thioarsenite 硫代亚砷酸盐 M_3AsS_3
thioarsenous acid(=thioarsenious acid) 硫代亚砷酸 H_3AsS_3
thioarsonous acid 硫代亚胂酸 $RAs(SH)_2$
thioaurite 硫金酸盐 $MAuS$
thiobarbituric acid(=malonyl thiourea) 硫代巴比妥酸；丙二酰硫脲 $C_4H_4O_2N_2S$
thiobarbituric acid reagent 硫代巴比妥酸试剂
thiobenzaldehyde 苯甲硫醛 C_6H_5CHS
thiobenzamide 硫代苯甲酰胺 $PhCSNH_2$
thiobenzanilide N-硫代苯酰苯胺 $C_6H_5NHCSC_6H_5$
thiobenzhydrol 硫代二苯甲醇
thiobenzimidazolone 亚苯基硫脲
thiobenzoic acid 硫代苯甲酸〔苯巯酸或苯疣酸〕 $C_6H_5COSH; C_6H_5CSOH$
thiobenzophenone 二苯甲硫酮 $(C_6H_5)_2CS$
thiobenzoyl chloride 硫代苯酰氯
thiobenzyl alcohol 硫代苄醇
4,4′-thiobis (6-t-butyl-m-cresol) 4,4′-硫代双(6-叔丁基间甲酚)〔防老剂〕
4,4′-thiobis(6-t-butyl-3-methyl-phenol) 4,4′-硫代双(6-叔丁基-3-甲苯酚)〔防老剂〕
4,4′-thiobis (di-sec-amyl phenol) 4,4′-硫代双(二仲戊基苯酚)〔防老剂〕
thiobis (3,5-di-t-butyl-4-benzyl phenol) 硫代双(3,5-二叔丁基-4-苄基苯酚)〔防老剂〕
3,3′-thiobis (2,6-di-t-butyl-4-ethyl propinoate) 3,3′-硫代双(2,6-二叔丁基-4-丙酸乙酯)〔稳定剂〕
2,2′-thiobis (4,6-dichlorophenol) 2,2′-硫代双(4,6-二氯苯酚)〔胶乳保存剂〕
4,4′-thiobis(4-methyl-6-t-butylphenol) 4,4′-硫代双(4-甲基-6-叔丁基苯酚)〔防老剂〕
2,2′-thiobis〔4-methyl-6-(α-methyl benzyl)phenol〕 2,2′-硫代双[4-甲基-6-(α-甲基苄基)苯酚]〔热稳定剂〕
1,1′-thiobis(2-naphthol) 1,1′-硫代双(2-萘酚)〔防老剂〕
thiocacodylate 硫代卡可酸盐
thiocarbamate 硫代氨基甲酸盐(酯) $CS(NH_2)OM$;

CS(NH$_2$)OR
thiocarbamic acid　硫代氨基甲酸　CS(NH$_2$)OH
thiocarbamic acid-O-ethyl ester　硫代氨基甲酸-O-乙酯；乙基黄原酰胺
thiocarbamic acid-S-ethyl ester　硫代氨基甲酸-S-乙酯　NH$_2$CO(SC$_2$H$_5$)
thiocarbamide(=thiourea)　硫脲　NH$_2$CSNH$_2$
thiocarbamidometry　硫脲滴定法
thiocarbamoyl(=thiocarbamyl)　氨基硫羰基；硫代氨甲酰　H$_2$NCS—
thiocarbamyl(=thiocarbamoyl)　硫代氨基甲酰；氨基硫羰基
thiocarbamyl sulfenamide　硫代氨基甲酰次磺酰胺〔促进剂〕
thiocarbanilide　对称二苯硫脲　(C$_6$H$_5$NH)$_2$CS
thiocarbazone　硫卡巴腙　—N=NCSNHNH—
thiocarbimide　硫代异硫氰酸
thiocarbin　硫卡比醇
thiocarbohydrazide　硫代对称二氨基脲　CS(NHNH$_2$)$_2$
thiocarbonate　硫代碳酸盐(酯)
thiocarbonated cellulose　硫代碳酸盐纤维素
thiocarbonic acid　硫代碳酸　HOCOSH;HOCSOH
thiocarbonic ester　硫代碳酸酯
thiocarbonyl　硫羰基　SC=
thiocarbonyl chloride(=thiocarburyl chloride)　二氯硫化碳　CSCl$_2$
thiocarbonyl group　硫代羰基
thiocarbonyl perchloride　高氯硫化碳〔防焦剂〕
thiocarboxylic acid　硫代羧酸
thiocarburyl chloride(=thiocarbonyl chloride)　二氯硫化碳
thiochroman　二氢苯并噻喃　C$_9$H$_{10}$S
thiochromanone　二氢苯并噻喃酮　C$_9$H$_8$OS
thiochrome　硫色素；脱氢硫胺素　C$_{12}$H$_{14}$ON$_4$S
thiochromene　1,2-苯并硫吡喃
thiochromone　1,4-苯并硫吡喃酮
thiochromonol　3-羟基-1,4-苯并硫吡喃酮　C$_9$H$_6$O$_2$S
thioclastic cleavage　硫解性断裂
thiocol　愈创酚磺酸钾
thio-compounds　硫代化合物
thiocoumarin　硫代香豆素
thiocresol　甲苯硫酚
thioctic acid　硫辛酸
thiocyanate(=thiocyanide)　①硫氰化物②硫氰酸盐(酯或根)
thiocyanate ceriometric determination　硫氰酸盐-硫酸铈滴定测定法
thiocyanate extraction system　硫氰酸盐萃取系
thiocyanate ion selective electrode　硫氰酸根离子选择电极

thiocyanate radical　硫(代)氰酸根
thiocyanate value　硫氰酸值
thiocyanation　硫氰化作用
thiocyanato(=thiocyano)　①氰硫基〔有机〕②硫(代)氰酸根合〔无机〕　N≡CS—
thiocyanic acid　硫氰酸　HSCN
thiocyanic ester　硫氰酸酯　RSCN
thiocyanide(=thiocyanate)　硫氰酸盐(酯或根)；硫氰化物
thiocyanimetry(=thiocyanometric titration; thiocyanometry)　硫氰酸盐滴定法
thiocyano(=thiocyanato)　①氰硫基〔有机〕②硫(代)氰酸根合〔无机〕　N≡CS—
thiocyanocarbons　硫氰(基)碳化合物
thiocyano dyestuffs　氰硫染料
thiocyanogen　硫化氰　(SCN)$_2$
thiocyanogen number(=thiocyanogen value)　硫氰值
thiocyanogen value(=thiocyanogen number)　硫氰值
thiocyanometric titration(=thiocyanometry;thiocyanimetry)　硫氰酸盐滴定法
thiocyanometry(=thiocyanometric titration; thiocyanimetry)　硫氰酸盐滴定(法)
thiocyanuric acid(=trithiocyanuric acid)　硫氰尿酸；三聚硫氰酸　C$_3$H$_3$N$_3$S$_3$
thiocyclohexanone　硫代环己酮
thiocyclopentanone　硫代环戊酮
thiocysteine　硫代半胱氨酸
thiodialkylamine　硫二烷基胺
thiodiazine　硫代二嗪〔促进剂〕
thiodiazolidine　四氢噻二唑
thiodiazoline　二氢噻二唑
thiodiethylene glycol　硫二甘醇；2,2′-二羟基二乙硫
thiodiglycol　硫二甘醇〔俗〕；2,2′-二羟基二乙硫　(HOCH$_2$CH$_2$)$_2$S
thiodiglycolic acid　亚硫基二乙酸　S(CH$_2$COOH)$_2$
thiodiglycolic anhydride　环硫二乙酸酐　OCOCH$_2$SCH$_2$CO

thiodiphenylamine(=phenothiazine)　吩噻嗪
thiodipropionate　硫代二丙酸盐(酯)
thioester　硫代酸酯*
thioester bond　硫酯键
thioethanolamine　硫代乙醇胺
thio-ether　硫醚　RSR
thioether acid salt grease　硫醚酸式盐润滑脂
thiofluorenone　硫代芴酮
thiofolicaine　硫叶卡因
thioform　碱式二硫代水杨酸铋

thioformaldehyde 硫甲醛；三聚甲硫醛 $CH_2(SCH_2)_2S$
thioformamide 硫代甲酰胺〔不宜叫甲硫酰胺或甲硫羰胺〕 $HCSNH_2$
thioformanilide N-硫代甲酰苯胺 $C_6H_5NHCH=S$
thioformin 硫甲酰素
thioformyl 硫醛基 HSC—
thiofuran 噻吩
thiogalactoside 硫代半乳糖苷
thiogenic dye 硫化染料
thioglucuronide 硫糖醛酸苷
thioglycerin 硫甘油（俗）；2,3-二羟基-1-丙硫醇 $HSCH_2C_2H_3(OH)_2$
thioglycerol 硫甘油
thioglycolate 硫基乙醇酸盐(酯)；巯基乙酸乙酯
thioglycollate 巯基乙酸盐(酯)
thioglycollic acid 巯基乙酸；氢硫基乙酸 $HSCH_2CO_2H$
1-thioguaiacol 1-硫代愈创木酚
6-thioguanine(6TG) 6-硫代鸟嘌呤；6-巯基鸟嘌呤
thiohistidine-betaine(=ergothioneine) 硫组氨酸甲基内盐；麦硫因
thiohydantoic 硫脲基酸
thiohydantoic acid 硫脲基乙酸
thiohydantoin 海硫因〔俗〕；乙内酰硫脲 $C_3H_4ON_2S$
thiohydantoin method 乙内酰硫脲法
thiohydracrylic acid 巯基丙酸 $HSCH_2CH_2CO_2H$
thiohydropyrimidine 硫氢嘧啶〔促进剂〕
thiohydroquinone 硫氢醌；对羟基苯硫酚 HSC_6H_4OH
thiohydroxy(=mercapto) 巯基；氢硫基
thiohypophosphate 全硫连二磷酸盐 $(M_2PS_3)_2$
thiohypophosphoric acid 全硫连二磷酸 $(H_2PS_3)_2$
thioic acid 硫代酸 RCSOH; RCOSH
thioimido ester 硫代亚氨酸酯 $RC(=NH)SR$
thioindigo 硫靛(蓝)
thioindigo blue 硫靛蓝
thioindigo bordeaux 硫靛枣红(瓮颜料)
thioindigo color 硫靛颜料；硫靛染料
thioindigo dye 硫靛染料
thioindigoid dye 硫准靛染料
thioindigoid maroon 硫准靛紫红
thioindigo pink 硫靛桃红
thioindigo scarlet 硫靛猩红
thioindigotic acid 硫靛蓝酸
thioindigo violet 硫靛紫
thioindigo white 硫靛白 $C_{16}H_{10}O_2S_2$
thioindoxyl 硫代-3-吲哚酚
thioindoxylic acid 硫代吲羟酸
thioketal 酮缩硫醇 $R_2C(SR)_2$
thio-ketone ①硫酮②丙硫酮
thiokol 聚硫橡胶；乙硫橡胶
thiokol-bendloop test 聚硫橡胶低温屈挠试验
thiokol latex 聚硫胶乳
thiokol-lined hose 聚硫橡胶衬里(软)管
thiokol lining 聚硫橡胶衬里
thiokol polymer 乙硫橡胶聚合物
thiokol polysulfide (rubber) 聚硫橡胶
thiol ①硫羟②硫醇
thiol-acetic acid 硫羟乙酸 CH_3COSH
thiol acid 硫羟酸*
thiolactic acid 硫羟乳酸 $CH_3CH(SH)COOH$
thiolate ①硫醇盐；烃硫基金属 RSM②硫羟酸盐 RCOSM
thiolation process 硫醇(羟)化法〔硫取代羟基中的氧〕
thiolcarbamic ethyl ester 硫羟氨基甲酸乙酯 $H_2NCOSC_2H_5$
thiolcarbonic acid 一硫羟碳酸 OHCOSH
thiolcarbonic ester 硫羟碳酸酯 RSCOOR
thiolester 硫羟酸酯 RCOSR
thiol ethylation 巯乙基化作用
thiolhistidine 巯(基)组氨酸
thiolic acid 硫羟酸；巯酸 —COSH
thiolignin 硫代木素
thiolinalool 硫代芳樟醇
α-thiolpropionic acid α-硫代丙酸；巯基丙酸
thiolysis 硫解(作用)
thiomalic acid 硫羟苹果酸 $HOOCCH(SH)CH_2COOH$
thiomersal(ate) 乙基汞硫代水杨酸钠
thion ①硫(的)②硫羰
thionalid 巯萘剂
thionamic acid 氨基磺酸 RNHSOOH
thionaphthene(=thianaphthene;benzothiophene) 硫茚；苯并噻吩
thionaphthol 萘硫酚
β-thionaphthol β-硫代萘酚
thionate 硫代硫酸盐
thionation(=sulfuration) 硫化作用
thioncarbonate 硫羰碳酸盐(酯) $CS(OM)_2$;$CS(OR)_2$
thioncarbonic acid 硫羰碳酸 $CS(OH)_2$
thioncarbonic ester 硫羰碳酸酯 $CS(OR)_2$
thion dye 硫化染料
-thione 硫酮 —CS—
thioneine 硫因；巯基组氨酸三甲(基)内盐
-thionic acid ①硫羰酸；砣酸 —CSOH②连多硫酸
thionine(=Lauth's violet) 硫堇；劳氏紫 $C_{12}H_9N_3S$
thionine blue 硫堇蓝
N-thionitrosodimethylamine N-硫代亚硝基二甲基胺

thionizer 脱硫塔
thiono- 硫羰
thionone 硫酮
thionone color 硫酮染料
thionophosphine sulphide 硫羰基膦化硫
thionothiolic acid 硫羧酸;硫羟羰酸
thionuric acid 硫尿酸;磺酸氨基巴比妥酸 $C_3H_3O_2N_2NHSO_3H$
thionyl(=sulfinyl) 亚硫酰 $=SO$
thionyl amines 亚硫酰胺(类)
thionyl aniline 亚硫酰苯胺
thionyl benzene 二苯亚砜 $(C_6H_5)_2=SO$
thionyl chloride 亚硫酰(二)氯
thionyl chlorobromide 亚硫酰氯溴
thionyl dialkylamine 亚硫酰二烷基胺
thionyl fluoride 亚硫酰(二)氟
thionyl hydrazine 亚硫酰肼
thionyl imide 亚硫酰亚胺
thionyl-p-phenylenediamine 亚硫酰对苯二胺〔防老剂〕
thionyl toluidines 亚硫酰甲苯胺(类)
thio-organo-tin compound 有机硫锡化合物
thio-oxamide 硫代草酰胺
thiooxine 8-巯基喹啉 C_9H_6NSH
thio-oxybiazoline 2-氧代-1-硫-3,4-二唑
thio-oxydiphenylamine 硫氧二苯胺
thio-ozone 硫代臭氧
thio-ozonide 硫代臭氧化物
thiopanic acid(=pantoyltaurine) 泛磺酸
thioparamizone 对乙酰胺基亚苄基缩氨基硫脲
thiopental sodium 5-乙基-5-(1-甲基丁基)-2-巴比妥酸钠;硫喷妥钠〔药〕
thiopentone sodium 5-乙基-5-(1-甲基丁基)-2-巴比妥酸钠
thiopentose 硫戊糖
thioperoxide 硫代过氧化物
thioperrhenate 硫代高铼酸盐 $MReO_3S$
thioperrhenic acid 硫代铼酸 $HReO_3S$
thiophane 四氢噻吩
thiophanone 噻吩烷酮
thiophanthrene 萘并[2,3-b]噻吩
thiophene 噻吩
thiophen(e) alcohol 噻吩甲醇
thiophene carboxylic acid 噻吩羧酸 C_4H_3SCOOH
thiophene-free benzene 无噻吩苯
thiophenic acid 噻吩甲酸
thiophenine 噻吩胺 C_8H_5NS
thiophenol(=phenthiol;phenylmercaptan) 苯硫酚 C_6H_5SH

thiophenoxide ion 苯硫氧化物离子
thiophenyl 苯硫基 PhS—
thiophenyl acetone 苯硫基丙酮
thiophilic 亲硫的
thiophilicity 亲硫性
thiophorase 辅酶 A 转移酶
thiophos(=parathion) 对硫磷;一六○五〔农药〕
thiophosgene(=thiocarbonyl chloride) 硫光气;二氯硫化碳 $Cl_2C=S$
thiophosphate 硫代磷酸盐 $M_3PO_3S;M_3PO_2S_2;M_3POS_3;M_3PS_4$
thiophosphoric acid 硫代磷酸 $H_3PO_3S;H_3PO_2S_2;H_3POS_3;H_3PS_4$
thiophosphoric anhydride 五硫化二磷 P_2S_5
thiophosphorous acid 硫代亚磷酸;全硫亚磷酸 H_3PS_3
thiophosphoryl 硫代磷酰 $PS\equiv$
thiophosphorylation 硫代磷酰化作用
thiophosphoryl bromide 三溴硫化磷
thiophosphoryl chloride 硫代磷酰氯 $PSCl_3$
thiophosphoryl triamide 硫代磷酰三胺 $PS(NH_2)_3$
thiophthalide 硫代苯酞
thiophthene 并噻吩 $C_6H_4S_2$
thiopicric acid 硫苦酸;2,4,6-三硝基苯硫酚 $C_6H_2(SH)(NO_2)_3$
thioplast 硫塑料
thiopropionamide 丙硫羧酰胺;硫代丙酰胺;丙硫羰胺 $C_2H_5CSNH_2$
thiopropionate 硫代丙酸酯(盐)
n-thiopropyl alcohol 正丙硫醇
thioproteose 硫胨
thiopyran(=thiapyran) 噻喃
thiopyronine 硫代焦宁
thiopyrophosphoryl bromide 硫代焦磷酰溴
thiopyrrolidone 硫代吡咯烷酮
thio resin 聚硫树脂
thioresorcinol 硫代间苯二酚;间巯基苯酚
thioridazine 硫利达嗪〔药〕
thio rubber 聚硫橡胶
thiosalicylamide 硫代水杨酰胺〔防焦剂〕
thiosalicylic acid 硫代水杨酸;邻巯基苯甲酸 $HSC_6H_4CO_2H$
thiosemicarbazide 氨基硫脲 $NH_2NHCSNH_2$
thiosemicarbazide colorimetry 氨基硫脲比色法
thiosemicarbazone 缩氨基硫脲 $R_2C=NNHCSNH_2$
thiosilicic acid 硫代硅酸;全硫硅酸 H_2SiS_3
thiosinamine 烯丙基硫脲
thiostannate 硫代锡酸盐;全硫锡酸盐 M_2SnS_3
thiostannic acid 硫代锡酸;全硫锡酸 H_2SnS_3

thiostannite 硫代亚锡酸盐；全硫亚锡酸盐 M_2SnS_2
thiostannous acid 硫代亚锡酸；全硫亚锡酸 H_2SnS_2
thiosuccimide 硫代琥珀酰亚胺
thio sugar 硫糖
thiosulfate 硫代硫酸盐(酯)
thiosulfinate 硫代亚磺酸酯
thiosulfite 硫代亚硫酸盐(酯)
thiosulfuric acid （一）硫代硫酸
thiotaurine 硫代牛磺酸
thiotepa 硫化三(环氧丙基)磷；塞替派〔药〕
thiotolene 甲基噻吩
thiouracil 硫尿嘧啶
thiouramil 硫代氨基丙二酰脲 $C_4H_3N_3O_2S$
thiourazole 硫氧代-1,2,4-三唑 $C_2H_3ON_3S$
thiourea(=thiocarbamide) 硫脲 NH_2CSNH_2
thiourea adduct 硫脲加合物
thiourea complex 硫脲配合物
thiourea cure 硫脲硫化
thiourea-formaldehyde resin 硫脲-甲醛树脂
thiourea inclusion compound 硫脲包结化合物
thiourea peroxide 过氧化硫脲；二氧化硫脲
thiourea resin 硫脲树脂；硫脲甲醛树脂
thioureido 硫脲基
thiourethane 硫代氨基甲酸乙酯 $NH_2COSC_2H_5$
thioureylene 1,3-亚硫脲基 —NH—CS—NH—
thiovanadate 硫代钒酸盐 M_3VS_4
thiovanadic acid 硫代钒酸；全硫钒酸
thiovioluric acid 硫代紫尿酸
thioxalic acid 硫草酸 HOCSCSOH
thioxane 噻㗁烷 C_4H_8OS
thioxanthamide(=O-ethyl thiocarbamate) 硫代氨基甲酸乙酯
thioxanthene 噻吨 $C_{13}H_{10}S$
thioxanthone 噻吨酮 $C_{13}H_8OS$
thioxene 二甲基噻吩
thioxo 硫代〔S 取代了 CH_2 中的两个 H〕
thioxylenol(=xylyl-mercaptan) 二甲苯基硫酚〔塑解剂〕
thiozon 臭硫 S_3
third ①第三(的)②三分之一
third electroviscous effect 第三电黏效应
third evaporator 第三级蒸发器
third grade gasoline(=third-structure gasoline) 三级汽油
third law of thermodynamics 热力学第三定律
third order aberration 第三级像差
third-order fluid 三阶流体
third-order meridional diffraction 三级子午线衍射
third order reaction 三级反应
third phase 第三相
third press 三榨油
third-structure gasoline(=third grade gasoline) 三级汽油
thirteen-membered ring(=thirteen ring) 十三元环
thirteen ring(=thirteen-membered ring) 十三元环
thistle funnel(=thistle tube) 蓟头漏斗；长梗漏斗
thistle-seed oil 蓟子油
thistle tube(=thistle funnel) 蓟头漏斗；长梗漏斗
thitsiol 缅漆酚
thiuram 秋兰姆〔1.二烃氨基硫羰(基) (R_2NCS—); 2.二硫化二甲氨基硫羰(Me_2NCSS)$_2$; 3.氨基硫羰(基) NH_2CS—〕
thiuram disulfide 二硫化四烷基秋兰姆
thiuram type accelerator 秋兰姆类促进剂
thiurea 硫脲 $CS(NH_2)_2$
thiuronium 硫脲鎓盐 $RCS(NH_2)_2^+X^-$
thixopectic liquid 触变胶液体
thixotrometer 触变仪
thixotrope 触变胶
thixotropic ①触变(的)；具有触变作用的②摇溶的
thixotropic agent 触变剂
thixotropic assistant 触变性助剂
thixotropic behaviour 触变行为
thixotropic binder ①触变性漆基②触变性黏结剂③触变性连接料〔油墨〕
thixotropic breakdown 触变破坏
thixotropic coatings 触变涂料
thixotropic coefficient 触变系数
thixotropic consistency 触变稠度
thixotropic dispersion 触变分散体
thixotropic effect 触变效应
thixotropic filler 触变性填料
thixotropic flow 触变型流动
thixotropic flow behavior 触变流动特性
thixotropic flow curve 触变型流动曲线
thixotropic fluid 触变液体
thixotropic fluid behaviour 触变性流体特性
thixotropic fluidity ①触变性流动度②触变流动性
thixotropic fluid substance 触变性流体物质
thixotropic gel 触变胶体
thixotropic index 触变指数
thixotropic interaction 交互触变作用；互生触变作用
thixotropic interior flat paint 触变性内用无光漆
thixotropic level 触变等级
thixotropic liquid 触变性液体
thixotropic loop 触变回线；触变环
thixotropic paint 触变漆
thixotropic plastic behaviour 触变塑性特性
thixotropic plastic-fluid behaviour 触变塑性流体特性

thixotropic plastic-fluid substance 触变塑性流体物质
thixotropic plastic substance 触变塑性物质
thixotropic propellant 触变喷气燃料
thixotropic ratio 触变比
thixotropic resetting time 触变再现(性)时间
thixotropic sol 触变性溶胶
thixotropic thickener 触变性增稠剂
thixotropic thickening 触变增稠过程(作用)
thixotropic vehicle 触变性漆料
thixotropic viscoplasticity 触变黏塑性
thixotropic viscosity 触变黏度
thixotropy 触变性*
thixotropy factor 触变系数
thixotropy index 触变指数
thiylation 引入含硫基
thiyl radical 含硫游离基
Thoman flap tray 索尔曼活门塔板
Thomas meal 托马斯碱性炉渣粉
Thomas meter 托马斯流量计
Thomas phosphate 托马斯磷肥
Thomas process 托马斯除磷炼铁法
Thomas slag 托马斯炉渣
Thompson process 汤普逊电弧焊接法
Thompson-Stewart process 汤普逊-斯特瓦特法〔碱式碳酸铅的改进制法之一〕
thomsenolite 方霜晶石
Thomsen process 汤姆森制纯碱及矾土法
Thomson-Fitzgerald furnace 汤姆孙-费兹格德电炉
thomsonite 杆沸石
Thomson's spectrograph 汤姆孙谱仪
thoracene 双(环辛四烯)合钍
Thorex pilot plant 钍雷克斯中间(试验)工厂
THOREX process 钍雷克斯流程
thoria 氧化钍 ThO_2
thoria effusion membrane 氧化钍渗流膜
thoria gel 氧化钍凝胶
thorianite 方钍石
thoriated tungsten 涂钍钨；镀钍钨
thoriated waste 含钍废物；被钍污染的废物
thorides 钍系元素
thorin 钍试剂*
thorite 钍石；硅酸钍矿
thorium 钍〔90 号元素, 化学符号 Th〕
thorium A 钍 A ^{216}Po
thorium anhydride 氧化钍
thorium antimonide 锑化钍
thorium arc lamp 钍弧灯
thorium B 钍 B ^{216}Po

thorium C 钍 C ^{212}Bi
thorium C′ 钍 C′ ^{212}Po
thorium C″ 钍 C″ ^{208}Tl
thorium chloranilate 氯冉酸钍
thorium chloride 氯化钍 $ThCl_4$
thorium D 钍 D；钍铅 ^{208}Pb
thorium decay series 钍衰变系
thorium deuteride 氘化钍
thorium dioxide 二氧化钍 ThO_2
thorium emanation(=thoron) 钍射气〔符号 Tn〕
thorium hydroxide 氢氧化钍 $Th(OH)_4$
thorium-lead age 钍-铅(法测)定年代法
thorium nitrate 硝酸钍 $Th(NO_3)_4$
thorium nitride 氮化钍；四氮化三钍 Th_3N_4
thorium oxide 氧化钍 ThO_2
thorium oxycarbonate 碳酸氧化钍
thorium series 钍系
thorium sulfate 硫酸钍 $Th(SO_4)_2$
thorium tetraisopropoxide 四异丙氧基钍 $Th[(CH_3)_2CHO]_4$
thorium X 钍 X ^{224}Ra
Thorman grid tray 索尔曼格条塔板
Thorman tray 索尔曼塔板；推液式条形泡罩塔板
thorn stone 钍石
thorny restharrow 刺芒柄花
thorocene 双环辛四烯合钍 $Th(C_8H_8)_2$
thorogummite 硅酸钍铀矿
thoron ①(=thorium emanation)钍射气②(=thorin)钍试剂
thoronol 1-(邻砷酸苯基偶氮)萘-2-酚-3,6-二磺酸
thorotrast 氧化钍胶体
thorough burning 完全燃烧
thorough-cut temperature ①热击穿温度②贯穿切割温度
thorough drying 充分干燥
thorough mixing 充分混合
thoroughness (除锈)彻底性；完全性；充分性
thorough washing 彻底洗涤；充分洗涤
thortveitite 钪钇石
thor-uraninite 钍铀矿
Thoulet's solution 杜氏碘化汞钾浓溶液
thraclene 乙酸-6-甲基庚烯-5-酯
thrapic acid 十七碳四烯酸
thrasher ①敲挞机②打谷机③松散机
thrashing 捶击
thrashing machine 捶击机
thread ①线②胶丝③螺纹
thread-advancing mechanism 绕丝装置
thread advancing reel 运丝辊

thread count　经纬密度；织物密度
thread counter　织物分析镜；织物密度镜
thread-cutting compound(=thread-cutting oil)　螺纹切削油
thread-cutting oil(=thread-cutting compound)　螺纹切削油
threaded bend　螺纹弯
threaded fitting　螺纹接口
threaded flange　螺纹法兰
threaded joint　螺纹接合
threaded pipe　螺纹管
threaded-plunger syringe　螺纹柱塞注射器
thread error　螺纹误差
thread fit　螺纹配合
thread-forming polyester　成纤聚酯
thread-forming property　成丝性质
thread fraying　缝线起毛
thread gauge　螺纹规
thread hole restoring　螺纹孔修复
thread imperfection tester　丝条疵点测试仪
threading speed　绕线速度
threading-up　生头
thread jaw chuck　三爪卡盘
thread-like molecular chain　线状分子链
thread-like molecule　线型分子
threadlike particle　线状粒子〔属纤维粒子〕
thread milling　螺纹铣削
three-acetate fiber　三醋酯纤维
three-arm rotational mo(u)lding machine　三臂式滚塑机
three-arm staker　裘皮拉软机
three attributes of color　色的三属性
three-axial cylinder structure　三维圆筒体结构
three-axial orthogonal structure　三维三向正则结构
three-blade marine type propeller agitator　三叶推进桨式搅拌器；三叶船用式螺旋桨搅拌器
three-body collision　三体碰撞
three-body recombination　三体再化合
three boiling system　三级蒸发系统〔制糖〕
three-bowl calender　三轮砑光机
three-brush bottle washer　三刷洗瓶机
three-button (filter) plate　三钮(滤)板
three carbon prototropic system　三碳质子移变异构系
three carbon rearrangement　三碳重排
three-cell furnace　三室炉
three center bond　三中心键
three center (type) reaction　三中心(型)反应
three-chamber mill　三仓磨
three-channel colorimeter　三道比色计
three-channel multipoint recorder　三道多点记录器
three-channel ultravioletvisible spectrophotometer　三道紫外-可见光分光光度计
three-color separation　三色分离
three-component system　三组分(物)系
three-coordinate complex　三配位复盐；三配位络(合)盐
three-cylinder washer　三筒洗机
three dimension　①三维空间②三维的；三度的；立体的
three-dimensional　三维的
three-dimensional arrangement　三维组合；立体结构
three dimensional chromatogram　三维色谱图
three dimensional (construction) polymer　立构聚合物
three-dimensional crosslinked network　三维交联网络；立体交联网络
three-dimensional crystalline structure　三维晶体结构
three-dimensional crystallinity　三维结晶性
three-dimensional cyclic structure　三维环状结构
three dimensional fluorescence spectrum　三维荧光光谱
three-dimensional highlight effect　立体辉光效应
three-dimensional irregularity　三维不规整性；三维不规整度
three-dimensionality of color　色的三维(三度)性；色的空间(立体)性
three-dimensional lattice　三维点阵
three dimensional model for spectrophotometry　分光光度法的三维模型；分光光度法的立体模型
three dimensional net structure　体型网格结构
three dimensional nuclear magnetic resonance　三维核磁共振
three-dimensional order　三维有序
three-dimensional packing　三维堆砌
three dimensional polycondensation　体型缩聚；三向缩聚
three-dimensional polymer　体型聚合物；网络聚合物；交联聚合物
three-dimensional polymeric molecule　体型聚合物分子；体型高分子
three dimensional radiofrequency quadrupole ion trap　三维射频四极离子阱
three dimensional scanner　三维扫描机；立体扫描机
three-dimensional space　三维空间
three dimensional structure　体型结构；三维结构
three dimensional type　三维型；体型
three dimensional weave　三向织物
three dimension polymerization　体型聚合
three dimension resolution surface　分离度曲面
three dimension scanning　三维扫描；主体扫描
three dimension total scanning fluorimetry　三维全扫描荧光法

three dimentional cross-linked molecule　三维交联分子；体型交联分子
three dimentional network structure　三维网格结构；体型网格结构
three dimentional polycondensation　体型缩聚
three dimentional structure　三维结构；体型结构
three-direction polymerization　三向缩合；体型聚合
three-drum system　三鼓法
three-electrode cell　三电极电解池
three electrode polarograph　三电极极谱仪
three-electrode potentiostat　三电极恒电位仪
three electrodes argon plasma　三电极氩等离子体
three electrode system　三电极系统；三电极体系
three-electrode (vacuum) tube　三级真空管
three-electron bond　三电子键
three-elementary weave　三原组织
three filter colorimeter　三色滤光式比色计
three-five compounds　Ⅲ-Ⅳ族化合物
threefold axis　三重轴
threefold barrier　三倍垒
three fragment fission　三裂变；三分裂
three-jaw self-centering chuck　自定心三爪卡盘
three-lens illumination　三透镜照明
three-level system　三能级系统
three line method　三线法
threeling　三连晶
three-lobed structure　三叶形结构
three-lobe oil pump　三圆突旋转泵
three-membered ring　三节环
three-neck distillation flask　三颈蒸馏瓶
three-necked bottle　三颈瓶；三颈玻璃瓶
three-necked flask　三颈烧瓶；三口烧瓶
three-number system　三参数值系统
three o'clock welding　横向自动焊
three-pan system　三锅系
three-pen recorder　三笔记录仪
three phase　三相的
three phase aerosol　三相(组成的)气溶胶制品
three-phase alternating current　三相交流电
three-phase current　三相电流
three-phase extraction　三相萃取
three phase region　三相区
three-pit system　三槽法
three-ply stock　三重纸
three-ply-wood　三合板；三层胶合板
three-point chiral recognition model　三点手性识别模型
three point interaction　三点相互作用
three prisms　三棱镜

three-quarters-filled shell effect　四分之三充满(电子)壳层效应
three-ring　三元环
three-roll coater　三辊涂漆机
three-roll grinder　三辊磨(颜)料机
three-roll mill　三辊滚轧机
three rolls plate bending machine　三辊卷板机
three-screw oil pump　三螺旋桨油泵
three slot burner　三缝燃烧器
three solvent mode　三元溶剂方式
three-stage compressor　三级压缩机
three-stage distillation　三级蒸馏
three stage fluidized bed dryer　三层(段)流化床干燥器
three-stage mass spectrometer　三级质谱计
three-stage mixing　三段混炼
three-stage nitration　三段硝化(法)
three-stage process　三段(硝化)法
three stages fluidized bed dryer　三层硫化床干燥器
three-stage steam ejector　三级蒸汽喷射器
three states of matter　物质的三态
three-throw pump　三程泵
three wavelength spectrophotometry　三波长分光光度法
three-way cock　三通活栓
three-way connection(=three-way piece)　三路管
three-way control valves　三路控制阀
three-way electrochromatography　三道电色谱法
three-way glass stopcock　三通玻璃栓
three-way key　三路电键
three-way piece(=three-way connection)　三路管
three-way pipe　三通管
three-way stop-cock　三路(活)栓；三通活栓
three way stopcock burette　三通滴定管
three-way tap　三通活栓
three-way valve　三路阀
three-wheel pump　三轮泵
three-zone hydrocracking　三段加氢裂化
threo-　苏〔词根〕
threo configuration　苏型构型*
threo-di-isotactic　对(映)双全同立构（是指双全同立构中两个不对称中心不是叠同的，而是镜(面对映)体即构型相反）
threo-diisotactic polymer　苏型双全同立构聚合物
threo-disyndiotactic　对(映)双间同立构
threo-disyndiotactic polymer　苏型双间同立构聚合物
threo form　苏型
threo isomer　苏型异构体
threomycin(=furanomycin)　苏氨霉素〔即：呋喃霉素〕
threonic acid　苏糖酸

threonine 苏氨酸 CH₃CH(OH)CH(NH₂)COOH
threonyl 苏氨酰 CH₃CH(OH)CH(NH₂)CO—
threose 苏糖
thresher 捶击机
threshing 捶击
threshing machine 捶击机
threshold ①阈；阈(值)*②(最低)限(度)③极限
threshold burst pressure 临界破裂压力；界限破裂压力
threshold concentration 阈浓度〔最低限度浓度〕；极限浓度；临界浓度
threshold dose 极限剂量；阈剂量
threshold drying 低限干燥
threshold effect 低限效应；阈效应
threshold energy 阈能
threshold fluence 阈值光流
threshold frequency 极限频率
threshold limit 阈限值；最大允许浓度
threshold limit dose 最高允许剂量
threshold limit value 阈限值
threshold of coagulation 凝结阈
threshold of reaction 反应极限
threshold signal 阈限信号
threshold temperature 阈温度
threshold treatment 极限处理
threshold value 阈值；临界值
thribble 三联管
throat 加料颈
throat sprayer 喷喉器〔可代替喷枪〕
throat stopper 炉喉栓
throat thickness 焊缝厚度
thrombase (=thrombin) 凝血酶；凝血因子Ⅱₐ
thrombelastogram 血栓弹力图
thrombelastography 血栓弹力描记术
thrombin (=thrombase) 凝血酶；凝血因子Ⅱₐ
thrombinogen 凝血酶原；凝血因子Ⅱ
thrombocyte 血小板
thromboplastid 血小板
thromboplastin (=factor Ⅲ) 组织促凝血酶原激酶；凝血因子Ⅲ
throttle 节流
throttle expansion valve 节流膨胀阀
throttle flow 节流
throttle plate 节流板
throttle valve 节流阀
throttle valve shaft 节流阀杆
throttling 节流；节气
throttling calorimeter 节流量热器
throttling controller 节流阀

throttling discharge 排出口节流；节流排出口
throttling expansion 节流膨胀
throttling flow meter 节流流量计；差压流量计
throttling set 节流装置
throttling valve 节流阀
through(=grinning) 露底〔面漆透出底漆颜色〕
through-air drum dryer 鼓式穿流干燥器
through-and-through coal 原煤
through-bond heteronuclear correlation 穿键杂核相关
through-drying 全干；干透〔指上下层漆完全干透〕
through-feed ironing machine 通过式熨平机
through-feed samming machine 通过式匀湿机
throughflow drying (=throughcirculation drying) 通风干燥；流通干燥
through-flow heater 通流加热器
through impregnation ①浸透②完全浸渍
through metal 金属支架
through plate girder 穿过式板梁
through-put ①处理量②生产量；生产能力
through-put capacity 物料通过额
through-put concentration 物料通过浓度
through-put of column 塔中蒸气通过速度
through-put rate 物料通过速率
through-put ratio 流量比；物料通过比率
through-put volume 物料通过体积
through-put weight 物料通过重量
through-resonance 贯穿共振
through-screen size 筛下产品粒度
through-station 中间站
through thermocouple 穿透式热电偶
through-tube 纱管
throwaway bio-degradable plastic bags 可降解一次性塑料袋
throw-away fuel cycle 一次通过式燃料循环
thrower 甩油环
throwing ①拉坯②捻〔生丝〕
throwing out ①气裂②沉淀③絮凝作用
throwing power 布散能力；深镀能力
thruflow analyzer 通流分析器
thrust 推力
thrust balancing device 推力平衡装置
thrust bearing 止推轴承
thrust load 推力负荷；纵向负荷；轴负荷
thrust washer 止推垫圈
thuja ①苧②金钟柏；崖柏
thuja acid(=thujic acid) 苧酸
4-thujaanol 4-苧醇
thujadiene 苧二烯；崖柏二烯

thujaketone 苧甲酮
thujaketonic acid 苧酮酸
thuja leaf oil(=white cedar leaf oil) 白柏叶油；侧柏叶油
thujane 苧烷；2-甲基-4-异丙基二环[3.1.0]已烷 $C_{10}H_{18}$
thuja oil 苧油；金钟柏油
thujaplicin 苧侧素；大叶崖柏素
α-thujaplicin α-苧侧素；大叶崖柏素
β-thujaplicin(=hinokitiol) β-苧侧素；4-异丙基苧酚酮；日扁柏素
γ-thujaplicin γ-苧侧素；4-异丙基苧酚酮
thujaplicine 苧侧醇
β-thujaplicine β-苧侧醇；4-异丙基苧酚酮
thujaplicinol 苧侧酚；侧柏酚
thujene ①苧烯②崖柏烯
thujetic acid 苧介酸 $C_{28}H_{22}O_{13}$
thujetin 氢化金钟柏配基
thujgenin 苧苷配基
thujic acid(=thuja acid) 苧酸
thujin 苧精；崖柏苷
thujoid 苧萃；侧柏浸提物
thujol ①苧醇②崖柏醇
thujone(=thalviol) 崖柏酮；艾菊酮
thujopsane 罗汉柏烷
thujopsanone 罗汉柏酮
thujopsene 罗汉柏烯
thujorhodin 玫红黄质
thujyl 苧基〔来自苧烷，指2位的基〕 $C_{10}H_{17}$—
thujyl acetate 乙酸苧酯；乙酸侧柏酯
thujyl alcohol 苧醇 $C_{10}H_{18}O$
thulia 氧化铥 Tu_2O_3
thulite 锰黝帘石
thulium 铥〔69号元素，化学符号Tu〕
thulium carbonate 碳酸铥 $Tu_2(CO_3)_3$
thulium chloride 氯化铥 $TuCl_3$
thulium hydroxide 氢氧化铥 $Tu(OH)_3$
thulium nitrate 硝酸铥 $Tu(NO_3)_3$
thulium oxalate 草酸铥
thulium oxide 氧化铥 Tu_2O_3
thulium sulfate 硫酸铥 $Tu_2(SO_4)_3$
thumb nut 蝶形螺帽
thumb-pressure can 指压注油器
thumb screw 脚螺丝〔化学天平的〕
thunbergene 黑松烯
thunbergin 寸拜精
Thunberg tube 通堡氏管
thunderbolt 闪电熔岩
thurberogenin 苏拜精宁
thurberogenone 苏拜精酮
thuringite 鳞绿泥石
Thurston friction machine 瑟斯顿摩擦试验机
Thurston tester 瑟斯顿试验机
Thury automatic regulator 瑟里自动校正器
Thury thread 细牙螺纹
thus 乳香
thuva 金钟柏
thuzic acid 侧柏酸
Th-X 射钍；钍X
Thylox process 泰洛克斯法气体硫代砷酸钠脱硫化氢过程〔湿法〕
thymacetin 百里酚 N-乙酰基乙氧苯胺
thyme 百里香；麝香草
thyme camphor 百里酚 $(CH_3)_2CHC_6H_3(CH_3)OH$
thymegol 邻硝基对百里酚磺酸汞钾
thyme lemon oil 百里柠檬油
thymene 百里烯 $C_{10}H_{16}$
thyme oil 百里香油
thymidine (dT; T) 胸(腺嘧啶脱氧核)苷
thymidine monophosphate(=thymidylic acid) 胸苷一磷酸；胸(腺嘧啶脱氧核)苷酸
thymidol 百里蓋醇〔百里酚和蓋醇(薄荷醇)的缩合产物〕
thymidylic acid (dTMP; TMP) 胸(腺嘧啶脱氧核)苷酸；胸苷一磷酸
thymin 胸腺激素
thymine (Thy) 胸腺嘧啶
thymine deoxyriboside kinase 脱氧胸苷激酶
thyminose 胸腺糖
thymodin(=thymol iodide) 百里碘酚
thymoform 百里仿 $C_{21}H_{28}O_2$
thymohydroquinone(=thymoquinol) 百里氢醌；5-甲基-2-异丙基-1,4-苯二酚 $C_{10}H_{14}O_2$
thymohydroquinone dimethyl ether 百里氢醌二甲醚
thymol(=5-methyl-2-isopropyl-phenol) 百里酚；麝香草酚；5-甲基-2-异丙基苯酚 $(CH_3)(C_3H_7)C_6H_3OH$
thymol acetate 乙酸百里酯 $CH_3CO_2C_{10}H_{13}$
thymol blue(=thymolsulfonphthalein) 百里酚磺酞；百里酚蓝 $C_{27}H_{30}O_5S$
thymol carbamate 氨基甲酸百里酚酯
thymol-carboxylic acid 百里酚羧酸 $(CH_3)(C_3H_7)C_6H_2=(OH)CO_2H$
thymol ethyl ether 百里酚乙醚 $C_2H_5OC_{10}H_{13}$
thymol formate 甲酸百里香酯
thymol iodide(=thymodin) 百里碘酚 $(C_{10}H_{12}OI)_2$
thymol methyl ether 百里酚甲醚 $CH_3OC_{10}H_{13}$
thymol phenacetin 百里酚非那西汀
thymol phenyl ether 百里酚苯醚 $C_6H_5OC_{10}H_{13}$
thymolphthalein 百里酚酞

thymolphthalein complexon (TPC) 百里酚酞(氨羧)络合剂
thymolphthalexone 百里酚酞(氨羧)配合剂
thymol reagent 百里酚试剂
thymol salicylate 水杨酸百里酚酯
thymolsulfonic acid 百里酚磺酸 $(CH_3)(C_3H_7)C_6H_2(OH)SO_3HH_2O$
thymolsulfonphthalein(=thymol blue) 百里酚磺酞；百里酚蓝
thymol type basil oil 百里香(酚型)罗勒油
thymol urethane 氨基甲酸百里酚酯
thymoquinol(=thymohydroquinone) 百里氢醌 $(C_3H_7)CH_3C_6H_2(OH)_2$
thymoquinone 百里醌 $O=C_6H_2(CH_3)(C_3H_7)=O$
thymoquinone-oxime 百里醌肟
thymosin 胸腺肽〔药〕
thymotal 氨基甲酸百里酚酯
thymotic acid 百里酸 $C_{11}H_{14}O_3$
thymotic alcohol 百里醇
thymotic anhydride 百里酸酐 $C_{11}H_{12}O_2$
thymotin alcohol 百里亭酚 $(CH_3)_2CHC_6H_3(CH_3)CH_2OH$
thymotinic acid 百里亭酸 $(C_3H_7)CH_3C_6H_2(OH)COOH$
thymotinic aldehyde 百里亭醛 $(C_3H_7)CH_3C_6H_2(OH)CHO$
thymotol 百里碘酚
thymovidin 胸维定
thymyl 百里基〔来自百里酚〕
$HC=C(CH_3)CH=CHC[CH(CH_3)_2]=C—$
thymylamine 百里基胺；5-甲基-2-异丙苯胺 $CH_3(NH_2)C_6H_3C_3H_7$
thymyl formate 甲酸百里香酯
thyratron 闸流管
thyresol O-甲基檀木萜
thyristor 可控硅
thyroid-releasing hormonestimulation test 甲状腺释放激素刺激实验
thyroid tablets 甲状腺片〔药〕
thyroxine 甲状腺素 $C_{15}H_{11}O_4NI_4$
tianeptine 噻奈普汀〔药〕
tiapride 硫必利〔药〕
tibolone 替勃龙〔药〕
tichromin 钛色敏
ticket 签条；票
ticket paper 票纸；券纸
tickling(=spin-tickling) 自旋挠痒法
ticresyl phosphite 亚磷酸三(甲苯酯) $(CH_3C_6H_4O)_3P$
tidal energy 潮汐能
tidal range exposure facility 潮汐带暴晒试验站

tidal wave 潮汐波；浪潮
tidewater exposure 潮水浸渍；潮汐浸渍
tie and dye leather 扎染革
tie coat(=block coat) 过渡(涂)层；中间(涂)层
tie dyeing 扎染
tie line 结线*
tiemannite 硒汞矿
tiered twisting machine 双层加捻机
tie rod(=draw rod) ①支柱②拉杆
tie water 结合水
tiferron 1,2-二羟基苯-3,5-二磺酸二钠
tiff ①方解石②重晶石
tigaldehyde 惕各醛 $MeCH=CMeCHO$
tiger powder 老虎火药
tight container 紧密容器
tight cure 充分硫化；彻底硫化；恰当硫化
tightener 收紧器
tightening 上紧；固定；紧密
tightening screw 紧固螺钉
tight filament 僵丝
tight fit 牢配合；紧(密)配合
tight gel(=macro-gel) 紧凝胶；大粒凝胶
tighthead oil can 密盖油听
tightness 严密度；紧密
tightness system 密封系统；密封装置
tight transition state 紧密过渡态
tight twist end 紧捻经丝
tiglaldehyde 惕各醛
tiglate 惕各酸酯
tiglic 巴豆的
tiglic acid 惕各酸；顺芷酸；顺式-2-甲基-2-丁烯酸 $CH_3CH=C(CH_3)CO_2H$
tiglic aldehyde 惕各醛；顺芷醛；顺式-2-甲基-2-丁烯-1-醛 $CH_3CH=C(CH_3)CHO$
tigloidine 惕各酰莨菪碱
tigloylmeteloidine 惕各酰陀罗碱
tiglyl 甲基巴豆酰(基)
tigogenin 惕告吉宁
tile ①瓦②砖；片砖
tile floor 砖地
tile flooring 砖铺地的
tile furnace 瓦窑
tile kiln 瓦窑
tile kilning 烧瓦
tile-like coating 仿瓷砖涂料
tile-lined 瓦衬〔炉〕
tile lining 衬瓦
tile-packed column 瓦片填充塔

tile packing 瓦填料
tile pipe 瓦管；瓦筒
tile press 压瓦机
tilery 瓦厂
tilestone 石板
tiling 铺瓦
till ①耕(作)；耕种②冰碛土
Tillman's reagent 提尔曼试剂
tilorone 双二乙氨乙基芴酮
tilted blade 倾刀
tilted-cone rheometer 斜(置圆)锥式流变仪
tilted-cylinder rheometer 斜(置圆)柱式流变仪
tilted mixer 倾斜式搅拌器
tilted plate 倾斜板〔测接触角〕
tilted-plate separator 斜板式分离器
tilter 倾架
tilting furnace 可倾炉
tilting gear 翻车装备
tilting head press 摇头印刷机；摇头压缩机
tilting hearth 可倾底
tilting pad 倾斜垫
tilting plate method 斜板法*
tilting slide method 斜板法〔测接触角〕
tilting tank 倾翻桶
tilt-pour process 倾注法
tilt trap 倾卸式阱
timber 木材；木料
timber dressing ①木材整饰②木材防腐处理
timbo 无患子皮
timbonine 无患子碱
time array detection 时间阵列检测
time averaging method 时间平均法
time-averaging operation 按时间取平均值(法)
time band method 时间区带法
time-basis sampling 定时取样；时基取样
time bomb 定时炸弹
time code 时码
time constant 时间常数
time controller(=timer) (自动)定时器
time course fitting 时程拟合法
time delay 时间延迟
time delay over current relay 延时过载继电器
time-dependent effect 含时效应；时间依赖效应
time derivative 时间导数
time domain 时域
time domain signal 时域信号
time efficiency 时间效率
time fuse ①定时信线②定时引信③定时信管；曳火线

time independent fluid 非时间依赖性流体；定常流体
time interval 时间间隔
time-lag action 延时作用
time-lag focusing 时间滞后聚焦
time-lag method for diffusion 扩散时滞法；扩散延时法
time marker 计时器
time normalization 时间归一化法
time of flight (TOF) 飞行时间
time-of-flight analysis 飞行时间分析
time of flight analyzer 飞行时间分析器
time of flight mass analyzer 飞行时间质量分析器
time of flight mass separator 飞行时间质量分离器
time-of-flight mass spectrometer (TOFMS) 飞行时间质谱仪
time-of-flight mass spectrometry 时间飞行质谱法
time of flight secondary ion mass spectrometry (TOFSIMS) 飞行时间二次离子质谱法
time-of-flight spectrometer 时间飞行质谱仪
time of half dying 半染时间
time of immersion 浸涂时间；浸渍时间
time of relaxation 松弛时间
time of run ①运转时间②展开时间
time-piece lubricant 时计润滑剂；钟表润滑剂
time proportional phase incrementation 时间与相位增量成比例的方法
time proportioning controller 时间比例控制器
timer(=time controller) (自动)定时器；记时器
time recorder 记时器；时间记录器
time resolved chemiluminescence 时间分辨化学发光
time-resolved emission spectrum 时间分辨发射光谱
time resolved fluorescence 时间分辨荧光
time-resolved fluorescence spectrometry 时间分辨荧光光谱法
time resolved fluorescence spectroscopy (TRFS) 时间分辨荧光光谱法
time resolved fluoroimmunoassay 时间分辨荧光免疫分析法
time resolved Fourier transform infrared spectroscopy 时间分辨傅里叶变换红外光谱法
time-resolved immuno-fluorescence spectrometry 时间分辨免疫荧光光谱法
time resolved laser induced fluorescence 时间分辨激光诱导荧光
time resolved spectra 时间分辨光谱
time resolved spectrochemical analysis 时间分辨分光分析
time resolved spectrometer 时间分辨光谱仪
time resolved spectrometry 时间分辨光谱法
time-resolved spectroscopy 时间分辨光谱学

time retardation 时间推迟
timer programmer 计时程序器
time-share 分时
timeshare computer 分时计算机
time sharing 时间共用
time sharing laser 时间分割激光器；分时激光器
time split 时间分流
time switch 定时开关
time-temperature exchange principle 时-温换算原则
time-temperature profile 时间温度分布图
time-temperature superposition principle 时-温叠加原理
time-to-amplitude converter 时间-振幅转换器
time to ignition 点燃时间
time weighted average (TWA) 时间加权平均
time window method 时间窗口法
timija mint 圆叶薄荷
timing 计时；校时
timing device 计时装置
timing gear 校时齿轮
timing pulse 定时脉冲
Timken film strength 泰姆肯膜强度
Timken index 泰姆肯指数
Timken machine 泰姆肯摩擦试验机
Timken test 泰姆肯试验
Timken test cup 泰姆肯试验杯
Timken tester 泰姆肯试验机
Timken wear test 泰姆肯磨耗试验
timnodonic acid 二十碳五烯酸
-tin 亭〔音译词根，指生物碱或生物化学物质〕
tin ①锡②镀锡；包锡③白铁罐
tin acetate 乙酸锡
tin alkyl 烷基锡 $SnR_2;SnR_4$
tin ammonium chloride 氯化铵锡
tin anhydride 锡酐；二氧化锡；氧化锡 SnO_2
tin aryl 芳基锡 $SnAr_2;SnAr_4$
tin ash 二氧化锡；氧化锡 SnO_2
tin bath 锡浴
tin bibromide 二溴化锡；溴化亚锡 $SnBr_2$
tin bichloride 二氯化锡；氯化亚锡 $SnCl_2$
tin bifluoride 二氟化锡；氟化亚锡 SnF_2
tin biiodide 二碘化锡；碘化亚锡 SnI_2
tin bisulfide 二硫化锡 SnS_2
tin bromide 溴化锡 $SnBr_2;SnBr_4$
tin bronze ①锡青铜②硫化锡〔别名，用作金粉颜料，代替真金粉〕
tincal 粗硼砂
tincalcite 钠硼钙石
tin chloride 氯化锡 $SnCl_2;SnCl_4$

tin chloride solution 氯化亚锡盐酸溶液
tin-cobalt coating 锡-钴涂层；锡钴电镀层
tin-cobalt electroplating tank 锡-钴电镀槽
tin-cobalt pyrophosphate 焦磷酸锡钴
tin-cobalt sulphate 硫酸锡钴
tin crystal 氯化亚锡；二氯化锡或氯化锡〔一种强酸材料；作皮革染色的媒染剂用〕
tinct 色泽
tinction 着(染)色
tinctorial power 着色本领
tinctorial property 着色性能
tinctorial quality 着色质量
tinctorial strength(=color strength) ①着色强度；着色力 ②着色颜料的有效体积浓度〔油墨〕
tinctorial value ①着色值；着色力②撤淡值〔油墨〕
tinctorial yield 着色率；给色量
tincture ①酊(剂)；药酒②着色；染色③色调④浸染
tincture of iodine 碘酒
Tindal ozonizer 廷德耳臭氧器
tin-decorating finish 印铁漆
tinder 火绒；火种
tindery 易燃的
tin dibromide 二溴化锡；溴化亚锡 $SnBr_2$
tin dichloride 二氯化锡；氯化亚锡 $SnCl_2$
tin diethyl 二乙基锡 $Sn(C_2H_5)_2$
tin diethyl dichloride 二氯化二乙基锡
tin diethyldithiocarbamate 二乙基二硫代氨基甲酸锡〔硫化剂〕
tin diethyl oxide 氧化二乙基锡
tin difluoride 二氟化锡；氟化亚锡 SnF_2
tin diiodide 二碘化锡；碘化亚锡 SnI_2
tin dimethyl 二甲基锡 $Sn(CH_3)_2$
tin dioxide 二氧化锡；氧化锡 SnO_2
tin diphenyl 二苯基锡 $Sn(C_6H_5)_2$
tin disulfide 二硫化锡 SnS_2
-tine 亭〔音译词根〕
tin-electroplated 用锡电镀过的
tin electroplating 锡的电镀
tin ethide 二乙基锡 $Sn(C_2H_5)_2$
tin family 锡族
tin family element 锡族元素
tin filling 灌锡
tin filling machine 灌锡机
tin flake (鳞)片状锡
tin fluoride 氟化锡 $SnF_2;SnF_4$
tin foil 锡箔；锡纸
Tinfos furnace 廷福斯电炉
tinge 色彩；色调；着色的；带色的

tingeing 着(染)色的
tingitamine 氨基嘧啶丙氨酸
tin hydroxide 氢氧化亚锡 Sn(OH)$_2$
tinidazole 替硝唑〔药〕
tin ingot 锡锭；锡块
tin iodide 碘化亚锡 SnI$_2$;SnI
tinkal 硼砂
tin leaf 锡箔；锡纸
tinman's scissors 白铁工剪刀；铅皮剪刀
tin methide 二甲基锡 Sn(CH$_3$)$_2$
tin milk 锡乳
tin monosulfide 一硫化锡 SnS
tin monoxide 一氧化锡 SnO
tin mordant 锡媒染剂
tin mordanting 用锡媒染
tinned 镀锡的；包锡的
tinned iron 白铁皮；马口铁
tinned joint 焊接；点焊接合
tinned plate(=tin plate) 马口铁
tinned sheet iron 白铁皮；马口铁
tinned tack 镀锡铁小钉
tinning 镀锡；包锡
tinning furnace 镀锡炉
tin oxide 氧化锡 SnO;SnO$_2$
tin oxychloride 二氯氧化锡 SnOCl$_2$
tin paint 罐用涂料
tin peroxide 二氧化锡；氧化锡 SnO$_2$
tin plate(=tinned plate) 马口铁
tin plating 镀锡
tin powder 锡粉
tin protobromide 溴化亚锡 SnBr$_2$
tin protochloride 氯化亚锡 SnCl$_2$
tin protofluoride 氟化亚锡 SnF$_2$
tin protoiodide 碘化亚锡 SnI$_2$
tin protosulfide 一硫化锡 SnS
tin protoxide 一氧化锡 SnO
tin pyrite 黄锡矿
tin roller 白铁滚筒
tin salt 锡盐（通常指二氯化锡或氯锡酸铵）
tinsel ①金属线②金属箔③金属丝交织物
tin soldering 锡焊
tin stabilizer 有机锡稳定剂
tinstone 锡石
tin sulfide 硫化锡 SnS; SnS$_2$
tint ①色调；色辉〔物〕②着色③浅色
tintage 上色；着色
tint base 调色浆
tinted lime wash ①着色石灰涂料②着色水胶涂料

tinted paint 浅色漆
tinted rubber 着色橡胶
tinted rubbing filler 可着色和打磨的腻子〔填孔剂〕
tinted stopper 着色填孔剂
tinter ①调色料；调色浆；色浆②着色器
tin test 锡试验
tin tetrabromide 四溴化锡 SnBr$_4$
tin tetrachloride 四氯化锡
tin tetraethide 四乙基锡 Sn(C$_2$H$_5$)$_4$
tin tetraethyl 四乙基锡
tin tetrafluoride 四氟化锡 SnF$_4$
tin tetraiodide 四碘化锡 SnI$_4$
tin tetramethide 四甲基锡 Sn(CH$_3$)$_4$
tin tetramethyl 四甲基锡
tin tetraphenyl 四苯基锡 Sn(C$_6$H$_5$)$_4$
tinting ①着色②染色③涂漆
tinting black 黑调色浆；调色用黑浆
tinting color 调色剂
tinting material 着色剂
tinting paste 调色浆；找色浆〔俗〕
tinting pigment 着色颜料
tinting power 着色力
tinting strength(=tint strength) 着色力
tinting system 调色系统；配色系统
tintless 无色的
tintmeter(=tintometer) 色调计；色辉计〔物〕
tintometer(=tintmeter) 色调计；色辉计〔物〕
tintone ①颜料着色(强)度②(颜料)撤淡(强度)〔油墨〕
tint permanence ①(颜料)冲淡能力②着色性
tint reduction 颜色冲淡
tint retention 保色性；固色作用
tin triethyl ①三乙基锡(基)②六乙二锡③三乙锡(基)
tin trimethyl ①三甲基锡(基)②六甲二锡③三甲锡(基)
tin trimethyl radical 三甲基锡基
tint stand oil 浅色熟油
tint strength(=tinting strength) 着色力
tinware 锡器
tin white 锡白；二氧化锡
tin white cobalt ①砷钴矿②方钴矿 (Co,Fe,Ni)As$_2$; CoAs$_2$
tiny balloons ①小气球②小球形瓶
tiodine 烯丙基硫脲合乙基碘
tioguanine 硫鸟嘌呤〔药〕
tiopronin 硫普罗宁〔药〕
tip ①管尖②尖(端)使倾斜
tip-cap 顶芽帽法·〔胶树〕
tip electrochemistry 针尖电化学
tip-lift type perforated plate 活动顶盖式多孔塔盘

tipped staple 尖状毛束
tipped tool 镶片刀具；镶刃刀具；焊接车刀
tipping car 倾卸车
tipping idler 倾斜惰轮
tipping-resistant coating composition 不倾出涂料；高触变性涂料
tipple ①翻车机②翻锭机
tip velocity 周缘速度
tirandamycin 提朗达霉素
tire (轮)胎；车胎
tire bead 胎圈；子口；胎口；胎耳；胎脚
tire chemistry 轮胎化学〔包括原材料、配方、混炼硫化工艺〕
tire compound 轮胎胶(料)
tire cord 帘子线
tire paint 轮胎漆
tire rubber 轮胎用橡胶
tire texture 轮胎帘布层
tire working load 轮胎工作负荷；轮胎安全负荷
tiron 钛试剂*；钛铁试剂；1, 2-二羟基苯-3, 5-二磺酸钠 $C_6H_2(OH)_2(SO_3Na)_2$
Tirril burner 提利灯
Tirrill burner 梯瑞尔煤气灯
tirucallol(=kanzuiol) 甘遂醇 $C_{30}H_{50}O$
Tiselius cell 蒂塞利乌斯电泳池
Tiselius electrophoresis apparatus 蒂塞利乌斯电泳仪
tissue ①组织；体素②织物③薄绸④(薄)纸
tissue compatibility 组织相容性
tissue electrode 活组织电极
tissue-equivalent ionization chamber 组织当量电离室
tissue homogenate 组织匀浆
tissue paper 薄(叶)纸；纱纸
tissue plasminogen activator 组织血纤维蛋白溶酶原激活剂
tissue slice 组织切片
tissue waxing-paper 涂蜡纤维纸
tissus sensor 组织传感器
titan(=titanium) 钛〔22号元素，化学符号 Ti〕
titan-alkoxide 烷氧基钛
titanate ①钛酸盐②钛酸酯
titanate chelate 钛酸酯螯合物
titanate coupling agent 钛酸酯偶联剂
titanated castor oil 钛酸酯化蓖麻油；钛酸丁酯改性蓖麻油
titanated fabric 钛酸纤维
titanated lithopone 钛钡白；含钛立德粉〔俗〕
titanate radical 钛酸根
titanate whisker 钛酸盐类晶须

titanaugite 钛辉石
titan-calcium white 钛钙白
titanclinohumite 钛斜硅镁石；钛橄榄石
titanellow 黄色氧化钛
titania 二氧化钛 TiO_2
titania ceramics 钛氧陶瓷
titania-opacified enamel 二氧化钛乳白搪瓷；钛白搪瓷
titania-opacified glaze 钛质乳白釉
titania porcelain 钛质瓷
titania whiteware 钛质瓷；钛白瓷(器)
titanic (正)钛；四价钛
titanic acid 钛酸
titanic acid anhydride 钛酸酐；二氧化钛
titanic acid ester 钛酸酯
titanic anhydride 钛酸酐
titanic chloride 四氯化钛
titanic hydroxide 氢氧化钛；原钛酸
titanic magnetite 钛磁铁矿
titanic schore(=titanic schorl) 钛金红石
titanic schorl(=titanic schore) 钛金红石
titaniferous 含钛的
titanite 楣石；钛铁矿
titanium 钛〔22号元素，化学符号 Ti〕
titanium acetylacetonate (四)乙酰丙酮络钛(螯合物)
titanium acetylacetone 乙酰丙酮钛
titanium acrylate isostearate oxyacetate 丙烯酰异硬脂酰钛羟乙酸酯
titanium acylate 酰化钛；酰基钛
titanium alkoxide polymer 烷氧基钛聚合物
titanium alum 钛矾〔主要成分为硫酸钛钾 $KTi(SO_4)_2 \cdot 12H_2O$〕
titanium aluminum chloride 氯化铝钛 Ti_3AlCl_{12}
titanium-aluminum polyisoprene 钛铝〔催化〕聚异戊二烯橡胶
titanium 4-aminobenzene sulfonate dodecyl benzene sulfonate oxyacetate 4-氨基苯磺酰十二烷基苯磺酸钛羟乙酸酯
titanium 2-amionbenzoate isostearate oxyacetate 2-氨基苯甲酸异硬脂酸钛羟乙酸酯
titanium benzoate 苯甲酸钛
titanium boride fibre 硼化钛纤维
titanium bromide 溴化钛
titanium brown 钛棕
titanium-calcium 钛钙白
titanium-calcium pigment 钛-钙(白)颜料
titanium carbide 碳化钛 TiC
titanium carbide fibre 碳化钛纤维
titanium carbonate 碳酸钛

titanium chelate 钛螯合物
titanium chloride 氯化钛 $TiCl_2$; $TiCl_3$; $TiCl_4$
titanium cyanide 氰化钛
titanium diamond 钛金刚石
titanium dianthranilate oxyacetate 二氨茴酸钛羟乙酸酯
titanium dichloride 二氯化钛 $TiCl_2$
titanium di(cumyl phenolate) oxyacetate 二(枯基苯(酚)氧基)钛羟乙酸酯
titanium di(dioctyl phosphate) oxyacetate 二(二辛基磷酸)钛羟乙酸酯
titanium di(dioctyl pyrophosphate) oxyacetate 二(二辛基焦磷酸)钛羟乙酸酯
titanium di(2-formyl phenoxy) oxyacetate 二(2-甲酰基苯氧)钛羟乙酸酯
titanium dihydroxide 氢氧化钛(Ⅱ);二羟基钛
titanium diisostearate oxyacetate 二异硬脂酸钛羟乙酸酯
titanium dimethacrylate oxyacetate 二甲基丙烯酸钛羟乙酸酯
titanium dioxide 二氧化钛;钛白
titanium dioxide dilution function 二氧化钛冲淡函数
titanium disilicide 二硅化钛 $TiSi_2$
titanium disulfide 二硫化钛 TiS_2
titanium ester 钛醚
titanium ethanolate 乙醇钛;钛酸乙酯
titanium ethylene glycolate 乙二醇钛(盐)
titanium family 钛族
titanium family element 钛族元素
titanium ferrocene(=titanocene dichloride) 二氯化二茂钛;二茂基二氯化钛
titanium green 钛绿
titanium hydride 氢化钛 TiH_2
titanium hydroxide 氢氧化钛 $Ti(OH)_4$
titanium isobutoxide 原钛酸四异丁酯;异丁醇钛;异丁氧钛
titanium isopropoxide 异丙氧基钛
titanium isopropylate(=tetraisopropyl titanate) 异丙醇钛;钛酸(四)异丙酯
titanium isostearate methacrylate oxyacetate 异硬脂酸甲基丙烯酸钛羟乙酸酯
titanium lining 钛衬里
titanium lithopone 钛钡白;含钛立德粉〔俗〕
titanium monoxide 一氧化钛 TiO
titanium nickel yellow 钛镍黄
titanium nitride 一氮化钛 TiN
titanium ore 钛矿
titanium oxalate 草酸钛 $Ti_2(C_2O_4)_3 \cdot 10H_2O$
titanium oxide(=titanium dioxide) 二氧化钛;钛白
titanium oxide ceramics 氧化钛陶瓷
titanium oxynitride 氮氧化钛
titanium oxyoxalate 草酸氧钛;碱式草酸钛
titanium peroxide 过(三)氧化钛 $TiO[O_2]$
titanium phosphate(TP) 磷酸钛
titanium pigment 钛白粉
titanium polymer 钛聚合物
titanium porcelain 钛瓷
titanium potassium fluoride 氟化钛钾
titanium potassium oxalate 草酸钛钾
titanium propanolate 丙醇钛;丙氧基钛;钛酸丙酯
titanium sesquioxide 三氧化二钛 Ti_2O_3
titanium sesquisulfate 倍半硫酸钛(Ⅲ);三硫酸二钛
titanium silicide 硅化钛 $TiSi_2$;Ti_2Si
titanium-silicon pigment 钛-硅颜料
titanium slag 高钛渣
titanium sponge 海绵(状)钛
titanium sublimation pump 钛升华泵
titanium sulfate 硫酸钛 $Ti_2(SO_3)_3$;$Ti(SO_4)_2$
titanium sulfide 硫化钛 TiS;TiS_2
titanium superoxide 过(三)氧化钛 $TiO[O_2]$
titanium tannage 钛鞣
titanium tetrachloride 四氯化钛
titanium tetra iodide 四碘化钛
titanium tetraisopropylate 四异丙醇钛
titanium tribromide 三溴化钛 $TiBr_3$
titanium trichloride(=titanous chloride) 氯化亚钛;三氯化钛 $TiCl_3$
titanium trifluoride 三氟化钛 TiF_3
titanium trihydroxide 三氢氧化钛
titanium triiodide 三碘化钛 TiI_3
titanium trioxide 三氧化钛 $TiO[O_2]$;TiO_3
titanium tritide 氚化钛
titanium white 钛白;(二)氧化钛钛(白)粉〔商〕
titanium yellow 钛黄
titanocene dichloride(=dicyclopentadienyl titanium dichloride;titanium ferrocene) 二氯化二茂钛;二环戊二烯基二氯化钛;二茂基二氯化钛
titanomagnetite 钛磁铁矿
titanometric titration 钛盐滴定(法)
titanometry 钛盐滴定法
titanosiloxane 钛-硅氧烷
titanous 三价钛;亚钛
titanous chloride(=titanium trichloride) 氯化亚钛;三氯化钛 $TiCl_3$
titanous oxalate(=titanium oxalate) 草酸亚钛
titanous sulfate 硫酸亚钛
titanox 钛白
titanoxane 钛氧烷

titan white(=titanium white)　钛白
titan yellow　太坦黄；钛黄
titanyl　钛氧(基)　TiO＝
titanyl acylate　钛氧酰化物
titanyl nitrate　硝酸氧钛　$TiO(NO_3)_2$
titanyl phthalate　邻苯二甲酸氧钛
titanyl sulfate　硫酸氧钛　$(TiO)SO_4$
titer(=titre)　①滴定率②滴定度③脂酸冻点测定(法)④脂酸冻点⑤当量溶液校准时的差⑥纤度
titer method(=titer system)　滴定浓度法
titer system(=titer method)　滴定浓度法
titer test　脂酸冻点测定(法)
titer tube　脂酸冻点试管
titer value　①滴定度②脂酸冻点数值
titone　钛铅钡白
titrable acidity　可滴定酸度
titrand　(被)滴定物*；(被)滴定液
titrant　滴定剂*；滴定(用)标准液
titrant-stream reference electrode　滴液参比电极
titratable acidity　可滴(定)酸度
titrate　①被滴定液②滴定
titrated solution　滴定过的溶液
titrating　滴定的
titrating coulometer　滴定库仑计
titrating head　滴定头
titration　滴定*
titration acidity　滴定酸度
titration analysis　滴定分析
titration apparatus　滴定仪器
titration bottle　滴瓶
titration cell　滴定池
titration constant　滴定常数
titration coulometer　滴定电量计
titration curve　滴定曲线
titration detector　滴定检测器
titration efficiency　滴定效率
titration end point　滴定终点
titration error　滴定误差
titration exponent　滴定指数
titration fraction　滴定分数
titration in nonaqueous solvent　非水溶剂滴定
titration jump　滴定突跃
titration level　滴定剂规度
titration of compensating anodic and cathodic waves　阴阳波对消滴定；补偿滴定；混液滴定
titration technique　滴定法；滴定技术
titration thief　滴定阱*
titration vessel　滴定容器

titrator　滴定器
titre (=titer)　①滴定率②滴定度③脂酸冻点测定(法)④脂酸冻点⑤当量溶液校准时的差⑥纤度
titre uniformity tester　纤度均匀度试验
titrimeter　滴定计
titrimetric　滴定分析的
titrimetric analysis　滴定分析(法)*
titrimetric-extraction method　滴定-萃取法
titrimetric factor　滴定因子；滴定系数
titrimetric method　滴定(分析)法
titrimetric standard　滴定标准
titrimetry　滴定(分析)法*
tjujamunite　钒钙铀矿
TLA process　三月桂胺萃取法
TLV(=threshold limit)　阈极限值；最大允许浓度
T matrix(=transition matrix)　跃迁矩阵
toaster　蒸(炒)缸
tobacco　烟草
tobacco essence　烟草香精
tobacco flavour　烟草香精
tobacco flower oil　烟草花油
tobacco leaf oil　烟叶油
tobacco pouch　烟草袋；烟荷包
tobacco-seed oil　烟草子油
tobacco smoke alkaloid　烟草烟(中的)碱
tobacco stems　烟草梗
tobacco waste　烟草屑
tobacco wax　烟草蜡
tobermorite　雪硅钙石；水化硅酸钙
Tobias acid　托拜厄斯酸；2-氨基-1-萘磺酸
tobinetin　刺槐亭
tobramycin　妥布霉素〔药〕
tocopherol (=vitamin E)　生育酚；抗不育维生素；维生素E
tocopheronic acid　生育酸
tocopherylamine　生育胺
toddaculin　飞龙掌血素
toddalia　飞龙掌血
toddalia leaf oil　飞龙掌血叶油
toddaline(=chelerythrine)　白屈菜赤碱
toddalolactone　飞龙掌血内酯；毛两面针素
toe-case　皮鞋包头
Toepler pump　托普勒泵；托氏泵
Tofranil　10,11-二氢-N,N-二甲基-5H-二苯并[b,f]氮䓬-5-丙胺；盐酸丙咪嗪
toggle　①肘节〔机工〕②绷节夹子〔皮革〕
toggle dryer　绷皮干燥器
toggle switch　搬扭开关；叉簧开关

toggle type injection mo(u)lding machine 肘节式注塑机
toggling 绷板(干燥)
tohi oil 橙皮油
toilet cream (盥洗用)香水
toilet perfume 盥洗用香水
toilet powder 爽身粉
toiletry 化妆用品
toilet soap 香皂
toilet water 花露水
tokamycin 东海霉素
tolamex 陶拉袂克斯〔炸药〕
tolamine 对甲苯磺酰氯胺钠；氯胺-T
tolane(=diphenylacetylene) 二苯乙炔 $C_6H_5C≡CC_6H_5$
tolane dibromide 二溴化二苯乙炔 $C_6H_5CBr=CBrC_6H_5$
tolane sulfide 硫化二苯乙炔
tolazoline(=peripherine) 陶拉唑啉
tolazoline hydrochloride 盐酸 4,5-二氢-2-苄基-1H-咪唑；2-苄基咪唑啉
tolbutamide 1-丁基-3-对甲苯磺酰基脲
tole ①铁皮；薄钢板；钢板；板材②涂漆的马口铁器皿
tolerable limit 许许范围；许许限度
tolerance 容许误差；冗差
tolerance deviation 容许偏差
tolerance dose 耐受剂量
tolerance error 容许误差
tolerance in color deviation 容许色差
tolerance interval 容许区间
tolerance level ①允许剂量(级)②允许的辐射量(级)③允许程度
tolerance limit 容许限*
tolerance test 耐药量试验
o-tolidine 邻联甲苯胺；邻二甲基二氨基联苯
tolidine(=dimethylbenzidine) 联甲苯胺；二甲基二氨基联苯 $(CH_3C_6H_3NH_2)_2$
3,3′-tolidine-4,4′-diisocyanate 3,3′-联甲苯胺-4,4′-二异氰酸酯
tolidine sulfate 硫酸联甲苯胺
tolil 甲苯偶酰 $(CH_3C_6H_3CO)_2$
tolilic acid 二甲苯基乙醇酸 $(CH_3C_6H_4)_2C(OH)COOH$
tolimidazole 甲基苯并咪唑
tolita(=TNT) 徒里达；三硝基甲苯
tolit(e)(=TNT) 徒里特；三硝基甲苯 $CH_3C_6H_2(NO_2)_3$
tollalyl sulfide 硫化二苯乙炔
toll enrichment 委托浓缩；收费浓缩
Tollens' reagent 土伦试剂
toloxy(=tolyloxy) 甲苯氧基
tolterodine 托特罗定〔药〕

tolu ①妥卢②妥卢香脂
tolualdehyde 甲苯甲醛 $CH_3C_6H_4CHO$
toluamide 甲苯甲酰胺
toluanilide N-对甲苯甲酰(基)苯胺
toluarsonic acid 妥卢胂酸；羧基甲苯胂酸 $HOOC(CH_3)C_6H_3AsO(OH)_2$
toluate 甲苯甲酸盐(酯) $CH_3C_6H_4COOM$
tolu balsam 妥卢香脂
tolu balsam oil 妥卢香脂油
toluene(=methylbenzene) 甲苯 $C_6H_5CH_3$
toluene azonaphthylamine 甲苯偶氮萘胺 $CH_3C_6H_4N=NC_{10}H_6NH_2$
toluene bromide 溴甲苯 $CH_3C_6H_4Br$
toluene chloride 氯甲苯 $CH_3C_6H_4Cl$
toluene diamine 甲苯二胺
toluene dichloride 二氯甲苯 $CH_3C_6H_3Cl_2$
toluene diisocyanate 甲苯二异氰酸酯
2,4-toluene diisocyanate 2,4-甲苯二异氰酸酯〔硫化剂〕
3,3′-p-toluene-4,4′-diisocyanate 3,3′-对甲苯-4,4′-二异氰酸酯〔硫化剂〕
toluene distillation method 甲苯蒸馏法
toluene disulfonate 甲苯二磺酸盐 $C_7H_6O_6S_2M_2$
toluene disulfonic acid 甲苯二磺酸 $CH_3C_6H_3(SO_3H)_2$
toluene-2,4-disulfonyl hydrazide 甲苯-2,4-二磺酰肼〔发泡剂〕
toluene-3,4-dithiol(TDT) 甲苯-3,4-二硫酚 $CH_3C_6H_3(SH)_2$
toluene equivalent 甲苯当量
toluene fluoride 氟甲苯 $CH_3C_6H_4F$
toluene fraction 甲苯馏分
toluene halide 卤甲苯 $CH_3C_6H_4X$
toluene iodide 碘甲苯 $CH_3C_6H_4I$
toluene monochloride 氯甲苯 $CH_3C_6H_4Cl$
toluene musk 甲苯麝香
toluene pentachloride 五氯甲苯 $CH_3C_6Cl_5$
toluene plant 甲苯工厂
toluene sulfamide 甲苯磺酰胺
toluenesulfinate 甲苯亚磺酸盐 $C_7H_7O_2SM$
toluenesulfinic acid 甲苯亚磺酸 $CH_3C_6H_4SO_2H$
toluene sulfochloride 甲苯磺酰氯 $CH_3C_6H_4SO_2Cl$
toluene sulfodichloramide 甲苯磺酰二氯胺 $CH_3C_6H_4SO_2NCl_2$
p-toluene sulfohydrazide 对甲苯磺酰肼〔发泡剂〕
p-toluene sulfonylsemicarbazide 对甲苯磺酰氨基脲〔发泡剂〕
p-toluidine method 对甲苯胺法
toluene sulfonamide 甲苯磺酰胺 $CH_3C_6H_4SO_2NH_2$
toluenesulfonanilide N-甲苯磺酰苯胺 $CH_3C_6SO_2NHC_6H_5$
toluenesulfonate 甲苯磺酸盐 $C_7H_7O_3SM$

toluene sulfonic acid　甲苯磺酸　$CH_3C_6H_4SO_3H$
toluenesulfonyl　甲苯磺酰　$CH_3C_6H_4SO_2-$
toluene-*p*-sulfonyl acetone hydrazone　对甲苯磺酰丙酮腙〔发泡剂〕
p-toluene sulfonyl azide　对甲苯磺酰叠氮〔发泡剂〕
toluenesulfonyl butylamine　*N*-甲苯磺酰丁胺　$CH_3C_6H_4SO_2NHC_4H_9$
toluene sulfonyl chloride　甲苯磺酰氯　$CH_3C_6H_4SO_2Cl$
toluenesulfonyl dibutylamine　*N*-甲苯磺酰二丁胺　$CH_3C_6H_4SO_2N=(C_4H_9)_2$
toluenesulfonyl dimethylamine　*N*-甲苯磺酰二甲胺　$CH_3C_6H_4SO_2N=(CH_3)_2$
toluenesulfonyl ethylamine　*N*-甲苯磺酰乙胺　$CH_3C_6H_4SO_2NHC_2H_5$
p-toluenesulfonyl hydrazide　对甲苯磺酰肼〔发泡剂〕
toluenesulfonyl methylamine　*N*-甲苯磺酰甲胺　$CH_3C_6H_4SO_2NHCH_3$
toluenesulfonyl methylaniline　*N*,*N*-甲苯磺酰基甲基苯胺　$CH_3C_6H_4SO_2N(CH_3)(C_6H_5)$
p-toluenesulfonyl semicarbazide　对甲苯磺酰氨基脲〔发泡剂〕
toluenesulfonyl toluidine　*N*-甲苯磺酰甲苯胺　$CH_3C_6H_4SO_2NHC_6H_4CH_3$
toluene tetrachloride　四氯甲苯　$CH_3C_6HCl_4$
toluene-*ω*-thiol　对甲苯硫酚
α-toluenethiol　*α*-苯甲硫醇
toluene tribromide　三溴甲苯　$CH_3C_6H_2Br_3$
toluene value(=toluene equivalent)　甲苯值
N-toluene-*N*′-xylene-*p*-phenylenediamine　*N*-甲苯基-*N*′-二甲苯基对苯二胺〔防老剂〕
toluenmeter　芳烃分析计
toluenone　2-甲基-2,4-环己二烯酮
toluenyl　亚苄基
toluic acid(=tolyl acid; toluyl acid)　甲基苯甲酸
toluic aldehyde　甲苯甲醛　$CH_3C_6H_4CHO$
toluic anhydride　甲(基)苯甲(酸)酐
toluic nitrile　甲苯(甲)腈
toluidide　*N*-某酰基甲苯胺　$CH_3C_6H_4NHCOR$
toluidine　甲苯胺
toluidine blue　甲苯胺蓝
toluidine hydrochloride　甲苯胺盐酸盐
toluidine maroon　甲苯胺紫红；甲苯胺栗红
toluidine red　甲苯胺红
toluidine sulfate　硫酸甲苯胺
toluidine toner　甲苯胺色原；甲苯胺色料
toluidino　甲苯氨基〔邻、间或对位〕　$CH_3C_6H_4NH-$
toluido(=toluino)　甲苯氨基　$CH_3C_6H_4NH-$
toluino(=toluido)　甲苯氨基

toluiquinone　甲苯醌
α-tolunitrile　*α*-苄基氰
tolunitrile　①〔专指〕苄基氰　$C_6H_5CH_2CN$②〔通指〕甲苯基氰　$CH_3C_6H_4CN$
toluol　甲苯　$C_6H_5CH_3$
toluol sulfonamide　甲苯磺酰胺
toluol sulfonanilide　甲苯磺酰苯胺
toluoyl(=toluyl)　甲苯酰〔邻、间或对〕　$CH_3C_6H_4CO-$
toluphenazine　甲基吩嗪
tolupiaselenole　甲基苯并硒二唑　$C_7H_6N_2Se$
toluquinoline　甲(基)喹啉　$C_{10}H_9N$
toluquinone(=2-methylbenzoquinone)　甲苯醌；2-甲基醌　$O=C_6H_3(CH_3)=O$
tolu-resinotannol　妥卢树脂鞣醇
toluresitannol　妥卢香醇
tolurin　维生素 B_1
toluyl(=toluoyl)　甲苯酰
toluyl acid(=tolyl acid; toluic acid)　甲基苯甲酸
toluyl aldehyde　甲苯甲醛
toluylate　①甲苯(甲)酸盐②甲苯酸酯
toluyl azo-*β*-naphthol　甲基苯甲酰偶氮 *β*-萘酚
toluylene　①甲代亚苯基　$C_6H_3CH_3=$②茋；二苯乙烯　$C_6H_5CH=CHC_6H_5$
toluylene blue　甲苯蓝
toluylene diamine　甲代苯二胺　$C_7H_{10}N_2$
toluylene diamine indophenol　甲苯二胺靛酚
toluylene diisocyanate　甲苯二异氰酸酯
toluylene dimaleimide　甲苯二(顺丁烯二酰亚胺)
toluylene hydrate　水合茋；苯(基)苄(基)甲醇
toluylene orange　甲苯橙
toluylene red　中性红
toluylene urea　亚甲苯基脲
m-toluylformic acid　间-甲苯酰基甲酸
toluylic acid　苯乙酸
m-toluyl peroxide　过氧化间甲苯酰
tolyantipyrine　甲苯基安替比林
tolyl　甲苯基〔邻、间或对〕　$CH_3C_6H_4-$
p-tolyl acetaldehyde　对甲基苯乙醛
p-tolyl acetate　乙酸对甲苯酯
tolyl-acetic acid　甲苯乙酸　$CH_3C_6H_4CH_2COOH$
tolyl acid(=toluyl acid;toluic acid)　甲基苯甲酸
tolyl acrylic acid　甲基丙烯酸；甲基肉桂酸
tolyl alcohol　甲苯基醇
tolyl aldehyde　甲苯醛
p-tolyl aldehyde　对甲基苯甲醛
o-tolyl alkyl ketimine　邻甲苯基烷基甲亚胺
tolylation　引入甲苯基
tolylazo-2-naphthol　甲苯偶氮-2-萘酚

o-tolyl-biguanide 邻甲苯基二胍
tolyl bromide 甲苯基溴
tolyl carbinol 甲苯基甲醇
tolyl chloride 甲苯基氯
p-tolyl diethanolamine 对甲苯二羟乙基胺〔乳化剂〕
tolylene ①(=toluylene)甲代亚苯基②亚甲代苯基—$CH_2C_6H_4$—③亚苄基 $C_6H_5CH=$
α-tolylene(=benzylidene) 苯亚甲基 $C_6H_5CH=$
tolylene diamine ①甲苯二胺 $CH_3C_6H_3(NH_2)_2$ ②亚苄基二胺 $NH_2CH_2C_6H_4NH_2$
tolylene diisocyanate 甲苯二异氰酸酯
2,4-tolylene diisocyanate dimer 2,4-甲苯二异氰酸酯二聚体
β-tolyl ethyl alcohol 甲基-β-苯乙醇
tolyl ethyl ether 甲苯基乙基醚
p-tolyl ethyl ether 对甲基苯乙醚
tolylhydrazine 甲苯肼
tolyl hydrazine hydrochloride 盐酸甲苯基肼
tolylhydroxylamine 甲苯胲
tolyl isorhodanate(=tolyl isothiocyanate) 异硫氰酸甲苯酯
tolyl isothiocyanate(=tolyl isorhodanate) 异硫氰酸甲苯酯
tolylmercuric chloride 甲苯基汞化氯 $CH_3C_6H_4HgCl$
p-tolyl methyl carbinol 对甲苯基甲基原醇
tolyl methyl ether 甲苯基甲基醚
m-tolyl methyl ether 间甲苯基甲醚
p-tolyl methyl ether 对甲苯基甲醚
tolyl mustard oil 甲苯基芥子油；异硫氰酸甲苯酯 $CH_3C_6H_4NCS$
tolyloxy(=toloxy) 甲苯氧基
p-tolyl phenylacetate 苯乙酸对甲苯酯
m-tolyl phenyl ether 间甲苯基苯醚
p-tolyl phenyl ether 对甲苯基苯醚
tolyl phenyl ketone 甲苯基苯基(甲)酮
N-o-tolyl-N'-phenyl-p-phenylenediamine N-邻甲苯基-N'-苯基对苯二胺〔防老剂〕
2-p-tolyl propene 2-对甲苯基丙烯
α-p-tolyl propionaldehyde α-对甲苯基丙醛
tolyl sulfone 甲苯基磺内酯
tolylsulfonyl(=tosyl) 甲苯磺酰基 $CH_3C_6H_4SO_2—$
p-(p-tolyl-sulfonylamide)-diphenylamine 对(对甲苯基亚磺酰氨基)二苯胺
tolylthiourea 甲苯基硫脲
tolylurea 甲苯基脲
tolypomycin 颗粒霉素
tomatidine 番茄碱；番茄苷配基
tomatin 番茄素

tomatine 番茄(碱糖)苷
tomato 番茄；西红柿〔俗〕
tomato flavour 番茄香精
tomato oil 番茄油
tomato pomace 番茄渣
tomato pulp 番茄纸浆
tomato-seed oil 番茄子油
tombak 铜锌合金
tomenin 毛樱桃苷
tomentosine(=othosenine) 奥索(千里光)碱
Tommasi lead-refining process 托马斯炼铅法
tomographic scintiphoto 断层闪烁照片
tomography 断层X射线照相法；断层显像法
tomography scanner 层析X射线扫描仪
tomoscan 断层扫描图谱
tonalide 吐纳麝香
tonalite 英云闪长岩
tonal paper 扩音纸
tonco beans(=tonka beans) 零陵香豆；黑香豆
tone ①音调②色泽；彩色③香韵；香调
tone colour 色调
tone intensity 彩色强度
tone-in-tone effect (不同类型纤维)同色深浅效应
tone light 色调浅
tone off 色泽渐浅直至消失
tone process 施彩法
toner 有机调色剂〔不含无机粒料的有机颜料〕
toner brown 色淀棕
toner pigment 调色颜料；有机颜料
toner yellow 色淀黄
tone scale 色调梯尺；色域
tongs 钳(子)
tongue 舌(子)
tongue and groove 榫槽面
tongue and groove joint 舌套式接头
tongue-and-groove labyrinth 舌槽迷路；密封迷路
tongued and grooved flanges 舌槽法兰；密封法兰
tongued and grooved surface 舌槽面；雌雄面
tongue flange 榫面法兰
tongue-groove seal face 准槽密封面
tongue of flame 火苗；火舌
tongue tear 切口撕裂
tonic ①强壮剂；补药②实的；增强的
tonicity 紧张性；张力
toning 调(匀颜)色
toning agent 调色剂
tonite 徒那特〔炸药〕
tonka 香豆

tonka bean camphor 黑香豆脑；香豆素
tonka beans 零陵香豆；黑香豆
tonka beans extract 黑香豆浸剂
tonkalide(=γ-hexalactone) γ-己内酯；香豆内酯
tonka resin 黑香豆树脂
tonka resinoid 黑香豆香树脂
tonnage 吨数
tonnage oxygen 工业用氧
tonometry 音调测量学
tonquim 零陵香；黑香豆
tonquinol 奎诺尔〔麝香代用品〕
tonquitone 东进麝香
Tooke gauge 托克测厚仪
tool ①工具；用具②(车)刀
tool bar 刀杆
tool bogie 工具车
tooling 手工烫花
tooling hole 工具孔
tool mechanism 工具机构
tool resistance 工具抗力〔机工〕
tool sharpening 磨锐刀具
tool steel 工(用)钢
toonalol 香椿脑
toonin 香椿素
toonol 香椿醇
toothed belt 锯齿形带
toothed wheel 齿轮
tooth paste 牙膏
tooth powder 牙粉
tooth rake 齿耙
tooth soap 牙用皂
tooth (type) coupling 齿式联轴器
toot poison 马桑毒苷；毒空木毒
top ①顶②上面③盖顶④截去顶端⑤最高的；最大的⑥陀螺
top-action hydraulic press 顶力水压泵
top-and-bottom press 压盖底机
top application 顶施
topaz 黄玉；黄精
topaz mine 黄玉矿
topazolite 黄榴石
top buffing 表面轻磨
top cap 顶盖
top cavity 上模腔
top-chrome dyeing 铬盐后处理染色；后铬媒染色；套铬媒染色
top-coat 面漆；面涂
topcoat delamination 面漆脱层

topcoat holdout 面漆不渗性
topcoating ①表面涂饰②表面涂料
topcoat paint 面漆
top color(=face color) ①面色②直视色
top dead centre 上死点
top deck(=main deck) 顶甲板；主甲板
top discharge 上出料(式)
top-down 由上而下的技术
top dressing 顶肥
top-driven centrifuge 顶动离心器
top feed(=top feeding) 顶部加料
top feeding(=top feed) 顶部加料
top felt 上毛布
top filler 炉顶加料器
top filling 炉顶加料
top finishing 表面涂饰
top firing 顶部加热
top force 上模塞；上阳模
top grinding 平磨
top guide 顶导架
tophan box 离心罐
tophan pot 离心罐
topical application 局部施用
topical magnetic resonance 定域磁共振
topical nuclear magnetic resonance spectrometry 定域核磁共振波谱法
top ignition oxygen index tester 顶燃式(需)氧指数测试仪
topiramate 托吡酯〔药〕
top lacquer 面漆
top loading 顶部加料；顶部装料
top-loading balance 上皿式天平；顶部加载天平
top-loading washing machine 顶装式洗涤机
topnotch ①最高质量的；第一流的②顶点
top note 顶香；头香
topochemical 局部化学的
topochemical polymerization 拓扑化学聚合
topochemical reaction 局部化学反应
topochemistry ①局部化学②拓扑化学
top of burner 灯头
top of can 听盖；桶盖
top of piston 活塞顶
topogon lens 弯月形透镜；小孔径宽视场镜头
topographical ①地形的②地形(学)的
topographical map 地形图
topography ①地形②地形学
topography contrast 形貌反差
topography copy 样品复型

topological entanglement　拓扑缠结
topological framework　拓扑结构
topological matrix　拓扑矩阵
topomineralogy　区域矿物学
topotactic polymerization　局部有规立构聚合
topotactic reaction　局部规整反应
topotecan　拓扑替康〔药〕
top overhaul　初步大修
topped　拔顶的
topped crude(=topped oil)　拔顶原油〔蒸去了轻油后所剩的原油〕
topped crude petroleum　拔顶原油
topped oil(=topped crude)　拔顶原油〔蒸去了轻油后所剩的原油〕
topped residuum　拔顶残油
top petrol tank　顶汽油箱
topping　①去梢②拔顶；蒸去轻油③去头
topping heater　拔顶加热器
topping of crude oil　原油拔顶；原油的直接蒸馏
topping of oil　石油拔顶
topping pipe still　拔顶管馏器
topping plant　拔顶厂；初馏装置
topping plant exchanger　拔顶车间换热器
topping plant pipe still　拔顶车间管馏器
topping processing　拔顶过程；蒸出轻馏分
topping reflux　塔顶回流
topping still　拔顶塔；粗馏塔
topping tower　拔顶塔
topping unit　拔顶装置；(常压)蒸馏装置
topping up　上油；添油
topping with gum　涂胶；上胶
top plate　顶板
top pressure　最高压力
top pressure plate　胎面样板
top-quality rubber　优质橡胶
top reflux　塔顶回流
top roll　上轴
top roller　顶辊
top roller arbor　皮辊芯子
tops　顶馏分；最初馏分；轻油
top seal　顶盖密封
top season　顶光；盖面光
top-side black　(水线以上)舷侧黑漆
topside painting　船壳外板涂装
top sizing　①上涂料②表面施胶
topsoil　①大气污染②耕作层土壤
top speed　最高速率
top-suspended basket centrifuge　上悬式离心机
top suspension　上悬式
top suspension centrifuge　上悬式离心机
top tank air　罐顶空气；油罐中最上层气体
top temperature　①塔顶温度②最高温度
top tower control　塔顶(温度)调节
top tray　顶板；顶盘
top up　上油；填油
top view　俯视图；顶视图
top wire　上网
top with gum　上胶；涂(橡)胶
toramin　三氯丁基丙二酸铵
torasemide　托拉塞米〔药〕
torbane　块煤
torbanite　块煤；藻烛煤
torbernite　铜铀云母
torch　火炬；炬管
torch assembly　火炬装置
torch atomizer　火炬喷射器
torch igniter　火炬点火器
torching　喷出火焰
torch lamp　火炬灯
torch oil　火炬油
toremifene　托瑞米芬〔药〕
toriconical closure　带折边锥形封头
toriconical head　(有)折边(的)锥形封头；准锥形封头
toriconical reducer　带折边锥形变径段
torilene　窃衣烯
tori seed oil　芥子油
torispherical head　碟形封头
tormentil tannic acid　委陵菜单宁酸
tornado dust collector　龙卷风式集尘器
torn grain　毛刺沟痕
torography　①地形②地形学③表面形状
toroid detector　环形检测器
Toronto arc　多伦多电弧
torpex　铝末混合炸药
torque　转(力)矩；扭(力)矩
torque breakdown test　转矩破裂试验
torr　托；毫米汞柱(压力)〔1torr＝1 mm Hg〕
torrefaction　焙烧；烘；烤
torrefying　焙烧
torreyal　榧烯醛
torreyene　榧烯
torreyol　香榧醇
torsilastic (rubber) spring　防扭振橡胶弹簧
torsiometer　扭力计
torsion　①扭转；扭力；扭矩②挠曲
torsional angle　扭转角

torsional bending　扭力弯曲
torsional braid　扭辫
torsional braid analysis(TBA)　扭辫分析
torsional braid tester　扭力带试验机
torsional creep　扭曲蠕变
torsional deflection　扭转变形
torsional deformation　扭(转)形变
torsional dynamometer　动力扭力计
torsional effect　扭转效应*
torsional elasticity　扭弹性
torsional flow　扭转流动
torsional laminar displacement　扭转层状位移
torsional load　扭力负载
torsionally oscillating rheometer　扭转振荡流变仪
torsional method　转扭法〔测黏度方法〕
torsional modulus　抗扭模数；刚性模数
torsional moment　扭(力)矩
torsional oscillation　扭(转)振(动)
torsional pendulum　扭摆
torsional relaxation　扭力松弛
torsional resistance　抗扭强度
torsional rigidity　扭转刚度
torsional rigidity modulus　扭曲刚性模量
torsional shear adhesion tester　扭转剪切黏合试验机
torsional shear strength　扭转剪切强度
torsional stiffness　扭转刚性度
torsional strain　扭应变
torsional strength　抗扭强度
torsional stress　扭转应力；扭应力〔物〕
torsional suspension　扭力式橡胶弹簧
torsional tension　扭应力
torsional test　扭力测试；扭力试验
torsional vibration　扭转振动
torsional vibration damper　防扭振器
torsional vibration rheometer　扭振流变仪
torsional viscometer　扭力黏度计
torsional viscosity　扭力(法)黏度
torsional yield strength　抗扭强度极限
torsion angle　扭转角*
torsion balance　扭力天平
torsion bar　扭力杆
torsion bearing　扭转轴承
torsion braid　扭辫
torsion braid analysis(TBA)　扭辫分析
torsion-braid analyzer　扭带分析仪〔对溶剂固着带加热，分析其扭转振动〕
torsion-cone consistometer　扭力-圆锥稠度计
torsion constant　抗扭模数
torsion couple　扭力偶
torsion dynamometer　扭力计
torsion galvanometer　扭转电流计
torsion impact test　扭力冲击试验
torsion joint　扭合接头
torsion meter　扭力计
torsion modulus　扭转模量
torsion moment　扭力矩
torsion pendulum　扭摆
torsion rod spring　扭杆弹簧
torsion spring　扭力弹簧
torsion suspension　防扭振橡胶
torsion tester　扭力试验仪
torsion thermobalance　扭力热天平
torsion-type balance　扭力天平；扭秤
torsion vibration　扭转振动
torsion visco(si)meter　扭力黏度计
torsion wire　扭力丝
tortuosity factor　曲折因数
torula yeast　酿母
torulin　①(=vitamin B_1)维生素 B_1 ②圆酵母素
torun metal　锡铜轴承合金
torus　纹孔塞；纹孔托
tosate　对甲苯磺酸基团
tossing tub　摇洗盆
tosyl(=tolylsulfonyl)　甲苯磺酰基　$CH_3C_6H_4SO_2-$
tosylate　甲苯磺酸盐(酯)
tosylation　甲苯磺酰化
tosyl cellulose　甲苯磺酰纤维素
tosylester of cellulose　纤维素对甲苯磺酸酯
tosyl group　甲苯磺酰基
total　①总数②总的；全部的③总计
total acidity　总酸度
total acid number(TAN)　总酸值
total acid value　总酸值
total adsorption capacity　①吸着剂的总活性②吸着剂的总活度
total alkali　总碱
total analysis　全分析
total angular momentum　总角动量
total area　总面积
total aromatics　总芳香物含量
total available alkali　总有效碱含量
total bag sampling system　总量取样法
total base number(TBN)　总碱值
total board thickness　印制板总厚度
total bond order　总键级
total calorific value　总发热量；总热值

total carbon 总碳量
total carbon analyzer 总碳分析仪
total cellulose 全纤维素
total chlorine free bleaching(TCF) 全无氯漂白
total color difference 总色差
total column volume 柱总体积
total combustion method 完全燃烧法
total condenser (完)全(冷)凝器
total consumption atomizer 全消耗型喷雾器
total consumption burner 全消耗型喷灯；全消耗型燃烧器
total content 总含量
total conversion 总转化率
total correlation coefficient 全相关系数*；总相关系数
total correlation spectroscopy 总相关谱
total current 总电流
total daily production 总日产量
total dissolved solid 总溶液固体量
total dynamic head 总压头
total effect 总效应
total efficiency 总效率
total emissive power 总发射能力
total encapsulation 全包膜；全包封
total energy balance 总能量衡算
total energy method 全能量法
total exchange capacity 总交换容量
total feed 总进料
total half-life 总半衰期
total hand value 织物综合手感值
total hardness 总硬度
total hardness of water 水的总硬度
total head 总高差
total heat(=integral heat) ①总热(量)②热函(数)；焓③积分热；变浓热
total heat content 总热函
total heat history 总热历程
total heating value 总热值
total heat of solution 总溶解热
total heat of steam 水蒸气总热
total heat transfer 总热传递
total hydrocarbon analysis 总烃分析
total hydrocarbon feed 混合烃原料
total hydrocarbon peak 总烃峰
total immersion jig 全浸式卷染机
total immersion thermometer 全浸温度计
total infrared absorbance reconstruction chromatogram 红外总吸光度重建色谱图
total internal reflection fluorescence microscopy 全内反射荧光显微镜
total ion chromatogram 总离子流色谱图
total ion current 总离子流
total ion current collector 总离子流收集器
total ionic chromatogram 总离子流色谱图
total ionic strength adjustment buffer(TISAB) 总离子强度缓冲剂
total ionization 总电离
total ionization cross section 总电离截面
total ion monitoring 总离子监测
total isolation method 全数分离法
totalizer 累加器；加法器；加法计算装置
total latent heat 总潜热
totalling 加总；总和
total liquid volume 液相总体积
total load 总负荷
total loss lubrication 全损耗性润滑
total luminescence spectrum 总发光光谱
totally cryostatic stabilized conductor 全低温稳定(超)导体
totally enclosed 全封闭式(的)
totally porous microbead 全多孔型微珠
totally porous packing 全多孔型填充剂*
totally porous support 全多孔载体
total make(=total output) 总出量；总产量
total-mixing cycle 总混炼周期〔包括装料排胶时间〕
total moisture 总水分
total molal heat content of gas 气体的总摩尔热函
total nitrogen analysis 全氮分析
total normal force 总法向力；总正交力
total obscuring power 总蒙蔽本领
total olefines 总烯含量
total organic carbon 总有机碳
total osmotic volume 总渗透体积
total output(=total make) 总出量；总产量
total overall reaction 全部反应
total oxygen demand 总需氧量
total percent aromatics 总芳香族百分率
total percent of ash 总灰分百分数
total phosphorus of soil 土壤中全磷
total plutonium 总钚〔所有钚同位素总和〕
total porosity 总孔率
total pressure 总压力
total pressure difference 总压力差
total quantum number 总量子数；主量子数
total rare earth(TRE) 总稀土(元素)
total recovery ①收回总量②收回总糖量
total reflection 全反射

total reflection refractometer 全反射折射计
total reflection X-ray fluorescence analysis 全反射 X 射线荧光分析
total reflection X-ray fluorescence spectrometer 全反射 X 射线荧光光谱仪
total reflectivity 全反射率
total reflux 全回流
total reflux operation 全回流操作
total reflux partial takeoff heads 全回流部分接取头
total residue on evaporation 蒸发总残渣
total resolution efficiency 总分辨效率
total response chromatogram 总响应色谱图
total retention time 总保留时间
total salt concentration 总盐浓度
total solid content 总固物含量
total solid matters 总固体物
total solids 总固体
total solids content 总固体含量
total solids test 总固体测定
total solubility parameter 总溶解度参数
total soluble matters 总溶质
total spectrum direct reading 全谱直读
total static head 总静力高差；总静压头
total sulfur 总硫量
total sulfur dioxide 总二氧化硫量
total sulfurous acid (=total sulfur dioxide) 总亚硫酸
total sum of squares 总平方和
total suspended matter 总悬浮物
total suspended substance 总悬浮物
total synthesis 全合成*
total systematization 总体配套系统化
total thickness 涂膜总厚度
total titratable alkali 总滴定碱
total transmittance 总透射比；总透射率
total treatment 完全处理
total unsaturates 总不饱和物
total vaporization method 完全弧烧法
total volatile alkaline 总挥发碱
total volatile content 总挥发分
total volatile organic compounds 总挥发性有机物
total volume 总体积
total water content 总水分
total weight 总重(量)
total weight loss 总失重
totaquina 金鸡纳混碱
totaquine 金鸡纳混碱制剂；全奎宁
totara tree 桃拓罗汉松
totarene(=rimuene) 桃柘烯；芮木烯；芮木泪柏烯

totarolone 桃拓酮
tote bin 运搬箱
tote box 运搬箱
totomycin 妥妥霉素；盐酸四环素
touch 接触；试验
touch dry 指触干燥〔涂料〕
touch roll 接触辊
touch sensitive glove 超薄手套
touchstone 试金石；标准
touch-up 补漆〔局部修补〕
touch-up coating (局部)修补用漆
tough ①(坚)韧的②黏着的
tough-brittle transition 韧脆转变
tough drum 油桶；槽形桶
toughened agent 增韧剂
toughened filter paper 硬质滤纸
toughened polystyrene(TPS) 韧性聚苯乙烯
toughener 增韧剂
toughening agent 增韧剂
toughening oil 黏稠油
tough film 坚韧膜
tough hardness 韧硬度
tough metal 韧性金属
toughness ①韧度②韧性③柔韧性；弹性；弯曲性
toughness and tenacity test 强韧性试验
toughness of fiber 纤维的强韧度
toughness test 韧性测试；韧性试验
tough rubber 硬橡胶
Toulet's solution 碘化汞钾的浓溶液
tourie(=tourill) (吸收)坛
tourill(=tourie) (吸收)坛
tourmaline 电气石
tourmaline gauge 电气石计
tourmaline pincette 电气石钳
tourmalinite 电气石
tourneforcine 紫丹碱
tournesol(=orchil) 苔藓红素；苔藓色素；石蕊素
tou-saghiz rubber 山胶草橡胶
tow ①麻屑②丝束；烟束
tow canning machine 丝束盘条器；丝束装筒器
tow converting unit 丝束叠丝装置
tower 塔
tower acid 塔酸
tower bottom 塔底残油
tower concentrator 塔式浓缩器
tower counter-flow 塔逆流
tower dryer 塔式干燥器
tower evaporation 塔式蒸发

tower evaporator 塔式蒸发器
tower filling 填塔料
tower gas cooler 塔式气体冷却器
tower hold-up 填塔液
tower mill 塔式粉碎机
tower packing 填塔料
tower scrubber 洗塔器
tower shell 塔壳体
tower skirt 塔裙；塔下部侧缘
tower sludge 塔内泥
tower still 塔式蒸馏釜
tower system 塔组
tower top temperature 塔顶温度
tower tray 塔盘；塔板
tower wall material 塔壁料
tower washer 塔式洗涤器
tower washing 塔式洗涤
town gas 民用煤气；民用燃气
town mains 城市(煤气)总管
Townsend cell (for saustic soda) 唐森德(制苛性钠)电池
tow piddler ①丝束运送器②丝束运送工
tow-stapling machine 丝束切断机
tow-to-sliver break spinning 丝束拉断成条纺纱(法)；丝束剪切成条纺纱法
toxacarol 灰毛豆酚
toxaphene(=octachloro-camphene) 八氯莰烯；毒莰烯
toxic ①(有)毒的；毒性的②毒剂；毒物
toxicant 毒(素)；毒物
toxicarol 毒灰叶酚
toxic dose 中毒(剂)量
toxic equivalent 毒当量
toxic fume 毒烟；毒气
toxicity test 毒性试验
toxicodendron 干毒漆树叶
toxicological analysis 毒物分析
toxicological detection 毒性检测
toxic pollutants 有毒污染物；毒性污染物
toxic substances 有毒物质
toxic waste disposal 毒性废水处理
toxiferine 毒马钱碱
toxins 毒素
toxisterol 毒甾醇
toy balloon 玩具气球
T-piece (=T-tube) T(形)管
T-pipe T(形)管
trabuk metal 锡镍锑铋合金
trace ①描；画②示踪；探索③痕量；微量④迹
traceability 溯源性
trace amount 痕量
trace analysis 痕量分析
trace catcher 痕量(成分)捕集剂
trace chemistry 痕量化学*
trace colloid 示踪胶体
trace components 痕量组分
trace constituent 痕量成分*
trace detection 痕量检测
trace element 痕量元素
trace element pollution 痕量元素污染
trace enrichment 痕量富集
trace impurity 微量杂质；痕量杂质
trace level 痕量级
trace metal 痕量金属；微迹金属
trace metal analysis of environment 环境中的痕量金属分析
trace of impurity 微量杂质
trace peroxidic impurity 微量过氧化物杂质
trace pollutant 痕量污染物
tracer 示踪剂*；示踪物
tracer analysis 示踪分析
tracer atom (=tagged atom) 标记原子；加有标记的原子；示迹原子；显迹原子
tracer bullet 示踪(子)弹；曳光(子)弹
tracer chemistry 示踪化学
tracer composition 曳光剂
tracer diffusion 示踪原子扩散
tracer element 示踪元素
tracer experiment 示踪实验；显迹实验
tracer-free 无指示剂的；无示踪物的
tracer isotope 示踪同位素
tracerlab 同位素示踪物实验室；放射化学实验室
tracer-labelling 同位素示踪物标记
tracer level 示踪量；痕迹量
tracer method 示踪法
tracer milling machine 靠模铣床
tracer needle 触针；描形针
tracer pulse chromatography 示踪脉冲色谱法
tracer scale 示踪量
tracer scale extraction 示踪量萃取
tracer shell 示踪(炮)弹；曳光(炮)弹
tracer study 示踪研究
tracer technique 示踪技术
tracer test of labelled drug 标记药物示踪实验
tracer yarn 标志纱
trace sediment 痕迹沉淀；痕迹沉积物
trachelanthamidine 颈花脒
trachelanthine 颈花碱

trachelogenin 络石配基
tracheloside 络石糖苷 $C_{36}H_{56}O_{18}$
trachreid 管胞
trachyte 粗面岩
tracing cloth 摹图布
tracing paper 摹图纸
tracing plan 底图
tracing point ①描迹针②针刺印迹；针迹〔用针沿图形花样剌小孔而后复印原图〕
tracing wheel 刻痕轮；划线轮；描绘轮
track ①径②径迹；跟踪
track aging 径迹老化
track annealing 径迹退火
track etch analysis 径迹蚀刻分析(法)
track etching 径迹蚀刻
track etching dosimeter 径迹蚀刻剂量计
track etch method 径迹蚀刻法
tracking accuracy 跟踪准确度
tracking errors 跟踪误差；循迹误差
tracking (heat) 电弧电阻(热)；伴热
tracking index 抗电弧径迹指数
tracking path 电弧径迹
tracking test 电弧径迹试验
track-sensitive target 径迹敏感靶
track tyre 拖拉机车胎；链式胎
traction 引力
traction force 牵(引)力
traction gas producer 牵机煤气发生器
tractive force 引力
tractor fuel 拖拉机燃料
tractor fuel oil 拖拉机燃料油
tractor oil 拖拉机油；拖拉机润滑油
tractor vapourizing oil 拖拉机挥发油
trade (brand) name 商标名；商品名
trade effluent 工业排出物；工业废水
trademark 商标
trade sales coating ①零售涂料；小包装涂料②(=architectural coating)建筑涂料〔美国涂料分类的一大类〕
trade sales paint ①零售漆；小包装漆②(美国)建筑漆
traditional paint 传统油漆
traffic line paint 交通线涂料
traffic marking paint 马路标志漆；马路划线漆
traffic noise 交通噪声
traffic paint 路标漆；路线漆
tragacanth 黄芪胶；龙胶
tragacantha shrub 胶黄芪
tragacanth gum 黄芪胶；龙须胶
tragacanthin 黄芪质；黄芪糖

tragasol gum 角豆树胶
tragon 角豆树胶
tragon seed gum 角豆子胶；酸豆胶
trail 拖尾
trail formation 拖尾
trailing ion 尾随离子
trails〔复〕 沟；刻槽
train ①列②导火线
training 整枝
training set 训练集
train oil 鲸骨脂
trajectory 轨迹；迹线
tram 加捻纬丝
tramadol 曲马多〔药〕
Tramex process 特拉美克斯过程〔从低酸度浓氯化锂溶液中用叔胺萃取分离锕系和锕系元素的过程〕
tramp 杂质
tramp material 外来杂质；混入物
tramp metal(=pick-up metal) 混入金属
tranexamic acid 氨甲环酸〔药〕
tranmission topography 透射形貌学〔晶体〕
tranquil flow 平静流动
tranquilizer 安神药〔药〕
trans- 〔拉丁字头〕 ①反(式)②超；跨；过；(以)
transacetalation(=exchange of acetal linkage) 缩醛(链)转移作用
transacetylation 乙酰转移作用
trans-acid 反式(脂肪)酸
trans-aconitic acid 反乌头酸
transactinide element 锕系后元素*
transactinium elements 锕后元素
transacylation 酰基移转(作用)
trans addition 反式加成*
transaldimination 转醛亚胺作用
transalkylation 烷基转移(作用)
transalkylation reaction 烷基转移反应
transamidation 转酰氨基作用
transaminase(=aminopherase) 转氨酶
trans and *gauche* 反式与旁式
transanhydrisation 酐交换作用
transannular 跨环
transannular bond(=transannular bridge) 跨环键；跨环桥〔指碳环或杂环的〕
transannular bridge(=transannular bond) 跨环桥；跨环键〔指碳环或杂环的〕
transannular effect 跨环效应
transannular hydrogen effect 跨环氢效应
transannular hydrogen transfer 跨环氢传递

transannular insertion　跨环插入*
transannular interaction　跨环相互作用*
transannular link(=transannular bridge)　跨环键；跨环桥
transannular linkage(=transannular bridge)　跨环键；跨环桥
transannular migration　跨环移位
transannular peroxide　跨环过氧化物〔指过氧基架在环上的〕
transannular reaction　跨环反应
transannular rearrangement(=transannular transposition)　跨环重排
transannular strain　跨环张力*
transannular transposition(=transannular rearrangement)　跨环重排
trans-azobenzene　反(式)偶氮苯
trans-azomethane　反(式)偶氮甲烷
trans-butadiene-piperylene rubber　反(式)丁二烯戊二烯橡胶
trans-2-butene　反(式)-2-丁烯
transcalent　透热的
transcalifornium element　锎后元素
transcolumn effect　传递柱效应
trans-communic acid　反式欧柏酸；反式瘿柏酸
trans-compound　反式化合物
trans-configuration　反式构型
transconformation　构象转移(现象)
transcrystallic cracking (=inner crystal crack)　穿晶裂纹
transcrystalline　横穿晶的；跨晶的
transcrystalline cracking　穿晶裂纹
transcrystallinity　跨晶结晶度
transcrystallization　穿晶(现象)；横结晶(作用)
trans-crystallization structure　晶变结构
transcurium element　锔后元素
trans-cyanoethylation　氰乙基转移作用
trans-derivative　反式衍生物
trans-double bond　交叉双键
transducer(s)　(能量)转换器
transductant　转导
trans effect　反位效应
trans elimination　反式消除*
trans-erythro〔词头〕反-赤藓式
transesterification　酯交换*
transesterification catalyst　酯基转移催化剂；酯交换催化剂
transesterification reaction　酯交换反应
transesterification reactor　酯交换反应釜
transesterification-type polymerization　酯基转移型聚合(反应)；酯交换型聚合(反应)
transetherification　醚交换；醚键转移(作用)

transfection　转染
transferability　转移能力；可移植性
transfer agent　转移剂
transferase　转移酶
transfer car　运搬车
transfer cell　转移池
transfer coating　移膜涂饰
transfer coefficient　传递系数
transfer dish　输送盘
transfer efficiency　①转移效率②传送(涂敷)效率
transference　①迁移②搬运；输送③输电④让与
transference apparatus　(离子)迁移器
transference cell　迁移电池
transference number　迁移数
transference tube　(离子)迁移管
transfer finishing　移膜涂饰(法)
transfer function　传递函数
transfer gantry　输送支架
transfer hydrogenation　转移氢化*
transfer hydrogenolysis　转移氢解
transfer-in box　转入箱
transfer lime　输送管；传送线
transfer-lime cooler　输送线冷却器
transfer limes　输送线
transfer-lime temperature　输送线温度
transfer-line exchanger　输送线换热器
transfer method　迁移(取样)法
transfermium element　超镄元素
transfer mould　传递模型模具
transfer moulding　传递模塑；压铸
transfer number　迁移数
transfer of gas　气体输送
transfer of heat　传热；热传递
transfer of populations via double resonance　双共振布居转移
transfer padding　转移轧染；转移轧液
transfer passage　输送通道
transfer pipe　输油管
transfer pipet(te)　移液吸(移)管；移液管
transfer process　移膜法
transfer pump　输送泵
transfer pumping unit　输送泵车间
transfer rate　输送率；输送速度
transfer ratio　转移系数；转换比
transfer reaction　(链位)转移反应
transferred electron generator　电子传递发生器
transferred heat　传递热
transfer resistance　转移电阻

transferring oil　输送油
transferritin saturation　血清运铁蛋白饱和度
transfer roller　①输送机；传送辊②送墨辊③转印辊
transfer speed　转移速度
transfer swith　换向开关
transfer system　传送系统
transfer unit　传递单位
transformation　①转变②换变；蜕变③变化④变换⑤相变
transformation coefficient　变换系数
transformation constant　变换常数
transformation of energy　能量转化
transformation range　转变范围
transformation reactions　变态反应
transformation region　变色区；变色范围
transformation series　(放射)蜕变系
transformation theory　变换论
transformation to linearity　线性化变换
transformed plywood　改形胶合板
transformer　变压器
transformer oil　变压器油；绝缘油
transformer winding　变压器线圈
transforming　转变；重排
transgenation　基因突变
trans-hexen-2-al　反式己烯-2-醛
trans-hexen-2-al-diethyl acetal　反式己烯-2-醛二乙缩醛
trans-2-hexen-1-ol　反式-2-己烯-1-醇
transient boiling range　过渡沸腾区
transient dipole moment　瞬间偶极矩
transient electroanalytical method　暂态电分析方法
transient equilibrium　瞬间平衡；暂时平衡
transient flow(=transition flow)　瞬变流动；不稳流动
transient ionic matrix effect　瞬时离子基体效应
transient isentropic tangential modulus　瞬态等熵切向模量
transient isothermal tangential modulus　瞬态等温切向模量
transient method　暂态法*
transient response　①瞬态特性②瞬变响应
transient stress-energy modulus　瞬态应力能模量
transient structure(=secondary structure)　暂时结构；二次结构〔炭黑〕
transient temperature　瞬变温度
transimission efficiency　传输效率
trans influence　反位影响
trans-interchange　反位转移作用
trans-isomer　反式异构体*
trans-isomeric　反式异构的

trans-isomeride　反式异构体
trans-isomerism　反式异构(现象)
transistor　晶体管；换能器
transistor amplifier　晶体管放大器
transistor grade reagent　晶体管级(纯)试剂
transistor ignition　晶体管点火
transistor ignitor　晶体管点火器
transistorized cotton colo(u)rimeter　晶体管棉花色泽仪
transistorized electronic thermoregulator　晶体管化电子调温器
transite plate　透明塑料板
transition　跃迁*
transitional　过渡的
transitional element (=transition element)　过渡元素
transitional pore　过渡型孔；中孔
transitional region　过渡区域
transition boundary layer　变迁层
transition cell　转变点电池
transition complex　过渡态复合物
transition curve　过渡曲线
transition distance　过渡距离
transition effect　跃迁效应
transition element(=transitional element)　过渡元素
transition energy　跃迁能*
transition fit　过渡配合
transition form　过渡型
transition heat　转化热；转变热；转换热
transition interval　变色区间*
transition knuckle　折边；多变接头
transition law　转变(定)律
transition layer　过渡层
transition matrix(=T matrix)　跃迁矩阵
transition metal　过渡金属元素
transition metal catalyst　过渡金属催化剂
transition moment　跃迁矩
transition of electron　电子跃迁；电子过渡
transition order　转变次序；跃迁次序
transition parameter in oriented crystallization　取向结晶作用中转变参数
transition period　过渡周期
transition phase　跃迁相
transition point　转变点
transition point temperature　转变点温度
transition polymerization　逐步聚合(作用)
transition potential　转变点电位
transition probability　跃迁概率
transition probability of spontaneous radiation　自发辐射跃迁概率

transition process 跃迁过程
transition range 转变范围
transition rate 跃迁率
transition region ①跃迁区域②过渡区③转变区
transition stage 过渡阶段
transition state 过渡态
transition-state signal 过渡性信号
transition state theory 过渡态理论
transition structure 过渡结构
transition temperature 转变温度
transition time 跃迁时间；过渡时间
transition time constant 跃迁时间常数
transition zone 过渡区域
transitron 负跨导管
transit site 过渡位置
transit site tank 另位油罐
transit time 过渡时〔物〕
transit time of droplets 小滴的过渡时间
translation 平移*
translational diffusion 平动扩散
translational motion 平移运动
translational oscillation 平移振动；直线振动
translational state 平动态
translational velocity 平动速度
translational vibration 平移振动；直线振动
translation group 平移群
translatory Brownian motion 平移布朗运动
translawrencium element 铹后元素
translucence ①半透明(性)②半透明度
translucent 半透明
translucent beads 半透明珠
translucent body 半透明体
translucent lighting 半透明照明
translucent rubber 半透明橡胶
translucent scale 透明尺
translucent soap 半透明皂
translucent white 半透明白
transmembrane pressure 跨膜压(透)
transmercuration 汞化转移作用
trans-metallation 金属转移作用
transmethylation 甲基移转(作用)
trans-migration 反式迁移(作用)
transmissibility ①可传性②可透性
transmission ①传递；传达②透光度③联动机件
transmission band 传频带；传输频带
transmission belt 传动带
transmission case 传动箱；齿轮箱
transmission chain 传动链
transmission coefficient ①传递系数②透射系数
transmission curve 传动曲线
transmission efficiency 传输效率
transmission electron 透射电子
transmission electron micrograph 电子透射显微照片
transmission electron microscope 电子透射显微镜；透射电镜
transmission electron microscopy 透射电子显微术
transmission electron microscopy-image analysis 透射电镜-图像分析法
transmission fabric 传动带用布
transmission filter ①透射滤波器②透射滤光片
transmission grating 透射光栅
transmission grease 润滑油
transmission interference filter 透射型干涉滤波器
transmission lime storage 管线中储存
transmission lubricant 传动装置润滑剂
transmission measurement 透射测定法
transmission mechanism 传动机关；传动机构
transmission of heat 热传递；传热
transmission of liquid 液体的传递
transmission oil 润滑油
transmission-scanning electron microscope 透射-扫描电子显微镜
transmission scattering electron 透射散射电子
transmission seal (液体)传动密封
transmission spectra 透射光谱
transmission topography 透射形貌术
transmission type microscope 透射式电子显微镜
transmissivity 透射率；透射系数
transmit 传送；透射
transmit interference 传输干扰
transmittance 透光度
transmittance ratio measurement method 透射比率测定法
transmittancy 透光度
transmitted beam 通过(光)束
transmitted flux 透射光通量
transmitted intensity 透射(光)强度
transmitted light 透射光
transmitter ①变送器②发送机
transmitting capacity 输送能力；传送能力
transmitting-type iridescent film 透射式闪光膜；透过式闪色膜
transmitting vacuum tube 发送真空管
transmodulator 转换器
transmountain line 越山管线
transmutation ①嬗变；换变；蜕变②点石成金

transmuted wood　变性木材
trans-nitration　亚硝基转移作用
transnobelium element　锘后元素
trans-2-cis-6-nonadienal　反-2-顺-6-壬二烯醛
trans-2-nonen-1-ol　反-2-壬烯-1-醇
transoccanic　越洋的〔物〕
transoid　反向
transoid conformation　反向构象*
trans-olefine　交叉结构烯烃
transparency　①透明(性)；透彻性②透明度；透彻(度)
transparent　透明的；透彻的
transparent color　透明色
transparent colored finishes　透明彩色面漆；透明彩色涂层
transparent cup grease　透明杯脂
transparent cutting oil　透明切削油
transparent emulsion　透明乳液(乳胶)
transparent film　透明膜
transparent lake　透明色淀
transparent layer　透明层
transparent liquid　透明液体
transparent paper　透明纸
transparent plastic model　透明塑料模型
transparent reactor　透明反应器
transparent soap　透明皂
transparent varnish　透明清漆
transparent vert emeraude　水合氧化铬绿；透明翡翠绿(绘画色)
transparent white　透明白〔矾土的别名〕
transparent window　透明窗
transparent zinc oxide(=zinc carbonate)　透明氧化锌；碳酸锌
transparent zinc white　透明锌白；透明氧化锌
transparit　玻璃纸
transparticle effect　传递颗粒效应
trans-passive region　过钝化区；超惰化区
transpiration cooling　蒸发冷却法
transpiration method　泻流法
transplutonics　钚后元素
transplutonium element　钚后元素；超钚元素
trans-1,4-polybutadiene　反(式)-1,4-聚丁二烯
trans-polybutadiene rubber　反式聚丁二烯橡胶
trans-1,4-polyisoprene　反(式)-1,4-聚异戊二烯
trans-1,4-polyisoprene rubber　反式-1,4-异戊二烯橡胶
trans-1,5-polypentenamer rubber　反式-1,5-聚戊烯橡胶
transport　①运输；运送②迁移具
transportable　可运输的；可迁移的
transportable container　移动式容器

transport apparatus　(离子)迁移器
transportation　①迁移②运输③运输工具
transportation cost　运输成本
transportation equipment　运输设备；运输装备
transportation leakage　运输漏损
transport band　运输带
transport detector　转移检测器
transporter　运输机
transport factor　运输因数
transport fuel　运输用燃料
transport-ionization detector　迁移电离检测器
transport lag　输送滞后
transport number　迁移数*
transport phenomenon　输运现象
transport properties of molecules　分子运动性质
transport ratio　位移率
transport selectivity　迁移选择性
transport tube　(离子)迁移管
transport velocity　转移速度
transposition　①换位；移位②移调③易位
transposition weighing　换位称法
trans-quantitative method　转化定量法
transship(ment)　换船
trans-stereoisomer　反式立体异构体
trans-tactic　有规反式构型
transtactic polymer　反式有规聚合物
transtat　可调变压器；自耦变压器
transtatic polymer　反式有规聚合物
trans-thioindigo　反式硫靛(蓝)
transthiolation　转硫醇作用
transthorium element　钍后元素
trans-2-trans-4-decadienal　反-2-反-4-癸二烯醛
trans-trans isomer　反-反式(同分)异构体
transudate　渗出液
transudation　渗出作用
transuranic　铀后的
transuranic element　铀后元素；超铀元素
transuranic waste　铀后废物
transuranide　铀后元素
transuranium　①铀后元素②铀后的
transuranium element　铀后元素*
transurface contactor　界面传递萃取器
transvaalin　海葱苷 A
transverse bending resilience　横弯曲弹性
transverse carcass cord　子午排列帘线
transverse contraction　横向收缩
transverse crack　横向裂纹
transverse current　涡流

transverse diffusion 横向扩散
transverse direction 横向
transverse elasticity 横向弹性
transverse flow 横流
transverse flow gating interface 横向流控界面
transverse heated graphite atomizer 石墨炉横向加热原子化器
transverse heating graphite atomizer 横向加热石墨原子化器
transverse heating-longitudinal Zeeman effect atomic absorption spectrophotometer 横向加热纵向塞曼原子吸收分光光度计
transversely excited atmospheric pressure 横向激励大气压
transverse magnetic field 横向磁场
transverse module of elasticity 横向弹性模量
transverse photothermal deflection technique 横向光热偏转法
transverse relaxation 横向弛豫*
transverse resilience 横弯曲弹性
transverse section 横剖面
transverse strain 横向应变；挠曲应变
transverse strength 抗挠强度；横向强度；扭曲强度
transverse superposed complex dynamic shear moduli 横向重合复数动态剪切模量
transverse test 横向(弯曲)试验
transverse test-piece 挠曲试样
transverse vibration 横向振动
transversion 换异(型碱)；颠换
trans-vinylation 乙烯基转移作用
trans-vinylene diisocyanate 反式亚乙烯基二异氰酸酯〔硫化剂〕
trap ①阱*；收集器②捕获；俘获
trap-ash(=trap-tufa; trap-tuff) 暗色岩灰
trapezium form 梯形；不规则四边形
trapezohedron 梯面体；偏方三八面体
trapezoid ①梯形的②梯形
trapezoidal tear 梯形撕破(试验)
trapezoidal thread 梯形螺纹
trapp 暗色岩
trapped air 残存空气；存气；内部气泡
trapped-air process 吹塑薄膜挤塑法
trapped electron (被)俘获电子
trapped exciton 俘获激发子
trapped ion 捕获(了)的离子
trapped ion analyzer cell 捕集离子分析室
trapped ion spectroscopy 离子阱谱法
trapped radical 截留基
trapped species 俘获组分；俘获粒子
trapped twisting method 加捻分股法
trapped vapour ①汽水分离的汽②(溶剂蒸发中的)残存气(雾)
trapping 捕集；截留
trapping agent 捕集剂
trapping efficiency 捕获效率
trapping electrode 阱电极
trapping parameter 阱参数
trapping system 捕集系统
trap-to-trap distillation 逐阱蒸馏
trap-tufa(=trap-ash) 暗色岩灰
trap-tuff(=trap-ash) 暗色岩灰
trashmeter 纤维杂质疵点测定仪
trash receptacle 废料桶
trash-type impeller 闭式叶轮
trass 火山土；粗面凝灰岩
trass cement 粗面凝灰岩水泥
trass concrete 粗面凝灰岩混凝土
Traube's rule 特劳勃规则〔在同系物的稀溶液中欲使表面张力降低得一样多所需的溶液浓度因分子中每增加一个 CH_2 而减少 1/3〕
traumatic acid 创伤酸；2-十二碳烯二酸
traumatin 愈创素
traumatol 碘甲酚
Trauton's coefficient(=coefficient of viscous traction) 拉伸黏性系数；黏性曳引系数
Trauzl (lead block) test 特劳泽(炸铅)试验
travas leaf oil 木姜子叶油
traveller 导丝钩
travelling band conveyor 移动式运输带
travelling bar grizzly 动格筛选
travelling belt 运输带
travelling belt screen 运输带筛
travelling counting glass 游动式织物密度镜
travelling crane 桥动起重机
travelling fabric loop 环形织物传送带
travelling furnace 活动炉；移动炉
travelling grate 转动炉箅
travelling grizzly 动格筛选
travelling hoist 天车；移动式起重机
travelling mixer 活动混合器
travelling oven 活动炉
travelling paddle mixer 活动桨混合器
travelling plate oven 活动盘炉
travelling scale 活动秤
travelling screen 活动筛
travelling speed 迁移速度
travelling spray booth 移动式喷漆橱(柜)

travelling stay　跟刀架；移动刀架
travelling steady　跟刀架
travelling tray oven　活动盘炉
travelling wave　行波
travelling wave tube amplifier　行波管放大器
travelling weigher　活动秤
travel switch　行程开关
traversal　粒子行径
traverse　①横(亘)的②横放③横过
traversellite　绿透辉石
traverse motion　(横向)往复移动(运动)
traverse motion threadguide　往复运动导丝器；横动导丝器
traverse relaxation time　横向弛豫时间
traverse spraying hcad　往复(返)式喷头；桥式喷头
travertine　石灰华；钙华
tray　①(浅)盘②塔盘③垫；座；托架
tray accumulator　盘式蓄电池
tray burner　盘架炉；搁架炉
tray cap　分馏塔盘的泡罩
tray classifier　槽式选粒机
tray column　板式塔
tray down-spout　塔盘下流管
tray drier　盘架干燥器
tray dyeing　浸染
tray efficiency　板效率
tray elevator　盘架升降机
tray flatness　塔板水平度
tray floor　塔板；塔盘
tray gradient　塔盘压力降
tray heater　(蒸馏塔)板框式加热器
tray method of dyeing　浸染(干燥)法
tray mold　盘式模型；敞口模
tray of column　塔盘；塔板
tray ring　塔盘环
tray riser　塔盘蒸气上升口
tray spacing　盘式距
tray support　塔盘支柱
tray thickener　多层增稠器
tray tower　盘式塔
tray water　网下白水
tray weir　塔盘堰
trazodone　曲唑酮〔药〕
treacle stage　黏液阶段〔塑〕
treacliness　①黏(滞)性②黏(滞)度
tread　①外胎面②棱〔车轮〕③轮距〔机工〕
tread concavity　胎面压扁；行驶变形
tread contour plate　胎面样模

tread cushioning layer　底胎面
tread design(=tread pattern)　胎面花纹
tread pattern(=tread design)　胎面花纹
tread pattern groove　胎面花纹槽
tread rubber　胎面胶
tread separation　胎面分层
treated　已经处理过的
treated carbonates　活性碳酸钙
treated clay(=treated earth)　活化白土
treated drying oil　处理干性油；精制干性油
treated earth(=treated clay)　活化白土
treated felt　油毡
treated gasoline　精制汽油
treated neutrals〔复〕　精制中性油
treated oil　精制油；加工油
treated paper　处理滤纸
treated particle　表面处理胶粉
treated pigment　(表面)处理的颜料
treated roofing　铺顶油毡
treated rubber　精制橡胶
treated surface　(已)处理的表面
treated tape　浸胶带
treated water　净水；已处理的水
treater　处理器；提纯器；净化器；纯净器；精制器
treating　①处理②浸润；精制
treating adsorbent　精制吸附剂
treating column(=treating tower)　精制塔
treating compound　浸润剂
treating plant　净化设备
treating process　精制过程；处理过程
treating solution　精制用溶液
treating tank　浸渍桶
treating tower(=treating column)　精制塔
treating water　处理水
treatment　①处理②医治③讨论
treatment losses　精制损失；处理损失
treatment tank　精制罐；处理罐
treatment time　处理时间
treatment transfer　处理物转移
treble　①三倍的；三重的②(使成)三倍
treble bond　三键
tree bark　树皮绉
tree basil oil　树罗勒油
tree elastomer　天然橡胶
tree graph　树图(法)
tree moss　粉肖扁枝衣；树苔〔俗〕
tree polymer　树状高分子
tree scraps〔复〕　树木搔取胶

trefle(=clover) 三叶草
trefle compound 三叶草香精
trefle perfume 三叶草香精
trefoil 三叶草
trehalosamine 海藻糖胺
trehalose(=mycose) 海藻糖 $C_{12}H_{22}O_{11} \cdot 2H_2O$
treibgas(=calorgas) 卡气；压缩混烃
tremelliform ①胶质型②银耳状
tremine 山黄麻碱
tremolite 透闪石
tren(=triaminotriethylamine) 三(氨乙基)胺
trench ①管沟；水渠；运河②挖掘
trench crossing 水沟交叉；掘沟横越
trencher(=trench-hoe) 挖沟机
trench-hoe(=trencher) 挖沟机
trend-analyzing control instrument 控制过程趋向的仪器
trepibutone 曲匹布通〔药〕
Tresca yield criterion 屈雷斯卡屈服判据
tretamine 2,4,6-三亚乙基亚氨基均三嗪
tretinoin 维 A 酸〔药〕
tretranitrophenol sulfonphthalein 四硝基苯酚磺酞
tri- 〔希腊及拉丁字头〕 三
triacanthine 三刺(皂荚)碱
triacetamide 三乙酰胺 $(CH_3CO)_3N$
triacetate ①三乙酸酯②〔专指〕甘油三乙酸酯
triacetate fibre 三醋酯纤维
triacetate membrane 三醋酯膜
triacetate staple fibre 三醋酯短纤维
triacetin(=glycerin triacetate) 三醋精；甘油三乙酸酯 $C_3H_5(CH_3CO_2)_3$
triacetonamine(=tetramethylpiperidone) 三丙酮胺；四甲基哌啶酮 $C_9H_{17}ON$
triacetonediamine 三丙酮二胺 $C_9H_{20}N_2O$
triacetoxy ethyl silane 三乙酰氧基乙基甲硅烷
triacetoxysilane 三乙酸基甲硅烷
triacetylacetonate aluminum 三乙酰丙酮合铝
triacetylated 三乙酰(基)的
triacetyl cellulose 三乙酰纤维素
triacetyl-glucose 三乙酰葡萄糖；葡萄糖三乙酸酯 $C_6H_9O_3(O_2CCH_3)_3$
triacetyl methyl hydroxynaphthoquinone 三乙酰基甲基羟基萘醌 $C_{17}H_{16}O_6$
triacid ①三(酸)价的②三元酸
triacid amide 三酰胺 $(RCO)_3N$
triacidic 三(酸)价的
triacid(ic) base 三(酸)价碱
triacid salt 三酸式盐
triacontane(=melissane) 三十烷；蜂花烷 $CH_3(CH_2)_{28}CH_3$
triacontanedron 三十面体
triacontanoic acid 三十烷酸
triacontanol 三十醇
triacontyl 三十(烷)基 $CH_3(CH_2)_{28}CH_2-$
triacontylene(=melene) 三十碳烯；蜂花烯
1,3,5-triacryl-hexahydrotriazine 1,3,5-三丙烯酰基-六氢化三嗪
triactic 三同立构
tri-active amylamine 三(旋性戊基)胺 $[C_2H_5CH(CH_3)CH_2]_3N$
triacylated 三酰(基)的
triacyl glycerol 三酰基甘油；三甘油酯
triad ①三单元组*②三素组③三价原子④三价基
triad prototropic system(=triad prototropy) 三质子互变体系
triad prototropy(=triad prototropic system) 三质子互变体系
triad stereosequence 三元立构规整序列
triage 筛余；筛余料
trial ①试；试用；试验②尝试③试验性
trial and error 反复试验放；试错法
trial and error method 试算法；累试法；近似法；逐次逼近法
trial cut 试切；试切割
trialkylaluminium 三烷基铝
trialkylamine 三烷基胺；叔胺 R_3N
trialkyl-arsine 三烃基胂；叔胂 R_3As
trialkyl-arsine cyanohalide 氰卤化三烃基胂 $R_3As(CN)X$
trialkyl-arsine dichloride 二氯化三烃基胂 R_3AsCl_2
trialkyl-arsine hydroxychloride 羟氯化三烃基胂 $R_3As(OH)Cl$
trialkyl-arsine oxide 氧化三烷基胂 $R_3As=O$
trialkylated 三烃基的
trialkylated benzene 三烃基苯
trialkylboron 三烷基硼
trialkylchlorosilane 三烷基氯甲硅烷；三烷基氯硅烷
trialkylphosphate 磷酸三烷基酯 $(RO)_3PO$
trialkyl phosphine 三烷基膦
trialkyl sulfonium iodide 碘化三烃基硫 R_3SI
trialkyl thiourea 三烷基硫脲〔促进剂〕
trialkyltin acrylate 丙烯酸三烷基锡
trialkyltin aminoacetate 氨基乙酸三烷基锡
trialkyltin aminopropionate α-氨基丙酸三烷基锡
trialkyltin butylacrylate 丁基丙烯酸三烷基锡
trialkyltin N-dialkylaminoalkoxide N-二烷基氨基烷氧基三烷基锡

trialkyltin morpholinoalkoxide 吗啉代烷氧基三烷基锡
trialkyltin piperazine alkoxide 哌嗪烷氧基三烷基锡
trialkyltin pyrrolidinyl alkoxide 吡咯烷烷氧基三烷基锡
trialkyltin thio-S-triazine derivative 三烷基锡硫代三嗪衍生物
triallylamine 三烯丙基胺
triallyl cyanurate 氰脲酸三烯丙酯
2,4,6-triallyl-3,5-dimethyl phenol 2,4,6-三烯丙基-3,5-二甲基苯酚〔防老剂〕
triallyl isocyanurate 异氰脲酸三烯丙酯
2,4,6-triallyl phenol 2,4,6-三烯丙基甲酚〔防老剂〕
triallyl phosphate 磷酸三烯丙酯
triallyl trimellitate 苯偏三酸三烯丙(基)酯
trial method 尝试法
trial mortar 火药(试验)研钵
trial production 试生产；产品试制
trial run ①试运转；试车②试生产；临时生产
trial test 探索性试验；预试验
triamcinolone 曲安西龙〔药〕
triamcinolone acetonide 曲安奈德〔药〕
triamido ①三酰氨基②(某)三氨基
triamine (某)三胺
triaminoazobenzene(=Bismarok brown) 三氨基偶氮苯；碱性棕；俾斯麦棕 $NH_2C_6H_4N=NC_6H_3(NH_2)_2$
triaminobenzene 三氨基苯
triaminopyrimidine 2,4,6-三氨基嘧啶
triaminotriethylamine(=tren) 三(氨乙基)胺
triaminotriphenyl-carbinol(=pararosaniline) 三氨基三苯甲醇；碱性副品红；副蔷薇苯胺 $HOC(C_6H_4NH_2)_3$
triaminotriphenyl-methane(=paraleucaniline) 三氨基三苯甲烷；副白苯胺 $(NH_2C_6H_4)_3CH$
triammonium phosphate 磷酸三铵 $(NH_4)_3PO_4$
triamorph (同质)三晶形物
triamterene 氨苯蝶啶〔药〕
triamyl amine 三戊胺 $(C_5H_{11})_3N$
triamyl borate 硼酸三戊酯 $(C_5H_{11}O)_3B$
triamyl orthoformate 原甲酸三戊酯 $HC(OC_5H_{11})_3$
triamyloxyboron 三戊氧基硼 $B(OC_5H_{11})_3$
triamylphosphine oxide(TAPO) 氧化三戊基膦 $(C_5H_{11})_3PO$
triangle ①三角(形)②(铁)三角
triangle moulding scraper 三角形造型刮板
triangleprogrammed titration 三角程序滴定法
triangle programmed titration technique 三角程控滴定法
triangle test 三点比较法
triangular 三角(形)的
triangular coordinate 三角坐标
triangular diagram 三角形图解
triangular file 三角锉
triangular phase diagram 三角形相图
triangular scraper 三角刮刀
triangular space 三角区
triangular voltage sweep 三角波(电压)扫描
triangular wave polarography 三角波极谱法*
triangular wave voltammetry 三角波伏安法
triangulation 成三角形
triangulation method (作)三角形法
triaralkyltin methacrylate 甲基丙烯酸三芳烷基锡
triaryl aluminum 三芳基铝
triaryl antimony 三芳基锑
triaryl-arsino dihydroxide 二氢氧化三芳基胂 $R_3As(OH)_2$
triarylated 三芳基的
triarylbismuth dihalide 三芳基二卤化铋
triarylmethane colouring matters 三芳基甲烷染料
triarylmethane dye 三芳甲烷染料
triaryl methyl 三芳基甲基
triaryl phosphate 磷酸三芳基酯 $(ArO)_3PO$
triaryl phosphine 三芳基膦
triarylphosphine oxide 氧化三芳基膦
triaryltin acrylate 丙烯酸三芳基锡
tri-aryltin methacrylate 甲基丙烯酸三芳基锡
triassic 三叠
triassic acid 三叠酸
triatomic ①三原子的②三羟基的
triatomic acid 三价酸
triatomic alcohol 三元醇
triatomic base 三价碱
triatomic molecule 三原子分子
triatomic phenol 三元酚
triatomic ring 三元环
triatomics 三原子物
1,4,7-triazaheptane 1,4,7-三氮杂庚烷；二乙撑三胺
triazanaphthalene 三氮杂萘
triazane 三氮烷 NH_2NHNH_2
triazanetriyl 三氮烷三基
triazene〔类名〕 三氮烯 $N_2H_2N=NH$
triazenediyl 1,3-三氮烯亚基 —NHN=N—；1,1-三氮烯亚基 >N—N=NH
triazene paper 三氮烯纸
triazeno 三氮烯基 $H_2NN=N—$
triazenyl 三氮烯基 $NH=NNH—；—N=NNH_2$
triazide 三叠氮化合物 $R(N_3)_3$

triazido 三叠氮基 (N₃—)₃
triazidotrinitrobenzene 三叠氮基三硝基苯
triazine 三嗪*
triazinedithiol 三嗪二硫酚
triazine-glyoxal adduct 三嗪-乙二醛加合物
s-triazine phase 均三嗪类固定相
triazine resin 三嗪系树脂
triazine ring 三嗪环
triazine triol 三聚氰酸
triazinyl 三嗪基 C₃H₂N₃—
triazo- 叠氮基 N₃—
triazoacetic acid 叠氮基乙酸 (N=N)=NCH₂CO₂H
triazobenzene(=phenylazide) 叠氮基苯 C₆H₅N₃
triazo-compound 叠氮化合物 RN₃
oso-triazole 1,2,5-三唑 NHN=CHCH=N
triazole 三唑*
1,2,4-triazole 1,2,4-三唑
2H-1,2,3-triazole(=osotriazole) 接三唑; 2H-1,2,3-三唑
triazolidine 三唑烷 C₂H₇N₃
s-triazolo- 均三唑并
triazolone 三唑酮 NHNHCON=CH
triazolyl 三唑基 C₂H₂N₃—
triazo-methane(=methyl azide) 叠氮基甲烷; 甲基叠氮 CH₃N₃
triazone 三嗪酮
tribasic ①三碱(价)的; 三元的②三代的
tribasic acid(=triprotic acid) 三元酸
tribasic alcohol 三元醇
tribasic calcium phosphate(=tertiary calcium phosphate) 三代磷酸钙; 正磷酸钙
tribasic carboxylic acid 三元羧酸 R(COOH)₃
tribasic ester 三元酸酯
tribasic lead phosphosilicate 三盐基磷硅酸铅
tribasic lead sulfate 三代硫酸铅; 三(碱)价硫酸铅
tribasic magnesium phosphate 正磷酸镁; 三代磷酸镁
tribenzal 三亚苄基
tribenzal-diamine(=hydrobenzamide) 三苯甲醛缩二胺
tribenzamide 三苯甲酰胺 (C₆H₅CO)₃N
tribenzo-p-diazine 三苯并哒嗪
tribenzoin 三苯精; 甘油三苯甲酸酯 (C₆H₅CO₂CH₂)₂CHOCOC₆H₅
tribenzoylmethane 三苯甲酰甲烷
tribenzyl 三苄基
tribenzylamine 三苄胺 (C₆H₅CH₂)₃N
tribenzyl-arsine oxide 氧化三苄胂 (C₆H₅CH₂)₃As=O
tribenzyl-benzene 三苄基苯

tribenzylchlorosilicane 三苄基氯化硅 (C₇H₇)₃SiCl
tribenzyl citrate 柠檬酸三苄酯 (C₆H₅CH₂)₃C₆H₅O₇
tribenzyl ethyl tin 三苄基乙基锡
tribenzylidene diamide 三亚苄二胺
tribenzylidene mannito 三亚苄基甘露醇; 三苄叉甘露醇
tribenzylidene sorbitol 三亚苄基山梨糖醇; 三苄叉山梨糖醇
tribenzylidene xyllitol 三亚苄基木糖醇; 三苄叉木糖醇
tribenzylphosphine oxide(TBzPO) 氧化三苄基膦 (C₆H₅CH₂)₃PO
tribenzyl phosphite 亚磷酸三苯甲酯
tribenzyltin chloride 三苄基氯化锡 (C₆H₅CH₂)SnCl
tribenzyl tin hydroxide 三苄基氢氧化锡
tribenzyl tin laurate 月桂酸三苄甲基锡
tribenzyltin methacrylate 甲基丙烯酸三苄基锡
tri-p-biphenyl phosphate 磷酸三对联苯酯 (C₆H₅C₆H₄O)₃PO
triblock copolymer 三嵌段共聚物
triblock (polymer) 三段型(共)聚合物
tribo- 〔希腊字头〕摩擦
tribochemistry 摩擦化学
tribochemistry corrosive wear 摩擦化学腐蚀磨损
tribochemistry fatigue wear 摩擦化学疲劳磨损
tribo-electric charging gun 摩擦起电喷枪
triboelectricity 摩擦电
triboelectric series 摩电序
triboelectrification 摩擦起电
tribology 摩擦学
triboluminescence 摩擦发光*
triboluminescent 摩擦发光的
tribometer 摩擦计
tribophysics 摩擦物理学
tribostatic powder spray 摩擦静电粉末喷涂
tribromated(=tribrominated) 三溴化的
tribromide 三溴化合物
tribrominated(=tribromated) 三溴化的
tribromo- 〔词头〕三溴
tribromoacetaldehyde 三溴乙醛
tribromoacetamide 三溴乙酰胺 Br₃CCONH₂
tribromoacetic acid 三溴乙酸 Br₃CCO₂H
tribromo-acetic fluoride 三溴乙酰氟 CBr₃COF
tribromo-acetyl bromide 三溴乙酰溴 CBr₃COBr
tribromo-acetyl chloride 三溴乙酰氯 CBr₃COCl
tribromoanthrarufin 三溴蒽绛酚
tribromo-benzene 三溴(代)苯 C₆H₃Rr₃
tribromo-compound 三溴化合物
tribromo-dichloroethane 三溴二氯乙烷
tribromo-ester 三溴代酯

1,1,2-tribromoethane(=vinyltribromide)　1,1,2-三溴乙烷　BrCH₂CHBr₂

tribromo-ethanol　2,2,2-三溴乙醇

tribromo-ether　三溴代醚

tribromoethyl alcohol　三溴乙醇　Br₃CCH₂OH

tribromo-ethylene　三溴乙烯　BrCH=CBr₂

tribromo-hydrin　1,2,3-三溴丙烷

tribromomesitylene　三溴米；三溴-1,3,5-三甲苯　Br₃C₆(CH₃)₃

tribromo-β-naphthol　三溴-β-萘酚

tribromophenol　三溴苯酚〔阻燃剂〕

tribromo-phenol bismuth　三溴酚铋

tribromophenyl acetate　乙酸三溴苯酯　CH₃CO₂C₆H₂Br₃

tribromophenyl salicylate　水杨酸三溴苯酯　HOC₆H₄CO₂C₆H₂Br₃

tribromopyrogallol　三溴连苯三酚；三溴焦棓酚

tribromoquinaldine　三溴喹哪啶；三溴-2-甲基氮杂萘　C₉H₆NCBr₃

tribromosalicylanilide　三溴水杨酰苯胺〔防霉剂〕

tribromo-salol　水杨酸三溴苯酯

tribrom pyranthrone　三溴代皮蒽酮

tribrom pyranthrone scarlet　三溴皮蒽酮猩红

tributoxy-boron　三丁氧基硼；正硼酸三丁酯　B(OC₄H₉)₃

tributoxyethyl phosphate　磷酸三丁氧基乙酯

tri-n-butoxytitanium monostearate　三正丁氧基钛单硬脂酸酯；单硬脂酸钛三正丁酯

tributyl acetylcitrate　乙酰柠檬酸三丁酯

tributyl aconitate　丙烯三羧酸三丁酯；乌头酸三丁酯〔增塑剂〕

tributyl amine　三丁胺　(C₄H₉)₃N

tributyl borate　硼酸三丁酯

tri-n-butyl borate　硼酸三正丁酯　(C₄H₉)₃BO

tributyl carbinol　三丁基甲醇　(C₄H₉)₃COH

tributyl citrate　柠檬酸三丁酯　(C₄H₉)₃C₆H₅O₇

tributyl lead acetate　乙酸三丁基铅

2,4,6-tri-t-butylnitrosobenzene　2,4,6-三-叔丁基亚硝基苯

tributyl orthoformate　原甲酸三丁酯　HC(OCH₂CH₂CH₃)₃

tributyloxyethyl phosphate　磷酸三丁氧乙酯〔阻燃剂、增塑剂〕

tributyl phenol　三丁基(苯)酚

tributyl phosphate　磷酸三丁酯　(C₄H₉O)₃PO

tri-n-butyl phosphine　三正丁基膦

tributylphosphine oxide(TBPO)　氧化三丁基膦　(C₄H₉)₃PO

tributyl thiourea　三丁基硫脲〔防老剂〕

tributyltin acetate　乙酸三丁基锡；三丁基乙酸锡

tributyltin chloride　三丁基氯化锡

tributyltin fumarate　富马酸三丁基锡；三丁基富马酸锡

tributyltin hydroxide　氢氧化三丁基锡；三丁基氢氧化锡

tributyltin isonicotinate　异烟酸三丁基锡；4-吡啶甲酸三丁基锡

tributyl tin laurate　月桂酸三丁基锡〔防霉剂〕

tributyltin monocarboxylate　单羧酸三丁基锡；三丁基单羧酸锡　(C₄H₉)₃SnOOCR

tributyltin oxide　氧化三丁基锡

tributyltin toluene sulfonate　甲苯磺酸三丁基锡

tributyrin(=glycerol tributyrate)　三丁精；甘油三丁酸酯；三丁酸甘油酯

tricalcium orthophosphate(=calcium orthophosphate)　正磷酸三钙；正磷酸钙

tricalcium phosphate　磷酸三钙　Ca₃(PO₄)₂

tricalcium saccharide　三钙蔗糖

tricalcium silicate　硅酸三钙〔水泥成分〕

tricaprin　三癸精；甘油三癸酸酯　C₃H₅(OOCC₉H₁₉)₃

tricaproin　三己精；甘油三己酸酯　(C₅H₁₁COO)₃C₃H₅

tricaprylin　三辛精；甘油三辛酸酯　(C₇H₁₅COO)₃C₃H₅

tricaprylmethylammonium nitrate(TCMAN)　硝酸三辛基甲基铵　(C₈H₁₇)₃(CH₃)NNO₃

tricapryl trimellitate　偏苯三酸三辛酯〔增塑剂〕

tricarballylic acid　丙三羧酸　(HOOCCH₂)₂CHCOOH

tricarboxylic acid　三羧酸　R(COOH)₃

tricarboxylic acid cycle　三羧酸循环

tricarboxylic cycle　三羧循环

tricarboxylic ester　三羧酸酯　R(COOR)₃

tricetin　3′,4′,5′,5,7-五羟黄酮

tricetylphosphine oxide(TCPO)　氧化三(十六烷基)膦　(C₁₆H₃₃)₃PO

trichlene　三氯乙烯

trichlene finishing system　三氯乙烯(配套)涂装法；三氯乙烯一条龙涂装法

trichlorated　三氯化的

trichloride　三氯化物

trichlorinated(=trichlorizated)　三氯化的

trichlorizated(=trichlorinated)　三氯化的

trichloroacetal　三氯乙缩醛；二乙醇缩三氯乙醛　Cl₃CCH(OC₂H₅)₂

trichloroacetaldehyde(=chloral)　三氯乙醛；氯醛

trichloroacetamide　三氯乙酰胺　Cl₃CCONH₂

trichloroacetamidophosphorus trichloride　三氯乙酰亚氨基三氯化磷　CCl₃CON=PCl₂

trichloroacetanilide　N-三氯苯基乙酰胺　C₈H₆Cl₃NO

trichloroacetic acid(TCA)　三氯乙酸

trichloro-acetic bromide　三氯乙酰溴　CCl₃COBr

trichloroacetic chloride 三氯乙酰氯 CCl₃COCl
trichloroacetone 三氯丙酮 CCl₃COCH₃
trichloro-acetyl bromide 三氯乙酰溴 CCl₃COBr
trichloro-acetyl chloride 三氯乙酰氯 CCl₃COCl
trichloro-acrylic acid 三氯代丙烯酸 Cl₂C＝CClCOOH
trichloro-aldehyde 三氯乙醛
trichloro-ammine platinite 三氯一氨合亚铂酸盐
2,3,4-trichloroaniline 2,3,4-三氯苯胺 Cl₂C₆H₂NH₂
2,4,6-trichloroanisole 2,4,6-三氯苯甲醚 C₇H₅Cl₃O
trichloro-benzene 三氯(代)苯 C₆H₃Cl₃
1,2,3-trichlorobenzene 1,2,3-三氯苯；连三氯苯 Cl₃C₆H₃
2,3,4-trichlorobenzoic acid 2,3,4-三氯苯甲酸；连三氯苯甲酸 Cl₃C₆H₂CO₂H
trichlorobromomethane 三氯溴甲烷 CBrCl₃
trichloro-butyl alcohol 三氯叔丁醇
trichloro-butyl malonate 丙二酸三氯丁基酯
trichloro-butyraldehyde 三氯丁醛 Cl₃C₃H₄CHO
trichlorobutyric acid 三氯丁酸
trichloro-compound 三氯化合物
trichloro-cupric acid 三氯铜酸 H[CuCl₃]
1,1,1-trichloro-2,2-dibromoethane 1,1,1-三氯-2,2-二溴乙烷 Br₂CHCCl₃
trichlorodivinylarsine 二乙烯基三氯化胂
trichloro ester 三氯代酯
1,1,1-trichloroethane(=methylchloroform) 1,1,1-三氯乙烷；甲基氯仿 CH₃CCl₃
trichloro-ethanol 2,2,2-三氯乙醇
trichloro ether 三氯代醚
trichloroether phosphate 磷酸三氯醚〔阻燃剂〕
trichloro-ethyl alcohol 2,2,2-三氯乙醇
trichloroethylene 三氯乙烯 ClCH＝CCl₂
trichloroethylene-based paint 三氯乙烯漆
trichloroethylene degreasing 三氯乙烯脱脂法
trichloroethylene phosphating solution 三氯乙烯磷化处理液
trichloroethylene process 三氯乙烯溶剂脱蜡过程
trichloroethylglucuronide 葡萄糖三氯乙基苷酸
trichloro-ethylideneimine 三氯乙亚胺 (Cl₃CCH＝NH)₃
tri-β-chloroethyl phosphate β-三氯乙基磷酸酯
trichlorofluoromethane 三氯氟甲烷〔发泡剂〕
trichloro-hydrin(=trichloropropane) 三氯丙烷
trichloro-hydroquinone 三氯氢醌
trichloro-iodomethane 三氯碘甲烷 Cl₃CI
1,2,3-trichloroisobutane 1,2,3-三氯异丁烷
trichloro-lactic acid 三氯乳酸 CCl₃CHOHCOOH
trichlorolactonitrile 三氯丙醇腈
trichloromelamine 三氯三聚氰胺〔防焦剂〕

trichloromethane 三氯甲烷
trichloromethane sulfonyl chloride 硫酰氯化三氯甲烷
trichloromethyl 三氯甲基 Cl₃C—
trichloromethyl chlorocarbonate 氯甲酸三氯甲酯；双光气 ClCO₂CCl₃
trichloromethyl chloroformate 氯甲酸三氯甲基酯
trichloromethyl phenyl carbinyl acetate 乙酸三氯甲苯基原酯
trichloro-methylthio 三氯甲硫基 Cl₃CS—
N-trichloromethylthio-N-phenyl-sulfamide N-三氯甲基硫代-N-苯基磺酰胺〔防焦剂〕
N-(trichloromethylthio) phthalimide N-(三氯甲基硫代)邻苯二甲酰亚胺〔防焦剂〕
N-(trichloromethylthio) tetrahydrophthalimide (=Captan) N-三氯甲基硫代四氢化邻苯二酰亚胺
trichloronaphthalene 三氯萘 C₁₀H₅Cl₃
trichloronitrile 三氯代腈(类)
trichloronitrobenzene 三氯硝基苯 C₆H₂(NO₂)Cl₃
trichloronitromethane(=chloropicrin) 三氯硝基甲烷；氯化苦
trichloronitrosobenzene 三氯代亚硝基苯
tri-β-chlorooctylphosphine oxide 氧化三(β-氯代辛基)膦 (ClC₈H₁₆)₃PO
trichlorophenol 三氯苯酚 Cl₃C₆H₂OH
trichloro-phenomalic acid 三氯乙酰丙烯酸 CCl₂COCH＝CHCOOH
N-2,4,5-trichlorophenylleuco auramine N-2,4,5-三氯苯基无色金胺
trichlorophenyl phosphate 磷酸三(氯苯酯) (ClC₆H₄O)₃PO
trichloropropane(=trichloro-hydrin) 三氯丙烷 CH₃CHClCHCl₂
trichloropropylene oxide 三氯氧化丙烷
trichloropropyl phosphate 磷酸三氯丙酯〔增塑剂、阻燃剂〕
trichloropurine 三氯嘌呤 C₅HCl₃N₄
trichloropyridine 三氯吡啶 C₅H₂NCl₃
trichloroquinone 三氯醌 O＝C₆HCl₃＝O
trichlororesorcinol 三氯间苯二酚 C₆HCl₃(OH)₂
trichlorosilane 三氯氢硅；三氯甲烷硅；硅氯仿 SiHCl₃
trichlorothiophenol 三氯硫酚〔塑解剂〕
trichlorotoluene 三氯甲苯 C₆H₂Cl₃(CH₃)
trichloro-tribromoethane 三氯三溴乙烷 Cl₂CBrCClBr₂
trichloro-triethylamine 三氯三乙胺
1,1,2-trichlorotrifluoroethane 1,1,2-三氯三氟乙烷
2,4,6-trichloro-1,3,5-trimethyl borazine 2,4,6-三氯代-1,3,5-三甲基环硼氮烷
trichlorotrivinylarsine 三氯三乙烯胂
trichocereine 仙影掌碱

trichocidin　杀滴虫青霉素
trichodesmine　毛束草碱
tricholomic acid　口蘑氨酸
trichosanic acid　栝楼酸
trichromat　具三色视觉者；色视觉正常者
trichromate　三铬酸盐
trichromatic analysis　三色分析
trichromatic coefficient　三原色系数
trichromatic colorimeter　三色比色仪；三色(激励)色度计
trichromatic colorimetry　三色定色法；三(原)色测色法
trichromatic coordinates　三色坐标
trichromatic specification　三色标
trichromatic system　三原色表色系统
tricin　麦黄酮；4′,5,7-三羟(基)-3′,5′-二甲氧黄酮
trick　换班；班次
trickle　细流
trickle bed　喷淋床
trickle bed reactor　滴流床反应器
trickle charge　连续补充充电
trickle cooler　水淋式冷却器
trickle cooling plant　滴流冷却装置
trickle dissolver　滴流式溶解器
trickle flow hydrodesulfurization　滴流式加氢脱硫(法)
trickle process　滴流法〔煤油、减压轻油的加氢脱硫法〕
trickling cooler　水淋冷却器
trickling filter　滴滤池
trickling internals　(塔)内部滴流填料
triclinic　三斜(晶)的
triclinic crystalline system　三斜晶系
triclinic lattice　三斜晶格
triclinic system　三斜晶系
triclofos sodium　磷酸三氯乙酯钠
tricosadienoic acid　二十三碳二烯酸　$C_{22}H_{41}COOH$
tricosandioic acid(=tricosanediacid)　二十三(碳)二酸
n-tricosane　正二十三(碳)烷　$CH_3(CH_2)_{21}CH_3$
tricosane diacid(=tricosandioic acid)　二十三烷二酸　$C_{21}H_{42}(COOH)_2$
tricosane dicarboxylic acid　二十五烷二酸　$C_{23}H_{46}(COOH)_2$
tricosanic acid　二十三(烷)酸　$C_{22}H_{45}(COOH$
tricosanol　二十三醇
12-tricosanone　12-二十三(烷)酮
tricosendioic acid(=tricosene diacid)　二十三碳烯二酸
tricosene diacid(=tricosendioic acid)　二十三碳烯二酸　$C_{21}H_{40}(COOH)_2$
tricosene dicarboxylic acid　二十三(碳)烯二羧酸　$C_{23}H_{44}(COOH)_2$
tricosenoic acid　二十三(碳)烯酸　$C_{22}H_{43}COOH$

tricosoic acid　二十三(烷)酸　$C_{22}H_{45}COOH$
tricosyl　二十三(烷)基　$CH_3(CH_2)_{21}CH_2-$
tricosylacetic acid　二十三基乙酸；二十五(碳)烷酸
tricot　经编织物
tricresol　邻、间、对三种甲酚混合物
tri-o-cresyl phosphate　磷酸三(邻甲苯酯)　$(CH_3C_6H_4O)_3PO$
tri-p-cresyl phosphate　磷酸三(对甲苯酯)
tricresyl phosphate　磷酸三甲酚酯
tri-p-cresyl phosphite　亚磷酸三(对甲苯酯)
tri-o-cresyl thiophosphate　硫代磷酸三(邻甲苯酯)
tricrotonylidene tetramine　三(亚丁烯基)四胺
triculamin　三底胺菌素
tricyanic acid　三聚氰酸
tricyano-　三聚氰基
tricyanoethane　三氰基乙烷　$CH_3C(CN)_3$
tricyanogen chloride　三聚氯化氰
tricyanomethide　三氰甲基化合物
tricyano-styrene　三氰基苯乙烯
tricyanovinylation　三氰基乙烯化(作用)
tricyclal　三环萜醛
tricyclamol chloride　氯化 1-(3-环己基-3-羟基-3-苯丙基)-1-甲基吡咯烷鎓
tricyclene　三环烯；三环萜　$C_{10}H_{16}$
tricyclenic acid(=dehydrocamphenilic acid)　三环萜酸
tricyclic　三环的
tricyclic aromatics　三环芳香化合物
tricyclic compound　三环化合物
tricyclic hydrocarbon　三环烃
tricyclic isochroman musk　三环异色满麝香
tricyclic naphthenes　三环环烷
tricyclic ring　三环核
tricycloalkene　三环链烯
tricyclo decenyl acetate　乙酸三环癸烯酯
tricyclo decenyl propionate　丙酸三环癸烯酯
tricyclo-ekasantalal　三环准檀香醛
tricyclohexyl borate　硼酸三环己酯
tricyclohexylmethylenephosphine oxide　氧化三(环己基次甲基)膦
tricyclohexyl phosphate　磷酸三环己酯　$(C_6H_{11}O)_3PO$
tricyclohexyltin acrylate　丙烯酸三环己基锡
tricyclohexyltin butylacrylate　丁基丙烯酸三环己基锡
tricyclopentadienide　三环戊二烯化物　$M(C_5H_5)_3$
tricyclopentadienyl-bromothorium　溴化三环戊二烯钍　$(C_5H_5)_3ThBr$
tricyclopentadienyl-butoxyuranium　三环戊二烯基丁氧基铀　$(C_5H_5)_3U(OC_4H_9)$
tricyclopentadienyl-californium　三环戊二烯锎

Cf(C_5H_5)$_3$

tricyclopentadienyl-chlorouranium 氯化三环戊二烯基铀 (C_5H_5)$_3$UCl

tricyclopentadienyl-cholesteryloxy-uranium 三环戊二烯基胆甾醇氧基铀 (C_5H_5)$_3$U(O$C_{27}H_{45}$)

tricyclopentadienyl-curium 三环戊二烯锔 Cm(C_5H_5)$_3$

tricyclopentadienyl-cyclohexyliloxy-uranium 三(环戊二烯)环己氧基铀 (C_5H_5)$_3$U(OC_6H_{11})

tricyclopentadienyl-cyclohexylisonitrile-uranium 三(环戊二烯)环己基异腈铀

tricyclopentadienyl-ethoxyuranium 三(环戊二烯)乙氧基铀 (C_5H_5)$_3$U(OC_2H_5)

tricyclopentadienyl-fluorothorium 氟化三(环戊二烯)钍 (C_5H_5)$_3$ThF

tricyclopentadienyl-iodouranium 碘化三(环戊二烯)合铀 (C_5H_5)$_3$UI

tricyclopentadienyl-isopropoxyuranium 三(环戊二烯)异丙氧基铀 (C_5H_5)$_3$U(OC_3H_7)

tricyclopentadienyl-metal alkoxide 三(环戊二烯)烷氧基合(某)金属 (C_5H_5)$_3$M(OR)

tricyclopentadienyl-methoxyuranium 三(环戊二烯)甲氧基合铀 (C_5H_5)$_3$U(OCH_3)

tricyclopentadienyl-nicotineplutonium 三(环戊二烯)烟碱钚 (C_5H_5)$_3$Pu($C_{10}H_{14}N_2$)

tricyclopentadienyl-octoxyuranium 三(环戊二烯)辛氧基铀 (C_5H_5)$_3$U(OC_8H_{17})

tricyclopentadienyl-plutonium 三(环戊二烯)钚

tricyclopentadienyl-tetrahydroborato-uranium 三(环戊二烯)四氢化硼基铀 (C_5H_5)$_3$U(BH_4)

tricyclopentadienyl-tetrahydrofuran-plutonium 三(环戊二烯)四氢呋喃钚 (C_5H_5)$_3$Pu(OC_4H_8)

tricyclopentadienyl-uranium chloride 三(环戊二烯)氯铀 (C_5H_5)$_3$UCl

tricyclovetivane 三环岩兰素

tricyclovetivene 三环岩兰烯

tricyclovetivenol 三环岩兰烯醇

tridecadienoic acid 十三碳二烯酸 $C_{12}H_{21}$COOH

tridecalactone 十三(烷)内酯

tridecamethylenediamine 十烷撑二胺

tridecanal(=tridecyl aldehyde) 十三醛

tridecandioic acid(=tridecane diacid) 十三烷二酸

tridecane ①(正)十三(碳)烷②十三(碳)(级)烷 $CH_3(CH_2)_{11}CH_3$

tridecane diacid(=tridecandioic acid) 十三烷二酸 $C_{11}H_{22}$(COOH)$_2$

tridecane dicarboxylic acid 十三烷二羧酸 $C_{13}H_{26}$(COOH)$_2$

tridecanoic acid 十三(烷)酸 $C_{12}H_{25}$COOH

1-tridecanol 1-十三(烷)醇 $CH_3(CH_2)_{11}CH_2OH$

2-tridecanone(=methyl n-undecyl ketone) 2-十三烷酮；甲基十一基(甲)酮

tridecanoyl 十三(烷)酰 $CH_3(CH_2)_{11}$CO—

2-tridecenal 2-十三烯醛

tridecendioic acid(=tridecene diacid) 十三碳烯二酸

tridecene diacid(=tridecendioic acid) 十三碳烯二酸 $C_{11}H_{20}$(COOH)$_2$

tridecene dicarboxylic acid 十三(碳)烯二羧酸 $C_{13}H_{24}$(COOH)$_2$

tridecene-2-nitrile 十三烯-2-腈

tridecenoic acid 十三(碳)烯酸 $C_{12}H_{23}$COOH

tridecoic acid 十三(烷)酸 $C_{12}H_{25}$COOH

tridecyl 十三(烷)基 $CH_3(CH_2)_{11}CH_2$—

tridecyl acrylate 丙烯酸十三(烷)酯

tridecyl alcohol 十三(烷)醇 $CH_3(CH_2)_{11}CH_2OH$

tridecyl aldehyde(=tridecanal) 十三醛

tridecyl amine 十三(烷)胺 $C_{13}H_{27}NH_2$

tridecyl benzene 十三烷基苯

tridecyl caproate 己酸十三(烷)酯 $C_5H_{11}CO_2C_{13}H_{37}$

tridecyl cyanide(=myristonitrile) 十四(烷)腈；肉豆蔻腈

tridecylendioic acid(=tridecylene diacid) 十三(碳)烯二酸

tridecylene 十三(碳)烯 $C_{13}H_{26}$

tridecylene diacid(=tridecylendioic acid) 十三(碳)烯二酸 $C_{11}H_{20}$(COOH)$_2$

tridecylene dicarboxylic acid 十三(碳)烯二羧酸 $C_{13}H_{24}$(COOH)$_2$

tridecylenic acid 十三(碳)烯酸 $C_{12}H_{23}$COOH

tridecylic acid 十三(烷)酸 $C_{12}H_{25}$COOH

tridecylic aldehyde 十三(烷)醛 $C_{12}H_{25}$CHO

tridecyl methacrylate(TDMA) 甲基丙烯酸十三烷酯

tridecyl phosphate(TDP) 磷酸三癸酯 ($C_{10}H_{21}O$)$_3$PO

tridecyl phosphine oxide(TDPO) 氧化三癸基氧膦 ($C_{10}H_{21}$)$_3$PO

tridecyl phosphite 亚磷酸三癸酯〔稳定剂〕

tri-dentate ①三(锯)齿状的②三配位基

tridentate ligand 三齿配位体(基)

tridepside 三缩酚酸

2,4,6-tri(dimethylaminomethyl) phenol 2,4,6-三(二甲基氨甲基)苯酚

tri(dimethylphenyl) phosphate 磷酸三(二甲苯酯)〔阻燃剂〕

tridione 三甲苯噁唑烷二酮

tridiphenylmethyl 三联苯(代)甲基

tri(dithiooxalato) cobaltate(Ⅲ) 三(二硫代草酸根)合钴(Ⅲ)酸盐

tridodecylamine(TDA) 三(十二烷基)胺；三月桂胺 ($C_{12}H_{25}$)$_3$N

tridodecylammonium nitrate 三(十二烷基)铵硝酸盐 TDA·HNO$_3$
tridodecyl borate 硼酸三(十二烷基)酯
tridodecyl phosphine oxide(TDPO) 氧化三(十二烷基)膦 $(C_{12}H_{25})_3PO$
tridymite 鳞石英
trielaidin (三)反油酸甘油酯
trielide 第三族元素化物
trieline 三氯乙烯
triels 第三族元素
trien(=triethylenetetramine) 三亚乙基四胺
triene 三烯
triene conjugation 三烯共轭化
trienic acid 三烯酸
trienol 三烯甘油酯
trienyl carbanion 三烯基碳负离子
trier ①试验机②试验者③取样器
triethanolamine(=trihydroxy ethylamine) 三乙醇胺〔俗〕；三(羟乙基)胺 $N(CH_2CH_2OH)_3$
triethanolamine borate 硼酸三乙醇胺〔环氧催化剂〕
triethanolamine linoleate 三乙醇胺亚油酸盐
triethanolamine monooleate ester 三乙醇胺单油酸酯
triethanolamine monostearate 三乙醇胺单硬脂酸酯
triethanol amine oleate 油酸合三乙醇胺〔乳化剂〕
triethanolamine orthotitanate 原钛酸三乙醇胺酯
triethanolamine stearate(=trihydroxy ethylamine stearate) 硬脂酸三乙醇胺
triethanolamine titanate 三乙醇胺络钛酸酯；钛酸三乙醇胺
triethenoid fatty acid 三烯脂肪酸
triethide 三乙基金属 $M(C_2H_5)_3$
triethoxy 三乙氧基
triethoxy-boron 三乙氧基硼；硼酸三乙酯 $B(OC_2H_5)_3$
triethoxy-silane(=triethoxy-silicane) 三乙氧基甲硅烷 $(C_2H_5O)_3SiH$
triethoxy-silicane(=triethoxy-silane) 三乙氧基甲硅烷
triethyl 三乙(烷)基
triethylacetic acid 三乙基乙酸 $(C_2H_5)_3CCOOH$
triethyl adipate 己二酸三乙酯〔增塑剂〕
triethyl aluminum 三乙基铝 $Al(C_2H_5)_3$
triethylamine 三乙胺 $(C_2H_5)_3N$
triethylamine hydrobromide 三乙胺氢溴酸盐 $(C_2H_5)_3NHBr$
triethylamine hydrochloride 盐酸三乙胺 $(C_2H_5)_3NHCl$
triethylaminoethyl cellulose 三乙氨基乙基纤维素〔含有 $(C_2H_5)_3N^+CH_2CH_2$—基置换的纤维素〕
triethyl ammonium allyl-sulphonate 烯丙基磺酸三乙铵
triethylantimony 三乙锑
triethyl arsenic 三乙砷 $(C_2H_5)_3As$
triethylarsine 三乙胂 $(C_2H_5)_3As$
triethyl-arsine cyanobromide 溴氰化三乙胂 $(C_2H_5)_3As(CN)Br$
triethyl-arsine hydroxybromide 溴羟化三乙胂 $(C_2H_5)_3As(OH)Br$
triethyl-benzene 三乙苯
triethyl benzyl ammonium calcium trichloride monohydrate 三乙基苄基铵合三氯化钙一水合物
triethyl-bismuth 三乙基铋 $Bi(C_2H_5)_3$
triethyl-bismuthine 三乙铋
triethylborane 三乙基甲硼烷
triethyl borate 硼酸三乙酯 $(C_2H_5O)_3B$
triethyl-borine 三乙基硼 $B(C_2H_5)_3$
triethyl-boron 三乙基硼 $(C_2H_5)_3B$
triethyl-carbinol 三乙基甲醇
triethylchloro-silicane 三乙基氯硅 $(C_2H_5)_3SiCl$
triethylcholine 三乙基胆碱
triethyl citrate 柠檬酸三乙酯 $(C_2H_5)_3C_6H_5O_7$
triethyl cyanurate 氰尿酸三乙酯；三聚氰酸三乙酯 $C_3N_3(OC_2H_5)_3$
triethylenediamine 三亚乙基二胺
triethylenediammonium 三亚乙基二铵
triethylene-glycol(=triglycol) 三甘醇；二缩三(乙二醇) $(HOCH_2CH_2OCH_2)_2$
triethylene glycol caprate 三甘醇癸酸酯〔增塑剂〕
triethylene glycol caprylatecaprate 三甘醇辛酸癸酸酯
triethylene-glycol diacetate 三甘醇二乙酸酯；二缩三(乙二醇二乙酸)酯 $(CH_3CO_2CH_2CH_2OCH_2)_2$
triethyleneglycol diacrylate 三甘醇二丙烯酸酯
triethylene glycol dibenzoate 三甘醇二苯甲酸酯
triethyleneglycol dimethacrylate 三甘醇二甲基丙烯酸酯
triethylene glycol dimethyl ether(=triglyme) 三甘醇二甲醚
triethylene glycol dipelargonate 三甘醇二壬酸酯〔增塑剂〕
triethylenephosphoramide 三亚乙基磷酰胺
triethylenetetraaminehexaacetic acid(TTHA) 三亚乙基四胺六乙酸
triethylene-tetramine 三亚乙基四胺；二缩三(乙二胺) $(H_2NCH_2CH_2NHCH_2)_2$
triethylenethiophosphoramide 三亚乙基硫代磷酰胺
triethyl-ethoxy-silane(=triethyl-ethoxy-silicane) 三乙基乙氧基硅
triethyl-ethoxy-silicane(=triethyl-ethoxy-silane) 三乙基乙氧基硅 $(C_2H_5)_3Si(OC_2H_5)$
triethyl-gallium 三乙(基)镓
tri-2-ethylhexyl phosphate(TEHP) 磷酸三(2-乙基己基)酯 $(C_8H_{17}O)_3PO$

tri-2-ethylhexyl phosphine oxide(TEHPO)　氧化三(2-乙基己基)膦　[C$_4$H$_9$CH(C$_2$H$_5$)CH$_2$]$_2$PO
triethylin　三乙灵；三乙基甘油醚　C$_3$H$_5$(OC$_2$H$_5$)$_3$
triethyl orthoacetate　原乙酸三乙酯　CH$_3$C(CO$_2$H$_5$)$_3$
triethyl orthoformate　原甲酸三乙酯　HC(OC$_2$H$_5$)$_3$
triethyl orthopropionate　原丙酸三乙酯　C$_2$H$_5$C(OC$_2$H$_5$)$_3$
triethyl phosphate　磷酸三乙酯　(C$_2$H$_5$O)$_3$PO
triethyl-phosphine　三乙膦　(C$_2$H$_5$)$_3$P
triethyl-phosphine oxide　氧化三乙膦　(C$_2$H$_5$)$_3$PO
triethyl-phosphine sulfide　硫化三乙膦　(C$_2$H$_5$)$_3$PS
triethyl-phosphite　亚磷酸三乙酯　(C$_2$H$_5$O)$_3$P
triethyl phosphonoformate　膦酸甲酸三乙酯
triethyl-silane ethyloxide　三乙基乙氧基硅　(C$_2$H$_5$)$_3$SiOC$_2$H$_5$
triethyl-silicane　三乙基甲硅烷　SiH(C$_2$H$_5$)$_3$
triethyl-silicoformate　三乙氧基甲硅烷　SiH(OC$_2$H$_5$)$_3$
triethyl-silicol ethyl ether　三乙基乙氧基硅　(C$_2$H$_5$)$_3$SiOC$_2$H$_5$
triethyl silicon　三乙基甲硅烷　SiH(C$_2$H$_5$)$_3$
triethyl-silicon hydroxide　三乙基氢氧化硅
triethyl-silicon oxide(=hexaethyl disiloxane)　双氧化(三乙基)硅　[(C$_2$H$_5$)$_3$Si]$_2$O
triethyl-stibine(=triethylantimony)　三乙基䏲；三乙锑　Sb(C$_2$H$_5$)$_3$
triethyltin(=hexaethylditin)　三乙基锡；六乙基二锡　[(C$_2$H$_5$)$_3$Sn]$_2$
triethyltin chloride　氯化三乙基锡　(C$_2$H$_5$)$_3$SnCl
triethyltin methacrylate　甲基丙烯酸三乙基锡
triethyl trimellitate　偏苯三酸三乙酯〔增塑剂〕
triethyl trimethylene triamine　三乙基三亚甲基三胺〔促进剂〕
triethylzincate anion　三乙基锌酸根阴离子
triferrin　仲核酸铁
trifid　三裂的
triflate　三氟甲磺酸盐
trifluoperazine　三氟拉嗪〔药〕
trifluoperazine hydrochloride　盐酸10-[3-(4-甲基哌嗪-1-)丙基]-2-三氟甲基吩噻嗪；三氟啦嗪〔安定药,治疗精神分裂症等〕
trifluorated　三氟化的
trifluoride　三氟化物
trifluorinated(=trifluorizated)　三氟化的
trifluorizated(=trifluorinated)　三氟化的
trifluoroacetic acid　三氟乙酸　CF$_3$COOH
trifluoro-acetic chloride　三氟乙酰氯　CF$_3$COCl
trifluoro-acetic fluoride　三氟乙酰氟　CF$_3$COF
trifluoroacetylacetone(TFA)　三氟乙酰丙酮　CF$_3$COCH$_2$COCH$_3$

trifluoro-acetyl chloride　三氟乙酰氯　CF$_3$COCl
trifluoroacetyleation　三氟乙酰化
trifluoro-benzene　三氟(代)苯　C$_6$H$_3$F$_3$
1,2,2-trifluoro-1-chloroethane　1,2,2-三氟-1-氯乙烷　F$_2$CHCHFCl
trifluorochloroethylene　三氟氯乙烯
trifluoro-compound　三氟化合物
trifluorodimethylhexanedione(TFDMHD)　三氟二甲基己二酮　CF$_3$COCH$_2$COC(CH$_3$)$_2$CH$_3$
trifluoroethanol　三氟乙醇
trifluoro ether　三氟(代)醚
trifluoro methane(=fluoroform)　三氟甲烷；氟仿　CHF$_3$
m-trifluoromethylphenol　间三氟甲基苯酚
trifluoromethylsulphone　三氟代甲砜
trifluoromethyl thionitrite　硫代亚硝酸三氟代甲酯；亚硝酸三氟代甲硫醇酯
trifluoropropanol　三氟丙醇
1,1,1-trifluoro propyl methyl siloxane polymer　1,1,1-三氟丙基甲基聚硅氧烷聚合物
trifluoropropyl siloxane　三氟丙基硅氧烷；氟硅橡胶
trifluoto ester　三氟(代)酯
trifolianol　三叶醇
triformin　三甲精；甘油三乙酸酯　C$_3$H$_5$(OOCH)$_3$
triformol　三聚甲醛
tri-functional　三官能的
trifunctional acid　三官能(度)酸
trifunctional branching agent　三官能支化剂
trifunctional end　三官能端基
trifunctional initiator　三官能引发剂*
trifunctional monomer　三官能(基)单体*
trifunctional organochlorosilane　三官能有机氯(甲)硅烷
trifyl(=trifluoromethanesulfonyl)　三氟甲磺酰基　CF$_3$SO$_2$—
trigalloyl　三(个)棓酰
trigalloyl acetone glucose　三棓酰丙酮葡萄糖
trigalloyl glucose　三棓酰葡萄糖
trigalloyl glycerol　三棓酰甘油
trigger　触发器
trigger mechanism　触发机理*
trigger sweep　触发(器)扫描
trigger temperature　诱发温度
trigly　二氯代三甘醇　Cl(C$_2$H$_4$O)$_2$CH$_4$Cl
triglyceride　甘油三酯；三酸甘油酯
triglycerin　三甘油；三缩丙三醇
triglycidyl isocyanurate(TGIC)　异氰脲酸三缩水甘油酯
triglycol(=triethylene glycol)　三甘醇　HO(CH$_2$CH$_2$O)$_3$H
triglycol dichloride　二氯化三甘醇
triglycol monoacetate　三甘醇单乙酸酯〔增塑剂〕

triglycylglycine 三甘氨酰甘氨酸；三缩四(乙氨酸) $NH_2(CH_2CONH)_3CH_2CO_2H$
triglyme(=triethylene glycol dimethyl ether) 三甘醇二甲醚
Trigly process 特里格里过程；二氯代三甘醇萃取法
trigonal 三角的
trigonal bipyramid 三角双锥
trigonal carbon 三角型碳*
trigonal crystal 三方结晶
trigonal hybridization 三角杂化*
trigonal hybrid orbital 正三角形杂化轨函数
trigonal system 三方晶系
trigonellinamide 胡芦巴酰胺
trigonelline 胡芦巴碱；N-甲基烟酸内盐
triguaiacyl 三(愈创木酚)基；三(邻甲氧苯酚)基
triguaiacyl phosphate 三(愈创木酚)磷酸酯
triguaiacyl phosphite 三(愈创木酚)亚磷酸酯
trihalide 三卤化合物
trihalo-acetyl derivative 三卤乙酰衍生物 CX_3COR
trihalogen acid 三卤酸；三卤代羧酸
trihalogenated 三卤代的
trihalogenated benzene(=trihalogeno-benzene) 三卤(代)苯
trihalogen ester 三卤代酯
trihalogen ether 三卤代醚
trihalogeno-benzene(=trihalogenated benzene) 三卤(代)苯 $C_6H_3X_3$
trihemellitic acid 苯偏三酸 $C_6H_3(COOH)_3$
triheptin 三庚精；甘油三庚酸酯
trihexosan 三聚己糖
tri-n-hexyl aluminum 三正己基铝
trihexylamine 三己胺 $(C_6H_{13})_3N$
trihexylene glycol biborate 二硼酸三(己二醇)酯
trihexyl naphthalene 三己基萘 $C_{28}H_{44}$
trihexylphenidyl 1-环己基-1-苯基-3-(3-哌啶基)丙醇
trihexyphenidyl 苯海索〔药〕
Tri-Homo mill 特赖-霍姆(胶体)磨
trihydrate 三水合物
trihydrazinotriazine 三肼基三嗪〔发泡剂〕
trihydric acid 三价酸
trihydric alcohol 三元醇
trihydric phenol 三元酚
trihydric salt 三酸式盐〔盐中仍含有三个酸性 H〕
trihydrocarbyl derivative 三烃基衍生物
trihydrocarbyl phosphine 三烃基膦
trihydrochloride 三盐酸化物
trihydrocyanic acid 三氢氰酸
trihydrol 三聚水

trihydroxy 三羟基
2,3,4-trihydroxyacetophenone(=alizarine yellow C) 2,3,4-三羟苯乙酮；茜素黄 C $(HO)_3C_6H_2COCH_3$
trihydroxy acid 三羟酸 $(HO)_3RCOOH$
trihydroxy alcohol 三元醇
1,2,3-trihydroxyanthraquinone(=anthragallol) 1,2,3-三羟基蒽醌；蒽棓酚 $C_6H_4(CO)_2C_6H(OH)_3$
trihydroxy benzene 三羟基苯 $C_6H_3(OH)_3$
1,2,3-trihydroxybenzene(=pyrogallol) 1,2,3-三羟基苯；焦棓酚 $(HO)_3C_6H_3$
2,3,4-trihydroxybenzoic acid 2,3,4-三羟基苯甲酸 $(HO)_3C_6H_2CO_2H$
2,3,4-trihydroxy-benzophenone(=allizarin yellow A) 2,3,4-三羟基苯基苯基(甲)酮；茜素黄 A $C_6H_5COC_6H_2(OH)_3$
trihydroxy-benzyl alcohol 三羟基苄醇 $(OH)_3C_6H_2CH_2OH$
trihydroxybutane 三羟基丁烷；丁三醇 $C_4H_7(OH)_3$
trihydroxy-butyraldehyde 三羟基丁醛 $C_3H_4(OH)_3CHO$
trihydroxy-butyric acid 三羟基丁酸 $C_3H_4(OH)_3COOH$
2,3,4-trihydroxybutyrophenone(=4-n-butyryl pyrogallol) 2,3,4-三羟基苯丁酮；4-丁酰焦棓酚 $(HO)_3C_6H_2COC_3H_7$
trihydroxy dibasic acid 三羟基二元酸 $(OH)_3R(COOH)_2$
trihydroxy ethylamine(=triethanol amine) 三羟基乙胺；三乙醇胺
trihydroxy ethylamine oleate 油酸合三羟基乙胺〔乳化剂〕
trihydroxy ethylamine stearate(=triethanolamine stearate) 硬脂酸三乙醇胺
5,6,7-trihydroxyflavone 5,6,7-三羟基黄酮
2,3,4-trihydroxyglutaric acid 2,3,4-三羟基戊二酸 $(CHOH)_3(COOH)_2$
trihydroxymethylaminomethane 三(羟甲基)甲胺 $H_2NC(CH_2OH)_3$
trihydroxy monobasic acid 三羟基一元酸 $(OH)_3RCOOH$
1,3,6-trihydroxynaphthalene 1,3,6-三羟基萘；1,3,6-萘三酚 $C_{10}H_5(OH)_3$
trihydroxy-oestrin 雌三醇
trihydroxy-pyridine 三羟吡啶
trihydroxy-stearic acid 三羟基硬脂酸
2,4,6-trihydroxytoluene 2,4,6-三羟基甲苯；甲苯间三酚 $CH_3C_6H_2(OH)_3$
trihydroxy tribasic acid 三羟基三元酸 $(OH)_3R(COOH)_3$
triindenyl samarium 三茚基钐 $Sm(C_9H_7)_3$
triiodated 三碘化的
triiodide 三碘化物

triiodinated(=triiodizated) 三碘化的
triiodizated(=triiodinated) 三碘化的
triiodoacetic acid 三碘乙酸；三碘醋酸 I_3CCO_2H
triiodo-acetic chloride 三碘乙酰氯 CI_3COCl
triiodo-acetic fluoride 三碘乙酰氟 CI_3COF
triiodo-acetyl chloride 三碘乙酰氯 CI_3COCl
2,4,6-triiodoaniline 2,4,6-三碘苯胺；间三碘苯胺 $C_6H_2I_3NH_2$
triiodo-benzene 三碘(代)苯 $C_6H_3I_3$
1,2,3-triiodobenzene 1,2,3-三碘苯；连三碘苯 $I_3C_6H_3$
2,3,5-triiodobenzoic acid 2,3,5-三碘苯甲酸 $I_3C_6H_2CO_2H$
triiodo-compound 三碘化合物
triiodo cresol 三碘甲酚
triiodo ester 三碘代酯
1,1,1-triiodoethane(=methyl iodoform) 1,1,1-三碘乙烷；甲基碘仿 CH_3CI_3
triiodo ether 三碘代醚
triiodomethane(=iodoform) 三碘甲烷；碘仿 HCI_3
2,4,6-triiodophenol 2,4,6-三碘苯酚；间三碘苯酚 $I_3C_6H_2OH$
triiodothyronine(T3) 三碘甲(状)腺原氨酸
triiodothyropyruvic acid 三碘甲(状)腺丙酮酸
tri-iron tetroxide 四氧化三铁；铁黑
tri-isoamylamine 三异戊胺 $(C_5H_{11})_3N$
tri-isoamyl-boron 三异戊基硼 $(C_5H_{11})_3B$
tri-isoamyl phosphate(TiAP) 磷酸三异戊酯 $(C_5H_{11}O)_3PO$
tri-isoamyltin chloride 三异戊基氯(化)锡 $[(CH_3)_2CHCH_2CH_2]_3SnCl$
triisobutene 三聚异丁烯
triisobutylaluminium(TIBAL) 三异丁基铝
tri-isobutylamine 三异丁胺 $(C_4H_9)_3N$
tri-isobutyl-boron 三异丁基硼 $(C_4H_9)_3B$
tri-isobutylene 三聚异丁烯 $(CH_3)_2C=C[C(CH_3)_3]_2$
triisocyanate 三异氰酸酯
triisodecyl trimellitate 偏苯三酸三异癸酯〔增塑剂〕
triisooctyl phosphate 磷酸三异辛酯〔增塑剂〕
triisooctyl trimellitate 偏苯三酸三异辛酯
tri-isopropanolamine 三异丙醇胺 $(HOC_3H_6)_3N$
tri-isopropyl amine 三异丙胺
triisopropylphenylsulfonyl chloride 三异丙基苯磺酰氯
triisopropyl phosphite 亚磷酸三异丙酯
triisopropylthiuram disulfide 二硫化三异丙基秋兰姆〔促进剂〕
triisovalerin 三异戊精；甘油三异戊酸酯 $(C_4H_9COO)_3C_3H_5$
triketohydrindene hydrate (水合)茚三酮

triketone 〔类名〕 三酮
triketopentane 戊(烷)三酮
trilactatozirconate 三乳酸锆酸盐
trilafon 奋乃静〔药〕
trilaminar(=trilaminate; trilaminated) 三层的
trilaminate(=trilaminar) 三层的
trilaminated(=trilaminar) 三层的
trilateral 三边的
trilateral cross section 三角形截面；三边形截面
trilaurin 三月桂精；甘油三月桂酸酯 $C_2H_5[OCO(CH_2)_{10}CH_3]_3$
trilaurylamine(TLA) 三月桂胺 $(C_{12}H_{25})_3N$
trilaurylamine oxide(TLAO) 三月桂胺氧化物 $(C_{12}H_{25})_3NO$
trilauryl phosphite 亚磷酸三月桂酯〔防老剂〕
trilauryl thiophosphite 硫代亚磷酸三月桂酯〔防老剂〕
tri-lead tetroxide 四氧化三铅；红丹〔俗〕
trilinear coordinates 三直线坐标
trilinear direct decomposition algorithm 三线性直接分解(算)法
trilinolein(=linolein) 三亚油精；甘油三亚油酸酯
trilinolenin 三亚麻精；甘油三亚麻酸酯
trilit(e)(=TNT) 三硝基甲苯
trillenoside 延龄草苷
trillion ①万亿；10^{12} ②$10^{18}$
trillion becquerel(=terabecquerel)(TBq) 万亿贝可勒尔
trilobal 三叶形(的)
trilobin 三叶素
trilobine 三叶(木防己)碱
trim ①修剪 ②密封面
trimebutine 曲美布汀〔药〕
trimelissin 甘油三蜂花酸酯
trimellitate (偏)苯三酸盐(酯)
trimellitic acid 1,2,4-苯三酸；偏苯三酸 $C_6H_3(COOH)_3$
trimellitic anhydride(TMA) 1,2,4-苯三酸酐
trimer 三聚体
trimerisation 三聚(作用)
trimesanthracenobenzene 三中蒽并苯
trimesic acid 1,3,5-苯三酸 $C_6H_3(COOH)_3$
trimesitinic acid 1,3,5-苯三酸 $C_9H_6O_6$
tri-meta-cresyl borate 硼酸三(间甲苯酯)
trimetallic catalyst 三金属催化剂
trimetazidine 曲美他嗪〔药〕
trimethadione 3,5,5-三甲基-2,4-二氧噁唑烷；三甲双酮〔抗癫痫药〕
trimethano- 〔词头〕 三(甲桥)
trimethide 三甲基金属 $M(CH_3)_3$
trimethoprim 三甲氧苄二氨嘧啶〔抗菌素用于治疗疟

疾、呼吸道或尿道感染〕
trimethoxy 三(甲氧基)
2,3,4-trimethoxybenzoic acid 2,3,4-三甲氧基苯甲酸 $(CH_3O)_3C_6H_2COOH$
trimethoxy-boron 三甲氧基硼；硼酸三甲酯 $B(OCH_3)_3$
trimethoxy methylol melamine 三甲氧基羟甲基蜜胺〔交联剂〕
2,4,5-trimethoxyphenyl 三甲氧苯基 $(CH_3O)_3C_6H_2$—
trimethoxyvinylsilane 三甲氧基乙烯基甲硅烷
trimethyl 三甲基
trimethyl-acetaldehyde 三甲基乙醛；叔戊醛 $(CH_3)_3CCHO$
trimethylacetaldoxime 三甲基乙醛肟；叔戊醛肟 $(CH_3)_3CCH=NOH$
trimethylacetic acid 三甲基乙酸；叔戊酸 $(CH_3)_3CCO_2H$
trimethylacetonitrile 三甲基乙腈；叔戊腈 $(CH_3)_3CCN$
2,4,6-trimethylacetophenone 2,4,6-三甲基苯乙酮 $(CH_3)_3C_6H_2COCH_3$
trimethyl-acetyl chloride 三甲基乙酰氯
trimethylalkylammonium 三甲基烷基铵
trimethyl aluminium 三甲基铝 $Al(CH_3)_3$
trimethylamine 三甲胺 $(CH_3)_3N$
trimethylamine hydrochloride 盐酸三甲胺 $(CH_3)_3NHCl$
trimethylamine methacrylate 三甲胺甲基丙烯酸酯
trimethylamine oxidase 三甲胺氧化酶
trimethyl-amine oxide 氧化三甲胺 $(CH_3)_3N=O$
trimethylammonium ion 三甲铵离子 $(CH_3)_3N^+H$
2,4,6-trimethylaniline(=mesidine) 2,4,6-三甲基苯胺；米胺
2,4,5-trimethylaniline(=pseudocumidine) 假枯胺；2,4,5-三甲基苯胺 $(CH_3)_3C_6H_2NH_2$
2,4,5-trimethylanilino(=pseudocumidino) 假枯氨基；2,4,5-三甲苯氨基
trimethyl-arsine 三甲胂 $As(CH_3)_3$
trimethyl-arsine dibromide 二溴化三甲胂 $(CH_3)_3AsBr_2$
trimethylarsine oxide 氧化三甲胂 $(CH_3)_3AsO$
3,4,5-trimethylbenzaldehyde 3,4,5-三甲基苯甲醛 $C_6H_2(CH_3)_3CHO$
1,2,3-trimethylbenzene 1,2,3-三甲基苯 $(CH_3)_3C_6H_3$
1,3,5-trimethylbenzene(=mesitylene) 1,3,5-三甲基苯；茉
1,2,4-trimethylbenzene(=pseudocumene) 假枯烯；1,2,4-三甲基苯 $C_6H_3(CH_3)_3$
2,3,4-trimethylbenzoic acid 2,3,4-三甲基苯甲酸 $(CH_3)_3C_6H_2CO_2H$
trimethyl-bismuthine 三甲基铋
trimethyl borate 硼酸三甲酯 $(CH_3O)_3B$
trimethyl borine 三甲基硼 $B(CH_3)_3$

trimethyl-boron 三甲基硼 $(CH_3)_3B$
2,3,3-trimethyl-1-butene 2,3,3-三甲基-1-丁烯 $(CH_3)_3CC(CH_3)=CH_2$
trimethyl carbinol 三甲基甲醇
trimethylcetylammonium bromide 三甲基十六(烷)基溴化铵
trimethylchlorosilane 三甲基氯硅烷
trimethylchloro-silicane 三甲基氯硅 $(CH_3)_3SiCl$
trimethyl citrate 柠檬酸三甲酯 $(CH_3)_3C_6H_5O_7$
trimethyl cyanurate 氰尿酸三甲酯；三聚氰酸三甲酯 $C_3N_3(OCH_3)_3$
trimethyl-cyclohexane 三甲基环己烷
3,5,5-trimethylcyclohexen-2-one-1(=isophorone) 3,5,5-三甲基-2-环己烯-1-酮；异佛尔酮
trimethyl cyclohexenyl butenone 三甲基环己烯基丁烯酮
trimethyl-cyclopentene 三甲基环戊烯
trimethyldihydroquinoline 三甲基二氢喹啉〔防老剂〕
trimethylene ①(=cyclopropane)环丙烷②1,3-亚丙基 —$CH_2CH_2CH_2$—
trimethylene acetal 1,3-丙二醇缩乙醛 $CH_3CHO(CH_2)_3O$
trimethylene bromide 亚丙基二溴；1,3-二溴代丙烷 $BrCH_2CH_2CH_2Br$
trimethylene bromohydrin 亚丙基溴醇；3-溴-1-丙醇 $Br(CH_2)_3OH$
trimethylene chlorobromide(=3-chloro-1-bromopropane) 3-氯-1-溴丙烷 $Cl(CH_2)_3Br$
trimethylene chlorohydrin 3-氯-1-丙醇 $Cl(CH_2)_3OH$
trimethylene cyanide 戊二腈
trimethylene diamine 亚丙基二胺；1,3-丙二胺 $NH_2(CH_2)_3NH_2$
trimethylene diamine-N,N'-diacetic-N,N'-dipropionic acid 丙二胺-N,N'-二乙酸-N,N'-二丙酸；丙二胺二乙二丙酸
trimethylene diisocyanate 亚丙基二异氰酸酯；丙撑二异氰酸酯
trimethylene dimercaptan 亚丙基二硫醇；1,3-丙二硫醇 $HS(CH_2)_3SH$
trimethylenedinitrilo-tetraacetic acid 丙二胺四乙酸
trimethylene-formal(=1,3-dioxan) 亚丙基甲缩醛；1,3-二噁烷；1,3-丙二醇缩甲醛；间二氧杂环己烷 $CH_2O(CH_2)_3O$
trimethylene glycol 亚丙基二醇；1,3-丙二醇 $HO(CH_2)_3OH$
trimethylene glycol diacetate 亚丙基二醇二乙酸酯；1,3-丙二醇二乙酸酯 $CH_2(CH_2O_2CCH_3)_2$
trimethylene glycol dibutyrate 1,3-丙二醇二丁酸酯

CH$_2$(CH$_2$O$_2$CC$_3$H$_7$)$_2$
trimethylene group　1,3-亚丙基
trimethylene iodohydrin　3-碘-1-丙醇　I(CH$_2$)$_3$OH
trimethylene oxide　氧杂环丁烷　(CH$_2$)$_3$O
trimethylene sulfide　亚丙基；硫杂丁环
trimethylene-trinitramine　三亚甲基三硝基胺
trimethyl-ethylene(=2-methyl-2-butene)　三甲基乙烯；2-甲基-2-丁烯　(CH$_3$)$_2$C=CHCH$_3$
trimethyl gallic acid　三甲基五味子酸
trimethyl-gallium　三甲基镓
trimethylgalloyl azide　叠氮三甲基棓酰
2,3,6-trimethyl glucose　2,3,6-三甲基葡萄糖
trimethyl-glycine　三甲铵基乙内盐；甜菜碱
tri-1-methylheptylphosphate(TMHP)　磷酸三(1-甲基庚基)酯　[C$_6$H$_{13}$CH(CH$_3$)O]$_3$PO
trimethylhexamethylenediamine　三甲基六亚甲基二胺；三甲己撑二胺
trimethyl hexamethylene diisocyanate　三甲基六亚甲基二异氰酸酯〔硫化剂〕
3,3,5-trimethyl hexanone peroxide　3,3,5-三甲基环己酮过氧化物〔不饱和聚酯树脂固化剂〕
trimethyl hexyl phthalate　邻苯二甲酸三甲基己酯〔增塑剂〕
trimethylin　三甲灵；甘油三甲基醚　C$_3$H$_5$(OCH$_3$)$_3$
trimethyl lauryl ammonium chloride　氯化三甲基十二烷基铵
trimethyl-methane　三甲基甲烷
trimethyl-naphthalene　三甲基萘
trimethyl octadecyl ammonium chloride　氯化三甲基十八烷基铵
trimethylol ethane　三羟甲基乙烷
trimethylol ethane trimethacrylate　三羟甲基乙烷三甲基丙烯酸酯
trimethylol propane　三羟甲基丙烷
trimethylolpropane polyglycidylether　三羟甲基丙烷多缩水甘油醚
trimethylolpropane polyglycidylether polyacrylate　三羟甲基丙烷多缩水甘油醚多丙烯酸酯
trimethylolpropane triester　三羟甲基丙烷三酯
trimethylolpropane trilaurate　三羟甲基丙烷三月桂酸酯
trimethylolpropane trimethacrylate　三羟甲基丙烷三甲基丙烯酸酯
trimethylolpropane tri(toluene diisocyanate)　三羟甲基丙烷三(甲苯二异氰酸酯)
trimethylol urea　三羟甲基脲
trimethylolpropane triacrylate(TMPTA)　三羟甲基丙烷三丙烯酸酯
trimethyl oxosulfonium iodide　碘化三甲氧硫鎓　(CH$_3$)$_2$S$^+$OI$^-$

2,2,3-trimethyl-pentane　2,2,3-三甲基戊烷　(CH$_3$)$_3$CCH(CH$_3$)C$_2$H$_5$
2,2,4-trimethyl-1,3-pentanediol(=neopentyl glycol)　三甲基戊二醇；新戊二醇
2,4,4-trimethyl-2-pentanol　2,4,4-三甲-2-戊醇　(CH$_3$)$_3$CCH$_2$C(OH)(CH$_3$)$_2$
2,3,3-trimethyl-1-pentene　2,3,3-三甲基-1-戊烯　C$_2$H$_5$C(CH$_3$)$_2$C(CH$_3$)=CH$_2$
2,4,5-trimethylphenol　2,4,5-三甲苯酚　(CH$_3$)$_3$C$_6$H$_2$OH
2,4,6-trimethyl phenol(=mesitol)　2,4,6-三甲苯酚；栾酚
trimethyl phenoxysilane　三甲基苯氧(甲)硅烷
2,3,5-trimethylphenyl　2,3,5-三甲苯基　(CH$_3$)$_3$C$_6$H$_2$—
trimethyl phenyl ammonium iodide　碘化三甲基苯基铵
2,4,6-trimethylphenyl-p-phenylenediamine　2,4,6-三甲苯基对苯二胺〔防老剂〕
trimethyl phosphate　磷酸三甲酯　(CH$_3$O)$_3$PO
trimethyl-phosphine　三甲基膦　(CH$_3$)$_3$P
trimethylphosphine oxide　氧化三甲基膦　(CH$_3$)$_3$PO
trimethyl phosphite　亚磷酸三甲酯
2,4,6-trimethyl-piperidine　2,4,6-三甲基哌啶　C$_8$H$_{11}$N
2,3,5-trimethylpyrazine　2,3,5-三甲基吡嗪
2,3,4-trimethylpyridine　2,3,4-三甲基吡啶
2,3,4-trimethylquinoline　2,3,4-三甲基喹啉　(CH$_3$)$_3$C$_9$H$_4$N
trimethyl silanol　三甲基硅醇；三甲基羟基硅烷
trimethylsiloxane group　三甲基硅氧基
trimethylsilyl　三甲代甲硅烷基
trimethyl silyl acetate　三甲基甲硅烷乙酸酯
trimethylsilylated polysilicic acid　三甲基甲硅烷基化聚硅酸
trimethyl silylation　三硅烷基化作用
trimethylsilyl-blocked polyorganosiloxane　三甲代甲硅烷基封端的聚有机硅氧烷
trimethylsilyl chloride　氯化三甲基硅烷
trimethylsilyl end-blocked　三甲代甲硅烷基封端的
trimethylsilyl ether　三甲基硅醚
trimethylsilyl isocyanate　三甲基甲硅烷(基)异氰酸酯
trimethylsilyl methacrylate　甲基丙烯酸三甲代甲硅酯
trimethyl-stibine(=trimethylantimony)　三甲䏲；三甲基锑　Sb(CH$_3$)$_3$
trimethyl-succinic acid　三甲基丁二酸　COOHCH(CH$_3$)C(CH$_3$)$_2$COOH
1,1,6-trimethyltetralin　1,1,6-三甲基四氢化萘；紫罗烯
tri-methylthiazine　三甲基噻嗪
2,4,5-trimethylthiazole　2,4,5-三甲基噻唑
trimethyl thiourea　三甲基硫脲〔促进剂〕
trimethyl tin　①三甲基锡(游基)②六甲二锡③三甲锡(基)
trimethyl tin bromide　溴化三甲锡　SnBr(CH$_3$)$_3$

trimethyltin chloride 氯化三甲基锡
trimethyl tin hydride 氢化三甲锡 $Sn(CH_3)_3H$
trimethyl tin hydroxide 氢氧化三甲锡 $Sn(OH)(CH_3)_3$
trimethyl tin oxide 氧化双三甲锡 $(CH_3)_3SnOSn(CH_3)_3$
trimethyl-tin-radical ①三甲基锡(游基)②三甲基锡基
trimethyl-tin sulfide 硫化三甲锡
trimethyl trimellitate 偏苯三酸三甲酯(增塑剂)
trimethyltrithiophosphine 三甲基硫磷(燃料)
trimethyl-tryptophane 三甲基色氨酸；刺桐子氨酸
trimethyl-urea 三甲基脲 $CH_3NHCON(CH_3)_2$
trimethyl-uric acid 三甲基尿酸 $C_5HN_4O_3(CH_3)_3$
trimethyl-xanthine 三甲基黄质；咖啡碱
trimetric 斜方(晶)的
trim joist 过梁
trim line 裁切线
trim materials 修整材料
trimmed size 成品规格；切边后规格
trimmer 修边机
trimming 修剪；修边
trimming and sorting 削整及分皮
trimming coil 塔顶旋管
trimming condenser 微调电容器
trimming machine 修整机
trimming oil 塔顶回流油
trimmings〔复〕 切屑；刨屑；碎皮
trimolecular 三分子的
trimolecular reaction 三分子反应
trimorphism 三晶(现象)
trimorphous 三晶形的
trim paint 门窗漆
trimyristicin(=glyceryl trimyristate) 三肉豆蔻酸甘油酯
trimyristin 三肉豆蔻精；甘油三(十四酸)酯 $(C_{13}H_{27}COO)_3C_3H_5$
trinaphthylene 联三萘
trineutron 三中子
Trinidad asphalt 特里尼达沥青
trinifer 三引发-转移剂
trinitrate 三硝酸酯(盐)
trinitration 三硝基化(作用)
trinitride 叠氮化物 RN_3
trinitrin 三硝基甘油；三硝酸甘油酯
1,2,5-trinitroacenaphthalene 1,2,5-三硝基苊 $(NO_2)_3C_{12}H_5$
trinitroacetonitrile 三硝基乙腈 $(NO_2)_3CCN$
2,4,6-trinitroaminophenol 2,4,6-三硝基氨基苯酚 $C_6H(NO_2)_3(NH_2)OH$
2,4,6-trinitroaniline 2,4,6-三硝基苯胺；苦基胺 $NH_2C_6H_2(NO_2)_3$
2,4,6-trinitroanisole 2,4,6-三硝基苯甲醚 $(NO_2)_3C_6H_2OCH_3$
2,4,6-trinitrobenzaldehyde 2,4,6-三硝基苯甲醛 $(NO_2)_3C_6H_2CHO$
1,2,3-trinitrobenzene 1,2,3-三硝基苯 $(NO_2)_3C_6H_3$
1,3,5-trinitrobenzene 1,3,5-三硝基苯
2,4,6-trinitrobenzoic acid 2,4,6-三硝基苯甲酸；三硝基安息香酸 $C_6H_2(NO_2)_3CO_2H$
trinitrobenzoic acid 三硝基苯甲酸
trinitro-t-butyl-toluene 三硝基叔丁基甲苯；甲苯麝香 $CH_3C_6HC(CH_3)_3(NO_2)_3$
trinitro-t-butylxylene 三硝基叔丁基二甲苯 $(NO_2)_3C_6(CH_3)_2C(CH_3)_3$
trinitro-cellulose 三硝基纤维素
trinitro-chlorobenzene 三硝基氯苯
trinitro-compound 三硝基化合物
trinitrocresol 三硝基甲酚
trinitro-m-cresol 三硝基间甲苯酚 $(NO_2)_3C_6H(CH_3)OH$
trinitrodichlorobenzene 三硝基二氯苯
1,1,1-trinitroethane 1,1,1-三硝基乙烷 $CH_3C(NO_2)_3$
2,4,7-trinitro fluorenone 2,4,7-三硝基芴酮
trinitro-glycerin 三硝基甘油
trinitrol 三硝油(俗)；季戊四醇四硝酸酯
trinitro-mesitylene 三硝基䓛；2,4,6-三硝基-1,3,5-三甲苯 $(O_2N)_3C_6(CH_3)_3$
trinitromethane(=nitroform) 三硝基甲烷；硝仿 $(NO_2)_3CH$
trinitro-methylamine 三硝基甲胺
trinitronaphthalene 三硝基萘
trinitro-$α$-naphthol 三硝基-$α$-萘酚 $(NO_2)_3C_{10}H_4OH$
trinitro-orcinol 三硝基苔黑酚；2,4,6-三硝基-5-甲基-1,3-苯二酚 $(NO_2)_3C_6(OH)_2CH_3$
2,4,6-trinitrophenol 2,4,6-三硝基苯酚；苦味酸
trinitrophenol 三硝基苯酚 $(NO_2)_3C_6H_2OH$
trinitro-phenoxide 三硝基酚盐 $C_6H_2(NO_2)_3OM$
trinitrophenyl-hydrazine 三硝苯基肼 $(NO_2)_3C_6H_2N_2H_3$
2,4,6-trinitrophenylmethyl nitramine(=tetryl) 特屈儿；2,4,6-三硝基苯甲硝胺 $(NO_2)_3C_6H_2N(CH_3)NO_2$
trinitroresorcin 三硝基间苯二酚
trinitro-resorcinol 三硝基间苯二酚；2,4,6-三硝基-1,3-苯二酚 $(NO_2)_3C_6H(OH)_2$
trinitroso-trimethylene triamine 三亚硝基三亚甲基三胺
trinitrosotrimethyl triamine 三亚硝基三甲基三胺〔发泡剂〕
trinitro-toluene 三硝基甲苯 $CH_3C_6H_2(NO_2)_3$
2,4,6-trinitrotoluene(=TNT) 2,4,6-三硝基甲苯；茶色炸药
2,3,4-trinitrotoluene(=$β$-trinitrotoluene) 2,3,4-三硝基甲苯；$β$-三硝基甲苯 $(NO_2)_3C_6H_2CH_3$

trinitro-triazidobenzene 三硝基三叠氮苯
2,4,6-trinitro-1,3,5-trimethylbenzene 2,4,6-三硝基-1,3,5-三甲苯 $(NO_2)_3C_6(CH_3)_3$
trinitrotriphenyl-carbinol 三(硝基苯基)甲醇 $(NO_2C_6H_4)_3COH$
trinitrotriphenyl-methane 三(硝基苯基)甲烷 $(NO_2C_6H_4)_3CH$
trinitro-xylene 三硝基二甲苯
2,3,6-trinitro-p-xylene 2,3,6-三硝基对二甲苯 $(NO_2)_3C_6H(CH_3)_2$
trinol(=TNT) 三硝基甲苯
tri(nonylphenyl) phosphite 亚磷酸三(壬基苯)酯〔防老剂〕
trinonyl phosphate(TNP) 磷酸三壬酯 $(C_9H_{19}O)_3PO$
trinuclear(=trinucleate; trinucleated) 三环的；三核的
trinuclear dye 三核染料
trinucleate(=trinuclear) 三环的；三核的
trinucleated(=trinuclear) 三环的；三核的
trioazole 三噁唑；三氧氮五环 CHO_3N
trioctadecyl amine 三(十八烷基)胺
trioctylamine(TOA) 三辛胺 $(C_8H_{17})_3N$
trioctyl (mono) methylammonium chloride(TOM-ACl) 氯化三辛基甲基铵 $(C_8H_{17})_3(CH_3)NCl$
trioctyl phosphate 磷酸三辛酯；磷酸辛酯
trioctylphosphine oxide(TOPO) 氧化三辛基膦 $(C_8H_{17})_3PO$
trioctylphosphine sulfide(TOPS) 硫化三辛基膦 $(C_8H_{17})_3PS$
trioctyl trimellitate 偏苯三酸三辛酯
triode 三极(真空)管
triode argon detector 三极管氩检测器
-triol〔词尾〕三醇
triol adduct 三醇加成物；三醇加合物
triolefin process 三烯法〔由丙烯制丁烯及乙烯法〕
triolein(=olein) 油精；三油精；三油酸甘油酯 $(C_{17}H_{33}COO)_3C_3H_5$
triol ester 三元醇酯
trioleyl phosphate 磷酸三油基酯
trioleyl phosphite 亚磷酸三油(醇)酯
trional(=methyl ethyl ketone disulfone) 台俄那〔药〕；三乙眠砜；2,2-二(乙砜基)丁烷 $(C_2H_5)(CH_3)C=(SO_2C_2H_5)_2$
-trione〔词尾〕三酮
triorganophosphite 亚磷酸三有机(基)酯
triorganotin compound 三有机锡化合物
triose 丙糖 $C_3H_6O_3$
triosephosphoric acid 磷酸三糖 $C_3H_5O_2OPO_3H_2$
trioxa- 三噁；三氧杂
trioxa-bicyclooctane 3,6,8-三噁二环[3.2.1]辛烷

trioxalatoferriate 三草酸根合铁酸盐
trioxan(e) 三噁烷；三氧杂环己烷 $(CH_2O)_3$
trioxide 三氧化物
trioxime 三肟
trioximido〔词头〕三肟基
trioximido-propane 三肟基丙烷
trioxin 三噁英；三聚甲醛
trioxy ①三氧代②三羟代
trioxysulfotungstate 三氧硫钨酸盐；一硫代钨酸盐 $M_2[WO_3S]$
tripalmitin(=palmitin) 棕榈精；三棕榈精；(三)棕榈酸甘油酯 $(C_{15}H_{31}COO)_3C_3H_5$
tripan red 锥虫红
tripelennamine 曲吡那敏〔药〕
tripelennamine hydrochloride 苄吡二胺；N-苄基-N',N'-二甲基-N-2(2-吡啶基)乙二胺盐酸盐〔抗组胺药〕
triperchromic acid 三过氧铬酸 H_3CrO_8
tripestone 弯硬石膏
triphane 锂辉石
triphasic 三相的
triphen〔词头〕三吩-；三苯-
tri-phenarsazine chloride 氯化三吩吡嗪
triphenol 三酚
triphenyl 三苯基
triphenylacetic acid 三苯基乙酸 $(C_6H_5)_3CCO_2H$
triphenyl aluminum 三苯基铝 $Al(C_6H_5)_3$
triphenylamine 三苯胺 $(C_6H_5)_3N$
triphenyl antimony 三苯锑
triphenylarsine 三苯胂 $As(C_6H_5)_3$
triphenyl-arsine cyanobromide 氰溴化三苯基胂 $(C_6H_5)_3AsBrCN$
triphenyl arsine methylene 亚甲基三苯胂 $(C_6H_5)_3As=CH_2$
triphenyl arsine oxide 氧化三苯基胂 $(C_6H_5)_3AsO$
triphenyl-arsine sulfide 硫化三苯基胂 $(C_6H_5)_3As=S$
1,3,5-triphenylbenzene 1,3,5-三苯基苯；间三苯基苯 $(C_6H_5)_3C_6H_3$
triphenyl bismuth 三苯基铋 $Bi(C_6H_5)_3$
triphenyl borate 硼酸三苯酯〔防老剂〕
triphenyl borine 三苯基硼 $B(C_6H_5)_3$
triphenyl-boron 三苯基硼 $(C_6H_5)_3B$
triphenylbromomethane 三苯基溴甲烷 $(C_6H_5)_3CBr$
triphenylcarbinol 三苯基甲醇 $(C_6H_5)_3COH$
triphenylcarbinol methyl ether 三苯基甲基甲基醚 $(C_6H_5)_3COCH_3$
triphenylcarbonium ion 三苯基碳离子
triphenylchloromethane 三苯基氯甲烷 $(C_6H_5)_3CCl$
triphenylchlorosilicane 三苯氯硅 $(C_6H_5)_3SiCl$

triphenylene 苯并[9,10]菲 $C_{18}H_{12}$
1,1,2-triphenylethane 1,1,2-三苯基乙烷 $CH(C_6H_5)_2CH_2(C_6H_5)$
α-triphenylguanidine α-三苯胍 $C_6H_5N=C(NHC_6H_5)_2$
triphenylhydrazine 三苯肼 $(C_6H_5)_2NNHC_6H_5$
triphenylmethane(=tritane) 三苯甲烷 $(C_6H_5)_3CH$
triphenyl lead acetate 三苯基乙酸铅〔有机铅防污毒料〕
triphenylmethane dye 三苯甲烷染料
triphenylmethane indicator 三苯甲烷指示剂
triphenylmethane lactone leuco dye 三苯甲烷内酯无色染料
triphenylmethane leuco-dyes 三苯甲烷无色染料
triphenylmethane triisocyanate 三苯甲烷三异氰酸酯〔硫化剂〕
triphenyl methide 三苯甲基化物
triphenyl methoxide 三苯基甲醇盐 $(C_6H_5)_3COM$
triphenylmethyl ①三苯甲(游)基②三苯甲基
triphenylmethylarsonium ion 三苯甲基砷鎓离子
triphenyl methylation(=tritylation) 三苯甲基化作用
triphenylmethyl cellulose 三苯甲基纤维素
triphenylmethyl chloride 三苯基氯甲烷
triphenyl orthoformate 原甲酸三苯酯 $HC(OC_6H_5)_3$
triphenyl-oxazole 三苯基㗁唑
triphenyl peroxide 过氧化三苯基 $[(C_6H_5)_3CO]_2$
triphenyl phosphate 磷酸三苯酯 $(C_6H_5O)_3PO$
triphenyl phosphine 三苯膦
triphenyl phosphine methylene 亚甲基三苯膦 $(C_6H_5)_3P=CH_2$
triphenyl phosphine oxide 氧化三苯膦 $(C_6H_5)_3P=O$
triphenylphosphine sulfide 硫化三苯膦 $(C_6H_5)_3PS$
triphenyl phosphite 亚磷酸三苯酯 $(C_6H_5O)_3P$
triphenyl (2-pyridylthio) tin N-oxide 三苯基(2-吡啶基硫代)锡 N-氧化物
triphenyl pyrylium perchlorate 三苯基吡喃鎓过氯酸盐
triphenylrosaniline sulfate 硫酸三苯蔷薇苯胺 $(C_{38}H_{32}N_3)_2SO_4$
triphenylselenonium 三苯硒
triphenylselenonium iodobismuthite 碘亚铋酸三苯硒(鎓)
triphenylsilyl 三苯甲硅烷基 $(C_6H_5)_3Si—$
triphenyl-stibine(=triphenylantimony) 三苯䏲；三苯锑 $Sb(C_6H_5)_3$
triphenyl-succinic anhydride 三苯基丁酸酐 $(C_6H_5)_3C_4HO_3$
triphenyltetrazolium chloride 氯化三苯基四唑(鎓)
triphenyl-thiophosphate 硫代磷酸三苯酯 $(C_6H_5O)_3PS$
triphenyl-thiopyrylium perchlorate 三苯基硫代吡喃鎓过氯酸盐
triphenyltin acetate 乙酸三苯基锡
triphenyltin acrylate 丙烯酸三苯基锡
triphenyltin aminoacetate 氨基乙酸三苯基锡
triphenyltin aminopropionate α-氨基丙酸三苯基锡
triphenyltin chloride 氯化三苯锡 $(C_6H_5)_3SnCl$
triphenyltin chloroacetate 氯乙酸三苯基锡
triphenyltin compound 三苯基锡化合物
triphenyltin α'-dibromosuccinate α'-二溴琥珀酸三苯基锡
triphenyltin dimethyl dithiocarbamate 二硫代二甲氨基甲酸三苯基锡
triphenyltin N-dimethylthiocarbamate 硫代二甲氨基甲酸三苯基锡
triphenyltin fluoride 氟化三苯基锡
triphenyltin hydride 氢化三苯基锡
triphenyltin methacrylate 甲基丙烯酸三苯基锡
triphenyltin monochloroacetate 一氯代乙酸三苯基锡
triphenyltin nicotinic acid 三苯基锡吡啶甲酸；三苯基锡烟酸
triphenyltin oxide 氧化三苯基锡
triphenyltin pyridine carboxalate 3-吡啶单羧酸三苯基锡
triphenyltin versatic acid 三苯基锡叔碳酸
triphosgene 三光气
triphosphane 三膦；三膦烷 H_2PPHPH_2
triphosphate 三磷酸盐(根) $M_5P_3O_{10}; P_3O_{10}^{5-}$
triphosphine 三膦；三膦烷 H_2PPHPH_2
triphosphopyridine nucleotide 三磷酸吡啶核苷酸；辅酶Ⅱ
triphylite 磷酸锂铁矿
triple ①三(倍)的②(重复)三分的③加三倍
triple beam balance 三梁天平；三梁秤
triple bond 三键
triple-bonded 三键的
triple carbon-to-carbon linkage 三重碳-碳键
triple chloride 三重氯化物
triple cloud points 三浊点
triple collision 三分子碰撞
triple conjugation 三共轭(化)
triple dye 三重染料
triple effect 三效(式)
triple effect evaporation 三效蒸发
triple effect evaporator 三效蒸发器
triple filament method 三灯丝方法
triple ion(=ion triplet) 三重离子；离子三重态；离子三聚体
triple-jet experiment 三(向)射流实验
triple link(=triple linkage) 三键
triple linkage(=triple link) 三键
triple-linked 三键的
triple nitrite reagent 三重亚硝酸盐试剂
triple O-ring shape seal 三重 O 形环密封

triple particle 三重态粒子
triple phosphate 三重磷酸盐
triple point 三相点*；三态点
triple pressed stearic acid 三压硬脂酸
triple quadrupole mass spectrometer 三级四极杆质谱仪
triple resonance 三共振
triple-roller mill 三辊磨
triple seal 三重密封
triple stage quadrupole 三级四极
triple stage quadrupole mass spectrometer 三级四极质谱仪
triple stage quadrupole mass spectrometry 三级四级质谱法
triple-substituted 三代的；三元取代的
triple substitution 三元取代
triple super phosphate 重过磷酸钙
triplet ①三(线)态*②三重峰*
triple tandem quadrupole system 三级串联式四极系统
triplet mechanism 三重态机理
triplet particle 三重态粒子
triplet state 三重(线)态
triple valve 三通阀
triplex ①三(倍)的；(有)三部的②三缸
triplex board 三层纸板
triplex pump 三缸泵
triplex reciprocating pump 三缸往复泵
triplicate ①三份*②三倍的
triplite 磷铁锰矿
tripod 三脚架；三角架
tripoli 硅藻土；磨石
tripoli earth 硅藻土
tripolite 硅藻土
tripotassium phosphate process 磷酸三钾法
tripper 倾料器
trippkeite 软砷铜矿
tripple segmental baffle 三重弓形折流板
triprolidine 曲普利啶〔药〕
triprolidine hydrochloride 盐酸曲普利啶；反(式)2-[3-(1-吡咯烷基)对甲苯丙烯基]吡啶盐酸盐；2-(1-(4-甲基苯基)-3-(1-吡咯烷基)-1-丙烯基)吡啶盐酸盐〔抗过敏药〕
tripropanolamine 三(羟丙基)胺；三丙醇胺 $N(CH_2CHOHCH_3)_3$
tripropionin 三丙酸甘油酯
tripropoxy-boron 三丙氧基硼；硼酸三丙酯 $B(OC_3H_7)_3$
tripropoxy-silane 三丙氧基甲硅烷
tripropoxy-silicane 三丙氧基甲硅烷 $(C_3H_7O)_3SiH$
tripropyl 三丙基 $(CH_3CH_2CH_2)_3$
tripropyl amine 三丙胺 $(C_3H_5CH_3)_3N$

tripropyl boron 三丙硼 $(C_3H_7)_3B$
tripropyl orthoformate 原甲酸三丙酯 $HC(OCH_2C_2H_5)_3$
tri-n-propyl phosphate 磷酸三-正丙酯
triprotic acid (=tribasic acid) 三元酸
triptane 2,2,3-三甲基丁烷
tripterygine 雷公藤碱；雷公藤红
tripterysium glucosides 雷公藤多苷〔药〕
triptorelin 曲普瑞林〔药〕
triptycene 三蝶烯
triptych silicon compound 三幅联硅化合物
triptyl radical 三蝶烯基
trip valve 切断阀
tripyrrole 三吡咯 $C_{12}H_{15}N_3$
triquinoyl ①环己六酮 $(CO)_6$ ②三醌基
trireactive glyceride 三反应性甘油酯
triricinoleidin(=triricinolein) 三蓖麻精；甘油三蓖麻酸酯
triricinolein(=triricinoleidin) 三蓖麻精；甘油三蓖麻酸酯
tris- 三(个)
trisaccharide 三糖类
trisaccharide maltose 麦芽三糖
tris-(alkylphenyl-*t*-butyltin)chlorid 氯化三(烷基苯叔丁基锡；三烷基苯叔丁基氯化锡
tris[2-(2-aminoethyl)-amino ethanolato](2-propanolato) titanium 三[2-(2-氨乙基)-氨基乙氧基](2-丙氧基)钛
tris(aralkyl) phosphine 三芳烷基膦
trisaturated glyceride 三饱和酸甘油酯
trisazo 三偶氮(基)
tris-azo compound 三偶氮化合物
trisazo dye 三偶氮染料
trisazo pigment 三偶氮颜料
tris buffer 三羟甲基氨基甲烷缓冲液
tris buffer solution 三(羟甲基)氨基甲烷缓冲液
tris(β-carboxyethyl) isocyanate 三(β-羧乙基)异氰酸酯
tris (chloroethyl) phosphate 磷酸三(氯乙基)酯
tris(2-chloroethyl) phosphate 磷酸三(2-氯乙酯)
tris(2-chloropropyl) phosphite 亚磷酸三(2-氯丙基)酯
tris (cyclopentadienyl) cerium 三(环戊二烯基)铈；三茂铈
tris-cyclopentadienyl-cyclohexyloxy-uranium 三茂环己氧基铀 $(C_5H_5)_3U(OC_6H_{11})$
tris-cyclopentadienyl-*n*-hexyloxy-uranium 三茂正己氧基铀 $(C_5H_5)_3U(OC_6H_{13})$
tris-cyclopentadienyl-*n*-neptunium fluoride 氟化三茂镎 $(C_5H_5)_3NpF$
tris-cyclopentadienyl-*n*-uranium chloride 氯化三茂铀 $(C_5H_5)_3UCl$
tris(2,3-dibromopropyl) phosphate 磷酸三(2,3-二溴丙基)酯

tris(diethyl malonato) alulninum 三(丙二酸二乙酯)合铝
tris[di-μ-hydroxo-bisethylene diamine chromium(Ⅲ)] chromium(Ⅲ) salt 三[二-μ-羟基-双乙二胺络铬(Ⅲ)]合铬(Ⅲ)盐
tris(ethylenediamine) cobalt(Ⅲ) chloride 氯化三(乙二胺)高钴
tris-(2-ethylhexyl) amine(TEHA) 三(2-乙基己基)胺 [$C_4H_9CH(C_2H_5)CH_2$]$_3$N
tris-β-hydroxyethyl-isocyanurate 三(β-羟乙基)异氰脲酸酯
tris (hydroxymethyl) aminomethane 三(羟甲基)氨基甲烷〔交联剂〕
tris(hydroxymethyl) phosphine 三(羟甲基)膦
trisilalkane 丙硅烷
trisilane 丙硅烷 Si_3H_8
trisilanyl 丙硅烷基
trisilanylene 1,3-亚丙硅烷基
trisilicic acid 聚三硅酸 ($H_6Si_3O_9$)$_n$
trisilthian 三甲硅二硫醚;三甲硅二硫烷
tris(p-isocyanate phenyl) thiophosphite 三(对异氰酸酯苯基)硫代亚磷酸酯
triskelion cross section 三弯叶形截面;三弯脚形截面
tris-methylcyclopentadienyl-neodymium 三甲基茂钕 $Nd(C_6H_7)_3$
trisnonyl phenyl phosphite 亚磷酸三壬基苯酯〔防老剂〕
tri-soap 三酸皂
trisodium cellulose 纤维素三钠
trisodium EDTA(=ethylenediaminetetra-acetic acid sodium salt) 乙二胺四乙酸三钠〔螯合剂〕
trisodium glycyrrhetinate 甘草酸三钠
trisodium hydroxyethylenediamine triacetate 羟乙基乙胺三乙酸三钠〔螯合剂〕
trisodium phosphate 磷酸三钠
trisome 三(染色)体性;三(染色)体细胞
trisorb 三碘甲状腺原氨酸(树脂)测定药箱
trisphenol 三苯酚
tris(polyhaloaliphatic) phosphate 磷酸三(多卤代脂族基)酯〔纤维用阻燃剂〕
trissalicylatotitanate(Ⅳ) 三水杨酸根合钛(Ⅳ)酸盐
tristearin(=stearin) 硬脂精;三硬脂精;(三)硬脂酸甘油酯 $C_3H_5(OOCC_{17}H_{35})_3$
tristimulus 三色激励的
tristimulus coefficient 三色激励系数;三色系数
tristimulus color 三刺激色;三色觉
tristimulus colorimeter 三色刺激值色度计;三刺激比色计
tristimulus colorimetry 三色激励测色法
tristimulus coordinates 三原色坐标;色度坐标;三刺激值坐标
tristimulus designation 三色标示
tristimulus diagram 三色图
tristimulus filter 三色激励滤色片
tristimulus integrator 三色刺激积分仪
tristimulus mask colorimeter 三色刺激掩膜色度计
tristimulus reading 三刺激值组合读数
tristimulus spectrophotometer 三原色分光光度计
tristimulus value 三色激励值
tris (triorganotin) borate 硼酸三有机锡
tris(tri-n-propyl stannyl) borate 硼酸三(三正丙基锡)盐
trisubstituted 三取代的;三元取代的
trisubstituted carbinol 三取代甲醇;叔醇
trisubstitution product 三取代产物
trisulfapyrimidines 三重磺胺嘧啶
trisulfide 三硫化合物
trisulfonate 三磺酸盐(酯) $R(SO_3M)_3$;$R(SO_2OR)_3$
trisulfonic acid 三磺酸 $R(SO_3H)_3$
tris-xanthogenato chromium(Ⅲ) 三(乙基黄原酸根)合铬(Ⅲ)
tritactic polymer 三(等)规(立构)聚合物
tritane 三苯甲烷 $(C_6H_5)_3CH$
tritane carboxylic acid 二苯甲基苯甲酸 $(C_6H_5)_2CHC_6H_4COOH$
tritanopia 蓝盲;第三色盲
tritartaric acid 焦三酒石酸
triterium 氚〔氢的同位素,符号 ^3H〕
triterpene 三萜(烯)
triterpene alcohol 三萜烯醇
triterpenic acid 三萜酸
triterpenoid 三萜系化合物
trithian 三噻烷;三硫杂环己烷
trithilite 垂塞拉特〔炸药〕
trithiocarbonate 三硫代碳酸盐(酯);全硫碳酸盐(酯)
trithiocarbonic acid 三硫代碳酸;全硫碳酸 HSCSSH
trithiocyanuric acid(=thiocyanuric acid) 硫氰尿酸;三聚硫氰酸 $C_3H_3N_3S_3$
trithiodibutylamine 三硫代二丁胺〔促进剂〕
trithioglycerin 三硫甘油;1,2,3-丙三硫醇 $C_3H_5(SH)_3$
trithionate 连三硫酸盐 $M_2S_3O_6$
trithionic acid 连三硫酸 $H_2S_3O_6$
trithioozone 臭硫 S_3
trithiophenyl phosphate 三硫代磷酸三苯酯 $(C_6H_5S)_3PO$
trithiophosphite 三硫代亚磷酸盐(酯)
trithiopyrophosphoric acid 三硫代焦磷酸
tritiated 氚化了的
tritiated compound 氚(标记)化合物
tritiated hexaoxyethylene dodecyl ether 氚化六氧乙烯十二烷基醚

tritiated hydrocarbon　氚化烃
tritiated titanium target　氚化的钛靶
tritiated waste　含氚废物
tritiated water standard　氚化水标准
tritiation　氚化*
tritide　氚化物
tritioboration　氚硼化
tritium　氚*
tritium dating　氚测定年龄
tritium dating method　氚测定年龄法
tritium foil electron capture detector　氚箔电子捕获检测器
tritium-free water　无氚水
tritium-generating reactor　产氚(反应)堆
tritium label autoradiography　氚示踪自动放射显影法
tritium-labelled　用氚示踪(标记)的
tritium labilization　氚活化；氚不稳定性
tritium oxide　氧化氚；氚水　$HTO; T_2O$
tritium ratio(TR)　氚比〔H^3/H^1=10^{-18}(原子比)为 1TR 相当于一克水中氚放射性为 7.2×10^{-3} 衰变/分〕
tritium target　氚靶
tritium unit(TU)　氚单位〔氚：氢原子比为 $1:10^{18}$ 为一氚单位；相当于 7.2 衰变/分/升水的放射性〕
tritol　三硝基甲苯
tritolylguanidine　三甲苯胍（促进剂）
tritolyl phosphate　磷酸三甲酚酯
tritopine　三陶品
tritriacontane　三十三(碳)烷　$C_{33}H_{68}$
tritriacontyl　三十三(烷)基　$CH_3(CH_2)_{31}CH_2-$
triturable　可研成粉的；可以粉化的
triturate　磨碎；磨碎物
triturating　研制过程
triturating machine　研制机
trituration　①研制剂②研制(作用)
trityl　三苯甲基　$(C_6H_5)_3C-$
trityl alcohol　三苯甲醇　$(C_6H_5)_3COH$
tritylation(=triphenyl methylation)　三苯甲基化作用
trityl bromide　三苯甲基溴　$(C_6H_5)_3CBr$
trityl cellulose　三苯甲基纤维素
trityl chloride　三苯甲基氯　$(C_6H_5)_3CCl$
trityl ether　三苯甲基醚
trityl magnesium chloride　氯化三苯甲基镁　$ClMgC(C_6H_5)_3$
trityrosine　三酪氨酸
triunsaturated glyceride　三不饱和酸甘油酯
triuranium octaoxide　八氧化三铀　U_3O_8
triuret　(二缩)三脲　$C_3H_6N_4O_3$
trivalence(=trivalency)　三价

trivalency(=trivalence)　三价
trivalent　三价的
trivalent alcohol　三元醇
trivalent carbon hypothesis　三价碳假说
trivalent element　三价元素
trivalent hydrocarbon radical　三价烃基
trivalent radical　①三价基②三价根
trivalerin　三戊精；甘油三戊酸酯　$(C_4H_9COO)_3C_3H_5$
trivariant system　三变物系
trivial formula　无意义分子式〔不大可能的式子〕
trivial name　俗名
trixylyl phosphate　磷酸三(个)二甲苯酯
trizirconium tetranitride　四氮化三锆　Zr_3N_4
troche　锭剂；片剂
trochoid　余摆线
trochoidal-focusing mass spectrometer　余摆线型聚焦质谱仪
trochoidal mass spectrometer　余摆线型质谱计
troctolite　橄长岩
troegerite　砷铀矿
troilite　陨硫铁；硫铁矿
trolley　空中吊运车
trolley scale　活动秤
trollixanthin　金莲花黄质；金梅草黄素
trolly oil　摇车油；电车油；吊车油
trombone cooler　蛇管冷却器
tromexan　二吡喃乙酸乙酯
trommel　转筒筛
trommelling　转筒筛选
Trommer's test　特罗麦尿糖试验
Trommsdorf effect　特罗姆斯多夫效应〔聚合体系黏度上升和聚合速度加快的现象〕
trona (=sesquicarbonate of soda)　天然碱；二碳酸氢三钠
tron(a) potash　氯钾天然碱
troop　群；团
troostite　锰硅锌矿
tropacocaine　托派可卡因
tropaeolum oil　旱金莲油
tropaic acid　3-羟基-2-苯基丙酸；托品酸　$C_6H_5CH(CH_2OH)COOH$
tropal　衬里织物
tropane　托烷；莨菪烷
tropate　托品酸盐(酯)　$C_6H_5CH(CH_2OH)COOM$；$C_6H_5CH(CH_2OH)COOR$
tropeine　(某)托品酯
tropentane　莨菪醇苯环戊酸酯
tropeolin D(=methyl orange)　金莲橙 D；甲基橙
tropeolin G　金莲橙 G；间胺黄

tropeolin O(=resorcine-azo-benzene-sulfonic acid)　金莲橙 O；偶氮苯间二酚磺酸
tropeolin OO(=orange IV;diphenylamino azo *p*-benzene sulfonic acid)　金莲橙 OO；二苯氨基偶氮对苯磺酸
tropeolin OOO(=sodium naphthol-azobenzene sulfonate)　金莲橙 OOO；酸性橙；萘酚-偶氮苯磺酸钠
tropic acid　托品酸；2-苯基-3-羟基丙酸；邻苯间羟基丙酸〔外消旋〕　$HOCH_2CH(C_6H_5)COOH$
tropical bleach　热带漂白粉；加氧化钙的漂白粉
tropical deterioration　热带劣化
tropical exposure testing　热带暴晒试验；热带暴露试验
tropicalization test　热带化试验
tropical point　回归点
tropical wood　热带木材
tropics　热带
tropide　托品交酯；托品酸交酯
tropidine　托品定；莨菪定
tropilidene　环庚(间)三烯
tropine　托品；莨菪碱；托品醇
tropine alkaloids　托品生物碱
tropine carboxylic acid　托品甲酸；芽子碱
tropine sulfate　硫酸托品
tropinone　托品酮；颠茄酮
tropisetron　托烷司琼〔药〕
tropolium ion　䓬；环庚三烯正离子
tropolone　环庚三烯酚酮
tropone　环庚三烯酮
troponoid　环庚三烯酮型化合物
tropopause　对流层顶
troposphere　对流层；运流层
tropoyl　托品酰；莨菪酰　$C_6H_5CH(CH_2OH)CO-$
troptometer　测扭计
tropyl　托品基
tropylium ion　䓬鎓离子
tropylium salt　䓬鎓盐
trotter oil　蹄爪油
trotyl(=TNT)　三硝基甲苯
trouble　①障碍；故障②(疾)病③激动
trouble-free　无故障的
trouble-free burning　正常燃烧；可靠燃烧
trouble-free operation　无故障运转；顺利运行
trouble-free running　无故障运转
trouble hunting　检查故障原因
trouble-locating　故障检查；故障探测
trouble point　故障点
trouble-proof　无故障的；不间断的
troubleshooting　①故障检查及排除②检修
troubleshooting equipment　故障排除设备(用具)

trough　(长)槽；池；沟；谷〔光谱〕
troughed belt　槽带
trough mixer　槽式混和器
trough-shaped container　槽形容器
trough tank　(薄层色谱)罐
trough truck　油槽车
trough washer　槽洗机
trough washery　槽洗(涤)
trouser's tear piece　裤形撕裂试片
trout liver oil　鲑鱼肝油
trout oil　鲑鱼油
Trouton viscosity　特鲁顿黏度
trowel　修平刀
trowel adhesive　高黏度黏合剂
trowelling　镘涂；抹涂〔涂料相当稠，可用瓦工抹子(镘刀)抹涂〕
troxerutin　三羟乙基芦丁；维生素 P_4；曲克芦丁
troxidone　三甲双酮；3,5,5-三甲基-2,4-二氧恶唑烷〔抗惊厥及抗癫痫药〕
troy weight　英国金衡制
trub　冷却残渣
trubenizing　挺平织物法
truck drier　小车干燥器
truck fill stand　装车台
truck haul(ing)　货车运输
truck-loading facilities　装车设备
truck-loading rack　装车架
truck manifold valves　货车多歧管阀
truck pipe line　管道干线
truck scale　车重计量
truck tank　油槽汽车
truck tyre　载重轮胎
true acidity　真酸度
true aryl nitro-compound　真芳基硝基化合物
true ash　真灰分
true boiling-point　真沸点
true boiling-point curve　真沸点曲线
true boiling-point distillation　真沸点蒸馏
true camphor wood　樟木
true chemical constant　真化学常数
true coefficient　真系数
true colloid　真胶体
true color of lubricating oil　润滑油的真色度
true combined sulfur(=technically bound sulfur)　有机结合硫
true copolymer　真共聚物
true counter-current flow　真逆流流动
true crystalline state　真晶态

true density 真密度
true electrolyte 实电解质*
true emulsion 真乳状液；稳定乳胶体
true equilibrium 真稳定平衡
true fluidity 真流动性
true hardness 真硬度
true heat capacity 真热容量
true levelling 真平整
true liquid 理想液体
true lubrication 真润滑；理想润滑
true mass flow 真质量流动；实际质量流动
trueness 真实性；确切性
true plasticizer 真增塑剂；单用增塑剂
true polymer 真聚合物
true polymerization of olefines 真烯烃聚合
true purity 真纯度
trueran 棉涤纶
true solution 真溶液
true specific gravity 真比重
true temperature drop 真温度降
true terpene 真萜烯
true total temperature 真总温度
true up 调整；校正
true value 真值*
true vapour pressure 真蒸气压
true yellow 正黄(色)；纯黄(色)
trug 灰浆槽
trumpet cooler 管式冷却器
trumpet honeysuckle 喇叭金银花
truncated cone 截锥
truncated octahedra 平截八面体
trunk piston 筒形活塞；裙式活塞
trunk polymer (被接枝的)主链聚合物
trunnion carrier (离心)管套座
trunnion cup 凸耳座
trunnion discharge mill 耳轴卸料磨
trunnion feed mill 耳轴加料磨
trunnion mill 耳轴磨
trunnion ring 耳轴环；管套环
truss 空间构架；捆
truth table 真值表
truxellic acid 吐雪酸
truxillic acid 古柯间二酸；2,4-二苯环丁烷二羧酸
truxilline 异托品基可卡因；吐昔灵 $C_{38}H_{46}N_2O_8$
α-truxilloyl chloride α-吐昔二酰氯
truxinic acid 古柯邻二酸；2,3-二苯环丁烷二羧酸
truxone 吐昔酮；双茚酮
try-and-error method 试探法；尝试法

trying out 炼脂；脂肪熔炼
try-out run 试运转；试车
trypaflavin(e) 盐酸 3,6-二氨基-10-甲基吖啶
trypan blue 台盼蓝；锥虫蓝
trypanocides 锥虫药；杀锥虫药
trypsin 胰蛋白酶
tryptamine 色胺；β-吲哚基乙胺
$$NHC_6H_4CH=CCH_2CH_2NH_2$$
tryptamine hydrochloride 盐酸色胺 $C_{10}H_{12}N_2HCl$
trypterygine(=celastrol) 雷公藤红；雷公藤碱
tryptophan(=tryptophane) 色氨酸；β-吲哚基丙氨酸
tryptophanase 色氨酸酶
tryptophane 色氨酸；β-吲哚基丙氨酸
$$C_6H_4NHCH=CC_2H_3(NH_2)CO_2H$$
tryptophol 色醇；β-吲哚乙醇
tryptophyl 色氨酰 $C_8H_6NCH_2CH(NH_2)CO—$
tsano oil 朝鲁油
tschermigite 铵明矾
Tschugajew's reaction 秋加耶夫检镍反应
T-square 丁字尺
t statistic t 统计
tsuduic acid 粗杜酸
tsugaresinol 铁杉树脂醇
tsuzuic acid(=tetradecenoic acid A) 粗租酸；十四碳烯酸 A
t test t 测验法〔统计〕
T titration T 滴定
T.T.T. curve 时间温度变态曲线
T-tube (=T-piece) T(形)管
tub 盆；槽
tubacurarine 土芭碱
tubaic acid 鱼藤酸；土芭酸
tubain 土芭树脂
tubanol 土芭酚
tuba root 土芭根
tubatoxin 鱼藤酮
tub coloring 槽法着色
tube ①管(子)②(真空)管③内胎④装于管中⑤管(形)的
tube-and-tank-process 管-罐裂化过程
tube-and-tank-unit 管-罐裂化车间
tube bender 弯管机
tube classifier 管式分粒器
tube cleaner 洗管机
tube clip 管夹
tube clipper 封管器
tube closing 管的封密

tube combustion method 管式燃烧法
tube cooling system 管冷却系统
tube cracking furnace 管式裂解炉
tube cutter 割管刀；割管机
tubed goods 压出制品
tube dryer 管式干燥器
tube electric furnace 管形电炉
tube electrometer amplifier 管式静电放大器
tube filling and closing machine 装管封管机
tube filling machine 装管机
tube fitting offset 傍绕连接管
tube fittings 管配件
tube funnel 管式漏斗
tube furnace 管式炉；环形炉
tube furnace pyrolyzer 管炉热解器
tube gauge 管式气压计
tube hanger 吊管架
tube heater 管式加热炉
tube heating furnace 管式加热炉
tube in tube condenser 套管式冷凝器
tubeless tyre 无管轮胎
tube-life charts 管使用时间图表〔裂化炉〕
tube machine 压出机
tube mill 管式磨
tube mill head 管式磨头
tube milling 管式磨
tube mill liner 管式磨衬里
tube mill lining （把)管式磨衬里
tube mill shell 管式磨外壳
tube mo(u)ld 内胎模；吹管模
tube pitch 管间距
tube plate(=tube sheet) 管板
tube plug 管塞
tube press 管压机；内胎硫化压热器
tuber ①制管机；制内胎的机器②块茎
tube reactor 管式反应器
tube reamer 扩管孔器
tuberone 晚香玉酮
tuberose 晚香玉
tuberose absolute 晚香玉净油
tuberose compound 晚香玉香精
tuberose concrete 晚香玉浸膏
tuberose oil 晚香玉油
tuberostemonine 块茎百部碱；对叶百部碱
tubers 块茎(类)
tuberyl acetate 乙酸二氢香芹酯
tuberyl alcohol 二氢香芹醇
tube sampler 管式取样器

tube scope 管内检查镜
tube screen 管状筛
tube sealing 封管
tube sealing machine 封管机
tube sheet(=tube plate) 管板
tube side 管程；管方
tube skin temperature 炉管表面温度
tube space 管际空间
tube spacing 管间距
tube steel 管钢
tube still 管式蒸馏器
tube-still duty 管式蒸馏器本领
tube still heater 列管加热器
tube-still process 管式炉裂化过程
tube stopper 管塞
tube support ①管支柱②管支承
tube-surface 管(加热)表面
tube target 管靶
tube viscometry 管式测黏法
tube wall 管壁
tube-wall atomization 管壁原子化
tube welding machine 焊管机
tubing ①管(子)②制管
tubing jumper 跨接管
tubing machine 制管机
tubing machine cylinder 压出机机筒
tubing properties 压出性能
tubocurarine 管箭毒碱
tubocurarine chloride 氯化管箭毒碱
tubomel 异烟肼
tub sizing 槽法上胶
tubular air cooler 管式降温器
tubular atomizer 管状喷雾器
tubular boiler(=fire tube boiler；multitubular boiler) 火管锅炉
tubular-bowl centrifuge 管式离心机
tubular-bowl clarifier 管式澄清器
tubular casing 管形套
tubular cooler 管式冷却器
tubular cyclones 管状旋风分离器；管状旋风除尘器
tubular drier 管式干燥器
tubular equipment 管状设备
tubular exchanger 管状换热器；管式交换器
tubular fibre 中空纤维
tubular float level indicator 管状浮标液面指示器
tubular-flow reactor 管道式流动反应器
tubular furnace 管式炉
tubular goods 管类材料

tubular heater 管式加热器；管式加热炉
tubular heat exchanger 管式换热器
tubular hosiery 圆筒形针织物
tubular oven 管式加热炉
tubular powder 管状(火)药
tubular reaction vessel 管式反应器；管式反应釜
tubular reactor 管状反应器
tubular rheostat 管状电阻
tubular sandstone 管状砂岩；筒状砂岩
tubular surface condenser 管式冷凝器
tubular type preheater 列管式预热器
tubular tyre 单管轮胎
tubular vapour cooler 管状蒸气冷却器
tubulate ①有管的②管状的
tubulated ①有管的②管状的
tubulation clamp 管式夹
tubulature 短导管
tubule 小(导)管
tubulous boiler 水管锅炉
tuck fabric 集圈织物
tucking (油鞣革)皂液处理
tucum oil 星实桐油
tufa 石灰华
tuff 凝灰岩
tuff lava 凝灰熔岩
tufting 栽绒法；簇绒法
tuft withdrawal tensometer 绒头拔出力试验仪
Tukon 屠康〔硬度单位〕
Tukon hardness tester 屠康硬度试验机
tulipiferine 鹅掌揪碱
tulipinolide 郁金香酯
Tulsa-type absorption process 突萨型气体汽油吸收过程
tumbago 铜、金、银合金
tumble 鼓转；辗转
tumble curing 转动硫化
tumble enamelling 转鼓涂漆
tumble pilling test 滚动起球试验；滚磨起球试验
tumble polishing 转鼓抛光
tumbler ①(大)杯②转鼓③齿轮换向器④颠动筒
tumbler gear 顺逆齿轮〔商〕
tumbler switch 起倒开关；转换开关
tumbler test 磨蚀试验
tumbler test (for coke) (焦炭)磨蚀试验
tumbler-type washing machine 转鼓式洗涤机
tumbling(=barrel finishing;rumbling) ①(小件)滚筒内抛光；转鼓内抛光②转筒中除毛刺
tumbling barrel 转鼓
tumbling barrel process 转涂法
tumbling coating 转筒浸涂法
tumbling cylinder 滚筒
tumbling drum 转鼓
tumbling polishing 滚筒抛光；转鼓抛光
tumor necrosis factor 肿瘤坏死因子
tumor specific antigen 肿瘤特异性抗原
tun 大桶
tunable 可调的
tunable dye-laser 可调谐染料激光器
tunable filter 可调谐滤光器
tunable laser 可调谐激光*
tunable laser source 可调谐激光光源
tunable organic dye laser 可调谐有机染料激光器
tunable probe head 可调谐探头
tundish 耐火材料槽
tune up ①调谐；配合②用化学溶剂清除发动机中沉积物
tune-up oil 清除(发动机沉积物用)油
tungar 钨氩管
tungar rectifier 钨氩管整流器
tungates 桐油金属皂催干剂
Tung distribution 董分布
tung oil (=china wood oil;mu oil) 桐油
tungoxyn 桐油氧化物
tungstate 钨酸盐 M_2WO_4
tungstated toners 钨酸盐色原(色料)
tungstate radicle 钨酸根
tungsten(=wolfram) 钨〔74号元素,化学符号 W〕
tungsten alloy 钨合金
tungsten blue reaction 钨蓝反应
tungsten blues(=tungsten bronzes) 钨青铜
tungsten bromide 溴化钨 WBr_2; WBr_5; WBr_6
tungsten bronze(=tungsten blues) 钨青铜
tungsten-bronze structure 钨青铜结构
tungsten carbide 碳化钨 WC; W_2C; W_3C
tungsten chloride 氯化钨 WCl_2; WCl_3; WCl_4; WCl_5; WCl_6
tungsten dibromide 二溴化钨 WBr_2
tungsten dichloride 二氯化钨 WCl_2
tungsten dioxide 二氧化钨 WO_2
tungsten dioxydichloride 二氯二氧化钨 WO_2Cl_2
tungsten disulfide 二硫化钨 WS_2
tungsten electrode 钨电极
tungsten filament 钨丝
tungsten filament lamp 钨丝灯
tungsten fluoride 氟化钨 WF_3; WF_6
tungsten hexachloride 六氯化钨 WCl_6
tungsten hexafluoride 六氟化钨 WF_6
tungsten iodide 碘化钨 WI_2; WI_4

tungsten-iodine lamp 碘钨灯
tungsten lake 钨色淀；磷钨酸色淀
tungsten minerals 钨矿物
tungsten ore 钨矿
tungsten oxide 氧化钨 $WO; W_2O_3; WO_2; WO_3; W_2O_5$
tungsten oxybromide 四溴氧化钨 $WOBr_4$
tungsten oxychloride 氯氧化钨 $WO_2Cl_2; WOCl_4$
tungsten oxyfluoride 四氟氧化钨；四氢化氧钨 WOF_4
tungsten pentabromide 五溴化钨 WBr_5
tungsten pentachloride 五氯化钨 WCl_5
tungsten pentoxide 五氧化二钨 W_2O_5
tungsten quartz infrared lamp 钨丝石英红外(线)灯
tungsten-rhenium alloy 铼钨合金
tungsten-rhenium filament 铼钨热丝
tungsten steel 钨钢
tungsten strip atomizer 钨带原子化器
tungsten sulfide 硫化钨 $WS_2; WS_3$
tungsten tanning 钨盐鞣制
tungsten tetrachloride 四氯化钨 WCl_4
tungsten tetrahydroxide 四羟化钨；四氢氧化钨 $W(OH)_4$
tungsten tetraiodide 四碘化钨 WI_4
tungsten trichloride 三氯化钨 WCl_3
tungsten trifluoride 三氟化钨 WF_3
tungsten trioxide 三氧化钨 WO_3
tungsten trisulfide 三硫化钨 WS_3
tungsten wire 钨丝
tungsten wire method 钨丝法
tungstic ①六价钨的；(正)钨的②五价钨的
tungstic acid 钨酸
tungstic ocher 钨赭石
tungstic oxide 三氧化钨 WO_3
tungstite 钨华
tungstomolybdic pigment 钨钼酸颜料
tungstophosphoric acid 钨磷酸
tungstosilicic acid 钨硅酸
tungstyl 钨氧基
tuning 调谐
T-union T形接管
tunnel ①隧道②烟道
tunnel bearing grease (船舶发动机)螺旋桨轴承润滑脂
tunnel block grease (船舶发动机)螺旋桨块状无水润滑脂
tunnel carbon black 烟道炭黑
tunnel cooler 隧道式冷却器
tunnel drier 隧(道)式干燥器
tunneling effect 隧道效应
tunneling magnetometer 隧道磁力计
tunneling mechanism 隧穿机制
tunneling type atomic force microscope 隧道原子力显微镜
tunnel kiln 隧道式窑
tunnel ling effect 隧道效应
tunnel oven 隧道式窑
tunnel-shaped atomizer 隧道形(式)雾化器
tunnel test ①风洞试验②隧道试验〔测阻燃性〕
tunnel-type tray cap 长方形泡罩
tuno gum 糖胶树胶
turanose 土冉糖；松二糖
turbid ①(混)浊的②纷乱的
turbidimeter 浊度计
turbidimetric and nephelometric procedure 浊度和比浊法
turbidimetric assay 浊度测定法
turbidimetric method 比浊法；浊度测定法
turbidimetric titration 比浊滴定(法)；浊度滴定
turbidimetry 比浊法；浊度测定法
turbidity 混浊度；浊性
turbidity indicator 混浊度指示剂
turbidity point 浊度点
turbidity scale 浊度标
turbidity standard 浊度标准
turbidity titration 浊度滴定
turbidity value 浊值
turbidness 混浊
turbidometer (=turbidimeter) 浊度计
turbinaceous 泥煤的
turbine 汽轮机；叶轮机；涡轮；透平机
turbine agitator 汽轮式搅拌器
turbine blower 汽轮鼓风机
turbine distributor 汽轮式分布器
turbine-driven compressor 涡轮压缩机
turbine fuel 汽轮机燃料
turbine gas absorber 汽轮气体吸收器
turbine gas purifier 汽轮气体净化器
turbine impeller 涡轮高速搅拌机
turbine meter 汽轮计
turbine mixer 汽轮混合机
turbine oil 汽轮(机)油
turbine pump(=turbo-pump) 汽轮泵
turbine reaction 汽轮反作用度
turbine sifter 汽轮筛
turbine stirrer 汽轮搅拌机
turbine type agitator 汽轮式搅拌器
turbine (type) mixer 汽轮(式)混合器
turbine wheel 汽轮
turbo-blower 汽轮鼓风机；离心鼓风机
turbocharger 汽轮增压器

turbo-compressor 汽轮压缩机；离心压缩机
turbo-compressor rotor 汽轮压缩机转子
turbo-disperser 汽轮分散机
turbo drier 涡轮式干燥器
turbo-dynamo 汽轮发电机
turboemulsifier 涡轮式乳化器
turbo-engine lube 汽轮机润滑油
turbo-engine lubrication 汽轮机润滑
turbo film evaporator 湍流膜式蒸发器
turbo-gas absorber 汽轮气体吸收器
turbo-generator 汽轮发电机
turbo-grid 汽轮式格子〔分馏塔盘的〕
turbo-grid tray （非溢流型）栅板塔盘
turbo-mix-contactor 汽轮式混合萃取器
turbo mixer 汽轮式混合器
turbomolecular pump 涡轮分子泵*
turbo-pump(=turbine pump) 汽轮泵
turbo-refrigerator 透平冷冻机
turbo-shelf drier 汽轮盘架干燥器
turbostratic 湍流；紊流
turbostratic kaolinite 乱层高岭石〔矿〕
turbo-viscosimeter 搅动黏度计
turbo-viscosity 搅动黏度
turbulence ①紊流(度)；湍流(度)②扰动；混乱；颠簸
　③湍流
turbulence boundary layer 湍流边界层
turbulence damping 湍流阻尼
turbulence damping and drag reduction(=Tom's effect)
　湍流降阻
turbulence intensity 紊流强度
turbulence number 湍流数
turbulence resistance 湍流阻力
turbulency(=turbulence) 湍动
turbulent 湍动的；湍流的；扰动的
turbulent air 湍动空气；湍气流
turbulent attack 紊流(湍流)腐蚀
turbulent bed 湍动床
turbulent boundary flow 湍动边界流动
turbulent boundary layer 湍流边界层
turbulent burner 紊流燃烧器；湍流燃烧器
turbulent contact absorber 湍动接触吸收器；湍流混合吸
　收器
turbulent diffusion 湍流扩散
turbulent energy spectrum 湍动能谱
turbulent-film vaporizer 湍流膜式蒸发器
turbulent flame 紊流火焰；湍流焰
turbulent flow burner 紊流燃烧器；湍流燃烧器*
turbulent fluidized bed 湍动流化床
turbulent foreflow 先行湍流
turbulent jet 湍流
turbulent mixing chamber 湍动混合室
turbulent motion 湍动；涡动；紊动
turbulent resistance 湍流阻力
turbulent separation 湍流分离
turbulent shear flow 湍性剪切流
turbulent transition 湍流过渡（转变）
turbulent wake 湍流尾流
turbulivity 紊流度；紊流系数
turbulizer ①高速换转连续混合机②紊流(湍流)增强器
Turek detonator 图雷克雷管
turf 泥煤；泥炭
turfary 沼泽；沼泽地；泥煤田
turgent 肿胀的
turgescence(=turgescency) 肿胀；紧张
turgescency(=turgescence) 肿胀；紧张
turgescent 肿胀的
turgid 肿胀的
turgidity(=turgidness) 肿胀；紧张
turgidness(=turgidity) 肿胀；紧张
turgite 水赤铁矿
turgor pressure 紧胀压；超扩散压
tur(i)cine 右旋水苏碱；右旋脯氨酸二甲内盐
turkey pea 灰叶
Turkey red 土耳其红〔染料工业〕
Turkey red oil 土耳其红油；磺化蓖麻油
Turkey stone 土耳其石；磨石
Turkish gallotannin 土耳其种没食子单宁
Turkish tannin 土耳其单宁
Turkish tobacco essence 土耳其型烟草香精
Turkish tobacco flavour 土耳其烟草香精
turmeric ①姜黄②郁金香
turmeric acid 姜黄酸
turmeric oil 姜黄油；郁金姜黄油
turmeric paper 姜黄试纸
turmeric paper test 姜黄纸试验
turmeric root 姜黄根
turmeric test paper 姜黄试纸
turmeric yellow(=curcumin) 姜黄；酸性黄
turmeron 姜黄倍半萜
turmerone ①姜黄酮②郁金酮
turn ①圈数；匝数②回转③转变
turnable conveyor belt 可转式运输带
turnaround 小修；预防修理
turnaround of unit 装置工作周期
turnaround plans 检修计划
turnaround speed 周转速度

turn-around time ①来回时间②周转时间③检修时间
turn-buckle ①紧线器②花篮螺丝；螺丝接头
Turnbull's blue 特恩布尔兰
turn coil reactor 盘管式反应器
turndown of gas burners 关闭气体燃烧器
turndown ratio 操作弹性；极限负荷比〔塔的〕
turner 车工；旋工
Turner's yellow 特纳黄〔碱式氯化铅的别名〕
turning ①镦坯②转动③车削
turning effect 转动效应
turning machine 车床
turning oil 循环油
turning organ 回转器
turning tool 车刀
turning vane 导向叶片
turnmeter 回转速度指示器
turnover ①颠复；翻转②周转
turnover job 大修工作
turnover number ①转换数*；周转率②酶变率
turnover rate 转换率；周转率
turn-ratio 匝(数)比(率)
turnsole ①石蕊②天芥菜
turnsole oil 天芥菜油
turnsole paper 石蕊试纸
turnstile 转盘；转台
turn table 转盘〔机工〕
turntable extruder 转台式压出机
turpentine ①松节油②松脂；松油脂
turpentine oil 松节油
turpentine oil-resisting test 耐松节油试验
turpentine substitute 松节油代用品
turpentine wood oil 松木油
turpentinic 松节油的
turpentining(=resin tapping) 采(割松)脂
turpeth 泻根
turpeth mineral 碱式硫酸汞
turpineol 萜品醇；松油醇
turps 松节油
turquois(e)(= agaphite) 绿松石
turquoise blue 甸子蓝
turquoise green 绿松石绿；绿色艳蓝
turret lathe 六角车床
turret type pump 塔式泵〔垂直安装马达的〕
turtle oil 鳖油
tus 放大器
Tuscan red 塔斯康红〔茜素红色淀颜料〕
tussah 野蚕丝
tussah silk 柞蚕丝

tussol 杏仁酸安替比林；美沙酮，美散痛〔用于戒除海洛因毒瘾〕
tussore 野蚕丝
tutia 碳酸锌
tutin 马桑苷
tutocaine 土透卡因〔药〕
tutocaine hydrochloride(=butamine) 盐酸土透卡因；布他明 $NH_2C_6H_4CO_2CH=(CH_3)CH(CH_3)CH_2N(CH_3)_2HCl$
Tutton's salts 塔顿盐
tutty 碳酸锌；未加工的氧化锌
Tutwiler burette 杜维勒滴定管
Tutwiler test 杜维勒试验〔测定油气中硫化氢的工业法〕
tuyere(=twyer) 吹风管嘴
tuyere cooling plate 风口冷凝板
tuyere stock 风口套管
T value T 值〔土壤碱交换能力〕
T-valve 三通阀
TVA thermogram TVA 热解曲线
Tw 特瓦德尔比重标度
Twaddell degree 特瓦德尔度
Twaddell scale 特瓦德尔标度
Twaddell's hydrometer 特瓦德尔比重计
Twaddle 特瓦德尔比重计标度
Twaddwell hydrometer 特瓦德维尔浮计
tween 吐温
tweer(=twere) 炉壁间；鼓风管
tweezers 镊(子)
twelve-membered ring(=twelve-ring) 十二元环
twelve-ring(=twelve-membered ring) 十二元环
twere 炉壁间；鼓风管
twice disk method 二次压片法
twice substituted 二取代了的
twill 斜纹织物
twill weave 斜织法；菱织法
twin 孪晶
twin-arc weatherometer 双碳弧灯人工老化试验机
twin axis 孪晶轴
twin-barrel extruder 双筒挤出机
twin-bladed cutter 双刃切器
twin boundaries 双晶间界；孪晶间界
twin calorimeter 孪形热量计
twin-carbon lamp 双碳弧灯
twin crystal 孪晶
twin-detector scanner 双探头扫描机
twin drum drier 双辊干燥机
twine ①细绳②捻；交织〔纺〕
twin ebulliometer 双式沸点升高计
twin elbow 双肘管

twin electrons　孪电子
twiner　捻线机
twin fuel pump　双燃料泵
twin grinder　双重碎木机；二袋式碎木机
twining　捻线
twining axis　孪晶轴；双晶轴
twining plane　双晶面；孪晶面
twin magnets　孪片磁体
twin naphtha-rerun units　双石脑油重蒸设备
twinned crystal　孪晶
twinned dendrite　双树枝状晶体
twinnig　①孪晶作用②孪生作用
twin nuclei　孪核；双环结构
twin-phase nozzle　双相喷嘴〔指气流式喷嘴〕
twin pipe line　双管线
twin plane　孪晶面
twin pump　双生泵；双缸泵
twin-rapier loom　双层剑杆织机
twin-roll- drum drier　双辊干燥机
twin (screw) compounder　双螺杆混料机
twin screw extruder machine　双螺杆压出机
twin-screw plodder　双螺杆压条机
twin-shell dry blender　双壁干式搅拌机
twin simplex pump(=twin single pump)　双单动泵
twin single pump(=twin simplex pump)　双单动泵
twin-start screw　双螺纹螺杆
twin tyres　双料轮胎
twin-worm plodder　双螺压条机
twist　①扭(转)②歪曲③捻度*
twist balance index　捻度平衡指数
twist boat　扭船式
twist boundaries　扭转晶界
twist chair　扭椅式
twist conformation　扭型构象*
twist counter　计捻器
twist drill　扭钻；钻子；麻花钻
twisted fiber　合股纤维
twisted internal charge transfer　扭曲分子内电荷转移
twisted molecular layer　扭曲分子层
twisted union yarn　异纤合股线
twisted warp yarn　弯曲纱线
twisted weave　扭织
twisted yarn　加捻纱〔纺〕
twist factor　扭转因子；卷曲因子
twist-heatset-untwist texturing process　加捻-热定型-解捻变形法
twisting　①扭转②加捻
twisting cohesion　卷附
twisting frame　捻丝机
twisting oil　加拈油
twisting paper　绉纸
twisting resistance　抗扭力；扭转阻力
twisting strain　扭应变
twisting stress　扭应力
twisting test　扭转试验
twisting vibration　扭转振动
twist irregularity　捻度不匀；捻度不匀率
twist loss　捻陷
twist moment　扭矩；转矩
twist of fiber　纤维的捻度
twist of yarn　捻度
twist regularity meter　捻度均匀度测定仪
twist stress relaxation　扭应力松弛
twist takeup　捻缩
twist thermo-hydrosetter　定捻锅
two-accelerator effect　双催速剂效应
two-acetate fiber　二醋酯纤维
two acid-two base systematic analysis　两酸两碱系统分析
two arm kneader　双臂搅拌机
two-bank stage pump　双排阶段泵
two-bath process　二浴法
two beam spectrometer　双光束光谱仪
two-bed hydrodesulfurization　双床加氢脱硫
two-boiling system　双沸系统设备
two bubble method　双泡法〔测定泡沫膜寿命的方法〕
two-button (filter) plate　双钮(滤)板
two carbon rearrangement　二碳(原子)重排反应
two catalytic process　两段催化法
two-cell heater　两室炉
two-chamber baling press　双箱打包机
two chamber vacuum furnace　双室真空炉
two-channel recorder　双道记录器
two circle goniometer　双圆测角器
two-coil flow　两旋管并流
two-coil selective cracking　两旋管选择裂化
two-color indicator　二色指示剂
two column photoionization gas chromatograph　双柱光电离子化气相色谱仪
two-column plot　双柱图
two-compartment oil tank　双层油罐
two component adhesive　双组分黏合剂
two-component copolymerization reaction　双组分共聚反应
two-component spraying　双组分喷涂
two-component system　双组分(物)系；二元物系

two container varnish 分装清漆；两罐装清漆
two-daylight press 双板压榨机
two-deck classifier 双台分粒器
two-deck drawplate oven 双层屉烤炉
two-deck oven 双层炉
two-dimensional chromatography 两向色谱(法)
two dimensional contour plot 二维等线图
two-dimensional correlated electron spin resonance 二维相关电子自旋共振谱
two-dimensional development 两向展开(法)
two-dimensional development method 双向展开法
two dimensional disc electrophoresis 双向盘状电泳
two-dimensional electron-electron double resonance 二维电子-电子双共振
two dimensional electron paramagnetic resonance 二维电子顺磁共振
two-dimensional electron spin echo envelope modulation 二维电子自旋回波包络线调制
two dimensional electrophoresis 二维凝胶电泳技术；双向电泳
two dimensional exchange spectroscopy 二维交换谱
two dimensional Fourier transform magnetic resonance 二维傅里叶变换核磁共振
two-dimensional gas 二维气体
two dimensional gel slab technique 双向凝胶平板法
two dimensional-incredible natural abundance double quantum transfer experiment 二维天然丰度双量子转移实验
two dimensional mass spectrometer 二维质谱仪
two dimensional method 二维法
two dimensional nuclear Overhauser effect spectroscopy 二维核欧沃豪斯效应谱
two-dimensional paper chromatography 双向纸色谱法
two-dimensional polymer 片型聚合物
two-dimensional pressure 二维压力
two-dimensional regularity 二向规整性
two-dimensional scanning technique 二维扫描技术；双向扫描技术
two-dimensional separation 两向分离
two dimensional spectral pattern 二维光谱图
two dimensional spin echo correlated spectroscopy 自旋回波相关二维谱
two dimensional stacked trace plot 二维堆积图
two dimensional thin layer chromatography 双向薄层色谱法
two-direction condensation polymerization 二向缩聚
two-electrode system 双电极系统
two-electrode tube 二极(真空)管
two-electron shift 二电子转移
two-electron system 双电子系(统)
two-ended living polystyrene 双端活性聚苯乙烯
two-ended oligomer 双端(活性)低聚物
two-film theory 双膜理论
two-flash system 两次闪蒸系统
two-flight screw 双螺纹螺杆
two-fluid cell 两液电池
two-fluid chromic acid cell 两液铬酸电池
two-fluid nozzle 双相喷嘴；二流喷嘴〔指气流式喷嘴〕
two-fluid theory 两流体(学)说
twofold axis 二重轴
twofold barrier 二倍垒
two-for-one twister 倍捻机
two grid ion source 双栅离子源
two-handled knife 双柄刮皮刀
two-line lubricating system 双线式润滑系统
two-line manifold 双道流路
two line method 二重线法
two line method of background correction 二线背景校正法
two-lipped drill 二缘钻头
two-liquid centrifuge 两液离心分离机
two-liquid manometer 两液气压计；两液测压计
two-liquid method 双液法；两液法
two-lobe rotary pump 双凸轮旋转泵
two-loop sampling valve 双环管进样阀
two-part cracking 两部分裂化
two-pass condenser 两通冷凝器；双路冷凝器
two-pen recorder 双笔记录仪
two-phase additive 两相添加剂
two-phase cleaning 两相净洗；浮液净洗
two-phase current 双相电流
two phase emf-titration technique 两相电动势滴定法
two-phase flow 两相流动
two-phase ion-exchange column 两相离子交换柱
two-phase polycondensation 两相缩聚
two-phase titration 两相滴定法
two-photon absorption 双光子吸收
two photon excited fluorescence 双光子激发荧光
two-photon excited fluorescent detection 双光子激发荧光检测
two-photon spectroscopy 双光子光谱
two photon transition 双光子跃迁
two-pi counter 2π计数管
two-piston reciprocating pump 双活塞往复泵
two-pit tannage 二槽鞣法
two-plate press 双板压榨机

two-ply 二层合板
two-ply stock 双重纸
two point examination 两点测定
two polarized electrode 双极化电极
two-position controller 自动控制器
two-ray oscillograph 双线示波器
two region reactor 双区型反应堆
two-roll crusher 双辊压碎机
two-roller mill 两辊磨
two-screw oil pump 双螺桨油泵
two-sided confidence interval 双侧置信区间
two-sided criterion of significance 双侧显著性检验
twosideness 两边性
two side test 双侧检验
two site immunochemiluminometric assay 双位点免疫化学发光分析
two-site immunoradiometric assay 双位免疫放射测定法
two-solvent process 双溶剂过程
two source method 双源法
two-speed feed 两种速度供料
two-speed gearbox 双速减速机
two-stage aeration tank 两级曝气池
two-stage air pump 两级空气泵
two-stage biofilter 两级生物滤池
two-stage catalytic hydrogenation 两段催化加氢
two-stage centrifugal pump 两级离心泵
two stage column 两级柱
two-stage combustion 两级燃烧
two-stage compounder 两段式混料机
two-stage compressor 两级压气机
two-stage deoiling 两段脱油法
two-stage desulfurization 两段脱硫
two-stage dewaxing 两段脱蜡
two-stage digester 两级消化器
two-stage ejector 两级喷射器
two-stage filter 两级过滤器
two-stage hydrocracking 两段加氢裂化
two-stage hydrogenation 两段加氢
two-stage liquification 两段液化
two-stage mass spectrometer 双级质谱计
two-stage nitration 二段硝化(法)
two-stage plasticator 两级压塑机
two-stage polymerization 两步聚合
two-stage process (酚醛树脂)两级制备法；二段(硝化)法
two-stage radiation induced polymerization 二步辐射诱导聚合(作用)
two-stage reforming 两段重整

two-stage resin 两级(酚醛)树脂
two-stage sampling 两级取样；二步采样
two stage separator 二级分离器
two-start screw 双螺纹螺杆；双头螺杆
two-step cure 两级硫化
two step method 两级法
two-story face 双层割面
two-story sedimentation tank 双层沉淀池
two-tone finish 双色调涂饰剂
two-tone leather 双色调革
two-toning 双色调涂饰
two-way classification 两种方式分组
two-way electrochromatography 两路电色谱(法)
two-way tumbler switch 双连开关〔商〕
two-zone hydrosulfurization 双区加氢脱硫
twyer(=tuyere) 吹风管嘴
Twyman interferometer 特外曼干涉仪
tychite 杂芒硝
tying-in 结经
tylan(=tylosin) 泰乐菌素
Tyler mesh 泰勒标准筛号
Tyler standard screen 泰勒标准筛
Tyler standard screen scale 泰勒标准筛制
Tyler standard sieve 泰勒标准筛
tyllithin 乙酰水杨酸锂
tylophorine 娃儿藤碱
tylophorinine 娃儿藤宁
tylose 甲基纤维素
tylosis 侵填体
Tylox process 泰勒克司过程
tympan 厚衬纸
tympan paper 印刷衬纸
Tyndall effect 丁铎尔效应
tyndallimetry 悬体测定法
Tyndall light 丁铎尔光(现象)
Tyndall meter 丁铎尔计
Tyndall phenomenon 丁铎尔现象
Tyndall scattering 丁铎尔散射
Tyne powder 泰恩(炸)药〔一种铵硝、钾硝、木屑炸药〕
type ①(典)型；类型；式样②型号
type analysis 结构族分析
type A standard uncertainty A类标准不确定度
type B standard uncertainty B类标准不确定度
type metal 铅字合金
type of compounds 化合物类型
type of decay 衰变形式；衰变方式
type of disintegration 蜕变类型；衰变形式；衰变方式
type of weave 组织类型

type reaction	典型反应
type sample	标(准)样(品)
π-type mass spectrometer	π型质谱仪
types of flow	流动类型
type solvent	典型溶剂
type test	典型试验
typewriting ink	打字机墨
typhasterol	香蒲甾醇
typhotoxin	伤寒毒
typhus vaccine	斑疹伤寒疫苗
typical	典型的
typical compound	典型化合物
typical element	典型元素
typical microstructure	典型显微结构
typical properties	典型性质
typical representative grease	典型润滑脂
typing	分型
typographical	印刷的
typographical printing	金属版印刷
typography	活字印刷术
typomorphic	类吗啡的
typophor brown	印刷棕；油溶棕
tyramine	酪胺；2-对羟苯基乙胺 $HOC_6H_4CH_2CH_2NH_2$
tyramine hydrochloride	盐酸酪胺
tyratol	氨基甲酸百里酚酯
tyre	(轮)胎；车胎
tyre assembling	外胎装配
tyre builder	轮胎成型床
tyre building	外胎成型
tyre building machine	轮胎成型床
tyre building room	轮胎成型室
tyre casing	外胎
tyre collapsion	轮胎的压缩
tyre cord	轮胎帘子布
tyre dynamic response research	轮胎动态响应分析
tyre fabric	轮胎布
tyre foundation	帘子布层；胎体；胎身
tyre heater	轮胎压热器
tyre iron	轮胎架
tyre mould	轮胎压模
tyre opener	轮胎撑开器
tyre recapping machine	轮胎翻新机
tyre reclaim	轮胎再生胶
tyre regrooving machine	轮胎重新刻纹机
tyre retreader	轮胎翻新机
tyre section	轮胎断面
tyre shoulder	胎肩
tyre soles process	胎面翻新
tyre spreader	轮胎撑开器
tyre stock	轮胎胶料
tyre tester(=tyre testing machine)	轮胎试验机
tyre testing machine(=tyre tester)	轮胎试验机
tyre tool	轮胎工具
tyre tread	轮间距离
tyre welder	轮胎热补器
Tyrian purple	泰尔红紫
tyroleucine	酪亮氨酸
tyrolite	铜泡石
tyrosal	水杨酸安替比林；苷元酚 $C_{18}H_{18}O_4N_2$
tyrosinase	酪氨酸酶
tyrosine	酪氨酸；3-对羟苯基丙氨酸
tyrosine carboxylase (=tyrosine decarboxylase)	酪氨酸脱羧酶
tyrosine decarboxylase(=tyrosine carboxylase)	酪氨酸脱羧酶
tyrosine iodinase	酪氨酸碘化酶
tyrosol	对羟苯基乙醇
tyrosyl	酪氨酰 $HOC_6H_4CH_2CH(NH_2)CO-$
tysonite	氟铈矿
tyvelose	泰威糖；3,6-二脱氧-D-甘露糖

U

Ubbelodhe drop point　乌伯娄德滴点
Ubbelodhe melting point　乌伯娄德熔点〔润滑脂〕
Ubbelodhe viscometer　乌氏黏度计
Ubbelodhe viscosimeter　乌伯娄德黏度计
U-bend　U形管
ubenimex　乌苯美司〔药〕
ubiquinone　泛醌；辅酶Q
U-bolt　U形螺栓
ucon oil　乌康油〔聚烯二醇油的商名〕
ucuhuba butter　肉豆蔻脂
ucuhuba oil　肉豆蔻油
ucuhyba fat　肉豆蔻脂
udell　(冷凝水气)接受器
Udex B-T-X plant　尤狄克斯苯-甲苯-二甲苯装置
Udex extraction process(=Udex process)　尤狄克斯抽提过程〔在逆流塔用二醇抽提芳烃〕
Udex process(=Udex extraction process)　尤狄克斯抽提过程〔在逆流塔用二醇抽提芳烃〕
U effect　U效应；发生交替电压
U-expansion joint　U形膨胀节
U-F　脲-甲醛树脂
U-form tube　U(形)管
ufuta　芝麻
UF value　滤渣值
Ugit process　乌吉特过程〔高温裂解〕
Uhde process　乌德法〔合成氨〕
uhligite　锆钙钛矿
UHP graphite electrodes　超高功率石墨电极
Uibricht sphere　乌布里希球
uintahite(=gilsonite)　硬沥青
uintaite(=gilsonite)　硬沥青
ukambine　乌坎宾〔一种箭毒碱〕
Ukena colo(u)rimeter　乌肯纳比色计
ulcer　溃疡
Ulco metal　乌耳科合金
U-leather ring　U形管圈
uleron (=uliron；diseptal)　乌利龙；磺酰磺胺二甲
ulexine　乌乐碱
ulexite　硼钠钙石；钠硼解石
ulinastatin　乌司他丁〔药〕
uliron (=uleron)　乌利龙；磺酰磺胺二甲$H_2NC_6H_4SO_2NHC_6H_4SO_2N(CH_3)_2$
ullage(=outage)　①罐空〔储罐或容器内液面上空高度或容积〕②根据罐空确定罐内液体容积
ullage reference-point　罐空计量的基准点
ullage rule　测量罐空的量尺
ullage table　测油尺〔根据罐空推算容器内液体量的表〕
ullaging　根据罐空计量罐内油料量
ullmannite　锑硫镍矿
ulmic acid　①滑榆酸②棕腐酸；赤榆酸
ulmin　①滑榆胶②棕腐质
ulmin compound　棕腐质化合物
ultimate　最后的；极限的
ultimate analysis　元素分析；最后分析
ultimate bearing strength　极限承载强度
ultimate bending strength　极限曲折强度
ultimate biodegradation　最终生物降解；极限生物降解〔转变成无机物的生物降解作用〕
ultimate composition　元素组成
ultimate compression strength　抗压强度
ultimate crushing strength　极限压碎强度
ultimate disposal　最终处置
ultimate disposal of radwaste　放射性废物最终处置
ultimate elongation　极限伸长
ultimate extension　极限延伸
ultimate flax fibre　亚麻单纤维
ultimate line　住留谱线
ultimate load　最大载荷；极限载荷
ultimate motor fuel　最优发动机燃料
ultimate Newtonian viscosity　极限牛顿黏度
ultimate oxidation　极限氧化(作用)
ultimate oxygen demand　总需氧量
ultimate particle　基本粒子
ultimate-precision　最佳精密度
ultimate-precision method　最佳精密法
ultimate precursor　极限前体；终极前体
ultimate production　总产量
ultimate property　极限性质
ultimate rational units　基本有理单位
ultimate resilience　极限回弹性
ultimate sensitivity　极限灵敏度
ultimate separation　极限分离
ultimate strain　极限应变
ultimate strength　极限强度*
ultimate stress　极限强度；极限应力
ultimate swelling　极限溶胀
ultimate temperature　极限温度
ultimate tensile elongation　①极限拉伸伸长②极限拉伸

伸长率
ultimate tensile strength　极限拉伸强度*
ultimate tension　抗拉强度极限
ultimate-use temperature　极限使用温度
ultimate vacuum　极度真空
ultimate viscosity number　最大黏度值
ultimate working unit　极限工作单元
ultimate yield　最终收率〔加工后产物收率〕
ultra-〔拉丁词头〕　①超；过；越；(以)外②极端；异常；过(度)
ultra accelerant　超催速剂；超促进剂
ultra accelerator　超促进剂
ultra-acidic kicker　酸性超促进助剂
ultrabasic rocks〔复〕　超基性岩
ultracat cracking　超催化裂化
ultracat regeneration　超催化再生
ultracentrifugal analysis　超(速)离心分析
ultracentrifugal characterization　超离心特性
ultracentrifugal sedimentation　超离心沉积
ultracentrifugal stability　(乳状液的)超离心稳定性
ultracentrifugation　超(速)离心法
ultracentrifugation sedimentation　超速离心沉降(法)
ultra-centrifugation sedimentation separation and analysis　超离心沉降分离分析方法
ultracentrifuge　超离心机*
ultracentrifuge method　超离心机法
ultracentrifuge sedimentation　超离心(机)淀积(法)
ultra chromatography (=fluorescence chromatography)　荧光色谱(法)
ultra-clean coal　特净煤
ultra-clean room　超净室
ultracold neutron　超冷中子
ultra-conjugate fiber　超共轭纤维
ultracrackate　超加氢裂化产物
ultracracking　超加氢裂化
ultracryotomy　冷冻超薄切片术；超冷冷切片
ultra-development(=progressive development; continuous running development)　流出展开
ultra fast accelerator　超速促进剂
ultrafast transient infrared spectroscopy　超快瞬变红外光谱
ultrafilter　超滤器
ultrafiltering balloon　超滤瓶
ultrafiltering crucible　超滤坩埚
ultrafiltrate　超滤液
ultrafiltration　超滤*
ultra-filtration funnel　超滤漏斗
ultrafiltration membrane　超滤膜
ultrafine　①超薄(的)〔薄膜〕②超精度(的)③超细度(的)
ultra fine calcium carbonate　超细碳酸钙
ultra-fine dust　特细粉末
ultra-fine fibre　超细纤维
ultra-fine filter　超滤器
ultra fine pulverizer　超微粉碎机
ultra fine talc pigment　超细粒子滑石粉
ultrafining　超加氢精制(法)
ultra-forming process　超重整过程
ultra-gamma ray (=cosmic ray)　超γ射线；宇宙线
ultrahertzian waves　赫(兹)外波
ultra high frequency　超高频率
ultra-high molecular weight polyethylene(UHMWPE)　超高分子量聚乙烯
ultra-high molecular weight polymer　超高分子量聚合物
ultra-high pressure mercury lamp　超高压水银灯
ultra-high purity　超高纯度
ultrahigh-strength steel　超高强度钢
ultra-high-structure black　超高结构炭黑
ultrahigh vacuum (HHV)　超高真空
ultra-high voltage electron microscope　超高压电子显微镜
ultraloft　超膨松
ultra-low ammonia latex　超低氨胶乳
ultra low density polyethylene(ULDPE)　超低密度聚乙烯
ultra low interfacial tension　超低界面张力
ultra-low viscosity oil　超低黏度油
ultramarine(=French blue)　群青；佛青
ultramarine blue　群青色
ultramarine brown　群青棕
ultramarine green　群青绿
ultramarine violet　群青紫
ultramarine yellow　群青黄
ultramicro-　①超微量②超微
ultramicro analysis　超微量分析
ultramicro (analytical) balance　超微量天平
ultramicro burette　超微量滴定管
ultramicrochemical manipulation　超微量化学操作
ultramicrochemistry　超微量化学
ultramicro-crystal　超微结晶
ultramicro-determination　超微量测定
ultramicro disk electrode　超微盘电极
ultramicroelectrode　超微电极
ultramicroelectrophoresis　超微量电泳
ultramicrometer　超测微计
ultramicro method　超微量法
ultramicron　超微粒子
ultramicropipette　超微量吸移管；超微量移液管
ultramicroscope　超显微镜

ultramicroscopic observation 高倍显微镜观察
ultramicroscopic organism 超显微微生物
ultramicroscopic particle 超显微粒子
ultramicroscopy 超显微术
ultra-micro sensor 超微传感器
ultramicrospectrophotometer 超显微分光光度计
ultramicrospectrophotometry 超显微分光光度测定法
ultramicrostructure 超微结构
ultramicrotechnique 超微技术
ultramicrotome 超薄切片机
ultraphonic emulsifier 超声波乳化器
ultraphotic 超视的
ultra powdery carbon black 超细粒子炭黑
ultra-precision machining 超精密机械加工
ultra-pure 超纯的
ultrapure water 纯水*
ultrapurification 超净化
ultra purity 超纯度
ultraquinine 超奎宁；类奎宁
ultra-radio frequency 超射频(率)
ultra-rapid vulcanization 超速硫化
ultra-red 红外的
ultra-red absorption spectrometry 红外线吸收光谱法
ultra-red ray 红外线
ultras 超促进剂
ultraselective cracking process 超选择性裂化法
ultrasensitive amino acid analyzer 超灵敏氨基酸分析仪
ultrashort pulse 超短脉冲
ultra short wave 超短波
ultrasonator 超声振荡器
ultrasonic 超声(波)的
ultrasonic absorption 超声波吸收
ultrasonic agglomeration 超声附聚；超声结块
ultrasonic agitation 超声搅拌
ultrasonically induced oxidation 超声诱发的氧化
ultrasonic annealing 超声退火；超声熟炼
ultrasonication 超声破碎
ultrasonic cleaning 超声波清洗
ultrasonic defectoscope 超声探伤器；超声检验器
ultrasonic degas 超声波脱气
ultrasonic degreaser 超声(波)脱脂装置
ultrasonic detector 超声(波)检测器
ultrasonic diffraction microscope 超声衍射显微镜
ultrasonic disintegrator 超声粉碎器；超声振荡器
ultra-sonic dispersion 超声分散
ultrasonic emulgator 超声波乳化器
ultrasonic emulsification 超声乳化(作用)
ultrasonic emulsion breaking 超声破乳(化)
ultrasonic examination 超声检验
ultrasonic extraction 超声萃取
ultrasonic extractor 超声萃取器
ultrasonic fault detector 超声探伤仪
ultrasonic ga(u)ge 超声波探伤仪
ultra-sonic generator 超声发生器
ultrasonic holography 超声全息照相法
ultrasonic inspection 超声探伤
ultrasonic interferometer 超声干扰仪
ultrasonic investigation 超声研究
ultrasonic leaching 超声浸出
ultrasonic machining 超声加工
ultrasonic metal welding 超声波金属焊接
ultrasonic nebulizer 超声(波)雾化器
ultrasonic radiation 超声辐射
ultrasonics 超声学〔物〕
ultrasonic separation 超声分离
ultrasonic slurry sampler 超声浆料进样器
ultrasonic test 超声检验
ultrasonic tester 超声探伤器；超声检验器
ultrasonic testing 超声波(探伤)试验
ultrasonic thickness gauge 超声厚度计
ultrasonic thickness meter 超声测厚仪
ultrasonic vibration dyeing 超声波振荡染色
ultrasonic viscometer 超声黏度计
ultrasonic washing 超声波洗涤法
ultrasonic washing machine 超声洗涤机
ultrasonic wave 超声波〔物〕
ultrasonic welding 超声波焊接
ultrasonimetric analysis 超声分析
ultrasono-cardio-tomography 超声心脏断层造影法；超声心动图描记法
ultrasonography 超声照相法；超声回声图描记法
ultrasonoscope 超声波(探测)仪
ultrasonotomography 超声断层造影法；超声断层显像图描记法
ultrasound 超声
ultrasound wave 超声波
ultrastructure 超显微镜结构分析
ultrasweetening 超级脱硫
ultrathermometer 限外温度计
ultrathin 超细的；超薄的
ultrathin coating by multiple polyelectrolyte adsorption/surface activation technique 聚电解质吸附-活化技术
ultra thin laminate 超薄型层压板
ultrathin membrane 超薄膜
ultrathin section 超薄切片
ultrathin sectioning 超薄切片

ultrathin window 超薄窗口
ultra trace 超痕量
ultratrace amount 超示踪量
ultra trace analysis 超痕量分析
ultraviolet ①紫外线的②紫外线
ultraviolet absorbent 紫外线吸收剂
ultraviolet absorber 紫外线吸收剂
ultraviolet absorption 紫外线吸收
ultraviolet absorption detector 紫外吸收检测器
ultraviolet absorption spectrometry 紫外线吸收光谱法
ultraviolet absorption test 紫外线吸收试验
ultraviolet activation 紫外线活化(作用)
ultraviolet analysis 紫外(线)分析
ultraviolet analyzer 紫外分析仪器
ultraviolet and visible light detector 紫外及可见光检测器
ultraviolet and visible spectrophotometry 紫外-可见分光光度法
ultra violet curing 紫外线硬化
ultraviolet degradation 紫外线降解(作用)
ultraviolet detector 紫外线检测器
ultraviolet fluorescence 紫外荧光
ultraviolet generator 紫外线发生器
ultraviolet-initiated 紫外(线)引发的
ultraviolet-initiated polymerization 紫外线引发聚合
ultraviolet ionizing laser 紫外电离激光器
ultraviolet irradiation 紫外线照射
ultraviolet lamp 紫外灯
ultraviolet laser 紫外激光器
ultraviolet laser resonance Raman spectroscopy 紫外激发共振拉曼光谱
ultraviolet light 紫外光
ultraviolet light rating 紫外等级
ultraviolet microscope 紫外线显微镜
ultraviolet monitor 紫外线监测器
ultraviolet photoelectron spectroscopy 紫外光电子能谱(法)
ultraviolet photometric titration 紫外光度滴定
ultraviolet plate 紫外感光板
ultraviolet radiation 紫外线照射(作用)
ultraviolet range(=ultraviolet region) 紫外线区域
ultraviolet ray 紫外线
ultraviolet ray absorber 紫外线吸收剂
ultraviolet ray absorbing glass 紫外吸收玻璃；防紫外线玻璃
ultraviolet ray intercepting glass 紫外吸收玻璃；防紫外线玻璃
ultraviolet ray lamp 紫外(线)灯
ultraviolet ray microscope 紫外显微镜
ultraviolet ray transmitting glass 透紫外线玻璃
ultraviolet recorder 紫外线记录器
ultraviolet region 紫外(光)区
ultraviolet resistance 耐紫外线性
ultraviolet screener 防紫外线剂
ultraviolet sensitive paper 紫外光敏纸
ultraviolet spectrogram 紫外线光谱图
ultraviolet spectrograph 紫外光谱
ultraviolet spectrometer 紫外光谱仪
ultraviolet spectrometry 紫外光谱法
ultraviolet spectrophotometer 紫外线分光光度计
ultraviolet spectrophotometric method 紫外分光光度法
ultraviolet spectrophotometry 紫外分光光度法*
ultraviolet spectroscope 紫外光分光镜
ultraviolet spectroscopy 紫外吸收光谱
ultraviolet spectrum 紫外光谱
ultraviolet stabilizer 紫外线稳定剂*
ultraviolet transmission ①紫外透射②紫外透射量
ultraviolet-visible colorimeter monitor 紫外-可见光比色监控器
ultraviolet-visible detector 紫外-可见光检测器
ultraviolet-visible effluent monitor 紫外-可见光流出物监控器
ultraviolet-visible light detector 紫外-可见光检测器
ultraviolet-visible-photoacoustic spectrometer 紫外-可见光声光谱仪
ultraviolet-visible spectrophotometer 紫外-可见分光光度计
ultraweak luminescence 超微弱发光
umangite 红硒铜矿
umbellaric acid 伞二酸 $C_8H_{12}O_4$
umbellate ①伞形酸盐②伞形酸酯 $(HO)_2C_6H_3CH=CHCOOM$; $(HO)_2C_6H_3CH=CHCOOR$
umbellatin 伞花碱
umbellic acid 伞形酸；2,4-二羟(基)肉桂酸
umbelliferone (=7-hydroxycoumarin) 伞形酮；7-羟基香豆素
umbelliferone methyl ether 伞形酮甲醚；7-甲氧基香豆素
umbelliprenin 伞形花醚
umbellonic acid 伞酮酸 $C_9H_{14}O_3$
umbellularic acid 伞柳二酸
umbellulic acid 加州月桂酸 $C_{11}H_{22}O_2$
umbellulone 加州月桂酮；伞形花酮 $C_{10}H_{14}O$
umber 棕土
umbilicaric acid 石耳酸
umbra 本影
umbrella collector 伞状收集器〔倒装伞状蒸气收集器〕
umbrella effect 伞效应*
umbrella reel 伞形丝框
umbrella roof 伞形顶〔油罐〕

U-mode　U型〔扩散信号〕
Umpherston beater　昂弗斯东打浆机
umpire analysis　裁判分析
umpolung　极反转
unaccelerated aging　自然老化
unaccelerated sulfur vulcanizate　无促进剂的硫化胶
unacceptable product　不合格品；废品
unaccepted product　不合格(产)品
unacclimated (activated sludge)　未驯化的(活性污泥)
unactivated state　非激发态；未活化态
unaged property value　老化前性能数值
unannealed track　未退火径迹；未熟炼径迹
unattackable electrode　抗蚀电极
unattended booster pumping station　自动增压泵站
unattended pumping station　自动泵油站
unavailability　不能利用；无效
unavailable　不能利用的；无效的；不可得的；不可用的
unavailable energy　不可用能源
unavailable fertilizer　无效肥料
unbaffled　①无阻板的②无挡板的
unbalance　①失衡②不平衡
unbalanced diaphragm　差压隔膜；单面受压隔膜
unbalanced gasoline　分馏不良的汽油
unbeaten pulp　生纸浆
unbiased　无偏性
unbiased estimate of population variance　总体方差的无偏估计
unbiased estimator　无偏估计值*；无偏估计量
unbiased sample　无偏样本
unbiased statistics　无偏统计
unbiased test　无偏检验
unbleached　未漂白的；原色的
unblended　未掺合的
unblended base stock　未掺油料〔1.未掺和的基本油料；2.无四乙铅的基本组分〕
unblended gasoline　未掺汽油
unblended octane number　掺和前辛烷值
unblended oil　未掺和油
unblocked-in article　注型制品
unbonded coatings　未黏合被覆层
unbound state　游离状态；未结合状态
unbound water　非结合水
unbranched　无支链的
unbranched chain　直链；无分支的链
unbranched-chain hydrocarbon　无支链烃
unbranched dicarboxylic acid　直链二羧酸
unbroken rubber　未塑炼生胶
unbuilt detergent　未复配洗涤剂〔无助洗剂的洗涤剂〕

unburned charge　未燃烧混合物
unburned fuel　未燃烧燃料
unburned gas　未燃烧气体
unburnt　未灼烧的
uncambered bottom　平底
uncap　开盖
uncarbonated lye　未碳酸化碱水
uncarine　钩藤碱
uncatalyzed　未催化的
uncatalyzed polymerization　无催化聚合
uncertain systematic error　未定系统误差
uncertainty　不确定度*；不确定(性)
uncertainty principle　测不准原理
uncertainty relation　测不准关系
uncharged acid　无荷(电)酸
uncharged carrier　中性载体
uncharged carrier electrode　中性载体电极
uncharged polymer　非荷电聚物
unclad　无外壳(的)
uncoated laser　不镀膜激光器
uncoated lens　不镀膜透镜
uncoiled molecule　非盘曲型的分子；伸直的分子
uncoiling　解卷；伸直
uncoiling of molecule　分子的伸直
uncolored　无色的；未着色的
uncolored cloth　无色布
uncombined sulfur　非化合硫
uncompatibility　①不相容性②非兼容性
uncompetitive inhibition　非竞争性抑制
uncompounded cement　纯胶浆
uncompounded rubber　无填料的胶料；纯胶料
uncondensable　不凝结的；不可冷凝的
uncondensed spark　非电容火花
unconditional transfer　无条件转移
unconfined blast　敞口吹风
unconfined mixture　无限制混合
unconjugated acid　非共轭(双键)酸
unconjugated double bond　非共轭双键
uncontaminated air　洁净空气
uncontrollable factor　不可控因素
uncontrolled flow　无控制流动
unconverted n-butane　未转化正丁烷
unconverted monomer　未聚合的单体
unconvertible hydrocarbon　不能转化的烃
uncooled　未冷凝的
uncork　未加塞的
uncorrected absorbance　未校正吸光度
uncorrected retention volume　未校正保留体积

uncorrelated 不相关的；无关的
uncorrelated sample 不相关样本
uncoupled bond 非偶联键
uncoupled electron 非偶(联)电子
uncracked asphalt 未裂化沥青
uncracked hydrocarbon 未裂化烃
uncracked residue 未裂化残油
uncrimping strain 解卷曲应变；未卷曲应变
uncrystallizable 不能结晶的
uncrystallized 非晶的；不可结晶的
unctuosity 油腻性；似油性
unctuous 油性的
unctuousness 油性
uncture 油膏
uncured ①未处治的②未硫化的
undark 夜明涂料
undarken 夜明化
undeca- 〔希腊字头〕十一
undecadienal 十一碳二烯醛
2,4-undecadienal 2,4-十一碳二烯醛
2,3-undecadione 2,3-十一碳二酮
undecalactone 十一(碳)烷酸内酯
γ-undecalactone γ-十一内酯；桃醛〔商〕；十四醛〔商〕
undecamethylenediamine 十一撑二胺
undecanal 十一醛
undecanate 十一酸酯
undecandienoic acid 十一碳二烯酸 $C_{10}H_{17}COOH$
undecandioic acid 十一烷二酸
undecane(=hendecane) 十一(碳)烷
undecane diacid 十一烷二酸 $COOH(CH_2)_9COOH$
undecane dicarboxylic acid 十一烷二羧酸；十三烷二酸 $C_{11}H_{22}(COOH)_2$
undecanoate 十一酸酯
undecanoic acid (=undecylic acid) 十一(烷)酸 $CH_3(CH_2)_9COOH$
undecanoic amide 十一(烷)酰胺 $CH_3(CH_2)_9CONH_2$
1-undecanol 1-十一(烷)醇 $CH_3(CH_2)_9CH_2OH$
undecanol 十一醇
2-undecanol 2-十一(烷)醇 $C_9H_{19}CHOHCH_3$
2-undecanone(=methyl n-nonyl ketone) 2-十一烷酮；甲基正壬基(甲)酮
undecanonitrile 十一(烷)腈 $CH_3(CH_2)_9CN$
undecanoyl 十一(烷)酰 $CH_3(CH_2)_9CO-$
undecen-2-al 十一烯-2-醛
undecenal(=undecylenic aldehyde) 十一(碳)烯醛
undecenate(=undecylenate) 十一烯酸盐(酯)
undecendioic acid 十一碳(一)烯二酸 $C_9H_{16}COOH$
undecene 十一碳烯 $C_{11}H_{22}$

undecene diacid 十一碳烯二酸 $C_9H_{16}(COOH)_2$
undecene dicarboxylic acid 十一碳烯二羧酸；十三碳(一)烯二酸 $C_{11}H_{20}(COOH)_2$
10-undecene-1-ol 10-十一碳烯-1-醇 $CH_2=CH(CH_2)_9OH$
undecenoate 十一酸酯
undecenoic acid 十一碳烯酸 $C_{10}H_{19}COOH$
10-undecenoic acid 10-十一烯酸
undecenol 十一(碳)烯醇
undecenyl 十一碳烯基
undecoic acid 十一(烷)酸 $C_{10}H_{21}COOH$
undecomposable 不可分解的
undecomposed 未分解的
undecyl 十一(烷)基
undecyl alcohol 十一醇
undecyl aldehyde 十一醛
undecylate 十一酸酯
undecyl dimethyl benzyl ammonium halide 十一烷基二甲基苄基卤化铵
undecylenate(=undecenate) 十一烯酸盐(酯)
undecylene 十一碳烯 $C_{11}H_{22}$
undecylene aldehyde 十一烯醛
undecylene diacid 十一碳(一)烯二酸 $C_9H_{16}(COOH)_2$
undecylene dicarboxylic acid 十三碳(一)烯二酸 $C_{11}H_{20}(COOH)_2$
undecylenic acid 十一烯酸
undecylenic alcohol 十一烯醇
undecylenic aldehyde(=undecenal) 十一(碳)烯醛
undecylenyl acetate 乙酸十一烯酯
undecylic acid(=undecanoic acid) 十一(烷)酸 $CH_3(CH_2)_9COOH$
undecylic alcohol 十一(烷)醇
undecylic aldehyde 十一(烷)醛 $CH_3(CH_2)_9CHO$
undecylic aldehyde oxime 十一(烷)醛肟 $CH_3(CH_2)_9CH=NOH$
undecylprodigiosin 十一烷基灵菌红素
undecyndioic acid (=undecyne diacid) 十一炔二酸
1-undecyne 1-十一(碳)炔 $CH_3(CH_2)_8C≡CH$
undecyne diacid(=undecyndioic acid) 十一炔二酸 $C_9H_{14}(COOH)_2$
undecyne dicarboxylic acid 十一炔二羧酸；十三碳(一)炔二酸 $C_{11}H_{18}(COOH)_2$
undecynic acid(=undecynoic acid) 十一碳(一)炔酸 $C_{10}H_{17}COOH$
undecynoic acid (=undecynic acid) 十一碳(一)炔酸
undegraded material 未降解物质
under ①在下；底下②不足③从属的
under-acetylation 未充分乙酰化；乙酰化不足

underbead cracking test 焊缝下裂纹试验
under bleached 欠漂的
underbody 物体下部；底部
underbody dipping(=slipper dipping) （车身)移动浸涂；(车身)下(身)浸涂
under-bound finish 基料不足的涂膜
underburnt 欠熟
undercoat ①衬底施②衬里衣
undercoat paint 头道漆
undercolored 欠染的〔染〕
undercompound winding 低复绕法
under cooked chips 夹生木片
undercooled 过冷的
undercooled liquid 过冷液体
undercooling 过冷；冷却过度
under cure 欠硫；欠固化
undercuring ①欠熟②欠硫化
undercut ①伐倒切口②伐采不足量③咬边
undercutting 侧蚀
underdraft furnace 下抽(风)炉
underdrawn 未充分拉伸的；拉伸不足的
underdriven 底动的
underdriven automatic batch centrifuge 三足式自动间歇卸料离心机
underdriven buhrstone mill 底动石磨
underdriven centrifuge 底动离心机
underdriven mixer 下传动式搅拌机
underfed 下给的；下送的
underfed stoker 下给加煤机
underfeed ①营养不足；饲养不足②下部进料
underfeeding 供料不足
underfired ①欠火的；欠煅的②下部燃烧的
underfiring 欠火；欠熟
underflow 底流
underflow of the classifier 分级机的下层流
underglaze 釉底的
underglaze blue （釉下)青花
underglaze color 釉底颜料〔染〕
underglaze decoration 釉底彩
underglazing 釉底彩
under-ground 地(面以)下的
underground coal gasification 煤的地下气化
underground corrosion 地下腐蚀
underground disposal 地下处理
underground gasification 地下气化
underground leaching 地下浸取；地下沥取
underground liquefied petroleum gas storage 液化石油气地下储存(库)

underground oil storage 地下储油库
underground storage pools 地下储油池
underground storage tank 地下储(油)罐
underground tank 地下储(油)罐
underground water 地下水
underheated 欠热的
underinflated tyre 缺气胎〔橡〕
underlayer 下层；底层
underlaying ①置于下②垫
under load ①起动；开动②在负荷下
under meter 流量计；流速计
undermethylation 甲基化不足
undermixing 混合不足
under-oxidized 氧化不足的；欠氧化的
underpressing ①压型不够②欠压榨
underpressure 真空计压力
underpressure distillation 减压蒸馏
underpriming 注油不足
underproof 不合格的；不合标准的；标准强度以下的
underproofed 不合格的
underproof spirit 不合格酒精；不合标准的酒精
underrate 低估的
underrated octane number 低估辛烷值
underrefining 精炼不足；欠精炼
undersaturated 欠饱和的
undersize ①筛底料②筛下
undersized ①欠浆的②欠胶的③筛下的
undersized chips 过小木片；筛下木片
undersized fabric 欠浆布
undersized paper 欠胶纸；欠填料纸
undersoil 底土；地面下土
understable 欠稳定的
under-stream period 开工期
under-sulfonate 磺化不足
undersurface filling 液面下灌装
undersurface loading 液面下灌装
under-tanned 未鞣透的
undertint 淡色；浅色；褪色
undertone ①底彩；淡色②底香〔香料〕
undertread 底胎面
undertrimmer 滚边机
under voltage 电压不足
undervulcanization 欠硫化
under-water cutting 水下切割
underwater granulator 水浸造粒机
underwater line 水下管线
underwater pelletizer 水浸切粒机
underwater tank 水下储罐

underwater welding 水下焊接
underwing refueling 翼下补充燃料
Underwood's equation 安德伍德方程
Underwood method 安德伍德法〔计算最小回流比〕
Underwood oxidation test 安德伍德氧化试验
undesulfured 未脱硫的
undeterminable losses〔复〕 不可测定的损失
undiluted 未稀释的
undiluted engine oil 未稀释的发动机油
undiluted liquefied petroleum gas 未(被空气)稀释的液化石油气体
undiluted oil 未稀释的油
undirectional motion 不定向运动
undissociated 未离解的
undissolved 不溶解的
undissolving 不溶解的
undistillable heavy residue 不能蒸馏的重残油
undisturbed 未扰乱的
undisturbed flow ①稳流②未扰乱流
undressed 未加工的；生的；未经处理的
undressed ore 原矿
undried pellets 未干燥的小片〔催化剂〕
undue 过分的；不当的；非常的
undue heat 过(度)热(量)；不适当的热量
undue toxicity 异常毒性
undulate 波动
undulation 波动；波荡
undulatory current 波动电流
undulatory motion 波荡运动
undung 山苍子脂
undyed fuel 未染色的燃料
unencapsulated source 非密封源；未封装源
unenriched uranium 未浓缩的铀
unequal angle bar 不等边角铁；不等山形铁
unequally accurate measurement 不等精度测量
unequal variances 不等方差
unesterified 未酯化的
unetherified 未醚化的
unevaporated 未蒸发的
uneven bottoming 打底不匀；底染不匀；预媒染不匀
uneven color(=uneven dye) 不匀染色
uneven distribution error 不均匀分布误差
uneven distribution of gum 胶质的不均匀分布
uneven dye(=uneven color) 不匀染色
uneven fracture 粗糙断面；不平断面
uneven front 不整齐前沿
uneven grain 不均匀颗粒
unevenness 不平坦性；不均匀性

unevenness coefficient of mean deviation 平均差不匀率
uneven raising 绒不齐
uneven register 合模不正
unexplosive 不爆炸的
unextended compound 无油胶料
unfamiliar oxidation state 希见氧化态；不常见氧化态
unfast color(=unfast dye) 不牢染色
unfast dye(=unfast color) 不牢染色
unfast dyeing 不牢染(色)
unfast dyestuff 不牢染料
unfermentable 不能发酵的
unfilled gum vulcanizate 纯硫化胶
unfiltered 未滤过的
unfinished goods 半制品；半成品
unfinished leather 未整饰的皮革
unfired 不烧的；不用火的
unfired pressure vessel 不用火加热的高压容器；热交换器
unfluxed asphalt 未稀释的沥青
unfolded 展开的；解开的
unfolded chain section 无折叠链段；伸展链段
unfolded structure 无折叠结构；伸展结构
unfolding 伸展
unfolding force 解折叠力
unformulated surfactant solution 单纯表面活性剂溶液
unfreezable gelatin dynamite 不冻胶(状)爆药
ungalvanized 未镀锌的
ungalvanized iron and steel 未镀锌钢铁
ungerade 反对称
ungerade states u 态；反对称态
ungerine 不老蒜碱；水仙花碱
unglazed ①未上釉的；素烧的②无光的
unglazed crucible 素烧坩埚
unglazed kraft paper(=pure kraft paper) 棕色平面包皮纸
unglazed porcelain ①素(烧)瓷②素烧瓷的
unglazed printing paper 无光道林纸
unglazed support 未釉化载体
ungraphitized carbon 非石墨化碳；无定形碳
ungrease 脱脂的
ungrindable sample 不能研碎的试样
unguent 软膏；油膏；药膏
unguentum 软膏(剂)
ungulic acid 蹄酸
unhair 去毛
unhaired hide 去毛皮；光皮
unhairing 去毛；脱毛
unhairing beam 刨皮砧板
unhairing cylinder 去毛机刀轴

unhairing knife 去毛刀
unhairing machine 去毛机
UNH conversion process 由六水合硝酸铀酰制备三氧化铀的转换过程
unhindered amine (=free amine) 未受阻胺；游离胺
unhomogeneity 不均一性
unhydrated 未水合的
unhydrolyzable 不能水解的
uni- 〔拉丁字头〕单；一
uniacid base 一(酸)价碱
uniaxial 单轴的
uniaxial compression modulus 单轴压缩模量
uniaxial crystal 单轴晶体
uniaxial drawing 单轴拉伸
uniaxial elongation 单轴拉伸
uniaxial extension 单轴延伸；单轴拉伸
uniaxially oriented polymer 单轴定向聚合物
uniaxial orientation 单轴取向
uniaxial tensile deformation 单轴拉伸形变
unicellular plastic 闭孔微孔塑料
unicellular rubber 闭孔泡沫橡胶
unicoil injection process 单旋管注入(裂化)法
unicolor 单色的
unicomponent peak 单组分峰
unicorn 角胡麻
unidentate ligand 单齿配体；一合配体
unidentified products 未鉴定的产物；组成未明的产物
unidimensional descending development 单向下行展开
unidimensional disc immunoelectrophoresis 单向盘状免疫电泳
unidirectional composite 单向复合材料
unidirectional coupler 单向波导管
unidirectional diffusion 单向扩散
unidirectional fabric 单向织물
unidirectional flow 单向流动
unidirectional prepreg 单向预浸料；单向预混料
unidirectional shear flow 单向剪切流
unidirectional weftless tape 无纬带
uni-drive gear box 单独传动齿轮箱
unified base 单一机座
unified chromatograph 多用色谱仪
unifining 加氢精制
unifining process 加氢精制过程〔脱去硫、氮、氧的化合物〕
uniflow type compressor 单流式压缩机；顺流式压缩机
uniflux fractionating tray 长条泡罩分馏塔盘
uniflux tray 单流式泡罩塔板；S形塔板
uniform ①(相)等的②(均)匀的

uniform acceleration 等加速度；均匀速度
uniform adsorption 均匀吸附
uniform apparent flow 均匀表观流动
uniform cantilever beam 等载悬臂深式试验
uniform circular motion 匀速圆周运动
uniform cross-section vessel 均匀横截面的容器
uniform degree 均匀度
uniform distribution 均匀分布*
uniform dyed fabric 匀染织物
uniform dyeing 均(匀)染(色)；素染〔染〕
uniform elongation 均匀伸长
uniform flow 匀流
uniform fluid 均匀流体
uniform fluid flow 均匀流体流动
uniform fraction 均匀馏分
uniform high efficiency open-width water mangle 均匀高效平幅轧水机
uniformity 均匀；统一；一致
uniformity coefficient 均匀系数；一致性系数
uniformity of fuel pulverization 燃料粉化的均匀性
uniformity of fuel spray 燃料喷射的均匀性
uniformity of mixture 混合物的均匀性
uniformly 均匀(性)；均一(性)
uniformly distributed load 均布载荷(重)
uniformly dyed fabric 素染织物；匀染织物
uniformly dyeing 均(匀)染(色)；素染
uniformly flowing plastic substance 均匀流动塑性物质
uniformly increasing deformation 匀增形变
uniformly labeled 均匀标记的
uniform magnetic field 匀强磁场
uniform mix 均匀混合物
uniform mixing 均匀混合
uniform motion 匀速运动
uniform plastic flow 均匀塑性流
uniform polymer 单分散聚合物
uniform rate of reaction 均匀反应速度
uniform simple shear 均匀简单剪切
uniform stream 匀流
uniform torsion 均匀扭力
uniform velocity 匀速度
unifunctional exchanger 单功能离子交换剂
unifunctional exchange resin 单功能交换树脂
unilateral 一边的；片面的
unilateral heating 单面加热
uniligand complex 单一配体配合物
unimer 单聚体
unimolecular 单分子的；一分子的
unimolecular acid-catalyzed acyl-oxygen cleavage 单分

子酸催化酰氧断裂
unimolecular base-catalyzed acyl-oxygen cleavage 单分子碱催化酰氧断裂
unimolecular bonded layer 单分子键合层
unimolecular decomposition 单分子分解
unimolecular electrophilic substitution 单分子亲电取代
unimolecular elimination 单分子消除*
unimolecular elimination reaction 单分子消除反应
unimolecular elimination through the conjugate base 单分子共轭碱消除
unimolecular film 单分子膜
unimolecular layer 单分子层
unimolecular mechanism 单分子(反应)机理
unimolecular nucleophilic substitution 单分子亲核取代*
unimolecular process 单分子过程
unimolecular reaction 单分子反应*
unimolecular termination 单分子终止*
uninflammability 不燃性
uninflammable 不易燃的
unintentional pollution 无意污染
uninteresting peak 不重要的峰；无用的峰
uninterrupted current ①连续电流②不间断流
uninterrupted run ①连续生产②连续运转
union ①活接头②联管节
union color 统一染料；万用染料
union colorimeter 联合比色计
union elbow 联管肘管
Unionfining 联合(石油公司)加氢精制(法)
unionization 不电离(作用)
unionized 未电离的
union melt welding 埋弧自动焊
union nut 活接头螺母
union splitting machine 联合剖(皮)机
union tee ①T形连接；三通连接②T形接头；三通
Union Town test method 爆震强度法辛烷值测定〔一种实际行车测定法〕
union twist yarn 混纺交织物染色
union yarn (棉毛)混纺纱
uniplanar orientation 单面取向
unipolar 单极的
unipolar conduction 单极传导(作用)
unipolar conductivity ①单向导电率②单向导电性
unipolar system 单极式
unipump 单一泵；摩托泵
unique ①唯一的；独特的②异常的
uniqueness 唯一(性)
uniqueness of compressible fluid motion 可压缩流体流动的唯一性

uniqueness of viscous fluid motion 黏性流体流动的唯一性
unique solution 唯一解
uniquinones 单醌类
unirradiated fuel 未照射过的燃料
Unisar 联合(石油公司)芳烃饱和(法)
Unisol (mercaptan extraction) process 尤尼索尔精制过程〔用含甲醇的氢氧化钠或钾的水溶液抽提油品中的硫醇和某些含氮化合物的再生式过程〕
unit ①单位②单元③个体④一(个)⑤整数⑥装置；设备
unit activity 单位活度
unit area 单位面积
unitary 单元的；一元的
unitary gas conversion process (=unitary polymerization process;unitary process) 单元气体转化过程
unitary matrix 酉(矩)阵
unitary polymerization process (=unitary gas conversion process) 单元气体转化过程
unitary process (=unitary gas conversion process) 单元气体转化过程
unitary reactor 单元反应器
unitary structure factor 单位结构因子
unitary system 单元系
unit capacity 单位流量；装置处理量
unit cell ①晶胞；格子单位②干电池③单元格子
unit cell constant 晶胞常数
unit cell dimension 晶胞大小；晶胞尺寸
unit cell parameter 晶胞参数
unit charge 单位电荷
unit coal 单位煤
unit construction mold 单构塑模
unit drive 单独传动；直接传动
unit dryer 单元干燥器
unite ①联合；(并)合(为)一②结合
unit equipment 单元设备
unit frame 整体式机架
unithiol (=sodium 2,3-dimercaptopropane sulfonate) 2,3-二巯基丙烷磺酸钠
unitization ①联合②联合经营
unit lattice 晶胞；单元格子
unit load 单位负荷
unit mass resolution 单位质量分辨(率)
unit matrix 单位矩阵
unit molecule 单分子
unit operation 单元操作
unit pressure 单位压力
unit process 单元作业；单元过程
unit pump 单位泵；电动泵

unit resistance 单位阻力
unit sample 单位样品
unit shaft system 基轴制
unit speed 单位转速
unit tenacity 单位强度
unit tension 单位拉力
unit tensor 单位张量
unit tyre vulcanizer 单元外胎硫化器
unit vector 单位向量
unit volume 单位体积
unit vulcanizer 单元硫化机
unit weight 单位重量；比重
univalence(=univalency) 一价；单价
univalency(=univalence) 一价；单价
univalent 一价的；单价的
univalent saturated hydrocarbon radical 一价饱和烃基
univalent unsaturated hydrocarbon radical 一价不饱和烃基
universal ①通用的；广用的②一般的③宇宙的
universal apparatus 通用仪器
universal buffer 广域缓冲剂
universal calibration 普适标定
universal calibration method 通用校正法；普适校正法
universal chromatograph 通用色谱仪
universal chuck 联动夹盘
universal clamp 广用夹
universal constant 普适常数
universal coupling 万向联轴节
universal detector 通用检测器
universal filter respirator 广用滤毒罐
universal frequency 通用频率
universal gas chromatograph 通用气相色谱仪
universal gas constant 通用气体常数
universal gear lubricant 通用齿轮润滑脂
universal grease 通用润滑脂
universal grinder 万能磨床
universal hydrometer 通用比重计
universal indicator 通用指示剂
universal joint 万向接头；万向节
universal joint assembly 万向接头装置
universal joint cross 万向十字(接)头；万向节十字头
universal kneading machine 通用捏和机
universal lubricating grease 通用润滑脂
universal microprobe analyzer 通用型微探针分析器
universal milling machine 通用铣床
universal mixer 通用混和机
universal primary standard substance 通用基准物(质)
universal series constant 普遍光谱系常数
universal stand 通用架；广用台
universal starching and drying machine 通用上浆干燥机
universal starch mangle 通用上浆机；多用途上浆机
universal strength tester 万能强力测试机
universal support 通用载体
universal testing machine 万能试验机
universal test machine 万能试验机
universal test paper 通用试纸
universe ①宇宙②整体
univis oil 尤尼维斯油；加有黏度指数增进剂的润滑油
unkitted 未成药箱的
unknown ①未知的②未知物
unknown losses 未知损失；其他损失
unknown sample 未知样品
unknown solution 未知(溶)液
unknown term 未知项〔数〕
unlabelled 未标记的
unlagged piping 未保温的管道
unleachable 不可浸出的
unleaded 未加铅的；未加四乙铅的
unleaded antiknock additive 无铅抗爆添加剂
unleaded blend 无铅掺和料；未加四乙铅的掺和燃料
unleaded gasoline 不加铅汽油；未加四乙铅的汽油
unleaded octane rating 无铅辛烷值
unleaded petrol 无铅汽油
unlevelling color(=unlevelling dye) 不匀染料
unlevelling dye(=unlevelling color) 不匀染料
unlimited deformation 无限形变
unlimited flow 无限流动
unlimited solvent reservoir 非限量溶剂储器
unloaded vulcanizate 纯硫化胶
unloader 卸料机
unloading 卸荷；卸去负荷；放空
unloading auger 卸荷钻
unloading device 卸料机；卸料装备
unloading line 空闲管路；卸油管路
unloading valve 卸荷阀；卸载阀
unlubricated 无润滑的
unlubricated friction 干摩；无润滑摩擦
unmendable 不可修缮的；不可改正的
unmethylated ①未甲基化的②未加甲醇的〔酒精〕
unmethylated base 未甲基化的碱
unmicellized 未胶束化的
unmixed 不混合的；未混合的
unmixedness factor 不混合性因数
unmodified 未改性的
unmodified resin 净树脂；原树脂
unneutralized mercaptan 未中和的硫醇

unoccupied surface state　空表面态
unofficial　非法定的；未入药典的
unorganic(=inorganic)　无机的
unorganized　①无组织②无定形的
unoriented bulk polymer　未取向本体聚合物
unoriented fibre　未取向纤维
unoriented region　非取向区；未取向区域
unox　过氧化聚烯烃类黏合剂
unoxidizable　不可氧化的
unoxidizable alloy　不锈合金；不可氧化的合金
unpaired electron　不成对电子
unperturbed　无扰
unperturbed chain dimension　无扰分子链尺寸
unperturbed dimension　无扰尺寸
unperturbed end-to-end distance　无扰末端距
unpigmented rubber　素橡胶
unplasticized　未增塑的
unpolarizable electrode　不极化电极
unpolarized　未极化的
unpolarized electrode　非极化电极
unprotected reversing thermometer　开端颠倒温度表
unprotonated　未质子化的
unpurified　未纯化的；未精制的
unreactable naphthenes　不反应的环烷
unreacted　未反应的
unreacted radical　未反应的基团
unreactiveness　(化学)惰性
unreactiveness of paraffins　烷烃的化学惰性
unreeling of tape　卷尺松开
unrefinable crude oil　不适于炼制的原油
unrefined　未精炼的；非精制的
unrefined oil　未精制油
unregulated polymer　无规聚合物
unresisted　自由的；未遭到阻碍的
unrestricted flow　无限制流动；自由流动
unrestricted Hartree-Fock(UHF) method　非限制(的)哈特莱-弗克法
unripened viscose　未熟成黏胶
unroasted　未经焙烧的
unroll bracket　退卷摇框
unsafe fuel　不安全燃料
unsaponifiable　不皂化的
unsaponifiable fraction　不皂化部分
unsaponifiable matter　不皂化物
unsaponifiable oil　不能皂化的油
unsaponifiable residue　不皂化残渣
unsaponified　未皂化的
unsaturated　不饱和的

unsaturated acid　不饱和酸
unsaturated acyclic hydrocarbon　不饱和无环烃；不饱和链烃
unsaturated affinity　不饱和亲和力
unsaturated alcohol　不饱和醇
unsaturated benzene hydrocarbon　不饱和苯烃
unsaturated bond　不饱和键
unsaturated coefficient　不饱和系数
unsaturated compound　不饱和化合物
unsaturated ester　不饱和酯
unsaturated fatty acid　不饱和脂肪酸
unsaturated glyceride　不饱和甘油酯
unsaturated hydrocarbon　不饱和烃
unsaturated link(=unsaturated linkage)　不饱和键
unsaturated linkage (=unsaturated link)　不饱和键
unsaturated materials　不饱和物质
unsaturated monomer　不饱和单体
unsaturated oligomer　不饱和低聚物
unsaturated polyester(UP)　不饱和聚酯
unsaturated polyester curing agent　不饱和聚酯固化剂
unsaturated rubber　不饱和橡胶
unsaturated side chain　不饱和支链
unsaturated solution　不饱和溶液
unsaturated synthetic rubber　不饱和合成橡胶
unsaturated value　不饱和值
unsaturation　不饱和(现象)
unsaturation test　不饱和度试验
unscreened　未筛的
unscreened coal　未筛煤
unscreened ore　原矿
unscreened stock　粗浆
unscrewing mo(u)ld　退扣式模具；旋转芯模具
unseeded　未加晶种的
unsettled soap　未静置皂
unshaded lobe　未蔽叶
unshared electron pair　未共享电子对
unsheeted pulp　散浆粕
unshielded　未掩蔽的；未防护的
unsized　①未分大小的②未筛分的；未过筛的③无浆的
unsized coal　未筛分煤
unsized fabric　无浆布
unsized paper　无胶纸
unslaked　未消的；生的〔石灰〕
unslaked lime　生石灰
unsoiled cotton　未染污的棉布
unsorted　未分级的；未选的
unsorted coal　原煤
unsound cement　无稳定体积水泥

unsound knot 活节
unsplit casing 整体气罐
unstability 不稳定性
unstabilized 不稳定(化)的；不能控制的
unstabilized crude 不稳定原油
unstable 不稳定的
unstable arc 不稳定电弧
unstable characteristic 不稳定特性
unstable combustion 不稳定燃烧
unstable component 不稳定组分
unstable compound 不稳定化合物
unstable constituents 不稳定成分
unstable equilibrium 不稳定平衡
unstable flow 不稳流动
unstable gasoline 不稳定汽油
unstable grade natural gasoline 不稳定天然汽油
unstable gum 不稳定胶质
unstable hydrocarbon 不稳定的烃
unstable ion 不稳定离子
unstable naphtha 不稳定石脑油
unstable radical 不稳定自由基；不稳定游离基
unstable salt pair 不稳定盐对
unstable state 不稳状态
unstained ①未染色的②纯净的
unsteady 不稳的
unsteady distillation 不稳定蒸馏
unsteady elastico-viscous flow 非稳态弹黏性流动
unsteady feed 不稳给料
unsteady flow 非稳流；不稳流动
unsteady fluid flow 非稳定流(体的流动)
unsteady heat transfer 不稳定传热
unsteady mass transfer 不稳定传质
unsteady motion 不稳运动
unsteady state 非恒稳态
unsteady-state diffusion 非稳态扩散
unstitched ①拆开缝线的②未装订的③未缝合的
unstopper 拔去塞子
unstrained ①未应变②未滤过的
unstripped gas 原料气体；富气；湿气
unstripped latex 原态胶乳
unsubstituted aromatics 未取代的芳香化合物
unsuccessive peak 不相连的峰
unsulfated cement 未加石膏的水泥
unsulfonated material 未磺化物
unsulfonated oil 未磺化油
unsulfonated residue 未磺化的残渣
unsulfonation value 不磺化值
unsupervised pattern recognition 无监督模式识别

unsupported catalyst 无载体催化剂
unsupported nickel catalyst 无载体镍催化剂
unsym(=unsymmetrical) ①不对称的②偏(位)；苯环的1,2,4位
unsymmetrical(=unsym) ①不对称的②偏(位)；苯环的1,2,4位
unsymmetrical carbon 不对称碳原子
unsymmetrical cooling 不对称冷却
unsymmetrical linear molecule 不对称线式分子
unsymmetrically quenching 不对称骤冷
unsymmetrical monomer 不对称单体
unsymmetrical peak 不对称峰
unsymmetrical structure 不对称结构
unsymmetrical titration 非对称滴定
unsymmetric molecule 不对称分子
unsymmetry 不对称(现象)
unsym-tetrachloroethane 不对称四氯乙烷；1,1,1,2-四氯乙烷
untest stock 未试验油料
unthickened fuel 非稠化燃料；液状油
untreated ①未处理的②未浸渍过的
untreated felt 未浸渍的毛毡
untreated fibre 未处理纤维
untreated gasoline 未处理的汽油
untreated oil 未处理的油
untreated rubber 生橡胶
untrimmed size 未切边的规格
untrivial solution 非零解
untuous 手套革的柔润丰满感
untwisted silk 未捻丝
untwist-retwist method 退捻加捻法；张力法
unused heat 未利用的热
unvented ①未放气的②没有排气口的
unvolatilized 不挥发的；非挥发的
unvulcanized 未硫化的
unvulcanized rubber 生胶；未硫化橡胶
unwashed coal 未洗煤
unweighed 未称量过的
unwindase 解链酶
unworked consistency of grease 润滑脂工作前的稠度〔未受机械作用〕
unworked grease 未使用过的润滑脂
unworked micropenetration 工作前微量针入度
unworked penetration 工作前针入度
unzippering 开链式解聚
unzipping(=depolymerization) 解聚(作用)
unzipping force 解拉链力
UOP alkylation UOP氢氟酸烷基化法

UOP copper sweetening　UOP 铜精制法
upas　见血封喉；含马钱子碱箭毒〔爪哇产〕
upconversion　上变频；向上转换
upcurve　上升曲线
up-draft　直焰的；上抽的；向上通风的
updraft furnace(=updraft furnace heater)　直焰炉；向上通风炉
updraft furnace heater(=updraft furnace)　直焰炉；向上通风炉
up-draft kiln　直焰窑
updraught flue　直焰烟道；上行烟道
updraught producer　直焰发生炉
up-ender　①调头装置②翻料机
upfield　高磁场
upflow　上流；向上流动
upflow filter　上流滤过器
upflow fluid-catalyst unit　上流流体催化装置
upflow principle　上流原理
up-flow system　上流系统
upflow tube　溢流管
upgrading　浓集；浓缩；加浓
upgrading hydrocarbon　改质烃
uphill line　上行管道
upholstery leather　装饰用皮革
U-pipe　U 形管子
upkeep　检修；修理；维护
upkeep cost　维护费
upland cotton　陆地棉
upper alarm limit　上警告限
upper cloud point　上浊点
upper control limit　上控制限
upper critical solution temperature(UCST)　最高临界共溶(溶解)温度
upper cylinder lubrication　上部汽缸润滑
upper dead center　上死点
upper end of evaporator　蒸发器的上端
upper explosion limit　爆炸上限
upper half mean length　上半部平均长度
upper header　上集管；上联箱
upper horizontal flue　上横气道
upper layer　表面层；上层
upper leather　面革
upper limit　上限
upper limit of variation　改动的上限
upper mold section　上半模；阳模
upper plastic limit(=liquid limit; sticky point)　液态极限
upper platen　上(压)板
upper pour point　高流动点；最高倾点

upper quartile length　上四分位长度
upper sample　上层(取的)试样
upper tank air　油罐上层空气蒸气
upper temperature limit　温度上限
upper warning limit　上警戒限
up punching type　上刺式
upright　①直立铁杆②直立柱
upright still　立式蒸馏釜
uprise　向上；升起；直立管
up run　上行〔水蒸气〕
upset　缩锻；镦锻；镦(粗)
upset operation　不正常操作
upsetting　缩锻；镦锻；镦(粗)
upsetting test of tubes　管子的压短试验
upshot fired furnace　火焰上喷式加热炉
upshot furnace(=upshot heater)　火焰上喷式加热炉
upshot heater(=upshot furnace)　火焰上喷式加热炉
upside-down batch　反序混炼〔密炼〕
upside-down load(ing)　反序加料
upside-down mixing　反序混炼〔密炼〕
upsilon　v〔希腊字母〕
up-sizing　放大尺寸；放宽规格
upspout　升流管
upstanding margin on a horizontal flange　水平缘外具直沿
upstream　逆流(的)；上游(的)
upstream line　上游管线；吸引管线
upstream pumping unit　上流泵站；逆流泵站
upstream side　上流一边；上流侧
up stroke hydraulic press　上推(柱)水压机
upstroke twisting　上行式捻线(工艺)
uptake　①摄取②保留③收集
uptake rate　摄入率
up travel stop　上行停止装置
upward cell　向上电池；氯气电池
upward deflection　向上偏转
upward displacement　向上排代(作用)
upward flow　上向流
uracil(=Ura)　尿嘧啶　$CH=CHCONHCONH$
uracil-4-carboxylic acid(=orotic acid)　4-羧基尿嘧啶；乳清酸
uracil riboside　尿苷；尿核苷；尿嘧啶核苷
uraconite　土硫铀矿
uradal(=adalin)　阿达林；3-溴-3-乙基丁酰脲 $(C_2H_5)_2CBrCONHCONH_2$
uralite　深绿纤维闪石
uralium　三氯乙醛合氨基甲酸乙脂

uralkyd 氨基甲酸酯改性醇酸树脂
uramido 脲基
uramido acetic acid 脲基乙酸
uramil 乌拉米尔；2-氨基巴比妥酸；2-氨基丙二酰脲
  (CONH)$_2$COCHNH$_2$

uramil diacetic acid (UDA) 2-氨基丙二酰脲二乙酸
  (CONH)$_2$COCHN(CH$_2$COOH)$_2$

uramine 胍
uramino- (=ureido-) 脲基
uranate 铀酸盐 M$_2$(UO$_4$); M$_2$(U$_2$O$_7$)
urandiol 尿甾二醇
urane 尿甾〔urethane 叫做尿烷〕
uranediol 马尿甾二醇
urania 二氧化铀
urania green(=Paris green) 巴黎绿
urania sol 二氧化铀溶胶
uranic 六价铀的；(正)铀的
uranic acid 铀酸
uranic fluoride 六氟化铀 UF$_6$
uranic oxide 三氧化铀 UO$_3$
uranide(s) 铀系元素*
uraniferous 含铀的
uranin 荧光素钠 Na$_2$C$_{20}$H$_{10}$O$_5$
uraninite 沥青铀矿
uranite 铀矿；云母铀
uranium 铀〔92号元素，化学符号U〕
uranium I 铀 I ^{238}U
uranium II 铀 II ^{234}U
uranium acetate 乙酸双氧铀 UO$_2$·(C$_2$H$_3$O$_2$)$_2$
uranium-actinium 锕铀 ^{235}U
uranium ammonium fluoride 氧化铀酰铵
  UO$_2$F$_2$·3NH$_4$F
uranium analyzer 铀分析器
uranium boride 硼化铀
uranium borohydride 硼氢化铀 U(BH$_4$)$_4$
uranium carbide 碳化铀 UC; UC$_2$
uranium chloride 氯化铀 UCl$_3$; UCl$_4$; UCl$_5$
uranium citrate 柠檬酸铀
uranium concentrate 铀浓缩物
uranium decay series 铀衰变系
uranium deuteride 氘化铀
uranium diethyldithiocarbamate 二乙基二硫代氨基甲酸
  铀 U[(C$_2$H$_5$)$_2$NCSS]$_4$
uranium dioxide 二氧化铀 UO$_2$
uranium ethoxide 乙氧基铀 U(C$_2$H$_5$O)$_n$
uranium ferrate 铁酸铀
uranium ferrocyanide 亚铁氰化铀 UFe(CN)$_6$
uranium fluoride 氟化铀 UF$_4$; UF$_6$
uranium formate 甲酸铀 U(HCOO)$_4$
uranium fuel 铀燃料
uranium glass 铀玻璃
uranium halide 卤化铀 UX$_n$
uranium hexacarbonyl 六羰基铀 U(CO)$_6$
uranium hexaethoxide 六乙氧基铀 U(C$_2$H$_5$O)$_6$
uranium hexafluoride 六氟化铀 UF$_6$
uranium hydride 氢化铀
uranium hydroxide 氢氧化铀
uranium isopropoxide 异丙氧基铀 U[OCH(CH$_3$)$_2$]$_n$
uranium-lead age 铀-铅定年代法
uranium mercaptide 硫醇铀 U(SR)$_4$
uranium mercuride 汞化铀 UHg$_n$〔n=2,3,4〕
uranium metaphosphate 偏磷酸铀 U(PO$_3$)$_4$
uranium mineral (含)铀矿物
uranium moluranite 水铀锰矿
uranium molybdate 钼酸铀
uranium niobate 铌酸铀
uranium ore 铀矿石
uranium oxalate 草酸双氧铀 UO$_2$C$_2$O$_4$
uranium oxide 氧化铀 U$_2$O$_3$; UO$_2$; UO$_3$
uranium oxide-chalcogenide 氧硫(或硒；碲)化铀 UOY
  〔Y=S; Se; Te〕
uranium oxychloride 二氯二氧化铀 UO$_2$Cl$_2$
uranium oxyfluoride 氟氧化铀 UOF$_2$; UO$_2$F$_2$
uranium pentachloride 五氯化铀 UCl$_5$
uranium pentamethoxide 五甲氧基铀 U(OCH$_3$)$_5$
uranium pentamethoxide monoethoxide 五甲氧基单乙氧
  基铀 U(OCH$_3$)$_5$(OC$_2$H$_5$)
uranium peroxide 四氧化铀 UO$_4$;UO$_2$[O]$_2$
uranium phosphide 磷化铀
uranium pile 铀堆
uranium pyrophosphate 焦磷酸铀
uranium-radium dating method 铀-镭定年代法
uranium-radium decay series 铀镭衰变系
uranium reagent 铀试剂
uranium reserve 铀储量
uranium series 铀系
uranium sesquioxide 三氧化二铀 U$_2$O$_3$
uranium silicate 硅酸铀
uranium sodium acetate 乙酸双氧铀钠
  UO$_2$(C$_2$H$_3$O$_2$)$_2$·2NaC$_2$H$_3$O$_2$
uranium sulfate 硫酸(四价)铀 U(SO$_4$)$_2$
uranium tetrabromide 四溴化铀 UBr$_4$
uranium tetrachloride 四氯化铀 UCl$_4$
uranium tetrafluoride 四氟化铀 UF$_4$

uranium tetraiodide 四碘化铀 UI_4
uranium tetraoxide 四氧化铀 $UO_4; UO_2[O_2]$
uranium tetraphenoxide 四苯氧基铀 $U(OC_6H_5)_4$
uranium thiocyanate 硫氰酸铀
uranium trichloride 三氯化铀 UCl_3
uranium trioxide 三氧化铀 UO_3
uranium tritide 氚化铀
uranium X_1 铀X_1 ^{234}Th
uranium X_2 铀X_2 ^{234}Pa
uranium Y 铀Y ^{231}Th
uranium yellow 铀黄；铀酸钠
uranium Z 铀Z ^{234}Pa
uranium-zirconium hydride 氢化铀锆
uranocene 双(环辛四烯)合铀 $U(C_8H_8)_2$
uranocircite 钡铀云母
uranoid 铀系元素
uranol 铀试剂
uranophane 硅钙铀矿
uranopilite 铀钙矿；水硫铀矿
uranopissite 方铀矿
uranospathite 水磷铀矿
uranospherite 纤铀铋矿
uranospinite 砷钙铀矿
uranotantalite 钽铀矿；铌钇矿
uranothallite 铀钙石
uranothorianite 方铀钍石
uranothorite 铀钍矿
uranotile 硅钙铀矿
uranotilite 硅钙铀矿
uranous 四价铀的；亚铀的
uranous chloride 四氯化铀 UCl_4
uranous dibutylphosphate 磷酸二丁酯亚铀盐 $U(DBP)_4$
uranous fluoride 四氟化铀 UF_4
uranous nitrate 硝酸亚铀 $U(NO_3)_4$
uranous oxide 二氧化铀 UO_2
uranous sulfate 硫酸(四价)铀 $U(SO_4)_2$
uranous sulfide 二硫化铀 US_2
uranous tetrafluoride 四氟化铀 UF_4
uranous-uranic oxide 八氧化三铀 $U_3O_8; UO_2 \cdot 2UO_3;$ $U(UO_4)_2; UO_2 \cdot UO_4$
uranphyllite 铜铀云母
uranul calcium phosphate 磷酸双氧铀钙
uranvitriol 铀铜矾
uranyl 铀酰
uranyl acetate 乙酸双氧铀
uranyl acetylacetone 乙酰丙酮铀酰
uranyl alcoholate 烷基醇铀酰 $UO_2(OR)_2$
uranyl ammonium carbonate 碳酸双氧铀铵

uranyl ammonium phosphate(UAP) 磷酸铀酰铵 $NH_4UO_2PO_4$
uranyl bromide 溴化铀酰 UO_2Br_2
uranyl carbonate 碳酸铀酰 $UO_2(CO_3)$
uranyl chloride 铀酰氯；氯化双氧铀
uranyl dibutylphosphate 磷酸二丁酯铀酰 $UO_2(DBP)_2$
uranyl fluoride 氟化铀酰 UO_2F_2
uranyl formate(=uranyl formiate) 甲酸双氧铀 $UO_2(HCO_2)_2$
uranyl formiate(=uranyl formate) 甲酸双氧铀
uranyl hydroxide 氢氧化双氧铀 $UO_2(OH)_2$
uranyl nickel acetate 乙酸双氧铀镍
uranyl nitrate 硝酸双氧铀
uranyl nitrate hexahydrate(UNH) 六水合硝酸铀酰 $UO_2(NO_3) \cdot 6H_2O$
uranyl oxalate 草酸双氧铀 $UO_2C_2O_4$
uranyl oxalate actinometer 草酸双氧铀光化线强度计
uranyl phosphate 磷酸氢双氧铀 $(UO_2)HOP_4$
uranyl potassium sulfate 硫酸双氧铀钾
uranyl propionate 丙酸铀酰 $UO_2(C_2H_5COO)_2$
uranyl sodium acetate 乙酸双氧铀钠 $UO_2(C_2H_3O_2) \cdot 2NaC_2H_3O_2$
uranyl sulfate 硫酸双氧铀 UO_2SO_4
uranyl sulfide 硫化双氧铀 UO_2S
uranyl tri-8-hydroxylquinolate 三(8-羟基喹啉)铀酰 $UO_2(C_9H_6NO)_2 \cdot C_9H_7NO$
uranyl uranate 铀酸双氧铀 $UO_2 \cdot UO_4; U(UO_4)_2;$ $U_3O_8; 2UO_2 \cdot 2UO_3$
uranyl zinc acetate 乙酸双氧铀锌
urao 天然碱
urapidil 乌拉地尔〔药〕
urari 箭毒
urate ①尿酸盐(或酯)②尿酸
urate oxidase 尿酸(氧化)酶
urazine 尿嗪；环二脲 $C_2H_4O_2N_4$
urazole 尿唑 $C_2H_3O_2N_3$
urban heat land effect 城市热岛效应
urbaninol 乌斑宁醇
urceoline 壶蒜碱
urdite 独居石
ψ-urea 异脲
urea(=carbamide) 尿素；脲* NH_2CONH_2
urea acetate 乙酸-脲
urea adduct(=urea complex) 尿素加合物
urea adduct method 尿素加合物法
urea apparatus 测脲仪
urea carboxylic acid 脲羧酸
urea citrate 柠檬酸脲

urea clathration 尿素笼形包含物形成作用
urea clearance test 尿素清除率试验
urea complex (=urea adduct) 尿素加合物
urea complex method 尿素配位法
urea complex process 尿素配位法
urea derivative 尿素衍生物
urea dewaxing 尿素脱蜡
urea extraction 尿素抽提
urea extractive crystallization 尿素抽提结晶作用
urea filtrate 尿素不配位滤液
urea-formaldehyde condensation product 脲甲醛缩合物
urea-formaldehyde plastic 脲甲醛塑料；尿素甲醛塑料
urea-formaldehyde resin 脲醛树脂
urea-formaldehyde resin fibre 脲醛树脂纤维
urea fractionation method 尿素分离法
urea glutarate 戊二酸脲
urea hydrochloride 盐酸脲 $CO(NH_2)_2 \cdot HCl$
ureameter(=ureometer) 尿素计
urea nitrate 硝酸脲 $CO(NH_2)_2 \cdot HNO_3$
urea nitrogen 脲氮
urea non-adduct-forming material 尿素不加合物
urea oxalate (=1-amino-naphthalene) 草酸脲 $CO(NH_2)_2 \cdot C_2H_2O_4$
urea oxalate dihydrate 二水合草酸脲 $CO(NH_2)_2C_2H_2O_4 \cdot 2H_2O$
urea peroxide 过氧化脲
urea-phosphate fire retardant 尿素-磷酸酯阻燃剂
urea quinine 盐酸脲合奎宁
urea resin 尿素树脂
ureas 尿素塑料(类)
urease 脲酶；尿素酶
urease electrode 尿素酶电极
urea sensor 尿素传感器
urea stibamine 脲脒胺〔药〕
urea synthesis 尿素合成
urea-thiourea-formaldehyde resin 脲-硫脲-甲醛树脂
urechitin 黄龙葵苷
urechitine 黄龙葵碱
ureide 酰脲
ureido- 脲基 $H_2NCONH-$
ureido-acid 脲基酸 $NH_2CONHCORCOOH$
β-ureidoisobutyric acid β-脲基异丁酸
β-ureidopropionic acid β-脲基丙酸
ureidosuccinic acid 脲基琥珀酸
ureogenesis 脲生成(作用)
ureometer(=ureameter) 尿素计
ureotelic 排尿素的
ureous acid(=xanthine) 黄嘌呤

urethane(=ethyl carbamate) 尿烷〔俗〕；氨基甲酸乙酯 $NH_2CO_2C_2H_5$
urethane block copolymer 氨基甲酸乙酯嵌段共聚物
urethane elastomer 聚氨基甲酸(乙)酯弹性材料
urethane foam 聚氨酯泡沫(塑料)
urethane leather by wet method 湿法(聚)氨基甲酸酯人造革
urethane rubber 聚氨酯橡胶
urethanes 聚氨酯橡胶
urethano 氨酯基
urethylan 尿基烷〔俗〕；氨基甲酸甲酯 $C_2H_5O_2N$
ureylene 1,3-亚脲基 $-NHCONH-$
uric acid 尿酸；2,6,8-三羟基嘌呤 $C_5H_4O_3N_4$
uricase 尿酸酶
uridine 尿苷；尿核苷；尿嘧啶核苷
uridine diphosphate 尿苷二磷酸
uridine diphosphate reductase 尿苷二磷酸还原酶
uridine monophosphate 尿苷一磷酸；尿苷酸
uridine triphosphate 尿苷三磷酸
uridylic acid 尿(嘧啶核)苷酸；尿苷一磷酸
urinalysis 尿分析
urine 尿
urine porphobilinogen qualitative test 尿卟胆定性实验
urinoid 尿型素；环己烯-3-酮
uriodone 3,5-二碘-4-吡啶酮-N-乙酸二乙醇(基)铵
urminine 朱红壶蒜碱
urobenzoic acid 马尿酸
urobilin 尿胆素
urobilinogen 尿胆素原
urocanic acid 尿刊酸；咪唑丙烯酸 $C_6H_6O_2N_2$
urocitral 可可碱柠檬酸钠
urokinase 尿激酶
uron 乌龙；1-氧杂-3,5-二氮杂环己烷-4-酮
uronic acid 糖醛酸 $CHO(CHOH)_xCOOH$
uronic anhydride 糖醛酐
uropac N-甲基-3,5-二碘-4-吡啶酮-2,6-二羧酸钠
uropherine 可可碱锂
uropherine benzoate 苯甲酸可可碱锂
uropherine salicylate 水杨酸可可碱锂
uropittin 尿色素树脂
uroporphyrin 尿卟啉
urosine 奎尼酸锂
urotropine (=hexamethylene-tetramine) 乌洛托品；(环)六亚甲基四胺
urotropin quinate 奎尼酸乌洛托品
urotropin salicylate 水杨酸乌洛托品
uroxameter value 紫外光强度值〔乙酸双氧铀存在下草酸分解量〕
uroxanthin 尿黄质；β-吲哚硫酸钾

ursaenate 乌萨烯酸盐(酯)
ursanic acid 乌散酸 $C_{30}H_{48}O_2$
12-ursen-3-beta-2,8-diol 熊果醇
ursin 熊果碱
ursodesoxycholic acid 乌索脱氧胆酸
ursol ①对苯二胺②乌索(染料)〔一种毛皮染料〕
ursolate 乌索酸盐
ursol B 乌索 B〔染〕；毛皮氨酚棕
ursolene 乌索蜡；越橘蜡〔取自大果越橘 Vaccinium macrocarpum 的果皮〕
ursol GG 乌索 GG〔染〕；毛皮氨酚黄
ursolic acid(=urson) 乌宋；乌索酸；熊果酸 $C_{30}H_{48}O_3$
ursol SC 乌索 SC〔染〕；毛皮二氨黑
urson (=ursolic acid) 乌宋；乌索酸；熊果酸
ursonic acid 熊果酸；乌索酸 $C_{30}H_{46}O_3$
urtica 荨麻
urumbrin N-甲基-3,5-二碘-4-吡啶酮-2,6-二羧酸钠
urunday 植物性鞣剂
urusene 漆烯 $C_{15}H_{28}$
urushi 〔日文源〕漆
urushic acid 漆酸 $C_{23}H_{36}O_2$
urushiol 漆酚 $C_6H_3(OH)_2C_{15}H_{27}$
urushi silk 漆丝
urushi tallow 漆蜡；漆脂
urushi wax 漆蜡
urushoil 漆油；漆酚
urylene ①1,3-亚脲基 —NHCONH—②对称取代脲 RNHCONHR
usability 使用性能
usable break-through capacity 实用漏出容量
usable life 适用期
usadhana(=vaidigavat) 香须芒草
USC process(=ultra selective cracking process) 超选择性裂解法
use ①(使)用；应用②用途
use characteristics ①使用性能〔指制品〕②工艺性能〔指原材料〕
used ①用过的；旧的②习惯于
used catalyst 用过的催化剂
used heat 废热；余热；用过的热
used lime 废石灰；用过的石灰
used liquid 废液；用过的液体
used oil 废油；用过的油
used oil reclaimer(=used oil regenerator) 废油再生器
used oil regenerator(=used oil reclaimer) 废油再生器
used up 用尽
useful capacity 有效容量；可用容量
useful life 有效期；使用寿命
useful range 有效测定范围；测量范围
use-proved product 使用合格的制品
use ratio 利用率
users 用户；使用者
use technology 使用技术
use-yellowing 使用泛黄(现象)
U-shaped column U 形柱
usnaric acid 地衣那酸 $C_{30}H_{22}O_{15}$
usnein 地衣酸
usnic acid 松萝酸；地衣酸 $C_{18}H_{16}O_7$
usninic acid 地衣酸
usoline 优琐林；液体石蜡油
USP acid test 美国药典酸性试验(法)
USP standard 美国药典标准
ussamycin 乌萨霉素
ussingite 紫脆云母；紫翠石
Uster cotton fibre length tester 乌斯特棉纤维长度仪
ustilagic acid 黑粉菌酸
ustilaginoidin 黑曲定；黑刺菌素
ustilic acid 三羟基十六(烷)酸
utilities 公用事业；辅助设备〔蒸汽、水电等〕
utilities air 公用气源
utility ①有用矿物②公用事业③有用
utility consumption 动力(水、电、汽)消耗量
utility design 实用设计
utility service 公用服务事业
utilization 利用；使有用
utilizational coefficient 利用系数
utilization factor 利用系数
utilization of heat 热的利用
utilization ratio 利用率
U-tube U(型)管
U-tube apparatus U 型苯胺点测定管
U-tube exchanger U 型管换热器
U-tube heat exchanger U 型管换热器
U-tube manometer U 管(流体)压力计
U-turn U 形转折
UV absorber 紫外吸收剂
uvanite 钒铀矿
uvaol(=12-ursen-3-beta-2,8-diol) 山楂醇；熊果醇；乌苏醇
uvarovite 钙铬榴石
uvate (=uvinate) 乌韦酸盐(酯) $(CH_3)_2C_4HOCOOM$; $(CH_3)_2C_4HOCOOR$
uva ursi 熊果叶
UV curing 辐射交联
uvic acid(=pyrotritaric acid;uvinic acid) 乌韦酸；2,5-二甲基-3-呋喃羧酸 $(CH_3)_2C_4HOCOOH$

uvinate(=uvate) 乌韦酸盐(酯) $(CH_3)_2C_4HOCOOM$; $(CH_3)_2C_4HOCOOR$
uvinic acid(=uvic acid) 乌韦酸；2,5-二甲基-3-呋喃羧酸 $(CH_3)_2C_4HOCOOH$
UV ink UV 油
uviol 透紫外线玻璃
uviol glass 透紫外线玻璃
uvioresistant 抗紫外的；不受紫外线作用的
UV-irradiation 紫外光照射
uvitic acid(=uvitinic acid) 乌韦特酸；5-甲基苯间二甲酸 $CH_3C_6H_3(COOH)_2$
uvitinic acid (=uvitic acid) 乌韦特酸；5-甲基苯间二甲酸 $CH_3C_6H_3(COOH)_2$
uvitonic acid 乌韦酮酸；6-甲基-2,4-吡啶二甲酸 $CH_3C_5H_2N(COOH)_2$
UV light stabilizer 紫外光稳定剂
uvoflavin 维生素 B_2
uwarowite 钙铬榴石
uzara (非洲)止泻萝藦

V

v- 连〔位次形容词〕
vaal water 无离子水
vacancy chromatography 空穴色谱(法)
vacancy cluster 空位簇
vacancy concentration 空位密度
vacancy defect 空位缺陷
vacancy element 空位元素
vacancy photometric detection 空位光度检测
vacant chromatography 空穴色谱法
vacant level 空能级
vacant peak 空穴峰
vaccenic acid 异油酸；11-十八碳烯酸
vaccinin 越橘酯；6-苯甲酰葡萄糖
vaccum embossing 真空压花
vac-dry sheliac 干燥漂白紫胶
vacocin(=vancomycin) 万古霉素
vacuity 真空；空间
vacuo 真空
vacuole ①液泡；空泡②析稀胶粒
vacuometer 真空计；低压计
vacuous 真空的
vacuum 真空
vacuum and pressure air pump 抽压空气泵
vacuum arc furnace 真空电弧炉
vacuum bag 真空加压袋
vacuum-bag molding 真空袋模制法；真空袋成型
vacuum bonding 真空黏合
vacuum booster pump 预(真)空泵
vacuum bottle 保温瓶
vacuum brake-hose 真空制动胶管
vacuum breaker ①真空断路器〔电〕②真空调节阀③单向阀；止回阀
vacuum casting 真空铸塑(法)；真空浇铸
vacuum chamber 真空箱；真空室
vacuum checker 真空检查器
vacuum cleaner 真空清洁器
vacuum coat 真空涂层；真空涂覆
vacuum cold trap 真空冷阱
vacuum column 真空蒸馏塔
vacuum cone dryer 真空锥形干燥器
vacuum-constant temperature drying apparatus 真空恒温干燥器
vacuum conveyor tube 真空输送管
vacuum cooling 真空冷却

vacuum coupling 真空联锁
vacuum crystallizer 真空结晶器
vacuum degasifer 真空脱气器
vacuum degasser 真空脱气装置
vacuum degassing 真空脱气
vacuum deposition 真空淀积；真空蒸镀
vacuum desiccator 真空保干器；真空干燥器
vacuum detector 真空检测器
vacuum diffusion pump 真空扩散泵
vacuum discharge 真空放电
vacuum discharge spectrum 真空放电光谱
vacuum distillation 真空蒸馏
vacuum distillation plant 真空蒸馏装置
vacuum distilling apparatus 真空蒸馏器
vacuum distilling column 真空蒸馏塔
vacuum drier 真空干燥器
vacuum drum dryer 真空鼓式干燥器；真空筒形干燥器
vacuum dryer 真空干燥器
vacuum drying 真空干燥
vacuum drying apparatus 真空干燥器；减压干燥器
vacuum drying oven 真空干燥箱
vacuum electric furnace 真空电炉
vacuum epitaxy(VE) 真空外延
vacuum evaporation 真空蒸发(作用)
vacuum evaporation coating 真空电镀
vacuum evaporator 真空蒸发器
vacuum exhaust 真空排气
vacuum expander 真空膨胀器
vacuum extraction still 真空(蒸馏)提取器
vacuum extract still 减压抽出器
vacuum fan 真空扇
vacuum filler(=vacuum filling machine) 真空填充器
vacuum filling machine(=vacuum filler) 真空填充器
vacuum filter 真空滤器；(空)吸滤器
vacuum filtration 真空过滤；减压过滤
vacuum filtration process 真空过滤法
vacuum flash distillation 真空一次挥发蒸馏；真空快速蒸馏
vacuum flasher 真空闪蒸器
vacuum flashing 减压闪蒸
vacuum flash vaporizer 真空快速蒸发器
vacuum flask (=Dewar (vacuum) flask) 真空瓶；杜瓦(真空)瓶
vacuum flotation process 真空浮选法

vacuum forming	真空成型；吸塑
vacuum fractionation	真空分馏；真空分级
vacuum freeze drying	真空冷冻干燥(器)
vacuum fuel feed	真空送给燃料；减压送给燃料
vacuum fusion analysis	真空熔融(微量气体)分析
vacuum fusion apparatus	真空熔融装置
vacuum fusion gas chromatography	真空融熔气相色谱(法)
vacuum fusion method	真空熔融法
vacuum gas oil	真空瓦斯油；减压瓦斯油
vacuum ga(u)ge	真空计
vacuum gauge pressure	真空计压力
vacuum grease	真空润滑油
vacuum-heating method	真空加热法
vacuum hydraulic lamination	真空液压法
vacuum hydro-extraction	真空脱水；真空萃取
vacuum impregnating	真空注入
vacuum injection	抽空进样
vacuum-injection process	真空注射法
vacuum intensifier	真空加强器
vacuum ionization detector	真空电离检测器
vacuum-jacketed column	真空套塔
vacuum kneader	真空捏和机
vacuum lamp	真空灯
vacuum level	真空能级
vacuum liquid chromatography	减压液相色谱
vacuum lock	真空锁
vacuum manometer	真空(压力)计
vacuum melting	真空熔化；真空熔融
vacuum metallizing	真空敷金属
vacuum meter	真空计
vacuum microelectronics	真空微电子
vacuum mo(u)lding	真空模塑(法)
vacuum multistage evaporator	真空分级蒸发器
vacuum oil	真空油
vacuumometer	真空计；低压计
vacuum overhead	减压蒸馏塔顶馏分
vacuum packing	真空填充
vacuum packing method	真空填充法
vacuum pan	真空锅
vacuum partial condenser	真空塔部分冷凝器
vacuum phototube	真空光电管
vacuum pickup device	真空吸取器；真空抽出装置
vacuum plasma spraying	真空等离子喷涂
vacuum plating(=vacuum evaporation coating)	真空电镀
vacuum plodder	真空压条机
vacuum polymerization	真空聚合(作用)
vacuum pressure valves	①真空压力阀②呼吸阀
vacuum production	建立真空；形成真空
vacuum pump	真空泵；真空抽机
vacuum pump oil	真空泵(用)油
vacuum receiver	真空(接)收器
vacuum rectifying apparatus	真空精馏装置
vacuum refining	真空精制
vacuum release valve	真空解除阀
vacuum relief valve	真空解除阀
vacuum rerun	真空再蒸馏；减压重蒸馏
vacuum residuum	减压渣油
vacuum retort	真空蒸馏甑
vacuum rotary filter	真空回转过滤器
vacuum sampler	真空取样器
vacuum seal	真空密封
vacuum shelf drier	真空盘架干器
vacuum shelf dryer	真空干燥柜
vacuum sintering	真空垂熔；减压熔结
vacuum sintering technique	真空熔结技术
vacuum slot	真空吸嘴
vacuum snap-back forming	真空反吸成型
vacuum space	真空空间
vacuum spark	真空火花
vacuum spark discharge	真空火花放电
vacuum spark ion source	真空火花离子源
vacuum spectrograph	真空摄谱仪
vacuum spectrography	真空摄谱法
vacuum spectrometer	真空分光计
vacuum spectroscopy	真空光谱学
vacuum still	真空蒸馏釜
vacuum stripper	真空解吸塔
vacuum sublimation	真空升华
vacuum suction filter	真空吸滤
vacuum system	真空系统
vacuum take-off	真空抽吸
vacuum tank	真空罐
vacuum tar	真空焦油
vacuum technology	真空工艺
vacuum test	真空检验
vacuum thermal balance	真空热天平
vacuum thermobalance	真空热天平
vacuum thermogravimetry	真空热重(量)分析法
vacuum tight	真空紧密的
vacuum tightness	真空度
vacuum topping	减压拔顶蒸馏
vacuum tower	真空蒸馏塔
vacuum transfer vessel	真空转移容器
vacuum trap	真空阱；真空凝气瓣
vacuum tray drier	真空盘架干燥器
vacuum truck	真空油槽车

vacuum tube 真空管
vacuum tube galvanometer (=electronic galvanometer) 真空管检流计
vacuum tube voltmeter (=electron tube voltmeter) 电子管伏特计
vacuum ultraviolet 真空紫外(区)
vacuum ultraviolet atomic emission detector 真空紫外原子发射检测器
vacuum ultraviolet radiation source 真空紫外源
vacuum ultraviolet ray 真空紫外线
vacuum ultraviolet spectral region 真空紫外光谱区
vacuum ultraviolet spectrograph 真空紫外摄谱仪
vacuum ultraviolet spectrometry 真空紫外光谱法
vacuum ultraviolet spectroscopy 真空紫外光谱学
vacuum ultraviolet spectrum 真空紫外光谱
vacuum unit 真空装置
vacuum valve 真空阀
vacuum vaporizing method 真空气化法
vacuum vapour deposition method 真空蒸镀法
vacuum viscometer 真空黏度计
vacuum voltmeter(=vacuum tube voltmeter) 电子管伏特计
vacuum welding 真空焊接
vacuum xanthate mixer 真空黄化混合机；真空黄化捏合机
vacuum X-ray spectrometer 真空X射线分光计
vacuum zone collector 真空区带收集器
vad 锰土
vadose 渗流的
vadose water 渗流水
vagusstoff 迷走神经释放物；乙酰胆碱
vaidigavat 香须芒草
valamin 水合戊烯戊酸酯
valence 化合价*
valence adjustment 调价；价态调节
valence analysis 价态分析
valence angle 价角
valence band 价带
valence band spectroscopy 价带谱
valence band structure 价带结构
valence bond (化合)价键
valence bond method 价键法*
valence bond structure 价键结构
valence bond theory 价键理论
valence bridge 价桥
valence electron (化合)价电子；(原子)价电子
valence electron approximation 价电子近似
valence electron pair 价电子对
valence fluctuation 价态起伏
valence force(=intratomic force) 原子间力；价力
valence-force field 价力场
valence group (化合)价组
valence-ionized polymer chain 价离子化的聚合链
valence isomer 价异构体
valence isomerisation 价异构化
valence isomerism 价异构
valence link(=valence linkage) 价键
valence linkage(=valence link) 价键
valence-matching principle 键价匹配原理
valence model 化合价模型；原子价模型
valencene 瓦伦烯；朱栾倍半萜
valence number (化合)价数；原子价数
valence shell 价电子层
valence-shell electron pair repulsion theory 价层电子对推斥理论
valence state 价态
valence state ionization potential 价态电离势
valence tautomerism 价互变异构
valence vibrations 价振动
valencia orange 巴伦西亚橘
valency (=valence) (化合)价；(原子)价
valency angle (化合)价角
valency bond (化合)价键
valency force (原子)价力
valency isomerism (原子)价异构(现象)
α-valene α-缬草烯
valenol 缬草醇
valent 化合价的；价的
valenta value (冰醋酸)油浊点
valent chain unit 价链单元
valentinite 锑华
valent weight 当量
valeral 戊醛
γ-valeralactone γ-戊内酯
valeraldehyde 戊醛 C_4H_9CHO
n-valeraldehyde (正)戊醛 $C_2H_5CH_2CH_2CHO$
valeraldehyde oxime 戊醛肟 $C_4H_9CH=NOH$
n-valeramide (正)戊酰胺 $CH_3(CH_2)_3CONH_2$
valerane 缬草烷
valeranone 缬草烷酮
valerate 戊酸盐(酯) $C_4H_9COOM;C_4H_9COOR$
valerene 戊烯
valerenic acid 缬草烯酸
valerenone 缬草烯酮
valerian 缬草
valeriana coreana oil 东北缬草油

valeriana jatamansi oil　马蹄香油
valerianate(=valerate)　戊酸盐(酯)　C_4H_9COOM; C_4H_9COOR
valerianic acid(=valeric acid)　戊酸　C_4H_9COOH
valerian oil　缬草油
valerian phenol　缬草酚
valeric acid(=valerianic acid)　戊酸　$C_2H_5CH_2CH_2COOH$
valeric aldehyde　戊醛　C_4H_9CHO
valeric anhydride　戊酸酐　$(C_4H_9CO)_2O$
valeric chloride(=valeryl chloride)　戊酰氯
valeridin　N-对乙氧苯基戊酰胺
valerin　甘油三戊酸酯
valeroidine　瓦勒洛定；异戊酸羟基莨菪酯
valerol　缬草油素
valerolactam　戊内酰胺
valerolactone　戊内酯
δ-valerolactone　δ-戊内酯；1,5-戊内酯
γ-valerolactone　γ-戊内酯；1,4-戊内酯
valerone　二异丁基(甲)酮；2,6-二甲基-4-庚酮
valeronitrile(=butylcyanide)　戊腈；丁基氰　$CH_3(CH_2)_3CN$
valerydin　N-对乙氧苯基戊酰胺
valeryl　戊酰　$CH_3(CH_2)_3CO—$
valeryl chloride(=valeric chloride)　戊酰氯　$CH_3(CH_2)_3COCl$
valeryl diethyl amide　N-二乙基戊酰胺
valerylene　2-戊炔
valeryl oxybutyrine　二甲氨基二甲异戊酰基丙酯氢溴酸盐
valeryl phenetidine　N-对乙氧苯基戊酰胺
valetin　凡立丁
valex　碗子浸膏粉
valex extract　碗子浸膏
validamycin　有效霉素
validol(=menthyl valerate)　戊酸蓝酯
V alignment　V 形排列
valine　缬氨酸；α-氨基异戊酸　$(CH_3)_2CHCHNH_2CO_2H$
valine aminotransferase　缬氨酸转氨酶
valinomycin(=aminomycin)　缬氨霉素〔即氨基霉素〕
vallesine　瓦蕾辛
Vallets mass　碳酸亚铁膏
valley　①波谷；最低点②(曲线的)凹部
valley die　谷形口型；中凹口型
valley-top ratio　谷峰比
Vallez filter　瓦利兹滤机
valonea dilactone　橡椀二内酯
valonea extract　橡椀栲胶
valoneaic acid　橡椀酸
valoneaic acid dilactone　橡椀酸二内酯
valonia　橡椀；碗子；橡碗子
valonia (acorn)　橡椀
valonia extract　碗子浸膏
valonia tannin　碗子鞣质；橡椀单宁
valsartan　缬沙坦〔药〕
γ-value　γ值；酯化度
δ-value　δ值*
value of expectation　期望值*
value of quantity　量值
values　测试值
valve　①阀；活门②真空管
valve-actuated pulse column　阀动脉冲柱
valve-and-sample loop system　阀及进样环管系统
valve effect　阀效应；单向导电性
valve electrode　阀电极；整流电极
valve grinding compound　磨阀物
valve grinding sand　磨阀砂
valve guard　阀片升程限制器
valve guide　阀杆导承
valve helmet mask　活门头兜面具
valve injection　阀注射；阀进样
valve leakage　阀漏失
valve leather　阀皮
valveless injection system　无阀注射系统
valve loop　阀环
valve position switch　阀定位开关
valve potentiometer　真空管电位计
valve-switching mechanism　阀转换机构
valve tower　浮阀塔
valve tray　浮阀塔板
valve voltmeter　电子管伏特计
valyl　缬氨酰；异戊氨酰　$(CH_3)_2CHCH(NH_2)CO—$
valylene　缬烯炔；2-甲基-1-丁烯-3-炔；异戊烯炔　$CH_2=C(CH_3)C≡CH$
valzin　对乙氧苯基脲
vamp split　裂面
van　①风扇②簸分机
vanadametry　钒酸盐滴定
vanadate　钒酸盐
vanadatimetric titration　钒酸盐滴定
vanadic　钒的
vanadic acid　钒酸
vanadic anhydride　钒酐；五氧化二钒
vanadic fluoride　五氟化钒　VF_5
vanadic oxide　五氧化二钒　V_2O_5
vanadic salts　钒盐
vanadic sulfate　硫酸钒

vanadinite 钒铅矿
vanadinu chloride 氯化钒 VCl_2; VCl_3; VCl_4
vanadite 亚钒酸盐 $M_2V_4O_9$; $M_2V_2O_5$
vanadium 钒〔23号元素，化学符号 V〕
vanadium benzohydroxamate 苯并异羟肟酸钒
vanadium carbide 一碳化钒 VC
vanadium contact process 钒接触法
vanadium dichloride 二氯化钒 VCl_2
vanadium difluoride 二氟化钒 VF_2
vanadium dioxide 二氧化钒 VO_2; V_2O_4
vanadium dioxymonochloride 氯化二氧钒
vanadium disulfide 二硫化钒
vanadium drier 钒干燥剂
vanadium family 钒族
vanadium family element 钒族元素
vanadium fluoride 氟化钒 VF_2; VF_3; VF_4; VF_5
vanadium hydroxide 氢氧化钒 $V(OH)_2$; $V(OH)_3$
vanadium minerals 钒矿物
vanadium monosulfide 一硫化钒 VS; V_2S_2
vanadium monoxide 一氧化钒 VO
vanadium nitride 一氮化钒 VN
vanadium oxide 氧化钒 V_2O; VO; V_2O_3; VO_2; V_2O_5
vanadium oxybromide 溴氧化钒 $VOBr_2$; $VOBr_3$
vanadium oxychloride 氯氧化钒 $(VO)_2Cl$; VOCl; $VOCl_2$; $VOCl_3$
vanadium oxydichloride 二氯(一)氧化钒 $VOCl_2$
vanadium oxydifluoride 二氟氧化钒 VOF_2
vanadium oxyfluoride 氟氧化钒
vanadium oxytribromide 三溴氧化钒；三溴化氧钒 $VOBr_3$
vanadium oxytrichloride 三氯氧化钒；三氯化氧钒 $VOCl_3$
vanadium oxytrifluoride 三氟氧化钒；三氟化氧钒 VOF_3
vanadium pentafluoride 五氟化钒 VF_5
vanadium pentasulfide 五硫化二钒 V_2S_5
vanadium pentoxide 五氧化二钒 V_2O_5
vanadium sesquioxide 三氧化二钒 V_2O_3
vanadium sesquisulfide 三硫化二钒 V_2S_3
vanadium silicide 硅化钒 V_2Si; VSi_2
vanadium suboxide 一氧化二钒 V_2O
vanadium sulfate 硫酸钒 VSO_4
vanadium sulfide 硫化钒 VS; V_2S_2; V_2S_3; V_2S_5
vanadium tetrachloride 四氯化钒 VCl_4
vanadium tetrafluoride 四氟化钒 VF_4
vanadium triacetylacetonate 三乙酰丙酮钒
vanadium trichloride 三氯化钒 VCl_3
vanadium trifluoride 三氟化钒 VF_3
vanadium trioxide 三氧化二钒 V_2O_3

vanadium trisulfide 三硫化二钒 V_2S_3
vanadol 双氧钒基
vanadometry 钒酸盐滴定法
vanadous 亚钒的
vanadous acid 亚钒酸 $H_2V_4O_9$
vanadous bromide 三溴化钒 VBr_3
vanadous chloride 二氯化钒；氯化亚钒
vanadous fluoride 三氟化钒 VF_3
vanadousometry 次钒酸盐滴定法
vanadous oxide 三氧化二钒；氧化亚钒 V_2O_3
vanadous sulfide 三硫化二钒；硫化亚钒 V_2S_3
vanadyl ①氧钒根②氧钒基
vanadyl bromide(s) 溴化氧钒(类)
vanadyl chloride 氯化氧钒 $(VO)_2Cl$; VOCl; $VOCl_2$; $VOCl_3$
vanadyl dibromide 二溴化氧钒
vanadyl dichloride 二氯化氧钒 $VOCl_2$
vanadyl difluoride 二氟化氧钒
vanadylic 氧钒根；三价氧钒根的 VO^{3+}
vanadylic bromide 三溴化氧钒
vanadylic chloride 三氯氧化钒；(三)氯化氧钒 $VOCl_3$
vanadyl monochloride 一氯化氧钒
vanadylous 亚氧钒根；一价氧钒根的 VO^+
vanadylous bromide 一溴化氧钒
vanadylous chloride 一氯化氧钒；氯化亚氧钒 VOCl
vanadyl semichloride 氯化二氧二钒
vanadyl sulfate 硫酸氧钒 $(VO)_2(SO_4)_3$
vanadyl tribromide 三溴化氧钒
vanadyl trifluoride 三氟化氧钒
vanadyl xanthate 黄原酸氧钒
vanaspati ①加维生素 A 的氢化植物油②花生油
vancancy-trapped electron 空穴捕获(的)电子
vancomycin 万古霉素
vancomycin hydrochloride 盐酸万古霉素
van Deemter's equation 范第德姆特方程式
Vandenberg catalyst 范登堡催化剂
Vandenberg reaction 范登堡氏反应
van der Graaf (electrostatic) accelerator 范德格拉夫静电加速器
van der Waals adsorption 范德华吸附(作用)
van der Waals'attractive force 范德华吸引力
van der Waals bond 范德华键
van der Waals bonding 范德华键合
van der Waals'calorimeter 范德华热量计
van der Waals constant 范德华常数
van der Waals distance 范德华距离
van der Waals equation 范德华方程式
van der Waals force 范德华力
van der Waals' isotherm 范德华等温线

van der Waals radius 范德华半径
van de Waals'attractive force 范德华吸引力
van de Waals' isotherm 范德华等温线
Vandyke brown(=Cassel brown) 天然棕土；铁棕
vane ①叶轮②风标
vane anemometer 叶轮风速计
vane angle 桨角；叶轮角
vane curvature 叶片曲率
vane diffuser 叶片扩压器
vane efficiency 叶片效率
vane exit angle 叶片出口角
vaneless diffuser 无叶扩压器
vane pump 叶片泵；滑片泵
vane solidity 叶片充实度
vane thickness 叶片厚度
vane twist 叶片扭曲度
vane valve 叶片阀
vane wheel 叶轮
Vanier bulb 范尼尔钾碱球管
Vanier's tube 范尼尔钾碱球管
vanilate 香兰酸盐(酯)
vanilione 香兰酮
vanilla absolute 香子兰净油
vanilla beans 香子兰(果)实
vanillacetone 香兰叉丙酮
vanilla concrete 香子兰浸膏
vanilla extract 香子兰浸液
vanillal(=vanillylidene) 亚香草基；4-羟-3-甲氧苯亚甲基 $CH_3OC_6H_5(OH)CH=$
vanillalacetone 亚香草基丙酮 $CH_3COCH=CHC_6H_3(OCH_3)OH$
vanilla oil 香子兰油
vanilla oleoresin 香子兰油树脂
vanilla resinoid 香子兰香树脂
vanilla tincture 香子兰酊
vanillic acid 香草酸；4-羟基-3-甲氧基苯甲酸 $CH_3OC_6H_3(OH)CO_2H$
vanillic alcohol 香草醇；3-甲氧基-4-羟基苄醇
vanillin 香草醛；3-甲氧基-4-羟基苯甲醛；香兰素 $(CH_3O)C_6H_3(OH)CHO$
vanillin acetate 乙酸香兰酯
vanillinal 香兰缩醛；香兰叉
vanillin ethyl ether 香草醛乙醚 $(CH_3O)C_6H_3=(OC_2H_5)CHO$
vanillyl alcohol 香草醇；4-羟基-3-甲氧基苯甲醇 $CH_3OC_6H_3(OH)CH_2OH$
vanillylidene 亚香草基；4-羟基-3-甲氧苯亚甲基
vanillylmandelic acid 香草扁桃酸；3-甲氧基-4-羟扁桃酸
vanillyl methyl ketone 香兰基甲酮
vanirom 3-乙氧基-4-羟基苯甲醛
vanishing cream 雪花膏
vanner 淘矿机
vanner edge belt 带挡边的运输带
vanning 淘矿
Van Slyke amino nitrogen method 范斯里克氨基氮测定法
Van Slyke blood gas apparatus 范斯里克血液气体检验器
Van Slyke buffer capacity 范斯里克缓冲容量
Van Slyke method (for amino nitrogen) 范斯里克(氨基氮)测定法
Van Slyke ninhydrin method 范斯里克茚三酮二氧化碳测定(法)
Van Slyke wet-ashing method 范斯里克湿灰化法
vanthoffite 钠镁矾
Van't Hoff's equilibrium box 范特霍夫平衡箱
Van't Hoff's factor 范特霍夫因数
Van't Hoff's law 范特霍夫定律
Van't Hoff solution 范特霍夫溶液
Van't Hoff theory 范特霍夫理论
vapocide 瓦波剂
vapodust 杀虫石油喷剂
vapometer 蒸气压力计
vapor ①蒸气②汽；水蒸气；蒸汽
vaporability 汽化性
vaporable 可汽化的；可蒸发的
vapor adsorption process 蒸气(汽)吸附法
vapor-air mixture 空气蒸气(汽)混合气
vapor-alkali reclaiming process 蒸气(汽)碱(再生)法
vaporation 汽化；蒸发
vapor balancer 蒸气(汽)平衡器
vapor balancing mechanism(=vapor balancing unit) 蒸气(汽)平衡器
vapor balancing unit (=vapor balancing mechanism) 蒸气(汽)平衡器
vapor barrier 蒸气(汽)防护栅
vapor bath 蒸气(汽)浴
vapor blasting 蒸气(汽)喷砂
vapor capacity 蒸气(汽)容量
vapor chimney 蒸气(汽)囱
vapor cleaning 蒸气(汽)净洗
vapor cloud 蒸气(汽)云
vapor composition 蒸气(汽)组成
vapor compression evaporator 蒸气(汽)压缩蒸发器
vapor compression machine 蒸气(汽)压缩机

vapor compression refrigerator （蒸气)压缩式制冷机
vapor compressor　蒸气(汽)压缩机
vapor condensation zone　蒸气(汽)冷凝区
vapor condenser　蒸气(汽)凝结器
vapor corrosion inhibitor　蒸气(汽)腐蚀抑制剂
vapor cure　蒸气(汽)处治
vapor curve　蒸气(汽)曲线
vapor degreasing　蒸气(汽)脱脂(法)
vapor delivery tube　蒸气(汽)配送管
vapor density　蒸气(汽)密度
vapor density apparatus　(测)蒸气(汽)密度器
vapor density bulb　蒸气(汽)密度球管
vapor density method　蒸气(汽)密度法
vapor deposition　汽相淀积；蒸镀
vapor deposition technique　气相沉积技术
vapor discharge lamp　蒸气(汽)放电灯
vapor disengagement　蒸气(汽)分离；蒸气(汽)释出
vapor distillation　蒸气(汽)蒸馏
vapor-dividing head　蒸气(汽)分配蒸馏头〔分馏柱〕
vapor dome　蒸气(汽)圆顶室
vapor eliminator　蒸气(汽)分离器
vapor enrichment　蒸气(汽)富集；重蒸馏〔蒸气(汽)中轻馏分的浓集〕
vapor expanded film　汽态扩张膜
vapor expansion　蒸气(汽)膨胀
vapor explosion　蒸气(汽)爆炸
vapor fractometer　汽相分离计
vapor-gas mixture　蒸气(汽)煤气混合气
vapor generation　蒸气(汽)发生
vapor heated　(用)蒸气(汽)加热的
vapor heat exchanger　蒸气(汽)换热器
vapor holdup　蒸气(汽)截留量
vaporimeter　挥发度计
vaporimetric method　蒸气(汽)测定法
vaporising fuel　汽化燃料
vaporizability　可汽化性
vaporizable(=vaporable)　可汽化的；可蒸发的
vaporization　汽化*
vaporization efficiency　蒸发效率；汽化效率
vaporization energy　汽化能
vaporization heat　蒸发(潜)热
vaporization losses　蒸发损失
vaporization method　蒸发方法
vaporization property　汽化性
vaporization temperature　汽化温度
vaporization zone　汽化区域
vaporized combustible　已汽化的可燃物
vaporizer　①气化室*②汽化器③喷雾器

vaporizing　汽化；蒸发；挥发
vaporizing oil　汽化油；煤油；重汽化器燃料
vaporizing tube　汽化管
vaporizing tube nipple　汽化管乳头
vaporizing velocity　蒸发速度
vapor-laden air　含蒸气(汽)的空气
vapor lamp　蒸气(汽)灯；煤油气灯〔灯塔用〕
vapor lift pump　蒸气(汽)提升泵
vapor line　蒸气(汽)管线
vapor-liquid equilibrium　汽液平衡
vapor-liquid extraction　汽液抽提
vapor-liquid nucleation　汽液成核
vapor-liquid ratio　气液比
vapor lock　气(汽)封；蒸气(汽)塞
vapor locking temperature　形成气(汽)塞的温度
vapor locking tendencies　形成气(汽)塞倾向
vapor losses　蒸气(汽)损失
vapor meters　蒸气(汽)仪表
vapor-nozzle　蒸气(汽)喷嘴
vapor-oil-water separator　汽-油-水分离器
vaporometer　蒸气(汽)压力计
vaporous　汽状的；多蒸气(汽)的
vaporous cloud　蒸气(汽)云
vapor phase　气相；蒸气(汽)相
vapor phase chromatography(=gas chromatography)　气相色谱(法)
vapor phase cracked gasoline　气相裂化汽油
vapor phase cracking　气相热裂
vapor phase diffusion　气相扩散
vapor-phase etching　气相腐蚀
vapor phase inhibitor　气相(氧化)抑制剂
vapor phase interference　蒸气(汽)相干扰
vapor phase nitration　气相硝化
vapor-phase operation　气相操作
vapor phase oxidation　气相氧化
vapor phase polymerization　气相聚合
vapor phase process　气相过程；气相作业
vapor phase pyrolysis　气相热解
vapor phase pyrolyzer　气相热解器
vapor phase rearrangement　气相重排
vapor-phase refining　气相精炼
vapor-phase system　气相系统
vapor-phase thermal cracking　气相热裂化
vapor-phase treatment　气相处理
vapor pipe　蒸气(汽)管
vapor polishing　蒸气(汽)抛光
vapor pressure　蒸气(汽)压(力)
vapor-pressure chart　蒸气(汽)压线图

vapor-pressure curve　蒸气(汽)压曲线
vapor pressure index　蒸气(汽)压指数
vapor-pressure inhibitor　蒸气(汽)压抑低剂
vapor pressure isotope effect(VPIE)　蒸气(汽)压同位素效应
vapor pressure lowering　蒸气(汽)压下降
vapor pressure osmometry(VPO)　蒸气(汽)压渗透法
vapor pressure ratio　蒸气(汽)压比
vapor-pressure test　蒸气(汽)压测定
vapor-pressure thermometer　蒸气(汽)压式温度计
vapor programmed　程序换蒸气(汽)(的)
vapor programming chamber　程序换蒸气(汽)展开槽
vapor-proof connection　气密连接；不漏蒸气(汽)的接头
vapor-proof curtain　气密幕〔在油桶、油罐内〕
vapor pyrolysis　蒸气(汽)热解
vapor recompression　蒸气(汽)再压缩(法)
vapor recovery　蒸气(汽)回收
vapor rectification process　蒸气(汽)分馏过程；选择冷凝过程；加压分馏〔分离丁烷〕
vapor-reflux contacting　蒸气(汽)与回流液的接触
vapor removal　蒸气(汽)分离
vapor return　蒸气(汽)再发
vapor riser　直蒸气(汽)管；竖式蒸气(汽)管
vapor sampling rod　蒸气(汽)进样杆
vapor seal　蒸气(汽)封闭
vapor-solid growth　气-固生长
vapor-space volume　蒸气(汽)空间的容积
vapor superheater　蒸气(汽)过热器
vapor temperature　蒸气(汽)温度
vapor tension　蒸气(汽)压(力)；蒸气(汽)张力
vapor tension meter　蒸气(汽)压计
vapor tension thermometer　蒸气(汽)温度计
vapor-testing apparatus　蒸气(汽)试验器〔试空气中可燃气体或蒸气(汽)用〕
vapor-tight tank　气密罐
vapor transmission property　蒸气(汽)透过性
vapor tube of flask　蒸馏瓶侧管
vapor uptake　蒸气(汽)上升管〔塔盘的〕
vapor velocity　蒸气(汽)速度
vapor volume　蒸气(汽)容积
vapotron　蒸发冷凝管
vapo(u)ration　气化；蒸发
vapo(u)r blasting　蒸气(汽)喷射
vapour cure　①蒸气(汽)硫化〔通常用一氯化硫气体〕②蒸气(汽)处理
vapour discharge lamp　蒸气(汽)放电灯
vapo(u)r generator　蒸气(汽)发生器
vapo(u)rimeter　挥发度计

vapourisation-condensation process　(须晶的)蒸发-冷凝(生长)法
vapourizing furnace　蒸发炉；气化炉
vapour leakage　蒸气(汽)泄漏
vapour-liquid chromatography　气液色谱分析
vapour-liquid contacting column　气液接触塔；气液混合塔
vapour-liquid-solid heterogeneous system　气-液-固非均匀系统
vapour permeability　蒸气(汽)渗透率；透气性
vapour-phase condensation polymerization　气相缩聚(作用)
vapour phase esterification　(蒸)气相酯化
vapour phase graft copolymerization　(蒸)气相接枝共聚
vapour phase hydrogenation　气相氢化
vapour-phase interference　蒸气(汽)相干扰
vapour phase matrix isolation fluorimetry　蒸气(汽)相基体隔离荧光法
vapour phase osmometer　气相渗透计
vapour phase radiation grafting　(蒸)气相辐射接枝
vapour plating　蒸气(汽)蒸镀
vapour pressure gradient　蒸气(汽)压力梯度
vapour-pressure osmometer　蒸气(汽)压渗透计
vapour pressure osmometry　蒸气(汽)渗透压(测定)法
vapour separator　蒸气(汽)分离器
vapour tension　蒸气(汽)压；汽压
vapour tight　气密的；蒸气(汽)密封的
vapour transmission　透(蒸)气(汽)性；透气性
varec　海藻；海藻灰
varek　海藻
Varga process　瓦尔加法〔原油残渣油加氢脱硫法〕
variability　变异性*
variable　①可变数；(可)变量②可变(更)的
variable anisotropy etching　可变各向异性腐蚀
variable-area flow meter　可变截面流量计
variable cell　可变槽；可变池
variable chemical composition　①可变化学成分②可变化学组成
variable coefficient　可变系数
variable condenser　可变电容器
variable delay　可变延迟时间
variable delivery pump　可变排量泵
variable-delivery rotary pump　可变排量旋转泵
variable depth method　平衡法；均衡法
variable dilution sampling system　可变稀释取样系统
variable displacement pump(=variable flow pump)　可变流量泵
variable feed　可变进料(量)
variable flow pump(=variable displacement pump)　可变流量泵

variable focus condenser 可变焦点聚光器
variable gap 可变狭缝；可变间隙
variable path-length cell 可变光程吸收池
variable pitch screw 收敛式螺杆
variable-rate sampling method 可变流量取样法
variable reference solution method 参比溶液变化法；对照溶液变化法
variable resistance 可变电阻
variable sine wave generator 可变正弦波发生器
variable specific heat 不定比热
variable speed box 变速装置；变速箱
variable speed motor 变速电动机
variable-speed pump 变速泵
variable step size 可变步长
variable temperature infrared spectroscopy 变温红外光谱
variable temperature probe 变温探头
variable temperature unit 变温单元
variable transformation 变数变换
variable valency 可变(化合)价
variable vapour-space tank 可变蒸气(汽)空间油罐
variable wavelength spectrophotometric detector 可变波长分光光度检测器
variacyclomycin 变环霉素
variamine blue 变胺蓝
variamycin 变易霉素
variance ①变更②变度③方差
variance analysis 方差分析
variance of peak 峰畸变
variance ratio 方差比
variance ratio test 方差比检验
variance within laboratory 室内方差
variate 统计数值
variate calibration 变量校正
variation ①变易②演变③变异④变差
variational method 变分法
variation between laboratories 室(组)间变差；室(组)间变异性
variation coefficient 变异系数
variation method 变分法
variation of tolerance 公差带
variation within laboratory 室(组)内变差；室(组)内变异性
variator ①变速器②聚束栅③伸缩缝
varied-angle glossmeter 变角光泽计
variegation 斑纹
variolaric acid 瓦拉酸；地衣缩酚羧酸
variscite 磷铝石

varispeed motor 变速电动机
varister 变阻器；可变电阻
varitran 自耦变压器
varnish ①(清)漆；假漆；(油)漆②(涂)上(清)漆
varnish base 清漆底子
varnish coat 清漆涂层
varnish coating ①清漆涂层②涂清漆
varnish colour 清漆色；清漆涂料
varnish formation 漆膜形成
varnishing 上清漆
varnishing machine 浸漆机；浸渍涂漆机；上光机；上蜡机
varnish kettle 清漆(熬炼)锅
varnish linseed oil 清漆亚麻仁油
varnish maker's naphtha 清漆用石脑油
varnish oil 清漆油
varnish paint (清)漆
varnish pot 清漆锅
varnish remover 除漆剂
varnish resin 清漆树脂
varnish stain ①着色漆②(清)漆(污)点
varnish test 清漆试验；积碳形成倾向试验
varve 纹泥；每年一度的冲积层
varying capacity 变动生产量；变动容量
varying duty 变动操作条件
varying load 变动负荷
varying underflow 变动底流
vasegen(=vasogen) 凡士精
vaseline(=petroleum jelly) 矿脂；凡士林；石油冻；石油膏
vaseline oil 凡士林油
vash oil 凡士林油；白油
vasicine(= peganine) 骆驼蓬碱
vasicinone 鸭嘴花碱酮
vasoconstrictor 血管收缩剂
vasodilator 血管扩张剂
vasogen(=vasegen) 凡士精
vasopressin 后叶加压素；抗利尿激素；垂体后叶荷尔蒙
vasopressin injection 加(血)压素注射
vasopressin tannic acid 鞣酸加压素〔药〕
vasothrombin 血管凝血酶
vasotocin 管催产素；8-精催产素
vasoxyl 二羟基去甲肾上腺素
vat ①瓮；(大)桶；大盆②瓮(染料)还原物③鞣池
vat blue 瓮蓝
vat brilliant violet 瓮亮紫
vat color 瓮染料
vat coloring 瓮染(法)

vat dark blue 瓮暗蓝
vat deep printing black 瓮深黑〔印刷用〕
vat dye 瓮染料;还原染料
vat dyeing 瓮染(法)
vat dyestuff 瓮染料
vateria fat 龙脑香脂〔取自 *Vateria*〕
vat ester dye 酯化还原染料〔可溶性还原染料〕
vat liquor ①浸提液②浸提碱
vatman 捞工
vat paper ①圆筒机制纸②手(工捞)制纸
vat pigment 瓮染料
vat powder(=sodium hydrosulfite) 保险粉
vattability 可还原性
vatting 瓮染
vat waste 沉积残渣;沉积废料
vat yellow 瓮黄
vauqueline 马钱子碱
vauquelinite 磷铬铜铅矿
Vautin cell (for caustic soda) 沃廷(制苛性钠)电池
vazadrine 异烟肼
V-bar mo(u)ld V 型镶条模具
V-belt 三角皮带
V-belt drive 三角带传动
V-board 高级耐水纸板
V-cillin(=penicillin V) 青霉素 V;苯氧甲基青霉素
V-cut V(形)割法;V 形槽
veals〔复〕 小牛革
veal tallow 小羊脂
veatchine 维特钦;维钦碱
vectograph 矢量图
vector ①矢(量);向量②媒介物
vector addition 矢量加法;向量加法
vector function 向量函数
vectorial calibration method 矢量校正法
vectorial summation 向量和
vector model 矢量模型
vector quantity 矢量;向量
vector treatment 向量处理
vecuronium bromide 维库溴铵〔药〕
veepa oil 楝树油
vegasite 黄铅铁矾
vegetable adhesive 植物胶黏剂
vegetable black(=lamp black) 灯黑;油烟
vegetable butter 植物脂
vegetable carbon 木炭
vegetable charcoal 木炭
vegetable fat 植物油脂
vegetable fatty acid 植物脂肪酸
vegetable filler 植物性填料
vegetable glue 植物胶
vegetable lipase 植物(解)脂酶
vegetable lubricant 植物性润滑剂
vegetable matter 植物性物质
vegetable parchment 植物羊皮纸
vegetable rubber 植物橡胶
vegetable silk 有光泽的籽纤维;黏胶纤维
vegetables pesticide residues rapid analysis 蔬菜农药残留物快速测定法
vegetable-syntan tannage 植物-合成鞣剂鞣法
vegetable tallow 植物脂
vegetable tanned 植物鞣的
vegetable tanned kips〔复〕 植物鞣小兽皮
vegetable tanned leather 植鞣革
vegetable tanned sole leather 植物鞣鞋底革
vegetable tannin extract 栲胶
vegetable tanning 植物鞣(法)
vegetable tanning material 植物鞣料
vegetable turpentine 植物质松节油
vegetable wax 植物蜡
vegetable wool 植物毛;假羊毛
vegetal 植物的
vegeto-alkali 植物碱;生物碱
vegifat 高植脂
vegol 浓缩维生素 E
vegolysen 六甲基-1,6-溴化己二胺
vehicle ①载色剂②媒介物③赋形剂
vehicle clearance 车身离地高度
vehicle emulsion 赋形剂乳状液
vehicle exhaust gas purification system 汽车尾气净化系统
vehicle fuel 车用燃料
veil 覆盖毡片
veiling 网纹涂装法;喷丝涂装法;流挂
vein ①静脉②矿脉③杂色条纹
veininess 血管腺
veinlet 细脉
vein wall ①脉壁②矿脉两侧黏(土)皮
veiny 有血管腺的
velbe(=vinblastine) 长春花碱;长春碱
vellum paper 仿羊皮纸
velocimeter 速度计
velocitron 直线型飞行质谱计
velocity 速度
velocity constant 速度常数
velocity dispersion 速度色散
velocity distribution 速度分布

velocity distribution parameter 速度分布参数
velocity field 速度场
velocity focusing 速度聚焦
velocity gauge 速度计
velocity gradient 速度梯度
velocity gradient tensor 速度梯度张量
velocity head 速度高差；速位差；速度头
velocity indicator 流速计；流速指示计
velocity meter 速度计
velocity of approach 趋近速度；进入速度
velocity of detonation 爆（轰）速（度）
velocity of diffusion 扩散速度
velocity of displacement 位移速度
velocity of flow 流动速度
velocity of initiation 引发速度
velocity of particle 质点速度；粒子速度
velocity of propagation 传播速度
velocity potential 速度势
velocity pressure 速度压力；动压力
velocity profile 速度分布图；速度变化图
velocity profile rearrangement 速度分布重排
velocity sedimentation 速度（梯度）沉降
velocity selector 选速器
velocity slip 速度滑脱；速度滑移
velocity strain 速度应变
velocity triangle 速度三角形
velocity vector 速度向量
velometer ①速度计②调速计
velour 绒面革
velsicol 1068(=chlordane) 氯丹；八氯化甲桥茚；一〇六八
velvet leather 正绒面革
vena contracta 缩脉；收缩断面〔射流最小截面断〕
veneer ①单板②薄（片）木
veneer clipper 单板截断机
veneer sewing machine 单板缝合机
veneer sheet 层板；胶合板；薄板
venetian blind interference 条状波形干扰
venetian red 铁红
venetian white 碱式碳酸铅与硫酸钡的混合物
Venice turpentine(=Larch turpentine) 威尼斯松节油
venlafaxine 文拉法辛〔药〕
venogram 静脉像；静脉造影照片
vent ①(通气)孔；通气口②出口③放气口④放空
vent coil 排气旋管
vent condenser 排气冷却器；排气冷凝器
vented-barrel injection machine 排气料
vented tank 通风储桶

vent fan 吸气器；排气器
vent gas 排出气
vent gas scrubber 排出气洗涤器
vent gas tower 排气塔
vent hole 通风孔；排气孔
ventilating duct 导风筒
ventilating fan 通风机；风扇
ventilating hole 通风孔
ventilating openings 通风孔〔煤油灯的〕
ventilation 通风(作用)；换气
ventilation breather 通气器；呼吸孔
ventilation property 透气性
ventilation valve 通风阀
ventilator 通风器
ventilization 排气
vent line 排气管线
vent scrubber 排气洗涤塔
vent stack 放空烟囱
Venturi 文丘里管；文丘里喷嘴；细腰管
Venturi absorber 文丘里吸收器
Venturi effect 文丘里效应
Venturi gas scrubber 文丘里涤气器
Venturi meter 文丘里速度计
Venturi scrubber 文丘里涤气器
Venturi thorat 文丘里喉管
Venturi tube 文丘里流量计；文丘里管
venus crystals 乙酸铜
venus-hair fern 铁绒蕨
verapamil 维拉帕米〔药〕
veratral(=veratrylidene) 亚藜芦基；3,4-二甲氧苯亚甲基
veratraldehyde 藜芦醛；3,4-二甲氧基苯醛 $(CH_3O)_2C_6H_3CHO$
veratramin(e) 藜芦胺
veratric acid(=3,4-dimethoxybenzoic acid) 藜芦酸；3,4-二甲氧基苯甲酸 $(CH_3O)_2C_6H_3CO_2H$
veratridine 藜芦定；藜芦碱 I
veratrine ①藜芦碱类②藜芦碱
veratrine alkaloids 藜芦碱类
veratrine hydrochloride 盐酸藜芦碱
veratrine nitrate 硝酸藜芦碱
veratrine sulfate 硫酸藜芦碱
veratroidine 无定形藜芦碱
veratrole 藜芦醚；邻二甲氧基苯 $(CH_3O)_2C_6H_4$
veratroyl 藜芦酰；3,4-二甲氧苯甲酰 $(CH_3O)_2C_6H_3CO-$
veratrum alkaloids 藜芦碱类
veratrum viride 绿藜芦
veratryl 藜芦基；3,4-二甲氧苄基；3,4-二甲氧苯甲基

(CH₃O)₂C₆H₃CH₂—
veratryl alcohol 藜芦基醇
veratryl aldehyde 藜芦醛
veratrylamine 藜芦基胺
verbanol 马鞭草烷醇
verbanone 马鞭草烷酮
verbascose 毛蕊花糖
verbena absolute 马鞭草净油；防臭木净油
verbena concrete 马鞭草浸膏
verbenal 马鞭草烯醛
verbenalin 马鞭草苷
verbenaloside 马鞭草苷 $C_{17}H_{25}O_{10}$
verbena oil 马鞭草油
verbenene 马鞭草烯
verbenol 马鞭草烯醇
verbenone 马鞭草烯酮
verdazulene 绿薁素
Verdet's equation 弗德特方程
verdigris 铜绿；碱式碳酸铜
verditer ①碳酸铜②氢氧化铜
verditer blue 铜蓝〔蓝色铜盐颜料〕
verditer green 铜绿〔绿色铜盐颜料〕
verdoflavin 核黄素；维生素 B_2
veridian 绿色颜料
verification 检定；校准；核验
verification by sampling 抽样检定
verine 藜芦因；维林〔解痉药〕
veritas ring 热量圈
verium silicotungstate 硅钨酸钫
verivate 衍生物
vermeil 银面用火涂金法；飞金
vermicide 杀肠寄生虫药
vermiculine 蠕形青霉素
vermiculite 蛭石
vermifuge 驱(肠寄生)虫药
vermilion ①银朱；硫化汞②朱红(色)③朱红(色)的
vermilionette 赛银朱
vermilion paint ①银朱涂料②朱红涂料
vermouth(=absinthium) 苦艾
vernadite 水合软锰矿
vernal grass note 鲜草香
vernalic acid 环氧-9-十八碳烯酸
Verneuil method 晶体生长焰熔法
vernier ①游标②游(标)尺
vernier caliper 游标卡尺
vernier scale 游标尺
vernolic acid 斑鸠菊酸；12,13-环氧油酸
vernonine 斑鸠菊苷

Verona green 佛罗纳绿
veronal(=diethyl barbituric acid) 佛罗那；二乙基巴比妥酸 $C_8H_{12}O_3N_2$
veronal buffered salt solution 佛罗那缓冲盐溶液
veronal sodium 溶性巴比妥；巴比妥钠
veronica 婆婆纳〔药〕
verrucarine 疣孢菌素
verrucose 疣状的
versalide 万山麝香
versamide 植物聚酰胺
versatile adhesives 万能黏合剂
versatile column oven 通用柱(加热)炉
versatile rotary rheometer 通用型旋转流变仪
versenic acid 依地烯酸
versenol 依地烯醇；羟乙基乙烯二胺三乙酸
vertex 极点；顶(点)
vertex flow 涡流
vertical assembler-dismantler 立式装卸装置
vertical blowing process 垂直喷吹法；立吹法
vertical compressor 立式压缩机
vertical continuous polymerizer 立式连续聚合管
vertical Couette flow 竖向库爱特流动
vertical decanter 立式倾析口
vertical deposition 垂直沉积
vertical diaphragm cell 竖隔膜电池
vertical drier 立式干燥器
vertical drill 立式钻床
vertical endless band buffer 立式环带磨革机
vertical feed opening 垂直进料口
vertical film 垂直膜
vertical flow reactor 直流型反应器；立式连续反应器
vertical flue 竖(式)烟道
vertical-flued oven 竖式烟道炉
vertical grinding dispersion machine 立式研磨分散机
vertical heating flue 竖(式)烟道
vertical incision method 垂直切割法
vertically split 垂直裂缝
vertically symmetrical plane grating mounting 垂直对称平面光栅装置
vertical microsection 垂直微切片
vertical mixer 立式混合机
vertical orientation 垂直定向
vertical paper electrophoresis 纵向纸电泳
vertical planning drawing 竖向布置图；垂直布置图；立面图
vertical Poiseuille flow 竖向泊肃叶流动
vertical position welding 立焊
vertical pump 立式泵

vertical reaction tower 立式反应塔
vertical retort ①竖式甑②立式(干馏)釜
vertical section 纵截面
vertical slab electrophoresis 垂直板电泳
vertical slab gel electrophoresis 垂直板凝胶电泳
vertical split casing 垂直剖分式机壳
vertical stacked extractor 垂直堆积式萃取器
vertical still 立式蒸馏
vertical support 立式支座
vertical transition 垂直跃迁
vertical transportable retort 立式移动甑
vertical tube evaporator 立管蒸发器
vertical tube-flash evaporator 立管闪蒸器；立管式急骤蒸发器
vertical tubular furnace 立式管式炉
vertical tubular heater 立式管式加热炉；立式管式加热器
vertical-type evaporator 立式蒸发器
vertical vessel 立式容器
vertical view 俯视图
vertical weld 立焊
vertical wicking test 垂直式芯吸试验
verticil 降压灵〔药〕
verticillol (=4,8,12,15,15-pentamethyl- bicycle [9.3.1] pentadeca-3,7-dien-12-ol) 日本金松醇
verticine 维尔惕僧；贝母碱
verticinone 去氢贝母碱
vertivert oil 须芒草油〔印度的一种野草〕
very close elements 极相近元素；很类似的元素
very high density 极高密度
very low density 极低密度
very soluble 极易溶(解)的
vescalagin 栎木鞣花素 $C_{41}H_{26}O_{26}$
vescalin 栎木素 $C_{27}H_{20}O_8$
vesicle 泡；囊
vesicular 起泡的
vesicular structure 多孔结构
vesipyrin (=phenyl o-acetylsalicylate) 2-(乙酰氧基)苯甲酸苯酯
vesorcinol 二羟(基)甲苯
vesotinic acid 羟基甲苯甲酸
vespel ferrule 聚氨酯卡套
vessel ①(容)器；器皿②(导)管
vessel barometer(=bulb barometer) 球管气压计
vessel catcher 容器捕集法
vessel course 筒节
vessel port 容器开口
vesuvian ①符山石②耐风火柴
vesuvianite 符山石；火山石

vesuvine R 维苏文R〔染〕；碱性棕
vesypin 2-(乙酰氧基)苯甲酸苯酯
vetacetyl 乙酸岩兰草酯；乙酸香根酯
vetivalene(=vetivazulene) 岩兰薁；4,8-二甲基-2-异丙薁；韦惕瓦烯
vetivane 岩兰草烷
vetivazulene(=vetivalene) 韦惕瓦烯；岩兰薁
vetivene 韦惕烯；岩兰烯
vetivenic acid 岩兰草酸
vetivenol 岩兰草醇；香根醇
vetivenol acetate 乙酸岩兰草酯；乙酸香根酯
vetivenyl 岩兰草基；香根基
vetivenyl isovalerate 异戊酸岩兰草酯；异戊酸香根酯
vetivenyl propionate 丙酸岩兰草酯；丙酸香根酯
vetiver 韦惕蔚；岩兰草
vetiver acetate 乙酸岩兰草酯
vetiver oil 岩兰草油；香根油
vetiverol 岩兰草醇
vetiverone 韦惕蔚酮
vetiveryl 岩兰草基；香根基
vetiveryl acetate 乙酸岩兰草酯；乙酸香根酯
vetiveryl formate 甲酸岩兰草酯；甲酸香根酯
vetiveryl isovalerate 异戊酸岩兰草酯；异戊酸香根酯
vetiveryl propionate 丙酸岩兰草酯；丙酸香根酯
vetivol 岩兰草醇
vetivone 韦惕酮；岩兰酮；香根(草)酮
β-vetivone β-岩兰草酮
vetrification 玻璃化
V-face die V型表面口模
"V" filament V形灯丝
viable but nonculturable state (微生物)活的非培养态
via hole 导通孔
vial(=phial) 管(形)瓶
viaminati 维胺酯〔药〕
vibracone 振动锥筛
vibramycin(=deoxytetracycline) 强力霉素
vibrating conveyor 振动输送器；振动加料器
vibrating dampener 振动减低器
vibrating dripping mercury electrode 振动滴汞电极
vibrating electrode 振动电极
vibrating feeder 振动加料器
vibrating jet method 振荡射流法〔测动表面张力〕
vibrating plate extractor 振动板式萃取器
vibrating platinum electrode 振动铂电极
vibrating platinum microelectrode 振动铂微电极
vibrating reed 振(荡)簧(片)
vibrating reed amplifier 振簧放大器
vibrating reed electrometer 振簧静电计

vibrating-reed indicator 振簧指示器
vibrating-reed method 振簧法
vibrating sample magnetometer(VSM) 振动样品磁强计
vibrating screen 摆(动)筛
vibrating screen dissolver 振动筛溶解器
vibration 振动；摆动
vibration absorber 减振器
vibrational band 振动谱带
vibrational correction 振动校正
vibrational coupling 振动耦合
vibrational energy 振动能
vibrational entropy 振动熵
vibrational fine structure 振动微细结构
vibrational flow 振动流动
vibrational frequency 振动频率
vibrational level 振动能级
vibrational partition function 振动配分函数
vibrational quantum number 振动量子数
vibrational relaxation 振动弛豫
vibrational relaxation time 振动弛豫时间
vibrational resilience 振动态回弹性
vibrational-rotational band 振动-转动谱带
vibrational-rotational spectra 振动-转动光谱；振转光谱
vibrational-rotational transition 振动-转动跃迁
vibrational spectroscopy 振动波谱法
vibrational spectrum 振动光谱
vibrational state 振动态
vibrational transition 振动跃迁
vibration amplitude 振幅
vibration curve 振动曲线
vibration dampener 防振器；减振器
vibration damper 减振器
vibration damping 振动阻尼
vibration frequency 振动频率
vibration mill 振动磨
vibration period 振动周期
vibration pick-up 振动传感器
vibration rotation spectrum 振动转动光谱
vibration rotation tunneling spectroscopy 振动旋转隧道光谱
vibration sieve 振动筛
vibration sift 振动筛
vibration spectrum analyzer 振动频谱分析器
vibration staking machine 振荡式刮软机
vibration state 振动态
vibrator ①振动器②振子
vibratory feeder 振动给料器
vibratory leacher 振动浸取器
vibratory motion 振动
vibro-compact fuel element 振动增实燃料元件
vibrometer 振动仪
vibron analysis 电子振动分析
vibronic spectrum 电子振动光谱
vibrostand 振动(试验)台
vibro washer 振荡洗涤剂
viburnitol 荚蒾醇；L-栎醇；环己五醇
viburnum dilatatum oil 荚蒾油
vicalloy 钴铁钒合金
Vicat needle 维卡针
Vicat needle apparatus 维卡针式测试器
Vicat needle tester 维卡针入测试仪
Vicat softening point test 维卡软化点试验
Vicat softening temperature 熔维卡软化点
vice 轧钳
vice bench 钳工台
vicianin 巢菜苷；毒蚕豆苷
vicianose 巢菜糖；蚕豆糖；荚豆二糖
vicinal 连〔1,2,3位 或 1,2,3,4位〕
vicinal compound 连位化合物
vicinal coupling 邻偶
vicinal effect 邻位效应
vicinal faces 邻晶面；邻近面
vicinal position 连位
vicine 蚕豆嘧啶葡糖苷
Vickers hardness 维氏硬度
Vickers hardness tester 维氏硬度计
Vickers image splitting eyepiece 维氏裂像目镜
victor bronze 铜锌铝铁钒合金
vicuna wool 骆马毛
vidarabine 阿糖腺苷〔药〕
videofluorometer 电视荧光计
video recording 显像记录
video recording device 显像记录装置
vidicon 光导摄像管
vidicon tube 光导摄像管
Vielle's manometric bomb 维里测压弹
vienna caustic 氢氧化钾-石灰混合物
Vienna lime 新华石灰；维也纳石灰
viferral 吡啶与三氯甲醛聚合物
vigna 豇豆
vigorous agitation 剧烈搅动
vigorous oxidation 剧烈氧化
Vigreux column 维格罗分馏柱
Villari effect 维拉里效应
Villavecchia test 维拉维琪亚试芝麻油法
vilmorrianine A 黄草乌碱A

vinaconic acid 酒康酸 $CH_2CH_2C(COOH)_2$
vinaline 牧豆树素
vinblastine 长春碱〔药〕
vincamajoreine 长春蔓碱
vincamedine 长春蔓美定
vincamine 长春蔓胺
vincanine 长春蔓宁
vincennite 三氯化砷及氰化氢混合剂
vinc fiber 藤纤维
vincristine 长春新碱〔药〕
vindesine 长春地辛〔药〕
vindoline 文朵灵；长春花朵灵
vindolinine 文朵宁；长春花朵宁
vine fiber 藤纤维
vinegar 醋
vinegar essence 醋精
vinegar naphtha 醋酸乙酯
vinegar of squill 海葱醋萃液
vine rubber 藤(本橡)胶
vinetine 尖刺(小檗)碱
vinic acid ①硫酸氢烷基酯 RSO_4H ②〔专指〕硫酸氢乙酯 $C_2H_5SO_4H$
vinic ether (二)乙醚 $C_2H_5OC_2H_5$
vinifera palm oil 竹油
vinine 文宁；长春花因
vinometer 酒精比重计
vinopyrine 酒石酸乙氧基苯胺
vinorelbine 长春瑞宾〔药〕
viny 葡萄树的
vinyl 乙烯基 $CH_2=CH-$
vinylacetate 乙烯基乙酸盐(酯) CH_3CHCH_2COOM
vinyl acetate-acrylamide copolymer 醋酸乙烯酯-丙烯酰胺共聚物
vinyl acetate polymer 醋酸乙烯(酯)聚合物
vinylacetic acid 乙烯基乙酸；丁烯酸 $CH_2=CHCH_2CO_2H$
vinylacetic ester 乙酸乙烯酯 $CH_2=CHOCOCH_3$
vinyl acetylene 乙烯基乙炔
vinylacrylic acid 乙烯基丙烯酸；2,4-戊二烯-1-酸 $CH_2=(CH_2)=CHCO_2H$
vinyl alcohol 乙烯醇
vinyl alcohol-N-methylolacrylamide copolymer 乙烯醇-N-羟甲基丙烯酰胺共聚物
vinylal fibre (聚)乙烯醇系纤维
vinylallyl type 戊间二烯型；乙烯基丙基型 $CH_2=CHCH=CHCH_2-$
vinyl-amine 乙烯胺 $CH_2=CHNH_2$

vinylation 乙烯化作用
vinyl benzene 乙烯基苯；苯乙烯
vinyl benzyl ether 乙烯基苄基醚
vinyl blend 乙烯掺和剂
vinyl bromide 乙烯基溴；溴代乙烯 $CH_2=CHBr$
vinyl butyl ether 乙烯基丁醚；丁氧基乙烯 $CH_2=CHOC_4H_9$
vinyl caproate 己酸乙烯酯
vinyl carbonate 碳酸乙烯酯
vinyl carborane 乙烯(基)碳硼烷
vinyl cellulose 乙烯基纤维素
vinyl chloride 乙烯基氯；氯乙烯 $CH_2=CHCl$
vinyl chloride-acetate copolymer 氯乙烯-乙酸乙酯共聚物
vinyl chloride acrylate latex 氯乙烯丙烯酸酯胶乳
vinyl chloride acrylonitrile copolymer 氯乙烯丙烯腈共聚物
vinyl chloride copolymer 氯乙烯共聚物
vinylchloride-ethylene copolymer 聚乙烯-乙烯共聚物
vinylchloride-ethylene-methylacrylate copolymer 聚乙烯-乙烯-丙烯酸甲酯共聚物
vinylchloride-ethylene resin 氯乙烯-乙烯树脂
vinylchloride-ethylene-vinyl 氯乙烯-乙烯/醋酸乙烯共聚物
vinylchloride-ethylene-vinylacetate copolymer 氯乙烯-乙烯-乙酸乙烯酯共聚物
vinyl chloride-methyl acrylate(VCMA) 氯乙烯-丙烯酸甲酯共聚物
vinyl chloride-methylmethacrylate(VCMMA) 氯乙烯-甲基丙烯酸甲酯共聚物
vinylchloride-methylmethylacrylate copolymer 氯乙烯-甲基丙烯酸甲酯共聚物
vinylchloride-octylacrylate copolymer 氯乙烯-丙烯酸辛酯共聚物
vinyl chloride-octyl acrylate resin(VCOA) 氯乙烯-丙烯酸辛酯树脂
vinyl chloride rubber (聚)氯乙烯(合成)橡胶
vinylchloride-vinylacetate copolymer 氯乙烯-乙酸乙烯酯共聚物
vinyl chloride-vinyl acetate copolymer fibre 氯乙烯-醋酸乙烯酯共聚(物)纤维
vinyl chloride-vinyl acetate resin(VCVAC) 氯乙烯-醋酸乙烯树脂
vinyl chloride-vinylidene chloride(VCVDC) 氯乙烯-偏氯乙烯共聚物
vinyl chloride-vinylidene chloride copolymer 氯乙烯-偏氯乙烯酯共聚物
vinyl chloroacetate 氯代乙酸乙烯(酯)

vinyl coating 乙烯基涂料
vinyl compound 乙烯系化合物
vinyl copolymer 乙烯系共聚物
vinyl cyanide 乙烯基氰；丙烯腈 $CH_2=CHCN$
vinyl cyclohexane 乙烯(基)环己烷
vinyl cyclohexene dioxide 乙烯基环己烯二酮
vinyl elastomer 乙烯系弹料
vinylene 1,2-亚乙烯基 —CH=CH—
vinylene carbonate 碳酸亚乙烯酯
vinylene chloride 1,2-二氯乙烯
vinylene monomer 1,2-亚乙烯单体；1,2-二取代乙烯单体
vinyl ester 乙烯基酯
vinyl ether ①乙烯醚 $(CH_2CH)_2O$ ②乙烯基醚 $CH_2=CHOR$
vinyl ethyl alcohol(=allyl carbinol) 乙烯基乙醇；烯丙基甲醇 $C_3H_5CH_2OH$
vinyl ethyl ether 乙烯基乙基醚；乙氧基乙烯 $CH_2=CHOC_2H_5$
vinyl fatty ester 脂肪酸乙烯酯
vinyl fiber 乙烯基纤维
vinyl fluoride 乙烯基氟；氟乙烯 $CH_2=CHF$
vinyl fluoride-vinylidine fluoride copolymer fibre 氟乙烯-偏氟乙烯共聚(物)纤维
vinyl fluorobenzene sulphonic acid 乙烯基氟代苯磺酸
vinyl foam 乙烯系泡沫体
vinyl formate 甲酸乙烯酯 $HCOOCH=CH_2$
vinylformic acid 丙烯酸；乙烯基甲酸
vinylglycollic acid 乙烯基乙醇酸；2-羟基-3-丁烯-1-酸 $CH_2=CHCHOHCO_2H$
vinyl guaiacol 乙烯基愈创木酚 $CH_2=CHC_6H_3(OH)(OCH_3)$
vinylic 乙烯的
vinylidene 亚乙烯基 $CH_2=C=$
vinylidene chloride 1,1-二氯乙烯；亚乙烯基二氯 $CH_2=CCl_2$
vinylidene chloride latex 偏(二)氯乙烯胶乳
vinylidene cyanide 亚乙烯基二氰 $CH_2=C(CN)_2$
vinylidene dicyanide 偏(二)氰乙烯
vinylidene dinitrile 偏(二)氰乙烯
vinylidene fibre (聚)偏(二)氯乙烯系纤维
vinylidene fluoride 1,1-二氟乙烯
vinylidene monomer 亚乙烯基单体*
vinylidene resin 亚乙烯树脂；1,1-二氯乙烯树脂
vinyl imidazole 乙烯基咪唑
vinyl iodide 乙烯基碘；碘代乙烯 $CH_2=CHI$
vinyl-β-ionol 乙烯基-β-紫罗兰醇
vinyl isoamyl ether 乙烯基异戊基醚 $CH_2=CHOC_5H_{11}$

vinyl isobutyl ether 乙烯基异丁基醚 $CH_2=CHOC_4H_9$
vinyl isopropyl ether 乙烯基异丙基醚 $CH_2=CHOC_3H_7$
vinyl ketones 乙烯基甲酮类 $CH_2=CHCOR$
vinyl laurate 月桂酸乙烯酯
vinyl leather 乙烯基革；乙烯基人造革
vinyl methyl ether 乙烯基甲基醚
vinyl methyl formamide polymer 甲基乙烯基甲酰胺聚合物
vinyl methyl ketone 乙烯基甲基(甲)酮
vinyl-modified polyamide 乙烯基改性聚酰胺
vinyl monochloroacetate 一氯乙酸乙烯酯
vinyl monomer 乙烯基单体*
vinylog 插烯物
vinylogue 插烯物；联乙烯物
vinylogy 插烯(作用)
vinyloxybenzene sulfonic acid 乙烯氧基苯磺酸
vinyloxy group 乙烯氧基
vinyloxy polyoxy alkylene ether 乙烯氧聚氧化亚烷基醚
vinyl paste 乙烯基树脂糊
vinylphenol 乙烯基苯酚
vinylphenyl acetate 乙酸乙烯苯酯 $CH_3CO_2C_6H_4CH=CH_2$
vinylphenyl ether 乙烯基苯基醚 $CH_2=CHOC_6H_5$
vinyl phosphonate 膦酸乙烯酯
vinylphosphonic acid 乙烯膦酸
N-vinyl phthalimide 乙烯替邻苯二酰亚胺；N-乙烯基邻苯二酰亚胺
vinyl plastic 乙烯基塑料
vinyl plastisol 乙烯塑料溶胶
vinyl polymer 烯类聚合物*；乙烯基聚合物
vinyl polymerization 烯类聚合*；乙烯基聚合
vinylpyrene 乙烯基芘
vinylpyridiene-styrene-butadiene rubber 丁苯吡橡胶
vinylpyridine-butadiene rubber 丁吡橡胶
vinyl pyrrolidinone 乙烯基吡咯烷酮
vinyl pyrrolidone 乙烯基吡咯烷酮
vinylquinoline 乙烯基喹啉
vinyl resin 乙烯基树脂
vinylsilane rubber 乙烯基硅橡胶
vinylsiloxane rubber 乙烯基硅橡胶
N-vinyl succinimide 乙烯基琥珀酰亚胺
vinyl sulfide 二乙烯基硫 $(CH_2=CH)_2S$
vinylsulfonate 乙烯基磺酸盐
vinyl sulfone dyestuff 乙烯砜染料
vinyl sulfonic acid 乙烯(基)磺酸 $CH_2=CHSO_3H$
vinyl tape (涂)聚氯乙烯绝缘带
vinyl thioether 二乙烯基硫醚 $(CH_2=CH)_2S$

vinyl thiolacetate 硫羟乙酸乙烯酯
vinyltriacetoxy silane 乙烯基三乙酰氧基硅烷〔偶联剂〕
vinyltribromide(=1,1,2-tribromoethane) 1,1,2-三溴乙烷 $BrCH_2CHBr_2$
vinyltri-*t*-butylperoxy silane 乙烯基三叔丁基过氧基硅烷〔偶联剂〕
vinyl trichloride 1,1,2-三氯乙烷
vinyltrichlorosilane 乙烯基三氯甲硅烷
vinyltriethoxysilane 乙烯基三乙氧基硅烷〔偶联剂〕
vinyl trifluoacetate 三氟乙酸乙烯酯
vinyltrimethoxy silane 乙烯基三甲氧基硅烷〔偶联剂〕
vinyl-type polymerization product 乙烯型聚合产品
viola crystallina 结晶紫；龙胆紫
violaguercitrin 芸香苷
violanin 堇菜苷；花翠素鼠李葡萄糖苷
violanthrene 紫蒽
violanthrole 二苯并羟蒽
violanthrone 紫蒽酮
violaxanthin(=zeaxanthin diepoxide) 堇菜黄质；紫黄质
violent acceleration 急剧加速
violent stirring 强烈搅拌
violet ①紫(色)②紫(色)的③紫罗兰
violet absolute 紫罗兰净油
violet acid 混合硫硝酸；混酸 $H_2SO_3 \cdot NO_2$
violet compound 紫罗兰香精
violet flower absolute 紫罗兰花净油
violet flower concrete 紫罗兰花浸膏
violet flower oil 紫罗兰花油
violet leaf alcohol 紫罗兰叶醇；2,6-壬二烯-1-醇
violet leaf aldehyde 紫罗兰叶醛；2,6-壬二烯-1-醛
violet leaf concrete 紫罗兰叶浸膏
violet leaf oil 紫罗兰叶油
violet-like odor 紫罗兰样香气
violet phosphorus 紫磷 P_4
violet root 堇菜根
violet root oil 紫罗兰根油
violin 堇叶素
violine 堇菜苷
violurate 紫尿酸盐 $C_4H_2O_4N_3M$
violuric acid 紫尿酸
viomycidine 胍基二氢吡咯甲酸
viomycin 紫霉素
viomycin sulfate 硫酸紫霉素
viosterol 紫甾醇〔用紫外线照射麦角甾醇而得〕
virgiline 灌豆碱
virgin ammonia liquor 粗氨水
virgin curve 起始曲线；初始曲线；新曲线
virgin dip 初割树脂

virgin fibre 原始纤维；未改性纤维；母体纤维
virgin gas oil 直馏粗柴油
virgin kerosene 直馏煤油
virgin material 新料
virgin naphtha 直馏石脑油
virgin oil ①直馏油（石油）②初榨橄榄油
virgin olive oil 初榨橄榄油
virgin resin ①纯树脂②无填料树脂
virgin rubber ①原胶；未干透的生胶②新鲜橡胶；新胶
virgin stock ①直馏油料②原浆
Virial coefficient 维里系数
Virial theorem 维里定理
viride nitens 亮绿〔染料〕
viridiflorine 绿花(倒提壶)碱
viridiflorol 绿花白千层醇
viridon ①媒染绿②毛皮绿〔染〕
viridon FF 媒染绿 FF；媒染绿
virosine ①长春素②夹竹桃碱
virtual coupling 虚假偶合
virtual image 虚像
virtual long-range coupling 虚假远程偶合*
virtual mechanism 假定机理
virtual orbital 虚(空)轨道
virtual state 虚态
virtual tautomerism 假互变异构(现象)
virtual value 有效值
virulence 毒性；毒力
virulence test 毒力试验
virulent ①(极)毒的②恶性的；致死的
visbreaker(=viscosity breaking furnace) 减黏裂化炉
visbreaker heater(=viscosity breaking furnace) 减黏裂化炉
visbreaking(=viscosity breaking) 减黏裂化；减黏轻度裂化；减低黏度
viscin 槲寄生素〔取自槲寄生 *viscum*〕
viscocurometer 黏度式固化仪
viscoelastic 能塑造变形的
visco-elastic behavior 黏弹特性
visco-elastic body 黏弹体
visco-elastic creep 黏-弹性蠕变
viscoelastic cross effect 黏弹性横向效应；黏弹性交叉效应
viscoelastic deformation 黏弹性形变
visco-elastic dispersion 黏弹性分散体
viscoelastic flow 黏弹性流动
viscoelastic fluid(=Maxwell fluid) 黏弹(性)流体
viscoelastic function ①黏弹函数②黏弹功能
viscoelasticity 黏弹性*
visco-elastic material 黏弹性物质

viscoelastic micropolar medium 微极性黏弹介质
viscoelastic modulus 黏弹模量
viscoelastic plasticity 黏弹性塑度；黏弹塑性
visco-elastic property 黏黏性质
viscoelastic region 黏弹区(域)
viscoelastic relaxation 黏弹松弛
viscoelastic response 黏弹响应
viscoelastic solid 黏弹固体
viscoelastic state 黏弹态
viscoelastic theory 黏弹性理论
viscoelastometer 黏弹计
viscogel 黏性凝胶
viscograph 黏度曲线仪
viscoid 黏性体
viscoinelastic fluid 黏性非弹性流体
viscol 硫化油〔经 S_2Cl_2 处理的〕
viscolizer 增稠剂；增黏剂
viscometer(=viscosimeter) 黏度计
viscometric average degree of polymerization 黏均聚合度
viscometric degree of polymerization 黏均聚合度
viscometric flow 测黏流动
viscometric function 测黏函数
viscometric molecular weight 黏度法测定的分子量
viscometric titration 黏度滴定
viscometry 黏度测定法
viscomill 高黏度磨机
viscoplastic flow 黏塑(性)流
viscoplasticity ①黏塑性②黏塑度
viscoplastic principal strain difference 黏塑主应变差
viscoplastic solid 黏塑性固体
viscoplastoelastic 黏塑弹性(的)
visco-plasto-elastomer 黏塑弹性体
viscorator 连续记录黏度计
viscorator viscometer 〔商〕浮子型黏度计
viscosaccharase 黏滞转化酶
viscose ①黏胶纤维②黏胶丝
viscose acetal fibre 黏胶纤维
viscose adhesive 黏胶胶黏剂
viscose-based carbon fibre 黏胶基碳纤维
viscose blowout 破皮纤维
viscose cellar 黏胶窖；黏胶窨
viscose cord fabric 黏胶帘布
viscose fiber 黏胶纤维
viscose filament spinning machine 黏胶长丝机
viscose film 黏胶薄膜
viscose high tenacity yarn 高强力黏胶丝
viscosel viscometer 〔商〕气动信号连续式黏度计

viscose paper 玻璃纸；黏胶纸
viscose process 黏胶法
viscose proteinised 蛋白质化黏胶纤维
viscose rayon 黏胶嫘萦；黏液丝
viscose ripening 黏胶熟成
viscose silk 黏胶(人造)丝
viscose solution 黏胶溶液
viscose sponge 黏胶海绵
viscose staple fiber 黏胶短纤维
viscose staple spinning machine 黏胶短丝机
viscose tank 黏胶桶；黏胶槽
viscosimeter(=viscometer) 黏度计
viscosimetric titration 黏度滴定
viscosimetry 黏度测定(法)
viscosine 黏渣油；暗色残油；精制残油
viscosity 黏度*
viscosity alarm recorder 黏度警报记录器
viscosity analyzer 黏度分析器
viscosity anomaly 黏度反常
viscosity apparatus 黏度仪
viscosity average 黏度平均值
viscosity-average molar mass 黏均分子量；黏均摩尔质量
viscosity-average molecular weight(=viscosity-average molar mass) 黏均分子量
viscosity blending chart 黏度掺和线图
viscosity breaker(=viscosity breaking furnace) 减黏裂化炉
viscosity breaking 减黏裂化
viscosity breaking furnace(=visbreaker;visbreaker heater) 减黏裂化炉
viscosity breaking plant(=viscosity breaking unit) 减黏裂化装置
viscosity breaking unit(=viscosity breaking plant) 减黏裂化装置
viscosity build-up 黏度上升
viscosity coefficient 黏度系数；黏度
viscosity constant 黏度常数
viscosity control agent 黏度控制剂
viscosity controller 黏度控制器
viscosity-controlling agent 黏度调节剂
viscosity conversion 黏度换算
viscosity cup 黏度测杯
viscosity curve 黏度曲线
viscosity degradation 黏度下降
viscosity density ratio 比密黏度
viscosity depressant ①减黏剂②黏度抑制剂
viscosity detector 黏度检测器

viscosity drop 黏度降
viscosity effect 黏度效应
viscosity effusion bridge 黏度泻流桥
viscosity equation 黏度方程
viscosity factor 黏性系数
viscosity fluctuation 黏性波动；黏度波动
viscosity function 黏度函数
viscosity gauge 黏度表；黏度规
viscosity gradient 黏度梯度
viscosity-gravity chart 黏度比重线图
viscosity-gravity constant 黏度-比重常数
viscosity increaser 增黏剂
viscosity index 黏度指数
viscosity index blending value 黏度指数掺和值
viscosity index chart 黏度指数线图
viscosity index constituent 影响黏度指数的成分
viscosity index improver 黏度指数改进剂
viscosity measurement 黏度测定
viscosity modifier 黏度改进剂*
viscosity-molecular weight dependence 黏度-分子量依赖性
viscosity-molecular weight equation 黏度-分子量方程
viscosity-molecular weight relationship 黏度-分子量关系
viscosity number 黏数
viscosity of latex 胶乳黏度
viscosity of polymer solution 高聚物溶液的黏度
viscosity-pressure property 黏度-压力特性
viscosity ratio 黏度比
viscosity sensor 黏度传感器
viscosity standard liquid 黏度(测定用)标准液
viscosity-temperature characteristic 黏温特性
viscosity-temperature property 黏度-温度特性
viscosity-temperature property tester 黏-温特性试验机
viscostatic lubricating oil 静黏态润滑油
viscous bitumen 黏沥青；半固体沥青
viscous compressible flow 可压缩黏性流
viscous damping 黏滞阻尼；黏性衰减；黏性阻尼
viscous deformation 黏性变形；塑性变形；塑变
viscous drag 黏性阻力
viscous-elastic behaviour 黏弹特性；黏弹表现
viscous-elastic material 黏弹性材料
viscous emulsion 黏性乳化
viscous filter 黏液过滤器
viscous flow 黏滞流
viscous flow leak 黏性流渗漏孔
viscous flow state 黏流态
viscous fluid 黏性流体
viscous fluid filter 黏液过滤器
viscous fluid flow 黏性流动；稳流；层流
viscous force 黏滞力
viscous fracture 黏性破坏
viscous friction 黏性摩擦；黏滞摩擦
viscous-gel formation 黏性凝胶形成
viscous heating 黏性发热
viscous in body 本体内黏度；厚层黏度
viscous Lame coefficient 拉梅黏性系数
viscous liquid 黏性液体
viscous liquid flow(=viscous fluid flow) 黏性液流动；稳流；层流
viscous liquid sample 黏性液体试样
viscous lubrication 黏滞润滑；厚层润滑
viscous melt-phase 黏熔相
viscous motion (黏)滞(运)动
viscous state 黏性状态；黏流态
viscous stress tensor 黏性应力张量
viscous thermoelastic material 黏性热弹(性)材料
viscous traction 黏性引力
viscurometer 黏度硫化仪
visibility ①可见性②可见度；能见度
visibility determination method 能见度测定方法
visible absorption spectrophotometry 可见吸收分光光度法
visible flame 可见火焰
visible light 可见光
visible light detector 可见光检测器
visible light source 可见光源
visible radiation 可见光辐射
visible-range monitor 可见光监测器
visible ray 可见光线
visible region 可见区
visible spectrometry 可见光谱法
visible spectrophotometer 可见分光光度计
visible spectroscopy 可见光谱法
visible spectrum 可见光谱
visigraph recorder 示波记录器
visnadin 氢吡豆素；维斯哪定
visnadine 氢吡豆素；维斯哪定
visor ①观察孔②护目镜
visual ①视(觉)的；可测的②目视的；可见的
visual colorimeter 目视比色计
visual colorimetric determination 目视比色(测定)
visual colorimetry 目视比色法
visual detection 目测(法)
visual difference 错视
visual digital display 目视数字显示
visual display monitor 目视显示监测器
visualization 目视观察；显像

visualizing agent　显色剂
visual light scattering photometer　目视式光散射光度计
visually inspect　目测检查
visual method　目测法
visual photometry　目视测光法
visual test　目测试验
visual titration　目视滴定(法)
vitaglass　维他玻璃；透紫外线玻璃
vitagonist　维生素拮抗物
vitamer　同效维生素
vitamin A(=axerophthol)　维生素 A；抗干眼醇
vitamin A$_1$(=retinol)　维生素 A$_1$；视网膜醇
vitamin A$_2$(=3-dehydroretinol)　维生素 A$_2$；3-脱氢视网膜醇
vitamin A acid(=retinoic acid)　视网膜酸；维生素
vitamin A ether　维生素 A 醚
vitamin A ketone　维生素 A 酮
vitamin B　维生素 B
vitamin B$_1$(=thiamine)　维生素 B$_1$；硫胺(素)；抗神经炎维生素
vitamin B$_2$(=riboflavin)　维生素 B$_2$；核黄素
vitamin B$_6$(=pyridoxine; pyridoxal; pyridoxamine)　维生素 B$_6$；抗皮肤炎维生素〔吡哆醇、吡哆醛及吡哆胺的总称〕
vitamin B$_{12}$(=cyanocobalamine)　维生素 B$_{12}$；(氰)钴胺(素)
vitamin B$_{12a}$(=hydroxocobalamine)　维生素 B$_{12a}$；羟钴胺(素)
vitamin B$_{13}$(=orotic acid)　维生素 B$_{13}$；乳清酸
vitamin B$_c$(=folic acid)　维生素 B$_c$；叶酸
vitamin B$_{15}$　维生素 B$_{15}$
vitamin B complex　复合维生素 B
vitamin B group　维生素 B 类
vitamin Bp　维生素 Bp
vitamin BT(=carnitine)　维生素 BT；肉碱
vitamin Bu　维生素 Bu
vitamin Bx　维生素 Bx
vitamin C(=ascorbie acid)　维生素 C；抗坏血酸
vitamin D　维生素 D
vitamin D$_2$(=ergocalciferol, calciferol)　维生素 D$_2$；麦角钙化(甾)醇
vitamin D$_3$(=cholecalciferol)　维生素 D$_3$；胆钙化(甾)醇
vitamin D$_4$　维生素 D$_4$
vitamin D palmitate　维生素 D 棕榈酸酯
vitamin D sulfate　维生素 D 硫酸酯
vitamin(e)　维生素
vitamin E(=tocopherol)　维生素 E；生育酚
vitamin E$_1$　维生素 E$_1$
vitamin F(=nicotinic acid)　维生素 F；烟酸
vitamin G　维生素 G
vitamin H(=biotin)　维生素 H；生物素　$C_{10}H_{16}O_3N_2S$
vitaminization　维生素化作用；加入维生素
vitaminized　维生素化了的；加了维生素的
vitamin K(=coagulation vitamin)　维生素 K；凝血维生素
vitamin K$_1$(=phylloquinone; 2-methyl-3-phytyl-1,4-naphthaquinone)　维生素 K$_1$；叶绿醌；2-甲基-3-植基-1,4-萘醌
vitamin K$_2$(=2-methyl-3-difarnesyl-1,4-naphthoquinone)　维生素 K$_2$；2-甲基-3-二法呢基-1,4-萘醌
vitamin L　维生素 L
vitamin L$_1$　维生素 L$_1$
vitamin M(=folic acid)　维生素 M；叶酸
vitaminosis　维生素过多症
vitamin P(=citrin)　维生素 P；柠檬素
vitamin PP(=nicotinic acid; nicotinamide)　维生素 PP；烟酸；烟酰胺；抗糙皮病维生素
vitamins A　A 族维生素
vitaminstoss　维生素大剂量治疗
vitamin T　维生素 T
vitamin U　维生素 U
vitamin units　维生素单位
vitamin value　维生素值
vitamycin　维他霉素
vitasterol　甾醇型维生素
vitavel-k　乙酸甲萘氢醌
vitexin　牡荆素；牡荆黄素(酮)
vitexine　牡荆色素
vitol　维生素
vitrain　镜煤
vitreosil　耐热石英
vitreous　①玻璃的②透明的
vitreous carbon　玻璃碳
vitreous carbon electrode　玻璃化碳电极
vitreous clinker　玻璃熔结块
vitreous copper　辉铜矿
vitreous enamel　①透明搪瓷②珐琅③釉瓷
vitreous fibre　玻璃质纤维；透明纤维
vitreous fluid　玻璃状液
vitreous fracture　玻璃状断面
vitreous luster　玻璃光泽
vitreousness　玻璃状态；透明性
vitreous sanitary ware　玻璃卫生器具
vitreous silica　透明石英
vitreous silica fibre　高硅氧玻璃纤维
vitreous silver ore　辉银矿
vitreous white arsenic　玻态砒霜　As_2O_3
vitrified borax　熔融硼砂
vitrified enamel　搪瓷
vitrified polymer　玻璃化聚合物；透明化聚合物

vitrifying 玻璃化
vitrifying point 玻璃化点；玻化点
vitrinite 煤榆胶
vitriol ①硫酸盐；矾②硫酸
vitriolate of soda 硫酸钠
vitriolate of tartar 硫酸钾
vitriol chamber 铅室
vitriol chamber plant 铅室(法制)硫酸厂
vitriol chamber process 铅室法
vitriolic 硫酸的
vitriolization ①硫酸处理②溶于硫酸
vitriol oil (浓)硫酸
vitriol plant 硫酸厂
vitriol salt 锌矾；皓矾；七水合硫化锌 $ZnSO_4 \cdot 7H_2O$
vitriolum veneris 硫酸铜
vitriol works〔复〕 硫酸厂
vitrite 氯化氰与三氯化砷的混合物
vittatine 条纹(小星蒜)碱
vivianite 蓝铁矿
vivo strength retention ①(生物)体内强度保持率②自然条件下强度保持率
VK-simplified continuous process VK 管简化连续(聚合)法
V-mode V 型〔吸收信号〕
V-notch weir 三角堰
voacamine(=voacanginine) 老刺木胺
voacangarine(=voacristine) 老刺木分亭
voacanginine(=voacamine) 老刺木胺
voacristine(=voacangarine) 老刺木分亭
vobasine 老刺木碱
Vohr ozonizer 沃尔臭氧器
void ①空隙②空(处)；真空③空(虚)的④无效的
voidage 空隙度；气泡量；空穴
void column 空柱
void concentration 孔隙密集度
void content 有孔度；气孔率
void density 孔隙(密)度
void distribution 孔隙分布
void factor 空隙因数
void fraction percent 空隙百分率；孔隙百分率
void-free 紧密的；密实的；无气孔的
void-free fibre 密实纤维；无孔隙纤维
void level 空隙度；空隙密度
void porosity ①空隙比；松度②孔积率
void ratio 空隙比；(滤床)空隙率
voids〔复〕 砂眼〔铸件〕；空隙
void-size distribution 空隙大小分布
void space 空隙空间
void spot 空斑

void time 死时间
void tower 空塔
void volume 空隙容积；空隙率；空体积；外水体积
void-volume marker 死体积标记物
Voigt body(=Kelvin body) 沃伊特体；开尔文体
Voigt model(=Kelvin model) 开尔文模型；沃伊特模型
voile yarn 巴里纱；华而纱
volatile 挥发组分
volatile acid 挥发性酸
volatile alkali 挥发性碱〔指：氨；氨气 NH_3〕
volatile ammonia 挥发性氨
volatile base nitrogen 挥发性碱性氮
volatile base value 挥发碱值
volatile carbon in oil 油中游离碳
volatile caustic 氢氧化铵 NH_4OH
volatile coal 挥发性煤
volatile combustible matter 挥发可燃物(质)
volatile component 挥发性组分
volatile compound 挥发性化合物
volatile constituent 挥发性组分
volatile content 挥发物含量
volatile crystals 挥发性结晶体
volatile distillates 挥发性馏分
volatile fatty acid 挥发性脂肪酸
volatile fatty acid number 挥发性脂肪酸值
volatile flux 挥发性助熔剂
volatile foreign matter 挥发性杂质
volatile fraction 挥发部分；挥发性馏分
volatile fuel 挥发性燃料
volatile grade 挥发度；蒸发度
volatile impurity 挥发性杂质
volatile inflammable liquid 挥发性可燃液体
volatile liquid 挥发性液体
volatile loss 挥发减量
volatile material 挥发物
volatile matter 挥发物
volatile oil 挥发油；香精油
volatile oil determination apparatus 挥发油测定器
volatile oil of mustard 芥子挥发油
volatile organic compounds 挥发性有机物
volatile poisons 挥发性毒物
volatile products 挥发性产物
volatile recovery system 挥发物回收装置
volatile resistance ①耐挥发性②耐挥发度
volatile salt 碳酸铵 $(NH_4)_2CO_3$
volatile separation 耐挥发分离
volatile solid 挥发性固体
volatile solvent 挥发性溶剂

volatile solvent-dispersed grease 用挥发性溶剂分散的润滑脂
volatile spirits 挥发精油〔汽油馏分的旧称〕
volatile with steam 随水蒸气挥发的
volatility ①挥发性②挥发度
volatility index(=volatility number) 挥发度指数；挥发度值
volatility number(=volatility index) 挥发度指数；挥发度值
volatility of oil 油的挥发度
volatility product 挥发度积
volatility range 挥发范围
volatility resistance 耐挥发性
volatility test 挥发度的测定
volatility with steam 随水蒸气挥发性
volatilizable 可挥发的
volatilization 挥发(作用)
volatilization curve 蒸发曲线
volatilization loss 挥发损失
volatilization rate profile 挥发速率分布图
volatilization test 挥发度测定
volatilizer 挥发剂；挥发器
volatilizer-premix burner system 挥发器-预混式燃烧器装置
volatilizing roasting 挥发煅烧
volborthite 钙钒铜矿
volcanic 火山的
volcanic ash 火山灰石
volcanic gas 火山气体
volcanic glass 火山玻璃；松脂石
volcanic mud 火山泥
volcanics 火山岩；含硅硫黄；火山物质；辉石
volcanic tuff 火山凝灰岩
volemite(=volemitol) 庚七醇
volemitol(=volemite) 庚七醇 $CH_2OH(CHOH)_5CH_2OH$
Volhard method 福尔哈德法
Volhard's solution 福尔哈德溶液
Volhard's volumetric method 福尔哈德容量测卤法
volkonskoite 铬膨润石
volleyball leather 排球革
voloxidation 氧化挥发
voloxidator 氧化挥发器
voloxidizer 氧化挥发(反应)器
volt 伏(特)
volta cell 伏打电池
volta couple 伏打电偶
volta effect 伏打效应
volta furnace 伏打电炉

voltage ①电压②伏特数
voltage adjuster 调压器
voltage contrast image 电压衬度像法
voltage-current dual 电压-电流对偶
voltage-dependent inactivation 电压依赖性失活
voltage divider 分压器
voltage drift 电压漂移
voltage drop 电压降
voltage effect 电压效应
voltage fluctuation 电压波动
voltage multiplier 电压倍增器
voltage rating 额定电压
voltage regulator 调电压器
voltage regulator tube 稳压管；恒电压放电管
voltage-scanning coulometry 电压-扫描库仑法
voltage-scan voltammetry 电压扫描伏安法
voltage sensitivity 电压灵敏度
voltage stabilizer 稳压器
voltage step 电压阶跃*
voltage sweep 电压扫描*
voltage to current converter 电压电流转换器
voltage to frequency conversion system 电压频率转换系统
voltage to frequency converter 电压频率转换器
voltaic battery (伏打)蓄电池
voltaic cell 伏打电池*
voltaic couple 伏打电偶
voltaic electricity 伏打电流
voltaic element 伏特电池
voltaic pile(=galvanic pile) 电堆
voltaic series 伏打系列
voltaite 绿镁铁矾；绿钾铁矾
voltameter(=coulometer) 库仑计；电量计
voltammeter(=galvano-voltammeter) 伏安计
voltammetric analysis 伏安分析
voltammetric cell 伏安池
voltammetric indicator electrode 伏安指示电极
voltammetric titration 伏(特)安(培)滴定；电压电流滴定
voltammetry 伏安法
voltammogram 伏安图*
voltamograph 伏安极谱仪
voltamoscope 伏安器
volt-ampere 伏(特)-安(培)
volta potential 伏打电位；外电位
volta series 置换次序
voltate regulator 调压器
voltate-to-frequency conversion system 电压频率转换系统

voltmeter 伏特计；电压表
voltmeter panel 伏特计板
voltohmmeter 伏特欧姆计
voltolizing 无声放电处理(法)
voltol oil 电聚油；高压电油
voltol process 电聚过程；高电压处理过程
voltzite 锌乳石；肝锌矿
volucrisporin 间羟基对苯基醌
volume ①体积②容积③强度
volume additivity 体积相加性
volume buret 量管
volume capacity 容量
volume change 体积变化
volume charge density 体电荷密度
volume concentration 体积浓度
volume conductance 体积电导率
volume conductivity 单位体积导电率
volume contraction 体积收缩；容积收缩
volume coulometer 体积电量计
volume crystalline fraction 体积结晶分数；体积结晶百分率
volume crystallinity 体积结晶度
volume deformation 体积形变
volume density 体积密度；视密度
volume dilatation (=cubical dilatation) 体积膨胀
volume dilatometer 体积膨胀计
volume drying 整体干燥；成堆干燥
volume elasticity 体积弹性
volume expansion 体积膨胀
volume expansion factor 体积膨胀系数
volume factor 体积因子
volume flow injection analysis 容量流动注射分析
volume flow rate ①体积流速；容速②体积流量
volume formal 体积克式浓度
volume formality 体积克式浓度
volume fraction 体积分数
volume increase 体积膨胀
volume indicator 容积指示器；体积指示器
volume integral 体积积分
volume limit 容积极限
volume loading 增量性填充；按体积份填充
volume marker 体积标记器
volume meter 容量计；量器
volume modulus of elasticity 弹性的体积模数
volumenometer 体积计；视密度计
volume percent 体积百分比
volume percentage 容积百分比
volume phase 体相

volume pipet(=volume pipette) 刻度吸移管
volume pipette(=volume pipet) 刻度吸移管
volume polarization 体积极化
volume porosity 体积孔隙度
volume-pound 容量磅
volume-pressure coefficient 体积-压力系数
volume-pressure master curve 体积-压力叠合曲线
volume quenching effect 体猝灭效应
volume ratio ①体积比②容积比
volume reading ①体积读数②容积读数
volume recovery ①体积回收率②容积回收率
volume reduction ratio 减容比
volume relaxation 体积弛豫
volume resistivity 体积电阻系数；体积电阻率
volumescope (气体)体积计
volume selective spectroscopy 区域选择波谱学
volume specific resistance 体积电阻率
volume susceptibility 体积敏感度
volume swell(ing) 体积溶胀
volume tank 洗气罐
volume-temperature coefficient 体积-温度系数
volumeter 体积计
volume theory 体积理论
volume-time yield (设备的)容积时间产量；空时产量
volumetric 容量的；容积的
volumetric-actuated filler 容量定量式加料器
volumetrical(=volumetric) ①体积的②容量的；容积的
volumetric analysis 容量分析(法)；体积分析(法)
volumetric apparatus 容量仪器
volumetric behavior of condensate 凝结液的容量性质
volumetric calorific value 容积热量
volumetric capacity 容量
volumetric chromatography 体积色谱(法)
volumetric coefficient 容积系数
volumetric contraction in blending of hydrocarbons 烃类混合时的容积收缩
volumetric correction factor 容量校准因素
volumetric cylinder 量筒
volumetric determination 容量测定
volumetric displacement meter 容量置换计
volumetric efficiency 容积效率
volumetric factor 容量(分析计算用)因数
volumetric feed 容积计量喂入；定容供料
volumetric feeder 定容喂入器；定容供料装置
volumetric filler(=volumetric filling machine) 容量装填机
volumetric filling machine(=volumetric filler) 容量装填机

volumetric flask(=measuring flask)　(容)量瓶
volumetric flow meter　体积流量计
volumetric flow totalizer　容积式流量累加器
volume(tric) fraction　体积分数；体积部分
volumetric gas analysis　气体体积分析
volumetric glass　量器
volumetric glass ware　容量玻璃器
volumetric heat capacity　体积热容
volumetric method　体积法*
volumetric molar concentration　体积摩尔浓度
volumetric oxygen transfer coefficient　容积传氧系数；体积传氧系数
volumetric pipette　(容量)吸移管
volumetric precipitation method　沉淀滴定法
volumetric proportions　容积比
volumetric rate of reaction coefficient　容量反应速度系数
volumetric solution　滴定液；滴定(用)液
volumetric standard　容量(分析)标准
volumetric utensil　容量测定器具
volumetric ware　容量仪器
volumetry　容量分析(法)
volume-unaltered modified simplex　体积不变改进单纯形
volume viscoelasticity(=bulk viscoelasticity)　体积黏弹性
volume voltameter　体积电量计
volume weight　容重〔单位容量中松放物体的重量〕
volume work　体积功
volume yield　体积得率
voluminal　①体积的②容量的
voluminal expansion　体积膨胀
voluminal resilience　体积回弹
voluminosimeter　体积仪
voluminosity(=rheological voluminosity)　容积度
voluminosity constant　容积(度)常数
voluminous　①容积大的②卷帙繁多的
volute　螺旋形；集气环
volute casing　蜗形机壳
volute chamber　环流室
volute collector　蜗壳式收集器；集气蜗壳
volute diffuser　蜗壳扩压器
volute pump　蜗壳泵
vomicin　吐素
vomicine　呕吐素；番木鳖次碱
vomic-nut　呕吐果；马钱子
vomipyrine　5-异丙基-7-甲基-7H-吡咯并[2,3-h]喹啉
vonedrine　汪尼君；仿泪准(药)；盐酸苯丙甲胺
von Muller's indicator(=tropeolin OOO)　金莲橙 OOO；酸性橙
von Richter reaction(=cine substitution)　变位取代；冯里席特反应〔取代基不在被代基位上〕
voracevine(=protocevine)　藜芦瑟文
vortex　旋涡；涡流
vortex agitator　旋涡搅拌器
vortex breaker　(容器底部流出口处)防涡器
vortexes〔复〕　涡旋漏斗
vortex field　旋涡场
vortex flow　涡流
vortex grit washer　涡流洗砂器
vortex lattice　涡旋点阵；涡串；涡栅
vortex line　旋涡线
vortex motion　涡流运动
vortex path　涡流轨迹
vortex pattern　涡流型；涡纹图案
vortex pump　涡动泵
vortex reactor　涡动反应器
vortex ring　涡(流)环
vortex shedding flowmeter　涡流流量计；旋涡流量计
vortex street　涡街
vortex tail　旋涡尾；尾涡流
vortex theory　涡流理论
vortex time　涡旋时间
vortex trail　旋涡尾
vortex trunk　旋涡螺丝
vortex tube　涡流管
vortex value　涡动值
vortex wake(=vortex tail)　旋涡尾；尾涡；尾涡流
vortices〔复〕(=vortexes)　涡旋漏斗
vorticity　涡动性
vorticity in isotropic turbulence　各向同性湍流的涡量
vorticity tensor　涡动性张量
Vortmann-Metzl's method　沃尔曼特-梅茨尔法〔测锑〕
Vosmaer ozonizer　伏斯梅尔臭氧器
votator　同心双管热交换器
Votator crystallizer　孚推多结晶器
V-packing　V 型轴封
vrbaite　硫砷锑铊矿
V-R relaxation　振动-转动弛豫
V-shaped gouge　V 形尖口凿
V-T relaxation　振动-平动弛豫
vug　小空窝；晶簇
vug crystal　窝内结晶
vulcacite　乌卡赛特〔硫化促进剂〕
vulcafor　乌卡福〔硫化促进剂〕
vulcalock process　异构化增强橡胶作业法
vulcameter　硫化仪
vulcamycin(=novobiocin)　新生霉素
vulcan fast red B　硫化坚牢红 B

vulcanite 硫化橡胶；硬质橡胶
vulcanizable plasticizer 可硫化的增塑剂
vulcanize 硫化橡胶；橡皮
vulcanization ①硫化②硬化
vulcanization accelerator 硫化促进剂
vulcanization activator 硫化活化(性)剂*
vulcanization aid 硫化助剂
vulcanization bond 硫化结合
vulcanization characteristic 硫化特性
vulcanization coefficient 硫化系数
vulcanization curve 硫化曲线
vulcanization cycle controller 硫化时序控制器
vulcanization leveller 硫化调节剂；硫化活性剂
vulcanization process 硫化工序
vulcanization rate 硫化速度
vulcanization reaction 硫化反应
vulcanization relationship 硫化关系曲线
vulcanization retarder 硫化延迟剂*
vulcanization system 硫化体系
vulcanizator 硫化剂
vulcanized ①硫化的②硬化的
vulcanized article 硫化成品
vulcanized asbestos 夹胶石棉制品〔如制动片〕
vulcanized bitumen 硬化沥青
vulcanized fatty oil 油膏〔软化剂、增量剂〕
vulcanized fiber 硬化纸板
vulcanized fiber paper 硬化纸
vulcanized latex 硫化橡浆
vulcanized oil 硫化油
vulcanized paper 氯化锌硬化的纸料

vulcanized rubber 硫化橡胶；橡皮
vulcanizer ①硫化罐②硫化机③硫化剂
vulcanizing 硫化的
vulcanizing agent 硫化剂
vulcanizing apparatus 硫化器
vulcanizing boiler 硫化罐
vulcanizing chamber 硫化室
vulcanizing group 硫化基团
vulcanizing heater ①加热硫化机②加热硫化罐
vulcanizing ingredient 硫化剂；硫化加入的成分
vulcanizing pan 硫化锅
vulcanizing plant 加硫设备；硫化装置
vulcanizing press 加压硫化机
vulcanizing properties 硫化性质
vulcan oil 硫化油；加硫动植物油
vulcanosol 乌卡洛塞尔〔混炼胶着色剂〕
vulcan powder 伏尔甘炸药〔30%硝化甘油，52.5% $NaNO_3$, 10.5%木炭，7%硫黄〕
vulex 硫化橡浆
vulkacit 乌卡西特〔硫化促进剂〕
vulkameter 硫化仪
vulnerability ①易损性；脆弱性②致命性；要害③弱点
vulnerable polymer 易损聚合物
vulneraria 创伤药
vulnerary ①疗创药②疗创的
vulpinic acid 枕酸甲酯
vulpinite 鳞硬石膏
vultex 硫化胶乳；硫化橡浆
V-value V 值；光学玻璃色散值
V-V relaxation 振动-振动弛豫

W

wabain 假虎刺苷
wacke 玄武土；玄武石
Wacker process 瓦克尔法〔乙烯直接氧化制乙醛法〕
WAC oxidation WAC 氧化试验
wad ①锰土②小块；小束③填絮④作成小团
wad clay 锰土
wadding (填)絮；填塞物
wadding filter 填塞过滤器；填絮过滤器
wade's balsam 复方安息香酊
Wadsworth prism system 华兹沃思棱镜系统
Waelz process 韦尔茨冶锌法
wafer ①片②(封信用的)干胶片③糯米纸
wafer ash 忽布根皮
wafer method 压片法
wagging (左右)摆动
wagging vibration 非平面摇摆振动
wagnerite 磷镁石
Wagner-Meerwein rearrangement 瓦-米重排作用
Wagner's reagent 瓦格纳试剂；碘-碘化钾溶液
wagofo 大毒芨
wagon retort 车(式)干馏釜；车(式)干馏炉
waist-belt leather 腰带革
wake 尾流；尾迹
WAK reprocessing plant (西德)卡尔斯鲁厄核燃料后处理厂
Walden inversion 瓦尔登反转*
Walden reductor 瓦尔登还原器；银还原器
wald oil 水梨子油
wale 线圈纵行
wale mark 凸痕
wall (器)壁
Wallace belt flexing machine 华莱士条形试样屈挠试验机
Wallaston prism 渥拉斯顿棱镜
wallboard ①壁板②建筑纸板
wallboard strip ①壁板压缝板条②壁板接缝胶带
wall coated open tubular column (WCOT) 涂壁空心柱
wall coating 墙壁漆
wall cooling 管壁冷却
wall correction factor (落球黏度计的)壁面校正因子
wall covering fabric 贴用布；壁布
wall effect ①壁效应*②管壁效应*
wallet 皮夹子
wall flow 壁流〔填料塔〕
wallflower concrete 罗兰花浸膏
wallflower-seed oil 罗兰花子油
wall friction 壁面摩擦
wall-insulation 墙壁保温
wall layer 壁层
wall paint 墙漆
wall paper 糊墙纸
wallpaper lacquer 墙纸漆
wall plaster 刷墙粉
wall primer 墙壁(打)底漆
wall radiant tubes 墙壁辐射管；壁暖气管
wall reaction 器壁反应
wall resistance 管壁热阻
wall rubber 墙橡胶
wall saltpetre 墙硝〔石墙上风化了的硝酸钙〕
wall sensor 壁式传感器
wall temperature 壁温度
wall thickness 壁厚
wall tile 墙面贴砖；面砖
wall tile press 瓷砖机
wall-treated open tubular column 壁处理开管柱
wall-type pyrometer 壁式高温计
wall white 吐渣
walnut 核桃；胡桃
walnut fatty alcohol 核桃脂肪醇
walnut oil 核桃油
walnut tree 胡桃
Walpole's buffer solution 沃波尔缓冲溶液
walrus hide 海象皮
Walsh diagram 沃尔什图
Walsrode powder 沃耳斯罗德(猎枪)火药
warbled hide 虻疮皮〔皮革〕
warble fly pest 虻害〔皮革〕
warble hole 虻疮孔〔皮革〕
warble lump 虻疮核〔皮革〕
Warburg finite diffusion process 沃伯格有限扩散过程
Warburg impedance 沃伯格阻抗
Warburg manometric apparatus 沃(伯)氏测压器
ware ①器皿②仪器
warehouse 仓库
warehouse space 仓库容积；堆栈空间
warehousing 入库；仓库业
warfare gas (战用)毒气
warfarin 杀鼠灵；华法令〔农药〕

war gas (战用)毒气
Waring blender 韦林氏搀合器
waring of cloth 布的耐磨性
warm air pig 热风生铁
warm color 暖色
warmer 加热辊
warm forging 热锻塑
warming 加热；热炼
warming mill 加热辊
warming-up 预热；加热
warm-operation 温试验运转〔中等放射性水平〕
warm-setting adhesive(=intermediate temperature adhesive) 中温硬化黏合剂
warm sludge 热泥；热渣
warm-solution tank 热溶液罐
warm sweat system 热发汗法
warm table 加热台
warm up 加热
warm-up temperature control 升温控制
warm-up time 预热时间
warm vat 温染槽
warm wash 热清洗
warm water 温水
Warner's kneader 瓦纳式捏合机
warning agent 警告剂
warning color 警告(颜)色
warpage 翘曲；扭曲
warp and filling 经纬线
warp and weft 经纬线
warp and woof 经纬线
warp density 经密；经纱密度
warp direction 经向
warped ①翘曲的②反卷的〔纺〕
warping 翘曲
warping-sizing machine 经浆联合机
warp knitted piezo-jacquard system 经编压电陶瓷提花装置
warp knitted two-directional elastic fabric 经编双向弹力织物
warp resistance 抗挠曲；耐扭曲
warp streak 反卷条花；经向条花〔纺〕
warp tear resistance 经向抗撕裂性
warp thread 经线
warp tying machine 经纱接头机；经纱打结机
warp-yarn 经纱
warrant 煤层下耐火泥
warrenite 针硫铅锑矿
wart 疣；瘤肿

war up mill(=warmer) 加热辊
warwickite 硼酸铁矿
wash ①洗(涤)②洗涤物③积胶面〔增强塑料〕
washability 可洗净性；耐洗刷性
washability of coal 煤的可洗性
washability tester 可洗性试验机
washable chamois leather 可洗麂皮
washable distemper ①(=washable water paint)耐擦洗水性漆②耐擦洗的胶性刷墙粉〔含水胶、酪素和乳化干性油〕
washable suede 耐洗绒面革
wash and wear fabrics 耐洗不皱织物；洗可穿织物〔免熨〕
washboard(=baseboard) ①护壁板；墙裙②踢脚板
wash bottle 洗瓶
wash box 洗涤箱
Washburn cell 瓦什伯恩测电导率电池
wash channel 洗涤孔道
wash coat 修补基面涂层
wash column 洗涤柱
wash drum 洗鼓
washed 洗(过)的；洗(净)的
washed clay 淘洗黏土；澄出黏土
washed coal 洗(净)煤
washed kaoline 洗陶土
washed metal 洗铁
washed ore 洗矿；精砂；精选矿
washed precipitate 洗过的沉淀
washed slack 洗净的碎煤
washed sulfur 洗过的硫
washer ①洗涤器；洗涤机②垫圈
washer box 洗涤箱
washery ①洗(涤)②洗涤间
washery coal 洗(净)煤
washery refuse 洗渣；洗涤废物；选矿渣
washes 洗涤物；洗涤废水；洗刷剂
wash fast 耐洗涤的
washfastness 耐洗性
wash filtrate vacuum receiver 洗涤滤液真空接受器
wash house 洗涤间
washing 洗(涤)
washing agent 洗涤剂
washing assistant 助洗剂；洗涤助剂
washing bath 水洗槽；洗涤浴
washing beater 洗涤打浆机
washing beck 洗涤桶
washing bottle 洗(涤)瓶
washing box 洗涤箱
washing chamber 冲洗(洗涤)室

washing chemistry　洗涤化学
washing column　洗涤柱
washing composition　洗涤剂
wash(ing) cycle　洗涤周期
washing cylinder　洗筒
washing device　洗涤器；洗涤装置
washing dolly　洗呢机
washing down　冲洗；冲刷
washing drum　洗鼓
washing fastness　耐洗性
washing fastness tester　水洗牢度机
washing filter　洗滤器
washing liquid　洗液
washing liquor　洗液
washing loss　洗涤损失
washing lye　洗用碱水
washing machine　洗涤机
washing mill　洗涤碾磨机
washing nozzle　清洗喷嘴
washing off　洗去；冲去
washing oil　洗涤油；吸收油
washing out　洗去；冲去
washing out method　洗出法；清洗法
washing powder　洗涤粉
washing power　洗涤本领
washing press　洗涤式压滤机
washing rate　淘洗率
washings　洗液
washing section　洗涤段
washing soda(=sal soda)　晶碱；洗涤碱；十水(合)碳酸钠
washing tank　洗涤槽
washing test　洗涤试验
washing tower　洗涤塔
washing trommel　洗涤转筒筛
washing-up　①洗涤；冲刷②墨辊清洗
washing water　洗水；洗液
wash leather　洗革；麂革
wash liquid(=wash liquor)　洗液
wash liquor(=wash liquid)　洗液
wash mill　淘泥机
wash off　洗去
wash oil　洗油；吸收油
washout　冲洗；冲净；清洗
wash-out hole　清洗孔
wash-out pipette　洗出移液管
wash pipe　冲水管
wash plate　挡水板
wash primer (=metal conditioner)　磷化底漆；洗涤底漆

wash-resistant　耐洗的
wash room　洗料室
wash still　初馏釜；酒精蒸馏锅
wash tank　洗涤槽
wash temperature　洗涤温度
wash thinner　洗涤用稀释剂〔洗去油漆用〕
wash time　洗涤时间
wash trommel　洗涤转筒筛
wash trough　洗槽
wash water　洗水；洗余水；洗液
wash-water settling tank　洗水澄清槽
wash-wear　耐洗
wash-wheel　洗皮转鼓
was moulding　蜡成型；蜡铸模
waste　①废物；废料；废(产)品②耗损③回丝〔纺〕
waste acid　废酸
waste alkali　废碱液
waste back-cycling　废液返回(循环)
waste burial ground　废物埋藏地
waste calcination facility(WCF)　放射性废液煅烧设施
waste coal　废煤
waste coal product　废煤产品
waste crude oil　废原油
waste disposal　废物处理
waste encapsulation and storage facility(WESF)　放射性废液封装储存设施
waste fibre　下脚纤维
waste fuel　废燃料；燃料渣
waste gas　废气
waste-gas burning　废气燃烧
waste-gas cleaning　废气净化
waste-gas desulfurization　废气脱硫
waste gas feed heater　用废气的加热器
waste gate　废料排出门
waste graveyard(=burial ground)　废物埋藏场；废物处置场所
waste grinder　废胶磨
waste grinding　碾废废胶
waste heat　废热；余热
waste heat boiler　废热锅炉
waste heat engine　废热发动机
waste heat exchanger　废热交换器
waste-heat flue　烟道；废热道
waste heating　废热加热；用废气加热
waste-heat oven　废热炉；余热炉
waste heat recovery　废热收回
waste immobilization plant　废物固化工厂
waste incinerator　废物焚烧炉

waste liquor 废液
waste lye 废碱液
waste oil 废油；用过的油
waste pack lubrication 废料填充润滑法
waste pack oil cup 废料填充油杯
waste paper 废纸；纸屑
waste pipe 废水管
waste processing 废物处理
waste product 废产物
waster 废铸品
waste reclamation 废物回收利用
waste residue 残渣
waste rubber 废橡皮；旧橡皮
waste solidification engineering prototype(WSEP) 放射性废液固化工程原型设备
waste stock 废料
waste strength 废物污染度
waste sulfite liquor(=waste sulfite lye) 亚硫酸盐废液
waste sulfite lye(=waste sulfite liquor) 亚硫酸盐废液
waste trap 废液分离器
waste treatment 废物处理
waste valve 废料排出阀
waste vault 废物库；废物窖
waste water 废水
waste-water emulsion 废水乳状液
waste water lagoon 废水氧化塘
waste water reclamation 废水回收
waste-water separator 废水分离器
waste water treatment 废水处理
waste wood 废木料
waste wool 废(羊)毛
watch dog 信号器
watch-glass (=watch crystal) 表面皿*；表(面)玻璃
watch glass clamp 表面玻璃夹
watch-glass test 表面玻璃试验〔测定油的干性或汽油中胶质〕
watching gauge glass 计液玻管；看水玻璃管
watch maker's oil(=watch oil) 钟表油
watch oil(=watch maker's oil) 钟表油
water ①水②浇(水)；洒水③冲淡；掺水
water absorbability 吸水性；吸湿性
water absorbent polymer 吸水(性)聚合物
water absorbing agent 水吸收剂；吸水剂
water-absorbing anionic polyelectrolyte 吸水性阴离子型聚电解质
water-absorbing polymer 吸水性高分子
water absorption ①吸水性②吸水率
water absorption ability 吸水能力

water absorption equilibrium 平均吸水率
water absorption test 吸水性试验
water absorption tube 吸水管
water affinity 亲水性
water analysis 水分析
water and oil-repellent agent 防水防油整理剂
water and sediment content 水及沉渣含量
water and sediment test 水及沉渣(含量)测定
water and soil conservation 水土保持
water and soil erosion 水土流失
water-anolone mixture 水-(环己)醇(环己)酮混合物
water aspirator 水泵
water avens root oil 丁字根油
water back ore hearth 水套冶矿炉
water bag ①硫化室；煮沸室②水袋
water balance ①水平衡②水量结算
water balance equation 水量结算方程式
water ballasting(=water bottom) 用水压舱
water barrier 防水层；防水材料
water-based adhesive 水基黏合剂；水性胶黏剂
water-based latex paint 水基胶乳漆
water based paint 水基涂料；水溶性漆料
water-based poultice 水基膏糊；水性膏糊
water-based rubbing filler 水性打磨填孔剂
water-based stoving paint 水基烘漆；水性烤漆
waterbased urethane 水基聚氨酯；水性聚氨酯
water base foam 水基泡沫
water bath ①水浴②水浴器③热水锅
water-bath battery 水浴组
water bath with regulator 带调节器的水浴
water-bearing concrete 含水混凝土
water-bearing layer 含水层
water blast 水流鼓风器
water blasting(=wet blast cleaning) ①(喷)水除锈；高压水除锈法②湿(法)喷砂除锈(法)
water bleeding from grease 润滑脂渗出的水
water blue 水溶蓝；酸性水溶青
water blush (ing) 水致发白〔漆病〕
water body 水体；水域
water body sediment 水体沉积物
water-borne adhesive 水基胶黏剂
water-borne coating 水性涂料
water-borne paint 水性漆；水基漆
water-borne preservative 水性防腐剂
water-borne radioactivity 水中放射性
water bosh 水封
water bosh generator(=water bosh producer) 水封式(气体)发生器

water bosh producer (=water bosh generator)　水封式(气体)发生器
water bottle　采水器
water bottom(=water ballasting)　用水压舱
water break　水膜残迹〔洗涤餐具、金属等硬表面物件，因效果不良，经漂清后在其外表上残存有小水珠的现象〕
water-break-free　(冲洗)水覆盖能力；水膜不破
water break-free surface　无水膜残迹表面
water budget　水量收支水分平衡
water calorimeter　水量热器
water capacity of bottle　比重瓶的水容量
water capacity of cooling system　冷却系统的水容量
water catchments　储水池
water cellulose　充水纤维素
water cement　水凝水泥
water cement ratio　水-水泥比；水灰比
water chestnut starch　马蹄粉；荸荠粉
water-circulating pipe　水循环管
water-circulating pump　水循环泵
water classification process　水力分级工艺(法)
water cleanser　水性净化剂；水性清洁剂
water collar　水冷夹套；水冷套
water collecting sump　蓄水坑
water color　水彩；水合颜料
water column　水柱
water column pressure　水柱压力
water compatibility of grease　润滑脂的调水性
water complex　水配合物
water concentration　湿选；水选；水力富集
water condensation　水冷凝
water condenser　水冷凝器
water conditioner　水处理剂；水质调节剂
water conditioning　水处理
water conditioning chemicals　水处理用化学品
water conservation　水源保护
water constants　水(的)常数
water contamination　水污染
water content　含水率；水分
water cooked reclaim　水煮回收〔胶〕
water cooled　水冷的；用水冷却的
water cooled condenser　水冷式冷凝器
water cooled cylinder　水冷式汽缸
water cooled rolls〔复〕　水冷辊
water-cooled torch tube　水冷炬管
water-cooled welding torch　水冷焊炬
water cooler　水冷(却)器；冷却器
water cooling　水冷却

water cooling pond　冷水池
water cooling tank　水冷却槽
water cooling tower　①水冷却塔②凉水塔
water coulometer　水解电量计
water-coverage test　水面遮盖力试验〔铝粉漂浮性遮盖力的检验方法〕
water-cress oil　水芹油
water crossing　越水；穿越水域障碍〔装设水管〕
water cure　①(热)水硫化②(热)水固化
water cured concrete　水熟混凝土；水处理过的混凝土
water curing　①水处理②热水硫化
water cushion　水垫
water-cyclohexane azeotrope　水-环己烷共沸(混合)物
water cylinder　①水缸②水筒
water demineralizing　水去矿物质；水软化
water deprivation　脱水(作用)
water desalination　水(的)脱盐(作用)
water digestion process　中性水油(再生)法
water digestion reclaim　水煮回收
water-dilutable coating　水稀释性涂料
water dip lacquer　带水(工件)浸涂漆；湿表面浸涂漆
water dispersibility method　水分散方法〔测表面活性剂的 HLB 值〕
water dispersible coating　水分散性涂料
water-dispersible hammer finish　水分散性锤纹漆
water dispersible oil　水分散性油
water dispersible zinc stearate　水分散性硬脂酸锌
water dispersion　①水分散法②水分散体
water displacement　(防锈油的)取代(排)水性能
water displacement test　排代水试验〔喷砂钢板在涂漆前检验防锈性的方法〕
water displacing chemicals　排代(取代)水的化学(药)品
water distillation　水的蒸馏
water distilling apparatus　水蒸馏器；蒸馏装置
water divide　分水岭；分水线；分水界
water divining　地下水源判断；地下水源探索
water dock　水酸模
water drain　①下水道②排水
water drip cooler　水淋冷却器；喷淋式冷却器
water drip test　滴水(腐蚀)试验
water driven centrifuge　水动离心机
water dropper　滴水器
water drop test　水滴试验
water-drying　水干；用水排代有机溶剂
water economizer　省水器
watered　加水的
watered column　充水塔；用水将油气赶尽的塔
watered silk sheen effect　(波纹墙纸的)波纹绸丝光效应

water electrode 水(成)电极
water electrolyzer 水电解器；水电解槽
water eliminator 挡水板；除水器
water emulsifiable 水溶性乳化的
water emulsifiable oil 水溶性乳化油
water emulsifiable paste 水溶性乳化脂膏
water emulsion 水乳胶；水(外相)乳状液
water emulsion inhibitor 水乳化抑制剂
water emulsion sludge 水乳化淤渣
water end efficiency 水缸效率
water environmental capacity of pollutant 污染物水环境容量
water equivalent 水当量
water-eroded alluvial gravel 水蚀冲积砾石
water examination 水质检查
water-extended polyester(WEP) 水扩张聚酯
water extract 水抽出物
water extractable content 水(可)提取物含量
water extractor 脱水机
water factor 水因数
water failure 停水事故
water fennel 水芹
water fennel oil 水芹油〔取自 Oenanthe phellandrium〕
water filler (木屑)水性填孔剂
water film coefficient 水膜系数
water filter 滤水器
water-finder 试水器；底部取样器
water-finding paper 试水纸
water-finding paste 试水糊
water finish 水纹面饰；水涂饰剂
water floation 水选；淘析
water flooding 水驱法；水攻法〔采油〕
water flooding additive 水驱添加剂；水攻添加剂
water flooding agent 水驱剂；水攻剂
water flood petroleum recovery 注水采油法
water-flow calorimeter 水流量热器
water flowers 水花
water fog 水雾
water fog test 水雾试验
water-fractionated clay 淘选陶土
water fractionated kaolin 水选高岭土；淘选高岭土
water-free 无水的
water from diffusion 渗滤水
water funnel 水漏斗
water gas 水煤气
water-gas catalyst 水煤气催化剂
water gas cell 水煤气电池
water gas condenser 水煤气冷凝器

water gas generator (=water-gas machine) 水煤气发生器
water-gas machine (=water-gas generator) 水煤气发生器
water gas pipe 水煤气管
water-gas pitch 水煤气沥青
water-gas process 水煤气过程
water-gas producer 水煤气发生器
water-gas reaction 水煤气反应
water-gas set 水煤气成套设备
water-gas shift 水煤气轮换
water-gas tar 水煤气焦油
water-gas tar emulsion 水煤气焦油乳状液
water-gas tar pitch 水煤气沥青
water gauge 水表
water gauge cock 水表旋塞
water glass 水玻璃；硅酸钠
water-glass color (=water-glass paint) 水玻璃颜料；硅酸盐颜料
water-glass enamel (=water-glass paint) 水玻璃颜料；硅酸盐颜料
water-glass paint(=water-glass color; water-glass enamel) 水玻璃颜料；硅酸盐颜料
water globules 水珠；水滴
water-ground 湿法粉碎；水磨(的)
water-ground limestone 水磨碳酸钙；重质碳酸钙
water ground mica 水磨云母
water hammer 水锤；水击作用
water hammering 水锤击；水击作用
water hammer in pipe line 管线中水锤
water hardening 水淬硬化
water hardness 水(的)硬度
water hardness ions 水硬离子
water head 水头；水位差
water-holding capacity 持水容量
water-holding power 持水能力
water horse power 有效功率；水马力
water hose 水龙带；输水胶管
water imbibition 水吸胀(作用)
water-imbibition value 吸水指标
water immersion 水浸
water immersion test 浸水试验〔过滤材料〕
water-immiscible medium 水不混溶的介质
water impermeability 不透水性
water-indicating paper(=waterfinding paper) 试水纸
watering 灌水；水洗；加水；给水；洒水
watering of tank 油罐的灌水〔赶出罐中气体〕
water-initiated polymerization 水引发聚合(作用)
water in oil 油包水；水在油中
water in oil emulsion 油包水乳状液*

water-in-oil test　油中水含量试验
water in oil type emulsion　油包水型乳浊液
water-insolubilized　(使)不溶于水的
water insoluble　水不溶性的
water-insoluble gum inhibitor　水不溶性胶质抑制剂
water insoluble inorganic substance　水不溶性无机物
water interferometer　水干涉计
water intoxication　水中毒
water jacket　①水夹套②外槽〔浸渍槽〕
water jacketed　水套的
water-jacketed condenser　水套冷凝器
water jacketed producer　水套(气体)发生器
water jacket furnace　水套(鼓风)炉；水套冶炼炉
water jet　水注〔物〕
water jet air pump　喷水空气泵
water jet aspirator　吸水泵
water jet condenser　喷水冷凝器
water jet pump　喷水泵
water jet scrubber　喷水洗涤器
water jet strip　水针板
water ladle　水杓
water layer　水层
water leaching　水浸沥
water leaf paper　手工水纸
waterless　无水的
waterless block grease　无水块状脂
waterless gasholder(=waterless holder)　无水储气器；干式储气柜
waterless holder(=waterless gasholder)　无水储气器；干式储气柜
water level　水平(面)；水准
water level indicator　水位指示器；水准指示器
water level regulator　水准调节器；水平面调节器
water-like　①亲水②似水
waterlike solvent　类水溶剂
water line　水线；供水管线；供水管路
water-line paint(=boot-topping paint)　水线涂料；水线漆
water loss　①失水量(率)②耗水量
water lute　水封；液封
water luted producer　水封发生器
water mangle　(压力)脱水机
Waterman method　华特曼(环分析)法
Waterman ring analysis　华特曼环分析〔石油的结构族组成分析〕
water mark　①水印②水线标志；水位标志
water measuring tank　量水槽
watermelon essence　西瓜香精
watermelon flavour　西瓜香精

water-melon oil　西瓜油
water metabolism　水代谢
water meter　水量计
water mill　水磨；水力磨粉机
water milling　水缩绒；水磨
water mint　水薄荷
water miscibility　水混溶性
water-miscible paint　水性漆
water-miscible solvent　水混溶性溶剂
water-mix filler　水调(干粉)填孔料；水调和填孔料
water-mixing nozzles　水混合喷嘴〔洗气时用〕
water of condensation　冷凝水
water of constituent　化合水；结构水
water of constitution　化合水；结构水
water of crystallization　结晶水
water of decomposition　分解水
water of hydration　①结合水；水化水②结晶水
water of hygroscopicity　吸湿水
water of plasticity　塑性水；增塑水
water-oil ratio　水油比
water-oil separator　水油分离器
water oven　(热)水烘箱
water paint　水性涂料
water pearl ash glass　水珠灰玻璃；硅酸钾玻璃；水钾玻璃
water penetration　透水性；水渗透
water penetration tester for leather　皮革透水性测定仪
water pepper　水蓼；辣蓼
water-pepper oil　蓼油
water phase　水相
water phase modifier　水相调节剂
water pipe　水管
water pollutant　水污染物
water pollution　水污染
water pollution chemistry　水污染化学
water pollution control　防止水质污染；水(质)污染控制
water pollution control engineering　水污染防治工程
water pollution control law　水质污染控制法；水质污染管理法
water pollution index　水(质)污染指数
water pollution monitor　水质污染监测
water power　水力〔机工〕
water-processed clay　湿法黏土；水选黏土
water-proof　防水的；耐水(性)的；不透水的；绝湿的
water-proof abrasive cloth　耐水砂布
water-proof abrasive paper　(耐)水砂纸
water-proof cement　防水水泥
water-proof cloth　防水布

water-proof compound　防水剂
water-proof concrete　防水混凝土
water proofer　防水剂
water-proof finish　①防水漆②防水饰面
water-proof grease(=water-repellent grease; water-resistant grease)　防水脂
water-proofing　防水的；不透水的
water-proofing additive　防水剂
water-proofing agent　防水剂
water-proofing masonry finish　防水圬工面漆；砖石结构防水面漆
water-proofing powder　防水粉
water-proof laminate　防水叠层
water-proof membrane　防水(隔)膜
water proofness　耐水性
water-proof sand paper　(耐)水砂纸
water-proof tannage　防水鞣法
water pump　水泵；抽水机
water-pump grease　水泵润滑脂
water pumps　水泵
water purificant　水净化剂
water purification　水净化；净水洗
water purification plant　净水厂
water purification station　净水站
water purification unit　净水设备
water purifier　净水器
water purifying　①净水(操作)②净水的
water quality analysis　水质分析
water quality analyzer　水质分析仪；水质计
water quality assessment index　水质评价指数
water quality automatic monitoring system　水质自动监测系统
water quality continuous and automatic monitoring system　水质连续自动监测系统
water quality criteria　水质标准
water quality index　水质指标
water quality model　水质模式
water quality monitoring　水质监测
water quality monitoring ship　水质监测船
water-quencher　水急冷器；水骤冷器
water-quench film　水骤冷(成型)薄膜
water quenching　水淬火
water ratio　含水率；水灰比
water reactive drier　水反应性催干剂
water-receptive　亲水的；吸水的
water recirculator　循环水冷却器；水循环器
water reclamation　水回收
water recovery　水回收
water-recovery apparatus　水回收设备
water reducer　减水剂
water-reducible　水稀释的；水稀释性的
water-reducible coating　①水稀释性涂料②(广义)水基涂料〔包括乳胶漆和水溶性树脂漆〕
water regain　吸水量；得水率
water-removal additive　脱水添加剂；除水剂
water-repellancy　防水性；拒水性；拨水性〔纺〕
water-repellent　防水(的)；拒水(的)
water repellent agent　抗水剂*
water-repellent grease(=water-proof grease)　防水脂
water-repellent silicone　防水性有机硅树脂
water repellent surface　斥水性表面；疏水性表面
water repellent wood preservative　疏水性木材防腐剂
water resistance　①防水性②防水度
water-resistant adhesive　耐水胶黏剂
water-resistant grease(=water-proof grease)　防水脂
water resisting　抗水的
water-resisting molding powder　抗水(性)塑(料)粉
water resources conservation　水资源保护
water-retaining capacity　保水量；吸水能力
water retention　保水作用；保水率
water-retention agent　保水助剂
water-retention value　保水值；含水值
water retentivity　保水性；水分保持性
water reticulation network　水网系统
water retted flax　水浸亚麻
water retting　水浸渍
water reuse　水再生；水再利用
water rheostat　水变阻器〔物〕
water-rich phase　富水相
water riding of tanks　油罐的浮水运送
water-rinsable paint remover　水刷洗型脱漆剂
water rinse　水漂洗
water roll　水辊
water sample bottle　采水瓶
water sampler　水取样器；采水器
water sampling　水采样
water scavenging agent　水净化剂
water scrubber　水洗器
water seal　水封
water seal arrangement　水封装置
water-sealed　防水的；隔水的
water-sealed gas holder　水封储气罐
water sealed gas meter　水封气表
water sealed producer　水封发生器
water-sealed tank　水封罐
water segregator　分水器

water sensitivity　水敏性
water separation index　水分离指数
water separator　脱水器
water shield gum　莼菜胶
water slinger　甩水器
water soda ash glass　水苏打灰玻璃；硅酸钠玻璃
water soda glass　水苏打玻璃；硅酸钠玻璃
water softener　①软水剂②软水器
water softening　水软化(作用)
water-softening agent　软水剂；水软化剂
water-softening chemicals　软水剂；软水化合物
water-softening power　软水能力
water-solubility　水溶性
water soluble　水溶(性)的
water soluble alkyd resin　水溶性醇酸树脂
water soluble amino resin　水溶性氨基树脂
water-soluble baking paint　水溶性烘漆
water-soluble catalyst　水溶(性)催化剂
water-soluble cellulose ether　水溶性纤维素醚
water-soluble film　水溶性薄膜
water-soluble grease　水溶性脂
water soluble group　水溶性基团
water-soluble gum　水溶性树胶
water-soluble lubricant　水溶性润滑剂
water-soluble oil　水溶性油
water-soluble paint　水溶性涂料
water-soluble paste　水溶性浆料(浆糊)
water soluble phenol resin　水溶性酚醛树脂
water soluble plant food　水溶性植物养料
water soluble polymer　水溶性高分子*
water soluble resin　水溶(性)树脂
water solubles　水溶物
water-soluble vehicle　水溶性漆料
water solution　水溶液
water solution of oil　油水乳化液
water sorption　①吸水作用②吸水率
water spot　水渍
water spotting(=rain spotting)　水渍；水斑；雨斑〔漆病〕
water spray　洒水；喷水；注水
water spray chamber　喷水箱；水雾箱
water spray condenser　洒水冷凝器
water-sprayed tank　洒水散热的油罐
water stain　水性着色剂
water stop joint　防水缝
water strainer　滤水网；滤水器
water suction pump　吸水泵
water supply　水供给
water supply pump　供水泵；冷水泵

water tank　水罐；水瓮；水桶；储水箱
water tap　(放)水龙头
water-tar emulsion　水-焦油乳浊液
water terminal　码头；港埠
water test　水压试验
water tex　水性可塑拉毛漆
water thinnable　水稀释(性)的；水冲淡性的
water thinnable paint　水稀释涂料
water-thinned masonry paint　水稀释性圬工漆；水稀释性砖石建筑漆
water thinned paint　水释涂料；水稀释漆
water-thinned stopper　水稀释性填孔剂
water thinned varnish(=aqueous varnish)　水性漆；水稀释性清漆
water-thinned vehicle　水稀释性漆料
water-tight　不透水的；不漏水的
water tightness　不透水(性)
water-tight seal　不透水密封
water titration (method)　水滴定法〔测水数〕
water to carbide acetylene generator　"加水到电石"式乙炔发生器
water tolerance　耐水性(或度)
water tolerance of fuel　燃料的耐水性
water tolerance test　耐水性试验
water-top tank　顶部盖水油罐
water trap　脱水器
water treatment　水处理；水的净化
water treatment plant　水处理厂
water treatment station　净水站；水处理站
water trough　水沟；水槽
water tube boiler　水管锅炉
water tube condenser　水管冷凝器
water tube gas condenser　水管气体冷凝器
water tunnel　水隧道
water turbine　水力涡轮机
water turbulence　水紊流；水涡流
water-type sludge　水型淤渣
water uptake　水的吸收
water value　水值
water vapor　水蒸气；水汽
water vapor distillation　水蒸气蒸馏
water vapor permeability　透湿性；透蒸气性
water vapor pressure　水蒸气压
water-vapor test　水蒸气试验〔如天然气中的〕
water vapor transmission　水蒸气透过性；透湿性
water vapour permeability　透湿性
water vein　水脉
water vulcanization　(过)热水硫化

water wash column　水洗塔
water washdown　水冲失
water washed soft clay　水漂软质瓷土(陶土)
water washing　水洗
water washout characteristics test　耐水洗性试验
water wave　水波
water wax　水乳化蜡
waterways　排水沟
water wettability　水(可)湿性；水润湿度
water-white　水白(色)的
water-white acid　①水白酸②水白盐酸
water-white color　水白色颜料；无色透明颜料
water-white distillate　水白馏分；精制前的煤油馏分
water-white kerosene　无色煤油
water-white mineral oil　水白矿物油；无色矿物油
water-white oil　水白油；无色油
water-white paraffin wax　水白石蜡〔赛波特色度为25的固体石蜡〕
water white transparency　无色透明
watery　①水的②有水的③似水的
watery distillate　①含水馏分②水状馏分
watery fusion　结晶熔化；低温熔融
Watson-Biemann interface　沃森-比曼接口
Watson-Biemann separator　①沃森-比曼分离器②沃森-比曼连接器
Watson formula　沃森经验式〔计算沸点〕
watt　瓦(特)〔功率单位〕
wattage　瓦(特)数
watt current　有效电流
Watten mixer　瓦腾(高速)搅拌器
watt-hour　瓦(特小)时
watt-hour meter　瓦(特小)时计
wattle bark extract　荆树皮浸膏
wattle extract　荆树(皮)栲胶
wattle gum　澳洲(树)胶
wattless current　无效电流
wattmeter　瓦特计
Waukesha motor　沃克沙马达〔电动机〕
wave　①波(浪)②波形曲线
wave amplitude　波幅
wave analysis　波形分析
wave clipper　削波计〔物〕
wave current　波动电流
wave damping　波阻尼
wave detector　验波器
waved plate　波纹板
waved shed weft insertion　波形梭口引纬
wave equation　波动方程(式)

wave filter　滤波器
wave form　波形
wave formation(=standing wave)　驻波
wave front　波阵面；波面
wave function　波函数
wave guide　导波器
wave height　波高
wave height concentration plot　波高浓度工作曲线法
wave length　波长
wavelength accuracy　波长精度
wavelength adjustment　波长调节
wave length calibration　波长校正
wavelength dispersion　波长色散
wavelength dispersion analysis　波长色散分析法
wavelength dispersion spectrometer　波长色散光谱仪
wavelength dispersion X-ray fluorescence spectrometer　波长色散X射线荧光光谱仪
wavelength dispersion X-ray fluorescence spectrometry　波长色散X射线荧光光谱法
wavelength dispersive spectrometer　波长色散光谱仪
wavelength exchange cam　波长变换凸轮
wavelength modulation　波长调制
wavelength plate　波长板
wavelength region　波长范围
wavelength reproducibility　波长重现性
wavelength resolution　波长分辨*
wavelength shift　波长移动
wave-length spectrometer　波长分光计；波长分光仪
wavelength standard　波长标准
wavelength table　波长表
wavelet　小波；子波
wavelet transformation　小波变换
wavelet transformation-multiple spectrophotometry　小波变换多元分光光度法
wavellite　银星石
wave mechanics　波动力学
wave meter　波长计
wave motion　波动
wave number　波数
wave number calibration　波数校准*
wave-number-wavelength conversion table　波长-波数转换表
wave of the first order　一级波
wave of the second order　二级波
wave of the zeroth order　零级波
wave-particle parallelism　波粒二象性
wave pattern　波型；波动图式
wave propagation　波传播

wave propagation method 波传播法
wave propagation test 波传播试验
wave property 波(动)性(质)
wave shed loom 波形开口织机；多梭口织机
wave spectrum 波谱
wave theory 波动(理)论
waviness 波纹
waviness phenomena 波纹现象
wavingness ①波纹②波动
wavy fiber 波状纤维
wavy flow 波式流动
wavy grown 多节的；纠曲的
wavy laminar flow 载波层流
wax 蜡
wax alcohol (由蜂)蜡(制得的高碳)醇
wax bayberry 蜡果杨梅脂
wax bean 蜡豆
wax-bearing crude 含蜡原油
wax-bearing distillate 含蜡馏分
wax-bearing fraction 含蜡馏分
wax-bearing oil 含蜡油
wax bloom 蜡霜
wax burnishing 蜡油抛光
wax cake 蜡饼
wax cake method 蜡饼法〔测定脂肪酸〕
wax centrifuging 离心分蜡
wax-chilling system 蜡冷冻结晶系统
wax conditioner 蜡调理剂
wax-containing mixture 含蜡混合物
wax content 蜡含量
wax crystal 蜡晶体
wax-deoiling process 蜡脱油法
wax dispersion 蜡分散液(体)
wax distillate 石蜡馏分
wax dope 降低石蜡凝固点用的添加剂
waxed 上过蜡的
waxed calf 上蜡犊皮
waxed flesh 上蜡肉面；上了蜡的皮里
waxed kips〔复〕 上蜡幼兽皮
waxed split 上蜡肉面；上了蜡的皮里
wax emulsion 蜡乳状液
wax emulsion compound 蜡乳浊液化合物
wax ester 蜡酯
wax-extension 蜡伸展
wax filter 蜡过滤器
wax filter drum 蜡过滤器转鼓
wax filter-pressing 蜡压滤
wax finish 蜡涂饰层

wax fraction 蜡馏分
wax fractionation 蜡分馏
wax fractionation process 蜡分馏法；脱蜡法
wax-free crude 无蜡原油
wax-free extract 去蜡浸制品
wax-free hydrocarbon mixture 无蜡烃混合物
wax-free layer 无蜡层
wax grained soap 蜡粒皂
wax haze 蜡朦翳；痕量蜡
wax-house man 蜡冷榨间工人
wax-impregnated graphite electrode 浸蜡石墨电极
waxing 涂蜡
waxing paper 蜡纸
wax ink 蜡质墨
wax lined 衬蜡的
wax matrix 蜡模
wax mixture 蜡混合物
wax-mixture flash section 蜡混合物闪蒸段
wax-mixture flow tank 蜡混合物中间罐
wax modifying agent 蜡的改进剂；降低含蜡油凝固点用的添加剂
wax moulding 蜡成型；蜡铸模
wax of bridelia leaf 土蜜树叶蜡
wax of kamala 粗糖柴蜡；菲岛桐蜡
wax oil 蜡油
wax-oil charge 蜡-油进料
wax-oil mixture 蜡-油混合物
wax-oil separation 蜡-油分离
wax-oil system 蜡-油系统
wax paper 蜡纸
wax pencil 蜡笔
wax phase 蜡相
wax phenol 聚亚甲基二酚型树脂〔内增塑聚甲苯醚树脂〕
wax plant 石蜡(制造)厂
wax plug(=wax wire) 蜡塞头
wax-polymer lubricant 蜡-聚合物润滑剂
wax pour point 蜡的倾点；蜡的凝固点
wax-precipitating solvent 沉淀蜡用的溶剂〔使蜡沉淀〕
wax precipitation 蜡析出
wax press 蜡压(滤)机
wax-producing catalyst 制蜡催化剂
wax separator 蜡分离器
wax slop 蜡馏分〔蜡浓缩馏分〕
wax-soap base 蜡-皂基〔润滑脂〕
wax stock 蜡馏液；蜡原料
wax stone 粗蜡；蜡块
wax storage 蜡储藏；储蜡罐
wax sweater 蜡发汗室

wax tailings〔复〕 蜡尾〔石油最后重馏分〕
wax thickener 蜡增稠剂
wax trap 蜡阱
wax wire(=wax plug) 蜡塞头
waxy ①蜡状的②蜡的
waxy component 蜡质组分
waxy crude 含蜡原油；多蜡原油
waxy distillate fraction 含蜡馏分
waxy fuel 含蜡燃料
waxy hydrocarbon 含蜡烃
waxy luster 蜡光
waxy oil 含蜡油
waxy petroleum distillate 含蜡石油馏分
waxy phase 蜡状相
waxy residue 含蜡残油
waxy sludge 蜡淤渣
waxy soap 蜡状皂
waxy stain 蜡性着色料
W-board 次级耐水纸板
weak ①淡(色)②弱
weak acid 弱酸
weak-acid cation exchange resin 弱酸性阳离子交换树脂
weak acid type ion exchanger 弱酸型离子交换剂
weak base 弱碱
weak-base anion exchange resin 弱碱性阴离子交换树脂
weak base type ion exchanger 弱碱型离子交换剂
weak bond 弱键
weak-bonded molecules 弱结合的分子
weak caustic liquor 稀碱液
weak caustic solution 弱苛性碱溶液
weak coal 脆煤；易碎煤
weak collision 弱碰撞
weak coupling system 弱耦合系统
weak electrolyte 弱(性)电解质
weakest mixture 最贫混合物
weak extinction of flame 弱化碱性
weak ferromagnetism 弱铁磁性
weak-field method 弱场法
weak link(=weak linkage) 弱键
weak linkage(=weak link) 弱键
weak-link theory 弱键理论
weakly acidic cation exchanger 弱酸性阳离子交换剂
weakly acidic cation-exchange resin 弱酸性阳离子交换树脂
weakly acidic resin 弱酸性树脂
weakly alkaline 弱碱(性)的
weakly alkaline washing agent 弱碱性洗涤剂
weakly basic anion exchange resin 弱碱性阴离子交换树脂
weakly basic resin 弱碱性阴离子交换树脂
weakly burned 略微焙烧过的
weakly coking coal 弱焦性煤
weak lye 稀碱液
weak magnetic nuclei 磁性弱核
weak mixture 贫混合物
weak nitrous acid 稀亚硝酸
weak pitch 弱煤
weak salt-cake 弱盐饼
weak segregation limit 弱分凝极限
weak segregation theory 弱分凝理论
weak solution 稀溶液
weak spot ①〔膜裂纤维〕薄点②缺陷
weak yarn 低强力纱；烂纱
weal bean method 弱束法
weale oil 鲸油
weapon ①武器②兵器
weapons-grade plutonium 武器级钚
wear ①磨耗；磨损②耐磨耗性③工作面损耗〔轴封〕
wearability 耐磨损性
wear and tear 磨耗；磨损；损耗
wear attrition of catalyst 催化剂磨耗
wear behaviour ①耐磨性②磨损行为；磨损表现
wear cycles 磨耗周期
wear factor 磨耗系数
wear hardness 耐磨性；磨耗性能
wear index 磨耗指数
wearing plate 耐擦板
wearing property 耐用性
wearing quality 耐用性
wearing quality of polymerized oil 聚合油抗磨耗性质
wearing resistance 抗磨力；磨损阻力；耐磨性
wearing ring 耐擦环
wearing surface 磨耗面；磨损面
wearing test 磨损试验
wear(ing) value 磨耗值
wear life 耐用时间
wear limit 磨损极限
wear loss 磨耗减量；磨耗量
wear machine 磨损试验机
wearometer 磨耗计
wear particle 磨粒
wear performance 耐磨性
wear prevention agent 防磨剂
wear-preventive additive 防磨添加剂
wear properties of lubricant 润滑剂的抗磨耗性质
wear rate 磨损速率

wear rating　相对磨耗率；磨耗等级
wear-reducing value of lubricant　润滑剂减磨值
wear reduction factor　磨耗减低因数
wear resistance　耐磨度；耐磨性
wear resistant　耐磨的
wear resistant coating　抗磨耗覆盖层
wear resistant filler　耐磨填料
wear ring　磨损环
wear test　磨耗试验；磨损试验
wear testing machine　磨耗试验机
weatherability　耐气候性；抗风化力
weatherable　耐候的；抗寒耐热的
weather aging　大气老化作用
weather cap　风帽
weather checking　天候老化龟裂
weather condition　气候条件
weathered crude oil(=weathered crude petroleum)　曝干的原油；吹干的粗石油；风蚀的原油
weathered crude petroleum(=weathered crude oil)　曝干的原油；吹干的粗石油；风蚀的原油
weathered galvanized iron　露天用镀锌铁
weathered gas　风蚀的气体
weathered paint film　①露天用漆膜②耐候性漆膜
weathered plastic　露天用塑料；耐候性塑料
weather exposure panels　天然暴晒样板；天然暴露样板
weather exposure test　风蚀暴露试验
weather fastness　耐气候性；耐天气性
weathering　①风蚀②老化
weathering agent　①防(天然)老化剂②防风化(蚀)剂
weathering aging　大气老化作用
weathering index　天候老化指数
weathering machine　风蚀性试验机；试耐气候器
weathering process　老化过程
weathering quality　耐风蚀性
weathering resistance　耐候性
weathering tank　风蚀罐
weathering test　老化试验；风蚀试验
weatherizing　使适应气候条件〔如机器、设备等〕
weatherometer　老化试验机；天候老化仪
weather proof　耐气候老化的
weatherproof adhesive　耐候性黏结剂〔胶黏剂〕
weather-proof color　①耐候性颜色②耐候性颜料
weather-proofing paint　耐候性漆
weather-proof joint　耐候性接缝(接头)
weatherproof tape　耐候性胶带
weather resistance　耐候性；耐天气性
weather stain　老化褪色；老化变色
weave　织物组织；织纹

weave exposure　露织物；纤纹显露
weavers broom　鹰爪豆
weave setting　织纹固定；定纹
weave texture　显布纹
web　①卷材②织物；网；料片
Webb furnace　韦布电炉
webbing　网纹；网纹涂装(法)；喷丝涂装(法)；成网
webbing-finish paint　网纹涂料；喷丝涂料〔一种装饰性美术漆〕
webbing lacquer　网纹漆；喷丝漆
web coating　网纹涂料；喷丝涂料
webcord process　帘布浸胶法
web cross-layer　交叉铺网机
weber　成网机
Weber evaporator　韦伯蒸发器
Weber number　韦伯数
Weber's law　韦伯定律
web-fed letterpress rotary machine　卷筒纸凸版轮转机
web formation　成网
webs　蹼状晶体
websterite　矾石
weckamine　中枢兴奋药
wedding agent　乳化剂；结合剂
wedding bristol　婚帖厚纸
wedding paper　婚帖纸
Wedekind cell　韦德金电池
wedge　①楔②楔入
wedge agent　楔入剂〔带长烃链的吸附表面活性剂〕
wedge bend test　(罐头漆)楔形弯曲试验
wedge brick　楔砖
Wedge burner(=Wedge kiln)　威吉煅烧炉〔活性白土再生用〕
wedge colorimeter　楔形比色计；楔管比色计
wedged chromatography　楔形色谱法
wedge diaphragm　楔形隔板
wedge distribution　楔形分布
wedged-shaped paper　楔形纸
wedged-shaped technique　楔形技术
wedged-tip technique　楔形技术
wedge effect　楔入效应
wedge filter　楔形滤光器
Wedge furnace　威吉炉；拱床炉
Wedge kiln(=Wedge burner)　威吉煅烧炉〔活性白土再生用〕
wedge method　(纸浆取样的)楔取法
wedge mount　楔形座
wedge phosphorus　楔形磷块
wedge-shaped film　楔形膜〔润滑剂〕
wedge-sharp strip　楔形纸条

wedge spectrogram 楔形光谱图
wedge spectrograph 楔形摄谱仪
wedge-strip chromatography 楔形条色谱(法)
wedge-type film 楔形漆膜〔指涂层断面形状为楔形，漆膜厚度逐渐变薄〕
wedging agent 楔入剂〔带长烃链的吸附表面活性剂〕
wedging compound 楔入化合物
wedging effect 楔入效应
wedging effect of oil film 油膜的楔入作用
weedicide 除草剂；除莠剂
weeding of grease 除去脂
weed killer 除草剂；除莠剂
weed oil 除草油
weed seed oil 黑麦草油
weeping 漏液；(塔板)滴漏现象
weeping out ①冷凝毛细管②小滴冷凝
weeping point 露液点
weep point 滴漏点；泄漏液点〔塔板〕
weft 纬(线)
weft laying carriage 铺纬滑架；敷纬滑架
weftless cord fabric 无纬织物
weid 黄木犀草
Weigert effect 胶体二色效应
weighable 可称量的
weighable amount 可称数量
weighable gum 可称胶质〔航空汽油指标〕
weigh batching 称量配料
weigh belt 称量带式运输器
weigh bridge 称量机；计量台
weigh center 称量中心
weighed inside oven (烘)箱内称重
weighed portion 称量〔准确称出的重量〕
weigher 秤(物)机
weigh feeder 定量供给装置
weigh hopper 称量给料斗
weighing 称量*
weighing accuracy 称量准确度
weighing area 配料间
weighing batcher 称量给料斗
weighing bottle 称量瓶
weighing burette(=weight burette) 称量滴定管
weighing by difference 减差称量法
weighing by substitution 替代称法
weighing by swings 摆动称法
weighing capacity (天平的)最大称量
weighing capillary for liquid sample 液体试样称量管
weighing controller 称量控制器
weighing cup 称瓶

weighing disk 称量皿
weighing form 称量形式
weighing funnel 称量漏斗
weighing machine 称量器；计重器；重量计；衡器
weighing pipette 称量吸移管
weighing scoop 称量勺
weighing station 称量台
weighing tube 称量管
weight ①重量②砝码③重力；(地心)引力④重体(物)⑤衡；计重单位⑥权重
weight-actuated filler 重量定量式加料器
weigh tank 称量槽；称量桶
weight-average chain length 重均链长
weight average degree of polymerization 重均聚合度
weight-average molar mass 重均分子量
weight-average molecular weight 重均分子量
weight average particle diameter 重均粒径
weight buret 称量滴定管
weight burette(=weighing burette) 称量滴定管
weight by volume 容重〔单位容量中松放物体的重量〕
weight checking scale 验重秤
weight coefficient 重量系数
weight concentration 重量浓度
weight conductivity 重量传导系数
weight conversion 重量转化
weight coulometer 重量电量计；重量库仑计
weight distribution curve 重量分布曲线
weight distribution function 重量分布函数
weighted 权(重)的；加权的
weighted average 权重平均；加权平均
weighted beaker 已称重的烧杯
weighted bottle 已称重的玻瓶
weighted cage 已称重的盒子
weighted centroid simplex method 加权形心单纯形法
weighted design 加权设计
weighted error 权重误差；加权误差；权差
weighted function treatment 权函数处理
weighted increment Kalman filter 加权增量卡尔曼滤波器
weighted least square method 加权最小二乘法
weighted mean 加权(平)均值*；权重平均值
weighted rating 加权等级
weighted regression 加权回归*
weight feed 定重加料
weight feeder 重量送料器
weight feeding 重量送料
weight-formal 重量克式量(浓度)的
weight-formality 重量克式浓度

weight-for-weight substitution 等重量代换
weight fraction 重量分数
weight gain 重量增加
weight hourly space velocity 重时空速
weight increase 增重；重量增加率
weightiness 重；重量
weighting ①加重②权重
weighting agent ①增重剂；加重剂②填充剂
weighting form 称量形式
weighting material 填充物；填料
weighting method 重量法；称重法
weighting of hydrometer 比重计称重
weighting scheme 加权方式
weight in vacuo 真空中重量
weight in wet base 湿重
weight-lever regulator 权杆调节器
weight loaded accumulator 荷权蓄力器
weight loading 重量负荷
weight loss 重量减轻；减量
weight loss on drying 干燥失重
weight loss on heating 加热失重；加热减量
weight mean molecular weight 重均分子量
weight method 称重法
weight-molal 重量摩尔(浓度)的；重模的
weight-molality 重量摩尔浓度
weight molar concentration 重量摩尔浓度
weightness 重；重量
weight-normal 重规的；重量当量浓度的
weight-normality 重(量)规(度)；重量当量浓度
weight of coated paint 涂料使用量
weight of oil 油重
weightometer 重量计
weight percent 重量百分数
weight per cent of catalyst 催化剂重量百分数
weight per epoxy equivalent(WPE) 环氧当量
weight per gallon 每加仑重量
weight per litre 每升重量〔克/升〕
weight pick-up 增重；重量增益
weight ratio of constituents 成分重量比
weights 砝码*
weight saving 重量减轻；重量节省
weight solid 重量固体分
weight spring 回综丝
weight strength ①单位重量爆炸力②重量强度
weight tank 称量桶
weight-to-capacity ratio 重量对容量的比
Weiland oxidation test 韦兰德氧化试验
Weimarn's formula 韦曼公式

weinschenkite 针磷钇铒矿
Wein's displacement law of radiation 维恩辐射位移定律
weir ①堰；坝；水口②鱼梁
weir flowmeter 堰式流量计
weir meter 溢流表
weir plate 堰板
Weir process 威尔脱蜡过程
weir tank 溢流罐
weiss 外斯〔原子磁矩单位〕
Weissenberg diffractometer 魏森贝格衍射仪
Weissenberg effect 魏森贝格效应*
Weissenberg extruder 魏森贝格挤出器
Weissenberg normal stress pattern 魏森贝格正应力图式
Weissenberg number 魏森贝格数
Weissenberg photograph 魏森贝格图
Weissenberg photography 魏森贝格照相法；魏森贝格图
Weissenberg postulate 魏森贝格假定
Weissenberg rheogoniometer 魏森贝格流变仪
weissite 碲铜矿
Weisz ring oven 维茨环炉
weld ①焊(接)；熔接；煅接②焊缝
weldability 可焊性；焊接性
weldability of cast-iron 铸铁的焊接性
weldability test 可焊性试验
weldable primer 可焊接底漆
weld(able) steel 焊接钢
weld crack 焊接裂纹
weld-decay 焊缝晶间腐蚀
weld-decay-free quality 无焊接侵蚀级；无焊缝晶间腐蚀级
weld defect 焊接缺陷
welded 焊接了的
welded bonds 熔合；熔合线
welded construction 焊接结构
welded flange 焊接法兰
welded joint 焊接接头
welded joint coefficient 焊接接头系数
welded pipe 焊制管
welded seam 焊缝
welded steel tank 金属焊接油罐
welded together 焊合
welded tube 焊接管
welded vessel 焊接容器
welder ①焊机②焊(接)工(人)；熔接工人
welder gloves 焊工手套
welder's health 焊工保健
welding 焊接；熔接
welding assembly 焊接构件

welding base metal　焊接金属；焊接母材
welding bead　焊道；焊蚕
welding booth(=welding shop)　焊接车间
welding condition　焊接规范
welding defect(=welding flaw)　焊接缺陷
welding deformation　焊接变形
welding direction　焊接方向
welding equipment　焊接设备
welding flaw(=welding defect)　焊接缺陷
welding flux　焊药；焊剂
welding head　烧焊枪
welding machine　焊接机；熔接机
welding mask　电焊面
welding metal cracking　焊缝金属裂纹
welding method　焊接方法
welding neck flange　焊颈法兰
welding plant　①焊接装置②焊接厂
welding position　焊接位置
welding positioner roller　焊接定位滚轮
welding process　焊接工艺
welding rod　焊条
welding roll　焊滚
welding seam　焊缝
welding sequence　焊接顺序
welding set　焊接设备
welding shop(=welding booth)　焊接车间
welding stress　焊接应力
welding symbol　焊接符号
welding technological parameter　焊接工艺参数
welding tip　焊嘴
welding together　焊接
welding tool　电焊工具
welding wire　焊丝
weld junction　熔合线
weldless tube　无缝管
weld mark　焊接痕(印)
weldment　焊件
weld metal　焊接金属
weld mignonette　淡黄木犀草
Weldon mud　韦耳登软泥
Weldon process　韦耳登(制氯气)法
weld penetration　熔深
weld steel　焊接钢
weld-through primer　可焊(点焊)接底漆
weld zone　焊接区
well　①井②油井③炉底
well base rim　凹陷式轮缘
well base tyre　凹缘轮胎

well-compounded paste　充分混合(混拌)的颜料浆
well-defined crystal　完好晶体
Weller's method　韦勒法〔钛比色分析〕
well-milled paint　充分研磨过的漆
well-mixed　充分混合的；良好混合的
well-ordered　序态良好的；良序的
well-ordered cluster　充分有序群集束；良序群集束
well pump　井泵
well-refined oil　精炼良好的油
wellsite　钡钙十字石
well strainer　油井粗滤器
well-tried technique　多次试验证明的技术；久经考验的技术
well-type counter　①井式计数管②井型计数器
well-type scintillator　井型闪烁器
well-ventilated　通风良好的
well water　井水
Welsbach (gas) mantle　韦耳斯巴奇纱罩
Welsh coal　韦尔希煤
Welsh method of copper smelting　韦尔希冶铜法
welt　①鞋底面间的革条②边；缘
Welter method of saponification　韦尔特皂化法
Welter's rule　韦尔特燃烧热法则
welting leather　镶边革；贴边革
Wenzel's law　文泽尔中和定律
Werner complex　维尔纳配合物
wernerite　方柱石
Werner mixer　维尔纳混合机
Werner-Pfleiderer mixer　维尔纳-普夫莱德雷尔混合机；维尔纳混合机
Werner's theory　维尔纳理论
Werner type chromium complex(=Volan)　维尔纳型铬配合(络合)物
Werner-type complex　维尔纳型配合物
Wesson tube　韦宋二氧化碳吸收管
Western blotting　免疫印迹法；蛋白质印迹法
western lubricating oil　西部润滑油〔指美国加里福尼亚润滑油〕
westfalite　外斯发拉特炸药
Westly cell　韦斯特利电池
Weston analyzer　韦斯顿分析器
Weston cadmium cell　韦斯顿镉电池
Weston (normal) cell (=cadmium (normal) cell)　韦斯顿(标准)电池
Weston standard cell　韦斯顿标准电池
Westphal balance method　韦氏比重称法
Westphal's specific gravity balance　韦斯特法尔比重天平

Westphal-type balance 韦斯特法尔型天平；韦氏比重天平
Westrumite oil 韦斯特鲁梅特油〔一种用来控制路面灰尘的铺路油〕
wet/dry tenacity ratio 干湿韧度比
wet abrasion resistance 耐湿磨性；抗湿摩擦性
wet abrasive paper （耐）水砂纸
wet abrasive resistance 耐湿磨性
wet-air pump 湿空气泵
wet analysis 湿分析(法)
wet-and dry-bulb psychrometer 干湿球湿度计
wet-and-dry bulb thermometer 干湿球温度计
wet and dry bulb thermometry 干湿球温度测定法
wet and dry hygrometer 干湿球温度计
wet ashing 湿法灰化；湿灰化
wet assay 湿分析(法)；湿验定
wet assaying 湿验定；湿分析
wet baffle chamber 湿挡板室
wet barrel tumbling 湿(法)转筒抛光
wet basis 湿基*
wet beating 湿打浆
wet beneficiation 湿法选矿；湿法富集(精选)
wet blacking 涂炭黑浆
wet blast ①湿鼓风②湿喷〔含有磨蚀剂的水〕
wet blast cleaning(=water blasting) ①(喷)水除锈；高压水除锈法②湿(法)喷砂除锈(法)
wet blast(ing) ①湿鼓风②湿喷〔含有磨蚀剂的水〕
wet blasting process 湿喷砂法
wet blast mold cleaning 喷湿砂洗模
wet blending 湿掺和
wet blue 湿铬鞣革
wet bonding 湿黏接
wet bottom producer 湿除灰发生炉；湿底发生炉
wet bulb depression 湿球降低〔湿度计〕
wet-bulb effect 湿球效应
wet bulb temperature 湿球温度
wet bulb temperature indicator 湿球温度指示计
wet bulb thermometer 湿球温度计；湿度计
wet burnishing ①湿罩光②湿磨光
wet calorific power 湿热值；湿卡值
wet calorific value(=wet calorific power) 湿热值；湿卡值
wet capacity 湿容量
wet carbonization 湿碳化(法)
wet cell 湿电池
wet cell washers for aerosols 气溶胶湿滤室
wet chemical method 湿(法)化学法
wet chemical oxidation 湿态化学氧化
wet chemical test 湿法化学试验方法
wet-chemistry automated analyzer 湿化学自动分析仪
wet churn 黄化溶解混合机；湿法黄化机
wet classification 湿式分级
wet classifier 湿式分级机
wet cleaner 湿法洗涤器；湿法除尘器
wet collector 湿式集尘器
wet collodion process 火棉胶湿板(照相)法
wet colour ①湿色料②未干色
wet column packing 湿法柱填充
wet combining 湿贴
wet combustion 湿(燃)烧
wet combustion method 湿热燃烧分解法
wet combustion process 湿燃烧法
wet concentration 湿选；水选
wet condenser 湿式冷凝器
wet corrosion 湿腐蚀
wet Cottrell 科特雷耳型湿式静电除尘器
wet crocking fastness rating 湿摩擦牢度等级
wet crushing 湿碎法
wet crushing mill 湿碎磨
wet cured polyurethane paint 湿固化聚氨酯涂料
wet cyclone 旋液分离器
wet cylinder liner 湿式汽缸套
wet deposition 湿沉降
wet digestion 湿法消化；湿法消解
wet distillation 水蒸气蒸馏
wet distillation process 水蒸气蒸馏过程
wet-dry sieving technique 湿法干法混合筛分技术；干湿(法)交替筛分技术
wet-dry tenacity ratio 干湿韧度比
wet-dry thermometer 干湿(球)温度计
wet-edge 边缘润湿
wet-edge agent 湿边剂
wet edge extender 湿边扩展剂
wet-edge retention 湿边保持性
wet-edge time 湿边时间
wet elutriation 洗涤；淘洗
wet elutriator 洗涤器
wet end 铜网部
wet etchant 湿法腐蚀
wet extrusion molding 湿压(出)模塑(法)
wetfast 耐湿的
wet fastness 耐湿性
wet-filling 湿法填装
wet filling packing technique 湿式填充技术
wet film 湿膜
wet film adapter ①湿膜附加器②湿膜接头

wet film thickness　湿膜厚度
wet film thickness ga(u)ge　湿膜厚度计；湿膜测厚仪
wet filter　湿滤过器
wet furfural condenser　含水糠醛冷凝器
wet gas　湿气〔含有大量汽油蒸气的天然气体〕；富气
wet gas-flow meter　湿球气流计
wet gas-holder　水封储气器
wet gas meter　湿球气体流量计
wet gas purifier　①湿煤气净化器②湿气净化器
wet gel strength　湿凝胶强度
wet grinder　湿磨机
wet grinding　湿(法研)磨
wet ground mica　水磨云母
wet ground muscovite mica　湿磨白云母粉
Wetherill grate　威载瑞尔法铸铁栅(炉箅)〔真空吸铸法〕
wet hiding power　湿膜遮盖力
wet ketone accumulator　湿酮收集器
wet lamination　湿膜压膜法
wet lay-up method　湿叠层法；湿铺叠法
wet litharge　一氧化铅　PbO
wet machine　浆纸机
wet method　湿法
wet mill　湿磨机
wet milling　湿磨法
wet mixing　湿式混合
wet modulus　湿模量
wet mortar　湿灰泥
wet natural gas　湿天然气
wetness　湿；潮湿；润湿
wet oil　含水石油
wet-on-wet coating　湿-湿涂装；湿-湿涂布
wet out　打湿；浸湿
wet-out rate　浸透速率
wet oxidation　湿式氧化
wet packing　湿法填装
wet packing method　湿装柱法
wet paint　湿漆；未干涂膜
wet pan mill　湿辊磨机
wet pan milling　湿辊磨机研磨
wet pans〔复〕　湿辊磨机
wet perimetric　润湿周边的
wet-pit pump　排水泵
wet point　润湿点
wet preparation　湿制剂
wet press　湿压机
wet pressed pulp　湿压纸浆
wet process　湿法
wet process beneficiation　湿法富集
wet processed clay　湿法黏土；湿法瓷土
wet-process enameling　湿法搪瓷
wet processing　湿处理
wet process rotary kiln　湿法旋转窑
wet puddling　湿搅炼；湿精炼
wet pulp　湿纸浆
wet purification　湿提纯(法)
wet quenching　湿淬火；湿法熄焦；湿猝灭
wet reaction　湿反应
wet reflux　湿回流
wet rendering　湿炼油法
wet-rub resistance　耐湿(摩)擦性
wet salted hide　盐湿皮
wet sand blast　湿式喷砂
wet sanding　水磨；湿砂光
wet saturated steam　含水饱和蒸汽
wet screening　湿筛法
wet scrubber　湿洗器
wet scrubbing　湿法洗涤
wet scrub resistance　耐湿擦洗性；耐水擦洗性
wet seal　液封
wet separation　湿选；水选
wet spinning　湿纺*
wet-spray　湿喷涂(法)
wet stamping　湿碎法
wet stamping mill　湿碎磨
wet steam　湿蒸汽
wet steam cure　①(直接)湿蒸硫化②(直接)湿蒸汽处治
wet steaming　湿蒸法
wet stock　①湿物料；湿材料②黏状浆
wet storage holders　湿式储气柜
wet-store brushes　漆刷湿洗存放罐
wet strength　湿强度
wet strength of adhesion　湿态黏合强度
wet strength paper　湿强纸
wet strength retention　①湿强度保留②湿强度保留率
wet sulphidization　湿法黄化
wet sump lubrication　湿槽润滑法
wet system of evacuating　湿体系抽空〔用喷水空气泵抽空的装置〕
wettability　湿润性；润湿性；可湿性
wettability power　吸湿能力；润湿率
wettable　可湿的
wetted surface column　湿面分馏塔；沾湿表面分馏塔
wetted wall column　湿壁塔
wetted wall tower　湿壁塔
wet tensile strength　湿拉力；湿抗拉强度
wetter-off　打口工〔玻〕

wet test 湿试法
wetting 焊料浸润；润湿
wetting ability 润湿能力
wetting action 润湿作用；润湿力
wetting agent 湿润剂*
wetting and penetrating agent 润湿渗透剂
wetting angle 润湿角
wetting-back 再润湿
wetting balance 润湿天平
wetting behavior 润湿行为
wetting capacity 湿润能力；浸润能力
wetting color 易润湿颜料
wetting effect 润湿效应
wetting efficiency 润湿效率
wetting error 润湿误差
wetting force 润湿力
wetting heat 润湿热
wetting index 润湿指数
wetting of cloth 布匹增湿
wetting-off 打掉；取下〔把玻璃吹制件自吹制管口打下〕
wetting out 浸湿；浸透；润湿
wetting out action 湿透作用
wetting penetration 润湿浸透
wetting point （吸油量测定时)润湿点
wetting power 润湿力*
wetting property 润湿性(质)
wetting quality 润湿性；濡湿性
wetting surface energetics 润湿表面能学
wetting tension （薄膜)润湿张力
wetting time 润湿时间
wet treatment ①湿处理②湿选
wet tube mill 筒式湿磨机
wet tumbling 湿(法)抛光
wet type air cooler 湿式空气冷却器
wet vacuum distillation 真空蒸气(汽)蒸馏
wet vacuum pump 湿气泵
wet vapor 湿蒸气(汽)
wet water 润湿水〔加有强润湿剂的水，灭火用〕
wet way 湿法
wet-way analysis 湿法分析
wet weight 湿重(量)
wet wheeling machine 湿磨机
wet wheel mill 湿轮碾机
wet winding 湿法缠绕
whale-bone oil 鲸骨油
whale grease 鲸脂膏
whale guano 鲸肥；鲸粉
whale oil 鲸油

whale quenching oil 淬火鲸油
whale shot 粗鲸蜡
whale tallow 鲸脂
wheat oil 麦油
wheat starch 麦淀粉
Wheatstone bridge 惠斯通电(阻)桥
wheat straw pulp 麦秆浆粕
wheel ①轮(子)②轮转烟火③砂轮；齿轮；飞轮④转动；旋动
wheel and axle 轮轴
wheel-bearing grease 轮轴承脂
wheel-bearing lubricator 轮轴承润滑器
wheeling 干滚
wheel lubricator 轮轴润滑器
wheel mill 轮碾机
wheel mill bed 碾机底板
wheel pump 轮泵
wheel stretcher(=wheel stretching machine) 轮压展机
wheel stretching machine(=wheel stretcher) 轮压展机
wheel stuffing 鼓中加脂(法)
whet-slate 砥石；磨石
whetstone 砥石；磨石；油石
whetstone-slate 砥石；磨石
whewellite 草酸钙石
whey 乳清
whipper ①搅打器②净毛机
whipping 搅打
whipping machine 打泡机
whipping method 挑焊运条法
whip resistance 耐搓；耐绞；耐缠绕
whirl brush 旋涡式磨刷法
whirling 涡转；旋涡
whirling motion 涡动
whirling runner ①旋滑除渣器②旋滑撇取浮皮器
whirlpool 涡流
whirl sintering 旋流烧结
whisker ①搅拌器②须晶
whisker column 晶须柱
whiskerising process 须晶生长法
whisker reinforced composite 须晶增强复合材料
whiskers 针状单晶
whisker-walled open tubular column 晶须开管柱
whisking machine 打泡机
white 白(色)
white acid 白酸〔1. 蚀玻酸；2. 白色的酸〕
white agate(=chalcedony) 白玛瑙；玉髓
white air-dry enamel 白色自干瓷漆
white alkali 白苏打；白碱

white alum 明矾;白矾
white analytical system 白色分析体系
white angola copal 白色安哥拉玬脂
white antimony 辉锑矿
white arsenic 砒霜*
white baking enamel 白色烘干瓷漆
white baking finish 白色烘干面漆
white baking primer 白色烘干底漆
white bauxite 白(色)铁矾土
white black 硅胶
white bloom 白花;白翳〔含硫化合物在煤油灯上沉积〕
white board 白纸板
white bole 白陶土;高岭土
white camphor oil 白樟油
white carbon 白炭墨〔二氧化硅的俗称〕
white carbon ink 白炭墨;白色涂改墨
white carrara glass 白色卡拉拉玻璃
white cast iron 白口铁
white cedar leaf oil(=thuja leaf oil) 白柏叶油;侧柏叶油
white cement 白水泥
white charcoal 白炭
white clover leaf oil 白三叶草油
white copperas 硫酸锌
white crepe rubber 白皱胶;白生胶
white cup grease 白杯脂
white damp 一氧化碳 CO
white dye(=fluorescent brightening agent) 荧光增白剂
white factice 白油膏
white factis 白硫化油膏
white fuming nitric acid 白色发烟硝酸
white gasoline 无铅汽油
white glass 白玻璃
white glaze 白釉
white gold 白金
white ground design 白底图案
white group 白色类〔载体〕
white gum 白藓叶桉
Whitehead process (for silver refining) 怀特赫德(精炼银)法
white hiding power 白色遮盖力
white India rubber substitute 白色代胶〔橡胶〕
white iron 白铁;马口铁
white iron pyrite 白铁矿
white kerosene 白色煤油
white lac(=bleached lac) 脱色紫胶
white lake 氢氧化铝;矾土白
white lead 铅白*
white lead carbonate 铅白;碱式碳酸铅
white lead-linseed oil paint 铅白亚麻油涂料;亚麻油铅白漆
white lead-ore 白铅矿
white lead paint 白铅漆
white lead paper 碳酸铅浸过的滤纸;铅白纸
white lead pigment 铅白颜料
white lead stiff paste 白铅油稠浆〔制油性填孔剂用〕
white lead sulfate 硫酸铅白;碱式硫酸铅〔俗名〕
white leather 白革;矾鞣革
white light 白光
white light interferometer 白光干涉仪
white light refractometer 白光折光计
white lime 白刷料;白涂料
white linseed oil paint 亚麻仁油白漆
white liquor 烧碱液
white lotion 硫酸锌-硫化钾溶液
white manifold paper 复印纸
white metal ①白色金属②白铜③以锡铅或锑为基的合金
white mica 白云母
white mineral oil(=paraffin oil) 石蜡油
white molybdate pigment 钼白颜料
white mundic 砷黄铁矿;毒砂 FeAsS
white mustard oil(=white mustard seed oil) 白芥子油
white mustard seed oil(=white mustard oil) 白芥子油
whitener 增白剂
whiteness ①白;白色②白度
whiteness degree 白度
whiteness index 白度指数
whiteness value 白度值
white nickel ore 白镍矿
whitening ①增白②干削
whitening agent 增白剂
whitening cylinder 干削机刀轴
whitening effect 增白效应
whitening machine 干削机
whitening stone 磨石
white non-opaque pigment 白色透光颜料;白色体质颜料
whiten soap 增白皂
white oil ①轻油②白油
white olivine 白橄榄石
white opaque pigment 白色不透明颜料
white paraffin 白石蜡
white paste 厚漆;铅漆
white petrolatum 白矿脂;白凡士林
white phosphorus 白磷;黄磷
white pickling 浸白

white pig iron 白口铁
white pigment 白色颜料
white pine oil 白松油
white plastic 塑性白漆〔船舱水箱内部用的热喷涂厚层涂料〕
white polish 白色虫胶漆
white pollution 白色污染
white portland cement 白色硅酸盐水泥
white precipitate 白降汞；氯化氨基汞
white products 透明石油产品
white pyrite 白铁矿
white reference standard 白色参比标准；白色对照标准
white resin 白树脂
white rock 滑石
white rouge 微细二氧化硅；白铁丹（俗）〔光学镜片抛光材料〕
whiteruss 液体石蜡
white rust ①白锈〔铝表面腐蚀〕②白膜〔镀锌层表面缺陷〕
white scale 白粗石蜡片
white seal 白印（级）〔氧化锌最优品级〕
white shellac 漂白紫胶
white shiner 有光纤维质
white simile paper 印书纸
white soda ash 煅苏打；碳酸钠
white-soil effect 白污效应
white soot 白炭黑〔补强剂〕
white spindle oil 白锭子油
white spirit 石油溶剂
white spruce oil 白云杉油
White's spherical cathode lamp 怀特球形空心阴极灯
white substitute 白色代胶
white sulfate of alumina 纯硫酸铝
white support 白色载体
white tanning agent 白鞣剂
white titanium 钛白
white tube enamel 白色管线瓷漆〔指铅、锡、铝管用漆〕
white varnish 白瓷漆用漆基
white vaseline 白矿脂；白凡士林
white vitriol 皓矾；七水(合)硫酸锌
white ware 白色陶器
whitewash ①白涂料②刷白③白垩灰浆
white water 白水
white water box 白水箱
white water pan 白水槽
white water tray 白水槽
white wax 白蜡
white wax grained soap 白蜡粒皂

white X-ray 连续 X 射线
white X-ray radiation 白色 X 射线
white zinc 锌白；氧化锌
white zinc paint 白锌漆
Whitfield's ointment 苯甲酸、水杨酸、聚乙二醇膏
whiting 白粉；研细的白垩
Whiting cell (for caustic soda) 惠廷(制苛性钠)电池
whitlockite 磷钙矿
whitneyite 淡砷铜矿
whizzed 旋离了的
whizzer 离心机
whizzer air separation 离心空气分离
whizzing 旋离；离心分离
whole body activation autoradiography 整体活化放射自显影法
whole body autoradiography(WBA) 整体放射自显影法
whole contract 整体压缩；整体收缩
whole contraction 整体收缩*
whole crude 全原油〔未曾提取汽油的石油〕
whole-current classifier 全流分级机
whole genome shotgun sequencing 全基因组鸟枪测序(技术)
whole latex 全胶乳
whole latex rubber 全浆橡胶
whole mask 全面罩
whole tire reclaim 轮胎再生橡胶
whole vision mask 全视面罩
wholly-aromatic polyamide 全芳基聚酰胺
Wiborgh phosphate 威博磷肥
wick 芯；灯芯
wickability 毛细管性；芯吸性
wick chromatogram 灯芯色谱(图)
wick effect 灯芯效应
wick feeder 灯芯式供料器
wick-felt oiler 毡芯给油器
wick holder 支芯管；芯挟
wicking 芯给〔油灯或加油器〕
wicking action 灯芯效应
wicking height 毛细升高值；吸水上升高度
wicking rate 毛细速率；吸芯速度
wicking type cell 芯吸型电解池
wick lubricator 芯润滑器
wick oiling 芯加油
widdrane 羽毛柏烷
widdrene(=thujopsene) 羽毛柏烯；罗汉柏烯
widdrenic acid 羽毛柏烯酸
widdrol 羽毛柏醇
widdrol epoxide 环氧羽毛柏醇

wide angle diffraction	广角 X 射线衍射
wide angle laser scattering photometer	广角激光散射光度计
wide angle X-ray diffraction	大角度 X 射线衍射
wide angle X-ray diffraction method	广角 X 射线衍射法
wide apron conveyor	宽型带式输送机
wide-band decoupling	宽带去偶
wide-belt sander	宽带砂光机
wide-boiling cut	宽沸点馏分；宽馏分
wide boiling sample	宽沸程试样
wide buffing machine	宽幅磨革机
wide-cut	宽区(域)切割
wide cut diesel fuel	宽馏分柴油
wide fraction	宽馏分
wide-meshed	粗筛孔的
wide meter	宽刻度仪表
wide mouth bottle	广口瓶
wide mouth bottle with ground stopper	广口磨塞瓶；带磨塞广口瓶
wide-mouthed flat-bottom extraction flask	广口平底抽提瓶
wide-necked bottle	广口瓶
wide neck flask	广口(烧)瓶
widening of the ring	环的扩大
wide plate test	宽板试验
wide range	宽量程
wide shaving machine	宽幅削匀机
widest boiling point range	最宽沸程范围
wide temperature lubricant	宽温度范围润滑剂
widia	渗碳碳化钨
Widmer column	卫得门分馏柱
width	幅宽
width of cut	切削宽度
width of mesh	网眼大小；筛号
width of panel	板宽
width of roller	漆辊宽度〔实际上是辊长〕；辊涂宽度
width of root face	钝边高度
width setting	定幅
Widtsoe-Tollens' test	盐酸反应
Wiedemann's additivity law	魏德曼加和定律
Wieldermann cell	维耳德曼电炉
Wielogaski furnace	维洛加斯基电炉
Wiese's reduced pressure desiccator	威斯减压干燥器
Wiggins gasholder	卫金司储气柜
Wijs method	威杰斯法〔测碘值法〕
Wijs solution	威杰斯一氯化碘溶液
Wijs value	威杰斯值；碘值
wild	①野的②氧化过头的
wild angelica	羌活；林白芷
wild basil oil(=calamintha oil)	风轮菜油
wild cherry	野黑樱桃
wild cinnamon	香叶众香树
wild coal	野煤；带煤层的片岩
wild currant (=red currant)	红茶藨子；红醋粟
wild gasoline	不稳汽油；高蒸气压汽油
wild grain	粗粒面
wild jasmin(=mock orange)	西洋山梅花
wild limonene tree	柠味树
wild mango oil	野芒果油；加蓬依苦木油〔取自 Irvingia gabonensis〕
wild marjoram oil	牛至油
wild natural gasoline	不稳自然汽油
wild pansy	三色堇
wild pea	野豌豆；巢菜
wild rosemary oil	喇叭茶油
wild rubber	野生橡胶
wild rubber plant	野生橡胶植物
wild rue	蒺藜
wild silk	野蚕丝
wild-tansy	野艾菊
wild thyme	野百里香
wild thyme oil	野百里香油〔取自 Thymus serpyllum〕
wild yellow root	姜黄
Wilfley table	韦尔弗利淘选台
wilfordic acid	雷公藤酸
Wilhelmy plate apparatus	威廉米悬片仪〔测表面张力〕
Wilhelmy plate method	威廉米悬片法
Wilhelmy slide method	威廉米悬片法
wilkinite(=bentonite)	膨润土；皂土；浆土
Wilkinson complex	威尔金森复合物
Wilk-Shapiro test for normality	正态性的威尔克-夏皮罗检验法
willardin	尿嘧啶丙氨酸
willemite	硅锌矿
Willesden goods	铜氨纸制品
William's gas apparatus	威廉气体分析器
Williams-Lendel-Ferry equation	WLF 公式
Williamson ether synthesis	威廉逊醚合成法
Williamson reaction	威廉逊制醚反应
Williamson's violet	威廉逊氏紫；铁氰化亚铁钾 $KFe[Fe(CN)_6] \cdot 2H_2O$
Williamson synthesis	威廉逊合成法
willow bark	柳皮
willow side	搓纹半张牛面革
wilnite	灰绿榴石
Wilson cloud chamber	威尔逊云(雾)室
Wilson fog method	威尔逊雾室法

Wilson tracks 威尔逊射线径迹
wilting water percentage 凋萎含水量
Wilzbach labeling 维茨巴赫标记
winch ①绞车②有柄曲拐
Winchester bushel 文切斯特蒲式耳〔容量名〕
Winchester gallon 文切斯特加仑〔液量名〕
wind ①风②通风③绕；卷
windage 空气阻力
wind bending moment 风弯矩
windborne particles of grit 风积砂粒
wind-borne salt ①升腾盐(分)〔空气中带的盐分〕②风成盐
wind-borne sediment 风成沉积层
winded (印刷)气流分页的；气流卷取
winder ①卷纸机②绕线器；卷缠机
wind furnace 风炉
wind gage 风速计
wind girder （储罐)防风梁
winding 卷绕
winding angle 缠绕角
winding central angle 缠绕中心角
winding-compression moulding 缠绕模压法
winding machine 缠绕机
winding pattern 缠绕线型
winding rate 缠绕速度
winding rate ratio 缠绕速比
winding tension 缠绕张力
winding up ①卷裹②绕紧
winding up roller 卷辊
wind load 风载荷
window factor analysis 窗口因子分析法
window function treatment 窗(口)函数处理
window glass 窗玻璃
windowless counter 无窗计数管
windowless presenter 无窗送料器
window material 窗(口)材料
window-pane paper 窗玻璃纸
wind pressure 风压
wind rate ①风量②风速
wind shield ①减阻帽②遮风屏
wind sifting action 风筛作用；气流筛分作用
wind stress 风压应力
wind tunnel turbulence 风道湍流；风洞湍流
windup drum 转鼓
wind vane 风标
wing 机翼
wing nut 蝶螺(钉)帽；螺旋套
wing pump 叶轮泵
wing top 鱼尾灯头
wing valve 圆盘导翼阀
wink-dry 瞬间干燥
Winkler burette 温克勒滴定管
Winkler-Koch unit 温克勒-科赫裂化装置〔无反应室〕
Winkler reagent for oxygen 温克勒氧吸收液〔没食子酸碱性溶液〕
Winkler's method 温克勒(溶解氧)测定法
Winkler's oxygen sampling bottle 温克勒采氧瓶
Winstein-Grunwald equation 温斯坦-格伦瓦尔德公式
winter barley 冬大麦
winter black oil 冬用黑油；低凝固点车油
winter capacity 冬季容量；冬季生产力
winterene 木兰烃
winter grade 冬用级
winter grade gasoline 冬用汽油
wintergreen 冬青
wintergreen oil 冬青油；水杨酸甲酯；鹿蹄草油
winterization 过冬准备
winterized stearin 氢化油；硬化油
winterized unit 过冬装置
winterizing of oil 油的冬化；油的冻凝
winter mushroom 冬菇
winter oil 冬用油
winter rape 蔓菁；芸苔
winter rue 冬芸香
winter sludge 冬季淤渣；低温淤渣
wipe ①擦；拭②条痕〔印刷故障〕③消磁
wipe cleaner 擦净剂
wiped film evaporator 刮板式薄膜蒸发器
wipe-dry 擦干
wiped wall still 扫壁蒸馏器；转膜蒸馏器
wipe-off type 擦拭型〔磷化清洗剂〕
wiper 擦具；涂油工具
wipe resistance 抗擦拭性
wiper lubricator 拭擦润滑器
wipe test 擦拭试验
wiping 黏合剂移位〔橡胶与金属结合〕
wiping action 纵向滑动
wiping cloth(=wiping rags) 抹布；拭布
wiping rags(=wiping cloth) 抹布；拭布
wiping solvent 刷涂用溶剂
wiping stain （木器)擦着色剂
wipla 拟白金；铬镍合金
wire 金属丝；金属线
wire-bar applicator(=wire-wound rod) 绕线棒控涂漆器；绕线式刮漆棒
wire bead 钢丝(橡皮)撑轮圈

wire bead core 钢丝撑轮圈
wire belt 钢丝胶带；钢丝运输带；三角带
wire brush 钢丝刷
wire brushing method 钢丝刷清洗法；钢丝刷除锈法(清理)
wire cloth ①金属丝布；筛布②铜丝布③钢丝布
wire cloth sieve 金属丝布筛；铜布(网)筛
wire coating 线缆涂料；线缆包皮
wire coil 金属螺旋线
wire-coil packed column 线圈填充塔
wire connection clamp 接线夹
wire cross 十字线
wire-cup brush 环形钢丝刷
wire cutter (切)钢丝钳〔商〕
wire damage 铁丝伤
wired edge tyre 直边外胎
wired glass(=wire glass) 络网玻璃
wire-drawing lubrication 拉线润滑；抽丝润滑
wire-drawing soap (金属)拉丝皂
wire drive unit 送丝装置
wire enamel 电缆瓷漆
wire enamel solvent 漆包线漆溶剂
wire explosion spraying 线爆喷涂
wire fabrication 线材制品
wire feeder 送丝装置
wire feed speed 送丝速度
wire feed system 送丝系统
wire fencing 铁丝栅栏
wire filter cloth 金属丝滤布
wire form cupric oxide 丝状氧化铜
wire frame 网架
wire ga(u)ge 线规〔物〕；量铁丝尺
wire gauze 线网；铁丝网
wire gauze diaphragm 线网隔层
wire gauze filter 线网过滤器
wire gauze packing 线网填充料
wire glass(=wired glass) 络网玻璃
wire-grid polarizer 网栅视偏振器
wire grill 钢丝护栅
wire-grommet V-belt 钢丝绳三角带
wire guide 导网
wire helix 金属螺旋线
wire inserted asbestos yarn 夹钢丝石棉股线
wireless 无线的；不用金属线的
wire loop 电线圈
wire mat 金属丝毡
wire mesh 线网
wire-mesh column 线网隔层分馏塔
wire-mesh demister 线网除雾沫器〔蒸馏塔内〕
wire mesh electrode (金属)网状电极
wire mesh filter panel 金属网式过滤板
wire mesh gauze 钢丝网
wire mesh oven 线网炉
wire mesh packing 金属丝网填料
wire mesh screen 金属网筛
wire mesh sections 线网分隔层
wire nail 铁丝圆钉
wire out cake machine 模烤压切机
wire part 铜网部(分)
wire repeated tester 金属丝往返试验机
wire-ribbon method 丝带法
wire rod applicator(=wire-wound rod) 绕线棒控涂漆器；绕线式刮漆棒
wire roll 铜网辊
wire rope 钢丝绳；铁丝绳
wire rope grease 钢丝绳润滑脂
wire rope oil 钢丝绳润滑油
wire screen (金属丝)网筛
wire shorts 铁丝段
wire showers 铜网喷水器
wire spiral 金属螺旋线
wire stirrer 线网搅拌器
wire table 网案；网架
wire transport detector 传动丝检测器
wire triangle 线三角；铁丝三角
wire weight 线状砝码；丝码
wire winding 绕丝
wire winding vessel 绕丝式容器
wire with flux-core 药芯焊丝
wire-wound doctor 绕线棒式刮漆器
wire-wound furnace 线绕电炉
wire wound resistor 线绕电阻
wire-wound rod (=wire-bar applicator;wire rod applicator) 绕线棒控涂漆器；绕线式刮漆棒
wire-wound vessel 线绕容器
wiriness 手感粗硬
wiring 线路；接(电)线
wiring diagram 接线图
wistarin 紫藤苷
witch hazel 金缕梅
witch hazel oil 金缕梅油
withania 睡茄
withanic acid 睡茄酸
withanine 睡茄碱
withaniol 睡茄醇
withdrawal rate (浸涂的)上提速度
withdrawing 卷绕

witherite 碳酸钡矿；毒重石
within laboratory error （实验）室内误差
within-package unevenness 内不匀率
withstand load 耐负荷；容许负荷
withstand paint-line temperature 涂装线容许温度
witness 证明管
Wittig reagent 维悌希试剂
WLF equation WLF 公式
Wobbe index 沃泊指数〔燃料气比重与热值关系〕
wobble hypothesis 摆动假说
wobble plate 摇摆板〔转动式活塞泵中的〕
wobble plate fuel pump 摇摆板燃料泵
wobbling 摇摆；摇动
wobbling drum 转鼓
Woburn diene value 沃本双烯值
woddnite 钛云母
Woehler's synthesis 沃勒合成法
WO emulsion(=W/O emulsion) 水油乳剂
wogonin 沃贡宁；汉黄芩素
wöhlerite 铌锆钠石
Wohl's reaction 伍耳水解乙酰基反应
Wohlwill gold-refining process 沃耳韦耳炼金法
Wolff bottle 三颈(玻璃)瓶
Wolffram's salt 二水合三氯化铂
Wolf process 伍尔夫浮选法
Wolfram (=tungsten) 钨〔74号元素，化学符号 W〕
wolframate (=tungstate) 钨酸盐
wolframate radical 钨酸根
wolframic acid 钨酸 H_2WO_4
wolframine 钨华
wolframite 黑钨矿*
wolfram ore 钨矿
wolfram steel 钨钢
wolfsbergite 硫铜锑矿
Wolf trap 乌尔夫吸尘器
wollastonite 硅灰石
wollastonite extender pigment 硅灰石体质颜料
Wollaston prism 沃拉斯顿棱镜
Wollaston wires 沃拉斯顿白金丝
Wolpert's air tester 沃尔珀特空气试验器
Wolters phosphate 沃尔特斯磷肥
wood 木(材)
wood alcohol 甲醇；木醇〔俗〕；木精〔俗〕
wood ash 木灰
wood beam-base 木梁基
wood block 木段；木块
wood block humidity controller 木块湿度控制器
wood boat-bottom paint 木船船底漆

wood cellulose 木纤维素
wood cement (=stick shellac) 紫胶柱；紫胶粒〔别名〕
wood char (=wood charcoal) 木炭
wood charcoal (=wood char) 木炭
wood cleaning ①木材的洁净②木材的清除
wood coal 木炭
wood coating (=wood product coating) 木器用涂料
wood color 材色
wood creosote 木杂酚油；山毛榉杂酚油
woodcut 木刻；木版
wood distillation 木材蒸馏(法)
wood distillation methanol 木馏(所得)甲醇
wood drier 木材干燥器
wood dye 植物染料；木材染料
wooden 木(制)的；木质的
wooden block ①木块②木托③木滑车
wooden-body lacquer 木胎漆器〔大漆〕
wooden stand 木架
wooden support 木架
wood fiber 木纤维
wood filler 木材填孔料；胶缝剂
wood filling 油灰
wood flour 木粉
wood former 木模
wood free 无(原料)木的；无木质的
wood-free printing paper 道林纸
wood-free writing paper 道林纸
wood gas 木(煤)气
woodgrain (=wood grain) 木纹；木材纹理
woodgrained laminate 木纹层压板
wood grain finish 木纹加工
wood grid packing ①木格填衬②木格包扎
wood gum 树胶；树脂
wood hydrolysis (=wood saccharification) 木材水解
wood lacquer 木器漆
wood lined pipe 木衬管
wood march 变豆菜
wood meal 木粉填料
wood molasses 木材糖浆
wood naphtha 粗木精
wood oil 木油；桐油
wood opal 木蛋白石；木化石
wood paper 木制纸
wood peg 木楔
wood pitch 木沥青
wood-plastic composite(WPC) 木塑料复合材料
wood polishing process 木器揩(擦)涂紫胶清漆(操作)法
wood preservation 木材防腐

wood preservative　木材防腐剂
wood preservative and stain　护木油〔商〕
wood preservative coatings　木材防腐涂料
wood product coating(=wood coating)　木器用涂料
wood pulp　木(纸)浆
wood pulp fiber　木浆纤维
wood ray(=xylary ray)　木射线
wood (retort) tar　甑制木焦油
wood rosin　木(蒸)松香
woodruff absolute　香车叶草净油
wood rule　木尺
wood-run　林场
wood saccharification(=wood hydrolysis)　木材水解
Wood's alloy　伍德合金
wood screw　木螺钉〔机工〕；木螺丝(钉)
Wood's glass　伍德玻璃
wood shavings〔复〕　刨花；木材碎片
wood slip　衬板
Wood's metal　伍德合金；铋铅锡镉合金
wood spirit　甲醇；木精
wood spirit of turpentine　汽馏松节油
wood stabilization　木材稳定化处理
wood stain　木材着色料
wood stave pipe　板桶管
wood stone　锯末、氧化镁和氯化镁的混合物
wood sugar(=xylose)　木糖
wood tar　木焦油
wood tar fraction　木焦油馏分
wood tar oil　木焦油
wood tar pitch　木焦油沥青
wood tin　木锡矿；纤锡矿
wood turpentine　木材松节油
wood vinegar　木醋酸
Woodward-Hoffmann's rule　伍德沃德-霍夫曼定则
woodwardite　木守矿；硫酸铝铜矿
Woodward rules　伍德沃德定则
wood waste　废木
woodwaxen　金雀花
Wood-Werkmans'reaction　伍德-瓦克曼氏反应〔生物体内丙酮酸到琥珀酸的反应〕
wood wool　刨花
wood wool filter　刨花滤器
wood work　①木工②木制品
woodwork coating　①木器涂装②木器涂料
wood working　(制造)木工的；木工(制造)的
woody aspect　木香格调
woody background　木香尾韵
woody note　木香

woody odor　木香
wool　羊毛；绒毛
wool alcohols　羊毛脂醇
wool chromaticity　毛纤维色度
wool-clip machine　剪毛机
wool degras　羊毛脂
wool dye　羊毛(用)染料
woolen　(羊)毛的
woolen fabric　毛织物(品)
woolen oil　粗纺油
wool fat　羊毛脂
wool fat ethanolamide　羊毛脂乙醇酰胺
wool fat pitch　羊毛脂沥青
wool grease　羊毛脂
wool grease pitch　羊毛脂沥青
wooling of viscose　黏胶的丝化作用
woolliness　棉性；羊毛性
wool oil　羊毛油
wool pitch　羊毛(脂)沥青
wool pulling　羊毛落纱
wool rinsing machine　洗毛机
wool-scouring soap　洗毛皂
wool skins〔复〕　(连毛)羊皮
wool soap　洗毛皂
wool stock　羊毛料
wool sweat　羊毛粗脂；羊毛油质
wool tops　羊毛条
wool violet　羊毛紫
wool washing machine　洗毛机
wool waste　羊毛废料
wool wax　羊毛蜡
wool wax acid　羊毛蜡酸
wool wax alcohol　羊毛蜡醇
Woolwich viscosity　伍尔维奇黏度
Woolwish test　伍耳维希试验
wool yarn grease　毛线脂；羊毛脂
woorara　箭毒
worenine　黄连宁；甲基黄连碱
wore paper　布纹纸
work　功*
workability　施工性能
work bench　①钳桌②工作台
work content　功函
work done factor　作功因数；耗功系数
worked consistency　用过润滑脂的稠度
worked grease　用过的润滑脂；工作过的润滑脂
worked grease penetration　用过的润滑脂的针入度
worked penetration　用过润滑脂的针入度

worked rubber 反复处理过的橡胶
worker ①打浆机②电铸版
worker-consistometer 打浆机型稠度计
work factor 工作因数；表示油使用性质的因素
work factor method 工作因数法〔在应用条件下评定润滑剂品质的方法〕
work factor test 工作因素试验
work function 功函
work hardening 功致硬化*
workhorse 多用途的(机器)
work improvement curve 工时减低曲线
working area 工作面积
working bin 工作仓
working capacity 工作容量
working condition 工作条件；使用条件
working cost 使用费时；工作费用
working curve 工作曲线
working cycle 工作循环；作业周期
working depth 工作面深度
working door (of furnace) (炉的)工作门
working electrode 工作电极
working face 工作面；作用面
working flow rate 工作流速
working fluid 工作流体；工作介质
working gas 工作气体
working hole 工作口；工作门
working instructions 操作规程
working life(=bench life) 使用寿命
working method 生产方法；加工方法；操作方法
working order 操作顺序；工作状态
working pressure 操作压力；有效压力
working rate 开工率
working resolving power 实际分辨本领
working scale 操作规模
working shaft 工作轴；工作矿井
working space 工作空间
working stability of grease 润滑脂的使用安定性
working standard 工作标准
working steam 工作蒸汽；新汽；活汽
working substance 有效物
working surface 工作面
working tank 工作油罐
working temperature 工作温度
working up(=lifting) 咬底（漆病）
working viscosity 施工黏度
working viscosity test(=draining test) 涂刷黏度试验
working volume 工作体积
working width 工作宽度；加工宽度

work-in-process 在制品
work-in-process inspection 服装半成品检验
work input 输入功
work lead 生铅（通常含有银）
workload 工作量
workmanship ①手艺；技术②精工
work of adhesion 黏附功*
work of cohesion 内聚功
work of expansion 膨胀功
work off ①除去②制造③改进
work of rupture 断裂功
work over the beam 架上操作
work piece 工作物；被加工件；半制品
workplace air 厂房空气
work point 工作地点；操作位置
work recovery 功恢复
works ①工厂②煤气厂③建筑物④机件
works bottling 工厂装瓶
workshop 车间；工厂
work softening 功致软化*
work study 作业研究
work test 工作试验〔美国海军部在工作条件下试验润滑油的方法〕
worm ①蜗杆②旋管
worm condenser 旋管冷凝器
worm conveyor 螺旋运输机
worm drive 螺杆传动
worm extruder 螺杆压出机
worm gear 蜗轮
worm-gear drive 蜗轮传动
worm gear grease 蜗轮润滑脂
worm gear oil 蜗轮油；传动装置用润滑油
worm hole 虫孔
worm knotter 螺旋式结筛
worm-like chain 蠕虫状链
worm-like polymer 似螺旋形聚合物
worm pipe 旋管；蜗管；蛇管
worm pitch 螺杆的螺距
worm press 螺杆压出机
worm reduction gear 蜗轮减速机
worm roll 麻花辊
wormseed(=absinthium) 苦艾
wormseed oil 美洲土荆芥油
worm shaft 蜗(轮)轴
worm wheel 蜗轮〔机工〕
worm wheel drive 蜗轮传动
wormwood 苦艾
wormwood oil 苦艾油

Worstall heat test　沃斯塔热试验法〔试验桐油胶化时间和温度关系的方法〕
worsted　绒布
worsted flannel　精纺法兰绒
worsted oil　精纺油
worsted yarn　精纺毛纱；精梳毛纱
W/O type emulsion　油包水型乳浊液
Woulf absorption bottle　渥耳夫吸收坛
Woulf bottle　渥耳夫(吸收)坛；渥耳夫瓶
Woulff's scrubbing bottle　沃尔夫洗气瓶
woundwort　疗伤绒毛花
woven asbesto　石棉布
woven canvas tyre　帆布轮胎
woven fabric　机织物
woven filament　长丝织物
woven glass fabric　玻璃纤维布
woven membrane　织物滤层
woven monofilament　单丝织物
woven roving　无捻粗纱织物；方格布
woven scrim　稀松织物
woven wire screen　金属线筛
wove paper　布纹纸
W-packing　W型轴封
wrap　包装纸
wrap angle　包角
wrap angle of the lobe　叶形轮的包角
wrap-around　环抱；环绕；包围(的)；包覆(的)
wrapped　①包装好的②包住的
wrapped pipe　包裹好的管道(绝热)
wrapped pipe machine　卷管机
wrapper　包装纸
wrapping　①包装②包装纸
wrapping angle　接触角；包容角；抱合角；包围角
wrapping foil　打包带
wrapping machine　包装机
wrapping paper　包装纸
W rays　W射线
wrench　①扳手；扳钳②偶单力组
wrightine　康丝碱
wringer　挤水机
wringer roll　脱水滚
wringing　扭；绞；挤
wringing fit　轻迫配合；轻打配合
wringing machine　挤水机
wring out　拧干；挤净；绞漆
wrinkle　①皱纹②起皱
wrinkle-chasing　消皱
wrinkled　皱的

wrinkled-ribbon model　折带式模型
wrinkle enamel　皱纹瓷漆
wrinkle finish(=ripple finish)　皱纹(罩面)漆；波纹面饰
wrinkle lacquer　皱纹清喷漆
wrinkle proof　不皱的；防皱的
wrinkle proofing　防皱整理
wrinkle resistance　抗皱性能
wrinkle varnish　皱纹清漆
wrinkling(=crinkling)　起皱；皱缩〔漆病〕
wrist-action shaker　肘(节运)摇摆机
wrist pin　肘节销；活塞销
wrist-watch strap　手表带
writing bond paper　二号纸
writing paper　写字纸
wrought iron　熟铁；锻铁
wrought iron of smithing quality　煅用熟铁
wrought iron pipe　熟铁管
wrought-iron pipe line　熟铁管线
wrought iron scrap　碎熟铁；废煅铁
wrought product　锻制件
wrought steel　焊接钢
wrought steel cylinder　熟钢滚筒
W-shaped peak　W形峰
wulfenite　彩钼铅矿
Wulff-Back crystallizer　伍尔夫-巴克结晶器
Wulff net　伍尔夫经纬圈
Wulff process　伍尔夫过程〔由烃类制乙炔〕
Wulff pyrolysis furnace　伍尔夫裂解炉
Wullner's law　伍尔纳水汽压定律
Wurlan process　沃兰(界面聚合)法
Wurster's blue　沃斯特氏蓝
Wurster's red　沃斯特氏红
Wurtz-Fitting reaction　武慈-菲提希反应
Wurtz flask　武慈蒸馏瓶
wurtzilite(=asphaltic pyrobitumens)　韧沥青；焦性石油沥青
wurtzite(=aegerite; aeonite)　纤锌矿
Wurtz reaction　武慈反应
wustite　铁酸盐
W-value detector　W值检测器
wyerone　蚕豆酮；5-(4-庚烯-2-炔酰基)-2-呋喃丙烯酸甲酯
Wyler's process(=urea method)　威勒法；尿素法〔铜酞制法之一〕
Wyoming bentonite　(美国)怀俄明(州)膨润土
wyomingite　金云斑白榴岩
Wysor machine　威索尔金属研磨机

X

X-alloy X 合金〔含高硼的镍铁〕
X-alloy-lined cylinder X 合金衬里的机筒
xanchromatic 黄色的
xanthaline(=papaveraldine) 鸦片黄；罂粟啶
xanthamide 黄原酰胺
xanthan 黄原胶
xanthane 苍耳烷
xanthan gum 黄原酸胶〔稳定剂〕
xanthate 黄原酸盐(酯) ROCS·SM;ROCS·SR
xanthate acid (纤维素)黄酸酯；(纤维素)黄酸(纤维素式黄酸酯)
xanthated 黄原酸化了的
xanthated cotton 黄化棉
xanthated gel 黄化凝胶
xanthate group 黄(原)酸盐基
xanthate group distribution 黄(原)酸盐基团分布
xanthate gum 黄原胶
xanthate ratio 黄化率；γ值〔纤维素黄酸盐酯化度〕
xanthate sulfur 黄酸酯硫
xanthate type accelerator 黄原酸类促进剂
xanthate viscosity (纤维素)黄(原)酸盐(碱液)黏度
xanthating 黄(原酸)化(作用)
xanthating baratte 黄化鼓
xanthating churn 黄原酸化搅拌器
xanthating crumb (纤维素)黄化屑子
xanthating drum 黄原酸化转鼓
xanthating resistance 黄原酸化阻力
xanthation 黄原酸化作用
xanthator 黄化鼓；黄化机
xanthein 花黄素
xanthene 呫吨 $C_6H_4CH_2 6H_4O$

xanthene color 呫吨染料
xanthene dye 呫吨染料
xanthenol 呫吨酚
xanthenone 呫吨酮
xanthenyl 呫吨基 $C_{13}H_9O-$
xanthenyl-carboxylic acid 呫吨羧酸
xanthic 黄的
xanthic acid 黄原酸；乙氧基二硫代甲酸
  $C_2H_5OCS·SH$
xanthic amide 黄原酰胺
xanthic degree 黄化度；酯化度
xanthic disulfide 二硫化(二)黄原酰 $ROCS·S_2CSOR$

-xanthin〔词尾〕 黄质〔与嘌呤一般无关，不应叫黄嘌呤也不能叫黄素，黄素是 flavin〕
xanthine (=2,6-dihydroxypurine;ureous acid) 黄嘌呤；2,6-二羟基嘌呤 $C_5H_4O_2N_4$
xanthin(e) bases 黄嘌呤碱类
xanthine nucleotide 黄苷酸
xanthinin 一种类似尿素的盐基性白色结晶；苍耳素
xantho- 〔希腊字头〕 黄
xanthochelidonic acid 黄白屈菜酸
  $CO(CH_2COCOOH)_2$
xanthochroism 黄化现象
xanthochromic 黄变的；黄色的
xanthochromium 黄铬基团
xanthogen 黄(原)酸基
xanthogenamide 乙黄原酰胺 $C_2H_5OCSNH_2$
xanthogenaminic acid(=thiocarbamic acid) 黄原酰胺酸；硫代氨基甲酸
xanthogenate 黄原酸盐(酯) ROCS·SM;ROCS·SR
xanthogenated 黄原酸化了的
xanthogenate gel 胶状黄原酸酯
xanthogenate reaction 黄原酸酯反应
xanthogenate solution 黄原酸酯溶液
xanthogenate sulfur 黄原酸酯硫
xanthogenate viscosity 黄原酸酯黏度
xanthogenating agent 黄化剂
xanthogenation 黄原酸化作用
xanthogenic acid ①黄原酸；乙荒酸 ROCSSH②〔专指〕乙基黄原酸 C_2H_5OCSSH
xanthoglobulin ①黄色(球)蛋白②次黄嘌呤
xanthohumol 黄腐酚
xantholine 山道年
xantholver 黄化溶解设备
xanthomat 真空黄化器
xanthonate 黄原酸盐(酯) ROCSSM; ROCSSR
xanthonated 黄原酸化了的
xanthonation 黄原酸化作用
xanthone(=xanthenone) 呫吨酮 $OC=(C_6H_4)_2O$
xanthonic acid 黄原酸；乙氧基二硫代甲酸
  C_2H_5OCSSH
xanthophyll 叶黄素；胡萝卜醇
xanthophyllite 绿脆云母
xanthopicrin 秦椒黄
xanthopone 乙基黄原酸锌
xanthoproteic reaction 黄色蛋白反应

xanthopterin 黄蝶呤；2-氨基-4,6-二羟基蝶呤
xanthopurpurin(=purpuroxanthene) 1,3-二羟基蒽醌
xanthoresinotannol 黄树脂单宁醇
xanthorhamnin 黄色鼠李苷
xanthorhodium 黄铑基团
xanthorrhiza 黄根
xanthorrhoea balsam 草树香膏
xanthoscillide 黄海葱苷
xanthosiderite 黄针铁矿
xanthosine (X,Xao) 黄(嘌呤核)苷
xanthosine monophosphate(=XMP) 黄苷一磷酸；黄苷酸
xanthostemone 黄梗酮；3-羟-4,4-二甲基-2-异丁基环己-2,5-二烯-1-酮
xanthostrumarin 欧龙牙草苷
xanthotoxol butyl ether 花椒毒酚丁醚
xanthoxylene 黄木萜；花椒萜
xanthoxyletin 黄木亭
xanthoxylin-N 黄木灵-N〔即：黄木亭〕
xanthoxylone 黄木素
xanthoxylum 黄木
xanthurenic acid 黄尿酸；4,8-二羟喹啉甲酸
xanthydrol (=9-hydroxyxanthene) 呫吨氢醇；9-羟基呫吨 HOCH=$(C_6H_4)_2$O
xanthydryl 呫吨氢基
xanthyl(=xanthenyl) 呫吨基
xanthylic acid (XMP) 黄苷酸
xanthylium 呫吨䓬
xanthylium uranyl chloride 夹氧杂蒽䓬氯化铀酰
xaxaquin 奎宁邻乙酰氧基苯甲酸
X-axis X 轴
X-band X 频带
xble. 坩埚
X-control chart 平均值控制图
X-coordinate X 坐标；横坐标
X-cut 十字裂口〔轮胎〕
xenate ①氙酸盐②氙酸根 XeO_4^-
xenene 联苯
xenic acid 氙酸
xenobiotic (药物、杀虫剂、致癌物等)异型生物质的
xenobiotics 生物异源物质；异生物质
xenocryst 异晶
xenoestrogens 外源雌激素
xenol 联苯酚；苯基苯酚 $C_6H_5C_6H_4OH$
xenolite 重夕线石
xenology 氙法定年代学
xenon 氙〔54 号元素，化学符号 Xe〕
xenon arc 氙弧
xenon arc lamp 氙弧光灯
xenon arc weatherometer 氙弧老化机；氙灯老化机
xenon ester 氙酯
xenon fadeometer 氙光耐晒牢度试验仪
xenon fluoride 氟化氙 XeF_2; XeF_4; XeF_6
xenon ionization detector 氙电离检测器
xenon lamp ①氙灯②氙气管
xenon-mercury lamp 汞氙灯
xenonology 氙法定年代学
xenon poisoning 氙中毒
xenon test 氙灯耐候试验
xenon tetrafluoride 四氟化氙
xenon trioxide 三氧化氙 XeO_3
xenotime 磷钇矿
xenyl (=biphenylyl) 联苯基 $C_6H_5C_6H_4-$
xenylamine 联苯基胺；苯基苯胺 $C_6H_5C_6H_4NH_2$
xeraphium 干燥粉
xero- 〔希腊字头〕 干
xeroform 三溴酚铋
xerogel 干凝胶
xerography(=zerography) 静电复印术
xeronic acid 干酮酸；二乙基丁烯二酸 COOHC(C_2H_5)=C(C_2H_5)COOH
xeropaste 干铅膏
xeroradiograph X 射线静电照相；静电复印法
xeroradiography 干板放射照相法
xestophanesin 没食子色苷
X-fracture 十字形裂口〔轮胎〕
ximenic acid 西门木烯酸；二十六碳-17-烯酸
ximenynic acid 西门木炔酸；11-十八碳烯-9-炔酸
Xin Hua lime 新华石灰；维也纳石灰
X-laboratory X 射线实验室
X-nucleus correlation with fixed evolution time 具有固定演化时间的 X 核相关实验
X-number X 数
xonotlite 硬硅钙石
X-radiation X 辐射；X 射线
X-radiography X 射线照相术
X-radioray X 射线
X-ray X 射线；X 光
X-ray absorptiometry X 射线吸收分析法
X-ray absorption X 射线吸收
X-ray absorption analysis X 射线吸收分析
X-ray absorption coefficient X 射线吸收系数
X-ray absorption detector X 射线吸收检测器
X-ray absorption edge X 射线吸收限
X-ray absorption edge spectrometry X 射线吸收限光谱法
X-ray absorption method X 射线吸收法
X-ray absorption near edge structure X 射线吸收近边

结构
X-ray absorption spectrometry   X 射线吸收光谱法
X-ray absorption spectrophotometry   单色 X 射线吸收分析法
X-ray activation analysis   X 射线活化分析
X-ray analysis   X 射线分析
X-ray analysis of crystal structure   X 射线(法)晶体结构分析
X-ray beam   X 射线束
X-ray camera   X 射线照相机
X-ray converter   X 射线转换器
X-ray crystal density   X 射线晶体密度
X-ray crystallography   X 射线晶体学*
X-ray defectoscopy   X 射线探伤法
X-ray density distribution curve   X 射线密度分布曲线
X-ray detection   X 射线检测
X-ray detector   X 射线检测器
X-ray diagnosis   X 射线诊断术
X-ray diagram   X 射线图；伦琴射线图
X-ray diffractiometry   X 射线衍射分析
X-ray diffraction   X 射线衍射
X-ray diffraction analysis   X 射线衍射分析
X-ray diffraction examination   X-射线衍射检验
X-ray diffraction intensity   X 射线衍射强度
X-ray diffraction method   X 射线衍射法
X-ray diffraction microscope   X 射线衍射显微镜
X-ray diffraction pattern   X 射线衍射图
X-ray diffraction photographic method   X 射线衍射照相法
X-ray diffraction studies   X 射线衍射研究
X-ray diffractometer   X 射线衍射仪
X-ray diffractometry(XRD)   X 射线衍射法
X-ray diffractometry line broadening method   X 射线衍射线宽化法
X-ray diffuse scattering   X 射线漫散射
X-ray dosimetry   X 射线剂量测定法
X-ray emission analysis   X 射线发射分析
X-ray emission energies   X 射线发射能量
X-ray emission spectrometry   X 射线发射光谱分析
X-ray energy spectrometer   X 射线能谱仪
X-ray energy spectrum   X 射线能谱
X-ray equipment   X 射线设备
X-ray examination   X 射线检查
X-ray excited Auger electron   X 射线激发俄歇电子
X-ray fault detector   X 射线探伤仪
X-ray fibre diagram   纤维的 X 射线图
X-ray film   X 射线照相；X 光胶片
X-ray filter   X 射线滤光片
X-ray flaw detector   X 射线探伤仪
X-ray fluorescence   X 射线荧光
X-ray fluorescence absorption   X 射线荧光吸收
X-ray fluorescence analysis   X 射线荧光分析
X-ray fluorescence method   X 射线荧光分析法
X-ray fluorescence spectra   X 射线荧光光谱
X-ray fluorescence spectrometer   X 射线荧光光谱仪*
X-ray fluorescence spectrometry   X 射线荧光光谱法*
X-ray fluorescence spectroscopy   X 射线荧光光谱学
X-ray fluorometry   X 射线荧光分析
X-ray fluoroscopic inspection   X 射线荧光检查
X-ray fluoroscopy   X 射线荧光检查(法)
X-ray ga(u)ging   X 射线测厚
X-ray generator   X 射线发生器
X-ray goniometer   X 射线测角器
X-ray identity period   X 射线恒等周期
X-ray image   X 射线照片
X-raying   X 射线分析
X-ray inspection   X 射线检查
X-ray intensity   X 射线强度
X-ray intensity unit   X 射线强度单位
X-ray interferometer   X 射线干涉仪
X-ray interferometry   X 射线干涉术
X-ray ionization   X 射线电离
X-ray laser (XRL)   X 射线激光器
X-ray long-period peak-intensity   X 射线长周期峰值强度
X-ray luminescence   X 射线发光
X-ray metallography   X 射线金相学
X-ray method   X 射线法
X-ray microanalysis   微区 X 射线分析
X-ray microanalyzer   X 射线微量分析仪
X-ray microradiography   X 射线显微放射照相法
X-ray microscope   X 射线显微镜
X-ray microscopy   X 射线显微术
X-ray microtopography   X 射线微区厚层断层摄影法
X-ray monochromator   X 射线单色器
X-rayogram   X 射线图式
X-ray pattern   X 射线花样
X-ray peak broadening   X 射线峰增宽
X-ray photoelectron spectrometry   X 射线光电子能谱法
X-ray photoelectron spectroscopic analysis   X 射线光电子能谱分析
X-ray photoelectron spectroscopy   X 射线光电子能谱法
X-ray photoelectron spectrum   X 射线电子能谱
X-ray photograph   X 射线照相
X-ray photometer   X 射线光度计
X-ray photometric analysis   X 射线光度分析
X-ray photometry   X 射线光度法

X-ray photon spectroscopy　X 射线光子光谱法
X-ray picture　X 射线照片
X-ray powder diffraction　X 射线粉末衍射
X-ray proof rubber　X 射线防护橡胶
X-ray radiographic inspection　X 射线照相检查
X-ray radiography　X 射线摄影术
X-ray region　X 射线区
X-ray resonance diffraction (XRRD)　X 射线共振衍射
X-ray rubber　X 射线橡胶
X-ray satellite subtraction　X 射线卫星峰扣除
X-ray small angle　小角 X 射线
X-ray small angle pattern　X 射线小角衍射图
X-ray source　X 射线光源
X-ray spectrochemical analysis　X 射线光谱化学分析
X-ray spectrogram　X 射线谱图
X-ray spectrograph　X 射线摄谱仪
X-ray spectrometer　X 射线分光计；X 射线光谱仪
X-ray spectrometric analysis　X 射线能谱分析；X 射线光谱分析
X-ray spectroscopy　X 射线光谱学
X-ray spectrum　X 射线谱
X-ray structural picture　X 射线结构图
X-ray structure　(晶体)X 射线结构
X-ray structure analysis　X 射线结构分析
X-ray structure diagram　X-射线结构图
X-ray take-off angle　X 射线取出角
X-ray target element　X 射线靶元素
X-ray test　X 射线试验
X-ray topography　①X 射线形貌学〔晶体〕②X 射线局部厚层断层摄影法
X-ray transmission method　X 射线透射法
X-ray tube　X 射线管；X 光管
X-ray unit　X 射线单位
X-ray wide angle diffractogram　X 射线广角衍射图
XRL (=X-ray laser)　X 射线激光器
X-shaped filament　X 形(截面)长丝
X-tube　X(形)管
X-unit　X 单位〔光波单位〕
xylan　木聚糖
xylanbassoric acid　木聚糖黄芪酸
xylene (=dimethylbenzene)　二甲苯
m-xylene　间二甲苯
o-xylene　邻二甲苯
p-xylene　对二甲苯
xylene azo-β-naphthol　二甲苯偶氮-β-萘酚
xylene bromide　溴代二甲苯　$BrC_6H_3(CH_3)_2$
xylene chloride　氯代二甲苯　$ClC_6H_3(CH_3)_2$
m-xylenediamine　间-二甲苯二胺〔环氧固化剂〕
xylene dichloride　二氯(代)二甲苯　$Cl_2C_6H_2(CH_3)_2$
xylene dihalide　二卤(代)二甲苯　$X_2C_6H_2(CH_3)_2$
xylene diisocyanate　二异氰酸二甲苯酯〔硫化剂〕
xylenediol　二羟甲基苯；苯二甲醇
xylene disulfonic acid　二甲苯二磺酸　$(CH_3)_2C_6H_2(SO_3H)_2$
xylene equivalent　二甲苯当量
xylene fluoride　氟代二甲苯　$FC_6H_3(CH_3)_2$
xylene halide　卤代二甲苯　$XC_6H_3(CH_3)_2$
xylene iodide　碘代二甲苯　$IC_6H_3(CH_3)_2$
xylene isomerization process　二甲苯异构化工艺
xylene isomerization-separation unit　二甲苯异构化分离装置
xylene monochloride　氯代二甲苯　$ClC_6H_3(CH_3)_2$
xylene monosulfonic acid　二甲苯磺酸　$(CH_3)_2C_6H_3SO_3H$
xylene musk(=xylol musk)　二甲苯麝香；三硝基二甲基叔丁苯　$(CH_3)_2C_6C(CH_3)_3(NO_2)_3$
xylenesulfonate　二甲苯磺酸盐　$(CH_3)_2C_6H_3SO_3M$
xylenesulfonic acid　二甲苯磺酸　$(CH_3)_2C_6H_3SO_3H$
xylenesulfonyl chloride　二甲苯磺酰氯
xylene tetrachloride　四氯代二甲苯　$Cl_4C_6(CH_3)_2$
xylenethiol　二甲苯基硫酚〔塑解剂〕
xylene trichloride　三氯代二甲苯　$Cl_3C_6H(CH_3)_2$
xylene value　二甲苯值
xylenol　二甲苯酚　$(CH_3)_2C_6H_3OH$
2,6-xylenol　2,6-二甲苯酚
xylenol blue　二甲苯酚蓝
xylenol carboxylic acid　二甲苯酚(羧)酸　$OHC_6H_2(CH_3)_2COOH$
xylenol-formaldehyde resin　二甲苯酚甲醛树脂
xylenol orange　二甲酚橙
xylenol resin　二甲(苯)酚树脂
p-xylenolsulfonephthalein　二甲磺酞
xylic acid　二甲苯甲酸　$(CH_3)_2C_6H_3COOH$
xylidene(s)　二甲苯胺
xylidic acid　4-甲基-1,3-苯二甲酸　$CH_3C_6H_3(COOH)_2$
xylidine　二甲代苯胺
xylidinic acid　甲基苯二甲酸
xylidino-　二甲代苯氨基　$(CH_3)_2C_6H_3NH—$
xylitan ester　失水木糖醇酯
xylitan monostearate　失水木糖醇单硬脂酸酯
xylite-phthalic resin　邻苯二甲酸木糖树脂
xylitol　木糖醇　$C_5H_{12}O_5$
xylitol ester　木糖醇酯
xylitol laurate　木糖醇月桂酸酯
xylitol stearate　木糖醇硬脂酸酯
xylo-　〔词头〕　木

xyloascorbic acid 木糖型抗坏血酸
xylobalsamum 几来香脂
xylocaine 利多卡因；木卡因
xylochloral 木糖三氯乙醛
xylocinnamomum 肉桂(树)
xylogen(=lignin) 木质；木质素
xyloid ①木质(制成)的②似木的
xyloidine 木炸药
xyloid material 木质物质
xyloketose (=xylulose) 木酮糖
xylol 混合二甲苯
xylolith 锯末、氧化镁和氯化镁混合物无机化学
xylol musk(=xylene musk) 二甲苯麝香
xylomannan 木甘露聚糖
xylon 木质；木纤维
xylonamide 木质酰胺
xylonic acid 木质酸 $CH_2OH(CHOH)_3COOH$
xylonite 赛璐珞
xylopine 木番荔枝碱
xylopinine(=l-norcoralydine) 番荔枝宁
xylopyranose 吡喃木糖；六环木糖
xyloquinone 二甲基醌
xylorcinol (=4,6-dimethylresorcinol) 木间二酚；4,6-二甲基苯间二酚 $(CH_3)_2C_6H_2(OH)_2$
xylose(=wood sugar) 木糖 $C_4H_9O_4CHO$
xylose absorption test 木糖吸收实验
xylosic acid 木糖酸 $HOC(CHOH)_3COOH$
xyloside 木糖苷
xylosidoglucose 木糖葡萄糖苷
xylosone 木酮(醛)糖；戊沙罗糖
xylostein 忍冬苷
xyloyl(=dimethylbenzoyl) 二甲苯酰 $(CH_3)_2C_6H_3CO—$
xylulose(=xyloketose) 木酮糖
xylyene carbinol 二甲苯基乙醇

xylyl ①二甲苯基 $(CH_3)_2C_6H_3—$②甲苄基 $CH_3C_6H_4CH_2—$
p-xylyl acetate 乙酸对甲苄酯 $CH_3C_6H_4CH_2O_2CCH_3$
xylyl alcohol ①苯二甲醇 $C_6H_4(CH_2OH)_2$②甲基苄醇 $CH_3C_6H_4CH_2OH$
2,3-xylylaldehyde 2,3-二甲基苯甲醛
xylylamine 甲苄胺
xylyl bromide 甲苄基溴
xylyl chloride 甲苄基氯
xylyl diphenyl phosphate 磷酸二甲苯二苯酯〔增塑剂〕
xylylene(=phenylenedimethylene) 亚二甲基苯 $—CH_2C_6H_4CH_2—$
xylylene alcohol 二甲苯基乙醇
xylylene amine 亚二甲苯基二胺
xylylene bromide 二溴甲基苯
xylylene chloride 二氯甲基苯
xylylene cyanide 二氰甲基苯
xylylene diamine 苯(撑)二甲(基二)胺
m-xylylenediamine 间-苯二甲胺
xylylene dichloride 二氯甲基苯
xylylene dihalide 二卤甲基苯
xylylene diisocyanate 苯二亚甲基二异氰酸酯
xylylene-glycol 亚二甲苯基二醇；1,4-苯二甲醇；对苯二甲醇
xylylenimine 二氢异吲哚
xylyl hydrazine 二甲苯肼
xylyl-mercaptan(=thioxylenol) 二甲苯基硫酚〔塑解剂〕
xylylol 二甲苯基二醇
xylyloxy 二甲苯氧基
xylylstearic acid 二甲苯基硬脂酸
X-Y plotter X-Y 绘图仪
X-Y recorder X-Y 记录器
xysmalobin 野棉根苷
XYZ tristimulus filter XYZ 三刺激(激励)滤色片

Y

yaba bark 甘兰树皮
yabine 甘兰树皮苷；药巴皮碱
yacca 禾木胶；草树(树)脂
yacca gum(=accaroid gum) 禾木胶
yageine(=harmine) 哈尔明；去氢骆驼蓬碱
yajeine 哈尔明；去氢骆驼蓬碱
yakusimycin 药师霉素
yam 薯蓣
Yankee machine 单烘缸造纸机；杨克式纸机
yara-yara β-萘基甲基醚
yardage 尺码
yarn 纱*
yarn breakage 断纱〔纺〕
yarn break detector 断丝检测器；纱线断头检测器
yarn cone 锥形纱筒〔纺〕
yarn cop 纱管〔纺〕
yarn dyeing 染纱〔纺〕
yarn grease 含绒毛脂
yarn joiner 纱线捻接器；长丝捻接器
yarn launching 纱线生头；丝条生头
yarnlike 仿纱型
yarn-like polyester 仿短纤纱涤纶
yarn mercerizing machine 纱线丝光机
yarn number 纱线支数；纱号〔纺〕
yarn path notation 垫纱运动图
yarn size 纱线支数；长丝支数
yarn spectrograph 条干不匀波长谱图；纱条频谱图；纱线条干波谱分析
yarn take-off speed 纱线送出速度；长丝退绕速度
yarn tight spots tester 丝束紧点检测仪
yarn twist 捻度〔纺〕
yarrow oil(=milfoil oil) 蓍草油
Yaryan evaporator 雅利安蒸发器
yatanoside 鸦胆子苷
yawable Pitot tube 可偏转的皮托(流速测定)管
Y-bend Y(形)管
Y-connection Y(形)接法；Y(形)接合
Y-coordinate Y 坐标；纵坐标
Y-cut Y(形)割法
Y-cut crystal Y 截晶体
yearly output 年产量
yearly production 年产量
yeast 酵母
yeast extract 酵母萃；酵母抽提物

yeatmanite 锑硅锰锌石
Yelinek evaporator 耶林尼克蒸发器
yellon 黄纶
yellow ①黄(色)的②黄(色)③蛋黄④使(变)黄；变黄
yellow accroides 黄禾木胶
yellow acid 1,3-二羟萘-5,7-二磺酸
yellow arsenic 雌黄；二硫化二砷
yellow barium chromate 钡铬黄；铬酸钡黄
yellow bark 金鸡纳树皮
yellow bean 黄豆
yellow birch 加拿大桦；黄桦
yellow brass 铜锌合金
yellow bugle 黄筋骨草
yellow chrome(=chrome yellow) 铬黄
yellow cinchona bark 黄金鸡纳皮
yellow copper 黄铜矿
yellow copperas 叶绿矾
yellow cross 黄十字(毒气)
yellow-cross ointment 防黄十字(气)软膏
yellow currant 金茶蔗子
yellow cypress 阿拉斯加花柏；黄柏
yellow dip 初割树脂
yellow discoloration ①变黄；黄变；泛黄②黄斑
yellow dock 酸模
yellow dye 黄色染料
yellow earth 赭石；赭黄土
yellow filter 黄色滤光片
yellow grass tree gum 戟叶黄胶树
yellow-guava 黄番石榴
yellow index(=yellowness index) ①黄色指数；黄度指数②泛黄指数
yellowing 泛黄；发黄；变黄
yellowing factor ①泛黄系数；变黄系数②变黄度
yellowing on aging 老化泛黄
yellowing resistance 抗黄变性；抗泛黄性
yellowing type isocyanate 泛黄型异氰酸酯
yellow iron oxide 黄色氧化铁
yellowish-brown 黄棕
yellowish litharge 黄相黄丹
yellowish-orange 黄橙色
yellow jasmine 酸模
yellow lake 黄色色淀
yellow lead 黄丹〔商〕；密陀僧〔俗〕；(一)氧化(一)铅
yellow lead ore 钼铅矿

yellow lead oxide 一氧化铅；密陀僧
yellow limonite 黄色褐铁矿；天然氧化铁黄
yellow liquor 黄(碱)液
yellow lithopone 黄立德粉
yellow lupine 黄羽扇豆
yellow mercuric oxide(=yellow mercury oxide) 黄色氧化汞 HgO
yellow mercury oxide(=yellow mercuric oxide) 黄色氧化汞
yellowness 泛黄度
yellowness factor 泛黄度系数；泛黄度
yellowness index(=yellow index) ①黄度指数；黄色指数 ②泛黄指数
yellow oak 栎；栎树
yellow ocher pigment 赭黄颜料
yellow oil 黄油〔木材化学〕
yellow ore 黄铜矿
yellow oxide 铁黄
yellow oxide of iron 氧化铁黄
yellow oxide of mercury 氧化汞黄
yellow ozokerite 黄地蜡
yellow petrolatum 黄凡士林；黄矿脂
yellow phosphorus(=white phosphorus) 黄磷；白磷
yellow pigment 黄色颜料
yellow pine oil 黄松油
yellow precipitate 黄沉淀；黄色氧化汞 HgO
yellow prussiate 亚铁氰化物；黄血盐
yellow prussiate of potash 黄血盐；亚铁氰化钾 $K_4[Fe(CN)_6]$
yellow prussiate of soda 黄血盐钠〔俗〕；亚铁氰化钠
yellow puccoon 白毛茛
yellow pulp 稻草纸浆
yellow resin 禾木树脂
yellow sandalwood 黄檀(香木)
yellow soap 黄皂
yellow soda ash 煅苏打；碳酸钠
yellow spot 黄斑
yellow stain ①银黄色料(陶瓷釉)②黄斑〔黄色污染〕
yellow straw pulp 稻草纸浆
yellow sweet clover oil 香草木犀油；避汊草油
yellow ultramarine ①柠檬黄②黄色颜料
yellow vaseline 黄矿脂；黄凡士林〔俗〕
yellow vat pigment (s) 黄瓮颜料(类)
yellow wax ①黄蜡；蜂蜡②黄石蜡
yellow wood 黄桑；黄色桑；盐肤木
yellow wood extract 黄桑提取物
yellow wort 黄木犀草
yellow zinc titanium ferrite 铁酸锌钛黄〔耐高温颜料〕
yenite 黑柱石
yerba ①草②冬青茶
yerba mate 冬青茶
yerbine 巴拉圭茶碱；冬青碱
Yerzley resilience 杨子尼回弹性
yew 浆果紫杉
Y factor (=vitamin B_6) Y 因素；维生素 B_6
Y-head Y 形机头；斜角机头
yield ①收率*②产量③产生
yield capacity 生产率
yield characteristic 收获率指标
yielding 屈服*
yielding material 流动物质
yielding point 屈服点
yield limit ①屈服极限②流动性范围
yield limit curve (蠕变试验)屈服极限曲线
yield locus 屈服轨迹
yield modulus 屈服模量
yield-octane relationship 收率-辛烷值关系
yield on charge 对原料收率
yield percentage 收率；收率百分数
yield per pass 每次收率；循环一次的收率
yield-point ①屈服点②流动点
yield point behaviour 屈服点特性
yield point test 屈服点试验
yield point value 塑变值；流动值；屈服值
yield polymer film 成膜率
yield ratio 屈强比
yield stimulant 刺激素；激长素
yield strain 屈服应变*
yield strength 屈服强度
yield stress 屈服应力*
yield temperature 屈服温度；流动温度
yield value 增益值*
yield value of stress 屈服应力值
yield zone 流动区；塑变区
Y-intercept Y 截距
-yl (某)基
ylang absolute 衣兰净油
α-ylangene α-衣兰烯
ylangol 衣兰醇
ylang ylang concrete 衣兰浸膏
ylang-ylang oil 衣兰油
-ylene〔词尾〕 ①某烯②二价某烃基
-ylene-〔词中〕 亚基〔如：ethylene diamine 亚乙基二胺〕
ylide 内鎓盐(式)〔如 $Ph_3—R_2$ 膦内鎓盐(式)〕
ynamine 炔胺*
-yne〔词尾〕 某炔

yohimbane 育亨烷
yohimbehe 育亨树皮
yohimbene 假育亨宾
yohimbenine 柯楠质
yohimbic acid 育亨酸 $C_{20}H_{24}O_3N_2$
yohimbine(=quebrachine) 育亨宾
yoke 轭
yoke (magnetizing) method 磁轭法〔磁粉探伤法〕
yolk ash 羊毛汗盐〔浸毛水经蒸发后所得的钾盐〕
yolk coal 不黏(结性)煤
yoloy 铜、镍、磷钢
yonkite 杨卡特〔炸药〕
York-Scheibel extractor 约克-沙伊贝尔萃取塔
Yorkshire grease 约夏羊毛脂〔指英国约克郡附近由洗涤羊毛所得的脂肪〕
yoshino 和纸
yoshino paper 和纸
yosimilon 二盐酸三甲氧苄嗪
Youden square 不同条件下实验设计法；挠丁方
youngberry 生浆果；未成熟浆果
Young-Laplace equation 杨-拉普拉斯公式
Young's equation 杨氏方程式
Young's modulus 杨氏模量；杨氏模数
Young's modulus in flexure 杨氏挠曲模量；杨氏揉曲模量
Young-Thollon prism system 杨-托伦棱镜系统
young viscose 低熟成度黏胶；未熟成黏胶
yperite 芥子气；双氯乙基硫 $ClC_2H_4SC_2H_4Cl$
Y-slit crack test Y型坡口抗裂试验
ysopol 牛膝草
ytterbia 氧化镱 Yb_2O_3
ytterbite 硅铍钇矿
ytterbium 镱〔70号元素，化学符号Yb,三价为正〕
ytterbium acetate 乙酸镱
ytterbium bromide 溴化镱 $YbBr_3$
ytterbium carbonate 碳酸镱 $Yb_2(CO_3)_3$
ytterbium chloride 氯化镱 $YbCl_3$
ytterbium hydroxide 氢氧化镱 $Yb(OH)_3$
ytterbium metaphosphate 偏磷酸镱 $Yb(PO_3)_3$
ytterbium nitrate 硝酸镱 $Yb(NO_3)_3$
ytterbium orthophosphate (正)磷酸镱 $YbPO_4$
ytterbium oxalate 草酸镱
ytterbium oxide 氧化镱 Yb_2O_3
ytterbium oxychloride 氯氧化镱 $YbOCl$
ytterbium sulfate 硫酸镱 $Yb_2(SO_4)_3$

yttergranate 钇榴石
yttria 三氧化二钇；氧化钇 Y_2O_3
yttrialite 硅钍钇矿
yttric 钇的
yttrium 钇〔39号元素，化学符号Y，三价为正〕
yttrium acetate 乙酸钇
yttrium aluminate laser 铝酸钇激光器
yttrium bromide 溴化钇 YBr_3
yttrium carbonate 碳酸钇 $Y_2(CO_3)_3$
yttrium chloride 氯化钇 YCl_3
yttrium fluoride 氟化钇 YF_3
yttrium hydrophosphate 磷酸氢钇 $Y_2(HPO_4)_3$
yttrium hydropyrophosphate 焦磷酸氢钇
yttrium hydrosulfate 硫酸氢钇 $Y(HSO_4)_3$
yttrium hydroxide 氢氧化钇 $Y(OH)_3$
yttrium iodide 碘化钇 YI_3
yttrium nitrate 硝酸钇 $Y(NO_3)_3$
yttrium oxide 三氧化二钇；氧化钇 Y_2O_3
yttrium pellet 钇片
yttrium peroxide 过氧化钇；九氧化四钇 Y_4O_9
yttrium phosphate 磷酸钇
yttrium sulfate 硫酸钇 $Y_2(SO_4)_3$
yttrium sulfide 硫化钇 Y_2S_3
yttrium superoxide 过氧化钇；九氧化四钇 Y_4O_9
yttrocerite 铈钇矿
yttrotantalite 钇钽矿
yttrotianite 钇楣石
Y-tube Y(形)管
yucatan elemi oil 犹加敦榄香油
yuccagenin 丝兰皂苷元
yucconin 丝兰皂苷
yuccoside 丝兰糖苷
yuenkanin 芫根苷〔黄酮类〕
yuh-tsu oil 柚皮油
Yukawa particle 汤川粒子；重电子
yukon 重电子
yulocrotine N-[2,6-二氧-1-(2-苯乙基)-3-氮环己基]-2-甲基丁酰胺
yunnannogenin 滇吉祥草皂苷元
yurotin(=eurotin) 败菌素
Yusho disease event 米糠油事件
yu-sho oil 油樟油
Y-valve Y型阀；角阀
Y zeolite Y沸石；Y型分子筛

Z

zaffre 钴蓝釉
zafirlukast 扎鲁司特〔药〕
Zahn cup 赞氏黏度杯；赞氏杯
Zaitsev rule 札依采夫规则*
zala 硼砂
zaleplon 扎来普隆〔药〕
zanaloin 桑芦荟素
Zanibar copal 非洲玷玴脂
zanthoxylum (fruit) essential oil 山椒油
Zanzibar copal 桑给巴尔玷玴脂
zanzibar gum 桑给巴尔树胶
zaomycin 沙阿霉素
zapon 硝基清漆
zapon lacquer 硝化棉(清)漆
zaratite 翠镍矿
Z-average chain length Z均链长
Z-average molecular weight Z均分子量
Z-blade mixer 曲拐式搅拌机
zdravetz oil 大根香叶油〔俗〕
zea ①玉蜀黍属②玉蜀黍花柱和柱头
zearalenone 玉米赫霉烯酮
zearin 地衣素
zeatin 玉米素；N^6-异戊烯腺嘌呤
zeaxanthin 玉米黄质
zeaxanthin epoxide (=antheraxanthin) 花药黄质；表氧化玉米黄质
zebra stripe 粗条纹
zebromal 二溴肉桂酸乙酯 PhCHBrCHBrCOOEt
zechstein 镁灰岩；蔡希斯坦统
zedalon 泽达龙〔药〕
zederone 蓬术环氧酮
zedoarone 蓬莪术酮
zedoary 蓬莪术
zedoary oil 蓬莪术油〔取自 Curcuma zedoaria〕
Zeeman atomic absorption spectrometry 塞曼原子吸收光谱法
Zeeman atomic absorption spectrophotometer 塞曼原子吸收分光光度计*
Zeeman background correction 塞曼背景校正
Zeeman background corrector 塞曼背景校正器
Zeeman broadening 塞曼(谱线)增宽；塞曼展宽
Zeeman effect 塞曼效应
Zeeman effect atomic absorption spectrometer 塞曼效应原子吸收光谱仪

Zeeman effect background correction 塞曼效应校正背景(法)
Zeeman energy levels 塞曼能级
Zeeman split 塞曼分裂
Zeeman splitting 塞曼分裂
Zeeman-tuned laser 塞曼调谐激光器
zein 玉米；玉蜀黍醇溶
Zeisel determination 蔡泽尔定量法〔测烷基〕
Zeisel method 蔡泽尔法〔测定乳化液中环氧乙烷、甘油单酸酯等含量的方法〕
Zeisel reaction 蔡泽尔碘甲烷反应
Zeise salt 蔡斯盐
Zeiss particle counter 蔡司粒子计数器
Zelan treatment 泽伦整理〔棉、麻、人造线、丝织物的防水和防污整理〕
zelio paste 铊糊
zellglas 玻璃纸
Zellner's paper 赛尔纳试纸
zellon 泽隆塑料；四氯乙烯
zellwolle 毛型黏胶短纤维；黏胶人造毛
Zener diode 稳压(二极)管
Zenker's solution 重铬酸钾、硫酸钠、氯化汞和乙酸溶液
zeolite 分子筛；泡沸石
zeolite catalysis 沸石催化
zeolite catalyst 沸石催化剂；分子筛催化剂
zeolite cavity 沸石空穴
zeolite conversion process 沸石转化法
zeolite ester 沸石酯
zeolite framework 沸石骨架
zeolite-metal catalyst 沸石-金属催化剂
zeolite molecular sieve 沸石分子筛
zeolite (molecular sieve) catalyst 沸石(分子筛)催化剂
zeolite synthesis 沸石合成(法)
zeolite water 沸石水
zeolite water softener 沸石水软化器
zeolitic cracking catalyst 沸石裂化催化剂
zeolum 沸石
zeophyllite 叶沸石
zeoscope 沸点检醇器
zeotin 半胱甲酯
zephiramine 氯化十四烷基二甲(基)苄(基)铵
zephirol 氯化十二烷基二甲基烷基铵
zephyr 席纹花布
zephyranthine 葱莲碱〔取自石蒜科〕

Zerien tower 泽林(吸收)塔
zerk 加油嘴
zero ①零②零度③零点；起点
zero absolute temperature 绝对零度
zero adjuster 零点调节器
zero adjustment 零点调整
zero-angle tire 子午(线)轮胎
zero burette 自满滴定管
zero calorie 零点卡〔物化〕
zero charge potential 零电荷电势
zero check (仪器的)调零；零点调整
zero compensation 零位补偿
zero concentration 零富集
zero contact angle 零接触角
zero creep 零蠕变
zero cure 起点硫化；零点硫化
zero current 零位电流
zero-current bipotentiometry 零电流双电位法
zero-current chronopotentiometry 零电流计时电位法
zero-current potentiometric titration 零电流电位滴定
zero-current potentiometry 零电流电位法
zero deflection 无偏差
zero detector 示零器
zero-differential overlap method(ZDO) 零微分重叠法
zero dimension 无因次
zero discharge 无出料；空转
zero discharge system 无出料系统
zero drift 零点漂移
zeroed 调到零处；调零点
zero emission 零排放
zero error 零位误差；零点误差
zero field 零场
zero-field nuclear quadrupole resonance 零场核四极共振
zero-field splitting 零场分裂
zero field splitting parameter 零场分裂参数
zero field splitting tensor 零场分裂张量
zero free volume state 零自由体积态
zero gas 零点调整用气体
zero-gauge 零隔距
zero graduation 零刻度
zerography(=xerography) 静电复印术
zero (hardness) water 零硬水；软水
zero hoopwise stress head contour 零周向应力封头曲面
zeroing 调整零点
zeroing of instrument 仪器调整零点
zero instrument 示零仪器
zero interfacial tension 零界面张力
zero level 零水准；零水平
zero line 基线；起点线；零线
zero load 零位负荷
zero mark 零度；零点标记
zero method(=null method) 零点法；衡消法
zero moisture absorption 零度吸湿
zero moment condition 零矩条件
zero Mooney processing aid 无黏度操作助剂
zero offset 零点偏移
zero oil 零度(倾点)油
zero-order reaction 零级反应
zero-order streamline 零级流线
zero phonon line 零声子线
zero point 零点
zero point adjustment screw 零点调节螺丝〔天平〕
zero point energy 零点能(量)
zero point shifting 零点移动；零点漂移
zero porosity ①无孔隙性②无孔隙
zero position 零位；起始位置
zero potential 零电势；零电位
zero potential point 零电位点
zero pour 零度倾流
zero pour point oil 零度倾点油
zero (ppm) water 零硬水；软水；蒸馏水
zero pressure molding 无压成型
zero-pressure pneumatic 零压轮胎〔橡〕
zero pressure resin 无压(固化)树脂；常压固化树脂
zero-quantum transition 零量子跃迁
zero reading 零点读数
zero release 零排放*
zero release plant 不排放废物的工厂
zero release reprocessing 不排放(放射性废物的)后处理(工厂)
zero setting 调到零点
zero shear viscosity 零切变速率黏度；零剪切黏度
zero shift 零点漂移
zero shrink 零收缩品
zero signal 零位信号
zero span extension 零档伸长
zero span tensile strength 零档断裂强度；零跨距拉伸强度
zero stability 零点稳定性
zero state 零态
zero state volume swell 基态体积溶胀
zero-stoichiometric method 零化学计量法；同位素交换分析法
zero strength temperature 零强度温度
zero strength time 零强度时间

zero sum rule　加和为零规律
zero-suppression　消零
zero-temperature　零度
zeroth law of thermodynamics　热力学第零定律
zeroth level　零级能级
zeroth order reaction　零级反应
zero tolerance　无误差
zero-twist yarn　无拈纱
zero valence(=zero valency)　零价
zero valency(=zero valence)　零价
zero valent　零价的
zerovalent compound　零价化合物
zero-valent nickel catalyst　零(化合)价镍催化剂
zero-valent state　零价态
zero voltage　零电压
zero wander　零点漂移
zero water　软水
zerumbone　球姜酮
zetameter　ζ仪
zeta-potential　ζ电势；ζ电位
Zettlemoyer's shortness　(油墨丝头)短度；蔡氏短度
zeugmatography　投影重建法
zewaphosphate　亚硫酸纤维素废液磷肥
zeyherine　刺桐碱
Z-form pool　Z形池
zibet　香猫香；麝猫香
zibetone(=civetone)　灵猫酮
zidovudine　齐多夫定〔药〕
Ziegler alcohol　齐格勒醇
Ziegler catalyst　齐格勒催化剂
Ziegler method　齐格勒法〔尤指聚乙烯的低压聚合法〕
Ziegler-Natta catalyst　齐格勒-纳塔催化剂*
Ziegler-Natta polymerization　齐格勒-纳塔聚合(反应)
Ziegler-Natta type polymerization　齐格勒-纳塔型聚合；有规立构聚合
Ziegler olefin　齐格勒法烯烃
Ziegler polyethylene　齐格勒法聚乙烯
Ziegler process　齐格勒法〔制低压聚乙烯法〕
Ziegler type catalyst　齐格勒催化剂
Ziehl's stain　齐耳染色剂
zierane　吉莉烷；茱萸烷
zierazulene　吉莉薁
zierin　吉莉苷
zierone　吉莉酮
Ziervogel process　齐尔福格提银法
zigzag　曲折的；锯齿形的；Z形的
zigzag baffle chamber　曲折挡板室
zigzag carbon chain　曲折碳链

zigzag chain　锯齿链*
zigzag chain configuration　曲折型链构型；锯齿型链构型
zigzag chain structure　曲折链结构；锯齿型链结构
zigzag configuration　锯齿构型
zigzag motion　①之字形往复运动②不规则运动
zigzag paint pattern　交错的油漆喷束〔指喷束边缘交错搭接〕
zigzag sampling　不规则取样；曲折取样
zigzag scanning　锯齿扫描
Zimmer's discontinuous disc-ring system　齐默式间歇盘环缩聚装置
Zimmermann-Reinhardt solution　齐默尔曼-莱因哈特溶液
Zimm plot　齐姆图
Zimm viscometer　齐姆黏度计
-zin　〔词尾〕嗪〔表示氮杂芳核〕
zinc　锌〔30号元素，化学符号 Zn〕
zinc accumulator　锌蓄电池
zinc acetate　乙酸锌　$Zn(C_2H_3O_2)_2$
zinc-air battery　锌-空气电池
zinc alkyl condensation　烷基锌缩合
zinc alkyl(s)　烷基锌
zinc alloy die casting　锌合金压(模)铸件
zincaluminite　锌明矾　$Zn_6Al_6(SO_4)_2(OH)_{26} \cdot 5H_2O$
zinc amalgam　锌汞合金
zincamide　氨基锌
zinc ammonium complex　锌铵络合物〔发泡剂〕
zinc ammonium complex salt process　锌铵络盐法；开撒姆法；中空注模法〔胶乳制品〕
zinc ammonium salts　锌铵盐类
zinc arc spray　电弧喷镀锌
zinc arsenate　砷酸锌
zinc arsenite　亚砷酸锌　$Zn(AsO_2)_2$
zinc ashes　镀锌浴面的氧化锌
zincate　锌酸盐
zincate accelerator　锌酸盐类促进剂
zinc baryta white(=lithopone)　锌钡白；立德粉
zinc-bearing accelerator　含锌促进剂
zinc 2-benzamidothiophenate　2-苯甲酰氨基硫酚锌〔塑解剂〕
zinc benzoate　苯甲酸锌
zinc blende　闪锌矿
zinc blende roaster　闪锌矿烤炉
zinc bloom　锌华；氧化锌　ZnO
zinc boiler plate　锌锅炉板
zinc borate　硼酸锌
zinc borate flame retardant　硼酸锌阻燃剂

zinc bromate 溴酸锌 Zn(BrO$_3$)$_2$
zinc bromide 溴化锌 ZnBr$_2$
zinc buffered magnesia 油膏状氧化锌镁
zinc butter 氯化锌
zinc 4-t-butyl-thiophenate 4-叔丁基硫酚锌〔塑解剂〕
zinc butyl xanthate 丁基黄原酸锌
zinc butyrate 丁酸锌 Zn(C$_4$H$_7$O$_2$)$_2$
zinc-cadmium sulfide 硫化锌镉〔荧光颜料〕
zinc carbide 碳化锌
zinc carbonate(=transparent zinc oxide) 透明氧化锌；碳酸锌
zinc carboxylate 羧酸锌
zinc chlorate 氯酸锌 Zn(ClO$_3$)$_2$
zinc chloride 氯化锌 ZnCl$_2$
zinc chloride process 氯化锌法〔精制裂化汽油〕
zinc chloroiodide solution 氯化锌碘化锌混合溶液
zinc chromate 铬酸锌 ZnCrO$_4$
zinc chrome 锌铬
zinc-chrome yellow 锌铬黄〔着色剂〕
zinc chromium 锌(铬)黄；铬酸锌
zinc citrate 柠檬酸锌
zinc coated(=galvanizing) 镀锌的
zinc copper couple 套淀积铜的锌片；锌铜偶
zinc Coslettising process 磷酸锌处理法〔Coslett 法的改进方法，用磷酸锌取代磷酸铁〕
zinc-crosslinked xanthate 锌交联(纤维素)黄酸盐
zinc crown 锌冕玻璃；上等锌玻璃
zinc cyanide 氰化锌 Zn(CN)$_2$
zinc dialkyl dithiophosphate 二烷基二硫代磷酸锌〔促进剂〕
zinc diamyldithiocarbamate 二戊基二硫代氨基甲酸锌〔促进剂〕
zinc dibenzamidothiophenate 二苯甲酰氨基硫酚锌〔塑解剂〕
zinc dibenzyldithiocarbamate 二苄基二硫代氨基甲酸锌〔促进剂〕
zinc dibutyl 二丁锌 Zn(CH$_2$CH$_2$CH$_2$CH$_3$)$_2$
zinc dibutyl dithiocarbamate 二丁基二硫代氨基甲酸锌
zinc dibutyldithiophosphate 二丁基二硫代磷酸锌〔促进剂〕
zinc dichromate 重铬酸锌 ZnCr$_2$O$_7$
zinc diethyl 二乙锌 Zn(C$_2$H$_5$)$_2$
zinc diethyldithiocarbamate 二乙基二硫代氨基甲酸锌〔促进剂〕
zinc diisobutyl 二异丁锌 Zn[CH$_2$CH(CH$_3$)$_2$]$_2$
zinc diisopropyl 二异丙锌 Zn[CH(CH$_3$)$_2$]$_2$
zinc diisopropyl dithiophosphate 二异丙基二硫代磷酸锌〔促进剂〕
zinc dimethacrylate 二甲基丙烯酸锌〔硫化剂〕
zinc dimethyl 二甲锌 Zn(CH$_3$)$_2$

zinc dimethyldithiocarbamate(=ziram) 二甲基二硫代氨基甲酸锌〔促进剂〕
zinc diphenyl 二苯锌 Zn(C$_6$H$_5$)$_2$
zinc di-n-propyldithiocarbamate 二正丙基二硫代氨基甲酸锌〔促进剂〕
zinc distillation furnace 蒸锌炉
zinc distillation retort 蒸锌甑
zinc dithiocarbamate 二硫代氨基甲酸锌〔促进剂〕
zinc dithiofurate 二硫化呋喃甲酸锌
zinc dithiofuroate 二硫代糠酸锌
zinc 1-dodecanesulfinate 1-十二(烷)亚磺酸锌
zinc dodecylisopropyldithiocarbamate 十二烷基异丙基二硫代氨基甲酸锌〔促进剂〕
zinc-doped titania 钛锌晶格颜料；钛锌黄
zinc drier 锌催干剂
zinc dust(=zinc powder) 锌粉
zinc dust distillation (用)锌粉蒸馏
zinc dust sacrificial anode 锌粉牺牲阳极
zinc ethide(=zinc ethyl) 二乙(基)锌 Zn(C$_2$H$_5$)$_2$
zinc ethyl (=zinc ethide) 二乙(基)锌
zinc ethylene bisdithiocarbamate (=zineb) 亚乙基双二代氨基甲酸锌
zinc ethyl hexanoate 乙基己酸锌〔促进剂〕
zinc-ethylphenyl dithiocarbamate 乙基苯基二硫代氨基甲酸锌
zinc ethyl sulfate 硫酸乙酯锌
zinc ethyl xanthate 乙基黄原酸锌
zinc ferrocyanide 氰亚铁酸锌 Zn$_2$[Fe(CN)$_6$]
zinc finger 锌指
zinc flake 片状锌粉；锌箔粉
zinc fluoride 氟化锌 ZnF$_2$
zinc fluorosilicate 氟硅酸锌；六氟络硅酸锌
zinc formaldehyde sulfoxylate 甲醛合次硫酸锌 Zn(OHCH$_2$SO$_2$)$_2$
zinc formate 甲酸锌 Zn(HCO$_2$)$_2$
zinc-free paint 无锌油漆
zinc fume ①锌蒸气②锌尘
zinc furnace 蒸锌炉
zinc glycerophosphate 甘油磷酸锌
zinc green 锌绿
zinc halide 卤化锌
zinc hydrogen phosphate 磷酸氢二锌
zinc hydroxide 氢氧化锌 Zn(OH)$_2$
zinc-hydroxocarboxylate complex 锌基羟羧酸盐配合物
zinc hydroxy carbonate 碱式碳酸锌
zinc hydroxy phosphite 碱式亚磷酸锌
zinc hypophosphite 次磷酸锌
zincic acid 锌酸 H$_2$ZnO$_2$

zinciferous 含锌的
zincing 镀锌
zinc iodate 碘酸锌 $Zn(IO_3)_2$
zinc iodide 碘化锌
zinc iodide-starch paper 碘化锌淀粉试纸
zinc iodide-starch solution 碘化锌淀粉溶液
zinc ionomer 锌离子(交联)聚合物
zinc iron cell 锌铁电池
zinc isopropyl xanthate 异丙基黄原酸锌
zincite 红锌矿
zincky(=zincous) ①锌的②含锌的③似锌的
zinc lactate 乳酸锌 $Zn(C_3H_5O_3)_2$
zinc laurate 月桂酸锌；十二烷基酸锌〔活性剂〕
zinc linoleate 亚油酸锌
zinc lupetidinedithiocarbamate 2,6-二甲基哌啶二硫代氨基甲酸锌；卢哌啶二硫代氨基甲酸锌〔促进剂〕
zinc manganate 锰酸锌 $ZnMnO_4$
zinc 2-mercaptobenzothiazole 2-硫醇基苯并噻唑锌〔促进剂〕
zinc metasilicate 硅酸锌 $ZnSiO_3$
zinc methide 二甲(基)锌 $Zn(CH_3)_2$
zinc methyl 二甲(基)锌
zinc methylphenyldithiocarbamate 甲基苯基二硫代氨基甲酸锌〔促进剂〕
zinc mill 锌磨
zinc molybdate 钼酸锌
zinc monochloroacetate 一氯乙酸锌
zinc mordant 锌媒染剂
zinc myristate 豆蔻酸锌
zinc naphthenate 环烷酸锌
zinc nitrate 硝酸锌 $Zn(NO_3)_2$
zinc nitride 二氮化三锌 Zn_3N_2
zinc octadecylcyclohexyldithiocarbamate 十八烷基环己基二硫代氨基甲酸锌〔促进剂〕
zinc octoate 辛酸锌
zinc octylisopropyldithiocarbamate 辛基异丙基二硫代氨基甲酸锌〔促进剂〕
zincograph ①锌版术②锌凸版③锌版画
zincography 锌版术
zinc ointment (氧化)锌油膏
zinc oleate 油酸锌
zincon 锌试剂*
zincous(=zincky) ①锌的②含锌的③似锌的
zinc oxalate 草酸锌 ZnC_2O_4
ZINCOXEN 锌可申〔碱性锌-乙二胺络合物，化学式为 $Zn(en)_3(OH)_2$，式中 en 代表乙二胺，可做纤维素溶剂〕
zinc oxide 氧化锌
zinc oxide cement 氧化锌接合剂

zinc oxide ointment (氧化)锌油膏
zinc oxide thickening value 氧化锌增稠试验值
zinc oxide viscosity test 氧化锌黏度试验
zinc palmitate 棕榈酸锌
zinc paraphenol sulfonate 对羟基苯磺酸锌；苯酚磺酸锌
zinc paste 氧化锌、淀粉、矿脂混合剂
zinc pentachlorophenate 五氯(苯)酚锌
zinc pentamethylene dithiocarbamate 五亚甲基二硫代氨基甲酸锌
zinc perhydrol(=zinc peroxide) 过氧化锌
zinc permanganate 高锰酸锌
zinc peroxide 过氧化锌 $Zn[O_2]$
zinc petroleum sulfonate 石油磺酸锌
zinc phenol sulfonate 酚磺酸锌
zinc phenylethyldithiocarbamate 苯基乙基二硫代氨基甲酸锌〔促进剂〕
zinc phosphate 磷酸锌 $Zn_3(PO_4)_2$
zinc phosphate coating 磷酸锌处理层(膜)
zinc phosphide 磷化锌
zinc phosphite 亚磷酸锌
zinc phthalocyanine 酞花菁锌
zinc picrate 苦味酸锌
zinc pigment 锌(系)颜料
zinc plate ①锌版〔印刷〕②锌版；锌片
zinc-plate process 镀锌法
zinc plating 镀锌
zinc potassium chromate 铬酸锌钾
zinc potassium cyanide 氰化锌钾；四氰锌酸钾 $K_2[Zn(CN)_4]$
zinc potassium iodide 碘化锌钾；四碘锌酸钾 $K_2[ZnI_4]$
zinc powder(=zinc dust) 锌粉
zinc pyrophosphate 焦磷酸锌 $Zn_2P_2O_7$
zinc resinate 树脂酸锌
zinc retort 蒸锌甑
zinc rhodanate 硫氰酸锌 $Zn(SCN)_2$
zinc rhodanide(=zinc rhodanage) 硫氰酸锌
zinc-rich mixture 富锌混合物
zinc rich paint 富锌漆
zinc rich primer 富锌底漆
zinc ricinoleate 蓖麻醇酸锌 $Zn(C_{18}H_{33}O_3)_2$
zinc rosin 含锌松香钙脂
zinc salt of mercaptobenzothiazole 巯基苯并噻唑锌盐
zinc salt of pentachlorothiophenol 五氯硫酚锌盐〔塑解剂〕
zinc sheet 锌片；锌板
zinc silicate 硅酸锌 $ZnSiO_3$; Zn_2SiO_4
zinc silicate glass 硅酸锌玻璃
zinc silicofluoride 氟硅酸锌
zinc sludge 锌矿泥

zinc smoke candle 锌烟罐
zinc soap 锌皂
zinc spar 菱锌矿
zinc spelter ①锌块②锌铜焊料
zinc spinel 锌尖晶石
zinc sponge 锌棉
zinc stability time test 锌稳定性时间试验
zinc stearate 硬脂酸锌
zinc storage battery 锌蓄电池
zinc styphnate 收敛酸锌
zinc succinate 琥珀酸锌；丁二酸锌
zinc sulfate 硫酸锌
zinc sulfhydrate 氢硫化锌 $Zn(HS)_2$
zinc sulfide 硫化锌 ZnS
zinc sulfide white 锌钡白；立德粉
zinc sulfite 亚硫酸锌 $ZnSO_3$
zinc sulfocarbolate 酚磺酸锌
zinc sulfocyanate(=zinc sulfocyanide) 硫氰酸锌 $Zn(SCN)_2$
zinc sulfocyanide(=zinc sulfocyanate) 硫氰酸锌
zinc tallate 松浆油酸锌
zinc tetraborate 四硼酸锌 ZnB_4O_7
zinc tetroxy chromate 四碱式铬酸锌
zinc thiobenzoate 硫代苯甲酸锌〔塑解剂〕
zinc thiocyanate(zinc thiocyanide) 硫氰酸锌 $Zn(SCN)_2$
zinc thiocyanide(=zinc thiocyanate) 硫氰酸锌
zinc thioxylenol 硫代二甲苯酚锌〔塑解剂〕
zinc thymotate 百里酸锌〔抗氧化剂〕
zinc undecylenate 十一烯酸锌
zinc uranyl acetate 乙酸铀酰锌
zinc vat 锌瓮
zinc vitriol 锌矾*
zinc white 锌白*
zinc white rust 锌白锈
zinc yellow 锌铬黄*；锌黄〔碱式铬酸锌与铬酸钠(或钾)的复盐〕
zinc yellow pigment 锌黄颜料
zineb 亚乙基双二硫代氨基甲酸锌；代森锌
zine fluosilicate 氟硅酸锌
zinethyl 二乙锌
zingerol 姜油酚
zingerone 姜油酮；β-(3-甲氧基-4-羟苯基)-2-丁酮 $HO(CH_3O)C_6H_3CH_2CH_2COCH_3$
zingiberene 姜烯 $C_{15}H_{24}$
zingiberol 姜醇 $C_{15}H_{26}O$
zingiberone 姜酮 $C_{15}H_{24}O$
zinin 氧化偶氮苯
zinkenite 辉锑铅矿
zinkite 红锌矿
zinkoxen 氢氧化锌乙二胺溶液
zinnkies 铜铁硫锡矿
zinnober 朱砂；辰砂
zinnwaldite 铁锂云母
zip length 断链长度；链节长度〔分子〕
zippeite 水铀矾
zipper closed conveyor belt 密闭式运输带
zipper conveyor 密闭式运输带
zippering (高密度聚乙烯薄膜)易撕裂现象
zippering pyrolysis 拉链裂解
zipping off(=chain depolymerization) 链状解聚
ziram 二甲二硫代氨基甲酸甲锌；福美锌
zircaloy 锆合金
Zircex process 泽尔塞克斯过程〔锆氢氯化为 $ZrCl_4$ 的过程，核燃料后处理首端过程之一〕
zircoaluminate coupling agent 锆铝酸盐偶联剂
zircocene polymer 二茂锆聚合
zircomium-tiatnium phosphate 磷酸锆钛
zircon 锆石
zircon alba 氧化锆
zirconate 锆酸盐 M_2ZrO_3; M_4ZrO_4
zirconia 氧化锆
zirconia tube furnace 氧化锆管电炉
zirconia whisker 氧化锆须晶
zirconic 锆的；含锆的
zirconic acid 锆酸 H_2ZrO_3; H_4ZrO_4
zirconic anhydride 锆酸酐；氧化锆
zirconin 锆试剂
zirconium 锆〔40号元素，化学符号 Zr〕
zirconium acetate 乙酸锆 $Zr(CH_3COO)_4$
zirconium acetylacetonate 乙酰基丙酮酸锆〔交联剂〕
zirconium alloy 锆合金
zirconium ammonium fluoride(=ammonium zirconifluoride) 氟化锆铵；锆氟酸铵
zirconium anhydride 二氧化锆；锆酸酐 ZrO_2
zirconium boride 硼化锆 ZrB
zirconium bromide 溴化锆 $ZrBr_4$
zirconium-n-butylate 正丁醇锆；锆酸正丁酯
zirconium carbide 一碳化锆 ZrC
zirconium carbonitride 碳氮化锆
zirconium chloride 氯化锆
zirconium crucible 锆(合金)坩埚
zirconium cyanonitride 氰氮化锆；碳氮化锆〔俗〕
zirconium dendrite 枝晶锆；树枝状锆(结晶)
zirconium deuteride 氘化锆
zirconium diboride 二硼化锆
zirconium dichloride 二氯化锆

zirconium dihydride 二氢化锆 ZrH_2
zirconium dioxide 二氧化锆 ZrO_2
zirconium diphosphide 二磷化锆 ZrP_2
zirconium diselenide 二硒化锆 $ZrSe_2$
zirconium disilicide 二硅化锆 $ZrSi_2$
zirconium dodeca-boride 十二硼化锆 ZrB_{12}
zirconium driers 锆催干剂；锆干料
zirconium fluoride 氟化锆 ZrF_4
zirconium hydroxide 氢氧化锆
zirconium-iron pink 锆铁粉红
zirconium lactate (三)乳酸锆
zirconium lake 锆深红色淀
zirconium naphthenate 环烷酸锆
zirconium nitrate 硝酸锆 $Zr(NO_3)_4 \cdot 5H_2O$
zirconium nitride 二氮化锆 ZrN_2
zirconium oxide 氧化锆 Zr_2O_3；ZrO_2
zirconium oxyacetate(=zirconyl acetate) 乙酸氧锆 $ZrO(C_2H_3O_2)_2$
zirconium oxybromide 二溴氧化锆 $ZrOBr_2$
zirconium oxycarbonate(=zirconyl carbonate) 碳酸氧锆；碱式碳酸锆
zirconium oxychloride 二氯氧化锆 $ZrOCl_2$
zirconium oxynitrate(=zirconyl nitrate) 硝酸氧锆 $ZrO(NO_3)_2 \cdot 4H_2O$
zirconium oxysulfate(=zirconyl sulfate) 硫酸氧锆
zirconium phenylarsonate 苯胂酸锆
zirconium phosphate 磷酸锆 $Zr_3(PO_4)_4$
zirconium phosphide 磷化锆 ZrP
zirconium-n-propylate 正丙醇锆；锆酸正丙酯；四丙氧基锆
zirconium pyrophosphate 焦磷酸锆
zirconium sesquioxide 三氧化二锆 Zr_2O_3
zirconium silicate(=zircon) 硅酸锆 $ZrSiO_4$
zirconium silicate fibre 锆石纤维；硅酸锆纤维
zirconium sodium lactate 乳酸锆钠
zirconium stearate 硬脂酸锆
zirconium sulfate 硫酸锆 $Zr(SO_4)_2$
zirconium sulfate dihydrate 二水合硫酸锆
zirconium tanned leather 锆鞣革
zirconium tanning 锆鞣
zirconium tanning agent 锆鞣剂
zirconium (tetra-acetylacetonate) 四乙酰基丙酮锆
zirconium tetrachloride 四氯化锆
zirconium tetrafluoride 四氟化锆 ZrF_4
zirconium tetraisopropoxide 四异丙氧基锆 $Zr(C_3H_7O)_4$
zirconium trifluoride 三氟化锆 ZrF_3
zirconium triiodide 三碘化锆 ZrI_3
zirconium white 锆白；氧化锆

zirconocene dichloride 二氯化二茂锆
zircon-opacified glaze (氧化)锆(乳)白釉
zircon oxide opacifier 氧化锆乳白釉；氧化锆瓷釉颜料
zircon porcelain 锆瓷 $ZrO_2 \cdot SiO_2$
zirconyl 氧锆基 $ZrO=$
zirconyl acetate 乙酸氧锆 $ZrO(CH_3COO)_2$
zirconyl bromide 二溴氧化锆 $ZrOBr_2$
zirconyl carbonate 碳酸氧锆 $(ZrO)CO_3$
zirconyl chloride 二氯氧化锆 $ZrOCl_2$
zirconyl hydroxide 氢氧化氧锆；羟基氧化锆
zirconyl hydroxychloride 碱式氯化氧锆 $ZrO(OH)Cl$
zirconyl hydroxynitrate 碱式硝酸锆 $ZrO(OH)NO_3$
zirconyl nitrate 硝酸氧锆 $ZrO(NO_3)_2$
zirconyl octoate 辛基氧锆 $ZrO(C_8H_{15}O_2)_2$
zirconyl stearate 硬脂酸氧锆
zirconyl sulfate 硫酸氧锆
zirc sponge 海绵(状)锆
Zirflex process 泽尔弗莱克斯过程〔锆合金溶解于 NH_4F 和 NH_4NO_3 溶液的过程,核燃料后处理首端过程之一〕
zirkite 氧锆石
zirlite 羟铝石
Z isomer Z 异构体*
Z-molar mass Z 均分子量
zoalene (=3,5-dinitro-orthotoluamide) 3,5-二硝基邻甲苯酰胺
zodiac ①黄道带②左迪阿克铜镍锌合金
zoisite 黝帘石
zolon-red 锆龙红
zolpidem 唑吡坦〔药〕
zonal 带(状)的；分区的；地区性的
zonal aberration 带像差；域像差
zonal centrifugation 区域离心
zonal centrifuge 区域离心机
zonal combustion 区域燃烧
zonal scanning 区带扫描
zonary 带状的
zone ①带；层；段；区②色区；色带③区带；谱带
zone-annealing method 区域热处理法
zone axis 晶带轴
zone broadening 区域宽化
zone collection 区带收集
zone detection 区带检测
zone electrophoresis 区带电泳*
zone leveling 区域致匀；区域匀化
zone location 区域定位
zone melting 区域熔化
zone melting apparatus 区熔法装置
zone melting chromatography 区熔色谱法

zone melting process　区域熔融法
zone melting purification　逐区熔化提纯；区熔纯化(法)
zone-melting technique　区域熔化技术
zone melt-out　区域熔融
zone migration　区域迁移
zone of combustion　燃烧层
zone of cracking　裂化层
zone of flame　火焰层
zone of heating　加热层
zone of high fatal temperature　致死高温带
zone of low temperature carbonization　低温碳化层
zone of negative pressure　负压层
zone of oxidation　氧化层
zone of semi-coking　半焦化层
zone penetration　区域渗透
zone purification　区域提纯
zone refiner　区域精炼器
zone refining　区域(熔融结晶)精制
zone sampling　区域采样法
zone sharpening　区带尖化；区域锐化
zone sharpening effect　区带锐化效应；区带尖化效应
zone spreading　区域扩展；区域扩散
zone tailing　色区拖尾
zone transfer technique　层递技术
zone velocity　区带速度
zoning　分区取样
zonography　厚层断层摄影法
zoom　图像放大
zoomaric acid　棕榈油酸；鲨油酸；9-十六碳烯酸
zooming　图像放大
zoonic acid　乙酸
zopiclone　佐匹克隆〔药〕
Zsigmondy's ultrafilter　基格蒙奇超过滤器
Z twist　Z 捻；反手捻
Z-type catalyst　齐格勒催化剂
Z-type 4-roll calender　Z 形四滚压延机
zunsober　硫化汞 HgS；朱砂〔俗〕
zwittergent　两性洗涤剂

zwitterion　两性离子*
zwitterion concept　两性离子概念
zwitter ion equilibrium　两性离子平衡
zwitterion exchanger　两性离子交换剂
zwitter ion hypothesis　两性离子假说
zwitterionic amino acid　两性离子氨基酸
zwitterionic chromogens　两性离子发色体
zwitterionic compound　两性离子化合物
zwitterionic detergent　两性离子洗涤剂
zwitterionic diazonium salt　两性离子重氮基盐
zwitterionic dyestuff　两性染料
zwitterionic intermediate　两性离子中间体
zwitterionics　两性离子表面活性剂
zwitterionic surfactant　两性离子表面活性剂*
zwitterion polymerization　两性离子聚合
zyclon B　青克郎 B〔炸药〕
zygadenine　棋盘花碱
zygofabagine　霸王精〔取自蒺藜科的生物碱〕
zylonite　赛璐珞
zyme　酶
zymin　①胰提出物②酶制剂③致病酶
zymine　胰酶(制剂)
zymochemistry　酶化学
zymo-exciter　促酶素
zymogen　酶原
zymogram　酶谱
zymohexase(=aldolase)　醛缩酶；二磷酸果糖酶
zymohydrolysis (=zymolysis)　①酶解(作用)②发酵
zymology　酶学
zymometer　发酵计；发酵检验器
zymosan　酵母聚糖
zymosimeter　发酵计；发酵检验器
zymosis　发酵(作用)
zymotechnic　酶技术；酶工艺学的
zymotechnique　发酵工艺；酿造术
zymotic　①发酵的；有关发酵的；由发酵引起的②发酵病的③传染病的
zymurgy　酿造